McGRAW-HILL ENCYCLOPEDIA OF

PHYSICS

McGRAW-HILL ENCYCLOPEDIA OF

PHYSICS

Sybil P. Parker
Editor in Chief

McGraw- Hill Book Company

New York   St. Louis   San Francisco
Auckland   Bogotá   Guatemala   Hamburg
Johannesburg   Lisbon   London   Madrid   Mexico
Montreal   New Delhi   Panama   Paris   San Juan
São Paulo   Singapore   Sydney   Tokyo   Toronto

1234567890   KPKP   89876543

ISBN 0-07-045253-9

Library of Congress Cataloging in Publication data

McGraw-Hill encyclopedia of science & technology.
  Selections.
  McGraw-Hill encyclopedia of physics.

  "All of the material in this volume has been published
previously in the McGraw-Hill encyclopedia of science
& techology, fifth edition" — T.p. verso.
  Includes bibliographies and index.
  1. Physics — Dictionaries.   I. Parker, Sybil P.
II. McGraw-Hill Book Company.   III. Title.   IV. Title:
Encyclopedia of physics.
QC5.M425   1983   530'.03'21   82-21721
ISBN 0-07-045253-9

Preface

Physics, the science of matter and energy, can be traced back to the ancients who attempted to describe the structure of the natural world. Efforts to explain the motion of celestial bodies are probably among the earliest of scientific endeavors and have occupied philosophers and scientists throughout the centuries.

On the basis of the celestial observations of Copernicus, Tycho Brahe, Johann Kepler, and Galileo, Sir Isaac Newton formulated his great treatise—*Philosophiae Naturalis Principia Mathematica* (1686–1687)—in which he used mathematical logic and elementary calculus to explain the laws of gravitation and motion, and thus initiated the development of newtonian or classical mechanics. This was a milestone in the history of physics; the field of mechanics evolved to a high level of perfection during the 18th and 19th centuries and still constitutes part of the theoretical framework for many subfields of physics as well as other basic and applied sciences.

By the beginning of the 20th century physics had reached the threshold of a new era marked by the revolutionary formulation of the mass-energy relation, $E = mc^2$, by Albert Einstein. He formulated the theory of relativity, giving an elegant mathematical description of the space-time structure of the universe, and resolved certain problems which had plagued newtonian mechanics.

Now it is believed that physicists are on the brink of yet another revolution as they attempt to unravel the mysteries of elementary particles and to express the fundamental interactions in a single unified theory. The understanding of nature provided by such a theory could lead to an explanation of the history and future of the cosmos as well as to an identification of the common origin of all known forces.

Beyond the realms of theoretical and particle physics, major discoveries in other fields during recent decades have found significant technological applications. For example, the discovery of optical pumping led to development of the laser. Superconductivity, which was discovered by H. Kamerlingh Onnes in 1911 but eluded theoretical explanation until 1957 when J. Bardeen, L. N. Cooper, and J. R. Schrieffer formulated their microscopic theory, opened the way for an array of superconducting devices, such as magnets. Discovery of the Josephson effect was the key to the development of extremely fast switching elements that will ultimately replace conventional semiconductors in faster and smaller computers.

All of these important developments, in addition to the basic principles of classical and modern physics, recent advances in theoretical and experimental research, and selected topics in mathematics, are covered in the *McGraw-Hill Encyclopedia of Phys-*

ics. The articles provide detailed information on all of the major branches of physics, including acoustics, atomic physics, particle physics, molecular physics, nuclear physics, classical mechanics, electricity, electromagnetism, fluid mechanics, heat and thermodynamics, low-temperature physics, optics, relativity, and solid-state physics.

This Encyclopedia is thoroughly comprehensive and up to date. The 760 alphabetically arranged articles, written by the leading international authorities on each of the subjects, were selected from the *McGraw-Hill Encyclopedia of Science and Technology* (5th ed., 1982). The text is supplemented by more than 1000 drawings, graphs, charts, and photographs. All information is readily accessible through a detailed analytical index and by the use of cross-references. Bibliographies provide lists of references for further reading. The Appendix includes International System (SI) conversion tables, a listing of mathematical notation, a table of fundamental constants, and a periodic table of the elements.

In preparing this Encyclopedia, Professors R. H. Good, Jr., and D. A. Bromley were most helpful and cooperative as Project Consultants. Moreover, the expertise they brought to the *Encyclopedia of Science and Technology* as Field Consultants is preserved in the present work. Other such Field Consultants deserving appreciation are Messrs. Bochner, Cowan, Hudson, Jacobs, Lapple, Lindsay, Seitz, and Steele.

This Encyclopedia will serve as an important source of information for scientists, engineers, students, librarians, science writers, and others interested in understanding the natural phenomena of the physical world.

Sybil P. Parker
EDITOR IN CHIEF

Aberration (optics)
Zeeman effect

Aberration (optics)

Deviation from perfect image formation. The deviation arising from the fact that light of different wavelengths follows different paths through an optical system is treated elsewhere. This article treats the monochromatic aberrations. *See* CHROMATIC ABERRATION; OPTICAL IMAGE.

The plotting and analyzing of the image errors of a given system are discussed. For a systematic discussion of the aberrations themselves and for definition of many of the terms used here *see* GEOMETRICAL OPTICS.

Image errors plot, meridional rays. After three (or more) rays are traced from an axis point, the intersection heights h'_i of the rays with the image plane can be plotted as functions of the height h' in the exit pupil. The curve thus obtained (Fig. 1a) can be approximated by an equation of the form of Eq. (1). It is symmetrical with respect to the zero

$$h'_i = bh'^2 + ch'^4 \qquad (1)$$

point. A shift of the image plane corresponds to a rotation of the curve around the axis. The quantities b and c are called the coefficients of the aperture errors (spherical aberration).

Plotting in the same way the intersection heights of a bundle of at least five meridional rays from one or more off-axis points with the image plane leads to a set of curves, as shown in Fig. 1b, which is for points 15° and 20° from the axis. The equation of such curves can be approximated by an equation of the form of Eq. (2).

$$h'_i = a + bh' + ch'^2 + dh'^3 + eh'^4 \qquad (2)$$

These curves can be considered to be the superposition of two sets of curves I and II in Fig. 2 having the relationships shown in Eq. (3).

$$\begin{aligned} h_I &= a + ch'^2 + eh'^4 \\ h_{II} &= (b + dh'^2)h' \end{aligned} \qquad (3)$$

The first set is symmetrical, with a giving the shift of focus and c and e the two aperture errors for the off-axis point. The second set is antisymmetric, b and d giving the first- and second-order asymmetric errors for the meridional rays.

The meridional rays do not give complete information about the image formation in an optical system. A knowledge of the skew rays is required, and this is obtained by analyzing as follows the intersection of skew rays with an image plane, the so-called spot diagrams.

A large number of rays uniformly distributed over the exit pupil are traced through the optical system from an object point off axis, either by direct tracing or by using an interpolation formula. The ray tracing gives the intersection points with the image plane, and the plot of these is the spot

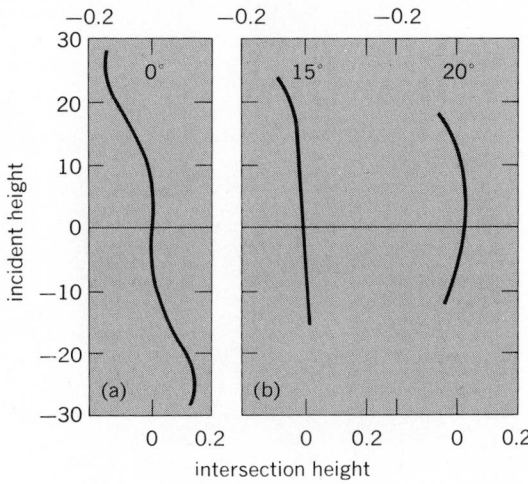

Fig. 1. Typical aperture-error (spherical-aberration) curves of a lens. (a) Field angle 0° (axis point); (b) field angles 15° and 20°.

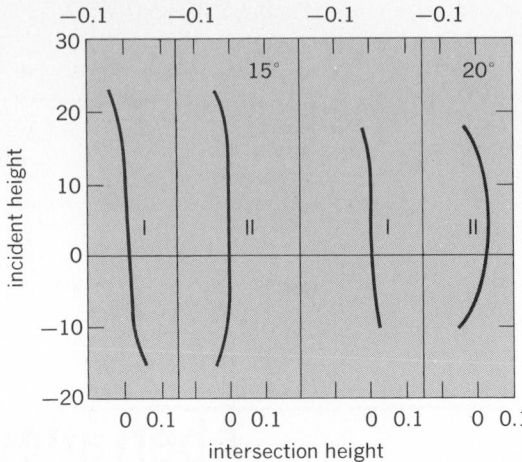

Fig. 2. Aperture-error curves of Fig. 1b divided into two parts according to Eq. (3).

diagram. In Fig. 3 these are shown for three field angles in the last line, indicated by T. These points also give two functions M and N which can be approximated by Eq. (4), where the coefficients are

$$
\begin{aligned}
M &= M_0 + M_1 y' + M_2(x'^2 + y'^2) + M_3 y'^2 \\
&\quad + M_4 y'(x'^2 + y'^2) + M_5(x'^2 + y'^2)^2 \\
N &= N_0 + N_1 y' + N_2(x'^2 + y'^2) + N_3 y'^2 \\
&\quad + N_4 y'(x'^2 + y'^2) + N_5(x'^2 + y'^2)^2
\end{aligned}
\tag{4}
$$

computed from the ray-tracing results by least-square methods.

Equations (5), where k' is the distance between

$$
\begin{aligned}
x'_i &= x'(1 + Nk') \\
&= x'_0 + x'_I + x'_{II} + x'_{III} + x'_{IV} + x'_V \\
y'_i &= y'(1 + Nk') + Mk' \\
&= y'_0 + y'_I + y'_{II} + y'_{III} + y'_{IV} + y'_V
\end{aligned}
\tag{5}
$$

image plane and exit pupil, can then be split up with respect to the power of the exit pupil. The zero-order terms of Eq. (6) and the deviation of the

$$
\begin{aligned}
x'_0 &= 0 \\
y'_0 &= M_0 k'
\end{aligned}
\tag{6}
$$

intersection point from the Gaussian image point give the distortion of the principal ray (the ray through the center of the exit pupil). The first-order terms of Eq. (7) are the first-order errors, resem-

$$
\begin{aligned}
x'_I &= x'(1 + Nk') \\
y'_I &= y'[1 + (N_0 + M_1)k']
\end{aligned}
\tag{7}
$$

bling the errors often called astigmatism. As shown in Fig. 3 by line I, the rays through a set of concentric circles in the pupil go through a set of concentric ellipses in the image plane.

The aberrations of second order are given by Eq. (8).

$$
\begin{aligned}
x'_{II} &= Nk'x'y' \\
y'_{II} &= y'^2(N_1 + M_2 + M_3)k' + M_2 k' x'^2
\end{aligned}
\tag{8}
$$

A set of rays through a concentric set of circles goes through a set of eccentric ellipses (line II, Fig. 3). This gives an asymmetric image point and the first-order asymmetry errors.

The third-order errors can be split in two forms, shown by Eqs. (9) and (10).

$$
\begin{aligned}
x'_{III} &= x'N_2 k'(x'^2 + y'^2) \\
y'_{III} &= y'N_2 k'(x'^2 + y'^2)
\end{aligned}
\tag{9}
$$

$$
\begin{aligned}
x'_{III}' &= x'N_3 k' y'^2 \\
y'_{III}' &= y'k'[M_4 x'^2 + (N_3 + M_4)y'^2]
\end{aligned}
\tag{10}
$$

The first is an aperture error, whereas the second may be classified as deformation errors. The author has suggested calling them Gullstrand errors. They are shown in lines IIIa and IIIb, respectively, in Fig. 3.

The fourth-order errors, given by Eq. (11), are

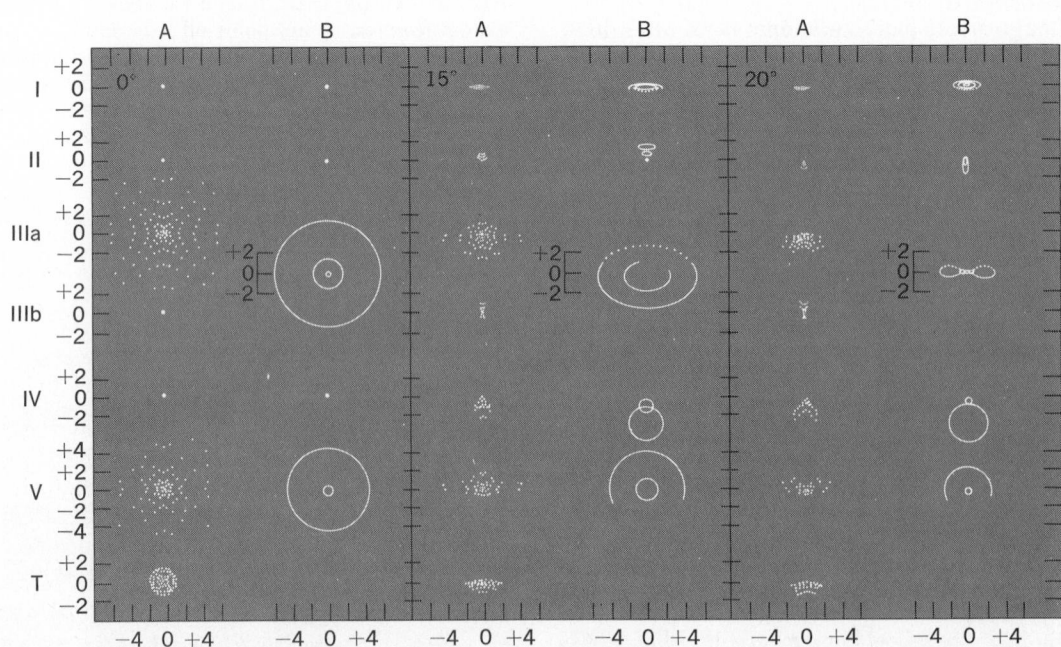

Fig. 3. An analysis of spot diagrams for three field angles 0°, 15°, 20°. A, spot diagrams found by ray tracing; B, theoretical patterns formed by rays passing through center and two zones of exit pupil; T, actual spot diagram; I–V, analytical diagrams corresponding to Eqs. (7) to (12). Diagrams in line T are the vector sums of the diagrams in lines I–V plus distortion corresponding to Eq. (6).

$$x'_{IV} = x'Nk'(x'^2 + y'^2)$$
$$y'_{IV} = [y'^2(N_4 + M_5)k' + M_5k'x'^2](x'^2 + y'^2) \quad (11)$$

asymmetry errors of the second order (line IV, Fig. 3), while Eq. (12) gives the second-order aperture errors (line V).

$$x'_V = x'N_5k'(x'^2 + y'^2)^2$$
$$y'_V = y'N_5k'(x'^2 + y'^2)^2 \quad (12)$$

This analysis of the image errors of a system permits one to obtain an insight into what happens if one of the system data is changed. While the spot diagram reacts to such a change in too complex a way, the variations of the parameters of the partial diagrams can be easily followed.

Even if all the errors are corrected for each point, one may still have the image points lying on a curved surface instead of a plane. These errors are called curvature of field. Even if the image points lie on a plane, the magnification may still vary with the position of the object point. These errors are called distortion. All the above errors are a function of the field, as seen in Fig. 3.

Previously, one tried to develop the characteristic functions as functions of aperture and field simultaneously. This still has theoretical interest, but seems impractical to describe systems with large field and aperture. A restriction to the first terms led to separation of the image errors into five groups: spherical aberration, coma, field errors (astigmatism and curvature), and distortion. One attempted to apply the above error types to finite field and aperture. However, this is not sufficient, since errors of second order in the field, like the Gullstrand error, are not covered. These generalized errors will nonetheless be discussed because of their historical interest, and because they are still used for testing optical systems.

Spherical aberrations. Aperture (spherical) aberrations have symmetry of rotation. They arise from the fact that the rays of different aperture generally do not come to the same focus. An axis point of a system with symmetry of rotation has only aperture errors. *See* MIRROR OPTICS.

A plot of image height against the square of the aperture gives for most systems a parabola, and the inclination of this parabola at the origin gives the Seidel coefficient of the aperture error. The trace of two meridional rays in addition to the calculation of the Gaussian focus is usually sufficient to give the aperture errors. In systems of very high aperture, such as microscopes, more rays should

be traced and more coefficients in the equation determined.

The aperture errors change with wavelength, and in a system designed for a wide range of wavelengths, the spherical and the chromatic aberrations have to be balanced against each other. A system in which the aperture aberrations do not change with wavelength is said to be spherochromatically corrected.

For an off-axis point, the asymmetry deformation errors must be separated from the aperture errors. The image analysis enables the aperture errors to be isolated and compared for different field angles. This procedure, again, enables the designer to balance out aperture aberrations as functions of the field angle by eventually introducing small aberrations of opposite order at the axis.

Coma. Coma is the popular name for the asymmetry errors in the image of a point. Coma occurs for one of two reasons: (1) The rays from the object point form a symmetrical image, but its appearance is unsymmetrical because the diaphragm vignettes the rays in an unsymmetrical manner; (2) the rays from the object point form an unsymmetrical image in the absence of vignetting.

The first kind of asymmetry can be easily corrected by shifting the stop (diaphragm) in such a way that the central ray of the imaging bundle goes through its center. The stop in this position is sometimes called the natural stop. The second kind of asymmetry, however, is intrinsic in the design of the system.

Seidel theory. Within the first-order approximation (Seidel theory), the asymmetry is always of the first kind. Unless the aperture error (spherical aberration) is corrected, one can always find a position of the stop to cut off the bundle coming from a near off-axis point in a symmetrical way. Figure 4 shows the shape of the image of a point in the presence of spherical aberration as the diaphragm is moved from the natural stop to a place where the bundle becomes extremely unsymmetrical.

For an axis point corrected for aperture errors, the rays from a nearby point through a set of concentric circles in the presence of Seidel coma go through a set of eccentric circles in the image plane, the common tangent of the circles forming an angle of 60°. This gives rise to the familiar comet-shaped figure from which the aberration derives its popular name.

Point image in outer field, small aperture. Here

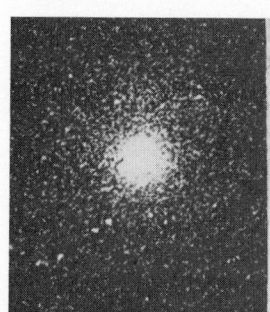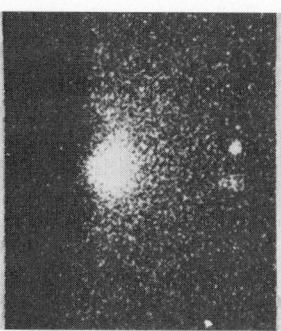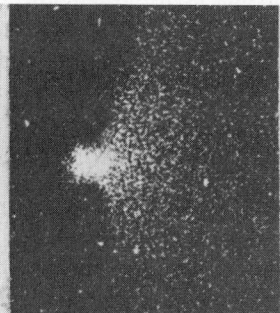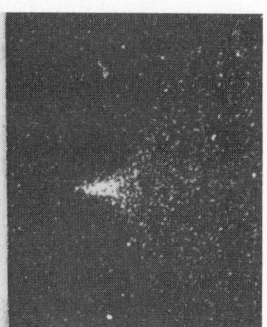

Fig. 4. Image of a point object, showing the increase in asymmetry in the presence of spherical aberration as the diaphragm is moved from the natural stop (left) to a position where the light bundle becomes extremely asymmetrical (right).

the presence of first-order asymmetry errors is characterized by the fact that the rays going through a set of concentric circles in the entrance pupil intersect the image in a set of eccentric ellipses, which have either two common tangents (outer coma) or a common secant (inner coma). The angle between the common tangents in the first case is not always 60°. Formulas have been derived for tracing data through an optical system to give these asymmetry errors, as astigmatism can be traced along a ray.

Point image near axis, large aperture. If the aperture errors are corrected, the system is, for a near axis point, free from all errors if, and only if, Abbe's sine condition is fulfilled. This means that, for all rays, Eq. (13) holds, where m is the Gaussian

$$mn' \sin u' = n \sin u \qquad (13)$$

magnification and u and u' are the angles which the ray forms with the axis in the object and image spaces, of index n and n', respectively. For an infinitely distant object, Eq. (13) is replaced by Eq. (14), where h is the entrance height of the rays par-

$$h = n'f' \; sin \; u' \qquad (14)$$

allel to the axis and f' is the (Gaussian) focal length.

In the presence of spherical aberration, the condition for symmetry, or freedom from coma, is given by Eq. (15), where Δ_s is the spherical aber-

$$\Delta_s/k' - \Delta_m/m' = 0 \qquad (15)$$

ration, k' the distance between exit pupil and Gaussian image, and Δ_m the difference between the aperture magnification given by Eq. (13) and the Gaussian magnification. This is the Staeble-Lihotzky isoplanasy condition. When the object is infinitely distant and the exit pupil is at the nodal point, this equation becomes Eq. (16), s being the back focus and f the focal length.

$$\Delta(s - f)/f = 0 \qquad (16)$$

Point image in outer field, general case. The fulfillment of the isoplanasy condition means that a point near the axis is symmetrically imaged. The condition covers only the coefficients of E which have the single index 2. Higher-order asymmetry errors must frequently be balanced out over the whole field to be imaged, by introducing a deviation from the sine condition. Asymmetry is the most disturbing error in an optical system because it makes the image extremely dissimilar to the object; therefore this balancing is of prime importance.

Astigmatism. The error which occurs because a wave surface in general has double curvature is called astigmatism. Even for a small circular stop, the rays from an object point do not come to a point focus but intersect a set of image planes in a set of ellipses, the diameters of which are proportional to the distances of the two foci from the image plane under consideration. Such an error exists even on the axis in systems which are not rotation symmetric, such as cylindrical and toric lenses and the astigmatic eye. In systems with rotation symmetry, it exists in general for the rays from an off-axis point going through a small pupil. Astigmatism on the axis is a common error in

the human eye arising from the fact that the refracting surfaces, especially the cornea, can have different powers in different meridians. It can be corrected by a spectacle lens in which at least one surface has different curvatures in different planes through the lens axis. The lens may be a cylinder, a torus, or a surface of second order with double symmetry. Since such a surface has different powers (a different number of diopters) in the two principal sections, it can be used to correct the different powers of the astigmatic eye.

The rays from an off-axis point through a small pupil, even in a system with rotation symmetry, envelop a surface (the caustic) to which they are tangent at two points that are usually separate. If the pupil is on the axis, the two points are the foci of the meridional rays and of the rays in the plane perpendicular to the plane of incidence, called the sagittal plane.

The distances s_t and s_s of the meridional and sagittal foci from the chosen image plane measure the meridional and sagittal astigmatism on the ray. The two corresponding Seidel errors are, for a small field, proportional to the astigmatic distances from the image plane for the principal or central ray (the ray through the center of the exit pupil). In lens design, $s_t - s_s$ is a measure of the astigmatism.

For a small field, the meridional astigmatism changes three times as fast as the sagittal as one of the lens parameters is varied. This means that for small changes the quantity in Eq. (17) remains

$$s_p = \tfrac{1}{2}(3s_s - s_t) \qquad (17)$$

unchanged. This quantity can be considered as a generalization of the Petzval sum or as Petzval field. Of course, the ratio of the rates of change is different from three in the outer part of the field when the field is large, and it is again important to introduce a certain amount of Petzval curvature into the system to balance it over the field.

Curvature of field. The best image of a plane object sometimes lies on a curved surface. An image formed on a flat screen will then be subject to the error known as curvature of field. If R_κ designates the curvature of the κth optical surface, the quantity shown in Eq. (18), where n_κ and n'_κ

$$R = \Sigma R_\kappa (1/n'_\kappa - 1/n_\kappa) \qquad (18)$$

are the refractive indices before and after the κth surface, is called the Petzval curvature or Petzval sum of the system. For thin-lens systems with finite distances, Eq. (18) is equivalent to Eq. (19), where ϕ_κ is the power of the κth lens.

$$R = \Sigma (\phi_\kappa/n_\kappa) \qquad (19)$$

When all the image-forming errors are corrected and the field is small, the Petzval curvature gives the axial curvature of the image of a plane object. The vanishing of R indicates that the Petzval condition is fulfilled. When a sizable field is to be covered, it is sometimes necessary to introduce a small amount of Petzval curvature to balance curvature errors of higher order.

The discovery of the significance of the Petzval sum enabled photographic lenses with plane fields to be constructed. These lenses were called anastigmats to distinguish them from aplanats, in which the meridional and sagittal fields were merely balanced against each other, one field having a positive curvature and the other a negative of nearly equal size. An anastigmat must contain at least one negative lens, as shown by Eq. (19). *See* LENS (OPTICS).

Distortion. Distortion is the error arising from the variation in magnification over the field of an optical system. It can occur in an optical system even if the system is perfectly corrected for image-forming errors. In an uncorrected system, it can be defined for any given angle as the difference between the Gaussian magnification and the magnification defined by the intersection point of the principal ray with the image plane (more generally, it can be defined as this quantity for an arbitrary ray).

Distortion that is positive, the magnification increasing with field angle, is called pincushion distortion because the image of a square has concave sides and thus looks like a pincushion (Fig. 5). The opposite type, negative distortion, is called barrel distortion because the image of a square has bulging sides. It is possible to balance distortion in an optical system by balancing higher-order distortion through introducing some third-order distortion of opposite sign. Distortion is sometimes intentionally introduced into wide-angle objectives to improve the uniformity of illumination.

The principal rays go through the centers of the exit and the entrance pupils; so, if the entrance pupil has no aperture error, freedom from distortion is achieved when $m = \tan u / \tan u'$, where m is the Gaussian lateral magnification and u and u' are corresponding field angles in object and image space, respectively. In the presence of aperture errors of the stop, this formula must be modified slightly.

Diapoint errors. M. J. Herzberger has succeeded in describing optical errors in a simple fashion. It was shown that the spot diagrams in the meridional plane, the diapoints, are easier analyzed than the ordinary spot diagrams. If all diapoints fall together, there is perfect correction; if all diapoints lie on a straight line, there is only spherical aberration; but if all diapoints lie on a curve, there are additional asymmetry errors. The deviations of the diapoint manifold from the best curve give the errors of deformation.

Thus, one could construct the center of gravity of the diapoints, the best straight line, and the best curve, and plot the deviations from it. Since for a small aperture the best point is the sagittal focus and the best ray is the principal ray, one calculates the spherical aberration as the longitudinal distance of a diapoint from the sagittal focus, and the aspherical error as the distance of the diapoint from the principal ray. The deformation error is given by the value of the functional determinant in Eq. (20), where x', y' are the coordinates of the in-

$$D = \begin{vmatrix} \dfrac{ds}{dx'} & \dfrac{ds}{dy'} \\ \dfrac{da}{dx'} & \dfrac{da}{dy'} \end{vmatrix} \quad (20)$$

tersection point of the ray with the exit pupil, which should be taken as the plane perpendicular to the axis at the intersection point of axis and principal ray. Here s is the derivation of the diapoint from the sagittal focus along the principal ray (spherical aberration), and a is the derivation orthogonal to the principal ray (asymmetry).

In calculating the diapoint spot diagram, one has again to be careful to see that the intersection points with the exit pupil of the rays, computed or interpolated, cover the exit pupil uniformly.

[MAX J. HERZBERGER]

Bibliography: G. A. Boutry, *Instrumental Optics*, 1962; E. U. Condon and H. Odishaw, *Handbook of Physics*, 1967; M. Herzberger, *Modern Geometrical Optics*, 1958; R. S. Longhurst, *Geometrical and Physical Optics*; H. G. Zimmer, *Geometrical Optics*, 1970.

Absolute zero

The lowest temperature on the scientific temperature scale. Like all other similar scalar quantities, the absolute temperature scale starts at zero and has arbitrary but convenient units going on up to high temperatures. Gross matter, which is in complete thermal equilibrium with all of its subdivided parts, has a property called its temperature, and this is measured in kelvins (K). A convenient and readily reproducible fixed point on this temperature scale is the triple point of pure water (an equilibrium mixture of ice, water, and water vapor) defined as 273.16 K. Some other values of interest (as constant-temperature baths) spread along the temperature scale are given in the table. There are several convenient thermometers to measure this low temperature range, but none can perform the impossible task of indicating the absolute zero of temperature, 0 K, because that temperature cannot be reached. It is possible to measure the temperature dependence of the magnetization of a system of electrons (or of nuclei) in gross matter; temperatures as low as 2×10^{-5} K have been reported.

The accepted standard thermometer for measuring the temperature scale is the gas thermometer, which uses the equation $PV = NkT +$ correction terms. In the equation P is the pressure in dynes per square centimeter, V is the volume in cubic centimeters, N is the number of molecules, k is Boltzmann's constant (1.38×10^{-16} erg/K), and T is the absolute temperature in degrees Kelvin. The correction terms, which are temperature-dependent, are caused by weak electric fields between the molecules. These intermolecular forces lead to gas liquefaction at low temperature; the pressure of the vapor then decreases exponentially with decrease in temperature. Even if the volume contained so little gas that the liquid or solid phase did not appear in macroscopic quantities, the vapor pressure would fall to zero as the temperature approaches 0 K because of adsorption of the molecules on the walls of the vessel.

The properties of matter at the absolute zero of

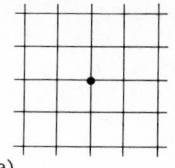

(a)

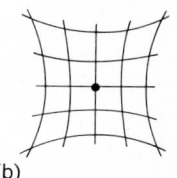

(b)

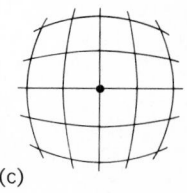

(c)

Fig. 5. Images of a rectangular object screen shown with (a) no distortion, (b) pincushion distortion, (c) barrel distortion. (*From F. A. Jenkins and H. E. White, Fundamentals of Optics, 3d ed., McGraw-Hill, 1957*)

Boiling points of some gases

Gas	Boiling point, K
He^4	4.2
H_2	20.4
N_2	77.3
O_2	90.1
C_2H_4	169.3
C_3H_6	226.1

temperature are always extrapolated from physical measurements made on them at available temperatures such as 1 K or even 0.01 K. Such studies have led to a statement of the third law of thermodynamics. The entropy of a system tends to a constant S_0 as the temperature of the system is made to approach 0 K. The value of S_0 may be set equal to zero for many systems. One popular error is to regard the absolute zero of temperature as characterized by the complete absence of motion or of energy of the system. The atoms in a solid, perhaps 10 cm³ in volume, have considerable energy locked into the lowest allowed energy states of vibration, even at 0 K. *See* CRYOGENICS; ENTROPY; KINETIC THEORY OF MATTER; LOW-TEMPERATURE PHYSICS; TEMPERATURE.

[CHARLES F. SQUIRE]

Bibliography: M. Kruisius and M. Vuorio, *Low Temperature Physics*, 1976; O. V. Lounasmas, *Experimental Principles and Methods below 1 K*, 1974; D. K. C. MacDonald, *Near Zero: The Physics of Low Temperature*, 1963; H. C. Wolfe (ed.), *Temperature: Its Measurement and Control in Science and Industry*, vol. 2, 1955.

Absorption of electromagnetic radiation

The process whereby the intensity of a beam of electromagnetic radiation is attenuated in passing through a material medium by conversion of the energy of the radiation to an equivalent amount of energy which appears within the medium; the radiant energy is converted into heat or some other form of molecular energy. A perfectly transparent medium permits the passage of a beam of radiation without any change in intensity other than that caused by the spread or convergence of the beam, and the total radiant energy emergent from such a medium equals that which entered it, whereas the emergent energy from an absorbing medium is less than that which enters, and, in the case of highly opaque media, is reduced practically to zero.

No known medium is opaque to all wavelengths of the electromagnetic spectrum, which extends from radio waves, whose wavelengths are measured in kilometers, through the infrared, visible, and ultraviolet spectral regions, to x- and γ- rays, of wavelengths down to 10^{-11} cm. Similarly, no material medium is transparent to the whole electromagnetic spectrum. A medium which absorbs a relatively wide range of wavelengths is said to exhibit general absorption, while a medium which absorbs only restricted wavelength regions of no great range exhibits selective absorption for those particular spectral regions. For example, the substance pitch shows general absorption for the visible region of the spectrum, but is relatively transparent to infrared radiation of long wavelength. Ordinary window glass is transparent to visible light, but shows general absorption for ultraviolet radiation of wavelengths below about 3100 A, while colored glasses show selective absorption for specific regions of the visible spectrum. The color of objects which are not self-luminous and which are seen by light reflected or transmitted by the object is usually the result of selective absorption of portions of the visible spectrum. Many colorless substances, such as benzene and similar hydrocarbons, selectively absorb within the ultraviolet region of the spectrum, as well as in the infrared. *See* COLOR; ELECTROMAGNETIC RADIATION.

Laws of absorption. The capacity of a medium to absorb radiation depends on a number of factors, mainly the electronic and nuclear constitution of the atoms and molecules of the medium, the wavelength of the radiation, the thickness of the absorbing layer, and the variables which determine the state of the medium, of which the most important are the temperature and the concentration of the absorbing agent. In special cases, absorption may be influenced by electric or magnetic fields. The state of polarization of the radiation influences the absorption of media containing certain oriented structures, such as crystals of other than cubic symmetry. *See* STARK EFFECT; ZEEMAN EFFECT.

Lambert's law. Lambert's law, also called Bouguer's law or the Lambert-Bouguer law, expresses the effect of the thickness of the absorbing medium on the absorption. If a homogeneous medium is thought of as being constituted of layers of uniform thickness set normally to the beam, each layer absorbs the same fraction of the radiation incident on it. If I is the intensity to which a monochromatic parallel beam is attenuated after traversing a thickness d of the medium, and I_0 is the intensity of the beam at the surface of incidence (corrected for loss by reflection from this surface), the variation of intensity throughout the medium is expressed by Eq. (1), in which α is a constant for

$$I = I_0 e^{-\alpha d} \tag{1}$$

the medium called the absorption coefficient. This exponential relation can be expressed in an equivalent logarithmic form as in Eq. (2), where

$$\log_{10}(I_0/I) = (\alpha/2.303)\, d = kd \tag{2}$$

$k = \alpha/2.303$ is called the extinction coefficient for radiation of the wavelength considered. The quantity $\log_{10}(I_0/I)$ is often called the optical density, or the absorbance of the medium.

Equation (2) shows that as monochromatic radiation penetrates the medium, the logarithm of the intensity decreases in direct proportion to the thickness of the layer traversed. If experimental values for the intensity of the light emerging from layers of the medium of different thicknesses are available (corrected for reflection losses at all reflecting surfaces), the value of the extinction coefficient can be readily computed from the slope of the straight line representing the logarithms of the emergent intensities as functions of the thickness of the layer.

Equations (1) and (2) show that the absorption and extinction coefficients have the dimensions of reciprocal length. The extinction coefficient is equal to the reciprocal of the thickness of the absorbing layer required to reduce the intensity to one-tenth of its incident value. Similarly, the absorption coefficient is the reciprocal of the thickness required to reduce the intensity to $1/e$ of the incident value, where e is the base of the natural logarithms, 2.718.

Beer's law. This law refers to the effect of the concentration of the absorbing medium, that is, the mass of absorbing material per unit of volume, on the absorption. This relation is of prime importance in describing the absorption of solutions of an absorbing solute, since the solute's concentration may be varied over wide limits, or the absorption of gases, the concentration of which depends on the pressure. According to Beer's law, each individual molecule of the absorbing material absorbs the same fraction of the radiation incident upon it, no matter whether the molecules are closely packed in a concentrated solution or highly dispersed in a dilute solution. The relation between the intensity of a parallel monochromatic beam which emerges from a plane parallel layer of absorbing solution of constant thickness and the concentration of the solution is an exponential one, of the same form as the relation between intensity and thickness expressed by Lambert's law. The effects of thickness d and concentration c on absorption of monochromatic radiation can therefore be combined in a single mathematical expression, given in Eq. (3), in which k' is a constant for a

$$I = I_0 e^{-k'cd} \qquad (3)$$

given absorbing substance (at constant wavelength and temperature), independent of the actual concentration of solute in the solution. In logarithms, the relation becomes Eq. (4). The values

$$\log_{10}(I_0/I) = (k'/2.303)\, cd = \epsilon cd \qquad (4)$$

of the constants k' and ϵ in Eqs. (3) and (4) depend on the units of concentration. If the concentration of the solute is expressed in moles per liter, the constant ϵ is called the molar extinction coefficient. Some authors employ the symbol a_M, which is called the molar absorbance index, instead of ϵ.

If Beer's law is adhered to, the molar extinction coefficient does not depend on the concentration of the absorbing solute, but usually changes with the wavelength of the radiation, with the temperature of the solution, and with the solvent.

The dimensions of the molar extinction coefficient are reciprocal concentration multiplied by reciprocal length, the usual units being liters/(mole) (cm). If Beer's law is true for a particular solution, the plot of $\log(I_0/I)$ against the concentrations for solutions of different concentrations, measured in cells of constant thickness, will yield a straight line, the slope of which is equal to the molar extinction coefficient.

While no true exceptions to Lambert's law are known, exceptions to Beer's law are not uncommon. Such exceptions arise whenever the molecular state of the absorbing solute depends on the concentration. For example, in solutions of weak electrolytes, whose ions and undissociated molecules absorb radiation differently, the changing ratio between ions and undissociated molecules brought about by changes in the total concentration prevents solutions of the electrolyte from obeying Beer's law. Aqueous solutions of dyes frequently deviate from the law because of dimerization and more complicated aggregate formation as the concentration of dye is increased.

Absorption measurement. The measurement of the absorption of homogeneous media is usually accomplished by absolute or comparative measurements of the intensities of the incident and transmitted beams, with corrections for any loss of radiant energy caused by processes other than absorption. The most important of these losses is by reflection at the various surfaces of the absorbing layer and of vessels which may contain the medium, if the medium is liquid or gaseous. Such

losses are usually automatically compensated for by the method of measurement employed. Losses by reflection not compensated for in this manner may be computed from Fresnel's laws of reflection. *See* REFLECTION OF ELECTROMAGNETIC RADIATION.

Scattering. Absorption of electromagnetic radiation should be distinguished from the phenomenon of scattering, which occurs during the passage of radiation through inhomogeneous media. Radiant energy which traverses media constituted of small regions of refractive index different from that of the rest of the medium is diverted laterally from the direction of the incident beam. The diverted radiation gives rise to the hazy or opalescent appearance characteristic of such media, exemplified by smoke, mist, and opal. If the centers of inhomogeneity are sufficiently dilute, the intensity of a parallel beam is diminished in its passage through the medium because of the sidewise scattering, according to a law of the same form as the Lambert-Bouguer law for absorption, given in Eq. (5), where I is the intensity of the pri-

$$I = I_0 e^{-\tau d} \qquad (5)$$

mary beam of initial intensity I_0, after it has traversed a distance d through the scattering medium. The coefficient τ, called the turbidity of the medium, plays the same part in weakening the primary beam by scattering as does the absorption coefficient in true absorption. However, in true scattering, no loss of total radiant energy takes place, energy lost in the direction of the primary beam appearing in the radiation scattered in other directions. In some inhomogeneous media, both absorption and scattering occur together. *See* SCATTERING OF ELECTROMAGNETIC RADIATION.

Physical nature. Absorption of radiation by matter always involves the loss of energy by the radiation and a corresponding gain in energy by the atoms or molecules of the medium.

The energy of an assembly of gaseous atoms consists partly of kinetic energy of the translational motion which determines the temperature of the gas (thermal energy), and partly of internal energy, associated with the binding of the extranuclear electrons to the nucleus, and with the binding of the particles within the nucleus itself. Molecules, composed of more than one atom, have, in addition, energy associated with periodic rotations of the molecule as a whole and with oscillations of the atoms within the molecule with respect to one another.

The energy absorbed from radiation appears as increased internal energy, or in increased vibrational and rotational energy of the atoms and molecules of the absorbing medium. As a general rule, translational energy is not directly increased by absorption of radiation, although it may be indirectly increased by degradation of electronic energy or by conversion of rotational or vibrational energy to that of translation by intermolecular collisions.

Quantum theory. In order to construct an adequate theoretical description of the energy relations between matter and radiation, it has been necessary to amplify the wave theory of radiation by the quantum theory, according to which the energy in radiation occurs in natural units called quanta. The value of the energy in these units, expressed in ergs or calories, for example, is the same for all radiation of the same wavelength, but differs for radiation of different wavelengths. The energy E in a quantum of radiation of frequency ν (where the frequency is equal to the velocity of the radiation in a given medium divided by its wavelength in the same medium) is directly proportional to the frequency, or inversely proportional to the wavelength, according to the relation given in Eq. (6), where h is a universal constant known as

$$E = h\nu \tag{6}$$

Planck's constant. The value of h is 6.63×10^{-27} erg-sec, and if ν is expressed in $\sec^{-1}$, E is given in ergs per quantum. *See* QUANTUM MECHANICS.

The most energetic type of change that can occur in an atom involves the nucleus, and increase of nuclear energy by absorption therefore requires quanta of very high energy, that is, of high frequency or low wavelength. Such rays are the γ-rays, whose wavelength varies downward from 10^{-9} cm. Next in energy are the electrons nearest to the nucleus and therefore the most tightly bound. These electrons can be excited to states of higher energy by absorption of x-rays, whose range in wavelength is from about 10^{-7} to 10^{-9} cm. Less energy is required to excite the more loosely bound valence electrons. Such excitation can be accomplished by the absorption of quanta of visible radiation (wavelength 7×10^{-5} cm for red light to 4×10^{-5} cm for blue) or of ultraviolet radiation, of wavelength down to about 10^{-5} cm. Absorption of ultraviolet radiation of shorter wavelengths, down to those on the border of the x-ray region, excites electrons bound to the nucleus with intermediate strength.

The absorption of relatively low-energy quanta of wavelength from about 10^{-3} to 10^{-4} cm suffices to excite vibrating atoms in molecules to higher vibrational states, while changes in rotational energy, which are of still smaller magnitude, may be excited by absorption of radiation of still longer wavelength, from the short-wavelength radio region of about 1 cm to long-wavelength infrared radiation, some hundredths of a centimeter long.

Gases. The absorption of gases composed of atoms is usually very selective. For example, monatomic sodium vapor absorbs very strongly over two narrow wavelength regions in the yellow part of the visible spectrum (the so-called D lines), and no further absorption by monatomic sodium vapor occurs until similar narrow lines appear in the near-ultraviolet. The valence electron of the sodium atom can exist only in one of a series of energy states separated by relatively large energy intervals between the permitted values, and the sharp-line absorption spectrum results from transitions of the valence electron from the lowest energy which it may possess in the atom to various excited levels. Line absorption spectra are characteristic of monatomic gases in general. *See* ATOMIC STRUCTURE AND SPECTRA.

The visible and ultraviolet absorption of vapors composed of diatomic or polyatomic molecules is much more complicated than that of atoms. As for atoms, the absorbed energy is utilized mainly in raising one of the more loosely bound electrons to a state of higher energy, but the electronic excitation of a molecule is almost always accompanied by simultaneous excitation of many modes of vibration of the atoms within the molecule and of rotation of the molecule as a whole. As a result, the absorption, which for an atom is concentrated in a very sharp absorption line, becomes spread over a considerable spectral region, often in the form of bands. Each band corresponds to excitation of a specific mode of vibration accompanying the electronic change, and each band may be composed of a number of very fine lines close together in wavelength, each of which corresponds to a specific rotational change of the molecule accompanying the electronic and vibrational changes. Band spectra are as characteristic of the absorption of molecules in the gaseous state, and frequently in the liquid state, as line spectra are of gaseous atoms. *See* MOLECULAR STRUCTURE AND SPECTRA.

Liquids. Liquids usually absorb radiation in the same general spectral region as the corresponding vapors. For example, liquid water, like water vapor, absorbs infrared radiation strongly (vibrational transitions), is largely transparent to visible and near-ultraviolet radiation, and begins to absorb strongly in the far-ultraviolet. A universal difference between liquids and gases is the disturbance in the energy states of the molecules in a liquid caused by the great number of intermolecular collisions; this has the effect of broadening the very fine lines observed in the absorption spectra of vapors, so that sharp-line structure disappears in the absorption bands of liquids.

Solids. Substances which can exist in solid, liquid, and vapor states without undergoing a temperature rise to very high values usually absorb in the same general spectral regions for all three states of aggregation, with differences in detail because of the intermolecular forces present in the liquid and solid. Crystalline solids, such as rock salt or silver chloride, absorb infrared radiation of long wavelength, which excites vibrations of the electrically charged ions of which these salts are composed; such solids are transparent to infrared radiations of shorter wavelengths. In colorless solids, the valence electrons are too tightly bound to the nuclei to be excited by visible radiation, but all solids absorb in the near- or far-ultraviolet region. *See* INTERMOLECULAR FORCES.

The use of solids as components of optical instruments is restricted by the spectral regions to which they are transparent. Crown glass, while showing excellent transparency for visible light and for ultraviolet radiation immediately adjoining the visible region, becomes opaque to radiation of wavelength about 3000 A and shorter, and is also opaque to infrared radiation longer than about 20,000 A in wavelength. Quartz is transparent down to wavelengths about 1800 A in the ultraviolet, and to about 40,000 A in the infrared. The most generally useful material for prisms and windows for the near-infrared region is rock salt, which is highly transparent out to about 150,000 A (15 μ). For a detailed discussion of the properties of optical glass *see* OPTICAL MATERIALS.

Fluorescence. The energy acquired by matter by absorption of visible or ultraviolet radiation, although primarily used to excite electrons to higher energy states, usually ultimately appears as increased kinetic energy of the molecules, that is, as heat. It may, however, under special circumstances, be reemitted as electromagnetic radiation. Fluorescence is the reemission, as radiant energy, of absorbed radiant energy, normally at wavelengths the same as or longer than those absorbed. The reemission, as ordinarily observed, ceases immediately when the exciting radiation is shut off. Refined measurements show that the fluorescent reemission persists, in different cases, for periods of the order of 10^{-9} to 10^{-6} sec. The simplest case of fluorescence is the resonance fluorescence of monatomic gases at low pressure, such as sodium or mercury vapors, in which the reemitted radiation is of the same wavelength as that absorbed. In this case, fluorescence is the converse of absorption: Absorption involves the excitation of an electron from its lowest energy state to a higher energy state by radiation, while fluorescence is produced by the return of the excited electron to the lower state, with the emission of the energy difference between the two states as radiation. The fluorescent radiation of molecular gases and of nearly all liquids, solids, and solutions contains a large component of wavelengths longer than those of the absorbed radiation, a relationship known as Stokes' law of fluorescence. In these cases, not all of the absorbed energy is reradiated, a portion remaining as heat in the absorbing material. The fluorescence of iodine vapor is easily seen on projecting an intense beam of visible light through an evacuated bulb containing a few crystals of iodine, but the most familiar examples are provided by certain organic compounds in solution—for instance, quinine sulfate, which absorbs ultraviolet radiation and reemits blue, or fluorescein, which absorbs blue-green light and fluoresces with an intense, bright-green color. *See* FLUORESCENCE.

Phosphorescence. The radiant reemission of absorbed radiant energy at wavelengths longer than those absorbed, for a readily observable interval after withdrawal of the exciting radiation, is called phosphorescence. The interval of persistence, determined by means of a phosphoroscope, usually varies from about 0.001 sec to several seconds, but some phosphors may be induced to phosphorescence by heat days or months after the exciting absorption. An important and useful class of phosphors is the impurity phosphors, solids such as the sulfides of zinc or calcium which are activated to the phosphorescent state by incorporating minute amounts of foreign material (called activators), such as salts of manganese or silver. So-called fluorescent lamps contain a coating of impurity phosphor on their inner wall which, after absorbing ultraviolet radiation produced by passage of an electrical discharge through mercury vapor in the lamp, reemits visible light. The receiving screen of a television tube contains a similar coating, excited not by radiant energy but by the impact of a stream of electrons on the surface. *See* PHOSPHORESCENCE.

Luminescence. Phosphorescence and fluorescence are special cases of luminescence, which is defined as light emission that cannot be attributed merely to the temperature of the emitting body. Luminescence may be excited by heat (thermoluminescence), by electricity (electroluminescence), by chemical reaction (chemiluminescence), or by friction (triboluminescence), as well as by radiation. *See* LUMINESCENCE.

Absorption and emission coefficients. The absorption and emission processes of atoms were examined from the quantum point of view by Albert Einstein in 1916, with some important results that have been realized practically in the invention of the maser and the laser. Consider an assembly of atoms undergoing absorption transitions of frequency ν sec^{-1} from the ground state 1 to an excited state 2 and emission transitions in the reverse direction, the atoms and radiation being at equilibrium at temperature T. The equilibrium between the excited and unexcited atoms is determined by the Boltzmann relation $N_2/N_1 = \exp(-h\nu/kT)$, where N_1 and N_2 are the equilibrium numbers of atoms in states 1 and 2, respectively, and the radiational equilibrium is determined by equality in the rate of absorption and emission of quanta. The number of quanta absorbed per second is $B_{12}N_1\rho(\nu)$, where $\rho(\nu)$ is the density of radiation of frequency ν (proportional to the intensity), and B_{12} is a proportionality constant called the Einstein coefficient for absorption. Atoms in state 2 will emit radiation spontaneously (fluorescence), after a certain mean life, at a rate of $A_{21}N_2$ per second, where A_{21} is the Einstein coefficient for spontaneous emission from state 2 to state 1. To achieve consistency between the density of radiation of frequency ν at equilibrium calculated from these considerations and the value calculated from Planck's radiation law, which is experimentally true, it is necessary to introduce, in addition to the spontaneous emission, an emission of intensity proportional to the radiation density of frequency ν in which the atoms are immersed. The radiational equilibrium is then determined by Eq. (7), where B_{21} is the Einstein

$$B_{12}N_1\rho(\nu) = A_{21}N_2 + B_{21}N_2\rho(\nu) \qquad (7)$$

coefficient of stimulated emission. The Einstein radiation coefficients are found to be related by Eqs. (8a) and (8b).

$$B_{12} = B_{21} \qquad (8a)$$

$$A_{21} = (8\pi h\nu^3/c^3) \cdot B_{21} \qquad (8b)$$

In the past when one considered radiation intensities available from terrestrial sources, stimulated emission was very feeble compared with the spontaneous process. Stimulated emission is, however, the fundamental emission process in the laser, a device in which a high concentration of excited molecules is produced by intense illumination from a "pumping" source, in an optical system in which excitation and emission are augmented by back-and-forth reflection until stimulated emission swamps the spontaneous process.

There are also important relations between the absorption characteristics of atoms and their mean lifetime τ in the excited state. Since A_{21} is the number of times per second that a given atom will emit a quantum spontaneously, the mean lifetime before emission in the excited state is $\tau = 1/A_{21}$. It can also be shown that A_{21} and τ are related, as shown in Eq. (9), to the f number or oscillator

$$A_{21} = 1/\tau = \frac{(8\pi^2\nu^2 e^2)}{mc^3} \cdot f$$
$$= 7.42 \times 10^{-22} f\nu^2 \qquad (\nu \text{ in sec}^{-1}) \qquad (9)$$

strength for the transition that occurs in the dispersion equations shown as Eqs. (13) to (17). The value of f can be calculated from the absorption integrated over the band according to Eq. (18).

Dispersion. A transparent material does not abstract energy from radiation which it transmits, but it always decreases the velocity of propagation of such radiation. In a vacuum, the velocity of radiation is the same for all wavelengths, but in a material medium, the velocity of propagation varies considerably with wavelength. The refractive index μ of a medium is the ratio of the velocity of light in vacuum to that in the medium, and the effect of the medium on the velocity of radiation which it transmits is expressed by the variation of refractive index with the wavelength λ of the radiation, $d\mu/d\lambda$. This variation is called the dispersion of the medium. For radiation of wavelengths far removed from those of absorption bands of the medium, the refractive index increases regularly with decreasing wavelength or increasing frequency; the dispersion is then said to be normal.

In regions of normal dispersion, the variation of refractive index with wavelength can be expressed with considerable accuracy by Eq. (10), known as

$$\mu = A + \frac{B}{\lambda^2} + \frac{C}{\lambda^4} \tag{10}$$

Cauchy's equation, in which A, B, and C are constants with positive values. As an approximation, C may be neglected in comparison with A and B, and the dispersion, $d\mu/d\lambda$, is then given by Eq. (11).

$$\frac{d\mu}{d\lambda} = \frac{-2B}{\lambda^3} \tag{11}$$

Thus, in regions of normal dispersion, the dispersion is approximately inversely proportional to the cube of the wavelength.

Dispersion by a prism. The refraction, or bending, of a ray of light which enters a material medium obliquely from vacuum or air (the refractive index of which for visible light is nearly unity) is the result of the diminished rate of advance of the wavefronts in the medium. Since, if the dispersion is normal, the refractive index of the medium is greater for violet than for red light, the wavefront of the violet light is retarded more than that of the red light. Hence, white light entering obliquely into the medium is converted within the medium to a continuously colored band, of which the red is least deviated from the direction of the incident beam, the violet most, with orange, yellow, green, and blue occupying intermediate positions. On emergence of the beam into air again, the colors remain separated. The action of the prism in resolving white light into its constituent colors is called color dispersion. *See* OPTICAL PRISM; REFRACTION OF WAVES.

The angular dispersion of a prism is the ratio, $d\theta/d\lambda$, of the difference in angular deviation $d\theta$ of two rays of slightly different wavelength which pass through the prism to the difference in wavelength $d\lambda$ when the prism is set for minimum deviation.

The angular dispersion of the prism given in Eq. (12) is the product of two factors, the variation,

$$\frac{d\theta}{d\lambda} = \frac{d\theta}{d\mu} \cdot \frac{d\mu}{d\lambda} \tag{12}$$

$d\theta/d\mu$, of the deviation θ with refractive index μ, and the variation of refractive index with wavelength, the dispersion of the material of which the prism is made. The latter depends solely on this material, while $d\theta/d\mu$ depends on the angle of incidence and the refracting angle of the prism. The greater the dispersion of the material of the prism, the greater is the angular separation between rays of two given wavelengths as they leave the prism. For example, the dispersion of quartz for visible light is lower than that of glass; hence the length of the spectrum from red to violet formed by a quartz prism is less than that formed by a glass prism of equal size and shape. Also, since the dispersion of colorless materials such as glass or quartz is greater for blue and violet light than for red, the red end of the spectrum formed by prisms is much more contracted than the blue.

The colors of the rainbow result from dispersion of sunlight which enters raindrops and is refracted and dispersed in passing through them to the rear surface, at which the dispersed rays are reflected and reenter the air on the side of the drop on which the light was incident.

Anomalous dispersion. The regular increase of refractive index with decreasing wavelength expressed by Cauchy's equation breaks down as the wavelengths approach those of strong absorption bands. As the absorption band is approached from the long-wavelength side, the refractive index becomes very large, then decreases within the band to assume abnormally small values on the short-wavelength side, values below those for radiation on the long-wavelength side. A hollow prism containing an alcoholic solution of the dye fuchsin, which absorbs green light strongly, forms a spectrum in which the violet rays are less deviated than the red, on account of the abnormally low refractive index of the medium for violet light. The dispersion of media for radiation of wavelengths near those of strong absorption bands is said to be anomalous, in the sense that the refractive index decreases with decreasing wavelength instead of showing the normal increase. The theory of dispersion, for which reference must be made to treatises such as those cited in the bibliography, shows, however, that both the normal and anomalous variation of refractive index with wavelength can be satisfactorily described as aspects of a unified phenomenon, so that there is nothing fundamentally anomalous about dispersion in the vicinity of an absorption band. *See* DISPERSION (RADIATION).

Normal and anomalous dispersion of quartz are illustrated in Fig. 1. Throughout the near-infrared, visible, and near-ultraviolet spectral regions (between P and R on the curve), the dispersion is normal and adheres closely to Cauchy's equation, but it becomes anomalous to the right of R. From S to T, Cauchy's equation is again valid.

Relation to absorption. Figure 1 shows there is an intimate connection between dispersion and absorption; the refractive index rises to high values as the absorption band is approached from the long-wavelength side and falls to low values on the short-wavelength side of the band. In fact, the theory of dispersion shows that the complete dispersion curve as a function of wavelength is governed by the absorption bands of the medium. In classical electromagnetic theory, electric charges are regarded as oscillating, each with its appropriate natural frequency ν_0, about positions of equilibrium within atoms or molecules. Placed in a radiation field of frequency ν per second, the oscillator in the atom is set into forced vibration,

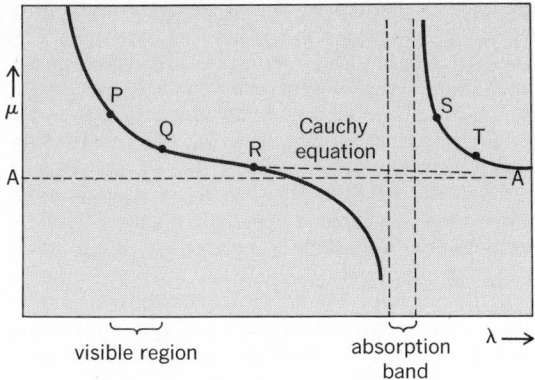

Fig. 1. Curve showing anomalous dispersion of quartz. A is limiting value of μ as λ approaches infinity. (*From F. A. Jenkins and H. E. White, Fundamentals of Optics, 3d ed., McGraw-Hill, 1957*)

with the same frequency as that of the radiation. When ν is much lower or higher than ν_0, the amplitude of the forced vibration is small, but the amplitude becomes large when the frequency of the radiation equals the natural frequency of the oscillator. In much the same way, a tuning fork is set into vibration by sound waves corresponding to the same note emitted by another fork vibrating at the same frequency. To account for the absorption of energy by the medium from the radiation, it is necessary to postulate that in the motion of the atomic oscillator some frictional force, proportional to the velocity of the oscillator, must be overcome. For small amplitudes of forced oscillation, when the frequency of the radiation is very different from the natural period of the oscillator, the frictional force and the absorption of energy are negligible. Near resonance between the radiation and the oscillator, the amplitude becomes large, with a correspondingly large absorption of energy to overcome the frictional resistance. Radiation of frequencies near the natural frequency therefore corresponds to an absorption band. *See* SYMPATHETIC VIBRATION.

To show that the velocity of the radiation within the medium is changed, it is necessary to consider the phase of the forced vibration, which the theory shows to depend on the frequency of the radiation. The oscillator itself becomes a source of secondary radiation waves within the medium which combine to form sets of waves moving parallel to the original waves. Interference between the secondary and primary waves takes place, and because the phase of the secondary waves, which is the same as that of the atomic oscillators, is not the same as that of the primary waves, the wave motion resulting from the interference between the two sets of waves is different in phase from that of the primary waves incident on the medium. But the velocity of propagation of the waves is the rate of advance of equal phase; hence the phase change effected by the medium, which is different for each frequency of radiation, is equivalent to a change in the velocity of the radiation within the medium. When the frequency of the radiation slightly exceeds the natural frequency of the oscillator, the radiation and the oscillator become 180° out of phase, which corresponds to an increase in the velocity of the radiation and accounts for the observed fall in re-

fractive index on the short-wavelength side of the absorption band.

The theory leads to Eqs. (13) through (17) for the refractive index of a material medium as a function of the frequency of the radiation. In the equations the frequency is expressed as angular frequency, $\omega = 2\pi\nu \ \mathrm{sec}^{-1} = 2\pi c/\lambda$, where c is the velocity of light. When the angular frequency ω of the radiation is not very near the characteristic frequency of the electronic oscillator, the refractive index of a homogeneous medium containing N molecules per cubic centimeter is given by Eq. (13a), where e and m are the charge and mass of the electron, and f is the number of oscillators per molecule of characteristic frequency ω_0. The f value is sometimes called the oscillator strength. If the molecule contains oscillators of different frequencies and mass (for example, electronic oscillators of frequency corresponding to ultraviolet radiation and ionic oscillators corresponding to infrared radiation), the frequency term becomes a summation, as in Eq. (13b), where ω_i is the characteristic frequency

$$\mu^2 = 1 + \frac{4\pi Ne^2}{m} \cdot \frac{f}{\omega_0^2 - \omega^2} \qquad (13a)$$

$$\mu^2 = 1 + 4\pi Ne^2 \sum_i \frac{f_i/m_i}{\omega_i^2 - \omega^2} \qquad (13b)$$

of the ith type of oscillator, and f_i and m_i are the corresponding f value and mass. In terms of wavelengths, this relation can be written as Eq. (14),

$$\mu^2 = 1 + \sum_i \frac{A_i \lambda^2}{\lambda^2 - \lambda_i^2} \qquad (14)$$

where A_i is a constant for the medium, λ is the wavelength of the radiation, and $\lambda_i = c/\nu_i$ is the wavelength corresponding to the characteristic frequency ν_i per second (Sellmeier's equation).

If the medium is a gas, for which the refractive index is only slightly greater then unity, the dispersion formula can be written as Eq. (15).

$$\mu = 1 + 2\pi Ne^2 \sum_i \frac{f_i/m_i}{\omega_i^2 - \omega^2} \qquad (15)$$

So long as the absorption remains negligible, these equations correctly describe the increase in refractive index as the frequency of the radiation begins to approach the absorption band determined by ω_i or λ_i. They fail when absorption becomes appreciable, since they predict infinitely large values of the refractive index when ω equals ω_i, whereas the refractive index remains finite throughout an absorption band.

The absorption of radiant energy of frequency very close to the characteristic frequency of the medium is formally regarded as the overcoming of a frictional force when the molecular oscillators are set into vibration, related by a proportionality constant g to the velocity of the oscillating particle; g is a damping coefficient for the oscillation. If the refractive index is determined by a single electronic oscillator, the dispersion equation for a gas at radiational frequencies within the absorption band becomes Eq. (16). At the same time an ab-

$$\mu = 1 + \frac{2\pi Ne^2}{m} \frac{f(\omega_0^2 - \omega^2)}{(\omega_0^2 - \omega^2)^2 + \omega^2 g^2} \qquad (16)$$

sorption constant κ enters the equations, related

to the absorption coefficient α of Eq. (1) by the expression $\kappa = \alpha c/2\omega\mu$. Equation (17) shows the

$$\kappa = \frac{2\pi Ne^2}{m} \frac{f\omega g}{(\omega_0{}^2 - \omega^2)^2 + \omega^2 g^2} \qquad (17)$$

relationship. For a monatomic vapor at low pressure, Nf is about 10^{17} per cubic centimeter, ω_0 is about 3×10^{15} per second, and g is about 10^{11} per second. These data show that, when the frequency of the radiation is not very near ω_0, ωg is very small in comparison with the denominator and the absorption is practically zero. As ω approaches ω_0, κ increases rapidly to a maximum at a radiational frequency very near ω_0 and then falls at frequencies greater than ω_0. When the absorption is relatively weak, the absorption maximum is directly proportional to the oscillator strength f. In terms of the molar extinction coefficient ϵ of Eq. (4), it can be shown that this direct relation holds, as seen in Eq. 18. The integration in Eq. (18) is carried out

$$f = 4.319 \times 10^{-9} \int \epsilon d\bar{\nu} \qquad (18)$$

over the whole absorption spectrum. The integral can be evaluated from the area under the curve of ϵ plotted as a function of wave number $\bar{\nu}$ cm^{-1} $= \nu(\sec^{-1})/c = 1/\lambda$.

The width of the absorption band for an atom is determined by the value of the damping coefficient g; the greater the damping, the greater is the spectral region over which absorption extends.

The general behavior of the refractive index through the absorption band is illustrated by the dotted portions of Fig. 2. The presence of the damping term $\omega^2 g^2$ in the denominator of Eq. (17) prevents the refractive index from becoming infinite when $\omega = \omega_0$. Its value increases to a maximum for a radiation frequency less than ω_0, then falls with increasing frequency in the center of the band (anomalous dispersion) and increases from a relatively low value on the high-frequency side of the band.

Figure 2 shows schematically how the dispersion curve is determined by the absorption bands throughout the whole electromagnetic spectrum. The dotted portions of the curve correspond to absorption bands, each associated with a distinct type of electrical oscillator. The oscillators excited by x-rays are tightly bound inner electrons; those excited by ultraviolet radiation are more loosely bound outer electrons which control the dispersion in the near-ultraviolet and visible regions, whereas those excited by the longer wavelengths are atoms or groups of atoms.

It will be observed in Fig. 2 that in regions of

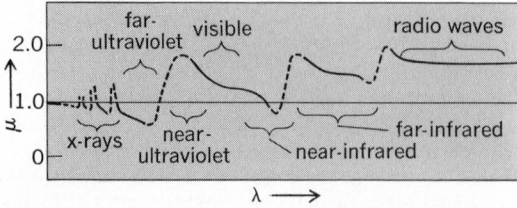

Fig. 2. Complete dispersion curve through the electromagnetic spectrum for a substance. (From F. A. Jenkins and H. E. White, Fundamentals of Optics, 3d ed., McGraw-Hill, 1957)

anomalous dispersion the refractive index of a substance may assume a value less than unity; the velocity of light in the medium is then greater than in vacuum. The velocity involved here is that with which the phase of the electromagnetic wave of a single frequency ω advances, for example, the velocity with which the crest of the wave advances through the medium. The theory of wave motion, however, shows that a signal propagated by electromagnetic radiation is carried by a group of waves of slightly different frequency, moving with a group velocity which, in a material medium, is always less than the velocity of light in vacuum. The existence of a refractive index less than unity in a material medium is therefore not in contradiction with the theory of relativity.

In quantum theory, absorption is associated not with the steady oscillation of a charge in an orbit but with transitions from one quantized state to another. The treatment of dispersion according to quantum theory is essentially similar to that outlined, with the difference that the natural frequencies ν_0 are now identified with the frequencies of radiation which the atom can absorb in undergoing quantum transitions. These transition frequencies are regarded as virtual classical oscillators, which react to radiation precisely as do the oscillators of classical electromagnetic theory.

Selective reflection. Nonmetallic substances which show very strong selective absorption also strongly reflect radiation of wavelengths near the absorption bands, although the maximum of reflection is not, in general, at the same wavelength as the maximum absorption. The infrared rays selectively reflected by ionic crystals are frequently referred to as reststrahlen, or residual rays. For additional information on selective reflection see REFLECTION OF ELECTROMAGNETIC RADIATION.

[WILLIAM WEST]

Bibliography: M. Born and E. Wolf, *Principles of Optics*, 5th ed., 1975; R. W. Ditchburn, *Light*, 2d ed., 1963; F. A. Jenkins *Fundamentals of Optics*, 4th ed., 1976; Optical Society of America, *Handbook of Optics*, 1978; A. Sommerfeld, *Lectures on Theoretical Physics*, vol. 4, 1954; J. A. Stratton, *Electromagnetic Theory*, 1941; J. Strong, *Concepts of Classical Optics*, 1958.

Abstract algebra

A term used synonymously with modern algebra and general algebra to describe the type of algebra which has been developed since the mid-1920s and has become a basic idiom of contemporary mathematics. In contrast with the earlier algebra, which was highly computational and was confined to the study of specific systems generally based on real and complex numbers, abstract algebra is conceptual and axiomatic and deals with systems which are arbitrary sets of elements of unspecified type, together with certain compositions satisfying prescribed lists of axioms.

A good insight into the difference between the older and the present approach can be obtained by comparing the older matrix theory with the more abstract linear algebra. Both deal with roughly the same portion of mathematics, the former from a direct approach which stresses calculations with matrices, while the latter from an axiomatic and geometric viewpoint which treats vector spaces and linear transformations as the basic notions, and matrices as secondary to these. *See* LINEAR ALGEBRA; MATRIX THEORY.

Features. Abstract algebra deals with a number of important algebraic structures, such as groups, rings, and lattices. *See* GROUP THEORY.

Such a structure consists of a set S of elements of unspecified nature endowed with a number of finitary compositions on S. If r is a positive integer, an r-ary composition associates with every r-tuple $(a_1, a_2, \ldots, a_r)$ of elements a_i in S a unique element $\omega(a_1, a_2, \ldots, a_r)$ in S. It is convenient to consider also nullary compositions which amount to the selection of particular elements of S. In the case of a group $S = G$ one has a single binary $(= 2$-ary) composition which is assumed to satisfy several simple conditions called the group axioms. In this case one usually writes ab for $\omega(a,b)$ or $a + b$, if the group is commutative—that is, $\omega(a,b) = \omega(b,a)$ for all a,b. In the case of a ring R one has two binary compositions, ab and $a + b$, which are subjected to a set of conditions known as the ring axioms.

In addition to the intrinsic study of algebraic structures, one is also interested in the study of the action of one algebraic structure on another. An important instance is the theory of modules and its special case of vector spaces. One defines a left module for a ring R as a commutative group M on which R acts on the left in the sense that, given a pair of elements (a,x) where a is in R and x is in M, then this determines a unique element ax in M. Moreover, the module product ax is assumed to satisfy the module axioms $a(x + y) = ax + ay$, $(a + b)x = ax + bx$, and $(ab)x = a(bx)$, where a,b are arbitrary elements of R and x,y are arbitrary elements of M.

A considerable part of the study of algebraic structures can be developed in a unified way which does not specify the particular structure. The deeper side of abstract algebra, however, does require specialization to the particular systems whose richness is to a large extent accounted for by their applicability to other areas of mathematics and physics. The general study of algebraic structures is called universal algebra. A basic notion here is that of homomorphism of one algebraic structure S into a second one S'. Here one has a $1-1$ correspondence $\omega \leftrightarrow \omega'$ between the sets of compositions on S and S', such that ω and ω' are both r-ary for the same $r = 0,1,2,3, \ldots$. Then a homomorphism f of S into S' is a mapping of S into S' such that $\omega(a_1, \ldots, a_r) = \omega'(f(a_1), \ldots, f(a_r))$ for all a_i in S and all the corresponding compositions ω, ω'. A substantial part of the theory of homomorphisms of groups and rings is valid in this general setting. Another basic notion of universal algebra is that of a variety of algebraic structures. This is defined to be the class of structures satisfying a given set of identities defined by means of the given compositions (that is, the associative or the commutative laws). Examples are the varieties of groups or rings.

New disciplines. A leading aspect of abstract algebra is its ability to encompass new disciplines. A notable example has been the rapid development of homological algebra, which had its inception around 1945. This includes a cohomology theory of groups and of modules and has important application to number theory and group theory. The basic

notions and results of this theory are too technical to discuss here.

Another new algebraic discipline which has been an outgrowth of homological algebra is that of category theory. The notion of a category is applicable throughout mathematics, for example, in set theory and topology. The notion is formulated in order to place on an equal footing a class of mathematical objects and the basic mappings between these. One defines a category to be a class of objects (generally mathematical structures) together with a class of morphisms (generally mappings of the structures) such that the following category axioms hold:

C1. With each ordered pair (a,b) of objects there is associated a unique set hom (a,b) of morphisms such that every morphism is contained in one and only one of the sets hom (a,b).

C2. If f is in hom (a,b) and g is in hom (b,c), then there is a uniquely determined element fg in hom (a,c).

C3. If f is in hom (a,b), g in hom (b,c), and h in hom (c,d), then one has the associativity $(fg)h = f(gh)$.

C4. To each object a there corresponds a morphism 1_a in hom (a,a) such that for any f in hom (a,b) and g in hom (c,a) one has $1_a f = f$, $g1_a = g$.

An example of a category is the class of all groups as objects and the class of all homomorphisms of pairs of groups as the class of morphisms. Another example is the class of all sets as objects and of mappings of one set into another as the class of morphisms.

If C and C' are categories, one defines a functor from C to C' as a mapping F of the objects of C into the objects of C', together with a mapping φ of the morphisms of C into the morphisms of C' such that the axioms listed below hold:

F1. For every a in C, $\varphi(1_a) = 1_{F(a)}$.

F2. If f is in hom (a,b) and g is in hom (b,c), where a,b,c are objects of C, then $\varphi(f)$ is in hom $(F(a),F(b))$, $\varphi(g)$ is in hom $(F(b),F(c))$, $\varphi(fg)$ is in hom $(F(a),F(c))$, and $\varphi(fg) = \varphi(f)\varphi(g)$.

The concepts of category and functor are turning up in many areas of mathematics, and it is conceivable that, like set theory, category theory will eventually find its way into elementary levels of mathematics too. [NATHAN JACOBSON]

Bibliography: J. R. Durbin, *Modern Algebra: An Introduction*, 1979; I. N. Herstein, *Topics in Algebra*, 2d ed., 1975; N. Jacobson, *Lectures in Abstract Algebra*, vols. 1–3, 1951–1964; A. G. Kurosh, *Lectures on General Algebra*, transl. from Russian, 1970; S. Lang, *Algebra*, 1965; S. MacLane, *Homology*, 1975; S. MacLane and G. Birkhoff, *Algebra*, 2d ed., 1979; L. W. Shapiro, *Introduction to Abstract Algebra*, 1975.

Acceleration

The time rate of change of velocity. Since velocity is a directed or vector quantity involving both magnitude and direction, a velocity may change by a change of magnitude (speed) or by a change of direction or both. It follows that acceleration is also a directed, or vector, quantity. If the magnitude of the velocity of a body changes from v_1 ft/sec to v_2 ft/sec in t sec, then the average acceleration a has a magnitude given by Eq. (1). To desig-

$$a = \frac{\text{velocity change}}{\text{elapsed time}} = \frac{v_2 - v_1}{t_2 - t_1} = \frac{\Delta v}{\Delta t} \quad (1)$$

nate it fully the direction should be given, as well as the magnitude. *See* VELOCITY.

Instantaneous acceleration is defined as the limit of the ratio of the velocity change to the elapsed time as the time interval approaches zero. When the acceleration is constant, the average acceleration and the instantaneous acceleration are equal.

If a body, moving along a straight line, is accelerated from a speed of 10 to 90 ft/sec in 4 sec, then the average change in speed per second is $(90 - 10)/4 = 20$ ft/sec in each second. This is written 20 ft per second per second or 20 ft/sec². Accelerations are also commonly expressed in meters per second per second (m/sec²), or in any similar units.

Whenever a body is acted upon by an unbalanced force, it will undergo acceleration. If it is moving in a constant direction, the acting force will produce a continuous change in speed. If it is moving with a constant speed, the acting force will produce an acceleration consisting of a continuous change of direction. In the general case, the acting force may produce both a change of speed and a change of direction.   [R. D. RUSK]

Angular acceleration. This is a vector quantity representing the rate of change of angular velocity of a body experiencing rotational motion. If, for example, at an instant t_1, a rigid body is rotating about an axis with an angular velocity ω_1, and at a later time t_2, it has an angular velocity ω_2, the average angular acceleration $\overline{\alpha}$ is given by Eq. (2),

$$\overline{\alpha} = \frac{\omega_2 - \omega_1}{t_2 - t_1} = \frac{\Delta\omega}{\Delta t} \quad (2)$$

expressed in radians per second per second. The instantaneous angular acceleration is given by $\alpha = d\omega/dt$.

Consequently, if a rigid body is rotating about a fixed axis with an angular acceleration of magnitude α and an angular speed of ω_0 at a given time, then at a later time t the angular speed is given by Eq. (3).

$$\omega = \omega_0 + \alpha t \quad (3)$$

A simple calculation shows that the angular distance θ traversed in this time is expressed by Eq. (4).

$$\theta = \overline{\omega}t = \left[\frac{\omega_0 + (\omega_0 + \alpha t)}{2}\right]t = \omega_0 t + \frac{1}{2}\alpha t^2 \quad (4)$$

In the figure a particle is shown moving in a circular path of radius R about a fixed axis through O with an angular velocity of ω radians/sec and an angular acceleration of α radians/sec². This particle is subject to a linear acceleration which, at any instant, may be considered to be composed of two components: a radial component $\mathbf{a}_r$ and a tangential component $\mathbf{a}_t$.

Radial acceleration. When a body moves in a circular path with constant linear speed at each point in its path, it is also being constantly acceler-

ated toward the center of the circle under the action of the force required to constrain it to move in its circular path. This acceleration toward the center of path is called radial acceleration. In the figure, the radial acceleration, sometimes called centripetal acceleration, is shown by the vector $\mathbf{a}_r$. The magnitude of its value is v^2/R, or ω^2/R, where v is the instantaneous linear velocity. This centrally directed acceleration is necessary to keep the particle moving in a circular path.

Tangential acceleration. The component of linear acceleration tangent to the path of a particle subject to an angular acceleration about the axis of rotation is called tangential acceleration. In the figure, the tangential acceleration is shown by the vector $\mathbf{a}_t$. The magnitude of its value is αR. *See* ROTATIONAL MOTION.

[C. E. HOWE/R. J. STEPHENSON]

Acoustic impedance

At a given surface, the complex ratio of effective sound pressure averaged over the surface to the effective flux (volume velocity or particle velocity multiplied by the surface area) through it. The unit is the newton-second/meter⁵, or the mks acoustic ohm. In the cgs system the unit is the dyne-second/centimeter⁵.

Specific acoustic impedance is the complex ratio of the effective sound pressure at a point to the effective particle velocity at a point. The unit is the newton-second/meter³, or the mks rayl. In the cgs system the unit is the dyne-second/centimeter³, or the rayl. The difference between specific acoustic impedance and acoustic impedance is in the specification of impedance at a point, as compared to the average over a surface. The specific acoustic impedance is generally employed in acoustical analyses, with the acoustic impedance being computed from it when required.

Characteristic acoustic impedance is the ratio of effective sound pressure at a point to the particle velocity at that point in a free, progressive wave. This ratio is equal to the product of the density of the medium ρ_0 times the speed of sound c in the medium. The characteristic impedance of a sound wave is analogous to the characteristic electrical impedance of an infinitely long, dissipationless transmission line. It is common in acoustical analyses to represent specific acoustic impedances in terms of their ratio to the characteristic impedance of air. For example, the specific acoustic impedance Z of a heavy drapery material may be written as $Z = 2\rho_0 c$, meaning that it is twice the characteristic impedance of air.

Impedance analogies. Acoustic impedance, being a complex quantity, can have real and imaginary components analogous to those in an electrical impedance. In applying this analogy, the real part of the acoustic impedance is termed acoustic resistance, and the imaginary part is termed acoustic reactance. *See* ELECTRICAL IMPEDANCE.

The analogy between acoustic and electrical impedances is useful in the solution of many acoustical problems because it permits the analysis to be conducted by the techniques of electrical circuit theory. In these analyses sound pressure is usually taken as analogous to voltage, and volume velocity as analogous to current. Various parts of acoustical circuits can be associated directly with

ACCELERATION

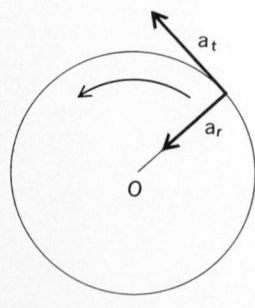

Radial and tangential accelerations in circular motion.

their electrical counterparts by employing these analogs.

Acoustic resistance is associated with the dissipative losses occurring when there is a viscous movement of a quantity of gas through a thin tube of mesh. It is analogous to electrical resistance.

Acoustic mass, associated with a mass of air accelerated by a net force which acts to displace the gas without appreciably compressing it, assumes the role of an inductance.

Acoustic compliance, associated with a volume of air that is compressed by a net force without an appreciable average displacement of the center of gravity of the air in the volume, acts as a capacitance. Both acoustic mass and acoustic compliance are reactive portions of acoustic impedance, in analogy with their electrical counterparts.

Examples of reactances. As simple examples of acoustical reactances, consider the low-frequency approximation to the impedances of an open-ended tube and that of a simple container having a given volume. If end corrections are neglected, the impedance of the tube is an acoustic mass M_a given by Eq. (1), where ρ_0 is the density of air, l is the

$$M_a = \frac{\rho_0 l}{S} \tag{1}$$

length of the tube, and S is its cross-sectional area. The impedance of the container is an acoustic compliance C_a given by Eq. (2), where V is the vol-

$$C_a = \frac{V}{\rho_0 c^2} \tag{2}$$

ume, ρ_0 the density of air, and c the velocity of sound.

The absorption of sound by a material is often described in terms of its acoustical impedance. For example, the absorption coefficient α of a material exposed to a normally incident plane wave of sound in air is given by Eq. (3), where Z is the

$$\alpha = 1 - \left(\frac{Z - \rho_0 c}{Z + \rho_0 c}\right)^2 \tag{3}$$

specific acoustic impedance at the surface of the material and $\rho_0 c$ is the characteristic impedance of air. *See* SOUND ABSORPTION.

[WILLIAM J. GALLOWAY]

Bibliography: L. L. Beranek, *Acoustics*, 1954; P. M. Morse and K. U. Ingard, *Theoretical Acoustics*, 1968; C. B. Officer, *Introduction to the Theory of Sound Transmission*, 1958; F. A. White, *Our Acoustic Environment*, 1975.

Acoustical holography

A technique that utilizes the interference pattern formed when two beams of sound intersect to form an image of an object placed in one of the sound beams.

The greatest interest in holography has resulted from its ability to generate an extremely realistic three-dimensional visual image. However, because the technique of holography involves the recording of a wave interference pattern, any process involving waves, such as sound waves or electromagnetic microwaves, can utilize the technique. Accordingly, numerous acoustical holography experiments have been made, and they have shown promise in the medical field (for locating cysts and tumors with x-ray-like acoustic pictures), in under-water sound (for detecting underwater objects), and as a supplement to present seismic procedures in locating offshore oil deposits. *See* HOLOGRAPHY; INTERFERENCE OF WAVES.

Liquid surface holography. Since a hologram is a photographic record of the interference pattern generated between a set of waves of interest and a set of reference waves, it is possible to make holograms using sound waves, provided the wave interference pattern can be formed.

One procedure which permits the real-time viewing of acoustic holograms, and thus bypasses the photographic recording process required in optical holography, uses two underwater quartz-crystal transducer sound sources (Fig. 1). The sound waves of interest (those originating at the object transducer and passing through the object) and the reference waves (those originating at the reference transducer) generate an interference pattern at the surface of the liquid, causing it to have extremely minute, stationary ripples (tiny ridges). These mechanical deformations of the liquid surface correspond exactly to the interference fringes of a hologram, so that immediate reconstruction of the acoustic hologram image is accomplished by causing laser light to be diffracted by these ridges as it is reflected off the liquid surface. This technique, as shown in Fig. 1, differs from the earlier reflection or sonar techniques, such as the procedure known as B-scan, in that it provides through-transmission images comparable to the images formed when x-rays are transmitted through the object. *See* DIFFRACTION.

Many parts of the human body transmit high-frequency sound waves in a manner different from x-rays. Accordingly, the application of hologram techniques as an alternative to x-rays has been examined, with the most rapid development occurring in liquid surface holography. Such equipment

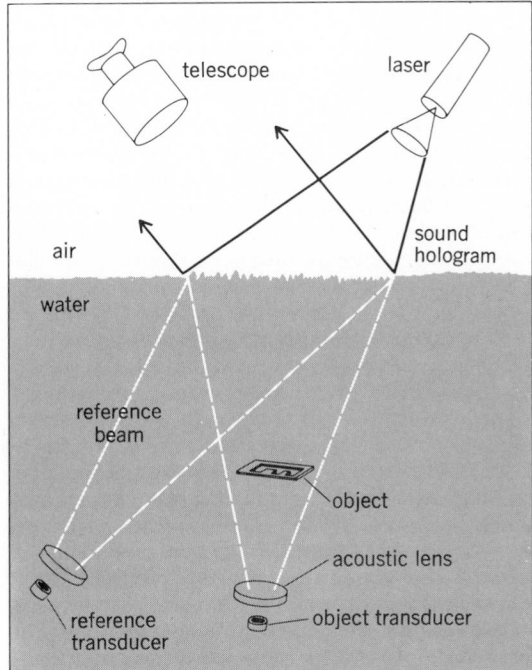

Fig. 1. Liquid-surface through-transmission holography system.

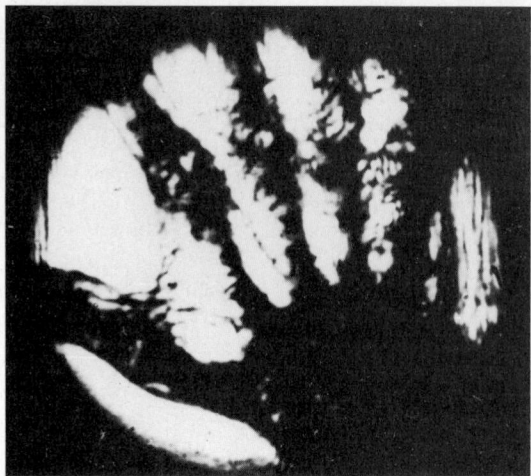

Fig. 2. A through-transmission, reconstructed, acoustical holographic image of a human hand (*Holosonics, Inc.*)

is now available commercially. The real-time through-transmission record obtained when a person's hand is immersed in a liquid and is made the object is shown in Fig. 2. Acoustical holography provides a differentiation between the various types of tissues, including delineation of the blood vessels, without any injection of dyes, as is required for x-rays. This ability to acoustically delineate soft tissue details has led to a fairly wide use of acoustical holography in the diagnosis of ordinary tumors. In particular, the small tumors and cysts occurring in the female breast can be detected by the through-transmission hologram technique.

Synthetic aperture holography. The synthetic aperture technique was first employed in radar. In it, a radiator emits successive pulses of a highly coherent (very-constant-frequency) signal. Because of this high coherence, the many echoes that return can be processed as though a single line-radiator having a length equal to the path had been used. The effective radiator length is thus quite large, and this large "synthetic" aperture provides records having extremely fine detail. This imaging technique has proved to be a very important addition to radar technology, and accordingly, extensive investigations have been undertaken to exploit its possibilities in various acoustical holography fields, including sonar, the seismic technologies, and the medical sonar technique called B-scan.

When synthetic aperture techniques are used in medical systems, they provide greater detail in body areas located near the acoustic transducer. Also, a number of acoustic planigram B-scan records (sectional images called tomograms) have been used to successively record, on the same hologram, planes at different depths (for example, in a series of different depth mammograms). Reconstruction of these optical holograms provides a three-dimensional image of the body volume under examination.

Holographic sonar. Sonar technology employing acoustics for detecting underwater objects has also benefited from acoustical holography developments. A holographic array of 400 acoustic transducers has been used in such a sonar, and its success resulted in a later version containing 5000

elements. The hologram synthetic aperture technique discussed above also appears to be valuable in sonar. Tests have shown that a sonar synthetic aperture having a dimension of at least ½ nautical mile (0.9 km) at 400 Hz is possible, corresponding to a coherence time of 7½ min. The ability of hologram sonars to delineate nearby (near-field) objects is one of their important advantages.

Geophysical exploration. Geophysical prospecting often employs acoustic waves which are sent into the earth, with the returning echoes then analyzed to appraise the likelihood of oil or gas being present in the substructure. Here too acoustical holography has proved useful, permitting special techniques (such as partial filtering) to suppress, in the records, artifacts which would otherwise obscure the desired information.

In the technique for locating offshore oil deposits, a cable which in practice would be 100 wavelengths or more in length is towed behind a ship equipped with a high-power transmitter emitting low-frequency coherent acoustic energy (Fig. 3).

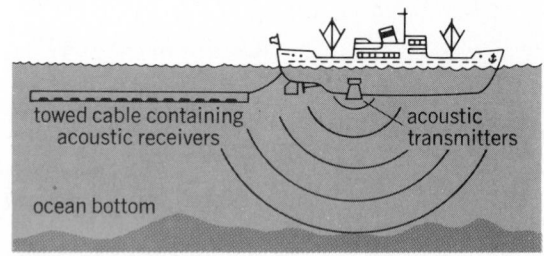

towed cable containing
acoustic receivers

acoustic
transmitters

ocean bottom

Fig. 3. Underwater viewing by acoustical holography.

Signals reflected or scattered from the ocean bottom and from geological layers below are picked up by the cable array, and holographic processing of the seismic data obtained permits the retrieval of useful information. One seismic holography test was carried out in the Gulf of Mexico over a known and well-defined salt dome having a diameter of approximately 2 mi (3.2 km), with the top of the dome being at a depth of 1156 ft (352 m). By using a phase quadrature technique, the exact outline of the dome in its expected position was observed in the hologram reconstruction. *See* SOUND; ULTRASONICS.

[WINSTON E. KOCK]

Bibliography: E. J. Barrakette et al. (eds.) *Optical Processing*, vol. 2, 1978; W. E. Kock, *The Creative Engineer*, 1978; W. E. Kock, *Engineering Application of Lasers and Holography*, 2d ed., 1977.

Acoustical image

If a point source of sound is placed on one side of an extended reflecting surface, it may be considered to have an "image" at an equivalent distance on the other side of the surface along a perpendicular projection, analogous to the familiar optical image. The effects of reflecting surfaces on acoustic waves frequently can be predicted by the use of such images. A source in front of a very reflective wall (for example, plaster, concrete, and masonry reflect 97–99% of incident sound energy) will have an image of "strength" almost equal to that of the source, vibrating in phase with the source. To obtain the total effect at any point due to the combined action of a source and such a reflecting surface, the effects due to the source itself and another source of equal strength placed at the image point are added. If there is a second reflecting surface present, this so-called first-order image will have an image, called the second-order image of the source, at an equivalent distance on the other side of the second reflecting surface. Similarly, an image of the second-order image is said to be a third-order image, and so on. Although this method is precise only when the reflecting surface is non-absorptive, it is often of considerable help in investigations of the action of sound waves in rooms. *See* OPTICAL IMAGE.

[CYRIL M. HARRIS]

Acoustics

The science of sound, which in its most general form endeavors to describe and interpret the phenomena associated with motional disturbances from equilibrium of elastic media. An elastic medium is one such that if any part of it is displaced from its original position with respect to the rest, as for example by an impact, it will return to its original state when the disturbing influence is removed. Acoustics was originally limited to the human experience produced by the stimulation of the human ear by sound incident from the surrounding air. Modern acoustics, however, deals with all sorts of sounds which have no relation to the human ear, for example, seismological disturbances and ultrasonics.

Basic acoustics may be divided into three branches, namely, production, transmission, and detection of sound.

Sound production. Any change of stress or pressure producing a local change in density or a local displacement from equilibrium in an elastic medium can serve as a source of sound. The human vocal mechanism is an obvious example. Other illustrations are provided by struck solids (as a drum, or violin or piano string), flow of air in a jet, and underwater explosions.

Modern acoustics as a science and technology has experienced enormous development by the invention of sources of sound which can be precisely controlled with respect to both frequency and intensity. The most important of these are called acoustic transducers, namely, devices by which any form of energy can be transformed into sound energy and vice versa. The most useful type is the electroacoustic transducer, in which electrical energy is transformed into the mechanical energy of sound. An example is the electrodynamic loudspeaker used in sound-reproducing systems, public address systems, and radio and television. Piezoelectric and magnetostrictive transducers are widely used in scientific and industrial applications of acoustics. An example of a nonelectric acoustic transducer is the siren, in which interrupted fluid flow results in the production of sound.

Sound transmission. Transmission of sound takes place through an elastic medium by means of wave motion. A wave is the motion through the medium of a disturbance as distinguished from the motion of the medium as a whole. An obvious example is a surface wave on water. Most sound waves of importance are transmitted compressional disturbances, that is, disturbances in which the pressure or density at any point in the medium is caused to vary from its equilibrium value. The change is then propagated through the medium with a velocity which depends on the type of wave in question, the nature of the medium, and the temperature. For example, the velocity of sound in still air of normal atmospheric composition at 0°C is 331.45 m/s. Another important property of sound transmission is intensity, measured by the average rate of flow of energy in the wave per unit of time and per unit area perpendicular to the direction of propagation. The intensity of all practical sound waves diminishes with distance from the source, a property known as attenuation. This is due either to the spreading of the sound-wave energy over larger and larger surfaces or to actual absorption of the energy by the medium with transformation into heat.

The most important sound waves are harmonic waves, defined as waves for which the propagated disturbance at any point in its path varies sinusoidally with time with a definite frequency or number of complete cycles per second (the unit being the hertz). For a harmonic wave the disturbance at any instant repeats itself in magnitude and phase at intervals along the direction of propagation equal to the so-called wavelength, which is, therefore, equal to the sound velocity divided by the frequency.

Acoustics deals with waves of all frequencies, but not all frequencies are audible by human beings, for whom the average range of audibility extends from 20 to 20,000 Hz. Sound below 20 Hz is referred to as infrasonic, whereas that above 20,000 Hz is ultrasonic.

Sound detection. The detection of sound is made possible by the incidence of transmitted sound energy or an appropriate acoustic transducer. For human beings with so-called normal hearing, the most important transducer is the ear, a remarkably sensitive organ able to detect a sound intensity as low as 10^{-16} watt per square centimeter. For modern applied acoustics, transducers such as the microphone, based on the piezoelectric effect, are widely used. Generally speaking, any transducer used as a source of sound is also available as a detector, though the sensitivity varies considerably with the type.

Applications. The practical applications of acoustics are multifarious. They include architectural acoustics, or the study of sound waves in closed rooms arranged for the satisfactory production and reception of speech and music; underwater acoustics, dealing specifically with all aspects of sound in the sea and its use for the detection of vessels and the exploration of the sea bed; engineering acoustics, including the whole technology of sound reproduction and recording, sound motion pictures, and radio and television, as well as the study of vibrations of solids and their control and the use of high-intensity ultrasonics in industrial processing (for example, in metallurgy). Noise control is also an important case of engineering acoustics. Further examples are provided by physiological and psychological acoustics, dealing with hearing in humans and animals. Communication acoustics considers the production and transmission of speech. Musical acoustics deals with the physics of musical instruments. Bioacoustics and medical acoustics take up the use of sound in medical diagnosis and therapy as well as its use in

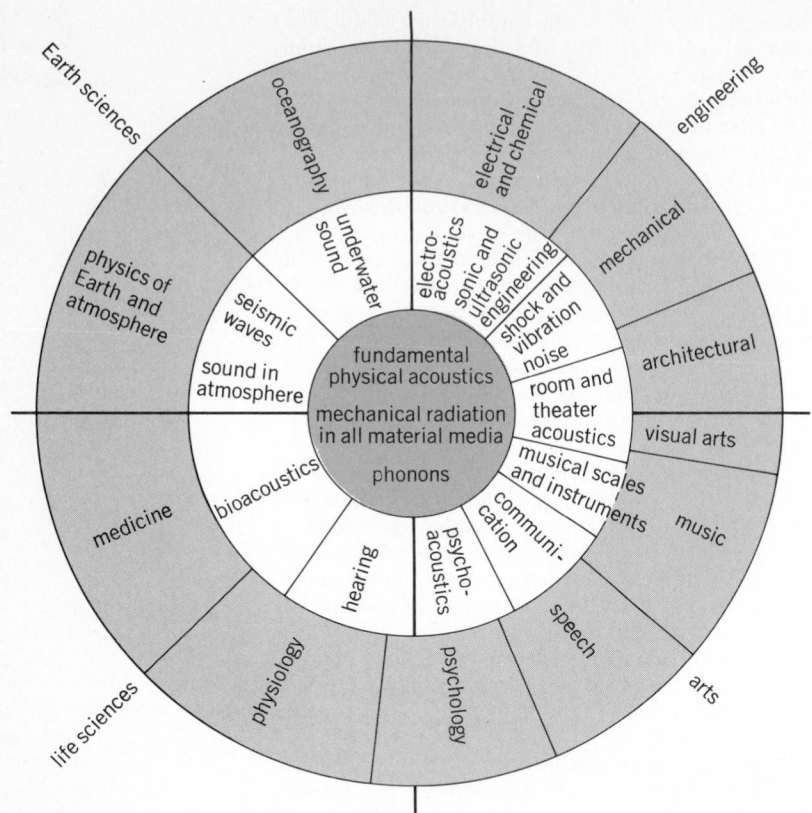

Relations among the various branches of acoustics and related fields of science and technology.

the study of the overall behavior of animals in general.

All applications of acoustics are based on the fundamental physical properties of sound waves. The extensive interdisciplinary range of acoustics as a science and technology is exhibited in the accompanying illustration. *See* ATMOSPHERIC ACOUSTICS; HYPERSONICS; MECHANICAL VIBRATION; SOUND; SOUND ABSORPTION; ULTRASONICS; VIBRATION.

[R. BRUCE LINDSAY]

Acoustooptics

That field of scientific investigation dealing with the interaction of acoustical and optical energy. The interaction usually occurs in a medium capable of supporting the propagation of both forms of energy, that is, one which is reasonably transparent to both. The chief acoustooptic effect is the change of the optical index of refraction of the material under the influence of strain caused by the acoustic wave. As the pressure on a material changes, so does its density and thus its index of refraction. *See* LIGHT; WAVE MOTION.

Although most acoustooptic effects were described and observed before 1935, their practical application awaited their successful union with lasers.

Magnitude of effect. In general, a small change Δn in the index of refraction n is proportional to $n^2 \Delta \rho$, where $\Delta \rho$ is the change in the density ρ. The relationship between the initial stress or pressure and the resulting changes in the index of refraction defines the photoelastic coefficient p. It can be shown that Δn equals the acoustic strain multi-

plied by $-n^3 p/2$. The ability of a low-intensity acoustic wave to interact with a light beam passing through it is proportional to the acoustic power and to the quantity $M = n^6 p^2 / v^3 \rho$, where v is the acoustic propagation velocity in the material. M is often referred to as a figure of merit since it depends on the properties of the medium alone. The properties which have the greatest influence on the effectiveness of acoustooptical materials are n and v. Liquids, in which v is relatively low, make good acoustooptical materials. However, due to the rapid increase in acoustic attenuation with frequency, liquids typically are useful only up to about 50 MHz. At the higher frequencies, solid materials, particularly those with high indexes of refraction, are used. Some crystalline materials such as sapphire and lithium tantalate have been used well into the microwave region, even at room temperature. *See* HYPERSONICS; ULTRASONICS.

Types of effect. In the interaction of a light and a sound beam, where each is reasonably well collimated and monochromatic, and they cross each other at right angles in an acoustooptical material, let the diameter of the light beam be small compared to the wavelength of the sound (its velocity v divided by its frequency). Under these conditions the light beam is alternately bent (refracted) back and forth from its original direction as the peaks and valleys of the acoustic pressure wave pass through the light beam. The acoustic wave acts as a time-varying prism for the light passing though it. This effect has been used in optical deflectors and control devices for lasers. It is useful only at relatively low frequencies, less than about 500 kHz. As the frequency increases and the wavelength becomes comparable to the diameter, the acoustic disturbance appears as a time-varying lens, alternately converging and diverging the light beam. This effect has also been used to control certain laser systems. *See* REFRACTION OF WAVES.

When the sound wavelength has decreased well below the diameter of the light beam, the light now must interact simultaneously with a series of periodic variations in the index of refraction as the sound wave propagates through it. The light beam is diffracted by this "acoustic diffraction grating" into one or more additional light beams propagating in new directions angularly displaced from the initial direction. An analogy can be made to the usual type of optical diffraction grating (consisting of a periodic array of slits), in which the angular displacement of the diffracted beams is proportional to the optical wavelength divided by the slit spacing. For acoustically diffracted light beams, the angular deflection is proportional to the optical wavelength divided by the acoustic wavelength. In this case, however, the "grating" is moving, and therefore causes a Doppler shift in the optical frequency of the diffracted light beams. This shift in optical frequency is, in most cases, exactly equal to the acoustic frequency. The intensity of the diffracted light beams is usually proportional to the intensity of the sound. *See* DIFFRACTION; DOPPLER EFFECT.

Applications. The ability of sound to control the amplitude, direction, and frequency of a light beam (particularly a laser beam) has led to a great variety of practical devices. The well-defined frequency and direction of a laser are used to full advantage in acoustooptical devices; and these devices are being increasingly used in image scanners, recorders, and projectors, as well as in a large number of signal-processing devices and laser controllers. *See* LASER.

Production of sound. In addition to the effect of acoustic disturbances on optical beams, the opposite effect occurs. That is, two light beams can be made to interact in an acoustooptical material to produce a sound wave. This experiment has been made possible only by the availability of very-high-power, monochromatic laser beams.

[MICHAEL J. BRIENZA]

Bibliography: R. Adler, Interaction between light and sound, *IEEE Spectrum*, 4:42–54, 1967; R. Y. Chiao, R. Y. Townes, and B. P. Stoiehoff, Stimulated Brillouin scattering and coherent generation of intense hypersound waves, *Phys. Rev. Lett.*, 12:592–595, 1964; E. U. Condon and H. Odishaw, *Handbook of Physics*, 2d ed., 1967; P. Debye and F. W. Sears, On the scattering of light by supersonic waves, *Proc. Nat. Acad. Sci.*, 18: 409–421, 1932; E. I. Gordon, A review of acoustooptical deflection and modulation devices, *Proc. IEEE*, 54:1391–1401, 1966; R. Lucas and P. Biquard, Optical properties of solids and liquids under ultrasonic vibrations, *J. Phys. Rad.*, 7th ser., 3:464–477, 1932; C. F. Quate, C. D. W. Wilkenson, and D. K. Winslow, Interaction of light and microwave sound, *Proc. IEEE*, 53:1604–1623, 1965; M. Ross, *Laser Applications*, vol. 2, 1974, vol. 3, 1977.

Action

An integral quantity associated with each possible motion of a system of particles, of fields, or both. For a particle system the action is given by Eq. (1),

$$S = \int \sum_{j=1}^{N} p_j(t)\dot{q}_j(t)\, dt \qquad (1)$$

where the q_js are generalized coordinates and the p_js their conjugate momenta. *See* HAMILTON-JACOBI THEORY; HAMILTON'S PRINCIPLE.

The action permits the following formulation of the dynamical equations, which is similar to but more restrictive than Hamilton's principle: Of all possible motions with a given energy from the initial values q_j to the final values $q_j{}'$, the dynamical motion is that for which the action is stationary (not always a minimum despite the name "principle of least action").

Hamilton's principle is sometimes loosely called an action principle also, especially in formulations of the field equations of classical and quantum field theory.

When the motion of a degree of freedom q_j, p_j is periodic, the separate integral shown in Eq. (2)

$$\int p_j \dot{q}_j\, dt = \int p_j\, dq_j \qquad (2)$$

holds. Taken over a period of the motion, it is called the action variable J_j. Action variables are useful as adiabatic invariants (quantities insensitive to slow variations in the external parameters of the system). This property was exploited in early quantum theory by assigning to the action variables fixed "quantized" values, namely, integral multiples of Planck's constant. *See* QUANTUM MECHANICS.

Action variables are likewise convenient in describing the behavior of charged particles in a magnetic field such as in a particle accelerator or in a plasma. In the latter case, for example, the J associated with the transverse motion around the magnetic field lines corresponds to a nearly constant magnetic moment opposing the field, with the result that the particles are repelled from strong-field regions and can be trapped in regions of low field. *See* PLASMA PHYSICS. [BERNARD GOODMAN]

Adiabatic demagnetization

The removal or diminution of a magnetic field applied to a magnetic substance when the latter has been thermally isolated from its surroundings. The process concerns paramagnetic substances almost exclusively, in which case a drop in temperature of the working substance is produced (magnetic cooling). *See* PARAMAGNETISM.

Thermodynamic principles. When a specimen is magnetized in a magnetic field **H** and a magnetic moment **M** is produced, there results a contribution to the energy of the specimen of magnitude $-\mathbf{M} \cdot \mathbf{H}$. By restricting the discussion to isotropic media for simplicity, and taking the ambient pressure to be modest and the compressibility of the substance to be very small, the thermodynamics of the situation may be taken over from the conventional results entirely by substituting $-MH$ for $+PV$ (where P is the pressure and V is the volume) everywhere. One particular result is the relation $(\partial T/\partial H)_S = (\partial M/\partial S)_H = -(T/C_H)(\partial M/\partial T)_H$, where T is the temperature, S is the entropy, and C_H is the heat capacity at constant magnetic field. As C_H is always positive, the sign of the magnetocaloric effect is determined by (and is opposite to the sign of) the quantity $(\partial M/\partial T)_H$. Where this latter is negative, as is the case for paramagnetics, $(\partial T/\partial H)_S$ is positive and magnetization produces heating, and vice versa. The maximum theoretical effect will not be achieved in practice unless the process is truly isentropic. In addition, it is obviously desirable that nonmagnetic contributions to C_H be kept small. In nearly all cases, this requires restricting the process to a region far below room temperature, where the lattice specific heat is very small in comparison with that of the assembly of magnetic ions (the spin system). Adiabatic demagnetization was employed from 1933 on to produce temperatures below those readily obtainable by using only liquid helium (that is, below 1 K). *See* LIQUID HELIUM; MAGNETOCALORIC EFFECT.

The process is most perspicuously discussed in terms of entropy rather than energy or heat capacity. The entropy S of a paramagnetic substance in zero magnetic field is depicted schematically in the illustration. The atomic magnets (electronic or nuclear), of moment μ, interact weakly with each other, but at sufficiently low temperatures ($T \sim U/k$, where k is Boltzmann's constant) this interaction energy U will restrict the motions of the atomic magnets and cause a drop in S, as seen at the left of the diagram. Well above this temperature region, the orientation of μ is practically random over the possible $(2J+1)$ orientations (where J is the quantum number appropriate to the situation under discussion), and the "infinte temperature" entropy of $R \cdot \ln (2J+1)$ per mole is approached, where R is the gas constant. At still higher temperatures, the lattice entropy begins to rise to nonnegligible values. Consider a starting condition represented by the point A, the tempera-

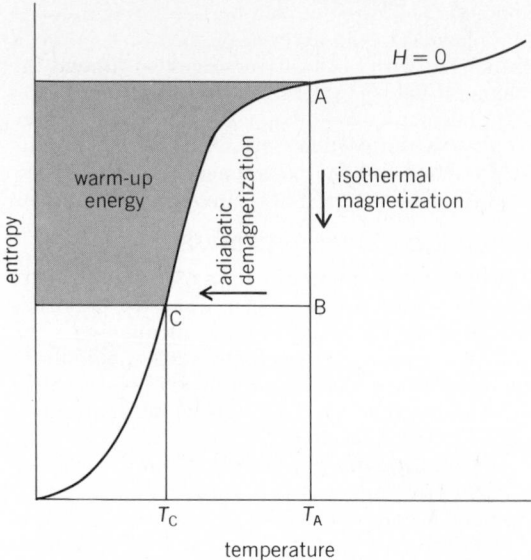

Temperature-entropy diagram of a paramagnetic substance, showing process of adiabatic demagnetization.

ture being T_A. Application of a magnetic field isothermally will reduce the entropy to a value corresponding to point B, say, governed by the thermodynamic relation $(\partial S/\partial H)_T = (\partial M/\partial T)_H$. (The energy which is released by the magnetization process is carried away by exchange gas, helium at low pressure, or a heat switch.) After reaching B, the substance is isolated from its surroundings and H is reduced to zero. Now, no energy exchange with the surroundings can occur, the system returns to the zero-field entropy curve along an isentrope (horizontal line BC), and it arrives at point C, where the temperature T_C is much lower than T_A.

Limiting temperatures. The thermodynamic properties are found to be functions of the quantity $(\mu H/kT)$. One may think of the situation as a competition between the magnetic (restraining) energy μH and the thermal (disruptive) energy kT. If one considers the interaction U as arising from an internal magnetic field h, then h at a temperature T_C and H at T_A are equally effective in holding the system at the diminished entropy represented by CB. Hence, given that S and, therefore, $(\mu H/kT)$ are constant, one may write (for a rough approximation) $\mu H/T_A = U/T_C = \mu h/T_C$ and $T_C/T_A = U/\mu H = h/H$. In order to produce a very low temperature, T_C, one requires h and T_A to be small and H large. Traditionally, low-temperature physicists have employed electronic paramagnets (paramagnetic salts) to reach limiting temperatures of 2 to 30 mK with $T_A \sim 1\ K$ and H lying in the range 10 to 60 kilooersteds or 0.8 to 4.8 MA/m (magnetic inductions of 10−60 kilogauss or 1−6 teslas). The lower the final temperature T_C, the more difficult it is to maintain the sample in that very-low-temperature region, given a fixed background heat influx. The "warm-up energy," represented in the illustration, will be obviously very small if the steep fall in S occurs at a very low temperature. Compromises may be effected by working with different salts or, if feasible, carrying out the adiabatic demagnetization to a nonzero final value of H, which results in a

higher final temperature but an enhanced heat capacity. *See* ENTROPY.

Nuclear adiabatic demagnetization. Nuclear magnetic moments are one or two thousand times smaller than their ionic (that is, electronic) counterparts, and the characteristic temperature U/k of their mutual interaction lies in the microkelvin rather than millikelvin region. Successful experiments in nuclear adiabatic demagnetization date from the mid-1950s. First, as will be readily deduced from the above discussion, the total thermal energy associated with a loss (or restoration) of a fixed amount of entropy at microkelvin temperatures is very small; hence special precautions for thermal isolation are required. It has, in fact, proved possible to routinely reduce heat leaks from the 1 erg s^{-1} (10^{-7} J s^{-1}) level to 1 erg min^{-1} (1.7×10^{-9} J s^{-1}) in a typical ensemble. Second, as the entropy leverage in the basic process is a function of $\mu H/T_A$, one must utilize very intense magnetic fields or very low starting temperatures (or a combination of both) to compensate for the thousandfold diminution in μ. As there are severe practical limits to substantially increasing H − a maximum induction of 15 teslas, say, using modern superconducting-solenoid technology − salvation is mainly sought in reducing T_A. A factor of 100 may be gained here, that is, from 1 K down to 0.01 K; the latter may be reached by a conventional electronic stage in a two-stage process or by other modern techniques, such as Pomeranchuk cooling or ^{3}He-^{4}He dilution refrigeration.

In moving from electronic to nuclear magnetic cooling, many difficulties and subtle features of design must be dealt with. Thermal conductivities, heat transfer coefficients, intersystem relaxation times, and irreversibilities must be understood and taken into account. (It may, for example, require several hours, rather than minutes, to remove the initial heat of magnetization.) Refinements have also entered through the utilization of the very intense, naturally occurring magnetic fields which permit one to substitute more subtle and more powerful techniques for the straightforward "brute force" technique discussed so far. Through the nuclear hyperfine interaction, the atomic nuclear moments may be subject to enormous magnetic fields arising from the same (or neighboring) atoms' electronic moments. In general, these hyperfine fields are constant and thus cannot be used for magnetic cooling. In certain systems, however, the electronic magnetic moment is suppressed, or quenched, by crystal electric field effects, and may be partially restored by an external magnetic field, as a result of quantum-mechanical "mixing" of excited states and the ground state of the atom. This induced hyperfine field may correspond to an enhancement of the external magnetic field by a factor of about 10 in praseodymium compounds and about 100 in thulium compounds, for example. The fact that these substances are metals also has practical advantages in energy transmission at very low temperatures. *See* ABSOLUTE ZERO; CRYOGENICS; LOW-TEMPERATURE PHYSICS.

[RALPH P. HUDSON]

Bibliography: R. P. Hudson, *Principles and Application of Magnetic Cooling,* 1972; O. V. Lounasmaa, *Experimental Principles and Methods below 1K,* 1974.

Adiabatic process

A thermodynamic process in which the system undergoing the change exchanges no heat with its surroundings. Reversible adiabatic processes are also isentropic; that is, they take place with no change in entropy. *See* ENTROPY; ISENTROPIC PROCESS.

By the first law of thermodynamics the change of the internal energy U in any process is equal to the sum of the heat Q gained and the work W done on the system. For an adiabatic process the change in internal energy when the system goes from state 1 to state 2 is equal to the external work performed *on* the system (which brings about the change). In the equation below, if U_2 is less than U_1, then W is negative and $-W$ is the work done *by* the system.

$$U_2 - U_1 = W$$

The events inside an engine cylinder are nearly adiabatic because the wide fluctuations in temperature take place rapidly compared to the speed with which the cylinder surfaces can conduct heat. Similarly, fluid flow through a nozzle may be so rapid that negligible exchange of heat between fluid and nozzle takes place. The compressions and rarefactions of a sound wave are rapid enough to be considered adiabatic. *See* THERMODYNAMIC PROCESSES.   [PHILIP E. BLOOMFIELD]

Admittance

The reciprocal of the impedance of an electric circuit. Admittance is expressed in the unit mho, coined from the inverse spelling of ohm, the unit of impedance. Admittance is used primarily in computations of parallel alternating-current circuits.

By using admittance Y, current I can be expressed as $I = EY$, where E is the voltage across the impedance Z. In terms of complex quantities where R is the total circuit resistance, X the total circuit reactance, G the conductance, and B the susceptance. However, these are not simple conductance $(1/R)$ and susceptance $(1/X)$. As seen from the equation, both G and B are combinations

$$Y = \frac{1}{Z} = \frac{1}{R \pm jX} = \frac{R}{R^2 + X^2} \mp j\frac{X}{R^2 + X^2} = G \pm jB$$

of resistance and reactance. *See* CONDUCTANCE; SUSCEPTANCE.   [BURTIS L. ROBERTSON]

Aerodynamics

The branch of aeromechanics dealing with the properties and characteristics of, and the forces exerted by, air and other gases in motion. The field of aerodynamics includes the science of a gas itself in motion and the science of bodies immersed in a gas between which there exists a relative motion. *See* GAS DYNAMICS.

Aerodynamics is a broad field with numerous specializations and applications, some of which extend into apparently unrelated fields of science and engineering. Perhaps the most frequently practiced function of the aerodynamicist is the analysis of the forces and moments exerted on a solid body in motion through the air.

Of fundamental significance in the term aerodynamics is the prefix, aero, which refers to the air of the Earth's atmosphere. Until man can

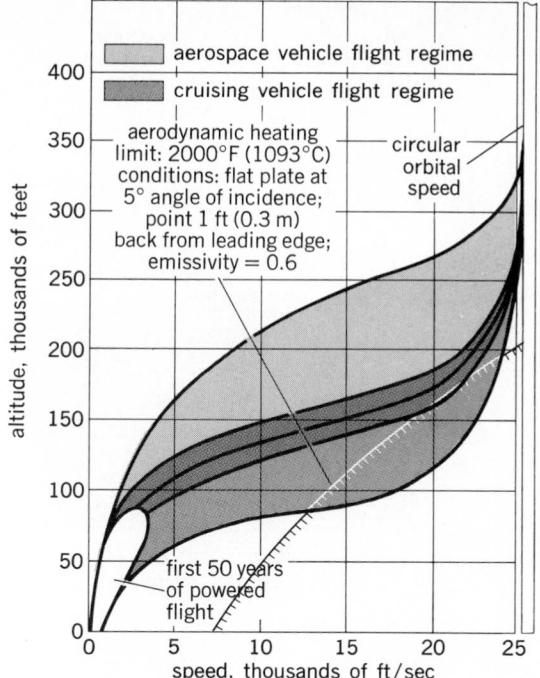

Fig. 1. Practical limits of aerodynamic flight within the atmosphere. 1 ft = 0.3048 m. (*Aero/Space Eng.*)

achieve flight within the atmospheres of other planets, the limits of aerodynamic flight and the majority of practical considerations will be confined to the limits of the Earth's atmosphere as defined in aerodynamic terms.

Figure 1 presents the practical limits of aerodynamic flight within the Earth's atmosphere based on a quantitative analysis of the governing factors. To a large extent the significant aerodynamic reactions of missiles in passing through the atmosphere also occur below the upper boundary shown in Fig. 1.

Two overlapping flight regimes are shown. The upper regime, defined as the aerospace-vehicle flight regime, indicates the operating region of reentry vehicles and space vehicles which use aerodynamic lift during their descent through the atmosphere. Its upper and lower boundaries are defined by a wing loading of 20 lb/ft² (1.0 kPa) and 40 lb/ft² (1.9 kPa), respectively. The lower flight regime in Fig. 1 is defined as the cruising-vehicle flight regime (sometimes referred to as the corridor of continuous flight). The upper and lower boundaries here are 10 lb/ft² (0.5 kPa) and 200 lb/ft² (10 kPa), respectively. The portion of this flight regime penetrated in the first 50 years of powered flight is also indicated.

The aerodynamic heating limit cuts off access to a large portion of the cruising flight regime between 12,000 and 24,000 ft/sec (3.7 and 7.3 km/sec). Actually, this temperature represents the approximate upper extreme of human engineering ability to penetrate what is sometimes called the thermal thicket; it cannot be accurately described as a barrier.

Strictly ballistic reentry vehicles could not reach the surface of the Earth wholly within the boundaries of the flight regimes just defined. The ballistic path of reentry in Fig. 1 would pass through

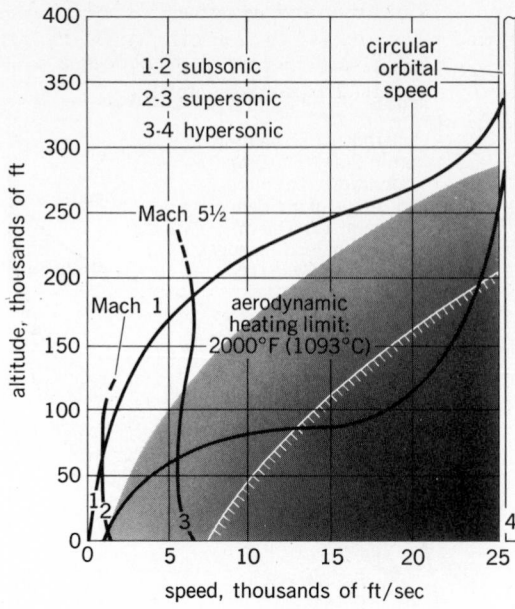

Fig. 2. The subsonic, supersonic, and hypersonic flight regimes within the atmosphere. 1 ft = 0.3048 m.

both regimes almost vertically at high speed, the greatest reduction in velocity occurring in the denser atmosphere below the cruising-vehicle flight regime. As a result, ballistic reentry vehicles suffer far greater extremes of aerodynamic heating during reentry as a result of air friction than do aerodynamic reentry vehicles.

Figure 2 is a plot similar to Fig. 1 except that the flight envelopes previously described have been subdivided into three speed regions: subsonic, supersonic, and hypersonic. The aerodynamic heating region or thermal thicket is shaded, the shades deepening with rising temperature.

At velocities below the speed of sound (Mach 1) air may be considered to be incompressible. The laws governing flow phenomena in this speed range comprise subsonic aerodynamics. Between Mach 1 and approximately Mach 5.5 is the region of supersonic aerodynamics. Supersonics differs markedly from subsonics because the air becomes compressible. The narrow transition region between subsonic and supersonic flight is characterized by a rapid drag rise often referred to as the sound barrier. *See* SONIC BARRIER.

Above a speed of about Mach 5.5 another branch of aerodynamics (hypersonics) has been defined to categorize phenomena which differ markedly from supersonic flow. At hypersonic speeds, properties of the gaseous medium in which a vehicle is immersed begin to differ significantly from those of air at lower speeds. High-temperature gas phenomena must now be included in the aerodynamic analysis.

[W. C. WALTER]

Bibliography: J. D. Anderson, Jr., *Introduction to Flight*, 1978; A. M. Kuethy and C. -Y. Chow, *Foundations of Aerodynamics*, 1976.

Aeromechanics

The science of air and other gases in motion or equilibrium. Aeromechanics has two branches, aerostatics and aerodynamics. Aeromechanics is a special case of the more general field of fluid mechanics, the science of fluids in motion or equilibrium. *See* FLUID MECHANICS.

Aerostatics is the branch of aeromechanics dealing with the equilibrium of air or other gases, and also with the equilibrium of bodies immersed in a gaseous medium. Examples of aerostatic phenomena are air being compressed in a closed container and the behavior of a dirigible or balloon. *See* AEROSTATICS.

Aerodynamics is the branch of aeromechanics dealing with the properties and characteristics of, and the forces exerted by, air and other gases in motion. The resistance and pressure of air flowing through a duct such as a wind tunnel, and the forces exerted by airflow over an airfoil-shaped compressor blade in a turbojet engine are aerodynamic in nature. *See* AERODYNAMICS.

[WILLIAM C. WALTER]

Aerostatics

The science of the equilibrium of gases and of solid bodies immersed in them when under the influence only of natural gravitational forces. Aerostatics is concerned with the balance between the weight of the gases and the weight of any object within them. Archimedes' law that an immersed body experiences a buoyancy force equal to the weight of the fluid displaced is the principal law of aerostatics, if the fluid is air, or of hydrostatics, if the fluid is water. Some phases of meteorology and the flight of balloons and dirigibles are based on aerostatics. In meteorology cloud and fog subsidence and simple pressure and temperature relations with altitude are predicted from aerostatic principles.

Strictly speaking, the air and the immersed body must be at rest for aerostatic principles to apply, but there are many problems where aerostatic forces essentially govern despite some movement. A convenient example of this is given by the motion of a dirigible through the air. Aerodynamic force (drag) limits the speed which the dirigible can achieve, yet the aerostatic forces essentially support the vehicle. This contrasts with the airplane, where aerodynamic forces provide both the lift and the drag. Another example is given by the atmosphere, where the pressure and density relations are determined to a first order by aerostatics, although some motion of the atmosphere takes place through winds and turbulence. *See* FLUID STATICS; HYDROSTATICS. [JOHN R. SELLARS]

Algebra

Classical algebra is a generalization of arithmetic, made possible by the use of symbols, usually letters such as x, y, a, b, α and β, for unknown numbers. The principal instrument for solving problems is the equation. The answers sought are denoted by letters, and statements in the form of equations are deduced about the quantities in a problem. To solve one or more equations, the value or values must be found which, when substituted for the unknowns, make the equation or equations true. Equations are said to be satisfied by such values. Most of the great developments in algebra were inspired and guided by the search for methods of solving equations, which is the central problem in algebra.

The extension of the number system from the whole numbers and fractions of arithmetic to a complete algebraic number system was influenced by the need for numbers to satisy equations. Thus, the equations $x^2 - 2 = 0$ and $x^2 + 1 = 0$ have no solutions among whole numbers and fractions. Even the extended number system does not provide solutions for all equations. And even though solutions can be shown to exist, they may be impossible to find. For example, these is no general method for finding the solutions of all fifth-degree equations.

Algebra is also concerned with inequalities, such as $x^2 + y^2 > 2xy$; with arrangements, such as the number of ways a set of elements can be chosen from a larger set (combinations), or the number of ways a set can be ordered (permutations); with the study of special sets of numbers such as the set of coefficients in the expansion of a binomial like $(a + b)^n$; and with manipulations of elements such as matrices, which have properties similar to, but different from, those of numbers. In modern algebra the relationship to arithmetic seems more remote than in classical algebra and the elements are often combined by rules at variance with those which stemmed from the arithmetic of experience.

Algebraic operations and expressions. The operations of arithmetic – addition, subtraction, multiplication, division, raising to powers, and extracting roots – are the algebraic operations. Applications of these operations to symbols for numbers, known and unknown, produce various kinds of algebraic expressions. A term is formed by multiplication: $4x^2y$. In a term the known multiplier of the literal part is the coefficient. The degree of the term is the sum of the exponents of unknowns. The degree of $4x^2y$ is 3. By multiplication and addition (which includes subtraction because of the rules for combining signed numbers), polynomials, or rational integral expressions, can be constructed. Polynomials are classified according to the number of terms into monomials, binomials, trinomials, and other multinomials. In polynomials the coefficients can be any fixed numbers:

$$5x^2y - 6y^3 + \tfrac{2}{3}x \qquad 3x^4 - \sqrt{2}x^2 + \tfrac{5}{3}$$

A fractional expression permits division by unknowns: $(x^2 + 1) / (x + 1)$. An expression is irrational if it denotes root extraction of an expression involving an inknown.

Whereas the operations in arithmetic are derived from practical experience, their generalizations in algebra accentuate the necessity of axioms. These axioms govern all operations.

Axioms. Algebra depends first upon the axioms of natural numbers (whole numbers of counting):

1. There exists a natural number 1.
2. Each number a has a successor $a+$
3. There is no natural number having 1 as its successor.
4. For each natural number a different from 1 there is precisely one number of which a is the successor.
5. If a set of natural numbers contains the number 1, and if for each number a in the set the successor $a+$ is in the set, then the set contains all natural numbers.

Definitions of sums and products are based on these properties. In order to provide differences in all cases, zero (0) and negative numbers are annexed to the system. The rules of signs which govern operations with negative numbers are as follows: If a and b are positive numbers,

1. $(-a) + (-b) = -(a + b)$
2. $-a + b = b - a = -(a - b)$
3. $(-a) \cdot b = -ab$
4. $(-a) \cdot (-b) = ab$

Ratios, quotients, or fractions are defined by ordered pairs,

$$a/b \quad \text{or} \quad \frac{a}{b}$$

subject to the restriction $b \neq 0$. Sums, products, and quotients of ratios are defined as follows:

$$\frac{a}{b} + \frac{c}{d} = \frac{ad + cb}{bd} \qquad \frac{a}{b} \cdot \frac{c}{d} = \frac{ac}{bd} \qquad \frac{a}{b} \div \frac{c}{d} = \frac{ad}{bc}$$

The algebraic operations are subject to the following axioms:

Addition	$a + b = b + a$	(commutative)
	$(a + b) + c = a + (b + c)$	(associative)
Multiplication	$ab = ba$	(commutative)
	$(ab)c = a(bc)$	(associative)
	$a(b + c) = ab + ac$	(distributive)
Equality	If $a = b$ and $b = c$, then $a = c$	

Operations with equations depend upon the axiom: If equals are added to, subracted from, multiplied by, or divided by equals (division by 0 is excluded), the results are equal. That is, if $a = b$, then $a + c = b + c$, $a - c = b - c$, $ac = bc$, and $a/c = b/c$, if $c \neq 0$. Transposition in an equation is simply adding or subtracting equals.

Some branches of modern algebra employ axioms that are different from those stated, and will not be covered here.

Number system of algebra. The positive and negative whole numbers, zero, and ratios of whole numbers constitute the rational number system. Addition, multiplication, and division are the rational operations. Root extraction is not a rational operation since its results are not usually rational numbers. For example, there is no rational number whose square is 2, nor any rational number whose square is -1. Two extensions of the number system provide irrational numbers and imaginary numbers.

1. A rational number is a number which can be expressed as the ratio of two integers, a/b, with $b \neq 0$. (An integer is a rational number, for example, $2 = 2/1$.)
2. An irrational number is a number which is not rational but which can be approximated as closely as desired by rational numbers. (Thus $\sqrt{2}$ is not rational, but for any $\epsilon > 0$ there are rational numbers a_r such that the difference between a_r and $\sqrt{2}$ is less than ϵ.)
3. The rational numbers and the irrational numbers make up the real number system.
4. A pure imaginary number is a number whose square is negative.
5. A complex number is the sum of a real number and a pure imaginary number.

The rational numbers are identical with the terminating and repeating decimals. All other infinite decimals are irrational numbers. A theory of real numbers, satisfactory to modern mathematicians, has been developed since about 1850.

A pure imaginary number can be expressed as a product $b\sqrt{-1}$, where b is a real number. The symbol i, replacing $\sqrt{-1}$, is the imaginary unit; hence all pure imaginary numbers have the form bi, and all complex numbers have the form $a + bi$, with a and b both real. The complex numbers include the real numbers (when $b = 0$) and the pure imaginary numbers (when $a = 0$). The term "imaginary" is applied to any complex number $a + bi$, with $b \neq 0$. The complex number system is sufficient for the needs of classical algebra. In modern algebra the number concept is more abstract.

Exponents and powers. The operation of repeated multiplication of like numbers is denoted by exponents and the results are called powers. Thus $2 \cdot 2 \cdot 2 \cdot 2 = 2^4 = 16$; the product of the four 2s is denoted by the exponent 4 applied to the base 2, and 16 is the fourth power of 2. From the definition of an exponent the following theorems are derived (a and b = any numbers other than 0 and n and m are positive integers):

1. $a^n \cdot a^m = a^{n+m}$. Example, $2^3 \cdot 2^4 = 2^7$.
2. $a^n \div a^m = a^{n-m}$ if $n > m$, and $1/a^{m-n}$ if $n < m$. Example, $2^4 \div 2^2 = 2^2 = 2^2$; $3^2 \div 3^3 = 1/3$.
3. $(a^n)^m = a^{nm}$. Example, $(2^2)^3 = 2^6$.
4. $(ab)^n = a^n b^n$. Example, $6^3 = (2^3 3^3) = 2^3 \cdot 3^3$.

This idea of an exponent is meaningful only for exponents which are positive integers. In order to extend the concept to other types of exponents, three additional definitions are made:

5. $a^0 = 1$ for all values of a other than 0.
6. $a^{-n} = 1/a^n$.
7. $a^{n/m} = \sqrt[m]{a^n} = (\sqrt[m]{a})^n$.

With these definitions the above rules hold for all rational exponents. For example,

$$3^{-4}3^2 = 3^{-4+2} = 3^{-2} = 1/3^2$$

$$a^{1/2} \cdot a^{1/3} = a^{1/2+1/3} = a^{5/6}$$

Radicals. The inverse operation to forming a power is that of extracting a root and is expressed by a radical sign. For example,

$$\sqrt{4} = 2 \qquad \sqrt[3]{8} = 2$$

In $\sqrt[n]{a}$, a is the radicand and n is the index. Similar radicals have the same radicand and the same index, and can be added by direct use of the distributive axiom. For example,

$$3\sqrt[3]{5} + 2\sqrt[3]{5} = 5\sqrt[3]{5}$$

Radicals with the same index can be multiplied or divided by multiplying or dividing radicands. For example,

$$\sqrt{6} \cdot \sqrt{5} = \sqrt{6 \cdot 5} = \sqrt{30}$$

Radicals can always be transformed to the same index by use of fractional exponents. For example,

$$\sqrt[3]{2} \cdot \sqrt{5} = 2^{1/3} \cdot 5^{1/2} = 2^{2/6} \cdot 5^{3/6}$$
$$= (2^2)^{1/6} \cdot (5^3)^{1/6} = \sqrt[6]{2^2} \cdot \sqrt[6]{5^3} = \sqrt[6]{500}$$

Because of the definition of a fractional exponent, a radical sign can be replaced by an exponent. For example,

$$\sqrt[3]{x} = x^{1/3}$$

Equations and identities. An equation is a statement that two expressions are equal. The statement may be false, for example, $x + 2 = x + 1$. It may be true for some values of the unknown or unknowns but not for all possible values. Such an equation is a conditional equation. To solve it is to find the values for which it holds true. For example, $3x - 12 = 0$ is true for $x = 4$ only. Or a statement of equality may be true for all values of the letters for which the expressions have meaning. For example, $3x - 2 - 2x + 6 = x + 4$. Such an equation is called an identity.

Solving conditional equations is the central problem of algebra. Identities are used for transforming and simplifying expressions, mainly to facilitate solving equations. Operations on algebraic expressions, including the factoring of polynomials and the combining of fractions and radicals, produce identities that make possible the reduction of conditional equations to manageable form.

Operations on polynomials. Operations on polynomials are deduced from the basic axioms. The distributive axiom is always applied. Others are used equally freely but with less awareness. For example,

1. $(5x + 3y) + (2x - 4y) = 5x + 2x + 3y - 4y$
   $= (5 + 2)x + (3 - 4)y = 7x - y$.
2. $(2x - y)(3x + 4y) = (2x - y)3x + (2x - y)4y$
   $= 2x \cdot 3x - y \cdot 3x + 2x \cdot 4y - y \cdot 4y$
   $= 6x^2 + 5xy - 4y^2$.

Long division is performed exactly as in arithmetic, after first arranging the terms of the dividend and the divisor in descending, or ascending, order of exponents of a letter. For example, $(2x^3 - 3y^3 - xy^2 - x^2y) \div (2x - 3y) = x^2 + xy + y^2$.

$$
\require{enclose}
\begin{array}{r}
x^2 + xy + y^2 \\[-2pt]
2x - 3y \enclose{longdiv}{2x^3 - x^2y - xy^2 - 3y^3} \\
\underline{2x^3 - 3x^2y} \\
2x^2y - xy^2 \\
\underline{2x^2y - 3xy^2} \\
2xy^2 - 3y^3 \\
\underline{2xy^2 - 3y^3} \\
\end{array}
$$

Factoring of polynomials. Factoring of polynomials is important in many procedures for solving equations. Ability to factor depends upon a knowledge of some basic special products, just as factoring of arithmetic numbers depends upon the multiplication table.

1. $ax + ay = a(x + y)$, by the distributive axiom.
2. $(x + y)(x - y) = x^2 - y^2$, whence by reading in reverse order, a rule for factoring the difference of squares is obtained. For example, $4a^2 - 9b^2 = (2a + 3b)(2a - 3b)$.
3. $(x \pm y)^2 = x^2 \pm 2xy + y^2$, whence a trinomial that is the square of a binomial can be identified. One term (the middle term as here written) is twice the product of the square roots of the other terms. For example, $x^2 + 6x + 9 = (x + 3)^2$. This knowledge is useful not only in factoring but also in discovering a general method of solving second-degree equations.

4. $(x+a)(x+b)=x^2+(a+b)x+ab$. Hence, trinomials of the form x^2+mx+n can be factored by finding, when this is possible, two numbers whose sum is m and whose product is n. For example, $x^2+5x+6=(x+2)(x+3)$.

5. By multiplication, $(x-y)(x^n+x^{n-1}y+x^{n-2}y^2+\cdots+xy^{n-1}+y^n)=x^{n+1}-y^{n+1}$. Hence, the difference of like powers of x and y always has the factor $x-y$. The form of the other factor is easily remembered, or it may be found by division. For example, $x^3-y^3=(x-y)(x^2+xy+y^2)$; $a^6-64=a^6-2^6=(a-2)(a^5+2a^4+4a^3+8a^2+16a+32)$.

6. By multiplication, $(x+y)(x^n-x^{n-1}y+x^{n-2}y^2-\cdots+y^n)=x^{n+1}+y^{n+1}$, when n is even so that the final term y^n is positive. It follows that the sum of like odd powers of x and y has the factor $x+y$. The other factor can be remembered or found by division. For example,

$$x^3+y^3=(x+y)(x^2-xy+y^2)$$
$$a^5+32=a^5+2^5$$
$$=(a+2)(a^4-2a^3+4a^2+4a^2-8a+16)$$

7. The factor theorem: If a polynomial in x is equal to zero when x is replaced by a, it has the factor $x-a$. For example, x^3+x^2-x-10 is equal to zero if $x=2$. Hence $x-2$ is a factor and the other factor is found by division to be x^2+3x+5.

Operations on fractions. Fractions are combined by applying the axioms to the definitions of sum, product, and quotient of fractions, as given above. Factoring plays an important role in carrying out these operations. From the definition of a product, it follows that

$$\frac{ab}{ac}=\frac{a}{a}\cdot\frac{b}{c}=1\cdot\frac{b}{c}$$

Hence, like factors can be canceled from the numerator and denominator since this is equivalent to division by 1. Examples are:

1. $\dfrac{3}{x+2}-\dfrac{2}{x+3}=\dfrac{3(x+3)-2(x+2)}{(x+2)(x+3)}$

$$=\frac{x+5}{(x+2)(x+3)}$$

2. $\dfrac{x^2-y^2}{x^2+5xy+6y^2}\cdot\dfrac{3x^2+7xy+2y^2}{3x^2-2xy-y^2}$

$$=\frac{(x-y)(x+y)}{(x+2y)(x+3y)}\cdot\frac{(x+2y)(3x+y)}{(3x+y)(x-y)}=\frac{x+y}{x+3y}$$

Conditional equations. Equations are classified first according to the kind of expressions in them. The most important class is that of rational integral equations, or equations containing only polynomials. Such equations are further classified according to degree and the number of unknowns.

An equation of the first degree is called a linear equation. A linear equation in one unknown presents no difficulties. One and only one solution exists and it is easily found. If $ax+b=0$, $ax=-b$, and $x=-b/a$.

An equation of the second degree is called a quadratic. In general, a quadratic in one unknown has two solutions, although these may coincide. The principal device for solving quadratics is a formula, found by the knowledge of the form of a trinomial perfect square.

Quadratic formula. This formula is derived by solving the general quadratic $ax^2+bx+c=0$, where $a\neq0$. After transposing and dividing by a, $x^2+(b/a)x=-c/a$. In factoring, $x^2+2kx+k^2=(x+k)^2$. If b/a is identified with $2k$, $k^2=b^2/4a^2$. Hence, adding $b^2/4a^2$ to both sides of the equation gives

$$x^2+\frac{b}{a}x+\frac{b^2}{4a^2}=\frac{b^2}{4a^2}-\frac{c}{a}$$

The trinomial on the left side is equal to $\left(x+\dfrac{b}{2a}\right)^2$.

Hence, $\left(x+\dfrac{b}{2a}\right)^2=\dfrac{b^2}{4a^2}-\dfrac{c}{a}=\dfrac{b^2-4ac}{4a^2}$

and $x+\dfrac{b}{2a}=\pm\sqrt{\dfrac{b^2-4ac}{4a^2}}=\pm\dfrac{\sqrt{b^2-4ac}}{2a}$

Finally, $x=-\dfrac{b}{2a}\pm\dfrac{\sqrt{b^2-4ac}}{2a}=\dfrac{-b\pm\sqrt{b^2-4ac}}{2a}$

This is the quadratic formula, which provides two solutions:

$$x_1=\frac{-b+\sqrt{b^2-4ac}}{2a}\qquad x_2=\frac{-b-\sqrt{b^2-4ac}}{2a}$$

If $b^2-4ac=0$, $x_1=x_2$. The quantity b^2-4ac is called the discriminant of the quadratic $ax^2+bs+c=0$. For example, the equation $2x^2-7x+3=0$ has $a=2$, $b=-7$, $c=3$. Hence

$$x_1=\frac{7+\sqrt{49-24}}{4}=\frac{7+5}{4}=3$$

and $x_2=\dfrac{7-\sqrt{49-24}}{4}=\dfrac{7-5}{4}=\dfrac{1}{2}$

The solutions of the quadratic are irrational if the discriminant is positive but not a perfect square, and are imaginary if the discriminant is negative. For example, $3x^2+5x+8=0$ has $a=3$, $b=5$, $c=8$, $b^2-4ac=25-96=-71$. The solutions are imaginary and are

$$x_1=\frac{-5+\sqrt{-71}}{6}\qquad x_2=\frac{-5-\sqrt{-71}}{6}$$

Ratio and proportion. The most useful comparison of two numbers is given by a ratio. Thus the ratio $2:3$, or $2/3$, denotes the relative sizes of 2 and 3. If two numbers a and b have the ratio of 2 to 3, it may be expressed as $a/b=2/3$. This equation expresses a proportion, which is simply the equality of two ratios.

The proportion $a/b=2/3$, or $a:b=2:3$, means that $3a=2b$. The extreme numbers, a and 3, in the array $a:b=2:3$ are called the extremes, whereas the intermediate numbers, b and 2, are the means. The equation $3a=2b$ results from multiplying both members of the proportion by $3b$ and states that, if four numbers are in proportion, the product of the extremes is equal to the product of the means. All fractional equations can be transformed to rational integral equations by the use of this principle.

Variation. Practical problems in proportion are often expressed in the language of variation. To say that the distance varies directly as the time is equivalent to saying that the ratio of the distance to the time is constant: $d/t=c=$ constant, or $d=ct$. Hence, if d_1 corresponds to t_1, and d_2 to t_2, then $d_1/t_1=d_2/t_2$ and $d_1/d_2=t_1/t_2$.

To say that for a perfect gas the pressure varies inversely as the volume means that the ratio of the pressure to the reciprocal of the volume (v inverted, to become $1/v$) is constant: $p/(1/v) = c$, or $pv = c$. If p_1 corresponds to v_1 and p_2 to v_2, $p_1v_1 = p_2v_2$, whence $p_1/p_2 = v_2/v_1$. In this instance, corresponding pressures and volumes are *inversely* proportional. *See* ANALYTIC GEOMETRY; EQUATIONS, THEORY OF.

[HOLLIS R. COOLEY]

Bibliography: P. M. Cohn, *Algebra*, vol. 1, 1974, vol. 2, 1977; H. R. Cooley et al., *Introduction to Mathematics*, 3d ed., 1967; N. Jacobson, *Basic Algebra One*, 1974, *Basic Algebra Two*, 1980; K. Knopp, *Infinite Sequences and Series*, 1956; R. R. Middlemiss, *Algebra for College Students*, 1953; M. A. Sobel and J. M. Banks, *Algebra: Its Elements and Structures*, 3d ed., book 1, 1976, book 2, 1977.

Alternating current

Electric current that reverses direction periodically, usually many times per second. Electrical energy is ordinarily generated by a public or a private utility organization and provided to a customer, whether industrial or domestic, as alternating current.

One complete period, with current flow first in one direction and then in the other, is called a cycle, and 60 cycles per second (60 hertz, or Hz) is the customary frequency of alternation in the United States and in all of North America. In Europe and in many other parts of the world, 50 Hz is the standard frequency. On aircraft a higher frequency, often 400 Hz, is used to make possible lighter electrical machines.

When the term alternating current is used as an adjective, it is commonly abbreviated to ac, as in ac motor. Similarly, direct current as an adjective is abbreviated dc.

Advantages. The voltage of an alternating current can be changed by a transformer. This simple, inexpensive, static device permits generation of electric power at moderate voltage, efficient transmission for many miles at high voltage, and distribution and consumption at a conveniently low voltage. With direct (unidirectional) current it is not possible to use a transformer to change voltage. On a few power lines, electric energy is transmitted for great distances as direct current, but the electric energy is generated as alternating current, transformed to a high voltage, then rectified to direct current and transmitted, then changed back to alternating current by an inverter, to be transformed down to a lower voltage for distribution and use.

In addition to permitting efficient transmission of energy, alternating current provides advantages in the design of generators and motors, and for some purposes gives better operating characteristics. Certain devices involving chokes and transformers could be operated only with difficulty, if at all, on direct current. Also, the operation of large switches (called circuit breakers) is facilitated because the instantaneous value of alternating current automatically becomes zero twice in each cycle and an opening circuit breaker need not interrupt the current but only prevent current from starting again after its instant of zero value.

Sinusoidal form. Alternating current is shown diagrammatically in Fig. 1. Time is measured horizontally (beginning at any arbitrary moment) and the current at each instant is measured vertically.

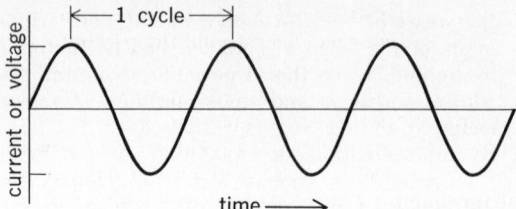

Fig. 1. Diagram of sinusoidal alternating current.

In this diagram it is assumed that the current is alternating sinusoidally; that is, the current i is described by Eq. (1), where I_m is the maximum in-

$$i = I_m \sin 2\pi ft \qquad (1)$$

stantaneous current, f is the frequency in cycles per second (hertz), and t is the time in seconds. *See* SINE WAVE.

A sinusoidal form of current, or voltage, is usually approximated on practical power systems because the sinusoidal form results in less expensive construction and greater efficiency of operation of electric generators, transformers, motors, and other machines.

Phase difference. Phase difference is a measure of the fraction of a cycle by which one sinusoidally alternating quantity leads or lags another. Figure 2 shows a voltage v which is described in Eq. (2) and a current i which is described in Eq. (3).

$$v = V_m \sin 2\pi ft \qquad (2)$$

$$i = I_m \sin (2\pi ft - \varphi) \qquad (3)$$

The angle φ is called the phase difference between the voltage and the current; this current is said to lag (behind this voltage) by the angle φ. It would be equally correct to say that the voltage leads the current by the phase angle φ. Phase difference can be expressed as a fraction of a cycle or in degrees of angle, or as in Eq. (3), in radians of angle, with corresponding minor changes in the equations.

If there is no phase difference, and $\varphi = 0$, voltage and current are in phase. If the phase difference is a quarter cycle, and $\varphi = \pm 90$ degrees, the quantities are in quadrature.

Power factor. Power factor is defined in terms of the phase angle. If the rms value of sinusoidal current from a power source to a load is I and the rms value of sinusoidal voltage between the two wires connecting the power source to the load is V, the average power P passing from the source to the load is shown as Eq. (4). The cosine of the phase

$$P = VI \cos \varphi \qquad (4)$$

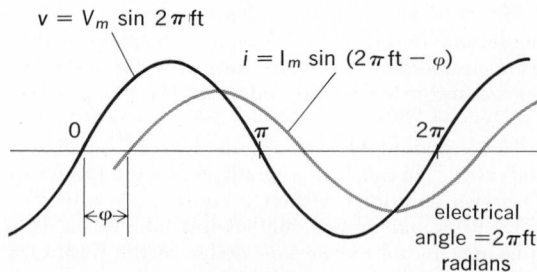

Fig. 2. The phase angle φ.

angle, cos φ, is called the power factor. Thus the rms voltage, the rms current, and the power factor are the components of power.

The foregoing definition of power factor has meaning only if voltage and current are sinusoidal. Whether they are sinusoidal or not, average power, rms voltage, and rms current can be measured, and a value for power factor is implicit in Eq. (5).

$$P = V I \text{ (power factor)} \qquad (5)$$

This gives a definition of power factor when V and I are not sinusoidal, but such a value for power factor has limited use.

If voltage and current are in phase (and of the same waveform), power factor equals 1. If voltage and current are out of phase, power factor is less than 1. If voltage and current are sinusoidal and in quadrature, power factor equals zero.

The phase angle and power factor of voltage and current in a circuit that supplies a load are determined by the load. Thus a load of pure resistance, as an electric heater, has unity power factor. An inductive load, such as an induction motor, has a power factor less than 1 and the current lags behind the applied voltage. A capacitive load, such as a bank of capacitors, also has a power factor less than 1, but the current leads the voltage, and the phase angle φ is a negative angle. *See* CAPACITANCE; CIRCUIT (ELECTRICITY); ELECTRIC CURRENT; ELECTRICAL IMPEDANCE; ELECTRICAL RESISTANCE; INDUCTANCE; JOULE'S LAW; OHM'S LAW; RESONANCE (ALTERNATING-CURRENT CIRCUITS).

[H. H. SKILLING]

Amorphous solid

A rigid material whose structure lacks crystalline periodicity; that is, the pattern of its constituent atoms or molecules does not repeat periodically in three dimensions. In the present terminology amorphous and noncrystalline are synonymous. A solid is distinguished from its other amorphous counterparts (liquids and gases) by its viscosity: a material is considered solid (rigid) if its shear viscosity exceeds $10^{14.6}$ poise ($10^{13.6}$ Pa·s). *See* CRYSTAL; SOLID-STATE PHYSICS; VISCOSITY.

Preparation. Techniques commonly used to prepare amorphous solids include vapor deposition, electrodeposition, anodization, evaporation of a solvent (gel, glue), and chemical reaction (often oxidation) of a crystalline solid. None of these techniques involves the liquid state of the material. A distinctive class of amorphous solids consists of glasses, which are defined as amorphous solids obtained by cooling of the melt. Upon continued cooling below the crystalline melting point, a liquid either crystallizes with a discontinuous change in volume, viscosity, entropy, and internal energy, or (if the crystallization kinetics are slow enough and the quenching rate is fast enough) forms a glass with a continuous change in these properties. The glass transition temperature is defined as the temperature at which the fluid becomes solid (that is, the viscosity $= 10^{14.6}$ poise $= 10^{13.6}$ Pa·s) and is generally marked by a change in the thermal expansion coefficient and heat capacity. [Silicon dioxide (SiO_2) and germanium dioxide (GeO_2) are exceptions.] It is intuitively appealing to consider a glass to be both structurally and thermodynamically related to its liquid; such a connection is more tenuous for amorphous solids prepared by the other techniques.

Types of solids. Oxide glasses, generally the silicates, are the most familiar amorphous solids. However, as a state of matter, amorphous solids are much more widespread than just the oxide glasses. There are both organic (for example, polyethylene and some hard candies) and inorganic (for example, the silicates) amorphous solids. Examples of glass formers exist for each of the bonding types: covalent [As_2S_3], ionic [$KNO_3 - Ca(NO_3)_2$], metallic [Pd_4Si], van der Waal's [o-terphenyl], and hydrogen [$KHSO_4$]. Glasses can be prepared which span a broad range of physical properties. Dielectrics (for example, SiO_2) have very low electrical conductivity and are optically transparent, hard, and brittle. Semiconductors (for example, As_2SeTe_2) have intermediate electrical conductivities and are optically opaque and brittle. Metallic glasses (for example, Pd_4Si) have high electrical and thermal conductivities, have metallic luster, and are ductile and strong.

Uses. The obvious uses for amorphous solids are as window glass, container glass, and the glassy polymers (plastics). Less widely recognized but nevertheless established technological uses include the dielectrics and protective coatings used in integrated circuits, and the active element in photocopying by xerography, which depends for its action upon photoconduction in an amorphous semiconductor. In optical communications a highly transparent dielectric glass in the form of a fiber is used as the transmission medium. In addition, metallic amorphous solids have been considered for uses that take advantage of their high strength, excellent corrosion resistance, extreme hardness and wear resistance, and unique magnetic properties.

Semiconductors. It is the changes in short-range order (on the scale of a localized electron), rather than the loss of long-range order alone, that have a profound effect on the properties of amorphous semiconductors. For example, the difference in resistivity between the crystalline and amorphous states for dielectrics and metals is always less than an order of magnitude and is generally less than a factor of 3. For semiconductors, however, resistivity changes of 10 orders of magnitude between the crystalline and amorphous states are not uncommon, and accompanying changes in optical properties can also be large.

Electronic structure. The model that has evolved for the electronic structure of an amorphous semiconductor is that the forbidden energy gap characteristic of the electronic states of a crystalline

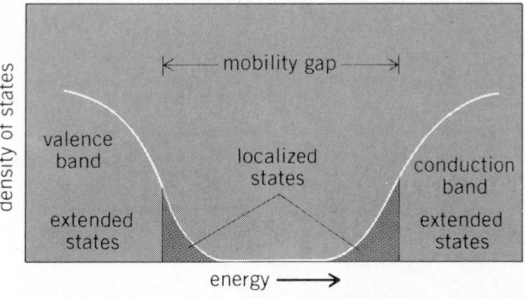

Fig. 1. Density of states versus energy for an amorphous semiconductor.

material is replaced in an amorphous semiconductor by a pseudogap. Within this pseudogap the density of states of the valence and conduction bands is sharply lower but tails off gradually and remains finite due to structural disorder (Fig. 1). The states in the tail region are localized; that is, their wave functions extend over small distances in contrast to the extended states that exist elsewhere in the energy spectrum. Because the localized states have low mobility (velocity per unit electric field), the extended states are separated by a mobility gap (Fig. 1) within which charge transport is markedly impeded. In each band, the energy at which the extended states meet the localized states is called the mobility edge. *See* BAND THEORY OF SOLIDS.

An ideal amorphous solid can be conceptually defined as having no unsatisfied bonds, a minimum of bond distortions (bond angles and lengths), and no internal surfaces associated with voids. Deviations from this ideality introduce localized states in the gap in addition to those in the band edge tails due to disorder alone. One important defect is called an unsatisfied, broken, or dangling bond. These dangling bonds create states deep in the gap which can act as recombination centers and markedly limit carrier lifetime and mobility. A large number of such states introduced, for example, during the deposition process will dominate the electrical properties.

Charge transport can occur by two mechanisms. The first is conduction of mobile extended-state carriers (analogous to that which occurs in crystalline semiconductors), for which the conductivity is proportional to $\exp(-E_g/2kT)$, where E_g is the gap width, T is the absolute temperature, and k is Boltzmann's constant. The second mechanism is hopping of the localized carriers, for which the conductivity is proportional to $\exp[-(T_0/T)^{1/4}]$, where T_0 is a constant (Mott's law). At low temperatures carriers hop from one localized trap to another, whereas at high temperatures they can be excited to the mobility edge.

Glassy chalcogenides. One class of amorphous semiconductors is the glassy chalcogenides, which contain one (or more) of the chalcogens sulfur, selenium, or tellurium as major constituents. These amorphous solids behave like intrinsic semiconductors, show no detectable unpaired spin states, and exhibit no doping effects. It is thought that essentially all atoms in these glasses assume a bonding configuration such that bonding requirements are satisfied; that is, the structure accommodates the coordination of any atom. These materials have application in switching and memory devices.

Tetrahedrally bonded solids. Another group is the tetrahedrally bonded amorphous solids, such as amorphous silicon and germanium. These materials cannot be formed by quenching from the melt (that is, as glasses) but must be prepared by one of the deposition techniques mentioned above. An amorphous to crystalline transformation in these materials is irreversible.

When amorphous silicon (or germanium) is prepared by evaporation, not all bonding requirements are satisfied, so a large number of dangling bonds are introduced into the material. These dangling bonds are easily detected by spin resonance or low-temperature magnetic susceptibility and create states deep in the gap which limit the transport properties. The number of dangling bonds can be reduced by a thermal anneal below the crystallization temperature, but the number cannot be reduced sufficiently to permit doping.

Amorphous silicon prepared by the decomposition of silane (SiH_4) in a plasma has been found to have a significantly lower density of defect states within the gap, and consequently the carrier lifetimes are expected to be longer. This material can be doped p- or n-type with boron or phosphorus (as examples) by the addition of B_2H_6 or PH_3 to the SiH_4 during deposition. This permits exploration of possible devices based on doping, which are analogous to devices based on doping of crystalline silicon.

One reason plasma-deposited silicon has a significantly lower density of defect states within the gap is that the process codeposits large amounts of hydrogen (typically 5–30% of the atoms, depending upon deposition conditions), and this hydrogen is very effective at terminating dangling bonds (Fig. 2). Other possible dangling-bond terminators (for example, fluorine) have been explored.

The ability to reduce the number of states deep in the gap and to dope amorphous silicon led directly to the development of an amorphous silicon photovoltaic solar cell. Intense effort has been devoted to improving the efficiency of these cells to the 8% level thought to be required for large-scale application. The appeal of amorphous silicon is that it holds promise for low-cost, easily fabricated, large-area cells.

Amorphous silicon solar cells have been constructed in heterojunction, *p-i-n*-junction, and Schottky-barrier device configurations and have been introduced for use in calculators and watches. The optical properties of amorphous silicon provide a better match to the solar spectrum than do those for crystalline silicon, but the transport properties of the crystalline material are better. Experiments indicate that hole transport in the amorphous material is the limiting factor in the conversion efficiency. *See* SEMICONDUCTOR.

[BRIAN G. BAGLEY]

Bibliography: R. H. Doremus, *Glass Science*, 1973; N. F. Mott and E. A. Davis, *Electronic Processes in Non-Crystalline Materials*, 2d ed., 1979; J. Tauc (ed.), *Amorphous and Liquid Semiconductors*, 1974.

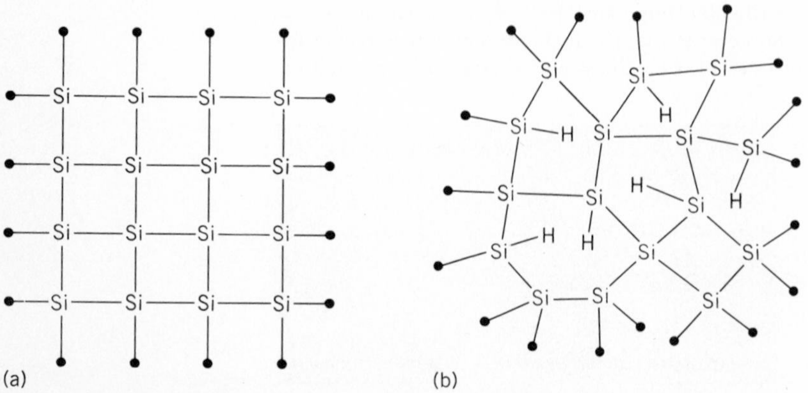

Fig. 2. Bonding of silicon. Bonds which continue the network are shown terminated by a dot. (*a*) Crystalline arrangement. (*b*) Amorphous structure with dangling bonds terminated by hydrogen.

Ampère's law

A law of electromagnetism which expresses the contribution of a current element of length dl to the magnetic induction (flux density) B at a point near the current. Ampère's law, sometimes called Laplace's law, was derived by A. M. Ampère after a series of experiments during 1820–1825.

Whenever an electric charge is in motion, there is a magnetic field associated with that motion. The flow of charges through a conductor sets up a magnetic field in the surrounding region. Any current may be considered to be broken up into infinitesimal elements of length dl, and each such element contributes to the magnetic induction at every point in the neighborhood. The contribution dB of the element is found to depend upon the current I, the length dl of the element, the distance r of the point P from the current element, and the angle θ between the current element and the line joining the element to the point P (see illustration). Ampère's law expresses the manner of the dependence by Eq. (1).

$$dB = k\frac{I\,dl\sin\theta}{r^2} \qquad (1)$$

Choice of units. The proportionality factor k depends upon the units used in Eq. (1) and upon the properties of the medium surrounding the current. As in other equations expressing observed relationships, there is an arbitrary choice as to which of the units is to be defined from Ampère's law, or by a relationship derived from the law. If values are assigned to B, I, and l, the factor k can be found by experiment. If, however, an arbitrary value is assigned to k, Ampère's law or an equation derived from Ampère's law may be used to define the unit of current since units of B are otherwise defined. In the rationalized meter-kilogram-second system of units, the latter choice is made, a value is assigned to k, and the ampere as a unit of current is defined from an equation derived from Ampère's law. When the current is in empty space, the factor k is assigned a value of 10^{-7} weber/amp-m.

As in other equations associated with electric and magnetic fields, for example Coulomb's law, it is convenient to replace k by a new factor μ_0 related to k as in Eq. (2). This substitution removes the

$$\mu_0 = 4\pi k \qquad (2)$$

factor 4π from many derived equations in which it would otherwise appear. With this substitution Ampère's law becomes Eq. (3). The factor μ_0 is

$$dB = \frac{\mu_0}{4\pi}\frac{I\,dl\sin\theta}{r^2} \qquad (3)$$

called the permeability of empty space. *See* ELECTRICAL UNITS AND STANDARDS; MAGNETIC PERMEABILITY.

Right-hand rule. The direction of dB is always perpendicular to the plane determined by the line tangent to the current element dl and the line joining dl to P. The sense of the lines is clockwise when looking in the direction of the current. The direction at each point may also be described in terms of a right-hand rule. If the current element is grasped by the right hand with the thumb pointing in the direction of the current, the fingers encircle the current in the direction of the magnetic induction. (If electron flow is used as the convention for current direction, the left hand is used in place of the right.)

Since the magnetic induction is everywhere perpendicular to the current element, it follows that the lines of induction or flux always form closed paths.

Field near a current. From Ampère's law, the field near a current may be calculated by finding the vector sum of the contributions of all the various elements that make up the current. This sum can be found (provided the integration can be carried out) by integrating over the whole length of the current, as in Eq. (4), where the limits are taken so

$$B = \int dB = \frac{\mu_0}{4\pi}\int\frac{I\,dl\sin\theta}{r^2} \qquad (4)$$

that all current elements are included and the integral represents the vector sum.

The experimental test of the validity of Ampère's law is not direct since experiments are made not upon the field due to individual elements, but upon the resultant field of the current as a whole. Thus, the applications of Ampère's law are in the computation of the field for known geometrical arrangements of the current. For simple current paths, the summation is easily carried out, as for the field at the center of a single circular conductor of radius a. For this conductor, $r = a$, and r is always perpendicular to dl, so that $\sin\theta = 1$. Furthermore, all contributions are perpendicular to the plane of the coil, so that the vector sum is the arithmetic sum as in Eq. (5). The field anywhere on

$$B = \frac{\mu_0}{4\pi}\int_0^{2\pi a}\frac{I\,dl}{a^2} = \frac{\mu_0}{4\pi}\frac{I}{a^2}\int_0^{2\pi a}dl = \frac{\mu_0 I}{2a} \qquad (5)$$

the axis of a flat coil, on the axis of a solenoid, or near a long straight conductor may be computed readily, and the result compared with experimental values. In general, the field of any current may be evaluated if dB can be expressed in Ampère's law, and the integration can be carried out. For the case of a long straight conductor *see* BIOT-SAVART LAW.

Ampère's law can be used as an alternative definition of magnetic induction and for defining magnetic field intensity. *See* MAGNETIC FIELD; MAGNETIC INDUCTION.

[KENNETH V. MANNING]

Bibliography: Berkeley Physics Course, vol. 2, *Electricity and Magnetism*, 1970; B. I. Bleany and B. Bleany, *Electricity and Magnetism*, 3d ed., 1976; J. C. Maxwell, *A Treatise on Electricity and Magnetism*, pt. 4, 3d ed., 1904; S. L. Oppenheimer and J. P. Borchers, *Direct and Alternating Currents*, 2d ed., 1973; R. Resnick and D. Halliday, *Physics*, 3d ed., 1977; F. W. Sears, *University Physics*, 5th ed., 1976.

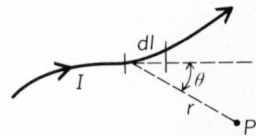

AMPERE'S LAW

Graphic representation of Ampère's law.

Amplitude (wave motion)

The maximum magnitude (value without regard to sign) of the disturbance of a wave. The term "disturbance" refers to that property of a wave which perturbs or alters its surroundings. It may mean, for example, the displacement of mechanical waves, the pressure variations of a sound wave, or the electric or magnetic field of light waves. Sometimes in older texts the word "amplitude" is used for the disturbance itself; in that case, amplitude as meant here is called peak amplitude. This is no longer common usage. *See* DISPLACEMENT (MECHANICS); LIGHT; SOUND; WAVE MOTION.

As an example, consider one-dimensional traveling waves in a linear, homogeneous medium. The wave disturbance y is a function of both a space coordinate x and time t. Frequently the disturbance may be expressed as $y(x,t) = f(x \pm vt)$, where v denotes the wave velocity. The plus or minus sign indicates the direction in which the wave moves, and the shape of the wave dictates the functional form symbolized by f. Then, the amplitude of the disturbance at some point x_0 is the maximum magnitude (that is, the maximum absolute value) achieved by f as time changes over the duration required for the wave to pass point x_0. A special case of this is the one-dimensional, simple harmonic wave $y(x,t) = A \sin [k(x \pm vt)]$, where k is a constant. The amplitude is A since the absolute maximum of the sine function is $+1$. The amplitude for such a wave is a constant. *See* HARMONIC MOTION; SINE WAVE.

If the medium which a wave disturbs dissipates the wave by some nonlinear behavior or other means, then the amplitude will, in general, depend upon position.　　　　[S. A. WILLIAMS]

Analog states

Certain states belonging to neighboring nuclear isobars and possessing identical structure except for the transformation of one or more neutrons into the same number of protons. Hundreds of examples of analog states (also known as isobaric analog states) have been observed throughout the periodic table, and their existence is considered to be a fundamental property of nuclear structure. *See* ISOBAR (ATOMIC PHYSICS); NUCLEAR STRUCTURE.

Starting with a nucleus of N number of neutrons and Z number of protons in a particular state of energy, angular momentum, parity, and so forth, an analog state may be constructed in the neighboring isobar with $N-1$ neutrons and $Z+1$ protons by replacing a neutron with a proton in the same orbit and then coupling the proton to the remaining nucleons in exactly the same fashion as the original neutron. Under the assumption of charge independence of nuclear forces, the only difference between a state and its analog should arise from the increase in the Coulomb interactions attributed to the increased number of protons in the nucleus. Since the total energy of a state has a Coulomb contribution, the most outstanding difference between analog states is their Coulomb energy. When this difference is taken into account, then the total energies of corresponding analog states become nearly equal. Another, smaller effect which must also be considered is the neutron-proton mass difference of 0.782 MeV. However, even then, all properties of analog states do not become precisely identical because there remain subtle differences in the behavior of neutrons and protons due to electromagnetic effects.

Isobaric spin. The simplest nucleus is hydrogen, which consists of a single proton; its analog is a free neutron. In the framework of the nuclear shell model, states of composite nuclei are built up from the individual neutrons and protons that are considered to be moving in orbits which can be calculated by assuming a suitable nuclear potential energy well. Such a framework proves ideal for understanding analog states when one introduces the concept of isobaric spin, also termed isotopic spin or isospin. The neutron and proton are endowed with a quantum number for isospin of $t = 1/2$ (just as they have $s = 1/2$ spin angular momentum). The z-projection of isospin is then defined as $t_z +1/2$ for neutrons and $t_z = -1/2$ for protons. The isospin of a nuclear state with N neutrons and Z protons is constructed by combining the individual neutron and proton isospins, so that the total z-projection is $T_z = \frac{(N-Z)}{2}$, and by analogy with angular momentum, the isospin T must be greater than or equal to T_z up to a maximum of $\frac{N+Z}{2}$. States which are analogs have the same total isospin T and differ only in their value of T_z. Interchange of a single neutron and proton either raises or lowers T_z by one unit. *See* ISOTOPIC SPIN.

Nuclei with N = Z. In nuclei with $N=Z$, the value of T_z is 0, and the ground and lowest excited states are characterized by $T = 0$. These states have no analogs, since replacement of a neutron by a proton, or the reverse, leads to $T_z = -1$ or $+1$, respectively, for which a value of $T = 0$ cannot occur. Physically, this means that when a neutron is transformed into a proton in a state with $T = 0$, the structure cannot be preserved merely by correcting for the difference in Coulomb energy. The neutron-proton exchange must also change one or more additional measurable properties of the nuclear state.

Figure 1 shows a portion of the energy level diagrams for three mass-12 nuclei: ^{12}B, ^{12}C, and ^{12}N. All of the states up to 15.11-MeV excitation in ^{12}C

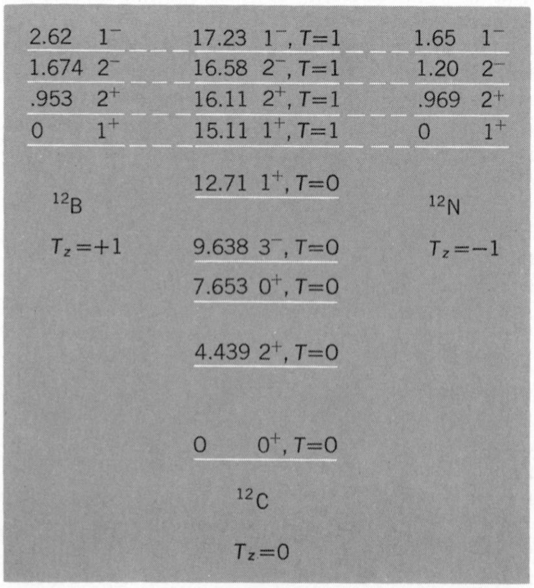

^{12}B		^{12}C	^{12}N	
2.62	1$^-$	17.23　1$^-$, T=1	1.65	1$^-$
1.674	2$^-$	16.58　2$^-$, T=1	1.20	2$^-$
.953	2$^+$	16.11　2$^+$, T=1	.969	2$^+$
0	1$^+$	15.11　1$^+$, T=1	0	1$^+$

^{12}B　　$T_z=+1$

12.71　1$^+$, T=0

9.638　3$^-$, T=0

7.653　0$^+$, T=0

4.439　2$^+$, T=0

0　　0$^+$, T=0

^{12}C

$T_z=0$

^{12}N　　$T_z=-1$

Fig. 1.　Partial level diagram of mass-12 isobars. Broken lines indicate analog states that are members of isospin triplets. Numbers at left of levels give energy in million-electron-volts, and symbols at right give spin and parity.

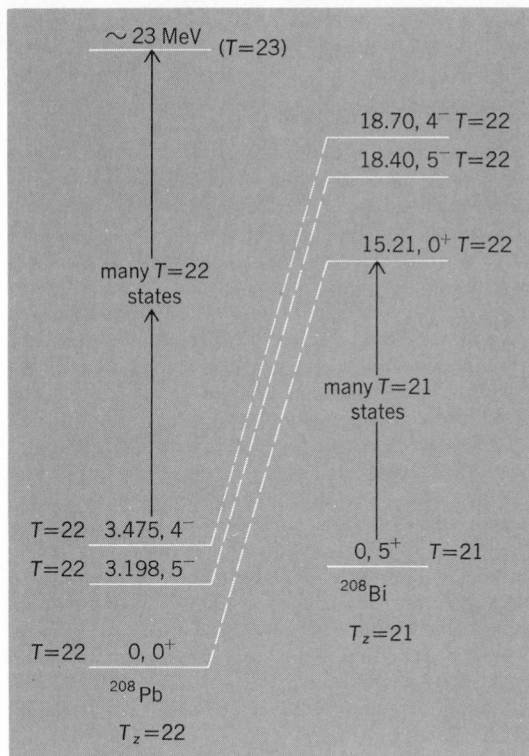

Fig. 2.   Partial level diagram of ^{208}Pb and ^{208}Bi. Broken lines indicate parent-analog state pairs. Numbers at left of levels give energy in million-electron-volts, and symbols at right give spin and parity.

have $T = 0$. The first $T = 1$ state at 15.11 MeV has spin 1 and positive parity. There should also exist two analog states with $T = 1$ and $T_z = +1$ and -1 respectively, which along with the 15.11-MeV state in ^{12}C form an isobaric analog triplet. These states are just the ground states of ^{12}B and ^{12}N, which are also known from experiments to have spin 1 and positive parity. In Fig. 1 the energy scale has been adjusted by the Coulomb energy differences of ^{12}B, ^{12}C, and ^{12}N and for the neutron-proton mass difference. It is assumed that all states of $T = 1$ in ^{12}C have corresponding analog states in ^{12}B and ^{12}N, forming a series of isobaric analog triplets. In a similar fashion at higher excitation energy, $T = 2$ states should occur in ^{12}C. Since $T = 2$ is compatible with $T_z = -2, -1, 0, +1$, and $+2$, such states should be members of isobaric analog quintets ranging over five isobars with mass 12. *See* PARITY (QUANTUM MECHANICS); SPIN (QUANTUM MECHANICS).

Nuclei with neutron excess (N > Z). Above ^{3}He, all stable nuclei have $N \geq Z$ because of the Coulomb repulsion of the protons. In cases such as ^{13}C and ^{27}Al where $N = Z + 1$, the isospin and its z-component have the value 1/2, and all states are members of isobaric analog doublets with the second member residing in the mirror nuclei ^{13}N and ^{27}Si respectively, both with $T_z = -1/2$.

Heavy nuclei. Until about 1961, it was thought that analog states were a property only of light-mass nuclei with mass number A less than about 40. This belief was based on the supposition that the increased Coulomb energy that results from the increased number of protons in heavy nuclei

would finally destroy the isobaric symmetry so that the analog states at high excitation energy would not remain as distinct measurable entities. However, in a number of experiments not specifically designed to search for analog states, dramatic evidence was found for their existence throughout the periodic table up to the heaviest known nuclei. This discovery was one of the most important in nuclear physics during the decade of the 1960s. It led to a large number of experiments and a great increase in knowledge of nuclear structure.

Mass 208 system. The mass 208 system provides an example of analog states in isobars with large neutron excess. Figure 2 shows a partial level diagram of ^{208}Pb and its neighbor ^{208}Bi. The z-component of isospin for ^{208}Pb is $T_z = 22$, and all states up to about 23 MeV of excitation also have $T = 22$. The value of T_z for ^{208}Bi is 21, which is also the value of T for all states up to 15.21 MeV. Just at that energy, however, an excitation is observed in the ^{208}Bi spectrum which possesses many of the properties of the ground state of ^{208}Pb. Moreover, if the Coulomb energy difference of 18.86 MeV is subtracted from ^{208}Bi, then the total energy of each of the two states is the same except for the neutron-proton mass difference. This leads to the identification of the excitation at 15.21 MeV in ^{208}Bi as the $T_z = 21$ analog of the $T = T_z = 22$ ground state of ^{208}Pb. Similarly, higher excitations of ^{208}Bi can be identified as analog states of excited states of ^{208}Pb, with nearly equal energies after correction for Coulomb energy and neutron-proton mass differences.

In neutron excess nuclei, the states with $T = T_z$ are often referred to as the parent states and the states with $T > T_z$ as the analog states. The microscopic structure of this parent-analog state pair in terms of the nuclear shell model is shown in Fig. 3. The ground state of ^{208}Pb is characterized by

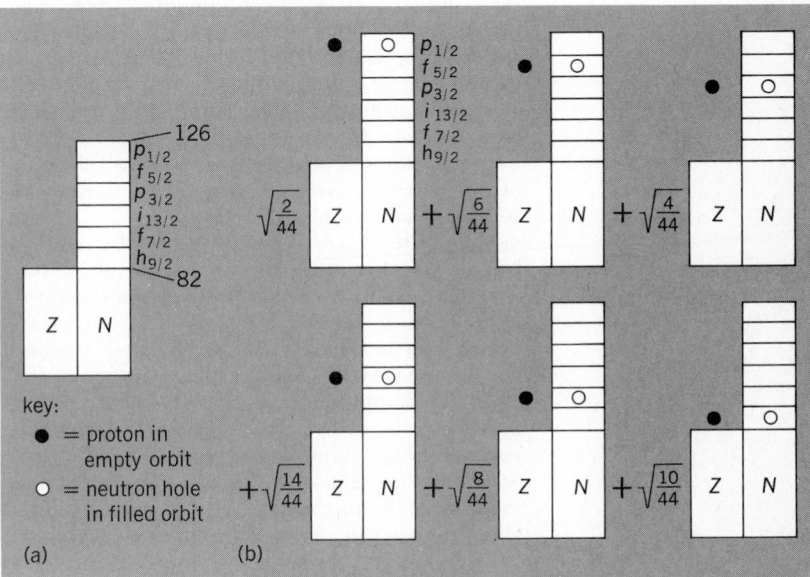

Fig. 3.   Shell model structure of a parent-analog state pair. Letters indicate orbital angular momentum of neutron orbits (s, p, d, f, g, . . . , for $l = 0$, 1, 2, 3, 4, . . .); the subscript gives the j value. (*a*) The parent state, the ground state of ^{208}Pb. (*b*) Its analog state in ^{208}Bi.

neutrons and protons filling all available orbits up to the closed shells at $Z = 82$ and $N = 126$. For the analog state in ^{208}Bi, one of the neutrons of ^{208}Pb is replaced by a proton in the same orbit. However, not all of the neutrons of ^{208}Pb are allowed to undergo this change. The Pauli exclusion principle limits the number of protons that may occupy a particular orbit, so that only those neutrons occupying orbits unfilled by protons are available for the interchange. Moreover, of those neutrons that are available, all have equal probability, so that the structure of the analog state consists of a superposition of many microscopic configurations (shown in Fig. 3b) that look almost like ^{208}Pb. Each microscopic configuration differs from ^{208}Pb by a proton occupying an unfilled orbit and a corresponding neutron hole in a neutron orbit. The numbers under the square root signs in Fig. 3b are the amplitudes which multiply each configuration. The amplitudes are determined by the number of neutrons that fill each orbit. Thus the structure of an analog state in a neutron excess nucleus is characterized by each neutron, which occupies an orbit unfilled by protons, appearing part of the time as a proton in the same orbit. The greater the neutron excess, the more neutrons available and the less time required for each to spend as a proton. *See* EXCLUSION PRINCIPLE; QUANTUM MECHANICS.

Width of states. An important distinction between analog states in light- and heavy-mass nuclei is their width. In nuclei with low mass, analog states are usually very narrow, not more than tens or hundreds of electron volts wide, and their widths are often comparable to those of their corresponding parent states. However, with increasing nuclear mass, analog state widths also increase, attaining values of about 200,000 eV in the lead region, while their corresponding parent states remain very narrow. Associated with the broadening of analog states in medium- and heavy-mass nuclei is the fact that they mix with, and their properties become fragmented over, many narrow states with one unit less of isospin which reside in the region of the analog state. In this way the analog state ceases to be an individual nuclear state, but rather it develops fine structure and is characterized by a distribution of its properties over the underlying states which span the region of the spectrum corresponding to the energy and width of the analog state.

Applications. The discovery of analog states in heavy nuclei has provided a powerful tool for studying nuclear structure that was previously unforeseen. First, it has enabled the study of Coulomb energies to be extended from low-mass nuclei to the heaviest nuclei; and second, because of their high excitation energy, analog states normally are able to decay by particle emission, unlike their corresponding parent states which usually decay only by γ-ray emission to the ground state. By measuring the properties of the particles that are emitted from the analog state, unique information is derived about the parent state which often cannot be obtained in any other fashion. *See* RADIOACTIVITY.   [NELSON STEIN]

Bibliography: J. D. Anderson et al., *Nuclear Isospin*, 1969; A. Bohr and B. R. Mottelson, *Nuclear Structure*, vol. 1, 1969, vol. 2, 1976, vol. 3, 1980; B. L. Cohen, *Concepts of Nuclear Physics*, 1971; D. H. Wilkinson (ed.), *Isospin in Nuclear Physics*, 1969.

Analysis of variance

Total variation in experimental data is partitioned into components assignable to specific sources by the analysis of variance. This statistical technique is applicable to data for which (1) effects of sources are additive, (2) uncontrolled or unexplained experimental variations (which are grouped as experimental errors) are independent of other sources of variation, (3) variance of experimental errors is homogeneous, and (4) experimental errors follow a normal distribution. When data depart from these assumptions, one must exercise extreme care in interpreting the results of an analysis of variance. Statistical tests indicate the contribution of the components to the observed variation.

In an illustrative experiment, t methods of treatment are under study, and n samples are measured for each treatment for a total of nt samples. Measurement X_{ij} of the ith sample that received the jth treatment records an overall effect μ, an effect β_j produced by the jth treatment, and an effect ϵ_{ij} produced by experimental error. The three effects are additive, so that Eq. (1) holds,

$$X_{ij} = \mu + \beta_j + \epsilon_{ij} \qquad (1)$$

where $i = 1, \ldots, n$; and $j = 1, \ldots, t$. The statistical problem is to test for the existence of these effects.

The analysis of variance in this example is presented in the table. Entries in the sum of squares

Illustrative example of the analysis of variance

Source of variation	Sum of squares	Degrees of freedom	Mean square
Between treatments	$T = n\Sigma_j (\overline{X}_j - \overline{X})^2$	$t - 1$	$T' = T/(t-1)$
Within treatments	$E = \Sigma_{i,j}(X_{ij} - \overline{X}_j)^2$	$t(n-1)$	$E' = E/t(n-1)$
Total	$Q = \Sigma_{i,j}(X_{ij} - \overline{X})^2$	$nt - 1$	

column represent that part of the total variation that is attributable to each source. Total sum of squares Q is the sum over all squared deviations of observations X_{ij} from the grand mean $\overline{X}$, Eq. (2).

$$\overline{X} = (\Sigma_i X_{ij})/nt \qquad (2)$$

Similarly, within treatments, sum of squares E is the sum over all squared deviations of observations X_{ij} within a treatment from the mean $\overline{X}_j$ of that treatment, Eq. (3). Also, between treatments,

$$\overline{X}_j = (\Sigma_i X_{ij})/n \qquad (3)$$

sum of squares T is n times the sum over all treatments of the squared deviations of treatment means $\overline{X}_j$ from grand mean $\overline{X}$ as defined by Eqs. (2) and (3). The sum of squares is generally computed more easily from the equivalent formulas (4)–(6).

$$Q = \Sigma_{i,j} X_{ij}^2 - \frac{(\Sigma_{i,j} X_{ij})^2}{nt} \qquad (4)$$

$$T = \Sigma_j \frac{(\Sigma_i X_{ij})^2}{n} - \frac{(\Sigma_{i,j} X_{ij})^2}{nt} \qquad (5)$$

$$E = Q - T \qquad (6)$$

The entries under degrees of freedom represent the number of independent comparisons upon which the sum of squares for the source of variation is based. In every case the linear restriction imposed by the relationship of the particular mean to the observations results in the loss of one degree of freedom. Therefore the number of degrees of freedom is always one less than the number of deviations used to compute the sum of squares.

The mean squares in the analysis of variance are obtained by dividing the sum of squares by the corresponding degrees of freedom. The within-treatments mean square is an estimate of σ^2, the variance of the error term ϵ_{ij} in the additive model. It represents the random or unexplained variation in the data. The between-treatments mean square is an estimate of $\sigma^2 + n\sigma_\beta^2$, where σ_β^2 is the variance of the treatment effects β_j.

If the treatment means differ substantially, the β_j effects estimated by $(\overline{X}_j - \overline{X})$ will differ correspondingly and will have a large variance σ_β^2. If on the other hand the means do not differ, the treatment effects β_j would be zero and σ_β^2 would be zero. In this case the treatment mean square would be equal to the error mean square and both would be independent estimates of σ^2. By comparing the ratio T'/E' of between-treatment mean square T' to within-treatment mean square E' with unity, the variation due to treatments is compared with the variation due to random or unexplained factors. If this ratio, called the F ratio, is close to unity, there is no evidence of a treatment effect. However, if ratio T'/E' is substantially greater than unity there may be a significant treatment effect.

To compare the mean squares objectively, one uses the F test of significance in which the statistical hypothesis is that $\sigma_\beta^2 = 0$. Under this hypothesis it can be concluded that the treatment effects are significantly different from zero at the significance level α if the calculated F ratio is greater than the value of F at the α point on the F distribution with $t - 1$ and $t(n - 1)$ degress of freedom. *See* STATISTICS.

[ROBERT L. BRICKLEY]

Bibliography: W. J. Dixon and F. J. Massey, *Introduction to Statistical Analysis*, 3d ed., 1969; D. L. Harnett, *Introduction to Statistical Methods*, 2d ed., 1975; P. G. Hoel, *Elementary Statistics*, 4th ed., 1976.

Analytic geometry

A branch of mathematics in which algebra is applied to the study of geometry. Because algebraic methods were first systematically applied to geometry in 1637 by the French philosopher-mathematician René Descartes, the subject is also called cartesian geometry. The basis for an algebraic treatment of geometry is provided by the existence of a one-to-one correspondence between the elements, "points" of a directed line g, and the elements, "numbers," that form the set of all real numbers. Such a correspondence establishes a coordinate system on g, and the number corresponding to a point of g is called its coordinate. The point O of g with coordinate zero is the origin of the coordinate system. A coordinate system on g is cartesian provided that for each point P of g, its coordinate is the directed distance $\overline{OP}$. Then all points of g on one side of O have positive coordinates (forming the positive half of g) and all points on the other side have negative coordinates. The point with coordinate 1 is called the unit point. Since the relation $\overline{OP} + \overline{PQ} = \overline{OQ}$ is clearly valid for each two points P, Q of directed line g, then $\overline{PQ} = \overline{OQ} - \overline{OP} = q - p$, where p and q are the coordinates of P and Q, respectively. Those points of g between P and Q, together with P, Q, form a line segment. In analytic geometry it is convenient to direct segments, writing PQ or QP accordingly as the segment is directed from P to Q or from Q to P, respectively. To find the coordinate of the point P that divides the segment $P_1 P_2$ in a given ratio r, put $\overline{P_1 P}/\overline{P_2 P} = r$. Then $(x - x_1)/(x - x_2) = r$ where x_1, x_2, x are the coordinates of P_1, P_2, P, respectively, and solving for x gives $x = (x_1 - rx_2)/(1 - r)$. Clearly r is negative for each point between P_1, P_2 and is positive for each point of g external to the segment. The midpoint of the segment divides it in the ratio -1, and hence its coordinate $x = (x_1 + x_2)/2$.

The plane. Choose any two intersecting lines g_1, g_2 of the plane, with cartesian coordinate systems selected on each so that the intersection point has coordinate O in each system. To each point P of the plane an ordered pair of numbers (x,y) is attached as coordinates, where x is the coordinate of the point of intersection of g_1 with the line through P parallel to g_2 and y is the coordinate of the point of intersection of g_2 with the line through P parallel to g_1. [If P lies on g, its coordinates are $(x,0)$ where x is the coordinate of P in the coordinate system on g_1. Similarly, each point of g_2 has coordinates $(0,y)$. The origin O has coordinates $(0,0)$. The lines g_1 and g_2 are called the x axis and y axis, respectively. It is usually convenient to take the same scale on each axis; that is, the segments joining the unit points on the two axes to the origin are congruent. The notation $P(x,y)$ denotes a point P with coordinates (x,y). If ω denotes the angle made by the positive halves of the two axes, and d the distance of points $P_1(x_1,y_1)$, $P_2(x_2,y_2)$, application of the law of cosines yields

$$d = [(x_1 - x_2)^2 + (y_1 - y_2)^2 + 2(x_1 - x_2)(y_1 - y_2) \cos \omega]^{1/2}$$

Since $\cos 90° = 0$, this important formula is simplified by taking the axes mutually perpendicular. Though it is occasionally useful to employ oblique axes (that is, $\omega \neq 90°$), the simplifications resulting from a rectangular cartesian coordinate system make it the usual choice. Such a cartesian coordinate system is assumed in what follows. Thus, the distance d of $P_1(x_1,y_1)$, $P_2(x_2,y_2)$ is given by $d = [(x_1 - x_2)^2 + (y_1 - y_2)^2]^{1/2}$.

Let θ denote the smaller of the two angles that a line g makes with the positive half of the x axis, measured in the (positive) direction of rotation (that is, from the positive half of the x axis to the positive half of the y axis). Angle θ is called the slope angle of g, and the number $\lambda = \tan \theta$, the slope of g, plays an important role in plane analytic geometry. Slope is not defined for any line perpendicular to the x axis. If $P_1(x_1,y_1)$, $P_2(x_2,y_2)$ are two distinct points of line g with slope angle $\theta \neq 90°$, and Q denotes the intersection of the line through P_1 perpendicular to the y axis with the line through P_2 perpendicular to the x axis, then $\lambda = \tan \theta = \overline{QP_2}/\overline{P_1Q} = (y_2 - y_1)/(x_2 - x_1)$.

Loci and equations. The correspondence between the geometric entity "point" and the arithmetic entity "pair of real numbers," upon which plane analytic geometry is based, results in associating with each geometric locus one or more equations that are satisfied by the coordinates of all those (and only those) points forming the locus (equations of the locus), and in associating with each system of equations in the variables x, y the figure (graph of the equations) whose points are determined by the pairs of numbers satisfying the equations. Thus the algebraic method of studying geometry is balanced by a geometric interpretation of algebra. A central problem in analytic geometry is that of finding equations of certain important figures among curves and surfaces.

Equations of lines. A line may be determined by data of various kinds, each yielding a different form for its equation. Thus an equation for a line g through $P_1(x_1,y_1)$ perpendicular to the x axis is evidently $x = x_1$. If g goes through $P_1(x_1,y_1)$, $P_2(x_2,y_2)$ and is not perpendicular to the x axis, let $P(x,y)$ be the coordinates of any other point of g. Then $(y-y_1)/(x-x_1) = \lambda = (y_2-y_1)/(x_2-x_1)$, whereas if (x,y) are the coordinates of a point not on g, $(y-y_1)/(x-x_1) \neq \lambda$. Hence an equation for g is shown as Eq. (1), from which follow Eqs. (2) and (3). The two-point form is given by Eq. (1),

$$(y-y_1)/(x-x_1) = (y_2-y_1)/(x_2-x_1) \qquad (1)$$
$$y-y_1 = \lambda(x-x_1) \qquad (2)$$
$$(x-x_1)/(x_2-x_1) = (y-y_1)/(y_2-y_1) \qquad (3)$$

the point-slope form by Eq. (2), and the symmetric form by Eq. (3). The validity of the determinant form

$$\begin{vmatrix} x & y & 1 \\ x_1 & y_1 & 1 \\ x_2 & y_2 & 1 \end{vmatrix} = 0$$

as well as the slope-intercept form, $y = \lambda x + b$, and the intercept form $x/a + y/b = 1$, $a \cdot b \neq 0$, also follow readily from Eq. (1).

A line g is determined by its distance p from the origin $O(p \geqq 0)$ and the angle β that the perpendicular to g from O makes with the x axis. (The perpendicular is directed from O to g in case g does not go through O, and so as to make $\beta < 180°$ in the contrary case.) The equation of g in terms of p and β (the so-called normal form) is $x \cos \beta + y \sin \beta - p = 0$. The directed distance from g to any point $P(x_0,y_0)$ is $x_0 \cos \beta + y_0 \sin \beta - p$. It follows that the equations of the bisectors of the angles formed by two lines

$$x \cos \beta_i + y \sin \beta_i - p_i = 0 \qquad (i = 1, 2)$$

are

$$x \cos \beta_1 + y \sin \beta_1 - p_1$$
$$= \pm (x \cos \beta_2 + y \sin \beta_2 - p_2)$$

The general form of an equation of a line is $Ax + By + C = 0$, where A, B are not both zero. The normal form is obtained from the general form

upon dividing the left-hand member by $\pm \sqrt{A^2 + B^2}$, choosing the sign of the radical opposite to that of C in case $C \neq 0$, the same as that of B in case $C = 0$ and $B \neq 0$, and the same as that of A in case $B = 0$ and $C = 0$. If $B = 0$, the general form reduces to $x = $ constant, a line perpendicular to the x axis, and if $B \neq 0$, one obtains $y = -(A/B)x + C/A$, a line with slope $-A/B$ and y intercept C/A. Thus to each line there corresponds a linear equation in x and y, and with each such equation is associated a line.

Angle between two lines. The angle ϕ ($0° < \phi < 180°$) from a line g_1 to another intersecting line g_2 is that through which g_1 must be rotated (about the point of intersection) to coincide with g_2. If θ_1, θ_2 are the slope angles of g_1, g_2, respectively, then

$$\phi = \theta_2 - \theta_1$$

and

$$\tan \phi = \frac{\tan \theta_2 - \tan \theta_1}{1 + \tan \theta_1 \tan \theta_2} = \frac{\lambda_2 - \lambda_1}{1 + \lambda_1 \cdot \lambda_2}$$

provided $\theta_1 \neq 90° \neq \theta_2$. Consequently, g_1 and g_2 are mutually perpendicular if and only if $1 + \lambda_1 \cdot \lambda_2 = 0$. The formula for $\tan \phi$ holds in case g_1, g_2 are parallel. Then $\phi = 0°$ and $\lambda_1 = \lambda_2$. It follows that two distinct lines $A_ix + B_iy + C_i = 0$, with $i = 1, 2$, are mutually perpendicular if and only if $A_1A_2 + B_1B_2 = 0$, and parallel provided $A_1B_2 - A_2B_1 = 0$.

Area of a triangle. Let $P_i(x_i,y_i)$, with $i = 1, 2, 3$, be vertices of a triangle whose area is denoted by A. Then $A = 1/2 \ d \cdot D$, where $d = [(x_1 - x_2)^2 + (y_1 - y_2)^2]^{1/2}$ and D is the distance of P_3 from the line joining P_1, P_2. Substituting the coordinates (x_3,y_3) of P_3 for (x,y) in the normal form of the equation of that line gives

$$D = \frac{\pm [(y_2 - y_1)x_3 - (x_2 - x_1)y_3 - x_1y_2 + x_2y_1]}{\sqrt{(x_2 - x_1)^2 + (y_2 - y_1)^2}}$$

and hence

$$A = \pm (1/2)(x_1y_2 + x_2y_3 + x_3y_1 - x_3y_2 - x_2y_1 - x_1y_3)$$

or

$$A = \pm(1/2) \cdot \begin{vmatrix} x_1 & y_1 & 1 \\ x_2 & y_2 & 1 \\ x_3 & y_3 & 1 \end{vmatrix}$$

The positive sign holds provided the vertices P_1, P_2, P_3 are in counterclockwise order, and the negative sign in the contrary case.

Linear combinations. If $u_1 = 0$, $u_2 = 0$ are equations of lines through a point P, for every choice of constants c_1, c_2 (except $c_1 = c_2 = 0$) the linear combination $c_1u_1 + c_2u_2 = 0$ is an equation of a line through P, and every line through P has an equation of that form. It follows that three lines $u_i = 0$, with $i = 1, 2, 3$, are concurrent provided there exist constants c_1, c_2, c_3 (not all zero) such that the linear combination $c_1u_1 + c_2u_2 + c_3u_3 = 0$ for every pair of numbers (x,y). Putting $u_i = A_ix + B_iy + C_i$, with $i = 1, 2, 3$, then c_1, c_2, c_3 are nontrivial solutions of the system of equations $c_1A_1 + c_2A_2 + c_3A_3 = 0$, $c_1B_1 + c_2B_2 + c_3B_3 = 0$, $c_1C_1 + c_2C_2 + c_3C_3 = 0$. Hence the lines $u_i = 0$, with $i = 1, 2, 3$, are concurrent if and only if

$$\begin{vmatrix} A_1 & B_1 & C_1 \\ A_2 & B_2 & C_2 \\ A_3 & B_3 & C_3 \end{vmatrix} = 0$$

Circle. By use of the formula for the distance of two points and the definition of a circle, an equation for a circle with center $C(x_0,y_0)$ and radius r ($r \geqq 0$) is found to be $(x-x_0)^2 + (y-y_0)^2 = r^2$. If $r=0$, the only (real) point of the locus is the center (x_0,y_0) and the circle is a point circle. The above equation is called the standard form. Expansion yields $x^2 + y^2 - 2x_0x - 2y_0y + x_0^2 + y_0^2 - r^2 = 0$, a quadratic in x, y with equal, nonzero, coefficients of x^2 and y^2, and with the product term xy lacking. Conversely, the locus of each such equation $A(x^2 + y^2) + 2Dx + 2Ey + F = 0$, $A \neq 0$, is a (real) circle with center $(-D/A, -E/A)$ and radius $r = (1/A)[D^2 + E^2 - AF]^{1/2}$ provided $D^2 + E^2 - AF \geqq 0$, for by "completing the square" the above equation may be put in the standard form. Let $P(x_1,y_1)$ be on the circle $(x-x_0)^2 + (y-y_0)^2 = r^2$. An equation for the tangent to the circle at P is easily seen to be $(y_1 - y_0)(y - y_1) + (x_1 - x_0)(x - x_1) = 0$. Adding $(x_1 - x_0)^2 + (y_1 - y_0)^2$ to the left side of this equation, and its equal r^2 [since $P(x_1,y_1)$ lies on the circle] to the right side, yields $(x_1 - x_0)(x - x_0) + (y_1 - y_0)(y - y_0) = r^2$ as an equation of the tangent. The formal process by which this equation may be obtained from the standard form of the equation of a circle is called polarization. If $P(x_1,y_1)$ is not on the circle, the line that is the locus of that equation is called the polar line of P with respect to the circle. If P is outside the circle, its polar line joins the contact points of the two tangents from P to the circle. *See* CIRCLE.

Polarization of the general equation of a circle gives $A(x_1x + y_1y) + D(x + x_1) + E(y + y_1) + F = 0$, as an equation of the tangent at (x_1,y_1) if that point is on the circle, and of the polar line of (x_1,y_1) otherwise. The tangential distance t from (x_1,y_1) to the circle $(x-x_0)^2 + (y-y_0)^2 = r^2$ is given by $t^2 = (x_1 - x_0)^2 + (y_1 - y_0)^2 - r^2$. Hence $(x - x_0)^2 + (y - y_0)^2 - r_0^2 = (x - x_1)^2 + (y - y_1)^2 - r_1^2$ is an equation of the locus of points with equal tangential distances from the two circles with centers (x_0,y_0), (x_1,y_1) and radii r_0, r_1, respectively. Since the equation is linear, it represents a line which evidently contains the common chord of the circles in case they intersect in two distinct points. If $u_i = 0$, with $i = 1, 2, 3$, denotes equations of three circles that intersect pairwise in two distinct points, then the three common chords are concurrent, since the equation $u_1 - u_2 = 0$ is a linear combination of the equations $u_2 - u_3 = 0$, $u_3 - u_1 = 0$. An equation of the circle through three non-collinear points $P_i(x_i,y_i)$, with $i = 1, 2, 3$, may be written

$$\begin{vmatrix} x^2 + y^2 & x & y & 1 \\ x_1^2 + y_1^2 & x_1 & y_1 & 1 \\ x_2^2 + y_2^2 & x_2 & y_2 & 1 \\ x_3^2 + y_3^2 & x_3 & y_3 & 1 \end{vmatrix}$$

Conic sections. Much of plane analytic geometry deals with a class of curves which (from the way in which they were first studied) are known as conic sections or conics. A conic is the locus of a point P that moves so that its distance from a fixed point F (the focus) is in a constant positive ratio ϵ (the eccentricity) to its distance from a fixed line (the directrix) not through F. Let $(c,0)$ be the coordinates of F, $c > 0$, and take the y axis as the directrix. Then $P(x,y)$ satisfies the equation $[(x-c)^2 + y^2]^{1/2} = \epsilon x$; that is, $(1-\epsilon^2)x^2 + y^2 - 2cx + c^2 = 0$, and it is easily seen that each point whose coordinates satisfy this equation is on the conic. Hence each conic is represented by a sec-

ond-degree equation in the cartesian coordinates (x,y). A conic is called a parabola, ellipse, or hyperbola accordingly as $\epsilon = 1$, $\epsilon < 1$, $\epsilon > 1$, respectively.

The parabola. When $\epsilon = 1$, the equation for the conic obtained above becomes $y^2 - 2cx + c^2 = 0$ or $y^2 = 2c(x - c/2)$. The curve has an axis of symmetry (the x axis when the focus and directrix are chosen as above). The point $(c/2,0)$, at which the axis intersects the curve is the vertex. Putting $x = c$ in the equation, it is seen that the chord through the focus that is perpendicular to the axis (the latus rectum) has length $2c$. All quadratics $x = Ay^2 + By + C$, $y = Ax^2 + Bx + C$, $A \neq 0$, are equations of parabolas (with axes perpendicular to the y axis, or the x axis, respectively). All other parabolas have equations $Ax^2 + 2Bxy + Cy^2 + 2Dx + 2Ey + F = 0$, with discriminant

$$\begin{vmatrix} A & B & D \\ B & C & E \\ D & E & F \end{vmatrix} \neq 0 \quad \text{and} \quad AC - B^2 = 0$$

(A quadratic with nonvanishing discriminant is called irreducible.) A simple standard form of an equation of a parabola is obtained by taking $(c/2,0)$ for focus ($c > 0$) and the line $x = -c/2$ for directrix, resulting in $y^2 = 2cx$. Polarizing gives $y_1y = c(x + x_1)$ for an equation of the tangent at the point $P(x_1,y_1)$ on the parabola. This tangent cuts the axis at the point $Q(-x_1,0)$ whose distance from the focus is $x_1 + c/2$. But this is the distance from the directrix to P, and consequently the triangle FPQ is isosceles with $\measuredangle FQP = \measuredangle FPQ$. It follows that the line joining F to any point P of the parabola makes the same angle with the tangent at P as does the line through P that is parallel to the axis of the parabola. Thus each light ray emanating from the focus is reflected parallel to the axis. *See* PARABOLA.

The ellipse. When $\epsilon < 1$, the equation of the general conic given above becomes

$$\left(\frac{x-c}{1-\epsilon^2}\right)^2 + \frac{y^2}{1-\epsilon^2} = \frac{\epsilon^2 c^2}{(1-\epsilon^2)^2} \qquad (\epsilon < 1)$$

This is an equation of the ellipse with focus $(c,0)$ and directrix the y axis. Putting $X = x - c/(1-\epsilon^2)$, $Y = y$, the standard form $X^2/A^2 + Y^2/B^2 = 1$ of the equation is obtained, where $A = \epsilon c/(1-\epsilon^2) > 0$, $B = \epsilon c/(1-\epsilon^2)^{1/2} > 0$. This is equivalent to referring the ellipse to a new set of coordinate axes obtained by translating the origin to the point $[c/(1-\epsilon^2),0]$. The coordinates of the focus F in the XY coordinate system are $[-c\epsilon^2/(1-\epsilon^2),0]$ and the equation of the directrix is $X = -c/(1-\epsilon^2)$. Since the standard form shows the curve to be symmetric with respect to each of the new axes and the origin, it is clear that $F'[c\epsilon^2/(1-\epsilon^2),0]$ and $X = c/(1-\epsilon^2)$ may be taken as focus and directrix. Putting $C = c\epsilon^2/(1-\epsilon^2)$, then $C^2 = A^2 - B^2$, and $X^2/A^2 + Y^2/B^2 = 1$ is seen to be the locus of points $P(X,Y)$ the sum of whose distances from $(C,0)$ and $(-C,0)$ is $2A$. This relation is frequently used to define an ellipse. The chord containing the foci is the major axis; its length is $2A$. The chord of length $2B$ through the midpoint of the foci, perpendicular to the major axis, is the minor axis. The tangent to $X^2/A^2 + Y^2/B^2 = 1$ at $P(X_1,Y_1)$ on the ellipse is $X_1X/A^2 + Y_1Y/B^2 = 1$. Lines PF, PF' make equal angles with the tangent; so sound or light that emanates from one focus is reflected to the other. An irreducible quadratic $Ax^2 + 2Bxy + Cy^2 + 2Dx + 2Ey + F = 0$ is an equation of an ellipse provided that $AC - B^2 > 0$.

The hyperbola. When $\epsilon > 1$, the equation for the conic is written $[x + c/(\epsilon^2 - 1)]^2 - y^2/(\epsilon^2 - 1) = \epsilon^2 c^2/(1 - \epsilon^2)^2$, $(\epsilon > 1)$. Putting $X = x + c/(\epsilon^2 - 1)$, $Y = y$ gives the standard form of a hyperbola $X^2/A^2 - Y^2/B^2 = 1$, where $A = \epsilon c/(\epsilon^2 - 1) > 0$, $B = \epsilon c/[\epsilon^2 - 1]^{1/2} > 0$. It is clear that the curve consists of two branches and is symmetric to the axes and the origin. Putting $C = c\epsilon^2(\epsilon^2 - 1)$, then both $F(C,0)$, $F'(-C,0)$ are foci of the hyperbola, with respective directrices $X = c/(\epsilon^2 - 1)$, $X = -c/(\epsilon^2 - 1)$, and $A^2 + B^2 = C^2$. Then $X^2/A^2 - Y^2/B^2 = 1$ is seen to be an equation of the locus of points $P(X,Y)$ such that $PF - PF' = 2A$. The lines $X/A \pm Y/B = 0$ are asymptotes of the hyperbola. As a point P traverses either branch of the hyperbola, its distances from an asymptote approach zero. An irreducible quadratic $Ax^2 + 2Bxy + Cy^2 + 2Dx + 2Ey + F = 0$ is an equation of an hyperbola provided $AC - B^2 < 0$.

Three-dimensional space. Let cartesian coordinate systems be established on each of three pairwise mutually perpendicular lines of three-space that intersect in O, the common origin of the systems. Suppose equal scales and call the lines the x axis, y axis, and z axis. To each point P of space an ordered triple (x,y,z) of real numbers is attached as rectangular cartesian coordinates, where x is the coordinate of the foot of the perpendicular from P to the x axis, and y and z are similarly defined. Thus every point of space has unique coordinates, and each ordered triple of real numbers is the coordinates of a point of space. If d denotes the distance of two points $P_1(x_1,y_1,z_1)$, $P_2(x_2,y_2,z_2)$, then $d = [(x_1 - x_2)^2 + (y_1 - y_2)^2 + (z_1 - z_2)^2]^{1/2}$. Let g denote any directed line and g' the line through O parallel to g and directed in the same sense. If α, β, γ denote the angles that g' makes with the x, y, and z axes, respectively (the direction angles of g), then $\cos \alpha$, $\cos \beta$, $\cos \gamma$ are the direction cosines of g. They satisfy the relation $\cos^2 \alpha + \cos^2 \beta + \cos^2 \gamma = 1$, and any three numbers λ, μ, ν such that $\lambda^2 + \mu^2 + \nu^2 = 1$ are the direction cosines of a (directed) line. Three numbers a, b, c proportional to the direction cosines λ, μ, ν of a directed line g are direction numbers of g. Clearly, direction numbers of parallel lines are proportional. If $P_1(x_1,y_1,x_1)$, $P_2(x_2,y_2,z_2)$ are distinct points of g, then $x_2 - x_1$, $y_2 - y_1$, $z_2 - z_1$ are direction numbers of g. It follows that $P(x,y,z)$ is on g if and only if $x - x_1 = t(x_2 - x_1)$, $y - y_1 = t(y_2 - y_1)$, $z - z_1 = t(z_2 - z_1)$, $-\infty > t > \infty$. These are parametric equations of g (t is the parameter), from which the symmetric equations $(x - x_1)/(x_2 - x_1) = (y - y_1)/(y_2 - y_1) = (z - z_1)/(z_2 - z_1)$ follow at once. The direction cosines of g, directed from P_1 to P_2, are $\cos \alpha = (x_2 - x_1)/d$, $\cos \beta = (y_2 - y_1)/d$, $\cos \gamma = (z_2 - z_1)/d$, where $d = [(x_2 - x_1)^2 + (y_2 - y_1)^2 + (z_2 - z_1)^2]^{1/2}$.

If $\alpha_1, \beta_1, \gamma_1$ and $\alpha_2, \beta_2, \gamma_2$ are direction angles of directed lines g_1, g_2, respectively, and θ denotes the angle between them, then $\cos \theta = \cos \alpha_1 \cos \alpha_2 + \cos \beta_1 \cos \beta_2 + \cos \gamma_1 \cos \gamma_2$. Hence g_1, g_2 are mutually perpendicular if and only if $a_1 a_2 + b_1 b_2 + c_1 c_2 = 0$, where a_1, b_1, c_1 and a_2, b_2, c_2 are direction numbers of g_1, g_2, respectively. Let the plane π go through $P_0(x_0,y_0,z_0)$ and be perpendicular to a line g with direction numbers a, b, c. Then $P(x,y,z)$ is in π if and only if $P = P_0$ or the line joining it to P_0 is at right angles to a line through P_0 that is parallel to

g; that is, if and only if $a(x - x_0) + b(y - y_0) + c(z - z_0) = 0$. Hence to each plane corresponds a linear equation. Conversely, if $P_0(x_0,y_0,z_0)$ satisfies the linear equation $Ax + By + Cz + D = 0$, with A, B, C not all zero, then $A(x - x_0) + B(y - y_0) + C(z - z_0) = 0$ is an equation of a plane through P_0, perpendicular to a line with direction numbers A, B, C, and so $Ax + By + Cz + D = 0$, with A, B, $C \neq 0, 0, 0$, is an equation of a plane. Clearly $x = 0$ is an equation for the plane determined by the y and z axes; equations of the other two coordinate planes are $y = 0$ and $z = 0$. If a directed perpendicular g from O to a plane meets the plane at P (g is directed from O to P in case the plane is not through O) the plane has equation $x \cos \alpha + y \cos \beta + z \cos \gamma = p$, where α, β, γ are the direction angles of g and $p = \overline{OP}$. This is the normal form of equation of a plane. The general form is reduced to it upon dividing by $\pm[A^2 + B^2 + C^2]^{1/2}$. The distance from $Ax + By + Cz + D = 0$ to $P(x_0,y_0)$ is $(Ax_0 + By_0 + Cz_0 + D)/\pm(A^2 + B^2 + C^2)^{1/2}$, where the sign is selected opposite that of D. (In case $D = 0$, other conventions are used.) Two planes $A_i x + B_i y + C_i z = 0$, with $i = 1, 2$, are parallel in case the number triples A_1, B_1, C_1 and A_2, B_2, C_2 are proportional, and are mutually perpendicular provided $A_1 A_2 + B_1 B_2 + C_1 C_2 = 0$ (since A_1, B_1, C_1 and A_2, B_2, C_2 are direction numbers of lines that are perpendicular to the respective planes). If $P_i(x_i,y_i,z_i)$, with $i = 1, 2, 3$, is not collinear, the plane determined has the equation

$$\begin{vmatrix} x & y & z & 1 \\ x_1 & y_1 & z_1 & 1 \\ x_2 & y_2 & z_2 & 1 \\ x_3 & y_3 & z_3 & 1 \end{vmatrix} = 0$$

Three planes $A_i x + B_i y + C_i z + D_i = 0$, with $i = 1, 2, 3$, intersect in one point if and only if

$$\begin{vmatrix} A_1 & B_2 & C_1 \\ A_2 & B_2 & C_2 \\ A_3 & B_3 & C_3 \end{vmatrix} = 0$$

The locus of points whose coordinates satisfy an equation $f(x,y,z) = 0$ is a surface. Curves may be thought of as intersections of two surfaces, and as such the coordinates of their points satisfy two equations $f(x,y,z) = 0$, $g(x,y,z) = 0$. Thus a line, considered as the intersection of two planes, is given by the two (simultaneous) equations $A_i x + B_i y + C_i z + D_i = 0$, with $i = 1, 2$. That line has direction numbers

$$\begin{vmatrix} B_1 & C_1 \\ B_2 & C_2 \end{vmatrix} \quad \begin{vmatrix} C_1 & A_1 \\ C_2 & A_2 \end{vmatrix} \quad \begin{vmatrix} A_1 & B_1 \\ A_2 & B_2 \end{vmatrix}$$

Curves are also represented parametrically by equations $x = f(t)$, $y = g(t)$, $z = h(t)$, the parameter t varying in an interval (a,b), finite or infinite. Parametric equations for the line have already been given. Additional examples are $x = r \cos t$, $y = r \sin t$, $z = 0$, with $0 \leq t \leq 2\pi$, parametric equations of the circle in the xy plane, with center at O and radius r; $x = a \cdot \cos t$, $y = a \cdot \sin t$, $z = kt$, with $k \neq 0$, $-\infty < t < \infty$, parametric equations of a circular helix—the curve of the thread of a machine (untapered) screw.

Special surfaces. It follows from the definition of a sphere and the formula for distance of two points that $(x - a)^2 + (y - b)^2 + (z - c)^2 = r^2$ is an equation for the sphere with center (a,b,c) and radius r, and by completing the squares of the x, y, and z terms in the equation $x^2 + y^2 + z^2 + 2Dx + 2Ey + 2Fz + G = 0$, it is seen that the locus of such an equation is a sphere with positive or zero radius, or there is no (real) locus.

Any equation in just two of the three coordinates is an equation of a cylinder whose elements are parallel to the axis of the missing variable. Thus the locus in 3-space of $x^2 + y^2 = r^2$ is a (right) circular cylinder whose elements are parallel to the z axis and which intersects the xy plane in the circle $x^2 + y^2 = r^2$, $z = 0$.

Any equation $f(x,y,z) = 0$, with $f(x,y,z)$ homogeneous in x, y, z (for example, $4xy - xz + yz = 0$, $x^3 - xy^2 + z^3 = 0$) has a cone with vertex O as locus.

A surface of revolution is obtained by rotating a plane curve C about a line g of its plane. If $f(x,y) = 0$, $z = 0$ are equations of C, and g is the x axis, the resulting surface of revolution has equation $f(x, \sqrt{y^2 + z^2}) = 0$. Thus the surface generated by revolving the circle $x_2 + (y - b)^2 = a^2$, $z = 0$, about the x axis (the torus or anchor ring, if $b > a$) has the equation $x^2 + (\sqrt{y^2 + z^2} - b)^2 = a^2$.

A quadric surface is the locus of points whose coordinates satisfy an equation of the form $Ax^2 + By^2 + Cz^2 + Dxy + Exz + Fyz + Gx + Hy + Jz + K = 0$, where at least one coefficient of a second-degree term is not zero. Some surfaces obtained by rotating conics about a line belong to this class, for example, spheres, prolate and oblate spheroids (given by rotating an ellipse about its major and minor axes, respectively), hyperboloids and paraboloids resulting from rotations of hyperbolas and parabolas about their axes of symmetry, and right circular cones and cylinders. Cylinders with conics for directrix curves are also members. Apart from degenerate cases, such as two planes, the remaining quadric surfaces and the standard forms of their equations are (1) ellipsoid, $x^2/a^2 + y^2/b^2 + z^2/c^2 = 1$; (2) hyperboloid of one sheet, $x^2/a^2 + y^2/b^2 - z^2/c^2 = 1$; (3) hyperboloid of two sheets, $x^2/a^2 - y^2/b^2 - z^2/c^2 = 1$; (4) elliptic paraboloid, $x^2/a^2 + y^2/b^2 = 2z$; and (5) hyperbolic paraboloid, $x^2/a^2 - y^2/b^2 = 2z$. Hyperboloids of one sheet and hyperbolic paraboloids are ruled surfaces; that is, each contains an infinity of straight lines, called generators. In fact, each of those surfaces contains two sets of generators.

n-Dimensions. Let cartesian coordinate systems with equal scales be established on each of n pairwise mutually perpendicular lines intersecting in the common origin O, and label the lines OX_1, OX_2, . . . , OX_n. To each point P of n space an ordered n-tuple $(x_1, x_2, . . . , x_n)$ of numbers is attached as coordinates, where x_i is the coordinate of the foot of the perpendicular from P to OX_i, with $i = 1, 2, . . . , n$. Two points $P(x_1, x_2, . . . , x_n)$, $Q(y_1, y_2, . . . , y_n)$ have distance $d = [(x_1 - y_1)^2 + (x_2 - y_2)^2 + \cdots + (x_n - y_n)^2]^{1/2}$. Direction angles $x_1, x_2, . . . , x_n$ of a directed line are defined as in three-dimensional analytic geometry, and the direction cosines satisfy the relation $\cos^2 \alpha_1 + \cos^2 \alpha_2 + \cdots + \cos^2 \alpha_n = 1$. Direction cosines of the line $g(P,Q)$, directed from P to Q, are $(y_1 - x_1)/d$, $(y_2 - x_2)/d$, . . . , $(y_n - x_n)/d$, and numbers proportional to them (for example, the numerators) are direction numbers of $g(P,Q)$. Hence $(X_1, X_2, . . . , X_n)$ are coordinates of a point on $g(P,Q)$ if and only if $(X_1 - x_1)/(y_1 - x_1) = (X_2 - x_2)/(y_2 - x_2) = \cdots = (X_n - x_n)/(y_n - x_n)$. These are symmetric

equations for the line determined by P, Q. Denoting the common value of the quotients by t gives the n parametric equations $X_i = x_i + t(y_i - x_i)$, with $i = 1, 2, . . . , n, -\infty < t < \infty$. Let g be a line through $C(c_i, c_2, . . . , c_n)$ with direction numbers $a_1, a_2, . . . , a_n$. Point C, together with all points X such that the line $g(C,X)$ is perpendicular to g, defines an $(n-1)$-dimensional subspace (hyperplane). Its equation is $a_1(X_1 - c_1) + a_2(X_2 - c_2) + \cdots + a_n(X_n - c_n) = 0$, since $X_i - c_i$, with $i = 1, 2, . . . , n$, are direction numbers of $g(C,X)$ and the equation is the condition that g and $g(C,X)$ be mutually perpendicular. It is readily seen that the locus of every linear equation $A_1X_1 + A_2X_2 + \cdots + A_nX_n + K = 0$ is a hyperplane perpendicular to a line with direction numbers A_1, A_2, . . . , A_n. It may be put in normal form by dividing by $\pm(A_1^2 + A_2^2 + \cdots + A_n^2)^{1/2}$, and the distance from it to a point $C(c_1, c_2, . . . , c_n)$ if found by substituting the coordinates of C for X_1, X_2, . . . , X_n. Subspaces of dimension $k(1 \leq k < n)$ are given by systems of $n - k$ linear equations. An equation of the hyperplane determined by n points is readily expressed in determinant form, as well as the condition that $n + 1$ points be on a hyperplane [substitute the coordinates of the $(n + 1)$-st point in the first row of the determinant equation of the hyperplane]. Discussion of loci of higher order is beyond the scope of this article.

This brief sketch of analytic geometry has dealt only with that coordinate system most frequently used. For other coordinate systems (polar), as well as a discussion of transformation of coordinates, *see* COORDINATE SYSTEMS. See also ALGEBRA; TRIGONOMETRY.   [LEONARD M. BLUMENTHAL]

Bibliography: A. A. Albert, *Solid Analytic Geometry*, 1966; G. Fuller, *Analytic Geometry*, 5th ed., 1979; W. F. Osgood and W. C. Graustein, *Plane and Solid Analytic Geometry*, 1921; M. H. Protter and C. B. Morrey, Jr., *Analytic Geometry*, 2d ed., 1975; G. Salmon, *A Treatise on Conic Sections*, 10th ed., 1896; P. H. Schoute, *Mehrdimensionale Geometrie*, 2 vols., 1902–1905.

Angular frequency

A measure of the rate of oscillation of sinusoidally varying phenomena. If a particle is moving with constant speed v on a circular path of radius r, as indicated in the figure, it will require a time $T = 2\pi r/v$ to complete one cycle and the frequency of the motion is $f = 1/T$. The angular frequency ω is then defined to be the number of radians traversed per unit time. Since there are 2π radians per cycle, $\omega = 2\pi/T = 2\pi f$.

The position of the particle in the diagram can be written in component form as $x = r \cos \theta$, $y = r \sin \theta$. If the particle is arbitrarily taken to have been at the point p of the figure at the zero of time, $t = 0$, then $\theta = \omega t$ and the position of the particle is given by $x = r \cos \omega t$, $y = r \sin \omega t$. These latter equations, indicating the sinusoidal or harmonic character of the position components, also illustrate the primary role of the angular frequency. Motion which is harmonic and phenomena which are describable in terms of harmonic motion depend upon the time via a trigonometric function whose argument is the product ωt. See FREQUENCY (WAVE MOTION); HARMONIC MOTION.

[K. L. KLIEWER]

ANGULAR FREQUENCY

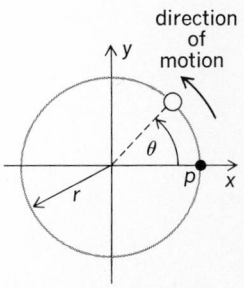

direction of motion

Illustration of angular frequency.

Angular momentum

In classical physics, angular momentum is the moment of momentum and is, conceptually, the momentum associated with rotation. A particle of mass m moving with velocity v and having position vector r with respect to a fixed coordinate origin is said to have an angular momentum L about this origin given by Eq. (1), where p is the

$$L = m(r \times v) = r \times p \qquad (1)$$

linear momentum. *See* CALCULUS OF VECTORS; MOMENTUM.

Angular momentum—like linear momentum—is an additive vectorial quantity. The angular momentum of a system of particles is obtained by summing the angular momenta of the individual particles.

In nonrelativistic mechanics, it is useful to express the angular momentum in terms of the angular velocity ω of the particle about the origin; the defining relation is $v = \omega \times r$. For a rigid body, rotating with angular velocity ω about a fixed point, one obtains from Eq. (1) the total angular momentum to be given by Eq. (2), where the sum is over

$$L = \sum_{i=1}^{N} m_i r^{(i)} \times (w \times r^{(i)}) \qquad (2)$$

the N mass points constituting the rigid body. Equation (2) may be written as Eq. (3), where I is

$$L = I \cdot \omega \qquad (3)$$

the moment of inertia, a tensor, and has the form given by Eqs. (4). *See* MOMENT OF INERTIA; RIGID-BODY DYNAMICS.

$$I_{xx} = \sum_{i=1}^{N} m_i(y_i^2 + z_i^2)$$
$$I_{xy} = -\sum_{i=1}^{N} m_i x_i y_i, \text{ etc.} \qquad (4)$$

Conservation. The importance of angular momentum in classical mechanics derives from the fact that it is conserved (remains constant in time) if no external torques are applied. Angular momentum conservation and rotation are of key importance, for example, in the structure, shape, and evolution of celestial bodies of all kinds. There is a close connection between conservation laws and invariance (symmetry) transformations (Noether's theorem). To be precise, if Lagrange's equations are invariant for spatial (temporal) translations, then linear momentum (respectively energy) is conserved. Similarly, if Lagrange's equations are invariant to rotations, then angular momentum is conserved. This deep connection between symmetry and conservation laws is not valid for Newton's laws. The requirement that Lagrange's equations apply constitutes a restriction on the forces. *See* CONSERVATION OF MOMENTUM; LAGRANGE'S EQUATIONS; SYMMETRY LAWS.

Quantum angular momentum. This connection between symmetry and conservation laws is valid in quantum physics, since quantum mechanics is based on hamiltonian (and hence lagrangian) equations. *See* NONRELATIVISTIC QUANTUM THEORY; QUANTUM MECHANICS.

In quantum physics, one associates to every classical observable an (hermitian) operator. For example, the momentum observable, p, is associ-

ated (Schrödinger realization) with the operator $-i\hbar\nabla$, where $i = \sqrt{-1}$, $\hbar$ is Planck's constant h divided by 2π, and ∇ is the gradient operator.

The operator associated in quantum physics to the angular momentum L of Eq. (1) is correspondingly defined by Eqs. (5)–(7). The vector operator

$$L_x = -i\hbar\left(y\frac{\partial}{\partial z} - z\frac{\partial}{\partial y}\right) \qquad (5)$$

$$L_y = -i\hbar\left(z\frac{\partial}{\partial x} - x\frac{\partial}{\partial z}\right) \qquad (6)$$

$$L_z = -i\hbar\left(x\frac{\partial}{\partial y} - y\frac{\partial}{\partial x}\right) \qquad (7)$$

L, whose components are defined above, is called the orbital angular momentum operator. Forming the vector cross product $L \times L$, one obtains Eq. (8),

$$L \times L = i\hbar L \qquad (8)$$

known as the angular momentum commutation law.

The total quantum angular momentum J is now defined as the most general hermitian operator satisfying the commutation rule given in Eq. (8), namely, $J \times J = i\hbar J$. In general, for a single particle, J is the sum of the two operators L and S shown in Eq. (9). Here L is the orbital angular momentum operator defined by Eqs. (5)–(7) and corresponds

$$J = L + S \qquad (9)$$

to the classical angular momentum, and S is an operator associated with the intrinsic, or spin, angular momentum of the particle and has no analog in classical theory. The spin angular momentum S is considered as intrinsic to the particle, since this angular momentum—unlike the orbital angular momentum L—is independent of the choice of origin of the coordinate frame. *See* SPIN (QUANTUM MECHANICS).

Eigenvalues and matrices. The components J_x, J_y, J_z of J, as well as the operator in Eq. (10), are hermitian and, moreover, J^2 commutes with

$$J^2 = J_x^2 + J_y^2 + J_z^2 \qquad (10)$$

all of the J_x, J_y, J_z; that is, $J^2 J_z - J_z J^2 = 0$, and similarly for J_x, J_y. This implies that the eigenvalue equations (11) and (12) have solutions for particular values of the real numbers α, β called

$$J^2\Psi = \alpha\Psi \qquad (11)$$

$$J_z\Psi = \beta\Psi \qquad (12)$$

eigenvalues. The functions Ψ are simultaneous eigenfunctions of J^2 and J_z. The particular values of α for which eigenfunctions Ψ exist are $j(j + 1)\hbar$, where $j = 0, \frac{1}{2}, 1, \frac{3}{2}, 2, \ldots$. For each such j there are $(2j + 1)$ permissible values $\beta = m\hbar$, where $m = -j, -j + 1, \ldots, j - 1, j$. Thus, the quantum angular momentum of a particle can have only values which are integral, or half-integral, multiples of $\hbar$.

The effect of the angular momentum operators J_x, J_y, J_z on the simultaneous eigenfunctions of J^2 and J_z can be expressed by a set of matrices called the matrices of angular momentum. Defining the operators $J_+ = J_x + iJ_y$ and $J_- = J_x - iJ_y$, one has, for each value of j, the three matrices as shown in Eqs. (13)–(15). Here $\delta_b{}^a$ is the Dirac delta function,

$$J_z^{(j)} = m\hbar\,\delta_{m'}^m \qquad (13)$$

$$J_+^{(j)} = [(j-m)(j+m+1)]^{1/2} \hbar \, \delta_{m'}^{m+1} \qquad (14)$$

$$J_-^{(j)} = [(j+m)(j-m+1)]^{1/2} \hbar \, \delta_{m'}^{m-1} \qquad (15)$$

which is 0 if $a \neq b$ and 1 if $a = b$, and the terms on the right-hand side represent the components of the matrices corresponding to the operators J_z, J_+, J_- for each j. The Condon-Shortley-Wigner phase convention has been used in Eqs. (13)–(15).

Rotations. Let f be a function of the spatial coordinates in the system used to describe the motion of a particle, and suppose that the coordinates are rotated through an angle θ about the z axis. The original function f will be transformed into a new function f', and one then writes Eq. (16), where $D(\theta)$ is the rotation operator transforming f

$$f' = D(\theta) \, f \qquad (16)$$

into f'. Equating the value of original function f at the original point with the value of the new function f' at the new point, one obtains Eq. (17); that is, the effect on a function of the spatial variables

$$D(\theta) = e^{-i\theta L_z/\hbar} \qquad (17)$$

x, y, z of a rotation of coordinates can be expressed in terms of the orbital angular momentum operators, Eqs. (5)–(7). A general rotation of the coordinate frame about any axis requires two generalizations: specifying an arbitrary rotation by the Euler angles (α, β, γ), and replacing the orbital angular momentum operator in Eq. (17) by the total angular momentum operator J so as to effect spin rotations as well). This yields the general rotation operator $D(\alpha\beta\gamma)$ of Eq. (18).

$$D(\alpha\beta\gamma) = e^{-i\alpha J_z/\hbar} \, e^{-i\beta J_y/\hbar} \, e^{-i\gamma J_z/\hbar} \qquad (18)$$

Since the operators J_x, J_y, J_z are associated with matrices, the operator $D(\alpha\beta\gamma)$ can also be expressed in terms of matrices. Specifically, for each j there is a $(2j+1) \times (2j+1)$ matrix whose component in the m'-th row and m-th column is given by Eq. (19), where Eq. (20) holds and $P_n^{(a,b)}(x)$ is the Jacobi polynomial, Eq. (21).

$$D_{m'm}^j(\alpha\beta\gamma) = e^{-im'\alpha} \, d_{m'm}^j(\beta) e^{-im\gamma} \qquad (19)$$

$$d_{m'm}^j(\beta) = \left[\frac{(j+m)!(j-m)!}{(j+m')!(j-m')!} \right]^{1/2} (\cos\beta/2)^{m'+m}$$
$$\cdot (\sin b/2)^{m'-m} P_{j-m}^{m-m',\,m+m'} (\cos\beta) \qquad (20)$$

$$P_n^{(a,b)}(x) \equiv 2^{-n} \sum_{\nu=0}^{n} \binom{n+a}{\nu} \binom{n+b}{n-\nu}$$
$$\cdot (x-1)^{n-\nu} (x+1)^\nu \qquad (21)$$

This connection between angular momentum and coordinate rotations is the basis for the close relationship between angular momentum theory and the symmetry properties of physical systems.

Quantal addition of angular momentum. The discussion above is concerned with the properties of a single angular momentum operator J. If the physical system is composed of two or more particles, or if one considers the angular momentum of a single particle as composed of two separate parts, for example, orbital angular momentum and intrinsic spin, then the problem of compounding the different angular momenta arises.

Specifically, two angular momenta J_1, J_2 which can be described independently at any instant of time are called kinematically independent mo-

menta. Mathematically, kinematic independence is expressed by Eq. (22), where the expression $[J_1, J_2]$ is the commutator of the operators J_1, J_2 and

$$[J_1, J_2] = 0 \qquad (22)$$

is defined in Eq. (23). The angular momentum $J = J_1 + J_2$ of a system composed of two kinematically

$$[J_1, J_2] = J_1 J_2 - J_2 J_1 \qquad (23)$$

independent momenta J_1, J_2 will have eigenfunctions Ψ_{jm} corresponding to the operators J^2, J_z, which are expressible in terms of the eigenfunctions $\Psi_{j_1 m_1}$ and $\Psi_{j_2 m_2}$ corresponding to the operators J_1^2, J_{1z} and J_2^2, J_{2z}, respectively. Specifically, the Ψ_{jm} are the linear combinations of $\Psi_{j_1 m_1}$, $\Psi_{j_2 m_2}$ shown in Eq. (24). The matrix coefficients $C_{m_1 m_2 m}^{j_1 j_2 j}$ characterize the addition of, or

$$\Psi_{jm} = \sum_{m_1 m_2} C_{m_1 m_2 m}^{j_1 j_2 j} \Psi_{j_1 m_1} \Psi_{j_2 m_2} \qquad (24)$$

or coupling between, J_1 and J_2 and are called (inaccurately) Clebsch-Gordan coefficients or, more properly, Wigner coefficients.

The Wigner coefficients will be zero unless the quantum numbers j_1, j_2, j, m_1, m_2, m satisfy Eq. (25) and (26).

$$m = m_1 + m_2 \qquad (25)$$

$$j = j_1 + j_2, j_1 + j_2 - 1, \ldots, |j_1 - j_2| + 1, |j_1 - j_2| \qquad (26)$$

In addition to being coupling coefficients, the matrices $C_{m_1 m_2 m}^{j_1 j_2 j}$ play a second fundamental role in the quantum theory of angular momentum. A tensor operator $\mathbf{T}$ is defined as a family of operators $\{T(JM)\}$ with $-J \leq M \leq J$, which satisfies the commutation rule shown in Eq. (27). Here the

$$[J_i, \mathbf{T}(JM)] = \sum_{M'} <JM'|J_i|JM> T(JM) \qquad (27)$$

bracket on the left is the commutator defined by Eq. (23), and $<JM'|J_i|JM>$ are the matrices of angular momentum defined by Eq. (13)–(15) with J_i equal to J_z, J_+ or J_- according as $i = 0, +1$, or -1. Each of the operators $T(JM)$ has a matrix representation whose matrix components are denoted by $<j'm'|T(JM)|jm>$. The statement of the fundamental Wigner-Eckart theorem is then given by Eq. (28). Here $<j'|T(J)|j>$ is a matrix called the reduced matrix element of the set of operators

$$<j'm'|\mathbf{T}(JM)|jm> = C_{mMm'}^{jJj'} <j|\mathbf{T}(J)|j> \qquad (28)$$

$\mathbf{T}(JM)$. Thus, according to this theorem, the matrices of a tensor operator $\mathbf{T}$ can be factored into two components, the first of which is a Wigner coefficient which can be calculated entirely from symmetry considerations, and the second of which is a matrix which contains information about the physical properties of the particular operator, and which is completely independent of the quantum numbers m, M, m' (hence the second term is orientation-independent). Conditions in Eq. (25) and (26) contain the conservation laws of angular momentum since they determine which of the tensor components are different from zero, and accordingly, the Wigner coefficients contain the selection rules which govern atomic spectra. That almost all the rules of spectroscopy follow from the symmetry of the problem is perhaps the most remarkable result of the quantum theory of angular momentum.

[L. C. BIEDENHARN]

Bibliography: L. C. Biedenharn and J. D. Louck, *Angular Momentum in Quantum Physics*, 1980; E. U. Condon and G. H. Shortley, *Theory of Atomic Spectra*, 1953; E. P. Wigner, *Group Theory and Its Application to Atomic Spectra*, 1959.

Anharmonic oscillator

A system that oscillates with a periodic motion that is not simple harmonic. An oscillator is anharmonic if the restoring force opposing a displacement from the position of equilibrium is a nonlinear function of the displacement. The free motion of such an oscillator may be a complicated function of time. It is periodic, but with a period that depends on the amplitude. A damping nonlinearity in the velocity can also give rise to anharmonicity in an oscillator.

If an anharmonic oscillator is driven by a force with a time dependence

$$F = F_0 \cos (2\pi f t)$$

the resulting steady-state motion will involve not only a response with frequency f, but also overtones whose frequencies are integral multiples of f, and in certain cases subharmonics with frequencies that are rational fractions of f.

Many oscillators that are approximately harmonic for sufficiently small amplitudes become anharmonic for larger motions. The thermal expansion of solids is attributed to the nonlinear forces between atoms (see illustration). Small atomic vibrations (at low temperatures) are harmonic, and are centered about the static equilibrium distance E. As the temperature rises, the amplitude of atomic oscillations increases, and the oscillations become anharmonic. One consequence is that the resulting motions are unsymmetric, and the mean separation between atoms increases beyond the static equilibrium distance. *See* HARMONIC OSCILLATOR; LATTICE VIBRATIONS; THERMAL EXPANSION.

[JOSEPH M. KELLER]

Bibliography: D. W. Jordan and P. Smith, *Nonlinear Ordinary Differential Equations*, 1977; A. H. Nayfeh and D. T. Mook, *Nonlinear Oscillations*, 1979; J. J. Stoker, *Non-linear Vibrations in Mechanical and Electrical Systems*, 1950.

Anisotropy (physics)

The quality of variation of a physical property with the direction in a body along which the property is measured. For example, the resistivity of certain single crystals measured with the electric field along a particular crystallographic direction may be higher than along directions perpendicular to it. Thus such crystals are anisotropic with respect to resistivity. Examples of bodies which are anisotropic in some of their properties are liquid "crystals," single crystals, and aggregates of polycrystals with a preferred orientation. *See* ELASTICITY; ISOTROPY.

[DAVID TURNBULL]

Antiferromagnetism

A property possessed by some metals, alloys, and salts of transition elements in which the atomic magnetic moments, at sufficiently low temperatures, form an ordered array which alternates or spirals so as to give no net total moment in zero applied magnetic field. Figure 1 shows the simple antiparallel arrangement of manganese moments

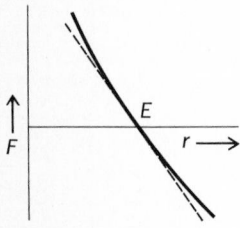

ANHARMONIC
OSCILLATOR

Interatomic force F as a function of the atomic separation r. Anharmonicity is produced by the departure of the actual force (solid curve) from the dashed line.

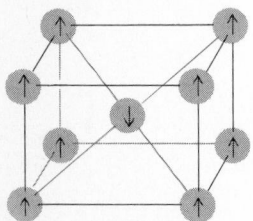

ANTIFERROMAGNETISM

Fig. 1.   The antiferromagnetism in manganese fluoride. Only manganese atoms are shown.

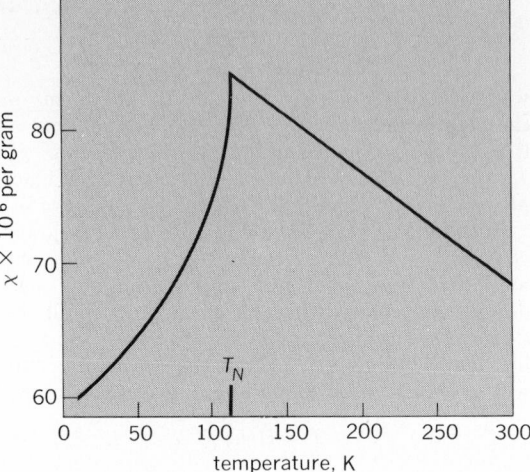

Fig. 2.   Magnetic susceptibility of powdered manganese oxide. (*After H. Bizette, C. F. Squire, and B. Tsai, 1938*)

at temperatures below 72 K in the unit cell of manganese fluoride (MnF_2). The most direct way of detecting such arrangements is by means of neutron diffraction. *See* NEUTRON DIFFRACTION.

Néel temperature. This is the transition temperature (L. Néel, 1932) below which the spontaneous antiparallel magnetic ordering takes place. A plot of the magnetic susceptibility of a typical antiferromagnetic powder sample versus temperature is shown in Fig. 2. Below the Néel point, which is characterized by the sharp kink in the susceptibility, the spontaneous ordering opposes the normal tendency of the magnetic moments to align parallel to the applied field. Above the Néel point, the substance is paramagnetic, and the susceptibility χ obeys the Curie-Weiss law, as in Eq. (1),

$$\chi = C/(T + \theta) \tag{1}$$

with a negative paramagnetic Curie temperature $-\theta$. The Néel temperature is similar to the Curie temperature in ferromagnetism. *See* CURIE TEMPERATURE; CURIE-WEISS LAW; MAGNETIC SUSCEPTIBILITY.

The cooperative transition that characterizes antiferromagnetism is thought to result from an interaction energy U of the form given in Eq. (2), where S_i and S_j are the spin angular momentum vectors associated with the magnetic moments of neighbor atoms i and j, and J_{ij} is an interaction constant which probably arises from the superexchange coupling discussed later, although formally Eq. (2) is identical to the Heisenberg exchange

$$U = -2\Sigma J_{ij} S \cdot S_j \tag{2}$$

energy. If all J_{ij} are positive, the lowest energy is achieved with all S_i and S_j parallel, that is, coupled ferromagnetically. Negative J_{ij} between nearest-neighbor pairs (i,j) may lead to simple antiparallel arrays, as in Fig. 1; if the distant neighbors also have sizable negative J_{ij}, a spiral array may have lowest total energy. *See* FERROMAGNETISM; HELIMAGNETISM.

A simple lattice like MnF_2 (Fig. 1) can be divided into sublattice 1, containing all corner atoms, and sublattice 2, containing all body-centered atoms. Nearest-neighbor interactions connect atoms on different sublattices. On the average, the interaction may be replaced by a single antiparallel coupling between the total magnetizations M_1 and M_2 of the two sublattices. Each sublattice acts as if it were in a large internal magnetic field (Weiss field) proportional to the negative magnetization of the other sublattice. This elementary approach was first given by Néel in 1932 and is analogous to the Weiss molecular field theory of ferromagnetism. A variety of more exact treatments of Eq. (2) have been made, but the basic features of antiferromagnetism appear in this simple model.

At high temperatures, the sublattice magnetizations obey the Curie law, as in Eqs. (3) and (4),

$$M_1 = (C'/T)(H_0 - \lambda M_2) \qquad (3)$$
$$M_2 = (C'/T)(H_0 - \lambda M_1) \qquad (4)$$

where C' is the Curie constant for a sublattice, H_0 is an applied external field, and $-\lambda$ is the proportionality constant of the internal Weiss field. From Eqs. (3) and (4), Eq. (5) is derived, which fits the Curie-Weiss law, Eq. (1), with the values shown in Eq. (6).

$$\chi = (M_1 + M_2)/H_0$$
$$= 2(C'/T)/[1 + \lambda(C'/T)] \qquad (5)$$
$$C = 2C' \qquad \theta = C'\lambda \qquad (6)$$

The condition that M_1 and M_2 can have finite values in the absence of H_0 (condition of spontaneous sublattice magnetization) is that the determinant of the coefficients of M_1 and M_2 in Eqs. (3) and (4) vanishes, which is satisfied at a temperature given by Eq. (7). This is the Néel temperature.

$$T_N = C'\lambda = \theta \qquad (7)$$

Equation (3) and (4) holds only for $T \geq T_N$; the Curie law takes a more complicated form for $T < T_N$. In this latter region, the sublattice magnetization varies with temperature essentially in the same manner as does the magnetization of ferromagnetism.

The preceding theory predicts $\theta/T_N = 1$; the experimental values (see table) range from 0.7 to 5. P. W. Anderson ascribes this disagreement to the oversimplified two-sublattice model. Anderson's multisublattice theory not only accounts for $\theta/T_N > 1$, but also predicts a variety of magnetic ordering arrangements, many of which have been confirmed by neutron diffraction. For example, the arrangement in MnO is shown in Fig. 3; the magnetic moments are all parallel in alternating planes.

Superexchange. This is an effective coupling between magnetic spins which is indirectly routed via nonmagnetic atoms in salts and probably via conduction electrons in metals. Consider the oxy-

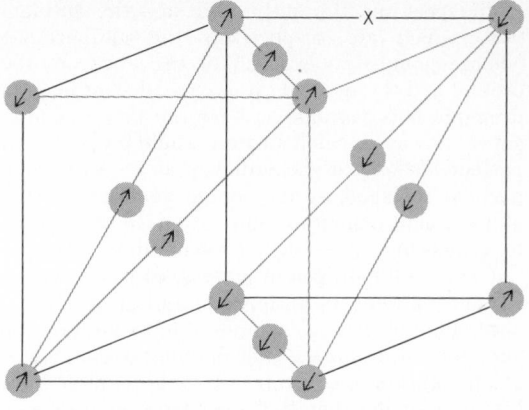

Fig. 3. The antiferromagnetism in manganese oxide. Only manganese atoms are shown.

gen atom at the position labeled X in Fig. 3. The three dumbbell-shaped electronic wave functions of the oxygen will each overlap a pair of manganese atoms (Fig. 4). Along any one of these dumbbells, the ground state is $Mn^{++}O^{--}Mn^{++}$, and the overlap mixes in the excited states $Mn^+O^-Mn^{++}$ and $Mn^{++}O^-Mn^+$, in which an electron "hops" from oxygen to manganese. The electron hops more easily if its magnetic moment is antiparallel to the manganese magnetic moment. Detailed consideration shows that there is an indirect tendency, from this mechanism, for the magnetic moments of the two manganese ions to be antiparallel; this can be expressed by an energy of the form $-J_{ij}S_i \cdot S_j$, with negative J_{ij}. This coupling aligns the moments of second-neighbor manganese ions in an antiparallel array, as in Fig. 3. First neighbors are coupled by "right-angled" superexchange from π-like bonding. This is probably comparable to the second-neighbor coupling in MnO but does not affect the ordering, primarily because it is geometrically impossible for all first neighbors to be antiparallel to one another.

In metals the conduction electrons may play the "hopping" role ascribed above to O^{--} electrons, or the antiferromagnetism may be related to periodic magnetic order in the electron energy bands.

Magnetic anisotropy. The magnetic moments are known to have preferred directions; these are shown in Figs. 1 and 3. In MnO, it is not known exactly in which direction the moments point, except that it is some direction in the (111) planes. Anisotropic effects come from magnetic dipole forces (predominant in MnF_2 and in MnO) and also from spin-orbit coupling combined with superexchange. Some nearly antiparallel arrays, such as Fe_2O_3, show a slight bending (called canting) and exhibit weak ferromagnetism. The anisotropy affects the susceptibility of powder samples and is of extreme importance in antiferromagnetic resonance. *See* MAGNETIC RESONANCE.

[E. ABRAHAMS; F. KEFFER]

Bibliography: C. Kittel, *Introduction to Solid State Physics*, 3d ed., 1966; G. T. Rado and H. Suhl (eds.), *Magnetism*, 4 vols., 1963–1967.

Antimatter

Substance consisting of atoms which are charge conjugates of atoms found in ordinary matter. Since physicists have demonstrated the existence

ANTIFERROMAGNETISM

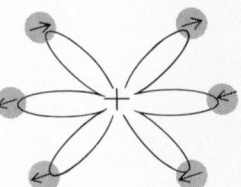

Fig. 4. Superexchange.

Some representative antiferromagnets

Substance	Crystal type	Néel temp., T_N in K	Paramagnetic Curie temp., θ in K
MnF_2	Rutile	67	80
MnO	NaCl	122	610
FeO	NaCl	198	507
$KMnF_3$	Perovskite	88	158
$CuCl_2 \cdot 2H_2O$	Orthorhombic	4.3	4.5
CrSb	NiAs	723	550
Cr_2O_3	Al_2O_3	307	485
$ZnFe_2O_4$	Spinel	9	
EuTe	NaCl	7.8	6
MnTe	Hexagonal close-packed	403	690

of the positron, the antiproton, and the antineutron, which are respectively the antiparticles (charge-conjugate particles) of the electron, the proton, and the neutron, it is clear that at least in principle it is possible to have the charge conjugate of any atom. Such an atom would be formed in perfect analogy to the ordinary atom, with each particle replaced by its charge conjugate. Such atoms would constitute antimatter. In the laboratory it has been possible to form antideuterons, the nuclei of antihydrogen of mass 2, composed of an antineutron and an antiproton, and an isotope of antihelium of mass 3, composed of an antineutron and two antiprotons. All of the charge conjugates of the known elementary particles which have lifetimes greater than 10^{-10} sec have been produced in the laboratory.

An atom of matter and its counterpart of antimatter, if brought in contact, would annihilate each other, giving rise to π-mesons and other particles; but all of these particles created would transform within microseconds into γ-rays, neutrinos, and electrons and their antiparticles.

Antimatter out of contact with ordinary matter would be stable, and there has been speculation about the presence in the cosmos of antiworlds, in which antimatter is prevalent. An extensive study of the possible role of antimatter in cosmology has been made by H. Alfvén. Astronomical observation cannot answer this question at present. *See* ELEMENTARY PARTICLE; POSITRON; SYMMETRY LAWS.

[JOSEPH LACH]

Bibliography: H. Alfvén, *Worlds-Antiworlds: Antimatter in Cosmology*, 1966.

Archimedes' principle

A body immersed in static fluid is acted upon by a vertical force equal to the weight of fluid displaced, and a body floating in the fluid displaces its own weight of fluid. For example, a balloon ascends because it displaces a volume of air which weighs more than the weight of the balloon. This principle was first stated by Archimedes (about 287–212 B.C.) and was used by him to determine the relative amounts of gold and silver in a crown. The principle can be proved by determining the difference in vertical components of fluid force acting on the lower and upper curved surfaces of the body. This force, called the buoyant force, acts vertically upward through the centroid of the displaced volume of fluid. *See* FLUID STATICS; HYDROSTATICS.

To find the specific gravity of a body, it is weighed separately in two fluids of specific weights γ_1 and γ_2 as illustrated. If its volume is V and its weight W, and it weighs F_1 in the fluid of specific weight γ_1 and F_2 in the fluid of specific weight γ_2, then Eq. (1) and (2) hold.

$$V = \frac{F_1 - F_2}{\gamma_2 - \gamma_1} \tag{1}$$

$$W = \frac{F_1\gamma_2 - F_2\gamma_1}{\gamma_2 - \gamma_1} \tag{2}$$

Its specific weight is then given by Eq. (3), and its

$$\gamma = \frac{W}{V} = \frac{F_1\gamma_2 - F_2\gamma_1}{F_1 - F_2} \tag{3}$$

specific gravity is the value of γ divided by the specific weight of water at standard conditions. That is, specific gravity is the ratio of the density of the substance to the density of water, or it can be given as the ratio of specific weight of the substance to specific weight of water, at standard conditions.

Specific gravity of a liquid may be related directly to its significant property. Thus, charge condition of an electrolyte in a storage battery, freezing temperature of a coolant, and energy density of a fuel such as kerosine are proportional to the specific gravities of these liquids. This correspondence provides a convenient and rapid means for measurement. The hydrometer uses Archimedes' principle to determine the specific gravity of liquids. It is a weighted body with a thin stem arranged so that it floats vertically with the liquid surface at some position along the stem, depending upon the specific gravity of the liquid. With a liquid of less specific gravity than water, more of the hydrometer is submerged, so that the weight of displaced liquid is the same in each case. By graduating the stem, specific gravities may be read directly from its depth of immersion in the liquid. Because specific gravity may vary rapidly with temperature, a temperature correction or measurement at a stated temperature is necessary.

[VICTOR L. STREETER]

Atmospheric acoustics

The science of sound in the atmosphere. The term is usually reserved for situations where departures of the atmosphere from an ideal homogeneous (having the same properties everywhere) medium affect propagation. Atmospheric acoustics is concerned with sound outdoors rather than indoors. Infrasound, as well as sound of audible frequencies, is within the scope of the subject.

Speed of sound. Sound travels relative to air with the speed given by Laplace's formula, Eq. (1),

$$c = (\gamma R_0 T/M)^{1/2} \tag{1}$$

resulting from the assumptions that air is an ideal gas and that negligible heat transfer accompanies the passage of a sound wave. Here T is absolute temperature in kelvins (temperature in degrees Celsius + 273.16), $R_0 = 8314$ J·kg^{-1} K^{-1} is the universal gas constant, γ is the specific heat ratio, and M is the average molecular weight. For dry air of normal composition, primarily 21% oxygen (O_2), 78% nitrogen (N_2), and 1% argon (A), γ is 1.4 and M is 29.0. Thus, c is 331 m/s at 0°C; this increases by approximately 0.6 m/s for each degree increase in temperature (Celsius).

The presence of water vapor causes both γ and M to decrease, such that Eq. (2) approximately

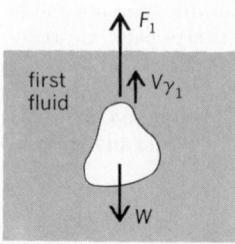
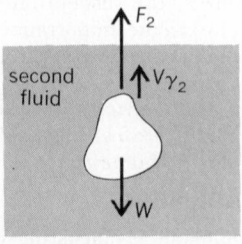

Free-body diagram for body suspended first in one fluid and then in a second fluid.

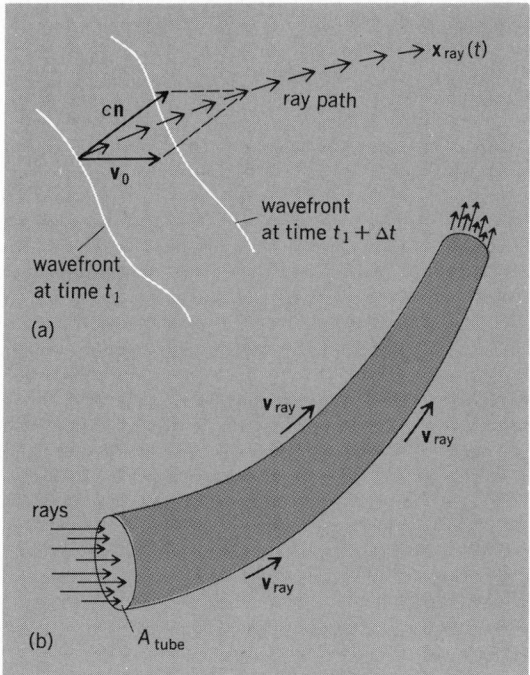

Fig. 1. Ray acoustics. (a) Definition of a ray path for a moving medium. (b) Ray tube formed by adjacent rays in an inhomogeneous medium.

$$c_{\text{wet}} = [1 + 0.16\, f(H_2O)]c_{\text{dry}} \qquad (2)$$

holds, where $f(H_2O)$ is the fraction of the air molecules that are water (H_2O) molecules. The water vapor correction is typically less than 1.5%, since $f(H_2O)$ rarely exceeds 0.07 (100% relative humidity at 40°C).

Linear acoustic approximations. Theoretical studies of the behavior of sound in the atmosphere generally employ the linear acoustic approximation to the partial differential equations governing sound in air. One of these equations is the fluid dynamic equation for conservation of mass. Another is Euler's generalization to a fluid of Newton's second law, with the gravity force taken into account. A third equation, with the neglect of thermal conduction and of irreversible entropy production, requires that the entropy of a fluid particle remain constant throughout its motion, or equivalently, that $p_T \rho_T^{-\gamma}$ remain constant for a fluid parti-

cle, where p_T and ρ_T are total pressure and density. *See* ENTROPY; EULER'S MOMENTUM THEOREM; FLUID-FLOW PRINCIPLES; FLUID MECHANICS; ISENTROPIC PROCESS.

The linear acoustic approximation results when one assumes that the deviations of pressure, density, and fluid velocity V_T from their equilibrium values, p_0, ρ_0, and V_0, are very small; that is, one sets $p_T = p_0 + p$, $\rho_T = \rho_0 + \rho$, $V_T = V_0 + V$, and subsequently neglects terms of higher order than first in p, ρ, and V. A first approximation to the actual atmosphere assumes that the ambient quantities p_0, ρ_0, V_0 depend only on height z above ground and that the ambient wind velocity V_0 is horizontal. The hydrostatic relation, Eq. (3), where $g = 9.8$ m/s is

$$dp_0/dz = -g\rho_0 \qquad (3)$$

the acceleration of gravity, results because the ambient quantities must themselves satisfy the fluid dynamic equations. These equations, together with the ideal gas equation $p_0 = \rho_0 R_0 T_0/M$ and Eq. (1), are used to derive the linear acoustic equations for the stratified atmosphere. *See* FLUID STATICS; GAS.

Ray acoustics. Audible sound and infrasound with wave periods somewhat shorter than 30 s propagate through the atmosphere primarily along rays. A ray is the trajectory (Fig. 1a) of a point that, relative to a person moving with the ambient wind, appears to move with speed c normal to a wavefront. Relative to a stationary coordinate system, the ray velocity $d\mathbf{x}_{\text{ray}}/dt$ is increased by the wind velocity, such that Eq. (4) results, where $\mathbf{n}$ is the

$$d\mathbf{x}_{\text{ray}}/dt = c\mathbf{n} + V_0 \qquad (4)$$

unit vector normal to a wavefront. If c and V_0 vary with position, as is typical in the actual atmosphere, then the directions of the normal vector $\mathbf{n}$ and of the ray velocity $d\mathbf{x}/dt$ will in general change along the path; the ray will be bent, or refracted. The general behavior of rays in the atmosphere can be deduced from ray-tracing equations that are derived from Eq. (4). These equations constitute a partial description of waves governed by the linear acoustic equations in the short-wavelength approximation. *See* REFRACTION OF WAVES.

Figure 2 shows typical long-range paths radiating from a source in the atmosphere. General features of such a plot can be explained qualitatively from a graph of effective sound velocity c_{eff}, sound speed plus wind component in plane of propaga-

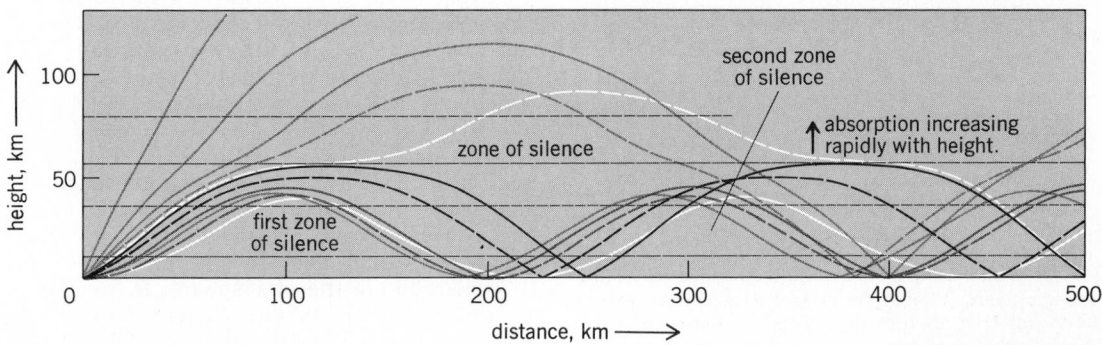

Fig. 2. Representative ray paths west to east in Northern Hemisphere in winter. (*From B. Gutenberg, Sound properties in the atmosphere, in T. F. Malone, ed., Compendium of Meteorology, American Meteorological Society, pp. 366–375, 1951*)

tion, versus height (Fig. 3). Rays can be trapped in either a low or an upper atmosphere sound channel where c_{eff} has a minimum. The greatest height attainable by a ray is its upper turning point, which is also the height for which c_{eff} is maximum along the ray path. If a path has a lower turning point, then at that altitude c_{eff} is the same as at the upper turning point. If there is some altitude at which c_{eff} is larger than at the ground, then it is possible for a ray proceeding obliquely upward from a source at an intermediate altitude to bend downward such that it eventually reaches the ground.

The magnitude and direction of upper atmosphere winds largely determine whether c_{eff} in the stratosphere can exceed its value at the ground. Since winds in the altitude range of 20–80 km reverse direction in spring and fall at temperate latitudes, long-range propagation east to west or west to east is markedly different in summer from that in winter. Local meteorological conditions such as inversions (temperature increasing with height near the ground) can also cause a bending back to the ground of rays initially proceeding obliquely upward.

A spectacular consequence of a c_{eff} profile that initially decreases with height, reaches a minimum, then increases to a value higher than that at the ground, is the existence of abnormal zones of audibility. Points at intermediate horizontal distances may receive no sound, while audible sound is received at greater distances. An example of this phenomenon is illustrated in Fig. 4, giving locations where sound was heard and not heard following an explosion. The anomalous zone of audibility to the east and south of the explosion, beyond 200 km, is explained if the stratospheric winds are blowing in these directions.

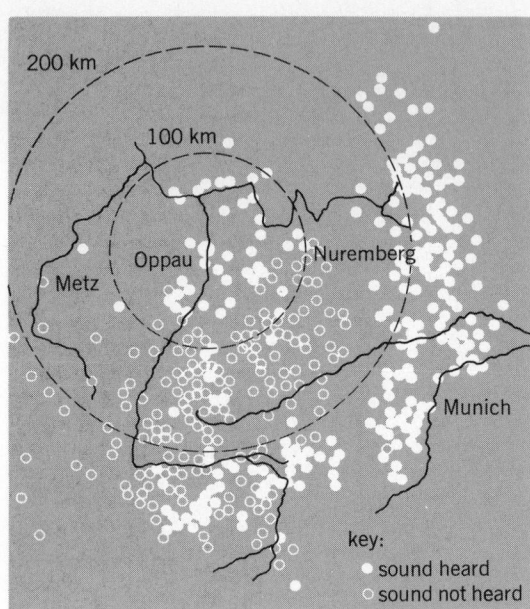

Fig. 4. Locations where sound was heard and not heard following an explosion at Oppau, Germany, on Sept. 21, 1921. (*From R. K. Cook, Strange sounds in the atmosphere, pt. I. Sound: Its Uses and Control, vol. 1, pp. 12–16. 1962*)

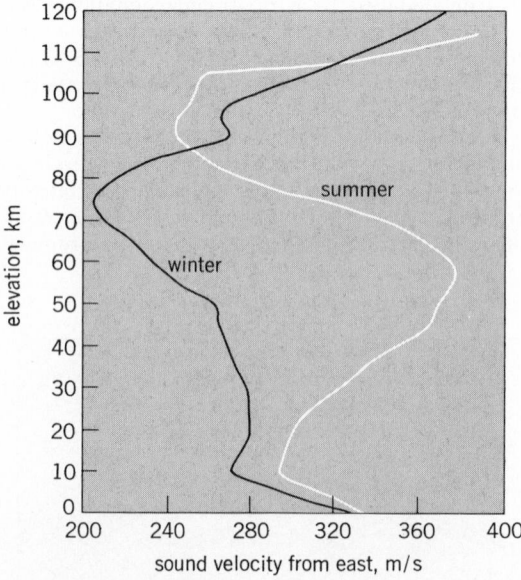

Fig. 3. Model atmospheric profiles of effective sound speed c_{eff} versus height for propagation east to west in northeastern United States. (*From D. Rind and W. E. Donn, Further use of natural infrasound as a continuous monitor of the upper atmosphere. J. Atmos. Sci., 32: 1694–1704, 1975*)

Amplitudes along ray paths. The linear acoustic equations in the short-wavelength approximation yield the prediction that a quantity known as the Blokhintzev invariant should be constant along a ray tube (Fig. 1b). This invariant is proportional to $(p^2)_{av}A_{tube}$, where $(p^2)_{av}$ is the average value of the square of the sound pressure and A_{tube} is the area of a cross section of the ray tube. Major variations in amplitude are thus associated with changes in A_{tube}. The geometrical acoustics or ray theory breaks down where A_{tube} vanishes (loci of such points are called caustics), but resumes validity some distance beyond the point where vanishing occurs, providing one associates an additional $\pi/2$ phase shift with the wave.

Shadow zones. Refraction can cause the existence of shadow zones within which no ray passes (Fig. 2). Since wind velocity near the ground increases with height, shadow zones are frequently found upwind of a source. Their presence and the additional ray tube spreading upwind explain why sound is generally more audible downwind of a source.

A small amount of acoustic energy penetrates into a shadow zone via creeping waves that propagate along the ground and that continually shed diffracted rays into the shadow zone. The amplitude decay with penetration distance is exponential, with a decay coefficient proportional to $f^{1/3}$ and to $(-dc_{eff}/dz)^{2/3}$, where f is frequency. The magnitude of the coefficient also varies with the ground impedance (discussed below). The dominant feature of shadow zone reception is the marked decrease in a sound's higher-frequency content

Attenuation in the atmosphere. Dissipation of acoustic energy in the atmosphere is caused by viscosity, thermal conduction, and molecular relaxation. The last arises because fluctuations in apparent molecular vibrational temperatures lag in

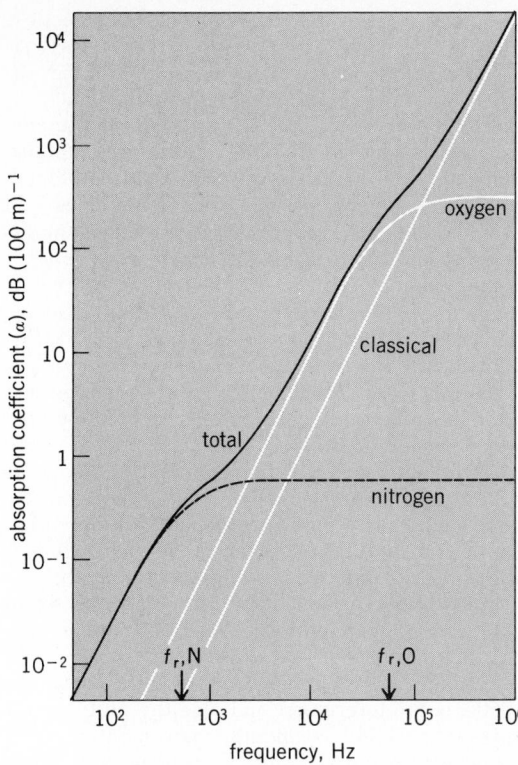

Fig. 5. Frequency dependence of attenuation coeffi-cient in units of 1 dB per 100 m (or of 1.15×10^{-3} neper/ m) for a pressure of 1 atm, temperature of 20°C, and a relative humidity of 70%; $f(H_2O = .016$. (*From J. E. Piercy, T. F. W. Embleton, and L. C. Sutherland, Review of noise propagation in the atmosphere, J. Acous. Soc. Amer., 61: 1403–1418, 1977*)

phase the fluctuations in translational tempera-tures. The vibrational temperatures of significance are those characterizing the relative populations of O_2 and N_2 molecules in ground and first-excited vibrational quantum states. Since collisions with H_2O molecules are much more likely to induce vi-brational state changes than are collisions with other O_2 or N_2 molecules, the sound attenuation varies markedly with absolute humidity.

The classical contribution due to viscosity and thermal conductions, Eqs. (5) and (6), of the

$$\alpha_{c1} = \frac{2\pi^2 f^2 \mu}{\rho_0 c^3} \cdot [(4/3) + (\mu_B/\mu) + (\gamma - 1)\kappa/(c_p\mu)] \quad (5)$$

$$(4/3) + (\mu_B/\mu) + (\gamma - 1)\kappa/(c_p\mu) \approx 2.5 \quad (6)$$

amplitude attenuation coefficient (nepers per me-ter) increases quadratically with frequency; the coefficient α_{c1}/f^2 at ground level at 27°C is of the order of 2×10^{-11} m^{-1} Hz^{-2}, so attenuation by classical mechanisms over distances less than 1 km is insignificant for frequencies less than 2 kHz. (Here μ, μ_B, κ, and c_p represent viscosity, bulk vis-cosity, thermal conductivity, and specific heat at constant pressure.)

The contributions α_1 and α_2 to the attenuation coefficient from N_2 and O_2 vibrational relaxations both also increase monotonically with frequency, but with a frequency dependence described by Eq.

(7) rather than as f^2. Here $\lambda = c/f$ if the nominal

$$\alpha_\nu = \frac{(\alpha_\nu\lambda)_{max}}{\lambda} \frac{2}{(f_\nu/f + f/f_\nu)} \quad (7)$$
$$\nu = 1, 2$$

wavelength of the sound and $(\alpha_\nu\lambda)_{max}$ is a frequen-cy-independent quantity that equals the maximum value that $\alpha_\nu\lambda$ will have when plotted versus f. This maximum occurs at the relaxation frequency f_ν. The values $(\alpha_1\lambda)_{max}$ and $(\alpha_2\lambda)_{max}$ are strongly dependent on temperature and derive from the molecular properties of N_2 and O_2; they are insen-sitive to the air's water vapor content. Representa-tive values for these at 20°C are .0002 and .0012, respectively.

For dry air at 20°C, the values of the relaxation frequencies f_1 and f_2 are of the order of 9 and 24 Hz, respectively, but these increase rapidly with $f(H_2O)$, such that f_1 becomes 360 Hz and f_2 be-comes 33 kHz when the fraction $f(H_2O)$ is 0.01. The values of α_{c1}/f^2, f_1, f_2, $(\alpha_1\lambda)_{max}$, and $(\alpha_2\lambda)_{max}$ are invariably such that N_2 vibrational relaxation dominates at low frequencies $f \ll f_1$, while O_2 vibrational relaxation dominates for $f_1 \ll f \ll f_2$. In the limit $f \gg f_2$, classical absorption dominates (Fig. 5).

Ground impedance. Experimental studies sug-gest that the ground behaves as a locally reacting surface for the reflection of audible sound such that, for any given frequency, the ratio of complex pressure amplitude to the into-ground component of the complex velocity amplitude **v**, namely v_z, is independent of the direction of the incident wave. This ratio, Eq. (8), is the specific acoustic im-

$$Z = p/(-v_z) \quad (8)$$

pedance of the ground; its real and imaginary parts are both positive, and typically decrease monotonically with increasing frequency. For grass-covered ground and within the middle range

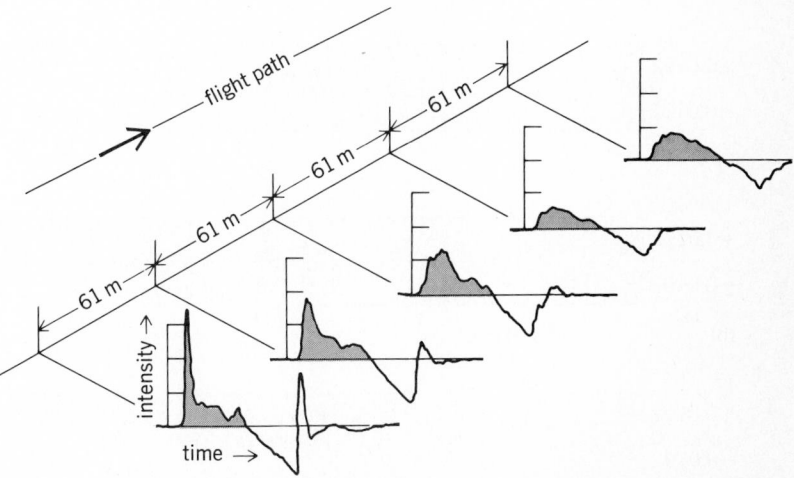

Fig. 6. Example of turbulence effects on sound in air: measured sonic boom pres-sure signatures at five microphones spaced at intervals of 61 m on the ground track of a fighter aircraft in steady-level supersonic flight (1.7 times speed of sound) at 8500 m altitude. (*From D. J. Maglieri, Sonic boom flight research: Some effects of air-plane operations and the atmosphere of sonic boom signatures, in A. R. Sebass, ed., Sonic Boom Research, NASA SP-147, 25–48, Apr. 12, 1967*)

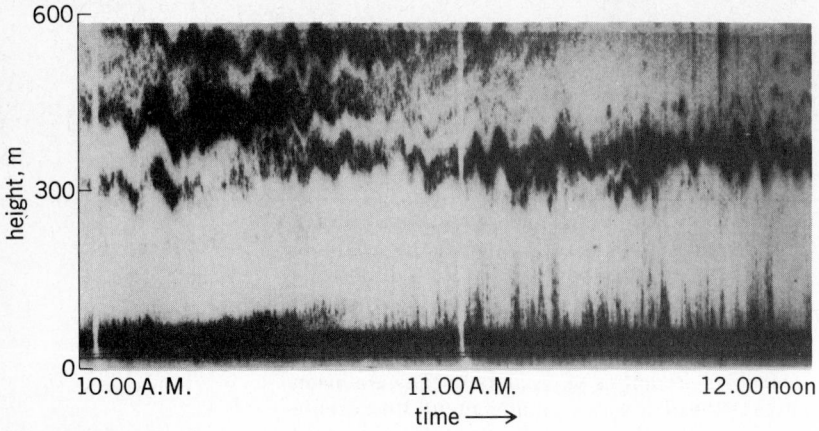

600

height, m

300

0

10.00 A.M.   11.00 A.M.   12.00 noon

time ⟶

Fig. 7. Facsimile record of acoustic echo sounding of atmosphere at the South Pole, Feb. 9, 1975. Dark regions above 300 m altitude correspond to temperature inhomogeneities associated with the passage of convective plumes over site. *(From E. H. Brown and F. F. Hall, Jr., Advances in atmospheric acoustics, Rev. Geophys. Space Phys., 16:47–110, 1978)*

of audible frequencies, both are of the order of $10\rho_0 c$.

The theory of reflection from a locally reacting surface yields an amplitude reflection coefficient R that goes to -1 when the angle of incidence θ_I (measured from the vertical) approaches grazing incidence. A consequence of this is that incident and reflected waves tend to cancel near the ground at larger distances from a source near the ground. Constructive reinforcement occurs at a point somewhat above the ground where the difference between direct and reflected path lengths is of the order of $\lambda/2$. The cancellation near the ground is offset to a minor extent by a surface wave which is decaying at a slow exponential rate with distance from the source and whose existence is caused by the boundary condition of Eq. (8) at the ground.

Turbulence effects. Larger-scale turbulence in the atmosphere causes the effective wave speed to fluctuate from point to point, so a nominally

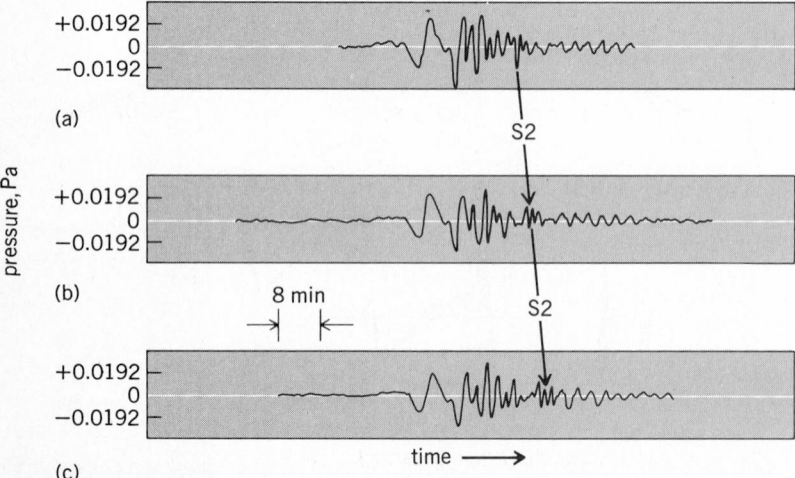

pressure, Pa

+0.0192
0
−0.0192

(a)

S2

+0.0192
0
−0.0192

(b)

8 min

S2

+0.0192
0
−0.0192

(c)

time ⟶

Fig. 8. Microbarograph of acoustic-gravity wave generated by a thermonuclear explosion of 5 megatons energy yield (2×10^{13} J), showing waveforms at the ground at great circle distances of (a) 7000 km, (b) 8000 km, (c) 9000 km. *(From D. G. Harkrider, Theoretical and observed acoustic-gravity waves from explosive sources in the atmosphere, J. Geophys. Res., 69:5295–5321, 1964)*

smooth wavefront develops ripples. One result is that the direction of a received ray may fluctuate with time in a random manner. Another is that adjacent ray tubes may focus or defocus because of the refraction associated with wavefront rippling. Consequently, the amplitude of the sound at a distant point will fluctuate with time and with small displacements in position. Lower frequencies tend to smear out such rippling effects, so turbulent distortions and fluctuations are most noticeable for higher frequencies. A demonstration of this is found in waveforms received on the ground during the overhead passage of a supersonic aircraft (Fig. 6). The turbulence in the atmosphere's boundary layer can distort the nominal N-wave shape into either a spiked waveform (focusing) or a rounded waveform (defocusing).

Turbulence and inhomogeneities in the atmosphere also scatter sound from its original direction, such that a collimated beam will spread in width and such that the amplitude along the axis will appear to undergo an additional attenuation.

Echosonde. Atmospheric acoustics has found extensive application in the study of meteorological disturbances in the lower atmosphere. The echosonde, invented by L. G. McAllister in the late 1960s, is the forerunner and also the prototype of current acoustic sounding instrumentation. A transducer pointing upward transmits a short acoustic pulse and then receives weak echoes over a relatively long period of time. Echoes received at time Δt after transmission have been reflected from inhomogeneities at height $c_0 \Delta t/2$. A facsimile display (Fig. 7) is produced by a side-by-side superposition of the echo histories for a long succession of such sounding experiments; this is interpreted as a picture of the time evolution of the height profile of the atmosphere's ability to backscatter acoustic waves. Darker regions on the display imply that, at the corresponding times and heights, the backscattering was strong. Meteorological interpretation of such records requires experience and an understanding of fundamental atmospheric processes. Refined systems in conjunction with the theory of sound propagation through a turbulent inhomogeneous atmosphere enable a quantitative determination of wind velocities above the ground and of the parameters characterizing the structure of turbulence.

Acoustic-gravity waves. Infrasound waves with periods longer than a minute are strongly affected by the Earth's gravity. The dominant features of some observed disturbances at stratospheric and ionospheric heights can be interpreted in terms of the model of planar sound waves in an isothermal atmosphere, for which Eq. (3) requires both p_0 and ρ_0 to decrease exponentially with height. The linear acoustic equations then predict that there is a forbidden range of angular frequencies ω in which sound waves cannot propagate, namely $\omega_b < \omega < \omega_a$, where ω_a and ω_b are given by Eqs. (9).

$$\omega_a = (\gamma/2)g/c \qquad (9a)$$

$$\omega_b = (\gamma - 1)^{1/2}g/c \qquad (9b)$$

(The wave period corresponding to ω_b is of the order of 5 to 10 min in the lower atmosphere.) Possible planar waves are divided into ordinary acoustic waves ($\omega > \omega_a$) and acoustic-gravity waves ($\omega < \omega_b$). If ω is comparable to or less than ω_a, then the phase velocity is not parallel to the group velocity.

Acoustic-gravity waves, in particular, are such that the vertical components of phase and group velocities have opposite signs, so a wave appearing to be propagating obliquely downward is actually transporting energy upward.

Acoustic-gravity waves can also be channeled in the lower atmosphere and along the ground such that they can carry infrasonic signals over large horizontal distances. Thus, for example, pressure waves generated by large explosions and traveling at slightly less than the sound speed at the ground can be detected at distances greater than halfway around the globe by microbarograph instrumentation. The dispersion of such waves with great-circle distance is illustrated in Fig. 8. The time origins of the microbarographs in the three parts of the figure have been shifted such that the early portions of the three waveforms are superposed. The early portion is channeled along the ground by gravity; the S2 arrival with slower speed is channeled at the height of sound speed minimum. Channeled acoustic-gravity waves are also generated by tornadoes, earthquakes, volcanic eruptions, and the flow of air over mountain ranges. The eruption of Krakatoa in 1883 generated waves that were detected even after a complete encirclement of the globe. *See* SOUND.

[ALLAN D. PIERCE]

Bibliography: E. H. Brown and F. F. Hall, Jr., Advances in atmospheric acoustics, *Rev. Geophys. Space Phys.*, 16:47–110, 1978; A. D. Pierce, *Acoustics: An Introduction to Its Physical Principles and Applications*, 1980; J. E. Piercy, T. F. W. Embleton, and L. C. Sutherland, Review of noise propagation in the atmosphere, *J. Acoust. Soc. Amer.*, 61:1403–1418, 1977; Special Issue on Infrasonics and Atmospheric Acoustics, *Geophys. J. Roy. Astron. Soc.*, vol. 26, December 1971.

Atom

The individual structure which constitutes the basic unit of any chemical element. This structure, consisting of a positively charged nucleus surrounded by a number of electrons of total negative charge equal to the positive charge on the nucleus, is essentially identical for all atoms of any one element. The nuclear charge, measured in units of the electronic charge, is called the atomic number and specifies the element. Masses of the atoms of stable elements range from 1.67×10^{-24} g to 3.95×10^{-22} g. Diameters of atoms are on the order of 10^{-8} cm; nuclear diameters are approximately 10^{-12} cm. *See* ATOMIC STRUCTURE AND SPECTRA; ISOTOPE; NUCLEAR STRUCTURE.

[F. A. JENKINS/W. W. WATSON]

Atomic beams

Unidirectional streams of neutral atoms passing through a vacuum. These atoms are virtually free from the influence of neighboring atoms but may be subjected to electric and magnetic fields so that their properties may be studied. The technique of atomic beams is identical to that of molecular beams. For historical reasons the latter term is most generally used to describe the method as applied to either atoms or molecules.

The method of atomic beams yields extremely accurate spectroscopic data about the energy levels of atoms, and hence detailed information about the interaction of electrons in the atom with each other and with the atomic nucleus, as well as information about the interaction of all components of the atom with external fields. For a detailed discussion *see* MOLECULAR BEAMS. [POLYKARP KUSCH]

Atomic constants

That group of physical constants which play a fundamental role in the basic theories of physics. These include the speed of light in vacuum, c; the magnitude of the charge on the electron, e, which is the fundamental unit of electric charge; the mass of the electron, m_e; Planck's constant, h; and the fine-structure constant, α.

These five quantities typify the different origins of the fundamental constants: c and h are examples of quantities which appear naturally in the mathematical formulation of certain physical theories—Einstein's theories of relativity, and quantum theory, respectively; e and m_e are examples of quantities which characterize the elementary particles of which all matter is constituted; and α, the fundamental constant of quantum electrodynamics (QED), is an example of quantities which are combinations of other fundamental constants, but are actually constants in their own right since the same combination always appears together in the basic equations of physics. (In the International System of Units, or SI, which is the unit system used throughout this article, $\alpha = \mu_0 ce^2/2h$, where μ_0, the so-called permeability of vacuum, is exactly $4\pi \times 10^{-7}$ henry/meter.)

Reasons for measurement. Reliable numerical values for the fundamental physical constants are required for two main reasons. First, they are necessary if quantitative predictions from physical theory are to be obtained. Second, and even more important, the self-consistency of the basic theories of physics can be critically tested by a careful intercomparison of the numerical values of fundamental constants obtained from experiments in the different fields of physics.

History of measurement. Although measurements of the fundamental constants date back to the 17th- and 18th-century determinations of c and G (the Newtonian gravitational constant), the field began to blossom only after 1900 with the onset of the modern era of physics. Great progress has occurred since World War II as a direct result of the technological advances made during the war in the fields of electronics and microwaves. In addition to c and G, other constants measured between 1900 and World War II include e by means of R. A. Millikan's oil drop experiment; the Faraday constant, F, using iodine- and silver-based coulometry; the Avogadro constant, N_A, by means of an x-ray technique; and the ratios e/m_e and h/e. Important postwar determinations, mostly related to atomic physics, include the proton gyromagnetic ratio, γ_p; the proton magnetic moment in nuclear magnetons, μ_p/μ_N; the free electron g-factor, g_e; and the fine structure and ground state hyperfine splitting of atomic hydrogen which, in combination with theory, leads to values of α.

In general, the accuracy of fundamental constants determinations has continually improved over the years. (By accuracy is meant the relative size of the uncertainty which must be assigned to the numerical value of the measured constant to indicate how far from the true but unknown value it may be. The uncertainty arises primarily from experimental limitations.) Whereas in the past, 100 ppm (0.01%) and even 1000 ppm (0.1%) measurements were commonplace, today 0.01 ppm and better determinations are not unusual (ppm = parts per million).

Impact of Josephson effect measurement. The fundamental constants field was given a new impetus in 1967 with the determination by W. H. Parker, B. N. Taylor, and D. N. Langenberg of the ratio $2e/h$ using the so-called ac Josephson effect in superconductors. The impact has largely been in the area of QED. To compare the theoretical predictions of QED with experiment requires an accurate value of α. Before the $2e/h$ measurement, the most accurate α values were obtained from experiment with the aid of theoretical equations containing significant contributions from QED. It was thus difficult to compare QED theory and experiment unambiguously since the theory had to be evaluated by using direct values of α derived from the experiments themselves. Such comparisons were therefore limited to the testing of internal consistency. Now, however, by combining the value of $2e/h$ with the measured values of certain other constants, a highly accurate indirect value of α can be obtained without any essential use of QED theory. Consequently, unambiguous comparisons can be made between QED theory and experiment. *See* Josephson effect; Quantum electrodynamics.

Hyperfine splitting of hydrogen. Among the quantities for which such a comparison was of critical importance in 1967 was the hyperfine splitting (hfs) in hydrogen. It can be measured experimentally to the phenomenal accuracy of 1 part in 10^{12} by using the hydrogen maser. In contrast, the theoretical QED equation for the hfs, which involves only well-known constants and α, is limited to an accuracy of a few parts per million because of the difficulty in calculating some of the terms in the equation from theory. *See* Maser.

One such term, called the proton polarizability correction, δ_N, arises from the fact that the proton in the hydrogen atom has an internal structure of its own. However, all calculations of δ_N show it to be rather small, 1 or 2 ppm at most. The small size of the theoretical value for δ_N was in marked contrast to that implied by the value of α accepted in 1967 (obtained from a measurement of the fine structure splitting in deuterium). When this value was used to calculate a theoretical value for the hydrogen hfs, when this was then compared with the experimental hydrogen maser value of the hfs, and when their difference was assumed to arise solely from the existence of a polarizability correction, it was found that $\delta_N = (43 \pm 9)$ ppm. This meant that the probability for δ_N to be as small as predicted by theory was only 1 in 20,000, a clear discrepancy. On the other hand, when the new value of α obtained from the Josephson effect measurement of $2e/h$ was used, it was found that $\delta_N = (2.5 \pm 4.0)$ ppm, consistent with the theoretical predictions. Thus the Josephson effect value of α removed the discrepancy and resulting challenge to QED.

This case is an excellent example of how fundamental constants experiments carried out in one field of physics can have important implications for other fields—a low-temperature solid-state physics experiment has given information about the excited states of the proton, a subject usually associated with high-energy physics. It is therefore a good example of the unity of physics as well as the role played by measurements of the fundamental constants. *See* Hyperfine structure.

Methods of obtaining α. In practice, there are several methods of obtaining indirect α values from $2e/h$. One involves F; μ_p/μ_N; the atomic mass of the proton, M_p; and the conversion factor relating the so-called as-maintained ampere in terms of which F was measured to the absolute or SI ampere. Another involves the Rydberg constant, R_∞; c; γ_p; the proton magnetic moment in units of the Bohr magneton, μ_p/μ_B; and the conversion factor relating the so-called as-maintained ohm associated with the electrical units in terms of which both $2e/h$ and γ_p are measured to the SI ohm. The ampere conversion factor and other similar conversion factors play an important role in the fundamental constants field because knowledge of their numerical values is necessary in order to express all measured quantities in the same system of units, that is, SI. Since the uncertainty in some of these factors is relatively large, they must in many cases be considered equal in importance to the particular fundamental constants with which they are associated, and separate experiments must be undertaken for their determination. *See* Electrical units and standards.

Least-squares method. This discussion illustrates the complex relationships which can exist among groups of constants and conversion factors, and that a particular constant may be determined either directly by measurement or indirectly by an appropriate combination of other directly measured constants. If the direct and indirect values have comparable accuracy, then both must be taken into account in order to arrive at a best value for that quantity. (By best value is meant that value believed to be closest to the true but unknown value.) Generally, each of the several routes which can be followed to a particular constant, both direct and indirect, will give a slightly different numerical value. Such a situation may be satisfactorily handled by the mathematical method known as least-squares. This technique provides a self-consistent procedure for calculating best "compromise" values of the constants from all of the available data. It automatically takes into account all possible routes and determines a single final value for each constant being calculated. It does this by weighting the different routes according to their relative uncertainties. The appropriate weights follow from the uncertainties assigned the individual measurements constituting the original set of data.

Least-squares studies of the constants were begun by R. T. Birge in the late 1920s, and continued by others, notably J. W. M. DuMond and E. R. Cohen. The two most recent studies were those of Taylor, Parker, and Langenberg in 1969, based on their Josephson effect determination of $2e/h$; and of Cohen and Taylor in 1973, carried out under the auspices of the CODATA Task Group on Fundamental Constants. (CODATA, the Committee on Data for Science and Technology, is under the jurisdiction of the International Council of Scientific Unions, ICSU.) The recommended values of Cohen and Taylor's 1973 adjustment were officially adopted for international use by the 8th CODATA

General Assembly in September 1973. Recommended values of selected constants from this adjustment are given in Table 1 and were those still in general use as of 1980.

To carry out a least-squares adjustment, the data are first divided into two groups: (1) the more precise data, or auxiliary constants, which have uncertainties sufficiently small that they can be considered as exactly known—for example, the speed of light, which has an uncertainty of only 0.004 ppm; and (2) the less precise, stochastic input data, which in the 1973 adjustment of Cohen and Taylor had uncertainties larger than about 0.2 ppm. A subset of constants is then chosen in terms of which all of the stochastic input data can be individually expressed, if necessary, with the aid of the auxiliary constants. It is actually the constants composing this subset which are directly subject to adjustment. In the 1973 work of Cohen and Taylor, these so-called adjustable constants were taken to be α, K, N_A, $\overline{R}$, Λ, and μ_μ/μ_p. Here, K is the conversion factor relating the ampere as maintained by the International Bureau of Weights and Measures (BIPM) in France to the SI ampere; $\overline{R}$ is the conversion factor relating the BIPM as-maintained ohm to the SI ohm; Λ is the conversion factor relating the kilo-X-unit, a unit of length used in the field of x-rays, to the SI unit of length, the meter; and μ_μ/μ_p is the ratio of the magnetic moment of the muon to that of the proton. With just these six quantities and the aid of selected auxiliary constants, a series of equations, generally known as the observational equations, were formed for all 27 separately available items of stochastic data. These 27 observational equations were then solved (with the aid of a computer) for the least-squares adjusted values of the six adjustable constants, and their uncertainties. Optimum values in the least-squares sense for other constants not directly subject to adjustment were then calculated from the six adjustable constants. (But this does not apply to the auxiliary constants since, for the purpose of the least-squares adjustment, they are taken to be exactly known.)

Critical analysis. Critical analysis of the input data, and deciding what uncertainty should be assigned each measurement, is the main problem in adjusting constants—the weight a particular stochastic datum carries in an adjustment is proportional to the reciprocal of the square of its uncertainty. Another important task is deciding how to handle "discrepant" data, that is, measurements

Table 1. Recommended values of selected fundamental constants as taken from the 1973 least-squares adjustment of Cohen and Taylor

Quantity	Symbol	Numerical value*	Uncertainty, ppm	Units†
Speed of light in vacuum	c	299792458(1.2)	0.004	$m \cdot s^{-1}$
Fine-structure constant,	α	7.2973506(60)	0.82	10^{-3}
$\mu_0 c e^2/2h$	α^{-1}	137.03604(11)	0.82	
Elementary charge	e	1.6021892(46)	2.9	10^{-19} C
Planck's constant	h	6.626176(36)	5.4	10^{-34} J $\cdot$ s
Avogadro constant	N_A	6.022045(31)	5.1	10^{23} mol^{-1}
Electron rest mass	m_e	9.109534(47)	5.1	10^{-31} kg
Proton rest mass	m_p	1.6726485(86)	5.1	10^{-27} kg
Ratio of proton mass to electron mass	m_p/m_e	1836.15152(70)	0.38	
Faraday constant, $N_A e$	F	9.648456(27)	2.8	10^4 C $\cdot$ mol^{-1}
Rydberg constant, $\mu_0{}^2 c^3 e^4 m_e/8h^3$	R_∞	1.097373177(83)	0.075	10^7 m^{-1}
Bohr radius, $\alpha/4\pi R_\infty$	a_0	5.2917706(44)	0.82	10^{-11} m
Free electron g-factor, or electron magnetic moment in Bohr magnetons	$g_e/2 = \mu_e/\mu_B$	1.0011596567(35)	0.0035	
Free muon g-factor, or muon magnetic moment in units of $eh/4\pi m_\mu$	$g_\mu/2$	1.00116616(31)	0.31	
Bohr magneton, $eh/4\pi m_e$	μ_B	9.274078(36)	3.9	10^{-24} J $\cdot$ T^{-1}
Proton gyromagnetic ratio	γ_p	2.6751987(75)	2.8	10^8 s^{-1} $\cdot$ T^{-1}
Proton magnetic moment in Bohr magnetons	μ_p/μ_B	1.521032209(16)	0.011	10^{-3}
Ratio of electron and proton magnetic moments	μ_e/μ_p	658.2106880(66)	0.010	
Proton magnetic moment in nuclear magnetons	μ_p/μ_N	2.7928456(11)	0.38	
Nuclear magneton, $eh/4\pi m_p$	μ_N	5.050824(20)	3.9	10^{-27} J $\cdot$ T^{-1}
Ratio of muon and proton magnetic moments	μ_μ/μ_p	3.1833402(72)	2.3	
Ratio of muon mass to electron mass	m_μ/m_e	206.76865(47)	2.3	
Muon rest mass	m_μ	1.883566(11)	5.6	10^{-28} kg
Compton wavelength of the electron, $h/m_e c$	λ_C	2.4263089(40)	1.6	10^{-12} m
Gravitational constant	G	6.6720(41)	615	10^{-11} m^3 $\cdot$ s^{-2} $\cdot$ kg^{-1}

*The numbers in parentheses are the one-standard-deviation uncertainties in the last digits of the quoted value, and the unified atomic mass scale ^{12}C $= 12$ has been used throughout.

†C $=$ coulomb, J $=$ joule, kg $=$ kilogram, m $=$ meter, mol $=$ mole, s $=$ second, T $=$ tesla.

Table 2. Comparison of the recommended values of selected fundamental constants as taken from the 1973 least-squares adjustment of Cohen and Taylor; the 1969 adjustment of Taylor, Parker, and Langenberg; and the 1963 adjustment of Cohen and DuMond

Quantity*	Value, 1973 adjustment, and ppm uncertainty		Value, 1969 adjustment, and ppm uncertainty		Change 1973–1969, ppm	Value, 1963 adjustment, and ppm uncertainty		Change 1973–1963, ppm
α^{-1}†	137.03604(11)	0.82	137.03602(21)	1.5	+0.15	137.0388(6)	4.4	−20
e	1.6021892(46)	2.9	1.6021917(70)	4.4	−1.6	1.60210(2)	12	+56
h	6.626176(36)	5.4	6.626196(50)	7.6	−3.0	6.62559(16)	24	+88
m_e	9.109534(47)	5.1	9.109558(54)	6.0	−2.6	9.10908(13)	14	+50
N_A	6.022045(31)	5.0	6.022169(40)	6.6	−21	6.02252(9)	15	−79
μ_p/μ_N	2.7928456(11)	0.38	2.792782(17)	6.2	+23	2.79276(2)	7.2	+34
F	9.648456(27)	2.8	9.648670(54)	5.5	−22	9.64870(5)	5.2	−25

*The units for e are 10^{-19} C; for h, 10^{-34} J · s; for m_e, 10^{-31} kg; for N_A, 10^{23} mol^{-1}; for F, 10^4 C · mol^{-1}.

†α^{-1}, the reciprocal of the fine structure constant, is given rather than α because it is a simpler number.

which differ from each other by statistically significant amounts in comparison with their assigned uncertainties. Such data cannot be included in an adjustment uncritically because the inconsistencies imply either incorrect uncertainty estimates or the presence of unknown measurement errors.

When confronted with such a situation, the constants adjuster can in general either (1) include the inconsistent data, but only after expanding (increasing) their assigned uncertainties so that they are no longer discrepant; or (2) decide on as sound a theoretical and experimental basis as feasible which of the inconsistent data are least reliable, and discard them, but expand no uncertainties. Thus, there are subjective factors in adjusting constants, and different reviewers may treat the same data differently, obtaining a somewhat different set of best values.

Comparison of 1969 and 1973 values. Some of the pitfalls of discarding data may be seen by comparing the 1969 adjustment of Taylor, Parker, and Langenberg with the 1973 adjustment of Cohen and Taylor. The most critical problem facing Taylor and colleagues was the internal inconsistency among the five available values of μ_p/μ_N, and an inconsistency between two of these values and the one available measurement of F. After much thought and analysis, Taylor and colleagues decided to discard the two "high" values of μ_p/μ_N and to retain the remaining three and F. The most difficult problem facing Cohen and Taylor also had to do with μ_p/μ_N and F. However, in the intervening 4 years, two new, very accurate (sub-ppm) μ_p/μ_N determinations were completed, which showed that the high values of μ_p/μ_N discarded in 1969 were more nearly correct than the three retained low values, and that it was F which was probably in error and should be the quantity discarded. This shift in outlook accounts for the large changes in the recommended values for certain constants as given in the 1969 and 1973 adjustments. These changes are readily apparent in Table 2, where the values of selected constants resulting from the 1973 adjustment are compared with their counterparts from the 1969 adjustment and the 1963 adjustment. (The changes occurring between the 1963 and 1969 values are primarily due to the change in α resulting from the measurement of $2e/h$ using the Josephson effect.) The table shows that (1) knowledge of the numerical values of the fundamental constants continually improves as new measurements become available; (2) the con-

stants are so intimately related to one another that a significant shift in the value of one will usually give rise to large shifts in the values of others; and (3) no set of recommended constants, such as is given in Table 1, should be taken as final and unalterable. Indeed, determinations of several constants since the 1973 adjustment, including N_A, R_∞, γ_p, and F, seem to indicate that there may be important changes in some of the 1973 recommended values when the next least-squares adjustment is carried out. *See* ATOMIC STRUCTURE AND SPECTRA; AVOGADRO NUMBER; ELECTRON; ELECTRON SPIN; ELEMENTARY PARTICLE; FINE STRUCTURE (SPECTRAL LINES); GRAVITATION; GYROMAGNETIC RATIO; LIGHT; MAGNETON; PLANCK'S CONSTANT; QUANTUM MECHANICS, RELATIVITY; RYDBERG CONSTANT.

[BARRY N. TAYLOR]

Bibliography: E. R. Cohen and B. N. Taylor, The 1973 least-squares adjustment of the fundamental constants, *J. Phys. Chem. Ref. Data*, 2(4): 663–734, 1973; D. N. Langenberg and B. N. Taylor (eds.), *Precision Measurement and Fundamental Constants*, Proceedings of the International Conference on Precision Measurement and Fundamental Constants, Washington, DC, August 1970, NBS SP–343, 1971; F. D. Rossini, *Fundamental Measures and Constants for Science and Technology*, 1974; J. H. Sanders and A. H. Wapstra (eds.), *Atomic Masses and Fundamental Constants 4*, Proceedings of the 4th International Conference on Atomic Masses and Fundamental Constants, Teddington, England, September 1971, published 1972; J. H. Sanders and A. H. Wapstra (eds.), *Atomic Masses and Fundamental Constants 5*, Proceedings of the 5th International Conference on Atomic Masses and Fundamental Constants, Paris, June 1975; B. N. Taylor, W. H. Parker, and D. N. Langenberg, *The Fundamental Constants and Quantum Electrodynamics*, 1969, also published in *Rev. Mod. Phys.*, 41(3):375–496, July 1969.

Atomic mass unit

An arbitrarily defined unit in terms of which the masses of individual atoms are expressed. One atomic mass unit is defined as exactly $^1/_{12}$ of the mass of an atom of the nuclide ^{12}C, the predominant isotope of carbon. The unit, also known as the dalton, is often abbreviated amu, and is designated by the symbol u. The relative atomic mass of a chemical element is the average mass of its atoms expressed in atomic mass units. *See* RELATIVE ATOMIC MASS.

Before 1961, two versions of the atomic mass unit were in use. The unit used by physicists was defined as $\frac{1}{16}$ of the mass of an atom of ^{16}O, the predominant isotope of oxygen. The unit used by chemists was defined as $\frac{1}{16}$ of the average mass of the atoms in naturally occurring oxygen, a mixture of the isotopes ^{16}O, ^{17}O, and ^{18}O. In 1961, by international agreement, the standard based on ^{12}C superseded both these older units. It is related to them by: 1 amu (international)$\cong$1.000 318 amu (physical)$\cong$1.000 043 amu (chemical). *See* ATOMIC WEIGHT.

[JONATHAN F. WEIL]

Atomic nucleus

The central region of an atom. Atoms are composed of negatively charged electrons, positively charged protons, and electrically neutral neutrons. The protons and neutrons (collectively known as nucleons) are located in a small central region known as the nucleus. The electrons move in orbits which are large in comparison with the dimensions of the nucleus itself. Protons and neutrons possess approximately equal masses, each roughly 1840 times that of an electron. The number of nucleons in a nucleus is given by the mass number A and the number of protons by the atomic number Z. Nuclear radii r are given approximately by $r = 1.4 \times 10^{-13} A^{1/3}$ cm. *See* NUCLEAR STRUCTURE.

[HENRY E. DUCKWORTH]

Atomic number

The number of elementary positive charges (protons) contained within the nucleus of an atom. It is denoted by the letter Z. For an electrically neutral atom, the number of planetary electrons is also given by the atomic number. Atoms with the same Z (isotopes) belong to the same element. The lightest element, hydrogen, has $Z = 1$. The heaviest naturally occurring element, uranium, has $Z = 92$. All elements up to and including $Z = 103$ (lawrencium) either occur in nature or have been created artificially. The atomic number of an atom is altered during radioactive decay: For α-emission, $Z \rightarrow Z - 2$; for β^{-}-emission, $Z \rightarrow Z + 1$; for β^{+}-emission or electron capture, $Z \rightarrow Z - 1$. When specifically written, the atomic number is usually placed before and below the elemental symbol, for example, $_1H$, $_{92}U$. *See* MASS NUMBER; RADIOACTIVITY.　　　　[HENRY E. DUCKWORTH]

Atomic physics

The study of the structure of the atom, its dynamical properties, including energy states, and its interactions with particles and fields. These are almost completely determined by the laws of quantum mechanics, with very refined corrections required by quantum electrodynamics. Despite the enormous complexity of most atomic systems, in which each electron interacts with both the nucleus and all the other orbiting electrons, the wavelike nature of particles, combined with the Pauli exclusion principle, results in an amazingly orderly array of atomic properties. These are systematized by the Mendeleev periodic table. In addition to their classification by chemical activity and atomic weight, the various elements of this table are characterized by a wide variety of observable proper-

ties. These include electron affinity, polarizability, angular momentum, multiple electric moments, and magnetism. *See* ATOMIC WEIGHT; QUANTUM ELECTRODYNAMICS; QUANTUM MECHANICS.

Each atomic element, normally found in its ground state (that is, with its electron configuration corresponding to the lowest state of total energy), can also exist in an infinite number of excited states. These are also ordered in accordance with relatively simple hierarchies determined by the laws of quantum mechanics. The most characteristic signature of these various excited states is the radiation emitted or absorbed when the atom undergoes a transition from one state to another. The systemization and classification of atomic energy levels (spectroscopy) has played a central role in developing an understanding of atomic structure. Control of populations of ensembles of excited atoms has led to the laser, which is itself now used to obtain even more refined information concerning atomic structure than has hitherto been possible. *See* LASER.

Atomic radiation represents one of nature's most refined probes for the study of a wide range of natural phenomena, such as the effects of fluctuations in empty space, of the anomalous magnetic moment of the electron, and even subtler high-order corrections to atomic energy levels caused by quantum electrodynamics. The isolated atom is one of the most reliable systems known in nature in terms of the reproducibility of its energy levels, and is used as a primary frequency standard for time calibration. *See* ATOMIC CONSTANTS; PARITY (QUANTUM MECHANICS).

The problem of the mutual interaction of atoms or fragments (for example, electrons and ions) is still more complicated than that of the isolated atom, but such interactions are important since they govern a wide variety of practical phenomena, including atmospheric physics, laser physics, plasma generation for controlled thermonuclear plasmas, materials research (including the influence of radiation on matter), and chemical reactions and molecular formation. This area is the domain of atomic collisions. *See* SCATTERING EXPERIMENTS (ATOMS AND MOLECULES).

Finally, the study of inner shell structure and interactions of heavier atoms at very high energies has become an active field of research. This is the domain of high-energy atomic physics, an area of increasing activity, which is leading to a better understanding of the structure and the dynamics of the inner, strongly bound atomic electrons. *See* ATOM; ATOMIC STRUCTURE AND SPECTRA; ELECTRON; NUCLEAR PHYSICS.

[BENJAMIN BEDERSON]

Atomic structure and spectra

The idea that matter is subdivided into discrete and further indivisible building blocks called atoms dates back to the Greek philosopher Democritus, whose teachings of the 5th century B.C. are commonly accepted as the earliest authenticated ones concerning what has come to be called atomism by students of Greek philosophy. The weaving of the philosophical thread of atomism into the analytical fabric of physics began in the late 18th and the 19th centuries. Robert Boyle is generally credited with introducing the concept of chemical elements, the irreducible units which are now recognized as individual atoms of a given element. In the early 19th century John Dalton developed his atomic theory, which postulated that matter consists of indivisible atoms as the irreducible units of Boyle's elements, that each atom of a given element has identical attributes, that differences among elements are due to fundamental differences among their constituent atoms, that chemical reactions proceed by simple rearrangement of indestructible atoms, and that chemical compounds consist of molecules which are reasonably stable aggregates of such indestructible atoms.

Electromagnetic nature of atoms. The work of J. J. Thomson in 1897 clearly demonstrated that atoms are electromagnetically constituted and that from them can be extracted fundamental material units bearing electric charge that are now called electrons. These pointlike charges have a mass of 9.110×10^{-31} kilogram and a negative electric charge of magnitude 1.602×10^{-19} coulomb. The electrons of an atom account for a negligible fraction of its mass. By virtue of overall electrical neutrality of every atom, the mass must therefore reside in a compensating, positively charged atomic component of equal charge magnitude but vastly greater mass. *See* ELECTRON.

Thomson's work was followed by the demonstration by Ernest Rutherford in 1911 that nearly all the mass and all of the positive electric charge of an atom are concentrated in a small nuclear core approximately 10,000 times smaller in extent than an atomic diameter. Niels Bohr in 1913 and others carried out some remarkably successful attempts to build solar system models of atoms containing planetary pointlike electrons orbiting around a positive core through mutual electrical attraction (though only certain "quantized" orbits were "permitted"). These models were ultimately superseded by nonparticulate, matter wave quantum theories of both electrons and atomic nuclei. *See* NONRELATIVISTIC QUANTUM THEORY; QUANTUM MECHANICS.

The modern picture of condensed matter (such as solid crystals) consists of an aggregate of atoms or molecules which respond to each other's proximity through attractive electrical interactions at separation distances of the order of 1 atomic diameter (approximately 10^{-10} m) and repulsive electrical interactions at much smaller distances. These interactions are mediated by the electrons, which are in some sense shared and exchanged by all atoms of a particular sample, and serve as a kind of interatomic glue which binds the mutually repulsive, heavy, positively charged atomic cores together. *See* SOLID-STATE PHYSICS.

Planetary atomic models. Fundamental to any planetary atomic model is a description of the forces which give rise to the attraction and repulsion of the constituents of an atom. Coulomb's law describes the fundamental interaction which, to a good approximation, is still the basis of modern theories of atomic structure and spectra: the force exerted by one charge on another is repulsive for charges of like sign and attractive for charges of unlike sign, is proportional to the magnitude of each electric charge, and diminishes as the square of the distance separating the charges. *See* COULOMB'S LAW.

Also fundamental to any planetary model is Thomson's discovery that electrically neutral atoms in some sense contain individual electrons whose charge is restricted to a unique quantized value. Moreover, Thomson's work suggested that nuclear charges are precisely equally but oppositely quantized to offset the sum of the constituent atomic electron charges. Boyle's elements, of which over a hundred have been discovered, may then be individually labeled by the number of quantized positive-charge units Z (the atomic number) residing within the atomic nucleus (unit charge $= +1.602 \times 10^{-19}$ C). Each nucleus is surrounded by a complement of Z electrons (with charges of -1.602×10^{-19} C each) to produce overall charge neutrality. Molecules containing two, three, . . . , atoms can then be thought of as binary, ternary, . . . , planetary systems consisting of heavy central bodies of atomic numbers Z_1, Z_2, . . . , sharing a supply of $Z_1 + Z_2 + \cdots$ electrons having the freedom to circulate throughout the aggregate and bond it together as would an electronic glue.

Atomic sizes. All atoms, whatever their atomic number Z, have roughly the same diameter (about 10^{-10} m), and molecular sizes tend to be the sum of the aggregate atomic sizes. It is easy to qualitatively, though only partially, account for this circumstance by using Coulomb's law. The innermost electrons of atoms orbit at small radii because of intense electrical attractions prevailing at small distances. Because electrons are more than 2000 times lighter than most nuclei and therefore move less sluggishly, the rapid orbital motion of inner electrons tends to clothe the slow-moving positive nuclear charge in a negative-charge cloud, which viewed from outside the cloud masks this positive nuclear charge to an ever-increasing extent. Thus, as intermediate and outer electrons are added, each experiences a diminishing attraction. The Zth, and last, electron sees $+Z$ electronic charge units within the nucleus compensated by about $(Z - 1)$ electronic charges in a surrounding cloud, so that in first approximation the Zth electron orbits as it would about a bare proton having $Z = 1$, at about the same radius. Crudely speaking, an atom of any atomic number Z thus has a size similar to that of the hydrogen atom. When the molecular aggregates of atoms of concern to chemistry are formed, they generally do so through extensive sharing or exchanging of a small number (often just one) of the outermost electrons of each atom. Hence, the interatomic spacing in both molecules and solids tends to be of the order of one to a very few atomic diameters.

Chemical reactions. In a microscopic planetary atomic model, chemical reactions in which rearrangements result in new molecular aggregates of atoms can be viewed as electrical phenomena, mediated by the electrical interactions of outer electrons. The reactions proceed by virtue of changing positions and shapes of the orbits of the binding electrons, either through their internal electrical interactions or through their being electrically "bumped" from outside (as by collisions with nearby molecules).

Difficulties with the models. Before continuing with a detailed account of the successes of detailed planetary atomic models, it is advisable to anticipate some severe difficulties associated with them. Some of these difficulties played a prominent role in the development of quantum theory, and others present as yet unsolved and profound challenges to classical as well as modern quantum theories of physics.

A classical planetary theory fails to account for several atomic enigmas. First, it is known that electrons can be made to execute fast circular orbits in large accelerating machines, of which modern synchrotrons are excellent examples. As they are then in accelerated motion, they radiate light energy, as predicted by Maxwell's theory, at a frequency equal to that of their orbital circular motions. Thus, a classical planetary theory fails to explain how atoms can be stable and why atomic electrons do not continuously radiate light, falling into the nucleus as they do so. Such a theory also does not account for the observation that light emitted by atomic electrons appears to contain only quantized frequencies or wavelengths. Furthermore, a classical planetary theory would lead one to expect that all atomic electrons of an atom would orbit very close to their parent nuclei at very small radii, instead of distributing themselves in shells of increasing orbital radius, which seem able to accommodate only small, fixed numbers of electrons per shell.

Any theory of atomic structure must deal with the question of whether the atom is mostly empty space, as a planetary system picture suggests, or whether the entire atomic volume is filled with electronic charges in smeared out, cloudlike fashion. Such a theory must also be concerned with whether electrons and nuclei are pointlike, structureless, and indivisible, or whether they can be subdivided further into smaller constituents, in analogy to the way in which Thomson was able to extract electrons as constituents of atoms. Questions which still have not received definitive answers concern how much energy is needed to construct electrons and nuclei, what their radii are, and why they do not fly apart under the explosive repulsion of half their charge distribution for the other half. A more advanced theory than the classical planetary model is also required to determine whether electrons really interact with each other and with nuclei instantaneously, as Coulomb's law would have it, or whether there is a finite interaction delay time. Finally, a fundamental question, which can be answered only by quantum theory, is concerned with whether electrons and nuclei behave as pointlike objects, as light does when it gives rise to sharp shadows, or whether they behave as extended waves which can exhibit diffraction (bending) phenomena, as when light or water waves bend past sharp edges and partially penetrate regions of shadow in undulatory intensity patterns.

Scattering experiments. A key experimental technique in answering many important questions, such as those just posed, is that of scattering, in which small, high-speed probe projectiles are used to interact with localized regions of more extended objects from which they rebound, or scatter. A study of the number of scattered projectiles, and their distributions in speed and angle, gives much information about the structure of target systems, and the internal distributions, speeds, and concentrations of constituent bodies or substructures. *See* SCATTERING EXPERIMENTS (ATOMS AND MOLECULES).

The scattering of x-rays by solids, for example,

was used by C. Barkla to indicate that the number of electrons in an atom is approximately half the atomic weight. The mechanism of the production of the secondary (scattered) x-rays is indicated in Fig. 1a. X-rays are electromagnetic waves of wavelength considerably smaller than the size of the atom. If an x-ray sets an atomic electron into vibration, there follows the emission of a wave of lesser amplitude which can be observed at directions outside the incident beam. *See* X-RAYS.

Rutherford's experiments on the scattering of alpha particles represented an important step in understanding atomic structure. Alpha particles are helium ($Z = 2$) nuclei emitted at high velocities by some radioactive materials. A beam of these particles was directed at thin foils of different metals, and the relative numbers scattered at various angles were observed. While most of the particles passed through the foil with small deflections, such as the lower particle in Fig. 1b, a considerable number of very large deflections occurred. The upper particle in Fig. 1b has undergone such a deflection. The precise results could be explained only if the positive electric charge of the atom is concentrated in the very small volume of the nucleus, which also contains almost all of the atom's mass. The diameter of the nucleus, found to depend on the atomic mass, was about 10^{-14} m for heavy nuclei. The nuclear charge was found to be the atomic number Z times the magnitude of the electron's charge.

The results of the scattering experiments therefore established the model of the atom as consisting of a small, massive, positively charged nucleus surrounded by a cloud of electrons to provide electrical neutrality. As noted above, such an atom should quickly collapse. The first step toward an explanation as to why it did not, came from the Bohr picture of the atom.

BOHR ATOM

The hydrogen atom is the simplest atom, and its spectrum (or pattern of light frequencies emitted) is also the simplest. The regularity of its spectrum had defied explanation until Bohr solved it with three postulates, these representing a model which is useful, but quite insufficient, for understanding the atom.

Postulate 1: The force that holds the electron to the nucleus is the Coulomb force between electrically charged bodies.

Postulate 2: Only certain stable, nonradiating orbits for the electron's motion are possible, those for which the angular momentum is an integral multiple of $h/2\pi$ (Bohr's quantum condition on the orbital angular momentum). Each stable orbit represents a discrete energy state.

Postulate 3: Emission or absorption of light occurs when the electron makes a transition from one stable orbit to another, and the frequency ν of the light is such that the difference in the orbital energies equals $h\nu$ (A. Einstein's frequency condition for the photon, the quantum of light).

Here the concept of angular momentum, a continuous measure of rotational motion in classical physics, has been asserted to have a discrete quantum behavior, so that its quantized size is related to Planck's constant h, a universal constant of nature. The orbital angular momentum of a point object of mass m and velocity v, in rotational mo-

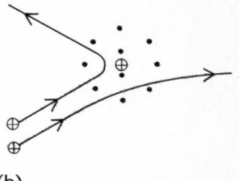

electrons

(a)   nucleus

(b)

Fig. 1. Scattering by an atom (a) of x-rays, (b) of alpha particles.

tion about a central body, is defined as the product of the component of linear momentum mv (expressing the inertial motion of the body) tangential to the orbit times the distance to the center of rotation. *See* ANGULAR MOMENTUM.

Modern quantum mechanics has provided justification for Bohr's quantum condition on the orbital angular momentum. It has also shown that the concept of definite orbits cannot be retained except in the limiting case of very large orbits. In this limit, the frequency, intensity, and polarization can be accurately calculated by applying the classical laws of electrodynamics to the radiation from the orbiting electron. This fact illustrates Bohr's correspondence principle, according to which the quantum results must agree with the classical ones for large dimensions. The deviation from classical theory that occurs when the orbits are smaller than the limiting case is such that one may no longer picture an accurately defined orbit. Bohr's other hypotheses are still valid.

Quantization of hydrogen atom. According to Bohr's theory, the energies of the hydrogen atom are quantized (that is, can take on only certain discrete values). These energies can be calculated from the electron orbits permitted by the quantized orbital angular momentum. The orbit may be circular or elliptical, so only the circular orbit is considered here for simplicity. Let the electron, of mass m and electric charge $-e$, describe a circular orbit of radius r around a nucleus of charge $+e$ and of infinite mass. With the electron velocity v, the angular momentum is mvr, and the second postulate becomes Eq. (1). The integer n is called the principal quantum number. The centripetal force required to hold the electron in its orbit is the electrostatic force described by Coulomb's law, as shown in Eq. (2). Here ϵ_0 is the permittivity of free

ATOMIC STRUCTURE
AND SPECTRA

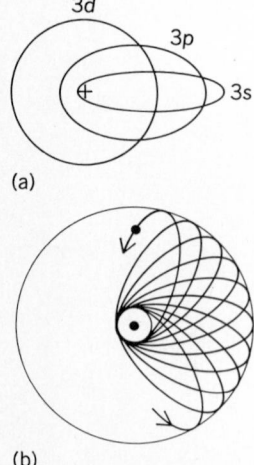

(a)

(b)

Fig. 2. Possible elliptical orbits, according to the Bohr-Sommerfeld theory. (*a*) The three permitted orbits for $n = 3$. (*b*) Precession of the 3s orbit caused by the relativistic variation of mass. (*From A. P. Arya, Fundamentals of Atomic Physics, Allyn and Bacon, pp. 281 and 286, 1971*)

$$mvr = n(h/2\pi) \qquad (n = 1, 2, 3, \ldots) \quad (1)$$

$$mv^2/r = e^2/4\pi\epsilon_0 r^2 \quad (2)$$

space, a constant included in order to give the correct units to the mks statement of Coulomb's law. *See* QUANTUM NUMBERS.

The energy of an electron in an orbit consists of both kinetic and potential energies. For these circular orbits, the potential energy is twice as large as the kinetic energy and has a negative sign, where the potential energy is taken as zero when the electron and nucleus are at rest and separated by a very large distance. The total energy, which is the sum of kinetic and potential energies, is given in Eq. (3). The negative sign means that the electron is bound to the nucleus and energy must be provided to separate them.

$$E = (mv^2/2) - mv^2 = -mv^2/2 \quad (3)$$

tron is bound to the nucleus and energy must be provided to separate them.

It is possible to eliminate v and r from these three equations. The result is that the possible energies of the nonradiating states of the atom are given by Eq. (4).

$$E = -\frac{me^4}{8\epsilon_0^2 h^2} \cdot \frac{1}{n^2} \quad (4)$$

The same equation for the hydrogen atom's energy levels, except for some small but significant corrections, is obtained from the solution of the Schrödinger equation for the hydrogen atom. *See* SCHRÖDINGER'S WAVE EQUATION.

The frequencies of electromagnetic radiation or light emitted or absorbed in transitions are given by Eq. (5) where E' and E'' are the energies of the

$$\nu = (E' - E'')/h \quad (5)$$

initial and final states of the atom. Spectroscopists usually express their measurements in wavelength λ or in wave number σ in order to obtain numbers of a convenient size. The frequency of a light wave can be thought of as the number of complete waves radiated per second of elapsed time. If each wave travels at a fixed velocity c (approximately 3×10^8 m/s in vacuum), then after t seconds, the distance traveled by the first wave in a train of waves is ct, the number of waves in the train is νt, and hence the length of each must be $ct/\nu t = c/\nu = \lambda$. The wave number, defined as the reciprocal of the wavelength, therefore equals ν/c. The quantization of energy $h\nu$ and of angular momentum in units of $h/2\pi$ does not materially alter this picture. The wave number of a transition is shown in Eq. (6). If $T = -E/hc$, then Eq. (7) results. Here T is called the spectral term.

$$\sigma = \frac{\nu}{c} = \frac{E'}{hc} - \frac{E''}{hc} \quad (6)$$

$$\sigma = T'' - T' \quad (7)$$

The allowed terms for hydrogen, from Eq. (4), are given by Eq. (8). The quantity R is the important Rydberg constant. Its value, which has been accurately measured by laser spectroscopy, is related to the values of other well-known atomic constants, as in Eq. (8). *See* RYDBERG CONSTANT.

$$T = \frac{me^4}{8\epsilon_0^2 ch^3} \cdot \frac{1}{n^2} = \frac{R}{n^2} \quad (8)$$

The effect of finite nuclear mass must be considered, since the nucleus does not actually remain at rest at the center of the atom. Instead, the electron and nucleus revolve about their common center of mass. This effect can be accurately accounted for and requires a small change in the value of the effective mass m in Eq. (8). The mass effect was first detected by comparing the spectrum of hydrogen with that of singly ionized helium, which is like hydrogen in having a single electron orbiting the nucleus. For this isoelectronic case, the factor Z^2 must be included in the numerator of Eqs. (2) and (8) to account for the greater nuclear charge. The mass effect was used by H. Urey to discover deuterium, one of three hydrogen isotopes, by the shift of lines in its spectrum because of the very small change in its Rydberg constant. *See* ISOELECTRONIC SEQUENCE; ISOTOPE SHIFT.

Elliptical orbits. In addition to the circular orbits already described, elliptical ones are also consistent with the requirement that the angular momentum be quantized. A. Sommerfeld showed that for each value of n there is a family of n permitted elliptical orbits, all having the same major axis but with different eccentricities. Figure 2*a* shows, for example, the Bohr-Sommerfeld orbits for $n = 3$. The orbits are labeled s, p, and d, indicating values of the azimuthal quantum number $l = 0$, 1, and 2. This number determines the shape of the orbit, since the ratio of the major to the minor axis is found to be $n/(l + 1)$. To a first approximation, the energies of all orbits of the same n are equal. In the case of the highly eccentric orbits, however, there is a slight lowering of the energy due to precession of the orbit (Fig. 2b). According to Einstein's theory of relativity, the mass increases somewhat in the inner part of the orbit, because of greater velocity. The velocity increase is greater as the eccentricity is greater, so the orbits of higher eccentricity have their energies lowered more. The quantity l is called the orbital angular momentum quantum number or the azimuthal quantum number. *See* RELATIVITY.

A selection rule limits the possible changes of l that give rise to spectrum lines (transitions of fixed frequency or wavelength) of appreciable intensity. The rule is that l may increase or decrease only by one unit. This is usually written as $\Delta l = \pm 1$ for an allowed transition. Transitions for which selection rules are not satisfied are called forbidden; these tend to have quite low intensities. The quantum number n may change by any amount. Selection rules for hydrogen may be derived from Bohr's correspondence principle. However, the selection rules, as well as the relation between n and l, arise much more naturally in quantum mechanics. *See* SELECTION RULES.

MULTIELECTRON ATOMS

In attempting to extend Bohr's model to atoms with more than one electron, it is logical to compare the experimentally observed terms of the alkali atoms, which contain only a single electron outside closed shells, with those of hydrogen. A definite similarity is found but with the striking difference that all terms with $l > 0$ are double. This fact was interpreted by S. A. Goudsmit and G. E. Uhlenbeck as due to the presence of an additional angular momentum of $1/2$ $(h/2\pi)$ attributed to the electron spinning about its axis. The spin quantum number of the electron is $s = 1/2$.

The relativistic quantum mechanics developed by P. A. M. Dirac provided the theoretical basis for this experimental observation. *See* ELECTRON SPIN.

Exclusion principle. Implicit in much of the following discussion is W. Pauli's exclusion principle, first enunciated in 1925, which when applied to atoms may be stated as follows: no more than one electron in a multielectron atom can possess precisely the same quantum numbers. In an independent, hydrogenic electron approximation to multielectron atoms, there are $2n^2$ possible independent choices of the principal (n), orbital (l), and magnetic (m_1, m_s) quantum numbers available for electrons belonging to a given n, and no more. Here m_1 and m_s refer to the quantized projections of I and s along some chosen direction. The organization of atomic electrons into shells of increasing radius (the Bohr radius scales as n^2) follows from this principle, answering the question raised above as to why all the electrons of a heavy atom do not collapse into the most tightly bound orbits. *See* EXCLUSION PRINCIPLE.

Examples are: helium $(Z = 2)$, two $n = 1$ electrons; neon $(Z = 10)$, two $n = 1$ electrons, eight $n = 2$ electrons; argon $(Z = 18)$, two $n = 1$ electrons, eight $n = 2$ electrons, eight $n = 3$ electrons. Actually, in elements of Z greater than 18, the $n = 3$ shell could in principle accommodate 10 more electrons but, for detailed reasons of binding energy economy rather than fundamental symmetry, contains full

$3s$ and $3p$ shells for a total of 8 $n = 3$ electrons, but often a partially empty $3d$ shell. The most chemically active elements, the alkalies, are those with just one outer orbital electron in an n state, one unit above that of a completely full shell or subshell. *See* ELECTRON CONFIGURATION.

Spin-orbit coupling. This is the name given to the energy of interaction of the electron's spin with its orbital angular momentum. The origin of this energy is magnetic.

A charge in motion through either "pure" electric or "pure" magnetic fields, that is, through fields perceived as "pure" in a static laboratory, actually experiences a combination of electric and magnetic fields, if viewed in the frame of reference of a moving observer with respect to whom the charge is momentarily at rest. For example, moving charges are well known to be deflected by magnetic fields. But in the rest frame of such a charge, there is no motion, and any acceleration of a charge must be due to the presence of a pure electric field from the point of view of an observer analyzing the motion in that reference frame. *See* RELATIVISTIC ELECTRODYNAMICS.

A spinning electron can crudely be pictured as a spinning ball of charge, imitating a circulating electric current (though Dirac electron theory assumes no finite electron radius—classical pictures fail). This circulating current gives rise to a magnetic field distribution very similar to that of a small bar magnet, with north and south magnetic poles symmetrically distributed along the spin axis above and below the spin equator. This representative bar magnet can interact with external magnetic fields, one source of which is the magnetic field experienced by an electron in its rest frame, owing to its orbital motion through the electric field established by the central nucleus of an atom. In multielectron atoms, there can be additional, though generally weaker, interactions arising from the magnetic interactions of each electron with its neighbors, as all are moving with respect to each other, and all have spin. The strength of the bar magnet equivalent to each electron spin, and its direction in space are characterized by a quantity called the magnetic moment, which also is quantized essentially because the spin itself is quantized. Studies of the effect of an external magnetic field on the states of atoms show that the magnetic moment associated with the electron spin is equal in magnitude to a unit called the Bohr magneton. *See* MAGNETIC MOMENT.

The energy of the interaction between the electron's magnetic moment and the magnetic field generated by its orbital motion is usually a small correction to the spectral term, and depends on the angle between the magnetic moment and the magnetic field or, equivalently, between the spin angular momentum vector and the orbital angular momentum vector (a vector perpendicular to the orbital plane whose magnitude is the size of the orbital angular momentum). Since quantum theory requires that the quantum number j of the electron's total angular momentum shall take values differing by integers, while l is always an integer, there are only two possible orientations for s relative to l: s must be either parallel or antiparallel to l. (This statement is convenient but not quite accurate. Actually, orbital angular momentum is a vector quantity represented by the quantum number l

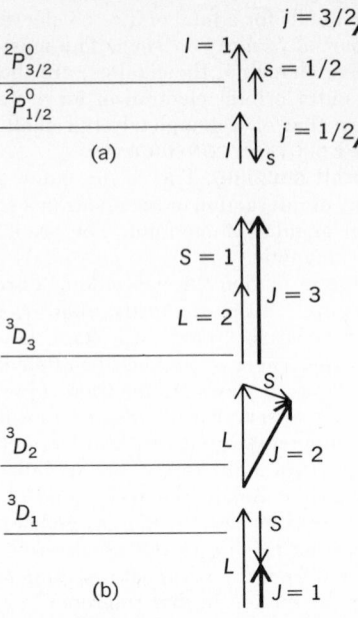

Fig. 3. Vector model for spectral terms arising from (a) a single p electron, and (b) two electrons, either two p electrons, or an s and a d electron.

and of magnitude $\sqrt{l(l+1)}\cdot(h/2\pi)$. There are similar relations for spin and total angular momentum. These statements all being true simultaneously, the spin vector cannot ever be exactly parallel to the orbital angular momentum vector. Only the quantum numbers themselves can be described as stated.) Figure 3a shows the relative orientations of these two vectors and of their resultant j for a p electron (one for which $l=1$). The corresponding spectral term designations are shown adjacent to the vector diagrams, labeled with the customary spectroscopic notation, to be explained later. *See* MAGNETON.

For the case of a single electron outside the nucleus, the Dirac theory gives Eq. (9) for the spin-

$$\Delta T = \frac{R\alpha^2 Z^4}{n^3} \cdot \frac{j(j+1) - l(l+1) - s(s+1)}{l(2l+1)\,(l+1)} \qquad (9)$$

orbit correction to the spectral terms. Here $\alpha = e^2/2\epsilon_0 hc \cong 1/137$ is called the fine structure constant. The fine structure splitting predicted by Eq. (9) is present in hydrogen, although its observation requires instruments of very high precision. A relativistic correction must be added.

In atoms having more than one electron, this fine structure becomes what is called the multiplet structure. The doublets in the alkali spectra, for example, are due to spin-orbit coupling; Eq. (9), with suitable modifications, can still be applied. These modifications may be attributed to penetration of the outer orbital electron within the closed shells of other electrons.

The various states of the atom are described by giving the quantum numbers n and l for each electron (the configuration) as well as the set of quantum numbers which depend on the manner in which the electrons interact with each other.

Coupling schemes. When more than one electron is present in the atom, there are various ways in which the spins and orbital angular momenta

can interact. Each spin may couple to its own orbit, as in the one-electron case; other possibilities are orbit–other orbit, spin-spin, and so on. The most common interaction in the light atoms, called LS coupling or Russell-Saunders coupling, is described schematically in Eq. (10). This notation

$$\{(l_1, l_2, l_3, \ldots)\,(s_1, s_2, s_3, \ldots)\} = \{L, S\} = J \qquad (10)$$

indicates that the l_i are coupled strongly together to form a resultant L, representing the total orbital angular momentum. The s_i are coupled strongly together to form a resultant S, the total spin angular momentum. The weakest coupling is that between L and S to form J, the total angular momentum of the electron system of the atom in this state. Suppose, for example, it is necessary to calculate the terms arising from a p ($l=1$) and a d ($l=2$) electron. The quantum numbers L, S, and J are never negative, so the only possible values of L are 1, 2, and 3. States with these values of L are designated P, D, and F. This notation, according to which $L=0, 1, 2, 3, 4, 5$, etc. correspond to S, P, D, F, G, H, etc. terms, survives from the early empirical designation of series of lines as sharp, principal, diffuse, and so on.

The spin $s=1/2$ for each electron, always, so the total spin $S=0$ or $S=1$ in this case. Consider only the term with $S=1$ and $L=2$. The coupling of L and S gives $J=1, 2$, and 3, as shown in Fig. 3b. These three values of J correspond to a triplet of different energy levels which lie rather close together, since the LS interaction is relatively weak. It is convenient to use the notation ^3D for this multiplet of levels, the individual levels being indicated with J values as a subscript, as $^3D_1$, $^3D_2$, and $^3D_3$. The superscript has the value $2S+1$, a quantity called the multiplicity. There will also be a term with $S=0$ and $L=2$ and the single value of $J=2$. This is a singlet level, written as ^1D or $^1D_2$. The entire group of terms arising from a p and d electron includes ^1P, ^3P, ^1D, ^3D, ^1F, ^3F, with all values of J possible for each term.

The ^2P state shown in Fig. 3a is derived from a single electron which, since $L=l=1$ and $S=s=1/2$, has J values of 3/2 and 1/2, forming a doublet. If there are three or more electrons, the number of possible terms becomes very large, but they are easily derived by a similar procedure. The resulting multiplicities are shown in the table. If two or more electrons are equivalent, that is, have the same n and l, the number of resulting terms is greatly reduced, because of the requirement of the Pauli exclusion principle that no two electrons in an atom may have all their quantum numbers alike. Two equivalent p electrons, for example, give only ^1S, ^3P, and ^1D terms, instead of the six terms ^1S, ^3S, ^1P, ^3P, ^1D, and ^3D possible if the

Possible multiplicities with different numbers of electrons

Number of electrons:	1	2	3	4
Values of S:	1/2	1, 0	3/2, 1/2	2, 1, 0
Multiplicities	Doublets	Singlets Triplets	Doublets Quartets	Singlets Triplets Quintets

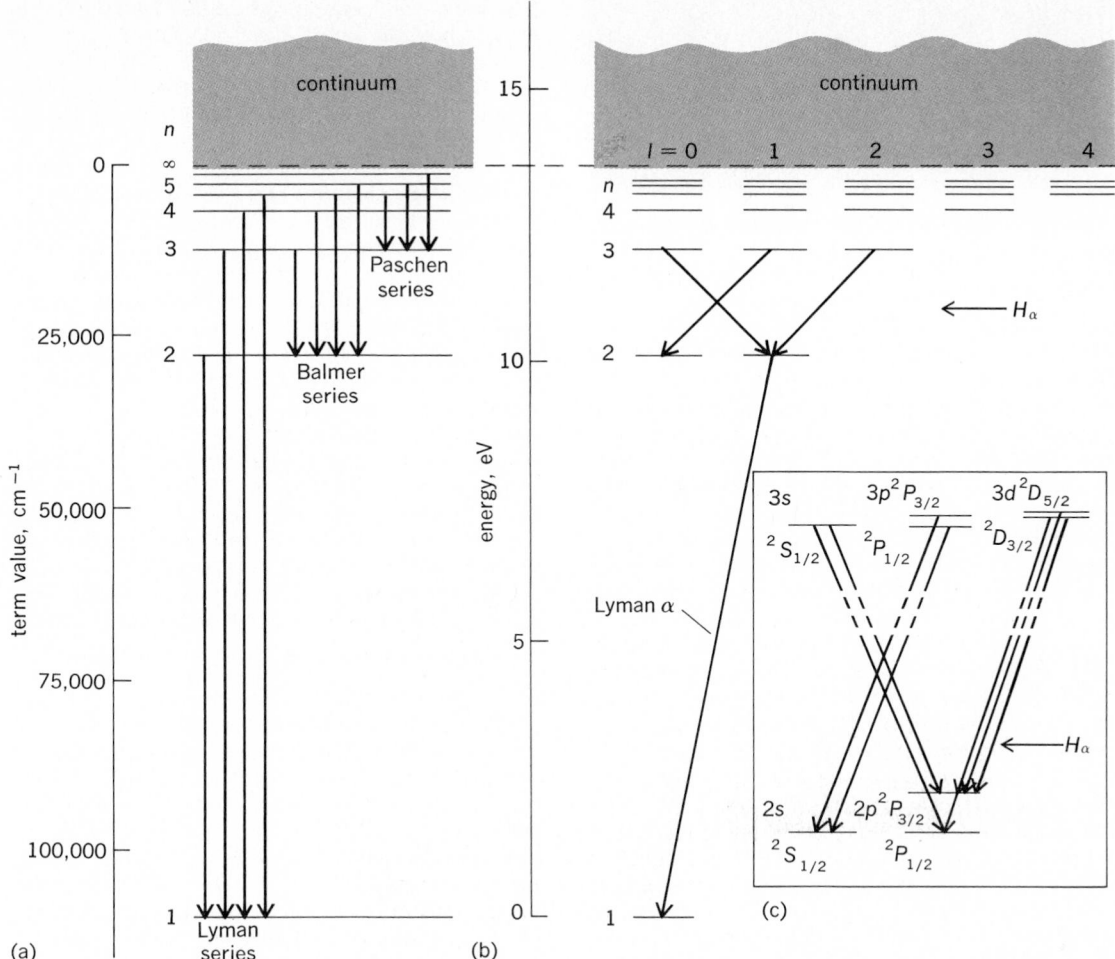

Fig. 4.   Terms and transitions for the hydrogen atom. (*a*) Bohr theory. (*b*) Bohr-Sommerfeld theory. (*c*) Dirac theory.

electrons are nonequivalent. The exclusion principle applied to equivalent electrons explains the observation of filled shells in atoms, where $L = S = J = 0$.

Coupling of the LS type is generally applicable to the low-energy states of the lighter atoms. The next commonest type is called jj coupling, represented in Eq. (11). Each electron has its spin cou-

$$\{(l_1, s_1)(l_2, s_2)(l_3, s_3) \ldots\}$$
$$= \{j_1, j_2, j_3, \ldots\} = J \quad (11)$$

pled to its own orbital angular momentum to form a j_i for that electron. The various j_i are then more weakly coupled together to give J. This type of coupling is seldom strictly observed. In the heavier atoms it is common to find a condition intermediate between LS and jj coupling; then either the LS or jj notation may be used to describe the levels, because the number of levels for a given electron configuration is independent of the coupling scheme.

SPECTRUM OF HYDROGEN

Figure 4*a* shows the terms R/n^2 in the spectrum of hydrogen resulting from the simple Bohr theory. These terms are obtained by dividing the numerical value of the Rydberg constant for hydrogen (109,678 cm^{-1}) by n^2, that is, by 1, 4, 9, 16, etc. The equivalent energies in electron volts (eV) may then

be found by using the conversion factor 1 eV = 8066 cm^{-1}. These energies, in contrast to the term values, are usually measured from zero at the lowest state, and increase for successive levels. They draw closer together until they converge at 13.598 eV. This corresponds to the orbit with $n = \infty$ and complete removal of the electron, that is, ionization of the atom. Above this ionization potential is a continuum of states representing the nucleus plus a free electron possessing a variable amount of kinetic energy.

The names of the hydrogen series are indicated in Fig. 4*a*. The spectral lines result from transitions between these various possible energy states. Each vertical arrow on the diagram connects two states and represents the emission of a particular line in the spectrum. The wave number of this line, according to Bohr's frequency condition, is equal to the difference in the term values for the lower and upper states and is therefore proportional to the length of the arrow. The wave numbers of all possible lines are given by Eq. (12), known as the

$$\sigma = T'' - T' = R\left(\frac{1}{n''^2} - \frac{1}{n'^2}\right) \quad (12)$$

Balmer formula, where the double primes refer to the lower energy state (larger term value) and the single primes to the upper state. Any particular series is characterized by a constant value of n'' and variable n'. Figure 5 shows the spectrum of

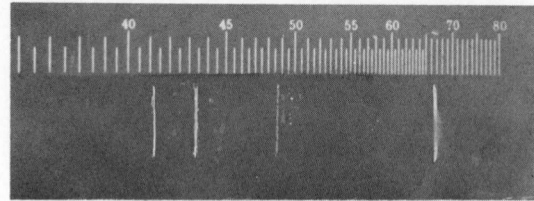

Fig. 5. Line emission spectrum of hydrogen. The scale gives the wavelengths in units of 10^{-8} m.

a hydrogen discharge tube taken with a quartz spectrograph of resolution insufficient to resolve the fine structure. The Balmer series, represented by Eq. (12) with $n'' = 2$, is the only one shown. Its first line, that for $n' = 3$, is the bright red line at the wavelength 656.3 nm and is called H_α. Succeeding lines H_β, H_γ, and so on, proceed toward the ultraviolet with decreasing spacing and intensity, eventually converging at 364.6 nm. Beyond this series limit there is a region of continuous spectrum. The other series, given by $n'' = 1, 3, 4$, etc., lie well outside the visible region. The Lyman series covers the wavelength range 121.5–91.2 nm in the vacuum ultraviolet, and the Paschen series 1875.1–820.6 nm in the infrared. Still other series lie at even longer wavelengths.

Since hydrogen is by far the most abundant element in the cosmos, its spectrum is extremely important from the astronomical standpoint. The Balmer series has been observed as far as H_{31} in the spectra of hot stars. The Lyman series appears as the strongest feature of the Sun's spectrum, as photographed by rockets and orbiting satellites such as Skylab above the Earth's atmosphere.

NUCLEAR MAGNETISM AND HYPERFINE STRUCTURE

Most atomic nuclei also possess spin, but rotate about 2000 times slower than electrons because their mass is on the order of 2000 or more times greater than that of electrons. Because of this, very weak nuclear magnetic fields, analogous to the electronic ones that produce fine structure in spectral lines, further split atomic energy levels. Consequently, spectral lines arising from them are split according to the relative orientations, and hence energies of interaction, of the nuclear magnetic moments with the electronic ones. The resulting pattern of energy levels and corresponding spectral-line components is referred to as hyperfine structure. *See* NUCLEAR MOMENTS.

The fine structure and hyperfine structure of the hydrogen terms are particularly important astronomically. In particular, the observation of a line of 21-cm wavelength, arising from a transition between the two hyperfine components of the $n = 1$ term, gave birth to the science of radio astronomy.

Investigations with tunable lasers. The enormous capabilities of tunable lasers have allowed observations which were impossible previously. For example, high-resolution saturation spectroscopy, utilizing a saturating beam and a probe beam from the same laser, has been used to measure the hyperfine structure of the sodium resonance lines (called the D_1 and D_2 lines). Each line was found to have components extending over about 0.017 cm^{-1}, while the separation of the D_1 and D_2 lines is 17 cm^{-1}. The smallest separation resolved was less than 0.001 cm^{-1}, which was far less than the Doppler width of the lines. *See* DOPPLER EFFECT; HYPERFINE STRUCTURE.

Isotope shift. Nuclear properties also affect atomic spectra through the isotope shift. This is the result of the difference in nuclear masses of two isotopes, which results in a slight change in the Rydberg constant. There is also sometimes a distortion of the nucleus. Isotope shifts are frequently measured by using an atomic beam as a light source. The element studied evaporates out from a heated oven into a cooled evacuated region which condenses all the material except those atoms passing out through a cooled slit. The atoms in this beam may then be excited by a beam of electrons at right angles; the emitted radiation is then viewed at right angles to the beam. The Doppler effect is reduced because of the low velocities of the atoms at right angles to the motion of the beam. Since it is generally desired to measure the separation of closely spaced wavelengths, the Doppler shift produced by the atomic beam, shifting all wavelengths the same amount, does not affect the measurement. *See* MOLECULAR BEAMS.

LAMB SHIFT AND QUANTUM ELECTRODYNAMICS

The Bohr-Sommerfeld theory, which permitted elliptical orbits with small relativistic shifts in energy, yielded a fine structure for H_α that did not agree with experiment. The selection rule $\Delta l = \pm 1$ permits three closely spaced transitions. Actually, seven components have been observed within an interval of 0.5 cm^{-1}. According to the Dirac theory, the spin-orbit and the relativistic correction to the energy can be combined in the single formula shown as Eq. (13), so that levels with the same n

$$\Delta T = \frac{R\alpha^2 Z^4}{n^3} \cdot \left(\frac{2}{2j+1} - \frac{3}{4n} \right) \qquad (13)$$

and j coincide exactly, as shown in Fig. 4c. The splittings of the levels are greatly exaggerated in the figure. Selection rules, described later, limit the transitions to $\Delta j = 0$ and ± 1, so there are just seven permitted transitions for the H_α line. Unfortunately for the theory, two pairs of these transitions coincide, if states of the same j coincide, so only five lines of the seven observed would be accounted for.

The final solution of this discrepancy came with the experimental discovery by W. Lamb, Jr., that the $2s^2S$ level is shifted upward by 0.033 cm^{-1}. The discovery was made by extremely sensitive microwave measurements, and it has subsequently been shown to be in accord with the general principles of quantum electrodynamics. The Lamb shift is present to a greater or lesser degree in all the hydrogen levels, and also in those of helium, but is largest for the levels of smallest l and n. Accurate optical measurements on the wavelength of the Lyman α line have given conclusive evidence of a shift of the $1s^2S$ level.

Though an extended discussion of the highly abstract field theory of quantum electrodynamics is inappropriate here, a simple picture of the lowest-order contribution to the Lamb shift in hydrogen can be given. No physical system is thought ever to be completely at rest; even at temperatures near absolute zero (about $-273°C$), there is always a zero-point oscillatory fluctuation in any observable physical quantity. Thus, electromagnetic fields have very low-level, zero-point fluctuations, even in vacuum, at absolute zero. Electrons respond to these fluctuations with a zero-point motion: in effect, even a point charge is smeared out over a small volume whose mean square radius, though tiny, is not exactly zero. The smearing out of the charge distribution changes the interaction energy of a charged particle with any additional external field in which it may reside, including that of an atomic nucleus. As s electrons spend more time near the nucleus, where the nuclear-field Coulomb electric field is most intense, than do, say, electrons having more extended orbits, the change in interaction energy is greater for s electrons than, say, p electrons, and is such as to raise the energy of s electrons above the exact equality with that for p electrons of the same n arising in Dirac's theory of the electron. *See* QUANTUM ELECTRODYNAMICS.

RADIATIONLESS TRANSITIONS

It would be misleading to think that the most probable fate of excited atomic electrons consists of transitions to lower orbits, accompanied by photon emission. In fact, for at least the first third of the periodic table, the preferred decay mode of most excited atomic systems in most states of excitation and ionization is the electron emission process first observed by P. Auger in 1925 and named after him. For example, a singly charged neon ion lacking a $1s$ electron is more than 50 times as likely to decay by electron emission as by photon emission. In the process, an outer atomic electron descends to fill an inner vacancy, while another is ejected from the atom to conserve both total energy and momentum in the atom. The ejection usually arises because of the interelectron Coulomb repulsion. *See* AUGER EFFECT.

The vast preponderance of data concerning level-electromagnetic atomic decays are observed, optically allowed, single-electron, outermost-shell excitations in low states of ionization. Since most of the mass in nature is found in stars, and most of the elements therein rarely occupy such ionization-excitation states, it can be argued that presently available data provide a very unrepresentative description of the commonly occurring excited atomic systems in nature. When the mean lives of excited atomic systems are considered, the relative rarity of lifetime measurements on Auger electron-emitting states is even more striking. The experimentally inconvenient typical lifetime range (10^{-12} to 10^{-16} second) accounts for this lack.

An effective means of creating such high ionization-excitation states is provided by beam-foil spectroscopy, in which ions accelerated by a Van de Graaff accelerator pass through a thin carbon foil. The resulting beam of ions can be analyzed so the charge states and velocities are accurately known. Highly excited states are produced by the interaction with the foil. The subsequent emission, detected with a spectrometer, allows the measurement of lifetimes of states which cannot be produced in any other source. When, for example, electromagnetic atomic decays are observed, optical or x-ray spectrometers may be useful (Fig. 6). When, as is most frequently the case, Auger processes are dominant, less familiar but comparably effective electron spectrometers are preferred. *See* BEAM-FOIL SPECTROSCOPY.

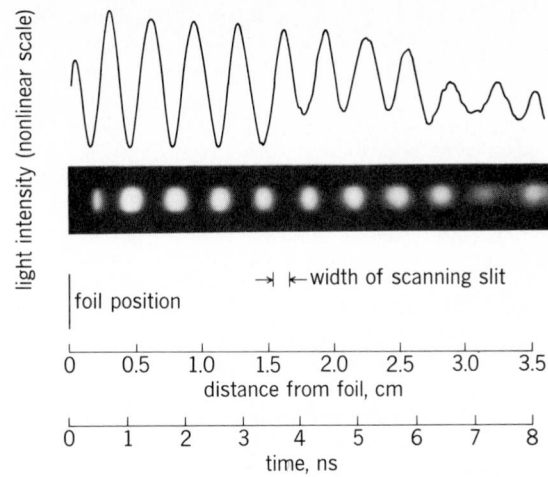

Fig. 6. Example of atomic process following excitation of a beam of ions in a thin solid target less than 1 μm in thickness. Quantum fluctuations in the light emitted by a 2-mm-diameter beam of such particles, traveling at a few percent of the speed of light, are shown as the atoms travel downstream from the target. (*From I. A. Sellin et al., Periodic intensity fluctuations of Balmer lines from single-foil excited fast hydrogen atoms, Phys. Rev., 184:56–63, 1969*)

DOPPLER SPREAD

In both cases, a common problem called Doppler broadening of the spectral lines arises, which can cause overlapping of spectral lines and make analysis difficult. The broadening arises from motion of the emitted atom with respect to a spectrometer. Just as in the familiar case of an automobile horn sounding higher pitched when the auto is approaching an observer than when receding, so also is light blue-shifted (to higher frequency) when approaching and red-shifted when receding from some detection apparatus. The percentage shift in frequency or wavelength is approximately $(v/c) \cos \theta$, where v is the emitter speed, c the speed of light, and θ the angle between $\vec{v}$ and the line of sight of the observer. Because emitting atoms normally are formed with some spread in v and θ values, there will be a corresponding spread in observed frequencies or wavelengths. Several ingenious ways of isolating only those atoms nearly at rest with respect to spectrometric apparatus have been devised. The most powerful employ lasers in the saturation spectroscopy mode mentioned above.

RECOIL ION SPECTROSCOPY

Because violent, high-velocity atomic collisions are needed to reach the highest ionization-excitation states possible, as for example in the beam-foil method, Doppler spread problems have distinctly hindered both the optical and electron spectroscopy of highly ionized matter. These problems have also hindered corresponding lifetime measurements obtained by monitoring the rate of decay in flight of atoms in necessarily well-identified upper states. Because highly ionized atoms are very important in stellar atmospheres, and in terrestrially produced highly ionized gases (plasmas) on which many potentially important schemes for thermonuclear fusion energy production schemes are based, it is important to overcome the Doppler spread problem. A discovery by I. A. Sellin and collaborators may represent an important step in this direction. They noted that for sufficiently highly ionized, fast, heavy projectiles, lighter target atoms in a gaseous target can be ionized to very high states of ionization and excitation under single-collision conditions, while incurring relatively small recoil velocities. Sample spectra indicate that a highly charged projectile can remove as many as $Z-1$ of the Z target electrons while exciting the last one remaining, all under single-collision-event conditions. Subsequent studies have shown that the struck-atom recoil velocities are characterized by a velocity distribution no worse than a few times thermal, possibly permitting a new field of precision spectroscopy on highly ionized atoms to be developed.

RELATIVISTIC DIRAC THEORY AND SUPERHEAVY ELEMENTS

Dirac's theory of a single electron bound to a point nucleus implies a catastrophe for hydrogen atoms of high Z, if Z could be made on the order of $137 \cong 1/\alpha$ or larger (in which $\alpha = e^2/2\epsilon_0 hc$ is the fine-structure constant). Dirac's expression for the electronic binding energy becomes unphysical at that point. Nuclear theorists have, however, contemplated the possible stable existence of superheavy elements, appreciably heavier than any observed heretofore, say, in the range $Z = 114$ to 126 or even larger. Though searches for such superheavy elements of either primordial or artificial origin have thus far proved unsuccessful, the advent of heavy-ion accelerators capable of bringing heavy particles to a distance of closest approach much smaller than the radius of the innermost shell in any atom, even a heavy one, raises the possibility of transient creation of a superheavy atom with a combined atomic number, $Z_{comb} = Z_1 + Z_2$, of about 180 or even greater (in which Z_1 and Z_2 are the individual atomic numbers of the colliding system). *See* SUPERTRANSURANICS.

The binding energy of a $1s$ electron in an atom increases rapidly with increasing atomic number.

As already noted, in the usual linear version of the Dirac equation, a catastrophe occurs when the nuclear charge reaches $(1/\alpha) \cong 137$. When $Z_1 + Z_2$ is sufficiently large, the $1s$ binding energy can reach and exceed twice the rest energy of the electron $m_e c^2$, in which m_e is the electron mass and c is the speed of light. By varying the target or projectile, one can trace the path of the electron energies as they dive toward the negative-energy sea, or, in the case of the $1s$ electron, into the negative-energy sea, to give rise to bound states degenerate with the negative-energy continuum (Fig. 7). The negative-energy sea at $-m_e c^2$ was introduced by Dirac and used to predict the existence of a new particle, the positron, represented by a hole in the sea. According to W. Greiner and colleagues, and independently, V. Popov, the $1s$ level acquires a width because of the admixture of negative-energy continuum states, and the $1s$-state width corresponds to a decaying state, provided there is nonzero final-state density. When a hole is present in the K shell, spontaneous positron production is predicted to take place. Production of an electron-positron pair is possible, since a final $1s$ ground state is available for the produced electron; the positron escapes with kinetic energy corresponding to overall energy balance.

Because of the effects of finite nuclear size in modifying the potential, $Z_1 + Z_2$ must reach some initial value Z_{CR} greater than 137 at which the "splash" into the continuum is made. According to nuclear-model-dependent estimates, the splash would occur at about 170. Some types of nonlinear additions to the Dirac equation may remove the diving into the negative-energy continuum, and these could be tested by observation of positron production as a function of $Z_1 + Z_2$. Other "limiting field" forms of nonlinear electrodynamics lead to larger values of Z_{CR}, and since the positron escape width turns out to be approximately proportional to the square of the difference between the effective quasi-nuclear charge and Z_{CR}, these nonlinearities could be tested too. Intensive searches for spontaneous positron production have been undertaken. *See* QUASIATOM.

UNCERTAINTY PRINCIPLE AND NATURAL WIDTH

A principle of universal applicability in quantum physics, and of special importance for atomic physics, is W. Heisenberg's 1927 uncertainty principle. In its time-energy formulation, it may be regarded as holding that the maximum precision with which the energy of any physical system (or the corresponding frequency ν from $E = h\nu$) may be determined is limited by the time interval Δt available for a measurement of the energy or frequency in question, and in no case can ΔE be less than approximately $h/2\pi\Delta t$. The immediate consequence is that the energy of an excited state of an atom cannot be determined with arbitrary precision, as such states have a finite lifetime available for measurement before decay to a lower state by photon or electron emission. It follows that only the energies of particularly long-lived excited states can be determined with great precision. The energy uncertainty ΔE, or corresponding frequency interval $\Delta \nu$, is referred to as the natural width of the level. For excited atoms radiating visible light, typical values of $\Delta \nu$ are on the order of 10^9 Hz. For inner electrons, which give rise to x-ray emission,

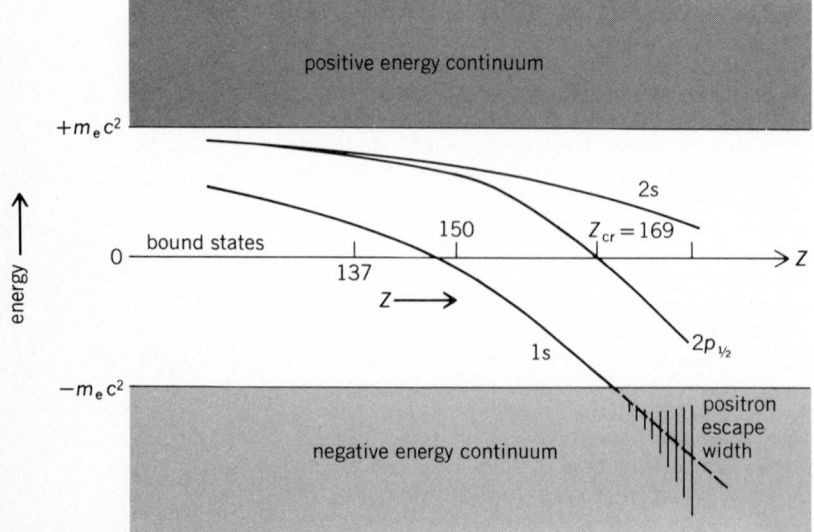

Fig. 7. Behavior of the binding energies of the $1s$, $2p_{1/2}$, and $2s$ electrons with increasing atomic number Z. (*From J. H. Hamilton and I. A. Sellin, Heavy ions: Looking ahead, Phys. Today, 26(4):42–49, April 1973*)

and also for Auger electron-emitting atoms, 10^{14} Hz is more typical. An energy interval $\Delta E = 1$ eV corresponds to about 2.4×10^{14} Hz. *See* UNCERTAINTY PRINCIPLE.

The time-energy uncertainty principle is perhaps best regarded as a manifestation of the basic wave behavior of electrons and other quantum objects. For example, piano tuners have been familiar with the time-frequency uncertainty principle for hundreds of years. Typically, they sound a vibrating tuning fork of accepted standard frequency in unison with a piano note of the same nominal frequency, and listen for a beat note between the tuning fork and the struck tone. Beats, which are intensity maxima in the sound, occur at a frequency equal to the difference in frequency between the two sound sources. For example, for a fork frequency of 440 Hz and a string frequency of 443 Hz, three beat notes per second will be heard. The piano tuner strives to reduce the number of beats per unit time to zero, or nearly zero. To guarantee a frequency good to $440 \pm .01$ Hz would require waiting for about 100 s to be sure no beat note had occurred ($\Delta\nu\Delta t \gtrsim 1$). *See* BEAT.

SUCCESSFUL EXPLANATIONS AND UNRESOLVED PROBLEMS

Many of the atomic enigmas referred to above have been explained through experimental discoveries and theoretical inspiration. Examples include Rutherford scattering, the Bohr model, the invention of the quantum theory, the Einstein quantum relation between energy and frequency $E = h\nu$, the Pauli exclusion and Heisenberg uncertainty principles, and many more. A few more are discussed below, but only in part, as the subject of atomic structure and spectra is by no means immune from the internal contradictions and defects in understanding that remain to be resolved.

In 1927 C. Davisson and L. Germer demonstrated that electrons have wave properties similar to that of light, which allows them to be diffracted, to exhibit interference phenomena, and to exhibit essentially all the properties with which waves are endowed. It turns out that the wavelengths of electrons in atoms are frequently comparable to the radius of the Bohr orbit in which they travel. Hence, a picture of a localized, point electron executing a circular or elliptical orbit is a poor and misleading one. Rather, the electrons are diffuse, cloudlike objects which surround nuclei, often in the form of standing waves, like those on a string. Crudely speaking, the average acceleration of an electron in such a pure standing wave, more commonly called a stationary state in the quantum theory, has no time-dependent value or definite initial phase. Since, as in Maxwell's theory, only accelerating charges radiate, atomic electrons in stationary states do not immediately collapse into the nucleus. Their freedom from this catastrophe can be interpreted as a proof that electrons may not be viewed as pointlike planetary objects orbiting atomic nuclei. *See* ELECTRON DIFFRACTION.

Atoms are by no means indivisible, as Thomson showed. Neither are their nuclei, whose constituents can be ejected in violent nucleus-nucleus collision events carried out with large accelerators. Nuclear radii and shapes can, for example, be measured by high-energy electron scattering experiments, in which the electrons serve as probes of small size. *See* NUCLEAR STRUCTURE.

However, a good theory of electron structure still is lacking, although small lower limits have been established on its radius. Zero-point fluctuations in the vacuum smear their physical locations in any event. There is still no generally accepted explanation for why electrons do not explode under the tremendous Coulomb repulsion forces in an object of small size. Estimates of the amount of energy required to "assemble" an electron are very large indeed. Electron structure is an unsolved mystery, but so is the structure of most of the other elementary objects in nature, such as protons, neutrons, neutrinos, and mesons. There are hundreds of such objects, many of which have electromagnetic properties, but some of which are endowed with other force fields as well. Beyond electromagnetism, there are the strong and weak forces in the realm of elementary particle physics, and there are gravitational forces. Electrons, and the atoms and molecules whose structure they determine, are only the most common and important objects in terrestrial life. *See* ELEMENTARY PARTICLE; FUNDAMENTAL INTERACTIONS.

Though the action-at-a-distance concept built into a Coulomb's-law description of the interactions of electrons with themselves and with nuclei is convenient and suprisingly accurate, even that model of internal atomic interactions has its limits. It turns out that electrons and nuclei do not interact instantaneously, but only after a delay comparable to the time needed for electromagnetic waves to travel between the two. Electrons and nuclei do not respond to where they are with respect to each other, but only to where they were at some small but measurable time in the past.

[IVAN A. SELLIN]

Bibliography: R. Eisberg and R. Resnick, *Quantum Physics*, 1974; T. W. Hänsch, A. L. Schawlow, and G. W. Series, The spectrum of atomic hydrogen, *Sci. Amer.*, 240(3):94–110, March 1979; H. Kuhn, *Atomic Spectra*, 2d ed., 1969; J. Slater, *Quantum Theory of Atomic Structure*, 1960.

Atomic weight

A number assigned to each chemical element which specifies the average mass of its atoms. The atomic weight scale is so chosen that the mass of an oxygen atom is exactly 16. Since an element may consist of two or more isotopes having atoms which differ in mass, the atomic weight of such an element depends on the relative proportions of its isotopes. The isotopic composition of all elements as they occur in nature, except those that are products of natural radioactivity, is practically constant. The atomic weight refers to this natural mixture. A list of recommended values for all elements is published biennially by the International Commission on Atomic Weights.

Two scales. Oxygen, formerly the basis of the atomic weight scale, itself possesses three stable isotopes. Two of these, O^{17} and O^{18}, are of low abundance, but they are nevertheless present in quantities sufficient to render the atomic weights as referred to the most abundant isotope, O^{16}, appreciably different from those referred to the natural mixture. Physicists determine atomic masses and abundances of individual isotopes with the mass spectrometer, and the atomic weights calculated from these data are expressed on a scale in which the mass of O^{16} equals 16. This so-called physical scale of atomic weights differs from the chemical scale, in which the natural mixture of oxygen isotopes has the value 16. To convert from the chemical to the physical scale, one multiplies by the factor 1.000275. Unfortunately, it appears that the isotopic composition of oxygen varies somewhat according to its source, so that the conversion factor would have to be 1.000278 for atmospheric oxygen and 1.000268 for oxygen derived from water. *See* ATOMIC MASS UNIT.

To avoid these difficulties, a new unit called relative nuclidic mass, defined as 1/12 the mass of carbon-12, was introduced in 1960. It is designated by the symbol u; thus, $C^{12} = 12u$. The table of relative atomic weights is now based on the atomic mass of $C^{12} = 12$. *See* RELATIVE ATOMIC MASS.

Determination of weights. There are four precision methods of determining atomic weights, all of which have comparable accuracy in the most favorable cases: (1) chemical combining weights, (2) limiting gas densities, (3) mass spectrometer measurements, and (4) nuclear reaction energies. The first two give results on the chemical scale, and the last two on the physical scale.

Combining weights. Method 1 is as old as atomic theory itself. If it is found, for example, that 1 g of silver (Ag), when converted into the chloride, always yields 1.32867 g of AgCl, one knows that the ratio of the atomic weights of chlorine and silver must be 0.32867. Taking the atomic weight of silver as 107.868 (one of the most accurately known values), this ratio yields for chlorine the value 35.453. Similar comparisons between other elements yields a set of atomic weights.

Gas density ratios. Method 2 is based on Avogadro's law, which states that equal volumes of two gases at the same temperature and pressure contain the same number of molecules. If this law were strictly accurate, the ratio of the weight of a liter of nitrogen (N_2) to that of a liter of oxygen (O_2), when multiplied by 16, would give the atomic weight of nitrogen. It is accurate, however, only for ideal gases. Hence, the ratio of densities must be extrapolated to the value it would have as these densities approach zero.

Mass spectrometer measurements. Method 3 compares the masses and abundances of different isotopic species by deflecting beams of their gaseous ions in electric and magnetic fields so that all ions having the same ratio of mass to electric charge are focused through a slit. The charge received on an electrode located behind the slit measures the relative abundance, while a knowledge of the deflection and the fields yields the mass per unit charge. Very accurate comparisons of masses can be made when beams of ions occur close together to form a doublet. Examples are $(C^{12}O^{16})^+$ and $(N_2^{14})^+$, or $(B^{10})^+$ and $(Ne^{20})^{++}$.

Nuclear reaction energies. Method 4 utilizes the Einstein mass-energy relation $\Delta E = \Delta M c^2$, where c is the velocity of light. If the energy of a bombarding particle and of the nuclear fragments resulting from a transmutation differ by a certain amount ΔE (the so-called Q value of the reaction), the total masses of all particles before and after the reaction must differ by an amount ΔM given by the Einstein relation, according to which 0.93114 Mev of energy is equivalent to 0.001 atomic weight units.

The results of such measurements, and of mass spectrometer work, are summarized in the packing-fraction curve, which displays the amounts by which individual nuclear masses differ from whole numbers when referred to $O^{16} = 16$. This curve is now well enough known so that an unknown atomic weight could be predicted by interpolation on the curve to better than 0.01 unit. *See* INERTIA OF ENERGY.

[F. A. JENKINS/W. W. WATSON]

Auger effect

An internal photoelectric process in an atom in which, for example, instead of the emission of a single characteristic x-ray from the filling of a vacancy in the K shell of electrons by an electron from a higher shell, an additional electron from the $L, M, \ldots$ shell is emitted with a kinetic energy equal to the energy of this x-ray minus the binding energy of this ejected electron. Such an electron is called an Auger electron. If the K-shell vacancy is in an atom of high atomic number Z, the return to the normal state may involve the emission of several x-rays and Auger electrons. That fraction of the vacancies in any electron shell that is filled by this Auger process is called the Auger yield, just as the fraction filled with accompanying x-ray emission is the fluorescence yield. The Auger effect is sometimes called autoionization.

[WILLIAM W. WATSON]

Avogadro number

The number of molecules N_0 in one mole of a substance, 6.02×10^{23}. Equal numbers of moles of all substances contain the same numbers of molecules. For a perfect gas, the molar volume contains a number of molecules equal to the Avogadro number. *See* MOLE.

Determinations of the Avogadro number by a variety of independent methods give results that agree very closely. The earliest estimates of its value were made during the latter half of the 19th century from the kinetic theory of gases. The viscosity of a gas can be shown by the kinetic theory to be proportional to $(1/N_0)\sigma^2$, where σ is the diameter of the molecules of the gas. If the molecules of a gas are considered as a number of elastic spheres of finite volume, then the expression relating the pressure p, temperature T, and volume V of a gram-molecule is $p(V - b) = RT$, where b is a constant representing the volume occupied by the molecules. It can be shown that b is proportional to $N_0\sigma^3$. If b and the viscosity are found experimentally, a rough value of N_0 can be calculated. The value of N_0 obtained by this method is 5×10^{23}, remarkably close to the accepted value. *See* KINETIC THEORY OF MATTER; VISCOSITY.

In 1909, J. Perrin determined values of N_0 from studies of the Brownian movement of colloidal particles and from the effect of gravity on their distribution with height. Albert Einstein showed that, for a particle moving completely randomly, as does a colloidal particle, the mean square of its displacement $\bar{x}^2$ in a given direction over a time t is related to the diffusion coefficient D by Eq. (1).

$$\bar{x}^2 = 2Dt \qquad (1)$$

If the particle obeys Stokes' law, Eq. (2) holds,

where η is the viscosity of the water and a is the diameter of the particle.

$$D = \frac{RT}{6\pi N_0 \eta a} \qquad (2)$$

Using these formulas and studying the mean square displacements of a very large number of colloidal particles over varying times through a microscope, Perrin obtained his N_0 value. *See* BROWNIAN MOVEMENT.

The numbers of particles n_1 and n_2, with energies E_1 and E_2, can be shown by Boltzmann's law to be related by Eq. (3). Now the potential energies

$$\frac{n_1}{n_2} = e^{[(E_2 - E_1)N_0]/RT} \qquad (3)$$

of particles at heights h_1 and h_2 in a colloidal suspension are given by $E_1 = Wh_1$ and $E_2 = Wh_2$, where W is the effective weight of the particles, allowing for the buoyancy of the water, calculated from their radius and density. Thus Eq. (4) follows.

$$N_0 = \frac{RT}{W(h_1 - h_2)} \ln\left(\frac{n_2}{n_1}\right) \qquad (4)$$

Perrin obtained particles of uniform size by fractional centrifuging and measured their radius microscopically. He measured the numbers of particles at different heights by direct counting of the number of particles in the field of view of the microscope. By this method, he obtained a value of 7.2×10^{23} for N_0. *See* BOLTZMANN CONSTANT.

By radioactivity measurements, a more accurate value of N_0 was obtained by B. B. Boltwood and E. Rutherford in 1911. They separated some radium salt from its decomposition products and measured the volume of helium produced after a known time and the rate of emission of α-particles per second per gram of radium. They also measured the decrease in the amount of radium during the course of the experiment. They obtained a value of 6.1×10^{23} for N_0. *See* RADIOACTIVITY.

In 1917, R. A. Millikan determined N_0 by a direct measurement of the charge on the electron e. He measured the rate of fall of electrically charged oil drops under gravity, and the rate of rise of the same drops when a vertical electric field was applied. From these measurements, a value for e was obtained. Now e and the Avogadro number N_0 are related by the expression $F = N_0 e$, where F is the faraday, the amount of electricity that will release 1 gram-equivalent on electrolysis. From Millikan's value of e and a knowledge of F, a value of 6.07×10^{23} was obtained for N_0. *See* ELECTRON.

The most accurate value of N_0 is obtained from x-ray measurements and density data. The wavelength of the x-rays is measured with a ruled diffraction grating, and the lattice spacing of a crystal d is determined by Bragg's relation, Eq. (5),

$$\lambda = 2d \sin \theta \qquad (5)$$

using x-rays of the same wavelength. The volume v per molecule is related to d by Eq. (6), where Φ is a

$$v = \Phi d^3/n \qquad (6)$$

geometrical factor and n is the number of mole-

cules in the unit cell. The density of crystal ρ must also be measured very accurately. Then the Avogadro number N_0 is equal to the ratio of the molecular weight M to the weight of a molecule m, as shown in Eq. (7), and so N_0 can be found. Crystals

$$N_0 = \frac{M}{m} = \frac{M}{\rho v} = \frac{nM}{\rho \Phi d^3} \qquad (7)$$

of calcite, $CaCO_3$, have been used in these determinations, and R. T. Birge gave a value of $N_0 = (6.0228_3 \pm 0.00011) \times 10^{23}$. T. Batuecas has pointed out that there is an uncertainty of 0.01 in the atomic weight of calcium, and he has used diamond instead of calcite to give a value of $(6.0236 \pm 0.00007) \times 10^{23}$ for N_0.

The most recent x-ray measurements give $6.0231_6 \times 10^{23}$ for N_0. The constants e, the charge on the electron; h, Planck's constant; F, the faraday; and N_0 are of course related, and in 1963 a committee of the National Academy of Sciences and the National Research Council of the United States recommended that the following values be adopted: $e = 1.60210 \times 10^{19}$ coulomb, $h = 6.6256 \times 10^{-34}$ joule sec, $F = 9.64870 \times 10^4$ coulomb mol^{-1}, and $N_0 = 6.02252 \times 10^{23}$ mol^{-1}. These values have been recommended internationally by the Twenty-third Conference of the International Union of Pure and Applied Chemistry. *See* ATOMIC CONSTANTS. [THOMAS C. WADDINGTON]

Avogadro's law

The principle that equal volumes of all gases and vapors, under the same conditions of temperature and pressure, contain identical number of molecules; also known as Avogadro's hypothesis. From Avogadro's law the converse follows that equal numbers of molecules of any gases under identical conditions occupy equal volumes. Therefore, under identical physical conditions the gram-molecular weights of all gases occupy equal volumes. Avogadro's law is not strictly obeyed by real gases at ordinary temperatures and pressures, although the deviations are only slight. At high pressures the deviations may be large. Avogadro's law can be shown to follow theoretically from the simple kinetic theory of gases. *See* GAS; KINETIC THEORY OF MATTER.

[THOMAS C. WADDINGTON]

Band spectrum

A spectrum consisting of groups or bands of closely spaced lines. Band spectra are characteristic of molecular gases or chemical compounds. When the light emitted or absorbed by molecules is viewed through a spectroscope with small dispersion, the spectrum appears to consist of very wide asymmetrical lines called bands. These bands usually have a maximum intensity near one edge, called a band head, and a gradually decreasing intensity on the other side. In some band systems the intensity shading is toward shorter waves, in others toward longer waves. Each band system consists of a series of nearly equally spaced bands called progressions; corresponding bands of different progressions form groups called sequences.

Six spectra of diatomic molecular fragments are illustrated in the photograph. The spectrum of a discharge tube containing air at low pressure is

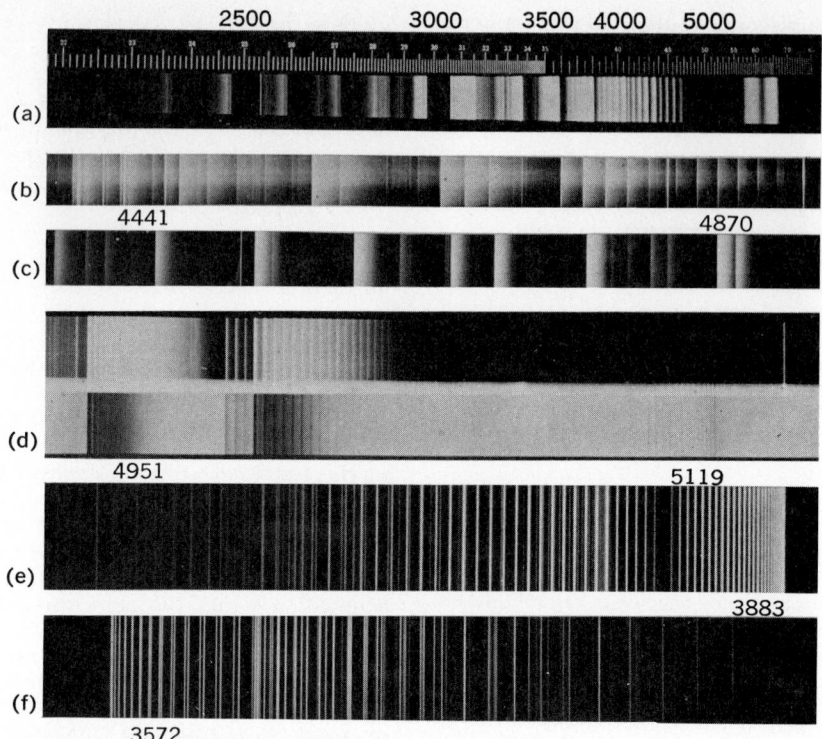

2500 3000 3500 4000 5000

(a)

(b)

4441 4870

(c)

(d)

4951 5119

(e)

3883

(f)

3572

Photographs of band spectra of (a) a discharge tube containing air at low pressure; (b) high-frequency discharge in lead fluoride vapor; (c) SbF (b and c taken with large quartz spectrograph, after Rochester); (d) BaF emission and absorption; (e) CN; and (f) NO. The measurements are in angstroms. (From F. A. Jenkins and H. E. White, *Fundamentals of Optics*, 3d ed., McGraw-Hill, 1957)

shown in *a*. It has four band systems: the γ-bands of nitrogen oxide (NO, 2300–2700 A), negative nitrogen bands (N₂⁺, 2900–3500 A), second-positive nitrogen bands (N₂, 2900–5000 A), and first-positive nitrogen bands (N₂, 5500–7000 A). The spectrum of high-frequency discharge in lead fluoride vapor in *b* has bands in prominent sequences. The spectrum in *c* shows part of one band system of SbF, and was obtained by vaporizing SbF into active nitrogen. Emission from a carbon arc cored with barium fluoride (BaF₂) and absorption of BaF vapor in an evacuated steel furnace are illustrated in *d*. These spectra were obtained in the second order of a diffraction grating, as were the spectra in *e* and *f*. The photograph *e* is that of the CN band at 3883 A from an argon discharge tube containing carbon and nitrogen impurities, and *f* is a band in ultraviolet spectrum of NO, obtained from glowing active nitrogen containing a small amount of oxygen.

When spectroscopes with adequate dispersion and resolving power are used, it is seen that most of the bands obtained from gaseous molecules actually consist of a very large number of lines whose spacing and relative intensities, if unresolved, explain the appearance of bands of continua (parts *e* and *f* of the figure). For the quantum mechanical explanations of the details of band spectra *see* MOLECULAR STRUCTURE AND SPECTRA.

[W. F. MEGGERS/W. W. WATSON]

Band theory of solids

A quantum-mechanical theory of the motion of electrons in solids. Its name comes from the fact that it predicts certain restricted ranges, or bands, for the energies of electrons in solids.

Suppose that the atoms of a solid are separated from each other to such a distance that they do not interact. The energy levels of the electrons of this system will then be those characteristic of the individual free atoms, and thus many electrons will have the same energy. Now imagine the atoms being slowly brought closer together. As the distance between atoms is decreased, the electrons in the outer shells begin to interact with each other. This interaction alters their energy and, in fact, broadens the sharp energy level out into a range of possible energy levels called a band. One would expect the process of band formation to be well advanced for the outer, or valence, electrons at the observed interatomic distances in solids. Once the atomic levels have spread into bands, the valence electrons are not confined to individual atoms, but may jump from atom to atom with an ease that increases with the increasing width of the band.

Although energy bands exist in all solids, the term energy band is usually used in reference only to ordered substances, that is, those having well-defined crystal lattices. In such a case, an electron energy state can be classified according to its crystal momentum **p** or its electron wave vector **k** = **p**/ℏ (where ℏ is Planck's constant *h* divided by 2π). If the electrons were free, the energy of an electron whose wave vector is **k** would be as shown in Eq. (1), where E_0 is the energy of the

$$E(\mathbf{k}) = E_0 + \hbar^2 \mathbf{k}^2 / 2m_0 \qquad (1)$$

lowest state of a valence electron and m_0 is the electron mass. In a crystal, however, the electrons are not free because of the effect of the crystal binding and the forces exerted on them by the atoms; consequently, the relation $E(\mathbf{k})$ between energy and wave vector is more complicated. The statement of this relationship constitutes the description of an energy band.

A knowledge of the energy levels of electrons is of fundamental importance in computing electrical, magnetic, optical, or thermal properties of solids.

Allowed and forbidden bands. The bands of possible electron energy levels in a solid are called allowed energy bands. It often happens that there are also bands of energy levels which it is impossible for an electron to have in a given crystal. Such bands are called forbidden bands, or gaps. The allowed energy bands sometimes overlap and sometimes are separated by forbidden bands. The presence of a forbidden band, or energy gap, immediately above the occupied allowed states (such as the region A to B in the illustration) is the princi-

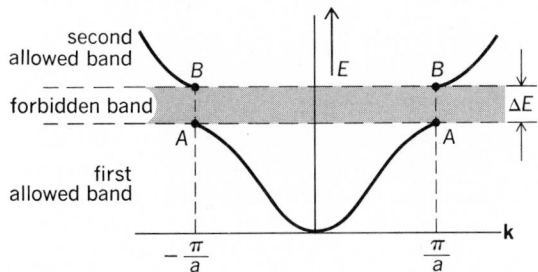

Electron energy E versus wave vector $\mathbf{k}$ for a monatomic linear lattice of lattice constant a. (After C. Kittel, Introduction to Solid State Physics, 2d ed., Wiley, 1956)

pal difference in the electronic structures of a semiconductor or insulator and a metal. In the first two substances there is a gap between the valence band or normally occupied states and the conduction band, which is normally unoccupied. In a metal there is no gap between occupied and unoccupied states.

The presence of a gap means that the electrons cannot easily be accelerated into higher energy states by an applied electric field. Thus, the substance cannot carry a current unless electrons are excited across the gap by thermal or optical means.

Effective mass. When an external electromagnetic field acts upon the electrons in a solid, the resultant motion is not what one would expect if the electrons were free. In fact, the electrons behave as though they were free but with a different mass, which is called the effective mass. This effective mass can be obtained from the dependence of electron energy on the wave vector, $E(\mathbf{k})$, in the following way.

Suppose there is a maximum or minimum of the function $E(\mathbf{k})$ at the point $\mathbf{k} = \mathbf{k}_0$. The function $E(\mathbf{k})$ can be expanded in a Taylor series about this point. For simplicity, assume that $E(\mathbf{k})$ is a function of the magnitude of $\mathbf{k}$ only, that is, is independent of the direction of $\mathbf{k}$. Then, by dropping terms higher than second order in the Taylor series, Eq. (2) results. By analogy with Eq. (1), a quantity

$$E(\mathbf{k}) = E(\mathbf{k}_0) + \tfrac{1}{2}(\mathbf{k} - \mathbf{k}_0)^2 \left(\frac{d^2E}{dk^2}\right)_{k_0} \quad (2)$$

m^* with the dimensions of a mass can be defined by the relation in Eq. (3).

$$\hbar^2/m^* = \left(\frac{d^2E}{dk^2}\right)_{k_0} \quad (3)$$

The quantity m^* is called the effective mass of electrons at $\mathbf{k}_0$. For many simple metals, the average effective mass is close to m_0, but smaller effective masses are not uncommon. In indium antimonide, a semiconductor, the effective mass of electrons in the conduction band is 0.013 m_0. In a semiclassical approximation, an electron in the solid responds to an external applied electric or magnetic field as though it were a particle of mass m^*. The equation of motion of an electron is shown in Eq. (4), where $\mathbf{v}$ is the electron velocity,

$$m^* \frac{d\mathbf{v}}{dt} = e(\mathbf{E} + \mathbf{v} \times \mathbf{B}) \quad (4)$$

$\mathbf{E}$ the electric field, $\mathbf{B}$ the magnetic induction, and e the charge of the electron.

It may happen that the energy $E(\mathbf{k})$ does depend upon the direction of $\mathbf{k}$. In such a case, the effective mass is a tensor whose components are defined by Eq. (5). Equation (4) remains valid with

$$\hbar^2/m_{ij}^* = \left(\frac{\partial^2E}{\partial k_i \, \partial k_j}\right)_{k_0} \quad (5)$$

a tensor m^*. Bismuth is an example of a metal in which the effective mass depends strongly on direction.

Transitions between states. Under external influences, such as irradiation, electrons can make transitions between states in the same band or in different bands. The interaction between the electrons and the vibrations of the crystal lattice can scatter the electrons in a given band with a substantial change in the electron momentum, but only a slight change in energy. This scattering is one of the principal causes of the electrical resistivity of metals. *See* ELECTRICAL RESISTIVITY.

An external electromagnetic field (for example, visible light) can cause transitions between different bands. In such a process, momentum must be conserved. Because the momentum of a photon $h\nu/c$ (where ν is the frequency of the light and c its velocity) is quite small, the momentum of the electron before and after collision is nearly the same. Such a transition is called vertical in reference to an energy band diagram. Conservation of energy must also hold in the transition, so absorption of light is possible only if there is an unoccupied state of energy $h\nu$ available at the same $\mathbf{k}$ as the initial state. These transitions are responsible for much of the absorption of light by semiconductors in the visible and near-infrared region of the spectrum.

Energy band calculation. As is the case for any quantum mechanical system, the energy levels of electrons in solids are determined in principle by the solution of the Schrödinger wave equation for the system. However, the enormous number of particles involved makes exact solution of this equation impossible. Approximate methods must be employed. The atoms are regarded as fixed in their equilibrium positions. Each electron is regarded as having an individual wave function $\psi_n(\mathbf{k}, \mathbf{r})$, in which $\mathbf{k}$ is the wave vector and the index n designates a particular band. The wave function $\psi_n(\mathbf{k}, \mathbf{r})$ is frequently called a Bloch function. The wave function of the many-electron system is written as a determinant of Bloch functions to satisfy the requirements of the Pauli exclusion principle. The general variational method of quantum mechanics may be employed to derive an equation for the individual Bloch functions. This equation, which is known as the Hartree-Fock equation, is similar to a one-electron Schrödinger equation in which each electron moves in the field of all the nuclei of the system and the average field of all the other electrons. An additional term, exchange, takes account of the reduction of the average electronic repulsion, because the probability of close encounters of electrons of parallel spin is reduced by the Pauli exclusion principle.

The Hartree-Fock equations are still quite complicated, and must be solved by the self-consistent field method. In this approach, some reasonable distribution of electrons is assumed to exist. An effective potential $V(\mathbf{r})$ can be calculated from this distribution, including the contribution from the nuclei. Usually it is an adequate approximation to include the exchange terms in this potential. Then the Bloch functions may be formed by solving Eq. (6), in which $V(\mathbf{r})$ is the potential described above

$$-\frac{\hbar^2}{2m_0}\nabla^2\psi_n(\mathbf{k},\mathbf{r})+V(\mathbf{r})\psi_n(\mathbf{k},\mathbf{r})=E_n(\mathbf{k})\psi_n(\mathbf{k},\mathbf{r}) \quad (6)$$

and $E_n(\mathbf{k})$ is the energy of an electron in band n having wave vector $\mathbf{k}$. The potential $V(\mathbf{r})$ is periodic in space with the periodicity of the crystal. The wave function $\psi_n(\mathbf{k},\mathbf{r})$ obtained from Eq. (6) yields a new electron distribution from which the potential $V(\mathbf{r})$ may be calculated again. The process is repeated until the potential used in the solution of Eq. (6) agrees with that obtained from the solution $\psi_n(\mathbf{k},\mathbf{r})$ to sufficient accuracy.

The local density approximation is frequently used to obtain this potential. In the simplest formula of this type, the Kohn-Sham potential, $V(\mathbf{r})$ is given by Eq. (7). The first term represents the po-

$$V(\mathbf{r})=-e^2\sum_\mu\frac{Z_\mu}{|\mathbf{r}-\mathbf{R}_\mu|}+e^2\int\frac{\rho(\mathbf{r}')d^3r'}{|\mathbf{r}-\mathbf{r}'|}-2e^2\left(\frac{3\rho(\mathbf{r})}{8\pi}\right)^{1/3} \quad (7)$$

tential energy of an electron in the field of all the nuclei of the system. These nuclei are located at sites $\mathbf{R}_\mu$ and have charge Z_μ. The second term contains the average electronic repulsion of the electron distribution, and $\rho(\mathbf{r})$ is the electron density. The third term approximately describes the exchange interaction.

The self-consistent field procedure is evidently quite complicated. Since it may not be practical to repeat the calculations to obtain a self-consistent solution, it is desirable to choose an initial $V(\mathbf{r})$ which approximates the actual physical situation as closely as possible. Choice of an adequate crystal potential is the chief physical problem in the calculation of energy levels of electrons in solids.

Several techniques are available for solving wave equation (6) with a given $V(\mathbf{r})$. Those in common use include the Green's function method, the augmented plane wave method, the orthogonalized plane wave method, and the linear combination of atomic orbitals method. It is also possible to use experimentally determined energy levels and effective masses to obtain a suitable potential for

use in Eq. (6). This procedure, which bypasses many of the difficulties of the self-consistent field approach, is known as the pseudopotential method and is now widely used. In simple metals (not transition metals) and wideband semiconductors, such as germanium and silicon, the pseudopotential is rather weak, and many properties can be calculated by using perturbation theory.

Density of states. Many properties of solids, including electron specific heat and optical and x-ray absorption and emission, are related rather simply to a basic function known as the density of states. This function, here denoted $G(E)$, is defined so that the number of electronic energy states in the range between E and $E+dE$ is $G(E)dE$. It can be shown that $G(E)$ is given by Eq. (8), in which $E(\mathbf{k})$ is the energy band function,

$$\begin{aligned}G(E)&=\frac{\Omega}{4\pi^3}\int\delta\bigl(E-E(\mathbf{k})\bigr)\,d^3k\\&=\frac{\Omega}{4\pi^3}\int\frac{dS_k(E)}{|\nabla_kE(\mathbf{k})|}\end{aligned} \quad (8)$$

δ is the Dirac delta function, Ω is the volume of a unit cell of the solid, and the constant multiplying the integral has been chosen so that the number of states is two (one for each spin direction) for each energy band. The first integral in Eq. (8) is taken over the Brillouin zone, the second over a surface of constant energy E in the zone. The density of states will show structure in the neighborhood of energies corresponding to points where $|\nabla_kE(\mathbf{k})|$ vanishes. Such points, which are required to exist by reasons of crystal symmetry and periodicity, are known as Van Hove singularities. The energies of states at which Van Hove singularities occur can be determined from optical and (sometimes) x-ray measurements.

Experimental information. A considerable amount of experimental information has been obtained concerning the band structures of the common metals and semiconductors. In metals experiments usually determine properties of the Fermi surface (the surface in $\mathbf{k}$ space bounding the occupied states at zero temperature), since only electrons in states near the top of the Fermi distribution are able to respond to electric and magnetic fields. For example, the magnetic susceptibility of a metal exhibits a component which is an oscillatory function of the reciprocal of the magnetic field strength when the field is strong (de Haas–Van Alphen effect). Measurement of the period of these oscillations determines the area of a cross section of the Fermi surface in a plane perpendicular to the direction of the magnetic field.

Other properties of the Fermi surface are obtained by studying the increase of the electrical resistance in a magnetic field, and from the magnetic field dependence of ultrasonic attenuation. The density of states (number of electron states per unit energy) at Fermi energy can be found from measurements of the electron contribution to the specific heat. Effective masses of electrons and holes can be determined by cyclotron resonance measurements in both semiconductors and metals.

Optical and x-ray measurements enable the determination of energy differences between states. The smallest width of the characteristic energy gap between valence and conduction bands in a semiconductor can be determined by measuring the wavelength at which the fundamental absorption begins. Application of a strong uniform magnetic field will produce an oscillatory energy dependence of this absorption, from which a precise determination of effective masses is possible.

In both semiconductors and metals, states removed in energy from the Fermi level can be studied optically. Measurements of the reflectivity will show structure at energies corresponding to transitions at symmetry points of the Brillouin zone where Van Hove singularities are likely to occur. To reveal such structure clearly against a background which frequently may be large, it is useful to subject the solid to a periodic perturbation, possibly an electric field, stress, or a change in temperature. This modifies the band structure slightly, particularly near Van Hove singularities, which tend to be sensitive to perturbations. As a result, the optical properties of the sample are slightly modulated, and this modulation is detected directly. It is possible to apply this technique in the soft x-ray wavelength region as well. Modulation of the soft x-ray emission by alternating stress (the piezo soft x-ray effect) enables the determination of the energies of transitions between band and deep core states.

Photoemission measurements can also yield information about energy bands below the Fermi energy. In these experiments, an electron is excited by a photon from a state of energy, E_i, below the Fermi energy to a state of high enough energy, E_f, such that the electron can escape from the crystal. The relation between the energies is $E_f = E_i + \hbar\omega$, where ω is the angular frequency associated of the light. The number of emitted electrons is studied as a function of the angle between the direction of the outgoing electrons and that of the incident photon (often chosen to be perpendicular to the surface). The polarization of the light can also be controlled. If the surface of a single crystal sample is plane, the component of the electron momentum parallel to the solid ($\mathbf{k}_\parallel$) is conserved. The measurement is then sensitive to the distribution of electrons in energy along a line in the Brillouin zone parallel to the surface normal and with the specified value of $\mathbf{k}_\parallel$. The results can be interpreted to give the positions of the energy band in the initial state. *See* BRILLOUIN ZONE; CRYSTAL STRUCTURE; EXCITON; FREE-ELECTRON THEORY OF METALS; HOLES IN SOLIDS; KRONIG-PENNEY MODEL; NONRELATIVISTIC QUANTUM THEORY; SEMICONDUCTOR; SPECIFIC HEAT OF SOLIDS.

[JOSEPH CALLAWAY]

Bibliography: P. W. Anderson, *Concepts in Solids*, 1964; N. W. Ashcroft and N. D. Mermin, *Solid State Physics*, 1976; J. Callaway, *Quantum Theory of the Solid State*, 1974; M. Cardona, *Modulation Spectroscopy*, Suppl. 11 of *Solid State Phys.*, 1969; G. C. Fletcher, *The Electron Band Theory of Solids*, 1971; W. Jones and W. H. March, *Theoretical Solid State Physics*, 1973; C. Kittel, *Quantum Theory of Solids*, 1963; J. C. Slater, *Quantum Theory of Molecules and Solids*, vol. 2: *Symmetry and Energy Bands in Crystals*, 1965; J. M. Ziman, *Principles of the Theory of Solids*, 2d ed., 1972.

Baryon

The generic name for any hadronic particle with baryon number $B = +1$. By far the most common baryons are the proton and neutron, the two states of the nucleon doublet $N = (p,n)$, whose intrinsic properties are listed in Table 1. The baryon number of any particular state may be deduced from its production or decay processes, or both, since the total baryon number is conserved (with possible rare exceptions discussed below) and $B = 0$ holds for all mesons and leptons. *See* LEPTON; MESON; NEUTRON; PROTON.

The scientific view of the hadrons changed greatly during the 1970s. It is now generally accepted that they are composite, consisting of spin-½ quarks (q), corresponding antiquarks ($\bar{q}$), and some number of gluons, the last being the quanta of the intermediate field which binds the quarks and antiquarks to form hadrons. $B = +\frac{1}{3}$ holds for a quark q, $B = -\frac{1}{3}$ for an antiquark $\bar{q}$, while $B = 0$ holds for a gluon. Thus, a baryon consists of three ("valence") quarks, together with some number of quark-antiquark ($q\bar{q}$) pairs (called the quark-antiquark sea) and of gluons. The known quarks are listed in Table 2. They must be assigned fractional charge values, relative to the proton charge. *See* GLUON; HADRON; QUARKS.

Color and quantum chromodynamics. This quark theory of the hadrons has been proposed in a quite specific form, known as quantum chromodynamics (QCD). It is a gauge theory based on a symmetry hypothesized for the hadronic interactions of the quarks, which says that these interactions are invariant with respect to an $SU(3)_C$ group of unitary transformations with modulus unity acting in an abstract three-dimensional space known (whimsically) as color space. Each quark type then has three color states, usually labeled by the suffixes r = red, g = green, and b = blue, corresponding to the three axes of this space. The gauge particle of this symmetry theory is the gluon, a neutral vector particle coupled universally with the currents of color, just as the photon, the gauge particle of quantum electrodynamics (QED), is coupled universally with the electromagnetic current. However, whereas the photon has no charge, the gluon has $(3^2 - 1) = 8$ color components, so that it is a color octet. Consequently, there is a gluon contribution to the color currents, and so the gluon field must interact with itself, introducing a nonlinearity into quantum chromodynamics which has no parallel in quantum electrodynamics. This nonlinearity has important implications for quantum chromodynamics, leading to its asymptotic freedom, the property that the coupling of gluon to the color current approaches zero at short distances, which is essential for even qualitative agreement between quantum chromodynamics predictions and the empirical data on high-energy collision processes. *See* COLOR (QUANTUM MECHANICS); QUANTUM CHROMODYNAMICS; QUANTUM ELECTRODYNAMICS; SYMMETRY LAWS (PHYSICS).

An important element in quantum chromodynamics is the confinement dogma, the assertion that only color singlet states have finite energy. This assertion implies that neither a quark nor a gluon can exist in a free state, since the former is a

Table 1. Known stable and semistable baryons and their properties

Baryon	Mass, MeV	Spin-parity	Strangeness (s)	Charm (C)	Lifetime, s	Dominant decay modes	Magnetic moment, n.m.†
p	938.280 ± 0.003	$\frac{1}{2}^+$	0	0	$>10^{36}$	—	2.7928
n	939.573 ± 0.003	$\frac{1}{2}^+$	0	0	917 ± 14	$p\bar{\nu}_e e^-$	-1.9130
Λ	1115.60 ± 0.05	$\frac{1}{2}^+$	-1	0	$2.63 \pm 0.02 \times 10^{-10}$	$p\pi^-$ (64%) $n\pi^0$ (36%) $p\bar{\nu}_e e^-$ (0.081 ± 0.003%)	-0.614 ± 0.005
Σ^+	1189.36 ± 0.06	$\frac{1}{2}^+$	-1	0	$8.0 \pm 0.04 \times 10^{-11}$	$p\pi^0$ (52%) $n\pi^+$ (48%) $p\gamma$ (0.12 ± 0.02%)	2.33 ± 0.13
Σ^0	1192.46 ± 0.08	$\frac{1}{2}^+$	-1	0	$6 \pm 1 \times 10^{-20}$	$\Lambda\gamma$	
Σ^-	1197.34 ± 0.05	$\frac{1}{2}^+$	-1	0	$1.48 \pm 0.01 \times 10^{-10}$	$n\pi^-$ $n\bar{\nu}_e e^-$ (0.11 ± 0.005%)	-1.41 ± 0.25
Ξ^0	1314.9 ± 0.6	$\frac{1}{2}^+$	-2	0	$2.9 \pm 0.1 \times 10^{-10}$	$\Lambda\pi^0$	-1.20 ± 0.06
Ξ^-	1321.3 ± 0.15	$\frac{1}{2}^+$	-2	0	$1.64 \pm 0.02 \times 10^{-10}$	$\Lambda\pi^-$ $\Lambda e^- \bar{\nu}_e$ (0.028 ± 0.012%)	-1.85 ± 0.75
Ω^-	1672.2 ± 0.3	$(\frac{3}{2}^+?)$	-3	0	$0.82 \pm 0.03 \times 10^{-10}$	$\Lambda\bar{K}^-$ (69%) $\Xi^0\pi^-$ (23%) $\Xi^-\pi^0$ (8%)	
Λ_c	2273 ± 6	$(\frac{1}{2}^+?)$	0	1	$\simeq 7 \times 10^{-13}$	$pK^-\pi^+$ (2.2 ± 1%)	

†The abbreviation n.m. denotes the unit $e\hbar/2M_p c$ (nuclear magneton).

color triplet and the latter a color octet. Zero mass is expected for a gauge particle (as is the case for the photon), hence for the gluon, and a mass of order 10 MeV is given for the (u,d) quarks in Table 2, but these values refer to the masses effective within hadronic states. In accord with this dogma, no observations of free gluons or quarks have yet been confirmed. Many theoreticians anticipate that this dogma will be deducible from quantum chromodynamics itself but, despite some favorable indications for this expectation, no rigorous proof that the dogma follows from quantum chromodynamics has been given.

Quantitative predictions of the properties of baryonic states are currently made using a simplified quark-quark (q-q) potential with the following features: (1) an attractive long-range potential, increasing with separation to ensure confinement, and (2) a spin-dependent potential representing one-gluon exchange, effective at small separation, where the regime of asymptotic freedom holds and perturbation theory is valid. Such predictions have had a great deal of success.

For a three-quark system, there is only one color-singlet (that is, scalar) wave function available, namely that given by Eq. (1), where the vector

$$\Phi_{\text{col.}}(1,2,3) = \mathbf{q}(1) \cdot \mathbf{q}(2) \times \mathbf{q}(3) \qquad (1)$$

$\mathbf{q} = (q_r, q_g, q_b)$ refers to color space. This is antisymmetric (A) with respect to permutation of the quark labels (1,2,3). The complete wave function for this system has the general form of Eq. (2), and

$$\psi(1,2,3) = \Sigma \Phi_{\text{col.}}(1,2,3)\, \psi_{\text{space}}(1,2,3)$$
$$\chi_{\text{spin}}(1,2,3)\, \Phi_{\text{flavor}}(1,2,3) \qquad (2)$$

must also have this permutation symmetry A, from general considerations (the Pauli spin-statistics theorem). These remarks imply that the space × spin × flavor wave function for a baryonic state must be permutation symmetric (S), as if the quarks obeyed Bose statistics, for those variables. Indeed, the achievement of this result was precisely the historical purpose behind the introduction of the color degree of freedom. Within this article, mesons are regarded as bound antiquark-quark ($\bar{q}$-q) systems, their color wave functions having the singlet form $\bar{\mathbf{q}}(1) \cdot \mathbf{q}(2)$. *See* QUANTUM STATISTICS.

Nucleons and isospin. The quark content of the nucleons is given by Eqs. (3). The nucleons are

$$p = (uud) \qquad n = (udd) \qquad (3)$$

ground states, so that their space wave functions ψ_{space} have $L = 0$ and are expected to be nodeless; this requires that these ψ_{space} wave functions have permutation symmetry S. The three quark spins sum to $S = \frac{1}{2}$, a spin state which has mixed (M) permutation symmetry. Antisymmetry for the wave function Eq. (2) then requires that the flavor wave functions (uud) and (udd) have mixed symmetry. (The product of two factors with M symmetry can be resolved into three terms, one with S, one with A, and one with M symmetry. Symbolically, M ⊗ M = S + A + M, whereas M ⊗ S = M and M ⊗ A = M.)

The u and d quarks are approximately degenerate in mass, their masses being small. For equal masses, their (kinetic + mass) energies would be the same. Their coupling with the gluon field has the simple form of Eq. (4), where the $g_{\mu\alpha}$ denote the

Table 2. Properties of established quarks†

Quark type	u (up)	d (down)	s (strange)	c (charmed)	b (bottom)		
Charge (Q/e_p)	$\frac{2}{3}$	$-\frac{1}{3}$	$-\frac{1}{3}$	$\frac{2}{3}$	$-\frac{1}{3}$	·	·
Mass, GeV	$\simeq 0.01$	$\simeq 0.01$	$\simeq 0.5$	$\simeq 1.5$	$\simeq 4.7$	·	·
Flavor	$I_3 = +\frac{1}{2}$	$I_3 = -\frac{1}{2}$	$s = -1$	$C = +1$	$b = +1$	·	·

†To each quark, there exists an antiquark with the opposite flavor values and with opposite intrinsic parity.

$$L_{\text{int.}} = f\Sigma g_{\mu\alpha}(\bar{u}\gamma_\mu\lambda_\alpha u + \bar{d}\gamma_\mu\lambda_\alpha d) \qquad (4)$$

components of the gluon field, the sum being over $\mu = 0, 1, 2, 3$ for time and space, and $\alpha = 1, 2, \ldots, 8$ for the components of octet color, and the $\{\lambda_\alpha\}$ are the infinitesimal operators of the $SU(3)_C$ group. In this case, the total energy remains unchanged for any $SU(2)$ transformation in the space (u,d). This invariance is the origin of the property known as charge independence for hadronic interactions. By analogy with the well-known group $SU(2)_\sigma$ for Pauli spin, this group is labeled $SU(2)_\tau$, its eigenvalues being known as the isospin I. The states (p,n) are then the $I_3 = (+\frac{1}{2}, -\frac{1}{2})$ components of an isospin doublet N. Isospin is well known in the classification of nuclear states; these occur as charge multiplets, a set of corresponding states in the nuclei with mass number A and the charge values $Z = I + A/2, I - 1 + A/2, \ldots, -I + A/2$, in which the nuclear interactions are the same. Thus, these $2I + 1$ states are said to have isospin component $I_3 = I, I - 1, \ldots, -I$, their energy being independent of I_3. The situation for systems of (u,d) quarks runs completely parallel with this.

This isospin symmetry is violated by the electromagnetic interactions, since the charge of the state depends on I_3 and the electromagnetic field couples with charge. This coupling will separate the masses of the u and d quarks, leading to terms in the energy which are not invariant under the $SU(2)_\tau$ transformations. Also, there will be electromagnetic interactions between the constituent quarks, which will violate this invariance. These charge-dependent effects are well known for the nuclear case, and the situation for (u,d) quarks is quite similar. *See* ISOTOPIC SPIN.

Baryon octet. The replacement of a d quark in the nucleon by an s quark produces a baryon state with spin parity $\frac{1}{2}^+$ and strangeness number $s = -1$, the latter being given by $[n(\bar{s}) - n(s)]$, where $n(q)$ denotes the number of quarks of type q in the system considered. The states thus reached have the quark structures of Eq. (5), the other factors in

$$(\Sigma^+, \Sigma^0, \Sigma^-) = (uus, (ud + du)s/\sqrt{2}, dds) \quad (5a)$$
$$\Lambda = (ud - du)s/\sqrt{2} \qquad (5b)$$

their wave functions Eq. (2) being identical with those for the nucleons; thus the isotriplet Σ and isosinglet Λ states are obtained. If two of the d quarks in Eq. (3) are replaced by s quarks, the isodoublet Ξ states of Eq. (6) are obtained. The s

$$(\Xi^0, \Xi^-) = (uss, dss) \qquad (6)$$

quark has $I = 0$, being unaffected by the $SU(2)_\tau$ transformations in the (u,d) space. The flavor wave function (sss) is necessarily symmetric and cannot occur with total spin $S = \frac{1}{2}$. Baryonic states with $s \neq 0$ are collectively termed hyperons. *See* HYPERON; STRANGE PARTICLES.

These eight baryon states $(p, n, \Sigma^+, \Sigma^0, \Sigma^-, \Lambda, \Xi^0, \Xi^-)$ all have the spin parity $\frac{1}{2}^+$ and the same internal wave functions. They are arrayed in Fig. 1, where the symmetry of their relationship is made evident; for this, it is helpful to use the quantum number $Y = (B + s)$, named hypercharge. In the approximation that the s quark has the same mass as the (u,d) quarks, the quark energies (kinetic + mass) are equal and the quark-gluon coupling has the form of Eq. (7), which is invariant under all

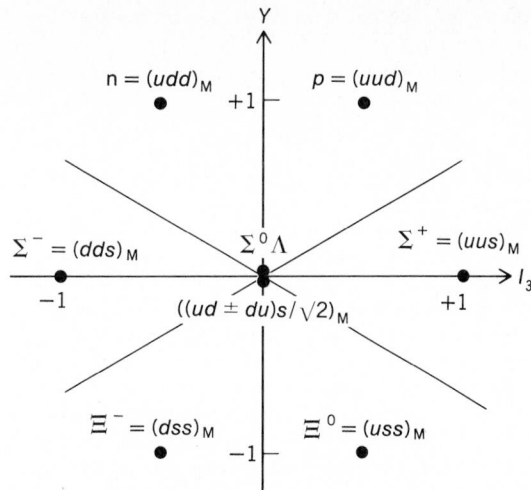

Fig. 1. The baryon octet states, arrayed with respect to I_3 as ordinate and $Y = (B + s)$ as abscissa. The charge number Q is given by $Q = I_3 + Y/2$. There are three axes of symmetry.

$$L_{\text{int.}} = f\Sigma g_{\mu\alpha}(\bar{u}\gamma_\mu\lambda_\alpha u + \bar{d}\gamma_\mu\lambda_\alpha d + \bar{s}\gamma_\mu\lambda_\alpha s) \quad (7)$$

$SU(3)_f$ transformations in the flavor space spanned by (u,d,s). The invariance of this expression for the pair-wise interchanges $(d \rightleftharpoons s)$, $(s \rightleftharpoons u)$, and $(u \rightleftharpoons d)$ is the origin of the threefold axes of symmetry shown in Fig. 1.

Mass differences. The mass difference $\delta m = (m(s) - m(u,d))$ is quite large, and so $SU(3)_f$ symmetry is much more strongly violated than $SU(2)_\tau$ symmetry. This is most apparent in the baryon mass values, which vary widely over the octet; the leading variation is that proportional to the strangeness s, which counts the s-quark content of each baryon. The approximately 75-MeV difference between the mean Σ mass $\bar{m}(\Sigma)$ and $m(\Lambda)$ has a more subtle origin, but is well accounted for on the basis of the quark-quark (qq) potential from quantum chromodynamics, described above. The small mass-differences within each isospin multiplet are believed due to electromagnetic effects. Assuming that these baryon states all have the same internal wave function, it follows that baryon octet states with the same charge Q have the same electromagnetic mass-shift $\delta m(Q)$; for example, $p = (uud)$ and $\Sigma^+ = (uus)$ have the same electromagnetic structure, since d and s have the same charge. From this observation, S. Coleman and S. Glashow pointed out in 1961 the remarkable $SU(3)_f$ relation of Eq. (8) connecting the mass dif-

$$m(\Xi^-) - m(\Xi^0)$$
$$= m(\Sigma^-) - m(\Sigma^+) + m(p) - m(n) \quad (8)$$

ferences within the N, Σ, and Ξ multiplets, which is rather well satisfied by the empirical masses. These remarks about electromagnetic structure also imply the magnetic moment equalities of Eq. (9), which are also in fair accord with the data.

$$\mu(\Sigma^+) = \mu(p)$$
$$\mu(\Sigma^-) = \mu(\Xi^-)$$
$$\mu(\Xi^0) = \mu(n) \qquad (9)$$

Production and reaction processes. The production and reaction processes observed for these baryons are governed by flavor selection rules. For hadronic processes (but not for weak processes) these selection rules are simply Eq. (10) for each

$$\Delta n(f) = 0 \tag{10}$$

flavor $f = B, Q, s, c, \ldots$; that is, each type of quark is independent of the others, and quarks do not change type. Quantum chromodynamics provides a ready understanding of these rules; for example, each term of Eq. (7) preserves quark flavor. The isospin symmetry $SU(2)_\tau$ implies some additional constraints going beyond Eq. (10), especially in relating quantitatively various reactions involving baryons within the same charge multiplets. Otherwise, which reactions occur or do not occur can be accounted for in terms of Eq. (10) and the empirical baryon and meson mass values. *See* SELECTION RULES (PHYSICS).

The Λ and Σ hyperons are formed by closely related interactions. Their formation in pion-proton collisions of sufficiently high energy, or in K^- absorption by protons, given by reactions such as (11)

$$\pi^- + p \rightarrow \begin{cases} K^0 + \Lambda \\ K^0 + \Sigma^0 \\ K^+ + \Sigma^- \end{cases}$$

$$K^- + p \rightarrow \begin{cases} \pi^- + \Sigma^+ \\ \pi^0 + \Sigma^0 \\ \pi^+ + \Sigma^- \\ \pi^0 + \Lambda \end{cases} \tag{11}$$

have been especially well studied. These reactions illustrate the selection rules for B, Q, and s; the (K^0, K^+) mesons have $s = +1$, while the $(K^-, \overline{K^0})$ mesons are their antiparticles, and have $s = -1$. It is illuminating to consider these reactions in terms of their constituent quarks, as shown in Fig. 2, where time progresses from left to right, and a quark line with backward-directed arrows represents an antiquark going forward in time. The simplest mechanisms for the reactions are equivalent to the exchange of $\bar{q}q$ (= mesons) or qqq (= baryon) systems between the meson and baryon. They generally involve intermediate $\bar{q}q$ creation and annihilation processes, but that of Fig. 2c involves only quark rearrangement. The simplest Ξ-production processes are given by reactions (12), where the

$$K^- + p \rightarrow \begin{cases} \Xi^- + K^+ \\ \Xi^0 + K^0 \end{cases} \tag{12}$$

$\Delta s = -2$ transition $\underline{N \rightarrow \Xi}$ is balanced by the $\Delta s = +2$ transition $\overline{K} \rightarrow K$.

Baryon-baryon interactions. The baryon-baryon interactions are of particular interest. Between nucleons, this interaction gives rise to the existence of atomic nuclei, and it has been particularly well studied, both empirically and theoretically, for the NN system. For large separations (greater than 0.8×10^{-15} m), the NN force is due to the exchange of pions and of other known mesons with masses less than about 1 GeV; for small separations (less than 0.4×10^{-15} m), a strong short-range repulsion is observed, possibly arising from the suppressive effects of the Pauli principle for quarks when the quark structures of the two nucleons overlap. At low energies, the outstanding feature of the NN interaction is its strong noncentral tensor component, which is due to one-pion exchange and is a direct consequence of the pseudoscalar nature of the pion. It also has a strong spin-orbit interaction, observed in NN interactions at higher energy and of much importance for the shell structure of nuclei.

Attractive ΛN forces of nuclear strength [with potentials $v(\Lambda n)$ of order 60% of $v(NN)$] are known to exist from the observation of Λ hypernuclei, each consisting of a Λ hyperon bound to an ordinary nucleus and denoted by $_\Lambda Z^A$. The $_\Lambda Z^A$ (ground state) systems remain stable until the Λ hyperon decays, since the conservation laws of Eq. (10) do not allow the deexcitation of the ΛN system by any hadronic process. Λ-proton elastic scattering has also been studied in the low-energy regime. Observations on excited Λ-hypernuclear states suggest that the ΛN spin-orbit interaction is relatively small.

The Σ hyperons also interact strongly with nucleons, as shown by the studies of $\Sigma^\pm$-proton elastic scattering at low energies. However, Σ hyperons also react strongly with nucleons of suitable charge, transforming into a Λ hyperon with release of about 75 MeV of kinetic energy; for example, reactions (13) are well known. These absorp-

$$\Sigma^- + p \rightarrow \Lambda + n \tag{13a}$$
$$\Sigma^+ + n \rightarrow \Lambda + p \tag{13b}$$

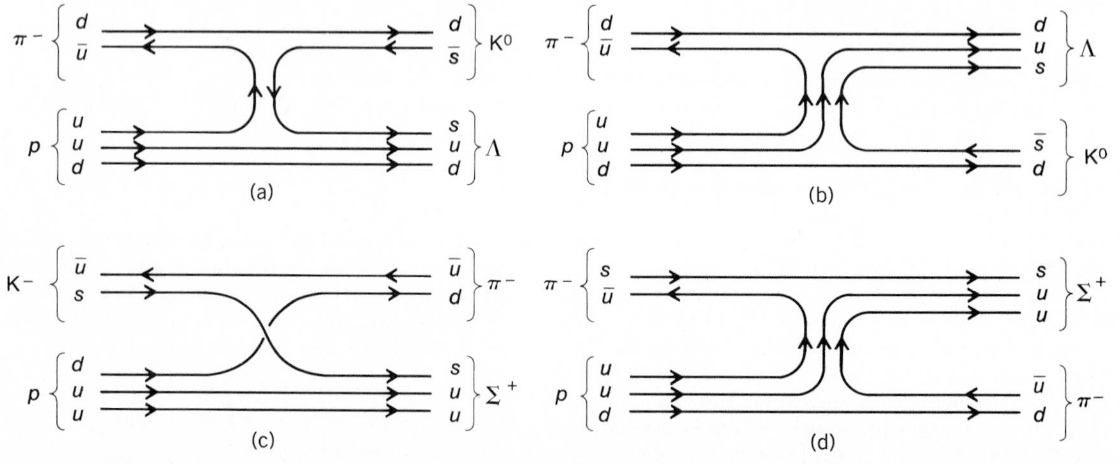

Fig. 2. Quark line graphs to illustrate meson-baryon reaction process. (a) $\pi^- + p \rightarrow K^0 + \Lambda$. (b) $\pi^- + p \rightarrow \Lambda + K^0$. (c) $K^- + p \rightarrow \pi^- + \Sigma^+$. (d) $\pi^- + p \rightarrow \Sigma^+ + \pi^-$. Time increases from left to right for these figures.

tive reactions preclude the existence of long-lived Σ hypernuclei, except for several special cases such as $n\Sigma^-$ or $nn\Sigma^-$ for which the selection rules of Eq. (10) allow no hadronic deexcitation process. There is no evidence for bound states of these last systems, but there is evidence suggesting that a class of exceptional Σ-hypernuclear levels may exist with lifetimes long on the nuclear scale.

The Ξ hyperons also interact strongly with nucleons; the little evidence available is consistent with theoretical expectation that processes (14)

$$\left.\begin{array}{c}\Xi^-+p\\\Xi^0+n\end{array}\right\}\to\Lambda+\Lambda \qquad (14)$$

should be strong, occurring through a rearrangement process. A striking example of this process was the observation of reaction (15) for a Ξ^- stopping in nuclear emulsion. From the total bind-

$$\Xi^-+C^{12}\to{}_{\Lambda\Lambda}He^6+Li^7 \qquad (15)$$

ing energy for the two Λ hyperons in ${}_{\Lambda\Lambda}He^6$, deduced from the energy released in its weak decay to ${}_\Lambda He^5 p\pi^-$, the ${}^1S_0$ $\Lambda\Lambda$ interation is known to have strength comparable with that for the ${}^1S_0$ ΛN interaction.

Baryon beta decay. The neutron beta-decay process (16) has been known for some decades. It

$$n\to p+e^-+\bar\nu_e \qquad (16)$$

was known quite early that the beta transition $n\to p$ involves two terms of opposite parity, vector (Fermi) and axial-vector (Gamow-Teller), with coupling amplitudes G_V and G_A. The empirical ratio $G_A/G_V=-1.253\pm0.007$ agrees well with the calculations of S. Adler (1965) and W. Weisberger (1966) based on the current algebra hypothesis (this reflects the property of chiral symmetry for the hadronic interactions). G_V is slightly smaller than the value G known for muon beta decay, $G_V/G=0.97\pm0.01$. The related process $\mu^-p\to n\nu_\mu$ is also well known, and its inverse, the neutrino-induced reaction $\nu_\mu n\to p\mu^-$, has been much studied for high-energy ν_μ beams (up to about 200 GeV) obtained from the decay $\pi^+\to\mu^+\nu_\mu$ of π^+ beams from high-energy proton accelerators. *See* NEUTRINO.

Analogous beta-decay processes are known for most of the hyperons in the baryon octet, but their rates are about an order of magnitude smaller than expected if their beta-decay amplitudes were the same as for the $n\to p$ transition. The Λ beta-decay process (17) is well known, with a branching ratio

$$\Lambda\to p+e^-+\bar\nu_e \qquad (17)$$

of $8.5\pm0.8\times10^{-4}$. The transition $\Lambda\to p$ involves $\Delta s=+1$, whereas the transition $n\to p$ of Eq. (16) involves $\Delta s=0$. For Σ hyperons, both types of transitions are known; $\Delta s=0$ holds for the decay modes (18) and $\Delta s=+1$ holds for the decay (19). The

$$\begin{array}{c}\Sigma^-\to\Lambda+e^-+\bar\nu_e\\\Sigma^+\to\Lambda+e^++\nu_e\end{array} \qquad (18)$$

$$\Sigma^-\to n+e^-+\bar\nu_e \qquad (19)$$

conceivable decay mode $\Sigma^+\to ne^+\nu_e$ is not observed, its branching ratio being less than 5×10^{-6}. For Ξ hyperons, the $\Delta s=+1$ decay mode $\Xi^-\to\Lambda e^-\bar\nu_e$ is known, but there is no evidence (branching ratio less than 3×10^{-3}) for the energetically favorable $\Delta s=+2$ mode $\Xi^-\to ne^-\bar\nu_e$. The only $\Delta s=+1$ decay mode for Ξ^0 would be $\Xi^0\to\Sigma^+e^-\bar\nu_e$, which

has not yet been observed (branching ratio less than 10^{-3}).

Parity and charge-conjugation violation. All of these beta-decay processes violate both parity conservation and charge-conjugation symmetry. Parity violation in beta decay was first observed in 1957, as an asymmetry in the angular distribution of beta electrons emitted from polarized Co^{60} nuclei. Later, this was demonstrated also for the beta decay of free neutrons, and subsequently also for the beta decay (17) of polarized Λ particles. *See* PARITY (QUANTUM MECHANICS).

Cabibbo theory. A remarkable synthesis of all the data on baryon beta-decay processes has been achieved, by the work of many theoreticians. This is based on the notion of a weak $\Delta Q=+1$ current J_μ^{wk+} which includes both vector and axial vector terms and which couples with itself. It consists of two parts, one for hadrons (meaning quarks) and the other for leptons, and so leads to the interaction amplitude (20). $(J_{lept}^{wk+})_\mu$ consists of a sum of terms

$$G\sum_\mu(J_{hadr}^{wk+}+J_{lept}^{wk+})_\mu{}^\dagger(J_{hadr}^{wk+}+J_{lept}^{wk+})_\mu \qquad (20)$$

$(e\nu_e)$, (μ,ν_μ), (τ,ν_τ), and so forth, for each lepton and its associated neutrino. The purely leptonic amplitude, which describes the muon decay $\mu^-\to\nu_\mu$ $(e\bar\nu_e)$ for example, is then given by the product of J_{lept}^{wk+} with its conjugate, as it occurs in expression (20), and has the universal amplitude G. The baryon beta-decay amplitudes stem from the hadron-lepton cross terms in Eq. (20). The crucial step forward was taken in 1963 by N. Cabibbo, who proposed that J_{hadr}^{wk+} for the quark flavors then known should be based on the single quark doublet

$$(u,d\cos\theta_C+s\sin\theta_C)$$

thus linking the $\Delta s=0$ and $\Delta s=+1$ transitions. This leads to expression (21) for the semileptonic

$$G\{\sum_\mu(J_{lept}^{wk+}(e^+\nu_e)+\dots)_\mu{}^\dagger(\cos\theta_C J_{hadr}^{wk+}(d\to u)$$
$$+\sin\theta_C J_{hadr}^{wk+}(s\to u))+\text{hermitian conjugate}\} \qquad (21)$$

weak transition processes, the $\Delta s=+1$ transition $s\to u$ having an intrinsic amplitude weaker by the factor $\tan\theta_C$ than that for the $\Delta s=0$ transition $d\to u$. Each J_{hadr}^{wk+} has two terms $(V_\mu+A_\mu)$, as was known for neutron beta decay. Equation (21) prescribes the relationship between the values for A_μ for the various baryonic transitions in terms of two numbers, one being (G_A/G_V) for $n\to p$ and the other being θ_C. These currents $(J_{hadr}^{wk+})_\mu$ for single quark transitions are automatically members of an octet, the property Cabibbo assumed; also the transitions induced by these currents necessarily have $\Delta Q=1$ for $\Delta s=+1$, or $\Delta Q=-1$ for $\Delta s=-1$, so that the interaction (21) forbids $\Sigma^+\to ne^+\nu_e$ while allowing $\Sigma^-\to ne^-\nu_e$. The value appropriate for the Cabibbo angle θ_C is given consistently by a wide variety of data, as $\theta_C=0.223\pm0.005$ radians. From Eq. (21), the value predicted for G_V/G is $\cos\theta_C=0.97$, consistent with the observed value given above.

$\Delta Q=0$ hadron transitions are also possible, based on neutral currents $(J_{hadr}^{wk0})_\mu$, but the data on mesonic transitions have shown clearly that there are no such transitions for $\Delta s=+1$, so they need not be considered for baryon decays. $\Delta Q=0$, $\Delta s=0$ transitions can and do occur, but are most accessible in the related inelastic neutrino scattering processes, of the type $\nu N \to \nu' +$ (hadrons). In 1970, S. Glashow, J. Iliopoulos, and L. Maiani proposed a specific mechanism to eliminate all $\Delta Q = 0$, $\Delta s = +1$ transitions, involving a new quark c, with the proposal that its contribution to $(J^{\text{wk}}_{\text{hadr}})_\mu$ should be based on a second quark doublet.

$$(c, -d \sin \theta_C + s \cos \theta_C)$$

Assuming the same strength for the neutral currents constructed from each of these two doublets, the two neutral currents of the type $J^{\text{wk0}}_{\text{hadr}}(d \to s)$ then formed cancel exactly. The dominant weak transition for the c quark is then the $\Delta Q = -1$ transition $c \to s$, with amplitude $G \cos \theta_C$.

Nonleptonic baryon decay. These are the dominant modes of hyperon decay, as shown in Table 1, typically having lifetimes of the order of 10^{-10} s. The decay $\Lambda \to p\pi^-$ has been especially well studied and shows a strong violation of parity conservation, as a large forward-backward asymmetry (coefficient $\alpha_\Lambda = 0.64$) in the pion distribution relative to the initial Λ-spin direction; this allows efficient measurements of Λ-spin polarization to be made for Λ-production processes. The same holds for the decay $\Sigma^+ \to p\pi^0$, where $\alpha^0_{\Sigma^+} = -0.98$. The modes $\Sigma^\pm \to n\pi^\pm$ are also well known but show only small asymmetries. The Ξ-hyperon decays are known, as $\Xi^- \to \Lambda\pi^-$ and $\Xi^0 \to \Lambda\pi^0$, both of which show a substantial asymmetry $\alpha_\Xi = -0.4$. All of these decays are weak, with $\Delta s = +1$.

The decay $\Sigma^0 \to \Lambda\gamma$ is not a weak process but an electromagnetic transition, allowed ($\Delta s = 0$) by the hadronic selection rules, with measured lifetime about 10^{-19} s. The decay $\Sigma^+ \to p\gamma$ is well known but involves $\Delta s = +1$, so that it is due to the joint action of weak and electromagnetic interactions. No $\Delta s = +2$ decay, such as $\Xi^0 \to p\pi^-$ or $\Xi^- \to p\pi^-\pi^-$, has been detected.

These weak nonleptonic decay processes are all implied by the terms $(J^{\text{wk}}_{\text{hadr}})_\mu{}^\dagger (J^{\text{wk}}_{\text{hadr}})_\mu$ of Eq. (21), including both the $\Delta Q = +1$ and the (not specified here) $\Delta Q = 0$ components. For example, the Λ decay results from the transition $s \to u(\bar{u}d)$, but

this is not simply related with the final state observed. Empirically, there is a $\Delta I = \frac{1}{2}$ rule, for $\Delta s = +1$ transitions, which means that the final state $B\pi$ is dominated by the isospins $|(I \pm \frac{1}{2})|$, where I is the initial baryon isospin. These nonleptonic hadronic terms of Eq. (21) also give rise to nonmesonic weak decay interactions; reaction (22a) represents a weak contribution to the NN forces, and its tiny effects have been identified in the study of nuclear processes, while reaction (22b)

$$NN \to NN \tag{22a}$$

$$\Lambda N \to NN \tag{22b}$$

is well known from the energetic nonmesonic decays observed for Λ hypernuclei, for example, $_\Lambda\text{He}^5 \to n\text{He}^4$ with approximately 173 MeV energy release. *See* WEAK NUCLEAR INTERACTIONS.

$3/2^+$ baryon decuplet. The three-quark state (uuu) with $S = \frac{3}{2}$ is flavor symmetric and spin symmetric. Antisymmetry for the full wave function [Eq. (2)] then requires the space wave function to have symmetry S, as holds for the nucleon states [Eq. (3)]. If the potentials $v(qq)$ are spin and flavor independent, this state will have the same $L = 0$ space wave function as the baryon octet, and it is therefore intimately connected with the latter. There are four such nonstrange states, denoted by Δ and having isospin $I = \frac{3}{2}$, whose quark structures are shown in Eqs. (23). The replacement of one,

$$\Delta^{++} = (uuu) \qquad \Delta^+ = (uud)_S$$
$$\Delta^0 = (udd)_S \qquad \Delta^- = (ddd) \tag{23}$$

two, or three of these d quarks by an s quark leads to an isotriplet Σ^*, an isodoublet Ξ^*, and an isosinglet Ω, respectively, forming the decuplet depicted in Fig. 3.

These Δ states are identified with the strong $I = \frac{3}{2}$, J^P (spin parity) $= \frac{3}{2}^+$ pion-nucleon resonances first observed in 1953, with mass about 1232 MeV. They are particle unstable, the breakup $\Delta \to N\pi$ being allowed by all the hadronic selection rules; their lifetimes are approximately 0.5×10^{-23} s. The Σ^* states have mass 1382 MeV and decay dominantly to $\Lambda\pi$. They were first established in 1960 from observations on $\Lambda\pi$ correlations in reaction (24), showing the occurrence

$$K^- + p \to \begin{Bmatrix} \Sigma^{*+} + \pi^- \\ \Sigma^{*-} + \pi^+ \end{Bmatrix} \to \Lambda + \pi^+ + \pi^- \tag{24}$$

of the intermediate states shown there. The Ξ^* states have mass 1532 MeV and decay to $\Xi\pi$, as observed in 1962 for reactions (25). The isosinglet

$$K^- + p \to \begin{Bmatrix} \Xi^{*-} + K^+ \to \Xi^- + \pi^0 + K^+ \\ \Xi^{*0} + K^0 \to \Xi^- + \pi^+ + K^0 \end{Bmatrix} \tag{25}$$

Ω^- was found in 1964, in reaction (26) depicted in Fig. 4, and has mass 1672 MeV. The Ω^- particle

$$K^- + p \to \Omega^- + K^+ + K^0 \tag{26}$$

is semistable; that is, it decays through the weak interactions, with lifetime on the order of 10^{-10} s; hadronic decay is forbidden, since its mass lies below the lowest threshold $m(\Xi^0 K^-) = 1808$ MeV for known hadrons with the same net strangeness.

It is this occurrence of the ^{28} and ^{410} multiplets (where the notation is $^{2S+1}\alpha$, α specifying the SU(3)$_f$ multiplicity) in association which indicates (Table 3) that their (spin × flavor) wave function has symmetry S. Without color, the spin-statistics

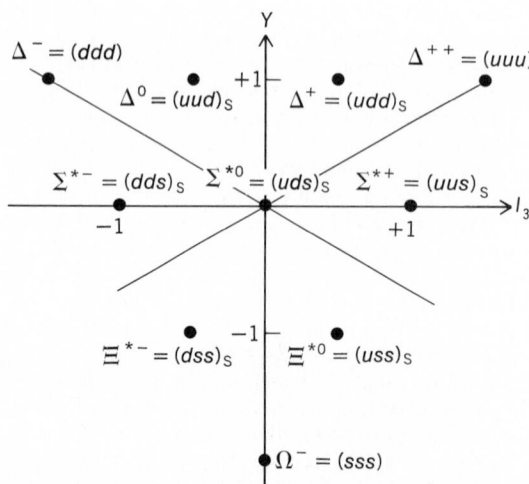

Fig. 3. The baryon decuplet states, arranged with respect to I_3 as ordinate and $Y = (B + s)$ as abscissa. The charge number Q is given by $Q = I_3 + Y/2$. There are three axes of symmetry.

Table 3. Permutation symmetries possible for × SU(3)$_f$ wave functions for three quarks

Permutation symmetry	SU(3) multiplets		
	Singlet	Octet	Decuplet
S		$S=\frac{1}{2}$	$S=\frac{3}{2}$
M	$S=\frac{1}{2}$	$S=\frac{1}{2}$ and $\frac{3}{2}$	$S=\frac{1}{2}$
A	$S=\frac{3}{2}$	$S=\frac{1}{2}$	

theorem would indicate that their Ψ_{space} wave function had symmetry A. Since these multiplets are ground states, it would be very difficult to achieve this, if it is indeed possible. It was these baryonic resonance observations which provided the empirical basis for the introduction of a three-dimensional color space for quarks.

With quantum chromodynamics, the mass difference $[m(\Delta) - m(N)] = 293$ MeV is attributed to the spin-spin component (Fermi-Breit term) of the one-gluon-exchange potential. Quantum chromodynamics also predicts approximately equal mass spacing between the four charge multiplets of the decuplet; the spacings observed from Δ to Ω are 150, 150, and 140 MeV, in turn. The existence of a semistable $\frac{3}{2}^+$ Ω^- hyperon with mass 1685 MeV had been predicted by M. Gell-Mann from SU(3)$_f$ symmetry in 1963, so the discovery of such a

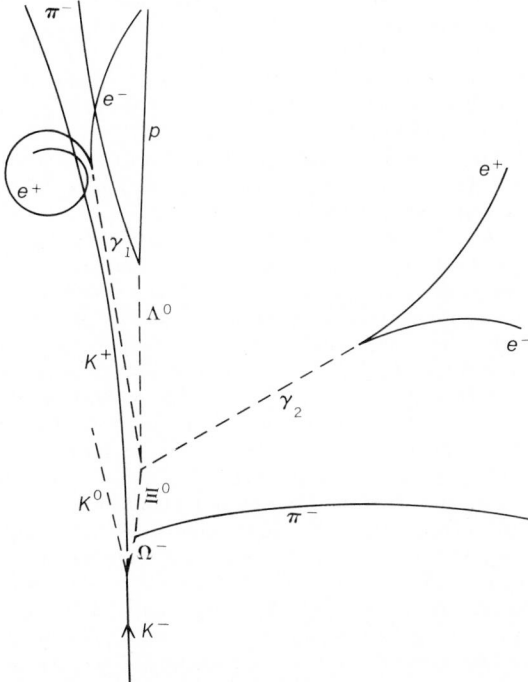

Fig. 4. Line drawing of hydrogen bubble chamber event involving production and decay of Ω^- particle. An incident K^- meson (momentum 5000 Mev/c) interacts with a proton giving Eq. (7). The K^0 meson is not observed but its presence is deduced from analysis of energy and momentum balance. $\Omega^- \to \Xi^0 \pi^-$ decay occurs, the (invisible) Ξ^0 travels some distance and then decays to neutral system $\Lambda \pi^0$; the Λ particle is detected from its decay $\Lambda \to p\pi^-$, the π^0 from the electron pair conversions of the two photons γ_1 and γ_2 resulting from the $\pi^0 \to \gamma\gamma$ decay. Dashed lines denote unobserved paths of neutral particles, deduced from measurements on visible particle tracks.

curious particle in 1964 quickly led to a general acceptance of SU(3)$_f$ symmetry.

Higher excited baryonic states. Many further particle-unstable baryon states, with lifetimes in the range 10^{-22} to 10^{-23} s, have become established. The $s = 0$ states have been explored most thoroughly, by the measurement of the angular and polarization angular distributions for pion-nucleon elastic and charge-exchange scattering and their partial-wave analysis to give phase shifts as a function of spin-parity and mass. Similarly, the $s = -1$ states have been studied, mostly using elastic scattering and reaction data for K^- on nucleons. Further evidence on the $s = -1$ states also comes from the study of final states in reactions $\bar{K}N \to (\Lambda$ or $\Sigma) +$ pions, just as the $\Sigma(1382)$ resonance (which lies below the $\bar{K}N$ threshold) was deduced from data on reaction (24). The $s = -2$ states can be studied only from the final state effects, for example, in $\bar{K}N \to K\Xi^*$, just as $\Xi(1530)$ was deduced from reaction (25), so that relatively little is known concerning Ξ^* states. The most prominent baryonic resonances established, up to mass values of order 2500 MeV, are listed in Table 4.

All of the resonances now established are consistent with the limitations of the three-quark model. For example, no resonances are established for $s = +1$; there is one candidate, for the $I = 0$ KN $\frac{1}{2}^-$ partial wave, deduced from elastic scattering and charge-exchange data, but despite many years of work the resonance interpretation is still not uniquely established for it. No resonances with $I = 2$, $s = -1$ are established for the $\pi\Sigma$ channels, nor with $I = \frac{3}{2}$, $s = -2$ for the $\pi\Xi$ channels, and there are indeed no qqq configurations which have these quantum numbers, so none is expected.

The density of excited baryon states per unit energy interval increases rapidly with increasing mass, as illustrated for the N* and Δ* states in Fig. 5. The validity of SU(3)$_f$ symmetry requires that there should exist Λ*, Σ*, and Ξ* states for each N*, and Σ*, Ξ*, and Ω* states for each Δ*, most of which have not yet been detected. Further exploration is going on, especially for high mass values, and there is every reason to expect the density of states to continue to increase in the mass region above 2500 MeV. The higher states already known indicate at least the following four mechanisms of excitation:

1. Rotational excitation of low-lying multiplets, such as the ground state multiplets with symmetry S, comprising the $\frac{1}{2}^+$ octet and the $\frac{3}{2}^+$ decuplet. This is well illustrated by the sequence of prominent Δ* states, for which Fig. 6 shows a plot of (mass)^2 versus spin J. The established points lie on a straight line, with intervals $\Delta J = 2$, which is generally known as a Regge (rotational) trajectory. *See* REGGE POLE.

2. Vibrational (radial) excitation of the ground state multiplets of a given type. In Fig. 5 the $\frac{1}{2}^+$ state N(1470) and $\frac{3}{2}^+$ state Δ(1690) constitute a repetition of the ground state multiplets with symmetry S. The precise nature of this radial excitation is still under investigation.

3. Excitations involving internal orbital angular momentum L with parity P. These are natural excitations to expect if baryons are composed of three quarks. Table 3 lists the SU(3)-spin combina-

Table 4. Known prominent and well-established excited baryonic states†

N states	J_P	Δ states	J_P	Λ states	J_P	Σ states	J_P	Ξ states	J_P
$N(1450)$	$\tfrac{1}{2}^+$	$\Delta(1232)$	$\tfrac{3}{2}^+$	$\Lambda(1405)$	$\tfrac{1}{2}^-$	$\Sigma(1385)$	$\tfrac{3}{2}^+$	$\Xi(1530)$	$\tfrac{3}{2}^+$
$N(1520)$	$\tfrac{3}{2}^-$	$\Delta(1620)$	$\tfrac{1}{2}^-$	$\Lambda(1520)$	$\tfrac{3}{2}^-$	$\Sigma(1660)$	$\tfrac{1}{2}^+$	$\Xi(1815)$	$\tfrac{3}{2}^-$
$N(1535)$	$\tfrac{1}{2}^-$	$\Delta(1640)$	$\tfrac{3}{2}^+$	$\Lambda(1670)$	$\tfrac{1}{2}^-$	$\Sigma(1670)$	$\tfrac{3}{2}^-$	$\Xi(1930)$	?
$N(1650)$	$\tfrac{1}{2}^-$	$\Delta(1700)$	$\tfrac{3}{2}^-$	$\Lambda(1690)$	$\tfrac{3}{2}^-$	$\Sigma(1750)$	$\tfrac{1}{2}^-$	$\Xi(2030)$	?
$N(1670)$	$\tfrac{3}{2}^-$	$\Delta(1900)$	$\tfrac{1}{2}^-$	$\Lambda(1800)$	$\tfrac{3}{2}^+$	$\Sigma(1765)$	$\tfrac{5}{2}^-$		
$N(1680)$	$\tfrac{5}{2}^-$	$\Delta(1910)$	$\tfrac{1}{2}^+$	$\Lambda(1815)$	$\tfrac{5}{2}^+$	$\Sigma(1915)$	$\tfrac{5}{2}^+$		
$N(1680)$	$\tfrac{5}{2}^+$	$\Delta(1920)$	$\tfrac{5}{2}^-$	$\Lambda(1830)$	$\tfrac{5}{2}^-$	$\Sigma(1940)$	$\tfrac{3}{2}^-$		
$N(1700)$	$\tfrac{3}{2}^-$	$\Delta(1920)$	$\tfrac{5}{2}^+$	$\Lambda(1870)$	$\tfrac{3}{2}^+$	$\Sigma(2030)$	$\tfrac{7}{2}^+$		
$N(1710)$	$\tfrac{1}{2}^+$	$\Delta(1950)$	$\tfrac{7}{2}^+$	$\Lambda(2100)$	$\tfrac{7}{2}^-$	$\Sigma(2250)$	?		
$N(1740)$	$\tfrac{3}{2}^+$	$\Delta(1960)$	$\tfrac{3}{2}^+$	$\Lambda(2110)$	$\tfrac{5}{2}^+$	$\Sigma(2455)$	?		
$N(1830)$	$\tfrac{3}{2}^-$	$\Delta(2010)$	$\tfrac{3}{2}^-$	$\Lambda(2350)$	$\tfrac{9}{2}^+$	$\Sigma(2595)$	?		
$N(1880)$	$\tfrac{3}{2}^+$	$\Delta(2420)$	$\tfrac{11}{2}^+$						
$N(1990)$	$\tfrac{7}{2}^+$	$\Delta(2850)$	$\tfrac{15}{2}^+$						
$N(2190)$	$\tfrac{7}{2}^-$	$\Delta(3230)$	?						
$N(2200)$	$\tfrac{9}{2}^-$								
$N(2220)$	$\tfrac{9}{2}^+$								
$N(2650)$	$\tfrac{11}{2}^-$								
$N(3030)$	?								

†There should be further columns for excited Ω states, made of three s quarks, and for charmed baryon states Λ_c and Σ_c, consisting of one c quark and two u or d quarks. Only $\Omega(1672)$ and $\Lambda_c(2273)$, both semistable, are known. There is some evidence for an unstable excited state $\Sigma_c(2500)$.

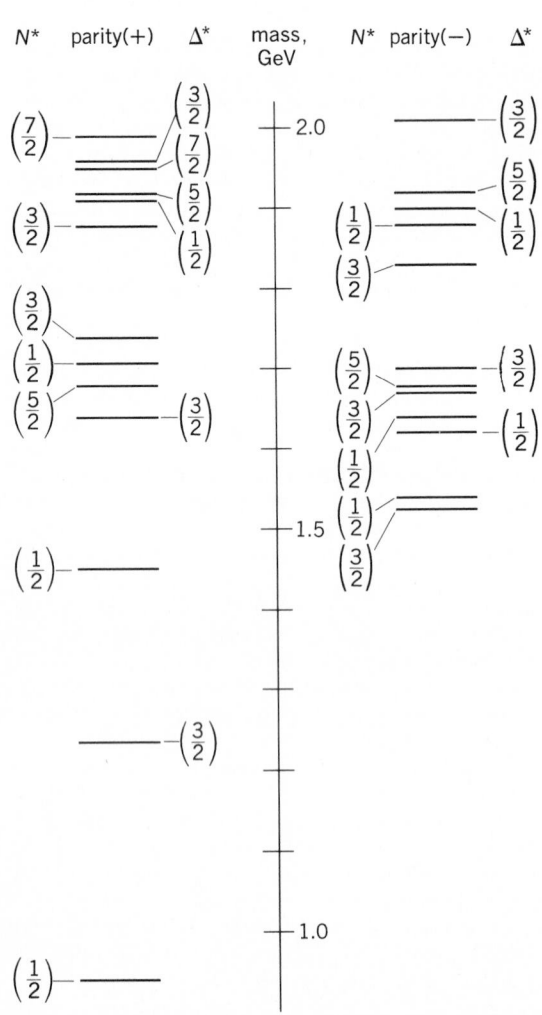

Fig. 5. Spectrum of known nucleonic excited states up to mass 2.0 GeV, with their spin values, those with positive parity being on the left half of the figure, those with negative parity being on the right half. N^* states are those with $I = 1/2$, Δ^* states those with $I = 3/2$.

tions which go together for a particular permutation symmetry of the space wave function. The observed states then have parity P and form submultiplets for each J obtained by the vector addition of L and S. It is convenient to define internal coordinates, $\boldsymbol{\rho}$ for the vector separation between two quarks and $\boldsymbol{\lambda}$ for the vector separation of the third quark from their center of mass, and to specify excited configurations by the quantum numbers $(n_\rho l_\rho;\, n_\lambda l_\lambda)$ for these internal degrees of freedom. Then $P = (-1)^{l\rho + l\tau}$ and $\boldsymbol{L} = \boldsymbol{l}_\rho + \boldsymbol{l}_\lambda$.

The ground state multiplets correspond to $(n_\rho l_\rho;\, n_\lambda l_\lambda) = (00;\, 00)$, which has symmetry S, assuming harmonic oscillator orbitals.

The first excited configuration is then a linear superposition of $(11;\, 00)$ and $(00;\, 11)$; its space wave function therefore has $L^P = 1^-$, with symmetry M. From Table 3, this comprises the multiplets $^2 10$, $^2 8$, and $^2 1$, together with $^4 8$, and so accounts for all the N^* and Δ^* states of negative parity lying below 1.8 GeV on Fig. 5. This supermultiplet also includes two $\mathrm{SU}(3)_\mathrm{f}$ singlets, $\Lambda(1520)$ with $\tfrac{3}{2}^-$ and $\Lambda(1405)$ with $\tfrac{1}{2}^-$. The various $\mathrm{SU}(3)_\mathrm{f}$ multiplets are not pure; in general, all states with the same s, I, J, and P in the same band $N = n_\rho + n_\lambda$ will mix strongly, and there can also be mixing between bands. With quantum chromodynamics, these mixings are prescribed, and considerable success has been achieved, especially by N. Isgur and G. Karl, in accounting for the data in this way.

The second excited configurations have even parity, consisting of the configurations $(22;\, 00)$, $(20;\, 00)$, $(11;\, 11)$, $(00;\, 20)$, and $(00;\, 22)$. These contain space wave functions with permutation symmetry S for $L = 2$ and 0, M for $L = 2$ and 0, and A for $L = 1$. The first group (S) includes the Regge excitations $(L = 2)$ of the ground state and the radial vibrations $(L = 0)$, as well as a number of $\mathrm{SU}(3)_\mathrm{f}$ multiplets with other values of J. All of the N^* and Δ^* states required by these multiplets have been found to exist. Some of the states required by the $\mathrm{SU}(3)_\mathrm{f}$ multiplets of the second group (M) are established, sufficient to suggest strongly

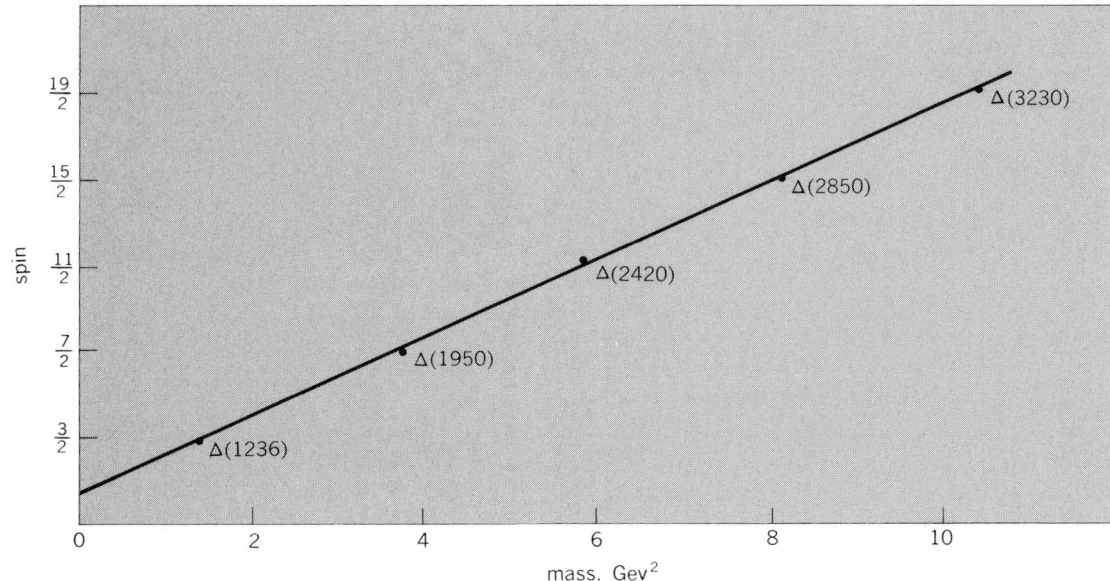

Fig. 6. (Mass)² for prominent $I = \frac{3}{2}$ nucleonic excited states plotted against their spin J. The uppermost state has been placed on the assumption that its spin-parity value continues the sequence $(\frac{3}{2}+)$, $(\frac{7}{2}+)$, $(\frac{11}{2}+)$, $(\frac{15}{2}+)$, ... The excellent straight-line fit is interpreted as a Regge trajectory for a sequence of rotational levels.

that these M supermultiplets exist. There is good reason to believe that states of the supermultiplet A will be difficult to produce and to detect, and there are no candidates for it yet.

The third excited configurations will have $P = -1$ and will be very numerous. The observed states include the $\frac{7}{2}^-$ Regge recurrence $N(2190)$ expected from the $\frac{3}{2}^-$ state $N(1520)$. A supermultiplet of symmetry S and $L = 1$ is also expected, and the five states above 1800 MeV, on the right half of Fig. 5, do have the (I,J,P) combinations to fill out this supermultiplet. No other assignment is available for the $\frac{5}{2}^-$ $\Delta(1920)$ state. It is unlikely that all the multiplets of the $N = 3$ band will ever be identified empirically, but at least no states have yet been found which are inconsistent with this model.

This qqq model for baryons is far from complete. It provides a good mnemonic for baryon spectroscopy, and even works well quantitatively, but it does not include any account of the quark sea or of gluonic components, even for the ground states. Other kinds of excitation are also possible; for example, if $s = +1$ resonances are established, they might correspond to multiquark states with the structure $(\bar{q}qqqq)$, and there might also exist excited states where the excitations are gluonic. However, there is no data calling for such excitations at present.

Charmed baryons and beyond. Further baryon states can be formed by replacing one or more of the u,d and s quarks of the states discussed above by a c quark. If the s and c quarks both had the same mass as the (u,d) quarks, the states formed would correspond to an SU(4)$_f$ symmetry. This is emphasized by Fig. 7, which shows the extensions of the $\frac{1}{2}^+$ baryon octet (Fig. 7a) and the $\frac{3}{2}^+$ baryon decuplet (Fig. 7b) obtained in this way. The lowest plane of each part of the figure consists of the charmless baryon states; those in Fig. 7a are in accord with the octet of Fig. 1, and those in

Fig. 7b with the decuplet of Fig. 3. If SU(4)$_f$ symmetry were exact, the states of the array in Fig. 7a or Fig. 7b would all have the same internal structure and would be the substates of a representation, ^{2}20$_M$ for Fig. 7a or ^{4}20$_S$ for Fig. 7b, of the SU(4)$_f$ group. These are two different 20-dimensional representations of this group; the suffix denotes the permutation symmetry of the flavor wave functions forming the basis of each SU(4)$_f$ representation. Figure 7a consists of four equal hexagons of side l put together to form a symmetric solid, whose outer surface consists of these hexagons plus four equilateral triangles of side l; Fig. 7b consists of four equilateral triangles of side $3l$, put together to form a pyramid.

In reality, the c-quark mass is so large that little quantitative detail of SU(4)$_f$ symmetry can survive in the physical situation, and the structure of Fig. 7 has value mainly for general comprehension and for the counting of states. This is clear from the quantum chromodynamics calculation of Isgur and his colleagues; the limitation of the space wave function to one particular symmetry and one particular band N is often a poor approximation. Their lowest calculated levels are given in Table 5; the states are labeled as for SU(3)$_f$, the suffix c meaning that the s quark has been replaced by a c quark (which increases the charge by +1). With these mass values, only the isosinglet $\Lambda_c^+(2260)$ will be semistable, and its weak decay will be due to the dominant c-quark transition $c \rightarrow s$, as remarked above.

The observed state $\Lambda_c^+(2273)$ has a number of

Table 5. Mass, spin, and parity values calculated from quantum chromodynamics for the low-lying C = +1 charmed baryonic states

$J^P =$	$\frac{1}{2}^+$	$\frac{3}{2}^+$	$\frac{1}{2}^-$	$\frac{3}{2}^-$
	$\Lambda_c(2260)$	$\Sigma_c(2510)$	$\Lambda_c(2510)$	$\Lambda_c(2590)$
	$\Sigma_c(2440)$			

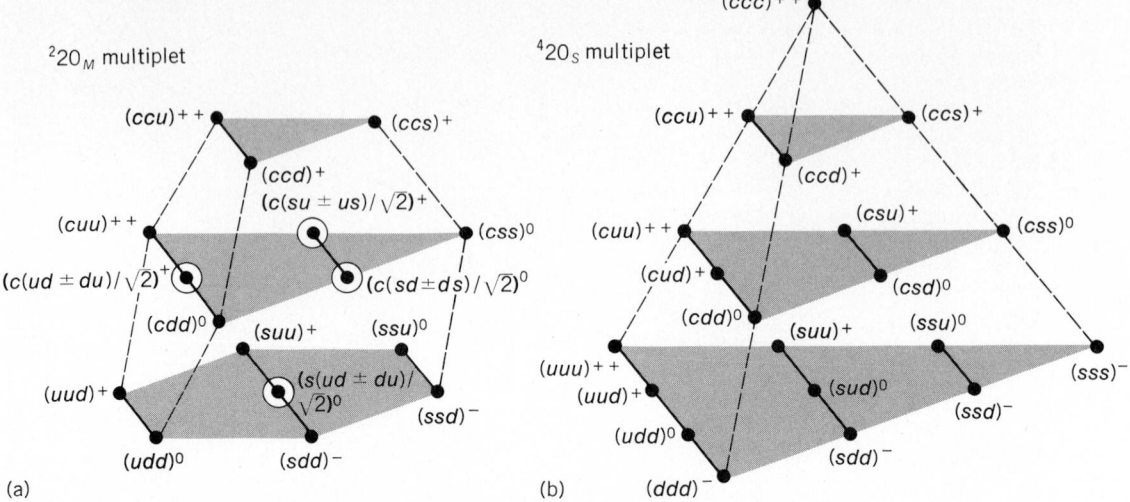

Fig. 7. Arrays of the states of baryons made from three quarks of the types *u*, *d*, *s*, and *c*, with no internal orbital angular momentum. The quark content and charge of each state is specified. The vertical axis specifies the number of *c* quarks. (*a*) ½⁺ baryons. (*b*) ³⁄₂⁺ baryons. The former have flavor symmetry M; the latter have symmetry S.

decay modes (Table 1), the strongest being to $pK^-\pi^+$, with branching ratio $2.2 \pm 1.0\%$. Its isospin has not been demonstrated to be $I = 0$, but it is reasonable to identify it with the lowest calculated state on Table 5. There is also some evidence for a Σ_c state of mass about 2500 MeV, decaying hadronically to $\pi\Lambda_c$. This may be identified with $\Sigma_c(2510)$, for this is the charmed analog to $\Sigma(1385)$, the decay $\Sigma_c \to \pi\Lambda_c$ being parallel with the decay $\Sigma \to \pi\Lambda$, but the spin-parity for the observed state $\Sigma_c(\sim 2500)$ is not yet known empirically. *See* CHARM.

The existence of a fifth quark *b* was demonstrated in 1978 in work with the PETRA electron-positron storage ring at the DESY laboratory, Hamburg, and theoreticians anticipate the existence of a sixth quark, to be named *t*, whose mass must exceed 19 GeV in order to account for the lack of evidence for this quark even at the top energy of PETRA. These quarks will add two further additive quantum numbers with conservation laws of the form of Eq. (10). Of course, there could well exist still further, heavier quarks at present unsuspected. Baryon states built of these heavier quarks, together with the lighter quarks, will certainly exist and will be observed in the course of time; no candidates to be identified with Λ_b have yet been brought forward. *See* PARTICLE ACCELERATOR.

Baryonic spectroscopy. It is clear that the baryonic levels present a spectroscopic situation not unlike, although more complicated than, that already explored for atomic nuclei and their excited states. There are electromagnetic and mesonic transitions between the baryonic levels whose systematics are just beginning to be studied. For example, the two mesonic transitions (27*a*) and (27*b*) are well known, and the radiative transition (28) has become established. Knowledge of

$$\Sigma(1765) \to \pi\Lambda(1520) \qquad (27a)$$
$$\Sigma(2030) \to \pi\Sigma(1385) \qquad (27b)$$

tion (28) has become established. Knowledge of

$$\Lambda(1520) \to \gamma\Lambda \qquad (28)$$

this baryon spectroscopy has been developing rapidly as has also theoretical understanding of it. *See* NUCLEAR SPECTRA.

Antibaryons. For every baryonic state mentioned above, there will exist an antibaryon state with opposite flavor quantum numbers, in particular with $B = -1$. The antiproton $\bar{p}$ was first identified in 1954, and antiproton beams have been available at high-energy proton accelerators for many years. Adaptation of the Super Proton Synchrotron (SPS) at CERN (Geneva) has been undertaken to serve as a proton-antiproton storage ring (named the Collider) capable of studying $\bar{p}$-p collisions up to 400 GeV center-of-mass energy in considerable detail.

Most of the expected antibaryon states have been detected and studied, in some detail, for example using $\bar{p}$ beams for reactions of the general type (29), where B_1 and B_2 denote any two baryonic

$$\bar{p} + p \to \bar{B}_1 + B_2 + (\text{mesons}) \qquad (29)$$

states (stable or unstable) and the mesons are such that the reaction conserves charge, strangeness, charm, and so forth. However, the reaction rates fall rapidly with increasing mass, strangeness, and charm for the final baryons. Electron-positron annihilation reactions can be studied efficiently using storage rings, and these suffer much less from this disadvantage; for example, at the storage ring SPEAR at SLAC (Stanford University), the energy available is well above the $\bar{\Lambda}_c\Lambda_c$ threshold, and reactions of the type (30) have given much of the knowledge of $\Lambda_c(2273)$.

$$e^+ + e^- \to \bar{\Lambda}_c + (\Lambda_c \text{ or } \Sigma_c) + \text{mesons} \qquad (30)$$

Antibaryon decay processes correspond directly to those for the corresponding baryons. They have the same lifetimes, within measurement accuracy. Several examples of the decay processes which have been observed are given in Eq. (31).

$$\begin{aligned}
\bar{\Lambda} &\to \bar{p}\pi^+ \\
\bar{\Sigma}^+ &\to \bar{n}\pi^+ \\
\bar{\Lambda}_c &\to \bar{p}K^+\pi^-
\end{aligned} \qquad (31)$$

Interaction with matter. When negatively charged particles with lifetimes longer than about 10^{-12} s come to rest in condensed matter, they are first captured into outer atomic orbits in the constituent elements and then cascade down to inner orbits by the ejection of Auger electrons and the emission of photons. Finally, they undergo absorption by the atomic nucleus, through reactions which release energy and cause nuclear breakup, giving rise to a nuclear star. *See* HADRONIC ATOM.

Σ^- stops in (nuclear) photographic emulsion often (about 75% of the time) do not lead to the emission of charged particles. The visible stars are small, with visible energy 10–20 MeV, and frequently (about 10% of Σ^- stops) emit a light Λ hypernucleus. These observations are consistent with the reaction mechanism of Eq. (13a), releasing two neutral particles of energy 75 MeV, the capture occurring from the nuclear periphery. Ξ^--capture stars result from absorptive reaction (14), with even less energy released, and frequently lead to the emission of two Λ hypernuclei. Antiproton stops in emulsion lead to large stars because the annihilation reaction $\bar{p}N \rightarrow$ (mesons) releases about 1880 MeV, with the emission of many mesons (mostly pions).

Proton stability. In grand unified theories (GUT), which attempt to account for both quarks and leptons, together with their strong, electromagnetic, and weak interactions, transitions $q \rightarrow l$ generally exist, at some level, since such theories assign leptons and quarks to common multiplets. This violation of baryon conservation opens the possibility that the lightest baryon, the proton, might not be absolutely stable but undergoes decay processes such as $p \rightarrow e^+\pi^0$ at a very low rate. Empirically, the proton is known to have a half-life exceeding 6×10^{35} s; its decay rate for $p \rightarrow \mu^+ +$ (mesons) is less than 3×10^{-38} s^{-1}. The simplest grand unified theories under discussion do predict a total proton decay rate of order $10^{-39\pm2}$ s^{-1}, while other theories may predict smaller decay rates, even zero. Since proton decay offers the possibility of discriminating between various grand unified theories, the detection of proton decay, or at least the improvement of the present empirical limits on its rate, is an important subject of investigation. *See* ELEMENTARY PARTICLE; FUNDAMENTAL INTERACTIONS.   [RICHARD H. DALITZ]

Bibliography: A. W. Hendry and D. B. Lichtenberg, The quark model, *Rep. Prog. Phys.*, 41: 1707–1780, 1978; G. Karl, in S. Homma, M. Kawaguchi, and H. Miyazawa (eds.), *Proceedings of the 19th International Conference on High Energy Physics, Tokyo, 1978*, pp. 135–137, 1979; Particle Data Group, *Rev. Mod. Phys.*, 52:S1–S286, 1980.

Beam-foil spectroscopy

A method of determining the energies and mean lives of excited electronic levels in monatomic ions and in atoms. Beams of particles (ions) of any element are energized in a particle accelerator and then sent in vacuum through a thin foil, usually of carbon. The particles emerge from the foil in various stages of ionization. Numerous energy levels are excited in each of those stages of ionization. The spontaneous loss of the energy of excitation takes place as the beam moves downstream from the foil. That loss is detected by means of the electromagnetic radiation or the electrons which the

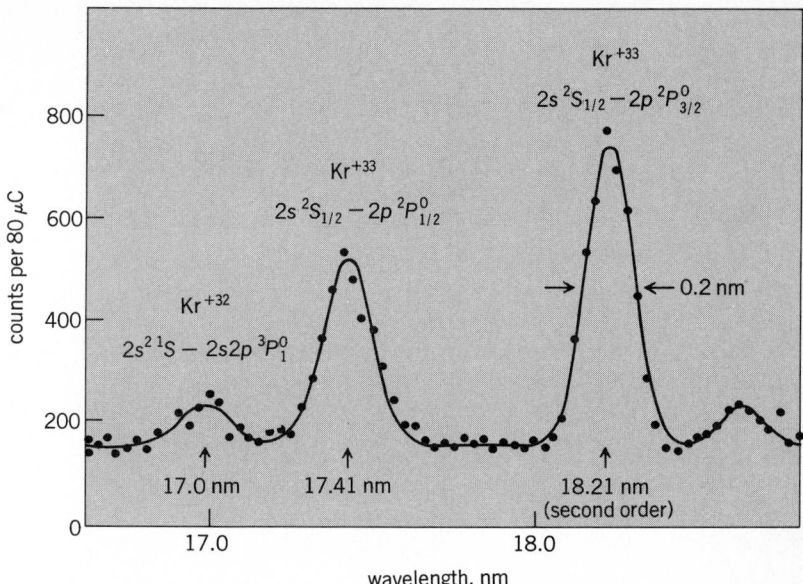

Fig. 1. Spectrum of krypton, excited by sending 714-MeV ions through a carbon foil with a thickness of 600 μg/cm^2. (*Data courtesy of J. A. Leavitt et al.*)

ions emit. Only experiments dealing with the light emission will be considered, since they represent the great majority of beam-foil researches. *See* PARTICLE ACCELERATOR.

Optical spectra. For optical radiations (4 to 700 nm), the wavelength is determined by a spectrometer employing a diffraction grating. More energetic forms of light (x-rays) are studied with energy-sensitive silicon crystals to which some lithium has been added. In both techniques, the radiation is transformed into an electrical pulse which is then amplified and counted, and the number of counts is plotted against the wavelength. Such a graph (a spectrum) contains peaks of various intensity occurring at wavelengths characteristic of the electronic structure of the excited ion. In the example shown in Fig. 1, three spectral lines are identified; the quantum-mechanical notation above each peak describes the origin of the lines.

Mean lives. The mean life of a level provides sensitive information about the quantum nature of the level itself and the mechanism whereby it connects, by either the absorption or emission of light, to some other level. In beam-foil spectroscopy, this mean life is found from observations on the intensity of a particular spectral line as a function of the separation of the emitter and the exciter foil. Beam-foil spectroscopy is presently the only known way of measuring mean lives for highly ionized systems. An example of the technique is shown in Fig. 2, in which the excitation mechanism is the same as in Fig. 1. The symbol τ stands for the mean life, the subscript identifying which electronic level is decaying. The 9.11-nm line in Fig. 2 is in first order, while the corresponding second-order line at 18.21 nm appears in Fig. 1.

Degree of ionization. The degree of ionization which can be achieved depends on the atomic number of the ions and its energy. Early work was restricted to a few elements at the low end of the periodic table and to energies of a few million electronvolts. Subsequently, more than 60 of the 94 naturally occurring elements were used, at ener-

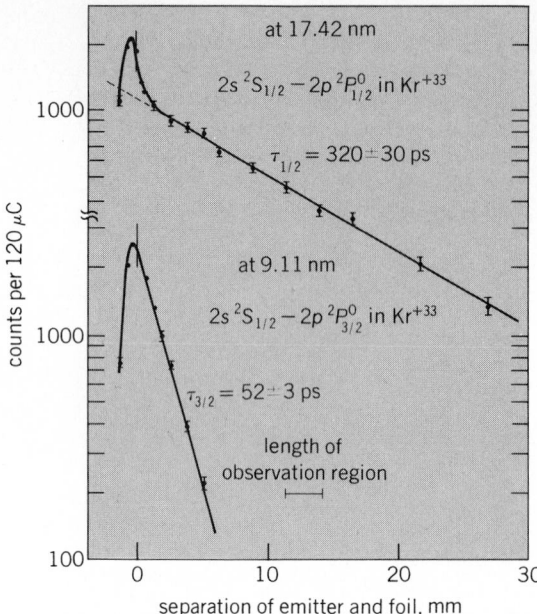

Fig. 2. Lifetime data for two levels in three-electron krypton ions, excited by the same mechanism as in Fig. 1. (*Data courtesy of J. A. Leavitt et al.*)

gies extending to nearly 900 MeV. In the 900-MeV work, ions of rhodium, element 45, were accelerated in the heavy-ion linear accelerator (HILAC) at the Lawrence Berkeley Laboratory of the University of California. Figures 1 and 2 are taken from a similar experiment on krypton, element 36, at 714 MeV. In both cases, the incident particles lost so many electrons that only three-electron (lithiumlike) or four-electron (berylliumlike) ions remained. Studies were made of the spectra (Fig. 1) and mean lives (Fig. 2) reflecting transitions of one of these remaining electrons. To reach the same degree of ionization in a hot plasma would require a temperature of $2.5 \times 10^{7}°C$.

Relativistic effects. The Coulomb force acting on the electrons of highly ionized atoms is so large that they move with speeds that are appreciable fractions of the speed of light; this means that the electronic motions must be treated relativistically. Since such theories are complicated, there is some uncertainty as to which kind of calculation is the best representation of the data; the beam-foil spectroscopy work on spectra and lifetimes is done partly to clarify the theory. Thus the krypton experiment and earlier work on iron, element 26, showed that the so-called dipole-length calculations are to be preferred over the dipole-velocity approach to the level lifetimes. An interesting relativistic effect is that electronic transitions which are forbidden under ordinary circumstances can become dominant over the usual allowed transitions. Such effects have been seen in the work at HILAC. *See* SELECTION RULES.

Lamb shift. A major advance in atomic theory was made when W. E. Lamb, Jr., showed that the Dirac treatment of the hydrogen atom had to be revised. The revision consisted of a small change in the energy levels of a one-electron system, the Lamb shift. Among other things, it is strongly dependent on the atomic number of the ion, and

beam-foil spectroscopy experiments have been done to see if current calculations are applicable to one-electron ions up to chlorine and argon, elements 17 and 18. Preliminary results suggest that the theory, quantum electrodynamics, is satisfactory. *See* QUANTUM ELECTRODYNAMICS.

The Lamb shift also occurs in two-electron ions, and that has been investigated up to oxygen, element 8.

Beam-foil interaction. The fundamental process which gives rise to the excitation phenomenon is still poorly understood. Efforts have been made to develop a semiempirical theory of beam-foil spectroscopy, the basic data being the relative populations of the excited levels as a function of particle energy and element. *See* ATOMIC STRUCTURE AND SPECTRA.   [STANLEY BASHKIN]

Bibliography: D. D. Dietrich et al., Oscillator strengths of the $2s\ ^{2}S_{112}-2p\ ^{2}P^{0}_{1/2,\ 3/2}$ transitions in Fe XXIV and the $2s\ ^{1}S_{0}-2s2p\ ^{3}P_{1}^{0}$ transition in Fe XXIII, *Phys. Rev. A*, 18:208–211, 1978; N. A. Jelley et al., Lamb shift and fine structure in $1s\ 3s\ ^{3}S - 1s\ 3p\ ^{3}P$ transitions in helium-like oxygen, *J. Phys. B: Atom. Molec. Phys.*, 12:2605–2611, 1979; Proceedings of the 5th International Conference on Beam-foil Spectroscopy, *J. Phys.*, C1:1–367, 1979.

Beat

A variation in the intensity of a composite wave which is formed from two distinct waves with different frequencies. Beats were first observed in sound waves, such as those produced by two tuning forks with different frequencies. Beats also can be produced by other waves. They can occur in the motion of two pendulums of different lengths and have been observed among the different-frequency phonons in a crystal lattice.

As a simple example, consider two waves of equal amplitudes A_1 and different frequencies ω_1 and ω_2 in hertz at the same spatial point given by $\psi_1(t) = A_1 \sin(\omega_1 t)$ and $\psi_2(t) = A_1 \sin(\omega_2 t)$. The sum of these waves at this point at an instantaneous time is given by Eq. (1), assuming that they are

$$\psi(t) = \psi_1(t) + \psi_2(t) = A_1 \{\sin \omega_1 t + \sin \omega_2 t\}$$
$$= \{2A_{-1} \cos [\tfrac{1}{2}(\omega_1 - \omega_2)t]\} \sin [\tfrac{1}{2}(\omega_1 + \omega_2)t] \quad (1)$$

coherent. The last form of the wave can be written as $\psi(t) = A \sin(\omega t)$, providing that the amplitude of the composite wave $\psi(t)$ is given by $A = 2A_1 \cos [\tfrac{1}{2}(\omega_1 - \omega_2)t]$ and the frequency is $\omega = \tfrac{1}{2}(\omega_1 + \omega_2)$. The illustration shows the case $A_1 = 1$, $\omega_1 = 8$ Hz, and $\omega_2 = 10$ Hz. The amplitude A of ψ is the enve-

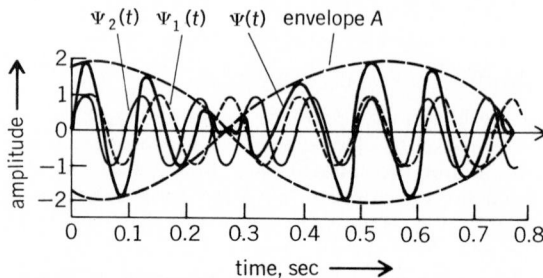

Production of beats. Two waves, $\psi_1(t)$ with frequency $\omega_1 = 8$ Hz and $\omega_2(t)$ with frequency $\omega_2 = 10$ Hz, produce composite wave $\psi(t)$ with amplitude A, an envelope that encloses ψ.

lope enclosing the curve of the total wave. The beat pulsations occur because of this low-frequency envelope.

There is a simple and interesting approximation to Eq. (1) when $\omega_1 - \omega_2 = \epsilon$ is small and short times are considered. The amplitude of $\psi(t)$ is approximately $B \simeq 2A_1 - \frac{1}{2} A_1 \epsilon^2 t^2$, and the composite wave is given by Eq. (2) for sufficiently short times.

$$\psi_{1f}(t) \approx \{2A - \frac{1}{2}A_1 \epsilon^2 t^2\} \sin [\frac{1}{2}(\omega_1 + \omega_2)t] \quad (2)$$

Of course, the beat phenomenon is lost in this limit. Similarly, the wave in Eq. (1) would reduce to Eq. (3), and only the beat frequency would appear

$$\psi_{hf} \approx A_1 \cos [\frac{1}{2}(\omega_1 - \omega_2)t] \quad (3)$$

if the sum frequency $\frac{1}{2}(\omega_1 + \omega_2)$ was too high to be observed.

The individual waves $\psi_1(t)$ and $\psi_2(t)$ could represent a model for monochromatic sound waves, the simple harmonic motion of two pendulums, two acoustic phonons in a crystal, or two electromagnetic waves at a point in space. For sound waves, the beats produce a throbbing sound at a frequency $(\omega_1 - \omega_2)/2$. In the example given in the illustration, $\omega_1 - \omega_2 = 2$ Hz, and one beat per second is produced. This is the envelope in the illustration. If the difference in frequencies is greater than 12–16 Hz, the beats are difficult to distinguish from one another unless the difference frequency is audible, and, in this case, a difference tone is produced. It is also possible to produce audible summation tones. *See* ELECTROMAGNETIC RADIATION; HARMONIC MOTION; LATTICE VIBRATIONS; SOUND.

One application of beat phenomena is to use one object with accurately known frequency ω_1 to determine the unknown frequency ω_2 of another such object. The beat-frequency or heterodyne oscillator also operates by producing beats from two frequencies. [BRIAN DE FACIO]

Bel

A logarithmic unit expressing the ratios of power, voltage, current, or sound intensity. The number of bels separating two power readings is the logarithm to the base 10 of their ratio (for example, two powers differ by 1 bel when their actual ratio is 10:1), while the number of bels separating two current readings or the sound pressures of an acoustical signal is twice the logarithm of the ratio of the currents.

It is convenient in acoustics to express sound intensity in logarithmic units because of the wide range of pressures to which the ear is sensitive. The strength of a sound is usually specified as the square root of the mean of the squares of the instantaneous pressures measured over a period of time. The sound intensity is proportional to the square of the sound pressure. The measure of the level in bels of a sound is $\log (I/I_R)$, where I is the intensity of the sound and I_R is a specified reference intensity. A smaller unit called the decibel, equal to 1/10 bel, is more commonly used. *See* DECIBEL.

[KARL D. KRYTER]

Bibliography: L. L. Beranek, *Acoustics*, 1954; W. A. Van Bergeijk et al., *Waves and the Ear*, 1960; F. A. White, *Our Acoustic Environment*, 1975.

Bernoulli's theorem

The relation between fluid pressure p and fluid velocity v along a streamline in the steady flow of an incompressible, inviscid fluid. Bernoulli's theorem may be written as in Eq. (1), where H is a con-

$$\frac{p}{\rho} + gz + \frac{v^2}{2} = H \quad (1)$$

stant along a streamline, ρ is the mass density of the fluid, g is acceleration due to gravity, and z is vertical height. Equation (1) is essentially an expression of conservation of energy because p/ρ is the pressure energy per unit mass of the fluid, gz is the potential energy per unit mass due to gravity, and $v^2/2$ is the kinetic energy per unit mass. It can also be regarded as an integral of momentum. *See* EULER'S MOMENTUM THEOREM.

If all streamlines come from a space where the fluid is at rest or in uniform motion in a straight line, the Bernoulli constant H is the same on all streamlines. Such flows are irrotational, implying the existence of a velocity potential ϕ such that the fluid velocity is given by $\mathbf{v} = \text{grad } \phi$. For irrotational flows, the unsteady Bernoulli's theorem may be written as in Eq. (2), which holds throughout

$$\frac{\partial \phi}{\partial t} + \frac{p}{\rho} + gz + \frac{v^2}{2} = H(t) \quad (2)$$

the fluid. The additional term $\partial \phi/\partial t$ in Eq. (2) is interpreted as the time rate of change of an impulsive pressure ϕ, which would be necessary to establish the instantaneous flow from rest. If the flow field is steady at some point (such as far away from a moving body in a large expanse of fluid), then H is constant; that is, it is not a function of time. For steady flows, $\partial \phi/\partial t = 0$. If the fluid is compressible, but still inviscid, the internal energy must be added to p/ρ, giving the enthalpy per unit mass, which for a perfect gas is given by Eq. (3),

$$\frac{\gamma}{\gamma - 1} \frac{p}{\rho} = C_p T \quad (3)$$

where $\gamma = C_p/C_v$, C_p is specific heat at constant pressure, C_v is specific heat at constant volume, and T is fluid temperature. It is assumed that $p = p(\rho)$; for a perfect gas this relation is $p = k\rho^\gamma$ where k is a constant. *See* FLUID-FLOW PRINCIPLES; LAPLACE'S IRROTATIONAL MOTION.

[ARTHUR E. BRYSON]

Bessel functions

By definition the solutions of Bessel's differential equation, Eq. (1). Bessel functions, also called cyl-

$$z^2 \, d^2y/dz^2 + z \, dy/dz + (z^2 - \nu^2)y = 0 \quad (1)$$

inder function, are examples of special functions which are introduced by a differential equation. Bessel functions are of great interest in purely mathematical concepts and in mathematical physics. They constitute additional functions which, like the elementary functions z^n, $\sin z$, e^z, can be used to express physical phenomena.

Applications of Bessel functions are found in such representative problems as heat conduction or diffusion in circular cylinders, oscillatory motion of a sphere in a viscous fluid, oscillations of a stretched circular membrane, diffraction of waves

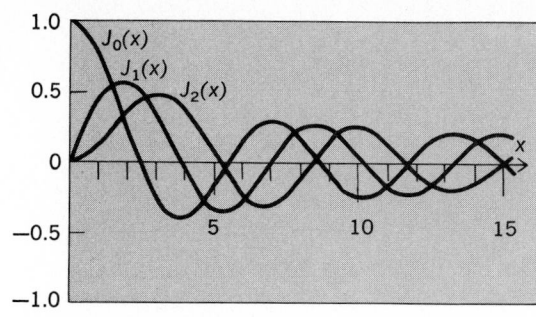

Fig. 1. Bessel functions $J_0(z)$, $J_1(z)$, $J_2(z)$ for $\nu=0,1,2$ and positive values of z.

There exist two relations between Bessel functions whose orders differ by one. If $C_\nu(z)$ stands for any one of the four kinds of functions $J_\nu(z)$, $Y_\nu(z)$, $H_\nu^{(1)}(z)$, $H_\nu^{(2)}(z)$, then Eq. (9) may be written,

$$(2\nu/z)C_\nu(z)=C_{\nu-1}(z)+C_{\nu+1}(z)$$
$$2dC_\nu(z)/dz=C_{\nu-1}(z)-C_{\nu+1}(z) \qquad (9)$$

and hence Eq. (10) holds.

$$z^\nu C_{\nu-1}(z)=d[z^\nu C_\nu(z)]/dz$$
$$z^{-\nu}C_{\nu+1}(z)=-d[z^{-\nu}C_\nu(z)]/dz \qquad (10)$$

Further properties. Some differential equations can be solved in terms of Bessel functions. The differential equation, Eq. (11), has the solutions

$$z^2\frac{d^2y}{dz^2}+(2\alpha-2\nu\beta+1)z\frac{dy}{dz}$$
$$+[\beta^2\gamma^2z^{2\beta}+\alpha(\alpha-2\nu\beta)]y=0 \quad (11)$$

$y=z^{\beta\nu-\alpha}C_\nu(\gamma z^\beta)$, while $d^2y/dz^2+(e^{2z}-\nu^2)y=0$ has the solutions $y=C_\nu(e^z)$. There are series with terms containing Bessel functions, whose sum can be expressed by elementary functions, for example, Eq. (12). In the last formula the functions

$$\exp(iz\cos\theta)=J_0(z)+2\sum_{n=1}^\infty i^nJ_n(z)\cos n\theta$$

$$1=J_0^2(z)+2\sum_{n=1}^\infty J_n^2(z) \qquad (12)$$

$$e^{iuz}=(\pi/2z)^{1/2}\sum_{n=0}^\infty i^n(2n+1)J_{n+1/2}(z)P_n(u)$$

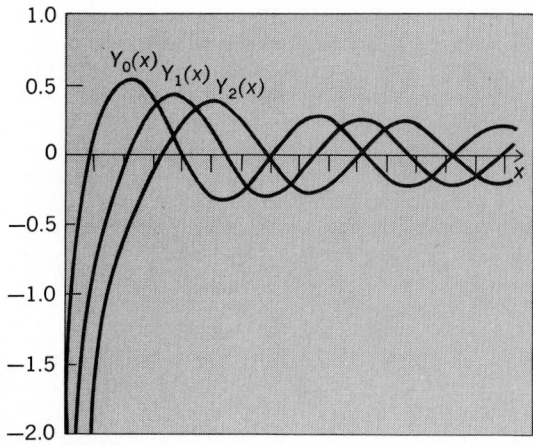

Fig. 2. Neumann functions $Y_0(z)$, $Y_1(z)$, $Y_2(z)$ for $\nu=0,1,2$ and for positive values of z.

$P_n(u)$ are the Legendre polynomials. There are also addition theorems of the type Eq. (13).

$$J_0(x\pm y)=J_0(x)J_0(y)$$
$$+2\sum_{n=1}^\infty(\mp1)^nJ_n(x)J_n(y) \quad (13)$$

Bessel functions are also useful because many definite integrals can be expressed in terms of Bessel functions. Examples are shown as Eq. (14).

$$\pi J_n(z)=i^{-n}\int_0^\pi e^{iz\cos\alpha}\cos n\alpha\,d\alpha \qquad (n=\text{integer})$$

$$\frac{\pi}{2}x^{1/2}[J_{1/3}(2x^{3/2})+J_{-1/3}(2x^{3/2})]$$
$$=\int_0^\infty\cos(t^3-3tx)\,dt \quad (14)$$

The last integral is called Airy's integral. Definite integrals containing Bessel functions can often be expressed by elementary functions, for example, Eq. (15), when the imaginary part of a is between -1 and $+1$.

$$\int_0^\infty e^{-t}J_0(at)\,dt=(1+a^2)^{-1/2} \qquad (15)$$

Expansions in terms of Bessel functions. In applications there is often the problem of expanding a function in terms of Bessel functions or expressing it as an integral with Bessel functions in the integrand.

The Fourier-Bessel series is a generalization of the Fourier series. Let $z_1, z_2, z_3, \ldots$ be the positive zeros of $J_\nu(z)$ for some real $\nu\geq-1/2$, arranged in increasing order of magnitude. Then a continuous function $f(x)$ defined in the interval $0\leq x\leq1$ can be expanded into the Fourier-Bessel series, Eq. (16), where the coefficients a_m are given by Eq. (17).

$$f(x)=\sum_{m=1}^\infty a_mJ_\nu(z_mx) \qquad (0<x<1) \quad (16)$$

$$[J_{\nu+1}(z_m)]^2a_m=\int_0^1 tf(t)J_\nu(z_mt)\,dt \qquad (17)$$

The Fourier-Bessel integral is a generalization of the Fourier integral. Let $f(x)$ be a continuous function, such that the integral $\int_0^\infty f(x)\,dx$ exists and is absolutely convergent. Furthermore, let ν be a real number $\geq-1/2$. Then the function $f(x)$ can be represented by an integral containing Bessel functions, the Fourier-Bessel integral, Eq. (18),

$$f(x)=\int_0^\infty t^{1/2}J_\nu(xt)g(t)\,dt \qquad (18)$$

where $g(t)$, the Fourier-Bessel transform of $f(x)$, is given by Eq. (19). Both representations hold also if $f(x)$ has a finite number of discontinuities, except at points of discontinuity themselves.

$$g(t)=\int_0^\infty x^{1/2}J_\nu(xt)f(x)\,dx \qquad (19)$$

Functions related to Bessel functions. In some applications, Bessel functions occur with a fixed value of arg $z\neq0$, for instance with arg $z=\pi/2$, that is, positive imaginary z. For convenience, special notations are used in such cases. For positive x the functions in Eq. (20) are defined. These func-

$$I_\nu(x) = \exp(-i\nu\pi/2) J_\nu(ix)$$
$$K_\nu(x) = (i\pi/2) \exp(i\nu\pi/2) H_\nu^{(1)}(ix) \qquad (20)$$

tions are called modified Bessel function and modified Hankel function or MacDonald functions of order ν. For real values of $\nu \geqq 0$, the functions $I_\nu(z)$ are monotonically increasing to infinity while the functions $K_\nu(z)$ are monotonically decreasing from infinity to zero as z goes through real values from 0 to ∞.

Particularly in problems for which the use of spherical coordinates is appropriate, there occur Bessel functions with a factor $z^{-1/2}$. They are called spherical Bessel, Neumann, and Hankel functions and defined by Eq. (21). These functions

$$\psi_\nu^{(1)}(z) = (\pi/2z)^{1/2} J_{\nu+1/2}(z)$$
$$\psi_\nu^{(2)}(z) = (\pi/2z)^{1/2} Y_{\nu+1/2}(z)$$
$$\psi_\nu^{(3)}(z) = (\pi/2z)^{1/2} H^{(1)}_{\nu+1/2}(z) \qquad (21)$$
$$\psi_\nu^{(4)}(z) = (\pi/2z)^{1/2} H^{(2)}_{\nu+1/2}(z)$$

can be expressed by elementary functions in finite form, if ν is an integer. In particular, Eq. (22)

$$\psi_0^{(1)}(z) = \frac{\sin z}{z} \qquad \psi_{-1}^{(1)}(z) = \frac{\cos z}{z}$$

$$\psi_1^{(1)}(z) = \frac{-\cos z}{z} + \frac{\sin z}{z^2} \qquad (22)$$

$$\psi_{-2}^{(1)}(z) = \frac{-\sin z}{z} - \frac{\cos z}{z^2}$$

may be written. *See* FOURIER SERIES AND INTEGRALS. [JOSEF MEIXNER]

Bibliography: M. Abramowitz and I. A. Stegun (eds.), *Handbook of Mathematical Functions,* 1964; A. Erdelyi (ed.), *Higher Transcendental Functions,* 3 vols., 1953–1955; A. Erdelyi (ed.), *Tables of Integral Transforms,* 2 vols., 1954; O. J. Farrell and B. Ross, *Solved Problems in Analysis: Gamma, Beta, Legendre and Bessel Functions,* 1963; E. Grosswald, *Bessel Polynomials,* 1979; E. Jahnke and F. Emde, *Tables of Higher Functions,* 4th ed., 1945; G. A. Korn and T. M. Korn, *Mathematical Handbook for Scientists and Engineers,* 2d ed., 1968; Y. L. Luke, *Special Functions and Their Approximations,* 1968–1969; W. Magnus and F. Oberhettinger, *Formulas and Theorems for the Special Functions of Mathematical Physics,* 3d ed., 1966; G. N. Watson, *Treatise on the Theory of Bessel Functions,* 2d ed., 1944.

Biot-Savart law

A law of physics which states that the magnetic flux density (magnetic induction) near a long, straight conductor is directly proportional to the current in the conductor and inversely proportional to the distance from the conductor. There is also a law in fluid dynamics bearing this name; it is concerned with vortex motion, and bears a close analogy to the law discussed in the present article.

The field near a straight conductor can be found by application of Ampère's law. Consider any element dl of the current-carrying conductor (see illustration). The contribution of this element to the flux density at point P is perpendicular to the plane of the paper and directed into the paper. *See* AMPERE'S LAW; MAGNETIC INDUCTION.

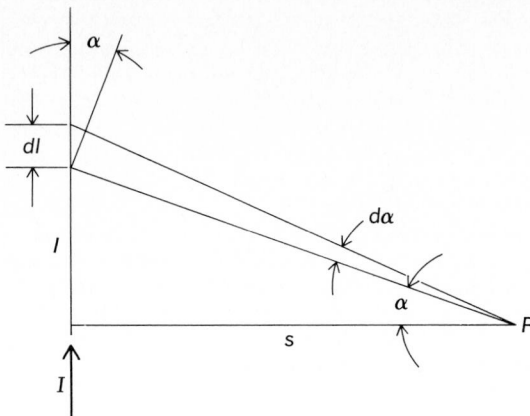

Diagram of the Biot-Savart law.

The flux density B at point P is found by summing up the contributions of all elements of the conductor from one end to the other. All these contributions are in the same direction at P, and hence the vector sum is the simple integral taken from one end of the wire to the other. The limits for the angle α are chosen to represent this sum. For a wire of any length, the lower end is represented by $\alpha = \alpha_1$ and the upper end by $\alpha = \alpha_2$. If I is the current, μ_0 the permeability of free space, and s the distance of P from the conductor, then the flux density is expressed by Eq. (1). For the special case

$$B = \int dB = \frac{\mu_0}{4\pi} \frac{I}{s} \int_{\alpha_1}^{\alpha_2} \cos\alpha \, d\alpha$$
$$= \frac{\mu_0}{4\pi} \frac{I}{s} (\sin\alpha_2 - \sin\alpha_1) \qquad (1)$$

of an infinite straight conductor, that is, one that is long in comparison to s, $\alpha_2 = 90°$ and $\alpha_1 = -90°$. Then Eq. (2) holds. This relation was originally de-

$$B = \frac{\mu_0}{4\pi} \frac{I}{s} [\sin 90° - \sin(-90°)]$$
$$= \frac{\mu_0}{4\pi} \frac{I}{s} (1+1) \quad \text{or} \quad B = \frac{\mu_0}{2\pi} \frac{I}{s} \qquad (2)$$

veloped from experimental observations by J. Biot and F. Savart, and was named after them as the Biot-Savart law.

The magnetic flux density B near the long, straight conductor is at every point perpendicular to the plane determined by the point and the line of the conductor. Therefore, the lines of induction are circles with their centers at the conductor. Furthermore, each line of induction is a closed line. This observation concerning flux about a straight conductor may be generalized to include lines of induction due to a conductor of any shape by the statement that every line of induction forms a closed path. [KENNETH V. MANNING]

Bibliography: W. M. Schwartz, *Intermediate Electromagnetic Theory,* 1964; F. W. Sears et al., *University Physics,* 5th ed., 1976; R. L. Weber et al., *College Physics,* 5th ed., 1974.

Birefringence

The splitting which a wavefront experiences when a wave disturbance is propagated in an anisotropic material; also called double refraction. In anisotropic substances the velocity of a wave is a function of displacement direction. Although the term birefringence could apply to transverse elastic waves, it is usually applied only to electromagnetic waves.

In birefringent materials either the separation between neighboring atomic structural units is different in different directions, or the bonds tying such units together have different characteristics in different directions. Many crystalline materials, such as calcite, quartz, and topaz, are birefringent. Diamonds, on the other hand, are isotropic and have no special effect on polarized light of different orientations. Plastics composed of long-chain molecules become anisotropic when stretched or compressed. Solutions of long-chain molecules become birefringent when they flow. This first phenomenon is called photoelasticity; the second, streaming birefringence.

For each propagation direction with linearly polarized electromagnetic waves, there are two principal displacement directions for which the velocity is different. These polarization directions are at right angles. The difference between the two indices of refraction is also called the birefringence. When the plane of polarization of a light beam does not coincide with one of the two principal displacement directions, the light vector will be split into components parallel to each direction. At the surface of such materials the angle of refraction is different for light polarized parallel to the two principal directions.

For additional information on birefringence and birefringent materials *see* Crystal optics; Polarized light; Refraction of waves.

[BRUCE H. BILLINGS]

Blackbody

An ideal energy radiator, which at any specified temperature emits in each part of the electromagnetic spectrum the maximum energy obtainable per unit time from any radiator due to its temperature alone. A blackbody also absorbs all the energy which falls upon it. The radiation properties of real radiators are limited by two extreme cases—a radiator which reflects all incident radiation, and a radiator which absorbs all incident radiation. Neither case is completely realized in nature. Carbon and soot are examples of radiators which, for practical purposes, absorb all radiation. Both appear black to the eye at room temperature, hence the name blackbody. Often a blackbody is also referred to as a total absorber. Such a total absorber constitutes a standard for the radiation of nonblackbodies, since Kirchhoff's law demands that the blackbody radiate the maximum amount of energy possible in any wavelength interval. For an extended discussion of blackbody radiation and Kirchhoff's law *see* Heat radiation. *See also* Graybody.   [HEINZ G. SELL; PETER J. WALSH]

Bloch theorem

The theorem which states that the wave function of an electron moving in a periodic potential (such as in a crystal lattice) has the form of a plane wave, modulated by a function which has the same periodicity as the potential. That is, the wave function $\psi_n(\mathbf{k},\mathbf{r})$ may be written as in the equation below,

$$\psi_n(\mathbf{k},\mathbf{r}) = e^{i\mathbf{k}\cdot\mathbf{r}}\, u_n(\mathbf{k},\mathbf{r})$$

where $\mathbf{k}$ is the wave vector, n is the band index, and $u_n(\mathbf{k},\mathbf{r})$ is a function which has the periodicity of the potential. *See* Band theory of solids.

[JOSEPH CALLAWAY]

Boltzmann constant

A constant occurring in practically all statistical formulas and having a numerical value of 1.3807×10^{-23} J/K. It is represented by the letter k. If the temperature T is measured from absolute zero, the quantity kT has the dimensions of an energy and is usually called the thermal energy. At 300 K (room temperature) $kT = 0.0259$ electron volt.

The value of the Boltzmann constant may be determined from the ideal gas law. For 1 mole of an ideal gas Eq. (1a) holds, where P is the pressure, V the volume, and R the universal gas constant. The value of R, 8.31 J/K mole, may be obtained from equation-of-state data. Statistical mechanics yields for the gas law Eq. (1b). Here N,

$$PV = RT \tag{1a}$$
$$PV = NkT \tag{1b}$$

the number of molecules in 1 mole, is called Avogadro's number and is equal to 6.02×10^{23} molecules/mole. Hence, comparing Eqs. (1a) and (1b), one obtains Eq. (2). Since k occurs explicitly in the

$$k = R/N = 1.3807 \times 10^{-23}\,\text{J/K} \tag{2}$$

distribution formula, Eq. (2), any quantity calculated using the Boltzmann distribution depends explicitly on k. Examples are specific heat, viscosity, conductivity, and the velocity of sound. Perhaps the most unusual relation involving k is the one between the probability of a state W and the entropy S, given by Eq. (3), which is obtained

$$S = k \ln W \tag{3}$$

by a process of identification similar to the one just described.

Almost any relation derived on the basis of the partition function or the Bose-Einstein, Fermi-Dirac, or Boltzmann distribution contains the Boltzmann constant. *See* Avogadro number; Boltzmann statistics; Bose-Einstein statistics; Fermi-Dirac statistics; Kinetic theory of matter; Statistical mechanics.

[MAX DRESDEN]

Boltzmann statistics

To describe a system consisting of a large number of particles in a physically useful manner, recourse must be had to so-called statistical procedures. If the mechanical laws operating in the system are those of classical mechanics, and if the system is sufficiently dilute, the resulting statistical treatment is referred to as Boltzmann or classical statistics. (Dilute in this instance means that the total volume available is much larger than the proper volume of the particles.) A gas is a typical example: The molecules interacting according to the laws of classical mechanics are the constituents of the system, and the pressure, temperature, and other parameters are the overall entities which determine the macroscopic behavior of the gas. In a case of this kind it is neither possible nor desirable to solve the complicated equations of motion of the molecules; one is not interested in the position and velocity of every molecule at any time. The purpose of the statistical description is to extract from the mechanical description just those features relevant for the determination of the macroscopic properties and to omit others.

Distribution function. The basic notion in the statistical description is that of a distribution function. Suppose a system of N molecules is contained in a volume V. The molecules are moving around, colliding with the walls and with each other. Construct the following geometrical representation of the mechanical system. Introduce a six-dimensional space (usually called the μ space), three of its coordinate axes being the spatial coordinates of the vessel x, y, z, and the other three indicating cartesian velocity components v_x, v_y, v_z. A molecule at a given time, having a specified position and velocity, may be represented by a point in this six-dimensional space. The state of the gas, a system of N molecules, may be represented by a cloud of N points in this space. In the course of time, this cloud of N points moves through the μ space.

Note that the μ space is actually finite; the coordinates x, y, z of the molecules' position are bounded by the finite size of the container, and the velocities are bounded by the total energy of the system. Imagine now that the space is divided into a large number of small cells, of sizes w_1, . . . , w_i, ... A certain specification of the state of the gas is obtained if, at a given time t, the numbers $n_1(t)$, . . . , $n_i(t)$, . . . of molecules in the cells 1, . . . , i, . . . are given. The change in the state of the system in the course of time is now determined by the behavior of the functions $n_i(t)$ as functions of time. Strictly speaking, these functions may change discontinuously in the course of time. Just how detailed the description so obtained is depends, of course, on the sizes chosen for the cells w_i. One gets the crudest possible description by having just one cell of size V in the coordinate subspace, with N particles in it all the time. A very detailed description is obtained if cells the size of a molecule are chosen. In that case even a small change in the state of the molecules will cause a profound alteration in the numbers n_i. To apply statistical methods, one must choose the cells such that on the one hand a cell size w is small compared to the macroscopic dimensions of the system, while on the other hand w must be large enough to allow a large number of molecules in one cell. That this is possible stems from the fact that the molecular dimensions (linear dimension about 10^{-8} cm) are small compared to macroscopic dimensions (typical linear dimension 10^2 cm). In this case it is possible to have about 10^{15} cells, each of linear dimension 10^{-3}, where in each cell there is "room" for 10^{15} molecules. If the cells are thus chosen, the numbers $n_i(t)$, the occupation numbers, will be slowly changing functions of time. The distribution functions $f_i(t)$ are defined by Eq. (1).

$$n_i(t) = f_i(t)w_i \tag{1}$$

The distribution function f_i describes the state of the gas, and f_i of course varies from cell to cell. Since a cell i is characterized by a given velocity range and position range, and since for appropriately chosen cells f should vary smoothly from cell to cell, f is often considered as a continuous function of the variables x, y, z, v_x, v_y, v_z. The cell size w then can be written as $dxdydzdv_x dv_y dv_z$.

In applications the continuous notation is often used; this should not obscure the fact that the cells are finite. L. Boltzmann called them physically infinitesimal.

Since a cell i determines both a position and a velocity range, one may associate an energy ϵ_i with a cell. This is the energy a single molecule possesses when it has a representative point in cell i. This assumes that, apart from instantaneous collisions, molecules exert no forces on each other. If this were not the case, the energy of a molecule would be determined by the positions of all other molecules.

Boltzmann equation; H theorem. Most of the physically interesting quantities follow from a knowledge of the distribution function; the main problem in Boltzmann statistics is to find out what this function is. It is clear that $n_i(t)$ changes in the course of time for three reasons: (1) Molecules located at the position of cell i change their positions and hence move out of cell i; (2) molecules under the influence of outside forces change their velocities and again leave the cell i; and (3) collisions between the molecules will generally cause a (discontinuous) change of the occupation numbers of the cells. Whereas the effect of (1) and (2) on the distribution function follows directly from the mechanics of the system, a separate assumption is needed to obtain the effect of collisions on the distribution function. This assumption, the collision-number assumption, asserts that the number of collisions per unit time, of type $(i,j) \rightarrow (k,l)$ (molecules from cells i and j collide to produce molecules of different velocities which belong to cells k and l), called $A_{ij}{}^{kl}$, is given by Eq. (2). Here

$$A_{ij}{}^{kl} = n_i n_j a_{ij}{}^{kl} \tag{2}$$

$a_{ij}{}^{kl}$ depends on the collision configuration and on the size and kind of the molecules but not on the occupation numbers. Furthermore, for a collision to be possible, the conservation laws of energy and momentum must be satisfied; so if $\epsilon_i' = 1/2m\mathbf{v}_i^2$ and $\mathbf{p}_i = m\mathbf{v}_i$, then Eqs. (3a) and (3b) hold. It is possible

$$\epsilon_i' + \epsilon_j' = \epsilon_k' + \epsilon_l' \tag{3a}$$

$$\mathbf{p}_i + \mathbf{p}_j = \mathbf{p}_k + \mathbf{p}_l \tag{3b}$$

to show that the geometrical factor $a_{ij}{}^{kl}$ has the property given by Eq. (4). Here $a_{kl}{}^{ij}$ is the geo-

$$w_i w_j a_{ij}{}^{kl} = w_k w_l a_{kl}{}^{ij} \tag{4}$$

metrical factor belonging to the collision which, starting from the final velocities (k and l), reproduces the initial ones (i and j). Gains and losses of the molecules in, say, cell i can now be observed. If the three factors causing gains and losses are combined, the Boltzmann equation, written as Eq. (5), is obtained. Here $\Delta_x f_i$ is the gradient of f

$$\frac{\partial f_i}{\partial t} + (\mathbf{v}_i \cdot \Delta_x f_i) + (\mathbf{X}_i \cdot \Delta_v f_i)$$
$$= \sum_{j,k,l} a_{ij}{}^{kl} w_j (f_k f_l - f_i f_j) \tag{5}$$

with respect to the positions, $\Delta_v f_i$ refers similarly to the velocities, and $\mathbf{X}_i$ is the outside force per unit mass at cell i. This nonlinear equation determines the temporal evolution of the distribution function. Exact solutions are difficult to obtain. Yet Eq. (5) forms the basis for the kinetic discussion of most transport processes. There is one remarkable general consequence, which follows from Eq. (5). If one defines $H(t)$ as in Eq. (6),

$$H(t) = \sum_i n_i \ln f_i \qquad (6)$$

one finds by straight manipulation from Eqs. (5) and (6) that Eq. (7) holds. Hence H is a function

$$\frac{dH}{dt} \le 0 \qquad \frac{dH}{dt} = 0 \qquad \text{if } f_i f_j = f_k f_l \qquad (7)$$

which in the course of time always decreases. This result is known as the H theorem. The special distribution which is characterized by Eq. (8) has

$$f_i f_j = f_k f_l \qquad (8)$$

the property that collisions do not change the distribution in the course of time; it is an equilibrium or stationary distribution.

Maxwell-Boltzmann distribution. It should be stressed that the form of the equilibrium distribution may be determined from Eq. (8), with the help of the relations given as Eqs. (3a) and (3b). For a gas which as a whole is at rest, it may be shown that the only solution to functional Eq. (8) is given by Eqs. (9a) or (9b). Here A and β are parameters,

$$f_i = A e^{-\beta \epsilon_i} \qquad (9a)$$

$$f(\mathbf{x}, \mathbf{v}) = A e^{(-1/2)\beta m v^2 - \beta U} \qquad (9b)$$

not determined by Eq. (8), and U is the potential energy at the point x, y, z. Equations (9a) and (9b) are the Maxwell-Boltzmann distribution. Actually A and β can be determined from the fact that the number of particles and the energy of the system are specified, as in Eqs. (10a) and (10b). From

$$\Sigma n_i = N \qquad (10a)$$
$$\Sigma n_i \epsilon_i = E \qquad (10b)$$

Eqs. (9) and (10) it can be shown immediately that Eqs. (11a) and (11b) hold. Therefore β is related

$$\frac{E}{N} = \frac{3}{2\beta} \qquad (11a)$$

$$A = \frac{N}{V} \left(\frac{\beta m}{2\pi} \right)^{3/2} \qquad (11b)$$

directly to the energy per particle while A is related to β and to the number density. Other physical quantities, such as pressure, must now be calculated in terms of A and β. Comparing such calculated entities with those observed, at a certain point one identifies an entity like β (a parameter in the distribution function) with a measured quantity, such as temperature. More precisely $\beta = 1/kT$, where k is the Boltzmann constant. This is the result of an identification. It is not a deduction. It is possible by specialization of Eq. (9b) to obtain various familiar forms of the distribution law. If $U = 0$, one finds immediately that the number of molecules whose absolute value of velocities lies between c and $c + dc$ is given approximately by Eq. (12). Equation (12) is useful in many applications.

$$4\pi V = A e^{-(1/2)\beta m c^2} c^2 \, dc \qquad (12)$$

If, on the other hand, a gas is in a uniform gravitational field so that $U = -mgz$, one finds again from Eqs. (9a) and (9b) that the number of molecules at the height z (irrespective of their veloci-

ties) is given by notation (13). If one uses the fact

$$\text{Constant} \times e^{-\beta mgz} \qquad (13)$$

that $\beta = 1/kT$, then notation (13) expresses the famous barometric density formula, the density distribution of an ideal gas in a constant gravitational field.

Statistical method; fluctuations. The indiscriminate use of the collision-number assumption leads, via the H theorem, to paradoxical results. The basic conflict stems from the irreversible results that appear to emerge as a consequence of a large number of reversible fundamental processes. Although a complete detailed reconciliation between the irreversible aspects of thermodynamics as a consequence of reversible dynamics remains to be given, it is now well understood that a careful treatment of the explicit and hidden probability assumptions is the key to the understanding of the apparent conflict. It is clear, for instance, that Eq. (2) is to be understood either in an average sense (the average number of collisions is given by $\overline{n_i n_j} a_{ij}^{kl}$, where $\overline{n_i n_j}$ is average of the product) or as a probability, but not as a definite causal law. The treatment usually given actually presupposes that $n_i n_j = \overline{n_i n_j}$; that is, it neglects fluctuations.

To introduce probability ideas in a more explicit manner, consider a distribution of N molecules over the cells w_i in μ space, as shown in notation (14). Let the total volume of the μ space be Ω.

$$w_1, w_2, \ldots, w_i, \ldots$$
$$n_1, n_2, \ldots, n_i, \ldots \qquad (14)$$
$$\epsilon_1, \epsilon_2, \ldots, \epsilon_i, \ldots$$

Suppose that N points are thrown into the μ space. What is the probability that just n_1 points will end up in cell 1, n_2 in cell 2, n_i in cell i, and so on? If all points in the space are equally likely (equal a priori probabilities), the probability of such an occurrence is shown in Eq. (15). One now really defines

$$W(n_1, \ldots, n_i, \ldots) = \frac{N!}{n_1! n_2! \cdots n_i! \cdots}$$
$$\left(\frac{w_1}{\Omega} \right)^{n_1} \cdots \left(\frac{w_i}{\Omega} \right)^{n_i} \cdots \qquad (15)$$

the probability of a physical state characterized by notation (14) as given by the essentially geometric probability shown in Eq. (15). The probability of a state so defined is intimately connected with the H function. If one takes the ln of Eq. (15) and applies Stirling's approximation for $\ln n!$, one obtains Eq. (16). The H theorem, which states that H decreas-

$$\ln W = -H \qquad (16)$$

es in the course of time, therefore may be rephrased to state that in the course of time the system evolves toward a more probable state. Actually the H function may be related (for equilibrium states only) to the thermodynamic entropy S by Eq. (17). Here k is, again, the Boltzmann constant.

$$S = -kH \qquad (17)$$

Equations (16) and (17) together yield an important result, Eq. (18), which relates the entropy of a

$$S = k \ln W \qquad (18)$$

state to the probability of that state, sometimes called Boltzmann's relation. *See* PROBABILITY.

The state of maximum probability, that is, the set of occupation numbers $n_1, \ldots, n_i, \ldots$, which maximizes Eq. (15), subject to the auxiliary conditions given in Eqs. (10a) and (10b), turns out to be given by the special distribution written as Eqs. (19a), (19b), and (19c). This again is the

$$n_i = A e^{-\beta \epsilon_i} \qquad (19a)$$

$$\Sigma n_i = N \qquad (19b)$$

$$\Sigma n_i \epsilon_i = E \qquad (19c)$$

Maxwell-Boltzmann distribution. Hence the equilibrium distribution may be thought of as the most probable state of a system. If a system is not in equilibrium, it will most likely (but not certainly) go there; if it is in equilibrium, it will most likely (but not certainly) stay there. By using such probability statements, it may be shown that the paradoxes and conflicts mentioned before may indeed be removed. The general situation is still not clear in all details, although much progress has been made, especially for simple models. A consequence of the probabilistic character of statistics is that the entities computed also possess this characteristic. For example, one cannot really speak definitively of the number of molecules hitting a section of the wall per second, but only about the probability that a given number will hit the wall, or about the average number hitting. In the same vein, the amount of momentum transferred to a unit area of the wall by the molecules per second (this, in fact, is precisely the pressure) is also to be understood as an average. This in particular means that the pressure is a fluctuating entity. In general, it may be shown that the fluctuations in a quantity Q are defined by expression (20), where

$$\frac{\overline{Q^2} - (\overline{Q})^2}{(\overline{Q})^2} \cong \frac{1}{N} \qquad (20)$$

$\overline{Q}=$ the average of Q, and $N=$ number of particles in the systems. When the fluctuations are small, the statistical description as well as the thermodynamic concept is useful. The fluctuations in pressure may be demonstrated by observing the motion of a mirror, suspended by a fiber, in a gas. On the average, as many gas molecules will hit the back as the front of the mirror, so that the average displacement will indeed be zero. However, it is easy to imagine a situation where more momentum is transferred in one direction than in another, resulting in a deflection of the mirror. From the knowledge of the distribution function the probabilities for such occurrences may indeed be computed; the calculated and observed behavior agree very well. This clearly demonstrates the essentially statistical character of the pressure. See BROWNIAN MOVEMENT.

For some important applications of Boltzmann statistics see KINETIC THEORY OF MATTER; for fundamental theory see STATISTICAL MECHANICS. See also BOLTZMANN TRANSPORT EQUATION; QUANTUM STATISTICS.

[MAX DRESDEN]

Bibliography: See STATISTICAL MECHANICS.

Boltzmann transport equation

An equation which is used to study the nonequilibrium behavior of a collection of particles. In a state of equilibrium a gas of particles has uniform composition and constant temperature and density. If the gas is subjected to a temperature difference or disturbed by externally applied electric, magnetic, or mechanical forces, it will be set in motion and the temperature, density, and composition may become functions of position and time; in other words, the gas moves out of equilibrium. The Boltzmann equation applies to a quantity known as the distribution function, which describes this nonequilibrium state mathematically and specifies how quickly and in what manner the state of the gas changes when the disturbing forces are varied. See KINETIC THEORY OF MATTER; STATISTICAL MECHANICS.

Equation (1) is the Boltzmann transport equation, where f is the unknown distribution func-

$$\frac{\partial f}{\partial t} = \left(\frac{\partial f}{\partial t}\right)_{\text{force}} + \left(\frac{\partial f}{\partial t}\right)_{\text{diff}} + \left(\frac{\partial f}{\partial t}\right)_{\text{coll}} \qquad (1)$$

tion which, in its most general form, depends on a position vector $\mathbf{r}$, a velocity vector $\mathbf{v}$, and the time t. The quantity $\partial f/\partial t$ on the left side of Eq. (1) is the rate of change of f at fixed values of $\mathbf{r}$ and $\mathbf{v}$. The equation expresses this rate of change as the sum of three contributions: first, $(\partial f/\partial t)_{\text{force}}$ arises when the velocities of the particles change with time as a result of external driving forces; second, $(\partial f/\partial t)_{\text{diff}}$ is the effect of the diffusion of the particles from one region in space to the other; and third, $(\partial f/\partial t)_{\text{coll}}$ is the effect of collisions of the particles with each other or with other kinds of particles.

The distribution function carries information about the positions and velocities of the particles at any time. The probable number of particles N at the time t within the spatial element $dxdydz$ located at (x,y,z) and with velocities in the element $dv_x dv_y dv_z$ at the point (v_x, v_y, v_z) is given by Eq. (2)

$$N = f(x,y,z,v_x,v_y,v_z,t) dxdydz dv_x dv_y dv_z \qquad (2)$$

or in vector notation, by Eq. (3). It is assumed that

$$N = f(\mathbf{r},\mathbf{v},t) d^3r d^3v \qquad (3)$$

the particles are identical; a different distribution function must be used for each species if several kinds of particles are present.

Specific expressions can be found for the terms on the right side of Eq. (1). Suppose that an external force F_x acts on each particle, producing the acceleration $a_x = F_x/m$ and hence changing the velocity by $\Delta v_x = a_x \Delta t$ in the time interval Δt. If a group of particles has the velocity v_x at time t, the same particles will have the velocity $v_x + \Delta v_x$ at the later time $t + \Delta t$. Therefore the distribution functions $f(v_x, t)$ and $f(v_x + \Delta v_x, t + \Delta t)$ satisfy the equality of Eq. (4), when only the effect of accelera-

$$f(v_x, t) = f(v_x + \Delta v_x, t + \Delta t) \qquad (4)$$

tion a_x is taken into account. Multiplying Eq. (4) by -1 and adding $f(v_x, t + \Delta t)$ to both sides yields Eq. (5).

$$f(v_x, t + \Delta t) - f(v_x, t)$$
$$= -[f(v_x + \Delta v_x, t + \Delta t) - f(v_x, t + \Delta t)] \qquad (5)$$

Multiplying the left side of Eq. (5) by $1/\Delta t$ and the right side by the equal quantity $a_x/\Delta v_x$, Eq. (6)

$$\frac{f(v_x, t + \Delta t) - f(v_x, t)}{\Delta t}$$

$$= -a_x \left[\frac{f(v_x + \Delta v_x, t + \Delta t) - f(v_x, t + \Delta t)}{\Delta v_x} \right] \quad (6)$$

is obtained. In the limit as $\Delta t \to 0$, Eq. (6) becomes Eq. (7). The generalization of this result for acceleration in an arbitrary direction is then given by

$$\left(\frac{\partial f}{\partial t} \right)_{\text{force}} = -a_x \frac{\partial f}{\partial v_x} \quad (7)$$

eration in an arbitrary direction is then given by Eq. (8). The quantity $(\partial f / \partial t)_{\text{force}}$ therefore depends

$$\left(\frac{\partial f}{\partial t} \right)_{\text{force}} = - \left[a_x \frac{\partial f}{\partial v_x} + a_y \frac{\partial f}{\partial v_y} + a_z \frac{\partial f}{\partial v_z} \right] \quad (8)$$

$$= -\mathbf{a} \cdot \frac{\partial f}{\partial \mathbf{v}}$$

on both $\mathbf{a}$, the rate of change of velocity of the particles, and $\partial f / \partial \mathbf{v}$, the variation of the distribution function with velocity.

In a similar way $(\partial f / \partial t)_{\text{diff}}$ depends on both $\mathbf{v}$, the rate of change of position of the particles, and $(\partial f / \partial \mathbf{r})$, the variation of the distribution function with position. One writes $\Delta x = v_x \Delta t$ in place of $\Delta v_x = a_x \Delta t$. Then the form of Eqs. (4)–(7) is unchanged, except that v_x is replaced by x and a_x by v_x. The final expression is given by Eq. (9).

$$\left(\frac{\partial f}{\partial t} \right)_{\text{diff}} = - \left[v_x \frac{\partial f}{\partial x} + v_y \frac{\partial f}{\partial y} + v_z \frac{\partial f}{\partial z} \right] \quad (9)$$

$$= -\mathbf{v} \cdot \frac{\partial f}{\partial \mathbf{r}}$$

If Eqs. (8) and (9) are substituted into Eq. (1), one gets Eq. (10), which is the usual form of the Boltz-

$$\frac{\partial f}{\partial t} + \mathbf{a} \cdot \frac{\partial f}{\partial \mathbf{v}} + \mathbf{v} \cdot \frac{\partial f}{\partial \mathbf{r}} = \left(\frac{\partial f}{\partial t} \right)_{\text{coll}} \quad (10)$$

mann equation. Before it can be solved for f, a specific expression must be found for $(\partial f / \partial t)_{\text{coll}}$, the rate of change of $f(\mathbf{r}, \mathbf{v}, \mathbf{t})$ due to collisions. The calculation of $(\partial f / \partial t)_{\text{coll}}$ begins with a mathematical description of the forces acting between particles. Knowing the type of statistics obeyed by the particles (that is, Fermi-Dirac, Bose-Einstein, or Maxwell-Boltzmann), the manner in which the velocities of the particles are changed by collisions can then be determined. The term $(\partial f / \partial t)_{\text{coll}}$ is expressed as the difference between the rate of scattering from all possible velocities $\mathbf{v}'$ to the velocity $\mathbf{v}$, and the rate of scattering from $\mathbf{v}$ to all possible $\mathbf{v}'$. For example, if the particles are electrons in a metal or semiconductor and if they are scattered elastically by imperfections in the solid, it is found that $(\partial f / \partial t)_{\text{coll}}$ obeys Eq. (11), where $W_{vv'}$ is the

$$\left(\frac{\partial f}{\partial t} \right)_{\text{coll}} = \sum_{\mathbf{v}'} W_{vv'} f(\mathbf{r}, \mathbf{v}', t) - \sum_{\mathbf{v}'} W_{vv'} f(\mathbf{r}, \mathbf{v}, t) \quad (11)$$

rate of scattering of one particle from $\mathbf{v}$ to $\mathbf{v}'$ or the reverse. A basic requirement for obtaining a meaningful expression for $(\partial f / \partial t)_{\text{coll}}$ in this way is that the duration of a collision be small compared to the time between collisions. For a gas of atoms or molecules this condition implies that the gas be dilute, and for electrons in a solid it implies

that the concentration of imperfections must not be too high.

Discussion. The Boltzmann expression, Eq. (10), with a collision term of the form in Eq. (11), is irreversible in time in the sense that if $f(\mathbf{r}, \mathbf{v}, t)$ is a solution then $f(\mathbf{r}, -\mathbf{v}, -t)$ is not a solution. Thus if an isolated system is initially not in equilibrium, it approaches equilibrium as time advances; the time-reversed performance, in which the system departs farther from equilibrium, does not occur. The Boltzmann equation therefore admits of solutions proceeding toward equilibrium but not of time-reversed solutions departing from equilibrium. This is paradoxical because actual physical systems are reversible in time when looked at on an atomic scale. For example, in a classical system the time enters the equations of motion only in the acceleration $d^2\mathbf{r}/dt^2$, so if t is replaced by $-t$ in a solution, a new solution is obtained. If the velocities of all particles were suddenly reversed, the system would retrace its previous behavior.

An actual system does not necessarily move toward equilibrium, although it is overwhelmingly probable that it does so. From a mathematical point of view it is puzzling that one can begin with the exact equations of motion, reversible in time, and by making reasonable approximations arrive at the irreversible Boltzmann equation. The resolution of this paradox lies in the statistical nature of the Boltzmann equation. It does not describe the behavior of a single system, but the average behavior of a large number of systems. Mathematically, the irreversibility arises from the collision term, Eq. (11), where the approximation has been made that the distribution function $f(\mathbf{r}, \mathbf{v}, t)$ or $f(\mathbf{r}, \mathbf{v}', t)$ applies to particles both immediately before and immediately after a collision.

A number of equations closely related to the Boltzmann equation are often useful for particular applications. If collisions between particles are disregarded, the right side of the Boltzmann equation, Eq. (10), is zero. The equation is then called the collisionless Boltzmann equation or Vlasov equation. This equation has been applied to a gas of charged particles, also known as a plasma. The coulomb forces between particles have such a long range that it is incorrect to consider the particles as free except when colliding. The Vlasov equation can be used by including the forces between particles in the term $\mathbf{a} \cdot (\partial f / \partial \mathbf{v})$. One takes $\mathbf{a} = q\mathbf{E}/m$, q being the charge of a particle and m the mass. The electric field $\mathbf{E}$ includes the field produced by the particles themselves, in addition to any externally produced field.

The Boltzmann equation is a starting point from which the equations of hydrodynamics, Eqs. (12a) and (12b), can be derived. The particles are now

$$\frac{\partial f}{\partial t} + \frac{\partial}{\partial \mathbf{r}} \cdot (\rho \mathbf{v}) = 0 \quad (12a)$$

$$\rho \left(\frac{\partial \mathbf{v}}{\partial t} + \mathbf{v} \cdot \frac{\partial \mathbf{v}}{\partial \mathbf{r}} \right) = \mathbf{F} - \frac{\partial}{\partial \mathbf{r}} \cdot \mathbf{p} \quad (12b)$$

considered as a continous fluid with density ρ and mean velocity $\mathbf{v}$. $\mathbf{F}$ is the external force per unit volume and $\mathbf{p}$ is the pressure tensor. Equation (12a) states mathematically that the mass of the fluid is conserved, while Eq. (12b) equates the rate of change of momentum of an element of fluid to the force on it. An energy conservation equation can also be derived. *See* HYDRODYNAMICS; NAVIER-STOKES EQUATIONS.

Applications. The Boltzmann equation can be used to calculate the electronic transport properties of metals and semiconductors. For example, if an electric field $\mathbf{E}$ is applied to a solid, one must solve the Boltzmann equation for the distribution function $f(\mathbf{r},\mathbf{v},t)$ of the $\mathbf{E}$ electrons, taking the acceleration in Eq. (10) as $\mathbf{a}=q\mathbf{E}/m$ and setting $\partial f/\partial \mathbf{r}=0$, corresponding to spatial uniformity of the electrons. If the electric field is constant, the distribution function is also constant and is displaced in velocity space in such a way that fewer electrons are moving in the direction of the field than in the opposite direction. This corresponds to a current flow in the direction of the field. The relationship between the current density $\mathbf{J}$ and the field $\mathbf{E}$ is given by Eq. (13), where σ, the electrical

$$\mathbf{J} = \sigma\,\mathbf{E} \tag{13}$$

conductivity, is the final quantity of interest. *See* FREE-ELECTRON THEORY OF METALS.

With the Boltzmann equation one can also calculate the heat current flowing in a solid as the result of a temperature difference, the constant of proportionality between the heat current per unit area and the temperature gradient being the thermal conductivity. In still more generality, both an electric field $\mathbf{E}$ and a temperature gradient $\partial T/\partial \mathbf{r}$ can be applied, where T is the temperature. Expressions are obtained for the electrical current density $\mathbf{J}$ and the heat current density $\mathbf{U}$ in the form of Eqs. (14a) and (14b). L_{11} is the electrical conductiv-

$$\mathbf{J} = L_{11}\,\mathbf{E} + L_{12}\partial T/\partial \mathbf{r} \tag{14a}$$

$$\mathbf{U} = L_{21}\,\mathbf{E} + L_{22}\partial T/\partial \mathbf{r} \tag{14b}$$

ity and $-L_{22}$ the thermal conductivity. Equations (14a) and (14b) also describe thermoelectric phenomena, such as the Peltier and Seebeck effects. For example, if a thermal gradient is applied to an electrically insulated solid so that no electric current can flow ($\mathbf{J}=0$), Eq. (14a) shows that an electric field given by Eq. (15) will appear. The quantity $-(L_{12}/L_{11})$ is called the Seebeck coefficient.

$$\mathbf{E} = -(L_{12}/L_{11})\,(\partial T/\partial \mathbf{r}) \tag{15}$$

Finally, if a constant magnetic field $\mathbf{B}$ is also applied, the coefficients L_{ij} in Eqs. (14a) and (14b) become functions of $\mathbf{B}$. It is found that the electrical conductivity usually decreases with increasing $\mathbf{B}$, a behavior known as magnetoresistance. These equations also describe the Hall effect, the appearance of an electric field in the y direction if there is an electric current in the x direction and a magnetic field in the z direction, as well as more complex thermomagnetic phenomena, such as the Ettingshausen and Nernst effects. The conductivity σ in Eq. (13) and the L_{ij} in Eqs. (14a) and (14b) are tensor quantities in many materials. If the field $\mathbf{E}$ and the temperature gradient $\partial T/\partial \mathbf{r}$ are in a given direction, the currents $\mathbf{U}$ and $\mathbf{J}$ are in such a case not necessarily in the same direction. *See* CONDUCTION (ELECTRICITY); CONDUCTION (HEAT); GALVANOMAGNETIC EFFECTS; HALL EFFECT; MAGNETORESISTANCE; THERMOELECTRICITY; THERMOMAGNETIC EFFECTS.

Nonequilibrium properties of atomic or molecular gases such as viscosity, thermal conduction, and diffusion have been treated with the Boltzmann equation. Although many useful results, such as the independence of the viscosity of a gas on pressure, can be obtained by simple approximate methods, the Boltzmann equation must be used in order to obtain quantitatively correct results. *See* VISCOSITY.

If one proceeds from a neutral gas to a charged gas or plasma, with the electrons partially removed from the atoms, a number of new phenomena appear. As a consequence of the long-range coulomb forces between the charges, the plasma can exhibit oscillations in which the free electrons move back and forth with respect to the relatively stationary heavy positive ions at the characteristic frequency known as the plasma frequency. This frequency is proportional to the square root of the particle density. If the propagation of electromagnetic waves through a plasma is studied, it is found that a plasma reflects an electromagnetic wave at a frequency lower than the plasma frequency, but transmits the wave at a higher frequency. This fact explains many characteristics of long-distance radio transmission, made possible by reflection of radio waves by the ionosphere, a low-density plasma surrounding the Earth at altitudes greater than 40 mi. A plasma also exhibits properties such as electrical conductivity, thermal conductivity, viscosity, and diffusion. *See* PLASMA PHYSICS.

If a magnetic field is applied to the plasma, its motion can become complex. A type of wave known as an Alfvén wave propagates in the direction of the magnetic field with a velocity proportional to the field strength. The magnetic field lines are not straight, however, but oscillate like stretched strings as the wave passes through the plasma. Waves that propagate in a direction perpendicular to the magnetic field have quite different properties.

When the plasma moves as a fluid, it tends to carry the magnetic field lines with it. The plasma becomes partially trapped by the magnetic field in such a way that it can move easily along the magnetic field lines, but only very slowly perpendicular to them. The outstanding problem in the attainment of a controlled thermonuclear reaction is to design a magnetic field configuration that can contain an extremely hot plasma long enough to allow nuclear reactions to take place. *See* MAGNETOHYDRODYNAMICS.

Plasmas in association with magnetic fields occur in many astronomical phenomena. Many of the events occurring on the surface of the Sun, such as sunspots and flares, as well as the properties of the solar wind, a dilute plasma streaming out from the Sun in all directions, are manifestations of the motion of a plasma in a magnetic field. It is believed that a plasma streaming out from a rapidly rotating neutron star will radiate electromagnetic energy; this is a possible explanation of pulsars, stars emitting regularly pulsating radio signals that have been detected with radio telescopes.

Many properties of plasmas can be calculated by studying the motion of individual particles in electric and magnetic fields, or by using the hydrodynamic equations, Eqs. (12a) and (12b), or the Vlasov equation, together with Maxwell's equations. However, subtle properties of plasmas, such as diffusion processes and the damping of waves, can best be understood by starting with the Boltzmann equation or the closely related Fokker-Planck equation. *See* MAXWELL'S EQUATIONS.

Relaxation time. It is usually difficult to solve the Boltzmann equation, Eq. (10), for the distribution function $f(\mathbf{r},\mathbf{v},t)$ because of the complicated form of $(\partial f/\partial t)_{\text{coll}}$. This term is simplified if one makes the relaxation time approximation shown in Eq. (16), where $f_0(v)$ is the equilibrium distribution

$$\left(\frac{\partial f}{\partial t}\right)_{\text{coll}} = -\frac{f(\mathbf{r},\mathbf{v},t) - f_0(\mathbf{v})}{\tau(v)} \quad (16)$$

function and $\tau(v)$, the relaxation time, is a characteristic time describing the return of the distribution function to equilibrium when external forces are removed. For some systems Eq. (16) follows rigorously from Eq. (11), but in general it is an approximation. If the relaxation time approximation is not made, the Boltzmann equation usually cannot be solved exactly, and approximate techniques such as variational, perturbation, and expansion methods must be used. An important situation in which an exact solution is possible occurs if the particles exert forces on each other that vary as the inverse fifth power of the distance between them. *See* RELAXATION TIME OF ELECTRONS.

Example. As an illustration of how the Boltzmann equation can be solved, consider a spatially uniform electron gas in a metal or a semiconductor. Let the scattering of electrons be described by Eq. (16). One can calculate the electric current which flows when a small, constant electric field E_x is applied. The spatial uniformity of the electron gas implies that $\partial f/\partial \mathbf{r} = 0$; since the electric field is constant, one requires also that the distribution function be constant ($\partial f/\partial t = 0$) so that the current does not depend on time. If one writes the acceleration of the electron as $a_x = F_x/m = -eE_x/m$, where $(-e)$ is the charge and m is the mass of the electron, Eq. (10) becomes Eq. (17). The difference

$$\frac{-eE_x}{m}\frac{\partial f}{\partial v_x} = -\frac{f(\mathbf{v}) - f_0(\mathbf{v})}{\tau(v)} \quad (17)$$

between f and f_0 is small; therefore f can be replaced by f_0 in the term containing E_x, since E_x itself is small. Thus Eq. (18) is obtained,

$$\partial f/\partial v_x \approx \partial f_0/\partial v_x = (\partial f_0/\partial v)(\partial v/\partial v_x)$$
$$= (\partial f_0/\partial v)(v_x/v) \quad (18)$$

where $v = (v_x^2 + v_y^2 + v_z^2)^{1/2}$ is the magnitude of the electron velocity; Eq. (17) then becomes Eq. (19).

$$f(\mathbf{v}) = f_0(\mathbf{v}) + \frac{eE_x\tau}{m}\frac{v_x}{v}\frac{\partial f_0}{\partial v} \quad (19)$$

Here $f_0(\mathbf{v})$ is the equilibrium distribution function describing the state of the electrons in the absence of an electric field. For a high electron density, as in metals, $f_0(\mathbf{v})$ is the Fermi-Dirac distribution function given by Eq. (20), where ζ is the Fermi

$$f_0(\mathbf{v}) \propto \frac{1}{e^{(mv^2/2 - \zeta)/kT} + 1} \quad (20)$$

energy, k is the Boltzmann constant, and T is the absolute temperature. In writing $f_0(\mathbf{v})$ in the form of Eq. (20), the simplifying assumption that the energy of an electron is $mv^2/2$ has been made. If the electron density is low, as it often is in semiconductors, $f_0(\mathbf{v})$ can be approximated by the Maxwell-

Boltzmann distribution given in Eq. (21). In either

$$f_0(\mathbf{v}) \propto e^{-mv^2/2kT} \quad (21)$$

case $f_0(\mathbf{v})$ depends only on the magnitude and not on the direction of the velocity. *See* FERMI-DIRAC STATISTICS; BOLTZMANN STATISTICS.

Therefore $f_0(\mathbf{v})$ is centered at $\mathbf{v} = 0$ in such a way that the net flow of electrons in every direction is zero. The nonequilibrium distribution function $f(\mathbf{v})$ is similar to $f_0(\mathbf{v})$, except that it is shifted in the $(-v_x)$ direction in velocity space. Consequently more electrons move in the $(-x)$ direction than in the $(+x)$ direction, producing a net current flow in the $(+x)$ direction. The current density J_x is calculated from $f(\mathbf{v})$ by Eq. (22), where the spatial inte-

$$J_x = -e \int v_x f(\mathbf{v}) \, d^3v\,d^3x \quad (22)$$

gral is to be taken over a unit volume, so that $\int d^3x = 1$. If $f(\mathbf{v})$, as given by Eq. (19) is substituted into Eq. (22), the term involving the equilibrium distribution function $f_0(\mathbf{v})$ by itself gives zero current, since no current flows at equilibrium. Hence one obtains Eq. (23).

$$J_x = \frac{e^2 E_x}{m} \int \frac{\tau(v) v_x^2}{v}\left(-\frac{\partial f_0}{\partial v}\right) d^3v \quad (23)$$

If n, the density of electrons, is introduced using the equation $n = \int f_0(\mathbf{v})d^3v$, Eq. (23) can be written as Eq. (24), where σ is given by Eq. (25), and $\bar{\tau}$, given by Eq. (26), is an average relaxation time.

$$J_x = \sigma E_x \quad (24)$$

$$\sigma = ne^2\bar{\tau}/m \quad (25)$$

$$\bar{\tau} = \frac{\int \tau \dfrac{v_x^2}{v}\left(-\dfrac{\partial f_0}{\partial v}\right) d^3v}{\int f_0 \, d^3v} \quad (26)$$

Equation (25) is the final expression for the electrical conductivity σ.

The above treatment is applicable only if the electrons in the solid behave as if they were nearly free. In general, the wave vector $\mathbf{k}$, rather than the velocity $\mathbf{v}$, must be used as a variable in the distribution function, and the final expression for σ depends on the electronic band structure of the solid. [RONALD FUCHS]

Bibliography: S. Chapman and T. G. Cowling, *The Mathematical Theory of Non-Uniform Gases*, 3d ed., 1970; W. B. Thompson, *An Introduction to Plasma Physics*, 2d ed., 1964; J. M. Ziman, *Electrons and Phonons*, 1960.

Boolean algebra

A branch of mathematics that was first developed systematically, because of its applications to logic, by the English mathematician George Boole, around 1850. Closely related are its applications to sets and probability.

Boolean algebra also underlies the theory of relations. A modern engineering application is to computer circuit design.

Set-theoretic interpretation. Most basic is the use of Boolean algebra to describe combinations of the subsets of a given set I of elements; its basic operations are those of taking the intersection or common part $S \cap T$ of two such subsets S and T, their union or sum $S \cup T$, and the complement S' of any one such subset S. These operations satisfy many laws, including those shown in Eqs. (1), (2), and (3).

$$S \cap S = S \qquad S \cap T = T \cap S$$
$$S \cap (T \cap U) = (S \cap T) \cap U \qquad (1)$$

$$S \cup S = S \qquad S \cup T = T \cup S$$
$$S \cup (T \cup U) = (S \cup T) \cup U \qquad (2)$$

$$S \cap (T \cup U) = (S \cap T) \cup (S \cap U)$$
$$S \cup (T \cap U) = (S \cup T) \cap (S \cup U) \qquad (3)$$

If O denotes the empty set, and I is the set of all elements being considered, then the laws set forth in Eq. (4) are also fundamental. Since these laws

$$O \cap S = O \qquad O \cup S = S \qquad I \cap S = S$$
$$I \cup S = I \qquad S \cap S' = O \qquad S \cup S' = I \qquad (4)$$

are fundamental, all other algebraic laws of subset combination can be deduced from them.

In applying Boolean algebra to logic, Boole observed that combinations of properties under the common logical connectives *and*, *or*, and *not* also satisfy the laws specified above. These laws also hold for propositions or assertions, when combined by the same logical connectives.

Boole stressed the analogies between Boolean algebra and ordinary algebra. If $S \cap T$ is regarded as playing the role of st in ordinary algebra, $S \cup T$ that of $s + t$, O of 0, I of 1, and S' as corresponding to $1 - s$, the laws listed above illustrate many such analogies. However, as first clearly shown by Marshall Stone, the proper analogy is somewhat different. Specifically, the proper Boolean analog of $s + t$ is $(S' \cap T) \cup (S \cap T')$, so that the ordinary analog of $S \cup T$ is $s + t - st$. Using Stone's analogy, Boolean algebra refers to Boolean rings in which $s^2 = s$, a condition implying $s + s = 0$. *See* SET THEORY.

Boolean algebra arises in other connections, as in the algebra of (binary) relations. Such relations $\rho, \sigma, \ldots$ refer to appropriate sets of elements I, $J, \ldots$. Any such ρ can be defined by describing the set of pairs (x, t), with x in I and y in J, that stand in the given relation—a fact symbolized $x \rho y$, just as its negation is written $x \rho' y$. Because of this set-theoretic interpretation, Boolean algebra obviously applies, with $x(\rho \cap \sigma)y$ meaning $x \rho y$ and $x \sigma y$, and $x(\rho \cup \sigma)y$ meaning $x \rho y$ or $x \sigma y$.

Abstract relationships. Before 1930, work on Boolean algebra dealt mainly with its postulate theory, and with the generalizations obtained by abandoning one or more postulates, such as $(p')' = p$ (Brouwerian logic). Since $a \cup b = (a' \cap b')'$, clearly one need consider $a \cap b$ and a' as undefined operations. In 1913 H. M. Sheffer showed one operation only $(a \mid b = a' \cap b')$ need be taken as undefined. In 1941 M. H. A. Newman developed a remarkable generalization which included Boolean algebras and Boolean rings. This generalization is based on the laws shown in Eqs. (5) and (6). From these assumptions, the idempo-

$$a(b+c) = ab + ac \qquad (a+b)c = ac + bc \qquad (5)$$

$$a1 = 1 \qquad a + 0 = 0 + a = a \qquad aa' = 0 \qquad (6)$$
$$a + a' = 1$$

tent, commutative, and associative laws (1) and (2) can be deduced.

Such studies lead naturally to the concept of an abstract Boolean algebra, defined as a collection of symbols combined by operations satisfying the identities listed in formulas (1) to (4). Ordinarily, the phrase Boolean algebra refers to such an abstract Boolean algebra, and this convention is adopted here.

The class of finite (abstract) Boolean algebras is easily described. Each such algebra has, for some nonnegative integer n, exactly 2^n elements and is algebraically equivalent (isomorphic) to the algebra of all subsets of the set of numbers $1, \ldots, n$, under the operations of intersection, union, and complement. Further, if m symbols $a_1, \ldots, a_m$ are combined symbolically through abstract operations $\cap$, $\cup$, and $'$ assumed to satisfy the identities of Eqs. (1) to (4), one gets a finite Boolean algebra with 2^{2m} elements—the free Boolean algebra with m generators.

Infinite relationships. The theory of infinite Boolean algebras is much deeper; it indirectly involves the whole theory of sets. One important result is Stone's representation theorem. Let a field of sets be defined as any family of subsets of a given set I, which contains with any two sets S and T their intersection $S \cap T$, union $S \cup T$, and complements S', T'. Considered abstractly, any such field of sets obviously defines a Boolean algebra. Stone's theorem asserts that, conversely, any finite or infinite abstract Boolean algebra is isomorphic to a suitable field of sets. His proof is based on the concepts of ideal and prime ideal, concepts which have been intensively studied for their own sake. Because ideal theory in Boolean algebra may be subsumed under the ideal theory of rings (via the correspondence between Boolean algebras and Boolean rings mentioned earlier), it will not be discussed here. A special property of Boolean rings (algebras) is the fact that, in this case, any prime ideal is maximal.

The study of infinite Boolean algebras leads naturally to the consideration of such infinite distributive laws as those in Eqs. (7) and (7').

$$x \cap \left(\bigcup_B y_\beta \right) = \bigcup_B (x \cap y_\beta)$$
$$x \cup \left(\bigcap_B y_\beta \right) = \bigcap_B (x \cup y_\beta) \qquad (7)$$

$$\bigcap_C \left[\bigcup_{A\gamma} u_{\gamma,\alpha} \right] = \bigcup_F \left[\bigcap_C u_{\gamma,\phi(\gamma)} \right]$$
$$\bigcup_C \left[\bigcap_{A\gamma} u_{\gamma,\alpha} \right] = \bigcap_F \left[\bigcup_C u_{\gamma,\phi(\gamma)} \right] \qquad (7')$$

For finite sets B of indices $\beta = 1, \ldots, n$, if $\bigcup_B y\beta$ means $y_1 \cup \cdots \cup y_n$, and so on, the laws (7) and (7') follow by induction from (1) to (3). Also, if the symbols x, y_β, and so on, in (7) and (7') refer to subsets of a given space I, and if $\bigcup_B y_\beta$ and $\bigcap_B y_\beta$ refer to the union and intersection of all y_β in B, respectively, then (7) and (7') are statements of general laws of formal logic. However, they fail in most infinite Boolean algebras. This is shown by the following result of Alfred Tarski: If a Boolean algebra A satisfies the generalized distributive laws (7) and (7'), then it is isomorphic with the algebra of all subsets of a suitable space I. A related result is the theorem of L. Loomis (1947):

Every σ-complete Boolean algebra is isomorphic with a σ-field of sets under countable intersection, countable union, and complement.

In general, such completely distributive Boolean algebras of subsets may be characterized by the properties of being complete and atomic. These properties may be defined roughly as the properties that (a) there exists a smallest element $\bigcup_B y_\beta$ containing any given set B of elements y_β, and (b) any element $y > 0$ contains an atom (or point) $p > 0$, such that $p > x > 0$ has no solution (from Euclid, "A point is that which has no parts"). Condition (b) is also implied by the "descending chain condition" of ideal theory.

Other forms. Nonatomic and incomplete Boolean algebras arise naturally in set theory. Thus, the algebra of measurable sets in the line or plane, ignoring sets of measure zero, is nonatomic but complete. The field of Borel sets of space is complete as regards countable families B of subsets S_β, but not for uncountable B. Analogous results hold for wide classes of other measure spaces and topological spaces, respectively. In any zero-dimensional compact space, the sets which are both open and closed (which "disconnect" the space) form a Boolean algebra; a fundamental result of Stone shows that the most general Boolean algebra can be obtained in this way.

Many other interesting facts about Boolean algebra are known. For instance, there is an obvious duality between the properties of $\cap$ and $\cup$ in the preceding discussion. However, so many such facts have natural generalizations to the wider context of lattice theory that the modern tendency is to consider Boolean algebra as it relates to such generalizations. For instance, the algebras of n-valued logic, intuitionist (Brouwerian) logic, and quantum logic are not Boolean algebras, but lattices of other types. The same is true of the closed sets in most topological spaces. *See* TOPOLOGY.

[GARRETT BIRKHOFF]

Bibliography: G. Birkhoff, *Lattice Theory*, 3d ed., Amer. Math. Soc. Colloq. Publ., vol. 25, 1967; G. Birkhoff and S. MacLane, *Survey of Modern Algebra*, 4th ed., 1977; G. Boole, *An Investigation of the Laws of Thought*, 1854; R. Sikorski, *Boolean Algebras*, 3d ed., 1969.

Bose-Einstein statistics

The statistical description of quantum mechanical systems in which there is no restriction on the way in which particles can be distributed over the individual energy levels. This description applies when the system has a symmetric wave function. This in turn has to be the case when the particles described are of integer spin.

Distribution probability. Suppose one describes a system by giving the number of particles n_i in an energy state ϵ_i, where the n_i are called occupation numbers and the index i labels the various states. The energy level ϵ_i is of finite width, being really a range of energies comprising, say, g_i individual (nondegenerate) quantum levels. If any arrangement of particles over individual energy levels is allowed, one obtains for the probability of a specific distribution, Eq. (1a). In Boltzmann statistics, this same probability would be written as Eq. (1b). *See* BOLTZMANN STATISTICS.

$$W = \prod_i \frac{(n_i + g_i - 1)!}{n_i!(g_i - 1)!} \tag{1a}$$

$$W = \prod_i \frac{g_i^{n_i}}{n_i!} \tag{1b}$$

The equilibrium state is defined as the most probable state of the system. To obtain it, one must maximize Eq. (1a) under the conditions given by Eqs. (2a) and (2b), which express the fact that

$$\Sigma n_i = N \tag{2a}$$
$$\Sigma \epsilon_i n_i = E \tag{2b}$$

the total number of particles N and the total energy E are fixed. One finds for the most probable distribution that Eq. (3) holds. Here, A and β are parameters to be determined from Eqs. (2a) and (2b);

$$n_i = \frac{g_i}{\dfrac{1}{A} e^{\beta \epsilon_i} - 1} \tag{3}$$

actually, $\beta = 1/kT$, where k is the Boltzmann constant and T is the absolute temperature.

In the classical case, that is, when Boltzmann statistics is employed, the equilibrium distribution may be obtained from a specific assumption about the number of collisions of a certain kind. One assumes that the number of collisions per second in which molecules with velocities in cells i and j in phase space produce molecules with velocities in cells k and l is given by Eq. (4), where $a_{ij}^{\,kl}$ is a

$$A_{ij}^{\,kl} = n_i n_j a_{ij}^{\,kl} \tag{4}$$

geometrical factor. In the Bose case, Eq. (3) may be obtained in a similar way from a collision number assumption, which is written as Eq. (5).

$$A_{ij}^{\,kl} = a_{ij}^{\,kl} n_i n_j \left(\frac{g_k + n_k}{g_k}\right)\left(\frac{g_l + n_l}{g_l}\right) \tag{5}$$

One observes the interesting fact that the number of collisions depends on the number of particles in the state to which the colliding particles are going. The more heavily populated these states are, the more likely a collision is. Quite often one defines f_i, the distribution function, by Eq. (6).

$$n_i = f_i g_i \tag{6}$$

Applications. An interesting and important result emerges when one applies Eq. (3) to a gas of photons, that is, a large number of photons in an enclosure. (Since photons have integer spin, this is legitimate.) For photons one has Eqs. (7a) and (7b),

$$\epsilon = h\nu \tag{7a}$$
$$p = \frac{h\nu}{c} \tag{7b}$$

where h is the Planck constant, ν the frequency of the photon, c the velocity of light, ϵ the energy of the photon, and p the momentum of the photon. The number of photons in a given energy or frequency range is given by Eq. (8), where V is the

$$g = \frac{8\pi}{c^3} V \nu^2 \, d\nu \tag{8}$$

volume of the enclosure containing the photons. Actually, for a gas of photons, Eq. (2a) is not necessary (the number of photons is not fixed), and thus the distribution function depends on just one parameter, β; A can be shown to be unity. If one also uses the fact that $\beta = 1/kT$, one obtains from Eq. (3), for the number of photons in the frequency range $d\nu$, Eq. (9). The energy density (energy per

$$n(\nu)\,d\nu = \frac{8\pi}{c^3} V \frac{\nu^2\,d\nu}{e^{h\nu/kT} - 1} \qquad (9)$$

unit volume) in the frequency range $d\nu$ is, by Eqs. (9) and (7a), Eq. (10). One recognizes Eq. (10) as the

$$\rho(\nu)\,d\nu = \frac{8\pi h}{c^3} \frac{\nu^3\,d\nu}{e^{h\nu/kT} - 1} \qquad (10)$$

celebrated Planck radiation formula for blackbody radiation. Thus, blackbody radiation must be considered as a photon gas, with the photons satisfying Bose-Einstein statistics. *See* HEAT RADIATION.

For material particles of mass m contained in a volume V, one may write Eq. (3) as Eq. (11), where

$$f(v_x v_y v_z)\,dv_x\,dv_y\,dv_z = \left(\frac{m}{h}\right)^3 V \frac{dv_x\,dv_y\,dv_z}{\frac{1}{A} e^{mv^2/2kT} - 1} \qquad (11)$$

$v = \sqrt{v_x{}^2 + v_y{}^2 + v_z{}^2}$ is the total velocity. Equations (2a) and (2b) may now be written as integrals. If the so-called virial theorem, written as Eq. (12),

$$PV = {}^2/{}_3 E \qquad (12)$$

is used where P is the pressure, V is the volume, and E the energy, Eqs. (2a) and (2b) yield a pair of implicit equations, which give the equation of state of an ideal Bose gas, Eqs. (13a) and (13b).

$$\frac{N}{V}\lambda^3 = B_{1/2}(A) \qquad (13a)$$

$$\lambda^3 \frac{P}{kT} = B_{3/2}(A) \qquad (13b)$$

Here λ^3 is defined as in Eq. (14) while Eq. (15)

$$\lambda^3 = \frac{h^2}{2\pi m kT} \qquad (14)$$

$$B_\rho(A) = \frac{1}{\Gamma(\rho + 1)} \int_0^\infty \frac{u^\rho\,du}{\frac{1}{A} e^u - 1} \qquad (15)$$

holds. In Eq. (15) Γ is the usual Γ function and $u = mv^2/2kT$. From Eqs. (13a) and (13b), one obtains the relation between P, V, and T by eliminating A.

Now if one develops the numerator of the integral in Eq. (15) in an infinite series, one obtains the so-called Einstein equations, Eqs. (16a) and (16b). It is easy to verify that the sums of Eqs. (16a)

$$\frac{N}{V} = \frac{1}{\lambda^3} \sum_{l=1}^\infty \frac{A^l}{l^{3/2}} \qquad (16a)$$

$$P = \frac{kT}{\lambda^3} \sum_{l=1}^\infty \frac{A^l}{l^{5/2}} \qquad (16b)$$

and (16b) diverge for $A > 1$; however, they still converge for $A = 1$. For $A = 1$, $N/V = 2.61/\lambda^3$ and $P = 1.34kT/\lambda^3$. Einstein interpreted $N/V = 2.61/\lambda^3$ as a maximum possible density. If the gas is compressed beyond this point, the superfluous particles will condense in a zero state, where they do not contribute to the density or the pressure. If the volume is decreased, this curious condensation phenomenon results, yielding the zero state which has the paradoxical properties of not contributing to the pressure, volume, or density. There is now

considerable evidence that many of the superfluid properties exhibited by liquid helium are in fact manifestations of an Einstein condensation. Particularly convincing is the observation that liquid He³, which has half integer spin and thus must satisfy Fermi-Dirac statistics, shows no superfluid properties. The discussion so far was for ideal Bose systems. For nonideal Bose systems the situation is currently under intensive study. It appears that a fairly complete description of the experimental facts of He⁴ is possible, thus providing direct evidence for the reality of the condensation phenomenon. *See* FERMI-DIRAC STATISTICS; NONRELATIVISTIC QUANTUM THEORY; QUANTUM STATISTICS; STATISTICAL MECHANICS.

[MAX DRESDEN]
Bibliography: See STATISTICAL MECHANICS.

Boundary-layer flow

The flow of that portion of a viscous fluid which is in the neighborhood of a body in contact with the fluid and in motion relative to the fluid. Wherever a viscous fluid flows past a boundary, the layers of the fluid nearest the boundary are subjected to shearing forces, which cause the velocity of these layers to be reduced. As the boundary is approached, the velocity continuously decreases until, immediately at the boundary, the fluid particles are at rest relative to the body. This region of retarded velocity is called the boundary layer, and a graph of the variation of velocity with distance from the wall or boundary describes a boundary-layer profile (Fig. 1). The primary effects of the viscosity of the fluid are concentrated in this boundary layer, whereas in the outer or free-stream flow the viscous forces are negligible. L. Prandtl founded modern boundary-layer research in 1904, when he recognized that the flow in the boundary layer could be treated separately from the free-stream flow and simplified the Navier-Stokes equations for use in the boundary layer. Because the friction forces due to the viscosity of the fluid are restricted to the boundary layer, this region is also often called the friction layer. *See* FLUID FLOW; NAVIER-STOKES EQUATIONS; VISCOSITY.

The velocity of the fluid within the boundary layer approaches the local free-stream velocity

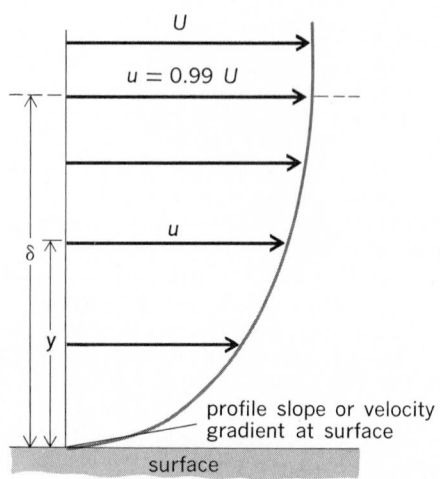

Fig. 1. Boundary-layer profile.

asymptotically as the distance from the wall is increased; thus the actual thickness of the boundary layer is difficult to determine. An arbitrary boundary-layer thickness δ is usually defined as that distance from the wall or boundary where the velocity of the flow in the boundary layer reaches 99% of the local free-stream velocity (Fig. 1). The thickness of a boundary layer depends upon the viscosity of the flowing fluid, the free-stream flow conditions, the roughness of the immersed surface or boundary, and the extent of the surface over which the fluid has passed. Each body moving through a viscous fluid is sheathed in a mantle of boundary-layer flow. The thicknesses of boundary layers surrounding vehicles moving through the air range from less than 0.1 in. near the front of a high-speed aircraft to more than 10 ft at the rear of a dirigible or an airship.

The viscous shearing force created within the boundary layer constitutes a major portion of the resistance or drag experienced by an airplane in flight or by fluids passing through pipes and channels; this shearing force opposes the relative motion of any viscous fluid past a bounding or immersed surface. The boundary-layer thickness generally increases with the extent of the surface over which the fluid has passed. In the flow of fluid through pipes, the boundary layer formed at the walls thickens until the pipe is entirely filled with boundary-layer flow (Fig. 2). When such a condi-

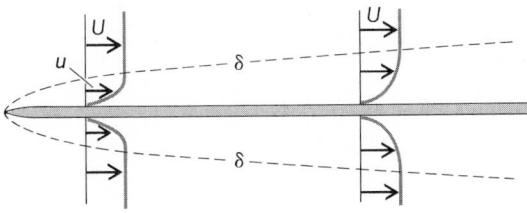

Fig. 2. Boundary-layer development in a pipe.

tion is reached, fully developed pipe flow is said to exist. In cases of unbounded flows, such as a thin, flat plate placed parallel to the flow direction, the boundary layer continues to increase in thickness as the fluid passes downstream (Fig. 3). This growth of the boundary layer results in a deflection or displacement of the free-stream flow away from the surface, thus apparently increasing the thickness of the plate.

Boundary-layer separation. In flowing over a surface of changing contour, the velocity and pressure of a fluid vary. As the flow velocity increases, the pressure must decrease. In the free stream, where there are negligible losses to viscosity, the flow is always able to proceed into regions of increasing pressure at the expense of its velocity, the opposing pressure forces being balanced by the forces that result from the change in momentum of the flow. However, within the boundary layer, where the momentum of the fluid has been reduced in overcoming viscous forces, the remaining momentum may be insufficient to allow the flow to proceed into regions of increasing pressure. At the position on the body where this condition exists, the boundary layer no longer continues to follow

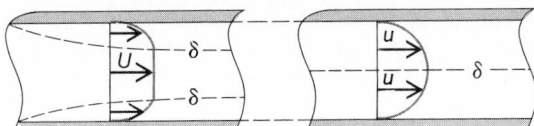

Fig. 3. Boundary-layer development on flat plate.

the contour of the boundary but instead leaves, or separates from, the surface. This phenomenon is known as boundary-layer separation (Fig. 4). Beyond the point of separation, between the surface and the separated boundary layer, a region of reversed or upstream flow exists. This reversed flow and the separated boundary layer subsequently coalesce to form a wake of low-velocity fluid behind the body. The separation of the flow in the boundary layer limits the lifting capabilities of wings and causes the stalling of airplanes. The efficiencies of fluid machines and mechanisms, including turbines, compressors, pumps, propelers, and so forth, are decreased by flow separation originating in the boundary layer. *See* BERNOULLI'S THEOREM.

Laminar and turbulent layer. The flow in the boundary layer formed on a body exists in one of two characteristic states: laminar or turbulent. If the flow is laminar, strata or laminae of fluid pass smoothly across the surface, whereas if it is turbulent, the flow is characterized by a rapid churning or mixing between layers of flow having different velocities. These two flow states are familiarly illustrated by the smoke rising from a cigarette in a still room. At first the smoke rises in smooth, continuous filaments, which after some distance begin to oscillate and finally erupt into boiling turbulent flow. This same phenomenon of transition from laminar to turbulent flow may occur within the boundary layer after the flow has passed over a surface for some distance. The flow on the forward or upstream portion of a body is generally laminar. However, as the boundary layer develops on the after or downstream part of the body, the flow may become turbulent. The flow in the boundary layer, at first smooth or laminar, eventually begins to fluctuate and finally bursts into random turbulence

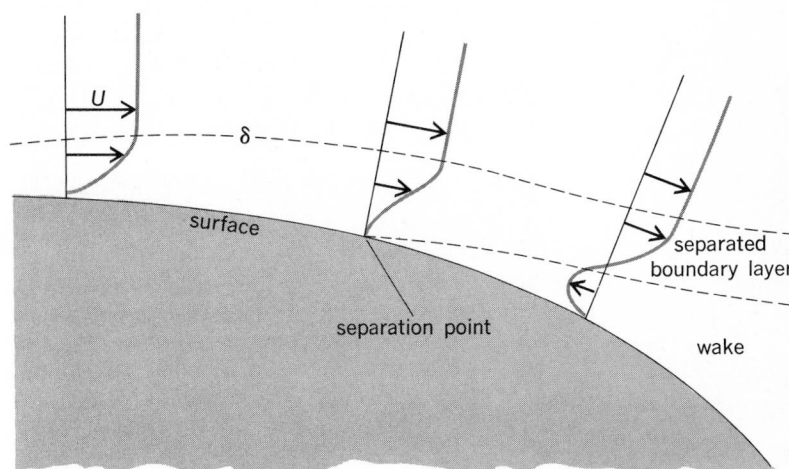

Fig. 4. Boundary-layer separation.

(Fig. 5). Turbulent flow may occur within the boundary layer even when, as is generally the case, the free-stream flow outside the boundary layer is laminar. Turbulent flow is readily distinguished from laminar by experiment: however, a complete and precise theoretical explanation of the processes which cause a laminar boundary layer to become turbulent has not yet been formulated.

Critical Reynolds number. Among the first to investigate the transition from laminar to turbulent flow, O. Reynolds suggested a dimensionless ratio, the Reynolds number, as a flow parameter useful in determining the extent of laminar boundary-layer flow over a surface. It has since been determined that for many specific shapes there exists a Reynolds number above which the flow in the boundary layer no longer remains laminar but becomes turbulent. This Reynolds number is known as the critical Reynolds number of the boundary layer. *See* REYNOLDS NUMBER.

In dealing with boundary-layer flows, the Reynolds number is usually written as in Eq. (1), where

$$R = U\delta^*/\nu \qquad (1)$$

U is the local free-stream velocity, ν is the kinematic viscosity of the fluid, and δ^* is the displacement thickness of the boundary layer, a characteristic parameter defined as Eq. (2), where u is the veloc-

$$\delta^* = \int_0^\delta \left(1 - \frac{u}{U}\right) dy \qquad (2)$$

ity at any distance y away from the surface. Transition generally will not occur below a Reynolds number $R = 575$.

The term critical Reynolds number is often used in a slightly different sense when bluff or blunt bodies, such as spheres or circular cylinders, placed perpendicular to the flow are being considered. In these cases, the critical Reynolds number is that value at which there occurs a sudden drop or decrease in the drag coefficient of the body. Here the Reynolds number is defined using the diameter D of the sphere or cylinder in place of the displacement thickness δ^* of the boundary layer. The critical Reynolds number of a sphere is about 325,000 and that for a circular cylinder about 450,000.

Stability of laminar layer. To predict the theoretical value of the critical Reynolds number of an arbitrary boundary layer, it has been found convenient to assume that the laminar boundary-layer flow is not absolutely smooth and laminar, but that it contains velocity fluctuations of infinitesimal amplitude. It is further assumed that these fluctuations cover a wide range of frequency of oscillation. Based upon these assumptions, the Tollmien-Schlichting stability theory has been developed. This theory shows that under certain conditions in the boundary layer, velocity fluctuations of particular frequencies are amplified, whereas fluctuations of all other frequencies are damped. The amplified fluctuations increase in magnitude and cause the flow within the boundary layer to oscillate at their particular frequency. These large oscillations within the boundary layer are known as Tollmien-Schlichting waves (Fig. 5b). The oscillations increase in amplitude and eventually cause turbulent flow. When the flow conditions are such that no fluctuations, regardless of frequency, are amplified, the boundary layer is said to be stable. If, however, the Reynolds number of the boundary layer is large enough, fluctuations of some frequency are amplified, the boundary layer is unstable, and the flow eventually becomes turbulent.

Transition to turbulent flow may also be caused by a finite disturbance to the boundary layer, such as surface roughness. A roughness element on the surface may result in localized separation of the boundary layer, causing turbulence to be shed from the element; frequently, the wake of the roughness element may cause a wedge of turbulent flow to be formed downstream, just as turbulent flow may be formed behind an object dipped into a smooth stream of water. It has been further postulated that such a transition can be excited even on a smooth surface by large velocity fluctuations within the boundary layer. Thus, amplified Tollmien-Schlichting waves can have the same effect in causing transition as disturbance or roughness elements on the surface. Both the Tollmien-Schlichting stability theory and Taylor's finite disturbance theory have been confirmed conclusively by experimentation.

Skin friction. The viscous shearing forces which retard the flow within the boundary layer stem

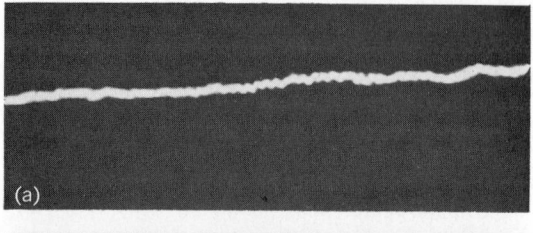

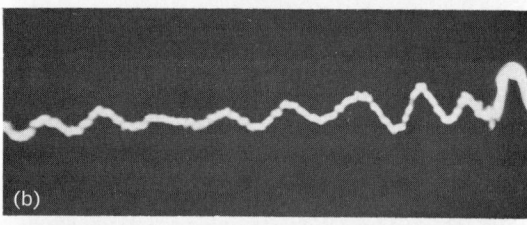

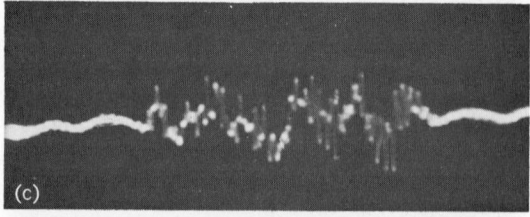

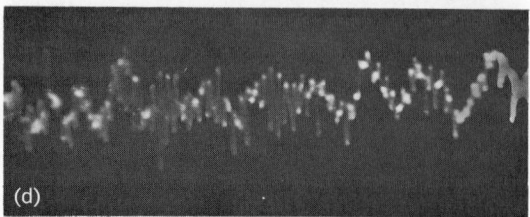

Fig. 5. Photographs of oscilloscope traces showing transitional velocity fluctuations in the boundary layer. (a) Laminar flow with small oscillations. (b) Oscillations amplified to form Tollmien-Schlichting waves. (c) Turbulent bursts indicating onset of transition. (d) Fully developed turbulent fluctuations.

from the friction developed between the moving fluid and the surface or skin of the immersed body. The accurate calculation and prediction of the surface shear or skin friction which will exist on a given body is the primary goal of boundary-layer theory. Because skin friction produces a direct shearing resistance or drag and also may result in separation of the boundary layer, a knowledge of its distribution and magnitude is useful in determining the performance of bodies moving through viscous fluids.

The shearing stresses existing within a fluid are proportional to the viscosity of the fluid and to the rate of change of strain within the fluid. Thus, for laminar flows, a relatively simple relation for the shearing stress may be written as in Eq. (3), where

$$\tau = \mu \frac{du}{dy} \qquad (3)$$

τ is the shearing or frictional stress, μ is the viscosity of the fluid, and du/dy is the gradient or rate of change of velocity normal to the surface. The skin friction or shearing stress at the surface is found by evaluating this relation at the surface where $y = 0$, so that Eq. (4) holds. Thus the skin friction of

$$\tau_0 = \mu \frac{du}{dy}\bigg|_0 \qquad (4)$$

a laminar boundary layer can be determined directly from its boundary-layer profile.

For turbulent flows, however, the shearing stress is further increased by turbulent mixing, which results in exchange of mass and momentum within the boundary layer. Laminar shear is caused only by molecular interaction within the fluid, whereas turbulent shear is determined mainly by the interaction of macroscopic portions or particles of fluid. In a turbulent flow, finite particles of fluid having different velocities are churned or mixed together by the turbulent fluctuations. In this manner, the exchange of momentum among the particles produces an increased resultant shearing stress. A particle moving from a low-velocity region into a region of higher velocity must be accelerated to the velocity of its new environment, and the force required to produce this acceleration reduces the momentum of the surrounding particles. The turbulent mixing and consequent momentum exchange within the boundary layer tend to produce a profile in which there is a more uniform distribution of velocity than in the case of laminar flow (Fig. 6). Prandtl's mixing-length theory treats the additional shearing stresses caused by turbulence as an apparent increase in the viscosity of the fluid, and gives the friction in the boundary layer as in Eq. (5), where ρ is the density

$$\tau = \rho \epsilon \frac{du}{dy} \qquad (5)$$

of the fluid and ϵ is the apparent kinematic viscosity, or eddy viscosity, of the fluid. The determination of ϵ is dependent on the scale of the turbulence and is indicated by the size of the turbulent eddies.

In the part of a turbulent boundary layer which is in the immediate proximity of the surface, the turbulent fluctuations are almost completely damped by the wall. This region is called the laminar sublayer, and the shear within this region, including the skin friction, follows the laws of laminar flow. This sublayer is, however, extremely thin and the velocity gradients are much larger; hence, the skin friction is much greater than for completely laminar boundary layers.

Layers in compressible flow. As the free-stream flow approaches sonic velocity, additional considerations complicate the behavior of the boundary layer. For example, the energy lost to skin friction produces sufficient heat to invalidate the assumptions of constant fluid density and viscosity. Thus the heat produced by skin friction results in a thermal boundary layer of varying temperature near the surface. At a Mach number $M = 3$, the surface temperature is raised about 50% above the temperature of the ambient fluid. The interaction of this variation of temperature with the variation of velocity in the boundary layer makes a rigorous theoretical analysis of the flow intractable. However, several parameters have been shown to influence the behavior of boundary layers in compressible flows. Among these is the Mach number $M = U/c$, where U is the velocity of the fluid and c is the sonic velocity. Also of importance is the Prandtl number $P = \nu/a$, where ν is the kinematic viscosity of the fluid and a is the thermal diffusivity.

The shock waves which may develop on a body also have a considerable and generally adverse effect upon the boundary layer. A shock wave almost inevitably results in premature transition from laminar to turbulent flow in the boundary layer and, if strong enough, will cause an early separation of the flow. Although theoretical treatments have been developed, the greater part of compressible boundary-layer technology is based upon empirical relations obtained from experimental results. *See* COMPRESSIBLE FLOW.

Boundary-layer control. The natural development of a boundary layer is affected by the contour and surface roughness of the body on which it is formed. Therefore, by the reduction of surface roughness and the choice of surface contours, some amount of control may be exerted upon the development of the boundary layer. Attention to surface roughness, waviness, and continuity is particularly important in the preservation of laminar boundary layers. Even the protuberance on a surface which is apparently most insignificant may produce localized transition to turbulent flow. Transition on aircraft wings is often produced by the remains of insects which have impinged upon the wings at high speeds. Rivet heads, skin laps and joints, and other surface discontinuities readily produce turbulent flow on airplanes.

The overall contour of the body is also an important consideration if laminar boundary-layer flow is to be maintained. Many laminar airfoil sections have been developed for high-speed aircraft. In fact, the term streamlining has come to describe the technique of designing shapes upon which the boundary-layer growth is minimized. The body shape also has a large influence upon the separation of the flow, and, to give low resistance or drag, shapes must be designed which prevent or delay boundary-layer separation. These techniques of influencing the development of the boundary layer by prescribing the geometry of bodies may be called geometric boundary-layer control.

Because the amount of control available by purely geometric means is limited, additional

BOUNDARY-LAYER FLOW

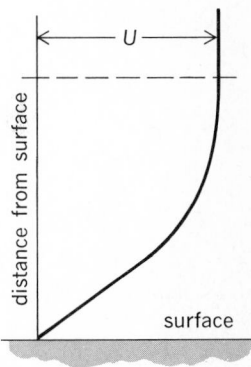

(a)

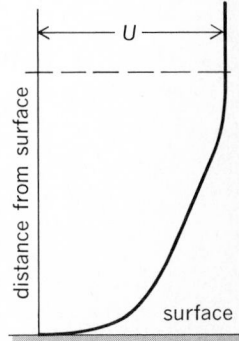

(b)

Fig. 6. Typical boundary-layer velocity profile.
(a) Laminar.
(b) Turbulent.

by a circular cylinder of infinite length or by a sphere, acoustic or electromagnetic oscillations in a circular cylinder of finite length or in a sphere, electromagnetic wave propagation in the waveguides of circular cross section, in coaxial cables, or along straight wires, and in skin effect in conducting wires of circular cross section. In these problems Bessel functions are used to represent such quantities as the temperature, the concentration, the displacements, the electric and magnetic field strengths, and the current density as function of space coordinates. The Bessel functions enter into all these problems because boundary values on circles (two-dimensional problems), on circular cylinders, or on spheres are prescribed, and the solutions of the respective problems are sought either inside or outside the boundary, or both.

Definition of Bessel functions. The independent variable z in Bessel's differential equation may in applications assume real or complex values. The parameter ν is, in general, also complex. Its value is called the order of the Bessel function. Since there are two linearly independent solutions of a linear differential equation of second order, there are two independent solutions of Bessel's differential equation. They cannot be expressed in finite form in terms of elementary functions such as z^n, $\sin z$, or e^z unless the parameter is one-half of an odd integer. They can, however, be expressed as power series with an exception for integer values of ν. The function defined by Eq. (2) is desig-

$$J_\nu(z) = \left(\frac{z}{2}\right)^\nu \sum_{l=0}^{\infty} \frac{(-z^2/4)^l}{l!\,\Gamma(\nu+l+1)} \qquad (2)$$

nated as Bessel's function of the first kind, or simply the Bessel function of order ν. $\Gamma(\nu+l+1)$ is the gamma function. The infinite series in Eq. (2) converges absolutely for all finite values, real or complex, of z. In particular, Eq. (3) may be ex-

$$J_0(z) = 1 - \frac{1}{1!1!}\left(\frac{z}{2}\right)^2 + \frac{1}{2!2!}\left(\frac{z}{2}\right)^4 - \frac{1}{3!3!}\left(\frac{z}{2}\right)^6 + \cdots$$
$$J_1(z) = \frac{z}{2} - \frac{1}{1!2!}\left(\frac{z}{2}\right)^3 + \frac{1}{2!3!}\left(\frac{z}{2}\right)^5 - \cdots \qquad (3)$$

pressed. Along with $J_\nu(z)$, there is a second solution $J_{-\nu}(z)$. It is linearly independent of $J_\nu(z)$ unless ν is an integer n. In this case $J_{-n}(z) = (-1)^n J_n(z)$. *See* GAMMA FUNCTION.

The Bessel function of the second kind, also called Neumann function, is defined by Eq. (4).

$$Y_\nu(z) = \frac{\cos \nu\pi J_\nu(z) - J_{-\nu}(z)}{\sin \nu\pi} \qquad (4)$$

If $\nu = n$, this expression is indeterminate, and the limit of the right member is to be taken. In particular, Eq. (5) may be expressed, with $\ln \gamma = 0.577215 \ldots$ (Euler's constant). There are two

$$Y_0(z) = \frac{2}{\pi} J_0(z) \ln\left(\frac{1}{2}\gamma z\right)$$
$$+ \frac{2}{\pi}\left[\frac{1}{1!1!}\left(\frac{z}{2}\right)^2 - \frac{1}{2!2!}\left(1+\frac{1}{2}\right)\left(\frac{z}{2}\right)^4\right.$$
$$\left. + \frac{1}{3!3!}\left(1+\frac{1}{2}+\frac{1}{3}\right)\left(\frac{z}{2}\right)^6 - \cdots\right] \qquad (5)$$

Bessel functions of the third kind, designated as first and second Hankel functions. They are

defined as Eq. (6).

$$H_\nu^{(1)}(z) = J_\nu(z) + iY_\nu(z)$$
$$H_\nu^{(2)}(z) = J_\nu(z) - iY_\nu(z) \qquad (6)$$

Elementary properties. For $z = 0$, then $J_0(0) = 1$, $J_n(0) = 0$ if $n = 1, 2, \ldots$, while $Y_n(0)$ and therefore also $H_n^{(1)}(0)$ and $H_n^{(2)}(0)$ are infinite for all n. The behavior of Bessel functions for small values of z is readily seen from the power series expansion in Eq. (2), if ν is not an integer. For integer $\nu = n \geq 0$, one has for small z the relations in notation (7) (the sign $\approx$ means approximately

$$J_n(z) \approx \left(\frac{z}{2}\right)^n \Big/ n! \qquad Y_0(z) \approx \frac{2}{\pi} \ln \frac{\gamma z}{2}$$
$$Y_n(z) \approx -\frac{1}{\pi}\left(\frac{z}{2}\right)^{-n}(n-1)! \qquad (n=1,2,\ldots) \qquad (7)$$

equal). Here $n! = 1 \cdot 2 \cdot 3 \ldots n$, $0! = 1$. The behavior of the Bessel functions for large values of $z(|z| \gg |\nu|^2)$ is given for all values of ν by the formulas in notation (8). They hold not only for real

$$J_\nu(z) \approx \left(\frac{2}{\pi z}\right)^{1/2} \cos\left(z - \frac{\nu\pi}{2} - \frac{\pi}{4}\right)$$
$$Y_\nu(z) \approx \left(\frac{2}{\pi z}\right)^{1/2} \sin\left(z - \frac{\nu\pi}{2} - \frac{\pi}{4}\right)$$
$$H_\nu^{(1)}(z) \approx \left(\frac{2}{\pi z}\right)^{1/2} \exp\left[i\left(z - \frac{\nu\pi}{2} - \frac{\pi}{4}\right)\right] \qquad (8)$$
$$H_\nu^{(2)}(z) \approx \left(\frac{2}{\pi z}\right)^{1/2} \exp\left[-i\left(z - \frac{\nu\pi}{2} - \frac{\pi}{4}\right)\right]$$

positive values of z but also for complex values in the angular domain $-\pi < \arg z < \pi$. These formulas are said to be asymptotic. This means that the difference of the left and right member, multiplied by a certain power of z, which is here $z^{3/2}$, is bounded as z goes to infinity along a straight line in the angular domain mentioned. The asymptotic formulas can be used to obtain approximate numerical values of the functions for sufficiently large $|z|$. They reveal that the Bessel functions are closely related to the functions $\cos z$, $\sin z$, $\exp(iz)$, $\exp(-iz)$, respectively.

The general behavior of Bessel functions can to a certain extent be inferred from the behavior near $z = 0$ and from the asymptotic behavior. The functions $J_\nu(z)$ and $Y_\nu(z)$ of real positive argument z and real order ν are oscillatory in z with a period that is not quite a constant but approaches 2π as $z \to \infty$. The oscillations have, however, an amplitude which decreases in accordance with $(2/\pi z)^{1/2}$ as $z \to \infty$. They are used in applications to represent standing cylindrical and spherical waves. The behavior of Hankel functions follows then from Eq. (6). They are used to represent progressing cylindrical and spherical waves.

Figures 1 and 2 give the Bessel and Neumann functions for $\nu = 0, 1, 2$ and positive values of z.

The large zeros $z_{\nu,s}$ of the function $J_\nu(z)\cos\alpha + Y_\nu(z)\sin\alpha$ with α a constant are given by the asymptotic expression $z_{\nu,s}^2 \approx [(s + \nu/2 - 1/4)\pi + \alpha]^2 + 1/4 - \nu^2$, in which s assumes large integer values. Even the first positive zero of $J_0(z)$ is very accurately given by this formula, namely, $z_{0,1} \approx 2.409$, as compared with the accurate value $z_{0,1} = 2.4048\ldots$

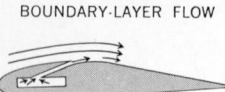

BOUNDARY-LAYER FLOW

(a)

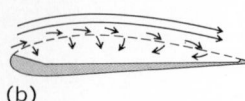

(b)

Fig. 7. Airfoils (a) with flowing boundary-layer control and (b) with suction boundary-layer control.

methods for influencing the development of the boundary layer have evolved. These methods in general are intended either to remove the low-momentum flow from the boundary layer or to restore the lost momentum. In the latter method, the retarded flow in the boundary layer is reenergized by supplying high-velocity flow through slots or jets in the surface of the body (Fig. 7a). This blowing boundary-layer control requires suitable pumps and ducts within the body to provide the necessary quantities of blown or ejected fluid. This technique has, as yet, not proven to be an economical method of control, although several aircraft utilizing variations of the principle have been flown. Blowing boundary-layer control is intended to suppress or delay the separation of the boundary layer rather than to preserve laminar flow and, as a result, is limited in its applications.

The method of controlling the boundary layer by the removal of the retarded flow, called suction boundary-layer control, may be applied both to maintain laminar flow and to prevent boundary-layer separation. By sucking away the flow in the lower regions of the boundary layer through slots or perforations in the surface, the development of the boundary layer can be readily influenced (Fig. 7b). The proper distribution of this suction, as determined by the requirements of the boundary-layer equations, allows a designer to tailor the development of the boundary layer to his particular demands. By maintaining the boundary-layer Reynolds number below its critical value, the flow can be kept in a stable condition and thus remain laminar. By thinning the boundary layer properly, it can be made to resist separation. Practical suction systems for maintaining laminar flow and for delaying flow separation have been demonstrated in flight.   [JOSEPH J. CORNISH, III]

Bibliography: S. Goldstein (ed.), *Modern Developments in Fluid Dynamics*, 2 vols., 1938; H. Schlichting, *Boundary Layer Theory*, 7th ed., 1979.

Boyle's law

A law of gases which states that at constant temperature the volume of a gas varies inversely with its pressure. This law, formulated by Robert Boyle (1627–1691), can also be stated thus: The product of the volume of a gas times the pressure exerted on it is a constant at a fixed temperature. The relation is approximately true for most gases, but is not followed at high pressures. The phenomenon was discovered independently by Edme Mariotte about 1650 and is known in Europe as Mariotte's law. *See* GAS; KINETIC THEORY OF MATTER.

[FRANK H. ROCKETT]

Bremsstrahlung

The electromagnetic radiation emitted by electrons when they pass through matter. Charged particles radiate when accelerated, and in this case the electric fields of the atomic nuclei provide the force which accelerates the electrons. The continuous spectrum of x-rays from an x-ray tube is that of the bremsstrahlung; in addition, there is a characteristic x-ray spectrum due to excitation of the target atoms by the incident electron beam. The major energy loss of high-energy (relativistic) electrons (energy $\gtrsim 10$ Mev, depending somewhat

upon material) occurs from the emission of bremsstrahlung, and this is the source of the γ-rays in a high-energy cosmic-ray shower. *See* COSMIC RAYS.

The spectrum of bremsstrahlung resulting from the collision of an electron with an atom is continuous and is roughly constant between $\nu = 0$ and $\nu = \nu_{max}$; ν_{max} is the maximum frequency of a photon which can be emitted; that is, $h\nu_{max} = T$, where T is the initial kinetic energy of the electron and h is Planck's constant. The angular distribution of bremsstrahlung is roughly isotropic at low (nonrelativistic) electron energies, but is largely restricted to the forward direction at high energies. Very little bremsstrahlung is emitted at an angle much larger than $\theta_c = m_e c^2 / T$ radians, where m_e is the electron mass and c the velocity of light. Bremsstrahlung emitted at the angle θ_c is polarized with the electric vector perpendicular to the plane containing the direction of radiation and the incident electron velocity. It is difficult to observe the polarization, because it is small except near the angle θ_c. A longitudinally polarized electron (that is, one with its spin parallel to its velocity) emits circularly polarized bremsstrahlung; this effect is not sensitive to angle, and has proved useful in analysis of the longitudinal polarization of electrons emitted in β-decay. *See* PLASMA PHYSICS.

[CHARLES GOEBEL]

Bibliography: J. J. Brandstatter, *Introduction to Waves, Rays, and Radiation in Plasma Media*, 1963; W. Heitler, *The Quantum Theory of Radiation*, 3d ed., 1954; J. D. Jackson, *Classical Electrodynamics*, 2d ed., 1975.

Brillouin zone

In the propagation of any type of wave motion through a crystal lattice, the frequency is a periodic function of wave vector **k**. This function may be complicated by being multivalued; that is, it may have more than one branch. Discontinuities may also occur. In order to simplify the treatment of wave motion in a crystal, a zone in **k**-space is defined which forms the fundamental periodic region, such that the frequency or energy for a **k** outside this region may be determined from one of those in it. This region is known as the Brillouin zone (sometimes called the first or the central Brillouin zone). It is usually possible to restrict attention to **k** values inside the zone. Discontinuities occur only on the boundaries. If the zone is repeated indefinitely, all **k**-space will be filled. Sometimes it is also convenient to define larger figures with similar properties which are combinations of the first zone and portions of those formed by replication. These are referred to as higher Brillouin zones.

The central Brillouin zone for a particular solid type is a solid which has the same volume as the primitive unit cell in reciprocal space, that is, the space of the reciprocal lattice vectors, and is of such a shape as to be invariant under as many as possible of the symmetry operations of the crystal. *See* CRYSTAL STRUCTURE; CRYSTALLOGRAPHY.

Zone construction. Let $\mathbf{a}_1, \mathbf{a}_2, \mathbf{a}_3$ be the primitive translation vectors for some crystal lattice. New vectors $\mathbf{b}_i$ with $i = 1, 2, 3$, are defined by Eq. (1),

$$\mathbf{a}_i \cdot \mathbf{b}_j = 2\pi \, \delta_{ij} \qquad (1)$$

where δ_{ij} is unity when $i = j$, and zero for other values of i. The vectors $\mathbf{b}_i$ are then given by Eq. (2).

$$\mathbf{b}_1 = \frac{2\pi \mathbf{a}_2 \times \mathbf{a}_3}{\mathbf{a}_1 \cdot \mathbf{a}_2 \times \mathbf{a}_3} \qquad \mathbf{b}_2 = \frac{2\pi \mathbf{a}_3 \times \mathbf{a}_1}{\mathbf{a}_1 \cdot \mathbf{a}_2 \times \mathbf{a}_3}$$

$$\mathbf{b}_3 = \frac{2\pi \mathbf{a}_1 \times \mathbf{a}_2}{\mathbf{a}_1 \cdot \mathbf{a}_2 \times \mathbf{a}_3} \qquad (2)$$

Now define vectors $\mathbf{K}_i$ by Eq. (3), where the h_i's

$$\mathbf{K}_i = h_1 \mathbf{b}_1 + h_2 \mathbf{b}_2 + h_3 \mathbf{b}_3 \qquad (3)$$

are arbitrary integers. The subscript i stands for some particular combination of the h. The end points of the vectors $\mathbf{K}_i$ form a lattice of points in reciprocal space. The vectors $\mathbf{K}_i$ have the property that plane waves of the form $e^{i\mathbf{K}_i \cdot \mathbf{r}}$ are periodic in the crystal lattice, since if some $\mathbf{r}' = \mathbf{r} + \mathbf{R}_n$ is considered, where $\mathbf{R}_n$ is a translation vector of the crystal, Eq. (4) holds.

$$e^{i\mathbf{K}_i \cdot \mathbf{r}'} = e^{i\mathbf{K}_i \cdot \mathbf{r}} e^{i\mathbf{K}_i \cdot \mathbf{R}_n} = e^{i\mathbf{K}_i \cdot \mathbf{r}} \qquad (4)$$

The last step follows since $\mathbf{K}_i \cdot \mathbf{R}_n$ is an integer times 2π. Consequently, the plane waves $e^{i\mathbf{K}_i \cdot \mathbf{r}}$ are suitable functions for the expansion of any function which is periodic in the lattice.

Unit cells can be constructed in the reciprocal space of the lattice of the ends of the vectors $\mathbf{K}_i$, just as is done in the real crystal space. The lines connecting one point with the other lattice sites are drawn, and the planes which are the perpendicular bisectors of these lines are constructed. The smallest enclosed solid figure is the first Brillouin zone. It is the smallest unit cell in the reciprocal lattice which has the symmetry of the entire lattice. Higher Brillouin zones are also formed. Brillouin zones for the body-centered cubic, face-centered cubic, and hexagonal close-packed lattices are shown in illustration $a-c$.

Application to band theory. Each electron wave function in the crystal can be classified according to some $\mathbf{k}$ inside the first Brillouin zone. For if $\mathbf{k}'$ is a vector in reciprocal space whose end point lies outside the zone, then it can be written as Eq. (5),

$$\mathbf{k}' = \mathbf{k} + \mathbf{K}_n \qquad (5)$$

where $\mathbf{k}$ lies inside the zone and $\mathbf{K}_n$ is a reciprocal lattice vector. According to the Bloch theorem, Eq. (6a) holds. But since $\mathbf{K}_n \cdot \mathbf{R}_j$ is an integer times 2π, Eq. (6b) is valid. Thus $\mathbf{k}'$ and $\mathbf{k}$ are equivalent in a certain sense. For this reason, it is possible to consider only the first zone in discussing the properties of solids.

$$\psi(\mathbf{k}', \mathbf{r} + \mathbf{R}_j) = e^{i\mathbf{k}' \cdot \mathbf{R}_j} \psi(\mathbf{k}', \mathbf{r}) \qquad (6a)$$

$$e^{i\mathbf{k}' \cdot \mathbf{R}_j} = e^{i\mathbf{k} \cdot \mathbf{R}_j} e^{i\mathbf{K}_n \cdot \mathbf{R}_j} = e^{i\mathbf{k} \cdot \mathbf{R}_j} \qquad (6b)$$

A fundamental point in the application of Brillouin zone theory to the study of the properties of metals and alloys is that the energy $E(\mathbf{k})$ of the states $\psi_{\mathbf{k}}$ must be a continuous function of $\mathbf{k}$ inside the zone (although it will be multivalued if the reduced zone scheme is employed). Discontinuities can occur only across the faces of the zone. The number of allowed states inside each Brillouin zone can be determined as follows: The number of allowed states per unit volume in the space of the vector $\mathbf{k}$ is $1/(2\pi)^3$ for each spin per unit volume of the crystal, or $2/(2\pi)^3$ altogether. Thus, the number of states in the zone for each atom in the crystal is $2V/(2\pi)^3 N$, where N is the number of atoms of the crystal, and V is the volume. A substance which has just enough electrons per atom to fill some zone may be an insulator or semiconductor. If there are not enough electrons to fill a zone, it must be a metallic conductor.

Alkali metals. These metals are body-centered cubic. The volume of the zone is $2(2\pi/a)^3$, where a is the lattice constant. The number of atoms per unit volume is $2/a^3$ so that the first zone can contain two electrons per atom. However, the atoms possess only one valence electron, so that these metals are good conductors.

Noble metals. Copper, silver, and gold are face-centered cubic, and again the zone can hold two electrons per atom. Thus, these also are good conductors. If the noble metals were adequately described by a free electron model, the Fermi surface (the surface bounding the occupied volume of reciprocal space) would be a sphere which would approach close to the surface of the zone. It is known, however, that the free electron model is not adequate for electrons near the Fermi surface, and that the Fermi surface touches the Brillouin zone boundary near the extremities of the (111) axes.

Divalent metals. Except for barium and mercury, the divalent metals have either hexagonal close-packed or face-centered cubic structures. The zones hold two valence electrons per atom, and on the basis of simple arguments, it would be expected that these materials should be insulators. However, it is likely that some of the electrons go over into the next Brillouin zone, leaving pockets of holes behind in the first zone.

Semiconductors and insulators. In diamond, silicon, and germanium, all of which have the diamond structure of two interpenetrating face-centered cubic (fcc) lattices, the Brillouin zone has the same form as for the fcc. It can hold four electrons per atom. Occupied states are separated from higher states by an energy gap, since these materials are semiconductors or insulators. Graphite, which has a hexagonal layer structure, can be regarded as a semiconductor with a vanishing energy gap. The Brillouin zone in a simple model is full at a temperature of 0 K, but there is no gap along the edges of the zone. In bismuth there is a higher Brillouin zone which holds five electrons per atom. There is a very small overlap of electrons across the faces of the zone, leading to the presence of a small number (about 10^{-5} per atom) of both electrons and holes at 0 K.

Binary alloys. Many of the properties of binary alloys which form substitutional solid solutions can be explained on the hypothesis that the solute contributes its valence electrons to form a composite system whose fundamental parameter is the electron per atom ratio. The electronic structure is assumed to be essentially unaltered. Many changes of the properties, such as the lattice parameter, the concentration at which the various phases appear, and the Hall coefficient, can be explained in terms of overlap or contact of the Fermi surface with the Brillouin zone. *See* BAND THEORY OF SOLIDS.

[JOSEPH CALLAWAY]

Bibliography: N. W. Ashcroft and N. D. Mermin, *Solid State Physics*, 1976; L. P. Bouckaert, R. Smoluchowski, and E. Wigner, Theory of Brillouin

BRILLOUIN ZONE

(a)

(b)

(c)

Brillouin zone for (*a*) body-centered cubic lattice, (*b*) face-centered cubic lattice, (*c*) hexagonal close-packed lattice.

zones and symmetry properties of wave functions in crystals, *Phys. Rev.*, 50:58–67, 1936; W. Hume-Rothery, *Electrons, Atoms, Metals, and Alloys*, 1955; H. Jones, *The Theory of Brillouin Zones and Electron States in Crystals*, 2d ed., 1975; C. Kittel, *Introduction to Solid State Physics*, 5th ed., 1976; R. S. Knox and A. Gold, *Symmetry in the Solid State*, 1964; F. Seitz and D. Turnbull (eds.), *Solid State Physics*, vol. 5, 1957, and vol. 6, 1958.

Brownian movement

The irregular motion of a body arising from the thermal motion of the molecules of the material in which the body is immersed. Such a body will of course suffer many collisions with the molecules, which will impart energy and momentum to it. Because, however, there will be fluctuations in the magnitude and direction of the average momentum transferred, the motion of the body will appear irregular and erratic, as shown in the diagram.

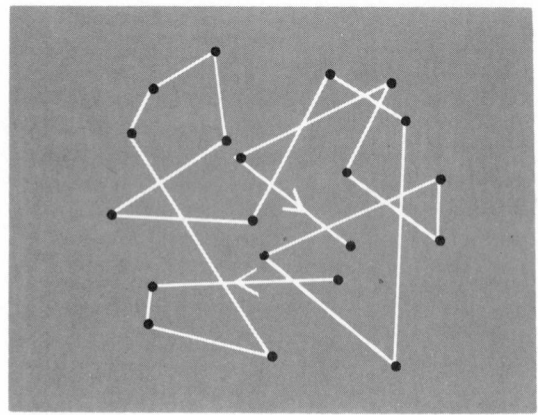

Random Brownian movement of a particle.

In principle, this motion exists for any foreign body suspended in gases, liquids, or solids. To observe it, one needs first of all a macroscopically visible body; however, the mass of the body cannot be too large. If its mass is M, one can estimate the root-mean-square velocity $\bar{v}$ by the equipartition law, written as Eq. (1). Here, k is the Boltzmann

$$\tfrac{1}{2}M\overline{v^2}=\tfrac{3}{2}kT \qquad \sqrt{\overline{v^2}}=\sqrt{\frac{3kT}{M}} \qquad (1)$$

constant and T is the absolute temperature. Hence, for a large mass, the velocity becomes small. For example, a 0.01-g mass has at 300 K a root-mean-square velocity of 3.5×10^{-6} cm/sec. *See* KINETIC THEORY OF MATTER.

Mirror in a gas. A simple observable example of Brownian motion is the motion of a mirror in a gas. If a mirror of moment of inertia I is suspended by a fine fiber, this mirror will execute a simple harmonic motion of natural frequency ω_0. Its mechanical equation of motion is written as Eq. (2), where

$$\frac{d^2\theta}{dt^2}+\omega_0{}^2\theta=0 \qquad (2)$$

θ is the angular displacement and $\omega=d\theta/dt$ the angular velocity. If this mirror is placed in the gas, the unequal forces on the two sides of it, which are due to the fluctuations in the momentum trans-

ferred by the molecules, will cause fluctuations around the harmonic oscillations. After a time, one will have equipartition of energy. For the rotational kinetic energy one has Eq. (3a). For the rotational potential energy, one has Eq. (3b).

$$\tfrac{1}{2}I\overline{\omega^2}=\tfrac{1}{2}kT \qquad (3a)$$

$$\tfrac{1}{2}I\omega_0{}^2\bar{\theta}^2=\tfrac{1}{2}kT \qquad (3b)$$

By careful measurement, relations such as Eqs. (3a) and (3b) may be checked. Brownian motion also yields a limit for the accuracy which may be obtained in a given experiment. *See* HARMONIC MOTION.

Langevin equation. If a small particle is suspended in a liquid, its displacement x in the course of time can be observed. The displacement will of course be erratic, but one can still calculate the average displacements, or root-mean-square displacements, at any time t. It is possible to compare these experimental results with theoretical calculations. If a small body moves through a liquid, it will experience the usual viscous drag, $-\beta Mv$, where M is the mass, v is the velocity, and β is a numerical coefficient depending on the viscosity. For a sphere of radius R, $M\beta$ has the value given by Stokes' law: $M\beta=6\pi R\eta$, where η is the coefficient of viscosity. In addition, a fluctuating force $MA(t)$, varying rapidly in the course of time, but of average value zero, acts on the particle. One may therefore write an equation of motion as Eq. (4). *See* VISCOSITY.

$$\frac{dv}{dt}=-\beta v+A(t) \qquad (4)$$

If one now multiplies Eq. (4), the Langevin equation, by x, averages over time, and applies the equipartition law, one obtains for a Stokes case Eq. (5). This result was originally derived by Albert

$$\overline{x^2}=\frac{kTt}{3\pi R\eta} \qquad (5)$$

Einstein; it expresses the root-mean-square displacement explicitly in terms of observable parameters. Equation (5) has been verified with an accuracy of about 0.5%.

[MAX DRESDEN]

Bibliography: R. B. Barnes and S. Silverman, Brownian motion as a natural limit to all measuring processes, *Rev. Mod. Phys.*, 6(3):162–192, 1934; S. Chandrasekhar, Stochastic problems in physics and astronomy, *Rev. Mod. Phys.*, 15(1):1–89, 1943; A. Einstein, *Investigations on the Theory of Brownian Movement*, 1926; R. Hersh and R. J. Griego, Brownian motion and potential theory, *Sci. Amer.*, 220(3):66–74, 1969; S. C. Port and C. J. Stone, *Brownian Motion and Classical Potential Theory*, 1978; N. C. Wang and G. E. Uhlenbeck, On the theory of the Brownian motion II, *Rev. Mod. Phys.*, 7(2,3):323–342, 1945.

Bubble chamber

A particle detector used in elementary particle physics research. Charged particles passing through liquid contained in the chamber leave visible tracks along their paths (Fig. 1). These tracks are then photographed for later study. Each curved track is actually a string of small bubbles marking the path of an elementary particle. Bubble chambers vary in size, in the liquid used, and in how fast photographs can be taken. They are particularly useful because they allow direct visualization of all the charged particles involved in a high-energy interaction, and they have been used to discover many new particles, such as the rho (ρ) and omega minus (Ω^-). *See* ELEMENTARY PARTICLE.

Principle of operation. The bubble chamber was invented by Donald Glaser in 1952. It was known that a pure liquid in a clean glass container can be raised above its boiling temperature without its starting to boil spontaneously. The boiling must be triggered by something. For example, a broken piece of glass dropped into a superheated liquid can make boiling violently erupt. As an electrically charged particle passes through any material (such as a liquid), it interacts with electrons in the material, leaving atoms ionized. Glaser discovered that in a superheated liquid this ionization energy is enough to trigger boiling along the path of the particle. The boiling bubbles grow rapidly, and so a photograph is quickly taken before the resolution of the individual tracks is destroyed. In order to make the bubble chamber sensitive to the arrival of another charged particle, the boiling must be stopped, and the liquid must be superheated again.

Operating cycle. A bubble chamber is cycled by the up and down movement of a piston (Fig. 2), causing pressure changes in the liquid which in turn cause a change in the liquid's boiling temperature. The piston first compresses the liquid so that the boiling point is above the temperature of the liquid. An accelerator produces the elementary particles to be studied and sends them as a beam toward the bubble chamber. Several milliseconds before the particles pass through the chamber, the piston expands the liquid, causing it to become superheated. The particles resulting from beam interaction with fluid nuclei in the chamber then produce bubbles, and about 1 ms later lights are flashed inside the chamber to record the tracks on film. Finally, the piston recompresses the liquid to stop the boiling and make the chamber ready for the next pulse of beam particles. This cycling can be faster than ten times a second for a rapid-cycling bubble chamber or as slow as four times a minute. Magnets are used to produce a magnetic field, which causes the charged particles to follow curved paths. Measurements of the curvature are then made to determine the particle's momentum. *See* PARTICLE ACCELERATOR.

Interactions and choice of liquid. Interest is focussed on the nuclear interactions that a beam particle has with the particles in the liquid. In Fig. 1, the beam particles are neutrinos which pass through the chamber from left to right. Neutrinos are electrically neutral, and so they do not leave tracks. When they interact, however, some of the particles produced will be charged, and the bubble chamber will record them. There are two interactions in Fig. 1: the first one with four produced particles in the upper left, and the second one in the lower center with five charged particles. A typical experiment comprises 50,000 analyzed photographs. Up to four cameras are used to photograph the tracks so that they can be reconstructed in three dimensions. Since the interactions occur

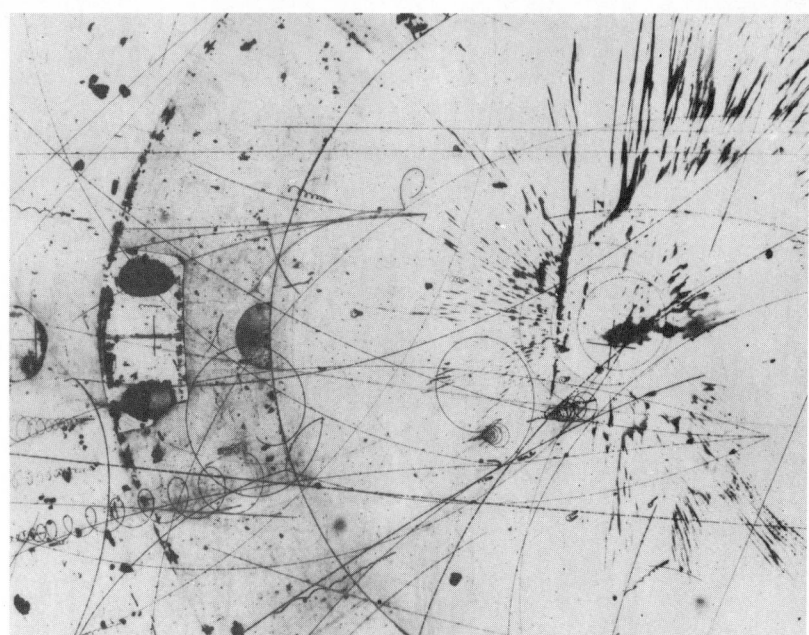

Fig. 1. Photograph of tracks produced by neutrinos in Fermilab's bubble chamber. (*Fermilab*)

with particles making up the liquid, the choice of liquid is important to the experimenter. Liquid hydrogen is used to study proton interactions because hydrogen is a simple atom made up of a sin-

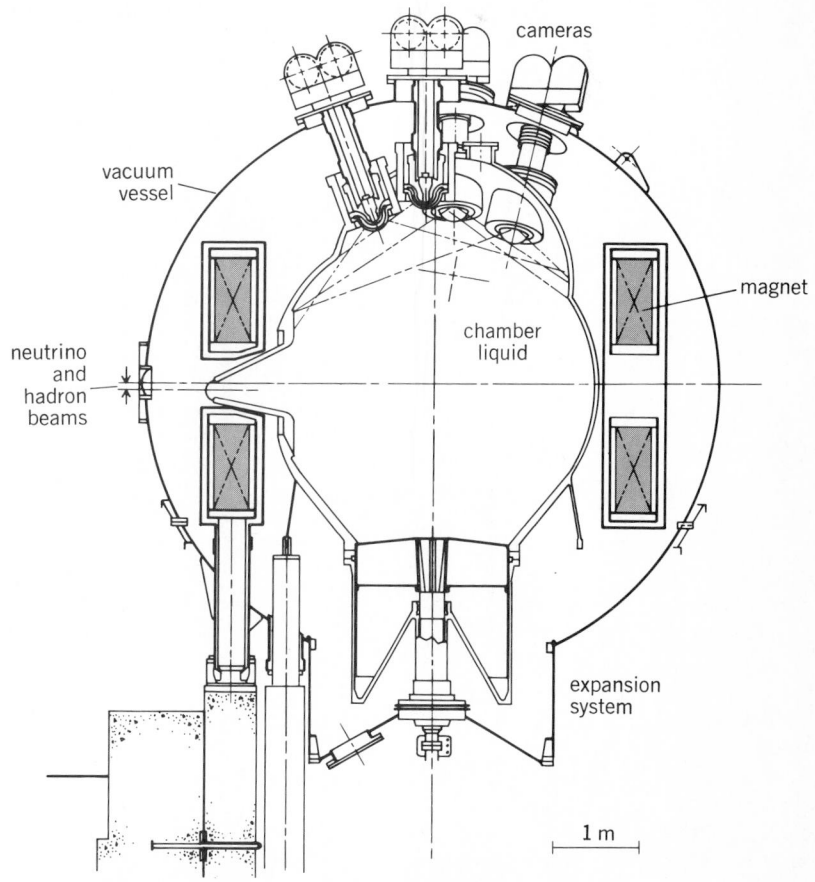

Fig. 2. Fermilab's 15-ft (4.6-m) bubble chamber.

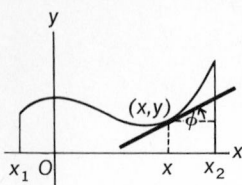

Fig. 1. Graphical representation of the derivative of $f(x)$.

gle proton and electron. Likewise deuterium, with one neutron and one proton, is used to study neutron and proton interactions. Liquid hydrogen and deuterium boil at a very low temperature; thus chambers using these liquids operate at about 27 K above absolute zero.

Photons are produced in most interactions, but they are hard to study directly since they too are neutral. However, a photon can convert into an electron-positron pair, which can be seen. The chances of this conversion process occurring increase as the density or the atomic number of the liquid increases. Thus some bubble chambers are built to use a heavy liquid such as propane or Freon. These heavy-liquid bubble chambers operate at much higher temperatures, around 50°C, but otherwise operate under the same principle as hydrogen bubble chambers.

Size. The first test bubble chamber was only a few centimeters across. Bubble chambers have steadily become larger and are now built as large as 15 ft (4.6 m) across. The main reason for making such a larger chamber is to be able to study neutrino interactions. Neutrinos interact only very weakly with other particles, and thus one uses a large volume of liquid to increase the chance of an interaction occurring.

Hybrid systems. Bubble chambers have been combined with other particle detectors to learn even more about the interactions occurring inside the chamber. In these hybrid systems, other particle detectors such as proportional wire chambers, spark chambers, lead glass counters, and Cerenkov counters are used to further study the particles before and after they leave the bubble chamber. *See* CERENKOV RADIATION; PARTICLE DETECTOR; SPARK CHAMBER.

[DONOVAN A. LJUNG]

Bibliography: B. Barish, Experiments with neutrino beams, *Sci. Amer.*, 229(2):30–38, August 1973; D. Glasser, The bubble chamber, *Sci. Amer.*, 192(2):46–50, February 1955; I. A. Pless, Bubble chambers, in D. M. Ritson (ed.), *Techniques of High Energy Physics*, pp. 87–114, 1961.

Calculus

The branch of mathematics dealing with two fundamental operations, differentiation and integration, which are carried out on functions. The subject, as traditionally developed in college textbooks, is partly an elementary development of the purely theoretical aspects of these operations and their interrelation, partly a development of rules and formulas for applying calculus to the standard functions which arise in algebra and trigonometry (with exponentials and logarithms included), and partly a collection of applications to problems of geometry, physics, chemistry, engineering, economics, and perhaps a few other subjects.

Derivative. The fundamental concept of differential calculus is that of the derivative of a function of one variable. The classical physical prototype of this concept is that of instantaneous velocity, which is the derivative of distance as a function of time. The derivative also has a highly significant geometrical realization which depends upon the graphical representation of a function in rectangular coordinates (x,y). If y is a differentiable function of x, perhaps as x increases from x_1 to x_2, the graph of the function is a continuous curve

with exactly one y for each x, and at each point the curve has a tangent line which is not parallel to the y axis. If ϕ is the angle, measured counterclockwise, from the positive x direction to the tangent (see Fig. 1), then $\tan \phi$ is equal to the derivative of y with respect to x. (This is on the supposition that the same unit of length is used along the two axes.) This $\tan \phi$ is also called the slope of the curve.

The standard notation for the derivative of y with respect to x is dy/dx. If the functional notation $y = f(x)$ is used, the derivative is often denoted by $f'(x)$. Modern practice is to use f for the function as an abstract entity, while $f(x)$ denotes the value of f at x. Then f' denotes the derivative as a function, and $f'(x)$ is the value of f' at x. For a precise definition of the derivative *see* DIFFERENTIATION.

Functions. To say that y is a function of x means that, as a consequence of some rule or definite agreement, there is a designated collection of permissible values of x (called the domain of the function) and an associated set of corresponding values of y (called the range of the function), of such a character that with each permissible value of x is paired a unique well-determined value of y. The function itself is the collection of all the pairs (x,y) which arise in this way. This collection exemplifies the rule or agreement. If f denotes the function, it is customary to write $y = f(x)$ to represent the dependence of y on x.

In calculus the domain of the function is usually composed of one or more intervals of the x axis.

Definite integral. If f is a function defined on the finite interval from x_1 to x_2 inclusive, the definite integral of f from x_1 to x_2, denoted by

$$\int_{x_1}^{x_2} f(x)\ dx$$

is defined by applying to f a rather intricate process which entails the consideration of what are called approximating sums. When the function f is subjected to certain restrictions, this process culminates in the determination of a number as the limit of the approximating sums, and this number is called the definite integral of f from x_1 to x_2. The integral is not defined unless the approximating sums do converge to a well-defined limit. A sufficient condition that this be so is that the function f be continuous.

There is a geometrical representation of the process of defining the definite integral, and it furnishes a plausible argument for the convergence of the approximating sums to a limit. Divide the interval from x_1 to x_2 into a finite number N of not necessarily equal parts. Let the lengths of these parts be $h_1, h_2, \ldots, h_N$, and let t_k be a value of x in the kth part (see Fig. 2). Then the expression

$$f(t_1)h_1 + f(t_2)h_2 + \cdots + f(t_N)h_N$$

is called an approximating sum. In Fig. 2, where the function is continuous and the function values are all positive, each term $f(t_k)h_k$ in the approximating sum is equal to the area of a certain shaded rectangle, and the whole sum is an approximation of the area between the graph of the function and the x axis, from x_1 to x_2 inclusive. The limiting process is carried on by increasing N and making the largest of the h_k's approach 0. It is then intui-

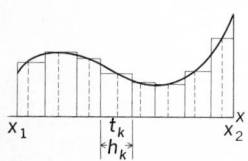

Fig. 2. The definite integral.

tively clear that the definite integral is the number which represents the exact area between the x axis and the graph. This geometrical interpretation of the integral is the basis of an important application of integral calculus, to the calculation of areas.

It would be tedious and difficult in practice to compute definite integrals by actually working out the limits of approximating sums. It is therefore fortunate that by purely mathematical reasoning it is possible to demonstrate a theorem which links derivatives and integrals and makes it possible, in many important instances, to compute definite integrals by an easier procedure. *See* INTEGRATION.

In the next paragraph are stated the two fundamental theorems of calculus. In these statements the adjective "continuous" appears. For a function $y = f(x)$ to be continuous means, roughly, that as x changes gradually, y must either change gradually, or not at all. Absolutely abrupt jumps are forbidden, and so are many more bizarre modes of irregular behavior. For a precise definition of continuity *see* DIFFERENTIATION.

Fundamental theorems. For the calculation of

$$\int_{x_1}^{x_2} f(x)\, dx$$

find, if possible, a function F with continuous derivative F' such that $F'(x) = f(x)$ when $x_1 \leqq x \leqq x_2$. Then Eq. (1) can be written. This is one of

$$\int_{x_1}^{x_2} f(x)\, dx = F(x_2) - F(x_1) \qquad (1)$$

the two central theorems. The other is stated as follows: Suppose f is continuous, and consider the function F defined by Eq. (2). Then F has a derivative given by $F'(x) = f(x)$.

$$F(x) = \int_{x_1}^{x} f(t)\, dt \qquad (2)$$

Law of the mean. The most important theorem about derivatives, exclusive of those which involve integrals, is variously called "the law of the mean" and "the mean-value theorem of differential calculus." It is stated in this form: Suppose f is continuous for x from x_1 to x_2 inclusive, and suppose f has a derivative for each x between x_1 and x_2. Then there is some x of this kind for which Eq. (3) holds

$$f'(x) = \frac{f(x_2) - f(x_1)}{x_2 - x_1} \qquad (3)$$

true. This theorem enables one to prove that, if $f'(x)$ is always positive, then $f(x)$ increases as x increases. Also that, if $f'(x)$ is always zero, then $f(x)$ remains constant as x changes. The first of these results is important for applications to the investigations of graphs. The second is virtually indispensable in the proof of the fundamental theorems of calculus stated above, and in establishing the uniqueness of the solutions of certain problems which involve antidifferentiation (the process of surmising what a function is from a knowledge of its derivative). Such problems abound in application of calculus to problems of motion.

Applications of derivatives. If y is a function of x, the derivative dy/dx is interpretable as the rate of change of y with respect to x. If y is distance and x is time, dy/dx is velocity. If y is work done by a force and x is time, dy/dx is power. Many motion problems, such as those arising in mechanics and formulated by Newton's second law, are expressible by posing an equation which a function and its derivative or derivatives must satisfy. These are called differential equations. The study of such equations is an extension and ramification of calculus. Differential equations are important in all branches of physics. They also arise in many problems of engineering and chemistry, for example, in the deformation of columns and beams, in the study of chemical reactions, and in radioactive decay. *See* DIFFERENTIAL EQUATION.

Problems of maximizing or minimizing a function usually involve derivatives. If a differentiable function f reaches a relative maximum or minimum at a point in the interior of an interval of its domain, then $f'(x)$ must be zero there. This helps in finding extreme values of f.

Applications of integrals. Integrals are used to compute areas bounded by curves in a plane. They are also used to compute other geometrical quantities, such as lengths of curves, areas of surfaces, and volumes of solids. For computing volumes and surface areas the natural tools are multiple integrals. Integrals are also used to compute quantities which occur in physics, such as center of mass, moment of inertia, work done by a variable force, and attraction due to gravitation. The concepts of first, second, and higher moments of a function, as used in statistics, are stated in terms of integrals.

Functions of several variables. To each pair (x_1, x_2) in a specified collection of number pairs, let there correspond a certain definite number y. This defines $y = f(x_1, x_2)$ as a function of two variables. Functions of three or more variables may also be considered. Differential calculus is extended to such functions through the study of partial derivatives with respect to the separate variables x_1, x_2, ... A concept of total differential is also relevant. For details *see* PARTIAL DIFFERENTIATION.

Integral calculus for functions of several variables is developed through the concept of a multiple integral. For instance, in the case of functions of two variables there is a concept of double integral. The theory of the subject deals with how the double integral is expressible in terms of two successive definite integrals of the sort occurring in calculus of functions of one variable. There is also a concept of a line integral, which is a kind of integral along a curve. There are relations between this concept, the concept of total differential, and the concept of double integral.

New developments. Calculus belongs to the branch of mathematics called analysis. One characteristic of analysis is its concern with infinite limiting processes. Current research in analysis is mostly on a level of development far beyond the elements of calculus. However, the ideas of calculus persist, though in much more generalized and abstract form, in the theory of functions and in functional analysis.

[ANGUS E. TAYLOR]

Bibliography: L. Bers and F. Karal, *Calculus*, 2d ed., 1976; R. Courant and F. John Fritz, *Introduction to Calculus and Analysis*, vols. 1 and 2, 1965, 1974; S. Lang, *A First Course in Calculus*, 4th ed., 1978; S. L. Salas and E. Hille, *Calculus*, 2d ed., 1974; S. K. Stein, *Calculus and Analytic Geometry*, 2d ed., 1977; A. E. Taylor and R. W. Mann, *Advanced Calculus*, 2d ed., 1972.

Calculus of tensors

The systematic study of tensors which led to an extension and generalization of vectors, begun in 1900 by two Italian mathematicians, G. Ricci and T. Levi-Civita, following G. F. B. Riemann's proposal concerning a generalization of euclidean geometry. The principal aim of the tensor calculus (absolute differential calculus) is to construct relationships which are generally covariant in the sense that these relationships or laws remain valid in all coordinate systems. The differential equations for the geodesics in a Riemannian space are covariant expressions; they yield a description of the geodesics which is valid for all coordinate systems. On the other hand, Newton's equations of motion require a preferred coordinate system for their description, namely, one for which force is proportional to acceleration (an inertial frame of reference). Thus Albert Einstein was led to a study of Riemannian geometry and the tensor calculus in order to construct the general theory of relativity.

Arithmetic, or vector, n-space. The coordinates of a point in a three-dimensional euclidean space are given by the triple of numbers (x,y,z) or (x^1,x^2,x^3), with $x=x^1$, $y=x^2$, $z=x^3$. The superscripts in x^i, $i=1, 2, 3$, are simply labels which enable one to distinguish and order the various elements in the triple of numbers. The totality of all number triples of the form (x^1,x^2,x^3) with the x^i, $i=1, 2, 3$, real, yields the arithmetic 3-space, designated as V_3. A simple generalization yields the arithmetic n-space, V_n. This space or manifold consists of all n-tuples of the form $(x^1,x^2, \ldots ,x^n)$, the x^i, $i=1, 2, \ldots , n$, taken as real numbers. Now a space of n dimensions is defined as any set of objects which can be put in 1:1 reciprocal correspondence with the arithmetic n-space. The 1:1 correspondence between the elements or points of the n-space and the arithmetic n-space can be chosen in many ways, and in general, the choice depends on the nature of the physical problem. In the special theory of relativity, an event is specified by the three space coordinates and the time, so that each event corresponds to a four-tuple $(x^1,x^2,x^3,x^4=ct)$.

Let a point P of an n-space correspond to the n-tuple $(x^1,x^2, \ldots ,x^n)$. Now consider the n equations, Eq. (1), and assume that one can solve for each x^i, yielding Eq. (2). It is assumed that

$$y^i=y^i(x^1,x^2, \ldots ,x^n) \qquad i=1,2, \ldots , n \qquad (1)$$

$$x^i=x^i(y^1,y^2, \ldots ,y^n) \qquad i=1,2, \ldots , n \qquad (2)$$

Eqs. (1) and (2) are single valued, and that the partial derivatives $\partial y^i/\partial x^j$, $\partial x^i/\partial y^j$, where $i, j= 1, 2, \ldots , n$, exist and are continuous. The n-space of which P is an element can also be put into 1:1 correspondence with the arithmetic n-space $(y^1,y^2, \ldots ,y^n)$. Thus, one has a new description of every point in the n-space. The points or elements of the n-space have not changed; one merely has a new description (coordinate system) of these points. Thus, Eq. (1) is called a transformation of coordinates.

The algebra of tensors is simplified by use of Einstein's summation convention. From the calculus, Eq. (1) yields Eq. (3). The index α occurs just twice in the last expression of Eq. (3), thus indicating a summation over this index. In a space of

$$\frac{\partial y^i}{\partial y^j}=\sum_{\alpha=1}^{n} \frac{\partial y^i}{\partial x^\alpha} \frac{\partial x^\alpha}{\partial y^j}=\frac{\partial y^i}{\partial x^\alpha} \frac{\partial x^\alpha}{\partial y^j} \qquad (3)$$

$$i,j=1,2, \ldots , n$$

three dimensions, the expression $S=a_{\alpha\beta}x^\alpha x^\beta$ requires a double summation because the indices α and β occur exactly twice. Thus,

$$S=\sum_{\beta=1}^{3} \sum_{\alpha=1}^{3} a_{\alpha\beta}x^\alpha x^\beta$$

and Einstein found it convenient to remove the summation signs. A further convenience occurs if the Kronecker delta is introduced. Thus, δ_j^i, $i,j= 1, 2, \ldots , n$, is a set of quantities whose numerical values are given by Eq. (4).

$$\delta_j^i=\begin{cases} 1 & \text{if} \quad i=j \\ 0 & \text{if} \quad i \neq j \end{cases} \qquad (4)$$

In particular

$$\delta_1^1=\delta_2^2= \cdots =\delta_n^n=1$$

$$\delta_2^1=\delta_3^1= \cdots =\delta_n^{n-1}=0$$

and for an n-space

$$\delta_\alpha^\alpha=\delta_1^1+\delta_2^2+ \cdots +\delta_n^n=n$$

Equation (3) may be written

$$\delta_j^i=\frac{\partial y^i}{\partial x^\alpha} \frac{\partial x^\alpha}{\partial y^i}$$

This expression yields

$$\left| \frac{\partial y^i}{\partial x^j} \right| \left| \frac{\partial x^\alpha}{\partial y^\beta} \right|=1$$

applying the rule for the product of two determinants, with

$$\left| \frac{\partial y^i}{\partial x^j} \right|$$

the Jacobian of the transformation in Eq. (1), and

$$\left| \frac{\partial x^i}{\partial y^j} \right|$$

the Jacobian of the inverse transformation in Eq. (2).

Contravariant vectors. In an n-space the locus of elements given by Eq. (5) represents a space

$$x^i=x^i(\lambda) \qquad i=1,2, \ldots , n \qquad \lambda_0 \leq \lambda \leq \lambda_1 \qquad (5)$$

curve Γ with λ a parameter. The n-tuple in notation (6), designated by $dx^i/d\lambda$, is defined to be a tangent

$$\left(\frac{dx^1}{d\lambda}, \frac{dx^2}{d\lambda}, \ldots , \frac{dx^n}{d\lambda} \right) \qquad (6)$$

element to the space curve, Eq. (5). Under an allowable coordinate transformation, Eq. (1), the space curve Γ can be represented by Eq. (7).

$$y^i=y^i(x^1(\lambda),x^2(\lambda), \ldots ,x^n(\lambda))=y^i(\lambda) \qquad (7)$$

$$i=1,2, \ldots , n$$

The components of the tangent vector for the $y=(y^1,y^2, \ldots ,y^n)$ coordinate system are given by notation (8), which represent the elements of the n-tuple, notation (6).

$$\frac{dy^i}{d\lambda} \qquad i=1,2, \ldots , n \qquad (8)$$

Certainly the x-coordinate system is no more important than the y-coordinate system as a description of the space curve Γ. One cannot say that $dx^i/d\lambda$ is the tangent vector; nor can one say that $dy^i/d\lambda$ is the tangent vector. If one considers all allowable coordinate systems, one obtains the entire class of tangent elements, each element (n-tuple) claiming to be the tangent vector for its particular coordinate system. It is the abstract collection of all these tangent elements that is specified as the tangent vector. To find what relationship exists between the components of the tangent vector when described by two different coordinate systems, Eq. (7) can be used to give Eq. (9). Note that at any point each component

$$\frac{dy^i}{d\lambda} = \frac{\partial y^i}{\partial x^\alpha} \frac{dx^\alpha}{d\lambda} \qquad i = 1, 2, \ldots, n \qquad (9)$$

$dy^i/d\lambda$ depends linearly on every component of $dx^i/d\lambda$. Moreover, Eq. (10) holds.

$$\frac{dx^i}{d\lambda} = \frac{\partial x^i}{\partial y^\alpha} \frac{dy^\alpha}{d\lambda} \qquad i = 1, 2, \ldots, n \qquad (10)$$

One can now form the following generalization: Any set of numbers $A^i(x^1, x^2, \ldots, x^n)$, $i = 1, 2, \ldots, n$, which transform according to the law shown by Eq. (11) under the coordinate transfor-

$$\overline{A}^i(\bar{x}^1, \bar{x}^2, \ldots, \bar{x}^n) = A^\alpha(x^1, x^2, \ldots, x^n) \frac{\partial \bar{x}^i}{\partial x^\alpha} \qquad (11)$$

$$i = 1, 2, \ldots, n$$

mation $\bar{x}^i = \bar{x}^i(x^1, x^2, \ldots, x^n)$, $i = 1, 2, \ldots, n$, is said to be a contravariant vector. The vector is not just the set of components in any coordinate system, but is rather the abstract element which is represented in each coordinate system x by the set of numbers $A^i(x)$. The $\overline{A}^i(\bar{x})$ depend linearly on the $A^\alpha(x)$; therefore the sum and difference of two vectors in an n-space are also vectors.

One immediately sees that the law of transformation for a contravariant vector is transitive. Let

$$\overline{A}^i = A^\alpha \frac{\partial \bar{x}^i}{\partial x^\alpha} \qquad \overline{\overline{A}}^i = \overline{A}^\alpha \frac{\partial \bar{\bar{x}}^i}{\partial \bar{x}^\alpha}$$

so that

$$\overline{\overline{A}}^i = \overline{A}^\beta \frac{\partial \bar{\bar{x}}^i}{\partial \bar{x}^\beta} = A^\alpha \frac{\partial \bar{x}^\beta}{\partial x^\alpha} \frac{\partial \bar{\bar{x}}^i}{\partial \bar{x}^\beta} = A^\alpha \frac{\partial \bar{\bar{x}}^i}{\partial x^\alpha}$$

which proves the statement.

If the components of a contravariant vector are known in one coordinate system, the components are known in all allowable coordinate systems from Eq. (11). A coordinate transformation does not yield a new vector; it merely changes the components of the same vector. Thus, a vector is said to be invariant under a coordinate transformation. Any object which is not changed under a coordinate transformation is called an invariant.

If $\phi(x^1, x^2, \ldots, x^n)$ is a scalar point function $\phi(x^1, x^2, \ldots, x^n) = \phi(x^1(\bar{x}), x^2(\bar{x}), \ldots, x^n(\bar{x})) = \phi(\bar{x}^1, \bar{x}^2, \ldots, \bar{x}^n)$, then Eq. (12) can be derived.

$$\frac{\partial \bar{\phi}}{\partial \bar{x}^i} = \frac{\partial \phi}{\partial x^\alpha} \frac{\partial x^\alpha}{\partial \bar{x}^i} \qquad (12)$$

The n-tuple

$$\left(\frac{\partial \phi}{\partial x^1}, \frac{\partial \phi}{\partial x^2}, \ldots, \frac{\partial \phi}{\partial x^n} \right)$$

yields the components of a covariant vector, called the gradient of ϕ.

More generally Eq. (13) holds, if one says that

$$\overline{A}_i(\bar{x}^1, \bar{x}^2, \ldots, \bar{x}^n) = A_\alpha(x^1, x^2, \ldots, x^n) \frac{\partial x^\alpha}{\partial \bar{x}^i} \qquad (13)$$

$$i = 1, 2, \ldots, n$$

the $A_\alpha(x)$ are the components of a covariant vector.

The law of transformation, Eq. (13), differs from that of Eq. (11) because in general,

$$\frac{\partial x^\alpha}{\partial \bar{x}^i} \neq \frac{\partial \bar{x}^i}{\partial x^\alpha}$$

However, for the group of orthogonal transformations $\bar{x}^i = a_\alpha{}^i x^\alpha$ such that $\mathbf{A} = \|a_j{}^i\|$ satisfies $\mathbf{A}^T = \mathbf{A}^{-1}$, one can show that $\partial \bar{x}^i/\partial x^\alpha = \partial x^\alpha/\partial \bar{x}^i$, and for this reason no distinction occurs between contravariant and covariant vectors in elementary vector analysis.

An important scalar invariant can be formed from a contravariant vector A^i and a covariant vector B_i. From their laws of transformation it follows that Eq. (14) holds.

$$\overline{A}^i \overline{B}_i = A^\alpha B_\beta \frac{\partial \bar{x}^i}{\partial x^\alpha} \frac{\partial x^\beta}{\partial \bar{x}^i} = A^\alpha B_\beta \delta_\alpha{}^\beta$$

$$= A^\alpha B_\alpha = A^i B_i \qquad (14)$$

Thus, the expression

$$A^i B_i = \sum_{i=1}^{n} A^i B_i$$

is invariant both in form and in its numerical value under a transformation of coordinates. One calls $A^\alpha B_\alpha$ the scalar, or inner, product of the two vectors A^i, B_i.

Tensors. The contravariant and covariant vectors discussed above are special cases of differential invariants called tensors. The components of a tensor are of the form

$$T^{a_1 a_2 \cdots a_r}_{b_1 b_2 \cdots b_s}(x^1, x^2, \ldots, x^n)$$

and the indices $a_1, a_2, \ldots, a_r, b_1, b_2, \ldots, b_s$ run through the integers $1, 2, \ldots, n$. The components transform according to the rule shown by Eq. (15).

$$\overline{T}^{a_1 a_2 \cdots a_r}_{b_1 b_2 \cdots b_s} = \left| \frac{\partial x}{\partial \bar{x}} \right|^N T^{a_1 a_2 \cdots a_r}_{\beta_1 \beta_2 \cdots \beta_s} \frac{\partial \bar{x}^{a_1}}{\partial x^{\alpha_1}}$$

$$\cdots \frac{\partial \bar{x}^{a_r}}{\partial x^{\alpha_r}} \frac{\partial x^{\beta_1}}{\partial \bar{x}^{b_1}} \cdots \frac{\partial x^{\beta_s}}{\partial \bar{x}^{b_s}} \qquad (15)$$

The exponent N of the Jacobian

$$\left| \frac{\partial x}{\partial \bar{x}} \right|$$

is called the weight of the tensor field. For $N = 0$ the tensor field is absolute; otherwise the tensor field is of weight N. A tensor density occurs for $N = 1$. The number of indices is $r + s$, the rank of the tensor. Vectors are tensors of rank one. The tensor of Eq. (15) is contravariant of order r and covariant of order s.

An important property of tensors is immediately evident from Eq. (15). If every component of a tensor vanishes in one coordinate system, the components vanish in all coordinate systems.

Two tensors are said to be of the same kind if they have the same order in their contravariant and covariant indices and if they are of the same weight. Additional tensors can be constructed as follows:

1. The sum and difference of two tensors of the same kind yields a new tensor of the same kind. This is apparent from the linear property of Eq. (15).

2. The product of two tensors is a new tensor. This can be shown for a special case. Let

$$\overline{T}_b{}^a = \left|\frac{\partial x}{\partial \bar{x}}\right| T_\alpha{}^\beta \frac{\partial x^\beta}{\partial \bar{x}^b} \frac{\partial \bar{x}^a}{\partial x^\alpha} \qquad \overline{S}^c = \left|\frac{\partial x}{\partial \bar{x}}\right|^3 S^\gamma \frac{\partial \bar{x}^c}{\partial x^\gamma}$$

Then $\overline{W}_b{}^{ac} \equiv \overline{T}_b{}^a \overline{S}^c = \left|\frac{\partial x}{\partial \bar{x}}\right|^4 (T_\beta{}^\alpha S^\gamma) \frac{\partial \bar{x}^a}{\partial x^\alpha} \frac{\partial \bar{x}^c}{\partial x^\gamma} \frac{\partial x^\beta}{\partial \bar{x}^b}$

$$= \left|\frac{\partial x}{\partial \bar{x}}\right|^4 W_\beta{}^{\alpha\gamma} \frac{\partial \bar{x}^a}{\partial x^\alpha} \frac{\partial \bar{x}^c}{\partial x^\gamma} \frac{\partial x^\beta}{\partial \bar{x}^b}$$

To illustrate the concept of contraction, consider the absolute tensor

$$\overline{A}_{kl}{}^{ij} = A_{\sigma\tau}{}^{\alpha\beta} \frac{\partial \bar{x}^i}{\partial x^\alpha} \frac{\partial \bar{x}^j}{\partial x^\beta} \frac{\partial x^\sigma}{\partial \bar{x}^k} \frac{\partial x^\tau}{\partial \bar{x}^l}$$

Replacing k by i and summing yields

$$\overline{B}_l{}^j \equiv \overline{A}_{il}{}^{ij} = A_{\sigma\tau}{}^{\alpha\beta} \frac{\partial \bar{x}^i}{\partial x^\alpha} \frac{\partial \bar{x}^j}{\partial x^\beta} \frac{\partial x^\sigma}{\partial \bar{x}^i} \frac{\partial x^\tau}{\partial \bar{x}^l}$$

$$= A_{\sigma\tau}{}^{\alpha\beta} \frac{\partial x^\sigma}{\partial x^\alpha} \frac{\partial \bar{x}^j}{\partial x^\beta} \frac{\partial x^\tau}{\partial \bar{x}^l}$$

$$= A_{\sigma\tau}{}^{\alpha\beta} \delta_\alpha{}^\sigma \frac{\partial \bar{x}^j}{\partial x^\beta} \frac{\partial x^\tau}{\partial \bar{x}^l}$$

$$= A_{\sigma\tau}{}^{\sigma\beta} \frac{\partial \bar{x}^j}{\partial x^\beta} \frac{\partial x^\tau}{\partial \bar{x}^l} = B_\tau{}^\beta \frac{\partial \bar{x}^j}{\partial x^\beta} \frac{\partial x^\tau}{\partial \bar{x}^l}$$

so that $B_\tau{}^\beta \equiv A_{\sigma\tau}{}^{\sigma\beta}$ is a mixed tensor. In general, one equates a certain covariant index with a contravariant index, sums on this repeated index, and obtains a new tensor whose rank is two less than that of the original tensor. This process of producing a new tensor is called the method of contraction.

A few examples of tensors can be listed.

1. The Kronecker delta is a mixed absolute tensor because

$$\delta_j{}^i \frac{\partial x^j}{\partial \bar{x}^\alpha} \frac{\partial \bar{x}^\beta}{\partial x^i} = \frac{\partial x^i}{\partial \bar{x}^\alpha} \frac{\partial \bar{x}^\beta}{\partial x^i} = \frac{\partial \bar{x}^\beta}{\partial \bar{x}^\alpha} = \bar{\delta}_\alpha{}^\beta$$

2. If A_i and B_i are absolute covariant vectors, then

$$\overline{C}_{ij} \equiv \overline{A}_i \overline{B}_j = A_\alpha B_\beta \frac{\partial x^\alpha}{\partial \bar{x}^i} \frac{\partial x^\beta}{\partial \bar{x}^j} = C_{\alpha\beta} \frac{\partial x^\alpha}{\partial \bar{x}^i} \frac{\partial x^\beta}{\partial \bar{x}^j}$$

so that $C_{\alpha\beta} \equiv A_\alpha B_\beta$ are the components of an absolute covariant tensor of rank two.

3. Let ϕ_i be the components of an absolute covariant vector. From $\overline{\phi}_i = \phi_\alpha (\partial x^\alpha / \partial \bar{x}^i)$, it' follows that

$$\overline{F}_{ij} \equiv \frac{\partial \overline{\phi}_i}{\partial \bar{x}^j} - \frac{\partial \overline{\phi}_j}{\partial \bar{x}^i} = \left(\frac{\partial \phi_\alpha}{\partial x^\beta} - \frac{\partial \phi_\beta}{\partial x^\alpha}\right) \frac{\partial x^\alpha}{\partial \bar{x}^i} \frac{\partial x^\beta}{\partial \bar{x}^j}$$

$$= F_{\alpha\beta} \frac{\partial x^\alpha}{\partial \bar{x}^i} \frac{\partial x^\beta}{\partial \bar{x}^j}$$

so that

$$F_{\alpha\beta} \equiv \frac{\partial \phi_\alpha}{\partial x^\beta} - \frac{\partial \phi_\beta}{\partial x^\alpha}$$

are the components of an absolute covariant

tensor. Note that $F_{\alpha\beta} = -F_{\beta\alpha}$, so that the tensor is skew symmetric. If the ϕ_i, $i = 1, 2, 3, 4$, are the components of the electromagnetic vector potential, the F_{ij} are the components of the electromagnetic field tensor.

Line element of Riemannian geometry. A simple generalization of the euclidean metric, or line element, $ds^2 = dx^2 + dy^2 + dz^2$, led G. F. B. Riemann to consider an n-space such that a metric is imposed on the space which yields the invariant distance ds between two nearby points whose coordinates differ by dx^i. The line element in a Riemannian n-space is given by Eq. (16).

$$ds^2 = g_{\alpha\beta}(x^1, x^2, \dots, x^n) \, dx^\alpha \, dx^\beta \qquad (16)$$

$$g_{\alpha\beta} = g_{\beta\alpha}$$

Under a coordinate transformation, $x^\alpha = x^\alpha(\bar{x})$, Eq. (16) becomes Eq. (17)

$$ds^2 = g_{\alpha\beta} \frac{\partial x^\alpha}{\partial \bar{x}^\sigma} \frac{\partial x^\beta}{\partial \bar{x}^\tau} d\bar{x}^\sigma \, d\bar{x}^\tau = \bar{g}_{\sigma\tau} d\bar{x}^\sigma \, d\bar{x}^\tau \quad (17)$$

$$\bar{g}_{\sigma\tau} = g_{\alpha\beta} \frac{\partial x^\alpha}{\partial \bar{x}^\sigma} \frac{\partial x^\beta}{\partial \bar{x}^\tau}$$

Hence the $g_{ij}(x)$ are the components of an absolute covariant tensor of rank two. One can show that the $g^{ij}(x)$ defined by $g^{ij} g_{jk} = \delta_k{}^i$ are the components of an absolute contravariant tensor, provided $|g_{ij}| \neq 0$. Furthermore, it can be shown that

$$|\bar{g}_{ij}| = |g_{ij}| \left|\frac{\partial x}{\partial \bar{x}}\right|^2$$

Thus, if A^i are the components of an absolute contravariant vector, then $|g_{\alpha\beta}|^{1/2} A^i$ are the components of a contravariant vector density.

Geodesics in a Riemannian space. If a space curve in a Riemannian n-space is given by $x^i = x^i(\lambda)$, $\lambda_0 \leqq \lambda \leqq \lambda_1$, the distance along the curve between the end points is given by Eq. (18).

$$L = \int_{\lambda_0}^{\lambda_1} \left(g_{\alpha\beta} \frac{dx^\alpha}{d\lambda} \frac{dx^\beta}{d\lambda}\right)^{1/2} d\lambda \qquad (18)$$

The geodesic path between two fixed points of a Riemannian space is that curve joining the two points which yields a minimum value for the integral of Eq. (18). The same path is obtained if one minimizes, as shown in notation (19).

$$\int_{P_0}^{P_1} \left(g_{\alpha\beta} \frac{dx^\alpha}{ds} \frac{dx^\beta}{ds}\right) ds \qquad (19)$$

Applying the Euler-Lagrange equations of the calculus of variations,

$$\frac{d}{ds}\left(\frac{\partial L}{\partial \dot{x}^i}\right) = \frac{\partial L}{\partial x^i}$$

with $L = g_{\alpha\beta} \dot{x}^\alpha \dot{x}^\beta$, yields the differential equations of the geodesics, Eq. (20), with Eq. (21) defining a term.

$$\frac{d^2 x^i}{ds^2} + \Gamma_{jk}{}^i(x) \frac{dx^j}{ds} \frac{dx^k}{ds} = 0 \qquad (20)$$

$$\Gamma_{jk}{}^i = \frac{1}{2} g^{i\sigma} \left(\frac{\partial g_{\sigma j}}{\partial x^k} + \frac{\partial g_{k\sigma}}{\partial x^j} - \frac{\partial g_{jk}}{\partial x^\sigma}\right) \qquad (21)$$

The important elements $\Gamma_{jk}{}^i$ are called the Christoffel symbols of the second kind. The elements $\{i, jk\} = g_{i\sigma} \Gamma_{jk}{}^\sigma$ are the Christoffel symbols of the first kind. Under a coordinate transformation, the Christoffel symbols transform according to the rule expressed in Eq. (22).

Thus, the Christoffel symbols are not the components of a mixed tensor. However, if the $\Gamma_{jk}{}^i(x)$ are given in an x-coordinate system, their values can be computed in an $\bar{x}$-coordinate system from Eq. (22).

$$\bar{\Gamma}_{jk}{}^i(\bar{x}) = \Gamma_{\beta\gamma}{}^\alpha(x)\frac{\partial x^\beta}{\partial \bar{x}^j}\frac{\partial x^\gamma}{\partial \bar{x}^k}\frac{\partial \bar{x}^i}{\partial x^\alpha} + \frac{\partial^2 x^\alpha}{\partial \bar{x}^j \partial \bar{x}^k}\frac{\partial \bar{x}^i}{\partial x^\alpha} \quad (22)$$

Covariant differentiation. From Eq. (11) follows Eq. (23), so that $\partial A^\alpha/\partial x^\beta$ are not the components of

$$\frac{\partial \bar{A}^i}{\partial \bar{x}^j} = \frac{\partial A^\alpha}{\partial x^\beta}\frac{\partial \bar{x}^i}{\partial x^\alpha}\frac{\partial x^\beta}{\partial \bar{x}^j} + A^\alpha \frac{\partial^2 \bar{x}^i}{\partial x^\beta \partial x^\alpha}\frac{\partial x^\beta}{\partial \bar{x}^j} \quad (23)$$

a mixed tensor. This is not surprising; in a euclidean space using coordinates other than rectangular coordinates, the unit vectors change directions from point to point. A differentiation process which yields a new tensor can be introduced by combining Eqs. (22) and (23). It can be shown, Eq. (24), that the terms are the components of a

$$A_{,j}{}^i \equiv \frac{\partial A^i}{\partial x^j} + A^\alpha \Gamma_{\alpha j}{}^i \quad (24)$$

mixed tensor. The term $A_{,j}{}^i$ is called the covariant derivative of A^i (the comma denoting covariant differentiation).

In a euclidean space using rectangular coordinates,

$$ds^2 = \sum_{i=1}^{n}(dx^i)^2$$

the $\Gamma_{jk}{}^i$ vanish from Eq. (21), and the covariant derivative reduces to the ordinary partial derivative.

The intrinsic derivative of A^i along a path $x^i = x^i(s)$, with s arc length, is given by Eq. (25).

$$\frac{\delta A^i}{\delta s} \equiv A_{,j}{}^i\frac{dx^j}{ds} = \frac{dA^i}{ds} + A^\alpha \Gamma_{\alpha j}{}^i\frac{dx^j}{ds} \quad (25)$$

The covariant derivative of the general tensor given by Eq. (15) is Eq. (26).

$$T_{\beta_1\beta_2\cdots\beta_s,j}^{\alpha_1\alpha_2\cdots\alpha_r} = \frac{\partial T_{\beta_1\beta_2\cdots\beta_s}^{\alpha_1\alpha_2\cdots\alpha_r}}{\partial x^j} + T_{\beta_1\beta_2\cdots\beta_s}^{\mu\alpha_2\cdots\alpha_r}\Gamma_{\mu j}^{\alpha_1}$$
$$+ \cdots + T_{\beta_1\beta_2\cdots\beta_s}^{\alpha_1\alpha_2\cdots\alpha_{r-1}\mu}\Gamma_{\mu j}^{\alpha_r} - T_{\mu\beta_2\cdots\beta_s}^{\alpha_1\alpha_2\cdots\alpha_r}\Gamma_{\beta_1 j}^\mu$$
$$- \cdots T_{\beta_1\beta_2\cdots\beta_{s-1}\mu}^{\alpha_1\alpha_2\cdots\alpha_r}\Gamma_{\beta_s j}^\mu - NT_{\beta_1\beta_2\cdots\beta_s}^{\alpha_1\alpha_2\cdots\alpha_r}\Gamma_{\mu j}^\mu \quad (26)$$

A few examples of covariant differentiation now follow:

1. Let A_i be an absolute covariant vector. The curl of A_i is defined by Eq. (27).

$$\text{curl } A_i = A_{i,j} - A_{j,i}$$
$$= \frac{\partial A_i}{\partial x^j} - A_\alpha\Gamma_{ij}{}^\alpha - \frac{\partial A_j}{\partial x^i} + A_\alpha\Gamma_{ji}{}^\alpha$$
$$= \frac{\partial A_i}{\partial x^j} - \frac{\partial A_j}{\partial x^i} \quad (27)$$

2. The divergence of A^i is defined by Eq. (28)

$$\text{div } A^i = A_{,\alpha}{}^\alpha = \frac{\partial A^\alpha}{\partial x^\alpha} + A^\alpha\Gamma_{\alpha\beta}{}^\beta$$
$$= \frac{1}{\sqrt{|g|}}\frac{\partial}{\partial x^\alpha}(\sqrt{|g|}\,A^\alpha) \quad (28)$$

with $|g| = |g|_{ij}|$. The proof of Eq. (28) is omitted.

3. If ϕ is an absolute scalar, the gradient of ϕ is defined by $\phi_{,i} = \partial\phi/\partial x^i$. The associated vector of $\phi_{,i}$ is $g^{\alpha\beta}\phi_{,\beta} = g^{\alpha\beta}(\partial\phi/\partial x^\beta) = A^\alpha$. Applying Eq. (28) to A^α yields the Laplacian of ϕ given by Eq. (29).

$$\text{Lap } \phi = \frac{1}{\sqrt{|g|}}\frac{\partial}{\partial x^\alpha}\left(\sqrt{|g|}\,g^{\alpha\beta}\frac{\partial\phi}{\partial x^\beta}\right) \quad (29)$$

See CALCULUS OF VECTORS; RIEMANNIAN GEOMETRY. [HARRY LASS]

Bibliography: R. M. Bowden and C. C. Wang, *Introduction to Vectors and Tensors*, 2 vols., 1975; H. Lass, *Vector and Tensor Analysis*, 1950; A. J. McConnell, *Applications of Tensor Analysis*, 1931; I. S. Sokolnikoff, *Tensor Analysis*, 2d ed., 1964; J. L. Synge and A. Schild, *Tensor Calculus*, 1969; T. Y. Thomas, *Concepts from Tensor Analysis and Differential Geometry*, 2d ed., 1965.

Calculus of variations

An extension of the part of differential calculus which deals with maxima and minima of functions of a single variable. The functions of the calculus of variations depend in an essential way upon infinitely many independent variables. Classically these functions are usually integrals whose integrand depends on a function whose specification by any finite number of parameters is impossible. For example, let R be a smooth bounded region of a space of m variables, $x_1, x_2, \ldots, x_m$, let y be any function of some smooth class on R and its boundary into real numbers or into n-tuples of real numbers and taking specified values on the boundary, and let $f(x,y,p)$ be a smooth function of $2m+1$ variables $x_1, x_2, \ldots, x_m, y, p_1, p_2, \ldots, p_m$. Then the integral, Eq. (1), is a function on the space

$$J = \int\cdots\int_R f(x,y,y_x) \quad (1)$$

of functions y to the real numbers, and this space of functions is infinitely dimensional unless excessive restrictions are placed on it. Here y_x denotes the derivatives $\partial y/\partial x$, and throughout this article subscripts will be used to denote derivatives and occasionally where the context is clear to denote particular values.

The calculus of variations studies such functions and their maxima and minima. The limitation of the competing functions is made realistically, and with sufficient restrictions it is possible to arrive at a rewarding theory; these restrictions do not always include the fixed boundary conditions stated above.

Principal applications may be to physical systems involving flexible components or time dependent orbits; equilibrium positions or orbits may be determined by minimizing energy or action integrals. The problems are of mathematical interest because of intrinsic difficulties (largely related to lack of topological compactness of bounded regions in spaces of infinitely many dimensions) and possibly because more progress with difficult nonlinear problems has been made here than elsewhere.

Theoretical basis. The classical theorems of differential calculus which are used and generalized are the following:

1. Necessary conditions for a function f of a single variable x to attain a local minimum at $x = x_0$ are $f'(x_0) = 0$ if the derivative exists at x_0 and if

there are neighboring points of definition of f for arguments on each side of x_0; $f'(x_0) \geqq 0$ if the derivative exists at x_0 and if there are neighboring points of definition of f for arguments larger than x_0.

2. If $f'(x_0) = 0$, if f is defined at some points neighboring x_0, and if $f''(x_0)$ exists, then a necessary condition for f to attain a local minimum at $x = x_0$ is $f''(x_0) \geqq 0$; conversely the conditions $f'(x_0) = 0$ and $f''(x_0) > 0$ guarantee a local minimum at x_0.

One standard technique of the calculus of variations is to derive necessary conditions for minima by restricting variations to a set of admissible variations depending smoothly on a single parameter and then to apply these theorems 1 and 2 from differential calculus.

For example, in the integral above, suppose that the function $f(x,y,p)$ is defined for all (x,y) in some region of $(m+1)$-space and for all p, and that f and all its first and second partial derivatives are continuous in this region. Suppose that the admissible functions y are restricted to functions which have continuous second derivatives in R and which take on the required boundary values. Let $g(x)$ be such an admissible function lying in the interior of the region over which $f(x,y,p)$ is defined, and let $z(x)$ be any function with continuous second derivatives on R and taking the value 0 at all points on the boundary of R. Then for e neighboring zero, the function $y = g(x) + ez(x)$ is admissible in the problem (it has continuous second derivatives and takes the prescribed values on the boundary). Then $z(x)$ is an admissible variation. If g does indeed afford a local minimum to J and if it is specified that the functions $y = g(x) + ez(x)$ are close to g for e neighboring zero, then propositions 1 and 2 above may be applied to the function $J(e)$ of the single variable e, where $J(e)$ is the value taken by J for y defined in terms of g and z as above. Under the conditions stated, the derivative $J'(0)$ does exist, and it may be computed by differentiating under the integral sign. The result is shown in Eq. (2).

$$J'(0) = \int \cdots \int_R [z(x) f_y(x,g,g_x) + \sum_j z_{x,j} f_{p,j}(x,g,g_x)] \quad (2)$$

This expression must be zero for any admissible $z(x)$ according to proposition 1. Under the restrictions set here each term of the expression to be summed may be integrated by parts, the integration of the term with factor $z_{x,j}$ being with respect to the single variable x_j; for the formula $\int u\,dv = uv - \int v\,du$ (with boundary values assigned to the term uv) one sets $dv = z_{x,j}\,dx_j$ and takes the remainder of the term as u. This yields, after due account is taken of the vanishing of z on the boundary, Eq. (3). A fundamental technique of the

$$J'(0) = \int \cdots \int_R z(x)[f_y(x,g,g_x) - \sum_j \frac{\partial}{\partial x_j} f_{p,j}(x,g,g_x)] \quad (3)$$

calculus of variations lies in exploiting the necessity of this last integral being zero independent of the admissible variation $z(x)$. The integral divided by the measure of R is the average over R of z times the expression in brackets, and the requirement that any function of a sufficiently wide class

times a given function averages to zero is met only if the given function is itself zero. Care must be taken to assure that this fundamental condition is met—that is, that the admissible variations are sufficiently general to permit invoking this lemma. In this case the bracketed expression must be zero, and the necessary condition becomes a differential equation, the Euler equation. For the function in Eq. (4), so that Eq. (5) holds, the Euler equation is just the Laplace equation, Eq. (6).

$$f(x,y,p) = p_1{}^2 + p_2{}^2 \quad (4)$$

$$J = \int \int \left[\left(\frac{\partial y}{\partial x_1} \right)^2 + \left(\frac{\partial y}{\partial x_2} \right)^2 \right] \quad (5)$$

$$\partial^2 y/\partial x_1{}^2 + \partial^2 y/\partial x_2{}^2 = 0 \quad (6)$$

Second variations are studied in a similar way, applying proposition 2 above to $J''(0)$. These give rise to conditions of Jacobi for a minimum. Generally, there are a set of conditions including Euler's condition above, a condition of Weierstrass, a condition of Legendre which may be derived from the Weierstrass condition, and the Jacobi condition, all known as necessary conditions for a minimum. These are generally set in terms of inequalities, and when the conditions are strengthened to demand strong inequalities (excluding equality) the conditions become sufficient for many interesting problems.

Multidimensional derivatives. Much of the work on the calculus of variations is devoted to meticulous detail with regard to the number of derivatives assumed to be available for various functions, particularly the competitive admissible functions $y(x)$. If too many derivatives are assumed, minima may not exist; if too few are assumed, the solution might not be sufficiently smooth to be acceptable in the light of the original statement of the problem. In an attempt to use fewer derivatives, different approaches are used depending on the number of independent variables x. For the multidimensional case, an approach has been presented by A. Haar; it leads to extended complications which are not amenable to description here. Haar introduced an additional function which has the effect of replacing the Laplace equation in the simple Dirichlet problem stated above by the Cauchy-Riemann equations.

For the case of single integrals, however, a lemma of du Bois-Reymond is applicable. This lemma states that a function must be a constant if the average of its product with every function of a sufficiently broad class is zero, where the class may be restricted to have average value zero. The proof takes the given function to be $u(x)$, its average over the interval to be some constant c, and requires that $u(x) - c$ be one of the admissible functions for comparison. This leads to a requirement that $[u(x) - c]^2$ average zero, and this means that $u(x)$ is essentially c.

Additional illustrative material of general results obtainable will now be presented in terms of single integral problems. Here take Eq. (7) and enlarge

$$J = \int_a^b f(x,y,y')\,dx \quad (7)$$

the class of admissible curves somewhat. The precise nature of this enlargement will not be important for the moment, but assume that among the admissible curves $y = g(x)$, $a(0) \leqq x \leqq b(0)$ affords a minimum to J. Assume also that a family of curves $y = z(x,e)$, $a(e) \leqq x \leqq b(e)$ is admissible and that the limiting value of $z(x,e)$ as e tends to zero is $g(x)$. Most particularly, however, it is desired to avoid the restriction that $z_x(x,e) \to g'(x)$, which was implied above by the variation $y = g(x) + ez(x)$. This allows variations more general than the ones above with regard to the variations of the end points of the curve, the limits of integration, and the behavior of the derivatives.

Curves of the type of $z(x,e)$ introduced here may be admitted as neighboring $y(x)$ for e neighboring zero either by using $\max_x |z(x,e) - y(x)|$ (possibly modified by some measure of disparity between the intervals of definition) or $\int [z(x,e) - y(x)]^2$ as a measure of distance between two points in the function space. In either case, $J(e)$ may be defined as the value of the integral J with f evaluated for x, $z(x,e)$ and $z_x(x,e)$, and it is again possible to study $J'(0)$ if it exists. However, it should be noted that there is no reason for expecting $J'(0)$ to exist in this case, for the integrand involves derivatives z_x which may not vary continuously. Actually a simple example of length shows what may happen; if J is a length integral, the z curves may be taken as zigzagging broken lines approaching a straight line segment but always making an angle of perhaps 45° with it. Then $J(e)$ for any e not zero is $\sqrt{2}$ times $J(0)$. There is illustrated here an important property known as lower semicontinuity. Although $J(e)$ is not continuous, its limit as the argument varies is never less than its value for the limiting value of the argument. This property, along with some compactness properties, suffices for all purposes of the calculus of variations, but it must be noted that it is reasonable to expect semicontinuity but not continuity in problems of the calculus of variations.

The method used before may be used to derive a formula for $J'(0)$; it is Eq. (8), where Eq. (9) holds.

$$J'(0) = f[b_0, g(b_0), z_x(b_0, 0)] \cdot b'(0) - f[a_0, g(a_0), z_x(a_0, 0)] a'(0) + \bar{J} \quad (8)$$

$$\bar{J} = \int_{a_0}^{b_0} \{ z_e j_y[x, g, z_x(x,0)] + z_{xe} J_p[x, g, z_x(x,0)] \} \quad (9)$$

This may be integrated by parts to give either Eq. (10) or Eq. (11), where H is a function of x such that Eq. (12) holds.

$$\bar{J} = \left\{ z_e J_p[x,g,z_x] \right\}_{a_0}^{b_0} + \int_{a_0}^{b_0} z_e \left\{ J_y - \frac{d}{dx} J_p \right\} \quad (10)$$

$$\bar{J} = \left\{ z_e H \right\}_{a_0}^{b_0} + \int_{a_0}^{b_0} z_{xe} \{ J_p - H \} \quad (11)$$

$$H' = J_y[x, g(x), g'(x)] \quad (12)$$

In all these formulas $z_z(x,0)$ has been retained instead of $g'(x)$ to indicate ambiguity in case of discontinuous derivatives.

End-point problems. If variations with fixed end points are admissible near g, it is possible to choose z with $z_e = 0$ at the end points, and so that a' and b' are zero for $e = 0$ and only $\bar{J}$ is left, so that $J'(0) = 0$ implies $\bar{J} = 0$. In the expression above for $\bar{J}$, Eq. (11), the first term is zero since $z_e = 0$ at the end points, the average of z_{xe} over the interval is zero, since $z_e = 0$ at both end points, and the lemma of du Bois-Reymond mentioned above may be applied. This means that the expression in curly brackets in the integrand is constant whether

$J_p(x,g,g')$ is known to be differentiable or not. Since H is obviously differentiable if g and z have first derivatives which are continuous on each of a finite set of closed intervals covering $[a,b]$, then the whole expression is differentiable, and the Euler equation, Eq. (6), follows under these weakened conditions. Furthermore, under some conditions of regularity, it is demonstrable that the solutions of this Euler equation have two continuous derivatives, so that minimizing functions in a wide class of functions are comfortably smooth.

The same formula may be used to get conditions which must exist in a corner. It is also applicable to problems with variable end conditions. It may be applied to a special variation to get the Weierstrass condition (see illustration).

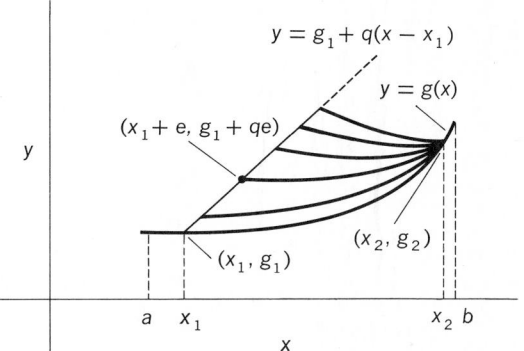

Weierstrass condition. Symbols explained in text.

These are curves with no general properties except that they vary slowly, approaching straightness at the top and conforming with the continuing curve at the bottom.

Here the lower curve $y = g(x)$ is assumed to be the extremal, and at a point $x = x_1$ on this curve a line with slope q is erected. The variations corresponding to e are taken by proceeding along g to x_1, then up the line to a point with abscissa $x_1 + e$, then along a curve of a smooth family rejoining g at a fixed point x_2, and then along g to b. The family between the line and the point of rejoining g is assumed to be smooth and to approach g with the $z_x(x,e)$ approaching g'. Thus the difficulty about limits of $z_x(x,0)$ are avoided except at the point $x = x_1$. Note that the variation is defined only for nonnegative values of e, and hence the value $J'(0)$ must be nonnegative if g is a minimum, but not necessarily zero. Intelligent straightforward application of the formulas developed just above yields this condition in the form shown in Eq. (13).

$$E(x,g,g',q)$$
$$= f(x,g,q) - f(x,g,q') - (q - g') F_p(x,q,g')$$
$$\geqq 0 \quad (13)$$

In problems with variable end points (which end points may, for example, be constrained to lie on prescribed curves) the end-point conditions which appear in these formulas must be taken into account. Here it is frequently possible to assume that $\bar{J}$ in the formula is zero, because of admissibility of sufficiently large classes of variations with fixed end points. The remaining conditions are called transversality conditions.

Single-integral problems. In classical single-integral problems it is frequently possible to integrate the Euler equations into a family of curves depending on a finite set of parameters (the initial conditions, for example). However, in other cases it is not possible to do this, and a solution must be arrived at or proved to exist in more direct ways. Frequently a minimizing sequence of functions is used; that is, a set of functions y_n is chosen so that the values of J corresponding to y_n approach their greatest lower bound. If there is a convergent subsequence of these y_n and if J itself is lower semicontinuous, then a limit y to the sequence must afford a minimum to J.

Other methods of solution and of proving the existence of solutions depend on general studies of Hilbert spaces. In these studies the integrals appear as operators in Hilbert space, and they are frequently reduced to quadratic forms by some majorizing process.

Problem of Bolza. Here the admissible class of functions is restricted by differential conditions. Problems of Bolza include the classical isoperimetric problems, in which the length of the admissible curves is specified. The problem of Bolza gives rise to multiplier rules which are based on the implicit function theorem. The theorem is applied by noting that if the rank of a matrix which can be caused to arise is maximal, then the restrictions imposed on admissible curves may be retained unchanged (in particular, or more generally they may be changed arbitrarily) and the function J may still be changed arbitrarily, hence reduced. A submaximal rank for this matrix is a necessary condition for a minimum; using this result the problem of Bolza may be transformed to an ordinary problem whose integral is a linear combination of the integrand of J and functions describing the original restrictions. The coefficients of this linear combination are initially undetermined; they are called Lagrange multipliers.

Finally, studies in the spirit of differential calculus but not restricted to local minima should be noted. Here the interest includes not only minima but all critical points (that is, all functions satisfying the Euler condition), and also other studies. It is true that classical problems in mechanics may have stable solutions corresponding to nonminimizing critical points. For sufficiently restricted problems it can be shown that the homology group of the space for which $J \leqq c$ for any c does not change as c varies until a critical level is reached; this is a level of J at which a critical point exists. The change in homology groups at critical points is related closely to the nature of the critical point. For example, in a finite dimensional space: on a sphere the number of pits minus the number of simple passes plus the number of peaks is two; on a torus the corresponding number is zero; a pass with three grooves running down and three ridges running up from it is to be counted as two simple passes for topological reasons. This is the Morse theory. *See* CONFORMAL MAPPING; DIFFERENTIATION; INTEGRATION; TOPOLOGY.

[CHARLES B. TOMPKINS]

Bibliography: G. A. Bliss, *Calculus of Variations*, 1925; G. A. Bliss, *Lectures on the Calculus of Variations*, 1946; O. Bolza, *Lectures on the Calculus of Variations*, 1931; J. W. Craggs, *Calculus of Variations*, 1973; L. E. Elsgole, *Calculus of Varia-*

tions, 1962; R. Hermann, *Differential Geometry and the Calculus of Variations*, 1973; L. A. Pars, *An Introduction to the Calculus of Variations*, 1963.

Calculus of vectors

In its simplest form, a vector is a directed line segment. Physical quantities, such as velocity, acceleration, force, and displacement, are vector quantities, or simply vectors, because they can be represented by directed line segments. The algebra of vectors was initiated principally through the works of W. R. Hamilton and H. G. Grassmann in the middle of the 19th century, and brought to the form presented here by the efforts of O. Heaviside and J. W. Gibbs in the late 19th century. Vector analysis is a tool of the mathematical physicist, because many physical laws can be expressed in vector form.

Addition of vectors. Two vectors **a** and **b** are added according to the parallelogram law (Fig. 1). An equivalent definition is as follows: From the end point of **a**, a vector is constructed parallel to **b**, of the same magnitude and direction as **b**. The vector from the origin of **a** to the end point of **b** yields the vector sum $\mathbf{s} = \mathbf{a} + \mathbf{b}$ (Fig. 2). Any number of vectors can be added by this rule.

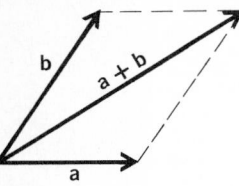

CALCULUS OF VECTORS

Fig. 1. Addition of two vectors.

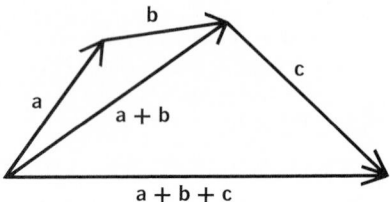

Fig. 2. Addition of three vectors.

Given a vector **a**, a class of vectors can be formed which are parallel to **a** but of different magnitudes. If x is a real number, the vector $x\mathbf{a}$ is defined to be parallel to **a** of magnitude $|x|$ times that of **a**. For $x > 0$, the two vectors **a** and $x\mathbf{a}$ have the same sense of direction, whereas for $x < 0$ the vector $x\mathbf{a}$ is in a reverse direction from that of **a**. The vector $-\mathbf{a}$ is the negative of the vector **a**, such that $\mathbf{a} + (-\mathbf{a}) = \mathbf{0}$, with **0** designated as the zero vector (a vector with zero magnitude). Subtraction of two vectors is defined by

$$\mathbf{a} - \mathbf{b} = \mathbf{a} + (-\mathbf{b})$$

The rules shown in notation (1), which conform to the rules of elementary arithmetic, can be readily deduced.

$$\begin{aligned}
\mathbf{a} + \mathbf{b} &= \mathbf{b} + \mathbf{a} \\
(\mathbf{a} + \mathbf{b}) + \mathbf{c} &= \mathbf{a} + (\mathbf{b} + \mathbf{c}) \\
x(\mathbf{a} + \mathbf{b}) &= x\mathbf{a} + x\mathbf{b} \\
x(y\mathbf{a}) &= (xy)\mathbf{a} \\
(x + y)\mathbf{a} &= x\mathbf{a} + y\mathbf{a} \\
0 \cdot \mathbf{a} &= \mathbf{0} \\
\mathbf{a} + \mathbf{0} &= \mathbf{a} \\
\mathbf{a} + \mathbf{b} = \mathbf{a} + \mathbf{c} \text{ implies } \mathbf{b} &= \mathbf{c} \\
\mathbf{a} = \mathbf{c}, \mathbf{b} = \mathbf{d} \text{ implies } \mathbf{a} + \mathbf{b} &= \mathbf{c} + \mathbf{d}
\end{aligned}$$

$|\mathbf{a} + \mathbf{b}| \leqq |\mathbf{a}| + |\mathbf{b}|$, with $|\mathbf{a}| =$ magnitude of **a**, etc.

(1)

Coordinate systems. The cartesian coordinate frame of analytic geometry is very useful for yield-

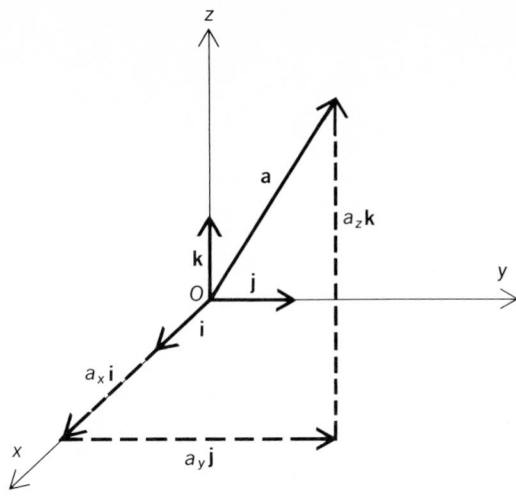

Fig. 3. Vectors in cartesian coordinate system.

ing a description of a vector (Fig. 3). The unit vectors **i**, **j**, **k** lie parallel to the positive x, y, and z axes, respectively. Any vector can be written as a linear combination of **i**, **j**, **k**. From Fig. 3, it is noted that Eq. (2) holds. Furthermore, the scalars

$$\mathbf{a} = a_x \mathbf{i} + a_y \mathbf{j} + a_z \mathbf{k} \qquad (2)$$

a_x, a_y, a_z are simply the projections of **a** on the x, y, and z axes, respectively, and are designated as the components of **a**. Thus, a_x is the x component of **a**, and so on. If the vector **b** is described by $\mathbf{b} = b_x \mathbf{i} + b_y \mathbf{j} + b_z \mathbf{k}$, then Eq. (3) holds.

$$\alpha \mathbf{a} + \beta \mathbf{b} = (\alpha a_x + \beta b_x) \mathbf{i} \\ + (\alpha a_y + \beta y_y) \mathbf{j} + (\alpha a_z + \beta b_z) \mathbf{k} \qquad (3)$$

In general, the components of a vector will be functions of the space coordinates and the time. To be more specific, consider a fluid in motion. At any time t the particle which is located at the point $P(x,y,z)$ will have velocity components which depend on the coordinates x, y, z, as well as the time t. Thus, the velocity field **v** of the fluid is represented by Eq. (4). A steady-state vector field

$$\mathbf{v} = v_x(x,y,z,t)\mathbf{i} + v_y(x,y,z,t)\mathbf{j} + v_z(x,y,z,t)\mathbf{k} \qquad (4)$$

exists if the components are time-independent. The force field of a fixed gravitating particle is of this type.

It is not necessary to describe a vector in terms of rectangular coordinates (Fig. 4). Let **a** be a

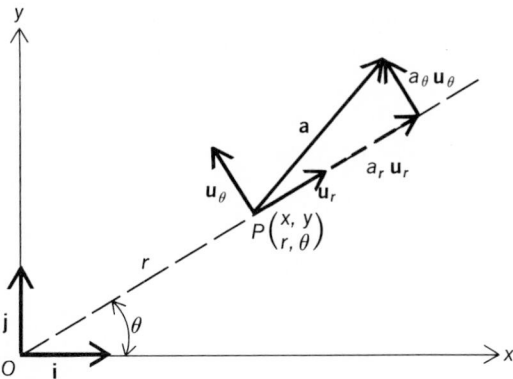

Fig. 4. Vectors in polar coordinate system.

vector in the xy plane with origin at the point $P(x,y)$. The point P can also be described in terms of polar coordinates (r,θ). Let $\mathbf{u}_r$, $\mathbf{u}_\theta$ be unit vectors in the directions of increasing r and θ, respectively. From Fig. 4 it follows that Eq. (5) holds

$$\mathbf{a} = a_x \mathbf{i} + a_y \mathbf{j} = a_r \mathbf{u}_r + a_\theta \mathbf{u}_\theta \qquad (5)$$

with

$$a_r = a_x \cos\theta + a_y \sin\theta \qquad a_\theta = -a_x \sin\theta + a_y \cos\theta$$

The components a_r, a_θ yield a description of the same vector **a**. Thus coordinate systems are simply a means of describing a vector. The vector is independent of the description.

Scalar or dot product of two vectors. From two vectors **a** and **b**, a scalar quantity is formed from the definition in Eq. (6), where θ is the angle between the two vectors when drawn from a common

$$\mathbf{a} \cdot \mathbf{b} = |\mathbf{a}| \cdot |\mathbf{b}| \cos\theta \qquad (6)$$

origin (Fig. 5). It is quickly verified that Eq. (7)

$$\mathbf{a} \cdot \mathbf{b} = \mathbf{b} \cdot \mathbf{a} \\ \mathbf{a} \cdot \mathbf{a} = |\mathbf{a}|^2 \\ (\mathbf{a}+\mathbf{b}) \cdot (\mathbf{c}+\mathbf{d}) = \mathbf{a} \cdot \mathbf{c} + \mathbf{a} \cdot \mathbf{d} + \mathbf{b} \cdot \mathbf{c} + \mathbf{b} \cdot \mathbf{d} \\ \mathbf{i} \cdot \mathbf{i} = \mathbf{j} \cdot \mathbf{j} = \mathbf{k} \cdot \mathbf{k} = 1 \\ \mathbf{i} \cdot \mathbf{j} = \mathbf{j} \cdot \mathbf{k} = \mathbf{k} \cdot \mathbf{i} = 0 \qquad (7)$$

holds. Here $\mathbf{a} \perp \mathbf{b}$ implies $\mathbf{a} \cdot \mathbf{b} = 0$, and conversely, provided $|\mathbf{a}| \cdot |\mathbf{b}| \neq 0$.

For $\mathbf{a} = a_x \mathbf{i} + a_y \mathbf{j} + a_z \mathbf{k}$, $\mathbf{b} = b_x \mathbf{i} + b_y \mathbf{j} + b_z \mathbf{k}$, it follows from Eq. (7) that Eq. (8) can be written.

$$\mathbf{a} \cdot \mathbf{b} = a_x b_x + a_y b_y + a_z b_z \qquad (8)$$

If **a** is a force field displaced along the vector **b**, then $\mathbf{a} \cdot \mathbf{b}$ represents the work performed by this force field.

Referring to Eq. (5), let $\mathbf{b} = b_x \mathbf{i} + b_y \mathbf{j} = b_r \mathbf{u}_r + b_\theta \mathbf{u}_\theta$. Then

$$\mathbf{a} \cdot \mathbf{b} = a_x b_x + a_y b_y = a_r b_r + a_\theta b_\theta \\ \equiv (a_x \cos\theta + a_y \sin\theta)(b_x \cos\theta + b_y \sin\theta) \\ + (-a_x \sin\theta + a_y \cos\theta)(-b_x \sin\theta + b_y \cos\theta)$$

Thus the scalar product of two vectors is independent of the descriptive coordinate system as is evident from the definition of the scalar product given by Eq. (6).

Vector or cross product of two vectors. In three-dimensional space, a vector can be formed from two vectors **a** and **b** in the following manner if they are nonparallel. Let **a** and **b** have a common origin defining a plane, and let **c** be that vector perpendicular to this plane of magnitude $|\mathbf{c}| = |\mathbf{a}||\mathbf{b}| \sin\theta$. If **a** is rotated into **b** through the angle θ, a right-hand screw will advance in the direction of **c** (Fig. 6). Thus Eq. (9) can be written.

$$\mathbf{c} = \mathbf{a} \times \mathbf{b} = |\mathbf{a}||\mathbf{b}| \sin\theta \, \mathbf{n} \qquad (9)$$

It follows that $\mathbf{a} \times \mathbf{b} = -(\mathbf{b} \times \mathbf{a})$, and that if **a** is parallel to **b**, $\mathbf{a} \times \mathbf{b} = 0$. Conversely, if $\mathbf{a} \times \mathbf{b} = 0$, then **a** is parallel to **b** provided $|\mathbf{a}||\mathbf{b}| \neq 0$.

The distributive law can be shown to hold for the vector product so that Eq. (10) holds.

$$(\mathbf{a}+\mathbf{b}) \times (\mathbf{c}+\mathbf{d}) \\ = \mathbf{a} \times \mathbf{c} + \mathbf{a} \times \mathbf{d} + \mathbf{b} \times \mathbf{c} + \mathbf{b} \times \mathbf{d} \qquad (10)$$

It follows from

$$\mathbf{i} \times \mathbf{i} = \mathbf{j} \times \mathbf{j} = \mathbf{k} \times \mathbf{k} = 0 \quad \mathbf{i} \times \mathbf{j} = \mathbf{k}, \mathbf{j} \times \mathbf{k} = \mathbf{i}, \mathbf{k} \times \mathbf{i} = \mathbf{j}$$

that for

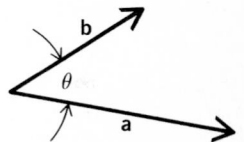

Fig. 5. Scalar product of two vectors.

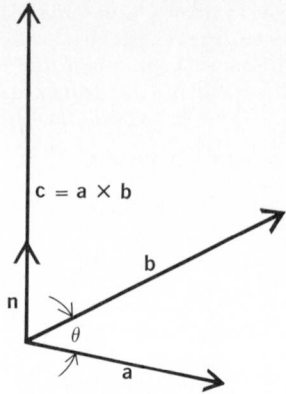

Fig. 6. Vector product of two vectors.

$$\mathbf{a} = a_x\mathbf{i} + a_y\mathbf{j} + a_z\mathbf{k} \qquad \mathbf{b} = b_x\mathbf{i} + b_y\mathbf{j} + b_z\mathbf{k}$$

Eq. (11) holds. The expression in Eq. (11) is to be

$$\mathbf{a} \times \mathbf{b} = \begin{vmatrix} \mathbf{i} & \mathbf{j} & \mathbf{k} \\ a_x & a_y & a_z \\ b_x & b_y & b_z \end{vmatrix} \qquad (11)$$

expanded by the ordinary rules governing determinants.

Multiple products involving vector and scalar products can be generated. The triple scalar product $\mathbf{a} \cdot (\mathbf{b} \times \mathbf{c})$ is given by Eq. (12). It can be shown

$$\mathbf{a} \cdot (\mathbf{b} \times \mathbf{c}) = \begin{vmatrix} a_x & a_y & a_z \\ b_x & b_y & b_z \\ c_x & c_y & c_z \end{vmatrix} \qquad (12)$$

that Eq. (13) is true. Geometrically, the scalar

$$\mathbf{a} \cdot (\mathbf{b} \times \mathbf{c}) = (\mathbf{a} \times \mathbf{b}) \cdot \mathbf{c} \equiv (\mathbf{abc}) \qquad (13)$$

triple product represents the volume of a parallelepiped formed with $\mathbf{a}$, $\mathbf{b}$, $\mathbf{c}$ as coterminous sides,

$$V = |\mathbf{a} \cdot (\mathbf{b} \times \mathbf{c})|$$

Of importance in the study of rigid-body motions is the triple vector product shown in Eq. (14).

$$\begin{aligned} \mathbf{a} \times (\mathbf{b} \times \mathbf{c}) &= (\mathbf{a} \cdot \mathbf{c})\mathbf{b} - (\mathbf{a} \cdot \mathbf{b})\mathbf{c} \\ (\mathbf{a} \times \mathbf{b}) \times \mathbf{c} &= (\mathbf{a} \cdot \mathbf{c})\mathbf{b} - (\mathbf{b} \cdot \mathbf{c})\mathbf{a} \end{aligned} \qquad (14)$$

From Eqs. (13) and (14) follow Eqs. (15) and (16).

$$\begin{aligned} (\mathbf{a} \times \mathbf{b}) \cdot (\mathbf{c} \times \mathbf{d}) &= (\mathbf{a} \cdot [\mathbf{b} \times (\mathbf{c} \times \mathbf{d})] \\ &= (\mathbf{a} \cdot \mathbf{c})(\mathbf{b} \cdot \mathbf{d}) - (\mathbf{a} \cdot \mathbf{d})(\mathbf{b} \cdot \mathbf{c}) \end{aligned} \qquad (15)$$

$$\begin{aligned} (\mathbf{a} \times \mathbf{b}) \times (\mathbf{c} \times \mathbf{d}) &= (\mathbf{acd})\mathbf{b} - (\mathbf{bcd})\mathbf{a} \\ &= (\mathbf{abd})\mathbf{c} - (\mathbf{abc})\mathbf{d} \end{aligned} \qquad (16)$$

Pseudovectors. It is easy to verify that a reflection of the space coordinates given by $x' = -x$, $y' = -y$, $z' = -z$, reverses the sign of the components of a vector. Under a space reflection, however, the components of the vector $\mathbf{a} \times \mathbf{b}$ do not change sign. This is seen from Eq. (11), for if a_x, a_y, a_z are replaced by $-a_x$, $-a_y$, $-a_z$, and if b_x, b_y, b_z are replaced by $-b_x$, $-b_y$, $-b_z$, the components of $\mathbf{a} \times \mathbf{b}$ remain invariant. Hence, $\mathbf{a} \times \mathbf{b}$ is not a true vector and therefore is given the title pseudovector. In electricity theory the magnetic field vector $\mathbf{B}$ is a pseudovector, whereas the electric field vector is a true vector provided the electric charge is a true scalar under a reflection of axes. The discussion of pseudovectors belongs properly to the domain of the tensor calculus. In general, vectors associated with rotations belong to the category of pseudovectors. In particular, the angular velocity vector associated with the motion of a rigid body is a pseudovector. *See* CALCULUS OF TENSORS.

Differentiation. There are three differentiation processes that are of conceptual value in the study of vectors; the gradient of a scalar, the divergence of a vector, and the curl of a vector.

Gradient of a scalar. The vector field $\mathbf{v}$ given by Eq. (17) has the property that the components of $\mathbf{v}$

$$\mathbf{v} = v_1(x,y,z)\mathbf{i} + v_2(x,y,z)\mathbf{j} + v_3(x,y,z)\mathbf{k} \qquad (17)$$

differ at individual points $P(x,y,z)$. The differential change in the individual components of $\mathbf{v}$ in moving from $P(x,y,z)$ to $Q(x+dx, y+dy, z+dz)$ is given by Eq. (18). It is suggestive to define the differential

$$dv_i = \frac{\partial v_i}{\partial x}dx + \frac{\partial v_i}{\partial y}dy + \frac{\partial v_i}{\partial z}dz \qquad i = 1, 2, 3 \quad (18)$$

of $\mathbf{v}$ by Eq. (19), because $\mathbf{i}$, $\mathbf{j}$, $\mathbf{k}$ are constant vectors (both in magnitude and direction).

$$d\mathbf{v} = dv_1\mathbf{i} + dv_2\mathbf{j} + dv_3\mathbf{k} \qquad (19)$$

If the components of a vector $\mathbf{v}$ are functions of a single parameter λ, the derivative of $\mathbf{v}$ with respect to λ is defined by Eq. (20).

$$\frac{d\mathbf{v}}{d\lambda} = \frac{dv_1}{d\lambda}\mathbf{i} + \frac{dv_2}{d\lambda}\mathbf{j} + \frac{dv_3}{d\lambda}\mathbf{k} \qquad (20)$$

The position vector of a particle is designated by $\mathbf{r} = x\mathbf{i} + y\mathbf{j} + z\mathbf{k}$. If the particle moves along a trajectory, the velocity and acceleration vectors associated with the motion of the particle are described by Eq. (21).

$$\mathbf{v} = \frac{d\mathbf{r}}{dt} = \frac{dx}{dt}\mathbf{i} + \frac{dy}{dt}\mathbf{j} + \frac{dz}{dt}\mathbf{k}$$

$$\mathbf{a} = \frac{d\mathbf{v}}{dt} = \frac{d^2\mathbf{r}}{dt^2} = \frac{d^2x}{dt^2}\mathbf{i} + \frac{d^2y}{dt^2}\mathbf{j} + \frac{d^2z}{dt^2}\mathbf{k}$$

$$(21)$$

The expression for the derivative of a vector described in curvilinear coordinates becomes more involved simply because the unit vectors in the curvilinear coordinate system need not remain fixed in space. Referring to Eq. (5),

$$\frac{d\mathbf{a}}{dt} = \frac{da_x}{dt}\mathbf{i} + \frac{da_y}{dt}\mathbf{j}$$

$$= \frac{da_r}{dt}\mathbf{u}_r + \frac{da_\theta}{dt}\mathbf{u}_\theta + a_r\frac{d\mathbf{u}_r}{dt} + a_\theta\frac{d\mathbf{u}_\theta}{dt}$$

From $\mathbf{u}_r = \mathbf{i}\cos\theta + \mathbf{j}\sin\theta$ and $\mathbf{u}_\theta = -\mathbf{i}\sin\theta + \mathbf{j}\cos\theta$, it follows that

$$\frac{d\mathbf{a}}{dt} = \left(\frac{da_r}{dt} - a_\theta\frac{d\theta}{dt}\right)\mathbf{u}_r + \left(\frac{da_\theta}{dt} + a_r\frac{d\theta}{dt}\right)\mathbf{u}_\theta$$

The differentiation rules shown in Eq. (22) exist.

$$\frac{d}{d\lambda}(\mathbf{u} \cdot \mathbf{v}) = \mathbf{u} \cdot \frac{d\mathbf{v}}{d\lambda} + \frac{d\mathbf{u}}{d\lambda} \cdot \mathbf{v}$$

$$\frac{d}{d\lambda}(\mathbf{u} \times \mathbf{v}) = \mathbf{u} \times \frac{d\mathbf{v}}{d\lambda} + \frac{d\mathbf{u}}{d\lambda} \times \mathbf{v} \qquad (22)$$

$$\frac{d}{d\lambda}(f\mathbf{v}) = f\frac{d\mathbf{v}}{d\lambda} + \frac{df}{d\lambda}\mathbf{v}$$

From the scalar $\phi(x,y,z)$, one can form the three partial derivatives $\partial\phi/\partial x$, $\partial\phi/\partial y$, $\partial\phi/\partial z$, from which the vector in Eq. (23) can be formed. The vector,

$$\left.\begin{array}{r}\text{gradient of }\phi\\\text{grad }\phi\\\text{del }\phi\equiv\nabla\phi\end{array}\right\}=\frac{\partial\phi}{\partial x}\mathbf{i}+\frac{\partial\phi}{\partial y}\mathbf{j}+\frac{\partial\phi}{\partial z}\mathbf{k}\quad(23)$$

grad ϕ, has two important properties. Grad ϕ is a vector field normal to the surface $\phi(x,y,z)=$ constant at every point of the surface. Moreover, grad ϕ yields that unique direction such that ϕ increases at its greatest rate. If $T(x,y,z)$ is the temperature at any point in space, then ∇T at any point yields that direction for which the temperature increases most rapidly. If $\phi(x,y,z)$ represents the electrostatic potential, then $\mathbf{E}=-\nabla\phi$ yields the electric field vector.

Divergence and curl of a vector. The del operator defined by Eq. (24) plays an important role in the

$$\nabla=\mathbf{i}\frac{\partial}{\partial x}+\mathbf{j}\frac{\partial}{\partial y}+\mathbf{k}\frac{\partial}{\partial z}\quad(24)$$

development of the differential vector calculus. For the vector field of Eq. (17), the divergence of $\mathbf{v}$ is defined by Eq. (25). The divergence of a vector is a scalar.

$$\operatorname{div}\mathbf{v}=\nabla\cdot\mathbf{v}=\frac{\partial v_1}{\partial x}+\frac{\partial v_2}{\partial y}+\frac{\partial v_3}{\partial z}\quad(25)$$

If ρ is the density of a fluid and $\mathbf{v}$ the velocity field of the fluid, then div $(\rho\mathbf{v})$ represents the rate of loss of mass of fluid per unit time per unit volume. The total loss of mass of fluid per unit time for a fixed volume V is given by the volume integral of notation (26).

$$\iiint_V \operatorname{div}(\rho\mathbf{v})\,dz\,dy\,dx\quad(26)$$

In electricity theory, the divergence of the displacement vector $\mathbf{D}$ is a measure of the charge density, $\nabla\cdot\mathbf{D}=\rho$.

By use of the del operator one obtains quite formally Eq. (27). The vector $\nabla\times\mathbf{v}$ is called the curl

$$\nabla\times\mathbf{v}=\begin{vmatrix}\mathbf{i}&\mathbf{j}&\mathbf{k}\\\frac{\partial}{\partial x}&\frac{\partial}{\partial y}&\frac{\partial}{\partial z}\\v_1&v_2&v_3\end{vmatrix}$$

$$=\left(\frac{\partial v_3}{\partial y}-\frac{\partial v_2}{\partial z}\right)\mathbf{i}+\left(\frac{\partial v_1}{\partial z}-\frac{\partial v_3}{\partial x}\right)\mathbf{j}$$

$$+\left(\frac{\partial v_2}{\partial x}-\frac{\partial v_1}{\partial y}\right)\mathbf{k}\quad(27)$$

of $\mathbf{v}$ (curl $\mathbf{v}$). Under a reflection of space coordinates, $\partial v_3/\partial y\rightarrow\partial(-v_3)/\partial(-y)=\partial v_3/\partial y$, and so on, so that the curl of $\mathbf{v}$ is a pseudovector. As has been noted previously, the class of pseudovectors is associated with rotations in space; thus, it is not strange that the curl of a velocity field is closely associated with an angular velocity vector field. The velocity of a fluid at a point $Q(x+dx,y+dy,z+dz)$ near $P(x,y,z)$ can be characterized as shown in Eq. (28), where $\mathbf{r}$ is the vector from P to Q and Eq. (29) holds. Now in general, $\boldsymbol{\omega}\times\mathbf{r}$ is the velocity

$$\mathbf{v}_Q=\mathbf{v}_P+\tfrac{1}{2}(\nabla\times\mathbf{v})_P\times\mathbf{r}+\tfrac{1}{2}\nabla(\mathbf{r}\cdot\mathbf{w})\quad(28)$$

$$\mathbf{w}=[(\mathbf{r}\cdot\nabla)\mathbf{v}]_P=\left(x\frac{\partial\mathbf{v}}{\partial x}+y\frac{\partial\mathbf{v}}{\partial y}+z\frac{\partial\mathbf{v}}{\partial z}\right)_P\quad(29)$$

of a point due to the angular velocity $\boldsymbol{\omega}$; thus, $\nabla\times\mathbf{v}=2\boldsymbol{\omega}$. The velocity at Q is simply the sum of a translatory velocity $\mathbf{v}_P$ plus a rigid-body rotational velocity $\boldsymbol{\omega}\times\mathbf{r}$, plus a nonrigid-body deformation, $\tfrac{1}{2}\nabla(\mathbf{r}\cdot\mathbf{w})$.

Formulas involving gradient, divergence, curl.

1. $\nabla(uv)=u\nabla v+v\nabla u$
2. $\nabla\cdot(\phi\mathbf{v})=\phi\,\nabla\cdot\mathbf{v}+(\nabla\phi)\cdot\mathbf{v}$
3. $\nabla\times(\nabla\phi)=\mathbf{0}$
4. $\nabla\cdot(\nabla\times\mathbf{v})=0$
5. $\nabla\times(\phi\mathbf{v})=\phi\,\nabla\times\mathbf{v}+(\nabla\phi)\times\mathbf{v}$
6. $\nabla\cdot(\mathbf{u}\times\mathbf{v})=(\nabla\times\mathbf{u})\cdot\mathbf{v}-(\nabla\times\mathbf{v})\cdot\mathbf{u}$
7. $\nabla\times(\mathbf{u}\times\mathbf{v})$
   $=(\mathbf{v}\cdot\nabla)\mathbf{u}-\mathbf{v}(\nabla\cdot\mathbf{u})+\mathbf{u}(\nabla\cdot\mathbf{v})-(\mathbf{u}\cdot\nabla)\mathbf{v}$
8. $\nabla(\mathbf{u}\cdot\mathbf{v})=\mathbf{u}\times(\nabla\times\mathbf{v})+\mathbf{v}\times(\nabla\times\mathbf{u})$
   $+(\mathbf{u}\cdot\nabla)\mathbf{v}+(\mathbf{v}\cdot\nabla)\mathbf{u}$
9. $\nabla\times(\nabla\times\mathbf{v})=\nabla(\nabla\cdot\mathbf{v})-\nabla^2\mathbf{v}$
10. $d\phi=d\mathbf{r}\cdot\nabla\phi+\frac{\partial\phi}{\partial t}dt\quad\text{for}\quad\phi=\phi(x,y,z,t)$
11. $\nabla\cdot(\nabla\phi)\equiv\nabla^2\phi=\frac{\partial^2\phi}{\partial x^2}+\frac{\partial^2\phi}{\partial y^2}+\frac{\partial^2\phi}{\partial z^2}$
12. If $\nabla\cdot\mathbf{f}=0$, then $\mathbf{f}=\nabla\times\mathbf{A}$ ($\mathbf{A}$ is called the vector potential)
13. If $\nabla\times\mathbf{f}=0$, then $\mathbf{f}=\nabla\phi$ (ϕ is called the scalar potential)

Integration.
If a closed surface is decomposed into a large number of small surfaces, a vector field normal to the surface can be constructed, each normal element being represented by $d\boldsymbol{\sigma}$. The magnitude of $d\boldsymbol{\sigma}$ is the area of the surface element dS, $d\boldsymbol{\sigma}=\mathbf{N}\,dS$.

If $\mathbf{f}$ is a vector field defined at every point of the surface, then notation (30) represents the total

$$\iint_S \mathbf{f}\cdot d\boldsymbol{\sigma}\quad(30)$$

flux of $\mathbf{f}$ through the surface S. The elements $d\boldsymbol{\sigma}$ point outward from the interior of S.

The divergence theorem of Gauss states that Eq. (31) is true, where R is the region enclosed by

$$\iint_S \mathbf{f}\cdot d\boldsymbol{\sigma}=\iiint_R (\nabla\cdot\mathbf{f})\,d\tau\quad(31)$$

S, $d\tau$ a volume element of R. For $\mathbf{f}=\rho\mathbf{v}$, the divergence theorem states that the net loss of fluid per unit time can be accounted for by measuring the total outward flux of the vector $\rho\mathbf{v}$ through the closed surface S bounding R.

The line integral of a vector field is described as follows: Let Γ be a space curve, and let $\mathbf{t}$ be the unit vector field tangent to Γ at every point of Γ in progressing from A to B, the initial and end points of the trajectory Γ.

The scalar integral in notation (32) is called the

$$\int_A^B (\mathbf{f}\cdot\mathbf{t})\,ds\quad(32)$$

line integral of $\mathbf{f}$ along Γ, with arc length s measured along Γ. If $\mathbf{f}$ is a force field, notation (32) represents the work performed by the force field if a unit test particle is taken from A to B along Γ.

The value of the integral of notation (32) will generally depend on the path from A to B. However, if $\mathbf{f}=\nabla\phi$, then

$$\int_A^B (\mathbf{f}\cdot\mathbf{t})\,ds=\int_A^B \nabla\phi\cdot d\mathbf{r}=\int_A^B d\phi=\phi(B)-\phi(A)$$

and the line integral is independent of the path of integration.

The theorem of Stokes states that Eq. (33) holds,

$$\oint_\Gamma \mathbf{f} \cdot d\mathbf{r} = \int\int_S (\nabla \times \mathbf{f}) \cdot d\boldsymbol{\sigma} \qquad (33)$$

where Γ is the boundary of the open surface S. If S is a closed surface, then

$$\oiint_S (\nabla \times \mathbf{f}) \cdot d\boldsymbol{\sigma} = 0$$

In electricity theory, a time-changing magnetic flux through an open surface induces a voltage in the boundary of the surface, so that

$$-\frac{\partial}{\partial t}\int\int_S \mathbf{B} \cdot d\boldsymbol{\sigma} = \oint_\Gamma \mathbf{E} \cdot d\mathbf{r}$$

Applying Stokes' theorem yields one of Maxwell's equations,

$$\nabla \times \mathbf{E} = -\frac{\partial \mathbf{B}}{\partial t}$$

See OPERATOR THEORY; POTENTIALS.

[HARRY LASS]

Bibliography: F. W. Bedford and T. D. Dwivedi, *Vector Calculus*, 1970; H. F. Davis and A. D. Snider, *Introduction to Vector Analysis*, 4th ed., 1979; O. D. Kellogg, *Foundations of Potential Theory*, 1929; N. Kemmer, *Vector Analysis*, 1977; H. Lass, *Vector and Tensor Analysis*, 1950; E. E. Wolstenholme, *Elementary Vectors: SI Units*, 3d ed., 1978.

Candlepower

Luminous intensity expressed in candelas. The term refers only to the intensity in a particular direction and by itself does not give an indication of the total light emitted. The candlepower in a given direction from a light source is equal to the illumination in footcandles falling on a surface normal to that direction, multiplied by the square of the distance from the light source in feet. The candlepower is also equal to the illumination of metercandles (lux) multiplied by the square of the distance in meters.

The apparent candlepower is the candlepower of a point source which will produce the same illumination at a given distance as produced by a given light source.

The mean horizontal candlepower is the average candlepower of a light source in the horizontal plane passing through the luminous center of the light source.

The mean spherical candlepower is the average candlepower in all directions from a light source as a center. Since there is a total solid angle of 4π (steradians) emanating from a point, the mean spherical candlepower is equal to the total luminous flux (in lumens) of a light source divided by 4π (steradians). *See* LUMINOUS INTENSITY; PHOTOMETRY.   [RUSSELL C. PUTNAM]

Canonical transformations

Transformations among the coordinates and momenta describing the state of a classical dynamical system which leave the canonical or Hamiltonian form of the equations of motion unchanged. *See*

HAMILTON'S EQUATIONS OF MOTION; HAMILTON'S PRINCIPLE.

General principles. The equations of motion of the system are derivable from Hamilton's principle once the Lagrangian of the system is given. Thus, Eq. (1) holds, where Δ is the variation in the inte-

$$\Delta \int_{t_1}^{t_2} L[q(t),\dot{q}(t),t]\,dt = 0 \qquad (1)$$

gral produced by a variation in the dependence of the q_j on t without varying the endpoints, implies Lagrange's equations of motion and hence also Hamilton's equations of motion. *See* LAGRANGE'S EQUATIONS; LAGRANGIAN FUNCTION.

This statement can also be written as Eq. (2),

$$\Delta \int_{t_1}^{t_2} \left(\sum_{j=1}^{f} p_j\dot{q}_j - H\right)dt = 0 \qquad (2)$$

where H is the Hamiltonian function. The content of Hamilton's principle is not changed if a total time derivative is added to the Lagrangian, since the variation of the integral of a total derivative vanishes identically. Thus, Eq. (2) is equivalent to Eq. (3), where ϕ is any differentiable function of the time and the old and new coordinates.

Equation (3) can be written as Eq. (4) provided that Eq. (5) is valid. Now, Eq. (4) is of the same form

$$\Delta \int_{t_1}^{t_2}\left[\sum_{j=1}^{f} p_j\dot{q}_j - H\right.$$
$$\left. - \frac{d}{dt}\phi\,(q_1,...,q_f;q'_1,...,q'_f;t)\right]dt = 0 \qquad (3)$$

$$\Delta \int_{t_1}^{t_2}\left[\sum_{j=1}^{f}(p'_j\dot{q}'_j) - H'\right]dt = 0 \qquad (4)$$

$$d\phi = \sum_{j=1}^{f}(p_j\,dq_j - p'_j\,dq'_j) - (H - H')\,dt \qquad (5)$$

as Eq. (2), but contains only primed variables, whereas Eq. (2) contains only unprimed variables. From Eq. (4), therefore, follow all the consequences that follow from Eqs. (2) or (1), but expressed in primed variables. The connection between the unprimed and the primed variables is given by Eq. (5), or equivalently by Eq. (6). Equa-

$$p_j = \frac{\partial\phi}{\partial q_j} \qquad p'_j = -\frac{\partial\phi}{\partial q'_j} \qquad H' = H + \frac{\partial\phi}{\partial t} \qquad (6)$$

tions (6) define a canonical transformation. In order to obtain the transformation explicitly, the first two sets of Eq. (6) must be solved for q', p', in terms of q, p, t. The quantity H' must be expressed as a function of primed variables before it can be used to get the equations of motion.

The function $\phi(q,q',t)$ is called the generator of the canonical transformation. The variables appearing in ϕ are the old and new coordinates and the time. The generator can be expressed as a function of any combination of old coordinates or momenta and of new coordinates or momenta. Thus with Eq. (7), it follows from Eqs. (5) and (7)

$$\psi(p,q',t) = \phi(q,q',t) - \sum_{j=1}^{f} p_j q_j \qquad (7)$$

that Eq. (8) holds. The same is true of Eq. (9) and

Eq. (10). There are thus four forms of generation for the same canonical transformation.

$$q_j = -\frac{\partial\psi}{\partial p_j} \qquad p'_j = -\frac{\partial\psi}{\partial q'_j} \qquad H' = H + \frac{\partial\psi}{\partial t} \quad (8)$$

$$\psi'(q,p',t) = \phi(q,q',t) + \sum_{j=1}^{f} p'_j q'_j \quad (9)$$

$$\phi'(p,p',t) = \phi(q,q',t) + \sum_{j=1}^{f} (p'_j q'_j - p_j q_j) \quad (10)$$

Canonical transformations constitute a group: The generator of the resultant of two successive canonical transformations is the sum of the generators of the separate transformations.

Thus, if Eqs. (11) and (12) hold, it follows that Eq. (13) is valid.

$$d\phi_1 = \sum_{j=1}^{f} (p_j dq_j - p'_j dq'_j) - (H - H') \, dt \quad (11)$$

$$d\phi_2 = \sum_{j=1}^{f} (p'_j dq'_j - p''_j dq''_j) - (H' - H'') \, dt \quad (12)$$

$$d(\phi_1 + \phi_2) = \sum_{j=1}^{f} (p_j dq_j - p''_j dq''_j) - (H - H'') \, dt \quad (13)$$

Applications. Canonical transformations have a number of important applications, of which the more important are briefly described.

Hamilton-Jacobi equation. A time-dependent canonical transformation changes the value of the Hamiltonian. If the transformation is properly chosen, the new Hamiltonian can be made to vanish, rendering the equations of motion trivial. A transformation which accomplishes this is generated by a complete integral of Eq. (14), which comes from

$$H\left(q, \frac{\partial\phi}{\partial q}, t\right) + \frac{\partial\phi}{\partial t} = 0 \quad (14)$$

Eq. (6) by setting $H' = 0$. Equation (14) is the Hamilton-Jacobi equation. A complete integral is a solution depending upon f constants in a nonadditive way. *See* HAMILTON-JACOBI THEORY.

Infinitesimal canonical transformations. If the new coordinates and momenta differ from the old only by small quantities, the transformation is called infinitesimal. Such a transformation is most easily described by a generator of the form of Eq. (7) as Eq. (15), where ϵ is an infinitesimal parameter.

$$\psi(p,q',t) = -\sum_{j=1}^{f} p_j q'_j + \epsilon X(q',p,t) \quad (15)$$

ter. Since ϵ^2 is regarded as negligible, this leads to Eq. (16). It is immaterial in Eq. (16) whether X is

$$q_k = q'_k - \epsilon\frac{\partial X}{\partial p_k} \qquad p'_k = p_k - \epsilon\frac{\partial X}{\partial q'_k} \quad (16)$$

regarded as a function of q', p, t or of q, p, t because X appears only in small terms, and q' differs from q by a small amount so that to first order in smallness, there is no difference. The quantity X is written now as $X(q,p,t)$. In this form, it is a function of the state of the system and is therefore a dynamical variable.

Let $q'_k - q_k = \delta q_k$, $p'_k - p_k = \delta p_k$. Then from Eq. (16), Eq. (17) is derived. Thus, if X is the z-compo-

$$\delta q_k = \epsilon\frac{\partial X}{\partial p_k} \qquad \delta p_k = -\epsilon\frac{\partial X}{\partial q_k} \quad (17)$$

nent of linear momentum of a particle, it generates an infinitesimal displacement in the z-direction. A component of angular momentum generates an infinitesimal rotation. The Hamiltonian generates the changes corresponding to a time-displacement, ϵ.

If $Y(q,p,t)$ is a second dynamical variable, the form of Y will change if its arguments are changed according to Eq. (17). Let $Y'(q', p', t)$ be expressed as Eq. (18). Then Eq. (19) holds.

$$Y'(q',p',t) = Y(q,p,t)$$
$$= Y(q',p',t) - \delta Y(q',p',t) \quad (18)$$

$$\delta Y = Y(q',p't) - Y(q,p,t)$$
$$= \sum_{j=1}^{f} \left(\frac{\partial Y}{\partial q_j}\delta q_j + \frac{\partial Y}{\partial p_j}\delta p_j\right)$$
$$= \sum_{j=1}^{f} \epsilon\left(\frac{\partial Y}{\partial q_j}\frac{\partial X}{\partial p_j} - \frac{\partial Y}{\partial p_j}\frac{\partial X}{\partial q_j}\right)$$
$$= \epsilon(Y,X) \quad (19)$$

Here (Y,X) is known as the Poisson bracket of the two dynamical variables Y and X. Clearly Eq. (20) is valid.

$$(Y,X) = -(X,Y) \quad (20)$$

If Y is the Hamiltonian H, then Eq. (21) can be

$$(X,Y) = (X,H) = \sum_{j=1}^{f} \left(\frac{\partial X}{\partial q_j}\frac{\partial H}{\partial p_j} - \frac{\partial X}{\partial p_j}\frac{\partial H}{\partial q_j}\right)$$
$$= \frac{dX}{dt} - \frac{\partial X}{\partial t} \quad (21)$$

written. Let $\partial X/\partial t = 0$. Equation (19) now reads as Eq. (22). If the Hamiltonian is invariant under the

$$\delta H = \epsilon(H,X) = -\epsilon(X,H) \quad (22)$$

infinitesimal canonical transformation generated by X, then X is a constant of the motion. This theorem connects the symmetry of a system directly with constants of motion.

Quantum mechanics is related to classical mechanics by defining a quantum mechanical commutator as $ih/2\pi$ times the Poisson bracket of the classically analogous dynamical variables. Thus, classically, the angular momentum components, j_x etc., satisfy the relation in Eq. (23). The corre-

$$(j_x,j_y) = j_z \quad (23)$$

sponding operators in quantum mechanics do not commute, but satisfy Eq. (24). *See* NONRELATIVISTIC QUANTUM THEORY.

$$j_x j_y - j_y j_x = \frac{ih}{2\pi}j_z \quad (24)$$

Perturbation theory. Let the Hamiltonian of a classical dynamical system be expressed as Eq. (25), where H_0 is simple and V is small. If $\phi_0(q,q_0,t)$

$$H = H_0 + V \quad (25)$$

is a complete integral of the Hamilton-Jacobi equation formed with H_0, Eq. (26), and $\phi(q,q',t)$ is a

$$H_0\left(q,\frac{\partial\phi_0}{\partial q},t\right)+\frac{\partial\phi_0}{\partial t}=0 \qquad (26)$$

complete integral of the Hamilton-Jacobi equation formed with H, then Eq. (27) holds, where ϕ_1 is small.

$$\phi=\phi_0+\phi_1 \qquad (27)$$

The exact Hamilton-Jacobi equation may be expanded in powers of ϕ_1, and in the first approximation, only linear terms in ϕ_1 and V retained, as in Eq. (28). From Eq. (26) and by use of Hamilton's

$$\begin{aligned}
&H_0\left(q,\frac{\partial\phi}{\partial q},t\right)+V\left(q,\frac{\partial\phi}{\partial q},t\right)\\
&=H_0\left(q,\frac{\partial\phi_0}{\partial q},t\right)+\left(\frac{\partial H_0}{\partial p}\right)_0\left(\frac{\partial\phi_1}{\partial q}\right)_0+V\left(q,\frac{\partial\phi_0}{\partial q},t\right)\\
&=-\frac{\partial\phi_0}{\partial t}-\frac{\partial\phi_1}{\partial t} \qquad (28)
\end{aligned}$$

equations of motion, Eq. (29) is obtained. The left

$$\left(\frac{\partial\phi_1}{\partial q}\right)_0\dot{q}_0+\frac{\partial\phi_1}{\partial t}=-V\left(q,\frac{\partial\phi_0}{\partial q},t\right) \qquad (29)$$

side of Eq. (29) is the total time derivative of ϕ_1 calculated on the unperturbed trajectory. To first order, Eq. (30) is written with the integral taken

$$\phi_1=-\int_{c_0}V\left(q,\frac{\partial\phi_0}{\partial q},t\right)dt \qquad (30)$$

along the unperturbed trajectory. Then, ϕ given by Eq. (27) is the generator of the canonical transformation correct to first order in V, or is the first approximation to the solution of the exact Hamilton-Jacobi equation. *See* PERTURBATION (MATHEMATICS); PERTURBATION (QUANTUM MECHANICS).

[PHILIP M. STEHLE]

Bibliography: See LAGRANGE'S EQUATIONS.

Capacitance

The ratio of the charge q on one of the plates of a capacitor (there being an equal and opposite charge on the other plate) to the potential difference v between the plates; that is, capacitance (formerly called capacity) is $C=q/v$.

In general, a capacitor, often called a condenser, consists of two metal plates insulated from each other by a dielectric. The capacitance of a capacitor depends on the geometry of the plates and the kind of dielectric used, since these factors determine the charge which can be put on the plates by a unit potential difference existing between the plates.

For a capacitor of fixed geometry and with constant properties of the dielectric between its plates, C is a constant independent of q or v, since as v changes, q changes with it in the same proportion. This statement assumes that the dielectric strength is not exceeded and thus that dielectric breakdown does not occur. (If it does occur, the device is no longer a capacitor.) If either the geometry or dielectric properties, or both, of a capacitor change with time, C will change with time.

In an ideal capacitor, no conduction current flows between the plates. A real capacitor of good quality is the circuit equivalent of an ideal capacitor with a very high resistance in parallel or, in alternating-current (ac) circuits, of an ideal capacitor with a low resistance in series.

Properties of capacitors. One classification system for capacitors follows from the physical state of their dielectrics.

Charging and discharging. These processes can occur for capacitors while the potential difference across the capacitor is changing if C is fixed; that is, q increases if v increases and q decreases if v decreases. If C and v both change with time, the rate of change of q with time is given by Eq. (1).

$$\frac{dq}{dt}=C\frac{dv}{dt}+v\frac{dC}{dt} \qquad (1)$$

Since the current i flowing in the wires leading to the capacitor plates is equal to dq/dt, Eq. (1) gives i in the wires. In many cases, C is constant so $i=C\,dv/dt$.

Energy of charged capacitor. This energy W_C is given by the formula $W_C=vq/2$, and is equal to the work the source must do in placing the charge on the capacitor. It is, in turn, the work the capacitor will do when it discharges.

Geometrical types. The geometry of a capacitor may take any one of several forms. The most common type is the parallel-plate capacitor whose capacitance C in farads is given in the ideal case by Eq. (2), where A is the area in square meters of

$$C=\frac{Ak\epsilon_0}{d} \qquad (2)$$

one of the plates, d is the distance in meters between the plates, ϵ_0 is the permittivity of empty space with the numerical value 8.85×10^{-12} coul^2/newton-m^2, and k is the relative dielectric constant of the dielectric between the plates. The value of k is unity for empty space and almost unity for gases. For other dielectrics, k ranges in value from one to several hundred. In order for Eq. (2) to give a good value of C for an actual capacitor, d must be very small compared to the linear dimensions of either plate. *See* DIELECTRIC CONSTANT; DIELECTRICS.

Each plate of a parallel-plate capacitor may be made up of many thin sheets of metal connected electrically with a corresponding number of thin sheets of metal making up the other plate. The sheets of metal and their intervening layers of dielectric are chosen and stacked in such a way that A will be large and d small without making the whole capacitor too bulky. The result is appreciable capacitance in a reasonable volume.

The spherical capacitor is made of two concentric metal spheres with a dielectric of relative dielectric constant k filling the space between the spheres. The capacitance in farads of such a capacitor is given by Eq. (3), where r_2 is the radius in

$$C=4\pi k\epsilon_0\frac{r_1r_2}{r_2-r_1} \qquad (3)$$

meters of the outer sphere and r_1 that of the inner sphere.

The cylindrical capacitor, as the name implies, is made of two concentric metal cylinders, each of length l in meters, with a dielectric filling the space between the cylinders. If r_2 and r_1 are the radii in meters of the outer and inner cylinders, respectively, and l is very large compared to $r_2 - r_1$, the capacitance C in farads is given by Eq. (4), where k is the relative dielectric constant

$$C = \frac{2\pi k \epsilon_0 l}{\ln (r_2/r_1)} \qquad (4)$$

of the dielectric and $\ln (r_2/r_1)$ indicates the natural logarithm of the ratio r_2/r_1.

Guard ring. This is often used with a standard parallel-plate capacitor, as shown in the diagram, in order that Eq. (2) shall more accurately represent its capacitance. It is the fringing of the electric lines of force which makes Eq. (2) inaccurate for an actual capacitor and, as shown in Fig. 2c, the fringing is nearly all at the outside edge of the guard ring, and thus is not associated with the charge Q which is put onto plate P_1 while the capacitor is being charged. It is the charge Q that determines the deflection of the ballistic galvanometer during the charging process. Then Eq. (2) gives the correct value of C that is needed to relate Q to the potential difference $\mathscr{E}$ across the plates by the equation $Q = C\mathscr{E}$. Thus, with C known from Eq. (2) and $\mathscr{E}$ known from a potentiometer measurement, Q may be computed and the ballistic galvanometer calibrated. This illustrates one use of a standard capacitor with a guard ring.

Body capacitance. When a part of the human body, say the hand, is brought near a high-impedance network, the body serves as one plate of a capacitor and the adjacent part of the network as the other plate. This situation is the equivalent of a capacitor of very low capacitance, in parallel between that part of the network and ground, since the human body can usually be considered as being a grounded conductor. This capacitance is known as body capacitance and enters as a part of the distributed capacitance of the network. A high-impedance network must be well shielded in order to eliminate the variable and undesirable effects of body capacitance. [RALPH P. WINCH]

Bibliography: Berkeley Physics Course, vol. 2: *Electricity and Magnetism*, 1970; R. Resnick and D. Halliday, *Physics*, 3d ed., 1977; F. W. Sears et al., *University Physics*, 5th ed., 1976.

Causality

In classical mechanics causality has been taken to mean that all the dynamical variables of a system can be precisely measured and their evolution in time is strictly determined by the forces. Thus, specifying the initial variables of position x, y, z and momentum p_x, p_y, p_z at time $t = t_0$ completely determines the future behavior of the system for $t > t_0$. Therefore, classically, one can repeatedly prepare a system in the same specific initial state with arbitrary precision and then measure the same final state after a given lapse in time.

In quantum mechanics there is also a rigorous law of causality. However, there are important differences. All the variables cannot be completely specified; there are inherent uncertainties δ such that, for example, $(\delta x)(\delta p_x) \geq$ Planck's constant. This Heisenberg uncertainty relation is completely built into the structure of quantum mechanics. Hence, one prepares an initial state and specifies a wave function $\psi(x,y,z,t_0)$ whose (absolute) square is interpreted as a probability distribution of finding the system at time t_0 at position x,y,z. The law of causality in quantum mechanics states that for an isolated system the initial state $\psi(x,y,z,t_0)$ completely determines, by the dynamics of the Schrödinger equation, the future state $\psi(x,y,z,t)$, which again leads to a probability distribution. To understand this physically, consider the scattering of protons off a target nucleus. Quantum mechanics can predict only an average distribution for the scattering process; thus it is the average of a large number of measured events that one accurately compares with theory. One also notes that in accord with the uncertainty principle there will always be some spread in momentum for the initial proton states. Furthermore, the detection of the protons disturbs the system (again because of the uncertainty principle), destroying the exact relation between ψ before and after the measurement. *See* QUANTUM MECHANICS.

Causality has also been used in reference to the principle that an event cannot precede its cause. Consider, for example, a packet of light (or electromagnetic radiation) having a sharp wave front incident on an obstacle. After the incident wave reaches the obstacle or scattering center (located at the origin) at time $t = 0$, scattered light moves out in all directions. Since light travels with the finite velocity c, causality requires that no scattered light be observed at any point P at a distance r from the scattering center until there is sufficient time for the incident light to reach P, that is, until $t > r/c$. This requirement places important restrictions on the behavior of the scattering process as a function of frequency ω. One writes the incident wave as

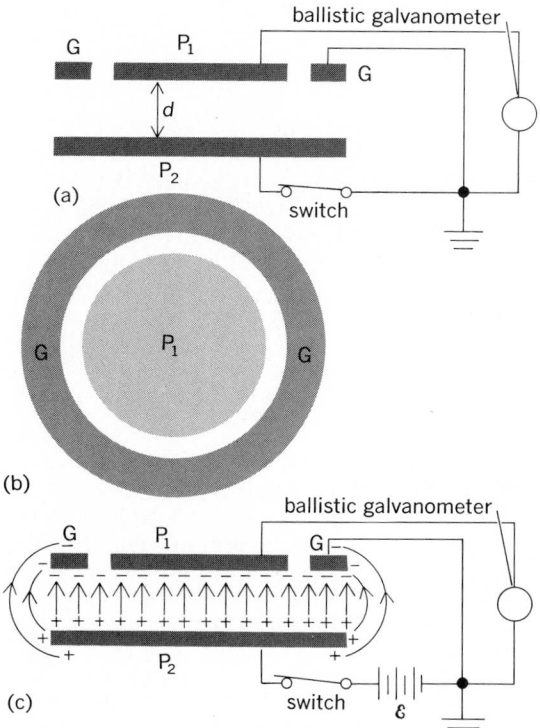

Parallel-plate capacitor P_1, P_2 with guard ring GG. Spacing between guard ring and plate is exaggerated. (*a*) Cross section with all parts of capacitor at ground potential. (*b*) Top view showing plate P_1 surrounded by guard ring. (*c*) Cross section of capacitor charged from battery whose electromotive force is $\mathscr{E}$.

expression (1). The statement that the incident wave does not reach the scattering center at the origin before $t = 0$ means that $A_I(0,t) = 0$ for $t < 0$.

Using Cauchy's residue theorem to evaluate (1) for $A_I(0,t)$ by contour integration, complet-

$$A_I(z,t) \propto \int_{-\infty}^{\infty} d\omega\, a(\omega)\, e^{i\omega(z/c - t)} \qquad (1)$$

ing the contour for $t < 0$ by an infinite semicircle in the upper-half complex ω plane where $Im\,\omega > 0$, it follows that $a(\omega)$ has no poles for $Im\,\omega > 0$. The scattered wave in the forward direction is given by expression (2), where $f(\omega)$ is the forward scatter-

$$A_S(z,t) \propto \int_{-\infty}^{\infty} d\omega\, f(\omega)\, a(\omega)\, e^{i\omega(z/c - t)} \qquad (2)$$

ing amplitude. Causality requires that $A_S(z,t) = 0$ for $(z/c - t) < 0$. It again follows (by a similar argument) that $f(\omega)$ has no poles for $Im\,\omega > 0$. This analyticity for $f(\omega)$ can be expressed in integral form, again using Cauchy's theorem, as in Eq. (3),

$$Re f(\omega) = \frac{1}{\pi} P \int_{-\infty}^{\infty} d\omega'\, \frac{Im f(\omega')}{\omega' - \omega} \qquad (3)$$

where P signifies that it is a principal value integral. Equation (3), which follows from causality [plus the assumption that $f(\omega) \to 0$ as $|\omega| \to \infty$], forms the basis of a large number of applications ranging from classical physics (for example, electric circuit theory) to relativistic quantum physics. (The causality principle in relativistic quantum field theory is expressed by the condition that the fields at different space-time points do not interfere with each other if the separation of the points is spacelike.) *See* RELATIVITY.

In particular, consider the scattering of light in an optical medium. Let $n(\omega)$ be the usual (real) geometrical index of refraction and $\gamma(\omega)$ the absorption coefficient of the medium. Here, one has for Eq. (3) the relationship shown as Eq. (4)

$$n(\omega) = 1 + \frac{c}{\pi} P \int_{0}^{\infty} \left(\frac{\omega'}{\omega}\right)^2 \frac{d\omega'}{\omega'^2 - \omega^2}\, \gamma(\omega') \qquad (4)$$

(Kramers-Kronig relation), relating the absorption process γ to the dispersive process n. Thus Eq. (4) is referred to as a dispersion relation, and the general equation, Eq. (3), is commonly called a dispersion relation even in its applications in high-energy physics. The forward dispersion relations for pion-nucleon scattering have been verified experimentally up to very high energies. *See* DISPERSION RELATIONS.

[GORDON L. SHAW]

Bibliography: J. Bjorken and S. Drell, *Relativistic Quantum Fields*, 1965; P. Fong, *Elementary Quantum Mechanics*, 1962; S. Gasiorowicz, *Elementary Particle Physics*, 1966; R. Newton, *Scattering Theory of Waves and Particles*, 1966.

Cayley-Klein parameters

A set of four complex numbers used to specify the orientation of a body, or equivalently, the rotation R which produces that orientation, starting from some reference orientation. They can be expressed in terms of the Euler angles ψ, θ, and ϕ, as in Eqs. (1).

$$\alpha = \cos\frac{\theta}{2}\, e^{-i(\psi - \phi)/2} \qquad \beta = -i\sin\frac{\theta}{2}\, e^{i(\psi - \phi)/2}$$
$$\gamma = -i\sin\frac{\theta}{2}\, e^{-i(\psi - \phi)/2} \qquad \delta = \cos\frac{\theta}{2}\, e^{i(\psi - \phi)/2} \qquad (1)$$

Often the set δ, $-\beta$, $-\gamma$, α is used with slightly different properties from those given here. The four complex numbers contain eight real numbers and satisfy relations in Eq. (2), where the bar de-

$$\delta = \bar{\alpha} \qquad \gamma = -\bar{\beta} \qquad \alpha\delta - \beta\gamma = \alpha\bar{\alpha} + \beta\bar{\beta} = 1 \qquad (2)$$

notes complex conjugate. These constitute five real conditions, leaving three independent parameters as required. *See* EULER ANGLES.

The Cayley-Klein parameters combine in a simple way under compound rotations when they are arranged in the square matrix array shown in (3).

$$\begin{pmatrix} \alpha & \beta \\ \gamma & \delta \end{pmatrix} \qquad (3)$$

If α, β, γ, δ correspond to the rotation R_1 and α', β', γ', δ', to R_2, then the parameters α'', β'', γ'', δ'' of the rotation $R_2 R_1$ (R_1 first, then R_2) are found by matrix multiplication. Thus Eq. (4) holds.

$$\begin{pmatrix} \alpha'' & \beta'' \\ \gamma'' & \delta'' \end{pmatrix} = \begin{pmatrix} \alpha' & \beta' \\ \gamma' & \delta' \end{pmatrix} \begin{pmatrix} \alpha & \beta \\ \gamma & \delta \end{pmatrix} \qquad (4)$$

A rotation $R(\psi,\theta,\phi)$ can be produced by three successive rotations. The corresponding decomposition of the Cayley-Klein parameter matrix is given by Eq. (5).

$$\begin{pmatrix} \alpha & \beta \\ \gamma & \delta \end{pmatrix} =$$
$$\begin{pmatrix} e^{-i\phi/2} & 0 \\ 0 & e^{i\phi/2} \end{pmatrix} \begin{pmatrix} \cos\frac{\theta}{2} & -i\sin\frac{\theta}{2} \\ -i\sin\frac{\theta}{2} & \cos\frac{\theta}{2} \end{pmatrix} \begin{pmatrix} e^{-i\psi/2} & 0 \\ 0 & e^{i\psi/2} \end{pmatrix} \qquad (5)$$

Although these parameters have been used to simplify somewhat the mathematics of spinning top motion, their main use is in quantum mechanics. There they are related to the Pauli spin matrices and represent the change in the spin state of an electron or other particle of half-integer spin under the space rotation $R(\psi,\theta,\phi)$. *See* MATRIX THEORY; SPIN (QUANTUM MECHANICS).

[BERNARD GOODMAN]

Bibliography: H. Goldstein, *Classical Mechanics*, 2d ed., 1980; W. Hauser, *Introduction to the Principles of Mechanics*, 1965.

Center of gravity

A fixed point in a material body through which the resultant force of gravitational attraction acts. The resultant of all forces or attractions produced by the Earth's gravity on a body constitutes its weight. This weight is considered to be concentrated at the center of gravity in mechanical studies of a rigid body. The location of the center of gravity for a body remains fixed in relation to the body regardless of the orientation of the body. If supported at its center of gravity, a body would remain balanced in its initial position. Coordinates, as for an aircraft, are conveniently chosen with origin at the center of gravity. *See* GRAVITY.

[NELSON S. FISK]

Center of mass

That point of a material body or system of bodies which moves as though the system's total mass existed at the point and all external forces were applied at the point. The Earth-Moon system moves in the Sun's gravitational field as though both masses were located at a center of mass some 3000 mi from the Earth's geometric center. The function of the center-of-mass concept is to permit analysis of the motion of an entire system as distinguished from that of its individual parts.

Consider a system of mass M composed of n bodies with masses $m_1, m_2, \ldots, m_n$ and radius vectors $r_1, r_2, \ldots, r_n$ measured from some common reference point. Separate the force on each body into internal and external components ${}^i f$ and ${}^e f$. The motion of the jth body is given by Eq. (1). The sum of the motions of all bodies is given by Eq. (2). By Newton's third law, $\sum_j {}^i f_j$ is zero, and sum of the external forces acts as a single resultant force as in Eq. (3).

$$m_j \frac{d^2 r_j}{dt^2} = {}^e f_j + {}^i f_j \qquad (1)$$

$$\sum_j m_j \frac{d^2 r_j}{dt^2} = \sum_j {}^e f_j + \sum_j {}^i f_j \qquad (2)$$

$$\sum_j {}^e f_j = F \qquad (3)$$

Define a point with radius vector R, such that Eq. (4) holds. Then Eq. (5) can be written, and by substitution from Eq. (2), Eqs. (6) and (7) are obtained.

$$MR = \sum_j m_j r_j \qquad (4)$$

$$\frac{d^2}{dt^2}(MR) = \frac{d^2}{dt^2}\sum_j m_j r_j = \sum_j m_j \frac{d^2 r_j}{dt^2} \qquad (5)$$

$$\frac{d^2}{dt^2}(MR) = F \qquad (6)$$

$$\frac{d^2 R}{dt^2} = \frac{F}{M} \qquad (7)$$

Equation (7), an expression of Newton's second law, states that the center of mass at R moves as though it possessed the total mass of the system and were acted upon by the total external force.

A simplification of the description of collisions can be obtained by using a coordinate system which moves with the velocity of the center of mass before collision. See COLLISION; RIGID-BODY DYNAMICS.

[JOHN P. HAGEN]

Center of pressure

A point on a plane surface through which the resultant force due to pressure passes. Such a surface can be supported by a single mounting fixture at its center of pressure if no other forces act. For example, a water gate in a dam can be supported by a single shaft at its center of pressure. For a nonvertical plane surface submerged in a motionless liquid, the center of pressure is vertically below the centroid of the volume of liquid above the surface. For a vertical submerged sur-

face, the distance from the surface of the liquid to the center of pressure is y_p given by $y_p = I/y_c A$, where A is the area of submerged surface, y_c is the distance below liquid surface to the centroid of area, and I is the moment of inertia of area A about the intersection line of the vertical surface with the liquid surface. See MOMENT OF INERTIA.

[NELSON S. FISK]

Centrifugal force

A fictitious or pseudo outward force on a particle rotating about an axis which by Newton's third law is equal and opposite to the centripetal force. Like all such action-reaction pairs of forces, they are equal and opposite but do not act on the same body and so do not cancel each other. Consider a mass M tied by a string of length R to a pin at the center of a smooth horizontal table and whirling around the pin with an angular velocity of ω radians per second. The mass rotates in a circular path because of the centripetal force $F_C = M\omega^2 R$ which is exerted on the mass by the string. The reaction force exerted by the rotating mass M, the so-called centrifugal force, is $M\omega^2 R$ in a direction away from the center of rotation. See CENTRIPETAL FORCE.

From another point of view, consider an experimenter in a windowless, circular laboratory that is rotating smoothly about a centrally located vertical axis. No object remains at rest on a smooth surface; all such objects move outward toward the wall of the laboratory as though an outward, centrifugal force were acting. To the experimenter partaking in the rotation, in a rotating frame of reference, the centrifugal force is real. An outside observer would realize that the inward force which the experimenter in the rotating laboratory must exert to keep the object at rest does not keep it at rest, but furnishes the centripetal force required to keep the object moving in a circular path. The concept of an outward, centrifugal force explains the action of a centrifuge.

[C. E. HOWE/R. J. STEPHENSON]

Centripetal force

The inward force required to keep a particle or an object moving in a circular path. It can be shown that a particle moving in a circular path has an acceleration toward the center of the circle along a radius. See ACCELERATION.

This radial acceleration, called the centripetal acceleration, is such that, if a particle has a linear or tangential velocity v when moving in a circular path of radius R, the centripetal acceleration is v^2/R. If the particle undergoing the centripetal acceleration has a mass M, then by Newton's second law of motion the centripetal force F_C is in the direction of the acceleration. This is expressed by Eq. (1), where ω is the constant angular velocity

$$F_C = Mv^2/R = MR\omega^2 \qquad (1)$$

and is equal to v/R. From the laws of motion given by Isaac Newton in his treatise *The Principia*, it follows that the natural motion of an object is one with constant speed in a straight line, and that a force is necessary if the object is to depart from this type of motion. Whenever an object moves in a curve, a centripetal force is necessary. In circular motion the tangential speed is constant but is changing direction at the constant rate of ω, so the

CENTRIPETAL FORCE

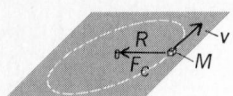

Fig. 1. Diagram of centripetal force in circular motion.

CENTRIPETAL FORCE

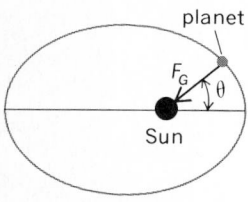

Fig. 2. Planet moving in an ellipse about the Sun.

centripetal force along the radius is the only force involved.

For example, a small heavy object tied by a cord of length R, with the other end of the cord held in the hand, can be whirled in a horizontal circular path. The person holding the cord has to exert an inward pull, the centripetal force, to maintain this circular motion (Fig. 1). When turning a corner in a car, the centripetal force necessary for the turn is provided by the frictional force between the road and the wheels. If the road is banked, then a component of the weight of the car can supply the necessary centripetal force. For a given angle of banking and radius of turn it is necessary to maintain a certain speed, if the component of weight is to exactly provide the necessary centripetal force. *See* CENTRIFUGAL FORCE.

As another example (Fig. 2), consider the motion of the Earth or other planet about the Sun. The planets move in elliptical paths about the Sun. This motion is maintained by the gravitational attraction of the planet by the Sun along a line joining the planet and Sun. In this case the radial distance between planet and Sun changes during the motion, so that the gravitational force is not in general equal to the centripetal force. If F_G is the gravitational force of attraction between the planet and Sun and R the distance between them, then Eq. (2)

$$F_G = M(a_R - R\omega^2) = M[d^2R/dt^2 - R\,(d\theta/dt)^2] \quad (2)$$

holds, where M is the mass of the planet, $a_R = d^2R/dt^2$ is the acceleration along the radial distance R, and $d\theta/dt$ is the instantaneous angular velocity. Since the path of a planet about the Sun is elliptical rather than circular, another force, besides the centripetal one, is required.

[R. J. STEPHENSON]

Bibliography: U. Haber-Schaim et al., *PSSC Physics*, 4th ed., 1976; F. Miller, *College Physics*, 4th ed., 1977; R. Resnick and D. Halliday, *Physics*, 3d ed., 1977; F. W. Sears et al., *University Physics*, 5th ed., 1976; R. J. Stephenson, *Mechanics and Properties of Matter*, 3d ed., 1969.

Cerenkov radiation

Light emitted by a high-speed charged particle when the particle passes through a transparent, nonconducting, solid material at a speed greater than the speed of light in the material. The blue glow observed in the water of a nuclear reactor, close to the active fuel elements, is radiation of this kind. The emission of Cerenkov radiation is analogous to the emission of a shock wave by a projectile moving faster than sound, since in both cases the velocity of the object passing through the medium exceeds the velocity of the resulting wave disturbance in the medium. This radiation, first predicted by P. A. Cerenkov in 1934 and later substantiated theoretically by I. Frank and I. Tamm, is used as a signal for the indication of high-speed particles and as a means for measuring their energy in devices known as Cerenkov counters.

Direction of emission. Cerenkov radiation is emitted at a fixed angle θ to the direction of motion of the particle, such that $\cos\theta = c/nv$, where v is the speed of the particle, c is the speed of light in vacuum, and n is the index of refraction of the medium. The light forms a cone of angle θ around the direction of motion. If this angle can be measured, and n is known for the medium, the speed of

the particle can be determined. The light consists of all frequencies for which n is large enough to give a real value of $\cos\theta$ in the preceding equation.

Cerenkov counters. Particle detectors which utilize Cerenkov radiation are called Cerenkov counters. They are important in the detection of particles with speeds approaching that of light, such as those produced in large accelerators and in cosmic rays, and are used with photomultiplier tubes to amplify the Cerenkov radiation. These counters can emit pulses with widths of $\sim 10^{-10}$ sec, and are therefore useful in time-of-flight measurements when very short times must be measured. They can also give direct information on the velocity of the passing particle. *See* PARTICLE DETECTOR.

Dielectrics such as glass, water, or clear plastic may be used in Cerenkov counters. Choice of the material depends on the velocity of the particles to be measured, since the values of n are different for the materials cited. By using two Cerenkov counters in coincidence, one after the other, with proper choice of dielectric, the combination will be sensitive to a given velocity range of particles.

The counters may be classified as nonfocusing or focusing. In the former type, the dielectric is surrounded by a light-reflecting substance except at the point where the photomultiplier is attached, and no use is made of the directional properties of the light emitted. In a focusing counter, lenses and mirrors may be used to select light emitted at a given angle and thus to give information on the velocity of the particle.

Cerenkov counters may be used as proportional counters, since the number of photons emitted in the light beam can be calculated as a function of the properties of the material, the frequency interval of the light measured, and the angle θ. Thus the number of photons which make up a certain size pulse gives information on the velocity of the particle.

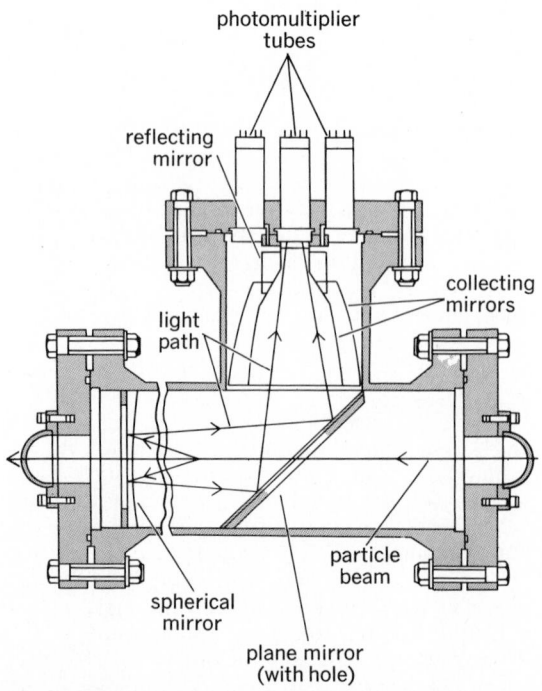

Differential gas Cerenkov counter.

Gas, notably CO_2 (see illustration), may also be used as the dielectric in Cerenkov counters. In such counters the intensity of light emitted is much smaller than in solid or liquid dielectric counters, but the velocity required to produce a count is much higher because of the low index of refraction of gas.

[WILLIAM B. FRETTER]

Bibliography: S. J. Lindenbaum and L. C. L. Yuan, Cerenkov counters, in L. C. L. Yuan and C. S. Wu (eds.), Nuclear Physics, pp. 162–194, vol. 5A of L. Marton (ed.), Methods of Experimental Physics, 1961; J. Litt and R. Meunier, Cerenkov counter technique in high-energy physics, Annu. Rev. Nucl. Sci., 23:1–43, 1973.

Chain reaction

A succession of generation after generation of acts of division (called fission) of certain heavy nuclei. The fission process releases about 200 MeV (3.2×10^{-4} erg $= 3.2 \times 10^{-11}$ J) in the form of energetic particles including two or three neutrons. Some of the neutrons from one generation are captured by fissile species (^{233}U, ^{235}U, ^{239}Pu) to cause the fissions of the next generations. The process is employed in nuclear reactors and nuclear explosive devices. See NUCLEAR FISSION.

The ratio of the number of fissions in one generation to the number in the previous generation is the multiplication factor k. The value of k can range from less than 1 to less than 2, and depends upon the type and amount of fissile material, the rate of neutron absorption in nonfissile material, the rate at which neutrons leak out of the system, and the average energy of the neutrons in the system. When $k = 1$, the fission rate remains constant and the system is said to be critical. When $k > 1$, the system is supercritical and the fission rate increases. See REACTOR PHYSICS.

A typical water-cooled power reactor contains an array of uranium rods (about 3% ^{235}U) surrounded by water. The uranium in the form of UO_2 is sealed into zirconium alloy tubes. The water removes the heat and also slows down (moderates) the neutrons by elastic collision with hydrogen nuclei. The slow neutrons have a much higher probability of causing fission in ^{235}U than faster (more energetic) neutrons do. In a fast reactor, no light nuclei are present in the system and the average neutron velocity is much higher. In such systems it is possible to use the excess neutrons to convert ^{238}U to ^{239}U. Then ^{239}U undergoes radioactive decay into ^{239}Pu, which is a fissile material capable of sustaining the chain reaction. If more than one ^{239}Pu atom is provided for each ^{235}U consumed, the system is said to breed (that is, make more fissile fuel than it consumes). In the breeder reactor, the isotope ^{238}U (which makes up 99.3% of natural uranium) becomes the fuel. This increases the energy yield from uranium deposits by more than a factor of 60 over a typical water-moderated reactor, which mostly employs the isotope ^{235}U as fuel.

A majority of the power reactors in the world today use water as both the moderator and the coolant. However, a limited number of reactors use heavy water instead of light water. The advantage of this system is that it is possible to use natural uranium as a fuel so that no uranium enrichment is needed. Some other power reactors are gas-cooled by either helium or carbon dioxide and are moderated with graphite.

[NORMAN C. RASMUSSEN]

Bibliography: R. D. Evans, The Atomic Nucleus, 1955; A. J. Henry, Nuclear-Reactor Analysis, 1975; J. R. Lamarsh, Introduction to Nuclear Engineering, 1975.

Charge-density wave

The ground state of a metal in which the conduction-electron charge density is sinusoidally modulated in space. The periodicity of this extra modulation is unrelated to the lattice periodicity. Instead it is determined by the dimensions of the conduction-electron Fermi surface in momentum space.

Description. The conduction-electron charge density $\rho_0(\vec{r})$ would ordinarily exhibit a dependence on position $\vec{r}$ having the same spatial periodicity as that of the positive-ion lattice. A metal with a charge-density wave (CDW) has instead a charge density given by Eq. (1). The amplitude of the

$$\rho(\vec{r}) = \rho_0(\vec{r})[1 + p \cos (\vec{Q} \cdot \vec{r} + \varphi)] \qquad (1)$$

charge-density wave is p, and typically has a value of approximately 0.1. The wave vector $\vec{Q}$ of the charge-density wave is determined by the conduction-electron Fermi surface. In a simple metal, having a spherical surface of radius p_F in momentum space, the wave vector is given by approximation (2), where h is Planck's constant. Although

$$Q \sim 4\pi p_F/h \qquad (2)$$

the length of a charge-density wave is comparable to a lattice constant, in general there is no integral relationship. The charge-density wave is then said to be incommensurate. In such a case, the total energy of the metal is independent of the phase φ in Eq. (1). See FERMI SURFACE.

Origin. Coulomb interactions between electrons are usually the cause of a charge-density wave instability. Two such contributions are the exchange energy, an effect of the Pauli exclusion principle, and the correlation energy, an effect of electron-electron scattering. Both energies are reduced in a charge-density wave structure. However, the classical, electrostatic energy opposes the instability and will suppress it unless the positive-ion lattice intervenes. See EXCLUSION PRINCIPLE.

Lattice distortion. Suppose that $\vec{u}(\vec{r})$ is the displacement of a positive ion from its lattice site at $\vec{r}$. Then a wavelike displacement given by Eq. (3)

$$\vec{u}(\vec{r}) = \vec{A} \sin [\vec{Q} \cdot \vec{r} + \varphi] \qquad (3)$$

will generate a positive-ion charge density that tends to cancel the electronic charge modulation of the charge-density wave. Typical values of the displacement amplitude are $A \sim 10^{-12} - 10^{-11}$ m. The ion-ion interactions must be small in order to permit this. Consequently charge-density waves are more likely to occur in metals with small elastic moduli.

Detection. The unambiguous signature of a charge-density wave is the observation of two satellites, on opposite sides of each Bragg reflection, in a single-crystal, x-ray scattering experiment. These are caused by the lattice displacement, Eq. (3). Charge-density wave satellites were first seen in metals like tantalum disulfide (TaS_2)

and tantalum diselenide ($TaSe_2$). These have three charge-density waves with $\vec{Q}$ directions separated by 120° in the hexagonal plane. At reduced temperature, transitions from incommensurate to commensurate $\vec{Q}$s are observed. The length of the charge-density wave is then an integral multiple of some lattice periodicity. *See* X-RAY DIFFRACTION.

Fermi surface effects. A charge-density wave is generated by, and generates in a bootstrap fashion, a periodic potential which modifies the conduction-electron energy spectrum. If the Fermi surface were originally simply connected—for example, spherical—it would become distorted and multiply connected. If a metal having cubic symmetry acquires one charge-density wave, its electronic properties will become very anisotropic and anomalous. A multiply connected Fermi surface can cause spectacular behavior in the low-temperature magnetoresistance. Such phenomena are observed in the alkali metals, for example, sodium and potassium. *See* MAGNETORESISTANCE.

$\vec{Q}$-domains. The $\vec{Q}$ direction in a single crystal, say a cubic one, will generally not be the same throughout the entire sample. There will be a domain structure analogous to magnetic domains in a ferromagnet. The $\vec{Q}$ direction will prefer some specific axis described, say, by direction cosines α, β, γ. In a cubic crystal there would be 24 equivalent axes and, therefore, 24 $\vec{Q}$-domain types. As a consequence, some physical properties of a sample will depend markedly on the orientation distribution of its $\vec{Q}$-domains. For example, the low-temperature resistivity of a potassium wire will increase by a factor of 4 if its $\vec{Q}$-domains are oriented parallel to the wire instead of perpendicular. Since the $\vec{Q}$-domain distribution can be altered by stress-induced domain regrowth, some physical properties will vary drastically from experiment to experiment, even on the same sample. Techniques for good control of $\vec{Q}$-domains have not yet been developed. *See* DOMAIN (CRYSTALLOGRAPHY); ELECTRICAL RESISTIVITY.

Phasons. Independence of the total energy on the phase φ, Eq. (1), leads to the existence of new low-energy, collective excitations. These correspond to a slowly varying phase modulation of the charge-density wave given by approximation (4).

$$\varphi \rightarrow \varphi(\vec{r}, t) \sim \sin(\vec{q} \cdot \vec{r} - \omega t) \qquad (4)$$

Quantized excitations of this type are called phasons. They exist only for small q, and their frequency $\omega(q)$ goes to zero with q.

Several physical phenomena caused by phasons have been observed. They are the cause of a low-temperature anomaly in the heat capacity of lanthanum digerminide ($LaGe_2$). Scattering of conduction electrons by phasons is the major contribution to the temperature-dependent electrical resistivity of potassium below 1.5 K. *See* BAND THEORY OF SOLIDS; CRYSTAL STRUCTURE; SPIN-DENSITY WAVE.   [ALBERT W. OVERHAUSER]

Bibliography: A. W. Overhauser, Charge-density waves and isotropic metals, *Adv. Phys.*, 27:343–363, 1978.

Charged particle beams

Unidirectional streams of charged particles traveling at high velocities. Charged particles can be accelerated to high velocities by electromagnetic fields. They are then able to travel through matter (termed an absorber), interacting with it, losing energy, and causing various effects important in many applications. The velocities under consideration in this article exceed 100,000 m/s (about 60 mi/s, or 200,000 mph), equivalent to an energy of 100 eV for a proton, and can approach the speed of light c (about 3×10^8 m/s, or 6.7×10^8 mph). Examples of charged particles are electrons, positrons, protons, antiprotons, alpha particles, and any ions (atoms with one or several electrons removed or added). In addition, some particles are produced artificially and may be short-lived (pions, muons).

Excluded from consideration are particles of, for example, cosmic dust (micrometeorites), which are clumps of thousands or millions of atoms. *See* ELECTRONVOLT; ELEMENTARY PARTICLE; PARTICLE ACCELERATOR.

Particle properties. Fast charged particles are described in terms of the following properties (values for some particles are given in Table 1):

Charge, z, in multiples of the electron charge $e = 1.6022 \times 10^{-19}$ coulomb. At small velocities the charge may be less than the charge of the nucleus because electrons may be present in some of the atomic shells.

Rest mass, M; usually the energy equivalent Mc^2 (in MeV) is given.

Rest mass, m, of electron; $mc^2 = 0.51104$ MeV.

Mass in atomic mass units, u; $A_m = Mc^2/931.481$ MeV.

Kinetic energy, T, in MeV.

Velocity, v, in cm/s or m/s.

$\beta = v/c$; $c = 299,792,458$ m/s = speed of light.

For absorbers, the information needed is:

Atomic number, Z (number of protons in nucleus).

Average atomic weight, A (usually in g/mole, but numerically equal to mass in u).

Absorber thickness, x, in cm or g/cm².

Physical state (solid, liquid, gas, plasma).

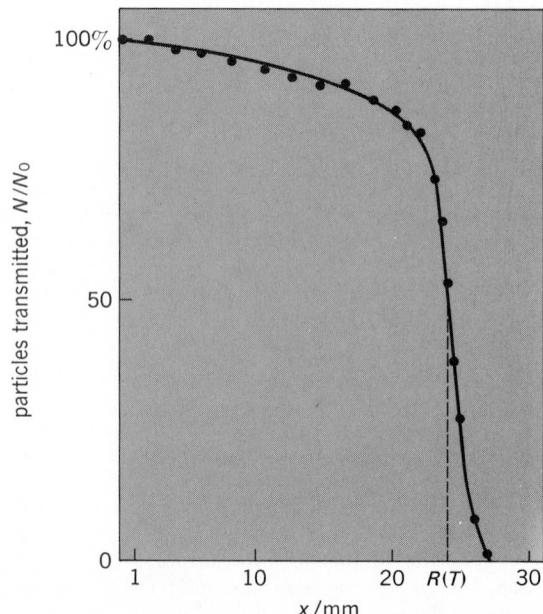

Fig. 1. Transmission curve for protons with kinetic energy $T = 144$ MeV through copper. (*From L. Koschmieder, Zur Energiebestimmung von Protonen aus Reichweitemessungen, Z. Naturforsch., 19a:1414–1416, 1964*)

Table 1. Properties of charged particles*

Ion	z	Lifetime, ns	Mass		
			10^{-24} g	u	MeV
Electron†	−1	Stable	0.910956	0.548593	511.004
Muon	1	2198.3	0.188357	0.113432	105.6598
Pion	1	26.04	0.248823	0.149846	139.578
Kaon	1	12.35	0.880322	0.530147	493.82
Sigma⁺	1	0.081	2.120318	1.276895	1189.40
Sigma⁻	−1	0.164	2.134436	1.285398	1197.32
		Mass excess,‡ MeV			
1N	0	8.0714	1.674920	1.0086652	939.553
1H	1	7.2890	1.672614	1.0072766	938.259
2H	1	13.1359	3.343569	2.0135536	1875.587
3H	1	14.9500	5.007334	3.0155011	2808.883
3He	2	14.9313	5.006390	3.0149325	2808.353
4He	2	2.4248	6.644626	4.0015059	3727.328
6Li	3	14.0884	9.985570	6.0134789	5601.443
7Li	3	14.9073	11.647561	7.0143581	6533.743
7Be	4	15.7689	11.648186	7.0147345	6534.093
9Be	4	11.3505	14.961372	9.0099911	8392.637
10B	5	12.0552	16.622243	10.0101958	9324.309
11B	5	8.6677	18.276741	11.0065623	10252.406
12C	6	0	19.920910	11.9967084	11174.708
13C	6	3.1246	21.587011	13.0000629	12109.314
14C	6	3.0198	23.247356	13.9999504	13040.691
14N	7	2.8637	23.246166	13.9992342	13040.024
15N	7	.1004	24.901771	14.9962676	13968.741
16O	8	−4.7365	26.552769	15.9905263	14894.875
17O	8	−.8077	28.220304	16.9947441	15830.285
18O	8	−.7824	29.880881	17.9947713	16761.791
19F	9	−1.4860	31.539247	18.9934674	17692.058
20Ne	10	−7.0415	33.188963	19.9869546	18617.472
21Ne	10	−5.7299	34.851833	20.9883627	19550.265
22Ne	10	−8.0249	36.508273	21.9858989	20479.451

*From *American Institute of Physics Handbook*, 3d ed., McGraw-Hill, 1972.

†Electron masses to be divided by 1000.

‡Mass excess given for neutral atoms; it is used to calculate nuclear reaction Q values.

See ATOMIC CONSTANTS; ATOMIC MASS UNIT; ENERGY; MASS.

Relation of velocity and energy. For small velocities ($\beta < 0.2$), the approximation $T = \frac{1}{2}Mc^2\beta^2$ is accurate to 3%. The expressions $\beta^2 = \zeta(\zeta+2)/(\zeta+1)^2$, and $T = Mc^2[(1/\sqrt{1-\beta^2})-1]$, with $\zeta = T/Mc^2$, are correct for all velocities. For $\zeta > 100$, the expression $\beta^2 = 1-(1/\kappa^2)$, with $\kappa = \zeta+1$, is more suitable to provide accurate values. *See* RELATIVISTIC MECHANICS.

Range. If a parallel beam of monoenergetic particles (that is, a beam in which all particles have exactly the same kinetic energy T) enters an absorber with a flat and smooth surface, it is found that all the particles (with the exception of electrons) travel along almost straight lines, slow down at approximately the same rate, and stop at approximately the same distance x from the surface (Fig. 1). The average distance traveled by the particles is called the mean range $R(T)$.

Energy loss and straggling. If the same beam travels through a thin absorber, it will emerge from it with a reduced average energy $\langle T_1 \rangle$ (Table 2). The difference between T and $\langle T_1 \rangle$ is called the average energy loss $\Delta = T - \langle T_1 \rangle$. Owing to the randomness of the number of collisions experienced by each particle, the range and the reduced energy fluctuate around the average values. This fluctuation is called straggling of energy loss or of range. (The shape of the curve in Fig. 1 for $x > 22$ mm is determined by range straggling.)

Charge state. At high velocities, an ion usually has the full charge ze of the nucleus. As soon as v drops below the velocity $u_K \cong zc/137$ of the K-shell electrons, electrons in the absorber will be attracted into K-shell orbits, thus reducing the total charge of the ion to a value $z^* = z-1$ or $z-2$. As the velocity drops further, more and more electrons will be attached to the ion; but since some of these electrons will be lost or gained as successive

Table 2. Calculated reduced average energy at various depths in copper of protons with T = 144 MeV*

x/mm	T_1/MeV
1	141
5	127
10	107
15	85
20	57

*From *American Institute of Physics Handbook*, 3d ed., McGraw-Hill, 1972.

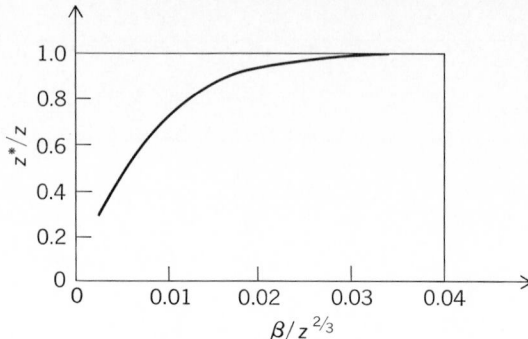

Fig. 2. Average charge z^* of a particle of velocity $v = \beta c$. (*From American Institute of Physics Handbook, 2d ed., McGraw-Hill, 1963*)

collisions take place, z^* must be considered as an average value which changes with v (see Fig. 2). *See* ELECTRON CONFIGURATION.

Interactions. In traveling through matter, charged particles interact with nuclei, producing nuclear reactions and elastic and inelastic collisions with the electrons (electronic collisions) and with entire atoms of the absorber (atomic collisions). Usually, in its travel through matter a charged particle makes few or no nuclear reactions or inelastic nuclear collisions, but many electronic and atomic collisions. The average distance between successive collisions is called the mean free path, λ. In solids, it is of the order of 10 cm for nuclear reactions. It ranges from the diameter of the atoms (about 10^{-10} m) to about 10^{-7} m for electronic collisions. The mean free path, λ, depends on the properties of the particle and, most importantly, on its velocity.

Nuclear interactions. In nuclear reactions [for example, $Be(d,n)B$] the incident particle is removed from the beam. Therefore, a reduction in the fluence (number of particles in the beam per cm²) will be observed. (The decrease in the number of particles in Fig. 1 for x less than about 22 mm is due to nuclear interactions.) Such an attenuation can be described by Eq. (1), where $e = 2.71828 \ldots$, $N_0 =$ initial particle fluence at $x = 0$,

$$\frac{N}{N_0} = e^{-x\Sigma} = e^{-x/\lambda} \tag{1}$$

$N =$ particle fluence at x, and $\Sigma =$ probability for an interaction to take place per centimeter of absorber. The mean free path is $\lambda = 1/\Sigma$; Σ is equal to $n\sigma$, where $\sigma =$ cross section per atom for nuclear reactions in cm²; and $n =$ number of atoms per cm³ of absorber material. *See* NUCLEAR REACTION; SCATTERING EXPERIMENTS (NUCLEI).

A rough estimate of the cross section for particles with $T/A_m > 10$ MeV can be obtained from Eq. (2). For thin absorbers, $N/N_0 \cong 1 - x/\lambda + \cdots$.

$$\sigma = 5 \times 10^{-26} \text{ cm}^2 \, (A_m^{1/3} + A^{1/3})^2 \tag{2}$$

An important nuclear interaction is Coulomb (or Rutherford) scattering: because both the incident particles and the nuclei of the absorber atoms have electric charge with values ze and Ze respectively, a change in the direction of motion of the particles will take place during the passage of the particles near the nuclei. The total cross section for this process is only slightly less than the cross-sectional area of the total atom. Usually, though, the angu-

lar deflection is much less than 0.01°, and only multiple scattering, the compounding of collisions with many atoms, will cause noticeable total deflections. If an observation of a very fine beam of particles were made along the direction of motion, the scattering events would be seen as small lateral displacements in random directions, and the final lateral displacement would be their vector sum. Although very few particles experience no deflection, the most probable location of the particles is still on the original line of the beam.

Bremsstrahlung. If a charged particle is accelerated, it can emit photons called bremsstrahlung. This process is of great importance for electrons as well as for heavy ions with $T \gg Mc^2$. It is used extensively for the production of x-rays in radiology. Electrons circulating in storage rings emit large numbers of photons with energies (100 eV – 1000 eV) not readily available from other sources. *See* ACCELERATION; BREMSSTRAHLUNG; SYNCHROTRON RADIATION.

Atomic collisions. At low velocities it may be convenient to consider separately collisions in which most of the energy loss is given as kinetic energy to a target atom. Usually, electronic excitation, electron rearrangements, and possibly ionization accompany this process. The term "nuclear collision" is used by some scientists. No simple quantitative description of atomic collisions is available. *See* SCATTERING EXPERIMENTS (ATOMS AND MOLECULES).

Electronic collisions. The interaction and energy transfer (see Fig. 3) between the charged particle and the electrons are caused by the Coulomb force. In general, except in tenuous plasmas, electrons are bound. In gases, all electrons are bound to individual atoms or molecules in well-defined orbits. For these isolated molecules (henceforth, atoms will be included with molecules), electrons can be moved into other bound orbits (excitation) requiring a well-defined energy ϵ_e. Another possibility is the complete removal of the electron from the atom (ionization) requiring an energy $\epsilon \geq I$, where I is the ionization energy for the particular electron. The secondary electron, which is called a delta ray, will have kinetic energy $K = \epsilon - I$. In both processes, the charged particle will lose energy; the energy loss is ϵ_e or ϵ, respectively. Also, it will be deflected very slightly. However, the change in direction is so small that it does not show in Fig. 3; the larger deflection caused by

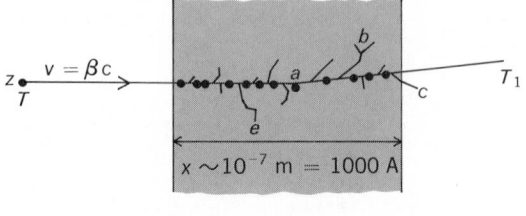

key:

⊤   ionization event

•—•   excitation event

Fig. 3. Energy loss by heavy charged particle. No details shown for energy losses of secondary electrons (delta rays). Rutherford scattering at point a. At point b, delta ray experiences collision that results in tertiary delta ray. At c, delta ray escapes from absorber.

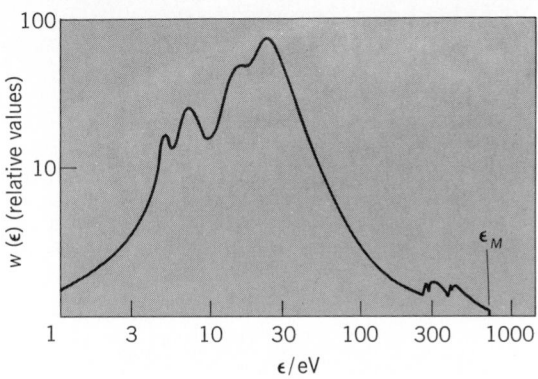

Fig. 4. Schematic single-collision spectrum $w(\epsilon)$ for heavy charged particles in adenine ($C_5N_5H_5$).

Rutherford scattering at point a is visible. *See* ATOMIC STRUCTURE AND SPECTRA; EXCITED STATE.

In liquids and solids, only the inner electrons are associated with a specific nucleus (in aluminum metal the K- and L-shell electrons). Excitation and ionization processes for these electrons are very similar to those in free molecules. The outer electrons are either associated with several neighboring nuclei (nonconducting materials) or, in metals (in Al, the three M-shell electrons), form a plasma-like cloud. Collective or plasma excitations ($\epsilon \cong 20$ eV) take place with high probability, but direct ionization ($\epsilon \gg 20$ eV) also occurs. At present, these processes are not well understood. Because of the requirements of momentum conservation, the maximum energy loss which can occur is given by $\epsilon_M \simeq 2mv^2 = 2mc^2\beta^2$ for particles heavier than electrons.

The probability for energy losses ϵ by the incident particle is described by the energy loss spectrum $w(\epsilon)$. Theoretical values have been calculated by H. Bethe. An energy loss spectrum for heavy charged particles in adenine ($C_5N_5H_5$) is shown in Fig. 4.

The structure between 3 and 30 eV relates to the outer electrons. Similar structures have been observed for many solids (including metals). Excitation and ionization of the K-shell of C (above 280

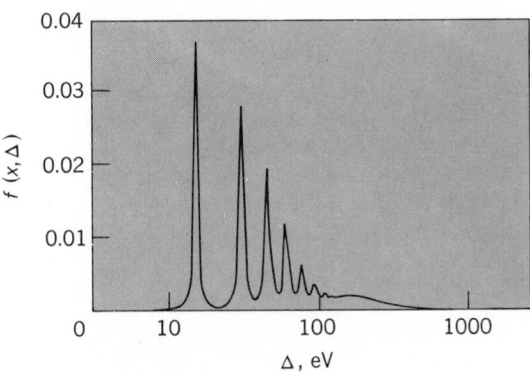

Fig. 5. Calculated straggling curve $f(x, \Delta)$ for 20-MeV protons incident on Al absorber of thickness $5.8 \cdot 10^{-8}$ m. Spikes represent multiples of the "plasma loss" at 15 eV. (*From H. Bichsel and R. Saxon, Comparison of calculational methods for straggling in thin absorbers, Phys. Rev., A11:1286–1296, 1975*)

eV) and N (above 400 eV) cause further structure. The average energy loss per collision is defined by Eq. (3).

$$\langle\epsilon\rangle = \int \epsilon \, w(\epsilon) \, d\epsilon \,/\, \int w(\epsilon) \, d\epsilon \quad (3)$$

Statistics of energy loss. The total energy loss δ of a particle traveling through matter is the sum of the energy losses ϵ_i in each collision: $\delta = \Sigma\epsilon_i = \epsilon_1 + \epsilon_2 + \epsilon_3 + \epsilon_4 + \ldots \epsilon_\nu$, where ν collisions have occurred, each with a probability given by $w(\epsilon)$ (see Fig. 4). If a large number of particles are observed, they will experience on the average $q = \langle\nu\rangle$ collisions (q is not an integer), and an average energy loss $\Delta = \langle\delta\rangle = q\langle\epsilon\rangle$, as long as collisions are uncorrelated with another. The number of collisions is distributed according to a Poisson distribution; the fraction $P(\nu)$ of particles having experienced exactly ν collisions is given by Eq. (4). The distribution function for

$$P(\nu) = \frac{q^\nu}{\nu!}e^{-q} \quad (4)$$

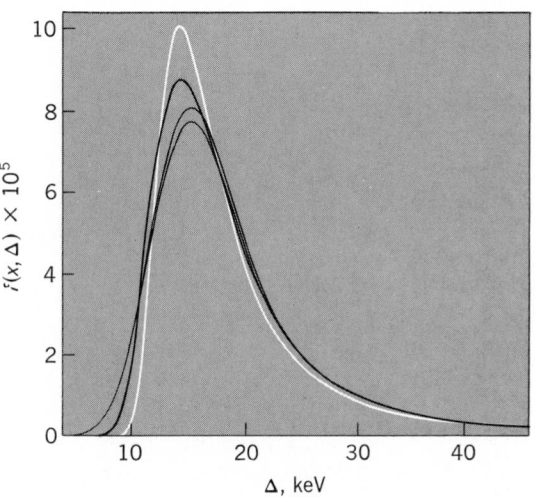

Fig. 6. Calculated straggling curve $f(x, \Delta)$ for 20-MeV protons incident on Al absorber of thickness $3.71 \cdot 10^{-6}$ m. Various theoretical calculations are presented. (*From H. Bichsel and R. Saxon, Comparison of calculational methods for straggling in thin absorbers, Phys. Rev., A11:1286–1296, 1975*)

energy losses is called a straggling function $f(\Delta,x)$. Examples are given in Figs. 5 and 6. For fairly thick absorbers, $f(\Delta,x)$ is approximately a gaussian of width proportional to $\sqrt{x}$. *See* DISTRIBUTION (PROBABILITY).

In many applications, the details of energy loss are not important, and a knowledge of the mean or average energy loss Δ is sufficient. If the total collision cross section per atom $w_t = \int w(\epsilon)d\epsilon$ is known (w_t in cm^2), the mean number of collisions is given by $q = xnw_t$, and the mean energy loss by Eq. (5). The quantity stopping power S thus is defined by Eq. (6), in MeV/cm. Since knowledge of w_t and

$$\Delta = xn\langle\epsilon\rangle w_t \equiv xS \quad (5)$$

$$S \equiv n\langle\epsilon\rangle w_t \quad (6)$$

$\langle\epsilon\rangle$ is not extensive, S is frequently determined in experimental measurements in which a beam passes through an absorber, and S is calculated from Eq. (7).

$$S = \lim_{x \to 0} \frac{\Delta}{x} = \lim_{x \to 0} \frac{T - \langle T_1 \rangle}{x} \tag{7}$$

Equation (5) is valid only if x is much smaller than the mean range, $R(T)$; otherwise, S varies significantly as the particle loses energy in the absorber. If x is not small the following procedure can be used to obtain $\langle T_1 \rangle$, provided that $R(T)$ has been tabulated: Find $R(T)$ in the range table, calculate $y = R(T) - x$, and then find the energy T_1 corresponding to y in the range table.

Stopping power. The stopping power is a function of the velocity v of the incident particle, its effective charge z^*, and the absorber material. Figure (7) shows $S(T)$ for heavy ions of various elements in aluminum. S is expressed in MeV cm²/g. This can be converted to MeV/cm by the formula $S(\text{MeV/cm}) = \rho S(\text{MeV cm}^2/\text{g})$, where ρ is the density of absorber material in g/cm³. Atomic collisions dominate in region I. For region III, S can be calculated by using the theoretical Bethe expression, Eq. (8), where I_A = average excitation energy

$$S = \frac{.30708}{\beta^2} \frac{Z}{A} \left[\ln \frac{2mc^2\beta^2}{I_A(1-\beta^2)} - \beta^2 - \frac{C}{\beta^2} - d(\beta) \right]$$
$$\cdot (z^*)^2 [1 + G(z^*, \beta)] \text{ MeV cm}^2/\text{g} \tag{8}$$

of absorber (approximately $Z \times 10^{-5}$ MeV); $C =$ shell correction constant; $d(\beta) =$ density correction, important for $T > Mc^2$; and $G(z^*, \beta) =$ correction due to the second Born approximation, important only for $\beta^2 < 0.01$. *See* PERTURBATION (QUANTUM MECHANICS).

A simpler approximate expression valid to about 10% for $z \le 10$ in the same region is given by Eq. (9).

$$S = \frac{2.6z^2}{\beta^{1.66}Z^{.25}} \text{ MeV cm}^2/\text{g} \qquad 0.1 < \beta < 0.88 \tag{9}$$

Range. A good approximation to the mean range $R(T)$ can be calculated from S using Eq. (10).

$$R(T) = \int_0^T \frac{d\tau}{S(\tau)} \tag{10}$$

Table 3. Range of validity of range formula

Range of kinetic energy T	Range of β	r	s
$10 \le T < 90$ MeV	$.145 < \beta < .4$	.116 g/cm²	1.84
$90 < T < 400$ MeV	$.4 < \beta < .7$	.275 g/cm²	2.33
$400 < T < 1000$ MeV	$.7 < \beta < .88$	.532 g/cm²	3.34

Usually, a numerical integration is performed to obtain R from a table of S or the Bethe formula. An approximation formula is Eq. (11). The range of validity for this range formula is given in Table 3.

$$R = r \cdot \beta^s \cdot \frac{Z^{.25}}{z^2} \quad z \le 10 \tag{11}$$

Channeling. In absorbers consisting of a single crystal, it has been found that the energy loss will be reduced if the direction of the particle beam coincides with certain preferred alignments of the crystal. It is believed that the particles travel through "open spaces" in the crystal, thus suffering a succession of collisions with relatively small energy losses and angular deflections, and tending to stay in the preferred direction (channel). Thus, even if the total number of collisions q were unchanged, the average energy loss per collision $\langle \epsilon \rangle$ would be reduced, and the total energy loss Δ would be less.

Ionization. The secondary electrons of energies K produced in ionizing electronic collisions will travel through the absorber and will also suffer various collisions, producing further energetic electrons, and so on. This process continues until the electrons have energy $K < I$. It has been found experimentally that the total average number j of ions produced in this way (in the distance x) is proportional to the energy Δ lost by the particle: $j = \Delta/\omega$.

The constant of proportionality ω introduced in this definition has values between 20 eV and 45 eV for gases, about 3.6 eV for silicon, and 2.96 eV for germanium. It is almost, but not exactly, independent of particle energy and type. If particles lose all their energy in the material, the total ionization J is related to the kinetic energy T: $J = T/W$. The relation between W and ω is: $T/W = \int dT/\omega$. $W = \omega$ only if ω is exactly independent of energy.

The ionization j along the path of a single particle increases with the distance traveled, approximately at the same rate as S. For a beam of particles, straggling also influences the total ionization at a given location. The function obtained from the combination (actually the convolution) of both effects is called the Bragg curve.

Electrons. Although the interactions discussed earlier all occur for electrons, there are some major differences between electron beams and beams of heavier particles. In general, the path of an electron will be a zigzag. Angular deflections in the collisions will frequently be large. Electron beams therefore tend to spread out laterally, and the number of primary electrons in the beam at a depth x in the absorber decreases rapidly.

Since it is not possible to distinguish individual electrons, it is customary in a collision between two electrons to consider the one emerging with the higher energy as the primary electron. The maximum energy loss in a collision therefore is $mv^2/4$ (for $T \ll mc^2$). The stopping power expression therefore is somewhat different for electrons.

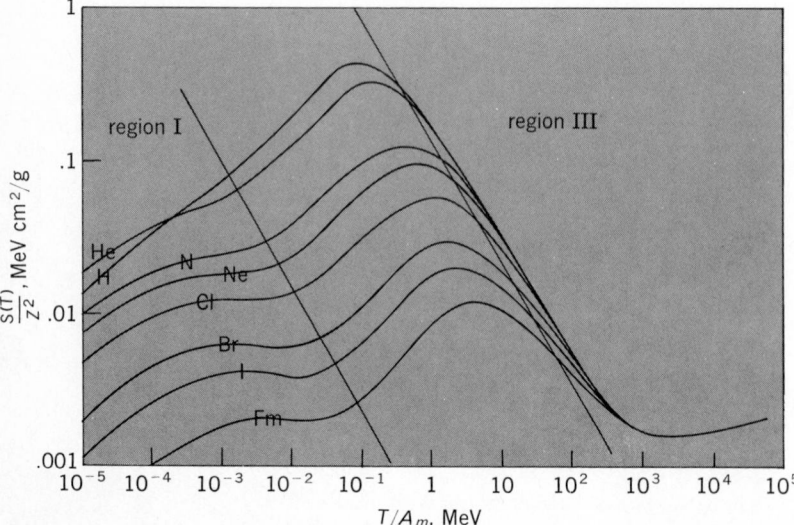

Fig. 7. *S(T)* for heavy ions in Al. (*From L. C. Northcliffe and R. F. Schilling, Range and stopping power tables for heavy ions, Nucl. Data Tables, A7:233, 1970*)

Biological effects. In general, for the same dose (the energy deposited per gram along the beam line) heavy charged particles will produce, because of their higher local ionization, larger biological effects than electrons (which frequently are produced by x-rays).

Observation. The most direct method of observing a beam of charged particles is to observe the electric current that they form (any flow of electric charges is an electric current). In all accelerators (such as cyclotrons, Van de Graaff generators, linear accelerators, and x-ray tubes) the beam current is measured as a primary monitor of the proper operation of the machine. It is not possible to identify the type of particles with current measurements (except for their electric charge). *See* PARTICLE DETECTOR.

Devices using ionization. If an electric field **E** is applied to an absorber irradiated with charged particles, the ions and electrons produced will travel in the direction **E**, and the resulting ionization current can be measured (electronic amplification usually is needed). If an oscilloscope is available, the ionization J associated with a single particle can be observed, and the energy (or energy loss) of the particle can be calculated from $T = JW$. Semiconductor detectors (chiefly silicon and germanium) are used extensively for this purpose, but gas-filled ionization chambers have also been used. Cloud chambers and bubble chambers use this principle, but individual ions or clumps of ionization are observed visually. Proportional counters, spark chambers, and Geiger-Müller tubes also operate on the same principle; but in the latter two only the presence of a particle is indicated, and J is not related to T. *See* BUBBLE CHAMBER; ELECTRIC FIELD; IONIZATION CHAMBER; SPARK CHAMBER.

Devices using excitation. The excited state of energy ϵ_e produced in excitation can decay with the emission of light (luminescence or scintillation). Early observations of radioactivity were made with this method (using ZnS screens and visual observation, usually with microscopes), and the method is used extensively with luminescent dials (for example, on wristwatches). The light emitted usually is detected and amplified with photomultipliers. Again, the energy T can be measured. Scintillators used are NaI(Tl), CsI, anthracene, stilbene, and various solids and liquids. *See* LIQUID SCINTILLATION DETECTOR; LUMINESCENCE; SCINTILLATION COUNTER.

A more indirect use of excitations and ionizations is in "chemical" devices (such as photographic emulsions, $FeSO_4$ solutions, thermoluminescence). *See* THERMOLUMINESCENCE.

Applications. Electron beams are used in the preservation of food. In medicine, electron beams are used extensively to produce x-rays for both diagnostic and therapeutic (cancer irradiation) purposes. Also, in radiation therapy, deuteron beams incident on Be and ^{3}H targets are used to produce beams of fast neutrons, which in turn produce fast protons, alpha particles, and carbon, nitrogen, and oxygen ions in the irradiated tissue. Energetic pion (~ 100 MeV), proton (~ 200 MeV), alpha (~ 1000 MeV), and heavier ion beams can possibly be used for cancer therapy. The existence of a Bragg peak for these particles promises improvements in the dose distribution within the human body.

The well-defined range of heavy ions permits their implantation at given depths in solids (this is useful in the production of integrated circuits). Radiation damage studies are performed with charged particles in relation to development work for nuclear fission and fusion reactors.

Charged particle beams are used in many methods of chemical and solid-state analysis. Nuclear activation analysis can be performed with heavy ions. *See* ELECTRON DIFFRACTION.

Isotopes can be produced with fast charged ions. *See* SUPERTRANSURANICS.

[HANS BICHSEL]

Bibliography: F. H. Attix and W. C. Roesch (eds.), *Radiation Dosimetry*, 1968; H. Bichsel, in *American Institute of Physics Handbook*, 3d ed., 1972; R. D. Evans, *The Atomic Nucleus*, 1955; M. Inokuti, *Rev. Mod. Phys.*, 43:297–347, 1971; M. Mladjenovic, *Radioisotope and Radiation Physics*, 1973; W. J. Price, *Nuclear Radiation Detection*, 2d ed., 1964.

Charles' law

A thermodynamic law, also known as Gay-Lussac's law, which states that at constant pressure the volume of a fixed mass or quantity of gas varies directly with the absolute temperature. Conversely, at constant volume the gas pressure varies directly with the absolute temperature. Jacques Alexandre Charles and Joseph Louis Gay-Lussac independently discovered the relation for an ideal gas. The relation is a useful and close approximation. *See* GAS.

[FRANK H. ROCKETT]

Charm

A term used in elementary particle physics to describe a new class of elementary particles.

Theory. Ordinary atoms of matter consist of a nucleus composed of neutrons and protons and surrounded by electrons. Over the years, however, a host of other particles with unexpected properties have been found, associated with both electrons (leptons) and protons (hadrons).

Leptons. The electron has as companions the mu meson (μ) and the tau meson (τ), approximately 200 times and 3700 times as heavy as the electron, respectively. These particles are similar to the electron in all respects except mass. In addition, there exist at least two distinctive neutrinos, one associated with the electron, ν_e, and another with the mu meson, ν_μ. It is still not known whether the τ also has its own neutrino. In all, there are five or six fundamental, distinct, structureless leptons. *See* LEPTON.

Hadrons. A similar but more complex situation exists with respect to the hadrons. These particles number in the hundreds, and unlike the leptons they cannot be thought of as fundamental. In fact, they can all be explained as composites of more fundamental constituents, called quarks. It is the quarks which now seem as fundamental as the leptons, and the number of quark types has also increased as new and unexpected particles have been experimentally uncovered. The originally simple situation of having an up quark (u; charge $+\frac{2}{3}$), and a down quark (d; charge $-\frac{1}{3}$) has evolved as several more varieties or flavors have had to be added. These are the strange quark (s;

charge $-\frac{1}{3}$), with the additional property or quantum number of strangeness ($S=-1$), to account for the unexpected characteristics of a family of strange particles; the charm quark (c; charge $+\frac{2}{3}$), possessing charm ($C=+1$) and no strangeness, to explain the discovery of the J/ψ particles, massive states three times heavier than the proton; and a fifth quark (b; charge $-\frac{1}{3}$) to explain the existence of the even more massive upsilon (Y) particles. *See* HADRON; J PARTICLE; QUARKS.

The quarks and leptons discovered so far appear to form a symmetric array (see table). Both the leptons and quarks come in pairs, although the anticipated partner (t) of the b quark has not yet been found. Whether more such pairs of leptons and quarks will be found is one of the outstanding questions in elementary particle physics.

Fundamental constituents of matter

Family of particles	Charge	Particles
Leptons	0	ν_e, ν_μ, $(\nu_\tau)^*$
	-1	e, μ, τ
Quarks	$+\frac{2}{3}$	u, c, $(t)^*$
	$-\frac{1}{3}$	d, s, b

*Existence uncertain.

Observations. The members of the family of particles associated with charm fall into two classes: those with hidden charm, where the states are a combination of charm and anticharm quarks ($c\bar{c}$), charmonium; and those where the charm property is clearly evident, such as the D^+ ($c\bar{d}$) meson and Λ_c^+ (cud) baryon.

Charmonium. In the charmonium family, six to seven states with various masses and decay modes have been identified. Although a detailed understanding of all these experimentally measured properties has not yet been achieved, everything seems to be in qualitative agreement with theoretical expectations.

Bare charm states. There are several identified bare charm states, including the D ($c\bar{d}$) mesons in both the $J^P = 0^-$ and $J^P = 1^-$ categories (where J is spin, and P is parity), and the Σ_c^{++} (cuu) and Λ_c^+ (cud) charmed baryons. Information about the lifetimes of these states has been derived from experiments utilizing the high resolution of emulsions to measure the finite distance traveled by charmed particles. The lifetimes of the Λ_c^+ and D^0 have been determined to be on the order of 10^{-13} s with that of the D^+ about a factor of 7 longer. These values are in good agreement with theoretical expectations.

The Λ_c^+ charmed baryon has been observed to be produced in a variety of interactions, including neutrino-proton, proton-proton, electron-positron, and neutrino-neon reactions. A large number of decay modes have been observed, including Kp, $Kp\pi$, $\Lambda\pi$, $\Lambda\pi\pi\pi$, and $Kp\pi\pi$. Precise mass values should result from studies of these production and decay modes when the systematic errors are better understood.

Prospects. Although reasonable progress has been made in the study of charmed states, only a handful of states has been observed, and much work remains to be done. Just as the basic SU(3) symmetry arose from a study of the numerous had-

ron strange and nonstrange resonances, the complete understanding of charm awaits the uncovering of additional states. *See* ELEMENTARY PARTICLE. [NICHOLAS P. SAMIOS]

Bibliography: W. Chinowsky, Psionic matter, *Ann. Rev. Nucl. Sci.*, 27:393–464, 1977; S. L. Glashow, Quarks with color and flavor, *Sci. Amer.*, 233(4):38–50, October 1975.

Chromatic aberration

The type of error in an optical system in which the formation of a series of colored images occurs, even though only white light enters the system. Chromatic aberrations are caused by the fact that the refraction law determining the path of light through an optical system contains the refractive index n, which is a function of wavelength λ. Thus the image position and the magnification of an optical system are not necessarily the same for all wavelengths, nor are the aberrations the same for all wavelengths.

In this article the chromatic aberrations of a lens system are discussed. For information on other types of aberration *see* ABERRATION (OPTICS). *See also* REFRACTION OF WAVES.

Dispersion formula. When the refractive index of glass or other transparent material is plotted as a function of the square of the wavelength, the result is a set of dispersion curves which appear to have an asymptote in the near-ultraviolet region and a straight portion in the near-infrared (up to 1 μ). In the glass catalogs the indices are given for selected wavelengths between 0.365 and 1.01 μ, which are the absorption (emission) bands of certain chemical elements. For an extended discussion of dispersion *see* ABSORPTION OF ELECTROMAGNETIC RADIATION. *See also* OPTICAL MATERIALS.

For the visible region, the Hartmann formula, Eq. (1), where a and b are constants for a given glass, has been much used. However, if an attempt is made to apply Eq. (1) to the near-ultraviolet or

$$n = a/(\lambda - b) \qquad (1)$$

the near-infrared, it proves to be insufficient. Over this more extended range, n can be given as a function of the wavelength by Eq. (2).

$$n = a + b\lambda^2 + \frac{c}{\lambda^2 - 0.028} + \frac{d}{(\lambda^2 - 0.028)^2} \qquad (2)$$

This formula, in which c and d are additional

Values of the four universal functions $a(\lambda)$ for various wavelengths

λ	a_1	a_2	a_3	a_4
1.0140	0.000000	0.000000	0.000000	+1.000000
0.7682	+0.031555	−0.276197	+1.051955	+0.192687
0.6563	0.000000	0.000000	+1.000000	0.000000
0.6438	−0.004774	+0.051075	+0.966511	−0.012813
0.5893	−0.024952	+0.326272	+0.744598	−0.045919
0.5876	−0.025492	+0.336338	+0.735480	−0.046326
0.5461	−0.033080	+0.597795	+0.479049	−0.043764
0.4861	0.000000	+1.000000	0.000000	0.000000
0.4800	+0.009340	+1.036699	−0.052699	+0.006659
0.4358	+0.143980	+1.193553	−0.394569	+0.057036
0.4047	+0.366779	+1.045064	−0.487634	+0.075791
0.3650	+1.000000	0.000000	0.000000	0.000000

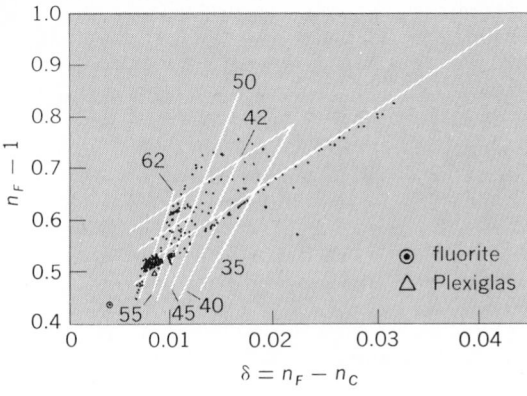

Fig. 1. Plot of (n_F-1) versus δ for selected glasses, largely from the Schott Catalog 350-E, and for fluorite and Plexiglas. The numbers on the lines dividing the glasses into groups represent ν values.

constants, enables one to compute n for any wavelength if it is given for four. An equivalent formula is shown by Eq. (3), where $a_1(\lambda)$, $a_2(\lambda)$, . . . are

$$n = n_1 a_1(\lambda) + n_2 a_2(\lambda) + n_3 a_3(\lambda) + n_4 a_4(\lambda) \quad (3)$$

functions of the form of Eq. (2), which assume for $\lambda_1, \lambda_2, \ldots$ the values 1, 0, 0, 0; 0, 1, 0, 0; 0, 0, 1, 0; and 0, 0, 0, 1, respectively. Choosing $\lambda_1 = 0.365 = \lambda^{**}$, $\lambda_2 = 0.4861 = \lambda_F$, $\lambda_3 = 0.6563 = \lambda_C$, $\lambda_4 = 1.014 = \lambda^*$, one finds in the table the four universal functions $a(\lambda)$ tabulated for different values of λ. Note that for any wavelength Eq. (4) holds.

$$a_1 + a_2 + a_3 + a_4 = 1 \quad (4)$$

Instead of giving the indices of a glass for the wavelengths specified, one can specify the glass by the refractive index n_F at one wavelength F, the dispersion $\delta = n_F - n_C$, and the partial dispersions in the ultraviolet and infrared, defined respectively by Eqs. (5). The partial dispersion for an arbitrary

$$P^{**} = \frac{n^{**} - n_F}{n_F - n_C} \qquad P^* = \frac{n_F - n^*}{n_F - n_C} \quad (5)$$

wavelength λ is then given by Eq. (6).

$$P_\lambda = P^{**} a_1(\lambda) + a_2(\lambda) + P^* a_4(\lambda) \quad (6)$$

In the literature the Abbe number is defined by Eq. (7), where n_D is the mean index for the D lines

$$\nu = (n_D - 1)/(n_F - n_C) \quad (7)$$

of sodium is frequently used to designate optical glass.

In Fig. 1, $(n_F - 1)$ is shown plotted against $(n_F - n_C)$ for various optical materials. One sees that in this kind of plot glasses of the same type lie on a straight line, while on the ordinary plot of n_D versus ν they lie on hyperbolas.

Color correction. A system of thin lenses in contact is corrected for two wavelengths λ_A and λ_B if the power of the combination is the same for both wavelengths as in Eq. (8), where the K_κs are the

$$\phi_A = \phi_B = \Sigma (n_A - 1) K_\kappa = \Sigma (n_B - 1) K_\kappa \quad (8)$$

differences between the first and last curvatures of the lenses, and the summation is over all the lenses, each with its particular value of n_A, n_B, and K_κ.

Two cemented lenses are corrected for wavelengths C and F for instance, if Eqs. (9) hold. *See* LENS (OPTICS).

$$\phi_1 + \phi_2 = \phi$$
$$\phi_1/\nu_1 + \phi_2/\nu_2 = \phi_C - \phi_F = 0 \quad (9)$$

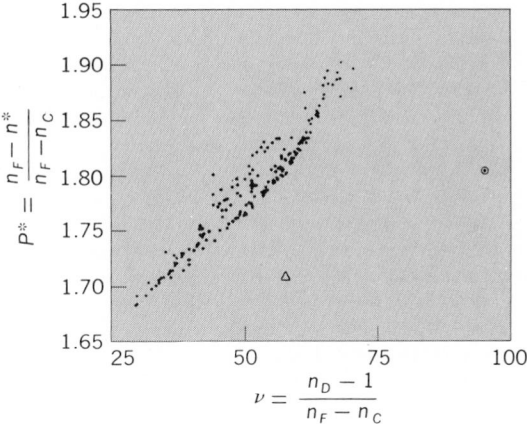

Fig. 3. Plot of P^* versus ν for materials of Fig. 1.

The two lenses are also corrected for a third wavelength if and only if, in addition, one has $P_1 = P_2$, where P_1 and P_2 are given by Eq. (6).

A system corrected for three colors is called an apochromat (in microscopy this term traditionally demands freedom from asymmetry in addition). An apochromat for the ultraviolet portion of the spectrum is possible only if the two glasses in Fig. 2 have the same P^{**} value. For the infrared, the glasses must have the same P^* value (Fig. 3). The $\nu - 1$ values for the two glasses should lie as far apart as possible to give low values for the powers.

Three lenses can be corrected for four wavelengths and therefore practically for the whole spectrum if the glasses lie on the straight line on the plot of P^{**} against P^* (Fig. 4). Such a system may be called a superachromat.

In lenses with finite thicknesses and distances, there are in Gaussian optics two errors to be corrected. One is a longitudinal aberration, which means that the Gaussian images do not lie in the same plane, and the other is a lateral aberration,

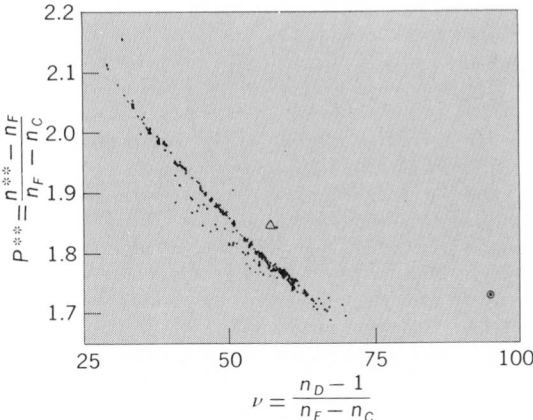

Fig. 2. Plot of P^{**} versus ν for materials of Fig. 1.

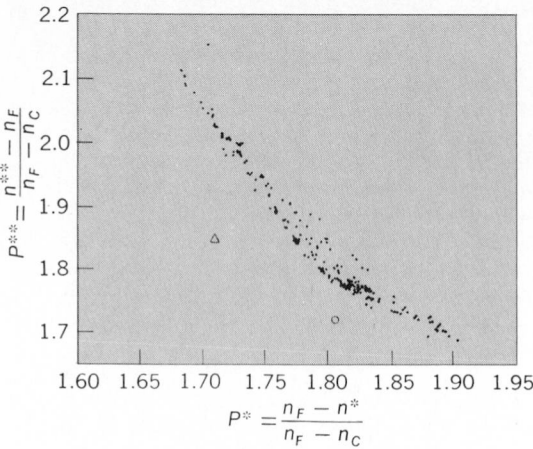

Fig. 4. Plot of P^{**} versus P^* for materials of Fig. 1.

which means that the images in different colors have different magnifications.

In the presence of longitudinal aberration, it is best to balance the lateral aberration so that the apparent sizes of the images as seen from the exit pupil coincide.

A system of two uncorrected lenses, such as a simple ocular, with one finite distance, cannot be corrected for both color errors. Two finite distances are needed to balance both color errors at the same time.

In a general system, all image errors are functions of the wavelength of light. However, in a lens system corrected for color, it is easily possible, with a small adjustment, to balance the correction of the aperture rays (spherical aberration) with respect to color by introducing a small amount of lateral color aberration.

If only two colors can be corrected, the choice of the colors depends on the wavelength sensitivity of the receiving instrument. For visual correction, the values for C and F are frequently brought together. Some optical systems contain filters to permit only a narrow spectral band to pass the instrument. This makes correction for color errors easier.

If light of a large band of wavelengths traverses the instrument, it is frequently desirable to use catadioptric systems, such as the Schmidt camera. The mirror or mirrors of these systems are used to obtain the necessary power without introducing color errors, and an afocal lens system can be added to correct monochromatic aberrations.

[MAX HERZBERGER]

Bibliography: E. U. Condon and H. Odishaw, *Handbook of Physics*, 1967; M. Herzberger, *Modern Geometrical Optics*, 1958; F. A. Jenkins and H. E. White, *Fundamentals of Optics*, 4th ed., 1976; Optical Society of America, *Handbook of Optics*, 1978; W. J. Smith, *Modern Optical Engineering*, 1976.

Circuit (electricity)

A general term referring to a system or part of a system of conducting parts and their interconnections through which an electric current is intended to flow. A circuit is made up of active and passive elements or parts and their interconnecting conducting paths. The active elements are the sources of electric energy for the circuit; they may be batteries, direct-current generators, or alternating-current generators. The passive elements are resistors, inductors, and capacitors. The electric circuit is described by a circuit diagram or map showing the active and passive elements and their connecting conducting paths.

Devices with an individual physical identity such as amplifiers, transistors, loudspeakers, and generators, are often represented by equivalent circuits for purposes of analysis. These equivalent circuits are made up of the basic passive and active elements listed above.

Electric circuits are used to transmit power as in high-voltage power lines and transformers or in low-voltage distribution circuits in factories and homes; to convert energy from or to its electrical form as in motors, generators, microphones, loudspeakers, and lamps; to communicate information as in telephone, telegraph, radio, and television systems; to process and store data and make logical decisions as in computers; and to form systems for automatic control of equipment.

Electric circuit theory. This includes the study of all aspects of electric circuits, including analysis, design, and application. In electric circuit theory the fundamental quantities are the potential differences (voltages) in volts between various points, the electric currents in amperes flowing in the several paths, and the parameters in ohms or mhos which describe the passive elements. Other important circuit quantities such as power, energy, and time constants may be calculated from the fundamental variables. For a discussion of these parameters *see* ADMITTANCE; CONDUCTANCE; ELECTRICAL IMPEDANCE; ELECTRICAL RESISTANCE; REACTANCE; SUSCEPTANCE.

Electric circuit theory is an extensive subject and is often divided into special topics. Division into topics may be made on the basis of how the voltages and currents in the circuit vary with time; examples are direct-current, alternating-current, nonsinusoidal, digital, and transient circuit theory. Another method of classifying circuits is by the arrangement or configuration of the electric current paths; examples are series circuits, parallel circuits, series-parallel circuits, networks, coupled circuits, open circuits, and short circuits. Circuit theory can also be divided into special topics according to the physical devices forming the circuit, or the application and use of the circuit. Examples are power, communication, electronic, solid-state, integrated, computer, and control circuits.

Direct-current circuits. In dc circuits the voltages and currents are constant in magnitude and do not vary with time (Fig. 1). Sources of direct current are batteries, dc generators, and rectifiers. Resistors are the principal passive element.

Magnetic circuits. Magnetic circuits are similar to electric circuits in their analysis and are often included in the general topic of circuit theory. Magnetic circuits are used in electromagnets, relays, magnetic brakes and clutches, computer memory devices, and many other devices. For a detailed treatment *see* MAGNETIC CIRCUITS.

Alternating-current circuits. In ac circuits the voltage and current periodically reverse direction with time. The time for one complete variation is known as the period. The number of periods in 1

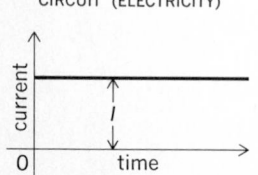

CIRCUIT (ELECTRICITY)

Fig. 1. Direct current.

sec is the frequency in cycles per second. A cycle per second has recently been named a hertz (in honor of Heinrich Rudolf Hertz's work on electromagnetic waves).

Most often the term ac circuit refers to sinusoidal variations. For example, the alternating current in Fig. 2 may be expressed by $i = I_m \sin \omega t$. Sinusoidal sources are ac generators and various types of electronic and solid-state oscillators; passive circuit elements include inductors and capacitors as well as resistors. The analysis of ac circuits requires a study of the phase relations between voltages and currents as well as their magnitudes. Complex numbers are often used for this purpose.

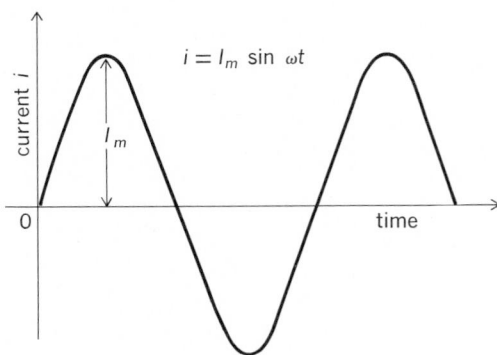

Fig. 2. Alternating current.

Nonsinusoidal waveforms. These voltage and current variations vary with time but not sinusoidally (Fig. 3). Such nonsinusoidal variations are usually caused by nonlinear devices, such as saturated magnetic circuits, electron tubes, and transistors. Circuits with nonsinusoidal waveforms are analyzed by breaking the waveform into a series of sinusoidal waves of different frequencies known as a Fourier series. Each frequency component is analyzed by ac circuit techniques. Results are combined by the principle of superposition to give the total response.

Electric transients. Transient voltage and current variations last for a short length of time and do not repeat continuously (Fig. 4). Transients occur when a change is made in the circuit, such as opening or closing a switch, or when a change is made in one of the sources or elements.

Series circuits. In a series circuit all the components or elements are connected end to end and carry the same current, as shown in Fig. 5.

Parallel circuits. Parallel circuits are connected so that each component of the circuit has the same potential difference (voltage) across its terminals, as shown in Fig. 6.

Series-parallel circuits. In a series-parallel circuit some of the components or elements are connected in parallel, and one or more of these parallel combinations are in series with other components of the circuit, as shown in Fig. 7.

Electric network. This is another term for electric circuit, but it is often reserved for the electric circuit that is more complicated than a simple series or parallel combination. A three-mesh electric network is shown in Fig. 8. *See* NETWORK THEORY.

Coupled circuits. A circuit is said to be coupled if two or more parts are related to each other through some common element. The coupling may be by means of a conducting path of resistors or capacitors or by a common magnetic linkage (inductive coupling), as shown in Fig. 9.

Open circuit. An open circuit is a condition in an electric circuit in which there is no path for cur-

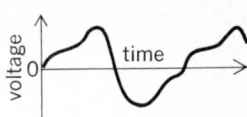
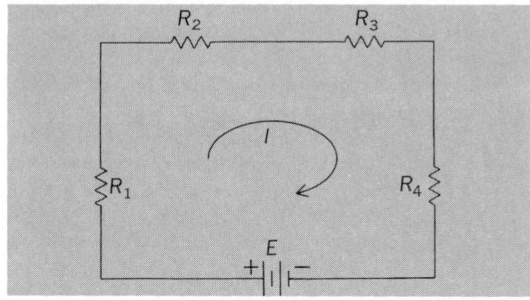

Fig. 5. Series circuit.

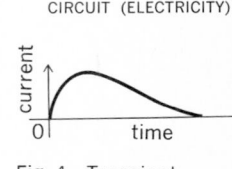

Fig. 3. Nonsinusoidal voltage wave.

Fig. 4. Transient electric current.

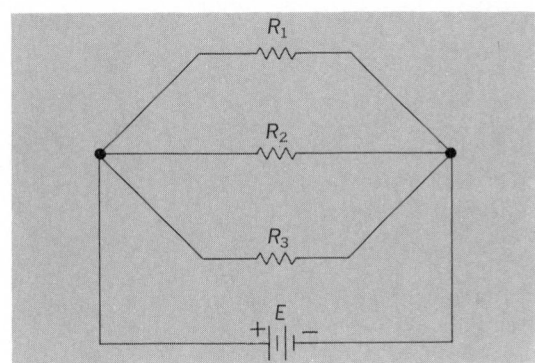

Fig. 6. Parallel circuit.

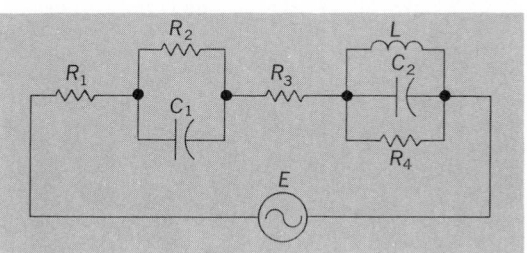

Fig. 7. Series-parallel circuit.

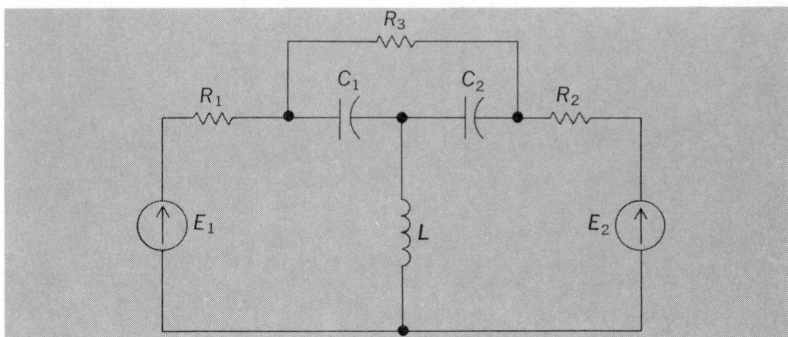

Fig. 8. A three-mesh electric network.

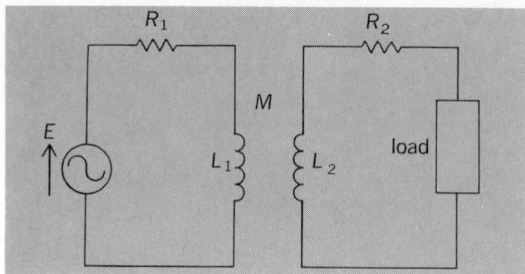

Fig. 9. Inductively coupled circuit.

rent flow between two points that are normally connected.

Short circuit. This term applies to the existence of a zero-impedance path between two points of an electric circuit.

Integrated circuit. The integrated circuit is a recent development in which the entire circuit is contained in a single piece of semiconductor material. Sometimes the term is also applied to circuits made up of deposited thin films on an insulating substrate.

[CLARENCE F. GOODHEART]

Bibliography: R. Boylestad, *Introductory Circuit Analysis*, 3d ed., 1977; E. Brenner and M. Javid, *Analysis of Electric Circuits*, 2d ed., 1967; P. Chirlian, *Electronic Circuits: Physical Principles, Analysis, and Design*, 1971; A. E. Fitzgerald et al., *Basic Electrical Engineering*, 4th ed., 1975; W. Hayt, Jr., and J. E. Kemmerly, *Engineering Circuit Analysis*, 3d ed., 1978; W. W. Lewis and C. F. Goodheart, *Basic Electric Circuit Theory*, 1957; R. E. Scott, *Elements of Linear Circuits*, 1965; R. Smith, *Circuits, Devices, and Systems*, 1966.

Classical field theory

More commonly known as continuum physics or continuum mechanics; the study of distributions of energy and matter under circumstances where the discrete nature of the latter is unimportant. This will generally be the case for a system of a (usually extremely) large number of particles. In this category is the study of the flow of fluids, heat, and other forms of energy; electromagnetic currents and waves, including optical phenomena; and macroscopic theories of the elastic and plastic deformation of solids. Thus, most of applied physical science, such as strength of materials, hydraulics, heat transfer, and aerodynamics, is included.

The quantities involved in these theories, for example, mass and energy densities, fluid velocities, and electric field strengths in a material medium, may be regarded as local averages of corresponding microscopic variables. In many cases, it has been possible to obtain the macroscopic field equations by an appropriate averaging of the microscopic motions over space and time intervals that are small macroscopically, but not microscopically. For example *see* STATISTICAL MECHANICS.

Mathematically, field theories are characterized by equations having partial derivatives with respect to position and time. These equations may express the requirements of conservation of energy, or linear and angular momentum, and also the geometrical properties of the particular type of field.

So-called constitutive equations must be added to the general equations. These define ideal materials by restricting the field variables in some way. Simple examples are the assumptions of incompressible or nonviscous fluids, of perfect gases, or of magnetic materials without hysteresis.

Many of the mathematical difficulties of continuum physics stem from the nonlinearity of the equations. In order to obtain tractable linear equations, the assumption is often made that the system departs from a simple type of motion by only a small amount.

Gravitational and unified field theories, although not a part of quantum theory at the present time, are not usually considered to be a part of continuum physics because of their ultimate concern with the structure of elementary particles. Quantum field theories arise from the dual wave-particle nature of matter. *See* GRAVITON; QUANTUM FIELD THEORY; THEORETICAL PHYSICS; UNIFIED FIELD THEORY.

[BERNARD GOODMAN]

Bibliography: L. D. Landau and E. M. Lifshitz, *Classical Theory of Fields*, 4th ed., 1976; P. Moon and D. E. Spenser, *Field Theory Handbook*, 2d ed., 1971; A Sommerfeld, *Lectures on Theoretical Physics*, vol. 2, 1950, vol. 3, 1952; D. E. Soper, *Classical Field Theory*, 1976.

Classical mechanics

The science dealing with the description of the positions of objects in space under the action of forces as a function of time. Some of the laws of mechanics were recognized at least as early as the time of Archimedes (287?–212 B.C.). In 1638, Galileo stated some of the fundamental concepts of mechanics, and in 1687, Isaac Newton published his *Principia*, which presents the basic laws of motion, the law of gravitation, the theory of tides, and the theory of the solar system. This monumental work and the writings of J. D'Alembert, J. L. Lagrange, P. S. Laplace, and others in the 18th century are recognized as classic works in the field of mechanics. Jointly they serve as the base of the broad field of study known as classical mechanics, or Newtonian mechanics. This field does not encompass the more recent developments in mechanics, such as statistical, relativistic, or quantum mechanics.

The general principles of classical mechanics are stated in mathematical form. With mathematical logic, one can deduce countless possible motions of bodies and then compare the predictions with experimental observations. Classical mechanics illustrates the essential nature of a physical theory, and it is usually an important ingredient in or a starting point for the various branches of modern physics. Its study offers one the opportunity to become acquainted with mathematical techniques and procedures which are useful in other fields.

In the broad sense, classical mechanics includes the study of motions of gases, liquids, and solids, but more commonly it is taken to refer only to solids. In the restricted reference to solids, classical mechanics is subdivided into statics, kinematics, and dynamics. Statics considers the action of forces that produce equilibrium or rest; kinematics deals with the description of motion without concern for the causes of motion; and dynamics involves the study of the motions of bodies under the actions of forces upon them. An important example of a force whose effect on bodies is often studied is the Earth's gravitational force. For some of the more important areas of classical mechanics *see* Collision; Dynamics; Energy; Force; Gravitation; Kinematics; Lagrange's equations; Mass; Motion; Precession; Rigid-body dynamics; Statics; Work.

[Newell S. Gingrich]

Clock paradox

The paradox produced by the use of incorrect arguments, concepts, and premises in the relativistic twin clock problem in which two initially synchronized perfect clocks, A and B, are first separated and then reunited. It is the true logical contradiction inherent in the assertion that, when reunited, the two clocks, side by side, are slower than each other. Now, two clocks, side by side, cannot be both slower than each other; those who infer this paradox to be a conclusion have, in some way, introduced premises or concepts incompatible with Albert Einstein's theory of relativity. The clock paradox is *not* a consequence of this theory.

When the relativistic twin clock problem is properly handled, no paradox occurs. Because of the long controversial history of the twin clock problem, some authors refer to it as the "clock paradox problem" or simply "clock paradox." Authors using the term "clock paradox" in this sense are not referring to a paradox at all, but to an ordinary relativistic problem for which the correct theoretical prediction is that, upon being reunited, one *and only one* clock is slower than the other.

In this article, "time dilation" and the twin clock problem are distinguished. Next, an example of the twin clock problem is presented. Common conceptual errors that lead to various forms of the clock paradox are discussed, particularly the false notions that (1) nature provides an obvious unique universal rule for determining the simultaneity of distant events, and (2) empty space is to be conceived as an unchangeable, passive constituent of the universe. Finally, brief mention is made of some of the experiments that have empirically confirmed the existence in nature of the twin clock effect, that is, the phenomenon of asymmetric natural aging as a path-dependent function of position and velocity.

Time dilation and twin clock effect. Time dilation is a symmetric artifact of the Lorentz transformations and the Einstein simultaneity convention. The twin clock effect, on the other hand, is an asymmetric physical phenomenon and the physical consequence of the fact that empty space has a metric structure.

Time dilation. Time dilation asserts that clocks at rest in different inertial frames symmetrically observe each other to "run slow." This assertion is no more paradoxical than the assertions that (1) a ship steams past an island and (2) the island floats past the ship. To a viewer on the ship, the island moves; to a viewer on the island, the ship moves. Since the island and the ship use different conventions for the rest, or stationary, state of motion, there is no paradox.

A similar situation arises in time dilation. In each inertial frame the rest clocks do indeed observe the rest clocks of other inertial frames to "run slow," but the Lorentz transformations leading to this result are derived from clocks that are

synchronized by a *convention* that does *not* have the same meaning in different inertial frames of reference.

Indeed, clocks synchronized within an inertial frame according to the Einstein simultaneity convention are *not* found to be synchronized when they are examined again in other inertial frames according to the "same" convention for distant simultaneity. Given a set of events $P_1, P_2, \ldots, P_n$ with time and space coordinates (t_1, x_1), (t_2, x_2), $\ldots$, (t_n, x_n) that are simultaneous in one frame, $t_i = t_j$, but spatially separated, $x_i \neq x_j$, then that same set of physical events will, in all other inertial frames, according to the "same" simultaneity convention, have coordinates (t_i', x_i') that are *not* simultaneous, $t_i' \neq t_j'$. This can easily be proved by mere inspection of the Lorentz transformations (Fig. 1) or by Lorentz-transforming the coordinates of two point events P_1 and P_2 for which, initially, $t_1 = t_2$ and $x_1 \neq x_2$. *See* Lorentz transformations.

$$t' = \gamma_v\left(t - \frac{vx}{c^2}\right)$$

$$x' = \gamma_v(x - vt)$$

(a)

$$t'' = \gamma_v\left(t + \frac{vx}{c^2}\right)$$

$$x'' = \gamma_v(x + vt)$$

(b)

$$\gamma_v \equiv \frac{1}{\sqrt{1 - \dfrac{v^2}{c^2}}}$$

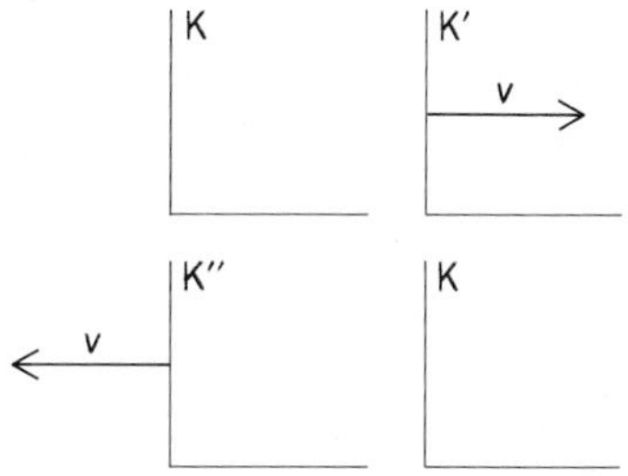

Fig. 1. Lorentz transformations. These expressions convert coordinates of events in the K inertial frame into their coordinates in (a) inertial frame K', moving to the right, with respect to frame K, at velocity v, and (b) frame K'', moving to the left, with respect to frame K, at velocity v.

Thus, unless additional information is supplied, the statement that "moving clocks run slow" is an artifact completely without physical significance. Time dilation is not a philosophical statement as to what is or is not real, and it tells nothing about the physical time differences that might exist between spatially separated clocks, were they (as is impossible) magically and instantaneously reunited. It is purely a symmetric, definitional artifact of measurement created by the adoption of the Einstein simultaneity convention, a "bookkeeping procedure" adopted to give empirical meaning to the statement that two or more spatially separated events occur simultaneously. This convention depends on the reference frame in the same way that the magnitudes of the components of an ordinary vector depend on the location and orientation of the axes of the reference frame in which the vector is to be resolved into components.

Twin clock effect. The twin clock effect is completely different from time dilation. The twin clock effect is the natural phenomenon that a clock having a past history of acceleration indicates a total elapsed time less than that indicated by an identical clock having no history of acceleration. It is a physical effect, and the amount by which the accelerated clock is found to be slow is the same regardless of the choice of inertial frame. The physical properties of empty space cause the twin clock effect.

In relativity the concept of empty space is not merely an abstract notion. Empty space is not empty, but has physical content. The physical reality of empty space, or space-time, is, in both the special and general theories, represented by a field whose components are functions of four parameters — three space parameters and one time parameter. In special relativity, the field functions representing the physical properties of empty space are all constant; in general relativity, however, they vary continuously with location and time. In special relativity, the existence of the twin clock effect is implied by the mathematical fact that the metric field functions representing the time-dependent properties of empty space have numerical signs opposite to those representing the spatial physical properties.

In classical mechanics, inertia is the property of massive bodies to resist acceleration. The entity relative to which acceleration occurs is "absolute space." By "absolute" is meant that, as indicated by the fact of inertia, empty space acts upon material objects but these objects do not, in turn, act upon empty space. In the special theory of relativity, empty space not only enables massive objects to resist acceleration but also modifies the rate at which the natural processes of these objects occur. Nevertheless, material objects still do not, in turn, physically influence empty space. In this sense, both acceleration and empty space are absolute in special relativity. *See* KINETICS (CLASSICAL MECHANICS); NEWTON'S LAWS OF MOTION.

However, in general relativity, massive objects warp, or "bend," empty space through the action of their active gravitational masses. To the extent that the metric field functions representing the physical properties of empty space are thereby also made to vary, the timekeeping processes of natural clocks are altered. Gravitational alteration of natural timekeeping processes may, in certain situations, significantly dominate the twin clock effect of special relativity, and it becomes possible for the traveling clock to be, at return, fast (older) rather than, as before, always slower (younger). *See* GRAVITATION.

In both the general and the special theory the twin clock effect is the consequence of the asymmetric physical metric structure of empty space.

From a purely formal standpoint, time dilation is related to the twin clock effect as a polygon is related to a triangle. Time dilation always involves, as Fig. 2 illustrates, a minimum of at least five dis-

tinct space-time events. Observers at rest in frame K compare space-time intervals P_1P_{2K} and P_1P_{3K}, while observers at rest in frame K' compare intervals $P_1P_{2K'}$ and $P_1P_{3K'}$. Since events P_{2K} and $P_{3K'}$ are simultaneous events in frame K, and events $P_{2K'}$ and P_{3K} are simultaneous in K', rest observers in frames K and K' each conclude that the rest clocks in the other frame are "running slow." On the other hand, the twin clock effect need involve not more than three distinct space-time events.

Twin clock problem. Let A and B be two perfect clocks initially synchronized at a common point of space and time. At some instant, let B depart from A and move along the positive x axis for 1000 sec with a constant velocity $v = (3/5)c$, where c is the speed of light (299,792,458 m/sec), turn around instantaneously, and take another 1000 sec to return to A with $v = -(3/5)c$. There are only three distinct physical events in this problem: B's departure, P_1; B's turnabout, P_{2B}; and B's return, P_3 (see Fig. 3).

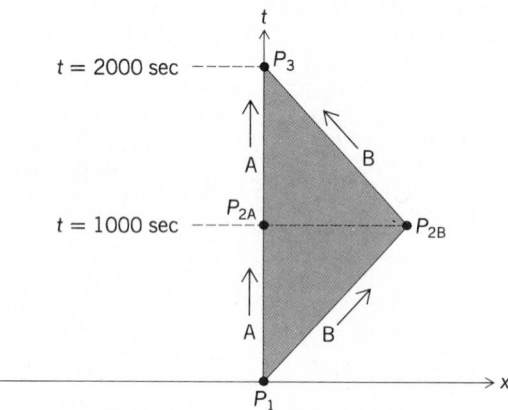

Fig. 3. Space-time diagram of the four events P_1, P_{2A}, P_{2B}, and P_3 in the inertial frame K. Clock A is at rest in this frame.

Let K be the inertial frame in which clock A is always at rest. Let K' be an inertial frame moving to the right with velocity $v = (3/5)c$; and let K'' be an inertial frame moving to the left with velocity $v = -(3/5)c$. For simplicity, the origins of K, K', and K'' are assumed to coincide when B departs from A.

The event P_{2A} is defined to be the event at A in K that, according to the Einstein simultaneity convention, is simultaneous with event P_{2B}. The coordinates of the four events P_1, P_{2A}, P_{2B}, and P_3 in frame K are given in the table. By using the Lorentz transformations in Fig. 1, the K frame coordinates of the four events are converted into their

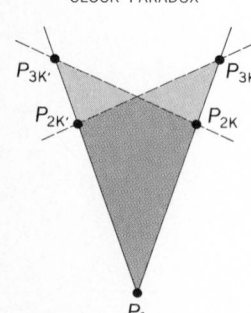

Fig. 2. The five distinct space-time point events that play a role in time dilation.

Coordinates of four space-time events, P_1, P_{2A}, P_{2B}, and P_3, in inertial frames K, K', and K''

	Frame K			Frame K'			Frame K''	
Event	t	x	Event	t'	x'	Event	t''	x''
P_1	0 sec	0 n_c* meters	P_1	0 sec	0 n_c meters	P_1	0 sec	0 n_c meters
P_{2A}	+1000	0 n_c	P_{2B}	+800	0 n_c	P_{2A}	+1250	+750 n_c
P_{2B}	+1000	+600 n_c	P_{2A}	+1250	−750 n_c	P_{2B}	+1700	+1500 n_c
P_3	+2000	0 n_c	P_3	+2500	−1500 n_c	P_3	+2500	+1500 n_c

*n_c is a dimensionless constant equal to 299,792,458; it is introduced for calculational convenience.

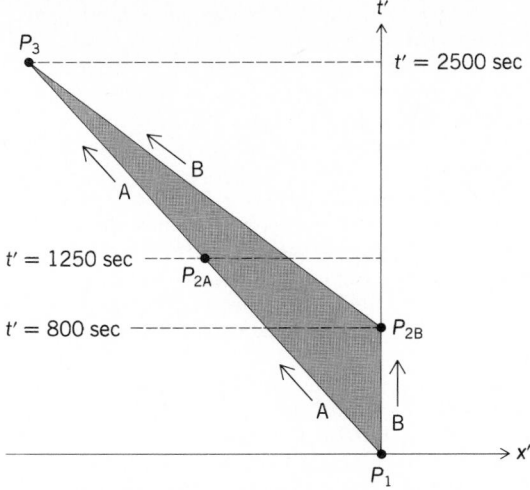

Fig. 4. Space-time diagram of the four events P_1, P_{2A}, P_{2B}, and P_3 in the inertial frame K'.

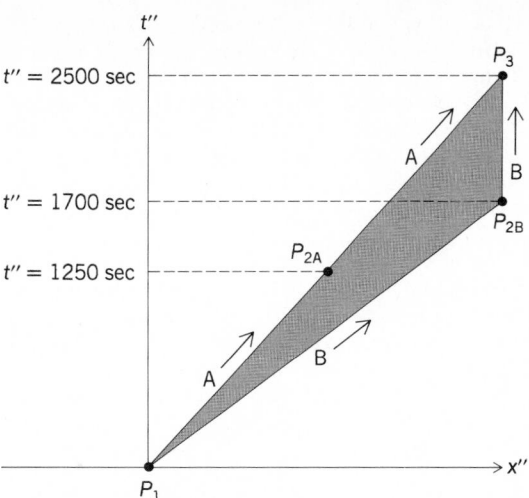

Fig. 5. Space-time diagram of the four events P_1, P_{2A}, P_{2B}, and P_3 in the inertial frame K''.

corresponding K' and K'' frame coordinates, also given in the table. Figures 3, 4, and 5 are space-time diagrams of the four events in frames K, K', and K'', respectively.

Starting with any one of the three frames, coordinates in the other two can be obtained by means of the appropriate Lorentz transformation. For transformations between K' and K'', the relativistic law for composition of velocities must be used to calculate the velocity of K'' with respect to K'.

Clock A's viewpoint. The table indicates that 2000 sec elapse between events P_1 and P_3 in frame K. Since clock A is always at rest in K, 2000 sec will have elapsed on it when it reaches P_3.

For clock B, however, it is impossible to find an inertial frame in which B is at rest for the entire timelike interval between events P_1 and P_3. Nevertheless, the time elapsed on clock B can be calculated using the fact that in K', between events P_1 and P_{2B}, clock B is at rest, and in K'', between events P_{2B} and P_3, clock B is again at rest. The table indicates that $800 - 0 = 800$ sec elapse between events P_{2B} and P_1 in frame K', and that $2500 - 1700 = 800$ sec elapse between events P_3 and P_{2B} in frame K''. The sum of elapsed times in the various frames in which clock B is at rest is $800 + 800 = 1600$ sec. Therefore, 1600 sec have elapsed on clock B when it reaches P_3.

Clock B is therefore slower (younger) than clock A by 400 sec when they reunite at P_3. This result is independent of the particular inertial frame used to obtain it.

Clock B's viewpoint. In B's view, clock A moves to the left, "turns around," and returns to B. Figures 6 and 7 show the space-time situation as viewed by B. Figure 6 is identical to Fig. 4 except that all space-time paths occurring after $t' = t'_{2B}$ have been drawn as dashed lines to emphasize that events P_{2A} and P_3 both occur *after* B departs from frame K'. Even though B is no longer in frame K' when P_{2A} and P_3 occur, both events still have definite space-time coordinates within frame K'. Figure 7 is identical to Fig. 5 except that all space-time paths occurring before $t'' = t''_{2B}$ have been dashed to indicate that events P_{2A} and P_1 occurred *before* clock B entered K''. The solid space-time paths in Fig. 6 show what B sees prior to A's "turn-

about"; and the solid space-time paths in Fig. 7 show what B sees after B adopts K'' as its new frame of reference.

As B departs from frame K' at $t' = t'_{2B}$, event P_{2A}, the event at which A "turns around," has yet to occur; but after B has entered frame K'' at $t'' = t''_{2B}$, event P_{2A} has already occurred! The reason is that distant simultaneity does not have the same physical realization in different inertial reference frames. When clock B adopts a new frame of reference, it gains access to an arrangement of physical information that would not have been available to it had B remained at rest in frame K'.

The sudden jump of event P_{2A} from a future event to a past event is not paradoxical; it merely results from the assumption that clock B is subjected to infinite forces when it transfers from frame K' to K''. If B is subjected to finite, gradual deceleration and acceleration, the time of the distant event P_{2A} changes smoothly and continuously from future event to past event. The "in-between" events which were "missed" because of the math-

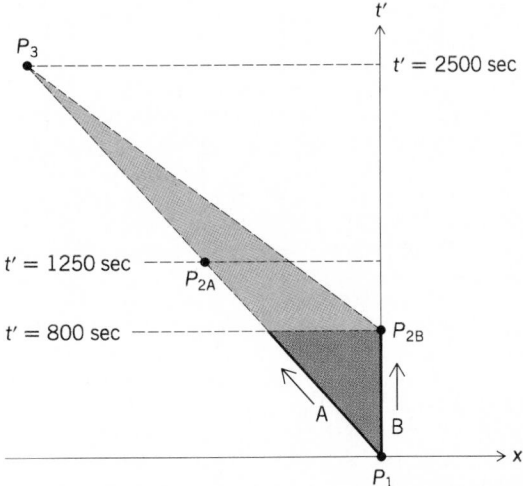

Fig. 6. Space-time diagram of the four events P_1, P_{2A}, P_{2B}, and P_3 in the inertial frame K' as determined by clock B while B is at rest in K'.

ematical discontinuity originally created by the assumption that B is subject to sharp sudden changes of direction and infinite accelerations will now be observed to occur continuously and rapidly in much the same way the relativistic Doppler shift smoothly changes from a red shift to a blue shift as B decelerates, stops and reverses direction, and accelerates back to speed. *See* DOPPLER EFFECT.

The fact that an event, for example, P_{2A}, can be converted, by mere adoption of a new frame of reference, from an event that has yet to occur into one that has already occurred, illustrates why it is impossible to find a common universal time system that all observers could agree to use.

In Fig. 6, $800 - 0 = 800$ sec elapse between events P_1 and P_{2B}; in Fig. 7, $2500 - 1700 = 800$ sec elapse between events P_{2B} and P_3. Hence in frames K′ and K″, in which clock B is alternately at rest, a total of $800 + 800 = 1600$ sec elapse on clock B between events P_1 and P_3.

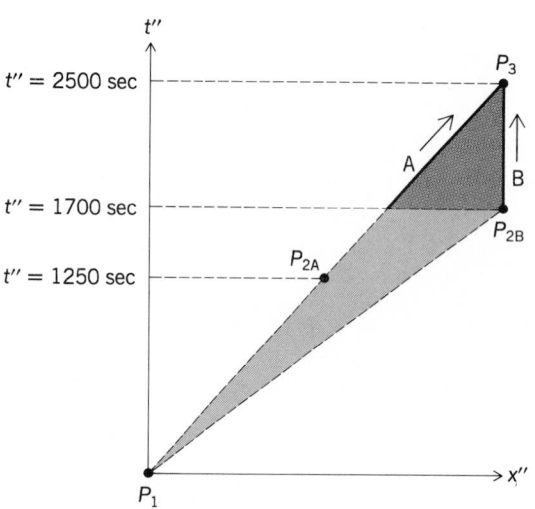

Fig. 7. Space-time diagram of the four events P_1, P_{2A}, P_{2B}, and P_3 in the inertial frame K″ as determined by clock B while B is at rest in K″.

In frame K′ (Fig. 6), clock A does not "turn around" until event P_{2A}, when $t' = 1250$ sec have elapsed in K′. Since clock B leaves K′ at $t' = 800$ sec, B misses the 450 sec that elapse in K′ between B's departure from K′, event P_{2B}, and the "turnaround" event P_{2A}. Nevertheless the 1250 sec that elapse in frame K′ between event P_1, A's departure from B, and event P_{2A}, A's "turnaround" in K′, still correspond to the elapse on clock A, in the frame K in which A is at rest, of $1000 - 0 = 1000$ sec. From B's viewpoint in frame K′, therefore, 1000 sec elapse on clock A between A's departure and A's "turnaround."

When B adopts frame K″ (Fig. 7), B discovers that A not only has already turned around but has been returning back toward B for the past $1700 - 1250 = 450$ sec. In frame K″, B misses the first 450 sec between A's "turnaround" at $t'' = 1250$ sec and A's reunion with B at $t'' = 2500$ sec. The $2500 - 1250 = 1250$ sec that elapse in frame K″ between events P_{2A} and P_3 correspond, in frame K, in which A is at rest, to the elapse on A of $2000 - 1000 = 1000$ sec.

As determined by clock B, therefore, 1000 sec

elapse on clock A on A's outward journey, and another 1000 sec on A's return journey, for a total of $1000 + 1000 = 2000$ sec elapsed in the frame in which A is at rest. Hence, even from clock B's point of view, clock A is still older than clock B when they are reunited at event P_3 by $2000 - 1600 = 400$ sec.

Significance of motion. One may object that B's viewpoint has not really been adopted, that the argument has been merely repeated from A's point of view. This objection is based on the notion that, in relativity, "only relative motions are significant." This idea is mistaken: in both classical mechanics and special relativity, one cannot, without introducing contradiction (or logical paradox), claim that the same body can move in an inertial, uniform, rectilinear fashion within one inertial frame and in an accelerated, nonuniform, curvilinear fashion in another, and one may not adopt the viewpoint of an accelerated observer for the purpose of pretending that this is possible. Isaac Newton *defined* an inertial frame of a reference as one in which a body subject to no forces moves in rectilinear path with constant velocity.

Even in general relativity, it is misleading to assert that only purely relative motions are meaningful. Mach's principle does not govern relativity theory in that sense. In general relativity, moving bodies are conceived as being in continuous interaction with empty space, a dynamic, physically active, continuous field endowed with an independent existence of its own. This field is initially created by gravitational mass but possesses thereafter an independent existence, in a manner analogous to the way electromagnetic fields initially created by moving electrons propagate and exist independently.

The relativistic concept of empty space as an existing, dynamic physical field is in common use in modern relativity. It is essential for speculations concerning geometrodynamical interpretations of elementary particles as stable metric "wormhole" structures, and for cosmological speculations concerning massive galactic stars that collapse into "black holes" that may connect into new universes as "white holes." *See* UNIFIED FIELD THEORY.

Once one understands that, in relativity, empty space is an existing reality with physical content, and is the immediate physical cause of twin clock effects, then the observable fact that natural clocks, moving about and interacting with empty space, alter their timekeeping rates should no longer seem enigmatic.

Empirical tests. A. H. Bucherer was the first (1909) to implicitly verify the twin clock effect when he experimentally found that high-speed electrons obey relativistic force laws rather than the classical Newtonian laws of motion. In 1938 H. E. Ives found that the frequencies of atomic spectral lines of rapidly moving atoms were decreased by an amount in agreement with the decrease predicted on the basis of relativistic time dilation. In 1941 B. Rossi, by counting the rates at which cosmic particles occur at various altitudes in the mountains of Colorado, found that cosmically created mesons survived, on the average, longer than would be considered possible on the basis of a purely classical Newtonian lifetime, and that the excess lifetime was in accord with relativistic time dilation.

During the 1950s many experiments were performed with meson lifetimes. It was repeatedly shown that the number of surviving mesons always exceeded, by a factor in agreement with relativistic time dilation, the number of mesons that were expected to survive on the basis of the classical Newtonian lifetime. All these meson experiments represent empirical confirmation of time dilation, and, at least to the extent that meson clocks were accelerated and the laboratory rest clocks were not, also a verification of the twin clock effect.

In 1960 the Mössbauer effect was used by R. V. Pound and G. A. Rebka to verify the general relativistic "red shift" or, better, "blue shift," the relativistic increase in the frequency of an atomic spectral line with the atom's increased distance from a heavy gravitating body. *See* MÖSSBAUER EFFECT.

In 1971 J. C. Hafele and R. E. Keating flew four atomically stabilized clocks around the world and directly observed both the special and general relativistic twin clock effects. This experiment excluded theoretical attempts to "explain" relativistic twin clock effects solely in terms of energy conservation and the principle of equivalence. The Hafele-Keating experiment clearly indicates that both the special and general twin clock effects have as their common physical cause and source the metric structure of the physical field properties of empty space.

The twin clock phenomena of special relativity are well understood, but the twin clock effects of general relativity are just beginning to be explored. Modern navigation systems routinely attain such high worldwide synchronization accuracies that gravitational twin clock effects are significant. Yet it is not known what is meant physically when a clock in Australia is said to be linked with a clock in the United States to an accuracy of 1 nanosecond (1 nsec $= 10^{-9}$ sec). The mere fact that the clocks at two such widely separated locations are on opposite sides of the Earth—and therefore alternately farther from and closer to the Sun—leads some scientists to expect that the two clocks have a time difference that varies periodically with a period of 1 day! However, this effect has not been detected. In any case, there is no reason to doubt that asymmetric natural aging as a path-dependent function of position and velocity is an empirical phenomenon exhibited by all known types of stable clocks. *See* RELATIVITY; SPACE-TIME.

[RICHARD E. KEATING]

Bibliography: A. H. Bucherer, The experimental confirmation of the relativity principle, *Ann. Phys.* (Leipzig), 28:513–536, 1909; R. Durbin, H. H. Loar, and W. W. Havens, The lifetime of the π^+ and π^- mesons, *Phys. Rev.*, 88:179–183, 1952; A. Einstein, *The Meaning of Relativity*, 5th ed., 1956; A. Einstein, On the electrodynamics of moving bodies, *Ann. Phys.* (Leipzig), 17:891–921, 1905 (available in a Dover reprint volume, *The Principle of Relativity*); T. T. Frankel, *Gravitational Curvature: An Introduction to Einstein's Theory*, 1979; R. Geroch, *General Relativity from A to B*, 1978; J. C. Graves, *The Conceptual Foundations of Contemporary Relativity Theory*, 1971; J. C. Hafele and R. E. Keating, Around-the-world atomic clocks, *Science*, 177:166–170, 1972; H. E. Ives and G. R. Stilwell, An experimental study of the rate of a moving atomic clock, *J. Opt. Soc. Amer.*, 28:215–226, 1938, and 31:369–374, 1941; L. Marder, *Time and the Space Traveller*, 1974; E. Martinelli and W. K. H. Panofsky, The lifetime of the positive π meson, *Phys. Rev.*, 77:465–469, 1950; C. W. Misner, K. S. Thorne, and J. A. Wheeler, *Gravitation*, 1973; R. V. Pound and G. A. Rebka, Jr., Apparent weight of photons, *Phys. Rev. Lett.*, 4:337–341, 1960; B. Rossi and D. B. Hall, Variation of the rate of mesotrons with momentum, *Phys. Rev.*, 59:223–228, 1941.

Coherence

The attribute of two or more waves, or parts of a wave, whose relative phase is constant during the resolving time of the observer. The concept has been developed most extensively in optics, but is applicable to all wave phenomena.

Coherence of two beams. Consider two waves, with the same mean angular frequency ω, given by Eqs. (1) and (2).

$$\Psi_A(x,t) = A \exp i [k(\omega)x - \omega t - \delta_A(t)] \qquad (1)$$
$$\Psi_B(x,t) = B \exp i [k(\omega)x - \omega t - \delta_B(t)] \qquad (2)$$

It is convenient, and no restriction, to choose both A and B real. These expressions as they stand could describe de Broglie waves in quantum mechanics. For real waves, such as components of the electric field in light or radio beams, or the pressure oscillations in sound, it is necessary to retain only the real parts of these and subsequent expressions. The frequency spectrum is assumed to be narrow, in the sense that a Fourier analysis of expressions (1) and (2) gives appreciable contributions only for angular frequencies close to ω. This assumption means that, on the average, $\delta_A(t)$ and $\delta_B(t)$ do not change much per period. *See* ELECTROMAGNETIC RADIATION; QUANTUM MECHANICS; SOUND.

Suppose that the waves are detected by an apparatus with resolving time T; that is, T is the shortest interval between two events for which the events do not seem to be simultaneous. For the human eye and ear, T is about 0.1 s, while a fast electronic device might have a T of 10^{-10} s. If the relative phase $\delta(t)$, given by Eq. (3),

$$\delta(t) = \delta_B(t) - \delta_A(t) \qquad (3)$$

does not, on the average, change noticeably during T, then the waves are coherent. If during T there are sufficient random fluctuations for all values of $\delta(t)$, modulus 2π, to be equally probable, then the waves are incoherent. If during T there are noticeable random fluctuations in $\delta(t)$, but not enough to make the waves completely incoherent, then the waves are partially coherent. These distinctions are not useful unless T is specified. On the one hand, only waves that have existed forever and that fill all of space can have absolutely fixed frequency and phase. On the other hand, two independent sound waves in which the phases change appreciably in 0.01 s would seem incoherent to the human ear, but would seem highly coherent to a fast electronic device.

The degree of coherence is related to the interference patterns that can be observed when the two beams are combined. The following variations of Young's two-slit interference experiment illustrate several possibilities. *See* INTERFERENCE OF WAVES.

Young's two-slit experiment. Figure 1 shows the usual Young's experiment. Near P, for linear wave phenomena, the resultant wave $\Psi(x,t)$ is the sum of $\Psi_A(x,t)$ and $\Psi_B(x,t)$. The observable intensity is proportional to $|\Psi|^2$, the square of the magnitude of Ψ. This observable intensity is the energy density or the mean photon density for electromagnetic waves, the energy density for acoustic waves, and the mean particle density for the wave functions of quantum mechanics. With the wave forms of Eqs. (1) and (2), it is given by Eq. (4).

$$|\Psi|^2 = |\Psi_A + \Psi_B|^2 = A^2 + B^2 + 2AB \cos \delta(t) \qquad (4)$$

The first term gives the intensity of beam A alone, the second gives the intensity of B alone, and the third term depends on the relative phase, given by Eq. (3).

The mean life of a typical excited atomic state is about 10^{-8} s. Collisions and thermal motion reduce the effective time of undisturbed emission to around 10^{-10} or 10^{-11} s in standard discharge tubes. The phase $\delta_A(t)$ of beam A therefore dances about erratically, with substantial changes occurring perhaps 10^{11} times per second. However, beam B comes from the same atoms and travels nearly the same distance, and the changes in $\delta_B(t)$ are the same as those in $\delta_A(t)$. The relative phase is

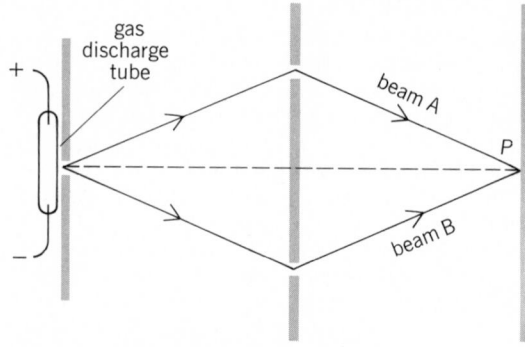

Fig. 1. Coherence. If beams A and B come from the same point source and traverse the same distance, they are coherent around P and produce there an interference pattern that has high visibility.

always zero at the exact midpoint P and takes on other time-independent values in the neighborhood of P. For each point in that neighborhood, $|\Psi|^2$ has a constant value that satisfies inequality (5).

$$|A - B|^2 \leq |\Psi|^2 \leq |A + B|^2 \qquad (5)$$

A clear interference pattern is observed even with a large-T detector such as a photographic plate.

Independent sources. Figure 2 shows an arrangement that uses two independent sources to produce the two beams. If the two sources are standard discharge tubes, $\delta_A(t)$ and $\delta_B(t)$ independently change erratically around 10^{11} times per second. Equation (4) is still valid, but the observed quantity is $|\Psi|^2$ averaged over the resolving time T, denoted by $\langle |\Psi|^2 \rangle_T$. When the average of Eq. (4) is formed, the last term on the right contributes nothing because $\cos \delta(t)$ randomly takes on values between $+1$ and -1 and thus obeys Eq. (6).

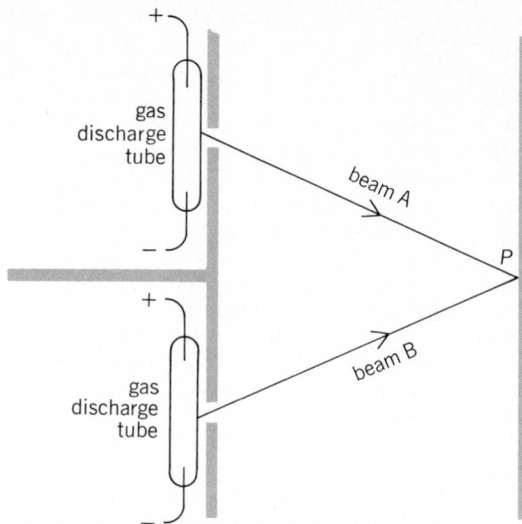

Fig. 2. Incoherence. Independent sources that change their phases frequently and randomly during the resolving time *T* are incoherent, and their superposition does not produce a visible interference pattern.

$$\langle \cos \delta(t) \rangle_T = 0 \qquad (6)$$

The two beams are incoherent, the observed intensity is simply the sum of the separate intensities, and the superposition does not produce a visible interference pattern. In general, if incoherent waves Ψ_A, Ψ_B, Ψ_C, . . . , are combined, much the same argument applies because the average over T of each cross term is zero, so that the observed intensity is given by Eq. (7).

$$\langle |\Psi|^2 \rangle_T = \langle |\Psi_A|^2 \rangle_T + \langle |\Psi_B|^2 \rangle_T + \langle |\Psi_C|^2 \rangle_T + \cdots \qquad (7)$$

The argument that results in Eq. (6) depends on there being random independent changes in the separate phases during times much shorter than T. If the discharge tubes in Fig. 2 are replaced by very well stabilized lasers, or by loudspeakers or radio transmitters, then the separate phases $\delta_A(t)$ and $\delta_B(t)$ can be made nearly constant during 0.1 s. The relative phase $\delta(t)$ is then of course also nearly constant, an interference pattern is visible, and the two waves can be called coherent. *See* LASER.

Effect of path length difference. Figure 3, like

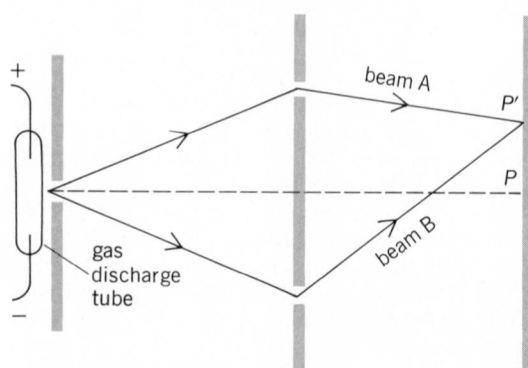

Fig. 3. Partial coherence. If the difference between the distances traversed by beams A and B is of the order of the coherence length, then the beams are partially coherent at *P'*.

Fig. 1, shows two beams that are emitted by the same point source and are therefore coherent at their origin. Here, however, the distance from the source to the point P' of wave addition is larger along beam B than along beam A by an amount l. In Fig. 3, the path length difference is due to the observation point P' not being near the central axis, but it could equally well be caused by the insertion of different optical devices in the two beams.

The waves that arrive at P' via path B at any instant must have left the source earlier by a time $\tau = l/c$, where c is the wave speed, than those that arrive via path A. Suppose again that the source is a standard discharge tube, so that the phase changes randomly about 10^{11} times per second. For $\tau \ll 10^{-11}$ s, the coherence between A and B at P' is not spoiled by the path length difference, and there is a visible interference pattern around P'. For $\tau \gg 10^{-11}$ s, the light that arrives via A left the source so much later than that which arrives via B at the same instant that the phase changed many times in the interim. The two beams are then incoherent, and their superposition does not produce a visible interference pattern around P'. For $\tau \simeq 10^{-11}$ s, perfect coherence is spoiled but there is less than complete incoherence; there is partial coherence and an interference pattern that is visible but not very sharp. The time difference that gives partial coherence is the coherence time Δt. The coherence time multiplied by the speed c is the coherence length Δl. In this example, Δl is given by Eq. (8).

$$\Delta l \simeq 10^{-11}\ \mathrm{s} \times 3.0 \times 10^{10}\ \mathrm{cm/s} = 0.3\ \mathrm{cm} \qquad (8)$$

Effect of extended source. Figure 4 differs from

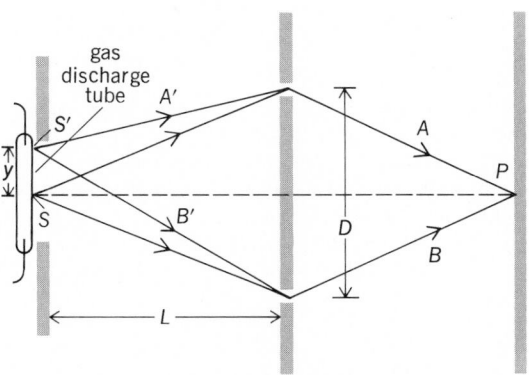

Fig. 4. Partial coherence. If the lateral extent y of the source is of the order of $L\lambda/D$, then the beams are partially coherent at P.

Fig. 1 in that it shows a large hole in the collimator in front of the discharge tube, so that there is an extended source rather than a point source. Consider two regions of the source, S near the center and S' near one end, separated by a distance y. During times longer than the coherence time, the difference between the phases of the radiation from the two regions changes randomly. Suppose that, at one moment, the radiation from S happens to have a large amplitude while that from S' happens to be negligible. The waves at the two slits are then in phase because S is equidistant from

them, and interference at P gives a maximum. Suppose that a few coherence times later the radiation from S happens to be negligible while that from S' happens to have a large amplitude. If the path length along A′ differs from that along B′ by half of the wavelength λ, then the waves at the two slits are out of phase. Now interference at P gives a minimum, and the pattern has shifted in a few coherence times. If the resolving time T is much larger than the coherence time Δt, the consequence is that beams A and B are not coherent at P.

The condition that there be a path length difference of $\lambda/2$ between B′ and A′ is given by Eq. (9),

$$\sqrt{L^2 + (y + D/2)^2} - \sqrt{L^2 + (y - D/2)^2} = \lambda/2 \qquad (9)$$

and for D and y both much smaller than L, expansion of the square roots gives Eq. (10).

$$2Dy = L\lambda \qquad (10)$$

This argument is rather rough, but it is clear that a source that satisfies Eq. (11)

$$y \simeq \frac{L\lambda}{D} \qquad (11)$$

is too large to give coherence and too small to give complete incoherence; it gives partial coherence.

Coherence of a single beam. Coherence is also used to describe relations between phases within the same beam. Suppose that a wave represented by Eq. (1) is passing a fixed observer characterized by a resolving time T. The phase δ_A may fluctuate, perhaps because the source of the wave contains many independent radiators. The coherence time Δt_W of the wave is defined to be the average time required for $\delta_A(t)$ to fluctuate appreciably at the position of the observer. If Δt_W is much greater than T, the wave is coherent; if Δt_W is of the order of T, the wave is partially coherent; and if Δt_W is much less than T, the wave is incoherent. These concepts are very close to those developed above. The two beams in Fig. 2 are incoherent with respect to each other if one or both have $\Delta t_W \ll T$, but are coherent with respect to each other if both have $\Delta t_W \gg T$.

The degree of coherence of a beam is of course determined by its source. Discharge tubes that emit beams with Δt_W small compared to any usual resolving time are therefore called incoherent sources, while well-stabilized lasers that give long coherence times are called coherent sources.

Consider two observers that measure the phase at the same time. If the observers are close to each other, they will usually measure the same phase. If they are far from each other, the phase difference between them may be entirely random. The observer separation that shows the onset of randomness is defined to be the coherence length of the wave, Δx_W. If this separation is in the direction of propagation, then Δx_W is given by Eq (12),

$$\Delta x_W = c\Delta t_W \qquad (12)$$

where c is the speed of the wave. Of course, the definition of Δx_W is also applicable for observer separations perpendicular to the propagation direction.

The coherence time Δt_W is related to the spectral purity of the beam, as is shown by the following argument. The wave can be viewed as a sequence of packets with spatial lengths around Δx_W. A typical packet requires Δt_W to pass a fixed point. Suppose that n periods of duration $2\pi/\omega$ occur during Δt_W, as expressed in Eq. (13).

$$\Delta t_W \frac{\omega}{2\pi} = n \qquad (13)$$

The packet can be viewed as a superposition of plane waves arranged to cancel each other outside the boundaries of the packet. To produce cancellation, waves must be mixed in which are in phase with the wave of angular frequency ω in the middle of the packet and which are half a period out of phase at both ends. In other words, waves must be mixed in with periods around $2\pi/\omega'$, where Eq. (14) holds.

$$\Delta t_W \frac{\omega'}{2\pi} = n \pm 1 \qquad (14)$$

The difference of Eqs. (13) and (14) gives Eq. (15), where the spread in angular frequencies $|\omega - \omega'|$ is called $\Delta\omega$.

$$\Delta t_W \Delta\omega = 2\pi \qquad (15)$$

This result should be stated as an approximate inequality (16).

$$\Delta t_W \Delta\omega \gtrsim 1 \qquad (16)$$

A large Δt_W permits a small $\Delta\omega$ and therefore a well-defined frequency, while a small Δt_W implies a large $\Delta\omega$ and a poorly defined frequency.

Quantitative definitions. To go beyond qualitative descriptions and order-of-magnitude relations, it is useful to define the fundamental quantities in terms of correlation functions.

Self-coherence function. As discussed above, there are two ways to view the random fluctuations in the phase of a wave. One can discuss splitting the wave into two beams that interfere after a difference τ in traversal time, or one can discuss one wave that is passing over a fixed observer who determines the phase fluctuations. Both approaches concern the correlation between $\Psi(x,t)$ and $\Psi(x, t + \tau)$. This correlation is described by the normalized self-coherence function given by Eq. (17).

$$\gamma(\tau) \equiv \frac{\langle \Psi(x, t+\tau)\, \Psi^*(x,t) \rangle}{\langle \Psi(x,t)\, \Psi^*(x,t) \rangle} \qquad (17)$$

The brackets $\langle\;\rangle$ indicate the average of the enclosed quantity over a time which is long compared to the resolving time of the observer. It is assumed here that the statistical character of the wave does not change during the time of interest, so that $\gamma(\tau)$ is not a function of t.

With the aid of the self-coherence function, the coherence time of a wave can be defined quantitatively by Eq. (18).

$$(\Delta t_W)^2 \equiv \frac{\int_{-\infty}^{+\infty} \tau^2 |\gamma(\tau)|^2\, d\tau}{\int_{-\infty}^{+\infty} |\gamma(\tau)|^2\, d\tau} \qquad (18)$$

That is, Δt_W is the root-mean-squared width of $|\gamma(\tau)|^2$. The approximate inequality (16) can also be made precise. The quantity $|g(\omega')|^2$ is proportional to the contribution to Ψ at the angular frequency ω' where $g(\omega')$ is the Fourier transform of Ψ, defined by Eq. (19).

$$g(\omega') \equiv \frac{1}{\sqrt{2\pi}} \int_{-\infty}^{+\infty} \Psi(0,t) e^{-i\omega' t}\, dt \qquad (19)$$

The root-mean-squared width $\Delta\omega$ of this distribution is given by Eq. (20),

$$(\Delta\omega)^2 = \frac{\int_0^\infty (\omega' - \omega)^2 |g(\omega')|^2\, d\omega'}{\int_0^\infty |g(\omega')|^2\, d\omega'} \qquad (20)$$

where ω is again the mean angular frequency. With these definitions, one can show that inequality (21) is satisfied.

$$\Delta t_W \Delta\omega \geq \tfrac{1}{2} \qquad (21)$$

The symbol $\gtrsim$ has become $\geq$; no situation can lead to a value of $\Delta t_W \Delta\omega$ that is less than $\frac{1}{2}$. Relation (21) multiplied by Planck's constant is the Heisenberg uncertainty principle of quantum physics, and can in fact be proved in virtually the same way. *See* NONRELATIVISTIC QUANTUM THEORY; UNCERTAINTY PRINCIPLE.

Complex degree of coherence. To describe the coherence between two different waves Ψ_A and Ψ_B, one can use the complex degree of coherence $\gamma_{AB}(\tau)$, defined by Eq. (22).

$$\gamma_{AB}(\tau) \equiv \frac{\langle \Psi_A(t+\tau)\, \Psi_B{}^*(t) \rangle}{[\langle \Psi_A{}^*(t)\, \Psi_A(t) \rangle \langle \Psi_B{}^*(t)\, \Psi_B(t) \rangle]^{1/2}} \qquad (22)$$

where the averaging process $\langle\;\rangle$ and the assumptions are the same as for the definition of $\gamma(\tau)$. If beams A and B are brought together, interference fringes may be formed where the time-averaged resultant intensity $\langle |\Psi|^2 \rangle$ has maxima and minima as a function of position. The visibility of these fringes, defined by Eq. (23).

$$V = \frac{\langle |\Psi|^2 \rangle_{\max} - \langle |\Psi|^2 \rangle_{\min}}{\langle |\Psi|^2 \rangle_{\max} + \langle |\Psi|^2 \rangle_{\min}} \qquad (23)$$

can be shown, for the usual interference studies, to be proportional to $|\gamma_{AB}(\tau)|$.

Examples. The concept of coherence occurs in a great variety of areas. The following are some applications and illustrations.

Astronomical applications. As is shown in the discussion of Fig. 4, extended sources give partial coherence and produce interference fringes with visibility V [Eq. (23)] less than unity. A. A. Michelson exploited this fact with his stellar interferometer (Fig. 5), a modified double-slit arrangement with movable mirrors that permit adjustment of the effective separation D' of the slits. According to Eq. (11), there is partial coherence if the angular size of the source is about λ/D'. It can be shown that if the source is a uniform disk of angular diameter Θ, then the smallest value of D' that gives zero V is $1.22\lambda/\Theta$. The interferometer can therefore be used in favorable cases to measure stellar diameters. The same approach has also been applied in radio astronomy. A different technique, developed by R. Hanbury Brown and R. Q. Twiss, measures the correlation between the intensities received by separated detectors with fast electronics. *See* INTERFEROMETRY.

Lasers and masers. Because they are highly coherent sources, lasers and masers provide very

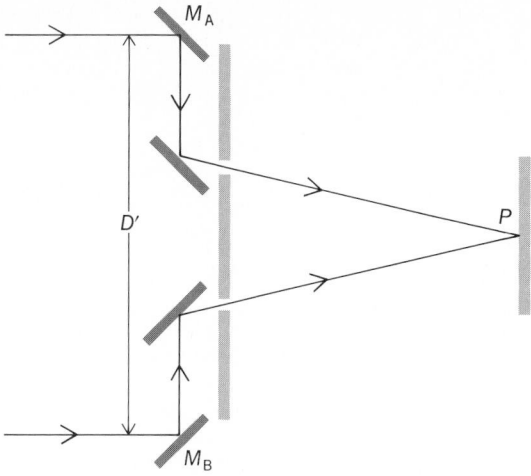

Fig. 5. Michelson's stellar interferometer. The effective slit separation D' can be varied with the movable mirrors M_A and M_B. The angular size of the source can be calculated from the resulting change in the visibility of the interference fringes around P.

large intensities per unit frequency, and double-photon absorption experiments have become possible. Atomic and molecular spectroscopy have as a consequence been revolutionized. One can selectively excite molecules that have negligible velocity components in the beam direction and thus improve resolution by reduction of Doppler broadening. States that are virtually impossible to excite with single-photon absorption can now be reached.

Musical acoustics. Fortunately, for most musical instruments, the coherence time is much less than human resolving time, so that the separate contributions in an orchestra are perceived incoherently. Equation (7) applies, and 20 violins seem to provide 20 times the intensity of 1 violin. If human resolving time were much shorter, the perceived intensity would drift between 0 and 400 times the intensity of a single violin.

Thermodynamic treatment of radiation. Coherence must be considered in thermodynamic treatments of radiation. As one should expect from the connection between entropy and order, entropy is reduced by an increase in coherence. The total entropy of two beams that are coherent with respect to each other is less than the sum of the entropies of the separate beams, because coherence means that the beams are not statistically independent. *See* ENTROPY; THERMODYNAMIC PRINCIPLES.

Quantum physics. In quantum physics, a beam of particles can for some purposes be described by Eq. (1) without the phase $\delta_A(t)$. If the wavelength is well defined, measurables such as barrier transmission coefficients and scattering cross sections of sufficiently small targets do not depend on phases and coherence. If, however, the scattering object is large, these aspects may have to be considered.

Electrons emitted by hot filaments show coherence times of roughly 10^{-14} s. In low-energy electron diffraction, where speeds of 10^8 to 10^9 cm/s are used, the coherence length is then about 10^{-5} cm, and the structure can be seen clearly only if the region being studied does not introduce path

length difference greater than this distance. *See* ELECTRON DIFFRACTION.

Most nuclear and particle physics experiments deal with energies so high that the wavelengths of the projectiles are less than 10^{-12} cm, far less than the amplitudes of motions of the scattering nuclei. The many nuclei therefore contribute to the scattered beam incoherently. Equation (7) applies, and the intensity of the scattered beam is proportional to the scattering cross section per nucleus times the number of nuclei.　　　　[ROLF G. WINTER]

Bibliography: M. Born and E. Wolf, *Principles of Optics*, 5th ed., 1975; L. Mandel and E. Wolf, Coherence properties of optical fields, *Rev. Mod. Phys.*, 37:231–287, 1965; J. I. Steinfeld (ed.), *Laser and Coherence Spectroscopy*, 1978; R. G. Winter, *Quantum Physics*, 1979.

Cohesion

The tendency of particles of a condensed phase (a solid or liquid) to stick together, or the processes or forces that lead to this tendency. The cohesive energy is the work necessary to pull the body apart into its constituent particles. For example, the cohesive energy of a metal is the work necessary to break the metal apart into individual atoms in their ground state. The cohesive energy of a molecular solid such as solid carbon dioxide is the energy necessary to separate the solid into isolated carbon dioxide molecules.

The study of cohesion is the study of the competing processes that tend to establish an equilibrium separation between molecules or atoms in a condensed phase. In normal matter all such forces are the result either of the Coulomb interaction between positive and negative charges or of quantum-mechanical kinetic energy. The Coulomb interaction lowers the energy (increases the bond strength) and tends to pull atoms together. This tendency toward higher density and stronger bonding is opposed by the quantum-mechanical kinetic energy. As a quantum-mechanical particle such as an electron is confined to a smaller volume, its kinetic energy increases. The cohesion of matter is a delicate balance between these two effects. *See* COULOMB'S LAW; EXCLUSION PRINCIPLE; UNCERTAINTY PRINCIPLE.

The study of cohesion is the analog in physics of the study of chemical bonding and intermolecular forces in chemistry. Cohesion is the binding of a macroscopic condensed body, rather than the binding of a small collection of atoms or the mutual interactions of a diffuse gas of particles. *See* INTERMOLECULAR FORCES.

There are a number of different classes of cohesion that can be distinguished by the nature of the balance between electrostatic and kinetic energies or by the spatial distribution of electronic charge. One such classification scheme distinguishes molecular, ionic, covalent, and metallic bonding. The classes represent arbitrary distinctions among systems in which the microscopic physics of cohesion are quite similar. It is possible that any individual system will display aspects of several of the following classes.

Molecular bonding. In a molecular crystal or liquid some of the atoms are strongly bound together in molecules, and these molecules are relatively weakly bound to the other molecules in a condensed phase. The intermolecular bond is usually due to the electrostatic dipole-dipole interaction. Molecules have no net electronic charge, but in some molecules, such as the water molecule (H_2O), the distribution of positive charge (due to the atomic nuclei) and negative charge (due to the electrons) is nonuniform, with one end of the molecule having a net positive charge, the other end a net negative charge. This charge distribution is called a permanent electric dipole moment. Two properly aligned dipoles attract each other as a result of the Coulomb interaction with a force that falls off as the inverse of the distance of separation raised to the fourth power.

Other molecules, such as the homonuclear diatomic molecules like oxygen, have no permanent dipole moment. The dominant cohesive interaction in these systems is the fluctuating dipole interaction studied by F. London. The random relative motion of electrons and nuclei in one molecule produces a fluctuating electrostatic dipole moment on that molecule. The electric field due to this fluctuating dipole acting on a second molecule induces a dipole moment on the second molecule. The induced dipole has a strength that is proportional to the inverse third power of the intermolecular separation. The resulting fluctuating dipole-induced dipole interaction is attractive, with a force that falls off as the seventh power of the separation between the molecules. When the molecules come close enough together that their electron clouds begin to overlap, the Pauli exclusion principle, which requires that no two identical particles such as electrons can be in the same quantum-mechanical state, forces a distortion of the electronic states and increases the electronic kinetic energy. This increase in kinetic energy reduces the bond strength. The competition between the attractive dipole-dipole interaction and the kinetic energy increase leads to the equilibrium intermolecular spacing.

Ionic bonding. In an ionic system, such as sodium chloride (NaCl), the constituent particles are ions with a net electronic charge, some positive and some negative. The Coulomb interaction between them leads to an attractive force. The system is kept from collapsing by an increase in kinetic energy due to the compression of charge as the system is compressed. This repulsive interaction is frequently described as core-core repulsion since most ionic systems are composed of closed-shell ions. Since the Coulomb interaction between charged particles is stronger than that between neutral molecules, the cohesive energy of ionic systems is significantly greater than that characteristic of molecular crystals.

Covalent bonding. Covalently bound crystals are a third major class of crystals, in which the atoms are held together by directional bonds. A bond forms because atoms are able to share their electrons. The shared electrons are subject to the Coulomb attraction of more than one nucleus, and thus the system of electrons and nuclei is more tightly bound than the separated atoms. Examples of covalent bonding are the two forms of carbon crystals, graphite and diamond, network-structure crystals such as the chalcogenides selenium and tellurium, cnd the tetrahedrally coordinated semiconductors silicon and germanium. The cohesive energy and density of these covalently bonded crystals are similar to those characteristic of ionic crystals. Since the covalent bonds are directional, covalent bonding is not well adapted to the cohesion of liquids. Silicon and germanium, for example, convert at their melting points from semiconducting crystals to metallic liquids.

Metallic bonding. The fourth class of cohesion is metallic bonding, which is characterized by relatively large cohesive energy and density, and by the fact that metals are good low-temperature electrical conductors in contrast to the other classes which are insulators. The concept of electron pair bonds is not applicable to these systems in which the coordination number, the number of atoms surrounding a given atom, is frequently much larger than the number of valence electrons, the electrons available for bonding. One of the most significant advances in the theory of cohesion was the explanation of the cohesive energy of alkali metals (lithium, sodium, potassium, rubidium, cesium, and francium) by E. Wigner and F. Seitz in 1934. Their theory provided the connection between the energy of an individual electronic state of a metal (the "energy bands") and its cohesive energy.

In its simplest form the Wigner-Seitz theory involves a two-step process for forming the solid from a collection of atoms. First, the boundary condition for the solution of the Schrödinger equation for the single-valence electronic wave function is changed. It is altered from its atomic form, where the wave function vanishes far from the atom, to the boundary condition appropriate to the most tightly bound valence state in the solid, a wave function whose slope vanishes on the boundary of the atomic cell. This second boundary condition leads to a state which is more tightly bound than the atomic valence state. Second, the electronic states are filled with electrons starting with the most tightly bound state. The Pauli exclusion principle implies that it is not possible to put all of the valence electrons from all of the atoms contained in a solid into this lowest state. It is necessary to fill higher-lying states, that is, states with increased kinetic energy. Wigner and Seitz worked out the quantitative implications of these steps and demonstrated that one can, within this model, accurately calculate the basic cohesive properties, the cohesive energy, the equilibrium density, and the compressibility.

Subsequent calculations have extended the Wigner and Seitz work to metals which possess more complicated electronic structures, such as the transition metals. When a condensed phase is formed, the atomic valence orbitals broaden into a band of states because of the change in boundary conditions afforded by the new environment. States are filled progressively, starting with those most tightly bound and satisfying the requirement that no two electrons occupy the same state. When a band of states formed from a particular atomic orbital begins to be filled, the bond strength per electron is large, but the number of electrons per atom is small, and thus the cohesive energy is relatively small. These bands of states effectively broaden symmetrically in energy about the atomic orbital from which they are formed. Therefore, in a completely filled band, as many states are filled that are less tightly bound as states that are more tightly bound, and again weak bonding results. It is in the intermediate situation, when only states that are more tightly bound than the atomic orbital are filled, that the strongest bonding occurs. This explains why the cohesive energy of transition metals is greatest near the center of a transition-metal row in the periodic table (that is, molybdenum and tungsten) and also why the ratio of the atomic size in a condensed phase to a measure of the radius of an isolated atom is smallest near the center of a row. *See* BAND THEORY OF SOLIDS; SOLID-STATE PHYSICS.

[C. D. GELATT, JR.]

Bibliography: N. W. Ashcroft and N. D. Mermin, *Solid State Physics*, 1976; W. Harrison, *Electronic Structure and the Properties of Solids: The Physics of the Chemical Bond*, 1980; C. Kittel, *Introduction to Solid State Physics*, 5th ed., 1976.

Collision

Any interaction between particles, aggregates of particles, or rigid bodies in which they come near enough to exert a mutual influence, generally with exchange of energy. The term collision, as used in physics, does not necessarily imply actual contact.

In classical mechanics, collision problems are concerned with the relation of the magnitudes and directions of the velocities of colliding bodies after collision to the velocity vectors of the bodies before collision. When the only forces on the colliding bodies are those exerted by the bodies themselves, the principle of conservation of momentum states that the total momentum of the system is unchanged in the collision process. This result is particularly useful when the forces between the colliding bodies act only during the instant of collision. The velocities can then change only during the collision process, which takes place in a short time interval. Under these conditions the forces can be treated as impulsive forces, the effects of which can be expressed in terms of an experimental parameter known as the coefficient of restitution, which is discussed later. *See* CONSERVATION OF MOMENTUM; IMPACT.

The study of collisions of molecules, atoms, and nuclear particles is an important field of physics. Here the object is usually to obtain information about the forces acting between the particles. The velocities of the particles are measured before and after collision. Although quantum mechanics instead of classical mechanics should be used to describe the motion of the particles, many of the conclusions of classical collision theory are valid. *See* SCATTERING EXPERIMENTS (ATOMS AND MOLECULES); SCATTERING EXPERIMENTS (NUCLEI).

Classification. Collisions can be classed as elastic and inelastic. In an elastic collision, mechanical energy is conserved; that is, the total kinetic energy of the system of particles after collision equals the total kinetic energy before collision. For inelastic collisions, however, the total kinetic energy after collision is different from the initial total kinetic energy.

In classical mechanics the total mechanical energy after an inelastic collision is ordinarily less than the initial total mechanical energy, and the mechanical energy which is lost is converted into heat. However, an inelastic collision in which the total energy after collision is greater than the initial total energy sometimes can occur in classical mechanics. For example, a collision can cause an explosion which converts chemical energy into mechanical energy. In molecular, atomic, and nuclear systems, which are governed by quantum mechanics, the energy levels of the particles can be changed during collisions. Thus these inelastic collisions can involve either a gain or loss in mechanical energy.

Consider a one-dimensional collision of two particles in which the particles have masses m_1 and m_2 and initial velocities u_1 and u_2. If they interact only during the collision, an application of the principle of conservation of momentum yields Eq. (1),

$$m_1 u_1 + m_2 u_2 = m_1 v_1 + m_2 v_2 \qquad (1)$$

where v_1 and v_2 are the velocities of m_1 and m_2, respectively, after collision.

Coefficient of restitution. It has been found experimentally that in collision processes Eq. (2)

$$e = \frac{v_2 - v_1}{u_1 - u_2} \qquad (2)$$

holds, where e is a constant known as the coefficient of restitution, the value of which depends on the properties of the colliding bodies. The magnitude of e varies from 0 to 1. A coefficient of restitution equal to 1 can be shown to be equivalent to an elastic collision, while a coefficient of restitution of zero is equivalent to what is sometimes called a perfectly inelastic collision. From the definition of e one can show that in a perfectly inelastic collision the colliding bodies stick together after collision, as two colliding balls of putty or a bullet fired into a wooden block would do. Equations (1) and (2) can be solved for the unknown velocities v_2 and v_1 in the one-dimensional collision of two particles.

The concept of coefficient of restitution can be generalized to treat collisions involving the plane motion of smooth bodies—both of particles and larger bodies for which rotation effects must be considered. For these collisions, experiments show that the velocity components to be used in Eq. (2) for e are the components along the common normal to the surfaces of the bodies at the point where they make contact in the collision. For smooth bodies the velocity components perpendicular to this direction are unchanged. Use of this result and the principle of conservation of momentum is sufficient to solve two-dimensional collision problems of smooth bodies. For collisions of smooth spheres the velocity components to be used in Eq. (2) for e are those on the line joining the centers of the spheres. Velocity components perpendicular to this direction are unchanged.

Center-of-mass coordinates. A simplification of the description of both classical and quantum mechanical collisions can be obtained by using a coordinate system which moves with the velocity of the center of mass before collision. (Since for an isolated system the center of mass of the system can be shown to be unaccelerated at all times, the velocity of the center of mass of the system of particles does not change during collision.) The coordinate system which moves with the center of mass is called the center-of-mass system, while the stationary system is the laboratory system.

The description of a collision in the center-of-mass system is simplified because in this coordinate system the total momentum is equal to zero, both before and after collision. In the case of a two-particle collision the particles therefore must be oppositely directed after collision, and the magnitude of one of the velocities in the center-of-mass system can be determined if the other magnitude is known. *See* CENTER OF MASS.

[PAUL W. SCHMIDT]

Bibliography: R. A. Becker, *Introduction to Theoretical Mechanics*, 1954; D. Halliday and R. Resnick, *Physics*, 3d ed., 1978; F. W. Sears et al., *University Physics*, 5th ed., 1976.

Color

That aspect of visual sensation enabling a human observer to distinguish differences between two structure-free fields of light having the same size, shape, and duration. Although luminance differences alone permit such discriminations to be made, the term color is usually restricted to a class of differences still perceived at equal luminance. These depend upon physical differences in the spectral compositions of the two fields, usually revealed to the observer as differences of hue or saturation.

Photoreceptors. Color discriminations are possible because the human eye contains three classes of cone photoreceptors that differ in the photopigments they contain and in their neural connections. Two of these, the R and G cones, are sensitive to all wavelengths of the visible spectrum from 380 to 700 nanometers. (Even longer or shorter wavelengths may be effective if sufficient energy is available.) R cones are maximally sensitive at about 570 nm, G cones at about 540 nm. The ratio R/G of cone sensitivities is minimal at 465 nm and increases monotonically for wavelengths both shorter and longer than this. This ratio is independent of intensity, and the red-green dimension of color variation is encoded in terms of it. The B cones, whose sensitivity peaks at about 440 nm, are not appreciably excited by wavelengths longer than 540 nm. The perception of blueness and yellowness depends upon the level of excitation of B cones in relation to that of R and G cones. No two wavelengths of light can produce equal excitations in all three kinds of cones. It follows that, provided they are sufficiently different to be discriminable, no two wavelengths can give rise to identical sensations.

The foregoing is not true for the comparison of two different complex spectral distributions. These usually, but not always, look different. Suitable amounts of short-, middle-, and long-wavelength lights, if additively mixed, can for example excite the R, G, and B cones exactly as does a light containing equal energy at all wavelengths. As a result, both stimuli look the same. This is an extreme example of the subjective identity of physically different stimuli known as chromatic metamerism. Additive mixture is achievable by optical superposition, rapid alternation at frequencies too high for the visual system to follow, or (as in color television) by the juxtaposition of very small elements which make up a field structure so fine as to exceed the limits of visual acuity. The integration of light takes place within each receptor, where photons are individually absorbed by single photopigment molecules, leading to receptor potentials that carry no information about the wavelength of the absorbed photons.

Colorimetry. Although colors are often defined by appeal to standard samples, the trivariant nature of color vision permits their specification in terms of three values. Ideally these might be the relative excitations of the R, G, and B cones. Because too little was known about cone action spectra in 1931, the International Commission on Illumination (CIE) adopted at that time a different but related system for the prediction of metamers (the CIE system of colorimetry). This widely used system permits the specification of tristimulus values X, Y, and Z, which make almost the same predictions about color matches as do calculations based upon cone action spectra. If, for fields 1 and 2, $X_1 = X_2$, $Y_1 = Y_2$, and $Z_1 = Z_2$, then the two stimuli are said to match (and therefore have the same color) whether they are physically the same (isometric) or different (metameric).

The use of the CIE system may be illustrated by a sample problem. Suppose it is necessary to describe quantitatively the color of a certain paint when viewed under illumination by a tungsten lamp of known color temperature. The first step is to measure the reflectance of the paint continuously across the visible spectrum with a spectrophotometer. The reflectance at a given wavelength is symbolized as R_λ. The next step is to multiply R_λ by the relative amount of light E_λ emitted by the lamp at the same wavelength. The product $E_\lambda R_\lambda$ describes the amount of light reflected from the paint at wavelength$_\lambda$. Next, $E_\lambda R_\lambda$ is multiplied by a value $\bar{x}_\lambda$, which is taken from a table of X tristimulus values for an equal-energy spectrum. The integral $\int E_\lambda R_\lambda \bar{x}_\lambda d_\lambda$ gives

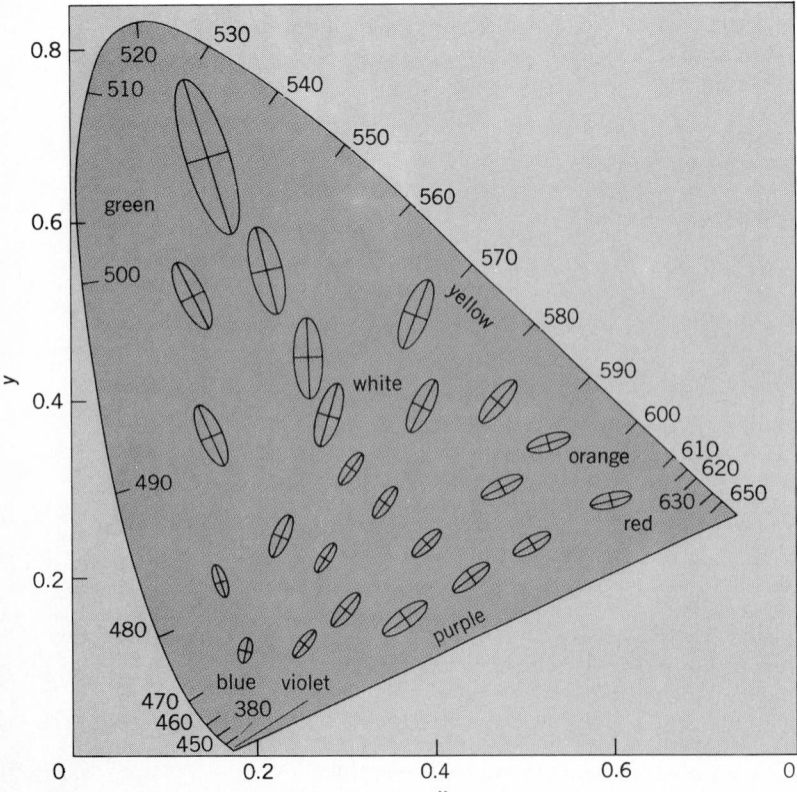

The 1931 CIE chromaticity diagram showing MacAdam's ellipses 10 times enlarged. (*From G. S. Wyszecki and W. S. Stiles, Color Science, copyright © 1967 by John Wiley and Sons; used with permission*)

the tristimulus value X for all of the light reflected from the paint. Similar computations using $\bar{y}_\lambda$ and $\bar{z}_\lambda$ yield tristimulus values Y and Z.

Tables of $\bar{x}_\lambda$, $\bar{y}_\lambda$, and $\bar{z}_\lambda$ are by convention carried to more decimal places than are warranted by the precision of the color matching data upon which they are based. As a result, colorimetric calculations of the type just described will almost never yield identical values, even for two physically different fields that are identical in appearance. For this and other reasons it is necessary to specify tolerances for color differences. Such differential colorimetry is primarily based upon experiments in which observers attempted color matches repeatedly, with the standard deviations of many such matches being taken as the discrimination unit.

Chromaticity diagram. Colors are often specified in a two-dimensional chart known as the CIE chromaticity diagram, which shows the relations among tristimulus values independently of luminance. In this plane, y is by convention plotted as a function of x, where $y = Y/(X + Y + Z)$ and $x = X/(X + Y + Z)$. [The value $z = Z/(X + Y + Z)$ also equals $1 - (x + y)$ and therefore carries no additional information.] Such a diagram is shown in the illustration, in which the continuous locus of spectrum colors is represented by the outermost contour. All nonspectral colors are contained within an area defined by this boundary and a straight line running from red to violet. The diagram also shows discrimination data for 25 regions, which plot as ellipses represented at 10 times their actual size. A discrimination unit is one-tenth the distance from the center of an ellipse to its perimeter. Predictive schemes for interpolation to other regions of the CIE diagram have been worked out.

If discrimination ellipses were all circles of equal size, then a discrimination unit would be represented by the same distance in any direction anywhere in the chart. Because this is dramatically untrue, other chromaticity diagrams have been developed as linear projections of the CIE chart. These represent discrimination units in a relatively more uniform way, but never perfectly so.

A chromaticity diagram has some very convenient properties. Chief among them is the fact that additive mixtures of colors plot along straight lines connecting the chromaticities of the colors being mixed. Although it is sometimes convenient to visualize colors in terms of the chromaticity chart, it is important to realize that this is not a psychological color diagram. Rather, the chromaticity diagram makes a statement about the results of metameric color matches, in the sense that a given point on the diagram represents the locus of all possible metamers plotting at chromaticity coordinates x, y. However, this does not specify the appearance of the color, which can be dramatically altered by preexposing the eye to colored lights (chromatic adaptation) or, in the complex scenes of real life, by other colors present in nearby or remote areas (color contrast and assimilation). Nevertheless, within limits, metamers whose color appearance is thereby changed continue to match.

For simple, directly fixated, and unstructured fields presented in an otherwise dark environment, there are consistent relations between the chromaticity coordinates of a color and the color sensations that are elicited. Therefore, regions of the chromaticity diagram are often denoted by color names, as shown in the illustration.

Although the CIE system works rather well in practice, there are important limitations. Normal human observers do not agree exactly about their color matches, chiefly because of the differential absorption of light by inert pigments in front of the photoreceptors. Much larger individual differences exist for differential colorimetry, and the system is overall inappropriate for the 4% of the population (mostly males) whose color vision is abnormal. The system works only for an intermediate range of luminances, below which rods (the receptors of night vision) intrude, and above which the bleaching of visual photopigments significantly alters the absorption spectra of the cones.

[ROBERT M. BOYNTON]

Bibliography: R. M. Boynton, *Human Color Vision*, 1979; R. W. Burnham, R. M. Hanes, and C. J. Bartleson, *Color: A Guide to Basic Facts and Concepts*, 1963; D. B. Judd and G. W. Wyszecki, *Color in Business, Science, and Industry*, 2d ed., 1963; G. W. Wyszecki and W. S. Stiles, *Color Science*, 1967.

Color (quantum mechanics)

A term used to describe a hypothetical quantum number carried by the quarks which are thought to make up the strongly interacting elementary particles. It has nothing to do with the ordinary, visual use of the word color.

The quarks which are thought to make up the strongly interacting particles have a spin angular momentum of one-half unit of $\hbar$ (Planck's constant). According to a fundamental theorem of relativity combined with quantum mechanics, they must therefore obey Fermi-Dirac statistics and be subject to the Pauli exclusion principle. No two quarks within a particular system can have exactly the same quantum numbers. *See* EXCLUSION PRINCIPLE; FERMI-DIRAC STATISTICS.

However, in making up a baryon, it often seemed necessary to violate this principle. The Ω^--particle, for example, is made of three strange quarks, and all three had to be in exactly the same state. O. W. Greenberg is responsible for the essential idea for the solution to this paradox. In 1964 he suggested that each quark type (u, d, and s) comes in three varieties identical in all measurable qualities but different in an additional property, which has come to be known as color. The exclusion principle could then be satisfied and quarks could remain fermions, because the quarks in the baryon would not all have the same quantum numbers. They would differ in color even if they were the same in all other respects.

The color hypothesis triples the number of quarks but does not increase the number of baryons and mesons. The rules for assembling them ensures this. Tripling the number of quarks does, however, have at least two experimental consequences. It triples the rate at which the neutral π-meson decays into two photons and brings the predicted rate into agreement with the observed rate.

The total production cross section for baryons and mesons in electron-positron annihilation is also tripled. The experimental result at energies between 2 BeV and 3 BeV is in reasonable agreement with the color hypothesis and completely incompatible with the simple quark model without color. *See* BARYON; ELEMENTARY PARTICLE; MESON; QUARKS. [THOMAS APPELQUIST]

Bibliography: O. W. Greenberg, Spin and unitary spin independence in a paraquark model of baryons and mesons, *Phys. Rev. Lett.* 13, 598–602 1964.

Color centers

Atomic and electronic defects of various types which produce optical absorption bands in otherwise transparent crystals such as the alkali halides, alkaline earth fluorides, or metal oxides. They are general phenomena found in a wide range of materials. Color centers are produced by gamma radiation or x-radiation, by addition of impurities or excess constituents, and sometimes through electrolysis. A well-known example is that of the *F*-center in alkali halides such as sodium chloride, NaCl. The designation *F*-center comes from the German word *Farbe*, which means color. *F*-centers in NaCl produce a band of optical absorption toward the blue end of the visible spectrum; thus the colored crystal appears yellow under transmitted light. On the other hand, KCl with *F*-centers appears magenta, and KBr appears blue. *See* CRYSTAL DEFECTS.

Color centers have been under investigation for many years. Theoretical studies guided by detailed experimental work have yielded a deep understanding of specific centers. The crystals in which color centers appear tend to be transparent to light and to microwaves. Consequently, experiments which can be carried out include optical spectroscopy, luminescence and Raman scattering, magnetic circular dichroism, magnetic resonance, and electromodulation. Color centers find practical application in radiation dosimeters, schemes have been proposed to use color centers in high-density memory devices, and tunable lasers have been made from crystals containing color centers. *See* ABSORPTION OF ELECTROMAGNETIC RADIATION.

Origin. Figure 1 shows the absorption bands due to color centers produced in potassium bromide by

exposure of the crystal at the temperature of liquid nitrogen (81 K) to intense penetrating x-rays. Several prominent bands appear as a result of the irradiation. The *F*-band appears at 600 mμ and the so-called *V*-bands appear in the ultraviolet. Uncolored alkali halide crystals may be grown readily from a melt with as few imperfections as one part in 10^5. In the unirradiated and therefore uncolored state, they show no appreciable absorption from the far-infrared through the visible, to the region of characteristic absorption in the far-ultraviolet (beginning at about 200 mμ in Fig. 1). Color centers may be introduced into such crystals by means other than exposure to penetrating radiation. The most important of these is a method known as additive coloration, in which the composition of the crystal is made to deviate from stoichiometric proportions by heating in the presence of the alkali metal vapor.

Color bands such as the *F*-band and the *V*-band arise because of light absorption at defects dispersed throughout the lattice. This absorption is caused by electronic transitions at the centers. On the other hand, colloidal particles, each consisting of many atoms, dispersed through an optical medium also produce color bands. In this case, if the particles are large enough, the extinction of light is due to both light scattering and light absorption. Colloidal gold is responsible for the color of some types of ruby glass. Colloids may also form in alkali halide crystals—for example, during heat treatment of an additively colored crystal which contains an excess of alkali metal.

Atomically dispersed centers such as *F*-centers are part of the general phenomena of trapped electrons and holes in solids. The accepted model of the *F*-center is an electron trapped at a negative ion vacancy. Many other combinations of electrons, holes, and clusters of lattice vacancies have been used to explain the various absorption bands observed in ionic crystals. The centers and models shown in Fig. 2 have been positively identified either by electron spin resonance or by a combination of other types of experimental evidence.

Impurities can play an important role in color-center phenomena. Certain impurities in ionic crystals produce color bands characteristic of the foreign ion. For example, hydrogen can be incorporated into the alkali halides with resultant appearance of an absorption band (the *U*-band) in the ultraviolet. In this case, the *U*-centers interact with other defects. The rate at which *F*-centers are produced by x-irradiation is greatly increased by the incorporation of hydrogen, the *U*-centers being converted into *F*-centers with high efficiency.

F-centers. *F*-centers may be produced in uncolored crystals by irradiation with ultraviolet, x-rays, γ-rays, and high-speed particles, and also by electrolysis. However, one of the most convenient methods is that of additive coloration. Alkali halide crystals may be additively colored by heating to several hundred degrees Celsius in the presence of the alkali metal vapor, then cooling rapidly to room temperature. The *F*-band that results is the same as that produced by irradiation and is dependent upon the particular alkali halide used, not upon the alkali metal vapor. For example, the coloration produced in KCl is the same whether the crystal is heated in potassium or in sodium vapor. This and

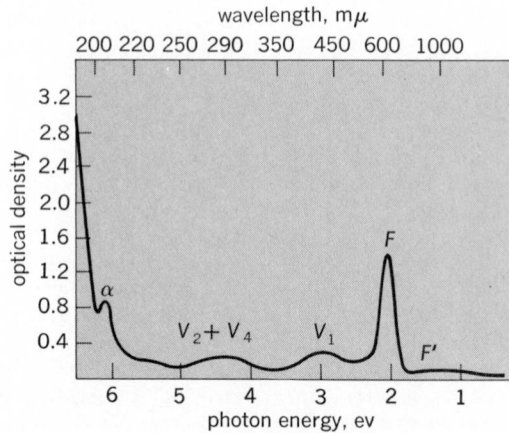

Fig. 1. Absorption bands produced in a KBr crystal by exposure to x-rays at 81 K. Optical density is equal to $\log_{10}(I_0/I)$, where I_0 is the intensity of incident light and I the intensity of transmitted light. Steep rise in optical density at far left is due to intrinsic absorption in crystal (modified slightly by existence of *F*-centers).

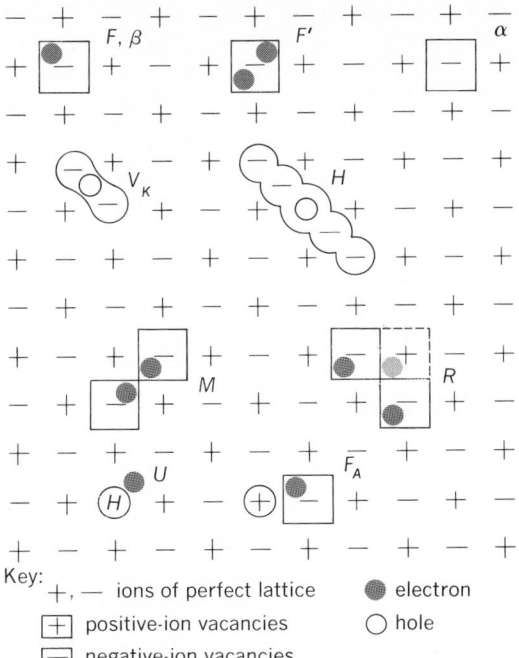

Key:
+, − ions of perfect lattice   ● electron
⊞ positive-ion vacancies   ○ hole
⊟ negative-ion vacancies

Fig. 2. Well-established models for several color centers in ionic crystal of NaCl structure. Designation β for F-center signifies disturbing influence of F-centers on tail of fundamental absorption band; characteristic absorption is also influenced by presence of vacancies, or α-centers (see Fig. 1). The V_K-center can be described as a self-trapped hole forming a Cl_2-molecule-ion, and the H-center can be described as a hole trapped at an interstitial halide ion. Both are stable at low temperature only. The R-center contains three negative-ion vacancies, one of which is in an adjacent lattice plane. H in circle at lower left represents hydrogen impurity.

other evidence indicate that excess alkali atoms enter the crystal as normal lattice ions by donating an electron to accompanying negative-ion vacancies to form F-centers. Figure 3 shows the position of the F-band in several different alkali halides.

Both the width and the exact position of the F-band change with temperature in the manner shown in Fig. 4. This behavior can be explained in terms of the thermal motion of the ions surrounding the center. Thermal motion is most important at high temperatures, whereas at low temperatures the width of the absorption band becomes less and approaches a constant value which prevails down to the very lowest temperatures.

Concentration of centers. The height of the absorption maximum and the width at half maximum may be used to calculate the concentration N of absorbing centers. For KCl, classical theory gives the concentration as $N = 1.3 \times 10^{16} AH$ cm^{-3}, where

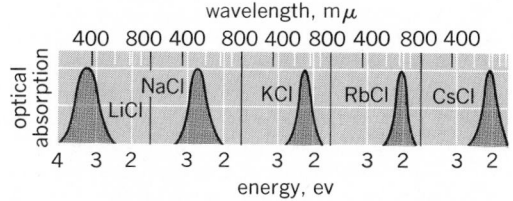

Fig. 3. F-bands in different alkali halide crystals.

A is the maximum absorption in cm^{-1}, and H is the half width of the absorption band in electronvolts. The numerical factor in this equation depends upon the material under consideration and is given for KCl. The general relation between absorption characteristics and defect concentration is slightly more complicated but is nevertheless useful to determine concentrations below the limit of detection by chemical means.

Shape of F-band. Refined measurements of the F-band at low temperatures show that it is not a simple bell-shaped curve as originally presented. In fact, it is found to contain a shoulder on the short-wavelength side, which is referred to as the K-band. It has been proposed that the K-band is in agreement with the energy-level scheme of the F-center. An electron in the vicinity of a negative-ion vacancy should have a ground state and one or more excited states below the bottom of the conduction band. The main absorption peak corresponds to excitation of the electron from the 1s

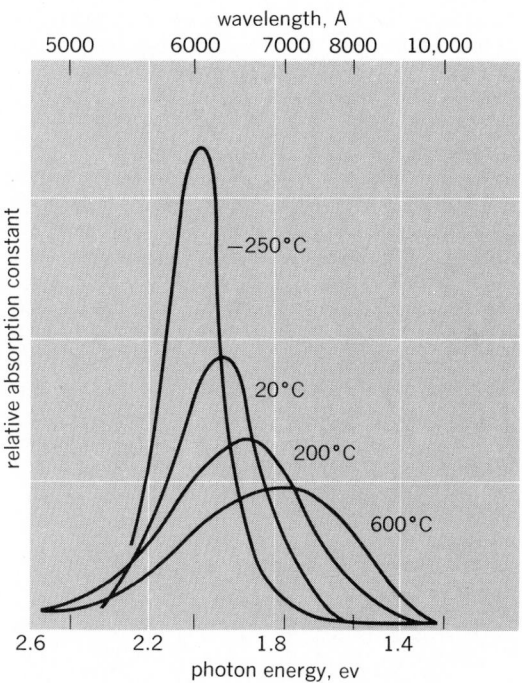

Fig. 4. Variation with temperature of the width of the F-band in a crystal of potassium bromide.

ground state to an excited 2p state. The K-band is thought to be due to transitions to the 3p and higher states which finally merge with the conduction band itself. F-center photoconductivity measurements tend to confirm the idea of discrete excited states, such as the 2p, which lie below the conduction band or continuum. *See* PHOTOCONDUCTIVITY.

Magnetic experiments. Some of the most detailed information on F-centers comes from electron spin resonance and from electron-nuclear double resonance. From these experiments the extent and character of the wave function of the electron trapped at the halogen-ion vacancy can be obtained. The electron even in the ground state is not highly localized but is spread out over the six neighboring alkali ions and to some extent over more distant neighbors. Faraday rotation studies indicated fine structure in the excited state of the F-center due to spin-orbit interaction. The spin-orbit splitting of the F-center is negative, in agreement with the theoretical prediction for an electron trapped at a vacancy. *See* MAGNETIC RESONANCE.

Lifetime of excited F-center. At low temperature an excited F-center decays spontaneously to the ground state with the emission of an infrared luminescent quantum. The radiative lifetime has been measured and found to be unexpectedly long, 0.57 μsec in KCl and as long as 2.0 μsec in KI. There are indications that this is due to a drastic change in the wave function following excitation and lattice relaxation. Lattice polarization must be taken into account in order to calculate the effect. The relaxed excited state is very spread out in the lattice, whereas the ground state is somewhat more localized.

F′-centers. When a crystal containing F-centers is irradiated at low temperature with light in the F-band itself, the F absorption decreases and a new band, known as the F' band, appears (Fig. 5). Experiments have shown that for every F'-center created, two F-centers are destroyed. The F'-center consists of an F-center which has captured an extra electron. Important photoelectric effects occur during conversion of F- to F'-centers and during irradiation of alkali halides with F-band light.

V-centers. As Fig. 1 shows, absorption bands arise in the ultraviolet as well as in the visible portion of the spectrum when KBr is irradiated while cold. These V-bands are apparently not present in crystals additively colored by the alkali metal vapor. Irradiation of a crystal produces both electrons and holes so that bands associated with trapped positive charges are to be expected in addition to electrons trapped at negative-ion vacancies (F-centers). V-bands can also be produced, at least in the case of KBr and KI, by introduction of an excess of halogen into the crystal.

Crystals containing V-centers have been studied by spin resonance techniques with the result that a center not previously observed optically has been identified. This center, sometimes referred to as the V_K-center (Fig. 2) is due to a hole trapped in the vicinity of two halogen ions which have become displaced from their normal lattice position to form a Cl_2-molecule-ion (in the case of KCl). It has been found in several different alkali halides, and its optical absorption and energy level structure have been correlated. The V_K-center is known to play an important role in the intrinsic luminescence of ionic crystals.

Other centers. Only the simpler defects in pure crystals, such as the alkali halides, are discussed above. Aggregate centers and centers associated with impurities may be important. The M- and R-centers shown in Fig. 2 are examples of the former; U-centers and F_A-centers, also shown in Fig. 2, are important examples of the latter. The alkali halide lattice is the classic matrix for color centers, which frequently can be readily formed by irradiation. An efficient photolytic process is involved. Analogous phenomena occur in crystals with the fluorite (CaF_2) structure and in oxides such as CaO and MgO. In these cases, the photolytic processes are different and apparently not as efficient as in an alkali halide. Crystal purity is also frequently a problem.

[FREDERICK C. BROWN]

Bibliography: F. C. Brown, *The Physics of Solids*, 1967; W. B. Fowler, *The Physics of Color Centers*, 1968; W. Hayes, *Crystals with the Fluorite Structure*, 1974; A. E. Hughes and B. Henderson, in J. Crawford and L. Slifkin (eds.), *Point Defects in Solids*, vol. 1, 1972; C. Kittel, *Introduction to Solid State Physics*, 4th ed., 1971; F. Lüty, *Surface Sci.*, 37:120, 1973; L. F. Mollenauer and D. H. Olson, *Appl. Phys. Lett.*, 24:386, 1974; J. H. Schulman and W. D. Compton, *Color Centers in Solids*, 1962; W. A. Sibley and D. Pooley, in H. Herman (ed.), *Treatise on Material Science and Technology*, vol. 5, 1974.

Combinatorial theory

The branch of mathematics which studies arrangements of elements (usually a finite number) into sets under certain prescribed constraints. Problems combinatorialists attempt to solve include the enumeration problem (how many such arrangements are there?), the structure problem (what are the properties of these arrangements and how efficiently can associated calculations be made?), and, when the constraints become more subtle, the existence problem (is there such an arrangement?).

Other names for the field include combinatorial mathematics, combinatorics, and combinatorial analysis, the last term having been coined by G. W. Leibniz in *Dissertatio de arte combinatoria* (1666), wherein he hoped for application of this field "in the entire sphere of science." Indeed, the availability today of high-speed computers which can deal with large but finite configurations seems to be bringing his prophecy near fulfillment. On the other hand, the study of combinatorial phenomena dates back to the very beginnings of mathematics. According to a Chinese legend recorded in

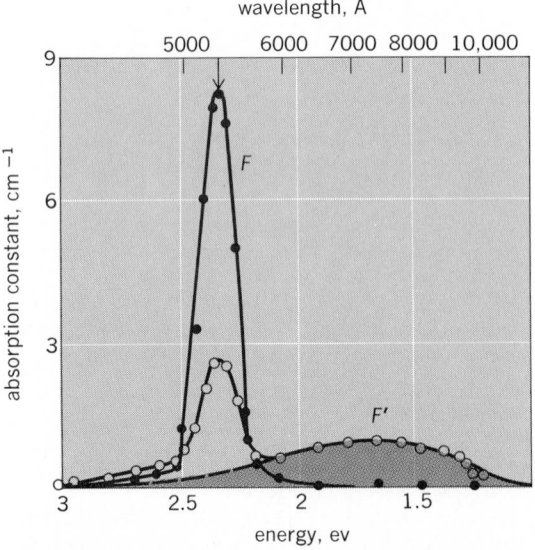

Fig. 5. F- and F'-bands in KCl at −235°C. F'-band was produced by irradiating crystal with light of wavelength shown by arrow at center of F-band. Such light causes the F-band to bleach as F'-band grows. (*After H. Pick*)

a divinatory book of the lesser Taoists (about 2200 B.C.), the magic square

$$
\begin{array}{ccc}
4 & 9 & 2 \\
3 & 5 & 7 \\
8 & 1 & 6
\end{array}
$$

was observed on the back of a divine tortoise. Apparently, permutations were considered as early as 1100 B.C., and Rabbi Ben Ezra, in the 12th century, probably knew the formula for the number of k-element subsets of an n-element set.

ENUMERATION

From the 17th well into the 20th century, combinatorial theory was used primarily to solve certain problems in probability. For example, to compute the probability that exactly three heads appear in five flips of a coin, one need only count the total number of sequences of heads and tails which may occur (32) and divide this number into the number of sequences which result in three heads and two tails (10). *See* PROBABILITY.

Permutations and combinations. In enumeration of arrangements it is always essential to know which arrangements are considered the same and which different. When k objects are to be selected from a set S and different orderings on the selected objects are to be counted separately, the arrangements are called permutations or relabelings of the set of k objects. But if order is to be disregarded, so that attention is focused only on the subset consisting of the k selected elements, then the choice is termed a combination. If elements of S are allowed to be selected more than once, then the arrangements are called permutations with repetition or combinations with repetition. If S has n different elements, Eqs. (1–4) enumerate these four basic

$$
P(n,k) = n(n-1)(n-2) \cdots (n-k+1)
$$
$$
= \frac{n!}{(n-k)!} \tag{1}
$$

$$
C(n,k) = \binom{n}{k} = \frac{P(n,k)}{k!} = C(n,n-k) \tag{2}
$$

$$
\overline{P}(n,k) = n^k \tag{3}
$$

$$
\bar{C}(n,k) = \binom{n+k-1}{k} \tag{4}
$$

cases. The number of k-permutations is shown in Eq. (1). In particular, $P(n,n) = n! = n(n-1) \cdots (2)(1)$ is the number of ways to order S. The number of k-combinations (or k-element subsets of S) is shown in Eq. (2). When repetition is allowed, Eq. (3) gives the number of permutations with repetition (or k-tuples) and Eq. (4) gives the number of combinations with repetition. Formula (4), although not as widely known as formula (2), recurs throughout mathematics: for instance, it gives the number of mixed partial derivatives of order k with n variables, or the number of solutions in nonnegative integers of the equation $x_1 + \cdots + x_n = k$. *See* CALCULUS.

Generating functions. The symbol $\binom{n}{k}$ in Eq. (2) is sometimes called a binomial coefficient, because it is the coefficient of x^k in the expansion of $(1+x)^n$, given in Eq. (5). When a finite or infinite

$$
C(n,x) = (1+x)^n = \sum_{k=0}^{n} \binom{n}{k} x^k \tag{5}
$$

sequence $(a_0, a_1, \ldots)$ is exhibited in this way as the coefficients of a polynomial or series $p(x) = \Sigma a_i x^i$, then $p(x)$ is called a generating function for the sequence. The general method of generating functions was developed by P. S. Laplace in 1812 but had already been used extensively in the 18th century by L. Euler. For example, substituting $x = -1$ in Eq. (5) shows that there are exactly as many even subsets of a set as odd subsets (and thus the probability of an even number of heads in n flips of a coin is 1/2). *See* SERIES.

Recursions. Algebraic identities involving generating functions are often proved from recursions (or difference equations) satisfied by the enumerating sequence. Thus the numbers $C(n,k)$ are uniquely determined by the boundary conditions $C(n,0) = 1$ for all n and $C(0,k) = 0$ for all $k > 0$, and by recursion (6). The generating function defined by Eq. (5)

$$
C(n,k) = C(n-1, k-1) + C(n-1, k) \tag{6}
$$

satisfies the algebraic identify $C(n,x) = (1+x) C(n-1,x)$.

Catalan numbers. The Catalan numbers are given in Eq. (7). They count the ways to insert paren-

$$
c_n = \frac{1}{n} \binom{2n-2}{n-1} \tag{7}
$$

theses in a string of n terms so that their product may be unambiguously carried out by multiplying two quantities at a time. Thus $c_4 = 5$, since $abcde$ may be computed by $(ab)(cd)$, $((ab)c)d$, $(a(bc))d$, $a((bc)d)$, or $a(b(cd))$. The Catalan numbers also enumerate the ways to place nonintersecting diameters in a convex polygon with $n+1$ vertices so that only triangular regions result. Further, c_n gives the number of ways to draw in succession all the balls from a container which originally contains $n-1$ white balls and $n-1$ black balls, so that at every stage there are at least as many black balls as white balls remaining in the container. Catalan numbers also appear in the theoretical measurement of preferences. Let A be a finite set, and let f be a function which assigns a real number to each element of A. Then x is said to be preferred to y if $f(x) \geq f(y) + 1$, and this induces a partial order on the elements of A. Thus, if $A = \{x,y,z\}$ and $f(x) = 0$, $f(y) = 1/2$, and $f(z) = 1$, the only preference is z to x. Assignments $\{1,2,3\}$, $\{0,1,1\}$, $\{0,0,1\}$, and $\{0,0,0\}$ give the four other essentially distinct preferential orders on A, and, in general, there are c_{n+1} such orders on an n-element set.

Catalan numbers obey the recursion $c_n = c_1 c_{n-1} + c^2 c_{n-2} + \cdots + c_{n-1} c_1$ and have the generating function $c(x) = 1/2 (1 - \sqrt{1 - 4x})$, where the root is to be expanded by means of the series form of the binomial expansion.

Stirling numbers. If instead of selecting elements from the set S, it is desired to distribute the elements of S into r indistinguishable cells so that no cell remains empty, there results a set-partition or partition of S into r blocks, and these are counted by the Stirling number of the second kind $S(n,r)$. The Stirling numbers satisfy the boundary conditions $S(n,1) = S(n,n) = 1$, and the recursion $S(n,r) = S(n-1,r-1) + rS(n-1,r)$ (which may be established by distinguishing within the family of partitions between the distributions for which the first $n-1$ elements of S leave a cell unoccupied which must be filled by the last element and those for which the first $n-1$ elements occupy every cell). In addition, they satisfy formula (8), where $(x)_r$

$$x^n = \sum_{r=1}^{n} S(n,r)(x)_r \qquad (8)$$

stands for the falling factorial $x(x-1)(x-2) \cdots (x-r+1)$. The Bell number, given by Eq. (9), is the

$$B(n) = \sum_{r=1}^{n} S(n,r) \qquad (9)$$

total number of partitions of an n-element set. These numbers also give the number of rhyme schemes for a poem of n lines. Thus, there are $B(4) = 15$ ways to rhyme a quatrain: *abcd, aabc, abac, . . . , aabb, . . . , abbb, aaaa.* [The Stirling numbers of the first kind, $s(n,r)$, are defined as the coefficients of the generating function of $(x)_n$.]

Fibonacci numbers. The Fibonacci numbers are defined by the elementary recursion $f_{n+2} = f_n + f_{n+1}$; $f_0 = 0$, $f_1 = 1$. Each f_n counts, for example, the number of sequences of $n-2$ zeros and ones with no consecutive zeros. The generating function $f(x)$ of the Fibonacci numbers satisfies the identity $(1 - x - x^2)f(x) = x$, from which the closed form $f_n = (\tau^n - (1 - \tau)^n)/\sqrt{5}$, where $\tau = (1 + \sqrt{5})/2$, may readily be deduced.

Asymptotic formulas. Occasionally, only a rough estimate of the number of arrangements of a certain kind is desired (for example, to estimate how long a computer will take to list all such arrangements), and then an asymptotic enumeration formula is sought. Thus, $n!$, which is the number of (complete) permutations $P(n,n)$, obeys for large n Stirling's formula (10), where $e = 2.71828 \cdots$

$$n! \sim \sqrt{2\pi n} \left(\frac{n}{e}\right)^n \qquad (10)$$

[here $f(n) \sim g(n)$ means that as n grows without bound the ratio $f(n)/g(n)$ approaches one].

Integer partitions. An integer partition of n is a nonincreasing sequence of positive integers whose sum is n; the number of these partitions is denoted by $p(n)$. Thus $p(4) = 5$, since $4, 3+1, 2+2, 2+1+1$, and $1+1+1+1$ all partition 4. No usable formula for $p(n)$ is known, but Euler discovered an elegant expression for the generating function: $p(x) = (1-x)^{-1}(1-x^2)^{-1} \cdots (1-x^m)^{-1} \cdots$ Asymptotically, $p(n)$ obeys formula (11), where e^x is the

$$p(n) \sim \left(\frac{1}{4n\sqrt{3}}\right) e^{\pi\sqrt{2n/3}} \qquad (11)$$

exponential function which uses the base e. Thus $p(n)$ is eventually greater than any polynomial function of n but is eventually less than a^n for any $a > 1$.

If the requirement that the sequence be nondecreasing is removed, so that, for example, the partition $1+3$ is considered distinct from the partition $3+1$, then the number of such partitions of n is much easier to count and is 2^{n-1}. If $\bar{p}(n)$ denotes the number of partitions of n into distinct parts, then $\bar{p}(x) = (1+x)(1+x^2)(1+x^3) \ldots$, and if $\bar{p}_k(n)$ is the number of partitions of n into distinct parts whose largest part is at most k, then $\bar{p}_k(x) = (1+x)(1+x^2) \cdots (1+x^k)$.

Partially ordered sets. Results in asymptotic enumeration include a formula for the number of partially ordered sets with n elements. The asymptotic estimate of approximately $2^{n^2/4}$ for the number of partially ordered sets with n elements results from the observation that as n becomes large the partially ordered sets which contain four elements w,x,y,z such that $w < x < y < z$ become asymptotically inconsequential (in the sense that the probability that a random partially ordered set contains a four-element chain approaches zero).

Probabilistic method. A nonconstructive proof technique termed the probabilistic method guarantees the existence of certain configurations by showing that the probability of their occurrence is positive. For any i and j, the existence of a graph all of whose circuits have at least i edges but whose chromatic number is at least j can be demonstrated by this technique. *See* GRAPH THEORY.

Ferrer's diagrams. Associated with an integer partition $n = a_1 + \cdots + a_k$ is an array of dots called its Ferrer's diagram, whose ith row contains a_i dots. Thus $4 = 3 + 1$ gives rise to the diagram $\bullet\!\!\bullet\!\!\bullet$.

A reflection of the diagram through its main diagonal interchanges rows and columns, converting $4 = 3 + 1$ into its dual partition $4 = 2 + 1 + 1$. This correspondence proves a number of identities, such as that the number of partitions of n into k parts equals the number of partitions of n which have greatest part k.

Möbius inversion. Another useful counting technique exploits the relationship between a pair of functions f and g defined on a lattice L. Suppose that g satisfies Eq. (12), and that the values of g are

$$g(s) = \sum_{s \leq t} f(t) \qquad (12)$$

known; then the values of f can be found if the calculation in Eq. (12) can be "inverted" to define f in terms of g. The principle of Möbius inversion states that there is a function $\mu(s,t)$, called the Möbius function of the lattice, defined for pairs of elements (s,t) in L and which depends only on the lattice, such that Eq. (13) holds.

$$f(s) = \sum_{s \leq t} \mu(s,t)g(t) \qquad (13)$$

For example, assume that the lattice in question is the Boolean algebra on the two points H and P. Let the subset $\{H\}$ stand for the property that a specific card in a deck is a heart, $\{P\}$ for the property that it is a picture card, $\{P,H\}$ for the property that it is both a heart and a picture card, and ϕ, the zero of the lattice, for the property that it is neither. As t varies over these four subsets, let $f(t)$ denote the number of cards in the deck which have precisely the property (or properties) t (and no others), and let $g(t)$ denote the number which have at least the property (or properties) t. Then clearly $g(\phi) = 52$, $g(\{H\}) = 13$, $g(\{P\}) = 12$, and $g(\{P,H\}) = 3$, and g and f stand in the relation given by Eq. (12). For Boolean algebras the Möbius function is given as follows: $\mu(s,t) = 0$ if s is not a subset of t, and $\mu(s,t) = (-1)^p$ otherwise, where p is the number of elements in t but not in s. [Thus, for all $s \subseteq t$, $\mu(s,t) = 1$ if the difference in sizes of s and t is even and $\mu(s,t) = -1$ if it is odd.] In our example, $f(\phi) = g(\phi) - g(\{H\}) - g(\{P\}) + g(\{P,H\}) = 30$ cards which are neither hearts nor picture cards. *See* BOOLEAN ALGEBRA.

For a general Boolean algebra, the Möbius inversion formula for $f(\phi)$ is more often called the principle of inclusion-exclusion, or sieve formula, because at various stages of the summation the correct number is alternately overcounted and then undercounted. Assume S has $n = g(0)$ elements each of which may have some of the properties P_1, ..., P_r. Further, let $g(i_1, i_2, \ldots, i_k)$ be the number of elements with (at least) properties P_{i_1}, $P_{i_2}, \ldots, P_{i_k}$. Again Eq. (12) is satisfied, so by Eq. (13), one arrives at formula (14) for $f(0) = f(\phi)$.

$$f(0) = n - \sum_{1 \leq i_1 \leq n} g(i_1) + \cdots$$
$$+ (-1)^p \sum_{1 \leq i_1 < \cdots < i_p \leq n} g(i_1, \ldots, i_p) + \cdots \quad (14)$$

Derangements. The derangement numbers D_n count complete permutations (without repetition) subject to the additional constraint that every element be "wrongly labeled." That is, if $S = \{1, \ldots, n\}$, then D_n is the number of ways to order S into sequences $a_1, \ldots, a_n$ such that $a_i \neq i$ for $i = 1, \ldots, n$. If the property P_i of a permutation is that $a_i = i$, then the sieve formula may be applied to prove Eq. (15). In probabilistic terms,

$$D_n = P(n,n) - C(n,1)P(n-1,n-1)$$
$$+ C(n,2)P(n-2,n-2) - \cdots$$
$$= n!\left(1 - 1 + \frac{1}{2!} - \frac{1}{3!} + \cdots + \frac{(-1)^n}{n!}\right) \sim \frac{n!}{e} \quad (15)$$

Eq. (15) implies that no matter what the size of the decks is, the probability is approximately $1/e$ that two identical and randomly shuffled decks of cards will never have the same card in the same place.

Gaussian coefficients. The Mobius function has been computed for other lattices which arise in enumeration, such as the lattice of all positive divisors of a given integer n (with elements ordered by divisibility), a lattice of integer partitions, the lattice of partitions of an n-element set (ordered by refinement: $s \leq t$ if every block of s is contained in a block of t), and the lattice of all subspaces of a vector space of dimension n over a finite field with q elements. The number of subspaces of dimension k in such a vector space is expressed in formula (16) for Gaussian coefficients. This apparent

$$\begin{bmatrix} n \\ k \end{bmatrix}_q = \frac{(q^n - 1)(q^{n-1} - 1) \cdots (q^{n-k+1} - 1)}{(q^k - 1)(q^{k-1} - 1) \cdots (q - 1)} \quad (16)$$

rational function of q is, in fact, a polynomial which is the generating function for the numbers $p_{m,k}(n)$ of integer partitions of n into at most m parts whose largest part is at most k, Eq. (17). An easy fact,

$$\begin{bmatrix} m + k \\ k \end{bmatrix}_q = \sum_{q=0}^{mk} p_{m,k}(n) q^n \quad (17)$$

established from the Ferrer's diagram, is that $p_{m,k}(n) = p_{m,k}(mk - n)$. A much harder result is that the sequence of coefficients is unimodal: $p_{m,k}(n + 1) \geq p_{m,k}(n)$ for all $n < mk/2$. *See* LINEAR ALGEBRA.

The lattices for Boolean algebras, subspaces, and set-partitions show striking similarities [such as the similarity between Eqs. (2) and (16)], which are studied in the general theory of geometric lattices or combinatorial geometries. For example, the methods applied to set partitions to study the four-color theorem of graph theory, when ap-

plied to the lattice of subspaces, provide a way to attack the problem of finding efficient error-correcting codes.

Magic squares. A (generalized) magic square is an n by n matrix of (not necessarily distinct) nonnegative integers all of whose row and column sums are equal to a prescribed number x. Thus $x = 15$ in the 3×3 magic square given earlier. The number of such magic squares is 1 when $n = 1$, $x + 1$ when $n = 2$, and in general is a polynomial of the form $a_m x^m + a_{m-1} x^{m-1} + \cdots + a_1 x + 1$, where $m = (n - 1)^2$. Although the coefficients a_i are not given explicitly in this formula, for each n they may be computed by the method of undetermined coefficients, so that a computer search of magic squares for small values of x will also determine the number of such squares for all values of x. *See* MATRIX THEORY.

Polya counting formula. If the number of ways to paint the faces of a cube with red and blue is to be computed, formula (3) would seem to give $2^6 = 64$; however, it is reasonable to assume that all six arrangements in which exactly one face is blue are identical, since for all such arrangements a rotation of the cube will make the top face blue. Under the condition that two coloring patterns are to be assumed equivalent if one can be transformed into the other by a rotation, the number of configurations is 10. This number can be derived by exhausting all cases or by the Pólya counting formula. The Pólya formula counts the number of functions from a set D (in the example, the faces of the cube) to a set R (the colors red and blue), with two functions f and g assumed to be the same if some element of a fixed group G of (complete) permutations of D (in this case, the 24 rotations of a cube) takes f into g. Actually, the Pólya formula provides a way to compute the generating function, with variables $x_1, \ldots, x_k$ corresponding to the elements of R, for which the coefficient of $x_1^{a_1} \cdots x_k^{a_k}$ is the number of functions which use each x_i value exactly a_i times. (For the painted cube, if $x_1 = $ red and $x_2 = $ blue, then the generating function is given by $x_1^6 + x_1^5 x_2 + 2x_1^4 x_2^2 + 2x_1^3 x_2^3 + 2x_1^2 x_2^4 + x_1 x_2^5 + x_2^6$.) The proof of the Pólya theorem depends on a special case proved earlier by W. Burnside: if G is a group consisting of m (complete) permutations of S, then the number of inequivalent permutations of S is

$$\frac{1}{m} \sum_{\sigma \in G} f(\sigma)$$

Here two permutations π_1 and π_2 are regarded as equivalent if there is some σ in G such that the relabeling that σ effects on the sequence π_1 gives the sequence π_2; and for each σ in G, $f(\sigma)$ denotes the number of permutations of S which do not change when relabeled by σ. *See* GROUP THEORY.

Application of Lefschetz theorem. A wide range of techniques is employed in enumeration theory. For example, the hard Lefschetz theorem from algebraic geometry was used to show that if A is any set of n distinct positive real numbers, and if for any subset B of A, $s(B)$ is the sum of its members, then $s(B)$ is constant on at most $g(n)$ subsets B, where $g(n) = \bar{p}_n([(n + 1)(n)/4])$, the number of partitions of $[(n + 1)(n)/4]$ into distinct parts whose largest part is at most n. Thus $g(n)$ is the middle coefficient of the polynomial $(1 + x)(1 + x^2)(1 + x^3) \cdots$

$(1 + x^n)$, and $g(n)$ is achieved by letting A be the arithmetic progression $\{1,2, \ldots , n\}$ and setting the sum equal to $[(n + 1)(n)/4]$.

PROPERTIES OF ARRANGEMENTS

An arrangement is primarily a family of subsets of a set S. An alternate way of formulating this concept is as an incidence system, also called a relation, which specifies when a particular element s of S is in a subset S_i of the family, for example, by means of the incidence matrix $M = [m_{ij}]$ of the family: the rows of the incidence matrix are indexed by the elements $s_1, \ldots , s_n$ of the set S, the columns of the matrix are indexed by the subsets $S_1, \ldots , S_m$ in the family, $m_{ij} = 1$ if s_i is in S_j, and $m_{ij} = 0$ otherwise. For example, if $S = \{s_1, s_2, s_3, s_4\}$, $S_1 = \{s_1, s_2\}$, $S_2 = S_4 = \{s_1, s_3, s_4\}$, and $S_3 = \{s_2\}$, the incidence of matrix of this arrangement is

$$\begin{bmatrix} 1 & 1 & 0 & 1 \\ 1 & 0 & 1 & 0 \\ 0 & 1 & 0 & 1 \\ 0 & 1 & 0 & 1 \end{bmatrix}$$

Systems of distinct representatives. If $n = m$ in an incidence system, a system of distinct representatives for the sets is a permutation $a_1, \ldots , a_n$ of S such that a_i is in S_i for $i = 1, 2, \ldots , n$. (In a congressional committee S, where each S_j might denote a particular cause or coalition, it might be desirable to find a system of distinct representatives so that each cause or coalition would have a different proponent.) In the example, s_1, s_3, s_2, s_4 is such a system.

Marriage theorem. A system of distinct representations certainly cannot exist unless any k of the subsets S_i together contain at least k distinct elements, and in fact it can be shown that this condition is also sufficient. This theorem is whimsically called the marriage theorem, since if S consists of a collection of n boys and each subset S_i consists of the boys acquainted with a particular girl g_i, then the conditions of the theorem guarantee that each girl could marry a boy of her acquaintance.

Algorithms for finding systems. Algorithms for computers exist which find systems of distinct representatives with reasonable efficiency: as n increases, the time a computer needs to find a system increases as the square of n.

Permanent. To enumerate the systems of distinct representatives, it is enough to compute the permanent of the incidence matrix defined by $\Sigma m_{1,a_1} m_{2,a_2} \cdots m_{n,a_n}$, where the sum is taken over all $n!$ permutations $(a_1, a_2, \ldots , a_n)$ of the set of column indices.

Maximal systems. When a complete system of distinct representatives does not exist, the size of a largest possible system is equal to the smallest possible value of the function $n - k + s_{i_1, \ldots , i_k}$ for all possible combinations $S_{i_1}, \ldots , S_{i_k}$ of subsets in the family, where $s_{i_1, \ldots , i_k}$ is the number of elements in $S_{i_1} \cup \cdots \cup S_{i_k}$. A related theorem asserts that the minimum number of blocks into which a partially ordered set may be partitioned subject to the constraint that any pair of elements x and y in same block be related ($x \leq y$ or $y \leq x$) is equal to the maximum possible size of a set S of pairwise unrelated elements. It is always

possible to find such a set S with the further property that any (complete) permutation π which preserves the partial order [if $x \leq y$ then $\pi(x) \leq \pi(y)$] must take members of S to other members of S. This fact can be used to show that in a Boolean algebra the maximum size of S is $C(n,[n/2])$. This number is the middle (largest) binomial coefficient and corresponds to the set S of all subsets of size $[n/2]$.

Assignment problem. Suppose it is desired to assign n workers to do n jobs, and entry a_{ij} of a matrix A measures how well the ith worker does the jth job. An assignment would be a permutation $j_1, \ldots , j_n$ of the jobs so that the ith worker performs the job j_i. The assignment problem requires that one find an assignment maximizing the total utility T, given by Eq. (18).

$$T = \sum_{i=1}^{n} a_{ij_i} \tag{18}$$

The problem may be solved by introducing auxiliary row and column numbers $u_1, \ldots , u_n$ and $v_1, \ldots , v_n$ subject to the condition $u_i + v_j \geq a_{ij}$ for all i and j. Then Eq. (19) is valid for any assign-

$$T = \sum_{i=1}^{n} a_{ij_i} \leq \sum_{i=1}^{n} u_i + \sum_{j=1}^{n} v_j = R \tag{19}$$

ment, since in an assignment each row and column is used exactly once. The marriage theorem shows that the maximum utility M of T is equal to the minimum m of R. Further, an algorithm is known which systematically changes the u's and v's to reduce R, arriving at numbers u_i and v_j for which the set of a_{ij}'s such that $a_{ij} = u_i + v_j$ admits a complete system of distinct representatives. For an assignment derived from such a system, $T = R$ and so M is achieved for this T. This method is practical in that the algorithm leads directly to a solution; and since the number of possible assignments is usually very large (for example if $n = 25$, this number is $25! > 10^{25}$), without some guidance there is no practical way to search for the optimal assignment or to recognize it when it is found. Another application of the marriage theorem is to prove topological fixed-point theorems. *See* TOPOLOGY.

Doubly stochastic matrices. A doubly stochastic matrix, a matrix of nonnegative real numbers such that every row sum and every column sum is equal to one, is a real-number analog of a magic square. The set of all n by n doubly stochastic matrices is convex in the sense that if $M_1, \ldots , M_k$ are all doubly stochastic and $a_1, \ldots , a_k$ are nonnegative numbers whose sum is one, then the convex combination

$$\sum_{i=1}^{k} a_i M_i$$

is also doubly stochastic. These matrices in fact form a convex polytope (a generalization of the notion of convex polygon) in Euclidean space $\mathbf{R}^{n^2}$, in that their entries are bounded and there are a finite number of vertices (a minimal set of matrices such that every doubly stochastic matrix is a convex combination of elements of this set). The marriage theorem can be used to prove that the $n!$ permutation matrices with exactly one 1 in each row and column are the vertices of the convex polytope of doubly stochastic matrices.

By a theorem in linear programming, linear functions are optimized on the vertices of a convex polytope. For the assignment problem this means that for a more general assignment in which the ith worker spends a fraction m_{ij} of his time on the jth job and each job is always being worked on (so that the matrix $M = [m_{ij}]$ is doubly stochastic), a maximum utility is obtained when each worker spends all his time at one job.

Upper-bound problem. In a typical problem in linear programming, polytopes in $\mathbf{R}^m$ occur which are defined by n linear constraints, such as those given by (20), where the x_j's are the variables

$$\sum_{j=1}^{m} a_{ij}x_j \leq b_i, \quad i = 1, \ldots, n \qquad (20)$$

(corresponding to vectors in $\mathbf{R}^m$). Algorithms to maximize linear functions on these polytopes usually involve proceeding from one vertex to another at which the function is greater. [An algorithm was given in 1979 using different (nonlinear) techniques, which guarantees that the number of steps needed to solve any such linear program is no more than a polynomial function of mn, but for most practical applications it has been found that a vertex-to-vertex method such as the simplex algorithm gives a faster answer.] It is important when allocating computer time to place an upper bound on the number of vertices. In 1970 it was shown that the number of vertices of a polytope in $\mathbf{R}^m$ defined by n linear constraints is at most $M(m,n)$, given by Eq. (21).

$$M(m,n) = \binom{n - \left[\frac{m+1}{2}\right]}{n - m} + \binom{n - \left[\frac{m+2}{2}\right]}{n - m} \qquad (21)$$

EXISTENCE AND CONSTRUCTION

In the 1920s R. A. Fisher noted that "the design and analysis of statistical experiments requires the construction of orthogonal latin squares and balanced incomplete block design". *See* STATISTICS.

Orthogonal Latin squares. A Latin square is a square n by n matrix with entries from the set $N = \{0, 1, \ldots, n-1\}$ so that each number occurs exactly once in each row and exactly once in each column. Two Latin squares are said to be orthogonal if when they are superposed the n^2 cells contain each of the n^2 pairs of numbers from N exactly once. For $n = 3$ one has the superposed orthogonal squares

00	11	22
12	20	01
21	02	10

where the first digits form the first square, and the second digits the second square. This square may be used to design an agricultural experiment which tests the interaction of three varieties of grain with three types of fertilizers. A field is divided into nine plots in each of which the choice of grain is made according to the first digit and the choice of fertilizer according to the second digit. The Latin squares assure the even distribution of the varieties of grain and fertilizer in both directions, so that effects such as the variation of the soil are mimimized, and the orthogonality allows the experimenter to try each fertilizer with each variety of grain.

For $n \leq 10$ each of these arrays can be viewed as a particular type of magic square whose entries are distinct two-digit numbers and such that each row and column sum is $11n(n-1)/2$ (33 for $n = 3$ and 495 for $n = 10$.).

If n is odd, it is easy to construct two orthogonal squares A and B. One takes $A = [a_{ij}]$ and $B = [b_{ij}]$ where $a_{ij} = i + j$, $b_{ij} = i + 2j$, reducing these sums by n or $2n$ if necessary to put them in the range $0, \ldots, n-1$. If n is a multiple of 4, it is not much more difficult to make a construction. But for n of the form $4m + 2$ it is considerably harder. It is clearly impossible when $n = 2$, and, by an exhaustive trial, G. Tarry showed it to be also impossible for $n = 6$. Euler had conjectured in 1782 that no pair of orthogonal Latin squares existed for any n of the form $4m + 2$, but in 1959 orthogonal Latin squares were found for every $4m + 2$ greater than or equal to 10. Two orthogonal squares for $n = 10$ are given in Fig. 1.

Fig. 1. Two orthogonal Latin squares for $n = 10$. (*Thomas Brylawski; Karl Petersen, photographer*)

Block designs. T. Kirkman posed the following puzzle in the 1850s: Is it possible for 15 schoolgirls to go for walks in 5 groups of 3 every afternoon, so that in seven afternoons every girl shall have walked with every other girl? This arrangement is indeed possible and the following is a solution, the girls being represented by letters $a, \ldots, o$:

Sun.	Mon.	Tue.	Wed.	Thur.	Fri.	Sat.
abi	*acj*	*adk*	*ael*	*afm*	*agn*	*aho*
cdf	*deg*	*efh*	*fgb*	*ghc*	*hbd*	*bce*
gjo	*hki*	*blj*	*cmk*	*dnl*	*eom*	*fin*
ekn	*flo*	*gmi*	*hnj*	*bok*	*cil*	*djm*
hlm	*bmn*	*cno*	*doi*	*eij*	*fjk*	*gkl*

The cited design is a special case of a balanced incomplete block design or (b,v,r,k,λ)-design: an arrangement of v elements or treatments x_i, $i = 1, \ldots, v$, into b subsets or blocks B_j, $j = 1, \ldots, b$, so that each B_j contains exactly k distinct elements, each element occurs in r blocks, and every combination of two elements $x_u x_v$ occurs together in exactly λ blocks.

Counting the total number of plots (incidences of

x_i in B_j or symbols in the above arrangement) in two ways—first summing over the blocks and then summing over the treatments, the constraint (22) is obtained. By considering all two-element subsets which contain x, constraint (23) is obtained. The two technical constraints (24) imply that the blocks are incomplete in that none of the blocks contains all or all but one of the treatments and none consists of a single treatment.

$$bk = vr \qquad (22)$$
$$r(k-1) = \lambda(v-1) \qquad (23)$$
$$2 < k < v - 1, \lambda > 0 \qquad (24)$$

Steiner triple systems. When $\lambda = 1$ and $k = 3$, such (b,v,r,k,λ) designs are known as Steiner triple systems, and in this case constraints (22) and (23) imply that v is of the form $6m + 1$ or $6m + 3$. Kirkman himself proved that triple systems exist for all such v.

The schoolgirl arrangement has the additional property that the blocks themselves are partitioned into r families called parallel classes of v/k blocks each such that every element occurs in exactly one block of each parallel class. Such designs are termed resolvable. A resolvable Steiner triple system is called a Kirkman triple system, and clearly in this case v must be divisible by three. In 1968 Kirkman triple systems (resolvable balanced incomplete block designs with block size three) were shown to exist for all $v = 6m + 3$.

An application of Steiner triple systems might arise in testing taste preferences among v products if every pair of products is to be compared, but each taster can efficiently taste only three products per day. A Kirkman system will, in addition, give an efficient schedule in which each product is used exactly once each day.

Fisher's inequality. Fisher's inequality states that $b \geq v$ in every design. If $b = v$, then necessarily $k = r$ and the design is called a symmetric (v,k,λ)-design. It is a property of these designs that every two blocks intersect in λ elements, and hence the transpose of the incidence matrix of a (v,k,λ)-design is itself a (v,k,λ)-design. Elementary matrix theory may be used to show that if v is even, then a symmetric design with parameters v,k,λ exists only if $k - \lambda$ is a perfect square; and deep results in number theory have been used to show that when v is odd, there must exist integers x, y, and z not all zero such that Eq. (25) holds.

$$x^2 = (k - \lambda)y^2 + (-1)^{(v-1)/2}\lambda z^2 \qquad (25)$$

Existence for fixed k. If k is fixed, then, for all but finitely many sets of parameters b,v,r,k,λ satisfying Eqs. (22), (23), and (24), there exists a block design having those parameters. When $k \leq 5$, all sets of parameters for which a block design exists have been determined. (A design is simple when $\lambda = 1$.)

Projective planes. When a symmetric design has $\lambda = 1$, condition (23) shows that if $k = n + 1$, then $v = n^2 + n + 1$. In analogy with the points (treatments) and lines (blocks) of a plane in projective geometry, such configurations are called projective planes. In this case, condition (25) states that if the parameter n, called the order of the plane, is of the form $4m + 1$ or $4m + 2$, then integers u and v exist with $n = u^2 + v^2$. Thus an infinite number of possible values of n are excluded, beginning with 6, 14, and 21. Whenever n is a power

of a prime (and thus the size of a finite field), a plane exists, namely, the projective plane coordinatized over the field (but these are not the only planes which exist for prime-power orders). In the early part of the 20th century, planes were constructed over many algebraic systems in which linear equations could be solved, but which were not associative. Planes coordinatized over such systems are different from those coordinatized by fields. The smallest of these planes has order 9 and thus has 91 points. *See* ALGEBRA.

Non-Desarguesian planes. The theorem of G. Desargues (which states that if corresponding vertices of two triangles lie on three concurrent lines, then the three points of intersection of respective sides of those triangles will lie on a line) holds in general only for those planes which arise from fields, and hence the term non-Desarguesian plane is used for the others. No (necessarily non-Desarguesian) projective plane has been found for any n that is not a prime power [and no n has been excluded except those not satisfying condition (19)]. Whether a projective plane exists for an n that is not a power of a prime is an important unsolved problem in existence theory and projective geometry. Orthogonal Latin squares and projective planes are related in that the existence of a projective plane of order n is equivalent to the existence of a set of $n - 1$ n by n Latin squares every pair of which are orthogonal. This would mean that for the first unresolved case, $n = 10$, nine pairwise orthogonal Latin squares must be found to produce a projective plane of order 10. No one has found more than two such squares.

Friendship theorem. Only for trivial projective planes can all the points be paired with all the lines in such a way that the incidence system is dualized, and each line is paired with a point not on that line. Stated differently, this is the friendship theorem: Among n people, if every pair of people has exactly one common friend, then there is someone who knows everyone else.

t-Designs. The concept of a block design in which every pair of elements is in the same number of blocks can be generalized to t-designs with parameters $t - (v,k,\lambda)$, in which each t-element subset of a v-element set is in a fixed number λ of k-element blocks. Every t-design is an s-design for all s such that $0 \leq s \leq t$ where the λ_s are given by Eq. (26), and these numbers must all be integers. Formulas (22) and (23) are special cases of Eq. (26), with $\lambda_0 = b$, the number of blocks, and

$$\lambda_s = \frac{\lambda P(v-s, t-s)}{P(k-s, t-s)} \qquad (26)$$

$\lambda_1 = r$, the replication number.

Designs are only known for relatively small values of t. For $t = 3$, there are the inversive planes which, like finite projective planes, combine classical geometry and finite fields. Let P be the points of a finite affine plane coordinatized by a field F with q elements. These points are in correspondence with the elements of an extension field F' with q^2 elements, where each point (r,s) is paired with the element $r + sx$, with x the root of an irreducible quadratic equation (like $\sqrt{-1}$ over the real numbers). An ideal point at infinity, ∞, is added giving $v = q^2 + 1$ points. The blocks are then the images of the set B_0 consisting of the line $\{(s,0):$

s element of F} along with ∞ under all linear fractional transformations, given by (27). Here ∞

$$y \to \frac{ay+b}{cy+d} \ (a,b,c,d \text{ elements of } F', \ ad = bc) \quad (27)$$

acts like $1/0$ so that Eq. (28) holds. The blocks (all

$$\frac{a^\infty + b}{c^\infty + d} = \frac{a}{c} \quad (28)$$

with $q + 1$ elements) are then called circles (with the original lines in the affine plane corresponding to circles through ∞). Circles have the property that every triple of points determines a unique circle ($\lambda = 1$), and further that if C is a circle which contains p but not q, there is a unique circle which contains q and is "tangent" to C at p.

The construction of t-designs has been closely related to the study of multiply transitive groups. The Mathieu groups give two designs for $t = 5$. One has parameters $5 - (12,6,1)$, and the other has parameters $5 - (24,8,1)$. In the first case, the blocks which contain a fixed three-element subset of points form a design isomorphic to the affine plane of order three, and in the second the blocks through three fixed points form a projective plane of order four.

At present no simple designs are known for $t > 5$. Steiner quadruple systems [designs with parameters $3 - (v,4,1)$] have been constructed for all v congruent to 2 or 4 modulo 6.

Error-correcting codes. The following matrix is the incidence system of a projective plane of order 2 (and symmetric Steiner triple system):

$$
\begin{array}{ccccccc}
1 & 0 & 0 & 0 & 1 & 0 & 1 \\
1 & 1 & 0 & 0 & 0 & 1 & 0 \\
0 & 1 & 1 & 0 & 0 & 0 & 1 \\
1 & 0 & 1 & 1 & 0 & 0 & 0 \\
0 & 1 & 0 & 1 & 1 & 0 & 0 \\
0 & 0 & 1 & 0 & 1 & 1 & 0 \\
0 & 0 & 0 & 1 & 0 & 1 & 1 \\
\end{array}
$$

Since $k = 3$ and $\lambda = 1$, each row has exactly three 1's and any two rows overlap in only one of these columns. Thus, any two rows differ in four of their entries. This matrix is an example of a binary block code of distance $m = 4$ and length $n = 7$: a set of codewords, each consisting of a sequence of n binary digits (zeros and ones), such that any two codewords differ in at least m places. If seven messages are coded by rows of the above matrix and one is transmitted to a receiver, the original message can be decoded—even if during transmission one of the 1's were changed to a 0 or conversely—by choosing the codeword closest to the received sequence. For example, if 1001011 were received, the last codeword was transmitted. On the other hand, if two digits were changed, resulting for example in a received message 1000110, the receiver could not tell whether the intended message corresponded to the first codeword or the second. Such a code will then "correct" one transmission error and decide if an intended codeword with up to three such errors has been correctly transmitted.

In general, a block code will correct up to

$[(m-1)/2]$ errors. The fundamental problem in coding theory is to construct codes with a large number of words and large distance but small length. In the example, the size of the code may be increased without decreasing the distance by adding a 1 to the end of all seven codewords, adjoining a new word of eight 1's, and then forming the eight additional words constructed from each of the others by interchanging all 0's and 1's.

Subdivision of square. An elegant combinatorial application of the theories of electrical networks and three-dimensional convex polytopes led in the 1940s to the construction of a square subdivided into n smaller squares no two of which are the same size (Fig. 2, where $n = 26$). In 1978 it was

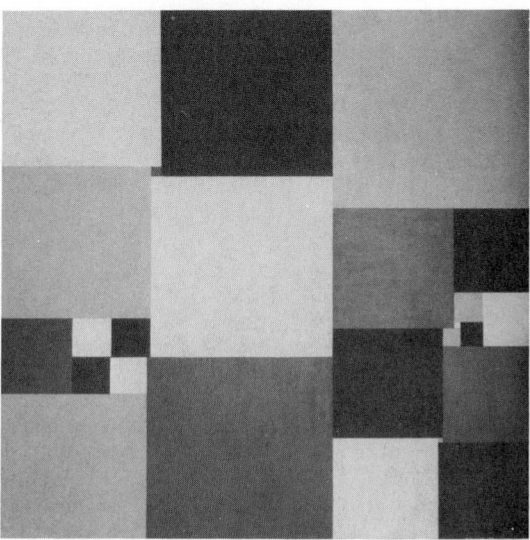

Fig. 2. Subdivision of a square into 26 smaller squares, no two of which are the same size. (*Thomas Brylawski; Karl Petersen, photographer*)

determined that the least number of unequal squares into which a square can be divided is 21, and that, up to symmetry, there is only the following subdivision. A square of length 112 is partitioned into squares of length 50, 35, 27, 8, 19, 15, 17, 11, 6, 24, 29, 25, 9, 2, 7, 18, 16, 42, 4, 37, and 33, respectively, where each square enters the big square as far north as possible and then as far west as possible.

Pigeonhole principle. The pigeonhole (or Dirichlet drawer) principle states that: if a very large set of elements is partitioned into a small number of blocks, then at least one block contains a rather large number of elements. From this result it follows, for example, that any permutation of the numbers from 1 to $mn + 1$ must contain either an increasing subsequence of length $m + 1$ or a decreasing subsequence of length $n + 1$. To see this, associate with each number in the permutation P the pair of numbers (i,d), where i (respectively d) is the length of the longest increasing (respectively decreasing) subsequence in which it is the first element. Then no two numbers in P are associated with the same pair (since if p precedes q, p initiates a longer increasing subsequence if $p < q$ and a longer decreasing subsequence if $p > q$), so that not every number can be associated with one of the mn pairs (i,d), $i=1, \ldots, m$; $d=1, \ldots, n$. One can do no better than $mn+1$, since if the numbers from 1 to mn are arranged sequentially in an n by m array and read off from left to right starting at the bottom, then the longest increasing subsequence has length m and the longest decreasing subsequence has length n. This construction belongs to the field of extremal combinatorial theory, a branch of combinatorics which constructs counterexamples of size n for theorems which hold for all $k \geq n+1$.

Ramsey's theorem. Ramsey's theorem, proved in the 1930s, generalizes the pigeonhole principle as follows: For any parameters $(r;r_1, \ldots, r_k)$ with each $r_i \geq r$, there exists a number N such that for all $n \geq N$, if S is an n-element set all $\binom{n}{r}$ of whose r-element subsets are partitioned into blocks $A_1, \ldots, A_k$, then there is some r_i-element subset S' of S such that all $\binom{r_i}{r}$ of the r-element subsets of S' are in the block A_i. For a given set of parameters, one then asks for the Ramsey number, or smallest N for which the theorem is true. When $r=1$, by the pigeonhole principle $N=r_1+\cdots+r_k-k+1$. For the parameters $(2;3,3)$, the theorem can be interpreted as follows: When the edges and diagonals of a regular polygon with a sufficiently large number N of vertices are all colored red or blue, there are three vertices of the polygon which are also the vertices of a triangle all of whose edges are the same color. One can always find a monochromatic triangle in a hexagon, and the extremal configuration consisting of a pentagon whose edges are red and whose five diagonals are blue shows that the Ramsey number for these parameters is 6.

Analogs of Ramsey's theorem have been proved for graphs, vector spaces, configurations in the Euclidean plane, and (with infinite cardinals as parameters) in set theory. However, very few Ramsey numbers have been computed. For example, when the edges and diagonals of a polygon are colored with four colors, the smallest N which guarantees a monochromatic triangle [that is for $(2;3,3,3,3)$] is known only to be between 51 and 63.

Many asymptotic results are known and have interpretations in graph theory. For example, the Ramsey number $(2;3,t)$ is known to be at most $ct^2/\log t$, where c is an absolute constant not depending on t. This means that in any graph with $ct^2/\log t$ vertices and no triangles, there is a subset of t vertices, no two of which are connected by an edge.

[THOMAS BRYLAWSKI]

Bibliography: E. F. Beckenbach (ed.), *Applied Combinatorial Mathematics*, 1964; C. Berge, *Principles of Combinatorics*, 1971; R. A. Brualdi, *Introductory Combinatorics*, 1977; L. Comtet, *Advanced Combinatorics*, 1974; M. Hall, *Combinatorial Theory*, 1967; C. L. Liu, *Topics in Combinatorial Mathematics*, 1972; J. Riordan, *Combinatorial Mathematics*, 1963; G.-C. Rota (ed.), *Studies in Combinatorial Theory*, 1978; H. J. Ryser, *Combinatorial Mathematics*, 1963; N. J. A. Sloane and F. J. MacWilliams, *The Theory of Error-Correcting Codes*, 1977.

Complex numbers and complex variables

A natural and extremely useful extension of the familiar real numbers. They can be introduced formally as follows. Consider the two-dimensional real vector space consisting of all ordered pairs (a_1, a_2) of real numbers. Geometrically this space can be identified with the ordinary euclidean plane, viewing the real numbers a_1, a_2 as the coordinates of a point in the plane.

The addition of vectors is defined by $(a_1,a_2) + (b_1,b_2) = (a_1 + b_1, a_2 + b_2)$ and is just the usual addition of vectors by the parallelogram law (Fig. 1). The multiplication of a vector (a_1,a_2) by a real number c is defined by $c(a_1,a_2) = (ca_1, ca_2)$, and is just the uniform dilation of the plane by the factor c. It may be asked whether it is possible to define a multiplication of one vector by another in such a manner that this multiplication is linear and satisfies the same formal rules as multiplication of real numbers. That is to say, it may be asked whether it is possible to associate to any two vectors (a_1,a_2) and (b_1,b_2) a product vector $(a_1,a_2) \cdot (b_1,b_2)$ in such a manner that:

(1) The multiplication is linear in the sense that

$$(a_1,a_2) \cdot [(b_1,b_2) + (c_1,c_2)]$$
$$= (a_1,a_2) \cdot (b_1,b_2) + (a_1,a_2) \cdot (c_1,c_2)$$

(2) It is associative,

$$(a_1,a_2) \cdot [(b_1,b_2) \cdot (c_1,c_2)] = [(a_1,a_2) \cdot (b_1,b_2)] \cdot (c_1,c_2)$$

(3) It is commutative,

$$(a_1,a_2) \cdot (b_1,b_2) = (b_1,b_2) \cdot (a_1,a_2)$$

(4) There is an identity vector (e_1,e_2) such that $(e_1,e_2) \cdot (a_1,a_2) = (a_1,a_2)$ for all vectors (a_1,a_2).

(5) For each vector (a_1,a_2) other than $(0,0)$ there is a vector (b_1,b_2) for which $(b_1,b_2) \cdot (a_1,a_2) = (e_1,e_2)$.

There does exist such a multiplication, given by Eq. (1), and it is essentially unique in the sense that

$$(a_1,a_2) \cdot (b_1,b_2) = (a_1b_1 - a_2b_2, a_1b_2 + a_2b_1) \quad (1)$$

any multiplication satisfying all the desired proper-

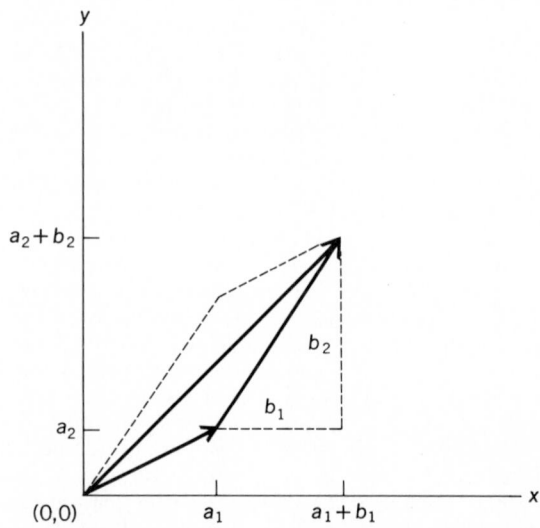

Fig. 1. Vector addition of two complex numbers by the parallelogram law.

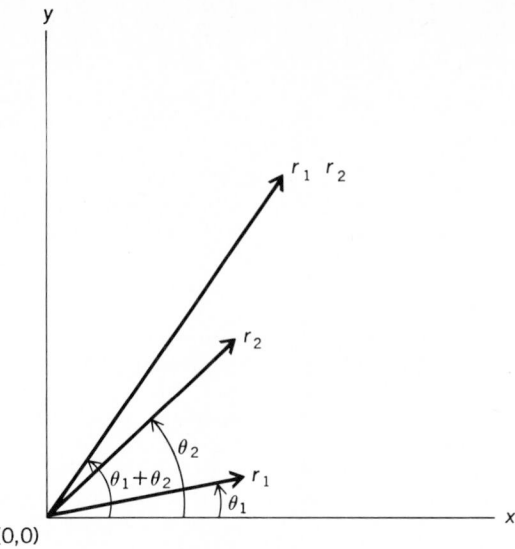

Fig. 2. Multiplication of complex numbers.

miliar addition formulas for the sine and cosine functions, that

$$(r_1 \cos \theta_1, r_1 \sin \theta_1) \cdot (r_2 \cos \theta_2, r_2 \sin \theta_2)$$
$$= [r_1 r_2 \cos (\theta_1 + \theta_2), r_1 r_2 \sin (\theta_1 + \theta_2)]$$

Thus multiplication of complex numbers amounts to multiplying their moduli and adding their arguments (Fig. 2).

Real and imaginary parts. Introducing the basis vectors $l = (1,0)$ and $i = (0,1)$, any vector (x,y) can be written uniquely as the sum $(x,y) = xl + yi$. Since the multiplication given by Eq. (1) is linear, the product of any two vectors is determined by giving the multiplication table for the basis vectors l and i. It follows readily from Eq. (1) that $l \cdot l = l$, $l \cdot i = i \cdot l = i$, and $i \cdot i = -l$, so that in particular the vector l is the identity element for the multiplication of complex numbers. Furthermore, the addition and multiplication of vectors of the form xl for all real numbers x reduce to the ordinary addition and multiplication of real numbers. There is consequently a natural imbedding of the real numbers into the complex numbers by associating to any real number x the complex number xl. The complex number l can be identified with the ordinary real number 1, and reflecting this identification the notation can be simplified by writing a complex number $z = (x,y) = xl + yi$ merely as $z = x + iy$. The component x is called the real part of the complex number z, and the component y is called the imaginary part. The modulus of the complex number z is denoted by $|z|$.

Complex conjugates. The vectors l and $-i$ are also a basis and satisfy exactly the same multiplication table as the vectors l and i, so the mapping which associates to any complex number $z = x + iy$ the complex number $\bar{z} = x - iy$ is a one-to-one mapping preserving the algebraic structure of the complex number system; thus $\overline{(z_1 + z_2)} = \bar{z}_1 + \bar{z}_2$ and $\overline{(z_1 \cdot z_2)} = \bar{z}_1 \cdot \bar{z}_2$. The complex number $\bar{z}$ is called the conjugate of z. It has the properties that $z = \bar{z}$ precisely when z is a real number, that $z \cdot \bar{z} = x^2 + y^2 = |z|^2$, and that in polar coordinates $\bar{z}$ is a complex number with the same modulus as z but with the negative argument (Fig. 3).

Polynomials. An ordinary real polynomial function $f(x) = x^n + a_1 x^{n-1} + \ldots + a_n$ can be extended to a function of a complex variable z in the obvious manner, merely setting $f(z) = z^n + a_1 z^{n-1} + \ldots + a_n$ and using the algebraic operations as defined for complex numbers. The advantage of this extension is that the fundamental theorem of algebra then holds; a nontrivial polynomial function over the complex numbers always has a root α, a complex value for which $f(\alpha) = 0$. The same result is true for polynomials with complex coefficients $a_1, \ldots, a_n$. A simple consequence of this result is that any polynomial function of degree n can be written as a product $f(z) = (z - \alpha_1)(z - \alpha_2) \ldots (z - \alpha_n)$ where the complex numbers $\alpha_1, \alpha_2, \ldots, \alpha_n$ are precisely the roots of this polynomial repeated according to multiplicity. Thus passing from the real to the complex number system simplifies the analysis of polynomial functions and clarifies many of their properties. This was in a sense the original motivation for introducing the complex number system.

ties can be reduced to Eq. (1) by a suitable choice of coordinates in the plane. The plane with the ordinary addition and scalar multiplication of vectors and with the vector multiplication in Eq. (1) is the complex number system. It is not possible to introduce a corresponding multiplication on vector spaces of higher dimensions, although, for example, if the commutativity condition is dropped, there is a multiplication satisfying the remaining conditions on the four-dimensional vector space (the quaternions). *See* QUATERNIONS.

Geometric interpretation. The multiplication in Eq. (1) can most easily be interpreted geometrically by introducing polar coordinates in the plane and writing $(a_1, a_2) = (r \cos \theta, r \sin \theta)$, where r is the length of the vector (a_1, a_2), or the modulus of the complex number (a_1, a_2), defined by $r^2 = a_1^2 + a_2^2$, and θ is the angle between the vector (a_1, a_2) and the first coordinate axis, or the argument of the complex number (a_1, a_2), defined by $\tan \theta = a_2 / a_1$. It follows immediately from Eq. (1), by using the fa-

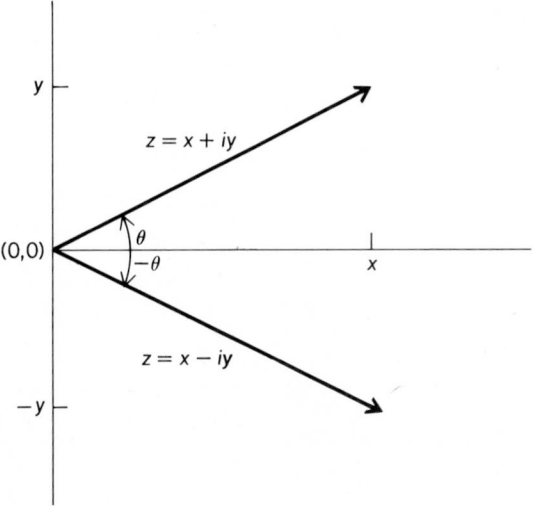

Fig. 3. Complex conjugate numbers.

Power series. In a somewhat similar manner, passing from the real to the complex number system simplifies and clarifies the analysis of more general functions than polynomials. Many of the familiar functions treated in elementary calculus have convergent Taylor series expansions and so can be written for all real numbers x sufficiently near a real number a in the form of a convergent infinite series

$$f(x) = \sum_{n=0}^{\infty} a_n(x-a)^n$$

where $n!\, a_n = f^{(n)}(a)$. If this series converges when the real number x is replaced by nearby complex numbers z, there results a natural extension of the function f to a complex-valued function of the complex variable z.

The question of convergence can be handled rather easily. A series of the form

$$\sum_{n=0}^{\infty} a_n(z-a)^n$$

where a_n, a, z are complex numbers is called a complex power series centered at the point a. For any such series there is a value r, a nonnegative real number or ∞, such that the power series converges for all complex numbers z satisfying $|z-a| < r$ and diverges for all complex numbers z satisfying $|z-a| > r$; r is called the radius of convergence of the power series. If $r = 0$, the power series diverges for all complex numbers z other than $z = a$, while if $r = \infty$, the power series converges for all complex numbers. Otherwise, the series converges for all complex numbers lying within a circle of radius r centered at the point a and diverges for all complex numbers outside that circle. The series may or may not converge at complex numbers on the circle itself. The radius of convergence can be determined by Hadamard's formula: $1/r = \lim \sup_{n \to \infty} |a_n|^{1/n}$. Alternatively and generally more simply the radius of convergence satisfies

$$\lim \inf_{n \to \infty} |a_n|/|a_{n+1}| \le r \le \lim \sup_{n \to \infty} |a_n|/|a_{n+1}|$$

If r is the radius of convergence of a power series, then that series converges absolutely and uniformly on the set of all complex numbers z satisfying $|z-a| \le \rho$ whenever $\rho < r$.

Using these observations it is easy to see, for example, that the exponential function can be extended to a complex-valued function of the complex variable z defined for any z by

$$e^z = \sum_{n=0}^{\infty} z^n/n!$$

Among the trigonometric functions the sine and cosine functions can also be defined for arbitrary complex numbers z by their Taylor series. Moreover, these functions satisfy the identity $e^{iz} = \cos z + i \sin z$ for all complex numbers z, an interesting relation among the elementary functions only apparent when passing to complex numbers. See SERIES.

Analytic functions. A complex-valued function f of a complex variable defined on some region Ω in the complex plane is analytic or holomorphic in Ω if for each point a in Ω the values of the function at all points z sufficiently near a are given by a power series expansion

$$f(z) = \sum_{n=0}^{\infty} a_n(z-a)^n$$

This power series must then be the Taylor series expansion of the function f at the point a and must converge in at least the largest circle centered at a and contained in Ω. An analytic function in a region Ω is necessarily a continuous function in Ω; moreover, as a function of the two real variables (x, y), an analytic function has partial derivatives of all orders.

Characterization by differentiability. There are a number of alternative characterizations of analytic functions, all of which are useful. The most common and perhaps the most surprising characterization, which is often taken as the definition of an analytic function, is that a function f is analytic in Ω precisely when the limit given by expression (2) exists at all points a in Ω. This limit is the com-

$$\lim_{z \to a} \frac{f(z) - f(a)}{z - a} \qquad (2)$$

plex analog of the ordinary derivative and is therefore called the complex derivative and written $f'(a)$. If

$$f(z) = \sum_{n=0}^{\infty} a_n(z-a)^n$$

where the power series converges in a disc $\triangle$ centered at a, then

$$f'(z) = \sum_{n=0}^{\infty} na_n(z-a)^{n-1}$$

and this series also converges in the disc $\triangle$. See CALCULUS.

Cauchy-Riemann equation. If f is analytic near a point a, then the complex derivative (2) can be calculated by letting z approach a through a sequence of points having the same imaginary value as a, hence $f'(a) = \partial f/\partial x$. Alternatively the complex derivative can be calculated by letting z approach a through a sequence of points having the same real value as a, hence $f'(a) = -i\,\partial f/\partial y$. Upon comparing these two results it follows that an analytic function f must satisfy the Cauchy-Riemann equation $\partial f/\partial x + i\,\partial f/\partial y = 0$, which is sometimes written $\partial f/\partial \bar{z} = 0$. Conversely if f is a complex-valued function defined in a region Ω of the complex plane, and if f has continuous first partial derivatives with respect to the real coordinates x and y, and if they satisfy the Cauchy-Riemann equation at all points of Ω, then f is analytic in Ω. It is not enough merely to require the existence of the first partial derivatives, and so this is not such a clean characterization of analytic functions as the preceding one, but it is nonetheless very useful. Weaker regularity conditions than the continuity of the first partial derivatives are possible. For instance, if the function f has first partial derivatives in the sense of distributions and if they satisfy the Cauchy-Riemann equation, then f is analytic.

Relation to harmonic functions. Setting $f(x,y) = u(x,y) + iv(x,y)$, where u and v are real-valued functions, the Cauchy-Riemann equation can be written equivalently as the pair of equations $\partial u/\partial x = \partial v/\partial y$, $\partial u/\partial y = -\partial v/\partial x$. It follows immediately from this that $\partial^2 u/\partial x^2 + \partial^2 v/\partial y^2 = 0$ and similarly for v, so that the real and imaginary parts of an analytic function are harmonic functions. Conversely if u is harmonic in a simply connected region Ω of the complex plane, the pair of real Cauchy-Riemann equations determines a harmonic function v such that $u + iv$ is analytic in Ω, and v is unique up to an additive constant; if Ω is not simply connected, the function v may not be single-valued. *See* DIFFERENTIAL EQUATION; LAPLACE'S DIFFERENTIAL EQUATION.

Morera's theorem. If f is an analytic function in a region Ω of the complex plane, then it follows from the Cauchy-Riemann equation that, using the terminology and notation of differential forms, the differential form $f dz = f dx + if dy$ is closed, since

$$d(f dx + if dy) = (i \partial f/\partial x - \partial f/\partial y) dx \wedge dy = 0$$

Therefore $\int_\gamma f dz = 0$ for any closed path γ homologous to zero in Ω. Conversely Morera's theorem asserts that if f is a continuous complex-valued function in Ω and if $\int_\gamma f dz = 0$ for all closed paths γ homologous to zero in Ω, then f is necessarily analytic in Ω. From this characterization of analyticity it follows immediately that the limit of a uniformly convergent sequence of analytic functions is also analytic.

Conformal mapping. An analytic function f in a region Ω of the complex plane can be viewed as a mapping from the plane region Ω into the complex plane. If $f(z) = u(z) + iv(z)$, this mapping is described in real terms by the pair of functions $u(x,y)$, $v(x,y)$. It follows readily from the Cauchy-Riemann equation that the jacobian determinant of this real mapping is given by Eq. (3). Since the jacobian de-

$$\det \begin{pmatrix} u_x & u_y \\ v_x & v_y \end{pmatrix} = u_x^2 + v_x^2 = |f'(z)|^2 \qquad (3)$$

terminant is thus always nonnegative, the mapping always preserves orientation. Moreover, the mapping is locally one-to-one whenever $f'(z) \neq 0$, and its inverse is readily seen also to satisfy the Cauchy-Riemann equation and therefore to be analytic. By analyzing further the condition imposed by the Cauchy-Riemann equation it follows that if $f'(a) \neq 0$, if γ_1 and γ_2 are two differentiable paths through the point a, and if the tangent vectors to γ_1 and γ_2 at the point a are at an angle θ apart, then the tangent vectors to the paths $f(\gamma_1)$ and $f(\gamma_2)$ at the point $f(a)$ are also at an angle θ apart. The mapping f thus preserves angles at a, and is therefore said to be conformal at a (Fig. 4). Conversely any continuously differentiable conformal mapping is an analytic function with a nonzero complex derivative, thus providing a very geometric characterization of analytic functions. *See* CONFORMAL MAPPING; PARTIAL DIFFERENTIATION.

For completeness, something should be said about the mapping property of an analytic function at a point at which the complex derivative is zero. The analytic function $f(z) = z^n$, for example, has a

zero derivative at $z = 0$ whenever $n > 1$. The mapping is not conformal but increases angles at the origin by a factor of n (Fig. 5) and so can be viewed as a mapping which wraps each circle about the origin in its domain n times about the origin in its range. Now if f is any analytic function at the origin with $f(0) = f'(0) = 0$, then this function can be written in the form $f(z) = g(z)^n$ for some integer $n > 1$, where g is analytic at the origin, $g(0) = 0$ but $g'(0) \neq 0$. Therefore the change of variable $w = g(z)$ reduces the function f to the form $f(w) = w^n$ just considered. It is clear from this that an analytic function takes open sets to open sets even when its derivative is zero and that any analytic function satisfies the maximum modulus theorem: if f is analytic at a and $|f(a)| \geq |f(z)|$ for all points z in some neighborhood of a, then f must actually be a constant. This seemingly simple property has a considerable range of applications.

Cauchy integral formula. An integral representation formula due to A. Cauchy plays a fundamental role in complex analysis and is traditionally the principal tool used in proving the equivalence of some of the alternative characterizations of analytic functions given above. The formula asserts that if γ is the smooth boundary curve of a finite region Ω in the complex plane (Fig. 6) and if f is analytic in an open neighborhood of the closed region $\Omega \cup \gamma$ (the union of Ω and γ), then Eq. (4) is valid. This shows, for example, that the

$$\frac{1}{2\pi i} \int_\gamma \frac{f(\zeta) d\zeta}{\zeta - z} = \begin{cases} f(z) & \text{whenever } z \text{ is a} \\ & \text{member of } \Omega \\ 0 & \text{whenever } z \text{ is not a} \\ & \text{member of } \Omega \cup \gamma \end{cases} \qquad (4)$$

values of the analytic function f in Ω are completely determined by the values of that function on the boundary of Ω. Equation (4) can be differentiated under the integral sign to yield the companion formula (5) for the derivatives of the

$$f^{(n)}(z) = \frac{n!}{2\pi i} \int_\gamma \frac{f(\zeta) d\zeta}{(\zeta - z)^{n+1}} \qquad (5)$$

function f at any point z in Ω. In particular, if Ω is a disc of radius r centered at a point a and if $|f(z)| \leq M$ on the boundary γ of Ω, then (5) implies that $|f^{(n)}(a)| \leq M r^{-n} n!$. Thus the derivatives of an analytic function cannot grow faster than $r^{-n} n!$ for some r as n tends to infinity. An immediate consequence of this estimate is Liouville's theorem, the assertion that the only functions analytic and bounded in the entire plane are the constant functions.

Relation to Green's integral theorem. The Cauchy integral formula (4) can be considered as a special case of a more general integral representation formula, following fairly directly from Green's integral theorem. If Ω and γ are as before and if f is any continuously differentiable complex-valued function of the real variables (x,y) in an open neighborhood of the closed region $\Omega \cup \gamma$, then Eq. (6) is valid for all z in Ω. Since analytic

$$f(z) = \frac{1}{2\pi i} \int_\gamma f(\zeta) \frac{d\zeta}{\zeta - z}$$
$$+ \frac{1}{2\pi i} \int \int_\gamma \frac{\partial f(\zeta)}{\partial \bar\zeta} \frac{d\zeta \wedge d\bar\zeta}{\zeta - z} \qquad (6)$$

functions are characterized by the Cauchy-Riemann equation $\partial f/\partial \bar z = 0$, it is evident that (6)

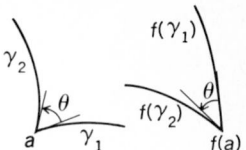

Fig. 4. Conformal mapping.

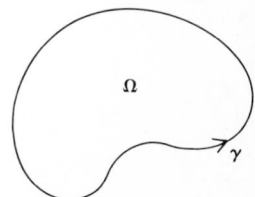

Fig. 6. The geometric situation in which the Cauchy integral formula holds.

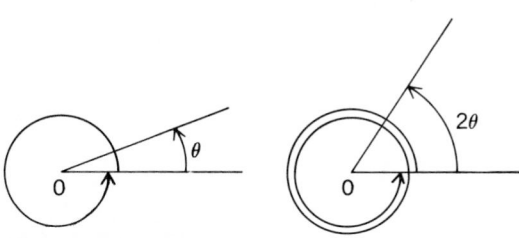

Fig. 5. The mapping z^2 at the origin.

reduces to (4) for analytic functions. The more general formula (6) is useful in investigating solutions of the partial differential equation $\partial f/\partial \bar{z} = g$ for a given function g. This equation has many applications in complex analysis.

Relation to Poisson integral formula. The Cauchy integral formula (4) is also closely related to the Poisson integral formula for harmonic functions, which can be derived from Eq. (4). The Poisson formula is useful in solving boundary-value problems for harmonic functions. For any continuous function on the smooth boundary γ of a bounded region Ω, there exists a harmonic function u in Ω with the given boundary values. If Ω is simply connected, the function u is the real part of an analytic function in Ω, so there thus exists a holomorphic function in Ω with an arbitrarily specified real part on γ. The imaginary part of this analytic function is determined uniquely up to an additive constant by the real part, so there cannot exist an analytic function in Ω with arbitrarily specified values on the boundary curve γ. On the other hand, if g is any sufficiently smooth function on γ, the integral $\int_{\gamma} g(\zeta) d\zeta/(\zeta - z)$ represents an analytic function $f_i(z)$ at all points z in Ω and another analytic function $f_0(z)$ at all points outside $\Omega \cup \gamma$. These functions have well-defined boundary values on γ, and $f_i(z) - f_0(z) = 2\pi i g(z)$ whenever z is on γ; the function $f_0(z)$ vanishes only when g is the boundary value on γ of an analytic function in Ω.

Isolated singularities. If a function f is analytic at all points of a disc $\triangle$ centered at a, except perhaps at the point a itself, then a is called an isolated singularity of the function f. There are three mutually exclusive possibilities for the nature of such a singularity:

(1) It may happen that $|f(z)|$ remains bounded near the point a. In that case the function can be extended to an analytic function even at the point a by suitably choosing the value of $f(a)$; the point a is then called a removable singularity.

(2) It may happen that $\lim_{z \to a} |f(z)| = \infty$. In that case a is a removable singularity for the function $g(z) = 1/f(z)$, so that $f(z) = 1/g(z)$ where $g(z)$ is an analytic function vanishing at the point a; and if $g(z) = (z - a)^k g_0(z)$ where $g_0(z)$ is analytic and $g_0(a) \neq 0$, then $f(z) = (z - a)^{-k} f_0(z)$ where $k > 0$ and $f_0(z)$ is analytic and $f_0(a) \neq 0$. The function $f(z)$ is said to have a pole of order k at the point a. The function $f_0(z)$ has a Taylor expansion at the point a as usual, so the function $f(z)$ has an expansion given by Eq. (7).

$$f(z) = a_{-k}(z-a)^{-k} + \ldots + a_{-1}(z-a)^{-1}$$
$$+ a_0 + a_1(z-a) + a_2(z-a)^2 + \ldots \quad (7)$$

The portion

$$\sum_{n=-k}^{-1} a_n(z-a)^n$$

of this expansion is called the principal part or singular part of the function $f(z)$ at the point a, while the remainder of the series represents an analytic function near a.

(3) If neither of the two preceding cases arises, the function $f(z)$ is said to have an essential singularity at the point a. A function having an essential singularity admits a Laurent series expansion near the point a of the form (8), in which

$$f(z) = \sum_{n=-\infty}^{+\infty} a_n(z-a)^n \quad (8)$$

infinitely many terms involving negative powers of the variable $(z - a)$ appear. Singularities of this type can occur and do so frequently. For example, whenever $f(z)$ has a pole at a point a, then $e^{f(z)}$ has an essential singularity at the point a. The behavior of a function near an essential singularity is rather complicated. A theorem of K. Weierstrass asserts that in any neighborhood of an essential singularity the values taken by the function come arbitrarily near any complex value. A much deeper theorem of E. Picard asserts that in any neighborhood of an isolated singularity a function actually takes all complex values with at most one possible exception.

Global properties. The zeros of an analytic function are isolated, in the sense that if f is analytic near a point a and $f(a) = 0$, then f has no other zeros in some neighborhood of a. That leads quite easily to the identity theorem: if $f(z)$ and $g(z)$ are analytic in a connected region Ω and if $f(a_n) = g(a_n)$ for an infinite sequence of points a_n having a limit point inside Ω, then $f(z) = g(z)$ at all points z in Ω. On the other hand, if a_n is any sequence of points in Ω having no limit point inside Ω, there are analytic functions in Ω having zeros precisely at the points a_n. This result is particularly easy in case that Ω is the entire complex plane. Indeed, in that case any analytic function having zeros at the points a_n can be written as a product given by Eq. (9), where $h(z)$ is an analytic function in the

$$f(z) = z^m e^{h(z)} \Pi_n \left(1 - \frac{z}{a_n}\right) e^{P_n(z)} \quad (9)$$

entire plane and $P_n(z)$ are suitably chosen polynomials ensuring that the product converges when there are infinitely many points a_n. There is a good deal of quite detailed information relating the growth of the function $f(z)$, the distribution of the zeros a_n, and the canonical product formula (9). This area of investigation is called the theory of entire functions. The infinite product expansions of even the elementary functions are quite interesting. For example, the expansion of $\sin \pi z$ is given by Eq. (10).

$$\sin \pi z = \pi z \, \Pi_{n=1}^{\infty} \left(1 - \frac{z^2}{n^2}\right) \quad (10)$$

A function analytic in a region Ω of the complex plane, except possibly for poles at some points of Ω, is called a meromorphic function in Ω. The points at which the function has poles can of course have no limit point inside Ω. Near any pole a function can be written as the quotient of two analytic functions, as already noted. There is an analogous global result: any meromorphic function in Ω can be written as the quotient of two functions each analytic in Ω. If $a_1, a_2, \ldots$ is any sequence of points in Ω having no limit point inside Ω and if $f_1(z), f_2(z), \ldots$ are principal parts of poles at these points, so that

$$f_\nu = \sum_{n=-k_\nu}^{-1} a_{\nu,n}(z-a_\nu)^n,$$

then there exists a meromorphic function $f(z)$ in Ω having poles with the specified principal parts at the specified points and no other singularities.

This is the analog of the familiar partial fraction expansion for quotients of polynomials. The resulting expansions of even the elementary functions are quite interesting, as for example the expansion given by Eq. (11).

$$\pi \cot \pi z = \frac{1}{z} + \sum_{n=1}^{\infty} \frac{2z}{z^2 - n^2} \qquad (11)$$

Analytic functions of several variables. A complex-valued function depending on several complex variables $f(z_1, \ldots, z_n)$ is called analytic at a point $(a_1, \ldots, a_n)$ if the values of that function at all points z_j sufficiently near a_j are given by a multiple power series expansion (12).

$$f(z_1, \ldots, z_n) =$$
$$\sum_{i_1=0}^{\infty} \cdots \sum_{i_n=0}^{\infty} a_{i_1 \cdots i_n}(z - a_1)^{i_1} \cdots (z - a_n)^{i_n} \quad (12)$$

If this series converges at a point $(z_1^0, \ldots, z_n^0)$ with $|z_i^0 - a_i| = r_i > 0$, then it is absolutely and uniformly convergent on the set of all points $(z_1, \ldots, z_n)$ for which $|z_i - a_i| \leq \rho_i$ whenever ρ_i are any positive constants such that $\rho_i < r_i$. This series thus converges to the same value for any ordering of the terms. For example, it can be rewritten as the power series (13) in the variable

$$f(z_1, \ldots, z_n) =$$
$$\sum_{i_n=0}^{\infty} a_{i_0}(z_1, \ldots, z_{n-1}) \cdot (z_n - a_n)^{i_n} \qquad (13)$$

z_n with coefficients which are analytic functions of the variables $z_1, \ldots, z_{n-1}$. Hence holding $z_1, \ldots, z_{n-1}$ constant, the function $f(z_1, \ldots, z_n)$ is analytic in the variable z_n alone, or more generally the function $f(z_1, \ldots, z_n)$ is analytic in each variable separately. Conversely a surprisingly nontrivial theorem of F. Hartogs asserts that a function $f(z_1, \ldots, z_n)$ that is analytic in each variable separately at all points of a region Ω is an analytic function of all n variables $z_1, \ldots, z_n$ in Ω.

Extension of results from one variable. Some results extend quite directly from one to n complex variables: the uniform limit of analytic functions is analytic; the maximum modulus theorem holds; and there is an extension of the Cauchy integral formula, asserting that if $f(z_1, \ldots, z_n)$ is analytic whenever z_j is a member of $\Omega_j \cup \gamma_j$ where Ω_j is a plane domain with smooth boundary γ_j, then Eq. (14) holds whenever z_j is a member of Ω_j for all indices j.

$$f(z_1, \ldots, z_n) = \left(\frac{1}{2\pi i}\right)^n \int_{\zeta_1 \epsilon \gamma_1} \cdots \int_{\zeta_n \epsilon \gamma_n}$$
$$\frac{f(\zeta_1, \ldots, \zeta_n)}{(\zeta_1 - z_1) \ldots (\zeta_n - z_n)} d\zeta_1 \ldots d\zeta_n \quad (14)$$

dices j. However, for the most part, the theory of analytic functions of several complex variables is far from being merely an extension of the standard results of the classical theory of functions of a single complex variable, by considering one variable at a time. Indeed, the theory of functions of several complex variables has quite a distinctive character with a considerable variety of results which perhaps are not the expected generalizations of the classical results but which shed new light on the familiar theory of analytic functions of a single variable.

Analytic continuation. An examination of the extended Cauchy integral formula (14) indicates some of the differences between the cases $n = 1$ and $n > 1$. When $n = 1$, either formula (14) or (4) expresses the values of an analytic function inside a domain Ω in terms of the values of the analytic function on the boundary of Ω, whereas when $n > 1$, formula (14) expresses the values of an analytic function inside a domain $\Omega_1 \times \ldots \times \Omega_n$ in terms of the values of the analytic function on a very small piece of the boundary. For example, when $n = 2$ the domain $\Omega_1 \times \Omega_2$ can be viewed as an open set of the four-dimensional euclidean space, its boundary is the three-dimensional subset $(\Omega_1 \times \gamma_2) \cup (\gamma_1 \times \Omega_2)$, but the integration in (14) is extended across the two-dimensional part $\gamma_1 \times \gamma_2$ of the boundary. This difference is not so superficial as one might at first imagine, but reflects the greater possibilities for analytic continuation of functions of $n > 1$ variables than of functions of a single variable.

For any domain Ω in the complex plane there are functions which are analytic in that domain but which cannot be extended as analytic functions across any part of the boundary of Ω. For example, there exist nontrivial analytic functions vanishing at any infinite sequence of points $a_1, a_2, \ldots$ of Ω having no limit point inside Ω, and if every boundary point of Ω is a limit of a subsequence of these points $a_1, a_2, \ldots$ then such an analytic function cannot be continued across any boundary point. However there are domains Ω in the space of $n > 1$ complex variables such that any function analytic in Ω extends to an analytic function in a properly larger domain. For example, it was shown by Hartogs that if Ω is an open neighborhood of the boundary of a compact subset K in the space of $n > 1$ complex variables, then every function analytic in Ω extends to a function analytic in the union of K and Ω (Fig. 7).

Thus in the study of functions of $n > 1$ complex variables there arise the problems of characterizing the natural domains of existence of analytic functions (called domains of holomorphy) and of determining for those domains which are not domains of holomorphy the largest domains to which all analytic functions can be extended (called the envelopes of holomorphy). Both problems have been extensively studied, the first one particularly so. Perhaps the most surprising result is that the domains of holomorphy can be characterized purely locally: a domain Ω in the space of n complex variables is a domain of holomorphy if and only if each point on the boundary of Ω has an open neighborhood $\cup$ such that the intersection of $\cup$ and Ω is a domain of holomorphy. The determination of the envelope of holomorphy of a domain is generally rather difficult, the situation being complicated by the fact that the envelope of holomorphy may not be realizable as a domain in the space of n complex variables.

Zeros and singularities. Another and not entirely unrelated difference between functions of $n > 1$ and functions of one complex variable is that in the case $n > 1$ the zeros and singularities of an analytic function are never isolated. The set of zeros of an analytic function of $n > 1$ variables is a set of topological dimension $2n - 2$ and is in most places locally euclidean but can have very complicated singularities. Strange spaces, such as the exotic

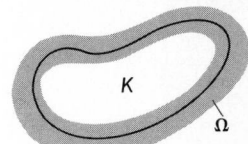

Fig. 7. Analytic continuation according to the theorem of Hartogs.

seven spheres, manifolds which are topologically but not differentiably equivalent to the ordinary seven-dimensional sphere, arise in this context. Even the purely local study of functions of several complex variables thus presents a considerable challenge. *See* TOPOLOGY.

[ROBERT C. GUNNING]

Bibliography: L. V. Ahlfors, *Complex Analysis*, 1953; H. Behnke and P. Thullen, *Theorie der Funktionen mehrer komplexer Veränderlichen*, 1970; J. B. Conway, *Functions of One Complex Variable*, 2d ed., 1979; E. Hille, *Analytic Function Theory*, 1959; L. Hormander, *An Introduction to Complex Analysis in Several Variables*, 1966.

Compressible flow

Flow in which the fluid density varies. In aerodynamic phenomena, when the flow velocity is large, it is necessary to consider that the fluid is compressible rather than to carry over from classical aerodynamics the assumption that the fluid has a constant density. Under this condition the speed of sound becomes an important factor. At relatively low speeds the changes of temperature and density of a fluid caused by the motion of a body in the fluid are almost negligible. However, if the body moves at high speed through the fluid, the motion can cause pronounced changes in density and temperature of the fluid. Hence, consideration of phenomena of this type involves not only classical fluid mechanics but thermodynamics as well.

The essential difference between an incompressible fluid and a compressible fluid is in the speed of sound. In an incompressible fluid the propagation of pressure change is essentially instantaneous; in a compressible fluid the propagation takes place with finite velocity. For example, if one strikes the surface of an incompressible fluid, the effect observed at great distances is, of course, less than at a smaller distance, but it reaches even infinite distance in essentially zero time. In a compressible fluid the effect propagates at finite velocity. A small disturbance propagates at the velocity of sound.

In aerodynamic phenomena the effects of the compressibility of the fluid can be important if the variation of fluid volume caused by the variation of pressure occurring in the flow is of the same order of magnitude as the variation in velocity which corresponds to the variation in pressure. The ratio of the fractional change in volume $\Delta v/v$ to the fractional change in velocity $\Delta u/u$ is equal to the ratio of the square of the velocity to the square of the velocity of sound $(u/a)^2$, where a is sonic velocity. Hence, when the velocity of the flow is the same order of magnitude as the velocity of sound in the flow, the variation of volume (or density) is the same order of magnitude as the variation in velocity. The large velocity variations occurring in high-speed flows therefore cause large changes in the fluid density.

The ratio of the local fluid velocity to the local sound velocity is the local Mach number. Mach number is an index of compressibility, serving as a measure of the relative importance of density changes in a fluid-flow field. In aerodynamic forces the error which results from the assumption of incompressibility in flow problems amounts to roughly $M_0^2/2\%$, when M_0 is the flight Mach number.   [JOHN E. SCOTT, JR.]

Compton effect

The increase in wavelength of electromagnetic radiation, observed mainly in the x-ray and gamma-ray region, on being scattered by material objects. This increase in wavelength, $\lambda_2 - \lambda_1 = \Delta\lambda$, of the scattered radiation, which is caused by the interaction of the radiation with the weakly bound electrons in the matter in which the scattering takes place, is given to good approximation by Eq. (1). Here λ_1 is the wavelength of the incident radia-

$$\lambda_2 - \lambda_1 = \Delta\lambda = (h/m_0 c)(1 - \cos\phi) \qquad (1)$$

tion, λ_2 is the wavelength of the radiation scattered at the angle ϕ, h is Planck's constant, m_0 is the rest mass the electron, c is the speed of light, and ϕ is the angle that the direction of the scattered radiation makes with the direction of the incident radiation.

The Compton effect illustrates one of the most fundamental interactions between radiation and matter and dislays in a very graphic way the true quantum nature of electromagnetic radiation. Together with the laws of atomic spectra, the photoelectric effect, and pair production, the Compton effect has provided the experimental basis for the quantum theory of electromagnetic radiation. For information on these and related topics *see* ANGULAR MOMENTUM; ATOMIC STRUCTURE AND SPECTRA; ELECTRON-POSITRON PAIR PRODUCTION; LIGHT; NONRELATIVISTIC QUANTUM THEORY; QUANTUM MECHANICS; UNCERTAINTY PRINCIPLE.

The Compton effect represents a great departure from earlier ideas concerning electromagnetic radiation. According to the original theory for the scattering of electromagnetic radiation by electrons in matter, which was developed by J. J. Thomson about 1900, the scattered radiation should have exactly the same wavelength as the incident radiation. This theory considers the incident radiation to have an oscillating electric field and shows that an electron would be forced by this electric field to oscillate with the same frequency as the field. The theory of electromagnetic radiation developed in the latter part of the 19th century by James Clerk Maxwell predicts that a point charge, such as an electron, when oscillating with a given frequency, will itself emit in all directions waves of electromagnetic radiation of exactly the same frequency. Therefore an increase in the wavelength, corresponding to a decrease in the frequency, of the scattered radiation is not to be expected if the scattering of x-rays takes place according to Thomson's theory. *See* ELECTROMAGNETIC RADIATION.

In a series of experiments, beginning in 1922, A. H. Compton confirmed the earlier conclusions of other scientists that the wavelengths of scattered x-rays increase, depending on the angle of scattering, a result in direct conflict with Thomson's theory. This discovery, along with its subsequent explanation by Compton, is regarded as one of the most significant contributions in physics. Compton showed that a beam of x-rays is composed of individual particles, called photons, each of which carries the energy $h\nu$ and also the linear momentum $h\nu/c$ in the direction of the beam, where h is Planck's constant, ν is the frequency of the radiation, and c is the speed of light. Moreover, the photon can impart energy and linear momentum to an individual electron in an elastic collision with the electron.

Experimental results. Using a crystal spectrometer, Compton made careful measurements of the wavelength spectrum of molybdenum K x-rays after they had been scattered at different angles from graphite. A diagram of the experimental apparatus for these measurements is shown in Fig. 1a. The spectrum of wavelengths of the molybdenum K x-rays after scattering from graphite at various angles is shown in Fig. 1b. While the incident radiation before scattering consists mainly of a fairly narrow range of wavelengths (that is, the molybdenum K-line), the spectrum observed after scattering consists of two peaks. One of these peaks P has essentially the same wavelength as the molybdenum K-line, but the second peak M has a longer wavelength. The wavelength of this second peak depends on the scattering angle, and is longer for larger scattering angles (Fig. 1b).

The dependence of the wavelength shift $\lambda_2 -$ $\lambda_1 = \Delta\lambda$ on scattering angle ϕ is plotted in Fig. 1c. The range of wavelengths due to the inhomogeneity in the scattering angle required by these experiments is also shown for the angles 45, 90, and 135°, and labeled $\delta\lambda$ in the figure. The first peak in Fig. 1b, which has the same wavelength as the incident radiation, is called the unmodified line. The longer-wavelength peak is called the modified line and is clearly due to a different type of scattering than Thomson predicted, since the wavelengths of the x-rays are increased in the scattering process. This is the type of scattering which yields the longer wavelengths in Compton scattering.

Theoretical explanation. Compton's explanation of the observed wavelength shift in x-ray scattering is based on developments which took place early in the 20th century. Max Planck provided an explanation for the observed intensity distribution with wavelength of electromagnetic radiation from a blackbody by introducing the idea that energy

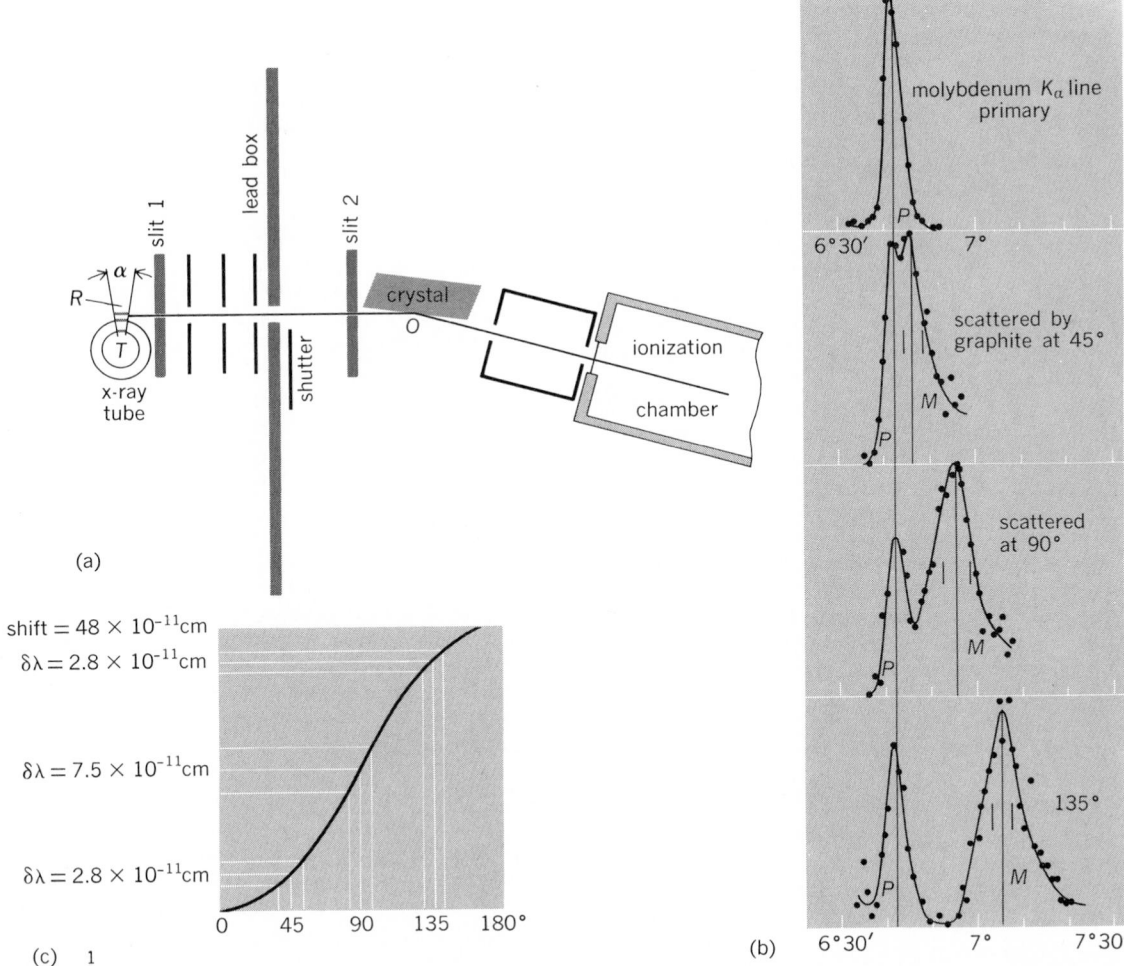

Fig. 1. Compton's experiment. (a) Characteristic K-lines from molybdenum target T of x-ray tube (shown in cross section) fall on graphite scatterer R. Scattered radiation is passed through slits 1 and 2, and analyzed spectrally by slow rotation of the calcite crystal and ionization chamber around pivot O. Longer wavelengths are observed by Bragg diffraction for larger angles of the calcite crystal with respect to the scattered beam defined by slits 1 and 2. Spread, or inhomogeneity, in scattering angle ϕ due to width of scatterer R is denoted by α. (b) Resulting spectra at three scattering angles. The ordinate is the intensity of the beam detected with ionization chamber and the abscissa is the angle that diffracting planes of calcite crystal make with scattered beam defined by slits 1 and 2. Larger angles of the calcite crystal spectrometer correspond to longer wavelengths in the spectrum of scattered radiation. (c) Graph showing dependence of shift $\lambda_2 - \lambda_1$ on ϕ. Effect of inhomogeneity α, about 0.31 radian, in scattering angle ϕ is shown in the graph. This spread gives rise to apparent spread $\delta\lambda$ in wavelength shift due to thickness of scatterer R, as shown on the ordinate of the graph.

could be emitted or absorbed in the blackbody only in discrete amounts equal to $h\nu$, where h is Planck's constant and ν is the frequency of the radiation. Planck's discovery, which marked the birth of quantum theory, was followed by Albert Einstein's explanation for photoemission. When light falls upon a surface, energetic electrons are emitted, and the energy of the electrons is found to be independent of the intensity but dependent on the frequency of the incident light which liberates the electrons. Einstein supplied the solution for this puzzle by postulating that the radiation field, that is, the beam of light, consists of particles called photons, each having energy $h\nu$, where h and ν are as Planck defined them. The photoelectric effect takes place when a photon is absorbed by an electron, so that the photon disappears and the electron assumes all of the energy $h\nu$ of the photon, less the energy required to bind the electron in its medium *see* HEAT RADIATION.

Compton's explanation of the wavelength shift assumes that, since x-rays are electromagnetic radiation, the picture given by Einstein of the quantum nature of the radiation field in the photoelectric effect describes the incident beam of x-rays. The scattering of the x-rays which gives rise to the peak of the modified line in the spectrum of the scattered x-rays (Fig. 1b) is due to photons or quanta of electromagnetic radiation scattering like material particles in collision with free or loosely bound electrons in the scattering material. Further, a photon, in addition to the energy $h\nu$, has a linear momentum $h\nu/c$ in the direction of travel of the beam. This momentum corresponds to that possessed by a massless particle having energy $h\nu$ and moving with the speed of light c, a fact which follows from Einstein's theory of special relativity. *See* RELATIVITY.

In other words, the x-ray scattering process, which gives rise to the increase in wavelength, is a process in which an x-ray photon of energy $h\nu$ and linear momentum $h\nu/c$ scatters as a mechanical particle would in colliding elastically with an electron which is at rest. In this type of collision, the x-ray particle transmits some of its energy and linear momentum to the electron which recoils. Therefore the x-ray photon after the collision has less energy $h\nu_2$ than before the collision. Since $\nu_2 = c/\lambda_2$ for a wave traveling with speed c, the wavelength λ_2 for the photon after the collision will be longer than λ_1, the wavelength it had before the scattering.

An important property of the elastic collision between two mechanical particles is that the energy and linear momentum will be conserved in the process. These two principles were used by Compton to calculate the increase in wavelength of the x-rays after scattering through a definite angle.

Scattering process. Figure 2 shows the Compton scattering process. The incident photon with linear momentum p_1 and wavelength λ_1 collides with the electron, which is initially at rest. After the collision, the photon scatters off at an angle ϕ with respect to the incident direction. The electron recoils and moves off at an angle θ with respect to the incident direction. A triangle resulting from the conservation of linear momentum of the particles involved in the collision is also shown in Fig. 2. This triangle shows the angular relationship which must be satisfied by ϕ and θ.

The principle of conservation of energy gives Eq. (2), where $h\nu_1$ is the energy of the incident photon,

$$h\nu_1 = h\nu_2 + m_0 c^2 \left[\frac{1}{\sqrt{1-\beta^2}} - 1 \right] \qquad (2)$$

ton, $h\nu_2$ is the energy of the scattered photon, and the second term on the right-hand side of the equation is the relativistic form of the kinetic energy of the recoiling electron.

In applying the principle of conservation of linear momentum, two additional equations are obtained. Conservation of the horizontal component of linear momentum (Fig. 2) gives Eq. (3), while

$$\frac{h\nu_1}{c} = \frac{h\nu_2}{c} \cos\phi + \frac{m_0 V}{\sqrt{1-\beta^2}} \cos\theta \qquad (3)$$

conservation of the vertical component yields Eq. (4). In these equations V is the velocity of the recoiling electron and β is V/c.

$$0 = \frac{h\nu_2}{c} \sin\phi - \frac{m_0 V}{\sqrt{1-\beta^2}} \sin\theta \qquad (4)$$

After making the substitution $\nu = c/\lambda$, squaring the equations, and combining the results, Eq. (1) is obtained. The quantity in Eq. (1), $h/m_0 c = 24.26 \times 10^{-11}$ cm, is called the Comptom wavelength of a free electron, and is the wavelength of a photon having energy $h\nu = m_0 c^2$, the rest energy of the electron. The Compton wavelength is the shift in the wavelength of a photon which is scattered through 90°, as seen in Eq. (1).

Two important physical results follow from Eq. (1). First, the wavelength shift $\Delta\lambda$ is independent of the wavelength of the incident photons and is, therefore, independent of the photon energy. Second, the shift $\Delta\lambda$ is independent of the type of material of the scatterer.

From Eq. (1) it follows that the energy of the scattered photon can be expressed as shown in Eq. (5). It then follows from Eqs. (2) and (5) that the

$$h\nu_2 = \frac{h\nu_1}{1 + \frac{h\nu_1}{m_0 c^2}(1 - \cos\phi)} \qquad (5)$$

kinetic energy E_R of the recoil electron can be expressed as in Eq. (6).

$$E_R = h\nu_1 - h\nu_2 = h\nu_1 \frac{\frac{h\nu_1}{m_0 c^2}(1 - \cos\phi)}{1 + \frac{h\nu_1}{m_0 c^2}(1 - \cos\phi)} \qquad (6)$$

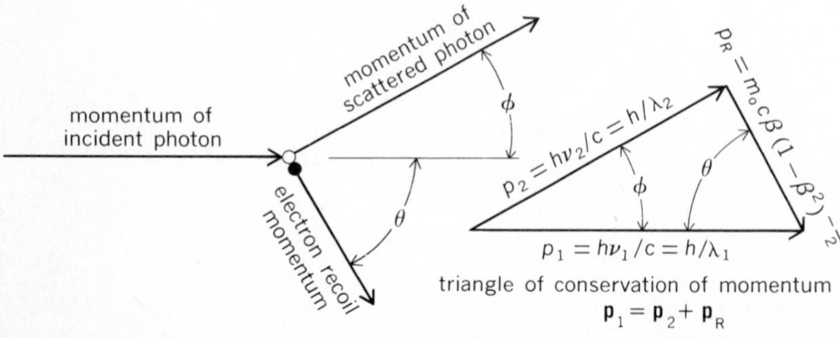

momentum of incident photon

momentum of scattered photon

electron recoil momentum

$p_R = m_0 c \beta (1 - \beta^2)^{-\frac{1}{2}}$

$p_2 = h\nu_2/c = h/\lambda_2$

$p_1 = h\nu_1/c = h/\lambda_1$

triangle of conservation of momentum
$\mathbf{p}_1 = \mathbf{p}_2 + \mathbf{p}_R$

Fig. 2. Diagram for derivation of wavelength shift, electron recoil energy, and angular correlation between scattering angle ϕ and electron recoil angle θ.

From the geometry of the momentum triangle of Fig. 2, Eq. (7) can be obtained. This is the required

$$\cot \theta = -\left(1 + \frac{h\nu_1}{m_0 c^2}\right) \tan \frac{\phi}{2} \qquad (7)$$

relationship for the angular correlation between the scattered photon and the recoil electron.

Compton-Debye effect. Peter Debye knew of Compton's published measurements of the wavelength shift of scattered x-rays, and independently developed the same theoretical explanation as Compton. His results were published at about the same time as Compton's; in Europe the effect has been known as the Compton-Debye effect.

Experimental verification. Verification of Compton's ideas appeared soon after his theoretical explanation for the shift. The recoil electrons, which are predicted by Compton's scattering theory, were detected independently by C. T. R. Wilson and Walther Bothe in Wilson cloud chambers. The momenta of the recoil electrons were measured by magnetic deflection and found to agree with Compton's theory by A. A. Bless. Bothe and Hans Geiger demonstrated the simultaneity of the appearance of the scattered photon and recoil electron. Then Compton and A. W. Simon, using a cloud chamber with partitions, were able to show that the predicted correlation in Eq. (7) between the direction of the scattered photon and that of the recoil electron is satisfied in individual Compton scattering processes. Later experiments with improved techniques showed the correctness of Compton's theory with greater accuracy.

The results of all these experiments led to the definite conclusion that electromagnetic radiation scattering from an electron, instead of spreading out in all directions around the scattering center as waves are expected to do, takes a definite direction in each individual process as a particle would.

Intensity distribution. Thomson's theory for the scattering of electromagnetic radiation predicts that the relative intensity of scattered radiation is symmetrical about a 90° scattering angle and proportional to the factor $1 + \cos^2 \phi$, where ϕ is the scattering angle. The angular distribution of scattered electromagnetic radiation as given by Thomson's theory is written as Eq. (8). Here I_ϕ/I_0 is the

$$I_\phi/I_0 = \frac{ne^4}{2r^2 m_0{}^2 c^4} (1 + \cos^2 \phi) \qquad (8)$$

ratio of the radiation scattered at an angle ϕ to the incident radiation intensity, n is the effective number of independently scattering electrons, e is the charge on the electron, and r is the distance from the scatterer.

Thomson's theory did not take into account the effect of the magnetic vector of the radiation. Including the effect of the magnetic vector would have provided the electron with a recoil in the scattering of radiation from the effect known as radiation pressure. Consideration of recoil of the electron from radiation pressure would have led to the prediction of an increase in the wavelength of the scattered radiation from the Doppler shift. Paul Dirac, Gregory Breit, and others took into account the effect of the relativistic Doppler shift of the photon in scattering from the electron, and found that the intensity distribution from the Thomson theory was to be multiplied by the Breit-Dirac re-

coil factor R, given by Eq. (9). *See* DOPPLER EFFECT; RADIATION PRESSURE.

$$R = \left[1 + \frac{h\nu_1}{m_0 c^2} (1 - \cos \phi)\right]^{-3} \qquad (9)$$

In addition to their charge and mass, electrons also have a quantized spin or angular momentum and an associated magnetic moment. Using Dirac's relativistic quantum mechanics and taking into account the interaction of the spin and magnetic moment of the electron with the electromagnetic radiation, O. Klein and Y. Nishina obtained a further factor which modifies the angular distribution of scattered x-rays. This factor can be written as notation (10). Combining this factor and the

$$1 + \frac{\left(\dfrac{h\nu_1}{m_0 c^2}\right)^2 (1 - \cos \phi)^2}{(1 + \cos^2 \phi)\left[1 + \dfrac{h\nu_1}{m_0 c^2} (1 - \cos \phi)\right]} \qquad (10)$$

Breit-Dirac recoil factor with the Thomson formula for the scattered x-ray angular distribution, the Klein-Nishina formula is obtained. This is written as Eq. (11). The quantities on the left-hand side of

$$I_\phi/I_0 = \frac{ne^4}{2r^2 m_0{}^2 c^4} \cdot \frac{1 + \cos^2 \phi}{\left[1 + \dfrac{h\nu_1}{m_0 c^2} (1 - \cos \phi)\right]^3}$$

$$\cdot \left\{1 + \frac{\left(\dfrac{h\nu_1}{m_0 c^2}\right)^2 (1 - \cos \phi)^2}{(1 + \cos^2 \phi)\left[1 + \dfrac{h\nu_1}{m_0 c^2} (1 - \cos \phi)\right]}\right\} \qquad (11)$$

the formula represent the same quantities as in the Thomson formula, Eq. (8). For the limiting case of low-energy x-rays ($h\nu_1 \to 0$), the Klein-Nishina formula reduces to the Thomson scattering law. But at photon energies above a few hundred kilovolts, the angular distribution departs significantly from the $1 + \cos^2 \phi$ distribution of the Thomson law. *See* ELECTRON SPIN; RELATIVISTIC QUANTUM MECHANICS.

Figure 3 shows the predicted results of the x-ray scattering angular distribution, comparing the

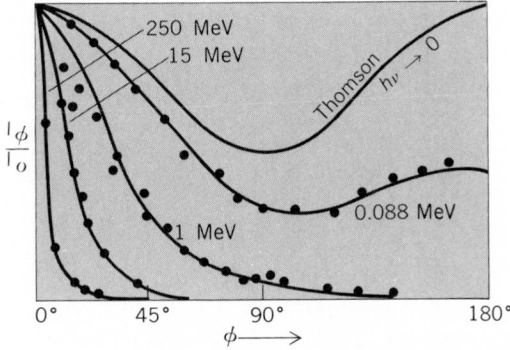

Fig. 3. Angular distribution of Compton scattered gamma rays. I_ϕ represents intensity of gamma rays scattered at angle ϕ with respect to incident beam, while I_0 represents intensity of incident beam before scattering. (*Adapted from R. S. Shankland, Atomic and Nuclear Physics, 2d ed., 1960*)

Klein-Nishina formula with the Thomson formula at various energies. The full lines in Fig. 3 are the theoretical predictions, and the points are experimental measurements.

If the modified line of Fig. 1b is due to scattering of an x-ray photon from an electron which recoils, the presence of the unmodified line in the spectrum of scattered radiation must be due to some other type of scattering. The fact that the wavelength changes in Compton scattering, giving rise to the modified line of Fig. 1b, is due to the decrease in energy of the scattered photon in imparting recoil energy to the electron. The unmodified line is due to photons that scatter from electrons which are too tightly held in the atom to recoil from the impact of the photon. That is, these interactions correspond to photon-electron interactions for which the conservation of momentum and energy would require the free-electron recoil energy to be comparable to, or less than, the electron binding energy in the atom. The momentum change accompanying scattering is communicated in such cases to the atom as a whole. Because of the atom's larger mass when compared to that of an electron, the photon transmits only a small contribution of linear momentum to the atom. Therefore the photon itself loses only a negligible amount of energy. The unmodified line is more prominent in scattering from atoms of higher atomic number because of the greater number of tightly bound electrons in these atoms. The theory giving the ratio of intensities of modified and unmodified scattering has been developed by I. Waller and D. R. Hartree.

There are many ways in which photons can interact with matter. In these interactions, usually photons either disappear completely or are scattered out of the initial beam of photons, so that the intensity of the beam is diminished as it moves through the scattering material. Of the many possible modes of interaction of photons with matter, there are four which predominate: Compton scattering (with a change in wavelength of the photon giving rise to the modified line in the scattered spectrum), elastic scattering (with no change in wavelength giving rise to the unmodified line), photoelectric effect or photoemission (when the visible and ultraviolet regions are most often involved), and electron-positron pair production (for photon energies above 1 MeV).

The Klein-Nishina formula, Eq. (11), can be integrated over the angle ϕ to obtain a formula which gives the relative intensity of photons scattered by the Compton effect compared with the initial intensity of the incident beam in the scattering material. This formula is related to the probability that Compton scattering will occur to a photon of a given energy $h\nu$. Formulas have been obtained giving the probability that the other three processes can occur. It has been demonstrated that, while all these processes can happen to some extent at most energies (except for electron-positron pair production which has a threshold at 1 MeV below which it cannot occur), the Compton effect is the predominant interaction for a large range of atomic numbers for photon energies between 1 and approximately 10 MeV.

Applications in science. The Compton effect has played a significant role in several diverse scientific areas. Compton scattering (often referred to as incoherent scattering, in contrast to Thomson scattering and also Rayleigh scattering, which are called coherent scattering) is important in nuclear engineering (radiation shielding), experimental and theoretical nuclear physics, atomic physics, plasma physics, x-ray crystallography, elementary particle physics, and astrophysics, to mention some of these areas.

In addition the Compton effect provides an important research tool in some branches of medicine, in molecular chemistry and solid-state physics, and in the use of high-energy electron accelerators and charged particle storage rings. Some examples of the role of the Compton effect in a few of these fields will be described below.

Attenuation of electromagnetic radiation. As mentioned above, if a beam of electromagnetic radiation traverses matter it is reduced in intensity by the absorption of photons and by the scattering of photons out of the beam by the matter. The relative importance of the principal processes in attenuating the beam depends in part on the energies of the photons in the beam. Considerable effort has been made to determine as accurately as possible the parameters determining the rate of attenuation of electromagnetic radiation as a function of energy through Compton scattering and through the other processes. For example, the contribution of Compton scattering (incoherent scattering) is an important correction in crystallographic studies through coherent x-ray scattering. *See* X-RAY CRYSTALLOGRAPHY.

Information on the attenuation properties of various materials is obviously important in the design of radiation shielding.

Medical applications. In addition to its role in the interpretation of x-ray photographs, the Compton effect has been used directly in the diagnoses of human diseases. One application is the use of Compton scattering at 90° of gamma rays from a radioactive ^{137}Cs source for the early detection of the bone disease osteoporosis. Since the intensity of the Compton-scattered gamma rays is proportional to n, the effective number of independently scattering electrons [see. Eq. (11)], and since the number of effective scattering electrons can be closely related to the density of the scattering material, the intensity of the Compton-scattered photons can be used as a measure of the density of the scattering material. Changes in density of bone are an indication of osteoporosis.

Study of electrons in matter. The Compton effect has been used to study the electronic structures of molecules and solid crystals. It was realized by J. W. M. DuMond that in addition to the Doppler shift from the recoil of the electron in the Compton-scattering process, there is also an additional Doppler shift of the photons which is due to the fact that the electrons from which the photons are scattered are not really at rest but are rapidly moving in their atoms. This second Doppler shift adds to or subtracts from the Compton shift, depending on whether the electron is moving toward or away from the incident photon before scattering. The result is a broadening of the modified line. Since the 1960s physicists and chemists have made extensive use of this Doppler broadening to study the momenta of electrons in atoms, molecules, and conducting and semiconducting crystals.

Production of gamma rays by inverse Compton effect. In 1948 E. Feenberg and H. Primakoff studied theoretically the possibility that the inverse Compton effect takes place in interstellar space and that it may be an astrophysical phenomenon important in the depletion of high-energy electrons. In the inverse Compton effect energetic electrons undergo elastic collisions with low-energy photons so that the electrons lose energy and the photons gain energy in the collision. The wavelength shift for the photon is to shorter wavelength. Feenberg and Primakoff suggested that energetic electrons would interact through the inverse Compton effect with starlight to reduce the energy of the electrons. In so doing, of course, the photons would often receive high energies and would then become energetic gamma rays.

Such a mechanism has been extensively considered as a source for the production of high-energy gamma rays which are observed in cosmic rays. The universal microwave radiation, discovered in 1965 by A. Penzias and R. Wilson has been considered as the source of initial photons for this process, since it provides a much higher density of photons in space than light could. The nature and source of the primary high-energy electrons which produce the gamma rays can be studied from the observed gamma-ray spectrum.

The inverse Compton effect can be used as a method for producing high-energy polarized gamma rays, by directing a homogeneous beam of photons of light from a laser into the energetic electron beam of an electron accelerator. Figure 4 shows the arrangement used by Richard Milburn and his colleagues at the Cambridge electron accelerator when this technique was being developed. In that arrangement the photons which are backscattered through the inverse Compton effect are increased uniformly in energy so that high-energy gamma rays are emitted from the accelerator and can be detected by a Cerenkov counter or some other type of gamma-ray detector. If photons are circularly polarized, the polarization state will be maintained and the resulting gamma rays will be circularly polarized in the same sense as the original laser photons. *See* PARTICLE ACCELERATOR; POLARIZED LIGHT.

Several laboratories have been active in studying and developing this method. At the electron accelerator at Frascati near Rome, for example, a beam of gamma-ray photons continuously variable in energy between 5 and 80 MeV has been obtained from a primary electron energy ranging between 370 and 1500 MeV. The photon intensities are between approximately 10^5 and 10^7 photons per second, the energy resolution is 1%, full width at half maximum, and the gamma-ray polarization is 99%. Such a beam of photons is valuable as a probe of nuclear structure in photo-nuclear reaction work.

The inverse Compton backscatter technique using photons from lasers is a useful method for determining the degree of spin polarization of the circulating electron and positron beams in high-energy storage rings. Under certain conditions there is a gradual buildup of this polarization through the effects of the synchrotron radiation emitted by the circulating beams. The calculation described earlier for the intensity of Compton scattered photons, the Klein-Nishina formula, Eq. (11), applies in the case of unpolarized electrons and unpolarized photons. When the calculation is carried out for the case of spin-polarized electrons and circularly polarized photons, it is found that there will be an asymmetry in the scattering depending on the degree of polarization of the electrons. This effect has been used to detect circularly polarized gamma rays scattered from polarized electrons in magnetized iron, in studying nonconservation of parity in beta decay. Thus, circularly polarized photons from a laser can be used to determine the extent of the polarization of the beams of circulating electrons and positrons in storage rings by examining the scattering asymmetry. *See* ELECTRON SPIN; PARITY (QUANTUM MECHANICS).

Another application of the inverse Compton

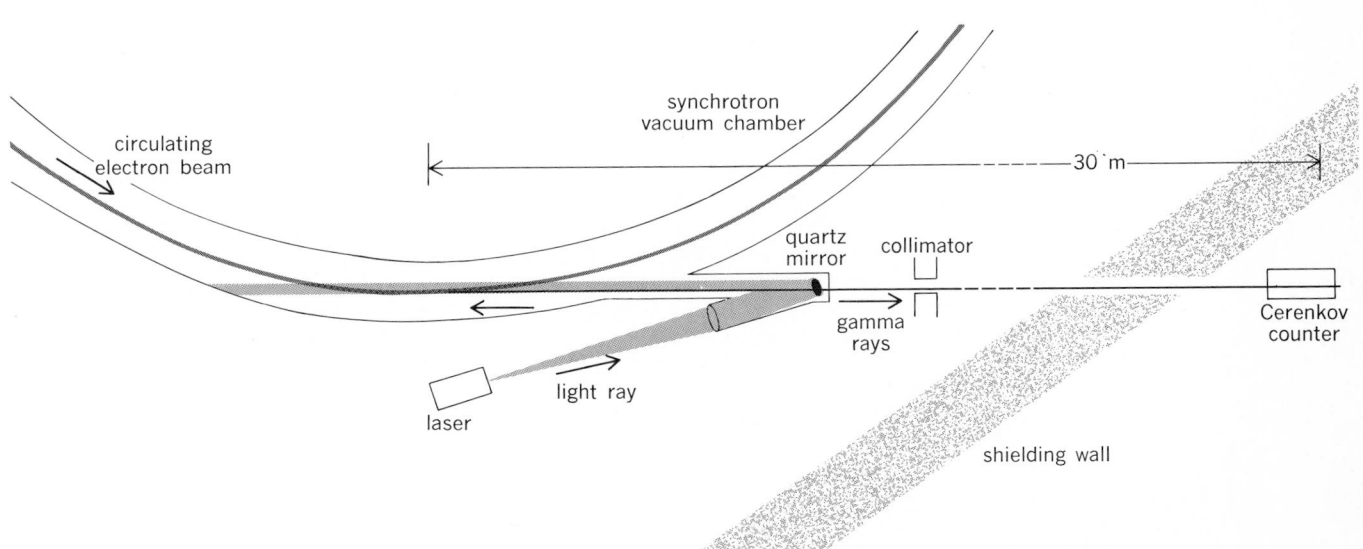

Fig. 4. Method for producing high-energy gamma rays at the Cambridge electron accelerator. Light from the ruby laser is reflected at the quartz mirror to meet 6-GeV electrons in the straight section of the machine. Backscattered gammas pass through the mirror and are counted in the total-absorption Cerenkov counter. (*Polarized gamma rays made by Compton scattering, Phys. Today, 21(5):77, 1968*)

effect using photons from a laser is to monitor the intensity of the charged particle beams, such as a proton beam, in high-energy accelerators. By monitoring the intensity of high-energy gamma rays produced through the inverse Compton backscattering from the beam, the intensity of the beam can be determined once the system has been calibrated.

Compton-like scattering from other particles. The Compton effect is usually associated with the elastic collision of a photon with an electron. However, a similar process occurs when a photon is scattered by a proton or some other elementary particle. Consideration has also been given to Compton-like scattering of neutrinos by nuclear matter in interstellar space.

It is believed that research on Compton scattering of high-energy gamma-ray photons from protons will provide details of the internal structure of the proton and possibly also of the photon itself. Quarks, believed to be the constituent particles of the proton, can be observed through their internal motions which give rise to a Doppler shift, similar to that observed in experiments on the motion of electrons in atoms. *See* QUARKS.

Conclusion. During the past 2 centuries an overwhelming amount of evidence has been obtained which shows that electromagnetic radiation, including light, x-rays, and gamma rays, has the nature of waves moving through space. The phenomena of diffraction and interference, which are exhibited by x-rays and gamma rays as well as by light, can be explained only if the radiation has a wave character.

Experiments involving the polarization of the electric vector of x-rays and gamma rays from scattering have shown that the waves are transverse vibrations, and that x-rays and gamma rays as well as light are linearly polarized with the electric vector of the beam predominantly in a given direction after scattering in matter. If the radiation is scattered at 90° with respect to its incident direction, the scattered radiation is completely linearly polarized. This statement is found true for both the unmodified and the Compton-scattered-modified line.

Perhaps the greatest significance of the Compton effect is that it demonstrates directly and clearly that in addition to its wave nature with transverse oscillations, electromagnetic radiation has a particle nature and that these particles, the photons, behave quite like material particles in collisions with electrons. This discovery by Compton and Debye led to the formulation of quantum mechanics by W. Heisenberg and E. Schrödinger and provided the basis for the beginning of the theory of quantum electrodynamics, the theory of the interactions of electrons with the electromagnetic field.

A finding also of fundamental importance is that material particles have a wave nature and exhibit interference effects. This dual nature of both electromagnetic radiation and material particles, like electrons and protons, lies at the very heart of modern quantum theory and is called the wave-particle duality. A striking example of this dual nature of material particles is that an electron will undergo a Compton-like effect, that is, there will be an increase in its De Broglie wavelength, when it scatters, say, from a proton, since some of the electron's momentum will be transmitted to the proton. However, the formula for the wavelength increase will be different from Eq. (1), since the electron is not a massless particle. *See* DE BROGLIE WAVELENGTH; ELECTRON DIFFRACTION.

The discovery and successful explanation of the Compton effect and the subsequent related experiments have provided the resolution of an important scientific controversy of 2000 years' duration: does electromagnetic radiation consist of waves or particles? The astonishing answer is that electromagnetic radiation is both.

[EASTMAN N. HATCH]

Bibliography: A. H. Compton, The scattering of x-rays as particles, *Amer. J. Phys.*, 29:817–820, 1961; R. D. Evans, The Compton effect, *Handbuch der Physik*, vol. 34, pp. 218–298, 1958; L. Federici et al., A new monochromatic and polarized photon beam at Frascati, in *Lecture Notes in Physics*, 108: 234–239 (Nuclear Physics with Electromagnetic Interactions Proceedings), Mainz, 1979; G. Hazan et al., The early detection of osteoporosis by Compton gamma ray spectroscopy, *Phys. Med. Biol.*, 22:1073–1084, 1977; J. H. Hubbell et al., Atomic form factors, incoherent scattering functions, and photon scattering cross sections, *J. Phys. Chem. Ref. Data*, 4:471–538, 1975, and 6: 615–616, 1977; I. Kaplan, *Nuclear Physics*, 2d ed., 1962; W. Niemann and K. O. Thielheim, Universal microwave radiation and extragalactic γ-radiation, *Nature*, 282:48–50, 1979; P. J. Schinder and S. L. Shapiro, Neutral currents and neutrino comptonization in high-temperature, nuclear matter, *Astrophys. J.*, 233:961–973, 1979; R. F. Schwitters, Experimental review of beam polarization in high-energy e^+e^- storage rings, *Amer. Inst. Phys. Conf. Proc.* no. 51, pp. 91–108, 1979; R. S. Shankland, *Atomic and Nuclear Physics*, 2d ed., 1960; R. H. Stuewer, *The Compton Effect: Turning Point in Physics*, 1975; B. G. Williams (ed.), *Compton Scattering*, 1976; R. Wilson, From the Compton effect to quarks and asymptotic freedom, *Amer. J. Phys.*, 45:1139–1147, 1977.

Compton wavelength

A convenient unit of length that is characteristic of any particle. By definition, the Compton wavelength λ_C of a particle of rest mass m is $\lambda_C = h/mc$, where h is Planck's constant and c is the velocity of light; this definition is analogous to that of the quantum-mechanical de Broglie wavelength λ_B, defined as $\lambda_B = h/p$, where p is the momentum classically given by $p = mv$, with m and v the particle mass and velocity respectively. *See* DE BROGLIE WAVELENGTH; PLANCK'S CONSTANT.

The Compton wavelength of the electron is 2.42631×10^{-12}m; of the proton 1.32141×10^{-15}m; of the muon 1.1734×10^{-14}m; and of the pion 8.883×10^{-15}m. The last is, by definition, the range of the strong nuclear force; in general, the Compton wavelength provides a convenient scale length in any given quantum-mechanical situation. *See* ELEMENTARY PARTICLE; QUANTUM FIELD THEORY.

The so-called reduced Compton wavelength, $\lambdabar_C = \lambda_C/2\pi$, is very frequently used instead of the Compton wavelength itself. *See* QUANTUM MECHANICS.

[D. ALLAN BROMLEY]

Conductance

For a series circuit the real, or in-phase, part of the complex representation for the admittance Y of a circuit. This is shown in Eq. (1), where B is the

$$Y = G \pm jB \qquad (1)$$

imaginary part, called the susceptance. Since $Y = 1/Z = 1/(R + jX)$, where R is the resistance and X is the total reactance $X_L - X_C$, Eq. (2) holds, and conductance G is given by Eq. (3). This is the gen-

$$Y = \frac{R}{R^2 + X^2} - j\frac{X}{R^2 + X^2} \qquad (2)$$

$$G = \frac{R}{R^2 + X^2} \qquad (3)$$

eral expression for conductance and shows that conductance is a function involving both resistance and reactance.

If reactance is negligible, then $G = R/R^2 = 1/R$, or conductance is the reciprocal of resistance. This is called simple conductance and is strictly correct only where the impedance contains no reactance. Thus in a dc circuit, conductance is the reciprocal of resistance. These reciprocal functions find application chiefly in computations of parallel circuits.

The total admittance of a number of parallel admittances, if expressed in complex quantities, is equal to the sum of the individual admittances. *See* ADMITTANCE.

[ARTHUR A. WELCH]

Conduction (electricity)

Electrical conduction may be defined as the passage of electric charge. This can occur by a variety of processes.

In metals the electric current is carried by free electrons. These are not bound to any particular atom and can wander throughout the metal. In general, the conductivity of metals is higher than that of other materials. At very low temperatures certain metals become superconductors, possessing infinite conductivity. The free electrons are able to move through the crystal lattice without any resistance whatsoever. *See* ELECTRICAL RESISTANCE; ELECTRICAL RESISTIVITY; FREE-ELECTRON THEORY OF METALS; SUPERCONDUCTIVITY.

In semiconductors (germanium, silicon, and so on) there are a limited number of free electrons and also "holes," which act as positive charges, available to carry current. The conductivity of semiconductors is much smaller than that of metals and, in contrast to most metals, increases with rising temperature. *See* HOLES IN SOLIDS; SEMICONDUCTOR.

Aqueous solutions of ionic crystals readily conduct electricity by means of the positive and negative ions present, for example, Na^+ and Cl^- in an ordinary solution of sodium chloride. Solid ionic crystals are themselves fair conductors. These crystals have sufficient lattice vacancies so that a few of the ions are able to migrate through the crystal under the influence of an applied electric field.

A strong electric field ionizes gas molecules, and thereby permits a flow of current through the gas in which the ions are the charge carriers. If sufficient ions are formed, there may be a spark.

Electric current can flow across a vacuum, for example, in a vacuum tube. The charge carriers are electrons emitted by the filament. The effective conductivity is low because of low available current densities at the normal temperatures of electron-emitting filaments. *See* ELECTRON EMISSION.

[JOHN W. STEWART]

Conduction (heat)

The flow of thermal energy through a substance from a higher- to a lower-temperature region. Heat conduction occurs by atomic or molecular interactions. Conduction is one of the three basic methods of heat transfer, the other two being convection and radiation. *See* CONVECTION (HEAT); HEAT RADIATION; HEAT TRANSFER.

Steady-state conduction is said to exist when the temperature at all locations in a substance is constant with time, as in the case of heat flow through a uniform wall. Examples of essentially pure transient or periodic heat conduction and simple or complex combinations of the two are encountered in the heat-treating of metals, air conditioning, food processing, and the pouring and curing of large concrete structures. Also, the daily and yearly temperature variations near the surface of the Earth can be predicted reasonably well by assuming a simple sinusoidal temperature variation at the surface and treating the Earth as a semi-infinite solid. The widespread importance of transient heat flow in particular has stimulated the development of a large variety of analytical solutions to many problems. The use of many of these has been facilitated by presentation in graphical form.

For an example of the conduction process, consider a gas such as nitrogen which normally consists of diatomic molecules. The temperature at any location can be interpreted as a quantitative specification of the mean kinetic and potential energy stored in the molecules or atoms at this location. This stored energy will be partly kinetic because of the random translational and rotational velocities of the molecules, partly potential because of internal vibrations, and partly ionic if the temperature (energy) level is high enough to cause dissociation. The flow of energy results from the random travel of high-temperature molecules into low-temperature regions and vice versa. In colliding with molecules in the low-temperature region, the high-temperature molecules give up some of their energy. The reverse occurs in the high-temperature region. These processes take place almost instantaneously in infinitesimal distances, the result being a quasi-equilibrium state with energy transfer. The mechanism for energy flow in liquids and solids is similar to that in gases in principle, but different in detail.

Fourier equation. The mathematical theory as well as the practical calculation of heat conduction is based on a macroscopic interpretation, as contrasted to the basic microscopic mechanism just described. From a physical point of view, it is reasoned that the steady heat flow from a surface (Fig. 1) at temperature t_1 to a parallel surface at t_2 is directly proportional to $(t_1 - t_2)$, the area A normal to

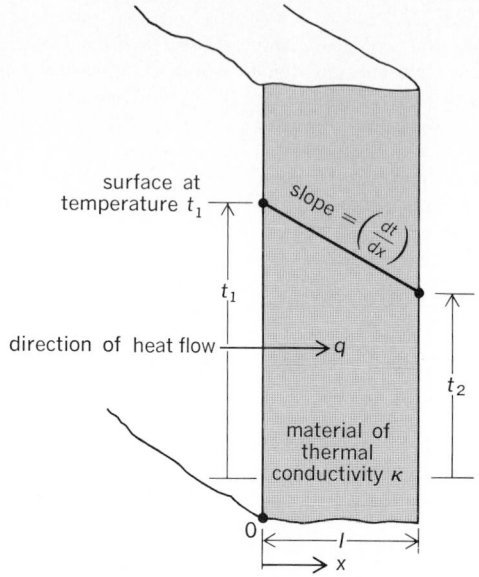

Fig. 1.   Heat flow by conduction.

heat-conducting ability of a substance, and depends not only on the particular substance involved, but also on the state of that substance. The Fourier equation is essentially a definition of κ which (Fig. 1) can be interpreted as the rate of heat flow per unit of area normal to the direction of flow when a unit temperature difference exists in unit length. Thus from Eq. (2) is derived Eq. (3), where

$$\kappa = \frac{-q/A}{(dt/dx)} = \frac{q/A}{(t_1 - t_2)/l} \qquad (3)$$

κ is seen to have the dimensions of a heat rate per unit area and per unit of temperature gradient. In the cgs system, it can be expressed in cal/(sec) (cm²)(°C/cm), which is equivalent to cal/(sec)(cm) (°C). In engineering, the units most frequently used are Btu/(hr)(ft)(°F).

Considerable progress has been made in the interpretation of thermal properties from theories of matter. This is particularly true for gases, where

the direction of flow, and the time of flow. τ, and inversely proportional to the distance l between the two planes. These factors are modified by a coefficient κ accounting for the heat-conducting nature of the particular substance between the two planes. Thus, the heat flow Q (in British thermal units, for example) is given by Eq. (1). In terms of

$$Q = \kappa A \frac{(t_1 - t_2)}{l} \tau \qquad (1)$$

the time rate of flow $q = Q/\tau$ through an infinitesimally thin layer dx, in which the temperature change is dt, this becomes Eq. (2). The minus

$$q = -\kappa A \left(\frac{dt}{dx}\right) \qquad (2)$$

sign is conventionally included to make q positive when heat flows in the increasing direction of x, since dt/dx is then negative. Although this equation was first proposed by J. Biot, it is named after J. Fourier in honor of Fourier's extensive contributions to the theory of heat conduction.

Thermal conductivity. The coefficient κ in Eqs. (1) and (2), called the thermal conductivity, is an important property of matter. It accounts for the

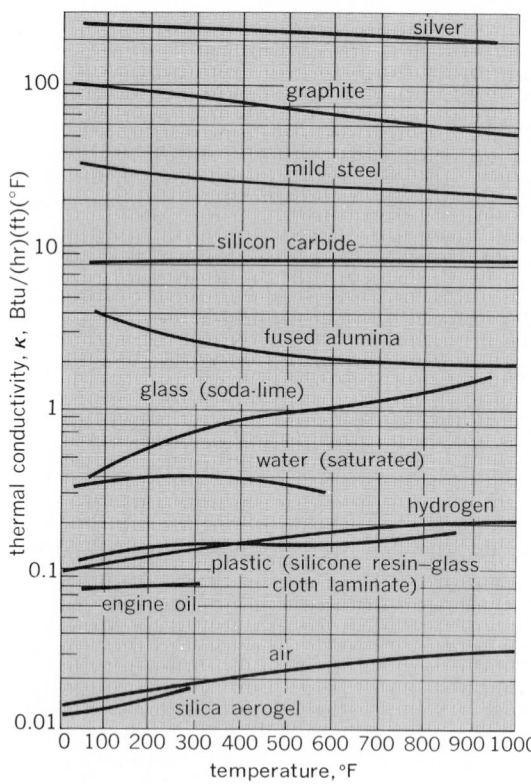

Fig. 3.   Graph of thermal conductivities of some typical examples of gases, liquids, and solids.

theory involving intermolecular forces has yielded very accurate results. The process of heat conduction in liquids is believed to be similar to that of sound transmission. In dielectric solids, energy is transmitted primarily by means of waves traveling through the atomic lattice; in metals, the electrons behave like an electron gas and provide for energy transfer as well as electrical conduction. This is the basis of the Wiedemann-Franz law, which states that $\kappa/\sigma T =$ constant (σ is the electrical conductivity and T the absolute temperature). *See* INTERMOLECULAR FORCES; KINETIC THEORY OF MATTER; THERMAL CONDUCTION IN SOLIDS.

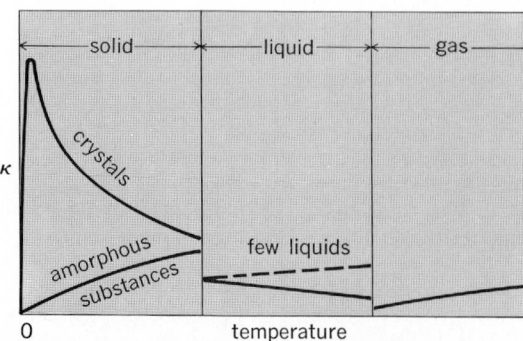

Fig. 2.   General variation of thermal conductivity with temperature throughout the three physical states. (*After L. S. Kowalczyk, Trans. ASME, 77:1021–1035, 1955*)

For materials occurring as crystalline or amorphous solids, the general trend of κ at atmospheric pressure throughout the three physical states is as shown in Fig. 2. The numerical value at the maximum, which occurs near absolute zero in crystalline substances, is comparatively high. For example, the κ of a copper crystal at 20°K has been found to be 7050 Btu/(hr)(ft)(°F)—more than 30 times its value at room temperature. Thermal conductivity of solids is discussed in a later section.

Because of the complexity and incomplete understanding of the mechanisms responsible for heat conduction, values of κ are usually determined experimentally. Results for typical gases, liquids, and solids throughout appropriate temperature ranges are shown in Fig. 3. The effect of pressure is significant primarily in gases.

Differential equation of conduction. The evidence of heat flow by conduction through a substance is the variation of the temperature with location and time. If the temperature as a function of the space coordinates and time is known or can be determined, the heat flow at any location and in any direction can be specified by appropriate differentiation. A given problem is normally attacked by solving the differential equation governing the temperature distribution in a homogeneous substance and making this solution fit the prescribed initial or boundary conditions. This differential equation, essentially an expression of the first law of thermodynamics applied to the heat flow, is derived by making a heat balance on an elemental volume in a medium (Fig. 4).

Considering first the x direction, the net heat flow into the element in time $\Delta\tau$ is the difference between that flowing in on the left minus that flowing out on the right; or, applying Eq. (1), Eq. (4)

$$\Delta Q_x = -\kappa\,\Delta y\,\Delta z\left(\frac{\partial t}{\partial x}\right)_x \Delta\tau$$
$$-\left[-\kappa\,\Delta y\,\Delta z\left(\frac{\partial t}{\partial x}\right)_{x+\Delta x}\Delta\tau\right] \quad (4)$$

is obtained. Accounting for the variation of κ with temperature is extremely difficult and may make the analytical solution of a problem impossible.

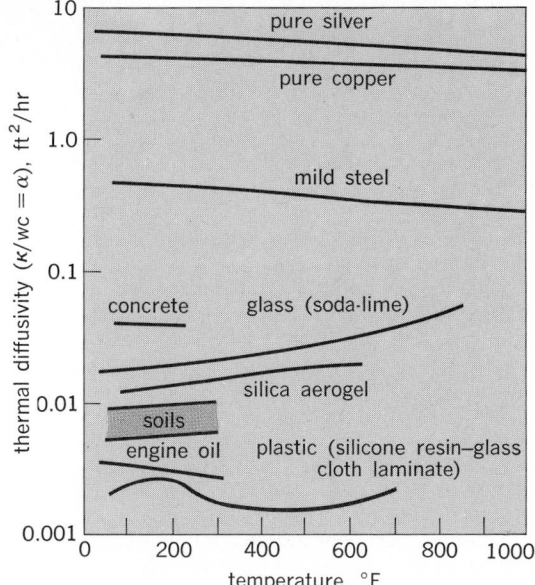

Fig. 5. Thermal diffusivities of some materials.

Because of this, it is customary to use an appropriate average value which is regarded as constant. Equation (4) can then be rearranged to read as Eq. (5a), which, as $\Delta x \rightarrow 0$, becomes Eq. (5b). Similar

$$\Delta Q_x = \kappa\,\Delta y\,\Delta z\,\Delta x\,\frac{\left(\frac{\partial t}{\partial x}\right)_{x+\Delta x} - \left(\frac{\partial t}{\partial x}\right)_x}{\Delta x}\,\Delta\tau \quad (5a)$$

$$\Delta Q_x = \kappa\,\Delta x\,\Delta y\,\Delta z\,\Delta\tau\,\frac{\partial^2 t}{\partial x^2} \quad (5b)$$

expressions apply for the y and z directions. Heat generated within the element at a uniform rate G per unit volume and time would add an amount $G\,\Delta x\,\Delta y\,\Delta z\,\Delta\tau$.

The net heat flow into the element would be manifest as stored energy and would be equal to notation (6), where w is the specific weight (weight

$$\Delta x\,\Delta y\,\Delta z\,wc\,\Delta t \quad (6)$$

per unit volume) of the medium, c its specific heat, and Δt the temperature rise in the time increment $\Delta\tau$. Equating the net flow into the element to that stored and letting Δx, Δy, Δz, and $\Delta\tau \rightarrow 0$ leads to Eq. (7). When no heat source is present, Eq. (7)

$$\kappa\left[\frac{\partial^2 t}{\partial x^2}+\frac{\partial^2 t}{\partial y^2}+\frac{\partial^2 t}{\partial z^2}\right]+G = wc\frac{\partial t}{\partial\tau} \quad (7)$$

becomes Eq. (8). The ratio $\kappa/wc = \alpha$ is defined as

$$\left[\frac{\partial^2 t}{\partial x^2}+\frac{\partial^2 t}{\partial y^2}+\frac{\partial^2 t}{\partial z^2}\right] = \frac{1}{\kappa/wc}\frac{\partial t}{\partial\tau} = \frac{1}{\alpha}\frac{\partial t}{\partial\tau} \quad (8)$$

the thermal diffusivity, and is the significant thermal property of a material for transient heat flow (Fig. 5). Equation (8) will be recognized as the equation governing a potential field written in terms of temperature. Similar equations are satisfied by other potential field phenomena, such as electricity, magnetism, diffusion, and ideal fluid flow. Because of this, solutions to problems in one field are applicable to analogous systems in the others. Also, an experimental solution of a problem

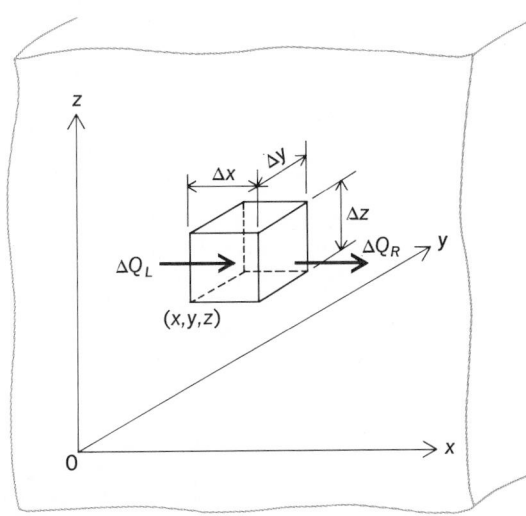

Fig. 4. Diagram of the heat flow through an elemental volume in a homogeneous medium.

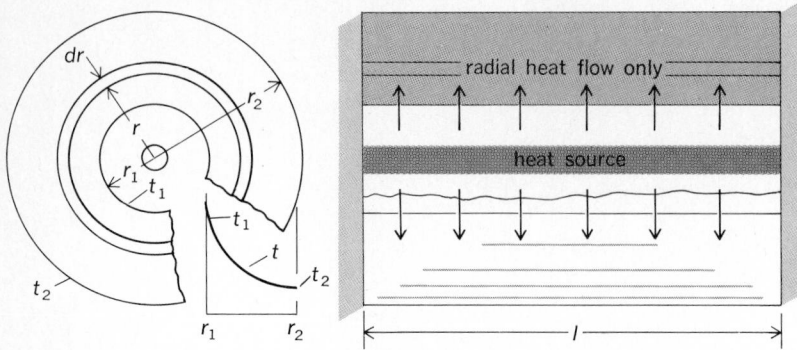
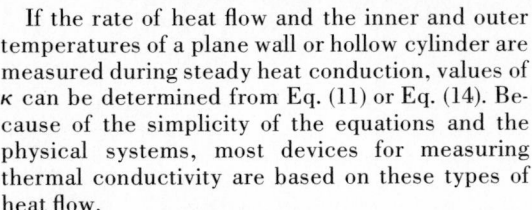

Fig. 6.   Steady-state conduction. Diagrams show the temperature distribution for steady radial heat flow through a circular cylinder wall.

in one field may be obtained from an analogous system in another. *See* POTENTIALS.

Steady-state conduction. When the temperature at all locations is constant with time or, mathematically, when $\partial t/\partial \tau = 0$, steady-state conduction exists. Many important practical problems fall in this category, the most familiar being heat flow through a wall and a hollow cylinder. Referring to Fig. 1 and assuming no heat generation within the wall, Eq. (8) reduces to Eq. (9). The terms $\partial^2 t/\partial y^2$

$$\frac{\partial^2 t}{\partial x^2} = 0 \qquad (9)$$

and $\partial t/\partial z^2$ are eliminated, since t is considered to vary only with x. The desired temperature distribution is obtained by integrating Eq. (9) twice and evaluating the two constants from the known temperatures at $x = 0$ and $x = l$. This leads to Eq. (10).

$$t = t_1 - (t_1 - t_2)\frac{x}{l} \qquad (10)$$

CONDUCTION (HEAT)

Application of Eq. (2) to obtain q yields Eq. (11).

$$q = \kappa A \frac{(t_1 - t_2)}{l} \qquad (11)$$

In the case of steady radial heat flow through a cylindrical wall (Fig. 6), Eq. (2) is applicable to any imaginary thin annular ring in the wall; thus Eq. (12) is obtained. Integration from $t = t_1$ to t, and r_1

$$q = -\kappa A \frac{dt}{dr} = -\kappa 2\pi rl \frac{dt}{dr} \qquad (12)$$

to r leads to Eq. (13), which indicates that the

$$t = t_1 - q\frac{\ln(r/r_1)}{2\pi\kappa l} \qquad (13)$$

effect of the increasing area for heat flow is to produce a logarithmic variation in the temperature, as shown by the curve in the left part of Fig. 6.

Heat flow through a cylindrical wall is usually expressed in the same form as Eq. (11), written as Eq. (14). Solving Eq. (13) for q and substituting in

$$q = \kappa A_m \frac{t_1 - t_2}{r_2 - r_1} \qquad (14)$$

Eq. (14) shows that the appropriate value for A_m must be as given in Eq. (15), which is called the logarithmic-mean area.

$$A_m = \frac{A_2 - A_1}{\ln(A_2/A_1)} \qquad (15)$$

If the rate of heat flow and the inner and outer temperatures of a plane wall or hollow cylinder are measured during steady heat conduction, values of κ can be determined from Eq. (11) or Eq. (14). Because of the simplicity of the equations and the physical systems, most devices for measuring thermal conductivity are based on these types of heat flow.

Interface resistance. Now consider steady-state heat flow through a wall composed of two or more layers of material, each with different uniform thermal properties. If the surfaces of the various layers are very smooth and in very good contact with each other, the temperature distribution will be continuous. At any interface (Fig. 7a), since q is constant, Eq. (16) holds. This shows that

$$\kappa_1\left(\frac{dt}{dx}\right)_1 = \kappa_2\left(\frac{dt}{dx}\right)_2 \qquad (16)$$

there is a discontinuity in the temperature gradient due to the change in κ.

Actual surfaces, even polished ones, are not smooth but have small projections and depressions. Consequently, when two surfaces are brought together, contact occurs primarily at projecting spots, as illustrated in Fig. 7b. The resulting contact area is only a small fraction of the nominal contact area. Plastic deformation at the points of contact of one or both materials usually occurs with the application of force to hold them together. Heat flows through both the small contact areas and the substance (usually a gas or liquid) filling the voids between the contacting protuberances.

The impairment to the heat-flow path caused by this imperfect contact is referred to as contact resistance. This is determined by extrapolating the measured temperature distribution in each material to the apparent interface location (Fig. 7b). The quotient of the resulting temperature difference Δt_i thus determined and the heat flux defines an interface resistance $R_i = \Delta t_i/q$. R_i depends on the roughnesses of the surfaces, the gas or liquid filling the voids, and the contact pressure. In general, the effect of contact resistance is significant only at low (for example, below 100 lb/in.²) interface pressures.

Internally generated heat. Conduction of heat generated internally occurs, for example, in the fuel elements used in nuclear reactors. Many of these elements are essentially long, flat plates over which a coolant flows. It is usually necessary to clad the fissionable material to prevent corrosion and keep radioactive particles from entering the coolant (Fig. 8). This cladding is undesirable from a heat-transfer standpoint, and is made as thin as possible and of the best adaptable heat-conducting material. Consequently, the temperature drop through it is usually small.

To illustrate the effect of the heat generation, assume that it is uniform with space and time in the radioactive material of Fig. 8 [that is, G in Eq. (7) is constant]. Heat flow from the ends of the elements and parallel to the flow direction of the coolant is negligible.

Therefore, Eq. (7) reduces to Eq. (17). A solution

$$\frac{d^2 t}{dx^2} + \frac{G}{\kappa} = 0 \qquad (17)$$

for t is obtained by integrating twice and applying the boundary conditions (considering cooling to be

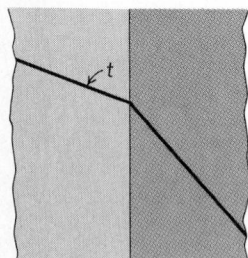

Fig. 7. Temperature distribution through composite wall. (*a*) With perfect interface contact. (*b*) For typical actual surfaces.

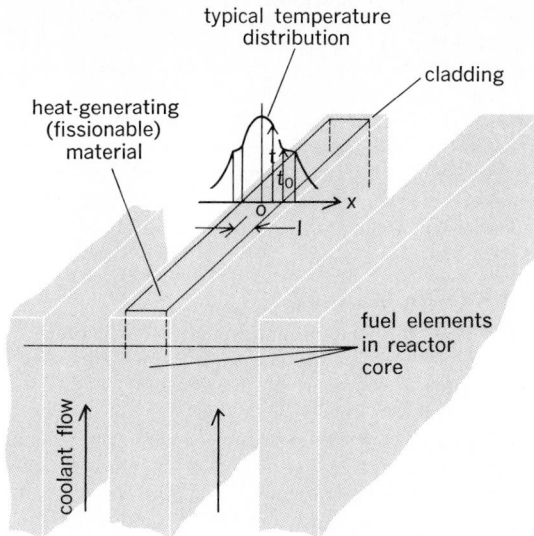

Fig. 8. Temperature distribution in reactor fuel plate.

the same on each side) $dt/dx = 0$ at $x = 0$ and $t = t_0$ at $x = l$. The result shows the temperature distribution to be parabolic, as in Eq. (18). The rate of

$$t = t_0 + \frac{G}{2\kappa}(l^2 - x^2) \qquad (18)$$

heat transfer to the coolant is obtained by differentiating Eq. (18) with respect to x, evaluating at $x = l$, and substituting in Eq. (2).

Since the heat generation rate is frequently dependent on the temperature, G will be more complex, possibly making an analytical solution impossible. Numerical methods of solution are then employed.

Periodic and transient conduction. These are the two kinds of non-steady-state heat flow. Periodic means a quasi-steady-state condition in which the temperature and heat flow at any location in a body vary continuously with time, but pass through the same series of values in a definite period of time, τ_0. A transient state results when the heat flow at any location is momentarily or permanently changed. The duration of the transient period is the time required for the system to return to its original or a new steady-state condition. A transient change may be superimposed on a periodic variation.

Restricting consideration to examples in which no heat generation is present, the fundamental differential equation to be satisfied for either periodic or transient heat flow is Eq. (8). An interesting application is the temperature variation in the Earth due to the diurnal temperature variation, or to a sudden change in surface temperature. Equation (8) reduces in this case to Eq. (19). Assume the

$$\alpha \frac{\partial^2 t}{\partial x^2} = \frac{\partial t}{\partial \tau} \qquad (19)$$

existence of a mean temperature of the Earth which is invariable with depth and that its surface temperature has been varying in a steady periodic manner long enough so that the original transient state due to starting the cyclic surface temperature has reached a steady periodic condition. If the surface temperature variation is given by Eq. (20),

$$t = t_0 \cos 2\pi \tau/\tau_0 \qquad (20)$$

the appropriate solution to Eq. (19) is given by Eq. (21). This result shows that the temperature distri-

$$t = t_0 e^{-\sqrt{\pi/2\alpha}\, x} \cos\left(2\pi\frac{\tau}{\tau_0} = \sqrt{\pi/2\alpha}\, x\right) \qquad (21)$$

bution looks like a wave traveling into the medium with the amplitude decreasing as the factor $e^{-\sqrt{\pi/2\alpha}\, x}$. Computation of the heat flow by determining $(\partial t/\partial t)_{x=0}$ from Eq. (21) indicates that heat flows in during one-half of the period τ_0 and out during the other half.

Calculations of the temperature distribution in the Earth over an 8-hr interval, using Eq. (21), are shown in Fig. 9. In this case the surface temperature varied from 29 to 41°F over a 24-hour period; the diffusivity of the soil was 0.0065 ft²/hr.

An illustration of a transient state is the temperature variation resulting from a sudden change of magnitude t_0 in the temperature at the surface of the Earth. When the initial temperature throughout is uniform and taken as the datum, the applicable solution to Eq. (19) is given by Eq. (22).

$$t = \frac{2t_0}{\sqrt{\pi}} \int_0^{\frac{x}{2\sqrt{\alpha\tau}}} e^{-\beta^2}\, d\beta \qquad (22)$$

If a body of soil in the Earth's surface is at a uniform temperature of 40°F and the surface temperature suddenly drops to 20°F, Eq. (22) can be applied to determine the depth at which the temperature will have dropped to freezing (32°F) in 12 hr. Taking $\alpha = 0.0065$ ft²/hr, $t = 32$°F after 12 hr is found to occur at $x = 0.27$ ft $= 2.5$ in.

Transient temperature changes are of fundamental importance in many industrial fields such as heat-treating, air conditioning, and food processing. This has stimulated the development of solutions to a large variety of problems, the use of many of which has been facilitated by presentation in graphical form.

Variable thermal conductivity. Most materials are sufficiently homogeneous so that their thermal properties are independent of position. Assuming that they vary only with temperature, the heat balance on the element of Fig. 4 leads to Eq. (23).

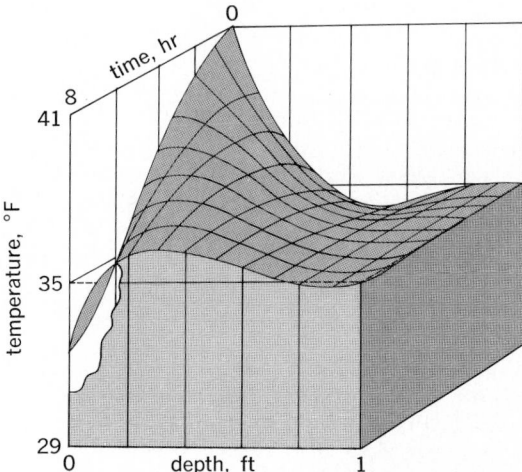

Fig. 9. Diagram of the temperature distribution in the ground for a daily periodic surface variation.

$$\frac{\partial}{\partial x}\left(\kappa\frac{\partial t}{\partial x}\right)+\frac{\partial}{\partial y}\left(\kappa\frac{\partial t}{\partial y}\right)+\frac{\partial}{\partial z}\left(\kappa\frac{\partial t}{\partial z}\right)+G=wc\frac{\partial t}{\partial\tau} \quad (23)$$

Upon carrying out the indicated differentiation, Eq. (23) becomes Eq. (24). By introducing a new

$$\kappa\left[\frac{\partial^2 t}{\partial x^2}+\frac{\partial^2 t}{\partial y^2}+\frac{\partial^2 t}{\partial z^2}\right]$$

$$+\frac{\partial\kappa}{\partial t}\left[\left(\frac{\partial t}{\partial x}\right)^2+\left(\frac{\partial t}{\partial y}\right)^2+\left(\frac{\partial t}{\partial z}\right)^2\right]+G=wc\frac{\partial t}{\partial\tau} \quad (24)$$

variable θ called the conductivity potential, notation (25), Eq. (24) can be simplified to Eq. (26).

$$\theta=\int_0^t \kappa\,dt \quad (25)$$

$$\frac{\partial^2\theta}{\partial x^2}+\frac{\partial^2\theta}{\partial y^2}+\frac{\partial^2\theta}{\partial z^2}+G=\frac{1}{\alpha}\frac{\partial\theta}{\partial\tau} \quad (26)$$

Eq. (26) has exactly the same form as Eq. (7). In many cases the variation of α with temperature (Fig. 5) is less than that of κ so that a mean constant value may be selected. This is true, for example, of metals at temperatures near absolute zero. In such cases, if G is not a function of t (or no heat source is present), solutions for constant α will apply with θ replacing t, provided that the boundary conditions are specified in terms of t or $\kappa(\partial t/\partial n)$ where n represents the variables x, y, z.

The utility of the conductivity potential is demonstrated in the calculation of the heat leak to a liquid-nitrogen tank through a support rod. Consider steady-state conditions with the exposed end of the rod at 300 K and the other at 77.3 K. Also assume there are no losses from the side of the rod which is 1 cm² in cross section, is 15 cm long, and is made of stainless steel (for which κ decreases from 0.15 at 300 K to 0.08 watt/cm K at 77.3 K). At any location q is given by Eq. (27). Since q is constant, Eq. (28) holds. Then Eq. (29) follows.

$$q=-A\,\kappa\,dt/dx \quad (27)$$

$$\frac{q}{A}\int_0^L dx=-\int_{t_1}^{t_2}\kappa\,dt=\int_0^{t_1}\kappa\,dt-\int_0^{t_2}\kappa\,dt \quad (28)$$

$$\frac{qL}{A}=\theta_1-\theta_2$$
$$q=A\frac{\theta_1-\theta_2}{L}=1\frac{30-3.3}{15}=1.78 \text{ watts} \quad (29)$$

[WARREN H. GIEDT]

Bibliography: V. S. Arpaci, *Conduction Heat Transfer,* 1966; R. Berman, *Thermal Conduction in Solids,* 1976; H. S. Carslaw and J. C. Jaeger, *Conduction of Heat in Solids,* 2d ed., 1959; J. P. Holman, *Heat Transfer,* 4th ed., 1976; M. N. Ozisik, *Heat Conduction,* 1979; W. M. Rohsenow and J. P. Hartnett (eds.), *Handbook of Heat Transfer,* 1973; J. R. Welty, *Engineering Heat Transfer, SI Version,* 1978.

Conductivity

A measure of the ability of a material to conduct electric current. It is the reciprocal of resistivity. Conductivity is commonly expressed as mhos per meter, since the unit of resistivity is the ohm-meter. The conductivity of metallic elements varies inversely with absolute temperature over the normal range of temperatures, but at temperatures approaching absolute zero the imperfections and impurities in the lattice structure of a material make the relationship more complicated. For thermal conductivity *see* CONDUCTION (HEAT).

The conductivity associated with conduction electrons in a semiconductor is known as *n*-type conductivity; that associated with the holes in an impurity semiconductor (equivalent to positive charges) is known as *p*-type conductivity. *See* SEMICONDUCTOR.   [JOHN MARKUS]

Conformal mapping

A special operation in mathematics in which a point set in one coordinate system is mapped or transformed into a corresponding point set in another coordinate system.

A mapping or transformation of a point set E in the xy plane is merely a correspondence $u=u(x,y)$, $v=v(x,y)$ which makes each point of E correspond to some point (u,v) of the uv plane. If distinct points (x,y) are transformed into distinct points (u,v), the mapping is one-to-one. A mapping is conformal if it is one-to-one and if small triangles are transformed into small triangles approximately similar to the original triangles, the approximation to similarity approaching perfection as the size of the triangles diminishes indefinitely. Since any small figure can be cut up into triangles, it follows that any small figure can be transformed into an approximately similar figure.

The following fact is decisive: The transformation defined by an analytic function whose derivative is different from zero is conformal. Let the transformation be $w=f(z)$, $z=x+iy$, $w=u+iv$, which is one-to-one in suitable neighborhoods of $z=z_0$ and $w=w_0=f(z_0)$, for example. Let two smooth curves C_1 and C_2 intersect at z_0; their images C'_1 and C'_2 intersect at w_0 (Fig. 1). If z_1 and z_2 lie on C_1 and C_2, respectively, their images w_1 and w_2 lie on C'_1 and C'_2. The angles of the rectilinear triangles $z_0z_1z_2$ and $w_0w_1w_2$ at z_0 and w_0 are, respectively,

$$\arg\frac{z_2-z_0}{z_1-z_0} \quad \text{and} \quad \arg\frac{w_2-w_0}{w_1-w_0}$$

where arg signifies the argument or amplitude of the complex number, and the ratios of the sides of the triangles adjoining z_0 and w_0 are

$$\left|\frac{z_2-z_0}{z_1-z_0}\right| \quad \text{and} \quad \left|\frac{w_2-w_0}{z_1-z_0}\right|$$

The assumed existence of $f'(z_0)$ asserts the relationship in Eq. (1). This determines Eq. (2). Thus,

$$\lim_{z_1\to z_0}\frac{w_1-w_0}{z_1-z_0}=\lim_{z_2\to z_0}\frac{w_2-w_0}{z_2-z_0}=f'(z_0)\neq 0 \quad (1)$$

$$\arg\frac{w_2-w_0}{w_1-w_0}-\arg\frac{z_2-z_0}{z_1-z_0}\to 0$$
$$\left|\frac{w_2-w_0}{w_1-w_0}\right|\bigg/\left|\frac{z_2-z_0}{z_1-z_0}\right|\to 1 \quad (2)$$

when the triangles are small, they are nearly similar in the sense that the angles at z_0 and w_0 are nearly equal, and the ratio of the adjacent sides of the one triangle is approximately equal to the corresponding ratio for the other triangle; these approximations approach perfection as the lengths of the sides of the triangles approach zero. *See* COM-

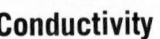

CONFORMAL MAPPING

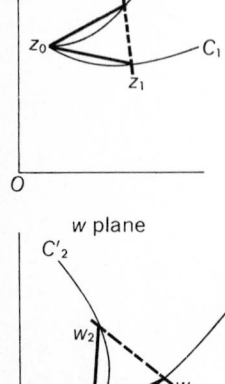

Fig. 1. Transformation of curves from z plane *to* w plane.

PLEX NUMBERS AND COMPLEX VARIABLES.

In the discussion just given, it follows that the angle between chords of C_1 and C_2 and the angle between chords of C'_1 and C'_2 approach equal limits; these limits are by definition the angles between the curves; so in a conformal transformation, angles between curves are preserved. Indeed, such angles are preserved algebraically as well as in magnitude.

It can also be shown that, conversely, a conformal map must be defined by an analytic function with nonvanishing derivative. This article deals only with the transformation of one plane region onto another; the term conformal applies also to the transformation of any surface onto another, such as the mapping of a portion of the Earth's surface onto a portion of a plane. The Mercator and stereographic projections are conformal in this same sense. Small regions are essentially unchanged in shape but not in relative size.

On the other hand, a map defined by an analytic function whose derivative is zero cannot be conformal. For instance, the transformation $w = z^2$ in the neighborhoods of $z = 0$ and $w = 0$ can be written $\Delta w = (\Delta z)^2$, so $\arg (\Delta w) = 2 \arg (\Delta z)$, $\arg (\Delta_1 w) = 2 \arg (\Delta_1 z)$; angles at $z = 0$ are doubled in transformation to the w plane.

Linear transformations. The transformation $w = \alpha z + \beta$, $\alpha \neq 0$ transforms every figure into a similar figure whose size is changed in the ratio $1 : |\alpha|$, for if $w_1 = \alpha z_1 + \beta$, $w_2 = \alpha z_2 + \beta$, then $w_2 - w_1 = \alpha(z_2 - z_1)$, whence $|w_2 - w_1| = |\alpha| \cdot |z_2 - z_1|$; also $\arg (w_2 - w_1) = \arg \alpha + \arg (z_2 - z_1)$, every line segment being rotated through the angle $\arg \alpha$.

The transformation $w = 1/\bar{z}$, where $\bar{z}$ indicates the conjugate of z, asserts essentially that $\arg w = \arg z$, $|w| = 1/|z|$ (Fig. 2). This transformation is not defined by an analytic function of z, and hence is not conformal. But the transformation is equivalent to the succession of transformations $z' = \bar{z}$, $w = 1/z'$. The first is merely a 180° rotation of the plane about the x axis and leaves all angles unchanged in magnitude but reversed in direction, and thus is inversely conformal; the second is conformal. Thus $w = 1/\bar{z}$ (called inversion in the unit circle) is also inversely conformal; it is one-to-one in the entire z plane and w plane except that $z = 0$ and $w = 0$ have no images. To avoid these exceptions, the plane may be "extended" by introducing ideal (or

fictitious) points "at infinity" $z = \infty$, $w = \infty$. When z is near 0, w is remote from $w = 0$; so $w = \infty$ can be considered the image of $z = 0$, and $w = 0$ the image of $z = \infty$. With this convention, the transformation is one-to-one in the extended plane. Angles between curves at infinity can be introduced either by studying limits of chords as one point of intersection recedes indefinitely or by projecting the plane stereographically onto a sphere by using a point of the sphere as center of projection (here the point at infinity in the plane projects into the center of projection). In either case it is also true that angles even at infinity are unchanged in magnitude by the transformations $w = 1/z$, $w = 1/\bar{z}$.

The general linear transformation of the complex variable is defined by Eq. (3), where α, β, γ, and δ

$$w = \frac{\alpha z + \beta}{\gamma z + \delta} \qquad \alpha \delta - \beta \gamma \neq 0 \qquad (3)$$

are constants; the purpose of the restriction on their values is merely to assure that the fraction does not reduce identically to a constant. This transformation is readily shown to be one-to-one in the extended plane. One of the important properties of such a transformation is that it leaves invariant the cross ratio of any four distinct points. If the points are z_1, z_2, z_3, z_4, the cross ratio is defined as in Eq. (4), with a suitable convention when

$$(z_1, z_2, z_3, z_4) \equiv \frac{(z_1 - z_2)(z_3 - z_4)}{(z_2 - z_3)(z_4 - z_1)} \qquad (4)$$

a point is at infinity; and if the image points are w_1, w_2, w_3, w_4 (any one of which may be at infinity), it is a matter of direct verification to prove $(w_1, w_2, w_3, w_4) = (z_1, z_2, z_3, z_4)$.

If four points lie on a circle, their cross ratio is real, as is expressed by Eq. (5), and is proved by

$$\arg \frac{z_2 - z_1}{z_2 - z_3} - \arg \frac{z_4 - z_1}{z_4 - z_3} = 0 \text{ or } \pi \qquad (5)$$

the euclidean properties of angles inscribed in a circle; conversely, if the cross ratio is real, the four points are either concyclic or collinear. For simplicity the concept of circle is enlarged to include straight line. If three distinct points z_1, z_2, z_3 are given, the point z is concyclic with them when, and only when, the cross ratio (z_1, z_2, z_3, z) is real; here is included the case that z may coincide with a z_k. Under a linear transformation the points z_1, z_2, z_3 go over into distinct points w_1, w_2, w_3, and z goes into w with preservation of the cross ratio. The latter is real when, and only when, z_1, z_2, z_3, z are concyclic, which occurs when, and only when, (w_1, w_2, w_3, w) is real, namely when, and only when, w_1, w_2, w_3, w are concyclic. Thus a linear transformation carries every "circle" into a "circle."

There exists a linear transformation which carries the unit disk into itself and any given point α of the disk into the origin, as shown in Eq. (6).

$$w = \lambda \frac{z - \alpha}{1 - \bar{\alpha} z} \qquad |\lambda| = 1 \qquad (6)$$

With $z\bar{z} = 1$, Eq. (7) holds. The point α with $|\alpha| <$

$$w\bar{w} = \frac{z - \alpha}{1 - \bar{\alpha} z} \frac{\bar{z} - \bar{\alpha}}{1 - \alpha \bar{z}} = \frac{z - \alpha}{1 - \bar{\alpha} z} \frac{1 - \bar{\alpha} z}{z - \alpha} = 1 \qquad (7)$$

1 is carried into $w = 0$, so by the continuity of the transformation $|z| < 1$ is carried into $|w| < 1$. It

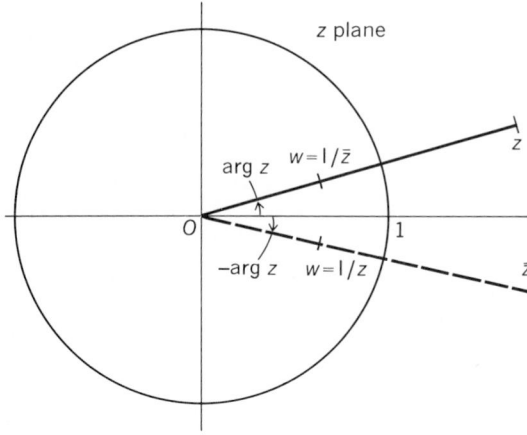

Fig. 2. An inversely conformal transformation.

can be shown that no other linear transformation has the property required.

Riemann's theorem. Trivial, simply connected regions of the plane are the plane of finite points and the extended plane; neither of these can be mapped one-to-one and conformally onto the unit disk. G. F. B. Riemann asserted (the fact was first proved by W. F. Osgood) that essentially any other simply connected region D (one having at least two boundary points) can be so mapped. The proof depends on extremal properties of univalent functions, namely, analytic functions assuming no value more than once in D. The most general of such maps is given by combination with the linear transformations just discussed. If, in particular, D is a Jordan region, the map may be defined so as to be one-to-one and continuous in the closed regions.

Multiply connected regions. If a region D is bounded by a finite number $p(>1)$ of mutually disjoint Jordan curves, it cannot be mapped one-to-one onto a circular disk. Various canonical regions have been proposed for the study of such regions and comparison of two such regions which may or may not be mappable onto each other. Such canonical regions may be bounded by p circles; by p parallel rectilinear slits; by slits of arcs of circles whose center is the origin; by a circle whose center is the origin plus such slits; by two circles whose common center is the origin plus such slits; by loci

$$\left| \frac{A(z-a_1)^{m_1}(z-a_2)^{m_2}\cdots(z-a_\mu)^{m_\mu}}{(z-b_1)^{n_1}(z-b_1)^{n_2}\cdots(z-b_\nu)^{n_\nu}} \right| = 1 \text{ and } B;$$

by regions bounded by arcs of logarithmic spirals; and by numerous other kinds of prescribed boundaries.

Riemann surfaces. When the map of a sector $0 < \arg z < \alpha$ is studied under the transformation $w = z^2$, it is seen that for small α the image is the sector $0 < \arg w < 2\alpha$, and that as α increases the image sector also increases (Fig. 3). When α passes through the value π, the boundary of the image sector passes through the value 2π and continues to increase beyond 2π, say, in a new sheet over the w plane, in the sense that as new points z occur it is natural to continue to adjoin new points w. But when the sector in the z plane reaches the angle 2π, it abuts on values of z already considered, so it is natural to "heal" the boundaries of the sector of angle 4π in the w plane by passing one sheet through another (say, along the positive half of Ox) and joining the edges. There is thus formed a Riemann surface over the w plane, which is useful for studying the behavior of $z = \sqrt{w}$ as a function of w with reference to the two branches of the function, their continuity, and their passing (along various w curves) continuously from one into the other.

There exists more generally a Riemann surface of n sheets for the inverse of any irreducible polynomial function $w = a_0 z^n + a_1 z^{n-1} + \cdots + a_n$. Moreover, the exponential function $w = e^z = e^{x+iy}$ can be defined as $e^x(\cos y + i \sin y)$ for complex z, whose analyticity for all z can be verified by the Cauchy-Riemann equations. The inverse function, Eq. (8), is infinitely many valued; its Riemann surface consists of infinitely many sheets over the w

$$z = \log w = \log[r(\cos\theta + i\sin\theta)] = \log r + i\theta \quad (8)$$

plane, each connected (for instance) along the positive half of Ox to its predecessor along one edge and to its successor along the other.

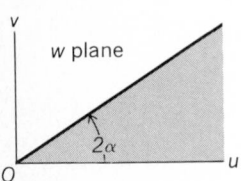

CONFORMAL MAPPING

Fig. 3. Formation of a Riemann surface.

Riemann surfaces may be defined abstractly also; their classification and properties have occupied mathematicians for more than a century, and unsolved problems remain abundant. *See* TOPOLOGY.

[JOSEPH L. WALSH]

Bibliography: L. Bieberback, *Conformal Mapping*, 1952; C. Caratheodory, *Conformal Representation*, 1952; Z. Nehari, *Conformal Mapping*, 1975.

Conservation of energy

The principle of conservation of energy states that energy cannot be created or destroyed, although it can be changed from one form to another. Thus in any isolated or closed system, the sum of all forms of energy remains constant. The energy of the system may be interconverted among many different forms—mechanical, electrical, magnetic, thermal, chemical, nuclear, and so on—and as time progresses, it tends to become less and less available; but within the limits of small experimental uncertainty, no change in total amount of energy has been observed in any situation in which it has been possible to ensure that energy has not entered or left the system in the form of work or heat. For a system that is both gaining and losing energy in the form of work and heat, as is true of any machine in operation, the energy principle asserts that the net gain of energy is equal to the total change of the system's internal energy. *See* THERMODYNAMIC PRINCIPLES.

Application to life processes. The energy principle as applied to life processes has also been studied. For instance, the quantity of heat obtained by burning food equivalent to the daily food intake of an animal is found to be equal to the daily amount of energy released by the animal in the forms of heat, work done, and energy in the waste products. (It is assumed that the animal is not gaining or losing weight.) Studies with similar results have also been made of photosynthesis, the process upon which the existence of practically all plant and animal life ultimately depends.

Conservation of mechanical energy. There are many other ways in which the principle of conservation of energy may be stated, depending on the intended application. Examples are the various methods of stating the first law of thermodynamics, the work-kinetic energy theorem, and the assertion that perpetual motion of the first kind is impossible. Of particular interest is the special form of the principle known as the principle of conservation of mechanical energy (kinetic E_k plus potential E_p) of any system of bodies connected together in any way is conserved, provided that the system is free of all frictional forces, including internal friction that could arise during collisions of the bodies of the system. Although frictional or other nonconservative forces are always present in any actual situation, their effects in many cases are so small that the principle of conservation of mechanical energy is a very useful approximation. Thus for a missile or satellite traveling high in space, the dissipative effects arising from such sources as the residual air and meteoric dust are so exceedingly small that the loss of mechanical energy $E_k + E_p$ of the body as it proceeds along its trajectory may, for many purposes, be disregarded. *See* ENERGY.

Mechanical equivalent of heat. The mechanical energy principle is very old, being directly derivable as a theorem from Newton's law of motion. Also very old are the notions that the disappearance of mechanical energy in actual situations is always accompanied by the production of heat and that heat itself is to be ascribed to the random motions of the particles of which matter is composed. But a really clear conception of heat as a form of energy came only near the middle of the 19th century, when J. P. Joule and others demonstrated the equivalence of heat and work by showing experimentally that for every definite amount of work done against friction there always appears a definite quantity of heat. The experiments usually were so arranged that the heat generated was absorbed by a given quantity of water, and it was observed that a given expenditure of mechanical energy always produced the same rise of temperature in the water. The resulting numerical relation between quantities of mechanical energy and heat is called the Joule equivalent, or mechanical equivalent of heat. The present accepted value is one $15°$ calorie $= 4.1855 \pm 0.0004$ joules.

Conservation of mass-energy. In view of the principle of equivalence of mass and energy in the restricted theory of relativity, the classical principle of conservation of energy must be regarded as a special case of the principle of conservation of mass-energy. However, this more general principle need be invoked only when dealing with certain nuclear phenomena or when speeds comparable with the speed of light (3×10^{10} cm/sec) are involved.

If the mass-energy relation, $E = mc^2$, where c is the speed of light, is considered as providing an equivalence between energy E and mass m in the same sense as the Joule equivalent provides an equivalence between mechanical energy and heat, there results the relation, 1 kg $= 9 \times 10^{16}$ joules. *See* INERTIA OF ENERGY.

Laws of motion. The law of conservation of energy has been established by thousands of meticulous measurements of gains and losses of all known forms of energy. It is now known that the total energy of a properly isolated system remains constant. Some parts or particles of the system may gain energy but others must lose just as much. The actual behavior of all the particles, and thus of the whole system, obeys certain laws of motion. These laws of motion must therefore be such that the energy of the total system is not changed by collisions or other interactions of its parts. It is a remarkable fact that one can test for this property of the laws of motion by a simple mathematical manipulation that is the same for all known laws: classical, relativistic, and quantum mechanical.

The mathematical test is as follows. Replace the variable t, which stands for time, by $t + a$, where a is a constant. If the equations of motion are not changed by such a substitution, it can be proved that the energy of any system governed by these equations is conserved. For example, if the only expression containing time is $t_2 - t_1$, changing t_2 to $t_2 + a$ and t to $t_1 + a$ leaves the expression unchanged. Such expressions are said to be invariant under time displacement. When daylight-saving time goes into effect, every t is changed to $t + 1$ hr. It is unnecessary to make this substitution in any known laws of nature, because they are all invariant under time displacement.

Without such invariance laws of nature would change with the passage of time, and repeating an experiment would have no clear-cut meaning. In fact, science, as it is known today, would not exist.

[DUANE E. ROLLER/LEO NEDELSKY]

Bibliography: K. R. Atkins, *Physics*, 3d ed., 1976; D. Halliday and R. Resnick, *Physics*, 3d ed., 1978; G. Laundry et al., *Physics: An Energy Introduction*, 1979; F. W. Sears et al. *University Physics*, 5th ed., 1976; E. P. Wigner, Symmetry and conservation laws, *Phys. Today*, 17(3):34–40, March 1964.

Conservation of mass

The notion that mass, or matter, can be neither created nor destroyed. According to conservation of mass, reactions and interactions which change the properties of substances leave unchanged their total mass; for instance, when charcoal burns, the mass of all of the products of combustion, such as ashes, soot, and gases, equals the original mass of charcoal and the oxygen with which it reacted.

The special theory of relativity of Albert Einstein, which has been verified by experiment, has shown, however, that the mass of a body changes as the energy possessed by the body changes. Such changes in mass are too small to be detected except in subatomic phenomena. Furthermore, matter may be created, for instance, by the materialization of a photon (quantum of electromagnetic energy) into an electron-positron pair; or it may be destroyed, by the annihilation of this pair of elementary particles to produce a pair of photons. *See* ELECTRON-POSITRON PAIR PRODUCTION; LIGHT; MASS; RELATIVITY.

[LEO NEDELSKY]

Conservation of momentum

The principle that, when a system of masses is subject only to forces that masses of the system exert on one another, the total vector momentum of the system is constant. Since vector momentum is conserved, in problems involving more than one dimension the component of momentum in any direction will remain constant. The principle of conservation of momentum holds generally and is applicable in all fields of physics. In particular, momentum is conserved even if the particles of a system exert forces on one another or if the total mechanical energy is not conserved.

Use of the principle of conservation of momentum is fundamental in the solution of collision problems. *See* COLLISION.

If a man standing on a well-lubricated cart steps forward, the cart moves backward. One can explain this result by momentum conservation, considering the system to consist of cart and man. If both man and cart are originally at rest, the momentum of the system is zero. If the man then acquires forward momentum by stepping forward, the cart must receive a backward momentum of equal magnitude in order for the system to retain a total momentum of zero.

When the principle of conservation of momentum is applied, care must be taken that the system under consideration is really isolated. For example, when a rough rock rolls down a hill, the isolated system would have to consist of the rock plus the earth, and not the rock alone, since momentum exchanges between the rock and the earth cannot be neglected.

Rocket propulsion. The propulsion of a rocket through space can be explained in terms of momentum conservation. Hot gases produced by the combustion of the fuel are expelled at high speed from the rear of the rocket. Although the total mass of these hot gases may not be large, the gases move with such a high velocity that the total momentum associated with them is appreciable. The momentum of the gases is directed backward. For momentum to be conserved, the rocket must acquire an equal momentum in the forward direction. If the rocket carries all the materials needed for the combustion of its fuel, its propulsion does not require air, and it can move through empty space.

Exploding bomb. An exploding bomb gives another application of the conservation of momentum. The total resultant vector momentum of all the pieces of the bomb immediately after explosion must equal the momentum of the unexploded bomb just before the explosion.

Proof of principle. The principle of conservation of momentum follows directly from Newton's second and third laws. While the principle will be proved here only for the straight-line motion of a two-particle system, it can be generalized to systems containing any number of particles. A particle is a mass with dimensions so small that rotational effects are negligible. Momentum will also be conserved for rigid bodies large enough that rotation must be considered, since rigid bodies can be treated as assemblies of many particles.

For the one-dimensional motion of an isolated two-particle system, Newton's third law states that the force F_{12} that particle 1 exerts on particle 2 is equal in magnitude and opposite in direction to the force F_{21} that particle 2 exerts on particle 1. Thus Eq. (1) holds.

$$F_{21} = -F_{12} \qquad (1)$$

By use of Newton's second law this equation can be expressed in terms of the momenta $m_1 v_1$ and $m_2 v_2$ of particles 1 and 2, respectively, where m_1, m_2, v_1, and v_2 are the masses and velocities of particles 1 and 2, respectively. Then Eq. (2) holds.

$$m_1 \frac{dv_1}{dt} = -m_2 \frac{dv_2}{dt} \qquad (2)$$

Integration gives Eq. (3), where c is a constant.

$$m_1 v_1 + m_2 v_2 = c \qquad (3)$$

This equation expresses the conservation of momentum for two particles moving in the same straight line.

Finally it should be mentioned that angular and linear momentum are independent quantities. A complete description of a system must include both quantities. The angular momentum of a system is conserved under quite general conditions. *See* ANGULAR MOMENTUM; CONSERVATION OF ENERGY; MOMENTUM. [PAUL W. SCHMIDT]

Constraint

A restriction on the natural degrees of freedom of a system. If n and m are the numbers of the natural and actual degrees of freedom, the difference $n - m$ is the number of constraints. In principle $n = 3N$, where N is the number of particles, for example, atoms. In practice n is determined by the number of effectively rigid components.

A holonomic system is one in which the n original coordinates can be expressed in terms of m independent coordinates and possibly also the time. It is characterized by frictionless contacts and inextensible linkages. The new coordinates, q_1, q_2, . . . , q_m, are called generalized coordinates. The equations of equilibrium and of motion may be expressed in terms of these coordinates. *See* LAGRANGE'S EQUATIONS.

Nonholonomic systems cannot be reduced to independent coordinates because the constraints are not on the n coordinate values themselves but on their possible changes. For example, an ice skate may point in all directions but at each position it must point along its path. This is a condition between changes (dx, dy) in the two position coordinates (x, y) and the direction angle θ, as in Eq. (1).

$$dy = \tan \theta \, dx \qquad (1)$$

This cannot be put in an integrated form, as in Eq. (2), since a skater can pass in different directions repeatedly over a point.

$$f(x, y, \theta) = 0 \qquad (2)$$

The static equilibrium conditions of a constrained system under impressed forces F_1, F_2, and so on are contained in a general statement, the principle of virtual work or virtual displacement. *See* STATICS; VIRTUAL WORK PRINCIPLE.

A moving constraint is one which changes with time, as in the case of a system on a moving platform. Moving constraints differ from the stationary constraints of static equilibrium in being able to do work. For example, a mass constrained to move on the floor of a rising elevator is carried also in the direction of the (vertical) force of constraint which does work on it. [BERNARD GOODMAN]

Bibliography: H. C. Corben and P. Stehle, *Classical Mechanics*, 2d ed., 1974; H. Goldstein, *Classical Mechanics*, 2d ed., 1980; D. T. Greenwood, *Classical Dynamics*, 1977; L. D. Landau and E. M. Lifshitz, *Mechanics*, 3d ed., 1976; K. R. Symon, *Mechanics*, 1971.

Convection (heat)

The transfer of thermal energy by actual physical movement from one location to another of a substance in which thermal energy is stored. A familiar example is the free or forced movement of warm air throughout a room to provide heating. Technically, convection denotes the nonradiant heat exchange between a surface and a fluid flowing over it. Although heat flow by conduction also occurs in this process, the controlling feature is the energy transfer by flow of the fluid—hence the name convection. Convection is one of the three basic methods of heat transfer, the other two being conduction and radiation. *See* CONDUCTION (HEAT); HEAT RADIATION; HEAT TRANSFER.

Natural convection. This mode of energy transfer is exemplified by the cooling of a vertical surface in a large quiescent body of air of temperature t_∞. As shown in Fig. 1a, the lower-density air next to a hot vertical surface moves upward because of the buoyant force of the higher-density cool air farther away from the surface. At any arbitrary vertical location x, the actual velocity variation parallel to the surface will be similar to that sketched in Fig. 1b, increasing from zero at the surface to a maximum, and then decreasing to zero as ambient surrounding conditions are reached. In contrast, the temperature of the air decreases from the heated wall value to the surrounding air temperature. These temperature and velocity distributions are clearly interrelated, and the distances from the wall through which they exist are coincident because, when the temperature approaches that of the surrounding air, the density difference causing the upward flow approaches zero.

The region in which these velocity and temperature changes occur is called the boundary layer. Because velocity and temperature gradients both approach zero at the outer edge, there will be no heat flow out of the boundary layer by conduction or convection. *See* BOUNDARY-LAYER FLOW.

An accounting of all of the energy streams entering and leaving a small volume in the boundary layer (Fig. 1c) during steady-state conditions yields Eq. (1), where κ, w and c_p are the thermal conduc-

$$\kappa\left(\frac{\partial^2 t}{\partial x^2}+\frac{\partial^2 t}{\partial y^2}\right)=wc_p\left(u\frac{\partial t}{\partial x}+v\frac{\partial t}{\partial y}\right) \qquad (1)$$

tivity, the weight density, and specific heat at constant pressure of the air, and u and v are the velocity components in the x and y directions, respectively. Equation (1) states that the net energy conducted into the element equals the increase in energy of the fluid leaving (convected) over what it had entering.

At the surface, $u=v=0$, and $\partial^2 t/\partial x^2=\partial t/\partial x=0$. Thus, $\kappa(\partial^2 t/\partial y^2)=0$ and Eq. (2) is obtained, where q

$$\kappa\frac{\partial t}{\partial y}=\text{constant}=q/A \qquad (2)$$

is the time rate of flow through an infinitesimally thin layer dx, and A is the cross-sectional area normal to the direction of flow. This shows that the heat transfer in the immediate vicinity of the wall is by conduction through a thin layer of air which does not move relative to the surface. At a very small distance from the surface, the velocity becomes finite and some of the energy conducted normal to the surface is convected parallel to it. This process causes the temperature gradient to decrease, eventually to zero. The solution of a problem requires determination of the tempera-

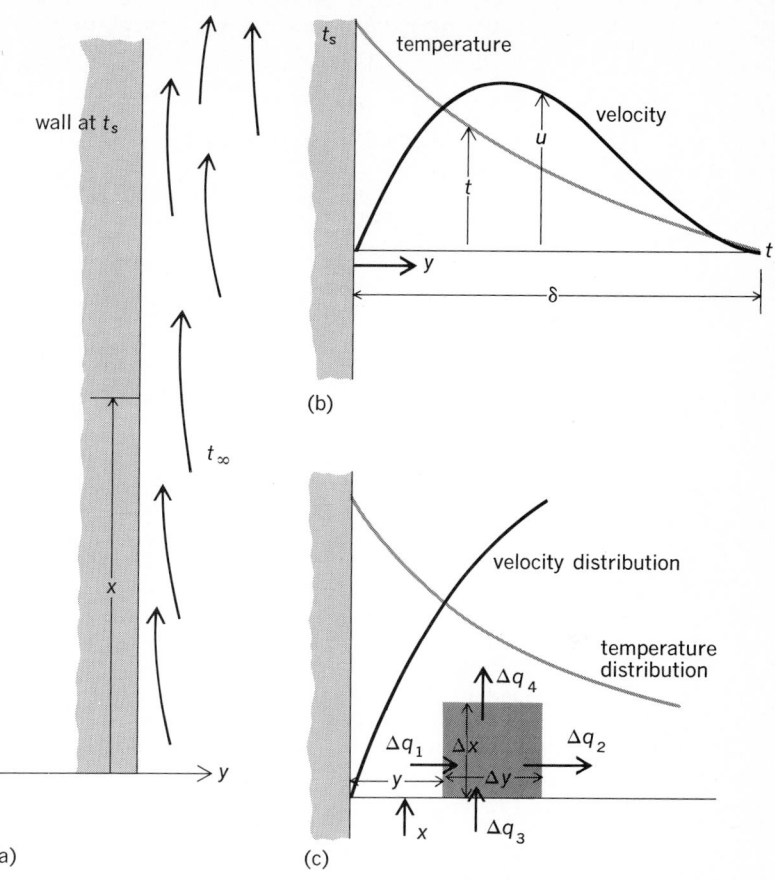

(a)   (b)   (c)

Fig. 1.   Temperature and velocity distributions in air near a heated vertical surface. (a) Upward movement of hot air. (b) Distributions at arbitrary vertical location. The distance δ is that distance at which the velocity and the temperature reach ambient surrounding conditions. (c) Distributions in boundary layer.

ture distribution throughout the boundary layer. From this, the temperature gradient at the wall and the rate of heat flow can be computed.

The effect of energy leaving a surface and remaining in the boundary layer is (1) a gradual increase in temperature of the air in this layer as it moves upward, and (2) diffusion of energy farther from the surface, entraining more air in (thickening) the boundary layer. This is shown in Fig. 2, where the interference fringes indicate lines of constant temperature. Figure 3 is a similar vis-

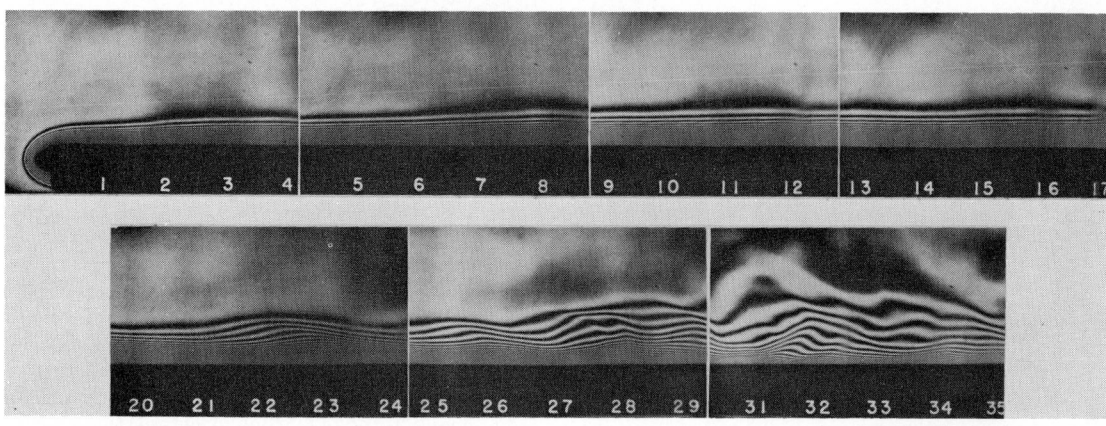

Fig. 2.   Laminar and turbulent natural convection flow along a vertical plate, as revealed by interference photographs. (*From E. R. G. Eckert and R. M. Drake, Jr., Heat and Mass Transfer, McGraw-Hill, 1959*)

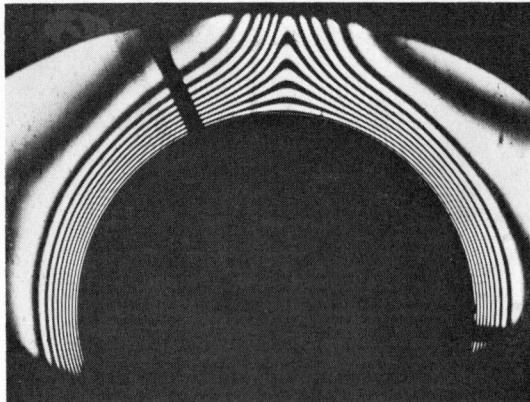

Fig. 3. Interference photograph of isotherms around heated horizontal cylinder (*From D. L. Doughty and W. H. Giedt, Proc. Instrum. Soc. Amer., 7:115–119, 1952*)

ualization of the free convection temperature field around a horizontal cylinder. The outer broad fringe indicates approximately the edge of the boundary layer which, by comparison with the cylinder (which measures 4 in. in diameter), is about 3/4 in. thick.

Forced convection. The effect of blowing air across the cylinder is shown in Fig. 4. Here the boundary layer on the forward half of the cylinder has become so thin that it is not possible to resolve the isotherms within it. Although the natural convection forces are still present in this latter case, they are clearly negligible compared with the imposed forces. The process of energy transfer from the heated surface to the air is not, however, different from that described for natural convection. The major distinguishing feature is that the maximum fluid velocity is at the outer edge of the boundary layer, as is illustrated in Fig. 5 for flow along a flat plate. This difference in velocity profile and the higher velocities provide more fluid near the surface to carry along the heat conducted normal to the surface. Consequently, boundary layers are very thin (greatly enlarged in Fig. 5 for clarity).

The properties of a fluid which influence its heat-convecting ability are the dynamic viscosity μ, heat capacity at constant pressure c_p, and ther-

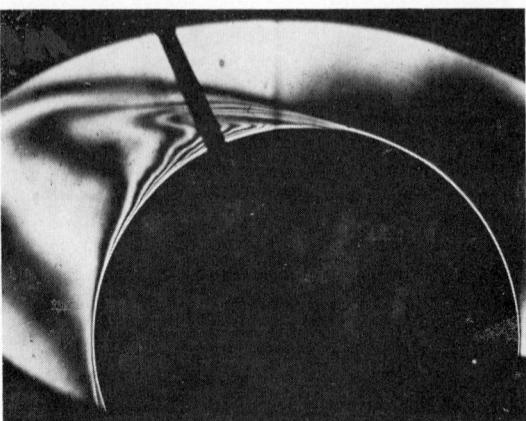

Fig. 4. Interference photograph showing isotherms around a heated cylinder normal to a 33 ft/sec airstream. (*From D. L. Doughty and W. H. Giedt, Proc. Instrum. Soc. Amer., 7:115–119, 1952*)

mal conductivity κ. These combine in a single significant property in the form $\mu c_p / \kappa = \mu / (\kappa / c_p)$, which is the ratio of the fluid viscosity to the quotient of its heat-conducting and -storage capacities. With proper units, this ratio is dimensionless and is called the Prandtl number, N_{Pr}. Fluids such as liquid metals, having low values of N_{Pr} (Fig. 6), are particularly effective for convective heat-transfer applications.

Heat-transfer coefficient. The convective heat-transfer coefficient h is a unit conductance used for calculation of convection heat transfer. It was introduced by Sir Isaac Newton, and until the mechanism of convection was properly interpreted, was thought to be a characteristic of the fluid flowing. To describe quantitatively the cooling of objects in air, Newton suggested Eq. (3). This

$$q = hA(t_{surface} - t_{fluid}) \qquad (3)$$

equation, known as Newton's law of cooling, is really a definition of h. This coefficient is determined by the slope of the fluid temperature distribution right at the surface, and the thermal conductivity κ of the fluid. Engineering units are Btu/(hr)(ft²)(°F).

Local and average coefficients. As the fluid in the heated or cooled boundary layer moves along an isothermal surface, it gradually approaches the temperature of the surface. This causes the temperature gradient in the fluid at the surface (and the rate of heat transfer) to decrease in the direction of flow. Taking, for example, an airstream at 80°F moving at 50 ft/sec over a flat plate at 30°F, the local heat-transfer coefficient h_x decreases in 1 ft to about 1/5 of its leading-edge value. For practical calculations, an average heat-transfer coefficient h is more useful. This is obtained by integrating h_x over the heat-transfer surface and dividing by the surface area. For this system, h is a function of the Prandtl number N_{Pr}, κ, ν, and u_∞ of the fluid. For comparing geometrically similar systems involving different fluids, dimensional analysis shows that the specific properties can be combined into dimensionless parameters, conveniently reducing the number of independent variables. For the flat plate, these are the Nusselt number, $N_{Nu} = hl/\kappa$, the Reynolds number, $N_{Re} = lu_\infty / \nu$, and the Prandtl number. For example, it can be shown that in this case Eq. (4) holds.

$$N_{Nu} = 0.664 N_{Pr}^{1/3} \sqrt{N_{Re}} \qquad (4)$$

In the case of free convection, the Grashof number, $N_{Gr} = \beta g l^3 \, \Delta t / \nu^2$ (where β is the coefficient of thermal expansion), replaces the Reynolds number. Other dimensionless numbers pertinent to convection include the Stanton number, $N_{St} = h/\rho u c_p = N_{Nu}/N_{Re}N_{Pr}$, which is related to the skin-friction coefficient. *See* DIMENSIONAL ANALYSIS; REYNOLDS NUMBER.

Turbulent flow. Heat convection in turbulent flow is interpreted similarly to that in laminar flow, which has been implied in previous paragraphs. Rates of heat transfer are higher for comparable velocities, however, because the fluctuating velocity components of the fluid in a turbulent flow stream provide a macroscopic exchange mechanism which greatly increases the transport of energy normal to the main flow direction. Because of the complexity of this type of flow, most of the information regarding heat transfer has been ob-

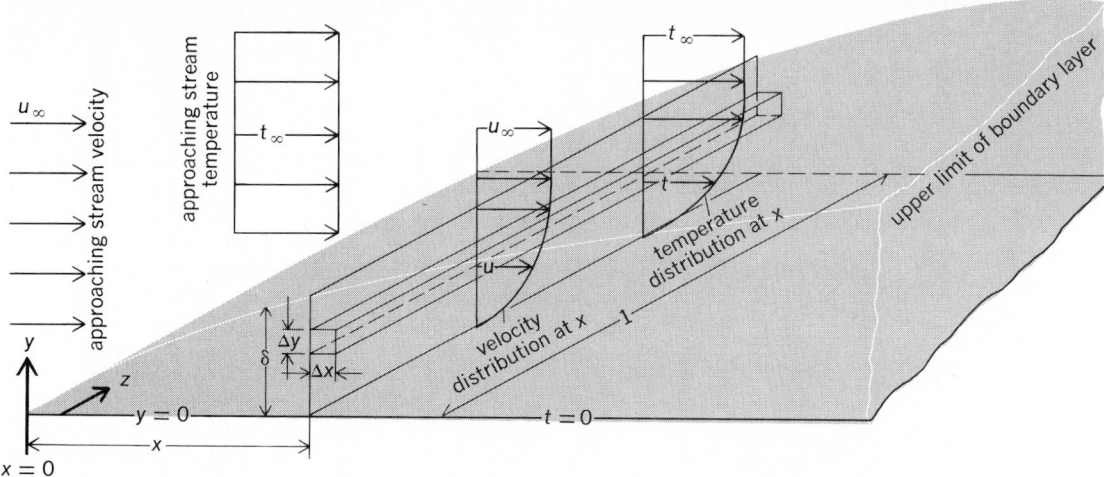

Fig. 5. Velocity and temperature distributions in a laminar boundary layer along a flat plate. The fluid is assumed to have a Prandtl number of 1, for which the velocity and temperature boundary layers will be of equal thickness. (*From W. H. Giedt, Principles of Engineering Heat Transfer, Van Nostrand, 1957*)

tained experimentally. Such results, combined with dimensional analysis, have yielded useful design equations, typical of which is Eq. (5), for predicting the heat transfer in a pipe, where d is the pipe diameter. *See* LAMINAR FLOW; TURBULENT FLOW.

$$\frac{hd}{\kappa} = 0.023 \left(\frac{du}{\nu}\right)^{0.8} \left(\frac{\mu c_p}{\kappa}\right)^{0.4} \quad (5)$$

Aerodynamic heating. Convection heat transfer which occurs during high-speed flight or high-velocity flow over a surface is known as aerodynamic heating. This heating effect results from the conversion of the kinetic energy of the fluid as it approaches a body to internal energy as it is slowed down next to the surface. In the case of a gas (Fig. 7), its temperature increases first, because of compression as it passes through a shock and approaches the stagnation region, and second, because of frictional dissipation of kinetic energy in the boundary layer along the surface. Typical velocity and temperature distributions near the surface of a missile are shown in Fig. 7. The maximum temperature possible would result from an adiabatic compression of the gas as it is slowed from the free stream velocity to zero. This maximum temperature depends on the square of the missile velocity, being, for example, around 1500°F for a speed of 4000 ft/sec (a Mach number of 4). Temperatures very close to the maximum may occur in the stagnation region; downstream maximum temperatures are lower because of decreasing pressure and heat convection toward the outer edge of the boundary layer.

Condensation and boiling. The phenomena of condensation and boiling are important phase-change processes involving heat release or absorption. Because vapor and liquid movement are present, the energy transfer is basically by convection. Local and average heat-transfer coefficients are determined and used in the Newton cooling-law equation for calculating heat rates which include the effects of the latent heat of vaporization.

Condensation. Consider a saturated or superheated vapor of a single substance in some region. When it comes in contact with a surface maintained at a temperature lower than the saturation temperature, heat flow results from the vapor, releasing its latent heat and condensing on the surface. The condensation process may proceed in two more or less distinct ways. If there are no impurities in the vapor or on the surface (which need not be smooth), the condensate will form a continuous liquid film. If, however, such contaminants as fatty acids or mercaptans are present, the vapor will condense in small droplets. These increase in size until their weight causes them to run down the surface. In doing so, they sweep the surface free

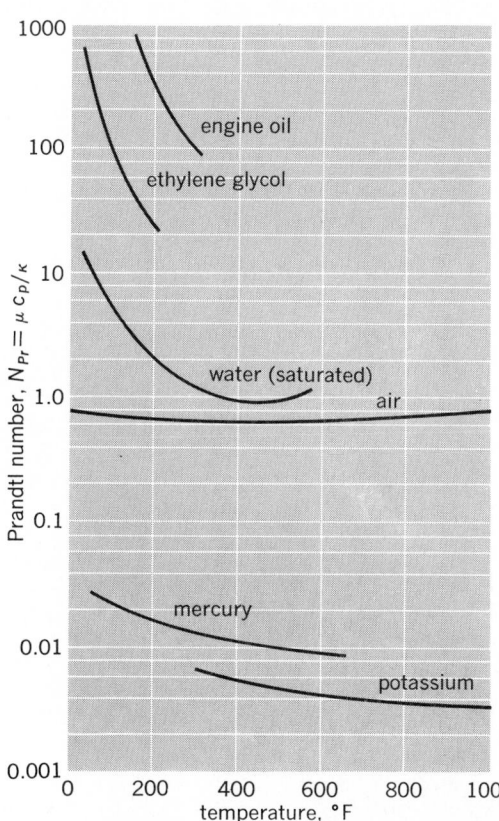

Fig. 6. Variation of Prandtl number with temperature.

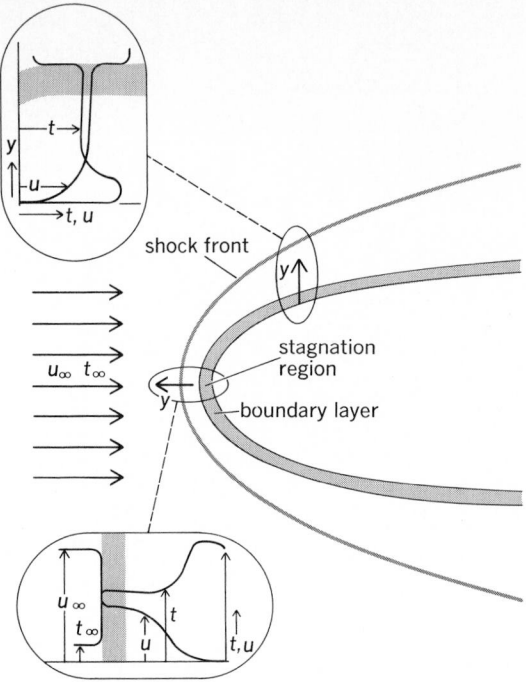

Fig. 7. Velocity and temperature distributions in the gas around a high-speed missile (thicknesses of shock wave and boundary layer exaggerated).

for formation of new droplets. For the same temperature difference between the vapor and surface, heat transfer with dropwise condensation may be 15–20 times greater than filmwise condensation. The dropwise type is therefore very desirable, but conditions under which it will occur are not predictable, and designs are limited to systems for which experimental results are available.

Boiling. In boiling, results indicate the existence of several regimes, as shown in Fig. 8. The important independent variable is the temperature difference between the hot surface and the fluid relatively far from the surface. For values of Δt up to approximately 10°F (A to B on the curve), the liquid is being superheated by natural convection, and q/A is proportional to $\Delta t^{5/4}$. With further increase in Δt(B to C), bubbles form at active nuclei

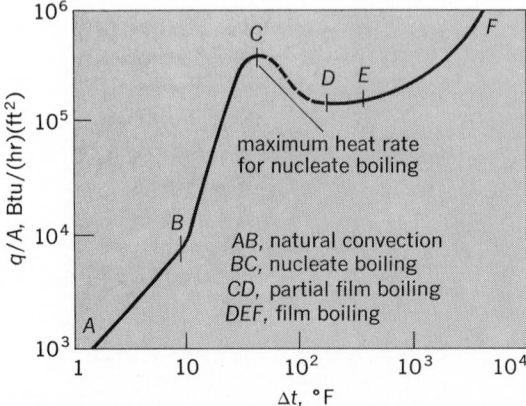

Fig. 8. Heat rate versus temperature difference during boiling of water at 212°F on an electrically heated platinum wire. (*After S. Nukiyama, J. Soc. Mech. Eng. Japan, 37(206):367–379, 1934*)

on the heated surface. These bubbles break away and rise through the pool, their stirring action causing the heat transfer to be much above that due to natural convection. This phenomenon is called nucleate boiling, and q/A varies as Δt^3 to Δt^4. When the rate of bubble formation becomes so rapid that the bubbles cannot get away before they tend to merge, a vapor film begins to form, through which heat must flow by conduction. The rate of heating then decreases with Δt until complete film boiling is reached and heat flows by radiation and conduction through the film. Point F corresponds to the melting point of the wire.

The high heat rates which occur during boiling make it a very effective means of absorbing the energy capable of being released in furnaces and nuclear reactors. This is also one reason why the vapor power-generating cycles have been successful.

Mass and momentum transfer. In a gas, heat conduction occurs by transfer of kinetic energy from high-temperature to lower-temperature molecules. This process requires a change in location of molecules, called mass transfer. If, in addition to their random velocities, the molecules have definite but different flow velocities, the molecular (mass) interchange will also result in a momentum transfer. This means that mass, momentum, and heat exchange are interrelated. This interrela-

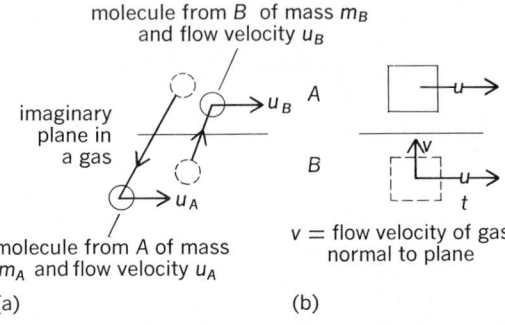

Fig. 9. Mass, momentum, and heat transfer. (a) By conduction: mass transfer $= m_B - m_A$; momentum transfer $= m_B u_B - m_A u_A$; heat transfer $= m_B c_B t_B - m_A c_A t_B$ (b) By convection: mass transfer (per unit time and area) $= \rho v$; momentum transfer $= (\rho v)u$; heat transfer $= (\rho v)c_p t$.

tionship exists in both conduction and convection phenomena (Fig. 9).

These three processes are described by similar equations, which are listed in the table. As mass conduction, W/A denotes the mass transfer per unit time and area of component A through B. The coefficient D_{AB} is a diffusion coefficient characterizing the diffusion of molecules of gas A through gas B, and C denotes the concentration of A (pounds per cubic foot) which is the driving potential for the process. The equation describing momentum transfer is basically an expression of Newton's second law of motion. The three coefficients, D_{AB}, μ, and κ, in these transport processes, are referred to as the transport properties.

In the case of convection, the same mass flux ρv carries with it the concentration of C of a given species, the velocity u, and the enthalpy $c_p t$.

The similarity between the fundamental equa-

Transfer equations

Phenomenon	Transfer of		
	Mass (diffusion)	Momentum (introducing shear stress)	Heat
By conduction	$\dfrac{W}{A} = D_{AB}\dfrac{\partial C}{\partial y}$	$\sigma = \dfrac{W}{A}\Delta u = \mu\dfrac{\partial u}{\partial y}$	$\dfrac{q}{A} = -\kappa\dfrac{\partial t}{\partial y}$
By convection	$\dfrac{W}{A} = (\rho v)\,C$	$\sigma = (\rho v)u$	$\dfrac{q}{A} = (\rho v)c_p t$

tions suggests that solutions for one process may be applicable to the others. The validity of this has been established for such cases as evaporation and condensation from a liquid surface into a gas, such as air, above it.

The applicability of these analogous solutions is, however, limited to conditions where one process can be regarded as independent of the others. When exchange rates are high and the properties of the species involved vary with temperature and pressure, it may be necessary to solve the differential equations governing each transport process simultaneously.

[WARREN H. GIEDT]

Bibliography: J. P. Holman, *Heat Transfer*, 4th ed., 1976; W. M. Kays, *Convective Heat and Mass Transfer*, 1966; W. H. McAdams, *Heat Transmission*, 3d ed., 1954; W. M. Rohsenow and J. P. Hartnett (ed.), *Handbook of Heat Transfer*, 1973; J. R. Welty, *Engineering Heat Transfer, SI Version*, 1978.

Coordinate systems

Schemes for locating points in a given space by means of numerical quantities specified with respect to some frame of reference. These quantities are the coordinates of a point. To each set of coordinates there corresponds just one point in any coordinate system, but there are useful coordinate systems in which to a given point there may correspond more than one set of coordinates.

A coordinate system is a mathematical language that is used to describe geometrical objects analytically; that is, if the coordinates of a set of points are known, their relationships and the properties of figures determined by them can be obtained by numerical calculations instead of by other descriptions. It is the province of analytic geometry, aided chiefly by calculus, to investigate the means for these calculations.

The most familiar spaces are the plane and the three-dimensional euclidean space. In the latter a point P is determined by three coordinates (x,y,z). The totality of points for which x has a fixed value constitutes a surface. The same is true for y and z so that through P there are three coordinate surfaces. The totality of points for which x and y are fixed is a curve and through each point there are three coordinate lines. If these lines are all straight, the system of coordinates is said to be rectilinear. If some or all of the coordinate lines are not straight, the system is curvilinear. If the angles between the coordinate lines at each point are right angles, the system is rectangular.

Cartesian coordinate system. This is one of the simplest and most useful systems of coordinates. It is constructed by choosing a point O designated as the origin. Through it three intersecting direct-

ed lines OX, OY, OZ, the coordinate axes, are constructed. The coordinates of a point P are x, the distance of P from the plane YOZ measured parallel to OX, and y and z, which are determined similarly (Fig. 1). In this system the coordinate lines through P are straight lines respectively parallel to the three coordinate axes. Usually the three axes are taken to be mutually perpendicular, in which case the system is a rectangular cartesian one. Obviously a similar construction can be made in the plane, in which case a point has two coordinates (x,y). It is this system that is used in the construction of graphs and charts of various data, whether observed or computed.

Polar coordinate system. This system is constructed in the plane by choosing a point O called the pole and through it a directed straight line, the initial line. A point P is located by specifying the directed distance OP and the angle through which the initial line must be turned to coincide with OP in position and direction (Fig. 2). The coordinates of P are (r,θ). The radius vector r is the directed line OP, and the vectorial angle θ is the angle through which the initial line was turned, $+$ if turned counterclockwise, $-$ if clockwise. To each pair of values (r,θ) there corresponds just one point, but any point has an endless number of sets of coordinates. The coordinate lines in this case are radial lines through the pole ($\theta = $ constant) and concentric circles with center at the pole ($r = $ constant). In spite of this lack of unique reciprocity between points and their coordinates, the polar

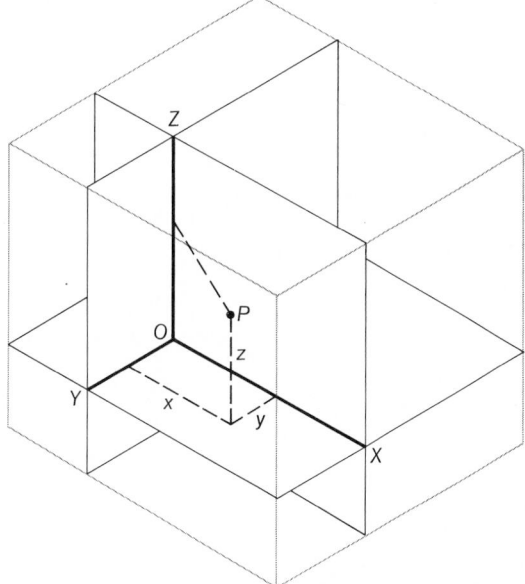

Fig. 1. Cartesian coordinate system.

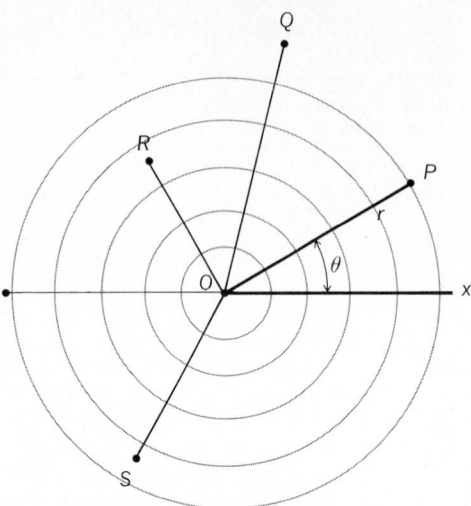

Fig. 2. Polar coordinate system.

system is useful in the study of spirals and rotations and in the investigation of motions under the action of central forces such as those of planets and comets.

Spherical coordinates. In three-dimensional euclidean space this system of coordinates is constructed by choosing a plane and in it constructing a polar coordinate system. At the pole O a polar axis OZ is constructed at right angles to the chosen plane. A point P, not on OZ, and OZ determine a plane. The spherical coordinates of P are then the directed distance OP denoted by ρ, the angle θ through which the initial line is turned to lie in ZOP and the angle $\phi = ZOP$ (Fig. 3). The coordinate lines are radial lines through O (θ and ϕ constant), meridian circles (ρ and θ constant) and circles of latitude (ρ and θ constant). This is an example of a curvilinear rectangle coordinate system. It is used in locating stars, in the study of spherical waves, and in problems in which there is spherical symmetry.

Cylindrical coordinates. These are constructed by choosing a plane with a pole O, an initial line in

it, and a polar axis OZ, as in spherical coordinates. A point P is projected onto the chosen plane. The cylindrical coordinates of P are (r,θ,z) where r and θ are the polar coordinates of Q and $z = QP$ (Fig. 3). This is also a curvilinear rectangular system, the coordinate lines being mutually perpendicular. This system is used in problems of fluid flow and in others in which there is axial symmetry.

What has been done above in the plane and in euclidean space of three dimensions can be extended to curved spaces of any number dimensions. For a space S of n dimensions with some known geometrical or physical properties, a coordinate system in which a point has coordinates $(u_1, u_2, \ldots , u_n)$ is chosen. The coordinate lines are curves along which only one coordinate varies. When the known properties of S are expressed in terms of these coordinates, it is then the province of differential geometry to investigate their consequences.

Transformation of coordinates. By means of a system of equations the description of a geometrical object in one coordinate system may be translated into an equivalent description in another coordinate system. Thus in a given space of n dimensions if there is a coordinate system A in which a general point P has coordinates $(u_1, u_2, \ldots , u_n)$ and a coordinate system B in which P has the coordinates $(v_1, v_2, \ldots , v_n)$, the transformation from system A to system B is the set of equations $u_i = f_i$ $(v_1 v_2, \ldots , v_n)$, for $i = 1, 2, \ldots , n$, which expresses each u in terms of the v's. These functions are obtainable from the relation between the two coordinate systems. They are not completely arbitrary, for they must be single-valued, independent, and such that if in the B system P has another set of coordinates, say $(v'_1, v'_2, \ldots , v'_n)$, then $f_i(v_1, v_2, \ldots , v_n) = f_i(v'_1, v'_2, \ldots , v'_n)$. Then if a geometrical locus in the space S is described by one or more equations of the form $F(u_1, u_2, \ldots , u_n) = 0$ in system A, the equivalent description in system B is one or more of the equations of the form

$$F[f_1(v), f_2(v), \ldots , f_n(v)] = 0$$

Other geometrical objects such as vectors and areas have more complicated laws of transformation, but each such law is expressible in terms of the equations of transformation. As remarked above, a coordinate system is a mathematical language; a transformation of coordinates plays the role of a dictionary that translates from one language to another.

The most important transformations of coordinates are between rectangular cartesian coordinate systems. One such set of coordinate axes is obtainable from another by a translation (in the physical sense) and a rotation. Consider a coordinate system A with coordinate axes OX, OY, OZ, and coordinates of a point $P(x,y,z)$. System B is obtained by moving the axes without turning to a point O' with coordinates $(a,b,c,)$. The new axes are $O'X'$, $O'Y'$, $O'Z'$ and the coordinates of P in this system are (x',y',z'); then the equations of transformation are $x = x' + a$, $y = y' + b$, $z = z' + c$. If the new system is obtained from A by a rotation of the coordinate axes, let $\alpha_1, \alpha_2, \alpha_3$; $\beta_1, \beta_2, \beta_3$; $\gamma_1, \gamma_2, \gamma_3$ be the angles which the new axes OX', OY', OZ' make with the original axes (only three of these angles are independent); then the equations

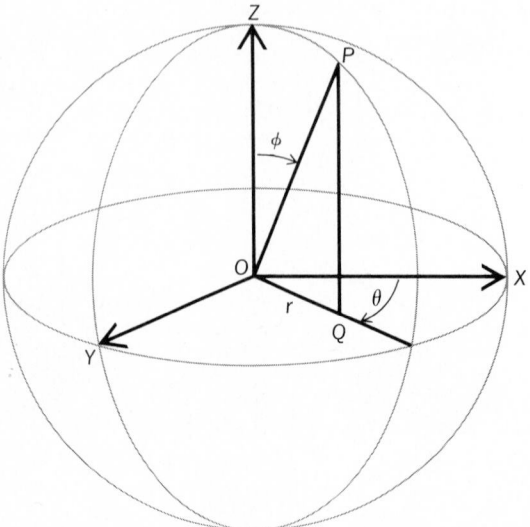

Fig. 3. Spherical coordinate system.

of transformation are

$$x = x' \cos \alpha_1 + y' \cos \beta_1 + z' \cos \gamma_1$$
$$y = x' \cos \alpha_2 + y' \cos \beta_2 + z' \cos \gamma_2$$
$$z = x' \cos \alpha_3 + y' \cos \beta_3 + z' \cos \gamma_3$$

The equations of transformation may be differently interpreted. The equations of transformation make the point P of coordinates (x,y,z), in the same coordinate system, correspond to a point P' of coordinates (x',y',z'). This constitutes a mapping of the space upon itself which maps a figure into some other figure. In applications one usually seeks a mapping that carries some pertinent loci into simpler loci while preserving some relevant properties. *See* ANALYTIC GEOMETRY; CALCULUS; CONFORMAL MAPPING; SPHERICAL HARMONICS.

[MORRIS S. KNEBELMAN]

Coordination number

The number of nearest neighbors of a point in a space lattice, of an atom or an ion in a structure, or of an anion or cation in a complex ion. The figure shows that the coordination number in the three cubic lattices P, I, and F is respectively 6, 8, and 12. In closely packed structures this number is 12. The structures of the subgroup elements are easily described by means of the coordination number of each atom, which is given by the rule of Hume-Rothery, or the $8-N$ rule. Here N is the number of the subgroup. Diamond, germanium, and silicon are in subgroup four; the coordination number is 4. Tellurium and selenium are in subgroup six, each atom having two nearest neighbors. Accordingly, the structure is an assembly of strings.

In ionic structures each ion is surrounded by a number of ions of opposite sign. This coordination is respectively 8/8, 6/6, 4/4, 8/4 in the cesium chloride, sodium chloride, zinc sulfide, and calcium fluoride structures. For a discussion of these structures *see* CRYSTAL STRUCTURE.

The anions are, in nearly all cases, larger than the cations. For any given coordination the distribution of the anions is as regular as possible. Each cation is in the center of a coordination polyhedron. The structures can therefore be described in terms of the way in which the anion polyhedrons are packed. As an example, in the rock salt structure, each cation is in the center of an octahedron formed by six anions. Two neighboring octahedrons have a common edge. Along these lines L. Pauling has deduced a number of rules which summarize general features of ionic compounds. The most important are listed here.

1. A coordinated polyhedron of anions is formed about each cation, the cation-anion distance being determined by the radius sum, and the coordination number of the cation by the radius ratio.

2. In a stable coordination structure, the total strength of the valence bonds which reach an anion from all the neighboring cations is equal to the charge of the anion.

3. The existence of edges and particularly of faces common to two anion polyhedrons decreases stability. This effect is large for cations with high valency and small coordination number, and is especially large when the radius ratio approaches the lower limit of stability of the polyhedron.

4. In a crystal containing different cations, those of high valency and small coordination number tend not to share polyhedron elements with each other.

[WILLY C. DE KEYSER]

Bibliography: M. J. Buerger, *Crystal Structure Analysis*, 1979; R. C. Evans, *Introduction to Crystal Chemistry*, 2d ed., 1964.

Coulomb excitation

A process in which two atomic nuclei that approach each other undergo transitions to excited states that are caused by the long-range Coulomb forces acting between the nuclei. The process can occur even when the nuclei do not come sufficiently close to allow the strong short-range nuclear forces to act.

That nuclei can be excited by this process is not so important in itself. What has proved to be enormously fruitful is the fact that experimental results based on this process have provided accurate values for electrical or electromagnetic properties of nuclei. To take an easily visualized example, Coulomb excitation measurements were instrumental in showing that many nuclear species are not spherical but are, instead, shaped like a football. *See* COULOMB'S LAW; NUCLEAR STRUCTURE.

Experiments. The field of experimental Coulomb excitation began in the mid-1950s. Early experiments involved bombarding targets with beams of protons or α-particles obtained from accelerators such as Van de Graaffs, cyclotrons, or linacs. Nuclei are made of protons and neutrons; the positively charged protons cause an intense electrostatic (Coulomb) field to surround nuclei. Positively charged bombarding particles, such as protons or α-particles, must have a large kinetic energy in order to overcome the strong Coulomb repulsion of nuclei and thereby reach the close distances required for the attractive nuclear forces to act.

Coulomb excitation was discovered when protons of much too small an energy to overcome this Coulomb repulsion of the target nuclei were observed to still cause nuclear excitation. The occurrence of nuclear excitation was confirmed by detecting the γ-rays emitted from the target when the nuclei decayed back down to the ground state. *See* PARTICLE ACCELERATOR; PROTON.

Although the first Coulomb excitation measurements were carried out with beams of protons or α-particles, it was soon realized that the use of beams of more highly charged projectiles would dramatically increase the probabilities for Coulomb excitation. Much work has been done with beams of ^{16}O, ^{20}Ne, and ^{40}Ar ions, and experiments have been reported using beams of ^{84}Kr and ^{136}Xe ions. Coulomb excitation with these heavy ions is characterized by the term "multiple excitation." Many more excited states of nuclei are appreciably populated with these heavy-ion beams.

Theory and interpretations. The theory of pure Coulomb excitation is well understood, but it is mathematically complicated. The modern approach is to use computer programs to generate specific theoretical results applicable to a particular experimental situation. However, with available computers, it is not feasible to obtain good quantitative theoretical results for the heaviest projectiles such as ^{136}Xe.

In order to interpret Coulomb excitation results precisely, it is important that the influence of nuclear forces be negligible so that a "pure" Cou-

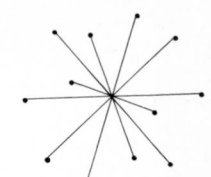

F

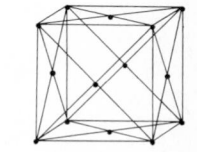

F

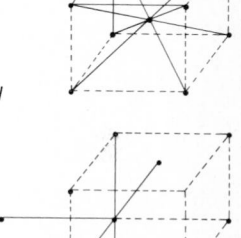

I

P

The coordination number in the three cubic lattices *F*, *I*, and *P* is shown to be 12, 8, and 6, respectively.

lomb excitation situation exists. Of course, at higher bombarding energies, where nuclear forces are an important part of the reaction, Coulomb excitation is still present; and, in fact, the combination of Coulomb and nuclear forces produces an interesting reaction situation which has been the subject of much research. *See* NUCLEAR REACTION; SCATTERING EXPERIMENTS (NUCLEI).

Measurement and applications. Coulomb excitation can be measured either by the direct analysis of the energy spectrum of the scattered projectiles or by the analysis of the γ-rays subsequently emitted by the excited nuclei. Both methods are used extensively. The chief advantage of a direct measurement of the spectrum of the scattered projectiles is the good accuracy with which excitation probabilities can be determined. A high degree of accuracy is valuable in determining such nuclear properties as static quadrupole moments of excited states, the existence and magnitude of hexadecapole or E4 moments, and the possibility of centrifugal stretching of nuclei.

The chief advantages offered by the detection of γ-rays following Coulomb excitation are: (1) The excellent energy resolution (approximately 2 keV) of Ge(Li) γ-ray detectors can be used to study the excited nuclear states; (2) good statistical accuracy (high counting rates) is achieved in much shorter times on the accelerator; and (3) more extensive information on the properties of the nuclear states can be extracted from γ-ray measurements. From measurements of angular distributions of the γ-rays, spin-parity values can be assigned to excited states, and information about the strength of magnetic dipole transitions in nuclei can be gained. *See* NUCLEAR SPECTRA.

Heavy-ion collisions. One of the important frontiers of nuclear physics research is the use of heavy-ion projectiles; producing collisions between large complex nuclei allows interesting questions to be asked about the behavior and structure of nuclei. In such collisions, the Coulomb interaction is always important. Thus it is likely that the Coulomb excitation process will continue to play a vital role in nuclear physics research.

[PAUL H. STELSON]

Bibliography: K. Adler and A. Winther, *Coulomb Excitation*, 1966; F. K. McGowan and P. H. Stelson, in J. Cerny (ed.), *Nuclear Spectroscopy and Reactions*, pt. C, 1974.

Coulomb's law

For electrostatics, Coulomb's law states that the direct force F of point charge q_1 on point charge q_2, when the charges are separated by a distance r, is given by $F = k_0 q_1 q_2/r^2$, where k_0 is a constant of proportionality whose value depends on the units used for measuring F, q, and r. It is the basic quantitative law of electrostatics. In the rationalized meter-kilogram-second (mks) system of units, $k_0 = 1/4\pi\epsilon_0$, where ϵ_0 is called the permittivity of empty space and has the value $\epsilon_0 = (1/4\pi \times 8.98776 \times 10^9)$ coul^2/newton-m$^2 \cong 8.85 \times 10^{-12}$ coul^2/newton-m^2. Thus, Coulomb's law in the rationalized mks system is as in the equation below, where q_1 and q_2 are

$$F = \frac{1}{4\pi\epsilon_0} \frac{q_1 q_2}{r^2}$$

expressed in coulombs, r is expressed in meters, and F is given in newtons. *See* ELECTRICAL UNITS AND STANDARDS.

The direction of F is along the line of centers of the point charges q_1 and q_2, and is one of attraction if the charges are opposite in sign and one of repulsion if the charges have the same sign. For a statement of Coulomb's law as applied to point magnet poles *see* MAGNETOSTATICS.

Experiments have shown that the exponent of r in the equation is very accurately the number 2. Lord Rutherford's experiments, in which he scattered α-particles by atomic nuclei, showed that the equation is valid for charged particles of nuclear dimensions down to separations of about 10^{-12} cm. Nuclear experiments have shown that the forces between charged particles do not obey the equation for separations smaller than this.

The direct force that one charged particle exerts on another is unaffected by the presence of additional charge and, in any electrostatic system, the equation gives this direct force between q_1 and q_2 under any conditions of charge configuration, including that in which intervening and surrounding matter is present and the molecules of the matter are polarized so that their charges contribute to this configuration. The total force on any one charge, say q_1, is the vector sum of the separate direct forces on q_1 due to q_2, q_3, q_4, and so on, each force computed separately by use of the equation as if all other charges were absent.

The permittivity ϵ of a medium is defined by $\epsilon = k\epsilon_0$, where k is the relative dielectric constant of the medium. Another name, specific inductive capacity, is also given to k. *See* DIELECTRIC CONSTANT.

If two free point charges q_1 and q_2 are immersed in an infinite homogeneous isotropic dielectric, the total force on one of them, say q_1, is given by $F = q_1 q_2/4\pi\epsilon r^2$ and the use of ϵ (in place of ϵ_0) takes proper account of the forces on q_1 due to the polarization charges of the dielectric molecules. It is only in a few special cases that this latter formulation of Coulomb's law is valid. *See* ELECTRIC CHARGE; ELECTROSTATICS.

[RALPH P. WINCH]

Bibliography: B. I. Bleaney and B. Bleaney, *Electricity and Magnetism*, 3d ed., 1976; Coulomb's Law Committee, *Amer. J. Phys.*, 18:6–11, 1950; E. M. Purcell, *Electricity and Magnetism*, Berkeley Physics Course, vol. 2, 1965; E. M. Pugh and E. W. Pugh, *Principles of Electricity and Magnetism*, 2d ed., 1976.

Creeping flow

Fluid flow in which the velocity of the flow is very small. For creeping flow, the Reynolds number $R_e = Ud/\nu$ is small (less than unity). Here U is the reference velocity, d is the reference length, and ν is the kinematic viscosity of the fluid. For low Reynolds number, the inertial force is negligible and the nonlinear terms in Navier-Stokes equations are neglected. The resultant flow is known as Stokes flow. One of the important applications of creeping flow is the motion of a tiny particle in a viscous flow, which is important in two-phase flow. If the particle is a sphere, its drag coefficient C_D is $C_D = 24/R_e$ if R_e is less than 1, where the reference length d is the diameter of the sphere.

When R_e is greater than 1, but not too large, the drag coefficient is a little larger than $24/R_e$. Such a flow is known as Oseen's flow, and may be considered as an upper limit of creeping flow. Another application of creeping flow is the lubrication problem which was initiated by O. Reynolds, who showed that two parallel or near-parallel surfaces can slide one over the other with only slight frictional resistance, even under great normal pressure, provided that a film of viscous flow is maintained. See FLUID FLOW; FLUID-FLOW PRINCIPLES; NAVIER-STOKES EQUATIONS; REYNOLDS NUMBER; VISCOSITY.

[S. I. PAI]

Bibliography: J. Happel and H. Brenner, *Low Reynolds Number Hydrodynamics*, 1965; S. I. Pai, *Two-Phase Flows*; S. I. Pai, *Viscous Flow Theory*, vol. 1: *Laminar Flow*, 1956.

Critical phenomena

The unusual physical properties displayed by substances near their critical points. The study of critical phenoma of different substances is directed toward a common theory.

Critical points. Ideally, if a certain amount of H_2O (water) is sealed inside a transparent cell and heated to a rather high temperature T, for instance, $T > 660$ K, the H_2O exists as a transparent homogeneous substance. When the cell is allowed to cool down gradually and reaches a particular temperature, namely the boiling point, the H_2O will go through a phase transition and separate into liquid and vapor phases. The liquid phase, being more dense, will settle into the bottom half of the cell. This sequence of events takes place for H_2O at most moderate densities. However, if the enclosed H_2O is at a density close to 322.2 kg m⁻³, rather extraordinary phenomena will be observed. As the cell is cooled toward 647 K, the originally transparent H_2O will become increasingly turbid and milky, indicating that visible light is being strongly scattered. Upon slight additional cooling, the turbidity disappears and two clear phases, water and vapor, are found. This phenomenon is called the critical opalescence, and the H_2O sample is said to have gone through the critical phase transition. The density, temperature, and pressure at which this transition happens specify the critical point and are called respectively the critical density ρ_c, the critical temperature T_c, and the critical pressure P_c. For H_2O, $\rho_c = 322.2$ kg m⁻³, $T_c = 647$ K, and $P_c = 2.21 \times 10^7$ Pa.

Different fluids, as expected, have different critical points. Although the critical point is the end point of the vapor pressure curve on the pressure-temperature (P-T) plane (Fig. 1), the critical phase transition is qualitatively different from that of the ordinary boiling phenomenon that happens along the vapor pressure curve. In addition to the critical opalescence, there are other highly unusual phenomena that are manifested near the critical point; for example, both the isothermal compressibility and heat capacity diverge to infinity as the fluid approaches T_c. See THERMODYNAMIC PROCESSES.

Many other systems, for example, ferromagnetic materials such as iron and nickel, also have critical points. The ferromagnetic critical point is also known as the Curie point. As in the case of fluids, a number of unusual phenomena take place near

the critical points of ferromagnets, including singular heat capacity and divergent magnetic susceptibility. The study of critical phenomena is directed toward describing the various anomalous and interesting types of behavior near the critical points of these diverse and different systems with a single common theory. See CURIE TEMPERATURE; FERROMAGNETISM.

Order parameters. One common feature of all critical phase transitions is the existence of a quantity called the order parameter. The net magnetization M is the order parameter for the ferromagnetic system. At temperatures T above T_c and under no external field, there is no net magnetization. However, as the temperature of the system is cooled slowly through T_c, a net magnetization M will appear precipitously. The increase in M is more gradual as temperature is reduced further. The nonzero magnetization is due to the partial alignment of the individual spins in the ferromagnetic substance. M is called the order parameter, since the state with partial alignment of spins is clearly more ordered than that with no alignment. The density difference in the liquid and vapor phases, $(\rho_l - \rho_v)/\rho_c$, is the proper order parameter in the fluid system; ρ_c in the denominator is the critical density. This order parameter has similar temperature dependence as the net magnetization M of a ferromagnetic system. For $T > T_c$ the order parameter is equal to zero, since there is only one homogeneous phase in the fluid. As the system is cooled through T_c, the fluid system phase separates with a precipitous increase in the difference of the liquid and vapor densities. A number of critical systems and their respective order parameters are listed in Table 1.

The order parameters assume power law behaviors at temperatures just below T_c. In the fluid and ferromagnetic systems, for $t < 0$, they follow Eqs. (1) and (2), and for $t > 0$ they obey Eqs. (3). In these

$$M = B(-t)^\beta \tag{1}$$

$$\frac{\rho_l - \rho_v}{\rho_c} = B(-t)^\beta \tag{2}$$

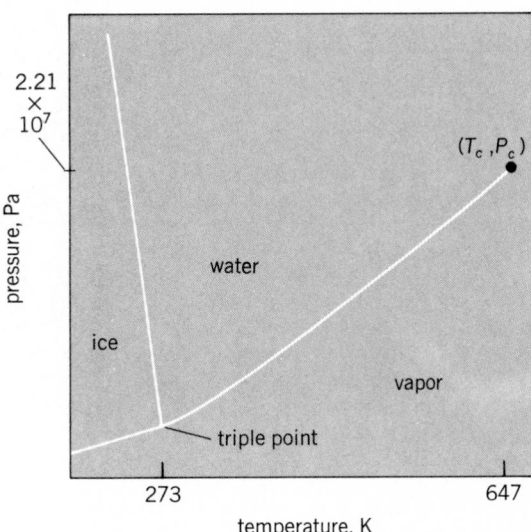

Fig. 1. Phase diagram of H_2O on pressure-temperature (P-T) plane.

Table 1. Order parameters, theoretical models, and classification according to universality class of various physical systems*

Universality class		Theoretical model	Physical system	Order parameter
$d=2$	$n=1$	Ising model in two dimensions	Adsorbed films	Surface density
	$n=2$	XY model in two dimensions	Helium-4 films	Amplitude of superfluid phase
	$n=3$	Heisenberg model in two dimensions		Magnetization
$d>2$	$n=\infty$	Spherical model	None	
$d=3$	$n=0$	Self-avoiding random walk	Conformation of long-chain polymers	Density of chain ends
	$n=1$	Ising model in three dimensions	Uniaxial ferromagnet	Magnetization
			Fluid near a critical point	Density difference between phases
			Mixture of liquids near consolute point	Concentration difference
			Alloy near order-disorder transition	Concentration difference
	$n=2$	XY model in three dimensions	Planar ferromagnet	Magnetization
			Helium 4 near super-fluid transition	Amplitude of superfluid phase
	$n=3$	Heisenberg model in three dimensions	Isotropic ferromagnet	Magnetization
$d\le4$	$n=-2$		None	
	$n=32$	Quantum chromo-dynamics	Quarks bound in protons, neutrons, etc.	

$$M=0; \frac{\rho_l-\rho_v}{\rho_c}=0 \qquad (3)$$

equations, $t=(T-T_c)/T_c$ is the reduced temperature, and β and B are respectively the critical exponent and amplitude for order parameter.

Measurement of the order parameter of fluid neon near the critical temperature ($T_c=44.54$ K, $\rho_c=484$ kg m^{-3}, $P_c=2.27\times10^6$ Pa) is shown in Fig. 2. Figure 2a shows data on the densities of liquid and vapor phases of neon, obtained by measuring the dielectric constant of fluid neon in the bottom and top halves of a cell. For T below T_c ($t<0$), the densities ρ_B and ρ_I in the bottom and top halves represent respectively the liquid and vapor densities ρ_l and ρ_v. For T above T_c ($t>0$), ρ_B is still greater than ρ_T because, due to the divergent isothermal compressibility, a density gradient is induced by the gravitational field. In this experiment, the vertical separation of the two capacitor plates is 2.68 mm and the average density of the

fluid $\bar{\rho}$ is 0.997 of the critical density ρ_c. In Fig. 2b, the order parameter $(\rho_l-\rho_v)/\rho_c$, as calculated from the data in Fig. 2a, is plotted versus the reduced temperature in a log-log scale. The plot shows the power law dependence of Eq. (2). The slope, or exponent β, is equal to 0.351 ± 0.004 in this experiment.

Power law behavior of other quantities. The anomalous behavior of other thermodynamic quantities near the critical point can also be expressed by power laws. These can be characterized by a set of critical exponents, labeled (besides β) α, γ, δ, ν, and η: α characterizes the divergent heat capacity; γ the susceptibility (of magnets) and isothermal compressibility (of fluids); δ the critical isotherm; ν the correlation length; and η the critical correlation function. The functional forms of these power laws for the fluid and magnet systems are shown in Table 2. The critical opalescence phenomenon discussed above is closely related to the strong density fluctuations induced

Table 2. Critical exponents and power laws in pure fluids and ferromagnets*

Thermodynamic quantity	Fluid	Ferromagnet	Power law
Specific heat	C_v	C_H	$\sim (t)^{-\alpha}$ for $t>0$ $\sim (-t)^{-\alpha'}$ for $t<0$
Order parameter $=S$	$(\rho_l-\rho_v)\rho_c^{-1}$	M	$=0$ for $t>0$ $\sim (-t)^{\beta}$ for $t<0$
Response function	$K_T=\frac{1}{\rho}\left(\frac{\partial P}{\partial \rho}\right)_T$	$X=\left(\frac{\partial M}{\partial H}\right)_T$	$\sim (t)^{-\gamma}$ for $t>0$ $\sim (-t)^{-\gamma'}$ for $t<0$
Critical isotherm	$P-P_c$	H	$\sim |S|^{\delta}\cdot$ sign of S
Correlation length	ξ	ξ	$\sim (t)^{-\nu}$ for $t>0$ $\sim (-t)^{-\nu'}$ for $t<0$
Critical correlation function of fluctuation	$G(r)$	$G(r)$	$\sim (r)^{-(d-2+\eta)}$

*$d=$ spatial dimensionality of the critical system; $H=$ magnetic field strength; $C_v=$ specific heat at constant volume; $C_H=$ specific heat at constant magnetic field strength; $K_T=$ isothermal compressibility; $X=$ susceptibility.

by the divergent isothermal compressibility near T_c. When the density fluctuations are correlated at lengths comparable to the wavelength of visible light, intense scattering occurs.

Mean field theories. The earliest attempts to understand the critical behavior were the van der Waals model for fluids (1873) and the Weiss model for ferromagnets (1907). These are mean field theories in the sense that the state of any particular particle in the system is assumed to be determined by the average properties of the system as a whole. In these models, all particles can be considered to contribute equally to the potential at each site. Therefore, the mean field theory essentially assumes the intermolecular interaction to be of infinite range at all temperatures. The mean field theories are qualitiatively quite successful in that they predict the existence of critical points and power law dependence of the various thermodynamic quantities near the critical point. They are not quantitively correct because the predicted values for the various exponents are not in agreement with exact model calculations or with experimental results (Table 3).

Scaling hypothesis. Theoretical efforts in the study of critical phenomena have been centered on predicting correctly the value of these critical exponents. One of the most important developments is the hypothesis of scaling. This hypothesis is model-independent and applicable to all critical systems. The underlying assumption is that the long-range correlation of the order parameter, such as the spin fluctuation in the ferromagnetic system and the density fluctuation in the fluid system near T_c, is responsible for all singular behavior. This assumption leads to a particular functional form for the equation of state near the critical point. With this simple assumption, it has been demonstrated that a number of inequalities among critical exponents that can be proved rigorously by thermodynamic arguments are changed into equalities. These equalities, or scaling laws, show that there are only two independent critical exponents; once two exponents in a system are given, all other exponents can be determined. What is truly impressive about this simple hypothesis is that the scaling laws, Eqs. (4–8), have been shown to be

$$\alpha = \alpha', \gamma = \gamma', \nu = \nu' \qquad (4)$$
$$2 = \alpha + \gamma + 2\beta \qquad (5)$$
$$2 = \alpha + 2\beta\delta - \gamma \qquad (6)$$
$$\nu d = 2 - \alpha \qquad (7)$$
$$\gamma = \nu(2 - \eta) \qquad (8)$$

correct in almost all real and theoretical model critical systems.

The meaning of these exponents are explained above and also in Table 2; d is not an exponent but the spatial dimensionality; the primed and unprimed exponents represent the value below and above T_c respectively. A large number of theoretical and experimental activities are concerned with putting the scaling hypothesis on a firm fundamental basis and understanding its universality and limitations.

Model systems. A great deal of insight has been gained by the construction and solution of model systems that can be solved exactly. The most famous one is the two-dimensional Ising model solved by L. Onsager in 1944. In this model, spins (little magnets) on a lattice are allowed to point in either the up or down directions, and it is assumed that only pairs of nearest-neighboring spins can interact. Onsager found a critical point for this system and calculated the values of the various critical exponents. The solution of this model is important because this is one of the very few model systems with exact solutions, and it is often used to check the validity of approximation techniques. There are many other model systems similar to the Ising model. They are distinguished from each other by the spatial (d) and spin (n) di-

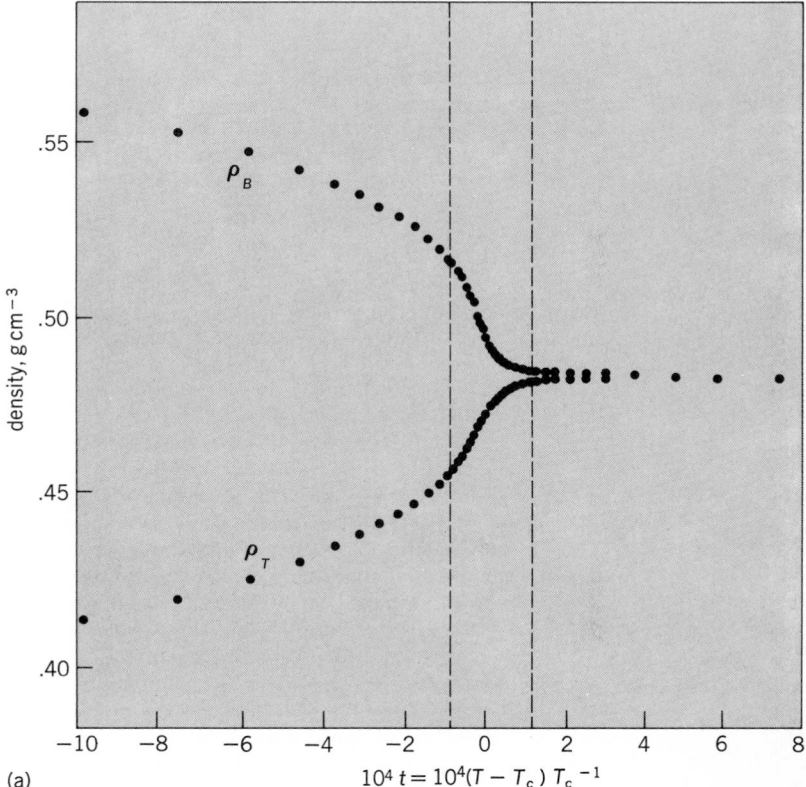

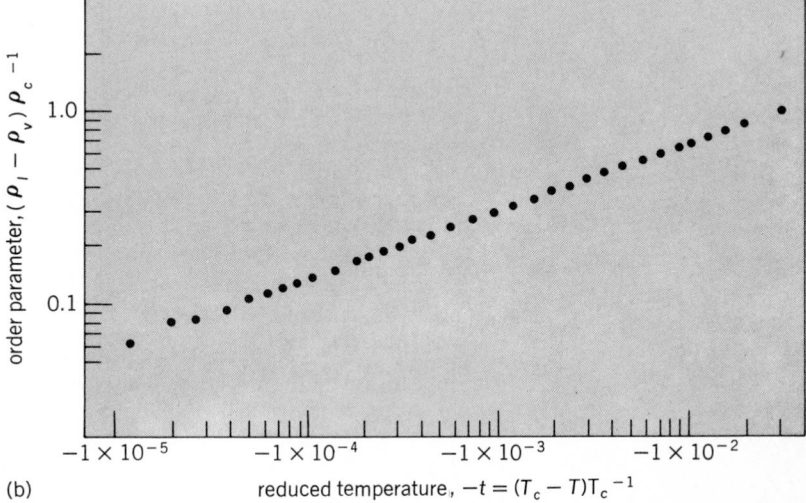

Fig. 2. Measurement of the order parameter of fluid neon near its critical point. (a) Densities of liquid and vapor phases of neon plotted against reduced temperature t. (b) Order parameter plotted against reduced temperature on log-log scale. (*Courtesy of M. Chan and M. Pestak; and M. Pestak, M.S. thesis, University of Toledo, 1979*)

Table 3. Values of critical exponents

Systems	α	β	γ	δ	ν	η
Mean field model	0 (discontinuity)	1/2	1	3	1/2	0
Ising model (exact) ($d=2, n=1$)	0 (logarithmic discontinuity)	1/8	7/4	15	1	0.25
Ising model (approximate) ($d=3, n=1$)	0.110	0.325	1.24	4.82	0.63	0.03
Heisenberg model (approximate) ($d=3, n=3$)	-0.10	0.36	1.38	4.80	0.705	0.03
Fluids*						
Xe	$0.08 \pm .02$	$0.325 - 0.337$	1.23 ± 0.03	4.40		
SF$_6$		$0.321 - 0.339$	1.25 ± 0.03			
Ne		$0.351 \pm .001$	1.24 ± 0.02			
^{3}He, ^{4}He		$0.355 \pm .005$	1.17 ± 0.03			
Ferromagnets* (isotropic)						
Iron, Fe	$-0.09 \pm .01$	$0.34 \pm .02$	1.33 ± 0.02			0.07 ± 0.07
Nickel, Ni	$-0.09 \pm .03$	$0.37 \pm .03$	1.34 ± 0.02	4.2 ± 0.1		

*Experimental data are the averaged measured values of a number of experiments.

mensionality. The spin can be oriented along an axis ($n = 1$, Ising model), or in any direction on a plane ($n = 2$, XY model), or in any direction in space ($n = 3$, Heisenberg model). *See* ISING MODEL.

These models are essentially simplified versions of real physical systems. The three-dimensional Heisenberg model ($d = 3$, $n = 3$), for example, clearly resembles the isotropic ferromagnets; the three-dimensional Ising model ($d = 3$, $n = 1$) can be found to correspond to the pure fluid system. The correspondence can be shown if the space accessible to the fluid is divided into lattice sites, and at each site the spin parameter is considered to be up if the site is occupied and down if it is not. There are other physical systems beside pure fluids that resemble the three-dimensional Ising model, for example, the binary fluid near its consolute mixing point, the uniaxial ferromagnet, and an alloy system near the order-disorder transition. A great deal of effort and ingenious mathematical techniques have been employed to obtain approximate solutions to these model systems.

Universality hypothesis. It has been observed that the measured values for the critical exponents are rather close to the calculated ones of the corresponding model system. This observation leads to the hypothesis of critical universality. According to this hypothesis, the details of the particle-particle interaction potential in the vicinity of the critical point are not important, and the critical behavior is determined entirely by the spatial dimensionality d and the spin dimensionality n. All systems, both model and real, that have the same value of d and n are said to be in the same universality class and have the same critical exponents. The hypotheses of scaling and universality are closely related: since the length of correlation between density or spin fluctuation diverges as one approaches the critical point, and any interaction potential between particles is finite in range, the details of interparticle potential are expected to be increasingly less important as one approaches the critical point. It has been shown that scaling laws can be derived from the universality hypothesis. Classification of model and physical systems according to universality classes is shown in Table 1.

Renormalization group. The renormalization group (RG) method, originally used in quantum field theory, has been introduced in the study of

critical phenomena. By employing a set of symmetry transformations, the ideas contained in the principle of universality and in the scaling hypothesis can be reformulated and incorporated much more economically. As a result, a fully operational formalism is obtained from which critical exponents can be calculated explicitly. Beside the success in critical phenomena, the renormalization group method is also found to be a very useful technique in many diverse areas of theoretical physics. *See* QUANTUM FIELD THEORY.

The renormalization group method, although extremely elegant, is not an exact and rigorous theory. There are many pieces of experimental evidence confirming the basic correctness of the renormalization group approach and the ideas of universality and scaling, but there are also a number of interesting discrepancies. For example, the critical exponents β and γ found for helium fluids appear to be distinctly different from that of room-temperature fluids and that of the three-dimensional Ising model values (see Table 3). One likely explanation is that the experimentally accessible temperature range for the helium experiment ($5 \times 10^{-5} < |t| < 1 \times 10^{-2}$), although very close to T_c, is not yet in the truly asymptotic critical regime. As a consequence, parameters other than the spatial and spin dimensionality also affect the value of the apparent critical exponent. This explanation is substantiated in a few experiments which show that the value of the exponents changes toward the universal value as the system approaches closer to T_c. However, there are other experiments which do not show this effect. Extremely precise and careful experiments will be needed to ascertain whether renormalization group theory, the principle of universality, and scaling laws give the complete picture of critical phenomena.

Dynamical effects. Besides the static critical phenomena, there are many interesting dynamical effects near the critical point, including critical slowing down, the dynamics of density and spin fluctuations, thermal and mass transport, and propagation and attenuation of sound. Understanding of these effects is far from complete.

Most critical phenomena experiments have been performed on three-dimensional systems, but a number have been done in two-dimensional and quasi-two-dimensional systems. The experiments include the order-disorder transition and the continuous melting transition of gases bound to a graphite surface and the superfluid transition of ⁴He films on substrates. These experiments provide interesting physical realizations of the various model systems in two dimensions. *See* PHASE TRANSITIONS; STATISTICAL MECHANICS.

[MOSES H. W. CHAN]

Bibliography: C. S. Domb and M. S. Green (eds.), *Phase Transitions and Critical Phenomena*, vols. 1–6, 1972–1979; S. K. Ma, *Modern Theory of Critical Phenomena*, 1976; H. E. Stanley, *Introduction to Phase Transition and Critical Phenomena*, 1971; K. Wilson, Problems in physics with many scales of length, *Sci. Amer.*, 241:(2)158–179, August 1979.

Cryogenics

The science of producing and maintaining very low temperatures. Before World War II, scientific research at very low temperatures was carried out at only a few specialized laboratories because of the large amount of time and money required for the construction, operation, and maintenance of the necessary gas liquefaction and associated facilities. Beginning in the prewar years, however, recognition of the economic advantages of applied cryogenics in certain industrial processes initiated a rapid expansion of both the scientific and industrial aspects of low-temperature technology. For example, liquid helium, which in 1939 was available in fewer than one dozen laboratories throughout the world, is now a readily available refrigerant in most large research institutes. Similarly, many prewar experimental applications of cryogenic engineering, such as tonnage oxygen plants to produce pure oxygen for metallurgical processes, are now in common use. In the vanguard of industrial cryogenics is the development of hydrogen and helium liquefaction plants capable of producing large quantities of these liquefied gases for both industrial and military use.

Production of low temperatures. From a practical standpoint, any given object may be cooled to and maintained at a low temperature most simply by placing it in thermal contact with a suitable liquefied gas held at a constant pressure. The temperature ranges available using baths of liquefied gases are given in the table. In principle, a liquefied gas can be used to provide constant bath temperatures from its triple point to its critical point. Bath temperature is varied by changing the pressure above the liquid, and, within the liquid range, any desired bath temperature can be maintained by removing precisely that amount of gas vaporized by heat leak into the bath liquid. In general, because heat leak cannot be excluded entirely, baths will eventually boil away unless replenished periodically.

As may be seen from the table, liquefied gases do not exist over the entire low-temperature range. The lowest temperature at which it is practical to use a liquid bath is about 0.3 K, and gaps exist between 5 and 14 K and between 44 and 55 K. If it is necessary to extend the accessible temperature range or realize certain ranges more conveniently, other methods of cold production are available.

Refrigerators can, in principle, be designed to produce and maintain a predetermined temperature at any point along the temperature scale. For refrigerators operating cyclically, the low temperature produced is not strictly constant because of the periodic withdrawal of heat from the low-temperature reservoir, but the temperature variation can be large or extremely small, depending on the cycle characteristics. When, however, the working substance changes phase in the low-temperature reservoir, its specific heat becomes infinite and the reservoir temperature remains constant. The low-temperature reservoir is simply a liquid bath in which the boil-off liquid is continuously replenished.

Refrigerator cycles operating above about 1 K generally involve compression and expansion of appropriately chosen gases. Because at lower temperatures it is impractical or impossible to use gases, liquid or solid working materials are employed along with different techniques to produce cooling. Adiabatic demagnetization of paramagnetic ions in solids, involving electron-spin paramagnets, is useful in the range 1 to 0.003 K, but could be applied at temperatures up to 300 K or more by appropriate choice of magnetic material. Other examples are: the ³He/⁴He dilution refrigerator (0.3 to 0.003 K); adiabatic compression of a liquid-solid mixture of ³He, made possible because of a slope reversal in the ³He melting curve, known as Pomeranchuk cooling (0.05 to 0.002 K); hyperfine enhanced nuclear refrigeration, which exploits a few select ions with magnetic moments that increase greatly with application of an external magnetic field (0.05 to 0.001 K); and adiabatic demagnetization of nuclear-spin paramagnets (10^{-3} to 10^{-6} K). *See* ADIABATIC DEMAGNETIZATION; LIQUID HELIUM.

Principles of cold production. The same basic principles of refrigeration may be used to explain

Selected physical constants of low-temperature liquids

Liquid	Triple point		Normal b.p., K	Critical point	
	Temp., K	Pressure, atm†		Temp., K	Pressure, atm†
Helium-3			3.2	3.3	1.2
Helium-4			4.2	5.2	2.3
n-Hydrogen	14.0	0.07	20.4	33.2	13.0
n-Deuterium	18.7	0.17	23.7	38.4	16.4
n-Tritium	20.6	0.21	25.0	42.5*	19.4*
Neon	24.6	0.43	27.1	44.7	26.9
Fluorine	53.5	<0.01	85.0	144.0	55.0
Oxygen	54.4	<0.01	90.2	154.8	50.1
Nitrogen	63.2	0.12	77.4	126.1	33.5
Carbon monoxide	68.1	0.15	81.6	132.9	34.5
Argon	83.8	0.68	87.3	150.7	48.0
Propane	85.5	<0.01	231.1	370.0	42.0
Propene	87.9	<0.01	226.1	365.0	45.6
Ethane	89.9	<0.01	184.5	305.4	48.2
Methane	90.7	0.12	111.7	191.1	45.8

*Calculated.   †1 atm = 1.01325×10^5 Pa.

gas liquefaction, the periodic temperature reduction of the refrigerator working substance, and the adiabatic demagnetization process. Cold production is related directly to the concept of temperature, which in turn is intimately connected with entropy S defined as $\int (C/T)\, dT$, where C is the specific heat of the substance being cooled. Entropy is a measure of the intrinsic thermal disorder associated with a substance, and, at absolute zero, the entropy of every substance in internal equilibrium is zero. Entropy is increased either by the introduction of thermal energy or by appropriately varying some other parameter upon which S depends. A conjugate relation exists such that, provided all other parameters are held constant, a change in T results in a change of S of the same sign. The significance of S for low temperatures stems from the concept of idealized processes, called isentropic (or constant S), which result in the maximum possible decrease in T for a given spontaneous parametric change. Because an isentropic process must be both adiabatic (thermally isolated) and reversible, in practice it can be only approximated. Nevertheless, the isentropic process provides a criterion against which the efficiency of any actual process may be gaged. From Fig. 1, in which the variation of S and T for an idealized substance is shown at two different constant values of a thermodynamic parameter X such as pressure or magnetization, it is seen that an isentropic parameter change from $A \to B$ produces the maximum reduction in T. (Spontaneous processes resulting in a decrease in S with a change in X are impossible.) See THERMODYNAMIC PRINCIPLES.

All real processes are irreversible to at least some extent, but these can also be used to lower the temperature of the substance to which they are applied, although the temperature drop for the same parameter change is, for an irreversible process, always less than that obtained from a reversible isentropic change. The cooling achievable by an irreversible process is represented on Fig. 1 by the path $A \to C$. (Under certain conditions, it is even possible in an irreversible process to produce, for the same parameter change, a heating instead of a cooling as shown by the path $A \to C'$.)

Cooling of a gas. In applying the above principles to a gas, let parameters X_1 and X_2 be the constant pressures P_1 and P_2, with $P_2 < P_1$. Then if the gas were expanded isentropically with the performance of external work (as in the driving of a turbine or piston expansion engine), one would obtain the maximum cooling as shown by the path $A \to B$. Actually, the temperature drop is somewhat less than the ideal because of frictional losses. An irreversible expansion of the gas from $P_1 \to P_2$, for example, through a throttle valve, would result in a smaller temperature drop, such as that given by the path $A \to C$.

Adiabatic demagnetization. If, in Fig. 1, the parameter curves are assumed to represent two different but constant magnetizations of a paramagnetic material, I_1 and I_2 with $I_2 < I_1$, and if initially the specimen is magnetized by an external magnetic field to a value I_1 at a temperature corresponding to that given by point A, then an isentropic demagnetization of the specimen to I_2 will result in the temperature drop $A \to B$. In practice, this is accomplished by thermally isolating the magnetized paramagnetic material before reducing the magnetic field. After the low temperature T_B has been reached, experiments may be carried out on the paramagnetic material itself if that is the object of interest, or alternatively, a second substance can be thermally connected with the paramagnetic material and cooled by it to some intermediate temperature dependent on the relative heat capacities of paramagnetic material and sample.

^{3}He/^{4}He dilution refrigerator. Another parameter X that may be varied to obtain cooling is the relative concentration c of two chemical species that form a nonideal solution. In such a solution, the mixing of the two components is accompanied by a heat of mixing ΔH_m and an excess entropy of mixing ΔS_m. Surprisingly, the liquid solutions of the two helium isotopes, ^{3}He and ^{4}He, are nonideal and, below about 0.86 K, separate into two phases, depending on c. In the region of miscibility, the mixing process absorbs heat from the surroundings and thereby forms the basis for a refrigeration system. Solutions with less than approximately 6% ^{3}He are miscible at all temperatures, so that very low temperatures can be obtained.

In practice, a continuously operating dilution refrigerator has as its principal element a mixing chamber containing a phase-separated mixture, with nearly pure ^{3}He (the light phase) on top and a dilute mixture of ^{3}He ($c = 6-7\%$) in the bottom phase. ^{3}He, circulated by a pump at ambient temperature, is continuously introduced into the upper phase, forced across the phase boundary into the dilute phase (where cooling occurs), and then withdrawn from the latter phase to continue the cycle. An object to be cooled can be attached to the bottom of the mixing chamber.

Cooling by refrigeration. If a suitable working

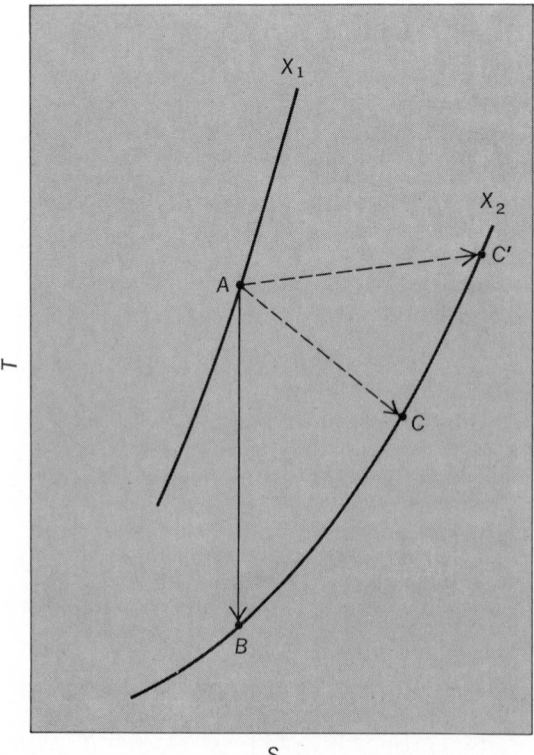

Fig. 1. Variation of entropy with temperature for two different constant values of thermodynamic parameter X.

substance is caused to undergo repeatedly one of the cooling procedures described above, a lower-than-ambient temperature may be maintained indefinitely. Machines capable of carrying out such processes are called refrigerators, and the closed sequence of states through which the working substance passes is called the refrigeration cycle. All reversible refrigeration cycles have the same efficiency, the maximum possible. One such cycle, called a Carnot cycle, is illustrated in Fig. 2. First a suitable substance, showing a large change in S with both T and some other parameter such as the pressure, is cooled isentropically ($A \rightarrow B$). It is then allowed to absorb heat isothermally ($B \rightarrow C$) by further changing the parameter as required to maintain constant temperature. Following this, a second isentropic process ($C \rightarrow D$) is carried out by thermally isolating the working substance and reversing the parametric change adiabatically. During this step, the working substance heats until it reaches the temperature of the hot reservoir, whereupon thermal contact is made between the working substance and hot reservoir. The net heat extracted from the low-temperature reservoir as well as the work done in going around the cycle is then transferred ($D \rightarrow A$) isothermally to the hot reservoir by further parameter changes until the working substance has returned to its original state, and the cycle is ready to repeat. As may be seen in Fig. 2, this sequence is represented by a rectangle on a T-S diagram. Actual refrigerators approach the efficiency of the Carnot cycle, but cannot exceed it.

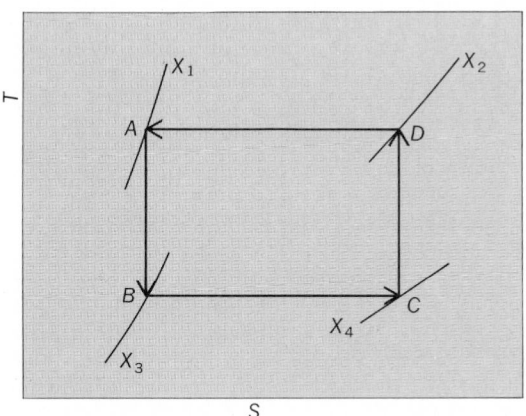

Fig. 2. An ideal refrigeration cycle on a temperature-entropy (T-S) diagram.

An application of the above principles is provided by the magnetic refrigerator, the low-temperature reservoir of which can operate as low as 0.003 K. The high-temperature reservoir consists of a liquid-helium bath held at 1.2 K (^{4}He), or 0.3 K (^{3}He), or 0.020 K (^{3}He/^{4}He dilution refrigerator), depending on what final temperature is required. The cycle is shown in Fig. 2, in which a paramagnetic salt is carried through the indicated sequence of isentropic and isothermal changes. In order to carry out adiabatic and isothermal changes at will, a heat switch is required between the working substance and the heat reservoirs. In the magnetic refrigerator the heat switch may be a lead wire, the

thermal conductivity of which can be changed by one or two factors of 10 when it is caused to pass from its superconducting to its "normal" state.

At higher temperatures, a gas may be used as the refrigerator working substance. The cycle is then located between two isobars on the T-S diagram, with the cooling phase occurring during the expansion of the compressed gas. The closeness of approach to a reversible expansion for a given design depends upon the importance of such factors as reliability, required simplicity of operation, and efficiency. For certain applications in which the refrigeration must be supplied at a constant low temperature, a portion of the gas is allowed to liquefy in the cold reservoir. The absorption of heat then occurs by evaporation of the liquid at constant pressure and hence constant T. If the heat absorption can occur over a range of low temperatures, however, the refrigeration is usually supplied directly by the expanded cold gas.

Although the refrigerators discussed above are based upon conventional refrigeration cycles, it should be remembered that any heat engine run in reverse constitutes a refrigerator. An example of a refrigerator utilizing a cycle quite different from the ones discussed above is the Stirling-cycle Norelco air liquefier developed in 1954. In this device, heat is absorbed and rejected to the heat reservoirs isothermally during expansion and compression of a gaseous working substance (hydrogen or helium), while the cooling and heating phases of the cycle can be considered to occur at either constant pressure or constant volume with the aid of a regenerator or heat accumulator.

Liquefaction of gases. In order to reduce the temperature of a gas and eventually liquefy it, any one of several procedures may be followed. The most frequently used are (1) cooling the gas at constant pressure, that is, constant parameter, by subjecting it to a series of progressively colder refrigerator reservoirs, or (2) allowing the gas itself to constitute the refrigerator working substance until it has cooled sufficiently to liquefy partially. Then the process is continued with the addition of make-up gas while simultaneously the liquid formed per cycle is withdrawn.

Maintenance of low temperatures. Once a temperature lower than ambient has been produced, it may be maintained indefinitely (provided the region itself contains no heat sources) if flow of heat to the cooled region can be prevented. Because no perfect thermal insulators exist, the problem resolves itself into minimizing the flow of heat to the refrigerated region. Devices designed to provide low-temperature environments in which operations may be carried out under controlled conditions are called cryostats.

Heat flows irreversibly from a high- to a low-temperature zone by three different mechanisms: radiation, convection, and conduction. Radiative heat losses are decreased by surfacing the cold region with material of the lowest possible emissivity. Because emissivities of 0.02 are practically attainable, a suitably coated surface can be made to reflect 98% of the radiant energy incident upon it. Radiation heat leak may be further diminished by the use of similarly surfaced radiation shields located between the hot and cold surfaces. Finally, the temperature of the hot radiating surface may often be reduced by cooling it to some temperature intermediate between ambient and the low temperature in question by some relatively inexpensive refrigerant. By the use of one or more of the above techniques, the heat leak from radiation can almost always be reduced to acceptable proportions. *See* EMISSIVITY; HEAT RADIATION; HEAT TRANSFER.

Heat leak from convection is also relatively easy to eliminate, either by evacuation of the space between the hot and cold surfaces to a very high vacuum, or by packing this intervening space with an insulating material such as a foamed plastic or a suitable powder. The gas is thus effectively displaced or localized sufficiently to prevent convection. In addition, the efficiency of a powder insulation is often enhanced significantly by evacuating it to a few millitorrs pressure (1 millitorr = 0.13 pascal). *See* CONVECTION (HEAT).

Conduction along the solid supports, piping, necessary openings, and the insulation itself constitutes the remaining mechanism by which heat can flow to the refrigerated region. As a general rule, all connecting material between hot and cold surfaces should have the lowest possible thermal conductivity, the smallest cross-sectional area, and the longest length consistent with the imposed requirements of mechanical stability. *See* CONDUCTION (HEAT).

These three heat-leak mechanisms are usually satisfactorily suppressed in a device known as a dewar (thermos bottle), a double-walled vessel with evacuated annular space and with walls fabricated from low-thermal-conductivity, thin material (glass, stainless steel, and so forth) and coated with a low-emissivity surface (such as silver or polished metal). Cryostats incorporate one or more (nested) dewars to contain cryogenic liquid baths; for example, a liquid nitrogen bath is usually used to cool the outside of the dewar containing a liquid helium bath.

Finding an optimum solution to any given problem requires an analysis of all the factors contributing to the heat leak. Beginning with the largest, the various heat leaks are minimized successively. Under these conditions, it frequently happens that a less than optimum, but also less expensive, material can be specified for a particular component when its fractional contribution to the total heat leak is small. *See* LOW-TEMPERATURE PHYSICS.

[F. J. EDESKUTY; K. D. WILLIAMSON; W. E. KELLER]

Bibliography: S. Fluegge (ed.), *Handbuch der Physik,* vols. 14 and 15, 1956; O. V. Lounasmaa, *Experimental Principles and Methods below 1 K,* 1974; K. D. Timmerhaus (ed.), *Advances in Cryogenic Engineering,* vols. 1–24, 1954–1979.

Crystal

This term, as used in science and technology, usually denotes a single crystal. A single crystal is a solid throughout which the atoms or molecules are arranged in a regularly repeating pattern. In electronics the term crystal is usually restricted to mean a single crystal which is piezoelectric. Examples of single crystals are most gems, piezoelectric quartz crystals used in controlling the frequencies of radio transmitters, and single crystals of galena (lead sulfide) used in crystal radios. *See* SINGLE CRYSTAL.

Most crystalline solids are made up of millions of tiny single crystals called grains and are said to be polycrystalline. These grains are oriented randomly with respect to each other. Any single crystal, however, no matter how large, is a single grain. Single crystals of metals many cubic centimeters in volume are relatively easy to prepare in the laboratory.

Single crystals differ from polycrystalline and amorphous substances in that their properties are anisotropic. Young's modulus, for example, is different for different directions in the crystal. Anisotropy is responsible for the fact that crystals will cleave (split) along very flat planes which are characteristic of the atomic stacking pattern. *See* CRYSTAL DEFECTS; CRYSTAL GROWTH; CRYSTAL OPTICS; CRYSTAL STRUCTURE; CRYSTALLOGRAPHY.

[HERMAN H. HOBBS]

Crystal absorption spectra

The wavelength or energy dependence of the attenuation of electromagnetic radiation as it passes through a crystal, due to its conversion to an equivalent amount of energy in the crystal. *See* ABSORPTION OF ELECTROMAGNETIC RADIATION.

When atoms are grouped into a regular array to form a crystal, their interaction with electromagnetic radiation is greatly modified. Absorption spectra of free atoms consist of a series of sharp lines at well-defined energies, owing to excitation of electrons between the discrete energy levels of the atoms. In a crystal, the sharp energy levels interact and are broadened into energy bands, and the cores of the atoms vibrate about their equilibrium positions. The ability of electromagnetic radiation to interact with these and other energy-storing processes of a crystal leads to a broad absorption spectrum that bears little resemblance to the spectra of the free parent atoms. *See* ATOMIC STRUCTURE AND SPECTRA.

The absorption spectrum of a crystal includes a number of distinct features, as indicated in the illustration. These absorption types are called lattice, intrinsic, extrinsic, or free-carrier, according to the physical process by which energy is extracted from the electromagnetic radiation. The illustration shows the absorption spectrum of a typical semiconducting crystal (GaAs). Absorption spectra of insulating crystals are similar, except that free-carrier absorption is negligible, and the absorption edge moves to higher energies. For metal crystals, the free-carrier contribution dominates, and the intrinsic absorption processes are much less pronounced. *See* SEMICONDUCTOR.

General properties. Absorption is measured in terms of the absorption coefficient, defined as the rate of attenuation of radiation per unit length, and it is expressed typically in units of cm^{-1}. This quantity varies from less than 3×10^{-5} cm^{-1} in the near infrared, for the exceedingly highly refined glasses used in fiber optics applications, to greater than 10^{6} cm^{-1}, for certain intrinsic absorption processes in crystals. Absorption is related to reflection through a more fundamental quantity called the dielectric function. The absorption coefficient is also related to the index of refraction by means of a frequency integral known as the Kramers-Kronig transform. *See* REFLECTION OF ELECTROMAGNETIC RADIATION.

Lattice absorption. Lattice absorption arises from the excitation of lattice vibrations (phonons) and is the equivalent of vibrational absorption in molecules. It occurs in the infrared and is responsible for most structure in the absorption spectra of crystals at energies below about 0.1 eV (wavelengths above about 10 μm). In a diatomic, partially ionic crystal such as GaAs, an electric field that forces the Ga cores one way simultaneously forces the As cores the other way. Near the frequency corresponding to a wavelength of 37 μm in GaAs, a resonance occurs which results in the generation of transverse optical (TO) phonons and a large increase in absorption. This is seen as the reststrahl peak in the spectrum of illustration *b*. Since the lattice restoring forces are anharmonic, several phonons can be created at the same time at the higher multiples of the reststrahl frequency. These multiphonon processes are responsible for the structures shown from 10 to 30 μm in the illustration. *See* LATTICE VIBRATIONS; MOLECULAR STRUCTURE AND SPECTRA; PHONON.

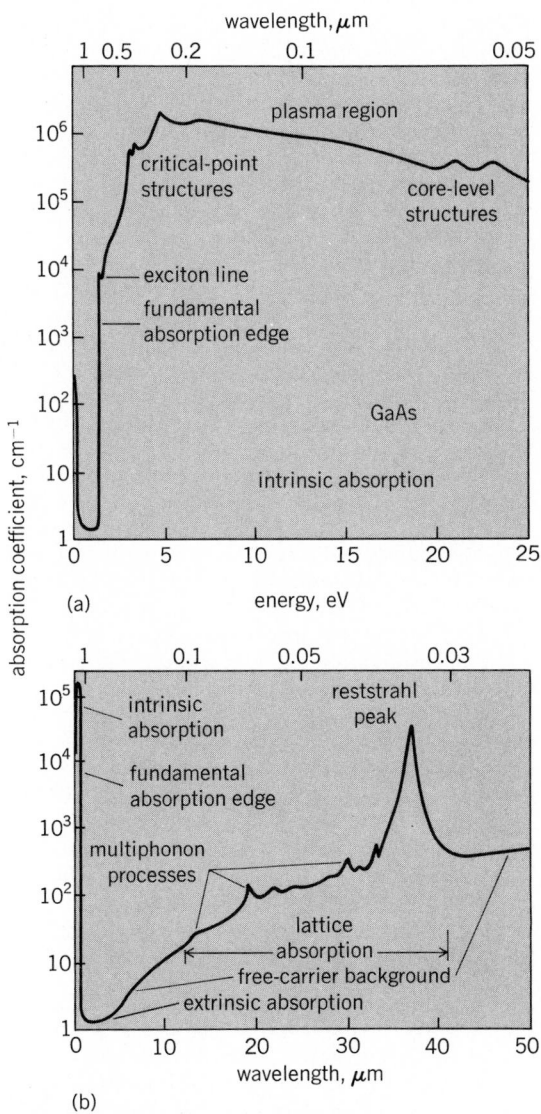

(a)

(b)

Absorption spectrum of GaAs. The region in which each type of absorption process is important is indicated. (*a*) Spectrum at higher energies. (*b*) Spectrum at lower energies.

Free-carrier absorption. If unbound charges are available to carry a current, the electromagnetic field causes motion of the charges directly, with a net energy transfer from the field to the medium as a result of charge-charge and charge-lattice collisions. Unbound charges are free carriers such as conduction (non-bonding-state) electrons and valence (bonding-state) vacancies or holes. The decreasing efficiency of energy dissipation by collision processes at high frequencies causes free-carrier absorption to fall off roughly as $1/E^2$, where $E = \hbar\omega$ is the energy of the electromagnetic radiation ($\hbar$ is Planck's constant divided by 2π, and ω is the frequency of radiation). For crystals with free-carrier concentrations less than about 10^{17} cm^{-3}, free-carrier absorption is significant only in the infrared. Since metallic free-carrier concentrations are much larger, of the order of 10^{23} cm^{-3}, free-carrier concentrations dominate the absorption properties of most metals throughout the visible and near-ultraviolet, producing their lustrous neutral gray appearance. But in a few spectacular exceptions, such as Cu and Au, both free-carrier and intrinsic absorption processes are of comparable importance, and the absorption spectra in the visible are modified significantly.

Intrinsic absorption. Electronic transitions between the bonding (filled-valence) energy bands and the antibonding (empty-conduction) energy bands produce the crystal equivalent of the line-absorption spectra of free atoms. These intrinsic absorption processes dominate the optical behavior of semiconductors and insulators in the visible and ultraviolet spectral regions. The variation of electron energy with electron momentum within the energy bands gives rise to structure in the absorption spectra of crystals in the visible and near ultraviolet. The most readily apparent intrinsic absorption feature is the fundamental absorption edge, which marks the boundary between the range of transparency at lower energies and the strong absorption that occurs at higher energies. The energy of the fundamental absorption edge is determined by the forbidden gap, the energy difference between the highest valence and lowest conduction-band state of the crystal. The absorption edge or forbidden gap may be nonexistent (in metals such as Pb, white tin; in semimetals such as Sb, Bi), zero (in semiconductors such as gray tin, one of the HgCdTe alloys), fall in the infrared (in semiconductors such as Ge, Si, GaAs), or occur in the visible (in semiconductors such as GaP, CdS) or well into the ultraviolet (in insulators such as C, diamond; NaCl; SiO$_2$; LiF). *See* BAND THEORY OF SOLIDS.

Intrinsic absorption spectra are of two types, indirect or direct, according to whether or not a phonon participates in the absorption process. Indirect absorption processes, which are much less probable events, give rise to absorption coefficients typically 100–1000 times less than those for direct transitions. They are important in certain crystals (Ge, Si, GaP) in which the highest valence and lowest conduction states have different values for the electron momentum, and in which an absorption process requires a phonon for momentum conservation. In other crystals (GaAs, ZnO) the highest valence and lowest conduction states occur at the same electron momentum, and the fundamental edge is direct.

For either direct or indirect transitions, the excited electron is attracted by the Coulomb interaction to the vacancy, or hole, that was created in the valence band by its excitation. There is a strong tendency for the electron and hole to bind together to form a hydrogenlike state called an exciton. Excitons greatly affect the shape of absorption spectra near the fundamental edge. The positive binding energy means that the energy needed to create an exciton is somewhat less than that of the forbidden gap. Creating excitons makes it possible to absorb radiation at energies less than that of the forbidden gap. The lowest possible exciton absorption process results in a single line seen at the absorption edges of most semiconductors and insulators. In addition, excitonic effects cause a strong enhancement of the absorption process immediately above the fundamental edge. *See* EXCITON.

Absorption structures above the fundamental edge are due to critical points, which—like the fundamental edge itself—represent energies at which new valence-conduction band-pair transitions become possible (an M_0 critical point at which the electron energy reaches a relative minimum as a function of electron momentum) or where formerly available band-pair transitions become no longer possible (an M_3 critical point at which electron energy reaches a relative maximum). Such singularities also occur at saddle points, where a relative maximum of electron energy is reached in one direction of momentum, and relative minima in the other two (an M_1 critical point), or where relative maxima are reached in two directions, and a relative minimum in the remainder (an M_2 critical point).

Above approximately 10 eV, the conduction-band states become more free-electron-like, and structure in absorption spectra dies out. This is the plasma region, as indicated in the illustration. Absorption structures beyond this region, in the far-ultraviolet and x-ray spectral regions, originate from core valence bands. Since the crystal potential is a minor perturbation on deep core levels, and the deep core levels contribute negligibly to bonding, they are very narrow in energy and can be considered more properly as atomic levels. Consequently, the source of any structure in core-level absorption spectra is primarily conduction bands. Core-level absorption spectra begin in the $20-25$ eV region for Ga-V compounds (GaP, GaAs, GaSe), near 100 eV for Si, and near 340 eV for C.

Extrinsic absorption. Extrinsic absorption processes involve states associated with deviations from crystal perfection, such as vacancies, interstitials, and impurities. Since these states occur in low (approximately $10^{14}-10^{18}$ cm^{-3}) concentrations relative to the host atoms (approximately 10^{23} cm^{-3}), extrinsic absorption is weak relative to intrinsic processes and is typically important only where the crystal is otherwise transparent in the forbidden gap. Nevertheless, the ability to obtain macroscopic lengths of crystals (on the order of a centimeter) can make extrinsic absorption readily apparent, as, for example, in the color centers of alkali halides, or in the poor optical quality of industrial diamonds. Although extrinsic processes are not so important in absorption, they are vitally important in luminescence, where the presence or absence of radiation may depend entirely on the types and concentrations of impurities in a crystal. *See* COLOR CENTERS; LUMINESCENCE.

[DAVID ASPNES]

Bibliography: J. N. Hodgson, *Optical Absorption and Dispersion in Solids*, 1970; T. S. Moss et al., *Semiconductor Opto-electronics*, 1973; J. I. Pankove, *Optical Processes in Semiconductors*, 1971; *Semiconductors and Semimetals*, vol. 3: Optical Properties of III-V Compounds, 1967.

Crystal counter

A detector of radiation in which the sensitive material is a high-resistivity crystal. The crystals are diced into wafers 0.5 to 2 mm in thickness and 4 to 100 mm^2 in surface area. Various techniques are used to contact the larger surfaces. A field of about 10^4 V/cm is used to collect the charge liberated by the radiation. Crystal and junction counters are very similar, but differ fundamentally in the way the high-resistivity material is created. In a junction counter, a reverse-biased junction creates a depletion region which is the high-resistivity region. At the same time, the reverse bias gives high charge-collection field. In the crystal counter, the high resistivity comes from the basic crystalline material. It is often difficult to decide whether a particular radiation detector should be categorized as a crystal counter or a junction counter.

Crystals that have been studied as possible crystal counters include diamond, silver chloride, zinc sulfide, thallium iodide, thallium chloride, cadmium telluride, and mercuric iodide. Cadmium telluride (CdTe) and mercury iodide (HgI$_2$) are considered the most preferable crystals. Their absorption for gamma rays is greater than that for silicon or germanium. Also, they have larger band gaps (1.45 eV for cadmium telluride and 2.15 eV for mercury iodide)—a characteristic which lowers leakage currents, and makes possible high-resolution room-temperature detection of gamma rays.

Circuit. The basic circuit for a crystal counter is shown in the illustration. When a charged particle enters the crystal, it loses energy by creating electron-hole pairs. A gamma ray entering the crystal interacts with an atom of the crystal. Then, depending upon the interaction process—photoelectric, Compton, or pair production—the gamma ray will transfer all or part of its energy to an electron or to an electron-positron pair. These charged particles, in turn, dissipate their energy by creating electron-hole pairs. The number of electron-hole pairs is proportional to the energy deposited in the crystal. *See* HOLES IN SOLIDS.

When an electron-hole pair is generated in the crystal, the electron and the hole induce equal and opposite charges on the electrodes (see illustration). As the electron moves toward the positive electrode, the positive charge it induces on this electrode increases uniformly. The operational amplifier provides this charge through the feedback capacitor, which becomes charged as shown. The positive charge on the collecting electrode reaches its maximum value and is neutralized when the electron is collected. Simultaneously, the hole induces a negative charge on the collecting electrode, and this charge decreases uniformly to zero as the hole moves to the negative electrode. This decreasing negative charge on the collecting

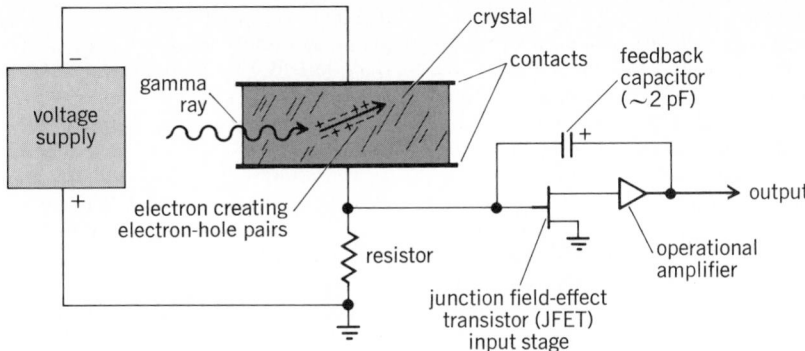

Counter circuit. Output signal depends on number of electron-hole pairs, which is proportional to energy deposited in crystal.

electrode is compensated by supplying positive charge from the operational amplifier and results in even more positive charge on the feedback capacitor.

Limitations and performance improvement. Only when both positive and negative carrier types are collected does a charge that is equal in magnitude to one of the carrier types appear on the feedback capacitor. This requirement of collecting both carriers places stringent demands on the crystal. That is, both carrier types should be similarly mobile to ensure an output charge signal proportional to the energy deposited in the crystal. If the carrier types are not similarly mobile, the charge signal is determined by the charge deposited and the location of that deposit. Hole trapping (or incomplete charge collection), excessive noise due to the high leakage current, the electronic noise of the preamplifier system, and the statistics of charge generation limit the resolution achieved with cadmium telluride and mercury iodide. Increasing the collecting field to minimize trapping causes an increase in the noise leakage current.

Other methods for reducing the effects of hole trapping include: (1) Differentiating the charge pulse. This gives a pulse which is proportional in magnitude to the number of moving carriers (in this case, electrons). (2) Searching for the spot in the crystal which gives the best resolution. (3) Introducing low-energy x-rays near the negative electrode. The electron-hole pairs are formed very near this electrode. Since the holes which become trapped are close to the negative electrode, their contribution to the total charge pulse is small.

Other methods such as cooling the detector, pulsing the bias, recording the spectra for only the first 20 seconds, and illuminating the crystal with infrared can be used to improve the performance.

Many of the preceding solutions may be unacceptable in the practical use of counters; however, they do show the inherent potential of good material. As methods of growing crystals improve to yield larger volumes that are highly uniform in purity, stoichiometry, and crystalline perfection, and as better detector fabrication methods are developed, crystal counters will achieve the efficiency of scintillators with the energy resolution of germanium. See PARTICLE DETECTOR; SCINTILLATION COUNTER.   [JAMES MC KENZIE]

Bibliography: R. C. Whited and M. M. Schieber, Cadmium telluride and mercuric iodide gamma radiation detectors, in D. A. Bromley (ed.), *Detectors in Nuclear Science*, Nuclear Instruments and Methods, vol. 162, pp. 113–123, 1979.

Crystal defects

Departures of a crystalline solid from a regular array of atoms or ions. A "perfect" crystal of NaCl, for example, would consist of alternating Na^+ and Cl^- ions on an infinite three-dimensional simple cubic lattice, and a simple defect (a vacancy) would be a missing Na^+ or Cl^- ion. There are many other kinds of possible defects, ranging from simple and microscopic, such as the vacancy and other structures illustrated in Fig. 1, to complex and macroscopic, such as the inclusion of another material, or a surface.

Natural crystals always contain defects, often in abundance, due to the uncontrolled conditions under which they were formed. The presence of defects which affect the color can make these crystals valuable as gems, as in ruby (Cr replacing a small fraction of the Al in Al_2O_3). Crystals prepared in the laboratory will also always contain defects, although considerable control may be exercised over their type, concentration, and distribution.

The importance of defects depends upon the material, type of defect, and properties which are being considered. Some properties, such as density and elastic constants, are proportional to the

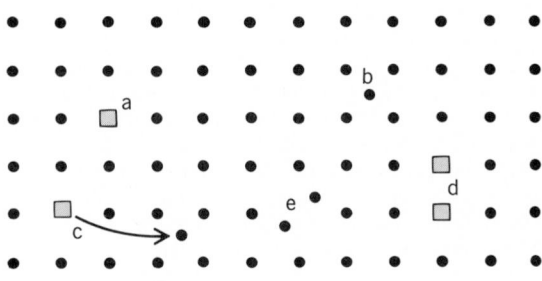

key:

a = vacancy (Schottky defect)
b = interstitial
c = vacancy—interstitial pair (Frenkel defect)
d = divacancy
e = split interstitial
▢ = vacant site

Fig. 1. Some simple defects in a lattice.

concentration of defects, and so a small defect concentration will have a very small effect on these. Other properties, such as the color of an insulating crystal or the conductivity of a semiconductor crystal, may be much more sensitive to the presence of small numbers of defects. Indeed, while the term defect carries with it the connotation of undesirable qualities, defects are responsible for many of the important properties of materials, and much of solid-state physics and chemistry and materials science involves the study and engineering of defects so that solids will have desired properties. A defect-free silicon crystal would be of little use in modern electronics; the use of silicon in devices is dependent upon small concentrations of chemical impurities such as phosphorus and arsenic which give it desired electronic properties.

This article will concentrate on the simple defects and their properties, after a brief mention of surfaces and interfaces. Certain types of defects are discussed in detail in other articles. *See* COLOR CENTERS; GRAIN BOUNDARIES; LATTICE VIBRATIONS; LUMINESCENCE; TWINNING (CRYSTALLOGRAPHY).

Surfaces and interfaces. The importance of surfaces depends very much upon the properties which are being considered and upon the geometry of the specimen. In a specimen no more than a few hundred atoms thick, a significant fraction of the atoms is close to the surface, while in a sample 10^{-2} m thick or greater, a much smaller fraction of the atoms is close to the surface. In the latter case, it makes sense to consider "bulk" properties which are surface-independent. Even in this case, surface effects may be important, as in the case of metals which reflect visible light from their surfaces, but they may generally be separated from bulk effects.

Surfaces are of great importance in determining the chemical properties of solids, since they are in contact with the environment. Such properties include chemical reactions involving the surface, such as corrosion; they also include the role of solids as catalysts in chemical reactions. It is often found that surface atoms reconstruct; that is, they displace and form bonds different from those existing in the bulk. Experimental studies of surfaces are difficult, due to the high vacuum which must be maintained and the relatively small number of atoms which are involved. *See* SURFACE PHYSICS.

An interface is the boundary between two dissimilar solids; it is in a sense an interior surface. For example, a simple metal-oxide-semiconductor (MOS) electronic device has two interfaces, one between metal and semiconductor, and the other between semiconductor and oxide. (In this case the oxide is not crystalline.) Interfaces tend to be chemically and electrically active, and to have large internal strains associated with structural mismatch.

Chemical impurities. An important class of crystal defect is the chemical impurity. The simplest case is the substitutional impurity, for example, a zinc atom in place of a copper atom in metallic copper. Impurities may also be interstitial; that is, they may be located where atoms or ions normally do not exist.

In metals, impurities usually lead to an increase in the electrical resistivity. Electrons which would

travel as unscattered waves in the perfect crystal are scattered by impurities. Thus zinc or phosphorus impurities in copper decrease the conductivity. Impurities can play an important role in determining the mechanical properties of solids. As discussed below, an impurity atom can interact with a structural defect called a dislocation to increase the strength of a solid; hydrogen, on the other hand, can lead to brittle fracture of some metals.

Impurities in semiconductors are responsible for the important electrical properties which lead to their widespread use. The electronic states of all solids fall into energy bands, quasicontinuous in energy. In nonmetals there is a band gap, a region of energy in which no states exist, between filled (valence) and empty (conduction) bands. Because electrons in filled bands do not conduct electricity, perfect nonmetals act as insulators. Impurities (and other defects) often introduce energy levels within the forbidden gap (Fig. 2). If these are close

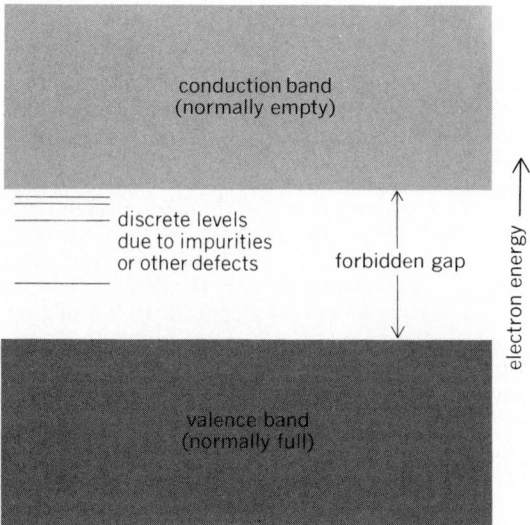

Fig. 2. Band structure of an imperfect nonmetallic crystal, including defect states which occur in the normally forbidden energy gap.

in energy to the valence or conduction band, electrons may be thermally excited from a nearby filled level (donor) into the conduction band, or from the valence band into a nearby empty level (acceptor), in the latter case leaving an unoccupied state or hole in the valence band. In both cases, conduction will occur in an applied electric field. Semiconductor devices depend on the deliberate addition (doping) of appropriate impurities. Other impurities may have energy levels which are not close in energy to either valence or conduction band (so-called deep traps). These may be undesirable if they trap the conducting particles. Thus careful purification is an important part of semiconductor technology. *See* BAND THEORY OF SOLIDS; SEMICONDUCTOR.

The energy levels associated with impurities and other defects in nonmetals may also lead to optical absorption in interesting regions of the spectrum. Ruby and other gems are examples of this; in addition, the phenomenon of light emission or luminescence is often impurity-related. Solid-state lasers generally involve impurity ions. *See* LASER.

Structural defects. Even in a chemically pure crystal, structural defects will occur. Thermal equilibrium exists when the Gibbs free energy G is minimized. G is given by Eq. (1), plus a pressure-

$$G = E - TS \qquad (1)$$

times-volume term which can often be neglected. Here E is the energy, T is the absolute temperature, and S is the entropy. A perfect crystal has a lower energy E than one containing structural defects, but the presence of defects increases the entropy S. Thus at $T = 0$ K, a perfect crystal is stable, but at higher values of T, G has its lowest value when structural defects exist.

Crystals may be grown by solidifying the molten substance. If a crystal is cooled slowly (annealed), the number of defects will decrease. However, as the temperature decreases it becomes more difficult for atoms to move in such a way as to "heal" the defects, and in practice one often has more defects than would be expected in thermal equilibrium. *See* FREE ENERGY.

In thermal equilibrium the number of defects at temperatures of interest may be nonnegligible. Application of the methods of statistical mechanics yields Eq. (2). Here n is the number of defects,

$$\frac{n}{N} = A e^{-BE/kT} \qquad (2)$$

N is the total number of atoms, E is the energy required to create a defect, and k is Boltzmann's constant. A and B are dimensionless constants, generally of order 1, whose actual value depends on the type of defect being considered. For example, for vacancies in monatomic crystals such as silicon or copper, $A = B = 1$. E is typically of order 1 eV; since $k = 8.62 \times 10^{-5}$ eV/K, at $T = 1000$ K $n/N = \exp[-1/(8.62 \times 10^{-5} \times 1000)] \simeq 10^{-5}$, or 10 parts per million. For many purposes this fraction would be intolerably large, although as mentioned this number may be reduced by slowly cooling the sample.

Simple defects. Besides the vacancy, other types of structural defects exist (Fig. 1). The atom which left a normal site to create a vacancy may end up in an interstitial position, a location not normally occupied. Or it may form a bond with a normal atom in such a way that neither atom is on the normal site, but the two are symmetrically displaced from it. This is called a split interstitial. The name Frenkel defect is given to a vacancy-interstitial pair, whereas an isolated vacancy is a Schottky defect. (In the latter case the missing atom may be thought of as sitting at a normal site at the surface.)

Simple defects are often combined. For example, when a small amount of alkaline-earth halide is melted with an alkali halide and the resulting mixture is recrystallized, the mixed solid is found to have a large number of alkali-ion vacancies, approximately one for each alkaline-earth ion. Since the alkaline-earth ions are divalent, the alkali-ion vacancies act as "charge compensators" to make the crystal neutral. Such a crystal has a relatively high electrical conductivity associated with the movement of positive ions into the vacancies (or, alternatively, movement of the vacancies

through the crystal). This general phenomenon of ionic conductivity is of practical interest, for example, in attempts to develop solid-state batteries.

Simple defects tend to aggregate if thermally or optically stimulated. The F-center in an alkali halide is an electron trapped at a halogen-ion vacancy. Under suitable conditions of optical or thermal stimulation, it is found that F-centers come together to form F_2- or M-centers (two nearest F-centers), F_3- or R-centers (three nearest F-centers), and so on.

Extended defects and mechanical properties. The simplest extended structural defect is the dislocation. An edge dislocation is a line defect which may be thought of as the result of adding or subtracting a half-plane of atoms (Fig. 3a). A screw dislocation is a line defect which can be thought of as the result of cutting partway through the crystal and displacing it parallel to the edge of the cut (Fig. 3b). The lattice displacement associated with a dislocation is called the Burgers vector b.

Dislocations generally have some edge and some screw character. Since dislocations cannot termi-

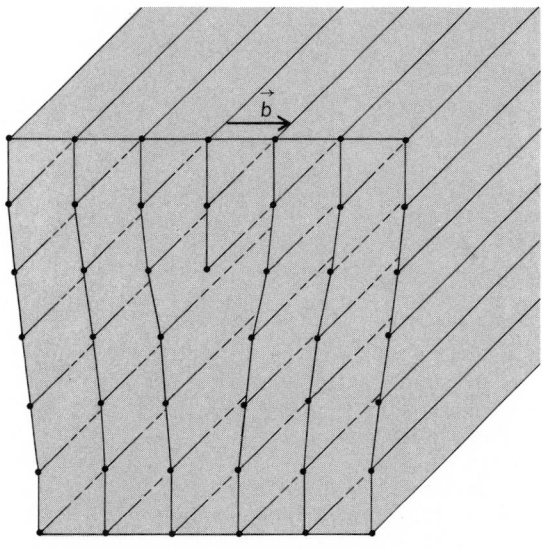

(a)

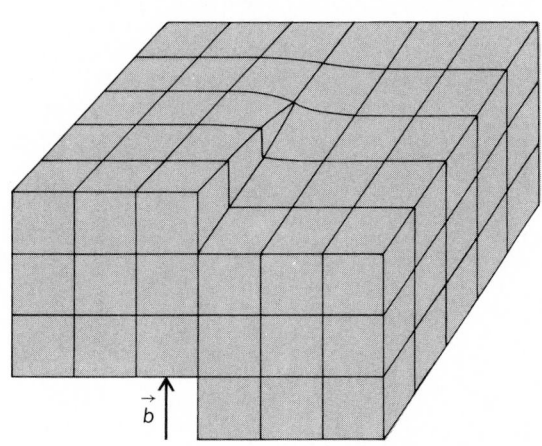

(b)

Fig. 3. Extended structural defects. (a) Edge dislocation. (b) Screw dislocation. In each case the Burgers vector is denoted by b.

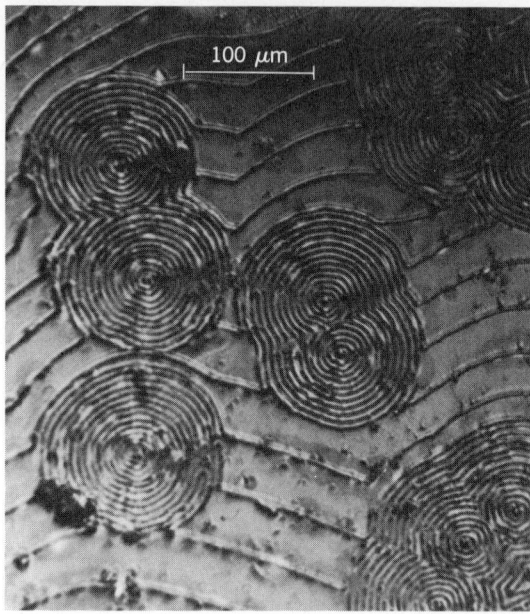

Fig. 4. Crystal-growth spirals on the surface of a silicon carbide crystal. Each spiral step originates in a screw dislocation defect. From the side the surface looks like an ascending ramp wound around a flat cone. (*General Electric*)

nate inside a crystal, they either intersect the surface or an internal interface (called a grain boundary) or form a closed loop within the crystal.

Dislocations are of great importance in determining the mechanical properties of crystals. A dislocation-free crystal is resistant to shear, because atoms must be displaced over high-potential-energy barriers from one equilibrium position to another. It takes relatively little energy to move a dislocation (and thereby shear the crystal), because the atoms at the dislocation are barely in stable equilibrium. Such plastic deformation is known as slip.

To extend the earlier argument about the difficulty in preparing perfect crystals, it should be anticipated that most crystals will contain dislocations in ample numbers and that special care would have to be taken to prepare a dislocation-free crystal. The latter is in most cases impractical (although in fact some electronic materials are dislocation-free). Although practical metallurgy developed centuries before dislocations were known

to exist, most strengthening processes are methods for dealing with dislocations.

Dislocations will not be mechanically deleterious if they cannot move. Three major ways of hindering dislocation motion are: "pinning" by dissolved impurities; "blocking" by inclusions; and tangling. Each of these is described briefly below.

Impurities tend to collect near dislocations if the solid is sufficiently hot that the impurities can move. The presence of these impurities along a dislocation removes the energetic equivalence of the dislocation in different positions; now it requires energy to move the dislocations unless the cloud of dissolved impurities can move as well; and to move them will involve overcoming a large energy barrier. This approach's success is lessened at higher temperatures, when the impurities tend to migrate away.

Small particles of a second substance can "block" dislocation motion. Steel is iron with small inclusions of iron carbide, whose major function is dislocation blocking.

A large, randomly oriented collection of dislocations can tangle and interfere with their respective motions. This is the principle behind work-hardening.

Dislocations have other important properties. The growth of crystals from the vapor phase tends to occur where screw dislocations intersect the surface, for in these regions there are edges rather than planes, and the attractive forces for atoms will be greater. Figure 4 shows spiral growth patterns of silicon carbide (SiC) derived from a screw dislocation. *See* CRYSTAL GROWTH.

A common planar structural defect is the stacking fault. This can be thought of as the result of slicing a crystal in half, twisting the two pieces with respect to each other, and then reattaching them. Another planar defect. the grain boundary, is the boundary between two crystallites. In some cases it can be described in terms of a parallel set of edge dislocations.

Alloys. A solid containing more than one element is called an alloy, especially when the solid is metallic. There are a vast number of possible alloys, and the present discussion will treat two-component or binary systems as examples.

The term stoichiometry is used to describe the chemical composition of compounds and the resulting crystal structure. The alkali halides are particularly simple, forming compounds with well-defined compositions (for example, NaCl). The situation becomes more complicated for compounds in which one or more components can take on more than one valence state. In particular, in the transition-metal oxides different compositions can occur, with different crystal structures and defects. For example, TiO_2 represents a well-defined perfect crystal, whereas TiO_{2-x} would have an oxygen deficiency whose value is described by the subscript x. In this case the deficit oxygens are associated with neither oxygen vacancies nor titanium interstitials, but rather with defect aggregates called shear planes.

If two elements are mixed uniformly throughout the solid, the atomic arrangement may be described as ordered or disordered. For example, an alloy composed of equal numbers of two metal

(a)   (b)

Fig. 5. Binary alloy. (a) Ordered lattice. (b) Disordered lattice.

atoms may form an ordered lattice with a regular periodic arrangement of the two atoms with respect to each other, or a disordered lattice in which some of the atoms are on the "wrong" site (Fig. 5). The ordered arrangement is most stable at very low temperatures. The temperature at which the structure becomes disordered is called the transition temperature of the order-disorder transformation.

Defect chemistry. Most of the preceding discussion has treated the simpler types of defects and their static properties, that is, their properties under conditions such that the nature and location of the defects do not change with time. Situations have also been alluded to in which the number, locations, or types of defects change with time, either because of thermal effects (for example, slowly cooling a hot sample and annealing vacancies) or because of external irradiation or defect interactions (causing, for example, the aggregation of alkali-halide F-centers).

For both scientific and practical reasons, much of the research on crystal defects is directed toward the dynamic properties of defects under particular conditions, or defect chemistry. Much of the motivation for this arises from the often undesirable effects of external influences on material properties, and a desire to minimize these effects. Examples of defect chemistry abound, including one as familiar as the photographic process, in which incident photons cause defect modifications in silver halides or other materials. Properties of materials in nuclear reactors is another important case. Only a few examples will be considered here.

Thermal effects. Thermal effects are the easiest to understand from a microscopic point of view: raising the temperature increases the amplitude of atomic vibrations, thereby making it easier for atoms to "hop" from place to place over potential-energy barriers. Almost all dynamic processes are temperature-dependent, occurring more readily at high temperatures.

Defect creation by irradiation. Radiation can have profound effects on materials in both direct and indirect ways. For example, a beam of high-energy electrons, neutrons, or atomic nuclei (such as alpha particles) can create defects by simple collision processes: the incident particle imparts a portion of its momentum and energy to a lattice atom, thereby releasing the atom from its normal position. If the released atom has enough energy, it may collide with and release a second lattice atom, thereby creating a second vacancy, and so on; thus one incident particle may lead to the production of a number of defects. Other processes also exist. *See* CHARGED PARTICLE BEAMS.

A beam of photons may also create defects through a variety of processes. If the photon energy equals or exceeds the band gap, ionization can occur; that is, electrons can be excited from core or valence bands into the conduction bands as the photons are absorbed. Very energetic photons (gamma rays) may be absorbed or scattered, generating fast electrons, which in turn eject atoms through collision processes such as those occurring with incident particles.

Lower-energy photons (x-rays, ultraviolet, visible, and so on) have insufficient energy to create defects by such processes. In a few cases simple ionization results in defect creation; for example, in alkali halides excitation of an electron on a halogen ion leads to a large potential energy for covalent bond formation between the excited halogen ion and a neighboring halogen ion. This large attractive force leads to large local lattice vibrations, and under suitable conditions a significant fraction of the potential energy is converted to kinetic energy of a halogen atom, leading to its ejection from a normal position and the creation of a halogen ion vacancy plus an electron, or F-center. A replacement sequence occurs (Fig. 6) in which each halogen displaces its neighbor in the indicated direction. Eventually a halogen interstitial some distance away bonds with an adjacent halogen ion and becomes a so-called H-center. The first stages of this process can occur in very short times (on the order of 10^{-12} s).

A laser beam is often observed to result in crystal destruction, even when the crystal is an insulator and the photon energy is well below the band gap. One way this can occur is first by ionization of defects by the laser, producing a few conduction electrons. These conduction electrons can in turn be accelerated, by the large electric field present in the laser beam, to sufficiently high energy to ionize normal atoms; thus more electrons are available to be accelerated, and a type of cascade process occurs which leads to destruction of the crystal.

Effects of radiation on preexisting defects. More subtle effects must often be considered when studying the effects of radiation on device performance. This involves the interaction of radiation with existing (and desirable) defects. Aggregation of F-centers occurs when alkali halides are irradiated in the F-absorption band. This is a several-step process: (1) The excited F-center loses its electron to the conduction band, leaving behind a positively charged vacancy. (2) The conduction electron is subsequently trapped at another F-center, forming a (negative) F'-center. (3) Under the attractive Coulomb force, the vacancy moves to-

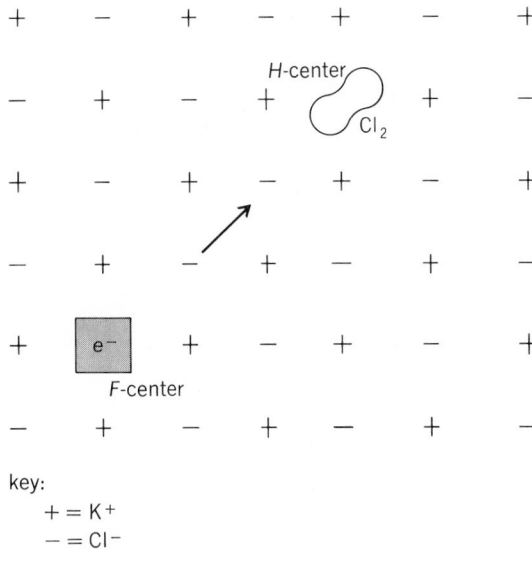

key:

  + = K⁺

  − = Cl⁻

Fig. 6. F- and H-center formation in KCl after a replacement sequence of halogen-halogen collisions along the direction indicated by the arrow.

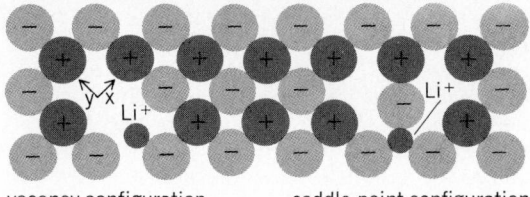

vacancy configuration   saddle-point configuration

(a)

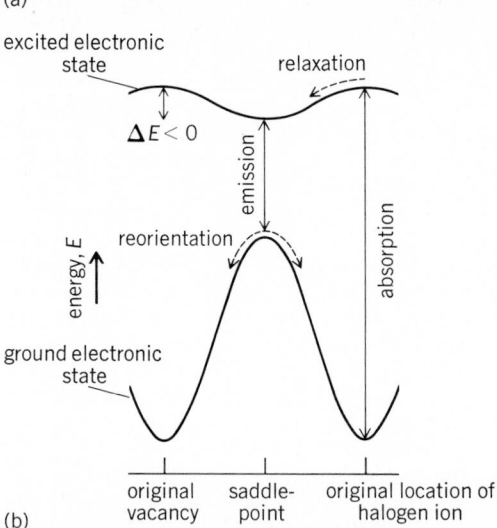

(b)

Fig. 7.   Reorientation of the type II F_A-center in alkali halides. (a) Vacancy and saddle-point configurations. (b) Schematic potential energy functions for motion of the neighboring (halogen)⁻ ion. *(From F. Luty, F_A-centers in alkali halide crystals, in W. Beall Fowler, ed., Physics of Color Centers, Academic, 1968)*

ward the F'-center. (4) The extra electron of the F'-center returns to the vacancy either by ionization or by tunneling, leading to two F-centers closer than normal. (5) The process repeats, until the two F-centers are adjacent.

Externally activated defect processes are also found in semiconductors, typically (but not solely) associated with irradiation. Charge-state processes involve the motion of preexisting defects following a change in the charge of the defect. For example, band-gap light generates electrons and holes in silicon, which can be trapped, for example, at a vacancy V in silicon. The doubly negative vacancy (V^{2-}) has a considerably smaller activation energy for motion than less negatively charged vacancies, and will consequently migrate more readily. Presumably differences in locations of the neighboring atoms are responsible for the different activation energies of motion.

Recombination processes may also occur. Here an electron and hole recombine at a defect and transfer their (electronic) energy to vibrational energy of the defect, leading to its motion. This effect has been observed, for example, at a recombination junction in GaAs, where defects introduced by proton bombardment were destroyed under charge injection.

A third possibility is excited-state effects. In this case a defect in an excited electronic state and its surroundings relax into a configuration in which defect motion is easier; in some instances, there may be a zero barrier energy to overcome. One of the best examples of this is the type II F_A-center in

alkali halide crystals, an F-center with an alkali impurity (usually Li) as one of its six nearest alkali neighbors (Fig. 7). After a photon is absorbed, the system relaxes into a "double-well" configuration with a Cl⁻ which was initially adjacent to both the Li⁺ and the vacancy moving into a "saddle-point" position half the way to the vacancy. The excited electron is then shared by the two small equivalent "wells." When the system returns to the ground state, there is equal probability of the Cl⁻ returning to its original site or to the original vacancy; in the latter case the vacancy will have moved. Similar processes are thought to occur in semiconductors. *See* CRYSTAL STRUCTURE; CRYSTALLOGRAPHY.   [W. BEALL FOWLER]

Bibliography: J. H. Albany (ed.), *Defects and Radiation Effects in Semiconductors, 1978,* 1979; J. H. Crawford, Jr., and L. M. Slifkin (eds.), *Point Defects in Solids,* vols. 1 and 2, 1972, 1975; N. B. Hannay (ed.), *Treatise on Solid State Chemistry,* vol. 2: *Defects in Solids,* 1975; C. Kittel, *Introduction to Solid State Physics,* 5th ed., 1976.

Crystal growth

The growth of crystals, of which all crystalline solids are composed, generally occurs by means of the following sequence of processes: (1) diffusion of the molecules of the crystallizing substance through the surrounding environment (or solution) to the surface of the crystal, (2) diffusion of these molecules over the surface of the crystal to special sites on the surface, (3) incorporation of molecules into the crystal at these sites, and (4) diffusion of the heat of crystallization away from the crystal surface. The rate of crystal growth may be limited by any of these four steps. The initial formation of the centers from which crystal growth proceeds is known as nucleation. *See* CRYSTALLIZATION.

Increasing the supersaturation of the crystallizing component or increasing the temperature independently increases the rate of crystal growth. However, in many physical situations the supersaturation is increased by decreasing the temperature. In these circumstances the rate of crystal growth increases with decreasing temperature at first, goes through a maximum, and then decreases. Often the growth is greatly retarded by traces of certain impurities.

After nucleation, the crystals in the medium grow isolated from one another for a time. However, if several differently oriented crystals are growing, they may finally impinge on one another, and intercrystalline (grain) boundaries will be formed. At relatively high temperatures, the average grain size in these polycrystalline aggregates increases with time by a process called grain growth, whereby the larger grains grow at the expense of the smaller.

Regular growth. During its growth into a fluid phase, a crystal often develops and maintains a definite polyhedral form which may reflect the characteristic symmetry of the molecular pattern of the crystal. The bounding faces of this form are those which are perpendicular to the directions of slowest growth. How this comes about is illustrated in Fig. 1, in which it is seen that the faces b, normal to the faster-growing direction, disappear, and the faces a, normal to the slower-growing directions, become predominant. Growth forms, like that shown in the figure, are not necessarily

CRYSTAL GROWTH

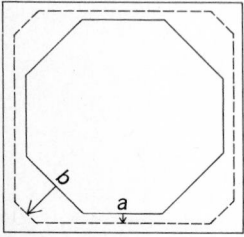

Fig. 1.   Schematic representation of cross section of crystal at three stages of growth.

equilibrium forms, but they are likely to be most regular when the departures from equilibrium are not large. *See* CRYSTAL STRUCTURE.

J. W. Gibbs pointed out that the molecular binding sites on the surface of a crystal can be of several kinds. Thus a molecule must be more weakly bound on a perfectly developed plane of molecules at the crystal surface than at a ledge formed by an incomplete plane one molecule thick (Fig. 2). Therefore, the binding of molecules in an island monolayer on the crystal surface will be less per molecule than it would be within a completed surface layer.

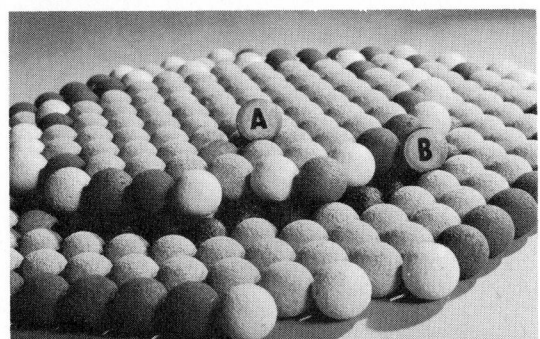

Fig. 2. A model of a crystal surface showing the inequivalent binding sites at which the atoms labeled *A* and *B* are located. (*General Electric Co.*)

The potential energy of a crystal is most likely to be minimum in forms containing the fewest possible ledge sites. This means that, in a regime of regular crystal growth, dilute fluid, and moderate departure from equilibrium, the crystal faces of the growth form are likely to be densely packed and molecularly smooth. There will be a critical size of monolayer, which will be a decreasing function of supersaturation, such that all monolayers smaller than the critical size will shrink out of existence, and those which are larger will grow to a complete layer. The critical monolayers form by a fluctuation process. Kinetic analyses along the lines initiated by M. Volmer and others indicate that the probability of critical fluctuations is so small that in finite systems perfect crystals will not grow, except at substantial departures from equilibrium. That, in ordinary experience, finite crystals do grow in a regular regime only at infinitesimal departures from equilibrium is explained by the screw dislocation theory of F. C. Frank. According to this theory, growth is sustained by indestructible surface ledges which result from the emergence of screw dislocations in the crystal face. *See* CRYSTAL DEFECTS.

The theory of interface structure and regular growth is not sufficiently developed to predict definitively the growth behavior and morphology of crystals growing into concentrated fluids or their own melts. However, a highly successful correlation, due to K. A. Jackson, indicates that the condition for the occurrence of a regular growth regime is that the entropy of crystallization be greater than $2k$ per molecule (k is Boltzmann's constant).

Irregular growth. When the departures from equilibrium (supersaturation or undercooling) are sufficiently large, the more regular growth shapes become unstable and dendritic (treelike), or cellular morphologies develop. Essentially, the development of protuberances on an initially regular crystal permits higher diffusive fluxes, but at the cost of higher interfacial area–crystal volume ratios. Theories which define the conditions for the onset of this morphological instability were developed by W. W. Mullins and R. F. Sekerka and by A. A. Chernov. These theories were generally confirmed by the studies of S. R. Coriell and coworkers, and revealed that a critical undercooling is indeed required for the development of protuberances on the edges of initially disk-shaped single ice crystals growing in water. *See* CRYSTAL WHISKERS; SINGLE CRYSTAL.

[DAVID TURNBULL]

Bibliography: W. K. Burton et al., *Phil. Roy. Soc. London*, 243A:299–358, 1951; A. A. Chernov, *Sov. Phys. Crystallogr.*, 16:734–753, 1972; R. H. Doremus et al. (eds.), *Growth and Perfection of Crystals*, 1958; R. L. Parker, in H. Ehrenreich (ed.), *Solid State Physics*, 1970.

Crystal optics

The study of the propagation of light, and associated phenomena, in crystalline solids. The propagation of light in crystals can actually be so complicated that not all the different phenomena are yet completely understood, and not all theoretically predicted phenomena have been demonstrated experimentally.

For a simple cubic crystal the atomic arrangement is such that in each direction through the crystal the crystal presents the same optical appearance. The atoms in anisotropic crystals are closer together in some planes through the material than in others. In anisotropic crystals the optical characteristics are different in different directions. In classical physics the progress of an electromagnetic wave through a material involves the periodic displacement of electrons. In anisotropic substances the forces resisting these displacements depend on the displacement direction. Thus the velocity of a light wave is different in different directions and for different states of polarization. The absorption of the wave may also be different in different directions. *See* DICHROISM; TRICHROISM.

In an isotropic medium the light from a point source spreads out in a spherical shell. The light from a point source embedded in an anisotropic crystal spreads out in two wave surfaces, one of which travels at a faster rate than the other. The polarization of the light varies from point to point over each wave surface, and in any particular direction from the source the polarization of the two surfaces is opposite. The characteristics of these surfaces can be determined experimentally by making measurements on a given crystal.

For a transparent crystal the theoretical optical behavior is well enough understood so that only a few measurements need to be made in order to predict the behavior of a light beam passing through the crystal in any direction. It is important to remember that the velocity through a crystal is not a function of position in the crystal but only of the direction through the lattice. For information closely related to the ensuing discussion *see* POLARIZED LIGHT. *See also* CRYSTAL STRUCTURE; REFRACTION OF WAVES.

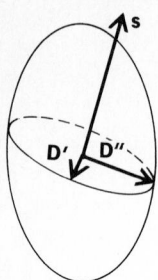

Fig. 1. Index ellipsoid, showing construction of directions of vibrations of **D** vectors belonging to a wave normal **s**. (*From M. Born and E. Wolf, Principles of Optics, Pergamon, 1959*)

Index ellipsoid. In the most general case of a transparent anisotropic medium, the dielectric constant is different along each of three orthogonal axes. This means that when the light vector is oriented along each direction, the velocity of light is different. One method for calculating the behavior of a transparent anisotropic material is through the use of the index ellipsoid, also called the reciprocal ellipsoid, optical indicatrix, or ellipsoid of wave normals. This is the surface obtained by plotting the value of the refractive index in each principal direction for a linearly polarized light vector lying in that direction (Fig. 1). The different indices of refraction, or wave velocities associated with a given propagation direction, are then given by sections through the origin of the coordinates in which the index ellipsoid is drawn. These sections are ellipses, and the major and minor axes of the ellipse represent the fast and slow axes for light proceeding along the normal to the plane of the ellipse. The length of the axes represents the refractive indices for the fast and slow wave, respectively. The most asymmetric type of ellipsoid has three unequal axes. It is a general rule in crystallography that no property of a crystal will have less symmetry than the class in which the crystal belongs. In other words, if a property of the crystal had lower symmetry, the crystal would belong in a different class. Accordingly, there are many crystals which, for example, have four- or sixfold rotation symmetry about an axis, and for these the index ellipsoid cannot have three unequal axes but is an ellipsoid of revolution. In such a crystal, light will be propagated along this axis as though the crystal were isotropic, and the velocity of propagation will be independent of the state of polarization. The section of the index ellipsoid at right angles to this direction is a circle. Such crystals are called uniaxial and the mathematics of their optical behavior is relatively straightforward. *See* CRYSTALLOGRAPHY.

Ray ellipsoid. The normal to a plane wavefront moves with the phase velocity. The Huygens wavelet, which is the light moving out from a point disturbance, will propagate with a ray velocity. Just as the index ellipsoid can be used to compute the phase or wave velocity, so can a ray ellipsoid be used to calculate the ray velocity. The length of the axes of this ellipsoid is given by the velocity of the linearly polarized ray whose electric vector lies in the axis direction. *See* PHASE VELOCITY.

The ray ellipsoid in the general case for an anisotropic crystal is given by Eq. (1), where α, β, and

$$\alpha^2 x^2 + \beta^2 y^2 + \gamma^2 z^2 = 1 \qquad (1)$$

γ are the three principal indices of refraction and where the velocity of light in a vacuum is taken to be unity. From this ellipsoid the ray velocity surfaces or Huygens wavelets can be calculated as just described. These surfaces are of the fourth degree and are given by Eq. (2). In the uniaxial

$$(x^2 + y^2 + z^2)\left(\frac{x^2}{\alpha^2} + \frac{y^2}{\beta^2} + \frac{z^2}{\gamma^2}\right) - \frac{1}{\alpha^2}\left(\frac{1}{\beta^2} + \frac{1}{\gamma^2}\right)x^2$$
$$- \frac{1}{\beta^2}\left(\frac{1}{\gamma^2} + \frac{1}{\alpha^2}\right)y^2 - \frac{1}{\gamma^2}\left(\frac{1}{\alpha^2} + \frac{1}{\beta^2}\right)z^2 + \frac{1}{\alpha^2\beta^2\gamma^2} = 0 \qquad (2)$$

case $\alpha = \beta$ and Eq. (2) factors into Eq. (3). These

$$x^2 + y^2 + z^2 = \frac{1}{\alpha^2}$$
$$\gamma^2(x^2 + y^2) + \alpha^2 z^2 = 1 \qquad (3)$$

are the equations of a sphere and an ellipsoid. The z axis of the ellipsoid is the optic axis of the crystal.

The refraction of a light ray on passing through the surface of an anisotropic uniaxial crystal can be calculated with Huygens wavelets in the same manner as in an isotropic material. For the ellipsoi-

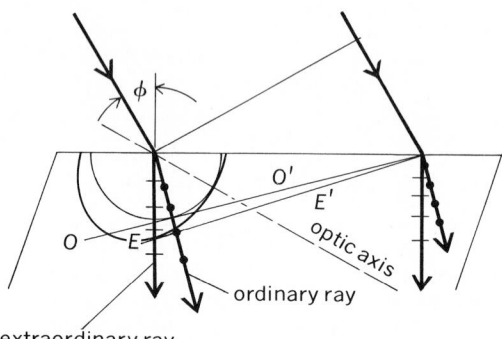

Fig. 3. Huygens construction when the optic axis lies in the plane of incidence. (*From F. A. Jenkins and H. E. White, Fundamentals of Optics, 3d ed., McGraw-Hill, 1957*)

dal wavelet this results in an optical behavior which is completely different from that normally associated with refraction. The ray associated with this behavior is termed the extraordinary ray. At a crystal surface where the optic axis is inclined at an angle, a ray of unpolarized light incident normally on the surface is split into two beams: the ordinary ray, which proceeds through the surface without deviation; and the extraordinary ray, which is deviated by an angle determined by a line drawn from the center of one of the Huygens ellipsoidal wavelets to the point at which the ellipsoid is tangent to a line parallel to the surface. The construction is shown in Fig. 2. The two beams are oppositely linearly polarized. When the incident beam is inclined at an angle ϕ to the normal, the ordinary ray is deviated by an amount determined by Snell's law of refraction, the extraordinary ray by an amount which can be determined in a manner similar to that of the normal incidence case already described. The plane wavefront in the crystal is first found by constructing Huygens wavelets as shown in Fig. 3. The line from the cen-

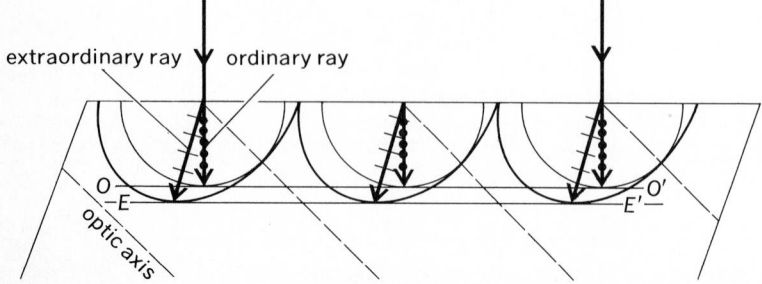

Fig. 2. Huygens construction for a plane wave incident normally on transparent calcite. If one proceeds to find the common tangents to the secondary wavelets shown, the two plane waves labeled *OO'* and *EE'* are obtained. (*From F. A. Jenkins and H. E. White, Fundamentals of Optics, 3d ed., McGraw-Hill, 1957*)

ter of the wavelet to the point of tangency gives the ray direction and velocity.

The relationship between the normal to the wavefront and the ray direction can be calculated algebraically in a relatively straightforward fashion. The extraordinary wave surface, or Huygens wavelet, is given by Eq. (3), which can be rewritten as Eq. (4), where ϵ is the extraordinary index of re-

$$\epsilon^2(x^2 + y^2) + \omega^2 z^2 = 1 \qquad (4)$$

fraction and ω the ordinary index. A line from the center of this ellipsoid to a point (x_1, y_1, z_1) on the surface gives the velocity of a ray in this direction. The wave normal corresponding to this ray is found by dropping a perpendicular from the center of the ellipsoid to the plane tangent at the point (x_1, y_1, z_1). For simplicity consider the point in the plane $y = 0$. The tangent at the point $x_1 z_1$ is given by Eq. (5). The slope of the normal to this line is given by Eq. (6). The tangent of the angle between

$$\epsilon^2 x x_1 + \omega^2 z z_1 = 1 \qquad (5)$$

$$\frac{z}{x} = \frac{\omega^2 z_1}{\epsilon^2 x_1} \qquad (6)$$

the optic axis and the wave normal is the reciprocal of this number, as shown in Eq. (7), where ψ is

$$\tan \varphi = \frac{\epsilon^2 x_1}{\omega^2 z_1} = \frac{\epsilon^2}{\omega^2} \tan \psi \qquad (7)$$

the angle between the ray and the optic axis. The difference between these two angles, τ, can be calculated from Eq. (8). This quantity is a maximum when Eq. (9) holds.

$$\tan \tau = \frac{\tan \varphi - \tan \psi}{1 + \tan \varphi \tan \psi} \qquad (8)$$

$$\tan \varphi = \pm \frac{\epsilon}{\omega} \qquad (9)$$

One of the first doubly refracting crystals to be discovered was a transparent variety of calcite called Iceland spar. This unaxial crystal cleaves into slabs in which the optic axis makes an angle of 45° with one pair of surfaces. An object in contact with or a few inches from such a slab will thus appear to be doubled. If the slab is rotated about a normal to the surface, one image rotates about the other. For the sodium D lines at 589 mμ the indices for calcite are given by Eq. (10). From these, the

$$\epsilon = 1.486 \qquad \omega = 1.659 \qquad (10)$$

maximum angle τ_{max} between the wave normal and the ray direction is computed to be 6° 16′. The wave normal at this value makes an angle of 41° 52′ with the optic axis. This is about equal to the angle which the axis makes with the surface in a cleaved slab. Accordingly, the natural rhomb gives nearly the extreme departure of the ray direction from the wave normal.

Interference in polarized light. One of the most interesting properties of plates of crystals is their appearance in convergent light between pairs of linear, circular, or elliptical polarizers. An examination of crystals in this fashion offers a means of rapid identification. It can be done with extremely small crystals by the use of a microscope in which the illuminating and viewing optical systems are equipped with polarizers. Such a polarizing micro-

scope is a common tool for the mineralogist and the organic chemist.

In convergent light the retardation through a birefringent plate is different for each direction. The slow and fast axes are also inclined at a different angle for each direction. Between crossed linear polarizers the plate will appear opaque at those angles for which the retardation is an integral number of waves. The locus of such points will be a series of curves which represent the characteristic interference pattern of the material and the angle at which the plate is cut. In order to calculate the interference pattern, it is necessary to know the two indices associated with different angles of plane wave propagation through the plate. This can be computed from the index ellipsoid. *See* INTERFERENCE OF WAVES.

Ordinary and extraordinary indices. For the uniaxial crystal there is one linear polarization direction for which the wave velocity is always the same. The wave propagated in this direction is called the ordinary wave. This can be seen directly by inspection of the index ellipsoid. Since it is an ellipsoid of revolution, each plane passing through the center will produce an ellipse which has one axis equal to the axis of the ellipsoid. The direction of polarization will always be at right angles to the plane of incidence and the refractive index will be constant. This constant index is called the ordinary index. The extraordinary index will be given by the other axis of the ellipse and will depend on the propagation direction. When the direction is along the axis of the ellipsoid, the extraordinary index will equal the ordinary index.

From the equation of the ellipsoid, Eq. (11), one

$$\frac{x^2}{\omega^2} + \frac{y^2}{\omega^2} + \frac{z^2}{\epsilon^2} = 1 \qquad (11)$$

can derive the expression for the ellipse and in turn Eq. (12) for the extraordinary index n_e as a

$$n_e = \frac{\omega \epsilon}{(\epsilon^2 \cos^2 r + \omega^2 \sin^2 r)^{1/2}} \qquad (12)$$

function of propagation direction. Since the ellipsoid is symmetrical it is necessary to define this direction only with respect to the ellipsoid axis. Here r is the angle in the material between the normal to the wavefront and the axis of the ellipsoid, n_e is the extraordinary index associated with this direction, ω is the ordinary index, and ϵ is the maximum value of the extraordinary index (usually referred to simply as the extraordinary index).

A slab cut normal to the optic axis is termed a Z-cut or C-cut plate (Fig. 4). When such a plate is placed between crossed linear polarizers, such as Nicol prisms, it gives a pattern in monochromatic light shown in Fig. 5. The explanation of this pattern is obtained from the equations given for the indices of refraction. To a first approximation the retardation for light of wavelength λ passing through the plate at an angle r can be written as Eq. (13), where d is the thickness of the plate. The

$$\Gamma = \frac{(n_e - \omega) d}{\lambda \cos r} \qquad (13)$$

axis of the equivalent retardation plate at an angle r will be in a plane containing the optic axis of the plate and the direction of light propagation. Wherever Γ is a whole number of waves, the light leaving the plate will have the same polarization as the

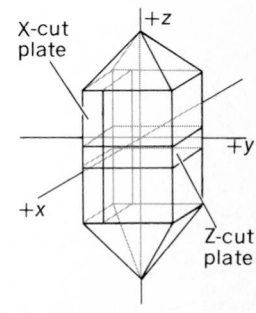

CRYSTAL OPTICS

X-cut plate

+z

+y

+x

Z-cut plate

Fig. 4. X-cut and Z-cut plates in NH$_4$H$_2$PO$_4$ crystal.

Fig. 5. Interference figure from fluorspar that has been cut perpendicular to the optic axis and placed between crossed Nicol prisms. (*From M. Born and E. Wolf, Principles of Optics, Pergamon, 1959*)

incident light. The ordinary index is constant. The extraordinary index is a function of r alone, as seen from Eq. (12). Accordingly, the locus of direction of constant whole wave retardation will be a series of cones. If a uniform white light background is observed through such a plate, a series of rings of constant whole wave retardation will be seen. Since the retardation is a function of wavelength, the rings will appear colored. The innermost ring will have the least amount of color. The outermost ring will begin to overlap and disappear as the blue end of the spectrum for one ring covers the red end of its neighbor. The crystal plate has no effect along the axes of the polarizers since here the light is polarized along the axis of the equivalent retardation plate. The family of circles is thus bisected by a dark cross. When the crystal is mounted between like circular polarizers, the dark cross is not present and the center of the system of rings is clear. During World World War II crystal plates were used in this fashion as gunsights. The rings appear at infinity even when the plate is close to the eye. Lateral movement of the plate causes no angular motion of the ring system. The crystal plate could be rigidly fastened to a gun mount and adjusted so as to show at all times the direction in space in which the gun was pointing.

Uniaxial crystal plates in which the axis lies in the plane of the plate are termed X-cut or Y-cut. When such plates are observed between crossed polarizers, a pattern of hyperbolas is observed.

When crystal plates are combined in series, the patterns become much more complex. In addition to the fringes resulting from the individual plates, a system of so-called moiré fringes appears.

Negative and positive crystals. In calcite the extraordinary wave travels faster than the ordinary wave. Calcite and other materials in which this occurs are termed negative crystals. In positive crystals the extraordinary wave travels slower than the ordinary wave. In the identification of uniaxial

minerals one of the steps is the determination of sign. This is most easily demonstrated in a section cut perpendicular to the optic axis. Between crossed linear polarizers the pattern in convergent light, as already mentioned, is a series of concentric circles which are bisected by a dark cross. When a quarter-wave plate is inserted between one polarizer and the crystal, with its axis at 45° to the polarizing axis, the dark rings are displaced outward or inward in alternate quadrants. If the rings are displaced outward along the slow axis of the quarter-wave plate, the crystal is positive. If they are displaced inward, it is negative.

If a Z-cut plate of a positive uniaxial crystal is put in series with a similar plate of a negative crystal, it is possible to cancel the birefringence so that the combination appears isotropic.

Biaxial crystals. In crystals of low symmetry the index ellipsoid has three unequal axes. These crystals are termed biaxial and have two directions along which the wave velocity is independent of the polarization direction. These correspond to the two sections of the ellipsoid which are circular. These sections are inclined at equal angles to the major axes of the ellipsoid, and their normals lie in a plane containing the major and intermediate axes. In convergent light between polarizers, these crystals show a pattern which is quite different from that which appears with uniaxial crystals. A plate cut normal to the major axis of the index ellipsoid has a pattern of a series of lemniscates and ovals. The directions corresponding to the optic axes appear as black spots between crossed circular polarizers. The interference pattern between crossed linear polarizers is shown in Fig. 6.

Angle between optic axes. One of the quantities used to describe a biaxial crystal or to identify a biaxial mineral is the angle between the optic axes. This can be calculated directly from the equation for the index ellipsoid, Eq. (14), where α, β, and γ

$$\frac{x^2}{\alpha^2} + \frac{y^2}{\beta^2} + \frac{z^2}{\gamma^2} = 1 \qquad (14)$$

Fig. 6. Interference figure from Brazil topaz. (*From M. Born and E. Wolf, Principles of Optics, Pergamon, 1959*)

are the three indices of refraction of the material. The circular sections of the ellipsoid must have the intermediate index as a radius. If the relative sizes of the indices are so related that $\alpha > \beta > \gamma$, the circular sections will have β as a radius and the normal to these sections will lie in the xz plane. The section of the ellipsoid cut by this plane will be the ellipse of Eq. (15). The radius of length β

$$\frac{x^2}{\alpha^2}+\frac{z^2}{\gamma^2}=1 \qquad (15)$$

will intersect this ellipse at a point x_1z_1 where Eq. (16) holds. The solution of these two equations

$$x_1{}^2+z_1{}^2=\beta^2 \qquad (16)$$

gives for the points x_1z_1 Eq. (17). These points and

$$x_1=\pm\sqrt{\frac{\alpha^2(\gamma^2-\beta^2)}{\gamma^2-\alpha^2}} \\ z_1=\pm\sqrt{\frac{\gamma^2(\alpha^2-\beta^2)}{\alpha^2-\gamma^2}} \qquad (17)$$

the origin define the lines in the xz plane which are also in the planes of the circular sections of the ellipsoid. Perpendiculars to these lines will define the optic axes. The angle Ω between the axes and the z direction will be given by Eq. (18).

$$\tan^2\Omega=\frac{z_1{}^2}{x_1{}^2}=\frac{(1/\alpha^2)-(1/\beta^2)}{(1/\beta^2)-(1/\gamma^2)} \qquad (18)$$

The polarization direction for light passing through a biaxial crystal is computed in the same way as for a uniaxial crystal. The section of the index normal to the propagation direction is ordinarily an ellipse. The directions of the axes of the ellipse represent the vibration direction and the lengths of the half axes represent the indices of refraction. One polarization direction will lie in a plane which bisects the angle made by the plane containing the propagation direction and one optic axis and the plane containing the propagation direction and the other optic axis. The other polarization direction will be at right angles to the first. When the two optic axes coincide, this situation reduces to that which was demonstrated earlier for the case of uniaxial crystals.

Conical refraction. In biaxial crystals there occurs a set of phenomena which have long been a classical example of the theoretical prediction of a physical characteristic before its experimental discovery. These are the phenomena of internal and external conical refraction. They were predicted theoretically in 1832 by Sir William Hamilton and experimentally demonstrated in 1833 by H. Lloyd.

As shown earlier, the Huygens wavelets in a biaxial crystal consist of two surfaces. One of these has its major axis at right angles to the major axis of the other. The two thus intersect. In making the geometrical construction to determine the ray direction, two directions are found where the two wavefronts coincide and the points of tangency on the two ellipsoids are multiple. In fact, the plane which represents the wavefront is found to be tangent to a circle which lies partly on one surface and partly on the other. The directions in which the wavefronts coincide are the optic axes. A ray

incident on the surface of a biaxial crystal in such a direction that the wavefront propagates along an axis will split into a family of rays which lie on a cone. If the crystal is a plane parallel slab, these rays will be refracted at the second surface and transmitted as a hollow cylinder parallel to the original ray direction. Similarly, if a ray is incident on the surface of a biaxial crystal plate at such an angle that it passes along the axes of equal ray velocity, it will leave the plate as a family of rays lying on the surface of a cone. The first of these phenomena is internal conical refraction; the second is external conical refraction. Equation (19)

$$\tan\psi=\frac{\beta}{\sqrt{\alpha\gamma}}\sqrt{(\beta-\alpha)(\gamma-\beta)} \qquad (19)$$

gives the half angle ψ of the external cone.

[BRUCE H. BILLINGS]

Bibliography: M. Born and E. Wolf, *Principles of Optics*, 5th ed., 1975; F. A. Jenkins and H. E. White, *Fundamentals of Optics*, 4th ed., 1976; E. E. Wahlstrom, *Optical Crystallography*, 5th ed., 1979; E. A. Wood, *Crystals and Light*, 1977.

Crystal structure

The arrangement of atoms or ions in a crystalline solid. Knowledge of the precise ways in which atoms and ions are distributed in crystals is of prime importance in solid-state physics, chemistry, metallurgy, mineralogy, geochemistry, and other fields. In 1912 M. von Laue, W. Friedrich, and P. Knipping first discovered that x-rays could be diffracted by crystals. Prior to this discovery, little was known about the arrangement of atoms and ions in solid materials. Knowledge of crystal structure is and has been obtained mainly from x-ray diffraction data, although electron diffraction and neutron diffraction have become important tools in crystal analysis.

This article discusses the important concepts and terminology involved in the structure of crystalline solids and describes the structure of various metals and some relatively simple crystalline compounds. For related information *see* CRYSTAL DEFECTS; CRYSTALLOGRAPHY; ELECTRON DIFFRACTION; FERROELECTRICS; MINERALOGY; NEUTRON DIFFRACTION; X-RAY CRYSTALLOGRAPHY; X-RAY DIFFRACTION.

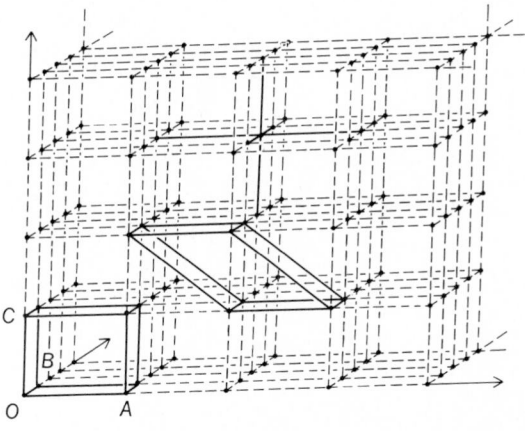

Fig. 1. A space lattice, two possible unit cells, and the environment of a point.

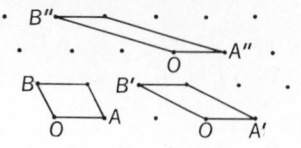

Fig. 2. Two-dimensional lattice, formed by the points in a lattice plane (*hkl*). All possible primitive parallelograms have the same surface area.

CONCEPTS AND TERMINOLOGY

In order to understand and describe crystal structures, several terms and concepts have been developed. Brief explanations of the most important are given in the following paragraphs.

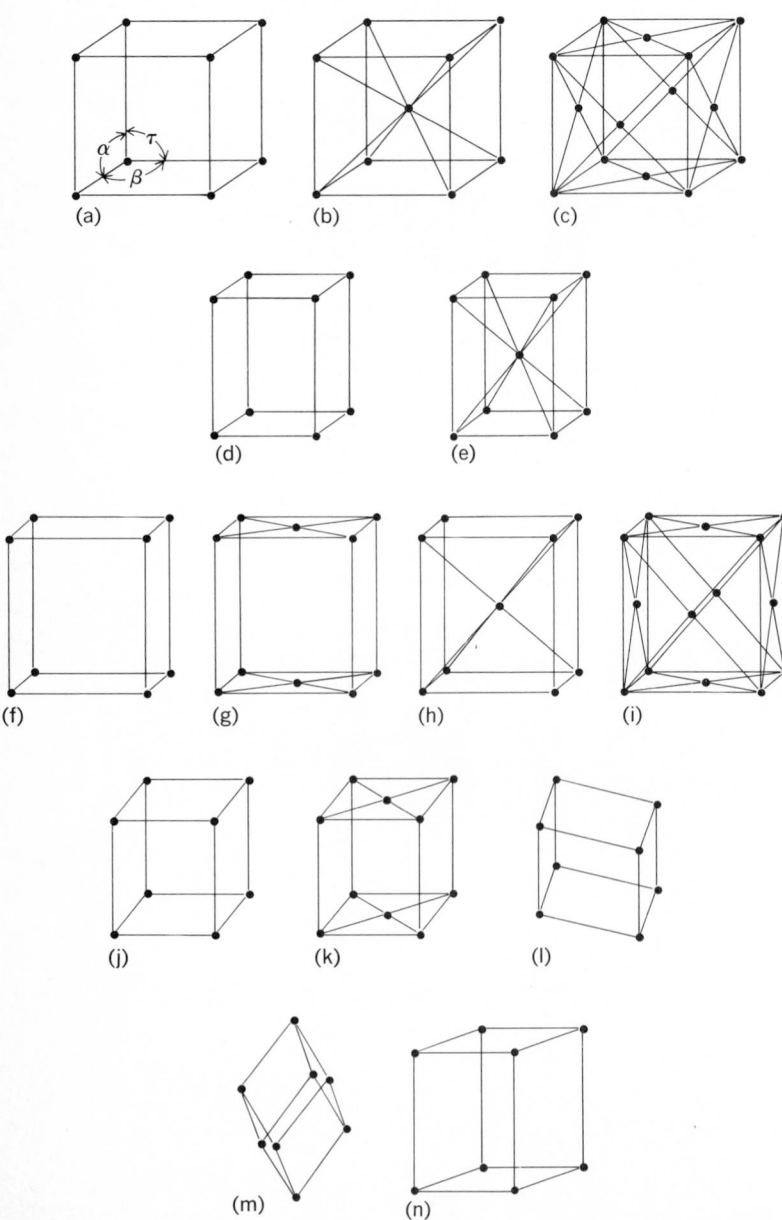

Fig. 3. The 14 Bravais space lattices. (*a*) Simple cubic. (*b*) Body-centered cubic. (*c*) Face-centered cubic. (*d*) Tetragonal. (*e*) Body-centered tetragonal. (*f*) Ortho-rhombic. (*g*) Base-centered orthorhombic. (*h*) Body-centered orthorhombic. (*i*) Face-centered orthorhombic. (*j*) Monoclinic. (*k*) Base-centered monoclinic. (*l*) Triclinic. (*m*) Trigonal. (*n*) Hexagonal.

Space lattices. A three-dimensional, indefi-nitely extended array of points, each of which is surrounded in an identical way by its neigh-bors, is known as a lattice or space lattice. The space lattice of a crystal is the representation of the periodicity with which matter is distributed in it. It is essential to distinguish a lattice from a crystal structure; a crystal structure is formed by associating with every lattice point an assembly of atoms identical in composition, arrangement, and orientation. The space-lattice concept was intro-duced by R. J. Haüy as an explanation for the spe-cial geometric properties of crystal polyhedrons. It was postulated that an elementary unit, having all the properties of the crystal, should exist, or con-versely that a crystal was built up by the juxtaposi-tion of such elementary units. If mathematical points forming the vertices of a parallelepiped $OABC$ (defined by three vectors $\overline{OA}, \overline{OB}, \overline{OC}$) are considered (Fig. 1), a space lattice is obtained by translations parallel to and equal to $\overline{OA}, \overline{OB}, \overline{OC}$. The parallelepiped is called the unit cell. The vec-tor **r** joining O to any lattice point can be written as $\mathbf{r} = m\overline{OA} + n\overline{OB} + r\overline{OC}$, where m, n, and r are integers.

In a space lattice two points define a row, and three define a lattice plane. Taking the directions of OA, OB, OC as axes and ABC as the unit plane, all planes and rows can be expressed by Miller indices.

In a row [*uvw*] two neighboring points are sepa-rated by a distance p_{uvw} which is characteristic for that row and all parallel ones. Two adjacent paral-lel rows of the family represented by [*uvw*] are a distance r_{uvw} apart. The plane (*hkl*) defines a fami-ly of lattice planes in which the points are identi-cally distributed and form a two-dimensional lat-tice. Figure 2 is an example.

This two-dimensional lattice can be deduced from two vectors such as $\overline{OA}, \overline{OB}$. The parallelo-gram OAB is the smallest unit from which the whole assembly can be obtained by parallel trans-lation. This statement applies equally well to the other two parallelograms $OA'B'$ and $OA''B''$ in Fig. 2 and, in fact, to an infinite number of such paral-lelograms, all of which have the same surface area $S_{hkl} = p_{uvw} r_{uvw}$. Thus the density of points in a giv-en row $1/p_{uvw}$ is proportional to the distance be-tween rows.

In the same way an infinity of unit cells can be considered in the space lattice. These cells all have the same volume $V = S_{hkl} d_{hkl}$, if d_{hkl} is the dis-tance between two adjacent planes (*hkl*). The dis-tance d_{hkl} is proportional to $1/S_{hkl}$, the latter prov-iding a measure for the density of points or the packing in (*hkl*). This remark, as well as the similar one made for the rows, is important when the prop-erties of the crystals are considered on an atomic scale. The faces and the edges of crystals are, re-spectively, densely packed lattice planes and rows. It is clear from a mathematical standpoint that from all possible cells the most convenient one should be chosen and that, whenever possible, preference should be given to a parallelepiped with mutually perpendicular edges.

The volume V of the unit cell is given by Eq. (1).

$$V = abc \sqrt{\begin{aligned} & 1 - \cos^2\alpha - \cos^2\beta - \cos^2\gamma \\ & \quad + 2\cos\alpha\cos\beta\cos\gamma \end{aligned}} \quad (1)$$

$$\frac{1}{d^2}=\frac{\dfrac{h}{a}\begin{vmatrix} h/a & \cos\gamma & \cos\beta \\ k/b & 1 & \cos\alpha \\ l/c & \cos\alpha & 1 \end{vmatrix}+\dfrac{k}{b}\begin{vmatrix} 1 & h/a & \cos\beta \\ \cos\gamma & k/b & \cos\alpha \\ \cos\beta & l/c & 1 \end{vmatrix}+\dfrac{l}{c}\begin{vmatrix} 1 & \cos\gamma & h/a \\ \cos\gamma & 1 & k/b \\ \cos\beta & \cos\alpha & l/c \end{vmatrix}}{\begin{vmatrix} 1 & \cos\gamma & \cos\beta \\ \cos\gamma & 1 & \cos\alpha \\ \cos\beta & \cos\alpha & 1 \end{vmatrix}} \quad (2)$$

where α, β, and γ are the angles defined by the crystal axes, as shown in Fig. 3a. The general expression for d_{hkl} (called d for convenience) is given by Eq. (2).

If $\alpha=\beta=\gamma=90°$, $\cos\alpha=\cos\beta=\cos\gamma=0$, and Eq. (2) reduces to Eq. (3), and if furthermore $a=b=c$, as in cubic lattices, this becomes $1/d^2=(h^2+k^2+l^2)/a^2$. When $a=b\neq c$, $\alpha=\beta=90°$, and

$$\frac{1}{d^2}=\frac{h^2}{a^2}+\frac{k^2}{b^2}+\frac{l^2}{c^2} \quad (3)$$

$\gamma=120°$ as in the hexagonal case, Eq. (4) holds.

$$1/d^2=(4/3a^2)(h^2+k^2+l^2)+l^2/c^2 \quad (4)$$

Symmetry can also be considered in a lattice. The lattice being indefinitely extended, there is no longer a point group but an array of regularly repeating symmetry elements. Each lattice point is a center of symmetry. Symmetry planes are parallel to lattice planes. Symmetry axes must be perpendicular to a lattice plane and coincide with, or be parallel to, a row.

Bravais lattices. These are the 14 different possible space lattices obtained on the basis that two lattices are different when the environment of their points is different. The 14 unit cells are represented in Fig. 3. If considered as solids, the combination of symmetry elements they exhibit can be determined. Seven point groups which are respectively the most symmetrical of each system are found. Accordingly the unit cells, and by extension the lattices, can be divided into seven groups corresponding to the seven crystallographic systems. Among the cells shown in Fig. 3, some have points at places other than corners. These cells are not primitive but are multiple cells chosen for convenience. As an example, the three primitive cells of the cubic lattices are, respectively, a cube, a rhombohedron with a plane angle of 109°28′, and a rhombohedron with an angle of 60°. The two rhombohedrons are extremely inconvenient to handle; consequently, the body-centered and face-centered cubes are adopted in their stead. Figure 4 shows the three cubic systems and the relationship of the last two to their primitive rhombohedrons. The letters P, I, and F stand for primitive, body-centered, and face-centered, respectively. A P cell contains one point, since each corner is used eight times in the formation of the complete assembly. An I cell contains two points, and an F cell four (one at the corners and six halves, as each point in the center of the faces belongs to two cells).

The external symmetry of a crystal is due to the periodically repeated arrangement of its atoms or ions. The Bravais lattices give the possible periodicities. A further step must consist in finding the number of possible periodic arrangements, two such arrangements being considered as different when they give rise to a different symmetry.

Two-dimensional structures are very well suited to illustrate this. In Figs. 5 and 6 a numeral 9 is chosen as the object; it is fully asymmetric and no false symmetry can therefore be introduced. It is multiplied by a periodic array of mirror planes and fourfold axes. Several interpenetrating identical lattices are formed. The two arrangements have the same Bravais lattice but not the same symmetry. In three-dimensional space the problem is similar, but other symmetry elements are considered.

Screw axes. These combine the rotation of an ordinary symmetry axis with a translation parallel to it and equal to a fraction of the unit distance in this direction. Figure 7 illustrates such an operation, which is symbolically denoted 3_1 and 3_2. The translation is respectively 1/3 and 2/3. The helices are added to help the visualization, and it is seen that they are respectively right- and left-handed. The projection on a plane perpendicular to the axis shows that the relationship about the axis remains in spite of the displacement. A similar type of arrangement can be considered around the other symmetry axes, and the following possibilities arise: 2_1, 3_1, 3_2, 4_1, 4_2, 4_3, 6_1, 6_2, 6_3, 6_4, and 6_5.

If screw axes are present in crystals, it is clear

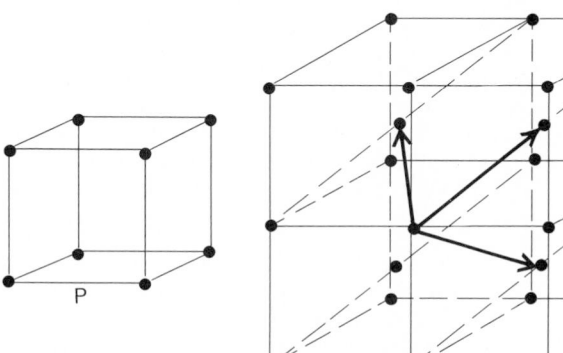

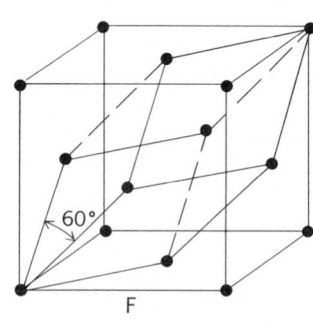

Fig. 4. The three cubic systems. The relationship of the F and I cells to their primitive rhombohedrons is demonstrated. (*After C. Kittel, Introduction to Solid State Physics, 2d ed., copyright © 1956 by John Wiley and Sons, Inc.; reprinted by permission)*

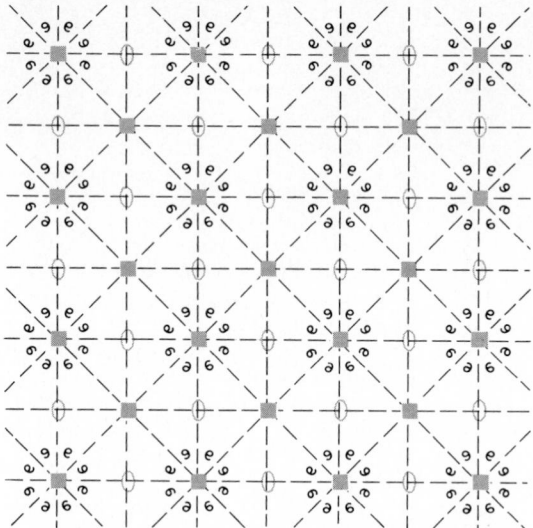

Fig. 5. Periodic pattern obtained by multiplication of asymmetric object by indicated array of symmetry elements. Both pattern and lattice have same symmetry elements. Mirror planes indicated by dotted lines.

that the displacements involved are of the order of a few angstroms and that they cannot be distinguished macroscopically from ordinary symmetry axes. The same is true for glide mirror planes.

Glide mirror planes. These combine the mirror image with a translation parallel to the mirror plane over a distance which is half the unit distance in the glide direction. This is illustrated in Fig. 8. Axial glide planes, denoted by a, b, c, have translations which are equal to $a/2$, $b/2$, $c/2$, respectively, where a, b, c are the lattice vectors. Diagonal glide planes, denoted by n, have translations of $(a + b)/2$, $(b + c)/2$, or $(c + a)/2$.

These new symmetry elements must be taken into account when the number of possible periodic arrangements is considered. This problem was solved by R. S. Federow (1885), A. M. Schoen-

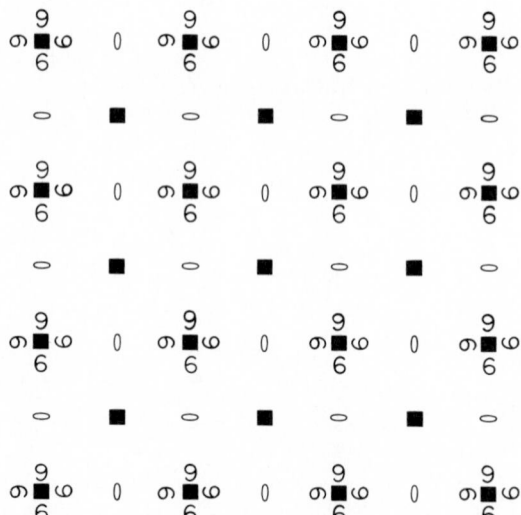

Fig. 6. Arrangement in which the periodic pattern has fewer symmetry elements than the lattice. This shows that the symmetry of the Bravais lattices can be lowered by the structure of the lattice points.

flies (1891), and W. Barlow (1894). A total of 230 arrangements or space groups is possible; of these, 32 can be distinguished macroscopically.

Space groups. These are indefinitely extended arrays of symmetry elements disposed on a space lattice. A space group acts as a three-dimensional kaleidoscope: An object submitted to its symmetry operations is multiplied and periodically repeated in such a way that it generates a number of interpenetrating identical space lattices. The fact that 230 space groups are possible means, of course, 230 kinds of periodic arrangement of objects in space. When only two dimensions are considered, 17 space groups are possible.

Space groups are denoted by the Hermann-Mauguin notation preceded by a letter indicating the Bravais lattice on which it is based. For example, P $2_12_12_1$ is an orthorhombic space group; the cell is primitive and three mutually perpendicular

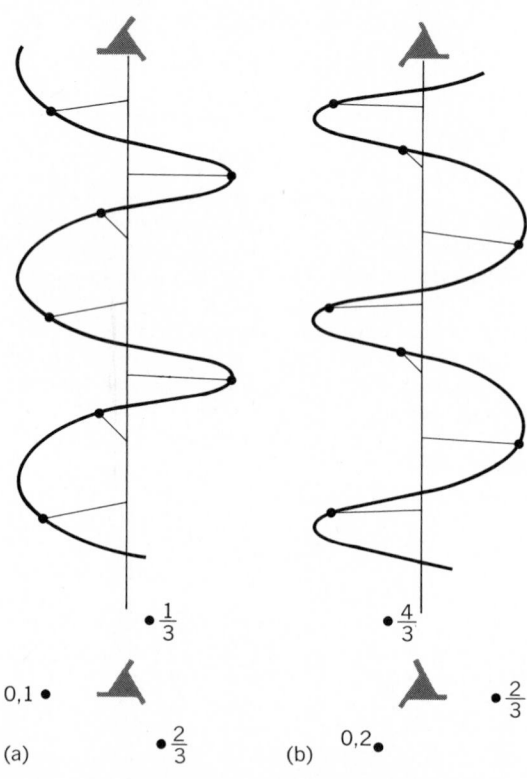

Fig. 7. Screw axes (a) 3_1 and (b) 3_2. Lower parts of figure are projections on a plane perpendicular to the axis. Numbers indicate heights of points above that plane.

screw axes are the symmetry elements. All space groups are listed in advanced textbooks.

J. D. H. Donnay and D. Harker have shown that it is possible to deduce the space group from a detailed study of the external morphology of crystals.

COMMON STRUCTURES

A discussion of the three structural systems found in metals and of the five systems found in crystalline compounds is given below.

Metals. In general, metallic structures are relatively simple, characterized by a dense packing and a high degree of symmetry. Manganese, gallium, mercury, and one form of tungsten are exceptions. Metallic elements situated in the sub-

Fig. 8. Symmetry elements involving (*a*) mirror and (*b*) glide plane.

groups of the periodic table gradually lose their metallic character and simple structure as the number of the subgroup increases. A characteristic of metallic structures is the frequent occurrence of allotropic forms; that is, the same metal can have two or more different structures which are most frequently stable in a different temperature range.

The forces which link the atoms together in metallic crystals are nondirectional. This means that each atom tends to surround itself by as many others as possible. This results in a dense packing, similar to that of spheres of equal radius, and yields three distinct systems: close-packed (face-centered) cubic, hexagonal close-packed, and body-centered cubic.

Close packing. For spheres of equal radius, close packing is interesting to consider in detail with respect to metal structures. Close packing is a way of arranging spheres of equal radius in such a manner that the volume of the interstices between the spheres is minimal. The problem has an infinity of solutions. The manner in which the spheres can be most closely packed in a plane *A* is shown in Fig. 9. Each sphere is in contact with six others; the centers form a regular pattern of equilateral triangles. The cavities between the spheres are numbered. A second, similar plane can be po-

sitioned in such a way that its spheres rest in the cavities 1, 3, 5 between those of the layer *A*. The new layer, *B*, has an arrangment similar to that of *A* but is shifted with respect to *A*. Two possibilities exist for adding a third layer. Its spheres can be put exactly above those of layer *A* (an assembly *ABA* is then formed), or they may come above the interstices 2, 4, 6. In the latter case the new layer is shifted with respect to *A* and *B* and is called *C*. For each further layer two possibilities exist and any sequence such as *ABCBABCACBA* . . . in which two successive layers have not the same denomination is a solution of the problem. They are all characterized by the fact that each sphere touches 12 others and all are equally densely packed. Such assemblies are rare in crystals. Complicated periodic sequences have, however, been observed, especially in carborundum. One structure is known where 89 layers form a sequence which is regularly repeated. In the vast majority of cases, periodic assemblies with a very short repeat distance occur. The cavities between the spheres, occupying 27% of the total volume, are of two types: tetrahedral cavities between four spheres, and octahedral ones between six spheres. For an assembly of *N* spheres, 3*N* cavities exist; 2*N* are tetrahedral and *N* are octahedral.

Face-centered cubic structure. This utilizes close packing characterized by the regular repetition of the sequence *ABC*. The centers of the spheres form a cubic lattice *F*, as shown in Fig. 10*a*. The densely packed planes of the type *A*, *B*, *C* are perpendicular to the threefold axis and can therefore be written as {111}. This form contains four sets of planes. These being the closest packed planes of the structure, d_{111} is greater than any other d_{hkl} of the lattice. The densest rows in these planes are <110>.

It is relatively easy to calculate the percentage of the volume occupied by the spheres. The unit cell has a volume a^3 and contains four spheres of radius *R*, their volume being $16\pi R^3/3$. The spheres touch each other along the face diagonal (Fig. 10*a*); $a\sqrt{2}$ is therefore equal to $4R$, or $R = a\sqrt{2}/4$. Substitution gives for the volume of the spheres $\pi\sqrt{2}\,a^3/6$, which is 73% of the volume of the cube.

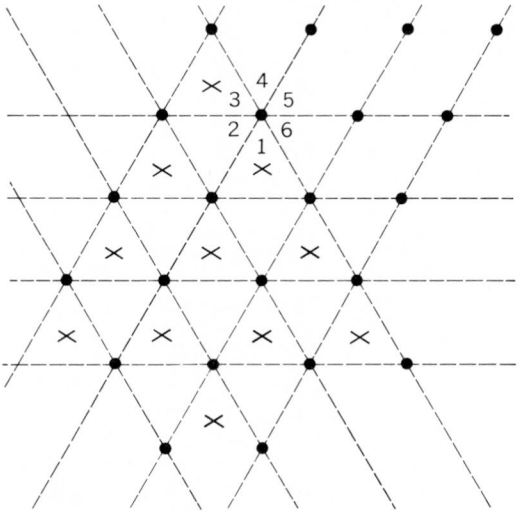

Fig. 9. Close packing of spheres of equal radius in a plane. The centers for the *A*, *B*, and *C* layers are indicat-

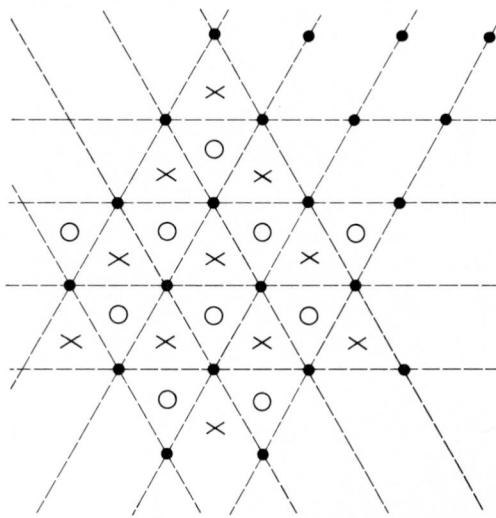

ed by a dot, a cross, and an open circle, respectively. Cavities between spheres are numbered.

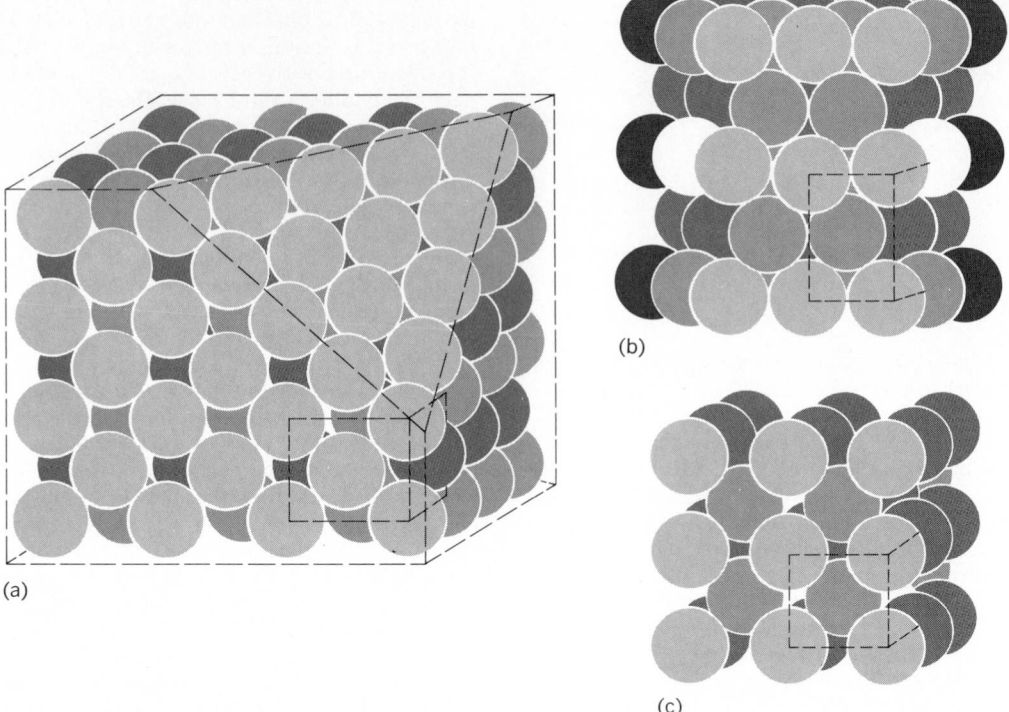

(a)

(b)

(c)

Fig. 10. Close packing of spheres in space. (a) Cubic close packing (face-centered cubic). One set of close-packed {111} planes (A, B, or C) is shown. (b) Hexagonal close packing. (c) Body-centered cubic arrangement.

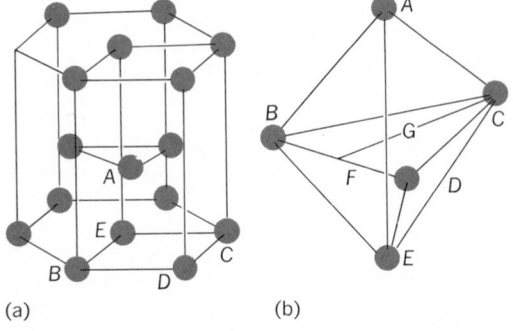

(a)   (b)

Fig. 11. Hexagonal close-packed structure. (a) Three unit cells, showing how the hexagonal axis results. One of the cells is fully outlined. (b) Calculation of the ratio c/a. Distance AE is equal to height of cell.

It is clear that the same percentage of the unit volume is filled in all other close-packed assemblies, that is, assemblies corresponding to other alternations of the planes A, B, C.

Hexagonal close-packed structure. This is a close packing characterized by the regular alternation of two layers, or $ABAB \ldots$. The assembly has hexagonal symmetry (Fig. 10b); the unit cell is shown in Fig. 11. Six spheres are at the corners of an orthogonal parallelepiped having a parallelogram as its base; another atom has as coordinates (1/3, 1/3, 1/2). The ratio c/a is easily calculated. The length $BD = a$, the edge of the unit cell. The height of the cell is $AE = 2AG = c$, defined in Eqs. (5) and (6).

$$AG = \sqrt{a^2 - \left(\frac{2}{3}\frac{a\sqrt{3}}{2}\right)^2} = a\sqrt{2/3} \qquad (5)$$

$$c = 2AG = a\sqrt{8/3} \qquad c/a = \sqrt{8/3} = 1.633 \qquad (6)$$

This latter value is important, for it permits determination of how closely an actual hexagonal structure approaches ideal close packing.

Body-centered cubic structure. This is an assembly of spheres in which each one is in contact with eight others, as shown in Fig. 10c.

The spheres of radius R touch each other along the diagonal of the cube, so that measuring from the centers of the two corner cubes, the length of

Metal structures

Metal	Modification	Stability range	Structure
Beryllium	α	To melting point	hcp
	?β	630°C to melting point	Hexagonal
Cadmium	α	To melting point	hcp
Calcium	α	To 450°C	fcc
	?β	300–450°C	Hexagonal
	γ	450°C to melting point	bcc
Cerium	α	To melting point	fcc
	?β	50°C	Hexagonal
Chromium	α	To melting point	bcc
	?β	Electrolytic form	Hexagonal
Cobalt	α	To 420°C	hcp mixed with fcc
	β	420°C to melting point	fcc
Gold	α	To melting point	fcc
Iron	α	To 909°C	bcc
	γ	909–1403°C	fcc
	δ	1403°C to melting point	bcc
Lead	α	To melting point	fcc
Magnesium	α	To melting point	Nearly hcp
Nickel	α	To melting point	fcc
	β	Electrolytic form	Hexagonal
Silver	α	To melting point	fcc
Tungsten	α	To melting point	bcc
Zirconium	α	To 862°C	hcp
	β	862°C to melting point	bcc
Zinc	α	To melting point	hcp

the cube diagonal is $4R$. The length of the diagonal is also equal to $a\sqrt{3}$, if a is the length of the cube edge, and thus $R = a\sqrt{3}/4$. The unit cell contains two spheres (one in the center and 1/8 in each corner) so that the total volume of the spheres in each cube is $8\pi R^3/3$. Substituting $R = a\sqrt{3}/4$ and dividing by a^3, the total volume of the cube, gives the percentage of filled space as 67%. Thus the structure is less dense than the two preceding cases.

The closest packed planes are {110}; this form contains six planes. They are, however, not as dense as the A, B, C planes considered in the preceding structures. The densest rows have the four <111> directions.

Tabulation of structures. The structures of various metals are listed in the table, in which the abbreviations fcc, hcp, and bcc, respectively, stand for face-centered cubic, hexagonal close-packed, and body-centered cubic. In this table the hexagonal structures classified as ? are still in doubt. The structures listed hcp are only roughly so; only magnesium has a c/a ratio (equal to 1.62) which is very nearly equal to the ratio (1.63) calculated earlier for the ideal hcp structure. For cadmium and zinc the ratios are respectively 1.89 and 1.86, which are significantly larger than 1.63. Strictly speaking, each zinc or cadmium atom has therefore not twelve nearest neighbors but only six. This departure from the ideal case for subgroup metals follows a general empirical rule, formulated by W. Hume-Rothery and known as the 8-N rule. It states that a subgroup metal has a structure in which each atom has 8-N nearest neighbors, N being the number of the subgroup. *See* COORDINATION NUMBER.

Crystalline compounds. Simple crystal structures are usually named after the compounds in which they were first discovered (diamond or zinc sulfide, cesium chloride, sodium chloride, and calcium fluoride). Many compounds of the type A^+X^-, $A^{++}X_2^-$ have such structures. They are highly symmetrical, the unit cell is cubic, and the atoms or ions are disposed at the corners of the unit cell and at points having coordinates which are combinations of 0, 1, 1/2, or 1/4.

Sodium chloride structure. This is an arrangement in which each positive ion is surrounded by six negative ions and vice versa. The arrangement is expressed by stating that the coordination is 6/6. The centers of the positive and the negative ions each form a face-centered cubic lattice. They are shifted one with respect to the other over a distance $a/2$, where a is the repeat distance (Fig. 12a). Systematic study of the dimensions of the unit cells of compounds having this structure has revealed that:

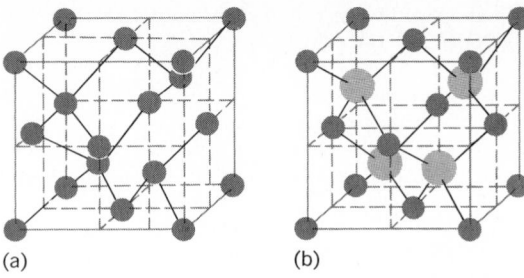

Fig. 13. Tetrahedral crystal compound structures. (a) Diamond. (b) Zinc blende.

1. Each ion can be assigned a definite radius. A positive ion is smaller than the corresponding atom and a negative ion is larger.

2. Each ion tends to surround itself by as many others as possible of the opposite sign because the binding forces are nondirectional.

On this basis the structure is determined by two factors: a geometrical factor involving the size of the two ions which behave in first approximation as hard spheres, and an energetical one involving electrical neutrality in the smallest possible volume. In the ideal case all ions will touch each other; therefore, if r_A and r_X are the radii of the ions, $4r_X = a\sqrt{2}$ and $2(r_A + r_X) = a$. Expressing a as a function of r_X gives $r_A/r_X = \sqrt{2} - 1 = 0.41$. When r_A/r_X becomes smaller than 0.41, the positive and negative ions are no longer in contact and the structure becomes unstable. When r_A/r_X is greater than 0.41, the positive ions are no longer in contact, but ions of different sign still touch each other. The structure is stable up to $r_A/r_X = 0.73$, which occurs in the cesium chloride structure.

Cesium chloride structure. This is characterized by a coordination 8/8 (Fig. 12b). Each of the centers of the positive and negative ions forms a primitive cubic lattice; the centers are mutually shifted over a distance $a\sqrt{3}/2$. The stability condition for this structure can be calculated as in the preceding case. Contact of the ions of opposite sign here is along the cube diagonal.

Diamond structure. In this arrangement each atom is in the center of a tetrahedron formed by its nearest neighbors. The 4-coordination follows from the well-known bonds of the carbon atoms. This structure is illustrated in Fig. 13a. The atoms are at the corners of the unit cell, the centers of the faces, and at four points having as coordinates (1/4, 1/4, 1/4), (3/4, 3/4, 1/4), (3/4, 1/4, 3/4), (1/4, 3/4, 3/4). The atoms can be divided into two groups, each forming a face-centered cubic lattice; the mutual shift is $a\sqrt{3}/4$.

Zinc blende structure. This structure, shown in Fig. 13b, has coordination 4/4 and is basically similar to the diamond structure. Each zinc atom (small circles in Fig. 13b) is in the center of a tetrahedron formed by sulfur atoms (large circles) and vice versa. The zinc atoms form a face-centered cubic lattice, as do the sulfur atoms.

Calcium fluoride structure. Figure 14 shows the calcium fluoride structure. If the unit cell is divided into eight equal cubelets, calcium ions are situated at corners and centers of the faces of the cell. The fluorine ions are at the centers of the eight cubelets. There exist three interpenetrating face-centered cubic lattices, one formed by the Ca

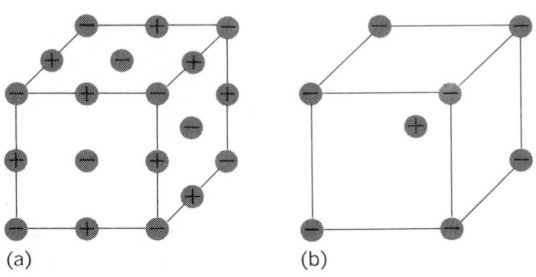

Fig. 12. Simple crystalline compound structures. (a) Sodium chloride. (b) Cesium chloride.

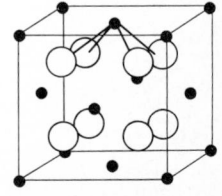

Fig. 14. Calcium fluoride structure.

ion and two by the F ions, the mutual shifts being (0, 0, 0), (1/4, 1/4, 1/4), (3/4, 3/4, 3/4).

[WILLY C. DEKEYSER]

Bibliography: Akademische Verlagsgesellschaft, *Strukturbericht*, vols. 1–7, *Z. Krist.*, suppl. vols., 1913–1939, reprint, 1943, and continued by International Union of Crystallography, *Structure Reports*, 1945–; P. J. Brown and J. E. Forsyth, *The Crystal Structure of Solids*, 1978; M. J. Buerger, *Crystal-Structure Analysis*, 1979; M. J. Buerger, *Introduction to Crystal Geometry*, 1977; J. D. H. Donnay and W. Nowacki, *Crystal Data*, Geol. Soc. Amer., Mem., no. 60, 1954; R. C. Evans, *Introduction to Crystal Chemistry*, 2d ed., 1964; N. F. M. Henry and K. Lonsdale, *International Tables for X-ray Crystallography*, vol. 1, 1952.

Crystal whiskers

Single crystals that have grown in a filamentary form. Such filamentary growths have been known for centuries; however, great interest in them developed only after it was discovered at Bell Telephone Laboratories in the 1950s that the strength exhibited by some whiskers in bend tests approaches that expected theoretically for perfect crystals. The strength of whiskers is sometimes as great as 0.06 G, where G is the shear modulus, while the strength of well-annealed single crystals of ordinary experience rarely exceeds 0.0001–0.001 G. *See* SINGLE CRYSTAL.

Whiskers have been grown by spontaneous extrusion or out of a supersaturated medium, the medium supersaturated by physical methods or chemical reaction. In all instances whiskers apparently form from singularities in the medium.

Bell Laboratories scientists discovered that crystal whiskers 1–2 μm in diameter and about 1 mm long are, under certain conditions, extruded from many different metals. It now appears that the reason for this spontaneous extrusion is relief of internal stress. It has not yet been proven that the spontaneously extruded whiskers, which are strong in bending, are exceptionally strong in tension.

Whiskers of a wide variety of substances, for example, mercury, graphite, sodium and potassium chlorides, copper, iron, and aluminum oxide, have been grown from supersaturated media. Whiskers grown in this way are usually a few microns in diameter and up to several centimeters long (see illustration). Some are exceptionally strong, both in bend tests and in tension tests.

In addition to exceptional strength, whiskers often have unique electrical, magnetic, or surface properties. This behavior can be interpreted to mean that the crystal structure of whiskers is virtually perfect, particularly with respect to line defects. Actually it appears that some whiskers contain line defects whereas others do not. However, no general correlations between whisker properties and whisker structure have as yet been established.

[DAVID TURNBULL]

Bibliography: S. S. Brenner, Growth and properties of "whiskers," *Science*, 128(3324):569–575, 1958; B. Chalmers (ed.), *Progress in Metal Physics*, vol. 6, 1956; C. Kittel, *Introduction to Solid State Physics*, 5th ed., 1976; B. R. Pamplin, *Crystal Growth*, 1975.

Crystallography

The branch of science that deals with the geometric description of crystals, their internal arrangement, and their properties. Long a part of mineralogy, crystallography is now a vast no-man's-land between physics, chemistry, physical metallurgy, mineralogy, and even biology.

This article deals with the geometric description of crystals; for other aspects of crystallography *see* CRYSTAL STRUCTURE and articles listed there.

Formal description of crystals. The anisotropic nature of crystals results in a freely growing crystal, bounded by flat faces. The following discussion presents a formalism for describing crystals exactly in terms of these faces.

Faces which are parallel to the same edge are said to form a zone; the direction of the edge is the zone axis. The number and relative importance of the faces bounding crystals of the same chemical compound can be very different. The crystals are then said to have a different habit. Faces and edges which are present as well as those which are geometrically possible must therefore be taken into account, and in the following discussion the term face is extended to any plane determined by two edges. An edge is, in the same sense, the intersection of two faces.

N. Steno found that the angle between similar faces is constant for different-sized crystals of the same substance. This means that a crystal grows by the parallel displacement of its faces. Therefore, only the direction of a normal to a face is of importance. As a consequence, the crystal is considered to be in the center of a sphere (Fig. 1). The normals to its faces from the center cut the sphere in points or poles whose position can be fixed by two coordinates φ and ρ (Fig. 2a). The constellation of faces is in this way replaced by a constellation of points on a sphere. The zone axes are replaced by great circles or zone circles; they are the intersections of the sphere with planes passing through the center of the sphere, perpendicular to

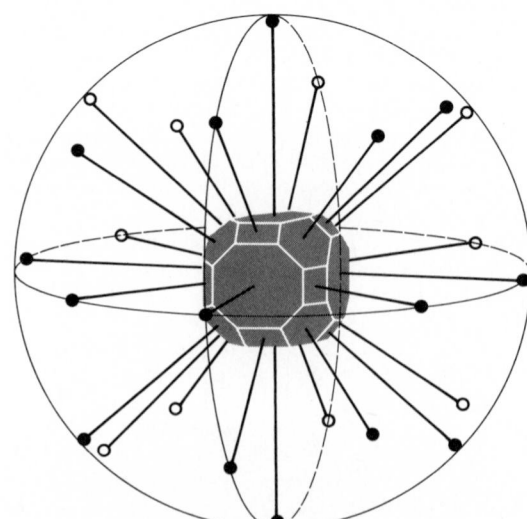

Fig. 1. Replacement of the constellation of faces by a constellation of poles on a sphere. Faces parallel to the same edge, that is, forming a zone, have their poles on a great circle called the zone circle.

the zone axis. In practice, φ and ρ values are obtained by "measuring" the crystal; two-circle goniometers are used for the purpose. All problems which can arise are connected with angles between faces and between zone axes; they can in principle be solved by spherical trigonometry once the φ and ρ values are known. Graphical constructions are, however, most helpful, and gnomonic and stereographic projections are used in such constructions. These projections are shown in Fig. 2. In stereographic projection, circles are projected as circles, and the angles between them are not altered. In gnomonic projection, great circles are projected as straight lines.

Crystal symmetry. Symmetry is a basic property of crystals. A crystal polyhedron (or better, the constellation of the normals) can be brought into successive indistinguishable positions by nontrivial operations. These are inversion, rotation around an axis over a specific angle, reflection into a plane, or combinations of these (Fig. 3). It is, however, convenient to consider all operations as rotations, and two types of axis can be considered: (1) a rotation or symmetry axis involving a rotation over an angle of $360°/n$ (n is the multiplicity of the axis) and (2) an inversion axis which combines the preceding operation with an inversion with respect to a point of the axis. It is found that only multiplicities of 1, 2, 3, 4, and 6 occur. This has a deeper meaning and is connected with the homogeneous filling of space. It can be understood intuitively by considering the two-dimensional example in Fig. 4. An assembly of identical polygons can completely cover a surface only when the polygons possess one of the following symmetry elements: 2, 3, 4, 6. The symmetry axes are represented by their multiplicity; inversion axes are written $\bar{1}, \bar{2}, \bar{3}, \bar{4}, \bar{6}$. Axis $\bar{1}$ is in fact an inversion center; $\bar{2}$ is equivalent to a mirror plane and is therefore written m. Not all of the other inversion axes are specific symmetry elements; that is, $\bar{3}$ is a combination of the operations 3 and $\bar{1}$, and $\bar{6}$ a combination of a threefold axis and a mirror plane perpendicular to it. Finally, the distinct symmetry elements are 1, 2, 3, 4, 6, $\bar{1}$, m, $\bar{4}$.

It can be proved that a symmetry axis, the perpendicular to a symmetry plane, the plane perpendicular to a symmetry axis, and a symmetry plane are all possible edges or faces of the crystal.

A crystal can exhibit a number of these symmetry elements, all going through a common point. A collection of symmetry elements applied about such a point is called a point group.

Combinations of symmetry elements are ruled by the following theorems:

1. When two out of the three elements $\bar{1}$, m, 2 are present, the third follows automatically.

2. The intersection of two mirror planes forming an angle φ is a symmetry axis with multiplicity $n = 360/2\varphi$.

3. When a mirror plane contains a symmetry axis of multiplicity n, this axis is the intersection of n mirror planes forming angles of $360°/n$.

4. When two twofold axes forming an angle φ intersect, the perpendicular at the intersection point on the plane they determine is a symmetry axis with multiplicity $n = 360/2\varphi$.

5. Any crystal polyhedron containing more than two axes with a multiplicity higher than two must

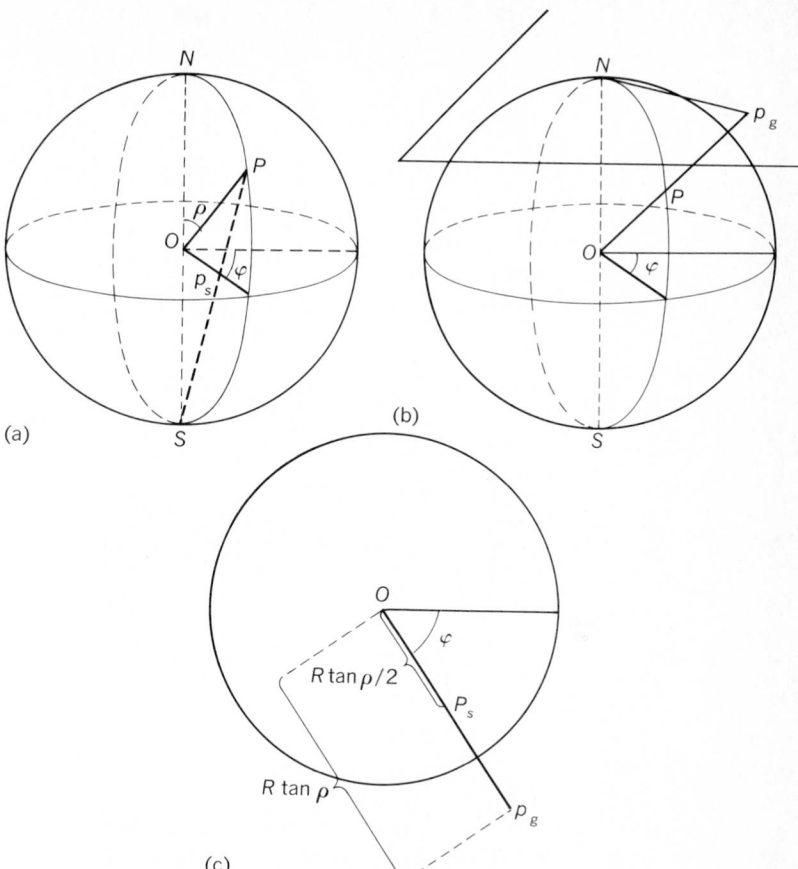

Fig. 2. Graphical constructions used in description of crystals. (*a*) Stereographic projection of *P*. Line joining pole *P*, defined by φ and ρ, to south pole *S* cuts equator at p_s. (*b*) Gnomonic projection of *P*. Line *OP* cuts the plane tangent to the sphere at *N* at point p_g. (*c*) Stereognomogram. Both projection planes are superimposed; p_s and p_g are on a radius making an angle φ with the origin and at distances respectively equal to $R\tan\rho/2$ and $R\tan\rho$.

necessarily present the combination of symmetry axes present in a cube or in a tetrahedron.

Taking these five theorems into account, it can be shown that 32 combinations or point groups are possible. These are conveniently represented by the Hermann-Mauguin notation, which provides the minimum information from which, with the aid of the five theorems, all the symmetry elements of a point group can be deduced. The code is as follows: If the axis of highest multiplicity is written first, the succeeding directions are perpendicular to it, and the succeeding planes are either perpendicular to it or pass through it. A mirror plane perpendicular to an axis is written $3/m$. The diagonal is not present when the plane passes through the axis; that is, one writes $3m$. The notation $m3$ means that the threefold axis is neither in the plane m nor perpendicular to it. A combination such as $3/mm$ represents a threefold axis, three planes passing through it and one perpendicular to it. The intersections of these mirror planes are twofold axes which are not written, but deduced.

All crystals can accordingly be divided into 32 classes if the point group is used as a criterion. Examples belonging to each class are known.

Crystal systems. The seven groups obtained when only the dominant symmetry elements of the

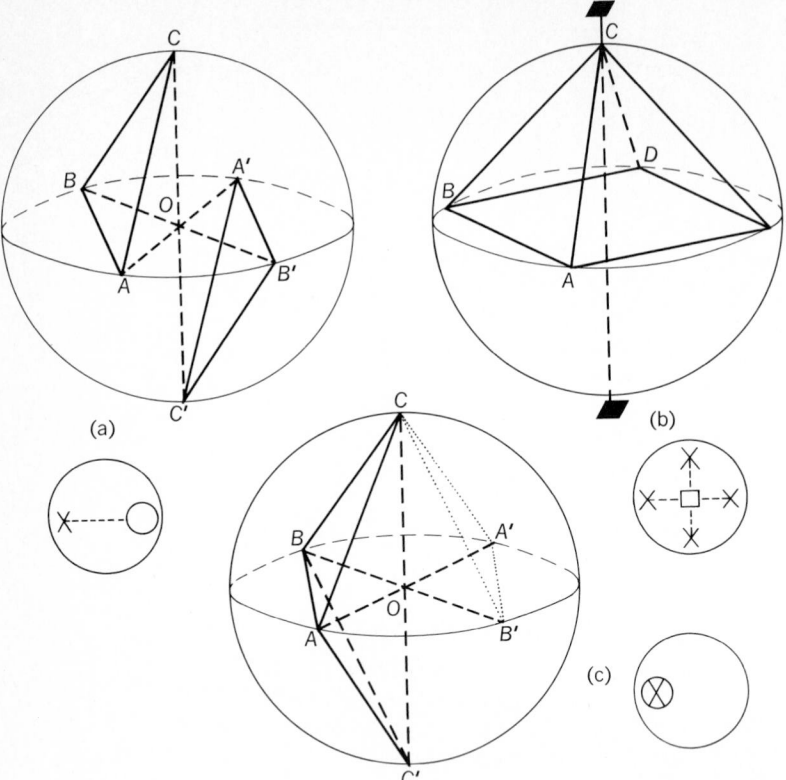

Fig. 3. Elementary symmetry operations in space and in projection. (a) O is in an inversion center. The pole of ABC (× in the small diagram) is projected stereographically from C′, that of A′B′C′ from C (○ in the diagram). (b) Fourfold axis and stereographic projection of the four planes related by its symmetry operation. (c) CC′ is a twofold inversion axis. ABC is first rotated over 180° into the intermediate position CA′B′ and by inversion to ABC′. This operation 2 gives the same result as a reflection of ABC into the equatorial plane, thus showing that $\overline{2} = m$.

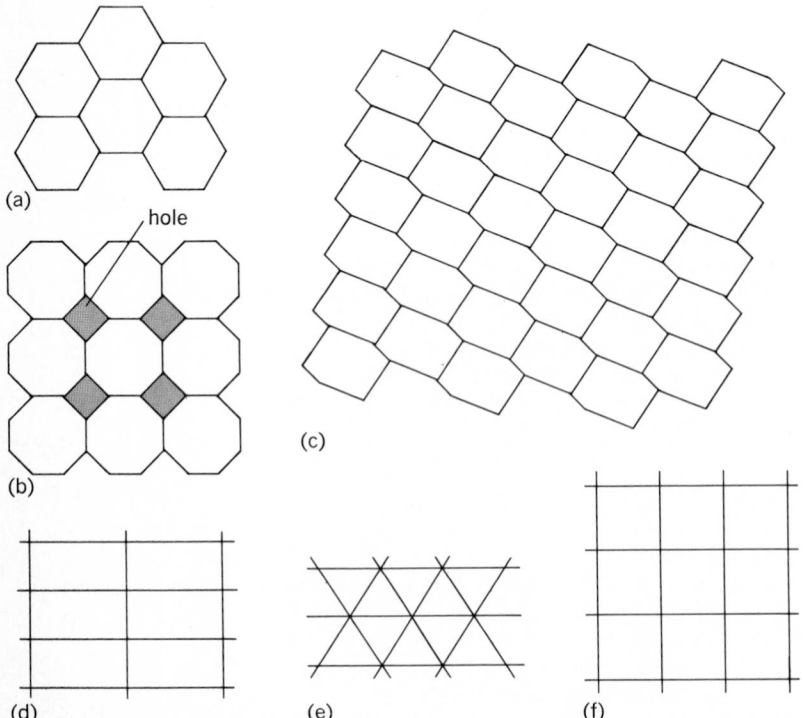

Fig. 4. Mosaics. Homogeneous filling of a surface with polygons is possible only when they have definite symmetry elements. Note the holes formed in b when octagons are used. Mosaics a, c, d, e, f fill space completely.

classes are taken into account are called crystal systems. In order of decreasing symmetry, they are called cubic, hexagonal, tetragonal, trigonal, orthorhombic, monoclinic, and triclinic. Sometimes the hexagonal and trigonal systems are considered as a single system. The systems and classes are summarized in Table 1, in which the Hermann-Mauguin notation is used. The expression "a cubic crystal" means that the crystal belongs to the cubic system.

Miller indices. These are used for the analytical description of crystal faces, and are shown in Fig. 5. A specially well-suited reference system used for that purpose consists of three nonparallel possible edges of the crystal which are chosen as axes and a possible plane of the crystal which intersects all three axes respectively at A, B, C. It is defined by the three angles α, β, γ and the ratio $OA:OB:OC$. Another plane, HKL, of the same crystal is defined by the ratio $OH:OK:OL$. The lengths OH, OK, OL are respectively measured with $OA = a$, $OB = b$, $OC = c$ as units, that is, by

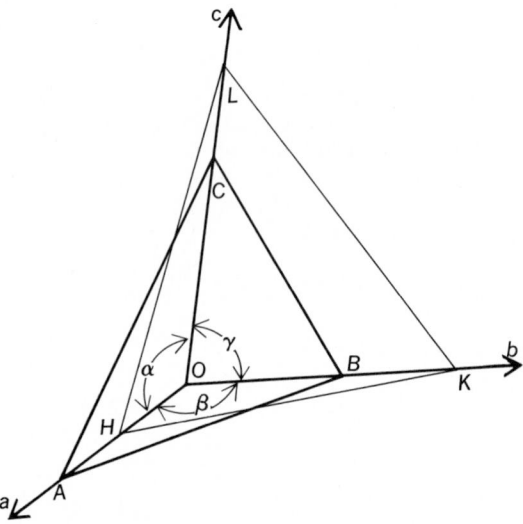

Fig. 5. The reference system used to define Miller indices; ABC is the unit plane.

the ratios a/OH, b/OK, c/OL. A measurement is a comparison (ratio) between what must be measured and a unit. The unit is generally written as the denominator in such ratios; the reverse is done here for a special reason which will be explained later. In Eq. (1), h, k, l are rational num-

$$(a/OH)/(b/OK)/(c/OL) = h/k/l \qquad (1)$$

bers which are small when the axes are well chosen (the law of rational indices). A different unit is used along each axis to take the anisotropy of the crystal into account. When, for instance, the temperature is raised, a, b, and c expand differently, but the ratios a/OH, b/OK, c/OL remain constant. The ratios $h:k:l$ define the plane HKL; h, k, l are called its Miller indices, and by convention, this is written (hkl), or ($\overline{hkl}$) when the indices are negative numbers. In geometrical crystallography, it is customary to precede the indices with a letter. The letters a, b, c are generally used for the axial

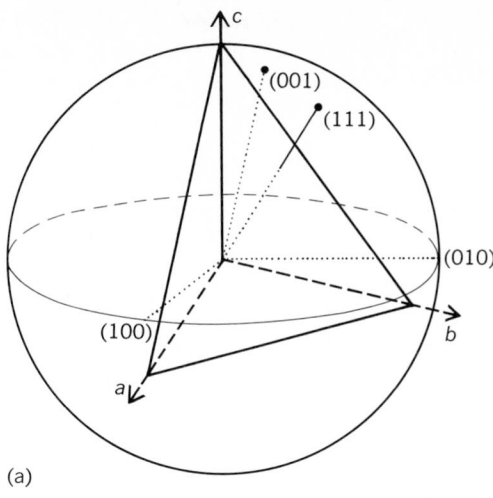

(a)

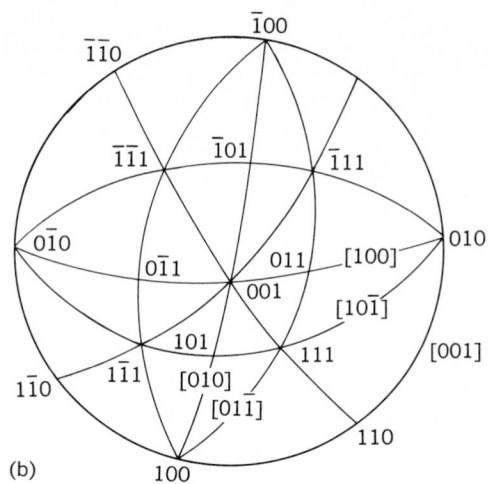

(b)

Fig. 6. Aspects of reference system. (a) Orientation of the reference system and poles of the axial and unit plane. (b) Projection. By the construction of great circles, four quadrants of type 001-010-100 are obtained.

planes, and the letter o is used for the plane ABC. The indices of these planes are respectively $a(100)$, $b(010)$, $c(001)$, $o(111)$. The last is therefore called unit plane. A plane (hkl) cuts from the axis segments OH, OK, OL whose ratios are, following the definition, equal to $(a/h)/(b/k)/(c/l)$. The equation of the plane in cartesian coordinates is therefore given by Eq. (2). The quantity d is a parameter, since

$$h\frac{x}{a} + k\frac{y}{b} + l\frac{z}{c} = d \qquad (2)$$

the plane can be shifted parallel to itself. The indices are respectively the coefficients in x/a, y/b, z/c in the equation of the plane. This is the special reason mentioned earlier.

Zone indices. Two planes $(h_1k_1l_1)$ and $(h_2k_2l_2)$ define an edge, the direction coefficients of which are determined by simultaneous solution of Eqs. (3a) and (3b). The direction coefficients are given

$$h_1\,x/a + k_1\,y/b + l_1\,z/c = 0 \qquad (3a)$$
$$h_2\,x/a + k_2\,y/b + l_2\,z/c = 0 \qquad (3b)$$

by notation (4) or by ua, vb, wc, when the determi-

$$a\begin{vmatrix} k_1 & l_1 \\ k_2 & l_2 \end{vmatrix} \qquad b\begin{vmatrix} l_1 & h_1 \\ l_2 & h_2 \end{vmatrix} \qquad c\begin{vmatrix} h_1 & k_1 \\ h_2 & k_2 \end{vmatrix} \qquad (4)$$

nants are respectively denoted by u, v, w. If a, b, c are taken as units along each of the axes, u, v, w are the coordinates of a point. The edge, or zone axis, determined by the two planes has the direction of the line joining the origin to the point with

crystallographic coordinates u, v, w; it is represented by $[uvw]$.

Conversely, two zones $[u_1v_1w_1]$ $[u_2v_2w_2]$ define a plane, whose indices are found by the same procedure. The condition that a plane (hkl) belongs to a zone $[uvw]$ is found by requiring that this plane be parallel to $[uvw]$, that is, that the plane passing through the origin must contain the point ua, vb, wc. This gives $hu + kv + lw = 0$.

Choice of reference system. It is useful to consider the stereographic and gnomonic projection of the reference system (Fig. 6). The c axis is vertical, that is, perpendicular to the projection plane. The poles $a(100)$, $b(010)$, $c(001)$, $o(111)$ define a number of zones, and the representative circles are easily constructed. Their indices, as well as those of the planes defined by their intersection, are found by the rules given earlier. In this way 26 planes having indices which are combinations of 0, 1, and $\bar{1}$ are obtained. The zone circles $[100]$, $[010]$, $[001]$ divide the projection into four quadrants. Each intersection of one of these with the reference circle as well as an interior point having indices formed by 1 or $\bar{1}$ is the intersection of three zone circles. Such a pattern is referred to as the basic pattern in the discussion that follows. The angles α, β, γ and the ratios $a:b:c$ can be determined graphically.

With the method of gnomonic projection (as in Fig. 7), indices can be assigned to a plane such as e by the following construction. Zones $[100]$ and $[010]$ are considered as axes of a coordinate sys-

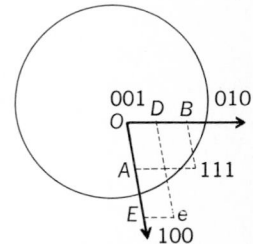

Fig. 7. Gnomonic projection of reference system, illustrating how indices can be graphically assigned to a plane e.

Table 1. Crystal systems and classes

Cubic	Hexagonal	Tetragonal	Trigonal	Ortho-rhombic	Mono-clinic	Triclinic
23	6	4	3	$2mm$	2	1
$(2/m)\bar{3}$	$\bar{6}$	$\bar{4}$	$\bar{3}$	222		$\bar{1}$
$\bar{4}3m$	$6/m$	$4/m$	$3m$	$(2/m)(2/m)(2/m)$	m	
432	$6mm$	$4mm$	$\bar{3}(2/m)$		$2/m$	
$(4/m)\bar{3}(2/m)$	$\bar{6}m2$	$\bar{4}2m$	32			
	622	422				
	$(6/m)(2/m)(2/m)$	$(4/m)(2/m)(2/m)$				

tem. The coordinates of the gnomonic projection of the plane *e* are measured with the coordinates of (111) as units, that is, by the ratio *OE/OA, OD/OB*. It can be shown that, provided the *c* axis is vertical, the indices of this plane are (*OE/OA, OD/OB,* 1) or in the case under consideration (2, 1/2, 1) or (4, 1, 2). This graphical construction only works for simple indices; 1/3 can be easily distinguished from 1/4, but not from 1/7 and 1/8. In such cases, the indices must be calculated, but such small fractions are highly improbable provided the axes are well chosen. This choosing of the axes is a major problem. Symmetry elements and the related elements which are possible faces or edges are usually chosen as axis and unit plane, as explained in the earlier discussion of crystal symmetry. The axis of highest symmetry is chosen as the *c* axis and is oriented vertically except in the monoclinic system, where *b* is usually the vertical axis. Inspection of the symmetry elements of the different classes indicates plainly that it is only in the cubic system that enough symmetry elements are present to permit an unambiguous determination of the reference system, that is, three mutually perpendicular twofold axes and a plane perpendicular to a threefold axis as unit plane; further inspection shows that mutually perpendicular axes can be chosen except in the monoclinic ($\beta \neq 90°$) and triclinic systems ($\alpha \neq \beta \neq \gamma \neq 90°$).

This means that two different observers describing the same noncubic crystal will probably choose a different reference system, that is, orient or set the crystal differently. The same face of the crystal will consequently have different indices in the two settings. Different procedures have been worked out to obtain uniformity. The Barker method is the simplest and best known. It consists of a number of convenient rules which allow two observers to come independently to an identical choice. This is based on the consideration that the basic pattern considered earlier is the generalization of what exists in the cubic system. The main idea is therefore to find in a stereographic projection a quadrant as in a basic pattern and to choose the corners and an inside intersection of three zones as elements of the reference system. Rules are provided to distinguish between them, that is, to find out which pole will be denoted as *a*, *b*, *c*, or *o*, and what sense will be assigned to the axis. At the same time, one has also a maximum chance that a maximum number of planes will have simple indices. It is clear that the number of rules grows as the indeterminacy becomes greater. It is also clear that the rules have no physical meaning.

Bravais indices. These are frequently used in the hexagonal and trigonal systems because they reveal the symmetry more clearly than do other indices. In the hexagonal system, the axes a_1, a_2, a_3 are perpendicular to the *c* axis (Fig. 8) and make angles of 120° with each other. The four indices are denoted (*hkil*), but it is easy to prove that $h + k = \bar{i}$; therefore one may write (*hk.l*). Thus, *i* is a redundant index but is used because of the additional clarity which it yields in this system. The unit plane is ($1\bar{1}01$).

The zone axes are found by considering the first two and the last indices as usual and inserting the sum of the first two with an opposite sign for the third one. As an example, the planes (0001) and ($1\bar{1}01$) define the zone [$11\bar{2}0$].

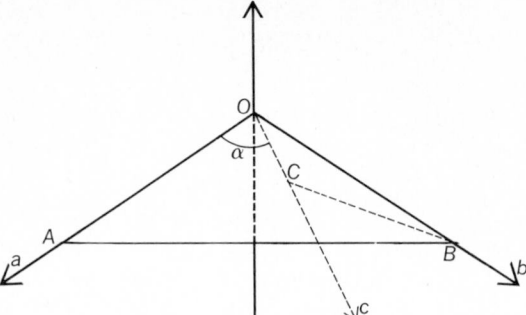

Fig. 9. Miller indices in trigonal system.

When Miller indices are used in the trigonal system, the three axes are chosen equally inclined around the threefold axis and making an equal angle α between them (Fig. 9). A plane normal to the symmetry axis is chosen as the (111) face. The system is therefore defined by α, *a:a:a*. Bravais indices (*hkil*) are easily transformed into Miller indices, here denoted (*pqr*), by the relations shown as Eq. (5).

$$\begin{aligned} p &= 2h + i - l \\ q &= -h + i + l \\ r &= -h - 2i + l \end{aligned} \qquad (5)$$

The reference system used in the seven crystal systems is shown in Table 2.

Table 2. Crystallographic reference systems

Crystal system	Angles formed by crystal axes	Defining ratio
Cubic	$\alpha = \beta = \gamma = 90°$	$a:a:a$
Hexagonal	$\alpha = \beta = 90°, \gamma = 120°$	$a:a:c$
Tetragonal	$\alpha = \beta = \gamma = 90°$	$a:a:c$
Trigonal	$\alpha = \beta = 90°, \gamma = 120°$	$a:a:c$
	$\alpha = \beta = \gamma \neq 90°$	$a:a:a$
Orthorhombic	$\alpha = \beta = \gamma = 90°$	$a:b:c$
Monoclinic	$\alpha = \gamma = 90°, \beta \neq 90°$	$a:b:c$
Triclinic	$\alpha \neq \beta \neq \gamma \neq 90°$	$a:b:c$

Crystal form. The set of planes obtained by submitting a plane (*hkl*) to the symmetry operations of a point group is called a crystal form. All these planes together are written as {*hkl*}; they all have indices which are cyclic permutations of *h, k, l, $\bar{h}$, $\bar{k}$, $\bar{l}$*. The plane (*hkl*) can occupy a general or special position with respect to the symmetry elements; accordingly, one obtains a general or special form.

A crystal polyhedron can be described by using zone axes [*uvw*]; the group obtained is written <*uvw*>.

Practical description of crystals. In practice, a crystal description is made as follows. The crystal is measured; that is, the φ and ρ values of the faces are determined and are plotted in a stereognomogram. Zone circles are drawn, and the intersections give possible faces. The Barker rules are applied to find the reference systems. The indices are found graphically, as are all other needed elements. Some of them are calculated afterward if more accuracy is wanted. A crystal drawing de-

CRYSTALLOGRAPHY

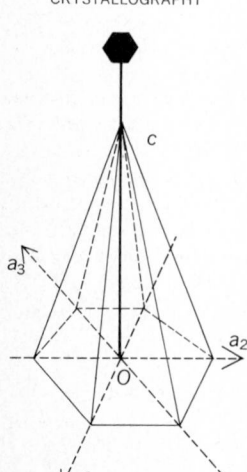

Fig. 8. Bravais indices in hexagonal system.

duced from the projection is also made. A description of the crystals measured is to be found in the book by P. H. Groth listed in the bibliography. These have been revised and completed in the Barker index.   [WILLY C. DEKEYSER]

Bibliography: L. V. Azaroff, *Introduction to Solids*, 1975; M. J. Buerger, *Contemporary Crystallography*, 1970; M. J. Buerger, *Elementary Crystallography*, paper 1978; F. C. Phillips, *Introduction to Crystallography*, 4th ed., 1971; R. W. G. Wyckoff, *Crystal Structures*, 2d ed., 1963–1971.

Curie temperature

The critical or ordering temperature for a ferromagnetic or a ferrimagnetic material. The Curie temperature T_c is the temperature below which there is a spontaneous magnetization M in the absence of an externally applied magnetic field, and above which the material is paramagnetic. In the disordered state above the Curie temperature, thermal energy overrides any interactions between the local magnetic moments of ions. Below the Curie temperature, these interactions are predominant and cause the local moments to order or align so that there is a net spontaneous magnetization.

In the ferromagnetic case, as temperature T increases from absolute zero, the spontaneous magnetization decreases from M_0, its value at $T = 0$. At first this occurs gradually, then with increasing rapidity until the magnetization disappears at the Curie temperature (see illustration). Many physical properties show an anomaly or a change in behavior at the Curie temperature. There is a peak in specific heat and in magnetic permeability, and there are changes in the behavior of such properties as electrical resistivity, elastic properties, and thermal expansivity. In ferrimagnetic materials the course of magnetization with temperature may be more complicated, but the spontaneous magnetization disappears at the Curie temperature. The Curie temperature can be determined from magnetization versus temperature measurements or from the related anomalies in other physical properties. The Curie temperature can be altered by changes in composition, pressure, and other thermodynamic parameters.

Well above the Curie temperature the magnetic susceptibility follows the Curie-Weiss law, and in fact the characteristic temperature θ in that law is called the paramagnetic Curie temperature. Generally θ is slightly above T_c. The susceptibility deviates from Curie-Weiss behavior as the temperature approaches T_c and θ. *See* CURIE-WEISS LAW; FERRIMAGNETISM; FERROMAGNETISM; MAGNETIC SUSCEPTIBILITY; MAGNETIZATION; PARAMAGNETISM.

In antiferromagnetic materials the corresponding ordering temperature is termed the Néel temperature (T_N). Below the Néel temperature the

magnetic sublattices have a spontaneous magnetization, though the net magnetization of the material is zero. Above the Néel temperature the material is paramagnetic. *See* ANTIFERROMAGNETISM.

The ordering temperatures for magnetic materials vary widely (see table). The ordering temperature for ferroelectrics is also termed the Curie temperature, below which the material shows a spontaneous electric moment. *See* FERROELECTRICS; PYROELECTRICITY.   [J. F. DILLON, JR.]

Bibliography: N. W. Ashcroft and N. D. Mermin, *Solid State Physics*, 1976; S. Chikazumi and S. H. Charap, *Physics of Magnetism*, 1964; A. H. Morrish, *Physical Principles of Magnetism*, 1966.

Curie-Weiss law

A relation between magnetic or electric susceptibilities and the absolute temperature which is followed by ferromagnets, antiferromagnets, nonpolar ferroelectrics and antiferroelectrics, and some paramagnets. The Curie-Weiss law is usually written as the following equation, where χ is the

$$\chi = C/(T - \theta)$$

susceptibility, C is a constant for each material, T is the absolute temperature, and θ is called the Curie temperature. Antiferromagnets and antiferroelectrics have a negative Curie temperature. The Curie-Weiss law refers to magnetic and electric behavior above the transition temperature of the material in question. It is not always precisely followed, and it breaks down in the region very close to the transition temperature. Often the susceptibility will behave according to a Curie-Weiss law in different temperature ranges with different values of C and θ. *See* CURIE TEMPERATURE; ELECTRIC SUSCEPTIBILITY; MAGNETIC SUSCEPTIBILITY.

The Curie-Weiss behavior is usually a result of an interaction between neighboring atoms which tends to make the permanent atomic magnetic (or induced electric) dipoles point in the same direction. The strength of this interaction determines the Curie temperature θ. In the magnetic case the interaction is caused by Heisenberg exchange coupling.

In many paramagnetic salts, especially those of the iron group, the Curie-Weiss behavior is due to crystalline distortions and their effect upon the atomic orbital angular momenta which contribute to the magnetization. For further discussion and derivation of the Curie-Weiss law *see* FERROMAGNETISM. *See also* ANTIFERROMAGNETISM; FERROELECTRICS. [ELIHU ABRAHAMS; FREDERIC KEFFER]

Current density

A vector quantity equal in magnitude to the instantaneous rate of flow of electric charge, perpendicular to a chosen surface, per unit area per unit time.

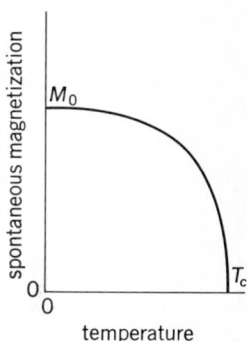

CURIE TEMPERATURE

Variation of spontaneous magnetization of a ferromagnetic material with temperature.

Ordering temperatures of magnetic materials.*

Ferromagnets	Curie temperature (T_c), K	Ferrimagnets	Curie temperature (T_c), K	Antiferromagnets	Néel temperature (T_N), K
Co	1388	Fe_3O_4	858	NiO	600
Fe	1043	$MnFe_2O_4$	573	Cr	311
$CrBr_3$	37	$Y_3Fe_5O_{12}$	560	$FeCl_2$	23.7
$GdCl_3$	2.2				

*From F. Keffer, *Handbuch der Physik*, vol. 18, pt. 2, Springer, 1966.

If a wire of cross-sectional area A carries a current I, the current density J is I/A. The units of J in the rationalized meter-kilogram-second system are amperes per square meter.

The concept of current density is useful in treating the flow of electricity through a conductor of nonuniform cross section, and in electromagnetic theory. In terms of current density Ohm's law can be written $J = E/\rho$, where E is electric field strength and ρ is resistivity. Current density is related to the number per unit volume n, the charge e, and the average velocity v of the effective charge carriers in a conductor by the formula $J = nev$.

[JOHN W. STEWART]

Cyclotron

A device which accelerates particles and heavy ions (deuterons, α-particles, molecular hydrogen ions, and others) in a constant magnetic field. The acceleration is produced by a pair of semicircular electrodes across which a high alternating voltage is produced. At a given frequency of excitation, depending on the value of the magnetic field and the type of particle to be accelerated, the particles spiral out from the center of the magnet at such a rate that their period of revolution in the magnetic field equals the period of the applied alternating voltage, thus maintaining synchronism with the accelerating field. For full discussion *see* PARTICLE ACCELERATOR.   [WOLFGANG K. H. PANOFSKY]

Cyclotron resonance experiments

The measurement of charge-to-mass ratios of electrically charged particles from the frequency of their helical motion in a magnetic field. Such experiments are particularly useful in the case of conducting crystals, such as semiconductors and metals, in which the motions of electrons and holes are strongly influenced by the periodic potential of the lattice through which they move. Under such circumstances the electrical carriers often have "effective masses" which differ greatly from the mass in free space; the effective mass is often different for motion in different directions in the crystal. Cyclotron resonance is also observed in gaseous plasma discharges and is the basis for a class of particle accelerators. *See* BAND THEORY OF SOLIDS; CYCLOTRON; PARTICLE ACCELERATOR.

The experiment is typically performed by placing the conducting medium in a uniform magnetic field H and irradiating it with electromagnetic waves of frequency ν. Selective absorption of the radiation is observed when the resonance condition $\nu = qH/2\pi m^*c$ is fulfilled, that is, when the radiation frequency equals the orbital frequency of motion of the particles of charge q and effective mass m^* (c is the velocity of light). The absorption results from the acceleration of the orbital motion by means of the electric field of the radiation. If circularly polarized radiation is used, the sign of the electric charge may be determined, a point of interest in crystals in which conduction may occur by means of either negatively charged electrons or positively charged holes. *See* HOLES IN SOLIDS.

For the resonance to be well defined, it is necessary that the mobile carriers complete at least $1/2\pi$ cycle of their cyclotron motion before being scattered from impurities or thermal vibrations of the crystal lattice. In practice, the experiment is usually performed in magnetic fields of 1000 to 100,000 oersteds (1 oersted $= 79.6$ amperes per meter) in order to make the cyclotron motion quite rapid ($\nu \sim 10-100$ GHz, that is, microwave and millimeter-wave ranges). Nevertheless, crystals with impurity concentrations of a few parts per million or less are required and must be observed at temperatures as low as 1 K in order to detect sharp and intense cyclotron resonances.

The resonance process manifests itself rather differently in semiconductors than in metals. Pure, very cold semiconductors have very few charge carriers; thus the microwave radiation penetrates the sample uniformly. The mobile charges are thus exposed to radiation over their entire orbits, and the resonance is a simple symmetrical absorption peak.

In metals, however, the very high density of conduction electrons present at all temperatures prevents penetration of the electromagnetic energy except for a thin surface region, the skin depth, where intense shielding currents flow. Cyclotron resonance is then observed most readily when the magnetic field is accurately parallel to the flat surface of the metal. Those conduction electrons (or holes) whose orbits pass through the skin depth without colliding with the surface receive a succession of pulsed excitations, like those produced in a particle accelerator. Under these circumstances cyclotron resonance consists of a series of resonances $n\nu = qH/2\pi m^*c$ ($n = 1,2,3 \ldots$) whose actual shapes may be quite complicated. The resonance can, however, also be observed with the magnetic field normal to the metal surface; it is in this geometry that circularly polarized exciting radiation can be applied to charge carriers even in a metal.

Cyclotron resonance is most easily understood as the response of an individual charged particle; but, in practice, the phenomenon involves excitation of large numbers of such particles. Their net response to the electromagnetic radiation may significantly affect the overall dielectric behavior of the material in which they move. Thus, a variety of new wave propagation mechanisms may be observed which are associated with the cyclotron motion, in which electromagnetic energy is carried through the solid by the spiraling carriers. These collective excitations are generally referred to as plasma waves. In general, for a fixed input frequency, the plasma waves are observed to travel through the conducting solid at magnetic fields higher than those required for cyclotron resonance. The most easily observed of these excitations is a circularly polarized wave, known as a helicon, which travels along the magnetic field lines. It has an analog in the ionospheric plasma, known as the whistler mode and frequently detected as radio interference. There is, in fact, a fairly complete correspondence between the resonances and waves observed in conducting solids and in gaseous plasmas. Cyclotron resonance is more easily observed in such low-density systems since collisions are much less frequent there than in solids. In such systems the resonance process offers a means of transferring large amounts of energy to the mobile ions, which is a necessary condition if nuclear fusion reactions are to occur. *See* NUCLEAR FUSION; PLASMA PHYSICS; WAVES AND INSTABILITIES IN PLASMAS.

[WALTER M. WALSH]

Bibliography: C. Kittel, *Introduction to Solid State Physics*, 4th ed., 1971; P. M. Platzman and P. A. Wolff, *Waves and Interactions in Solid State Plasmas*, 1973.

D'Alembert's principle

The principle that the resultant of the external forces **F** and the kinetic reaction acting on a body equals zero. The kinetic reaction is defined as the negative of the product of the mass m and the acceleration **a**. The principle is therefore stated as $\mathbf{F} - m\mathbf{a} = 0$. While D'Alembert's principle is merely another way of writing Newton's second law, it has the advantage of changing a problem in kinetics into a problem in statics. The techniques used in solving statics problems may then provide relatively simple solutions to some problems in dynamics; D'Alembert's principle is especially useful in problems involving constraints. *See* CONSTRAINT.

If D'Alembert's principle is applied to the plane motion of a rigid body, the techniques of plane statics can be used. The principal advantage is that in a dynamics problem the torques must be calculated about a fixed point or about the center of mass, while in statics the torques can be calculated about any point.

[PAUL W. SCHMIDT]

Bibliography: H. Goldstein, *Classical Mechanics*, 2d ed., 1980; R. J. Stephenson, *Mechanics and Properties of Matter*, 3d ed., 1969.

Dalitz plot

Pictorial representation in high-energy nuclear physics for data on the distribution of certain three-particle configurations. Many elementary-particle decay processes and high-energy nuclear reactions lead to final states consisting of three particles (which may be denoted by a,b,c, with

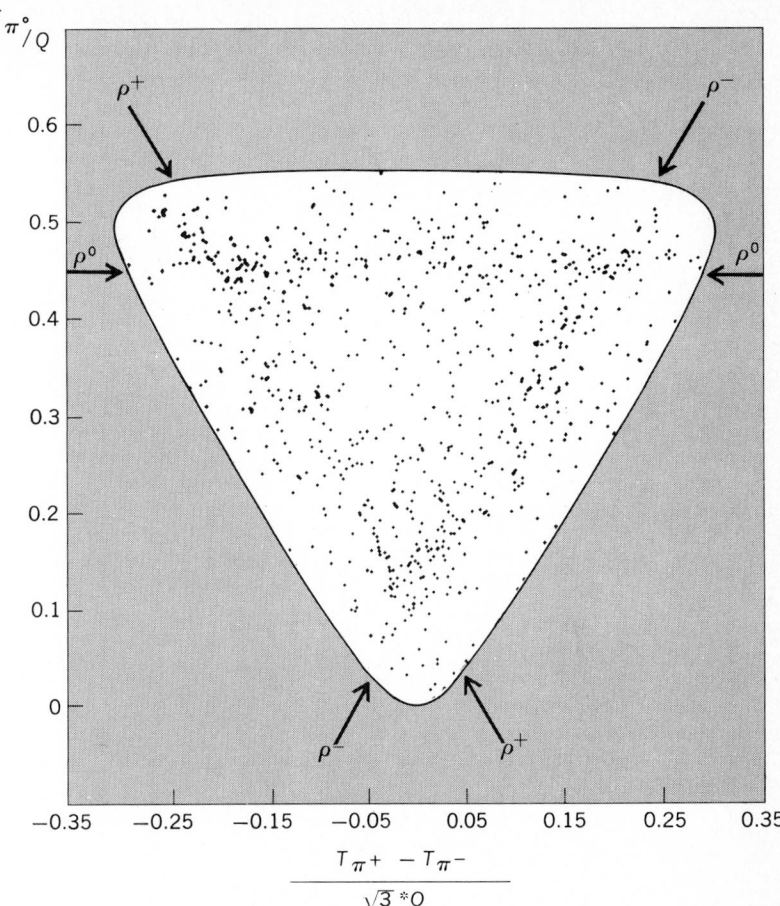

Fig. 2. Dalitz plot for 823 examples of the antiproton annihilation process, $p + \bar{p} \to \pi^+ + \pi^+ + \pi^0$, as reported by C. Baltay and coworkers. Arrows show positions expected for $\rho(765)$-meson resonance bands, which appear clearly for the $(\pi^+\pi^-)$, $(\pi^-\pi^0)$, and $(\pi^0\pi^+)$ systems. Distribution is symmetrical between the six sectors obtained by drawing the three axes of symmetry. The authors interpret this distribution as being due to (1) the reaction $\bar{p}p \to \rho\pi$ occurring in the $I = 0$ S_1^3 initial state, (2) the reaction $pp \to 3\pi$ (s-wave pions) occurring in the $I = 1$ S_0^1 initial state, with roughly equal intensities. (*After C. Baltay et al., Annihilation of Antiprotons in Hydrogen at Rest, Phys. Rev., 140:B1039, 1965*)

mass values m_a, m_b, m_c). Well-known examples are provided by the K-meson decay processes, Eqs. (1) and (2), and by the K- and $\bar{K}$-meson reactions with

$$K^+ \to \pi^+ + \pi^+ + \pi^- \qquad (1)$$
$$K^+ \to \pi^0 + \mu^+ + \nu \qquad (2)$$

hydrogen, given in Eqs. (3) and (4). For definite to-

$$K^+ + p \to K^0 + \pi^+ + p \qquad (3)$$
$$K^- + p \to \Lambda + \pi^+ + \pi^- \qquad (4)$$

tal energy E (measured in the barycentric frame), these final states have a continuous distribution of configurations, each specified by the way this energy E is shared among the three particles. (The barycentric frame is the reference frame in which the observer finds zero for the vector sum of the momenta of all the particles of the system considered.) *See* ELEMENTARY PARTICLE.

Equal mass representation. If the three particles have kinetic energies T_a, T_b, and T_c (in the barycentric frame), Eq. (5) is obtained. As shown

$$T_a + T_b + T_c = E - m_a c^2 - m_b c^2 - m_c c^2 = Q \qquad (5)$$

in Fig. 1, this energy sharing may be represented

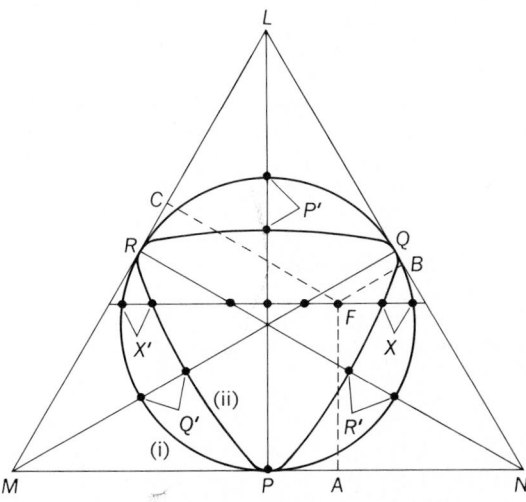

Fig. 1. Configuration of a three-particle system (abc) in its barycentric frame is specified by a point F such that the three perpendiculars FA, FB, and FC to the sides of an equilateral triangle LMN (of height Q) are equal in magnitude to the kinetic energies T_a, T_b, T_c, where Q denotes their sum. See Eq. (6). Curve (i) encloses all points F which correspond to physically allowed configurations, for equal masses and nonrelativistic kinematics; curve (ii) corresponds to curve (i) when relativistic kinematics are used, appropriate to the decay process $\omega(785\ Mev) \to \pi^+\pi^-\pi^0$.

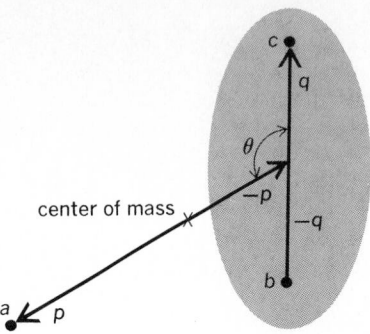

Fig. 3. Illustration of the coordinate system for relativistic three-particle system. For given total energy E, the two momenta, q and p, are related in magnitude by the following equations: $E = \sqrt{(m_a^2 + p^2)} + \sqrt{(M_{bc}^2 + p^2)}$ and $M_{bc} = \sqrt{(m_b^2 + q^2)} + \sqrt{(m_c^2 + q^2)}$.

uniquely by a point F within an equilateral triangle LMN of side $2Q/\sqrt{3}$, such that the perpendiculars FA, FB, and FC to its sides are equal in magnitude to the kinetic energies T_a, T_b, and T_c. This exploits the property of the equilateral triangle that $(FA + FB + FC)$ has the same value (Q, the height of the triangle) for all points F within it. The most important property of this representation is that the area occupied within this triangle by any set of configurations is directly proportional to their volume in phase space. In other words, a plot or empirical data on this diagram gives at once a

picture of the dependence of the square of the matrix element for this process on the a,b,c energies.

Not all points F within the triangle LMN correspond to configurations realizable physically, since the a,b,c energies must be consistent with zero total momentum for the three-particle system. With nonrelativistic kinematics (that is, $T_a = p_a^2/2m_a$, etc.) and with equal masses m for a,b,c, the only allowed configurations are those corresponding to points F lying within the circle inscribed within the triangle, shown as (i) in Fig. 1. With unequal masses the allowed domain becomes an inscribed ellipse, touching the side MN such that $MP:PN$ equals $m_b:m_c$ (and cyclical for NL and LM). More generally, with relativistic kinematics ($T_a = \sqrt{(m_a^2 c^4 + p_a^2 c^2)} - m_a c^2$, etc.), the limiting boundary is distorted from a simple ellipse or circle. This is illustrated in Fig. 1 by the boundary curve (ii), drawn for the $\omega \to 3\pi$ decay process, where the final masses are equal. This curve was also calculated by E. Fabri for Eq. (1), and this plot is sometimes referred to as the Dalitz-Fabri plot. In the high-energy limit $E \to \infty$, where the final particle masses may be neglected, the boundary curve approaches a triangle inscribed in LMN.

The following points (and the regions near them) are of particular interest:

1. All points on the boundary curve. These correspond to collinear configurations, where a,b,c have parallel momenta.

2. The three points of contact with the triangle LMN. For example, point P corresponds to the situation where particle c is at rest (and therefore carries zero orbital angular momentum).

3. The three points which are each farthest from the corresponding side of the triangle LMN. For example, point P' on Fig. 1 corresponds to the situation where b and c have the same velocity (hence zero relative momentum, and zero orbital angular momentum in the bc rest frame).

If the process occurs strongly through an intermediate resonance state, say $a + (bc)^*$ where $(bc)^* \to b + c$, there will be observed a "resonance band" of events for which T_a has the value appropriate to this intermediate two-body system. Such a resonance band runs parallel to the appropriate side of the triangle [the side MN for the case $(bc)^*$, and cyclically] and has a breadth related with the lifetime width for the resonance state.

Antiproton annihilation. The Dalitz plot shown for equal masses in Fig. 1 has been especially useful for three-pion systems, since it treats the three particles on precisely the same footing. Points placed symmetrically with respect to the symmetry axis PL represent configurations related by the interchange of π_b and π_c. The symmetry axes PL, QM, and RN divide the allowed region into six sectors; the configurations in each sector can be obtained from those corresponding to one chosen sector (for example, the sector such that $T_a \geqq T_b \geqq T_c$) by the six operations of the permutation group on three objects. These operations are of particular interest for three-pion systems, since pions obey Bose statistics; the intensities in the six sectors are related with the permutation symmetry of the orbital motion in the three-pion final state. The Dalitz plot shown in Fig. 2, for the antiproton capture reaction $\bar{p}p \to \pi^+\pi^-\pi^0$, illustrates these points. Three ρ-meson bands [$\rho(765 \text{ Mev}) \to \pi\pi$]

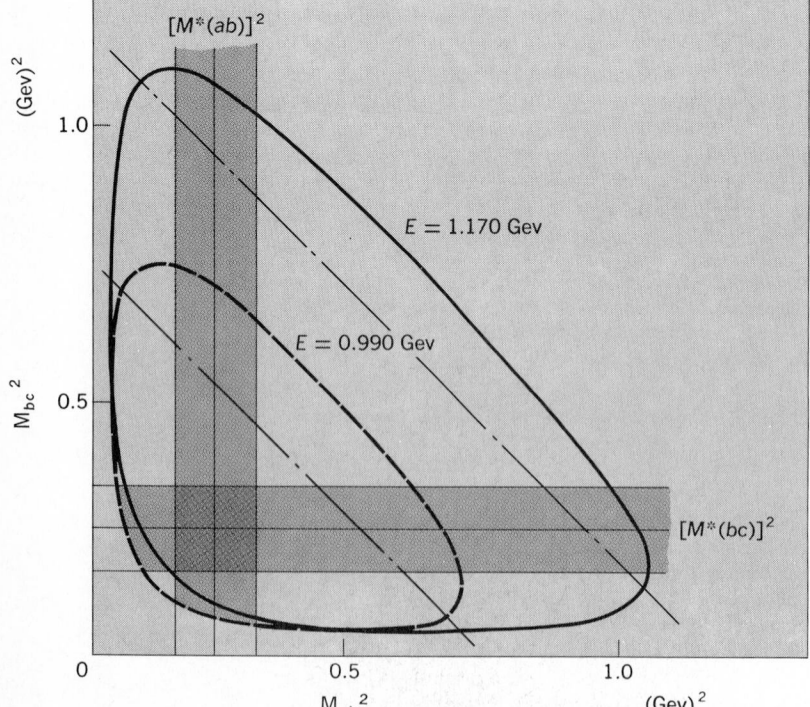

Fig. 4. An unsymmetrical Dalitz plot. Configuration of system (abc) is specified by a point (M_{ab}^2, M_{bc}^2) in a rectangular coordinate system, where M_{ij} denotes the barycentric energy of the two-particle system (ij). Kinematic boundaries have been drawn for equal masses $m_a = m_b = m_c = 0.14$ Gev and for two values of total energy E, appropriate to a three-pion system ($\pi^+ \pi^- \pi^+$). Resonance bands are drawn for states (ab) and (bc) corresponding to a (fictitious) π-π resonance mass 0.5 Gev and full width 0.2 Gev. The dot-dash lines show the locations a (ca) resonance band would have for this mass of 0.5 Gev, for the two values of the total energy E.

are seen, corresponding to intermediate states $\pi^+\rho^-$, $\pi^0\rho^0$, and $\pi^-\rho^0$; the six sectors have equal intensity. *See* BOSE-EINSTEIN STATISTICS.

Relativistic three-particle system. Figure 3 depicts a less symmetric specification for a three-particle system. The momentum of c in the (bc) barycentric frame is denoted by $\boldsymbol{q}$, the momentum of a in the (abc) barycentric frame by $\boldsymbol{p}$, and the angle between $\boldsymbol{p}$ and $\boldsymbol{q}$ by θ. For fixed energy T_a, the points F on Fig. 1 lie on a line parallel to MN; as $\cos\theta$ varies from $+1$ to -1, the point F representing the configuration moves uniformly from the left boundary X' to the right boundary X. If cartesian coordinates are used for F, with origin P and y axis along NM (as has frequently been found useful in the literature), then Eqs. (6) hold. Note

$$x = T_a/Q \qquad y = (T_b - T_c)/Q\sqrt{3} \qquad (6)$$

that, for fixed x, y is linearly related with $\cos\theta$.

Unsymmetrical plot. The Dalitz plot most commonly used is a distorted plot in which each configuration is specified by a point with coordinates $(M_{ab}{}^2, M_{bc}{}^2)$ with respect to right-angled axes. This depends on the relationship given in Eq. (7)

$$M_{bc}{}^2 = (E - m_a c^2)^2 - 2ET_a \qquad (7)$$

and its cyclic permutations, for the total barycentric energy M_{bc} of the two-particle system bc. This plot may be obtained from Fig. 1 by shearing it to the left until LM is perpendicular to MN, and then contracting it by the factor $\sqrt{3}/2$ parallel to MN [which leads to a cartesian coordinate system (T_c, T_a)], and finally reversing the direction of the axes [required by the minus sign in Eq. (7)] and moving the origin to the point $M_{ab}{}^2 = M_{bc}{}^2 = 0$. The plots shown in Fig. 4 correspond in this way to the relativistic curve (ii) in Fig. 1 for two values of the total energy E. This distorted plot retains the property that phase-space volume is directly proportional to the area on the plot. As shown in Fig. 4, the $(ab)^*$ and $(bc)^*$ resonance bands have a fixed location on this plot; data from experiments at different energies E can then be combined on the same plot to give a stronger test concerning the existence of some intermediate resonance state. On the other hand, the $(ca)^*$ resonance bands run across the plot at 135° and move as E varies, so that a different choice of axes [say $(M_{ca}{}^2, m_{ab}{}^2)$] is more suitable for their presentation.

Conclusion. It must be emphasized that the Dalitz plot is concerned only with the internal variables for the system (abc). In general, especially for reaction processes such as Eqs. (3) and (4), there are other variables such as the Euler angles which describe the orientation of the plane of (abc) relative to the initial spin direction or the incident momentum, which may carry additional physical information about the mechanism for formation of the system (abc). The Dalitz plots usually presented average over all these other variables (because of the limited statistics available); this sometimes leads to a clearer picture in that there are then no interference terms between states with different values for the total spin-parity. However, it is quite possible to consider Dalitz plots for fixed values of these external variables, or for definite domains for them. *See* EULER ANGLES; GOLDHABER TRIANGLE.

[RICHARD H. DALITZ]

Bibliography: R. H. Dalitz, *K*-mesons and hyperons: Their strong and weak interactions, *Rep.*

Progr. Phys. 20:163, 1957; G. Kallen, *Elementary Particle Physics*, 1964; C. Zemach, Three-pion decays of unstable particles, *Phys. Rev.*, 133: B1201, 1964.

Dalton's law

The total pressure that a mixture of gases exerts is equal to the sum of the separate pressures which each of the gases would exert if it alone occupied the whole volume. This law was observed by John Dalton (1766–1844) and is also called the law of additive pressures. It may be stated mathematically as $p_m = p_a + p_b + p_c + \cdots$, where p_m is the pressure produced by the mixture and p_a, p_b, p_c, $\cdots$ are the partial pressures of the several gases in the mixture. The partial pressure is the pressure exerted by a constituent gas that occupies the whole volume occupied by the mixture at the same temperature and pressure in the absence of the other gases. *See* AVOGADRO'S LAW; GAS; THERMODYNAMIC PRINCIPLES.

[GEORGE A. HAWKINS]

Bibliography: T. L. Brown, *General Chemistry*, 2d ed., 1968; M. J. Sienko and R. A. Plane, *Chemistry*, 5th ed., 1975.

Damping

A term broadly used to denote either the dissipation of energy in, and the consequent decay of, oscillations of all types or the extent of the dissipation and decay. The energy losses arise from frictional (or analogous) forces which are unavoidable in any system or from the radiation of energy to space or to other systems. For sufficiently small oscillations, the analogous forces are proportional to the velocity of the vibrating member and oppositely directed thereto; the ratio of force to velocity is $-R$, the mechanical resistance. For the role of damping in the case of forced oscillations, where it is decisive for the frequency response, *see* FORCED OSCILLATION; RESONANCE (ACOUSTICS AND MECHANICS). *See also* HARMONIC MOTION; MECHANICAL VIBRATION; OSCILLATION; VIBRATION.

Damped oscillations. An undamped system of mass m and stiffness s oscillates at an angular frequency $\omega_0 = (s/m)^{1/2}$. The effect of a mechanical resistance R is twofold: It produces a change in the frequency of oscillation, and it causes the oscillations to decay with time. If u is one of the oscillating quantities (displacement, velocity, acceleration) of amplitude A, then Eq. (1) holds in the

$$u = Ae^{-\alpha t}\cos\omega_d t \qquad (1)$$

damped case, whereas in the undamped case Eq. (2) holds. The reciprocal time $1/\alpha$ in Eq. (1) may be called the damping constant.

$$u = A\cos\omega_0 t \qquad (2)$$

In Eqs. (1) and (2), the origin for the time t is chosen so that $t = 0$ when $u = A$. The damped angular frequency ω_d in Eq. (1) is always less than ω_0; its value will be given later. According to Eq. (1), the amplitude of the oscillation decays exponentially; the time, given in Eq. (3), is that required for

$$1/\alpha = 2m/R \qquad (3)$$

the amplitude to decrease to the fraction $1/e$ of its initial value.

A common measure of the damping is the logarithmic decrement δ, defined as the natural logarithm of the ratio of two successive maxima of the decaying sinusoid. If T is the period of the oscillation, then Eq. (4) holds, so that Eq. (1) becomes Eq. (5). Thus $1/\delta$ is the number of cycles required for

$$\delta = \alpha T \tag{4}$$

$$u = Ae^{-\delta t/T} \cos \omega_d t \tag{5}$$

the amplitude to decrease by the factor $1/e$ in the same way that $1/\alpha$ is the time required.

The Q of a system is a measure of damping usually defined from energy considerations. In the present case, the stored energy is partly kinetic and partly potential; when the displacement is a maximum, the velocity is zero and the stored energy is wholly potential, while at zero displacement, the energy is wholly kinetic. The Q is π times the ratio of peak energy stored to energy dissipated per cycle. In the present example, this reduces to Eq. (6). The damped frequency ω_d of

$$Q = \omega_0 m/R = \pi/\delta \tag{6}$$

Eq. (1) is related to the undamped frequency ω_0 of Eq. (2) by Eq. (7), so that for high-Q (lightly

$$(\omega_d/\omega_0)^2 = 1 - (1/2Q^2) \tag{7}$$

damped) systems, it is only slightly less than ω_0. *See* ENERGY.

Overdamping; critical damping. If α in Eq. (1) exceeds ω_0, then the system is not oscillatory and is said to be overdamped. If the mass is displaced, it returns to its equilibrium position without overshoot, and the return is slower as the ratio α/ω_0 increases. If $\alpha = \omega_0$ (that is, $Q = 1/2$), the oscillator is critically damped. In this case, the motion is again nonoscillatory, but the return to equilibrium is faster than for any overdamped case.

Distributed systems. An undamped, one-dimensional wave of frequency $\omega/2\pi$ propagated in the positive direction of x is represented by Eq. (8),

$$u = A \cos \omega(t - x/c) \tag{8}$$

c being the velocity of the wave. If the vibration is maintained at $x = 0$ at the value $u = A \cos \omega t$, then the damping manifests itself as an exponential decrease of amplitude with distance x. Equation (8) is replaced by Eq. (9). The attenuation α' may

$$u = Ae^{-\alpha' x} \cos [\omega(t - x/c)] \tag{9}$$

depend on frequency. If the medium is terminated, the wave will be reflected from the ends, and a system of standing waves will be set up. Examples are a rod carrying sound waves, a piece of electrical transmission line or waveguide, and a vibrating violin string. Such a system has a number of natural frequencies $\omega_n/2\pi$, at each of which it behaves like the lumped system of the previous sections. The decay of a vibration is characterized by Q in Eq. (10), where $\lambda_n = 2\pi c/\omega_n$ is the wavelength. *See* STANDING WAVE.

$$Q = \omega_n/2\alpha' c = \pi/\alpha' \lambda_n \tag{10}$$

Hysteresis damping. At a given instant, the elongation (strain) of a metal bar which is under periodic, alternating stress is not determined exactly by the instantaneous value of the stress existing at that time. For example, the elongation is less at a given stress value when the stress is increasing than when it is decreasing. This phenomenon, which is known as mechanical hysteresis, causes

an undesirable energy loss. A vibration problem of serious nature exists in the blades of jet engines and other steam and gas turbines. The blade material itself exhibits a mechanical hysteresis damping which holds the vibrations in check. When the stress is small, the hysteresis damping is very small in all metals, but it rises suddenly when the stress reaches a certain value. Unfortunately, in most metals the stress at which hysteresis damping becomes large and that at which the metal fails because of fatigue are very close together. However, a much higher hysteresis damping at safe stresses than that of ordinary steel is exhibited by certain alloys.

Oscillating electrical circuits. A simple series electrical circuit consisting of an inductance L, resistance R, and capacitance C is exactly analogous to the mechanical system described by Eqs. (1)–(6). The inductance, resistance, and elastance ($1/C$) correspond to the mass, mechanical resistance, and stiffness, respectively. A distributed electrical circuit, such as a section of transmission line or wave guide, is analogous to a vibrating rod or disk.

In the ordinary electrical oscillator, the frequency is controlled by an electrical resonator (tank) lumped at the lower, and distributed at the higher, frequencies. Good frequency stability is associated with a high-Q tank. For frequencies of tens of megahertz (MHz) and below, mechanical resonators can be constructed which have a much higher Q than the equivalent electrical tanks. Thus, very stable electrical oscillators have mechanical resonators as their frequency-determining elements. Such an electromechanical system can operate only if there is some coupling between the electrical aspects and mechanical aspects of the system.

The coupling can be arranged in various ways. In some materials, such as quartz, the constitutive relations involve the mechanical and electrical variables jointly; thus, for example, an electric field may produce a strain in the absence of any stress. Thus, the coupling is inherent in the quartz itself. Quartz crystals are used for frequency control of oscillators in the range from kilohertz to perhaps 100 MHz; the Q of a high-frequency crystal may be several million. Some low-frequency oscillators are controlled by tuning forks having a Q of several hundred thousand, the action being similar to that of the electric bell or buzzer. *See* PIEZOELECTRICITY.

The electrostatic motor-generator effect provides the coupling in such mixed systems as the condenser microphone and electrostatic loudspeaker, and the electromagnetic motor-generator effect plays the same role in the dynamic microphone, dynamic loudspeaker, and in various electrical instruments.

Reading of meters. Galvanometers and other electrical indicating instruments are examples of damped electromechanical systems. The free period depends on the moment of inertia of the rotating system (for example, of the coil in a galvonometer) and on the stiffness of the suspension or spring. If an electrical input is suddenly applied, the indicator will, if the system is highly underdamped, overshoot its equilibrium value and then execute a damped sinusoidal oscillation about it; if the system is highly overdamped, the indicator will approach its final reading sluggishly. If the reading time is taken as the time required to reach the equilibrium value ±1%, then the minimum reading time is obtained if the logarithmic decrement is 83% of the critical value (relative damping = 0.83) and is equal to 67% of the free period.

In portable and switchboard instruments, the damping is either viscous or magnetic or both. Viscous damping is achieved with vanes attached to the movement which move in a narrow, air-filled space; magnetic damping is an eddy-current effect. The eddy currents are generated in the coil frame or in a metal plate moving between magnetic poles; this latter arrangement is used in magnetically damped analytical balances. *See* EDDY CURRENT.

In the d'Arsonval galvanometer, the damping is largely due to the generator action of the moving coil, and it can be adjusted by varying the external circuit resistance. The same is true, to a lesser extent, of sensitive microammeters. *See* RESONANCE (ALTERNATING-CURRENT CIRCUITS).

[MARTIN GREENSPAN]

Bibliography: F. K. Harris, *Electrical Measurements*, 1952, reprint 1975; L. E. Kinsler and A. R. Frey, *Fundamentals of Acoustics*, 2d ed., 1962; C. Kittel, W. D. Knight, and M. A. Ruderman, *Mechanics*, Berkeley Physics Course, vol. 1, 2d ed., 1973; L. Meirovitch, *Elements of Vibration Analysis*, 1975; B. Pippard, *The Physics of Vibration*, vol. 1, 1978; R. F. Steidel, *An Introduction to Mechanical Vibrations*, 2d ed., 1979.

De Broglie wavelength

The wavelength $\lambda = h/p$ associated with a beam of particles (or with a single particle) of momentum p; $h = 6.63 \times 10^{-27}$ erg-sec is Planck's constant. The same formula gives the momentum of an individual photon associated with a light wave of wavelength λ. This formula, along with the profound proposition that all matter has wavelike properties, was first put forth by Louis de Broglie in 1924, and is fundamental to the modern theory of matter and its interaction with electromagnetic radiation. *See* NONRELATIVISTIC QUANTUM THEORY; QUANTUM MECHANICS. [EDWARD GERJUOY]

Decibel

A logarithmic unit used to express the magnitude of a change in level of power, voltage, current, or sound intensity. A decibel (dB) is 1/10 bel.

In acoustics a step of 1 bel is too large for most uses. It is therefore the practice to express sound intensity in decibels. The level of a sound of intensity I in decibels relative to a reference intensity I_R is given by notation (1). Because sound intensity is

$$10 \log_{10} \frac{I}{I_R} \qquad (1)$$

proportional to the square of sound pressure P, the level in decibels is given by Eq. (2). The reference

$$10 \log_{10} \frac{P^2}{P_R{}^2} = 20 \log_{10} \frac{P}{P_R} \qquad (2)$$

pressure is usually taken as 0.0002 dynes/cm² or 0.0002 microbar. (The pressure of the Earth's atmosphere at sea level is approximately 1 bar.) A sinusoidal alternation in pressure at a frequency of 1000 Hz is barely audible to the average person when it has a root-mean-square sound pressure of 0.0002 microbar. By this definition such a tone has a sound pressure level of 0 dB. *See* BEL.

The neper is similar to the decibel but is based upon natural (Napierian) logarithms. One neper is equal to 8.686 dB.

[KARL D. KRYTER]

Deep inelastic collisions

A type of nuclear reaction in which the two nuclei interact strongly while their surfaces overlap. Deep inelastic collisions of one nucleus with another represent a class of reaction mechanism with interaction times intermediate between direct and compound nucleus processes. They often occur in collisions of heavy nuclei at center-of-mass energies less than 5 MeV per nucleon above the Coulomb barrier. During the brief encounter of the two nuclei, energy and mass flow from one nucleus into the other. After a short time, corresponding to the time it takes for the intermediate dinuclear complex to make a partial rotation (10^{-22} to 5×10^{-21} s), the two nuclei separate again. Of the characteristic features of these collisions, two of the most important properties are the sizable amounts of energy dissipation and nucleon (neutron and proton) exchange.

Light-particle emission serves as a probe of the dissipation and deexcitation mechanisms in deep inelastic collisions. Such studies with neutrons have shown that the nuclear temperatures of the two fragments at the time of separation are equal. From this result it can be inferred that thermal equilibrium is established during the deep inelastic collision time. Furthermore, it can be concluded from neutron multiplicity measurements that most of the kinetic energy dissipated in a deep inelastic collision goes into the intrinsic heat of the two reaction fragments, resulting in temperatures of the order of 2 to 3 MeV ($2-3 \times 10^{10}$ K).

Studies of the variance in the charge distribution for fixed mass asymmetry have shown that the charge equilibration degree of freedom, like the excitation energy, is relaxed in less than 2×10^{-22} s. In contrast to the short relaxation times for the above two intrinsic degrees of freedom, the mass asymmetry degree of freedom relaxes more slowly. This results in final fragments with average masses near those of the target and projectile. However, each mass peak has a width associated with nucleon exchange. From gamma-ray and sequential fission measurements, it can be inferred that the fragments from a deep inelastic collision are rotating rapidly.

The observed mass distributions and their evolution with kinetic energy loss are in accord with the assumption of a successive exchange of single nucleons proceeding simultaneously with dissipation of relative kinetic energy in many small steps. Models based on the idea that nucleons transport momentum through a window between the colliding nuclei, leading to the dissipation of energy and orbital angular momentum and fluctuations in the charge and mass number, have been very successful. The calculation of the current of nucleons transferred from one nucleus to the other requires that account be taken of the occupancy of the single-particle orbits in accordance with the Pauli principle. In such models, the motion of the heavy ions in the presence of conservative and dissipative forces is approximated by newtonian trajectories. *See* NUCLEAR REACTION; NUCLEAR STRUCTURE.   [JOHN R. HUIZENGA]

Bibliography: M. Lefort and C. Ngô, Deep inelastic reactions with heavy ions: A probe for nuclear macrophysics studies, *Ann. Phys.*, 3:5–114, 1978; J. Randrup, Theory of transfer-induced transport in nuclear collisions, *Nucl. Phys.*, 327:490–516, 1979; W. V. Schröder and J. R. Huizenga, Damped heavy-ion collisions, *Annu. Rev. Nucl. Sci.*, 27:465–547, 1977.

Degeneracy (quantum mechanics)

A term referring to the fact that two or more stationary states of the same quantum-mechanical system may have the same energy even though their wave functions are not the same. In this case the common energy level of the stationary states is degenerate. The statistical weight of the level is proportional to the order of degeneracy, that is, to the number of states with the same energy; this number is predicted from Schrödinger's equation.

Except for so-called accidental degeneracy, degeneracy is associated with special symmetries of the physical system, and can be removed by destroying this symmetry. For example an energy level of total angular momentum $\sqrt{J(J+1)}\, h/2\pi$ has a $(2J+1)$-fold degeneracy that results from the circumstance that the $2J+1$ allowed values of J_z, the z-component of total angular momentum, all have the same energy. In a magnetic field **H** along z, this level splits into $2J+1$ energy levels, since the energy now depends on J_z, that is, on the angle between **H** and the total angular momentum.

In quantum mechanics and in other branches of mathematical physics, the term degeneracy is employed also to characterize the eigenvalues of operators other than the energy operator. *See* EIGENVALUE (QUANTUM MECHANICS); NONRELATIVISTIC QUANTUM THEORY.   [EDWARD GERJUOY]

Degree of freedom

Any one of the number of independent ways in which the space configuration of a mechanical system may change. A material particle confined to a line in space can be displaced only along the line, and therefore has one degree of freedom. A particle confined to a surface can be displaced in two perpendicular directions and accordingly has two degrees of freedom. A particle free in physical space has three degrees of freedom corresponding to three possible perpendicular displacements. A system composed of two free particles has six degrees of freedom, and one composed of N free particles has $3N$ degrees. If a system of two particles is subject to a requirement that the particles remain a constant distance apart, the number of degrees of freedom becomes five. Any requirement which diminishes by one the degrees of freedom of a system is called a holonomic constraint. Each such constraint is expressible by an equation of condition which relates the system's coordinates to a constant, and may also involve the time. When applied to systems of particles, a holonomic constraint frequently has the geometrical significance of confining a particle to a specified surface, which may be time-dependent. *See* CONSTRAINT.

A mechanical system may be conceived as a set of N particles in space subject to a certain number

K of holonomic constraints which reduce the number of degrees of freedom to $3N - K$. In the case of a rigid structure consisting of more than two non-collinear particles, the number of degrees of freedom is independent of the number of particles and is equal to six.   [RUSSELL A. FISHER]

Bibliography: H. Goldstein, *Classical Mechanics*, 2d ed., 1980; K. R. Symon, *Mechanics*, 3d ed., 1971; H. D. Young, *Fundamentals of Mechanics and Heat*, 2d ed., 1973.

De Haas–van Alphen effect

An oscillatory behavior of the magnetic moment of a pure metal crystal with changes in the applied magnetic field B, at very low temperatures. It is named after its discoverers W. J. de Haas and P. M. van Alphen. This effect has its origin in the Bohr-Sommerfeld quantization of the orbits of conduction electrons under the influence of the magnetic field.

Origin. A free electron in a magnetic field moves in a circular orbit with an angular frequency given by the cyclotron resonance frequency, Eq. (1),

$$\omega_c = \frac{eB}{m} \tag{1}$$

where e and m are the mass and charge of the electron. But the energy E of this simple harmonic motion must be quantized according to the Planck prescription, Eq. (2), where $\hbar$ is Planck's constant

$$E_n = \left(n + \frac{1}{2}\right)\hbar\omega_c \tag{2}$$

divided by 2π, and n numbers the quantized energy levels.

In a metal at low temperatures at zero magnetic field, electrons are not at rest, but have a continuous distribution of energies up to some limiting energy E_F called the Fermi energy; this arises because the Pauli exclusion principle keeps the electrons from all falling into the lowest energy state. Those electrons with energy close to E_F dominate the behavior of the metal. In a magnetic field the allowed electron energies are not continuous, but must correspond to one of those given by Eq. (2). Suppose for some value of n, B_n, an electron energy level matches the Fermi energy E_F, that is, $E_n = E_F$ (see illustration); from Eqs. (1) and (2), this corresponds to Eq. (3).

$$E_F = \frac{\left(n + \frac{1}{2}\right)\hbar e}{m} B_n \tag{3}$$

See CYCLOTRON RESONANCE EXPERIMENTS; EXCLUSION PRINCIPLE; FREE-ELECTRON THEORY OF METALS.

In order for n to increase by 1, the magnetic field must change so that Eq. (4) is satisfied. Eliminating

$$E_F = \frac{\left(n + \frac{3}{2}\right)\hbar e}{m} B_{n+1} \tag{4}$$

nating n between Eqs. (3) and (4) yields Eq. (5) for

$$\triangle\left(\frac{1}{B}\right) \equiv \frac{1}{B_{n+1}} - \frac{1}{B_n} = \frac{e\hbar}{mE_F} \tag{5}$$

the amount by which the magnetic field must change in order for n to increase by 1. The quantity $\triangle(1/B)$ determines the "period" of the de Haas–van Alphen oscillations (it is a period in

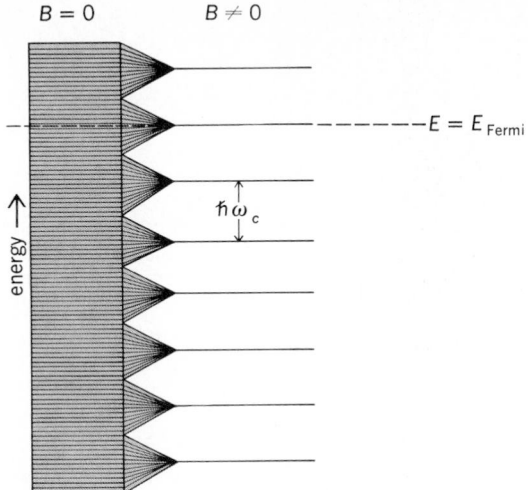

$B = 0$ $B \neq 0$

$--- E = E_{\text{Fermi}}$

energy

$\hbar\omega_c$

Distribution of electron energy in a metal at zero and finite magnetic field B, showing the condition that an electron energy level matches the Fermi energy.

magnetic field, not time). Since $E = p^2/2m$, where p is the electron momentum, Eq. (6) is valid, where

$$\Delta\left(\frac{1}{B}\right) = \frac{2eh}{p_F^2} \qquad (6)$$

p_F is the limiting momentum corresponding to the energy E_F. Thus, a measurement of the period of the oscillations permits a determination of the limiting or Fermi momentum of the electrons. A detailed quantum statistical-mechanical treatment shows that many of the properties of a metal, including the magnetization, oscillate with field according to Eq. (6). See MAGNETIZATION.

Measurement of effective mass. The assumption that the electrons occupy only states with E_n equal to or less than E_F involves the implicit assumption that the metal is cooled to absolute zero. At any finite temperature, states with E_n greater than E_F will also be occupied. The ratio of the number of electrons in two neighboring energy levels E_n and E_{n+1} is governed by the Boltzmann occupation factor e $(E_{n+1} - E_n)/k_BT$, where k_B is Boltzmann's constant. If T is large enough such that k_BT is much greater than $\hbar\omega_c$, this factor becomes unity, and in this case there is nothing particularly significant about the condition $E_F = E_n$; that is, the amplitude of the oscillations is expected to be greatly attenuated. On the other hand, at very low temperatures where $\hbar\omega_c$ is greater than k_BT, strong oscillations are expected. Thus, a measurement of the temperature dependence of the oscillation amplitude permits a determination of the cyclotron frequency, or equivalently the electron mass. In a metal, as in a semiconductor, electrons behave with an effective mass m^* rather than the free electron mass, and the de Haas–van Alphen effect thus allows a measurement of m^* in metals. See KINETIC THEORY OF MATTER.

Determination of Fermi surface. When electrons move in a metal, the energy-momentum relation in general differs radically from that of a free particle. The momenta $\vec{p}_F$ which correspond to the Fermi energy depend on the direction $\hat{p}_F$ within the crystal; that is, the shape of the Fermi surface differs from a sphere.

Application of the Bohr-Sommerfeld quantiza-

tion rules shows that Eq. (6) for the de Haas–van Alphen period must be generalized to Eq. (7),

$$\Delta\left(\frac{1}{B}\right) = \frac{2\pi\hbar e}{a_p} \qquad (7)$$

where a_p is the area of the orbit in momentum space, that is, the cross section of the Fermi surface.

For a given Fermi surface, the cross-sectional area is a function of the momentum parallel to the field and ranges between various extremal values. In principle, all of these sections make an oscillatory contribution. However, for all sections other than extremal sections, the oscillatory contributions at neighboring sections tend to cancel. Thus in Eq. (7) a_p is the area of an extremal section (or sections for metals whose Fermi surface supports multiple extremal cross sections). By studying the dependence of a_p on the direction $\hat{B}$ of the magnetic field, sufficient information can be obtained to construct the detailed shape of the Fermi surfaces of a metal. A further condition for observation, in addition to low temperatures, is sample purity: the electron-scattering time τ must be longer than, or at least comparable to, a cyclotron period $2\pi/\omega_c$.

Virtually all metals available in sufficient purity have been studied by the de Haas–van Alphen effect. The effect is also the most powerful probe of Fermi surface properties in alloys and intermetallic compounds.

Prescriptions for deconvolution of extremal areas to Fermi momenta are well developed. Furthermore, measurements of angular-dependent effective mass m^* can be used to determine the velocity of electrons at the Fermi surface. See BAND THEORY OF SOLIDS; FERMI SURFACE.

[J. B. KETTERSON]

Bibliography: G. W. Crabtree et al., Anisotropic many-body effects in the quasiparticle velocity of Nb, *Phys. Rev. Lett.*, 42:390–393, 1979; D. P. Karim, J. B. Ketterson, and G. W. Crabtree, *J. Low Temp. Phys.*, 30:389, 1978; I. M. Lifshitz and A. M. Kosevich, *Zh. Eksperim Theor. Fiz.*, 29:730, 1955; L. F. Mattheiss, *Phys. Rev.*, 31:373, 1970.

Demagnetization

A process for reducing a magnetized material to a neutral state, that is, to a state when both the magnetic induction B and the magnetic field strength H are equal to zero.

The relationship between B and H for a ferromagnetic material is complex because it depends upon the previous magnetic history of the sample. The state of the sample is not uniquely specified by a point on a magnetization curve because the value of B corresponding to a given value of H depends on the sequence of events by which the value of H is reached. For the state of the sample to be independent of the previous history, it must be demagnetized by a standard process. For a discussion of magnetization curves see MAGNETIZATION.

The standard demagnetization process involves applying a magnetizing force above a critical value for the material used, and successively reversing H in direction as the value of H is steadily reduced to a value less than those to be used in subsequent experiments. If the final value of H is essentially zero, the specimen is demagnetized. The reversal may be carried out by mechanically reversing a direct current, or by using an alternating current that is steadily reduced or from which the sample is gradually removed.

The adiabatic demagnetization of paramagnetic salts is a technique used to produce temperatures very near the absolute zero. *See* ADIABATIC DE-MAGNETIZATION.   [KENNETH V. MANNING]

Bibliography: B. Bleaney and B. I. Bleaney, *Electricity and Magnetism*, 3d ed., 1976; E. M. Pugh and E. W. Pugh, *Principles of Electricity and Magnetism*, 2d ed., 1970.

Density

The mass per unit volume of a material. The term is applicable to mixtures and pure substances and to matter in the solid, liquid, gaseous, or plasma state. Density of all matter depends on temperature; the density of a mixture may depend on its composition, and the density of a gas on its pressure. Common units of density are grams per cubic centimeter, and slugs or pounds per cubic foot. The specific gravity of a material is defined as the ratio of its density to the density of some standard material, such as water at a specified temperature, for example, 60°F, or, for gases the basis may be air at standard temperature and pressure. Another related concept is weight density, which is defined as the weight of a unit volume of the material. *See* MASS; WEIGHT.

[LEO NEDELSKY]

Density matrix

A matrix which is constructed as the most general statistical description of the states of a many-particle quantum-mechanical system. The state of a quantum system is described by a normalized wave function $\psi(x,t)$, (where x stands for all coordinates of the system, and t for the time), which satisfies the Schrödinger equation, Eq. (1), where

$$H\psi(x,t) = \frac{h}{i}\frac{\partial \psi(x,t)}{\partial t} \qquad (1)$$

H is the Hamiltonian of the system, and $\hbar$ is Planck's constant divided by 2π. Furthermore, $\psi(x,t)$ may be expanded in terms of a complete orthonormal set $\{\varphi(x)\}$, as in Eq. (2). Then, the density matrix is defined by Eq. (3), and this density matrix describes a pure state.

$$\psi(x,t) = \sum_n a_n(t)\varphi_n(x) \qquad (2)$$

$$\rho_{mn}(t) = a_n{}^*(t)a_m(t) \qquad (3)$$

sity matrix describes a pure state. Examples of pure states are a beam of polarized electrons and the photons in a coherent beam emitted from a laser. *See* LASER; NONRELATIVISTIC QUANTUM THEORY.

In quantum statistics, one deals with an ensemble of N systems which have the same Hamiltonian. If the αth member of the ensemble is in the state ψ^α in Eq. (4), the density matrix is defined as the ensemble average, Eq. (5). In general, this

$$\psi^\alpha(x,t) = \sum_n a_n{}^\alpha(t)\varphi_n(x) \qquad (4)$$

$$\rho_{mn}(t) = \frac{1}{N}\sum_\alpha [a_n{}^\alpha(t)]^*a_m{}^\alpha(t) \qquad (5)$$

density matrix describes a mixed state, for ex-

ample, a beam of unpolarized electrons or the photons emitted from an incoherent source such as an incandescent lamp. The pure state is a special case of the mixed state when all members of the ensemble are in the same state. *See* STATISTICAL MECHANICS.

Diagonal elements. The diagonal element $\rho_{nn}(t) = |a_n(t)|^2$ of the density matrix for a pure state gives the probability of finding the quantum system in the state $\varphi_n(x)$ at time t. The set $\{\varphi(x)\}$ is complete; consequently, the normalization property of the probabilities implies that $\sum_n \rho_{nn}(t) = \text{Tr}(\rho) = 1$. The trace is unity also for a mixed state. Since all diagonal elements are nonnegative, it also follows that $0 \le \rho_{nn} \le 1$.

Observables. The ensemble average of an observable A (an operator that represents an observable quantity) can be written as Eq. (6). Since $\bar{\bar{A}}$

$$\bar{\bar{A}} = \frac{1}{N}\sum_\alpha \sum_{mn}(a_n{}^\alpha)^*(a_m{}^\alpha)\int \varphi_n{}^*A\varphi_m\,dx$$
$$= \sum_{mn}\rho_{mn}A_{nm} = \text{Tr}(\rho A) \qquad (6)$$

is the trace of a matrix, its value is independent of the choice of the orthonormal set $\{\varphi(x)\}$.

Pure and mixed states. For a pure state, Eq. (7) holds, or, equivalently, $\rho^2 = \rho$ or $\text{Tr}\,\rho^2 = 1$. But,

$$(\rho^2)_{mn} = \sum_p a_p{}^*a_m a_n{}^*a_p = a_n{}^*a_m = \rho_{mn} \qquad (7)$$

for a mixed state, $\text{Tr}\,\rho^2 < 1$ always holds. The necessary and sufficient condition for a state to be a pure state is $\rho^2 = \rho$.

Eigenvalues and eigenvectors. The density matrix is Hermitian, $\rho_{mn}{}^* = \rho_{nm}$, and is diagonalizable. The eigenvalues of ρ for a pure state are all zero except one, which has the value unity. The eigenvector corresponding to the nonzero eigenvalue is the pure state the system is in. For a mixed state, the eigenvectors are the possible pure states the system may have, and the eigenvalues are the probabilities of these states. Thus all eigenvalues lie between 0 and 1. The extreme case of a mixed state is the random state for which all N members of the ensemble are in different states. All eigenvalues corresponding to the occupied states have the value $1/N$. *See* MATRIX THEORY.

Liouville equation. The density matrix satisfies the Liouville equation, Eq. (8), where H is the

$$i\hbar\frac{\partial \rho}{\partial t} = [H,\rho] \qquad (8)$$

Hamiltonian of the system. In a state of equilibrium $\partial\rho/\partial t = 0$. Then, Eq. (8) implies that ρ is a function of the Hamiltonian H alone. For an ensemble in thermal equilibrium, ρ is related to temperature T according to Eq. (9), where k is the Boltzmann

$$\rho = \frac{e^{-H/kT}}{\text{Tr}\,e^{-H/kT}} \qquad (9)$$

constant. This result, together with Eq. (6), allows a complete determination of the equilibrium thermodynamic properties of the system. For example, the entropy S is proportional to the expectation value of $\ln\rho$ as given in Eq. (10).

$$S = -k\overline{\overline{\ln\rho}} = -k\,\text{Tr}(\rho\ln\rho) \qquad (10)$$

The relationship in Eq. (10) between the entropy

and the density matrix holds for nonequilibrium systems as well. On the basis of knowledge of the eigenvalues of the density matrix, it can be deduced that the entropy of a pure state is zero, that of a mixed state is nonnegative, and that of a random state has the highest possible value of $k \ln N$ for an ensemble of N systems. Hence, entropy is a measure of the randomness of the state of an ensemble. *See* ENTROPY.

Effect of measurement. Another consequence of the Liouville equation, Eq. (8), is that, for an isolated system whose Hamiltonian is time-independent, a pure state cannot evolve into a mixed state, or vice versa, as time goes on. The act of a physical measurement, however, can interrupt the time development of a system so that a pure state can be turned into a mixed state. Consider a Stern-Gerlach experiment which employs a non-uniform magnetic field to separate a beam of electrons into two beams which have spins parallel and antiparallel to the field. The direction of the field is chosen as the z axis. The incoming electrons are assumed to be polarized along an axis different from the z axis. The wave function of the incoming beam is then $\psi = a\psi_+ + b\psi_-$, where ψ_+, ψ_- are the wave functions of the spin-up and spin-down states and $|a|^2 + |b|^2 = 1$. The density matrix is shown in Eq. (11). This matrix satisfies the requirement $\rho^2 =$

$$\rho = \begin{pmatrix} |a|^2 & ab^* \\ a^*b & |b|^2 \end{pmatrix} \tag{11}$$

ρ for a pure state. After passing through the Stern-Gerlach apparatus, the density matrix becomes Eq. (12), which represents a mixed state. The

$$\rho' = \begin{pmatrix} |a|^2 & 0 \\ 0 & |b|^2 \end{pmatrix} \tag{12}$$

measurement has resulted in an increase in the entropy or randomness of the electron beam.

Perturbations. The Liouville equation is also very useful in determining the effect of a perturbation on the density matrix and, hence, on the properties of the system. Let the unperturbed Hamiltonian be H_0 and the perturbing Hamiltonian be H_1. The density matrix is similarly written as the sum of the unperturbed part ρ_0 and a correction ρ_1. Furthermore, it is assumed that the system is in equilibrium state before the perturbation is applied. Then Eq. (8) takes the form of Eq. (13). If the perturbation is weak, the last term of Eq. (13)

$$i\hbar \frac{\partial \rho_1}{\partial t} = [H_0, \rho_1] + [H_1, \rho_0] + [H_1, \rho_1] \tag{13}$$

can be ignored because it is of the second order of the perturbation. When this is done, the correction term can be solved to obtain Eq. (14). The Heisenberg operator $H_1(t - t')$ is defined in Eq. (15). The

$$\rho_1 = -\frac{i}{\hbar} \int_{t_0}^{t_1} [H_1(t - t'), \rho_0] dt' \tag{14}$$

$$H_1(t - t') = \exp\left[\frac{i}{\hbar} H_0(t - t')\right] H_1 \exp\left[-\frac{i}{\hbar} H_1(t - t')\right] \tag{15}$$

lower limit of the integration in Eq. (14) is the time at which the perturbation is turned on. *See* PERTURBATION (QUANTUM MECHANICS).

From this result the first-order correction to the expectation value of an observable A can be calcu-

lated as in Eq. (16). Notice that ρ_1 involves only the

$$\overline{\overline{A}}_1 = \mathrm{Tr}\,(\rho_1 A) \tag{16}$$

perturbation and the equilibrium density matrix. Thus, the first-order correction, called the linear response of the system to the perturbation, can be completely calculated from the knowledge of the equilibrium state. This is the starting point of the modern theory of transport phenomena. *See* KINETIC THEORY OF MATTER.   [S. H. LIU]

Bibliography: C. Cohen-Tannoudji et al., *Quantum Mechanics*, 1978; L. D. Landau and E. M. Lifshitz, *Quantum Mechanics: Non-Relativistic Theory*, 3d ed., 1977; L. D. Landau and E. M. Lifshitz, *Statistical Physics*, 2d ed., 1969; P. Roman, *Advanced Quantum Theory*, 1965; D. ter Haar, *Reports on Progress in Physics*, 24:304, 1961; J. von Neumann, *Mathematical Foundations of Quantum Mechanics*, 1955.

Deuteron

The nucleus of the atom of heavy hydrogen, H^2 (deuterium). The deuteron d is composed of a proton and a neutron. As the simplest multinucleon nucleus, the deuteron has been the subject of extensive study. Its binding energy is 2.227 Mev; that is, this is the amount of energy which must be added to a deuteron for it to dissociate into a proton and a neutron. The accurate determination of this dissociation energy provides the means of calculating the mass of the neutron, the mass of the deuteron (2.014187 amu) and proton being known from other experiments.

The intrinsic angular momenta, or spins, of the proton and neutron combine to produce a deuteron spin of unity; hence, the deuteron obeys the type of quantum statistics known as Bose-Einstein statistics. The deuteron possesses a magnetic moment (0.857407 nuclear magnetons) and an electric quadrupole moment (2.738×10^{-27} cm^2).

Deuterons are much used as projectiles in nuclear bombardment experiments, especially to produce (d,p), (d,n), and (d,α) reactions. In the first two reactions, because of the low binding energy of the deuteron, the neutron n or proton p is stripped from it and captured by the target nucleus. Meanwhile, the other half of the deuteron (that is, the proton or neutron) carries away the excess energy. The H^1/H^2 abundance ratio in nature is 6700. *See* NUCLEAR REACTION.

[HENRY E. DUCKWORTH]

Diamagnetism

That branch of magnetism which treats of diamagnetic phenomena and of the properties of diamagnetic bodies. Diamagnetism is a property exhibited by substances with a negative magnetic susceptibility, that is, by substances which magnetize in a direction opposite to that of an applied magnetic field. A diamagnetic substance has a magnetic permeability less than 1, and is repelled when placed near a magnet. The magnetization of diamagnetic substances is associated with the currents induced on application of a magnetic field. According to Lenz's law, the flow of an induced current is in such a direction as to oppose the change of flux of the inducing field; this accounts for the negative susceptibility. The diamagnetic susceptibility is invariably small, of the order of

-10^{-5} cm^3/mole. *See* LENZ'S LAW; MAGNETIC PERMEABILITY; MAGNETIC SUSCEPTIBILITY.

All matter responds to applied fields in this diamagnetic fashion. However, some substances also have net electronic orbital or spin magnetic moments, or both, which can be aligned by an applied magnetic field in a direction along (not opposite to) the field; this property is called paramagnetism. For these substances, as shown in Eq. (1), the observed

$$\chi = \chi_d + \chi_p \qquad (1)$$

served susceptibility χ is the sum of diamagnetic and paramagnetic terms. Under ordinary conditions χ_d is temperature-independent. Hence, if χ_p follows the inverse temperature dependence of Curie's law, one can experimentally determine the separate contributions of χ_d and χ_p by measuring χ as a function of temperature. However, if χ_p is also temperature-independent, as in metals and in Van Vleck paramagnetism, the problem of determining the separate contributions becomes quite difficult. *See* PARAMAGNETISM.

Although all matter exhibits diamagnetism, only those substances in which $\chi_p = 0$ are referred to as diamagnetic. This is because paramagnetism, if present, usually predominates, and the gross magnetic response of the material is paramagnetic. Important exceptions are the alkali and alkaline earth metals, where χ_p is unusually small and of the order of χ_d, and salts or solutions containing only a small fraction of paramagnetic atoms. The condition for $\chi_p = 0$, and hence pure diamagnetism, is that all electronic spins be paired and all orbital moments either be zero or effectively cancel one another. Nearly all molecules with an even number of electrons satisfy this condition; an important exception is O_2. The condition is also satisfied by most nonmetallic solids, except compounds containing atoms with incomplete inner-shell electron groups, such as the transition, rare-earth, and actinide elements. *See* ELECTRON SPIN.

As stated previously, the diamagnetic response of a substance is small; only a very small fraction of the applied magnetic field is shielded from the interior of the substance by the induced diamagnetic currents. There is one case, however, in which the inducing field is completely shielded (except for small surface effects). This is the perfect diamagnetism exhibited by superconductors, and is known as the Meissner effect. *See* MEISSNER EFFECT; SUPERCONDUCTIVITY.

Langevin theory. The Langevin theory of diamagnetism (P. Langevin, 1905) is based on the idea of inducing an electronic current inside an atom. The theory employs the Larmor theorem. *See* LARMOR PRECESSION.

In a magnetic field H, the precession of the Z electrons within the atom is equivalent to a current equal to $-Z(e/c)(\omega_L/2\pi)$ in electromagnetic units. Here e/c is the magnitude of the electronic charge in emu, and ω_L is the angular Larmor frequency as shown in Eq. (2), where m is the electronic mass.

$$\omega_L = -eH/2mc \qquad (2)$$

The magnetic moment μ arising from this induced current is equal to the product of the current and the area of the current loop, as in Eq. (3), where $\overline{\rho^2}$

$$\mu = -Z(e/c)(\omega_L/2\pi)(\overline{\rho^2}) \qquad (3)$$

is the statistical average, over a large number of

atoms, of the square of the perpendicular distance of an electron from the field axis. This average is equivalent to $\overline{x^2} + \overline{y^2}$ if H is along z. For a random assembly of atoms, since $\overline{x^2} = \overline{y^2} = \overline{z^2}$, one may write Eq. (4), where $\overline{r^2}$ is the mean square distance

$$\overline{\rho^2} = (2/3)(\overline{x^2} + \overline{y^2} + \overline{z^2}) = (2/3)\,\overline{r^2} \qquad (4)$$

of the electron from the nucleus. Thus, the diamagnetic susceptibility of N atoms is given by Eq. (5). This is P. Langevin's result, as corrected by

$$\chi_d = N\mu/H = -(Ze^2N/6mc^2)\,\overline{r^2} \qquad (5)$$

W. Pauli. The molar susceptibility χ_M is obtained by replacing N in Eq. (5) by Avogadro's number. Numerically, it may be expressed as Eq. (6), where

$$\chi_M = (-2.83 \times 10^{10}\ \text{cm/mole})\,\Sigma\overline{r^2} \qquad (6)$$

the summation is to be taken over all the electron orbits in the atom. Since $\overline{r^2}$ is of the order of 10^{-15} cm^2, this gives $\chi_M \sim -10^{-5}$ cm^3/mole.

Langevin's formula is not modified by quantum mechanics, and the problem becomes that of determining $\overline{r^2}$ of the electronic wave function. The calculation for many-electron atoms is quite involved, and experimental values of $\overline{r^2}$ as determined by using Eq. (5) or (6) give a very useful check of the nature of the wave function for large r. This complements x-ray and electron diffraction data which give information, for the most part, only for small r.

Ionic crystals. In the case of diamagnetism in ionic crystals, the Larmor theorem holds for the individual ions. The diamagnetic susceptibility may be computed with reasonable accuracy from the sum of the individual ion susceptibilities.

Diamagnetism in rare gases and in rare-gas configurations of ions in ionic crystals is shown in the table.

Molar diamagnetic susceptibilities (all $\times 10^{-6}$ cm^3/mole)

He	-2.0	Li	-0.7	Mg	-4.3	F	-9.4
Ne	-7.0	Na	-6.1	Ca	-10.7	Cl	-24.2
Ar	-19.2	K	-14.6	Sr	-18.0	Br	-34.5
Kr	-28	Rb	-22.0	Ba	-29.0	I	-50.6
Xe	-43	Cs	-35.0				

Molecules. In most molecules the electrons are not moving in a single field of force, and the Larmor theorem breaks down. The total susceptibility of H_2 is given by the sum of Eq. (5) and the Van Vleck paramagnetism, the latter corresponding to the presence of a mean-square magnetic moment, although the mean moment vanishes. Calculations for more complicated molecules are exceedingly difficult. In aromatic ring molecules, such as benzene, with H normal to the ring, the electrons can precess around the ring, or at least in partial ring-like orbits about many nuclei. This gives rise to a much larger susceptibility than is possible when H is parallel to the ring. Crystals with layerlike structures, such as antimony, bismuth, and graphite, also exhibit large anisotropies in diamagnetic susceptibilities.

Bohr–van Leeuwen theorem. This theorem (N. Bohr, 1911; J. H. van Leeuwen, 1919) proves the complete absence of magnetism in classical theory. For bound electrons this comes from a cancellation of paramagnetic and diamagnetic susceptibilities, providing one does not (as did Langevin) implicitly quantize the paramagnetic moments by setting them all equal to μ. For a system of free electrons confined to a box, the induced diamagnetic currents in the interior of the box are just cancelled by the currents from electrons which bounce in cuspoidal paths off the walls. Thus, magnetism is inexplicable in classical physics and is a quantum phenomenon.

Free electrons. The diamagnetism of free electrons, which vanishes classically, was calculated quantum-mechanically by L. Landau (1930). For particles obeying Fermi-Dirac statistics, such as the electrons in a metal, the numerical value of the Landau diamagnetism is exactly one-third of the Pauli spin paramagnetism.

Bound electrons. The diamagnetism of bound electrons in other than ionic crystals is difficult to calculate. In metals the diamagnetism is a sum of contributions from the nonconducting core electrons, for which the Langevin theory is adequate, and from the conduction electrons, for which the Landau theory must be modified to take account of the periodic potential from the ion cores. Metals in the bismuth group have unusually high diamagnetic susceptibilities ($\sim -10^{-4}$ cm³/mole) coming from conduction electrons. These metals also show the de Haas–van Alphen effect, a quasi-periodic variation of susceptibility when plotted against $1/H$ at low temperatures. The susceptibility may even oscillate between diamagnetism and paramagnetism.   [ELIHU ABRAHAMS; FREDERIC KEFFER]

Bibliography: A. G. Morrish, *The Physical Principles of Magnetism*, 1965, reprint 1979; L. N. Mulay and E. A. Boudreaux, *Theory and Application of Molecular Diamagnetism*, 1976; J. H. Van Vleck, *The Theory of Electric and Magnetic Susceptibilities*, 1932.

Dichroism

In certain anisotropic materials, the property of having different absorption coefficients for light polarized in different directions.

There are few natural materials which exhibit strong dichroism. One of the first to be discovered was tourmaline. Light transmitted by thin plates of dark forms of tourmaline is almost completely polarized. *See* POLARIZED LIGHT.

In isotropic optical materials the optical density is defined as in Eq. (1), where I_0 is the intensity of

$$d = \log \frac{I_0}{I} \tag{1}$$

the incident light, and I that of the transmitted light. In anisotropic materials that are dichroic, the value of d can vary as a function of the vibration direction of the electric vector of the light wave. Just as the index ellipsoid is used to define the birefringence of a material, a density surface can be used to define the dichroism. *See* CRYSTAL OPTICS.

Compared to the literature on birefringence and optical activity, there has been relatively little material on dichroism. This is partly because of the difficulty in making measurements. The Kramers-Kronig relationship shows that any material whose refractive index is different from unity and varies as a function of wavelength will absorb radiation at some wavelength. From the Kramers-Kronig relationship it is apparent that all optically anisotropic materials should be dichroic. From the values of the refractive index at different wavelengths, the spectral positions and intensity of the absorption can be calculated. In a linear birefringent material the refractive index depends on the polarizing direction or electric vector of the radiation. For each direction of propagation there are two perpendicular vibration directions with different refractive indices. It can be inferred that, at some wavelength, the absorption for light vibrating in these two directions will be different. For transparent materials this wavelength is in the ultraviolet and the absorption is thus difficult to measure. The absorption difference is also frequently so small that it cannot be detected. In other words, the absorption is apparently the same in each direction. Furthermore, the dichroic band may coincide with a region of isotropic absorption.

If the absorption in a dichroic material is different for different linear states of polarization, the material is termed linear dichroic. If it is different for right and left circularly polarized light, it is termed circular dichroic. Similarly, there can be elliptically dichroic crystals. In biaxial crystals there are three different refractive indices corresponding to an electric vector lying along each of the three orthogonal axes of the index ellipsoid. Such a crystal has dichrosim which is different for light traveling along each of these three principal axes.

Most dichroic materials exist in the form of relatively thin sheets. Here one is dealing with a section of the dichroic surface. Associated with the sheet of material will be a direction of maximum absorption and one of minimum absorption. The density equation can be rewritten as Eqs. (2),

$$d_\parallel = \log \frac{I_{0\parallel}}{I_\parallel} \qquad d_\perp = \log \frac{I_{0\perp}}{I_\perp} \tag{2}$$

where the densities are now for light vibrating parallel or perpendicular to the axis of the section of dichroic surface. *See* TRICHROISM.

[BRUCE H. BILLINGS]

Dielectric constant

For a given dielectric material, the ratio of electrical capacitance of a dielectric-filled capacitor to a vacuum capacitor of identical dimensions. This is defined by Eq. (1), where C is the capacitance of

$$\kappa = \frac{C}{C_0} \tag{1}$$

the dielectric-filled capacitor and C_0 is the capacitance of the empty capacitor.

The dielectric constant κ is also known as the specific inductive capacity or as the relative permittivity. It is perhaps most familiar as the proportionality constant in Coulomb's law of electrostatics. For a given charge distribution, the dielectric constant expresses the ratio of electric field strength in vacuum to that in a dielectric, the latter field being reduced by the polarization of the dielectric medium. *See* CAPACITANCE; COULOMB'S LAW; DIELECTRICS; ELECTRIC FIELD; PERMITTIVITY.

The values of κ for low frequency or static fields

Selected dielectric constants

Substance	Temperature, °C	Frequency, Hz	κ
Dry air, CO_2 free	20		1.00054
NaCl	25	10^2	5.9
MgO	25	10^2	9.65
Al_2O_3	25	10^2	10.55*
			8.6†
TiO_2	25	10^5	170*
			86†
$BaTiO_3$	25	10^5	180*
			2000†
Polyethylene and paraffin	25	10^2	2.25
Rubber, vulcanized	25	10^2	2.94
Quartz, fused	25	10^2	3.78
Mica	25	10^4	7.3‡
			6.9§
Water	25	10^5	78.2
Ice	−12	10^5	4.8
HCN	20		114.9
CH_3OH	25	10^6	31

*Electric field parallel to principal axis of crystal.
†Electric field perpendicular to principal axis of crystal.
‡Electric field parallel to sheet of material.
§Electric field perpendicular to sheet of material.

range from 1 to more than 10,000 for typical dielectrics. The dielectric constants of gases are only slightly greater than unity, while high values occur for many polar liquids and certain ionic solids. The table lists the dielectric constants of some selected dielectric materials.

Measurement. The experimental methods of measuring dielectric constants depend on the frequency range under investigation. For frequencies below about 10^9 Hz, the permittivity or impedance of a dielectric sample inserted in a parallel-plate capacitor may be measured in suitable circuits; a Schering bridge arrangement is commonly employed up to 10^7 Hz, and resonant circuits in the range 10^4–10^9 Hz. For frequencies above 10^8 Hz, the dielectric constant may be determined by measuring the interaction of electromagnetic waves with the medium. From about 10^8 to 10^{11} Hz, the material is usually inserted in wave guides or coaxial lines, and the standing-wave patterns measured. At still higher frequencies, optical techniques involving reflection and transmission measurements are employed. Measurement techniques employed in analytical chemistry are discussed later.

Macroscopic theory. The dielectric constant κ in Eq. (2) is a dimensionless parameter relating the

$$\kappa = \frac{\epsilon}{\epsilon_0} = \frac{\mathbf{D}}{\epsilon_0 \mathbf{E}} = 1 + \frac{\gamma \mathbf{P}}{\epsilon_0 \mathbf{E}} = 1 + \gamma\chi \qquad (2)$$

macroscopic quantities displacement $\mathbf{D}$, and polarization $\mathbf{P}$, with electric field $\mathbf{E}$. In Eq. (2) ϵ and ϵ_0 are the permittivities of dielectric and vacuum, respectively, χ is the electric susceptibility, and γ is a geometrical factor ($\epsilon_0 = 1$ and $\gamma = 4\pi$ in cgs electrostatic units; $\epsilon_0 = 8.854 \times 10^{-12}$ farad/m and $\gamma = 1$ in rationalized mks units). The quantities κ and χ are scalar or tensor quantities for isotropic or anisotropic dielectrics, respectively. *See* ELECTRIC SUSCEPTIBILITY.

For electric fields varying sinusoidally with time, where the phase of the displacement may be retarded with respect to the field, Eq. (2) may be employed using complex number notation. Thus, $\kappa^* = \kappa' - i\kappa''$, where κ' and κ'' designate the components of the permittivity ratio in phase with $\mathbf{E}$ and retarded 1/4 cycle, respectively. The term dielectric constant is usually restricted to the real part, κ', while κ^* is designated as the complex relative permittivity.

Microscopic theory. The dielectric constant of a material depends on its polarization in an applied field or, microscopically, on the relative displacements, in the field direction, of the electrons and nuclei composing the molecules of the dielectric. These displacements are associated with changes in rotational and vibrational motions of the electrons and nuclei upon application of an electric field. This leads to a frequency dependence and temperature dependence resulting from the inertial characteristics of the motions and the initial state of excitation of the system. The dielectric constant also depends on field strength, since for sufficiently high fields the polarization will no longer be proportional to the field because saturation or breakdown phenomena occur. In these cases, the nonlinear portions of the polarization lead to harmonic frequency generation and interaction between two or more electromagnetic waves.

Theory of static polarization. A molecule in an electric field develops an average dipole moment, $(\boldsymbol{\mu})_{\mathrm{av}} = \alpha\mathbf{E}$, where α is the polarizability. For a polar molecule, α may be expressed as $\alpha = \alpha_o + \alpha_p$, where α_p is the polarizability arising from the partial orientation of the permanent dipole moment $\boldsymbol{\mu}_p$, and α_o is the polarizability due to all other processes. Then $(\boldsymbol{\mu})_{\mathrm{av}} = \alpha_o\mathbf{E} + (\boldsymbol{\mu}_p \cos\theta)_{\mathrm{av}}$ where $(\boldsymbol{\mu}_p \cos\theta)_{\mathrm{av}}$ is the average value of the component of $\boldsymbol{\mu}_p$ in the field direction, θ being the angle between $\boldsymbol{\mu}_p$ and $\mathbf{E}$.

The potential energy U of the permanent dipole is given by $U = -|\boldsymbol{\mu}_p||\mathbf{E}|\cos\theta$. For a system in thermodynamic equilibrium, with the probabilities of the possible states given by the Boltzmann distribution, it can be shown that $(\boldsymbol{\mu}_p \cos\theta)_{\mathrm{av}} = \boldsymbol{\mu}_p L(x)$, where the Langevin function $L(x) = \coth x - (1/x)$ and $x = |\boldsymbol{\mu}_p||\mathbf{E}|/kT$ (where k is the Boltzmann constant, and T is the absolute temperature). For $x \ll 1$, $L(x) \cong x/3$; in this approximation the microscopic parameters have been multiplied by N, the number of molecules per unit volume, to give the polarization, Eq. (3). Equation (3) is a

$$\mathbf{P} = N(\boldsymbol{\mu})_{\mathrm{av}} = N(\alpha_o + |\boldsymbol{\mu}_p|^2/3kT)\mathbf{E} \qquad (3)$$

form of the Langevin-Debye formula, from which the permanent dipole moments of polar molecules may be obtained from the temperature variation of measured values of polarization. For additional information *see* MOLECULAR STRUCTURE AND SPECTRA.

Local field. In the previous section, the mutual interaction of the molecules was not considered explicitly. For gases at low densities, molecular interaction can usually be neglected; but for matter at higher densities, the interaction becomes important and is usually treated by introducing the concept of local field $\mathbf{E}_l$, the average electric field strength at a molecular site. $\mathbf{E}_l$ is conveniently divided into two parts, $\mathbf{E}_s$ and $\mathbf{E}_r$, where $\mathbf{E}_s$ is the field due to polarization by molecules inside a small sphere centered at the molecular site, and $\mathbf{E}_r$ is the field due to external charges and polarization by molecules outside the sphere; the sphere is taken sufficiently large that the medium outside may be considered continuous. $\mathbf{E}_r$ is obtained by subtracting the field $\mathbf{E}_p$ at the center of a uniformly polarized isolated sphere from the macroscopic field $\mathbf{E}$ in the dielectric. Thus, a sphere of uniform polarization, $\mathbf{P}$, has a surface charge density $s = |\mathbf{P}| \cos \theta$, where θ is the angle between $\mathbf{P}$ and the normal to the surface. The field at the center contributed by a unit area of the surface is equal to $-\gamma s/4\pi\epsilon_0 R^2$, where R is the radius of the sphere, and where its component in the direction of $\mathbf{P}$ is $-\gamma s \cos \theta/4\pi\epsilon_0 R^2$. Integrating over the surface gives the net field $\mathbf{E}_p = \gamma \mathbf{P}/3\epsilon_0$. Finally, $\mathbf{E}_r$ is expressed by Eq. (4).

$$\mathbf{E}_r = \mathbf{E} - \mathbf{E}_p = \mathbf{E} + \frac{\gamma \mathbf{P}}{3\epsilon_0} = \frac{(\mathbf{D} - 2\gamma\mathbf{P}/3)}{\epsilon_0}$$
$$= \frac{\mathbf{E}(\kappa' + 2)}{3} = \mathbf{E}(1 + \gamma\chi/3) \qquad (4)$$

By assuming that $\mathbf{E}_s = 0$, one obtains the Lorentz local field $\mathbf{E}_l = \mathbf{E}_r$ which leads to the expressions in Eqs. (5)–(7) for P, χ, and κ'. The last equa-

$$\mathbf{P} = N(\boldsymbol{\mu})_{av} = N\alpha\mathbf{E}_l = N\alpha(\mathbf{E} + \gamma\mathbf{P}/3\epsilon_0)$$
$$= N\alpha\mathbf{E}/(1 - \gamma N\alpha/3\epsilon_0) \qquad (5)$$

$$\chi = \mathbf{P}/\epsilon_0\mathbf{E} = N\alpha/\epsilon_0(1 - \gamma N\alpha/3\epsilon_0) \qquad (6)$$

$$\kappa' = 1 + \gamma\chi = 1 + \gamma N\alpha/(\epsilon_0 - \gamma N\alpha/3) \qquad (7)$$

tion may be solved for α, giving $\gamma N\alpha/3\epsilon_0 = (\kappa' - 1)/(\kappa' + 2)$ which is a form of the Clausius-Mosotti equation. This formula gives fairly good agreement with experiments up to moderate densities for nonpolar molecules, but fails to account for the behavior of strongly polar molecules.

In the Lorentz field approximation, the susceptibility becomes infinite if $\gamma N\alpha = 3\epsilon_0$. Since α for polar molecules is given by $\alpha = \alpha_o + |\boldsymbol{\mu}|^2/3kT$, there exists a critical temperature below which $\alpha > 3\epsilon_0/\gamma N$ and the material should undergo spontaneous polarization. This "polarization catastrophe" is not confirmed by experiment except for a class of crystals known as ferroelectrics. *See* FERROELECTRICS.

Onsager theory. The polarization catastrophe is avoided in a dielectric theory of polar molecules due to L. Onsager. In his treatment, the local field is calculated for an actual spherical cavity of molecular size in the dielectric using Laplace's equation, which gives $\mathbf{E}_l = 3_{\kappa}'\mathbf{E}/(2_{\kappa}' + 1)$. A polar molecule with dipole moment $\boldsymbol{\mu}_p$ inserted in the cavity induces an additional reaction field as given by Eq. (8), where V is the volume of the cavity. This reac-

$$\mathbf{E}_R = \frac{\gamma\boldsymbol{\mu}_p 2(\kappa' - 1)}{\epsilon_0 V 3(2\kappa' + 1)} \cdot \qquad (8)$$

tion field is parallel to $\boldsymbol{\mu}_p$ and thus exerts no aligning torque on the molecule. For this theory, the polarization is given by Eq. (9). Since $\gamma\mathbf{P} = (\kappa' - 1) \cdot \epsilon_0\mathbf{E}$, Eqs. (10) and (11) may be written, where $z = $

$$\mathbf{P} = N\alpha\mathbf{E}_l = N\alpha \frac{3\kappa'\mathbf{E}}{2\kappa' + 1} \qquad (9)$$

$$\frac{\gamma N\alpha}{\epsilon_0} = \frac{(2\kappa' + 1)(\kappa' - 1)}{3\kappa'} \qquad (10)$$

$$\kappa' = \frac{1}{4}\left[1 + 3z + 3\left(1 + \frac{2z}{3} + z^2\right)^{1/2}\right] \qquad (11)$$

$\gamma N\alpha/\epsilon_0$. The agreement with theory is satisfactory for most polar liquids, but is inadequate for systems with hydrogen bonds.

[ROBERT D. WALDRON]

Applications in analysis. Because the dielectric constant is related to chemical structure, it can be used for both qualitative and quantitative analysis. Alone, it is a nonspecific indication of qualitative constitution. Unless the sample is known to be a pure material, the dielectric constant is of very little value since for mixtures it is generally not a simple additive function of composition. Even for pure samples, it is seldom used because there are generally more sensitive methods available. On the other hand, from measurements of the dielectric constant and refractive index taken together, it is possible to compute the dipole moment of a species. This quantity is a measure of the separation of charge within the molecule, and thus often gives explicit clues concerning structure when composition is known from other means. Azobenzene ($C_{12}H_{10}N_2$), diiodoacetylene (C_2I_2), and carbon suboxide (C_3O_2) were first assigned their symmetrical molecular structures unambiguously because dielectric measurements indicated a lack of dipole moment. The α and β isomers of benzene hexachloride ($C_6H_6Cl_6$) used in insecticides were first characterized in the same fashion.

The dielectric constant is a nonadditive function for mixtures, and so for quantitative analysis it is necessary to prepare calibration curves relating measurements to composition for standard samples. After this has been done, however, the method is rapid and efficient. It is most applicable to two-component mixtures. When the ratio of the dielectric constants of the components of a two-component mixture is 2:1, the accuracy of the determination is 0.2–2%. The method can be applied to three-component mixtures also if some independent method is available for determining one of the components. For more complex systems, the errors mount rapidly, except in special cases. If the dielectric constants for all but one constituent in a multicomponent system are similar, and there is little interaction between them in solution, then the unique component can often be determined. This is the situation in the analysis for toluene in the presence of complex mixtures of aliphatic hydrocarbons in petroleum refining. Determination of moisture in cereal grains and other solids is based on a similar principle.

Experimentally, the dielectric constant is determined by the ratio of the capacitances of a capacitor measured with and without the sample between its plates. Measurements can be made at low (1000–10,000 Hz) or high (1 MHz) frequencies. The equipment needed at low frequencies is simpler to operate than that needed at high frequencies. However, direct contact between sample and electrodes is unnecessary at high frequencies. *See* DIPOLE MOMENT; POLARIZATION OF DIELECTRICS.

[WILLIAM H. REINMUTH]

Bibliography: C. J. F. Bottcher, *Theory of Dielectric Polarization*, vol. 1: *Dielectrics in Static Fields*, 2d ed., rev. by O. C. Van Belle et al., 1973; C. J. F. Bottcher and P. Bordewijk, *Theory of Dielectric Polarization*, vol. 2: *Dielectrics in Time Dependent Fields*, 2d ed., 1978.

Dielectrics

Materials which are electrical insulators or in which an electric field can be sustained with a minimum dissipation of power. In a more general sense, dielectrics include all materials except condensed states of metals.

Dielectrics are employed as insulation for wires, cables, and electrical equipment, as polarizable media for capacitors, in devices used for the propagation or reflection of electromagnetic waves, and for a variety of dielectric devices, such as rectifiers and semiconductor devices, piezoelectric transducers, dielectric amplifiers, and memory elements.

The electrical response of a normal dielectric can be described by its dielectric or breakdown strength, conductivity or dielectric loss, and dielectric constant. The behavior of nonlinear dielectrics depends also on the amplitude and time variation of the electric field.

Dielectric strength. This is defined as the maximum electric field which can be applied to a dielectric without causing breakdown, the abrupt irreversible drop in resistivity at high fields often accompanied by destruction of the material. Dielectric strengths of most insulating materials lie in the range from 10^4 to 10^7 volts/in. at room temperature and low frequencies and decrease at higher temperatures. The breakdown strengths of gases increase nearly linearly with pressure over a considerable range, but at very low pressures the values also increase, and the dielectric strength of a high vacuum is superior to gases at atmospheric pressure. Dielectric breakdown is caused by an enormous increase in the number of charge carriers because of collisions or thermal ionization and field emission. *See* FIELD EMISSION.

Dielectric loss. This is the power dissipated in a dielectric because of conduction processes. This power loss results from thermal dissipation of the electrical energy expended by the field. It is caused by molecular collisions. It can be described by any of the following related parameters: the conductivity σ, the loss factor ϵ'', the power factor $\cos\theta$, and the loss tangent or dissipation factor $\tan\delta$. Of these, only σ is applicable to direct current problems. The conductivity σ is the current density $\mathbf{I}$ per unit field strength $\mathbf{E}$ in phase with the applied voltage. The loss factor ϵ'', which is the imaginary part of the permittivity, is related to the conductivity by $\sigma = \omega\epsilon''/\gamma$, where ω equals 2π times the frequency. The power factor, $\cos\theta$, is the ratio of conduction or loss current in phase with the applied voltage to the total current in any circuit, and θ is the phase angle between current

and voltage. The dissipation factor, $\tan\delta$, is the ratio of loss current to reactive or charging current, where $\delta = 90° - \theta$. This is expressed in terms of permittivity as in Eqs. (1), where $|\epsilon^*| = \sqrt{\epsilon'^2 + \epsilon''^2}$

$$\epsilon^* = \epsilon' - i\epsilon'' \qquad \cos\theta = \epsilon''/|\epsilon^*| \qquad (1)$$

and $\tan\delta = \epsilon''/\epsilon'$. For low-loss materials, $\cos\theta$ $\tan\delta$ are nearly equal. *See* PERMITTIVITY.

The power dissipated per unit volume is $p = \sigma|\mathbf{E}|^2 = |\mathbf{I}||\mathbf{E}|\cos\theta$. This power loss increases at high temperatures and, in many substances, at high frequencies for a given field strength. This effect is commercially employed in dielectric heating equipment for industrial and therapeutic purposes.

Dielectric constant. The dielectric constant or permittivity relative to vacuum is important in many applications. Materials with high dielectric constants are desirable for capacitors, since they permit a reduction in size for a given capacitance, while low dielectric constants are usually preferred for cable and transformer insulation. For an extended discussion *see* DIELECTRIC CONSTANT.

Dispersion. The conductivity and dielectric constant, or alternatively the permittivity, have a frequency dependence determined by the molecular mechanisms of polarization in the dielectric. Classically, one may describe the action of an electric field $\mathbf{E}$ on a charged particle by Eq. (2), where

$$m\frac{d^2\mathbf{r}}{dt^2} + f\frac{d\mathbf{r}}{dt} + k\mathbf{r} = e\mathbf{E} \qquad (2)$$

e, m, and $\mathbf{r}$ are the charge, mass, and position of the particle and f and k are the frictional and restoring force constants. (The first term represents an acceleration, and the remaining terms represent the net force acting on the particle.) The solution for a sinusoidal field $\mathbf{E} = \mathbf{E}_0\exp(i\omega t)$ is given in Eq. (3),

$$\mathbf{r} = \frac{e\mathbf{E}}{m}\frac{1}{\omega_0^2 - \omega^2 + i\omega f/m} \qquad (3)$$

where $\omega_0^2 = k/m$. The polarization is expressed by Eq. (4), where N is the number of particles per unit

$$\mathbf{P} = (\epsilon^* - \epsilon_0)\mathbf{E}/\gamma = N\mathbf{r}e \qquad (4)$$

volume and γ is a geometrical factor equal to 4π or 1 in the cgs or mks systems, respectively. Thus, Eq. (5) holds. For $f/m < \omega_0$, the frequency depend-

$$\epsilon^* = \epsilon_0 + \frac{\gamma Ne^2}{m}\frac{1}{\omega_0^2 - \omega^2 + i\omega f/m} \qquad (5)$$

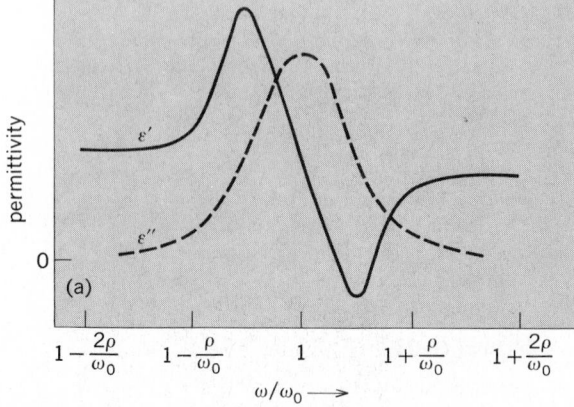
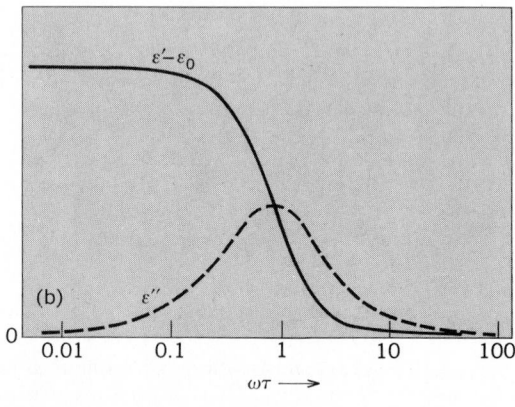

Fig. 1. Dispersion of permittivity. (*a*) Resonance spectrum ($\rho = f/m$). (*b*) Relaxation spectrum ($\tau = f/m\omega_0^2$).

ence of ϵ^* is described as the resonance spectrum of a damped harmonic oscillator shown in Fig. 1a. For $f/m \gg \omega_0$, the frequency dependence approaches that of a relaxation circuit, as shown in Fig. 1b.

The experimentally observed frequency dependences of the dielectric constant and permittivity can be satisfactorily accounted for in terms of resonance and relaxation processes. Resonance dispersion is associated with changes in the electronic or vibrational energy of molecules; the resonance frequencies usually occur at frequencies greater than 10^{12} Hz. Relaxation spectra occur for polar molecules in viscous media and for solids exhibiting interfacial polarization; the time constant, $\tau = f/m\omega_0^2$, is usually greater than 10^{-12} sec and may extend to very long periods. *See* Molecular structure and spectra; Polarization of dielectrics.

The presence of conduction phenomena characterized by a long relaxation time τ leads to the occurrence of dielectric hysteresis and absorption.

Dielectric hysteresis is analogous to magnetic hysteresis. It is the dependence of dielectric polarization on the previously applied electric fields, that is, the previous electrical history of the sample. For an applied field $\mathbf{E} = \mathbf{E}_0 \exp(i\omega t)$, where $\omega > 1/\tau$, the polarization describes an ellipse when plotted versus $\mathbf{E}$ as shown in curve A, Fig. 2. For $\omega \ll 1/\tau$, $\mathbf{P}$ approaches a single valued function of $\mathbf{E}$ given by $\mathbf{P} = \chi\epsilon_0\mathbf{E}$, where χ is the electric susceptibility. Ferroelectric materials exhibit spontaneous polarization and show hysteresis even for nearly static fields (curve B, Fig. 2). *See* Ferroelectrics; Magnetic hysteresis.

Dielectric absorption or the dielectric aftereffect is the charging current or polarization which builds up or decays slowly when the field applied to a dielectric is changed. It is usually caused by space-charge polarization and may in exceptional cases persist for months or years. *See* Electret.

Dielectrics whose polarization varies nonlinearly with the applied field generate harmonic waves when exposed to sinusoidal fields. This effect permits frequency multiplication and also signal mixing of independent waves which are used in parametric amplifiers and laser harmonic generators.

Dielectric materials. Vacuum or gaseous dielectrics other than air have had relatively little dielectric application except in electron tube devices, voltage regulators, and lightning arresters.

Dielectric liquids are principally employed for impregnating porous insulation in high-voltage cables and capacitors and as insulating media for transformers and circuit breakers. In the latter applications, the heat transfer properties of the liquid are also important. Mineral oils, halogenated hydrocarbons, and silicone oils are the most important commercial dielectric liquids.

Solid dielectrics are employed for the vast majority of commercial applications. Important solid dielectrics include many ceramics and glasses; plastics and rubber; minerals such as quartz, mica, magnesia, and asbestos; and paper and fibrous products. The mechanical and thermal properties as well as the electrical response are important in the choice of a dielectric for a particular product. For high mechanical strength and temperature resistance, ceramic and mineral insulators are preferred, while plastics and rubber are

employed where flexibility is desired. Low-loss, nonpolar dielectrics, such as polyethylene or polystyrene, are necessary for many ultrahigh-frequency applications.

The material requirements for dielectric devices are usually determined by the specific electrical characteristics desired for the operating frequencies selected. Semiconductors such as silicon and germanium and piezo- or ferroelectric ceramics including heavy metal titanates, zirconates, and niobates have found considerable application.

[Robert D. Waldron]

Bibliography: J. C. Burfoot and G. W. Taylor, *Polar Dielectrics*, 1979; N. E. Hill et al., *Dielectric Properties and Molecular Behavior*, 1969; A. R. von Hippel, *Dielectric Materials and Applications*, 1966; A. A. Zaky and R. Hawley, *Dielectric Solids*, 1970; I. S. Zheludev, *Physics of Crystalline Dielectrics*, 2 vols, 1971.

Differential equation

A relationship between a function and its derivatives.

Definitions. If there is one independent variable, the differential equation is called an ordinary differential equation. The general form of such an equation is shown in Eq. (1), where t is the in-

$$F(t,u,u', \ldots, u^{(n)}) = 0 \tag{1}$$

dependent variable; u is a function of t; $u' = du/dt$, $u'' = d^2u/dt^2, \ldots, u^{(n)} = d^nu/dt^n$ are the derivatives of u; and $F(t,u_0,u_1, \ldots, u_n)$ is a given function of the $n+2$ variables $t,u_0, \ldots, u_n$. The positive integer n is called the order of the differential equation; that is, the order of the differential equation is the order of the highest derivative that occurs in the equation. As an example, consider Eq. (2), which is an ordinary differential equation of order 2. A function u is said to be a solution to Eq. (1) if, when u and its derivatives up to order n are substituted into F, the identity expressed in Eq. (1) is valid for t in some interval $a < t < b$. In Eq. (2),

$$u'' + u = 0 \tag{2}$$

$u = \cos t$ and $u = \sin t$ are solutions for $-\infty < t < \infty$, as can be immediately verified.

If there are two or more independent variables, the equation is called a partial differential equation. Again, the order of the equation is the order of the highest partial derivative that occurs in the equation. Thus, the general form of a partial differential equation of order 2, with two independent variables, is given by Eq. (3), where x and y are the

$$F(x,y,u,u_x,u_y,u_{xx},u_{xy},u_{yy}) = 0 \tag{3}$$

independent variables; $u = u(x,y)$ is a function: $u_x = \partial u/\partial x$, $u_y = \partial u/\partial y$, $u_{xx} = \partial^2 u/\partial x^2$, $u_{xy} = \partial^2 u/\partial x\partial y$, and $u_{yy} = \partial^2 u/\partial y^2$ are the derivatives of u; and $F(x, y, u, p, q, r, s, t)$ is a given function of eight variables. For example, Eq. (4) is a partial

$$u_{xy} = 0 \tag{4}$$

differential equation of order 2. It is easily verified that $u = h(x) + g(y)$ is a solution of Eq. (4) for any choice of functions h and g.

A differential equation is linear if the function

DIELECTRICS

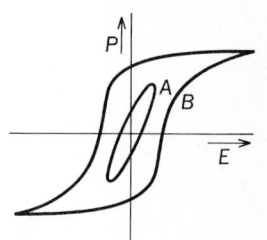

Fig. 2. Dielectric hysteresis. Curve A shows polarization for $\omega > 1/\tau$. Curve B shows quasistatic polarization for a ferroelectric crystal.

u and its derivatives appear linearly in the equation, and is nonlinear otherwise. Thus, Eqs. (2) and (4) are linear equations, whereas Eqs. (5), (6), and (7) are nonlinear.

$$u' = \sin u \qquad (5)$$

$$(1 + u_y^2)u_{xx} - 2u_x u_y u_{xy} + (1 + u_x^2)u_{yy} = 0 \qquad (6)$$

$$u_t + uu_x + u_{xxx} = 0 \qquad (7)$$

Partial differential equations occur with more than two independent variables; see Eqs. (33), (35), and (36). Another possibility is a system of differential equations. This consists of two or more equations involving one or more unknown functions which are to be solved simultaneously. For example, Eqs. (8) are a first-order linear system of

$$u' = v$$
$$\qquad (8)$$
$$v' = -u$$

ordinary differential equations. The Cauchy-Riemann equations, Eqs. (9), are a first-order linear

$$u_x - v_y = 0$$
$$\qquad (9)$$
$$u_y + v_x = 0$$

system of partial differential equations. *See* COMPLEX NUMBERS AND COMPLEX VARIABLES.

Ordinary differential equations. The process of solving differential equations is called integration, and is analogous to the solution of the simple equation $u'(t) = h(t)$, which is actually the indefinite integral of h. The value of an indefinite integral is unique except for an arbitrary constant of integration. In general, the solution of an ordinary differential equation of order n involves n arbitrary constants of integration. For example, the general solution of Eq. (2) is $u(t) = A \cos t + B \sin t$, where A and B are arbitrary constants; and the solution of Eq. (5) is $u(t) = 2 \tan^{-1}(ke^t)$, where k is an arbitrary constant. *See* INTEGRATION.

In the full generality of Eq. (1), it is not possible to prove that solutions exist. For example, Eq. (10)

$$1 + \left|\frac{du}{dt}\right|^2 = 0 \qquad (10)$$

has no solution. However, if an equation is of the type of Eq. (11), and if $f(t, u_0, \ldots, u_{n-1})$ has con-

$$u^{(n)} = f(t, u, \ldots, u^{(n-1)}) \qquad (11)$$

tinuous first derivatives on the set defined by $|t - t_0| < d$, $|u_0 - a_0| < d$, $|u_1 - a_1| < d$, $\ldots$, $|u_{n-1} - a_{n-1}| < d$, where $a_1, \ldots, a_n$, and d are constants, then there is one and only one function $u(t)$ defined and continuously differentiable on the interval $|t - t_0| < \epsilon$, for some $\epsilon > 0$, which is a solution of Eq. (11) and which satisfies the initial conditions $u(t_0) = a_0$, $u'(t_0) = a_1$, $\ldots$, $u^{(n-1)}(t_0) = a_{n-1}$. This result is purely an existence result. That is, it says a solution exists but does not say how to find it. Indeed, it may easily be impossible to find a solution which is expressible in terms of elementary functions.

Equations of first order. It is usual in the case of equations of first order to use x as the independent variable and to denote the unknown function by

$y = y(x)$. Then Eq. (11) reduces to $dy/dx = f(x, y)$. This equation is usually written in differential form, Eq. (12), where $f = -P/Q$. In this notation, Eq. (5) would be rewritten as Eq. (13). For a given

$$P(x,y)dx + Q(x,y)dy = 0 \qquad (12)$$

$$\sin y\, dx - dy = 0 \qquad (13)$$

function $v(x,y)$ the differential of v is defined by $dv = v_x dx + v_y dy$. The first-order equation is said to be exact if there is a function v such that $dv = P dx + Q dy$, that is, if $\partial v/\partial x = P$ and $\partial v/\partial y = Q$. In this case the solution to Eq. (12) is given implicitly by the equation $v(x,y) = c$, with c an arbitrary constant. For example, Eq. (14) is exact, since

$$y\, dx + x\, dy = 0 \qquad (14)$$

$d(xy) = y\, dx + x\, dy$, and the solution is $xy = c$. A general example of exact equations is given by the case when, in Eq. (12), P is a function of x only and Q is a function of y only. Then, if $v(x,y) = \int P(x)dx + \int Q(y)dy$, the solution is $v(x,y) = c$. For example, Eq. (15) is exact and has the solution $x - \log \tan(y/2) = c$.

$$dx - \frac{dy}{\sin y} = 0 \qquad (15)$$

When Eq. (12) is not exact, it may be possible to find an integrating factor, that is, a function $\mu(x,y)$ such that the equation $\mu P dx + \mu Q dy$ is exact. If $\mu P dx + \mu Q dy = dv$, then, once more, the solution is given by $v(x,y) = c$. Theoretically it is always possible to find an integrating factor, provided P and Q are not both equal to zero at some point. In practice, however, it may be difficult. For Eq. (16),

$$y\, dx - x\, dy = 0 \qquad (16)$$

which is not exact, an integrating factor is $1/y^2$, since $d(x/y) = (y\, dx - x\, dy)/y^2$, and the solution is given by $x/y = c$. A general example is provided by the case of separable variables when, in Eq. (12), $P(x,y) = A(x)B(y)$ and $Q(x,y) = C(x)D(y)$. Then $\mu = 1/C(x)B(y)$ is an integrating factor, since $[A(x)/C(x)]dx + [D(y)/B(y)]dy = 0$ is exact. Using this method, Eq. (13)—and hence Eq. (5)—can be reduced to Eq. (15). As another example, consider Eq. (17), a linear first-order differential equation,

$$\frac{dy}{dx} + a(x)y = z(x) \qquad (17)$$

or in differential form, Eq. (18). This equation is not exact, but $\mu(x) = \exp\left(\int a(x)dx\right)$ is an integrating factor. After multiplication by μ, Eq. (18)

$$dy + [a(x)y - z(x)]dx = 0 \qquad (18)$$

becomes $dv = 0$, where $v(x,y) = y \exp\left(\int a(x)dx - \int z(x) \exp\left(\int a(x)dx\right)dx\right)$. Thus the general solution of Eq. (17) is given by Eq. (19), where c is an arbitrary constant.

$$y(x) = \exp\left[-\int a(x)dx\right]$$
$$\left\{\int z(x)\exp\left[\int a(x)dx\right]dx + c\right\} \qquad (19)$$

Second-order linear equations. The study of oscillation in mechanical systems and of simple electric circuits leads to linear ordinary differential equations of second order, typified by Eq. (20).

$$u'' + a(t)u' + b(t)u = v \qquad (20)$$

This equation is said to be homogeneous when v is identically equal to zero; otherwise, it is nonhomogeneous. The linearity of the equation has important implications for the sets of solutions to Eq. (20). If u_1 and u_2 are two linearly independent solutions to the homogeneous equation, then $Au_1 + Bu_2$ is also a solution for any choice of constants A and B. Furthermore, any solution of the homogeneous equation is of this form. Since the difference of two solutions to the nonhomogeneous equation is a solution to the homogeneous equation, it follows that the general solution to the nonhomogeneous equation has the form of expression (21), where u_p is a particular solution to the non-

$$u_p + Au_1 + Bu_2 \qquad (21)$$

homogeneous equation, u_1 and u_2 are linearly independent solutions to the homogeneous equation, and A and B are arbitrary constants. As an example, consider Eq. (22). The corresponding homogeneous equation is Eq. (2). The functions $u_1(t) = \cos t$ and $u_2(t) = \sin t$ are linearly independent solutions to Eq. (2); thus the general solution to Eq. (2) is $u(t) = A \cos t + B \sin t$. It is easy to find a particular solution to Eq. (22), namely, $u_p(t) = t$. Hence the general solution to Eq. (22) is $u(t) = t + A \cos t + B \sin t$.

$$u'' + u = t \qquad (22)$$

When the coefficients $a(t)$ and $b(t)$ in Eq. (20) are constants, solutions to the homogeneous equation exist of the form $u(t) = e^{rt}$, where r is a constant which may be complex. Substitution into Eq. (20) (with $v = 0$) yields the quadratic equation $r^2 + ar + b = 0$, which may be solved for r. If $a^2 - 4b > 0$, there are two distinct real roots, and thus two linearly independent solutions to the homogeneous equation exist. If $a^2 - 4b = 0$, there is only one root. A second solution in this case is $u(t) = te^{rt}$. If $a^2 - 4b < 0$, the roots are complex conjugate, $r_1 = \alpha + i\beta$, $r_2 = \alpha - i\beta$. Then two real linearly independent solutions to the homogeneous equation are $e^{\alpha t} \cos \beta t$ and $e^{\alpha t} \sin \beta t$. *See* DAMPING; FORCED OSCILLATION; HARMONIC MOTION.

If the coefficients of Eq. (20) are nonconstant, it is sometimes difficult to find solutions to the homogeneous equation. However, if two linearly independent solutions u_1 and u_2 to the homogeneous equation are known, the method of variation of parameters yields solutions to the nonhomogeneous equation. The idea is to look for a solution to Eq. (20) of the form $A(t)u_1(t) + B(t)u_2(t)$, subject to the condition that $A'u_1 + B'u_2 = 0$. After solving for A and B, formula (23) is obtained for a particular solution to Eq. (20), where the kernel $k(t,s)$ is given by Eq. (24).

$$u_p(t) = \int_{t_0}^{t} k(t,s)v(s)\,ds \qquad (23)$$

$$k(t,s) = \frac{u_2(t)u_1(s) - u_1(t)u_2(s)}{u_1(s)u_2'(s) - u_2(s)u_1'(s)} \qquad (24)$$

If the coefficients $a(t)$ and $b(t)$ of Eq. (20) are analytic functions (that is, if a and b have power series expansions

$$a(t) = \sum_{n=0}^{\infty} a_n t^n, \; b(t) = \sum_{n=0}^{\infty} b_n t^n$$

which converge for t near zero), then power series solutions to the homogeneous equation can be found. The idea is to set

$$u(t) = \sum_{n=0}^{\infty} u_n t^n$$

Then

$$u'(t) = \sum_{n=0}^{\infty} nu_n t^{n-1} \; \text{ and } \; u''(t) = \sum_{n=0}^{\infty} n(n-1)u_n t^{n-2}$$

The series for a, b, u, u', and u'' are substituted into the homogeneous equation, resulting in a power series which is identically equal to zero. Setting the coefficients equal to zero results in a sequence of recursion relations for the coefficients u_n. This method can be extended to the case for which not $a(t)$ and $b(t)$ themselves but $ta(t)$ and $t^2 b(t)$ are analytic functions. This extension is essential since many important equations have this feature. For example, the solution of Bessel's equation, Eq. (25), is Bessel's function of order n, shown in Eq. (26), where Γ denotes the gamma function of Euler. Another example is the hypergeometric equation of Gauss, Eq. (27). If c is not zero or a negative integer, a solution is the hypergeometric function shown in Eq. (28). *See* BESSEL FUNCTIONS; GAMMA FUNCTION.

$$t^2 u'' + tu' + (t^2 - n^2)u = 0 \qquad (25)$$

$$J_n(t) = \left(\frac{t}{2}\right)^n \sum_{k=0}^{\infty} \frac{(-1)^k (t/2)^{2k}}{\Gamma(k+1)\Gamma(n+k+1)} \qquad (26)$$

$$t(1-t)u'' + [c - (a+b+1)t]u' - abu = 0 \qquad (27)$$

$$F(a,b,c,t) = 1 + \frac{a \cdot b}{1 \cdot c} t + \frac{a(a+1)b(b+1)}{1 \cdot 2 \cdot c(c+1)} t^2 + \cdots \qquad (28)$$

Partial differential equations. The analysis of many physical problems leads to linear partial differential equations of second order. When there are two independent variables x and y, the most general equation of this sort is Eq. (29), where the

$$au_{xx} + 2bu_{xy} + cu_{xx} + du_x + eu_y + fu = g \qquad (29)$$

coefficients $a, \ldots, g$ are given functions of x and y. It will be assumed that $a^2 + b^2 + c^2 \neq 0$. Then Eq. (29) is called elliptic if $b^2 - ac < 0$, parabolic if $b^2 - ac = 0$, and hyperbolic if $b^2 - ac > 0$. For each of these types there are certain typical boundary value problems which arise in physical applications and which are of mathematical interest. When there are more than two independent variables, or when the order of the equation is larger than two, there are equations which have the same qualitative properties, as do these three types, and for which similar boundary value problems exist. These equations are accordingly called elliptic, parabolic, or hyperbolic. However, there are also equations which do not belong to any of these three categories.

Elliptic equations. The prototype is Laplace's equation, Eq. (30), for two independent variables, and Eq. (31) for three independent variables. The

$$u_{xx} + u_{yy} = 0 \qquad (30)$$

$$u_{xx} + u_{yy} + u_{zz} = 0 \qquad (31)$$

solutions are called harmonic functions. A typical boundary value problem is the Dirichlet problem. It consists of finding a harmonic function $u(x,y)$ in a region in the plane [say, the disk $D = \{(x,y)|x^2 + y^2 < 1\}$] with prescribed values on the boundary of the region [$u(x,y) = g(x,y)$ for $x^2 + y^2 = 1$, where g is a given function). This corresponds to the physical problem of finding the steady-state temperature distribution $u(x,y)$ in a plate when the temperature distribution on the edge of the plate is known. *See* LAPLACE'S DIFFERENTIAL EQUATION.

Parabolic equations. The prototype is the heat equation, shown in one space variable x in Eq. (32), and in two space variables x and y in Eq. (33).

$$\frac{\partial u}{\partial t} - \frac{\partial^2 u}{\partial x^2} = 0 \qquad (32)$$

$$\frac{\partial u}{\partial t} - \frac{\partial^2 u}{\partial x^2} - \frac{\partial^2 u}{\partial y^2} = 0 \qquad (33)$$

The variable t represents time. If $u(t,x)$ is the temperature in a rod modeled by the interval $a \le x \le b$, and if there is no source of heat in the rod, then u is a solution to Eq. (32). A typical boundary value problem is the initial value problem, which consists of finding a solution to Eq. (32) for $0 < t < \infty$ and $-\infty < x < \infty$ with $u(0,x) = f(x)$ given. The physical interpretation consists of finding the temperature distribution in an infinite rod for all time when the initial temperature distribution is known. *See* CONDUCTION (HEAT).

Hyperbolic equations. The prototype is the wave equation, shown for one, two, and three space variables in Eqs. (34), (35), and (36), respectively. Again,

$$\frac{\partial^2 u}{\partial t^2} - \frac{\partial^2 u}{\partial x^2} = 0 \qquad (34)$$

$$\frac{\partial^2 u}{\partial t^2} - \frac{\partial^2 u}{\partial x^2} - \frac{\partial^2 u}{\partial y^2} = 0 \qquad (35)$$

$$\frac{\partial^2 u}{\partial t^2} - \frac{\partial^2 u}{\partial x^2} - \frac{\partial^2 u}{\partial y^2} - \frac{\partial^2 u}{\partial z^2} = 0 \qquad (36)$$

t represents time. Equation (34) is also called the equation of the vibrating string; Eq. (35) is called the equation of the vibrating membrane. As these names indicate, a large variety of oscillatory phenomena lead to hyperbolic equations. A typical boundary value problem is the Cauchy problem, which involves finding a solution $u(t,x)$ to Eq. (34) for $0 < t < \infty$ and $-\infty < x < \infty$, with $u(0,x) = f(x)$ and $u_t(0,x) = g(x)$ given. If $u(t,x)$ is interpreted as the displacement of an infinite string, then the Cauchy problem consists of deriving $u(t,x)$ for all time when the initial displacement $u(0,x) = f(x)$ and the initial velocity $u_t(0,x) = g(x)$ are both known. *See* VIBRATION; WAVE EQUATION; WAVE MOTION.

Other equations. An example of a partial differential equation which is not of the types previously considered is the Korteweg-deVries (KdV) equation, Eq. (7), which describes shallow-water waves in a one-dimensional channel. The KdV equation is a nonlinear equation of third order. Another nonlinear equation of interest is Eq. (6), which is called the minimal surface equation; its solution represents surfaces of minimal surface area (for example, soap films spanning wire frames have minimal surface area). *See* CALCULUS; DIFFERENTIATION.   [JOHN C. POLKING]

Bibliography: C. R. Chester, *Techniques in Partial Differential Equations*, 1971; R. Courant and D. Hilbert, *Methods of Mathematical Physics*, vol. 1, 1953, and vol. 2, 1962; G. F. Simmons, *Differential Equations, with Applications and Historical Notes*, 1972; E. C. Young, *Partial Differential Equations: An Introduction*, 1972.

Differentiation

A mathematical operation performed on a function to determine the effect of a change in the value of the independent variable. Functions of one variable are considered in this article. For differentiation of functions of several variables *see* PARTIAL DIFFERENTIATION. If f is a function of x, defined on an interval containing x_0, the derivative at x_0 is, by definition, that shown in Eq. (1). For

$$f'(x_0) = \lim_{x \to x_0} \frac{f(x) - f(x_0)}{x - x_0} \qquad (1)$$

generalities about derivatives and calculus *see* CALCULUS.

In the quotient on the right, in the definition of $f'(x_0)$, x is restricted to the interval on which f is defined. If $y = f(x)$, $f'(x_0)$ is also denoted by

$$\left(\frac{dy}{dx}\right)_{x = x_0}$$

The limit defining $f'(x_0)$ may not exist. If it does, f is called differentiable at x_0.

The derivative of $f'(x)$, called the second derivative, is denoted by

$$f''(x) \quad \text{or} \quad \frac{d^2 y}{dx^2}$$

A function f is called continuous at x_0 if x_0 is in the domain of f and

$$\lim_{x \to x_0} f(x) = f(x_0)$$

The precise formulation of the limit concept used here is the following: Let g be a function and let A be a number. Then Eq. (2) means that to each posi-

$$\lim_{x \to x_0} g(x) = A \qquad (2)$$

tive number ϵ corresponds some positive number δ such that $|g(x) - A| < \epsilon$ whenever x is a number in the domain of g such that $x \ne x_0$ and $|x - x_0| < \delta$. It is required of g that its domain shall contain numbers x as close to x_0 as desired, but different from x_0. The domain of g may also contain x_0, but this is irrelevant.

It is a theorem that, if f is differentiable at x_0, then f is continuous at x_0. However, a function can be continuous at x_0, but not differentiable there. An example is $f(x) = |x|$ at $x_0 = 0$. It is even possible for a function to be continuous on an interval and yet not differentiable at any point of this interval.

The chief elementary applications of differentiation are (1) in expressing rates of change (velocity, acceleration), and in solving problems where, through functional relationship, the rate of change of one variable is calculated when the rate of change of another variable is known; (2) in studying graphs of functions and, more generally, in studying curves in the plane or in space of three dimensions; (3) in expressing scientific laws or principles in the form of differential equations; and (4) in the expression of various extensions and applications of the law of the mean, including such topics as l'Hospital's rule and Taylor's formula or series.

Principles. The general technique of differentiation is built upon the rules for differentiating combinations of differentiable functions. If $u = f(x)$ and $v = g(x)$ are functions differentiable for the same values of x, then $u + v$ and uv are differentiable and so is u/v if $v \neq 0$. The formulas are shown in Eqs. (3). If u is constant in

$$\frac{d}{dx}(u+v) = \frac{du}{dx} + \frac{dv}{dx}$$

$$\frac{d}{dx}(uv) = u\frac{dv}{dx} + v\frac{du}{dx} \qquad (3)$$

$$\frac{d}{dx}\left(\frac{u}{v}\right) = \frac{v\dfrac{du}{dx} - u\dfrac{dv}{dx}}{v^2}$$

value, then $du/dx = 0$. A very powerful instrument of technique is furnished in the chain rule for composite functions. If y is a differentiable function of u and u is a differentiable function of x, then y is a differentiable function of x, and

$$\frac{dy}{dx} = \frac{dy}{du}\frac{du}{dx}$$

In functional notation, if $y = f(u)$ and $u = h(x)$, then $y = F(x)$, where $F(x) = f[h(x)]$, and then $F'(x) = f'[h(x)]h'(x)$.

The technique of differentiation also leans upon facts about inverse functions. If $y = f(x)$, where f is a differentiable function and $f'(x)$ is either always positive or always negative on an interval, for example, when $a \leq x \leq b$, then for each y from $f(a)$ to $f(b)$ inclusive there is just one x such that $a \leq x \leq b$ and $y = f(x)$. Thus there is defined a function g such that $x = g(y)$ is equivalent to $y = f(x)$ with x and y restricted as indicated. This function g is differentiable, and

$$g'(y) = \frac{1}{f'(x)} = \frac{1}{f'[g(y)]}$$

An example: $y = \sin x$, $x = \sin^{-1} y$ (sometimes written $x = \arcsin y$), where

$$-\frac{\pi}{2} \leq x \leq \frac{\pi}{2} \quad \text{and} \quad -1 \leq y \leq 1$$

Here $f'(x) = \cos x$,

$$\frac{d}{dy}(\sin^{-1} y) = \frac{1}{\cos x} = \frac{1}{\sqrt{1-\sin^2 x}} = \frac{1}{\sqrt{1-y^2}}$$

Algebraic and transcendental functions. The functions studied in elementary calculus are of two kinds, algebraic and transcendental. The basic differentiation formula for algebraic functions is

$$\frac{dy}{dx} = nx^{n-1} \quad \text{if} \quad y = x^n$$

where n is any rational number. This rule may be combined with the chain rule and used in connection with the rules for dealing with sums, products, and quotients. The differentiation of algebraic functions in general may require use of implicit

function theorems. The elementary transcendental functions are the trigonometric functions, the logarithm functions, and their inverses. For a discussion of implicit function theorems *see* PARTIAL DIFFERENTIATION.

Trigonometric functions. In calculus the trigonometric functions are defined on the assumption that angles are measured in radians, so that $\sin x$ means the sine of x radians. Differentiation of $f(x) = \sin x$ is based on the fact that

$$\frac{\sin x}{x} \to 1 \quad \text{as} \quad x \to 0$$

This is equivalent to $f'(0) = 1$. By combining this result with trigonometric identities, the two basic formulas shown in Eqs. (4) are derived. Then the

$$\frac{d}{dx}\sin x = \cos x$$
$$\frac{d}{dx}\cos x = -\sin x \qquad (4)$$

derivatives of the other functions are worked out by using the rule for quotients. The results are shown in Eqs. (5). These formulas may also be

$$\frac{d}{dx}\tan x = \sec^2 x \qquad \frac{d}{dx}\cot x = -\csc^2 x$$
$$\frac{d}{dx}\sec x = \sec x \tan x \qquad \frac{d}{dx}\csc x = -\csc x \cot x \qquad (5)$$

combined with the chain rule. For example,

$$\frac{d}{dx}\sin u = \cos u \frac{du}{dx}$$

The inverse trigonometric functions are differentiated by the methods for inverse functions, as explained earlier. The definitions are

$$y = \sin^{-1} x \text{ means } x = \sin y \text{ and } -\frac{\pi}{2} \leq y \leq \frac{\pi}{2}$$
$$y = \cos^{-1} x \text{ means } x = \cos y \text{ and } 0 \leq y \leq \pi$$
$$y = \tan^{-1} x \text{ means } x = \tan y \text{ and } -\frac{\pi}{2} < y < \frac{\pi}{2}$$
$$y = \cot^{-1} x \text{ means } x = \cot y \text{ and } 0 < y < \pi$$

The differentiation formulas are shown in Eqs. (6).

$$\frac{d}{dx}\sin^{-1} x = \frac{1}{\sqrt{1-x^2}} \qquad \frac{d}{dx}\cos^{-1} x = \frac{-1}{\sqrt{1-x^2}}$$
$$\frac{d}{dx}\tan^{-1} x = \frac{1}{1+x^2} \qquad \frac{d}{dx}\text{ctn}^{-1} x = \frac{-1}{1+x^2} \qquad (6)$$

The inverses of the secant and cosecant functions are little used, and there is no standard usage about the definitions needed to make them single-valued.

Exponentials and logarithms. These two functions go together, a logarithm function being the inverse of an exponential function, or vice versa. The traditional treatment of these functions in differential calculus was for many years as follows: The nature of exponentials was assumed as known from algebra, and the definition $y = \log_a x$ if $x = a^y$ (where $a > 0$, $a \neq 1$, and $x > 0$), as well as the algebraic properties of logarithms, was also assumed as known. As a first step toward the differentiation of a logarithm function, it was traditional to show that $(1 + t)^{1/t}$ approaches a limit, denoted by e, as t approaches 0. Moreover, $2 < e < 3$, and $e = 2.718$, approximately. It can then be shown that

$$\frac{d}{dx}\log_a x = \frac{1}{x}\log_a e$$

This formula suggests the advantage of choosing $a = e$ as the base for logarithms. The base e logarithm of x, called the natural logarithm of x, is denoted by $\log x$ with no subscript, or by $\ln x$. (The latter notation is favored by engineers and many physical scientists.) Then $y = \ln x$ is equivalent to $x = e^y$, and $y = e^x$ is equivalent to $x = \ln y$. The derivative of e^x can be worked out by the rule for inverse functions. The simple formulas are

$$\frac{d}{dx}\ln x = \frac{1}{x} \quad \text{and} \quad \frac{d}{dx}e^x = e^x$$

For an arbitrary base a, where $a \neq 1$ and $0 < a$, the formulas are

$$\frac{d}{dx}\log_a x = \frac{\log_a e}{x} \quad \text{and} \quad \frac{d}{dx}a^x = a^x \log_e a$$

It is often convenient to know that

$$\log_a x = (\log_a b)(\log_b x)$$

In particular, with $a = 1$, one obtains

$$\log_a b = (\log_b a)^{-1}$$

The modern trend is to recognize that neither exponentials nor logarithms are adequately defined and studied by students before they come to calculus. In any case, satisfactory discussion of these functions requires the theory of limits. Hence it is reasonable to use calculus itself to define logarithms and develop their properties. When this is done, the development proceeds as follows: The "natural logarithm function" is defined when $x > 0$ by Eq. (7). This means that $\ln x$

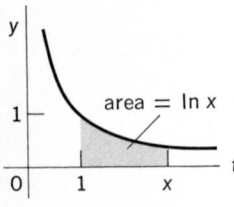

DIFFERENTIATION

The natural logarithm function.

$$\ln x = \int_1^x \frac{dt}{t} \tag{7}$$

is the area between the curve $y = 1/t$ and the t axis from $t = 1$ to $t = x$, reckoned as positive if $x > 1$ and reckoned as negative if $0 < x < 1$ (see illustration).

By use of the fundamental properties of integrals and the relation between differentiation and integration, it can be shown quite directly that

$$\ln 1 = 0, \quad \frac{d}{dx}\ln x = \frac{1}{x}$$

and that $\ln(AB) = \ln A + \ln B$. Moreover, $\ln x$ is a continuous function of x which increases as x increases and is such that $\ln x \to +\infty$ as $x \to +\infty$ and $\ln x \to -\infty$ as $x \to 0$. There is then a well-defined inverse function (call it the E function) such that $y = \ln x$ is equivalent to $x = E(y)$ if $x > 0$. The number $E(1)$ is denoted by e; that is, e is the unique positive number such that $\ln e = 1$. The rules about inverse functions show that $E'(x) = E(x)$. If $a > 0$ and x is an integer, it is possible to show easily that

$$a^x = E(x \ln a)$$

This formula serves as the general definition of a^x when x is not an integer, and the results are consistent with what is expected, so that the usual exponent laws are valid. The general definition of logarithms, from this point of view, is Eq. (8). In particular, $\log_e x = \ln x$, because $\ln e = 1$.

$$\log_a x = \frac{\ln x}{\ln a} \tag{8}$$

In this method of developing properties of exponentials and logarithms by calculus, the fact that $(1 + t)^{1/t} \to e$ as $t \to 0$ is not needed, but it can be proved, more easily than in the older traditional development, after everything else has been worked out.

The number e enters naturally into the concept of continuous compounding of interest. If the sum $\$P$ is placed at interest at the nominal rate of $100r\%$, compounded n times a year, the accumulated sum $\$S$ after t years is shown in Eq. (9). If

$$S = P\left(1 + \frac{r}{n}\right)^{nt} \tag{9}$$

now n is increased indefinitely, so that interest is compounded more and more frequently, the fact that

$$\lim_{n \to \infty}\left(1 + \frac{r}{n}\right)^{n/r} = e$$

shows that in the limit of continuously compounded interest, the formula for the accumulated sum after t years is $S = Pe^{rt}$. This formula is characterized by the differential equation, Eq. (10), which

$$\frac{dS}{dt} = rS \tag{10}$$

expresses what is sometimes called the law of natural growth. In general, if y depends on x in such a way that $dy/dx = ky$, where k is a nonzero constant, then $y = y_0 e^{kx}$, where $y = y_0$ when $x = 0$. This situation occurs in radioactive decay and in many types of growth and diminution processes in chemistry and natural science.

Applications. Several important applications of differentiation are discussed in this section.

Velocity and acceleration. If a point moves on the x axis, with coordinate x at time t, its velocity is dx/dt and its acceleration is d^2x/dt^2. When the point with coordinates (x,y) moves in the xy plane, its velocity and acceleration are vectors, the x and y components of velocity are dx/dt and dy/dt, respectively, and the corresponding components of acceleration are d^2x/dt^2, d^2y/dt^2. These results are extended naturally for a point (x,y,z) moving in three-dimensional space.

For the plane case, if the point has polar coordinates (r,θ), the velocity and acceleration can be resolved into components in the r and θ directions. The r direction at (r,θ) is the direction directly away from the origin (r increasing and θ constant). The θ direction is $90°$ from the r direction in the counterclockwise sense. The r and θ components of velocity are, respectively, dr/dt and $r(d\theta/dt)$, assuming that $r > 0$. The corresponding components of acceleration are Eqs. (11) and (12). These

$$\frac{d^2r}{dt^2} - r\left(\frac{d\theta}{dt}\right)^2 \tag{11}$$

$$r\frac{d^2\theta}{dt^2} + 2\frac{dr}{dt}\frac{d\theta}{dt} \tag{12}$$

expressions are useful in discussing the motion of a particle under the influence of a central force, as in the case of a single mass moving in the gravitational field of a fixed center of attraction according to the inverse-square law.

Still another useful way of resolving acceleration into components involves the tangential and normal directions to the path. Here one needs to deal with arc length and curvature. If s is arc length measured along a plane curve in a preassigned positive sense and if K is curvature, then the velocity is a vector whose component along the curve is ds/dt, the component at right angles to the curve being zero. The component of acceleration along the curve in the positive sense is d^2s/dt^2, and the component at right angles to the curve (90° counterclockwise from the positive sense along the curve) is v^2K, where $v = ds/dt$. The curvature of a circle is the reciprocal of its radius R, and so a point moving around a circle of radius R with speed v experiences an acceleration toward the center of amount v^2/R. This is called centripetal acceleration. If the speed is not uniform, there is also an acceleration along the curve.

Curve tracing. If $f'(x) > 0$, $f(x)$ increases as x increases, whereas $f(x)$ decreases as x increases when $f'(x) < 0$. Points where $f'(x) = 0$ are called critical points. With x and y axes in the usual position (x positive to the right, y positive toward the top of the page), the graph of $y = f(x)$ is called concave upward over an interval of x values if the tangent line turns counterclockwise as x increases. If the tangent line turns clockwise as x increases, the curve is called concave downward. If $f''(x) > 0$, the curve is concave upward, and it is concave downward if $f''(x) < 0$. It follows that y is at a relative maximum if $f'(x) = 0$ and $f''(x) < 0$, at a relative minimum if $f'(x) = 0$ and $f''(x) > 0$. A point where the concavity changes from one sense to the other is called a point of inflection. Such a point is a point of relative maximum or minimum for $f'(x)$. A sufficient condition for a point of inflection is that $f''(x) = 0$, $f'''(x) \neq 0$.

Simple harmonic motion. A point moving on the x axis in such a way that

$$\frac{d^2x}{dt^2} = -\omega^2 x$$

where $\omega > 0$ is executing what is called simple harmonic motion. If a point travels around a circle with constant speed, its orthogonal projection on a diameter executes simple harmonic motion with the center of the circle as the origin of coordinates on the diameter. The methods of calculus enable one to infer from the foregoing differential equation that x depends on t by a formula

$$x = A \cos (\omega t + \alpha)$$

where A and α are constants. The moving point completes one cycle of its motion in time $2\pi/\omega$.

Law of the mean. For a discussion of the simple case of this law *see* CALCULUS.

An extended version, often called Cauchy's formula, is the following: Suppose F and G are continuous when $a \leq x \leq b$, differentiable when $a < x < b$, and suppose $G(b) \neq G(a)$. Suppose also that $F'(x)$ and $G'(x)$ are never both zero together. Then there is some number X such that $a < X < b$ and

$$\frac{F(b) - F(a)}{G(b) - G(a)} = \frac{F'(X)}{G'(X)}$$

Rule of l'Hospital. This rule for finding the limit of a quotient of two functions in certain circumstances is stated as follows: Subject to certain general conditions on f and g, if either $f(x) \to 0$ and $g(x) \to 0$ as $x \to a$ or $g(x) \to +\infty$ or $g(x) \to -\infty$ as $x \to a$, then

$$\lim_{x \to a} \frac{f(x)}{g(x)} = \lim_{x \to a} \frac{f'(x)}{g'(x)}$$

provided the limit on the right exists as a finite limit or as either $+\infty$ or $-\infty$. The general conditions are that f and g are differentiable as $x \to a$ and neither $g(x)$ nor $g'(x)$ is 0 as $x \to a$.

Taylor's formula. This formula, Eq. (13), is a

$$f(x) = f(a) + \frac{f'(a)}{1!}(x - a) + \cdots$$
$$+ \frac{f^{(n)}(a)}{n!}(x - a)^n + R_n \qquad (13)$$

finite sum expression for $f(x)$ where R_n is called the remainder. The two most useful formulas for R_n are Eq. (14), integral form, and Eq. (15), Lagrange's

$$R_n = \frac{1}{n!} \int_a^x f^{(n+1)}(t)(x - t)^n \, dt \qquad (14)$$

$$R_n = \frac{f^{(n+1)}(X)}{(n+1)!}(x - a)^{n+1} \qquad (15)$$

form. In the latter form, X is some number between x and a. If, for fixed a and x, $R_n \to 0$ as $n \to \infty$, $f(x)$ is said to be represented by Taylor's infinite series expansion. *See* INTEGRATION; OPERATOR THEORY; SERIES.

[ANGUS E. TAYLOR]

Bibliography: M. H. Protter and C. B. Morrey, *College Calculus with Analytic Geometry*, 3d ed., 1977; S. K. Stein, *Calculus and Analytic Geometry*, 2d ed., 1977.

Diffraction

The bending of light, or other waves, into the region of the geometrical shadow of an obstacle. More exactly, diffraction refers to any redistribution in space of the intensity of waves that results from the presence of an object that causes variations of either the amplitude or phase of the waves. Most diffraction gratings cause a periodic modulation of the phase across the wavefront rather than a modulation of the amplitude. Although diffraction is an effect exhibited by all types of wave motion, this article will deal only with electromagnetic waves, especially those of visible light. Some important differences that occur with microwaves will also be mentioned. For discussion of the phenomenon as encountered in other types of waves *see* ELECTRON DIFFRACTION; NEUTRON DIFFRACTION; SOUND.

Diffraction is a phenomenon of all electromagnetic radiation, including radio waves; microwaves; infrared, visible, and ultraviolet light; and x-rays. The effects for light are important in connection with the resolving power of optical instruments. *See* X-RAY DIFFRACTION.

There are two main classes of diffraction, which are known as Fraunhofer diffraction and Fresnel diffraction. The former concerns beams of parallel light, and is distinguished by the simplicity of the mathematical treatment required and also by its practical importance. The latter class includes the effects in divergent light, and is the simplest to observe experimentally. A complete

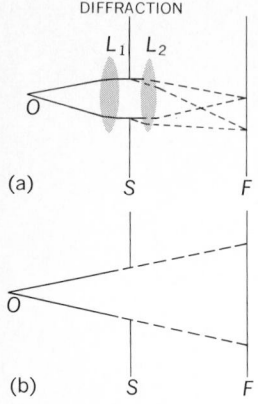

DIFFRACTION

L_1 L_2

(a)

(b)

Fig. 1. Observation of the two principal types of diffraction, in the case of a circular aperture. (a) Fraunhofer and (b) Fresnel diffraction.

explanation of Fresnel diffraction has challenged the most able physicists, although a satisfactory approximate account of its main features was given by A. Fresnel in 1814. At that time, it played an important part in establishing the wave theory of light.

To illustrate the difference between the methods of observation of the two types of diffraction, Fig. 1 shows the experimental arrangements required to observe them for a circular hole in a screen S. The light originates at a very small source O, which can conveniently be a pinhole illuminated by sunlight. In Fraunhofer diffraction, the source lies at the principal focus of a lens L_1, which renders the light parallel as it falls on the aperture. A second lens L_2 focuses parallel diffracted beams on the observing screen F, situated in the principal focal plane of L_2. In Fresnel diffraction, no lenses intervene. The diffraction effects occur chiefly near the borders of the geometrical shadow, indicated by the broken lines. An alternative way of distinguishing the two classes, therefore, is to say that Fraunhofer diffraction concerns the effects near the focal point of a lens or mirror, while Fresnel diffraction concerns those effects near the edges of shadows. Photographs of some diffraction patterns of each class are shown in Fig. 2. All of these may be demonstrated especially well by using the light beam from a neon-helium laser. *See* LASER.

FRAUNHOFER DIFFRACTION

This class of diffraction is characterized by a linear variation of the phases of the Huygens secondary waves with distance across the wavefront, as they arrive at a given point on the observing screen. At the instant that the incident plane wave

occupies the plane of the diffracting screen, it may be regarded as sending out, from each element of its surface, a multitude of secondary waves, the joint effect of which is to be evaluated in the focal plane of the lens L_2. The analysis of these secondary waves involves taking account of both their amplitudes and their phases. The simplest way to do this is to use a graphical method, the method of the so-called vibration curve, which can readily be extended to cases of Fresnel diffraction. *See* HUYGENS' PRINCIPLE.

Vibration curve. The basis of the graphical method is the representation of the amplitude and phase of a wave arriving at any point by a vector, the length of which gives the magnitude of the amplitude, and the slope of which gives the value of the phase. In Fig. 3 are shown two vectors of amplitudes a_1 and a_2, pertaining to two waves having a phase difference δ of 60°. That is, the waves differ in phase by one-sixth of a complete vibration. The resultant amplitude A and phase θ (relative to the phase of the first wave) are then found from the vector sum of a_1 and a_2, as indicated. A mathematical proof shows that this proposition is rigorously correct and that it may be extended to cover the addition of any number of waves.

The vibration curve results from the addition of a large (really infinite) number of infinitesimal vectors, each representing the contribution of the Huygens secondary waves from an element of surface of the wavefront. If these elements are assumed to be of equal area, the magnitudes of the amplitudes to be added will all be equal. They will, however, generally differ in phase, so that if the elements were small but finite each would be drawn at a small angle with the preceding one, as

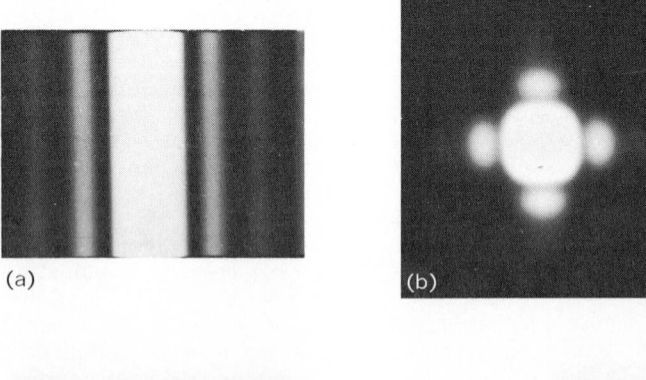

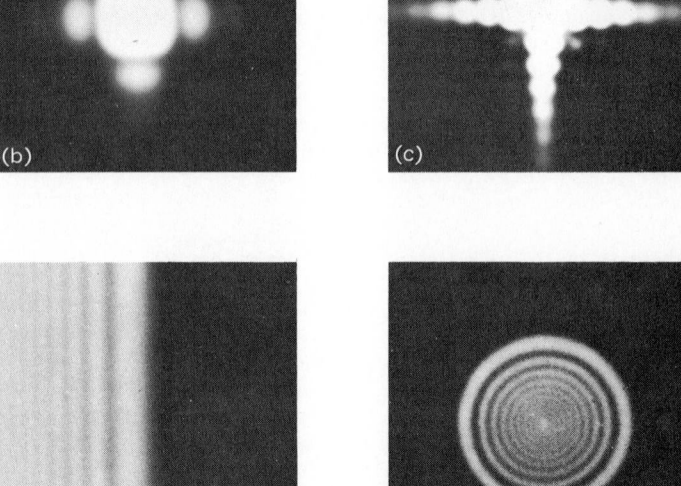

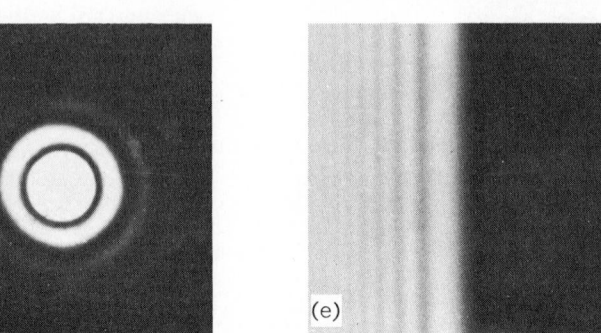

Fig. 2. Diffraction patterns, photographed with visible light. (a) Fraunhofer pattern, for a slit; (b) Fraunhofer pattern, square aperture, short exposure; (c) Fraunhofer pattern, square aperture, long exposure (F. S. Harris); (d) Fraunhofer pattern, circular aperture (R. W. Ditchburn); (e) Fresnel pattern, straight edge; (f) Fresnel pattern, circular aperture.

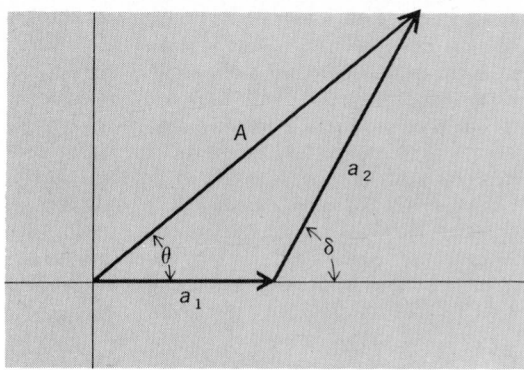

Fig. 3. Graphical addition of two amplitudes.

shown in Fig. 4a. The resultant of all elements would be the vector A. When the individual vectors represent the contributions from infinitesimal surface elements (as they must for the Huygens wavelets), the diagram becomes a smooth curve, the vibration curve, shown in Fig. 4b. The intensity on the screen is then proportional to the square of this resultant amplitude. In this way, the distribution of the intensity of light in any Fraunhofer diffraction pattern may be determined.

The vibration curve for Fraunhofer diffraction by screens having slits with parallel, straight edges is a circle. Consider, for example, the case of a slit of width b illustrated in Fig. 5. The edges of the slit extend perpendicular to the plane of the figure, and the slit is illuminated by plane waves of light coming from the left. If s is the distance from O of a surface element ds (ds actually being a strip extending perpendicular to the figure), the extra distance that the wavelet from ds must travel in reaching a point on the screen lying at the angle θ from the center is $s \sin \theta$. Since this extra distance determines the phase difference, the latter varies linearly with s. This condition necessitates that the vibration curve be a circle.

The intensity distribution for Fraunhofer diffraction by a slit as a function of the angle θ may be simply calculated as follows. The extra distances traveled by the wavelets from the upper and lower edges of the slit, as compared with those from the center, are $+(b/2) \sin \theta$ and $-(b/2) \sin \theta$. The corresponding phase differences are $2\pi/\lambda$ times these quantities, λ being the wavelength of the light. Using the symbol β for $(\pi b \sin \theta)/\lambda$, it is seen that the end points of the effective part of the vibration curve must differ in slope by $\pm\beta$ from the slope at its center, where it is taken as zero. Figure 6 shows the form of the vibration curve for $\beta = \pi/4$, that is, for $\sin \theta = \frac{1}{4}\lambda/b$. The resultant is $A = 2r \sin \beta$, where r is the radius of the arc. The amplitude A_0 that would be obtained if all the secondary waves were in phase at the center of the diffraction pattern, where $\theta = 0$, is the length of the arc. Thus, Eq. (1) holds. The intensity at any angle is given by Eq. (2), where I_0 is the intensity at the center of the pattern. Figure 7 shows a graph of this function.

$$\frac{A}{A_0} = \frac{\text{chord}}{\text{arc}} = \frac{2r \sin \beta}{2r\beta} = \frac{\sin \beta}{\beta} \tag{1}$$

$$I = I_0 \frac{\sin^2\beta}{\beta^2} \tag{2}$$

The central maximum is twice as wide as the subsidiary ones, and is about 21 times as intense as the strongest of these. A photograph of this pattern is shown in Fig. 2a.

The dimensions of the pattern are important, since they determine the angular spread of the light behind the slit. The first zeros occur at values $\beta = (\pi b \sin \theta)/\lambda = \pm\pi$. In most cases the angle θ is extremely small, so that Eq. (3) holds. For a slit 1

$$\sin \theta \approx \theta = \pm\frac{\lambda}{b} \tag{3}$$

mm wide, for example, and green light of wavelength 5×10^{-5} cm, Eq. (3) gives the angle as only 0.0005 radian, or 1.72 minutes of arc. The slit would have to be much narrower than this, or the wavelength much longer, for the approximation to cease to be valid.

The main features of Fraunhofer diffraction patterns of other shapes can be understood with the aid of the vibration curve. Thus for a rectangular or square aperture, the wavefront may be subdivided into elements parallel to either of two adjacent sides, giving an intensity distribution which follows the curve of Fig. 7 in the directions parallel

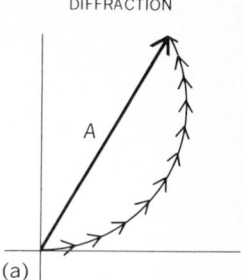

(a)

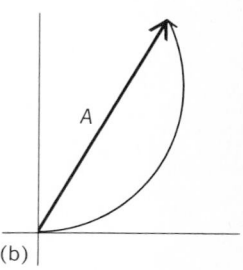

(b)

Fig. 4. Vibration curves. (a) Addition of many equal amplitudes differing in phase by equal amounts. (b) Equivalent curve, when amplitudes and phase differences become infinitesimal.

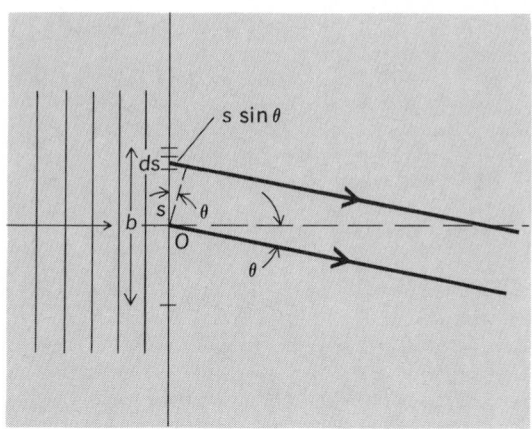

Fig. 5. Analysis of Fraunhofer diffraction by a slit.

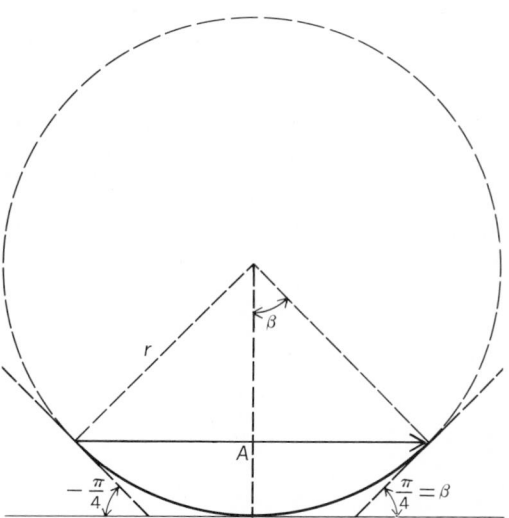

Fig. 6. Vibration curve and resultant amplitude for a particular point in Fraunhofer diffraction by a slit.

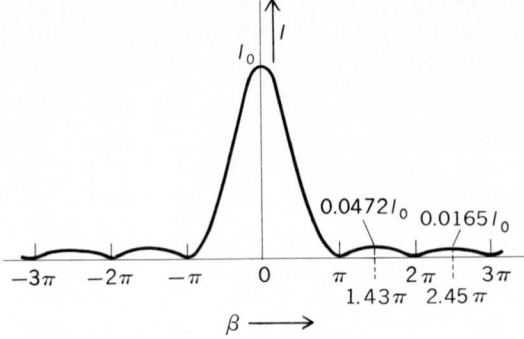

Fig. 7. Intensity distribution curve for Fraunhofer diffraction by a slit.

DIFFRACTION

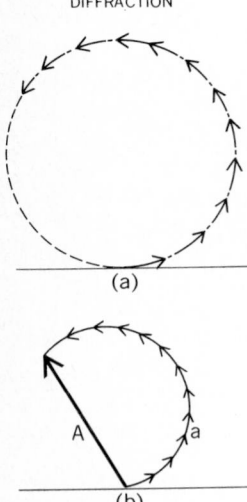

Fig. 8. Graphical analysis of diffraction grating. (a) Vibration curve for grating of 12 slits. (b) Resultant amplitude A formed by adding amplitudes a from individual slits. Each a represents the chord of one of the short arcs in part a.

to the two sides. Photographs of such patterns appear in Fig. 2b and c. In Fig. 2c it will be seen that there are also faint subsidiary maxima lying off the two principal directions. These have intensities proportional to the products of the intensities of the side maxima in the slit pattern. The fact that these subsidiary intensities are extremely low compared with that of the central maximum has an important application to the apodization of lenses, to be discussed later.

Diffraction grating. An idealized diffraction grating consists of a large number of similar slits, equally spaced. Equal segments of the vibration curve are therefore effective, as shown in Fig. 8a. The resultants a of each segment are then to be added to give A, the amplitude due to the whole grating, as shown in Fig. 8b. The phase difference between successive elements is here assumed to be very small. As it is increased, by going to a larger angle θ, the resultant A first goes to zero at an angle corresponding to λ/W, where W is the total width of the grating. After going through numerous low-intensity maxima, A again rises to a high value when the phase difference between the successive vectors for the individual slits approaches a whole vibration. These small vectors a are then all lined up again, as they were at the center of the pattern ($\theta = 0$). The resulting strong maximum represents the "first-order spectrum," since its position depends on the wavelength. A similar condition occurs when the phase difference becomes two, three, or more whole vibrations, giving the higher-order spectra. By means of this diagram, it is possible not only to predict the intensities of successive orders for an ideal grating, but also to find the sharpness of the maxima which represent the spectrum lines. *See* DIFFRACTION GRATING.

Determination of resolving power. Fraunhofer diffraction by a circular aperture determines the resolving power of instruments such as telescopes, cameras, and microscopes, in which the width of the light beam is usually limited by the rim of one of the lenses. The method of the vibration curve may be extended to find the angular width of the central diffraction maximum for this case. Figure 9 compares the treatments of square and circular apertures by showing, above, the elements of equal phase difference into which the wavefront may be divided, and, below, the corresponding vibration curves. For the square aperture shown in Fig. 9a, the areas of the surface elements are

equal, and the curve forms a complete circle at the first zero of intensity. In Fig. 9b these areas, and hence the lengths of the successive vectors, are not equal, but increase as the center of the curve is approached, and then decrease again. The result is that the curve must show somewhat greater phase differences at its extremes in order to form a closed figure. An exact construction of the curve or, better, a mathematical calculation shows that the extreme phase differences required are $\pm 1.220\pi$, yielding Eq. (4) for the angle θ at the first

$$\sin \theta \approx \theta = \pm \frac{1.220\lambda}{d} \qquad (4)$$

zero of intensity. Here d is the diameter of the circular aperture. When this result is compared with that of Eq. (3), it is apparent that, relative to the pattern for a rectangle of side $b = d$, this pattern is spread out by 22%. Obviously it now has circular symmetry and consists of a diffuse central disk, called the Airy disk, surrounded by faint rings (Fig. 2d). The angular radius of the disk, given by Eq. (4), may be extremely small for an actual optical instrument, but it sets the ultimate limit to the sharpness of the image, that is, to the resolving power. *See* RESOLVING POWER (OPTICS).

Other applications. Among the applications of Fraunhofer diffraction, other than the calculation of resolving power, are its use in (1) the theory of certain types of interferometer in which the interfering light beams are brought together by diffraction, (2) the theory of microscopic imaging, and (3) the design of apodizing screens for lenses.

Michelson's stellar interferometer. This instrument was devised by A. A. Michelson to overcome the limitation expressed by Eq. (4). In front of a telescope are placed two fixed mirrors, shown at M_2 and M_3 in Fig. 10, and two others, M_1 and M_4, movable so as to vary their separation D. The light of a star is directed by these mirrors into a telescope as shown, and the diffraction patterns, of size corresponding to the aperture of the mirrors, are superimposed in the focal plane P of the lens. When D is small the resulting pattern is crossed by interference fringes. If the light source has a finite width, increasing D causes these fringes to be-

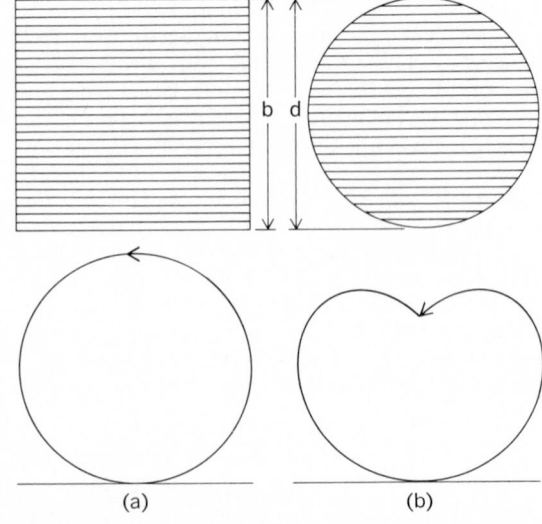

Fig. 9. Condition of vibration curves at first minimum for (a) a square aperture and (b) a circular aperture.

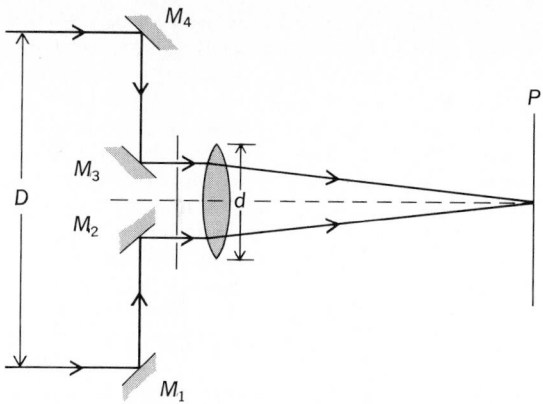

Fig. 10. Michelson stellar interferometer, which is used for measuring the diameters of stars.

come indistinct and eventually to disappear. For a circular disk source, such as a star, the angle subtended by the disk must be $1.22\lambda/D$ for disappearance of the fringes. The resolving power of the telescope has thus effectively been increased in the ratio D/d. In this way angular diameters as small as 0.02 second of arc have been measured, corresponding to $D=24$ ft. See INTERFERENCE OF WAVES; INTERFEROMETRY.

Microscopic imaging. E. Abbe's theory of the microscope evaluates the resolving power of this instrument by considering not only the diffraction caused by the limited aperture of the objective, but also that caused by the object itself. If this object is illuminated by coherent light, such as a parallel beam from a point source, its Fraunhofer diffraction pattern is formed in the rear focal plane of the microscope objective. Abbe took as an object a diffraction grating, and in this case the pattern consists of a series of sharp maxima representing the various orders $m=0,\pm1,\pm2,\ldots$. If the light of all these orders is collected by the objective, a perfect image of the grating can be formed where the light is reunited in the image plane. In practice, however, the objective can include only a limited number of them, as is indicated in Fig. 11. Here only the orders $+2$ to -2 are shown entering the objective. The higher orders, involving greater values of θ, would miss the lens.

The final image must be produced by interference in the image plane of the Huygens secondary waves coming from the various orders. To obtain a periodic variation of intensity in that plane, at least two orders must be accepted by the objective. The angle θ_1 at which the first order occurs is given by $\sin\theta_1=\lambda/d$, where d is the spacing of the lines in the grating. The limit of resolution of the microscope, that is, the smallest value of d that will produce an indication of separated lines in the image, may thus be found from the angular aperture of the objective. In order for it to accept only the orders $0,\pm1$, the lens aperture 2α must equal at least $2\theta_1$, giving $d=\lambda/\sin\alpha$. This resolving limit may be decreased by illuminating the grating from one side, so that the zero-order light falls at one edge of the lens, and that of one of the first orders at the other edge. Then $\theta_1=2\alpha$, and the limit of resolution is approximately $0.5\lambda/\sin\alpha$.

Apodization. This is the name given to a procedure by which the effect of subsidiary maxima (such as those shown in Fig. 7) may be partially suppressed. Such a suppression is desirable when one wishes to observe the image of a very faint object adjacent to a strong one. If the fainter object has, for example, only 1/1000 of the intensity of the stronger one, the two images will have to be far enough apart so that the principal maximum for the fainter one is at least comparable in intensity to the secondary maxima for the stronger object at that point. In the pattern of a rectangular aperture it is not until the tenth secondary maximum that the intensity of these falls below 1/1000 of the intensity of the principal maximum. Here it has been assumed, however, that the fainter image lies along one of the two principal directions of diffraction of the rectangular aperture, perpendicular to two of its adjacent sides. At 45° to these directions the subsidiary maxima are much fainter (Fig. 2c), and even the second one has an intensity of only 1/3700, the square of the value of the second subsidiary maximum indicated in Fig. 7.

The simplest apodizing screen is a square aperture placed over a lens, the diagonal of the square being equal to the diameter of the lens. If the lens is the objective of an astronomical telescope, for example, the presence of a fainter companion in a double-star system can often be detected by turning the square aperture until the image of the companion star lies along its diagonal. Apodizing screens may be of various shapes, depending on the purpose to be achieved. It has been found that a screen of graded density, which shades the lens

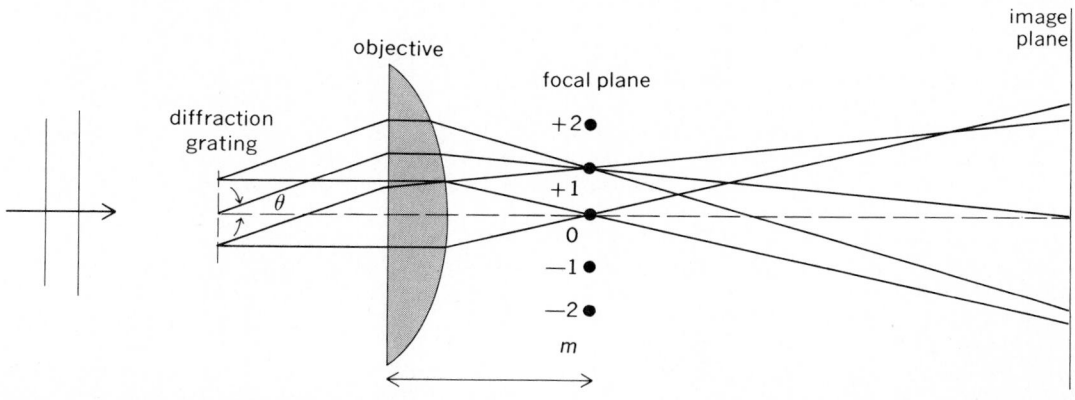

Fig. 11. Abbe's method of treating the resolving power of a microscope.

from complete opacity at the rim to complete transparency at a small distance inward, is effective in suppressing the circular diffraction rings surrounding the Airy disk. In all types of apodization there is some sacrifice of true resolving power, so that it would not be used if the two images to be resolved were of equal intensity.

FRESNEL DIFFRACTION

The diffraction effects obtained when the source of light or the observing screen are at a finite distance from the diffracting aperture or obstacle come under the classification of Fresnel diffraction. This type of diffraction requires for its observation only a point source, a diffracting screen of some sort, and an observing screen. The latter is often advantageously replaced by a magnifier or a low-power microscope. The observed diffraction patterns generally differ according to the radius of curvature of the wave and the distance of the point of observation behind the screen. If the diffracting screen has circular symmetry, such as that of an opaque disk or a round hole, a point source of light must be used. If it has straight, parallel edges, it is desirable from the standpoint of brightness to use an illuminated slit parallel to these edges. In the latter case, it is possible to regard the wave emanating from the slit as a cylindrical one. For the purpose of deriving the vibration curve, the appropriate way of dividing the wavefront into infinitesimal elements is to use annular rings in the first case, and strips parallel to the axis of the cylinder in the second case.

Figure 12 illustrates the way in which the radii of the rings, or the distances to the edges of the strips, must be chosen in order that the phase difference may increase by an equal amount from one element to the next. Figure 12a shows a section of the wavefront diverging from the source S, and the paths to the screen of two secondary wavelets. The shortest possible path is b, while r is that for another wavelet originating at a distance s above the "pole" O. Since all points on the wavefront are at the same distance a from S, the path difference between the two routes from S to P is r − b. When this is evaluated, to terms of the first order in s/a and s/b, the phase difference is given by Eq. (5). The phase difference across an elemen-

$$\delta = \frac{2\pi}{\lambda}(r-b) \approx \frac{\pi(a+b)}{ab\lambda}s^2 = Cs^2 \qquad (5)$$

tary zone of radius s and width ds then becomes $d\delta = 2Csds$, so that for equal increments of δ the increment of s must be proportional to 1/s. The annular zones and the strips drawn in this way on the spherical or cylindrical wave, respectively, looking toward the pole from the direction of P, are shown in Fig. 12b and c.

For the annular zonal elements, the areas $2\pi sds$ are all equal, and hence the amplitude elements of the vibration curve should have the same magnitude. Actually, they must be regarded as falling off slowly, due to the influence of the "obliquity factor" of Huygens' principle. The resulting vibration curve is nearly, but not quite, circular, and is illustrated in Fig. 13a. It spirals in toward the center C, at a rate that has been considerably exaggerated in the figure. The intensity at any point P on the axis of a circular screen centered on O can now be determined as the square of the resultant amplitude

A for the appropriate part or parts of this curve. The curve shown in Fig. 13a is for a circular aperture which exposes five Fresnel zones, each one represented by a half-turn of the spiral. The resultant amplitude is almost twice as great (and the intensity four times as great) as it would be if the whole wave were exposed, in which case the vector would terminate at C. The diffraction by other circular screens may be determined in this same manner.

Zone plate. This is a special screen designed to block off the light from every other half-period zone, and represents an interesting application of Fresnel diffraction. The Fresnel half-period zones are drawn, with radii proportional to the square roots of whole numbers, and alternate ones are blackened. The drawing is then photographed on a reduced scale. When light from a point source is sent through the negative, an intense point image is produced, much like that formed by a lens. The zone plate has the effect of removing alternate half-turns of the spiral, the resultants of the others all adding in the same phase. By putting $\delta = \pi$ and $a = \infty$ in Eq. (5), it is found that the "focal length" b of a zone plate is s_1^2/λ, where s_1 is the radius of the first zone.

Cornu's spiral. This is the vibration curve for a cylindrical wavefront, and is illustrated in Fig. 13b. The areas of the elementary zones, and hence the magnitudes of the component vectors of the vibration curve, decrease rapidly, being proportional to ds, and hence to 1/s (Fig. 12c). The definition of Cornu's spiral requires that its slope δ at any point

DIFFRACTION

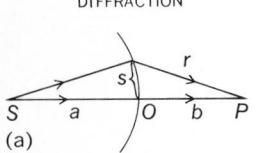

(a)

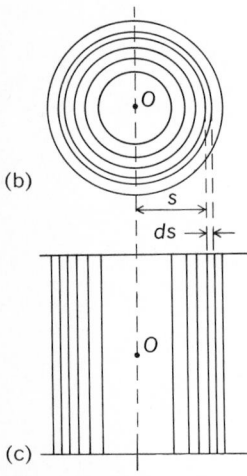

(b)

(c)

Fig. 12. Division of wavefront for constructing vibration curve for Fresnel diffraction. (a) Section of wave diverging from S and paths to screen of two secondary wavelets. (b) Annular zones. (c) Strips.

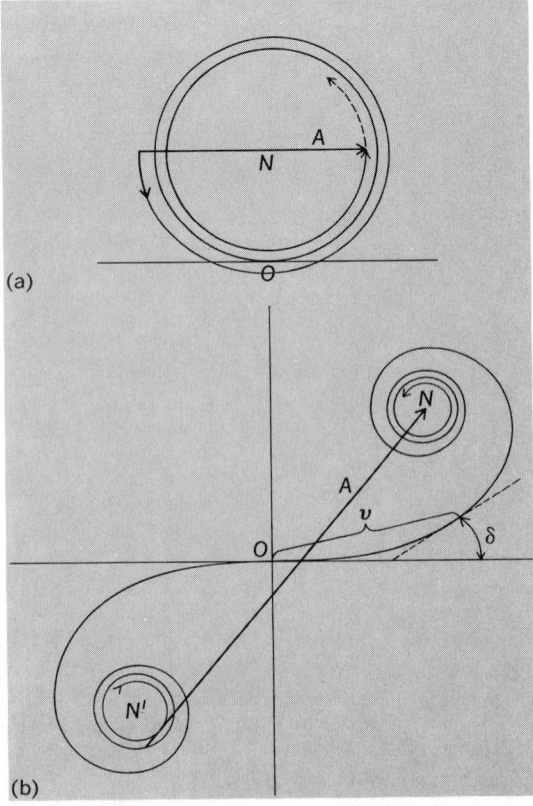

(a)

(b)

Fig. 13. Vibration curves for Fresnel diffraction patterns. (a) Circular division of wavefront. (b) Strip division (Cornu's spiral).

be proportional to the square of the corresponding distance s measured up the wavefront [Fig. 12a and Eq. (5)]. The length of the spiral from the origin is proportional to s, but it is usually drawn in terms of the dimensionless variable v, defined by $v = s\sqrt{2C/\pi}$ in the notation of Eq. (5). The coordinates of any point on the curve may then be found from tables of Fresnel's integrals.

As an example of the application of Cornu's spiral, consider the diffraction of an opaque straight edge, such as a razor edge, illuminated by light from a narrow slit parallel to the edge. At some point outside the edge of the shadow, say one which exposes three half-period strips beyond the pole, the resultant amplitude will be that labeled A in Fig. 13b. The intensity will be greater than that given by the amplitude NN', which represents the amplitude for the whole (unobstructed) wave. On going further away from the edge of the shadow, the tail of the vector A will move along the spiral inward toward N', and the intensity will pass through maxima and minima. At the edge of the geometrical shadow the amplitude is ON, and the intensity is just one-fourth of that due to the unobstructed wave. Further into the shadow the intensity approaches zero regularly, without fluctuations, as the tail of the vector moves up toward N. A photograph of the straight-edge pattern is shown in Fig. 2e.

Babinet's principle. This states that the diffraction patterns produced by complementary screens are identical. Two screens are said to be complementary when the opaque parts of one correspond to the transparent parts of the other, and vice versa. Babinet's principle is not very useful in dealing with Fresnel diffraction, except that it may furnish a short-cut method in obtaining the pattern for a particular screen from that of its complement. The principle has an important application for Fraunhofer diffraction, in parts of the field where there is zero intensity without any screen. Under this condition the amplitudes produced by the complementary screens must be equal and opposite, since the sum of the effects of their exposed parts gives no light. The intensities, being proportional to the squares of the amplitudes, must therefore be equal. In Fraunhofer diffraction the pattern due to a disk is the same as that due to a circular hole of the same size.

DIFFRACTION OF MICROWAVES

The diffraction of microwaves, which have wave-lengths in the range of millimeters to centimeters, has been intensively studied since World War II because of its importance in radar work. Many of the characteristics of optical diffraction can be strikingly demonstrated by the use of microwaves. Microwave diffraction shows certain features, however, that are not in agreement with the Huygens-Fresnel theory, because the approximations made in that theory are no longer valid. Most of these approximations, for example, that made in deriving Eq. (5), depend for their validity on the assumption that the wavelength is small compared to the dimensions of the apparatus. Furthermore, it is not legitimate to postulate that the wave has a constant amplitude across an opening, and zero intensity behind the opaque parts, except for the very minute waves of light.

As an example of the failure of classical diffraction theory when applied to microwaves, the results of the diffraction by a circular hole in a metal screen may be mentioned. The observed patterns begin to show deviations from the Fresnel theory, and even from the more rigorous Kirchoff theory, when the point of observation is within a few wavelengths' distance from the plane of the aperture. Even in this plane itself there are detectable variations of intensity. There are also polarization effects that could not have been predicted from the earlier theories, which treat light as a scalar, rather than a vector, wave motion. An exact vector theory of diffraction, developed by A. Sommerfeld, has been applied only in a few simple cases, but the measurements at microwave frequencies agree with it wherever it has been tested. *See* LIGHT.

[F. A. JENKINS/W. W. WATSON]

Bibliography: C. J. Ball, *Introduction to the Theory of Diffraction*, 1971; M. Born and E. Wolf, *Principles of Optics*, 5th ed., 1975; F. A. Jenkins and H. E. White, *Fundamentals of Optics*, 4th ed., 1976; A. Sommerfeld, *Lectures on Theoretical Physics*, vol. 4: *Optics*, 1954; W. T. Welford, *Optics*, 1977.

Diffraction grating

An optical device consisting of an assembly of narrow slits or grooves, which by diffracting light produces a large number of beams which can interfere in such a way as to produce spectra. Since the angles at which constructive interference patterns are produced by a grating depend on the lengths of the waves being diffracted, the waves of various lengths in a beam of light striking the grating will be separated into a number of spectra, produced in various orders of interference on either side of an undeviated central image. By controlling the shape and size of the diffracting grooves when producing a grating and by illuminating the grating at suitable angles, a beam of light can be thrown into a single spectrum whose purity and brightness may exceed that produced by a prism. Gratings can now be made with much larger apertures than prisms, and in such form that they waste less light and give higher intrinsic dispersion and resolving power. A single grating can be used over a much broader range of spectrum than can any single prism, and its dispersion will vary less rapidly with wavelength. Gratings are being used increasingly in large spectrographs and for highly precise spectroscopic work, as well as in monochromators and analytical spectrographs. *See* DIFFRACTION; INTERFERENCE OF WAVES; OPTICAL PRISM.

Transmission gratings consist of a large number of narrow transparent and opaque slits alternating side by side in regular order and with uniform separation, through which a beam of light will appear as a series of spectra in various orders of interference. Such gratings are conveniently used in small spectroscopes and spectrometers, but only for visible light, since they are usually not transparent to ultraviolet or infrared radiation. They are commonly made by contact molding from a master grating.

Reflection gratings, either plane or concave, are used in most spectrographs. Such a grating may consist of an original ruling or of a metal-coated replica from an original. Large grating replicas practically indistinguishable in performance or permanence from an original can now be made.

Production of gratings. Gratings are engraved by highly precise ruling engines, which use a diamond tool to press into a highly polished mirror surface a series of many thousands of fine shallow burnished grooves. Gratings for the range 1500–10,000 A are commonly ruled with 5000–30,000 grooves per inch (the usual value is near 15,000), on a thin layer of aluminum deposited on glass by evaporation in vacuum. Gratings for the infrared region are often ruled on gold, silver, copper, lead, or tin mirrors, with coarser groove spacings.

If a grating is to give resolution approaching the theoretical limit, its grooves must be ruled straight, parallel, and equally spaced to within a few tenths of the shortest incident wavelength. The proper overall spacing of grooves must also be maintained if changes in focal properties are not to result. Scattered light and false images may arise from local spacing error and groove shape variations of only a few hundredths of the diffracted wavelength.

Among the false lines produced by imperfect gratings are Rowland ghosts, which arise from periodic errors in groove position; Lyman ghosts, which come from a combination of periodicities in ruling; and satellites, caused by sets of irregularly placed grooves, which may seriously reduce resolution. Target pattern arises from unequal contribution of light from all parts of a grating, and is especially prevalent in concave gratings, in which the shape of the grooves may change as the cutting angle of the diamond changes.

Gratings of 2-, 4-, or 6-in. ruled width are commonly used in commercial spectrographs, with projection distances of 20–180 in. In large research instruments, gratings of 6- to 10-in. ruled width are used with projection distances of 10–50 ft or more. The largest modern gratings, used in their highest orders, show resolving power $\lambda/\delta\lambda$ in excess of 900,000 in the green region of the spectrum, and in excess of 1.5×10^6 at shorter wavelengths. Here λ is the mean wavelength of two closely spaced, just resolvable spectral lines, and $\delta\lambda$ is their wavelength difference. Such gratings give resolution equal to that of most interferometers, and in addition provide greater photographic speed, are easier to adjust, follow more simple laws of wavelength distribution, and permit a wider range of wavelengths to be photgraphed at one time without crossed dispersion. *See* RESOLVING POWER (OPTICS).

Properties of gratings. A grating illuminated at angle α (measured from the normal) will direct wavelength λ toward angle β in accordance with the formula $m\lambda = d(\sin\alpha \pm \sin\beta)$, where m is an integral order of interference, d is the grating constant, or distance between consecutive grooves, and the $+$ and $-$ signs refer to orders on opposite sides of the normal. The linear dispersion produced by a grating on a photographic plate depends on its intrinsic angular dispersion multiplied by the distance P from grating to plate. The intrinsic angular dispersion is given by the formula shown below. Theoretically the resolving power of

$$\frac{d\beta}{d\lambda} = \frac{1}{\lambda}\left(\frac{\sin\alpha}{\cos\beta} + \tan\beta\right)$$

a grating is mN, where N is the number of grooves in the grating. Resolving power is not directly de-

pendent on the number of grooves, since for gratings of a given size m and N are inversely related. It is basically dependent on the number of wavelengths of optical retardation the grating introduces between the extreme rays leaving it. Another useful concept is the resolving limit $d\sigma$, the smallest wave number difference the grating can resolve, which remains essentially constant for a given angle of illumination of a given grating at all wavelengths, except as errors in groove spacing become more important for shorter wavelengths.

The manner in which incident light will be distributed among the various orders of interference depends upon the shape and orientation of the groove sides and on the relation of wavelength to groove separation. When $d \gtrsim \lambda$, diffraction effects predominate in controlling the intensity distribution among orders, but when $d > \lambda$, optical reflection from the sides of the grooves is more strongly involved. It is possible to "blaze" a grating by ruling its grooves so that their sides reflect a large fraction of the incoming light of suitably short wavelengths in one general direction. Controlled groove shape is especially important in the gratings known as echelettes and echelles, in which as much as 80% of the incoming light may be sent into one particular order for a given wavelength. Many ordinary gratings are blazed.

Grating spectroscopes. These consist usually of a slit, a lens or mirror to collimate the light sent through the slit into a parallel beam, a transmission or reflection grating to disperse the light, a lens or mirror to focus the light into spectrum lines (which are monochromatic images of the slit in the light of each wavelength passing through it), and an eyepiece for viewing the spectrum (Fig. 1). If a

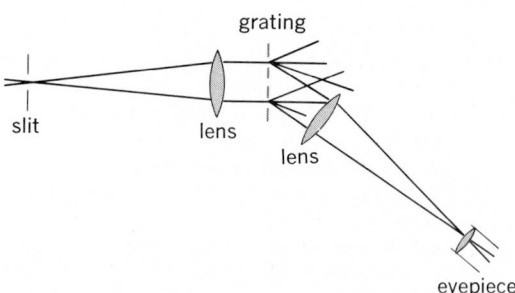

Fig. 1. Transmission grating spectroscope.

camera is substituted for the telescope, the instrument becomes a grating spectrograph. If a photoelectric cell, a thermocouple, or other radiation-detecting device is used instead of a camera or telescope, the device becomes a grating spectrometer.

Echelette grating. This has coarse groove spacing and is designed for the infrared region, the grooves being so shaped and of such size that most of the radiation is concentrated by reflection into a small angular coverage. Radiation of any given wavelength is thus concentrated largely into one order by shaping the point of the ruling diamond to give grooves with comparatively flat sides and by choosing a groove separation which minimizes diffraction effects.

Echelle grating. This is designed for use in high orders and at angles of illumination greater than

45° to obtain high dispersion and resolving power by the use of high orders of interference. Echelles have properties lying midway between those of plane gratings of the ordinary type and interferometers of the reflection echelon type, orders of interference ranging from 100 to 1000 being used. Overlapping orders are separated by using crossed dispersion.

Concave grating. This is a widely used form of reflection grating with which a spectrograph can be formed that has no auxiliary optical parts except a slit and a camera. Being ruled on a concave mirror, this type of grating can both collimate and focus the light that falls upon it. It is made by spacing straight grooves equally along the chord (rather than the arc) of a spherical or paraboloidal mirror surface. Light which passes through a slit and falls on such a grating is dispersed by it into spectra which are in focus on the Rowland circle, a circle drawn tangent to the face of the grating at its midpoint, having a diameter equal to the radius of curvature of the grating surface.

A great advantage of the concave grating is that it provides a dispersing and focusing system free of refracting material, so that it can be used with ultraviolet, visible, or infrared radiation interchangeably so long as its grooves diffract radiation and its surface has adequate reflecting power. A disadvantage is its astigmatism at high angles of incidence or reflection, which can, however, be diminished with various optical devices. A plane reflection grating used with two concave mirrors avoids this difficulty.

Grating mountings. The slit, grating, and camera of a concave grating spectrograph can be placed anywhere on the Rowland circle so that any desired wavelength range can be photographed in the desired order (Fig. 2).

The various possible combinations of fixed and moving parts give rise to a number of different grating mountings. In the Rowland mounting, camera and grating are connected by a bar forming a diameter of the Rowland circle, the two running on tracks placed at right angles with the slit fixed at their junction. A spectrum of limited extent having uniform dispersion is then produced at the camera, and camera and grating can be moved on the tracks to shift wavelength coverage. In the Paschen-Runge mounting, slit and grating are fixed, and photographic plates can be clamped to a fixed track almost anywhere on the Rowland circle. In the Eagle mounting, most suited to long, narrow housing, the grating can be rotated and moved toward or away from the slit. The slit is placed close to the camera, which is arranged to rotate so that it can be kept on the Rowland circle.

All these mountings suffer from astigmatism arising from using the grating off-axis. Although this does not markedly reduce the sharpness of the spectrum lines, it may result in a great loss of light intensity when the grating is used at high angles, and it makes difficult the sharp focusing of step filters, sector disks, and interferometer patterns that are placed at the slit.

Astigmatism is greatly reduced in the Wadsworth mounting, in which the slit is placed at the principal focus of a concave mirror, so that the light falling on the grating is in a parallel beam. The grating can be illuminated at any desired angle up to about 40°, and light is taken off along the grating normal, the spectrum being focused on a photographic plate at only half the usual distance. The usual dispersion of the grating is thus cut in half, but the speed of the spectrograph is increased fourfold, and at high angles the speed is increased much more as a result of reduction of astigmatism.

Most modern commercial grating spectrographs, because of the need for portability, are based either on the Eagle mounting, with which a rather limited portion of the spectrum can be photographed at one time, or the Wadsworth mounting, which can give greater spectral coverage without resetting but is bulkier and cannot be used at such high values of $m\lambda$. In order to obtain more complete spectrum coverage in a single exposure, an echelle with crossed dispersion or a grating with some other device for separating the orders may be used.

Plane reflection gratings are ordinarily used in the Littrow mounting (Fig. 3), in which a single

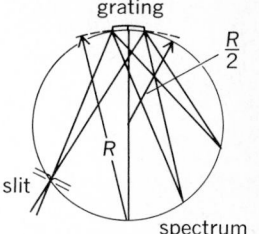

DIFFRACTION GRATING

Fig. 2. Rowland circle.

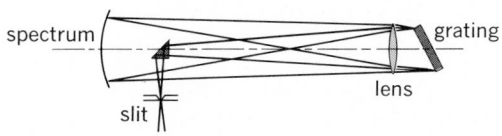

Fig. 3. Littrow mounting of a plane grating.

lens serves for both collimating and focusing, or in the Ebert mounting in which there are two concave mirrors.   [GEORGE R. HARRISON]

Bibliography: R. Chang, *Basic Principles of Spectoscopy*, 1978; D. G. Childers, *Modern Spectrum Analysis*, 1978; F. A. Jenkins and H. E. White, *Fundamentals of Optics*, 4th ed., 1976; D. A. Ramsay, *Spectroscopy*, 1976; B. P. Straughan and S. Walker, *Spectroscopy*, 2d ed., 1978.

Dimensional analysis

A technique that involves the study of dimensions of physical quantities. Dimensional analysis is used primarily as a tool for obtaining information about physical systems too complicated for full mathematical solutions to be feasible. It enables one to predict the behavior of large systems from a study of small-scale models. It affords a convenient means of checking mathematical equations. Finally, dimensional formulas provide a useful cataloging system for physical quantities.

Theory. All the commonly used systems of units in physical science have the property that the number representing the magnitude of any quantity (other than purely numerical ratios) varies inversely with the size of the unit chosen. Thus, if the length of a given piece of land is 300 ft, its length in yards is 100. The ratio of the magnitude of 1 yd to the magnitude of 1 ft is the same as that of any length in feet to the same length in yards, that is, 3. The ratio of two different lengths measured in yards is the same as the ratio of the same two lengths measured in feet, inches, miles, or any other length units. This universal property of unit systems, often known as the absolute significance of relative magnitude, determines the structure of all dimensional formulas. *See* UNITS OF MEASUREMENT.

In defining a system of units for a branch of science such as mechanics or electricity, certain quantities are chosen as fundamental and others as secondary, or derived. The choice of the fundamental units is always arbitrary and is usually made on the basis of convenience in maintaining standards. In mechanics the fundamental units most often chosen are mass, length, and time. Standards of mass (the standard kilogram) and of length (the standard meter) are readily manufactured and preserved, while the rotation of the Earth gives a sufficiently reproducible standard of time. Secondary quantities such as velocity, force, and momentum are obtained from the primary set of quantities according to a definite set of rules.

Assume that there are three primary, or fundamental, quantities α, β, and γ (the following discussion, however, is not limited to there being exactly three fundamental quantities). Consider a particular secondary quantity, expressed in terms of the primaries as $F(\alpha,\beta,\gamma)$ where F represents some mathematical function. For example, if $\alpha =$ mass, $\beta =$ length, and $\gamma =$ time, the derived quantity velocity would be $F(\alpha,\beta,\gamma) = \beta/\gamma$.

Now, if it is assumed that the sizes of the units measuring α, β, and γ are changed in the proportions $1/x$, $1/y$, and $1/z$, respectively, then the numbers measuring the primary quantities become $x\alpha$, $y\beta$, and $z\gamma$, and the secondary quantity in question becomes $F(x\alpha, y\beta, z\gamma)$. Merely changing the sizes of the units must not change the rule for obtaining a particular secondary quantity.

Consider two separate values of the secondary quantity $F(\alpha_1,\beta_1,\gamma_1)$ and $F(\alpha_2,\beta_2,\gamma_2)$. Then, according to the principle of the absolute significance of relative magnitude, Eqs. (1a) and (1b) hold.

$$\frac{F(x\alpha_1,y\beta_1,z\gamma_1)}{F(x\alpha_2,y\beta_2,z\gamma_2)} = \frac{F(\alpha_1,\beta_1,\gamma_1)}{F(\alpha_2,\beta_2,\gamma_2)} \quad (1a)$$

$$F(x\alpha_1,y\beta_1,z\gamma_1) = F(x\alpha_2,y\beta_2,z\gamma_2)\frac{F(\alpha_1,\beta_1,\gamma_1)}{F(\alpha_2,\beta_2,\gamma_2)} \quad (1b)$$

Differentiating partially with respect to x, holding y and z constant, gives Eq. (2), where F' represents the total derivative of the function F with respect to its first argument.

$$\alpha_1 F'(x\alpha_1,y\beta_1,z\gamma_1)$$
$$= \alpha_2 F'(x\alpha_2,y\beta_2,z\gamma_2)\frac{F(\alpha_1,\beta_1,\gamma_1)}{F(\alpha_2,\beta_2,\gamma_2)} \quad (2)$$

Next set the coefficients x, y, $z = 1$. This gives Eq. (3). This relation must hold for all values of the

$$\frac{\alpha_1 F'(\alpha_1,\beta_1,\gamma_1)}{F(\alpha_1,\beta_1,\gamma_1)} = \frac{\alpha_2 F'(\alpha_2,\beta_2,\gamma_2)}{F(\alpha_2,\beta_2,\gamma_2)} \quad (3)$$

arguments α, β, and γ and hence is equal to a constant. The subscripts can now be dropped, giving Eq. (4). The general solution of this differential

$$\frac{\alpha}{F}\frac{dF}{d\alpha} = \text{constant} \quad (4)$$

equation if $F = C_1\alpha^a$, where a is a constant and C_1 is in general a function of β and γ.

The above analysis can now be repeated for the parameters β and γ, leading to the results given in Eqs. (5a) and (5b). These solutions are consistent

$$F = C_2(\alpha,\gamma)\beta^b \quad (5a)$$
$$F = C_3(\alpha,\beta)\gamma^c \quad (5b)$$

only if $F = C\ \alpha^a\beta^b\gamma^c$, where C, a, b, and c are constants. Thus every secondary quantity which satisfies the condition of the absolute significance of relative magnitude is expressible as a product of powers of the primary quantities. Such an expression is known as the dimensional formula of the secondary quantity. There is no requirement that the exponents a,b,c be integral.

Examples of dimensional formulas. Table 1 gives the dimensional formulas of a number of mechanical quantities in terms of mass M, length L, and time T.

Table 1. Dimensional formulas of common quantities

Quantity	Definition	Dimensional formula
Mass	Fundamental	M
Length	Fundamental	L
Time	Fundamental	T
Velocity	Distance/time	LT^{-1}
Acceleration	Velocity/time	LT^{-2}
Force	Mass × acceleration	MLT^{-2}
Momentum	Mass × velocity	MLT^{-1}
Energy	Force × distance	ML^2T^{-2}
Angle	Arc/radius	0
Angular velocity	Angle/time	T^{-1}
Angular acceleration	Angular velocity/time	T^{-2}
Torque	Force × lever arm	ML^2T^{-2}
Angular momentum	Momentum × lever arm	ML^2T^{-1}
Moment of inertia	Mass × radius squared	ML^2
Area	Length squared	L^2
Volume	Length cubed	L^3
Density	Mass/volume	ML^{-3}
Pressure	Force/area	$ML^{-1}T^{-2}$
Action	Energy × time	ML^2T^{-1}
Viscosity	Force per unit area per unit velocity gradient	$ML^{-1}T^{-1}$

In order to extend this list to include the dimensional formulas of quantities from other branches of physics, such as electricity and magnetism, one may do either of the following:

1. Obtain the dimensional formulas in terms of a particular unit system without introducing any new fundamental quantities. Thus, if one uses Coulomb's law in the centimeter-gram-second electrostatic system in empty space (dielectric constant = 1), one obtains definitions (6a) and (6b).

$$\text{Force} = \frac{(\text{charge})^2}{(\text{distance})^2} \quad (6a)$$

$$\text{Charge} = \text{distance} \times \text{square root of force} \quad (6b)$$

Thus, the dimensions of charge are $M^{1/2}L^{3/2}T^{-1}$. The dimensional formulas of other electrical quantities in the electrostatic system follow directly from definitions (7) and so forth. *See* ELECTRICAL UNITS AND STANDARDS.

$$\text{Potential} = \text{energy/charge} = M^{1/2}L^{1/2}T^{-1}$$
$$\text{Current} = \text{charge/time} = M^{1/2}L^{3/2}T^{-2} \quad (7)$$
$$\text{Electric field} = \text{force/charge} = M^{1/2}L^{-1/2}T^{-1}$$

2. Introduce an additional fundamental quantity to take account of the fact that electricity and magnetism encompass phenomena not treated in mechanics. All electrical quantities can be defined in terms of M, L, T, and one other, without resorting to fractional exponents and without the artificial assumption of unit (and dimensionless) dielectric constant as in the electrostatic system. The usual choice for the fourth fundamental

quantity is charge Q, even though charge is not a preservable electrical standard. Then definitions (8) and so forth hold.

$$\text{Potential} = \text{energy/charge} = ML^2T^{-2}Q^{-1}$$
$$\text{Current} = \text{charge/time} = QT^{-1} \qquad (8)$$

It must be realized that the choice of fundamental quantities is entirely arbitrary. For example, a system of units for mechanics has been proposed in which the velocity of light is taken as a dimensionless quantity, equal to unity in free space. All velocities are then dimensionless. All mechanical quantities can then be specified in terms of just two fundamental quantities, mass and time. This is analogous to the reduction of the number of fundamental quantities needed for electricity from four to three by taking the dielectric constant (permittivity) of empty space as unity.

Furthermore, one could increase the number of fundamental quantities in mechanics from three to four by adding force F to the list and rewriting Newton's second law as $F = Kma$, where K is a constant of dimensions $M^{-1}L^{-1}T^2F$. One could then define a system of units for which K was not numerically equal to 1.

In the past there has been considerable controversy as to the absolute significance, if any, of dimensional formulas. The significant fact is that dimensional formulas always consist of products of powers of the fundamental quantities.

Applications. The important uses of the technique of dimensional analysis are considered in the following sections.

Checking of equations. It is intuitively obvious that only terms whose dimensions are the same can be equated. The equation 10 kg = 10 m/sec, for example, makes no sense. A necessary condition for the correctness of any equation is that the two sides have the same dimensions. This is often a help in the verification of complicated analytic expressions. Of course, an equation can be correct dimensionally and still be wrong by a purely numerical factor.

A corollary of this is that one can add or subtract only quantities which have the same dimensions (except for the trivial case which arises when two different equations are added together). Furthermore, the arguments of trigonometric and logarithmic functions must be dimensionless; otherwise their power series expansions would involve sums of terms with different dimensions. There is no restriction on the multiplication and division of terms whose dimensions are different.

π Theorem. The application of dimensional analysis to the derivation of unknown relations depends upon the concept of completeness of equations. An expression which remains formally true no matter how the sizes of the fundamental units are changed is said to be complete. If changing the units makes the expression wrong, it is incomplete. For a body starting from rest and falling freely under gravity, $s = 16t^2$ is a correct expression only so long as the distance fallen s is measured in feet and the time t in seconds. If s is in meters and t in minutes, the equation is wrong. Thus, $s = 16t^2$ is an incomplete equation. The constant in the equation depends upon the units chosen. To make the expression complete, the numerical factor must be replaced by a dimensional constant, the

acceleration of gravity g. Then $s = 1/2gt^2$ is valid no matter how the units of length and time are changed, since the numerical value of g can be changed accordingly.

Assume a group of n physical quantities x_1, x_2, . . . , x_n, for which there exists one and only one complete mathematical expression connecting them, namely, $\phi(x_1, x_2, . . . , x_n) = 0$. Some of the quantities x_1, x_2, . . . , x_n may be dimensional constants. Assume further that the dimensional formulas of the n quantities are expressed in terms of m fundamental quantities α, β, γ, ... Then it will always be found that this single relation ϕ can be expressed in terms of some arbitrary function F of $n - m$ independent dimensionless products π_1, π_2, . . . , π_{n-m}, made up from among the n variables, as in Eq. (9). This is known as the π

$$F(\pi_1, \pi_2, . . . , \pi_{n-m}) = 0 \qquad (9)$$

theorem. It was first rigorously proved by E. Buckingham. The proof is straightforward but long. The only restriction on the πs is that they be independent (no one expressible as products of powers of the others). The number m of fundamental quantities chosen in a particular case is immaterial so long as they also are dimensionally independent. Increasing m by 1 always increases the number of dimensional constants (and hence n) by 1, leaving $n - m$ the same.

The main usefulness of the π theorem is in the deduction of the form of unknown relations. The successful application of the theorem to a particular problem requires a certain amount of shrewd guesswork as to which variables x_1, x_2, . . . , x_n are significant and which not. If $\phi(x_1, x_2, . . . , x_n)$ is not known one can often still deduce the structure of $F(\tau\tau_1, \tau\tau_2, . . . , \tau\tau_{n-m}) = 0$ and so obtain useful information about the system in question.

An example is the swinging of a simple pendulum. Assume that the analytic expression for its period of vibration is unknown. Choose mass, length, and time as the fundamental quantities. Thus $m = 3$. One must make a list of all the parameters pertaining to the pendulum which might be significant. If the list is incomplete, no useful information will be obtained. If too many quantities are included, the derived information is less specific than it might otherwise be. The quantities given in Table 2 would appear to be adequate.

Table 2. Dimensional formulas for a simple pendulum

Quantity	Symbol	Dimensional formula
Mass of bob	m	M
Length of string	l	L
Acceleration of gravity	g	LT^{-2}
Period of swing	τ	T
Angular amplitude	θ	0

The microscopic properties of the bob and string are not considered, and air resistance and the mass of the string are likewise neglected. These would be expected to have only a small effect on the period. Thus $n = 5$, $n - m = 2$, and therefore one expects to find two independent dimensionless products. These can always be found by trial and error. There may be more than one possible set of independent πs. In this case one π is simply θ, the angular amplitude. Another is $l/\tau^2 g$. Since no other of the n variables contains M in its dimensional formula, the mass of the bob m cannot occur in any dimensionless product. The π theorem gives $F(\theta, l/\tau^2 g) = 0$. Therefore the period of vibration does not depend upon the mass of the bob.

Because τ appears in only one of the dimensionless products, this expression can be explicitly solved for τ to give $\tau = G(\theta)\sqrt{l/g}$, where $G(\theta)$ is an arbitrary function of θ. Now make the further assumption that θ is small enough to be neglected. Then $n = 4$, $n - m = 1$, and $F(l/\tau^2 g) = 0$. Thus $l/\tau^2 g = $ constant, or $\tau \propto \sqrt{l/g}$. The π theorem thus leads to the conclusion that τ varies directly as the square root of the length of the pendulum and inversely as the square root of the acceleration of gravity. The magnitude of the dimensionless constant (actually it is 2π) cannot be obtained from dimensional analysis.

In more complicated cases, where a direct solution is not feasible, this method can give information on how certain variables enter a particular problem even where $F(\pi_1, \pi_2, \ldots, \pi_{n-m}) = 0$ cannot be explicitly solved for one variable. The procedure is particularly useful in hydraulics and aeronautical engineering, where detailed solutions are often extremely complicated.

Model theory. A further application of dimensional analysis is in model design. Often the behavior of large complex systems can be deduced from studies of small-scale models at a great saving in cost. In the model each parameter is reduced in the same proportion relative to its value in the original system.

Once again the case of the simple pendulum is a good example. It was found from the π theorem that $F(\theta, l/\tau^2 g) = 0$. If the magnitudes of θ, l, τ, and g are now changed in such a way that neither argument of F is changed in numerical value, the system will behave exactly as the original system and is said to be physically similar. Evidently θ cannot be changed without altering any of the arguments of F, but l, g, and τ can be varied. Suppose that it was desired to build a very large and expensive pendulum which was to swing with finite amplitude. One could build a small model of, say, 1/100 the length and time its swing for an amplitude equal to that for the desired pendulum. The acceleration of gravity g would be the same for the model as for the large pendulum. The period for the model would then be just 1/10 that for the large pendulum. Thus the period of the large pendulum could be deduced before the pendulum was ever built. In practice one would never bother with the π theorem in cases as simple as this where a full analytic solution is possible. In many situations where such a solution is not feasible, models are built and extensively studied before the full-scale device is constructed. This technique is standard in wind tunnel studies of aircraft design. *See* DYNAMIC SIMILARITY.

Cataloging of physical quantities. Dimensional formulas provide a convenient shorthand notation for representing the definitions of secondary quantities. These definitions depend upon the choice of primary quantities. The π theorem is applicable no matter what the choice of primary quantities is.

Changing units. Dimensional formulas are help-ful in changing units from one system to another. For example, the acceleration of gravity in the centimeter-gram-second system of units is 980 cm/sec^2. The dimensional formula for acceleration is LT^{-2}. To find the magnitude of g in mi/hr^2, one would proceed as in Eq. (10).

$$980 \frac{\text{cm}}{\text{sec}^2} \times \frac{\begin{array}{c}\text{(conversion factor} \\ \text{for length)}\end{array}}{\begin{array}{c}\text{(conversion factor} \\ \text{for time)}^2\end{array}}$$

$$= 980 \frac{\text{cm}}{\text{sec}^2} \times \frac{\dfrac{1}{1.609 \times 10^5} \dfrac{\text{mi}}{\text{cm}}}{(1/3600)^2 (\text{hr}^2/\text{sec}^2)} \quad (10)$$

$$= 7.89 \times 10^5 \text{ mi/hr}^2$$

In the past the subject of dimensions has been quite controversial. For years unsuccessful attempts were made to find "ultimate rational quantities" in terms of which to express all dimensional formulas. It is now universally agreed that there is no one "absolute" set of dimensional formulas. Some systems are more symmetrical than others and for this reason are perhaps preferable. The representation of electrical quantities in terms of M, L, and T alone through the electrostatic form of Coulomb's law leads to somewhat awkward fractional exponents, but nevertheless is just as "correct" as a representation in which charge is used as a fourth fundamental unit.

A highly symmetrical pattern results if energy, linear displacement, and linear momentum are chosen as the fundamental quantities in mechanics. In electricity one can use energy, charge, and magnetic flux. The corresponding quantities for a vibrating mass on a spring and the analogous alternating-current circuit with inductance and capacitance have similar dimensional formulas. In this analogy energy is invariant, charge corresponds to displacement, and magnetic flux corresponds to linear momentum. This correspondence is not displayed in conventional dimensional formulas.

[JOHN W. STEWART]

Bibliography: P. W. Bridgman, *Dimensional Analysis*, 1931, reprint 1976; E. Isaacson and M. Isaacson, *Dimensional Methods in Physics*, 1975; H. L. Langhaar, *Dimensional Analysis and Theory of Models*, 1951, reprint 1979; K. Nagami, *Dimension Theory*, 1970.

Dimensions (mechanics)

Length, mass, time, or combinations of these quantities serving as an indication of the nature of a physical quantity. Quantities with the same dimensions can be expressed in the same units. For example, although speed can be expressed in various units such as miles/hour, feet/second, and meters/second, all these speed units involve the ratio of a length unit to a time unit; hence, the dimensions of speed are the ratio of length L to time T, usually stated as LT^{-1}. The dimensions of all mechanical quantities can be expressed in terms of L, T, and mass M. The validity of algebraic equations involving physical quantities can be tested by a process called dimensional analysis; the terms on the two sides of any valid equation must have the same dimensions. *See* DIMENSIONAL ANALYSIS; UNITS OF MEASUREMENT.

[DUDLEY WILLIAMS]

Dipole

Any object or system that is oppositely charged at two points or poles, such as a magnet or a polar molecule. The properties of a dipole are determined by its dipole moment, that is, the product of one of the charges by their separation directed along an axis through the centers of charge. *See* DIPOLE MOMENT.

An electric dipole consists of two electric charges of equal magnitude but opposite polarity, separated by a short distance (see figure); or more generally, a localized distribution of positive and negative electricity without net charge whose mean positions of positive and negative charge do not coincide. For a discussion of electric dipole radiation *see* ELECTROMAGNETIC RADIATION.

Molecular dipoles which exist in the absence of an applied field are called permanent dipoles, while those produced by the action of a field are called induced dipoles. *See* POLAR MOLECULE.

[ROBERT D. WALDRON]

The term magnetic dipole originally referred to the fact that a magnet has two poles and, because of these two poles, experiences a torque in a magnetic field if its axis is not along a magnetic flux line of the field. It is now generalized to include electric circuits which, because of the current, also experience torques in magnetic fields. *See* MAGNET.

[RALPH P. WINCH]

Dipole moment

A mathematical quantity characteristic of a dipole unit equal to the product of one of its charges times the vector distance separating the charges. The dipole moment $\boldsymbol{\mu}$ associated with a distribution of electric charges q_i is given by Eq. (1), where

$$\boldsymbol{\mu} = \sum_i q_i \mathbf{r}_i \tag{1}$$

$\mathbf{r}_i$ is the vector to the charge q_i. For systems with a net charge (for example, positive), the origin is taken at the mean position of the positive charges (and vice versa). Dipole moments have the dimensions coulomb-meters in the rationalized mks system, and statcoulomb-centimeters in the cgs electrostatic system. Molecular dipole moments are often expressed in debye units, where 1 debye $= 10^{-18}$ statcoulomb-cm. *See* DIPOLE.

The electric potential Φ of a dipole at a long distance $\mathbf{R}$ from the dipole is given by Eq. (2), where θ

$$\Phi = \gamma \frac{|\boldsymbol{\mu}| \cos \theta}{4\pi \epsilon_0 |\mathbf{R}|^2} \tag{2}$$

is the angle between $\boldsymbol{\mu}$ and $\mathbf{R}$, ϵ_0 is the permittivity of vacuum, and γ is a geometrical factor. ($\epsilon_0 = 1$ and $\gamma = 4\pi$ in cgs electrostatic units; $\epsilon_0 = 8.854 \times 10^{-12}$ farad/m and $\gamma = 1$ in rationalized mks units.)

The potential energy U of a dipole in a uniform electric field $\mathbf{E}$ is $U = -|\boldsymbol{\mu}||\mathbf{E}| \cos \phi$, where ϕ is the angle between $\boldsymbol{\mu}$ and $\mathbf{E}$.

The induced dipole moment $\boldsymbol{\mu}_i$ of a molecule may be expressed in terms of molecular parameters by the equation $\boldsymbol{\mu}_i = \alpha \mathbf{E}_L$, where α is the polarizability and $\mathbf{E}_L$ is the local field strength acting at the molecular site. This relation permits the interpretation of the macroscopic polarization and hence dipole moments in terms of molecular processes. *See* DIELECTRIC CONSTANT: POLARIZATION OF DIELECTRICS.   [ROBERT D. WALDRON]

Direct current

Electric current which flows in one direction only through a circuit or equipment. The associated direct voltages, in contrast to alternating voltages, are of unchanging polarity. Direct current corresponds to a drift or displacement of electric charge in one unvarying direction around the closed loop or loops of an electric circuit. Direct currents and voltages may be of constant magnitude or may vary with time.

Batteries and rotating generators produce direct voltages of nominally constant magnitude. Direct voltages of time-varying magnitude are produced by rectifiers, which convert alternating voltage to pulsating direct voltage.

Direct current ordinarily is not widely distributed for general use by electric utility customers. Instead, direct-current (dc) power is obtained at the site where it is needed by the rectification of commercially available alternating current (ac) power to dc power.   [D. D. ROBB]

Discriminant

For a polynomial $f(x) = a_n x^n + a_{n-1} x^{n-1} + \cdots + a_1 x + a_0$, the discriminant is given by the expression $D = a_n^{2n-1} (x_1 - x_2)^2 (x_1 - x_3)^2 \cdots (x_1 - x_n)^2 (x_2 - x_3)^2 \cdots (x_2 - x_n)^2 \cdots (x_{n-1} - x_n)^2$, where $x_1, x_2, \ldots, x_n$ are the roots of the equation $f(x) = 0$. There are $n(n-1)/2$ terms $(x_i - x_j)^2$ in the product corresponding to all possible selections of two indices i less than j from the numbers 1, 2, $\ldots, n$.

The importance of the discriminant D lies in the fact that D vanishes if, and only if, the equation $f(x) = 0$ has equal roots. Since the value of the discriminant is unchanged if any two letters x_i and x_j are interchanged, it is a symmetric function of the roots, and can be expressed in terms of the coefficients of $f(x)$. Such an expression for D is most easily obtained, using the result $D = [(-1)^{n(n-1)/2}/a_n]\ R_x(f,f')$, where $f' = f'(x)$ is the derivative of $f(x)$ and $R_x(f,f')$ is the resultant of $f(x)$ and $f'(x)$. For example, with $f(x) = a_2 x^2 + a_1 x + a_0$, and $f'(x) = 2a_2 x + a_1$, notation (1) is obtained,

$$R_x(f,f') = \begin{vmatrix} a_2 & a_1 & a_0 \\ 2a_2 & a_1 & 0 \\ 0 & 2a_2 & a_1 \end{vmatrix} = -a_2 a_1^2 + 4a_2^2 a_0 \tag{1}$$

and $D = a_1^2 - 4a_2 a_0$, which is the discriminant of a quadratic polynomial. The discriminant of a cubic polynomial $f(x) = a_3 x^3 + a_2 x^2 + a_1 x + a_0$ is shown in notation (2).

$$D = 18a_3 a_2 a_1 a_0 - 4a_2^3 a_0 + a_2^2 a_1^2 \\ - 4a_3 a_1^3 - 27a_3^2 a_0^2 \tag{2}$$

See EQUATIONS, THEORY OF; POLYNOMIAL SYSTEMS OF EQUATIONS.

[ROSS A. BEAUMONT]

Dispersion

The separation of a complex of electromagnetic or sound waves into its various frequency components. For example, a beam of white light can be separated into its monochromatic components by virtue of the different velocities of rays of different wavelength of the beam as it passes through a prism. The dispersion of a material, such as glass

DIPOLE

Electric dipole with moment $\mu = Qd$.

or water, at a given wavelength in the electromagnetic spectrum is defined as the rate of change of refractive index with wavelength at the wavelength in question. For an extended discussion of the dispersion of light *see* ABSORPTION OF ELECTROMAGNETIC RADIATION.

[WILLIAM WEST]

Dispersion relations

Relations between the real and imaginary parts of a response function. The term dispersion refers to the fact that the index of refraction of a medium is a function of frequency. In 1926 H. A. Kramers and R. Kronig showed that the imaginary part of an index of refraction (that is, the absorptivity) determines the real part (that is, the refractivity); this is called the Kramers-Kronig relation. The term dispersion relation is now used for the analogous relation between the real and imaginary parts of any response function, such as Eq. (14) below.

Response function. Consider a system in which a cause $C(t)$ (for example, a force) and its effect $E(t)$ (for example, a displacement) are related by Eq. (1), where $G(t)$ is called the response function.

$$E(t) = \int_{-\infty}^{t} dt' \, G(t - t') C(t') \qquad (1)$$

Because this relation is linear in both $C(t)$ and $E(t)$, the response is said to be linear: The superposition of two causes results in the sum of their effects. Because the response function G is a function only of $t - t'$, the response is said to be time-independent: The effect of time t of a cause at time t' depends only on the time difference $t - t'$. Because the upper limit of integration is t [or equivalently, $G(t)$ can be said to vanish for negative argument, Eq. (2)], the response is said to be causal: the cause has no effect at earlier times. *See* CAUSALITY.

$$G(t) = 0 \qquad \text{if } t < 0 \qquad (2)$$

Many examples can be given. A few pairs of cause and effect are: force and spatial displacement, electric field and polarization (G is electrical susceptibility), and incident wave and scattered wave (G is scattering amplitude).

Pure-tone response. The Fourier transform $g(\omega)$ of the response function $G(t)$ is defined in Eq. (3)

$$g(\omega) = \int_{0}^{\infty} dt \, e^{i\omega t} G(t) \qquad (3)$$

and the inverse relation in Eq. (4). The function

$$G(t) = \frac{1}{2\pi} \int_{-\infty}^{\infty} d\omega \, e^{-i\omega t} g(\omega) \qquad (4)$$

$g(\omega)$ is complex-valued; however, if $G(t)$ is real [which it must be if $C(t)$ and $E(t)$ of Eq. (1) are real], the crossing relation, Eq. (5), is valid, where an

$$g^*(\omega) = \int_{0}^{\infty} dt \, e^{-i\omega^* t} G(t) = g(-\omega^*) \qquad (5)$$

asterisk indicates complex conjugation. The function $g(\omega)$ is called the frequency-dependent response; it describes the response to a simple harmonic (pure tone) cause: if the cause is given by Eq. (6), then Eq. (1) implies that its effect is

$$C(t) = \cos(\omega t - \eta), \quad \omega \text{ real} \qquad (6)$$

given by Eq. (7), where c.c. means complex conju-

gate and $|g|$ and δ are given by Eq. (8). One sees

$$
\begin{aligned}
E(t) &= \int_{-\infty}^{\infty} dt' \, G(t - t') \cos(\omega t' - \eta) \\
&= \tfrac{1}{2} \int_{-\infty}^{\infty} dt' \, G(t - t') e^{-i(\omega t' - \eta)} + \text{c.c.} \\
&= \tfrac{1}{2} e^{-i(\omega t - \eta)} \int_{-\infty}^{\infty} dt' \, G(t - t') e^{i\omega(t - t')} + \text{c.c.} \\
&= \tfrac{1}{2} e^{-i(\omega t - \eta)} g(\omega) + \text{c.c.} \\
&= \tfrac{1}{2} |g(\omega)| e^{-i(\omega t - \eta - \delta(\omega))} + \text{c.c.} \\
&= |g(\omega)| \cos(\omega t - \eta - \sigma(\omega))
\end{aligned}
\qquad (7)
$$

$$g(\omega) = |g(\omega)| e^{i\delta(\omega)} \qquad (8)$$

that the magnitude $|g|$ and phase δ of $g(\omega)$ are the amplitude and phase shift, respectively, of the simple harmonic effect relative to the cause. *See* FOURIER SERIES AND INTEGRALS; INTEGRAL TRANSFORM.

Causality. Causality, Eq. (2), has an important consequence for $g(\omega)$, namely the dispersion relations given in Eq. (14) below. Their derivation will be outlined, using Cauchy's formula and other properties of analytic functions of a complex variable. Let ω take on complex values, $\omega = \text{Re } \omega + i \text{ Im } \omega$. Then $g(\omega)$ as given by Eq. (3) is an analytic function of ω as long as the integral, and its derivative with respect to ω, exist (that is, are finite). For any stable system, $G(t)$ is bounded in magnitude for all values of t. It follows that the integral in Eq. (3) exists for all ω such that $\text{Im } \omega > 0$ ("in the upper half ω-plane") then the magnitude of the factor $e^{i\omega t}$ of the integrand falls exponentially with increasing t according to Eq. (9).

$$|e^{i\omega t}| = e^{-(\text{Im } \omega)t} \qquad (9)$$

Since, on the other hand, for decreasing t this factor rises exponentially, the existence of the integral depended on causality, Eq. (2). So, causality implies that $g(\omega)$ is an analytic function in the upper half ω-plane. *See* COMPLEX NUMBERS AND COMPLEX VARIABLES.

Consequently one may write Cauchy's formula for $g(\omega)$, Eq. (10), where the integration contour C

$$g(\omega) = \frac{1}{2\pi i} \int_{C} d\omega' \, g(\omega')/(\omega' - \omega) \qquad (10)$$

is any simple loop, traversed counterclockwise, which surrounds the point ω and which lies in the upper half ω'-plane (the region where $g(\omega)$ is an analytic function). Letting the point ω approach the contour C, the Cauchy formula comes to the form Eq. (11), where P means principal value.

$$g(\omega) = \frac{1}{2\pi i} \left[\text{P} \int_{C} \frac{d\omega'}{\omega' - \omega} g(\omega') + i\pi g(\omega) \right] \qquad (11)$$

[Roughly speaking, it means: omit from the range of integration the interval of C which lies within a distance ϵ on either side of the singular point $\omega' = \omega$, and then take the limit $\epsilon \to 0$.] Solving for $g(\omega)$ gives Eq. (12), and then taking real and imaginary parts gives Eqs. (13). HT means Hilbert transform

$$g(\omega) = \frac{-i}{\pi} \text{P} \int_{C} \frac{d\omega'}{\omega' - \omega} g(\omega') \equiv -i\text{HT}[g] \qquad (12)$$

$$\text{Re } g = \text{HT}[\text{Im } g] \qquad \text{Im } g = -\text{HT}[\text{Re } g] \qquad (13)$$

(using the contour C), and Re g and Im g are the real and imaginary parts of g; these relations hold for any function $g(\omega)$ which is analytic inside the contour C. Let now the contour C be expanded until it encloses the entire upper half ω'-plane. If $g(\omega)$ vanishes at large ω the only part of the contour which contributes to the HT integral runs along the real ω'-axis. Finally, using crossing, Eq. (5), the integral can be expressed as an integral over just positive ω', Eqs. (14). These are the disper-

$$\operatorname{Re} g(\omega) = \frac{2}{\pi} P \int_0^\infty d\omega'\, \omega'\, \operatorname{Im} g(\omega')/\omega'^2 - \omega^2)$$

$$\operatorname{Im} g(\omega) = \frac{-2\omega}{\pi} P \int_0^\infty d\omega'\, \operatorname{Re} g(\omega')/(\omega'^2 - \omega^2) \qquad (14)$$

sion relations for the frequency-dependent response function $g(\omega)$. They say that $\operatorname{Re} g$ and $\operatorname{Im} g$ are not independent functions; given one of them, the other is determined.

Resonance. Although a causal $g(\omega)$ is analytic in the upper ω-plane, it can, and usually does, have singularites in the lower ω-plane. For instance, $g(\omega)$ may have a pole at $\omega = \omega_0$ (and to satisfy Eq. (5) likewise at $\omega = -\omega_0^*$); causality requires that $\operatorname{Im} \omega_0 < 0$. In Eq. (15) such a pole term of $g(\omega)$ is

$$g(\omega) = \frac{A}{\omega - \omega_0} + \frac{A}{-\omega - \omega_0^*} + \cdots$$

$$= \frac{\operatorname{Re} A + i \operatorname{Im} A}{\omega - \omega_R + i^{1/2}\gamma} + \frac{\operatorname{Re} A - i \operatorname{Im} A}{-\omega - \omega_R - i^{1/2}\gamma} + \cdots \qquad (15)$$

written out, with the notation $\operatorname{Re} \omega_0 = \omega_R$, $\operatorname{Im} \omega_0 = -\frac{1}{2}\gamma$. If γ is small, then $|g(\omega)|^2$ as a function of real ω peaks sharply at $\omega = \omega_R$ with a width (defined as the interval of ω between the half height points, where $|g(\omega)|^2$ is half as big as the peak) equal to γ. This is called a resonance of the response function $g(\omega)$; ω_R is its position and γ is its width. *See* RESONANCE (ACOUSTICS AND MECHANICS); RESONANCE (ALTERNATING-CURRENT CIRCUITS).

The consequence of a resonance for the effect $E(t)$ is found by using Eq. (15) in Eqs. (1) and (4). If the cause $C(t)$ turns off rapidly enough, then at large t the exponential factor in the integrand of Eqs. (16) decreases into the lower ω-plane, and so $E(t)$

$$= \int dt'\, C(t') \frac{1}{2\pi} \int d\omega\, e^{-i\omega(t-t')} \frac{A}{\omega - \omega_0} + \text{c.c.} + \cdots$$

$$\xrightarrow{t \to \infty} -iA \int dt'\, C(t') e^{-i\omega_0(t-t')} + \text{c.c.} + \cdots$$

$$= [-iA e^{-i\omega_R t} \int dt'\, C(t') e^{i\omega_0 t'} + \text{c.c.}] e^{-1/2\gamma t} + \cdots$$

$$= C_1 \cos(\omega_R t + C_2) e^{-1/2\gamma t} + \cdots \qquad (16)$$

where C_1, C_2 are constants

the ω-intergration can be done by residues. The result is that the effect $E(t)$ has an exponentially decaying term whose time constant (for E^2) is γ^{-1}; that is, the width of the resonance peak of $|g(\omega)|^2$ and the decay constant of $E(t)^2$ are reciprocals of one another. In more physical terms, the decaying oscillation of $E(t)$ reflects a (decaying) normal mode of oscillation, a "ringing," of the system.

If $\operatorname{Im} \omega_0$ were positive, contrary to causality, the calculation of Eqs. (16) would show the effect $E(t)$ to have an exponentially growing term at an early time, preceding the cause. As for $g(\omega)$, the

qualitative effect of causality is that at the resonance, where the magnitude $|g(\omega)|$ has a narrow peak, the phase of $g(\omega)$ [$\delta(\omega)$ in Eq. (7)] rises rapidly; if $\operatorname{Im} \omega_0$ were positive, contrary to causality, the phase would instead fall rapidly.

Dielectric constant and refractive index. The response of a macroscopically homogeneous dielectric (polarizable) medium to an electromagnetic plane wave of frequency ω is described by the dielectric constants $\epsilon_{||}(\omega)$ and $\epsilon_\perp(\omega)$. The subscripts distinguish whether the electric field is parallel or perpendicular to the direction of propagation of the plane wave. The discussion will be limited to $\epsilon_\perp(\omega)$, which will be referred to as simply $\epsilon(\omega)$. The refractive index $n(\omega)$ is simply related: $n^2 = \epsilon$. But $n(\omega)$ itself will not in general satisfy dispersion relations because if $\epsilon(\omega)$ has a zero in the upper ω-plane, $\sqrt{\epsilon} = n$ will have a square root singularity. *See* DIELECTRIC CONSTANT.

At $\omega = \infty$, $\epsilon(\omega)$ has the value 1 (the vacuum value). At $\omega = 0$, $\epsilon(\omega)$ is singular if the medium is conductive (that is, contains charges which are free to move), according to Eq. (17), where σ_0 is the

$$\epsilon(\omega) \xrightarrow{\omega \to 0} \frac{4\pi i \sigma_0}{\omega} - \frac{\omega_0^2}{\omega^2} \quad \omega_0^2 = n_v 4\pi e^2/m \qquad (17)$$

static conductivity and ω_0 is the plasma frequency; n_v is the number density of free charges of charge e and mass m. Thus the function given in Eq. (18)

$$g(\omega) = \epsilon(\omega) - 1 - \frac{4\pi i \sigma_0}{\omega} + \frac{\omega_0}{\omega^2} \qquad (18)$$

has the properties assumed for $g(\omega)$ above and can be substituted for it in Eq. (14), resulting in the dispersion relations, Eqs. (19). In Eqs. (19), terms

$$\operatorname{Re} \epsilon(\omega) = 1 - \frac{\omega_0^2}{\omega^2} + \frac{2}{\pi} \int_0^\infty d\omega'$$

$$\frac{\omega' \operatorname{Im} \epsilon(\omega') - \omega \operatorname{Im} \epsilon(\omega)}{\omega'^2 - \omega^2} \qquad (19a)$$

$$\operatorname{Im} \epsilon(\omega) = \frac{4\pi\sigma_0}{\omega} - \frac{2}{\pi}\omega \int_0^\infty d\omega'$$

$$\frac{\operatorname{Re} \epsilon(\omega') + \omega_0^2/\omega'^2 - \operatorname{Re} \epsilon(\omega) - \omega_0^2/\omega^2}{\omega'^2 - \omega^2} \qquad (19b)$$

have been added to the integrands in order to make them nonsingular; these terms do not alter the integrals because $P \int d\omega'/(\omega'^2 - \omega^2) = 0$.

By setting ω equal to special values, interesting results are obtained; for simplicity these will be given for the case of a nonconductor, where $\sigma_0 = \omega_0 = 0$. Setting $\omega = 0$ in Eq. (19a) results in an expression for the static dielectric constant, Eq. (20). Taking ω large in Eqs. (19) gives Eqs. (21),

$$\epsilon(0) = 1 + \frac{2}{\pi} \int_0^\infty d\omega\, \operatorname{Im} \epsilon(\omega)/\omega \qquad (20)$$

$$\operatorname{Re} \epsilon(\omega) \longrightarrow 1 - \omega_\infty^2/\omega^2 \qquad (21a)$$

where $\omega_\infty^2 = \frac{2}{\pi} \int_0^\infty d\omega\, \omega\, \operatorname{Im} \epsilon(\omega)$

$$\operatorname{Im} \epsilon(\omega) \longrightarrow 4\pi\sigma_\infty/\omega \qquad (21b)$$

where $4\pi\sigma_\infty = \frac{2}{\pi} \int_0^\infty d\omega\, [\operatorname{Re} \epsilon(\omega) - 1]$

which are sum rules. The first of these relates an integral over all frequencies of the absorptivity, $\operatorname{Im} \epsilon$, to the quantity ω_∞^2, for which a direct calculation gives the value $\omega_\infty^2 = \Sigma\, n_v 4\pi e^2/m$; where the sum is over all kinds of charges in the medium, in contrast to the formula for the plasma frequency ω_0, Eq. (17); at infinite frequency all of the particles of the medium behave as though free. *See* ABSORPTION OF ELECTROMAGNETIC RADIATION.

Forward scattering amplitude. It can be shown by arguments like those used above that the forward scattering amplitude $f(\omega)$ of light incident on an arbitrary body satisfies the dispersion relation given in Eq. (22) [the less interesting relation

$$\mathrm{Re}\, f(\omega) = f(0) + \frac{\omega^2}{2\pi^2 c}\int_0^\infty d\omega'\, \frac{\sigma_T(\omega') - \sigma_T(\omega)}{\omega'^2 - \omega^2} \quad (22)$$

where $f(0) = -Q^2/Mc^2$

giving $\mathrm{Im}\, f$ in terms of $\mathrm{Re}\, f$ has been omitted], where Q is the charge, M is the mass of the body, and c is the speed of light. In Eq. (22) $\mathrm{Im}\, f(\omega)$ has been expressed in terms of the total cross section σ_T by the optical theorem, Eq. (23); this formula

$$\mathrm{Im}\, f(\omega) = (k/4\pi)\sigma_T(\omega)$$

$$(23)$$

where $k =$ wave number
is true for the scattering of waves of any sort.

Equations (22) and (14) can be related by using the "optical potential" formula, Eq. (24), which

$k^2(\text{in medium})/k^2(\text{in vacuum})$

$$= n^2 - 1 + 4\pi n_v f(\omega)/k^2 \quad (24)$$

relates the forward scattering amplitude $f(\omega)$ of a wave (of any sort) of frequency ω and wave number k incident on a body to the index of refraction n of that wave in a medium consisting of a random arrangement of such bodies with mean number density n_v. However, Eq. (24) is valid only for a dilute medium, whereas Eqs. (14) and (22) are both exact if causality is true.

An interesting consequence of Eq. (22) is that a system consisting of the electromagnetic field and point charges, obeying classical mechanics, cannot be causal. For such a system it can be shown that $f(\omega)$ vanishes for $\omega = \infty$; using this in Eq. (22) gives Eq. (25), which is untrue since both

$$0 = f(0) - (2\pi^2 c)^{-1}\int_0^\infty d\omega\, \sigma_T \quad (25)$$

terms on the right-hand side are negative. In the quantum theory, by contrast, $\sigma_T > 0(1/\omega)$ and so no formula like Eq. (25) follows from causality. *See* NONRELATIVISTIC QUANTUM THEORY.

Elementary particle theory. In the quantum theory the scattering amplitudes of matter waves satisfy dispersion relations similar to Eq. (22) if causality holds. This gives the possibility of experimentally testing causality up to very high frequencies (corresponding to very short time intervals) by using, for instance, protons of energy up to 500 GeV, incident on protons in a target, at Fermilab. Since the right-hand side of Eq. (22) requires knowledge of σ_T at all ω, one can strictly say only that the observed $f(\omega)$ is consistent with causality and a reasonable extrapolation of σ_T beyond observed energies.

Not only is there no experimental evidence against causality, but causality has always been a property of relativistic quantum field theory (elementary particle theory). The following is essentially equivalent to causality: Considered as

functions of the Lorentz invariant scalars formed from the 4-momenta of the interacting particles (for instance, the total energy and the momentum transfer in the center-of-mass frame, in the case of a two-body scattering amplitude), the S-matrix elements (scattering amplitudes) of quantum field theory are analytic functions. More precisely, they are analytic except for singularities on Landau surfaces in the space of the (complex) arguments, the Lorentz scalars. These singularities are branch points, and if the amplitude is analytically continued around one, the amplitude becomes changed by a certain amount.

On the one hand, knowing this change allows the amplitude itself to be determined, by use of Cauchy's formula [a dispersion relation like Eq. (22) is a simple example of this]. On the other hand, the change is given in terms of other S-matrix elements. [The optical theorem, Eq. (23), is a simple example of this; the point is that σ_T is a sum (and integral) over products of two amplitudes.] These S-matrix equations, resulting from the combination of the analyticity and the unitarity of the S-matrix, are the subject of S-matrix theory. *See* RELATIVITY.

These equations can to some extent replace the equations of motion of quantum field theory. They have the advantage over the only known systematic solution of the equations of motion, namely perturbation theory (Feynman diagrams), that they involve only S-matrix elements, that is, the amplitudes of real, observable processes. However, there is no known way to find the solution of the S-matrix equations which corresponds to a given set of field theory equations of motion.

It was the original hope of many workers in S-matrix theory that the equations of motion were in fact irrelevant and that the S-matrix was determined by a few properties of low-mass particles (in fact perhaps by none, in the extreme view termed bootstrapping). However, the successes of quantum chromodynamics (QCD) in describing and predicting many phenomena in high-energy physics seems to show that a specific field theory, whose elementary quanta (quarks and gluons) are not observable as particles, underlies the S-matrix of "elementary" particle interactions. Of course, the S-matrix equations are true in any case, and will remain an important tool in the description and correlation of experimental data. *See* ELEMENTARY PARTICLE; QUANTUM CHROMODYNAMICS; QUANTUM FIELD THEORY; SCATTERING EXPERIMENTS (NUCLEI); SCATTERING MATRIX.

[CHARLES J. GOEBEL]

Bibliography: G. F. Chew, *The Analytic S Matrix*, 1966; R. Churchill, *Complex Variables and Applications*, 1960; R. J. Eden, *High Energy Collisions of Elementary Particles*, 1967; R. J. Eden et al., *The Analytic S Matrix*, 1966; R. Good and T. Nelson, *Classical Theory of Electric and Magnetic Fields*, 1971; A. Martin and T. Spearman, *Elementary Particle Theory*, 1970; P. Roman, *Advanced Quantum Theory*, 1965.

Displacement

When an object is moved from one position to another, it is said to be displaced, and the linear distance from the initial to the final position, regardless of the length of path followed, is called the displacement. The displacement is always in a particular direction, and consequently displacement is a vector quantity involving direction as well as magnitude. In rectilinear motion the magnitude of the displacement is also the length of path or the distance traversed, but since length of path does not involve direction, it is a scalar quantity. For instance, the shortest distance or length of path is the same from Washington to Philadelphia as it is from Philadelphia to Washington, but the displacement of an object in one direction is entirely different from its displacement in the opposite direction.

When a body is rotated about any axis, it is said to undergo angular displacement. Angular displacement is commonly measured in radians or degrees, one radian being equal to 57.3°. *See* ROTATIONAL MOTION.   [R. D. RUSK]

Displacement current

The name given by J. C. Maxwell to the term $\partial \mathbf{D}/\partial t$ which must be added to the current density $\mathbf{i}$ to extend to time-varying fields A. M. Ampère's magnetostatic result that $\mathbf{i}$ equals the curl of the magnetic intensity $\mathbf{H}$. In integral form this result is given by Eq. (1), where the unit vector $\mathbf{n}$ is perpen-

$$\oint \mathbf{H} \cdot d\mathbf{s} = \int_S \left(\mathbf{i} + \frac{\partial \mathbf{D}}{\partial t} \right) \cdot \mathbf{n}\, dS \qquad (1)$$

dicular to the surface dS. The concept of displacement current has important consequences for insulators and for free space where $\mathbf{i}$ vanishes. For conductors, however, the difference between Eq. (1) and Ampère's result is negligible. *See* AMPERE'S LAW; MAXWELL'S EQUATIONS.

In order to show that the displacement term is essential, consider a parallel-plate capacitor charged by a circuit carrying an alternating current. Let a closed curve s encircle one of the charging wires and be the boundary of two surfaces, S_1 which passes through the capacitor gap and S_2 which cuts the charging wire. By Gauss' electric flux theorem, the charge Q on the plate and wire between S_1 and S_2 is given by Eq. (2), where the

$$Q = \int_{S_1} \mathbf{D} \cdot \mathbf{n}\, dS - \int_{S_2} \mathbf{D} \cdot \mathbf{n}\, dS \qquad (2)$$

normal is taken in the direction of current flow in both S_1 and S_2. The current I or $\int \mathbf{i} \cdot \mathbf{n}\, dS$ equals $\partial Q/\partial t$ so that Eq. (3) holds. Thus, the inclusion of the displacement current is needed to make Eq. (1) valid for any surface S bounded by s.

$$\int_{S_1} \frac{\partial \mathbf{D}}{\partial t} \cdot \mathbf{n}\, dS = \int_{S_2} \left(\mathbf{i} + \frac{\partial \mathbf{D}}{\partial t} \right) \cdot \mathbf{n}\, dS = \oint \mathbf{H} \cdot d\mathbf{s} \qquad (3)$$

If one defines current as a transport of charge, the term displacement current is certainly a misnomer when applied to a vacuum where no charges exist. If, however, current is defined in terms of the magnetic fields it produces, the expression is legitimate. In a dielectric, where an electric field produces a displacement of the negative charges with respect to the positive ones, the name has meaning. Maxwell had this sort of picture, even for a vacuum, where he postulated a polarizable ether. *See* ETHER HYPOTHESIS.

[WILLIAM R. SMYTHE]

Bibliography: *See* MAXWELL'S EQUATIONS.

Distribution (probability)

The results of a series of independent trials, random variables, or errors often occur in fairly regular and predictable patterns. These patterns can be expressed mathematically, and the most important of them are called the binomial, normal, and Poisson distributions.

Binomial distribution. Consider n independent trials, each of which results in success S or failure F, with corresponding probabilities p and $q = 1 - p$. Denote by S_n the number of successes. Because there are $\binom{n}{k}$ possible ways to select k places for S and $n - k$ places for F, the probability distribution of the random variable S_n is given by

$$P\{S_n = k\} = \binom{n}{k} p^k q^{n-k}$$

where $k = 0, 1, \ldots, n$. This is the binomial distribution. Its expectation is np, its variance npq. *See* PROBABILITY.

If one lets a random variable X_k equal 1 or 0 according as the kth trial results in S or F, then $S_n = X_1 + \cdots + X_n$. The binomial distribution can therefore be approximated by the normal distribution in accordance with the central limit theorem. This special case is known as the DeMoivre-Laplace theorem; putting

$$x_k = (k - np)(npq)^{-1/2}$$

it asserts that

$$P\{S_n = k\} \sim (2\pi)^{-1/2} \exp(-x_k^2/2)$$

$$P\{a < S_n < b\} \sim (2\pi)^{-1/2} \int_{x_{a-1/2}}^{x_{b+1/2}} \exp(-t^2/2)\, dt$$

with a percentage error tending to 0 as $n \to \infty$ provided x_a and x_b are restricted to a fixed interval. (With $p = q = 1/2$ these formulas are useful for the evaluation of binomial coefficients and their sums.)

Normal distribution. The standard normal density is defined for $-\infty < x < \infty$ by

$$\phi(x) = (2\pi)^{-1/2} e^{-x^2/2}$$

The standard normal distribution function or error function $\Phi(x)$ is its integral from $-\infty$ to x. The normal distribution function with mean m and variance s^2 is $\Phi[(x - m)/s]$; its density is $(2\pi)^{-1/2} s^{-1} e^{-(x - m)^2/(2s^2)}$. As x goes from $-\infty$ to ∞, the function increases from 0 to 1. It plays an important role in many fields. In particular, $u(t, x; \xi) = 2\pi^{-1/2} t^{-1/2} e^{-(x - \xi)^2/rt}$ is the fundamental solution of the heat (or diffusion) equation $u_t = u_{xx}$ and represents the heat distribution on the x axis at time t caused by a unit heat source initially concentrated at $x = \xi$. Probabilistically, this represents the transition probabilities in the Wiener process.

A random variable whose distribution is normal is called normal or Gaussian. The sum of two independent Gaussian variables is again Gaussian; hereby the means and variances add. Analytically this means that the convolution of two normal distributions is again normal. This property characterizes the normal distributions among distributions with finite variances.

Central limit theorem and error theory. These are the best-known and most important applications of the normal distribution. The nature of the central limit theorem is best seen from the simplest special case: If $X_1, X_2, \ldots$ are independent random variables having a common distribution with mean m and variance s^2, their sum $S_n = X_1 + \cdots + X_n$ has mean $\mu = nm$ and variance $\sigma^2 = ns^2$; the corresponding standardized variable $S_n^* = (S_n - \mu)\sigma^{-1}$ has a probability distribution $P\{S_n^* \leq x\}$ tending to $\Phi(x)$ as $n \to \infty$; in other words, the distribution of S_n gets close to $\Phi[(x - \mu)\sigma^{-1}]$.

The striking feature of this result is that an essential property of a sum of many independent components is independent of the nature of these components. The general central limit theorem shows this to be true under much wider conditions: The distribution of the sum S_n tends to normality even if the distributions of the components X_k vary with k, provided only that the components are likely to be of the same order of magnitude (so that no component has individually a noticeable effect on the sum). It is not even necessary that the X_k be independent.

Many empirical quantities (for example, the amount of water in a reservoir, certain inherited characteristics such as height, and the experimental error of physical measurements) represent the cumulative effect of many small components, and the statistical fluctuations of such quantities may be expected to follow the normal distribution. In particular, under the authority of K. F. Gauss, it has been assumed for a long time that experimental errors are approximately normally distributed. Modern research has shown the limitations of this assertion. Many distributions appear to the untrained eye as nearly normal, but refined statistical tests prove that even the finest physical measurements depart considerably from normality, and that assignable causes (that is, large contribution components) can be discovered statistically. This discovery shows the questionable character of the classical methods to predict the probable experimental error. In industrial quality control the departures from normality are used efficiently to discover assignable causes and thus to spot coming trouble at an early stage.

Misinterpretation of theory. Much harm has been caused by the widespread misinterpretation of the limit theorems in probability and of the meaning of statistical equilibrium in stochastic processes. The situation may be explained in terms of a coin-tossing game in which H and T count 1 and -1, respectively. Here $X_k = \pm 1$ with probability 1/2, and $m = 0$, $s^2 = 1$. The operational meaning of the central limit theorem in this case is as follows. Fix a large n. Repeat the same game of n tossings independently many times. One would then expect that in about 50% of the cases $S_n > 0$, in about 25% of the cases $S_n > 0.67n^{1/2}$, in about 16% of the cases $S_n > 2n^{1/2}$, and so on. What the central limit does not say is that in one game about half of the sums $S_1, \ldots, S_n$ will be positive. In fact, the arcsine law shows the opposite to be true: It is much more probable that all $S_j > 0$ than that there be equally many positive and negative ones.

Multivariate normal distribution. The previous theory carries over without essential changes to n dimensions. The n-dimensional normal density is defined by $(2\pi)^{-n/2} D^{1/2} e^{-Q(x_1, \ldots, x_n)/2}$, where Q is a positive definite quadratic form with determinant D; the matrix of variances and covariances is the reciprocal of the matrix of Q. If the n-dimensional joint distribution of the random variables $X_1, \ldots, X_n$ is normal, then each X_j is normal. The converse assertion is false, even though found in textbooks. The multivariate normal distribution is important for stationary stochastic processes. *See* STOCHASTIC PROCESS.

Poisson distribution. The Poisson distribution with parameter λ is the probability distribution of a random variable assuming the values 0, 1, 2, ... with probabilities $p_k(\lambda) = e^{-\lambda} \lambda^k/k!$. Both its expectation and variance equal λ. This is one of the most important distributions; it plays a basic role in the theory of stochastic processes and in many applications. A full understanding of its character can be obtained only from a postulational derivation and from the consideration of its many generalizations. However, much can be gained by the following elementary approach starting from the binomial distribution.

Consider a large number n of independent trials, each of which results in success or failure with probabilities p and $q = 1 - p$. Ordinarily, interest is restricted to the case where p is very small, but the expected number of successes $np = \lambda$ is of moderate size. Typical examples may be obtained by considering centenarians, color-blind people, or triplets in a large population; the defectives in an allotment of screws or fuses; or the wrong calls among all calls arriving during a day at a busy telephone. The number of successes in the n trials is a random variable with the binomial distribution, but under the present circumstances the binomial is close to the Poisson distribution and may be replaced by it. In fact, the probability of no success is $q^n = (1 - \lambda/n)^n$, which is close to $e^{-\lambda}$, the first term of the Poisson distribution. Now the ratio of the kth to the $(k-1)$th term in the binomial distribution approaches that of the Poisson distribution.

This reasoning explains why the statistical fluctuations in the phenomena cited above follow approximately the Poisson distribution. In other circumstances the Poisson distribution appears, not as an approximation, but as the exact expression of a law of nature. This is true in particular of processes where certain events, such as radioactive disintegrations, mutations, power failures, and accidents, occur in time in such a way that (1) the probability that an event occurs during any given time interval of length dt is, asymptotically, λdt; and (2) there is no interaction or aftereffect between nonoverlapping time intervals. Under these assumptions a time interval of length t can be divided into $n = t/dt$ subintervals, each representing a trial in which the probability of success is close to λdt; a change to the limit $dt \to 0$ gives the Poisson expression $p_k(\lambda t)$ as exact probability for the occurrence of k events during a time interval of length t. A similar argument applies to random distributions of points in space, with t interpreted as volume; typical examples are stars, flaws of material, raisins in a cake, and animal litters in a field. Thus "perfect randomness" of a chance aggregate of points in space or time usually means that the fluctuations obey the Poisson law. *See* ANALYSIS OF VARIANCE; STATISTICS.

[WILLIAM FELLER]

Bibliography: S. A. Book, *Essentials of Statis-*

tics, 1978; W. Feller, *An Introduction to Probability Theory and Its Applications*, vol. 1, 3d ed., 1968, vol. 2, 2d ed., 1971; R. L. Wine, *Beginning Statistics*, 1976.

Domain (solids)

A region in a solid within which elementary atomic or molecular magnetic or electric moments are uniformly arrayed.

Ferromagnetic domains are regions of parallel-aligned magnetic moments. Each domain may be thought of as a tiny magnet pointing in a certain direction. A ferromagnet is generally composed of many such domains, pointing in many different directions. For the origin and fundamental properties of ferromagnetic domains *see* FERROMAGNETISM; for the effect of domains on magnetization *see* MAGNETIZATION.

Antiferromagnetic domains are regions of antiparallel-aligned magnetic moments. They are associated with the presence of grain boundaries, twinning, and other crystal inhomogeneities.

Ferroelectric domains are electrical analogs of ferromagnetic domains. See FERROELECTRICS.

[ELIHU ABRAHAMS; FREDERIC KEFFER]

Doppler effect

A change in the observed frequency of sound, light, or other waves, caused by motion of the source or of the observer. A familiar example for sound waves is the increase (decrease) in pitch of a train whistle as the train approaches (passes). The optical phenomenon is shown in the altered frequencies of spectral lines in the light emitted from a moving star. If the star and the Earth are moving closer to one another, more light pulses are received in a given time interval, and the color emitted from the star appears to be shifted toward the violet end of the spectrum. When the distance between the Earth and the star is increasing, the observed light is shifted toward the red end of the spectrum. The color shifts of remote galaxies are taken as evidence that the universe is expanding.

In astronomy, color differences between the approaching sides and receding sides have been used to compute the rotation of the Sun and planets.

Acoustical Doppler effect. Acoustical observations of a moving source emitting sound at a constant frequency make its pitch appear greater when the source is approaching the listener, and smaller when the source-to-listener distance is increasing. The effect is based on the fact that the listener perceives as frequency the number of sound waves arriving per second.

The acoustical Doppler effect deals with cases of relative motion between the listener and the source, and includes the effects of motion of the medium itself relative to both the source and the listener. The wave velocity u of the sound in the medium is a property of the medium and its value is referred to that medium. The wavelength λ, frequency f, and velocity u are related in wave propagation by the equation $u = f\lambda$.

A distinction needs to be drawn between the case in which the source moves relative to the listener fixed in the medium, and the case in which the listener moves with respect to the source fixed in the medium.

In the first case, if the source moves toward the fixed observer with a velocity v_S, waves emitted with a frequency f_S appear to have their wavelength shortened in the ratio $(u - v_S)/u$, because of a crowding of the waves (Fig. 1) which, however, still arrive at the listener with a velocity u.

In the second case, if the listener moves toward the fixed source, the waves appear to him to arrive with a velocity $(u + v_L)$. The wavelength of the sound in the medium is unchanged in this case and is equal to that measured when both the listener and the source are fixed in the medium.

Consider now the effect of the velocity of the medium relative to the listener and the source. If v_M is the component of this velocity taken positive in the direction from the listener to the source, and if v_L and v_S are the velocity components along the line joining the listener to the source and are now taken to be positive in the direction from the listener to the source, then the general equation relating the observed frequency f_L and the source frequency f_S is given by Eq. (1).

$$\frac{f_L}{u + v_L - v_M} = \frac{f_S}{u + v_S - v_M} \quad (1)$$

Optical Doppler effect. This phenomenon seems at first to be analogous to the acoustical Doppler effect, but the causes, detailed effects, and explanation of the optical Doppler effect are fundamentally different and result from the relativistic behavior of light. *See* LIGHT; RELATIVITY.

Differences between the two effects. Three fundamental differences exist between the acoustical and the optical Doppler effects.

1. The optical frequency change does not depend upon whether it is the source or the observer that is moving with respect to the other, whereas the acoustical frequency change is different in the two cases.

2. No effect is observable in the acoustical case when the source, or the observer, moves at right angles to the line connecting the source and the observer. An optical frequency change is observable under such conditions.

3. The motion of the medium through which the waves are propagated does not affect the observed optical frequency, whereas it does affect the observed acoustical frequency.

Light source in motion. The mathematical expressions of the observable effects involving light and other electromagnetic waves are arrived at by noting that the propagation of a given plane wave must be described by the same law in the source frame and the observer frame, according to the relativistic principle of equivalence. Accordingly, the equation of propagation of the plane wave written for the source frame of coordinates is transformed to the observer frame of coordinates with the help of the well-known Lorentz transformations, and the relevant factors on the two sides of the resulting equation, identifying the descriptions in the two frames, are identified. The result is expressed in Eqs. (2) and (3). These equations re-

$$f_O = \frac{f_S \sqrt{1 - (v^2/c^2)}}{1 - (v/c)\cos\theta_O} \quad (2)$$

$$\cos\theta_O = \frac{\cos\theta_S + (v/c)}{1 + (v/c)\cos\theta_S} \quad (3)$$

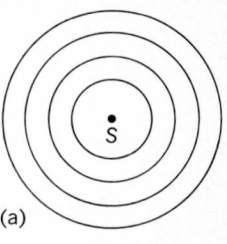

(a)

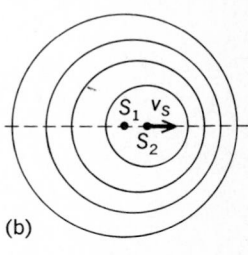

(b)

Fig. 1. Spherical waves from a point source. (*a*) At rest; (*b*) in motion.

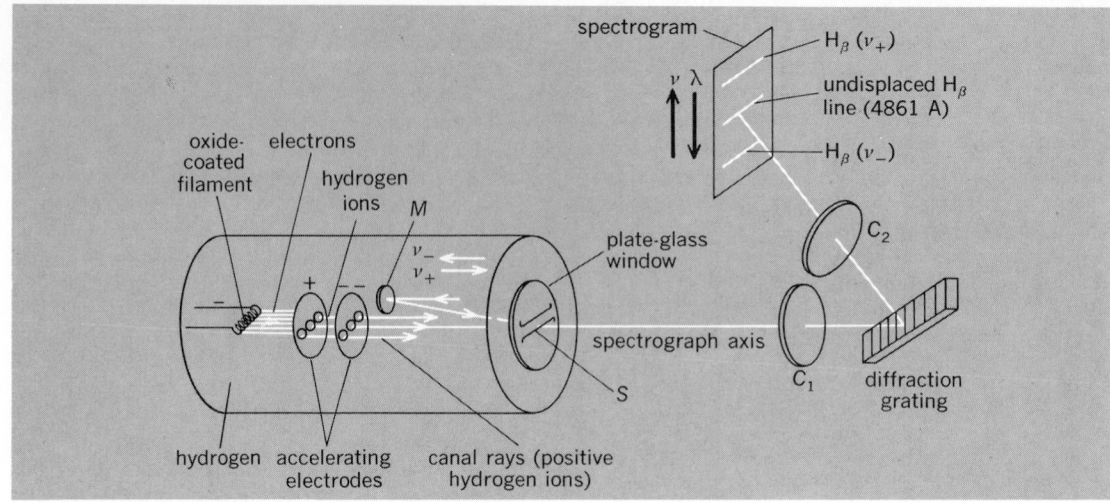

Fig. 2. Ives and Stilwell experiment. C_1 and C_2, collimating and camera lenses; S, spectrograph entrance slit; M, concave mirror focused on slit, observing light from hydrogen ions moving from slit. (*G. W. Stroke*)

late the frequency f_O and angle θ_O measured in the observer frame to the frequency f_S and angle θ_S that would be measured in the source frame, under the conditions in which the source frame is measured (in the observer frame) to move with a velocity v relative to the observer frame; c is the velocity of light in free space.

Examination of the frequency relation shows that it incorporates two factors: a purely relativistic direction-independent factor, $f_O \sim f_S \cdot \sqrt{1 - (v^2/c^2)}$, according to which the observed frequency will be smaller than the source frequency regardless of the apparent direction of motion of the source (transverse Doppler effect); and a direction-dependent factor, $f_O \sim f_S/[1 - (v/c) \cos \theta_O]$, showing a further dependence on the direction of relative motion. Like the acoustical Doppler effect, this factor is understandable on the basis of classical arguments.

The part involving the direction of relative velocity (radial Doppler effect) can be derived by counting as the observed frequency f_O the number of waves arriving in a time interval dt_O, corresponding to the difference in the times of arrival of a first wave and of a last wave traveling with a velocity c toward the observer. The waves are emitted at a frequency f_S by a source traveling with a velocity v at an angle θ_O with respect to the observer. During a given time interval dt_S the source emits a total of $f_S \, dt_S$ waves. The relativistic velocity-dependent part is then included by noting that the source frequency will appear to be $f_S\sqrt{1 - (v^2/c^2)}$ according to the special theory of relativity.

H. E. Ives and G. R. Stilwell (1938), skeptical as to the conclusions of the special theory of relativity, set out to verify the velocity-dependent part of the frequency shift (transverse Doppler effect) observed at zero angle ($\theta_O = 0$). They measured the wavelengths of the H_β line in the direction of motion ($\theta_O = 90°$) of hydrogen canal rays at 18,000 volts (Fig. 2) and in the opposite direction ($\theta_O = -90°$) for which the frequencies are respectively $f_{O1} = f_S\sqrt{1 - \beta^2}/1 - \beta)$ and $f_{O2} = f_S\sqrt{1 - \beta^2}/(1 + \beta)$, where $\beta = v/c$. They determined the average $f_O = (f_{O1} + f_{O2})/2 = f_S/(\sqrt{1 - \beta^2})$. This result was found to be in accord with the theoretical value $f_O = f_S/\sqrt{1 - \beta^2}$, thus providing a direct proof of the "dilatation of time," according to which the observer thinks that the source period T_S is $T_O/\sqrt{1 - \beta^2}$ and therefore greater than the observer period T_O.

[GEORGE W. STROKE]

Bibliography: A. Einstein et al., *The Principle of Relativity*, 1923; T. P. Gill, *The Doppler Effect*, 1965; W. C. Michels, Phase shifts and the Doppler effect, *Amer. J. Phys.*, 24(2):51–53, 1956; R. Resnick and D. Halliday, *Physics*, 3d ed., 1978; A. J. W. Sommerfeld, *Lectures on Theoretical Physics*, vol. 4, 1954; F. W. Sears et al., *University Physics*, 5th ed., 1976.

Dynamic nuclear polarization

The creation of assemblies of nuclei whose spin axes are not oriented at random, and which are in a steady state that is not a state of thermal equilibrium. Under commonly occurring conditions, the spin axes of nuclei (with nonzero spin) are oriented at random; where this is not so, the nuclei are said to be polarized. Assemblies of polarized nuclei are not in a state of thermal equilibrium except under rather extreme conditions (for example, temperatures below 10 millikelvins and magnetic fields greater than several teslas), and therefore schemes have been devised to produce polarized assemblies, in a steady state which is not a state of thermal equilibrium, under less extreme conditions of temperature and so forth. Such schemes constitute dynamic nuclear polarization.

Among the many applications of polarized nuclei are the following. Nuclear forces are spin-dependent, and although the spin-dependent part can be found by using unpolarized assemblies, the experiments are simpler and their interpretation is clearer if polarized nuclei are used. Assemblies of polarized nuclei have a lower geometrical symmetry than assemblies of randomly oriented nuclei, and so these have been used to investigate the fundamental symmetries of nature. Polarized nuclei have been used to enhance the signal in free precession magnetometers and similar instruments, and the use of an assembly of polarized nuclei as a gyroscope has also been suggested. *See* NUCLEAR ORIENTATION; PARITY (QUANTUM MECHANICS); SPIN (QUANTUM MECHANICS).

The simplest and most common parameter used to define the state of polarization is called the "polarization" or vector polarization **P**. It is a vector whose component along any given axis (for example, the z axis) is the average value of the projection on that axis of a unit vector parallel to the nuclear spin axis. Often, nuclei lie with their spins pointing along the z axis in equal numbers in both directions, but none lies in the x-y plane. Such a system is not random, but **P** = 0. The parameter used to quantify this arrangement is the average value of $(3/2 \cos^2 \Theta - 1/2)$ and is one component of the so-called tensor polarization. In general, the average value of any zonal harmonic is a measure of a kind of polarization. It is a property of nuclear spin operators that harmonics of order greater than $2J$ vanish identically, where J is the nuclear spin. In particular, spin 0 nuclei cannot be polarized and spin 1/2 nuclei can have a vector polarization only. *See* ANGULAR MOMENTUM; SPHERICAL HARMONICS.

There are, in general, two common features of all dynamic nuclear polarization schemes. There is some external system or interaction, for example, an oscillating magnetic field, which disturbs the system; this is called the pumping mechanism. There are also processes which tend to restore the system to thermal equilibrium; these are called relaxation processes. However, the functions of these two processes are not easily distinguished, for, in many schemes, the relaxation mechanisms actually play a part in setting up the nuclear polarization, this being one of the stages back to thermal equilibrium.

Microwave pumping. Consider an electron spin and a nucleus of spin 1/2 (for example, a proton) in close association. Typically, the electron spin is part of a paramagnetic ion, and the proton is a hydrogen nucleus in a nearby water molecule. In a magnetic field of about 1 T, the energy levels of this combination are as shown in the illustration. M_e and m_p are, respectively, the electron and proton spin magnetic quantum numbers which specify

whether the electron and proton spins are oriented in the same direction as the magnetic field or in the opposite direction. An oscillating magnetic field is applied to induce transitions between pairs of levels. The two transitions A and B, in which only an electron spin is reversed, constitute the "allowed" transition. The two weaker transitions C and D, in which a nuclear spin is also reversed, occurring one at a higher and the other at a lower field (or frequency) than the allowed transition, are called forbidden transitions. *See* MAGNETIC RESONANCE; SELECTION RULES.

In the most common and most successful mechanism of this type for polarizing protons, one of the forbidden transitions (C, for instance) is excited strongly enough to saturate it. Then the populations of levels β and γ are equal, and are kept equal by the oscillating magnetic field. The strongest relaxation processes, which try to bring the system into thermal equilibrium, are associated with the strongest transitions A and B. As a result of these processes, fewer systems occupy level α than level γ, and similarly, fewer systems occupy level β than level δ. Thus more proton spins are oriented in one direction ($m_p = +1/2$) than in the opposite direction ($m_p = -1/2$), and a nuclear polarization results. By saturating the other forbidden transition, an equal and opposite polarization results. *See* MAGNETIC RELAXATION.

In another method the allowed transition is saturated, and so the populations of levels α and γ are equal, and similarly the populations of levels β and δ are equal. The strong relaxation processes, which involve only the inversion of an electron spin, are ineffective since they are swamped by the radio-frequency field. The "cross relaxation" processes, which connect level α with level δ (called the flip-flop relaxation), and which connect level β with level γ (called the flip-flip relaxation), are of different strengths. This is because in the flip-flop transition, spin angular momentum is conserved, but not so in the flip-flip transition. The relative populations of the levels are thus determined by the relative strengths of the two processes. If, as in the case of metals, the flip-flip transition can be neglected, the flip-flop process results in fewer systems occupying level α than level δ, and a nuclear polarization results. If the electron spin and the nuclear spin are coupled by magnetic dipole interaction, and direct nuclear spin lattice relaxation can be neglected, the flip-flip process is twice as strong as the flip-flop process, and a smaller nuclear polarization of the opposite sign results.

Although this explanation describes very effectively what happens, it is oversimplified. Nuclei do not interact in turn with individual electron spins, and the latter are not independent. Especially when protons are polarized in organic materials, the effect of the oscillating magnetic field is to cool the assembly of interacting electron spins, which then cool the nuclei by thermal contact. These processes are known by the acronym DONKEY effects. *See* PARAMAGNETISM.

In order to produce large polarizations, the external magnetic field should be of the order of 2 T, the frequency of the oscillating field should be about 70 GHz, and the temperature should be 1 K or less. Values of polarization of 0.8 or more (corresponding to 90% of the spins oriented in one direction and only 10% in the opposite direction)

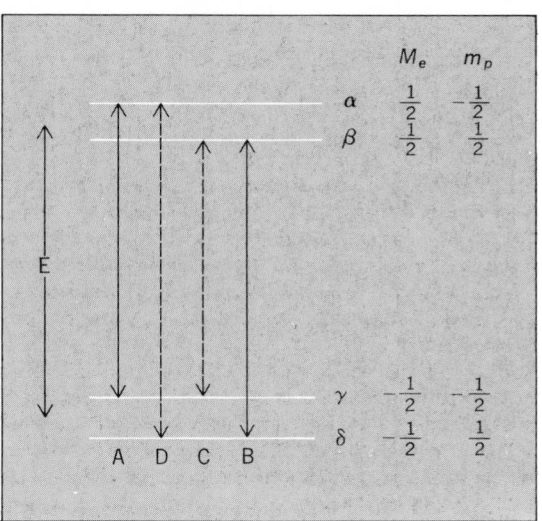

The energy levels of an electron and a proton in a magnetic field.

are regularly obtainable for protons polarized in $La_2Mg_3(NO_3)_{12}\cdot24H_2O$ in which 1% of the lanthanum, La, is replaced by neodymium, Nd, or in an empirically determined mixture consisting mainly of butanol, C_4H_9OH, and about 1% of porphyrexide, a free radical. Deuteron polarizations of 0.4 have been obtained by using C_4D_9OD and porphyrexide at temperatures of 0.3 K.

Polarization by rotation. This method was most spectacularly demonstrated by rotating a crystal of yttrium ethylsulfate, $Y(C_2H_5SO_4)_3\cdot9H_2O$, in which a few percent of yttrium, Y, was replaced by ytterbium, Yb, in a magnetic field of 1 T at a temperature of 1.4 K, at a rate of 60 revolutions per second. The proton polarization was found to be 0.19 (corresponding to 59.5% of the spins oriented in one direction and 40.5% in the other direction); higher values can be obtained by modifying the experimental conditions. The splitting of the two energy levels of the protons in the external magnetic field is independent of the orientation of the crystal, but the splitting of the two energy levels of the Yb^{3+} ion depends strongly on the orientation of the magnetic field relative to the c axis of the crystal, being quite large when these are parallel and almost zero when they are perpendicular. The Yb^{3+} ions are strongly coupled to the lattice when this angle is 45°, but almost isolated when it is 0 or 90°. Thus, as the crystal is rotated, when it reaches the 45° position, the Yb^{3+} ions quickly come to equilibrium at the temperature of the lattice. Then, as the crystal is rotated to the 90° position, the energy levels of the Yb^{3+} ions move together; this is equivalent to adiabatic demagnetization, and the spin temperature of the Yb^{3+} ions drops. Near the 90° position, the Yb^{3+} ions can exchange energy with the protons, because the two splittings are the same, and a flip-flop transition involving a Yb^{3+} ion and a proton conserves energy. The protons and the Yb^{3+} ions thus come to thermal equilibrium at the temperature of the Yb^{3+} ions after demagnetization. Continuous rotation repeats the cycle, and the ideal steady state is one in which the proton polarization is the same as the polarization of the Yb^{3+} spins at the 45° position. *See* ADIABATIC DEMAGNETIZATION; MAGNETIC SUSCEPTIBILITY.

Optical pumping. This method, pioneered by Alfred Kastler and Jean Brossel, makes use of the fact that circularly polarized light carries angular momentum, and when it is absorbed, that angular momentum is given to the absorber. Typically circularly polarized resonance radiation (such as sodium-D radiation) is incident on, for example, sodium vapor at room temperature (at a pressure of about 10^{-6} torr or 10^{-4} Pa) in a small magnetic "guide" field directed along the direction of the light beam. It excites only the $\sigma+$ transitions in which the magnetic quantum number of the absorbing atom, which specifies its angular momentum along the direction of the magnetic field, increases by 1. When the excited state decays, on the average there is no change in the magnetic quantum number. Thus an equilibrium is set up in which the ground-state atom spins are polarized. Polarizations of almost 1.0 can easily be obtained for the alkali metals.

The absorbing atoms are usually mixed with a gas such as argon, called a buffer gas, at a pressure of $1-100$ torr (10^2-10^4 Pa) to delay their diffusion to the walls of the container where they may be depolarized. If some other gas is present instead of, or in addition to, the buffer gas, the angular momentum of the absorber is transferred in collisions to the atoms of this gas and can end up as a nuclear polarization.

This method has been very successfully applied to ^3He, which acts as a buffer gas. The absorbing atoms are metastable 2^3S_1 ^3He atoms produced by striking a weak electrodeless discharge in the ^3He gas. Polarizations of up to 0.4 have been achieved, and under suitable conditions, when the polarizing mechanism is turned off, the polarization decays with a time constant of the order of several days. *See* ATOMIC STRUCTURE AND SPECTRA; OPTICAL PUMPING; ZEEMAN EFFECT. [JAMES M. DANIELS]

Bibliography: J. M. Daniels, *Oriented Nuclei: Polarized Targets and Beams*, 1965; *Proceedings of the International Conference on Polarized Targets and Ion Sources*, C. E. N. Saclay, 1966; M. E. Rose (ed.), *Nuclear Orientation*, ISRS 6, 1963; G. Shapiro (ed.), *Proceedings of the 2d International Conference on Polarized Targets*, Lawrence Berkeley Laboratory, 1971.

Dynamic similarity

A relationship existing between two homologous fluid-flow systems such that corresponding parts of the systems experience similar net forces. Dynamically similar flows about geometrically similar bodies will themselves be geometrically similar (see illustration). Consequently, this concept is basic to the meaningful extrapolation of model results to full-scale performance. However, geometrically similar flows are not necessarily dynamically similar.

Dynamic similarity between two flow systems will occur if certain nondimensional parameters formed from the flow variables have the same values for both systems.

One important parameter for establishing dy-

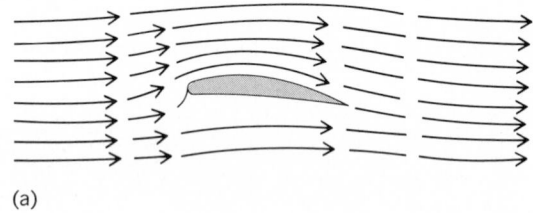

(a)

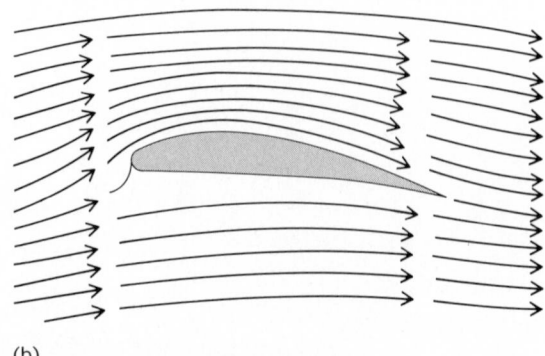

(b)

Dynamically similar flows about (a) airfoil profile and (b) airfoil profile geometrically similar to a.

Dynamic similarity parameters

Nondimensional parameter*	Name	Physical effect
$\rho VL/\mu$	Reynolds number	Viscosity
V/c	Mach number	Compressibility
V^2/Lg	Froude number	Gravity
x_m/L	Knudsen number	Pressure
$\rho V^2 L/\sigma$	Weber number	Surface tension
$c_p\mu/\kappa$	Prandtl number	Heat conduction
$\beta TgL^3\rho^2/\mu^2$	Grashof number	Free convection

*The reference variables are defined as follows: L, length; μ, coefficient of viscosity; c, speed of sound; g, acceleration of gravity; σ, surface tension; κ, coefficient of thermal conductivity; β, coefficient of thermal expansion; T, temperature; x_m, mean free path of molecule; and c_p, specific heat at constant pressure.

namically similar flows is pressure coefficient $p/\rho V^2$, where p, ρ, and V are, respectively, a reference pressure, density, and velocity. Other important nondimensional parameters are given in the table along with associated physical effects which characterize the parameters.

The parameters in the table can be determined analytically by dimensional analysis or by examining the invariance of the differential equations and boundary conditions for the flow systems under scalar transformations of length, time, and mass. Other useful parameters can also be defined. Often these can be obtained as ratios of the tabulated parameters.

In practice, it is often difficult to establish equality of all similarity parameters simultaneously for two flows. Equality of parameters corresponding to dominant flow properties is usually sufficient. Given low-speed viscous flows, for example, the Reynolds numbers of the two flows would be equated, but the Mach numbers might be ignored. *See* DIMENSIONAL ANALYSIS; FLUID MECHANICS; MACH NUMBER.

[ARTHUR G. HANSEN]

Bibliography: H. L. Langhaar, *Dimensional Analysis and Theory of Modes*, 1951, reprint 1979; S. Pai, *Viscous Flow Theory*, vol. 1, 1965; R. H. Sabersky et al., *Fluid Flow*, 2d ed., 1971.

Dynamics

That branch of mechanics which deals with the motion of a system of material particles under the influence of forces, especially those which originate outside of the system under consideration. From Newton's third law of motion, namely, to every action there is an equal and opposite reaction, the internal forces cancel in pairs and do not contribute to the motion of the system as a whole, although they determine the relative motion, if any, of the several parts.

Particle dynamics refers to the motion of a single particle under the influence of external forces, particularly electromagnetic and gravitational forces. The dynamics of a rigid body is the study of the motion, under given forces, of a system of particles, the distances between which are postulated to be constant throughout the motion.

In classical dynamics the basic relation that enables the motion to be determined once the force is known is Newton's second law of motion, which states that the resultant force on a particle is equal to the product of the mass of the particle times its acceleration. For a many-particle system it becomes impracticable to write and solve this equation for each individual particle and, in general, the motion may be computed only on a statistical basis (that is, by the methods of statistical mechanics) unless, as for a few particles or a rigid body, the number of degrees of freedom is sufficiently small. *See* DEGREE OF FREEDOM (MECHANICS); KINEMATICS; KINETICS (CLASSICAL MECHANICS); NEWTON'S LAWS OF MOTION; RIGID-BODY DYNAMICS; STATISTICAL MECHANICS.

[HERBERT C. CORBEN/BERNARD GOODMAN]

Eddy current

An electric current induced within the body of a conductor when that conductor either moves through a nonuniform magnetic field or is in a region where there is a change in magnetic flux. It is sometimes called Foucault current. Although eddy currents can be induced in any electrical conductor, the effect is most pronounced in solid metallic conductors. Eddy currents are utilized in induction heating and to damp out oscillations in various devices.

Causes. If a solid conductor is moving through a nonuniform magnetic field, electromotive forces (emfs) are set up that are greater in that part of the conductor that is moving through the strong part of the field than in the part moving through the weaker part of the field. Therefore, at any one time in the motion, there are many closed paths within the body of the conductor in which the net emf is not zero. There are thus induced circulatory currents that are called eddy currents (see illustration). In accordance with Lenz's law, these eddy currents circulate in such a manner as to oppose the motion of the conductor through the magnetic field. The motion is damped by the opposing force. For example, if a sheet of aluminum is dropped between the poles of an electromagnet, it does not fall freely, but is retarded by the force due to the eddy currents set up in the sheet. If an aluminum plate oscillates between the poles of the electromagnet, it will be stopped quickly when the switch is closed and the field set up. The energy of motion of the aluminum plate is converted into heat energy in the plate. *See* LENZ'S LAW.

Eddy currents are also set up within the body of

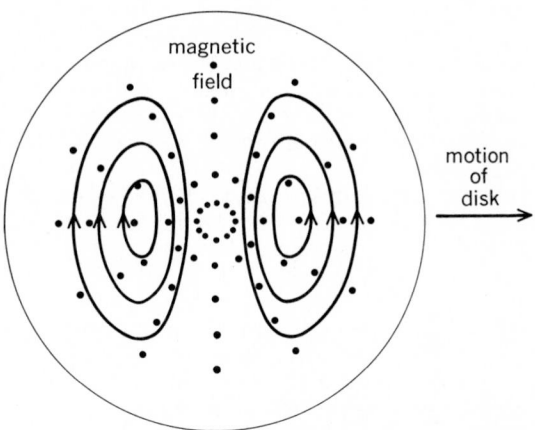

Eddy currents which are induced in a disk moving through a nonuniform magnetic field.

a material when it is in a region in which the magnetic flux is changing rapidly, as in the core of a transformer. As the alternating current changes rapidly, there is also an alternating flux that induces an emf in the secondary coil and at the same time induces emfs in the iron core. The emfs in the core cause eddy currents that are undesirable because of the heat developed in the core (which results in high energy losses) and because of an undesirable rise in temperature of the core. Another undesirable effect is the magnetic flux set up by the eddy currents. This flux is always in such a direction as to oppose the change that caused it, and thus it produces a demagnetizing effect in the core. The flux never reaches as high a value in the core as it would if there were no eddy currents.

Laminations. Induced emfs are always present in conductors that move in magnetic fields or are present in fields that are changing. However, it is possible to reduce the eddy currents caused by these emfs by laminating the conductor, that is, by building the conductor of many thin sheets that are insulated from each other rather than making it of a single solid piece. In an iron core the thin iron sheets are insulated by oxides on the surface or by thin coats of varnish. The laminations do not reduce the induced emfs, but if they are properly oriented to cut across the paths of the eddy currents, they confine the currents largely to single laminae, where the paths are long, making higher resistance; the resulting net emf in the possible closed path is small. Bundles of iron wires or powdered iron formed into a core by high pressure are also used to break up the current paths and reduce the eddy currents.

[KENNETH V. MANNING]

Bibliography: B. Bleaney and B. I. Bleaney, *Electricity and Magnetism*, 3d ed., 1976; E. M. Pugh and E. W. Pugh, *Principles of Electricity and Magnetism*, 2d ed., 1970.

Eigenfunction

If an equation containing a variable parameter possesses nontrivial solutions only for certain special values of the parameter, these solutions are called eigenfunctions and the special values are called eigenvalues. In older books these are referred to as proper, or characteristic, functions and values; the more fashionable terms are half-translations of the German *Eigenfunktion* and *Eigenwert*. The equation of the problem may be a matrix equation, that is, a set of simultaneous linear algebraic equations, a differential or integral equation, or occasionally something more complicated, such as a combination of these kinds. This article shows some typical situations, chosen for their mathematical simplicity, in which these problems arise.

Matrix equations. Consider the eigenvalue equation given as Eq. (1), where M_{ij} is a 3×3 ma-

$$\sum_j M_{ij} u_j = \lambda u_i \qquad (i = 1,2,3) \qquad (1)$$

trix assumed real and symmetrical ($M_{ij} = M_{ji}$), and λ is the free parameter. (Here one considers vectors in three dimensions, but the extension to n dimensions is immediate.) The equation has the trivial solution $u_i = 0$ ($i = 1,2,3$). To find the nontrivial solutions, one writes the equations in detail, as in Eqs. (2). They are soluble if, and only if, the

$$(M_{11} - \lambda)u_1 + M_{12}u_2 + M_{13}u_3 = 0$$
$$M_{21}u_1 + (M_{22} - \lambda)u_2 + M_{23}u_3 = 0 \qquad (2)$$
$$M_{31}u_1 + M_{32}u_2 + (M_{33} - \lambda)u_3 = 0$$

determinant of the coefficients vanishes, as in Eq. (3). Multiplied out, this gives a cubic equation for λ

$$\begin{vmatrix} M_{11} - \lambda & M_{12} & M_{13} \\ M_{21} & M_{22} - \lambda & M_{23} \\ M_{31} & M_{32} & M_{33} - \lambda \end{vmatrix} = 0 \qquad (3)$$

that can be shown to have three real roots, which may be denoted by $\lambda_1, \lambda_2, \lambda_3$. These are the eigenvalues of the matrix M_{ij}, and together they constitute its spectrum. If two of them are equal, the spectrum is said to be degenerate; this case will not be discussed here.

Corresponding to the jth eigenvalue, let the three components of the vector **u** obtained by solving Eqs. (2) be u_{1j}, u_{2j}, u_{3j} or, in general, u_{ij} ($i,j = 1,2,3$). From Eq. (1), Eq. (4) is obtained, and multiplication by u_{ik} and summation give Eq. (5).

$$\sum_j M_{ij} u_{jl} = \lambda_l u_{il} \qquad (i,l = 1,2,3) \qquad (4)$$

$$\sum_{i,j} u_{ik} M_{ij} u_{jl} = \lambda_l \sum_i u_{ik} u_{il} \qquad (5)$$

Interchanging the roles of k and l gives Eq. (6),

$$\sum_{i,j} u_{il} M_{ij} u_{jk} = \lambda_k \sum_i u_{il} u_{ik} \qquad (6)$$

and by further interchanging the labels i and j on the left one can write this as Eq. (7). Because M_{ij} is

$$\sum_{i,j} u_{jl} M_{ji} u_{ik} = \lambda_k \sum_i u_{ik} u_{il} \qquad (7)$$

assumed symmetrical, the left sides of Eqs. (5) and (7) are equal, and one has Eq. (8). This means that,

$$(\lambda_k - \lambda_l) \sum_i u_{ik} u_{il} = 0 \qquad (k,l = 1,2,3) \qquad (8)$$

if $k \neq l$ so that, assuming nondegeneracy, $\lambda_k \neq \lambda_l$, then Eq. (9) holds. The second form is the first

$$\sum_i u_{ik} u_{il} = 0 \quad \text{or} \quad \mathbf{u}_k \cdot \mathbf{u}_l = 0 \qquad (9)$$

written in usual vector notation. Since the **u**'s are not zero, the eigenvectors of a symmetrical matrix belonging to different eigenvalues are orthogonal. One can also, by suitable choice of scale, normalize $\mathbf{u}_j$ to unit length, so that Eq. (10) holds, where δ_{jl} is the Kronecker delta.

$$\sum_i u_{ij} u_{il} = \mathbf{u}_j \cdot \mathbf{u}_l = \delta_{jl} \qquad (10)$$

Since $\mathbf{u}_1$, $\mathbf{u}_2$, and $\mathbf{u}_3$ are three orthogonal vectors in a 3-space, they span the space, and any other vector a can be written with them as a basis, as in Eq. (11). Thus, Eq. (12) is obtained. In terms of

$$\mathbf{a} = \sum_k a_k \mathbf{u}_k \text{ with } a_k = \mathbf{a} \cdot \mathbf{u}_k \qquad (11)$$

$$\mathbf{a} = \sum_k (\mathbf{a} \cdot \mathbf{u}_k) \mathbf{u}_k \qquad (12)$$

components, this is written as Eq. (13), and since a_i

$$a_i = \sum_j a_j \sum_k u_{jk} u_{ik} \qquad (13)$$

is arbitrary, Eq. (14) is valid.

$$\sum_k u_{ik}u_{jk} = \delta_{ij} \qquad (14)$$

Although Eqs. (10) and (14) look alike, their contents are quite different, since the first index labels axes and the second labels eigenvalues. Thus Eq. (14) refers to sum of products corresponding to different eigenvalues; it is called a completeness relation. The foregoing argument is of course valid for a space of n dimensions, but the extension to infinite n, or to cases in which n varies continuously, is more complicated.

The following is one example among many of the utility of these considerations. In determining the motion of a rigid body subject to arbitrary forces (or no forces), it turns out to be best to describe the motion in terms of a set of axes fixed in the body. The inertial properties of the body are characterized by a tensor I_{ij} called the inertial tensor, which in general has the symmetrical form shown in Eq. (15).

$$I_{ij} = \begin{pmatrix} I_{11} & I_{12} & I_{13} \\ I_{12} & I_{22} & I_{23} \\ I_{13} & I_{23} & I_{33} \end{pmatrix} \qquad (15)$$

The equations of motion are much easier to solve if all the off-diagonal terms are zero and only I_{11}, I_{22}, and I_{33} remain. This can always be accomplished by orienting the axes properly in the body, as follows: Solve the eigenvalue problem, Eq. (1), for the matrix I_{ij}. Consider the u_{ij} as a matrix and introduce instead of the original basis vectors a new basis chosen so that any vector x_i now has components y_i, with x_j being defined as in Eq. (16).

$$x_j = \sum_l u_{jl}y_l \qquad (16)$$

Consider the quadratic form $F = \sum_{ij} I_{ij}x_ix_j$. In the new basis it becomes Eq. (17). Thus in the new

$$\begin{aligned} F &= \sum_{i,j,k,l} I_{ij}u_{ik}u_{jl}y_ky_l \\ &= \sum_{i,k,l} u_{ik}\lambda_l u_{il}y_ky_l \quad \text{[by Eq. (1)]} \\ &= \sum_{k,l} \delta_{kl}\lambda_l y_ky_l = \sum_l \lambda_l y_l^2 \quad \text{[by Eq. (10)]} \end{aligned} \qquad (17)$$

coordinate system the general tensor I_{ij} is replaced by a diagonal tensor whose nonvanishing elements are the three eigenvalues λ_i. This example illustrates the direct utility of the eigenvectors and eigenvalues of a matrix.

Differential equations. A different kind of eigenvalue problem arises in connection with certain differential equations. The simplest is that of a vibrating uniform string fixed at the ends, which may be located at $x=0$ and L. The equation for the displacement $y(x,t)$ is Eq. (18), where T is the tension

$$T\frac{\partial^2 y(x,t)}{\partial x^2} - \rho\frac{\partial^2 y(x,t)}{\partial t^2} = 0 \qquad (18)$$

and ρ is the mass per unit length. The periodic motions have $y(x,t) = y(x)\cos\omega(t-t_0)$, so that Eq. (19) holds. Here $\rho\omega^2/T$ is the eigenvalue, and the

$$\frac{\partial^2 y(x)}{\partial x^2} + \frac{\rho}{T}\omega^2 y(x) = 0 \qquad (19)$$

nontrivial solutions are those for which $y(0)=$

$y(L)=0$ but $y(x)\neq 0$. Clearly, they are of the form given in Eq. (20). (The negative n gives no new

$$y(x) = A\sin\frac{n\pi x}{L} \qquad (n=1,2,\ldots) \qquad (20)$$

solutions.) Substitution into Eq. (19) gives Eq. (21),

$$\omega = \frac{n\pi}{L}\sqrt{\frac{T}{\rho}} \qquad (21)$$

which are the proper frequencies of the vibrating string.

The same mathematical problem occurs in quantum mechanics if one asks for the eigenfunctions of a particle in a one-dimensional region of length L bounded by impenetrable walls. In this case the equation for the Schrödinger wave function $\psi(x,t)$ is Eq. (22), in which m is the particle's

$$\frac{\hbar^2}{2m}\frac{\partial^2\psi(x,t)}{\partial x^2} + i\hbar\frac{\partial\psi(x,t)}{\partial t} = 0 \qquad (22)$$
$$\psi(0,t) = \psi(L,t) = 0$$

mass and $\hbar$ is Planck's constant divided by 2π. Since $\psi(x,t) = \psi(x)e^{-iEt/\hbar}$, Eq. (23) holds, which is formally the same as Eq. (19).

$$-\frac{\hbar^2}{2m}\frac{\partial^2\psi(x)}{\partial x^2} = E\psi(x) \qquad \psi(0) = \psi(L) = 0 \qquad (23)$$

A particle bound only by a harmonic restoring force $F = -kx$ obeys the Schrödinger equation, Eq. (24). Here the boundary conditions are at infinity

$$-\frac{\hbar^2}{2m}\frac{\partial^2\psi(x)}{\partial x^2} + \tfrac{1}{2}kx^2\,\psi(x) = E\psi(x) \qquad (24)$$
$$\psi(\pm\infty) = 0$$

but the general character of the problem is unchanged. The solutions are written in terms of the Hermite polynomials, and the eigenvalues are given in Eq. (25).

$$E = (n+\tfrac{1}{2})\hbar\sqrt{\frac{k}{m}} \qquad (n=0,1,2,\ldots) \qquad (25)$$

Continuous distributions of eigenvalues arise when the boundary conditions do not significantly limit the eigenfunction, as for example if one has a particle free to move anywhere in infinite space. Here the equation is again (23), but without the boundary conditions, and the solutions can be written as Eqs. (26) for all positive and negative values

$$\psi(x) = Ae^{ikx} \qquad E = \frac{(\hbar k)^2}{2m} \qquad (26)$$

of k. It is recognized that the particle's momentum is $\hbar k$ and that such a free particle can have any momentum at all, just as in classical theory.

Other differential eigenvalue problems in quantum mechanics involve the possible values of angular momentum. The eigenfunctions in this case are the spherical harmonics. The fields of electrical and mechanical engineering give rise to many other kinds of functions often of great complexity.

The eigenfunctions of differential eigenvalue problems have an orthogonality property similar to that of the eigenvectors of a matrix. Let the equation be, for example, Eq. (27), where $\mathscr{D}y(x)$ might

$$\mathscr{D}y(x) = \lambda y(x) \qquad (27)$$

be the left side of Eq. (23) or (24), with $y(a)=$

$y(b)=0$, and suppose for simplicity that the solutions are nondegenerate (different λ's correspond to different eigenfunctions). Let y_i correspond to λ_i and y_j to λ_j. Take the equation for y_i, $\mathscr{D}y_i(x)=\lambda_iy_i(x)$, and multiply it by y_j. Then multiply the equation for y_j by y_i and subtract. Equation (28) results.

$$y_j\mathscr{D}y_i - y_i\mathscr{D}y_j = (\lambda_i - \lambda_j)y_iy_j \qquad (28)$$

Now integrate from a to b. It is easily seen that for the operators in Eqs. (23) and (24), Eq. (29) holds.

$$\int_a^b y_i\mathscr{D}y_j\,dx = \int_a^b y_j\mathscr{D}y_i\,dx \qquad (29)$$

Operators for which this is true are called self-adjoint. If $\mathscr{D}$ is self-adjoint and $\lambda_i \neq \lambda_j$, Eq. (30) is obtained.

$$\int_a^b y_iy_j\,dx = 0 \qquad (i \neq j) \qquad (30)$$

tained. Furthermore, one can normalize y so that $\int_a^b y_i^2\,dx = 1$ for all i and in general, Eq. (31) holds.

$$\int_a^b y_iy_j\,dx = \delta_{ij} \qquad (31)$$

This is the analog of the matrix orthogonality relation (10), the self-adjoint operator being analogous to the symmetrical matrix. The completeness relation analogous to Eq. (24) is Eq. (32), where

$$\sum_k y_k(x)\,y_k(x') = \delta(x-x') \qquad (32)$$

$\delta(x-x')$ is Dirac's delta-function and, if all or part of the range of eigenvalues is continuous, the sum must be wholly or partially replaced by an integral. A rigorous proof of Eq. (32) in the most general case has not been given, but it is known to be true in most cases of interest. The importance of this property is that it enables one to show that an arbitrary function can be expanded in terms of the eigenfunctions of a self-adjoint operator, as in Eq. (33a). The term C_n is defined in Eq. (33b). The

$$f(x) = \sum_n C_ny_n(x) \qquad (a \leq x \leq b) \qquad (33a)$$

$$C_n = \int_a^b f(x)\,y_n(x)\,dx \qquad (33b)$$

sense of these equations is that, whenever the sum converges, it converges to the function $f(x)$. *See* INTEGRAL TRANSFORM.

Differential-to-matrix relation. Suppose one wishes to solve the differential eigenvalue problem, Eq. (34), where $\mathscr{D}$ is self-adjoint in the region

$$\mathscr{D}Y_n(x) = \Lambda_nY_n(x) \qquad (34)$$

$a < x < b$, and suppose one knows a complete set of eigenfunctions $y_j(x)$, of some other operator, orthonormal in the same region. It is then possible to expand the unknown function Y_n in terms of the known y_j's as in Eq. (35), where the $u_j^{(n)}$ are constant coefficients to be found. The eigenvalue equation becomes Eq. (36). Multiply by $y_i(x)$ and

$$Y_n(x) = \sum_j u_j^{(n)}y_j(x) \qquad (35)$$

stant coefficients to be found. The eigenvalue equation becomes Eq. (36). Multiply by $y_i(x)$ and

$$\sum_j u_j^{(n)}\mathscr{D}y_j(x) = \Lambda_n\sum_j u_j^{(n)}y_j(x) \qquad (36)$$

integrate from $x = a$ to b. Then, from Eq. (31), Eq.

(37) is obtained. Writing the integral as D_{ij}, Eq. (38)

$$\sum_j u_j^{(n)}\int_a^b y_i(x)\mathscr{D}y_j(x)\,dx = \Lambda_nu_i^{(n)} \qquad (37)$$

$$\sum_j D_{ij}u_j^{(n)} = \Lambda_nu_i^{(n)} \qquad (38)$$

holds. This is exactly in the form of Eq. (1), except that the matrix has now, in general, an infinite number of rows and columns, so that one must be careful about questions of convergence in carrying out sums. Furthermore, it can be seen from Eq. (29) that, if $\mathscr{D}$ is self-adjoint, then D_{ij} is symmetrical, which explains the correspondence noted above.

For some further applications, including generalizations to functions which take on complex values, *see* EIGENVALUE (QUANTUM MECHANICS). *See also* NONRELATIVISTIC QUANTUM THEORY; QUANTUM MECHANICS.

[DAVID PARK]

Bibliography: G. Arfken, *Mathematical Methods for Physicists*, 1970; E. A. Kraut, *Fundamentals of Mathematical Physics*, 1967.

Eigenvalue (quantum mechanics)

If an equation containing a variable parameter possesses nontrivial solutions only for certain special values of the parameter, these solutions are called eigenfunctions and the special values are called eigenvalues.

The eigenfunction-eigenvalue relation is of particular importance in quantum mechanics because of its prominence in the equations which relate the mathematical formalism of the theory with physical results. *See* EIGENFUNCTION.

In quantum mechanics, dynamical variables f are represented by operators $\hat{f}$ operating on a wave function $\psi_m(x,t)$, which describes the system under discussion in some state, which may be denoted by m. (Here x represents all the coordinates used to describe the system.) The expectation value of the quantity f is given by Eq. (1),

$$\langle f \rangle = \int \psi_m^*(x,t)\,\hat{f}\psi_m(x,t)\,(dx) \qquad (1)$$

where $\int(dx)$ represents integration over all coordinates. The requirement that dynamical variables take on only real values leads to a restriction on the operators representing them. Let $f_{mn}(t)$ be defined as in Eq. (2). The restriction is that for all m and n, Eq. (3) holds. This is an immediate exten-

$$f_{mn}(t) = \int \psi_m^*(x,t)\,\hat{f}\psi_n(x,t)\,(dx) \qquad (2)$$

$$f_{mn}(t) = f_{nm}^*(t) \qquad (3)$$

sion of the self-adjoint operators introduced in eigenfunction. Operators of this kind are called hermitian operators. The properties of completeness and orthogonality, as well as the equivalence of matrix and differential formulations shown in eigenfunction, all have their counterparts with hermitian operators. The eigenfunction-eigenvalue relation is especially important here because of the postulate that measurements of a dynamical variable f give definite and predictable values only for those states which are eigenfunctions of the corresponding operator $\hat{f}$. These values are the eigenvalues of $\hat{f}$ as in Eq. (4), where the number n, called a

$$\hat{f}\psi_n = f_n\psi_n \qquad (4)$$

quantum number, labels the different eigenfunctions. (The eigenvalues f_n with different subscripts are not necessarily different; two different states with the same eigenvalue are said to be degenerate.) If f enters the theory as a hermitian differential operator, Eq. (4) can readily be transformed into a matrix equation, Eq. (5). With the

$$\sum_m f_{lm} u_m^{(n)} = f_n u_l^{(n)} \qquad (5)$$

advent of high-speed computers this is often the easiest approach to the determination of eigenvalues. It is also shown in eigenfunction that the eigenvectors of a matrix define a transformation to a basis with respect to which the matrix is diagonal. Thus the process of finding eigenvalues and eigenfunctions is often called diagonalization, and the matrix written with respect to such a basis is diagonal. The question of how many operators can be diagonalized at the same time is answered by the following theorem: If two hermitian operators $\hat{f}$ and $\hat{g}$ commute, there exists a complete set of eigenfunctions that diagonalize both of them. In practice, one of the operators is usually the Hamiltonian, $\hat{H}$. Since any operator that commutes with $\hat{H}$ is constant, it follows that the eigenvalues of all other operators that commute with $\hat{H}$ and with each other are constant, so that the quantum numbers labeling them identify constant properties of the system. It is fundamental to the theory that a system be characterized as completely as possible by the quantum numbers corresponding to the largest possible set of independent operators that commute with each other and with the Hamiltonian. *See* DEGENERACY (QUANTUM MECHANICS); NONRELATIVISTIC QUANTUM THEORY; QUANTUM MECHANICS.

[DAVID PARK]

Bibliography: C. Cohen-Tannoudji et al., *Quantum Mechanics*, vol. 1, 1978; S. G. Gasiorwicz, *Quantum Physics*, 1974; D. A. Park, *Introduction to the Quantum Theory*, 2d ed., 1974; D. Saxon, *Elementary Quantum Mechanics*, 1968; E. L. White, *Basic Quantum Mechanics*, 1966.

Elasticity

The property whereby a solid material changes its shape and size under the action of opposing forces, but recovers its original configuration when the forces are removed. The theory of elasticity deals with the relations between the forces acting on a body and the resulting changes in configuration, and is important in many branches of science and technology, for instance, in the design of structures, in the theory of vibration and sound, and in the study of the forces between atoms in crystal lattices.

Elastic constants. The forces acting on a body are expressed as stresses and measured as force per unit area. Thus if a bar $ABCD$ of square cross section (Fig. 1a) is fixed at one end and subjected to a force F uniformly distributed over the other end DC, the stress is $F/(DC)^2$. This stress causes the bar to become longer and thinner and to assume the shape $A'B'C'D'$. The strain is measured by the ratio (change in length)/(original length),

that is, by $(B'C' - BC)/(BC)$. According to Hooke's law, stress is proportional to strain, and the ratio of stress to strain is therefore a constant, in this case the Young's modulus, denoted by E, so that $E = F(BC)/(DC)^2(B'C' - BC)$.

Poisson's ratio σ is defined as the ratio of lateral strain to longitudinal strain, so that $\sigma = BC(DC - D'C')/DC(B'C' - BC)$. The bar of Fig 1a is in a state of tension, and the stress is tensile; if the force F were reversed in direction, the stress would be compressive. Stresses of this type are called direct or normal stresses; a second type of stress, known as tangential or shear stress, is illustrated in Fig. 1b. In this case, the configuration $ABCD$ becomes $ABC'D'$, with the shear forces F acting in the directions AB and CD. The shear strain is measured by the angle θ, and if the body is originally a cube, the shear stress is $F/(DC)^2$. The ratio of stress to strain, $F/(DC)^2\theta$, is the shear or rigidity modulus G, which measures the resistance of the material to change in shape without change in volume.

A further elastic constant, the bulk modulus k, measures the resistance to change in volume without changes in shape, and is illustrated in Fig. 1c. The original configuration is represented by the circle AB, and under a hydrostatic (uniform) pressure P, the circle AB becomes the circle $A'B'$. The bulk modulus is then $k = Pv/\Delta v$, where $\Delta v/v$ is the volumetric strain. The reciprocal of the bulk modulus is the compressibility.

Determination of values. The elastic constants may be determined directly in the way suggested by their definitions; for instance, Young's modulus can be determined by measuring the relative extension of a rod or wire subjected to a known tensile stress. Less direct methods are, however, usually more convenient and accurate. Prominent among these are the dynamic methods involving frequency of vibration and velocity of sound propagation. The elastic constants can be expressed in terms of frequency of (or velocity in) regularly shaped specimens, together with the dimensions and density, and by measuring these quantities, the elastic constants can be found. *See* ULTRASONICS.

The elastic constants can also be determined from the flexure and torsion of bars. As an illustration, consider a bar AB (Fig. 2) of breadth b (in the x_1 direction) and depth d (in the x_2 direction) supported by forces F at the ends, and loaded symmetrically by forces F at points C and D. Over the portion CD there is a uniform bending moment $M = Fl_2$, and the theory of bending shows that the

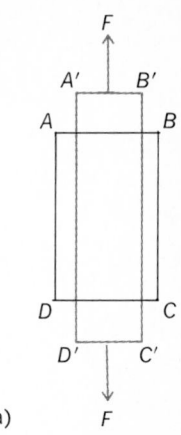

(a)

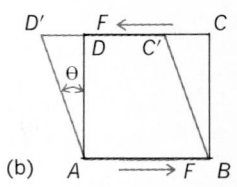

(b)

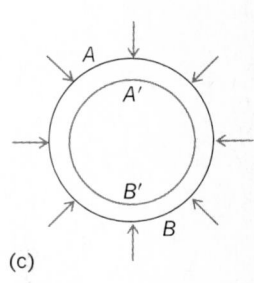

(c)

Fig. 1. Stresses on a bar. (a) Direct or normal stress. (b) Tangential or shear stress. (c) Change in volume with no change in shape. (All deformations are exaggerated.)

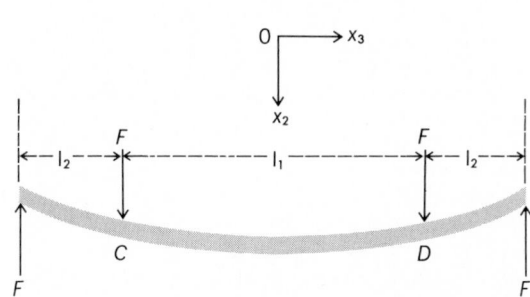

Fig. 2. Flexure and torsion in a bar.

portion CD is bent into the arc of a circle such that Eq. (1) applies, where R is the radius of curvature,

$$R = EI/M \tag{1}$$

E is Young's modulus and I is the moment of inertia of cross section, equal to $bd^3/12$ for a rectangular cross section. The longitudinal stress at the lower face of the bar is tensile, and at the upper face, compressive. The middle plane of the bar is free of stress, and is the neutral axis. The stress at a distance x_2 from the neutral axis is shown in Eq. (2).

$$T = Ex_2/R = Mx_2/I \tag{2}$$

It is thus possible to determine E from Eq. (1) by measuring I, R, and M; conversely, if E is known, the stress may be determined from Eq. (2).

Practical limitations. In practice, stress is only proportional to strain, and the strain is only completely recoverable within certain limits called the elastic limits of the material. The stress below which the strain is completely recoverable is sometimes called the limit of perfect elasticity, and the stress up to which Hooke's law is obeyed is sometimes called the proportional limit or limit of linear elasticity. Above the elastic limits, the material is subject to time-dependent effects, and as the stress is further increased, the ultimate strength of the material is approached.

Theory of elasticity. In classical elasticity theory, it is assumed that the strains are always small; Hooke's law is therefore obeyed; the strains are completely recoverable and, moreover, are superposable, so that the strain produced by the joint action of two or more stresses is the sum of the strains produced by them individually.

In order to develop the theory, it is necessary to specify the stresses and strains more closely. Figure 3 shows the stress components T_{ij} (where i, j may take the values 1, 2, or 3) acting on the faces of a cube, parallel to the coordinate axes x_1, x_2, x_3. The first suffix indicates the direction of the stress component and the second the direction of the normal to the plane under consideration. Stresses of the type T_{11} are normal stresses, and of the type T_{12}, shear stresses. The conditions for zero rotation of the cube are $T_{12} = T_{21}$, $T_{13} = T_{31}$, $T_{23} = T_{32}$, and there are therefore six independent stress components.

In addition to the stresses T_{ij}, body forces proportional to volume (for instance, forces due to the weight of the body) may also be acting. If the stresses T_{ij} vary with position, application of Newton's second law leads to Eq. (3) for the x_1 direction

$$\frac{\partial T_{11}}{\partial x_1} + \frac{\partial T_{12}}{\partial x_2} + \frac{\partial T_{13}}{\partial x_3} + X_1 = \rho f_1 \tag{3}$$

where ρ is the density, f_1 is the acceleration, and X_1 the body force component per unit volume along x_1, together with two similar equations for the x_2 and x_3 directions. If $f_1 = f_2 = f_3 = 0$, these equations become the equations of equilibrium, and if, further, $X_1 = X_2 = X_3 = 0$, they become the equations of equilibrium in the absence of body forces. The preceding equations are important in many branches of elastic theory and, for example, provide a starting point in the study of vibrating bodies and of the twisting of cylinders and prisms with cross sections of various shapes.

The components of strain are specified in a simi-

lar way to the stresses. There are six independent strain components: S_{11}, S_{22}, S_{33}, S_{23}, S_{13}, and S_{12}. If, as a result of strain, the coordinates of a point x_1, x_2, x_3 become $x_1 + u_1$, $x_2 + u_2$, $x_3 + u_3$, the quantities u_1, u_2, and u_3 are the components of the displacement vector, and the strain components are as displayed in Eq. (4), so that, for example the re-

$$S_{ij} = \tfrac{1}{2}\left(\frac{\partial u_i}{\partial x_j} + \frac{\partial u_j}{\partial x_i}\right) \tag{4}$$

lations shown by Eq. (5) would hold true.

$$S_{11} = \frac{\partial u_1}{\partial x_1} \qquad S_{12} = \tfrac{1}{2}\left(\frac{\partial u_1}{\partial x_2} + \frac{\partial u_2}{\partial x_1}\right) \tag{5}$$

By eliminating the displacements from these equations, the so-called compatibility equations are obtained with three of the type shown in Eq. (6)

$$\frac{\partial^2 S_{22}}{\partial x_3{}^2} + \frac{\partial^2 S_{33}}{\partial x_2{}^2} = 2\,\frac{\partial^2 S_{23}}{\partial x_2\,\partial x_3} \tag{6}$$

and three of the type shown in Eq. (7).

$$\frac{\partial^2 S_{11}}{\partial x_2\,\partial x_3} = \frac{\partial}{\partial x_1}\left(-\frac{\partial S_{23}}{\partial x_1} + \frac{\partial S_{13}}{\partial x_2} + \frac{\partial S_{12}}{\partial x_3}\right) \tag{7}$$

The stresses and strains have so far been denoted by two suffixes. This is essential if the methods of tensor analysis are to be applied to elasticity problems, but for many purposes a single suffix notation is adequate. The change from a two- to a one-suffix notation for the stresses is simply $T_{11} = T_1$, $T_{22} = T_2$, $T_{33} = T_3$, $T_{23} = T_4$, $T_{13} = T_5$, $T_{12} = T_6$. The change of notation for the strains is $S_{11} = S_1$, $S_{22} = S_2$, $S_{33} = S_3$, $2S_{23} = S_4$, $2S_{13} = S_5$, $2S_{12} = S_6$: the factor 2 is required to make the strains S_4, S_5, and S_6 conform with the usual definition of shear strain (Fig. 1b).

Hooke's law generalized. Hooke's law may be generalized to the statement that each stress component is proportional to each strain component, equivalent to the six equations

$$T_1 = c_{11}S_1 + c_{12}S_2 + c_{13}S_3 + c_{14}S_4 + c_{15}S_5 + c_{16}S_6$$
$$\cdots\cdots\cdots\cdots\cdots\cdots\cdots$$
$$T_6 = c_{61}S_1 + c_{62}S_2 + c_{63}S_3 + c_{64}S_4 + c_{65}S_5 + c_{66}S_6$$

which may be written more concisely as Eq. (8),

$$T_q = \sum_r c_{qr}S_r \tag{8}$$

where the summation extends over $r = 1, 2, 3, 4, 5$, and 6. The elastic constants c_{qr} are termed the stiffnesses; there are altogether 36 of them but they are subject to the reciprocal relations $c_{qr} = c_{rq}$ imposed by thermodynamic requirements, and the number is thus reduced to 21.

Additional relations can be derived from the three assumptions that the interatomic forces act along the lines joining the centers of atoms in the lattice, that the atoms are situated at centers of symmetry, and that the lattice is initially at zero stress. These relations, called Cauchy relations, are $c_{23} = c_{44}$, $c_{13} = c_{55}$, $c_{12} = c_{66}$, $c_{14} = c_{56}$, $c_{25} = c_{46}$, $c_{45} = c_{36}$ and, if true, would reduce the number of stiffnesses to 15. Experiment shows, however, that they are not true in general; nevertheless, their investigation provides an indication of the extent to which these three assumptions hold in any particular case.

The generalized Hooke's law can also be written

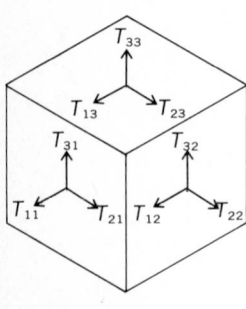

ELASTICITY

Fig. 3. Stress components.

Stress-strain relations

Stress	Strain			
	Orthorhombic	Hexagonal	Cubic	Isotropic
$T_1 =$	$c_{11}S_1 + c_{12}S_2 + c_{13}S_3$	$c_{11}S_1 + c_{12}S_2 + c_{13}S_3$	$c_{11}S_1 + c_{12}S_2 + c_{12}S_3$	$c_{11}S_1 + c_{12}S_2 + c_{12}S_3$
$T_2 =$	$c_{12}S_1 + c_{22}S_2 + c_{23}S_3$	$c_{12}S_1 + c_{11}S_2 + c_{13}S_3$	$c_{12}S_1 + c_{11}S_2 + c_{12}S_3$	$c_{12}S_1 + c_{11}S_2 + c_{12}S_3$
$T_3 =$	$c_{13}S_1 + c_{23}S_2 + c_{33}S_3$	$c_{13}S_1 + c_{13}S_2 + c_{33}S_3$	$c_{12}S_1 + c_{12}S_2 + c_{11}S_3$	$c_{12}S_1 + c_{12}S_2 + c_{11}S_3$
$T_4 =$	$c_{44}S_4$	$c_{44}S_4$	$c_{44}S_4$	$(c_{11}-c_{12})S_4/2$
$T_5 =$	$c_{55}S_5$	$c_{44}S_5$	$c_{44}S_5$	$(c_{11}-c_{12})S_5/2$
$T_6 =$	$c_{66}S_6$	$(c_{11}-c_{12})S_6/2$	$c_{44}S_6$	$(c_{11}-c_{12})S_6/2$

to express the strains in terms of the stresses given in Eq. (9), in which the quantities s_{qr} are the elastic

$$S_q = \sum_r s_{qr} T_r \qquad (r = 1, 2, 3, 4, 5, 6) \qquad (9)$$

compliances. If the six simultaneous equations of Eq. (8) are solved for the strains, the compliances are obtained in terms of the stiffnesses as in Eq. (10), where Δc is the determinant shown in Eq. (11)

$$s_{qr} = \Delta c_{qr}/\Delta c \qquad (10)$$

$$\Delta c = \begin{vmatrix} c_{11} & c_{12} & c_{13} & c_{14} & c_{15} & c_{16} \\ c_{12} & c_{22} & c_{23} & c_{24} & c_{25} & c_{26} \\ c_{13} & c_{23} & c_{33} & c_{34} & c_{35} & c_{36} \\ c_{14} & c_{24} & c_{34} & c_{44} & c_{45} & c_{46} \\ c_{15} & c_{25} & c_{35} & c_{45} & c_{55} & c_{56} \\ c_{16} & c_{26} & c_{36} & c_{46} & c_{56} & c_{66} \end{vmatrix} \qquad (11)$$

and Δc_{qr} is the cofactor obtained by deleting the row and column containing c_{qr} from the determinant Δc.

The 21 stiffnesses (or compliances) of the generalized Hooke's law describe the elastic behavior of a material belonging to the triclinic crystal system. The existence of symmetry elements reduces the number of independent elastic constants in the other crystal systems to the following numbers: monoclinic, 13; orthorhombic, 9; tetragonal, 7 or 6; trigonal, 7 or 6; hexagonal, 5; and cubic, 3. Materials belonging to all of these systems are anisotropic, and the elastic properties depend upon direction within the material. If the properties are independent of direction, the material is isotropic and its elastic behavior is completely described by two independent stiffnesses (or compliances). *See* CRYSTALLOGRAPHY.

The stress-strain relations, referred to the principal axes in the orthorhombic, hexagonal, cubic, and isotropic systems, are given in the table. The equations involving the compliances are completely analogous, with S and T interchanged, and s_{qr} written for c_{qr}, except where $T_q = \frac{1}{2}(c_{11} - c_{12}) S_q$, in which case $S_q = 2(s_{11} - s_{12}) T_q$.

Rochelle salt is an example of an orthorhombic crystal; materials which, although not crystalline, possess the same symmetry and matrix of elastic constants as orthorhombic crystals are said to be orthotropic. Wood and plywood are materials of this description, and orthotropic elastic theory has also been applied to laminated plastics and reinforced concrete.

Single-crystal zinc, cobalt, magnesium, and ice are hexagonal materials; they are transversely isotropic because the properties are independent of direction in all of the planes normal to the hexagonal axis.

Single-crystal copper, gold, silver, nickel, and the alkali halides (for example, sodium chloride) are important cubic materials. The stress-strain equations are derived from those of the orthorhombic system by superimposing the condition that the three principal directions are all equivalent. This does not mean that the properties are independent of direction; for example, the compliance s'_{11} in an arbitrary direction is given by Eq. (12), where a_1,

$$s'_{11} = s_{11} - 2(s_{11} - s_{12} - s_{44}/2)$$
$$\cdot (a_1^2 a_2^2 + a_2^2 a_3^2 + a_3^2 a_1^2) \qquad (12)$$

a_2, a_3 are the cosines of the angles between the arbitrary direction and the cubic axes. This equation shows that s'_{11} depends on orientation unless $s_{44}/2 = (s_{11} - s_{12})$.

[R. F. S. HEARMON]

Bibliography: A. P. Boresi and P. P. Lynn, *Elasticity in Engineering Mechanics: With an Introduction to Numerical Stress Analysis*, 1974; S. F. Borg, *Fundamentals of Engineering Elasticity*, 1962, reprint 1973; R. F. S. Hearmon, *An Introduction to Applied Avisotropic Elasticity*, 1961; J. F. Nye, *Physical Properties of Crystals: Their Representation by Tensors and Matrices*, 1957; S. P. Timoshenko and J. M. Gere, *Theory of Elastic Stability*, 2d ed., 1961; S. P. Timoshenko and J. N. Goodier, *Theory of Elasticity*, 3d ed., 1969.

Electret

A solid dielectric possessing persistent dielectric polarization. An electret is the analog of a magnet. Electrets are made by cooling suitable dielectrics from elevated temperatures in strong electric fields. A special class called photoelectrets is produced by the removal of light from an illuminated photoconductor in an electric field. *See* POLARIZATION OF DIELECTRICS.

Electrets can be prepared from certain organic waxes and resins (for example, carnauba wax) or from ferroelectric crystals or ceramics such as barium titanate. Photoelectrets have been prepared from sulfur, cadmium and zinc sulfides, and anthracene. Electrets are metastable; their polarizations decay slowly after removal of the applied field and more rapidly with increasing temperature. Space-charge polarization is the principal mechanism involved in electret formation, except for ferroelectric substances. *See* DIELECTRICS; FERROELECTRICS; MAGNET. [ROBERT D. WALDRON]

Electric charge

A basic property of elementary particles of matter. One does not define charge but takes it as a basic experimental quantity and defines other quantities in terms of it. The early Greek philosophers were aware that rubbing amber with fur produced properties in each that were not possessed before the rubbing. For example, the amber attracted the fur after rubbing, but not before. These new properties were later said to be due to "charge." The amber was assigned a negative charge and the fur was assigned a positive charge.

According to modern atomic theory, the nucleus of an atom has a positive charge because of its protons, and in the normal atom there are enough extranuclear electrons to balance the nuclear charge so that the normal atom as a whole is neutral. Generally, when the word charge is used in electricity, it means the unbalanced charge (excess or deficiency of electrons), so that physically there are enough "nonnormal" atoms to account for the positive charge on a "positively charged body" or enough unneutralized electrons to account for the negative charge on a "negatively charged body."

The rubbing process mentioned "rubs" electrons off the fur onto the amber, thus giving the amber a surplus of electrons, and it leaves the fur with a deficiency of electrons. *See* ELECTROSTATICS.

In line with the previously mentioned usage, the total charge q on a body is the total unbalanced charge possessed by the body. For example, if a sphere has a negative charge of 1×10^{-10} coulomb, it has 6.24×10^8 electrons more than are needed to neutralize its atoms. The coulomb is the unit of charge in the meter-kilogram-second (mks) system of units. *See* COULOMB'S LAW; ELECTRICAL UNITS AND STANDARDS.

The surface charge density σ on a body is the charge per unit surface area of the charged body. Generally, the charge on the surface is not uniformly distributed, so a small area ΔA which has a magnitude of charge Δq on it must be considered. Then σ at a point on the surface is defined by the equation below.

$$\sigma = \lim_{\Delta A \to 0} \frac{\Delta q}{\Delta A}$$

The subject of electrostatics concerns itself with properties of charges at rest, while circuit analysis, electromagnetism, and most of electronics concern themselves with the properties of charges in motion. *See* CAPACITANCE.   [RALPH P. WINCH]

Electric current

The net transfer of electric charge per unit time. It is usually measured in amperes. The passage of electric current involves a transfer of energy. Except in the case of superconductivity, a current always heats the medium through which it passes. For a discussion of the heating effect of a current *see* JOULE'S LAW.

On the other hand, a stream of electrons or ions in a vacuum, which also may be regarded as an electric current, produces no local heating. Measurable currents range in magnitude from the nearly instantaneous 10^5 or so amperes in lightning strokes to values of the order of 10^{-16} amp, which occur in research applications.

All matter may be classified as conducting, semiconducting, or insulating, depending upon the ease with which electric current is transmitted through it. Most metals, electrolytic solutions, and highly ionized gases are conductors. Transition elements, such as silicon and germanium, are semiconductors, while most other substances are insulators.

Electric current may be direct or alternating. Direct current (dc) is necessarily unidirectional but may be either steady or varying in magnitude. By convention it is assumed to flow in the direction of motion of positive charges, opposite to the actual flow of electrons. Alternating current (ac) periodically reverses in direction.

Conduction current. This is defined as the transfer of charge by the actual motion of charged particles in a medium. In metals the current is carried by free electrons which migrate through the spaces between the atoms under the influence of an applied electric field. Although the propagation of energy is a very rapid process, the drift rate of the individual electrons in metals is only of the order of a few centimeters per second. In a superconducting metal or alloy the free electrons continue to flow in the absence of an electric field after once having been started. In electrolytic solutions and ionized gases the current is carried by both positive and negative ions. In semiconductors the carriers are the limited number of electrons which are free to move, and the "holes" which act as positive charges.

Displacement current. When alternating current traverses a condenser, there is no physical flow of charge through the dielectric (insulating material), but the effect on the rest of the circuit is as if there were a continuous flow. Energy can pass through the condenser by means of the so-called displacement current. James Clerk Maxwell introduced the concept of displacement current in order to make complete his theory of electromagnetic waves. *See* ALTERNATING CURRENT; CONDUCTION (ELECTRICITY); DIELECTRICS; DIRECT CURRENT; DISPLACEMENT CURRENT; ELECTRICAL RESISTANCE; FREE-ELECTRON THEORY OF METALS; SEMICONDUCTOR; SUPERCONDUCTIVITY.

[JOHN W. STEWART]

Electric field

A condition in space in the vicinity of an electrically charged body such that the forces due to the charge are detectable. An electric field (or electrostatic field) exists in a region if an electric charge at rest in the region experiences a force of electrical origin. Since an electric charge experiences a force if it is in the vicinity of a charged body, there is an electric field surrounding any charged body.

Field strength. The electric field intensity (or field strength) E at a point in an electric field has a magnitude given by the quotient obtained when the force acting on a test charge q' placed at that point is divided by the magnitude of the test charge q'. Thus, it is force per unit charge. A test charge q' is one whose magnitude is small enough so it does not alter the field in which it is placed. The direction of E at the point is the direction of the force F on a positive test charge placed at the point. Thus, E is a vector point function, since it has a definite magnitude and direction at every point in the field, and its defining equation is Eq. (1).

$$\mathbf{E} = \frac{\mathbf{F}}{q'} \tag{1}$$

Principle of superposition. As applied to electric fields, this principle states that the total **E** at a point P due to the combined influence of a distribution of point charges is the vector sum of the electric field intensities that the individual point charges would produce at P if each acted alone. Thus, using the rationalized mks system of units, Eq. (2) holds, where $\epsilon_0 \cong 8.85 \times 10^{-12}$

$$E = \frac{1}{4\pi\epsilon_0} \sum_{i=1}^{n} \frac{q_i}{r_i^2} \quad \text{vector sum} \quad (2)$$

coulomb^2/newton-m^2 is the permittivity of empty space, q_i is the ith charge (in coulombs) in the distribution, and r_i is the distance in meters from q_i to P. The units of **E** in the mks system are newtons/coulomb, which are the same as volts/meter. A common method of solving for **E** in a particular known distribution of charges is to evaluate the vector sum in Eq. (2). In many cases, however, Gauss' theorem affords a more powerful and convenient method.

Electric displacement. Electric flux density or electric displacement **D** in a dielectric (insulating) material is related to **E** by either of the equivalent equations shown as Eqs. (3), where **P** is the polari-

$$D = \epsilon_0 E + P \qquad D = \epsilon E \qquad (3)$$

zation of the medium, and ϵ is the permittivity of the dielectric which is related to ϵ_0 by the equation $\epsilon = k\epsilon_0$, k being the relative dielectric constant of the dielectric. In empty space, $D = \epsilon_0 E$. The units of **D** are coulombs/meter^2.

In addition to electrostatic fields produced by separations of electric charges, an electric field is also produced by a changing magnetic field. The relationship between the **E** produced and the rate of change of magnetic flux density $d\mathbf{B}/dt$ which produces it is given by Faraday's law of induced electromotive forces (emfs) in Eq. (4), where $d\mathbf{s}$ is a

$$\oint E \cdot ds = -\int_A \frac{dB}{dt} \cdot dA \qquad (4)$$

vector element of path length directed along the path of integration in the general sense of **E**. Thus $\oint E \cdot ds$ is the emf induced in this closed path of integration. The area of the surface bounded by the path of integration is A and the direction of $d\mathbf{A}$, an infinitesimal vector element of this area, is the direction of the thumb of the right hand when the fingers encircle the path of integration in the general sense of **E**. The right side of Eq. (4) is seen to be the negative of the time rate of change of the magnetic flux linking the path of integration chosen for the left side.

In an electrostatic field, $\oint E \cdot ds$ is always zero. *See* ELECTRIC CHARGE; ELECTROMAGNETIC INDUCTION; POTENTIALS.   [RALPH P. WINCH]

Bibliography: Berkeley Physics Course, vol. 2: *Electricity and Magnetism*, 1970; R. Resnick and D. Halliday, *Physics*, 3d ed., 1978; F. W. Sears et al., *University Physics*, 5th ed., 1976.

Electric susceptibility

A dimensionless parameter measuring the ease of polarization of a dielectric. The susceptibility χ is equal to the ratio of polarization **P** to the product of electric field strength **E** and vacuum permittivity ϵ_0, as in Eq. (1) ($\epsilon_0 = 1$ in cgs electrostatic units, and 8.854×10^{-12} farad/m in mks units).

$$\chi = P/\epsilon_0 E \qquad (1)$$

The electric susceptibility is related to the dielectric constant κ' by Eq. (2), where γ is a geomet-

$$\chi = (\kappa' - 1)/\gamma \qquad (2)$$

rical factor equal to 4π or 1 in the cgs or mks systems, respectively. This ambiguity in the definition of χ is unfortunate and must be considered in evaluating published data. *See* ELECTRICAL UNITS AND STANDARDS.

The electric susceptibility can be related to the polarizability α by expressing the polarization in terms of molecular parameters. Thus Eqs. (3) hold,

$$P = N\langle\boldsymbol{\mu}\rangle_{avg} = N\alpha E_L \qquad \chi = \frac{N\alpha E_L}{\epsilon_0 E} \qquad (3)$$

where N is the number of molecules per unit volume, $\langle\boldsymbol{\mu}\rangle_{avg}$ is their average dipole moment, and E_L is the local electric field strength at a molecular site. At low concentrations, E_L approaches **E**, and the susceptibility is proportional to the concentration N. For a discussion of the properties and measurement of electric susceptibility *see* DIELECTRIC CONSTANT; POLARIZATION OF DIELECTRICS.

[ROBERT D. WALDRON]

Electrical impedance

The total opposition that a circuit presents to an alternating current. Impedance, measured in ohms, may include resistance R, inductive reactance X_L, and capacitive reactance X_C. *See* REACTANCE.

The impedance of the series RLC circuit is given by Eq. (1).

$$Z = \sqrt{R^2 + (X_L - X_C)^2} \text{ ohms (magnitude)} \quad (1)$$

In terms of complex quantities, this impedance is given by Eq. (2). The two components of Z are

$$Z = R + j(X_L - X_C) \qquad (2)$$

at right angles to each other in an impedance diagram. Therefore, impedance also has an associated angle, given by Eq. (3). The angle is called the

$$\theta = \arctan \frac{X_L - X_C}{R} \qquad (3)$$

phase, or power-factor, angle of the circuit. The current lags or leads the voltage by angle θ depending on whether X_L is greater or less than X_C.

Impedance may also be defined as the ratio of the rms voltage to the rms current, $Z = E/I$. This is a form of Ohm's law for ac circuits.

[BURTIS L. ROBERTSON]

Electrical resistance

That property of an electrically conductive material that causes a portion of the energy of an electric current flowing in a circuit to be converted into heat. In 1774 A. Henley showed that current flowing in a wire produced heat, but it was not until 1840 that J. P. Joule determined that the rate of conversion of electrical energy into heat in a conductor, that is, power dissipation, could be expressed by the relation given in notation (1).

$$H/t \propto I^2 R \qquad (1)$$

The day-to-day determination of resistance by measuring the rate of heat dissipation is not practical. However, this rate of energy conversion is also VI, where V is the voltage drop across the element in question and I the current through the element, as in Eq. (2), from which the more conventional

$$H/t \propto I^2R = VI \qquad (2)$$

relationship implied by Ohm's law, Eq. (3), is apparent. *See* ELECTRICAL RESISTIVITY; OHM'S LAW.

$$R = V/I \qquad (3)$$

[CHARLES E. APPLEGATE]

Electrical resistivity

The electrical resistance offered by a homogeneous unit cube of material to the flow of a direct current of uniform density between opposite faces of the cube. Also called specific resistance, it is an intrinsic, bulk (not thin-film) property of a material. Resistivity is usually determined by calculation from the measurement of electrical resistance of samples having a known length and uniform cross section according to Eq. (1), where ρ is the resistiv-

$$\rho = RA/l \qquad (1)$$

ity, R the measured resistance, A the cross sectional area, and l the length. In the mks system, the unit of resistivity is the ohm-meter. Therefore, in Eq. (1), resistance is expressed in ohms and the sample dimensions in meters.

The room-temperature resistivity of pure metals extends from approximately 1.5×10^{-8} ohm-meter for silver, the best conductor, to 135×10^{-8} ohm-meter for manganese, the poorest pure metallic conductor. Most metallic alloys also fall within the same range. Insulators have resistivities within the approximate range of 10^8 to 10^{16} ohm-meters. The resistivity of semiconductor materials, such as silicon and germanium, depends not only on the basic material but to a considerable extent on the type and amount of impurities in the base material. Large variations result from small changes in composition, particularly at very low concentrations of impurities. Values typically range from 10^{-4} to 10^5 ohm-meters. *See* ELECTRICAL RESISTANCE; SEMICONDUCTOR.

Temperature effects. The temperature coefficients (changes with temperature) of resistivity of pure metallic conductors are positive. Resistivity increases by about 0.4%/°K at room temperature and is nearly proportional to the absolute temperature over wide temperature ranges. As the temperature is decreased toward absolute zero, resistivity decreases to a very low residual value, ρ_0, for some metals. The resistivity of other metals abruptly changes to zero at some temperature above absolute zero, and they become superconductors. For example, lead becomes superconducting at 7.26°K and aluminum at 1.1°K. No simple theory has been developed to explain all of the effects observed experimentally. However, a first approximation is given by Matthiessen's rule, Eq. (2), where ρ is the bulk resistivity of a metal, ρ_0 is

$$\rho = \rho_0 + \rho(T) \qquad (2)$$

the resistivity at absolute zero and is dependent on impurities and crystal imperfections in the material, and $\rho(T)$ is a temperature-dependent factor related to the scattering of conduction electrons in the material by crystal lattice vibrations. *See* LATTICE VIBRATIONS; MATTHIESSEN'S RULE.

The dependence of resistivity on temperature is the basis on which resistance thermometers operate. Strain-free platinum thermometers provide the secondary standard for temperature measurement over a wide range of temperatures.

Certain alloys, such as constantan, manganin, and one of gold-chromium, which have very small or negligible temperature coefficients have been developed. These materials are used in the construction of resistance standards.

Most insulating materials and semiconductors have a negative temperature coefficient of resistivity. The resistivity decreases with increasing temperature and usually at a more rapid rate than with metals. This property is used to advantage for more sensitive, but less stable, resistance thermometers. However, it also limits the maximum operating temperature of semiconductor devices and the choice of suitable electrical insulation for high-temperature applications.

Magnetic effects. Metals, and some semiconductors in particular, exhibit a change in resistivity when placed in a magnetic field. Theoretical relations to explain the observed phenomena have not been well developed. For most metals, the change is approximately 0.01%/10 kgauss. Bismuth exhibits the unusual coefficient of about 100%/10 kgauss. Some semiconductors have coefficients approaching 5000%/10 kgauss. Transducers based on this effect are used to measure both dc and high-frequency magnetic field strength.

Elastic deformation effects. The resistivity of materials is sensitive to both tensile and compressive stress. The effect is reproducible for deformations within the elastic limit of the material, but there are many anomalous experimental observations for which no explanation has been developed. For instance, a sample being elongated in tension would be expected to change its resistivity in accordance with dimensional changes and the relationship of Eq. (1), but in fact, the change can be between 1/2 and 10 times that expected, depending on the material. Coefficients can be positive or negative, large or small. This phenomenon, even though not understood, has practical application in the design of electrical resistance strain gages and the measurement of ultrahigh pressures, that is, pressures of thousands of atmospheres.

[CHARLES E. APPLEGATE]

Bibliography: J. S. Dugdale, *Electrical Properties of Metals and Alloys*, 1978; D. E. Gray (ed.), American Institute of Physics *Handbook*, 3d ed., 1972.

Electrical units and standards

The standard in terms of which electrical quantities are evaluated, the quantities so adopted being known as units. The ohm, for example, is a unit of electrical resistance. The electrical units in practical use today, and also in extensive theoretical use, were designated by the Eleventh General Conference of Weights and Measures in 1960 as members of the International System of Units (Système International d'Unités, abbreviated SI in all languages). This action by the General Conference was the culmination of an effort initiated by A. Giorgi at the beginning of this century to bring the practical electrical units into a coherent system with appropriate mechanical units of the metric system.

ELECTRICAL UNITS

To accomplish the above objective, the base units for mechanical quantities were arbitrarily selected: the meter for the unit of length, the kilogram for the unit of mass, and the second for the unit of time. Units for other mechanical quantities are derived from these units in accordance with physical laws and concepts such as the unit of speed, the meter per second, and the unit for acceleration, the meter per second per second.

This system was originally called the mks system to distinguish it from the cgs system (based on the centimeter, gram, and second).

Meter-kilogram-second system. Acting under authority given it by the Eighth General Conference of Weights and Measures, the International Committee of Weights and Measures in 1937 proceeded to define a unit for force (now called the newton, N) and units for energy and power in mechanical terms. The theoretical magnitudes of these units are given below.

Unit of force. The force which gives to a mass of 1 kilogram an acceleration of 1 meter per second per second.

Joule (J). The work done when the point of application of the mks unit of force is displaced a distance of 1 meter in the direction of the force.

Watt (W). The power which gives rise to the production of energy at the rate of 1 joule per second.

The Committee then proceeded to define electric and magnetic units in terms of these mechanical units. The revised units were to replace the definitions which had been in effect for many years such as the "mercury ohm" and the "silver ampere." The revised definitions of electrical and magnetic units which have been accepted since 1948 were given by the Committee as follows.

Ampere (A). The constant current which, if maintained in two straight parallel conductors of infinite length, of negligible circular sections, and placed 1 meter apart in a vacuum, would produce between these conductors a force equal to 2×10^{-7} mks unit of force per meter of length.

Volt (V). The difference of electric potential between two points of a conducting wire carrying a constant current of 1 ampere, when the power dissipated between these points is equal to 1 watt.

Ohm (Ω). The electric resistance between two points of a conductor when a constant difference of potential of 1 volt, applied between these two points, produces in the conductor a current of 1 ampere, the conductor not being the seat of any electromotive force.

Coulomb (C). The quantity of electricity transported in 1 second by a current of 1 ampere.

Farad (F). The capacitance of a capacitor between the plates of which there appears a difference of potential of 1 volt when it is charged by a quantity of electricity equal to 1 coulomb.

Henry (H). The inductance of a closed circuit in which an electromotive force of 1 volt is produced when the electric current in the circuit varies uniformly at a rate of 1 ampere per second.

Weber (Wb). The magnetic flux which, linking a circuit of 1 turn, produces in it an electromotive force of 1 volt as it is reduced to zero at a uniform rate in 1 second.

The revised definitions were intended solely to fix the magnitudes of the units and not the methods to be followed for their practical realization. This realization is effected in accord with the well-known laws of electromagnetism. For example, the definition of the ampere represents only a particular case of the general formula expressing the forces which are developed between conductors carrying electric currents, chosen for the simplicity of its verbal expression. It serves to fix the constants in the general formula which has to be used for the realization of the unit.

A special name was added to the list by the Eleventh General Conference of Weights and Measures in 1960, the tesla (T) for the unit of magnetic flux density (one weber per square meter).

Centimeter-gram-second systems. Two systems of electric and magnetic units have been in use in scientific circles for a long time but both are rapidly giving way to the International System. They are the electrostatic system of units (esu) and the electromagnetic system of units (emu).

The electrostatic system defines a unit charge as that charge which exerts 1 cgs unit of force (1 dyne, which is equivalent to 10^{-5} newton) on another unit charge when separated from it by a distance of 1 centimeter in a vacuum. All other units of the system are derived from this definition by assigning unit coefficients in equations relating electric and magnetic quantities to each other. The units so derived are often referred to in terms of the SI units with the prefix "stat," for example, statvolts, statohms, and statamperes.

The electromagnetic system defines a unit magnetic pole (a highly fictitious concept) as that pole which exerts 1 cgs unit of force on another unit pole when separated from it by a distance of 1 centimeter in a vacuum. All other units of the system are derived from this definition in accord with the principles set forth above for the electrostatic system. Units so derived are often referred to in terms of the SI units with the prefix "ab," for example, abvolts, abohms, and abamperes. Special names are given to some magnetic units of the emu system such as the maxwell and the gauss, which correspond, respectively, to the weber and the tesla of SI, although differing from them in magnitude.

The magnitudes of corresponding units of the electrostatic and electromagnetic systems differ from each other by a factor theoretically equal to the speed of light, c, (3×10^{10} centimeters per second, approximately) or its square. Thus, 1 abampere is equal to 3×10^{10} statamperes; 1 statvolt is equal to 3×10^{10} abvolts; and hence 1 statohm is equal to 9×10^{20} abohms.

The esu system is found convenient for handling purely electrostatic problems. A combination of the two systems in which electrostatic quantities are expressed in esu and magnetic and electromagnetic quantities in emu, with appropriate use of the conversion constant c between the two systems, is called the Gaussian system. All of these systems are rapidly giving way to use of SI units in treatment of electric, magnetic, and electromagnetic phenomena.

ELECTRICAL STANDARDS

Electrical standards are the physical embodiments by means of which the electrical units are realized and maintained. The ampere is unique in this system, since an arbitrary constant, other than

(a)

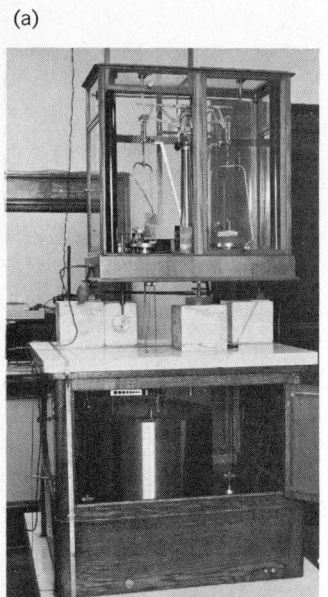

(b)

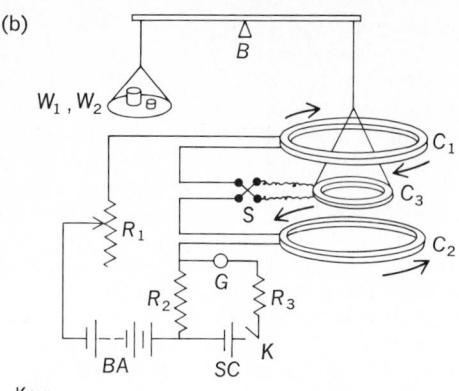

Key:

B	= weight beam	R	= protective resistor
W_1	= tare weight	BA	= battery
W_2	= balancing weight	S	= reversing switch
C_1, C_2	= fixed coils	SC	= standard cell
C_3	= moving coil	G	= galvanometer
R_1	= adjusting resistor	K	= key
R_2	= standard resistor		

Fig. 1. Current balance: (a) exterior and (b) schematic.

unity is employed in its definition to bring the entire system into agreement with the mechanical units while still adhering substantially to the old value for the unit.

Not all standards for electrical quantities are maintained in the national standards laboratories, such as the National Bureau of Standards; the only

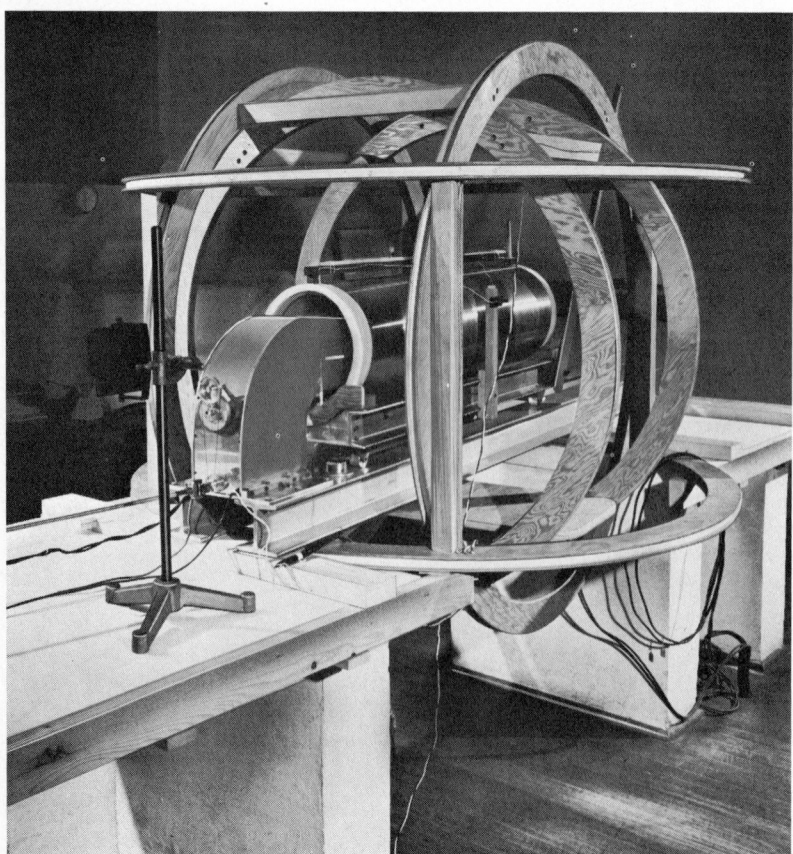

Fig. 2. Electrodynamometer, the current balance used in absolute determination of the ampere at the National Bureau of Standards.

standards maintained are those for the volt, the ohm, the farad, and sometimes the henry, since very stable standards for these quantities can be produced. The other electrical quantities are determined from suitable combinations of these.

Determination of the ampere. Since the ampere is defined in terms of the force between two current-carrying wires, the conventional means of determining the ampere is by some kind of current balance in which the force between the current-carrying elements is compared with the force of gravity on a known mass. One form of current balance consisting of two fixed coils and a movable coil supported by one arm of an equal arm balance is shown in Fig. 1. An electric current, supplied by a battery and controlled by a rheostat, is sent through the fixed and movable coils. The current flows through the fixed coils in opposite directions so that the magnetic fields produced by them are in opposition in the region of the movable coil. Here the magnetic field is directed horizontally and radially with respect to the axis of the coil system. The direction of the current in the movable coil is controlled by a reversing switch. When the current flowing through the movable coil is in one direction, the force exerted on it by the currents in the other coils is downward; but when the current is reversed, the force is upward. The force of gravity on the mass of the movable coil is balanced by a tare weight on the other arm of the balance.

The currents and the balancing weights are adjusted so that when the force due to the current is upward on the coil, the tare weight just balances the coil; but when the current is reversed, the additional weight must be added to achieve balance.

When balance is achieved, the current through the coil is given by Eq. (1), in which C is the con-

$$2\,Ci^2 = mg \tag{1}$$

stant computed from the dimensions of the coil assembly, i is the current in amperes, m is the mass of the small weight, and g is the acceleration of gravity at the place where the experiment is performed.

Current circulating in the coil system is passed through a standard resistor having a resistance of 1 ohm, approximately, but its value must be accurately known.

A standard cell, that is, an electrochemical cell which produces a constant emf, with a protective resistor and a galvanometer and key in series with it, is placed across the known resistor. If no deflection of the galvanometer is observed when the key is closed in this circuit, then the emf of the standard cell is given by Eq. (2), where r is the re-

$$ir = v \tag{2}$$

sistance of the known resistor. Thus the experiment for the determination of the ampere is actually an experiment for determining the emf of a standard cell.

Several such standard cells are calibrated by means of the current balance, and they in turn preserve the unit of voltage when the current balance is not in use. A current of 1 ampere may thus be established at anytime by sending a current of such strength through a 1-ohm resistor that it gives rise to exactly 1-volt drop across it.

In modern versions of the current balance, both the fixed coils and the movable coils are wound in a single layer on cylinders of marble, pyrex, or fused silica, in which accurate grooves have been lapped to maintain the wires in fixed positions so that the dimensions of the coils can be accurately measured.

Another form of the current balance, the Pellat balance, is called an electrodynamometer (Fig. 2). It consists of a long solenoid within which another cylindrical coil is balanced on knife edges so that it is free to rotate. The axis of the inner coil is at right angles to the axis of the solenoid. When the electric current is sent through both coils, a torque is exerted on the movable coil which is balanced by a weight suspended at the end of an arm extending out horizontally from the rotatable coil. When balance is attained, the current through the coil system is given by Eq. (3), where C is the computed

$$Ci^2 = mgl \qquad (3)$$

constant of the coil system, and l the length of the lever arm on which the weight is supported. When the current in the rotatable coil is reversed, the torque is reversed and the balancing weight is moved to an arm on the opposite side of the coil. The electrical circuit used with the Pellat balance is essentially identical with that used with the other type. Experiments at the National Bureau of Standards with the two forms of current balance gave agreement to within about 1 part in a million.

Similar experiments in the national standardizing laboratories of other countries have served to establish the value for the unit of voltage for those countries, but there were known differences in values assigned to the standards in various countries, as demonstrated by international comparisons. Increased accuracy with which such experiments have been conducted since World War II has permitted assignment of values to the electrical standards in much closer conformity with the theoretically defined values of the units. By international agreement, decision was made that electrical standards in use throughout the world should be referred to a uniform basis after Jan. 1, 1969, bringing them into closer agreement with the units they embody. The new volt at the National Bureau of Standards is 8.4 parts in a million smaller than the old volt. Thus, a standard cell which was assigned a value of, say, 1.0183000 volts before the changeover would now be assigned a value of 1.0183086 volts.

Determination of the ohm. Several methods have been employed for the determination of the ohm. The impedance of ohms of an electric circuit containing only resistive elements for direct or alternating current is given by Eq. (4), where r is the

$$z = r \qquad (4)$$

resistance in ohms. But for a circuit containing only inductive or capacitive elements the impedance for alternating current is, respectively, given by Eqs. (5) and (6), where L is the inductance in

$$z = \omega L \qquad (5)$$
$$z = 1/\omega C \qquad (6)$$

henries, C is the capacitance in farads, and ω is the angular frequency of the alternating current in radians per second.

The inductance of an arrangement of current-carrying elements or the capacitance of an arrangement of charge-bearing elements can be cal-culated readily from electromagnetic principles and the geometry of the arrangement in units of inductance (henries) or in units of capacitance (farads). The impedance in ohms of either of these arrangements may be calculated for given alternating current frequencies from Eqs. (5) or (6). If one of these calculated inductors or capacitors is placed in one arm of an alternating-current bridge, its impedance can be compared with that of a resistor in another arm of the bridge, thus establishing the value of the resistor in ohms (Fig. 3).

For many years the values of resistors were obtained from the values of computable inductors in most cases. The discovery of the Thompson-Lampard theorem in electrostatics in 1956 led to a great improvement in the art. In comprehending this theorem, a new form of capacitor should be visualized, consisting of a long metal tube divided into four segments by longitudinal cuts coplanar with the axis of the tube. If C_1 is the capacitance per unit length between one opposite pair of segments when the other pair is grounded, and C_2 is the capacitance per unit length between the other pair of opposite segments when the first pair is grounded, then Eq. (7) holds, where ϵ_0 is the electric constant.

$$e^{-\Pi C_1/\epsilon_0} + e^{-\Pi C_2/\epsilon_0} = 1 \qquad (7)$$

If C_1 and C_2 are nearly equal, the tube length is reduced as shown in Eq. (8), where $\Delta C = C_1 - C_2$.

$$\overline{C} = \frac{C_1 + C_2}{2} = \epsilon_0 \frac{\ln 2}{\Pi}\left(1 + 0.087\frac{\Delta C^2}{\overline{C}^2}\right) \qquad (8)$$

Fig. 3. Computable capacitance standard used in absolute determination of ohm.

The tube need not be of cylindrical form. It is possible to replicate this arrangement with carefully machined parts such as cylindrical gage blocks insulated from each other, each block corresponding to one segment of the tube. With this arrangement C_1 and C_2 can be made nearly equal and ΔC becomes vanishingly small.

Since the length of cylindrical gage blocks may be measured with very great accuracy, the capacitance of this type of capacitor can be calculated from the length measurement with correspondingly great accuracy. The greatest uncertainty in calculation of the capacitance arises from uncertainty of the knowledge of the speed of light because the constant ϵ_0 is implicitly defined in the International System by Eq. (9), where c is the speed of light

$$c^2 = 1/\epsilon_0\mu_0 \tag{9}$$

and μ_0 has the arbitrarily assigned numerical value $4\Pi \times 10^{-7}$.

The impedance of a practical-size capacitor of this type is very great, about 10^8 ohms. However, its impedance can be compared with that of a 1-ohm resistor in successive steps so that nearly the full accuracy of the calculation can be realized. *See* PHYSICAL MEASUREMENT; UNITS OF MEASUREMENT.   [ALVIN G. MC NISH]

Bibliography: F. L. Hermech and R. F. Dziuba, *Precision Measurement and Calibration*: *Electrical*, Nat. Bur. Stand. Spec. Publ. no. 300, vol. 3, 1968; F. D. Rossini, *Fundamental Measures and Constants for Science and Technology*, 1974.

Electricity

Electricity comprises those physical phenomena involving electric charges and their effects when at rest and when in motion. Electricity is manifested as a force of attraction, independent of gravitational and short-range nuclear attraction, when two oppositely charged bodies are brought close to one another. It is now known that the elementary (nondivisible) electric charges are possessed by electrons and protons. The charge of the electron is equal in magnitude to that of the proton, but is electrically opposite. The electron's charge is arbitrarily termed negative, and that of the proton, positive. Magnetism, those physical phenomena involving magnetic fields and their effects upon materials, manifests itself in the presence of moving electric charge. For this reason, magnetism was originally considered to be a part of electricity. *See* ELECTRIC CHARGE; MAGNETISM.

Historical development. The earliest observations of electric effects were made on naturally occurring substances. Magnetism was observed in the attraction of metallic iron by the iron ore magnetite. The natural resin amber was found to become electrified when rubbed (triboelectrification) and to attract lightweight objects. Both of these phenomena were known to Thales of Miletus (640–546 B.C.). Jerome Cardan in 1551 first clearly distinguished the difference between the attractive properties of amber and magnetite, thus presaging the division of electric and magnetic effects. He also envisioned electricity as a type of fluid, a viewpoint that was developed more extensively in the late 18th and early 19th centuries. In 1600 W. Gilbert observed variations in the amounts of electrification of various substances. He divided

substances into two classes, according to whether they did or did not electrify by rubbing. The division actually is into poor and good conductors, respectively. A two-fluid theory was first proposed by C. F. duFay in 1733. A one-fluid theory of electricity was propounded in 1747 by Benjamin Franklin, who called an excess of the fluid positive electrification, and a deficiency of fluid negative electrification. This theory fell into disrepute, but the choice of positive and negative remains. Although fluid theories of electricity were superseded at the end of the 19th century, the concept of electricity as a substance persists.

The quantitative development of electricity began late in the 18th century. J. B. Priestley in 1767 and C. A. Coulomb in 1785 discovered independently the inverse-square law for stationary charges. This law serves as a foundation for electrostatics. *See* COULOMB'S LAW; ELECTROSTATICS.

In 1800 A. Volta constructed and experimented with the voltaic pile, the predecessor of modern batteries. It provided the first continuous source of electricity. In 1820 H. C. Oersted demonstrated magnetic effects arising from electric currents. The production of induced electric currents by changing magnetic fields was demonstrated by M. Faraday in 1831. In 1851 he also proposed giving physical reality to the concept of lines of force. This was the first step in the direction of shifting the emphasis away from the charges and onto the associated fields. *See* ELECTROMAGNETIC RADIATION; ELECTROMAGNETISM.

In 1865 J. C. Maxwell presented his mathematical theory of the electromagnetic field. This theory proposed a continuous electric fluid. It remains valid today in the large realm of electromagnetic phenomena where atomic effects can be neglected. Its most radical prediction, the propagation of electromagnetic radiation, was convincingly demonstrated by H. Hertz in 1887. Thus Maxwell's theory not only synthesized a unified theory of electricity and magnetism, but also showed optics to be a branch of electromagnetism. *See* MAXWELL'S EQUATIONS.

The developments of theories about electricity subsequent to Maxwell have all been concerned with the microscopic realm. Faraday's experiments on electrolysis in 1833 had indicated a natural unit of electric charge, thus pointing toward a discrete rather than continuous charge. Thus, the groundwork for exceptions to Maxwell's theory of electromagnetism was laid even before the theory was developed. H. A. Lorentz began the attempt to reconcile these viewpoints with his electron theory in 1895. He postulated discrete charges, called electrons. The interactions between the electrons were to be determined by the fields as given by Maxwell's equations. The existence of electrons, negatively charged particles, was demonstrated by J. J. Thomson in 1897 using a Crookes tube. The existence of positively charged particles (protons) was shown shortly afterward (1898) by W. Wien, who observed the deflection of canal rays. Since that time, many particles have been found having charges numerically equal to that of the electron. The question of the fundamental nature of these particles remains unsolved, but the concept of a single elementary charge unit is apparently still valid. Of these many particles only two, the electron and the proton, exist in a stable condition on Earth. *See* BARYON; ELECTRON; ELEMENTARY PARTICLE; HYPERON; MESON; PROTON; QUARKS.

A second departure from classical Maxwell theory was brought on by M. Planck's studies of the electromagnetic radiation emitted by "black" bodies. These studies led Planck to postulate that electromagnetic radiation was emitted in discrete amounts, called quanta. This quantum hypothesis ultimately led to the formulation of modern quantum mechanics. The most satisfactory fusion of electromagnetic theory and quantum mechanics was achieved in 1948 with the work of J. Schwinger and R. Feynman in quantum electrodynamics, which suppressed the particle aspect and emphasized the field. *See* HEAT RADIATION; QUANTUM ELECTRODYNAMICS; QUANTUM MECHANICS.

Sources. The sources of electricity in modern technology depend strongly on the application for which they are intended.

The principal use of static electricity today is in the production of high electric fields. Such fields are used in industry for testing the ability of components such as insulators and condensers to withstand high voltages, and as accelerating fields for charged-particle accelerators. The principal source of such fields today is the Van de Graaff generator. *See* PARTICLE ACCELERATOR.

The major use of electricity today arises in devices using electric currents alternating at low or zero frequency. The use of alternating current, introduced by S. Z. de Ferranti in 1885–1890, allows power transmission over long distances at very high voltages with a resulting low percentage power loss followed by highly efficient conversion to lower voltages for the consumer through the use of transformers. Large amounts of zero-frequency current, that is, direct current, are used in the electrodeposition of metals, both in plating and in metal production, for example, in the reduction of aluminum ore. To avoid power transmission difficulties, such facilities are frequently located near sources of abundant power. *See* ALTERNATING CURRENT; DIRECT CURRENT; ELECTRIC CURRENT.

The principal sources of low-frequency electricity are rotary generators whose operation is based on the Faraday induction principle. The force to drive such generators derives from the flow of water or the expansion of gases, as in steam and internal combustion engines. The primary heat has been derived principally from fossil fuels. Economic considerations, particularly the cost of natural gas and oil and the need to conserve these for petrochemical purposes, are leading to increased reliance on nuclear reactors as the heat source. In addition, the use of coal is reemerging, and intensive efforts are being made for the discovery and development of geothermal sources. Other sources, such as fusion, solar, and oceanic, appear several decades away from significant application.

A more direct method of using fission or fusion reactors is the direct conversion of the energy released in the nuclear process into electricity. This has been achieved on a laboratory scale in the case of fission reactors. *See* NUCLEAR FUSION.

Many high-frequency devices, such as communications equipment, television, and radar, involve the consumption of only moderate amounts of power, generally derived from low-frequency sources. If the power requirements are moderate and portability is needed, the use of ordinary chemical batteries is possible. Ion-permeable membrane batteries are a later development in this line. Fuel cells, particularly hydrogen-oxygen systems, are being developed. They have already found extensive application in Earth satellite and other space systems. The successful use of thermoelectric generators based on the Seebeck effect in semiconductors has been reported in the Soviet Union and in the United States. In a particularly compact low-power device constructed in the United States, the heat needed for the operation of such a generator has been supplied by the energy release in the radioactive decay of suitably encapsulated isotopes produced in fission reactors. *See* THERMOELECTRICITY.

The Bell solar battery, also a semiconductor device, has been used to provide charging current for storage batteries in telephone service and in communications equipment in artificial satellites.

There are a number of other effects which might also serve to convert various forms of energy into electrical energy, but they do not seem generally practicable.

The changing magnetic flux required for the Faraday induction may be produced by an oscillating (rather than rotary) mechanism or by varying the temperature of a magnetic circuit whose components are made of a substance with a highly temperature-dependent permeability. It has been proposed to extract the energy of the fission (or possibly the fusion) reaction directly by inducing currents in external circuits by the changing magnetic field of bursts of ions from the reaction.

Direct conversion of mechanical energy into electrical energy is possible by utilizing the phenomena of piezoelectricity and magnetostriction. These have some application in acoustics and stress measurements. Pyroelectricity is a thermodynamic corollary of piezoelectricity. *See* MAGNETOSTRICTION; PIEZOELECTRICITY; PYROELECTRICITY.

Some other sources of electricity are those in which charged particles are released with some energy and collected in some manner. Charged particles are suitably released in radioactive decay, in the photoelectric effect, and in thermionic emission, among other ways. The photovoltaic effect may also be in this group. *See* PHOTOELECTRICITY.

The differences of work functions of various materials can be used for energy conversion. The contact potential difference may be used to convert heat directly to electricity or to provide improved collection for currents arising from some other source such as radioactivity. *See* WORK FUNCTION (THERMODYNAMICS).

Other possible sources of electricity arise from the existence of electrokinetic potentials in flowing fluids and of phase-transition potentials such as occur in the Workman-Reynolds effect. The possibilities of combining several effects also exist as exemplified in thermogalvanic potentials. It also appears that organic materials (as distinguished from the inorganic materials for which most of the work already described was done) merit investigation. A primitive type of organic solar battery has been developed. *See* CIRCUIT (ELECTRICITY); CONDUCTION (ELECTRICITY); ELECTRICAL UNITS AND STANDARDS.

[WALTER ARON]

Bibliography: P. H. Abelson (ed.), *Energy: Use, Conservation and Supply*, American Association for the Advancement of Science, 1974; B. I. Bleaney and B. Bleaney, *Electricity and Magnetism*, 3d ed., 1976; R. P. Feynman et al., *Feynman Lectures on Physics*, vol. 2, 1964; E. M. Pugh and E. W. Pugh, *Principles of Electricity and Magnetism*, 2d ed., 1970.

Electrodynamics

The study of the relations between electrical, magnetic, and mechanical phenomena. This includes considerations of the magnetic fields produced by currents, the electromotive forces induced by changing magnetic fields, the forces on currents in magnetic fields, the propagation of electromagnetic waves, and the behavior of charged particles in electric and magnetic fields. Classical electrodynamics deals with fields and charged particles in the manner first systematically described by J. C. Maxwell, whereas quantum electrodynamics applies the principles of quantum mechanics to electrical and magnetic phenomena. Relativistic electrodynamics is concerned with the behavior of charged particles and fields when the velocities of the particles approach that of light. Cosmic electrodynamics is concerned with electromagnetic phenomena occuring on celestial bodies and in space. *See* ELECTROMAGNETISM; MAXWELL'S EQUATIONS; QUANTUM ELECTRODYNAMICS; RELATIVISTIC ELECTRODYNAMICS.

[JOHN W. STEWART]

Electromagnetic field

A changing magnetic field always produces an electric field, and conversely, a changing electric field always produces a magnetic field. This interaction of electric and magnetic forces gives rise to a condition in space known as an electromagnetic field. The characteristics of an electromagnetic field are expressed mathematically by Maxwell's equation. *See* ELECTRIC FIELD; ELECTROMAGNETIC RADIATION; ELECTROMAGNETIC WAVE; MAGNETIC FIELD; MAXWELL'S EQUATIONS.

[JESSE W. BEAMS]

Electromagnetic induction

The production of an electromotive force either by motion of a conductor through a magnetic field in such a manner as to cut across the magnetic flux or by a change in the magnetic flux that threads a conductor. *See* MAGNETIC FLUX.

Motional electromotive force. A charge moving perpendicular to a magnetic field experiences a force that is perpendicular to both the direction of the field and the direction of motion of the charge. In any metallic conductor, there are free electrons, electrons that have been temporarily detached from their parent atoms.

If a conducting bar (Fig. 1) moves through a magnetic field, each free electron experiences a force due to its motion through the field. If the direction of the motion is such that a component of the force on the electrons is parallel to the conduc-

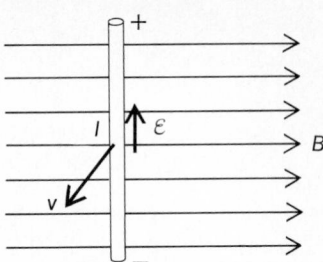

Fig. 1. Flux density *B*, motion *v*, and induced emf$_e$ when a conductor of length *l* moves in a uniform field. (*From R. L. Weber, M. W. White, and K. V. Manning, Physics for Science and Engineering, McGraw-Hill, 1957*)

tor, the electrons will move along the conductor. The electrons will move until the forces due to the motion of the conductor through the magnetic field are balanced by electrostatic forces that arise because electrons collect at one end of the conductor, leaving a deficit of electrons at the other. There is thus an electric field along the rod, and hence a potential difference between the ends of the rod while the motion continues. As soon as the motion stops, the electrostatic forces will cause the electrons to return to their normal distribution.

From the definition of magnetic induction (flux density) *B*, the force on a charge *q* due to the motion of the charge through a magnetic field is given by Eq. (1), where the force *F* is at right angles to a

$$F = Bqv \sin \theta \qquad (1)$$

plane determined by the direction of the field, and the component $v \sin \theta$ of the velocity is perpendicular to the field. When *B* is in webers/m^2, *q* is in coulombs, and *v* is in meters/sec, the force is in newtons. *See* MAGNETIC INDUCTION.

The electric field intensity *E* due to this force is given in magnitude and direction by the force per unit positive charge. The electric field intensity is equal to the negative of the potential gradient along the rod. In motional electromotive force (emf), the charge being considered is negative. Thus, Eqs. (2) hold. Here *l* is the length of the con-

$$E = \frac{F}{-q} = -Bv \sin \theta = -\frac{\mathscr{E}}{l}$$

$$\mathscr{E} = Blv \sin \theta \qquad (2)$$

ductor in a direction perpendicular to the field, and $v \sin \theta$ is the component of the velocity that is perpendicular to the field. If *B* is in webers/m^2, *l* is in meters, and *v* is in meters/sec, the emf $\mathscr{E}$ is in volts.

This emf exists in the conductor as it moves through the field whether or not there is a closed circuit. A current would not be set up unless there were a closed circuit, and then only if the rest of the circuit does not move through the field in exactly the same manner as the rod. For example, if the rod slides along stationary tracks that are connected together, there will be a current in the closed circuit. However, if the two ends of the rod were connected by a wire that moved through the field with the rod, there would be an emf induced in the wire that would be equal to that in the rod and opposite in sense in the circuit. Therefore, the net emf in the circuit would be zero, and there would be no current.

$\Phi_2(x,y)$ depends on the boundary conditions. *See* WAVE EQUATION.

Spherical waves. A wave is spherical when the instantaneous value of any field element such as **E** or **B** is constant in phase over a sphere. The radiation from any source of finite dimensions becomes spherical at great distances in an unbounded, isotropic, homogeneous medium. The equation for an undamped spherical wave is Eq. (5). The first term represents a diverging and the

$$F = r^{-1}\Phi_1(\theta,\varphi)\,f(r-vt) \\ + r^{-1}\Phi_2(\theta,\varphi)\,f(r+vt) \quad (5)$$

second a converging wave. Again, the form of $\Phi_1(\theta,\varphi)$ and $\Phi_2(\theta,\varphi)$ depends on the nature of the source and other boundary conditions.

Damped waves. If there are energy losses which are proportional to the square of the amplitude, as in the case of a medium of conductivity γ which

obeys Ohm's law, then the wave is exponentially damped, and Eq. (1) becomes Eq. (6). The symbol

$$E_x = E_0 e^{-\alpha z}\cos(\omega t - \beta z) \quad (6)$$

α is called the attenuation constant, and β the wave number or phase constant which equals ω/v', where v' is the damped-wave velocity. The electric wave amplitude at the origin has been taken as E_0. The ratio of E_0 to B_0, as well as that of α to β, depends on the permeability μ, the capacitivity ϵ, and the conductivity γ of the medium. In terms of the phasor $\check{E}_x$, Eq. (6) may be written as the real part of Eq. (7). This is exactly the form for the cur-

$$E_x = \check{E}_x e^{j\omega t} = E_0 e^{-(\alpha+j\beta)z}e^{j\omega t} \quad (7)$$

rent on a transmission line. (Phasors are complex numbers of form such that, when multiplied by $e^{j\omega t}$, the real part of the product gives the amplitude, phase, and time dependence.)

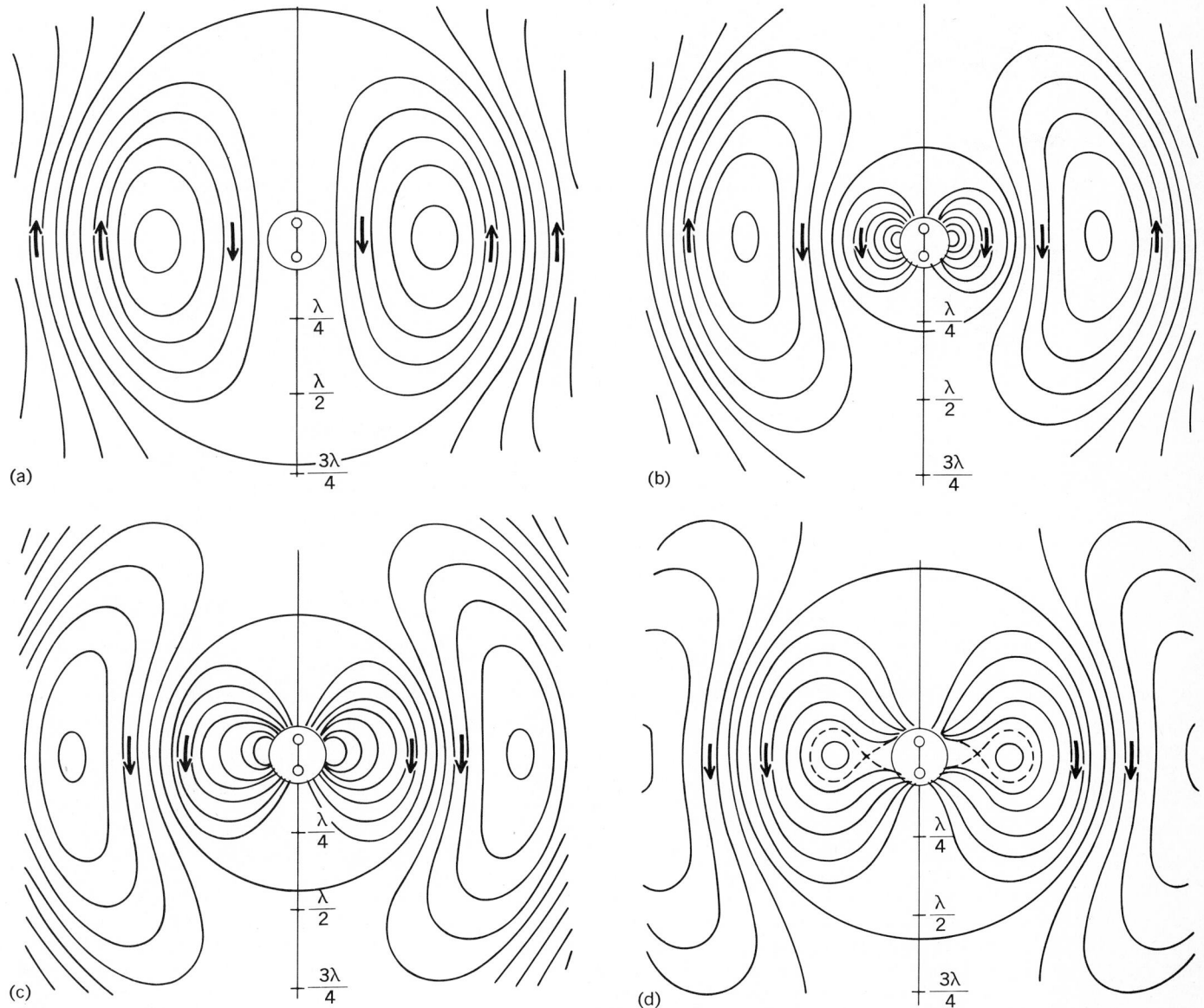

Diagrams of electric dipole. The outward moving electric field lines generated by the Hertzian oscillator are shown at successive eighth-period intervals. (*a*) $t=0$; (*b*) $t=T/8$; (*c*) $t=T/4$; (*d*) $t=3T/8$.

Wave impedance. Those trained in transmission line theory find it useful to apply the same techniques to wave theory. Consider an isolated tubular section of the wave in Eq. (1) bounded by $x=0$, $x=1$, and $y=0$, $y=1$ as a transmission line. The potential across the line between $x=0$ and $x=1$ is E. The line integral of B around the $x=0$ boundary from $y=0$ to $y=1$ is μI by Ampère's law and equals B because B is zero on the negative side. Thus, the impedance of the line is, making use of Eqs. (1) and (2), given by Eq. (8). This de-

$$\check{Z}_k = \frac{V}{I} = \frac{\mu E}{B} = \frac{E}{H} = \left(\frac{\mu}{\epsilon}\right)^{1/2} = \eta \qquad (8)$$

pends only on the properties of the medium and is known as the wave impedance. In transmission line theory the ratio μ/ϵ would be replaced by the ratio of the series impedance $\check{Z}_L = j\omega L$ to the shunt admittance $\check{Y} = j\omega C$, where L is the inductance per unit length, and C the capacitance per unit length across the line. If there is a resistance R per unit length across the line, then $1/R$ must be added to Y. This resistance is $1/\gamma$ for the tubular section. Thus, for a conducting medium, Eq. (8) becomes Eq. (9). The last term is a common transmission

$$\check{Z}_k = \left(\frac{j\omega\mu}{\gamma + j\omega\epsilon}\right)^{1/2} = \frac{j\omega\mu}{\alpha + j\beta} \qquad (9)$$

line form. The reflection and refraction of plane waves at plane boundaries separating different mediums may be calculated by transmission line formulas with the aid of Eqs. (8) and (9).

Electric dipole. A charge undergoing simple harmonic motion in free space is a dipole source when the amplitude of the motion is small compared with the wavelength. The term is loosely applied to the Hertzian oscillator, usually pictured as a dumbbell-shaped conductor in which the electrons oscillate from one end to the other, leaving the opposite end periodically positive. An electric dipole of moment M is defined as the product qa when two large, equal and opposite charges, $+q$ and $-q$, are placed a small distance a apart. A dipole is oscillating when M is periodic in time and is the simplest source of spherical waves. Much can be learned by a study of H. Hertz's picture of the outward moving electric field lines at successive time intervals of one-eighth period in a plane which passes through the Hertzian oscillator axis, shown in the figure. The most striking feature of the pictures is that, after breaking loose from the dipole, all electric field lines are closed, which means that the divergence of E is zero. This is true of all unbounded waves. It is also noteworthy that the waves become truly spherical with a fixed wavelength λ only in a direction perpendicular to the dipole and at a distance which greatly exceeds the dipole dimensions. This distance is beyond the edges of the picture. Lengths $\lambda/4$, $\lambda/2$, and $3\lambda/4$ are marked off on the axis for comparison. The magnetic field lines are circles coaxial with the oscillator, so they intersect the plane of the diagram normally. They are most dense where the electric lines are closely spaced. The radiant energy emitted by atoms and molecules is essentially radiation of the dipole type. *See* ABSORPTION OF ELECTROMAGNETIC RADIATION; DIFFRACTION; HEAT RADIATION; INFRARED RADIATION; INTERFERENCE OF WAVES; LIGHT; MAXWELL'S EQUATIONS; PO-LARIZATION OF WAVES; RADIATION; REFLECTION OF ELECTROMAGNETIC RADIATION; REFRACTION OF WAVES; SCATTERING OF ELECTROMAGNETIC RADIATION; ULTRAVIOLET RADIATION; WAVE MOTION; X-RAYS.

[WILLIAM R. SMYTHE]

Bibliography: Berkeley Physics Course, vol. 2: *Electricity and Magnetism*, 1970; M. Born and E. Wolf, *Principles of Optics*, 5th ed., 1975; J. G. Brown, *X-rays and Their Applications*, 1975; D. Dearholt and W. McSpadden, *Electromagnetic Wave Propagation*, 1973; F. A. Jenkins and H. E. White, *Fundamental of Optics*, 4th ed., 1976; P. Lorrain and D. R. Corson, *Electromagnetic Fields and Waves*, 2d ed., 1970; S. Ramo, J. R. Whinnery, and T. Van Duzer, *Fields and Waves in Communications Electronics*, 1965; V. Rojansky, *Electromagnetic Fields and Waves*, 1979; A. Shadowitz, *The Electromagnetic Field*, 1974; W. R. Smythe, *Static and Dynamic Electricity*, 3d ed., 1968.

Electromagnetic wave

A disturbance, produced by the acceleration or oscillation of an electric charge, which has the characteristic time and spatial relations associated with progressive wave motion. A system of electric and magnetic fields moves outward from a region where electric charges are accelerated, such as an oscillating circuit or the target of an x-ray tube. The wide wavelength range over which such waves are observed is shown by the electromagnetic spectrum. The term electric wave, or Hertzian wave, is often applied to electromagnetic waves in the radar and radio range. Electromagnetic waves may be confined in tubes, such as wave guides, or guided by transmission lines. They were predicted by J. C. Maxwell in 1864 and verified experimentally by H. Hertz in 1884. *See* ELECTROMAGNETIC RADIATION.   [WILLIAM R. SMYTHE]

Electromagnetism

The branch of science dealing with the observations and laws relating electricity to magnetism. Electromagnetism is based upon the fundamental observations that a moving electric charge produces a magnetic field and that a charge moving in a magnetic field will experience a force.

The magnetic field produced by a current is related to the current, the shape of the conductor, and the magentic properties of the medium around it by Ampère's law. *See* AMPERE'S LAW.

The magnetic field at any point is described in terms of the force that it exerts upon a moving charge at that point. The electrical and magnetic units are defined in terms of the ampere, which in turn is defined from the force of one current upon another.

The association of electricity and magnetism is also shown by electromagnetic induction, in which a changing magnetic field sets up an electric field within a conductor and causes the charges to move in the conductor. *See* EDDY CURRENT; ELECTRICITY; ELECTROMAGNETIC INDUCTION; FARADAY'S LAW OF INDUCTION; HALL EFFECT; INDUCTANCE; LENZ'S LAW; MAGNETIC CIRCUITS; MAGNETIC FIELD; MAGNETIC FLUX; MAGNETIC INDUCTION; MAGNETISM; MAGNETOMOTIVE FORCE; MAXWELL'S EQUATIONS. RELUCTANCE.

[KENNETH V. MANNING]

Electromotive force (emf)

The electromotive force, represented by the symbol ε, around a closed path in an electric field is the work per unit charge required to carry a small positive charge around the path. It may also be defined as the line integral of the electric intensity around a closed path in the field. The abbreviation emf is preferred to the full expression since emf, also called electromotance, is not really a force. The term emf is applied to sources of electric energy such as batteries, generators, and inductors in which current is changing.

The magnitude of the emf of a source is defined as the electrical energy converted inside the source to some other form of energy (exclusive of electrical energy converted irreversibly into heat), or the amount of some other form of energy converted in the source into electrical energy, when a unit charge flows around the circuit containing the source. In an electric circuit, except for the case where an electric current is flowing through resistance and thus electrical energy is changed irreversibly into heat energy, electrical energy is converted into another form of energy only when current flows against an emf. On the other hand, some other form of energy is converted into electrical energy only when current flows in the same sense as an emf.

For a discussion of motional emf *see* ELECTRO-MAGNETIC INDUCTION. *See also* MAGNETOMOTIVE FORCE.　　　　　　　　　[RALPH P. WINCH]

Bibliography: E. M. Purcell, *Electricity and Magnetism*, Berkeley Physics Course, vol. 2, 1965; R. Resnick and D. Halliday, *Physics*, 3d ed., 1978; F. W. Sears et al., *University Physics*, 5th ed., 1976.

Electron

An elementary particle which is the negatively charged constituent of ordinary matter. The electron is the lightest known particle which possesses an electric charge. Its rest mass is $m_e \cong 9.1 \times 10^{-28}$ g, about 1/1836 of the mass of the proton or neutron, which are, respectively, the positively charged and neutral constituents of ordinary matter. Discovered in 1895 by J. J. Thomson in the form of cathode rays, the electron was the first elementary particle to be identified. *See* ATOMIC STRUCTURE AND SPECTRA; ELECTRIC CHARGE; ELEMENTARY PARTICLE; NUCLEAR STRUCTURE.

Charge. The charge of the electron is $-e \cong -4.8 \times 10^{-10}$ esu $= -1.6 \times 10^{-19}$ coulomb. The sign of the electron's charge is negative by convention, and that of the equally charged proton is positive. This is a somewhat unfortunate convention, because the flow of electrons in a conductor is thus opposite to the conventional direction of the current. *See* CONDUCTION (ELECTRICITY).

The most accurate direct measurement of e is the celebrated oil drop experiment of R. A. Millikan in 1909, in which the charges of droplets of oil in air are measured by finding the electric field which balances each drop against its weight. The weight of each drop is determined by observing its rate of free fall through the air, and using Stokes' formula for the viscous drag on a slowly moving sphere. The charges thus measured are integral multiples of e. For more precise values of e and m_e *see* ATOMIC CONSTANTS.

Electrons and matter. Electrons are emitted in radioactivity (as beta rays) and in many other decay processes; for instance, the ultimate decay products of all mesons are electrons, neutrinos, and photons, the meson's charge being carried away by the electrons. The electron itself is completely stable, according to all available evidence. Electrons contribute the bulk to ordinary matter; the volume of an atom is nearly all occupied by the cloud of electrons surrounding the nucleus, which occupies only about 10^{-13} of the atom's volume. The chemical properties of ordinary matter are determined by the electron cloud. *See* MESON; RADIOACTIVITY.

The electron obeys the Fermi-Dirac statistics, and for this reason is often called a fermion. One of the primary attributes of matter, impenetrability, results from the fact that the electron, being a fermion, obeys the Pauli exclusion principle; the world would be completely different if the lightest charged particle were a boson, that is, a particle that obeys Bose-Einstein statistics. *See* BOSE-EINSTEIN STATISTICS; EXCLUSION PRINCIPLE; FERMI-DIRAC STATISTICS.

Spin. Every elementary particle possesses an intrinsic angular momentum called its spin. The spin of the electron has the magnitude $1/2\ \hbar$, where $\hbar$ is Planck's constant h divided by 2π. An electron thus has two spin states: spin up and spin down. To describe this, the nonrelativistic wave function is a two-component function, that is, a vector in a two-dimensional spin-space; the two linearly independent vectors represent the two possible spin states. In 1928 P. A. M. Dirac derived the corresponding relativistic wave equation (Dirac equation). Here, the electron wave function must have four components; correspondingly, for a wave of given momentum, there are four internal states. In addition to the two-valued spin coordinate, there is an energy coordinate; that is, for a momentum p, the energy can be $\pm\sqrt{(m_e c^2)^2 + (pc)^2}$, where c is the velocity of light. *See* ELECTRON SPIN; RELATIVISTIC QUANTUM THEORY.

Positron. The negative energy states were at first an embarrassment, for they extend downward indefinitely; an electron would cascade indefinitely downward in energy, radiating photons. Electrons obey the exclusion principle, however, and therefore one can avoid this conclusion by assuming that in empty space all the negative energy states are already occupied, and so exclude any more electrons. A new process is possible now; such a negative energy electron, by absorbing energy, can go to a positive energy state. This leaves behind a hole in the sea of negative-energy electrons; the hole has a positive energy, because it represents a missing negative energy. In fact, such a hole has all the properties of an electron except that it appears to have a positive charge (because it represents a missing negative charge). This particle is the positron, first discovered in 1932 by C. D. Anderson in a cloud-chamber study of cosmic radiation. *See* POSITRON.

The electron (sometimes called a negatron) and the positron are on an equal footing; if one started with a Dirac wave equation for the positron, identifying electrons with holes in the negative-energy positron states, one would get an equivalent theory. The apparent dissymmetry inherent in the construction of the hole theory disappears from the results when the total charge, energy, and momentum of empty space is defined to be zero; actually, the dissymmetry can be avoided at all stages in the formalism of quantum field theory. *See* QUANTUM FIELD THEORY; SYMMETRY LAWS.

Magnetic moment. The electron has magnetic properties by virtue of its orbital motion about the nucleus of its atom and its rotation (spin) about its own axis. The magnetic moment of the electron is predicted by the Dirac equation to be the value shown in the equation below. The actual moment

$$\mu_D = \frac{e\hbar}{2m_e}$$

μ differs from μ_D by a small amount (anomalous magnetic moment) due to electromagnetic radiative corrections: $\mu = 1.0011\ \mu_D$. This theoretical value, calculated using renormalized quantum field theory, agrees with the experimental value. *See* QUANTUM ELECTRODYNAMICS.

Other information. It would be difficult to list all the articles wherein the electron forms an integral part of the discussion. The articles listed in this paragraph and in the preceding discussion are intended to be merely a representative sample. *See* BAND THEORY OF SOLIDS; COMPTON EFFECT; ELECTRICITY; ELECTROMAGNETIC RADIATION; ELECTRON CAPTURE; ELECTRON CONFIGURATION; ELECTRON DIFFRACTION; FREE-ELECTRON THEORY OF METALS; MAGNETISM; NONRELATIVISTIC QUANTUM THEORY; PARTICLE ACCELERATOR; QUANTUM MECHANICS; RELATIVISTIC ELECTRODYNAMICS.

[CHARLES J. GOEBEL]

Electron affinity

The amount of energy release when an electron at rest is captured by a species M, producing the negative ion M⁻. The electron affinity of a species M can also be thought of as the ionization potential of the negative ion M⁻. Stated in terms of a chemical equation, the electron affinity of a species M is equal to the exothermocity of the reaction $e + M \rightarrow M^-$, where the negative ion M⁻ is left in its lowest electronic, vibrational, and rotational state.

If the electron affinity of M is negative, the M⁻ ion is unstable with respect to decomposition into $M + e$. Most atoms have positive electron affinities, even though there is no net Coulomb attraction between the electron and the atom until the electron is close enough to be "a part of the atom." The simple rules of chemical valency provide a qualitative guide to the magnitude of electron affinities. Thus the noble gases, which have a filled outer electronic shell and are chemically inert, are not capable of binding an additional electron to form a negative ion. The largest electron affinities are possessed by the halogens, atoms which require only one additional electron to fill the valence shell.

The major exception to this concept is that multiply charged negative ions—for example, O⁻⁻, one of many multiply charged negative ions which are stable in solution—are not stable in the gas phase. The ability to place more than one additional electron in the valence shell of a neutral atom or molecule appears to come from the medium; the solvent shell surrounding the ion in liquid solutions and the amorphous or crystalline region surrounding the ion in solids.

Experimental methods. While accurate ionization potentials of the elements have been known for a number of years, comparable data for electron affinities of the elements have become available only more recently. In order to determine the ionization potential of an element, one can simply make a vapor of the element, place it in an optical spectrometer, and look for the onset of absorption corresponding to the photoionization process $h\nu + A \rightarrow A^+ + e$. The energy of the photon corresponding to the threshold wavelength for this process is the ionization energy of the species A. The analogous method for determination of an electron affinity is through observation of the threshold of the very similar photodetachment reaction $h\nu + A^- \rightarrow A + e$. Again, the threshold wavelength for this process corresponds to the electron affinity of the species A. Unfortunately, it has not proved possible to produce sufficiently large densities of negative ions to be able to observe directly the threshold for the photodetachment process in a photoabsorption measurement; consequently, determination of accurate electron affinities has lagged far behind the determination of accurate ionization potentials.

The major experimental advances which now enable accurate electron affinity determinations have been the development of ion-beam techniques and the availability of intense light sources in the form of lasers. In modern experiments to determine electron affinities, negative ions are formed in an electrical discharge, extracted through a small aperture into a high-vacuum region, formed into a negative ion beam, mass-analyzed, and intersected by an intense laser beam. The laser-beam−negative-ion-beam intersection takes place in a high-vacuum region where no collisions are likely. The occurrence of a photodetachment event is determined by detection of the photodetached electron.

Two experimental methods have evolved which can produce accurate electron affinities. In the first method the laser is a tunable laser, and one searches for the wavelength corresponding to the threshold for the photodetachment process. In this case the electron affinity is given directly by the threshold wavelength, and is in principle determinable to accuracies of 10^{-6} eV. In the second type of experiment, called photoelectron spectroscopy, a fixed-frequency laser (of known photon energy) is employed, and electrostatic fields are used to determine the kinetic energy of the ejected electron. From simple energy conservation arguments, the electron affinity is then given by the photon energy less the kinetic energy of the ejected electron. This latter technique is quite general, but is limited in accuracy by the resolution of the electron energy analyzer (typically 10^{-2} eV). *See* LASER.

Periodic trends. These laser photodetachment studies dramatically improved knowledge of the electron affinities of the elements. The illustration is a periodic table showing the current best values of the electron affinities of the elements. Most of the data shown here were obtained by using laser photodetachment methods. The periodic trends in electron affinities and the qualitative effects described earlier are immediately apparent. In addition, a number of more subtle trends are observable. For example, while one expects that filled-shell species such as the rare gases will

1 H 0.7542							2 He <0
3 Li 0.620	4 Be <0	5 B 0.282	6 C 1.268	7 N ≤0	8 O 1.462	9 F 3.399	10 Ne <0
11 Na 0.546	12 Mg <0	13 Al 0.442	14 Si 1.385	15 P 0.743	16 S 2.0772	17 Cl 3.615	18 Ar <0
19 K 0.5012	20 Ca <0	31 Ga 0.3	32 Ge 1.2	33 As 0.80	34 Se 2.0206	35 Br 3.364	36 Kr <0
37 Rb 0.4860	38 Sr <0	49 In 0.3	50 Sn 1.25	51 Sb 1.05	52 Te 1.9708	53 I 3.061	54 Xe <0
55 Cs 0.4715	56 Ba <0	81 Tl 0.3	82 Pb 0.349	83 Bi 0.947	84 Po 1.9	85 At 2.8	86 Rn <0

21 Sc <0	22 Ti 0.2	23 V 0.526	24 Cr 0.667	25 Mn <0	26 Fe 0.164	27 Co 0.667	28 Ni 1.157	29 Cu 1.226	30 Zn <0
39 Y ≈0	40 Zr 0.429	41 Nb 0.886	42 Mo 0.747	43 Tc 0.7	44 Ru 1.1	45 Rh 1.138	46 Pd 0.558	47 Ag 1.303	48 Cd <0
57 La 0.5	72 Hf <0	73 Ta 0.323	74 W 0.816	75 Re 0.15	76 Os 1.1	77 Ir 1.566	78 Pt 2.128	79 Au 2.3086	80 Hg <0

Periodic table showing the best values for the electron affinities of the elements. All values are reported in electronvolts. The value <0 implies that the negative ion is unstable with respect to decomposition into an electron and a neutral atom. The solid bar below represents the relative uncertainty in the electron affinity. (*From H. Hotop and W. C. Lineberger, Binding energies in atomic negative ions, J. Phys. Chem. Ref. Data, 4:539–576, 1975*)

not be capable of binding an additional electron, the illustration shows that half-filled shells (for example, N and P) also exhibit small or negative electron affinities. Again, this effect is the result of the fact that a half-filled valence shell is spherically symmetric and behaves somewhat as though it were a filled shell. A similar situation is also seen for half-filled *d*-shells, as in the transition metals.

These same techniques are beginning to provide a number of accurate electron affinity determinations for molecules and free radicals, and new insight into the structural and chemical properties of ions in the gas phase. *See* IONIZATION POTENTIAL.   [W. C. LINEBERGER]

Bibliography: R. R. Corderman and W. C. Lineberger, Negative ion spectroscopy, *Annu. Rev. Phys. Chem.*, 30:347–376, 1979; H. Hotop and W. C. Lineberger, Binding energies in atomic negative ions, *J. Phys. Chem. Ref. Data*, 4:539–576, 1975; B. K. Janousek and J. I. Brauman, Electron affinities, in M. T. Bowers (ed.), *Gas Phase Ion Chemistry*, 1980; H. S. W. Massey, *Negative Ions*, 3d ed., 1976.

Electron capture

The process in which an atom or ion passing through a material medium either loses or gains one or more orbital electrons. In the passage of charged particles (defined here as nuclei having more or less than Z atomic electrons, where Z is the atomic number) through matter, the capture (and loss) of electrons is an important process in the slowing down of the particles and therefore has a strong influence on their range. Thus a neutral hydrogen atom loses only about half as much energy per centimeter as the positively charged proton in passing through matter consisting of light elements.

For the ordinary charged particles (α-particles and protons) the capture process is important only at low energies, when the particle velocity is of the order of electron velocities in the stopping material, and thus is important at the end of the range. For fission fragments, however, which initially have a large excess of positive charges, electron capture occurs immediately and continues throughout the slowing-down process. This fact causes the energy-loss mechanisms at the latter part of the range to be different for fission fragments and protons or α-particles. The heavy ions (nuclei of oxygen, argon, and so forth with all atomic electrons stripped away) now available are intermediate in mass between fission fragments and the light particles and have higher velocity than fission fragments. Thus their energy loss is relatively unaffected by electron capture in the early part of the range, but in later stages this process has important consequences. *See* NUCLEAR FISSION.

The nuclear capture of electrons (*K*-capture) occurs by a process quite different from atomic capture and is in fact a consequence of the general β-interaction. This general interaction includes β⁻-decay (the oldest known β-transformation and hence the name), β⁺-decay (or positron decay), and *K*-capture, the latter so called because the electron captured by the nucleus is taken from the *K*-shell (the shell nearest the nucleus) of atomic electrons. The probability of occurrence of electron capture by the nucleus obviously depends on the amount of time the electrons spend at the nucleus, that is, on the size of the electron wave function at the nuclear center. Since to a very good approximation only electrons with zero orbital angular momentum have a wave function that is finite at the center, capture is not expected from any but the *K*-shell. However, a second-order process can occur, in which (to speak pictorially and thus somewhat imprecisely) an *s* electron (from the *K*-shell) is captured with the simultaneous transition of a *p* electron (from the *L*-shell) to the *K*-shell with the emission of γ-radiation. This differs from ordinary x-radiation following *K*-capture by the fact that energy is not conserved in the transition (it must, of course, be conserved in the whole process). Since *K*-electrons spend more time in the nucleus for large-*Z* nuclei than for small, *K*-capture is more probable in heavier nuclei. The second-order process (called *L*-capture) is even more strongly *Z*-dependent and actually controls the shape of the γ-ray spectrum for very heavy nuclei. *See* RADIOACTIVITY.

What has been concluded so far depends on the atomic (that is, electromagnetic) interactions. But the process itself is a result of the β-interaction, which is between the electron field and the nucleon field. This weak interaction (so called because the processes involving the interaction take place in times that are long on the nuclear time scale), which couples electrons (or positrons) with nuclei, γ-rays, and neutrinos, has been the subject of increased study because it has been shown to demonstrate nonconservation of parity. The electron and positron are identical except for electromagnetic interactions; thus a nucleus which is energetically capable of *K*-capture will also usually be capable of positron emission, and the two processes do indeed compete in several nuclei. *See* AUGER EFFECT; PARITY; QUANTUM FIELD THEORY.

[MC ALLISTER H. HULL]

Electron configuration

The orbital arrangement of an atom's electrons. Negatively charged electrons are attracted to a positively charged nucleus to form an atom or ion. Although such bound electrons exhibit a high degree of quantum-mechanical wavelike behavior, there still remain particle aspects to their motion. Bound electrons occupy orbitals that are somewhat concentrated in spatial shells lying at different distances from the nucleus. As the set of electron energies allowed by quantum mechanics is discrete, so is the set of mean shell radii. Both these quantized physical quantities are primarily specified by integral values of the principal, or total, quantum number n. The full electron configuration of an atom is correlated with a set of values for all the quantum numbers of each and every electron. In addition to n, another important quantum number is l, an integer representing the orbital angular momentum of an electron in units of $h/2\pi$, where h is Planck's constant. The values 1, 2, 3, 4, 5, 6, 7 for n and 0, 1, 2, 3 for l together suffice to describe the electron configurations of all known normal atoms and ions, that is, those that have their lowest possible values of total electronic energy. The first seven shells are also given the letter designations K, L, M, N, O, P, and Q respec-

Distribution of electrons in the atoms

Element and atomic number	K 1,0 $1s$	L 2,0 $2s$	L 2,1 $2p$	M 3,0 $3s$	M 3,1 $3p$	M 3,2 $3d$	N 4,0 $4s$	N 4,1 $4p$	N 4,2 $4d$	N 4,3 $4f$	O 5,0 $5s$	O 5,1 $5p$	O 5,2 $5d$	O 5,3 $5f$	Ground term	Ionization potential, eV
H 1	1	—	—	—	—	—	—	—	—	—	—	—	—	—	$^2S_{1/2}$	13.5981
He 2	2	—	—	—	—	—	—	—	—	—	—	—	—	—	$^1S_0$	24.5868
Li 3	2	1	—	—	—	—	—	—	—	—	—	—	—	—	$^2S_{1/2}$	5.3916
Be 4	2	2	—	—	—	—	—	—	—	—	—	—	—	—	$^1S_0$	9.322
B 5	2	2	1	—	—	—	—	—	—	—	—	—	—	—	$^2P^0_{1/2}$	8.298
C 6	2	2	2	—	—	—	—	—	—	—	—	—	—	—	$^3P_0$	11.260
N 7	2	2	3	—	—	—	—	—	—	—	—	—	—	—	$^4S^0_{3/2}$	14.534
O 8	2	2	4	—	—	—	—	—	—	—	—	—	—	—	$^3P_2$	13.618
F 9	2	2	5	—	—	—	—	—	—	—	—	—	—	—	$^2P^0_{3/2}$	17.422
Ne 10	2	2	6	—	—	—	—	—	—	—	—	—	—	—	$^1S_0$	21.564
Na 11		Neon configuration		1	—	—	—	—	—	—	—	—	—	—	$^2S_{1/2}$	5.139
Mg 12				2	—	—	—	—	—	—	—	—	—	—	$^1S_0$	7.646
Al 13				2	1	—	—	—	—	—	—	—	—	—	$^2P^0_{1/2}$	5.986
Si 14				2	2	—	—	—	—	—	—	—	—	—	$^3P_0$	8.151
P 15				2	3	—	—	—	—	—	—	—	—	—	$^4S^0_{3/2}$	10.486
S 16				2	4	—	—	—	—	—	—	—	—	—	$^3P_2$	10.360
Cl 17				2	5	—	—	—	—	—	—	—	—	—	$^2P^0_{3/2}$	12.967
Ar 18				2	6	—	—	—	—	—	—	—	—	—	$^1S_0$	15.759
K 19			Argon configuration			—	1	—	—	—	—	—	—	—	$^2S_{1/2}$	4.341
Ca 20						—	2	—	—	—	—	—	—	—	$^1S_0$	6.113
Sc 21						1	2	—	—	—	—	—	—	—	$^2D_{3/2}$	6.54
Ti 22						2	2	—	—	—	—	—	—	—	$^3F_2$	6.82
V 23						3	2	—	—	—	—	—	—	—	$^4F_{3/2}$	6.74
Cr 24						5	1	—	—	—	—	—	—	—	$^7S_3$	6.765
Mn 25						5	2	—	—	—	—	—	—	—	$^6S_{5/2}$	7.432
Fe 26						6	2	—	—	—	—	—	—	—	$^5D_4$	7.870
Co 27						7	2	—	—	—	—	—	—	—	$^4F_{9/2}$	7.86
Ni 28						8	2	—	—	—	—	—	—	—	$^3F_4$	7.635
Cu 29						10	1	—	—	—	—	—	—	—	$^2S_{1/2}$	7.726
Zn 30						10	2	—	—	—	—	—	—	—	$^1S_0$	9.394
Ga 31						10	2	1	—	—	—	—	—	—	$^2P^0_{1/2}$	5.999
Ge 32						10	2	2	—	—	—	—	—	—	$^3P_0$	7.899
As 33						10	2	3	—	—	—	—	—	—	$^4S^0_{3/2}$	9.81
Se 34						10	2	4	—	—	—	—	—	—	$^3P_2$	9.752
Br 35						10	2	5	—	—	—	—	—	—	$^2P^0_{3/2}$	11.814
Kr 36						10	2	6	—	—	—	—	—	—	$^1S_0$	13.999
Rb 37				Krypton configuration					—	—	1	—	—	—	$^2S_{1/2}$	4.177
Sr 38									—	—	2	—	—	—	$^1S_0$	5.693
Y 39									1	—	2	—	—	—	$^2D_{3/2}$	6.38
Zr 40									2	—	2	—	—	—	$^3F_2$	6.84
Nb 41									4	—	1	—	—	—	$^6D_{1/2}$	6.88
Mo 42									5	—	1	—	—	—	$^7S_3$	7.10
Tc 43									5	—	2	—	—	—	$^6S_{5/2}$	7.28
Ru 44									7	—	1	—	—	—	$^5F_5$	7.366
Rh 45									8	—	1	—	—	—	$^4F_{9/2}$	7.46
Pd 46									10	—	—	—	—	—	$^1S_0$	8.33

Distribution of electrons in the atoms (cont.)

Element and atomic number		Configuration of inner shells	N — 4,3 / 4f	O — 5,0 / 5s	O — 5,1 / 5p	O — 5,2 / 5d	O — 5,3 / 5f	P — 6,0 / 6s	P — 6,1 / 6p	P — 6,2 / 6d	Q — 7,0 / 7s	Ground term	Ionization potential, eV
Ag	47	Palladium configuration	−	1	−	−	−	−	−	−	−	$^2S_{1/2}$	7.576
Cd	48		−	2	−	−	−	−	−	−	−	$^1S_0$	8.993
In	49		−	2	1	−	−	−	−	−	−	$^2P^0_{1/2}$	5.786
Sn	50		−	2	2	−	−	−	−	−	−	$^3P_0$	7.344
Sb	51		−	2	3	−	−	−	−	−	−	$^4S^0_{3/2}$	8.641
Te	52		−	2	4	−	−	−	−	−	−	$^3P_2$	9.01
I	53		−	2	5	−	−	−	−	−	−	$^2P^0_{3/2}$	10.457
Xe	54		−	2	6	−	−	−	−	−	−	$^1S_0$	12.130
Cs	55	The shells 1s to 4d contain 46 electrons	−	The shells 5s to 5p contain 8 electrons		−	−	1	−	−	−	$^2S_{1/2}$	3.894
Ba	56		−			−	−	2	−	−	−	$^1S_0$	5.211
La	57		−			1	−	2	−	−	−	$^2D_{3/2}$	5.5770
Ce	58		1			1	−	2	−	−	−	$^1G^0_4$	5.466
Pr	59		3			−	−	2	−	−	−	$^4I^0_{9/2}$	5.422
Nd	60		4			−	−	2	−	−	−	$^5I_4$	5.489
Pm	61		5			−	−	2	−	−	−	$^6H^0_{5/2}$	5.554
Sm	62		6			−	−	2	−	−	−	$^7F_0$	5.631
Eu	63		7			−	−	2	−	−	−	$^8S^0_{7/2}$	5.666
Gd	64		7			1	−	2	−	−	−	$^9D^0_2$	6.141
Tb	65		(8)			(1)	−	(2)	−	−	−	$(^8G_{13/2})$	5.852
Dy	66		10			−	−	2	−	−	−	$^5I_8$	5.927
Ho	67		11			−	−	2	−	−	−	$^4I^0_{15/2}$	6.018
Er	68		12			−	−	2	−	−	−	$^3H_6$	6.101
Tm	69		13			−	−	2	−	−	−	$^2F^0_{7/2}$	6.184
Yb	70		14			−	−	2	−	−	−	$^1S_0$	6.254
Lu	71		14			1	−	2	−	−	−	$^2D_{3/2}$	5.426
Hf	72	The shells 1s to 5p contain 68 electrons				2	−	2	−	−	−	$^3F_2$	6.865
Ta	73					3	−	2	−	−	−	$^4F_{3/2}$	7.88
W	74					4	−	2	−	−	−	$^5D_0$	7.98
Re	75					5	−	2	−	−	−	$^6S_{5/2}$	7.87
Os	76					6	−	2	−	−	−	$^5D_4$	8.5
Ir	77					7	−	2	−	−	−	$^4F_{9/2}$	9.1
Pt	78					9	−	1	−	−	−	$^3D_3$	9.0
Au	79	The shells 1s to 5d contain 78 electrons					−	1	−	−	−	$^2S_{1/2}$	9.22
Hg	80						−	2	−	−	−	$^1S_0$	10.43
Tl	81						−	2	1	−	−	$^2P^0_{1/2}$	6.108
Pb	82						−	2	2	−	−	$^3P_0$	7.417
Bi	83						−	2	3	−	−	$^4S^0_{3/2}$	7.289
Po	84						−	2	4	−	−	$^3P_2$	8.43
At	85						−	2	5	−	−	$^2P^0_{3/2}$	
Rn	86						−	2	6	−	−	$^1S_0$	10.749
Fr	87						−	2	6	−	(1)		
Ra	88						−	2	6	−	2	$^1S_0$	5.278
Ac	89						−	2	6	1	2	$^2D_{3/2}$	5.17
Th	90						−	2	6	2	2	$^3F_2$	6.08
Pa	91						2	2	6	1	2	$^4K_{11/2}$	5.89
U	92						3	2	6	1	2	$^5L_6$	6.05
Np	93						4	2	6	1	2	$^6L_{11/2}$	6.19
Pu	94						6	2	6	−	2	$^7F_0$	6.06
Am	95						7	2	6	−	2	$^8S^0_{7/2}$	5.993
Cm	96						7	2	6	1	2	$^9D^0_2$	6.02
Bk	97						(9)	2	6	(0)	(2)	$^6H^0_{5/2}$	6.23
Cf	98						(10)	2	6	(0)	(2)	$^5I_8$	6.30
Es	99						(11)	2	6	(0)	(2)	$^4I^0_{15/2}$	6.42
Fm	100						(12)	2	6	(0)	(2)	$^3H_6$	6.50
Md	101						(13)	2	6	(0)	(2)	$^2F^0_{7/2}$	6.58
No	102						(14)	2	6	(0)	(2)	$^1S_0$	6.65
Lw	103						(14)	2	6	(1)	(2)		

tively. Electrons with l equal to 0, 1, 2, and 3 are desigrated s, p, d, and f, respectively. *See* QUANTUM MECHANICS; QUANTUM NUMBERS.

In any configuration the number of equivalent electrons (same n and l) is indicated by an integral exponent (not a quantum number) attached to the letters s, p, d, and f. According to the Pauli exclusion principle, the maximum is s^2, p^6, d^{10}, and f^{14}. For example, the configuration $1s^2 2s^2 2p^6 3s$ of the ground or lowest energy state of sodium means that there are two electrons with $n = 1$, $l = 0$; two with $n = 2$, $l = 0$; six with $n = 2$, $l = 1$; and one with

$n = 3$, $l = 0$. Higher values of n and l can be achieved by excitation of normal atoms, either through photon absorption or by particle impact, temporarily moving one or more electrons to unfilled shells of larger radii and increasing the atom's electronic energy. Such excited electronic configurations are inherently unstable. Excited electrons ultimately fall back down to their normal locations closer to the nucleus, a process accompanied by the emission of one or more quanta of electromagnetic radiation (photons). While the atom is excited, the partially depleted shells are said to possess vacancies. *See* EXCLUSION PRINCIPLE.

An electron configuration is categorized as having even or odd parity, according to whether the sum of p and f electrons is even or odd. Strong spectral lines result only from transitions between configurations of unlike parity. *See* PARITY (QUANTUM MECHANICS).

Insofar as they are known from spectroscopic investigations, the electron configurations characteristic of the normal or ground states of the first 103 chemical elements are shown in the table. In the next-to-last column of the table, the spectral term of the energy level with lowest total electronic energy is shown. The main part of the term symbol is a capital letter, S, P, D, F, and so on, that represents the total electronic orbital angular momentum. Attached to this is a superior prefix, 1, 2, 3, 4, and so on, that indicates the multiplicity, and an anterior suffix, 0, $\frac{1}{2}$, 1, $\frac{3}{2}$, 2, $\frac{5}{2}$, and so on, that shows the total angular momentum, or J value, of the atom in the given state. A sign $^\circ$ above the J value signifies that the spectral term and electron configuration have odd parity.

The last column of the table presents the first ionization potential of the atom when this has been derived from spectroscopic observations. In any atomic spectrum, two or more spectral lines with certain similar properties may form a series such that the reciprocal wavelengths $1/\lambda$ (number of waves per centimeter $= \sigma$) can be closely represented by a formula of the Rydberg type, $\sigma = L - R/(n + \mu)^2$, in which L is the limit of the series. R is called the Rydberg constant, and the principal quantum number n has successive integral values to which a constant fractional part μ is added. The second term vanishes when n approaches infinity, and the series limit is thus evaluated. This limit is usually coincident with the ground state of the ion, and is thus a measure (in wave-number units) of the energy required to remove from an atom its least firmly bound electron and transform a neutral atom into a singly charged ion. The energy required to ionize an atom is usually expressed in electronvolts (1 eV = 8065.48 wave numbers) and is called its first ionization potential. *See* ATOMIC STRUCTURE AND SPECTRA; IONIZATION POTENTIAL; RYDBERG ATOM; RYDBERG CONSTANT.

Molecules consist of two or more atoms that at least partially share one or more electrons with one another. Sets of quantum numbers specify molecular electron configurations in much the same way as for atoms. However, the physical meaning of some molecular quantum numbers is different than that for the atomic case, as is the term notation. *See* MOLECULAR STRUCTURE AND SPECTRA.

[JAMES E. BAYFIELD]

Electron diffraction

The phenomenon associated with interference processes that occur when electrons are scattered by atoms to form diffraction patterns. The wave character of electrons is shown most strikingly, and doubtless most conclusively, by the phenomena of interference. For this reason, the diffraction of electrons presents the most obvious confirmation of quantum mechanics. Because of the dependence of the diffraction pattern on the distances between the atoms, electron diffraction is also an important tool for the study of the structure of crystals and of free molecules, analogous to the use of x-rays for these purposes. *See* X-RAY CRYSTALLOGRAPHY; X-RAY DIFFRACTION.

According to quantum theory, any particle moving with momentum mv has a wavelength $\lambda = h/mv$, where h is Planck's constant. If the particle is an electron, and its velocity is the result of having fallen through the potential difference V, this formula becomes Eq. (1), where m_0 and e are the

$$\lambda = \frac{h}{(2m_0 eV)^{1/2}(1 + eV/2m_0 c^2)^{1/2}} \qquad (1)$$

rest mass and charge of the electron, and c is the velocity of light. The last factor in the denominator represents the relativity correction, which is negligible at low voltages and amounts to only 5% at 100,000 V. For V in volts and λ in nanometers, Eq. (2) is a good approximation to Eq. (1) at non-

$$\lambda = \left(\frac{1.5}{V}\right)^{1/2} \qquad (2)$$

relativistic energies. According to energy $E = eV$, two major techniques of structure analysis are distinguished: low-energy electron diffraction (LEED) ($E \simeq 5 - 500$ eV) and high-energy electron diffraction (HEED) ($E \simeq 5 - 500$ keV); medium-energy electron diffraction (MEED) ($E \simeq 500\text{eV} - 5\text{keV}$) is of little importance. Unlike neutrons and x-rays, electrons penetrate matter only for a very short distance before they lose energy (by inelastic scattering) or are scattered elastically (diffracted). Due to the energy loss, the wavelength of an electron changes according to Eq. (2) so that it can no longer interfere with the other incident electrons. This loss of coherence occurs in condensed matter typically after mean free paths for inelastic scattering ranging from several tenths of nanometers at low energies ($E \simeq 50\text{eV}$) to several tens of nanometers at high energies, which determines the application range of LEED and HEED. *See* COHESION; DE BROGLIE WAVELENGTH; DIFFRACTION; INTERFERENCE OF WAVES; MEAN FREE PATH; NONRELATIVISTIC QUANTUM THEORY; QUANTUM MECHANICS.

LOW-ENERGY ELECTRON DIFFRACTION

LEED is used mainly for the study of the structure of single crystal surfaces and of processes on such surfaces that are associated with changes in the lateral periodicity of the surface. A monochromatic, nearly parallel electron beam of 10^{-4} to 10^{-3}m in diameter strikes the surface, usually at normal incidence. The elastically backscattered electrons are separated from all other electrons by a retarding field and detected with a suitable mov-

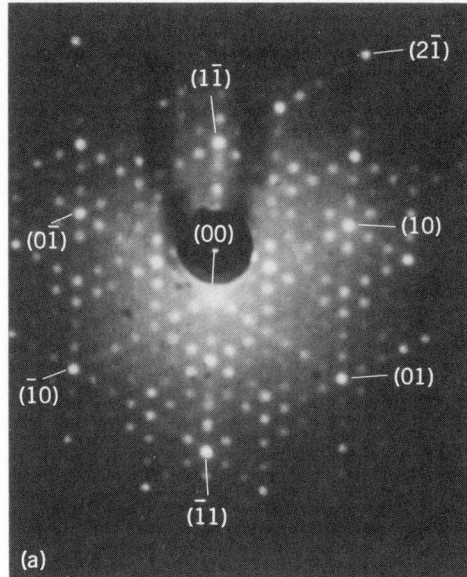

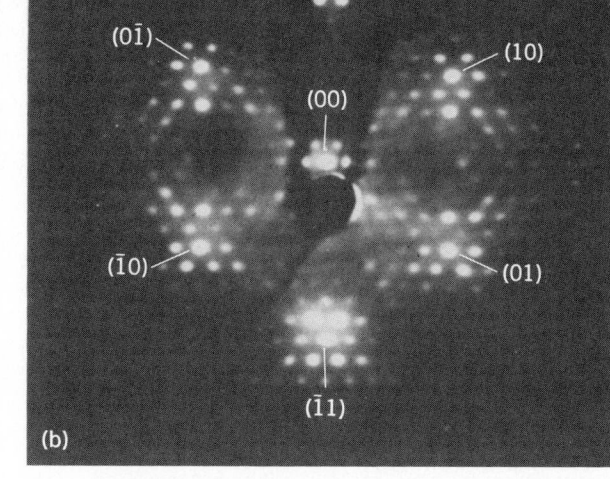

Fig. 1. LEED patterns from single-crystal surfaces. (a) Clean silicon (111) surface (7 × 7 structure), $E = 85$ eV.

(b) Ordered carbon monoxide adsorption layer on a tungsten (110) surface, $E = 120$ eV. (E. Bauer)

able collector or, more frequently—after acceleration to about 5 keV energy—on a hemispherical fluorescent screen with the crystal in its center. Typical LEED patterns obtained in the second detection mode are shown in Fig. 1. The numbers are indices (h,k) labeling diffraction spots of the surfaces. Typical "$I(V)$ curves" obtained in the first mode are shown in Fig. 2. Here the intensity of the diffraction spots is measured as a function of the energy $E = eV$ of the incident electrons.

To obtain such results, the surface must be carefully cleaned and kept in ultrahigh vacuum (UHV, pressure $\simeq 10^{-12}$ N m^{-2}) during the experiment and must be exposed to well-defined amounts of gases or vapors at low pressures if adsorption, condensation, or corrosion is to be studied. This requirement, as well as the fact that many surfaces are changed by the electron beam—for example, due to dissociation in the case of ionic crystals or due to desorption of adsorbed gases— limits the applicability of LEED. Nevertheless, the number of surfaces and surface processes that can and have been studied is very large, and the main limitation of LEED is not caused by difficulties in obtaining LEED patterns but rather in evaluating them.

Interpretation of patterns. The difficulties in interpreting LEED patterns are caused by the strong elastic scattering of slow electrons by atoms. As a consequence of this, the electrons observed in the LEED pattern have been scattered not only once in the crystal but several times on the average, in contrast to x-ray and neutron diffraction, in which single scattering is a good approximation. Thus evaluation of the spot intensities of LEED patterns cannot be done in the first Born approximation (known as "kinematical" theory) as in x-ray diffraction but requires a multiple-scattering or "dynamical" theory. Such an evaluation, with the additional complications arising in the scattering of slow electrons by atoms and in the

surface region of a crystal, requires large-scale computers. As a consequence, only the structures of surfaces with small unit meshes (two-dimensional unit cells) have been evaluated by LEED. For this reason, alternative evaluation methods have been developed in which dynamical effects are reduced by proper averaging procedures so that a kinematical interpretation becomes possible. It is expected that significant improvements in the speed of computers and further developments of the averaging procedures will also allow a meaningful interpretation of complicated superstructure patterns, such as those shown in Fig. 1.

The interpretation difficulties are considerably reduced if only the lateral periodicity of the surface, but not the location of the atoms in the unit mesh, is to be determined. The determination of

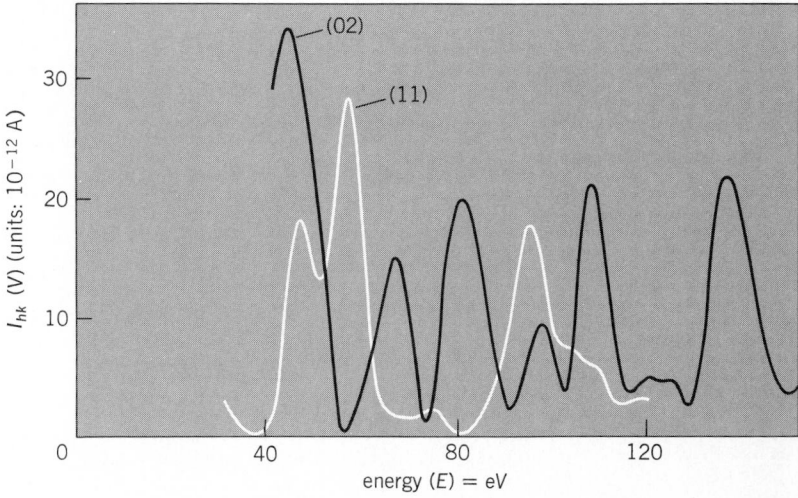

Fig. 2. I(V) curves of spots (hk) in the LEED pattern from a clean tungsten (110) surface. (From J. O. Porteus, in G. A. Somorjar, The Structure and Chemistry of Solid Surfaces, Wiley, 1969)

the lateral periodicity of the surface structure and of the size and shape of domains with a given structure requires only the evaluation of the geometry of the pattern, that is, of the position, size, and shape of the diffraction spots but not of their intensity. Because of the simple relation between LEED pattern, which is essentially the reciprocal lattice of the surface, and the real surface lattice, this evaluation is easy, although not unique, without additional information, such as the coverage of adsorption layers. Thus the most important application of LEED has been the study of surface processes related to changes in lateral periodicity.

Surface structure. Clean surfaces of crystals usually have the lateral periodicity that is expected from their bulk crystal structure. The analysis of $I(V)$ curves of such surfaces (see Fig. 2) in general reveals that also the distribution of the atoms normal to the surface is the same as in the bulk; this means that the crystal has the structure of the bulk up to the very surface (that is, it has an "ideal surface"). There are, however, several exceptions: gold, platinum, and iridium (100) surfaces and various germanium and silicon surfaces (see Fig. 1a) show complicated superstructures (known as "reconstructed surfaces"). Several models for the structures such as a hexagonal overlayer on the (100) surfaces or vacancies on the semiconductor surfaces have been proposed, but a full structure determination has not been made because of the difficulties in interpretation. On ideal surfaces, the decrease of the spot intensities with temperature has shown that the amplitudes of the thermal vibrations of surface atoms normal to the surface

can be much larger than in the bulk. *See* CRYSTAL STRUCTURE; LATTICE VIBRATIONS.

Adsorption. Foreign atoms can interact with a clean surface in a variety of ways: they can either adsorb on top of the surface (nonreconstructive adsorption), or mix with the atoms of the first or several substrate layers (reconstruction, alloying), or roughen the surface (faceting, formation of three-dimensional crystals). They can also be adsorbed only temporarily, reacting with other adsorbed atoms or molecules to form intermediate complexes, then can desorb with or without further reaction.

The last case is very important in heterogeneous catalysis, but it has been rather elusive in LEED for two reasons: weak adsorption (1) requires high gas pressures to maintain a coverage sufficient to obtain a LEED pattern and (2) frequently does not lead to ordered adsorption, particularly in gas mixtures. Nevertheless, a number of surface reactions such as $CO + O_2$ on platinum have been studied, especially in connection with Auger electron spectroscopy (AES). Physisorption, which also represents a weak interaction with the substrate, is increasingly studied with LEED. An example is the two-dimensional phase transition of xenon on graphite.

The most important contribution of LEED, however, is to the understanding of chemisorption, which precedes corrosion and, in many cases, epitaxy. Here, not only the structure of many adsorption systems—mainly of gases such as H_2, O_2, N_2, and CO on metals such as wolfram, nickel, copper, and platinum, or metals on other metals

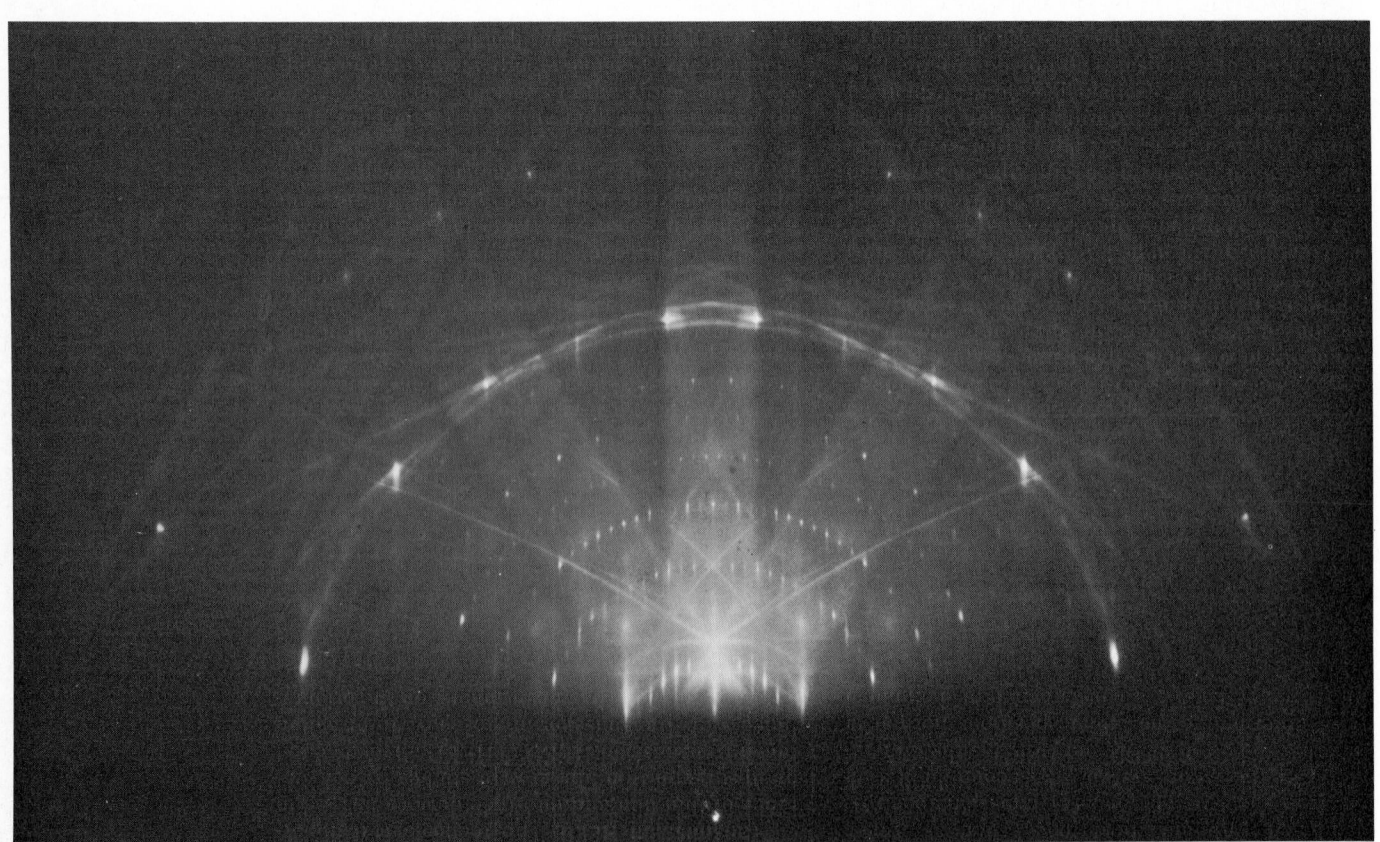

Fig. 3. RHEED pattern from a flat (111) surface of a silicon single crystal. (*Varian Associates*)

and semiconductors—has been studied, but also the kinetics of the adsorption and desorption process as well as changes in the adsorption layer upon heating. The combination of LEED with AES and with work-function measurements has proven particularly powerful in these studies, because such methods give the coverage and information on the location of the adsorbed atoms normal to the surface. Combining LEED with other complementary techniques such as ion scattering spectroscopy, electron energy loss spectroscopy, or photoelectron spectroscopy has become increasingly popular and can enable the elimination of ambiguities in the interpretation of many LEED results. *See* SURFACE PHYSICS.

HIGH-ENERGY ELECTRON DIFFRACTION

HEED is used mainly for the study of the structure of thin foils, films, and small particles (thickness or diameter of 10^{-9} to 10^{-6} m), of molecules, and also of the surfaces of crystalline materials. A monochromatic, usually nearly parallel, electron beam with a diameter of 10^{-3} to 10^{-8} m is incident on the target. The forward-scattered electrons (backscattering is negligible) are detected by means of a fluorescent screen, a photoplate, or some other current-sensitive detector, usually without the inelastically scattered electrons being eliminated.

Reflection HEED. The availability of UHV instruments has made it possible to study the structure of surfaces under clean conditions by reflection HEED (RHEED). Many experiments have confirmed the theoretical expectation that RHEED has a sensitivity comparable to that of LEED. Therefore, similar to LEED, RHEED can be used for the determination of the lateral arrangement of the atoms in the topmost layers of the surface, including the structure of adsorbed layers. Although it is more convenient to deduce the periodicity of the atomic arrangement parallel to the surface from LEED patterns than from RHEED patterns, LEED frequently becomes inapplicable when the surface is rough. This usually occurs in the later stages of corrosion or in precipitation, for example, of silicon carbide on silicon, when small crystals grow on the surface. In such investigations RHEED is far superior to LEED. Figure 3 shows a RHEED pattern of a flat silicon (111) single-crystal surface, a sample for which LEED is equally well suited (Fig. 1*a*). It is a Kikuchi diagram characterized by bands and lines of varying intensity onto which the short streaks of the Laue diagram of the reconstructed surface layer are superimposed. Figure 4 is the RHEED pattern of a rough surface of a polycrystalline evaporated film of CaF_2 with strong fiber texture. In this sample, the surface consists of many small facets distributed at random and inclined against the mean surface so that no LEED pattern could be obtained.

Like LEED, RHEED gives little information on the chemical nature of the atoms that produce the diffraction pattern. Therefore, RHEED has been combined with analytical tools, such as Auger electron spectroscopy or x-ray emission spectroscopy, in the same system. A second limitation inherent to both RHEED and LEED is the difficulty of determining the atomic positions normal to the surface. This is a consequence of dynamical

Fig. 4. RHEED pattern from an evaporated calcium fluoride film. (*E. Bauer*)

effects and of absorption. These limitations have led to a number of controversial interpretations, such as the question of reconstruction upon gas adsorption or of the nature of reconstructed "clean" surfaces with superstructure. Nevertheless, both techniques have given valuable information on corrosion, precipitation, adsorption, condensation, film growth, and other surface and interface phenomena that could not be obtained by other methods.

Scanning HEED. In scanning HEED (SHEED) the diffracted electrons are not recorded on photographic film but are directly measured electronically with sensitive detectors. By moving the detector across the diffraction pattern or by deflecting the diffracted electrons across a stationary detector (scanning), the intensity distribution in the diffraction pattern can be displayed quantitatively on an XY recorder. If an energy filter is put in front of the detector, the inelastically scattered electrons, which are usually not taken into account in the quantitative intensity evaluation, may be filtered out so that only the elastically scattered electrons are measured. This technique is particularly useful in transmission through polycrystalline samples, which produce a ring pattern (Debye-Scherrer diagram). It has also been used in single-crystal samples producing a spot pattern (Laue diagram) and even in reflection diffraction from surfaces. The main application of SHEED is in the study of processes which are accompanied by changes of the intensity distribution, such as the growth of thin films and annealing and corrosion processes.

Transmission HEED. The technological importance of thin film and interface devices has led to an upsurge of thin film growth studies by conventional transmission HEED (THEED), usually combined with transmission microscopy. Information obtained this way has been mainly on the orienta-

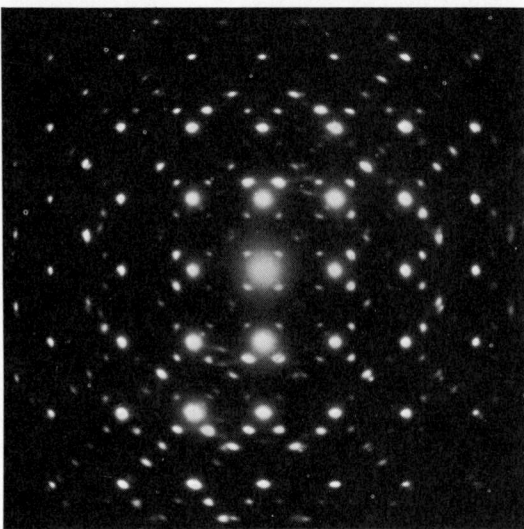

Fig. 5. THEED pattern from epitaxial gold particles on a lead sulfide single-crystal film. (*From A. K. Green, J. Dancy, and E. Bauer, Growth of Au on PbS, PbSe, PbTe and SnTe thin film substrates, J. Vac. Sci. Technol., 10: 494–502, 1973*)

tion of the crystallites composing the film. Figure 5 shows an example, the diffraction pattern of an epitaxial gold film on a single-crystal PbS layer.

Diffraction electron microscopy. In transmission electron microscopy of crystalline specimens the main contrast mechanism is the diffraction contrast. To understand it, knowledge of the diffraction pattern and of diffraction theory is necessary. High-voltage electron microscopes (with electron energies of about 1 MeV) have become available. As a consequence, samples with thicknesses of the order 10^{-6} m can now be studied, as compared to the former upper thickness limit of about 10^{-7} m. This has significance for metallurgy and the physics of metals and semiconductors. It is now possible to study the structure, distribution, and behavior of imperfections in metals, alloys, and semiconductors in bulk with little influence from the boundaries of the material. Other electron-microscope techniques that rely heavily on diffraction are the use of illumination systems producing tilted or conical illumination of the specimen and of objective apertures which transmit only diffracted electrons. With these techniques it has become possible to produce, by diffraction contrast, images of crystal planes 1.5×10^{-10} m apart. [E. BAUER]

DIFFRACTION IN GASES AND LIQUIDS

Electron diffraction in gases and liquids is similar in principle to that in solids; the differences arise from the lack in gases and liquids of any highly regular arrangement of the component atoms. In gases the low density makes it possible to study diffraction by individual atoms and molecules. The results obtained from monatomic gases represent the density of electronic charge in the atom as a function of the distance from the nucleus. The results from gaseous polyatomic molecules represent the equilibrium distances between the atomic nuclei and the average amplitudes of vibration associated with these distances. Liquids have

been studied much less thoroughly, both in theory and in practice, than have gases.

It should be remembered that the structures of molecules in the gaseous state may be different from those in the liquid or solid states; for example, hydrofluoric acid in the gas forms an $(HF)_6$ ring, while in solution or liquid it forms an infinite chain (—FHFHFHFHFHFHF—); biphenyl in the gas is twisted with conformational equilibria, whereas in the solid state it is coplanar in one frozen conformer.

Applications. Typical questions of molecular structure studied by electron diffraction include those of configuration and size in many molecules, with special interest in the variation of chemical bond distances in related molecules (such as the 0.068-A decrease in C—F distance in the series CH_3F, CH_2F_2, CHF_3, CF_4), the variation in the angles between chemical bonds, the distinction between geometric isomers, the degree of restricted rotation around chemical bonds, and in general the relations between the geometry of molecules and their energy and chemical behavior.

The application of electron diffraction in liquids has been restricted because of the less precise information obtainable from molecules which lie in close contact but in irregular arrangements.

The liquids studied so far are certain metals (mercury, tin, tin-bismuth, and tin-aluminum alloys), paraffin oils, and thin layers of water. For monatomic molecules in the liquid state the distances between nearest neighbors are fairly uniform, whereas the separation between more distant neighbors varies considerably. Evaluation of the closest distance of approach can be made to about 0.1 A; major developments in the theory are still required.

The most interesting application of electron diffraction to a liquid or amorphous condensed phase has been in the examination of the surface layers on polished solids. The diffuse diffraction patterns often obtained may be the result of an amorphous arrangement in the surface or of the poor resolving power of very tiny crystalline particles. It is probable that both these states are produced in the polishing of different solids.

Theory and techniques. Structural information about gaseous molecules is obtained by having a fine beam of electrons pass through the gas and strike a photographic plate. The electrons in the beam interact with the charged particles in the atoms (electrons and nuclei) and are bent away from the original direction through varying angles, as registered in the pattern on the plate. The observed variation in the number of scattered electrons with increasing angle is interpreted as an interference effect; that is, electron scattering can be described in the language of diffracted waves for which the resultant is the sum of the component wavelets whose amplitudes and phases are influenced by the wavelength of the incident radiation and the relative positions of the scattering centers. The equivalent wavelength of the electrons, λ, is determined by their energy and is given by Eqs. (1) or (2). The values of λ commonly used in diffraction by gases are in the range $7-5 \times 10^{-12}$ m.

The intensity I of electrons scattered by a spherically symmetrical atom is computed by the Schrödinger wave equation; the result is given by Eq. (3),

$$I = k|f|^2 \tag{3}$$

where k is a constant and f is a complex function called the atomic scattering factor. If the atomic number Z is not too large, Eq. (4) holds, where a is

$$|f|^2 = [2(Z-F)/as^2]^2 \tag{4}$$

the Bohr radius and F is given by Eq. (5). Here $\rho(r)$

$$F = 4\pi \int_0^\infty r^2 \rho(r) [(\sin sr)/sr]\, dr \tag{5}$$

is density of electronic charge at distance r from the nucleus, $s = 4\pi (\sin \theta)/\lambda$ with 2θ equal to the angle between the scattered electron and the original beam, and λ the electron wavelength. Experimental observations on I as a function of s in monatomic gases lead to the determination of $\rho(r)$.

The intensity of electrons scattered by a collection of independent molecules having all possible orientations is given by Eq. (6), where the summa-

$$I(s) = k \sum_i \left[|f_i|^2 + (2/as^2)^2 S_i \right]$$
$$+ \sum_i \sum_j f_i^* f_j \int_0^\infty P_{ij}(r) [(\sin sr)/sr]\, dr \tag{6}$$

tions are taken over all the atoms in the molecule and $(2/as^2)^2 S_i$ represents the intensity of inelastic scattering by the ith atom. The double summation has a term for each pair of atoms i and j. The relative probability of finding the distance between the atom i and j at various values is represented by $P_{ij}(r)$. In principle the use of the observed intensity to determine P_{ij} for each pair of atoms in the molecule constitutes a structure determination for the molecule. The expression is simpler when the atomic motions are nearly harmonic, as in the case of CCl_4. For this molecule the double summation has only two distinct terms. The first is given by expression (7); the second for Cl—Cl is similar.

$$8 f_C f_{Cl} \exp(-l^2_{CCl} s^2)(\sin sr_{CCl})/sr_{CCl} \tag{7}$$

The four parameters which describe the structure are the equilibrium distances, r_{CCl} and r_{ClCl}, and the average amplitudes of vibration, l_{CCl} and l_{ClCl}. Molecular parameters such as these can be determined with high precision when a proper treatment of the scattering factors is included, and when accurate experimental intensities are obtained.

Special equipment is required to meet the conditions assumed in the scattering theory. The electron beam is accelerated with a steady voltage between 30,000 and 70,000 volts, which should be constant within about 0.01%. The beam is focused by electrostatic and magnetic lenses so that it has a diameter of no more than 0.1 mm at the photographic plate. The whole path of the beam is enclosed in a high-vacuum chamber (pressure of about 10^{-5} mm or 10^{-3} Pa of mercury) so that no appreciable electron scattering will occur in the residual air. The gas specimen is introduced in a fine jet so that the volume in which the electrons meet the gas is no more than 1.0 mm in diameter; high-speed pumping is required to remove the gas as rapidly as possible. In front of the photographic plate a rotating sector is mounted to modify the intensity of electrons reaching the plate so that the normally rapid decrease of intensity with increas-

ing angle of scattering is leveled off in a known way and the emulsion is able to register the incident electrons over a wide range of angle.

Interpretation of results. The pattern observed is a set of light and dark circular bands whose spacings and relative intensities depend on the composition and structure of the specimen molecules. The pattern is scanned by a recording microphotometer which yields (after calibration of the photographic emulsion) a tracing or a digital output of the experimental data of I as a function of s. These data are interpreted with the aid of a Fourier transform of the intensity expression of Eq. (6) to give the P_{ij} functions for the pairs of atoms, as illustrated for CCl_4 in Fig. 6. The two prominent peaks represent the C—Cl and Cl—Cl distances; the small deviations from the background line are caused by errors in the experimental data and in the method of interpretation. From the positions and widths of the peaks it is determined that Eqs. (8) hold (1 A = 10^{-10} m).

$$\begin{aligned}
r_{CCl} &= 1.766 \pm 0.003 \text{ A}\\
l_{CCl} &= 0.060 \pm 0.005 \text{ A}\\
r_{ClCl} &= 2.887 \pm 0.004 \text{ A}\\
l_{ClCl} &= 0.068 \pm 0.003 \text{ A}
\end{aligned} \tag{8}$$

This relatively high precision in determining the distances in gas molecules has been achieved since about 1950 with the development of new instrumentation and the application of high-speed computing methods. The relative intensities of scattered electrons registered on the photographic plate can be measured to better than 0.1%; the time saved by the use of electronic computing methods in interpreting the data permits a more rigid application of the criteria for satisfactory agreement between experiment and theory.

Through 1975 over a thousand gases had been studied by the methods giving internuclear distances within 0.004 A or better. In favorable cases precisions exceeding 0.001 A have been obtained; when these can be compared with spectroscopic results, good agreement is found if allowance is made for the difference in the nature of the distances measured. The molecular structures of more than 500 gases have been reported by earlier electron diffraction procedures with uncertainties 10 times larger or more.

The applicability of the gas diffraction method alone is limited to simple molecules. Only when

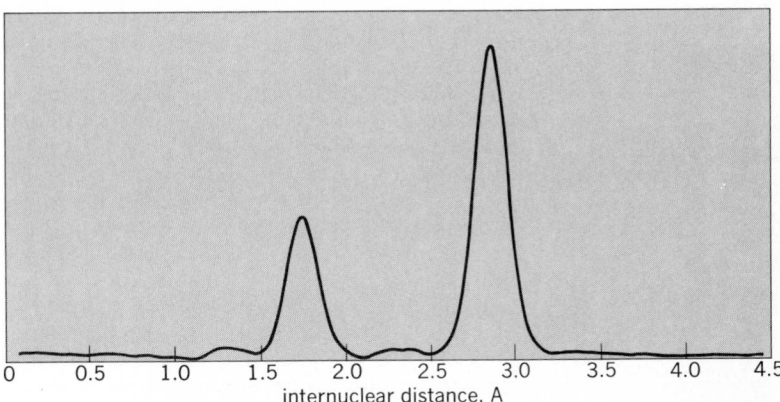

Fig. 6. Experimental distribution of internuclear distances in CCl_4.

the three-dimensional structure can be derived uniquely from a one-dimensional spectrum of internuclear distances blurred by thermal motion is a complete structure determination possible from electron diffraction data alone; light atoms in the presence of heavy ones in the same molecule are less precisely located; distances as close as 0.03 A can barely be resolved. The range of the method is greatly increased, however, when some structural features in the molecule can be assumed from the results of other methods of investigation. For example, if a six-membered ring of atoms is known to have trigonal symmetry, the number of structural parameters is decreased from 12 to 2. *See* NEUTRON DIFFRACTION; SCATTERING EXPERIMENTS (ATOMS AND MOLECULES); SCATTERING EXPERIMENTS (NUCLEI).   [LAWRENCE O. BROCKWAY]

Bibliography: L. S. Bartell, in A. Weissberger and B. W. Rossiter (eds.), *Physical Methods in Chemistry*, 4th ed., vol. 1, 1971; E. Bauer, in R. F. Bunshah (ed.), *Techniques of Metals Research*, vol. 2, pt. 2, 1969; E. Bauer, in R. Gomer (ed.), *Interactions on Metal Surfaces*, 1975; S. H. Bauer in D. Henderson (ed.), *Physical Chemistry: An Advanced Treatise*, vol. 4, ch. 14, 1970; R. A. Bonham and M. Fink, *High Energy Electron Scattering*, ACS Monogr., 1974; L. H. Germer, Low-energy electron diffraction, *Phys. Today*, 17:19, 1964; L. H. Germer, The structure of crystal surfaces, *Sci. Amer.*, 212:32–41, March 1965; R. L. Hilderbrandt and R. A. Bonham, Structure determination by gas electron diffraction, *Annu. Rev. Phys. Chem.*, 22:279–312, 1971; J. B. Pendry, *Low Energy Electron Diffraction*, 1974; T. B. Rymer, *Electron Diffraction*, 1970; J. A. Strozier, in J. M. Blakely (ed.), *Surface Physics of Crystalline Solids*, 1974; L. E. Sutton (ed.), *Interatomic Distances*, 1958, suppl., 1964; G. Thomas and W. L. Bell, in R. F. Bunshah (ed.), *Techniques of Metals Research*, vol. 2B, 1970; M. F. Tompsett, SHEED in materials science, *J. Mater. Sci.* 7:1069–1079, 1972.

Electron magnetic moment

The electron has magnetic properties by virtue of (1) its orbital motion about the nucleus of its parent atom and (2) its rotation about its own axis. The magnetic properties are best described through the magnetic dipole moment associated with 1 and 2. The classical analog of the orbital magnetic dipole moment is the dipole moment of a small current-carrying circuit. The electron spin magnetic dipole moment may be thought of as arising from the circulation of charge, that is, a current, about the electron axis; but a classical analog to this moment has much less meaning than that to the orbital magnetic dipole moment. The magnetic moments of the electrons in the atoms that make up a solid give rise to the bulk magnetism of the solid.

The magnetic moment of the electron has been measured to 40 parts per trillion and, based on the theory of quantum electrodynamics, its value has been calculated to 150 parts per trillion. The two values are in agreement. This experimental verification of the theoretical calculation of the electron magnetic moment constitutes the most accurate and direct test of the theory of quantum electrodynamics which has been made to date (1980). It also constitutes the most precise confrontation of

any experiment with a theoretical prediction in all of science. For a detailed discussion of magnetic moments of electrons *see* ELECTRON SPIN. *See also* ATOMIC STRUCTURE AND SPECTRA; MAGNETISM; QUANTUM ELECTRODYNAMICS.   [ARTHUR RICH]

Electron-positron pair production

A process in which a negative electron (negatron) and a positive electron (positron) are simultaneously created in the vicinity of a nucleus or an elementary particle. In external pair production, an electromagnetic wave (photon) is absorbed and creates an electron pair. The absorption of high-energy γ-rays is due mainly to this effect (see diagram). Internal pair production is not associated with observable electromagnetic radiation and may occur when an excited nucleus releases some of its internal energy.

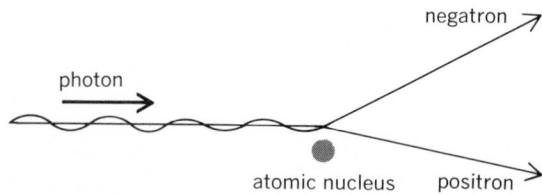

External pair (negatron-positron) production.

Pair production is of considerable theoretical interest, not only as an example of the materialization of energy, but also as a striking confirmation of the relativistic quantum theory proposed by P. A. M. Dirac. This theory has made possible quantitative predictions of production probability, differential electron distribution, and kinetic energy partition. The results are in satisfactory agreement with experimental findings. *See* RELATIVISTIC QUANTUM THEORY.

External pair production can take place only if the energy of the photon exceeds $2mc^2$ ($m=$ electron mass, $c=$ velocity of light) or 1.02 Mev, which is the energy required for production of an electron pair at rest. The energy excess, $h\nu - 2mc^2$ (ν is the frequency of the photon, h is Planck's constant), appears as kinetic energy of the created particles. The sharing of this energy between the electrons takes place in a random way, such that the positron, for example, may assume any energy between zero and $h\nu - 2mc^2$ with about the same probability. Because of the electrostatic repulsion of the nucleus, the positron actually obtains a higher energy on the average than the negatron.

Conservation laws require that the momentum of the initial photon be transferred to the product particles. Simple calculations show that this can be fulfilled only if a third particle or system of particles takes part in the process. It may be a nucleus, as is usually the case, but in principle any charged particle may restore the momentum balance. For a given energy division between the electrons, the nucleus may recoil in any direction; consequently, the direction in which the electrons are emitted is not fixed, but is randomly distributed. As a consequence of its large mass, the nucleus receives only a vanishingly small part of the initial photon energy. For a discussion of conservation laws *see* NUCLEAR REACTION.

Internal pairs are often emitted from radioactive substances. After radioactive decay, the daughter nucleus may be left with excess energy. Although this energy is usually released as electromagnetic radiation, pair production may compete when the energy exceeds $2mc^2$, the probability increasing with higher energy release. The angular correlation of the pairs and the production probability also depend on the multipole order of the transition. *See* MULTIPOLE RADIATION; POSITRON; QUANTUM FIELD THEORY.

[GUNNAR BACKSTROM]

Bibliography: R. D. Evans, *The Atomic Nucleus*, 1955; S. Gasiorowicz, *Structure of Matter: A Survey of Modern Physics*, 1979.

Electron spin

That property of an electron which gives rise to its angular momentum about an axis within the electron. Spin is one of the permanent and basic properties of the electron. Both the spin and the associated magnetic dipole moment of the electron were postulated by G. E. Uhlenbeck and S. Goudsmit in 1925 as necessary to allow the interpretation of many observed effects, among them the so-called anomalous Zeeman effect, the existence of doublets (pairs of closely spaced lines) in the spectra of the alkali atoms, and certain features of x-ray spectra. *See* SPIN (QUANTUM MECHANICS).

All theory that concerns itself with electronic, nuclear, atomic, and molecular phenomena includes the electron spin in its formulation to obtain a theoretical structure consistent with experimental observation. The electron thus possesses the intrinsic property of spin angular momentum (rotational motion about an axis), in addition to the intrinsic properties of charge and mass. *See* ELECTRON.

The spin quantum number is s, where s is always 1/2. This means that the component of spin angular momentum along a preferred direction, such as the direction of a magnetic field, is $\pm\frac{1}{2}\hbar$ where $\hbar = h/2\pi$ and h is Planck's constant. The spin angular momentum of the electron is not to be confused with the orbital angular momentum of the electron associated with its motion about the nucleus. In the latter case the maximum component of angular momentum along a preferred direction is $l\hbar$, where l is the angular momentum quantum number and may be any positive integer or zero. The total orbital angular momentum is $\sqrt{l(l+1)}\,\hbar$. In this discussion the terms angular momentum or magnetic dipole moment describe the maximum component of these quantities along a field direction. *See* QUANTUM NUMBERS.

Electron magnetic moment. The electron has a magnetic dipole moment by virtue of its spin. The approximate value of the dipole moment is the Bohr magneton μ_0 which is equal to $eh/4\pi mc = 9.27 \times 10^{-21}$ erg/oersted, where e is the electron charge measured in electrostatic units, m is the mass of the electron, and c is the velocity of light. (In SI units, $\mu_0 = 9.27 \times 10^{-24}$ J/T.) The orbital motion of the electron also gives rise to a magnetic dipole moment μ_l that is equal to μ_0 when $l=1$ (Fig. 1). In Fig. 1 is shown the simple case of an electron in circular motion in a magnetic field. The electron is shown with a spin angular momentum parallel to the orbital angular momentum. Physically it could equally well be in the opposite

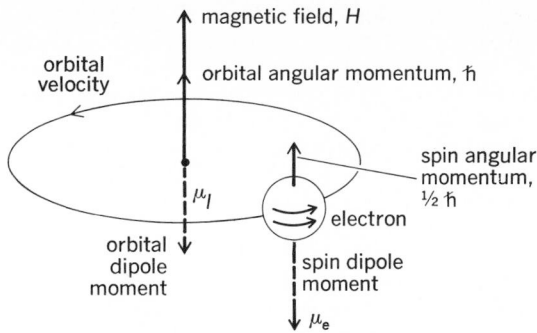

Fig. 1. Diagram of orbital angular momentum G_l and orbital magnetic moment μ_l due to a negative charge revolving in circle of radius a.

direction. The direction of the dipole angular momentum (both are vectors), and the magnetic moments are therefore negative. For a positron (a positively charged particle having the same mass and magnitude of charge as the negatively charged electron) the magnetic moments are positive, that is, in the same direction as the angular momentum. *See* MAGNETON; POSITRON.

The orbital magnetic moment of an electron can readily be deduced with the use of the classical statements of electromagnetic theory in quantum-mechanical theory; the simple classical analog of a current flowing in a loop of wire describes the magnetic effects of an electron moving in an orbit. The spin of an electron and the magnetic properties associated with it are, however, not possible to understand from a classical point of view. The classical radius of the electron is $e^2/(2mc^2) = 1.41 \times 10^{-13}$ cm; a reasonable distribution of mass and electric charge within this radius which would lead to a magnetic moment μ_0 leads to the calculation of a peripheral velocity of the electron far greater than the velocity of light. This is, of course, wholly precluded by the special theory of relativity. No theory of the structure of the electron has been formulated which makes the spin of the electron amenable to simple pictorial understanding. Nevertheless, the interpretation of the spectra of atoms and molecules, the magnetic properties of materials, and other phenomena on an atomic scale unambiguously require that the electron have the property of spin. To the extent that physical theory assumes these properties and does not concern itself with questions about the structure of the electron, it is quite adequate to deal with a large range of physical phenomena in a highly quantitative way.

In the Landé g-factor, g is defined as the negative ratio of the magnetic moment, in units of μ_0, to the angular momentum, in units of $\hbar$. For the orbital motion of an electron, $g_l = 1$. For the spin of the electron the appropriate g-value is $g_s \cong 2$; that is, unit spin angular momentum produces twice the magnetic moment that unit orbital angular momentum produces.

The following discussion is limited to atoms which have a single electron outside of closed electron shells. Both the orbital and spin angular momenta of the electrons within closed shells add up in such a way that their net angular momentum is zero. The single electron outside closed shells may have $l = 0, 1, 2, \ldots$ By the usual rules developed from quantum mechanics the total angular momentum quantum number of the electron, which is called j, is $l \pm s$ (Fig. 1) except when $l = 0$, in which case the total angular momentum is s. For instance, when $l = 1$, $j = \frac{1}{2}$ or $\frac{3}{2}$. These relations may be represented by vector diagrams, as shown in Fig. 2. Since the revolution of the electron about the nucleus produces a magnetic field at the electron, and since the electron has a magnetic dipole moment, the energy of the atom is different when the vector s is parallel to l $(j = l + \frac{1}{2})$ and when s is antiparallel (180°) to l $(j = l - \frac{1}{2})$. Thus the spin of the electron causes a doubling of the energy levels in all single-electron atoms except when $l = 0$, in which case level is single. In the case of sodium, the familiar yellow lines (the D lines) comprise a closely spaced doublet that arises from a transition from a state of $l = 1$ to one of $l = 0$. The doubling of the lines is a direct consequence of the existence of the electron spin.

Energy-level splitting. The total electronic magnetic moment of an atom depends on the state of coupling between the orbital and spin angular momenta of the electron (Fig. 2). In the single-electron case, an atom in the state for which $l = 0$ has only the magnetic moment associated with the spin, and this moment may be oriented in either of two directions with respect to an externally applied magnetic field. However, when the atom is in a state for which $l = 1$, j can be $\frac{1}{2}$ or $\frac{3}{2}$ and the Landé g-factors for these two states are $\frac{2}{3}$ and $\frac{4}{3}$ respectively. That is, the magnetic moment per unit angular momentum is equal neither to that which characterizes the orbital motion nor to that which characterizes the spin motion. In a magnetic field, a single energy level characterized by l and j is split into several energy levels, each described by the component of j, m_j, along a magnetic field, where $m_j = j, j-1, \ldots, -(j-1), -j$. The energy of the level is the zero field energy plus the term $m_j g_j \mu_0 H$. In a transition between two energy levels that gives rise to a spectral line, the energy of the emitted or absorbed photon is the zero field energy difference between the two levels plus the difference between two magnetic energy terms as given previously. The resultant splitting of the line, which is single at zero magnetic field, into a line of two or more components is called the Zeeman effect. The Zeeman effect has been called anomalous when it is not explicable purely in terms of the

orbital motion of the electron. The introduction of the electron spin allows the interpretation of all observed Zeeman effects. *See* ZEEMAN EFFECT.

Atomic beam measurements. Prior to 1940 spectroscopic measurements had been made only on optical spectra, and Zeeman effects were small effects superimposed on lines of considerable natural width. Exact measurements of the splitting of lines in a magnetic field, therefore, could not be made on such lines, and all data on the Zeeman effect were consistent with the statement that $g_s = 2g_l$. With the development of spectroscopy by the atomic beam method, a new order of precision in the measurement of the frequencies of spectral lines became possible. By using the atomic-beam techniques, it became possible to measure g_s/g_l directly, with the result $g_s/g_l = 2$ (1.001168 ± 0.000005). The magnetic moment of the electron therefore is not μ_0 but $1.001168\mu_0$, or equivalently the g factor of the electron departs from 2 by the so-called g-factor anomaly defined as $a = (g_2 - 2)/2$ so that $\mu = (1 + a)\mu_0$. Thus the first molecular beam work gave $a = 0.001168$. *See* MOLECULAR BEAMS.

Calculation of g-factor anomaly. It is not possible to give a qualitative description of the effects which give rise to the g-factor anomaly of the electron. The detailed theoretical calculation of the quantity is in the domain of quantum electrodynamics, and involves the interaction of the zero-point oscillation of the electromagnetic field with the electron. Comparison of theoretical determination of a with its experimental measurement constitutes the most accurate and direct existing test of the theory of quantum electrodynamics.

Theoretical work on the g-factor anomaly, based on the principles of quantum electrodynamics, began simultaneously with its experimental discovery. The initial prediction of the value of the anomaly was made by J. Schwinger in 1948, who showed that the anomaly could be written as $a = 0.5$ (α/π), where α, the fine-structure constant, is given by $\alpha = e^2/\hbar c \simeq 1/137$. Shortly thereafter it was shown that a could be expressed as a power series in α (more customarily, α/π), that is, one can write $a = A(\alpha/\pi) + B(\alpha/\pi)^2 + C(\alpha/\pi)^3 + D(\alpha/\pi)^4 + \ldots$. Calculation of B and C from quantum electrodynamics has proven to be very difficult, with errors occurring at various stages of the work. However, it is now generally agreed that $A = 0.500$, $B = -0.328478$, and $C = 1.184 \pm 0.007$. In order to find a theoretical value for a, one also needs an accurate value of α. Such a value is given by work at the National Bureau of Standards where, from a number of precise measurements of quantities which enter the definition of α, it has been found that $\alpha^{-1} = 137.035963 \pm 0.000015$. The result of substituting this value of α into the expression for a gives the most recent theoretical value of a as $a(\text{theoretical}) = 0.001\ 159\ 652\ 570 \pm 0.000\ 000\ 000\ 150$. The uncertainty quoted for this number is due primarily to error in α and also to error in the coefficient. It is possible that both of these sources of error will be reduced dramatically. A calculation of D has been undertaken. This is of importance since, if $D = 1$, this term contributes $0.000\ 000\ 000\ 027$ to $a(\text{theoretical})$, and such a contribution, as small as it is, must be considered in future research. *See* ATOMIC CONSTANTS; QUANTUM ELECTRODYNAMICS.

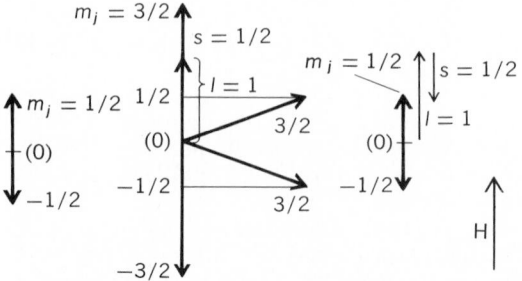

Fig. 2. Diagram illustrating addition of l and s vectors into a resultant j vector. The allowed values of m_j are also shown. In no case does the (0), in parentheses, give an allowed value.

Measurement of g-factor anomaly. The initial molecular-beam method for measuring a was based on measurement of g_s/g_l, which yields $1 + a$ rather than a itself. Such experiments have been superseded by two new types of experiments which have the major advantages of measuring a directly, and of measuring a for electrons trapped in electric and magnetic fields but not bound to atoms. The second feature means that various corrections due to the presence of an atom are now not necessary.

The new experiments may be readily understood, qualitatively, by noting that if an electron (spin angular momentum $\hbar/2$) moves at a low velocity v ($v/c \ll 1$) perpendicular to a magnetic field B, the particle will rotate in the field at a frequency (the cyclotron frequency) given by $f_C = (1/2\pi)(eB/mc)$, while its magnetic moment (and spin) will precess about the field at a frequency (the spin precession frequency) given by $f_S = (1 + a)f_C$. The difference between these frequencies, often called the g-2 or difference frequency, is given by $f_D = f_S - f_C = af_C$. It represents the rate at which the spin of the particle rotates relative to the particle's velocity.

One technique for measuring a consists of trapping very-low-energy electrons in a magnetic field and measuring f_D and f_C simultaneously, with a being given by the relation $a = f_D/f_C$.

A slightly different version of the same technique consists of trapping electrons whose velocity is about half the speed of light (electrons accelerated to several hundred thousand volts) in a magnetic field and determining the same ratio. Relativistic effects manifest themselves in both the expressions for f_S and f_C, but cancel exactly when the ratio f_D/f_C is formed. Consequently comparison of f_D/f_C, using both the high- and low-energy techniques, constitutes an excellent check of the separate measurements, as well as one of the most precise tests extant of the special theory of relativity, which predicts that the frequency ratio should be independent of velocity.

The most accurate experimental result has been obtained by using the low-energy technique. This result is a(experimental) = 0.001 159 652 200 ± 0.000 000 000 040, and it is in agreement, within the limits of error, with a(theoretical). The comparison of a(experimental) with a(theoretical) presented above constitutes the most precise confrontation of any experiment with a theoretical prediction in the history of science, a confrontation at the level of 150 parts per billion. Even more accurate experiments at both high and low energy and, as mentioned above, improved theoretical predictions have been undertaken. *See* ATOMIC STRUCTURE AND SPECTRA; GYROMAGNETIC RATIO; QUANTUM MECHANICS.　　[ARTHUR RICH]

Bibliography: E. Merzbacher, *Quantum Mechanics*, 1970; A. Rich and J. C. Wesley, The current status of the lepton g-factors, *Rev. Mod. Phys.*, 44(2):250–283, 1972; F. K. Richtmyer, E. H. Kennard, and T. Lauritsen, *Introduction to Modern Physics*, 5th ed., 1969; V. F. Weisskopf, Recent developments in the theory of the electron, *Rev. Mod. Phys.*, 21(2):305–315, 1949.

Electronvolt

A unit of energy used for convenience in atomic systems. Specifically, it is the change in energy of an electron, or of any particle having a charge numerically equal to that of an electron, when it is moved through a difference of potential of 1 mks volt. Its value (in mks units) is obtained from the equation $W = qV$, where W is energy in joules, q the charge in coulombs, and V the potential difference in volts. For a potential difference of 1 volt and the electronic charge of 1.6022×10^{-19} coulomb, the electronvolt is 1.6022×10^{-19} joule. *See* ELECTRON; IONIZATION POTENTIAL.

[GLENN H. MILLER]

Electrooptics

The branch of physics which deals with the influence of an electric field on the optical properties of matter, especially in its crystalline form. These properties include transmission, emission, and absorption of light.

An electric field applied to a transparent crystal can change its refractive indexes and, therefore, alter the state of polarization of light propagating through it. When the refractive-index changes are directly proportional to the applied field, the phenomenon is termed the Pockels effect. When they are proportional to the square of the applied field, it is called the Kerr effect. *See* KERR EFFECT.

The Pockels effect is used in a light modulator called the Pockels cell. This device (see Fig. 1)

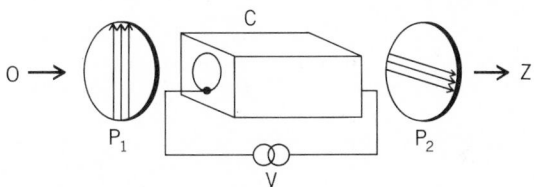

Fig. 1. Pockels cell light modulator. The arrows represent a light beam.

consists of a crystal C placed between two polarizers P_1 and P_2 whose axes are crossed. A crystal often used for this application is potassium dihydrogen phosphate (KH_2PO_4), also known as KDP. Ring electrodes bonded to two crystal faces allow an electric field to be applied parallel to the axis OZ, along which a light beam (for example, a laser beam) is made to propagate. The crystal has been cut and oriented in such a way that, in the absence of an electric field, the polarization of light propagating along OZ does not change, and therefore no light is transmitted past P_2. However, when an electronic driver V applies an electric field of the proper magnitude (2000 V/cm is a typical value)

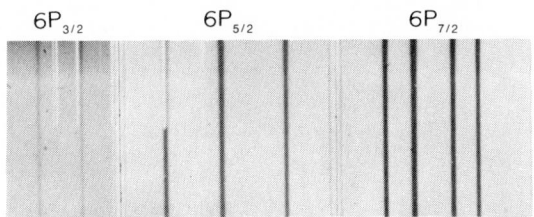

Fig. 2. Splitting of absorption lines by the electric field in a crystal ($GdAc_3 \cdot 6H_2O$). $6P_{3/2}$, $6P_{5/2}$ and $6P_{7/2}$ are the spectral term symbols. (*Johns Hopkins University*)

across the crystal, the crystal becomes birefringent, and light that has been vertically polarized by P_1 undergoes a 90° change in polarization in the crystal, and is now well transmitted past P_2. Pockels cells can be switched on and off in well under 1 ns (one-billionth of a second). *See* LASER.

Electric fields which are already present inside some crystals can change the light-emission and absorption characteristics of ions in the crystal. Electric fields cause a shift and a splitting of certain rare-earth ion-absorption lines which can be clearly seen at low temperatures (Fig. 2).

[MICHEL A. DUGUAY]

Bibliography: A. Yariv, *Introduction to Optical Electronics*, 1971.

Electroscope

An instrument for detecting the presence and sign of an electric charge. It is the simplest type of ionization chamber. *See* IONIZATION CHAMBER.

The illustration shows a common type of simple gold-leaf electroscope. Gold leaf, shown as L, is used because it is an extremely thin conducting foil which has low mass per unit area and is very flexible. Hence, it responds quickly and vigorously to small electrostatic forces. Aluminum foil is almost as satisfactory as gold foil. In the illustration P is a metal support post for L; I is an insulator of high quality through which P passes and terminates in a metal knob K; and H is a cylindrical metal housing with flat ends and windows so located that the motion and final position of L are visible. H serves as a grounded electrostatic shield, as well as a shield against air currents. The base B supports the electroscope.

The hard-rubber rod R (illustration *a*) with its negative charge has set up the charge distribution by the process of electrostatic induction. The response shown is a test for the fact that R has a charge. *See* ELECTROSTATICS.

To leave the electroscope with a net charge, a grounded conductor is touched to K so that the surplus electrons on P and L go off to ground, leaving the bound positive charge on K. The ground connection is then broken and R is removed. At this stage (illustration *b*) the electroscope is said to have a positive charge because there is a positive charge on its leaf system.

If an electroscope has a charge of known sign, as illustration *b*, it can be used to test the sign of an unknown charge, as in illustration *c*, where the metal test ball T, with its insulating handle J, has the unknown charge. In the situation pictured, L moves farther away from P as T is brought slowly up toward K, showing that T has a positive charge. If T had a negative charge, L would move toward P, as T slowly approaches K. The converse situation, if the leaf system in illustration *c* had a negative charge initially, can be readily visualized.

Although electroscopes have been built with a wide variety of geometries, the principle of operation is essentially the same for all. If an electroscope has a scale, permitting quantitative measurements, it is called an electrometer or electrostatic voltmeter.

[RALPH P. WINCH]

Bibliography: A. Smith and J. H. Cooper, *Elements of Physics*, 9th ed., 1979; R. L. Weber et al., *College Physics*, 5th ed., 1974.

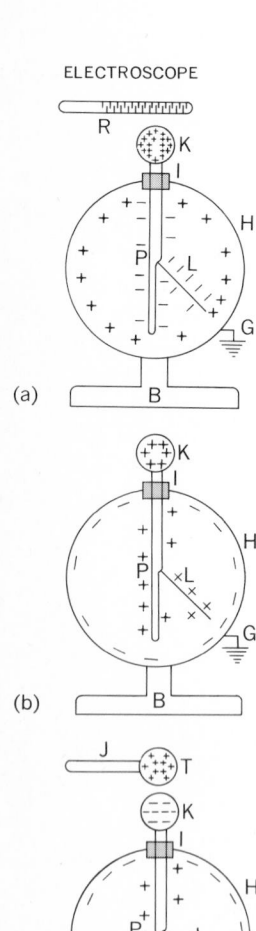

ELECTROSCOPE

Electroscope. (*a*) Being charged by induction by negative charge on hard-rubber rod R. (*b*) Positive charge left on its leaf after induction process is complete. (*c*) Testing the sign of an unknown charge on test ball T.

Electrostatic induction

A method whereby an electrical conductor becomes electrified when near a charged body. Its usefulness arises from the fact that for every electric charge there is somewhere an equal and opposite induced charge. For a detailed discussion *see* ELECTROSTATICS.

[RALPH P. WINCH]

Electrostatics

The study of electric charges at rest under conditions where the positive and negative charges are separated from each other. The term static electricity is often used to refer to electric charges at rest.

Electrification. Electrification by friction or rubbing occurs when one substance is rubbed with another and a surplus of electrons, each with its negative charge, is rubbed onto one of the bodies, leaving a deficiency of electrons, that is, a positive charge, on the other body. The most commonly used materials for electrification by friction are hard rubber and fur. The hard rubber has a negative charge after being rubbed with cat fur. Only good insulators show a net charge after rubbing, because electrical conduction permits neutralization of the charge very rapidly in other materials. This effect is evident if the fur is held in the experimenter's hand during the rubbing process, because electrical conduction through the experimenter's body will neutralize the positive charge nearly as fast as it appears on the fur. One can get a net positive charge on a Lucite (Plexiglas) plate by rubbing it with a polyethylene plastic film. The lucite will retain the charge because it is a good insulator. *See* ELECTRIC CHARGE.

Electrostatic induction. This process affords another way of charging a body, as illustrated in Fig. 1, where the charged, hard-rubber rod R is responsible for the charging process. Since like charges repel each other and unlike charges attract each other, the negative charge on R repels some electrons to the opposite side of the metal sphere M. This leaves an equal positive charge on the part of the sphere near R. If a grounded lead (for example, the experimenter's finger) is touched to the metal sphere, the unbound electrons flow off to ground leaving the positive charge bound by the presence of the charge on R. Then if the ground lead is removed and next R is removed, the metal sphere is left with a net positive charge which, in the end, is uniformly distributed over the surface of the sphere. When the charge on a conductor is static, that is, after it has completed its redistribution, it is in any case all on the surface of the conductor.

The induced charge is always equal in magnitude, as well as opposite in sign, to the inducing charge. In Fig. 1, not all the induced charge is on the sphere M, since the sphere does not surround the inducing charge on R, and there are other adjacent objects. Some of the induced charge is located on these nearby objects (the floor and walls and the experimenter's body).

Electrostatic generators. These devices, sometimes called static machines, depend on electrification by friction and by induction for their operation. The simplest electrostatic generator is the electrophorus shown in Fig. 2. The hard-rubber plate R has been given a negative charge by rubbing it with cat fur. The metal plate D has an insulating handle H. When D is set on R, it touches it at only a few high points because R is microscopically rough, even though it appears polished to the eye. A positive charge is induced on the lower surface of D, and an equal, negative charge appears on the upper surface of D in the induction process. Now if a grounded conductor is touched to D, the electrons go off to ground and the ground-connection is then removed. Next, by using the insulating handle H, D is lifted from the rubber plate, and there is a net positive charge on D which can be shared with a receiver and used for experimentation.

Since no charge is removed from R while charging by induction, D can be recharged from R as many times as the above process is repeated, and a large charge can be built up on the receiver. A rotating machine which would carry out the repetition of the induction process might be built from the electrophorous as follows. In Fig. 2, imagine an insulating wheel with a series of metal disks like D mounted on its rim and insulated from each other. As the wheel rotates, each disk in turn comes to the position of D in Fig. 2, where a positive charge is induced on its lower surface, and a stationary ground wire removes the equal negative charge from its upper surface. When D moves away from its position over R, the stationary ground wire loses contact, and D leaves with its net positive charge which it delivers to the receiver at some other place along the path of its rotation. As D moves away, another metal disk comes into position over R, and the process is repeated. Thus, charge will be transferred to the receiver as long as the wheel is turned. For this purpose, R could just as well be a metal plate which has been given a negative (or positive) charge by conduction, because this will serve as well as the hard-rubber plate as long as D does not come close enough to establish electrical contact between D and R. The Wimshurst machine is an electrostatic generator based on this method of operation. For information on the important electrostatic devices used in nuclear physics *see* PARTICLE ACCELERATOR.

It should be pointed out that there is actually no motion of positive charges in metals during current flow. The description as given here is a convenient fiction which has been in common usage since Benjamin Franklin's time. In metals, it is the motion of the free electrons which constitutes the electric current; however, motion of electrons in one direction is equivalent to motion of positive charge in the opposite direction. Thus the preceding description is entirely valid as long as this equivalence is kept in mind.

Conservation of charge. The law of conservation of electric charge, an empirical law of experience, says that the total net electric charge in the universe remains constant. As has been explained, when electrification is produced by friction or by induction, there is merely a separation of equal quantities of negative and positive charge; that is, charge is not created in either process. When energy is changed to particles of matter which have electric charges, as in the creation of an electron-positron pair from a γ-ray photon, one particle (the

electron) has a negative charge and the other particle (the positron) has an equal positive charge. Thus, no net charge has been created. Conversely, when annihilation of charged particles occurs, as in electron-positron annihilation to produce a pair of γ-ray photons, equal quantities of positive and negative charge disappear, and thus, no net charge has been destroyed.

Method of electric images. This is one of the schemes for solving certain types of electrostatic problems easily by the substitution of a physically simple, but mathematically equivalent, situation. The types of problems where it is useful are those in which fixed free charges (such as q of Fig. 3a) have set up an induced charge on the surface of a conductor (such as the grounded infinite metal plane AB of Fig. 3a), and a complete description of the electrostatic field in the region of the charge and external to the conductor is desired (for example, the region to the right of AB in Fig. 3a). The equivalent situation used is one in which the conductor with its complicated distribution of induced charges is removed and replaced by a simple distribution of one or more fictitious charges which will produce the same electric field in the region under study. The fictitious charges, which for the purpose of the solution replace the conductor with its induced charge, are called the images of the fixed free charges.

As an illustration, the method of images will be used to solve the special problem posed in Fig. 3a, where q is a positive point charge considered to be on the x axis at a distance $x = a$ from the origin of coordinates located at the surface of the grounded infinite metal plane AB. First, however, the problem in Fig. 3b will be solved where there are two point charges $+q$ and $-q$, as shown, and nothing else in the universe. By using the principle of superposition for potential and the fact that electric potential is a scalar point function, the potential at point P is given by Eq. (1) or, in the problem of Fig.

$$V = \frac{1}{4\pi\epsilon_0} \sum_{i=1}^{n} q_i/r_i \qquad (1)$$

3b by Eq. (2). Here $\epsilon_0 = 8.85 \times 10^{-12}$ coulomb²/newton-m² is called the permittivity of empty space.

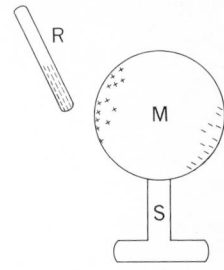

Fig. 1. Charging by electrostatic induction. Metal sphere M, on insulating stand S, is charged by induction because of presence of negative charge on hard-rubber rod R.

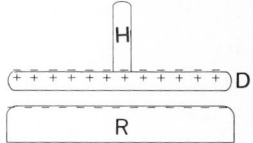

Fig. 2. Electrophorus. When the metal plate D with insulating handle H is placed on the rubber plate R, charge is induced as shown.

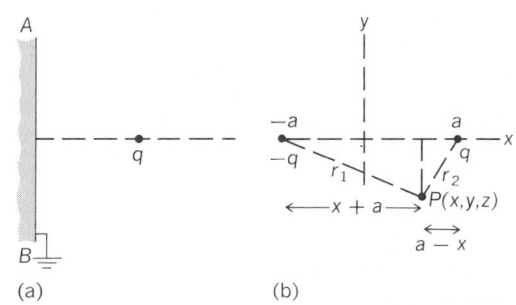

Fig. 3. Illustration of the method of electric images. (*a*) Point charge q at a distance a in front of an infinite grounded conducting plane AB. The problem is to find the potential function and equation for electric field intensity in the region to the right of AB, the field being caused by point charge q, and its equal distributed induced charge on the conducting plane. (*b*) Physically simpler but mathematically equivalent situation for the region of interest.

$$V = \frac{q}{4\pi\epsilon_0[(x-a)^2 + y^2 + z^2]^{1/2}}$$

$$-\frac{q}{4\pi\epsilon_0[(x+a)^2 + y^2 + z^2]^{1/2}} \quad (2)$$

From Eq. (2), the equipotential surfaces can be determined by assigning specific values to V. The particular equipotential surface for which $V = 0$ is the surface where $x = 0$, because the two terms on the right in Eq. (2) then become equal in magnitude as well as opposite in sign. Hence, the yz plane is the equipotential surface for $V = 0$ and lies midway between the two point charges. Thus, the field would be the same if a grounded infinite plane conductor were placed in the yz plane, and this is just the situation in Fig. 3a, the problem to be solved. Therefore, the field is the same to the right of the yz plane in Fig. 3a and 3b, and Eq. (2) is the potential function in this region in both cases (it does not, of course, apply on the left of the conducting plane AB in Fig. 3a). The charge on the left of the yz plane is said to be the electric image of the charge on the right. The gradient of the potential function gives the vector equation for the electric field $\mathbf{E}$, and with equations for $\mathbf{E}$ and V known, most questions with regard to the physical problem can be answered. Also, the image force on q due to the distributed charge induced on the infinite plane can be seen to be given by Coulomb's law applied to the two point charges of Fig. 3b. *See* CAPACITANCE; COULOMB'S LAW; DIELECTRICS; ELECTRIC FIELD; ELECTROSCOPE; POTENTIALS.   [RALPH P. WINCH]

Bibliography: Berkeley Physics Course, vol. 2: *Electricity and Magnetism*, 1970; B. I. Bleaney and B. Bleaney, *Electricity and Magnetism*, 3d ed., 1976; E. M. Pugh and E. W. Pugh, *Principles of Electricity and Magnetism*, 2d ed., 1970.

Electrostriction

A form of elastic deformation of a dielectric induced by an electric field; specifically, the term applies to those components of strain which are independent of reversal of the field direction. Electrostriction is a property of all dielectrics and is thus distinguished from the converse piezoelectric effect, a field-induced strain which changes sign upon field reversal and which occurs only in piezoelectric materials. *See* PIEZOELECTRICITY.

Electrostrictive strain is approximately proportional to the electric susceptibility, elastic compliance, and the square of the field strength, and is extremely small for most materials.

The electrostrictive effect in certain ceramics is employed for commercial purposes in electromechanical transducers for sonic and ultrasonic applications.

[ROBERT D. WALDRON]

Elementary particle

A particle which cannot be described as compound, in the present state of knowledge. The elementary particles are thus the fundamental constituents of matter.

The known elementary particles are listed in Table 1. The graviton, the quantum of the gravitational field, has been omitted from Table 1 since it plays no role in high-energy particle physics; it is firmly predicted by theory, but the prospect of direct observation is exceedingly remote. The heavy vector bosons $W^\pm$ and Z^0 are expected to be observed during the 1980s; their properties have been deduced from the weak interactions, for which they are responsible. Gluons and quarks are never seen as free particles; this phenomenon is known as confinement. Hadrons, strongly interacting particles (see Table 2), are compounds of quarks and gluons; essentially, mesons are composed of a quark-antiquark pair, $q\bar{q}$, and baryons are three quarks, qqq, bound together by the exchange of gluons. *See* BARYON; GLUONS; GRAVITON; HADRON; MESON; QUARKS.

The gauge bosons, γ, g, $W^\pm$, and Z^0, are the quanta of gauge fields. Their fundamental couplings are indicated in Table 1. (The couplings have each been stated in one form, but by the principle of line reversal there may be other equivalent forms; for instance, $U \Rightarrow W^+D$ means also $W^-U \Rightarrow D$, $W^- \Rightarrow \overline{U}D$, and so forth.) There are also couplings between three or four gauge bosons. The photon γ, the quantum of the electromagnetic field, is coupled to charge; that is, the coupling $A \Rightarrow \gamma A$, the amplitude of a process in which a γ is emitted by the elementary particle A, is proportional to Q_A, the charge of A. Similarly, the gluon g, the quantum of the gluon (or color) field, is coupled to color, discussed below. Of the elementary particles, only the quarks and gluons carry color. The different kinds of quarks, called flavors of quark, all have the same color (that is, they are all color SU_3 triplets) and hence are coupled equally to the gluon. This flavor independence of the gluon coupling results in the flavor SU_N symmetries of the hadrons and their strong interactions, discussed below. The coupling $A \Rightarrow gA'$ vanishes unless A and A' belong to the same color multiplet, that is, unless A and A' are both gluons or both the same flavor of quark. Thus, flavor is conserved in strong interactions. The term flavor may be used in a broader sense in which, for example, ν_e and e are two lepton flavors. In this broader sense, the flavor of a quark or lepton changes when it emits or absorbs a $W^\pm$, $U \Rightarrow W^+D$, as discussed below. The exchange of a $W^\pm$ between two particles, changing their flavors, is a weak interaction. The exchange of a Z^0 is a neutral-current weak interaction. *See* COLOR (QUANTUM MECHANICS); FLAVOR; LEPTON; PHOTON.

Experimental evidence for confinement is found, for example, in inelastic electron-proton scattering at high energy, in particular, in deep inelastic scattering, in which the electron loses a sizable fraction of its energy. The results show that the charge of the proton is carried by pointlike (radius less than 10^{-1} femtometer) particles of small mass. But no such particles are seen in the final state of this process, or indeed of any other high-energy collision. What is seen is a narrow shower of hadrons. The interpretation is that the electron scatters off one of the quarks in the proton and gives it a large energy and momentum, the quark responding as though it were a free particle of mass much less than 100 MeV (consistent with the masses of the u and d quarks in Table 1). Later, through the production of quark-antiquark pairs, the energy and momentum of the struck quark is divided up among a number of hadrons, mostly pions, a process called hadronization or fragmentation of the quark, which is to be distinguished from the decay

Table 1. The elementary particles[a]

Gauge bosons $J_C^P = 1_-^-$ Self-conjugate except $\overline{W^+} = W^-$.

Name	Symbol	Charge[b]	Mass, GeV	Couplings
Photon	γ	0	0	$A \Rightarrow \gamma A$
Gluon[c]	g	0	0	$A \Rightarrow gA'$
Weak bosons[d]				
Charged	$W^\pm$	± 1	80	$U \Rightarrow W + D$
Neutral	Z^0	0	90	$A \Rightarrow Z^0 A$

Fermions $J = \frac{1}{2}$ All have distinct antiparticles, except perhaps the neutrinos.

Name	Charge[b]	Symbol	Mass, GeV	Symbol	Mass, GeV	Symbol	Mass, GeV
Leptons Neutrinos	0	ν_e	$<10^{-7}$	ν_μ	$<.0006$	ν_τ	$<.5$
Charged leptons[e]	-1	e	.0005	μ	.106[f]	τ	1.78[f]
Quarks[c] Up type	$\frac{2}{3}$	u	.005	c	1.4	(t	>18)
Down type	$-\frac{1}{3}$	d	.01	s	.15	b	4.8

[a]The graviton, with $J_C^P = 2_+^+$, has been omitted, since it plays no role in high-energy particle physics.
[b]In units of the proton charge.
[c]The gluon is a color SU_3 octet $\{8\}$; each quark is a color triplet $\{3\}$. These colored particles are confined constituents of hadrons; they do not appear as free particles.
[d]The weak bosons have not yet been observed.
[e]Any further charged leptons have mass greater than 15 GeV.
[f]The μ and τ leptons are unstable, with the following mean life and principal decay modes (branching ratios in %):

$$\mu \qquad \tau_\mu = 2.2 \times 10^{-6} \text{ s} \qquad e\bar{\nu}_e \nu_\mu \ 100$$

$$\tau \qquad \tau_\tau = 5 \times 10^{-13} \text{ s} \qquad \begin{array}{l} \mu\bar{\nu}_\mu\nu_\tau \ 18 \\ e\bar{\nu}_e\nu_\tau \ 17 \\ (\text{hadrons})^-\nu_\tau \ 65 \end{array}$$

of a free particle. The string model of confinement, discussed below, gives a concrete picture of this process. The resulting shower of hadrons whose total momentum vector is roughly that of the original quark is called a hadronic jet (like a jet of water which breaks up into a spray of droplets). Such jets are the closest available phenomenon to the actual observation of a quark as a free particle.

Interactions. The interactions of elementary particles are responsible for their scattering and transformations (decays and reactions). They conform to the same symmetry groups revealed by the particle mass spectrum. Because of interactions, an isolated particle will decay into other particles if this does not violate the selection rules. Two elementary particles passing near each other may transform, perhaps into the same particles but with changed momenta (elastic scattering) or into other particles (inelastic scattering). The rates or cross sections of these transformations, and so also the interactions responsible for them, fall into three groups: strong (typical decay rates of $10^{21} - 10^{23}$ s^{-1}), electromagnetic ($10^{16} - 10^{19}$ s^{-1}), and weak ($< 10^{15}$ s^{-1}). Strong interactions occur only between hadrons and have the largest symmetry group. Electromagnetic interactions result from the coupling of charge to the electromagnetic field. They are the best-understood interactions.

Weak interactions are unobservable in competition with strong or electromagnetic interactions. They are observable only when they do something which those much stronger interactions cannot do (forbidden by the selection rules); for instance, by changing flavors they can make a particle decay which would otherwise be stable, and by making parity-violating transition amplitudes they can produce an otherwise absent asymmetry in the angular distribution of a reaction. *See* FUNDAMENTAL INTERACTIONS; STRONG NUCLEAR INTERACTIONS; WEAK NUCLEAR INTERACTIONS.

Stability. Most particles are unstable and decay into smaller-mass particles. The only particles which appear to be stable are the massless particles (graviton, photon), the neutrinos (possibly massless), the electron, the proton, and the ground states of stable nuclei, atoms, and molecules. It is speculated that some or all of the neutrinos may be massive and unstable, and that the proton (and therefore all nuclei) may be unstable. The present view is that the only massive particles which are strictly stable are the electron and the lightest neutrino(s). The electron is the lightest charged particle; its decay would be into neutral particles and could not conserve charge. Likewise, the lightest neutrino is the lightest fermion; its decay would be into bosons and could not conserve angular momentum. *See* ELECTRON; NEUTRON; PROTON.

The unstable elementary particles must be studied within a short time of their creation, which occurs in the collision of a fast (high-energy) particle with another particle. Such fast particles exist in nature, namely the cosmic rays, but their flux is small; thus most elementary particle research is based on high-energy particle accelerators. *See* COSMIC RAYS; NUCLEAR REACTION; PARTICLE ACCELERATOR; PARTICLE DETECTOR; REST MASS.

Hadrons can be divided into the quasistable (or semistable) and the unstable. The quasistable hadrons are simply those that are too light to decay into other hadrons by way of the strong interactions, such decays being restricted by the requirement that isotopic spin I, strangeness s, charm c, and any other flavors, be conserved. The quasistable hadrons that decay through weak interactions have long mean lives—greater than 10^{10} times the characteristic time of strong interactions, $\hbar/m_\pi c^2 = 0.5 \times 10^{-23}$ s where $\hbar$ is Planck's constant h divided by 2π, m_π is the mass of the π meson, or pion, and c is the speed of light. Three hadrons, π^0, η, and Σ^0, can decay by way of the electromagnetic interaction, which conserves I_3 but not I. These three have mean lives of the order of $10^5 \times \hbar/m_\pi c^2$. Figure 1 shows masses and primary decay modes of the quasistable baryons.

The important practical distinctions in the experimental study of interactions are among (1) the stable massive particles (electrons and nuclei), which can be used as target particles; (2) the particles with mean lives greater than 10^{-8} s (γ, ν, n, $\mu^\pm$,

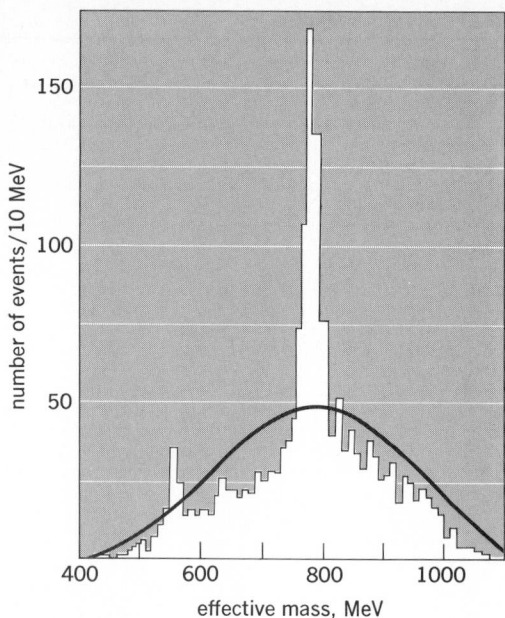

Fig. 2. Observation of the η meson ($m = 549$ MeV) and the ω meson ($m = 783$ MeV) as resonances in the reaction $\pi^+ p \rightarrow \pi^+ p \pi^+ \pi^- \pi^0$. Effective mass is the total relativistic energy of three of the emerging pions in their center-of-mass coordinate system. Peak in curve indicates that collision can create a short-lived particle of corresponding mass, which decays into three pions.

$\pi^\pm$, K_L, $K^\pm$), which can only be used in beams as incident particles; (3) the quasistable hadrons with mean lives of the order of 10^{-10} s ($\Xi^{0,-}$, Λ, $\Sigma^\pm$, K_S, Ω^-), which have only a small, but usable, chance of interacting in matter before decaying; and (4) the remaining hadrons, which have a vanishingly small chance of reinteracting except when produced within a nucleus.

Resonances. The unstable hadrons are also called particle resonances or excited hadrons. Their lifetimes, of the order of $\hbar/m_\pi c^2$, are much too short to be observed directly. Instead they appear, through the uncertainty principle, as spreads in the masses of the particles—that is, in their widths—just as in the case of nuclear resonances (Fig. 2). *See* UNCERTAINTY PRINCIPLE.

The first unstable hadron observed was the "3,3" resonance [in current notation $\Delta(1232)$] in pion-nucleon scattering. That process, $\pi N \rightarrow \Delta \rightarrow \pi N$, is an example of making a resonance by formation. In principle, any of the decay channels of a resonance can be used as an initial channel to form the resonance, for example, $\omega\pi\pi \rightarrow A_2 \rightarrow \rho\pi$. But practically, the initial channel must be composed of just two particles, at least one of which is stable so that it can be used as a target, and the other sufficiently stable so that it can be used as a beam particle. In many cases the resonance decays only weakly, or not at all, into such channels; then the resonance can only be made in production along with other particles. An example of this is the production of the ρ^0 meson in the reaction $\pi^- p \rightarrow n\rho^0 \rightarrow n\pi^+\pi^-$. Production is a less effective way of observing a resonance than is formation. The only resonances which can be made by formation are the neutral $J_C^P = 1_-^-$ mesons (where J_C^P is a combined symbol for spin J, parity P, and charge parity C), which can be formed in e^+e^- collisions,

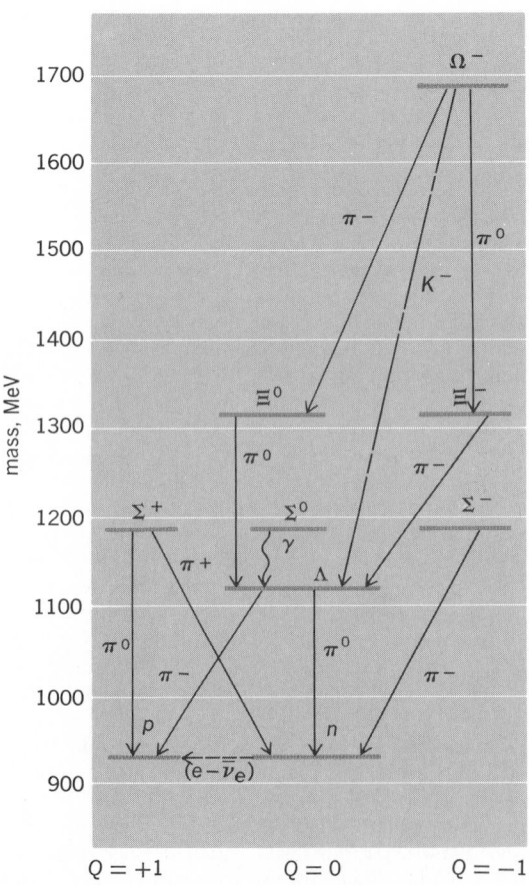

Fig. 1. Decay modes of quasi-stable baryons.

Table 2. The hadrons (strongly interacting particles)

Hadronic quantum numbers‡	Symbol (mass, MeV)	J_C^{P}†	Width, MeV	Decay products and branching ratios, %
Mesons				
	$\pi(138)$	0^-	Hadronically stable	
	$\rho(775)$	1^-	160	$\pi\pi 100$
	$\delta(980)$	0^+_+	50	$\eta\pi, K\bar{K}$
	$A_1(1100-1300)$	1^+_+	300	$\rho\pi 100?$
$I=1$	$B(1230)$	1^+	130	$\omega\pi 100$
	$A_2(1315)$	2^+_+	100	$\rho\pi 70, \eta\pi\pi 15, \omega\pi\pi 10, K\bar{K}5$
	$\rho'(1600)$	1^-	300	$4\pi 85, \pi\pi 15$
	$A_3(1660)$	2^-	200	$f\pi 60, \rho\pi 30$
	$g(1690)$	3^-	200	$\pi\pi 25, \omega\pi, A_2\pi, \rho\rho, \ldots$
	$\eta(549)$	0^-_+	Hadronically stable	
	$\omega(782)$	1^-_-	10	$\pi\pi\pi 90, \pi\gamma 9, \pi\pi 1$
	$\eta'(958)$	0^-_+	0.3	$\eta\pi\pi 66, \rho^0\gamma 30, \omega\gamma 3, \gamma\gamma 2$
	$S^*(980)$	0^+_+	40	$K\bar{K}, \pi\pi$
	$\phi(1020)$	1^-_-	4	$K^+K^- 49, K_L K_S 35, \rho\pi 15, \eta\gamma 2$
	$f(1275)$	2^+_+	180	$\pi\pi 83, K\bar{K}3, \pi\pi\pi\pi 3$
	$\zeta(1275)$	0^-_+	70	$\eta\pi\pi$
	$D(1285)$	1^+_+	30	$\eta\pi\pi 50, \rho\pi\pi 40, K\bar{K}10$
	$\epsilon(1300)$	0^+_+	200-400	$\pi\pi 90, K\bar{K}10$
	$E(1420)$	1^+_+	50	$K^*\bar{K} + \bar{K}^*K, \delta\pi?$
	$\iota(1440)$	0^-_+	50	$\delta\pi$
	$f'(1515)$	2^+_+	70	$K\bar{K}100?$
	$\theta(1650)$	2^+_+	200	$\eta\eta$
	$\omega(1665)$	3^-_-	165	$\rho\pi, B\pi$
	$h(2040)$	4^+_+	150	$\pi\pi, K\bar{K}$
$I=0$	$\eta_c(2980)$	0^-_+		$\eta\pi\pi, K\bar{K}\pi, 4\pi, K\bar{K}\pi\pi, \ldots$
	$\psi(3097)=\psi/J$	1^-_-	0.06	mesons 86, $\mu\bar{\mu}7, e\bar{e}7$
	$\chi(3413)=\chi_0$	0^+_+		$\pi\pi, K\bar{K}, 4\pi, \ldots, J\gamma?$
	$\chi(3508)=\chi_1$	1^+_+		$J\gamma 30, 4\pi, \ldots$
	$\chi(3554)=\chi_2$	2^+_+		$J\gamma 15, \pi\pi, K\bar{K}, 4\pi, \ldots$
	$\eta(3590)=\eta_c'$	0^-_+		$\eta\pi\pi, K\bar{K}\pi, 4\pi, K\bar{K}\pi\pi, \ldots$
	$\psi(3686)=\psi$	1^-_-	0.2	$J\pi\pi 50, \chi(3413)\gamma 7,$ $\chi(3508)\gamma 7, \chi(3554)\gamma 7,$ $J\eta 4, \mu\bar{\mu}1, e\bar{e}1$
	$\psi(3770)=\psi''$	1^-	25	$D\bar{D}100$
	$\psi(4030)$	1^-	50	hadrons
	$\psi(4160)$	1^-	80	hadrons
	$\psi(4415)$	1^-	~40	hadrons
	$Y(9450)=Y$	1^-_-	0.06	$\mu^+\mu^- 2, e^+e^- 2$, hadrons
	$Y(10,010)=Y'$	1^-_-	0.03	$\mu^+\mu^-, e^+e^-$, hadrons
	$Y(10,340)=Y''$	1^-_-		$\mu^+\mu^-, e^+e^-$, hadrons
	$Y(10,560)=Y'''$	1^-	10-20	$B\bar{B}$ dominant?

‡For footnotes see page 303.

(continued)

the ordinary (nonstrange, noncharmed, and so forth) baryons, which can be formed in πN collisions, and the singly strange ($s = -1$) baryons, which can be formed in $\bar{K}N$ collisions.

General properties. In relativistic quantum mechanics, a particle is a system of definite mass and spin. Thus an O_2 molecule, say, in a definite energy level is just as much a particle as is an electron or a pi meson; the concept of particle has nothing to do with elementarity or structure. An unstable particle has a complex mass m, with Im $m = -\frac{1}{2}\Gamma$, where $\Gamma = \hbar/\tau$, where τ is the mean life of the particle. The spin, the intrinsic part of the angular momentum of a particle (the part which does not vanish in the rest frame), is a half integer, in units of $\hbar$. A particle is also characterized by other quantum numbers, such as parity and lepton number, which are eigenvalues of further symmetry operators that commute with mass and spin. *See* ANGULAR MOMENTUM; EIGENVALUE (QUANTUM MECHANICS); NONRELATIVISTIC QUANTUM THEORY; QUANTUM MECHANICS; QUANTUM NUMBERS; REST MASS; SPIN; SYMMETRY LAWS.

An individual particle is never unique; that is, any number of identical copies can exist as well. This means that a field whose quantum is the particle can be defined. There is thus no distinction between kinds of fields and kinds of particles: one can speak interchangeably of the electromagnetic field or of the photon particle, and of the electron or of the electron field. Each particle is either a boson or a fermion, according to whether its spin is an integral or a half-odd-integral multiple of $\hbar$. In the macroscopic world, where classical physics holds, fermions can only appear as particles, but bosons can appear either as particles or as fields if many identical bosons are put into the same mode. *See* QUANTUM FIELD THEORY; QUANTUM STATISTICS.

Antiparticles. To each kind of particle there corresponds an antiparticle, or conjugate particle, which has the same mass and spin, also has the same space parity and charge parity (quantum numbers which have the values $+$ or $-$ and are conserved multiplicatively), belongs to the conjugate representation (multiplet) of internal symmetry

Table 2. The hadrons (strongly interacting particles)[cont.]

Hadronic quantum numbers‡	Symbol (mass, MeV)	J_c^P†	Width, MeV	Decay products and branching ratios, %
Mesons (cont.)				
$s=+1$ $I=\frac{1}{2}$	$K(495)$	0^-	Hadronically stable	
	$K^*(892)=K^*$	1^-	50	$K\pi\,100$
	$Q_1(1280)$	1^+	120	$K\pi\pi(K\rho, K^*\pi?), K\omega?$
	$Q_2(1400)$	1^+	150	$K\pi\pi(K^*\pi, K\rho?)$
	$K^*(1435)$	2^+	100	$K\pi\,49, K^*\pi\,27, K^*\pi\pi\,11, K\rho\,7, K\omega\,4, K\eta\,3$
	$\kappa(1500)$	0^+	250	$K\pi$
	$K^*(1785)$	3^-	125	$K\pi\pi(K\rho, K^*\pi), K\pi\,20$
$I=\frac{1}{2}$ $c=+1$	$D(1866)$	0^-	Hadronically stable	
	$D(2008)=D^*$	1^-	<2	$D\pi, D\gamma$
$I=0$ $s=+1, c=+1$	$F(2030)=F$	$0^-?$	Hadronically stable	
	$F(2140)=F^*$	$1^-?$		$F\gamma$
$I=\frac{1}{2}$ $b=+1$	$B(5200)=B$	0^-	Hadronically stable	

Baryons				$N\pi$	$\Delta\pi$	$N\rho$	$N\epsilon$	$N\eta$
	$N(939)=N$	$\frac{1}{2}^+$	Hadronically stable					
$I=\frac{1}{2}$	$N(1450)$	$\frac{1}{2}^+$	200	$50-65,$	$23,$	$7,$	$7,$	18
	$N(1520)$	$\frac{3}{2}^-$	125	$55,$	$23,$	$19,$	<5	
	$N(1535)$	$\frac{1}{2}^-$	150	$40,$	$1,$	$3,$	$2,$	55
	$N(1650)$	$\frac{1}{2}^-$	150	$60,$	$4-15,$	$7-21,$	$<10,$	$\Lambda K\,10, \Sigma K\,2-7$
	$N(1680)$	$\frac{5}{2}^-$	155	$40,$	$50,$	5		
	$N(1680)$	$\frac{5}{2}^+$	130	$60,$	$18,$	$13,$	22	
	$N(1700)$	$\frac{3}{2}^-$	120	$10,$	$15-40,$	$<5,$	$<40,$	4
	$N(1710)$	$\frac{1}{2}^+$	120	$20,$	$10-20,$	$40-65,$	$15-40,$	$2-20, \Sigma K\,10$
	$N(1810)$	$\frac{3}{2}^+$	200	$17,$	$20,$	$45-70,$	20	
	$N(1990)$	$\frac{7}{2}^+$	250	$5,$				3
	$N(2190)$	$\frac{7}{2}^-$	250	$15,$				2
	$N(2200)$	$\frac{9}{2}^-$	250	$10,$				2
	$N(2220)$	$\frac{9}{2}^+$	300	$20,$				1
	$N(2650)$	$\frac{11}{2}^-$	400	5				
	$N(3030)$	$?$	400					
$I=\frac{3}{2}$	$\Delta(1232)=\Delta$	$\frac{3}{2}^+$	115	100				
	$\Delta(1620)$	$\frac{1}{2}^-$	140	$32,$	$40,$	<50		
	$\Delta(1640)$	$\frac{3}{2}^+$	250	$20,$	$30-45,$	<10		
	$\Delta(1700)$	$\frac{3}{2}^-$	200	$15,$	$<50,$	40		
	$\Delta(1910)$	$\frac{1}{2}^+$	220	$20-25,$	$<40,$			$\Sigma K\,2-20$
	$\Delta(1920)$	$\frac{5}{2}^-$	200	$4-12$				
	$\Delta(1920)$	$\frac{5}{2}^+$	250	$15,$	$10-30,$	60		
	$\Delta(1950)$	$\frac{7}{2}^+$	240	$40,$	$30,$	20		
	$\Delta(2420)$	$\frac{11}{2}^+$	300	10				
	$\Delta(2850)$	$\frac{15}{2}^+?$	400					
	$\Delta(3230)$	$\frac{19}{2}^+?$	440					

(continued)

(for example, an antiquark belongs to an antitriplet $\{\bar{3}\}$ of color SU_3), and has opposite values of charge, I_3, strangeness, and so forth (quantum numbers which are conserved additively). For instance, the antielectron is the positron. If the antiparticle is the same as the particle, it is called self-conjugate; examples are the photon γ and the neutral pion π^0. The equality of masses implies the equality of lifetimes of particle and antiparticle. Thus the positron is stable; however, in the presence of ordinary matter it soon annihilates with an electron, and thus is not a component of ordinary matter. *See* ANTIPROTON; PARITY; POSITRON.

The conjugation operator, which transforms a particle to an equal-mass antiparticle, is a symmetry operator. A self-conjugate particle is an eigenstate of the operator, and so the eigenvalue of the latter is a quantum number of the particle. If weak interactions are neglected, the conjugation operator can be taken to be C, the charge conjugation operator. The γ and π^0 have the C values $-$ and $+$ respectively. When weak interactions are taken into account, C is not a symmetry operator, but CP, the product of the charge conjugation and space inversion operators, is. Usually the effect of the weak interactions is negligible; for instance, they mix into the π^0 unmeasurably small amounts of channels with $C=-$ and $P=+$. But there are exceptions. Neutrinos interact only through the weak interactions; the effect of C on a neutrino is to make a particle which is uncoupled from the weak interactions, which means that it is unobservable. The operation which produces the observed antineutrino is CP. Another example is the K^0: the K^0 is strange and therefore not self-conjugate; that is, $\overline{K^0} \neq K^0$ (the K^0 has quark content $d\bar{s}$, so the $\overline{K^0}$ is $s\bar{d}$). Here, weak interactions cannot be neglected, because in their absence the K^0 and $\overline{K^0}$ are degenerate; the weak interactions

Table 2. The hadrons (strongly interacting particles)[cont.]

Hadronic quantum numbers‡	Symbol (mass, MeV)	J_C^P†	Width, MeV	Decay products and branching ratios, %			
Baryons (cont.)	$\Lambda(1116)=\Lambda$	$\frac{1}{2}^+\{8\}$	Hadronically stable				
				$N\overline{K}$	$\Sigma\pi$		
	$\Lambda'(1405)$	$\frac{1}{2}^-\{1\}$	40		100		
	$\Lambda'(1520)$	$\frac{3}{2}^-\{1\}$	16	46,	42,	$\Lambda\pi\pi10$	
	$\Lambda(1670)$	$\frac{1}{2}^-\{8\}$	40	20,	$20-60$,	$\Lambda\eta\,15-35$	
	$\Lambda(1690)$	$\frac{3}{2}^-\{8\}$	60	25,	$20-40$,	$\Lambda\pi\pi25, \Sigma\pi\pi20$	
$I=0$	$\Lambda(1800)$	$\frac{1}{2}^-\{8\}$	300	$25-40$			
	$\Lambda(1815)$	$\frac{5}{2}^+\{8\}$	80	60,	10,	$\Sigma^*\pi5-10$	
	$\Lambda(1830)$	$\frac{5}{2}^-\{8\}$	95	<10,	$35-75$,	$\Sigma^*\pi>15$	
	$\Lambda(1870)$	$\frac{3}{2}^+\{8\}$	100	$15-40$,	$3-10$		
	$\Lambda'(2100)$	$\frac{7}{2}^-\{1\}$	250	30,	5		
	$\Lambda(2110)$	$\frac{5}{2}^+$	200	$5-25$,	<40,	$N\overline{K}^*(892)\,20-60$	
	$\Lambda(2350)$	$\frac{9}{2}^+$	120	12,	10		
$s=-1$	$\Sigma(1193)=\Sigma$	$\frac{1}{2}^+\{8\}$	Hadronically stable				
				$N\overline{K}$	$\Lambda\pi$	$\Sigma\pi$	
	$\Sigma(1385)=\Sigma^*$	$\frac{3}{2}^+\{10\}$	35		88,	12	
	$\Sigma(1660)$	$\frac{1}{2}^+\{8\}$	100	<30			
	$\Sigma(1670)$	$\frac{3}{2}^-\{8\}$	50	10,	<20,	$20-60$	
	$\Sigma(1750)$	$\frac{1}{2}^-\{8\}$	75	$10-40$,	$5-20$,	<8,	$\Sigma\eta15-55$
$I=1$	$\Sigma(1765)$	$\frac{5}{2}^-\{8\}$	120	41,	14,	1,	$\Lambda(1520)\pi19, \Sigma^*\pi9$
	$\Sigma(1915)$	$\frac{5}{2}^+\{8\}$	100	10,	15		
	$\Sigma(1940)$	$\frac{3}{2}^-$	220	<20			
	$\Sigma(2030)$	$\frac{7}{2}^+\{10\}$	180	20,	20,	$5-10$,	$\Lambda(1520)\pi15, \Sigma^*\pi10, \Delta\overline{K}15$
	$\Sigma(2250)$	$\frac{7}{2}^-?$	100	<10			
	$\Sigma(2455)$	?	120				
	$\Sigma(2595)$	?	200				
$s=-2$ $I=\frac{1}{2}$	$\Xi(1318)=\Xi$	$\frac{1}{2}^+\{8\}$	Hadronically stable				
	$\Xi(1530)$	$\frac{3}{2}^+\{10\}$	10	$\Xi\pi100$			
	$\Xi(1820)$	$\frac{3}{2}^-?\{8\}$	20	$\Lambda\overline{K}45, \Xi(1530)\pi45, \Sigma\overline{K}10$			
	$\Xi(2030)$	?	16	$\Sigma\overline{K}80, \Lambda\overline{K}20$			
$s=-3$ $I=0$	$\Omega^-(1672)$	$\frac{3}{2}^+$	Hadronically stable				
$c=+1$ $\begin{cases}I=0\\I=1\end{cases}$	$\Lambda_c^+(2273)$	$\frac{1}{2}^+$	Hadronically stable				
	$\Sigma_c(2440)$	$\frac{1}{2}^+$	$\Lambda_c^+\pi$				

†Numbers following J_C^P in baryon entries give multiplicity of dominant SU(3) supermultiplet.

‡The hadronic quantum numbers are i-spin magnitude I, the third component of i-spin I_3, strangeness s, charm c, bottomness b, and so forth. Each entry in the table represents an i-spin multiplet of $2I+1$ states with $I_3=-I, -I+1, \ldots, I$. The values of s, c, b, and so on are specified in the table only if nonzero.

determine the eigenstates according to degenerate perturbation theory. The mass eigenstates if nondegenerate must be CP eigenstates (neglecting T violation). They *are* nondegenerate because the states to which they can decay through the weak interaction are different: a $J=0$ $\pi\pi$ system has $CP=+$; a $J=0$ $\pi\pi\pi$ system has $CP=-$. The mass eigenstate, that is, particle, called K_S (short-lived neutral K) decays almost always into $\pi\pi$; the particle called K_L (long-lived neutral K) decays into $\pi\pi\pi$, $\pi e\nu$, and $\pi\mu\nu$ about equally.

Strictly, the conjugation operator is CPT, that is, the product of charge conjugation, space inversion, and time reversal; this is because CPT is always a symmetry operator according to the CPT theorem of Pauli and Lüders, but experiment shows that none of the factors of CPT are. The violation of the CP symmetry (or equivalently violation of T, according to the CPT theorem) is observed only in K_L decay: the K_L decays into $\pi\pi$ with a small branching ratio, 0.3%; therefore it is not exactly a $CP=-$ eigenstate. CP violation is also seen in the unequalness of the branching ratios of $K_L \rightarrow \pi^+l^-\bar{\nu}$ and $\pi^-l^+\nu$ ($l=e$ or μ); they differ by 0.3%.

Multiplets. A characteristic of the hadrons is that they are grouped into isotopic-spin (i-spin) multiplets (for example, n, p; π^-, π^0, π^+); the masses of the particles in each multiplet differ by only a few MeV. Each multiplet can be assigned a certain magnitude I of the isotopic-spin vector **I** (sometimes denoted by **T**). Just as the spin projection states of a particle with spin S form a multiplet with $2S+1$ members, the charge projection states of a hadron with isotopic spin I form a multiplet with $2I+1$ members. The charge Q (in units of the fundamental charge e) of each member of a multiplet is linearly related to its value of I_3, the third component of isotopic spin, by $Q=I_3+\frac{1}{2}Y$, where Y is an integer characteristic of the multiplet, termed hypercharge. *See* ISOTOPIC SPIN.

According to the quark model, i-spin symmetry results from the smallness of the mass difference of the lightest kinds (flavors) of quarks, the u and the d, together with the flavor independence of the glue force which binds quarks together to form hadrons. The fundamental i-spin doublet is (u,d), with $I_3 = (\frac{1}{2}, -\frac{1}{2})$; all heavier quarks are i-spin singlets, with $I = I_3 = 0$. Since their charges are $Q_u = \frac{2}{3}$, $Q_d = -\frac{1}{3}$ (in units of the proton charge) the charge of a hadron is $Q = \frac{2}{3}N_u - \frac{1}{3}N_d + Q' = I_3 + \frac{1}{6}(N_u + N_d) + Q'$, where N_u is the net number of u quarks (number of u minus number of $\bar{u}$) in the hadron, N_d is the net number of d quarks, and Q' is the charge carried by heavier quarks (if any) in the hadron. Only the term I_3 in this formula for Q varies in a given multiplet, and so the formula agrees with $Q = I_3 + \frac{1}{2}Y$. For "ordinary" hadrons (nonstrange, noncharmed, nonbottom, and so forth), which are made of the "ordinary" quarks u and d, the hypercharge Y is seen to be 0 for a meson ($q\bar{q}$) and 1 for a baryon (qqq).

Supermultiplets. The hadrons exhibit a further grouping into supermultiplets of i-spin multiplets, the masses of the multiplets in a supermultiplet differing by a few hundred MeV. These supermultiplets appear to be the consequence of a symmetry that is less exact than i-spin symmetry. This symmetry is with respect to a group of transformations SU_3. The regular representation of the group SU_3 is an octet, consisting of i-spin multiplets with $I = 0, \frac{1}{2}, \frac{1}{2}$, and 1 (hence $1 + 2 + 2 + 3 = 8$ members); the lightest mesons (η, K, $\overline{K}$, π, all with $J^P = 0^-$) and baryons (Λ, N, Ξ, Σ, all with $J^P = \frac{1}{2}^+$) in fact form just such octets.

The vector ($J^P = 1^-$) mesons ω, ϕ, K^*, $\overline{K}^*$, and ρ form another octet plus a singlet, and the $J^P = \frac{3}{2}^+$ baryon resonances $\Delta(1232)$, $\Sigma(1385)$, $\Xi(1530)$, and $\Omega(1672)$ form a deket ($4 + 3 + 2 + 1 = 10$), which is a representation of SU_3.

According to the quark model, this SU_3 symmetry and the pattern of charges in the SU_3 multiplets results from the existence of a third kind (flavor) of quark, the s (strange) quark, with charge the same as the d quark, namely $-\frac{1}{3}$, together with the flavor independence of the glue force; that is, all three quarks u, d, and s have the same interaction with the glue field. The resulting flavor SU_3 symmetry is broken by the relatively large mass of the s, approximately 150 MeV. The three quarks make up the fundamental triplet, $\{3\}$, representation of SU_3; it follows from the vector addition rules that mesons, with the composition $q\bar{q}$, occur only as singlets or octets, $\{1\}$ or $\{8\}$, and baryons, qqq, occur only as singlets, octets, or dekets, $\{1\}$, $\{8\}$, or $\{10\}$. This is consistent with the nonobservation of so-called exotic hadrons which other SU_3 multiplets would contain, such as doubly charged mesons or positive strangeness baryons. *See* GROUP THEORY.

Hadrons are known since 1974 which contain yet more massive quarks, the c and the b (see Table 1). The resulting symmetry is badly broken, and the supermultiplets hardly recognizable. The hadronic quantum numbers I_3, strangeness s, charm c, bottomness b, and so forth, listed in Table 2, specify the net flavor content of a hadron according to the scheme $I_3 = \frac{1}{2}(N_u - N_d)$, $s = -N_s$, $c = N_c$, $b = -N_b$, and so forth, where, as before, N_u is the net number of u quarks, and so on.

Nomenclature. Nomenclature is closely tied to knowledge. When i-spin symmetry became apparent, the term nucleon came into use to mean the multiplet whose two members are the proton and the neutron. Similarly, all other i-spin multiplets have been given names, for example, pi

meson π and sigma hyperon Σ. The members of a multiplet are distinguished by writing charge as a superscript, for example, π^+ and Σ^0. (But p and n are usually written instead of N^+ and N^0.) Antiparticles are denoted by putting a bar over the symbol of the corresponding particle: antiproton $\bar{p}$, antisigma-plus [= anti(sigma-minus)] hyperon $\overline{\Sigma^+}$ ($= \overline{\Sigma}^-$), and so forth. (But the antielectron $\bar{e}$, the positron, is usually written e^+; likewise $\bar{\mu} = \mu^+$.) The equality $\overline{\pi} = \pi$ is valid in the sense that conjugation permutes only the members of the multiplet: $\overline{\pi^+} = \pi^-$, and so forth.

With the discovery of large numbers of unstable hadrons, it has become impractical to give each hadron an individual name; thus the custom has arisen of calling it by the name of a similar hadron of lower mass, the symbol (sometimes asterisked) being followed by its mass (approximate, in MeV). Baryons are named after the lowest-mass baryon with the same I and s, for example, $\Lambda(1520)$ and $\Xi(1530)$.

Baryons composed only of ordinary and strange quarks can only have six combinations of I and s, so there are just six names: N, Δ, Λ, Σ, Ξ, Ω. Further, the only flavor SU_3 multiplets which these baryons form are $\{1\}:\Lambda$, $\{8\}:N$, Λ, Σ, Ξ, and $\{10\}:\Delta$, Σ, Ξ, Ω. Therefore an N belongs to an $\{8\}$, and a Δ or an Ω belongs to a $\{10\}$, uniquely, but a Λ belongs to either a $\{1\}$ or an $\{8\}$, and a Σ or a Ξ belongs to an $\{8\}$ or a $\{10\}$. In fact, a Λ is never a pure $\{1\}$ or $\{8\}$, because the mass of the s quark mixes a Λ in a $\{1\}$ with a Λ of the same J^P in an $\{8\}$; this mixing also occurs for Σ or Ξ hyperons. If the masses of the members of the multiplets concerned are known, one can estimate the relative amount of $\{1\}$ and $\{8\}$ in a given Λ. In Table 2 the dominant SU_3 multiplet is given. The singly charmed analogs of the strange baryons Λ and Σ (the baryons in which the s quark is replaced by a c quark) are called Λ_c and Σ_c respectively.

Most established mesons have their own proper names. But, in the manner of baryons, some are given the same name as the lowest-mass meson of the same I and s (and C, if applicable). Irregularly, D is the name of two unrelated mesons, as is B. The lightest charmonium ($c\bar{c}$) pseudoscalar is named η_c, in the manner of the Λ_c, though not very logically, because the η is not close to 100% $s\bar{s}$. (Following the same scheme, it would have been logical to call the $D(1866)$ "K_c", and the $B(5200)$ "K_b.")

In contrast to the i-spin multiplets, there are no accepted names or symbols for the supermultiplets. One must denote them by their $J^P_{(C)}$ and their rank in mass. For instance, the "lowest (or ground state) $J^P = \frac{1}{2}^+$" specifies the baryon supermultiplet consisting of N, Λ, Σ, Ξ, Λ_c, Σ_c, and so forth.

Regge recurrences. It is known from the interactions of hadrons that they are not point particles, and so it could be expected that hadrons would have rotationally excited states, as do molecules, for instance. Such excited states would be a sequence of hadrons with increasing spins (J_0, $J_0 + 1$, . . .) and masses, but with the same values of other quantum numbers (except for parities: P, C, and G, which would alternate in sign). For historical reasons, such hadrons are called Regge recurrences. The relation between their spin and mass is termed a Regge trajectory. For a rigid body, this would be of the form $m = m_0 + J^2/2\mathscr{I}$

($\mathscr{I}$ = moment of inertia), but for hadrons, it is found empirically to be the form $m^2(J) = b(J - \alpha_0)$, where $b \cong 1$ GeV^2. For example, the $I = 1$ mesons ρ, A_2, and g are recurrences with $J^P(G) = 1^-(+)$, $2^+(-)$, and $3^-(+)$ respectively. There is no $0^-(-)$ member of this trajectory; according to the empirical mass-spin relation, it would have had an imaginary mass.

Because of the exchange character of many forces, it is expected that alternate Regge recurrences will actually fall on separate trajectories (if the two trajectories happen to coincide, as in the case of the mesons ρ, A_2, and g, one says that there is exchange degeneracy). For instance, the $I = \frac{1}{2}$ baryons $N(939)$ and $N(1680)$, with $J^P = \frac{1}{2}^+$ and $\frac{5}{2}^+$ respectively, are recurrences on whose trajectory a $\frac{3}{2}^-$ baryon would have a mass 1370 MeV; in fact, there is such a baryon but with a different mass, namely 1520. Thus this baryon, together with a recurrence, the $N(2190)$, $\frac{7}{2}^-$, lies on a different trajectory. *See* REGGE POLE.

Quantum chromodynamics. It appears that the "glue" field which binds quarks together to make hadrons is a Yang-Mills (that is, a non-Abelian) gauge field of an SU$_3$ symmetry group, color SU$_3$. The quanta of the field are called gluons, and its quantum theory is called quantum chromodynamics (QCD). The gluon field resembles the electromagnetic field, but has an internal symmetry index (octet index) which runs over eight values; that is, there are really eight fields, corresponding to the eight parameters needed to specify an SU$_3$ transformation. (Technically, the field transforms under the group is the same manner as the generators, that is, as the adjoint representation.) Just as the electromagnetic field is coupled to (that is, photons are emitted and absorbed by) the density and current of a conserved quantity, charge, the gluon field is coupled to color. This is a shorthand way of saying the following: The octet of gluon fields is coupled to an octet of color charges; the matrix element of the color charge a of a particle belonging to the representation (multiplet) $\mathfrak{R}$ of color SU$_3$ (all fields and particles belong to color SU$_3$ multiplets because color SU$_3$ is a good symmetry) which makes a transition between color states A and B (A and B label members of the multiplet $\mathfrak{R}$) is given in notation (1), where $(T_a)^{\mathfrak{R}}$ is

$$g(T_a)^{\mathfrak{R}}_{BA} \tag{1}$$

the matrix of the generator T_a of SU$_3$ in the representation $\mathfrak{R}$, and g is a constant which is the same for all particles. The important thing is that the coupling of the gluon to a particle is fixed by the color of the particle (that is, what member of what color multiplet) and just one universal coupling constant g, analogous to the electronic unit of charge e. (The analogy breaks down in quantum theory, as discussed below; the quantity g is no longer constant but it is still universal.)

A color singlet particle does not couple to gluons because $(T_a)^1 = 0$; such colorless particles include the photon, leptons, and any collections of colored particles (quarks and gluons) which are vector-coupled to form a color singlet. Gluons are not colorless, and therefore they are coupled to themselves. This situation is very different from electromagnetism, where the photon does not carry charge. The consequence of this self-coupling of massless particles is a severe infrared (small

momentum transfer or large distance) divergence of perturbation theory. In particular, the interaction between two colored particles through the gluon field, which in lowest order is an inverse-square Coulomb force, proportional to g^2/r^2 (where r is the distance between the particles), becomes stronger than this inverse-square force at larger r. A way of describing this is to say that the coupling constant g is effectively larger at larger r; this defines the so-called running coupling constant $g(r)$. According to the first-order radiative correction, $g(r)$ becomes infinite at a certain distance r_c. As r is raised toward r_c, $g(r)$ rises, and the perturbation series becomes less reliable; all that can be said is that the interaction is very different from Coulomb for $r \gtrsim r_c$.

This situation is very different from that in quantum electrodynamics (QED). There, the interaction between, say, two μ^+ leptons becomes precisely of the inverse-square form at larger distance, with coefficient equal to the product of the leptons' charges, e^2. In dimensionless form, this is $e^2/\hbar c \approx 1/137$, the dimensionless parameter α^{-1} of QED. At smaller distance the interaction becomes stronger (vacuum polarization), but this is a very small effect at distances larger than the Compton wavelength of the lightest charged particle (the electron). In QCD the lightest colored particle (the gluon) is massless, and so vacuum polarization, the deviation from the inverse-square law, occurs at all r. There is also a sign difference: in QED, vacuum polarization shields a charge—just as in any polarizable medium—resulting in an increase in the apparent charge as one gets closer, where the shielding is less effective; in QCD, vacuum polarization—the effect of the gluon self-coupling—is antishielding, resulting in an apparent decrease in the color strength as one gets closer.

There is no way to define a coupling *constant* g^2 in QCD analogous to e^2 in QED. Instead of such a dimensionless parameter, QCD has the scale parameter r_c. At small r, or equivalently at large momentum transfer, the running coupling becomes small (asymptotic freedom); consequently, perturbation theory becomes reliable. Calculations in perturbative QCD, such as for the cross section for gluon bremsstahlung or the gluonic radiative corrections to the cross section for $e^+e^- \rightarrow$ hadrons, are in good agreement with experiment. This is an important reason for the present view that QCD is the correct theory of the hadron glue. *See* QUANTUM CHROMODYNAMICS; QUANTUM ELECTRODYNAMICS.

String model of confinement. A specific form for the gluonic force between two colored particles, at large r, greater than r_c, namely that it falls to a nonzero constant value λ, of the order of $\hbar c r_c^{-2}$, is suggested by a model, the superconductor analogy. This force is confining. (Evidence for this form of the gluonic force also comes from a nonperturbative method of calculating QCD, lattice QCD.) A superconductor excludes a magnetic field from its interior, below a penetration depth D, except in the form of flux bundles, which have a diameter of order D, and one unit of flux, namely $\Phi_0 = 2\pi/Q_{sc}$, where Q_{sc} is the charge of the carriers of the supercurrent. (For a real superconductor, the carriers are Cooper pairs of electrons, so $Q_{sc} = 2e$.) Superconductivity is destroyed along the core of the bundle. If a magnetic monopole were available and put into a superconductor, the magnetic flux leaving the monopole would gather into a flux bundle. That is, closer to the monopole than D, the magnetic field would be roughly that of an isolated monopole; farther away, it would be in the form of a bundle of diameter D. A flux bundle has a certain energy per unit length, of order $\Phi_0^2 D^2 \equiv \lambda$; hence if a pair of monopoles of equal but opposite magnetic charge were put into the superconductor, the lowest energy state would be one in which the flux bundle went straight from one to the other. This means that the monopoles are confined by a long-distance force of constant magnitude λ. (The energy per unit length of the bundle is the same as the tension it exerts, just as the energy per unit area of a surface is its surface tension.) At distances much less than D, the force is of the usual inverse-square form. *See* MAGNETIC MONOPOLES; SUPERCONDUCTIVITY.

The conjecture is that the vacuum is like a superconductor with respect to color, with the interchange, however, of electric and magnetic quantities. That is, the vacuum acts like a color magnetic superconductor which confines color flux into bundles which have a diameter of order r_c and an energy per unit length equal to λ of order $\hbar c r_c^{-2}$. The color flux bundles run between colored particles; they can also form closed loops. These flux bundles are often idealized as having vanishing diameter and are then called strings. This idealization is obviously good only if the flux bundles are long compared to r_c, and if their local radius of curvature is always much larger than r_c.

The motion, according to classical mechanics, of such a string with a particle (quark) at each end, which has the lowest energy for a given angular momentum J, is as usual a rigid rotation. The string is straight, rotating in a plane; centrifugal force balances the tension of the string. At large J, the masses of the quarks at the ends are irrelevant; most of the energy and angular momentum of the system is in the string. The result in this limit is that $E^2 = 2\pi\lambda J$, where E is the total rest energy, that is, mass. Such a linear relation between E^2 and J is just what is found experimentally for the relation between the mass and spin of hadrons of the lowest mass for given spin (the leading Regge trajectory), aside from a constant term, which can be viewed as a quantum effect. The value of λ is 0.185 GeV^2/$\hbar c$ = 1.5×10^4 newtons = 15 metric tons weight. An idealization of the hadron S matrix, the Veneziano or dual model, which embodied the feature of linear Regge trajectories, turned out to have a spectrum of hadrons which was just the spectrum of a string (with nothing at its ends); this was the first appearance of strings in the theory of hadrons. In view of the idealization of the string model and the neglect of the quark masses, it is a mystery that the leading Regge trajectories are so accurately linear.

The string (flux bundle) model also gives a picture for the process of fragmentation: If two quarks, tied with a string, move apart at high velocity, the string is stretched. When it is longer than L, where $\lambda L = 2m_q$, it is unstable to the process of breaking. Breaking involves the creation of a pair of quarks of mass m_q to cap the new string ends (just as breaking a bar magnet creates

a pair of north-south poles). The energy required for the pair creation comes from the loss of a length L of string (that is, L is the length of the created gap); hence the formula above for L. If the new lengths of string are still being stretched enough, they too will break, and so on. Thus the process of hadronization of quarks is really the fragmentation of strings. This process in various reactions is shown schematically in Fig. 3.

Electron-positron annihilation. Electron-positron annihilation at high energies is observed in colliding-beam machines. To lowest order in the fine-structure constant, $\alpha^{-1} = e^2/\hbar c \approx 1/137$, there are two reactions. One is $e^+e^- \to \gamma\gamma$, which is completely described by quantum electrodynamics and will not be discussed further. The other is $e^+e^- \to \gamma^* \to x\bar{x}$, where γ^* is a virtual photon. If x is a lepton, the cross section at a total center-of-mass energy $\sqrt{s}(= 2E_{e+} = 2E_{e-})$ well above threshold (that is, $\sqrt{s} >> 2m_x$) is independent of the lepton's mass m_x and proportional to the square of the charge, Q_x^2. If x is a quark, the only difference at energies s greater than the order of 1 GeV is that its coupling to the gluon field makes a small correction to the cross section and "fragments" the quarks into hadrons (Fig. 3a). The consequence of this behavior is (Eq. 2) for the ratio R

$$R \equiv \sigma(\text{hadrons})/\sigma(\mu\bar{\mu}) = 3\sum_f Q_f^2 \qquad (2)$$

of the cross sections for $e^+e^- \to$ hadrons and $e^+e^- \to \mu^+\mu^-$, where the sum is over all the flavors of quarks whose mass is less than the beam energy and the factor of three comes from the sum over the color. For example, at $\sqrt{s} = 8$ GeV, the flavors are u, d, s, and c, and so $R = 3\,[(2/3)^2 + (1/3)^2 + (1/3)^2 + (2/3)^2] = 10/3$. The reaction $e^+e^- \to$ hadrons has the great advantage in studying hadron properties that the hadronic final state is created as a single high-energy $q\bar{q}$ pair, whereas in ep, pp, πp, and other reactions more quarks are involved.

Equation (2) for R is not valid near the threshold for making a quark pair, for instance in the case of the c quark near $s = 2m_c = 2.8$ GeV. (The zero-point energy of the quarks, due to confinement, moves the effective threshold energy a few hundred MeV higher.) This is because there is little if any energy for fragmentation; consequently the cross section is strongly modulated, with peaks at $c\bar{c}$ resonant states. These states all have $J_C^P = 1^-$, the quantum numbers of the γ^*. Below threshold for fragmenting into the lightest hadrons which contain a c or $\bar{c}$, that is, $\sqrt{s} < 2m_{D(1866)}$, these states namely $\psi(3097)$ and $\psi(3686)$, can decay only hadronically through annihilation of the $c\bar{c}$ pair into gluons. The selection rules require that the number of gluons is at least three (Fig. 3d), so that the "fragmentation" of these gluons does not make the back-to-back double jets seen in $e^+e^- \to \gamma^* \to q\bar{q} \to$ hadrons. This contrast in the distribution of hadrons between events seen on-resonance and off-resonance is more marked at higher energy, namely at the Υ mesons. At still higher energy, namely the expected $t\bar{t}$ mesons at some mass greater than 36 GeV, the hadronic decays of these mesons should show clearly three jets, from the fragmentation of the three gluons. *See* J PARTICLE; UPSILON PARTICLES.

The process $e^+e^- \to \gamma^* \to x\bar{x}$ can also proceed through the Z^0: $e^+e^- \to Z^{0*} \to x\bar{x}$. At low energy

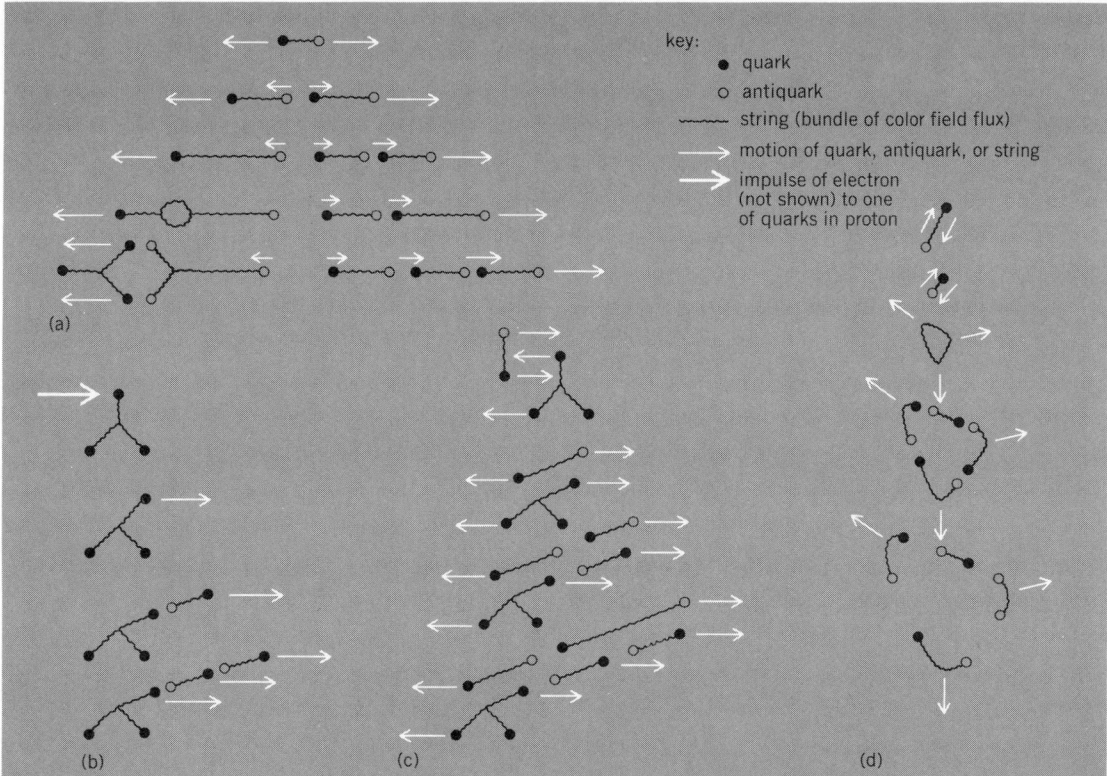

Fig. 3. Development of hadronic jets in high-energy collisions according to the string model. (*a*) Annihilation of *e*⁺*e*⁻ into hadrons. First picture is just after the virtual photon has materialized into a quark pair. (*b*) Hard *ep* collision (deep inelastic scattering). (*c*) Soft *πp* collision. String exchange occurs between first and second pictures. (*d*) Zweig's rule-violating hadronic decay of a heavy quarkonium $J_C^P = 1_-^-$ meson.

this additional amplitude is negligible, but at higher energy it is larger, relative to the γ^* amplitude, and at the mass of the Z^0, $\sqrt{s} = m_{z^0} = 90$ GeV, it will dominate. This will be the experimental discovery of the Z^0. Already at $\sqrt{s} \approx 35$ GeV, the interference of the parity violating Z^{0*} amplitude with the γ^* amplitude makes an easily observable forward-backward asymmetry in the angular distribution of the x and $\bar{x}$.

High-energy hadron-hadron collisions. According to the quark model (QCD), the dominant process in high-energy hadron-hadron collisions is the exchange of a gluon when the two hadrons overlap. This leaves each hadron in a color octet state (by the vector-coupling rule, singlet + octet = octet), and hence they are tied (confined) to one another. Described in more detail, the effect of the exchange of a gluon between a quark in one hadron and a quark in the other hadron is to exchange the strings which were attached to the quarks, so that there are now two strings between the hadrons (Fig. 3*c*). This color transfer interaction between the hadrons is usually soft, transferring little momentum. The strings stretch as the particles continue on their way and then fragment, just as in deep inelastic *ep* scattering (Fig. 3*b*) or in *e*⁺*e*⁻ annihilation into hadrons (Fig. 3*a*). Because in the present case there are two fragmenting strings, twice as many hadrons are expected to be produced for a given energy put into the quark-string systems, and this is roughly so. The model also agrees with the rough constancy of the cross section with energy.

Hadron structure. A generally good semiquantitative understanding of the properties of hadrons follows from assuming that the force between quarks in a hadron is described by a potential of Coulomb form at small r, less than $r_c \approx 1$ femtometer, and linear in r, namely λr, at large r. The description of hadrons composed of heavy quarks (the $c\bar{c}$ and $b\bar{b}$ systems) is particularly good, because their motion in their low-lying bound states is nonrelativistic, so the eigenfunction of a Schrödinger equation is a good approximation to the wave function. (For light quarks, which move relativistically in their bound state, the approximation of the interaction by a static potential is not very good.)

Mesons. The simplest hadrons are the mesons, composed of a quark and an antiquark (called quarkonium, after positronium). The colors of the quark and the antiquark, a color triplet and antitriplet respectively, must be vector-coupled to a singlet. Since the spins of the q and $\bar{q}$ can be vector-coupled to either singlet or triplet, the possible states are (in spectroscopic notation) ^1L, with $J = L$ and $P = (-)^{L+1}$, $C = (-)^L$ and ^3L, with $J = L - 1$, L, or $L + 1$, $C = (-)^{L+1}$. (The values of charge parity C apply only when the quark and antiquark are conjugates of one another, that is, of conjugated flavors.) Thus there are sequences of states with $J_C^P = 0_+^-, 1_-^+, \ldots, 0_+^+, 1_-^-, \ldots, 1_-^-, 2_+^+, \ldots,$ and $1_+^+, 2_-^-, \ldots.$ The state 0_+^- and the sequence $0_+^+, 1_-^-, \ldots$ are missing; in fact no such mesons have been found.

The lowest states of a given quark pair are

expected to be the S states, $^1S_0$ and $^3S_1$, with $J^P_C = 0^-_+$ and 1^-_- respectively. They would be degenerate in mass if it were not for the spin-spin interaction which is the color analog of the interaction of magnetic moments through the magnetic field, that is, the hyperfine interaction. This makes the pseudoscalar, 0^-_+, lighter than the vector, 1^-_-.

In each spin-space state, the four mesons composed of light quarks, $u\bar{d}$, $u\bar{u}$, $d\bar{d}$, and $d\bar{u}$, constitute an i-spin triplet $[u\bar{d}, (u\bar{u} - d\bar{d})/\sqrt{2}, d\bar{u}]$ and a singlet $[(u\bar{u} + d\bar{d})/\sqrt{2}]$. One would expect these mesons to have about the same mass (of course the $u\bar{d}$ and $d\bar{u}$ have exactly the same mass, being mutual charge conjugates) to the accuracy to which the u and d quarks have the same mass, namely about 5 MeV. The four mesons $u\bar{s}$, $d\bar{s}$, $s\bar{d}$, and $s\bar{u}$ constitute an i-spin doublet $[u\bar{s}, d\bar{s}]$ and its conjugate $[s\bar{u}, s\bar{d}]$, with strangeness $+1$ and -1 respectively. These four mesons have about the same mass, which is higher than the mass of the first four, nonstrange, mesons by about 150 MeV, the mass of the s quark. The meson with composition $s\bar{s}$, an i-spin singlet, has a mass which exceeds the strange mesons again by about m_s. A set of mesons with the just-described properties is called an ideal nonet. The lowest 1^-_- mesons form such a nonet, since $m_\rho \approx m_\omega$ ($775 \approx 782$) and $m_{K^*} - m_\omega \approx m_\phi - m_{K^*}$ ($110 \approx 128$). Also the $s\bar{s}$ composition of the ϕ is shown by the large branching ratio of its decay into $K\bar{K}$, 84% (despite the small phase space available), since that decay proceeds merely by the breaking of the string between the s and $\bar{s}$ with creation of a pair of light quarks, whereas a decay such as $\phi \rightarrow \pi\rho$ requires the annihilation of the $s\bar{s}$ and the creation of two pairs of light quarks. The observation that the latter reaction is weaker than the former is known as Zweig's rule.

The ideal nonet can be extended to an ideal 16-plet by letting the quarks be u, d, s, or c. The largeness of m_c causes the wave function of the $c\bar{c}$ ground state, the ψ, to be concentrated at small distance (this is an elementary property of the nonrelativistic Schrödinger equation) where the potential is negative, and consequently the mass of the ψ exceeds the rest mass $2m_c$ by noticeably less than in the other mesons where at least one of the quarks is light. A consequence is that the ψ is lighter than the lightest pair of mesons containing c quarks, that is, $m_\psi < 2m_D$, so that all of its hadronic decays must violate Zweig's rule (Fig. 3d), resulting in the quite small total width of the ψ.

There is a nonet of pseudoscalar, 0^-_+, mesons below 1 GeV, but it is far from ideal: the lowest i-spin singlet $\eta(549)$ is much heavier than the triplet $\pi(138)$. On the evidence of their masses, the π, K, $\bar{K}$, and η form roughly an SU$_3$ octet, having the form $(u\bar{u} + d\bar{d} - 2s\bar{s})/\sqrt{6}$, so that $m_\eta \approx (4m_K - m_\pi)/3$. This is because there is a large interaction matrix element in the 0^- state between the quark pair channel and the two gluon channel: $q\bar{q} \leftrightarrow gg$; since gluons are flavorless, this can happen only in the flavor singlet state, which is thus shifted away in mass from the octet state. (The largeness of m_c, m_b, and so forth, means that $c\bar{c}$, $b\bar{b}$, and so forth mix very little with the glue.) By contrast, in the 1^- state, selection rules force the glue state to have at least three gluons: $q\bar{q} \leftrightarrow ggg$; the "glue" mixing effect is therefore weaker. The foregoing explanation of the low-lying pseudoscalar mesons has a glaring defect, however:

the effect of mixing the $q\bar{q}$ channel with "glue" should drive down the lowest i-spin singlet state below the π mass (according to the rule that two energy levels are repelled when coupled by an interaction); this is in fact observed in the 2^+_+ mesons, where $m_f = 1275 < 1315 = m_{A_2}$. The nonobservation of an isosinglet 0^- meson lighter than the π has been explained in sophisticated ways, but not so far in any simple convincing way.

Baryons. Baryons, according to the quark model, are composed of three quarks, qqq, in a color singlet; this is antisymmetric in the color indices of the quarks, so the baryon wave function is symmetric in the remaining coordinates: space, spin, and flavor. The spins of the three quarks can be added (vector-coupled) to give $S = \frac{3}{2}$, symmetric, or $S = \frac{1}{2}$, of mixed symmetry. The flavors of the quarks can be vector-coupled to a total flavor state (multiplet) which is symmetric, mixed, or antisymmetric. For instance, for three flavors (the quark is then a flavor SU$_3$ triplet, $\{3\}$), these three total flavor states are $\{10\}$, $\{8\}$, and $\{1\}$ respectively. It is useful to couple the spin and flavor into states of overall symmetry (these are states of the group SU$_{2N}$, where N is the number of flavors), because if the spin dependence of the quark-quark interaction is neglected, all members of one such state have the same spatial wave function and the same mass.

The ground-state spatial wave function is symmetric and has vanishing angular momentum, $L = 0$, and even parity. Hence its states have $J = S$, and are a $\frac{3}{2}^+$ $\{10\}$ and a $\frac{1}{2}^+$ $\{8\}$. The spin-spin (hyperfine) interaction raises the mass of the $\frac{3}{2}^+$ and lowers the mass of the $\frac{1}{2}^+$. The states of $\{8\}$ containing 0, 1, or 2 s quarks are i-spin multiplets with $I = \frac{1}{2}$, 0 and 1, and $\frac{1}{2}$ respectively. The observed lightest $\frac{1}{2}^+$ baryons are $N(939)$, $\Lambda(1116)$, $\Sigma(1193)$, and $\Xi(1318)$. Although the Λ and Σ have the same number of s quarks, their masses are different. The following is the explanation for this mass difference: The two light quarks in the Λ form an isosinglet, and hence (because the spin-flavor wave function is symmetric) a spin singlet, and so there is no spin-spin interaction between the strange quark and the light quarks; the entire spin-spin interaction is between the light quarks. This makes the hyperfine energy greater in magnitude (and hence lowers the mass more) in the Λ than in the Σ, because the (color) magnetic moment is inversely proportional to the quark mass. Because the lowest members of the $\frac{1}{2}^+$ octet are all hadronically stable, they are long-lived enough that their magnetic moments can be measured; these are all consistent with the quark model.

For any reasonable quark-quark interaction, the first excited orbital state of qqq is 1^-, of mixed symmetry. The spin-spin and spin-orbit interactions mix states of the same J^P and i-spin. (Thus, for instance, there are three $\frac{1}{2}^-$ Λ states which are mixed.) The calculation of the mixing shows a remarkable agreement with the observed masses and couplings of the low-lying negative-parity baryons (Fig. 4). (This is not a calculation from first principles; several parameters, representing the matrix elements of the spin-dependent interactions, are chosen to best fit the data. But the number of data compared to the number of parameters is very large.) The only notable dis-

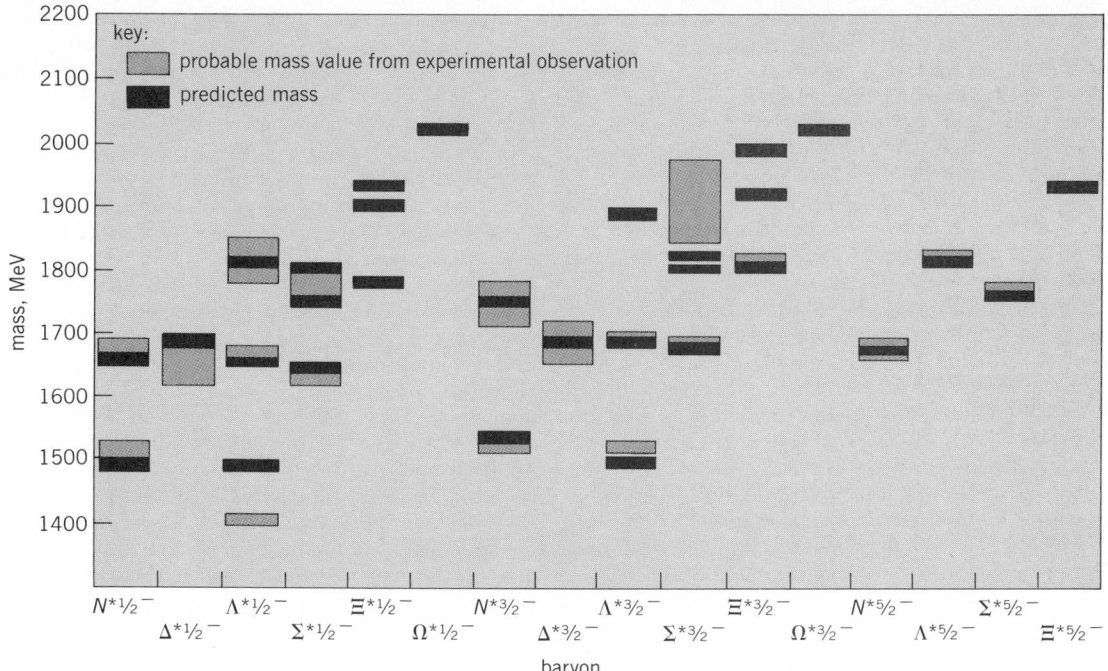

Fig. 4. Comparison of the predicted and observed spectrum of negative-parity baryons. (*After N. Isgur and G. Karl, P-wave baryons in the quark model, Phys. Rev., D18:4187–4205,1978*)

crepancy is that the lowest $\frac{1}{2}^-$ Λ, $\Lambda(1405)$, is predicted about 100 MeV too high; presumably its mass has been depressed by coupling to $\overline{K}N$ continuum states lying higher in energy.

Calculations of the second excited orbital level, using a harmonic oscillator model, yield a large number of positive-parity states, far more than are experimentally observed. The discrepancy has been explained by calculations showing that most of these baryons are weakly coupled to pseudoscalar mesons, so that they cannot be seen in the usual way, using incident π or K beams to form them, or looking for their decays into π, K, or η plus a baryon.

Exotic hadrons. In addition to the "conventional" hadrons with the structure $q\bar{q}$ or qqq, there are other assortments of quarks and/or gluons which can form a color singlet. It is not known whether such hadrons can be long-lived enough to be seen as a resonance peak. If such an object were observed, there would be two possible ways of knowing that it was not a "conventional" hadron: one is that the spectrum of "conventional" hadrons are well enough understood that an interloper would be noticeable; the other would be that the unconventional hadron was an exotic, that is, with quantum numbers which were impossible for a conventional hadron. The most discussed unconventional hadrons are "glueballs" and baryonium.

A glueball is made only of gluons. (In the string picture, it is an endless string, a loop.) It is a flavorless meson, therefore self-conjugate, and with $I = 0$. There are two candidates for glueballs which seem to be interlopers, the $\iota(1440)$, with $J_C^P = 0_+^-$ and the $\theta(1640)$ with $J_C^P = 2_+^+$. Neither, however, has the expected decay branching ratios for a glueball. Since these mesons are not exotics, they can have $q\bar{q}$ channels mixed in, which will change the branching ratios. Conversely, this would mean that the conventional 0_+^- and 2_+^+ mesons have pure gluon channels mixed in; this is certainly not ruled out. The lightest exotic glueball is a three-gluon system with $J_C^P = 1_+^-$; no such meson has been seen.

Baryonium is the name given to a meson with the structure $qq\bar{q}\bar{q}$ for the reason that by merely making a quark pair (in the string model, breaking one string) it can decay (if heavy enough) into a baryon-antibaryon pair. (The name is rather misleading since the baryon pair has nothing to do with the basic structure of the meson, in contrast to the meaning of positronium or quarkonium.) It is not known whether it can more easily decay into a pair of mesons, without having even to make one pair of quarks. An example of an unmistakably exotic baryonium would be doubly charged meson.

Weak interaction. From the quark point of view, the so-called charged-current weak interactions are interactions which are flavor-changing; here flavor is meant in a broad sense which includes leptons. In the older language, these interactions violate I_3 conservation ($I_3 = \pm\frac{1}{2}$ for the u and d quarks respectively, 0 for all others), and/or s (strangeness), and so forth. All experimental results to date are consistent with the weak reactions being point four-Fermi interactions, that is, occurring when the four fermions are at one point in space-time. But such an interaction is not renormalizable and thus not acceptable as an elementary interaction. The properties of the known charged-current four-Fermi interactions are consistent with being the result of the exchange of a massive charged vector (spin 1) particle, called the $W^\pm$ (W^+ and W^- are antiparticles of one another), which is emitted and absorbed by the so-called charged current. It turns out that this particle can be part of a renormalizable field theory, which is a spontaneously broken gauge theory (by the Higgs mechanism). This is the electro-weak theory of S. Weinberg, S. Glashow, and A. Salam. It has four gauge fields, whose quanta, after the symmetry breaking, are the W^+, W^-, a massive neutral vector particle Z^0, and the photon. The Z^0 is coupled to all particles. The strengths of these couplings are controlled by a parameter, the so-called Weinberg angle θ_W, whose value is determined by observation of the neutral-current weak interactions which exchange of the Z^0 yields. These were unknown before this theory suggested their existence. They have now been observed in νe scattering, high-energy e^+e^- scattering, deep inelastic ep scattering, and the low-energy e-nucleus interaction in the atom. In many of these, the Z^0 exchange is observable only because its coupling is partly axial vector, and so interferes with photon exchange to produce a parity-violating effect.

The property of the Z^0 coupling, that it is diagonal (that is, the particle which emits a Z^0 remains unchanged, as with the photon), is deduced from the nonobservation of decays such as $K^+ \rightarrow \pi^+ \nu \bar{\nu}$ and $\mu \rightarrow ee\bar{e}$, which show the absence of a $Z^0 d\bar{s}$ coupling, or a $Z^0 e\bar{\mu}$ coupling respectively. Another place where a $Z^0 d\bar{s}$ coupling would be noticeable is in the $K^0 - \overline{K}^0$ system: exchange of a Z^0 would produce the transition $d\bar{s} \leftrightarrow s\bar{d}$, that is, $K^0 \leftrightarrow \overline{K}^0$, which would make a much larger difference in the masses of K_S and K_L than is observed. In fact, this mass difference is even smaller than what is produced by the second-order effect of two exchanges of a $W^\pm$: $d\bar{s} \leftrightarrow u\bar{u} \leftrightarrow s\bar{d}$; this discrepancy led to the prediction of the charmed quark c, since if properly coupled, its contribution in $d\bar{s} \leftrightarrow c\bar{c} \leftrightarrow s\bar{d}$ can cancel against the previous one.

An important property of the weak interaction is universality. This was first observed in the fact that the interactions responsible for the beta decays $\mu \rightarrow e\bar{\nu}_e \nu_\mu$ and $n \rightarrow pe\bar{\nu}_e$ are nearly equal; that is, the couplings of the $W^\pm$, $W\nu_\mu\bar{\mu}$, and $Wu\bar{d}$, are nearly equal. (Likewise it is found that the couplings $W\mu\bar{\nu}_\mu$ and $We\bar{\nu}_e$ are equal.) This points to a deep similarity of leptons and quarks. Despite the fact that the strangeness-changing-coupling Wus is about $\frac{1}{4}$ the size of the previously mentioned couplings, universality is maintained by Cabbibo's scheme: the couplings Wev_e and Wud_C are exactly equal (and like Wev_μ, the coupling Wus_C vanishes). Here, d_C and s_C are a mixture of d and s, given by Eq. (3), where θ_C is called the Cabibbo angle; in

$$d_C = \cos\theta_C d + \sin\theta_C s \\ s_C = -\sin\theta_C d + \cos\theta_C s \tag{3}$$

words, the doublets (d_C, s_C) and (d, s) are related by an SU_2 transformation, the Cabibbo rotation. With the discovery of the b quark (and expectation of the t) the scheme of Eq. (3) has been extended to all three $Q = -\frac{1}{3}$ quarks; that is, an SU_3 rotation relates (d_C, s_C, b_C) and (d, s, b). The precise statement of universality is that the couplings of the $W^\pm$ are of the form WUD, that is, $U \Rightarrow W^+D$, where the pair (U, D) stands for any of the pairs (ν_e, e), (ν_μ, μ), (ν_τ, τ), (u, d_C), (c, s_C), and (t, b_C). These pairs are called weak i-spin doublets. (The coupling is V-A, that is, to the left-hand helicity component only.) It is at present unknown why the members of the weak i-spin doublets are not the mass eigenstates; a related mystery is the scheme of values of the quarks and leptons.

Grand unified theories. Since the fields responsible for the fundamental forces (glue, electromagnetic and weak) are all of one type, namely gauge fields, an obvious possibility is that they are parts of one large gauge field, the gauge field of a large symmetry group SU_5 or larger. The theories of this are called grand unified theories (GUTs). For interactions with momentum transfers greater than 100 GeV, that is, involving distances less than 10^{-3} femtometer, the masses of the $W^\pm$ and Z^0 are irrelevant, but the gauge fields (gluon or color and electroweak) still do not constitute parts of one gauge field, because they have different coupling constants $g_{Strong}(= g)$, g_{Weak}, and $g_{EM}(= e)$. (The ratio of g_{Weak} to e is described by the weak angle θ_W.) However, these couplings vary with momentum transfer (running coupling constants, as in QCD). With increasing momentum transfer, g and g_{Weak} decrease whereas e increases, and so they become more nearly equal. By a remarkable coincidence, all three become equal (within errors) at the same momentum transfer (this means that the value of θ_W is correctly given), namely 10^{14} GeV, the grand unification energy. In grand unified theories, this energy is the mass of superheavy leptoquark gauge bosons, analogous to the $W^\pm$ and Z^0 bosons of the electroweak subtheory. At momentum transfers larger than this, the mass of the leptoquark bosons is irrelevant and the large symmetry is unbroken; for smaller momentum transfers, the symmetry is broken and the three couplings g, g_{Weak}, and e become different, the more so the lower the momentum transfer.

In these theories, the leptons and quarks occur together in multiplets of the large symmetry group (this is how universality of the weak boson coupling comes about naturally). The couplings of the leptoquark gauge bosons turn leptons into quarks or vice versa (this is the reason for the name leptoquark), or quarks into antiquarks. The exchange of a leptoquark boson can therefore result in the transformation $qqq \rightarrow l\bar{q}q$, for example, $p \rightarrow e^+\pi^0$. This baryon and lepton-number-violating interaction is a much weaker interaction than the analogous ordinary weak interaction, because leptoquark bosons are much heavier than weak bosons. The predicted ratio of the lifetime of the proton to that of a hyperon is the order of the fourth power of the ratio of the boson masses:

$$\tau_p \approx (m_{l-q}/m_W)^4 \tau_\Lambda \\ \approx (10^{14} \text{ GeV}/10^2 \text{ GeV})^4 10^{-10} \text{ s} \approx 10^{38} \text{ s}$$

Remarkably, this is only slightly longer than the present experimental lower limit, and experiments have been undertaken to extend this limit. Another consequence of such interactions that violate baryon number is that if they also violate T and C symmetries they can create a net number of baryons in the big bang, and thus explain the present nonvanishing density of baryons.

[CHARLES J. GOEBEL]

Bibliography: A. W. Hendry and D. B. Lichtenbert, The quark model, *Rep. Prog. Phys.*, 41: 1707–1780, 1978; Y. Nambu, The confinement of quarks, *Sci. Amer.*, 235(5):48–60, 1976; G. K. O'Neill and D. Cheng, *Elementary Particle Physics*, 1979; Particle Data Group: Review of particle properties, *Rev. Mod. Phys.*, 52(2), part 2, 1980; R. F. Schwitters, Fundamental particles with charm, *Sci. Amer.*, 237(4):56–70, 1977.

Elliptic function and integral

In a certain sense, elliptic integrals are the simplest integrals not expressible in terms of elementary functions; elliptic functions arise as the inverse functions of certain elliptic integrals.

Let R be a rational function of x and y, and set $I = \int R(x,y) \, dx$. I can be expressed in terms of elementary functions if y^2 is a polynomial of degree 2 or less in x. If y^2 is a polynomial of degree 3 or 4 in x, I cannot in general be expressed in terms of elementary functions and is called an elliptic integral. (If y^2 is a polynomial of degree 5 or higher, I is called a hyperelliptic integral, and if y satisfies an algebraic equation whose coefficients are polynomials in x, I is called an abelian integral.) The standard elliptic functions are analogous to trigonometric functions. Trigonometric functions may be defined as the inverse functions of certain integrals of the form I; they satisfy differential equations, are periodic functions, and may alternatively be obtained as the "simplest" periodic functions. The standard elliptic functions are the inverse functions of certain elliptic integrals; they satisfy differential equations of order 1 and degree 2, are doubly periodic functions, and may alternatively be obtained as the "simplest" doubly periodic functions.

Applications. In geometry elliptic functions or integrals arise in determining the length of an arc of an ellipse, hyperbola, or lemniscate, the surface of an ellipsoid, geodesics on quadrics of revolution, parametric representations of plane cubic curves or, more generally, curves of genus 1, conformal representation problems, and other problems. In analysis there are applications to differential equations (Lamé's equation, diffusion equation, and others); in number theory, in a great variety of problems. In the physical sciences elliptic functions or integrals appear in potential theory both through conformal representations and in the potential of an ellipsoid, in the theory of elastica, the pendulum, in rigid body motion, in Green's functions in heat conduction and diffusion theory, and many other problems.

Reduction of elliptic integrals. By suitable homographic substitution, $x' = (ax + b)/(cx + d)$, $ad - bc \neq 0$, the elliptic integral I can be reduced to an elliptic integral in which the polynomial y^2 appears in normal form. The two customary normal forms are: Legendre's normal form, $y^2 = (1 - x^2)(1 - k^2 x^2)$ where k, the modulus, is a real or complex number, $|k| \leq 1$ and $k^2 \neq 1$, and it is usual to set $x = \sin \phi$; and the Weierstrass canonical form, $y^2 = 4x^3 - g_2 x - g_3$ where g_2 and g_3, the invariants, are real or complex numbers. I can be expressed as a linear combination of an integral of rational functions and the elliptic integrals of the first, second, and third kinds defined in Legendre's normal form, respectively, as Eqs. (1)–(3), with

$$F(\phi,k) = \int_0^\phi dt / \Delta(t,k) \qquad (1)$$

$$E(\phi,k) = \int_0^\phi \Delta(t,k) \, dt \qquad (2)$$

$$\Pi(\phi,n,k) = \int_0^\phi \frac{dt}{(1 + n \sin^2 t) \, \Delta(t,k)} \qquad (3)$$

$\Delta(t,k) = (1 - k^2 \sin^2 t)^{1/2}$, and in the Weierstrass canonical form as expressions (4) with $y = (4x^3 - g_2 x - g_3)^{1/2}$.

$$\int \frac{dx}{y} \qquad \int \frac{x \, dx}{y} \qquad \int \frac{dx}{(x - c) y} \qquad (4)$$

In Legendre's theory, $\mathbf{K} = \mathbf{K}(k) = F(\pi/2,k)$ and $\mathbf{E} = \mathbf{E}(k) = E(\pi/2,k)$ are called the complete elliptic integrals of the first and second kinds, respectively, $k' = (1 - k^2)^{1/2}$ is the complementary modulus, and $\mathbf{K}' = \mathbf{K}'(k) = F(\pi/2,k')$, $\mathbf{E}' = \mathbf{E}'(k) = \mathbf{E}(\pi/2,k')$. Complete elliptic integrals as functions of k satisfy linear differential equations of the second order and are hypergeometric functions of k^2. They also satisfy Legendre's relation, $\mathbf{KE}' + \mathbf{K}'\mathbf{E} - \mathbf{KK}' = \pi/2$ identically in k.

Periods and singularities. Elliptic integrals are many-valued functions. Any two determinations of I differ by a sum of integral multiples of certain real or complex numbers, the periods. E, F, and Π are many-valued functions of the complex variable $x = \sin \phi$. All three functions have branch points at $x = \pm 1, \pm k^{-1}$, and Π has branch points also at $x = \pm in^{-1/2}$. The periods of F are $4\mathbf{K}$ and $2i\mathbf{K}'$, those of E are $4\mathbf{E}$ and $2i(\mathbf{K}' - \mathbf{E}')$. Since the complete elliptic integrals are real when $0 \leq k \leq 1$, the first (second) of these periods is called the real (imaginary) period. Although E and F are many-valued functions of x, E is a single-valued function of F provided that corresponding values of E and F are obtained by integration over the same path and using the same determination of $\Delta(t,k)$.

Inversion of elliptic integrals. Jacobian elliptic functions are determined by inversion of the functional relation $u = F(\phi,k)$. It is usual to write Eqs. (5).

$$\begin{aligned}
\phi &= \operatorname{am} u = \operatorname{am}(u,k) \\
\sin \phi &= \operatorname{sn} u = \operatorname{sn}(u,k) \\
\cos \phi &= \operatorname{cn} u = \operatorname{cn}(u,k) \\
\Delta(\phi,k) &= \operatorname{dn} u = \operatorname{dn}(u,k)
\end{aligned} \qquad (5)$$

With the additional conditions $\operatorname{sn} 0 = 0$, $\operatorname{cn} 0 = \operatorname{dn} 0 = 1$, it turns out that $\operatorname{sn} u$, $\operatorname{cn} u$, $\operatorname{dn} u$ are single-valued analytic functions of the complex variable u, and that they are doubly periodic and also regular except for simple poles. Nine additional functions are introduced by the notations $1/\operatorname{pn} u = \operatorname{np} u$, $\operatorname{pn} u/\operatorname{qn} u = \operatorname{pq} u$ where p and q stand for any of the letters s, c, d. Similarly, the Weierstrass $\wp$-*function* is introduced by writing relation, Eq. (6), in the form $w = \wp(z) = \wp(z;g_2,g_3)$,

$$z = \int_\infty^w (4t^3 - g_2 t - g_3)^{-1/2} \, dt \qquad (6)$$

and $\wp(z)$ turns out to be single-valued, doubly periodic, and regular except for double poles.

Doubly periodic functions. The term p is called a period of a single-valued analytic function $f(z)$, regular except for isolated singularities, if $f(z + p) = f(z)$. A nonconstant periodic function is either simply periodic, when all its periods are integral multiples of a single period, or else doubly periodic, when its periods are $2m\omega + 2n\omega'$ where m and n are integers, 2ω and $2\omega'$ are primitive periods, and $\operatorname{Im} \omega'/\omega > 0$. Two points of the z plane are congruent if they differ by a period. A parallelogram in the z plane is a period parallelogram if every point in the plane is congruent to exactly one point of the parallelogram. If no singularity or zero of $f(z)$ lies on the boundary of the period parallelogram, the parallelogram is called a cell. An elliptic function is a doubly periodic function which is regular except for poles. An elliptic function has a finite number of poles in every cell; the sum of the residues at these poles is zero, and the sum of the orders of these poles is called the order of the function.

Every elliptic function of order 0 is a constant. There is no elliptic function of order 1. An elliptic function of order $r > 1$ assumes in every cell each complex value r times (counting multiplicity). The difference of two elliptic functions with the same periods, same poles, and the same principal parts at each pole is a constant. The quotient of two elliptic functions with the same periods, poles, and zeros (including multiplicities), is a constant. All elliptic functions with a given pair of primitive periods form an algebraic field, and any two such functions are connected by an algebraic relation. Every elliptic function satisfies an algebraic differential equation of the first order. Every elliptic function possesses an algebraic addition theorem, that is, an algebraic relation connecting $f(u)$, $f(v)$, and $f(u+v)$. Conversely, any single-valued analytic function of z which is regular except for poles and possesses an algebraic addition theorem is either a rational function of z, or a rational function of e^{az} for some a, or else an elliptic function.

The simplest nontrivial elliptic functions are those of order 2. Choice of a basic function with two simple poles in a cell leads to Jacobi's functions, choice of a function with a double pole, to Weierstrass's functions.

Jacobian elliptic functions. Write $s = \text{sn } u$, $c = \text{cn } u$, $d = \text{dn } u$, $s' = ds/du$, and so on. The periods, zeros, and poles are given in Table 1 in which m and n are integers. These functions possess symmetry properties around the points $u = 0$, $\mathbf{K}$, $i\mathbf{K}'$ which are set out in Table 2 and on account of which it is sufficient to study the functions in the parallelogram whose vertices are 0, $\mathbf{K}$, $\mathbf{K} + i\mathbf{K}'$, and $i\mathbf{K}'$. There is a very large number of identities for these functions. Some of the basic ones are shown in Eqs. (7). The addition theorem for

$$
\begin{aligned}
&s^2 + c^2 = 1, \ k^2 s^2 + d^2 = 1, \ d^2 - k^2 c^2 = k'^2 \\
&s' = cd, \ c' = -sd, \ d' = -k^2 sc, \\
&s'^2 = (1 - s^2)(1 - k^2 s^2), \text{ etc.} \\
&\text{sn } (iu,k) = i \text{ sc } (u,k'), \ \text{cn } (iu,k) = \text{nc } (u,k') \\
&\text{dn } (iu,k) = \text{dc } (u,k')
\end{aligned} \tag{7}
$$

Table 1. Properties of Jacobian elliptic functions

Function	Primitive periods	Zeros	Poles
sn u	$4\mathbf{K}, 2i\mathbf{K}'$	$2m\mathbf{K} + 2ni\mathbf{K}'$	
cn u	$4\mathbf{K}, 2\mathbf{K} + 2i\mathbf{K}'$	$(2m+1)\mathbf{K} + 2ni\mathbf{K}'$	$2m\mathbf{K} + (2n+1)i\mathbf{K}'$
dn u	$2\mathbf{K}, 4i\mathbf{K}'$	$(2m+1)\mathbf{K} + (2n+1)i\mathbf{K}'$	

Table 2. Symmetries of Jacobian elliptic functions

v	$-u$	$2\mathbf{K} - u$	$2i\mathbf{K}' - u$
sn v	$-\text{sn } u$	sn u	$-\text{sn } u$
cn v	cn u	$-\text{cn } u$	$-\text{cn } u$
dn v	dn u	dn u	$-\text{dn } u$

s is given in Eq. (8). There are similar addition

$$
\text{sn } (u+v) = \frac{\text{sn } u \text{ cn } v \text{ dn } v + \text{sn } v \text{ cn } u \text{ dn } u}{1 - k^2 \text{ sn}^2 u \text{ sn}^2 v} \tag{8}
$$

theorems for c and d. These, in combination with the formulas for sn (iu), for example, serve to express sn $(u + iv)$ in terms of elliptic functions of u and v; they provide formulas for sn $(2u)$, sn $(u/2)$, and so on. By means of these formulas the values of the Jacobian functions at the points $m\mathbf{K}/2 + ni\mathbf{K}'/2$ (m, n integers) and at the points $m\mathbf{K} + in\mathbf{K}' + u$ can be obtained.

The elliptic functions reduce to elementary functions if one or both of the periods become infinite (degenerate elliptic functions, see Table 3).

Table 3. Degenerate elliptic functions

$\mathbf{K}$	$\mathbf{K}'$	k	k'	sn u	cn u	dn u
∞	$\frac{\pi}{2}$	1	0	tanh u	sech u	sech u
$\frac{\pi}{2}$	∞	0	1	sin u	cos u	1
∞	∞		0	1	1

Weierstrass's functions. With $w = 2m\omega + 2n\omega'$, sums and products running over all pairs of integers m, n except $m = n = 0$, there are these functions: Weierstrass sigma function, Eq. (9), which is

$$
\sigma(z) = z\Pi\left\{\left(1 - \frac{z}{w}\right) \exp\left[\frac{z}{w} + \frac{1}{2}\left(\frac{z}{w}\right)^2\right]\right\} \tag{9}
$$

an entire function; Weierstrass zeta function, $\zeta(z) = \sigma'(z)/\sigma(z)$, which is a meromorphic function; and Weierstrass $\wp$-function, $\wp(z) = \zeta'(z)$, which is an elliptic function of order 2 with double poles at $z = 0$ and congruent points. The invariants are $g_2 = 60\Sigma w^{-4}$ and $g_3 = 140\Sigma w^{-6}$. Legendre's relation becomes $\omega'\zeta(\omega) - \omega\zeta(\omega') = \pi i/2$. The $\wp$-function satisfies the differential equation, Eq. (10), and

$$
\wp'^2(z) = 4\wp^3(z) - g_2\wp(z) - g_3 \tag{10}
$$

possesses the addition theorem given as Eq. (11). It

$$
\wp(u+v) = \frac{1}{4}\left[\frac{\wp'(u) - \wp'(v)}{\wp(u) - \wp(v)}\right]^2 - \wp(u) - \wp(v) \tag{11}
$$

is a homogeneous function of degree -2 in z, ω, ω'. Every elliptic function with primitive periods 2ω, $2\omega'$ can be expressed in the form $R_1[\wp(z)] + \wp'(z)R_2[\wp(z)]$, where $R_1(w)$ and $R_2(w)$ are rational functions of w; and there are also representations in terms of zeta and sigma functions. Degenerate cases lead to elementary functions.

Theta functions. The function in Eq. (12), with a

$$
\theta(v|\tau) = \sum_{r=-\infty}^{\infty} e^{i\pi r^2\tau + i\pi r(2v+1)} \tag{12}
$$

fixed τ and Im $\tau > 0$, is an even entire function of v. It has period 1, it is multiplied by $e^{-i\pi(2v+\tau)}$ when v is increased by τ, and it has simple zeros at the points $v = m + (n + 1/2)\tau$ (m, n integers). It is usual to consider four theta functions, Eqs. (13).

$\theta(x/2, i\pi t)$ satisfies the partial differential equation $\partial^2 y/\partial x^2 = \partial y/\partial t$ and has a simple Laplace transform. Elliptic functions and elliptic integrals can

$$\theta_1(v) = -ie^{i\pi(v+\tau/4)}\theta\left(v+\frac{\tau}{2}\right)$$

$$\theta_2(v) = e^{i\pi(v+\tau/4)}\theta\left(v+\frac{1+\tau}{2}\right) \qquad (13)$$

$$\theta_3(v) = \theta(v+1/2)$$

$$\theta_4(v) = \theta(v)$$

be expressed in terms of theta functions: $\tau = \omega'/\omega$ in the case of Weierstrass functions and $\tau = i\mathbf{K}'/\mathbf{K}$ in the case of Jacobian functions or Legendre's normal form of elliptic integrals.

Transformation theory. The set of periods of an elliptic function may be described by various pairs of primitive periods. The change from one pair of primitive periods to another pair is called a transformation of the elliptic function or integral. The quotient of the primitive periods, τ, undergoes a homographic substitution, $\tau = (a\tau + b)/(c\tau + d)$, where a, b, c, d are integers and $D = ad - bc$ is positive and is called the degree of the transformation. All transformations of degree 1 form the modular group. The study of these transformations is of great theoretical interest, has applications to number theory, and is also used for the numerical computation of elliptic functions. It is also connected with the study of elliptic modular functions, or analytic functions, $f(\tau)$, with the property that $f(\tau)$ and $f(\dot\tau)$ are algebraically connected whenever τ and $\dot\tau$ are connected by a transformation of the modular group. *See* FOURIER SERIES AND INTEGRALS.

[A. ERDELYI]

Bibliography: P. F. Byrd and M. D. Friedman, *Handbook of Elliptic Integrals for Engineers and Scientists*, 2d ed., 1971; P. DuVal, *Elliptic Functions and Elliptic Curves*, 1972; A. Erdélyi et al., *Higher Transcendental Functions*, vol. 2, 1953; S. Lang, *Elliptic Functions*, 1973; National Bureau of Standards, *Handbook of Mathematical Functions*, 1964; E. H. Neville, *Elliptic Functions: A Primer*, 1971; H. E. Rauch and A. Leibowitz, *Elliptic Functions, Theta Functions and Riemann Surfaces*, 1973; E. T. Whittaker and G. N. Watson, *A Course of Modern Analysis*, 4th ed., 1940.

Emissivity

The ratio of the radiation intensity of a nonblackbody to the radiation intensity of a blackbody. This ratio, which is usually designated by the Greek letter ϵ, is always less than or just equal to one. The emissivity characterizes the radiation or absorption quality of nonblack bodies. Published values are readily available for most substances. Emissivities vary with temperature and also vary throughout the spectrum. For an extended discussion of blackbody radiation and related information *see* HEAT RADIATION.

There are several methods by means of which the emissivity can be determined. The one most commonly used is the cavity method. In this technique a fine hole is provided in a radiating surface, and the ratio of the radiation intensity from the surface to the radiation intensity from the hole yields the emissivity directly. This method is quite accurate. One can also use an optical pyrometer to determine the emissivity from the brightness temperatures of the hole and the surface in conjunction with Wien's law of radiation.

The total emissivity when introduced into the Stefan-Boltzmann law gives the total radiated energy W in joules per square centimeter of the real heat radiator as $W = \epsilon\sigma T^4$. Here T represents the absolute temperature and σ, the radiation constant, has the value 5.67×10^{-12} joule cm^{-2} K^{-4}. This energy is always smaller than the energy radiated by the blackbody, since ϵ is less than 1. For example, the total emissivity for tungsten is 0.32 at 2500°C, which means that at the same temperature tungsten radiates approximately one-third the energy of a blackbody.

The spectral emissivity ϵ_λ (the subscript λ denotes the wavelength) provides information on the energy distribution. Any spectral emissivity value is valid only for a narrow wavelength interval. The wavelength at which ϵ_λ has been determined is indicated by a subscript, for instance, $\epsilon_{0.655}$. A spectral emissivity of zero means that the heat radiator emits no radiation at this wavelength. Strongly selective radiators, such as insulators or ceramics, have spectral emissivities close to 1 in some parts of the spectrum, and close to zero in other parts. Carbon has a high spectral emissivity throughout the visible and infrared spectrum, exceeding 0.90 in certain portions; thus carbon is a good blackbody radiator. Tantalum is the only metal with a spectral emissivity greater than 0.5 in the visible spectrum. All other metals have a lower spectral emissivity. Tungsten is a relatively good emitter, with a spectral emissivity of 0.43–0.47 within the visible region of the spectrum.

[HEINZ G. SELL; PETER J. WALSH]

Energy

The ability of one system to do work on another system. There are many kinds of energy: chemical energy from fossil fuels, electrical energy distributed by a utility company, radiant energy from the Sun, and nuclear energy from a reactor. The units of energy include ergs, joules, kilowatts, foot-pounds, and foot-poundals. Work and heat have the same units as energy, but are entirely different physical concepts.

Work. If a particle is moved from a point in space $\vec{r}_1$ to a $\vec{r}_2$ along a curve C by a force $\vec{F}$, the force does an amount of work $W_{12}(C)$ given by Eq. (1), where $d\vec{R}$ parameterizes the path along C and

$$W_{12}(C) = \int_{\vec{r}_1}^{\vec{r}_2} \vec{F} \cdot d\vec{R} \qquad (1)$$

$\vec{F} \cdot d\vec{R}$ is the dot product of $\vec{F}$ and $d\vec{R}$, that is, the component of $\vec{F}$ along $d\vec{R}$. Since forces cause changes in motion through Newton's laws, the work in Eq. (1) could impart motion to a stationary object, stop a moving object, or change the magnitude or direction of the object's velocity. Thus work is a technical term, and requires both a force and a displacement. A person traveling around the Earth at constant speed would undergo a displacement of approximately 40,000 km, but would do no work since this displacement is perpendicular to the gravitational force of the Earth. Similarly, a person who pushes on a wall without moving it does no work because the wall has no displacement. *See* FORCE; WORK.

A force is called a conservative force if the work which it does in displacing a particle from one point to another is independent of the path of the

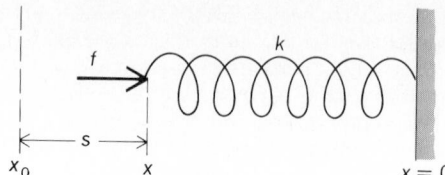

Fig. 1. Potential energy of a compressed spring. Here $s = x - x_0$ is the distance that the spring has been compressed from its equilibrium length x_0. The spring constant is k, and the external force f compresses the spring.

particle for all paths. Conservative forces can be expressed as the gradient of a scalar function $V_\alpha(\vec{r})$, which is called the potential energy function for the conservative force. Each type of conservative force corresponds to a kind of interaction in nature. Only five potentials, or interactions, are required for all present understanding in fundamental physics: gravitational potential energy; electromagnetic potential energy; weak, or neutron decay, potential energy; superweak potential energy; and nuclear potential energy. The first two potential energy functions are fairly well understood. S. Weinberg and A. Salam have done work toward unifying electromagnetism with weak interactions. *See* ELECTROMAGNETISM; FUNDAMENTAL INTERACTIONS; GRAVITATION; STRONG NUCLEAR INTERACTIONS; WEAK NUCLEAR INTERACTIONS.

Any particle or system of particles subject to conservative forces has two kinds of energy, potential energy and kinetic energy. Potential energy is the energy due to position or configuration, and kinetic energy is the energy due to motion.

Potential energy. The electric field $\vec{E}$ and the magnetic field $\vec{B}$ in a region of space can be calculated from an electrostatic potential and a magnetic vector potential. Then the force on a charged particle can be calculated, and the work done on the particle along a specific path by the force can be obtained. The electromagnetic potentials in Schrödinger's equation account for the atomic structure of matter. Thus, macroscopic changes in potential energy that result from pushing an object from one place to another with a stick or pulling it with a string have their microscopic origin in the electromagnetic potential energy which holds the atoms in place in the stick or string together. However, there are macroscopic forces which produce different potential energy functions, such as the elastic spring (Fig. 1). The potential energy func-

tion for a spring is given by Eq. (2), where k is in

$$V(x) = \tfrac{1}{2} k (x - x_0)^2 \tag{2}$$

units of N/m and x_0 is the equilibrium spacing. The hookean force law $F = -k(x - x_0)$ can be derived by taking the negative derivative, $-d/dx$, of $V(x)$ in Eq. (2). There are also simple functional forms for the five fundamental interactions. Each potential energy function specifies the dependence of the force on position, and each one contains a constant, such as the x_0 in Eq. (2), which is a reference potential energy. This constant part of the potential energy simply specifies a reference point and does not contribute to the force (Fig. 2). *See* POTENTIALS.

Kinetic energy. The kinetic energy of a body is due to its motion. A point particle with mass m and velocity v has kinetic energy given by Eq. (3). Any

$$E = \tfrac{1}{2} m v^2 \tag{3}$$

body in motion has Eq. (3) as its translational kinetic energy. The velocity must be specified with respect to some inertial reference frame.

A rigid body can move as in Eq. (3), but in addition it can rotate about its center. The rotational kinetic energy E_R, for a body with moment of inertia I and angular frequency ω, in units of radians per second is given by Eq. (4). The most general

$$E_R = \tfrac{1}{2} I \omega^2 \tag{4}$$

kinetic energy for a solid body in motion is given by Eq. (5). Understanding of the significance of

$$E_k = \tfrac{1}{2} m v^2 + \tfrac{1}{2} I \omega^2 \tag{5}$$

Eqs. (3)–(5) is aided by consideration of the order of magnitude of the quantities appearing there. The mass of a large automobile is about 900 kg. A translational speed of 55 mph is about 25 m/s. The translational kinetic energy of this automobile is 275,000 J. The moment of inertia of this auto is $I = mk^2$, where k is the radius of gyration ($k \sim 1.5$ m). If the auto rotates about its center at $\omega = 16.7$ rad/s, which is a little more than 2.5 revolutions per second, its rotational kinetic energy will also equal 275,000 J. *See* ROTATIONAL MOTION.

Reference frames. The galilean principle of relativity is that all of the mechanical properties of systems are the same in all inertial systems which move at constant velocity with respect to each other. Thus the velocity must be measured in some particular inertial frame. A person in an auto traveling at a constant velocity of 55 mph (or 25 m/s) as measured from the ground is at rest in an inertial frame traveling at 25 m/s attached to the car. However, if the car collides with a large tree, the driver gains no advantage from his or her zero velocity because the tree is racing toward the car at 55 mph in this inertial frame.

Conservation of energy. Energy is conserved for all isolated mechanical systems. This is because if a system A is isolated, there is no other system B that it can give any energy to, and its total energy must remain constant. This system A can convert kinetic energy to potential energy, and it can convert one form of potential energy to another, but the total energy must remain the same. The meaning of conserved total energy is that the system has the same value of total energy at all times. The definition of time translational invariance is that the system as a whole does not change

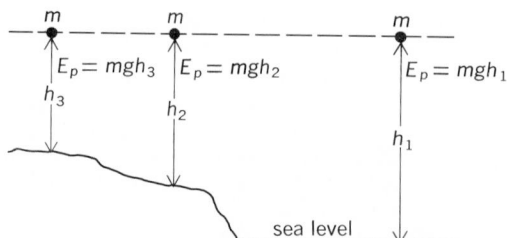

Fig. 2. Gravitational potential energy $E_p = mgh$ for an object of mass m and weight mg shown at the same height h for different reference levels. The different values of E_p correspond to the additive constant discussed in the text.

in time, that is, the system is isolated. *See* CONSER-VATION OF ENERGY.

Internal energy. Macroscopic bodies and processes are not isolated in reality, so that some small energy transfer is present. In the laboratory, the isolation of a gas, solid, or fluid sample can be approximated. Usually, one object is in contact with another one, such as a book on a table or a pan of water on a stove burner. Energy can be transferred in these cases, and dissipations can occur. A new variable, the temperature T, is required to describe equilibrium of a new physical quantity, heat. If two objects have the same temperature, they are in thermal equilibrium and no heat will flow between them. If their temperatures are not equal, heat will flow from the body with a higher temperature to the body with a lower temperature. Heat has the same units as work and energy, but is physically different. *See* HEAT.

The first law of thermodynamics is a generalized conservation-of-energy law for all macroscopic processes. It may be expressed by Eq. (6), where

$$dU = dW + dQ \qquad (6)$$

dU is the internal energy of the system, dW is the work done on the system, and dQ is the heat added to the system. The symbol d emphasizes that the work and heat are path-dependent, or are inexact differentials, whereas the internal energy $U(S,V)$ is an exact or path-independent differential, where S is the entropy and V is the volume. The interpretation of the first law of thermodynamics, Eq. (6), is that of generalized conservation of energy, generalized to include dissipations, dQ. The microscopic origin of the macroscopic internal energy is that of the average energy $\bar{\epsilon}$ given by Eq. (7), where $P(\epsilon_i)$ is

$$U = \bar{\epsilon} = \sum_i \epsilon_i P(\epsilon_i) \qquad (7)$$

the probability that the N molecules of the system have energy ϵ_i. It is the large number of particles in macroscopic systems, N of order equal to or greater than 10^{28}, which sharpens the statistical distribution of energies and makes the internal energy a single number. The large numbers also allow pressure, volume, temperature, and entropy to be treated as simple variables when they are really statistical distributions. The average statistical energy is path-independent, and is equal to the sum of a path-dependent work term which involves only forces and displacements (or pressure and volume) and a path-dependent heat term which involves temperature and entropy. The work term includes all mechanical effects, and the heat term involves only thermal variables. *See* INTERNAL ENERGY; THERMODYNAMIC PRINCIPLES.

Mass-energy. In 1905 A. Einstein showed that at high velocities near the speed of light important modifications must be made in physical concepts. One particularly radical idea which he advanced was that space and time are not independent, but rather are two aspects of the same object, a space-time manifold. This necessitated a reexamination of the concept of energy and led to the conclusion that the inertia, or mass m, depends upon its energy through the mass-energy relation, Eq. (8), where

$$E = mc^2 \qquad (8)$$

c is the speed of light in vacuum. Furthermore, energy and momentum conservation become joined in a single four-momentum conservation law in special relativity. Equation (8) is valid for all known physical and chemical processes. *See* INERTIA OF ENERGY; RELATIVITY.

Nuclear energy, first in the atomic bomb and then in peaceful applications such as energy-generating reactors, was developed in part due to Eq. (8). Since it was experimentally known that large mass changes Δm occur in nuclear reactions, Eq. (8) implies that the energy released is $\Delta E = \Delta mc^2$. In the nuclear reaction $^2H + {}^3H \rightarrow {}^1n + {}^4He + Q$ among the light nuclei, the energy released, Q, is almost 20 MeV. This reaction is prominent in fusion technology because 20 MeV, approximately 20,000,000 times the energy released in one TNT or dynamite process, is especially large. In one fission process of ^{235}U, over 200 MeV are released, but are distributed among two or three neutrons and one or more heavier nuclei, so that "gathering" the energy for use is more difficult. *See* NUCLEAR FISSION; NUCLEAR FUSION.

Field energy. Whereas the particles and rigid bodies of mechanics are localized objects which have positions and velocities, there are quantities called fields which are defined throughout a region, such as fluids, the electromagnetic field, and sound and water waves. Fields are recognizable disturbances. Each field satisfies some field equation which determines what sort of disturbance is created under various conditions. A fluid or water wave is a disturbance in the fluid or water medium. A sound wave is a pressure wave in a solid, liquid, or gas. The electromagnetic wave consists of an electric vector $\vec{E}(\vec{r}, t)$ and a magnetic vector $\vec{B}(\vec{r}, t)$, and can propagate through a vacuum; that is, no medium is needed. Maxwell's field equations for the electromagnetic field require that $\vec{E}$ and $\vec{B}$ are mutually perpendicular and are orthogonal to $\vec{k}$, the direction of propagation of the wave. If a field is set up in an isolated region of space Ω, an energy density $e(\vec{r})$ exists such that the total energy E_T is given by Eq. (9), and is conserved (since Ω is iso-

$$E_T = \int_\Omega e(\vec{r}) d^3 r \qquad (9)$$

lated from external agents once the field is established). For an electromagnetic field, Eq. (9) becomes Eq. (10), where ϵ and μ are the permittivity

$$E_T = \frac{1}{2} \int_\Omega \left(\epsilon \, \vec{E}^2 + \frac{1}{\mu} \, \vec{B}^2 \right) d^3 r \qquad (10)$$

and permeability of the region and $\vec{E}$ and $\vec{B}$ are the electromagnetic field vectors. In the case of fields, the conserved total energy E_T is constant because of time translational invariance of the field equations. *See* CLASSICAL FIELD THEORY; ELECTROMAGNETIC RADIATION; MAXWELL'S EQUATIONS; SOUND; WAVE MOTION.

Atomic energy. An atomic or subatomic particle subject to the potential energy $V(\vec{r})$ is described by a function $\psi(\vec{r})$ whose behavior is governed by Schrödinger's equation. For a one-dimensional particle of mass m in a potential $V(x)$, Schrödinger's equation is given by Eq. (11), where $\hbar = 1.054 \times$

$$H\Psi(x, t) = \left\{ -\frac{\hbar^2}{2m} \frac{d^2\Psi}{dx^2} + V(x)\Psi(x, t) \right\}$$
$$= i\hbar \left(\frac{\partial \Psi}{\partial t} \right) \qquad (11)$$

10^{-34} J-s is Planck's constant, x is the space coordinate, t is the time parameter, and H is the hamiltonian operator for the particle. If $\psi(x, t) = \phi(x)T(t)$, it is straightforward to show that $T(t) = e^{-iEt/\hbar}$ and Eq. (11) reduces to Eq. (12), where E_n is the value

$$H\phi_n(x) = E_n\phi_n(x) \qquad (12)$$

of the total energy of the particle. Depending upon which H and V are used, a form of Eq. (12) describes electronic states in atoms, proton and neutron states in a nucleus, or atoms in a molecule. In all of these cases the hamiltonian acts as an energy operator, and the time translational invariance of the system requires that the energy be conserved in Schrödinger's equation too. *See* ENERGY LEVEL; NONRELATIVISTIC QUANTUM THEORY; QUANTUM MECHANICS.

Time translational invariance. For each case discussed in this article, there are systems or states of systems which do not change in time; and each of these systems has a conserved total energy. Conversely, when systems or states of systems change in time, the total energy must change. Thus a single concept, time translational invariance, unifies many kinds of energy in many theories. *See* UNITS OF MEASUREMENT.

[BRIAN DE FACIO]

Bibliography: P. C. W. Davies, *The Forces of Nature*, 1979; D. Halliday and R. Resnick, *Physics*, 3d ed., 1978; F. A. Kaempffer, *The Elements of Physics: A New Approach*, 1967; Physical Science Study Committee, *Physics*, 1960.

Energy level

Self-contained physical systems of molecular size or smaller are found to have states of internal motion or excitation in which their energies take on definite stationary values. The energies of these states are called energy levels.

Atomic spectra. The existence of stationary states was first clearly seen in the study of atomic spectra. In 1900 Johannes Rydberg remarked that the frequencies of many observed spectral lines could be expressed in terms of the differences between pairs of numbers in a certain set called spectral terms; in 1908 Walter Ritz proposed this as a fundamental physical law, which later became known as the combination principle. Five years later Niels Bohr explained the combination principle in dynamical terms: Atoms exist in stationary states; when an atom passes from one stationary state to another of lower energy, it delivers the extra energy to a single quantum of light. Since according to Max Planck the energy of a quantum is proportional to its frequency, it follows that the spectral terms, written in appropriate units, are the energy values of the stationary states and that Ritz's combination principle merely expresses the conservation of energy. *See* ATOMIC STRUCTURE AND SPECTRA.

Experimentally it is found that most terms consist of several energy levels spaced close together. This multiplet structure is now explained as largely an effect of magnetic interactions associated with spinning electrons. It has further been discovered that certain transitions between levels are far more likely to occur than others. This is an effect of the conservation of angular momentum. Like electrons, photons (light quanta) carry the intrinsic angular momentum known as spin, but photons carry a full unit of $\hbar$ ($\hbar$ is Planck's constant divided

by 2π), whereas electrons have only half as much. Thus, ordinarily, the angular momentum of an atom changes by only one unit in a radiative transition. Occasionally a photon may be emitted from a point off-center in the atom; if so, it carries orbital angular momentum as well as its spin, and the atom's angular momentum may change by more than one unit. Rules like this which state that certain transitions are excluded, or at least are much less probable than others, are called selection rules. Normally, as here, they have their origin in the requirement that some dynamical quantity be conserved, but certain of the selection rules governing transitions among elementary particles have not yet been understood in this way. *See* SELECTION RULES.

The simplest of all atoms is hydrogen. Its energy levels, ignoring multiplet structure, are shown in Fig. 1. The obvious regularity of the terms can be summarized in a formula first guessed in 1885 by Johann Jakob Balmer on the basis of the four spectral lines then known. As derived by Bohr from the first primitive version of quantum mechanics in

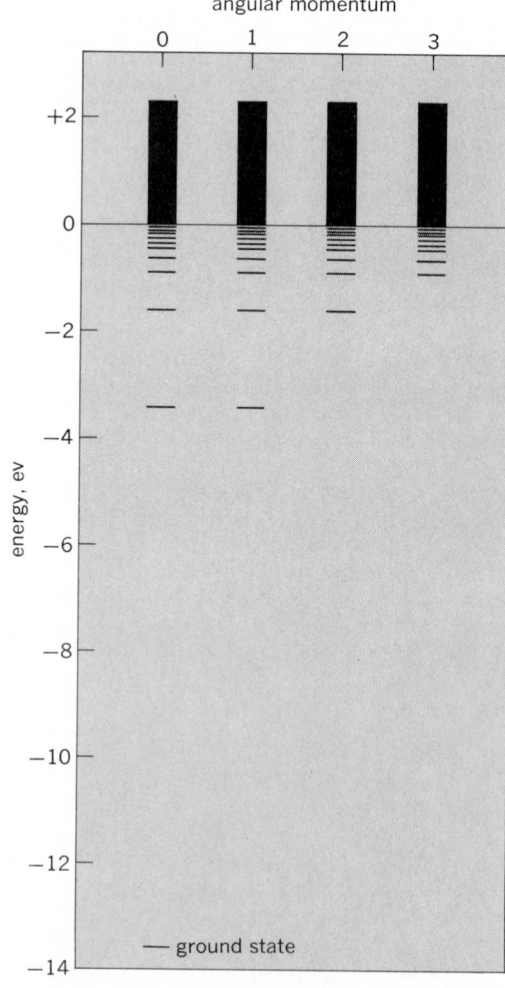

Fig. 1. Energy levels of the hydrogen atom, classified by the orbital angular momentum of the electron, expressed in units of $\hbar$. Lines show energy levels of bound states; shading shows continuous spectrum corresponding to positive energies. Allowed radiative transitions correspond to changes of one unit in the angular momentum of the electron.

1913, the formula is stated below, where m is the

$$E_n = -\frac{me^4}{2n^2\hbar^3} \qquad n = 1,2,3,\ldots$$

electron's mass (or, more accurately, the reduced mass of an electron and a proton), e is the electronic charge, and n is an integer called the principal quantum number. The same formula is given by modern quantum mechanics, and in hydrogen the multiplet structure is explained quantitatively as the effect of two perturbations—the moving electron's intrinsic magnetic moment interacting with the Coulomb field of the proton, and the relativistic variation of the electron's mass with velocity. It should be noted that only the states with negative energy have discrete levels. Positive energies correspond to a proton and an electron with too much energy to be bound together. For them any energy is allowed, so that the levels above zero form a continuum. *See* QUANTUM MECHANICS.

Figure 2 shows for contrast the energy levels of cesium. This is an atom in which one electron, cloud of electrons, does all the radiating. The differences from the hydrogen spectrum are explainable by the Pauli exclusion principle and the

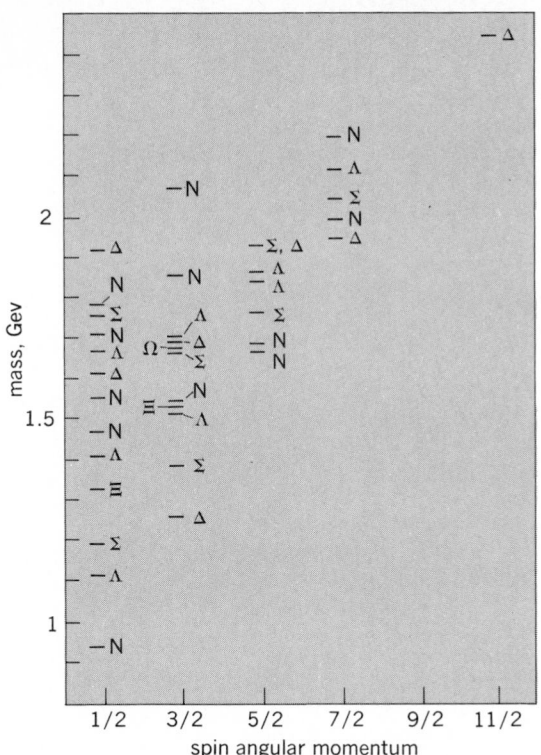

Fig. 3. Mass levels (expressed in energy units) of the heavy particles (baryons). Multiplet structure is as follows: N nucleon, 0, 1; Λ lambda, 0; Σ sigma, −1, 0, 1; Δ delta, −1, 0, 1, 2; Ξ xi, −1, 0; Ω omega, −1. Numbers give the electric charges in multiplets; particles in each multiplet have masses which differ by a few million electron volts.

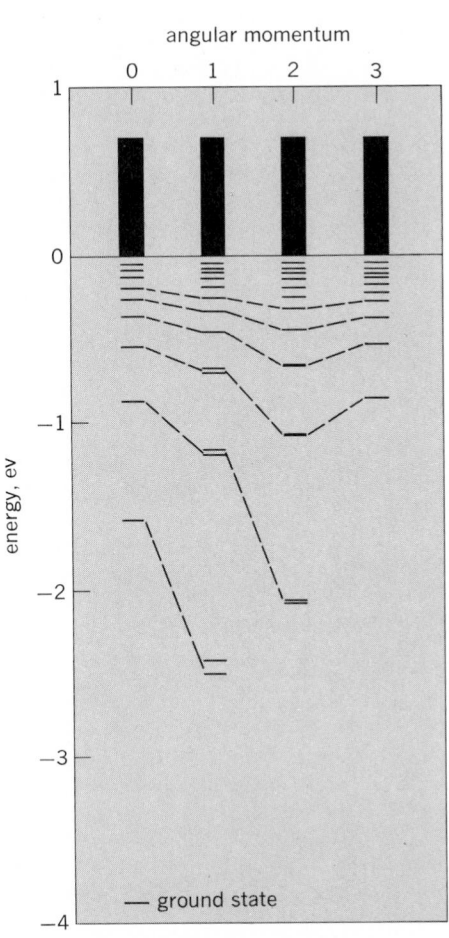

Fig. 2. Energy levels of cesium, showing some of the multiplet structure. All levels with orbital angular momentum greater than zero are double, with the states in which the electron's spin is in the same direction as its orbital motion having higher energy. Dashed lines connect corresponding levels, which, in absence of the inner core of electrons, would lie at the same energy.

perturbations due to the central cloud.

As one might expect, the energy levels of molecules are far more complex in their structure than those of atoms, since in addition to electronic transitions like those in atoms, the molecule as a whole can rotate, stretch, and twist in various ways, and the stationary states of these motions give rise to sets of energy levels related by selection rules similar to those governing atomic transitions. *See* MOLECULAR STRUCTURE AND SPECTRA.

Nuclear and particle physics. The experimental study of nuclear energy levels is made difficult by the fact that the continuous spectrum begins rather low, that is, typical level spacings are a few Mev (a million times greater than those of atomic levels), and a few more Mev disrupt a nucleus. Further, there are several ways in which a nucleus in an excited energy level can lose energy in addition to the radiation of a photon: It can lose an electron, a positron, or more rarely one or two protons or neutrons, a deuteron, or an alpha particle. The emission of anything but a photon changes the identity of the nucleus, and new selection rules associated with the dynamics of these processes come into play.

In reactions involving the so-called elementary particles, the same considerations enter, but often in simpler form because the internal structure of these particles, whatever it may be, is presumably simpler than those of all but the simplest nuclei. The situation with particles is however complicated by the existence of transitions and selection rules that have no counterpart among nuclei. *See* ELEMENTARY PARTICLE.

For many years, as long as only a very few particles were known, each was regarded as essentially independent of the others, but as their number has grown and knowledge about their transmutations and interactions has accumulated, it has become clear that the most fruitful way to think about them is as the energy levels of some kind of dynamical system whose nature is still not clear. In this view, the neutron and the proton are considered as belonging in first approximation to the same energy level of a system whose other levels include the lambda (Λ), the sigmas (Σ^-, Σ^0, and Σ^+), the cascade particles (Ξ^- and Ξ^0), and perhaps further particles as well. The mass differences between neutron and proton, the three Σs, and the two Ξs represent multiplet structures analogous to those encountered in atomic spectra. Figure 3 shows a few of the known levels of the mesons and baryons. The present (1969) task of particle physics is to explain these levels as the physicists of 50 years ago showed that the stationary states of atoms could be explained, but the work is still far from complete. [DAVID PARK]

Bibliography: A. Beiser, *Concepts of Modern Physics*, 2d ed., 1973; H. D. Young, *Fundamentals of Waves, Optics and Modern Physics*, 1975.

Enthalpy

For any system, that is, the volume of substance under discussion, enthalpy is the sum of the internal energy of the system plus the system's volume multiplied by the pressure exerted by the system on its surroundings. This may be expressed as $U + PV = H$, where U is the system's internal energy, P the pressure of the system, V the system's volume, and H the enthalpy of the system. The sum of $U + PV$ is given the special symbol H primarily as a matter of convenience because this sum appears repeatedly in thermodynamic discussion. Consistent units must, of course, be used in expressing the terms in the above equation. Previously, enthalpy was referred to as total heat or heat content, but these terms are misleading and should be avoided. Enthalpy is, from the viewpoint of mathematics, a point function, as contrasted with heat and work, which are path functions. Point functions depend only on the initial and final states of the system undergoing a change; they are independent of the paths or character of the change. Mathematically, the differential of a point function is a complete or perfect differential. *See* CALCULUS; MAXWELL'S EQUATIONS.

Because the absolute value of internal energy of even a simple system is usually unknown, recorded values of enthalpy are relative values measured above some convenient but arbitrarily chosen datum. Thus in the steam tables of Keenan and Keyes, the datum is liquid water at 32°F (0°C) and under its own vapor pressure. At this state water is assumed to have an enthalpy equal to zero. Under this assumption the internal energy of water in this state is a negative quantity equal to PV. No complication is introduced by this fact, although visualization of negative energies of this kind may be disturbing to some. There is limited utility for absolute enthalpies because only the changes in enthalpy are measurable. It is instructive to examine the utility of the enthalpy function in terms of some simple but important thermodynamic processes.

The first law of thermodynamics is merely a statement of the law of conservation of energy. The first law alone indicates that:

1. For a chemical reaction carried out at constant pressure and temperature with no work performed except that resulting from keeping the internal and external pressure equal to each other as the volume changes, the change in enthalpy of the system (the material taking part in the chemical reaction) is numerically equal to the heat that must be transferred to maintain the above-mentioned conditions. This heat is often loosely referred to as the heat of reaction. More properly, it is the enthalpy change for the reaction.

2. So-called heat balances on heat exchangers, furnaces, and similar industrial equipment that operate under steady flow conditions are really enthalpy balances.

3. The work developed in a steadily running adiabatic engine or turbine is equivalent to the enthalpy change of the fluid passing through the engine.

4. The adiabatic, irreversible, steady flow of a stream of materials through a porous plug or a partially opened valve under circumstances where the change in kinetic energy of flow is negligible (a Joule-Thomson process) results in no change in enthalpy of the flowing stream. Although no change in enthalpy results from this process, there is a loss in the energy available for doing work as a result of the pressure drop across the plug or valve. For change in enthalpy with pressure or temperature *see* THERMODYNAMIC PRINCIPLES. *See also* ENTROPY; THERMODYNAMIC PROCESSES.

[HAROLD C. WEBER; WILLIAM A. STEELE]

Entropy

A function first introduced in classical thermodynamics to provide a quantitative basis for the common observation that naturally occurring processes have a particular direction. Subsequently, in statistical thermodynamics, entropy was shown to be a measure of the number of microstates a system could assume. Finally, in communication theory, entropy is a measure of information. Each of these aspects will be considered in turn. Before the entropy function is introduced, it is necessary to discuss reversible processes.

Reversible processes. Any system under constant external conditions is observed to change in such a way as to approach a particularly simple final state called an equilibrium state. For example, two bodies initially at different temperatures are connected by a metal wire. Heat flows from the hot to the cold body until the temperatures of both bodies are the same. As another example, a vessel containing a gas is connected through a stopcock to an evacuated vessel. When the stopcock is opened, the gas expands to fill the whole of the available space uniformly. It is common experience that the reverse processes never occur if the systems are left to themselves; that is, heat is never observed to flow from the cold to the hot body, nor will the gas compress itself into one of the vessels. Max Planck classified all elementary processes into three categories: natural, unnatural, and reversible.

Natural processes do occur, and proceed in a

direction toward equilibrium. Unnatural processes move away from equilibrium and never occur. If A → B is a natural process between states A and B, then B → A is an unnatural process. A reversible process is an idealized natural process that passes through a continuous sequence of equilibrium states. Consider the evaporation of a liquid in the presence of its vapor at a pressure P. Let the equilibrium vapor pressure of the liquid be p. If $P < p$, liquid evaporates as a natural process. If $P > p$, evaporation is an unnatural process and will not occur; indeed, the opposite process—condensation—will take place. Finally, if $P = p$, both processes of condensation and evaporation are reversible and can be initiated by a very slight increase or decrease in the external pressure P.

A useful idea is that a reversible process may be exactly reversed by an infinitesimal change in the external conditions. If a hot object is placed adjacent to a much colder object, the heat-flow direction cannot be reversed by small changes in the temperature of either object. In reversible processes, work is accomplished through small pressure differences, and heat transfer occurs through small temperature differences.

Entropy function. The state function entropy S puts the foregoing discussion on a quantitative basis. The function is not derived in this article; but, rather, some of its properties are stated, and its implications are discussed mainly by example. Entropy is related to q, the heat flowing into the system from its surroundings and to T, the absolute temperature of the system. The important properties for this discussion are:

1. $dS > q/T$ for a natural change.

$dS = q/T$ for a reversible change.

It is necessary to introduce both S and T together. A formal derivation would show T^{-1} as an integrating factor leading to the complete differential dS.

2. The entropy of the system S is made up of the sum of all the parts of the system so that $S = S_1 + S_2 + S_3 \cdots$. *See* HEAT; TEMPERATURE; THERMODYNAMIC PRINCIPLES.

Heat flow. Consider two bodies, α and β, at different temperatures separated by an adiabatic (no heat transfer) wall. If the two bodies are connected by a fine wire that allows a small heat flow q from α to β, then $dS_\alpha = -q/T_\alpha$ and $dS_\beta = q/T_\beta$.

For the whole system, Eq. (1) holds. If $T_\alpha > T_\beta$,

$$dS = dS_\alpha + dS_\beta = q\left(\frac{1}{T_\beta} - \frac{1}{T_\alpha}\right) \tag{1}$$

$dS > 0$, and heat flows from α to β as a natural process. The process could be continued until $T_\alpha = T_\beta$ and $dS = 0$.

Once the constraint of the adiabatic wall is abrogated, the entropy increases to a maximum value, and T_α becomes equal to T_β. This is a special case of the most important notion in thermodynamics; that is, the system will assume that equilibrium state which maximizes the entropy at constant energy, consistent with the constraints.

Nonconservation of entropy. In his study of the first law of thermodynamics, J. P. Joule caused work to be expended by rubbing metal blocks together in a large mass of water. By this and similar experiments, he established numerical relationships between heat and work. When the experiment was completed, the apparatus remained unchanged except for a slight increase in the water temperature. Work (W) had been converted into heat (Q) with 100% efficiency. Provided the process was carried out slowly, the temperature difference between the blocks and the water would be small, and heat transfer could be considered a reversible process. The entropy increase of the water at its temperature T is $\Delta S = (Q/T) = (W/T)$.

Since everything but the water is unchanged, this equation also represents the total entropy increase. The entropy has been created from the work input, and this process could be continued indefinitely, creating more and more entropy. Unlike energy, entropy is not conserved. *See* CONSERVATION OF ENERGY.

Although the heat transfer is considered to be reversible in order to calculate the entropy increase, the overall process of converting work into heat is irreversible. The frictional process that converts kinetic energy into the heat of the metal blocks is a natural process. In fact, the impossibility of the reverse process is Lord Kelvin's statement of the second law of thermodynamics. Heat cannot be completely converted into work without other changes occurring in the surroundings. For example, a gas in a cylinder can be expanded reversibly by extracting heat from a large constant-temperature bath. All of the heat extracted from the bath is converted into work, but eventually the pressure of the gas system would be reduced to an unusable level. The system has changed, and the process cannot continue indefinitely. If one tries to convert heat into work through a system undergoing a cycle so that the system will return to its initial state, one finds that only a portion of the heat input does work and that the remainder must be rejected to a lower temperature; this is just the process which takes place in a heat engine. *See* THERMODYNAMIC PROCESSES.

Degradation of energy. Energy is never destroyed. But in the Joule friction experiment and in heat transfer between bodies, as in any natural process, something is lost. In the Joule experiment, the energy expended in work now resides in the water bath. But if this energy is reused, less useful work is obtained than was originally put in. The original energy input has been degraded to a less useful form. The energy transferred from a high-temperature body to a lower-temperature body is also in a less useful form. If another system is used to restore this degraded energy to its original form, it is found that the restoring system has degraded the energy even more than the original system had. Thus, every process occurring in the world results in an overall increase in entropy and a corresponding degradation in energy. R. Clausius stated the first two laws of thermodynamics as: "The energy of the world is constant. The entropy of the world tends toward a maximum."

Increasing entropy and mixing. Once the atomic theory of matter is accepted, the entropy concept can be made much clearer. It is then found through statistical thermodynamics that the increase of entropy toward its maximum value at equilibrium corresponds to the change of the system toward its most probable state consistent with the constraints. The most probable state represents the most mixed or most random state. Mixing must be given a broad interpretation which includes particle or configurational mixing, and spreading of energy over the particles or thermal mixing. Diffusion of one gas into another represents obvious configurational mixing and increased entropy. Irreversible expansion of a gas represents configurational mixing of the molecules over the available space. Heat flow represents spreading of the kinetic energy between the particles. Friction spreads the kinetic energy of the body over the constituent particles. Sometimes the energy-spread entropy increase and the configurational entropy increase are not compatible, and a compromise is struck. A subcooled liquid adiabatically crystallizes to a lower configurational entropy but gains even more entropy through the additional energy levels made available. The same sort of behavior occurs in partially miscible liquids—some configurational entropy is sacrificed in order to gain a large amount of energy-spread entropy. *See* STATISTICAL MECHANICS.

Absolute entropy. The third law of thermodynamics (Nernst's heat theorem) refers to the vanishing of entropy at zero temperature. In 1912 Planck proposed that the theorem applied to pure crystalline solids. However, the theorem is now known to be applicable to gases and, by all reasonable expectation, is applicable to any system. Thus, any substance at finite temperatures has an absolute entropy, the value of which can be determined from either calorimetric or spectroscopic data. Absolute entropies, together with thermochemical data, are very useful in the calculation of equilibrium compositions of reaction systems.

The statistical viewpoint is that a thermodynamic state at finite temperatures corresponds to many microstates. During an observation the microstates of a system undergo continuous rapid transitions. Since entropy is proportional to the logarithm of the number of available microstates, the Nernst theorem implies that the thermodynamic state at zero temperature corresponds to a single microstate. Thus, at zero temperature, even a ferromagnetic material should exist in a single state, fully magnetized in a direction determined by its inevitable interactions with the environment.   [WILLIAM F. JAEP]

Measure of information. The probability characteristic of entropy leads to its use in communication theory as a measure of information. The absence of information about a situation is equivalent to an uncertainty associated with the nature of the situation. This uncertainty, designated H, is the entropy of the information about the particular situation, Eq. (2), where $p_1, p_2, \ldots, p_n$ are the proba-

$$H(p_1, p_2, \ldots, p_n) = -\sum_{k=1}^{n} p_k \log p_k \qquad (2)$$

bilities of mutually exclusive events, the logarithms are taken to an arbitrary but fixed base, and $p_k \log p_k$ always equals zero if $p_k = 0$. For example, if $p_1 = 1$ and all others ps are zero, the situation is completely predictable beforehand; there is no uncertainty and so the entropy is zero. In all other cases the entropy is positive.

In introducing entropy of an information space, C. E. Shannon described a source of information by its entropy H in bits per symbol. The ratio of the entropy of a source to the maximum rate of signaling that it could achieve with the same symbols is its relative entropy. One minus relative entropy is the redundancy of the source.

                 [FRANK H. ROCKETT]

Bibliography: J. Aczel and Z. Daroczy, *Measures of Information and Their Characterizations*, 1975; H. B. Callen, *Thermodynamics*, 1960; K. G. Denbigh, *Principles of Chemical Equilibrium*, 3d ed., 1971; J. D. Fast, *Entropy: The Significance of the Concept of Entropy and Its Applications in Science and Technology*, 2d ed., 1968; A. I. Khinchin, *Mathematical Foundations of Information Theory*, 1957; C. E. Shannon, A mathematical theory of communication, *Bell Syst. Tech. J.*, 27(3): 379–423, 27(4):623–656, 1948; R. C. Tolman, *The Principles of Statistical Mechanics*, 1980; K. Wark, *Thermodynamics*, 3d ed., 1977.

Epicycloid

A curve traced by a point on a circle that rolls on the convex side of a fixed circle. The term is also occasionally applied to the curve generated by a point on the prolongation of the radius of a circle as the circle rolls on a straight line. If the rolling circle has radius r and the fixed circle radius R, as in the illustration, parametric equations of the epi-

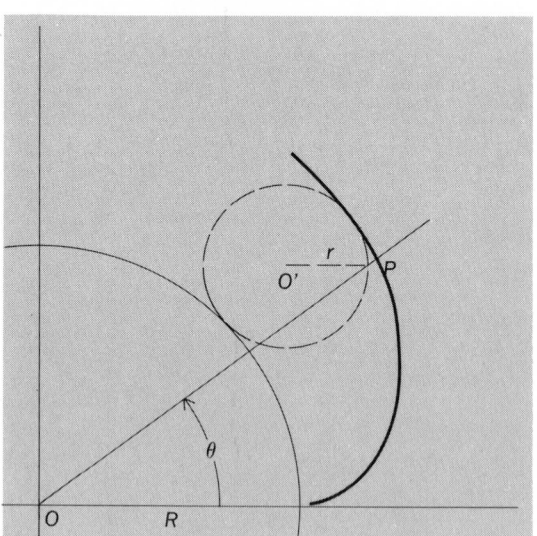

Diagram of an epicycloid.

cycloid are as shown below. The curve is closed

$$x = (R + r) \cos \theta - r \cos (1 + R/r)\theta$$
$$y = (R + r) \sin \theta - r \sin (1 + R/r)\theta$$

provided R/r is a rational number. It is a cardioid when $r = R$. The curve was known to the Greek astronomer Hipparchus about 140 B.C., and was investigated by Gerard Desargues in 1639 and Leonhard Euler in 1781. Desargues made the first applications of epicycloids to the design of gear teeth.

                 [LEONARD M. BLUMENTHAL]

Equation of continuity

An equation of continuity appears in many branches of physics. In dynamic field theory, it is essentially a statement that charge is conserved or that the rate of increase of charge in any region equals the current $\mathbf{i}$ flowing into that region. If v is the volume enclosed by the surface S, this statement may be expressed in integral form, as given in Eq. (1), where ρ is the charge density and $\mathbf{n}$ is a unit

$$\int_S \mathbf{i} \cdot \mathbf{n} \, dS = \int_v \boldsymbol{\nabla} \cdot \mathbf{i} \, dv = -\frac{\partial}{\partial t} \int_v \rho \, dv \qquad (1)$$

vector normal to S. The second integral comes from the first by Gauss' divergence theorem. When currents and charges are confined to a surface S bounded by the curve s, this becomes Eq. (2), where θ is the angle between ds and $\mathbf{i}$ which is

$$\oint i \sin \theta \, ds = -\frac{\partial}{\partial t} \int_S \rho \, dS \qquad (2)$$

directed into the area S. This states that the current crossing the boundary of an area equals the rate of increase of charge in the area. The volume in Eq. (1) is arbitrary, so the integrands in the last two integrals are equal. This leads to the differential form in Eq. (3), where i_x, i_y, and i_z are

$$\boldsymbol{\nabla} \cdot \mathbf{i} = \frac{\partial i_x}{\partial x} + \frac{\partial i_y}{\partial y} + \frac{\partial i_z}{\partial z} = -\frac{\partial \rho}{\partial t} \qquad (3)$$

the rectangular components of $\mathbf{i}$. Maxwell's equations satisfy Eq. (3). For application to the motion of charged particles, Eq. (3) would be written as Eq. (4). When $\mathbf{i}$ and ρ vary sinusoidally with time,

$$\boldsymbol{\nabla} \cdot \rho\mathbf{v} = \rho\left(\frac{\partial v_x}{\partial x} + \frac{\partial v_y}{\partial y} + \frac{\partial v_z}{\partial z}\right) = -\frac{\partial \rho}{\partial t} \qquad (4)$$

they may be written as phasors, which are complex numbers such that when multiplied by $e^{j\omega t}$, the real part of the product gives the amplitude, phase, and time dependence. The only change in the preceding equations when $\mathbf{i}$ and ρ are phasor quantities is the replacement of $\partial/\partial t$ by $j\omega$. *See* FLUID-FLOW PRINCIPLES; MAXWELL'S EQUATIONS.

[WILLIAM R. SMYTHE]

Bibliography: See WAVE EQUATION.

Equations, theory of

The branch of mathematics concerned with finding facts concerning the roots of algebraic equations and finding methods for obtaining them. The most important type of algebraic equation is the polynomial equation in one unknown which is an expression of the form $f(x) = a_n x^n + a_{n-1} x^{n-1} + \cdots + a_1 x + a_0 = 0$, where x is called the unknown, or variable; n is a positive whole number; and the a_i, with $i = 0, 1, \ldots, n$, are constants, or fixed numbers, called coefficients of the equation. The left member of the equation is called a polynomial in one variable of degree n. A root of such an equation is a number which, when substituted for the variable x, makes the left member zero. For example, 3 is a root of the equation $x^3 + 2x^2 - 13x - 6 = 0$. In addition, systems of equations in one or more variables are considered, and here the problem is to find values for the variables which simultaneously satisfy each equation of the

system. *See* POLYNOMIAL SYSTEMS OF EQUATIONS.

The topics covered in a systematic study of the theory of equations can be placed in the following principal subdivisions: properties of a polynomial which do not depend on the particular number system containing the coefficients of the polynomial; factorization of polynomials; equations with coefficients which are rational, real, or complex numbers; determination of bounds for real roots, and systematic methods for approximating real roots of equations; the solution of quadratic, cubic, and quartic equations by radicals.

This article is limited to polynomials and to equations which have rational, real, or complex numbers as coefficients. Each of these number systems constitutes a number field.

Let $f(x) = a_n x^n + a_{n-1} x^{n-1} + \cdots + a_1 x + a_0$ and $g(x) = b_m x^m + b_{m-1} x^{m-1} + \cdots + b_1 x + b_0 \neq 0$ be polynomials with coefficients in a number field F. Then the division algorithm, which is a formal statement of the ordinary division process for polynomials, states that there exist unique polynomials $q(x)$ and $r(x)$, with coefficients in F, such that $f(x) = q(x)g(x) + r(x)$, where either $r(x) = 0$ or $r(x)$ has degree less than $g(x)$. If $r(x) = 0$, then $g(x)$ divides $f(x)$. Immediate corollaries of the division algorithm are the remainder and factor theorems. If $g(x)$ is the linear polynomial $(x - a)$, then $f(x) = q(x)(x - a) + r$, where r is a constant. Then, substituting $x = a$, the result is $f(a) = r$. This is the remainder theorem. If $f(a) = r = 0$, then a is a root of the equation $f(x) = 0$, giving the factor theorem, which states that a is a root of $f(x) = 0$ if, and only if, $(x - a)$ is a factor of $f(x)$. It follows from the factor theorem that the equation $f(x) = 0$ has at most n roots in F, where $f(x)$ is of degree n.

The greatest common divisor (gcd) of two polynomials $f(x)$ and $g(x)$ is a polynomial $d(x)$ which divides both $f(x)$ and $g(x)$, and is divisible by every other polynomial which also divides $f(x)$ and $g(x)$. A process called the euclidean algorithm, based on successive application of the division algorithm, is used to find the gcd $d(x)$ having coefficients in the smallest number field that contains the coefficients of $f(x)$ and $g(x)$. The polynomials $f(x)$ and $g(x)$ are called relatively prime if their gcd is a constant.

The factorization of a polynomial $f(x)$ depends on the particular number field F under consideration. For example, the polynomial $x^5 - 1/2 x^4 - x^3 + 1/2 x^2 - 2x + 1$, having coefficients which are rational numbers, factors as

$$(x^2 + 1)(x^2 - 2)(x - 1/2)$$

over the rational numbers, as

$$(x^2 + 1)(x - \sqrt{2})(x + \sqrt{2})(x - 1/2)$$

over the real numbers, and as

$$(x + i)(x - i)(x - \sqrt{2})(x + \sqrt{2})(x - 1/2)$$

over the complex numbers. A polynomial $f(x)$ with coefficients in F is irreducible over F if it cannot be expressed as a product of polynomials of lower degree. Every polynomial can be expressed in essentially one way as a product of irreducible factors, although there is no general algorithm which enables one to obtain this expression. For polynomials with rational coefficients, there is such an algorithm, devised by L. Kronecker. There are methods for finding the repeated factors of a polynomial which are often useful as a first step in factoring a polynomial.

If a polynomial equation $f(x) = 0$ with rational coefficients is multiplied by a suitable whole number, an equation with whole number coefficients is obtained. If r/s, a rational number in lowest terms, is a root of such an equation, then r divides the constant term and s divides the leading coefficient. Hence, the rational roots of $f(x) = 0$ can be found by a finite number of trials.

The fundamental theorem of algebra states that a polynomial equation with complex coefficients has a complex root. From this it follows immediately that a polynomial of degree n with complex coefficients factors into n linear factors over the complex numbers.

If $f(x) = 0$ has real coefficients, then it can be shown that, if the complex number $a + bi$ is a root of $f(x)$, the conjugate complex number $a - bi$ is also a root. Thus, the real, irreducible factors of a polynomial with real coefficients are linear or quadratic. A further consequence is that, if $f(x)$ with real coefficients has odd degree, then $f(x) = 0$ has at least one real root.

The property of a polynomial $f(x)$ with real coefficients, which is basic for the study of the equation $f(x) = 0$, is that $f(x)$ defines a continuous real function. The location principle which follows from this states that, if there exist real numbers $a < b$ such that $f(a)$ and $f(b)$ have opposite signs, then $f(x) = 0$ has a real root r such that $a < r < b$. The location principle is used to isolate the real roots and is basic in systematic schemes such as Horner's method and Newton's method for approximating the real roots. A numerical method, known as Graeffe's method, can be used to approximate both the real and the complex roots.

Important results used in finding the real roots of $f(x) = 0$ with real coefficients include the following: Rolles' theorem from differential calculus, which states that, between two real roots of $f(x) = 0$, there is at least one real root of the derivative $f'(x) = 0$; Sturm's theorem, which gives an exact count of the number of real roots in an interval between two real numbers $a < b$; and Descartes' rule of signs, which states that the number of positive roots of $f(x) = 0$ equals the number of variations in sign of the coefficients of $f(x)$ minus a nonnegative even number. The negative roots of $f(x) = 0$ are positive roots of $f(-x) = 0$.

Results on the bounds for the real roots of $f(x) = 0$ are based on the fact that, for sufficiently large values of x, the sign of $f(x)$ is the same as the sign of the leading term $a_n x^n$ of $f(x)$. In particular, $f(x) = 0$ has no real roots for $|x| \geq |a_k/a_0| + 1$, where a_k is the coefficient of $f(x)$ with the greatest numerical value.

Polynomial equations of degree 2, 3, and 4 are solvable by radicals. This means that there are formulas which give the roots in terms of the coefficients of the equation and that these formulas involve only the rational operations and the operation of extraction of roots. The principal methods for the solution of the cubic and quartic equations were devised by J. Cardan and L. Ferrari, respectively. By use of the Galois theory of equations, it can be proved that for $n > 4$ there cannot exist a formula involving only rational oper-

ations and root extractions for expressing the roots of every polynomial equation of degree n in terms of the coefficients. *See* MATRIX THEORY.

[ROSS A. BEAUMONT]

Bibliography: R. A. Barnett, *Elementary Algebra: Structure and Use*, 3d ed., 1980; P. K. Rees, F. W. Sparks, and C. S. Rees, *College Algebra*, 7th ed., 1976; J. V. Uspensky, *Theory of Equations*, 1948, paper 1958.

Ether hypothesis

James Clerk Maxwell and his contemporaries in the 19th century found it inconceivable that a wave motion should propagate in empty space. They therefore postulated a medium, which they called the ether, that filled all space and transmitted electromagnetic vibrations.

During the last half of the 19th century, dozens of models were tried, but all broke down at some point. Direct experimental attempts to establish the existence of an absolute ether frame of reference, in which Maxwell's equations hold and light has the velocity c, have failed. The best known of these is the Michelson-Morley experiment, in which an attempt was made to measure the velocity of the Earth relative to the ether. *See* LIGHT.

Every hypothesis (ether drag, Lorentz contraction, and so on) invented to reconcile some experiment with the ether concept has been disproved by some other experiment. At present, there is no evidence whatever that the ether exists. *See* MAXWELL'S EQUATIONS.

[WILLIAM R. SMYTHE]

Bibliography: E. L. Hill, Optics and relativity theory, in E. U. Condon and H. Odishaw (eds.), *Handbook of Physics*, 2d ed., 1967; J. D. Jackson, *Classical Electrodynamics*, 2d ed., 1975; W. K. H. Panofsky and M. Phillips, *Classical Electricity and Magnetism*, 2d ed., 1962; E. T. Whittaker, *A History of the Theories of Aether and Electricity*, 2 vols., 1960.

Euler angles

Three angular parameters that specify the orientation of a body with respect to reference axes. They are used for describing rotating systems such as gyroscopes, tops, molecules, and nonspherical nuclei. They are not symmetrical in the three angles but are simpler to use than other rotational parameters.

Unfortunately, different definitions of Euler's angles are used, and therefore it is confusing to compare equations in different references. The definition given here is the majority convention according to H. Margenau and G. Murphy.

Let $OXYZ$ be a right-handed cartesian (right-angled) set of fixed coordinate axes and $Oxyz$ a set attached to the rotating body.

The orientation of $Oxyz$ can be produced by three successive rotations about the fixed axes starting with $Oxyz$ parallel to $OXYZ$. Rotate through (1) the angle ψ counterclockwise about OZ, (2) the angle θ counterclockwise about OX, and (3) the angle ϕ counterclockwise about OZ again. The line of intersection OK of the xy and XY planes is called the line of nodes.

Denote a rotation about OZ, for example, by Z (angle). Then the complete rotation is, symbolically, given by Eq. (1) where the rightmost operation is done first.

Direction cosines stated in terms of Euler angles

	X	Y	Z
x	$\cos\psi\cos\phi - \sin\psi\sin\phi\cos\theta$	$\cos\psi\sin\phi + \sin\psi\cos\phi\cos\theta$	$\sin\psi\sin\theta$
y	$-\sin\psi\cos\phi - \cos\psi\sin\phi\cos\theta$	$-\sin\psi\sin\phi + \cos\psi\cos\phi\cos\theta$	$\cos\psi\sin\theta$
z	$\sin\phi\sin\theta$	$-\cos\phi\sin\theta$	$\cos\theta$

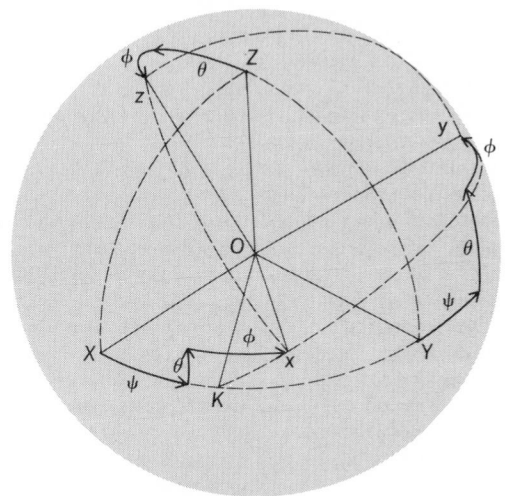

Euler angles. The successive movements of the axes on a unit sphere described in the text are shown by arrows. The complete rotation may also be obtained by a different sequence of rotations, namely, first through ϕ about *OZ*, then through θ about the displaced x axis (which is *OK*), then through ψ about *OZ*.

$$R(\psi,\theta,\phi) = Z(\phi)X(\theta)Z(\psi) \qquad (1)$$

A point P will have coordinates (x,y,z) with respect to the body axes and (X,Y,Z) with respect to the fixed axes. These are related by linear Eqs. (2)–(4), where (x,X) is the angle between the axes

$$x = X\cos(x,X) + Y\cos(x,Y) + Z\cos(x,Z) \qquad (2)$$
$$y = X\cos(y,X) + Y\cos(y,Y) + Z\cos(y,Z) \qquad (3)$$
$$z = X\cos(z,X) + Y\cos(z,Y) + Z\cos(z,Z) \qquad (4)$$

Ox and OX, and so forth. The nine direction cosines are expressed in terms of the three Euler angles in the table.

Inspection of the figure makes it apparent that no operation in $R(\psi, \theta, \phi)$ can be replaced by a combination of the other two. Therefore, three parameters are needed to specify the orientation, and the amounts of the angles are unique (barring additional 360° rotations). In dynamical problems of rotating bodies, ψ, θ, and ϕ can be used as independent angular coordinates.

Molecules and nuclei undergo oscillatory changes in shape while rotating. The body axes apply to the average shape. For another set of rotational parameters *see* CAYLEY-KLEIN PARAMETERS.

[BERNARD GOODMAN]

Bibliography: A. L. Fetter and J. D. Walecka, *Theoretical Mechanics of Particles and Continua*, 1980; H. Goldstein, *Classical Mechanics*, 2d ed., 1980; W. Hauser, *Introduction to the Principles of Mechanics*, 1965; H. Margenau and G. M. Murphy, *The Mathematics of Physics and Chemistry*, 2d ed., 1956, reprint 1976.

Euler's equations of motion

A set of three differential equations expressing relations between the force moments, angular velocities, and angular accelerations of a rotating rigid body. They are equations of motion in the usual dynamical sense, having the form of Eqs. (1)–(3).

$$I_1(d\omega_1/dt) + (I_3 - I_2)\omega_2\omega_3 = M_1 \qquad (1)$$
$$I_2(d\omega_2/dt) + (I_1 - I_3)\omega_3\omega_1 = M_2 \qquad (2)$$
$$I_3(d\omega_3/dt) + (I_2 - I_1)\omega_1\omega_2 = M_3 \qquad (3)$$

The formulation employs as coordinate axes the three principal axes of rotational inertia of the body that can rotate about a body-fixed point, which is the center of mass if constraints are absent. These reference axes, which form a right-hand set, are indicated by subscripts 1, 2, and 3 in the equations, where I_1, I_2, and I_3 represent the principal moments of inertia; ω_1, ω_2, and ω_3 the angular velocities about the axes; M_1, M_2, and M_3 the corresponding force moments; and t the time.

In the general case, these equations cannot be integrated, but solution is possible in special cases. Soluble problems of interest include that in which force moments are absent, the resulting complex behavior being called Poinsot motion, and that in which two of the principal moments of inertia are identical and only one force moment is present. The latter case includes spinning tops and gyroscopes. *See* RIGID-BODY DYNAMICS.

[RUSSELL A. FISHER]

Bibliography: H. C. Corben and P. Stehle, *Classical Mechanics*, 2d ed., 1960, reprint 1974; A. L. Fetter and J. D. Walecka, *Theoretical Mechanics of Particles and Continua*, 1980; H. Goldstein, *Classical Mechanics*, 2d ed., 1980.

Euler's momentum theorem

A principle of fluid mechanics which states that momentum of the particles in a moving frictionless, or inviscid, fluid is conserved. This theorem is expressed by Euler's hydrodynamical equations, a set of nonlinear partial differential equations. When viscous shear forces are included, the momentum equations are called the Navier-Stokes equations. The Eulerian method of viewing fluid motion is to consider it as a velocity-pressure field; that is, the velocity of the fluid particles is considered as a vector function of position in space and time, $\mathbf{v} = \mathbf{v}(\mathbf{r},t)$, and the pressure as a scalar function of position and time, $p = p(\mathbf{r},t)$, where $\mathbf{r}$ is the position vector locating a point in space and t is time. Euler's equations are essentially a restatement of Newton's single-particle law, force equals mass times acceleration, adapted to the many-particle continuum concept of fluid motion. In such a flowing continuum the acceleration of a particle of fluid at a given point in the fluid is given by Eq. (1),

$$\mathbf{a} = d\mathbf{v}/dt = \partial\mathbf{v}/\partial t + (\mathbf{v} \cdot \nabla)\mathbf{v} \qquad (1)$$

where $(\partial \mathbf{V}/\partial t)\, dt$ represents the velocity change in the time interval dt at a fixed point and $(\mathbf{v}\, dt \cdot \nabla)\, \mathbf{v}$ represents the velocity change of the particle in traveling the distance $\mathbf{v}\, dt$ from one point in the fluid to another point during time dt. Consider a volume V in the fluid with a surface S and let $\mathbf{n}$ be a unit vector normal to the surface, positive outward. One of the forces acting on this volume is the resultant of the pressure forces given by expression (2) which, by the analog to Gauss' diver-

$$-\iint_S p\mathbf{n}\, dS \qquad (2)$$

gence theorem, is the equivalent of expression (3).

$$-\iiint_V \nabla p\, dV \qquad (3)$$

Thus, the negative of pressure gradient $-\Delta p$ may be regarded as a force per unit volume acting on a fluid particle. There may also be body forces such as gravity and electromagnetic forces (if the fluid is electrically conducting). Let the amount of body force per unit mass be $\mathbf{F}$; then the force per unit volume is $\rho\mathbf{F}$, where ρ is the density. Euler's equations (as one vector equation) then become Eq. (4),

$$p[(\partial \mathbf{v}/\partial t) + (\mathbf{v}\cdot\nabla)\mathbf{v}] = -\nabla p + p\mathbf{F} \qquad (4)$$

where the left-hand side is (mass per unit volume) times acceleration and the right-hand side represents the forces per unit volume. In rectangular cartesian coordinates (x,y,z) with velocity components $(u,v,w,)$ and body force components (X,Y,Z), the relations are Eqs. (5). An integral of these

$$\rho\left[\frac{\partial u}{\partial t} + u\frac{\partial u}{\partial x} + v\frac{\partial u}{\partial y} + w\frac{\partial u}{\partial z}\right] = -\frac{\partial p}{\partial x} + \rho X \qquad (5a)$$

$$\rho\left[\frac{\partial v}{\partial t} + u\frac{\partial v}{\partial x} + v\frac{\partial v}{\partial y} + w\frac{\partial v}{\partial z}\right] = -\frac{\partial p}{\partial y} + \rho Y \qquad (5b)$$

$$\rho\left[\frac{\partial w}{\partial t} + u\frac{\partial w}{\partial x} + v\frac{\partial w}{\partial y} + w\frac{\partial w}{\partial z}\right] = -\frac{\partial p}{\partial z} + \rho Z \qquad (5c)$$

equations can be found for flows starting from rest or uniform motion. In this way the problem of determining the velocity field is reduced to purely kinematic considerations. *See* BERNOULLI'S THEOREM; GAS DYNAMICS; LAPLACE'S IRROTATIONAL MOTION; NAVIER-STOKES EQUATIONS.

[ARTHUR E. BRYSON]

Excited state

In quantum mechanics, a stationary state of higher energy than the lowest stationary state or ground state of a particle or a system of particles. Customarily, only bound stationary states, which generally are at most denumerably infinite in number, are spoken of as excited, although the formal quantum theory often treats the noncountable unbound stationary states on an equal footing with the bound states. Conventionally, the excited states are ranked in order of increasing energy; that is, the second excited state has higher energy than the first, which lies higher than the zeroth or ground state. Unlike the ground state, excited states frequently are degenerate. *See* DEGENERACY (QUANTUM MECHANICS); GROUND STATE; METASTABLE STATE; STATIONARY STATE.

[EDWARD GERJUOY]

Exciton

In a nonmetallic crystal, a quantum of electronic excitation which transports energy but not charge. There are two customary ways of picturing an exciton, termed the Frenkel model and the Wannier model.

Models. The Frenkel model applies to a solid consisting of weakly interacting atoms or molecules, a prime example being organic crystals such as anthracene and napthalene. The excitons are very much like the excited electronic states of the isolated molecules which make up the crystal. Although largely localized on a single molecule, an exciton can hop readily from one molecule to another in the crystal.

The Wannier model begins at the opposite extreme, with an electron in the conduction band and a hole in the valence band which are bound weakly to each other by the Coulomb force. This exciton overlaps many atoms in the crystal and does not resemble the excitation of any constituent atom. The dielectric screening of the crystal serves to weaken the attraction between electron and hole, and thus to increase their average separation. The Wannier model is therefore most appropriate for materials with a high dielectric constant—semiconductors such as Ge and Si, in particular. The internal energy of a Wannier exciton, relative to that of a free electron and hole, is given approximately in units of electron volts (eV) by Eq. (1),

$$E = -13.6\mu/\kappa^2 n^2 \qquad (1)$$

where $n = 1, 2, \ldots$ is a quantum number specifying the particular exciton state, κ is the dielectric constant, and μ is a factor which accounts for the reduced mass of the exciton being, in general, not the same as that of a free electron. Without κ and μ, Eq. (1) would be recognized as the formula for the energy levels of a hydrogen atom. For the lowest ($n = 1$) exciton level in Ge, for example, experiments give $E \approx 0.004$ eV. The corresponding average separation of the electron and hole is about 50 nearest-neighbor Ge distances. By contrast, the average radius of an exciton in its lowest state in an alkali halide or rare gas crystal is not much greater than the radius of a single constituent ion or atom. *See* BAND THEORY OF SOLIDS; DIELECTRIC CONSTANT; HOLES IN SOLIDS; SEMICONDUCTOR.

For many crystals, including rare gas crystals and most ionic crystals, the appropriate description of an exciton lies intermediate between the Frenkel and Wannier models. It is apparent that simple excitons cannot exist in metals, because the high electronic conductivity serves to cancel any Coulomb force holding the electron and hole together.

Kinetic energy. In addition to its internal energy, an exciton possesses kinetic energy associated with its motion through the crystal. In thermal equilibrium the average kinetic energy is related to the temperature T by Eq. (2), where k is Boltz-

$$1/2\, Mv^2 = 3/2\, kT \qquad (2)$$

mann's constant, and v is the exciton velocity. For an exciton whose effective mass M happened to coincide with that of a free electron, v would be

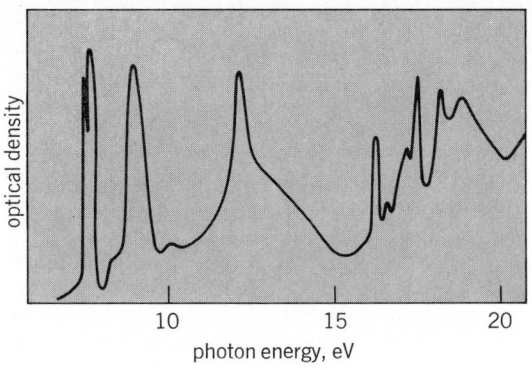

(a)

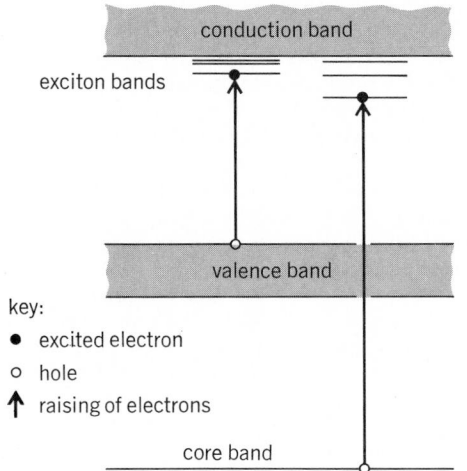

conduction band

exciton bands

valence band

key:
- ● excited electron
- ○ hole
- ↑ raising of electrons

core band

(b)

Fig. 1. Optical absorption of RbCl. (*a*) Absorption spectrum. (*b*) Schematic energy-band diagram illustrating the creation of excitons from two different bands. Exciton consists of an electron and a hole that is left behind when the electron is raised from the filled band.

roughly 10^5 m/s at $T = 300$ K. *See* BOLTZMANN CONSTANT.

Optical spectra. Excitons are the dominant feature in the visible and ultraviolet optical absorption and reflection spectra of many crystals. Figure 1*a* is an asorption spectrum for the ionic crystal RbCl. Because every ion in the crystal contributes to the absorption, all but the thinnest samples are opaque for photon energies above 7 eV. Most of the peaks in the spectrum can be attributed to excitons, although they arise from different valence and conduction bands and not from different values of n in Eq. (1). In particular, the peaks near 7.5 eV consist of excitons created when electrons are raised from the highest filled band or valence band, whereas the peaks in the 16–18 eV range arise from a deep core level which is, in fact, the filled $4p$ atomic level of the Rb$^+$ ion (Fig. 1*b*). Other features in the spectrum relate to the complex internal structures of the energy bands.

Because of line broadening, few materials exhibit distinct absorption peaks corresponding to $n > 2$ in Eq. (1). An exception is Cu_2O, in which lines up to $n = 11$ have been observed. In materials with higher dielectric constants, optical absorption

due to excitons is generally weak relative to the band-to-band transitions. Excitons can be created not only by light but also by any form of ionizing radiation. *See* CRYSTAL ABSORPTION SPECTRA.

Decay and trapping. An exciton is not stable against electron-hole recombination and can thus exist for only a brief interval before it decays by emitting a photon (recombination luminescence) or by giving up its energy to the crystal as heat or as stored energy. In most materials, the lifetime is long enough that the exciton becomes trapped before it decays. The trap is often an impurity ion, and the recombination luminescence is then characteristic of the impurity as well as the host crystal. An example is the green recombination luminescence in GaP that contains a small concentration of nitrogen atoms. *See* LUMINESCENCE; TRAPS IN SOLIDS.

In the alkali or alkaline-earth halides, the exciton actually traps itself spontaneously without the aid of crystal imperfections. This self-trapping comes about because of changes in the electrostatic forces between ions at the site of the exciton. The self-trapping process converts a substantial fraction of the energy of the exciton into heat, and thus the energy of the photon emitted when the electron and hole recombine can be less than half the energy of the original exciton.

Electron-hole droplets. The production of a high concentration of excitons in Ge and Si crystals at very low temperatures leads to the phenomenon of exciton condensation, which is similar in many respects to the condensation of water molecules in air to form fog droplets. The condensed phase is referred to as an electron-hole droplet, since the excitons lose their individual character, and the droplet becomes an electrically conducting plasma.

Droplets decay through exciton evaporation or by electron-hole recombination, which produces characteristic infrared luminescence having a photon energy slightly lower than that of the isolated exciton. Droplets may be maintained if the

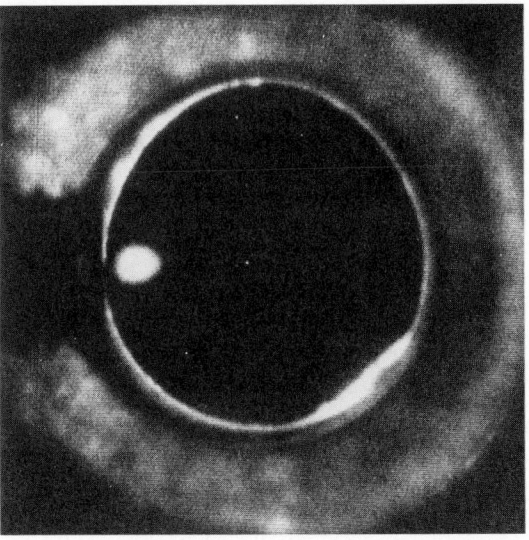

Fig. 2. An electron-hole droplet (light spot) consisting of condensed excitons in a circular sample of Ge. (*From J. P. Wolfe et al., Photograph of an electron-hole drop in germanium, Phys. Rev. Lett., 34:1292–1293, 1975*).

exciton concentration is replenished, continuous laser illumination being a good method of doing so. In Ge, electron-hole droplets typically have an electron (or hole) concentration of 10^{17} per cubic centimeter and a radius of 5 μm, and are stable against evaporation below about 6 K. When the exciting light is turned off, the droplets disappear in the 10^{-5} to 10^{-4}-s time range. Figure 2 shows a 4-mm-diameter Ge crystal containing a single large electron-hole droplet. This photograph was taken using the infrared recombination luminescence from the droplet itself. An elastic strain was applied to this crystal to facilitate the production of a droplet of exceptional size.

Excitons are intermediate states in many photochemical reactions, providing transport or conversion of energy, or both. In many halides an exciton can convert itself into stored energy in the form of a stable crystal defect; that is, it can give up its energy by forcing a halide ion from its lattice site to a nearby interstitial position. The exciton concept is not confined to simple crystalline materials but is descriptive of energy-transfer phenomena in liquids, glasses, polymers, and biological systems. In particular, the transfer of light energy among chlorophyll molecules during the initial stage of photosynthesis may be regarded as an exciton process. *See* COLOR CENTERS.

[MILTON N. KABLER]

Bibliography: F. Abelès (ed), *Optical Properties of Solids*, 1972; D. L. Dexter and R. S. Knox, *Excitons*, 1965; R. J. Elliot and A. F. Gibson, *An Introduction to Solid State Physics and Its Applications*, 1974; C. D. Jeffries, Electron-hole condensation in semiconductors, *Science*, 189(4207):955–964, 1975.

Exclusion principle

No two electrons may simultaneously occupy the same quantum state. This principle, often called the Pauli principle, was first formulated by Wolfgang Pauli in 1925 and, for time-independent quantum states, it means that no two electrons may be described by state functions which are characterized by exactly the same quantum numbers. In addition to electrons, all known particles having half-integer intrinsic angular momentum, or spin, obey the exclusion principle. It plays a central role in the understanding of many diverse phenomena, including the periodic table of the elements and their chemical activities, the electron contribution to the specific heat of metals, the shell structure in the atomic nucleus analogous to that of electrons in atoms, and certain symmetries in the scattering of identical particles. *See* ANGULAR MOMENTUM; NONRELATIVISTIC QUANTUM THEORY; QUANTUM NUMBERS; SPIN.

When a system of identical particles is described by a wave function, the indistinguishability of the particles implies that the wave function must have certain symmetry properties when the coordinates of any two of the particles are interchanged. Specifically, the wave function either remains unchanged or is changed only in sign when such a coordinate interchange is made. For the two-particle system, Eq. (1) holds, where $\mathbf{x}_i$

$$\Psi(\mathbf{x}_1\mathbf{x}_2) = \pm\Psi(\mathbf{x}_2\mathbf{x}_1) \qquad (1)$$

denotes all the coordinates of the particle i, such as space coordinates, spin, and any others necessary, and Ψ is the wave function.

When the plus sign applies, the wave function is said to be symmetric; the particles then are termed bosons and obey Bose-Einstein statistics. When the minus sign applies, the wave function is antisymmetric under particle exchange; the particles then are termed fermions and obey Fermi-Dirac statistics. Bosons are particles with integer spin, while fermions have half-integer spin, measured in units of Planck's constant. This connection between spin and statistical laws obeyed by the particles may be proved mathematically under appropriate assumptions. *See* BOSE-EINSTEIN STATISTICS; FERMI-DIRAC STATISTICS.

The Pauli principle follows from the symmetry of the many-fermion wave function in the special case when this wave function may be written as a product of single-particle wave functions. For example, consider two identical fermions and two single particle states. Let $\psi_\alpha(1)$ denote particle 1 in state α, $\psi_\beta(2)$ particle 2 in state β, and so on, where α and β represent all quantum numbers labeling the states. Then a possible state of the two-particle system is shown by Eq. (2), which obviously

$$\Psi(1,2) = \psi_\alpha(1)\,\psi_\beta(2) - \psi_\alpha(2)\,\psi_\beta(1) \qquad (2)$$

changes sign if 1 and 2 are interchanged. If $\alpha = \beta$, then Ψ is identically zero; that is, no such state exists which is exactly the statement of the Pauli principle. Note that Ψ may be expressed as a determinant, as in Eq. (3).

$$\Psi = \begin{vmatrix} \psi_\alpha(1) & \psi_\beta(1) \\ \psi_\alpha(2) & \psi_\beta(2) \end{vmatrix} \qquad (3)$$

For more than two particles, one may still form product wave functions which are also determinants. The exchange of a pair of particles is equivalent to the interchange of two rows of the determinant which is known to change its sign. Furthermore, if any two of the column labels, the α and β above, are the same, the determinant vanishes identically, which again is the Pauli principle. The most general wave function is a linear combination of all possible determinant wave functions, and the antisymmetry still holds, but the Pauli principle does not, since it applies only if there is a one-to-one correspondence between the number of single-particle states and the number of particles.

In the example above a single function was used to denote a single particle state. In many physical applications it is more convenient to express the wave function of a single particle as a product of a function which describes its spatial properties and one which describes its spin orientation. For spin s there are $2s + 1$ possible spin orientations ranging from $-s$ to s in integer steps. Electrons, protons, and neutrons all have $s = \frac{1}{2}$. The wave function for a collection of identical particles may then also be expressed as the product of a spatial wave function for the collection times a spin wave function for the collection. For the case of two particles, the Pauli principle dictates that, if the spatial wave function is symmetric, the spin function must be antisymmetric and vice versa. For more than two identical fermions, mixed symmetries in each part may occur, but they must be complimentary in the sense that the overall wave function obeys the Pauli principle.

It is often convenient to regard protons and neutrons as simply two different states of the same fermion, termed a nucleon. This approach takes into account that the proton is charged and the neutron is not, but ignores the small mass excess of the neutron. The additional quantum number needed to describe the two possible charge states is called isotopic spin. In this case the wave function of a single nucleon may be taken to be the product of a spatial part, an ordinary spin-orientation part, and a part to describe its charge state, that is, isotopic spin. The ideas about the symmetry apply similarly for a collection of nucleons. In this case one speaks of the generalized Pauli principle, to reflect the fact that the overall wave function is antisymmetric. For two nucleons this is attained by choosing the symmetries of the parts of the wave function as indicated in the table where + denotes symmetric and − antisymmetric. Again, for more than two nucleons mixed symmetries occur, but the overall wave function obeys the generalized Pauli principle. See ISOTOPIC SPIN; SUPERMULTIPLET.

Choice of symmetries for two nucleons

Space	Spin	Isotopic spin
+	+	−
+	−	+
−	+	+
−	−	−

One important consequence of the exclusion principle is that, if there are N different single-particle states corresponding to a given energy level, then, at most, N identical fermions may have that same energy. For example, in the atomic case the single-electron states have allowed energies labeled by the quantum number $n = 1, 2, 3, \cdots$ corresponding to the lowest energy, next lowest, and so on. For a given value of n, the orbital angular momentum l may have any one of the values $l = 0, 1, 2 \cdots n - 1$. For each l value there are $2l + 1$ states labelled by m_l, which range in integer steps from $-l$ to l, and specify the orbit orientation. In addition, there are the two possible spin orientations, $\pm\frac{1}{2}$. For $n = 1$, then, there are two different single-particle states, while for $n = 2$, there are eight. In general, $2n^2$ is the maximum number of electrons that may have an energy specified by n.

Using the fact that a system will try to occupy the state of lowest possible energy, the electron configuration of atoms may be understood by simply filling the single-particle energy levels according to the Pauli principle. This is the basis of Niels Bohr's explanation of the periodic table. For example, the one electron of H goes into the $n = 1$ level, as do both of the electrons of He. The $n = 1$ level is then said to be full. The next element in the periodic table is Li, and its third electron must go into the $n = 2$ level. For Ne, of atomic number 10, two of its 10 electrons go into the $n = 1$ level and eight into the $n = 2$ level, so that both are full. Now, the chemical activity of an element is determined mainly by its outermost (largest n) electrons, and a filled level is essentially inert. One sees therefore why He and Ne are noble gases and have very low

chemical activity, whereas Li, which has one more electron than the very stable He configuration, is very active chemically. By extending these ideas, such phenomena as binding energies and valences may be understood. Even if the Pauli principle applied only to this one area, it would still be a very important contribution to physics. See ATOMIC STRUCTURE AND SPECTRA.   [S. A. WILLIAMS]

Bibliography: R. Eisberg and R. Resnick, *Quantum Physics of Atoms, Molecules, Solids, Nuclei and Particles,* 1974; S. Gasiorowicz, *Quantum Physics,* 1974; S. Gasiorowicz, *Structure of Matter: A Survey of Modern Physics,* 1979; D. A. Park, *Introduction to Quantum Theory,* 2d ed., 1974; R. F. Streater and A. S. Wightman, *PCT, Spin and Statistics and All That,* 1964.

Extrapolation

A process in mathematics used to find the value of a function outside its tabulated values. This is done as in interpolation by assuming that over a small range of x the function may be closely approximated by a polynomial or some other readily computed function. See INTERPOLATION.

Formulas. Any of the interpolation formulas can be used, therefore, and the desired value of x substituted in them. Thus, for example, if $y = f(x)$ has been tabulated at $x = x_{-N}, x_{-N+1}, \ldots, x_{-1}, x_0$, the Gregory-Newton interpolation formula Eq. (1),

$$y = y_0 + u\,\delta y_{-1/2} + \frac{1}{2!}\,u(u+1)\delta^2 y_{-1}$$

$$+ \frac{1}{3!}\,u(u+1)(u+2)\delta^3 y_{-3/2} + \cdots$$

$$+ \frac{1}{m!}\,u(u+1)\cdots(u+m-1)\delta^m y_{-m/2} \quad (1)$$

may be used to determine a polynomial equation passing through the $m + 1$ ordinates $y_{-m}, y_{-m+1}, \ldots, y_{-1}, y_0$. These differences give Eq. (2),

$$u = \frac{x - x_0}{h} \quad (2)$$

and the differences $\delta^k y_{-k/2}$, $k = 1, 2, 3, \ldots$, are the same as those used in interpolation.

If $-1 < u < 0$, then $x_{-1} < x < x_0$ and the formula is used to interpolate. On the other hand, substitution of positive values of u permits its use for extrapolation for y beyond y_0, the last value tabulated.

If the function $y = f(x)$ is known, the error introduced by using a polynomial to extrapolate for the value of the function can be expressed by adding to Eq. (1) a remainder term, shown as notation (3),

$$\frac{1}{(m+1)!}\,u(u+1)\cdots(u+m)\,f^{(m+1)}(\xi) \quad (3)$$

where now ξ is any value of x lying between the smallest and the largest of the numbers $x_{-m}, x_{-m+1}, \ldots, x_{-1}, x_0$, and x. This term will be larger for extrapolation, $u > 0$, than for interpolation, $u < 0$, for two reasons. First, since u is positive for extrapolation, the coefficient of $f^{(m+1)}(\xi)$ will be larger. Second, since the range of values permitted ξ is larger, $|f^{(m+1)}(\xi)|$ must be assumed to be larger. It is necessary, in calculating the error, to take the largest absolute value of this $(m+1)$th derivative of $f(x)$ in the above range of ξ. If there is a singular-

Difference table

x	$y = \log x$	δy	$\delta^2 y$	$\delta^3 y$
1.00	0.0000 000			
		43 214		
1.01	0.0043 214		−426	
		42 788		8
1.02	0.0086 002		−418	
		42 370		9
1.03	0.0128 372		−409	
		41 961		8
1.04	0.0170 333		−401	
		41 560		7
1.05	0.0211 893		−394	
		41 166		$7 + \epsilon$
1.06	0.0253 059		$−387 + \epsilon$	
		$40\,779 + \epsilon$		$7 + \epsilon$
1.07	$0.0293\,838 + \epsilon \times 10^{-7}$		$−380 + 2\epsilon$	
		$40399 + 3\epsilon$		$7 + \epsilon$
1.08	$0.0334\,237 + 4\epsilon \times 10^{-7}$		$−373 + 3\epsilon$	
		$40026 + 6\epsilon$		
1.09	$0.0374\,263 + 10\epsilon \times 10^{-7}$			

ity of $f(x)$ or of its derivatives near the value of x required in the extrapolation, the remainder term in Eq. (3) would indicate that the extrapolation could involve a large error.

If it is necessary to extrapolate a distance greater than h, the interval of the table, beyond the limits of a table, it may be helpful to proceed by first extending the entries in the table by extrapolation. This may be done by assuming, for instance, that some order of difference remains constant, or by letting u take on positive integral values in Eq. (1). An estimate of the errors introduced by extending the table in this way can be made by attempting to extrapolate for the last few entries in the table from the earlier entries. For values not near a singularity of the function tabulated, these errors should be of about the same size as those introduced in extending the table. Of course, the same number of entries should be added in the two cases.

Having extended the entries in the table by extrapolation, one can look upon the problem of finding y for an intermediate value of x as just the problem of interpolation. If the degree of the interpolating polynomial is large enough, the error of this interpolation can be ignored in comparison with the error in extrapolating for the additional entries.

Example. From the portion of the difference table lying above the line, find by extrapolation the logarithms of 1.07, 1.08, and 1.09 and determine a probable limit for the error.

If it is assumed that the third difference stays constant at seven units in the seventh decimal place, the difference table can be extended, as shown below the line, by working from right to left. For this purpose the ϵs added to the number are ignored.

To determine the maximum error in the extrapolated values it must be recognized that the true third differences would not necessarily be equal to 7×10^{-7}. The maximum error would occur if all these third differences assumed were too high or too low. Therefore the true values of logarithms, rounded off to seven decimal places, will differ from those computed above by less than the ϵs attached to the numbers in the table.

Since a reasonable value for ϵ is ±1, reasonable **upper** limits for the errors in the extrapolation for the logarithms of 1.07, 1.08, and 1.09 are, respectively, 1, 4, and 10 units in the last decimal place.

Comparison of these values with the true values reveals that the errors are actually 0, −1, and −2 units in the last decimal place. It should be clear from this example, however, that the error in extrapolating beyond the limits of a table can be expected to grow rapidly for each new entry added.

[KAISER S. KUNZ]

Faraday effect

Rotation of the plane of polarization of a beam of linearly polarized light when the light passes through matter in the direction of the lines of force of an applied magnetic field. Discovered by M. Faraday in 1846, the effect is often called magnetic rotation. The magnitude α of the rotation depends on the strength of the magnetic field H, the nature of the transmitting substance, the frequency v of the light, the temperature, and other parameters. In general, $\alpha = VxH$, where x is the length of the light path in the magnetized substance and V the so-called Verdet constant. The constant V is a property of the transmitting substance, its temperature, and the frequency of the light.

The Faraday effect is particularly simple in substances having sharp absorption lines, that is, in gases and in certain crystals, particularly at low temperatures. Here the effect can be fully explained from the fundamental properties of the atoms and molecules involved. In other substances the situation may be more complex, but the same principles furnish the explanation.

Rotation of the plane of polarization occurs when there is a difference between the indices of refraction n^+ for right-handed polarized light and n^- for left-handed polarized light. Most substances do not show such a difference without a magnetic field, except optically active substances such as crystalline quartz or a sugar solution. It should be noted that the index of refraction in the vicinity of

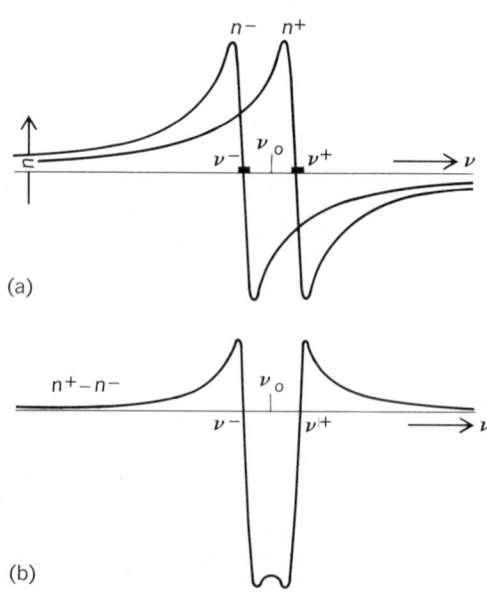

Fig. 1. Curves used in explaining the Faraday effect. (a) Index of refraction for left-handed circularly polarized light (n^-) and right-handed light (n^+) in the vicinity of an absorption line split into a doublet (v^-, v^+) in a magnetic field. (b) Difference between two curves, $n^+ - n^-$. Magnetic rotation is proportional to this difference.

an absorption line changes with the frequency (Fig. 1a). *See* POLARIZED LIGHT.

When the light travels parallel to the lines of force in a magnetic field, an absorption line splits up into two components which are circularly polarized in opposite directions; this is the normal Zeeman effect. This means that, for one line, only right-handed circularly polarized light is absorbed, and for the other one, only left-handed light. The indices of refraction n^- and n^+ bear to their respective absorption frequencies the same relation as indicated in Fig. 1a; that is, they are identical in shape but displaced by the frequency difference between the two Zeeman components. It is evident that $n^+ - n^-$ is different from zero (Fig. 1b), and the magnetic rotation is proportional to this difference. The magnitude of the rotation is largest in the immediate vicinity of the absorption line and falls off rapidly as the frequency of the light increases or decreases. *See* ZEEMAN EFFECT.

The Faraday effect may be complicated by the fact that a particular absorption line splits into more than two components or that there are several original absorption lines in a particular region of the spectrum.

The case represented in Fig. 1 is independent of the temperature, and the rotation is symmetric on both sides of an absorption line. This case is called, not quite correctly, the diamagnetic Faraday effect. It occurs when the substance is diamagnetic, which means the splitting of the absorption line is due to the splitting of the upper level only (Fig. 2a), and the lower level of the line is not split.

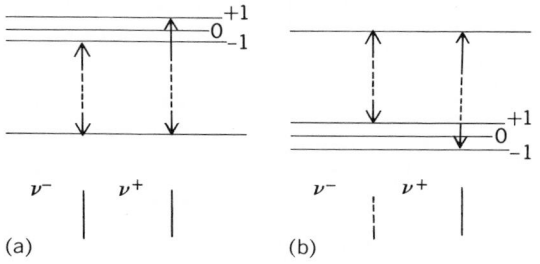

(a)   (b)

Fig. 2. Two cases of Faraday effect. (a) Diamagnetic case, which is temperature-independent, and (b) paramagnetic case, which is temperature-dependent.

The same situation prevails in general when the intensity of the two Zeeman components is equal. This holds for all substances except paramagnetic salts at very low temperatures (Fig. 2b), in which case the v^- component is absent.

In the latter case at high temperatures there are an equal number of ions in the +1 and −1 states, and the two Zeeman components have equal intensities, as in the previously discussed situation. When the temperature is lowered, however, the ions concentrate more and more in the lower level (−1), and therefore absorption by ions in the upper level (+1) disappears. At very low temperatures only the high-frequency component in the Zeeman pattern is left. In this case the n^- refraction coefficient is not affected by the presence of the absorption line and is a constant. The difference $n^+ - n^-$ will therefore have the shape of the n^+ curve in Fig. 1a. Here the rotation is not symmetric with respect to the absorption line. If it

is right-handed on one side of the line, it will be left-handed on the other. As the temperature is raised, the other line comes in with increasing strength until the two are nearly equal. In the transition region the Faraday effect depends strongly on the temperature. This case has been called the paramagnetic Faraday effect.

It is possible to modulate laser light by use of the Faraday effect with a cylinder of flint glass wrapped with an exciting coil. But since this coil must produce a high magnetic field of about 19,000 gauss, this method of modulation has been little used. For a discussion of other phenomena related to the Faraday effect *see* MAGNETOOPTICS.

[G. H. DIEKE/W. W. WATSON]

Faraday's law of induction

A statement relating an induced electromotive force (emf) to the change in magnetic flux that produces it. For any flux change that takes place in a circuit, Faraday's law states that the magnitude of the emf ε induced in the circuit is proportional to the rate of change of flux as in expression (1).

$$\varepsilon \propto -\frac{d\Phi}{dt} \qquad (1)$$

The time rate of change of flux in this expression may refer to any kind of flux change that takes place. If the change is motion of a conductor through a field, $d\Phi/dt$ refers to the rate of cutting flux. If the change is an increase or decrease in flux linking a coil, $d\Phi/dt$ refers to the rate of such change. It may refer to a motion or to a change that involves no motion.

Faraday's law of induction may be expressed in terms of the flux density over the area of a coil. The flux Φ linking the coil is given by Eq. (2),

$$\Phi = \int B \cos \alpha \, dA \qquad (2)$$

where α is the angle between the normal to the plane of the coil and the magnetic induction B. The integral is taken over the area A enclosed by the coil. Then, for a coil of N turns, Eq. (3) holds.

$$\varepsilon = -N\frac{d\Phi}{dt} = -N \int \frac{d(B \cos \alpha)}{dt} A \qquad (3)$$

See ELECTROMAGNETIC INDUCTION; MAGNETIC FLUX.   [KENNETH V. MANNING]

Fermi-Dirac statistics

The statistical description of particles or systems of particles that satisfy the Pauli exclusion principle. This description was first given by E. Fermi, who applied the Pauli exclusion principle to the translational energy levels of a system of electrons. It was later shown by P. A. M. Dirac that this form of statistics is also obtained when the total wave function of the system is antisymmetrical. *See* EXCLUSION PRINCIPLE; NONRELATIVISTIC QUANTUM THEORY.

Distribution function. Such a system is described by a set of occupation numbers $\{n_i\}$ which specify the number of particles in energy levels ϵ_i. It is important to keep in mind that ϵ_i represents a finite range of energies, which in general contains a number, say g_i, of nondegenerate quantum states. In the Fermi statistics, at most one particle is allowed in a nondegenerate state. (If spin is taken into account, two particles may be contained in such a state.) This is simply a restatement of the

Pauli exclusion principle, and means that $n_i \leqq g_i$. The probability of having a set $\{n_i\}$ distributed over the levels ϵ_i, which contain g_i nondegenerate levels, is described by Eq. (1), which gives just the

$$W = \prod_i \frac{g_i!}{(g_i - n_i)!\, n_i!} \qquad (1)$$

number of ways that n_i can be picked out of g_i, which is intuitively what one expects for such a probability. In Boltzmann statistics this same probability is given by Eq. (2). The equilibrium

$$W' = \prod_i \frac{g_i{}^{ni}}{n_i!} \qquad (2)$$

state which actually exists is the set of ns that makes W a maximum, under the auxiliary conditions given in Eqs. (3a) and (3b). These conditions

$$\Sigma n_i = N \qquad (3a)$$
$$\Sigma n_i \epsilon_i = E \qquad (3b)$$

express the fact that the total energy E and the total number of particles N are given. *See* BOLTZMANN STATISTICS.

Equation (4) holds for this most probable distri-

$$n_i = \frac{g_i}{\dfrac{1}{A} e^{\beta \epsilon_i} + 1} \qquad (4)$$

bution. Here A and β are parameters, to be determined from Eq. (3); in fact, $\beta = 1/kT$, where k is Boltzmann's constant and T is the absolute temperature. When the 1 in the denominator may be neglected, Eq. (4) goes over into the Boltzmann distribution; this provides a procedure for identifying β. It is known that in classical statistics the Boltzmann distribution may be obtained if specific assumptions are made as to the number of collisions taking place. It is there assumed that the number of collisions per second in which molecules with velocities in cells i and j in phase space produce molecules with velocities in cells k and l is given by Eq. (5). Here $A_{ij}{}^{kl}$ is a geometrical factor.

$$A_{ij}{}^{kl} = n_i n_j a_{ij}{}^{kl} \qquad (5)$$

This leads to the Boltzmann distribution. The Fermi distribution, Eq. (4), may be obtained if instead of Eq. (5) one assumes Eq. (6) for the number

$$A_{ij}{}^{kl} = a_{ij}{}^{kl} n_i n_j \left(\frac{g_k - n_k}{g_k} \right)\left(\frac{g_l - n_l}{g_l} \right) \qquad (6)$$

of collisions. One observes from Eq. (6) the interesting quantum theoretical feature that the probability for a collision depends on the occupation numbers of the states into which the colliding particles will go. In particular, if these final states are filled up ($n_k = g_k$), no collision with that state as a final state can occur.

The distribution f_i is often used; it is defined by Eq. (7).

$$n_i = f_i g_i \qquad (7)$$

Applications. For a system of N electrons, each of mass m in a volume V, Eq. (4) may be written as Eq. (8), where h is Planck's constant. Equations

$$f(v_x v_y v_z)\, dv_x\, dv_y\, dv_z = 2\left(\frac{m}{h}\right)^3 V \frac{dv_x\, dv_y\, dv_z}{\dfrac{1}{A} e^{mv^2/2kT} + 1} \qquad (8)$$

(3a) and (3b) may now be transformed into integrals, thus yielding Eqs. (9a) and (9b). Here $\lambda =$

$$\frac{N}{V}\frac{\lambda^3}{2} = U_{1/2}(A) \qquad (9a)$$

$$\frac{E}{3kT}\frac{\lambda^3}{V} = U_{3/2}(A) \qquad (9b)$$

$(h^2/2\pi mkT)^{1/2}$ is the thermal de Broglie wavelength, and $U_\rho(A)$ is the Sommerfeld integral defined by Eqs. (10a)–(10c), where $u = mv^2/2kt$ and

$$U_\rho(A) = \frac{1}{\Gamma(\rho+1)} \int_0^\infty \frac{u^\rho\, du}{\dfrac{1}{A} e^u + 1} \qquad (10a)$$

$$U_\rho(A) \cong A \quad (A \ll 1) \qquad (10b)$$

$$U_\rho(A) \cong \frac{(\ln A)^{\rho+1}}{\Gamma(\rho+1)} \quad (A \gg 1) \qquad (10c)$$

Γ is the usual Γ-function. Very often one writes instead of A in the Fermi distribution a quantity μ defined by Eq. (11). It may be shown (for instance,

$$\mu = kT \ln A \qquad A = e^{\mu/kT} \qquad (11)$$

by going to the classical limit) that μ is the chemical potential. It may be seen that if ϵ is large $(1/kT)$ $(\epsilon - \mu) \gg 1$; hence the Fermi distribution goes over into a Maxwell-Boltzmann distribution It is easy to verify that this inequality may be transcribed so as to state expression (12). Physically

$$\frac{V}{N} \gg \lambda^3 \qquad (12)$$

this is reasonable because expression (12) says that classical conditions pertain when the volume per particle is much larger than the volume associated with the de Broglie wavelength of a particle. For example, $V/N\lambda^3$ is about 7.5 for helium gas (He³) at 4°K and 1 atm. This indicates that classical statistics may perhaps be applied, although quantum effects surely play a role. For electrons in a metal at 300°K, $V/N\lambda^3$ has the value 10^{-4}, showing that classical statistics fail altogether for electrons in metals. When the classical distribution fails, a degenerate Fermi distribution results. Numerically, if $A \gg 1$, a degenerate Fermi distribution results; if $A \ll 1$, the classical results are again obtained. For example, Eqs. (9a), (9b), and (10b) show that $E = (3/2)NkT$ and the specific heat should be $(3/2)R = (3/2)Nk$. However for an electron gas this does not apply since such a system is degenerate for normal temperatures and $A \gg 1$. A somewhat lengthy calculation yields the result that, in the case $A \gg 1$, the contribution to the specific heat is negligible. This resolves an old paradox, for, according to the classical equipartition law, the electronic specific heat C should be $(3/2)Nk$, whereas in reality it is very small; in fact, $C = \gamma T$, where γ is a very small constant. This is a consequence of the fact that an electron gas at normal temperature is a degenerate Fermi gas. The electrical resistance of a metal can be understood on a classical picture, but the lack of a specific heat is a pure quantum effect. *See* BOSE-EINSTEIN STATISTICS; KINETIC THEORY OF MATTER; QUANTUM STATISTICS; STATISTICAL MECHANICS. [MAX DRESDEN]

Fermi surface

A surface of constant energy in a space defined by the components of the wave vectors of a system of half-integral spin particles. This is not a real surface, but is a geometrical description of the dynamical behavior of conduction electrons in solids. Its name comes from the fact that half-integral spin particles obey Fermi-Dirac statistics and at the zero of temperature T fill energy levels up to a maximum energy, the Fermi energy, and no energy levels are occupied above this energy. The wave vectors of electrons having the Fermi energy at $T = 0$ define the Fermi surface. *See* FERMI-DIRAC STATISTICS.

Free-electron approximation. Fermi surfaces are used in the discussion of the electrical conduction properties of solids. As a first approximation to consider, assume that the electrons responsible for electrical conduction in a solid constitute a free-electron gas. In this approximation the energy of the conduction electrons is entirely kinetic, with vanishingly small potential interactions with either the atomic electrons or other conduction electrons. The only interaction is the fact that they obey Fermi-Dirac statistics. Thus, the energy of electrons at the Fermi energy is given by Eq. (1), where

$$E_F = \frac{\hbar^2 k_F^2}{2m} \qquad (1)$$

k_F is the wave vector of electrons of this energy and m is the electronic mass. At $T = 0$ all states with k less than k_F are filled, and those with k greater than k_F are empty. This situation can be represented geometrically by a sphere of radius k_F, as shown in Fig. 1. All states inside the sphere are filled, and states outside are empty. The surface of this sphere is a surface of constant energy E_F — the Fermi surface. *See* FREE-ELECTRON THEORY OF METALS.

Shapes in real metals. The Fermi energy for electrons in metals is very large compared to ordinary thermal energies or to energies the electrons acquire due to electrical forces in electrical conduction processes. Because of the Pauli exclusion principle, only a small number of electrons having energies near the Fermi energy can contribute to conduction processes, and this makes the details

of the properties of the Fermi surface of prime importance to understanding electrical conduction in solids. Two phenomena not included in this free-electron approximation cause real metals to have Fermi surfaces which are geometrical shapes other than spheres: conduction electrons experience potential interactions with the atoms in the crystal, and they experience Bragg reflections at the boundaries of the Brillouin zone. *See* BAND THEORY OF SOLIDS; BRILLOUIN ZONE.

In metals which are formed from multivalent atoms, the volume of the Fermi sphere is larger than the volume of the Brillouin zone. In these cases the Fermi surface will intersect Brillouin zone boundaries. Electrons moving on the Fermi surface experience Bragg reflection when their wave vector magnitude and direction is such that it coincides with a Brillouin zone boundary. Because each Brillouin zone is a closed geometrical surface and electrons are reflected at each surface, an electron is confined to paths in wave vector space which cause them to remain inside the particular Brillouin zone in which they started.

Potential interactions with the atomic core states cause the Fermi surface to deviate from a sphere. The potential interactions also produce gaps in the energy spectrum. In addition to the energy gaps determining whether a solid is an insulator or conductor, potential interactions also distort shapes of Fermi surfaces.

Measurement of shapes. With few exceptions, measurements of the shapes and dimensions of a Fermi surface involve the application of a magnetic field. In the presence of a magnetic field $\vec{B}$, an electron experiences a force (the Lorentz force) given by Eq. (2), where $\vec{k}$ is the time derivative of

$$\vec{F} = \hbar \dot{\vec{k}} = -\frac{e}{c}(\vec{v} \times \vec{B}) \qquad (2)$$

the electron's wave vector $\vec{k}$, $\vec{v}$ is the velocity of the electron, e is the magnitude of the electron charge, and c is the speed of light. This force causes the wave vector of an electron to change its direction

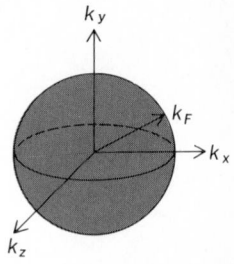

Fig. 1. Free-electron-gas Fermi surface in the space defined by the components of the wave vectors of the electrons.

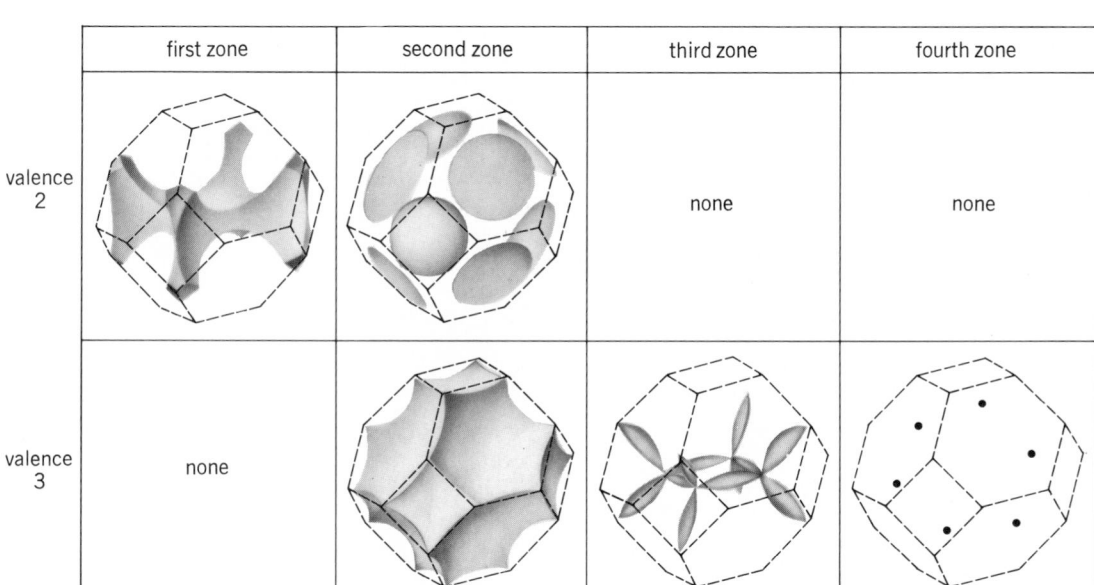

	first zone	second zone	third zone	fourth zone
valence 2			none	none
valence 3	none			

Fig. 2. Free-electron Fermi surfaces for face-centered cubic metals of valence 2 and 3. (*From N. W. Ashcroft and N. D. Mermin, Solid State Physics, Holt, Rinehart and Winston, 1976*)

in a plane perpendicular to both $\vec{v}$ and $\vec{B}$. The resultant motion of an electron in wave vector space is an orbit in a plane perpendiuclar to the direction of $\vec{B}$. If a real space coordinate system is superimposed on the k space coordinate system, it can be seen by taking a time integral of Eq. (2) that the real space orbit of an electron has the same path shape as the k space orbit, but is rotated by $\pi/2$ radians about the direction of $\vec{B}$. Thus, a determination of electronic paths on the Fermi surface also determines their real space paths.

There are many techniques for experimentally determining the dimensions and other detailed properties of Fermi surfaces of metals which involve the application of a magnetic field. Some of these techniques include the de Haas–van Alphen effect, the radio-frequency size effect, cyclotron resonance, acoustical geometrical resonance, and magnetoresistance measurements. All of these measuring techniques involve an input to the electron system of energies which are very small compared to the Fermi energy. The application of a magnetic field produces a force on an electron which is perpendicular to its velocity and therefore produces no work on it. Consequently, the electron's energy is not changed by the magnetic field. Therefore, these techniques measure properties of the Fermi surface. *See* CYCLOTRON RESONANCE EXPERIMENTS; DE HAAS–VAN ALPHEN EFFECT; MAGNETORESISTANCE; ULTRASONICS.

The path shapes which are measured include the effects of distortion due to potential interactions and the effect of Bragg reflection at the zone boundaries. Thus, a different set of paths is determined for electrons in each Brillouin zone which intersects the Fermi surface; when all possible paths are determined, different geometrical shapes for portions of the Fermi surface are obtained for each zone. It is also possible to have more than one section of Fermi surface per Brillouin zone due to multiple intersections of the Fermi surface with the zone boundary. Examples of Fermi surfaces for particular Brillouin zones of free-electronlike metals are shown in Figs. 2 and 3. Since the valency of atoms which make up a conductor determines the volume of the Fermi surface, different Fermi surfaces occur for conductors of the same crystal structure (the same Brillouin zone). The Fermi surface of conductors consisting of atoms of the same valency depends on the crystal structure in which the conductor is formed. Distortions from these free-electron Fermi surfaces occur because of conduction electron–atomic potential interactions and symmetry requirements on spin-orbit coupling imposed by the crystal lattice. *See* CRYSTAL STRUCTURE.

[R. G. GOODRICH]

Bibliography: N. W. Ashcroft and N. D. Mermin, *Solid State Physics*, 1976; A. P. Cracknell and K. C. Wong, *The Fermi Surface: Its Concept, Determination, and Use in the Physics of Metals*, 1973; C. Kittel, *Introduction to Solid State Physics*, 5th ed., 1976; P. B. Visscher and L. M. Falicov, Fermi surface properties of metals, *Phys. Status Solidi*, B54:9–51, 1972.

Ferrimagnetism

A specific type of ordering in a system of magnetic moments or the magnetic behavior resulting from such order. In some magnetic materials the magnetic ions in a crystal unit cell may differ in their magnetic properties. This is clearly so when some of the ions are of different species. It is also true for similar ions occupying crystallographically inequivalent sites. Such ions differ in their interactions with other ions, because the dominant exchange interaction is mediated by the neighboring nonmagnetic ions. They also experience different crystal electric fields, and these affect the magnetic anisotropy of the ion. A collection of all the magnetic sites in a crystal with identical behavior is referred to as a magnetic sublattice. A material is said to exhibit ferrimagnetic order when, first, all moments on a given sublattice point in a single direction and, second, the resultant moments of the sublattices lie parallel or antiparallel to one another. The notion of such an order is due to L. Nèel, who showed in 1948 that its existence would explain many of the properties of the magnetic ferrites. *See* FERROMAGNETISM.

At high temperatures all magnetic systems are disordered. As the temperature of a potentially ferrimagnetic system is lowered, there comes a point, the Curie temperature, at which all sublattices simultaneously acquire a moment and arrange themselves in a definite set of orientations. The Curie temperature and the temperature dependence of the sublattice moments depend in a

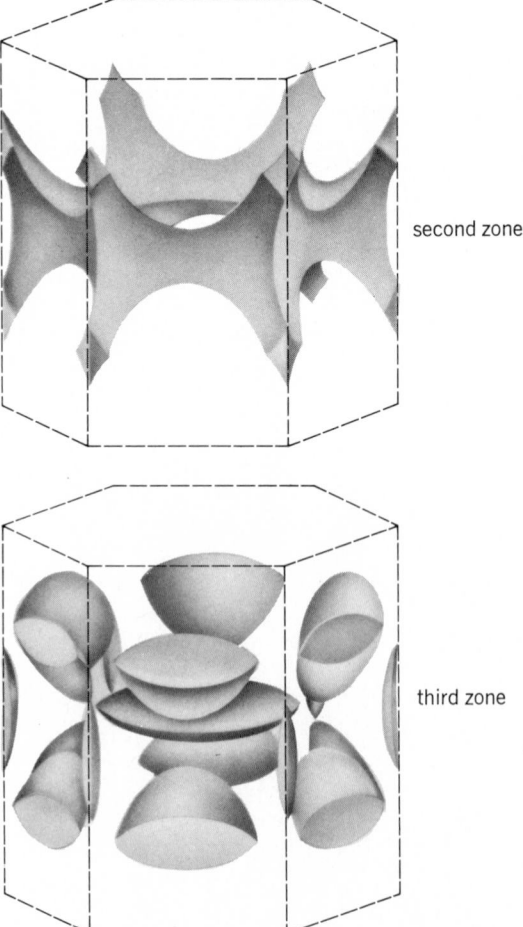

second zone

third zone

Fig. 3. Free-electron Fermi surface for a divalent hexagonal–close-packed metal which ignores spin-orbit splitting. *(From N. W. Ashcroft and N. D. Mermin, Solid State Physics, Holt, Rinehart and Winston, 1976)*

complicated way on the magnetic properties of the individual ions and upon the interactions between them. In general, there is a net moment, the algebraic sum of the sublattice moments, just as for a normal ferromagnet. However, its variation with temperature rarely exhibits the very simple behavior of the normal ferromagnet. For example, in some materials, as the temperature is raised over a certain range, the magnetization may first decrease to zero and then increase again. Ferrimagnets can be expected, in their bulk properties, measured statically or at low frequencies, to resemble ferromagnets with unusual temperature characteristics. *See* CURIE TEMPERATURE.

The most versatile of ferrimagnetic systems are the rare-earth iron garnets. The garnet unit cell has three sets of inequivalent magnetic sites, differing in their coordination to neighboring oxygen ions. Two of the sets, with 24 and 16 sites respectively per unit cell, are each occupied by Fe^{3+} ions, and the corresponding sublattices orient antiparallel. The remaining set of 24 sites may be occupied by nonmagnetic ions, such as Y, or by a magnetic rare-earth ion. This sublattice, when magnetic, usually lies parallel to the 24-site Fe^{3+} lattice. Rare-earth ions of various species have widely different magnetic moments, exchange interactions, and crystalline anisotropies. By a suitable choice of rare-earth ions, it is possible to design ferrimagnetic systems with prescribed magnetizations and temperature behavior.

[LAURENCE R. WALKER]

Bibliography: D. J. Craik, Intrinsic and technical properties of oxides, in D. J. Craik (ed.), *Magnetic Oxides*, pp. 1–96, 1975; A. H. Morrish, *Physical Principles of Magnetism*, 1979.

Ferroelectrics

Crystalline substances which have a permanent spontaneous electric polarization (electric dipole moment per cubic centimeter) that can be reversed by an electric field. In a sense, ferroelectrics are the electrical analog of the ferromagnets, hence the name. The spontaneous polarization is the so-called order parameter of the ferroelectric state, just as the spontaneous magnetization is the order parameter of the ferromagnetic state. The names Seignette-electrics or Rochelle-electrics, which are also widely used, are derived from the name of the first substance found to have this property, Seignette salt or Rochelle salt. *See* FERROMAGNETISM.

The reversibility of the spontaneous polarization is due to the fact that the structure of a ferroelectric crystal can be derived from a nonpolarized structure by small displacements of ions. In most ferroelectric crystals, this nonpolarized structure becomes stable if the crystal is heated above a critical temperature, the ferroelectric Curie temperature; that is, the crystal undergoes a phase transition from the polarized phase (ferroelectric phase) into an unpolarized phase (paraelectric phase). The change of the spontaneous polarization at the Curie temperature can be continuous or discontinuous. The Curie temperature of different types of ferroelectric crystals range from a few degrees absolute to a few hundred degrees absolute. As a rule, the ferroelectric phase is the low-temperature phase; however, there are crystals which are ferroelectric in a relatively narrow temperature range

only, and others stay polarized up to the temperature of decomposition or melting.

Classification. From a practical standpoint ferroelectrics can be divided into two classes. In ferroelectrics of the first class, spontaneous polarization can occur only along one crystal axis; that is, the ferroelectric axis is already a unique axis when the material is in the paraelectric phase. Typical representatives of this class are Rochelle salt, KH_2PO_4, $(NH_4)_2SO_4$, guanidine aluminum sulfate hexahydrate, glycine sulfate, colemanite, and thiourea.

In ferroelectrics of the second class, spontaneous polarization can occur along several axes that are equivalent in the paraelectric phase. The following substances, which are all cubic above the Curie point, belong to this class: $BaTiO_3$-type (or perovskite-type) ferroelectrics; $Cd_2Nb_2O_7$; $PbNb_2O_6$; certain alums, such as methyl ammonium alum; and $(NH_4)_2Cd_2(SO_4)_3$. Some of the $BaTiO_3$-type ferroelectrics have, below the Curie temperature, additional transition temperatures at which the spontaneous polarization switches from one crystal axis to another crystal axis. For example, $BaTiO_3$ and $KNbO_3$ polarize with decreasing temperature first along a [100] axis, then the polarization switches into a [110] axis, and finally into a [111] axis.

From a scientific standpoint, one can distinguish proper ferroelectrics and improper ferroelectrics. In proper ferroelectrics, for example, $BaTiO_3$, KH_2PO_4, and Rochelle salt, the spontaneous polarization is the order parameter. The structure change at the Curie temperature can be considered a consequence of the spontaneous polarization. The unit cell of the crystal in the ferroelectric phase contains the same number of chemical formula units as the unit cell in the paraelectric phase. In improper ferroelectrics, the spontaneous polarization can be considered a by-product of another structural phase transition. The unit cell in the ferroelectric phase is an integer multiple of the unit cell in the paraelectric phase. Examples of such systems are $Gd(MoO_4)$ and boracites. The dielectric, elastic, and electromechanical behavior of the two types of ferroelectrics differ significantly.

Ferroelectric domains. The spontaneous polarization can occur in at least two equivalent crystal directions; thus, a ferroelectric crystal consists in general of regions of homogeneous polarization that differ only in the direction of polarization. These regions are called ferroelectric domains. Ferroelectrics of the first class consist of domains with parallel and antiparallel polarization (Fig. 1a),

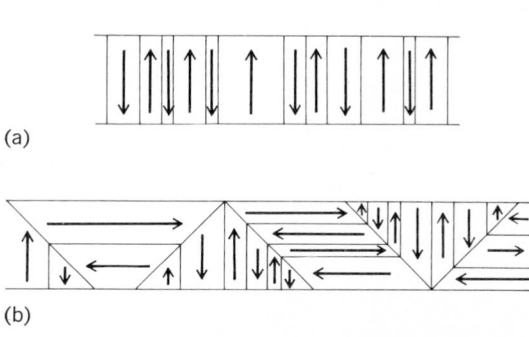

(a)

(b)

Fig. 1. Domain configurations (simplified) encountered in ferroelectric crystals. (a) First class. (b) Second class.

Fig. 2.   Ferroelectric domains in $BaTiO_3$ photographed through a polarizing microscope. Ferroelectric domains range from macroscopic to submicroscopic size.

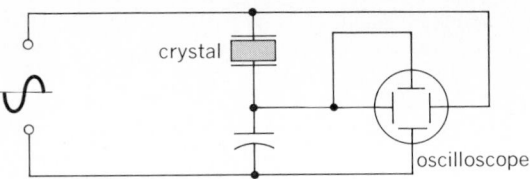

Fig. 4.   Circuit for the display of ferroelectric hysteresis loops on an oscilloscope.

whereas ferroelectrics of the second class can assume much more complicated domain configurations (Fig. 1b). The region between two adjacent domains is called a domain wall. Within this wall, the spontaneous polarization changes its direction. The wall between antiparallel domains is probably only a few lattice spacings thick, whereas the wall between domains that are polarized at a right angle to each other is probably thicker. Ferroelectric domains can be observed in a number of substances by means of the polarizing microscope (Fig. 2) because of their birefringence, or double refraction. The ferroelectric domains range in size from macroscopic (millimeters) to submicroscopic. *See* BIREFRINGENCE.

Ferroelectric hysteresis. When an electric field is applied to a ferroelectric crystal, the domains that are favorably oriented with respect to this field grow at the expense of the others, for example, by sidewise motion of domain walls. In addition, favorably oriented domains can nucleate and grow until the whole crystal becomes one single domain. When the field is reversed, the polarization reverses through the same processes. The relation between the resulting polarization P of the whole

crystal and the externally applied electric field E is given by a hysteresis loop (Fig. 3). The shape of the hysteresis loop depends strongly upon the perfection of the crystal as well as upon the rate of change of the externally applied field E. A simple circuit that permits the observation of ferroelectric hysteresis loops by means of an oscilloscope is shown in Fig. 4. In some ferroelectrics, the polarization can be reversed within a fraction of 1 μsec.

Spontaneous polarization. The magnitude of the permanent or spontaneous polarization P_s of a domain can be obtained from the hysteresis loop by extrapolating the saturation branch to zero external field (Fig. 3). For most ferroelectrics, the values of P_s are between 10^{-7} and 10^{-4} coulomb/cm² (Fig. 5). In nonferroelectric dielectrics, electric fields between 10^5 and 10^8 volts/cm would be necessary in order to achieve such large polarizations.

Dielectric properties. As a rule, the dielectric constant ϵ measured along a ferroelectric axis increases in the paraelectric phase when the Curie temperature is approached. In many ferroelectrics, this increase can be approximated by the Curie-Weiss law, shown in the equation below.

$$\epsilon = \frac{C}{T - T_0}$$

Here T designates the temperature of the crystal, and T_0 is equal to or somewhat smaller than the transition temperature. C is the so-called Curie constant. For $BaTiO_3$, this law holds unaltered up to frequencies of 2.4×10^{10} Hz. Dispersion sets in in the far-infrared. The dielectric constant drops when the crystal becomes spontaneously polarized (Fig. 6). In the ferroelectric phase, the dielectric constant has two components. The first component is the dielectric constant of the individual domains. It is independent of the frequency and of the electric field generally up to far-infrared frequencies. The second component is due to domain wall motions, that is, to partial reversal of the spontaneous polarization. This process can give rise to large dielectric losses, and it depends strongly upon the frequency, the electric field strength, the domain structure, and the temperature. In uniaxial ferroelectrics, the dielectric constant measured perpendicular to the ferroelectric axis generally does not show a very pronounced anomaly near the Curie temperature, and in some cases it has even the same order of magnitude and temperature dependence as for any normally behaving dielectric crystal. *See* CURIE-WEISS LAW; DIELECTRIC CONSTANT; DIELECTRICS.

Piezoelectric properties. Ferroelectrics can be divided into two groups according to their piezoelectric behavior.

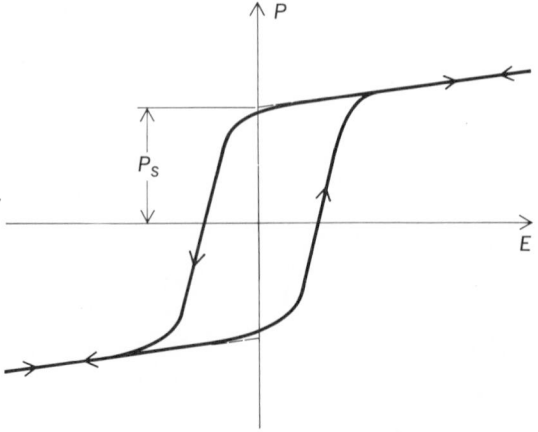

Fig. 3.   Net polarization P of a ferroelectric crystal versus externally applied electric field E.

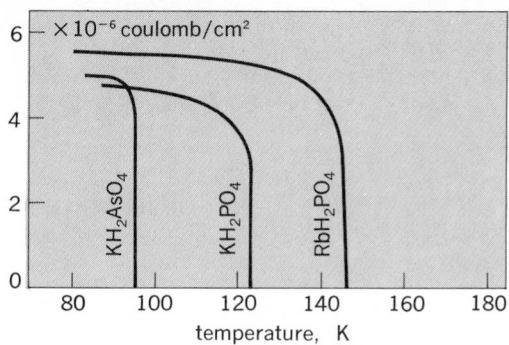

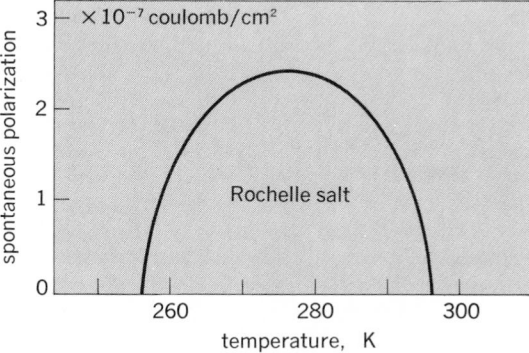

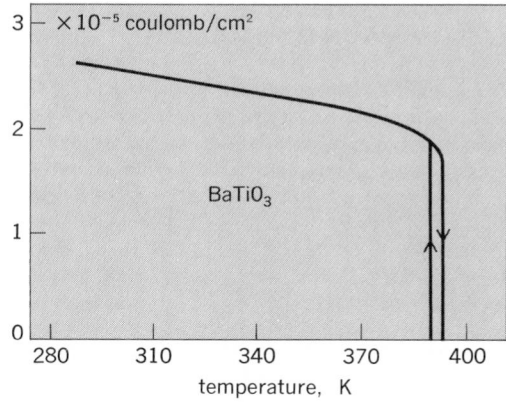

Fig. 5. Dependence upon temperature of the spontaneous polarization of some ferroelectrics.

The ferroelectrics in the first group are already piezoelectric in the unpolarized phase. Those piezoelectric moduli which relate stresses to polarization along the ferroelectric axis have essentially the same temperature dependence as the dielectric constant along this axis, and hence become very large near the Curie point. The spontaneous polarization gives rise to a large spontaneous piezoelectric strain which is proportional to the spontaneous polarization. In KH_2PO_4-type ferroelectrics and in Rochelle salt, for example, this strain is a shear in the plane perpendicular to the axis of polarization. It reaches 27 min of arc in KH_2PO_4 and about 1.8 min of arc in Rochelle salt. The piezoelectric modulus decreases as the spontaneous polarization increases. But with sufficiently large stresses, it is possible to align the domains and reverse the spontaneous polarization (Fig. 7). The relation between the resulting polarization of the whole crystal and the mechanical stress is given by a hysteresis loop analogous to the loop of Fig. 3 (piezoelectric hysteresis). This effect can simulate a very large piezoelectric modulus.

The ferroelectrics in the second group are not piezoelectric when they are in the paraelectric phase. However, the spontaneous polarization lowers the symmetry so that they become piezoelectric in the polarized phase. This piezoelectric activity is often hidden because the piezoelectric effects of the various domains can cancel. However, strong piezoelectric activity of a macroscopic crystal or even of a polycrystalline sample occurs when the domains have been aligned by an electric field. The spontaneous strain is proportional to the square of the spontaneous polarization. In $BaTiO_3$, for example, the crystal (which has cubic symmetry in the unpolarized phase) expands along the axis of polarization and contracts at right angles to it. The strain is of the order of magnitude of 1%. The spontaneous polarization cannot be reversed by a mechanical stress in ferroelectrics of this group. *See* PIEZOELECTRICITY.

Crystal structure. The structures of different types of ferroelectrics are entirely different, and it is not possible to establish a general rule for the occurrence of ferroelectricity. The structures of a

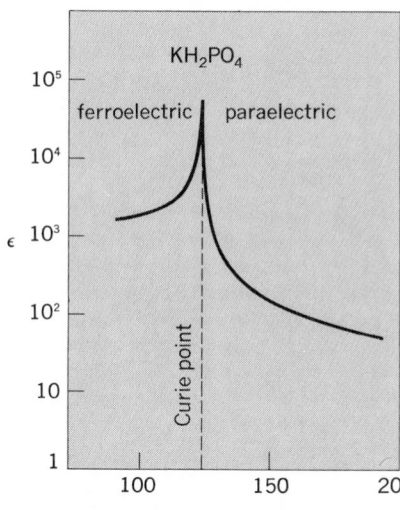

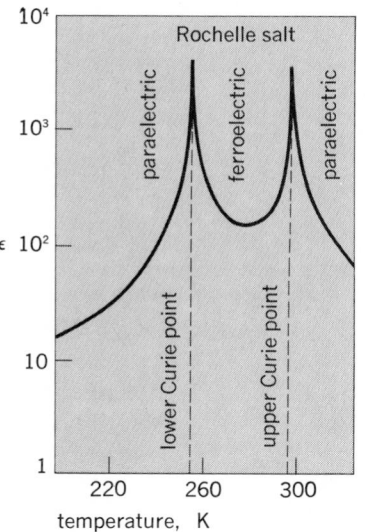

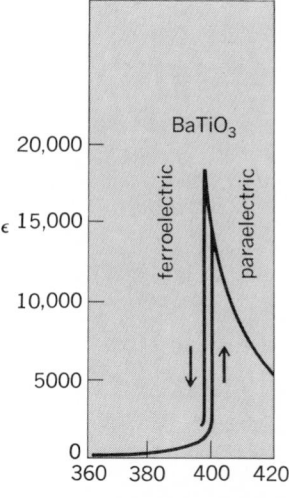

Fig. 6. Anomalous temperature dependence of relative dielectric constant of ferroelectrics at transition temperature.

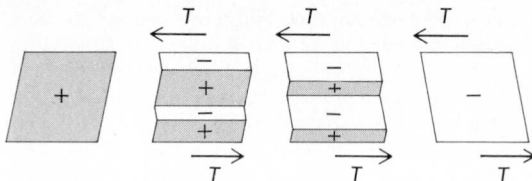

Fig. 7. Schematic representation of the reversal of the spontaneous polarization by a mechanical shear stress T in KH_2PO_4 and Rochelle salt.

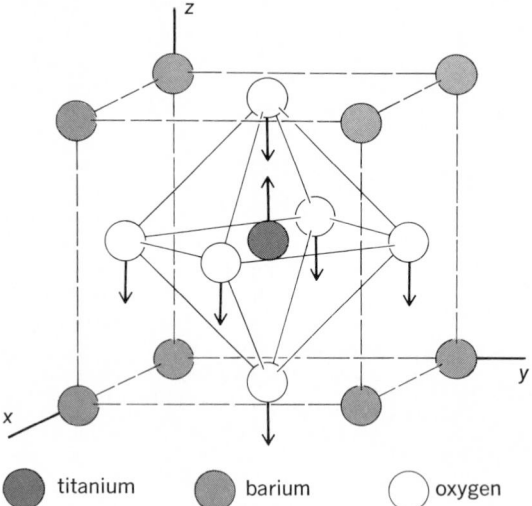

titanium   barium   oxygen

Fig. 8. Crystal lattice of $BaTiO_3$. Arrows indicate displacements of the ions when crystal becomes polarized.

number of ferroelectrics and the minute changes that they undergo when spontaneous polarization occurs are known in great detail from x-ray diffraction and neutron diffraction studies. In a qualitative way, the process of polarization is best understood for ferroelectrics of the $BaTiO_3$ type. Figure 8 shows schematically the structure of the unit cell of a $BaTiO_3$ crystal in the unpolarized state, and the arrows indicate the direction in which the ions are slightly displaced when the lattice becomes spontaneously polarized along the

axis z. The order of magnitude of the displacements is 1% of the unit cell dimension. However, these displacements do not account quantitatively for the observed polarization, because other changes of the electronic structure occur as well. *See* CRYSTAL STRUCTURE.

In KH_2PO_4-type ferroelectrics, hydrogen bonds $O—H \cdots O$ play an important part in the ferroelectric effect. Above the Curie temperature, the hydrogen ions are statistically distributed over the two possibilities $O—H \cdots O$ and $O \cdots H—O$, whereas below the Curie point, one or the other of these two possibilities is strongly favored, depending upon the sign of the spontaneous polarization.

Antiferroelectric crystals. These materials are characterized by a phase transition from a state of lower symmetry (generally low-temperature phase) to a state of higher symmetry (generally high-temperature phase). The low-symmetry state can be regarded as a slightly distorted high-symmetry state. It has no permanent electric polarization, in contrast to ferroelectric crystals. The crystal lattice can be regarded as consisting of two interpenetrating sublattices with equal but opposite electric polarization. This state is referred to as the antipolarized state.

In a certain sense, an antiferroelectric crystal is the electrical analog of an antiferromagnetic crystal. In the high-symmetry phase, the sublattices are unpolarized and indistinguishable. In general, antiferroelectric crystals have more than one axis along which the sublattices can polarize. Therefore, the low-symmetry phase consists of regions of homogeneous antipolarization which differ only in the orientation of the axis along which antipolarization has occurred. These regions are called antiferroelectric domains and can be observed by the polarizing microscope. Because these domains have no permanent electric dipole moment, an electric field generally has little influence on domain structure. *See* ANTIFERROMAGNETISM.

The dielectric constant of antiferroelectric crystals is generally larger than it is for nonferroelectric crystals and has an anomalous temperature dependence. It increases as the transition temperature is approached and drops when antipolarization occurs (Fig. 9). In some antiferroelectrics the

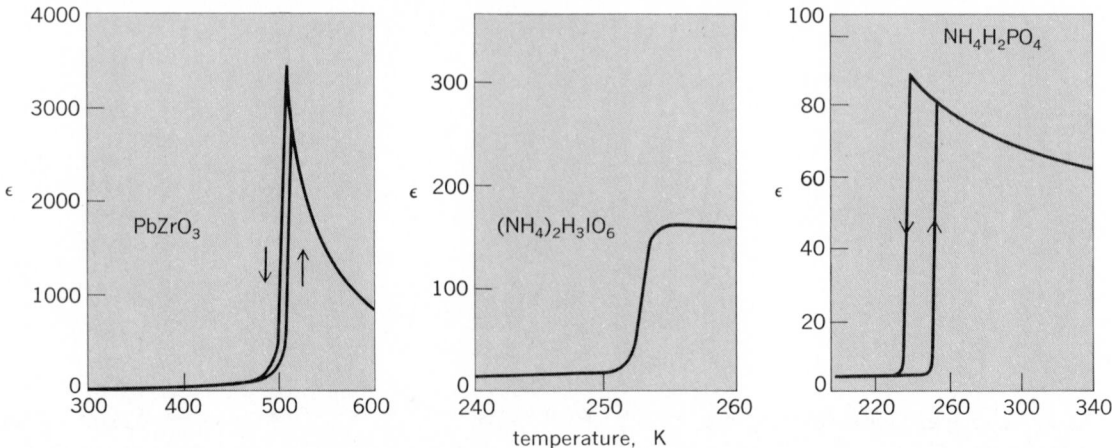

temperature, K

Fig. 9. Anomalous temperature dependence of relative dielectric constant of antiferroelectric crystals. Note increase as transition temperature is approached and drop when antipolarization occurs.

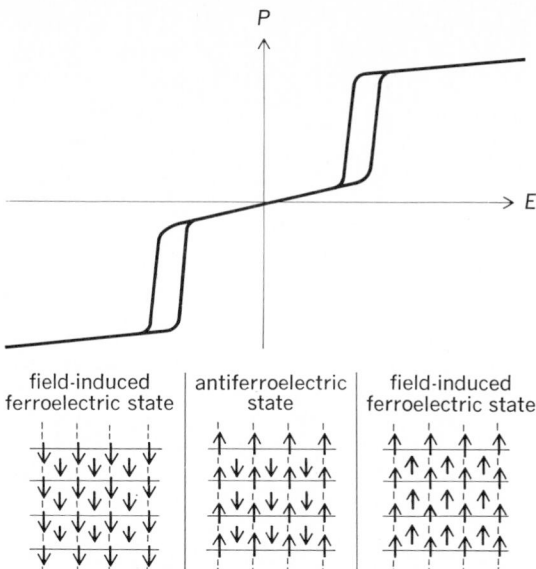

Fig. 10. Polarization *P* of antiferroelectric PbZrO₃ versus externally applied electric field *E*. Strong fields "switch" the antiferroelectric crystal into a ferroelectric state, as shown here schematically.

phase transition is discontinuous; in others it is continuous.

The structure of antiferroelectric crystals is generally closely related to the structure of ferroelectric crystals. Some antiferroelectrics even undergo phase transitions from an antipolarized state into a spontaneously polarized, ferroelectric state; in others a sufficiently strong electric field applied along an antiferroelectric axis reverses the polarity of one of the sublattices so that a ferroelectric state results. The crystal reverts, however, to the antiferroelectric state when the electric field is removed. Figure 10 shows net polarization versus externally applied field for such a case.

Compounds with antiferroelectric properties are PbZrO₃, PbHfO₃, NaNbO₃ (isomorphous with ferroelectric BaTiO₃), WO₃ (structure related to BaTiO₃), NH₄H₂PO₄ and isomorphous NH₄ salts (isomorphous with ferroelectric KH₂PO₄), (NH₄)₂-H₃IO₆, Ag₂H₃IO₆, and certain alums.

Origin of phase transition. The ferroelectric phase transition results from an instability of one of the normal lattice vibration modes. On approaching the transition temperature, the frequency of the relevant normal mode decreases (soft mode). The restoring force of the mode displacements tends to zero. When the stability limit is reached, the displacements corresponding to the soft mode freeze in, and the ferroelectric phase results. The ferroelectric soft mode is polar (infrared-active) and of infinite wavelength. The antiferroelectric phase transition, on the other hand, emerges from a soft lattice mode with a finite wavelength equal to an integer multiple of a lattice period.

Applications. The piezoelectric effect of ferroelectrics (and certain antiferroelectrics) finds numerous applications in electromechanical transducers. The large electrooptical effect (birefringence induced by an electric field) is used in light modulators. In certain ferroelectrics (for example, BaTiO₃, LiNbO₃, KTaNbO₃, and LiTaO₃), light can induce changes of the refractive indices. These substances can be used for optical information storage and in real-time optical processors. The temperature dependence of the spontaneous polarization corresponds to a strong pyroelectric effect which can be exploited in thermal and infrared sensors. [WERNER KANZIG]

Bibliography: R. Blinc and B. Žekš, *Soft Modes in Ferroelectrics and Antiferroelectrics,* 1974; V. M. Fridkin, *Ferroelectric Semiconductors,* 1980; V. M. Fridkin, *Photoferroelectrics,* 1979; M. E. Lines and A. M. Glass, *Principles and Applications of Ferroelectrics and Related Materials,* 1977; T. Mitsui, *An Introduction to the Physics of Ferroelectrics,* 1976; T. Mitsui et al., *Ferro- and Antiferroelectric Substances,* in K. M. Hellwege and A. M. Hellwege (eds.), Landolt-Bornstein Series, group 3, vol. 9, 1975.

Ferromagnetism

A property exhibited by certain metals, alloys, and compounds of the transition (iron group), rare-earth, and actinide elements in which, below a certain temperature called the Curie temperature, the atomic magnetic moments tend to line up in a common direction. Ferromagnetism is characterized by the strong attraction of one magnetized body for another, a phenomenon known before 600 B.C.

Atomic magnetic moments arise when the electrons of an atom possess a net magnetic moment as a result of their angular momentum. The combined effect of the atomic magnetic moments gives rise to a relatively large magnetization, or magnetic moment per unit volume, for a given applied field. Above the Curie temperature, a ferromagnetic substance behaves as if it were paramagnetic: Its susceptibility approaches the Curie-Weiss law. The Curie temperature marks a transition between order and disorder of the alignment of the atomic magnetic moments. Some materials exhibit a special form of ferromagnetism below the Curie temperature called ferrimagnetism. *See* CURIE TEMPERATURE; CURIE-WEISS LAW; ELECTRON SPIN; FERRIMAGNETISM; MAGNETIC SUSCEPTIBILITY; PARAMAGNETISM.

The characteristic property of a ferromagnet is that, below the Curie temperature, it can possess a spontaneous magnetization in the absence of an applied magnetic field. Upon application of a weak magnetic field, the magnetization increases rapidly to a high value called the saturation magnetization, which is in general a function of temperature. For typical ferromagnetic materials, their saturation magnetizations, and Curie temperatures *see* MAGNETIZATION.

The tasks of a theory of ferromagnetism are to account for the spontaneous magnetization below the Curie point, the temperature dependence of the saturation magnetization, and the nature of the magnetization process, or magnetization curve.

Weiss theory. The Weiss molecular field theory of ferromagnetism (P. Weiss, 1907) represents the first realistic attempt to account for the properties of a ferromagnet. This theory rests on two hypotheses:

1. Below the Curie point, a ferromagnetic substance is composed of small, spontaneously magnetized regions called domains. The total magnetic

moment of the material is the vector sum of the magnetic moments of the individual domains. It is now known that these assumed domains really exist and are usually between 0.1 and 0.01 cm across.

2. Each domain is spontaneously magnetized because a strong molecular (magnetic) field tends to align the individual atomic magnetic moments within the domain.

The consequence of these assumptions is that, while each domain is spontaneously magnetized, the directions of magnetization of the domains do not coincide; therefore the overall magnetization of the sample may be much smaller than if it were composed of a single domain. Application of a relatively weak field of the order of 100–1000 oersteds (and often very much less) is sufficient to align the directions of magnetization of the domains, thereby achieving a large magnetization.

The second hypothesis of the Weiss theory leads to the existence of a Curie temperature below which a domain may be spontaneously magnetized in the absence of an applied magnetic field. This comes about in the following way. If the domain is spontaneously magnetized, there must be some sort of interaction between the atomic magnetic moments which tends to align them. Otherwise the domain would behave paramagnetically. The average strength of this interaction may be represented by an internal magnetic field, the Weiss molecular field, which is proportional to the magnetization of the domain. Thus the effective field acting on any atomic magnetic moment within the domain may be written as Eq. (1), where H_0 is an externally ap-

$$H = H_o + \lambda M \tag{1}$$

plied magnetic field and λM (M = magnetization) is the Weiss molecular field whose order of magnitude in iron is 10^7 oersteds. It is relatively easy to deduce the magnetic susceptibility above the Curie point since it is known that the ferromagnetic substance behaves like a paramagnet above the Curie temperature. If the Curie law holds, Eq. (2)

$$M/H = C/T \tag{2}$$

is valid, where C is the Curie constant. In the Curie law, H is taken to be the effective magnetic field acting on an atomic magnetic moment, Eq. (1). From Eqs. (1) and (2), Eq. (3) is obtained. The susceptibility χ is the magnetization per unit applied field H_0, so that from Eq. (3), in the electromagnetic system of units, Eq. (4) is derived. The Curie

$$M/(H + \lambda M) = C/T \tag{3}$$

ceptibility χ is the magnetization per unit applied field H_0, so that from Eq. (3), in the electromagnetic system of units, Eq. (4) is derived. The Curie

$$\chi = M/H_0 = C/(T - C\lambda) = C/(T - T_c) \tag{4}$$

temperature is defined by $T_c = C\lambda$, and the susceptibility follows the Curie-Weiss law above the Curie temperature. The form of the Curie-Weiss law leads to a nonzero magnetization when $H_0 = 0$ at T_c.

Below the Curie point, where the Curie-Weiss law breaks down, the Weiss theory leads to a spontaneous magnetization and also predicts the temperature dependence of the saturation magnetization. This comes about if it is taken into account that the Curie law is only an approximation valid for weak fields or high temperatures; it does not allow for saturation effects. The correct quantum

mechanical expression to replace Eq. (2) is given by Eqs. (5) and (6), where N is the number of atoms

$$M = NJg\mu_B B_J(a^*) \tag{5}$$

$$a^* = g\mu_B(H_0 + \lambda M)J/kT \tag{6}$$

per unit volume, each with angular momentum quantum number J, g is the spectroscopic splitting factor (the measure of the energy level splittings), μ_B is the Bohr magneton, k is Boltzmann's constant, T is the absolute temperature, and $B_J(a^*)$ is the Brillouin function. For $T < T_c$, the value of the magnetization for an applied field of H_0 is obtained by solving Eqs. (5) and (6) simultaneously. In particular, the spontaneous magnetization is obtained by setting $H_0 = 0$ in Eq. (6). See GYROMAGNETIC EFFECT; GYROMAGNETIC RATIO; MAGNETON.

The solution for the spontaneous magnetization ($H_0 = 0$) may be obtained graphically for any temperature T, as shown in Fig. 1. The dashed line of

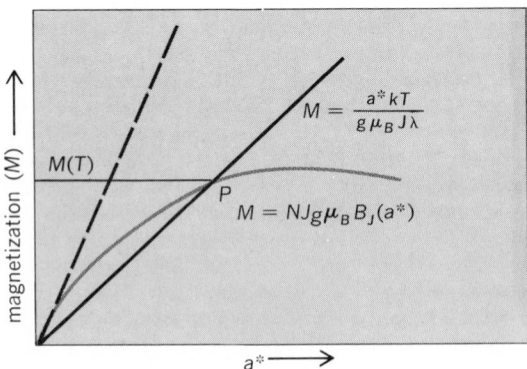

Fig. 1. Spontaneous magnetization below the Curie point. Intersection P determines $M(T)$.

Fig. 1 represents the line $M = a^*kT_c/g\mu_B J$ and is the line for the largest value of T for which there is a solution. This is the Curie point, and for $T > T_c$ there is no spontaneous magnetization. The value of T_c obtained in this way [the slope of Eq. (5) at $a^* = 0$] is $T_c = Ng^2\mu_B^2 J(J + 1)\lambda/3k$, a result which is consistent with the earlier definition $T_c = C\lambda$. The results obtained from Fig. 1 are plotted as a function of temperature in Fig. 2. The value of J chosen is $\frac{1}{2}$.

The plot of Fig. 2 is in overall general agreement with the experimental results for iron, nickel, and cobalt. The low-temperature behavior, however, is better described by magnon theory. See MAGNON.

Heisenberg theory. The Heisenberg theory of ferromagnetism (W. Heisenberg, 1928) treats the origin of the Weiss molecular field on an atomic basis. It may be remarked at the outset that ordinary dipole-dipole interactions among atomic magnetic moments are much too small to account for the Weiss field. The foundation of the Heisenberg theory is the Pauli exclusion principle. Consider two adjacent atoms of a ferromagnetic material. Each has an atomic magnetic moment arising from electron spin angular momentum. The Pauli principle requires that electrons with parallel spin on adjacent atoms keep out of each other's way. If electrons with parallel spin are farther apart than

FERROMAGNETISM

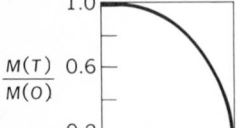

Fig. 2. Temperature dependence of spontaneous magnetization for $J = \frac{1}{2}$.

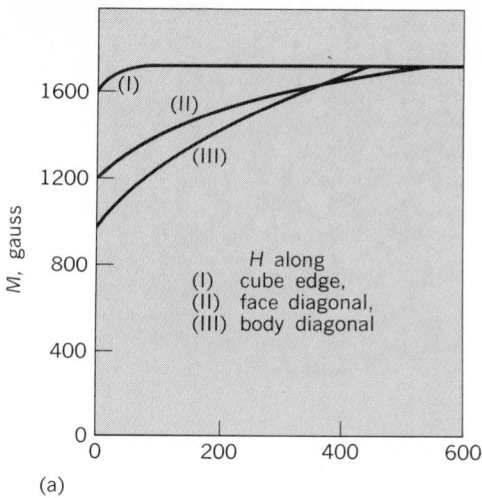

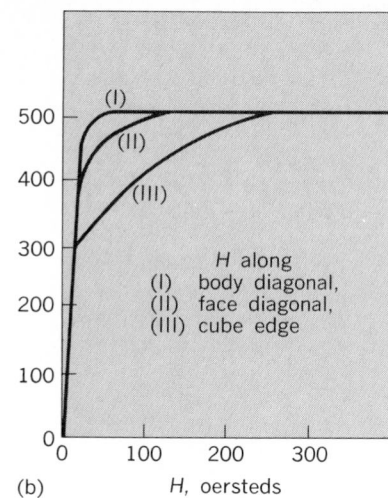

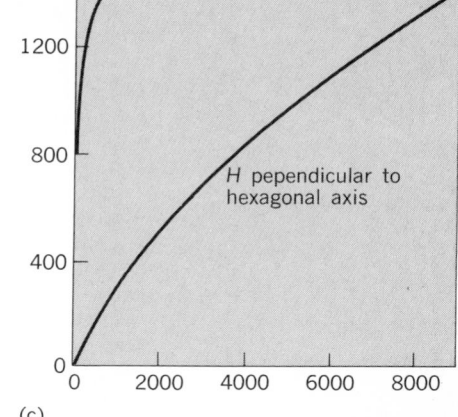

Fig. 3. The magnetization curves, for various directions of applied field H, indicated for single crystals (a) of iron, (b) of nickel, and (c) of cobalt. (*After K. Honda and S. Kaya, 1926–1928*)

they would be if their spins were antiparallel, then the electrostatic Coulomb repulsion between them will be less in the former case. The lowest energy state of the system will thus occur when the electron spins are parallel, and this arrangement will be favored. However, if the electrons are kept apart, their kinetic energies will increase and this will tend to make the antiparallel arrangement the state of lowest energy. It is the competition between these two effects which determines whether or not a substance is ferromagnetic. Heisenberg showed that the whole situation could be expressed in terms of an effective interaction between the electron spins, the exchange interaction energy, which had first been derived by P. A. M. Dirac. In Eq. (7), $\mathbf{S}_i$ and $\mathbf{S}_j$ are the spin angular

$$\text{Exchange energy} = -2J_{ij}\mathbf{S}_i \cdot \mathbf{S}_j \qquad (7)$$

momentum vectors of the two electrons i and j, and J_{ij} is the so-called exchange integral between the electrons. The exchange integral decreases rapidly with distance between the electrons and depends in a complicated way upon the spatial distribution (wave function) of the electrons. It is extremely difficult to compute. If the exchange integral is positive, the parallel arrangement is favored, and if J is large enough, ferromagnetism should result. If J is large and negative, antiferromagnetism or ferrimagnetism supposedly arises. The order of magnitude of J is given by $J \sim kT_c \sim 10^{-13}$ erg. See EXCLUSION PRINCIPLE.

There seems to be no question that the Heisenberg theory correctly accounts for the tendency of electrons in the same ferromagnetic atom to exhibit parallel spins. However, whether or not it leads to the correct explanation of the interatomic alignment of spins is still a subject of much controversy. In insulators it usually proves possible to express the coupling between atomic spins by Eq. (7), provided J_{ij} is interpreted as an effective exchange integral. In metals the problem is much more complicated. There is little doubt, however, that the basic Heisenberg idea is correct and that the molecular field arises from the interplay between electrical forces and the effects of Pauli exclusion.

Crystalline anisotropy energy. This accounts for the experimental fact that ferromagnets tend to magnetize along certain crystallographic axes, called directions of easy magnetization. For example, a single crystal of iron, which is made up of a cubic array of iron atoms, tends to magnetize in the directions of the cube edges. It requires about 1.4×10^5 ergs/cm³ (at room temperature) to move the magnetization into a hard direction along a cubic body diagonal. Single-crystal magnetization curves, for various directions of applied field H, are shown in Fig. 3.

N. S. Akulov (1929–1931) showed that the anisotropy energy U_A could be expressed conveniently in an ascending power series of the direction cosines between the magnetization and the crystal axes. For cubic crystals the lowest-order terms take the form of Eq. (8), where α_1, α_2, and α_3 are

$$U_A = K_1(\alpha_1{}^2\alpha_2{}^2 + \alpha_2{}^2\alpha_3{}^2 + \alpha_3{}^2\alpha_1{}^2) + K_2(\alpha_1{}^2\alpha_2{}^2\alpha_3{}^2) \qquad (8)$$

direction cosines with respect to the three cube edges, and K_1 and K_2 are temperature-dependent parameters characteristic of the material, called anisotropy constants. In general, $|K_1| > |K_2|$, and further terms are unnecessary. In iron, K_1 is positive, and therefore U_A is a minimum when any single direction cosine $\alpha_i = 1$ and the other $\alpha_j = 0$. That is, the cube edges are easy directions. In

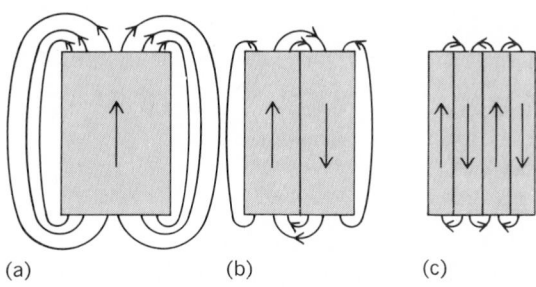

Fig. 4. Lowering of magnetic field energy by domains. (a) Lines of force for a single domain. (b) Shortening of lines of force by division into two domains. (c) Reduction of field energy by further subdivision.

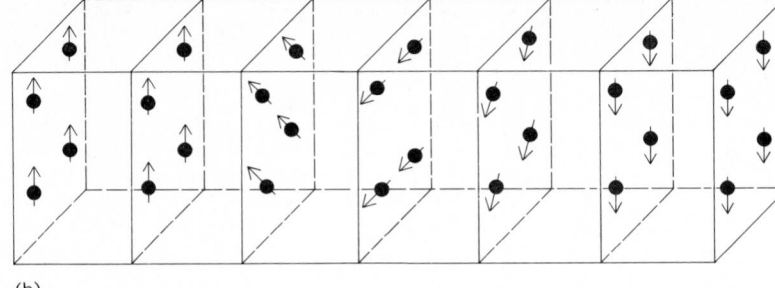

(a)   (b)

Fig. 5. Lowering of exchange energy by the transition zone known as Bloch wall. The reversal of magnetization between domains does not take place abruptly as is shown in (a) but by degrees as is illustrated in (b).

nickel, K_1 is negative and hence the body diagonals are easy directions.

For crystals of other than cubic symmetry the energy U_A must be expressed in a form which is different from Eq. (8) and appropriate to the particular symmetry.

Anisotropy constants can be determined from (1) analysis of magnetization curves, (2) the torque on single crystals in a large applied field, and (3) single-crystal ferromagnetic resonance. *See* MAGNETIC RESONANCE.

The Heisenberg exchange energy, Eq. (7), is isotropic and cannot account for the observed anisotropy, which probably has its origin in a complicated interplay of spin-orbit coupling, crystalline electric fields, and overlap of orbital wave functions. Anisotropy energy depends on the state of strain of the crystal, giving rise to magnetostriction, that is, changes in length of a substance when it is magnetized. *See* MAGNETOSTRICTION.

Ferromagnetic domains. Small regions of spontaneous magnetization, formed at temperatures below the Curie point, are known as domains. As shown in Fig. 4, domains originate in order to lower the magnetic energy. In Fig. 4b it is shown that two domains will reduce the extent of the external magnetic field, since the magnetic lines of force are shortened. On further subdivision, as in Fig. 4c, this field is still further reduced.

An alternate way to describe the energy reduction is to note that the interior demagnetizing fields, coming from surface poles, are much smaller in the long, thin domains of Fig. 4c than in the "fat" domain of 4a.

The question arises as to how long this subdivision process continues. With each subdivision there is a decrease in field energy, but there is also an increase in Heisenberg exchange energy, since more and more magnetic moments are aligning antiparallel. Finally a state is reached in which further subdivision would cause a greater increase in exchange energy than it would cause decrease in field energy, and the ferromagnet will assume this state of minimum total energy.

Bloch wall. Also because of exchange energy, the reversal of magnetization between domains does not occur abruptly but takes place gradually through a transition zone called the Bloch wall (F. Bloch, 1932) (Fig. 5). To understand the reason for this wall, consider the exchange energy involved. Let the angle between the magnetization of neighbor planes be ϕ and assume that, as is usual, the exchange energy is appreciable only between neighbor atoms. According to Eq. (7), the exchange energy between atoms on neighbor planes varies as $-\cos\phi$, or as a constant $+\phi^2/2$ if ϕ is small. If total reversal takes place in N planes, $\phi = \pi/N$, and the total exchange energy of the wall will be given by expression (9).

$$U_E \text{ (wall)} \propto \text{constant} + N\frac{1}{2}(\pi/N)^2 \qquad (9)$$

The question now arises as to why N should not continue to increase until the entire crystal becomes a single Bloch wall. This does not happen because the decrease in exchange energy is accompanied by an increase in anisotropy energy. Many of the intermediary planes of Fig. 5b must of necessity have their magnetization along hard directions. The larger the value of N, the greater must be the number of such planes. The thickness of the wall can be determined by finding that value of N which makes the sum of exchange and anisotropy energies a minimum. In iron the wall is ~500 A thick and has the total energy ~2 erg/cm².

Domain arrangement. The orientation of domains in a crystal is determined primarily by the need to minimize the magnetic energy (Fig. 4). It is possible to eliminate all surface magnetic poles by forming flux-closure domains (Fig. 6). Here the normal component of magnetization is continuous across all domain boundaries. The demagnetizing fields are zero everywhere, except for the trivial effect of surface poles in the Bloch walls. In a uniaxial crystal, that is, a crystal with a single easy direction, an arrangement as shown in Fig. 6b will be preferred since it has a lower density of magnetization normal to the easy direction, or in the hard direction. Even in cubic crystals, in which all directions of magnetization in Fig. 6 may be easy, 6b will be preferred because of magnetostriction. In iron, for example, each domain increases in length along the direction of magnetization by a fraction $\sim 2 \times 10^{-5}$. Thus the domains of Fig. 6 can be fitted smoothly together only by straining them elastically against this magnetostriction, and the required elastic strain energy will be smaller in 6b than in 6a.

In polycrystals the domain structure is much more complicated, depending upon such variables as grain orientation and grain boundaries. It is possible, however, for domains to cross grain boundaries. *See* GRAIN BOUNDARIES.

On minimizing the total contributions from (1) magnetic, or demagnetizing, energy, (2) anisotropy, (3) magnetostriction, (4) elastic strain, and (5) Bloch wall energy, it is found that, depending upon the composition and shape of the crystal, the

FERROMAGNETISM

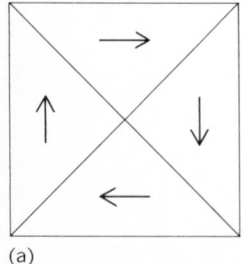

(a)

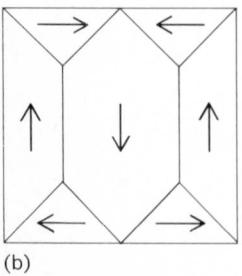

(b)

Fig. 6. Flux-closure domains. (a) Large domains at right angles. (b) Reduction in their size, causing reduction of anisotropy energy of uniaxial crystals or of strain energy of cubic crystals.

theoretical domain thickness should vary from about 0.1 to 0.001 cm.

Bitter powder patterns. Direct experimental evidence of the existence of domains is furnished by Bitter powder patterns (F. Bitter, 1931). First a surface is prepared by careful polishing with fine abrasive, followed (for metallic samples) by electropolish. Then a drop of a colloidal suspension of magnetic Fe_2O_3 particles is placed on the surface and covered by a microscope cover glass. These particles gather at surface regions where the magnetic field is largest and can be observed easily with a microscope. It is seen from Fig. 5 that the surface field is largest at the center of the Bloch wall; that is, lines of surface poles will demarcate the wall. Thus the colloidal particles concentrate along domain boundaries, as shown in Fig. 7. If the surface deviates by even a few degrees of angle from a simple lattice plane, the pattern becomes very complicated. The beginnings of this may be seen in the "fir tree" patterns at the lower right of Fig. 7.

The powder patterns only delineate the boundaries and do not themselves show the direction of magnetization. This direction may, however, be inferred from a study of how particles gather around a scratch made on the surface, coupled with observation of domain growth when the sample is placed in a magnetic field.

The Kerr magnetooptic effect (rotation of the plane of polarization of light reflected from a magnetic surface) has also been used to study domains. J. F. Dillon has shown that single-crystal slabs of ferromagnetic yttrium iron garnet (YIG) are transparent to visible light and that the Faraday effect (rotation of the plane of polarization of light transmitted along a magnetic field) can be used to observe domains. The effect is very striking, and the growth and diminution of domains are readily visible. *See* FARADAY EFFECT; KERR EFFECT.

It is also possible to observe domains by means of transmission or reflection of electrons in an electron microscope. This technique, and also the optical techniques mentioned above, are particularly useful if the Bloch walls are extremely thick (because of small magnetocrystalline anisotropy) and the resultant Bitter powder patterns very blurred.

Bloch wall motion. This accounts for the initial portion of the magnetization curve. As the wall passes through regions of crystal strain, or over a number of foreign atoms, it may suddenly go over a potential energy hump and into a minimum. There will be a small, almost discontinuous jump in the magnetization, easily detected by a search coil (a device for measuring change of flux density) and amplifier. This is called the Barkhausen effect. It was once thought erroneously that this effect comes from a sudden, complete reversal of magnetization within a domain. However, H. J. Williams and W. Shockley have shown that many Barkhausen jumps accompany the motion of a single domain wall.

Soft magnetic materials. Materials easily magnetized and demagnetized are called "soft"; these are used in alternating-current machinery. The problem of making cheap soft materials is complicated by the fact that readily fabricated metals usually have many crystalline boundaries and crystal grains oriented in many directions. In such metals the magnetization process is accompanied by much irreversible Bloch wall motion and by much rotation against anisotropy, which is usually irreversible.

The ideal cheap soft material would be an iron alloy fabricated by some inexpensive technique which results in all crystal grains being oriented in the same or nearly the same direction. Various complicated rolling and annealing methods have been discovered in the continued search for better grain-oriented or "cube-textured" steels.

In some situations the cost of fabrication is secondary, and alloys with the smallest possible crystalline anisotropy and magnetostriction are demanded. The first such need appears to have been for the inductive loading of submarine telegraph cables. This gave rise to the perfection of permalloy, an alloy of 21.5% Fe and 78.5% Ni. The constant K_1 of Eq. (8) takes opposite signs in pure Fe and pure Ni, and at the permalloy composition the resultant anisotropy goes to zero. More precisely, the composition for zero anisotropy is not quite the same as for zero magnetostriction, and permalloy represents a compromise. Addition of a third element, generally molybdenum, can simultaneously drive anisotropy and magnetostriction to zero and, as a bonus, decrease resistivity.

Hard magnetic materials. Materials which neither magnetize nor demagnetize easily are called "hard"; these are used in permanent magnets. Early attempts to fabricate hard magnets centered around the introduction of various inclusions to impede Bloch wall motion. Martensitic alloys are of this type.

The fine-particle precipitates have dominated the market. These alloys consist basically of small and elongated magnetic particles embedded in a nonmagnetic matrix. L. Néel first pointed out that an isolated, sufficiently small magnetic particle (of radius roughly 10^{-6} cm, of iron) will not energetically support a Bloch wall and hence can contain only a single domain. Elongated single domain magnets (ESD) are fabricated from metallic alloys

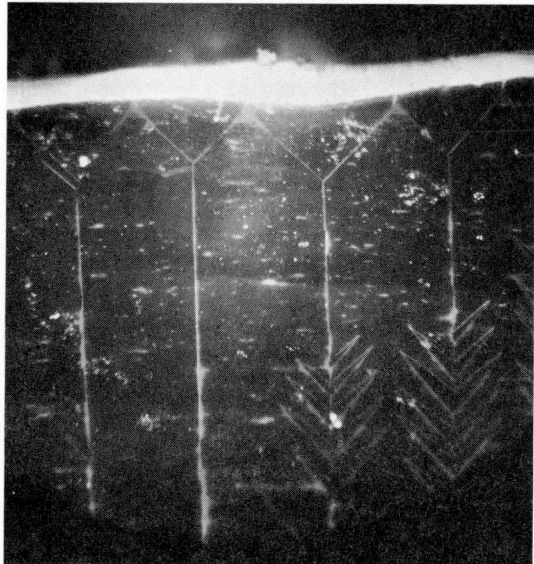

Fig. 7. Bitter powder patterns on a (100) surface of silicon-iron. (*Photograph by H. J. Williams*)

which have been cooled into elongated precipitates (Alnico) or electrodeposited into a soft lead matrix (Lodex). They are also made from pressed ceramic materials of huge magnetocrystalline anisotropy, such as strontium ferrite.

[ELIHU ABRAHAMS; FREDERIC KEFFER]

Bibliography: S. Chikazumi and S. H. Charap, *Physics of Magnetism,* 1964; reprint 1978; B. D. Cullity, *Introduction to Magnetic Materials,* 1972; A. Pekalski et al., *Magnetism in Metals and Metallic Compounds,* 1975; G. T. Rado and H. Suhl (eds.), *Magnetism,* 5 vols., 1963–1973.

Feynman integral

A technique, also called the sum over histories, which is basic to understanding and analyzing the dynamics of quantum systems. It is named after fundamental work of Richard Feynman. The crucial formula gives the quantum probability density for transition from a point q_0 to a point q_1 in time t as expression (1), where $S(\text{path})$ is the classical me-

$$\int \exp\left[iS(\text{path})/\hbar\right] d(\text{path}) \tag{1}$$

chanical action of a trial path, and $\hbar$ is the rationalized Planck's constant. The integral is a formal one over the infinite-dimensional space of all paths which go from q_0 to q_1 in time t. Feynman defines it by a limiting procedure using approximation by piecewise linear paths. *See* ACTION.

Feynman integral ideas are especially important in quantum field theory, where they not only are a useful device in analyzing perturbation series but are also one of the few nonperturbative tools available.

An especially attractive element of the Feynman integral formulation of quantum dynamics is the classical limit, $\hbar \to 0$. Formal application of the method of stationary phase to expression (1) says that the significant paths for small $\hbar$ will be the paths of stationary action. One thereby recovers classical mechanics in the hamiltonian stationary action formulation. *See* LEAST-ACTION PRINCIPLE.

Feynman-Kac formula. Mark Kac realized that if one replaces time by a formal purely imaginary time (that is, replaces the Schrödinger equation by a heat equation), then one can make precise mathematical sense out of Feynman's integral in terms of the stochastic process introduced by Norbert Wiener in his theory of Brownian motion. The resulting formula for the semigroup e^{-tH}, with $H = -\frac{1}{2}\Delta + V$ the quantum hamiltonian (Δ is the laplacian operator and V is the potential energy), is given by Eq. (2). Here E_x is the expectation with

$$(e^{-tH}f)(x) = E_x\left(\exp\left\{-\int_0^t V[b(s)]ds\right\}f[b(t)]\right) \tag{2}$$

respect to Wiener's process, $b(s)$, starting at the point x, and f is an arbitrary bounded measurable function to which e^{-tH} is applied. Equation (2) is called the Feynman-Kac formula. It makes a wide variety of probabilistic ideas applicable to a rigorous mathematical analysis of quantum theory. *See* BROWNIAN MOVEMENT; STOCHASTIC PROCESS.

Euclidean quantum field theories. The passage to imaginary time initiated by Kac is especially useful in relativistic quantum field theory. Since the Lorentz covariance of relativity is replaced by euclidean group covariance under this replace-

ment, the new theory is often called euclidean field theory. The corresponding Feynman integral is often called a euclidean path integral and has become the standard formulation of Feynman integrals in most modern discussions of quantum field theory. The analogs of stationary action solutions are called instantons. *See* INSTANTON.

Statistical mechanics analogy. The appearance of the exponential in Eq. (2) suggests an analogy between quantum dynamics and classical statistical mechanics. The analog of the total space-time dimension in a euclidean quantum field theory is then the dimension of space in statistical mechanics. Just as the phenomenon of phase transitions occurs in statistical mechanics, the analogous phenomena of spontaneously broken symmetries and dynamical instability occur in quantum field theory. These ideas are a basic element of all modern theories of elementary particle physics. They are usually analyzed within a Feynman integral framework. *See* ELEMENTARY PARTICLE; NONRELATIVISTIC QUANTUM THEORY; PHASE TRANSITIONS; QUANTUM FIELD THEORY; STATISTICAL MECHANICS; SYMMETRY LAWS.

[BARRY SIMON]

Bibliography: R. P. Feynman, Space-time approach to nonrelativistic quantum mechanics, *Rev. Mod. Phys.,* 20:367–387, 1948; R. P. Feynman and A. Hibbs, *Quantum Mechanics and Path Integrals,* 1965; M. Kac, On some connections between probability theory and differential equations, *Proceedings of the 2d Berkeley Symposium on Mathematical Statistics Problems,* pp. 189–215, 1950; B. Simon, *Functional Integration and Quantum Physics,* 1979.

Field emission

The emission of electrons from a metal or semiconductor into vacuum (or a dielectric) under the influence of a strong electric field. In field emission, electrons tunnel through a potential barrier, rather than escaping over it as in thermionic or photoemission. The effect is purely quantum-mechanical, with no classical analog. It occurs because the wave function of an electron does not vanish at the classical turning point, but decays exponentially into the barrier (where the electron's total energy is less than the potential energy). Thus there is a finite probability that the electron will be found on the outside of the barrier. This probability varies as $e^{-cA^{1/2}}$, where c is a constant and A the area under the barrier. *See* NONRELATIVISTIC QUANTUM THEORY.

For a metal at low temperature, the process can be understood in terms of the illustration. The metal can be considered a potential box, filled with electrons to the Fermi level, which lies below the vacuum level by several electronvolts. The distance from Fermi to vacuum level is called the work function, ϕ. The vacuum level represents the potential energy of an electron at rest outside the metal, in the absence of an external field. In the presence of a strong field F, the potential outside the metal will be deformed along the line AB, so that a triangular barrier is formed, through which electrons can tunnel. Most of the emission will occur from the vicinity of the Fermi level where the barrier is thinnest. Since the electron distribution in the metal is not strongly temperature-dependent,

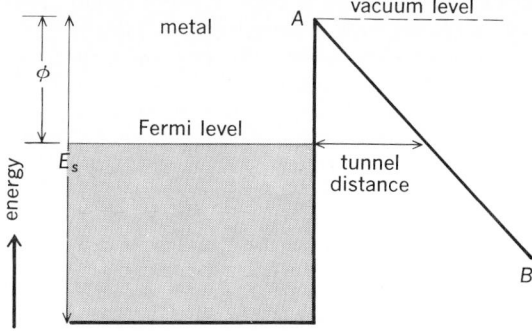

Diagram of the energy-level scheme for field emission from a metal at absolute zero temperature.

field emission is only weakly temperature-dependent and would occur even at the absolute zero of temperature. The current density J is given by the Fowler-Nordheim equation below, where B is a

$$J = BF^2 e^{-6.8 \times 10^7 \phi^{3/2}/F}$$

field-independent constant of dimensions A/V² (A = amperes or state), ϕ is work function in electronvolts, and F is applied field in V/cm. The factor $\phi^{3/2}/F$ is proportional to the square root of the area under the barrier at the Fermi level. Appreciable emission requires fields of $4 - 7 \times 10^7$ V/cm, depending on ϕ. See FREE-ELECTRON THEORY OF METALS.

Field emission is most easily obtained from sharply pointed metal needles whose ends have been smoothed into nearly hemispherical shape by heating. Tip radii r_t equal to or less than 100 nm can be obtained in this way; because of its small size, the emitter is generally a single crystal. If an emitter is surrounded by a hemispherical anode raised to a voltage V, then the field F is approximately $V/5r_t$ at the emitter. Thus, modest voltages suffice for emission. The electric lines of force diverge radially from the tip; since the electron trajectories initially follow the lines of force, they also diverge, and a highly magnified emission map of the emitter surface can be obtained, for instance, by making the anode a fluorescent screen. This constitutes a field-emission microscope, invented by E. W. Müller in 1936. Since work function and hence emission are affected by adsorbed layers, the field-emission microscope is very useful for studying adsorption, particularly surface diffusion. Field emitters are widely used in "ordinary" and scanning electron microscopes as high-brightness quasi-point sources of electrons, since emission occurs as if it originated from the center of the emitter cap.

Field emission can also occur from electrode asperities into insulating liquids, and thus initiates electrical breakdown. It can occur from isolated atoms or molecules in high fields, and is then called field ionization. Field ionization forms the basis of the field-ion microscope, and is a useful method of generating ions in analytical mass spectrometry. Internal field emission can occur from the valence to the conduction band of a semiconductor in a high field, and is then known as Zener breakdown.

[ROBERT GOMER]

Fine structure

A term referring to the closely spaced groups of lines observed in the spectra of the lightest elements, notably hydrogen and helium. The components of any one such group are characterized by identical values of the principal quantum number n, but different values of the azimuthal quantum number l and the angular momentum quantum number j.

According to P. A. M. Dirac's relativistic quantum mechanics, those energy levels of a one-electron atom that have the same n and j coincide exactly, but are displaced from the values predicted by the simple Bohr theory by an amount proportional to the square of the fine-structure constant α. The constant α is dimensionless, and nearly equal to 1/137. Its value is actually 0.007297351 ± 0.000000006. In 1947 deviations from Dirac's theory were found, indicating that the level having $l = 0$ does not coincide with that having $l = 1$, but is shifted appreciably upward. This is the celebrated Lamb shift named for its discoverer, Willis Lamb, Jr. Modern quantum electrodynamics accounts for this shift as being due to the interaction of the electron with the zero-point fluctuations of the electromagnetic field. See ATOMIC STRUCTURE AND SPECTRA; QUANTUM ELECTRODYNAMICS; QUANTUM NUMBERS; RELATIVISTIC QUANTUM THEORY.

In atoms having several electrons, this fine structure becomes the multiplet structure resulting from spin-orbit coupling. This gives splittings of the terms and the spectral lines that are "fine" for the lightest elements but that are very large, of the order of an electronvolt, for the heavy elements.

[F. A. JENKINS/W. W. WATSON]

First-order transition

A change in state of aggregation of a system accompanied by a discontinuous change in enthalpy, entropy, and volume at a single temperature and pressure. This transition may be between liquid and gas, between solid and gas, between solid and liquid, or between two solid phases. For the differences between first- and second-order transition see SECOND-ORDER TRANSITION. See also TRANSITION POINT.

[ROBERT L. SCOTT]

Flavor

A generic term referring to quarks, the most elementary constituents of matter. The evidence is conclusive that all matter is constituted of quarks of at least five different types of flavors, with a sixth predicted. They are labeled u, d, s, c, b, and t. The last, the t quark, is predicted from symmetry considerations. Corresponding mnemonics are up, down, strange, charmed, bottom, and top. The u, c, and t flavors each carry a positive electric charge equal in magnitude to two-thirds that of the electron, and the d, s, and b flavors have a negative charge one-third that of the electron.

The proton is composed of two u- and one d-flavored quarks; the neutron of two d and one u. Strange particles contain an s; charmed particles contain a c-flavored quark. The strange lambda particle is similar to a neutron except that one of the d quarks has been replaced by an s quark. See

BARYON; CHARM; NEUTRON; PROTON; STRANGE PARTICLES.

Mesons are composed of quark-antiquark ($q,\bar{q}$) combinations. For example, the ρ meson is made of $u\bar{u}$ or $d\bar{d}$, the ϕ is $s\bar{s}$, the ψ/J is $c\bar{c}$, and the Υ is $b\bar{b}$. The particle corresponding to $t\bar{t}$ has yet to be discovered. It is presumed that this is because its rest mass is so great as to preclude its production at the highest-energy accelerators currently available. *See* J PARTICLE; MESON; UPSILON PARTICLES.

Interquark forces are produced by the exchange of massless, spin-one bosons called gluons. Evidence exists that these forces are flavor-independent. *See* GLUONS.

A subdivision of flavors into colors is required from statistical considerations. Each quark flavor occurs in three colors. The color of the quarks in particles is such that the particles are colorless. *See* COLOR (QUANTUM MECHANICS).

Whether quarks can exist in isolation is an open question. Various reports of their detection remain to be confirmed. A major theoretical problem lies in understanding why quarks appear to be always confined, and always appear in combination with other quarks to compose particles which have integral multiples 0, ±1, ±2, and so forth, of the electron charge. *See* ELEMENTARY PARTICLE; QUARKS.

[VAL FITCH]

Fluid dynamics

The science of fluids in motion. Fluid dynamics attempts to describe the motion of a fluid as it is displaced and deformed by the action of moving or fixed boundaries. Fluid dynamics may be divided into two parts, hydrodynamics and aerodynamics. For low Mach numbers (ratio of velocity of fluid to local acoustic velocity) both hydro- and aerodynamics may be treated in the same manner, but for Mach numbers over about 1/2, compressibility must be taken into account.

Fluid dynamics makes use of both theoretical developments and experimental results. Simplifying assumptions are generally made in theoretical studies in order to make the equations manageable. The extent to which the results differ from the flow of real fluids must be determined by experiment, and corrections applied to the theoretical treatment to obtain practical results. *See* FLUID MECHANICS.

Theoretical hydrodynamics has been made a useful tool in solution of flow problems by use of the Prandtl hypothesis, which states that with fluids of low viscosity the effects of viscosity are limited to a narrow region along the boundaries. The problem may be solved as if the fluid were frictionless (nonviscous) to determine the velocity and pressure intensity throughout the fluid except at the boundaries. The flow near the boundaries is called boundary-layer flow, and takes into account viscous effects and the fact that the velocity at the boundary relative to the boundary is zero. From a study of the boundary layer, its growth may be computed, as well as the tangential shear force it exerts on the boundary. Under certain conditions of adverse pressure distribution, the boundary-layer film immediately adjacent to the wall comes to rest, and the bounding streamline separates from the wall. This phenomenon is known as separation and results in turbulence and formation of a wake downstream from the separation point, which increases the energy losses. Beyond the separation point, the fluid does not follow the boundary and does not regain pressure as velocity is reduced. There remains an additional drag or pressure force on a body immersed in a flowing fluid, known as form drag. *See* BOUNDARY-LAYER FLOW; HYDRODYNAMICS.

At very high altitude or low pressure instead of the Prandtl hypothesis, it is assumed that individual molecules slip along the surface of a flight vehicle in slip flow. If the fluid is ionized and reacts with electric or magnetic fields as well as with thermal and mechanical boundary conditions, additional behavior is obtained. *See* SUPERAERODYNAMICS. [VICTOR L. STREETER]

Bibliography: I. G. Currie, *Fundamental Mechanics of Fluids*, 1974; H. Lamb, *Hydrodynamics*, 6th ed., 1945; A. J. H. Rouse (ed.), *Advanced Mechanics of Fluids*, 1959, reprint 1976; V. L. Streeter, *Fluid Dynamics*, 1948.

Fluid flow

Motion of a fluid as a continuum. Fluids flow whereas solids move as bodies. In flow, the individual particles of a substance move relative to each other as well as to their surroundings. A fluid is a substance that flows under the slightest stress. Thus, the pressure of a gas is sufficient to cause it to flow throughout a container to which it is admitted. The weight of a liquid is sufficient to cause it to flow in a container but without significant change in volume. Considerable external force is required to cause a solid to flow.

By analogy, electricity, heat, and other forms of energy are said to flow. This article discusses only flow of fluids. *See* HEAT TRANSFER.

Fluids. A fluid may be liquid, vapor, or gas. A liquid will fill the container which holds it, but it may have a free surface, that is, a surface from which all pressure is removed except that of its own vapor. All liquids are relatively incompressible. *See* FLUIDS.

A vapor is a gas whose temperature and pressure are such that it is very near the liquid phase. Thus, steam is considered to be a vapor because its state is not far from that of water.

A gas may be defined as a highly superheated vapor; that is, its state is far removed from the liquid phase. Thus. air is considered to be a gas because its state is normally very far from that of liquid air. A gas is very compressible, and when all external pressure is removed, it tends to expand indefinitely. A gas is therefore in equilibrium only when it is completely enclosed.

The volume of a liquid is altered only slightly by changes in either pressure or temperature unless the temperature change is considerable or unless the initial temperature is near the critical temperature, but the volume of a gas or a vapor is greatly affected by changes in either pressure or temperature. *See* GAS; LIQUID.

Ideal and real flow. When there is no friction between adjacent moving particles, flow is termed ideal; that is, viscosity is zero. In ideal flow, internal forces at any section are always normal to the section. Forces are purely pressure forces. Such flow is approached but never achieved in reality.

In a real fluid, tangential or shearing forces always come into being whenever motion takes

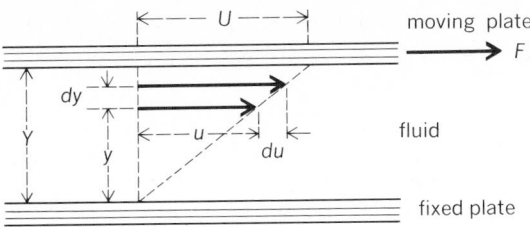

Fig. 1. Viscous flow.

place, thus giving rise to fluid friction, because these forces oppose the sliding of one particle past another. These friction forces are due to a property called viscosity.

Viscosity. The viscosity of a fluid is a measure of its resistance to shear or to angular deformation. Consider two parallel plates large enough that edge conditions may be neglected, and placed a small distance apart, and assume that the space between is filled with the fluid (Fig. 1). Assume that the lower surface is stationary while the upper one is moved relative to it with a velocity U by the application of a force F corresponding to some area A of the moving plate. Particles of the fluid in contact with each plate will adhere to it, and if the distance Y is not too great and the velocity U not too high, the velocity gradient will be a straight line. The action of the fluid may be likened to a series of thin sheets, each of which slips a little relative to the next. Experiment has shown that for a large class of fluids the shear stress τ between any two thin sheets of fluid may be expressed as Eq. (1), where μ is the coefficient of viscosity. This is

$$\tau = \frac{F}{A} = \mu \frac{U}{Y} = \mu \frac{du}{dy} \qquad (1)$$

also called the absolute or dynamic viscosity. *See* NEWTONIAN FLUID.

In a Newtonian fluid, the viscosity does not change with the rate of deformation and is represented by the straight line (Fig. 2). The slope of this line is determined by the magnitude of the viscosity. In a non-Newtonian fluid, μ varies with the rate of deformation. Printer's ink, pastes, paints, greases, and most suspensions in water, such as coal slurries and drilling muds, are non-Newtonian fluids. *See* NON-NEWTONIAN FLUID.

Absolute viscosity. In the metric system, the unit

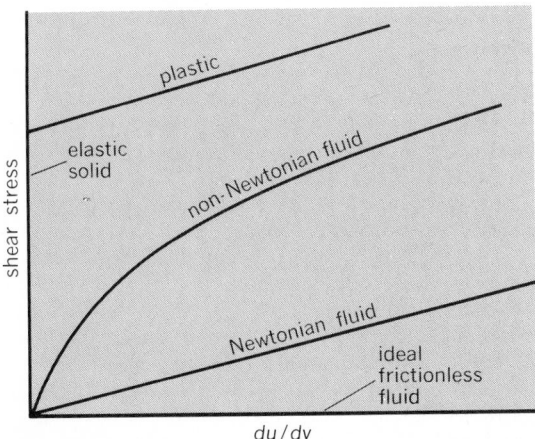

Fig. 2. Types of frictional flow.

of absolute viscosity is the poise (1 poise = 1 dyne-sec/cm²). Because most fluids have low viscosities, the centipoise (0.01 poise) is frequently a more convenient unit. It has the further advantage that the viscosity of water at 68.4°F (1 centipoise) provides a convenient reference.

In the English gravitational, or engineers', system, the viscosity unit in pounds (force) is 1 lb-sec/ft² or 1 slug/ft-sec. In the English absolute system, viscosity is expressed in terms of poundal-sec/ft² or lb/ft-sec. There are no names for the English units.

The conversion factors are 1 poise ≡ 100 centipoises ≡ 0.00209 lb-sec/ft² ≡ 0.0672 poundal-sec/ft².

The viscosities of all liquids decrease with an increase in temperature, whereas those of all gases and vapors increase. The absolute viscosity of both liquids and gases is practically independent of pressure except for extremely high pressures, at which there is a small increase.

Kinematic viscosity. In many fluid-flow problems there occurs the value of viscosity divided by density. This is called kinematic viscosity, and is $\nu = \mu/\rho$. In the metric system the unit of kinematic viscosity is the stoke (1 stoke = 1 poise divided by density, g/cm³). A convenient unit is the centistoke (0.01 stoke). The dimensions of kinematic viscosity are cm²/sec.

There is no name for the unit of kinematic viscosity in the English system, but the dimensions are ft²/sec. It is necessary to be consistent in the units employed. Thus, viscosity in pounds (force)-sec/ft² should be divided by density in slugs/ft³, or viscosity in poundal-sec/ft² should be divided by density in lb/ft³, which is numerically the same as specific weight.

The conversion factors are 1 stoke ≡ 100 centistokes ≡ 0.001076 ft²/sec and 1 ft²/sec ≡ 929 cm²/sec. *See* VISCOSITY.

Density and specific weight. Density is mass per unit volume and is normally indicated by ρ. In the metric system the dimensions of ρ are g/cm³. In the English systems they are either slugs/ft³ or lb/ft³ according to whether the engineers' system or the absolute system is used. The value of density is, however, the same for any location.

Specific weight is weight per unit volume and is commonly designated by either w or γ. In the English engineers' system dimensions are lb/ft³. In the absolute system they are slugs/ft³ where 1 slug ≡ 32.174 lb (mass). Specific weight in the engineers' system varies with the value of gravity and hence varies with location.

The relation between density and specific weight is $\rho = w/g$, where g is the acceleration of gravity. The standard sea-level values of g are 980.66 cm/sec² of 32.174 ft/sec². See DENSITY; SPECIFIC GRAVITY.

Compressibility. All liquids are relatively incompressible, and for most purposes water in particular can be treated as incompressible; yet it is 10 times as compressible as steel. The passage through water of a sound wave, which is really a pressure wave, is a result of the water's compressibility.

In dealing with the flow of air or other gases when the change in pressure is small, so that the change in density is negligible, even gases can be treated as incompressible. For an airplane flying at speeds of less than 250 mph the air can be considered to be of constant density. However, as the speed approaches that of sound in air, which is of the order of 700 mph, the pressure of the air adjacent to the body becomes materially different from that at some distance away, and air must then be considered to be compressible. *See* COMPRESSIBLE FLOW.

The change in volume from v to v_1 of a fluid as the result of a change in pressure Δp can be determined by $\Delta v/v_1 = -\Delta p/E_v$, where E_v is the mean value of the volume modulus of elasticity for the pressure range. For water the value of the volume modulus varies with both temperature and pressure, but a typical value for most conditions is 300,000 psi (Δp and E_v must both be in the same units). For isothermal compression of a perfect gas, $E_v = p$ and for an adiabatic compression $E_v = kp$, where k is the ratio of the specific heat at constant pressure to that at constant volume. For air the value of k is normally about 1.4, and p is the average pressure range. For low pressures air is many thousand times as compressible as water.

For practical use the change in volume of a gas is better determined by means of the perfect gas equation of state.

Perfect gas. A perfect gas is one whose equation of state is $pV = WRT$ or $pv = RT$, where p is absolute pressure, V is total volume, v is volume per unit weight, W is total weight, T is absolute temperature, and R is a gas constant. A consistent system of units must be used. Thus p may be in lb/ft^2, V in ft^3, v in ft^3/lb, W in lb, and T in °F $+459.7°$. The value of R in this system is 1546/molecular weight.

There is no perfect gas, but air and other so-called permanent gases (gases at conditions far from their liquid states) may usually be so considered, and for such real gases the product of R times molecular weight may range from 1485 to 1786. For air the value of R is 53.3. Even water vapor in the air follows the above law closely with $R = 1541/18 = 85.6$. For normal steam pressures, this does not hold, and vapor tables must be used. Also, for real gases when high pressure or low temperatures are involved the equation of state is more complicated. *See* VIRIAL EQUATION.

Real gases. For real gases, in which the conditions are such that the perfect gas law is not sufficiently accurate, a compressibility factor Z may be used in the equation of state $pv = ZRT$, where Z may range from 0.2 to 3 in extreme cases.

The van der Waals equation of state is sometimes useful at high pressures. *See* VAN DER WAALS EQUATION.

Ideal fluid flow. The flow of an ideal fluid is also termed inviscid flow, because it assumes an imaginary fluid of zero viscosity which therefore has no fluid friction. In such flow all particles of the fluid would move along individual streamlines and with equal velocities (Fig. 3).

Because there is no fluid friction, the total en-

ergy is the same throughout. For an incompressible liquid Bernoulli's equation, Eq. (2), can be used

$$\frac{p}{w} + z + \frac{V^2}{2g} = \text{constant} \qquad (2)$$

from one streamline to the next and also along any streamline. In Eq. (2) p/w is pressure head in height of the fluid. If p is pressure in lb/ft^2 and w is specific weight in lb/ft^3, p/w is a linear quantity in feet of the fluid. Potential energy of a unit weight of fluid is represented by z, which is a height above any arbitrary datum plane. The kinetic energy per unit weight is represented by $V^2/2g$, which is a linear quantity with V the velocity in ft/sec and g the acceleration of gravity in ft/sec^2.

Each term in Bernoulli's equation is a linear quantity that also represents energy per unit weight. Strictly speaking, the energy possessed by a particle of fluid is the sum of its potential and kinetic energy; that is, $z + V^2/2g$, whereas p/w represents the work done as a result of pressure and motion, but work and energy are measured in the same units.

In many real cases fluid friction is so small that results obtained by Bernoulli's theorem are sufficiently accurate for practical purposes or in many cases need only be modified slightly. Thus, the velocity of a jet issuing from a tank under a head h can be obtained by introducing a velocity coefficient into the equation such that $V = C_v\sqrt{2gh}$, where C_v may have a value as high as 0.98 or 0.99. Thus, the actual value of V is little less than the ideal value.

However, caution must be used in neglecting the effect of viscous friction; in many cases, the fluid friction is large, and the actual result may be very different from the ideal frictionless value.

The effect of viscous friction really originates where the fluid is in contact with a solid surface, and at a considerable distance from such surface the effect of friction is much less. Thus, fluid friction resulting from viscosity produces significant results near a body, such as an airplane, but at some distance away the relative velocity of the air may be considered as uniform (Fig. 3). *See* BOUNDARY-LAYER FLOW.

Real fluid flow. Real fluid flow and viscous flow are synonymous terms; all real fluids are more or less viscous. Because of fluid friction, Bernoulli's equation must be modified by the introduction of a term h_f, which represents the loss of head, or energy. Thus, an energy equation between two points is written as Eq. (3), which shows that the total

$$\frac{p_1}{w} + z_1 + \frac{V_1^2}{2g} = \frac{p_2}{w} + z_2 + \frac{V_2^2}{2g} + h_f \qquad (3)$$

head always decreases in the direction of flow. This equation applies to an incompressible fluid. *See* BERNOULLI'S THEOREM.

In some real cases the loss of head may be small and can be disregarded with slight error, but in other cases it may be large. Values of h_f are obtained for specific cases by equations that are more or less empirical and are based upon experimental data.

Laminar flow. Laminar flow may also be called streamline flow, because all particles of the fluid move in distinct and separate lines. It is called laminar flow because the action is as if layers or lamina of fluid slide relative to each other. In the

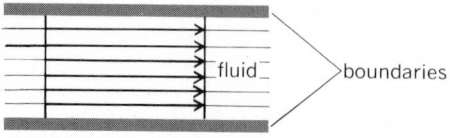

Fig. 3. Ideal fluid flow.

case of a laminar flow in a circular pipe, the velocity adjacent to the wall is zero and increases to a maximum in the center of the pipe (Fig. 4). The velocity profile in a circular pipe is a parabola, and the average velocity is 0.5 times the maximum velocity in the center.

In the case of laminar flow in a circular pipe, the loss of head due to fluid friction is not given by any empirical equation but is given by a rational equation known as the Hagen-Poiseuille law, Eq. (4),

$$h_f = \frac{p_1 - p_2}{w} = 32 \frac{\mu}{w} \frac{L}{D^2} V = 32\nu \frac{L}{gD^2} V \qquad (4)$$

where h_f is linear head in feet of the fluid, p is pressure in lb/ft³, μ is absolute viscosity in lb/ft-sec, w is specific weight in lb/ft³, ν is kinematic viscosity in ft²/sec, L is distance between pressure levels (1) and (2) in ft, D is diameter in ft, and V is velocity in ft/sec.

Laminar flow in a circular pipe will be found when the Reynolds number, $DV\rho/\mu = DV/\nu$, is less than 2000. *See* LAMINAR FLOW.

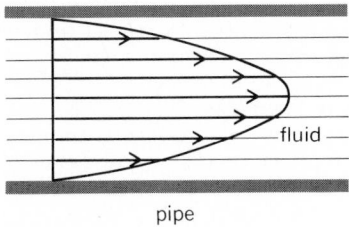

Fig. 4. Laminar flow in a circular pipe.

Turbulent flow. In turbulent flow no distinct streamlines are found. Instead, the fluid consists of a mass of eddies (Fig. 5a). No two particles can follow the same or similar paths (Fig. 5b). The velocity profile shows a maximum velocity in the center, while near the wall the velocity is about one-half the center velocity. The profile is flatter for a smooth pipe than it is for a rough one, and the ratio of average velocity to maximum velocity in the center ranges from about 0.74 for a very rough pipe to about 0.88 for a very smooth one.

In a circular pipe the loss of head due to fluid friction is given by $h_f = f(L/D)V^2/2g$, where f is a function of both Reynolds number and the wall roughness, L/D is the ratio of length to diameter, V is the average velocity in ft/sec, and g is the acceleration of gravity, normally taken as 32.2 ft/sec².

Turbulent flow occurs in a circular pipe when Reynolds number has values greater than 2000.

In the case of a solid body immersed in a stream, there is a turbulent wake in the rear (Fig. 6a and b). This produces a pressure difference which results in a force on the body known as form or pressure drag. *See* TURBULENT FLOW; WAKE FLOW.

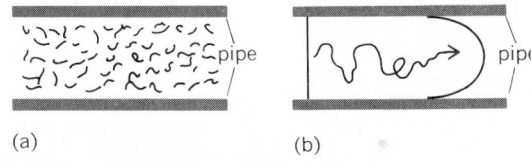

Fig. 5. Turbulent flow in a circular pipe. (a) General turbulence. (b) Path of a single particle.

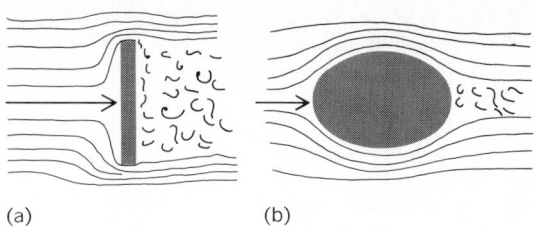

Fig. 6. Turbulent wake of solid body immersed in a stream. (a) Rectangular body. (b) Oval body.

Uniform flow. If at a given instant the velocity is the same in both magnitude and direction at every point in space, the flow is uniform. This strict definition can have little meaning for the flow of a real fluid when the velocity varies across a section. However, when the size and shape of cross section are the same in any given length, the flow is said to be uniform. Specifically, the flow in a pipe of constant diameter is uniform; the flow in a pipe of varying size is not. Also, the flow in an open canal is uniform if the size and shape of the cross section of the stream are the same at different locations along the canal.

Steady flow. By steady flow is meant that all conditions at any one point are constant with respect to time. True steady flow is found only with laminar flow. In turbulent flow there are continual fluctuations in velocity and pressure at every point. However, if the values fluctuate on both sides of an average value that is constant, then the flow is mean steady flow.

In steady flow, conditions are usually constant in time from one section to another, although not necessarily the same at different sections. Thus, along the line of flow the equation of continuity applies, which is written as Eq. (5), or for a fluid of constant specific weight, as Eq. (6).

$$W = w_1 A_1 V_1 = w_2 A_2 V_2 = \text{constant} \qquad (5)$$

$$Q = A_1 V_1 = A_2 V_2 = \text{constant} \qquad (6)$$

Here W is flow in lb/sec, w is specific weight in lb/ft³, A is cross-sectional area in ft² normal to the velocity, V is average velocity across the section in ft/sec, and Q is flow in ft³/sec.

Unsteady flow. This means that conditions are changing with respect to time and ultimately may become either steady flow or zero flow. This changing rate of flow may take place slowly, as when the action of a valve in a channel produces a gradual change in the rate, or it may take place rapidly as a result of a sudden closure, which produces a phenomenon known as water hammer. It is also found in such a case as the flow from one reservoir to another, in which equilibrium is approached as the two levels approach each other.

Unsteady flow also includes periodic motion such as that of waves on beaches, tidal motion in estuaries, and other oscillations. The difference between such cases and mean steady flow in turbulent flow is that the deviations from the mean are much greater and the time scale is also much longer. *See* WAVE MOTION IN LIQUIDS.

Compressible flow. For compressible fluids such as gases and vapors it is necessary to add

FLUID FLOW

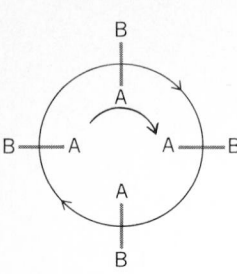

(a)

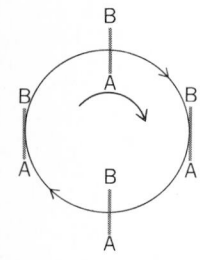

(b)

Fig. 7. Rotational and irrotational flow. (a) Particles rotate on their axes as they move around. (b) Particles retain absolute orientation during motion.

thermal terms to the Bernoulli equation. The energy equation between station 1 and station 2 then becomes Eq. (7), where q is Btu/lb of fluid flowing,

$$\frac{p_1}{w_1} + JI_1 + z_1 + \frac{V_1{}^2}{2g} - Jq = \frac{p_2}{w_2} + JI_2 + z_2 + \frac{V_2{}^2}{2g} \quad (7)$$

which may be transferred from the fluid to the surroundings. If the heat flow is into the fluid, then q is negative, as the equation is written. Internal energy is thermal energy and is represented by I in Btu/lb, and J is 778 ft-lb/Btu. For a perfect gas $\Delta I = c_v \Delta T$ for each unit of weight. For a gas or a vapor the quantity p/w is usually large relative to $z_1 - z_2$, because of the small value of w in general. Therefore the z terms are usually, but not always, negligible.

Because p/w, which equals pv, and I are usually associated for gases and vapors, it is customary to replace them by a single term called enthalpy H, which may be described algebraically as $H = I + pv/J$ in Btu/lb. Therefore the energy equation becomes Eq. (8). See ENTHALPY.

$$JH_1 + \frac{V_1{}^2}{2g} - Jq = JH_2 + \frac{V_2{}^2}{2g} \quad (8)$$

Isothermal flow of a gas. For a perfect gas $I_1 = I_2$ if the temperature is unchanged, and $p_1/w_1 = p_2/w_2$. Hence for this special case Eq. (9) holds, which

$$V_2{}^2 - V_1{}^2 = -2gJq \quad (9)$$

shows that an isothermal flow is accompanied by an absorption of heat when there is an increase in kinetic energy. This equation is true either with or without friction, because friction supplies some of the heat necessary to maintain constant temperature; however, less heat is absorbed from the surrounding medium. Consequently q is less, and the resultant increase in kinetic energy is less.

Adiabatic flow. In adiabatic flow, heat transfer is zero; hence the energy equation becomes Eq. (10),

$$JH_1 + \frac{V_1{}^2}{2g} = JH_2 + \frac{V_2{}^2}{2g} \quad (10)$$

and because $778 \times 2g$ is practically 50,000, this may be written as Eq. (11).

$$V_2{}^2 - V_1{}^2 = 50,000 (H_1 - H_2) \quad (11)$$

For a perfect gas $\Delta H = c_p \Delta T$, where c_p is specific heat per pound at constant pressure. Therefore for gases Eq. (12) holds.

$$V_2{}^2 - V_1{}^2 = 50,000 \, c_p \, (T_1 - T_2) \quad (12)$$

Also for gases $c_p = kR/(k-1)J$, where $k = c_p/c_v$ (ratio of specific heats at constant pressure and at constant volume) and R is the gas constant. Then, because $pv = RT$, Eq. (13) holds.

$$\frac{V_2{}^2 - V_1{}^2}{2g} = \frac{k}{k-1} (p_1 v_1 - p_2 v_2) \quad (13)$$

The preceding equations are valid with or without fluid friction. For a frictionless flow the values of H_2, T_2, and v_2 are all determined as the result of an isentropic expansion from p_1 to p_2. Fluid friction increases the numerical value of these quantities and thus decreases the numerical values of the right-hand sides of the equations. See ISENTROPIC FLOW.

Rotational and irrotational flow. In rotational flow, each minute particle of fluid rotates about its own axis. A specific case is a forced vortex in which a fluid rotates like a solid body. Such a case might be obtained with a fluid in a cylindrical vessel rotating about its central axis, assuming no motion of the fluid relative to the container. A more common case is that in which the container is fitted with vanes, as the impeller of a centrifugal pump. As a small element rotates about the central axis, it also rotates about its own axis (Fig. 7a).

In irrotational flow, each infinitesimal particle or element of the fluid preserves its original orientation (Fig. 7b). Because an element of fluid can be caused to rotate about its axis only by the application of viscous forces, rotational flow is possible only with a real or viscous fluid, whereas irrotational flow is possible only for an ideal or nonviscous fluid. For fluids of low viscosity, such as air or water, irrotational flow may be approached in a free vortex. In a free vortex, a body of fluid rotates without the application of external torque because of angular momentum previously imparted to it. Examples are the rotation of fluid after leaving the impeller of a centrifugal pump, a tornado, or the rotation of water entering the drain of a tub. See VORTEX.

Boundary-layer flow. For an ideal frictionless fluid the velocity of flow adjacent to a surface would be the same as that at a distance. In actuality, adhesion between fluid and boundary surface tends to make the velocity of the fluid at the surface equal to the velocity of the surface body. For a very small distance from the surface, the velocity increases with distance at a very rapid rate because of viscosity within the fluid. The flow in this thin layer is laminar in character. This thin layer is known as the laminar boundary layer.

There is then a transition zone, the boundaries of which are indefinite, and beyond this the flow is fully turbulent. Much farther from the surface, the effect of the surface vanishes and the flow is undisturbed. The layer between the laminar one and the undisturbed field is known as the turbulent boundary layer. It is thick and there is no sharp line of demarcation between it and the undisturbed uniform flow. Such a case may be found with an airplane at a considerable height above the ground or a submarine deeply submerged beneath the surface. In the case of flow in a pipe or other conduit the turbulent boundary layers from opposite sides may meet so that there is no zone of undisturbed flow. Viscosity effects are most pronounced at or near a solid boundary and diminish rapidly with distance from the boundary. See FLUID-FLOW PRINCIPLES; FLUID MECHANICS; GAS DYNAMICS; HYDRODYNAMICS.

[ROBERT L. DAUGHERTY]

Bibliography: J. W. Daily and D. R. F. Harleman, *Fluid Dynamics*, 1966; R. L. Daugherty and J. B. Franzini, *Fluid Mechanics*, 7th ed., 1977; V. L. Streeter, *Handbook of Fluid Dynamics*, 1961; W. L. Streeter and E. B. Wylie, *Fluid Mechanics*, 7th ed., 1979.

Fluid-flow principles

Fundamentals of fluid dynamics that govern motion phenomena. Fluid dynamics may be divided into hydrodynamics and gas dynamics. Hydrodynamics relates to the motion of liquids and gases where density changes are negligible. In general, gases flowing at speeds much less than the speed of sound may be approximated as incompressible. Only the mechanical principles of (1) mass conservation and (2) momentum conservation are necessary to determine hydrodynamic fluid motions. Gas dynamics treats the motion of gases where significant density changes occur, and it requires, in addition to the principles used in hydrodynamics, the principles of thermodynamics, namely, (3) the conservation of energy (first law of thermodynamics), (4) the equation of state of the gas, and in some cases also (5) the no-perpetual-motion-machines law (second law of thermodynamics).

For chemically reacting fluids, flows involving both the liquid and gas phases of a substance, and inhomogeneous mixtures of different fluids, concepts and principles from physical chemistry must be added. For electrically conducting fluids in the presence of electromagnetic fields, electromagnetic body forces occur, requiring the introduction of concepts and principles from electromagnetic theory. In general, only homogeneous nonreacting fluids with zero electrical conductivity are considered here. *See* GAS DYNAMICS; HYDRODYNAMICS; MAGNETOHYDRODYNAMICS; PLASMA PHYSICS; THERMODYNAMIC PRINCIPLES.

Flow continuity. The most common way of describing a fluid-flow field is by giving velocity **v**, pressure p, density ρ, and temperature T as functions of position in space and time.

Mass. The principle of mass conservation states that the rate of increase of fluid mass in any volume V, fixed in space, equals the net rate of mass flow into this volume. If S is the surface of this volume with **n** a unit vector normal to the surface, this principle is stated analytically as Eq. (1). By the

$$\frac{\partial}{\partial t}\int\int\int_V \rho\,dV = -\int\int_S \rho\mathbf{v}\cdot\mathbf{n}\,dS \qquad (1)$$

use of Gauss' divergence theorem the surface integral can be written as a volume integral as in Eq. (2), which implies that Eq. (3) holds for any

$$\int\int_S \rho\mathbf{v}\cdot\mathbf{n}\,dS = \int\int\int_V \boldsymbol{\nabla}\cdot(\rho\mathbf{v})\,dV \qquad (2)$$

$$\int\int_V\int\left[\frac{\partial\rho}{\partial t} + \boldsymbol{\nabla}\cdot(\rho\mathbf{v})\right]dV = 0 \qquad (3)$$

volume V. This in turn implies that the integrand itself must vanish everywhere so that Eq. (4) holds. This is called the equation of continuity.

$$\frac{\partial\rho}{\partial t} + \boldsymbol{\nabla}\cdot(\rho\mathbf{v}) = 0 \qquad (4)$$

Momentum. The principle of momentum conservation is a statement of Newton's particle laws applied to the many-particle fluid system. The rate of increase of momentum in a fixed volume must equal the net rate at which momentum is convected into the volume plus the forces acting on the volume. The analytical statement for an inviscid fluid is given in Eq. (5), where **F** is body force per

$$\rho\frac{\partial\mathbf{v}}{\partial t} + \rho(\mathbf{v}\cdot\boldsymbol{\nabla})\mathbf{v} = -\nabla p + \rho\mathbf{F} \qquad (5)$$

unit mass. For irrotational hydrodynamic flows it is necessary only to find a velocity field such that the flow passes around solid boundaries and the pressure is determined post facto by Bernoulli's theorem. *See* BERNOULLI'S THEOREM; EULER'S MOMENTUM THEOREM; NAVIER-STOKES EQUATIONS.

Energy conservation. The principle of energy conservation (first law of thermodynamics) brings in the concepts of temperature, heat, and internal energy. Applied to a fixed volume in the fluid flow it states that the rate of increase of internal energy in any fixed volume in space is equal to the net rate of flow of internal energy into the volume, plus the net rate of heat flow into the volume, plus the rate at which work is being done on the volume. In general, it is only near the boundaries of solid objects in the fluid flow or in shock waves that the heat conductivity and viscosity of fluids must be considered, for these are the regions where high temperature and velocity gradients occur. *See* BOUNDARY-LAYER FLOW; SHOCK WAVE.

Elsewhere in the fluid flow, the changes of state of the fluid particles are closely adiabatic and reversible. If all fluid particles originate from a zone of uniform state, that part of the fluid flow that does not pass through shock waves or enter into boundary layers is isentropic. Many of the flows encountered in gas dynamics are well approximated as isentropic flows. In boundary layers and shock waves entropy production occurs through the processes of viscous dissipation (work done by viscous shear forces) and heat addition by conduction, which are phenomena outside the considerations of classical thermodynamics. *See* ISENTROPIC FLOW.

As mentioned previously, considerations of internal energy and heat are, in general, not necessary to determine hydrodynamic flow fields. The reason for this is that the fluid kinetic energy per unit mass is so small compared to the internal energy per unit mass that the internal energy remains nearly constant, even if the whole kinetic energy is transformed into heat. On the other hand, transfer of heat by a fluid necessarily involves the use of the energy conservation principle and the finite heat conductivity of the fluid. In forced-convection heat transfer at low speeds, the temperature field can be determined from the velocity field but does not affect the velocity field. Natural-convection heat transfer involves buoyant forces due to density changes and yet involves speeds much less than sound speed; hence it is an exception to the earlier statement that low-speed velocity fields are determined only by mass and momentum considerations. *See* HEAT TRANSFER.

The thermal equation of state of a fluid is a relationship among the variables, pressure p, density ρ, and temperature T, which is valid when the fluid is in thermodynamic equilibrium. Similarly, the caloric equation of state relates the internal energy e to ρ and T. These relationships are derived from experiments or, where possible, statistical mechanics or kinetic theory. The assumption of local thermodynamic equilibrium in the fluid is usually sufficiently accurate, although changes from one state to another necessarily involve small relaxation times which, in some cases, may not be negligible, for instance, the flow of a gas such as carbon dioxide through a shock wave. *See* KINETIC THEORY OF MATTER.

The second law of thermodynamics introduces the concept of entropy and is necessary in some gas-dynamic flows to determine the direction in which certain flow changes can or cannot occur. For example, the flow through a stationary normal shock wave appears to be possible both from subsonic to supersonic and from supersonic to subsonic until the entropy changes are considered; then it is clear that the subsonic to supersonic shock cannot occur spontaneously in nature. The second law is also the basis of the law of mass action, which is essential in determining equilibrium concentrations of various gases in a mixture of gases; a partially dissociated gas is such a mixture (molecular gas and atomic gas), and an electrically neutral, partially ionized gas or plasma is a mixture of atomic species, ions, and electrons. *See* ENTROPY.

[ARTHUR E. BRYSON]

Fluid mechanics

The science concerned with fluids, either at rest or in motion. It deals with pressures, velocities, and accelerations in the fluid, including fluid deformation and compression or expansion. Fluid mechanics may be divided into two branches, fluid statics and fluid dynamics; the first deals with pressure intensities and forces exerted by a fluid at rest, and the second with forces exerted on fluids and their resulting motions. *See* FLUID DYNAMICS; FLUID STATICS.

The laws of fluid mechanics control a great portion of natural phenomena. The flight of an insect or bird, the motion of a fish through water, the relative movement of air masses as in frontal weather systems, and the eruption of a volcano are examples of flow that follow the laws of fluid mechanics. The science of fluid mechanics is involved in many phases of aeronautical, chemical, civil, and mechanical engineering. It requires the combination of theoretical analysis and orderly experimentation. In the theoretical approach, assumptions are made as needed to keep the resulting mathematical expressions manageable, such as assuming an incompressible fluid, or one without viscosity. The experimental work must be planned in such a way that similitude relations are used to a maximum advantage.

Definitions. Flow may be classified in many ways, such as turbulent, laminar; real, ideal; isothermal, isentropic; steady, unsteady; and uniform or nonuniform.

Turbulent-flow situations are most prevalent in engineering practice. The fluid particles move in irregular paths, causing an exchange of momentum from one portion of fluid to another. The fluid regions involved in the transfer of momentum due to turbulence can range in size from large-scale turbulence with many cubic miles of fluid participating in a single tornadic eddy to a small-scale turbulence with only a few thousand molecules interacting. Turbulent flow causes the conversion of mechanical energy into thermal energy at a rate varying roughly as the square of the velocity. *See* TURBULENT FLOW.

In laminar flow, particles move along smooth paths in layers, or laminae, with one layer gliding over an adjacent one. Laminar flow is governed by Newton's law of viscosity. It may be considered as

flow in which all turbulence has been damped out by the action of viscosity. *See* NEWTONIAN FLUID.

A real fluid always has viscosity and, whether a liquid or a gas, has compressibility. An ideal fluid is considered to be frictionless (nonviscous) and incompressible. There is no means by which an ideal fluid can convert mechanical energy into thermal energy. When ideal fluid particles have no rotation, the fluid is irrotational and a velocity potential exists.

When gas flows without change in temperature, the flow is isothermal. When flow occurs such that no heat is added or subtracted at the boundaries, the flow is adiabatic. Reversible adiabatic (frictionless adiabatic) flow is called isentropic flow. *See* ISENTROPIC FLOW.

Steady flow occurs when conditions at any point in the fluid do not change with time. In unsteady flow one or more quantities, such as density or velocity, change with time at a point. Uniform flow occurs when the velocity vector throughout the fluid is everywhere the same at any instant, and nonuniform flow occurs when the velocity vector has varying values throughout the fluid at any instant. The strict definitions of steady flow and uniform flow are relaxed in practical flow situations, due to turbulent fluctuations in the first case and to variations over a cross section in the second case. For example, liquid flow through a long pipe at constant rate is steady-uniform flow; liquid flow through a long pipe at a decreasing rate is unsteady-uniform flow; flow through an expanding tube at a constant rate is steady-nonuniform flow; and flow through an expanding tube at an increasing rate is unsteady-nonuniform flow.

A streamline is a continuous line through the fluid that has the direction of the local velocity vector at every point. A stream tube is composed of all streamlines through a small closed curve.

Basic equations of fluid flow. In any fluid-flow situation, three conditions exist: (1) Newton's laws of motion hold for every particle at every instant; (2) the continuity relationship holds, that is, net mass inflow into any small volume per unit time equals its time rate of increase of mass; (3) at a boundary, the velocity component normal to the boundary equals the velocity component of the boundary normal to itself. For real fluids, in addition, the tangential component of fluid velocity at the boundary is zero relative to the boundary. *See* BOUNDARY-LAYER FLOW; EULER'S MOMENTUM THEOREM; HYDRODYNAMICS.

Integration of Newton's second law of motion may lead either to the Bernoulli equation or to the momentum equation. By considering the steady flow of a frictionless, incompressible fluid, the Bernoulli equation states that the mechanical energy remains constant along a streamline. By considering steady, irrotational flow of a frictionless incompressible fluid, a form of the Bernoulli equation is obtained that shows that the energy is constant everywhere throughout the fluid. For steady flow the momentum equation states that the resultant force acting on any free body of fluid is just equal to its time rate of change of momentum or, for a fixed control volume, that the resultant force equals the difference between the momentum per unit time leaving and the momentum per unit time entering. *See* BERNOULLI'S THEOREM.

The continuity equation for steady flow states

that the mass per unit time flowing along a stream tube is everywhere constant.

Use of the basic equations permits many fluid-flow situations to be analyzed, provided that energy losses are small and can be neglected. In certain special situations, application of the continuity, Bernoulli, and momentum equations permits the energy losses to be approximately computed. *See* SHOCK WAVE.

For the vast majority of real fluid-flow situations, experimental information is required to determine the amount of mechanical energy converted to thermal energy. The effect of losses in steady flow of a fluid along a streamline may be expressed by including one or more loss terms in the energy equation. The equation states that the energy per unit weight at one point is equal to the energy remaining at a downstream point plus all the losses between the two points. For example, in turbulent pipe flow the losses due to wall friction tend to vary almost directly as the length of pipe, inversely as the diameter, and directly as the square of the velocity. Fluid properties and condition of the pipe-wall surface enter in a more complicated manner. Experimentally the losses may be determined and expressed in a dimensionless form by a chart so that the results apply to other fluids and to other sizes of geometrically similar pipe. Losses in pipe flow due to changes in cross section or direction, such as elbows and valves, tend to vary about as the square of the velocity, or as a constant times the kinetic energy per unit weight.

Similitude. The performance of hydraulic structures or fluid machines may be studied by utilizing similitude relations. Two systems are said to be dynamically similar if (1) they are geometrically similar, (2) they have the same boundary conditions, and (3) either their streamline configurations are geometrically similar or their dynamic pressure ratios at corresponding points are the same.

The particular method of interpreting the data from model studies depends upon the types of forces that predominate. For example, with hydraulic structures such as dams, spillways, and canal transitions, the important forces are those due to gravity and inertia of the liquid, with viscous forces of lesser importance. The ratio of inertial forces to gravity forces is a dimensionless parameter, known as the Froude number. By adjusting flows and depths so that the Froude number is the same in model and prototype, measurements in the model can be made and converted to corresponding prototype values. If the ratio of linear dimensions of the prototype to the model is λ, then the velocity in the prototype is $\sqrt{\lambda}$ times the corresponding velocity in the model, and (for the same liquids in model and prototype) the pressure intensity in the prototype is λ times the corresponding pressure intensity in the model. *See* DYNAMIC SIMILARITY.

Model studies of fluid machinery. In making a test on a model of a turbomachine, special relationships must be observed. For geometrically similar streamlines, the ratio of velocity of flow at some point in the machine to the peripheral velocity of the runner or rotor must be the same in model and prototype. Also, since inertial forces are of great importance in the machine, there must be a definite relation between head on the machine and velocity at some point within the machine, this re-

lation being the same for model and prototype. In addition, for compressible flow the ratio of velocity of fluid to local acoustic velocity must be the same at corresponding points in model and prototype. The above conditions, together with geometric similitude, permit tests on a model to be used to predict performance of the prototype. The above relations, referred to as homologous relationships, do not permit viscous forces to be scaled properly; hence there is a slight difference in efficiency of the various sizes. The larger the machine, the more efficient it is, but with the change being usually not more than 2 or 3%.

In flow through pipes and other closed conduits, the controlling forces are inertial and viscous for velocities that are small compared with acoustic velocity. The ratio of inertial to viscous forces is expressed by a dimensionless parameter known as Reynolds number. When two geometrically similar closed-flow systems have the same Reynolds number at corresponding points, the dimensionless flow and loss coefficients will be the same. When closed-channel flow occurs at velocities near acoustic velocity, the Mach number (ratio of velocity to acoustic velocity) becomes a controlling parameter. *See* REYNOLDS NUMBER.

[VICTOR L. STREETER]

Bibliography: R. W. Fox and A. T. McDonald, *Introduction to Fluid Mechanics*, 2d ed., 1978; A. J. H. Rouse (ed.), *Advanced Mechanics of Fluids*, 1959, reprint 1976; V. L. Streeter, and E. B. Wylie, *Fluid Mechanics*, 7th ed., 1979; J. K. Vennard and R. L. Street, *Elementary Fluid Mechanics*, 5th ed., 1975.

Fluid statics

The determination of pressure intensities and forces exerted by liquids and gases at rest. Hydrostatics, although implying the statics of water alone, applies to liquids in general. By definition, a fluid at rest cannot sustain a shear stress; therefore, a fluid force exerted on an element of boundary area must act normal to the area. For example, consider the small free body of fluid shown in the illustration. Equilibrium requires that $dp = -\gamma\,dz$, in which p is the absolute pressure, γ is the specific weight (weight of a unit volume of fluid), and z is the elevation, measured vertically upward. For liquids, γ is substantially constant, and the equation shows that pressure decreases linearly as the elevation increases. *See* HYDROSTATICS.

With gases, the variation of γ with pressure or elevation must be known in order to integrate the equation. For an isothermal gas ($p/\gamma = p_0/\gamma_0 =$ constant), the pressure variation with elevation is given by Eq. (1), in which p_0 and γ_0 are the values of p and γ at elevation z_0.

$$p = p_0 \exp\left[-(z - z_0)/(p_0/\gamma_0)\right] \qquad (1)$$

On the average, the absolute temperature T of the atmosphere tends to decrease linearly with elevation within the troposphere, $T = T_0 - \beta z$. By using the general gas law $p/\gamma = RT$, in which R is the gas constant, the specific weight is given by Eq. (2). Use of this expression for γ in terms of p

$$\gamma = p/R(T_0 - \beta z) \qquad (2)$$

and z yields the variation of pressure with elevation in a standard atmosphere, as in Eq. (3), in which p_0 and T_0 are the values at $z = 0$.

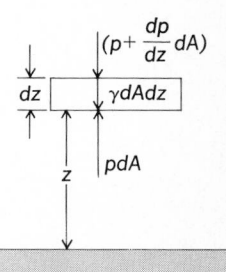

Free-body diagram for vertical forces acting on fluid element.

$$p = p_0(1 - \beta z/T_0)^{1/R\beta} \qquad (3)$$

The determination of fluid pressure is of importance in many flow-measuring devices. The manometer is one method which makes use of transparent tubes with one or more liquids contained in them. By measuring the difference in elevation of the fluid menisci, the desired pressure is determined from the laws of fluid statics.

Fluid forces on plane surfaces may be determined by integration of $p\,dA$ over the surface, with p the pressure intensity and dA an element of the surface area. For liquids, the magnitude of the force is the product of the area and the pressure at the centroid of the area. The line of action of the resultant force is normal to the surface and acts at a point termed the center of pressure. For horizontal submerged surfaces, the centroid of the area and the center of pressure coincide, but for all other orientations of the surface, the center of pressure is below the centroid (at less elevation).

To determine the resultant fluid force exerted on a curved surface, it is convenient to consider horizontal and vertical components of the force. The horizontal component of force exerted on a curved surface is equal to the force exerted on a projection of the surface onto a plane normal to the direction of the component. The line of action for the horizontal force will be through the center of pressure of the projected area. The vertical component of force exerted on a curved surface by the fluid pressure equals the weight of liquid vertically above the curved surface, and acts through the centroid of this volume of liquid. The resultant force is thus the sum of three vectors, two horizontal components at right angles and the vertical component. *See* Archimedes' principle; Fluid mechanics.

[VICTOR L. STREETER]

Fluids

Substances with no reference configuration of permanent significance. Aggregates of matter in which the molecules are able to flow past each other without limit and without fracture planes forming are usually classified as fluids. The subdivisions of fluids known as gases, vapors, and liquids, each of which exhibits successively closer association of the molecules and is distinguished by different thermodynamic and mechanical properties, compose a group of easily distinguishable fluids, some having mainly Newtonian and the rest mainly non-Newtonian flow properties. *See* Non-Newtonian fluid.

Solid aggregates of matter, whether continuous or divided in the form of powders, can usually be distinguished from fluids on the molecular scale in that long-term and long-range order are apparent in the structure.

There are some substances, the semisolids, which appear to be able to flow without fracture and yet which also have some of the attributes of solids such as the ability to form free-standing figures. Butter in a temperate environment is an excellent example of a material with this dual nature.

Similarly, some fluids possess elastic properties, the so-called viscoelastic fluids. Such substances are not easy to classify in one or another

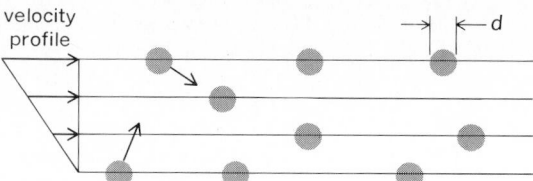

Fig. 1. Maxwell's molecular system for gases.

category, and their properties are not well understood in depth.

Gases. Gases at low density are quite amenable to the theoretical treatment of J. C. Maxwell in which the molecules are visualized as rigid, nonattracting spheres in relative motion, as shown in Fig. 1.

The theoretical expression for the viscosity μ of a gas, derived from considerations of momentum exchange between the spheres in different planes, is given by Eq. (1), where m is the mass of

$$\mu = \frac{2}{3\pi^{3/2}} \frac{\sqrt{mkT}}{d} \qquad (1)$$

one molecular sphere, k is the Boltzmann constant, T is the absolute temperature, and d is the sphere diameter.

An important consequence of Eq. (1) is that the viscosity of a low-density gas is proportional to the square root of the absolute temperature but is independent of pressure. This is radically different from the behavior of liquids.

As condensation conditions are approached, the viscosity starts to differ significantly from that predicted by Eq. (1). The temperature exponent of 0.5 increases to between 0.6 and 1.0, and the gas viscosity is *no longer independent of pressure*. The general behavior of the viscosity as the temperature and pressure are varied is shown in Fig. 2. The viscosity of gas mixtures is generally not simply related to the viscosity of the components and the mole fractions present. *See* Gas.

Liquids. An approximate theory which enables the viscosity of liquids to be estimated from other physical properties is based on an application of the theory of rate processes developed by E. Eyring and coworkers. In this theory it is assumed that migration of liquid molecules occurs by their achieving enough vibrational energy to surmount the barrier surrounding their current site. *See* Liquid.

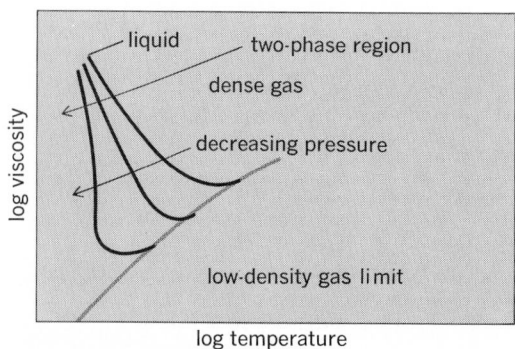

Fig. 2. The effect of varying temperature and pressure upon the fluid viscosity.

In a state of rest the energy barrier is assumed symmetrical, but under the action of applied stress the barrier is distorted so that the molecule is biased toward jumping in the forward direction (Fig. 3). The relationship between the shear stress τ and the shear rate $\dot{\gamma}$ derived by Eyring and co-workers is given by Eq. (2), where ΔG_0 is a free

$$\dot{\gamma} = \frac{a}{\delta} \left(\frac{kT}{h} \exp -\frac{\Delta G_0}{RT} \right) \left(2 \sinh \frac{a\tau\bar{v}}{2\delta RT} \right) \quad (2)$$

energy of activation indicated in Fig. 3, h is Planck's constant, k is the Boltzmann constant, $\bar{v}$ is the molar volume, R is the gas constant, T is the absolute temperature, and a and δ are molar spacings defined in Fig. 3.

Note that generally the shear stress – shear rate of relation (2) is non-Newtonian. For fluids of *small molecular weight* it is known from the amassed experimental data that Newton's viscous law, Eq. (3), is generally appropriate for all isothermal

$$\tau = \mu \dot{\gamma} \quad (3)$$

flows. The exceptional condition of periodic stresses of megahertz frequency may find Eq. (3) inadequate even for small molecules.

The Eyring formula, Eq. (2), predicts that the stress will be in phase with the shear rate, *but this is not generally found in non-Newtonian flow.*

The temperature dependence of the viscosity can often be described over limited temperature ranges by the exponential relation shown in Eq. (4).

$$\mu = A \exp \frac{E}{RT} \quad (4)$$

The effect of high pressure can also often be described by the same type of relation, with pressure replacing temperature in Eq. (4). This is an important consideration in the field of lubrication.

Particle suspensions. The addition of rigid, non-interacting spheres to a Newtonian fluid increases the viscosity according to the Einstein equation, Eq. (5), where c is the volume concentration of the

$$\mu = \mu' (1 + 2.5c) \quad (5)$$

spheres, μ' is the viscosity of the suspending

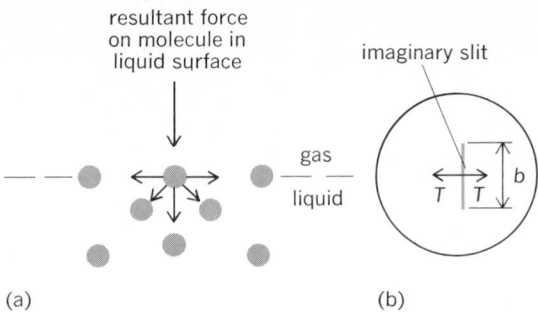

(a)   (b)

Fig. 4.   Surface tension. (a) Forces (attractive on one hemisphere only) on a liquid molecule at a gas-liquid interface. (b) Tensile force per unit length of slit.

liquid, and μ is the viscosity of the suspension. For concentrated solutions a power series in the volume concentration is required.

Fluid surfaces. A molecule of liquid in a gas-liquid interface finds itself in a different force field from those in the core of the gas or the liquid. This results from the surface molecule having attractive forces only on one hemisphere, as shown in Fig. 4a. The resultant effect is that the liquid surface at any interface appears to be in a state of tension, like the surface of an inflated balloon.

An imaginary slit in the surface of Fig. 4b would have a tensile force per unit length, given by Eq. (6).

$$t = \frac{T}{b} \quad (6)$$

The well-known capillary rise in a fine bore tube (Fig. 5) results from surface tension effects and is a common method for measuring surface tension from Eq. (7), where ρ is the liquid density, g is the

$$t = \frac{\rho g H r}{2 \cos \theta} \quad (7)$$

gravitational acceleration, H is the capillary meniscus elevation, and r is the capillary tube radius.

Equation (7) yields a static surface tension, but if there is shearing in the surface, then surface viscosity will be exhibited as a two-dimensional analog of bulk viscosity. Furthermore, if macromolecules are present in the surface, shear elasticity will exhibit itself as a surface phenomenon.

Diffusion. The tendency for a substance to achieve uniform distribution throughout the space available to it is known as diffusion. It is exhibited by all classes of matter, whether gas, liquid, or solid, but is most vigorous in the case of gases. The basic demonstration is the placing together of the mouths of two jars, one containing a heavy gas, say nitrogen which is placed in the bottom jar, and the other a light gas, say hydrogen which is placed in the top jar. After a time has elapsed, the hydrogen and nitrogen will be found to be evenly distributed throughout the two jars.

Graham's law of diffusion, which is an approximation, states that the rate of diffusion of a gas is inversely proportional to the square root of the density, as in expression (8). Diffusion separation is

$$D \propto \left(\frac{1}{\rho} \right) \quad (8)$$

FLUIDS

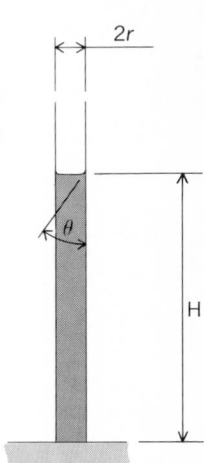

Fig. 5.   Meniscus rise due to surface tension.

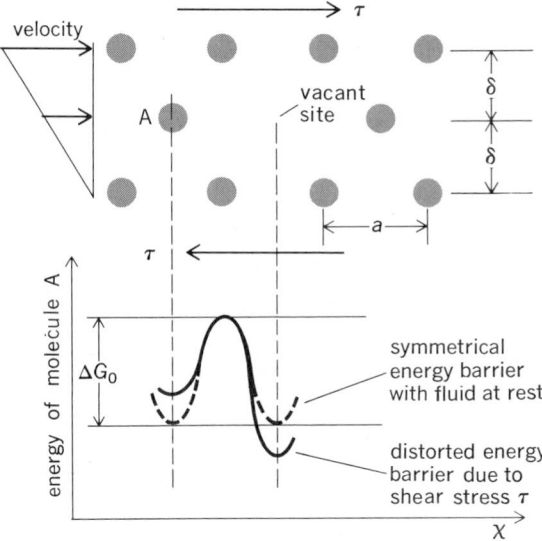

Fig. 3.   Eyring molecular system for liquids.

an important process in the field of chemical engineering.　[J. HARRIS; W. L. WILKINSON]

Bibliography: R. A. Alberty and F. Daniels, *Physical Chemistry*, 5th ed., 1979; G. Barrow, *Physical Chemistry*, 4th ed., 1979.

Fluorescence

Fluorescence is generally defined as a luminescence emission that is caused by the flow of some form of energy into the emitting body, this emission ceasing abruptly when the exciting energy is shut off. In attempts to make this definition more meaningful it is often stated, somewhat arbitrarily, that the decay time, or afterglow, of the emission must be of the order of the natural lifetime for allowed radiative transitions in an atom or a molecule, which is about 10^{-8} s for transitions involving visible light. Perhaps a better distinction between fluorescence and its counterpart, phosphorescence, rests not on the magnitude of the decay time per se, but on the criterion that the fluorescence decay is temperature-independent. If this latter criterion is adopted, the luminescence emission from such materials as the uranyl compounds and rare-earth-activated solids would be called slow fluorescence rather than phosphorescence. The decay of their luminescence takes place in milliseconds to seconds, rather than in 10^{-8} s, showing that the optical transitions are somewhat "forbidden"; but the decay is temperature-independent over a considerable range of temperature, and it follows an exponential decay law, shown in the equation below, that is to be expected for sponta-

$$I = I_0 \exp\left(-t/\tau_1\right)$$

neous transitions of electrons from an excited state of an atom to the ground state when the atom has a transition probability per unit time $1/\tau_1$. (τ_1 is called the natural radiative lifetime or fluorescence lifetime). In this equation I is the luminescence intensity at a time t, and I_0 is the intensity when $t = 0$. *See* SELECTION RULES.

In applying this criterion, one should take note of the following restriction which arises because all luminescent systems ultimately lose efficiency or are "quenched" at elevated temperatures, each system having its own characteristic temperature for the onset of this so-called thermal quenching. Quenching sets in because increase of temperature makes available other competing atomic or molecular transitions that can depopulate the excited state and dissipate the excitation energy nonradiatively. Hence, even for a fluorescence, a temperature dependence of decay time will be observed at temperatures where thermal quenching becomes operative, because $1/\tau_{obs} = 1/\tau_1 + 1/\tau_{diss}$, where τ_{obs} is the observed fluorescence lifetime and $1/\tau_{diss}$ is the probability for a competing dissipative transition.

In the literature of organic luminescence, the term fluorescence is used exclusively to denote a luminescence which occurs when a molecule makes an allowed optical transition. Luminescence with a longer exponential decay time, corresponding to an optically forbidden transition, is called phosphorescence, and it has a different spectral distribution from the fluorescence. *See* PHOSPHORESCENCE.

The decay time of fluorescent materials varies widely, from the order of 5×10^{-9} s for many organ-

ic crystalline materials up to 2 s for the europium-activated strontium silicate phosphor. Fluorescent materials with decay times between 10^{-9} and 10^{-7} s are used to detect and measure high-energy radiations, such as x-rays and gamma rays, and high-energy particles such as alpha particles, beta particles, and neutrons. These agents produce light flashes (scintillations) in certain crystalline solids, in solutions of many polynuclear aromatic hydrocarbons, or in plastics impregnated with these hydrocarbons. *See* SCINTILLATION COUNTER.

The so-called fluorescent lamps employ the luminescence of gases and solids in combination to produce visible light. A fluorescent lamp consists of a glass tube filled with a low-pressure mixture of argon gas and mercury vapor, coated on the inside surface with a luminescent powder or blend of such powders, and having an electrode at each end. An electrical discharge is passed through the gas between the two electrodes, exciting the mercury atoms to luminesce. This results in the emission of both visible and ultraviolet light. At the low pressures employed, approximately half of the electrical energy input to the lamp is converted into the 253.7-nm radiation characteristic of the mercury atom. The phosphor coating is chosen for the efficiency with which it is excited by this wavelength of ultraviolet and for the color of visible luminescence that is desired. The small amount of visible light generated by the discharge itself is largely transmitted by the phosphor coating and adds slightly to the luminous output of the lamp. Luminous efficiencies of fluorescent lamps are considerably higher than those of incandescent lamps, which unavoidably convert into heat the major portion of the electrical energy supplied to them. *See* ABSORPTION OF ELECTROMAGNETIC RADIATION; LUMINESCENCE.

[JAMES H. SCHULMAN; CLIFFORD C. KLICK]

Focal length

A measure of the collecting or diverging power of a lens or an optical system. Focal length, usually designated f' in formulas, is measured by the distance of the focal point (the point where the image of a parallel entering bundle of light rays is formed) from the lens, or more exactly by the distance from the principal point to the focal point. *See* GEOMETRICAL OPTICS.

The power of a lens system is equal to n'/f', where n' is the refractive index in the image space (n' is usually equal to unity). A lens of zero power is said to be afocal. Telescopes are afocal lens systems. *See* LENS.

[MAX HERZBERGER]

Force

Force may be briefly described as that influence on a body which causes it to accelerate. In this way, force is defined through Newton's second law of motion.

This law states in part that the acceleration of a body is proportional to the resultant force exerted on the body and is inversely proportional to the mass of the body. An alternative procedure is to try to formulate a definition in terms of a standard force, for example, that necessary to stretch a particular spring a certain amount, or the gravitational attraction which the Earth exerts on a standard object. Even so, Newton's second law inextricably links mass and force. *See* ACCELERATION; MASS.

Many elementary books in physics seem to expect the beginning student to bring to his study the same kind of intuitive notion concerning force which Isaac Newton possessed. One readily thinks of an object's weight, or of pushing it or pulling it, and from this one gains a "feeling" for force. Such intuition, while undeniably helpful, is hardly an adequate foundation for the quantitative science of mechanics.

Newton's dilemma in logic, which did not trouble him greatly, was that, in stating his second law as a relation between certain physical quantities, he presumably needed to begin with their definitions. But he did not actually have definitions of both mass and force which were independent of the second law. The procedure which today seems most free of pitfalls in logic is in fact to use Newton's second law as a defining relation.

First, one supposes length to be defined in terms of the distance between marks on a standard object, or perhaps in terms of the wavelength of a particular spectral line. Time can be supposed similarly related to the period of a standard motion (for example, the rotation of the Earth about the Sun, the oscillations of the balance wheel of a clock, or perhaps a particular vibration of a molecule). Although applying these definitions to actual measurements may be a practical matter requiring some effort, a reasonably logical definition of velocity and acceleration, as the first and second time derivatives of vector displacement, follows readily in principle.

Absolute standards. Having chosen a unit for length and a unit for time, one may then select a standard particle or object. At this juncture one may choose either the absolute or the gravitational approach. In the so-called absolute systems of units, it is said that the standard object has a mass of one unit. Then the second law of Newton defines unit force as that force which gives unit acceleration to the unit mass. Any other mass may in principle be compared with the standard mass (m) by subjecting it to unit force and measuring the acceleration ($\mathbf{a}$), with which it varies inversely. By suitable appeal to experiment, it is possible to conclude that masses are scalar quantities and that forces are vector quantities which may be superimposed or resolved by the rules of vector addition and resolution.

In the absolute scheme, then, Eq. (1) is written

$$\mathbf{F} = m\mathbf{a} \qquad (1)$$

for nonrelativistic mechanics; here boldface type denotes vector quantities. The quantities on the right of Eq. (1) are previously known, and this statement of the second law of Newton is in fact the definition of force. In the absolute system, mass is taken as a fundamental quantity and force is a derived unit of dimensions MLT^{-2} ($M=$ mass, $L=$ length, $T=$ time).

Gravitational standards. The gravitational system of units uses the attraction of the Earth for the standard object as the standard force. Newton's second law still couples force and mass, but since force is here taken as the fundamental quantity, mass becomes the derived factor of proportionality between force and the acceleration it produces. In particular, the standard force (the Earth's attraction for the standard object) produces in free fall what one measures as the gravitational acceleration, a vector quantity proportional to the standard force (weight) for any object. It follows from the use of Newton's second law as a defining relation that the mass of that object is $m = w/g$, with g the magnitude of the gravitational acceleration and w the magnitude of the weight. The derived quantity mass has dimensions FT^2L^{-1}.

Because the gravitational acceleration varies slightly over the surface of the Earth, it may be objected that the force standard will also vary. This may be avoided by specifying a point on the Earth's surface at which the standard object has standard weight. In principle, then, the gravitational system becomes no less absolute than the so-called absolute system.

Composition of forces. By experiment one finds that two forces of, for example, 3 units and 4 units acting at right angles to one another at point 0 produce an acceleration of a particular object which is identical to that produced by a single 5-unit force inclined at arccos 0.6 to the 3-unit force, and arccos 0.8 to the 4-unit force (see figure). The laws of vector addition thus apply to the superposition of forces.

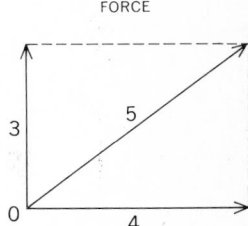

Vector addition of forces.

Conversely, a single force may be considered as equivalent to two or more forces whose vector sum equals the single force. In this way one may select the component of a particular force which may be especially relevant to the physical problem. An example is the component of a railroad car's weight along the direction of the track on a hill.

Statics is the branch of mechanics which treats forces in nonaccelerated systems. Hence, the resultant of all forces is zero, and critical problems are the determination of the component forces on the object or its structural parts in static equilibrium. Practical questions concern the ability of structural members to support the forces or tensions. *See* STATICS.

Specially designated forces. If a force is defined for every point of a region and if this so-called vector field is irrotational, the force is designated conservative. Physically, it is shown in the development of mechanics that this property requires that the work done by this force field on a particle traversing a closed path is zero. Mathematically, such a force field can be shown to be expressible as the (conventionally negative) gradient of a scalar function of position V, Eq. (2).

$$\mathbf{F} = -\nabla V \qquad (2)$$

A force which extracts energy irreversibly from a mechanical system is called dissipative, or nonconservative. Familiar examples are frictional forces, including those of air resistance. Dissipative forces are of great practical interest, although they are often very difficult to take into account precisely in phenomena of mechanics.

The force which must be directed toward the center of curvature to cause a particle to move in a curved path is called centripetal force. For example, if one rotates a stone on the end of a string, the force with which the string pulls radially inward on the stone is centripetal force. The reaction to centripetal force (namely, the force of the stone on the string) is called centrifugal force. *See* CENTRIFUGAL FORCE; CENTRIPETAL FORCE.

Methods of measuring forces. Direct force measurements in mechanics usually reduce ultimately to a weight comparison. Even when the elastic distortion of a spring or of a torsion fiber is used, the calibration of the elastic property will often be through a balance which compares the pull of the spring with a calibrated weight or the torsion of the fiber with a torque arising from a calibrated weight on a moment arm.

In dynamic systems, any means of measuring acceleration—for example, through photographic methods or radar tracking—allows one to calculate the force acting on an object of known mass.

Units of force. In addition to use of the absolute or the gravitational approach, one must contend with two sets of standard objects and lengths, the British and the metric standards. All systems use the second as the unit of time. In the metric absolute system, the units of force are the newton and the dyne. The newton, the unit of force in the International System (SI), is that force which, when applied to a body having a mass of 1 kilogram, gives it an acceleration of 1 m/s². The poundal is the force unit in the British absolute system, whereas the British gravitational system uses the pound. Metric gravitational systems are rarely used. Occasionally one encounters terms such as gram-force or kilogram-force, but no corresponding mass unit has been named. *See* UNITS OF MEASUREMENT.   [GEORGE E. PAKE]

Bibliography: F. Bueche, *Principles of Physics*, 3d ed., 1977; M. L. Bullock, Systems of units in mechanics: A summary, *Amer. J. Phys.*, 22:291–299, 1954; C. Kittel, W. D. Knight, and M. A. Ruderman, *Berkeley Physics Course*, vol. 1: *Mechanics*, 2d ed., 1973; R. Resnick and D. Halliday, *Physics*, 3d ed., 1977.

Forced oscillation

An oscillation produced in a simple oscillator or equivalent mechanical system by an external periodic driving force. Forced oscillations are to be contrasted with free oscillations which occur when the system is displaced from equilibrium and released. *See* HARMONIC MOTION; MECHANICAL VIBRATION; OSCILLATION; VIBRATION.

Application of the driving force results at first in two simultaneous oscillations, one at the frequency of the free oscillations, and one at the frequency of the impressed driving force. The former, the so-called transient, eventually decays, whereas the latter builds up to the steady-state value. Small damping is associated with slow decay of the transient and slow establishment of the steady state, whereas large damping is associated with fast decay. *See* DAMPING.

Consider a simple oscillator having mass m, stiffness (ratio of restoring force to displacement) s, and mechanical resistance (the negative of the ratio of resisting force to velocity) R. The undamped (that is, calculated on the basis $R = 0$) natural angular frequency is $\omega_0 = (s/m)^{1/2}$.

If a sinusoidal driving force of amplitude F and angular frequency $\omega = 2\pi f$ is applied, then after the steady state is reached, the velocity amplitude is given by Eq. (1), where Z is the complex me-

$$V = F/|Z| \tag{1}$$

chanical impedance and the vertical bars denote absolute value, or magnitude. The quantity Z depends on m, R, and s. *See* MECHANICAL IMPEDANCE.

The maximum velocity amplitude occurs for $\omega = \omega_0$ and is given by Eq. (2). The ratio of V to V_m

$$V_m = F/R \tag{2}$$

is dependent on the frequency only through the variable of Eq. (3), so that the amplitude response

$$X = \frac{\omega}{\omega_0} - \frac{\omega_0}{\omega} \tag{3}$$

at, for example, twice the resonant frequency is the same as that at half the resonant frequency (geometric symmetry), and on the damping only through the so-called quality factor given in Eq. (4). In these terms the response is given by

$$Q = \omega_0 m/R \tag{4}$$

Eq. (5) and the phase, ϕ, of the velocity, relative

$$V/V_m = (1 + Q^2 X^2)^{1/2} \tag{5}$$

to the driving force, is given by Eq. (6).

$$\phi = \arctan(-QX) \tag{6}$$

Equation (5) shows that if Q is large, the response is appreciable only if X is small, so that ω is not too different from ω_0. Thus, high Q corresponds to narrow bandwidths. Large bandwidths, centered on ω_0, are obtained with low Qs. According to Eq. (6), the velocity leads, is in phase with, or lags behind the force for frequencies below, on, or above resonance, respectively. *See* RESONANCE (ACOUSTICS AND MECHANICS).

In some cases the displacement response is the one of interest. The displacement amplitude at zero frequency is $d_0 = F/s$, whereas that at $\omega = \omega_0$, with $X = 0$, is exactly Q times as great. The maximum displacement amplitude occurs at an angular frequency somewhat below ω_0, which is also the natural frequency of the damped system. The difference is small if Q is large. Flat displacement response occurs at frequencies well below resonance.

[MARTIN GREENSPAN]

Bibliography: R. E. D. Bishop, *Vibration*, 2d ed., 1979; J. P. Den Hartog, *Mechanical Vibrations*, 4th ed., 1956; L. E. Kinsler and A. R. Frey, *Fundamentals of Acoustics*, 2d ed., 1962.

Foucault pendulum

A pendulum or swinging weight, supported by a long wire, by which J. B. L. Foucault demonstrated in 1851 the rotation of Earth on its axis. Foucault used a 28-kg iron ball suspended on about a 60-m wire in the Pantheon in Paris. The upper support of the wire restrains the wire only in the vertical direction. The bob is set swinging along a meridian in pure translation (no lateral or circular motion). In the Northern Hemisphere the plane of swing appears to turn clockwise; in the Southern Hemisphere it appears to turn counterclockwise, the rate being 15 degrees times the sine of the local latitude per sidereal hour. Thus, at the Equator the plane of swing is carried around by Earth and the pendulum shows no apparent rotation; at either pole the plane of swing remains fixed in space while Earth completes one rotation each sidereal day. *See* PENDULUM.

[FRANK H. ROCKETT]

Bibliography: G. O. Abell, *Exploration of the Universe*, 3d ed., 1975; C. Kittel, W. D. Knight, and M. A. Ruderman, *Mechanics*, Berkeley Physics Course, vol. 1, 2d ed., 1973; O. Struve, B. Lynds, and H. Pillans, *Elementary Astronomy*, 1959.

Fourier series and integrals

Mathematical tools for the study of periodic phenomena, indispensable in the theory of wave motion (for example, light waves and sound waves), oscillatory mechanical systems (such as vibrating strings), and celestial orbits. Fourier series and the related topics discussed below have important applications to other branches of mathematics, of which the theories of probability and partial differential equations are worthy of special mention. Finally, the disciplines motivated by the subject itself enjoy a prominent position in pure mathematical research.

A function f of a real variable is said to be periodic with period T if $f(t+T)=f(t)$ for each t. The simplest examples of functions with a given period T are pure harmonics, that is, functions of the form $f(t)=a_n \cos n\omega t + b_n \sin n\omega t$, where $\omega = 2\pi T^{-1}$ is the fundamental frequency and a_n, b_n are constants. The basic idea for the applications of Fourier series is that any function f of period T which satisfies rather mild restrictions can be represented as a superposition of pure harmonics, as in relation (1), or, in a more convenient form, using complex exponentials, as relation (2).

$$f(t) \sim \sum_{n=0}^{\infty} (a_n \cos n\omega t + b_n \sin n\omega t) \tag{1}$$

$$f(t) \sim \sum_{-\infty}^{\infty} c_n e^{in\omega t} \tag{2}$$

If term-by-term integration of relation (2) is assumed to be legitimate, an easy calculation shows that Eq. (3) can be formed. (The interval of

$$c_n = T^{-1} \int_{-T/2}^{T/2} f(t) e^{-in\omega t} dt \tag{3}$$

integration could be any interval of length T). This motivates the formal definition of Fourier series. Suppose that f is a function of period T such that expression (4) exists and is finite. The coefficients

$$\int_{-T/2}^{T/2} |f(t)| \, dt \tag{4}$$

$\{c_n\}$ defined by Eq. (3) are the Fourier coefficients of f, and the series in relation (2), the Fourier series expansion of f. The coefficients uniquely determine the function; that is, if $c_n = 0$ for each n, then f is essentially the zero function. In addition, many formal operations in which the series is substituted term by term for the function can be justified. Two important questions present themselves immediately. Let Eq. (5) by the Nth partial sum of

$$s_N(t) = \sum_{-N}^{N} c_n e^{in\omega t} \tag{5}$$

the Fourier series of f. The first question is, does $s_N(t)$ converge to $f(t)$ as N approaches ∞? The second question is, given a sequence $\{c_n\}$, is it the sequence of Fourier coefficients of some function?

The Fourier series of a continuous function need not converge everywhere. If t_0 is a given point, the convergence of $s_N(t_0)$ to $f(t_0)$ depends on the behav-

ior of $f(t)$ for t in the neighborhood of t_0. However, if one takes the averaged partial sums, Eq. (6), then

$$\sigma_N = (N+1)^{-1} \sum_{k=0}^{N} s_k \tag{6}$$

for continuous f, $\sigma_N \to f$ uniformly. The knowledge alone of the ordinary convergence of a Fourier series is of little importance in applications because, for purposes of calculation, it is necessary to know something about the rapidity of convergence. An example of a theorem treating this is the following: Suppose $|df/dt| \leqq M$ everywhere; then $|f(t) - s_N(t)| \leqq 1/2\pi M(N+1)^{-1}$.

The Riemann-Lebesgue lemma asserts that, if $\{c_n\}$ is the sequence of Fourier coefficients of an integrable function, then $c_N \to 0$ as $n \to \pm\infty$. The converse is false: Not all trigonometric series with coefficients tending to zero, expression (7), are Fourier series.

$$\sum_{-\infty}^{\infty} c_n e^{in\omega t} \tag{7}$$

To obtain a really satisfactory theory, one modifies the above questions and restricts attention to a special class of functions, namely, the class L^2 of those functions f of period T for which the integral shown in expression (8) exists and is

$$\int_{-T/2}^{T/2} |f(t)|^2 \, dt \tag{8}$$

finite. L^2 is a Hilbert space with the inner product of two functions f, g being as in Eq. (9), where the

$$(f,g) = T^{-1} \int_{-T/2}^{T/2} f(t) \bar{g}(t) \, dt \tag{9}$$

bar denotes complex conjugation. In lieu of ordinary convergence, convergence in the mean exists. An example of this is shown in Eq. (10).

$$\lim_{N \to \infty} \int_{-T/2}^{T/2} |f(t) - s_N(t)|^2 \, dt = 0 \tag{10}$$

The partial sums s_N can be characterized by an extremal property: Among all functions p of the form shown in Eq. (11), the minimum of the quadratic deviation, expression (12), is achieved

$$p(t) = \sum_{-N}^{N} a_n e^{in\omega t} \tag{11}$$

$$\int_{-T/2}^{T/2} |f(t) - p(t)|^2 \, dt \tag{12}$$

uniquely by $p = s_N$, the Nth partial sum of the Fourier series of f. For functions of the class L^2, there is the Parseval identity, Eq. (13).

$$T^{-1} \int_{-T/2}^{T/2} |f(t)|^2 \, dt = \sum_{-\infty}^{\infty} |c_n|^2 \tag{13}$$

If f and g are two functions of L^2 with Fourier coefficients $\{c_n\}$ and $\{d_n\}$, respectively, then Eq. (14) holds. Let l^2 denote the ensemble of se-

$$T^{-1} \int_{-T/2}^{T/2} f(t) \bar{g}(t) \, dt = \sum_{-\infty}^{\infty} c_n \bar{d}_n \tag{14}$$

quences of complex numbers $\{c_n\}$ such that expression (15) is finite. One of the implications of

$$\sum_{-\infty}^{\infty} |c_n|^2 \tag{15}$$

Eq. (13) is that the Fourier coefficients of a function of class L^2 belong to l^2. The truth of the converse is the Riesz-Fischer theorem: If $\{c_n\}$ is a sequence of class l^2, then the trigonometric series shown in expression (16) is the Fourier series of a

$$\sum_{-\infty}^{\infty} c_n e^{in\omega t} \qquad (16)$$

function of period T belonging to the class L^2. These statements are not peculiar to Fourier series, but rather they describe the general situation in a series expansion in terms of orthogonal functions. The set of functions $\{e^{in\omega t}\}$, in which n ranges over all integers, forms a complete orthonormal set in the L^2-space of functions on an interval of length T.

Almost periodic functions. These are a generalization of periodic functions. Suppose f is periodic with period T_0; then $f(t + nT_0) = f(t)$ for each integer n. In particular, in any interval of length greater than T_0, there is at least one number T of the form $T = nT_0$ so that $f(t + T) = f(t)$. A continuous function f is said to be uniformly almost periodic if, given a positive number ε, there is another positive number T_ε such that in any interval of length greater than T_ε there is at least one number T for which $|f(t + T) - f(t)| < \varepsilon$, for all t. An almost periodic function f has an almost periodic Fourier series shown by relation (17); see also relation (2).

$$f(t) \sim \sum_{\nu} c_\nu e^{i\omega_\nu t} \qquad (17)$$

If f is actually periodic with fundamental frequency $\omega_0 = 2\pi T_0^{-1}$, then each ω_ν for which $c_\nu \neq 0$ will be an integral multiple of ω_0; in the general case, the series in relation (17) is to be regarded as a superposition of pure harmonics without a common period. The generalization of Eq. (3) giving the formula for the almost periodic Fourier coefficients is Eq. (18). In Eq. (18) the interval of integration

$$c_\nu = \lim_{T \to \infty} T^{-1} \int_{-T/2}^{T/2} f(t) e^{-i\omega_\nu t} \, dt \qquad (18)$$

can be taken to be any interval of length T, the limit always exists, and the limit is different from zero only for a countable number of frequencies ω_ν. If f is uniformly almost periodic, suitable combinations of the partial sums in relation (17) converge uniformly on the whole real line to f. Conversely, a uniform limit of trigonometric polynomials, that is, finite sums of the form given in relation (17), is a uniformly almost periodic function. There are also classes of discontinuous almost periodic functions, for example, the generalization of L^2 to the almost periodic case, which will not be discussed here. However, Eq. (13) has a simple analog, represented by Eq. (19).

$$\lim_{T \to \infty} T^{-1} \int_{-T/2}^{T/2} |f(t)|^2 \, dt = \Sigma |c_\nu|^2 \qquad (19)$$

Trigonometric integrals. The class of functions which admit a representation in the form (17) is still too narrow for many purposes. It is necessary to admit continuous sums, or integrals, as well. A generalization is given by using Stieltjes integrals, relation (20). However, to get a useful theory, one

$$f(t) \sim \int_{-\infty}^{\infty} e^{i\omega t} \, d\sigma(\omega) \qquad (20)$$

has to admit more general quantities than $d\sigma$ in Eq. (20). Often, expressions such as Eq. (53), involving

tempered distributions, suffice. The representation of functions by generalized Fourier transforms continues to present challenging problems in mathematical research.

If σ is of bounded variation on the whole real line, the integral above exists in the ordinary sense and σ is uniquely determined by its Fourier-Stieltjes transform f. A case of particular importance in probability is when σ is a distribution function, that is, σ is an increasing function with $\sigma(-\infty) = 0$ and $\sigma(+\infty) = 1$. If X is a real-valued random variable, then $\sigma(\omega) = $ probability $\{X < \omega\}$ is a distribution function. In this situation, f is called the characteristic function of the probability distribution. An important property of characteristic functions is that they are positive definite, by which is meant that, for each finite collection $(c_1, t_1), \ldots, (c_n, t_n)$ of pairs of complex numbers c_k and points t_k, Eq. (21) holds.

$$\sum_{j,k=1}^{n} c_j \bar{c}_k f(t_j - t_k) \geq 0 \qquad (21)$$

Bochner's theorem states: A function f of a real variable is a characteristic function, that is, it is representable in the form (20) with σ a distribution function, if and only if f is continuous, $f(0) = 1$, and f is positive definite. Characteristic functions are very useful in probability; the techniques involved are valid whenever one is dealing with Fourier-Stieltjes transforms. These techniques are described below.

If X_1 and X_2 are two random variables with distribution functions σ_1 and σ_2, respectively, the random variable $X_3 = X_1 + X_2$ has the distribution function σ_3 defined by Eq. (22), the convolution of

$$\sigma_3(\omega) = \int_{-\infty}^{\infty} \sigma_2(\omega - \xi) \, d\sigma_1(\xi) \qquad (22)$$

σ_1 and σ_2. More generally, if σ_1 and σ_2 are two functions of bounded variation, their convolution σ_3, as defined in Eq. (22), is also of bounded variation The Fourier-Stieltjes transform of a convolution is the product of the Fourier-Stieltjes transforms, that is, $f_3 = f_1 f_2$. Thus, the rather complicated operation of convolution can be replaced by the simpler operation of multiplication. Another useful principle concerns what are called limit theorems. Suppose $\{X_n\}$ is an infinite sequence of random variables with X_n having the distribution function σ_n and the characteristic function f_n. The question arises as to whether there is a limiting distribution. Not only are the calculations generally simplified by using characteristic functions, but also rigorous criteria are available; for example, if the sequence $\{f_n\}$ converges uniformly in some interval containing 0, then Eq. (23) exists

$$\lim_{n \to \infty} f_n(t) = f(t) \qquad (23)$$

for every t, f is also a characteristic function, corresponding to a distribution function σ, and Eq. (24) holds at each point of continuity of σ.

$$\lim_{n \to \infty} \sigma_n(\omega) = \sigma(\omega) \qquad (24)$$

Proceeding purely formally from the representation in relation (20), differentiating n times gives relation (25). Assuming that the appropriate inte-

$$\frac{d^n}{dt^n} f(t) \sim i^n \int e^{i\omega t} \omega^n \, d\sigma(\omega) \qquad (25)$$

grals are absolutely convergent, one can calculate the moments of the function of bounded variation σ from Eq. (26).

$$\int \omega^n \, d\sigma(\omega) = i^{-n} \frac{d^n}{dt^n} f(0) \qquad (26)$$

However, a more significant application of the device is that, if L is a differential operator with constant coefficients, then at least heuristically, relation (27) holds, where λ is the polynomial determined by Eq. (28).

$$L f(t) \sim \int e^{i\omega t} \lambda(\omega) \, d\sigma(\omega) \qquad (27)$$

$$L_t e^{i\omega t} = \lambda(\omega) e^{i\omega t} \qquad (28)$$

More generally, if L is any linear operation which commutes with translations, that is, $LU(\tau) = U(\tau)L$ for each τ, where $U(\tau)$ is defined by $(U(\tau)f)(t) = f(t+\tau)$, then $L_t e^{i\omega t} = \lambda(\omega) e^{i\omega t}$, and the situation is analogous. The idea is that the solution f of the equation $Lf = g$, where $g(t) \sim \int e^{i\omega t} \, d\rho(\omega)$, should be representable in the form (20), where $\lambda(\omega) \, d\sigma(\omega) = d\rho(\omega)$, and thus the problem is formally solved by division: $d\sigma(\omega) = d\rho(\omega)/\lambda(\omega)$. However, to get a useful theory, one has to admit generalized functions σ which are not of bounded variation and may not even be functions in the ordinary sense. A rigorous discussion of the representation of functions by generalized Fourier transforms is not only beyond the scope of this article but also a subject which presents challenging problems in mathematical research.

Fourier integrals. There is one class of functions for which a completely satisfactory theory of Fourier transforms exists. That is the generalization of the L^2-theory of Fourier series to Fourier integrals. It is customary to use L^2 to denote the Hilbert space of functions f such that expression (29) exists and is finite with the inner product shown in Eq. (30).

$$\int_{-\infty}^{\infty} |f(t)|^2 \, dt \qquad (29)$$

$$(f,g) = (2\pi)^{-1} \int_{-\infty}^{\infty} f(t)\bar{g}(t) \, dt \qquad (30)$$

For a function $f(t)$ in L^2, it is possible to form a Fourier transform $F(\omega)$ which is again a function in L^2. The function f will then have the representation (20), with $d\sigma(\omega) = F(\omega) \, d\omega$. That is, relation (31) holds. F is computed from Eq. (32), where the

$$f(t) \sim \int_{-\infty}^{\infty} e^{i\omega t} F(\omega) \, d\omega \qquad (31)$$

$$F(\omega) = \lim_{T \to \infty} (2\pi)^{-1} \int_{-T}^{T} f(t) e^{-i\omega t} \, dt \qquad (32)$$

limit need not converge in the ordinary sense, but only in the mean. Otherwise, the theorems about Fourier series usually have appropriate generalizations. For example, if one forms the partial sums shown in Eq. (33), one again has convergence in the mean, as shown in Eq. (34).

$$s_N(t) = \int_{-N}^{N} e^{i\omega t} F(\omega) \, d\omega \qquad (33)$$

$$\lim_{N \to \infty} \int_{-\infty}^{\infty} |f(t) - s_N(t)|^2 \, dt = 0 \qquad (34)$$

The Parseval identity takes the form of Eq. (35).

$$(2\pi)^{-1} \int_{-\infty}^{\infty} |f(t)|^2 \, dt = \int_{-\infty}^{\infty} |F(\omega)|^2 \, d\omega \qquad (35)$$

If f and g are two functions of L^2 with Fourier transforms F and G, respectively, then Eq. (36) holds and, in particular, for the convolution shown in Eq. (37).

$$(2\pi)^{-1} \int_{-\infty}^{\infty} f(t)\bar{g}(t) \, dt = \int_{-\infty}^{\infty} F(\omega)\bar{G}(\omega) \, d\omega \qquad (36)$$

$$(2\pi)^{-1} \int_{-\infty}^{\infty} f(t-s)g(s) \, ds$$
$$= \int_{-\infty}^{\infty} e^{i\omega t} F(\omega) G(\omega) \, d\omega \qquad (37)$$

Theory of distributions. The study of trigonometric integrals, as well as differential equations and other related topics, can be greatly simplified by considering generalized functions or, more properly, distributions.

The most important properties of an ordinary function φ can be deduced from a knowledge of its behavior as an integrator in definite integrals of the form shown by expression (38). The functions f

$$\int_{-\infty}^{\infty} f(t) \, \varphi(t) \, dt \qquad (38)$$

which occur in these integrals are called test functions, and to determine φ one needs to know the value of expression (38) for sufficiently many test functions. The most important class of test functions is **D**, the class of infinitely differentiable functions which vanish outside a bounded interval. For continuous functions φ, it is clear that the integral of expression (38) has a definite, complex numerical value for all f in **D** and, in addition, the expressions shown in (38) have certain linearity and continuity properties. The linearity property is shown in Eq. (39), which holds for all f and g of

$$\int_{-\infty}^{\infty} \{af(t) + bg(t)\} \varphi(t) \, dt$$
$$= a \int_{-\infty}^{\infty} f(t) \, \varphi(t) \, dt + b \int_{-\infty}^{\infty} g(t) \, \varphi(t) \, dt \qquad (39)$$

the class **D** and all complex constants a and b. The continuity condition is this: If $\{f_n\}$ is a sequence of functions of the class **D**, all of which vanish outside of some fixed interval $(-T,T)$, such that the sequence of functions and all their derivatives are uniform for all t as in Eq. (40), then Eq. (41) holds.

$$\lim_{n \to \infty} \frac{d^k}{dt^k} f_n(t) = \frac{d^k}{dt^k} f(t) \qquad (40)$$

$$\lim_{n \to \infty} \int_{-\infty}^{\infty} f_n(t) \, \varphi(t) \, dt = \int_{-\infty}^{\infty} f(t) \, \varphi(t) \, dt \qquad (41)$$

In physical problems where φ corresponds to a density of mass, electric charge, or a similar quantity, there is a great gain in generality in considering Stieltjes integrals $\int_{-\infty}^{\infty} f(t) \, d\mu(t)$, where now μ is thought of as a distribution of mass, charge, and so on. In case μ has a density φ, Eq. (42) holds. Even

$$\int_{-\infty}^{\infty} f(t) \, d\mu(t) = \int_{-\infty}^{\infty} f(t) \, \varphi(t) \, dt \qquad (42)$$

if μ does not possess a density in the ordinary sense, it is often convenient to replace $d\mu(t)$ by the symbol $\varphi(t)\,dt$. There is no implication that φ signifies a function; all that is intended is that $\varphi(t)\,dt$ symbolize a meaningful integrator.

A distribution of the class $\mathbf{D}'$ is a quantity, usually written in the form $\varphi(t)\,dt$, such that the expressions shown in (38) have definite values for each f of the class $\mathbf{D}$ and such that the linearity and continuity conditions hold. Various formal manipulations with distributions are legitimate once careful definitions have been made. For a differentiable function φ, the integration by parts formula gives Eq. (43). This formula is taken as the

$$\int_{-\infty}^{\infty} f\frac{d\varphi}{dt}\,dt = -\int_{-\infty}^{\infty}\frac{df}{dt}\varphi\,dt \qquad (43)$$

definition of the derivative of a distribution; the meaning of d/dt is defined by the right-hand side of Eq. (43). The product of a distribution φ by an infinitely differentiable function h is defined by Eq. (44). In general, the product $h\varphi$ is not defined

$$\int_{-\infty}^{\infty} f(t)\,[h(t)\varphi(t)]\,dt$$
$$= \int_{-\infty}^{\infty} [f(t)h(t)]\varphi(t)\,dt \quad (44)$$

if h is not infinitely differentiable; thus the product of two distributions usually does not have a meaning.

A notable example of a distribution is the delta function, δ, defined by Eq. (45). What has been

$$\int_{-\infty}^{\infty} f(t)\,\delta(t)\,dt = f(0) \qquad (45)$$

done is to write $f(0)$, a linear and continuous functional of the function f, in the form shown in (38). The expression $\delta(t)$ does not symbolize any function, but $\delta(t)\,dt$ can be viewed as a substitute for $d\mu$, where μ represents a unit mass or charge concentrated at the origin. Derivatives of distributions have been defined above; an example which has a physical realization as an electric dipole is $d\delta/dt$. The effect of this distribution on a test function is given by Eq. (46). The calculation

$$\int_{-\infty}^{\infty} F\frac{d\delta}{dt}\,dt = -\frac{df}{dt}(0) \qquad (46)$$

is made by applying Eq. (43) with $\varphi = \delta$ and then using the definition of δ. As an example of multiplication of a distribution by a function, one has $t\delta(t) = 0$. The reasoning is this: $\int_{-\infty}^{\infty} f[t\delta]\,dt = \int_{-\infty}^{\infty} [tf]\,\delta\,dt$, and the right-hand side is, by definition, the value of $tf(t)$ at $t=0$, namely, 0. Note that δ^2 has no meaning.

Another interesting example is provided by the principal value of an apparently divergent integral. As a special case, consider $\int_{-\infty}^{\infty} f(t)t^{-1}\,dt$. The singularity of t^{-1} at the origin makes it appear that the integral diverges if $f(0) \neq 0$, but one can define a meaningful distribution, P.V. t^{-1}, as in Eq. (47),

$$\text{P.V.}\int_{-\infty}^{\infty} f(t)t^{-1}\,dt$$
$$= \int_{-\infty}^{\infty} \{f(t) - f(0)u(t)\}t^{-1}\,dt \quad (47)$$

where u is any even function in $\mathbf{D}$ with $u(0) = 1$. The right-hand side of Eq. (47) makes sense as an ordinary integral, and its value is independent of the choice of u. Observe that t P.V. $t^{-1} = 1$.

The distributions considered as examples above can also be defined as derivatives by use of Eq. (43). The relevant formulas are Eqs. (48) and (49).

$$\frac{d^2}{dt^2}|t| = \frac{d}{dt}\,sgn(t) = 2\delta(t) \qquad (48)$$

$$\frac{d^2}{dt^2}\{t\log|t| - t\} = \frac{d}{dt}\log|t| = \text{P.V. }t^{-1} \quad (49)$$

In general, for any distribution $\varphi(t)\,dt$ on a finite interval, there exists a continuous function ψ and an integer k such that $\varphi(t) = d^k/dt^k\,\psi(t)$. Thus distributions may be viewed as sums of derivatives, in a generalized sense, of continuous functions.

Tempered distributions. The Fourier integral can be generalized to distributions, but it is simpler to restrict one's attention to tempered distributions. These are derivatives of continuous functions of polynomial growth or, alternatively, distributions of the class $\mathbf{S}'$, defined by the same procedure as used above for distributions of the class $\mathbf{D}'$ except that a larger class of test functions is used, namely, the class $\mathbf{S}$ of infinitely differentiable functions f with the property that $t^m\,d^n/dt^n\,f(t)$ is bounded for all m and n. The Fourier transform of a function f of the class $\mathbf{S}$ is the function F defined by the absolutely convergent integral shown in Eq. (50).

$$F(\omega) = (2\pi)^{-1}\int_{-\infty}^{\infty} f(t)e^{-i\omega t}\,dt \qquad (50)$$

It is clear that F is also a function of the class $\mathbf{S}$ and that the inverse Fourier transform is given by the absolutely convergent integral shown in Eq. (51). Indeed, the Fourier transform of $t^m\,d^n/dt^n\,f(t)$ is $(i\,d/d\omega)^m\,(i\omega)^n F(\omega)$.

$$f(t) = \int_{-\infty}^{\infty} e^{i\omega t}F(\omega)\,d\omega \qquad (51)$$

The Fourier transform of a tempered distribution $\varphi(t)\,dt$ is the distribution $\Phi(\omega)\,d\omega$ of the class $\mathbf{S}'$, such that for all f of class $\mathbf{S}$, Eq. (52) exists.

$$(2\pi)^{-1}\int_{-\infty}^{\infty} f(t)\bar{\varphi}(t)\,dt = \int_{-\infty}^{\infty} F(\omega)\overline{\Phi}(\omega)\,d\omega \quad (52)$$

One may write symbolically relation (53), but this must be understood as a convenient shorthand for the relationship given by Eq. (52).

$$\varphi(t) \sim \int_{-\infty}^{\infty} e^{i\omega t}\Phi(\omega)\,d\omega \qquad (53)$$

With proper care, the theory of Fourier transforms of tempered distributions can be developed in close analogy with the L^2 theory of Fourier integrals.

The Fourier transform changes differentiation into multiplication, and vice versa: The Fourier transform of $d\varphi/dt$ is $i\omega\Phi(\omega)$ and that of $t\varphi(t)$ is $i(d\Phi/d\omega)$. This is important for applications to differential equations, and it is helpful in computing Fourier transforms. The transform of a polynomial $P(t)$ is $P(i\,d/d\omega)\delta$. The transform of P.V. t^{-1} is $(2i)^{-1}\,sgn(\omega)$. The transform of δ is the constant 1, and that of $\delta(t+T)$ is $e^{i\omega T}$. Finally, if α and β are positive numbers with $\alpha\beta=2\pi$, one has the Poisson summation formula, giving the Fourier transform represented by Eq. (54). With this last

$$\alpha \sum_{m=-\infty}^{\infty} \delta(t+\alpha m) = \beta \sum_{n=-\infty}^{\infty} \delta(\omega+\beta n) \quad (54)$$

formula one can treat Fourier series, as well as Fourier integrals, as special cases of the Fourier transform of tempered distributions.

The preceding discussion has been confined to functions of one variable, but the theory can be extended to functions of several variables. The results persist except for trivial modifications, if one considers t and ω as n-dimensional vectors, $\mathrm{t} = (t^{(1)}, \ldots, t^{(n)})$, $\omega = (\omega^{(1)}, \ldots, \omega^{(n)})$, and ωt is replaced in the exponentials by the inner product $\omega^{(1)}t^{(1)} + \cdots + \omega^{(n)}t^{(n)}$. Here one is dealing with multiple Fourier series and multiple Fourier transforms. The discussion of representations of the form for functions of several variables is of particular interest in the study of partial differential equations with constant coefficients. *See* ORTHOGONAL POLYNOMIALS; PROBABILITY; SERIES.

[C. S. HERZ]

Bibliography: R. V. Churchill and J. W. Brown, *Fourier Series and Boundary Value Problems*, 3d ed., 1978; R. Edwards, *Fourier Series: A Modern Introduction*, 1979; M. J. Lighthill, *Introduction to Fourier Analysis and Generalized Functions*, 1958; E. C. Titchmarsh, *Introduction to the Theory of Fourier Integrals*, 2d ed., 1950; A. Zygmund, *Trigonometrical Series*, 1969.

Frame of reference

A base to which to refer physical events. A physical event occurs at a point in space and at an instant of time. Each reference frame must have an observer to record events, as well as a coordinate system for the purpose of assigning locations to each event. The latter is usually a three-dimensional space coordinate system and a set of standardized clocks to give the local time of each event. For a discussion of the geometrical properties of space-time coordinate systems *see* SPACE-TIME. *See also* RELATIVITY.

Newtonian reference frames. In the ordinary range of experience, where light signals, for all practical purposes, propagate instantaneously, the time of an event is quite distinct from its space coordinates, since a single clock suffices for all observers, regardless of their state of relative motion. The set of reference frames which have a common clock or time is called newtonian, since Isaac Newton regarded time as having invariable significance for all observers.

Inertial reference frames. In these a body moves with constant velocity when free of impressed forces. It follows that one inertial frame cannot be accelerating with respect to another. If x_1, y_1, z_1 and x_2, y_2, z_2 are the coordinates of a particle P in parallel inertial frames 1 and 2, where 2 is moving with respect to 1 with velocity components v_x, v_y, v_z, then Eq. (1) holds. (Newtonian frames are

$$\begin{aligned} x_2 &= x_1 - v_x t \\ y_2 &= y_1 - v_y t \\ z_2 &= z_1 - v_z t \end{aligned} \quad (1)$$

understood here so that $t_2 = t_1$.) This is called a Galilean transformation. Since the vs are independent of time, the component accelerations of P in 1 and 2 are the same, that is, Eq. (2) is valid.

$$A_{x2} = d^2x_2/dt^2 = d^2x_1/dt^2 = A_{x1} \text{ etc.} \quad (2)$$

Accelerated reference frames. A noninertial frame of reference is called an accelerated frame. A point fixed in such a frame will be accelerated with respect to inertial frames and so must be held by a force. An observer in an accelerated frame will see accelerations that are not proportional to the impressed forces, since a force is required to keep the particle at rest. Newton's second law does not hold without modification.

Rotating reference frames. A rotating frame of reference is one whose coordinate axes are rotating with respect to some inertial system. A point P fixed in the rotating frame moves uniformly about a circle in the inertial frame and is accelerated toward the axis of rotation. There is an additional acceleration (with respect to the inertial system) if P moves in the rotating system. For example, if the point moves away from the axis along a moving radius, it gains circular speed also. *See* CENTRIPETAL FORCE.

Newton's second law can be modified to take account of these two accelerations that are not observed in the rotating system. Let A_{Ac}, A_C, A_{Co} be the observed, centripetal, and Coriolis accelerations, respectively. The total acceleration in the inertial frame is given by Eq. (3), where the accel-

$$A = A_{Ac} + A_C + A_{Co} \quad (3)$$

erations must be added vectorially. Defining the centrifugal force, $Fc = -mAc$, and the Coriolis force, $F_{Co} = -mA_{Co}$, permits Newton's second law to be written as Eq. (4). In rotating systems a con-

$$mA_{Ac} = F + F_C + F_{Co} \quad (4)$$

sistent mechanical theory is obtained if the two additional forces are added to the impressed forces. Conversely, the presence of such forces will enable the observer to determine the motion of his reference frame.

Astronomical reference frames. These are fixed relative to celestial objects. The most convenient one for observational purposes is, of course, a geocentric system. The Earth's axis of rotation is used as a coordinate axis. A second reference direction is chosen as the intersection of the Earth's equatorial plane and the plane of the Earth's orbit (the ecliptic). The basic time unit is the Earth's rotation period, the sidereal day. The positions and times of distant astronomical events are determined indirectly.

Reference frames are regarded as more basic as more phenomena appear regular with respect to them. Thus the planetary motions are described naturally in an ecliptic coordinate system centered on the Sun, and the motions of stars as viewed from the Earth appear regular in a coordinate system with its origin at the center of the galaxy.

Reference frames based on the mean positions of large enough celestial objects, for example, a group of galaxies, approximate inertial frames closely. Between two such reference frames, the distances and relative velocities may be so large that transmission of light signals takes considerable time. Newtonian reference frames with a single universal time are no longer meaningful, and each reference frame must have its own system of time. Inertial frames of this type are called Galilean reference frames, a somewhat misleading name since the transformation of the coordinates of an event from one such frame to another is not the Galilean transformation given earlier (with $t_1 = t_2$) but one which involves the time directly. It is known as a Lorentz transformation. Galilean reference frames are often called Lorentz frames. *See* LORENTZ TRANSFORMATIONS.

[BERNARD GOODMAN]

Bibliography: G. O. Abell, *Exploration of the Universe*, 3d ed., 1975; A. Einstein, *The Meaning of Relativity*, rev. ed., 1956; C. Kittel, W. D. Knight, and M. A. Ruderman, *Mechanics*, Berkeley Physics Course, vol. 1, 2d ed., 1973.

Franck-Condon principle

In any molecular system the transition from one energy state to another occurs so near to instantaneously that the nuclei of the atoms involved can be considered to be stationary during the transition. This principle, proposed by J. Franck in 1925 and developed on the basis of quantum mechanics by E. U. Condon in 1928, is important in discussing systems where more than one atom is involved. It is therefore valuable in problems of molecular spectroscopy and in the interpretation of the optical properties of liquids and solids.

An example of the principle's operation involves luminescent centers in solids. In the state of lowest energy (ground state), the distance from a given atom in the center to its nearest neighboring atoms adjusts itself so that the total energy of the system assumes a minimum value. For an ionic crystal, such as sodium chloride, there is a balance between the electrostatic and other forces tending to pull the ions together and the repulsive forces tending to keep the ions apart. In a state of higher energy (excited state), the system also has a minimum energy position, but this is generally different from that of the ground state because of a different balance of the forces involved. If the luminescent system is in its ground state and absorbs light which raises it to an excited state, the Franck-Condon principle specifies that immediately after the transition the nuclei involved are still in the equilibrium configuration for the ground state. This configuration does not represent the condition of minimum energy for the excited state, and the system rapidly changes to its new minimum, giving up the excess energy as heat. A similar process occurs in the luminescence transition to the ground state, and heat energy is again given off when the luminescent center readjusts to the ground-state equilibrium position. The center has now returned to its inital state. However, it should be noted that the absorption transition was achieved by the action of light energy only, whereas the system returned to the ground state by giving up heat energy twice and emitting the remaining energy as luminescence. As a result, the luminescent emission is of lower energy than the absorbed light, and since the wavelength varies inversely as the energy, the emission is of longer wavelength than is the absorbed light. This general result, known as the Stokes shift in luminescence, is thus explained on the basis of the Franck-Condon principle. *See* LUMINESCENCE; MOLECULAR STRUCTURE AND SPECTRA.

[CLIFFORD C. KLICK; JAMES H. SCHULMAN]

Bibliography: J. H. Crawford and L. M. Slifkin (eds.), *Point Defects in Solids*, vol. 1, 1972; E. U. Condon, The Franck-Condon principle and related topics, *Amer. J. Phys.*, 15:365–375, 1947; R. J. Elliott and A. F. Gibson, *An Introduction to Solid State Physics*, 1974.

Free-electron theory of metals

A model of a metal, originally proposed by H. A. Lorentz and improved by A. Sommerfeld in 1928, in which the free electrons, that is, those giving rise to the conductivity, are regarded as moving in a potential—the sum of the average potential due to the metal ions in the lattices and to all the remaining free electrons—which is approximated as constant everywhere inside the metal. The free electrons are thus assumed to move independently of one another throughout the space bounded by the surfaces of the metal. Interaction of the free electrons with the ion cores is assumed to be negligible. The free electrons are taken to be identical to the valence electrons of the free metal atoms; thus, alkali metals contain one free electron per atom, and aluminum contains three. The free electron theory should actually be termed the quasi-free-electron theory, to distinguish between electrons in a metal which behave nearly as do free electrons, and truly free electrons, such as those in a vacuum tube.

A number of important physical properties of some metals, in particular the simple monovalent metals, are explained satisfactorily in terms of the quasi-free-electron model. Among these are the electronic specific heat, the magnetic susceptibility, and the electrical and thermal conductivity. To understand the behavior of the more complex metals and of semiconductors, it became necessary to introduce the more complicated band theory of solids. *See* BAND THEORY OF SOLIDS.

Effective mass. In reality, the true crystalline potential seen by a valence electron is not constant nor independent of the position and motion of the remaining valence electrons. In fact, the potential that is due to ions located at regular lattice points is periodic in space; in the modern version of the quasi-free-electron model the effect of this periodic variation is taken into account by assigning to the conduction electrons an effective mass m^* which is generally different from the mass m of a free electron. The correlation of the motion of a given electron with that of the remaining conduction electrons may be treated theoretically and leads to certain corrections which are of considerable importance in some instances but apparently do not greatly influence the results of the theory of conductivity based on a model which neglects such correlations.

The effective mass m^* is a tensor which relates the force **F** to the acceleration **a** through the equation $F = m^*a$. In a coordinate system in which m^* is diagonal the components of the tensor generally are unequal, and some, or all, of the components may be negative. If the components of m^* differ along different crystallographic directions, it is called an anisotropic effective mass. Such anisotropy is the rule rather than the exception and is a consequence of the physical fact that the crystalline potential in which the conduction electrons move is anisotropic. The quasi-free-eletron model assumes, however, that m^* is a scalar.

Fermi energy. Electrons are elementary particles of half-integral spin; hence they obey Fermi-Dirac statistics. They carry a charge $e = -4.8 \times 10^{-10}$ esu and have an intrinsic magnetic moment μ. According to Fermi-Dirac statistics if there is an assembly of electrons in thermal equilibrium at temperature T, the probability that an allowed quantum state of energy ϵ is occupied is given by the Fermi distribution

$$f_0(\epsilon) = \frac{1}{1 + \exp\left[\,(\epsilon - \eta)/kT\,\right]} \qquad (1)$$

where k is Boltzmann's constant. The parameter η is called the Fermi energy. It is that energy for which the probability of occupancy is one-half. It is also common practice to speak of the assembly of quasi-free electrons as an electron gas. *See* FERMI-DIRAC STATISTICS; KINETIC THEORY OF MATTER.

Wave vector. In quantum mechanics, free electrons are described by wave functions of the form

$$\varphi_\mathbf{k}(r) = e^{i\mathbf{k} \cdot \mathbf{r}} \qquad (2)$$

that is, by plane waves. Here $\mathbf{k}$ is called the wave vector. Its allowed values are determined by the boundary conditions of the problem. For present purposes, $\mathbf{k}$ may be regarded as a continuous variable. The energy of a free electron is given by

$$\epsilon(\mathbf{k}) = \mathbf{p}_\mathbf{k}^2/2m = \hbar^2\mathbf{k}^2/2m \qquad (3)$$

where $\mathbf{p}_\mathbf{k}$ is the momentum of the electron in the state $\mathbf{k}$, and $\hbar$ is Planck's constant h divided by 2π. The correct wave function for an electron in a crystalline lattice is given not by Eq. (2) but by

$$\psi_\mathbf{k}(r) = U_\mathbf{k}(\mathbf{r})\,e^{i\mathbf{k} \cdot \mathbf{r}} \qquad (4)$$

known as a Bloch function. Here $U_\mathbf{k}(\mathbf{r})$ is a function which is periodic in the lattice. To a first approximation the relationship between the energy and the wave vector is still given by Eq. (3) but with the mass of the free electron replaced by the effective mass m^*.

Density of states. An important quantity which enters into nearly all calculations of electronic properties of metals is the density of states, $N(\epsilon)$; $N(\epsilon)\,d\epsilon$ is the number of quantum states per unit volume, for a specified electron spin orientation, in the energy range between ϵ and $\epsilon + d\epsilon$. For quasi-free electrons, $N(\epsilon)$ is given by

$$N(\epsilon) = \frac{(2m^*)^{3/2}}{4\pi^2\hbar^3}\,\epsilon^{1/2} \qquad (5)$$

The total number of electrons per unit volume is thus

$$n = 2\int_0^\infty N(\epsilon)f_0(\epsilon)\,d\epsilon$$

$$= \frac{(2m^*)^{3/2}}{2\pi^2\hbar^3}\int_0^\infty \epsilon^{1/2}f_0(\epsilon)\,d\epsilon \qquad (6)$$

The factor 2 appears in Eq. (6) because each state $\mathbf{k}$ can accommodate two electrons, the two electrons having opposite spin orientation. The integral in Eq. (6) must be evaluated numerically in

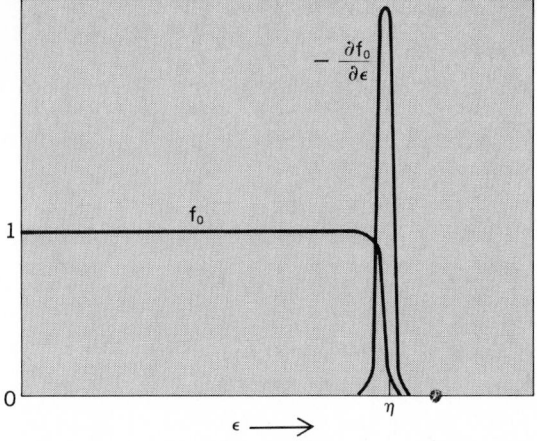

Functions f_0 and $-\partial f_0/\partial \epsilon$ plotted against energy ϵ. (*From F. Seitz, The Modern Theory of Solids, McGraw-Hill, 1940*)

the general case. In metals, however, n is so large that $f_0(\epsilon)$ has the form shown in the figure. Nearly all states below the energy η are fully occupied, and the energy range in which $f_0(\epsilon)$ differs significantly from unity and from zero is relatively narrow (of the order kT) compared to the region in which $f_0(\epsilon) = 1$. An electron gas for which this situation holds is said to be highly degenerate. In that case $\partial f_0/\partial \epsilon$ is nonvanishing only over a small energy range, as shown in the figure, and an integral of the form $\int_0^\infty g(\epsilon)\dfrac{\partial f_0}{\partial \epsilon}\,d\epsilon$ can be expressed in terms of a rapidly converging series:

$$\int_0^\infty g(\epsilon)\frac{\partial f_0}{\partial \epsilon}\,d\epsilon$$

$$= -g(\eta) - \frac{\pi^2}{6}(kT)^2\left(\frac{\partial^2 g}{\partial \epsilon^2}\right)_{\epsilon = \eta} + \cdots \qquad (7)$$

The integral in Eq. (6) and similar integrals which occur in the evaluation of properties of quasi-free electrons in metals can be put into the form of Eq. (7) by an integration by parts.

One then finds that

$$\eta = \frac{(2m^*)^{3/2}}{3\pi^2\hbar^3}\,\eta^{3/2}\left[1 + \frac{\pi^2}{8}\left(\frac{kT}{\eta}\right)^2 + \cdots\right] \qquad (8)$$

and

$$\eta = \eta_0\left[1 - \frac{\pi^2}{12}\left(\frac{kT}{\eta_0}\right)^2 + \cdots\right] \qquad (9)$$

where

$$\eta_0 = \frac{\pi^2\hbar^2}{2m^*}\left(\frac{3\eta}{\pi}\right)^{2/3} \qquad (10)$$

is the Fermi energy at $T = 0$ K. The series expansion, Eq. (7), is useful only if $\eta \gg kT$. The dependence of the Fermi energy on the number of electrons per unit volume is apparent from Eq. (10). The fundamental reason for this dependence is found in the Pauli exclusion principle. An additional electron which is added to an electron gas that is confined within a fixed volume will tend to occupy the lowest possible state. If, however, the electrons already present occupy all available states between $\epsilon = 0$ and $\epsilon = \eta_0$, the exclusion principle requires that the additional electrons go into a state whose energy is greater than η_0. Thus as *n* increases, the energy of that state for which $f_0(\epsilon) = \frac{1}{2}$ also increases. *See* EXCLUSION PRINCIPLE.

Electronic specific heat. It is now a relatively simple matter to calculate the equilibrium properties of an electron gas. For example, the average energy per electron is the total energy divided by η, namely

$$\bar{\epsilon} = \frac{2\int \epsilon N(\epsilon) f_0(\epsilon)\, d\epsilon}{\eta}$$

$$= \frac{3}{5}\eta\left[1 + \frac{5}{12}\pi^2\left(\frac{kT}{\eta_0}\right)^2 + \cdots\right] \quad (11)$$

Equation (11) shows that the contribution to the specific heat per electron, $\partial\epsilon/\partial T = C_v$, is given by

$$C_v = k\frac{\pi^2}{2}\frac{kT}{\eta_0} \quad (12)$$

Attention is directed to two features of Eq. (12). First $C_v \ll 3/2k$, the classical expression for the specific heat of a free particle. Second, C_v is proportional to the absolute temperature, also in contrast to the classical result. Both of these features are consequences of the degeneracy of the electron gas. Only a small fraction of the electrons in a metal, namely those whose energies are within the small energy range of width equal to approximately $2kT$ and centered at $\epsilon = \eta$, change their energy as a result of thermal excitation. Moreover, the number of electrons which can be excited thermally is proportional to the width of this energy region, and hence proportional to T. Thus, C_v is small, and is proportional to T. *See* SPECIFIC HEAT OF SOLIDS.

The function $f_0(\epsilon)$ gives the equilibrium distribution. Under the influence of applied forces, such as electric and magnetic fields, the steady state distribution $f(\mathbf{k},\mathbf{r})$ of electrons among the allowed energy states differs from f_0. It is convenient to normalize $f(\mathbf{k},\mathbf{r})$ in such a manner that the time average of the number of electrons in the volume element $dxdydzdk_x dk_y dk_z$ is

$$\frac{1}{4\pi^3}f(\mathbf{k},\mathbf{r})\, dxdydzdk_x dk_y dk_z \quad (13)$$

At equilibrium it then follows

$$f(\mathbf{k},\mathbf{r}) = f_0(\epsilon) \quad (14)$$

Conductivity. In the theory of conductivity, the central problem is that of finding the function $f(\mathbf{k},\mathbf{r})$ in the presence of applied fields and temperature gradients. Once $f(\mathbf{k},\mathbf{r})$ is known, the calculation of electric and thermal currents, for example, is formally trivial. Since the current contributed by one electron is $e\mathbf{v}$, where $\mathbf{v}$ is the velocity of the free electron, the current density due to an electron gas whose distribution is $f(\mathbf{k},\mathbf{r})$ is given by

$$\mathbf{J}(\mathbf{r}) = \frac{e}{4\pi^3}\int \mathbf{v}_k f(\mathbf{k},\mathbf{r})\, d\mathbf{k} \quad (15)$$

The thermal current density $\mathbf{Q}(\mathbf{r})$ is obtained by replacing e in Eq. (15) by the energy ϵ:

$$\mathbf{Q}(\mathbf{r}) = \frac{1}{4\pi^3}\int \epsilon\mathbf{v}_k f(\mathbf{k},\mathbf{r})\, d\mathbf{k} \quad (16)$$

Evaluation of Eqs. (15) and (16) leads to expressions

$$\sigma = \frac{\eta e^2\tau}{m^*} \qquad \kappa = \frac{\eta\pi^2 k^2\tau T}{3m^*} \quad (17)$$

for the electrical and thermal conductivities.

Here σ is the electrical conductivity, κ is the thermal conductivity, and τ is called the relaxation time of the electrons. *See* BOLTZMANN TRANSPORT EQUATION; CONDUCTION (HEAT); RELAXATION TIME OF ELECTRONS.

[FRANK J. BLATT]

Bibliography: F. J. Blatt, *Physics of Electronic Conduction in Solids*, 1968; C. Kittel, *Introduction to Solid State Physics*, 5th ed., 1976; F. Seitz and D. Turnbull (eds.), *Solid State Physics*, vol. 1, 1955, vol. 4, 1957; J. M. Ziman, *Principles of the Theory of Solids*, 2d ed., 1972.

Free energy

A term in thermodynamics which in different treatments may designate either of two functions defined in terms of the internal energy E or enthalpy H, and the temperature-entropy product TS.

The function $(E - TS)$ is the Helmholtz free energy and is the function ordinarily meant by free energy in European references. The Gibbs free energy is the function $(H - TS)$. For the Lewis and Randall school of American chemical thermodynamics, this is the function meant by the free energy F. To avoid confusion with the symbol F as applied elsewhere to the Helmholtz free energy, the symbol G has also been used. A recent development was the introduction of the name free enthalpy, with symbol G, for the Gibbs function. *See* WORK FUNCTION (THERMODYNAMICS).

Theory. For a closed system (no transfer of matter across its boundaries), the work which can be done in a reversible isothermal process is given by the series shown in Eq. (1). For these conditions,

$$W_{\text{rev}} = -\Delta A = -\Delta(E - TS) = -(\Delta E - T\,\Delta S) \quad (1)$$

$T\,\Delta S$ represents the heat given up to the surroundings. Should the process be exothermal, $T\Delta S < 0$, then actual work done on the surroundings is less than the decrease in the internal energy of the system. The quantity $(\Delta E - T\,\Delta S)$ can then be thought of as a change in free energy, that is, as that part of the internal energy change which can be converted into work under the specified conditions. This then is the origin of the name free energy. Such an interpretation of thermodynamic quantities can be misleading, however; for the case in which $T\,\Delta S$ is positive, Eq. (1) shows that the decrease in "free" energy is greater than the decrease in internal energy.

For constant temperature and pressure in a reversible process the decrease in the Gibbs function G for the system again corresponds to a free-energy change in the above sense, since it is equal to the work which can be done by the closed system other than that associated with its change in volume ΔV under the given constant pressure P. The relations shown in Eq. (2) can be formed since $\Delta H = \Delta E + P\Delta V$.

$$\Delta G = -(\Delta H - T\,\Delta S) = W_{\text{net}} = W_{\text{rev}} - P\,\Delta V \quad (2)$$

Each of these free-energy functions is an extensive property of the state of the thermodynamic system. For a specified change in state, both ΔA and ΔG are independent of the path by which the change is accomplished. Only changes in these functions can be measured, not values for a single state.

The thermodynamic criteria for reversibility, irreversibility, and equilibrium for processes in closed systems at constant temperature and pressure are expressed naturally in terms of the function G. For any infinitesimal process at constant temperature and pressure, $-dG \geqq \delta w_{net}$. If δw_{net} is never negative, that is, if the surroundings do no net work on the system, then the change dG must be negative or zero. For a reversible differential process, $-dG > \delta w_{net}$; for an irreversible process, $-dG > \delta w_{net}$. The free energy G thus decreases to a minimum value characteristic of the equilibrium state at the given temperature and pressure. At equilibrium, $dG = 0$ for any differential process taking place, for example, an infinitesimal change in the degree of completion of a chemical reaction. A parallel role is played by the work function A for conditions of constant temperature and volume. Because temperature and pressure constitute more convenient working variables than temperature and volume, it is the Gibbs free energy which is the more commonly used in thermodynamics.

Partial molal quantities. For a particular homogeneous phase in the absence of surface, gravitational, and magnetic forces, the free energy G depends on the numbers of moles of the constituents present, the temperature T, and the pressure P. Let Ω represent the total number of constituents, n_i the number of moles of typical constituent i, and designate by subscript n constant composition, by subscript n_j constancy of the number of moles of all constituents except n_i; then Eq. (3) is formed.

$$dG(T,P,n_1, \ldots ,n_\Omega)$$
$$= \left(\frac{\partial G}{\partial T}\right)_{P,n} dt + \left(\frac{\partial G}{\partial p}\right)_{T,n} + \sum_{i=1}^{\Omega} \left(\frac{\partial G}{\partial n_i}\right)_{T,P,n_j} dn_i \quad (3)$$

In Eq. (3) the term $\left(\frac{\partial G}{\partial n_i}\right)_{T,P,n_i}$ is the chemical potential μ_i of the ith constituent. It is identical to the partial molal free energy $\overline{G}_i$ of Lewis and Randall. It then follows that Eq. (4) holds.

$$dG = -S\, dT + V\, dP + \sum_{i=1}^{\Omega} \mu_i\, dn_i \quad (4)$$

Because the chemical potentials at constant T,P are intensive variables whose values are fixed, like that of the density, by the relative number of moles of the various constituents present, and are independent of the total mass of the phase, this equation can be integrated for constant T,P and relative composition starting from $n_i = 0$ to obtain Eq. (5).

$$G(T,P,n_1, \ldots ,n_\Omega) = \sum_{i=1}^{\Omega} n_i\mu_i \quad (5)$$

This yields Eq. (6). Consistency with the expression for dG in Eq. (4) requires that Eq. (7) hold.

$$dG = \sum_{i=1}^{\Omega} \mu_i\, dn_i + \sum_{i=1}^{\Omega} n_i\, d\mu_i \quad (6)$$

$$S\, dT - V\, dP + \sum_{i=1}^{\Omega} n_i\, d\mu_i = 0 \quad (7)$$

This is the Gibbs-Duhem equation. For constant temperature and pressure, this relation imposes a condition on the composition variation of the set of chemical potentials.

Heterogeneous systems. The free energy of a closed, heterogeneous system is the sum of the free energies of its various phases. In the absence of such a constraint as provided by the subdivision

of the system by a rigid, semipermeable membrane, the general thermodynamic criterion of equilibrium requires that the temperature and pressure be uniform throughout the system and that the chemical potential of each constituent have a common value for all phases in which it is present. Further, if any of the constituents can be formed from others, the chemical potentials of the reactants and products are related in accordance with the stoichiometry of the reaction equation. Thus, for the reaction in Eq. (8), at equilibrium Eq.

$$A + 2B \rightleftharpoons 3C + 4D \quad (8)$$

(9) can be formed. Expressing each chemical po-

$$\mu_A + 2\mu_B = 3\mu_C + 4\mu_D \quad (9)$$

tential μ_i in terms of the standard value μ_i^0 and its associated activity term $RT \ln a_i$ results in Eq. (10).

$$RT \ln \left(\frac{a_C{}^3 a_D{}^4}{a_A a_B{}^2}\right)_{equil} = -(3\mu_C{}^0 + 4\mu_D{}^0 - \mu_A{}^0 - 2\mu_B{}^0)$$
$$= -\Delta G^0 \quad (10)$$

In Eq. (10) ΔG^0 is called the standard free-energy change for the reaction. Its value depends on the standard states chosen, but for a given temperature and pressure, it is a constant characteristic of the reaction involved. A true equilibrium constant K then results as shown by Eqs. (11) and (12). If the

$$K = \left(\frac{a_C{}^3 a_D{}^4}{a_A a_B{}^2}\right)_{equil} \quad (11)$$

$$RT \ln K = -\Delta G^0 \quad (12)$$

pressure for each standard state is fixed and independent of the pressure of the reaction system, ΔG^0 and hence K are functions of temperature only. This is the conventional approach in treating gas-phase equilibria, but not ordinarily for condensed phases.

Since the activities can be correlated with partial pressures or concentrations through fugacity coefficients or activity coefficients, this thermodynamic approach eliminates the uncertainties otherwise associated with equilibrium calculations based on the law of mass action.

The prediction of an equilibrium constant then requires the calculation of ΔG^0 for the reaction. The so-called third-law method involves calculation for the reaction at 25°C of the value of ΔH^0, the standard heat of reaction, from tabulated standard heat of formation data and of ΔS^0 from tabulated third-law entropies. These are combined in the sense of $\Delta G^0 = \Delta H^0 - T\, \Delta S^0$ to permit calculation of the equilibrium constant for 25°C. This in turn is used for evaluation of the integration constant in the integration of the relation in Eq. (13).

$$\frac{d \ln K}{dT} = \frac{\Delta H^0}{RT^2} \quad (13)$$

The integration requires expression of ΔH^0 as a function of temperature, which necessitates a knowledge of the heat capacities $C^0_{P(i)}$ for the various reactants over the temperature range involved.

Alternatively, if values of the free-energy function $(G^0 - H^0_{298}/T)$ are available, either from experimental measurement or from statistical thermodynamical computations, they can be combined with the standard heat of reaction at 25°C to give the desired result, Eq. (14).

$$\Delta G^0 = \Delta \left(\frac{G^0 - H^0{}_{298}}{T} \right) + \frac{\Delta H^0{}_{298}}{T} \tag{14}$$

See ENTROPY; HEAT CAPACITY.

[PAUL BENDER]

Bibliography: K. G. Denbigh, *Principles of Chemical Equilibrium*, 2d ed., 1968; E. Fermi, *Thermodynamics*, 1956; J. W. Gibbs, *Collected Works*, vol. 1, 1948; K. S. Pitzer and L. Brewer, *Thermodynamics*, rev. ed., 1961; F. T. Wall, *Chemical Thermodynamics*, 2d ed., 1965.

Free fall

The accelerated motion toward the center of the Earth of a body acted on by the Earth's gravitational attraction and by no other force. If a body falls freely from rest near the surface of the Earth, it gains a velocity of approximately 9.8 m/s every second. Thus, the acceleration of gravity g equals 9.8 m/s² or 32.16 ft/s². This acceleration is independent of the mass or nature of the falling body. For short distances of free fall, the value of g may be considered constant. After t seconds the velocity v_t of a body falling from rest near the Earth is given by Eq. (1).

$$v_t = gt \tag{1}$$

If a falling body has an initial constant velocity in any direction, it retains that velocity if no other forces are present. If other forces are present, they may change the observed direction and rate of fall of the body, but they do not change the Earth's gravitational pull; therefore a body may still be thought of as freely "falling" even though the resultant observed motion is upward. In the case of an initial downward velocity v_0, the downward velocity v_t after t seconds is given by Eq. (2). In the

$$v_t(\text{down}) = v_0 + gt \tag{2}$$

case of a body projected upward with an initial velocity v_0, the resultant upward velocity after t seconds is given by Eq. (3), and the body loses

$$v_t(\text{up}) = v_0 - gt \tag{3}$$

upward velocity; that is, it is accelerated downward at the rate of free fall, just as it would be if it were falling. When its resultant upward velocity has been reduced to zero, the rising body has reached its maximum height.

For a body falling a very large distance from the Earth, the acceleration of gravity can no longer be considered constant. According to Newton's law of gravitation, the force between any two bodies varies inversely with the square of the distance between them; therefore with increasing distance between any body and the Earth, the acceleration of the body toward the Earth decreases rapidly. The final velocity v_f, attained when a body falls freely from a very large distance h, involves integration of the force equation as a function of distance giving as a result Eq. (4), where R is the radius of the Earth.

$$v_f = \sqrt{\frac{2ghR^2}{R^2 + Rh}} \tag{4}$$

In free fall from an infinite distance to the surface of the Earth, Eq. (4) reduces to $v_f = \sqrt{2gR}$, which gives a numerical value of 11.3 km/s or 7 mi/s. This is consequently the "escape velocity,"

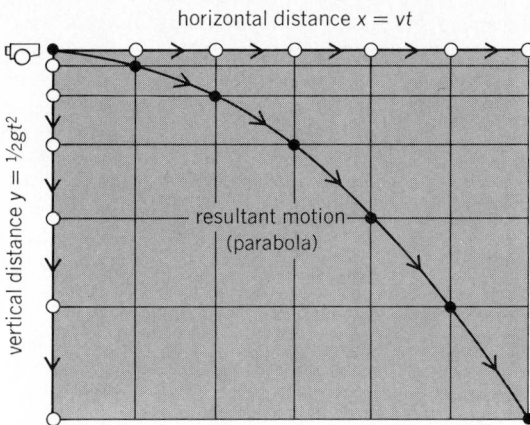

Fig. 1. Path of projectile fired horizontally, illustrating independence of horizontal and vertical components of motion.

the initial upward velocity for a rising body to completely overcome the Earth's attraction.

Because of the independent action of the forces involved, a ball thrown horizontally or a projectile fired horizontally with velocity v will be accelerated downward at the same rate as a body falling from rest, regardless of the horizontal motion (Fig. 1). If an observer could travel with the same initial horizontal velocity as the ball or the projectile, the observer would remain above it and in the observer's frame of reference the observed motion would be only that of free fall downward.

At a sufficiently large horizontal velocity, a projectile would fall from the horizontal only at the same rate that the surface of the Earth curves away beneath it (Fig. 2). The projectile would thus remain at the same elevation above the Earth and in effect become an earth satellite. The critical horizontal velocity v_h at which this would occur is that for which the required inward acceleration v^2/r for a body to be in uniform circular motion at velocity v about a circle of radius r is exactly supplied by the free-fall acceleration of gravity g at that elevation. Thus, for the special case of horizontal motion near the Earth's surface, where $v_h = v$ and $r = R$ (the radius of the Earth), the required orbital velocity for the projectile to go into satellite motion is given by Eq. (5), and is close to

$$v_h = \sqrt{gR} \tag{5}$$

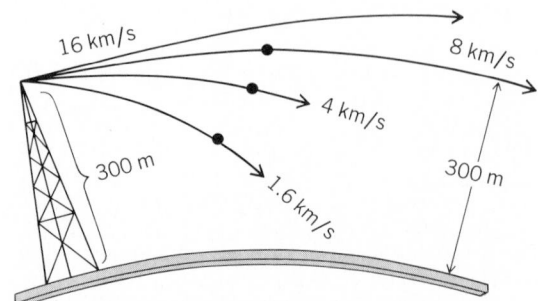

Fig. 2. Paths of projectiles fired horizontally at high velocities over the curved surface of the Earth. A projectile fired horizontally with a velocity of 8 km/s would circle the Earth like a satellite at constant altitude.

8 km/s or 5 mi/s. At a higher velocity the satellite would rise to a higher orbit.

It was Newton's perception that for the Moon to be a satellite of the Earth the Moon must be in continual free fall toward the Earth which led him to propose his universal law of gravitation. *See* GRAVITATION.

[ROGERS D. RUSK]

Frequency

The number of times which sound pressure, electrical intensity, or other quantities specifying a wave vary from their equilibrium value through a complete cycle in unit time. The most common unit of frequency is the hertz (Hz); 1 Hz is equal to 1 cycle per second. In one cycle there is a positive variation from equilibrium, a return to equilibrium, then a negative variation, and return to equilibrium. This relationship is often described in terms of the sine wave, and the frequency referred to is that of an equivalent sine-wave variation in the parameter under discussion. *See* SINE WAVE.

Frequency is a convenient means for describing the various ranges of interest in wave motion. For example, audible sound is between approximately 20 and 20,000 Hz. Infrasonic frequencies are below approximately 20 Hz; sound having frequencies above the audible is termed ultrasonic. Electromagnetic waves vary in frequency from less than 1 Hz for commutated direct current up to 10^{23} Hz for the most energetic γ-rays that have been observed. Within this range, typical approximate frequency ranges are: AM radio in the United States, 550 to 1700 kHz; FM radio, 88 to 108 MHz; visible light, 4×10^{14} to 7.5×10^{14} Hz. *See* ANGULAR FREQUENCY; WAVE MOTION.

[WILLIAM J. GALLOWAY]

Fringe

The part of optics using the light or dark bands produced by interference or diffraction of light. Distances between fringes are usually very small, because of the short wavelength of light. Fringes are clearer and more numerous when produced with light of a single color.

Diffraction fringes are formed when light from a point source, or from a narrow slit, passes by an opaque object of any shape. The Fraunhofer diffraction fringes produced when the light approaches and leaves the obstacle in essentially plane waves are especially important in the theory of optical instruments. *See* DIFFRACTION; RESOLVING POWER (OPTICS).

Interference fringes are obtained by bringing together two or more beams of light that have originated from a common source. This is usually accomplished by means of an apparatus especially designed for the purpose called an interferometer, although interference fringes may also be seen in nature. Examples are the colors in a soap film and in an oil film on water. When the fringes are controllable, for example, by changing the paths traversed by the two beams in an interferometer, they are valuable for accurate measurement of small distances and of slight differences in refractive index. *See* INTERFERENCE OF WAVES; INTERFEROMETRY.

If the light from a laser is passed through an interferometer, features such as coherence and phase fluctuations may be checked by observing the sharpness and drift of the fringes. *See* LASER.

[F. A. JENKINS/W. W. WATSON]

Fundamental frequency

The lowest frequency at which a system vibrates freely. The mode of vibration having this frequency is called the fundamental mode. For a complex wave consisting of a sum of sinusoidal components, that component having the lowest frequency is the fundamental and its frequency is the fundamental frequency. *See* MODE OF VIBRATION.

[ROBERT W. YOUNG]

Fundamental interactions

Fundamental forces that act between the elementary particles of which all matter is assumed to be composed.

Four fundamental interactions. At present, four fundamental interactions are distinguished.

Gravitational interaction. This interaction manifests itself as a long-range force of attraction between all elementary particles. The force law between two particles of masses m_1 and m_2 separated by a distance r is well approximated by the Newtonian expression $G_N(m_1 m_2/r^2)$, where G_N is the Newtonian constant ($G_N = 6.6720 \pm 0.0041 \times 10^{-8}$ cm^3 g^{-1} s^{-2}). The dimensionless constant $(G_N m_e m_p)/\hbar c$ is usually taken as the constant characterizing the gravitational interaction (m_e, m_p are electron and proton masses; $2\pi\hbar =$ Planck's constant, $c =$ velocity of light).

Electromagnetic interaction. This interaction is responsible for the long-range force of repulsion of like and attraction of unlike electric charges. The dimensionless constant characterizing the strength of electromagnetic interaction is the fine-structure constant $\alpha = e^2/\hbar c = 1/(137.03604 \pm 0.00011)$, where e is the (unrationalized electrostatic) electron charge. At comparable distances, the ratio of gravitational to electromagnetic interactions (as determined by the strength of respective forces between an electron and a proton) is given by the ratio of the constants $G_N m_e m_p/e^2 \approx 4 \times 10^{-37}$.

In modern quantum field theory, the electromagnetic interaction and the forces of attraction or repulsion between charged particles are pictured as arising secondarily as a consequence of the primary process of emission of one or more photons (quanta of light) by an accelerating electric charge (in accordance with Maxwell's equations) and the subsequent reabsorption of these quanta by a second charged particle. The space-time diagram (first introduced by R. F. Feynman) for this exchange is shown in Fig. 1. The same picture may also be valid for the gravitational interaction (in accordance with the quantum version of A. Einstein's gravitational equations), however, with the exchange of a zero-rest-mass gravitons (*g*) rather than the zero-rest-mass photons. (The physical existence of the graviton has, however, not yet been demonstrated.)

In accordance with this picture, the electromagnetic interaction is usually represented by reaction (1), where γ is the photon, emitted by the

$$e + P \rightarrow (e + \gamma) + P \rightarrow e + (P + \gamma) \rightarrow e + P \quad (1)$$

electron and reabsorbed by the proton. For this in-

FUNDAMENTAL
INTERACTIONS

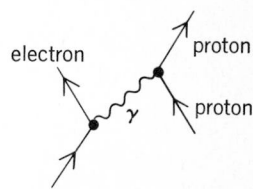

Fig. 1. Feynman diagram of electromagnetic interaction between an electron and a proton.

teraction, and also for the gravitational interaction $e+P \rightarrow (e+g)+P \rightarrow e+(P+g) \rightarrow e+P$, the nature of the participating particles (electron e and proton P) is the same, before and after the interaction. *See* LIGHT; MAXWELL'S EQUATIONS; PHOTON; QUANTUM ELECTRODYNAMICS; QUANTUM FIELD THEORY; QUANTUM MECHANICS.

Weak nuclear interactions. The third fundamental interaction is the weak nuclear interaction, whose characteristic strength for low-energy phenomena, measured by the so-called Fermi constant, G_F ($G_F = 1.026 \times 10^{-5}\ m_P{}^{-2}\hbar^3/c$), is a thousand times weaker than electromagnetic. Unlike electromagnetism and gravitation, weak interactions are short-range interactions, with a force law of the type $(1/r^2)e^{-Kr}$, the range of the force K^{-1} being much smaller than 10^{-15} cm. Also, unlike electromagnetism and gravitation, weak interactions, until 1973, appeared always to change the nature of the interacting particles, as in reactions (2), where P = proton, N = neutron, μ^- = muon, ν_e and ν_μ are the electron and muon neutrinos, and $\bar{\nu}_e$ and $\bar{\nu}_\mu$ are the corresponding antineutrinos. In reaction (2a), for example, the weak interaction transforms a proton into a neutron and at the same time an electron into a neutrino.

$$P + e^- \xrightarrow{\text{Weak}} N + \nu_e \quad \text{(this reaction is equivalent to}$$
$$\beta\text{-decay of the neutron } N \rightarrow P + e^- + \bar{\nu}_e) \quad (2a)$$

$$P + \mu^- \xrightarrow{\text{Weak}} N + \nu_\mu \quad \text{(muon capture by a proton}$$
$$\text{with the emission of a neutrino)} \quad (2b)$$

$$\mu^- + \nu_e \xrightarrow{\text{Weak}} e^- + \nu_\mu \quad \text{(this reaction is equivalent}$$
$$\text{to muon decay } \mu^- \rightarrow e^- + \bar{\nu}_e + \nu_\mu) \quad (2c)$$

One of the most crucial discoveries in particle physics was the discovery in 1973 of the so-called neutral currents, which manifest themselves through weak interactions of the type where the nature of the interacting particles is not changed during the interaction, as in reactions (3).

$$\nu_\mu + e^- \xrightarrow{\text{Weak}} \nu_\mu + e^-$$
$$\nu_\mu + P \xrightarrow{\text{Weak}} \nu_\mu + P$$
$$\nu_e + N \xrightarrow{\text{Weak}} \nu_e + N \qquad (3)$$
$$e + P \xrightarrow{\text{Weak}} e + P$$

In contrast to gravitation, electromagnetism, and strong nuclear interactions, weak interactions violate conservation of left-right and particle-antiparticle symmetries. *See* PARITY (QUANTUM MECHANICS); SYMMETRY LAWS; WEAK NUCLEAR INTERACTIONS.

Strong nuclear interaction. The fourth fundamen-

tal interaction is the strong nuclear interaction between protons and neutrons, which resembles the weak nuclear interaction in being short-range, though the range is approximately 10^{-13} cm rather than $\ll 10^{-15}$ cm. As the name implies, within this range of distances the strong force overshadows all other forces between these particles, with a characteristic strength parameter of the order of unity compared with the electromagnetic strength parameter $\alpha \approx 1/137$.

Protons and neutrons are themselves believed to be made up of yet more basic entities, the "up" (u) and "down" (d) quarks. Each quark is assumed to be endowed with one of three color quantum numbers [conventionally labeled red (r), yellow (y), and blue (b)]. The strong nuclear force can be pictured as arising through an exchange of gluons (G), color-carrying quanta of the strong force, analogous to photons in electromagnetism, which are exchanged between quarks (contained inside protons and neutrons) as in reaction (4). Since neu-

$$\text{Quark} + \text{quark} \rightarrow (\text{quark} + \text{gluon}) + \text{quark}$$
$$\rightarrow \text{quark} + (\text{gluon} + \text{quark}) \quad (4)$$
$$\rightarrow \text{quark} + \text{quark}$$

trinos, electrons, and muons do not contain quarks, their interactions among themselves or with protons and neutrons do not exhibit the strong nuclear force. *See* COLOR (QUANTUM MECHANICS); GLUONS; LEPTON; QUANTUM CHROMODYNAMICS; QUARKS; STRONG NUCLEAR INTERACTIONS.

Properties of the four fundamental interactions are summarized in the table.

Unification of interactions. Ever since the discovery and clear classification of these four interactions, particle physicists have attempted to unify these interactions as aspects of one basic interaction between all matter. The work of M. Faraday and J. C. Maxwell in the 19th century, which united the distinct forces of electricity and magnetism as aspects of one single interaction, is a model for the unification idea.

Gravitation and electromagnetism. The first attempt in this direction was that of Einstein, who, having succeeded in understanding gravitation as a manifestation of the curvature of space-time, tried to comprehend electromagnetism as another geometrical manifestation of the properties of space-time, thus achieving a unification between these forces. In this attempt, to which he devoted all his later years, he is considered to have failed. *See* UNIFIED FIELD THEORY.

Weak and electromagnetic interactions. A totally different type of unification of weak and electromagnetic interactions was suggested (employing the gauge principle of H. Weyl, C. N. Yang, R. Mills and R. Shaw) by S. Glashow and by Abdus

Properties of interactions

Interaction	Characteristic strength	Range	Exchanged quanta
Gravitational	10^{-39}	Long-range	Gravitons
Electromagnetic	10^{-2}	Long-range	Photons
Weak nuclear	10^{-5}	Short-range, 10^{-15} cm	W^+, Z^0, W
Strong nuclear	1	Short-range, 10^{-13} cm	Gluons

Salam and J. C. Ward. This followed a parallel between these two interactions, pointed out by J. S. Schwinger, providing one assumed that the then known classes of weak interactions (2) were mediated by exchanges of W^+ and W^-, quanta of weak interactions, just as electromagnetism is mediated by the exchange of quanta of light, the photons. The empirical properties of weak interaction phenomena, extensively experimented on during 1957, had revealed that if $W^\pm$ existed, these quanta of the weak interaction must carry an intrinsic spin of magnitude $\hbar$, just like the photon, the quantum of electromagnetism, which also carries spin $\hbar$. If the assumption is made that the intrinsic coupling strength of weak interactions is the same as for electromagnetism, that is, the characteristic coupling parameter for both these interactions is of the order of the fine-structure constant $\alpha \approx 1/137$, one may then deduce that the effective constant G_F for weak interactions is approximately $10^{-5}\, m_p^{-2}\hbar^3/c$, provided the masses of the $W^\pm$ particles are in excess of $\sqrt{\alpha\hbar^3/G_F c}$, that is, 37 GeV or greater.

One important consequence of this unification is the prediction of the existence of weak interactions represented by reactions (3), or, equivalently, the existence of a new heavy intermediate weak-quantum (called Z^0) which, unlike the W^+ and W^- (which carry the same electric charges as the proton and the electron), would be electrically neutral.

This unified theory was further elaborated on by Weinberg and Salam, who gave estimates of the strength of the predicted new weak interactions (3), mediated by Z^0 [such as $\nu + P \to (\nu + Z^0) + P \to \nu + (Z^0 + P) \to \nu + P$] and the precise masses of $W^\pm$ and Z^0, using ideas of spontaneous symmetry breaking.

In summary, the unified theory of weak and electromagnetic interactions of Salam and Weinberg predicts that, first, the $W^\pm$ bosons must exist with masses of the order of 80 GeV, and second, that weak current phenomena of reactions shown in (3) must be observed with strengths comparable to other weak phenomena of reactions shown in (2). The predicted mass of the Z^0 particle mediating these interactions is of the order of 90 GeV. The theory is based on a deep fundamental principle: the gauge principle.

Since 1973, the verification of the prediction about neutral-current phenomena of reactions shown in (3) at various accelerator laboratories has given decisive support to the notion of a basic unity between weak and electromagnetic interactions into a single so-called electroweak interaction. The most crucial experiment in this respect, carried out at Stanford during 1978, exhibited interference effects between the photon (γ) and the Z^0 particle in scattering of electrons from protons (Fig. 2). One of the major predictions for experiments still to be carried out is the actual production of $W^\pm$ and Z^0 particles in the laboratory when sufficiently energetic beams of electrons and positrons or protons and antiprotons are available for experimentation (through reactions such as $P + \bar{P} \to Z^0 \to W^+ + W^-$). For energies in excess of $W^\pm$ and Z^0 masses, weak interactions are expected to exhibit their primitive strength of the order of α. The weakness of weak interactions is apparently a transitory phenomenon, related to the fact that all experimentation that has so far been carried out is

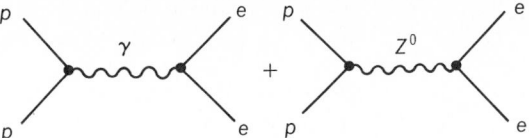

Fig. 2. Feynman diagrams corresponding to the Stanford (1978) experiment, which showed the interference effects of protons and Z^0 quanta in electron-proton scattering.

basically exploring a relatively low-energy regime, much below the masses of $W^\pm$ and Z^0.

Inclusion of strong interactions. The gauge unification of weak and electromagnetic interactions, which started with the observation that the mediating quanta ($W^\pm$, Z^0, γ) for these interactions possess intrinsic spin $\hbar$, can be carried further to include strong nuclear interactions as well, if strong interactions are also mediated through quanta (gluons) carrying spin $\hbar$. Indirect experimental evidence exists for the existence of gluons, and it should be possible to verify if their spin is indeed $\hbar$. If this is so, a complete gauge unification of all three forces, electromagnetic, weak-nuclear, and strong-nuclear, into a single electronuclear force seems plausible. Such a unification necessarily means that the distinction between quarks on the one hand and neutrinos, electrons, and muons on the other must disappear and that all interactions (weak, electromagnetic, and strong) are facets of a universal force, with primitive strength of order of α, this universality manifesting itself for experiments carried out at sufficiently high energies. Another consequence of this disappearance of distinction between quarks and neutrinos and electrons appears to be the possiblity, first predicted by J. C. Pati and Salam and elaborated afterwards by Glashow and H. Georgi, of proton decays (with a lifetime of approximately 10^{31} years) into neutrinos and pions or positrons and pions. The discovery of proton instability against decay into these particles would be an epic discovery and a direct confirmation of the unification ideas.

Prospects. Future work in particle physics is likely to be concerned with the detailed elaboration and experimental verification of these unification ideas, between strong, weak, and electromagnetic interactions. It is possible that the gravitational interaction will also be unified with the other three interactions by extensions of gauge ideas to the so-called supersymmetry between particles (e, ν, μ, quarks) and quanta ($\gamma, W^\pm, Z^0, G, g$). A clue that such a unification does take place may lie in the empirical and theoretical expression (5) which

$$\alpha^{-1} \approx \log(G_N m_e^2/c\hbar) \qquad (5)$$

relates the fine structure constant α to the newtonian constant G_N. *See* ELEMENTARY PARTICLE.

[ABDUS SALAM]

Bibliography: S. Glashow, Partial symmetries of weak interactions, *Nucl. Phys.*, 22:579–588, 1961; F. J. Hasert et al., Search for elastic muon-neutrino electron scattering, *Phys. Lett.*, 46B:121–124, 1973; J. C. Pati and A. Salam, Is baryon number conserved?, *Phys. Rev. Lett.*, 31:661–664, 1973; C. Prescott et al., Parity non-conservation in inelastic electron scattering, *Phys. Lett.*, 77B:347–352, 1978; A. Salam, Weak and electromagnetic interactions, *Nobel Symposium*, Gothenburg, pp. 367–377, 1968; A. Salam and J. Strathdee, Quantum gravity and infinities in quantum electrodynamics, *Lett. Nuovo Cimento*, 4:101–108, 1970; A. Salam and J. C. Ward, Weak and electromagnetic interactions, *Nuovo Cimento*, 11:568–577, 1959; A. Salam and J. C. Ward, Weak and electromagnetic interactions, *Phys. Lett.*, 13:168–171, 1964; S. Weinberg, A model of leptons, *Phys. Rev. Lett.*, 19: 1264–1266, 1967.

Galilean transformations

The family of mathematical transformations used in Newtonian mechanics to relate the space and time variables of uniformly moving (inertial) reference systems.

In the simple case of two similarly oriented cartesian reference frames, moving along their common (x,x') axis, the transformation equations can be put in the form of Eqs. (1), where x, y, z and

$$x' = x - vt \qquad y' = y \qquad z' = z \qquad t' = t \qquad (1)$$

x', y', z' are the space coordinates of a given particle, and v is the speed of one system relative to the other.

The transformation equations for cartesian reference frames having arbitrary displacements and orientations take the more general form ($x_1 = x$, $x_2 = y$, $x_3 = z$) of Eqs. (2), where a_1, a_2, a_3, a_4 and v_1,

$$x'_j = \sum_{k=1}^{3} c_{jk}(x_k - a_k - v_k t) \qquad (2)$$

$$t' = t - a_4$$

v_2, v_3 are arbitrary real numbers and the coefficients (c_{jk}) are constants. The matrix $C = [c_{jk}]$ is real and orthogonal, so that it satisfies the condition $C^{-1} = C_t$, where C^{-1} and C_t are the inverse and transposed matrices of C.

The Galilean transformations form a 10-parameter group which can be generated from translations of the space and time coordinates, rotations of the space coordinate frame, and transformations to moving reference frames. *See* FRAME OF REFERENCE.

[E. L. HILL]

Galvanomagnetic effects

Electrical and thermal phenomena occurring when a current-carrying conductor or semiconductor is placed in a magnetic field. The galvanomagnetic effects are closely related to the thermomagnetic effects. Both sets of effects yield important information on the band structure of metals and semiconductors and on the nature of the conductivity process. *See* THERMOMAGNETIC EFFECTS.

Let the electric current density j be transverse to the magnetic field H_z, for example, along x. Then the following transverse-transverse effects are observed:

1. Hall effect, an electric field along y, as shown in Eq. (1), where R is the Hall coefficient. According to experimental conditions, R may be the adiabatic or the isothermal Hall coefficient. *See* HALL EFFECT.

$$E_y = Rj_x H_z \qquad (1)$$

ing to experimental conditions, R may be the adiabatic or the isothermal Hall coefficient. *See* HALL EFFECT.

2. Ettingshausen effect, a temperature gradient

along y, as in Eq. (2), where P is the Ettingshausen coefficient.

$$\frac{\partial T}{\partial y} = Pj_x H_z \qquad (2)$$

Also the following transverse-longitudinal effects are observed:

3. Transverse magnetoresistance, an electrical potential change along x. *See* MAGNETORESISTANCE.

4. Nernst effect, a temperature gradient along x.

Let the electric current density j be along H. Then, the most important effect is longitudinal magnetoresistance, or an electrical potential change along H.

These effects arise from the Lorentz force on moving charges in a magnetic field. They are complicated by interaction between these charges and the lattice in which they move. For this reason, the signs of R and P may be negative or positive, depending upon composition and in many cases on temperature; the magnitudes show tremendous range.   [ELIHU ABRAHAMS; FREDERIC KEFFER]

Bibliography: N. W. Ashcroft and N. D. Mermin, *Solid State Physics*, 1976; S. Fluegge (ed.), *Handbuch der Physik*, vol. 20, 1957; C. Kittel, *Introduction to Solid State Physics*, 5th ed., 1976; F. Seitz and D. Turnbull (eds.), *Solid State Physics*, vol 5, 1957, and suppl. 4, 1963; H. Weiss, *Structure and Application of Galvanomagnetic Devices*, 1969.

Gamma function

A particular mathematical function that can be used to express many definite integrals. There are, however, no significant applications where the gamma function by itself constitutes the essence of the solution. Instead it occurs usually in connection with other functions, such as Bessel functions and hypergeometric functions.

A special case of the gamma function is the factorial $n! = 1 \cdot 2 \cdot 3 \cdot \cdots \cdot n$ (for example, $1! = 1$, $2! = 2$, $3! = 6$, $4! = 24$). It is defined only for integral positive values of n. The factorial occurs, for instance, in the expansion

$$\exp z = 1 + z/1! + z^2/2! + z^3/3! + \cdots$$

The binomial coefficient (N/n) can be expressed in terms of factorials as $N!/[n!(N-n)!]$. Many occurrences of the factorial are found in combinatorial theory (for instance, $n!$ is the number of permutations of n different elements), in probability theory, and in the applications of this theory to statistical mechanics. For large values of n, the factorial can be easily, although only approximately, computed with Stirling's formula,

$$n! \approx (n/e)^n (2\pi n)^{1/2}$$

This formula is in error by a factor, which lies, for all $n \geq 1$, between 1 and $1/(11n)$. For $n = 10$, Stirling's formula gives approximately $3598695.6 \cdots$ instead of the accurate value $10! = 3628800$.

The gamma function can be considered as a certain interpolation of the factorial to nonintegral values of n. It is defined by the definite integral

$$\Gamma(z) = \int_0^\infty u^{z-1} e^{-u} \, du.$$ This definition is applicable to all complex values of z with positive real part. The value of $\Gamma(n+1)$ coincides with $n!$ if $n = 1, 2,$

3, ... The definition of $\Gamma(z)$ can be extended to all complex values of z by the principle of analytic continuation.

Other notations are in use: $z!$ or $\Pi(z)$ for $\Gamma(z+1)$. Sometimes the phrase factorial function is used for $z!$ or $\Pi(z)$ when z is arbitrary.

From its definition follow the fundamental properties of the gamma function:

1. $\Gamma(z+1)=z\Gamma(z)$ (difference equation or functional equation).
2. $\Gamma(z)\Gamma(1-z)=\pi/\sin(\pi z)$,
   $\Gamma(1/2+z)\,\Gamma(1/2-z)=\pi/\cos(\pi z)$.
3. $\Gamma(z)\,\Gamma(z+1/2)=\pi^{1/2}2^{1-2z}\Gamma(2z)$ (Legendre's duplication formula).
4. $\Gamma(z)$ is a regular function in the complex z plane except for the points $z=-m$ $(m=0, 1, 2, \ldots)$, where it goes to infinity, as does $[(-1)^m m!\,(z+m)]^{-1}$, if z is close to $-m$.
5. As shown in the figure, $\Gamma(z)$ is real for real values of z, $\Gamma(1)=0!=1$, $\Gamma(1/2)=\pi^{1/2}$.

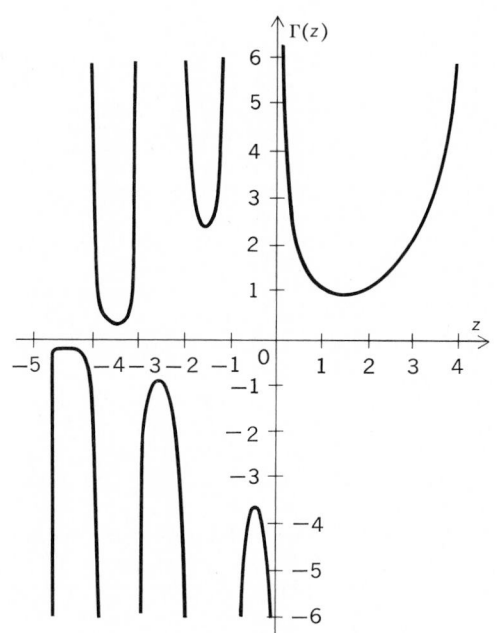

Depiction of gamma function $\Gamma(z)$ for real values of z.

In the angular domain $-\pi < \arg z < \pi$, there exists an expansion of the gamma function, namely,

$$\Gamma(z+1)$$
$$= z^{z+1/2}e^{-z}(2\pi)^{1/2}\left[1+\frac{1}{12z}+\frac{1}{288z^2}+\cdots\right]$$

which is not convergent but nevertheless useful for the numerical computation of $\Gamma(z)$ for large z. It gives approximate numerical values of the gamma function for large values of z if one takes only a finite number of terms into account. In particular for real positive values of z, the error is less than the last term taken into account. This expansion generalizes Stirling's formula.

Psi function. The logarithmic derivative of the gamma function $d\ln\Gamma(z)/dz = \Gamma'(z)/\Gamma(z)$ is denoted by $\psi(z)$. This function has the properties:

$$\psi(z+1)-\psi(z)=\frac{1}{z}$$

$$\psi(1)=-C=-0.5772157\ldots$$

$$\psi(n+1)=-C+1+\frac{1}{2}+\frac{1}{3}+\cdots+\frac{1}{n}$$
$$\text{for } n=1,2,3,\ldots$$

$$\psi(1/2)=-C-2\ln 2$$

$$\psi(z)=-C+\sum_{n=0}^{\infty}\left(\frac{1}{1+n}-\frac{1}{z+n}\right)$$

Beta function. The beta function generalizes the binomial coefficient (N/n) to noninteger values of N and n. The beta function $B(x,y)$ is by definition equal to $\Gamma(x)\Gamma(y)/\Gamma(x+y)$. Therefore one has $(N/n)^{-1}=(N+1)B(n+1, N-n+1)$.

Some of the definite integrals, whose values can be expressed by the gamma function, the psi function, and the beta function are:

$$\int_0^\infty \exp(-at^n)t^z\,dt$$

$$=(z+1)^{-1}a^{-(z+1)/n}\Gamma\left(\frac{z+n+1}{n}\right)$$
$$(\operatorname{Re} a>0, \operatorname{Re} z>-1)$$

$$\psi(z)=\int_0^\infty\left[\frac{e^{-t}}{t}-\frac{e^{-zt}}{1-e^{-t}}\right]dt$$

$$=-C+\int_0^1\frac{1-t^{z-1}}{1-t}\,dt \qquad (\operatorname{Re} z>0)$$

$$B(x,y)=\int_0^1 t^{x-1}(1-t)^{y-1}\,dt$$

$$=2\int_0^{\pi/2}\sin^{2x-1}\zeta\cos^{2y-1}\zeta\,d\zeta$$

$$(\operatorname{Re} x>0, \operatorname{Re} y>0)$$

See BESSEL FUNCTIONS. [JOSEF MEIXNER]

Bibliography: E. Artin, *The Gamma Function*, 1964; R. Askey, *Orthogonal Polynomials and Special Functions*, 1975; A. Erdelyi (ed.), *Higher Transcendental Functions*, vol. 1, 1953; W. Magnus et al., *Formulas and Theorems for the Special Functions of Mathematical Physics*, 3d ed., 1966; I. N. Sneddon, *Special Functions of Mathematical Physics and Chemistry*, 1980; E. T. Whittaker and G. N. Watson, *Course of Modern Analysis*, 4th ed., 1940, reprint 1976.

Gas

A phase of matter characterized by relatively low density, high fluidity, and lack of rigidity. A gas expands readily to fill any containing vessel. Usually a small change of pressure or temperature produces a large change in the volume of the gas. The equation of state describes the relation between the pressure, volume, and temperature of the gas. In contrast to a crystal, the molecules in a gas have no long-range order.

At sufficiently high temperatures and sufficiently low pressures, all substances obey the ideal gas, or perfect gas, equation of state, shown as Eq. (1), where p is the pressure, T is the absolute temperature, v is the molar volume, and R is the gas constant. Absolute temperature T expressed on the Kelvin scale is related to temperature t expressed on the Celsius scale as in Eq. (2).

$$pv = RT \qquad (1)$$

$$T = t + 273.16 \qquad (2)$$

The gas constant is

$$R = 82.0567 \text{ cm}^3\text{-atm/(mole)(K)}$$
$$= 82.0544 \text{ ml-atm/(mole)(K)}$$

The molar volume is the molecular weight divided by the gas density.

Empirical equations of state. At lower temperatures and higher pressures, the equation of state of a real gas deviates from that of a perfect gas. Various empirical relations have been proposed to explain the behavior of real gases. The equations of J. van der Waals (1899), Eq. (3), of P. E. M. Berthelot (1907), Eq. (4), and F. Dieterici (1899), Eq. (5),

$$\left(p + \frac{a}{v^2}\right)(v - b) = RT \qquad (3)$$

$$\left(p + \frac{a}{Tv_2}\right)(v - b) = RT \qquad (4)$$

$$pe^{a/vRT}(v - b) = RT \qquad (5)$$

are frequently used. In these equations, a and b are constants characteristic of the particular substance under consideration. In a qualitative sense, b is the excluded volume due to the finite size of the molecules and roughly equal to four times the volume of 1 mole of molecules. The constant a represents the effect of the forces of attraction between the molecules. In particular, the internal energy of a van der Waals gas is $-a/v$. None of these relations gives a good representation of the compressibility of real gases over a wide range of temperature and pressure. However, they reproduce qualitatively the leading features of experimental pressure-volume-temperature surfaces.

Schematic isotherms of a real gas, or curves showing the pressure as a function of the volume

for fixed values of the temperature, are shown in Fig. 1. Here T_1 is a very high temperature and its isotherm deviates only slightly from that of a perfect gas; T_2 is a somewhat lower temperature where the deviations from the perfect gas equation are quite large; and T_c is the critical temperature. The critical temperature is the highest temperature at which a liquid can exist. That is, at temperatures equal to or greater than the critical temperature, the gas phase is the only phase that can exist (at equilibrium) regardless of the pressure. Along the isotherm for T_c lies the critical point, C, which is characterized by zero first and second partial derivatives of the pressure with respect to the volume. This is expressed as Eq. (6). At temperatures

$$(\partial p/\partial v)_c = (\partial^2 p/\partial v^2)_c = 0 \qquad (6)$$

lower than the critical, such as T_3 or T_4, the equilibrium isotherms have a discontinuous slope at the vapor pressure. At pressures less than the vapor pressure, the substance is gaseous; at pressures greater than the vapor pressure, the substance is liquid; at the vapor pressure, the gas and liquid phases (separated by an interface) exist in equilibrium.

Along one of the isotherms of the empirical equations of state discussed above, the first and second derivatives of the pressure with respect to the volume are zero. The location of this critical point in terms of the constants a and b is shown below; p_c and v_c are the pressure and volume at the critical temperature.

	Van der Waals	*Berthelot*	*Dieterici*
p_c	$\dfrac{a}{27b^2}$	$\left(\dfrac{aR}{216b^3}\right)^{1/2}$	$\dfrac{a}{4e^2b^2}$
v_c	$3b$	$3b$	$2b$
T_c	$\dfrac{8a}{27Rb}$	$\left(\dfrac{8a}{27Rb}\right)^{1/2}$	$\dfrac{a}{4Rb}$
$\dfrac{p_c v_c}{RT_c}$	0.3750	0.3750	0.2706

Some typical values of $p_c v_c/RT_c$ for real gases are as follows: 0.30 for the noble gases, 0.27 for most of the hydrocarbons, 0.243 for ammonia, and 0.232 for water. The van der Waals and Berthelot equations of state, Eqs. (3) and (4), cannot quantitatively reproduce the critical behavior of real gases because no substance has a value of $p_c v_c/RT_c$ as large as 0.375. The Dieterici equation, Eq. (5), gives a good representation of the critical region for the light hydrocarbons but does not represent well the noble gases or water.

At temperatures lower than the critical point, the analytical equations of state, such as the van der Waals, Berthelot, or Dieterici equations, give S-shaped isotherms as shown in Fig. 2. From thermodynamic considerations, the vapor pressure is determined by the requirement that the cross-hatched area DEO be equal to the cross-hatched area AOB. Under equilibrium conditions, the portion of this isotherm lying between A and D cannot occur. However, if a gas is suddenly compressed, points along the segment AB may be realized for a short period until enough condensation nuclei form to create the liquid phase. Similarly, if a liquid is suddenly overexpanded, points along DE may occur for a short time. For low temperatures, the

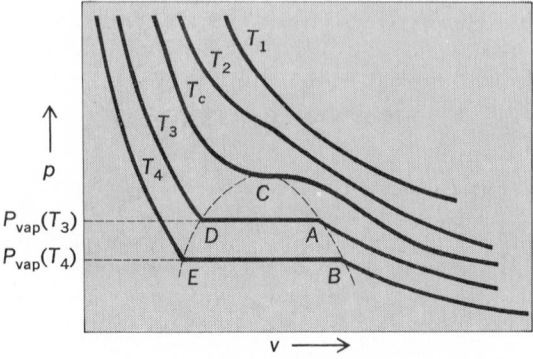

Fig. 1. Schematic isotherms of a real gas. C is the critical point. Points A and B give the volume of gas in equilibrium with the liquid phase at their respective vapor pressures. Similarly, D and E are the volumes of liquid in equilibrium with the gas phase.

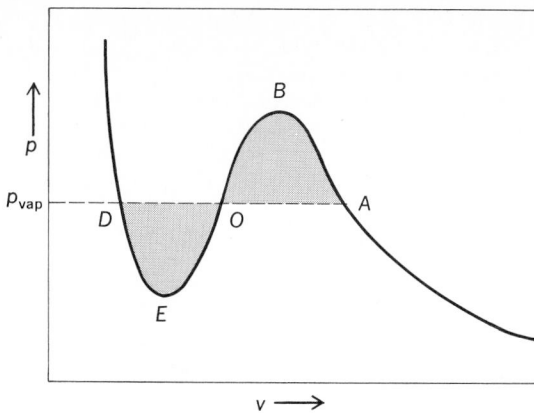

Fig. 2. Schematic low-temperature isotherm as given by van der Waals, Berthelot, or Dieterici equations of state. Here the line *DOA* corresponds to the vapor pressure. The point *A* gives the volume of the gas in equilibrium with the liquid phase, and *D* gives the volume of the liquid. The segment of the curve *DE* represents overexpansion of the liquid. The segment *AB* corresponds to supersaturation of the vapor. However, the segment *EOB* could not be attained experimentally.

point E may represent a negative pressure corresponding to the tensile strength of the liquid. However, the simple analytical equations of state can not be used for a quantitative estimate of these transient phenomena. Actually, it is easy to show that the van der Waals, Berthelot, and Dieterici equations give poor representations of the liquid phase since the volume of most liquids (near their freezing point) is considerably less than the constant b.

Principle of corresponding states. In the early studies, it was observed that the equations of state of many substances are qualitatively similar and can be correlated by a simple scaling of the variables. To describe this result, the reduced or dimensionless variables, indicated by a subscript r, are defined by dividing each variable by its value at the critical point. These variables are given in Eqs. (7)–(9).

$$p_r = p/p_c \qquad (7)$$
$$T_r = T/T_c \qquad (8)$$
$$v_r = v/v_c \qquad (9)$$

In its most elementary form, the principle of corresponding states asserts that the reduced pressure, p_r, is the same function of the reduced volume and temperature, v_r and T_r, for all substances. An immediate consequence of this statement is the statement that the compressibility factor, expressed as Eq. (10), is a universal function of

$$z = pv/RT \qquad (10)$$

the reduced pressure and the reduced temperature. This principle is the basis of the generalized compressibility chart of O. A. Hougen and K. M. Watson (Fig. 3). This chart was derived from data on the equation-of-state behavior of a number of common gases.

It follows directly from the principle of corresponding states that the compressibility factor at the critical point z_c should be a universal constant. It is found experimentally that this constant varies somewhat from one substance to another. On this

account, A. L. Lydersen, R. A. Greenkorn, and Hougen (see Hougen in bibliography) have developed empirical tables of the compressibility factor and other thermodynamic properties of gases as functions of the reduced pressure and reduced temperature for a range of values of z_c. Such generalized corresponding-states treatments are very useful in predicting the behavior of a substance on the basis of scant experimental data.

Theoretical considerations. The equation-of-state behavior of a substance is closely related to the manner in which the constituent molecules interact. Through statistical mechanical considerations, it is possible to obtain some information about this relationship. If the molecules are spherically symmetrical, the force acting between a pair of molecules depends only on r, the distance between them. It is then convenient to describe this interaction by means of the intermolecular potential $\varphi(r)$ defined so that the force is the negative of the derivative of $\varphi(r)$ with respect to r. *See* INTERMOLECULAR FORCES.

Two theoretical approaches to the equation of state have been developed. In one of these approaches, the pressure is expressed in terms of the partition function Z and the total volume V of the container in the manner of Eq. (11). Here k is the

$$p = kT(\partial \ln Z/\partial V) \qquad (11)$$

Boltzmann constant or the gas constant divided by the Avogadro number N_0, $k = R/N_0$. For a gas made up of spherical molecules or atoms with no internal structure, the partition function is given as Eq. (12). In this expression, φ_{ij} is the energy of

$$Z = \frac{1}{N!}\left(\frac{2\pi mkT}{h^2}\right)^{3N/2}$$
$$\times \int \exp\left(-\sum_{i>j}\frac{\varphi_{ij}}{kT}\right) dv_1\, dv_2 \cdots dv_N \qquad (12)$$

interaction of molecules i and j and the summation is over all pairs of molecules, h is Planck's constant, N is the total number of molecules, and the integration is over the three cartesian coordinates of each of the N molecules. The expression for the partition function may easily be generalized

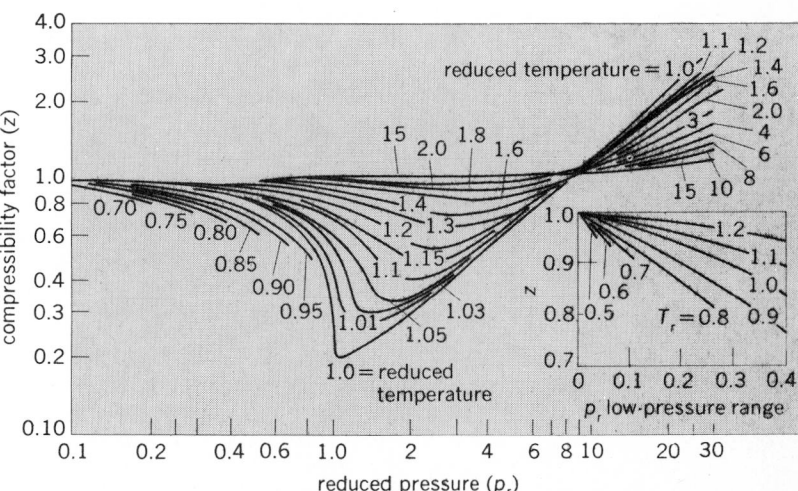

Fig. 3. The compressibility factor pv/RT as a function of the reduced pressure $p_r = p/p_c$, and reduced temperature $T_r = T/T_c$. (*From O. A. Hougen, K. M. Watson, and R. A. Ragatz, Chemical Process Principles, pt. 2, Wiley, 1959*)

to include the effects of the structure of the molecules and the effects of quantum mechanics.

In another theoretical approach to the equation of state, the pressure may be written as Eq. (13),

$$p = \frac{NkT}{V} - \frac{2\pi N^2}{3V^2} \int g(r) \frac{d\varphi}{dr} r^3 \, dr \qquad (13)$$

where $g(r)$ is the radial distribution function. This function is defined by the statement that $2\pi(N^2/V) g(r)r^2 \, dr$ is the number of pairs of molecules in the gas for which the separation distance lies between r and $r + dr$. The radial distribution function may be determined experimentally by the scattering of x-rays. Theoretical expressions for $g(r)$ are being developed.

The compressibility factor $z = pV/NkT$ may be considered as a function of the temperature, T, and the molar volume, v. In the virial form of the equation of state, z is expressed as a series expansion in inverse powers of v, as in Eq. (14). The

$$z = 1 + B(T)/v + C(T)/v^2 + \cdots \qquad (14)$$

coefficients $B(T), C(T), \ldots$, which are functions of the temperature, are referred to as the second, third, $\ldots$, virial coefficients. This expansion is an important method of representing the deviations from ideal gas behavior. From statistical mechanics, the virial coefficients can be expressed in terms of the intermolecular potential. In particular, the second virial coefficient is Eq. (15). If the

$$B(T) = 2\pi N_0 \int (1 - e^{-\varphi/kT}) \, r^2 \, dr \qquad (15)$$

intermolecular potential is known, Eq. (15) provides a convenient method of predicting the first-order deviation of the gas from perfect gas behavior. The relation has often been used in the reverse manner to obtain information about the intermolecular potential. Often $\varphi(r)$ is expressed in the Lennard-Jones (6–12) form, Eq. (16), where ϵ and

$$\varphi(r) = 4\epsilon \left[\left(\frac{\sigma}{r} \right)^{12} - \left(\frac{\sigma}{r} \right)^{6} \right] \qquad (16)$$

σ are constants characteristic of a particular substance. Values of these constants for many substances have been tabulated. In terms of these constants, the second virial coefficient has the form of Eq. (17), where $B^*(kT/\epsilon)$ is a universal func-

$$B(T) = (2/3)\pi N_0 \sigma^3 B^*(kT/\epsilon) \qquad (17)$$

tion. If all substances obeyed this Lennard-Jones (6–12) potential, the simple form of the law of corresponding states would be rigorously correct. *See* KINETIC THEORY OF MATTER.

[C. F. CURTISS; J. O. HIRSCHFELDER]
Bibliography: J. O. Hirshfelder et al., *Molecular Theory of Gases and Liquids*, 1964; R. Holub and P. Vonka, *The Chemical Equilibrium of Gaseous Systems*, 1976; O. A. Hougen, K. M. Watson, and R. A. Ragatz, *Chemical Process Principles*, pt. 2, 1959.

Gas constant

Boyle's law and Charles' law may be combined into a single expression, $pV/T = $ a constant, showing how the volume V of a given mass of gas depends upon its temperature T and pressure p. If the mass of gas chosen is 1 g-mole, then the constant, known as the gas constant, is written as R. Hence, $pV = RT$ for 1 g-mole. *See* BOYLE'S LAW; CHARLES' LAW; GAS.

The numerical value of the gas constant R is obtained by dividing the molar or gram-molecular volume of a perfect gas at the melting point of ice and a pressure of 760 mm of mercury at the same temperature (the standard atmosphere) by the absolute temperature at the ice point, $R = V_0/T_0$. The best values available for R, in various units, are given below.

$R = 8.31433 \pm 0.00034 \times 10^7$ ergs/(deg)(mole)
$\quad = 1.98717$ cal/(deg)(mole)
$\quad = 8.20544 \pm 0.00037 \times 10^{-2}$ liter-atm/(deg)(mole)

The kinetic theory of gases relates R to the specific heats of a perfect monatomic gas at constant pressure C_p and constant volume C_v. C_p is equal to $\frac{5}{2}R$ and C_v to $\frac{3}{2}R$. Maxwell's law of the equipartition of energy shows that any degree of freedom of a system possesses an energy of $\frac{1}{2}RT$ per mole, making a specific heat contribution of $\frac{1}{2}R$. A perfect monatomic gas has three degrees of freedom, three independent directions of molecular motion, and a specific heat C_v of $\frac{3}{2}R$. In a perfect diatomic gas molecules rotate in two independent directions; their specific heat C_v is $\frac{5}{2}R$. *See* KINETIC THEORY OF MATTER.

[THOMAS C. WADDINGTON]
Bibliography: R. H. Perry and C. H. Chilton (eds.), *Chemical Engineers' Handbook*, 5th ed., 1973.

Gas dynamics

The study of the motion of gases which takes into account thermal effects generated by the motion. Gas dynamics combines fluid mechanics and thermodynamics and differs from gas statistics in that there is motion and from gas kinematics in that the forces exerted on or by the gas are considered.

Scope of subject. Several other names are used to define the subject. The most important ones are aerothermodynamics, aerothermochemistry, fluid dynamics, compressible fluid flow, and supersonic aerodynamics. This terminology reflects the fact that in each particular case different aspects of gas dynamics are emphasized. Magnetogasdynamics, which includes effects due to magnetism and electricity, has applications in rocket reentry, propulsion, and astrophysics. *See* COMPRESSIBLE FLOW; FLUID DYNAMICS.

Gas dynamics deals with the motion of a continuous medium and is not concerned with the behavior of individual atoms or molecules which constitute the gas. At low pressures, however, such as may be encountered at very high altitudes, the particle mean free path is very large and continuum considerations may not be applicable. The Knudsen number, kn, which is defined as the ratio of the mean free path to a characteristic length of interest, characterizes the regimes of continuum and free molecule flow, respectively. Thus, if $kn = 0.01$, the continuum gas dynamic equations are usually valid. On the other hand, when kn is 10 or greater, the flow must be described by the Boltzmann equation. The transition region when $kn = 1$ is referred to as slip flow. *See* BOLTZMANN TRANSPORT EQUATION; SUPERAERODYNAMICS.

Fundamental relations. The fundamental conservation principles of mechanics and thermodynamics constitute the theoretical basis of gas dynamics. The conservation laws can be derived in Lagrangian form with respect to a specific mass of flowing gas or in Eulerian form with respect to the rate at which gas enters and leaves a fixed control volume in space. The Eulerian conservation laws are expressed by (1) the continuity equation (conservation of mass); (2) the Navier-Stokes equations (conservation of momentum); and (3) the energy equation (conservation of energy).

A list follows of the principal notations used in the field.

A = cross-sectional area
a = acoustic velocity
c = speed of light
c_p = specific heat at constant pressure
c_v = specific heat at constant volume
f = friction factor, as in Eq. (53)
G = mass velocity, $G = \rho V$
G' = mass flow per unit time
h = enthalpy
k = thermal conductivity
L = characteristic length
m = molecular mass
n_i = number of particles of component i per unit volume
p = pressure
p_0 = total or stagnation pressure
Q_m = heat
R, R' = gas constant
r_h = hydraulic radius
T = temperature
T_0 = total or stagnation temperature
t = time
u = internal energy
V = average gas velocity
$\mathbf{V}$ = gas vector velocity
$\mathbf{v}_i$ = particle vector velocity
w_i = rate of production of species i
x = fractional dissociation
α = Mach angle
γ = specific heat ratio, c_p/c_v
ν = kinematic viscosity
ρ = density
∇ = gradient operator
Super and subscripts
$(\ )^*$ = critical values
$(\ )_o$ = reservoir condition

Continuity equation. Two types of continuity equation can be written: global and species conservation. The global continuity equation is expressed by Eq. (1), where the first term defines the

$$\nabla \cdot (\rho \mathbf{V}) + \frac{\partial \rho}{\partial t} = 0 \qquad (1)$$

rate of change of the mass flow with respect to the space coordinates, whereas the second term, usually called the source term, indicates changes with respect to time within the control volume. *See* CALCULUS OF VECTORS.

If the flow is steady and there are no sources present, the continuity equation reduces to Eq. (2).

$$\nabla \cdot (\rho \mathbf{V}) = 0 \qquad (2)$$

For incompressible flow, it is given simply by Eq. (3).

$$\nabla \cdot \mathbf{V} = 0 \qquad (3)$$

It is often convenient in steady irrotational flow to introduce a function φ satisfying Eq. (4).

$$\nabla \varphi = \mathbf{V} \qquad (4)$$

The incompressible continuity equation, Eq. (3), becomes Eq. (5). The identity $\nabla \times (\nabla \varphi) = 0$ shows that the flow is irrotational. Equation (5) is known

$$\nabla^2 \varphi = 0 \qquad (5)$$

as Laplace's equation. Solutions to Laplace's equation which satisfy the appropriate boundary conditions are solutions of gas kinematics since they describe a flow without regard to the forces maintaining it.

If the gas undergoes chemical changes, the concentration of species is altered, which affects the spatial distribution of energy since each species has its particular velocity $\mathbf{v}_i$. Thus, in order to properly keep track of the distribution of energy, one must account for the rate of change of each specie or component by a continuity equation of the form shown in Eq. (6). The term on the right-hand side

$$\frac{\partial n_i}{\partial t} + \nabla \cdot (n_i \mathbf{v}_i) = w_i \qquad (6)$$

represents the rate of specie production. When Eq. (6) is summed for all species, one arrives at the continuity equation, Eq. (1).

Frequently a one-dimensional approach is followed in gas dynamics, in which case the properties are assumed to vary mainly in the flow direction. The integral of the steady global continuity equation is then given by Eq. (7).

$$\rho AV = \text{constant} = G' \qquad (7)$$

Momentum equation. The momentum equation expresses the conservation of momentum. It must take into consideration the effects of friction and of external body forces such as gravity, magnetism, and possibly others. Written in vector form, it is given by Eq. (8). The term D/Dt represents the mobile operator which is defined by Eq. (9). The first term on the right-hand side of Eq. (8) repre-

$$D\mathbf{V}/Dt = -\frac{1}{\rho} \nabla p + \nu \nabla^2 \mathbf{V}$$

$$+ (\tfrac{1}{3})\nu \nabla (\nabla \cdot \mathbf{V}) + \sum_{l=1}^{N} \mathbf{F}_l \qquad (8)$$

$$D/Dt = (\partial/\partial t) + \mathbf{V} \cdot \nabla \qquad (9)$$

sents the forces exerted on the fluid at the control volume boundaries; the second and third terms on the right-hand side are the viscous stresses; and F_l stands for any body force such as gravity, electric, and magnetic forces. Equation (8) in its vectorial form is quite general and applies to classical gas dynamics and aerothermochemistry as well as to magnetohydrodynamics. The specialization to particular coordinate systems, for example, cartesian coordinates, is obtained by standard vector manipulation methods.

It may not be always necessary, however, to consider all terms. For example, if electromagnetic effects are not present but viscous effects are included, one obtains the Navier-Stokes equation. *See* NAVIER-STOKES EQUATIONS.

A general solution for the Navier-Stokes equation has not been found, and only a few particular solutions exist. When the momentum equation is simplified to exclude viscous effects as well as forces on the body, one obtains Euler's equation, shown as Eq. (10). Its one-dimensional form is given by Eq. (11).

$$DV/Dt = -\frac{1}{\rho}\nabla p \qquad (10)$$

$$\frac{dp}{\rho} + V\,dV = 0 \qquad (11)$$

Equation (11) can be integrated for incompressible flow (that is, $\rho = $ constant) to yield Bernoulli's equation, shown as Eq. (12).

$$\frac{p}{\rho} + \frac{V^2}{2} = \text{constant} \qquad (12)$$

For compressible flow, one needs to consider the thermodynamics of the flow, which is done through the energy equation.

Energy equation. The energy equation expresses the principle of conservation of energy (first law of thermodynamics) as applied to a flowing gas. There are a large number of equivalent forms in which this equation can be written. The apparent differences arise from the use of such subsidiary thermodynamic relations as, for example, the equation of state $p = \rho RT$ for a perfect gas, or $p/\rho^\gamma = $ constant for an isentropic process, to express one state variable in terms of another. A fairly common form of the energy equation is given in Eq. (13). The left-hand side of Eq. (13) gives the

$$\frac{\rho D(u + \frac{1}{2}V^2)}{Dt} = \nabla \cdot p\mathbf{V}$$
$$\text{(I)}$$
$$+ \rho \sum_{l=1}^{N} \mathbf{F}_l \cdot \mathbf{V} + \nabla \cdot k\nabla T + \sum_{m=1}^{M} Q_m \qquad (13)$$
$$\text{(II)} \qquad\qquad \text{(III)} \qquad \text{(IV)}$$

change in internal and kinetic energy, which is balanced on the right-hand side by the rate at which work is done by (I) the pressure forces and (II) the body forces and (III) by the heat conducted across the boundary. Term (IV) accounts for any other energy-transfer mechanism, such as the transfer of heat generated by the dissipative action of viscosity and electrical conductivity or radiative heat transfer or transfer of electromagnetic energy.

If the flow is steady, one obtains Eq. (14). Furthermore, if the flow is adiabatic, then the work

$$\rho\mathbf{V}\cdot\nabla\left(u + \frac{1}{2}V^2 + \frac{p}{\rho}\right)$$
$$= \nabla \cdot k\nabla T + \rho\Sigma\mathbf{F}\cdot\mathbf{V} + \Sigma Q \qquad (14)$$

thermore, if the flow is adiabatic, then the work done by the shear forces Φ does not leave the system but simply raises the internal energy of the gas and is, therefore, accounted for by u. In the absence of body forces, all terms on the right-hand side of Eq. (14) are zero, and the energy equation can be integrated to yield Eq. (15). Moreover, if the gas is perfect, one obtains Eq. (16).

$$u + \frac{1}{2}V^2 + \frac{p}{\rho} = \text{constant} \qquad (15)$$

$$\frac{\gamma}{\gamma - 1}\frac{p}{\rho} + \frac{V^2}{2} = c_p T + \frac{V^2}{2} = \text{constant} \qquad (16)$$

The idealizations introduced by the concept of a perfect gas are approximately satisfied at moder-

ate temperatures and pressures. At very high pressures or temperatures or both, the real nature of the gas molecules must be taken into consideration. In gas dynamics, one is not ordinarily concerned with very high pressure; but high temperatures, and the attendant effects on the gas, are regularly encountered in missile flight, combustion, nuclear reactors, and many other technological applications. A very simple correction to the perfect equation of state can be made if the gas molecules dissociate. The equation of state then can be written as Eq. (17), where $R'/m = R$ and x is

$$p = R'Tn(1 + x) \qquad (17)$$

the fraction of the gas which is dissociated—the bookkeeping is done by the conservation of species equations. *See* VIRIAL EQUATION.

At sufficiently high temperatures, the electrons may become so energetic as to leave their orbits and become free electrons. When this happens, the gas is said to be ionized and is called a plasma. *See* PLASMA PHYSICS.

In a fairly complex situation where the gas can consist of molecules, atoms, ions, and electrons, the equation of state may have to be modified to account for more than just one dissociating species, as in Eq. (18). Another consequence of the

$$p = R'Tn(1 + \Sigma x_i) \qquad (18)$$

nonideal nature of the gas is the fact that it takes time for a molecule to dissociate, to ionize, and to combine with another species. The time that it takes to effect such a change is called relaxation time. When the relaxation time is short compared to a characteristic flow time, which might be the time it takes a fluid element to pass through a shock, one speaks of thermodynamic equilibrium; this state is characterized by Damköhler's first ratio, given by Eq. (19).

$$\text{Da}_\text{I} = \frac{t_\text{transit}}{t_\text{relaxation}} \gg 1 \qquad (19)$$

At equilibrium, the degree of ionization X is given by the Saha equation, Eq. (20), where E_0 is a characteristic parameter of the gas.

$$\frac{x^2}{1 - x^2} = \text{constant}\,\frac{T^{5/2}}{p}\exp\;-(E_0/RT)] \qquad (20)$$

For the case $\text{Da}_\text{I} \ll 1$, one speaks of frozen flow. Here the relaxation time is so large that the gas behaves nearly like a perfect gas.

When relaxation and transit times are comparable, the necessary accounting of all the possible species depends on a detailed knowledge of all possible chemical processes and the controlling rates; these rates have been established for a large number of reactions. However, the details of even a relatively simple phenomenon such as the burning of a Bunsen burner are still not completely understood.

Real gases deviate in other important aspects from a perfect gas. In the latter, the specific heat at constant pressure c_p, and at constant volume c_v are constants. In a real gas, the manner in which the molecules store, so to say, the heat must be considered. A gas that obeys the perfect gas thermal equation of state $p = \rho RT$ has the caloric equation of state given by Eq. (21), where c_{p0} and γ are

$$\frac{c_p}{c_{p0}} = 1 + \frac{\gamma - 1}{\gamma}\left[\left(\frac{\theta}{T}\right)^2 \frac{e^{\theta/T}}{(e^{\theta/T}-1)^2}\right] \qquad (21)$$

constant reference values. For $\gamma = 1.4$, the correction at 3000 K is approximately 26%. θ refers to molecular vibrational-energy constant. For air between 300 and 2500 K, the value of $\theta \approx 2800$ K.

Dimensionless parameters. Of the large number of dimensionless parameters that can be formed, there are a few that are particularly useful in effecting simplifications in the equations of motion. The simplifications are generally the consequence of one or more parameters being very large or very small.

The Knudsen number and Damköhler's ratio have been mentioned already. The Reynolds number $Re = VL/\nu$ is the dominating parameter when the effects of viscosity and the inertia of the fluid both contribute to the gas motion. The phenomena which are peculiar to gas dynamics, however, can best be categorized in terms of the Mach number. The Mach number is the ratio of the flow speed to the speed of sound. The speed of sound is the propagation velocity not only of audible sound but of any weak pressure disturbance. *See* DIMENSIONAL ANALYSIS; DYNAMIC SIMILARITY.

Gas-dynamic flow regimes can then be classified as follows.

$$
\begin{array}{ll}
M < 1 & \text{subsonic flow} \\
M = 1 & \text{sonic flow} \\
0.9 < M < 1.1 & \text{transonic flow} \\
M > 1 & \text{supersonic flow} \\
M > 5 & \text{hypersonic flow}
\end{array}
$$

Speed of sound. Consider a tube with insulated walls filled with a compressible nonviscous gas. Neglecting all external forces, Euler's equation in one dimension, Eq. (22), and continuity, Eq. (23), describe the motion.

$$\frac{\partial V}{\partial t} + V\frac{\partial V}{\partial x} = -\frac{1}{\rho}\frac{\partial p}{\partial x} \qquad (22)$$

$$\frac{\partial \rho}{\partial t} + \frac{\partial (\rho V)}{\partial x} = 0 \qquad (23)$$

For the case of small perturbations, the equation of state is given by Eq. (24), where ρ_0 is a reference density.

$$\rho = \rho_0(1 + \epsilon) \qquad (24)$$

Substituting Eq. (24) into Eqs. (22) and (23), one obtains Eqs. (25) and (26).

$$\frac{\partial V}{\partial t} = -\frac{\rho}{\rho_0}\frac{dp}{d\rho}\frac{\partial \epsilon}{\partial x} \qquad (25)$$

$$\frac{\partial \epsilon}{\partial t} = -\frac{\partial V}{\partial x} \qquad (26)$$

Eliminating ϵ between Eqs. (25) and (26), one obtains the one-dimensional wave equation, Eq. (27), where Eq. (28) defines the velocity of propagation of weak disturbances in general and sound in particular. If one assumes the transformation to be

$$\frac{\partial^2 V}{\partial t^2} = a^2\frac{\partial^2 V}{\partial x^2} \qquad (27)$$

$$a^2 = \frac{dp}{d\rho}\bigg|_{\rho = \rho_0} \qquad (28)$$

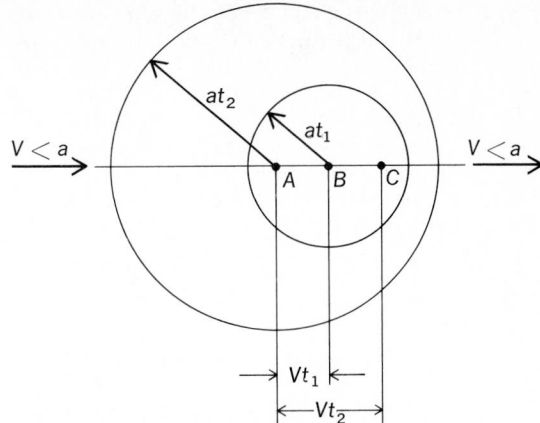

Fig. 1. Wavefronts produced by a point source moving at subsonic velocity. (*A. B. Cambel and B. H. Jennings, Gas Dynamics, McGraw-Hill, 1958*)

isentropic, which is quite reasonable, one obtains Eq. (29).

$$a^2 = \gamma R T \qquad (29)$$

The derivation of the speed of sound is illustrative of a whole class of problems that can be treated by the simple wave equation. The term waves is used in gas dynamics not only in the classical sense but also to denote wavefronts.

Consider a pulsating pressure source moving with velocity V. This source starts pulsating at time $t = 0$. In a time interval t_1, the source travels a distance Vt_1, while the signal reaches the surface of a sphere of radius at_1 (the surface of this sphere constitutes the wavefront at time t_1). At a later time t_2, the source point is at Vt_2, while the signal front is at at_2. Thus, as long as $V/a = M < 1$, the source point is always inside the outermost wavefront (Fig. 1).

Still another use of the term wave refers to a wave envelope. If the point source moves so fast that $V > a$, the wavefront spheres will no longer contain the source. The condition is shown in Fig. 2. The envelope to this family of spheres is a cone, known as the Mach cone, and the Mach angle is such that $\sin \alpha = 1/M$.

The Mach line constitutes a demarcation. In Fig. 2 the fluid outside the Mach line will receive no signal from the source. T. von Kármán has appro-

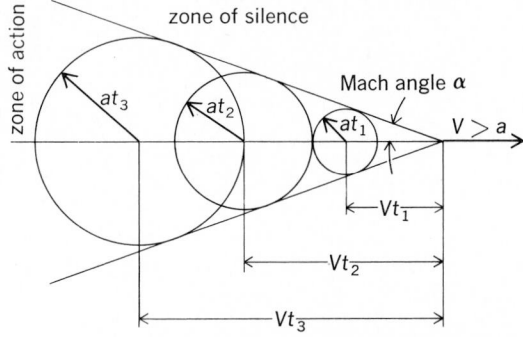

Fig. 2. Rule of forbidden signals from a point source moving at supersonic velocity. (*A. B. Cambel and B. H. Jennings, Gas Dynamics, McGraw-Hill, 1958*)

priately called this phenomenon the rule of forbidden signals and designated the region ahead of the Mach line the zone of silence and the region inside the Mach line the zone of action.

Shock waves. In the same manner in which a Mach wave is the envelope of infinitesimal disturbances, a shock wave is the envelope of finite disturbances. The steady conditions on either side of a standing shock wave can be obtained by applying the conservation laws (Fig. 3) expressed by Eqs. (30), (31), and (32), where $h = u + p/\rho$ is termed

$$\rho_1 V_1 = \rho_2 V_2 \quad \text{(continuity)} \quad (30)$$

$$p_1 + \rho_1 V_1^2 = p_2 + \rho_2 V_2^2 \quad \text{(momentum)} \quad (31)$$

$$h_1 + \frac{V_1^2}{2} = h_2 + \frac{V_2^2}{2} \quad \text{(energy)} \quad (32)$$

enthalpy, which for a perfect gas is given by $c_p T$. By a simple rearrangement of these equations, one obtains for a perfect gas the approximate expressions for the shock Mach number, Eq. (33) for

$$M_s \approx 1 + \frac{\gamma + 1}{4\gamma} \frac{p_2 - p_1}{p_1} \quad (33)$$

weak shocks and Eq. (34) for strong shocks.

$$M_s \approx \left(\frac{\gamma + 1}{2\gamma} \frac{p_2}{p_1}\right)^{1/2} \quad (34)$$

In a sound wave, $p_2 - p_1 \cong 0$ and, therefore, $M_s = 1$; for $p_2/p_1 = 4$, the shock speed is roughly twice the speed of sound, as shown in Eq. (34).

Detonation and deflagration waves. Other interesting gas dynamic waves are characterized by the same continuity and momentum equations. The energy equation, however, is modified to include a term which accounts for chemical heat release; such waves are either detonations or deflagrations.

Eliminating the kinetic energy from the energy equation by the use of the momentum equation yields Eq. (35).

$$h_1 - h_2 + Q = \tfrac{1}{2}(p_1 - p_2)\left(\frac{1}{\rho_2} + \frac{1}{\rho_1}\right) \quad (35)$$

For a given Q, zero or nonzero, and given p_1 and ρ_1 (which through the equation of state gives h_1) and one additional variable behind the wave, for example, V_2 or h_2, the locus of all possible combinations of ρ_2 and p_2 can be plotted on a so-called Hugoniot diagram (Fig. 4). The lines OJ and OK are tangent to the Hugoniot curve; O is the point $p_1, 1/\rho_1$. Points J and K separate strong and weak

	M_1	M_2	p_2/p_1	V_2/V_1	ρ_2/ρ_1	T_2/T_1
detonation	>1	≦1	>1	<1	>1	>1
deflagration	<1	<1	<1	>1	<1	>1

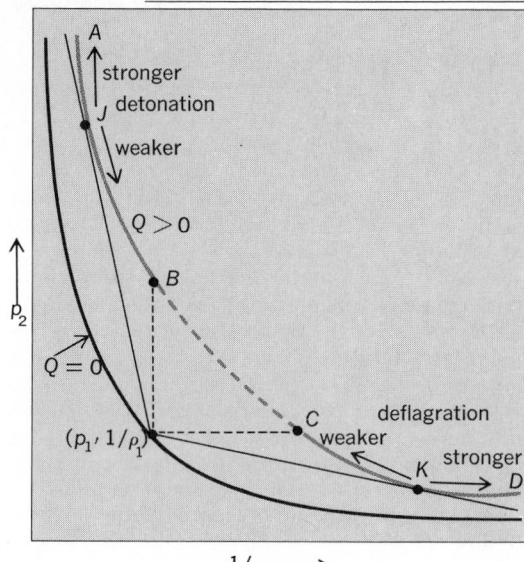

Fig. 4.   Hugoniot diagram and Chapman-Jouget conditions. (*A. B. Cambel and B. H. Jennings, Gas Dynamics, McGraw-Hill, 1958*)

waves. Flows corresponding to J and K are characterized by the fact that all the thermodynamic and fluid-mechanic variables have an extremum. Transitions from B to C or vice versa involve a decrease in entropy and are therefore forbidden. The slope of the tangent is connected through the momentum equation to the flow velocity by Eq. (36).

$$\frac{p_2 - p_1}{\dfrac{1}{\rho_1} - \dfrac{1}{\rho_2}} = (\rho_2 V_2)^2 = \frac{\Delta p}{\Delta\left(\dfrac{1}{\rho}\right)}\bigg|_2 \quad (36)$$

The derivative at constant entropy (since the entropy is stationary near J) is given by Eq. (37).

$$\rho^2 \frac{dp}{d\rho}\bigg|_2 = \rho_2{}^2 a_2{}^2 \quad (37)$$

In other words, the wave has a Mach number of unity with respect to the gas behind it. The points are appropriately named Chapman-Jouget points after the men who discovered their unique properties. The significance of these points lies in the fact that a stable detonation will eventually reach the point J and a deflagration point K.

Hydromagnetic (Alfvén) waves. Illustrative of the interaction of an electromagnetic field with a flowing plasma is the hydromagnetic, or Alfvén, wave. It is assumed that the velocity has only one component, for example, V_y, and the applied magnetic field B_x is perpendicular to it; the electric field E_z is perpendicular to B_x, and the current density j_z is parallel to the electric field. In order to simplify matters, it is assumed that the medium is incompressible so that the energy equation need not be considered. The fluid, moreover, is assumed

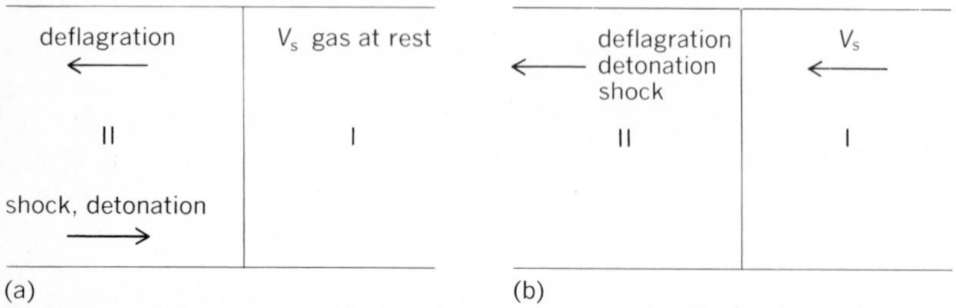

deflagration ←	V_s gas at rest	deflagration detonation shock ←	V_s ←
‖	Ⅰ	‖	Ⅰ
shock, detonation →			

| (a) | | (b) | |

Fig. 3.   Gas dynamic discontinuity. (*a*) Moving discontinuity, (*b*) Stationary discontinuity. (*A. B. Cambel and B. H. Jennings, Gas Dynamics, McGraw-Hill, 1958*)

to be inviscid and possesses infinite electrical conductivity. The equations of motion are then given by Eqs. (38) and (39).

$$\rho \frac{\partial V_y}{\partial t} = j_z B_x \qquad (38)$$

$$E_z - V_y B_x = 0 \qquad (39)$$

Substituting these equations into Maxwell's equations yields a wave, Eq. (40), with the propaga-

$$\frac{\partial^2 E_z}{\partial x^2} = \left(1 + \frac{4\pi\rho c^2}{B_x{}^2}\right)\frac{1}{c^2}\frac{\partial^2 E_z}{\partial t^2} \qquad (40)$$

tion velocity given by expression (41), called the Alfvén speed, where c is the speed of light. *See* MAXWELL'S EQUATIONS.

$$\frac{c}{(1 + 4\pi\rho c^2/B_x{}^2)^{1/2}} \qquad (41)$$

Mach number functions. In many applications it is reasonable to assume that the gas is perfect—both thermally and calorically—and that the flow is adiabatic. It then becomes very useful to express all the dependent variables in terms of the Mach number.

One usually starts with the energy equation and defines a stagnation temperature T_0 by expression (42). Physically, T_0 represents the temperature the gas would have if all its kinetic energy were transformed into thermodynamic enthalpy. The stagnation enthalpy can be measured by a thermometer immersed into a gas stream.

From Eq. (42), one simply obtains Eq. (43),

$$\frac{V^2}{2} + c_p T = c_p T_0 \qquad (42)$$

$$\frac{T_0}{T} = 1 + \frac{\gamma - 1}{2} M^2 \qquad (43)$$

where T is now called the static temperature. This temperature is measured with a thermometer at rest with respect to the gas. The stagnation temperature is constant in any adiabatic flow, even through a shock, and thus provides an excellent reference parameter. By using the isentropic relation $p = \rho^\gamma$ and the perfect gas law, one can define a reference stagnation pressure and density by means of Eqs. (44) and (45).

$$\frac{p_0}{p} = \left(1 + \frac{\gamma - 1}{2} M^2\right)^{\gamma/\gamma - 1} \qquad (44)$$

$$\frac{\rho_0}{\rho} = \left(1 + \frac{\gamma - 1}{2} M^2\right)^{1/\gamma - 1} \qquad (45)$$

When $M^2 \ll 1$, one may expand Eq. (44) by the binomial theorem to obtain Eq. (46).

$$p_0 = p + \frac{1}{2}\rho V^2 \left(1 + \frac{M^2}{4} + \ldots\right) \qquad (46)$$

The deviation from Bernoulli's equation, Eq. (12), due to compressibility at $M = 0.5$ is only 6%.

Flows can be classified as internal flow and external flow. Internal flow refers to the cases where the gas is constrained by a duct of some sort. Characteristically external flow is flow over an airplane or missile.

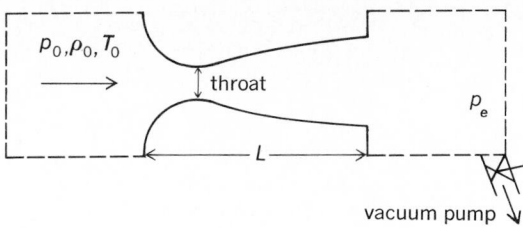

Fig. 5. Convergent-divergent nozzle. (*A. B. Cambel and B. H. Jennings, Gas Dynamics, McGraw-Hill, 1958*)

Internal one-dimensional flow. Internal flows are conveniently characterized by (1) the shape of the duct and its variation, (2) the heat transfer through the walls of the duct and internal heat sources, and (3) frictional effects. By varying one of these characteristics at a time, the essential features of internal flow can be discussed most simply.

Variable area flow. A device to accelerate the flow of a gas or liquid is termed a nozzle. In most engineering applications the contour of the nozzle is first converging and then diverging; it thus has a minimum cross section, called a throat.

For isentropic flow in a convergent-divergent nozzle in which the flow is supersonic in the divergent section, the velocity at the throat is sonic, that is, $M^* = 1$. The throat pressure is then said to be critical p^* and is given by Eq. (47). Velocity and

$$p^* = p_0 \left(\frac{2}{\gamma + 1}\right)^{\gamma/(\gamma-1)} \qquad (47)$$

pressure are related in this case by Eq. (48).

$$V = \frac{2\gamma R}{\gamma - 1} T_0 \left[1 - (p/p_0)^{(\gamma-1)/\gamma}\right] \qquad (48)$$

Consider a convergent-divergent deLaval nozzle inserted between two reservoirs as in Fig. 5. There will be no flow if the ratio of exit pressure p_e to reservoir pressure p_0 is $p_e/p_0 = 1$ (case a in Fig. 6). If p_e is reduced so that it is slightly less than the entrance pressure, the nozzle will act like a conventional venturi, as represented by curve b in Fig. 6. For this case, the flow is always subsonic and resembles incompressible flow. When the exit pressure is reduced further, the critical pressure can

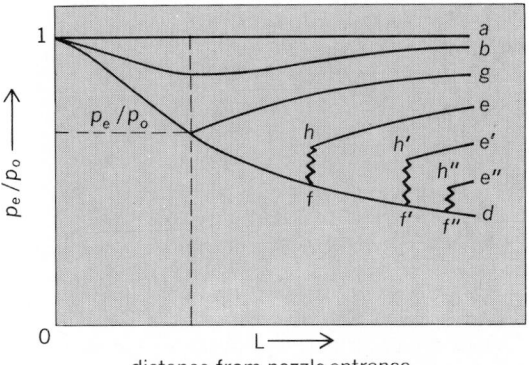

Fig. 6. Pressure distribution in convergent-divergent nozzle between two reservoirs. (*A. B. Cambel and B. H. Jennings, Gas Dynamics, McGraw-Hill, 1958*)

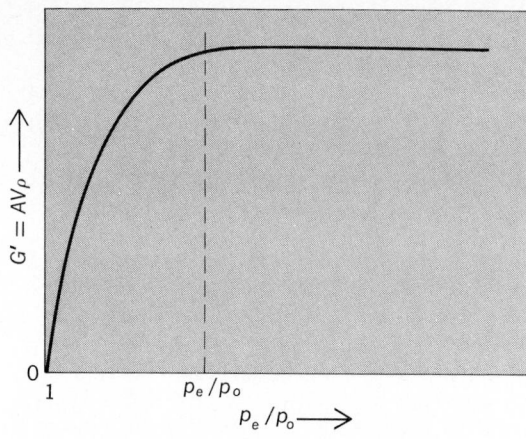

Fig. 7. Flow rate for given p_0, through convergent-divergent nozzle between two reservoirs. (*A. B. Cambel and B. H. Jennings, Gas Dynamics, McGraw-Hill, 1958*)

be reached at the throat, as curve g shows. In this case, the velocity is sonic at the throat but is never supersonic within the nozzle, even though the pressure at the throat corresponds to the critical. The minimum pressure which can exist in the nozzle outlet is depicted by point d, for which the pressure at the throat will be the critical. Here the velocity in the converging section is subsonic, in the diverging section it is supersonic, and at the throat it is sonic. For the range of exit pressures from p_d to p_g, the rate-of-flow curve is the same, for a given reservoir pressure p_0, and is plotted in Fig. 7. The flow rate reaches a maximum value and remains there over this wide range of exit pressures.

Even in the absence of friction, isentropic flow can exist only for the range of exhaust pressures from p_a to p_g and at the pressure reached along curve d, but not at intermediate pressures. The pressures p_g and p_d are the significant design pressures for a given nozzle. For exhaust pressures in the range between p_d and p_g, shocks will occur in the nozzle, raising the pressure from f to h (or f' to h'), followed by a pressure rise after the shock points to an exit pressure such as p_e. If p_e is less than p_d, the jet leaving the nozzle is said to be underexpanded and will drop in pressure after leaving the mouth of the nozzle. The velocity V_1 in front of a shock is supersonic, and the velocity V_2 behind a normal in contrast to an oblique shock is always subsonic. For a normal shock $V_1 V_2/a^{*2} = 1$, where a^* is the critical speed of sound corresponding to $M = 1$. Thus for $V_1 > V_2$, $V_1/a^* > 1$ and therefore $V_2/a^* < 1$. See ISENTROPIC FLOW.

Diabatic flow. Heat exchangers and combustion chambers are devices in which heat transfer occurs. The equations describing nonadiabatic or diabatic processes are complicated; consequently, certain limiting assumptions are usually required to make possible analytical solutions of the equations.

These assumptions are that (1) the flow takes place in a constant-area section, (2) there is no friction, (3) the gas is perfect and has constant specific heats, (4) the composition of the gas does not change, (5) there are no devices in the system which deliver or receive mechanical work, and (6) the flow is steady.

Equations which conform to these requirements are called Rayleigh equations, and the associated flow is designated as Rayleigh flow. Designating by $Q_{1 \to 2}$ the quantity of heat introduced between stations 1 and 2, one obtains (for the energy equation) Eq. (49).

$$Q_{1 \to 2} = c_p(T_2 - T_1) + (V_2{}^2 - V_1{}^2)/2$$
$$= h_2 - h_1 + (V_2{}^2 - V_1{}^2)/2 \tag{49}$$

If the stagnation enthalpy is introduced, then $Q_{1 \to 2}$ is given by Eq. (50), which can be expressed in terms of stagnation temperatures by Eq. (51).

$$Q_{1 \to 2} = h_{02} - h_{01} \tag{50}$$
$$Q_{1 \to 2} = c_p(T_{02} - T_{01}) \tag{51}$$

Because $Q_{1 \to 2} \neq 0$ for diabatic flow and because $c_p > 0$ always, it follows that $T_{02} \neq T_{01}$. This inequality states that in diabatic flow the stagnation temperature is not solely determined by the reservoir conditions, as is the case with adiabatic flow. Heating raises the stagnation temperature; cooling lowers it. *See* ADIABATIC PROCESS; ENTHALPY.

The locus of points of properties during a constant-area, frictionless flow with heat exchange is called the Rayleigh line. By definition, along the Rayleigh line the continuity equation and the momentum equation must apply. Equation (30) applies to steady flow in a constant-area duct. Mass velocity by definition is $G = \rho V$ and, from Eq. (31), the momentum relation is $p + \rho V^2 = C$, where C is a constant. Consequently, one obtains Eq. (52), which is one of the many Rayleigh-line equations.

$$p + \frac{G^2}{\rho} = C \tag{52}$$

The variations of pressure, temperature, and density with Mach number for Rayleigh flow are plotted in Fig. 8. The fact that the curve for $T_0/T_0{}^*$ in Fig. 8 reaches a maximum at a Mach number of unity indicates that it is impossible to pass from one flow domain into the other by the same heat-transfer process. Thus, if heat is added to a subsonic flow, the flow can be accelerated only until its Mach number becomes unity. Further addition of heat will not further accelerate the gas but will result in choking of the flow. As a consequence,

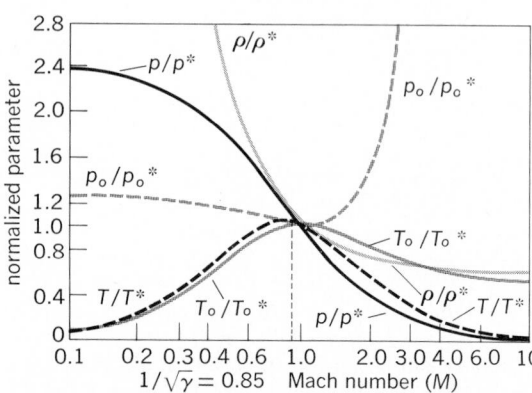

Fig. 8. Diabatic flow parameters for specific heat ratio of 1.4. Asterisk where $M = 1$. (*A. B. Cambel and B. H. Jennings, Gas Dynamics, McGraw-Hill, 1958*)

Table 1. Variation of flow properties for Rayleigh flow

Property	Heating		Cooling	
	$M > 1$	$M < 1$	$M > 1$	$M < 1$
T_0	Increases	Increases	Decreases	Decreases
p	Increases	Decreases	Decreases	Increases
p_0	Decreases	Decreases	Increases	Increases
V	Decreases	Increases	Increases	Decreases
T	Increases	Increases when $M < 1/\sqrt{\gamma}$ Decreases when $M > 1/\sqrt{\gamma}$	Decreases	Decreases when $M < 1/\sqrt{\gamma}$ Decreases when $M > 1/\sqrt{\gamma}$

the flow must readjust itself, which it will do by lowering its initial Mach number. Table 1 summarizes some of the Rayleigh flow phenomena.

Flow with friction. In long pipes the effects of friction may result in a significant pressure drop. Over a length dx, this pressure drop dp is given by the Fanning equation, Eq. (53), where f is a friction

$$dp = -f \frac{\rho V^2}{2r_h} dx \qquad (53)$$

factor that must be determined experimentally. To solve friction-flow problems analytically, certain simplifying assumptions are made and the resulting hypothetical flow is called Fanno flow. The Fanno flow assumptions are the same as those for Rayleigh flow except that the assumption that there is no friction is replaced by the requirement that the flow be adiabatic. Numerous Fanno flow equations may be written by combining the energy and the continuity equations in accordance with these assumptions. In Fig. 9 may be seen the variation of properties during Fanno flow. Table 2 summarizes the trends of the most important properties during subsonic and supersonic flow.

Table 2. Fanno flow phenomena

Property	Initial flow is subsonic	Initial flow is supersonic
M	Increases	Decreases
V	Increases	Decreases
p	Decreases	Increases
T	Decreases	Increases
ρ	Decreases	Increases

When the Fanno and Rayleigh lines are plotted for the same constant mass velocity $G = \rho V$, the curves appear as in Fig. 10. The Rayleigh and Fanno lines have two points of intersection, denoted by a and b; a normal shock connects these two points. The flow through a shock wave is irreversible; thus, associated with it is an increase in entropy. Point b, thus, always lies to the right of point a.

There is another interesting point about the Fanno curve. If frictional flow continues along the subsonic portion of the Fanno line, the Mach number tends to increase toward unity, whereas if it continues along the supersonic portion, the Mach number decreases toward unity. As in the case of Rayleigh flow, it is impossible, by virtue of the second law of thermodynamics, to pass from one flow regime to the other (subsonic into supersonic or conversely) unless the mass velocity is readjusted.

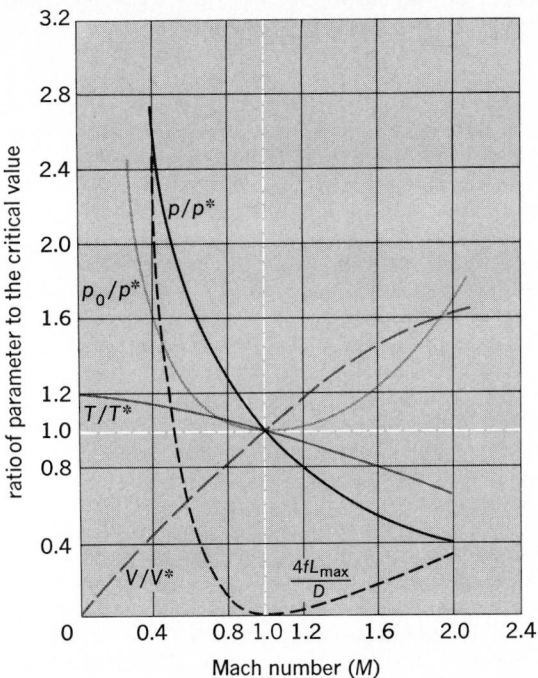

Fig. 9. Functions for constant-area flow with friction ($k = 1.4$). Asterisk where $M = 1$. (*A. B. Cambel and B. H. Jennings, Gas Dynamics, McGraw-Hill, 1958*)

External flow. Boundary layers and wakes are the centers of interest in external flows. Here the effects of compressibility are substantially more difficult to analyze than in internal flows, if for no other reason than the inapplicability of a one-dimensional approach.

Boundary layers. Ballistic missiles and space vehicles enter the Earth's atmosphere with velocities typically of 6–7 km/sec. The corresponding Mach number, depending on the altitude, is of the order of 20. The energy that maintains the bow shock and the work done to overcome the viscous shear stresses, Φ, reduce the kinetic energy of the vehicle, and it decelerates. This loss of kinetic

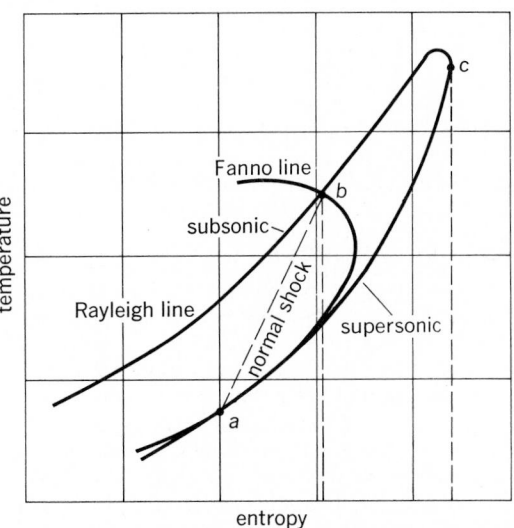

Fig. 10. Rayleigh and Fanno lines. (*A. B. Cambel and B. H. Jennings, Gas Dynamics, McGraw-Hill, 1958*)

energy, which is the work of the drag forces, reappears in part as the increased enthalpy and temperature of the fluid near the vehicle surface. The rate of heat transfer driven by the large enthalpy gradient $(h_g - h_s)/\Delta$, where h_g is the gas enthalpy at some suitably defined distance Δ from the surface and h_s is the enthalpy right at the surface, is so large that special protection must be afforded the vehicle. To this purpose, the vehicle can be covered with a material designed to char, melt, or gasify and in so doing absorb much of the heat that would otherwise penetrate into the structure. This "ablation" process, as it is called, introduces large amounts of material (some of it chemically active, some of it ionized, and some of it radiating) into what ordinarily would be called a boundary layer. Unfortunately, many of the assumptions that permit one to introduce the boundary-layer simplifications are violated here. Very complex computer programs have been developed, however, that yield reasonably accurate estimates of these effects. *See* BOUNDARY-LAYER FLOW.

Wakes. Hypersonic wakes were observed long before the space age. The tails of "shooting stars" are the luminous wakes of meteors as they enter the atmosphere and burn up. Meteor velocities range between 20 and 70 km/sec, and temperatures as high as 3500 K have been estimated to be necessary for vaporization. Since meteor trails, as well as reentry-vehicle wakes, contain electrons which scatter electromagnetic energy, the trails can be observed with radar. The estimation of the decay rate of these electrons provides a simple example of wake chemistry. *See* WAKE FLOW.

The probability of capture per second of an electron by an ion is $\beta_i n_e$, where β_i is the recombination coefficient of the particular ion and n_e is the electron density. The capture rate dn_e/dt between an electron and an ion is then given by Eq. (54) for a single ionized species.

$$\frac{dn_e}{dt} = n_e \beta_i n_i \tag{54}$$

In a neutral plasma $n_e = n_i$, and Eq. (54) can be integrated to yield Eq. (55), where n_0 is the ion

$$n_e = \left(\frac{1}{n_0} + \beta_i t\right)^{-1} \tag{55}$$

density at time zero.

[JOSHUA MENKES; ALI B. CAMBEL]

Bibliography: H. W. Liepmann and A. Roshko, *Elements of Gasdynamics*, 1957; A. H. Shapiro, *The Dynamics and Thermodynamics of Compressible Fluid Flow*, 2 vols., 1953 and 1954; R. Zucker, *Fundamentals of Gas Dynamics*, 1977.

Geometrical optics

The geometry of light rays and their imagery through optical systems. (The phenomena of diffraction due to the finite apertures of the lens systems are neglected in geometrical optics.)

Reflection and refraction laws. Light moves in straight lines through homogeneous media and changes its direction at the surface separating two such media, for instance, air and glass. An incident ray at the bounding surface is divided into two; one is reflected back into the first medium and the other penetrates the second medium after being bent or refracted. The incident, reflected, and refracted rays all lie in one plane containing the surface normal and form angles i, i_r, and i' with the surface normal such that Eqs. (1) and (2) hold where n and n' are the refractive indices of the media separated by the refracting surface. In order to obtain a solution of Eq. (2), i and i' must

$$i_r = \pi - i \tag{1}$$

$$n \sin i = n' \sin i' \tag{2}$$

be chosen so that they are in the same quadrant. *See* LIGHT; REFLECTION OF ELECTROMAGNETIC RADIATION; REFRACTION OF WAVES.

These formulas together with pure geometry make it possible to trace a ray through a system of lenses. The specific form of the ray-tracing formula should be adapted to the tools of the lens designer; that is, it will be different for a person using logarithm tables and for one using an electric desk machine or an electronic computer.

Point source. A point source is either an artificial light source which is so small that it appears to a given optical system as a point, or a luminous object, such as a star, which is so far away that it sends out coherent light.

All physical objects have finite areas. However, because of diffraction at the aperture of an optical system, or of the eye, an object which is small compared with the Airy disk will be imaged as an Airy disk; that is, it will be indistinguishable from the theoretical image of a mathematical point. Such an object can be given as the definition of a physical point. For a discussion of the Airy disk *see* DIFFRACTION.

Characteristic function. The aim of the optical designer is to see what happens to all the rays coming from every point of the object to be imaged. Moreover, he wants to direct the rays so that all of them coming from a fixed object point are collected at a fixed image point (freedom from aberration). Mostly, he wants all these image points to lie on a plane (freedom from field curvature), and he wants the image to be similar in shape to the object (freedom from distortion). Finally, correction should be achieved for light of different wavelengths (freedom from chromatic aberrations). *See* ABERRATION (OPTICS); CHROMATIC ABERRATION.

The basic tool for investigating all these problems is the characteristic function. If a coordinate system is chosen (origin O and x, y, z axes in object space and origin O' and x', y', z' axes in image space), a ray in object space is specified by the coordinates x, y, 0 of its intersection point with the plane $z = 0$ and by the optical direction cosines ξ and η formed by the ray with the x and y axes. The ray in image space is specified in the same way with primed coordinates. (An optical direction cosine is a direction cosine multiplied by the refractive index of the respective medium.)

The eight quantities x, y, x', y', ξ, η, ξ', and η' are, however, not independent. There exists a characteristic function E of, for instance, the quantities x, y, x', and y', from which the other four quantities ξ, η, ξ', and η' can be computed.

It is found that Eqs. (3) hold, where E_x, . . . ,

$$-\xi = \frac{\partial E}{\partial x} = E_x \qquad \xi' = \frac{\partial E}{\partial x'} = E_{x'}$$

$$-\eta = \frac{\partial E}{\partial y} = E_y \qquad \eta' = \frac{\partial E}{\partial y'} = E_{y'} \tag{3}$$

are introduced as abbreviations for $\partial E/\partial x, \ldots$.

The characteristic function has a physical meaning. It is the optical path—the sum of the paths in each medium multiplied by the corresponding refractive indices—from starting point (coordinates $x, y, 0$) to final point (coordinates $x', y', 0$). The validity of Eq. (3) presupposes that x, y, x', and y' determine a single ray, that is, that no two rays from a point in the plane $z = 0$ go through the same point in the plane $z' = 0$.

In case the optical system has an axis of rotation, as do most optical systems, the origins O and O' are best chosen on the axis of rotation, which will be the $z(z')$ axis. Then the x' axis may be chosen parallel to the x axis and in the same direction, and the y' axis parallel to the y axis. In this case, the characteristic function depends only on three parameters, for instance, Eqs. (4); and Eqs. (3)

$$e_1 = \tfrac{1}{2}(x^2 + y^2)$$
$$e_2 = xx' + yy' \qquad (4)$$
$$e_3 = \tfrac{1}{2}(x'^2 + y'^2)$$

transform to Eqs. (5), where $E_i = \partial E/\partial e_i$ is introduced as an abbreviation.

$$-\xi = E_1 x + E_2 x' \qquad \xi' = E_2 x + E_3 x'$$
$$-\eta = E_1 y + E_2 y' \qquad \eta' = E_2 y + E_3 y' \qquad (5)$$

When ξ' and η' as well as x' and y' are known, the intersection point of the rays with an arbitrary plane at the distance z' from the image origin can be computed and thus the image formation on any plane or curved surface investigated.

The characteristic function for any special image formation can be given in explicit form. For instance, the characteristic function for imaging the plane $z = 0$ onto a plane at the distance z'_0 from the origin with constant magnification m_0 and without distortion is given by Eq. (6), where f is an arbitrary function of e_1. The characteristic function

$$E = n'z'_0\left[1 + \frac{2}{(m_0 z'_0)^2}(e_1 + m_0 e_2 + m_0^2 e_3)\right]^{1/2} + f(e_1) \qquad (6)$$

for imaging the plane $z = 0$ onto the surface given by Eqs. (7), with m and therefore z' being given

$$z' = \phi\,(x'^2 + y'^2)$$
$$y' = my \qquad x' = mx \qquad (7)$$

functions of e_1 leads to Eq. (8). This leads to a

$$E = n'z'\left[1 + \frac{2}{(mz')^2}(e_1 + me_2 + m^2 e_3)\right]^{1/2} \qquad (8)$$

sharp image of the points of a plane with field curvature and distortion present.

The existence of the characteristic function can be used to prove that it is impossible to image sharply more than one plane except in a trivial case. The only such image formation possible would be an image comparable to that formed by a plane mirror, in which each object is imaged sharply and undistorted with a magnification which is equal to the ratio n/n' of object and image space. *See* MIRROR OPTICS.

Two surfaces can be imaged sharply only if the object and image surfaces are specific second-order surfaces which are imaged undistorted. The magnification m_1 and m_2 of the first and of the sec-

ond surface respectively must obey the condition of Eq. (9).

$$m_1 m_2 = n^2/n'^2 \qquad (9)$$

Gaussian optics. The first approximation to optical image formation is called Gaussian optics. It describes the rays which are so near the axis that one can assume $x, y, \xi, \eta, x', y', \xi'$, and η' to be so small that only linear terms of their Taylor series need be considered. This gives the position and magnification of the image for small apertures and, if the image is fairly sharp, plane, and undistorted, it also gives, at least approximately, the corresponding data for the image of a finite object with finite field.

Equations (10) describe the image formation if α,

$$x' = \alpha x + \beta\xi \qquad \xi' = \gamma x + \delta\xi$$
$$y' = \alpha y + \beta\eta \qquad \eta' = \gamma y + \delta\eta \qquad (10)$$

β, γ, and δ are assumed to be constant and connected by relation (11).

$$\alpha\delta - \beta\gamma = 1 \qquad (11)$$

Equations (10) and (11) give the image coordinates as functions of the object coordinates. Equations of this type, which correspond to the ray-tracing formulas, are called direct equations. Shifting the origin on object and image side by the amounts z and z' respectively changes the coefficients α, β, γ, and δ, as in Eqs. (12).

$$\bar\alpha = \alpha + \gamma\frac{z'}{n'}$$
$$\bar\beta = \beta - \alpha\frac{z}{n} + \delta\frac{z'}{n'} - \gamma\frac{zz'}{nn'} \qquad (12)$$
$$\bar\gamma = \gamma$$
$$\bar\delta = \delta - \gamma\frac{z}{n}$$

The quantity $(-\gamma)$, which is independent of the shift (invariant), is called the power of the system. A system for which $(-\gamma)$ is zero is called an afocal system.

Conjugate points. Shifting the image origin so that $\bar\beta$ vanishes leads to Eqs. (13).

$$x' = \bar\alpha x \qquad \xi' = \gamma x + \bar\delta\xi$$
$$y' = \bar\alpha y \qquad \eta' = \gamma y + \bar\delta\eta \qquad (13)$$

The rays from the object origin $x = y = z = 0$ meet at the image origin $x' = y' = z' = 0$. The image origin thus obtained is said to be conjugate to the object origin. All the rays from a point $x = x_0$, $y = y_0$ meet at a point $x'_0 = \bar\alpha x_0$, $y'_0 = \bar\alpha y_0$. This means that an object in the plane $z = 0$ is, within the limits of validity of Gaussian optics, imaged sharply with a magnification $m = \bar\alpha$ in the plane $z' = 0$. In the special case that $m = \bar\alpha = 1$, the two conjugate points are called *principal points*. If $m = \bar\alpha = n/n'$, the points are called *nodal points*. In case the object is at one of the principal points, it is imaged at the other principal point with unit magnification; in case the ray from the object origin ($x = y = z = 0$) passes through one of the nodal points, it leaves the system parallel to its original direction, or $\xi/n = \xi'/n'$, $\eta/n = \eta'/n'$.

Focal point and focal plane. Shifting the image origin the distance $z'/n' = -\alpha/\gamma$ makes $\bar\alpha = 0$; that is, since $\bar\beta\bar\gamma = -1$, Eqs. (14) hold. In this case a sys-

$$x' = -\xi/\gamma \qquad \xi' = \gamma x + \bar{\bar{\delta}}\xi$$
$$y' = -\eta/\gamma \qquad \eta' = \gamma y + \bar{\bar{\delta}}\eta \qquad (14)$$

tem of parallel rays coming from an "infinite axis point" (in the language of optics) meets at the image origin. This point is called the image focal point. A system of parallel rays (direction ξ_0, η_0) is imaged sharply at a point of the focal plane ($x'_0 = -\xi_0/\gamma, y'_0 = -\eta_0/\gamma$).

Shifting the object origin the distance $z/n = \delta/\gamma$ makes $\bar{\delta} = 0$; that is, since $\bar{\beta}\bar{\gamma} = -1$, Eqs. (15) hold.

$$x' = \bar{\alpha}x - \xi/\gamma \qquad \xi' = \gamma x$$
$$y' = \bar{\alpha}y - \eta/\gamma \qquad \eta' = \gamma y \qquad (15)$$

The rays from the object origin (which is called the object focal point) emerge parallel to the axis. The rays from an arbitrary point x_0, y_0 of the object focal plane $z = 0$ emerge parallel to one another ($\xi' = \gamma x_0, \eta' = \gamma y_0$).

Afocal systems. The power ($-\gamma$) is invariant against a shift of object and image origin. Until now $\gamma \neq 0$ had to be assumed. The case in which $\gamma \neq 0$ leads to a system which is called afocal, since it has no finite focal point. When object or image is shifted so that β vanishes Eqs. (16) are obtained, since $\alpha\delta = 1$.

$$x' = \alpha x \qquad \xi' = \xi/\alpha$$
$$y' = \alpha y \qquad \eta' = \eta/\alpha \qquad (16)$$

Any object is imaged with the constant magnification α, and a parallel bundle of object rays emerges as a parallel bundle with an angular magnification $n/n'\alpha$ for all rays.

Types of optical systems. It is customary to say that an object-side parallel bundle is a bundle that comes from an object point at infinity, and a parallel emerging bundle is a bundle that is said to be focused at an infinite image point. Therefore, there are four kinds of optical systems corresponding to the four choices of origins just considered: (1) enlarging systems—object and image are at finite conjugates; (2) photographic objectives—a distant object is imaged in the focal plane of the optical system; (3) eyepieces and microscope objectives—a near object is imaged at infinity to be seen by the relaxed eye; and (4) telescopes—a distant object is imaged at infinity to be seen by the relaxed eye. *See* LENS.

In an afocal system, the magnification remains constant for all object distances. When the origin is chosen at conjugate points, $\beta = \gamma = 0$, $\delta = 1/\alpha$ and the distances of z and z' of object and image respectively from the origin and the magnification m are given by Eqs. (17)

$$z'/n' = \alpha^2 z/n$$
$$m = \alpha \qquad (17)$$

On the other hand, in a system of finite power, the magnification changes from object distance to object distance.

When the origins are chosen at the two focal points, $\alpha = \delta = 0$ and the distances z and z' of object and image and their magnification m are given by Eqs. (18).

$$zz' = -nn'/\gamma^2$$
$$m = z'\gamma/n' = -n/z\gamma \qquad (18)$$

The distance from the principal point ($m = 1$)

to the focal point is given by Eqs. (19).

$$z' = -n'/\gamma = f'$$
$$z = n/\gamma = f \qquad (19)$$

The quantity f' is called the *focal length* of the optical system. *See* FOCAL LENGTH.

The distance from the nodal point to the focal point is found, by setting $m = n/n'$ in Eqs. (18), to be as shown in Eqs. (20).

$$z' = -n/\gamma = -f$$
$$z = n'/\gamma = -f' \qquad (20)$$

The nodal points and the principal points coincide for $n = n'$. For a single refracting spherical surface, the nodal points coincide with the center of the refracting surface and the principal points coincide with the vertex of the refracting surface.

When the origins are chosen at the principal points ($\alpha = \delta = 1$, $\beta = 0$), Eqs. (12) give Eqs. (21)

$$\frac{n'}{s'} - \frac{n}{s} = -\gamma = \frac{n'}{f'} = -\frac{n}{f}$$
$$m = 1 - \frac{s'}{f'} = \frac{1}{1 - s/f} = \frac{s'}{s} \cdot \frac{n}{n'} \qquad (21)$$

for the distances s and s' of conjugate points and their magnification.

When the origins are chosen at the nodal points ($\alpha = n/n'$, $\delta = n'/n$, $\beta = 0$), the distances c, c' of conjugate points and their magnification are given by Eqs. (22).

$$m = \frac{n}{n'} - \frac{c'}{f'} = \frac{1}{\frac{n'}{n} - c/f} = \frac{c'}{c}$$
$$\frac{1}{n'c'} - \frac{1}{nc} = -\frac{\gamma}{nn'} = \frac{1}{nf'} = -\frac{1}{n'f} \qquad (22)$$

Off-axis points. Gaussian optics considers only the image of points near the axis, but it is used as an approximation for the trace of finite rays, especially to compute the amount of light which the optical system transmits. Most optical systems, especially photographic lenses, contain a diaphragm which can be stopped down. In general the diaphragm determines the smallest aperture for the object point on the axis, since the other lenses can be made big enough to avoid cutting out any

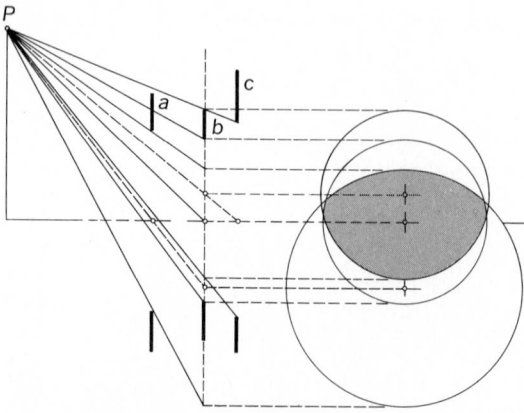

Fig. 1. Projection of apertures from finite point P onto entrance pupil. (*From M. Herzberger, Modern Geometrical Optics, Interscience, 1958*)

light. However, for off-axis points, the first and last surfaces may also cut off some light. This is called *vignetting*. To obtain an idea of the amount of light going through the system (Fig. 1), one must construct in object (or image) space the Gaussian image of the first lens, (*a*), the last lens, (*c*), and the diaphragm, (*b*), considering not only the position of each but also its magnification.

When these three apertures are projected from the object point onto a plane—for instance, the image of the diaphragm in object or image space, called the entrance pupil or the exit pupil, respectively—three eccentric circles in the plane of the entrance (exit) pupil are obtained. The rays from the object point through the region common to the three circles give a measure for the vignetting of the light if the object point moves away from the axis (Fig. 2).

Image-error theory. For a system with finite aperture and field (beyond the Gaussian domain), all the rays from a given object point do not generally meet at the Gaussian image point. Such a system is said to have image errors.

Let the object and image origins be chosen at the axis point of the object to be imaged and at the axis point of the exit pupil (the Gaussian image of the aperture stop) respectively. Then E, the characteristic path between the two planes $z = 0$ and $z' = 0$, is a function of e_1, e_2, e_3 [Eqs. (23)]. The

$$
\begin{aligned}
e_1 &= \tfrac{1}{2}(x^2 + y^2) \\
e_2 &= xx' + yy' \\
e_3 &= \tfrac{1}{2}(x'^2 + y'^2)
\end{aligned}
\tag{23}
$$

quantity e_1 depends only on the position of the object point (it is a field coordinate) and e_3 depends on the aperture in which the ray intersects the exit pupil (it is an aperture coordinate), while e_2 is a mixed coordinate (linear in both field and aperture).

The image errors are then given by two functions M and N having the property that the intersection point (coordinates $\bar{x}'$, $\bar{y}'$) of the rays with a plane at the distance z' from the exit pupil is designated by Eqs. (24).

$$
\begin{aligned}
\bar{x}' &= (1 + Nz')x' + Mz'x \\
\bar{y}' &= (1 + Nz')y' + Mz'y
\end{aligned}
\tag{24}
$$

If M and N are constant and z' is chosen equal to $-1/N_0$, Eqs. (24) give Eqs. (25), That is, the

$$
\bar{x}' = -\frac{M_0}{N_0} x
$$

$$
\bar{y}' = -\frac{M_0}{N_0} y
\tag{25}
$$

$$
\bar{z}' = -\frac{1}{N_0}
$$

points of the object plane are imaged sharply and without distortion onto the points of the plane at the distance $z' = -1/N_0$ from the exit pupil. The coefficients of the Taylor series expansion of M and N can therefore be considered as image errors.

Functions M and N can be computed from the characteristic function E and are given by Eqs. (26).

$$
M = E_2/\zeta' \qquad N = E_3/\zeta'
$$

$$
\zeta' = n'\left[1 - \frac{2}{n'^2}(E_2^2 e_1 + E_2 E_3 e_2 + E_3^2 e_3)\right]^{1/2}
\tag{26}
$$

There is a differential relation between M and N, given by Eq. (27).

$$
\begin{aligned}
M_3 - N_2 &= (2Me_1 + Ne_2)(MN_2 - NM_2) \\
&+ (Me_2 + 2Ne_3)(MN_3 - NM_3)
\end{aligned}
\tag{27}
$$

If E, M, and N are developed in a Taylor series with respect to the e_i, as in Eqs. (28), it can be

$$
\begin{aligned}
E &= E_0 + \Sigma \bar{E}_i e_i + \tfrac{1}{2}\Sigma \bar{E}_{ik} e_i e_k + \tfrac{1}{6}\Sigma \bar{E}_{ik\lambda} e_i e_k e_\lambda + \cdots \\
M &= \bar{M}_0 + \Sigma \bar{M}_i e_i + \tfrac{1}{2}\Sigma M_{ik} e_i e_k \\
&\qquad + \tfrac{1}{6}\Sigma \bar{M}_{ik\lambda} e_i e_k e_\lambda + \cdots \\
N &= \bar{N}_0 + \Sigma \bar{N}_i e_i + \tfrac{1}{2}\Sigma \bar{N}_{ik} e_i e_k \\
&\qquad + \tfrac{1}{6}\Sigma \bar{N}_{ik\lambda} e_i e_k e_\lambda + \cdots
\end{aligned}
\tag{28}
$$

shown that Eqs. (29) are valid, whereby the lower-

$$
\begin{aligned}
n' \cdot \bar{M}_{ik\lambda} \cdots &= \bar{E}_{2ik\lambda} \cdots \\
&\qquad + \text{lower-order terms} \\
n \cdot \bar{N}_{ik\lambda} \cdots &= \bar{E}_{3ik\lambda} \cdots \\
&\qquad + \text{lower-order terms}
\end{aligned}
\tag{29}
$$

order terms of $n' \cdot \bar{M}_{3k\lambda} \cdots$ and $n' \cdot \bar{N}_{2k\lambda} \cdots$ in general will not be the same though they both start with $\bar{E}_{23k\lambda}$, which has led to controversies about the number of image errors. The independent image errors are given by the derivatives of E_2 and E_3. Thus, if object and aperture order are considered to be equivalent, that is, if a larger and larger area surrounding the axis is considered, there are five errors of the first order (frequently called third order) given by so-called error coefficient (30) and nine errors of the second order (fre-

$$
\bar{E}_{21}, \bar{E}_{22}, \bar{E}_{23}, \bar{E}_{31}, \bar{E}_{33}
\tag{30}
$$

quently called fifth order) given as (31). In short,

$$
\begin{aligned}
&\bar{E}_{211}, \bar{E}_{212}, \bar{E}_{213}, \bar{E}_{222}, \bar{E}_{223}, \\
&\bar{E}_{233}, \bar{E}_{311}, \bar{E}_{313}, \bar{E}_{333}
\end{aligned}
\tag{31}
$$

there are $\binom{n+3}{n+1} - 1$ image-error coefficients of order n corresponding to $2\binom{n+2}{n}$ not all independent image errors of order n; namely, notations (32)

$$
\bar{M}_1, \bar{M}_2, \bar{M}_3 \qquad \bar{N}_1, \bar{N}_2, \bar{N}_3
\tag{32}
$$

for first-order errors and notations (33) for second-order errors.

$$
\begin{aligned}
&\bar{M}_{11}, \bar{M}_{12}, \bar{M}_{13}, \bar{M}_{22}, \bar{M}_{23}, \bar{M}_{33} \\
&\bar{N}_{11}, \bar{N}_{12}, \bar{N}_{13}, \bar{N}_{22}, \bar{N}_{23}, \bar{N}_{33}
\end{aligned}
\tag{33}
$$

When a system with sharp image formation for every point is investigated, it is found that M and N are functions of e_1 alone.

Coefficients (34) determine the curvature of the

$$
\bar{N}_1, \bar{N}_{11}, \ldots, \text{ or } \bar{E}_{31}, \bar{E}_{311}, \bar{E}_{3111}, \ldots
\tag{34}
$$

image and coefficients (35) determine the change

$$
\bar{M}_1, \bar{M}_{11}, \ldots, \text{ or } \bar{E}_{21}, \bar{E}_{211}, \bar{E}_{2111}, \ldots
\tag{35}
$$

of magnification or the errors of distortion. These errors may exist even if every point is sharply imaged.

For an axis point ($x = y = 0$), N becomes a function of e_3 alone. Then the rays through a set of concentric circles in the exit pupil go through a set of concentric circles in the image plane, and the image of the object point is concentric. It has been

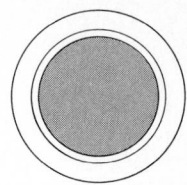

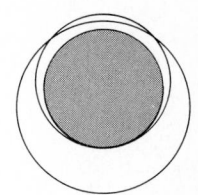

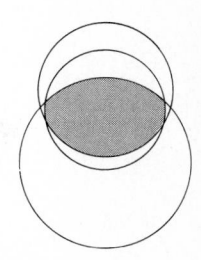

Fig. 2. Change in vignetting diagram as point moves off axis. (*From M. Herzberger, Modern Geometrical Optics, Interscience, 1958*)

suggested that these errors be called aperture errors. The name spherical aberration is used in the literature.

A point on the axis of a cylindrical torus or an ellipsoid has two planes of symmetry. Thus the rays through a set of concentric circles in the exit pupil go through a set of curves with two axes of symmetry and the same center of symmetry in the image plane. The corresponding configuration in the image plane is a set of deformed circles. The corresponding errors may be called deformation errors.

The rays from an off-axis point have a plane of symmetry, the meridional plane, through the object point and the axis. The image in the plane at the distance z' from the exit pupil thus has only one symmetry axis. The deviation from double symmetry is caused by the coefficients which belong to odd powers in e_2. These errors may be called asymmetry or coma errors.

Thus a specific error coefficient may be said to have an order with respect to aperture and field, a field rank and an aperture rank, and a degree of deformation, of coma, or of both. For instance, the coefficient (36) has order 7, field rank 7, aperture rank 9, and coma degree 3 ($o_7 f_7 a_9 c_3$).

$$\overline{E}_{11222333\cdots} \tag{36}$$

In this case the image-error coefficients (and the corresponding image errors of first order or third order in common practice) can be characterized as notation (37). The error usually called astigma-

$$
\begin{array}{llll}
\overline{E}_{12} : \overline{M}_1 & o_1 f_3 a_1 c_1 & \text{(Distortion)} \\
\overline{E}_{13} : \overline{N}_1 & o_1 f_2 a_2 & \text{(Curvature, sagittal)} \\
\overline{E}_{22} : \overline{M}_2 & o_1 f_2 a_2 d_1 & \text{(Astigmatism)} \\
\overline{E}_{23} : \overline{M}_3, \overline{N}_2 & o_1 f_1 a_3 c_1 & \text{(Coma)} \\
\overline{E}_{33} : \overline{N}_3 & o_1 f_0 a_4 & \text{(Spherical aberration)}
\end{array} \tag{37}
$$

tism is in this nomenclature a deformation error; the rays through an aperture circle go through an ellipse in the image plane.

The fifth-order errors can correspondingly be characterized as at notation (38).

$$
\begin{array}{lll}
\overline{E}_{112} : \overline{M}_{11} & o_2 f_5 a_1 c_1 & \text{(Field coefficient of} \\
& & \text{distortion)} \\
\overline{E}_{113} : \overline{N}_{11} & o_2 f_4 a_2 & \text{(Field coefficient of} \\
& & \text{curvature)} \\
\overline{E}_{122} : \overline{M}_{12} & o_2 f_4 a_2 d_1 & \text{(Field coefficient of} \\
& & \text{astigmatism)} \\
\overline{E}_{123} : \overline{M}_{13}, \overline{N}_{12} & o_2 f_3 a_3 c_1 & \text{(Field coefficient of} \\
& & \text{first-order coma)} \\
\overline{E}_{133} : \overline{N}_{13} & o_2 f_2 a_4 & \text{(Field coefficient of} \\
& & \text{aperture error)} \\
\overline{E}_{222} : \overline{M}_{22} & o_2 f_3 a_3 c_3 & \text{(Third-order coma)} \\
\overline{E}_{223} : \overline{N}_{22}, \overline{M}_{23} & o_2 f_2 a_4 d_1 & \text{(Aperture coefficient} \\
& & \text{of astigmatism)} \\
\overline{E}_{233} : \overline{N}_{23}, \overline{M}_{33} & o_2 f_1 a_5 c_1 & \text{(Aperture coefficient} \\
& & \text{of first-order coma)} \\
\overline{E}_{333} : \overline{N}_{33} & o_2 a_6 & \text{(Aperture coefficient} \\
& & \text{of aperture error)}
\end{array} \tag{38}
$$

If the image errors of a curved object are considered, the error coefficients vary. However, it can be shown that the errors not containing the index 1 are unchanged. They are invariant with respect to curvature geof object and image surface. In case they are zero, there can exist an object that is sharply imaged.

Since E gives the coordination for any object and image ray, a knowledge of E must suffice to give the image errors for any object. If the coefficients (39), which are the aperture errors of the

$$\overline{E}_1, \overline{E}_{11}, \overline{E}_{111}, \cdots \tag{39}$$

stop, are added to the image-error coefficients, formulas can be obtained for investigating the image errors that arise when object and stop are moved. Here another division of errors is suitable. It is obvious that the errors containing 2,2 and 1,3 have the same rank with respect to aperture and field. Combinations of these errors can designate the degree of skewness of the errors, which can be divided into meridional errors and skew errors of first, second, etc. types. The third- and fifth-order errors are then given by coefficients (40) and (41).

Zero type (meridional errors):

First order: $\overline{E}_{12}, \overline{E}_{13} + 2\overline{E}_{22}, \overline{E}_{23}, \overline{E}_{33}$

$$
\begin{aligned}
&\text{Second order: } \overline{E}_{112}, \overline{E}_{113} + 4\overline{E}_{122}, 3\overline{E}_{123} \\
&\qquad\qquad + 2\overline{E}_{222}, \overline{E}_{133} + 4\overline{E}_{223}
\end{aligned} \tag{40}
$$

First type:

First order: $\overline{E}_{22} - \overline{E}_{12}$

$$
\begin{aligned}
&\text{Second order: } \overline{E}_{113} - \overline{E}_{122}, \overline{E}_{123} - \overline{E}_{222}, \\
&\qquad\qquad \overline{E}_{133} - \overline{E}_{223}
\end{aligned} \tag{41}
$$

The coefficients of zero type are the meridional coefficients, which transform by themselves, and the coefficients of each type can then be chosen so that they transform by themselves if the object (or stop) position is changed. This means that, for each type, there exists a coefficient which is invariant against the position of both the object and the stop. The first such invariant was found by Josef Petzval and equals $\overline{E}_{22} - \overline{E}_{13}$, the Petzval condition; the next would be $\overline{E}_{2222} - 2\overline{E}_{2213} + \overline{E}_{1313}$, and so on.

Another analysis of the image errors can be made by considering the diapoint configuration. The diapoint of the object point for a ray is defined as the point where the ray intersects the meridional plane. The coordinates x'_p, y'_p, and z'_p of this point are given by Eqs. (42).

$$
\begin{aligned}
x'_p &= -(M/N)x \\
y'_p &= -(M/N)y \\
z'_p &= -1/N
\end{aligned} \tag{42}
$$

Thus the three-dimensional problem is transformed into a plane problem because one can set $x = 0$ without loss of generality. The coefficients of the development of M/N can be considered as lateral errors and those of $1/N$ as longitudinal errors.

Interpolation theory. In analyzing an optical system, it is not suitable to develop E into a Taylor series since such a series converges slowly and does not give a good enough approximation for the rays from an object point that is distant from the axis or for a system having a large aperture. It is, however, possible to derive an interpolation formula which gives a very good approximation. A number of rays, for instance nine, are traced from a point $x = 0$, $y = y_0$ (for an axis point $y_0 = 0$) through the optical system, and the intersection points $\bar{x}'$, $\bar{y}'$ with a plane at the fixed distance z'_0 from the exit pupil are determined. From Eqs. (43) a series

$$\bar{x}' = (1 + Nz'_0)x'$$
$$\bar{y}' = (1 + Nz'_0)y' + Mz'_0 y_0 \qquad (43)$$

of values for M and N as functions of e_2 and e_3 is obtained. These values are fitted by least squares with formulas (44).

$$M = M_0 + M_2 e_2 + M_3 e_3 + \tfrac{1}{2}M_{22}e_2^2$$
$$+ M_{23}e_2 e_3 + \tfrac{1}{2}M_{33}e_3^2$$
$$N = N_0 + N_2 e_2 + N_3 e_3 + \tfrac{1}{2}N_{22}e_2^2 \qquad (44)$$
$$+ N_{23}e_2 e_3 + \tfrac{1}{2}N_{33}e_3^2$$

Having found the coefficients of formulas (44), it is possible to compute M and N and therefore $\bar{x}'$ and $\bar{y}'$ for a large number of rays from the object point going through the vignetted exit pupil. If the exit pupil is uniformly illuminated, these points should be chosen so that they uniformly fill the (vignetted) exit pupil. The plot of the intersecting points x', y' with the plane at the distance z' then gives a measure of the intensity of the light distribution in the image.

As a measure of the quality of the image, the reciprocal of the radius of the circle that contains a certain percentage of the rays (for instance, 70, 80, or 90%) may be taken. The image also can be dissected into its aperture, comatic, and deformation errors with respect to aperture (keeping the object point, that is, the field, constant). This assists the designer in comparing different design stages since the corresponding figures are easy to analyze, in contrast to the complicated image figures.

Moreover, by integrating in the x or y direction, the spread function can be obtained; that is, the image of a line in the meridional direction or the sagittal direction can be investigated.

A simple mathematical consideration shows that a small sinusoidal test object at the object point is imaged sinusoidally but with a different amplitude and phase. (A sinusoidal test object is a pattern in which the intensity varies like a sine wave in the lateral direction, being kept constant in the longitudinal direction.) A series of sinusoidal test objects varying in the number of "waves" per millimeter is imaged.

The variation in amplitude gives a measure of the deterioration of the image with respect to resolution of gratings, and the change of phase gives information about the asymmetry of the image of the sinusoidal test object.

The plot of the sine-wave response as a function of the frequency is regarded as giving information sufficient to compare objectives of similar construction.

Diffraction. It is possible to compute diffraction for an off-axis point from geometrical optics. If M and N are known, Eqs. (45) hold.

$$E_2 = Mn'/[1 + 2\,(M^2 e_1 + MN e_2 + N^2 e_3)]^{1/2}$$
$$E_3 = Nn'/[1 + 2\,(M^2 e_1 + MN e_2 + N^2 e_3)]^{1/2} \qquad (45)$$

In view of the integrability condition, Eqs. (27), these equations when integrated give E as a function of e_2 and e_3 and thus give the phase difference at every point of the exit pupil. Integration of the exponential e^{iks}, where s is the sum of E and the optical distance to a fixed point over the exit pupil,

enables the (relative) light intensity at the point in question to be computed.

[MAX HERZBERGER]

Bibliography: E. U. Condon and H. Odishaw, *Handbook of Physics*, 1967; M. Herzberger, *Modern Geometrical Optics*, 1958; reprint 1978; M. Herzberger and H. Pulvermacher, Finite image-error theory based on the diapoint configuration, *J. Opt. Soc. Amer.*, vol. 58, no. 8, 1968; F. A. Jenkins and H. E. White, *Fundamentals of Optics*, 4th ed., 1976; R. Kingslake, *Applied Optics and Optical Engineering*, vol. 3, 1965; R. S. Longhurst, *Geometrical and Physical Optics*, 1974; Optical Society of America, *Handbook of Optics*, 1978.

Geometry

A branch of mathematics concerned with the properties of space, including points, lines, curves, planes and surfaces in space, and figures bounded by them.

Euclidean geometry. Geometry as a high school subject is largely based on the *Elements* of Euclid of Alexandria, a 13-volume work written about 300 B.C., of which the first 6 volumes present plane geometry, the next 4 are concerned with numbers and length, and the last 3 develop solid geometry. Euclid's *Elements* present in a logical order a sequence of over 400 mathematical propositions or theorems, proved on the basis of a set of 10 axioms or postulates which are assumed to be "self-evident" or true without proof—for example: "two points determine a line." Many theorems involve properties of triangles, circles, and other geometric figures, concerning lengths, angles, and congruence.

The most famous of Euclid's assumptions is the fifth, called the parallel postulate, which asserts that: given a line l and a point P not on l, there is one and only one line through P in the plane containing l and P that does not intersect l. Many unsuccessful attempts were made over the centuries to deduce Euclid's parallel postulate from the other axioms. Equivalent to this postulate is the assumption that the sum of the interior angles in a triangle is 180°. In a triangle bounded by arcs of three great circles on a sphere, the sum of the angles always exceeds 180°.

Noneuclidean geometries. K. F. Gauss is credited with discovering an "elliptic" noneuclidean geometry in which there are no parallels, but the other euclidean axioms are satisfied. One way to model it is to call each diameter of a sphere a "point," and call each plane through the center (intersecting the sphere in a great circle) a "line." Then each two points determine a unique line and each two lines determine one point.

Another type of noneuclidean geometry, called hyperbolic geometry, was discovered about 1830 by J. Bolyai and N. I. Lobachevski. Through a given point P not on a line l, there are many lines that do not meet l. This geometry may be modeled by defining as "lines" those circular arcs that meet a large circle C at right angles and restricting the term "points" to those within the absolute circle C. Then two points determine a unique line, but there are many lines through a point P not on a line l that do not intersect l. Since, therefore, consistent geometries exist in which the parallel postulate does not hold, this postulate is independent of the others. *See* NONEUCLIDEAN GEOMETRY.

Projective geometry. Projective geometry studies geometric properties invariant under projections (as in photography) which may distort angles and preserve neither lengths nor ratios of lengths, whereas euclidean geometry is confined to properties of figures such as angles and ratios of lengths that are invariant under rigid motions, reflections, and similarity transformations. If A, B, C, and D are four collinear points, the so-called cross ratio $(BA/BC) \div (DA/DC)$ is preserved under projection. To each set of parallel lines in euclidean geometry is added a "point at infinity" or "vanishing point" in projective geometry, so that each two lines meet in just one point. Lines joining a point P not on l to four points A, B, C, and D on l have the same cross ratio as the four points. A comprehensive theory is derived from the invariance of cross ratio, including a complete theory of conic sections. Hyperbolic, elliptic, and euclidean geometries may be studied in terms of projective transformations that fix respectively a real nondegenerate conic, or an imaginary nondegenerate or degenerate line conic, the latter consisting of two imaginary points at infinity that lie on all circles, but on no other conics.

Algebraic geometry. Algebraic geometry is a study of solutions of systems of polynomial equations $f_j(x) = 0$ in several variables x_i, thought of as coordinates of a point $x = (x_1, x_2, \ldots, x_n)$ in n-dimensional space. Classical studies emphasize properties of an irreducible algebraic plane curve $f(x_1, x_2) = 0$ (f an irreducible polynomial), such as singular and multiple points and genus. The curve is rational (of genus 0) if x_1 and x_2 can be expressed as rational functions of a parameter t. Two curves $f(x_1, x_2) = 0$ and $g(u_1, u_2) = 0$ are birationally equivalent if both the coordinates of each curve can be expressed as rational functions of the coordinates of the other. In modern studies, the coefficients in the polynomials f_j are chosen from an arbitrary field k, and solutions x are sought in an algebraically closed field K containing k. The set X of all points x at which all the f_j vanish is called an algebraic closed set, or variety. If X is not empty, then the set of all polynomials $f(x)$ that vanish on X form an ideal with a finite basis in the ring of polynomials. Modern algebraic geometry studies these ideals and corresponding varieties. *See* POLYNOMIAL SYSTEMS OF EQUATIONS.

Differential geometry. Differential geometry uses the tools of the calculus to study properties of curves and surfaces, usually in a euclidean space where the squared distance between nearby points is given in rectangular coordinates by the pythagorean formula $ds^2 = dx^2 + dy^2 + dz^2$. Arc length s is defined along "smooth" curves. Geodesics on a surface, like great circles on a sphere, are curves of shortest length between points not too far apart. Curvature is defined at each point of a smooth space curve, and torsion (twisting) at points not belonging to a straight portion of the curve; both are constant on a helix (similar to a coiled spring). Curvatures of surfaces play an important role.

Riemannian geometry. Riemannian geometry, named for G. F. B. Riemann, generalizes the concepts of differential geometry to noneuclidean spaces of any number n of dimensions in which there may or may not be any "straight" lines of infinite extent. Points P are specified by coordinates x^i like longitude and latitude on the Earth's surface. Squared distance is expressed by a quadratic form $ds^2 = \Sigma \, g_{ij} dx^i dx^j$, summed over i and j from 1 to n, where the functions g_{ij} are components of a "metric tensor" that is a scalar function of n vectors, varying in value from point to point. Under a differentiable change of coordinates (somewhat more general than the change from rectangular to spherical coordinates in Euclidean analytic geometry), the components g_{ij} are changed. But geodesic paths and a certain curvature obtained from second derivatives of the g_{ij} are independent of the choice of coordinates x^i and describe intrinsic properties of the space. *See* RIEMANNIAN GEOMETRY.

Concepts of riemannian geometry are employed in Albert Einstein's theory of relativity. In his special theory (1905), the invariant interval between events in space time is $ds^2 = dt^2 - (dx^2 + dy^2 + dz^2)/c^2$, where c is the velocity of light, about 3×10^8 m/s. Thus neither the apparent time interval dt between events nor the space interval but a combination of the two is the same for all observers. In his general theory (1916), Einstein introduces a more general riemannian metric with g_{ij}'s representing the local gravitational potential, and explains the acceleration ascribed by Newton to a gravitational force as acceleration due to motion along curved world lines in space-time that are bent by the gravitational field. *See* RELATIVITY; SPACE-TIME.

[J. SUTHERLAND FRAME]

Bibliography: H. S. M. Coxeter, *Non-Euclidean Geometry*, 5th ed., 1965; L. P. Eisenhart, *Riemannian Geometry*, 1966; W. C. Graustein, *Differential Geometry*, 1966; I. G. Macdonald, *Algebraic Geometry*, 1968; G. deB. Robinson, *The Foundations of Geometry*, 4th ed., 1959; V. Snyder et al., *Selected Topics in Algebraic Geometry*, 1970.

Giant nuclear resonances

Systematic excitations of the atomic nucleus which occur with great strength in a concentrated energy region. When high-energy electromagnetic radiation (gamma radiation) impinges on a nucleus, it can be strongly absorbed into a number of high-lying resonances: one in which the nucleus is excited in a dipole mode of oscillation that is electric in nature (E1), a second dipole mode that is magnetic in nature (M1), and a third excitation that can be identified as an electric quadrupole oscillation (E2). When excited into these resonances, the nucleus can deexcite by emitting gamma radiation or, if energetically allowed, by emitting particles, especially protons and neutrons. If proton and neutron emissions are allowed, it is possible to study these resonances by the inverse process in which the proton or the neutron is captured in the giant resonance and gamma radiation is emitted. These giant resonances can also be excited by inelastic scattering of electrons and other particles such as protons, deuterons, ^{3}He, and α-particles. For inelastic excitation with particles, greater momentum can be imparted to the nucleus, and higher modes such as octupole oscillations can be excited. The particles can also excite the electric monopole vibration (E0), which corresponds to a "breathing" mode. *See* MULTIPOLE RADIATION; NUCLEAR MOMENTS.

Giant E1 resonances. The giant E1 resonance has long been the object of intensive study. The

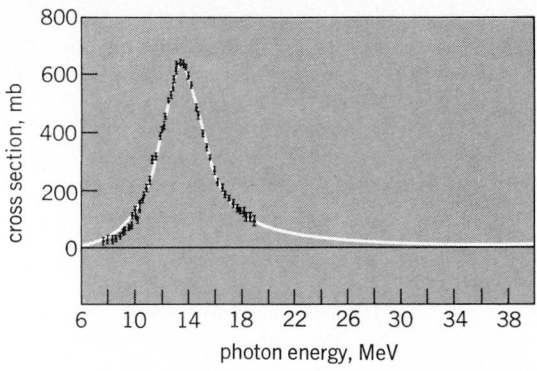

Fig. 1. Giant E1 resonance in ^{208}Pb observed with the photonuclear reaction, 1 mb $= 10^{-31}$ m². (From B. L. Berman and S. C. Fultz, Measurements of the giant dipole resonance with monoenergetic photons, Rev. Mod. Phys., 47:713–761, 1975)

Summary of giant resonance properties*

Property	E1	M1†	E2
E_x	$77/A^{1/3}$	$40/A^{1/3}$	$63/A^{1/3}$
Strength	Theoretical limit	$\simeq$Theoretical limit	$\leq$ Theoretical limit
Γ/E_x	0.2	≤ 0.2	$\simeq 0.2$

*E_x is the excitation energy in MeV and Γ is the width of the resonance. A is the number of protons and neutrons in the nucleus.

†Properties established for light nuclei.

three important properties which characterize it are its systematic occurrence in all nuclei, its great strength, and its localized nature. The reactions that have been used to study the giant E1 resonances are the following:

1. The photoneutron process (γ,n). In this case, a γ-ray is absorbed by the nucleus, and the subsequent neutron emission from the nucleus is measured as a function of the γ-ray energy. A typical E1 giant resonance obtained with this method is shown in Fig. 1.

2. The inverse proton capture reaction (p,γ). In this type of experiment, the γ-ray yield is measured as a function of the incident proton energy. A typical giant resonance is shown in Fig. 2.

3. Inelastic electron scattering (e,e'). In this case, the energy distribution of scattered electrons is measured at a fixed energy of the incident electrons. In general, a detailed analysis is required to separate the giant E1 resonance from other giant excitations.

The dominant features of the E1 resonance may be summarized as follows (see the table):

1. In nuclei of medium and heavy mass, the E1 resonance occurs at an excitation energy E_x of $77/A^{1/3}$ MeV, where A is the mass number of the nucleus. However, in the light nuclei, below ^{40}Ca, the energy of the resonance falls off to values of $50/A^{1/3}$ for the very lightest nuclei.

2. The giant E1 resonance "exhausts" the theoretically allowed limit for absorption of gamma radiation by an E1 mode.

3. Perhaps the most impressive feature of the E1 resonance is its localized nature, despite the fact that it occurs at a high excitation energy where the nucleus can decay in many ways. From the lightest to the heaviest nuclei, the width Γ is given by $\Gamma/E_x \simeq 1/5$, with some important exceptions. The best-established broadening of the resonance is that caused by the deformation of the nucleus.

The giant E1 resonance can be described in terms of characteristic single-particle excitations in the nucleus which absorb most of the E1 strength. Alternatively, it can be attributed to collective oscillations of the nucleus. In the latter picture, the basic mode is one in which all the protons in the nucleus vibrate against all the neutrons. Nuclear theory has been successful in showing the equivalence of these two pictures and in accounting for the prominent features of the E1 resonance.

Giant M1 resonances. Information on the giant M1 resonances has become rather extensive and exists for nuclei from mass 8 to 208. The methods that have been used to study the M1 resonance can exists for nuclei from mass 8 to 208. However, the resonance and its properties have been well established only for the light nuclei. The methods that have been used to study the M1 resonance can be summarized as follows:

1. The inverse proton capture reaction (p,γ). Some resonances have also been studied by reactions of the type (X,$Y\gamma$) where X and Y stand for nuclear particles.

2. Gamma-ray fluorescence (γ,γ'). In these experiments the nucleus is excited by means of inelastic γ-ray scattering.

3. Inelastic electron scattering at 180°. The use of 180° scattering is necessary to sort out magnetic from electric multipoles. A typical case of an M1 resonance observed with electron scattering is shown in Fig. 3.

4. The photoneutron process (γ,n). This process has been used to give valuable information just above the neutron threshold in heavy nuclei. Information comes also from (n,γ) results.

The properties of the M1 resonances are summarized in the table. The giant M1 resonance can be described in terms of single-particle excitations in the nucleus which produce a change in the direction of the particle spin. In the collective picture, the oscillation can be thought of as one in which particles with one spin direction vibrate against those with the opposite spin direction.

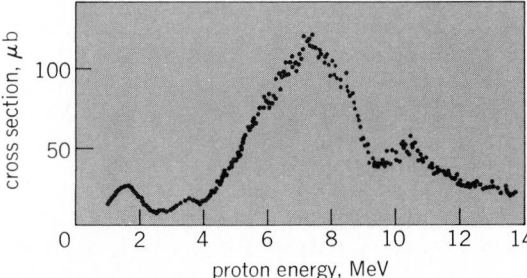

Fig. 2. Giant E1 resonance at 7.2 MeV in ^{12}C observed with the "inverse" proton capture reaction, ^{11}B(p,γ)^{12}C. 1 μb $= 10^{-34}$ m². (From R. G. Allas et al., Radiative capture of protons by B^{11} and the giant dipole resonance in C^{12}, Nucl. Phys., 58:122–144, 1964)

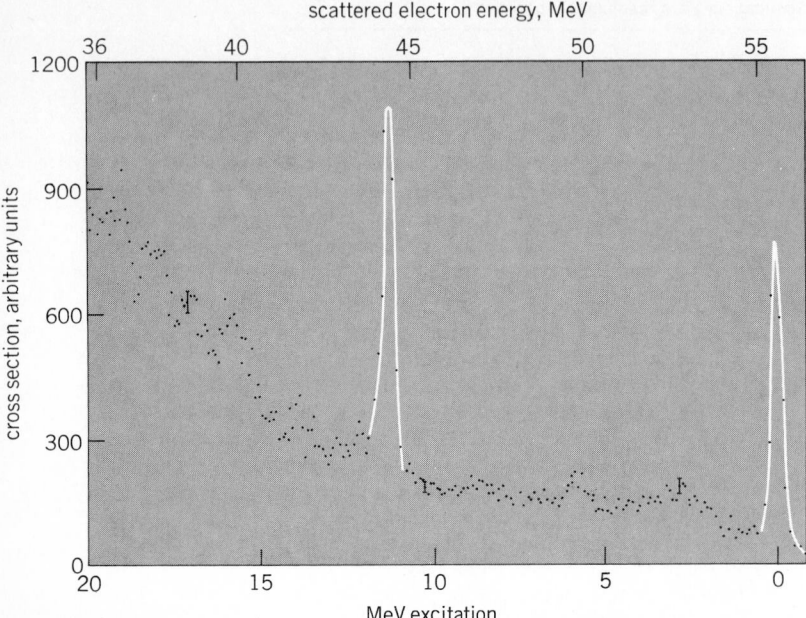

scattered electron energy, MeV

Fig. 3. Giant M1 resonance at 11.3 MeV in ^{20}Ne observed with inelastic electron scattering at 180°. Energy of incident electrons is 56.0 MeV. (*From L. W. Fagg, Electroexcitation of nuclear magnetic dipole transitions, Rev. Mod. Phys., 47:683–711, 1975*)

Giant E2 resonances. Interest in the study of the giant E2 resonances has stemmed from the observation (in 1971) of these resonances below the giant E1 resonance in electron scattering and their identification in inelastic proton scattering. E2 strength had been seen earlier in proton capture experiments in the lighter nuclei. However, the new observations established the E2 resonance as a compact, systematic excitation occurring in all nuclei. The methods which have been used to study E2 strength can be classified as follows:

1. Inelastic electron scattering (*e,e′*). Since the momentum imparted to the nucleus can be easily varied, electron scattering provides a sensitive method for studying the excitations.

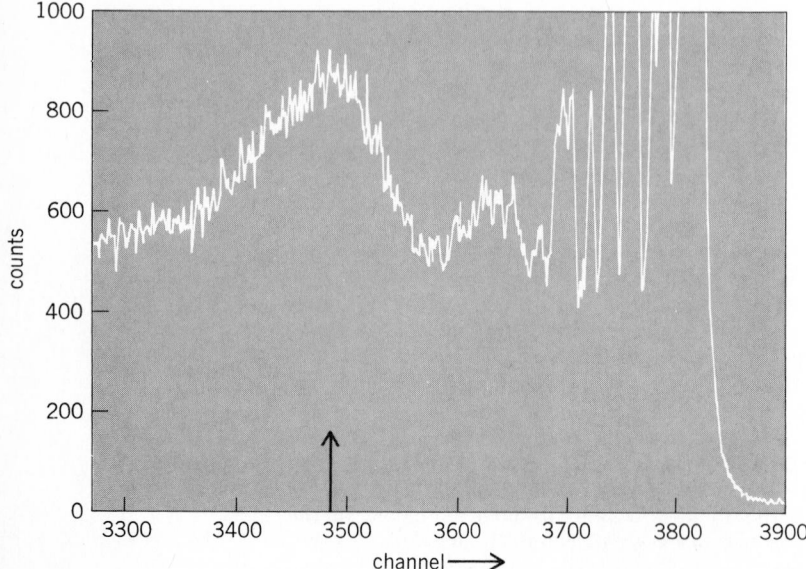

Fig. 4. Giant E2 resonance at channel 3490 (equal to an excitation energy of 13.3 MeV) observed in the inelastic scattering of 152-MeV α-particles from ^{120}Sn.

2. Inelastic scattering by nuclear particles (X,X'). The E2 resonances have been observed in the inelastic scattering of a variety of particles, such as protons, deuterons, ^{3}He, α-particles, heavy ions, and pions. Systematic properties have been developed throughout the whole nuclear table of isotopes, as given in the table. A typical E2 resonance is shown in Fig. 4.

3. The inverse capture reaction (X,γ). The (p,γ) work has become much more definitive by the use of polarized protons. Important information has been obtained from the (γ,p) process. A great deal of evidence has also been accumulated from the (α,γ) reaction.

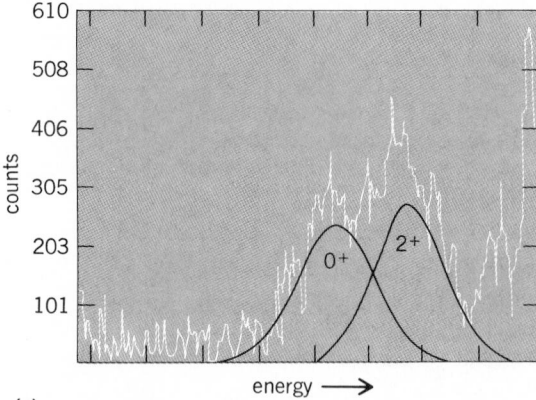

(a)

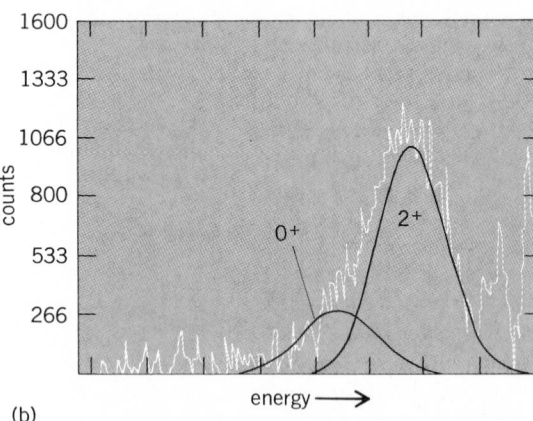

(b)

Fig. 5. Giant E0 resonance, labeled 0$^+$, observed in the inelastic scattering of 127-MeV α-particles from ^{208}Pb. (a) Scattering angle $\theta = 0°$. (b) $\theta = 4°$. (*From D. H. Youngblood, The giant monopole resonance: An experimental review, in F. E. Bertrand, ed., Giant Multipole Resonances, vol. 1, pp. 113–137, Harwood Academic Publishers, 1980*)

As for the dipole resonances, the E2 resonances can be successfully described by means of single-particle excitations or with a collective picture in which the nucleus undergoes shape oscillations of a quadrupole nature.

Giant E0 resonances. The importance of these resonances lies in the fact that the energy at which they occur determines the incompressibility of a nucleus, a property that is basic to understanding the force between neutrons and protons. These resonances became well established through a series of measurements on the inelastic excitation of a nucleus by α-particles. It was necessary to observe the scattered α-particles at very small angles in order to separate the E0 resonance from the other resonances discussed above. An illustration of this separation is shown in Fig. 5, where there is a marked increase in the E0 (0^+) peak relative to the E2 (2^+) peak in going from a scattering angle $\theta = 4°$ to $\theta = 0°$. Many of the properties of the E0 resonance have been established, but active research has continued to extend and quantify the results. The resonances are being studied by inelastic excitation by nuclear particles and electrons, and their decay properties are being investigated. *See* NUCLEAR REACTION; NUCLEAR SPECTRA; NUCLEAR STRUCTURE. [STANLEY S. HANNA]

Bibliography: B. L. Berman and S. C. Fultz, Measurements of the giant dipole resonance with monoenergetic photons, *Rev. Mod. Phys.*, 47: 713–761, 1975; F. E. Bertrand, Excitation of giant multipole resonances through inelastic scattering, *Annu. Rev. Nucl. Sci.*, 26:457–509, 1976; L. W. Fagg, Electroexcitation of nuclear magnetic dipole transitions, *Rev. Mod. Phys.*, 47:683–711, 1975; S. S. Hanna, Giant multipole resonances, in *Photonuclear Reactions*, International School on Electroand Photonuclear Reactions, vol. 1, pp. 275–339, 1976; D. H. Youngblood, The giant monopole resonance: An experimental review, in *Topical Conference on Giant Multipole Resonances*, Oak Ridge, TN, 1979.

Gibbs function

The Gibbs function G, also known as Gibbs free energy or free enthalpy, is defined in the equation shown, where E is the internal energy, p is the pressure, v is the volume, T is the absolute temper-

$$G = E + pv - TS$$

ature and S is the entropy. The Gibbs function is most useful in analyzing systems held at constant temperature and pressure. Under these conditions, the change in the Gibbs function ΔG of a system is a measure of the maximum attainable work, not including the work of displacing the environment. For example, ΔG represents the maximum electrical work obtainable from a galvanic cell. When the only work done by the system at constant temperature and pressure is displacing the environment, the equilibrium state is characterized by G having reached its minimum value. Since chemical processes frequently occur at constant temperature and pressure, the Gibbs function is extensively used in chemical engineering for calculating phase equilibrium and reaction equilibrium. *See* FREE ENERGY.

[WILLIAM F. JAEP]

Bibliography: R. W. Haywood, *Equilibrium Thermodynamics for Engineers and Scientists*, 1980; J. B. Jones and G. A. Hawkins, *Engineering Thermodynamics*, 1960; S. L. Kittsley, *Physical Chemistry*, 3d ed., 1969; M. Mark and A. R. Foster, *Thermodynamic Principles and Applications*, 1979; W. J. Moore, *Physical Chemistry*, 3d ed., 1962; K. Wark, *Thermodynamics*, 3d ed., 1977.

Gluons

The hypothetical force particles which are believed to bind quarks into "elementary" particles. Although theoretical models in which the strong interactions of quarks are mediated by gluons have been successful in predicting, interpreting, and understanding many phenomena in particle physics, free gluons remain undetected in experiments (as do free quarks). According to prevailing opinion, an individual gluon cannot be isolated.

Color. In 1961 M. Gell-Mann and Y. Ne'eman independently suggested that the strong (nuclear) interaction respected the unitary symmetry SU(3) and that the strongly interacting particles called hadrons could be classified according to the patterns prescribed by SU(3). The family groups or supermultiplets that emerged were confined to a few of the simplest possibilities permitted under SU(3) symmetry. Mesons, the hadrons with integral spin in units of $\hbar$ (Planck's constant h divided by 2π), occur only in families with one or eight members. The baryons, which possess half-integral spin, fit into groups with 1, 8, or 10 members. Gell-Mann and G. Zweig separately showed in 1963 that this circumstance could be explained by the hypothesis that hadrons were composites of fundamental constituents that have come to be called quarks. In this quark model of hadrons, a meson is composed of one quark and one antiquark, and a baryon is composed of three quarks. All the hadrons then known could be built out of three different varieties (or flavors) of quarks, denoted up, down, and strange. To account for the observed pattern of mesons and baryons, quarks must be spin-$\frac{1}{2}$ particles. *See* UNITARY SYMMETRY.

Although these rules reproduce the properties of the observed hadron states, they lead to a theoretical inconsistency. The characteristics of the unstable hadron resonance known as Δ^{++} (1232 MeV/c^2), which decays into a proton and a positively charged pi meson, require that it be composed of three up quarks in a configuration which is symmetric under the interchange of any pair of quarks. However, according to the Pauli exclusion principle (which first emerged in the description of atomic structure), identical spin-$\frac{1}{2}$ particles cannot occupy the same quantum state. The quark model could be brought into agreement with the Pauli principle, without compromising any of its successes, if a new attribute were ascribed to the quarks which would make the three up quarks distinguishable. For fanciful reasons, this new attribute is now known as color, though it has no connection with the color of visible light. Quarks are said to come in three colors, most frequently given the arbitrary labels red, blue, and green. A Δ^{++} resonance composed of one red up quark, one blue up quark, and one green up quark will then have the observed properties and be consistent with the laws of quantum mechanics. In this picture the antiparticle of a red up quark is an anti-red anti-up quark, so that the mesons are described as colorless quark-antiquark pairs.

Support for the idea that each quark flavor comes in three distinguishable colors has come from the lifetime of the neutral pi meson, and from the rate at which strongly interacting particles are produced in electron-positron annihilations. Theoretical predictions for these observables are sensitive to the number of distinct quark species, and thus to the number of colors.

The fundamental particles that do not experience strong interactions are the leptons, which like the quarks are spin-$\frac{1}{2}$ particles that are structureless at the current limits of resolution. The most familiar examples of leptons are the electron, the muon, and the neutrinos. Each lepton flavor comes in but a single species, which is to say that leptons are colorless. It is therefore appealing to regard color as the strong-interaction analog of the electric charge. Like electric charge, color cannot be created or destroyed in any of the known interactions; it is said to be conserved. *See* COLOR (QUANTUM MECHANICS); LEPTON.

Gauge symmetry. The existence of a conserved quantity is quite generally a consequence of a continuous group of symmetry transformations which leave the laws of physics unchanged in form. For example, the conservation of energy follows from the fact that physical laws depend upon the time interval between occurrences, and not upon an absolute time measured on some master clock. Translation in time (that is, the resetting of clocks) is a symmetry of the equations of physics. Symmetries relating to internal properties of particles, like electric charge, are known as gauge symmetries. In the case of conservation of the color charge, a natural choice is the unitary group SU(3), now applied to color rather than flavor. *See* SYMMETRY LAWS.

Yang-Mills theory. It frequently happens that the symmetries respected by a phenomenon are recognized before a complete theory has been developed. The question thus arises as to whether a complete theory of nuclear forces could be deduced from a knowledge of the symmetry. If the equations of physics are required to be invariant in form under local symmetry transformations which may be different at every point in space and time, the interactions related to the symmetry are completely fixed. The manner in which this could be accomplished for any continuous symmetry was indicated by C. N. Yang and R. L. Mills in 1954. Nuclear forces had long been known not to distinguish between the proton and neutron. From the point of view of nuclear forces, the designations proton and neutron are purely conventional. This symmetry among protons and neutrons is called isospin invariance. Yang and Mills investigated the consequences of the hypothesis that the nuclear force among protons and neutrons could be derived by imposing local isospin invariance (which is to say that the convention could be chosen independently at every point of space and time). In general, the requirement of local gauge invariance implies that the interaction must occur through the exchange of massless spin-1 bosons. One species of force particle corresponds to each conserved quantity. This made the Yang-Mills theory unacceptable as a description of nuclear forces. It predicted that nuclear forces were mediated by three massless "gauge bosons," whereas the short range (on the order of 10^{-13} cm) over which nuclear forces are observed to act demands that the force particles be massive, as proposed by H. Yukawa. *See* ISOTOPIC SPIN; QUANTUM FIELD THEORY.

Quantum chromodynamics. Applying similar reasoning to the idea that a local color gauge symmetry should prescribe the strong interaction among quarks leads to the gauge theory of strong interactions which has been called quantum chromodynamics (QCD). The mediators of the strong interaction are eight massless vector bosons, which are named gluons because they make up the "glue" which binds quarks together. It is

hoped that the infinite range of the forces mediated by the gluons may help to explain why free quarks have not been isolated. The gluons themselves carry color. Hence, strong interactions among gluons will also occur through the exchange of gluons. It is therefore believed that gluons, as well as quarks, may be permanently confined. According to this view, only colorless objects may exist in isolation.

Experimental evidence. No evidence has been reported for isolated or free gluons. As indicated above, the current interpretation of quantum chromodynamics is that free gluons cannot exist. Therefore it is necessary to devise indirect means to test the idea that gluons exist with all the desired attributes.

Inelastic electron-proton scattering. Early support for the existence of an electrically neutral glue within the proton came from 1968 experiments on inelastic electron-proton scattering carried out at the Stanford Linear Accelerator Center (SLAC). These experiments indicated that the electrons were not scattered electromagnetically from the proton as a whole, but from individual pointlike charged objects subsequently identified with the quarks. They also showed that only about half of the energy of a rapidly moving proton is carried by its charged constituents. The remainder must then be borne by neutral constituents which do not interact electromagnetically. This role would naturally be played by the gluons.

Charmonium lifetimes. Further evidence for the utility of the gluon concept was provided by the unusually long lifetime of the strongly decaying charmonium state J/ψ. In quantum electrodynamics the atom composed of an electron and an antielectron (positron) is known as positronium. Positronium occurs in two forms: orthopositronium, in which the electron and positron spins are aligned, and parapositronium, in which the spins are opposite. The electron and positron may annihilate into photons. The spinless parapositronium state may decay into two photons (Fig. 1*a*), but the spin-1 orthopositronium state must decay into three photons (Fig. 1*b*). The difficulty of radiating an additional

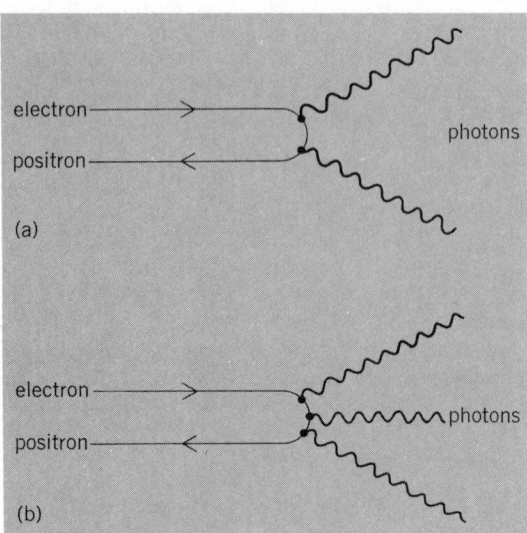

Fig. 1. Positronium decay. (*a*) Decay of parapositronium into two photons. (*b*) Decay of orthopositronium into three photons.

photon is reflected in the fact that orthopositronium lives 1120 times longer than parapositronium. In similar fashion, charmonium, the strong-interaction "atom" composed of a charmed quark and a charmed antiquark, decays by the annihilation of the quark and antiquark into gluons. The gluons materialize into the observed hadrons through the action of the confinement mechanism, with unit probability. For the pseudoscalar para-charmonium level, designated η_c, the semifinal state is composed of two gluons (Fig. 2a). The vector particle J/ψ, which corresponds to orthocharmonium, must decay into three gluons (Fig. 2b). The remarkably long lifetime of J/ψ and the large ratio of J/ψ to η_c lifetimes (approximately 500) argue for the aptness of the analogy. Decays of the still heavier quarkonium state upsilon also support this. *See* CHARM; J PARTICLE; POSITRONIUM.

Three-jet pattern. The most unambiguous evidence for the existence of gluons was reported in 1979 by a number of experimental groups working at the high-energy electron-positron storage ring PETRA at the Deutsches Elektron-Synchrotron (DESY) in Hamburg. It had earlier been established in experiments at SLAC and DESY that the dominant mechanism for hadron production in electron-positron annihilations is electron + positron → quark + antiquark, with the quarks materializing into hadrons. This interpretation explains the rate of particle production and the characteristic angular distribution of the sprays or jets of hadrons that emerge from the collisions (Fig. 3a). If quantum chromodynamics is correct, one of the outgoing quarks may occasionally radiate an energetic gluon, just as a fast electron may radiate an energetic photon. When this happens, the hadrons may be expected to emerge in a three-jet pattern (Fig. 3b). A number of examples of this behavior have been reported in electron-positron annihilations at center-of-mass energies exceeding 24 GeV. By far the most graceful interpretation of these findings is that gluon radiation is being observed indirectly.

Other implications. The existence of gluons with the properties implied by quantum chromodynamics has additional consequences. The interactions of quarks and gluons specified by quantum chromodynamics suggest the existence of a number of new species of hadrons. The most important of these from the gluon perspective are quarkless mesons, composed entirely of gluons. The simplest of these glueballs, as they are sometimes called, may be searched for in the radiative decay $J/\psi \rightarrow$ photon + 2 gluons, or in the decays of heavier quarkonium states.

It is to be expected that further experimental tests of the properties which have been imputed to gluons will be forthcoming. The angular distribution of hadron jets from vector states of heavy quarkonia may exhibit a three-lobed pattern that reflects the three-gluon semifinal state. Detailed characteristics of the three-jet events in electron + positron → quark + antiquark + gluon → hadrons may reveal which jet of hadrons has emerged from the gluon, and test the spin-1 nature of the gluon. Other tests of the gluon spin are to be had from analysis of the pattern of scaling violations in inelastic lepton scattering, and from comparisons of the hadronic decay rates of various quarkonium states.

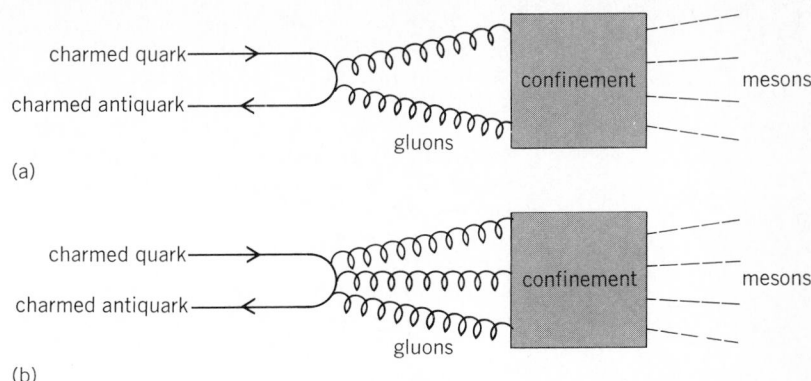

(a)

(b)

Fig. 2. Charmonium decay. (a) Decay of paracharmonium through a two-gluon semifinal state. (b) Decay of orthocharmonium through a three-gluon semifinal state.

Proof of the existence of gluons with the canonical properties would vindicate the idea of color gauge symmetry and provide strong encouragement for quantum chromodynamics and, by deri-

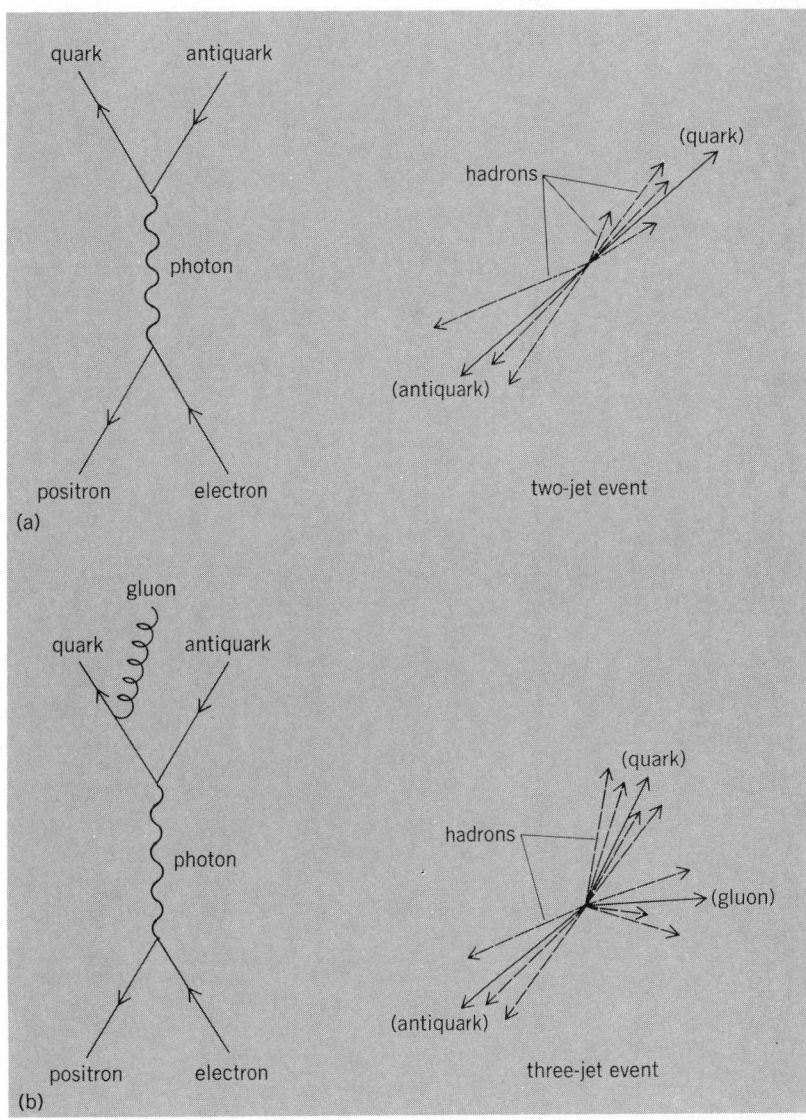

Fig. 3. Mechanisms for hadron production in electron-positron annihilation. (a) Two-jet event produced by the mechanism electron + positron → quark + antiquark. (b) Three-jet event produced by the mechanism electron + positron → quark + antiquark + gluon. The observed hadrons are represented by broken lines.

vation, for grand unified theories of the strong, weak, and electromagnetic interactions. *See* ELEMENTARY PARTICLE; FUNDAMENTAL INTERACTIONS; QUANTUM CHROMODYNAMICS; QUARKS.

[C. QUIGG]

Bibliography: N. Calder, *The Key to the Universe*, 1977; S. L. Glashow, Quarks with color and flavor, *Sci. Amer.*, 233(4):38–50, October 1975; T. B. W. Kirk and H. D. I. Abarbanel (eds.), *Proceedings of the 1979 International Symposium on Lepton and Photon Interactions at High Energies, Fermilab, Batavia, IL*, 1980; Y. Nambu, The confinement of quarks, *Sci. Amer.*, 235(5):48–60, November 1976.

Goldhaber triangle

The phase space triangle, or Goldhaber triangle, corresponds to the kinematically allowed boundary for a high-energy reaction leading to four or more particles. In a high-energy reaction between two particles a and b yielding four particles 1, 2, 3, and 4 in the final state ($a + b \to 1 + 2 + 3 + 4$), it is convenient to consider the reaction in terms of the production of two intermediate-state quasi-particle composites x and y, which then decay into two particles each, as in expression (1).

$$a + b \to x \quad + y$$
$$\quad \downarrow \to 1 + 2 \quad \downarrow \to 3 + 4 \qquad (1)$$

Most high-energy interactions indeed proceed through such intermediate steps, in which, for specific values of the invariant masses $m_x = M_x^*$ and $m_y = M_y^*$, the quasi-particle composites may form resonances. However, the description in

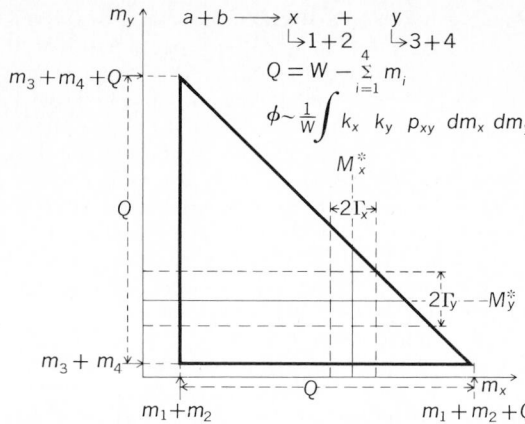

Fig. 1. Definition of the kinematical boundary of the Goldhaber triangle for four particles.

terms of the composites x and y, with the invariant masses m_x and m_y as variables, is valid irrespective of whether or not these composites form resonances.

Kinematical limits. The kinematical limits in this representation are particularly simple, namely, they form a right-angle isosceles triangle. A Goldhaber triangle plot corresponds to plotting a point (m_x, m_y) for each event occurring in the above high-energy reaction. Because of the kinematical constraints, these points must all lie inside the triangle.

If one considers the general reaction given in Eq. (1), then the length of each of the two equal sides of the triangle is Q, defined in Eq. (2). Here W

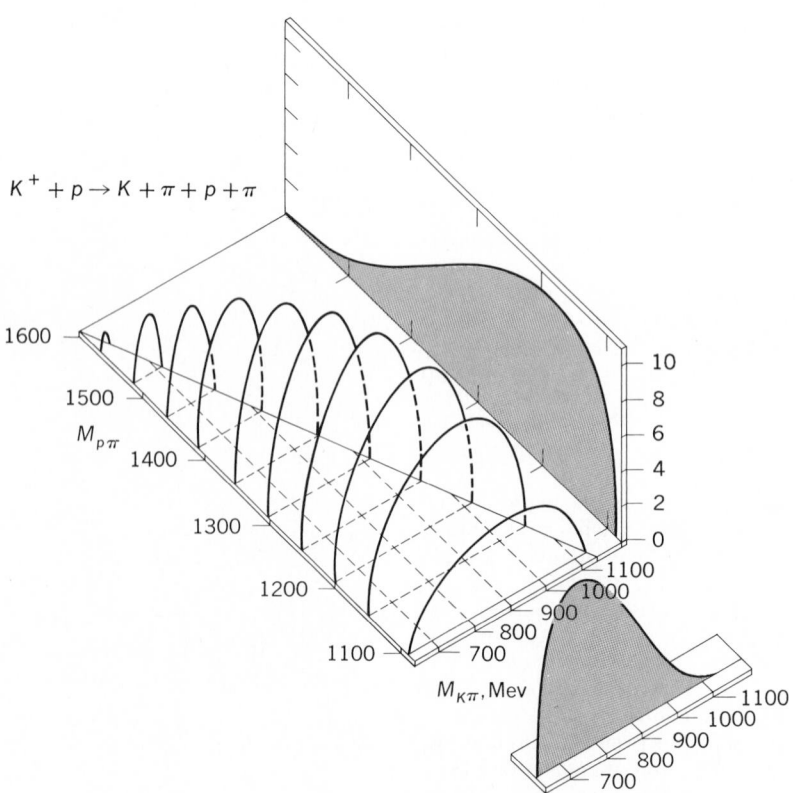

Fig. 2. Phase space distribution over the Goldhaber triangle for the example $K^+p \to (K^+\pi^-) + (p\pi^+)$ at 2 Gev/c incident laboratory momentum. The shaded areas cor-respond to the phase space projections onto the two mass axes.

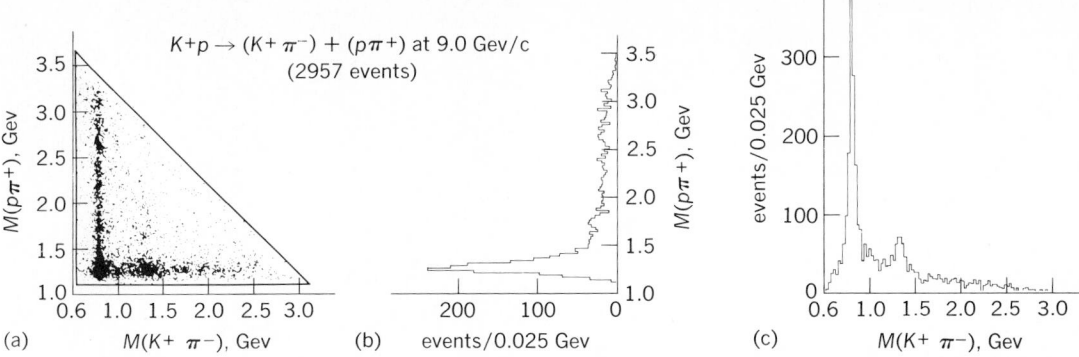

Fig. 3. Goldhaber triangle plot. (*a*) Example of a triangle plot. (*b*) Projections for $M(p\pi^+)$. (*c*) Projection for $M(K^+\pi^-)$. (*From A. Firestone, G. Goldhaber, and B. C. Shen, Lawrence Radiation Laboratory, Berkeley, Calif.*)

$$Q = W - \sum_{i=1}^{4} m_i \qquad (2)$$

is the total energy in the center of mass of particles *a* and *b*, and m_i, for $i = 1$ to 4, is the mass of the particles 1 to 4. Hence Q corresponds to the total kinetic energy available in the reaction. All quantities are in energy units of millions or billions of electronvolts.

The values of m_x and m_y run over the intervals $m_1 + m_2 \leqq m_x \leqq m_1 + m_2 + Q$ and $m_3 + m_4 \leqq m_y \leqq m_3 + m_4 + Q$, respectively. The effect of changing the incident momentum, and thus Q, is then simply to move the hypotenuse of the triangle, leaving the two sides, as well as the location of any resonances which may occur for certain mass values of the composites *x* and *y*, fixed.

In the triangle corresponding to the general reaction (Fig. 1), the vertical and horizontal bands indicate resonances at masses M_x^* and M_y^* with full width at half-maximum height or Γ_x and Γ_y, respectively. The bands shown of width 2Γ represent the regions usually chosen if the events corresponding to a given resonance are selected.

Phase space distribution. The phase space is given by $\Phi \propto 1/W \int k_x k_y p_{xy} \, dm_x \, dm_y$ where the integral extends over the triangle. Here k_x and k_y are the momenta in the center of mass of the composites *x* and *y*, respectively, and p_{xy} is the outgoing momentum of each of the composites in the overall center of mass of particles *a* and *b*. It is noteworthy that, along each of the three sides of the triangle, one of the factors in the integrand vanishes. This corresponds to the fact that, along the m_x axis, $m_y = m_3 + m_4$; thus there is no internal kinetic energy in the *y* composite and hence $k_y = 0$. Similarly $k_x = 0$ along the m_y axis. The hypotenuse corresponds to the situation for which the entire available energy is in the mass of the two composites; thus $W = m_x + m_y$. Hence here one gets $p_{xy} = 0$. The phase space distribution over the triangle is illustrated in Fig. 2 for the example $K^+ p \rightarrow K^+ \pi^- p\pi^+$ for an incident laboratory momentum of 2 Gev/c. The shaded areas correspond to the projections on the two mass axes.

Comparison with Dalitz plot. Superficially there is a great similarity between the Dalitz plot and the Goldhaber triangle plot. There are, however, several important differences: (1) The Dalitz plot applies to three particles in the final state and corresponds to plotting m_y^2 versus m_x^2; however, in this case the two composites *x* and *y* have one particle in common. (2) The Dalitz plot has the advantage that the phase space distribution is uniform but the kinematical boundary is a more complicated function of the variables. In the Goldhaber triangle plot the phase space distribution is more complicated, as shown in Fig. 2, but the kinematical boundary is very simple—a triangle. (3) If two resonances overlap on the plot the interpretation is completely different. In the Dalitz plot the overlap corresponds to interference between the two resonances; in the Goldhaber triangle plot the overlap corresponds to double resonance formation, that is, both the *x* and *y* composites form essentially independent resonances at the same time. *See* DALITZ PLOT.

Example of triangle plot. To illustrate the appearance of the Goldhaber triangle plot, Fig. 3 shows an example from the reaction $K^+ p \rightarrow (K^+ \pi^-) + (p\pi^+)$ at a laboratory momentum of 9 Gev/c. The two sets of particles in parentheses correspond to the choice of *x* and *y* composites, respectively. The clear-cut vertical band corresponds to the $K^{*0}(890)$ meson resonance with $K^+\pi^-$ decay. The horizontal band corresponds to the $\triangle^{++}(1238)$ baryon resonance with $p\pi^+$ decay. The overlap region of these two correspond to $K^*(890) + \triangle^{++}(1238)$ double resonance formation. A second faint vertical band corresponds to the $K^{*0}(1420)$ meson resonance again with $K^+\pi^-$ decay. This resonance is most apparent in the $K^{*0}(1420) + \triangle^{++}(1238)$ double-resonance region. The projections on the two mass axes are also shown.

Extension to five particles. The concepts discussed here can be extended to five particles in the final state; here the composite *x* corresponds to three particles: 1, 2, and 3; and the composite *y* to two particles: 4 and 5. The kinematical limits will still be determined, as before, by an isosceles triangle where the generalization is straightforward. For example, in reaction (3) a double reso-

$$\pi^+ p \rightarrow (\pi^+ \pi^- \pi^0) + (p\pi^+) \qquad (3)$$

nance formation occurs for ω^0 and $\triangle^{++}(1238)$ production according to reaction (4).

$$\pi^+ p \rightarrow \omega^0(890) \quad + \triangle^{++}(1238)$$
$$ \hookrightarrow \pi^+ \pi^- \pi^0 \quad \hookrightarrow p\pi^+ \qquad (4)$$

[GERSON GOLDHABER]

Grain boundaries

The surfaces separating individual crystals, or grains, in a crystalline solid. Many solids, especially metallic and ceramic materials, are characterized by a periodic internal structure. This structure is defined by arrangement of atoms (or ions) in a precise three-dimensional array to form a crystal. The extent of the arrangement is large with respect to the size of the atom or ion. Although solids may be a single crystal, it is much more common for solids to be composed of many crystals, which retain their characteristic structure but are misoriented with respect to neighboring crystals by a relative rotation of their principal axes. The boundary separating these individual grains is known as a grain boundary and is characterized by a two-dimensional structure which may be related to the structure of neighboring grains. Therefore, grain boundaries represent planar defects in an otherwise perfect crystalline material. Grains and grain boundaries are ubiquitous in nature; for example, a polycrystalline cube, measuring 1 cm on an edge, could easily contain over 1,000,000,000 grains and 1 m² of grain boundary area. *See* CRYSTAL STRUCTURE.

Microscopic structure. It was first believed that grain boundaries were composed of an amorphous (structureless) layer measuring many atomic diameters in thickness. However, mounting experimental and theoretical evidence indicates that grain boundaries not only are nearer to a single atomic diameter in thickness, but also are characterized by a periodic structure, regardless of the degree of misorientation between adjacent grains. One special type of structure is illustrated in Fig. 1, for two grains, characterized by cubic crystal structures and rotated with respect to one another by an angle of 36.9° about a common cube axis. This particular grain boundary merely provides an example of one possible boundary structure; the structure of other boundaries may vary considerably, depending on degree of misorientation between adjacent grains and inclination of the boundary plane with respect to these grains.

The structure of grain boundaries may conveniently be divided into three categories; low-angle boundaries, random high-angle boundaries, and special high-angle boundaries. Low-angle grain boundaries separate two adjacent grains which have been rotated (tilted or twisted) by a small angle (generally less than 15°) with respect to one another. For low-angle grain boundaries, the transition from one grain to another is only slightly per-

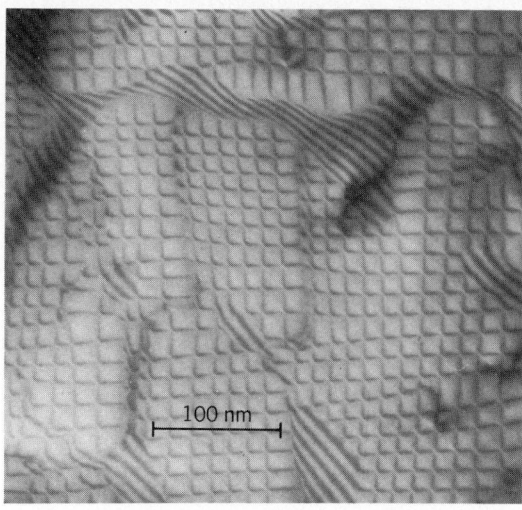

Fig. 2. High-resolution transmission electron micrograph of a low-angle grain boundary in gold, wherein a cross-grid of dislocations, characterized by spacing of about 15 nm, is resolved. (*Courtesy of F. Cosandey*)

turbed, usually by a series of line defects aligned in the plane of the boundary (Fig. 2). These line defects are known as dislocations and consist of an elastic stress field and a (noncrystalline) core region. As the angle of misorientation between adjacent grains increases, the spacing between dislocations decreases until the dislocation cores overlap, thus defining a high-angle grain boundary. These boundaries may be composed of a disordered (random) atomic array due to close proximity of adjacent dislocation cores, or may be composed of an ordered (special) atomic array, wherein the boundary assumes a certain two-dimensional periodic structure. Such special high-angle boundaries often are assigned special names, such as twin boundaries, stacking faults, and coincidence boundaries. One example of a (special) coincidence boundary is illustrated in Fig. 1. It is evident that this misorientation (36.9° about a cube axis) represents a special case, resulting in a periodic grain boundary structure and an extension of a sublattice of the original lattice (solid circles) across the boundary. *See* CRYSTAL DEFECTS; TWINNING.

Effects on macroscopic properties. The microscopic structure of grain boundaries is intimately related to many macroscopic properties, such as electrical resistivity, corrosion resistance, tensile strength, and fracture toughness. Moreover, in a macroscopic sense, grain boundaries may be characterized by a surface tension, which requires a certain balance of forces at intersections of grain boundaries with other boundaries, with external (free) surfaces, or even with boundaries separating dissimilar crystallographic materials (phases). The behavior of grain boundaries closely resembles that of a soap froth, since both are governed by capillarity. Unlike soap films, however, surface tension of grain boundaries is not only a function of the degree of misorientation between adjacent grains, but also a function of inclination of the boundary plane with respect to these grains. Therefore, many properties associated with grain boundaries are anisotropic. *See* SURFACE TENSION.

Segregation of impurities. The fact that grain

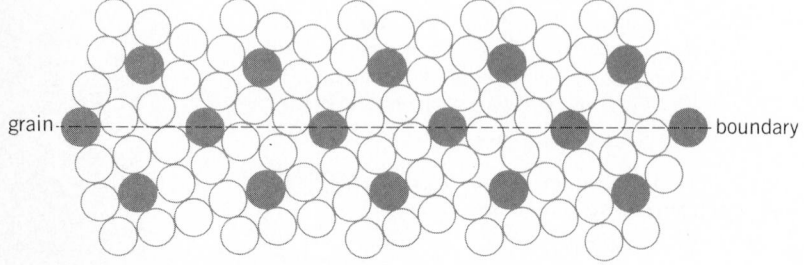

Fig. 1. A grain boundary in a cubic structure, produced by rotation of one grain with respect to the other by 36.9° about a cube axis. In this particular case, a sublattice (solid circles) extends across the boundary without interruption.

boundaries are characterized by a more open structure than their neighboring grains encourages segregation of impurities, especially if the diameter of the impurity atom differs significantly from that of the parent material. Accordingly, a material containing a small fraction of an impurity, on the average, could actually contain a large fraction of this impurity localized at grain boundaries. Localization of impurities at grain boundaries often imparts undesirable properties to the overall material, such as susceptibility to corrosion and fracture. Moreover, the degree of segregation at grain boundaries depends on structure of the boundary, which, in turn, depends on degree of misorientation between adjacent grains and inclination of the boundary plane with respect to these grains. Therefore, effects of segregation of impurities at grain boundaries in polycrystalline materials depend on both amount and character of existing boundaries, as well as characteristics of the impurity itself.

Mass transport. In contrast to segregation of impurities at grain boundaries, which is a dynamic equilibrium effect, impurities can also diffuse much more rapidly along grain boundaries, which is a kinetic effect. Accordingly, mass transport may occur either through grain interiors or along grain boundaries in a given (polycrystalline) material. Mass transport is especially severe in thin films, since not only is the film thickness small, but also the grain size is small, thus providing many short-circuit (grain boundary) paths for diffusion. Grain boundaries also play an important role in promoting solid-state phase transformations by providing internal surfaces for (heterogeneous) nucleation of a new phase.

Mobility of boundaries. At sufficiently high temperatures, grain boundary movement can occur by removal of atoms from one grain and reattachment to the adjacent grain. Since the case of disattachment and reattachment of atoms depends on local grain boundary structure, grain boundaries are characterized by a spectrum of mobilities; that is, velocity per unit pressure. Accordingly, special and low-angle grain boundaries are relatively immobile, whereas random high-angle boundaries are relatively mobile. These particular factors are important during recrystallization, grain growth, and texture formation. In these processes, numerous grains are nucleated in a deformed matrix and grow in such a manner as to convert the entire matrix to its original state. The resultant polycrystalline material is often characterized by preferential orientation or texture of existing grains. Conversely, proper choice of deformation and annealing conditions may be used to control (refine) grain size. At elevated temperatures, grains may continue to grow due to imbalances of surface tension. *See* CRYSTAL GROWTH.

Mechanical properties. Perhaps the effect of grain boundaries on mechanical properties is more important than on any other class of properties. In general, strength and ductility are controlled by the motion of dislocations under the influence of applied forces, which is responsible for plastic deformation. Grain boundaries provide an effective barrier to this motion so that strength increases and ductility decreases with decreasing grain size. Curiously, the role of grain boundaries is reversed at elevated temperatures, at which

small grain size, that is, a large amount of grain boundary area, facilitates plastic deformation by mass transport of constituent atoms along these boundaries. This phenomenon, known as creep, often limits usefulness of materials at elevated temperatures. Fine-grained polycrystalline materials sometimes exhibit extraordinarily large degrees of plasticity at elevated temperatures. This large degree of plasticity, termed superplasticity, actually is an accelerated form of creep involving rapid diffusion of constituent atoms along grain boundaries in order to relieve the applied stress.

Even when motion of dislocations is sufficiently suppressed to severely limit plastic deformation, for example, at low temperatures, failure may occur by fracture. This phenomenon may occur either by nucleation of cracks and subsequent propagation along grain boundaries (intergranular) or by subsequent propagation through the interior of grains (transgranular). A very important technical problem that is associated with the presence of grain boundaries is stress-corrosion cracking. This phenomenon involves selective corrosion at grain boundaries under a mechanical stress until eventual failure occurs.　　[CHARLES L. BAUER]

Bibliography: G. A. Chadwick and D. A. Smith (eds.), *Grain Boundary Structure and Properties*, 1976; P. Chaudhari and J. W. Matthews (eds.), *Grain Boundaries and Interfaces*, Surf. Sci. Spec. Vol. no 31 (1972); D. McLean, *Grain Boundaries in Metals*, 1957; L. E. Murr, *Interfacial Phenomena in Metals and Alloys*, 1975.

Graph theory

A branch of mathematics that belongs partly to combinatorial analysis and partly to topology. Its applications occur (sometimes under other names) in electrical network theory, operations research, organic chemistry, theoretical physics, and statistical mechanics, and in sociological and behavioral research. Both in pure mathematical inquiry and in applications, a graph is customarily depicted as a topological configuration of points and lines, but usually is studied with combinatorial methods. *See* COMBINATORIAL THEORY; TOPOLOGY.

Origin of graph theory. Graph theory and topology are said to have started simultaneously in 1736 when L. Euler settled the celebrated Königsberg bridge problem. In Königsberg, there were two islands linked to each other and to the banks of the Pregel River by seven bridges. Figure 1 illustrates both this setting and its topological abstraction as a graph. The points *a*, *b*, *c*, *d* correspond to land areas, and the connecting lines to bridges. The problem is to start at one of the land areas and to cross all seven bridges without ever recrossing a bridge. Euler proved that there is no solution, and he established a rule that applies to any connected

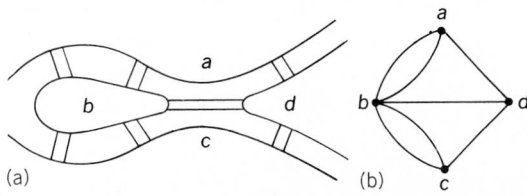

Fig. 1. Königsberg bridge problem. (*a*) The seven bridges of Königsberg. (*b*) Corresponding graph.

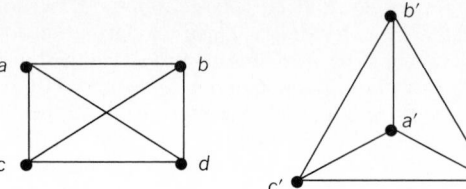

Fig. 2. Two isomorphic graphs.

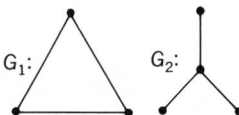

Fig. 3. Two graphs for which either a 120° rotation or a vertical reflection is an automorphism.

· GRAPH THEORY

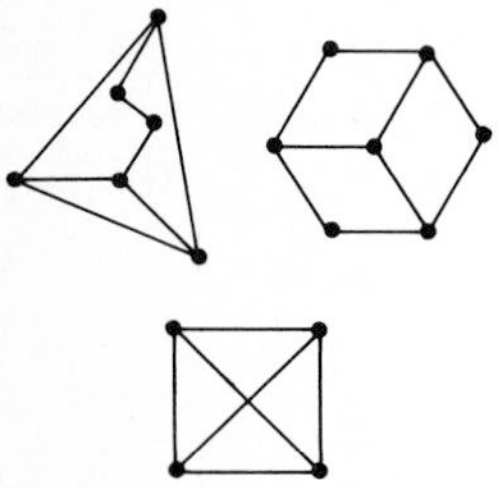

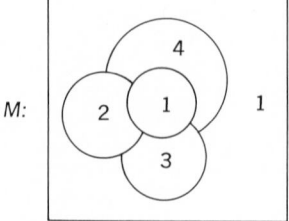

Fig. 4. Homeomorphic but nonisomorphic graphs.

graph: such a traversal is possible if and only if at most two points are odd, that is, each is the terminus for an odd number of lines. Euler also proved that the number of odd points in a graph is always an even number. Thus, a complete traversal without recrossing any lines is possible if the number of odd points is zero or two. If zero, the complete traversal ends at the starting point.

In geometry a graph might arise as the set of vertices and edges of a convex, three-dimensional polyhedron, such as a pyramid or a prism. Euler derived an important property of all such polyhedra. Let V, E, and F be the numbers of vertices, edges, and faces of such a polyhedron. Euler proved that $V - E + F = 2$, which is now called the Euler equation. For instance, a cube has $V = 8$, $E = 12$, and $F = 6$, so that $8 - 12 + 6 = 2$. Euler's observations have been extended to a theorem about imbeddings of graphs in surfaces and to the Euler-Poincaré characteristic for cell complexes in combinatorial topology.

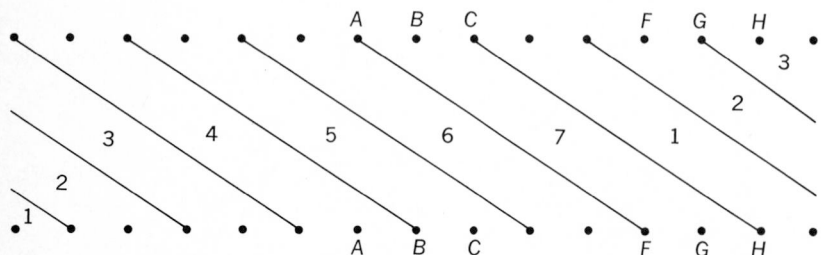

Fig. 5. Plane map requiring four colors.

Fig. 6. Map on a torus (doughnut) that requires seven colors. To form the torus, paste opposite sides of the rectangle together.

Definitions. A graph consists of a set of points, a set of lines, and an incidence relation that designates the end points of each line. In many applications no line starts and ends at the same point. (Such a line would be called a loop.) Also, no two lines have the same pair of end points. A graph whose lines satisfy these conditions is called simplicial. The valence of a point is the number of lines incident on it, calculated so that a loop is twice incident on its only end point. Two graphs are isomorphic if there is one-to-one correspondence from the point set and line set of one onto the point set and line set, respectively, of the other that preserves the incidence relation. The point correspondence $a \rightarrow a'$, $b \rightarrow b'$, $c \rightarrow c'$, $d \rightarrow d'$ indicates an isomorphism between the two graphs of Fig. 2.

An automorphism of a graph is an isomorphism of a graph with itself. For instance, a plane rotation of 120° would yield an automorphism of either of the two graphs in Fig. 3. A plane reflection through a vertical axis would also yield an automorphism of either of them. The set of all automorphisms of a graph G forms the automorphism group of G. R. Frucht proved in 1938 that every finite group is the automorphism group of some graph. Two graphs are homeomorphic (Fig. 4) if, after smoothing over all points of valence 2, the resulting graphs are isomorphic. *See* GROUP THEORY.

Map coloring problems. Drawing a graph on a surface decomposes the surface into regions. One colors the regions so that no two adjacent regions have the same color, rather like a political map of the world. It is a remarkable fact that for a given surface, there is a single number of colors that will always be enough no matter how many regions occur in a decomposition of the surface. The smallest such number is called the chromatic number of that surface. It is easy to draw a plane map, as in Fig. 5, that requires four colors. In 1976 K. Appel and W. Haken settled a problem dating back to about 1850, by showing that four colors are always enough for plane maps.

Some maps on more complicated surfaces require more than four colors. For instance, Fig. 6 illustrates a map on a torus (the surface of a doughnut) that needs seven. To obtain the toroidal map from the rectangular drawing, first match the top to the bottom to get a cylindrical tube. Then match the left end of the cylinder to the right end to complete the torus. Whereas, before this matching, region 7 meets only regions 1 and 6, after the matching it also meets region 2 along FG, region 3 along GH, region 4 along AB, and region 5 along BC. In fact, after the matching, each of the seven regions borders every other region. It follows that seven colors are necessary. No map on the torus needs more than seven colors, as P. J. Heawood proved in 1890. G. Ringel and J. W. T. Youngs completed a calculation in 1968 of the chromatic numbers of all the surfaces except the plane or sphere.

Planarity. A graph is planar if it can be drawn in the plane so that none of its lines cross each other. Neither of the two graphs in Fig. 7 can be drawn in the plate. K. Kuratowski proved in 1930 that a graph is planar if and only if it contains no subgraph homeomorphic to either of those two graphs. Testing all the subgraphs might be a very tedious process, even on a fast computer. In 1974 J. Hop-

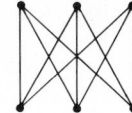

Fig. 7. Prototypes of all nonplanar graphs.

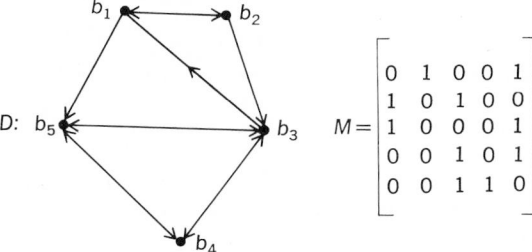

Fig. 8. A digraph and its adjacency matrix.

croft and R. Tarjan obtained an extremely fast alternative planarity test. The time it takes a computer to perform the Hopcroft-Tarjan test is linearly proportional to the time it takes to read its point set into the computer.

There are methods to decide for any graph and any surface whether the graph can be drawn on the surface without edge crossings. The time to execute such methods is unfeasibly large for most graphs and most surfaces except the plane or the sphere. Ringel has constructed many important special drawings on higher-genus surfaces.

Variations. In a directed graph, or digraph, each line ab is directed from one end point a to the other end point b. There is at most one line from a to b. The adjacency matrix $M = (m_{ij})$ of a digraph D with points $b_1, b_2, \ldots, b_n$ has the entry $m_{ij} = 1$ if the line $b_i b_j$ occurs in D; otherwise $m_{ij} = 0$ (Fig. 8).

An oriented graph is obtained from an ordinary graph by assigning a unique direction to every line. If there is one line between every pair of points and no loops, an ordinary graph is called complete. An oriented complete graph is called a tournament (Fig. 9).

Applications. A. Cayley reformulated the problem of counting the number of isomers of saturated hydrocarbons ($C_n H_{2n+2}$) in graphical language (Fig. 10). Each isomer is a tree all of whose vertices have valence 1 for hydrogen, or 4 for carbon. G. Polya devised a general theorem in 1937 for enumeration to provide a solution to such problems. F. Harary and others have solved many related problems by applying and extending Polya's theorem. Extremely effective use of Polya's theorem occurs in theoretical physics, where G. Ford and G. Uhlenbeck solved several graphical enumeration problems arising in statistal mechanics.

Suppose that some of the points of a graph correspond to workers $x_1, \ldots, x_m$, that the rest of the points correspond to jobs $y_1, \ldots, y_m$, and the presence of a line between x_i and y_j means that worker x_i is capable of performing job y_j. The personnel assignment problem is to find m lines so that each worker x_i is matched to exactly one possible job. In the optimal assignment problem, labels on the lines tell how well a worker can do a particular job. An algorithm due to H. Kuhn and

J. Munkres solves the optimal assignment problem.

If the points of a graph represent cities and the lines between them are labeled with distances, one might want to find the shortest path from one point to another. An efficient method to determine a shortest path was developed by E. Dijkstrain in 1959. K. Menger proved in 1927 that if A and B are disjoint sets of a connected graph G, then the minimum number of points whose deletion separates A from B equals the maximum number of disjoint paths between A and B. L. Ford and D. Fulkerson have generalized Menger's theorem into a method for solving network flow problems.

According to the physical laws of G. Kirchhoff and G. Ohm, any set of voltages applied to the input nodes of an electrical network determines the voltages at all other nodes and the currents on every branch. Kirchhoff also proved the result known as the matrix-tree theorem: Let G be a connected graph with points $b_1, \ldots, b_n$, and let $A = (a_{ij})$ be the matrix such that a_{ij} is the valence of b_i and, for $i \neq j$, $a_{ij} = -1$ if b_i is adjacent to b_j or 0 otherwise. Then the cofactor of each entry a_{ij} equals the number of spanning trees of G, that is, the number of trees in G that includes every point of G (Fig. 11).

Numerous applications of graph theory to social and behavioral science have been developed, many by Harary and his coauthors. If points represent persons and lines represent such interrelationships as communication, linking, or power, then a graph may depict various aspects of social organization. For instance, anthropologists use graphs to describe kinship, and management scientists use them to display corporate hierarchy.

Graph theory is presently in a phase of rapid growth. Two of the major branches not described here are external graph theory, founded by P. Turan and developed by P. Erdös, and hypergraph theory, developed by C. Berge. Material theory, originated by H. Whitney and expanded by W. Tutte, is closely related to graph theory. Tutte is also responsible for important results in many

GRAPH THEORY

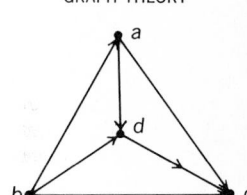

Fig. 9. A tournament.

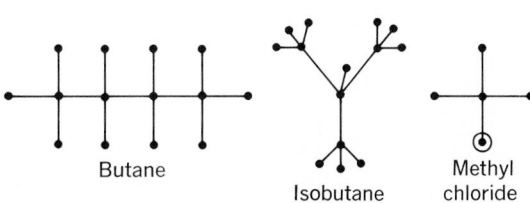

Fig. 10. The two isomers of a saturated hydrocarbon.

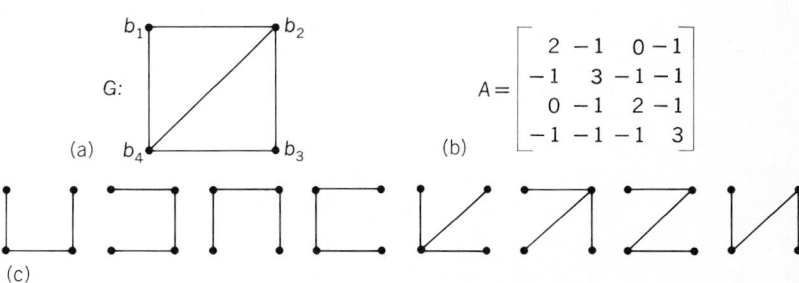

Fig. 11. Matrix-tree theorem. (a) Graph G. (b) Corresponding matrix A. (c) Spanning trees of G.

other areas of graph theory and combinatorial research, including connectivity, decomposition, and chromatic numbers.

[JONATHAN L. GROSS]

Bibliography: L. Beineke and R. Wilson (eds.), *Selected Topics in Graph Theory*, 1978; C. Berge, *Graphs and Hypergraphs*, 1973; N. Biggs, E. Lloyd, and R. Wilson, *Graph Theory 1736–1936*, 1976; B. Bollobas, *Extremal Graph Theory*, 1978; A. Bondy and R. Murty, *Graph Theory with Applications*, 1976; F. Harary, *Graph Theory*, 1969; F. Harary and E. Palmer, *Graphical Enumeration*, 1973; G. Ringel, *Map Color Theorem*, 1974; W. Tutte, *Introduction to the Theory of Matroids*, 1971.

Gravitation

The mutual attraction between all masses and particles of matter in the universe. In a sense this is one of the best-known physical phenomena. During the 18th and 19th centuries gravitational astronomy, based on Newton's laws, attracted many of the leading mathematicians and was brought to such a pitch that it seemed that only extra numerical refinements would be needed in order to account in detail for the motions of all celestial bodies. In the present century, however, Albert Einstein shattered this complacency, and the subject is currently in a healthy state of flux.

Until the 17th century, the sole recognized evidence of this phenomenon was the gravitational attraction at the surface of the Earth. Only vague speculation existed that some force emanating from the Sun kept the planets in their orbits. Such a view was expressed by Johannes Kepler (1571–1630), the author of the laws of planetary motion. But a proper formulation for such a force had to wait until Isaac Newton (1643–1727) founded Newtonian mechanics, with his three laws of motion, and discovered, in calculus, the necessary mathematical tool. *See* NEWTON'S LAWS OF MOTION.

Newton's law of gravitation. Newton's law of universal gravitation states that every two particles of matter in the universe attract each other with a force that acts in the line joining them, the intensity of which varies as the product of their masses and inversely as the square of the distance between them. Put into symbols, the gravitational force F exerted between two particles with masses m_1 and m_2 separated by a distance d is given by Eq. (1), where G is called the constant of gravitation.

$$F = Gm_1m_2/d^2 \qquad (1)$$

A force varying with the inverse square power of the distance from the Sun had been already suggested—notably by Robert Hooke (1635–1703) but also by other contemporaries of Newton, such as Edmund Halley (1656–1742) and Christopher Wren (1632–1723)—but this had only been applied to circular planetary motion. The credit for accounting for, and partially correcting, Kepler's laws and for setting gravitational astronomy on a proper mathematical basis is wholly Newton's.

Newton's theory was first published in *The Principia* in 1686. According to Newton, it was formulated in principle in 1666 when the problem of elliptic motion in the inverse-square force field was solved. But publication was delayed in part because of the difficulty of proceeding from the "particles" of the law to extended bodies such as the Earth. This difficulty was overcome when Newton established that, under his law, bodies having spherically symmetrical distribution of mass attract each other as if all their mass were concentrated at their respective centers.

Newton verified that the gravitational force between the Earth and the Moon, necessary to maintain the Moon in its orbit, and the gravitational attraction on the surface of the Earth were related by an inverse square law of force. Let E be the mass of the Earth, assumed to be spherically symmetrical with radius R. Then the force exerted by the Earth on a small mass m near the Earth's surface is given by Eq. (2), and the acceleration of gravity on the Earth's surface, g, by Eq. (3).

$$F = GEm/R^2 \qquad (2)$$
$$g = GE/R^2 \qquad (3)$$

Let a be the mean distance of the Moon from the Earth, M the Moon's mass, and P the Moon's sidereal period of revolution around the Earth. If the motions in the Earth-Moon system are considered to be unaffected by external forces (principally those caused by the Sun's attraction), Kepler's third law applied to this system is given by Eq. (4).

$$4\pi^2 a^3/P^2 = G(E + M) \qquad (4)$$

Equations (3) and (4), on elimination of G, give Eq. (5).

$$g = 4\pi^2 \frac{E}{E+M} \frac{a^2}{R^2} \frac{a}{P^2} \qquad (5)$$

Now the Moon's mean distance from the Earth is $a = 60.27R = 3.84 \times 10^{10}$ cm and the sidereal period of revolution is $P = 27.32$ days $= 2.361 \times 10^6$ sec. These data give, with $E/M = 81.35$, $g = 977$ cm sec^{-2}, which is close to the observed value.

This calculation corresponds in essence to that made by Newton in 1666. At that time the ratio a/R was known to be about 60, but the Moon's distance in miles was not well known because the Earth's radius R was erroneously taken to correspond to 60 mi per degree of latitude instead of 69 mi. As a consequence, the first test was unsatisfactory. But the discordance was removed in 1671 when J. Picard's measurement of an arc of meridian in France provided a reliable value for the Earth's radius.

Gravitational constant. Equation (3) shows that the measurement of the acceleration due to gravity at the surface of the Earth is equivalent to finding the product G and the mass of the Earth. Determining the gravitational constant by a suitable experiment is therefore equivalent to "weighing the Earth."

In 1774 N. Maskelyne determined G by measuring the deflection of the vertical by the attraction of a mountain. This method is much inferior to the laboratory method in which the gravitational force between known masses is measured. In the torsion balance two small spheres, each of mass m, are connected by a light rod, suspended in the middle by a thin wire. The deflection caused by bringing two large spheres each of mass M near the small ones on opposite sides of the rod is measured, and the force is evaluated by observing the period of

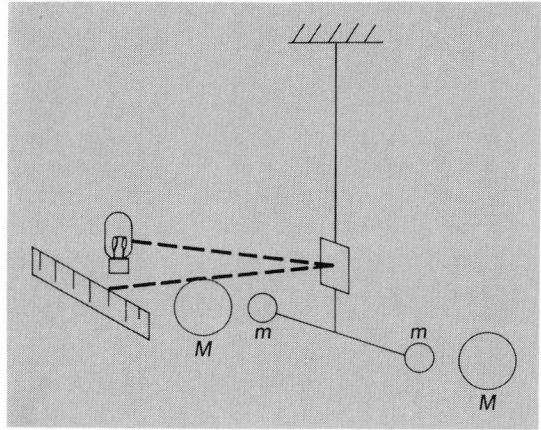

The torsion balance.

oscillation of the rod under the influence of the torsion of the wire (see illustration.) This is known as the Cavendish experiment, in honor of H. Cavendish who achieved the first reliable results by this method in 1797–1798. More recent determinations using various refinements yield the results: constant of gravitation $G = 6.67 \times 10^{-11}$ mks units; mass of Earth $= 5.98 \times 10^{27}$g. The uncertainty of these results is probably about one-half unit of the last place given.

In Newtonian gravitation G is an absolute constant, independent of time, place, and the chemical composition of the masses involved. Partial confirmation of this was provided before Newton's time by the experiment attributed to Galileo when different weights released simultaneously from the top of the tower of Pisa reached the ground at the same time. Newton found further confirmation, experimenting with pendulums made out of different materials. Early in this century, R. Eötvös found that different materials fall with the same acceleration to within 1 part in 10^7. The accuracy of this figure was extended by R. H. Dicke to 1 part in 10^{11}, using aluminum and gold, and by V. B. Braginskii and V. I. Panov to 0.9×10^{-12} with a confidence of 95%, using aluminum and platinum.

With the discovery of antimatter, there was speculation that matter and antimatter would exert a mutual gravitational repulsion. But experimental results indicate that they attract one another according to the same laws as apply to matter of the same kind.

Mass and weight. A cosmology with changing physical "constants" was first proposed in 1937 by P. A. M. Dirac. Field theories applying this principle have since been proposed by P. Jordan and D. W. Sciama and, in 1961, by C. Brans and R. H. Dicke. In these theories G is diminishing; for instance Brans and Dicke suggest a change of about 2×10^{-11} per year. This would have profound effects on phenomena ranging from the evolution of the universe to the evolution of the Earth. For instance, stars evolve faster if G is greater, so that stellar evolutionary ages computed with constant G at its present value would be too great. The Earth, compressed by gravitation, would expand, having a profound effect on surface features. Planetary orbits would gradually be increasing in

size. About 3×10^9 years ago the Sun would have been hotter than it is now, and the Earth and its orbit would have been smaller, so that the temperature on the Earth's surface might have approached the boiling point of water; this would be important for the origin of life on the Earth. Astronomical observations of the planets over the past few hundred years are not accurate enough for the predicted change to be detected; but T. C. Van Flandern has reported that observations of the motion of the Moon are consistent with the predicted change in G. He suggests a change of the order of 1 part in 10^{-10} per year. From radar ranging experiments I. I. Shapiro gives a limit of no greater than 4×10^{-10}. However, using cosmological arguments, J. D. Barrow has argued that the figure can be no greater than 1.5×10^{-12}.

In the equations of motion of Newtonian mechanics, the mass of body appears as inertial mass as a factor of the acceleration, and as gravitational mass in the expression of the gravitational force. The equality of these masses is confirmed by the Eotvos experiment. It justifies the assumption that the motion of a particle in a gravitational field does not depend on its physical composition. In Newton's theory the equality can be said to be a coincidence, but not in Einstein's theory, where inertia and gravitation are unified.

While mass in Newtonian mechanics is an intrinsic property of a body, its weight depends on certain forces acting on it. For example, the weight of a body on the Earth depends on the gravitational attraction of the Earth on the body and also on the centrifugal forces due to the Earth's rotation. The body would have lower weight on the Moon, even though its mass would remain the same. *See* CENTRIFUGAL FORCE.

Gravity. This should not be confused with the term gravitation. Gravity is the older term, meaning the quality of having weight, and so came to be applied to the tendency of downward motion on the Earth. Gravity or the force of gravity are today used to describe the intensity of gravitational forces, usually on the surface of the Earth or another celestial body. So gravitation refers to a universal phenomenon, while gravity refers to its local manifestation.

A rotating planet is oblate (or flattened at the poles) to a degree depending on the ratio of the centrifugal to the gravitational forces on its surface and on the distribution of mass in its interior. The variation of gravity on the surface of the Earth depends on these factors and is further complicated by irregular features such as oceans, continents, and mountains. It is investigated by gravity surveys and also through the analysis of the motion of artificial satellites. Because of the irregularities, no mathematical formula has been found that satisfactorily represents the gravitational field of the Earth, even though formulas involving hundreds of terms are used. The problem of representing the gravitational field of the Moon is even harder because the surface irregularities are proportionately much larger.

In describing gravity on the surface of the Earth, a smoothed out theoretical model is used, to which are added gravity anomalies, produced in the main by the surface irregularities.

Gravity waves are waves in the oceans or atmosphere of the Earth whose motion is dynamically governed by the Earth's gravitational field. They should not be confused with gravitational waves, which are discussed below.

Gravitational potential energy. This describes the energy that a body has by virtue of its position in a gravitational field. If two particles with masses m_1 and m_2 are a distance r apart and if this distance is slightly increased to $r + \triangle r$, then the work done against the gravitational attraction is $Gm_1 m_2 \triangle r/r^2$. If the distance is increased by a finite amount, say from r_1 to r_2, the work done is given by Eq. (6). If $r_2 \to \infty$, Eq. (7) holds.

$$W_{r_1, r_2} = Gm_1 m_2 \int_{r_1}^{r_2} dr/r^2$$
$$= Gm_1 m_2 (1/r_1 - 1/r_2) \qquad (6)$$

$$W_{r_1, \infty} = Gm_1 m_2/r_1 \qquad (7)$$

If one particle is kept fixed and the other brought to a distance r from a very great distance (infinity), then the work done is given by Eq. (8).

$$-U = -Gm_1 m_2/r \qquad (8)$$

This is called the gravitational potential energy; it is (arbitrarily) put to zero for infinite separation between the particles. Similarly, for a system of n particles with masses $m_1, m_2, \ldots, m_n$ and mutual distance r_{ij} between m_i and m_j, the gravitational potential energy $-U$ is the work done to assemble the system from infinite separation (or the negative of the work done to bring about an infinite separation), as shown in Eq. (9).

$$-U = -G \sum_{i<j} m_i m_j/r_{ij} \qquad (9)$$

A closely related quantity is gravitational potential. The gravitational potential of a particle of mass m is given by Eq. (10), where r is distance

$$V = -Gm/r \qquad (10)$$

measured from the mass. The gravitational force exerted on another mass M is M times the gradient of V. If the first body is extended or irregular, the formula for V may be extremely complicated, but the latter relation still applies.

A good illustration of gravitational potential energy occurs in the motion of an artificial satellite in a nearly circular orbit around the Earth which is affected by atmospheric drag. Because of the frictional drag the total energy of the satellite in its orbit is reduced, but the satellite actually moves faster. The explanation for this is that it moves closer to the Earth and loses more in gravitational potential energy than it gains in kinetic energy. *See* ENERGY.

Similarly, in its early evolution a star contracts, with the gravitational potential energy being transformed partly into radiation, so that it shines, and partly into kinetic energy of the atoms, so that the star heats up until it is hot enough for thermonuclear reactions to start.

Another related phenomenon is that of speed of escape. A projectile launched from the surface of the Earth with speed less than the speed of escape will return to the surface of the Earth; but it will not return if its initial speed is greater (atmospheric drag is neglected). For a spherical body with mass

M and radius R, the speed of escape from its surface is given by Eq. (11). For the Earth V_e is 11.2

$$V_e = (2MG/R)^{1/2} \qquad (11)$$

km/sec; for the Moon it is 2.4 km/sec, which explains why the Moon cannot retain an atmosphere such as the Earth's. By analogy, a black hole can be considered a body for which the speed of escape from the surface is greater than the speed of light, so that light cannot escape. P. S. Laplace speculated along these lines; but the analogy is not really exact since Newtonian mechanics is not valid. The question as to whether the universe will continue to expand can be considered in the same way. If the density of matter in the universe is great enough, then expansion will eventually cease and the universe will start to contract. At present the density cannot be found with sufficient accuracy to decide the question.

Application of Newton's law. In modern times Eq. (5), in a modified form with appropriate refinements to allow for the Earth's oblateness and for external forces acting on the Earth-Moon system, has been used to compute the distance to the Moon. The results have only been superseded in accuracy by radar measurements and observations of corner reflectors placed on the lunar surface.

Newton's theory passed a much more stringent test than the one described above when he was able to account for the principal departures from Kepler's laws in the motion of the Moon. Such departures are called perturbations. One of the most notable triumphs of the theory occurred when the observed perturbations in the motion of the planet Uranus enabled J. C. Adams in 1845 and U. J. Leverrier in 1846 independently to predict the existence and calculate the position of a hitherto unobserved planet, later called Neptune. When yet another planet, Pluto, was discovered in 1930, its position and orbit were strikingly similar to predictions based on the method used to discover Neptune. But the discovery of Pluto must be ascribed to the perseverance of the observing astronomers; it is not massive enough to have revealed itself through the perturbations of Uranus and Neptune.

F. W. Bessel observed nonuniform proper motions of Sirius and Procyon and inferred that each was gravitationally deflected by an unseen companion. It was only after his death that these bodies were telescopically observed, and they both later proved to be white dwarfs. More recently P. van de Kamp has accumulated evidence for the existence of some planetary masses around stars. The discovery of black holes (which will never be directly observed) hinges in part on a visible star showing evidence for having a companion of sufficiently high mass (so that its gravitational collapse can never be arrested).

Newton's theory supplies the link between the observed motion of celestial bodies and certain physical properties, such as mass and sometimes shape. Knowledge of stellar masses depends basically on the application of the theory to binary-star systems. Analysis of the motions of artificial satellites placed in orbit around the Earth has revealed refined information about the gravitational field of the Earth and of the Earth's atmosphere. Similarly, satellites placed in orbit around the Moon have yielded information about its gravitational field, and other space vehicles have yielded the best information to date on the masses and gravitational fields of other planets.

Newtonian gravitation has been applied without apparent difficulty to the motion in distant star systems. But over very great distance (or over very small distances, when gravitation is swamped by other forces) it has not been confirmed or disproved.

Accuracy of Newtonian gravitation. Newton was the first to doubt the accuracy of his law when he was unable to account fully for the motion of the perigee in the motion of the Moon. In this case he eventually found that the discrepancy was largely removed if the solution of the equations were more accurately developed. Further difficulties to do with the motion of the Moon were noted in the 19th century, but these were eventually resolved when it was found that there were appreciable fluctuations in the rate of rotation of the Earth, so that it was the system of timekeeping and not the gravitational theory that was at fault.

A more serious discrepancy was discovered by Leverrier in the orbit of Mercury. Because of the action of the other planets, the perihelion of Mercury's orbit advances. But allowance for all known gravitational effects still left an observed motion of about 43 seconds of arc per century unaccounted for by Newton's theory. Attempts to account for this by adding an unknown planet or by drag with an interplanetary medium were unsatisfactory, and a very small change was suggested in the exponent of the inverse square of force. This particular discordance was accounted for by Albert Einstein's general theory of relativity in 1916, but the final word on the subject has yet to be said. *See* RELATIVITY.

Gravitational lens. Light is deflected when it passes through a gravitational field, and an analogy can be made to the refraction of light passing through a lens. It has been suggested that a galaxy situated between an observer and a more distant source might have a focusing effect, and that this might account for some of the observed properties of quasi-stellar objects. The multiple images of one quasar (QSO, 0957 + 567 A,B) are almost certainly caused by the light from a single body passing through a gravitational lens.

Testing of gravitational theories. One of the greatest difficulties in investigating gravitational theories is the weakness of the gravitational coupling of matter. For instance, the gravitational interaction between a proton and an electron is weaker by a factor of about 5×10^{-40} than the electrostatic interaction. (If gravitation alone bound the hydrogen atom, then the radius of the first Bohr orbit would be 10^{13} light-years, or about 1000 times the radius of the Hubble universe!) The contrast between the accuracy achieved in the laboratory when measuring G and the accuracy required if the possible inconstancy of G is to be investigated shows that gravitation is not a laboratory subject. But the astronomical universe provides a wealth of situations for investigating gravitation. Their main drawback is that they must be taken as they occur: with the exception of some experiments included in the space program, the situations cannot be controlled, modified, or repeated for experimental convenience.

In the solar system there are planetary orbits around the Sun and satellite orbits around the planets. Gravitational effects on electromagnetic radiation as well as those on orbiting bodies can be observed and tested against theory. There are stars that rotate, oscillate, explode, and collapse, including white dwarfs, neutron stars (observed as pulsars), and probably black holes, some with enormous masses. There are binary systems, some with orbital periods as low as 1000 sec. There are star clusters, galaxies (some of them exploding), clusters of galaxies, and finally there is the universe itself. Of special interest is the "binary pulsar" PSR 1913+16. The observed star is a pulsar, and it has been speculated that the unseen companion is also a highly condensed object. The period is almost 8h, the orbital eccentricity is 0.6, and the motion of periastron is 4.2° per year.

In the solar system, increased observational accuracy is needed to detect small departures from Newtonian gravitational theory. In other situations, these departures may be very great, but observable effects are attenuated due to great distances. Of greatest interest are systems where there is both a high concentration of matter and rapid motion, involving rotation, revolution, or collapse.

Relativistic theories. Before Newton, detailed descriptions were available of the motions of celestial bodies—not just Kepler's laws but also empirical formulas capable of representing with fair accuracy, for their times, the motion of the Moon. Newton replaced description by theory, but in spite of his success and the absence of a reasonable alternative, the theory was heavily criticized, not least with regard to its requirement of "action at a distance" (that is, through a vacuum). Newton himself considered this to be "an absurdity," and he recognized the weaknesses in postulating in his system of mechanics the existence of preferred reference systems (that is, inertial reference systems) and an absolute time. Newton's theory is a superb mathematical one that represents the observed phenomena with remarkable accuracy.

Einstein showed in his special theory of relativity that these postulates were physically unacceptable. In his general theory of relativity he incorporated gravitational phenomena in such a way that there was no longer any preferred reference system. He treated the phenomenon of gravitation as a consequence of the geometric properties of space-time, this geometry being affected by the presence of gravitating matter. The acceleration of a body is determined by the local geometry, and the Newtonian concepts of gravitational force and action at a distance are abandoned. Mathematically the theory is far more complicated than Newton's. Instead of the single potential described above, Einstein worked with 10 quantities that are members of a tensor.

Principle of equivalence. An important step in Einstein's reasoning is his "principle of equivalence": that a uniformly accelerated reference system imitates completely the behavior of a uniform gravitational field. Imagine, for instance, a scientist in a space capsule infinitely far out in empty space so that the gravitational force on the capsule is negligible. Everything would be weightless; bodies would not fall; and a pendulum clock would not work. But now imagine the capsule to be accelerated by some agency at the uniform rate of 981 cm/sec². Everything in the capsule would then behave as if the capsule were stationary on its launching pad on the surface of the Earth and therefore subject to the Earth's gravitational field. But after its original launching, when the capsule is in free flight under the action of gravitational forces exerted by the various bodies in the solar system, its contents will behave as if it were in the complete isolation suggested above. Note that this principle requires that all bodies fall in a gravitational field with precisely the same acceleration and that this is confirmed by the Eotvos experiment mentioned earlier. Also, if matter and antimatter were to repel one another, it would be a violation of the principle. *See* FREE FALL.

Einstein's theory requires that experiments should have the same results irrespective of the location or time. This has been said to amount to the "strong" principle of equivalence.

Classical tests. The ordinary differential equations of motion of Newtonian gravitation are replaced in general relativity by a nonlinear system of partial differential equations for which general solutions are not known. Apart from a few special cases, knowledge of solutions comes from methods of approximation. For instance, in the solar system, speeds are low so that the quantity v/c (v is the orbital speed and c is the speed of light) will be small (about 10^{-4} for the Earth). The equations and solutions are expanded in powers of this quantity; for instance, the relativistic correction for the motion of the perihelion of Mercury's orbit is adequately found by considering no terms smaller than $(v/c)^2$. This is called the post-Newtonian approximation. (Another approach is the weak-field approximation.)

Einstein's theory has appeared to pass three famous tests. First, it accounted for the full motion of the perihelion of the orbit of Mercury. (Mercury is the most suitable planet, because it is the fastest-moving of the major planets and has a high eccentricity, so that its perihelion is relatively easily studied.) Second, the prediction that light passing a massive body would be deflected has been confirmed with an accuracy of about 5%. Third, Einstein's theory predicted that clocks would run more slowly in strong gravitational fields compared to weak ones; interpreting atoms as clocks, spectral lines would be shifted to the red in a gravitational field. This, again, has been confirmed with moderate accuracy.

I. I. Shapiro has confirmed predictions of the theory in an experiment in which radar waves were bounced off Mercury; the theory predicts a delay of about 2×10^{-4} sec in the arrival time of a radar echo when Mercury is on the far side of the Sun and close to the solar limb. Tests, similar in principle, have been conducted using observations of the Mariner space vehicles, the accuracy of confirmation being in the region of 4%. E. Fomalont and R. Stramek observed the deflection of microwave radiation passing close to the Sun, using radio interferometry with a baseline of 35 km. The amount of bending they found is 1.015 ± 0.011 times the amount predicted by general relativity. In another test, the precession of a gyroscope in orbit around the Earth is to be studied for evidence of the so-called geodetic precession. The motion of

a perihelion is suitable for study since its effects continue to accumulate. Other periodic (noncumulative) orbital effects have until recently been too small to observe. But the current revolution in observational techniques and accuracy has changed the situation; post-Newtonian terms are now routinely included in many calculations of the orbits of planets and space vehicles, and comparison with observations will furnish tests of the theory.

The observation and analysis of gravitation waves, discussed below, will constitute further tests.

Mach's principle. One of the most penetrating critiques of mechanics is due to E. Mach, toward the end of the 19th century. Some of his ideas can be traced back to Bishop G. Berkeley early in the 18th century. Out of Mach's work there has arisen what is known as Mach's principle; this is philosophical in nature and cannot be stated in precise terms. The idea is that the motion of a particle is only meaningful when referred to the rest of the matter in the universe. Geometrical and inertial properties are meaningless for an empty space, and the motion of a particle in such space is devoid of physical significance. Thus the behavior of a test particle should be determined by the total matter distribution in the universe and should not appear as an intrinsic property of an absolute space. If this is so, then the quantitative aspects of physical laws (that is, the various constants involved) should be dependent on position.

Brans-Dicke theory. The field theory developed in 1961 by Brans and Dicke is perhaps the best-known theory in conformity with Mach's principle. For instance, the expansion of the universe causes G to diminish in time. In this theory the gravitational field is described by a tensor and a scalar, the equations of motion being the same as those in general relativity. The addition of a scalar field leads to the appearance of an arbitrary constant, whose value is not known exactly. The Brans-Dicke theory predicts that the relativistic motion of the perihelion of Mercury's orbit is reduced compared with Einstein's value, and also that the light deflection should be less. With regard to the orbit of Mercury, Dicke pointed out that if the Sun were oblate, this might account for some of the motion of the perihelion. In 1967 he announced that measurements showed a solar oblateness of about 5 parts in 100,000 (or a difference in the polar and equatorial radii of about 34 km). His observations and discussion are still subject to some controversy. The difference between the theory and that of general relativity can be parameterized by the number ω, where $\gamma = (1 + \omega)/(2 + \omega)$; for general relativity $\gamma = 1$. Dicke has proposed $\omega \sim 7.5$; but the results of Fomalont and Stramek indicate a value of ω greater than 23, for which the predictions of the two theories would not be greatly different.

There are, of course, many other theories not mentioned here.

Supergravity. This is the term applied to a highly mathematical theory of gravitation forming part of a unified field theory in which all types of forces are included. *See* FUNDAMENTAL INTERACTIONS; UNIFIED FIELD THEORY.

Gravitational waves. The existence of gravitational waves, or gravitational "radiation," was predicted by Einstein shortly after he formulated his general theory of relativity. They are now a feature of any relativity theory. Gravitational waves are "ripples in the curvature of space-time." In other words, they are propagating gravitational fields, or propagating patterns of strain, traveling at the speed of light. They carry energy and can exert forces on matter in their path, producing, for instance, very small vibrations in elastic bodies. The gravitational wave is produced by change in the distribution of some matter. It is not produced by a rotating sphere, but would result from a rotating body not having symmetry about its axis of rotation: a pulsar, perhaps. In spite of the relatively weak interaction between gravitational radiation and matter, the measurement of this radiation is now technically possible and may already have been achieved. This is due to the work of Joseph Weber, whose original and pioneering work has led to a very exciting situation in science. The present situation contains some uncertainties; but gravitational-wave astronomy has been added to other branches of astronomy, and a new window is opening to the universe.

A classical problem, solved by Einstein, concerns the gravitational radiation from a rod spinning about a perpendicular axis through its center. If the rod has moment of inertia about the axis of spin I ($I = Md^2/3$, where M is the mass of the rod in kilograms and $2d$ its length in meters) and angular velocity ω, the power of the radiation in watts (1 W = 10^7 ergs/sec) is given by Eq. (12), where G

$$P = \frac{32GI^2\omega^6}{5c^5} = 1.73 \times 10^{-52} I^2 \omega^6 \qquad (12)$$

is the constant of gravitation in mks units and c is the speed of light in meters per second. A calculation using a steel rod of mass 4.9×10^5 kg (490 metric tons), length 20 m, and angular velocity $\omega = 28$ rad/sec, limited by the balance between centrifugal force and tensile strength, gives 2.2×10^{-29} W. So the problem of the generation and detection of gravitational waves in the laboratory is at present somewhat academic.

In electromagnetic theory, electric-dipole radiation is dominant. The gravitational analog of the electric dipole is the mass dipole moment whose time rate of change is the total momentum of the system; since this is constant, there is no gravitational dipole radiation; the principal power is in quadrupole radiation. The radiation has fairly elaborate polarization properties.

Binary systems. Consider a binary star system having period P hr and masses m_1, m_2, where the relative orbit is circular. If $M = m_1 + m_2$ and $\mu = m_1 m_2/M$, the power output by gravitational radiation is given by Eq. (13), where $M_\odot$ is the mass of

$$P_B = \left(\frac{\mu}{M_\odot}\right)^2 \left(\frac{M}{M_\odot}\right)^{4/3} P^{-10/3} \, 3.0 \times 10^{26} \text{ W} \qquad (13)$$

the Sun. For the orbit of the Earth around the Sun, P_B is about 200 W. The gravitation radiation extracts energy from the system. If a binary system has a relative elliptic orbit, then most of the energy is extracted at the closest point of approach and the orbit approaches a circle; then the orbit will gradually shrink, with the bodies colliding after a "spiral time" given by Eq. (14), where a_0 is the

$$\tau_0 = \frac{5c^5}{256G^3} \frac{a_0^4}{\mu M^2} \qquad (14)$$

initial radius of the relative orbit. Under this mechanism the Earth would have fallen toward the Sun less than a centimeter in the lifetime of the solar system!

Clearly one must look outside the solar system for promising sources. Ordinary binaries are not helpful. Sirius and its companion, with a spiral time of 7×10^{21} years, radiate at 10^8 W, the flux received at the Earth being 10^{-31} W. The closer the members of the system are to each other, the more promising they are; some eclipsing binaries can generate power that would be observed on the Earth at about 10^{-20} W, and have spiral times of the order of 10^{10} years. The shortest periods known are for close pairs consisting of a white dwarf and a main sequence star; here spiral times can be as low as 10^9 years, and the predicted flux at the Earth for the most promising candidate, ι Boo, is 18×10^{-18} W. For these binaries, gravitational radiation appears to play an important part in their physical characteristics. Matter flows toward the white dwarf from the companion star, causing flickering and occasional nova outbursts. The stars are very close, and it seems that the contraction of the orbit, caused by gravitational radiation, plays a crucial part in instigating the flow of matter. Closer binaries can at present only be generated by hypothesis. Two neutron stars, having solar masses and with 10^4-km separation, radiate at 3×10^{34} W and have a spiral time of 3 years. With such a system the formulas given above show that the evolution becomes increasingly rapid and the power input increases, so that the final stages of collapse constitute a burst of radiation. The period of the "binary pulsar" has been observed with high precision over several years. J. H. Taylor, L. A. Fowler, and P. M. McCulloch have reported that the period is diminishing at the annual rate of 1 part in 10^9, or 3 parts in 10^{12} per revolution. This is consistent with the prediction of general relativity, and encourages belief in the reality of gravitational radiation, although this particular radiation would not be directly observable.

Pulsars. The most rapidly rotating single objects that have been observed are pulsars. These are neutron stars rotating with periods mostly less than 1 sec. From their irregular light curves it is reasonable to suppose that they do not possess symmetry about the axis of rotation. Suppose that they are assumed homogeneous and the equatorial section is an ellipse with axes a and b, and that the ellipticity of the equator is $\epsilon = (a - b)/a$. If the star rotates with angular velocity ω, then the power radiated is given by Eq. (15), where I is the moment

$$P_R = \frac{32G\omega^6 I^2 \epsilon^2}{5c^5} \qquad (15)$$

of inertia about the axis of rotation. A promising candidate here is the pulsar in the Crab Nebula, remnant of a supernova; the period of rotation is 0.033 sec; the moment of inertia is likely to be of the order of 4×10^{37} kg m^2, and the power output can be estimated by writing Eq. (15) as Eq. (16),

$$P_R = \left(\frac{I}{4 \times 10^{37} \text{ kg m}^2}\right)^2 \left(\frac{P}{0.033 \text{ sec}}\right)^{-6} \left(\frac{\epsilon}{10^{-3}}\right)^2 10^{31} \text{ W} \qquad (16)$$

where P is the period. Clearly it is important to estimate ϵ. The periods of rotation are known to be increasing, and this puts an upper limit on ϵ. It is estimated that the flux received from the Crab pulsar would be less than 3×10^{-20} W. Some pulsars occasionally show sudden changes or glitches in their rotational period. These could be due to starquakes (neutron stars have solid surfaces) and might lead to strong bursts of gravitational radiation.

Explosive events. The gravitational collapse involved in a supernova explosion might produce the strongest radiation that can be observed. The processes involved can only be tentatively estimated, and unfortunately supernova occur about once every 100 years in the Galaxy. But their radiation, probably in short bursts, could be sufficiently powerful for them to be observed from other galaxies. It is possible that stellar collapse takes place without the display of a supernova, so estimates of frequency may be much too low.

Many galaxies show evidence of explosive activity. For quasars, gravitational radiation at 10^{38} W has been suggested, and for explosions in galactic centers, 10^{30} W; but these estimates are not at all definitive.

As matter falls into a black hole, it will release a burst of gravitational radiation; the energy released is proportional to the square of the mass captured and inversely proportional to the mass of the black hole; the time of the outburst is proportional to the mass of the black hole. It has been suggested that there might be a large black hole at the center of the galaxy; if its mass were 10^8 solar masses and it captured a star of 1 solar mass, a burst of energy 10^{37} W might be produced. If the black hole were rotating or the infalling star somehow had a speed greater than that acquired from falling from infinity, then the energy could be greater.

Nature of radiation. The radiation discussed could be continuous or in bursts. The radiation would have a spectrum that might be discrete, as in the case of rotation or orbital revolution, where the fundamental frequency is 2ω (there will also be harmonics), or broadband in the case of explosive events. The longest wavelengths suggested are from the primordial history of the universe, when they could be greater than the size of a galaxy; the shortest, in supernovae and stellar collapse.

Dirac worked out a quantum theory for this radiation; the graviton is a theoretical particle postulated as the quantum of the gravitational field. *See* GRAVITON; QUANTUM FIELD THEORY.

Detection of gravitational waves. When a gravitational wave interacts with a system of particles, the particles wiggle slightly; in the case of a solid body, strains are set up in the body; what is actually measured is a sort of tidal effect. Most of the detectors currently under consideration involve strains in solid bodies and they involve the principle of resonance; that is, they react much more strongly to radiation of a given frequency than to other frequencies. When radiation of the correct frequency impinges on the detector it oscillates, as if ringing; radiation at other frequencies is essentially ignored.

Weber's experiment. Weber's detectors principally consist of cylinders suspended in vacuum. They are typically of aluminum, 66 cm in diameter and 153 cm long, weighing 1.5×10^3 kg and reso-

nant at 1661 Hz. They are directional, being most strongly sensitive to radiation traveling perpendicular to the axis of the cylinder. The strains in the cylinder are converted into measurable voltages by quartz strain gages: piezoelectric crystals bonded around the girth of the cylinder. Strains of the order of 1 part in 10^{17} can be detected. There will be continued background thermal "noise," a random effect due to the thermal motion of individual molecules. Since the gravitational signals are of the same order of magnitude as this noise, it is necessary to have two independent receivers, and to look for coincidences in the signals received.

In the principal experiments, Weber analysed signals received 1000 km apart. Since 1969 Weber has reported coincidences at the average rate of about three times a day; typical displacements are around 5×10^{-17} m. He has looked for possible correlations between his coincidences and solar flares, electric storms, surges in the interstate power grid, network television broadcasting, seismic events, and cosmic rays, with no success, and so concludes that the observations are consistent with the detection of gravitational radiation.

Interpretation of Weber's results. In 1970 Weber reported that the coincident pulses were strongly correlated with sidereal time in a manner consistent with the antenna pattern if the gravitational radiation were coming from the center of the Galaxy. This leads to an astrophysically satisfying (in principle) source for the radiation and also a distance for the source: the galactic center. It therefore makes possible the calculation of the energies at the source that would produce the flux observed at Earth. Immediately problems arise. Each burst is what would be expected from a supernova or stellar collapse; if so, these events are more than 1000 times more frequent that had been expected. By using the formula $E = mc^2$, the mass loss that would be the equivalent to this energy can be calculated, and it appears that the Galaxy loses at least 500 solar masses each year; since Weber cannot be observing all events, perhaps this figure should be 10 times as great! But this is not acceptable; the Galaxy would be used up in a time of about one-hundredth of its known age.

The calculation of total energy uses the assumption that the energy is beamed equally in all directions and over a broad band. The band must be at least wide enough to include 1580 Hz, since Weber has observed coincidences at this frequency also. Many suggestions have been made to ease the situation: (1) Gravitational waves have been emitted at this rate for only a small fraction of the age of the Galaxy; this would imply that there is something special about the present age, and that there is something even more special about gravitational waves, since no other observed phenomenon shows the same preference for this era. (2) The source might be much nearer to Earth than the center of the Galaxy. (3) The radiation is strongly focused; into the galactic plane, for instance. (4) There may be some mechanism by which the radiation is magnified between the source and the Earth. (5) There is also the possibility that Weber's events are not caused by gravitational waves. Theories for the origin of the radiation include the possible presence of a very massive black hole at the galactic center, perhaps even disk-shaped, and the possibility of synchrotron-type gravitation radiation.

In another experiment Weber used a disk-shaped antenna. This was designed to search for scalar gravitation radiation. The result was negative, but the disk reacted in a way that was consistent with a source of tensor gravitational waves at the galactic center. So Weber claims similar results from two quite different antennas.

Other experiments. No other experimenter has reproduced Weber's results, although many have tried. A second generation of instruments is in operation or is being devised; greater sensitivity has been sought by making the detector more massive, cooler, or purer (that is, using large crystals such as silicon or sapphire that naturally have less noise). In another approach a laser beam is split into two parts along freely suspended perpendicular arms, each beam being reflected many times by mirrors at the ends; if the length of one arm changes relative to the other, this will be shown through interference when the beams are recombined. It has been suggested that the very accurate tracking of spacecraft could be exploited to detect the effect of gravitational waves on Doppler shifts. *See* DOPPLER EFFECT.

[J. M. A. DANBY]

Bibliography: R. H. Dicke, *Gravitation and the Universe*, 1970; G. Gamow, *Gravity*, 1962; J. L. Logan, Gravitational waves: A progress report, *Phys. Today*, 26:44–52, March 1973. C. W. Misner, K. S. Thorne, and J. A. Wheeler. *Gravitation*, 1973; R. Pinheiro, Gravity wave astronomy, *Astronomy*, 7(6):6–14, 1979; W. H. Press and K. S. Thorne, Gravitational wave astronomy, *Annu. Rev. Astron. Astrophys.*, 10:335–374, 1972; T. J. Sejnowski, Sources of gravity waves, *Phys. Today*, 27:40–48, January 1974; K. S. Thorne, The future of gravitational wave astronomy, *Mercury*, 7(3):58–61, 1979; S. Weinberg, The forces of nature, *Amer. Sci.*, 65:171–176, 1977; D. T. Whiteside, *J. Hist. Astron.*, 1:5–19, 1970; E. T. Whittaker, *From Euclid to Eddington: A Study of Conception of the External World*, 1949; E. T. Whittaker, *A History of the Theories of the Aether and Electricity*, 1954.

Graviton

A theoretically deduced particle postulated as the quantum of the gravitational field. According to Einstein's theory of general relativity, accelerated masses (or other distributions of energy) should emit gravitational waves, just as accelerated charges emit electromagnetic waves. And according to quantum field theory, such a radiation field should be quantized; that is, its energy should appear in discrete quanta, called gravitons, just as the energy of light appears in discrete quanta, namely photons.

The properties of the graviton follow from the properties of the classical gravitational field. That is, its rest mass and charge are zero (like the photon); it has spin 2 in units of $h/2\pi$, where h is Planck's constant, and is therefore a boson, which is a particle that obeys Bose-Einstein statistics. Because of its vanishing rest mass, its spin is restricted to be parallel to its motion, so that a graviton has only two independent spin states (again like a photon).

Later experiments, which may be observations of gravitational radiation, are discussed in the article on gravitation. Such observations are difficult, because matter is very weakly coupled to the gravitational field (the gravitational force between an electron and a proton is only 10^{-39} times the electrical force between them), so that the rate of emission and absorption of gravitational radiation is very small. *See* ELEMENTARY PARTICLE; GRAVITATION; QUANTUM FIELD THEORY; RELATIVITY.   [CHARLES J. GOEBEL]

Bibliography: G. K. O'Neill and D. Cheng, *Elementary Particle Physics*, 1979; C. E. Swartz, *The Fundamental Particles*, 1965.

Gravity

The gravitational attraction at the surface of a planet or other celestial body. The quantity g is often referred to simply as "gravity" or "the force of gravity" of Earth, both of which are incorrect. The force of gravity means the force with which a celestial body attracts an object, that is, the weight of the object. The letter g represents the acceleration caused by the gravitational force and, of course, has the dimensions of acceleration. *See* GRAVITATION.

[DIRK BROUWER/G. M. CLEMENCE]

Graybody

An energy radiator which has a blackbody energy distribution, reduced by a constant factor, throughout the radiation spectrum or within a certain wavelength interval. The designation "gray" has no relation to the visual appearance of a body but only to its similarity in energy distribution to a blackbody. Most metals, for example, have a constant emissivity within the visible region of the spectrum and thus are graybodies in that region. The graybody concept allows the calculation of the total radiation intensity of certain substances by multiplying the total radiated energy (as given by the Stefan-Boltzmann law) by the emissivity. The concept is also quite useful in determining the true temperatures of bodies by measuring the color temperature. For a discussion of the Stefan-Boltzmann law and color temperature *see* HEAT RADIATION. *See also* BLACKBODY.

[HEINZ G. SELL; PETER J. WALSH]

Ground state

In quantum mechanics, the stationary state of lowest energy of a particle or a system of particles. The ground state may be bound or unbound; when bound, its energy generally is a finite amount less than the energy of the next higher or first excited state. In the typical circumstance that the potential energy is zero at infinite separation, the magnitude of the negative ground-state energy is the binding energy, that is, the energy required to separate all the particles infinitely. *See* ENERGY LEVEL; EXCITED STATE; NONRELATIVISTIC QUANTUM THEORY; NUCLEAR BINDING ENERGY; STATIONARY STATE.

[EDWARD GERJUOY]

Group theory

Any set of elements which is equipped with an operation (called multiplication) satisfying the requirements (1), (2), and (3) is called a group.

Group theory is the branch of mathematics devoted to the properties of groups. Group theory has applications in the theories of relativity, quantum mechanics, and crystallography, and also in some branches of algebra and in analytic function theory.

Requirements on the group operation. The group operation is supposed to give for every pair of elements, for example, g and h, in the group, another element gh (called their product). Multiplication is supposed to satisfy the following requirements.

Associative law. If g_1, g_2, and g_3 are any elements of the group, then

$$(g_1 g_2) g_3 = g_1 (g_2 g_3) \tag{1}$$

that is, in forming a product of three elements one obtains the same result by first multiplying g_1 and g_2 and then multiplying the result and g_3, as first multiplying g_2 and g_3 and then multiplying g_1 and the result.

Existence of an identity element. There is in the group an element e (called the identity) with the property

$$eg = ge = g \tag{2}$$

for every element g of the group.

Existence of inverses. For each element g of the group there is an element g^{-1} (called the inverse of g) which satisfies

$$g^{-1}g = gg^{-1} = e \tag{3}$$

Examples. An example of a group is the set of positive real numbers equipped with the operation of ordinary multiplication. The identity element is then the number 1 and the inverse of an element is its reciprocal. Another example of a group is the set of all real numbers equipped with the operation of ordinary addition. The identity element is then the number 0, and the inverse of an element is its negative.

Both these examples are commutative groups (also called abelian groups after the mathematician Niels Abel) because their multiplication law satisfies

$$gh = hg \tag{4}$$

for every g and h in the group. An example of a noncommutative group (also called nonabelian) is the group of rotations of a three-dimensional rigid body around a point. Here an element of the group is a rotation, that is, the act of rotating the body around a certain axis in space by a certain angle. The group operation means merely successive transformation; $g_1 g_2$ stands for the act of rotation obtained by carrying out the acts of rotation g_2 and g_1 successively in that order. The identity element is the act of rotating through zero angle or making no rotation. The inverse of an element is the rotation around the same axis by the same angle but in the opposite sense. It is easy to check that if g_1 is a rotation around the vertical by 90° and g_2 a rotation around some horizontal axis by 90°, that $g_1 g_2 \neq g_2 g_1$, so the group is noncommutative.

These three examples are all of groups with an infinite number of elements, that is, infinite groups. There are also finite groups. A simple example is the group with two elements, the identity e and another e_1 satisfying $e_1 e_1 = e$.

Topological groups. For infinite groups, it often occurs that a group has a natural geometry. For example, in the group of positive real numbers described above one has the geometry of the real numbers. This geometry is compatible with the group operations in the sense that the product gh is a continuous function of g and h, and the inverse g^{-1} is a continuous function of g. Generalizing this scheme to arbitrary groups, one arrives at the notion of a topological group (or less frequently a continuous group) which is a set of elements equipped not only with a group operation but also with a topology, the two notions being required to be compatible in the above sense. A topology can be specified in a number of ways; the net effect of any of them is to give meaning to the phrase: Two group elements are close to one another. For example, in the case of the rotation group discussed above one can define a topology by specifying that two rotations are close to one another when their axes have nearly the same direction and their angles of rotation differ by very little. *See* TOPOLOGY.

There is a particular class of topological groups which is of great importance. These are the so-called Lie groups (named after the mathematician Sophus Lie). These are topological groups in which it is possible to label the group elements by a finite number of coordinates in such a way that the coordinates of gh and g^{-1} are analytic functions of the coordinates of g and h and of g, respectively. That means that they should be representable as convergent power series in those coordinates. Most of the topological groups occurring in applications are Lie groups. One important example of a group which is not a Lie group (because its elements cannot even be labeled by a finite number of coordinates in such a way as to make the group operation continuous) is the group of all permissible coordinate transformations in the general theory of relativity. The significant transformations in the special theory of relativity, the so-called Lorentz transformations, do form a Lie group. *See* RELATIVITY.

Applications to physical sciences. The principal applications of group theory outside mathematics itself have to do with the classification and exploitation of symmetries in physical systems.

Crystallography. An important case is the problem of classifying crystal structures. To each crystal one associates two groups: its point group and its space group. The point group is the set of all those rotations of three-dimensional space with center in a fixed atom of the crystal, which carry atoms into identical atoms. The space group is the set of motions of three-dimension space which carry atoms of the crystal into identical atoms. The classification of the different possible symmetries of crystals reduces to the problem of finding all the different point and space groups. When the point and space groups of a crystal are known, one can say a good deal about the physical properties of the crystal, because the symmetry imposes restrictions on the possible form of such quantities as the elastic and dielectric constants. *See* CRYSTALLOGRAPHY.

Quantum mechanics. The classification of crystal structure can be carried out within the framework of classical physics (as well as that of quantum theory), but it is in quantum mechanics that the most extensive and significant applications of group theory exist. In quantum mechanics, the state of a physical system is described by a wave function, and a symmetry transformation (for example, a rotation or translation) of the state yields a new wave function related to the old by a linear transformation. For the particular case of a stationary state of a system, symmetric under the symmetry transformations considered, the new wave function must also be a stationary state of the same energy as the old: A symmetry transformation then carries wave functions belonging to a given eigenvalue of the energy into linear combinations of themselves. In any case, one is led to the problem of classifying the different possible transformation laws of a set of states under the given symmetry group. In mathematical terminology, the problem turns out to be that of classifying all representations of the symmetry group, that is, of finding all correspondences between group elements g and linear operators $M(g)$ such that

$$M(g_1)M(g_2) = M(g_1 g_2) \qquad (5)$$

When the symmetry group is the rotation group, the classification of the representations can be made in terms of angular momentum; for the translation group in space, it is in terms of linear momentum. The classification leads to characteristic orthogonality relations and selection rules, the best known of which are the conservation laws of angular momentum and linear momentum. Another consequence is the theorem which says that a particle of spin s can have electromagnetic moments of order at most $2s$. For practical calculations of approximate wave functions of atomic nuclei, the techniques afforded by the theory of the representations of the rotation group have become nearly indispensable.

When the symmetry group of the physical system is the inhomogeneous Lorentz group, as is the case when the theory in question satisfies the requirements of the special theory of relativity, the classification of the representations is by parameters whose physical meanings are mass and spin. It leads to a classification of the possible wave equations for elementary particles. Although this theory is one of the best grounded and most satisfactory parts of existing elementary particle theory, the representations of the inhomogeneous Lorentz group, unlike those of the rotation group, have as yet been of little use in practical calculations.

There are significant applications of the theory of Lie groups to the description of the internal degrees of freedom of elementary particles (internal, in this context, means not describing space-time symmetry properties). Supermultiplets of elementary particles have been recognized which transform approximately according to irreducible representations of the group SU(3) of all 3×3 unitary matrices of determinant one. *See* ELEMENTARY PARTICLE; SYMMETRY LAWS.

Applications within mathematics. The applications of group theory within mathematics itself are numerous and important. The part of mathematics in which the notion of group was first clearly isolated was the theory of algebraic equations.

Algebra. Here, the basic idea is to associate a group (the so-called Galois group, named after the mathematician E. Galois) with the algebraic equation

$$a_0 + a_1 x + \cdots + a_n x^n = 0$$

in such a way that the structure of the group reflects the type of operations necessary to compute the roots from the coefficients $a_0, \ldots, a_n$. For example, using this method one obtains a simple proof that for $n > 4$, no solution in roots of the general equation exists.

Topology. In topology, groups are used to classify the structure of various geometric objects. The homology and homotopy groups serve this purpose for a manifold. As an example consider the homotopy group. To define it, one chooses a point P of the manifold and considers all continuous closed paths starting and ending at P. A notion of product of such paths is defined: If x_1 and x_2 are two paths, the product $x_1 x_2$ is the closed path which is obtained by tracing out x_2 and then x_1. The inverse of a path is then the same curve traversed in the opposite direction and the identity path consists of the point P alone. Paths are divided into equivalence classes, two paths lying in the same equivalence class if one can be deformed continuously into the other. The notion of product then extends to equivalence classes, the product of two equivalence classes being obtained as the equivalence class of the product of any two representative paths. With this definition of product, the set of equivalence classes of continuous closed paths starting at P is a group, the (one-dimensional) homotopy group (also called fundamental group). Provided the manifold under discussion is connected (that is, provided any two points of the manifold can be connected by a continuous curve lying in the manifold), the fundamental group is essentially the same, independent of which point is taken as starting point for the paths. If the fundamental group of the manifold consists of one element only, the manifold is said to be simply connected.

Analytic functions. Group theory plays a fundamental role in the theory of analytic functions. This is true not only for the thoroughly studied case of one complex variable, but also for the case of several complex variables, which is less well understood. Only the former will be discussed here. The starting point is the fact that each analytic function has a natural domain of definition, its Riemann surface. There is another Riemann surface associated with it, its universal covering surface, which is simply connected. The universal covering surface may have several sheets for each sheet of its underlying Riemann surface, each point of the latter being replaced by several "lying above" it. Any function which is analytic on the underlying Riemann surface can be regarded as analytic on the universal covering surface; it will take the same values at those points which lie above a given point. The analytic function regarded as defined on the universal covering surface is invariant under transformations which carry the points lying above any point of the Riemann surface into themselves. These transformations form a group. Now a second basic fact is that any simply connected Riemann surface can be mapped in a one-to-one and analytic fashion onto one of three regions: the interior of the unit circle, the complex plane, or the complex plane closed by adding a point at infinity. Thus, an analytic function on an arbitrary Riemann surface can be regarded as an analytic function defined in one of these three regions and invariant under a certain group of transformations. Such functions are called automorphic functions. *See* COMPLEX NUMBERS AND COMPLEX VARIABLES.

Structure of groups. If G is a group and a subset G' of its elements is a group under the same law of multiplication, then G' is called a subgroup of G. Especially significant are the invariant subgroups which have the property that if g is any element of G' and h any element of G then hgh^{-1} is an element of G'. Let G and H be two groups and suppose that f is a mapping of G into H (that is, a function which for each element, g, of G yields an element $f(g)$ in H) with the property $f(gh)=f(g)\,f(h)$. Then f is called a homomorphism of G into H and G is homomorphic to H. The subset of G consisting of those elements which f maps onto the identity in H form an invariant subgroup, the kernel of the homomorphism f. If $H = G$ so that f maps G into itself, it is called an endomorphism. If f maps G one-to-one onto H then it is called an isomorphism. Finally, if f is an isomorphism of G onto G, it is called an automorphism of G. The idea of isomorphism gives precise meaning to the vague notion that two groups have the same structure. Evidently, a representation of a group, defined in Eq. (5), is a homomorphism of the group into the group of linear transformations of a vector space.

For topological groups it is natural to require of a homomorphism that it preserve not only the group structure but also the topological structure; for a topological group, a homomorphism is defined as a continuous mapping, f, of G into H satisfying $f(gh) = f(g)f(h)$. The group structure and the topological structure of a topological group impose severe restrictions on each other. Evidence for this statement is the solution of Hilbert's fifth problem (posed in 1900 and solved about 50 years later), which may be stated roughly: Every topological group whose elements can be labeled by a finite number of coordinates so that the group operations are continuous is isomorphic to a Lie group.

For some topological groups it is possible to define a notion of integration on the group which has invariance properties under group multiplication. Namely, if F is a complex-valued function defined on the group and $\int F(g)\,d\mu(g)$ is the integral of F over the group with respect to a measure $d\mu(g)$, then $d\mu(g)$ is said to be left-invariant if

$$\int F(g)\ d\mu(g) = \int F(hg)\ d\mu(g)$$

holds for all integrable F and all h in the group. Analogously, right-invariance is defined by

$$\int F(g)\ d\mu(g) = \int F(gh)\ d\mu(g)$$

When a left-invariant measure exists so does a right-invariant, and they are unique up to a multiplicative factor. For a compact group, that is, one for which $\int d\mu(g) < \infty$, left- and right-invariant measures coincide.

Invariant integration is a powerful tool in the study of representations of groups. Consider, for example, two irreducible representations of a compact group, G, by unitary matrices. (A representation is irreducible if there is no proper subspace of vectors carried into itself by all matrices of the representation. It is unitary if all the matrices are unitary.) Then

$$\int M^{(1)}(g)_{\kappa\lambda}\,\overline{M^{(2)}(g)}_{\mu\nu}\,d\mu(g) =$$

$$\begin{Bmatrix}1\\0\end{Bmatrix}\frac{\delta_{\kappa\mu}\,\delta_{\lambda\nu}\int d\mu(g)}{\dim M^{(1)}(g)} \qquad (6)$$

where the first alternative holds if $M^{(1)} = M^{(2)}$ and the second if the y are not equivalent ($M^{(1)}$ and $M^{(2)}$ are equivalent if there exists a unitary operator U such that $M^{(1)}(y) = UM^{(2)}(y)\,U^{-1}$ for all g in G); $\dim M^{(1)}(g)$ stands for the number of rows (or columns) in $M^{(1)}(g)$. The angular momentum selection rules referred to above are consequences of Eq. (6) in the case that G is the three-dimensional rotation group.

For Lie groups there is an important method of analysis, the so-called infinitesimal method. Consider, for example, a Lie group of matrices with matrix multiplication as group multiplication. (Not all Lie groups are isomorphic to such groups, but many are.) A one-parameter subgroup is a subset of elements, $g(t)$, labeled continuously by a real parameter, t, and satisfying

$$g(t_1)g(t_2) = g(t_1 + t_2)$$

for all real t_1 and t_2. Such a subgroup can be written in the form $g(t) = \exp tx$. The matrix x is called the infinitesimal element of the one-parameter subgroup. If the group elements are labeled by n parameters one can choose n distinct one-parameter subgroups with linearly independent infinitesimal elements $x_1, \ldots, x_n$. These matrices will then satisfy commutation relations.

$$x_j x_k - x_k x_j = \sum_{l=1}^{n} C_{jk}{}^{l} x_l$$

The constants $C_{jk}{}^{l}$ are called the structure constants of the Lie group, and they largely determine the structure of the group. *See* GRAPH THEORY; SET THEORY.

[ARTHUR S. WIGHTMAN]

Bibliography: G. Birkhoff and S. MacLane, *Survey of Modern Algebra*, 4th ed., 1977; G. Burns, *Introduction to Group Theory with Applications*, 1977; A. W. Joshi, *Elements of Group Theory for Physicists*, 2d ed., 1977; H. Lipkin, *Lie Groups for Pedestrians*, 2d ed., 1966; E. P. Wigner, *Group Theory and Its Application to the Quantum Mechanics of Atomic Spectra*, 1959.

Group velocity

The velocity of propagation of a group of waves forming a wave packet; also, the velocity of energy flow in a traveling wave or wave packet. The pure sine waves used to define phase velocity v_p do not ever really exist, for they would require infinite extent. What do exist are groups of waves, wave packets, which are combined disturbances of a group of sine waves having a range of frequencies and wavelengths. Good approximations to pure sine waves exist, provided the extent of the media is very large in comparison with the wavelength of the sine wave. In nondispersive media, pure sine waves of different frequencies all travel at the same speed v_p, and any wave packet retains its shape as it propagates. In this case, the group velocity v_g is the same as v_p. But if there is dispersion, the wave packet changes shape as it moves, because each different frequency which makes up the packet moves with a different phase velocity. If v_p is frequency-dependent, then v_g is not equal to v_p. *See* PHASE VELOCITY; SINE WAVE; WAVE MOTION; WAVE PACKET.

The relationship between v_g and v_p is very easily derived in one dimension. Consider a wave packet made of two waves: the first has wavelength λ and phase velocity v_p; the second has wavelength $\lambda + \Delta\lambda$ and phase velocity $v_p + \Delta v_p$. The wave packet is the combined disturbance which is just the sum of the two waves taken over their infinite extent. The "position" of the wave packet may be taken as the position of maximum disturbance which occurs at some point x_0 at time $t = 0$, as shown in illustration *a* at the position of the crests labeled 2. For clarity, the waves are drawn separated. After some time t_0, crest 2 of the second wave will have moved ahead of crest 2 of the first wave, and the crests labeled 1 will be aligned as shown in illustration *b*. The position of the wave packet has moved forward a distance D in time t_0, so the group velocity is $v_g = D/t_0$. But $D = v_p t_0 - \lambda$ is evident from the figure, as is $\lambda + \Delta\lambda - \lambda = \Delta\lambda = (v_p + \Delta v_p)t_0 - v_p t_0 = \Delta v_p t_0$. Therefore, $t_0 = \Delta\lambda/\Delta v_p$, and upon combining these one has $v_g = v_p - \lambda(\Delta v_p/\Delta\lambda)$. A calculus derivation would yield Eq. (1), which is usually

$$v_g = v_p - \lambda \frac{dv_p}{d\lambda} \qquad (1)$$

given as the basic relationship. A more convenient from utilizes the equation $v_p = \lambda f$, where f is the frequency associated with wavelength λ. By the chain rule of differentiation, $dv_p/d\lambda = f + \lambda(df/d\lambda)$ so $v_g = v_p - \lambda f - \lambda^2(df/d\lambda) = -\lambda^2(df/d\lambda)$. But then one uses the wave number $k = 2\pi/\lambda$ to finally write Eq. (2), where $\omega = 2\pi f$ is the angular fre-

$$v_g = \frac{d\omega}{dk} \qquad (2)$$

quency and $\omega/k = v_p$. This last form is readily generalized to more than one dimension.

For waves in deep water, $v_p = ck^{-1/2}$, where c is a constant. Thus $\omega = ck^{1/2}$ and $v_g = d\omega/dk = \frac{1}{2} v_p$. These waves are very strongly dispersive, and the individual wave crests slide through the group at twice the group speed. *See* LIGHT; SOUND; WAVE EQUATION; WAVE MOTION IN LIQUIDS.

[S. A. WILLIAMS]

Bibliography: A. P. Fench, *Vibrations and Waves*, 1971; F. Lokkowicz and A. C. Melissinos, *Physics for Scientists and Engineers*, 1975.

Gyromagnetic effect

An effect arising from the relation between the angular momentum and the magnetization of a magnetic substance. It is the effect which is exploited in the measurement of the gyromagnetic ratio of magnetic materials. The gyromagnetic effect is demonstrated by a simple experiment in which a freely suspended magnetic substance is subjected to a magnetic field. Upon a change in direction of the magnetic field, the magnetization of the substance must change. In order for this to happen, the atoms must change their angular momentum. Since there are no external torques acting on the system, the total angular momentum must remain constant. Thus the sample must acquire a mass rotation which may be measured. In this way, the gyromagnetic ratio may be determined. Two common methods of determination are the Einstein-de Haas method and the Barnett method. *See* GYROMAGNETIC RATIO.

Einstein-de Haas method. This is usually used to determine the gyromagnetic ratio of ferromagnetic materials. Imagine a ferromagnetic substance in the shape of a cylinder suspended on one end by a torsion fiber, forming thereby a torsional pendulum. A magnetic field is applied parallel to the axis of the cylinder, for example, by means of a coil surrounding the cylinder. The ferromagnetic cylinder is now magnetized in the direction of the magnetic field. If the field is suddenly reversed, the magnetization will reverse, and the accompanying change in angular momentum of the atoms will be balanced by a mass rotation of the cylinder which can be measured by noting the change in amplitude of displacement of the torsional pendulum. A measurement is made of the change in magnetization by means of a magnetometer. The ratio of magnetization change to angular momentum change is the magnetomechanical ratio, the reciprocal of the gyromagnetic ratio.

Barnett method. This is an alternative to the Einstein-de Haas technique. In this experiment, a

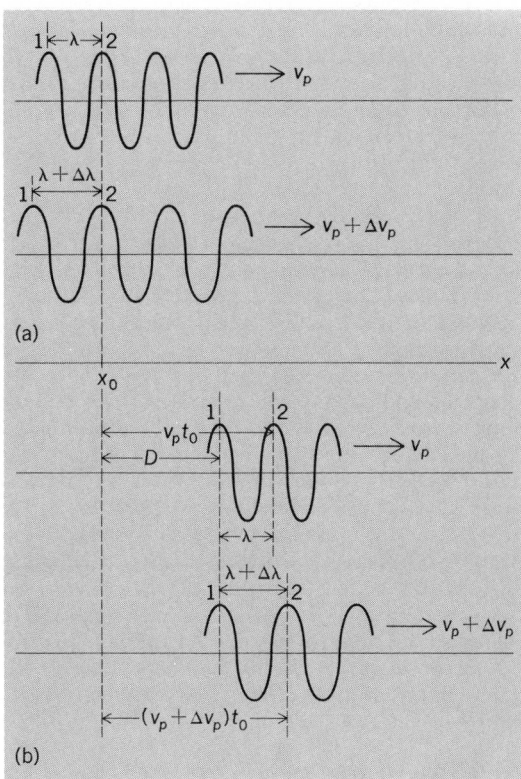

One-dimensional wave packet made of two waves. (*a*) Maximum disturbance of wave packet at time t_0 is located at x_0, where crests labeled 2 coincide. (*b*) After time t_0, wave packet has moved distance D to a position at which crests labeled 1 now coincide.

ferromagnetic bar with a coil surrounding it is spun rapidly about its axis. The atoms, therefore, acquire an angular momentum in the direction of rotation. The magnetization thereby produced is measured by stopping the rotation abruptly and measuring the voltage induced in the surrounding coil which is part of a fluxmeter circuit. Now the same increase in magnetization in the direction of the rotation could be achieved by applying a magnetic field in that direction instead of rotating the bar. In the first case, if an angular momentum L makes an angle θ with the axis of rotation, it will experience a torque $L\omega \sin \theta$, where ω is the angular velocity of rotation, tending to turn it into the axis of rotation. In the second case, the magnetic field H would produce a torque on L in the same direction of $(1/\gamma)LH \sin \theta$, where $1/\gamma$ is the magnetomechanical ratio (L/γ = magnetic moment). If the magnetizations in the two cases are the same, then the torques are the same also, and Eq. (1) holds. Thus Eq. (2) is obtained.

$$L\omega \sin \theta = (1/\gamma)LH \sin \theta \qquad (1)$$

$$\gamma = H/\omega \qquad (2)$$

In this way the magnetomechanical ratio can be measured. In an experiment similar to the one described, S. J. Barnett showed in 1914 that the magnetomechanical ratio for a ferromagnet was close to e/mc, the ratio of charge to mass of an electron. [ELIHU ABRAHAMS; FREDERIC KEFFER]

Bibliography: S. J. Barnett, *Proc. Amer. Acad. Arts Sci.*, 75:109, 1944; C. Kittel, *Introduction to Solid State Physics*, 5th ed., 1976.

Gyromagnetic ratio

The ratio of angular momentum to magnetic moment for atomic systems. This ratio is usually expressed in terms of the magnetomechanical factor g', as in Eq. (1). The ratio is written here in electromagnetic units; thus, e/c and m are the charge and

$$\frac{\text{Angular momentum}}{\text{Magnetic moment}} = \frac{2mc}{g'e} \qquad (1)$$

mass of the electron. The factor g' is sometimes loosely called the gyromagnetic ratio.

Magnetomechanical ratio. This quantity is the inverse of the gyromagnetic ratio. It is usually denoted by γ and is equal to $g'e/2mc$.

The magnetomechanical ratio of a substance identifies the origin of the magnetic moment. For example, for electron spin the angular momentum is $1/2\hbar$, where $\hbar$ is Planck's constant divided by 2π. The magnetic moment is the Bohr magneton $e\hbar/2mc$. Thus, the magnetomechanical ratio is given by Eq. (2). Since $\gamma = g'e/2mc$, for electron spin

$$\gamma = \frac{e\hbar/2mc}{\hbar/2} = \frac{e}{mc} \qquad (2)$$

$g' = 2$. For orbital angular momentum, $\gamma = e/mc$ and $g' = 1$. The experimental values of g' for most ferromagnetic materials are in the neighborhood of 2, showing that the major contribution to the magnetization comes from the electron spin. Deviations of g' from 2 show the extent to which the orbital motion contributes to the magnetization. In superconductors, on the other hand, the fact that

$g' = 1$ shows that the diamagnetic currents which cause the Meissner effect are caused by electrons. For discussion of the measurement of the magnetomechanical ratio *see* GYROMAGNETIC EFFECT. *See also* ELECTRON SPIN; MEISSNER EFFECT; SUPERCONDUCTIVITY.

Spectroscopic splitting factor. This is the measure of the enery-level splittings of an atomic system in a magnetic field. For a free electron spinning in a magnetic field H, the energy levels are split according to $\Delta E = g\mu_B H$ where μ_B is the Bohr magneton and g is the spectroscopic splitting factor, or g factor, and is equal to 2.00. For electron spins in paramagnetic salts, there are complicated energy-level schemes because of the spin-orbit coupling and crystalline field interactions; the g factor is different from case to case and may depend upon the orientation of the magnetic field with respect to the crystal axes. Under these circumstances, the g factor is defined quantum mechanically. For ferromagnetic materials, the spectroscopic splitting factor is defined similarly; it is the factor in the ferromagnetic resonance condition which gives the splitting of the energy levels in the magnetic field. It is not generally equal to the magnetomechanical factor g', as the two are affected differently by the effects of orbital angular momentum. For free atoms, the spectroscopic splitting factor is identical to the Landé g factor. For further discussion of g factors *see* MAGNETIC RESONANCE.

[ELIHU ABRAHAMS; FREDERIC KEFFER]

Hadron

The generic name of a class of particles which interact strongly with one another. Examples of hadrons are protons, neutrons, the π-, K-, and D-mesons, and their antiparticles. Protons and neutrons, which are the constituents of ordinary nuclei, are members of a hadronic subclass called baryons, as are strange and charmed baryons, for example, Λ_s^0 and Λ_c^0. Baryons have half-integral spin, obey Fermi-Dirac statistics, and are known as fermions. Mesons, the other subclass of hadrons, have zero or integral spin, obey Bose-Einstein statistics, and are known as bosons. The electric charges of baryons and mesons are either zero or ± 1 times the charge on the electron. Masses of the known mesons and baryons cover a wide range, extending from the pi-meson, with a mass approximately one-seventh that of the proton, to the upsilon-meson, with a mass about 10 times the proton mass. The spectrum of meson and baryon masses is not fully understood. *See* BARYON; BOSE-EINSTEIN STATISTICS; FERMI-DIRAC STATISTICS; MESON; NEUTRON; PROTON.

It has been believed that the net number of baryons in the universe is a conserved quantity. In making this count, baryons and antibaryons are arbitrarily assigned baryon numbers $+1$ and -1, respectively, so that production or annihilation of a baryon-antibaryon pair in a given reaction has no effect on the net baryon number. (Mesons are assigned baryon number zero.) In this scheme, the least massive baryon, the proton, would be stable, and all other baryons unstable. However, it has been suggested, as a result of some theoretical attempts to unify the strong, weak, and electromagnetic interactions, that protons may be unstable with a lifetime many orders of magnitude greater than the age of the universe; experimental searches for such an instability have been undertaken. *See* FUNDAMENTAL INTERACTIONS; SYMMETRY LAWS.

In addition to baryon number, hadrons and their antiparticles are assigned quantum numbers to represent other properties of hadronic matter. Among these are isobaric (*i*-) spin (related to electric charge), strangeness, charm, and quite probably others, all of which denote collateral families within the main family of hadrons. Strong interactions conserve *i*-spin, and hence hadrons may form *i*-spin multiplets whose members differ in mass by only a very small fraction of the proton mass, for example, the proton-neutron doublet. Hadrons with nonzero values of the strangeness and charm quantum numbers, as well as hadrons without a decay mode that conserves *i*-spin, are quasi-stable; their decays take place through the weak interaction, which does not conserve strangeness or charm, or through the electromagnetic interaction, which does not conserve *i*-spin. The quasistable hadrons have a lifetime many orders of magnitude longer than that of nonstrange, noncharmed hadrons, which decay with the conservation of *i*-spin; because of their very short lifetimes (less than 10^{-20} s), the latter are often called elementary particle-particle scattering resonances. *See* CHARM; ISOTOPIC SPIN; STRANGE PARTICLES; WEAK NUCLEAR INTERACTIONS.

Much of the data relating to the properties of hadrons (both baryons and mesons) can be interpreted as if hadrons consist of more elementary constituents known as quarks and gluons. This conception of hadrons has widespread appeal in elementary particle physics, and will be the subject of much further exploration and testing. *See* ELEMENTARY PARTICLE; GLUONS; QUARKS.

[A. K. MANN]

Hadronic atom

A hydrogenlike system that consists of a strongly interacting particle (hadron) bound in the Coulomb field and in orbit around any ordinary nucleus. The kinds of hadronic atoms that have been made and the years in which they were first identified include pionic (1952), kaonic (1966), Σ^--hyperonic (1968), and antiprotonic (1970). They were made by stopping beams of negatively charged hadrons in suitable targets of various elements, for example, potassium, zinc, or lead. The lifetime of these atoms is of the order of 10^{-12} sec, but this is long enough to identify them and study their characteristics by means of their x-ray spectra. They are available for study only in the beams of particle accelerators. Pionic atoms can be made by synchrocyclotrons and linear accelerators in the 500-MeV range. The others can be generated only at accelerators where the energies are greater than about 6 GeV. *See* ELEMENTARY PARTICLE; HADRON; PARTICLE ACCELERATOR.

The hadronic atoms are smaller in size than their electronic counterparts by the ratio of electron to hadron mass. For example, in pionic calcium, atomic number $Z = 20$, the Bohr radius of the ground state is about 10 fermis (1 fermi = 1 femtometer = 10^{-15} m) and in ordinary calcium it is about 2500 fermis. Thus the atomic electrons are practically not involved in the hadronic atoms and the equations of the hydrogen atom are applicable.

The close approach of the hadrons to their host nuclei suggests that hadron-nucleon and hadron-nucleus forces will be in evidence, and this is one of the motivations for studying these new types of atoms.

X-ray emissions. Negative hadrons are captured into orbits of large principal quantum number n by the attraction of the positively charged nuclei. As the hadrons fall through successively smaller Bohr orbits (cascade), electrons are ejected from the cloud of atomic electrons (Auger effect). When a hadron has reached about the same radius as that of the electronic atom ground state, x-ray emission becomes the dominant method for the system to shed its excitation energy. X-rays, whose energy increases with each successive jump, are emitted until the hadron reaches the ground state ($n = 1$) of the hadronic atom or is absorbed by the nucleus in a strong interaction. *See* ATOMIC STRUCTURE AND SPECTRA; AUGER EFFECT; X-RAYS.

The lines of the spectra of special interest are due to transitions between the lowest quantum levels because the hadrons are then closest to the nuclei and the nuclear forces perturb the orbits. The effects expected and observed are that some of the lines are slightly broadened (energy indefinite) and the average transition energy is different from that predicted solely on the basis of Coulomb effects.

The series of x-ray lines generally cuts off rather abruptly at $n > 1$. However, in light pionic atoms the ground state is reached. Kaons, Σ^-, and antiprotons ($\bar{p}$) can probably reach the ground state only when the nucleus is singly charged. In kaonic chlorine atoms, for example, the series ends at $n = 3$, and in kaonic lead the series ends at $n = 7$.

Experimentally the hadrons are generated by a beam of protons incident on a metallic target. A secondary beam is used to transport the particles to the target in which the atoms are to be made and studied. The arrival of a hadron is signaled by a set of scintillation counters as it is slowed down by passage through a moderator of carbon or beryllium. The thickness of the moderator is adjusted so that a maximum number of hadrons stops in the target under investigation. The targets are usually sheets of metal or disk-shaped boxes of powder or liquid. The x-ray detectors are semiconductors of silicon or germanium. Efficiencies for detecting an x-ray that comes from within the target are around 5×10^{-3}, including factors of solid angle and target self-absorption. The energy resolution of the detectors is of paramount importance. Detectors are in use whose line widths at half maximum height are around 500 eV for energies below 100 keV and about 1 keV for energies around 1 MeV. The illustration is an example of a kaonic x-ray spectrum of chlorine obtained by an ultra-pure germanium detector. The lines are labeled according to their hadronic transitions. The intensity of the lines average about 0.3 x-ray per stopped kaon for the principal lines ($\Delta n = -1$). About 5,000,000 kaons were stopped to obtain the spectrum. *See* PARTICLE DETECTOR.

The interpretation of the x-ray spectra of pionic atoms is complicated by the necessity for the pion to react with two nucleons rather than one nucleon alone, as in reaction (1), where N stands for either a

$$\pi + N + N \rightarrow N + N + \text{Kinetic energy} \qquad (1)$$

proton or a neutron. The two-nucleon final state is required for the reaction to conserve momentum.

Through the use of the line width and energy shift data of all the measurements on many elements throughout the periodic table, a calculation was made to determine whether pionic x-rays would show a difference between the root mean square (rms) radii of the distributions of protons and neutrons in the nuclei. The result was that no significant difference was seen.

Targets of chemical compounds have been employed to determine the probability that the pions land on nucleus Z_1 (or nucleus Z_2). It had been predicted that the probability of formation of a pionic atom would be proportional to Z. The intensities of the x-ray lines indicate that in general the number of pionic atoms found are proportional to Z, but there are exceptions occurring in some oxides.

The pions' decision is made in the outer regions of the atoms where the energies are those governing chemical effects. A more comprehensive study of the "chemistry" of pionic atoms could lead to increased understanding of some solid-state structures.

Kaonic atoms. Kaonic atoms were expected to yield valuable information concerning the surface of nuclei because a kaon reacts very strongly with either a single neutron or a single proton, as in reaction (2).

$$K^- + N \rightarrow \pi + \text{Hyperon} \qquad (2)$$

On the basis of theory it had been predicted that more neutrons than protons would be found on the surfaces of nuclei. One of the first interpretations of the behavior of the series of x-ray lines of various elements ranging from $Z=3$ through $Z=92$ suggested that the neutron dominance of nuclear surfaces was verified. The reasoning went along the following lines: Electron scattering experiments measure to a high accuracy the charge (proton) distributions in nuclei. If one assumed the same distribution was applicable to neutrons and

if one used the known kaon-nucleon interaction strengths, it would be possible to predict at which Z the x-ray series of the kaonic atoms would terminate. The experiments indicated a higher capture rate than expected, and this led to the idea that nuclear matter (neutrons) was encountered by the kaons in a region above the conventional nuclear surface. It was later pointed out that a resonance between K^- and proton, $Y^*(0)1405$, would probably enhance the affinity of kaons for protons on the nuclear outskirts, and this could account for the increased capture rate observed experimentally. Even though extensive data have been gathered on lines shifted in energy and broadened by strong interactions, at the end of 1974 the question of the distribution of nucleons in nuclear matter had not been resolved. A fully satisfactory model that related the observed x-ray data to nucleon density distributions has not been devised. *See* NUCLEAR STRUCTURE.

It has been found that kaonic x-ray intensities exhibit a remarkable and unexpected dependence on the atomic number Z of the elements in which kaons are stopped. The effect probably occurs with other hadronic atoms. Kaonic x-ray intensities rise and fall with peaks apparently near Z with closed shells (noble gases). For example, at $Z=19$ and 35 the intensities are 0.5 x-ray per stopped kaon, but for $Z=24$ and 25 the intensities are about 0.1. The reasons for this behavior are not known. Apparently the atomic electron configurations influence the cascade processes in such a way as to send more or fewer hadrons into orbits of low angular momentum where they undergo nuclear absorption even at large n.

Σ^--hyperonic atoms. When kaons are absorbed by nucleons, about 20% of the hyperons produced are Σ^--particles. In the light elements most of the Σ^--hyperons are ejected from the nucleus in which they are generated. Some of them are captured by target nuclei; Σ^--hyperonic atoms are formed and emit characteristic x-rays, just as do the kaonic atoms. Weak x-ray lines due to the hyperonic atoms are found along with the kaonic x-ray lines (see illustration). The hyperonic lines are of special

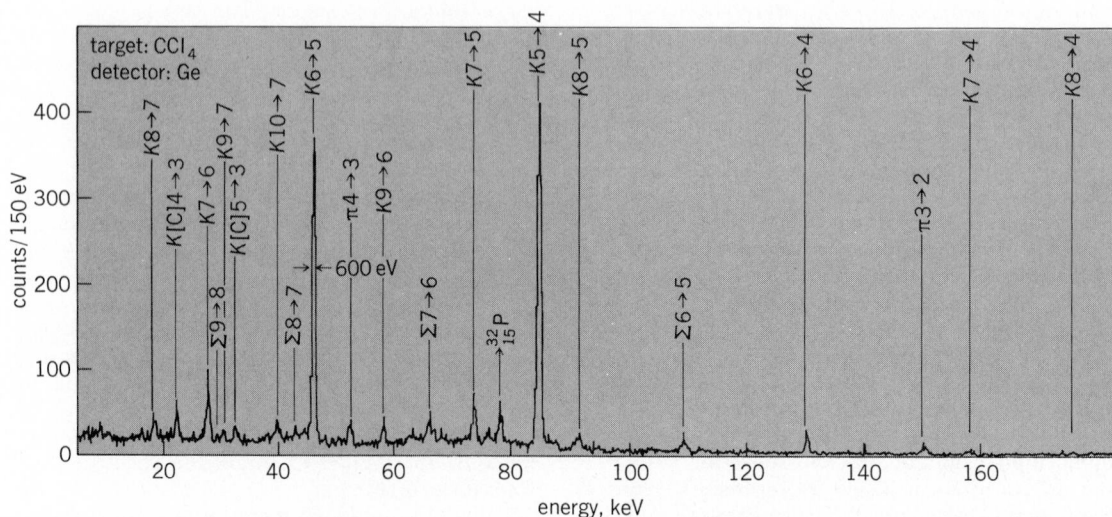

X-ray spectrum resulting from kaons stopped in carbon tetrachloride. Lines from Σ^--hyperonic atoms are seen along the kaonic x-rays. The pionic lines came from decay products. Nuclear γ-rays of the first excited state of phosphorus-32 are seen at 78 keV. (*Lawrence Berkeley Laboratory*)

interest because they are actually doublets due to the magnetic moment of the Σ^-. X-ray lines of Σ^--atoms in some heavy elements have been measured and limits set on the Σ^- magnetic moment:

$$\mu = -1.40{}^{+0.41}_{-0.28} \text{ or } 0.65{}^{-0.40}_{+0.28} \text{ nuclear magnetons}$$

Perhaps Σ^- will turn out to be a suitable probe of the nucleus. No resonances are known that would complicate the interpretation of the x-ray spectra, but future experiments depend upon the availability of more intense beams of kaons.

Antiproton atoms. Antiproton atoms are the latest in the series of hadronic atoms to be observed. Their x-ray lines are doublets due to the magnetic moment of $\bar{p}$. The splitting has been measured and μ found to be -2.790 ± 0.022 nuclear magnetons, which agrees very well with the proton magnetic moment as expected.

There are two more hadrons with lifetimes long enough to be candidates for hadronic atom formation: the negative xi (Ξ^-) and the negative omega (Ω^-), but even at the largest accelerators, these particles are too scarce for their atoms to be detected. [CLYDE E. WIEGAND]

Bibliography: E. H. S. Burhop, in E. H. S. Burhop (ed.), *High Energy Physics*, vol. 3, 1969; C. E. Wiegand, *Phys. Rev. Lett.*, 22:1235, 1969; C. E. Wiegand and G. L. Godfrey, *Phys. Rev.*, A9:2282, 1974.

Half-life

The time required for one-half of a given material to undergo chemical reactions; also, the average time interval required for one-half of any quantity of identical radioactive atoms to undergo radioactive decay.

Chemical reactions. The concept of the time required for all of the material to react is meaningless, because the reaction goes very slowly when only a small amount of the reacting material is left and theoretically an infinite time would be required. The time for half completion of the reaction is a definite and useful way of describing the rate of a reaction.

The specific rate constant k provides another way of describing the rate of a chemical reaction. This is shown in a first-order reaction, Eq. (1),

$$k = \frac{2.303}{t} \log \frac{c_0}{c} \tag{1}$$

where c_0 is the initial concentration and c is the concentration at time t. The relation between specific rate constant and period of half-life, $t_{1/2}$, in a first-order reaction is given by Eq. (2). In a

$$t_{1/2} = \frac{2.303}{k} \log \frac{1}{1/2} = \frac{0.693}{k} \tag{2}$$

$$t_{1/2} = \frac{1}{k c_0} \tag{3}$$

first-order reaction, the period of half-life is independent of the initial concentration, but in a second-order reaction it does depend on the initial concentration according to Eq. (3).

[FARRINGTON DANIELS]

Radioactive decay. The activity of a source of any single radioactive substance decreases to one-half in 1 half-period, because the activity is always proportional to the number of radioactive atoms present. For example, the half-period of Co^{60} (cobalt-60) is $t_{1/2} = 5.3$ years. Then a Co^{60} source whose initial activity was 100 curies will decrease to 50 curies in 5.3 years. The activity of any radioactive source decreases exponentially with time t, in proportion to exp $-0.693\, t/t_{1/2}$. After 1 half-period (when $t = t_{1/2}$) the activity will be reduced by the factor $e^{-0.693} = 1/2$. In 1 additional half-period this activity will be further reduced by the factor $1/2$. Thus, the fraction of the initial activity which remains is $1/2$ after 1 half-period, $1/4$ after 2 half-periods, $1/8$ after 3 half-periods, $1/16$ after 4 half-periods, and so on.

The half-period is sometimes also called the half-value time or, with less justification, but frequently, the half-life. The half-period is 0.693 times the mean life or average life of a group of identical radioactive atoms. The probability is exactly $1/2$ that the actual life-span of one individual radioactive atom will exceed its half-period. *See* RADIOACTIVITY. [ROBLEY D. EVANS]

Hall effect

The development of a transverse electric potential gradient in a current-carrying conductor placed in a magnetic field when the conductor is positioned so that the direction of the magnetic field is perpendicular to the direction of current flow. Analysis of the Hall effect yields important information on the band structure of metals and semiconductors and on the nature of the conductivity process. In semiconductor research, the magnitude of the Hall effect in simple cases provides a direct estimate of the concentration of charge carriers. The Hall effect is one of the so-called galvanomagnetic effects. *See* GALVANOMAGNETIC EFFECTS.

The experimental arrangement for observing the Hall effect is shown schematically in Fig. 1. A voltage V_H appears between the sides of the specimen whenever the current density J_x and the magnetic field B_z are nonvanishing. The electric field $E_H = V_H/d = E_y$, using the coordinates of Fig. 1, is found to be proportional to the product of J_x and B_z. The quantities V_H and E_H are called the Hall voltage and Hall field, respectively.

Physical interpretation. A simple interpretation of the Hall effect may be given by the following argument. Each charge carrier in the solid is assumed to move with a drift velocity $v_{dx} = J_x/ne$. Here n is the number of charge carriers per unit

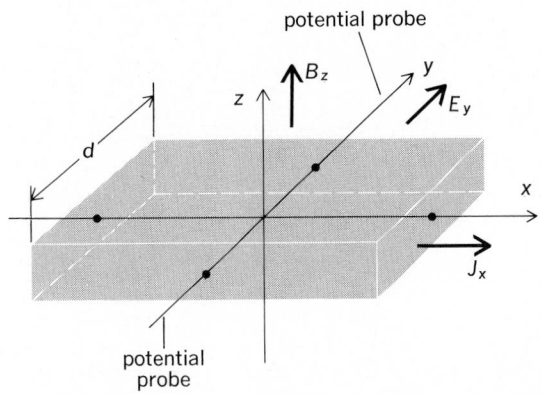

Fig. 1. Hall effect.

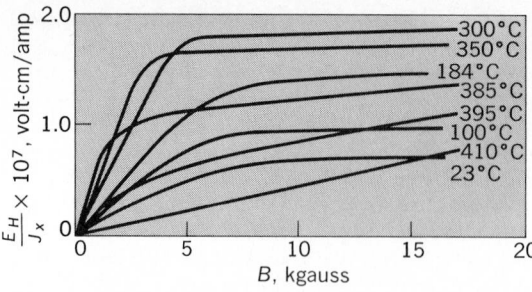

Fig. 2. Diagram of the Hall field per unit current density in nickel as a function of magnetic induction between room temperature and 410°C. (*After A. W. Smith, Phys. Rev., vol. 30, no. 1, 1910*)

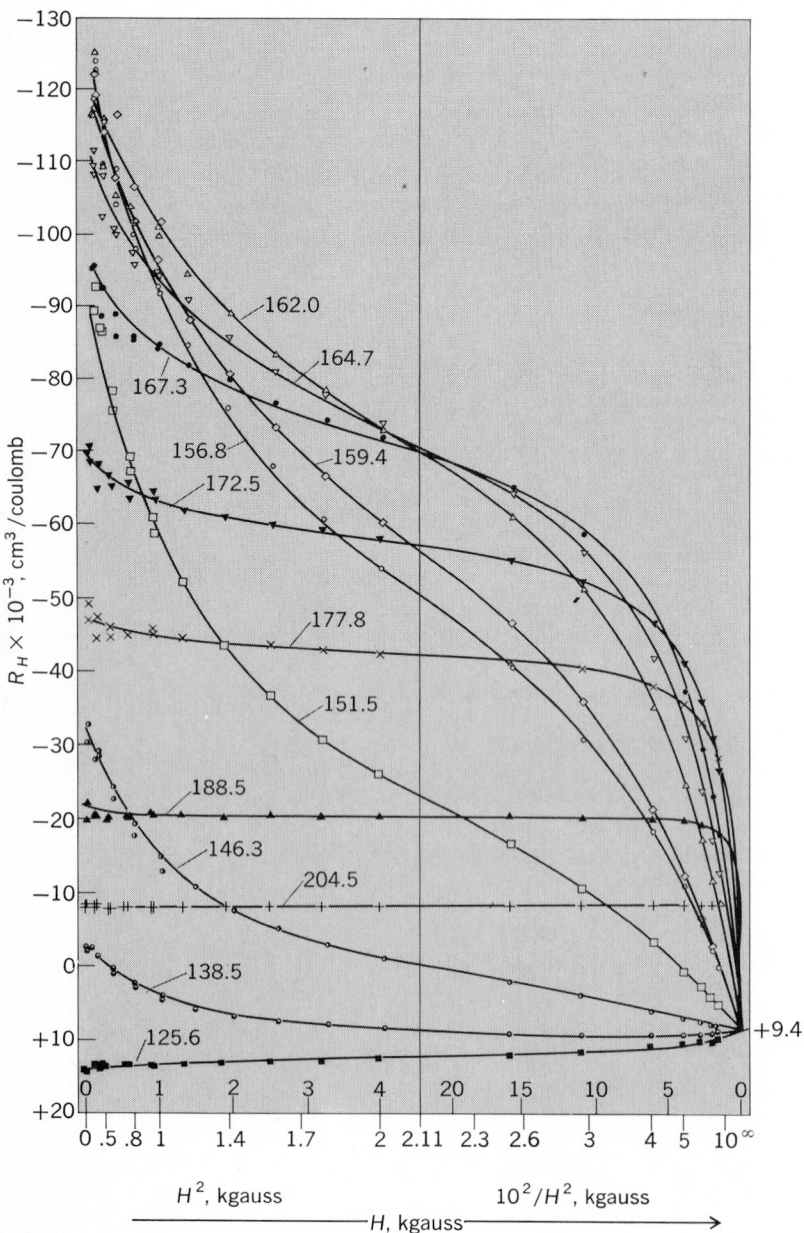

Fig. 3. Hall coefficient of *p*-type InSb as a function of magnetic field at various temperatures. Note change in scale on abscissa. Temperatures are in kelvins. (*From G. Fischer, Helv. Phys. Acta, 33:472, 1960*)

volume and e is the charge of each carrier. Each charge carrier experiences a Lorentz force

$$F_{yB} = -\frac{e}{c} v_{dx} B_z \qquad (1)$$

where c is the velocity of light. Unless this force is compensated by some other force, the charge carriers will acquire a drift velocity in the negative y direction, giving rise to a component of current in the y direction. The experimental arrangement, however, requires that $J_y = 0$. The necessary compensating force is provided by the Hall field E_y, whose magnitude and direction is such that

$$F_y = F_{yB} + F_{yE} = 0 = -\frac{e}{c} v_{dx} B_z + eE_y \qquad (2)$$

Hence

$$E_y = \frac{v_{dx} B_z}{c} = \frac{J_x B_z}{nec} \qquad (3)$$

The Hall coefficient R is defined

$$R = E_y / J_x B_z \qquad (4)$$

Thus

$$R = 1/nec \qquad (5)$$

A similar result can be obtained by solving the Boltzmann transport equation. The Hall coefficient of a metal provides the most direct information of both sign and number of charge carriers, provided the free electron model is valid. The sign of R is the same as the sign of the charge e, and the magnitude of R is inversely proportional to n. The number of conduction electrons per unit volume calculated from measured Hall coefficients agrees reasonably well with the product of valence and number of atoms per unit volume for the monovalent metals. Anomalous results are often obtained for polyvalent metals, as is apparent from the table. *See* FREE-ELECTRON THEORY OF METALS.

In these cases a two-band model is frequently invoked in the interpretation of the Hall effect and other transport phenomena. Because n is almost independent of the temperature T, R also changes only slightly with temperature.

At high magnetic fields and low temperatures ($T < 4$ K), the Hall effect in a metal single crystal sample generally depends sensitively on the orientations of the field and current relative to the crystal axes. This anisotropy is intimately related to the topology of the Fermi surface; hence, such measurements are useful in the study of the Fermi surface of metals. *See* BRILLOUIN ZONE.

Thermal side effects. It can be shown that an isothermal condition can be maintained in a specimen only if a thermal current can flow freely in the transverse direction. Unless the sample is immersed in a constant temperature bath, $\partial T/\partial y \neq 0$. Under nonisothermal conditions, the finite temperature difference between the sides of the sample may give rise to spurious thermal emfs at the potential probes. The generation of a transverse temperature gradient by a current flowing in the presence of a transverse magnetic field is known as the Ettingshausen effect, and proper correction for this effect should be made. A convenient way of avoiding these thermal difficulties is to use an alternating current as the primary current J_x, thereby preventing the establishment of a significant temperature gradient in the sample.

Effect in ferromagnetic metals. Ferromagnetic metals exhibit anomalously large Hall effects at temperatures below the ferromagnetic Curie temperature. Moreover, the Hall voltages are not linear functions of B. Typical curves are shown in Fig. 2. The knees of the curves occur at saturation, and the shapes of the curves suggest that E_H be written in the form $E_H = R_0 H + R_1 M$. Here M is the magnetization, and R_0 and R_1 are designated the ordinary and extraordinary—or anomalous—Hall coefficients, respectively. The coefficient R_0 is not a sensitive function of temperature; R_1 changes by orders of magnitude as the temperature is increased from 4 K to the Curie temperature.

Effect in semiconductors. As in metals, the sign and magnitude of R determine the sign and number of charge carriers, provided only one type of carrier (electrons or holes) is present. At high temperatures (intrinsic region) R decreases approximately exponentially with increasing temperature, largely because of the exponential increase in carrier density with temperature. At low temperatures (extrinsic region) R is approximately constant. In the intrinsic, and in some cases also in the extrinsic region, the Hall coefficient must be calculated with the aid of a two-band model because more than one type of carrier is responsible for charge transport. The Hall coefficient, especially in a multicarrier situation, is field-dependent (Fig. 3). This field dependence provides valuable information on mobility ratios and relaxation times. *See* RELAXATION TIME OF ELECTRONS; SEMICONDUCTOR.

Effect in ionic crystals. Alkali halides become electronic conductors only under the influence of light that will excite electrons from the valence band to the conduction band. Because the photoelectrons are few in number, they form a classical electron gas. Measurements of the Hall effect and of photoconductivity provide information on the mobility of these electrons. The scattering mechanisms and the effective masses of conduction electrons in ionic crystals are greatly influenced by the local polarization of the crystal by these charge carriers, and therefore the usual theories of mobility are not directly applicable here. For this reason, Hall effect and conductivity measurements in ionic crystals are of considerable interest. The measurements require rather special techniques and are more difficult than in metals or semiconductors. The data that are available indicate that current theories correctly describe the interaction between conduction electrons and lattice vibrations in alkali halides.   [FRANK J. BLATT]

Bibliography: N. W. Ashcroft and N. D. Mermin, *Solid State Physics*, 1976; *Effects in Semiconductors*, 1963; F. J. Blatt, *Physics of Electronic Conduction in Solids*, 1968; F. C. Brown, *Physics of Solids*, 1967; E. Fawcett, High-field galvanomagnetic properties of metals, *Advan. Phys.*, 13:139, 1964; C. Kittel, *Introduction to Solid State Physics*, 5th ed., 1976; F. Seitz and D. Turnbull (eds.), *Solid State Physics*, vol. 1, 1955, and vols. 4 and 5, 1957.

Hamilton-Jacobi theory

A theory that provides a means for discussing the motion of a dynamical system in terms of a single partial differential equation of the first order. It rests on the fact that Hamilton's principle, Eq. (1),

$$\Delta \int_{t_1}^{t_2} \left[\sum_{j=1}^{f} (p_j \dot{q}_j) - H \right] dt = 0 \qquad (1)$$

is unaffected by adding a total time derivative to the integrand. If Eq. (2) is substracted from the

$$\frac{d}{dt} \phi(q, \dot{q}, t) = \sum_{j=1}^{f} (p_j \dot{q}_j - p'_j \dot{q}'_j) - (H - H') \quad (2)$$

integrand, Eq. (1) is reproduced with all quantities replaced by primed quantities. Equation (2) is equivalent to Eqs. (3). *See* CANONICAL TRANSFORMATIONS; HAMILTON'S PRINCIPLE.

$$p_j = \frac{\partial \phi}{\partial q_j} \qquad p'_j = -\frac{\partial \phi}{\partial q'_j} \qquad H' = H + \frac{\partial \phi}{\partial t} \quad (3)$$

Equations (3) show that it is possible to choose ϕ so that $H' = 0$. Then the equations of motion in the primed variables are simply written as Eqs. 4. The

$$q'_j = \text{constant} \qquad p'_j = \text{constant} \qquad (4)$$

description of the motion of the system lies in the function $\phi(q, q', t)$ which generates this transformation. Equations (3) can be solved for q_j, p_j in terms of q'_j, p'_j, and t. The q'_j, p'_j may be chosen as the values of the q_j, p_j at time t_0; then ϕ generates the q_j, p_j at time t from those at time t_0.

The generator of this transformation must be a complete integral of the Hamilton-Jacobi equation, Eq. (5), which results from Eqs. (3) with $H' = 0$. A

$$H\left(q_i, \frac{\partial \phi}{\partial q_i}, t\right) + \frac{\partial \phi}{\partial t} = 0 \qquad (5)$$

complete integral of such a partial differential equation is a solution containing f nonadditive constants of integration which may be taken as the q'_j.

If $H(q, p, t)$ is independent of t so that it may be written $H(q, p)$, then the Hamilton-Jacobi equation is separable with respect to t. Let $\phi(q, q', t)$ be expressed as in Eq. (6). Then Eq. (5) becomes Eq. (7).

$$\phi(q, q', t) = S(q, q') - Et \qquad (6)$$

$$H\left(q_i, \frac{\partial S}{\partial q_i}\right) = E \qquad (7)$$

A complete integral of this equation is required, depending on f nonadditive constants including E.

Schrödinger equation. Equation (5) or (7) bears a close formal resemblance to the Schrödinger wave equation. This resemblance is, in fact, much more than formal. The Hamilton-Jacobi equation is the equation determining the canonical transformation from a set of initial coordinates and momenta to those at time t.

In quantum mechanics the state vector at time t is found from that at time 0 by a unitary transformation, as in Eq. (8). If the coordinates are taken

$$\psi(t) = U(t)\,\psi(0) \qquad (8)$$

as the quantum numbers describing the state vector $\psi(t)$; then U satisfies the Schrödinger equation, Eq. (9). *See* NONRELATIVISTIC QUANTUM THEORY.

$$H\left(q, \frac{ih}{2\pi}\frac{\partial}{\partial q}, t\right)U - \frac{ih}{2\pi}\frac{\partial U}{\partial t} = 0 \qquad (9)$$

Separation of variables. The integration of a partial differential equation is usually more difficult than the solution of a set of ordinary differential equations. It frequently happens, however, that the Hamilton-Jacobi equation is separable in a suitably chosen coordinate system, in which case it may be soluble in practice. The general solution is not required, but only a complete integral. Even this is not explicitly required, as only derivatives of ϕ appear in the equation and in the transformation equations which are to be obtained ultimately. In separable multiply periodic systems, the frequencies of the motion (as distinct from the trajectory) can be obtained without constructing the transformation function for all times, but just for the periodic times.

The complete integral of the Hamilton-Jacobi equation required is the time integral of the Lagrangian of the system along the trajectory of the system passing through the point q. Denoting by primes quantities in the transformed coordinate system in phase space, Eqs. (10) and (11) hold.

$$L' = \sum_{j=1}^{f} p'_j q'_j - H' = 0 = L - \frac{d\phi}{dt} \qquad (10)$$

$$\phi = \int_{t_0}^{t} L\,dt = \int_{t_0}^{t}\left(\sum_{j=1}^{f} p_j \dot{q}_j - H\right)dt \qquad (11)$$

If H is independent of time, so that $H = E$, Eqs. (12) and (13) hold.

$$\phi = \int_{t_0}^{t}\sum_{j=1}^{f} p_j\,dq_j - E(t - t_0) \qquad (12)$$

$$S = \sum_{j=1}^{f}\int_{t_0}^{t} p_j\,dq_j \qquad (13)$$

Equation (12) is especially useful if the system is further separable in the sense that each p_j can be expressed as a function of the conjugate q_j only. Then the problem has been reduced to the evaluation of the integrals occurring in Eq. (12) with arbitrary upper limits (indefinite integrals).

Action variables. If the system is separable and multiply periodic, by which is meant that either each p_k, q_k is a periodic function of the time or that p_k is a periodic function of q_k, it is useful to introduce the action variables J_k. These are defined by Eq. (14), where the integration is over one period of

$$J_k = \oint p_k\,dq_k \qquad (14)$$

p_k and q_k or is over one period of p_k as a function of q_k. In a separable system, the energy is expressible

as a function of the action variables, as in Eq. (15).

$$E = H(J_1, \ldots, J_f) \qquad (15)$$

Consider now that Eq. (12) is integrated over a period of τ_k of the variable p_k. The function ϕ generates the contact transformation from the initial variables to the variables τ_k later. For the kth degree of freedom, this is given by expression (16a) or (16b). In either case, $p_k\,dq_k$ is the same before

$$p_k \rightarrow p_k \qquad q_k \rightarrow q_k \qquad (16a)$$
$$p_k \rightarrow p_k \qquad q_k \rightarrow q_k + \text{constant} \qquad (16b)$$

and after the transformation, and hence ϕ, according to Eq. (2), is independent of the initial and final q_k and thus contains no reference to the kth degree of freedom; in particular, it is independent of J_k. Equation (12) becomes Eq. (17). The integrals in

$$\phi = J_k + \sum_{\substack{j=1 \\ j \neq k}}^{f}\int_{t_1}^{t_1 + \tau k} p_j\,dq_j = H(J)\tau_k \qquad (17)$$

Eq. (17) are independent of J_k, the term $j = k$ being explicitly missing. Hence, differentiation of Eq. (17) with respect to J_k leads to Eqs. (18).

$$0 = 1 - \frac{\partial H}{\partial J_k}\tau_k \qquad \nu_k = \frac{1}{\tau_k} = \frac{\partial H}{\partial J_k} \qquad (18)$$

To calculate the frequencies ν_k of a separable multiply periodic system, find the energy as a function of the action variables. Equations (18) then give the frequencies. [PHILIP M. STEHLE]

Bibliography: *See* LAGRANGE'S EQUATIONS.

Hamilton's equations of motion

The motion of a mechanical system may be described by a set of first-order ordinary differential equations known as Hamilton's equations. Because of their remarkably symmetrical form, they are often referred to as the canonical equations of motion of a system. They are equivalent to Lagrange's equations, but the fact that they are of first order and highly symmetrical makes them advantageous for general discussions of the motion of systems. *See* LAGRANGE'S EQUATIONS.

Definitions. Hamilton's equations can be derived from Lagrange's equations. Let the coordinates of the system be q_j ($j = 1, 2, \ldots, f$), and let the dynamical description of the system be given by the Lagrangian $L(q,\dot{q},t)$, where q denotes all the coordinates and a dot denotes total time derivative. Lagrange's equations are then given by Eq. (1). The momentum p_j canonically conjugate to q_j

$$\frac{d}{dt}\frac{\partial L}{\partial \dot{q}_j} - \frac{\partial L}{\partial q_j} = 0 \qquad (1)$$

is defined by Eq. (2). It is assumed that the equa-

$$p_j = \frac{\partial L}{\partial \dot{q}_j} \qquad (2)$$

tions from Eq. (2) are soluble for the velocities $\dot{q}_j$ in terms of the coordinates and momenta. If, as most commonly occurs, the only part of L containing the velocities is the kinetic energy, then the canonical momenta are the usual momenta. For example, if $L = m\dot{x}^2/2 - V(x)$, then $p_x = \partial L/\partial \dot{x} = m\dot{x}$. *See* LAGRANGIAN FUNCTION.

The Hamiltonian H is defined by Eq. (3). Differentiating Eq. (3) leads to Eqs. (4), where use

$$H = \sum_{j=1}^{f} p_j \dot{q}_j - L(q, \dot{q}, t) \qquad (3)$$

$$dH =$$

$$\sum_{j=1}^{f} \left[\left(p_j - \frac{\partial L}{\partial \dot{q}_j} \right) d\dot{q}_j + \dot{q}_j dp_j - \frac{\partial L}{\partial q_j} dq_j \right] - \frac{\partial L}{\partial t} dt \qquad (4a)$$

$$dH = \sum_{j=1}^{f} (\dot{q}_j dp_j - \dot{p}_j dp_j) - \frac{\partial L}{\partial t} dt \qquad (4b)$$

has been made of Eqs. (1) and (2) in going from (4a) to (4b). This shows that H must be a function of the qs, the ps, and t, as the differentials of only these quantities appear in dH. This being so, Eq. (5) follows.

$$dH(q,p,t) = \sum_{j=1}^{f} \left(\frac{\partial H}{\partial q_j} dq_j + \frac{\partial H}{\partial p_j} dp_j \right) + \frac{\partial H}{\partial t} dt \qquad (5)$$

lows. Hamilton's canonical equations, Eqs. (6),

$$\dot{q}_j = \frac{\partial H(q,p,t)}{\partial p_j} \qquad \dot{p}_j = -\frac{\partial H(q,p,t)}{\partial q_j} \qquad (6)$$

result from identifying coefficients of differentials of the independent quantities q_j, p_j, in Eqs. (4) and (5). It is essential that H be written as a function of q, p, t only, and that no velocities appear.

Phase-space coordinates. Hamilton's equations are most easily visualized by introducing the phase space of the system. This is a space of $2f$ dimensions in which the coordinates and momenta of the system serve as the coordinates of a point representing the state (the position and velocity of each mass point) of the system. Hamilton's equations give the velocity of this phase point as a function of its position in phase space. An important theorem named for J. Liouville follows directly from Hamilton's equations. Consider several points in a neighborhood of phase space at a given time. These points represent several possible sets of initial conditions for the system. As time proceeds, these points will move. If a large number of points is considered, one may speak of the density of phase points. Liouville's theorem states that the density of phase points in the neighborhood of a given point in phase space does not change as this point moves. The proof is merely the demonstration that the phase points move like an incompressible fluid, that is, the divergence of their velocity vanishes, as in Eq. (7). For additional information on Liouville's theorem and phase space *see* STATISTICAL MECHANICS.

$$\sum_{j=1}^{f} \left(\frac{\partial \dot{q}_j}{\partial q_j} + \frac{\partial \dot{p}_j}{\partial p_j} \right) = \sum_{j=1}^{f} \left(\frac{\partial^2 H}{\partial q_j \partial p_j} - \frac{\partial^2 H}{\partial p_j \partial q_j} \right) = 0 \qquad (7)$$

on Liouville's theorem and phase space *see* STATISTICAL MECHANICS.

Applications. As they stand, Hamilton's equations, Eqs. (6), are no easier to integrate directly than Lagrange's. Elimination of the ps from Eqs. (6) leads back to Lagrange's equations. Hamilton's equations are of great advantage in more general discussions, and they permit the making of canonical transformations which can lead to simplifications. They also lend themselves to numerical integration in some cases where Liouville's theorem is important, as in ion and electron optics. *See* CANONICAL TRANSFORMATIONS.

The Hamiltonian function H of classical me-

chanics is used to form the quantum mechanical Hamiltonian operator. *See* NONRELATIVISTIC QUANTUM THEORY. [PHILIP M. STEHLE]

Bibliography: See LAGRANGE'S EQUATIONS.

Hamilton's principle

A variational statement known as Hamilton's principle forms a basis from which the equations of motion of a classical dynamical system may be deduced. Consider a mechanical system whose configuration is specified by f independent generalized coordinates $q_1, \ldots, q_f$. Let the configuration at time t_1 be given by the values $q_j(t_1)$ and that at time t_2 by the values $q_j(t_2)$. If the system is described dynamically by a Lagrangian function $L(q, \dot{q}t)$ where q stands for all the f coordinates, then, according to Hamilton's principle, the trajectory or path of the system between the two given times is that which makes the value of the integral in Eq. (1) stationary

$$\Phi = \int_{t_1}^{t_2} L[q(t), \dot{q}(t), t] dt \qquad (1)$$

relative to nearby paths between the same end points and taking the same time.

Lagrange's equations may be derived from Hamilton's principle as follows: Let $q_j(t)$ represent the actual trajectory, and let $q_j(t) + \delta q_j(t)$ represent a nearby path. Because the end points are fixed, $\delta q_j(t_1) = \delta q_j(t_2) = 0$. If ϕ is the value of Φ for the trajectory and $\phi + \Delta\phi$ is the value of Φ for the nearby path, then Eq. (2) holds. Integrating

$$\Delta\phi = \int_{t_1}^{t_2} \sum_j \left(\frac{\partial L}{\partial q_j} \delta q_j + \frac{\partial L}{\partial \dot{q}_j} \delta \dot{q}_j \right) dt \qquad (2)$$

by parts, and noting that term (3) vanishes at t_1 and

$$\sum_j \frac{\partial L}{\partial \dot{q}_j} \delta q_j \qquad (3)$$

t_2, one obtains Eq. (4). Hamilton's principle asserts

$$\Delta\phi = \int_{t_1}^{t_2} \sum_j \left(\frac{\partial L}{\partial q_j} - \frac{d}{dt} \frac{\partial L}{\partial \dot{q}_j} \right) \delta q_j dt \qquad (4)$$

that $\Delta\phi$ vanishes for arbitrary δq_j. This can happen only if the coefficient of each δq_j vanishes at all times, as shown in Eq. (5). Hence, Lagrange's

$$\frac{\partial L}{\partial q_j} - \frac{d}{dt} \frac{\partial L}{\partial \dot{q}_j} = 0 \qquad (5)$$

equations must be satisfied by the coordinates specifying the trajectory of the system.

Hamilton's principle is invariant under any transformation of coordinates. It is also unaffected by the addition of a total time derivative to the Lagrangian, because the integral of such a total derivative is independent of the path of integration. This allows the coordinates q_j and the momenta $p_j = \partial L/\partial \dot{q}_j$ to be subjected to arbitrary canonical transformations without changing the form of Lagrange's equations. *See* CANONICAL TRANSFORMATIONS; LAGRANGE'S EQUATIONS; LAGRANGIAN FUNCTION; MINIMAL PRINCIPLES.

[PHILIP M. STEHLE]

Harmonic motion

A periodic motion that is a sinusoidal function of time. It is often called simple harmonic motion (SHM). It is the simplest possible type of vibratory motion. The motion is symmetric about its midpoint, at which the velocity is greatest and the acceleration is zero. At the extreme displacements or turning points, the velocity is zero, and the acceleration is a maximum. The motion is characterized by a unique frequency (without overtones).

Harmonic motion may be present in very simple mechanisms. For example, if a wheel is rotating at constant speed about a fixed axis, the projection on any fixed line of the motion of a point on the wheel is simple harmonic. Harmonic motion may also result from the response of a vibrating system to a periodic—in particular a sinusoidal—force. Harmonic motion is the typical motion of most simple systems that have been displaced from a position of stable equilibrium and then released, provided that the damping is negligible. The motion of a pendulum is approximately simple harmonic for small amplitudes. *See* PENDULUM.

If x represents the displacement measured from the midposition and t the time, then harmonic motion can be described by either of the forms

$$x = A \cos(\omega t) + B \sin(\omega t)$$
$$\text{or} \quad x = C \sin(\omega t - \delta) \qquad (1)$$

The constants A, B, C, and δ are not all independent, but have the following relations:

$$A = -C \sin \delta \qquad B = C \cos \delta$$

The amplitude C represents maximum displacement in one direction from the center (half the total motion between extreme positions); δ is a phase angle whose value depends on the precise instant at which the oscillation was started, or alternatively on the phase of the motion when $t = 0$. These quantities are illustrated in Fig. 1.

The remaining constant, ω, is known as the angular frequency. Dimensionally, ω is the reciprocal of time. Thus the product ωt is a pure number, to be interpreted as an angle measured in radians. When ωt increases by 2π, the motion repeats. Thus the angular frequency ω is related to ordinary frequency f (number of complete oscillations in unit time) and period T (duration of one complete oscillation) by Eq. (2).

$$\omega = 2\pi f = 2\pi/T \qquad (2)$$

The velocity v and acceleration a, obtained by differentiating Eq. (1), are given by

$$v = dx/dt = \omega C \cos(\omega t - \delta) = \omega \sqrt{C^2 - x^2} \qquad (3)$$
$$a = dv/dt = -\omega^2 C \sin(\omega t - \delta) = -\omega^2 x \qquad (4)$$

Because the net force acting on a body is equal to

the mass of the body multiplied by its acceleration, Eq. (5) shows that in SHM the force must be proportional to the displacement.

$$F = ma = -m\omega^2 x \qquad (5)$$

Conversely, SHM occurs whenever, for a body that is displaced from an equilibrium position, there is a net restoring force proportional to the displacement.

Energy considerations. The importance of harmonic motion lies in the simplicity of its time dependence, as given by Eqs. (1), and in the frequent occurrence of linear restoring forces. For sufficiently small displacements of almost any mechanical system from equilibrium, the restoring force or torque is always approximately proportional to the displacement.

This can be explained by some considerations about potential energy. At a point of stable equilibrium, potential energy is necessarily a minimum. If potential energy is a well-behaved function of position, it may be expanded as a series of powers of the distance from any point. (An important exception is electrostatic potential, which varies inversely as the distance from the charge and so is not well-behaved in the immediate neighborhood of the charge.) In the expansion of potential energy about a point of stable equilibrium, the special minimum property guarantees that the linear term must vanish. The next term, quadratic in the distance from the equilibrium point, does not ordinarily vanish and provides a good approximation of the potential energy for sufficiently small displacements. A quadratic, or parabolic, variation of potential with distance is equivalent to a force that varies linearly.

In a system oscillating freely about an equilibrium position, the energy changes from potential to kinetic and back again. For SHM the potential energy, proportional to the square of the displacement, is greatest at the extremities of the motion. Conversely, the kinetic energy $1/2 mv^2$ is zero at the turning points and maximum when the body is going past its equilibrium position. The total energy (kinetic plus potential) is constant. In SHM it is proportional to the square of the amplitude of the motion, and the average potential energy just equals the average kinetic energy.

The frequency of a freely oscillating system is determined by the stiffness and inertia of the system. If the motion is harmonic, it is also isochronous, which means that the frequency is independent of the amplitude of the motion.

Weight on elastic spring. If an elastic spring is stretched a distance x beyond its natural length, or compressed a distance x short of its natural length, it exerts a restoring force equal to $-kx$. The stiffness of the spring is measured by the spring constant k. Consider a spring which is suspended vertically and set into vertical oscillations, with a mass m attached to its free end (Fig. 2). If m is large compared to the mass of the spring, the latter can be neglected.

The equation of motion for the mass m is

$$mg - kx = ma = md^2x/dt^2$$

where g is the acceleration of gravity. The solution of this equation can be written in the form

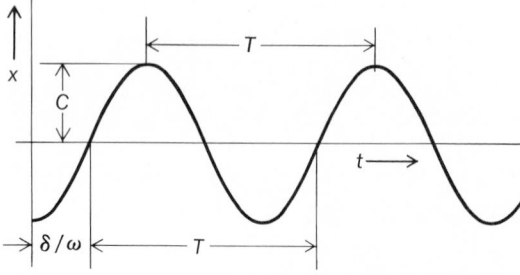

Fig. 1. Representation of simple harmonic motion.

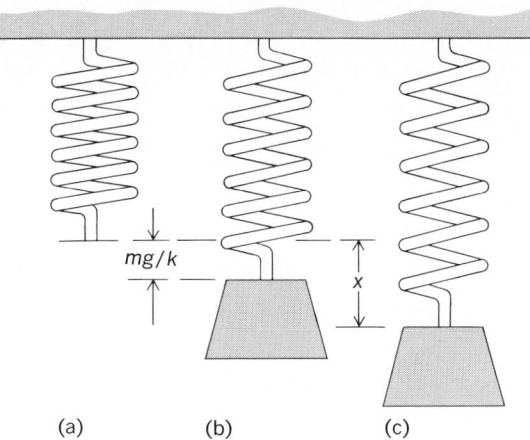

Fig. 2. Weight on an elastic spring. (*a*) Unloaded. (*b*) Statically loaded. (*c*) Weight and spring oscillating.

$$x = mg/k + C \sin (\omega t - \delta)$$
with $\qquad \omega^2 = k/m$

The term mg/k is simply the extension of the spring due to the static weight, and marks the loaded equilibrium position. Displacement from this equilibrium position calls forth a linear restoring force, and the resulting motion is simple harmonic.

Potential energy (PE) for this example is the sum of an elastic energy $\frac{1}{2}kx^2$ and a gravitational energy $-mgx$. Then

$$\begin{aligned} PE &= \tfrac{1}{2}kx^2 - mgx \\ &= \tfrac{1}{2}k(x - mg/k)^2 - m^2g^2/2k \end{aligned}$$

showing the quadratic dependence on displacement from equilibrium. The constant term $-m^2g^2/2k$ is of no importance to the motion; it can always be removed by a new choice for the zero of potential energy.

Rotational harmonic motion. Rotational as well as translational SHM may occur. In this case the angular displacement is a sinusoidal function of time. A free angular vibration will be simple harmonic if the angular displacement from an equilibrium orientation produces a restoring torque proportional to the displacement. An example is the torsional pendulum (Fig. 3). One end of a torsionally flexible elastic rod is held fixed. To the other end is fastened a disk, or another body of large moment

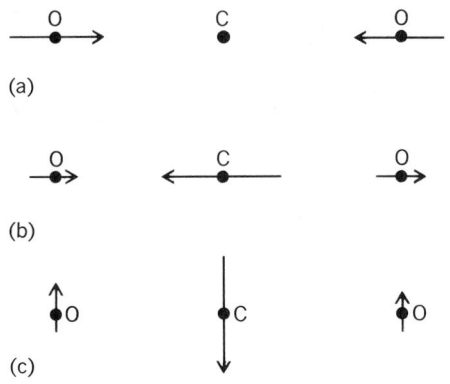

(a)

(b)

(c)

Fig. 4. Illustration of three normal modes of the CO_2 molecule. (*a*) Lower-frequency in-line mode. (*b*) Higher-frequency in-line mode. (*c*) Bending mode.

of inertia. If the rod is twisted and then released, the rod and disk will undergo angular SHM, provided that the torque in the rod is proportional to the angle of twist.

Atomic vibrations. The realization that atoms are continually vibrating in motions that are nearly harmonic is essential for understanding many properties of matter, including molecular spectra, heat capacity, and heat conduction.

In a diatomic molecule, the distance between the atoms is not precisely fixed. There is an equilibrium separation corresponding to a minimum in potential energy. The actual magnitude of the interatomic distance oscillates about the equilibrium distance with a motion that is approximately simple harmonic, if the energy of oscillation is small.

In a polyatomic molecule or in a crystalline solid, the atoms also vibrate about equilibrium positions. Because of the large number of degrees of freedom, however, the situation is more complicated. For example, in the carbon dioxide (CO_2) molecule neither oxygen atom by itself moves harmonically, or even periodically. On the other hand, the motion of the CO_2 molecule can be analyzed into a number of independent motions, called normal modes, each of which is by itself simple harmonic. In one such motion, for example, the two oxygen atoms move in phase toward or away from the carbon atom (Fig. 4*a*). The actual motion of the atoms in a molecule is a superposition of the various normal modes and a rotation of the molecule as a whole. *See* DAMPING; FORCED OSCILLATION; HARMONIC OSCILLATOR; LATTICE VIBRATIONS; MOLECULAR STRUCTURE AND SPECTRA; PERIODIC MOTION; VIBRATION; WAVE MOTION.

[JOSEPH M. KELLER]

Bibliography: G. R. Fowles, *Analytical Mechanics*, 3d ed., 1977; K. R. Symon, *Mechanics*, 3d ed., 1971.

Harmonic oscillator

Any physical system that is bound to a position of stable equilibrium by a restoring force or torque proportional to the linear or angular displacement from this position. If such a body is disturbed from its equilibrium position and released, and if damping can be neglected, the resulting vibration will be simple harmonic motion, with no overtones. The frequency of vibration is the natural frequency of the oscillator, determined by its inertia (mass) and the stiffness of its restoring force.

The harmonic oscillator is not restricted to a mechanical system, but might, for example, be electric. Typical electronic oscillators, however, are only approximately harmonic.

If a harmonic oscillator, instead of vibrating freely, is driven by a periodic force, it will vibrate harmonically with the period of the force; initially the natural frequency will also be present, but any damping will eventually remove the natural motion. The response of a harmonic oscillator driven by a general force $f(t)$, an arbitrary function of time, is described by a linear differential equation. If $x(t)$ represents the displacement as a function of time t, then the equation of motion is

$$m\, d^2x/dt^2 = -c\, dx/dt - kx + f(t) \qquad (1)$$

The left side of Eq. (1) represents the mass m

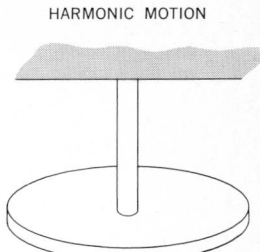

Fig. 3. Torsional pendulum.

multiplied by the acceleration. The force on the right side of the equation includes, in addition to the restoring force $-kx$ and the driving force, a "viscous damping" force $-c\,dx/dt$ that is proportional to the velocity. (This damping may vanish.) The explicit solution of Eq. (1) is

$$x = e^{-bt}\{A\sin(\omega t - \delta)$$

$$+ \frac{1}{m\omega}\int_{-\infty}^{t} e^{bT}\sin[\omega(t-T)]f(T)\,dT\} \quad (2)$$

where $b = c/2m$, $\omega^2 = (k/m) - (c/2m)^2$, and A and δ are arbitrary constants which can be set equal to zero if the oscillator displacement and velocity were zero before application of the force $f(t)$. The variable of integration T is the time at which the force $f(T)$ is considered to act. *See* DAMPING; FORCED OSCILLATION; HARMONIC MOTION.

In both quantum mechanics and classical mechanics, the harmonic oscillator is an important problem. It is one of the few rigorously soluble problems of quantum mechanics. The quantum mechanical description of electromagnetic, electronic, mesonic, and other fields is usually carried out in terms of a (time) Fourier analysis. The individual Fourier components of noninteracting fields are independent harmonic oscillators.

The Hamiltonian for a harmonic oscillator is

$$H = p^2/2m + m\omega^2 q^2/2 \quad (3a)$$

$$= P^2/2 + \frac{1}{2}\omega^2 Q^2 \quad (3b)$$

where ω is the angular frequency characteristic of the oscillator. The mass m in Eq. (3a) is made to disappear by replacing the coordinate q and the conjugate momentum p by

$$Q = m^{1/2} q$$
$$P = m^{-1/2} p$$

in Eq. (3b). (For explanation of the term conjugate momentum *see* LAGRANGE'S EQUATIONS.) The corresponding Schrödinger equation for the wave function ψ is then

$$-(\hbar^2/2)\,d^2\psi/dQ^2 + (\omega^2 Q^2/2)\psi = E\psi \quad (4)$$

where E is the energy, and $\hbar$ is Planck's constant divided by 2π. This equation possesses quadratically integrable solutions only for the characteristic energy values

$$E = (n + \frac{1}{2})\hbar\omega \quad (5)$$

where n is any positive integer. For these cases, ψ can be expressed in terms of the Hermite polynomial $H_n(y)$ of degree n, as

$$= C_n e^{(-y^2/2)} H_n(y)$$
where $\quad y = (\omega/\hbar)^{1/2}Q \quad (6)$

is dimensionless, and

$$C_n^2 = \left(\frac{\omega}{\pi\hbar}\right)^{1/2}\frac{1}{2^n n!}$$

See ANHARMONIC OSCILLATOR; NONRELATIVISTIC QUANTUM THEORY.

[JOSEPH M. KELLER]

Bibliography: J. P. Den Hartog, *Mechanical Vibrations*, 4th ed., 1956; R. M. Eisberg, *Fundamentals of Modern Physics*, 1961; S. Gasiorowicz, *Quantum Physics*, 1974; H. Goldstein, *Classical Mechanics*, 2d. ed., 1980; C. Kittel, W. D. Knight, and M. A. Ruderman, *Mechanics*, Berkeley Physics Course, vol. 1, 2d ed., 1973; J. W. S. Rayleigh, *The Theory of Sound*, 2d ed., 1894, reprint 1945.

Heat

For the purposes of thermodynamics, it is convenient to define all energy while in transit, but unassociated with matter, as either heat or work. Heat is that form of energy in transit due to a temperature difference between the source from which the energy is coming and the sink toward which the energy is going. The energy is not called heat before it starts to flow or after it has ceased to flow. A hot object does contain energy, but calling this energy heat as it resides in the hot object can lead to widespread confusion. *See* ENERGY; INTERNAL ENERGY.

Heat flow is a result of a potential difference between the source and sink which is called temperature. Work is energy in transit as a result of a difference in any other potential such as height. Work may be thought of as that which can be completely used for lifting weights. Heat differs from work, the other type of energy in transit, in that its conversion to work is limited by the fundamental second law of thermodynamics, or Carnot efficiency. This natural law is that the fraction of the heat Q convertible to work is determined by the relation $dW = Q(dI/T)$ for processes where the source and sink are differentially different in temperature, or by the relation $dW = dQ(T_1 - T_2)/T_1$ where the source (at T_1) and the sink (at T_2) differ by a finite temperature interval. *See* WORK.

For the above relations to be valid, temperature must be expressed on a thermodynamic temperature scale. Conversely, any temperature scale for which the above relations are valid, irrespective of the substance or material under investigation, is a thermodynamic temperature scale. The perfect gas law defines a scale in which the temperature is proportional to the thermodynamic temperature. In order to make the two scales be identical, the triple point of water (temperature and pressure at which ice, water, and vapor are in equilibrium) is defined to be at 273.16 kelvins on both the ideal-gas and the thermodynamic scales. *See* TEMPERATURE; THERMODYNAMIC PRINCIPLES.

[HAROLD C. WEBER; WILLIAM A. STEELE]

Bibliography: C. O. Bennett and J. E. Myers, *Momentum, Heat, and Mass Transfer*, 2d ed., 1974; P. W. Bridgman, *The Nature of Thermodynamics*, 1941.

Heat capacity

The quantity of heat required to raise a unit mass of homogeneous material one unit in temperature along a specified path, provided that during the process no phase or chemical changes occur is known as the heat capacity of the material in question. The unit mass may be 1 g, 1 lb, or 1 gram-molecular weight (1 mole). Moreover, the path is so restricted that the only work effects are those necessarily done on the surroundings to cause the change to conform to the specified path. The path, except as noted later, is at either constant pressure or constant volume. This definition conforms to an average heat capacity for the chosen unit change in temperature.

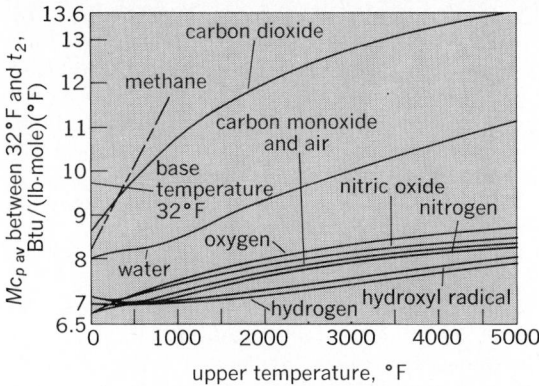

Fig. 1. Variation of average molar heat capacity with temperature for several common gases. (*Based on data of H. C. Hottel, MIT*)

Instantaneous heat capacity at a particular temperature is defined as the rate of heat addition relative to the temperature change at the temperature in question; that is, on a plot of heat addition Q as a function of temperature T, instantaneous heat capacity is given by the slope of the curve at the temperature in question. Units of heat capacity are energy units per unit mass of material per unit change in temperature.

In accordance with the first law of thermodynamics, heat capacity at constant pressure C_p is equal to the rate of change of enthalpy with temperature at constant pressure $(\partial H/\partial T)_p$. Heat capacity at constant volume C_v is the rate of change of internal energy with temperature at constant volume $(\partial U/\partial T)_v$. Moreover, for any material, the first law yields the relation in Eq. (1). *See* ENTHALPY; INTERNAL ENERGY.

$$C_p - C_v = \left[P + \left(\frac{\partial U}{\partial V} \right)_T \right] \left(\frac{\partial U}{\partial T} \right)_P \quad (1)$$

Gases. For one mole of a perfect gas, the preceding relation becomes $MC_p - MC_v = R$, where M is the molecular weight of the gas under discussion and R is the perfect gas law constant. For gases the ratio C_p/C_v is usually designated by the symbol K.

For monatomic gases at moderate pressures, MC_v is about 3, K is about 1.67, and the heat capacity changes but little with temperature. For diatomic gases, MC_v is approximately 5 at 20°C and moderate pressures. Change of heat capacity with temperature is usually small. The value of K is between 1.40 and 1.42. For triatomic gases at moderate pressures, MC_v varies from 6 to 7 and changes rapidly with temperature. The value of K varies but is always smaller than that for the less complex molecules at the same conditions of pressure and temperature.

For gases with more than three atoms per molecule, no generalizations are reliable. However, as molecular complexity increases, heat capacity increases, the influence of temperature on heat capacity increases, and K decreases. Figure 1 shows average MC_p for several common gases. Up to pressures of a few atmospheres, the effect of pressure on heat capacity of gases is small and is usually neglected.

Solids. For solids, the atomic heat capacity (heat capacity when the unit mass under discussion is 1 at.wt) may be closely approximated by an equation of type (2), where $n = 1$ for elements of simple crys-

$$C_v = J \left(\frac{T}{\theta} \right)^n \quad (2)$$

talline form, but has a smaller value for those of more complex structures; θ is characteristic of each element; J is a function that is the same for all substances; T is absolute temperature. Figure 2 compares measured with calculated values.

For all solid elements at room temperature, C_v is about 6.4 calories per gram atom per degree Celsius. This approximation may be used when no experimental data are available, but errors may be considerable, particularly for elements with atomic weights less than 39. Kopp's law states that for solids the molal heat capacity of a compound at room temperature and pressure approximately equals the sum of heat capacities of the elements in the compound. Errors are considerable but may be reduced by judicious choice of atomic heat capacities for the lighter elements. Recommended values for some of these are given in the table of constants for Kopp's law. Use of Kopp's law is justified only when no experimental data are available.

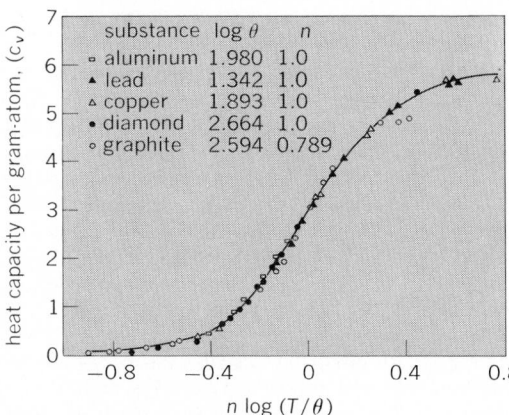

Fig. 2. Atomic heat capacity as a function of temperature (K). (*Based on data of G. N. Lewis and G. E. Gibson, J. Amer. Chem. Soc., 39:2554–2581, 1917*)

Heat capacities of some elements

substance	log θ	n
aluminum	1.980	1.0
lead	1.342	1.0
copper	1.893	1.0
diamond	2.664	1.0
graphite	2.594	0.789

Heat capacities of some elements

Element	Heat capacity
All heavy elements	6.4
Boron	2.7
Carbon	1.8
Fluorine	5.0
Hydrogen	2.3
Oxygen	4.0
Phosphorus and sulfur	5.4
Silicon	3.5

Figures 3 and 4 give instantaneous heat capacities for some industrially important solids.

Liquids. For liquids and solutions no useful generally applicable approximations are available. For aqueous solutions of inorganic salts the approximate heat capacity of the solution may be estimat-

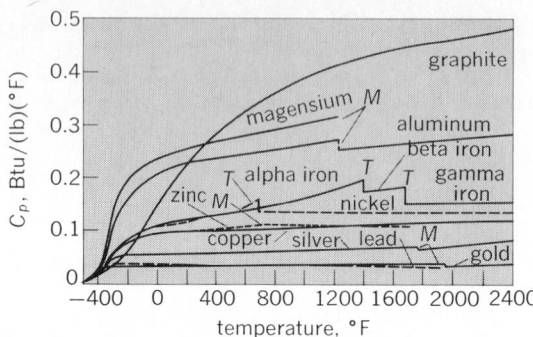

Fig. 3. Change in heat capacity of some industrially important solids with temperature. *M*= melting point; *T*= transition temperature. (*Based on data of K. K. Kelly, U.S. Bur. Mines Bull., no. 371, 1934*)

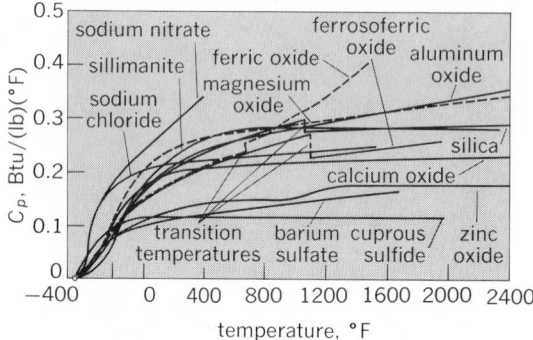

Fig. 4. Change in heat capacity of compounds with temperature. (*Based on data of K. K. Kelly, U.S. Bur. Mines Bull., no. 371, 1934*)

ed by assuming the dissolved salt to have negligible heat capacity. Thus, in a 20% by weight solution of any salt in water 0.8 would be the estimated heat capacity.

Effect of pressure on heat capacities at any temperature may be calculated by the relations in Eqs. (3a) and (3b).

$$\left(\frac{\partial MC_p}{\partial P}\right)_T = -T\left(\frac{\partial^2 V}{\partial T^2}\right)_P \qquad (3a)$$

$$\left(\frac{\partial MC_v}{\partial V}\right)_T = T\left(\frac{\partial^2 P}{\partial T^2}\right)_V \qquad (3b)$$

Constant temperature. Not so familiar as C_p and C_v are the heat necessary to cause unit change in pressure in a unit mass of material at constant temperature and the heat required to cause unit change in volume at constant temperature. These are designated $\partial Q_p/\partial P$ and $\partial Q_T/\partial V$. Similarly, $\partial Q_V/\partial P$ and $\partial Q_p/\partial V$ may be called heat capacities. *See* SPECIFIC HEAT; THERMODYNAMIC PRINCIPLES.

[HAROLD C. WEBER]

Bibliography: W. H. Brown, *Thermodynamics and Heat Engines*, 1964; K. E. Bett et al., *Thermodynamics for Chemical Engineers*, 1975.

Heat radiation

The energy radiated by solids, liquids, and gases as a result of their temperature. Such radiant energy is in the form of electromagnetic waves and covers the entire electromagnetic spectrum, extending from the radio-wave portion of the spectrum through the infrared, visible, ultraviolet, x-ray, and γ-ray portions. From most hot bodies on Earth this radiant energy lies largely in the infrared region. *See* ELECTROMAGNETIC RADIATION; INFRARED RADIATION.

Radiation is one of the three basic methods of heat transfer, the other two methods being conduction and convection. *See* CONDUCTION (HEAT); CONVECTION (HEAT); HEAT TRANSFER.

A hot plate at 400 K (261°F) may show no visible glow; but a hand which is held over it senses the warming rays emitted by the plate. A temperature of more than 1000 K is required to produce a perceptible amount of visible light. At this temperature a hot plate glows red and the sensation of warmth increases considerably, demonstrating that the higher the temperature of the hot plate the greater the amount of radiated energy. Part of this energy is visible radiation, and the amount of this visible radiation increases with increasing temperature. A steel furnace at 1800 K shows a strong yellow glow. If a tungsten wire (used as the filament in incandescent lamps) is raised by resistance heating to a temperature of 2800 K, it emits a bright white light. As the temperature of a substance increases, additional colors of the visible portion of the spectrum appear, the sequence being first red, then yellow, green, blue, and finally violet. The violet radiation is of shorter wavelength than the red radiation, and it is also of higher quantum energy.

In order to produce strong violet radiation, a temperature of almost 3000 K is required. Ultraviolet radiation necessitates even higher temperatures, and there is no solid on Earth which can withstand such temperatures without melting. The Sun emits considerable ultraviolet radiation, as evidenced by the sunburn it produces. The spectral distribution of the Sun's radiation has been measured, and the temperature of the Sun's surface has been determined from Wien's displacement law and corresponds to about 6000 K (Wien's law is discussed later). Such temperatures have been produced on Earth in gases ionized by electrical discharges. The mercury-vapor lamp used on highways, the fluorescent lamp used in offices, and the xenon compact-arc lamp used in searchlights are good examples of such gas discharges. They emit large amounts of ultraviolet radiation. Temperatures up to 20,000 K, however, are still much too low to produce x-rays or γ-radiation. Approaches to the utilization of nuclear energy have made use of the fusion of deuterons in magnetically constricted arcs at extremely high currents. By this means, temperatures above 1×10^6 K have been obtained for small fractions of a second. These devices require enormous amounts of energy to produce such high temperatures. A gas maintained at such temperatures emits x-rays and γ-rays. *See* NUCLEAR FUSION; ULTRAVIOLET RADIATION.

Theory. The emission of radiation is explained in terms of excited atoms and nuclei. For example, electrons in an atom can be ejected from their normal orbits around the atom into those farther from the nucleus. When this happens the atom is said to be in an excited state. This occurs when energy, supplied from outside a substance, is converted into thermal motion and finally into excitation. A short time after excitation, the electrons return to their normal orbits and give off their excess energy ΔE in the form of radiation of a particular frequency ν. This wavelength may be determined by the relation $\Delta E = h\nu$, where h is Planck's constant. *See* ATOMIC STRUCTURE AND SPECTRA; PLANCK'S CONSTANT.

In a gas, the thermal motion consists of substantially unhindered movement of the individual particles with different velocities. In a solid, on the other hand, the thermal motion is an oscillating movement of the particles, with varying displacements, about their fixed positions. The extent of the thermal motion depends upon the temperature. The hotter the substance, the greater the thermal motion and the higher the intensity and energy of the radiation. An energy distribution of the radiation intensity results, for example, from the distribution of velocities of the particles in a gas or from the distribution of displacements of the particles about their positions in a solid.

Further, the maximum available energy (excitation energy) depends upon temperature, and this explains why the energy of emitted radiation shifts to shorter wavelength (that is, higher energy) as the temperature is increased. For instance, a temperature of 1000 K produces just enough excitation energy for the dark red glow which contains the longest wavelengths within the visible portion of the spectrum. As explained before, higher temperatures or greater excitation energies are necessary to excite measurable quantities of the shorter wavelength regions. It is obvious that with decreasing temperatures, less excitation energy is available, and the amount of heat radiation decreases until finally at the absolute zero of temperature (0 K) substances radiate no energy because all atomic motion has ceased. However, for a definition of so-called zero point energy *see* QUANTUM MECHANICS.

The radiated energy per second is commonly expressed in terms of joules per second, or watts. Other units often used are ergs per second or calories per second. These are related to each other as follows: 1 watt = 1 joule/sec = 10^7 erg/sec = 0.239 cal/sec. For instance, the Sun radiates onto 1 cm² of the Earth's surface 2 cal/min or $\frac{1}{30}$ cal/sec or about $\frac{1}{7}$ watt. The total energy radiated from 1 cm² of a tungsten wire in an incandescent lamp at 2800 K is 112 watts. The same wire at room temperature emits only 0.0015 watt.

Energy distribution curves. In order to evaluate the usefulness of a heat radiator, energy distribution curves are used. These are graphs of relative or absolute radiated energy versus the wavelength of radiation (expressed in micrometers, 10^{-6} m, or in angstroms, 10^{-10} m) or the frequency (velocity of light/wavelength) expressed in hertz (cycles per second).

Such graphs show how the energy radiated from a substance at a certain temperature is distributed over the various portions of the spectrum. The usefulness of these graphs lies in the fact that they provide information, for example, on the effectiveness of a radiator as a light source or as a heating element. Furthermore, the area under the energy distribution curve is equivalent to the total

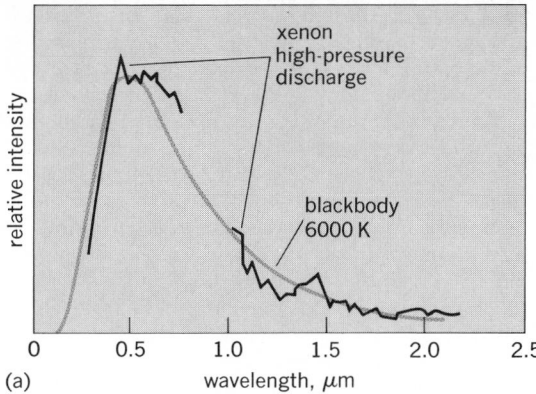

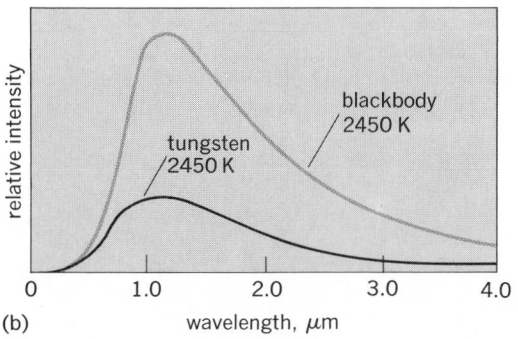

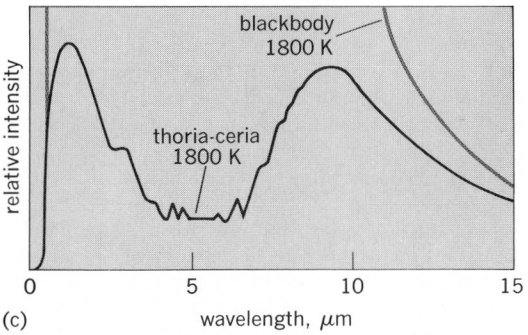

Fig. 1.   Energy distribution curves for (a) xenon high-pressure discharge (*from W. Meyer, ed., Technischwissenschaftliche Abhandlungen aus dem Osramkonzern, vol. 6, Springer, 1953*); (b) tungsten (*from Amer. Inst. Mining Met. Eng., Pyrometry: The Papers and Discussions of a Symposium on Pyrometry, 1920*); (c) a typical ceramic (*from R. W. Pohl, Einführung in die Optik, Springer, 1948*).

radiated energy. The energy distribution curves for tungsten metal, thoria plus 1% ceria (a ceramic), and the xenon high-pressure electrical gas discharge are illustrated in Fig. 1.

The energy distribution of various substances differs because of their internal properties and their surface condition. As a common rule, which holds well above 3 μm, substances with good electrical conductivity, especially metals, are poor emitters of radiation and are good reflectors (for example, silver or aluminum). Insulators radiate strongly in the infrared region of the spectrum and have gaps of low radiation intensity near the visible portion of the spectrum, as shown in Fig. 1c. These gaps are due to the electronic band structure of insulators. Roughening the surface of all radiators increases the emitted energy. This is true because tiny holes in the surface act as cavity radiators, radiating almost blackbody energy. (Cavity radiators and blackbody radiation are discussed later.)

Energy distribution curves are obtained by passing white (heterochromatic) light through a monochromator (quartz prism, grating, and the like) and measuring the spectral intensity of radiation with a phototube or a thermopile. The measured intensities at the various wavelengths are then plotted either as percent of the maximum intensity (relative energy) or as absolute intensity (absolute energy) versus the wavelength. *See* RADIOMETRY.

A radiator used in heating rooms should produce much infrared radiation (heat) and no light, whereas much visible light and little heat is desired from a light source. Unfortunately, an energy distribution curve gives a true picture of the radiator for one particular temperature only. If more information is needed, a set of such curves would have to be provided for the temperature range of interest.

Blackbody radiation. Because of the tedious experimental work involved in determining such curves, a different and more fruitful approach is generally taken. Two quantities characterize a heat radiator completely: the total emissivity and the spectral emissivity, which are designated by ϵ and ϵ_λ, respectively, where the subscript λ designates wavelength. Both emissivities, in conjunction with the radiation properties of a blackbody, describe fully the behavior of a real heat radiator. The radiation properties of a blackbody are completely stated by Planck's radiation law.

Planck's law and the concept of blackbody radiation are of utmost importance for the understanding of heat radiation. The blackbody signifies in the domain of heat radiation what any other standard, such as the standard meter, signifies in its own domain. A blackbody is defined as a body which emits the maximum amount of heat radiation. Although there exists no perfect blackbody radiator in nature, it is possible to construct one on the principle of cavity radiation. *See* BLACKBODY.

A cavity radiator is usually understood to be a heated enclosure with a small opening which allows some radiation to escape or enter. The escaping radiation from such a cavity has the same characteristics as blackbody radiation.

As can be seen from Fig. 2, radiation energy which enters the cavity is almost completely absorbed because of the multiple reflections it encounters. This follows because at each reflecting point some of the energy is absorbed by the walls. The absorptivity of the cavity hole is essentially unity, independent of the wall material. As a consequence of Kirchhoff's law, the emissivity of the cavity is also unity, and this fulfills the definition of a blackbody radiator. In practice, the cavity is approximated by a small hole or even a wedge cut into a surface.

Kirchhoff's law. This law correlates mathematically the heat radiation properties of materials at thermal equilibrium. It is often called the second law of thermodynamics for radiating systems.

Kirchhoff's law can be expressed as follows: The ratio of the emissivity of a heat radiator to the absorptivity of the same radiator is a function of frequency and temperature alone. This function is the same for all bodies, and it is equal to the emissivity of a blackbody. When ϵ is the emissivity of a real radiator, α its absorptivity, and $E = 1$ the emissivity of a blackbody, Kirchhoff's law takes the form of Eq. (1).

$$\epsilon/\alpha = E = 1 \qquad (1)$$

A substance, when brought without contact into an evacuated enclosure the walls of which are at a constant but higher temperature than the body, will assume the wall temperature after some time. However, it will not exceed it. Under these conditions, the exchange of energy can take place only by radiation. As the test body receives radiation from the walls it will absorb some of it, transforming it into motion of its elementary particles, and thereby raising its own temperature. Thermal equilibrium is obtained when the temperature of the walls and the test body is the same; in this case the test body must emit as much energy as it receives. If it absorbs all the impinging radiation, it is a blackbody. If it absorbs only a fraction of the impinging radiation, the other part must be reflected in order to maintain the equilibrium. These statements require that the absorptivity be equal to the emissivity. This is the form in which Kirchhoff's law is often stated. For opaque bodies, absorptivity plus reflectivity must be equal to unity, and therefore the emissivity and the absorptivity respectively must be unity minus the reflectivity. A consequence of Kirchhoff's law is the postulate that a blackbody has an emissivity which is greater than that of any other body. *See* KIRCHHOFF'S LAWS OF ELECTRIC CIRCUITS.

Planck's radiation law. This celebrated law represents mathematically the energy distribution of the heat radiation from 1 cm^2 of surface area of a blackbody at any temperature. It is the only heat radiation law which is accurate throughout the entire spectrum. The basis of Planck's radiation law was experimental data obtained from measurements on cavity radiators.

Planck's radiation law has great importance. Formulated by Max Planck early in the 20th century, it laid the foundation for the advance of modern physics and the advent of quantum theory. In determining the heat radiation of hot bodies Planck's radiation law is a basic tool in research and development, both in science and industry. The radiation law can be used to predict light output of incandescent lamps, the cooling time of molten steel, heat dissipation of nuclear reactors, the energy radiated from the Sun, the temperature of

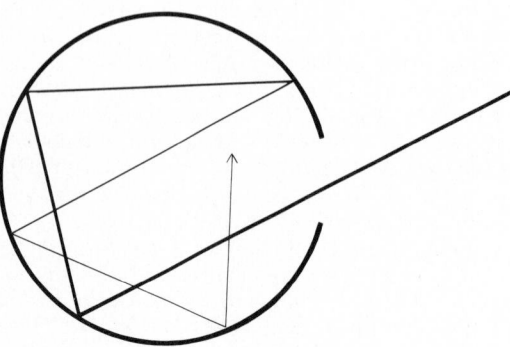

Fig. 2. Cavity radiator.

the stars, and many other important applications.

Although Planck's law can be derived on theoretical grounds alone, it was deduced from experiment. Prior attempts to calculate the heat radiation of a blackbody had described the radiation as consisting of electromagnetic waves whose energy content could vary continuously. Those attempts did not match the experimental results. Planck replaced the concept of continuous energy with the idea that the energy existed in bundles; that is, the energy was quantized. *See* QUANTUM MECHANICS.

This concept was a drastic innovation at that time. However, upon calculating the radiation, Planck found that the expression in Eq. (2) described the experimental results completely. This is the mathematical expression of Planck's radiation law, where R_λ is the total energy radiated from the body measured in watts per square centimeter per unit wavelength, at the wavelength λ. The wavelength in this formula is measured in micrometers. The quantity T is the temperature in degrees Kelvin, and e is the base of the natural logarithms. Figure 3 presents graphs of Planck's law for various temperatures and shows substances which attain these temperatures. It should be noted that these substances will not radiate as predicted by Planck's law since they are not blackbodies themselves.

$$R_\lambda = 37{,}418/\lambda^5(e^{14{,}388/\lambda T} - 1) \qquad (2)$$

As can be seen from Fig. 3, the radiation increases at every point of the energy spectrum as the temperature is increased. At all temperatures, the energy radiated at the extremes of the energy spectrum approaches zero and has a maximum at some place in between. The total area under any of the curves, when plotted as in Fig. 1, measures the total energy radiated by the body at the temperature represented by the curve.

Three important aspects of Planck's radiation law can be examined. First, the behavior of the law at the extremes of the energy spectrum leads to a discussion of Wien's radiation law, the Rayleigh-Jeans radiation law, and the so-called ultraviolet catastrophe. Second, the shift of the wavelength at which the maximum energy is radiated can be studied as the temperature is changed. This leads to Wien's displacement law. Finally, the total amount of energy radiated at any temperature can be investigated. This leads to the Stefan-Boltzmann law. The four laws mentioned were well known prior to the formulation of Planck's law. It is the fact that the Planck law so neatly sums up the four earlier laws and introduces the implication of energy quantization which made it of such importance in the development of modern physics during the 20th century.

Rayleigh-Jeans law. The heat radiation from a blackbody at long wavelengths is adequately described by the Rayleigh-Jeans radiation law. For larger values of λT, Planck's law simplifies to the Rayleigh-Jeans law, as shown in Eq. (3). This law

$$R_\lambda = 2.6007\, T/\lambda^4 \qquad (3)$$

states that the energy radiated at any temperature increases without limit as the wavelength decreases. As can be seen from Figs. 1 and 3, this law can be accurate only for wavelengths much larger than that at which the maximum occurs. For wavelengths shorter than this maximum, the energy radiated from a blackbody actually decreases again. If a blackbody acted as predicted by the Rayleigh-Jeans law, then the energy radiated at very short wavelengths, in the ultraviolet region, would become extremely large and the total energy radiated would be infinite. This is known as the ultraviolet catastrophe, and would be valid at any temperature, no matter how low.

Wien's radiation law. This law is valid at short wavelengths and is obtained from Planck's law by taking λT as very small. Planck's formula then becomes Eq. (4). This law is accurate in the visible region of the spectrum below 3000 K.

$$R_\lambda = \frac{37{,}418}{\lambda^5} e^{-14{,}388/\lambda T} \qquad (4)$$

Wien's displacement law. This law is obtained from Planck's law by the process of differentiation. It describes the shift with temperature of the wavelength at which the maximum amount of radiation occurs by Eq. (5). Thus, the product of the

$$\lambda_{\max} T = 2898 \qquad \text{(micrometer-degrees)} \qquad (5)$$

temperature of a blackbody and the wavelength at which the maximum amount of radiation occurs is a constant. Wien's law has wider significance than this, however. Dividing Planck's law by T^5 results in Eq. (6). On the right-hand side of Eq. (6) the

$$R_\lambda/T^5 = 37{,}418/(\lambda T)^5(e^{14{,}388/\lambda T} - 1) \qquad (6)$$

wavelength and temperature always appear multiplying each other. This means that only one curve is needed to express Planck's law, for all tempera-

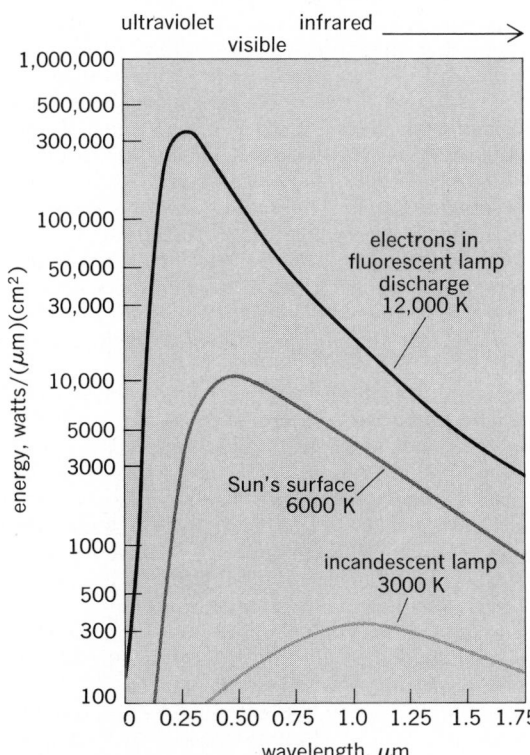

Fig. 3. Graphs of Planck's law for various temperatures.

tures, if a graph is used in which the radiation energy divided by the fifth power of the temperature is plotted versus the product of wavelength and temperature. This is illustrated in Fig. 4. It is helpful here to measure temperature in thousands of degrees. For comparison, the Rayleigh-Jeans law is also illustrated.

Wien's displacement law is helpful in determining the temperature of hot bodies. If λ_{max} can be measured, the temperature is immediately obtained from the displacement law. This is how an astronomer measures the temperature of a star. As an example, the radiation from the Sun's surface has a maximum in the green region of the energy spectrum in the vicinity of $\lambda = 0.5$ μm (Fig. 3). From Wien's law the surface temperature of the Sun must then be about 6000 K.

Stefan-Boltzmann law. This law states that the total energy radiated from a hot body increases with the fourth power of the temperature of the body. This law can be derived from Planck's law by the process of integration and is expressed mathematically as Eq. (7), where R_T is the total

$$R_T = 5.670 \times 10^{-12}T^4 \qquad (7)$$

amount of energy radiated from a blackbody in watts per square centimeter. When R_T is multiplied by the total emissivity, the total energy radiated from a real heat radiator is obtained.

The rapid increase in heat radiation with temperature is quite evident from the Stefan-Boltzmann law. If the absolute temperature is doubled, say from 273 to 546 K (32 to 524°F), then the energy radiated increases sixteenfold. Thus, the attainment of very high temperatures requires large amounts of energy to overcome the loss of energy by heat radiation. Temperatures greater than 3×10^7 K are encountered in a hydrogen bomb explosion. Such temperatures are 100,000 times higher than room temperature. Therefore, the energy radiated by 1 cm² of a substance at this high temperature will be 1×10^{20} times as much as that radiated at room temperature by the same substance. This energy, if radiated by a blackbody, would boil 2×10^7 tons of ice water in 1 sec.

Temperature determination. The apparent temperature of a real heat radiator can be deter-

mined by comparison with a blackbody whose temperature is known. This is done in any one of three customary ways, based upon the radiation laws described.

1. The radiation temperature of a surface is the temperature of the blackbody which radiates the same total energy per unit area in 1 sec as does the surface. This temperature is based upon the Stefan-Boltzmann law.

2. The brightness temperature of a surface is the temperature of the blackbody which has the same brightness at a certain wavelength as has the surface. The wavelength is normally taken as 0.655 μm, and the brightness temperature thus measured can be used in conjunction with the spectral emissivity $\epsilon_{0.655}$ and Wien's radiation law to calculate the true temperature of the body.

3. The color temperature of a surface is the temperature of a blackbody whose radiation has the same or approximately the same energy distribution as the surface. This is illustrated in Fig. 1*a*, where it is seen that the xenon high-pressure discharge has a color temperature approximating 6000 K. For a graybody, that is, a body whose emissivity does not vary throughout the spectrum, the color temperature and true temperature are the same. *See* INCANDESCENCE.

[HEINZ G. SELL; PETER J. WALSH]

Bibliography: M. Born, *Atomic Physics*, 8th ed., 1969; S. Fluegge (ed.), *Handbuch der Physik*, vol. 26, 1958; A. L. King, *Thermophysics*, 1962; R. Siegel and J. R. Howell, *Thermal Radiation Heat Transfer*, 1971; M. R. Wehr et al., *Physics of the Atom*, 3d ed., 1978.

Heat transfer

Heat, a form of kinetic energy, is transferred in three ways: conduction, convection, and radiation. Heat can be transferred only if a temperature difference exists, and then only in the direction of decreasing temperature. Beyond this, the mechanisms and laws governing each of these ways are quite different. This article gives introductory information on the three types of heat transfer (also called thermal transfer) and on important industrial devices called heat exchangers.

Conduction. Heat conduction involves the transfer of heat from one molecule to an adjacent one as an inelastic impact in the case of fluids, as oscillations in solid nonconductors of electricity, and as motions of electrons in conducting solids such as metals. Heat flows by conduction from the soldering iron to the work, through the brick wall of a furnace, through the wall of a house, or through the wall of a cooking utensil. Conduction is the only mechanism for the transfer of heat through an opaque solid. Some heat may be transferred through transparent solids, such as glass, quartz, and certain plastics, by radiation. In fluids, the conduction is supplemented by convection, and if the fluid is transparent, by radiation.

The conductivities of materials vary widely, being greatest for metals, less for nonmetals, still less for liquids, and least for gases. Any material which has a low conductivity may be considered to be an insulator. Solids which have a large conductivity may be used as insulators if they are distributed in the form of granules or powder, as fibers, or as a foam. This increases the length of

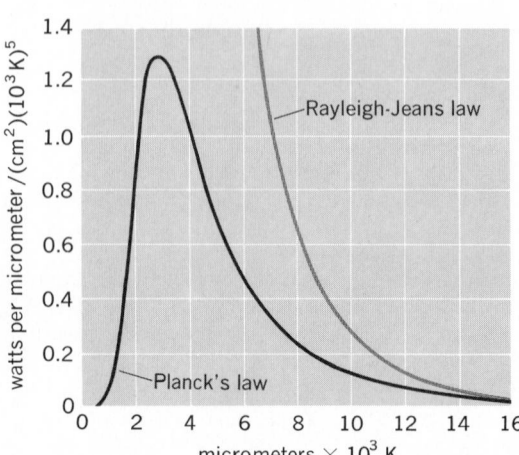

Fig. 4. Planck's law expressed as Wien's displacement law. Rayleigh-Jeans law is shown for comparison.

path for heat flow and at the same time reduces the effective cross-sectional area, both of which decrease the heat flow. Mineral wool, glass fiber, diatomaceous earth, glass foam, Styrofoam, corkboard, Celotex, and magnesia are all examples of such materials. *See* CONDUCTION (HEAT).

Convection. Heat convection involves the transfer of heat by the mixing of molecules of a fluid with the body of the fluid after they have either gained or lost heat by intimate contact with a hot or cold surface. The transfer of heat at the hot or cold surface is by conduction. For this reason, heat transfer by convection cannot occur without conduction. The motion of the fluid to bring about mixing may be entirely due to differences in density resulting from temperature differences, as in natural convection, or it may be brought about by mechanical means, as in forced convection.

Most of the heat supplied to a room from a steam or hot-water radiator is transferred by convection. In fact, the heat from the fire in the furnace heating the hot water or steam is transferred to the boiler wall by convection, and the hot water or steam transfers heat from the boiler to the radiator by convection. Iced tea is cooled and soup heated by convection. *See* CONVECTION (HEAT).

Radiation. Solid material, regardless of temperature, emits radiations in all directions. These radiations may be, to varying degrees, absorbed, reflected, or transmitted. The net energy that is transferred by radiation is equal to the difference between the radiations emitted and those absorbed.

The radiations from solids form a continuous spectrum of considerable width, increasing in intensity from a minimum at a short wavelength through a maximum and then decreasing to a minimum at a long wavelength. As the temperature of the object is increased, the entire emitted spectrum decreases in wavelength. As the temperature of an iron bar, for example, is raised to about 1000°F, the radiations become visible as a dark red glow. As the temperature is increased further, the intensity of the radiation increases and the color becomes more blue. This process is quite apparent in the filament of a light bulb. When the bulb is operated at less than normal voltage, the light appears quite red. As the voltage is increased, the filament temperature increases and the light progressively appears more blue.

Liquids and gases only partially absorb or emit these radiations, and do so in a selective fashion. Many liquids, especially organic liquids, have selective absorption bands in the infrared and ultraviolet regions. *See* ABSORPTION OF ELECTROMAGNETIC RADIATION.

Transfer of energy by radiation is unique in that no conducting substance is necessary, as with conduction and convection. It is this unique property that makes possible the transfer of large amounts of energy from the Sun to the Earth, or the transfer of heat from a radiant heater in the home. It is the ready transfer of heat by radiation from a California orange grove to outer space on a clear night that sometimes results in a frost. The presence of a shield of clouds will tend to prevent this loss of heat and often prevent the frost. By means of heat lamps and gold-plated reflectors, heat may be transferred deep into the layer of enamel on a car body, with resultant hardening of the enamel from the inside out. It is also the transfer over great distance of quantities of radiant energy that makes the atomic bomb so destructive. *See* HEAT RADIATION.

Design considerations. By utilizing a knowledge of the principles governing the three methods of heat transfer and by a proper selection and fabrication of materials, the designer atttempts to obtain the heat flow required for his purposes. This may involve the flow of large amounts of heat to some point in a process or the reduction in flow in others. It is possible to employ all three methods of heat transfer in one process. In fact, all three methods operate in processes that are commonplace. In summer, the roof on a house becomes quite hot because of radiation from the Sun, even though the wind is carrying some of the heat away by convection. Conduction carries the heat through the roof where it is distributed to the attic by convection. The prudent householder attempts to reduce the heat that enters the rooms beneath by reducing the heat that is absorbed in the roof by painting the roof white. He may apply insulation to the underside of the roof to reduce the flow of heat through the roof. Further, heated air in the attic may be vented through louvers in the roof.

Heat transferred by convection may be transferred as heat of the convecting fluid or, if a phase change is involved, as latent heat of vaporization, solidification, sublimation, or crystallization. The human body can be cooled to less than ambient temperature by evaporation of sweat from the skin. Dry ice absorbs heat by sublimating the carbon dioxide. Heat extracted from the products of combustion in the boiler flows through the gas film and the metal tube wall and converts the water inside the tube to steam, all without greatly changing the temperature of the water.

Heat exchangers. In industry it is generally desired to extract heat from one fluid stream and add it to another. Devices used for this purpose have passages for each of the two streams separated by a heat-exchange surface in the form of plates or tubes and are known as heat exchangers. Needless to say, the automobile radiator, the hot-water heater, the steam or hot-water radiator in a house, the steam boiler, the condenser and evaporator on either the household refrigerator or air conditioner, and even the ordinary cooking utensils in everyday use are all heat exchangers. In power plants, oil refineries, and chemical plants, two commonly used heat exchangers are the tube-and-shell and the double-pipe exchangers. The first consists of a bundle of tubes inside a cylindrical shell. One fluid flows inside the tubes and the other between the tubes and the shell. The double-pipe type consists of one tube inside another, one fluid flowing inside the inner tube and the other flowing in the annular space between tubes. In both cases, the tube walls serve as the heat-exchange surface. Heat exchangers consisting of spaced flat plates with the hot and cold fluids flowing between alternate plates are also in use. Each of these exchangers essentially depends upon convection heat flow through a film on each side of the heat-exchange surface and conduction through the surface. Countless special modifications, often also utilizing radiation for heat transfer, are in use in industry.

In these exchangers, the fluid streams may flow parallel concurrently or in mixed flow. In most cases, the temperatures of the various streams remain essentially constant at a given point, and the process is said to be a steady-state process. As the streams move through the exchangers, unless there is a phase change, the fluids are continuously changing in temperature, and the temperature gradient from one stream to the other may be continuously varying. To determine the amount of surface needed for a given process, the designer must evaluate the effective temperature gradient for the particular condition and exchanger.

With extremely high temperatures, or with gas streams carrying suspended solids, the use of conventional heat exchangers becomes impractical. Under these conditions, the transfer of heat from one stream to another becomes more economical by the alternate heating and cooling of refractory solids or by checkerwork as in the blast-furnace hot stove, in the glass-furnace regenerator, or in the Royster stove. At lower temperatures, metal packing is frequently employed, as in the Ljungstrom preheaters or in regenerators for liquid-air production. In petroleum refining and in the metallurgical industry, exchangers are being employed in which one or more of the streams are fluidized beds of solids, the large area of the solids tending to produce very high rates of heat exchange. In some of these devices and also in nuclear power reactors, large quantities of heat are being generated in the exchangers. Here one of the principal problems involves the rapid removal of this heat before the temperature rises to the point where the equipment is damaged or destroyed.

Often the heating or cooling of a body is desired. In this case, the body representing the second stream does not remain at constant temperature, the heat being transferred representing a change in the heat content of the body. Such a process is known as an unsteady-state process. The heating or cooling of food and canned products in utensils, refrigerators, and sterilizers; the heating of steel billets in metallurgical furnaces; the burning of brick in a kiln; and the calcination of gypsum are examples of this type of process. *See* HEAT.

[RALPH H. LUEBBERS]

Bibliography: S. Banerjee and J. T. Rogers (eds.), *Heat Transfer Nineteen Seventy-Eight*: *Proceedings*, 8 vols., 1979; G. M. Dusinberre,. *Numerical Analysis of Heat Flow*, 1949; E. R. G. Eckert, *Introduction to Heat and Mass Transfer*, 1963; J. P. Holman, *Heat Transfer*, 3d ed., 1978; M. Jakob, *Heat Transfer*, vol. 1, 1949; D. Q. Kern, *Process Heat Transfer*, 1950.

Helicity

A fundamental quantized variable used in quantum mechanics to specify the relative orientations of spin and linear momentum of massless particles. It is a requirement of fundamental Dirac quantum mechanics that such particles have their spins aligned either parallel or antiparallel to their linear momentum. Particles having parallel alignment are arbitrarily assigned helicity $+1$; those have antiparallel alignment, -1. *See* MOMENTUM; SPIN.

In a classic experiment on K electron capture by ^{152}Eu, M. Goldhaber, L. Grodzins, and A. Sun-

yar first showed that the neutrino emitted in the weak nuclear interaction had negative helicity—that its spin was aligned antiparallel to its momentum. An equivalent description of this situation is that these neutrinos are left-handed. Symmetry requires that antineutrinos be right-handed and have positive helicity. *See* ELECTRON CAPTURE; NEUTRINO; SYMMETRY LAWS.

This discovery also requires that the Gamow-Teller beta decay proceed predominantly by means of an axial vector interaction. *See* SELECTION RULES; WEAK NUCLEAR INTERACTIONS.

The questions of whether the neutrino is truly massless and of helicity and lepton number conservation are intimately related. One of the main points at issue is whether the lepton number of the neutrino is uniquely related to its helicity; usually it is taken as the negative of the helicity. Other theories have been suggested wherein the two are completely dissociated and wherein the usual additive lepton number conservation law is replaced by a multiplicative one. *See* ELEMENTARY PARTICLE; QUANTUM MECHANICS.

[D. ALLAN BROMLEY]

Bibliography: E. Commins, *Weak Interactions*, 1973; M. Goldhaber, L. Grodzins, and A. Sunyar, *Phys. Rev.*, 109:1015–1017, 1958.

Helimagnetism

A property possessed by some metals, alloys, and salts of transition elements in which the atomic magnetic moments, at sufficiently low temperatures, are arranged in a spiral or helix. It may be seen from the illustration that simple antiferromagnets and ferromagnets can be considered as nonconical helimagnets with helical angles ϕ of 180 and 0°, respectively. In the same way, nonconical helimagnets may be considered as conical helimagnets with cone angle θ of 0°. Some typical helimagnets are listed in the table. The magnetic structures have been detected by neutron diffraction.

Helimagnetism arises when the exchange coupling parameters J_{ij} between the spins $\mathbf{S}_i$ and $\mathbf{S}_j$ of the magnetic atoms lie within a certain range of values. The exchange energy is given by Eq. (1).

$$E_{ij} = -2J_{ij}\mathbf{S}_i \cdot \mathbf{S}_j \tag{1}$$

which is summed over all pairs of atoms (i,j). *See* FERROMAGNETISM.

Consider an axial structure with strong ferromagnetic coupling (positive J_{ij}) between atoms which are in the same plane, and weaker ferromagnetic or antiferromagnetic coupling between atoms which are in neighbor planes. Let this latter coupling between nearest neighbor planes be J_1, be-

Some representative helimagnets

Substance	Magnetic structure	Temperature, K
MnO_2	Nonconical helix	$0 < T < 84$
$MnAu_2$	Nonconical helix	$0 < T < 363$
Dy	Nonconical helix	$85 < T < 179$
	Ferromagnet	$0 < T < 85$
$MnCr_2O_4$	Simple ferrimagnet	$18 < T < 43$
	Complex conical helix	$0 < T < 18$
Er	Conical helix	$0 < T < 20$
	Complex oscillation	$20 < T < 53$
	Sinusoidal antiferromagnet	$53 < T < 85$

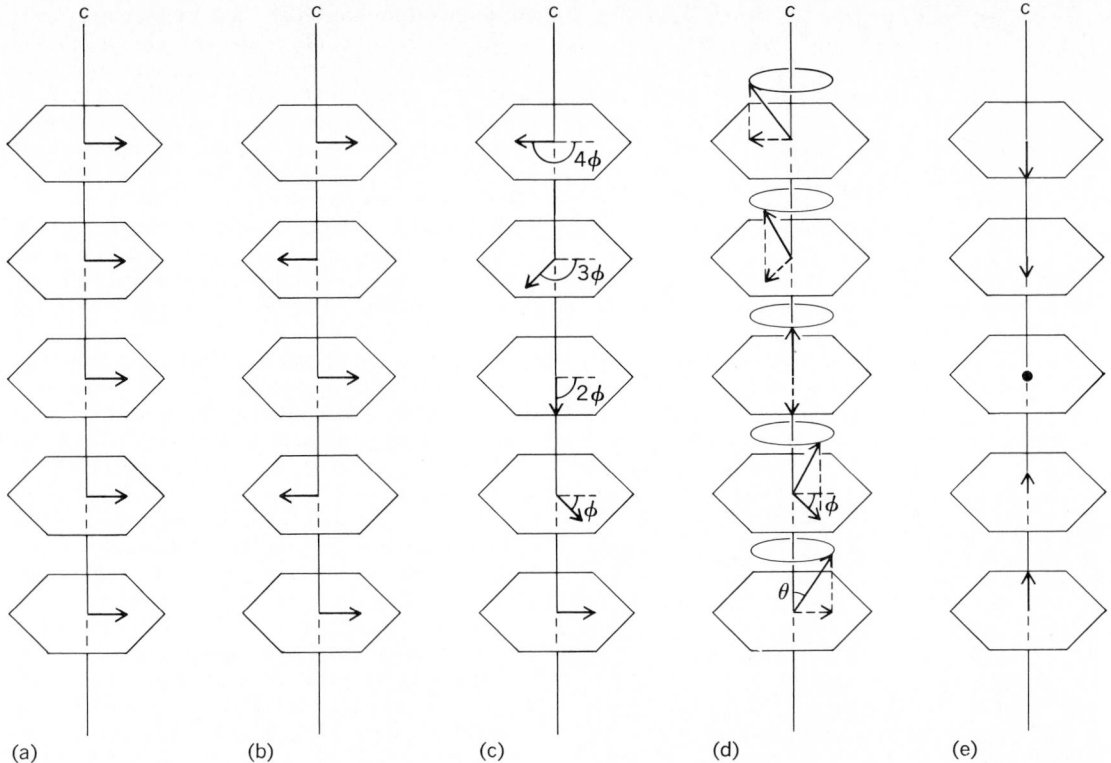

Some magnetically ordered arrays. Shown are directions taken by magnetic moments of consecutive planes along the *c* axes of hexagonal structures. (*a*) Simple ferromagnet; (*b*) simple antiferromagnet; (*c*) nonconical helimagnet of helical angle ϕ; (*d*) ferromagnetic conical helix of cone angle θ and helical angle ϕ; (*e*) sinusoidal antiferromagnet. In *a, b,* and *c* the moments all lie in the planes. In *d* there is a net ferromagnetic component along the *c* axis and in *e* there is an oscillating moment along the *c* axis.

tween next neighbor planes be J_2, and so on. Let the angle between the ferromagnetically coupled spins of the nth plane and those of the first plane be ϕ_n. The exchange energy is then given by expression (2), where J is the intraplanar exchange.

$$E \propto -\Sigma_n[J + 2J_1 \cos (\phi_{n+1} - \phi_n) + 2J_2 \cos (\phi_{n+2} - \phi_n) + \cdot \cdot \cdot] \quad (2)$$

If one assumes a helical array, then Eq. (3)

$$\phi_n = nka + \text{constant} \quad (3)$$

holds, where ka, which is the ϕ of array (*c*) of the figure, is the phase angle between neighbor planes. With a equal to the spacing between planes, k becomes the wave number describing the pitch of the helix.

On substituting Eq. (3) into Eq. (2), one obtains Eq. (4) where J_k is the Fourier transform of the

$$E \propto -[J + 2J_1 \cos ka + 2J_2 \cos 2ka + \cdot \cdot \cdot] = -J_k \quad (4)$$

exchange parameter J_{ij}, when the latter is considered a function of distance between planes only.

It is seen that the exchange energy is minimized by that value of k which maximizes J_k. In this way the pitch of the helix is determined.

Ferromagnets have maximum J_k at $k = 0$. This cannot be achieved if the interplanar J_n are mostly negative, that is, antiferromagnetic. In the latter case what happens physically is that the planes try to order antiferromagnetically with respect to one

another. It is impossible for them all to do this, and a helical compromise results.

Conical helimagnets have maximum J_k at $k \neq 0$, and also anisotropic coupling which favors spin alignment along the crystal axis.

Helimagnetism is found in most of the rare-earth metals and their alloys and in a few salts of the iron group. Some helimagnets have quite complicated patterns, such as manganese sulphate which, below 10 K, orders into two antiferromagnetically coupled conical helices.

[FREDERIC KEFFER]

Bibliography: B. D. Cullity, *Introduction to Magnetic Materials*, 1972; S. Fluegge (ed.), *Encyclopedia of Physics*, vol. 18, pt. 2, 1966; A. Pekalski et al., *Magnetism in Metals and Metallic Compounds*, 1975; D. Wagner, *Introduction to the Theory of Magnetism*, 1972; K. Yosida, *Progress in Low Temperature Physics*, vol. 4, 1964.

Helmholtz coils

A pair of flat circular coils with equal numbers of turns and equal diameters, arranged with a common axis, and connected in series to have a common current (Fig. 1). The purpose of the arrangement is to obtain a magnetic field H_{AB} that is more nearly uniform than that of a single coil (Fig. 2) without the use of a long solenoid.

The optimum arrangement is that in which the distance between the two coils is equal to the radius of one of the coils (Fig. 1). For this arrangement, the variation of the field strength near the center of the apparatus is a minimum, and the field is

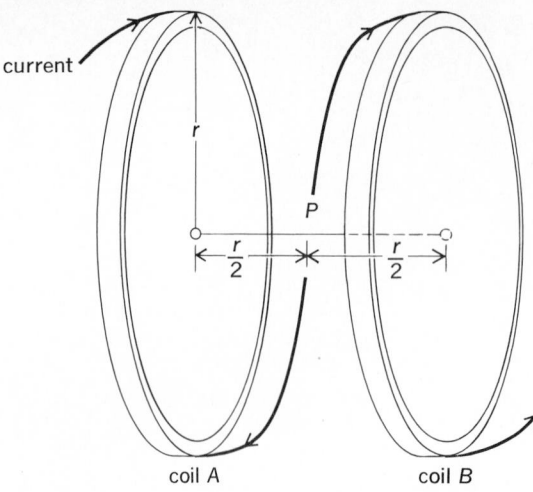

Fig. 1. Helmholtz coils, arranged with common axis and connected in series. (From L. B. Loeb, *Fundamentals of Electricity and Magnetism*, 3d ed., copyright © 1947 by John Wiley and Sons, Inc.; reprinted by permission)

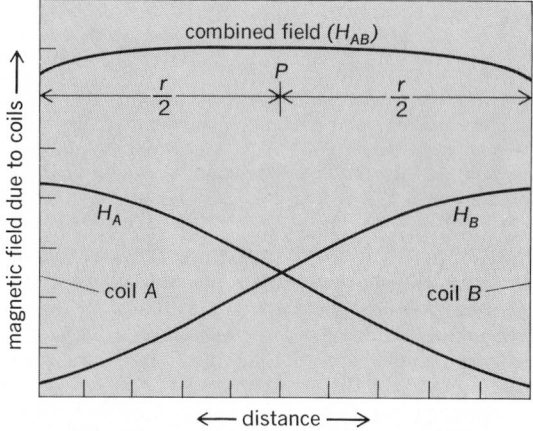

Fig. 2. Separate and combined fields of Helmholtz coils. Note the region of constant resultant field at point *P*. (From L. B. Loeb, *Fundamentals of Electricity and Magnetism*, 3d ed., copyright © 1947 by John Wiley and Sons, Inc.; reprinted by permission)

nearly uniform near the center. The field is the sum of the fields produced there by the individual coils. [KENNETH V. MANNING]

Bibliography: F. J. Bueche, *Introduction to Physics for Scientists and Engineers*, 2d ed., 1975; C. H. Durney and C. C. Johnson, *Introduction to Modern Electromagnetics*, 1969; A. F. Kip, *Fundamentals of Electricity and Magnetism*, 2d ed., 1968; J. D. Kraus and K. R. Carver, *Electromagnetics*, 2d ed., 1973; A. Shadowitz, *The Electromagnetic Field*, 1975.

High-pressure physics

High-pressure physics is concerned with the effects of high pressure on the properties of matter. Since most properties of matter are modified by pressure, the field of high-pressure physics encompasses virtually all branches of physics. The justification for classifying high-pressure physics as a separate field is that rather specialized and often ingenious techniques are need-

ed, both to produce high pressures and to make measurements of changes of physical properties of a material at high pressure. This field is therefore analogous to the fields of low- and high-temperature physics. Indeed, high-pressure experiments have been performed at temperatures approaching absolute zero and at temperatures as high as 5000°C. *See* LOW-TEMPERATURE PHYSICS.

Pressure is defined as force per unit area; commonly used units of pressure include pounds per square inch (psi), kilograms per square centimeter (kg/cm²), atmospheres (atm), bars and pascals (N/m²). The bar, equal to 1,000,000 dynes/cm² or about 14.5 psi (that is, slightly less than 1 atm), is the basic unit that has been used by most researchers, although the approved unit of pressure in the International System (SI) of Units is now the pascal (Pa), which is equal to 10^{-5} bar. *See* PRESSURE.

The "high" of high pressure physics connotes experimental difficulty. At liquid-helium temperatures, pressures of several hundred bars are considered high. In general, however, the high-pressure range may be arbitrarily regarded as extending from about 1 kbar upward to the present experimental limit. Prolonged static high pressures of as much as 500 kbars (50 GPa) at room temperature have been achieved in certain types of apparatus. Sample sizes are small—about 50 micrograms (μg)—but the observational techniques employed—x-ray diffraction and electrical resistance measurements—are ideally suited to small samples. If simultaneous prolonged high temperatures in the range from 1000 to 2000°C are required, the limit of static high pressures attainable is reduced to about 200 kbars (20 GPa). Additional experimental requirements, such as the need for nonmagnetic pressure vessels in magnetic resonance studies, further limit maximum attainable pressures.

Transient pressures as high as about 10,000,000 bars (1000 GPa) have been attained in shock waves produced by high explosives or by projectile impact. This article is primarily concerned with static high-pressure experimentation, but it should be noted that the results of shock-wave experiments generally complement the results of prolonged static high-pressure experiments.

High-pressure effects. The major effects of high pressure on matter include diminution of volume, phase transitions, changes in electrical, optical, magnetic, and chemical properties, increases in viscosity of liquids, and increases in the strength of most solids. The magnitudes of these pressure effects may be illustrated by some examples, both general and specific. If a gas, initially at low pressure and a temperature below critical, is compressed, the first effect of pressure is to reduce the free space separating the atoms or molecules. At pressures on the order of several hundred bars, the volume will be about one-thousandth of the initial volume and a change of phase from gas to liquid will occur. Further compression of the liquid, by pressures up to about 50 kbars (5 GPa), results in an additional volume decrement of only 20−50%. For liquids the greatest effect of pressure may be the increase in viscosity, which can be as high as a millionfold for a pressure increase of 10 kbars (1 GPa).

For liquids that freeze to a solid phase of greater

density than the liquid, the effect of pressure is to raise the freezing temperature. In the case of substances, such as water, having a solid phase of lower density than the liquid, the freezing point is lowered by pressure. The freezing point of water is lowered to $-20°C$ by a pressure of about 2 kbars (0.2 GPa). Above pressures of 2 kbars, however, water crystallizes in new solid forms which are denser than the liquid, and the freezing point is raised by further increases of pressure. At 45 kbars (4.5 GPa) the freezing point of water is $190°C$.

In general solids are less compressible than liquids, and the compressibility of both solids and liquids decreases with increasing pressure. However, solids and liquids show wide individual variations in compressibility. At a pressure of 200 kbars, solid sodium is reduced to about half its initial volume, but diamond is reduced in volume by only a few percent at the same pressure. The electrical conductivity of metals is generally increased by pressure, but there are numerous exceptions to this rule. Changes in the electrical conductivity of metals are typically on the order of 10% for a 10-kbar pressure change. Phase transitions in metals generally result in discontinuities in the pressure-versus-resistance curves. These resistance discontinuities in various metals, for example, bismuth, iron, and lead, may be used as calibration points for high-pressure apparatus.

At high pressure many solids exhibit polymorphic phase changes, that is, a rearrangement of the atoms or molecules in the solid. There are no universally applicable rules governing the number of phase changes or the kind of phase change to be expected at high pressure, but there is a thermodynamic requirement that the phase that is stable at high pressure must have a smaller volume than the phase that is stable at low pressure. Camphor has 11 probable solid phases, and seven solid phases of water are known to exist in the pressure range extending to 45 kbars. On the other hand, many elements and compounds are known to exist in only one solid phase up to pressures of over 1000 kbars (100 GPa). *See* THERMODYNAMIC PRINCIPLES.

Frequently, dramatic changes in physical properties result from phase changes. Ferromagnetic iron transforms to a paramagnetic form at pressures somewhat above 100 kbars. In the same pressure range, the semiconducting element germanium transforms into a metallic phase that has an electrical conductivity greater than a million times that of the semiconductor. Similar semiconductor-to-metal transitions at high pressure have been observed in the cases of silicon, indium arsenide, gallium antimonide, indium phosphide, aluminum antimonide, and gallium arsenide. *See* SEMICONDUCTOR.

Many phases that form at high pressure transform back to low-pressure phases as the pressure is released. However, some high-pressure phases may be retained in a metastable condition at low pressures, and some low-pressure phases can persist metastably at high pressure. Diamond, the high-pressure form of carbon, is thermodynamically unstable at room temperature and pressures below about 12 kbars (1.2 GPa). Nonetheless diamond persists indefinitely as a metastable phase at low temperatures; it transforms to the stable form, graphite, only when heated to temperature in excess of 1000°C at low pressure. Graphite can be exposed to pressures in excess of 200 kbars without transforming to diamond. Simultaneous high temperatures are required before diamond can form. Through use of a molten metal catalyst to lower the activation energy for the transformation, diamond is commercially synthesized at pressures of about 40 kbars (4 GPa) and temperatures of about 1500°C. Without the catalyst, pressures in excess of 100 kbars (10 GPa) and temperatures above 2000°C are required for diamond synthesis. *See* METASTABLE STATE.

An important area of research is the study of electronic transitions at high pressure. Mössbauer spectral measurements of iron compounds have demonstrated the reversible reduction of Fe^{3+} (ferric) ions to Fe^{2+} (ferrous) ions at high pressures. Similar reductions of Mn^{3+} to Mn^{2+} and of Cu^{2+} to Cu^+ have been observed by optical spectroscopy at high pressures. This research is relevant to the problem of estimating the possible chemical composition of the interior of the Earth. Some researchers now hypothesize that only a negligible amount of ferric iron (Fe^{3+}) is present at depths below 1000 km.

Experimental methods and problems. The simplest type of high-pressure apparatus is a thick-walled, hollow cylinder closed by sliding pistons to form a cavity. The sample is contained in the cylinder and force is applied to the pistons. Both the force and the change in volume can be measured by ordinary instruments. However, large corrections must be made for the pressure-induced distortion of the sample cavity and for the friction between pistons and cylinder. For liquids, the stresses will be hydrostatic, and the corrections for sample-cavity distortion may be calculated from elasticity theory. However, additional corrections, for example, for friction and for compression of the seals used to prevent leakage, must be determined empirically. Electrical leads may be introduced into the pressure cavity; the liquid may then serve as a pressure-transmitting fluid during measurements of the electrical properties of various materials. The electrical leads can also be connected to a coil of wire made from a metal or alloy having a well-known pressure variation of resistivity. The pressure in the liquid may then be determined independently by measuring the resistance change in the wire. Measurement of the compressibility of solid samples can best be made if the sample is immersed in a fluid of known compressibility.

At pressures above about 25 kbars (2.5 GPa) at room temperature, however, most liquids have solidified, and solid pressure-transmitting media must be used. At these higher pressures, an important part of the experimental problem is to ensure that the nonhydrostatic, or shearing, components of stress are as small as possible in comparison with the hydrostatic component. The pressure-transmitting medium must be a soft solid, chosen empirically as one which does not become prohibitively stiff at high pressure.

The strength of materials presently available limits the simple piston-and-cylinder apparatus to a maximum pressure of about 50 kbars. Apparatus designed for higher pressures takes advantage of the fact that the strength of many materials is greatly enhanced by pressure. Multistage apparatus has been built in which the sample is contained in a pressure vessel, which is in turn contained in a larger pressure vessel. Pressures up to about 90 kbars (9 GPa) have been reached in two-stage apparatus, at the cost of greatly increased complexity of the apparatus and marked reduction in the accuracy with which measurements can be made. Through ingenious use of geometrical factors, the benefits of multistaging can be realized without having to construct concentric pressure vessels. However, the problem of measurement is enormously complicated, because only a small and imprecisely known fraction of the total force applied to the system actually serves to compress the sample. In some types of apparatus, over 98% of the total applied force is used to provide support for the highly stressed regions of the apparatus.

Temperature may be varied in high-pressure experiments by externally heating or cooling the entire pressure vessel. The temperature limit for externally heated pressure vessels is about 1000°C, a limit determined by the reduction in strength of the vessel material at high temperatures. Higher temperatures may be obtained in internally heated pressure chambers in which heat is generated by passage of current through a resistance element; alternatively, the heat of a chemical reaction may be used. Temperature is measured by thermocouples within the pressure cavity, and suitable corrections must be made for the effect of pressure on thermocouple output. The effect of pressure on thermocouple output varies greatly according to the combination of metals or alloys chosen for the thermocouple junction. If two different thermocouple junctions are subjected to the same temperature and pressure, their outputs may be used to determine both temperature and pressure.

X-ray diffraction has become an important tool for studying volume changes and phase changes at high pressure. A portion of the pressure chamber must be constructed of a material of low atomic number, such as boron, beryllium, or diamond, to provide a "window" that is relatively transparent to x-rays. If an x-ray beam of suitable wavelength is passed through the sample, a portion of this beam is diffracted by the sample. The x-ray diffraction pattern, which may be recorded on film, is characteristic of the crystal structure and the atomic spacings of the material under investigation. If a calibrant (that is, a material that behaves in a known manner under pressure) is mixed with the specimen, the diffraction patterns of calibrant and specimen may be simultaneously recorded at high pressure. The diffraction pattern of the calibrant then serves as an internal standard of pressure. *See* X-RAY DIFFRACTION.

Other ingenious techniques have been used for high-pressure studies of the Mössbauer effect, nuclear magnetic resonance, diffusion in solids, and many other effects of current interest in solid-state physics research. The primary problem of high-pressure physics continues to be accuracy of measurement.

No article on high-pressure research would be complete without mention of the late Percy W. Bridgman, who won the 1946 Nobel Prize in physics. For 40 years, beginning in 1908, Bridgman's work completely dominated the field of high-pressure physics. Virtually all high-pressure research

depends heavily on apparatus designed by Bridgman and on his pioneering measurements of changes in physical properties at high pressure.

[R. K. LINDE; P. S. DE CARLI]

Bibliography: G. W. Boehm, The alchemy of ultrahigh pressure, *Fortune,* October 1964; P. W. Bridgman, *The Physics of High Pressure,* reprint, 1949; H. G. Drickamer and C. W. Frank, *Electronic Transitions and the High Pressure Chemistry and Physics of Solids,* 1973; E. C. Lloyd (ed.), *Accurate Characterization of the High-Pressure Environment,* NBS Spec. Publ. 326, 1971.

Holes in solids

Vacant electron energy states near the top of an energy band in a solid are called holes. A full band cannot carry electric current; a band nearly full with only a few unoccupied states near its maximum energy can carry current, but the current behaves as though the charge carriers are positively charged. The situation can be understood in terms of the definition of the effective mass: if the energy band is specified by a function $E(k)$, where k is the magnitude of the wave vector **k**, the effective mass for a spherical band is given by the equation here, where $\hbar$ is Planck's constant divided by 2π. Near a maximum of the band, the

$$m^* = \hbar^2 \left(\frac{\partial^2 E}{\partial k^2} \right)^{-1}$$

vided by 2π. Near a maximum of the band, the

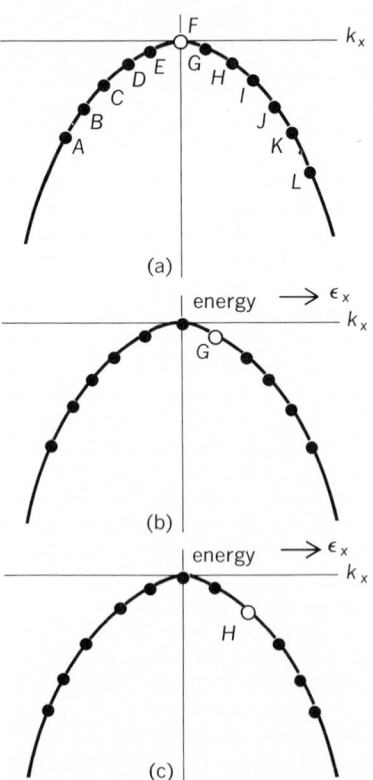

Process of hole conduction. (*a*) At time $t = 0$, energy states *A* through *L* are filled except *F*. (*b*) An electric field ϵ_x is applied in the +*x* direction. The force on the electrons is in the $-k_x$ direction, and all electrons make transitions in the $-k_x$ direction, moving the hole to *G*. (*c*) After a further interval, the electrons move farther along, and the hole is now at *H*. (*After C. Kittel, Introduction to Solid State Physics, 3d ed., copyright © 1966 by John Wiley and Sons, Inc.; used with permission*)

second derivative of the energy is negative, so the effective mass is negative. States for which the effective mass is negative are defined as hole states. Carriers in such states behave under the influence of an external electromagnetic field as though they carry positive charge. *See* BAND THEORY OF SOLIDS.

The process of conduction in such a system may be visualized in the following way. An electron moves against an applied electric field by jumping into a vacant state. This transfers the position of the vacant state, or propagates the hole, in the direction of the field, as shown in the diagram. Whether conduction occurs by electrons or holes is determined experimentally from the sign of the Hall emf. If a current is carried in the presence of a magnetic field perpendicular to the current, an emf is developed perpendicular to the current and to the field. The sign of this emf depends on the charge on the carriers. *See* HALL EFFECT.

Hole conduction is important in many semiconductors, notably germanium and silicon. The occurrence of hole conduction in semiconductors can be favored by alloying with a material of lower valence than the "host." Semiconductors in which the conduction is primarily due to holes are called *p* type. Hole conduction is also observed in some metals, including iron and chromium. In other metals, including aluminum and bismuth, both holes and electrons may be present in equilibrium. *See* SEMICONDUCTOR.

[JOSEPH CALLAWAY]

Bibliography: N. W. Ashcroft and N. D. Mermin, *Solid State Physics*, 1976; C. Kittel, *Introduction to Solid State Physics*, 5th ed., 1976; W. Shockley, *Electrons and Holes in Semiconductors*, 1950, reprint 1976.

Holography

A technique for recording, and later reconstructing, the amplitude and phase distributions of a coherent wave disturbance. Invented by Dennis Gabor in 1948, the process was originally envisioned as a possible method for improving the resolution of electron microscopes. While this original application has not proved feasible, the technique is widely used as a method of optical image formation, and in addition has been successfully used with acoustical and radio waves. *See* ACOUSTICAL HOLOGRAPHY.

Fundamentals of the technique. The technique is accomplished by recording the pattern of interference between the unknown "object" wave of interest and a known "reference" wave (Fig. 1). In general, the object wave is generated by illuminating the (possibly three-dimensional) subject of concern with a highly coherent beam of light, such as supplied by a laser source. The waves reflected from the object strike a light-sensitive recording medium, such as photographic film or plate. Simultaneously a portion of the light is allowed to bypass the object, and is sent directly to the recording plane, typically by means of a mirror placed next to the object. Thus incident on the recording medium is the sum of the light from the object and a mutually coherent "reference" wave. *See* LASER.

While all light-sensitive recording media respond only to light intensity (that is, power), nonetheless in the pattern of interference between reference and object waves there is preserved a

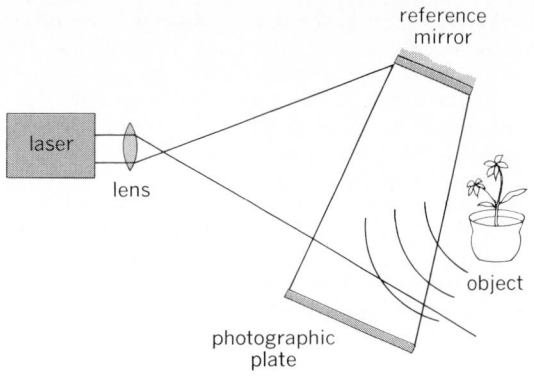

Fig. 1. Recording a hologram.

complete record of both the amplitude and the phase distributions of the object wave. Amplitude information is preserved as a modulation of the depth of the interference fringes, while phase information is preserved as variations of the position of the fringes.

The photographic recording obtained is known as a hologram (meaning a "total recording"); this record generally bears no resemblance to the original object, but rather is a collection of many fine fringes which appear in rather irregular patterns (Fig. 2). Nonetheless, when this photographic transparency is illuminated by coherent light, one of the transmitted wave components is an exact duplication of the original object wave (Fig. 3). This wave component therefore appears to originate from the object (although the object has long since been removed) and accordingly generates a virtual image of it, which appears to an observer to exist in three-dimensional space behind the transparency. The image is truly three-dimensional in the sense that the observer must refocus his eyes to examine foreground and background, and indeed can "look behind" objects in the foreground simply by moving his head laterally.

Also generated are several other wave components, some of which are extraneous, but one of which focuses of its own accord to form a real image in space between the observer and the transparency. This image is generally of less utility than the virtual image because its parallax relations are opposite to those of the original object.

Applications. The holographic technique has a number of unique properties which make it of great value as a scientific tool. Although the field is young, and new applications are continually emerging, certain important areas can be identified.

Microscopy. Historically, microscopy is the potential application of holography that has motivat-

Fig. 2. Typical appearance of a hologram (under magnification).

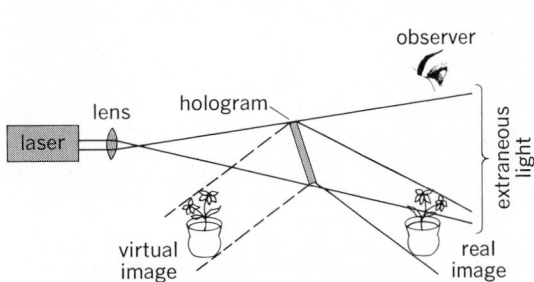

Fig. 3. Obtaining images from a hologram.

ed much of the early work, including the original work of Dennis Gabor. The use of holography for optical microscopy has been amply demonstrated, but it is generally agreed that these techniques are not serious competitors with more conventional microscopes in ordinary microscopy.

Nonetheless, there is one area in which holography offers a unique potential for optical microscopy. This area might be called "high-resolution volume imagery." In conventional microscopy, high transverse resolution is achieved only at the price of a very limited depth of focus; that is, only a limited portion of the object volume can be brought into focus at one time. It is possible, of course, to explore a large volume in sequence by continuously refocusing to examine new regions of the object volume, but such an approach is often unsatisfactory, particularly if the object is a dynamic one, continuously in motion. A solution to this problem is to record a hologram of the object by using a pulsed laser. The dynamic object is then "frozen" in time, but the recording contains all information necessary to explore the full object volume with an auxiliary optical system. Sequential observation is acceptable because the object (that is, the holographic image) is no longer dynamic. This approach has been fruitfully applied to the microscopy of three-dimensional volumes of living biological specimens and to the measurement of particle-size distributions in aerosols.

Interferometry. Holography has been demonstrated to offer the capability of several unique kinds of interferometry. This capability is a consequence of the fact that holographic images are coherent; that is, they have well-defined amplitude and phase distributions. Any use of holography to achieve the superposition of two coherent images will result in a potential method of interferometry.

The most powerful holographic interferometry techniques are based on the following property: When a photographic emulsion is multiply exposed to form several superimposed holograms, upon reconstruction the several corresponding virtual images are formed simultaneously and therefore interfere. Likewise the various real images interfere.

The most dramatic demonstrations of this type of interferometry were performed by R. E. Brooks,

L. O. Heflinger, and R. F. Wuerker, using a pulsed ruby laser. Two laser pulses were used to record two separate holograms on the same transparency. Any changes of the object between pulses resulted in well-defined fringes of the interference in the reconstructed image (Fig. 4). The technique is particularly well suited for performing interferometry through imperfect optical elements (for example, windows of poor quality), thus making possible certain kinds of interferometry that could not be achieved by any classical means.

Memories. Optical memories for storing large volumes of binary data in the form of holograms have been developed for commercial use. Such a memory consists of an array of small holograms, each capable of reconstructing a different "page" of binary data. When one of these holograms is illuminated by coherent light, it generates a real image consisting of an array of bright or dark spots, each spot representing a binary digit. This image falls on a detector array, with one detector element for each binary digit. Thus to read a single binary digit at a specific location in the memory, a beam deflector causes light to illuminate the appropriate hologram page, and the output of the proper detector element is interrogated to determine whether a bright spot of light exists at that particular location in the image.

There are two advantages of such a holographic memory over a more conventional "bit-by-bit" memory, in which the binary digits are directly recorded as transparent or opaque spots on film. First, when a bit-by-bit memory is recorded with high packing density (large numbers of bits per unit area), certain practical problems become serious. To determine the value of a single bit in the memory, an image of the recorded data must be formed on a detector element or array. This image must be formed with high magnification, but simultaneously there must be extremely precise registration between the image and the detectors to assure that the desired bit falls in its proper location on the detector array. Such registration is difficult and costly to achieve. The use of a holographic memory greatly alleviates this registration problem, for if the proper recording and readout geometries are used, once the correct hologram is illuminated, the real image is produced in perfect registration with the detector array.

Second, again at high packing densities, imperfections of the recorded memory (for example, dust specks or scratches on the film) can obliterate certain areas of a bit-by-bit memory, causing serious errors in the recovered data. However, because of the high degree of redundancy present in each page of a holographic memory, such defects on the memory have relatively little effect on the image retrieved. Thus with a holographic memory, highly accurate data can be retrieved under conditions that might make a more conventional bit-by-bit memory inoperable.

Display. There has been interest in the use of holography for purposes of display of three-dimensional images. Applications have been found in the field of advertising, and there is increased use of holography as a medium for artistic expression. A significant technical development in this area has been the perfection of a type of recording known as a multiplex hologram. Such a recording typically consists of a large number of separate holograms,

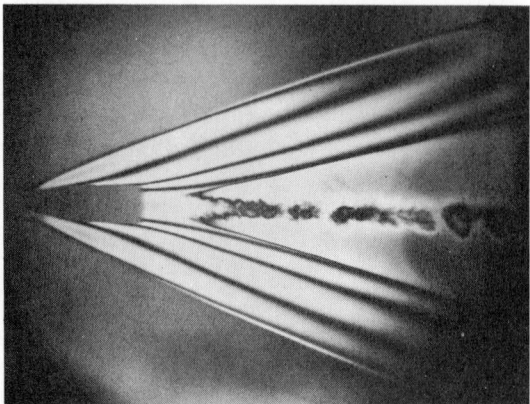

Fig. 4. Image taken by the technique of holographic interferometry, showing the compressional waves generated by a high-speed rifle bullet. (*Courtesy of R. E. Brooks, L. O. Heflinger, and R. F. Wuerker*)

all in the form of thin, contiguous, vertical strips on a single piece of film. Each of these holograms produces a virtual image of a different ordinary photograph of the subject of interest. In turn, each such photograph was originally taken from a slightly different angle. Thus when the observer examines the virtual image produced by the entire set of holograms, each eye looks through a different hologram and sees the subject from a different angle. The resulting stereo effect produces a nearly perfect illusion of three-dimensionality. Furthermore, as an observer moves the head horizontally, or as the collection of holograms is rotated, the observer's two eyes continuously see a changing pair of images. If the original set of photographs is properly chosen, the image can be made to move or dance about in nearly any desired fashion. Very dramatic three-dimensional displays of animated subjects can thus be constructed from a series of ordinary photographs.

Other applications. A variety of other applications of holography has been proposed and demonstrated, including the analysis of modes of vibration of complicated objects, measurement of strain of objects under stress, generation of very precise depth contours on three-dimensional objects, and high-resolution imagery through aberrating media. These and other applications of holography will be useful in future scientific and engineering problems. *See* INTERFERENCE OF WAVES; INTERFEROMETRY. [JOSEPH W. GOODMAN]

Bibliography: R. J. Collier, C. B. Burkhardt, and L. H. Lin, *Holography*, 1971; J. B. DeVelis and G. O. Reynolds, *Theory and Applications of Holography*, 1967; J. W. Goodman, *Introduction to Fourier Optics*, 1968; E. N. Leith and J. Upatnieks, Photography by Laser, *Sci. Amer.*, vol. 212, no. 6, 1965; H. M. Smith, *Principles of Holography*, 1969.

Huygens' principle

An assumption regarding the behavior of light waves, originally proposed by C. Huygens in the 17th century to explain the fact that light travels in straight lines and casts sharp shadows. Large-scale waves, such as sound waves or water waves, bend appreciably into the shadow. The special behavior of light may be explained by Huygens' principle, which states that "each point on a wavefront may be regarded as a source of secondary waves, and the position of the wavefront at a later time is determined by the envelope of these secondary waves at that time." Thus a wave WW originating at S is shown in illustration *a* at the instant it passes through an aperture. If a large number of circular secondary waves, originating at various points on WW, are drawn with the radius r representing the distance the wave would travel in time t, the envelope of these secondary waves is the heavily drawn circular arc $W'W'$. This represents the wave after t. If, as Huygens' principle requires, the disturbance is confined to the envelope, it will be 0 outside the limits indicated by points W'.

Careful observation shows that there is a small amount of light beyond these points, decreasing rapidly with distance into the geometrical shadow. This is called diffraction. *See* DIFFRACTION.

The Huygens-Fresnel principle, a modification of Huygens' original formulation, is capable of explaining diffraction. A. Fresnel in 1814 postulated that the amplitude of any secondary wave decreas-

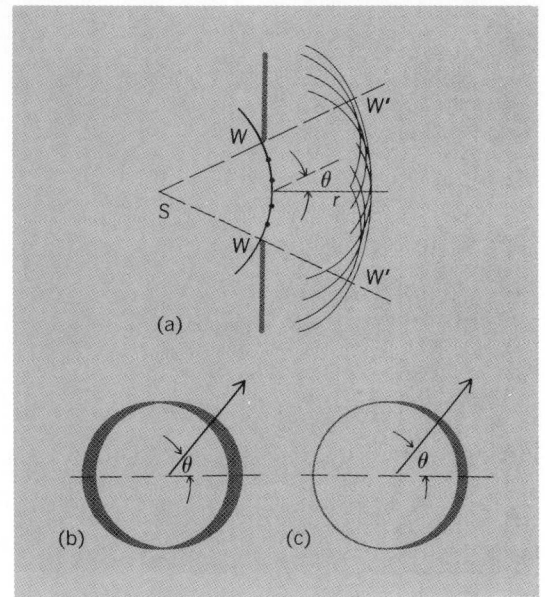

Huygens' principle. (*a*) The construction for a spherical wave. (*b*, *c*) Amplitude of the secondary wave according to Fresnel and Kirchhoff, respectively.

es in proportion to cos θ, when θ is the angle between the normal to the original wavefront and any point on the secondary wave (see illustration *b*, where the thickness of the arc indicates the amplitude). Fresnel then modified Huygens' requirement that the disturbance be confined to the envelope, by specifying that at any point the disturbance was the resultant of all displacements due to secondary waves reaching that point. In this way Fresnel was able to explain the complicated diffraction patterns that are produced by sending light through small apertures. Subsequent theoretical investigations by G. Kirchhoff showed that the correct obliquity factor should be $1 + \cos \theta$ instead of cos θ (illustration *c*). Approximations made by Fresnel had compensated for this. A discrepancy in the phase of the resultant wave of one-quarter period was also explained by Kirchhoff.

[FRANCIS A. JENKINS/WILLIAM W. WATSON]

Hydrodynamics

That branch of continuum mechanics which treats of the laws of motion of an incompressible fluid and of the interactions of the fluid with its boundaries. Partly because of the age-old interest in systems of water works and water-borne vehicles, and partly because of scientific curiosity, the subject has achieved a high state of both practical and analytical development. The present article presents a brief rational sketch of the state of this subject. For more detail on specific parts of the subject *see* FLUID FLOW; FLUID-FLOW PRINCIPLES; FLUID MECHANICS.

Flow field. Fluids have many physical properties, including, besides thermodynamic and electromagnetic properties, the dynamical ones of density, viscosity, adhesion, and cohesion. Only the dynamical properties will be considered in this article, and furthermore, because capillary effects will not be treated, the last two properties will be of interest only insofar as they affect the inception of cavitation.

A flow field can be studied from either of two points of view, the Lagrangian, in which one is concerned with the history of each fluid particle, and the Eulerian, in which interest is focused, rather, on the velocity vectors associated with each point. Although the Lagrangian method is convenient for demonstrating certain kinematical properties of a flow field, the alternative approach will be used, because it has been found, on the whole, to be simpler and more powerful. *See* EULER'S EQUATIONS OF MOTION; LAGRANGE'S EQUATIONS.

A fruitful concept in hydrodynamics is that of the ideal or inviscid fluid. In such a fluid, across any element of area, the fluid on one side exerts a normal stress or pressure on the fluid on the other side, the magnitude of which is independent of the orientation of the element. Thus there is a scalar pressure field associated with the vector velocity field. A principle problem of hydrodynamics is to determine these fields corresponding to prescribed boundary conditions.

When viscosity is taken into account, the normal stresses across an area element are not, in general, independent of orientation, and in addition, tangential stresses are present. Instead of a scalar pressure field one must consider now a tensor stress field. These stresses, or their integrals, must be determined in order to solve the important problem of obtaining the force and moment acting on a body moving through the fluid. *See* NEWTONIAN FLUID; VISCOSITY.

Hydrokinematics. For the present purpose, the molecular structure of a fluid can be ignored and it can be treated as a continuum. Thus various properties of the fluid will be considered as continuous functions of space and time. If the position of a point is represented in terms of its coordinates x,y,z with respect to a rectangular cartesian coordinate system, the velocity vector field may be expressed in the form $\mathbf{U}\,(x,y,z,t)$, with components $u(x,y,z,t)$, $v(x,y,z,t)$, $w(x,y,z,t)$, magnitude as in Eq. (1), and the mass density of the fluid (mass per unit volume) as the function $\rho(x,y,z,t)$.

$$V = \sqrt{u^2 + v^2 + w^2} \qquad (1)$$

A basic relation, which expresses the law of conservation of mass, is the equation of continuity which, in regions containing no fluid sources or sinks, may be written in the form of Eq. (2).

$$\operatorname{div}(\rho\mathbf{U}) = \frac{\partial(\rho u)}{\partial x} + \frac{\partial(\rho v)}{\partial y} + \frac{\partial(\rho w)}{\partial z} = -\frac{\partial\rho}{\partial t} \quad (2)$$

If ρ is constant, as is assumed hereafter, the equation becomes Eq. (3).

$$\operatorname{div}\mathbf{U} = 0 \qquad (3)$$

From the latter equation can be deduced the existence of two families of stream surfaces, as in Eqs. (4), defined by constant values of a and b, the

$$\psi(x,y,z) = a \qquad \chi(x,y,z) = b \qquad (4)$$

mutual intersections of which define the streamlines of the flow pattern, that is, lines tangent at a given instant to the velocity vectors.

According to this definition, the differential equations of a streamline are as shown in Eq. (5).

$$\frac{dx}{u(x,y,z,t_0)} = \frac{dy}{v(x,y,z,t_0)} = \frac{dz}{w(x,y,z,t_0)} \qquad (5)$$

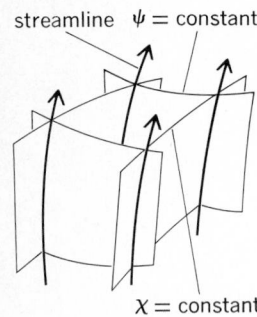

HYDRODYNAMICS

streamline $\psi = \text{constant}$

$\chi = \text{constant}$

Fig. 1. Stream channel formed by stream surfaces.

This should be distinguished from the equations of a path line, Eq. (6), which gives the path followed by a fluid particle. If the flow is steady (time-independent) the two sets of lines are coincident. In any case they are tangent at the location of a particle.

$$\frac{dx}{u(x,y,z,t)} = \frac{dy}{v(x,y,z,t)} = \frac{dz}{w(x,y,z,t)} \qquad (6)$$

If the stream functions ψ and χ have been determined for a flow problem, the velocity vector would be given by the vector cross product, as in Eq. (7).

$$\mathbf{U} = \operatorname{grad}\psi \times \operatorname{grad}\chi \qquad (7)$$

These stream functions also have the significant property that the rate of flow Q through a stream channel bounded by the four surfaces $\psi_1, \chi_1, \psi_2, \chi_2$ (Fig. 1) is as in Eq. (8).

$$Q = (\psi_2 - \psi_1)(\chi_2 - \chi_1) \qquad (8)$$

In two important special cases the stream functions reduce to a single one. For two-dimensional flow there is the Lagrange stream function $\psi(x,y)$, in terms of which the velocity components are as given by Eqs. (9).

$$u = \frac{\partial\psi}{\partial y} \qquad v = -\frac{\partial\psi}{\partial x} \qquad (9)$$

Constant values of ψ give the streamlines, and the rate of flow q between a pair of streamlines ψ_1 and ψ_2 is given by Eq. (10).

$$q = \psi_2 - \psi_1 \qquad (10)$$

The other case is that of axisymmetric flow, for which there is the Stokes stream function $\psi(r,z)$, where the z axis is coincident with the axis of symmetry and r denotes distance perpendicular to it. *See* STOKES STREAM FUNCTION.

The velocity components are now given by Eqs. (11) and the rate of flow between two stream surfaces ψ_1 and ψ_2 is given by Eq. (12).

$$u = -\frac{1}{r}\frac{\partial\psi}{\partial z} \qquad w = \frac{1}{r}\frac{\partial\psi}{\partial r} \qquad (11)$$

$$Q = 2\pi(\psi_2 - \psi_1) \qquad (12)$$

A useful concept in hydrodynamics is that of sources or sinks at various points of a fluid at which fluid is entering or leaving a region.

If $4\pi M$ (where M is called the source strength) denotes the volume rate of entry at a source (or sink, if negative), it is seen, by considering the discharge through a sphere of radius r with its center at the source, that the fluid velocity due to the source is radial and of magnitude $V = M/r^2$. If there are many sources in a region, Gauss' theorem states that the flux of fluid through the surface of this region is the product of the sum of the source strengths by 4π.

If the sources are distributed continuously through a region with a density m per unit volume, so that $md\tau$ is the strength of the sources in a volume element $d\tau$, it is necessary to generalize the equation of continuity to become Eq. (13).

$$\operatorname{div}\mathbf{U} = 4\pi m \qquad (13)$$

Analysis of the behavior of a fluid element in a

stream shows that it is rotating at an angular velocity $\boldsymbol{\omega}/2$, where $\boldsymbol{\omega}$ is defined by Eq. (14) and is called

$$\boldsymbol{\omega} = \text{curl } \mathbf{U} \qquad (14)$$

the vorticity vector. This vector is closely related to the circulation, defined as the line integral around a closed curve of the tangential component of the velocity, as in Eq. (15). By application of

$$\Gamma = \oint \mathbf{U} \cdot d\mathbf{s} \qquad (15)$$

Stokes' theorem it is readily shown that the circulation is also given by the surface integral of the vorticity over a surface bounded by the curve, as in Eq. (16), where $\mathbf{n}$ is the unit normal vector at a

$$\Gamma = \int_s \boldsymbol{\omega} \cdot \mathbf{n} \, dS \qquad (16)$$

point of the surface, positive in the sense of advance of a right-hand screw relative to the positive sense of describing the closed curve. *See* VORTEX.

Lines tangent to the vorticity vector at every point are called vortex filaments and a group of vortex lines can form a vortex tube. These have the property that a vortex tube can begin or end only at a boundary, unless it is in the form of a ring. Furthermore, the circulation is constant at all sections of a vortex tube and is called its strength. Two additional laws of vortex motion, for an inviscid fluid on which only conservative forces such as gravitational attraction are acting, are (1) vortex filaments are always composed of the same fluid particles, and (2) the strength of a vortex tube is constant with respect to time.

A straight-line vortex filament of strength Γ induces a velocity $V = \Gamma/r$ tangent to a circle of radius r with center on the line, in the plane perpendicular to it.

A flow field can be determined when its distributions of sources and vortices are prescribed. The problem is to determine the velocity $\mathbf{U}$ from Eqs. (17).

$$\text{div } \mathbf{U} = 4\pi m \qquad \text{curl } \mathbf{U} = \boldsymbol{\omega} \qquad (17)$$

This is accomplished by putting $\mathbf{U} = \mathbf{U}_1 + \mathbf{U}_2$, as defined by Eqs. (18).

$$\begin{aligned}\text{div } \mathbf{U}_1 &= 4\pi m & \text{curl } \mathbf{U}_1 &= 0 \\ \text{div } \mathbf{U}_2 &= 0 & \text{curl } \mathbf{U}_2 &= \boldsymbol{\omega}\end{aligned} \qquad (18)$$

The velocity field $\mathbf{U}_1$ is said to be irrotational, $\mathbf{U}_2$ to be solenoidal. Then Eqs. (19)–(22) hold. The

$$\mathbf{U}_1 = \text{grad } \phi \qquad (19)$$

$$\phi = -\int \frac{m(\xi,\eta,\zeta) \, d\xi \, d\eta \, d\zeta}{[(x-\xi)^2 + (y-\eta)^2 + (z-\zeta)^2]^{1/2}} \qquad (20)$$

$$\mathbf{U}_2 = \text{curl } \mathbf{A} \qquad (21)$$

$$\mathbf{A} = \frac{1}{4\pi} \int \frac{\boldsymbol{\omega}(\xi,\eta,\zeta) \, d\xi \, d\eta \, d\zeta}{[(x-\xi)^2 + (y-\eta)^2 + (z-\zeta)^2]^{1/2}} \qquad (22)$$

functions ϕ and $\mathbf{A}$ are called the scalar and vector potentials.

Hydrokinetics. A fluid element is acted upon by body forces (forces per unit mass) such as gravitational attraction, and stresses on its surface (forces per unit area). Let $\mathbf{B}$, with components B_x, B_y, B_z, denote the body force. The stresses can be represented by the symmetric tensor τ_{ij}, where the index i indicates that the stress is acting on a plane

with positive normal in the direction of increasing x, y, or z according as $i = 1, 2,$ or 3; and $j = 1, 2,$ or 3 indicates the x, y, or z component of the stress (Fig. 2). In an inviscid fluid this stress tensor reduces to the scalar pressure field.

The force $\mathbf{F}$ and moment $\mathbf{M}$ on a body in a fluid then assume the forms given in Eqs. (23) for an inviscid fluid. For a viscous fluid, Eqs. (24) hold.

$$\begin{aligned}\mathbf{F} &= -\int p\mathbf{n} \, dS + \rho \int \mathbf{B} \, d\tau \\ \mathbf{M} &= -\int p\mathbf{r} \times \mathbf{n} \, dS + \rho \int \mathbf{r} \times \mathbf{B} \, d\tau\end{aligned} \qquad (23)$$

$$\begin{aligned}X &= \sum_{i=1}^{3} \int \tau_{1i} n_i \, dS + \rho \int B_x \, d\tau \\ Y &= \sum_i \int \tau_{2i} n_i \, dS + \rho \int B_y \, d\tau \\ Z &= \sum_i \int \tau_{3i} n_i \, dS + \rho \int B_z \, d\tau \\ L &= \sum_i \int (y\tau_{3i} - z\tau_{2i}) n_i \, dS + \rho \int (yB_z - zB_y) \, d\tau \\ M &= \sum_i \int (z\tau_{1i} - x\tau_{3i}) n_i \, dS + \rho \int (zB_x - xB_z) \, d\tau \\ N &= \sum_i \int (x\tau_{2i} - y\tau_{1i}) n_i \, dS + \rho \int (xB_y - yB_x) \, d\tau\end{aligned} \qquad (24)$$

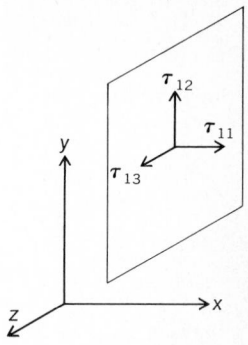

Fig. 2. Normal and tangential stresses.

Application of Newton's laws of motion to an element of an inviscid fluid leads to the Euler equations, Eqs. (25).

$$\begin{aligned}\frac{\partial u}{\partial t} + u\frac{\partial u}{\partial x} + v\frac{\partial u}{\partial y} + w\frac{\partial u}{\partial z} &= B_x - \frac{1}{\rho}\frac{\partial p}{\partial x} \\ \frac{\partial v}{\partial t} + u\frac{\partial v}{\partial x} + v\frac{\partial v}{\partial y} + w\frac{\partial v}{\partial z} &= B_y - \frac{1}{\rho}\frac{\partial p}{\partial y} \\ \frac{\partial w}{\partial t} + u\frac{\partial w}{\partial x} + v\frac{\partial w}{\partial y} + w\frac{\partial w}{\partial z} &= B_z - \frac{1}{\rho}\frac{\partial p}{\partial z}\end{aligned} \qquad (25)$$

Equations (25) may also be written in the vector form as Eq. (26). The form of Eq. (26) is useful

$$\frac{\partial \mathbf{U}}{\partial t} + \boldsymbol{\omega} \times \mathbf{U} + \tfrac{1}{2}\,\text{grad}\,(V^2) = \mathbf{B} - \frac{1}{\rho}\,\text{grad}\,p \qquad (26)$$

when it is desired to transform the equations of motion to other coordinate systems.

It will be supposed that the body force is a conservative one so that it can be expressed in the form $\mathbf{B} = \text{grad }\Omega$ where Ω is a scalar potential function. For example, if z is taken positive upwards at a point on the surface of the Earth, the scalar potential for gravitational attraction is $\Omega = -gz$, where g is the acceleration of gravity.

Probably the most frequently applied result of hydrodynamics is the Bernoulli equation which can be derived as a first integral of the Euler equations. This assumes different forms for various assumed conditions:

1. If the flow is steady, then for points along a streamline or along a vortex line, Eq. (27) holds.

$$p + \tfrac{1}{2}\rho V^2 + \rho\Omega = \text{constant} \qquad (27)$$

2. In a region in which the flow is irrotational the equation is Eq. (28).

$$p + \tfrac{1}{2}\rho V^2 + \rho\Omega + \frac{\partial \phi}{\partial t} = \text{constant} \qquad (28)$$

3. If the flow is steady and the streamlines and vortex lines are parallel, the equation is Eq. (29).

$$p + \tfrac{1}{2}\rho V^2 + \rho\Omega = \text{constant} \qquad (29)$$

4. If the coordinate axes are in motion, the origin having the velocity components U, V, W, and the coordinate system the angular velocity Λ, then Eq. (30) holds, in which $\Lambda \cdot \mathbf{r} \times \mathbf{U}$ denotes the triple

$$p + \tfrac{1}{2}\rho\left[(u-U)^2 + (v-V)^2 + (w-W)^2\right]$$

$$+ \rho\Omega + \Lambda \cdot \mathbf{r} \times \mathbf{U} + \frac{\partial\phi}{\partial t} = \text{constant} \qquad (30)$$

scalar product of the indicated vectors, and $\mathbf{U}$ and its components u, v, w are velocities of the fluid. For elaboration *see* BERNOULLI'S THEOREM.

If a body is immersed in a steady stream of velocity and pressure V_∞ and p_∞ at infinity, and if V and p denote the velocity and pressure at a point on the body, the Bernoulli equation may be written in the dimensionless form, Eq. (31).

$$\frac{p - p_\infty}{\tfrac{1}{2}\rho V_\infty^{\,2}} = 1 - \left(\frac{V}{V_\infty}\right)^2 \qquad (31)$$

At the point of maximum velocity on the body the pressure has a minimum value, denoted by p_m. Because ratio $(p_m - p_\infty)/(1/2\rho V^2)$ is a constant for the body, the value of p_m can be reduced either by reducing the ambient pressure p_∞ (as can be done in a variable-pressure water tunnel) or by increasing V_∞. If by either means p_m is reduced to the vapor pressure of a liquid p_v, and the liquid begins to vaporize, the liquid is said to undergo cavitation, and Eq. (32) holds. This is called the vapor-pres-

$$\sigma_v = \frac{p_\infty - p_v}{\tfrac{1}{2}\rho V_\infty^{\,2}} \qquad (32)$$

sure cavitation number. The phenomenon of cavitation is of great technical importance in the design of ship propellers, turbines, and other hydraulic structures because of the erosion caused by the collapsing cavitation bubbles. The inception pressure may be greater than the vapor pressure if a considerable amount of entrained air is present; on the other hand, a specimen of liquid may be able to withstand tensions of thousands of atmospheres if special care has been taken to remove gas nuclei from it.

When viscosity is taken into account, the equations of motion become the Navier-Stokes equations, Eqs. (33). *See* NAVIER-STOKES EQUATIONS.

$$\frac{\partial u}{\partial t} + u\frac{\partial u}{\partial x} + v\frac{\partial u}{\partial y} + w\frac{\partial u}{\partial z}$$

$$= B_x - \frac{1}{\rho}\frac{\partial p}{\partial x} + \nu\left(\frac{\partial^2 u}{\partial x^2} + \frac{\partial^2 u}{\partial y^2} + \frac{\partial^2 u}{\partial z^2}\right)$$

$$\frac{\partial v}{\partial t} + u\frac{\partial v}{\partial x} + v\frac{\partial v}{\partial y} + w\frac{\partial v}{\partial z}$$

$$= B_y - \frac{1}{\rho}\frac{\partial p}{\partial y} + \nu\left(\frac{\partial^2 v}{\partial x^2} + \frac{\partial^2 v}{\partial y^2} + \frac{\partial^2 v}{\partial z^2}\right) \qquad (33)$$

$$\frac{\partial w}{\partial t} + u\frac{\partial w}{\partial x} + v\frac{\partial w}{\partial y} + w\frac{\partial w}{\partial z}$$

$$= B_z - \frac{1}{\rho}\frac{\partial p}{\partial z} + \nu\left(\frac{\partial^2 w}{\partial x^2} + \frac{\partial^2 w}{\partial y^2} + \frac{\partial^2 w}{\partial z^2}\right)$$

Equations (33) may also be written in the vector form as Eq. (34) in which ν is the kinematic viscosity.

$$\frac{\partial \mathbf{U}}{\partial t} + \boldsymbol{\omega} \times \mathbf{U} + \tfrac{1}{2}\,\text{grad}\,(V^2)$$

$$= \mathbf{B} - \frac{1}{\rho}\,\text{grad}\,p - \nu\,\text{curl}\,\boldsymbol{\omega} \qquad (34)$$

These and the equation of continuity give four equations for solving for u, v, w, and p. If the equations can be solved, the stresses can then be obtained from the fundamental relations between stresses and rates of strain, as shown in Eqs. (35), where μ is the coefficient of dynamic viscosity.

$$\tau_{11} = -p + 2\mu\frac{\partial u}{\partial x} \qquad \tau_{22} = -p + 2\mu\frac{\partial v}{\partial x}$$

$$\tau_{33} = -p + 2\mu\frac{\partial w}{\partial x} \qquad \tau_{23} = \mu\left(\frac{\partial w}{\partial y} + \frac{\partial v}{\partial z}\right) \quad (35)$$

$$\tau_{31} = \mu\left(\frac{\partial u}{\partial z} + \frac{\partial w}{\partial x}\right) \qquad \tau_{12} = \mu\left(\frac{\partial v}{\partial x} + \frac{\partial u}{\partial y}\right)$$

Two immediate and important consequences of the Navier-Stokes equations are (1) the circulation in a closed circuit moving with the fluid diminishes at a time rate given by $v\oint (\text{curl}\,\boldsymbol{\omega}) \cdot ds$, and (2) the energy per unit volume of a fluid diminishes at a time rate given by expression (36). These furnish a

$$\mu\left[2\left(\frac{\partial u}{\partial x}\right)^2 + 2\left(\frac{\partial v}{\partial y}\right)^2 + 2\left(\frac{\partial w}{\partial z}\right)^2 + \left(\frac{\partial w}{\partial y} + \frac{\partial v}{\partial z}\right)^2\right.$$

$$\left. + \left(\frac{\partial u}{\partial z} + \frac{\partial w}{\partial x}\right)^2 + \left(\frac{\partial v}{\partial x} + \frac{\partial u}{\partial y}\right)^2\right] \qquad (36)$$

measure of the effect of viscosity on the rates of dissipation of both vorticity and energy.

Irrotational flow. In irrotational flow the vorticity is zero and the velocity is expressible as the gradient of a scalar potential $\mathbf{U} = \text{grad}\,\phi$. The equation of continuity in rectangular coordinates then assumes the form of Laplace's equation, Eq. (37). When the flow is irrotational, the terms

$$\frac{\partial^2\phi}{\partial x^2} + \frac{\partial^2\phi}{\partial y^2} + \frac{\partial^2\phi}{\partial z^2} = 0 \qquad (37)$$

due to viscosity in the Navier-Stokes equations vanish so that they become formally identical with the Euler equations. Thus, in either viscous or inviscid irrotational flow, if Laplace's equation has been solved for the velocity potential, the pressure field can be obtained from the Bernoulli equation.

Elementary solutions of Laplace's equation are: for uniform flow in the direction of a line having direction cosines l, m, n, Eq. (38); for a source of strength M at the origin, Eq. (39); and for a doublet of strength Δ oriented along the x axis, Eq. (40).

$$\phi = V(lx + my + nz) \qquad (38)$$

$$\phi = -\frac{M}{(x^2 + y^2 + z^2)^{1/2}} \qquad (39)$$

$$\phi = -\frac{x\Delta}{(x^2 + y^2 + z^2)^{3/2}} \qquad (40)$$

A doublet may be obtained by letting a source and sink of equal strength approach each other, holding constant the product of the source strength by the distance between them (Fig. 3).

By combining these elementary solutions given in Eqs. (38)–(40), many interesting flow patterns can be obtained; for example, Eq. (41), formed by

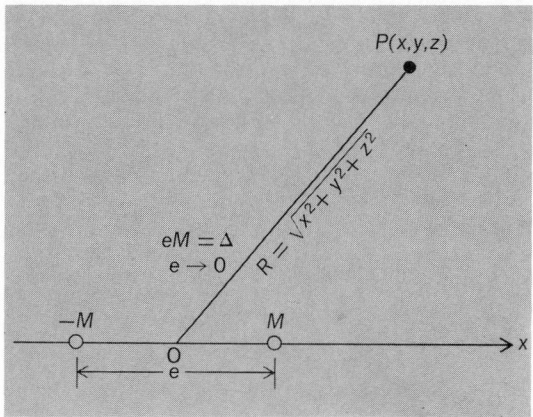

Fig. 3. Definition sketch of doublet.

$$\phi = Vx\left[1 + \frac{a^3}{2(x^2+y^2+z^2)^{3/2}}\right] \qquad (41)$$

adding the potentials of a uniform stream and a doublet, gives the flow about a sphere. Bodies obtained as stream surfaces by trying various combinations of sources, sinks, and doublets in a uniform stream are called Rankine bodies. A necessary condition for such bodies to be closed is that the algebraic sum of the strengths of the sources and sinks be zero.

Methods of solving the direct problem, of finding the flow subject to prescribed boundary conditions, have also been devised. These methods may be classified in five categories.

Separation of variables. This reduces Laplace's equation to several ordinary differential equations, the solutions of which are obtained as sets of orthogonal functions. Combinations of these are then found which satisfy the boundary conditions.

Method of images. This method is suitable when the boundaries are planes, spheres, or circular cylinders. This technique has reached its culmination in the discovery of the so-called sphere and circle theorems which immediately give the modification of the flow when a sphere or circle is introduced into a preexisting flow pattern.

Method of integral equations. Two important classes of problems, the Neumann problem, in which the normal component of the velocity is prescribed on the boundary, and the Dirichlet problem, in which the values of the potential are given on the boundary, can be formulated as Fredholm integral equations. Although it is tedious to solve these equations numerically by hand, programs are available for solving them on moderate- or high-speed automatic computers.

Solution by relaxation. This is a numerical method in which the flow region is divided into a fine network, at the corners of which the values of ϕ are assumed as an initial approximation. The finite-difference form of Laplace's equation and the boundary conditions are then used to adjust the assumed values by trial and error. This method is also tedious.

Conformal mapping. The theory of functions of a complex variable furnishes a remarkably powerful tool for solving two-dimensional irrotational flow problems. Riemann's mapping theorem gives assurance that boundaries of regions can be mapped into circles by means of analytic functions. If the proper function is found, then it also transforms the flow about these boundaries into the much simpler problem of the flow about the circles, which can be solved. The difficulty in conformal mapping lies in finding the appropriate mapping functions. Many flow problems with simple boundaries can be solved by applying the properties of elementary functions of complex variables. For arbitrary shapes, several numerical procedures have been developed.

Viscous flow. Since the viscosity of the most common fluids, air and water, is very small, it is an excellent approximation to treat their flows as inviscid except at very low Reynolds numbers ($Vl/v < 10$) and in the neighborhood of the boundaries. Because the velocity of a fluid at a wall is zero relative to that of the boundary (nonslip condition), a so-called boundary layer in which viscous effects are important is present, within which appreciable shear stresses occur. Outside of this boundary layer the flow may be treated as inviscid and to a good approximation, the pressure across the boundary layer may be assumed to be that in the inviscid flow at the outer edge of the boundary layer.

A simplified form of the Navier-Stokes equations, called the boundary-layer equations, yields laminar-flow solutions for the velocity profiles. It is known, however, that boundary-layer flows are often turbulent. A theory of the stability of the laminar boundary layer has been developed and current research is shedding considerable light on the processes whereby a disturbed laminar flow eventually breaks down into the typical random motions of turbulence, but the mechanisms are not yet clear.

Although turbulent flow is believed to be governed by the Navier-Stokes equations, it has not yet been possible to derive any turbulent-flow solutions from these equations. The difficulty lies in the fact that the Reynolds equations for turbulent flow, which are derived from the Navier-Stokes equations by an averaging process, introduce new unknowns for which additional relations are required. The well-developed theory of homogeneous isotropic turbulence, and the body of accumulated experimental results and hypotheses concerning turbulence in shear flows, have not yet supplied these missing relations.

A phenomenon of real flow that is of great practical concern is that of separation of flow. This occurs in a boundary layer when the fluid in the

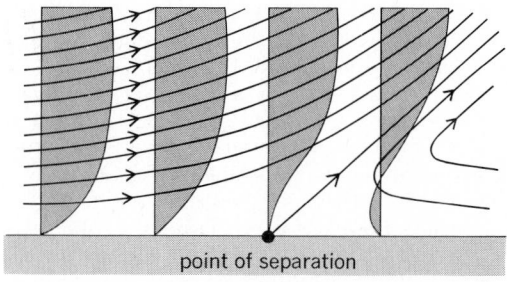

point of separation

Fig. 4. Boundary-layer separation.

neighborhood of a wall is brought to rest and caused to reverse its direction by the action of an adverse pressure gradient, that is, increasing pressures in the downstream direction (Fig. 4). When the boundary layer is turbulent, separation occurs farther downstream (or not at all) than when it is laminar, because the momentum of the layers near the wall is reinforced by turbulent interchange with layers of higher mean velocity due to the random motions. *See* BOUNDARY-LAYER FLOW; LAMINAR FLOW; TURBULENT FLOW.

[LOUIS LANDWEBER]

Bibliography: S. Goldstein (ed.), *Modern Developments in Fluid Dynamics*, 1938, reprint 1965; H. Lamb, *Hydrodynamics*, 6th ed., 1945, reprint 1965; B. LaMeharte, *An Introduction to Hydrodynamics and Water Waves*, 1975; H. Rouse (ed.), *Advanced Mechanics of Fluids*, 1959, reprint 1976; V. Streeter and E. B. Wylie, *Fluid Mechanics*, 7th ed., 1979.

Hydrostatics

The study of liquids at rest. In the absence of motion, there are no shear stresses; the internal state of stress at any point is determined by pressure alone. Hence, the pressure at a point is the same in all directions. Pressure acts normally to all boundary surfaces. For equilibrium under gravity, regardless of the shape of the containing vessel, the pressure is uniform over any horizontal cross section. Pressure varies with height or depth in accordance with the relation $dp = wdz$, where w is the specific weight of the liquid (pounds per cubic foot), and z is the height or depth in feet measured positive downward. Two different reference levels are used in measuring pressure. For many engineering purposes, gage pressure is used with pressure measured relative to atmospheric pressure as zero. For most scientific purposes, pressure is referred to true zero. Normal atmospheric pressure at sea level caused by the weight of the air above is approximately 14.7 psia.

Applications of hydrostatics. Storage tanks, underwater tunnels, gates for hydraulic structures, walls, dams, sheetpiling and bulkheads, pressure-measuring devices, hydraulic presses, and other pressure-actuated systems are applications of hydrostatics.

Hydrostatic forces on immersed surfaces. Force F is the pressure p multiplied by the area A on which it acts. Its magnitude is in pounds and its direction is normal to the area. It is a vector quantity and may be broken into components usually taken horizontally and vertically. On all surfaces, plane or curved, forces acting on elementary areas are evaluated as $dF = pdA$.

On a plane surface all elementary force vectors are parallel and the total force is the sum of the elementary forces. Hence, the total force is the product of the total area and the average pressure acting on it. The average pressure is that at the centroid of the area. The total force is independent of the orientation of the plane surface it acts on as long as the depth to its centroid is the same.

Pressure volume. In calculating the total hydrostatic force acting on a plane surface, a solid is imagined with the plane surface area as its base and the fluid pressure at each point on the base erected

there as an altitude. The total force is the volume of the solid. If the total force were imagined to act at a single point, the center of pressure of the surface, it would pass through the center of gravity of the solid and be normal to the surface.

On a nonplane surface, the elementary force vectors are not parallel. The summation of their horizontal and vertical components gives respectively the horizontal and vertical components of the total force. The total force is determined from these components by vector addition. The horizontal component of the total force is that which would be exerted on the vertical plane projected area of the nonplane surface. The vertical component of the total force is the weight of the liquid volume extending vertically from the nonplane surface up to the free surface of the liquid. As single forces, these components would pass through the centers of gravity of their respective volumes. When combined, they give the magnitude, line of action, and point of application of the total force on the nonplane surface.

Buoyant force. This is the force exerted vertically upward by a fluid on a body wholly or partly immersed in it. Its magnitude is equal to the weight of the fluid displaced by the body. This value is also the vertical component of the fluid pressure force acting upward against the bottom of the body minus the fluid pressure force component (if any) acting vertically downward against the top of the body. If this buoyant force equals the weight of the body, the body will remain at the given level. If it exceeds the weight of the body, the latter will rise, and vice versa. The buoyant force as a single magnitude acts vertically upward through the center of buoyancy which is the center of gravity of the displaced fluid. *See* ARCHIMEDES' PRINCIPLE.

Stability. The stability of a wholly or partly immersed body is determined by the relative positions of its center of gravity G and center of buoyancy B. The position of G depends upon the distribution of the mass within the body; the position of B depends upon the shape of the submerged portion of the body. If G lies directly below B, the body will be stable; under an angular displacement, a righting moment will tend to restore the body to its original position.

A floating body, depending upon its shape, may be stable even if G lies above B. An angular rotation or heel will not change the volume of the displaced fluid but may change its shape and the lateral position of B so that a righting moment through B and G may exist to restore the body to its original position. The point of intersection of the vertical line through the displaced position of B with the line drawn through G and the original position of B is M. If M lies above G for a given angle of heel, a righting moment will exist; if M lies below G, an overturning moment will arise to capsize the body. In most ships, for angles of heel up to 10–15°, M remains in a practically constant position—the metacenter.

Pressure-measuring instruments. Two types of instruments measure pressure—gages and manometers.

Gages. A metallic element such as a curved tube or a flexible diaphragm which deforms under liquid pressure is the usual sensing element of a gage. The deformation is changed mechanically or electrically into a calibrated dial reading. The bourdon gage, used for measurement of static or slowly changing pressure, converts mechanically the deformation of a curved metal tube into a reading in pressure units. Each gage is designed to be accurate for a selected range of pressures but does not give accurate readings of short-time pressure fluctuations.

A pressure transducer converts deformation of a metallic diaphragm into an electric current differential, which is calibrated to read in pressure units. Each transducer is designed to be accurate for a selected range of pressures and reacts to fluctuations of microsecond duration. The cost is high and accessory instrumentation elaborate; application is generally to the measurement of rapidly fluctuating pressures.

Manometers. Glass tubes in which the height of a liquid is a measure of the pressure being sought are called manometers. A simple manometer is open to the atmosphere at one end and connected to the pressure source at the other; it is called a piezometer if its liquid is the same as that in the tank, pipe, or other device to which it is connected. Measured above a selected datum, the height to the top of the liquid in the open-ended tube is the piezometer head. If the pressure source is a pipe, the height of the piezometer liquid level above the pipe center is the pressure head in the pipe. The pressure in pounds per square foot is obtained by multiplying this height in feet by the specific weight of the liquid.

A differential manometer measures the difference in pressure between two sources. The glass tube is usually an erect or inverted U partly filled with a liquid other than those liquids (or gases) of which the pressure difference is desired. One leg of the U is connected to one pressure source, the other leg to the other source. If the pressure sources are at the same elevation and contain the same liquid, the pressure difference is the vertical distance (displacement) between the tops of the manometer liquid in the two legs of the U multiplied by the difference between the specific weight of the manometer liquid and that of the displacing liquid. Measurement of small differences can be magnified by using a manometer liquid of low specific gravity; for large pressure differences mercury is commonly used.

Pressure transmission. Pressure applied to a confined liquid is transmitted with equal intensity throughout the liquid and by it to all surfaces of the confining vessel or piping. Hence, a small force applied to a small area of a confined liquid can create a large force against a large area. If the small and large areas are pistons the device may be a hydraulic press or jack. Because the transmitting liquid is practically incompressible and its volume virtually constant, the linear movement of the large piston will be to that of the small piston in inverse proportion to their areas. The principle of multiplying a force by means of liquid pressure applies also to hydraulic brakes, power steering, control systems, and the like; the actuating force may be a pump instead of a small piston.

[WILLIAM ALLAN]

Bibliography: R. Resnick and D. Halliday, *Physics,* 3d ed., 1977; J. U. Thoma, *Hydrostatic Power Transmission,* 1979.

Hypercharge

A quantized attribute, analogous to electric charge, introduced in the classification of a subset of elementary particles—the so-called baryons—including the proton and neutron as its lightest members. As far as is known, electric charge is absolutely conserved in all physical processes. Hypercharge was introduced to formalize the observation that certain decay modes of baryons expected to proceed by means of the strong nuclear force simply were not observed. *See* ELECTRIC CHARGE.

Unlike electric charge, however, the postulated hypercharge was found not to be conserved absolutely; the weak nuclear interactions do not conserve hypercharge—and indeed can change hypercharge by ± 1 or 0 units. These observations establish certain relationships but do not define the hypercharge. The simplicity of the relationships suggested a hypercharge (symbol Y) as follows: for the proton and neutron, $Y = +1$; for the pion, $Y = 0$.

For example, consider the heavier Λ^0 baryon, which in principle is energetically unstable against decay into a proton and a pion in a characteristic time of about 10^{-23} sec. But it is observed that the Λ^0 lives about 10^{13} times longer than this characteristic time, suggesting that, for this baryon, $Y = -1$, and thus it can decay into a proton and a pion only by means of the weak nuclear interaction. Systematic observations of this kind have permitted assignment of hypercharge values to all the baryons and antibaryons.

When the known baryons are classified according to their electric charge and their hypercharge, they naturally group into octets in the scheme first proposed by M. Gell-Mann and K. Nishijima. The quarks, hypothesized as the fundamental building blocks of matter, must have fractional hypercharge as well as electrical charge; the simplest quark model suggests values of $1/3$ and $2/3$, respectively. *See* BARYON; ELEMENTARY PARTICLE; QUANTUM MECHANICS; QUARKS; SYMMETRY LAWS; UNITARY SYMMETRY.

[D. ALLAN BROMLEY]

Hyperfine structure

A closely spaced structure of the spectrum lines forming a multiplet component in the spectrum of an atom or molecule, or of a liquid or solid. In the emission spectrum for an atom, when a multiplet component is examined at the highest resolution, this component may be seen to be resolved, or split, into a group of spectrum lines which are extremely close together. This hyperfine structure may be due to a nuclear isotope effect, to effects related to nuclear spin, or to both.

Isotope effect. The element zinc, for example, has three relatively abundant naturally occurring nuclear isotopes, ^{64}Zn, ^{66}Zn, and ^{68}Zn. The radius of a nucleus increases with the nuclear mass and, for a given element, the Coulomb interaction of the nucleus with the atomic s-electrons will be slightly weaker when the nuclear size is larger. This nuclear size effect causes a slight shift of certain of the spectrum lines, and this shift will be different for each isotope. For a mixture of ^{64}Zn, ^{66}Zn, and ^{68}Zn, certain of the multiplet components will thus consist of three closely spaced lines, one line for each isotope. A study of the isotope effect for an element leads, for example, to information about the dependence of the nuclear size on isotope mass, that is, on the number of neutrons in the isotope. *See* ISOTOPE SHIFT.

Structure due to nuclear spin. For the zinc isotopes discussed above, the nuclear spin $I = 0$, and these nuclei will be nonmagnetic and, in effect, have a spherical shape. If $I \neq 0$, however, two new nuclear properties may be observed. The nucleus may have a magnetic moment, and the shape of the nucleus may not be spherical but rather may be that of a prolate or oblate spheroid; that is, it may have a quadrupole moment. *See* SPIN.

Atoms and molecules. If the electrons in an atom or a molecule have an angular momentum, the electron system may likewise have a magnetic moment. An electron quadrupole moment may also exist. The magnetic moment of the nucleus may interact with the magnetic moment of the electrons to produce a magnetic hyperfine structure. The quadrupole moments of the nucleus and of the electrons may couple to give an electric quadrupole hyperfine structure. In a simple example, the magnetic and the electric quadrupole hyperfine structure may be described by an energy operator H_{hfs} which has the form below, where $\bar{S}$ is

$$H_{\text{hfs}} = A\bar{I} \cdot \bar{S} + P(\bar{I}_z^2 - 3I(I+1))$$

an operator describing electron spin, $\bar{I}$ and $\bar{I}_z$ are operators describing the nuclear spin and its z component, and I gives the magnitude of the nuclear spin. A and P are coupling constants which may take positive or negative values, and may range in magnitude from zero to a few hundred meters^{-1}. The term in A describes a magnetic, and the term in P a quadrupole, hyperfine structure.

The measurement of a hyperfine structure spectrum for a gaseous atomic or molecular system can lead to information about the values for A and P. These values may be interpreted to obtain information about the nuclear magnetic and quadrupole moments, and about the atomic or molecular electron configuration.

Important methods for the measurement of hyperfine structure for gaseous systems may employ an interferometer, or use atomic beams, electron spin resonance. or nuclear spin resonance. *See* INTERFEROMETRY; MAGNETIC RESONANCE; MOLECULAR BEAMS.

Liquid and solid systems. Hyperfine structure coupling may also occur and may be measured for liquid and solid systems. For liquids and solids, measurements are often made by electron spin or nuclear spin resonance methods. For solids, and for radioactive nuclei, one may, for example, also employ the Mössbauer effect or the angular correlation of nuclear gamma rays. *See* MÖSSBAUER EFFECT.

For a diamagnetic solid, $A = 0$ in the equation above, and if the crystalline environment of an atom is cubic, $P = 0$ also. If this environment is not cubic, P may have a finite measurable value. *See* DIAMAGNETISM.

If the solid is paramagnetic, ferromagnetic, or antiferromagnetic, A may be finite and measurable, and again P may or may not be zero depending on whether the atomic environment is cubic

or not. *See* ANTIFERROMAGNETISM; FERROMAGNETISM; PARAMAGNETISM.

One may gain information about the nuclear moments and about electron bonding and magnetic structure from measurements of hyperfine structure for liquids and solids. Such measurements are extensively used, for example, in atomic and condensed matter physics, chemistry, and biology. *See* ATOMIC STRUCTURE AND SPECTRA; NUCLEAR MOMENTS.

[LOUIS D. ROBERTS]

Bibliography: A. Abragam, *The Principles of Nuclear Magnetism*, 1961; R. S. Raghavan and D. E. Murnick (eds.), *Hyperfine Interactions IV*, 1978; M. E. Rose, *Elementary Theory of Angular Momentum*, 1957.

Hypergeometric functions

The analytic continuation of the function defined by the series in Eq. (1), where the shifted factorial $(a)_n$ is defined by Eq. (2). It satisfies differential equation (3), and for $|z| < 1$, Re c > Re b > 0, is given by the integral representation of Eq. (4),

$$_2F_1(a,b;c;z) = \sum_{n=0}^{\infty} \frac{(a)_n(b)_n}{(c)_n n!} z^n$$
$$|z| < 1 \tag{1}$$

$$(a)_n = a(a+1) \cdots (a+n-1)$$
$$n = 1,2, \ldots ,(a)_0 = 1 \tag{2}$$

$$z(1-z)y'' + [c - (a+b+1)z]y' - aby = 0 \tag{3}$$

$$_2F_1(a,b;c;z) = \frac{\Gamma(c)}{\Gamma(b)\Gamma(c-b)}$$
$$\int_0^1 (1-zt)^{-a} t^{b-1}(1-t)^{c-b-1} dt \tag{4}$$

where Γ represents the gamma function. *See* COMPLEX NUMBERS AND COMPLEX VARIABLES; GAMMA FUNCTION; SERIES.

The interest in hypergeometric functions comes from the many important functions which are special cases of the general hypergeometric function, the rich theory which has been developed for the general hypergeometric function, and the many times they occur in applications. Classically hypergeometric functions have arisen in science as solutions to differential equations. As discrete, rather than continuous, models of physical phenomena have become increasingly useful, hypergeometric functions have continued to arise as solutions to the equations governing these models. *See* DIFFERENTIAL EQUATION.

Elementary functions. Among the special cases are the elementary functions in Eqs. (5).

$$\log(1-z) = z \,_2F_1(1,1;2;z)$$
$$\sin^{-1}z = z \,_2F_1\left(\frac{1}{2},\frac{1}{2};\frac{3}{2};z^2\right)$$
$$\tan^{-1}z = z \,_2F_1\left(\frac{1}{2},1;\frac{3}{2};-z^2\right) \tag{5}$$
$$(1-z)^{-a} = \,_2F_1(a,b;b;z)$$

Continued fraction expansions. An example of a result which is useful, far from obvious, and yet is a simple consequence of the general theory of hypergeometric functions is the continued fraction expansion, first given by Johann Lambert, of

$\tan^{-1}z$ in Eq. (6). Since $\tan^{-1} 1 = \pi/4$, the continued

$$\tan^{-1}z = \cfrac{z}{1 + \cfrac{\frac{1.1}{1.3}z^2}{1 + \cfrac{\frac{2.2}{3.5}z^2}{1 + \cfrac{\frac{3.3}{5.7}z^2}{1 + \cdots}}}} \tag{6}$$

fraction in Eq. (6) gives an explicit expression for π.

C. F. Gauss found a continued fraction for the ratio $_2F_1(a,b+1;c+1;z)/_2F_1(a,b;c;z)$, which reduces to Eq. (6) in the special case $a = \frac{1}{2}$, $b = 0$, $c = \frac{1}{2}$. To obtain this continued fraction, Gauss derived three-term recurrence relations which connect three hypergeometric functions, two of which are contiguous to the third. Two functions are contiguous if two of the three parameters, "a," "b," and "c," are equal and the third differs by one. These contiguous relations can be thought of as three-term difference equations which are a discrete analog of differential equation (3). Most of the occurrences of hypergeometric functions arise because they satisfy a second-order differential or difference equation. *See* INTERPOLATION.

Transformations. A simple change of variables $(t = 1 - s)$ in the integral of Eq. (4) gives linear (fractional) transformation (7). Iterating this trans-

$$_2F_1(a,b;c;z) = (1-z)^{-a}{}_2F_1(a,c-b;c;z/(z-1)) \tag{7}$$

formation, after using the symmetry in "a" and "b," gives Eq. (8). *See* CONFORMAL MAPPING.

$$_2F_1(a,b;c;z) = (1-z)^{c-a-b}{}_2F_1(c-a,c-b;c;z) \tag{8}$$

There is a very important subclass of hypergeometric functions, depending on two, rather than three, parameters, which has a "quadratic transformation." The two basic ones are Eqs. (9).

$$\begin{aligned}
&_2F_1(2a,2b;a+b+\tfrac{1}{2};z) \\
&\quad = {}_2F_1(a,b;a+b+\tfrac{1}{2};4z(1-z)) \\
&_2F_1(2b,a;2a;z) \\
&\quad = (1-z)^{-b}{}_2F_1(b,a-b;a+\tfrac{1}{2};z^2/(4z-4))
\end{aligned} \tag{9}$$

This class is important because each function in it can be multiplied by an algebraic function to give a Legendre function, and all Legendre functions arise in this way.

Confluent hypergeometric functions. Differential equation (3) has regular singular points at the points 0, 1, and ∞. These can be moved to arbitrary points z_1, z_2, z_3 by a linear fractional transformation, and the resulting equation exhibits the symmetries given in linear and quadratic transformations (7)–(9), in a more transparent fashion. Also the resulting singular points can be made to coalesce. For the ordinary hypergeometric function this procedure is called confluence. The function $_2F_1(a,b; c;z/b)$ satisfies a differential equation whose regular singular points are at 0, b, and ∞, and if b is allowed to become large, the resulting function is called a confluent hypergeometric function. Explicitly, it is given by Eq. (10). Transformation

$$_1F_1(a;c;z) = \sum_{n=0}^{\infty} \frac{(a)_n}{(c)_n} \frac{z^n}{n!} \tag{10}$$

formula (8) becomes Eq. (11). Quadratic trans-

$$_1F_1(a;c;z) = e^z{}_1F_1(c-a;c;-z) \tag{11}$$

formation (9) implies Eq. (12). The function on the

$$e^{-z}{}_1F_1(a;2a;2z) = {}_0F_1(-;a+\tfrac{1}{2};z^2/4) \tag{12}$$

right-hand side is a simple multiple of a Bessel function of imaginary argument, or $I_{a-(1/2)}(z)$. Using this limit, differential equation (3) becomes Eq. (13).

$$zy'' + (c-z)y' - ay = 0 \tag{13}$$

It has a regular singular point at 0 and an irregular singular point at infinity, and one of its solutions is $_1F_1(a;c;z)$. *See* BESSEL FUNCTIONS.

Generalized hypergeometric functions. Since a number of sums, such as $_2F_1$, $_1F_1$, and $_0F_1$, arise in a very natural way, it is useful to consider the more general hypergeometric function defined by Eq. (14) where $a_p = a_1, \ldots, a_p; b_q = b_1, \ldots, b_q$.

$$_pF_q(a_p;b_q;z) = \sum_{n=0}^{\infty} \frac{(a_1)_n \cdots (a_p)_n}{(b_1)_n \cdots (b_q)_n} \frac{z^n}{n!} \tag{14}$$

All of these generalized hypergeometric functions satisfy differential equations of order $\max(p, q+1)$, where if $p > q+1$ one of the a's is assumed to be a negative integer, since the series does not converge except when $z = 0$ without this assumption.

Orthogonal polynomials. These functions also satisfy difference equations in the parameters a_i and b_j. Some of these difference equations give rise to polynomials which are orthogonal with respect to many of the important distribution functions in statistics. Some of these are listed in the table.

These polynominals are useful in studying statistical and probabilistic problems. Meixner and Laguerre polynomials play an essential role in birth and death processes when the birth and death parameters are linear functions of the size of the population. Jacobi polynomials for certain special values of the parameters α and β are the zonal spherical harmonics on spheres and projective spaces. For the case $\alpha = \beta = 0$, the polynomials reduce to Legendre polynomials and the sphere is the unit sphere in three dimensions. This point of view leads to some of the deepest formulas known about hypergeometric functions, especially those known as addition formulas. For example, Jacobi polynomials satisfy formula (15) where notation (16) applies, and $c(m,k,\alpha,\beta,n)$ is a product of

$$\begin{aligned}
&P_n^{(\alpha,\beta)}\Big(\frac{(1+x)(1+y)}{2} + \frac{r^2(1-x)(1-y)}{2} \\
&\quad + (1-x^2)^{1/2}(1-y^2)^{1/2}r\cos\varphi - 1\Big) \\
&= \sum_{k=0}^{n}\sum_{m=0}^{k} c(m,k,\alpha,\beta,n) f_{n,m,k}(x)f_{n,m,k}(y) \\
&\quad \cdot P_m^{(\alpha-\beta-1,\beta+k-m)}(2r^2-1)r^{k-m} \\
&\quad \cdot P_{k-m}^{(\beta-\frac{1}{2},\beta-\frac{1}{2})}(\cos\varphi)f_{n,m,k}(x) \tag{15} \\
&= (1-x)^{(k+m)/2}(1+x)^{(k-m)/2}P_{n-k}^{(\alpha+k+m,\beta+k-m)}(x) \tag{16}
\end{aligned}$$

shifted factorials. *See* PROBABILITY; SPHERICAL HARMONICS; STATISTICS.

Krawtchouk polynomials when $p = \frac{1}{2}$ are the zonal spherical harmonics on the space consisting

Orthogonal polynomials expressed as generalized hypergeometric functions

Polynomial	Hypergeometric representation	Distribution	
Jacobi $P_n^{(\alpha,\beta)}(x)$	$\dfrac{(\alpha+1)_n}{n!}\,{}_2F_1(-n,n+\alpha+\beta+1;\alpha+1;(1-x)/2)$	$(1-x)^\alpha(1+x)^\beta,$	$-1 < x < 1,$
Laguerre $L_n^\alpha(x)$	$\dfrac{(\alpha+1)_n}{n!}\,{}_1F_1(-n;\alpha+1;x)$	$x^\alpha e^{-x},$	$x > 0,$
Hermite $H_n(x)$	$(2x)^n\,{}_2F_0\left(-\dfrac{n}{2},\dfrac{1-n}{2};-;-\dfrac{1}{x^2}\right)$	$e^{-x^2},$	$-\infty < x < \infty,$
Hahn $Q_n(x;\alpha,\beta,N)$	${}_3F_2\left(\begin{matrix}-x,-n,n+\alpha+\beta+1\\-N,\alpha+1\end{matrix};1\right)$	$\dfrac{(\alpha+1)_x}{x!}\dfrac{(\beta+1)_{N-x}}{(N-x)!},$	$x = 0,1,\ \ldots\ ,N,$
Meixner $M_n(x;\beta,c)$	${}_2F_1(-n,-x;\beta;1-c^{-1})$	$\dfrac{(\beta)_x c^x}{x!}$	$x = 0,1,\ \ldots\ ,$
Krawtchouk $K_n(x;p,N)$	${}_2F_1(-n,-x;-N;p^{-1})$	$\dbinom{N}{x}p^x(1-p)^{N-x},$	$x = 0,1,\ \ldots\ ,N,$
Charlier $c_n(x;a)$	${}_2F_0(-n;-x;-;-a^{-1})$	$a^x/x!,$	$x = 0,1,\ \ldots\ .$

of the vertices of the unit cube in N-space and there is a corresponding addition formula. The vertices of the unit cube can be considered as a message of zeros and ones, and Krawtchouk polynomials have been shown to play an important role in coding theory. *See* ORTHOGONAL POLYNOMIALS.

Special values of the argument. A number of hypergeometric functions can be evaluated as quotients of gamma functions when the argument z takes on special values. Gauss' sum, Eq. (17),

$$_2F_1(a,b;c;1) = \frac{\Gamma(c)\Gamma(c-a-b)}{\Gamma(c-a)\Gamma(c-b)} \quad (17)$$

$$c > a + b$$

occurs often in applications. For the generalized hypergeometric series, conditions have to be placed on the parameters before the series can be explicitly summed. Two general classes of sums often arise. One, called well-poised, occurs when $p = q+1$ and the parameters can be paired so that $a_1+1 = a_2+b_1 = \cdots = a_{q+1}+b_q$. A. C. Dixon summed the general well-poised $_3F_2$ at $z=1$. A series is called very-well-poised if it is well-poised and $a_2 = b_1+1$. In the second type, called k-balanced, $p = q+1$, $a_1+\cdots+a_{q+1}+k = b_1+\cdots+b_q$ for some integer k, and one of the a_i's is a negative integer. J. F. Pfaff summed the 1-balanced $_3F_2$ at $z=1$. The most complicated sum of this type which has been discovered is John Dougall's sum of the very well-poised, 2-balanced $_7F_6$ at $z=1$.

These sums, and related transformation formulas, are fundamental and occur in many applications. Among these is G. Racah's work on complex spectra, T. Regge's symmetries of the Clebsch-Gordan coefficients, and E. Sparre Andersen's work on fluctuation theory of random walks. *See* STOCHASTIC PROCESS.

Generalizations. There are many generalizations of hypergeometric functions which are also very useful. One is to replace the shifted factorials by the quantities in Eq. (18), and to introduce appropriate powers of q as multiplicative factors. The

$$(q^a)_{q,n} = (1-q^a)(1-q^{a+1})\cdots(1-q^{a+n-1}) \quad (18)$$

propriate powers of q as multiplicative factors. The resulting series are connected with theta functions, and with a pair of formulas of L. J. Rogers and S. Ramanujan which are important in number theory and combinatorial analysis. One of these is

Eq. (19) where $\prod\limits_{n=0}^{\infty} a_n = a_0 a_1 \cdots a_n \cdots$ is an infinite

$$\sum_{n=0}^{\infty} \frac{q^{n^2}}{(q)_{q,n}} = \frac{1}{\prod\limits_{n=0}^{\infty}(1-q^{5n+1})(1-q^{5n+4})} \quad (19)$$

product. In terms of partitions of positive integers as the sum of positive integers, this says that the number of ways of writing a positive integer as the sum of positive integers so that the difference between any two of the parts is at least two is equinumerous with the number of ways of writing the same integer as the sum of positive integers so that each of the parts is either one more or one less than an integer divisible by five. An infinite number of sums similar to Eq. (19) have been found by George Andrews, but the sums are now multiple sums. *See* COMBINATORIAL THEORY; ELLIPTIC FUNCTION AND INTEGRAL.

Jay Goldman and G.-C. Rota have used the known observation that the coefficients $[(q)_{q,n}]/[(q)_{q,k}(q)_{q,n-k}]$ arise when counting the subspaces of dimension k of an n-dimensional vector space over a field of q elements to obtain some identities for q-hypergeometric functions. The results they found were known, but their observations opened up new fields for applications of this generalization of hypergeometric functions.

In 1880 Paul Appell introduced four hypergeometric functions of two variables. Willard Miller has shown that these functions arise very naturally from a group theoretic point of view, and thus have possible applications in the study of elementary particles by methods employing group theory. *See* ELEMENTARY PARTICLE; GROUP THEORY.

A final generalization is to hypergeometric functions of matrices. These functions have arisen in statistical multivariate analysis and in number theory. *See* MATRIX THEORY. [RICHARD ASKEY]

Bibliography: R. Askey, *Orthogonal Polynomials and Special Functions*, 1975; R. Askey (ed.), *Theory and Application of Special Functions*, 1975; A. Erdélyi et al., *Higher Transcendental Functions*, 3 vols. 1953–1955; W. Miller, *Lie Theory and Special Functions*, 1968; L. J. Slater, *Generalized Hypergeometric Functions*, 1966; N. Ja. Vilenkin, *Special Functions and the Theory of Group Representations*, 1968.

Hyperon

A collective name for any baryon with nonzero strangeness number s. The name hyperon has generally been limited to particles which are semistable, that is, which have long lifetimes relative to 10^{-22} s and which decay by photon emission or through weaker decay interactions. Hyperonic particles which are unstable (that is, with lifetimes shorter than 10^{-22} s) are commonly referred to as excited hyperons. The known hyperons with spin $1/2\hbar$ (where $\hbar$ is Planck's constant divided by 2π) are Λ, Σ^-, Σ^0, and Σ^+, with $s=-1$, and Ξ^- and Ξ^0, with $s=-2$, together with the Ω^- particle, which has spin $3/2\hbar$ and $s=-3$. The corresponding antihyperons have baryon number $B=-1$, opposite strangeness s, and charge Q; they are all known empirically, except for $\overline{\Omega}^+$.

The first excited hyperon was reported in 1960. The state $\Sigma(1385)^+$ was observed as a $\pi\Lambda$ resonance in the final state of reaction (1). The symbol $\Sigma(m)$

$$K^- + p \rightarrow \pi^- + \Sigma(1385)^+ \rightarrow \pi^- + \pi^+ + \Lambda \quad (1)$$

or $\Lambda(m)$ indicates that the strangeness is $s=-1$ and that the isospin is $I=1$ or 0, respectively. The superscript gives the charge, and m is the mass in million electronvolts (MeV); if no m is given, the symbol refers to the ground state, for example, Λ means $\Lambda(1115.5)$. *See* ISOTOPIC SPIN.

Reaction (1) is an example of an excited hyperon production reaction. Formation reactions are also possible for most excited hyperons with $s=-1$. For example, the properties of $\Lambda(1520)$ are particularly well known from observations on its formation and decay, for K^- mesons incident on hydrogen, reaction (2). $\Sigma(1385)$ cannot be formed in

$$K^- p \rightarrow \Lambda(1520) \rightarrow \begin{cases} K^- + p \\ \pi^\pm + \Sigma^\pm, \text{etc.} \end{cases} \quad (2)$$

this way, because its mass lies below the $K^- p$ threshold energy, $m_K + m_p \simeq 1432$ MeV.

There is no deep distinction between hyperons and excited hyperons, beyond the phenomenological definition above. Indeed, the hyperon $\Omega(1672)^-$ and the excited hyperons $\Xi(1530)$ and $\Sigma(1385)$, together with the unstable nucleonic states $\Delta(1236)$, are known to form a unitary decuplet of states with spin $3/2\hbar$. *See* BARYON; ELEMENTARY PARTICLE; SYMMETRY LAWS; UNITARY SYMMETRY.

[RICHARD H. DALITZ]

Hypersonics

A development of ultrasonics (also known as pretersonics) dealing with the production and utilization of sound waves of frequencies above 500 MHz. These frequencies are sufficiently high to approximate thermal waves, and they can interact directly with phonons (quantized sound waves which carry the thermal energy in an electrically nonconducting solid). They also interact with electrons and holes (which carry the electrical energy and much of the thermal energy of a metal or semiconductor) and with spin waves in ferromagnetic materials. *See* PHONON; ULTRASONICS; WAVE MOTION.

A considerable development has occurred in the field of hypersonics due to the production of thin-film transducers of cadmium sulphide or zinc oxide, the use of lasers in Brillouin scattering or Bragg reflection from sound waves, and the use of thermal pulses. They have provided considerable evidence on the interaction of acoustic waves with other thermal and electrical waves in the media. Frequencies as high as 2.5 THz have been obtained.

Transducer methods. By evaporating cadmium and sulfur under controlled conditions, thin films of hexagonal single-crystal CdS can be produced on various crystal substrates. These are high-resistivity crystals and are oriented with the c or hexagonal axis perpendicular to the substrate. By putting a field normal to the transducer, a longitudinal wave can be sent into the specimen, while a field parallel to the transducer face produces a shear wave. Microwave cavities can be used to generate these types of fields in the transducers, and both longitudinal- and shear-wave measurements can be made in a variety of materials.

Figure 1 shows typical measurements of the attenuation of the two shear waves and the longitudinal wave in a single crystal of aluminum oxide, Al_2O_3. Below $20-30$ K the attenuation is independent of the temperature. This region is assumed to be controlled by scattering losses due to imperfections in the crystal and transducers; the loss is a good measure of the imperfections in the crystal. Above this region the attenuation for the slow shear wave increases as the fourth power of the temperature from 20 to 80 K. The increase in attenuation is in agreement with the theory of L. Landau and G. Rumer which considers the direct interactions of the acoustic waves with the thermal phonons. This formula can be put into the form of Eq. (1), where α is the attenuation in

$$\alpha = 60\gamma^2 \, kT/Mv^2 \, (T/\theta)^3 \, 2\pi/\lambda \quad (1)$$

nepers per centimeter, γ the Grueneisen constant, k the Boltzmann constant, M the average atomic mass, v an average sound velocity, T the absolute temperature, θ the Debye temperature, and λ the acoustic wavelength. The agreement with the formula is quite good. The fast shear wave and the

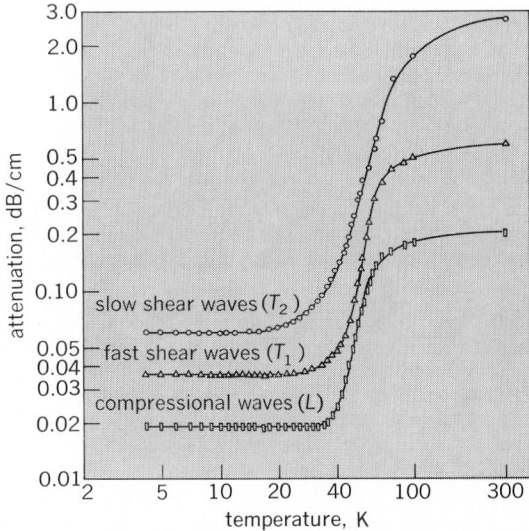

Fig. 1. Attenuation measurements for two shear and one longitudinal waves in sapphire, Al_2O_3. Measurements made at 1 GHz. (*After J. De Klerk*)

longitudinal wave behave in a different manner with slopes proportional to T^7 and T^9, respectively. Explanations for these values have not been obtained.

For higher temperatures, when the product of the angular frequency ω times the thermal relaxation time τ is much less than unity, individual interactions between sound waves and phonons can no longer be followed. In this region two effects cause the attenuation. These are the thermoelastic effect, which depends on the propagation of heat from the compressed part of the wave to the expanded part, and the Akheiser effect, which causes an instantaneous separation of the phonon modes followed by an equilibration of these temperatures which occurs with a relaxation time τ. The Akheiser effect produces a loss about 40 times the thermoelastic loss for insulators. According to a newer theory, this loss is given by Eq. (2), where

$$\alpha_{\text{(nepers/cm)}} = \omega^2 D (E_o K / C_v \overline{V}^2)/2\rho V^3(1+\omega^2\tau^2) \quad (2)$$
$$\tau = 3K/C_v\overline{V}^2$$

the ratio of the total thermal energy E_o to the specific heat C_v is proportional to a factor F times the absolute temperature T. F varies from 0.25 at very low temperatures to unity above the Debye temperature. D is a nonlinear constant which can be calculated when the third-order moduli of the material are known, K the thermal conductivity, ρ the crystal density, V the sound velocity, and $\overline{V}$ the Debye average velocity. A number of third-order moduli have been measured for at least six crystals and the agreement with Eq. (2) is good.

The transducer method of measurement has been extended to 118 GHz and measurements have been made for quartz at low temperatures.

Optical methods. The oldest method for measuring the properties of liquids and solids at high frequencies is the Brillouin light-scattering method. L. Brillouin showed that thermal sound waves produce scattered light of frequencies $f_l \pm f_s$, where f_s is the sound frequency and f_l the light frequency. The frequency difference f_s is related to the angle of observation φ and the relative wavelengths by either of the forms in Eqs. (3), where ω_s is 2π times

$$2\lambda_s \sin\frac{\varphi}{2} = \lambda_l$$
$$\omega_s = 2\left(\frac{vn}{c}\right)\omega_l \sin\frac{\varphi}{2} \quad (3)$$

the sound frequency f_s, $\omega_l = 2\pi f_l$, and λ_s and λ_l are the sound and light wavelengths, respectively. The velocity of light in the medium is c/n, where n is the refractive index and v the sound velocity. For the 253.7-nm mercury line, an angle $\varphi = 90°$, and an index of refraction in fused silica of 1.5057, the sound wavelength becomes 1.19×10^{-5} cm. Since the velocities of longitudinal and shear waves are, respectively, 5.97×10^5 and 3.76×10^5 cm/s, the corresponding frequencies are 5×10^{10} (50 GHz) and 31.9 GHz. For sound velocities from 10^5 to 10^6 cm/s, the corresponding frequencies are 10 GHz and 83 GHz. *See* BRILLOUIN ZONE.

The Brillouin method has been considerably improved by the use of lasers which produce a continuous light of a much higher intensity. The exposure time is reduced to very reasonable values, and the method is used considerably for ultrasonic measurements. This system has shown that all liquids can propagate shear waves at sufficiently high frequencies and that longitudinal waves suffer a dispersion at frequencies for which the shear modulus becomes operative. No disper-

The first experiments on the Brillouin scattering of laser light were performed on organic liquids,

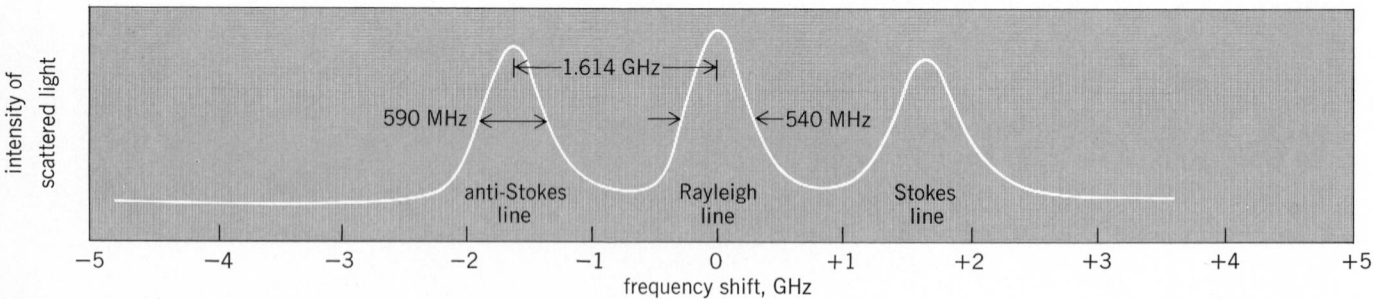

(a)

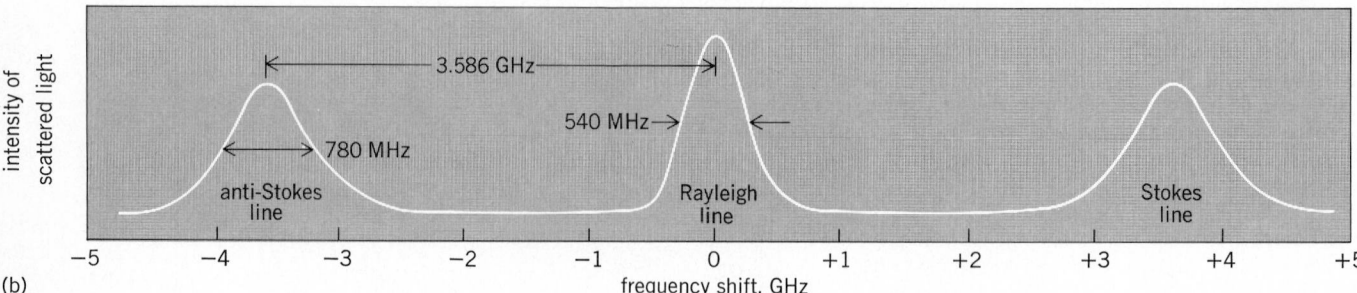

(b)

Fig. 2. Brillouin spectra in toluene. (*a*) Scattering angle 30.0°, temperature 22.4°C, giving a phonon frequency of 1.614 GHz. (*b*) Scattering angle 70.0°, temperature 22.8°C, giving a phonon frequency of 3.586 GHz, (*From P. Fleury, Light scattering, in W. P. Mason and R. Thurston, eds., Physical Acoustics, vol. 6, pp. 1–64, Academic Press, 1970*)

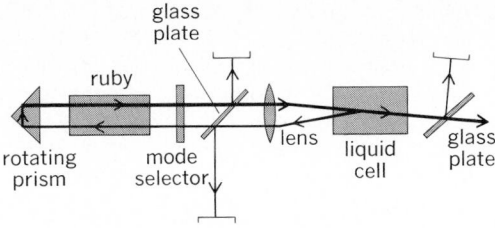

Fig. 3. Experimental arrangement showing path of stimulated Brillouin scattering. Thickness of line shows the relative strengths of the light scattered and amplified. (*After E. Garmire and C. H. Townes*)

and a large fraction of subsequent Brillouin studies have been devoted to such materials. Light scattering as an acoustic tool is most useful, for in room-temperature liquids acoustic phonons of wavelength in the 10^{-5}-cm range have frequencies in the $10^9 - 10^{10}$-Hz range. It is in this frequency range that many of the most striking acoustic effects occur in liquids. Among the most important of these are the large dispersions in velocity and absorption of sound associated with relaxation phenomena. Since the acoustic frequencies in a Brillouin experiment can be easily changed by varying the scattering angle, rather thorough investigations of relaxations are possible.

Figure 2 shows Brillouin scattering spectra in toluene. The central or Rayleigh peak in each spectrum is due to nonpropagating entropy fluctuations. Since the wavelength of the He-Ne laser is 632.8 nm, while the index of refraction for toluene is 1.4911, it follows from Eq. (3) that the velocity of sound at 1.614 GHz and 22.4°C (Fig. 2a) is 1.32×10^5 cm/s, while the velocity at 3.586 GHz (Fig. 2b)—corrected to 22.4°C from 22.8°C—is 1.347×10^5 cm/s. The velocity at ultrasonic frequencies, corrected to the same temperature, 22.4°C, is 1.319×10^5 cm/s. Hence it is evident that a relaxation effect is being observed. This is obvious also from the increased width of the Brillouin peaks with higher frequencies: the instrumental width contributes just over 500 MHz to the line width, but a definite increase in broadening at higher frequencies is nevertheless evident in Fig. 2b.

Such liquids as benzene, carbon tetrachloride, cyclohexane, hexylene, nitrobenzene, and acetic acid exhibit relaxations that can be measured by Brillouin scattering. Other liquids, such as water, acetone, methanol, and ethanol, have not exhibited dispersion in the Brillouin range.

For solids, no dispersion has been found for insulating them, and the elastic moduli at the highest-measure frequencies are the same as for low frequencies. For metals, a dispersion can occur due to the motion of dislocations.

Brillouin scattering has been applied in the study of phase transitions. For some crystals, such as ammonium chloride, there is a big change in the hypersonic and ultrasonic velocities at a phase transition, whereas for potassium dihydrogen phosphate there is no observable change, *See* PHASE TRANSITIONS.

The Brillouin light-scattering mechanism has been coupled to a laser amplifier to produce very intense sound waves. By passing the scattered light from a crystal or liquid through the laser (Fig. 3), the beam is amplified and returned to the cell along with the fundamental laser beam. The system becomes self-sustaining (lases) at a definite threshold of power. The two light waves act through the electro-strictive effect in the liquid or crystal to produce a very intense sound wave. For crystals this may be so large as to fracture the crystal. It has been suggested that, by resorting to low temperatures, the level for lasing can be made low enough to produce sound waves which do not fracture the crystal, possibly proving a new source of intense sound waves. *See* LASER.

Thermal pulse method. Thermal pulses of very short duration have also been used to measure the properties of insulating and conducting crystals. A thermal pulse of very short duration can be considered as a superposition of a wide band of randomly phased acoustic waves centered around the frequency given by Eq. (4), where h is Planck's con-

$$f = kT/h = 2.08 \times 10^{10}T \qquad (4)$$

stant. Since meaningful measurements have been made up to temperatures of 25 K, this corresponds to a frequency of 500 GHz.

The heat-pulse experiments require as basic elements a small heater or thermal transducer, to produce an excitation of a known pulse width, and

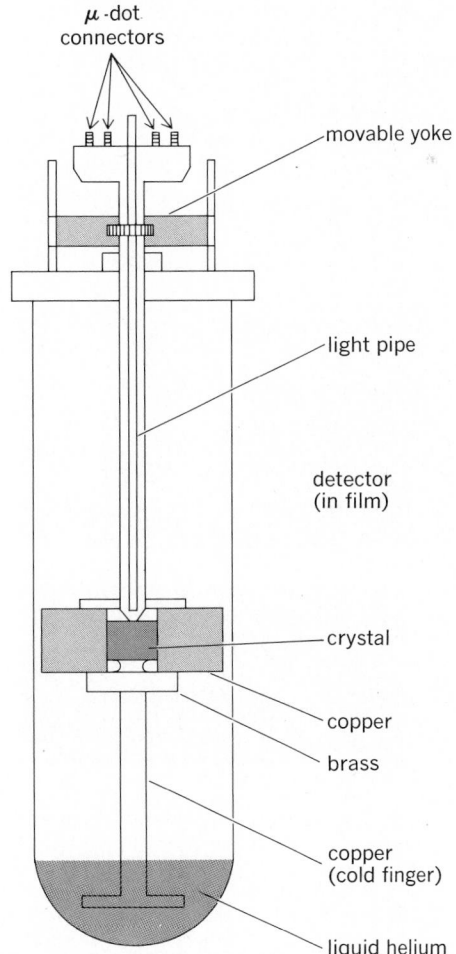

Fig. 4. Apparatus for measuring heat pulses as a function of the temperature. (*From R. J. von Gutfeld and A. H. Nethercot, in W. P. Mason and R. Thurston, eds., Physical Acoustics, vol. 5, Academic Press, 1969*)

a thermal receiver, whose response is proportional to the incident thermal flux. Both of these requirements can be satisfied by the evaporation of thin-film heater and detector circuits on opposite polished faces of the crystal. The sensitivity of such receiving bolometers is determined by the temperature coefficient of the resistivity. At temperatures between 2 and 8 K, the sensitivity can be maximized by the use of various superconducting films near their transition temperatures.

Figure 4 shows a typical experimental arrangement for producing and detecting heat pulses. A thin (200 nm) evaporated indium-tin alloy covers the entire front face. A laser pulse of 40 nanoseconds, that is, 40×10^{-9} s, produces the increase in temperature. Because of the large amount of energy in the pulse, instantaneous temperature rises up to 38 K are produced. The receiving bolometer is a pure indium film, which has a relatively high temperature-dependent resistance.

Figure 5 shows a series of responses at tempera-

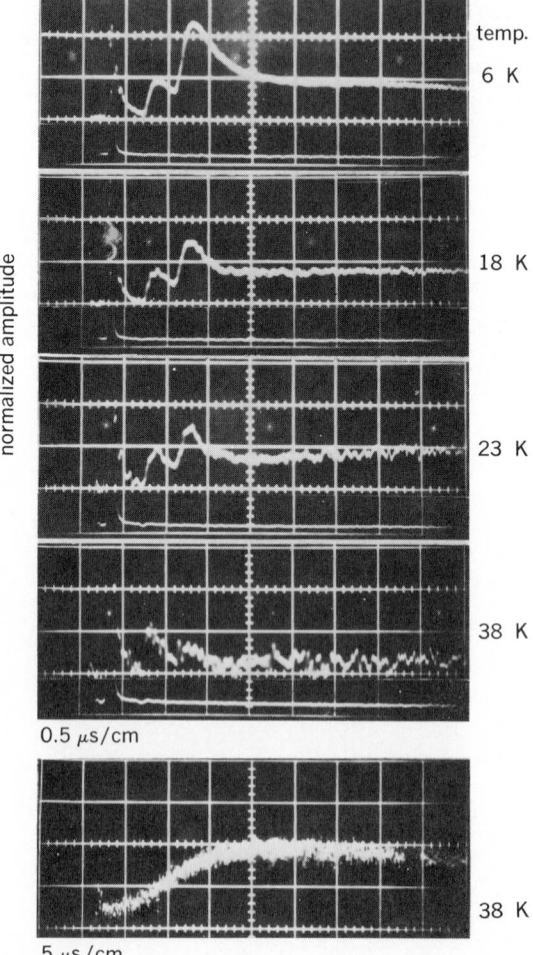

Fig. 5. Observed heat pulses (upper trace) after propagating through 1/2-cm Z-cut sapphire crystal at four different temperatures. The initial pulse is due to laser light falling directly on the detector; the second and third pulses represent heat arriving with the longitudinal and transverse phonon velocities. The lower trace represents the direct response of a phototube to the laser light. (*R. J. von Gutfeld and A. H. Nethercot, Temperature dependence of heat-pulse propagation in sapphire, Phys. Rev. Lett., 17:868, 1966*)

tures from 6 up to 38 K for sapphire, Al_2O_3, along the hexagonal axes. The observed discrete pulses for the lowest three temperatures are consistent with the velocity of longitudinal and shear waves along the hexagonal axis. The observed velocities are very close to those obtained from ultrasonic wave measurements. At 38 K these waves are no longer observable as pulses, and the response is similar to that expected when the waves interact with thermal phonons and produce the expected heat-wave response. The complete heat-flow equation, taking account of acoustic wave transmission, is given by Eq. (5), where T is the absolute temper-

$$\tau \partial^2 T/\partial t^2 + \partial T/\partial t = \kappa \partial^2 T/\partial x^2 \qquad (5)$$

ature, t the time variable, x the space variable, τ the effective phonon relaxation time, and κ the thermal diffusivity. Equation (5) contains a second time derivative which is not usually included in diffusive heat-flow problems in which the rate of change of temperature is slow compared to the relaxation rates. It is this term that prevents a temperature pulse from propagating with an infinite velocity.

Thermal pulse methods have also been used to investigate the transmission of heat in conducting materials. Here the largest share of the heat is carried by electrons and, in the absence of collisions of electrons and phonons, this heat would be carried with the Fermi velocity. The most complete measurements have been made for gallium by R. J. von Gutfeld and A. H. Nethercot. They measured the Fermi velocity to be 6×10^7 cm/s at 2 K.

Other effects. In addition to the effects mentioned above, hypersonic methods have been used to study the interaction of acoustic waves with electrons and holes. This interaction in metals produces a large attenuation at low temperatures. In metals a variation of attenuation with magnetic fields at low temperatures has delineated the Fermi surface of many metals. In semiconductors measurements of attenuation and velocity changes have determined the relaxation time for going from one part of the energy surface to another. For n-type material this relaxation time is determined by collisions of phonons with electrons or by electrostatic scattering of electrons by the charged impurity atoms. In p-type silicon the mechanism for change of momentum from one part of the surface to another is the capture of the hole in an excited state and the subsequent ejection by thermal vibrations. This process produces a very high loss at low temperatures. *See* RELAXATION TIME OF ELECTRONS.

Shear hypersonic waves also couple to spin waves in ferromagnetic material and produce a dispersion in a nonuniform magnetic field. This effect is of interest in nondispersive delay lines. Ultrasonic amplification of sound waves has been shown in piezoelectric semiconductors to which a dc field is applied. Many other effects are being investigated. A wide and fruitful field of endeavor thus appears to be opened up by the new hypersonic wave technique.

Acoustic microscopes. Another application of hypersonics is the acoustic microscope. This was suggested as early as 1936 by S. Sokolov, but it was not until sound waves with very high frequen-

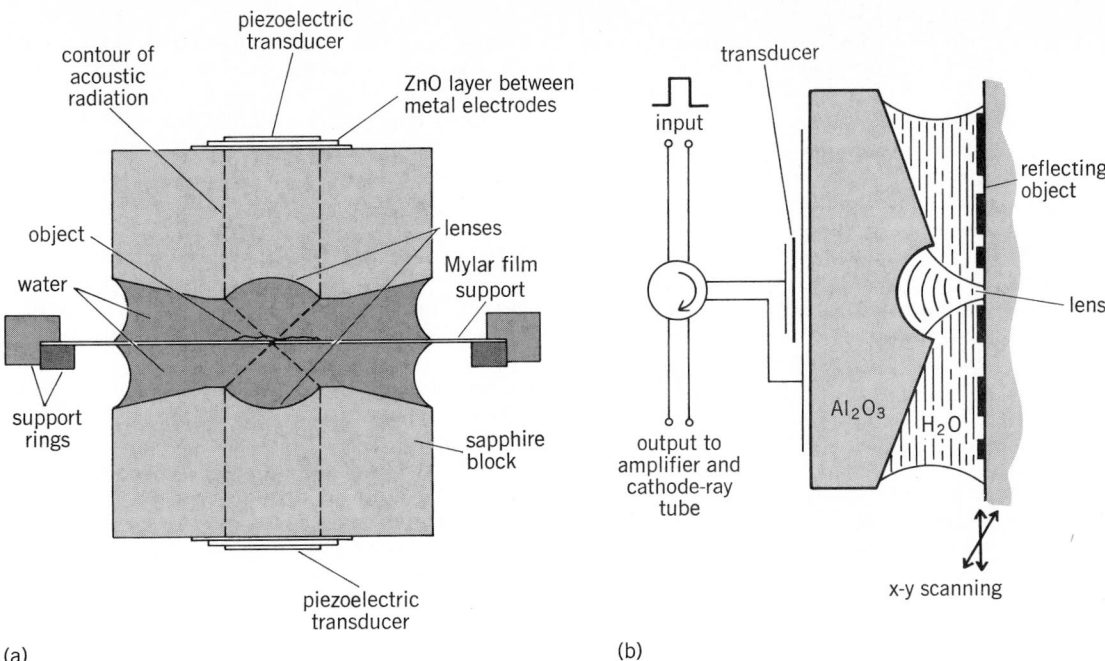

(a)

(b)

Fig. 6. Acoustic microscope, (a) Transmission type. (b) Reflection type. (From C. F. Quate. A. Atalar. and H. K. Wickramsinghe. Acoustic microscopy with mechanical scanning: A review. Proc. IEEE, 67:1092–1114, 1979)

cies with wavelengths on the order of those for optical wavelengths were obtained that acoustic microscopes were able to compete with optical means.

There are two types of microscopes: the transmission and the reflecting types (Fig. 6). In the transmission type (Fig. 6a), a high-frequency longitudinal wave, on the order of 1–3 GHz, is generated by a thin layer of zinc oxide and is transmitted through a sapphire crystal having a very low acoustic attenuation, as shown by Fig. 1. The beam is focused on the sample by means of a concave lens ground in the crystal, having a radius of 190 μm, to a focused point having a diameter of about 1 micrometer. In order to cut down on the reflection loss between the sapphire crystal and the water, a quarter-wave section of a glass, with an impedance intermediate to the sapphire and the water, is evaporated on the concave lens. The beam is transmitted through the object and is refocused on the piezoelectric transducer. The response picked up is proportional to the attenuation and reflection loss of the object. The object is scanned mechanically by means of a piezoelectric transducer, and the output is displayed as a picture on a cathode-ray tube. Figure 7 shows optical and acoustic images of normal red blood cells.

In the reflection-type microscope (Fig. 6b), the wave is focused on the reflecting surface and sent back to the transducer. A pulse of the high frequency, about 20 nanoseconds wide, is sent into the transducer and is then sent to the reflecting object. The returned pulse is separated from the transmitting pulse by a time-gating circuit called a circulator. The output can be displayed on a cathode-ray tube in a similar manner after being scanned mechanically. This type of microscope is useful in scanning metallurgical specimens and

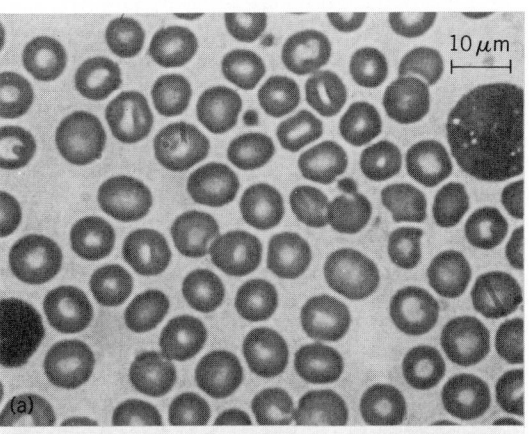

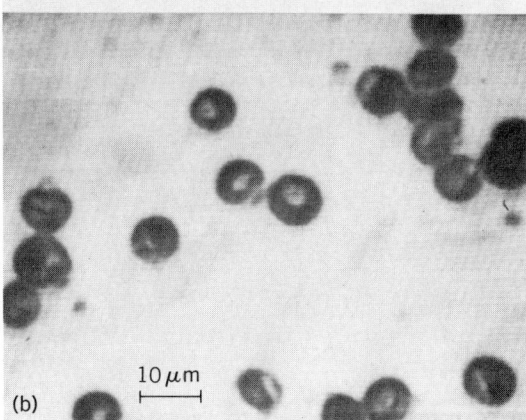

Fig. 7. Normal red blood cells. (a) Optical image of cells fixed in methanol with Leishman's stain. (b) Acoustic image (1100 MHz) of unstained cells fixed in methanol. (From C. F. Quate. A. Atalar. and H. K. Wickramsinghe. Acoustic microscopy with mechanical scanning: A review. Proc. IEEE. 67:1092–1114, 1979)

integrated circuit specimens. Biological material can also be placed on the reflecting surface, as in the transmission microscope.

[WARREN P. MASON]

Bibliography: L. Brillouin, Diffusion of light and x-rays by a transparent homogeneous body, *Ann. Phys. Paris*, 17(9):88–122, 1922; R. Y. Chiao and C. H. Townes, Stimulated Brillouin scattering and coherent generation of intense hypersonic waves, *Phys. Rev. Lett.*, 12:592, 1964; R. J. von Gutfeld and A. H. Nethercot, Temperature dependence of heat-pulse propagation in sapphire, *Phys. Rev. Lett.*, 17:868, 1966; W. P. Mason (ed.), *Physical Acoustics*, vols. 1–4, 1965–1968; W. P. Mason and R. Thurston (eds.), *Physical Acoustics*, vols. 5–14, 1969–1979.

Hysteresis

A phenomenon wherein two (or more) physical quantities bear a relationship which depends on prior history. More specifically, the response Y takes on different values for an increasing input X than for a decreasing X.

If one cycles X over an appropriate range, the plot of Y versus X gives a closed curve which is referred to as the hysteresis loop. The response Y appears to be lagging the input X.

Hysteresis occurs in many fields of science. Perhaps the primary example is of magnetic materials where the input variable H (magnetic field) and response variable B (magnetic induction) are traditionally chosen. For such a choice of conjugate variables, the area of the hysteresis loop takes on a

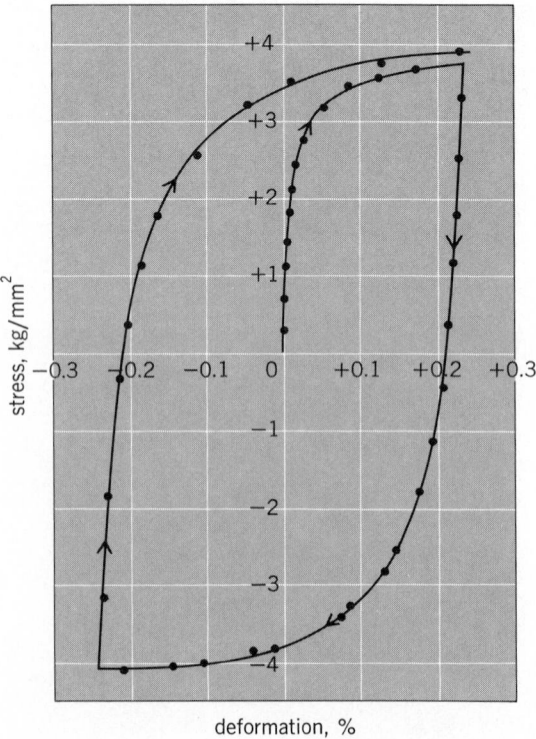

Plot of stress versus strain (deformation) for a single crystal of brass. The positive and negative strains correspond to elongation and compression, respectively. (*From G. Sachs and H. Shoji, Zug-Druckversuche an Messing-Kristallen (Bauschinger effect), Z. Physik, 45:776–796, 1927*)

special significance, namely the conversion of energy per unit volume to heat per cycle. In a similar way the dielectric energy loss per unit volume per cycle can be obtained by using the variables E (electric field) and D (electric displacement). For mechanical hysteresis, it is customary to take the variables stress and strain (see illustration), where the energy density loss per cycle is related to the internal friction. In the framework of particle rather than continuum mechanics, the energy loss per cycle is given by the hysteresis loop for the applied force and the particle displacement. Thermal hysteresis is characteristic of many systems, particularly those involving phase changes, but here the hysteresis loops are not usually related to energy loss. *See* FERROELECTRICS; MAGNETIC HYSTERESIS; THERMAL HYSTERESIS.

Hysteresis is intimately associated with microscopic irreversibility. In the electric and magnetic situations the motion of domain walls in ordered structures is impeded by microscopic inhomogeneities in the matrix which catch and then let go under increasing stress. Such effects are clearly irreversible and in the magnetic case are evident as the Barkhausen noise. On the other hand, any domain rotation is largely reversible and does not contribute to the hysteresis loss. In the mechanical situation, the pinning and unpinning of dislocations gives rise to the hysteresis. *See* CRYSTAL DEFECTS; DOMAIN.

Time-dependent effects, where the final equilibrium value of the history-independent response Y occurs some time after the input X is applied, are complex and are not usually referred to as hysteresis effects. However, cycled experiments (resonance and relaxation) whose frequencies are comparable with the natural frequencies of the system will display an X-versus-Y dependence of the hysteretic form. The largest values of the hysteresis area will occur when the applied frequency coincides closely with natural frequencies of resonance or relaxation.

[H. B. HUNTINGTON; R. K. MAC CRONE]

Illuminance

A term expressing the density of luminous flux incident on a surface. This word has been proposed by the Colorimetry Committee of the Optical Society of America to replace the term illumination. The definitions are the same. The symbol of illumination is E, and the equation is $E = dF/dA$, where A is the area of the illuminated surface and F is the luminous flux. *See* PHOTOMETRY.

[RUSSELL C. PUTNAM]

Impact

A force which acts only during a short time interval but which is sufficiently large to cause an appreciable change in the momentum of the system on which it acts. The momentum change produced by the impulsive force is described by the momentum-impulse relation. For a discussion of this relation *see* IMPULSE (MECHANICS).

The concept of impulsive force is most useful when the time in which the force acts is so short that the system which it acts on does not move appreciably during this time. Under these conditions the momentum of the system is changed rapidly by a finite amount. The details of the way in which the force varies with time are unimportant, since only the impulse determines the momentum change.

Ordinarily the forces occurring in collisions are impulsive. In the processes in which impulsive forces occur, mechanical energy can be dissipated, and attempts to apply the conservation of mechanical energy may lead to incorrect results. *See* COLLISION; CONSERVATION OF ENERGY.

A phenomenon known as impulsive loading occurs when materials are subjected to high-speed impacts or explosive charges. The study of failure of materials under impulsive loads is increasing in technological importance.   [PAUL W. SCHMIDT]

Bibliography: R. A. Becker, *Introduction to Theoretical Mechanics*, 1954; R. Resnick and D. Halliday, *Physics*, 3d ed., 1977; J. S. Rinehart, *Stress Transients in Solids*, 1975.

Impulse (mechanics)

The integral of a force over an interval of time. For a force $\mathbf{F}$, the impulse $\mathbf{J}$ over the interval from t_0 to t_1 can be written as Eq. (1). The impulse thus rep-

$$\mathbf{J} = \int_{t_0}^{t_1} \mathbf{F}\, dt \tag{1}$$

resents the product of the time interval and the average force acting during the interval. Impulse is a vector quantity with the units of momentum.

The momentum-impulse relation states that the change in momentum over a given time interval equals the impulse of the resultant force acting during that interval. This relation can be proved by integration of Newton's second law over the time interval from t_0 to t_1. Let $\mathbf{P}$ represent the momentum at time t, with $\mathbf{P}_0$ and $\mathbf{P}_1$ being the values of $\mathbf{P}$ at times t_0 and t_1, respectively. Then Eq. (2) holds.

$$\mathbf{J} = \int_{t_0}^{t_1} \mathbf{F}\, dt = \int_{t_0}^{t_1} (d\mathbf{P}/dt)\, dt$$
$$= \int_{\mathbf{P}_0}^{\mathbf{P}_1} d\mathbf{P} = \mathbf{P}_1 - \mathbf{P}_0 \tag{2}$$

If, as is ordinarily true, the mass m is constant, the momentum change can be expressed in terms of the velocities $\mathbf{v}_1$ and $\mathbf{v}_0$ at times t_1 and t_0, respectively, giving Eq. (3).

$$\mathbf{J} = m(\mathbf{v}_1 - \mathbf{v}_0) \tag{3}$$

The concept of impulse is ordinarily most useful when the forces are large but act only for a short period. In most of these cases it is necessary to know only the momentum change, which is determined by the impulse. The relation between momentum and impulse thus has the advantage of eliminating the need for a detailed knowledge of how the forces, which can be very complicated, change with time. Forces which occur during collisions are of this type. *See* COLLISION.

A further simplification of this type of system is obtained if the time interval is short enough for the system to be considered essentially stationary during the time of action of the force, so that the momentum change occurs almost instantaneously. Except for the changed momentum, the motion of the system is treated as it would have been if no impulse had occurred. The effect of the impulse can thus be considered to provide a set of initial

conditions for the motion. In a ballistic galvanometer, for example, there is an electric current for only a short time, during which the galvanometer coil, originally at rest, is given an impulse by the force associated with the current. The only effect of the current on the later motion of the coil, which can be considered stationary during the time there is a current, is to provide an initial velocity for the subsequent motion. *See* IMPACT; MOMENTUM.

[PAUL W. SCHMIDT]

Bibliography: F. Bueche, *Introduction to Physics for Scientists and Engineers*, 3d ed., 1980; R. Resnick and D. Halliday, *Physics*, 3d ed., 1977.

Incandescence

The emission of visible radiation by a hot body. A theoretically perfect radiator, called a blackbody, will emit radiant energy according to Planck's radiation law at any temperature. Prediction of the visual brightness requires additional consideration of the sensitivity of the eye, and the radiation will be visible only for temperatures of the blackbody which are above some minimum. The relation between brightness and temperature is plotted in the figure. As shown, the minimum temperature for incandescence for the dark-adapted eye is about 390°C. Under these ideal observing conditions, the incandescence appears as a colorless glow. The dull red light commonly associated with incandescence of objects in a lighted room requires a temperature of about 500°C. *See* BLACKBODY; HEAT RADIATION.

Not all sources of light are incandescent. A cold gas under electrical excitation may emit light, as in the so-called neon tube or the low-pressure mercury-vapor lamp. Ultraviolet light from mercury vapor may excite visible light from a cold solid, as in the fluorescent lamp. Luminescence is the term used to refer to the emission of light due to causes other than high temperature, and includes thermoluminescence, in which emission of previously trapped energy occurs on moderate heat-

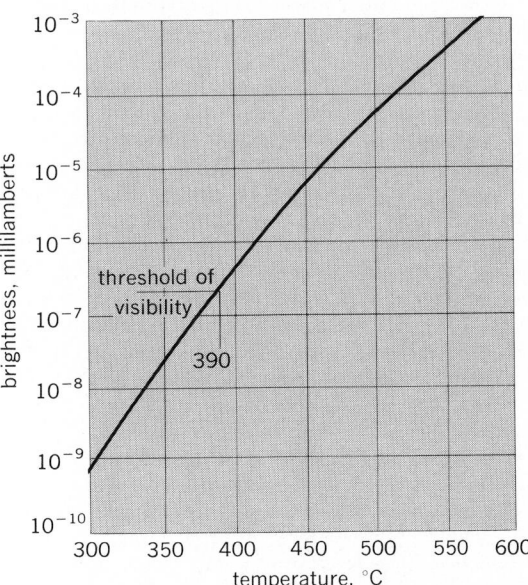

Graph showing the relation between the brightness of a blackbody and temperature.

Approximate color temperatures of common light sources

Source	Color temperature, K
Candle	1925
Kerosine lamp	2000
Common tungsten-filament 100-watt electric light bulb	2800
Carbon arc	4000
Sun	5800

ing. *See* THERMOLUMINESCENCE.

Flames are made luminous by incandescent particles of carbon. Gas flames can be made to produce intense light by the use of a gas mantle of thoria containing a small amount of ceria. This mantle is a good emitter of visible light, but a poor emitter of infrared radiation. As less heat is lost in the long waves, the mantle operates at a higher temperature than a blackbody would and, hence, produces more intense visible light.

A useful criterion of an incandescent source is its color temperature, the temperature at which a blackbody has the same color, although not necessarily the same brightness. The color temperature of common light sources depends upon operating conditions. Approximate values are given in the accompanying table.

[H. W. RUSSELL/GEORGE R. HARRISON]

Bibliography: M. M. Benarie, Optical pyrometry below red heat, *J. Opt. Soc. Amer.*, 47:1005–1009, 1957; F. A. Jenkins and H. E. White, *Fundamentals of Optics*, 4th ed., 1976.

Incompressible flow

Fluid motion without change in density. For practical purposes liquids are assumed to flow incompressibly. At low velocity this is nearly the case; however, even for liquids, abrupt changes in velocity produce compression or rarefaction. Usually a liquid flows under the action of gravity to occupy the lower portion of an open vessel. This property is taken as the distinguishing feature of a liquid. A gas, in contrast, flows compressibly to occupy any closed space to which it is confined, regardless of the initial volumes of gas and space. This property distinguishes a gas. As with liquids, slow flow of a gas is closely approximated by assuming it to be incompressible. In general, whenever fluid velocities are low (less than a fourth) relative to the rate of propagation of a pressure wave in the fluid, density variations during flow will be negligible and the flow will be effectively incompressible.

Analysis. Incompressible flow is quite frequently analyzed by supplementing the solution for an inviscid or "perfect" fluid with the effects of fluid viscosity. Simple flows such as uniform flow, sources, sinks, and vortexes may be represented by mathematical expressions defining flow velocity. These solutions may be superimposed to form realistic representation of complex inviscid flows, such as an airfoil moving through air or a boat hull through water. The results are mathematical expressions for velocity magnitude and direction at all points of the flow field. Pressure (p) may then be related to velocity (V) at a point in the flow by means of Bernoulli's equation, given below.

$$p + \tfrac{1}{2}\rho V^2 = \text{constant}$$

Here ρ is the constant fluid density. Forces on the boundary due to pressure may then be calculated. It remains then to determine the manner in which viscosity affects the flow field, hence the pressure distribution, and the additional force tangential to the boundary caused by fluid friction.

In this regime of incompressible flow, viscosity plays an important role, since it determines the behavior of the fluid adjacent to the boundaries of the flow (the boundary layer) and in the regions in which the fluid flow does not follow the boundaries (separated regions). Reynolds number, a nondimensional ratio of the inertia and viscous force in the fluid, provides a measure of flow characteristic which is quite useful in correlating experimental data with theory. *See* VISCOSITY.

Applications. There is a vast variety of practical problems which may be evaluated through use of both inviscid and viscous incompressible flow theory and experimental data. One might first think of aircraft moving at low airspeeds, hovercraft (air-cushion vehicles), helicopters and balloons passing through the atmosphere, boats of various types passing through water (here only the flow beneath the surface is pertinent to this regime), surface vehicles such as cars and trains, and wind effects on structures causing unusual loads or oscillations. Other equally important applications of incompressible flow theory include heating and air conditioning design, conveying of solid particles and liquid droplets, and airflow in various industrial processes such as steelmaking. *See* COMPRESSIBLE FLOW; GAS DYNAMICS; MACH NUMBER; REYNOLDS NUMBER.

[JAMES E. MAY]

Bibliography: T. von Karman, *Aerodynamics*, 1954, reprint 1964; A. M. Kuethe and C.-Y. Chow, *Foundations of Aerodynamics: Bases of Aerodynamic Design*, 3d ed., 1976; V. L. Streeter, *Fluid Mechanics*, 7th ed., 1979.

Inductance

That property of an electric circuit or of two neighboring circuits whereby an electromotive force is induced (by the process of electromagnetic induction) in one of the circuits by a change of current in either of them. The term inductance coil is sometimes used as a synonym for inductor, a device possessing the property of inductance. *See* ELECTROMAGNETIC INDUCTION; ELECTROMOTIVE FORCE (EMF).

Self-inductance. For a given coil, the ratio of the electromotive force of induction to the rate of change of current in the coil is called the self-inductance L of the coil, given in Eq. (1), where e is

$$L = -\frac{e}{dI/dt} \qquad (1)$$

the electromotive force at any instant and dI/dt is the rate of change of the current at that instant. The negative sign indicates that the induced electromotive force is opposite in direction to the current when the current is increasing (dI/dt positive) and in the same direction as the current when the current is decreasing (dI/dt negative). The self-inductance is in henrys when the electromotive force is in volts, and the rate of change of current is in amperes per second.

An alternative definition of self-inductance is the number of flux linkages per unit current. Flux linkage is the product of the flux Φ and the number of turns in the coil N. Then Eq. (2) holds. Both

$$L = \frac{N\Phi}{I} \qquad (2)$$

sides of Eq. (2) may be multiplied by I to obtain Eq. (3), which may be differentiated with respect to t,

$$LI = N\Phi \qquad (3)$$

as in Eqs. (4). Hence the second definition is equivalent to the first.

$$L\frac{dI}{dt} = N\frac{d\Phi}{dt} = -e \qquad (4)$$

or
$$L = -\frac{e}{dI/dt}$$

Self-inductance does not affect a circuit in which the current is unchanging; however, it is of great importance when there is a changing current, since there is an induced emf during the time that the change takes place. For example, in an alternating-current circuit, the current is constantly changing and the inductance is an important factor. Also, in transient phenomena at the beginning or end of a steady unidirectional current, the self-inductance plays a part.

Consider a circuit of resistance R and inductance L connected in series to a constant source of potential difference V. The current in the circuit does not reach a final steady value instantly, but rises toward the final value $I = V/R$ in a manner that depends upon R and L. At every instant after the switch is closed the applied potential difference is the sum of the iR drop in potential and the back emf $L\,di/dt$, as in Eq. (5), where i is

$$V = iR + L\frac{di}{dt} \qquad (5)$$

the instantaneous value of the current. Separating the variables i and t, one obtains Eq. (6). The solution of Eq. (6) is given in Eq. (7).

$$\frac{di}{(V/R) - i} = \frac{R}{L}\,dt \qquad (6)$$

$$i = \frac{V}{R}\left(1 - e^{-(R/L)t}\right) \qquad (7)$$

The current rises exponentially to a final steady value V/R. The rate of growth is rapid at first, then less and less rapid as the current approaches the final value.

The power p supplied to the circuit at every instant during the rise of current is given by Eq. (8).

$$p = iV = i^2R + Li\,di/dt \qquad (8)$$

The first term i^2R is the power that goes into heating the circuit. The second term $Li\,di/dt$ is the power that goes into building up the magnetic field in the inductor. The total energy W used in building up the magnetic field is given by Eq. (9). The

$$W = \int_0^t p\,dt = \int_0^t Li\frac{di}{dt}\,dt = \int_0^I Li\,di = \tfrac{1}{2}LI^2 \qquad (9)$$

energy used in building up the magnetic field re-

mains as energy of the magnetic field. When the switch is opened, the magnetic field collapses and the energy of the field is returned to the circuit, resulting in an induced emf. The arc that is often seen when a switch is opened is a result of this emf, and the energy to maintain the arc is supplied by the decreasing magnetic field.

Mutual inductance. The mutual inductance M of two neighboring circuits A and B is defined as the ratio of the emf induced in one circuit $\mathscr{E}$ to the rate of change of current in the other circuit, as in Eq. (10).

$$M = -\frac{\mathscr{E}_B}{(dI/dt)_A} \qquad (10)$$

The mks unit of mutual inductance is the henry, the same as the unit of self-inductance. The same value is obtained for a pair of coils, regardless of which coil is taken as the starting point.

The mutual inductance of two circuits may also be expressed as the ratio of the flux linkages produced in circuit B by the current in circuit A to the current in circuit A. If Φ_{BA} is the flux threading B as a result of the current in circuit A, Eqs. (11) hold. Integration leads to Eq. (12).

$$\mathscr{E}_B = -N_B\frac{d\Phi_{BA}}{dt} = -M\frac{dI_A}{dt} \qquad (11)$$

or
$$N_B\,d\Phi_{BA} = M\,dI_A$$

$$M = \frac{N_B\Phi_{BA}}{I_A} \qquad (12)$$

[KENNETH V. MANNING]

Inertia

That property of matter which manifests itself as a resistance to any change in the motion of a body. Thus when no external force is acting, a body at rest remains at rest and a body in motion continues moving in a straight line with a uniform speed (Newton's first law of motion). The mass of a body is a measure of its inertia. *See* MASS.

[LEO NEDELSKY]

Inertia of energy

The principle of inertia of energy states that the inertial properties of matter determine, and are determined by, its total energy content. If E is the total energy content, and m_0 the rest mass of a piece of matter, c being the speed of light, the mass-energy relation is $E = m_0c^2$. This formula was proposed on general grounds by H. Poincaré in 1900 and was deduced from the special theory of relativity by Albert Einstein in 1905.

If the mass of a body changes by an amount Δm, the corresponding energy change is $\Delta E = c^2 \cdot \Delta m$. The existence of disintegration processes in atomic nuclei has made it possible to test this formula with great accuracy. Energy released from nuclear reactions provides the power source in nuclear reactors, as well as the principal energy source in the Sun and other stars. The radiation emitted from the Sun is equivalent to a loss of mass of about 4,000,000 tons/sec.

The statement that the rest mass of matter is determined by its total energy content is not susceptible of a simple test since there exists no independent measure of the latter quantity. The validity of the general principle was adopted by Einstein as a cornerstone of his theory of gravitation. According to this theory, the gravitational properties of matter are determined by the distribution of energy in the universe. Matter and energy are used as interchangeable terms, all forms of energy being subject to gravitational action. Physical predictions made by Einstein on the basis of this principle are (1) light passing near a star should be deviated by the gravitational field of the star, and (2) light emitted from a massive star should lose energy in escaping from the star and consequently should appear to be slightly reddened with respect to a terrestrial source (gravitational red shift). Both these effects have been subjected repeatedly to physical test; results favor Einstein's predictions. *See* MASS; RELATIVITY.

[E. L. HILL]

Infrared radiation

Electromagnetic radiation in which wavelengths lie in the range from about 1 μm to 1 mm. This radiation therefore has wavelengths just a little longer than those of visible light and cannot be seen with the unaided eye. The radiation was discovered in 1800 by William Hershel, who used a prism to refract the light of the Sun onto mercury-in-glass thermometers placed just past the red end of the visible spectrum generated by the prism. Because the techniques and materials used to collect, focus, detect, and display infrared radiation are different from those of the visible, and because many of the applications of infrared radiation are also quite different, a technology has arisen, and many scientists and engineers have specialized in its application. *See* ELECTROMAGNETIC RADIATION.

Infrared techniques. A complete infrared system consists of a source, background, intervening medium, optical system, detector, electronics, and display.

Sources. The source can be described by the spectral distribution of power emitted by an ideal body (a blackbody curve). This distribution is characteristic of the temperature of the body. A real body is related to it by a radiation efficiency factor or emissivity which is the ratio at every wavelength of the emission of a real body to that of a blackbody under identical conditions. Figure 1 shows curves for these ideal blackbodies radiating at a number of different temperatures. The higher the temperature, the greater the total amount of radiation. The total number of watts per square meter is given by $5.67 \times 10^{-8}T^4$, where T is the absolute temperature in kelvins (K). Higher temperatures also provide more radiation at shorter wavelengths. This is evidenced by the maxima of these curves moving to shorter wavelengths with higher temperatures. *See* EMISSIVITY; HEAT RADIATION.

The sources can be either cooperative or uncooperative. Some examples of the former include tungsten bulbs (sometimes with special envelopes), globars, and Nernst glowers. These are made of rare-earth oxides and carbons. They closely approximate blackbodies and are used mostly for spectroscopy. Lasers have been used in special applications. Although they provide very intense, monochromatic, and coherent radiation, they are limited in their spectral coverage. The principal infrared lasers have been carbon dioxide (CO_2) and

carbon monoxide (CO) gas lasers and lead-tin-tellurium (PbSnTe) diodes. *See* LASER.

Transmitting medium. The radiation of one of these sources propagates from the source to the optical collector. This path may be through the vacuum of outer space, 1 m of laboratory air, or some arbitrary path through the atmosphere. Figure 2 shows a low-resolution transmission spectrum of the atmosphere and the transmissions of the different atmospheric constituents. Two of the main features are the broad, high transmission regions between 3 and 5 μm and between 8 and 12 μm. These are the spectral regions chosen for most terrestrial applications. Figure 3 shows a small section of this spectrum, illustrating its complexity. The radiation from the source is filtered by the atmosphere in its propagation so that the flux on the optical system is related to both the source spectrum and the transmission spectrum of the atmosphere.

Optical system. A lens, mirror, or a combination

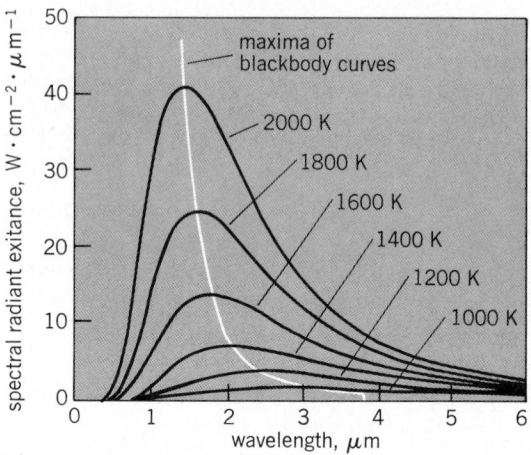

(a)

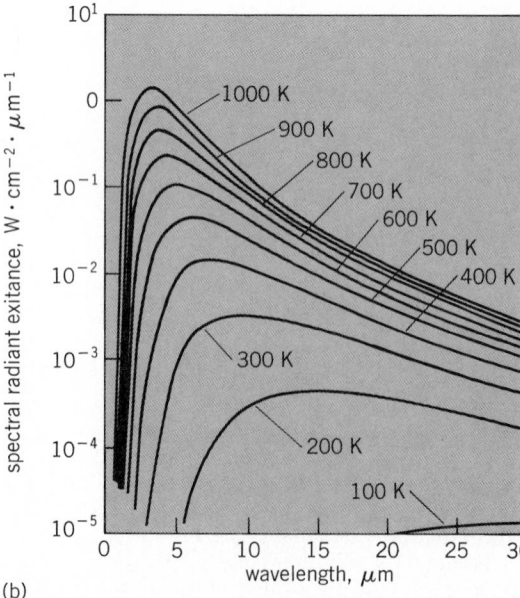

(b)

Fig. 1. Radiation from blackbodies at different temperatures, shown on (a) a linear scale and (b) a logarithmic scale.

of them is used to focus the radiation onto a detector. Since glass is opaque in any reasonable thickness for radiation of wavelengths longer than 2 μm, special materials must be used. The table lists the properties of some of the most useful materials for infrared instrumentation. In general, these are not as effective as glass is in the visible, so many infrared optical systems use mirrors instead. The mirrors are characterized by the blank and by its coating. Blanks are usually made of aluminum, beryllium, or special silica materials. The choices are based on high strength, light weight, and good thermal and mechanical stability. They are then coated with thin evaporated layers of aluminum, silver, or gold. The reflectivities of thin-film metallic coatings increase with wavelength, and the requirements for surface smoothness are also less stringent with increasing wavelength. *See* MIRROR OPTICS; OPTICAL MATERIALS.

Detectors. Photographic film is not useful for most of the infrared spectrum. The response of the silver halide in the film, even when specially processed, is only sensitive to about 1.2 μm. It cannot respond to radiation in the important 3−5-μm and 8−12-μm atmospheric transmission bands. If there were a film for the infrared, it would have to be kept cold and dark before and after exposure. Accordingly, infrared systems have used point (elemental) detectors or arrays of them. These detectors are based either on the generation of a change in voltage due to a change in the detector temperature resulting from the power focused on it, or on the generation of a change in voltage due to some photon-electron interaction in the detector material. This latter effect is sometimes called the internal photoelectric effect. Electrons which are bound to the different atomic sites in the crystal lattice receive a quantum of photon energy. They are freed from their bound lattice positions and can contribute to current flow. The energy in electronvolts required to do this is 1.24/λ (with λ in micrometers). Thus only a very small binding energy, about 0.1 eV, is permitted in photon detectors. The thermal agitation of the lattice could cause spontaneous "emission," so most infrared photodetectors are cooled to temperatures from 10 to 100 K. This does not affect the speed of response of photodetectors, which depends upon photon-electron interactions, but it does slow down thermal detectors. These are cooled because they generally respond to a relative change in temperature (and for a lower temperature, a small absolute change gives a larger relative change), and thermal noise is also reduced at low temperatures. *See* PHOTOELECTRICITY; RADIOMETRY.

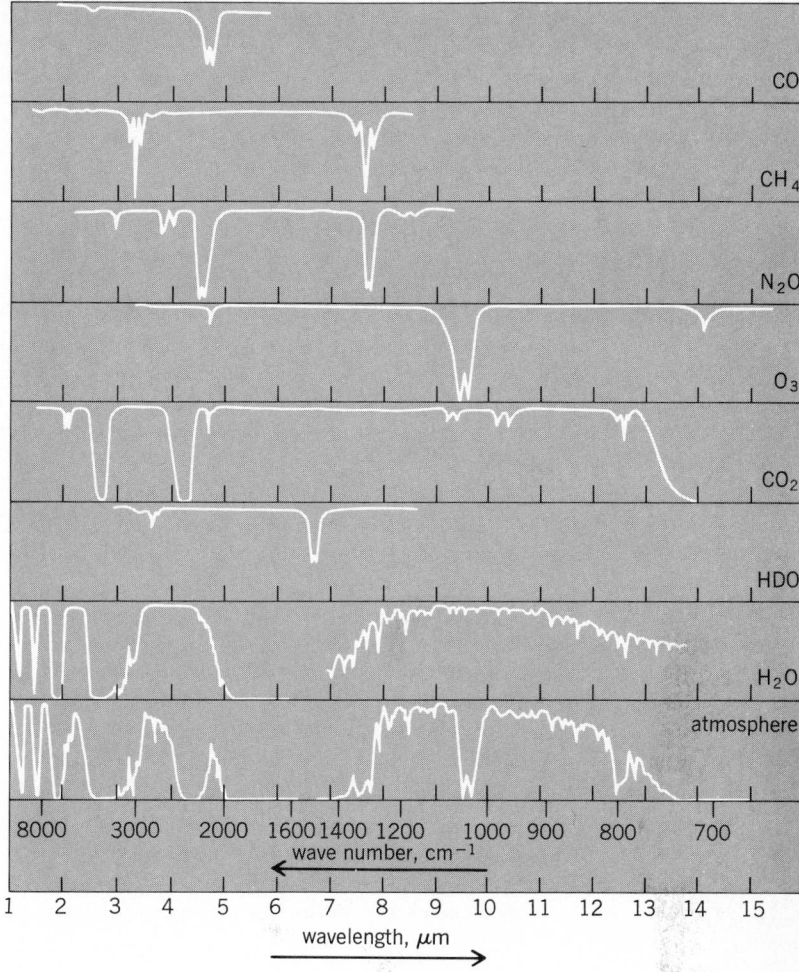

Fig. 2. Low-resolution transmission spectra of the atmosphere and of constituent gases.

Electronic circuitry. The voltage or current from the detector or detector array is amplified by special circuitry, usually consisting of metal oxide semiconductor field-effect transistors (MOSFETs) which are designed for low-temperature operations. The amplified signals are then handled very much like television signals. One important feature of most of these systems is that the system does not yield a direct response; only changes are recorded. Thus a "dc restore" or absolute level must be established with a thermal calibration source. The black level of the display can then be chosen by the operator. *See* TRANSISTOR.

Reticle. A reticle or chopper is an important feature of nonimaging systems. A typical reticle and

Properties of materials for infrared instrumentation

Material	Transmission region, μm	Approximate refractive index	Comment
Germanium	2−20	4	Opaque when heated
Silicon	2−15	3.5	Opaque when heated
Fused silica	0.3−2.5	2.2	Strong, hard
Zinc selenide	0.7−15	2.4	Expensive
Magnesium fluoride	0.7−14	1.6	Not very strong
Arsenic sulfur glass	0.7−12	2.2	Not always homogeneous
Diamond	0.3−50	1.7	Small sizes only
Salt	0.4−15	1.5	Attacked by moisture

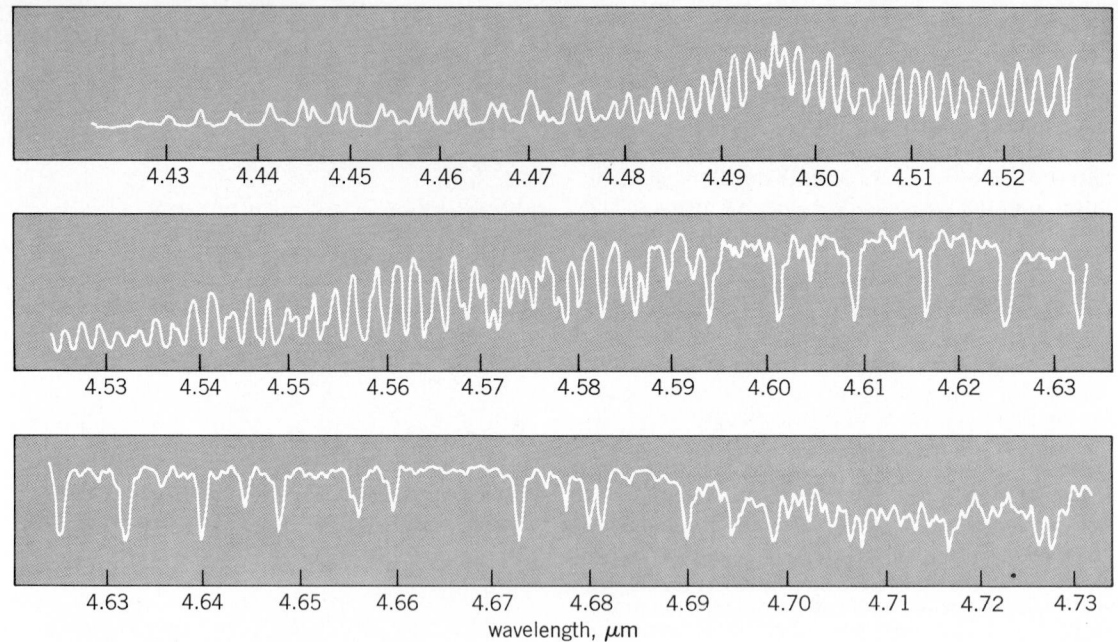

Fig. 3. High-resolution atmospheric transmission spectrum between 4.43 and 4.73 μm.

its use are shown in Figure 4. An image of the scene is portrayed on a relatively large fraction of the reticle—anywhere from 10 to 100%. The lens just behind it collects all the light that passes through the reticle and focuses it on a detector. The reticle is rotated. If the scene is uniform, there will be no change in the detector output. However,

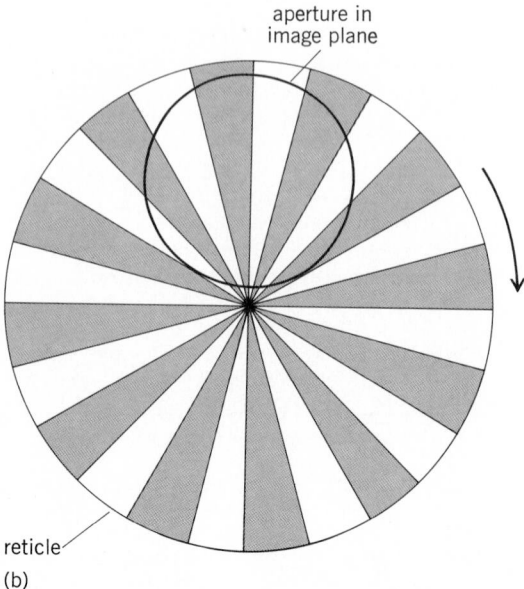

(a)

(b)

reticle

Fig. 4. Reticle system. (a) Configuration of components of system. (b) Reticle, showing area covered by image.

if the scene has a point source (like the image of the hot exhaust of an aircraft engine), then a point image is formed on the reticle and a periodic detector output is generated. The phase and frequency of this output can be used with properly designed reticles to obtain the angular coordinates of the point source. The reticle pattern can also be used to reduce the modulation obtained from almost uniform scenes which radiate from large areas. Reticles are used in most infrared tracking systems, although other schemes are sometimes employed.

Applications. Infrared techniques have been applied in military, medical, industrial, meteorological, ecological, forestry, agricultural, chemical, and other disciplines.

Meteorological applications. Weather satellites use infrared imaging devices to map cloud patterns and provide the imagery seen in many weather reports. Substances at temperatures between 200 and 300 K emit copious amounts of infrared radiation, but are not hot enough to emit in the visible. The Earth has a temperature of approximately 300 K, and high-altitude clouds are colder (about 250 K). An infrared sensor placed on a synchronous-orbit satellite can easily sense this difference and present a picture of these temperature patterns. Although the technique is not very widely known, radiometric sensors on lower-altitude satellites can determine the vertical temperature distribution along any path to the Earth. The infrared emission of the Earth's atmosphere is a function of wavelength; a space-borne radiometer senses the emitted radiation at several different wavelengths. The different wavelength bands correspond to greater atmospheric transmission and some "look" deeper into the atmosphere than others. Calculations determine the atmospheric temperature distribution that is required to produce the observed wavelength distribution based on these facts. Although there can be problems with the uniqueness and conversion of such an inversion or fitting process, the results obtained have been accurate to within a few degrees and about a kilometer of altitude.

Medical applications. Infrared imaging devices have also been used for breast cancer screening and other medical diagnostic applications. In most of these applications, the underlying principle is that pathology produces inflammation, and these locations of increased temperature can be found with an infrared imager. The major disadvantage is that localized increases of about 1 K usually need to be detected, and a fair number of nonpathological factors (or at least not the ones in question) can cause equivalent heating. In addition to breast cancer detection (which detects 80% or more cases and has about a 15% false-alarm rate), infrared techniques have been used in the analysis of severe burns, faulty circulation, back problems, and sinus ailments, and has even been proposed to test for medical malingerers.

Airborne infrared imagers. Airborne infrared imagers have been used to locate the edge of burning areas in forest fires. Typically, a forest fire in the western United States is ignited by the lightning of a late afternoon thunderstorm; the valley becomes filled with smoke by the next morning when the crews arrive. The longer wavelengths of the emitted infrared radiation penetrate the smoke better than the visible wavelengths, so the edges of fire are better delineated.

Thermal pollution contained in the power-plant effluent into rivers has been detected in various locations of the United States, and the viability of crops by assessment of the moisture content has also been accomplished with some degree of success.

Infrared spectroscopy. Infrared spectrometers have long been a powerful tool in the hands of the analytical chemist. The spectrum of a substance in either absorption or emission provides an unmistakable "fingerprint." The only problem is to find the spectrum of the substance that matches the unknown.

Military and space applications. The best-known military techniques are probably the Sidewinder air-to-air missile and the systems which detect the extremely large amounts of radiation from the plume of launching intercontinental rockets at great distances. Infrared systems have also been used for stabilizing satellites by sensing the Earth's horizon, for night vision applications, and for perimeter warning and gas detection.

[WILLIAM L. WOLFE]

Bibliography: S. S. Ballard (ed.), *Proceedings of the Institute of Radio Engineers: Infrared*, special issue, September 1959; R. D. Hudson, Jr., *Infrared System Engineering*, 1969; R. Vanzetti, *Practical Applications of Infrared Techniques*, 1972; W. L. Wolfe and G. J. Zissis, *The Infrared Handbook*, 1978.

Infrasound

Sound waves, particularly in the atmosphere, whose frequencies of pressure variation and of vibration are below the audible range, that is, lower than about 20 Hz. Earthquake and seismic waves are elastic waves which occur at infrasonic frequencies in the Earth's crust and in the oceans and seas. The physical laws of propagation in the atmosphere are essentially the same as for audible sound. This article discusses the physics of generation and propagation, the results of measurement of infrasound, the atmospheric global propagation of infrasound at very low frequencies, and the effects of infrasound on people.

The local speed of infrasound in air at ambient temperatures near 20°C is about 340 m/s, the same as for audible sound. The wavelengths for all frequencies less than 20 Hz are therefore greater than 17 m, and so the scattering by buildings, trees, hills, and so forth outdoors is much less than for audible sound. There are practically no acoustical shadows cast by infrasound.

Origin and global propagation. At frequencies less than about 1.0 Hz, infrasound propagates through the atmosphere for distances of thousands of kilometers without substantial loss of energy. Sounds at these frequencies are almost always present at measurable intensities. Those of natural origin have many causes, including tornadoes, volcanic explosions, earthquakes, the aurora borealis, waves on the seas, large meteorites, and lightning discharges. A specific example is the infrasound from the explosion in 1883 of the volcano Krakatoa in the East Indies. The absorption of infrasound from the explosion was low enough so that the waves were still detectable after traveling around the Earth several times. The somewhat weaker sound waves from the explosion in 1963 of the volcano Mount Agung, located on the island of Bali, were detected and measured with modern infrasonic microphones at several places in North America.

Waves having high sound pressure levels, lasting sometimes for hours, are (1) oscillations of the atmospheric jet stream, with periods over 300 s and a sound pressure level (SPL) of about 115 dB above a reference pressure of 2×10^{-5} pascals; (2) fluctuations in pressure on a windy day at sound pressure levels of about 125 dB; (3) radiation of auroral infrasound downward to the Earth's surface from the ionosphere, with periods of 20–200 s

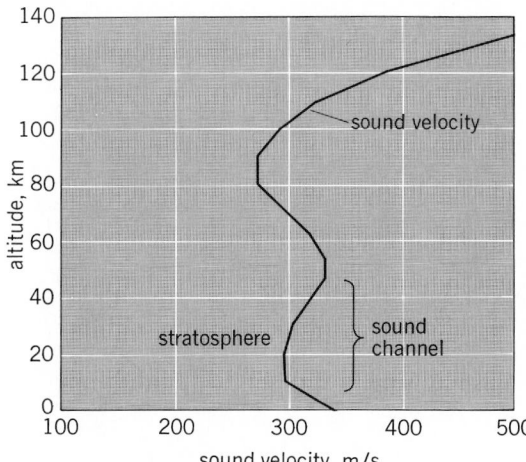

Fig. 1. Local sound velocity in the 1962 U.S. Standard Atmosphere. Details of the real atmosphere vary with location on the Earth's surface and with the seasons of the year. (*From R. K. Cook, Atmospheric sound propagation, in Proceedings of the Scientific Meetings of the Panel on Remote Atmospheric Probing, vol. 2, pp. 633–669, Committee on Atmospheric Sciences, National Academy of Sciences/National Research Council, January 1969*)

and sound pressure levels of about 90 dB; (4) microbaroms radiated landward from storms over the oceans, with periods of 4–7 s and sound pressure levels of about 85 dB.

Effects of air temperature on infrasound. At a frequency near 0.03 Hz (corresponding to a period of oscillation T of about 30 s) the wavelength of infrasound is almost 10 km. The infrasonic wave field then extends well up into the atmosphere. The variation of local sound speed with altitude is a gross feature which has a substantial effect on propagation (Fig. 1). The variation is due to the variation of atmospheric temperature with altitude. The speed minimum in the stratosphere causes the waves to be channeled between the ground and the layer of relatively high sound speed at a 50-km altitude. Loosely speaking, the layer serves as a reflector, albeit a poor one. For the shorter waves, where T is less than approximately 15 s, sound-ray trajectories are useful for studying propagation. In general, the rays from a source at ground level are alternately reflected between the layer at a 50-km altitude and the surface of the ground. Since the waves spend much of their propagation time in the relatively cool stratosphere, the speed over the Earth's surface is on the average of about 300 m/s.

Effects of gravity. At still lower frequencies, the gravitational field of the Earth influences the propagation of sound, particularly the phase velocity. The reason is that the potential energy of the air moving under gravity then becomes comparable to the usual potential energy of elastic compression and rarefaction in a sound wave. Such waves are called acoustic-gravity waves (Fig. 2). The approximately exponential decrease of atmospheric density with altitude z above the Earth's surface also influences sound propagation at all frequencies. For an isothermal atmosphere with a sound velocity $c = 333$ m/s, the density will decrease as exp $(-z/H)$, where H is the scale height of the atmosphere, approximately 8.1 km. Theory shows that the combined effect of gravity and decrease in

density causes the sound pressure for a plane wave of sound sent vertically upward to decrease as exp $(-z/2H)$, whereas the particle velocity will increase as exp $(+z/2H)$, so that the sound intensity remains constant. If the frequency of oscillation is decreased until the period $T_R = 4\pi c/\gamma g = 305$ s (where $\gamma = 1.40$ is the ratio of specific heats, and $g = 9.8$ m/s² is the acceleration of gravity), then the phase velocity of the upward traveling wave becomes infinite (Fig. 2). T_R is the resonant period for vertical oscillations of the atmosphere. But for waves of period less than about 100 s (with frequencies greater than about 10^{-2} Hz), gravity effects on sound speed are not significant.

Effects of infrasound on people. Pressure fluctuations at infrasonic frequencies arising from unsteady air flow are included as infrasound, even though they do not result in significant radiated sound energy. At a normal walking speed—two steps per second, for instance—a person experiences body motions at a frequency of 2 Hz and is aware of these motions through the kinesthetic and touch sensations. A running person has body motions at appreciably higher frequencies, perhaps 4–6 Hz. When a person's head moves vertically over a range of 3 cm while walking, the head is subjected to atmospheric pressure fluctuations of about a 0.4-Pa amplitude (86 dB above 2×10^{-5} Pa) at infrasonic frequencies. These arise from the vertical pressure gradient of the atmosphere caused by gravity. Often there is infrasound in the atmosphere having sinusoidal pressures of about 80–85 dB, in the range 0.10–0.20 Hz. These are the ubiquitous microbaroms, originating with surface waves on the seas and oceans. When the wind blows, turbulent pressure fluctuations in the atmosphere occur at amplitudes up to tens of pascals, at infrasonic frequencies. People are unaware of these pressures via the sensation of hearing.

But sufficiently strong infrasound is "audible," contrary to simple acoustic tradition. The threshold sound pressure level (the least intensity for audibility) is about 92 dB at 16 Hz, and increases 12 dB per octave to about 140 dB at 1.0 Hz. However, there is no sensation of tone. Listeners variously describe audible infrasound as "pumping," "popping effect," or "chugging."

With the advent of civilization and particularly in modern times, people have been subjected to increasing amounts of infrasound and vibrations in the range of frequencies 1–20 Hz. The physiological effects are commingled with, and difficult to separate from, the effects of whole-body oscillations in the same range of frequencies.

Motion sickness of people in boats must have been one of the earliest noticeable effects of vibration at very low frequencies. The human body is particularly sensitive to vibrations and infrasound near 7 Hz, at which frequency there is an overall mechanical resonance of organs in the abdominal and chest cavities. Unpleasant physiological effects with pathological overtones occur when human subjects and animals are exposed to infrasound in the laboratory at sound pressure levels of 130 dB and greater. The equipment and machinery of modern technology can expose people to infrasound which is much stronger and occurs more often than in the natural environment. For example, substantial pressure oscillations at a frequen-

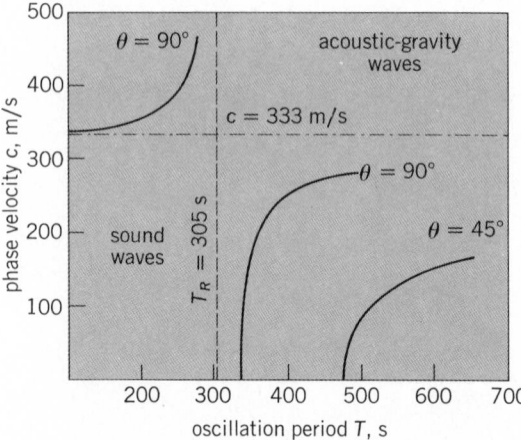

Fig. 2. Phase velocities for sound waves in the atmosphere at infrasonic frequencies. $\theta =$ angle between propagation direction and the horizontal plane. (*From R. K. Cook, Atmospheric sound propagation, in Proceedings of the Scientific Meetings of the Panel on Remote Atmospheric Probing, vol. 2, pp. 633–669, Committee on Atmospheric Sciences, National Academy of Sciences/National Research Council, January 1969*)

cy of a few hertz can occur in an automobile driven at a moderate speed with windows open. Artificial sources include powerful explosions and the sonic booms from supersonic aircraft. However, scientific data have refuted earlier speculations on the hazards of infrasound to people. *See* ATMOSPHERIC ACOUSTICS; SOUND.   [RICHARD K. COOK]

Bibliography: R. K. Cook, Atmospheric sound propagation, in *Proceedings of the Scientific Meetings of the Panel on Remote Atmospheric Probing*, vol. 2, pp. 633–669, Committee on Atmospheric Sciences, National Academy of Sciences/National Research Council, January 1969; Special issue on infrasonics and atmospheric acoustics, *Geophys. J. Roy. Astron. Soc.*, vol. 26, nos. 1–4, December 1971; W. Tempest (ed.), *Infrasound and Low Frequency Vibration*, 1976.

Instanton

A solution to euclidean (imaginary time) equations of motion that interpolates between two classical ground states. Consider, for example, a particle of mass m in a "double-well" potential $V(x)$ (illustration *a*). Classical ground states correspond to

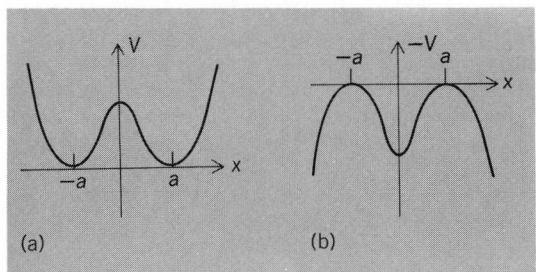

Potentials for the example of an instanton. (*a*) "Double-well" potential with classical ground states at $x = \pm a$. (*b*) "Double-hill" potential for the euclidean equation of motion, which has a solution that is an instanton.

the particle sitting at $x = \pm a$. Replacing τ by it (where t is the time coordinate and $i = \sqrt{-1}$) in the equation of motion

$$m d^2 x / dt^2 = -dV/dx \qquad (1)$$

one obtains the euclidean equation

$$m d^2 x / d\tau^2 = -dV_E/dx \qquad (2)$$

where the potential $V_E = -V$ is the "double-hill" potential (illustration *b*). The solution of Eq. (2),

$$x(\tau) = a \tanh (2a\tau/m) \qquad (3)$$

corresponding to the particle leaving $x = -a$ at $\tau = -\infty$, rolling down the valley and reaching $x = a$ at $\tau = \infty$, is an instanton by the above definition.

The most widely used example is the BPST (Belavin-Polyakov-Schwartz-Tyupkin) instanton from the Yang-Mills theory, which is a kind of generalized electromagnetic theory with many mutually interacting electromagnetic fields. (These fields are believed to bind the quarks to form particles such as the proton and neutron. The ground states which the BPST instanton connects have zero field strength; that is, they correspond to free space. Since this field configuration differs (appreciably)

from free space only for a finite time τ, G. 'tHooft named it the instanton. He also pointed out that, at the quantum level, instantons mediate tunneling between the approximate ground states with wave functions peaked at the classical minima ($x = \pm a$ in the above example). *See* GLUONS; NONRELATIVISTIC QUANTUM THEORY; QUANTUM CHROMODYNAMICS.

[R. SHANKAR]

Bibliography: A. A. Belavin et al., Pseudoparticle solutions of the Yang-Mills equations, *Phys. Lett.*, 59B:85–87, 1975; G. 'tHooft, Symmetry breaking through Bell-Jackiw anomalies, *Phys. Rev. Lett.*, 37:8–11, 1976.

Integral equation

An equation of the form typyfied by Eq. (1). The major problem is to decide when there is a function $\phi(x)$ which is a solution to the equation. Equations such as (1) arise from the analysis of ordinary differential equations. For example, it is easy to verify that the function $\phi(t)$ is a solution of differential equation (2) if and only if ϕ is a solution of integral equation (3). This equation is the same type as Eq. (1), with the function K given by Eq. (4).

$$\int_a^b K(x,y,\phi(y))dy + f(x) = a(x)\phi(x) \qquad (1)$$

$$\frac{d\phi}{dt} = F(t,\phi) \qquad \phi(0) = A \qquad (2)$$

$$\phi(t) = A + \int_0^t F(s,\phi(s))ds \qquad (3)$$

$$K(t,s,\phi) = \begin{cases} F(s,\phi) & 0 \le s \le t \\ 0 & t < s \end{cases} \qquad (4)$$

Integral equation (1) is an equation of the first kind if $a(x) \equiv 1$, and an equation of the second kind if $a(x) \equiv 0$.

The domain of integration, which in Eq. (1) is the interval $[a,b]$, can be a domain in two or more variables, a curve, or even a surface. The unknown function ϕ and the kernel K can be vector-valued functions, in which case Eq. (1) is called a system of integral equations.

Volterra equations. If $K(x,y,\phi) = 0$ for $x \le y \le b$, Eq. (1) is called a Volterra equation. Thus Eq. (3) is an example of a Volterra equation. Under mild assumptions about the smoothness of $K(x,y,\phi)$, for example, if K is continuous and if $|K(x,y,\phi) - K(x,y,\psi)| \le A|\phi - \psi|$, where A is a constant, for all x,y in the domain of integration and for all ϕ and ψ—then the Volterra equation of the second kind, Eq. (5), can be solved by the method of successive

$$\phi(x) = \int_a^x K(x,y,\phi(y))dy + f(x) \qquad (5)$$

approximations (also known as the method of Picard). The idea is to define a sequence of functions $\phi_0, \phi_1, \ldots$ according to the scheme of Eqs. (6). Under the assumed conditions, it can be shown

$$\phi_0(x) = f(x)$$
$$\phi_1(x) = \int_0^x K(x,y,\phi_0(y))dy + f(x)$$
$$\cdots \qquad (6)$$
$$\phi_{n+1}(x) = \int_0^x K(x,y,\phi_n(y))dy + f(x)$$

that this sequence converges to a function $\phi(x)$ which is the only solution to Eq. (5). The method of Picard is fundamental to the study of integral equations, but occurs also in the study of many other functional equations.

Linear integral equations. A special case of some interest is that of linear integral equations. The function $K(x,y,\phi)$ is a linear function of ϕ. The linear equations of the first and second kind are shown in Eqs. (7) and (8) respectively. The function

$$f(x) = \int_a^b K(x,y)\phi(x)\,dy \tag{7}$$

$$\phi(x) = f(x) + \lambda \int_a^b K(x,y)\phi(y)\,dy \tag{8}$$

$K(x,y)$ is called the kernel, and the complex number λ is called the parameter. The equation of the second kind, Eq. (8), is homogeneous if $f(x) \equiv 0$. In typical cases, the homogeneous equations will have only the trivial solution $\phi(x) \equiv 0$. For some values of the parameter λ, however, there will be nontrivial solutions. Such a value of the parameter λ is called a characteristic value for K, and the corresponding function ϕ is called a characteristic function for K. These concepts are related to corresponding concepts in linear algebra and operator theory. The operator L_K acting on the function ϕ is defined by Eq. (9). Then L_K is an example of a lin-

$$L_K\phi(x) = \int_a^b K(x,y)\phi(y)\,dy \tag{9}$$

ear operator, that is, if ϕ and ψ are functions and α and β are complex constants, then $L_K(\alpha\phi + \beta\psi) = \alpha L_K\phi + \beta L_K\psi$. If the function ϕ is a characteristic function for K with corresponding characteristic value λ, then the homogeneous equation can be rewritten as Eq. (10). In terms of linear operator

$$L_K\phi = \frac{1}{\lambda}\phi \tag{10}$$

theory, Eq. (10) means that $1/\lambda$ is an eigenvalue (proper value) for the linear operator L_K, and ϕ is a corresponding eigenvector (proper vector), or in this case, eigenfunction. A linear Volterra equation has no characteristic values. *See* EIGENFUNCTION; MATRIX CALCULUS; MATRIX THEORY; OPERATOR THEORY.

Fredholm equations. The integral equation, Eq. (7) or (8), is called a Fredholm equation, and $K(x,y)$ is called a Fredholm kernel if $\|K\| < \infty$, where $\|K\|$, the norm of K, is defined by Eq. (11).

$$\|K\|^2 = \int_a^b \int_a^b |K(x,y)|^2 dx\,dy \tag{11}$$

This is certainly true if K is continuous or even bounded. Even some infinite discontinuities are allowable: for example, if $|K(x,y)| \leq M|x - y|^{-\alpha}$, where M is a constant and $\alpha < 1/2$, then K is a Fredholm kernel.

Fredholm equations of the second kind for which the parameter satisfies the inequality $|\lambda|\,\|K\| < 1$ can be shown to have unique solutions by the method of Picard. The result is Eq. (12) where the resolvent $R_K(x,y;\lambda)$ is given by the Neumann series, Eqs. (13) and (14). Equation (12) can be rewritten in terms of the operator L_K as Eq. (15), when $L_K^2 f = L_K(L_K f), \ldots, L_K^{n+1}(f) = L_K(L_K^n f)$.

$$\phi(x) = f(x) + \lambda \int R_K(x,y;\lambda)f(y)\,dy \tag{12}$$

$$R_K(x,y;\lambda) = \sum_{n=1}^{\infty} \lambda^{n-1}K^{(n)}(x,y) \tag{13}$$

$$K^{(1)}(x,y) = K(x,y) \tag{14}$$

$$K^{(n+1)}(x,y) = \int_a^b K(x,z)K^{(n)}(z,y)\,dz$$

$$\phi = \sum_{n=0}^{\infty} \lambda^n L_K^n f \tag{15}$$

As a result there are no characteristic values for which $|\lambda| \leq 1/\|K\|$. In general, for any constant $M > 0$ there are only finitely many characteristic values which satisfy $|\lambda| \leq M$. Further, only finitely many linearly independent characteristic functions are associated with each characteristic value.

For the kernel K the conjugate kernel is defined to be $K^*(x,y) = \overline{K(y,x)}$, where the bar denotes complex conjugation. It can be shown that λ is a characteristic value for the kernel K if and only if $\bar{\lambda}$ is a characteristic function for the conjugate kernel K^*. Furthermore, the number of linearly independent corresponding characteristic functions is the same for λ and K as for $\bar{\lambda}$ and K^*. Finally, inhomogeneous equation (16) has a solution if and only if f satisfies Eq. (17) for every solution ψ of conjugate homogeneous equation (18). These results are a

$$\phi(x) = f(x) + \lambda \int_a^b K(x,y)\phi(y)\,dy \tag{16}$$

$$\int_a^b f(x)\,\overline{\psi(x)}\,dx = 0 \tag{17}$$

$$\psi(x) = \lambda \int_a^b K^*(x,y)\psi(y)\,dy \tag{18}$$

statement of the Fredholm alternative which may be stated more succinctly as follows: For a given value of λ, either λ is a characteristic value and the homogeneous equation has a nontrivial solution, or λ is not a characteristic value and the inhomogeneous equation has a solution for every choice of the function f. *See* DIFFERENTIAL EQUATION; INTEGRAL TRANSFORM; INTEGRATION.

[JOHN POLKING]

Bibliography: H. Hochstadt, *Integral Equations*, 1973; B. L. Moiseiwitch, *Integral Equations*, 1977.

Integral transform

An integral relation between two classes of functions. For example, a relation such as Eq. (1) is said to define an integral transform. More generally, the integral may be a multiple integral, and the functions f, G, ϕ may depend on a larger number of variables. Equation (1) is thought of as transforming a

$$f(x) = \int_{-\infty}^{\infty} G(x,y)\phi(y)\,dy \tag{1}$$

whole class, or space, of functions $\phi(y)$ into another class of functions $f(x)$. The function $G(x,y)$ is the kernel of the transform. One of the important uses of such a transform is based on the fact that a problem posed in one of the two spaces in question may be more easily solved in the other. For example, a differential equation to be solved for the function $\phi(y)$ may become an algebraic equation for the unknown function $f(x)$. *See* LAPLACE TRANSFORM.

The two basic problems for any integral transform are inversion and representation. In inversion the aim is to recover $\phi(y)$ from $f(x)$, the kernel $G(x,y)$ being known. That is, Eq. (1) is thought of as an integral equation (of the first kind) to be solved for the unknown function $\phi(y)$. A means of calculating $\phi(y)$ from $f(x)$ is called an inversion formula, and in its presence the transform achieves maximum utility, since explicit passage from each space to the other is thus assured. In representation the question is which functions $f(x)$ may be written or represented in the form (1). That is, one asks which functions $f(x)$ will make Eq. (1) solvable for $\phi(y)$. Usually this problem becomes more tractable when the solutions $\phi(y)$ are restricted to some subspace such as the class of positive or bounded functions.

Inversion theory. In many cases where the inversion problem has been solved, an inversion operator, such as a differential or integral operator, has been found, depending perhaps on a parameter, which accomplishes the inversion. It may be denoted by $O_t[f(x)]$. This means that for each value of the parameter t some operation O on $f(x)$ is defined. For example, t might be 3 and the operation might consist of computing the third derivative of $f(x)$. Suppose that O_t applied to Eq. (1) produces Eq. (2), where $G_t(x,y)$ is a new kernel for

$$O_t[f(x)] = \int_{-\infty}^{\infty} G_t(x,y)\phi(y)\,dy \qquad (2)$$

each t. In favorable cases the integral (2) approaches $\phi(x)$ as t approaches a suitable limit, for example, when G_t approaches the Dirac δ-function $\delta(x-y)$.

An important special case of Eq. (1) is the convolution transform, when the kernel is a function of $(x-y)$, Eq. (3). An equivalent form of Eq. (3) is Eq. (4), since the change of variable $x = e^t$, $y = e^{-u}$ carries Eq. (4) into Eq. (3) after a suitable change in notation.

$$f(x) = \int_{-\infty}^{\infty} G(x-y)\phi(y)\,dy \qquad (3)$$

$$F(x) = \int_{0}^{\infty} K(xy)\Phi(y)\,dy \qquad (4)$$

ries Eq. (4) into Eq. (3) after a suitable change in notation.

Named transforms. Many of the classical transforms have been named for the original or principal investigator. A few of the more important ones can be listed with the names ordinarily attached to them.

For Eq. (1):
A. Bilateral Laplace $G(x,y) = e^{-xy}$
B. Fourier $G(x,y) = e^{ixy}$
C. Mellin $G(x,y) = y^{x-1}$ $y > 0$
           $= 0$ $y < 0$
D. Stieltjes $G(x,y) = (x+y)^{-1}$ $y > 0$
           $= 0$ $y < 0$

For Eq. (3):
E. Dirichlet $G(x) = x^{-1} \sin x$
F. Fejér $G(x) = x^{-2} (\sin x)^2$
G. Fractional (Weyl form) $G(x) = (-x)^{\mu-1}$ $x < 0$
           $= 0$ $x > 0$

H. Hilbert $G(x) = x^{-1}$
I. Picard $G(x) = e^{-|x|}$
J. Poisson $G(x) = (1+x^2)^{-1}$
K. Laplace (unilateral) $G(x) = e^x e^{-e^x}$
L. Stieltjes $G(x) = \operatorname{sech}(x/2)$
M. Weierstrass or Gauss $G(x) = e^{-x^2}$

For Eq. (4):
N. Fourier cosine $K(x) = \cos x$
O. Fourier sine $K(x) = \sin x$
P. Hankel $K(x) = \sqrt{x} J_\nu(x)$
Q. Laplace (unilateral) $K(x) = e^{-x}$
R. Meijer $K(x) = \sqrt{x} K_\nu(x)$

Here $J_\nu(x)$ is a Bessel function and $K_\nu(x)$ a modified Bessel function. Certain transforms are listed twice because of their frequent appearance in two equivalent forms. In fact, all of the above are convolution transforms except A, B, and C. It can be seen also that A and C are the same transform (set $y = e^{-t}$ in C).

Convolution inversion. The inversion theory for a large class of convolution transforms has been completed. Operational calculus is a guide to the method. If the symbol D stands for differentiation with respect to x but is nonetheless treated as a number in the familiar series represented by Eq. (5), one is led to make the definition in Eq. (6),

$$e^{tD} = \sum_{k=0}^{\infty} \frac{t^k}{k!} D^k \qquad (5)$$

$$e^{tD} f(x) = \sum_{k=0}^{\infty} \frac{t^k}{k!} D^k f(x) = f(x+t) \qquad (6)$$

since the series (6) is precisely the Maclaurin expansion of $f(x+t)$ if D^k is allowed to mean a kth derivative. *See* OPERATOR THEORY.

Now consider the bilateral Laplace transform of $G(t)$ in Eq. (1), denoting it by $1/E(s)$, as shown in Eq. (7). Replacing s by D and using Eq. (6), one has Eq.

$$\frac{1}{E(s)} = \int_{-\infty}^{\infty} e^{-st} G(t)\,dt \qquad (7)$$

(8). If $E(D)$ were a number, one could solve Eq. (8)

$$\begin{aligned}
\frac{1}{E(D)}\phi(x) &= \int_{-\infty}^{\infty} e^{-tD}\,\phi(x)G(t)\,dt \\
&= \int_{-\infty}^{\infty} \phi(x-t)G(t)\,dt \\
&= \int_{-\infty}^{\infty} G(x-y)\phi(y)\,dy \\
&= f(x) \qquad x-t = y \qquad (8)
\end{aligned}$$

to obtain Eq. (9). Finally, reverting to the original meaning of D as a derivative, one has in Eq. (9) an

$$\phi(x) = E(D)f(x) \qquad (9)$$

inversion of Eq. (1) by means of a differential operator $E(D)$.

This argument is meant to be exploratory only, but the result is accurate for a large class of kernels G and their corresponding inversion functions

E. In summary, the inversion function is the reciprocal of the bilateral Laplace transform of the kernel. It has been shown that the result is correct if, for example, $E(s)$ is the infinite product in Eq. (10),

$$E(s) = e^{bs - cs^2} \prod_{k=1}^{\infty} \left(1 - \frac{s}{a_k}\right) e^{s/a_k} \qquad (10)$$

where $c \geqq 0$ and the series of real constants $\sum_{k=1}^{\infty} a_k$ converges.

For example, if $K(x) = e^{-x}$, then Eq. (4) is the Laplace transform. Expressed as a convolution transform as in Eq. (3), it becomes Eq. (11), where G is

$$e^x F(e^x) = \int_{-\infty}^{\infty} G(x-y) \Phi(e^{-y}) \, dy \qquad (11)$$

given in the above list as entry K. The bilateral Laplace transform of this kernel is the familiar gamma function, Eq. (12), whose reciprocal has a

$$\Gamma(1-s) = \int_{-\infty}^{\infty} e^{-st} G(t) \, dt = \int_{0}^{\infty} e^{-t} t^{-s} \, dt \quad (12)$$

well-known expansion in the form of Eq. (10). In Eq. (13) γ is Euler's constant. In the present example Eq. (9) becomes Eq. (14), or if $e^{-x} = t$, Eq. (15)

$$E(s) = \frac{1}{\Gamma(1-s)} = e^{-\gamma s} \prod_{k=1}^{\infty} \left(1 - \frac{s}{k}\right) e^{s/k} \quad (13)$$

$$e^{-\gamma D} \prod_{k=1}^{\infty} \left(1 - \frac{D}{k}\right) e^{D/k} e^x F(e^x) = \Phi(e^{-x}) \quad (14)$$

$$\lim_{k \to \infty} \frac{(-1)^k}{k!} F^{(k)} \left(\frac{k}{t}\right) \left(\frac{k}{t}\right)^{k+1} = \Phi(t) \qquad (15)$$

may be written. This familiar inversion formula also serves to illustrate the operator O_t appearing in Eq. (2). In the present case the operator is a differential one, and the parameter t is an integer k which tends to ∞. *See* CONFORMAL MAPPING; INTEGRATION.   [DAVID V. WIDDER]

Bibliography: B. Davies, *Integral Transforms and Their Applications*, 1978; A. Erdelyi (ed.), *Tables of Integral Transforms*, 2 vols., 1954; I. N. Sneddon, *The Use of Integral Transforms*, 1972; D. V. Widder, *Transform Theory*, 1971; K. B. Wolf, *Integral Transforms in Science and Engineering*, 1978.

Integrated optics

The study of optical devices, singly and in combinations, that are based on light transmission in waveguides, that is, structures that confine the propagating light to a region with one very small dimension (or sometimes two), of the order of the wavelength of the light. The principal motivation for these studies is to enable the combination of individual devices thus miniaturized, through waveguides or other means, into a functional optical system mounted on a small substrate. The resulting system is called an integrated optical circuit (IOC) by analogy with the semiconductor type of integrated circuit. An integrated optical circuit could, for example, include a laser, switches, polarizers, modulators, detectors, and so forth. An important use envisioned for integrated optical circuits is in connection with optical communica-

tions through the medium of glass fibers, which are themselves waveguides. Integrated optical circuits could be used in such a system as optical transmitters, repeaters, and receivers. *See* OPTICAL FIBERS.

The advantages of having an optical system in the form of an integrated optical circuit rather than a conventional series of components on an optical bench, include, apart from miniaturization, the reduced sensitivity to ambient-temperature gradients and changes, to airborne acoustical effects, and to mechanical vibrations of the separately mounted parts. As in the case of semiconductor integrated circuits, the integrated optical circuit might be made of essentially one material, modified for the different components by incorporating suitable substituents or dopants, or of many different materials. The latter option, called hybrid, has the advantage that each component could be optimized, for example, by using a gallium arsenide or neodymium yttrium-aluminum-garnet (YAG) laser, a lithium niobate modulator or switch, a silicon detector, and so forth. In the former case, the integrated optical circuit is called monolithic and is expected to have the advantage of ease of processing, similar to the situation for monolithic semiconductor integrated circuits. The most promising material for monolithic integrated optical circuits is gallium arsenide, since with suitable substituents this material may be made into a laser, switch, modulator, detector, and so forth.

Guided waves. The simplest type of waveguide is a three-layer or sandwich structure with the index of refraction largest in the middle layer, designated "film" in Fig. 1. The top layer or "cover" is very frequently air, and it will be assumed that is the case in what follows. In a guided wave the light is not distributed uniformly across the guide, but in a pattern that depends on the indices of refraction of all three layers and the height of the film. The pattern may be a simple one, such as one of those shown in Fig. 1a, or higher numbers in that sequence, or it may be a superposition of such patterns. In the simplest possible guided waves, called modes of the guide, a pattern with m zeroes, such as one of those in Fig. 1a, is swept down the guide with a speed characteristic for that particular pattern. The mode may also be described by a characteristic wave vector k_m for the mth mode, directed along the path of the light, that is, along the guide.

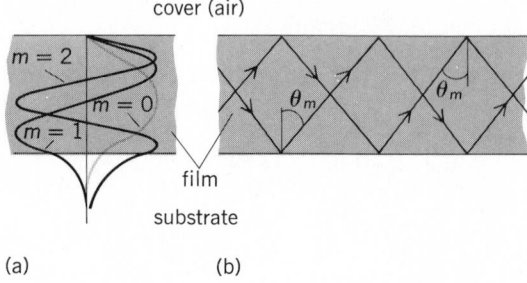

Fig. 1. Optical waveguide. (*a*) Light amplitude variation across waveguide for first few guided modes. Each mode is assigned a number *m* equal to number of times the light amplitude goes to zero inside film. (*b*) Ray picture of guided waves. (*From E. Conwell, Integrated optics, Physics Today, 29(5):48–56, 1976.*)

A pattern such as one of those in Fig. 1*a* represents a standing wave, which is equivalent to two progressing waves moving in opposite directions. To form the guided wave, one superimposes on these the appropriate uniform speed along the guide. The guided wave is thus equivalent to two rays bouncing back and forth with a characteristic angle θ_m, as indicated in Fig. 1*b*. From this point of view it is readily seen that the larger index of the middle layer keeps the wave confined because there is total internal reflection of the two rays every time they strike the surfaces.

Materials and fabrication. Waveguides have been made of many different materials with many different techniques. The film has been made of sputtered glass, sputtered oxides of tantalum or zinc, epitaxial gallium arsenide, ion-bombarded gallium arsenide, epitaxial garnets, sputtered and epitaxial lithium niobate, nitrobenzene liquid, nematic liquid crystals, and a number of other organics and polymers.

Most of the materials with desirable electrooptic or nonlinear properties for active waveguides or waveguide devices are single crystals, high-quality thin films of which are difficult to obtain. A good way to obtain waveguiding in such a case is to create a thin layer of higher refractive index at the top surface of a suitable single crystal by diffusion or ion exchange. With lithium niobate, for example, satisfactory guides may be made by heating in vacuum to diffuse out lithium oxide, or by diffusing in various metallic impurities, such as titanium. Although this type of treatment yields a region in which the refractive index varies with distance in a thin layer, rather than being constant as in the middle layer of the sandwich guide, guided waves can be obtained provided the refractive index decreases with distance below the surface. The modes are rather similar to those shown in Fig. 1.

Waveguides that confine light in two dimensions, rather than one, consist of a channel or strip of higher refractive index than its surroundings. A channel guide can be made by ion implantation or diffusion through a mask into a thin film, a raised strip guide by masking the desired strip region and removing the surrounding film by sputtering or etching.

Coupling of external light beams. An external light beam may be coupled into a waveguide by focusing the light on the end of the guide, but this is an inefficient process because the height of the film is much less than the width of the usual laser beam. It is clear, however, that light cannot be coupled in by illuminating the top of the guide from any angle because, as noted earlier, the guided light is kept inside by total internal reflection. (A light ray must be able to travel the same path in either direction.) It can, however, be coupled in from the top if it first goes through a prism made of higher-refractive-index material (Fig. 2). If the prism is close enough to the surface of the guide, within a wavelength of light, light may leak through the air gap into the guide. The exact condition to be satisfied is that the component of the propagation vector in the prism parallel to the guide be equal to the characteristic wave vector k_m of some guide mode. This is achieved by varying the angle of the incident beam. Prism coupling is capable of over 90% efficiency, but does not ap-

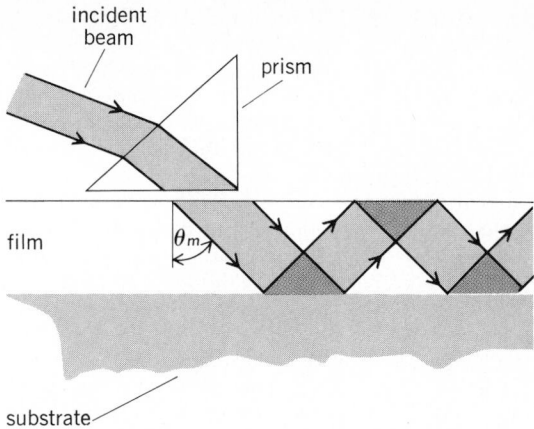

Fig. 2. Prism coupler. (*From E. Conwell, Integrated optics, Phys. Today, 29(5):48–56, 1976*)

pear to lend itself well to integration. A coupler that is generally less efficient but more suitable for integration is the grating coupler (Fig. 3). The grating is usually made of a thin layer of photoresist that has been exposed to an optical interference pattern and then developed. Its action can be described by saying that, as a periodic structure (with period *d*), it exchanges momentum with the light wave, making it possible to obtain a match with some wave vector k_m. Output coupling is provided by the same type of couplers.

Modulators. Modulators are used to impress information on the guided light wave. Many integrated optics modulators are based on the same principles as many bulk modulators, notably the change in indices of refraction caused by an electric field in materials such as lithium niobate. The main difference is that the light is confined in the small area of the waveguide, providing the advantage of a much smaller power requirement. Another type of modulator, which does not have a bulk analog, is based on the principle of the directional coupler used in microwave systems. This acts by controlling electrically the switching of a guided wave from one channel waveguide to a parallel and identical one, spaced within about a wavelength of light from the first.

Lasers. The gallium arsenide (GaAs) diode laser is already an integrated optics device in a sense, since its efficiency is due partly to waveguiding arising from refractive index differences between

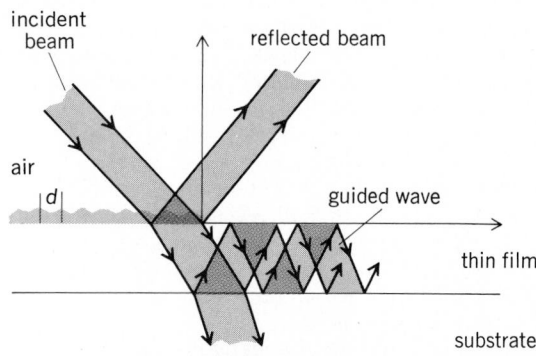

Fig. 3. Grating coupler. (*From E. Conwell, Integrated optics, Phys. Today, 29(5):48–56, 1976*)

regions of different impurity content. More deliberate progress has been made toward integrated-optics types of lasers by using waveguides with a periodic variation in their properties. This is most easily achieved as a periodic variation in width, which may be created by the technique described for the grating coupler. The periodic variation results in a strong reflection of waves with wave vector equal to π/d, d being the spatial period. Thus, by using for the active medium a periodic waveguide with d chosen so that the laser frequency is strongly reflected, it is possible to dispense with end mirrors. The feedback provided by the end mirrors, in other words, is replaced by distributed feedback. Alternatively, periodic structures of this kind can be used at the ends of the active medium to act as mirrors. *See* LASER.

[ESTHER M. CONWELL]

Bibliography: E. M. Conwell and R. D. Burnham, Materials for integrated optics: GaAs, in *Annu. Rev. Mater.*, 8:135–179, 1978; Special issue on integrated optics, *IEEE Trans. Microwave Theory Tech.*, MTT-23(1):1–180, 1975; T. Tamir (ed.), *Integrated Optics*, 1975; P. K. Tien, Integrated optics and new wave phenomena in optical waveguides, *Rev. Mod Phys.*, 49(2):361–420, 1977.

Integration

An operation of the infinitesimal calculus which has two aspects. The roots of one go back to antiquity, for Archimedes and other Greek mathematicians used the "method of exhaustion" to compute areas and volumes. A simple example of this is the approximation to the area of a circle obtained by inscribing a regular polygon of known area, and then repeatedly doubling the number of sides. The areas of the successive polygons are computable with the help of elementary geometry. The limit of the sequence of these areas gives the area of the circle. The area of each polygon can be regarded as being made up of the sum of the areas of triangles with vertices at the center of the circle, and so the process described is a constructive definition of an integral which is the limit of a sum. Modern definitions of integrals as limits of sums are discussed in this article.

The other aspect of integration is the process of finding antiderivatives, that is, for a given function $f(x)$ to find another function $g(x)$ whose derivative is $f(x)$. This aspect is related to the first by the fundamental theorem of integral calculus, so both processes are called integration.

Sir Isaac Newton emphasized the antiderivative aspect of integration, and his work shows how much can be done in the applications of integral calculus without introducing limits of sums. However, limits of sums lead to very fruitful theoretical developments in the theory of integration, as in the notion of multiple integrals, for example, and hence lead to a wider variety of applications. Leibnitz, a 17th-century mathematician, inspired the development in this direction, but many years elapsed before the theory was given a firm logical foundation. In the early 19th century A. L. Cauchy gave a clear-cut definition of the definite integral for continuous functions and a proof of its existence. Later, G. F. B. Riemann discussed the integral for discontinuous functions and gave a necessary and sufficient condition for its existence. Thus the most generally used definition of the integral as

the limit of a sum has come to be called the Riemann integral.

Riemann integral. The precise definition of the Riemann integral for a real function f of one real variable x on a finite interval $a \leqq x \leqq b$ may be formulated as follows. Let P be a partition of the interval $[a,b]$ into n subintervals by points t_i, where $t_{i-1} < t_i, t_0 = a$, $t_n = b$, and consider a sum S of the form of Eq. (1), where $t_{i-1} \leqq x_i \leqq t_i$. The sum S

$$S = \sum_{i=1}^{n} f(x_i)(t_i - t_{i-1}) \tag{1}$$

depends not only on the partition P but on the choice of the intermediate points x_i. It may happen that the sum S approaches a definite limit I when the maximum of the numbers $(t_i - t_{i-1})$ tends to zero, and in this case I is called the Riemann integral (or the definite integral) of f from a to b, and is denoted by the Leibnitzian symbol (2).

$$\int_a^b f(x)\, dx \tag{2}$$

Also, f is said to be integrable on $[a,b]$. When f is a continuous function with positive values on the interval $[a,b]$, the integral has a simple geometrical interpretation as the area bounded by the x axis, the ordinates $x = a$ and $x = b$, and the graph of $y = f(x)$ (Fig. 1). It is convenient to use Eq. (3) as a definition of the symbol on the left.

$$\int_b^a f(x)\, dx = -\int_a^b f(x)\, dx \tag{3}$$

For functions f and g which are integrable on $[a,b]$, the integral has the following properties:

(i) f is integrable on every subinterval of $[a,b]$.
(ii) For every triple of points c, d, and e in $[a,b]$,

$$\int_c^d f(x)\, dx + \int_d^e f(x)\, dx = \int_c^e f(x)\, dx$$

(iii) $f(x) + g(x)$ is integrable, and

$$\int_a^b [f(x) + g(x)]\, dx = \int_a^b f(x)\, dx + \int_a^b g(x)\, dx$$

(iv) $f(x)g(x)$ is integrable, and in particular $cf(x)$ is integrable for every real number c, and

$$\int_a^b cf(x)\, dx = c \int_a^b f(x)\, dx$$

(v) If $f(x) \leqq g(x)$ on $[a,b]$, then

$$\int_a^b f(x)\, dx \leqq \int_a^b g(x)\, dx$$

(vi) $|f(x)|$ is integrable, and

$$\left| \int_a^b f(x)\, dx \right| \leqq \int_a^b |f(x)|\, dx$$

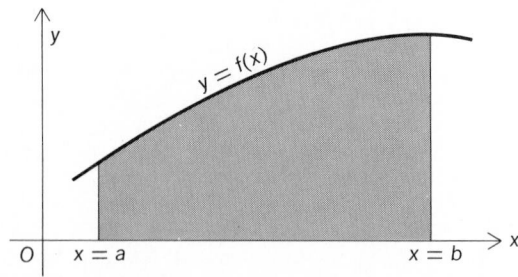

Fig. 1. Graph of $y = f(x)$.

It can be proved that a function f is integrable on $[a,b]$ if and only if the following two conditions are satisfied: f is bounded on $[a,b]$; and the set of points where f is discontinuous can be enclosed in a series (possibly infinite) of intervals, the sum of whose lengths is arbitrarily near to zero.

Antiderivatives. To develop the fundamental theorem of integral calculus let u be a variable in the interval $[a,b]$, on which f is integrable; then formula (4) defines a function of u which may be

$$\int_a^u f(x)\ dx \qquad (4)$$

noted by $I(u)$. If f is continuous on $[a,b]$, then f is integrable, and it is also true that $I(u)$ has a derivative $I'(u)=f(u)$. Now let h be any antiderivative of f, that is, $h'(u)=f(u)$ on $[a,b]$. Then $I'(u)-h'(u)=0$, so $I(u)-h(u)=$ constant $=-h(a)$, by the theorem of the mean for derivatives and the fact that $I(a)=0$, so relation (5) can be written. This is the funda-

$$I=I(b)=h(b)-h(a) \qquad (5)$$

mental theorem of integral calculus, and it shows that definite integrals may be calculated by the process of finding antiderivatives. For this reason antiderivatives are frequently called indefinite integrals and denoted by $\int f(x)\,dx$, and special methods of finding indefinite integrals for frequently occurring functions occupy a large part of elementary calculus. The principal methods are outlined in the next section. The standard notation for an indefinite integral of $f(x)$ is formula (6).

$$\int f(x)\ dx \qquad (6)$$

Elementary methods of integration. Obviously, each formula for differentiation yields a formula for indefinite integration.

Method of substitution. From these indefinite integrals, additional formulas can be obtained by the method of substitution, which is based on the chain rule for differentiating composite functions. For example, from formula (7) one obtains Eq. (8), by the substitution $v=a^2+x^2$, and Eq. (9), by the substitution $v=\sin x$.

$$\int x^m\ dx = \frac{x^{m+1}}{m+1} \qquad (7)$$

$$\int (a^2+x^2)^m x\ dx = \frac{(a^2+x^2)^{m+1}}{2(m+1)} \qquad (8)$$

$$\int \sin^m x \cos x\ dx = \frac{\sin^{m+1} x}{m+1} \qquad (9)$$

Method of partial fractions. In case $f(x)$ is a quotient of two polynomials in x, $f(x)$ may be represented as a sum of a polynomial and terms of the form, shown as notations (10) and (11), where m is a posi-

$$\frac{1}{(ax+b)^m} \qquad a \neq 0 \qquad (10)$$

$$\frac{Ax+B}{(cx^2+dx+e)^m} \qquad c \neq 0, d^2-4ce<0 \qquad (11)$$

tive integer and the coefficients are real. This is called the method of partial fractions. Its actual execution requires a knowledge of the factors of the denominator of $f(x)$. Terms of the form (10) have known indefinite integrals. Terms of the form (11) with $m=1$ can also be integrated by elementary formulas. When $m>1$, a term of the form (11) can be written as a sum of terms of the forms (12)

and (13), where $\alpha=(2dA-4cB)/(d^2-4ce)$, $\beta=$

$$\frac{\alpha}{(\gamma x^2+dx+e)^{m-1}} \qquad (12)$$

$$\frac{(\beta x+\gamma)\ (2cx+d)}{(cx^2+dx+e)^m} \qquad (13)$$

$-\alpha/2$ and $\gamma=(dB-2eA)/(d^2-4ce)$. By the process of integration by parts, the indefinite integral of a term of the form (13) can be expressed in terms of the integral of a term such as (11), with m replaced by $m-1$. Reduction formulas of this type are given in standard tables of integrals.

A rational function of $\sin x$ and $\cos x$ can be integrated by the method just described after the substitution $u=\tan (x/2)$ has been applied. This substitution reduces the problem to the integration of a rational function of u. Various functions involving radicals can be integrated by means of trigonometric substitutions or by other substitutions which reduce the problem to the integration of a rational function. For example, by setting $u=a+bx$, one finds Eq. (14), and when m is a nonnegative integer

$$\int x^m(a+bx)^p\ dx = \int \left(\frac{u-a}{b}\right)^m u^p \frac{du}{b} \qquad (14)$$

the right side can be multiplied out and integrated by the power formula, even though p is a fraction.

Integration by parts. This is a very powerful and important method. It follows from the formula for differentiating a product, namely Eq. (15). After

$$d(fg)=f\,dg+g\,df \qquad (15)$$

the terms are rearranged and integrated, Eq. (16)

$$\int f\,dg=fg-\int g\,df \qquad (16)$$

or (17) may be written, where the arbitrary con-

$$\int fg'\ dx=f(x)g(x)-\int gf'\ dx \qquad (17)$$

stant, as usual, has been omitted. As an example, consider Eq. (18), which is a special case of the

$$I=\int \frac{2x^2\ dx}{(x^2+1)^m} \qquad (18)$$

form (13). Let $f=x$, $g'=2x/(x^2+1)^m$. Then $f'=1$ and Eq. (19) can be written, and by Eq. (17), Eq. (20) follows.

$$g=\frac{1}{(1-m)(x^2+1)^{m-1}} \qquad (19)$$

$$I=\frac{x}{(1-m)(x^2+1)^{m-1}}$$

$$-\frac{1}{(1-m)}\int \frac{dx}{(x^2+1)^{m-1}} \qquad (20)$$

The indefinite integrals of many of the commonly occuring functions are given in tables of integrals in handbooks and textbooks.

The elementary functions are those expressible by means of a finite number of algebraic operations and trigonometric and exponential function and their inverses. The integrals of many elementary functions are known to be not elementary, so they define new functions. A number of these are sufficiently important that their values have been tabulated. Examples are Eqs. (21) and (22). $F(k,x)$

$$F(k,x)=\int_0^x \frac{dx}{\sqrt{(1-x^2)(1-k^2x^2)}} \qquad (k^2 \neq 1) \qquad (21)$$

$$\text{Si } x = \int_0^x \frac{\sin x \, dx}{x} \tag{22}$$

is called an elliptic integral of the first kind. The inverse of $u = F(k,x)$ is an elliptic function called the sine amplitude of u and denoted by $x = \sin u$. Some nonelementary functions have been included in tables of integrals. *See* ELLIPTIC FUNCTION AND INTEGRAL.

Improper integrals. This term is used to refer to an extension of the notion of definite integral to cases in which the integrand is unbounded or the domain of integration is unbounded. Consider first the case when $f(x)$ is integrable (in the sense defined above) and hence bounded on every interval $[a + \epsilon, b]$ for $\epsilon > 0$, but is unbounded on $[a,b]$. Then by definition, Eq. (23) is written, provided the

$$\int_a^b f(x) \, dx = \lim_{\epsilon \to 0} \int_{a+\epsilon}^b f(x) \, dx \tag{23}$$

limit on the right exists. For example, if $f(x) = x^{-2/3}$, relation (24) follows. Similarly, Eq. (25) follows.

$$\int_0^1 x^{-2/3} \, dx = \lim_{\epsilon \to 0} \int_\epsilon^1 x^{-2/3} \, dx$$
$$= \lim_{\epsilon \to 0} 3[1 - \epsilon^{1/3}] = 3 \tag{24}$$

$$\int_{-1}^0 x^{-2/3} \, dx = \lim_{\epsilon \to 0} \int_{-1}^{-\epsilon} x^{-2/3} \, dx$$
$$= \lim_{\epsilon \to 0} 3[(-\epsilon)^{1/3} + 1] = 3 \tag{25}$$

When $f(x)$ is integrable on every finite subinterval of the real axis, by definition Eqs. (26) and (27) hold, provided the limits on the right exist.

$$\int_a^{+\infty} f(x) \, dx = \lim_{b \to +\infty} \int_a^b f(x) \, dx \tag{26}$$

$$\int_{-\infty}^a f(x) \, dx = \lim_{b \to -\infty} \int_b^a f(x) \, dx \tag{27}$$

More general cases are treated by dividing the real axis into pieces, each of which satisfies one of the conditions just specified. Equation (28) is an example.

$$\int_{-\infty}^{+\infty} x^{-5/3} \, dx = \lim_{a \to -\infty} \int_a^{-1} x^{-5/3} \, dx$$
$$+ \lim_{\delta \to 0} \int_{-1}^{-\delta} x^{-5/3} \, dx + \lim_{\epsilon \to 0} \int_\epsilon^1 x^{-5/3} \, dx$$
$$+ \lim_{b \to +\infty} \int_1^b x^{-5/3} \, dx \tag{28}$$

ple. Because the second and third of the limits on the right do not exist, integral (29) does not exist.

$$\int_{-\infty}^{+\infty} x^{-5/3} \, dx \tag{29}$$

However, it is sometimes useful to assign a value to it, called the Cauchy principal value, by replacing definition (28) by Eq. (30). In this particular case the value is zero.

$$\int_{-\infty}^{+\infty} x^{-5/3} \, dx = \lim_{e \to 0} \left[\int_{-1}^{-\epsilon} x^{-5/3} \, dx + \int_\epsilon^1 x^{-5/3} \, dx \right]$$
$$+ \lim_{b \to +\infty} \left[\int_{-b}^{-1} x^{-5/3} \, dx + \int_1^b x^{-5/3} \, dx \right] \tag{30}$$

The preceding definitions apply to cases when the integrand $f(x)$ is bounded, except in arbitrarily small neighborhoods of a finite set of points. In the

closing years of the 19th century various extensions were made to more general cases. Then Henri Lebesgue produced a comparatively simple general theory for the case of absolutely convergent integrals. The integral of Lebesgue will be discussed below. The integral of Denjoy includes both nonabsolutely convergent integrals and the integral of Lebesgue.

Multiple integrals. The concept called the Riemann integral can be extended to functions of several variables. The case of a function of two variables illustrates sufficiently the additional features which arise. To begin with, let $f(x,y)$ denote a real function defined on a rectangle of the form (31).

$$R: a \le x \le b \qquad c \le y \le d \tag{31}$$

Let P be a partition of R into n nonoverlapping rectangles R_i with areas A_i, and let (x_i, y_i) be a point of R_i. Define S by Eq. (32). In case the sum S tends to a definite limit I when the maximum diagonal of a rectangle R_i tends to zero, then f is said to be in-

$$S = \sum_{i=1}^n f(x_i, y_i) A_i \tag{32}$$

tegrable over R, and the limit I is called the Riemann integral of f over R. It will be denoted here by the abbreviated symbol $\int \int_R f$. Properties corresponding to those numbered (i) to (iv) can be proved for these double integrals, except that property (ii) should now read as follows:

(ii') If the rectangle R is the union of two rectangles S and T, then Eq. (33) can be written.

$$\iint_R f = \iint_S f + \iint_T f \tag{33}$$

Necessary and sufficient conditions for a function $f(x,y)$ to be integrable over R can be stated as follows: f is bounded on R; and the set of points where f is discontinuous can be enclosed in a series of rectangles, the sum of whose areas is arbitrarily near to zero.

To define the integral of a function $f(x,y)$ over a more general domain D where it is defined, suppose that D is enclosed in a rectangle R, and define $F(x,y)$ by Eqs. (34). Then f is said to be integrable over D in case F is integrable over R, and Eq. (35) holds, by definition.

$$F(x,y) = f(x,y) \text{ in } D$$
$$F(x,y) = 0 \text{ outside } D \tag{34}$$

$$\iint_D f = \iint_R F \tag{35}$$

When the function $f(x,y)$ is continuous on D, and D is defined by inequalities of the form (36), where

$$a \le x \le b \qquad \alpha(x) \le y \le \beta(x) \tag{36}$$

the functions $\alpha(x)$ and $\beta(x)$ are continuous, the double integral of f over D always exists, and may be represented in terms of two simple integrals by Eq. (37). In many cases, this formula makes possible the evaluation of the double integral.

$$\iint_D f = \int_a^b \left[\int_{\alpha(x)}^{\beta(x)} f(x,y) \, dy \right] dx \tag{37}$$

Improper multiple integrals have been defined in a variety of ways. There is not space to discuss these definitions and their relations here.

Line, surface, and volume integrals. A general discussion of curves and surfaces requires an extended treatise. The following outline is restricted to the simplest cases.

A curve may be defined as a continuous image of an open interval. Thus a curve C in the xy plane is given by a pair of continuous functions of one variable, Eq. (38), whereas a curve in space is given by

$$x = f(u) \quad y = g(u) \quad a < u < b \qquad (38)$$

a triple of such functions. The mapping (38) is more properly called a representation of a curve, and a curve is then defined as a suitable class of such representations. Note that a curve such as Eq. (38) is a path rather than a locus. For example, notation (39) is the path going twice around the

$$x = \cos u \quad y = \sin u \quad 0 < u < 4\pi \qquad (39)$$

unit circle, Eq. (40), in the counterclockwise direc-

$$x^2 + y^2 = 1 \qquad (40)$$

tion. A curve is called smooth in case the function f and g have continuous derivatives $f'(u)$ and $g'(u)$ which are never simultaneously zero.

The length of a smooth curve C may be defined by Eq. (41) when this integral exists (as a proper or

$$L(C) = \int_a^b \sqrt{[f'(u)]^2 + [g'(u)]^2}\, du \qquad (41)$$

improper integral). When the integral does not exist, the length of C is said to be infinite. This definition of length in terms of the integral may be shown to coincide with the geometric definition in a much larger class than the class of smooth curves. When C has finite length, the position vector $[f(u),g(u)]$ approaches definite limits when u tends to a or to b, which are the ends of the curve C. A smooth curve has at every point a nonzero tangent vector $\mathbf{T}(u)$, with components $f'(u)$ and $g'(u)$, so that the integrand of $L(C)$ is the length $|\mathbf{T}(u)|$ of this vector. Thus one may write $ds = |\mathbf{T}(u)|\, du$, where s is the arc length measured from some convenient point on C.

Let $A(x,y)$ be a bounded continuous function defined on a set containing a plane curve C having finite length. Then A determines three functions of C called line integrals, whose symbols and definitions, Eqs. (42), follow. If $-C$ denotes the

$$\int_C A\, dx = \int_a^b A[f(u),g(u)]f'(u)\, du$$

$$\int_C A\, dy = \int_a^b A[f(u),g(u)]g'(u)\, du \qquad (42)$$

$$\int_C A\, ds = \int_a^b A[f(u),g(u)]|T(u)|\, du$$

curve C traversed in the opposite direction, as shown in Eqs. (43), a reversal of the orientation of

$$\int_{-C} A\, dx = -\int_C A\, dx$$

$$\int_{-C} A\, dy = -\int_C A\, dy \qquad (43)$$

$$\int_{-C} A\, ds = \int_C A\, ds$$

C changes the signs of the first two but not of the third of these functions.

The extension of the preceding definitions to curves in three-space is made in an obvious way. An important application of line integrals is to express the work done by a force field on a moving particle. Thus, if a force field $\mathbf{F}(x,y,z)$ has components $F_1(x,y,z)$, $F_2(x,y,z)$, and $F_3(x,y,z)$ in the directions of the x, y, and z axes, then the work done by the field on a particle moving on a curve C is given by formula (44). When $\mathbf{T}$ is the tangent vector of C, expressed in terms of a parameter u, the work W can be expressed in terms of the dot product by Eq. (45). If $\mathbf{T}_1$ denotes the tangent vector of unit

$$W = \int_C F_1\, dx + F_2\, dy + F_3\, dz \qquad (44)$$

$$W = \int_a^b \mathbf{F} \cdot \mathbf{T}\, du \qquad (45)$$

length, notation (46) then can be written.

$$W = \int_C \mathbf{F} \cdot \mathbf{T}_1\, ds \qquad (46)$$

To pass from curves to surfaces replace the open interval $a < u < b$ by a bounded connected open set D in a uv plane, which may be called the parameter plane. The domain D can be restricted to be of sufficiently simple shape so that every function which is continuous and bounded in D is integrable over D, as a multiple integral. For example, D may be the interior of a circle, or of a rectangle. Then a surface S in three-space is defined by a triple of functions continuous on D, Eqs. (47). Such

$$\begin{aligned} x &= f(u,v) \\ y &= g(u,v) \\ z &= h(u,v) \end{aligned} \qquad (47)$$

a surface is called smooth in case the functions f, g, and h have continuous first partial derivatives in D, and the three Jacobians listed in formulas (48),

$$J_1 = \frac{\partial(y,z)}{\partial(u,v)} \qquad J_2 = \frac{\partial(z,x)}{\partial(u,v)} \qquad J_3 = \frac{\partial(x,y)}{\partial(u,v)} \qquad (48)$$

are never simultaneously zero. A smooth surface has at every point a nonzero normal vector $\mathbf{J}(u,v)$ with components J_1, J_2, J_3, and length $|\mathbf{J}|$.

The area of a smooth surface S may be defined by the integral (49), and it is always finite when the vector $\mathbf{J}$ has bounded length.

$$\sigma(S) = \iint_D |\mathbf{J}(u,v)| \qquad (49)$$

If $A(x,y,z)$ is a bounded continuous function defined on a set containing a surface S with finite area, four surface integrals can be defined as in notation (50). If $\mathbf{F}(x,y,z)$ is a vector field with com-

$$\iint_S A\, dy\, dz = \iint_D A[f(u,v),g(u,v),h(u,v)]J_1(u,v)$$

$$\iint_S A\, dz\, dx = \iint_D A[f(u,v),g(u,v),h(u,v)]J_2(u,v)$$

$$\iint_S A\, dx\, dy = \iint_D A[f(u,v),g(u,v),h(u,v)]J_3(u,v)$$

$$\iint_S A\, d\sigma = \iint_D A[f(u,v),g(u,v),h(u,v)]|\mathbf{J}(u,v)|$$

$$(50)$$

ponents F_1, F_2, F_3, and $\mathbf{n}$ denotes the unit vector normal to S, with components $J_1/|\mathbf{J}|$, $J_2/|\mathbf{J}|$, $J_3/|\mathbf{J}|$, then a combination, formula (51), of the first three

$$\iint_S F_1 \, dy \, dz + F_2 \, dz \, dx + F_3 \, dx \, dy \qquad (51)$$

kinds of surface integral may also be written in the form of Eq. (52), where $\mathbf{F \cdot n}$ is the dot product. Ex-

$$\iint_S \mathbf{F \cdot n} \, d\sigma \qquad (52)$$

pression (51) or (52) is referred to as the integral of the vector $\mathbf{F}$ over the side of the surface S to which the vector $\mathbf{n}$ points. The integral over the oposite side of S is denoted by formula (53), whose value

$$\iint_S F_1 \, dz \, dy + F_2 \, dx \, dz + F_3 \, dy \, dx \qquad (53)$$

is the negative of that in (51), by Eq. (54). The last

$$\frac{\partial(z,y)}{\partial(u,v)} = -J_1, \text{ and so on} \qquad (54)$$

integral in Eq. (50), however, is independent of a choice of side of the surface.

The important formulas of Stokes and of Gauss, Eqs. (55), are expressed in terms of line and sur-

$$\int_C \mathbf{F \cdot T} \, ds = \iint_S \text{curl } \mathbf{F \cdot n} \, d\sigma \quad \text{(Stokes)}$$
$$\qquad (55)$$
$$\iint_S \mathbf{F \cdot n} \, d\sigma = \iiint_V \text{div } \mathbf{F} \quad \text{(Gauss)}$$

face integrals and the triple or volume integral analogous to the double integral. In the formula of Stokes, $\mathbf{F}$ is a vector field, S is a smooth surface having a piecewise smooth boundary curve C, directed so that S lies to the left of an observer proceeding along C on the side of S on which the unit normal vector $\mathbf{n}$ is chosen, and $\mathbf{T}$ is the unit tangent vector to C. In terms of the components of F, Stokes' formula may be written as Eq. (56), where

$$\int_C F_1 \, dx + F_2 \, dy + F_3 \, dz = \iint_S (F_{3y} - F_{2z}) \, dy \, dz$$
$$+ (F_{1z} - F_{3x}) \, dz \, dx + (F_{2x} - F_{1y}) \, dx \, dy \quad (56)$$

$F_{3y} = \partial F_3/\partial y$, and so on. When the surface S lies in the xy plane, this equation reduces to Green's formula, Eq. (57). The latter is readily proved for re-

$$\int_C F_1 \, dx + F_2 \, dy = \iint_S (F_{2x} - F_{1y}) \, dx \, dy \quad (57)$$

gions S of simple shape, and from it Stokes' formula may be deduced.

In the formula of Gauss (also called the divergence theorem), V is a space domain bounded by a smooth or piecewise smooth surface S, and $\mathbf{n}$ is the exterior normal of S. In terms of the components of F, Gauss' formula may be written as Eq. (58). See CALCULUS OF VECTORS.

$$\iint_S F_1 \, dy \, dz + F_2 \, dz \, dx + F_3 \, dx \, dy$$
$$= \iiint_V (F_{1x} + F_{2y} + F_{3z}) \quad (58)$$

Functions defined by integrals. Mention has already been made of elliptic functions, which may be defined as the inverse functions of elliptic integrals. Certain definite integrals may also be used to define important nonelementary functions which occur frequently in applications. For example, the gamma function may be defined by Eq. (59) for $x > 0$. The Bessel functions of integral or-

$$\Gamma(x) = \int_0^\infty e^{-t} t^{x-1} \, dt \qquad (59)$$

der may be defined by Eq. (60) for all values of x.

$$J_n(x) = \frac{1}{\pi} \int_0^\pi \cos(nt - x \sin t) \, dt \qquad (60)$$

The properties of these functions may be derived from these expressions with the help of various methods of advanced analysis, such as the theory of functions of a complex variable. See BESSEL FUNCTIONS; GAMMA FUNCTION.

In general, if $f(x,t)$ is an integrable function of t on $a \le t \le b$ for each x, then Eq. (61) is a well-

$$g(x) = \int_a^b f(x,t) \, dt \qquad (61)$$

defined function of x. When suitable conditions are satisfied by the function f, the derivative $g'(x)$ may be calculated by formula (62). This process is called

$$g'(x) = \int_a^b \frac{\partial}{\partial x} f(x,t) \, dt \qquad (62)$$

differentiation under the integral sign. It is valid in case f and $\partial f/\partial x$ are continuous in (x,t) for $c \le x \le d$, $a \le t \le b$, where a and b are finite, and even in more general cases. However, in the simple example given in Eqs. (63), formula (62) leads to a wrong result.

$$f(x,t) = x^3 e^{-tx^2} \qquad g(x) = \int_0^\infty f(x,t) \, dt \quad (63)$$

Approximate and mechanical integration. The definition of the definite integral itself gives a means of calculating its value approximately. There remains the question of a suitable selection of the functional values $f(x_i)$ for use in formula (1). It is usual to divide the interval $[a,b]$ into a number n of equal parts of length h (Fig. 2). If the points x_i are taken at the midpoints of the subintervals, one obtains the midpoint formula, Eq. (64). For a curve

$$S = h \sum_{i=1}^n f[a + (i - 1/2)h] \qquad (64)$$

that is concave downward, this gives too large a value.

Another formula, called the trapezoidal rule, is derived by calculating the area below a polygon inscribed in the graph of $f(x)$ (Fig. 3). This formula is Eq. (65). It gives too small a value for a curve that is concave downward.

$$T = \frac{h}{2} \left[f(a) + f(b) + 2 \sum_{i=1}^{n-1} f(a + ih) \right] \quad (65)$$

The error in the approximation may sometimes be reduced without increasing the number of subintervals by use of the parabolic rule (Simpson's rule). To obtain the formula, the subarcs of the graph of $f(x)$ are replaced by arcs of parabolas rather than by line segments. Since three points determine a parabola (with vertical axis), the interval $[a,b]$ is divided into an even number n of subintervals. The area under the parabola passing through the points on the graph of $f(x)$ having abscissas given in formulas (66) is given by formula (67).

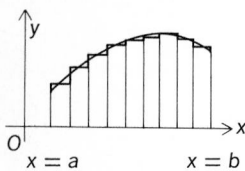

INTEGRATION

Fig. 2. Division of interval $[a,b]$ into parts.

INTEGRATION

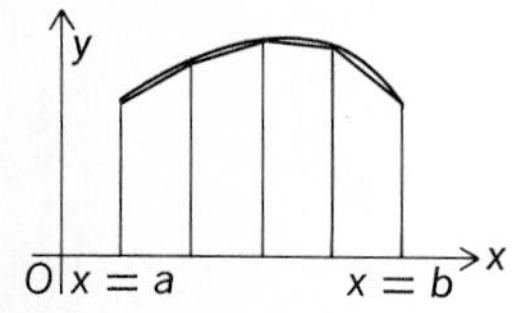

Fig. 3. Polygon inscribed in graph of $f(x)$.

Therefore, the parabolic rule gives approximation (68).

$$a+(2i-2)h \quad a+(2i-1)h \quad a+2ih \quad (66)$$

$$\frac{h}{3}\left\{f[a+(2i-2)h]\right.$$
$$\left.+4f[a+(2i-1)h]+f(a+2ih)\right\} \quad (67)$$

$$P=\frac{h}{3}\left\{f(a)+f(b)+2\sum_{i=1}^{\frac{n}{2}-1}f(a+2ih)\right.$$
$$\left.+4\sum_{i=1}^{n/2}f[a+(2i-1)h]\right\} \quad (68)$$

When a function $f(x)$ is given only by a table, its integral is most simply computed by one of these formulas or a similar formula.

When a function $f(x)$ is given by a graph, mechanical means may be used to calculate associated areas. One such means is the polar planimeter, which registers on a rotating wheel the area enclosed by a closed curve around which a tracing point is passed. The integraph, invented by Abdank Abakanowicz, is designed to draw the graph of an indefinite integral of $f(x)$ when a tracing point is passed over the graph of $f(x)$. These simple devices were the forerunners of more complex machines designed to solve differential equations, such as the differential analyzer of Vannevar Bush. For large-scale computations involving formulas such as (64), (65), or (68), a digital computer may be preferred.

Other methods of integration. The method of differentiation under the integral sign is often convenient for the evaluation of definite integrals, even when other methods are available. For example, Eq. (69) leads to Eq. (70). Or, if one has Eq. (71), Eq. (72) follows.

$$g(x)=\int_0^a e^{xt}\,dt=\frac{e^{xa}-1}{x} \quad (69)$$

$$g'(x)=\int_0^a te^{xt}\,dt=\frac{ae^{xa}}{x}-\frac{e^{xa}-1}{x^2} \quad (70)$$

$$g(x)=\int_0^a \frac{dt}{t^2+x^2}=\frac{1}{x}\arctan(a/x) \quad (71)$$

$$g'(x)=-2x\int_0^a \frac{dt}{(t^2+x^2)^2}$$
$$=-\frac{1}{x^2}\arctan(a/x)-\frac{a}{x(a^2+x^2)} \quad (72)$$

Another example in which the conditions for validity of the method still hold, though they are more difficult to verify, is given by Eq. (73), by means of which Eq. (74) can be derived, by integration by

$$g(x)=\int_0^\infty e^{-xt}\frac{\sin t}{t}\,dt \qquad 0\le x<\infty \quad (73)$$

$$g'(x)=-\int_0^\infty e^{-xt}\sin t\,dt=\frac{-1}{1+x^2} \quad (74)$$

parts twice. Hence $g(x)=-\arctan x+C$, and $C=\pi/2$, since Eq. (75) can be proved.

$$\lim_{x\to\infty} g(x)=0 \quad (75)$$

In particular, Eq. (76) holds.

$$g(0)=\int_0^\infty \frac{\sin t}{t}\,dt=\pi/2 \quad (76)$$

In case an indefinite integral is not elementary but the integrand is representable by an infinite series, the method of integrating term by term gives a representation for the integral which may be used for purposes of approximation. This can be seen in Eq. (77).

$$\text{Si}(x)=\int_0^x \frac{\sin t\,dt}{t}=\sum_0^\infty \frac{(-1)^n x^{2n+1}}{(2n+1)^2(2n)!} \quad (77)$$

Certain definite integrals may be readily evaluated by means of contour integrals, that is, integrals of analytic functions taken along curves in the complex plane.

Integral of Lebesgue. In the study of the "space" of real functions defined, for example, on the interval $a\le x\le b$, it is frequently useful to take formula (78) as the distance between the func-

$$\int_a^b |f(x)-g(x)|\,dx \quad (78)$$

tions f and g. This distance already has a meaning when f and g are Riemann-integrable, that is, bounded and not too discontinuous, in the sense specified for the Riemann integral. There is no generally useful extension of the concept of integral to apply to all real functions on $[a,b]$, but it is desirable to extend it to apply to the functions obtained from the continuous ones by certain limiting processes. In particular, it is desirable to have correspond to each sequence (f_n) of functions satisfying the Cauchy condition for convergence in terms of the distance (78), namely Eq. (79), a func-

$$\lim_{\substack{m=\infty\\n=\infty}} \int_a^b |f_m(x)-f_n(x)|\,dx=0 \quad (79)$$

tion g which is integrable (in the extended sense), and for which Eq. (80) holds. An extended

$$\lim_{n=\infty} \int_a^b |f_n(x)-g(x)|\,dx=0 \quad (80)$$

definition of integral having this property was given by H. L. Lebesgue in his thesis. It made obsolete many of the complicated extensions of the Riemann integral which had been previously proposed. Following Lebesgue, various mathematicians have proposed other ways of defining the integral which are equivalent to that of Lebesgue.

F. Riesz devised a definition which can be stated quite simply, at least for the case of a bounded function $g(x)$. As a first definition, a point set S in the interval $[a,b]$ has measure zero in case it can be enclosed in a sequence (finite or infinite) of intervals, the sum of whose lengths is arbitrarily small. Also, a step function $f(x)$ is defined as one which is constant on each interval of a partition of $[a,b]$, as in Fig. 4. The Riemann integral of a step function is expressible as a finite sum. Then a bounded function $g(x)$ is integrable in Lebesgue's sense in case it is the limit of a uniformly bounded sequence of step functions $f_n(x)$, at each point of $[a,b]$ except those in a set S with measure zero, and by definition Eq. (81) can be written. In case the step

$$\int_a^b g(x)\,dx=\lim_n \int_a^b f_n(x)\,dx \quad (81)$$

functions f_n in Eq. (81) are replaced by Lebesgue-integrable functions forming a uniformly bounded sequence, no new functions g are obtained.

The integral of Lebesgue is also defined for

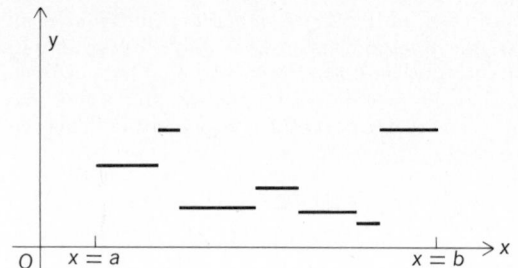

Fig. 4.   Step function $f(x)$.

unbounded functions, but the points of infinite discontinuity do not need to be considered one by one. For each function $g(x)$ and each positive integer N let $g_N(x)$ denote the lesser of $g(x)$ and N. If each $g_N(x)$ is Lebesgue-integrable in the sense already defined, expression (82) forms a nondecreasing sequence, and so if expression (82) is bounded,

$$\int_a^b g_N(x)\,dx \qquad (82)$$

it tends to a finite limit, which is taken as the value of integral (83). An arbitrary function $g(x)$ is the

$$\int_a^b g(x)\,dx \qquad (83)$$

difference of two nonnegative functions, Eqs. (84)

$$g^+(x) = \frac{|g(x)| + g(x)}{2} \qquad (84)$$

and (85), and one may write Eq. (86) whenever both

$$g^-(x) = \frac{|g(x)| - g(x)}{2} \qquad (85)$$

$$\int_a^b g(x)\,dx = \int_a^b g^+(x)\,dx - \int_a^b g^-(x)\,dx \qquad (86)$$

terms on the right have a meaning according to the definition just given. It is provable that the class of all functions for which integrals exist according to the definitions just given, does indeed have the property of completeness with respect to the Cauchy condition (79). The theory of Lebesgue yields many other useful results, including an extension of the fundamental theorem of integral calculus. One important restriction on the class of Lebesgue-integrable functions which is implicit in the definition is that when $g(x)$ is integrable so is $|g(x)|$. This restriction does not hold for the improper integrals which were described earlier.

The various methods of defining Lebesgue integrals are extensible also to functions of several variables. This extension throws light on the properties of multiple Riemann integrals and on the reduction of multiple integrals to repeated integrals.

Other definitions of integration. If $\alpha(x)$ is a fixed function defined on the interval $[a,b]$, the sum S defined by Eq. (1) may be replaced by that in Eq. (87). Then if the limit I of S exists in this case,

$$S = \sum_{i=1}^n f(x_i)[\alpha(t_i) - \alpha(t_{i-1})] \qquad (87)$$

it is called the Stieltjes integral of f with respect to

α, and is denoted by integral (88). It has many of

$$\int_a^b f(x)\,d\alpha(x) \qquad (88)$$

the properties of the Riemann integral, especially in case the function α is nondecreasing. In addition, it may take special account of the values of f at a finite or countable set of points. For example, if $\alpha(x) = c_i$ on $u_{i-1} < x < u_i$, where $u_0 < a$, $u_n > b$, $a < u_i < b$ for $i = 1, \ldots, n-1$, and if $f(x)$ is continuous at each u_i, then Eq. (89) may be written.

$$\int_a^b f(x)\,d\alpha(x) = \sum_{i=1}^{n-1} f(u_i)[c_{i+1} - c_i] \qquad (89)$$

ten. With some restrictions to ensure convergence, this may be extended to the case in which α has infinitely many discontinuities.

In the case in which α is nondecreasing, the Stieltjes integral has an extension which is similar to that of the Lebesgue integral and is called the Lebesgue-Stieltjes integral.

Other extensions of the Lebesgue integral are the integrals of Denjoy. They apply to certain functions f for which $|f|$ is not integrable, and include the improper integrals, but not the Cauchy principal value. Numerous other types of integrals have been defined. In particular the integral of Lebesgue has been extended to cases in which the independent variable lies in a suitable space of infinitely many dimensions, or in which the functional values of f lie in such a space. *See* CALCULUS; FOURIER SERIES AND INTEGRALS; SERIES.

[LAWRENCE M. GRAVES]

Bibliography: S. K. Berberian, *Measure and Integration*, 1965, reprint 1970; H. B. Dwight, *Tables of Integrals and Other Mathematical Data*, 4th ed., 1961; P. R. Halmos, *Measure Theory*, 1950, reprint 1974; K. Jacobs, *Measure and Integral*, 1978; W. F. Pfeffer, *Integrals and Measures*, 1977; W. Rudin, *Principles of Mathematical Analysis*, 3d ed., 1976; A. J. Weir, *General Integration and Measure*, 1974.

Interference of waves

The process whereby two or more waves of the same frequency or wavelength combine to form a wave whose amplitude is the sum of the amplitudes of the interfering waves. The interfering waves can be electromagnetic, acoustic, or water waves, or in fact any periodic disturbance.

The most striking feature of interference is the effect of adding two waves in which the trough of one wave coincides with the peak of another. If the two waves are of equal amplitude, they can cancel each other out so that the resulting amplitude is zero. This is perhaps most dramatic in sound waves; it is possible to generate acoustic waves to arrive at a person's ear so as to cancel out noise that is disturbing him. In optics, this cancellation can occur for particular wavelengths in a situation where white light is a source. The resulting light will appear colored. This gives rise to the iridescent colors of beetles' wings and mother-of-pearl, where the substances involved are actually colorless or transparent.

Two-beam interference. The quantitative features of the phenomenon can be demonstrated most easily by considering two interfering waves. The amplitude of the first wave at a particular point in space can be written as Eq. (1), where A_0 is

$$A = A_0 \sin (\omega t + \varphi_1) \qquad (1)$$

the peak amplitude, and ω is 2π times the frequency. For the second wave Eq. (2) holds, where

$$B = B_0 \sin (\omega t + \varphi_2) \qquad (2)$$

$\varphi_1 - \varphi_2$ is the phase difference between the two waves. In interference, the two waves are superimposed, and the resulting wave can be written as Eq. (3).

$$A + B = A_0 \sin (\omega t + \varphi_1) + B_0 \sin (\omega t + \varphi_2) \qquad (3)$$

Equation (3) can be expanded to give Eq. (4).

$$A + B = (A_0 \sin \varphi_1 + B_0 \sin \varphi_2) \cos \omega t$$
$$+ (A_0 \cos \varphi_1 + B_0 \cos \varphi_2) \sin \omega t \qquad (4)$$

By writing Eqs. (5) and (6), Eq. (4) becomes Eq. (7), where C^2 is defined in Eq. (8). When C is less

$$A_0 \sin \varphi_1 + B_0 \sin \varphi_2 = C \sin \varphi_3 \qquad (5)$$

$$A_0 \cos \varphi_1 + B_0 \cos \varphi_2 = C \cos \varphi_3 \qquad (6)$$

$$A + B = C \sin (\omega t + \varphi_3) \qquad (7)$$

$$C^2 = A_0{}^2 + B_0{}^2 + 2A_0B_0 \cos (\varphi_2 - \varphi_1) \qquad (8)$$

than A or B, the interference is called destructive. When it is greater, it is called constructive. For electromagnetic radiation, such as light, the amplitude in Eq. (7) represents an electric field strength. This field is a vector quantity and is associated with a particular direction in space, the direction being generally at right angles to the direction in which the wave is moving. These electric vectors can be added even when they are not parallel. For a discussion of the resulting interference phenomena *see* POLARIZED LIGHT. *See also* SUPERPOSITION PRINCIPLE.

In the case of radio waves or microwaves which are generated with vacuum tube or solid-state oscillators, the frequency requirement for interference is easily met. In the case of light waves, it is more difficult. Here the sources are generally radiating atoms. The smallest frequency spread from such a light source will still have a bandwidth of the order of 10^7 Hz. Such a bandwidth occurs in a single spectrum line, and can be considered a result of the existence of wave trains no longer than 10^{-8} sec. The frequency spread associated with such a pulse can be written as notation (9), where

$$\Delta f \approx \frac{1}{2\pi t} \qquad (9)$$

t is the pulse length. This means that the amplitude and phase of the wave which is the sum of the waves from two such sources will shift at random in times shorter than 10^{-8} sec. In addition, the direction of the electric vector will shift in these same time intervals. Light which has such a random direction for the electric vector is termed unpolarized. When the phase shifts and direction changes of the light vectors from two sources are identical, the sources are termed coherent.

Splitting of light sources. To observe interference with waves generated by atomic or molecular transitions, it is necessary to use a single source and to split the light from the source into parts which can then be recombined. In this case, the

amplitude and phase changes occur simultaneously in each of the parts at the same time.

Young's two-slit experiment. The simplest technique for producing a splitting from a single source was done by T. Young in 1801 and was one of the first demonstrations of the wave nature of light. In this experiment, a narrow slit is illuminated by a source, and the light from this slit is caused to illuminate two adjacent slits. The light from these two parallel slits can interfere, and the interference can be seen by letting the light from the two slits fall on a white screen. The screen will be covered with a series of parallel fringes. The location of these fringes can be derived approximately as follows: In Fig. 1, S_1 and S_2 are the two slits separated by a distance d. Their plane is a distance l from the screen. Since the slit S_0 is equidistant from S_1 and S_2, the intensity and phase of the light at each slit will be the same. The light falling on p from slit S_1 can be represented by Eq. (10) and

$$A = A_0 \sin 2\pi f \left(t - \frac{x_1}{c} \right) \qquad (10)$$

from S_2 by Eq. (11), where f is the frequency, t the

$$B = A_0 \sin 2\pi f \left(t - \frac{x_2}{c} \right) \qquad (11)$$

time, c the velocity of light; x_1 and x_2 are the distances of P from S_1 and S_2, and A_0 is the amplitude. This amplitude is assumed to be the same for each wave since the slits are close together, and x_1 and x_2 are thus nearly the same. These equations are the same as Eqs. (1) and (2), with $\varphi_1 = x_1/c$ and $\varphi_2 = x_2/c$. Accordingly, the square of the amplitude or the intensity at P can be written as Eq. (12).

In general, l is very much larger than y so that Eq. (12) can be simplified to Eq. (13).

$$I = 4A_0{}^2 \cos^2 \frac{2\pi f}{c} (x_1 - x_2) \qquad (12)$$

$$I = 4A_0{}^2 \cos^2 \pi \left(\frac{yd}{l\lambda} \right) \qquad (13)$$

Equation (13) is a maximum when Eq. (14) holds and a minimum when Eq. (15) holds, where n is an integer.

$$y = n\lambda \frac{l}{d} \qquad (14)$$

$$y = (n + \tfrac{1}{2})\lambda \frac{l}{d} \qquad (15)$$

Accordingly, the screen is covered with a series of light and dark bands called interference fringes. If the source behind slit S_0 is white light and thus has wavelengths varying perhaps from 4000 to

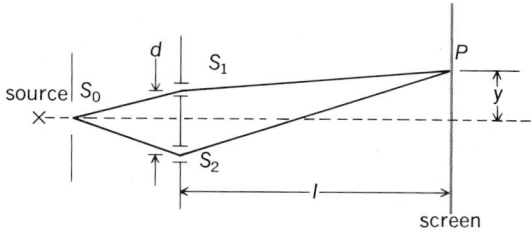

Fig. 1. Young's two-slit interference.

7000 A, the fringes are visible only where $x_1 - x_2$ is a few wavelengths, that is, where n is small. At large values of n, the position of the nth fringe for red light will be very different from the position for blue light, and the fringes will blend together and be washed out. With monochromatic light, the fringes will be visible out of values of n which are determined by the diffraction pattern of the slits. For an explanation of this *see* DIFFRACTION.

The energy carried by a wave is measured by the intensity, which is equal to the square of the amplitude. In the preceding example of the superposition of two waves, the intensity of the individual waves in Eqs. (1) and (2) is A^2 and B^2, respectively. When the phase shift between them is zero, the intensity of the resulting wave is given by Eq. (16).

$$(A + B)^2 = A^2 + 2AB + B^2 \qquad (16)$$

This would seem to be a violation of the first law of thermodynamics, since this is greater than the sum of the individual intensities. In any specific experiment, however, it turns out that the energy from the source is merely redistributed in space. The excess energy which appears where the interference is constructive will disappear in those places where the energy is destructive. This is illustrated by the fringe pattern in the Young two-slit experiment. The energy on the screen from each slit alone is given by Eq. (17), where A_0^2 is the

$$E_1 = \int_0^\infty A_0^2 \, dy \qquad (17)$$

intensity of the light from each slit as given by Eq. (10). The intensity from the two slits without interference would be twice this value. The intensity with interference is given by Eq. (18). The com-

$$E_3 = \int_0^\infty 4A^2 \cos^2 \left[2\pi \left(\frac{yd}{l\lambda} \right) \right] dy \qquad (18)$$

parison between $2E_1$ and E_3 need be made only over a range corresponding to one full cycle of fringes. This means that the argument of the cosine in Eq. (18) need be taken only from zero to π. This corresponds to a section of screen going from the center to a distance $y = l\lambda/2d$. From the two slits individually, the energy in this section of screen can be written as Eq. (19).

$$2E_1 = 2 \int_0^{l\lambda/2d} A_0^2 \, dy = \frac{A_0^2 l\lambda}{d} \qquad (19)$$

With interference, the energy is given by Eq. (20). Equation (20) can be written as Eq. (21).

$$E_3 = \int_0^{l\lambda/2d} 4A_0^2 \cos^2 \left[2\pi \left(\frac{yd}{l\lambda} \right) \right] dy \qquad (20)$$

$$E_3 = \frac{l\lambda}{2\pi d} \int_0^\pi 4A_0^2 \cos^2 \varphi \, d\varphi = \frac{A_0^2 l\lambda}{d} \qquad (21)$$

Thus, the total energy falling on the screen is not changed by the presence of interference. The energy density at a particular point is, however, drastically changed. This fact is most important for those waves of the electromagnetic spectrum which can be generated by vacuum-tube oscillators. The sources of radiation or antennas can be made to emit coherent waves which will undergo interference. This makes possible a redistribu-

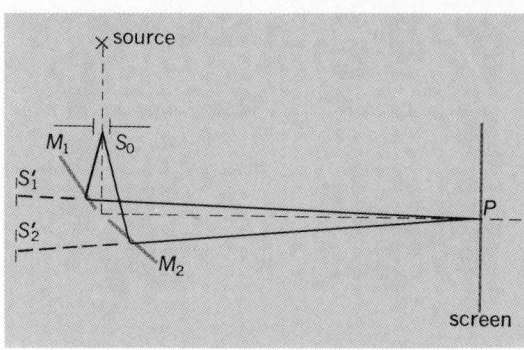

Fig. 2. Fresnel's double-mirror interference.

tion of the radiated energy. Quite narrow beams of radiation can be produced by the proper phasing of a linear antenna array.

The double-slit experiment also provides a good illustration of Niels Bohr's principle of complementarity. For detailed information on this *see* QUANTUM MECHANICS.

Fresnel double mirror. Another way of splitting the light from the source is the Fresnel double mirror (Fig. 2). Light from the slit S_0 falls on two mirrors M_1 and M_2 which are inclined to each other at an angle of the order of a degree. On a screen where the illumination from the two mirrors overlaps, there will appear a set of interference fringes. These are the same as the fringes produced in the two-slit experiment, since the light on the screen comes from the images of the slits S'_1 and S'_2 formed by the two mirrors, and these two images are the equivalent of two slits.

Fresnel biprism. A third way of splitting the source is the Fresnel biprism. A sketch of a cross section of this device is shown in Fig. 3. The light from the slit at S_0 is transmitted through the two halves of the prism to the screen. The beam from each half will strike the screen at a different angle and will appear to come from a source which is slightly displaced from the original slit. These two virtual slits are shown in the sketch at S'_1 and S'_2. Their separation will depend on the distance of the prism from the slit S_0 and on the angle θ and index of refraction of the prism material. In Fig. 3, a is the distance of the slit from the biprism, and l the distance of the biprism from the screen. The distance of the two virtual slits from the screen is thus $a + l$. The separation of the two virtual slits is given by Eq. (22), where μ is the refractive index of

$$d = 2a(\mu - 1)\theta \qquad (22)$$

the prism material. This can be put in Eq. (14) for the two-slit interference pattern to give Eq. (23) for the position of a bright fringe.

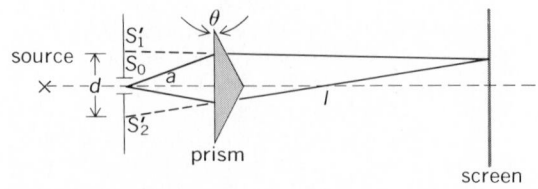

Fig. 3. Fresnel biprism interference.

$$y = n\lambda \frac{a+l}{2a(\mu-1)\theta} \qquad (23)$$

A photograph of the experimental equipment for demonstrating interference with the Fresnel biprism is shown in Fig. 4. A typical fringe pattern is shown in Fig. 5. This pattern was obtained with a mercury-arc source, which has several strong spectrum lines, accounting in part for the intensity variation in the pattern. The pattern is also modified by diffraction at the apex of the prism.

Billet split lens. The source can also be split with the Billet split lens (Fig. 6). Here a simple lens is sawed into two parts which are slightly separated.

Lloyd's mirror. An important technique of splitting the source is with Lloyd's mirror (Fig. 7). The slit S_1 and its virtual image S'_2 constitute the double source. Part of the light falls directly on the screen, and part is reflected at grazing incidence from a plane mirror. This experiment differs from the previously discussed experiments in that the two beams are no longer identical. If the screen is moved to a point where it is nearly in contact with the mirror, the fringe of zero path difference will lie on the intersection of the mirror plane with the screen. This fringe turns out to be dark rather than light, as in the case of the previous interference experiments. The only explanation for this result is that light experiences a 180° phase shift on reflection from a material of higher refractive index than its surrounding medium. The equation for maximum and minimum light intensity at the screen must thus be interchanged for Lloyd's mirror fringes.

Amplitude splitting. The interference experiments discussed have all been done by splitting the wavefront of the light coming from the source. The energy from the source can also be split in amplitude. With such amplitude-splitting techniques, the light from the source falls on a surface which is partially reflecting. Part of the light is transmitted, part is reflected, and after further manipulation these parts are recombined to give the interference. In one type of experiment, the light transmitted through the surface is reflected from a second surface back through the partially reflecting surface, where it combines with the wave reflected from the first surface (Fig. 8). Here the arrows represent the normal to the wavefront of the light passing through surface S_1 to surface S_2. The wave is incident at A and C. The section at A is partially transmitted to B, where it is again partially reflected to C. The wave leaving C now consists of two parts, one of which has traveled a longer distance than the other. These two waves will interfere. Let AD be the perpendicular from the ray at A to the ray going to C. The path difference will be given by Eq. (24), where μ is the

$$\Delta = 2\mu(AB) - (CD) \qquad (24)$$

refractive index of the medium between the surfaces S_1 and S_2, and AB and CD are defined in Eqs. (25) and (26).

$$(AB) = d/\cos r \qquad (25)$$

$$(CD) = 2(AB)\sin r \cos i \qquad (26)$$

From Snell's law, Eq. (27) is obtained and thus Eq. (28) holds.

Fig. 4. Equipment for demonstrating Fresnel biprism interference.

$$\sin i = \mu \sin r \qquad (27)$$

$$\Delta = \frac{2\mu d}{\cos r} - \frac{2\mu d}{\cos r}\sin^2 r = 2\mu d \cos r \qquad (28)$$

The difference in terms of wavelength and the phase difference are, respectively, given by Eqs. (29) and (30).

$$\Delta' = \frac{2\mu d \cos r}{\lambda} \qquad (29)$$

$$\Delta\varphi = \frac{4\pi\mu d \cos r}{\lambda} + \pi \qquad (30)$$

The phase difference of π radians is added because of the phase shift experienced by the light reflected at S_1. The experimental proof of this 180° phase shift was shown in the description of interference with Lloyd's mirror. If the plate of material has a lower index than the medium in which it is immersed, the π radians must still be added in Eq.

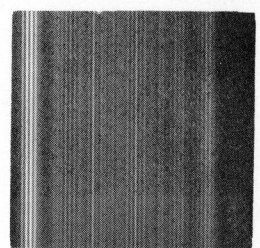

Fig. 5. Interference fringes formed with Fresnel biprism and mercury-arc light source.

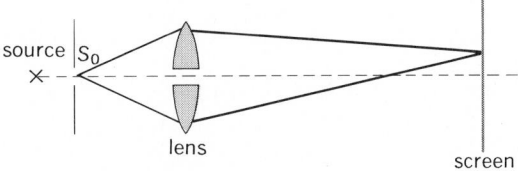

Fig. 6. Billet split-lens interference.

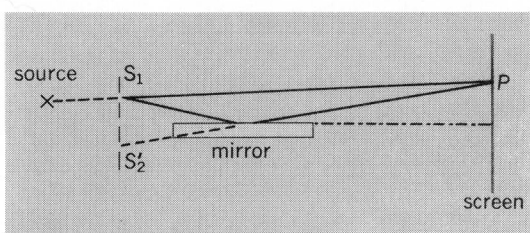

Fig. 7. Lloyd's mirror interference.

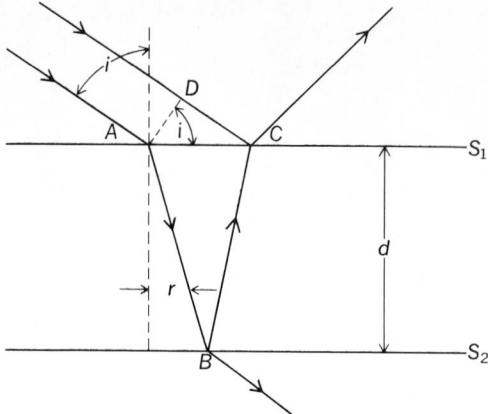

Fig. 8. Dielectric-plate-reflection interference.

(30), since now the beam reflected at S_2 will experience this extra phase shift. The purely pragmatic necessity of such an additional phase shift can be seen by considering the intensity of the reflected light when the surfaces S_1 and S_2 almost coincide. Without the extra phase shift, the two reflected beams would be in phase and the reflection would be strong. This is certainly not proper for a film of vanishing thickness. Constructive interference will take place at wavelengths for which $\Delta\varphi = 2m\pi$, where m is an integer. If the surfaces S_1 and S_2 are parallel, the fringes will be located optically at infinity. If they are not parallel, d will be a function of position along the surfaces and the fringes will be located near the surface. The intensity of the fringes will depend on the value of the partial reflectivity of the surfaces.

Testing of optical surfaces. Observation of fringes of this type can be used to determine the contour of a surface. The surface to be tested is put close to an optically flat plate. Monochromatic light is reflected from the two surfaces and examined as in Fig. 8. One of the first experiments with fringes of this type was performed by Isaac Newton. A convex lens is pressed against a glass plate and illuminated with monochromatic light. A series of circular interference fringes known as Newton's rings appear around the point of contact. From the separation between the fringes, it is possible to determine the radius of curvature of the lens.

Thin films. Interference fringes of this two-surface type are responsible for the colors which appear in oil films floating on water. Here the two surfaces are the oil-air interface and the oil-water interface. The films are close to a visible light wavelength in thickness. If the thickness is such that, in a particular direction, destructive interference occurs for green light, red and blue will still be reflected and the film will have a strong purple appearance. This same general phenomenon is responsible for the colors of beetles' wings.

Channeled spectrum. Amplitude splitting shows clearly another condition that must be satisfied for interference to take place. The beams from the source must not only come from identical points, but they must also originate from these points at nearly the same time. The light which is reflected from C in Fig. 8 originates from the source later

than the light which makes a double traversal between S_1 and S_2. If the surfaces are too far apart, the spectral regions of constructive and destructive interference become so close together that they cannot be resolved. In the case of interference by wavefront splitting, the light from different parts of a source could only be considered coherent if examined over a sufficiently short time interval. In the case of amplitude splitting, the interference when surfaces are widely separated can only be seen if examined over a sufficiently narrow frequency interval. If the two surfaces are illuminated with white light and the eye is used as the analyzer, interference cannot be seen when the separation is more than a few wavelengths. The interval between successive wavelengths of constructive interference becomes so small that each spectral region to which the eye is sensitive is illuminated, and no color is seen. In this case, the interference can again be seen by examining the reflected light with a spectroscope. The spectrum will be crossed with a set of dark fringes at those wavelengths for which there is destructive interference. This is called a channeled spectrum. For large separations of the surfaces, the separation between the wavelengths of destructive interference becomes smaller than the resolution of the spectrometer, and the fringes are no longer visible.

Fresnel coefficient. The amplitude of the light reflected at normal incidence from a dielectric surface is given by the Fresnel coefficient, Eq. (31),

$$A = A_0 \frac{n_1 - n_2}{n_1 + n_2} \qquad (31)$$

where A_0 is the amplitude of the incident wave and n_1 and n_2 are the refractive indices of the materials in the order in which they are encountered by the light. In the simple case of a dielectric sheet, the intensity of the light reflected normally will be given by Eq. (32), where B is the amplitude of the

$$C^2 = A^2 + B^2 + 2AB \cos\varphi \qquad (32)$$

wave which has passed through the sheet and is reflected from the second surface and back through the sheet to join A. The value of B is given by Eq. (33), where the approximation is made that

$$B = \frac{n_2 - n_3}{n_2 + n_3} \qquad (33)$$

the intensity of the light is unchanged by passing through the first surface and where n_3 is the index of the material at the boundary of the far side of the sheet.

Nonreflecting film. An interesting application of Eq. (32) is the nonreflecting film. A single dielectric layer is evaporated onto a glass surface to reduce the reflectivity of the surface to the smallest possible value. From Eq. (32) it is clear that this takes place when $\cos\varphi = -1$. If the surface is used in an instrument with a broad spectral range, such as a visual device, the film thickness should be adjusted to put the interference minimum in the first order and in the middle of the desired spectral range. For the eye, this wavelength is approximately in the yellow so that such films reflect in the red and blue and appear purple. The index of the film should be chosen to make $C^2 = 0$. At this point Eqs. (34)–(36) hold.

$$(A - B)^2 = 0 \qquad (34)$$

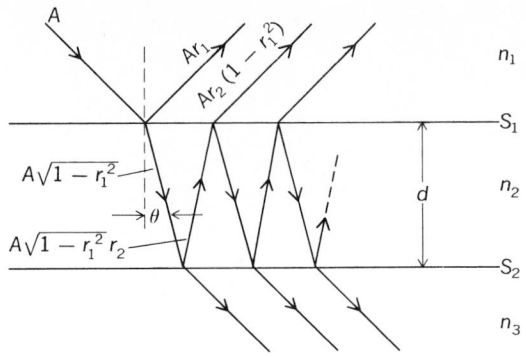

Fig. 9. Multiple reflection of wave between two surfaces.

$$\frac{n_1 - n_2}{n_1 + n_2} = \frac{n_2 - n_3}{n_2 + n_3} \qquad (35)$$

$$n_1 n_2 - n_2{}^2 + n_1 n_3 - n_2 n_3$$
$$= n_1 n_2 - n_1 n_3 + n_2{}^2 - n_2 n_3 \qquad (36)$$

Equation (36) can be reduced to Eq. (37). In the case of a glass surface in air, $n_1 = 1$ and $n_3 \cong 1.5$. Magnesium fluoride is a substance which is frequently used as a nonreflective coating, since it is hard and approximately satisfies the relationship of Eq. (37). The purpose of reducing the reflection

$$n_2 = \sqrt{n_1 n_3} \qquad (37)$$

from an optical element is to increase its transmission, since the energy which is not reflected is transmitted. In the case of a single element, this increase is not particularly important. Some optical instruments may have 15−20 air-glass surfaces, however, and the coating of these surfaces gives a tremendous increase in transmission.

Haidinger fringes. When the second surface in two-surface interference is partially reflecting, interference can also be observed in the wave transmitted through both surfaces. The interference fringes will be complementary to those appearing in reflection. Their location will depend on the parallelism of the surfaces. For plane parallel surfaces, the fringes will appear at infinity and will be concentric rings. These were first observed by W. K. Haidinger and are called Haidinger fringes.

Multiple-beam interference. If the surfaces S_1 and S_2 are strongly reflecting, it is necessary to consider multiple reflections between them. For air-glass surfaces, this does not apply since the reflectivity is of the order of 4%, and the twice-reflected beam is much reduced in intensity.

In Fig. 9 the situation in which the surfaces S_1 and S_2 have reflectivities r_1 and r_2 is shown. The space between the surfaces has an index n_2 and thickness d. An incident light beam of amplitude A is partially reflected at the first surface. The transmitted component is reflected at S_2 and is reflected back to S_1 where a second splitting takes place. This is repeated. Each successive component of the waves leaving S_1 is retarded with respect to the next. The amount of each retardation is given by Eq. (38).

$$\varphi = \frac{4\pi n d}{\lambda} \cos \theta \qquad (38)$$

Equation (7) was derived for the superposition of two waves. It is possible to derive a similar expression for the superposition of many waves. From Fig. 9, the different waves at a plane somewhere above S_1 can be represented by the following expressions:

Incoming wave $= A \sin \omega t$
First reflected wave $= A r_1 \sin \omega t$
Second reflected wave $= A(1 - r_1{}^2) r_2 \sin (\omega t + \varphi)$
Third reflected wave $= -A(1 - r_1{}^2) r_1 r_2{}^2$
          $\sin (\omega t + 2\varphi)$

By inspection of these terms, one can write down the complete series. As in Eq. (3), the sine terms can be broken down and coefficients collected. A simpler method is to multiply each term by $i = \sqrt{-1}$ and add a cosine term with the same coefficient and argument. The individual terms then are all of the form of expression (39), where m is an integer.

$$B e^{-i\omega t} e^{-im\varphi} \qquad (39)$$

The individual terms of expression (39) can be easily summed. For the reflected wave one obtains Eq. (40). Again, as in the two-beam case, the

$$R = \frac{r_1 + r_2 e^{-i\varphi}}{1 + r_1 r_2 e^{-i\varphi}} \qquad (40)$$

minimum in the reflectivity R is obtained when $\varphi = N\pi$, where N is an odd integer and $r_1 = r_2$. The fringe shape, however, can be quite different from the earlier case, depending on the values of the reflectivities r_1 and r_2. The greater these values, the sharper become the fringes.

It was shown earlier how two-beam interference could be used to measure the contour of a surface. In this technique, a flat glass test plate was placed over the surface to be examined and monochromatic interference fringes were formed between the test surface and the surface of the plate. These two-beam fringes have intensities which vary as the cosine squared of the path difference. It is very difficult with such fringes to detect variations in fringe straightness or, in other terms, variations of

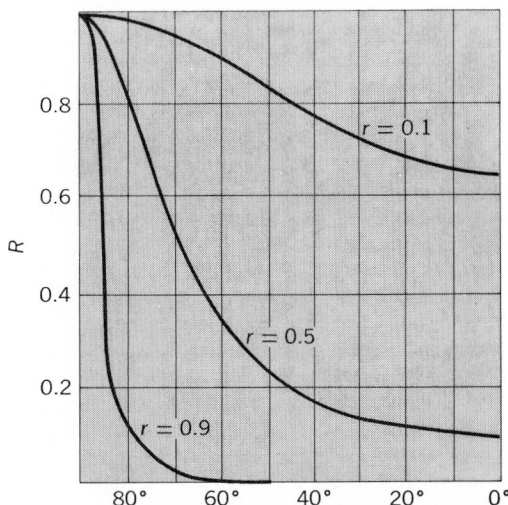

Fig. 10. The shape of multiple-beam fringes for different values of surface reflectivity.

surface planarity that are smaller than 1/20 wavelength. If the surface to be examined is coated with silver and the test surface is also coated with a partially transmitting metallic coat, the reflectivity increases to a point where many beams are involved in the formation of the interference fringes. The shape of the fringes is given by Eq. (40). The shape of fringes for different values of r is shown in Fig. 10. With high-reflectivity fringes, the sensitivity to a departure from planarity is increased far beyond 1/20 wavelength.

It is thus possible with partially silvered surfaces to get a much better picture of small irregularities than with uncoated surfaces. The increase in sensitivity is such that steps in cleaved mica as small as 10 A in height can be seen by examining the monochromatic interference fringes produced between a silvered mica surface and a partially silvered glass flat. *See* INTERFEROMETRY.

[BRUCE H. BILLINGS]

Bibliography: M. Born and E. Wolf, *Principles of Optics,* 5th ed., 1975; A. M. Cook, *Interference of Electromagnetic Waves,* 1971; O. S. Heavens, *Optical Properties of Thin Solid Films,* 1955; F. A. Jenkins and H. E. White, *Fundamentals of Optics,* 4th ed., 1976; J. Strong, *Concepts of Classical Optics,* 1958.

Interferometry

The design and use of optical interferometers. Optical interferometers based on both two-beam interference and multiple-beam interference of light are extremely powerful tools for metrology and spectroscopy. A wide variety of measurements can be performed, ranging from determining the shape of a surface to an accuracy of less than a millionth of an inch (25 nm) to determining the separation, by millions of kilometers, of binary stars.

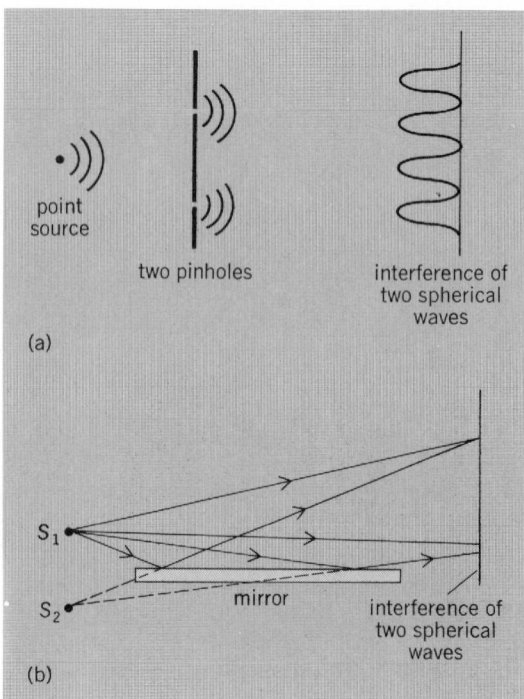

Fig. 1. Interference produced by division of wavefront. (a) Young's two-pinhole interferometer. (b) Lloyd's mirror.

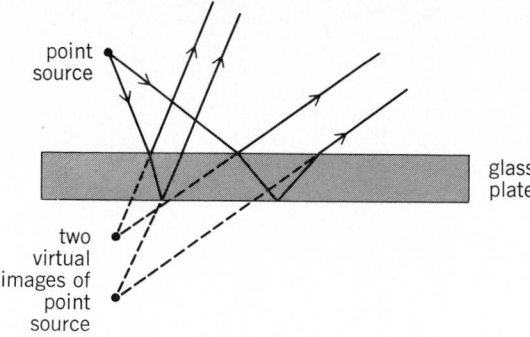

Fig. 2. Division of amplitude.

In spectroscopy, interferometry can be used to determine the hyperfine structure of spectrum lines. By using lasers in classical interferometers as well as holographic interferometers and speckle interferometers, it is possible to perform deformation, vibration, and contour measurements of diffuse objects that could not previously be performed.

Basic classes of interferometers. There are two basic classes of interferometers: division of wavefront and division of amplitude. Figure 1 shows two arrangements for obtaining division of wavefront. For the Young's double pinhole interferometer (Fig. 1*a*), the light from a point source illuminates two pinholes. The light diffracted by these pinholes gives the interference of two point sources. For the Lloyd's mirror experiment (Fig. 1*b*), a mirror is used to provide a second image S_2 of the point source S_1, and in the region of overlap of the two beams the interference of two spherical beams can be observed. There are many other ways of obtaining division of wavefront; however, in each case the light leaving the source is spatially split, and then by use of diffraction, mirrors, prisms, or lenses the two spatially separated beams are superimposed.

Figure 2 shows one technique for obtaining division of amplitude. For division-of-amplitude interferometers a beam splitter of some type is used to pick off a portion of the amplitude of the radiation which is then combined with a second portion of the amplitude. The visibility of the resulting interference fringes is a maximum when the amplitudes of the two interfering beams are equal. *See* INTERFERENCE OF WAVES.

Michelson interferometer. The Michelson interferometer (Fig. 3) is based on division of amplitude. Light from an extended source S is incident on a partially reflecting plate (beam splitter) P_1. The light transmitted through P_1 reflects off mirror M_1 back to plate P_1. The light which is reflected proceeds to M_2 which reflects it back to P_1. At P_1, the two waves are again partially reflected and partially transmitted, and a portion of each wave proceeds to the receiver R, which may be a screen, a photocell, or a human eye. Depending on the difference between the distances from the beam splitter to the mirrors M_1 and M_2, the two beams will interfere constructively or destructively. Plate P_2 compensates for the thickness of P_1. Often when a quasi-monochromatic light source is used with the interferometer, compensating plate P_2 is omitted.

The function of the beam splitter is to superimpose (image) one mirror onto the other. When the mirrors' images are completely parallel, the interference fringes appear circular. If the mirrors are slightly inclined about a vertical axis, vertical fringes are formed across the field of view. These fringes can be formed in white light if the path difference in part of the field of view is made zero. Just as in other interference experiments, only a few fringes will appear in white light, because the difference in path will be different for wavelengths of different colors. Accordingly, the fringes will appear colored close to zero path difference, and will disappear at larger path differences where the fringe maxima and minima for the different wavelengths overlap. If light reflected off the beam splitter experiences a one-half-cycle relative phase shift, the fringe of zero path difference is black, and can be easily distinguished from the neighboring fringes. This makes use of the instrument relatively easy.

The Michelson interferometer can be used as a spectroscope. Consider first the case of two close

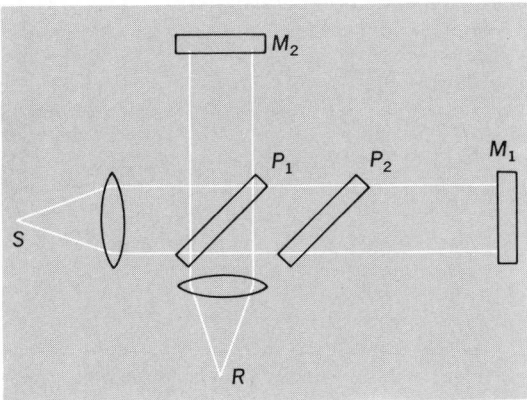

Fig. 3. Michelson interferometer.

spectrum lines as a light source for the instrument. As the mirror M_1 is shifted, fringes from each spectral line will cross the field. At certain path differences between M_1 and M_2, the fringes for the two spectral lines will be out of phase and will essentially disappear; at other points they will be in phase and will be reinforced. By measuring the distance between successive maxima in fringe contrast, it is possible to determine the wavelength difference between the lines.

This is a simple illustration of a very broad use for any two-beam interferometer. As the path length L is changed, the variation in intensity $I(L)$ of the light coming from an interferometer gives information on the basis of which the spectrum of the input light can be derived. The equation for the intensity of the emergent energy can be written as Eq. (1), where β is a constant, and $I(\lambda)$ is the intensity of the incident light at different wavelengths λ.

$$I(L) = \int_0^\infty I(\lambda) \cos^2\left(\frac{\beta L}{\lambda}\right) d\lambda \qquad (1)$$

sity of the incident light at different wavelengths λ. This equation applies when the mirror M_1 is moved linearly with time from the position where the path difference with M_2 is zero, to a position which depends on the longest wavelength in the spectrum

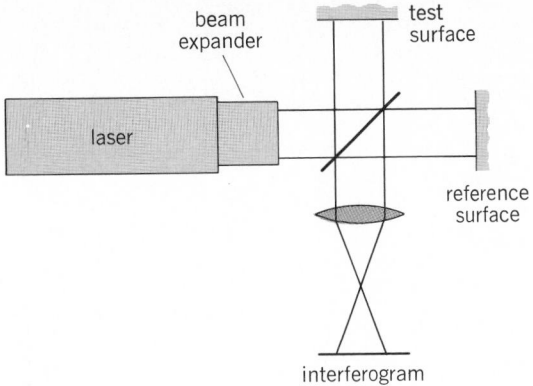

Fig. 4. Twyman-Green interferometer for testing flat surfaces.

to be examined. From Eq. (1), it is possible mathematically to recover the spectrum $I(\lambda)$. In certain situations, such as in the infrared beyond the wavelength region of 1.5 μm, this technique offers a large advantage over conventional spectroscopy in that its utilization of light is extremely efficient.

Twyman-Green interferometer. If the Michelson interferometer is used with a point source instead of an extended source, it is called a Twyman-Green interferometer. The use of the laser as the light source for the Twyman-Green interferometer has made it an extremely useful instrument for testing optical components. The great advantage of a laser source is that it makes it possible to obtain bright, good-contrast, interference fringes even if the path lengths for the two arms of the interferometer are quite different. *See* LASER.

Figure 4 shows a Twyman-Green interferometer for testing a flat mirror. The laser beam is expanded to match the size of the sample being tested. Part of the laser light is transmitted to the reference surface, and part is reflected by the beam splitter to the flat surface being tested. Both beams are reflected back to the beam splitter, where they are combined to form interference fringes. An imaging lens projects the surface under test onto the observation plane.

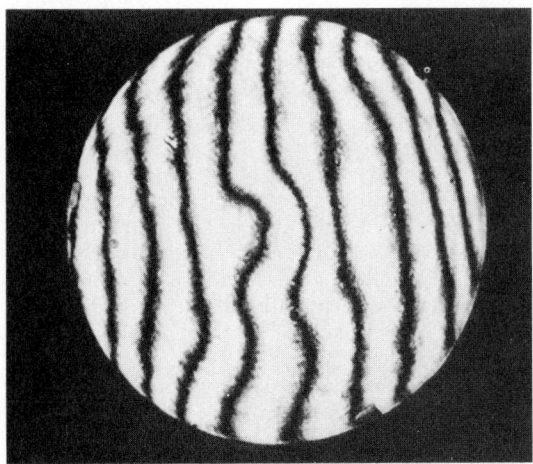

Fig. 5. Interferogram obtained with the use of a Twyman-Green interferometer to test a flat surface.

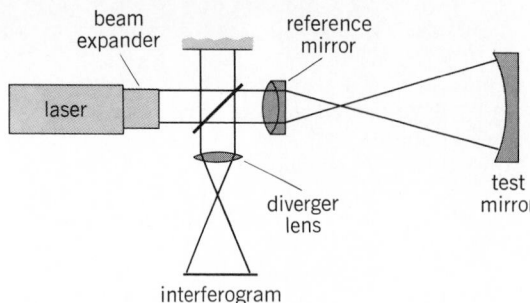

Fig. 6. Twyman-Green interferometer for testing spherical mirrors or lenses.

Fringes (Fig. 5) show defects in the surface being tested. If the surface is perfectly flat, then straight, equally spaced fringes are obtained. Departure from the straight, equally spaced condition shows directly how the surface differs from being perfectly flat. For a given fringe, the difference in optical path between light going from laser to reference surface to observation plane and the light going from laser to test surface to observation plane is a constant. (The optical path is equal to the product of the geometrical path times the refractive index.) Between adjacent fringes (Fig. 5) the optical path difference changes by one wavelength, which for a helium-neon laser corresponds to 633 nm. The number of straight, equally spaced fringes and their orientation depend upon the tip-tilt of the reference mirror. That is, by tipping or tilting the reference mirror the difference in optical path can be made to vary linearly with distance across the laser beam.

Deviations from flatness of the test mirror also cause optical path variations. A height change of half a wavelength will cause an optical path change of one wavelength and a deviation from fringe straightness of one fringe. Thus, the fringes give surface height information, just as a topographical map gives height or contour information.

The existence of the essentially straight fringes provides a means of measuring surface contours relative to a tilted plane. This tilt is generally introduced to indicate the sign of the surface error, that is, whether the errors correspond to a hill or a valley. One way to get this sign information is to push in on the piece being tested when it is in the interferometer. If the fringes move toward the right when the test piece is pushed toward the beam splitter, then fringe deviations from straightness toward the right correspond to high points (hills) on the test surface and deviations to the left correspond to low points (valleys).

The basic Twyman-Green interferometer (Fig. 4) can be modified (Fig. 6) to test concave-spherical mirrors. In the interferometer, the center of curvature of the surface under test is placed at the focus of a high-quality diverger lens so that the wavefront is reflected back onto itself. After this retroreflected wavefront passes through the diverger lens, it will be essentially a plane wave, which, when it interferes with the plane reference wave, will give interference fringes similar to those shown in Fig. 5 for testing flat surfaces. In this case it indicates how the concave-spherical mirror differs from the desired shape. Likewise, a convex-

spherical mirror can be tested. Also, if a high-quality spherical mirror is used, the high-quality diverger lens can be replaced with the lens to be tested.

Fizeau interferometer. One of the most commonly used interferometers in optical metrology is the Fizeau interferometer, which can be thought of as a folded Twyman-Green interferometer. In the Fizeau, the two surfaces being compared, which can be flat, spherical, or aspherical, are placed in close contact. The light reflected off these two surfaces produces interference fringes. For each fringe, the separation between the two surfaces is a constant. If the two surfaces match, straight, equally spaced fringes result. Surface height variations between the two surfaces cause the fringes to deviate from straightness or equal separation, where one fringe deviation from straightness corresponds to a variation in separation between the two surfaces by an amount equal to one-half of the wavelength of the light source used in the interferometer. The wavelength of a helium source, which is often used in a Fizeau interferometer, is 587.56 nm; hence one fringe corresponds to a height variation of approximately 0.3 μm.

Mach-Zehnder interferometer. The Mach-Zehnder interferometer (Fig. 7) is a variation of the Michelson interferometer and, like the Michelson interferometer, depends on amplitude splitting of the wavefront. Light enters the instrument and is reflected and transmitted by the semitransparent mirror M_1. The reflected portion proceeds to M_3, where it is reflected through the cell C_2 to the semitransparent mirror M_4. Here it combines with the light transmitted by M_1 to produce interference. The light transmitted by M_1 passes through a cell C_1, which is similar to C_2, and is used to compensate for the windows of C_1.

The major application of this instrument is in studying airflow around models of aircraft, missiles, or projectiles. The object and associated airstream are placed in one arm of the interferometer. Because the air pressure varies as it flows over the model, the index of refraction varies, and thus the effective path length of the light in this beam is a function of position. When the variation is an odd number of half-waves, the light will inter-

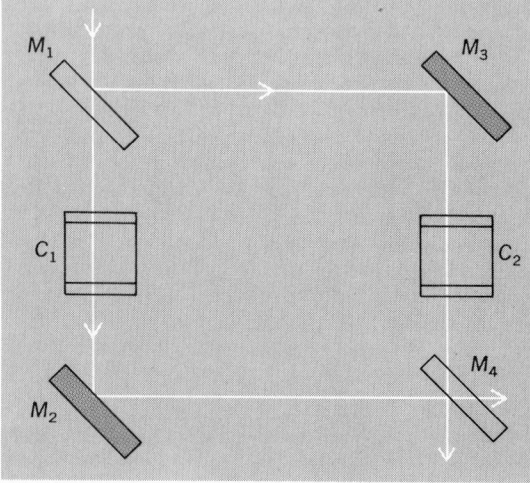

Fig. 7. Mach-Zehnder interferometer.

fere destructively and a dark fringe will appear in the field of view. From a photograph of the fringes, the flow pattern can be mathematically derived.

A major difference between the Mach-Zehnder and the Michelson interferometer is that in the Mach-Zehnder the light goes through each path in the instrument only once, whereas in the Michelson the light traverses each path twice. This double traversal makes the Michelson interferometer extremely difficult to use in applications where spatial location of index variations is desired. The incoming and outgoing beams tend to travel over slightly different paths, and this lowers the resolution because of the index gradient across the field.

Shearing interferometers. In a lateral-shear interferometer, an example of which is shown in Fig. 8, a wavefront is interfered with a shifted version of itself. A bright fringe is obtained at the points where the slope of the wavefront times the shift between the two wavefronts is equal to an integer number of wavelengths. That is, for a given fringe the slope or derivative of the wavefront is a constant. For this reason a lateral-shear interferometer is often called a differential interferometer.

Another type of shearing interferometer is a radial-shear interferometer. Here, a wavefront is interfered with an expanded version of itself. This interferometer is sensitive to radial slopes.

The advantages of shearing interferometers are that they are relatively simple and inexpensive, and since the reference wavefront is self-generated, an external wavefront is not needed. Since an external reference beam is not required, the source requirements are reduced from those of an interferometer such as a Twyman-Green. For this reason, shearing interferometers, in particular lateral-shear interferometers, are finding much use in applications such as adaptive optics systems for correction of atmospheric turbulence where the light source has to be a star, or planet, or perhaps just reflected sunlight.

Michelson stellar interferometer. A Michelson stellar interferometer can be used to measure the diameter of stars which are as small as 0.01 second of arc. This task is impossible with a ground-based optical telescope since the atmosphere limits the resolution of the largest telescope to not much better than 1 second of arc.

The Michelson stellar interferometer is a simple adaptation of Young's two-slit experiment. In its first form, two slits were placed over the aperture of a telescope. If the object being observed were a true point source, the image would be crossed with a set of interference bands. A second point source separated by a small angle from the first would produce a second set of fringes. At certain values of this angle, the bright fringes in one set will coincide with the dark fringes in the second set. The smallest angle α at which the coincidence occurs will be that angle subtended at the slits by the separation of the peak of the central bright fringe from the nearest dark fringe. This angle is given by Eq. (2), where d is the separation of the slits, λ the dominant wavelength of the two sources, and α their angular separation. The measurement of the separation of the sources is performed by adjusting the

$$\frac{\lambda}{2d} = \alpha \qquad (2)$$

inant wavelength of the two sources, and α their angular separation. The measurement of the separation of the sources is performed by adjusting the

separation d between the slits until the fringes vanish.

Consider now a single source in the shape of a slit of finite width. If the slit subtends an angle at the telescope aperture which is larger than α, the interference fringes will be reduced in contrast. For various line elements at one side of the slit, there will be elements of angle α away which will cancel the fringes from the first element. By induction, it is clear that for a separation d' such that the slit source subtends an angle as given by Eq. (3)

$$\alpha' = \frac{\lambda}{d'} \qquad (3)$$

the fringes from a single slit will vanish completely. For additional information on the Michelson stellar interferometer *see* DIFFRACTION.

Fabry-Perot interferometer. All the interferometers discussed above are two-beam interferometers. The Fabry-Perot interferometer (Fig. 9) is a multiple-beam interferometer since the two glass plates are partially silvered on the inner surfaces, and the incoming wave is multiply reflected between the two surfaces. The position of the fringe maxima is the same for multiple-beam interference as two-beam interference; however, as the reflectivity of the two surfaces increases and the number of interfering beams increases, the fringes become sharper.

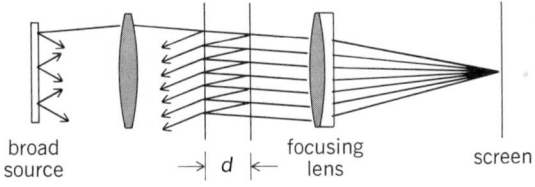

Fig. 9. Fabry-Perot interferometer.

A quantity of particular interest in a Fabry-Perot is the ratio of the separation of adjacent maxima to the half-width of the fringes. It can be shown that this ratio, known as the finesse, is given by Eq. (4),

$$\mathscr{F} = \frac{\pi \sqrt{R}}{1 - R} \qquad (4)$$

where R is the reflectivity of the silvered surfaces.

The multiple-beam Fabry-Perot interferometer is of considerable importance in modern optics for spectroscopy. All the light rays incident on the Fabry-Perot at a given angle will result in a single circular fringe of uniform irradiance. With a broad diffuse source, the interference fringes will be narrow concentric rings, corresponding to the multiple-beam transmission pattern. The position of the fringes depends upon the wavelength. That is, each wavelength gives a separate fringe pattern. The minimum resolvable wavelength difference is determined by the ability to resolve close fringes. The ratio of the wavelength λ to the least resolvable wavelength difference $\Delta\lambda$ is known as the chromatic resolving power $\mathscr{R}$. At nearly normal incidence it is given by Eq. (5),

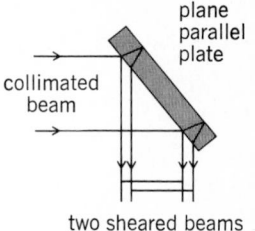

Fig. 8. Lateral shear interferometer.

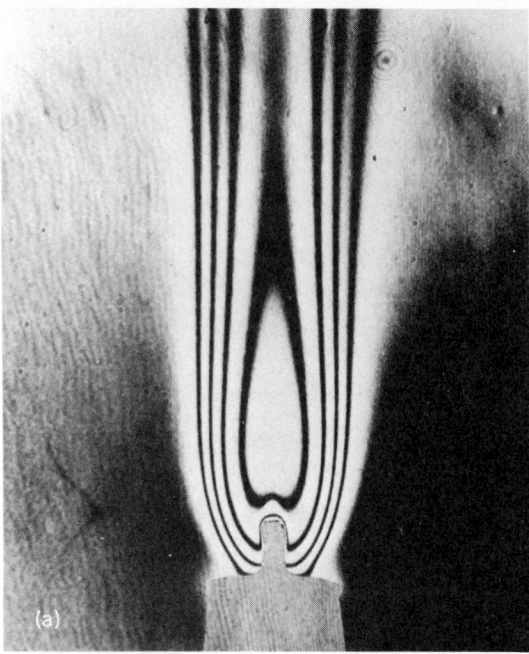

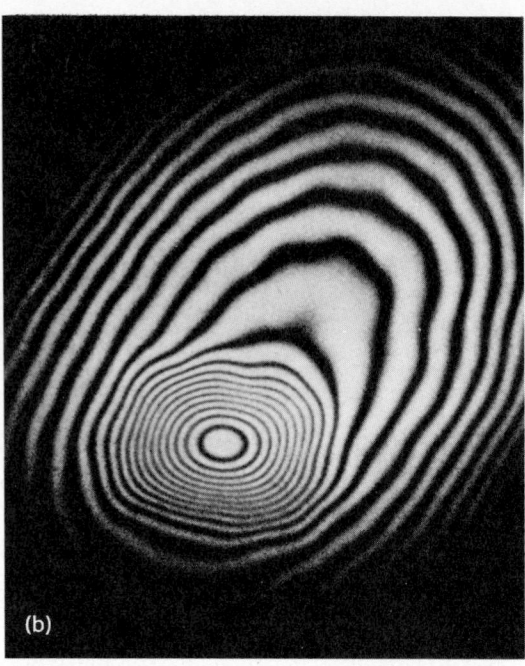

Fig. 10. Double-exposure holographic interferograms. (a) Interferogram of candle flame. (b) Interferogram of debanded region of honeycomb construction panel.

(From C. M. Vest, Holographic Interferometry, Copyright © 1978 by John Wiley and Sons, Inc.; used with permission)

$$\mathscr{R} = \frac{\lambda}{(\Delta\lambda)_{min}} = \mathscr{F} \frac{2nd}{\lambda} \tag{5}$$

where n is the refractive index between the two mirrors separated a distance d. For a wavelength of 500 nm, $nd = 10$ mm, and $R = 90\%$, the resolving power is well over 10^6. *See* REFLECTION OF ELECTROMAGNETIC RADIATION; RESOLVING POWER.

When Fabry-Perot interferometers are used with lasers, they are generally used in the central spot scanning mode. The interferometer is illuminated with a collimated laser beam and all the light transmitted through the Fabry-Perot is focused onto a detector, whose output is displayed on an oscilloscope. Often one of the mirrors is on a piezoelectric mirror mount. As the voltage to the piezoelectric crystal is varied, the mirror separation is varied. The light output as a function of mirror separation gives the spectral frequency content of the laser source.

Holographic interferometry. A wave recorded in a hologram is effectively stored for future reconstruction and use. Holographic interferometry is concerned with the formation and interpretation of the fringe pattern which appears when a wave, generated at some earlier time and stored in a hologram, is later reconstructed and caused to interfere with a comparison wave. It is the storage or time-delay aspect which gives the holographic method a unique advantage over conventional optical interferometry.

A hologram can be made of an arbitrarily shaped, rough scattering surface, and after suitable processing if the hologram is illuminated with the same reference wavefront used in recording the hologram, the hologram will produce the original object wavefront. If the hologram is placed back into its original position, a person looking through the hologram will see both the original object and the image of the object stored in the hologram. If the object is now slightly deformed, interference fringes will be produced which tell how much the surface is deformed. Between adjacent fringes the optical path between the source and viewer has changed by one wavelength. While the actual shape of the object is not determined,

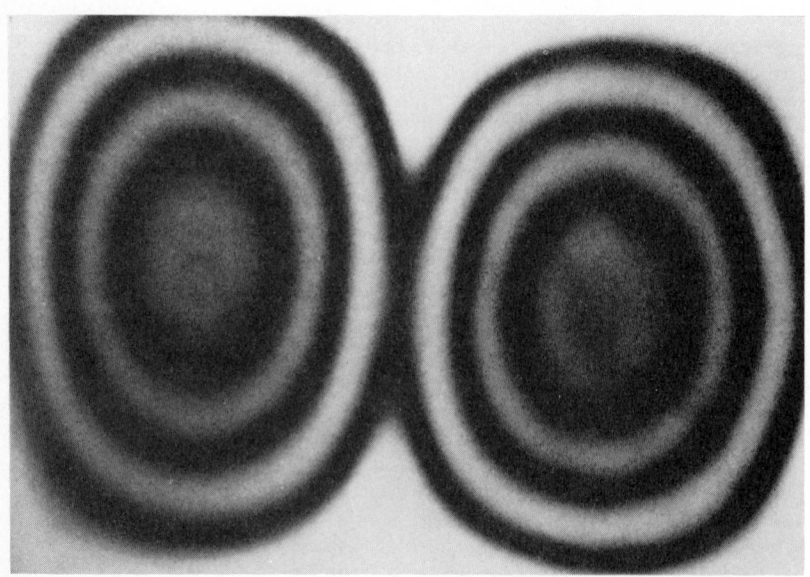

Fig. 11. Photograph of time-average holographic interferogram. *(From C. M. Vest. Holographic Interferometry, copyright © 1978 by John Wiley and Sons, Inc.; used with permission)*

the change in the shape of the object is measured to within a small fraction of a wavelength, even though the object's surface is rough compared to the wavelength of light.

Double-exposure. Double-exposure holographic interferometry (Fig. 10) is similar to real-time holographic interferometry described above, except now two exposures are made before processing: one exposure with the object in the undeformed state and a second exposure after deformation. When the hologram reconstruction is viewed, interference fringes will be seen which show how much the object was deformed between exposures. The advantage of the double-exposure technique over the real-time technique is that there is no critical replacement of the hologram after processing. The disadvantage is that continuous comparison of surface displacement relative to an initial state cannot be made, but rather only the difference between two states is determined.

Time-average. In time-average holographic interferometry (Fig. 11) a time-average hologram of a vibrating surface is recorded. If the maximum amplitude of the vibration is limited to some tens of light wavelengths, illumination of the hologram yields an image of the surface on which is superimposed several interference fringes which are contour lines of equal displacement of the surface. Time-average holography enables the vibrational amplitudes of diffusely reflecting surfaces to be measured with interferometric precision. *See* HOLOGRAPHY.

Speckle interferometry. A random intensity distribution, called a speckle pattern, is generated when light from a highly coherent source, such as a laser, is scattered by a rough surface. The use of speckle patterns in the study of object displacements, vibration, and distortion is becoming of more importance in the nondestructive testing of mechanical components. For example, time-averaged speckle photographs can be used to analyze the vibrations of an object in its plane. In physical terms the speckles in the image are drawn out into a line as the surface vibrates, instead of being double as in the double-exposure technique. The diffraction pattern of this smeared-out speckle-pattern recording is related to the relative time spent by the speckle at each point of its trajectory (Fig. 12).

Speckle interferometry can be used to perform astronomical measurements similar to those performed by the Michelson stellar interferometer. Stellar speckle interferometry is a technique for obtaining diffraction-limited resolution of stellar objects despite the presence of the turbulent atmosphere that limits the resolution of ground-based telescopes to approximately 1 arc-second. For example, the diffraction limit of the 5-m-diameter Palomar Mountain telescope is approximately 0.02 arc-second, 1/50 the resolution limit set by the atmosphere.

The first step of the process is to take a large number, perhaps 100, of short exposures of the object, where each photo is taken for a different realization of the atmosphere. Next the optical diffraction pattern, that is, the squared modulus of the Fourier transform of all the short-exposure photographs is added. By taking a further Fourier transform of each ensemble average diffraction

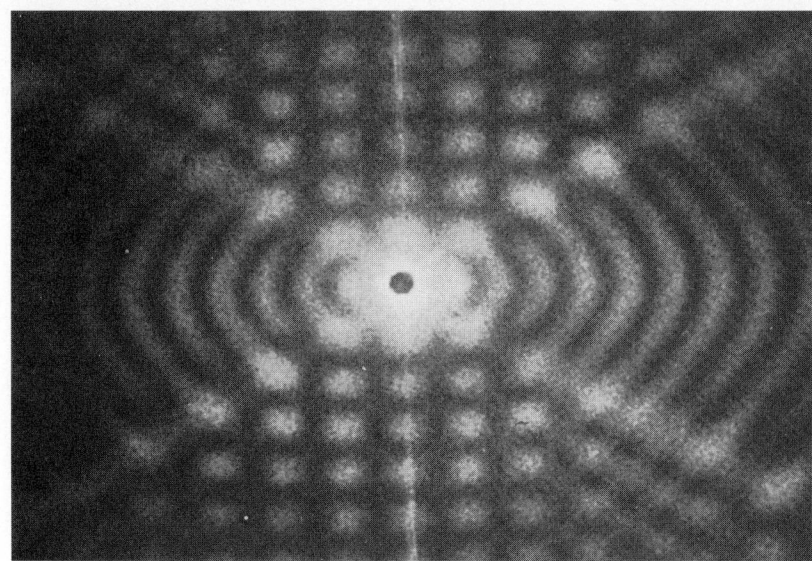

Fig. 12. Diffraction patterns from time-averaged speckle interferogram of a surface vibrating in its own plane with a figure-eight motion. (*From J. C. Dainty, Laser Speckle and Related Phenomena, Springer-Verlag, 1975*)

pattern, the ensemble average of the spatial autocorrelation of the diffraction-limited images of each object is obtained. *See* SPECKLE.

[JAMES C. WYANT]

Bibliography: J. C. Dainty (ed.), *Laser Speckle and Related Phenomena*, 1975; R. K. Erf (ed.), *Speckle Metrology*, 1978; E. Hecht and A. Zajac, *Optics*, 1974; D. Malacara (ed.), *Optical Shop Testing*, 1978; C. Vest, *Holographic Interferometry*, 1979.

Intermolecular forces

Attractive or repulsive interactions that occur between all atoms and molecules. These forces, which become significant at molecular separations of about 1 nm or less, are much weaker than forces associated with chemical bonds or electrostatic interactions of charged particles. They are important, however, since they are responsible for many of the physical properties of solids, liquids, and pressurized gases. Intermolecular forces also determine to an important extent the three-dimensional arrangement of biological molecules, polymers, and even smaller molecules.

Description. A simple description of intermolecular forces can begin with the example of two interacting argon atoms. The atoms are electrically neutral and do not undergo chemical bond formation.

Figure 1 shows the potential energy of two argon atoms as a function of their separation. At distances of about 1 nm or greater this energy is essentially zero and the atoms exert no forces on each other. (The force is the negative gradient, or slope, of the potential energy.) Between 0.4 and 0.8 nm the potential energy decreases and the atoms experience forces of attraction. For distances less than 0.3 nm the potential energy rises sharply as the atoms repel each other. At a distance of 0.38 nm the forces of attraction and repulsion balance each other. The potential energy (and correspond-

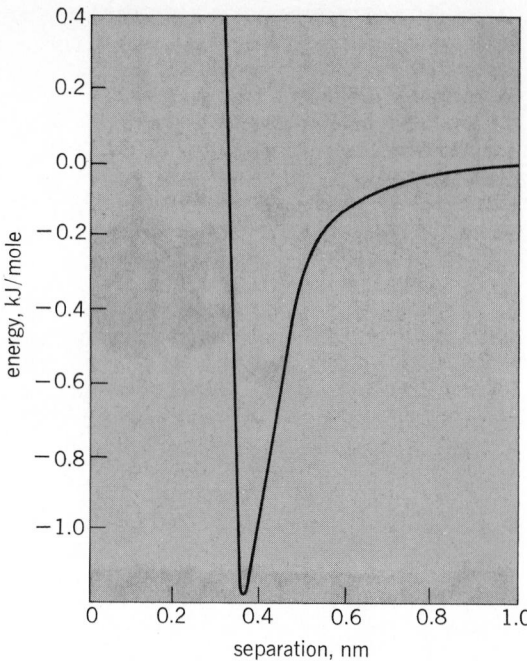

Fig. 1. The intermolecular potential energy of two argon atoms. (*From J. M. Parson, P. E. Siska, and Y. T. Lee, Intermolecular potentials from cross-beam differential elastic scattering measurements, IV:Ar + Ar, J. Chem. Phys., 56: 1511–1516, 1972.*)

ing intermolecular forces) between other pairs of atoms exhibits the same general shape as shown in Fig. 1, although the quantitative values of energy and separation are somewhat different. For intermolecular forces between molecules the relative orientations as well as distances are important and the description is more complex. In general, for either atoms or molecules at separations of 0.3 nm or less, the intermolecular forces are repulsive. At longer range, usually greater than 0.3 nm, the intermolecular forces are attractive. And at some intermediate distance, usually 0.3–0.4 nm (which depends on orientation in the case of molecules), the intermolecular forces of attraction and repulsion just balance.

Origin. The origin of intermolecular forces is again most simply discussed by considering two interacting atoms. Quantum mechanics indicates that the rapid motion of the electrons causes instantaneous fluctuations in the charge density around the nucleus. For atoms far apart the electrons in one atom move independently of electrons in the other atom, and on the average the charge distribution is symmetric as shown in Fig. 2a. At distances where attractive forces become important, the average charge distribution is still symmetric. However, an instantaneous fluctuation in the electron distribution in one atom can now affect its neighbor nearby. A charge separation in one atom occurs when the electron cloud shifts toward one side of the atom, barring its nucleus to a slight extent. In the other atom the electrons have moved in concert toward this barred nucleus, and an electrostatic attraction is set up. This is illustrated schematically in an exaggerated fashion in Fig. 2b. At another instant the electron clouds may shift in the opposite direction, and the other

atom has its nuclear charge partially exposed to the electrons of its neighbor. The electron motions in both atoms are correlated so that an attractive electrostatic force is maintained while the averaged motions assure a symmetric distribution about each atom. These attractive forces are often called London or dispersion forces.

At small separations the electron clouds can overlap, and repulsive forces are set up. These are called Pauli or exchange forces and are also explained by quantum mechanics. They are essentially a consequence of the reluctance of electrons to be confined into the same small region of space. Atoms or molecules brought close together will respond to exchange forces by a permanent distortion of their electron distribution as shown in Fig. 2c.

All atoms and molecules experience dispersion and exchange forces, which thus are a common component of intermolecular forces. Neutral molecules, in addition, may interact with each other because they possess permanent electrical polarity expressed as a dipole, quadrupole, or higher multipole moments. The electrostatic forces associated with these interacting multipole moments depend on the orientation of the molecules and may be either attractive or repulsive. The corre-

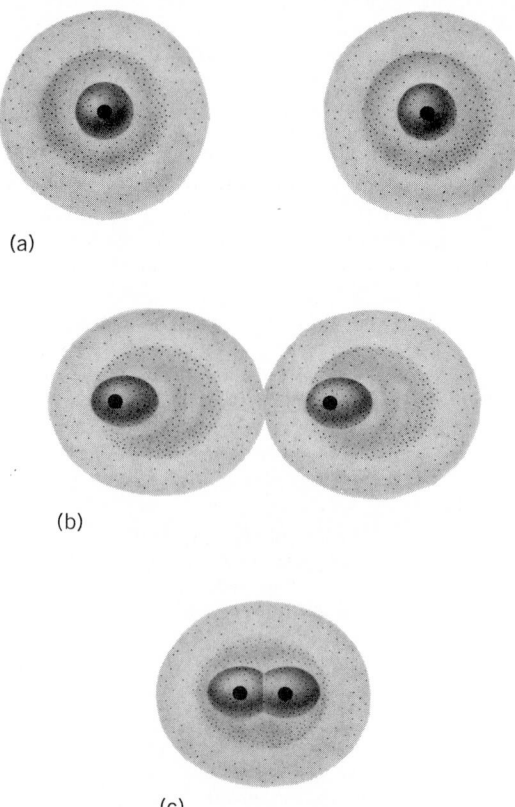

Fig. 2. Schematic illustration of intermolecular interaction. (a) There is no interaction between the atoms that are 1 nm or more apart. (b) For atoms separated by about 0.8 nm or less, dispersion forces which are attractive result from correlated fluctuations of the electron charge distribution in the atoms. (Distribution shown is greatly exaggerated.) (c) For the atoms closer together, 0.3 nm or less, exchange forces which are repulsive cause a permanent distortion of the electron charge distribution. (Distribution shown is greatly exaggerated.)

sponding energies are usually somewhat less than dispersion or exchange energies. The dispersion, exchange, and permanent multipole electrostatic forces taken together are usually called van der Waals forces. Energies associated with the formation of hydrogen bonds (that is, between two HF or H_2O molecules) are somewhat larger than van der Waals energies.

Interactions considerably stronger than those just discussed sometimes occur between atoms or molecules. The energies of chemical bond formation are hundreds of times greater than that shown by the intermolecular potential well of Fig. 1. Electrostatic interactions between charged particles are likewise relatively strong. These interactions are usually not classified as intermolecular forces.

Occurrence. Intermolecular forces are responsible for many of the bulk properties of matter in all its phases. A realistic description of the relationship among pressure, volume, and temperature of a gas must include the effects of attractive and repulsive forces between molecules. At increased pressures and sufficiently low temperatures the attractive forces between molecules in the gas will cause it to liquefy. The viscosity, surface tension, and diffusion of liquids are examples of physical properties which are a consequence of intermolecular forces. Repulsive forces prevent the molecules from approaching one another too closely and account for the high compressibility of liquids. Intermolecular forces between near and distant neighbors dictate the ordered molecular arrangements in crystalline solids. These forces also account for the elasticity of solids. A detailed accounting of the intermolecular forces in the condensed phase is complex since it must include the interactions of each molecule with many of its neighbors. Nevertheless, the energy of each pair of atom interactions is approximately described by an intermolecular potential of the sort shown in Fig. 1.

Intermolecular forces are also important between atoms within a molecule. Even for a molecule as small as ethane, CH_3CH_3, they direct important details in the molecular structure. Chemical bonds dictate the arrangement of the hydrogen atoms about each carbon atom as well as the distance between carbon atoms. However, intermolecular forces mold the final structure, which keeps the hydrogen atoms on one CH_3 group staggered with respect to those on the other CH_3 group. Thus the staggered rather than the eclipsed configuration for ethane as shown in Fig. 3 is the most stable three-dimensional structure. In an analogous way for larger molecules, proteins, and other biological molecules, the complex spatial ar-

rangement assumed is determined in part by the balance of attractive and repulsive intermolecular forces between atoms that are chemically bonded within the molecule.

Atoms and molecules may be held to a solid surface by intermolecular or van der Waals forces. This weak bonding, called physisorption, has many important applications. The trapping out of molecules from the gas phase onto cooled surfaces is the basis of pumps for producing vacuums. Undesirable odors or colors in food or water may sometimes be removed by use of filters which capture by physisorption the offending contamination. The selective adsorption of molecules by surfaces is a useful method for separation of mixtures of molecules.

Study methods. The importance of intermolecular forces has been responsible for their extensive study for many decades. In the early 1970s most of the information on intermolecular forces was inferred from the study of matter in bulk. Measurements of the viscosity of gases, or crystal structure of solids, for example, were used to estimate the quantitative nature of the intermolecular interactions that must produce these physical properties. However, it has since been found that studies of individual molecular interactions yield the information more directly.

In molecular beam experiments, low-density streams of atoms or molecules are directed so that individual particles collide. The way in which the molecules rebound as a result of their collision is determined by their initial velocities which can be controlled. Intermolecular forces can be extracted from the experimental data. The intermolecular potential energy curve shown in Fig. 1 was obtained from studies of the collision dynamics of argon atoms. Mappings of the potential energy surfaces of other atoms and molecules are being obtained by this technique. *See* MOLECULAR BEAMS.

Another approach is to study van der Waals molecules. In these experiments clusters of atoms or molecules are formed at low temperatures in the gas phase because of their intermolecular attractions for each other. Clusters of two or three atoms or molecules are called van der Waals molecules. For example, gaseous argon at the temperature of the boiling liquid ($-186°C$) contains about 98% Ar atoms, and the remaining 2% are Ar_2 van der Waals molecules. The ultraviolet spectrum of the gas at low temperatures reveals features due to Ar_2 which can be used to characterize the bond strength and the intermolecular forces between the argon atoms in the van der Waals molecule.

Spectroscopy of van der Waals molecules formed by clusters of chemically bonded molecules has also revealed much about intermolecular forces which depend on the orientation of the molecules within the cluster. Gaseous H_2, O_2, or HF contains small concentrations of $(H_2)_2$, $(O_2)_2$, or $(HF)_2$. The structures of these van der Waals molecules are shown in Fig. 4. The chemical bonds in H_2, O_2, or HF are about 0.1 nm long and not affected by the formation of the $0.3-0.4$ nm intermolecular bond of the van der Waals molecule. In $(H_2)_2$ the intermolecular forces do not depend much on the orientation of either H_2, and as a consequence each H_2 molecule, while weakly bound to its neigh-

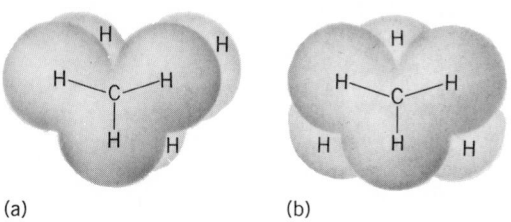

Fig. 3. Intermolecular forces and ethane. (*a*) Eclipsed configuration. (*b*) Staggered configuration.

INTERMOLECULAR FORCES

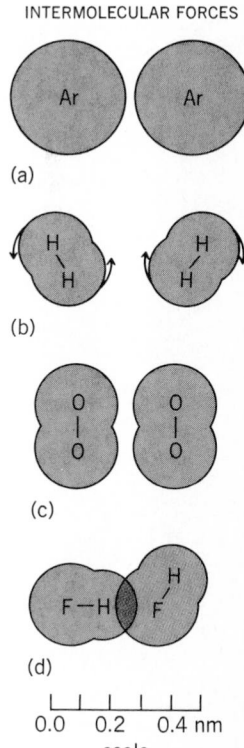

Fig. 4. The structures of some van de Waals molecules. (*a*) Argon. (*b*) Hydrogen. (*c*) Oxygen. (*d*) Hydrogen fluoride. (*From G. Ewing, Structure and properties of van der Waals molecules, Accounts Chem. Res., 8:185–192, 1975*)

bor, rotates freely within the cluster. The arrows shown in Fig. 4 are meant to represent this freedom of internal rotation. The $(O_2)_2$ van der Waals molecule appears to reside in a rectangular configuration, while $(HF)_2$ exhibits a bent structure characteristic of hydrogen bond formation. While chemical bonds produce rigid molecules with well-defined geometries, intermolecular forces maintain rather floppy structures of the van der Waals molecules. Internal motions in $(O_2)_2$ or $(HF)_2$ produce considerable distortions of the static structure representations in Fig. 4. The structures of several dozen van der Waals molecules are now known. The determination of properties of this new class of compounds promises to provide a deeper insight into the nature of intermolecular forces. See MOLECULAR STRUCTURE AND SPECTRA.

Theoretical approaches to intermolecular interactions have taken two directions. Detailed quantum-mechanical calculations have been performed on the interactions of very simple systems, for example, two He atoms. These calculations seek to determine the wave functions, importance of the correlated motions of the electrons, and the precise nature of the energy of the interaction. This theoretical approach then seeks a deeper understanding of the quantum-mechanical origin of intermolecular forces. A more pragmatic approach uses the electron distribution of the isolated molecule from previous calculations. This distribution is treated as an "electron gas" with an associated electric field. It is the response of the interacting molecules to these electric fields that is responsible for intermolecular forces. Calculations of the electron gas model appear to produce reliable intermolecular energies for both interacting atoms and molecules, with a modest amount of computational effort. See NONRELATIVISTIC QUANTUM THEORY; STATISTICAL MECHANICS.

[GEORGE E. EWING]

Bibliography: B. L. Blaney and G. E. Ewing, Van der Waals molecules, *Ann. Rev. Phys. Chem.*, 27:553–586, 1976; S. T. Ceyer and G. A. Somojai, Surface scattering, *Annu. Rev. Phys. Chem.*, 28:477–499, 1977; J. O. Hirschfelder, C. F. Curtiss, and R. B. Bird, *Molecular Theory of Gases and Liquids*, 1954; T. Kihara, *Intermolecular Forces*, 1978; Y. S. Kim and R. G. Gordon, Unified theory for intermolecular forces between closed shell atoms and ions, *J. Chem. Phys.* 61:1–16, 1974.

Internal energy

A characteristic property of the state of a thermodynamic system, introduced in the first law of thermodynamics. For a static, closed system (no bulk motion, no transfer of matter across its boundaries), the change ΔE in internal energy for a process is equal to the heat Q absorbed by the system from its surroundings minus the work w done by the system on its surroundings. Only a change in internal energy can be measured, not its value for any single state. For a given process, the change in internal energy is fixed by the initial and final states and is independent of the path by which the change in state is accomplished.

The internal energy includes the intrinsic energies of the individual molecules of which the system is composed and contributions from the interactions among them. It does not include contributions from the potential energy or kinetic energy of the system as a whole; these changes must be accounted for explicitly in the treatment of flow systems. Because it is more convenient to use an independent variable (the pressure P for the system instead of its volume V), the working equations of practical thermodynamics are usually written in terms of such functions as the enthalpy $H = E + PV$, instead of the internal energy itself. See ENTHALPY.

[PAUL J. BENDER]

Interpolation

A process in mathematics used to estimate an intermediate value of one (dependent) variable which is a function of a second (independent) variable when values of the dependent variable corresponding to several discrete values of the independent variable are known.

Suppose, as is often the case, that it is desired to describe graphically the results of an experiment in which some quantity Q is measured, for example, the electrical resistance of a wire, for each of a set of N values of a second variable v representing, perhaps, the temperature of the wire. Let the numbers Q_i, $i=1, 2, \ldots, N$, be the measurements made of Q and the numbers v_i be those of the variable v. These numbers representing the raw data from the experiment are usually given in the form of a table with each Q_i listed opposite the corresponding v_i. The problem of interpolation is to use the above discrete data to predict the value of Q corresponding to any value of v lying between the above v_i. If the value of v is permitted to lie outside these v_i, the somewhat more risky process of extrapolation is used. See EXTRAPOLATION.

Graphical interpolation. The above experimental data may be expressed in graphical forms by plotting a point on a sheet of paper for each pair of values (v_i, Q_i) of the variables. One establishes suitable scales by letting 1 in. represent a given number of units of v and of Q. If v is considered the independent variable, the horizontal displacement of the ith point usually represents v_i and its vertical displacement represents Q_i.

If, for simplicity, it is assumed that the experimental errors in the data can be neglected, then the problem of interpolation becomes that of drawing a curve through the N data points P_i having coordinates (x_i, y_i) that are proportional to the numbers v_i and Q_i, respectively, so as to give an accurate prediction of the value Q for all intermediate values of v. Since it is at once clear that the N measurements made would be consistent with any curve passing through the points, some additional assumptions are necessary in order to justify drawing any particular curve through the points. Usually one assumes that the v_i are close enough together that a smooth curve with as simple a variation as possible should be drawn through the points.

In practice the numbers v_i and Q_i will contain some experimental error, and, therefore, one should not require that the curve pass exactly through the points. The greater the experimental uncertainty the farther one can expect the true curve to deviate from the individual points. In some cases one uses the points only to suggest the type of simple curve to be drawn and then adjusts this type of curve to pass as near the individual points as possible. This may be done by choosing a function that contains a few arbitrary parameters that may be so adjusted as to make the plot of the function miss the points by as small a margin as possible.

For many purposes, however, one uses a French curve and orients it so that one of its edges passes very near a group of the points. Having drawn in this portion of the curve, one moves the edge of the French curve so as to approximate the next group of points. An attempt is made to join these portions of the curve so that there is no discontinuity of slope or curvature at any point on the curve.

Tabular interpolation. This includes methods for finding from a table the values of the dependent variable for intermediate values of the independent variable. Its purpose is the same as graphical interpolation, but one seeks a formula for calculating the value of the dependent variable rather than relying on a measurement of the ordinate of a curve.

In this discussion it will be assumed that x_i and y_i ($i = 1, 2, \ldots, N$), which represent tabulated values of the independent and dependent variables, respectively, are accurate to the full number of figures given. Interpolation then involves finding an interpolating function $P(x)$ satisfying the requirement that, to the number of figures given, the plot of Eq. (1) pass through a selected number of

$$y = P(x) \qquad (1)$$

points of the set having coordinates (x_i, y_i). The interpolating function $P(x)$ should be of such a form that it is made to pass readily through the selected points and is easily calculated for any intermediate value of x. Since many schemes are known for determining quickly the unique nth degree polynomial that satisfies Eq. (1) at any $n + 1$ of the tabulated values and since the value of such a polynomial may be computed using only n multiplications and n additions, polynomials are the most common form of interpolating function.

If the subscripts on x and y are reassigned so that the points through which Eq. (1) passes are now (x_0, y_0), (x_1, y_1), $\ldots$, (x_n, y_n), the polynomial needed in Eq. (1) may be written down by inspection, and one has Eq. (2), where Eq. (3) applies.

$$y = \sum_{k=0}^{n} \frac{L_k(x)}{L_k(x_k)} y_k \qquad (2)$$

$$L_k(x) = \frac{(x - x_0)(x - x_1) \cdots (x - x_n)}{(x - x_k)} \qquad (3)$$

Equation (3) is Lagrange's interpolation formula for unequally spaced ordinates. Since $L_k(x)$ vanishes for all x_s in the set $x_0, x_1, \ldots, x_n$ except x_k, substituting $x = x_s$ in the right-hand side of Eq. (1) gives rise to only one nonzero term. This term has the value y_s, as required.

For $n = 1$ Eqs. (2) and (3) give rise to Eq. (4),

$$y = \frac{x - x_1}{x_0 - x_1} y_0 + \frac{x - x_0}{x_1 - x_1} y_1 \qquad (4)$$

whose plot is a straight line connecting the points (x_0, y_0) and (x_1, y_1). Such an interpolation is referred to as linear interpolation and is used in all elemen-

tary discussions of interpolation; however, another equivalent form of this equation, given below in Eq. (12), is more often used in these cases.

Suppose the table were obtained from the equation $y = f(x)$, in which $f(x)$ is some mathematical function having continuous derivatives of all order up to and including the $(n + 1)$th. It is then possible to obtain an accurate expression for intermediate values of $f(x)$ by adding to the right-hand side of Eq. (2) a so-called remainder term, expression (5),

$$\frac{(x - x_0)(x - x_1) \cdots (x - x_n)}{(n+1)!} f^{(n+1)}(\xi) \qquad (5)$$

where $f^{(n+1)}(\xi)$ is the $(n + 1)$th derivative of $f(x)$ at some point $x = \xi$ lying between the smallest and largest of the values $x_0, x_1, \ldots, x_n$. Since the value of ξ is not known, the remainder term is used merely to set an upper limit on the truncation error introduced by using Lagrange's interpolation formula.

If the ordinates are equally spaced, that is, $x_s = x_0 + sh$ where h is the interval of tabulation, Lagrange's interpolation formula simplifies considerably and may be written as Eq. (6), where n is

$$y = \sum_{k=-p}^{n-p} A_k(u) y_k = A_{-p}(u) y_{-p} + \cdots$$
$$+ A_0(u) y_0 + \cdots + A_{n-p}(u) y_{n-p} \qquad (6)$$

the degree of interpolating polynomial that now passes through the $n + 1$ points (x_{-p}, y_{-p}), $\ldots$, (x_{n-p}, y_{n-p}). Here p is the largest integer less than or equal to $n/2$, and the $A_{-p}(u)$, $\ldots$, $A_{n-p}(u)$ are polynomials in the variable shown as Eq. (7). The

$$u = \frac{x - x_0}{h} \qquad (7)$$

polynomials of the variable in Eq. (7) have been tabulated as functions of u.

Inverse interpolation. If the value of y is known and the value of the corresponding independent variable x is desired, one has the problem of inverse interpolation. Since the polynomials $A_k(u)$ in Eq. (6) are known functions of u, and the values of y and y_k are also known, the only unknown in this equation is u. Thus the problem reduces to that of finding a real root u of an nth-degree polynomial in the range $0 < u < 1$. Having found u, one may find x from Eq. (7). For a discussion of numerical methods for solving for such a root and for more information on interpolation *see* NUMERICAL ANALYSIS.

One may also perform an inverse interpolation by treating x as a function of y. Since, however, the intervals between the y_i are not equal, it is necessary to employ the general interpolation formula of Eq. (2) with the xs and ys interchanged.

Round-off errors. In the tabulated values, round-off errors ε_i resulting from the need to express the entries y_i of the table as finite decimals will cause an additional error in the interpolated value of y that must be added to the truncation error discussed before. The effect of these errors on the application of Lagrange's interpolation formula is seen by Eq. (6) to be a total error ε_T in y given by Eq. (8). Letting e be the smallest positive

$$\varepsilon_T = \sum_k A_k(u) \varepsilon_k \qquad (8)$$

number satisfying the condition $e > |\varepsilon_k|$ for all k, one knows from Eq. (8) that relation (9) holds.

$$|\varepsilon_T| \leqq \sum_k |A_k(u)||\varepsilon_k| \leqq e \sum_k |A_k(u)| \qquad (9)$$

Since the sum of the $A_k(u)$ is equal to 1, the factor

$$\sum_k |A_k(u)|$$

in Eq. (9) is usually not much larger than 2 or 3 and thus the interpolated value of y has about the same round-off error as the individual entries.

Use of finite differences. For some purposes it is more convenient to use an interpolating formula based not so much on the entries y_i of a central difference table as upon their differences.

x_{-2}	y_{-2}				
		$\delta y_{-3/2}$			
x_{-1}	y_{-1}		$\delta^2 y_{-1}$		
		$\delta y_{-1/2}$		$\delta^3 y_{-1/2}$	
x_0	y_0		$\delta^2 y_0$		$\delta^4 y_0$
		$\delta y_{1/2}$		$\delta^3 y_{1/2}$	
x_1	y_1		$\delta^2 y_1$		$\delta^4 y_1$
		$\delta y_{3/2}$		$\delta^3 y_{3/2}$	
x_2	y_2		$\delta^2 y_2$		
		$\delta y_{5/2}$			
x_3	y_3				

Each difference $\delta^k y_s$ is obtained by subtracting the quantity immediately above and to the left of it from the quantity immediately below and to the left; thus, Eq. (10) can be written, where k and $2s$

$$\delta^k y_s = \delta^{k-1} y_{s+1/2} - \delta^{k-1} y_{s-1/2} \qquad (10)$$

are required to be integers. For example,

$$\delta y_{1/2} = y_1 - y_0 \quad \text{and} \quad \delta^2 y_0 = \delta y_{1/2} - \delta y_{-1/2}$$

An interesting property of a difference table is that, if y, the dependent variable tabulated, is a polynomial of the nth degree in x, its kth difference column will represent a polynomial of degree $n - k$. In particular, its nth differences will all be equal and all higher differences will be zero. For example, consider a table of cubes and the difference table formed from it by the rule given above.

x	$y = x^3$			
0	0			
		1		
1	1		6	
		7		6
2	8		12	0
		19		6
3	27		18	0
		37		6
4	64		24	0
		61		6
5	125		30	
		91		
6	216			

Most functions $f(x)$, when tabulated at a small enough interval $\Delta x = h$, behave approximately as polynomials and therefore give rise to a difference table in which some order of difference is nearly constant. Consider, for example, the difference table of $\log x$, in which the third differences fluctuate between 7 and 9 times 10^{-7}.

x	$y = \log x$	δy	$\delta^2 y$	$\delta^3 y$
1.00	0.0000 000			
		43 214		
1.01	0.0043 214		−426	
		42 788		8
1.02	0.0086 002		−418	
		42 370		9
1.03	0.0128 372		−409	
		41 961		8
1.04	0.0170 333		−401	
		41 560		7
1.05	0.0211 893		−394	
		41 166		
1.06	0.0253 059			

Experimental data, if taken at small enough interval of the independent variable, would be expected to exhibit much the same behavior as a mathematical function except for the presence of experimental error. The presence of the latter will cause the differences to have a more or less random fluctuation. The size of the fluctuation may, in fact, be used to indicate the number of significant figures in the data.

The constancy of the third differences for $\log x$ indicates that for the accuracy and the interval used, a third-degree polynomial may be employed as an interpolating function. Since such a polynomial is determined by the choice of four coefficients, one would expect the interpolation formula to involve four numbers derivable from the difference table. Thus the forward-interpolation formula of Gauss, Eq. (11), can be written. If

$$y = y_0 + u\,\delta y_{1/2} + \frac{1}{2} u(u-1)\,\delta^2 y_0$$

$$+ \frac{1}{3!} u(u^2 - 1)\,\delta^3 y_{1/2} + \frac{1}{4!} u(u^2 - 1)(u - 2)\,\delta^4 y_0$$

$$+ \frac{1}{5!} u(u^2 - 1)(u^2 - 4)\,\delta^5 y_{1/2} + \cdots \qquad (11)$$

terminated after the fourth term, it represents a third-degree polynomial in $u = (x - x_0)/h$, and hence in x. It involves the four constants y_0, $\delta y_{1/2}$, $\delta^2 y_0$, and $\delta^3 y_{1/2}$. Since any one of the entries in the y column may be chosen as y_0, the differences required are picked from a central difference table, for example, in relationship to this entry. The interpolating polynomial obtained passes through the four points (x_{-1}, y_{-1}), (x_0, y_0), (x_1, y_1), and (x_2, y_2). In general, the interpolating polynomial will pass through only those points whose y coordinate is needed to form the differences used in the formula.

If one terminates the series in Eq. (11) after the second term, one obtains Eq. (12). This is the lin-

$$y = y_0 + u\delta y_{1/2} = y_0 + u(y_1 - y_0) \qquad (12)$$

ear interpolation formula most often used when making a simple interpolation in a table.

There are a great variety of interpolation formulas, such as Gregory-Newton's, Stirling's, and Bessel's, that differ mainly in the choice of differences used to specify the interpolating polynomial.

Difference equations. Repeated application of Eq. (10) may be used to express any difference in terms of the tabulated values, for example, Eqs. (13). Expressed in a more general form Eqs. (13)

$$\delta y_{1/2} = y_1 - y_0$$
$$\delta^2 y_0 = y_1 - 2y_0 + y_{-1} \tag{13}$$

become Eqs. (14). If one sets the second difference

$$\delta f(x) = f\left(x + \frac{h}{2}\right) - f\left(x - \frac{h}{2}\right) \tag{14}$$
$$\delta^2 f(x) = f(x+h) - 2f(x) + f(x-h)$$

equal to zero, one obtains a so-called difference equation for $f(x)$. In general, a difference equation is any equation relating the values of $f(x)$ at discrete values of x.

Difference equations play much the same role in analytical work as differential equations. Because they can be interpreted in terms of only those values of $f(x)$ tabulated at some interval $\Delta x = h$, they are admirably adapted to numerical computations. For this reason most numerical solutions of differential equations involve approximating the equation by a suitable difference equation.

For ordinary differential equations the transformation to a difference equation can be made by replacing each derivative in the equation by the appropriate difference expression according to formula (15), where $f^{(n)}(x)$ designates the nth derivative

$$f^{(2k)}(x) \rightarrow \frac{1}{h^{2k}} \delta^{2k} f(x)$$

$$f^{(2k+1)}(x) \rightarrow \tag{15}$$

$$\frac{1}{2h^{2k+1}}\left[\delta^{2k+1}f\left(x + \frac{h}{2}\right) + \delta^{2k+1}f\left(x - \frac{h}{2}\right)\right]$$

ative of $f(x)$. The difference equation resulting can then, as mentioned before, be used to express the relationship between the values of $f(x)$ at the discrete points $x_s = x_0 + sh$, $s = 0, 1, 2, \ldots, n$.

Partial difference equations. Suppose one chooses to specify a function of two variables $f(x,y)$ by giving its value at some regular array of points in the xy plane having coordinates (x_m, y_n). Then, in place of a linear partial differential equation for $f(x,y)$ one has a linear partial difference equation, Eq. (16), where $g(x,y)$ is a known function and i

$$\sum_{s,t} A_{st} f(x_{i+s}, y_{j+t}) = g(x_i, y_j) \tag{16}$$

and j any of the set of integers for which the difference equation has significance. A difference equation in which some of the $f(x_{i+s}, y_{j+t})$ occur to a power other than the first is termed a nonlinear difference equation.

If one employs a square lattice makeup of the points, Eqs. (17), then Laplace's differential equa-

$$x_m = x_0 + mh$$
$$y_n = y_0 + nh \tag{17}$$

tion is approximated by difference equation (18),

$$f_{i+1,j} + f_{i,j+1} + f_{i-1,j} + f_{i,j-1} - 4f_{ij} = 0 \tag{18}$$

$$f_{mn} = f(x_m, y_n) \tag{19}$$

where, for simplicity, Eq. (19) holds.

[KAISER S. KUNZ]

Bibliography: S. Conte and C. de Boor, *Ele-*

mentary Numerical Analysis, 3d ed., 1980; P. J. Davis, *Interpolation and Approximation*, 1963, reprint 1975; A. Ralston and P. Rabinowitz, *A First Course in Numerical Analysis*, 2d ed., 1978.

Inverse-square law

Any law in which a physical quantity varies with distance from a source inversely as the square of that distance. When energy is being radiated by a point source (see illustration), such a law holds, provided the space between source and receiver is filled with a nondissipative, homogeneous, isotropic, unbounded medium. All unbounded waves become spherical at distances r, which are large compared with source dimensions so that the angular intensity distribution on the expanding wave surface, whose area is proportional to r^2, is fixed. Hence emerges the inverse-square law. *See* POINT SOURCE.

Similar reasoning shows that the same law applies to mechanical shear waves in elastic media and to compressional sound waves. It holds statistically for particle sources in a vacuum, such as radioactive atoms, provided there are no electromagnetic fields and no mutual interactions. The term is also used for static field laws such as the law of gravitation and Coulomb's law in electrostatics.

[WILLIAM R. SMYTHE]

Ionization chamber

An instrument for detecting ionizing radiation by measuring the amount of charge liberated by the interaction of ionizing radiation with suitable gases, liquids, or solids. These radiation detectors have played an important part in the development of modern physics and continue to find new applications in basic scientific research, in industry, and in medicine.

Principle of operation. While the gold leaf electroscope (Fig. 1) is the oldest form of ionization chamber, instruments of this type are still widely used as monitors of radiations by workers in the nuclear or radiomedical professions. In this device, two thin flexible pieces of gold leaf are suspended in a gas-filled chamber. When these are electrically charged, as in Fig. 1, the electrostatic repulsion causes the two leaves to spread apart. If ionizing radiation is incident in the gas, however, electrons are liberated from the gas atoms. These

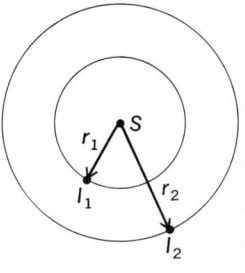

INVERSE-SQUARE LAW

Point source S emitting energy of intensity I. The inverse-square law states that $I_1/I_2 = r_2{}^2/r_1{}^2$.

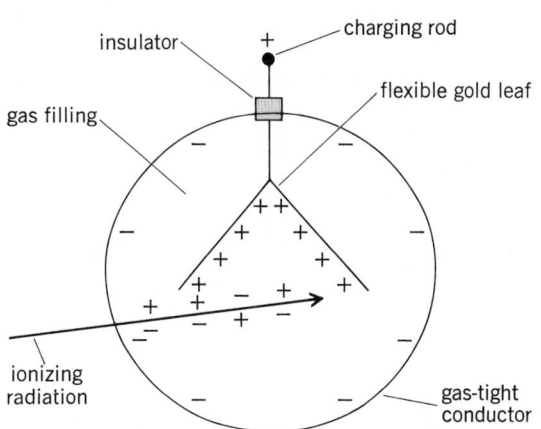

Fig. 1. Gold leaf electroscope used as a radiation detector.

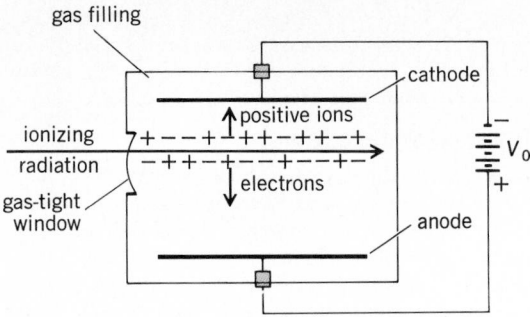

Fig. 2. Parallel-plate ionization chamber.

electrons then drift toward the positive charge on the gold leaf, neutralizing some of this charge. As the charge on the gold leaves decreases, the electrostatic repulsion decreases, and hence the separation between the leaves decreases. By measuring this change in separation, a measure is obtained of the amount of radiation incident on the gas volume. While this integrated measurement may be convenient for applications such as monitoring the total radiation exposure of humans, for many purposes it is useful to measure the ionization pulse produced by a single ionizing particle. *See* ELECTROSCOPE.

The simplest form of a pulse ionization chamber consists of two conducting electrodes in a container filled with gas (Fig. 2). A battery, or other power supply, maintains an electric field between the positive anode and the negative cathode. When ionizing radiation penetrates the gas in the chamber—entering, for example, through a thin gastight window—this radiation liberates electrons from the gas atoms leaving positively charged ions. The electric field present in the gas sweeps these electrons and ions out of the gas, the electrons going to the anode and the positive ions to the cathode.

The basic ion chamber signal consists of the current pulse observed to flow as a result of this ionization process. Because the formation of each

electron-ion pair requires approximately 30 eV of energy on the average, this signal is proportional to the total energy deposited in the gas by the ionizing radiation.

Because the charge liberated by a single particle penetrating the chamber is small, very-low-noise high-gain amplifiers are needed to measure this charge. In the early days, this was a severe problem, but such amplifiers have become readily available with the development of modern solid-state electronics.

In a chamber, such as that represented in Fig. 2, the current begins to flow as soon as the electrons and ions begin to separate under the influence of the applied electric field. The time it takes for the full current pulse to be observed depends on the drift velocity of the electrons and ions in the gas. These drift velocities are complicated functions of gas type, voltage, and chamber geometry. However, because the ions are thousands of times more massive than the electrons, the electrons always travel several orders of magnitude faster than the ions. As a result, virtually all pulse ionization chambers make use of only the relatively fast electron signal. The electron drift velocities for a few gases are given in Fig. 3. Using one of the most common ion chamber gases—argon with a small amount of methane—with electrode spaces of a few centimeters and voltages of a few hundred volts, the electron drift time is of order a microsecond, while the positive-ion drift time is of order milliseconds. By using narrow-bandpass amplifiers sensitive only to signals with rise times of order a microsecond, only the electron signals are observed.

Energy spectrum. One of the most important uses of an ionization chamber is to measure the total energy of a particle or, if the particle does not stop in the ionization chamber, the energy lost by the particle in the chamber. When such an energy-sensitive application is needed, a simple chamber geometry such as that shown in Fig. 2 is not suitable because the fast electron signal charge is a function of the relative distance that the ionization event occurred from the anode and cathode. If an ionization event occurs very near the cathode, the electrons drift across the full electric potential V_o between the chamber electrodes, and a full electron current pulse is recorded; if an ionization count occurs very near the anode, the electrons drift across a very small electric potential, and a small electron pulse is recorded. This geometrical sensitivity is a result of image charges induced by the very slowly moving positive ions. It can be shown that if the electrons drift through a potential difference ΔV, the fast electron charge pulse is $q' = (\Delta V/V_o)q$, where q is the total ionization charge liberated in the gas.

This geometrical dependence can be eliminated by introducing a Frisch grid as indicated in Fig. 4. This grid shields the anode from the positive ions and, hence, removes the effects of the image charges. By biasing the anode positively, relative to the grid, the electrons are pulled through the grid and collected on the anode. Now no signal is observed on the anode until the electrons drift through the grid, but the signal charge which is then observed is the full ionization charge q.

While the ionization chamber generates only small quantities of signal charge for incident parti-

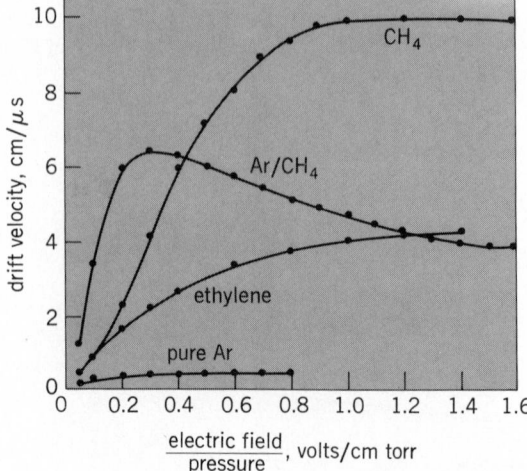

Fig. 3. Electron drift velocity in four different gases as a function of the ratio of the applied electric field strength in volts/centimeter to gas pressure in torrs. 1 torr = 133 Pa. (*From H. W. Fulbright, Ionization chambers, Nucl. Instr. Meth., 162(1979):21–28, 1979*)

cles or photons of MeV energies, the resulting signals are nevertheless well above the noise level of modern low-noise electronic amplifiers. When the signals generated by many incident particles of the same energy are individually measured, and a histogram is plotted representing the magnitude of a signal pulse versus the total number of pulses with the magnitude, then an energy spectrum results. Such a spectrum, smoothed out, consists of an essentially gaussian distribution with standard deviation σ (Fig. 5). Assuming a negligible contribution from amplifier noise, it might at first sight appear that σ should correspond to the square root of the average number of electron-ion pairs produced per incident particle. In fact, σ is usually found to be less than this by a substantial amount, usually designated F, where F is the Fano factor. *See* PROBABILITY; STATISTICS.

It is usual to express the width of an energy distribution such as that of Fig. 5 not in terms of σ but in terms of the "full width at half maximum," usually designated FWHM, or Δ. It can be shown that, for situations in which the width of the energy spectrum is governed by statistics alone, the FWHM is given by the equation $\Delta = 2.36\sqrt{F\epsilon E}$,

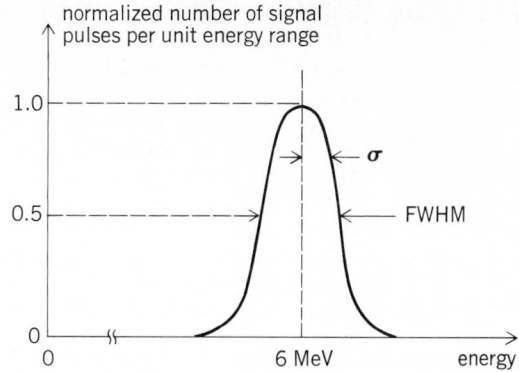

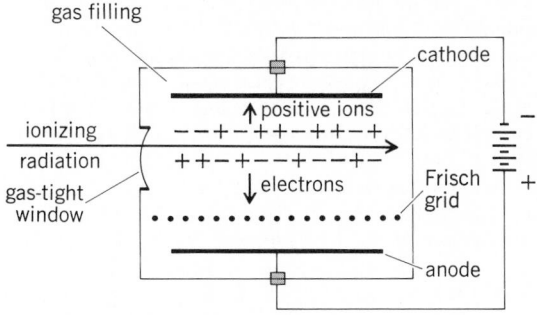

Fig. 4. Frisch grid parallel-plate ionization chamber.

where ϵ is the average energy required to create an electron-ion pair and E is the energy deposited in the chamber by each incident particle or photon. Values of the Fano factor F as low as 0.1 have been observed for certain gases.

Gaseous ionization chambers. Because of the very few basic requirements needed to make an ionization chamber (namely, an appropriate gas with an electric field), a wide variety of different ionization chamber designs are possible in order to suit special applications. In addition to energy information, ionization chambers are now routinely built to give information about the position within the gas volume where the initial ionization event occurred. This information can be important not only in experiments in nuclear and high-energy physics where these position-sensitive detectors were first developed, but also in medical and industrial applications.

This position-sensitive capability results from the fact that, to a good approximation, electrons liberated in an ionizing event drift along the electric field line connecting the anode and cathode, and they drift with uniform velocity (Fig. 3). Hence, a measure of the drift time is a measure of the distance from the anode that the ionization

Fig. 5. Idealized energy spectrum produced by monoenergetic (6-MeV) alpha particles incident on an ideal gridded ionization chamber. The spectrum consists of a single gaussian peak with standard deviation σ.

occurred. A simple illustration of this use is a heavy-ion nuclear physics detector "telescope" used in the basic study of nuclear reactions (Fig. 6). Ionizing charged particles (such as ^{1}H, ^{4}He, and ^{12}C) produced in nuclear collisions enter the detector telescope through a thin gas-tight window at the left, pass through two Frisch grid ionization chambers, and then stop in a solid-state ionization detector. Measurement of the ionization produced in the gas versus the total energy of the particle as determined by the solid-state ionization detector gives sufficient information to uniquely identify the mass and atomic charge of the incident particle. Because the response of the solid-state detector is fast relative to the electron drift time, the difference in time of the signals from the solid-state detector and the anode determines the electron drift time and, hence, the distance above the grid that the particle entered the ionization chamber. This distance can be used to determine the nuclear scattering angle. Hence, a very simple device can be designed which gives several pieces of useful information. While this example illustrates the principles, very complex ionization chambers are now routinely used in heavy-ion and high-energy physics where tens or a hundred signals are recorded (using a computer) for a single ionization event. Position-sensitive heavy-ion detectors with active surfaces as large as a square meter have been developed.

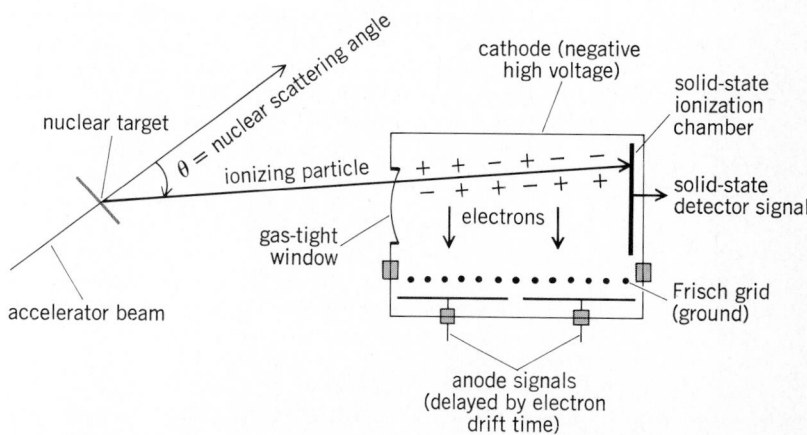

Fig. 6. Heavy-ion detector telescope used to study nuclear reactions.

Aside from applications in basic nuclear physics reseach, ionization chambers are now widely used in other applications. Foremost among these is the use of gas ionization chambers for radiation monitoring. Portable instruments of this type usually employ a detector containing approximately 1 liter of gas, and operate by integrating the current produced by the ambient radiation. They are calibrated to read out in convenient units such as milliroentgens per hour.

Another application of ionization chambers is the use of air-filled chambers as domestic fire alarms. These employ a small ionization chamber containing a low-level radioactive source, such as ^{241}Am, which generates ionization at a constant rate, the resulting current being monitored by a small solid-state electronic amplifier. On the introduction of smoke into the chamber (which is open to the ambient air), the drifting charged ions tend to attach themselves to the smoke particles. This reduces the ionization chamber current, since the moving charge carriers are now much more massive than the initial ions and therefore exhibit correspondingly reduced mobilities. The observed reduction in ion current is used to trigger the alarm.

Another development in ion chamber usage is that of two-dimensional imaging in x-ray medical applications to replace the use of photographic plates. This imaging depends on the fact that if a large flat parallel-plate gas ionization chamber is illuminated with x-rays (perpendicular to its plane), the resulting charges will drift to the plates and thereby form an "image" in electrical charge of the point-by-point intensity of the incident x-rays. This image can be recorded xerographically by arranging for one plate to be a suitably charged insulator. This insulator is preferentially discharged by the collected ions. The resulting charge pattern is recorded by dusting the insulator with a fine powder and transferring this image to paper as in the usual xerographic technique. Alternatively, the xerographic insulator may be a photoconductor, such as selenium, which is preferentially discharged by the ionization produced in the solid material. This is then an example of a solid ionization chamber, and its action closely parallels the operation of the optical xerographic copying machines. Such x-ray imaging detectors provide exceedingly high-quality images at a dosage to the patient substantially less than when photographic plates are used.

Gaseous ionization chambers have also found application as total-energy monitors for high-energy accelerators. Such applications involve the use of a very large number of interleaved thin parallel metal plates immersed in a gas inside a large container. An incident pulse of radiation, due for example to the beam from a large accelerator, will produce a shower of radiation and ionization inside the detector. If the detector is large enough, essentially all of the incident energy will be dissipated inside the detector (mostly in the metal plates) and will produce a corresponding proportional quantity of charge in the gas. By arranging that the plates are alternately biased at a positive and negative potential, the entire device operates like a large interleaved gas ion chamber. The total collected charge is then a measure of the total energy in the initial incident pulse of radiation.

Solid ionization chambers. Ionization chambers can be made where the initial ionization occurs, not in gases, but in suitable liquids or solids. In fact, the discovery of extremely successful solid-state ionization detectors in the early 1960s temporarily diverted interest from further developments of gas-filled chambers.

In the solid-state ionization chamber (or solid-state detector) the gas filling is replaced by a large single crystal of suitably chosen solid material. In this case the incident radiation creates electron-hole pairs in the crystal, and this constitutes the signal charge. In practice, it has been found that only very few materials can be produced with a sufficiently high degree of crystalline perfection to allow this signal charge to be swept out of the crystal and collected. Although many attempts were made in this direction in the 1940s in crystal counters, using such materials as AgCl, CdS, and diamond, these were all failures due to the crystals not having adequate carrier transport properties. In the late 1950s, however, new attempts were made in this direction using single crystals of the semiconductors silicon and germanium. These were highly successful and have led to detectors that have revolutionized low-energy nuclear spectroscopy.

There are two important differences between solid and gas-filled ionization chambers. First, it takes much less energy to create an electron-hole pair in a solid than it does to ionize gas atoms. Hence, the intrinsic energy resolution obtainable with solid-state detectors is better than with gas counters. Gamma-ray detectors with resolutions better than 180 eV are commerically available. Second, in the case of solid semiconductors, the positive charge is carried by electron "holes" whose mobilities are similar to those of electrons. Hence, both the electrons and holes are rapidly swept away by the electric field and, as a result, no Frisch grid is needed to electrically shield the anode from the image charge effects of slow-moving positive ions as in the case of gas- or liquid-filled ionization chambers. *See* CRYSTAL COUNTER; SEMICONDUCTOR.

Liquid ionization chambers. The use of a liquid in an ionization chamber combines many of the advantages of both solid and gas-filled ionization chambers; most importantly, such devices have the flexibility in design of gas chambers with the high density of solid chambers. The high density is especially important for highly penetrating particles such as gamma rays. Unfortunately, until the 1970s the difficulties of obtaining suitable high-purity liquids effectively stopped development of these detectors.

During the 1970s, however, a number of groups built liquid argon ionization chambers and demonstrated their feasibility. A Frisch grid liquid argon chamber achieved a resolution of 34 keV (FWHM). Significant development of large-area position-sensitive liquid ionization chambers is likely. Such chambers would be valuable as gamma-ray detectors for live time imaging in medical and industrial applications as well as in basic nuclear science.

Proportional counters. If the electric field is increased beyond a certain point in a gas ionization chamber, a situation is reached in which the free electrons are able to create additional electron-ion pairs by collisions with neutral gas atoms. For this to occur, the electric field must be sufficiently high so that between collisions an electron can pick up an energy that exceeds the ionization potential of the neutral gas atoms. Under these circumstances gas multiplication, or avalanche gain, occurs, thereby providing additional signal charge from the detector.

A variety of electrode structures have been employed to provide proportional gas gain of this type. The most widely used is shown in Fig. 7. Here a fine central wire acts as the anode, and the avalanche gain takes place in the high field region immediately surrounding this wire. In practice, under suitable circumstances, it is possible to operate at gas gains of up to approximately 10^6.

The gas gain is a function of the bias voltage applied to the proportional counter and takes the general form shown in Fig. 8.

Similar avalanche multiplication effects can occur in semiconductor junction detectors, although there the situation is less favorable, and such devices have not found very widespread use except as optical detectors. *See* JUNCTION DETECTOR.

The large gas gains realizable with proportional counters have made them extremely useful for research applications involving very-low-energy radiation. In addition, their flexibility in terms of geometry has made it possible to construct large area detectors, of the order of 1 m², suitable for use as x-ray detectors in space. Essentially all that has been learned to date regarding x-ray astronomy has involved the use of such detectors aboard space vehicles.

Further exceedingly useful applications of gas proportional counters involve their use as position-sensitive detectors. In Fig. 7, for example, if the anode wire is grounded at both ends, then the signal charge generated at a point will split and flow to ground in the ratio of the resistances of the center wire between the point of origin and the two ends of the wire. This device therefore comprises a one-dimensional position-sensitive detector. Such devices are widely used as focal plane detectors in magnetic spectrographs. Similar position-sensitive operation can be obtained by taking account of the rise time of the signals seen at each end of the wire. Further extension of such methods allows two-dimensional detectors to be produced, a wide variety of which are under investigation for medical and other imaging uses.

The relatively large signals obtainable from gas

proportional counters simplifies the requirements of the subsequent amplifiers and signal handling systems. This has made it economically feasible to employ very large arrays, of the order of thousands, of such devices in multidimensional arrays in high-energy physics experiments. By exploiting refinements of technique, it has proved possible to locate the tracks of charged particles to within a fraction of a millimeter in distances measured in meters. Such proportional counter arrays can operate at megahertz counting rates since they do not exhibit the long dead-time effects associated with spark chambers. *See* SPARK CHAMBER.

Geiger counters. If the bias voltage across a proportional counter is increased sufficiently, the device enters a new mode of operation in which the gas gain is no longer proportional to the initial signal charge but saturates at a very large, and constant, value. This provides a very economical method of generating signals so large that they need no subsequent amplification.

The most widespread use of Geiger counters continues to be in radiation monitoring, where their large output signals simplify the readout problem. They have also found extensive use in cosmic-ray research, where again their large signals have made it feasible to use arrays of substantial numbers of detectors without excessive expenditures on signal-processing electronics. *See* PARTICLE DETECTOR.

[WILLIAM A. LANFORD]

Bibliography: D. A. Bromley (ed.), *Detectors in Nuclear Science*, 1979; P. W. Nicholson, *Nuclear Electronics*, 1974; W. J. Price, *Nuclear Radiation Detection*, 2d ed., 1964.

Ionization potential

The potential difference through which a bound electron must be raised to free it from the atom or molecule to which it is attached. In particular, the ionization potential is the difference in potential between the initial state, in which the electron is bound, and the final state, in which it is at rest at infinity.

The concept of ionization potential is closely associated with the Bohr theory of the atom. Although the simple theory is applicable only to hydrogenlike atoms, the picture furnished by it conveys the idea quite well. In this theory, the allowed energy levels for the electron are given by the equation below, where E_n is the energy of the state

$$E_n = -k/n^2 \quad n = 1, 2, 3, \ldots$$

described by n. The constant k is about 13.6 ev for atomic hydrogen. The energy approaches zero as n becomes infinite. Thus zero energy is associated with the free electron. On the other hand, the most tightly bound case is given by setting n equal to unity. By the definition given above, the ionization potential for the most tightly bound, or ground, state is then 13.6 ev. The ionization potential for any excited state is obtained by evaluating E_n for the particular value of n associated with that state. For a further discussion of the energy levels of an atom *see* ATOMIC STRUCTURE AND SPECTRA; ELECTRONVOLT.

The ionization potential for the removal of an electron from a neutral atom other than hydrogen is more correctly designated as the first ionization potential. The potential associated with the re-

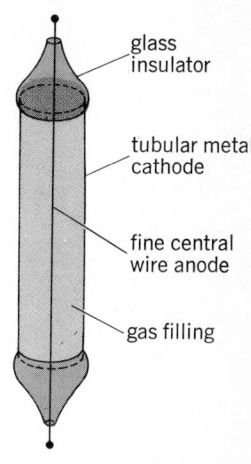

glass insulator

tubular metal cathode

fine central wire anode

gas filling

Fig. 7. Basic form of a simple single-wire gas proportional counter.

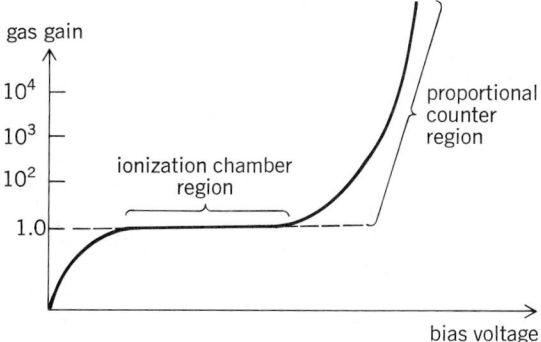

gas gain

10^4

10^3

10^2

1.0

ionization chamber region

proportional counter region

bias voltage

Fig. 8. Plot of gas gain versus applied voltage for a gas-filled radiation detector.

moval of a second electron from a singly ionized atom or molecule is then the second ionization potential, and so on.

Ionization potentials may be measured in a number of ways. The most accurate measurement is obtained from spectroscopic methods. The transitions between energy states are accompanied by the emission or absorption of radiation. The wavelength of this radiation is a measure of the energy difference. The particular transitions that have a common final energy state are called a series. The series limit represents the transition from the free electron state to the particular state common to the series. The energy associated with the series limit transition is the ionization energy.

Another method of measuring ionization potentials is by electron impact. Here the minimum energy needed for a free electron to ionize in a collision is determined. The accuracy of this type of measurement cannot approach that of the spectroscopic method. *See* ELECTRON CONFIGURATION.

[GLENN H. MILLER]

Bibliography: S. Gasiorowicz, *Structure of Matter: A Survey of Modern Physics*, 1979; F. K. Richtmyer et al., *Introduction to Modern Physics*, 6th ed., 1969; M. R. Wehr et al., *Physics of the Atom*, 3d ed., 1978.

Isentropic flow

The flow of a fluid is isentropic when its entropy is identical at all points in the flow. Isentropic flow can be approached for fluids flowing either in a duct or over the outside of a body. Because the entropy of the fluid is a thermodynamic property, similar to the enthalpy or energy of a fluid, the value of the entropy is fixed by the state of the fluid. For a pure substance, in the absence of external forces, entropy is a function of two independent properties. For example, in the absence of gravity, capillarity, electricity, and magnetism, the entropy of a single-phase fluid is a function of the pressure and temperature. *See* ENTHALPY; ENTROPY; ISENTROPIC PROCESS; THERMODYNAMIC PRINCIPLES.

One of the simplest examples of isentropic flow is the flow of a fluid through a nozzle wherein the fluid is accelerated by means of a pressure gradient. This flow can be easily computed for the situation shown in the illustration from the conservation of energy (first law of thermodynamics). Per unit mass of fluid, Eq. (1) holds, where v_1 and

ISENTROPIC FLOW

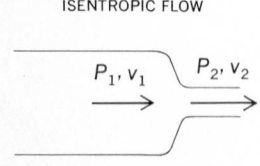

Flow through a nozzle.

$$\frac{1}{2}v_1{}^2 + h_1 = \frac{1}{2}v_2{}^2 + h_2 \tag{1}$$

v_2 are the entering and exiting velocities of the fluid and h_1 and h_2 are the corresponding fluid enthalpies. Here the cross-sectional areas and pressures are assumed constant in each half of the nozzle. Thus Eq. (2) holds, where P_1 is the

$$h_1 = P_1 V_1 + U_1 \tag{2}$$

fluid pressure, V_1 is the volume per unit mass, and U_1 is the internal energy per unit mass.

In an actual nozzle, the fluid flow is not completely isentropic because (1) the fluid shear stress at the walls is not zero, thereby introducing some friction; (2) a significant rate of heat transfer can occur between fluid and walls, as in a rocket nozzle; (3) a significant rate of mass transfer or diffusion may occur normal to the streamlines, thus producing local changes of entropy in the real

flow; and (4) chemical reactions can occur in the flow, thus causing local changes of entropy.

Isentropic flow is often used as a basis of comparison of the real flow with the ideal flow. The figure of merit for flow in a nozzle is defined by Eq. (3), where v_a is the actual measured velocity issu-

$$\text{Nozzle efficiency} \equiv v_a/v_s \tag{3}$$

ing from the nozzle, and v_s is a hypothetical velocity for isentropic flow of the same fluid from the same initial state to the same exit pressure as the real flow. The concept of isentropic flow is useful for fluid flow inside ducts and outside of variously shaped bodies. Isentropic flow is also used for predicting such flows as those of perfect gases; real gases; dissociating and chemically reacting systems; liquids; two-phase, single-, and multicomponent systems; and plasmas.

Isentropic fluid flow can be obtained in irreversible processes by selecting a process in which the local entropy could increase and then providing sufficient heat transfer to maintain the entropy constant at all points. *See* FLUID-FLOW PRINCIPLES; GAS DYNAMICS.

[PHILIP E. BLOOMFIELD]

Bibliography: F. W. Sears and G. L. Salinger, *Thermodynamics, the Kinetic Theory of Gases and Statistical Mechanics*, 3d ed., 1975; V. L. Streeter (ed.), *Handbook of Fluid Dynamics*, 1961.

Isentropic process

In thermodynamics a change that is accomplished without any increase or decrease of entropy is referred to as isentropic. Since the entropy always increases in a spontaneous process, one must consider reversible or quasistatic processes. During a reversible process the quantity of heat transferred, dQ, is directly proportional to the system's entropy change, dS, as in Eq. (1), where T is the absolute

$$dQ = TdS \tag{1}$$

temperature of the system. Systems which are thermally insulated from their surroundings undergo processes without any heat transfer; such processes are called adiabatic. Thus during an isentropic process there are no dissipative effects and, from Eq. (1), the system neither absorbs nor gives off heat. For this reason the isentropic process is sometimes called the reversible adiabatic process. *See* ADIABATIC PROCESS; ENTROPY; THERMODYNAMIC PROCESSES.

Work done during an isentropic process is produced at the expense of the amount of internal energy stored in the nonflow or closed system. Thus, the useful expansion of a gas is accompanied by a marked decrease in temperature, tangibly demonstrating the decrease of internal energy stored in the system. For ideal gases the isentropic process can be expressed by Eq. (2), where P is the

$$P_1 V_1{}^k = P_2 V_2{}^k = \text{constant} \tag{2}$$

pressure in pounds per square foot, V is the volume in cubic feet, and k is the ratio between the specific heat at constant pressure and the specific heat at constant volume for the given gas. It can be closely approximated by the values of 1.67 and 1.40 for dilute monatomic and diatomic gases, respectively. For a comparison of various processes involving a gas *see* POLYTROPIC PROCESS.

[PHILIP E. BLOOMFIELD]

Ising model

A model which consists of a lattice of "spin" variables with two characteristic properties: (1) each of the spin variables independently takes on either the value +1 or the value −1; and (2) only pairs of nearest neighboring spins can interact. The study of this model (introduced by Ernst Ising in 1925) in two dimensions, where many exact calculations have been carried out explicitly, forms the basis of the modern theory of phase transitions and, more generally, of cooperative phenomena.

The two-dimensional Ising model was shown to have a phase transition by R. E. Peierls in 1936, and the critical temperature or Curie temperature, that is, the temperature at which this phase transition takes place, was calculated by H. A. Kramers and G. H. Wannier and by E. W. Montroll in 1941. Major breakthroughs were accomplished by Lars Onsager in 1944, by Bruria Kaufman and Onsager in 1949, and by Chen Ning Yang in 1952. Onsager first obtained the free energy and showed that the specific heat diverges as $-\ln|1 - T/T_e|$ when the temperature T is near the critical temperature T_c; Kaufman and Onsager computed the short-range order; and Yang calculated the spontaneous magnetization. Since then several other properties have been obtained, and since 1974 connections with relativistic quantum field theory have been made. *See* QUANTUM FIELD THEORY.

Cooperative phenomena. A macroscopic piece of material consists of a large number of atoms, the number being of the order of Avogadro's number (approximately 6×10^{23}). Thermodynamic phenomena all depend on the participation of such a large number of atoms. Even though the fundamental interaction between atoms is short-ranged, the presence of this large number of atoms can, under suitable conditions, lead to an effective interaction between widely separated atoms. Phenomena due to such effective long-range interactions are referred to as cooperative phenomena. The simplest examples of cooperative phenomena are phase transitions. The most familiar phase transition is either the condensation of steam into water or the freezing of water into ice. Only slightly less familiar is the ferromagnetic phase transition that takes place at the Curie temperature, which, for example, is roughly 1043 K for iron.

Of the several models which exhibit a phase transition, the Ising model is the best known. In three dimensions the model is so complicated that no exact computation has ever been made, while in one dimension the Ising model does not undergo a phase transition. However, in two dimensions the Ising model not only has a ferromagnetic phase transition but also has very many physical properties which may be exactly computed. Indeed, despite the restriction on dimensionality, the two-dimensional Ising model exhibits all of the phenomena peculiar to magnetic systems near the Curie temperature. *See* CURIE TEMPERATURE; FERROMAGNETISM; MAGNETISM; SECOND-ORDER TRANSITION.

Definition of model. The mutual interaction energy of the pair of spins σ_α and $\sigma_{\alpha'}$ when α and α' are nearest neighbors may be written as $-E(\alpha,\alpha') \cdot \sigma_\alpha \sigma_{\alpha'}$. The meaning of this is that the interaction energy is $-E(\alpha,\alpha')$ when σ_α and $\sigma_{\alpha'}$ are both +1 or −1, and is $+E(\alpha,\alpha')$ when $\sigma_\alpha=+1$, $\sigma_{\alpha'}=-1$, or

$\sigma_\alpha=-1$, $\sigma_{\alpha'}=1$. In addition, a spin may interact with an external magnetic field H with energy $-H\sigma_\alpha$. From these two basic interactions the total interaction energy for the square lattice may be written as Eq. (1), where j specifies the row and k

$$E = -\sum_j \sum_k \left[E_1(j,k)\sigma_{j,k}\sigma_{j,k+1} + E_2(j,k)\sigma_{j,k}\sigma_{j+1,k} + H\sigma_{j,k} \right] \quad (1)$$

specifies the column of the lattice. In this form the interaction energies $E_1(j,k)$ and $E_2(j,k)$ are allowed to vary arbitrarily throughout the lattice. A special case of great importance is the translationality invariant case (E_1 and E_2 independent of j and k) which was studied by Onsager in 1944. This is the model needed to study a pure ferromagnet without impurities. *See* MAGNETIC FIELD.

Several generalizations of Ising's original model have been considered. For example, σ can be allowed to take on more values than just ± 1, and interactions other than nearest neighbor can be used. For these generalizations no exact calculations have been performed in two or three dimensions. However, various approximate calculations indicate that the phase transition properties of these models are the same as those of the Onsager lattice.

The extension to the nontranslationally invariant case where $E_1(j,k)$ and $E_2(j,k)$ are treated as independent random variables is important for studying the effects of impurities in ferromagnetics. Some extensions in this direction were made in 1968 by Barry McCoy and Tai Tsun Wu.

Thermodynamic properties. The basic simplification in framing the definition of the Ising model of the preceding section is the choosing of the fundamental variables to be the numbers $\sigma_{j,k}$ which can be only +1 or −1. Because of this choice there can be no terms in the interaction energy which refer to kinetic energy or to angular momentum. Consequently, the $\sigma_{j,k}$ do not change with time, and study of the system is, by necessity, confined to those physical properties which depend only on the distribution of energy levels of the system. When the number of energy levels is large, this study requires the use of statistical mechanics.

Statistical mechanics allows the calculation of average macroscopic properties from the microscopic interaction E. If A is some property of the spins σ of the system, then the thermodynamic average of A is given by Eq. (2), where T is the tem-

$$\langle A \rangle = \lim_{N \to \infty} \frac{1}{Z} \sum_{\{\sigma\}} A e^{-E(\sigma)/kT} \quad (2)$$

perature, k is Boltzmann's constant, Z is given by Eq. (3), the sums are over all values of $\sigma_{j,k}=\pm 1$,

$$Z = \sum e^{-E(\sigma)/kT} \quad (3)$$

and N is the number of rows and the number of columns. It is mandatory that the thermodynamic limit $N \to \infty$ be taken for these thermodynamic averages to have a precise meaning. *See* BOLTZMANN CONSTANT.

The most important thermodynamic properties of a ferromagnet are the internal energy per site $u = \langle E/N^2 \rangle$, the specific heat $c = \partial u/\partial T$, the magnetization per site $M = \langle \sigma \rangle$, and the magnetic sus-

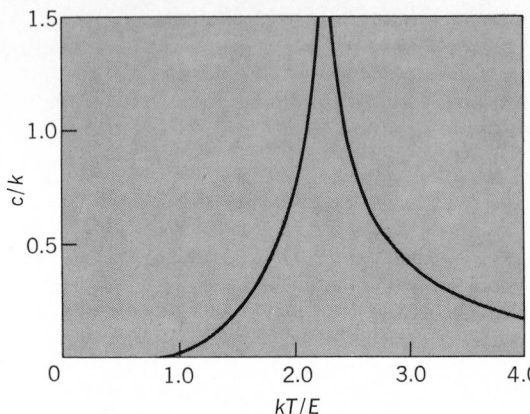

Fig. 1. Specific heat of Onsager's lattice for $E_2/E_1 = 1$.

ceptibility $\chi = \partial M/\partial H$. These quantities have all been computed for the two-dimensional Ising model $E_1 \neq E_2$, but for convenience this discussion is restricted to $E_1 = E_2 = E$.

Onsager studied the two-dimensional square lattice at $H = 0$ and computed the specific heat exactly. From that calculation he found that the specific heat was infinite at the critical temperature of Kramers and Wannier given as the solution of Eq. (4). When T is close to the critical temperature T_c, the specific heat is approximated by Eq. (5).

$$\sinh 2E/kT = 1 \qquad (4)$$

$$c \sim -\frac{8E^2}{kT_c^2\pi} \ln|1 - T/T_c| \qquad (5)$$

The behavior of the specific heat for any temperature is plotted in Fig. 1.

The spontaneous magnetization is defined as $M(0) = \lim_{H \to 0^+} M(H)$. For $T > T_c$, $M(0) = 0$. For $T < T_c$, Yang found that $M(0)$ is given by Eq. (6). When T is

$$M(0) = [1 - (\sinh 2E/kT)^{-4}]^{1/8} \qquad (6)$$

near T_c, $M(0)$ is approximated by Eq. (7). The be-

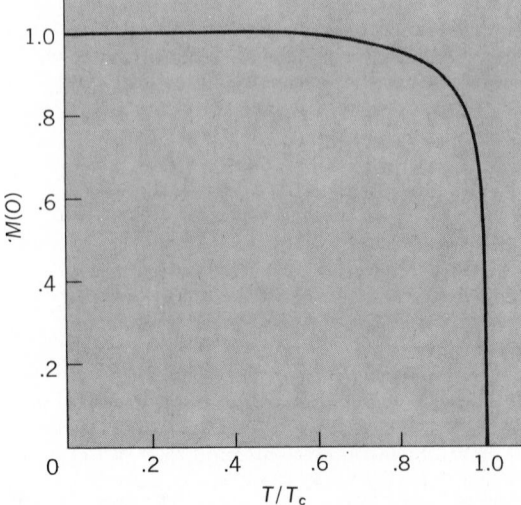

Fig. 2. Spontaneous magnetization $M(0)$ of Onsager's lattice for $E_2/E_1 = 1$.

$$M(0) \sim \left[\frac{8\sqrt{2}}{kT_c}(1 - T/T_c)\right]^{1/8} \qquad (7)$$

havior $M(0)$ as a function of T is plotted in Fig. 2.

The magnetic susceptibility χ at $H = 0$ is much more difficult to compute than either the specific heat or the spontaneous magnetization. Indeed, no closed form expression is known for χ over the entire range of temperature. However, near T_c it is known that as $T \to T_c^+$, χ is approximated by Eq. (8), and, as $T \to T_c^-$, is approximated by Eq. (9):

$$\chi(T) \sim C_0^+|1 - T_c/T|^{-7/4} + C_1^+|1 - T_c/T|^{-3/4} + C_2$$
$$(8)$$

$$\chi(T) \sim C_0^-|1 - T_c/T|^{-7/4} + C_1^-|1 - T_c/T|^{-3/4} + C_2$$
$$(9)$$

where $C_0^+ = 0.96258\ 17322\cdots$
$\quad\quad C_0^- = 0.02553\ 69719\cdots$
$\quad\quad C_1^+ = 0.07498\ 81538\cdots$
$\quad\quad C_1^- = -0.0198\ 94107\cdots$

See INTERNAL ENERGY; MAGNETIC SUSCEPTIBILITY; MAGNETIZATION; SPECIFIC HEAT OF SOLIDS; STATISTICAL MECHANICS; THERMODYNAMIC PRINCIPLES.

Random impurities. A question which can be very usefully studied in the Ising model is the generalization of statistical mechanics to deal with the experimental situation in which the interaction energy of the system is not completely known because of the presence of impurities. The term "impurity" refers not only to the presence of foreign material in a sample but to any physical property such as defects or isotopic composition which makes lattice sites different from one another. The distribution of these impurities is governed by spin-independent forces. At least two different situations can be distinguished.

1. As the temperature changes, the distribution of impurities may change; such a situation will occur, for example, near the melting point of a lattice.

2. The distribution of impurities may be independent of temperature, at least on the time scale of laboratory measurements; such a distribution will obtain when the temperature of a lattice is well below the melting temperature. Impurities of this sort are said to be frozen in. *See* CRYSTAL DEFECTS.

For a study of frozen-in impurities to be realistic, the impurities must be distributed at random throughout the lattice. The translational invariance of the system is now totally destroyed, and it is not at all clear that the phase transition behavior of the pure and random system should be related to each other at all.

These problems were studied for a special case of the Ising model by McCoy and Wu in 1968. They let $E_2(j,k)$ depend on j but not k and kept $E_1(j,k)$ independent of both j and k. Then the variables $E_2(j)$ were chosen with a probability distribution $P(E_2)$. When $P(E_2)$ was of narrow width, they showed that logarithmic divergence of Onsager's specific heat is smoothed out into an infinitely differentiable essential singularity. Such a smoothing out of sharp phase transition behavior may in fact have been observed. The results of one such ex-

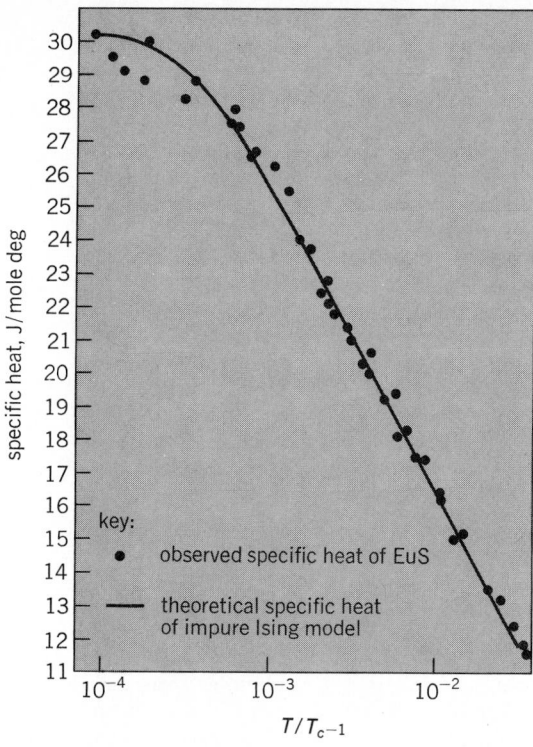

specific heat, J/mole deg

key:

● observed specific heat of EuS

—— theoretical specific heat of impure Ising model

T/T_{c-1}

Fig. 3. Comparison of the impure Ising model specific heat with the observed specific heat of EuS for $T > T_c$.

periment, carried out by B. J. C. van der Hoeven and colleagues, are compared with the result of the Ising model random specific heat calculation in Fig. 3. [BARRY M. MC COY; TAI TSUN WU]

Bibliography: C. Domb and M. Green, *Phase Transitions and Critical Phenomena*, 1972; B. McCoy and T. T. Wu, *The Two Dimensional Ising Model*, 1973; L. Onsager, Crystal statistics 1: A two-dimensional model with order-disorder transition, *Phys. Rev.*, 65:117–149, 1944; T. T. Wu et al., Spin-spin correlation functions for the two-dimensional Ising model: Exact theory in the scaling region, *Phys. Rev.*, B, 13:316–374, 1976; C. N. Yang, The spontaneous magnetization of two-dimensional Ising model, *Phys. Rev.*, 85:808–816, 1952.

Isobar (atomic physics)

One of two or more atoms which have a common mass number A but which differ in atomic number Z. Thus, although isobars possess approximately equal masses, they differ in chemical properties; they are atoms of different elements. Isobars whose atomic numbers differ by unity cannot both be stable; one will inevitably decay into the other by β^--emission $(Z \to Z+1)$, β^+-emission $(Z \to Z-1)$, or electron capture $(Z \to Z-1)$. There are many examples of stable isobaric pairs, such as Ti^{50} $(Z = 24)$ and Cr^{50} $(Z = 26)$, and four examples of stable isobaric triplets. At most values of A the number of known radioactive isobars exceeds the number of stable ones. *See* ELECTRON CAPTURE; RADIOACTIVITY. [HENRY E. DUCKWORTH]

Isobaric process

A thermodynamic process in which the heat transfer to or from the gaseous system causes a volume change at constant pressure. This process can be illustrated by the expansion of a substance when it is heated. The system is then capable of doing an amount of work on its surroundings. The maximum work will be done when the external pressure P_{ext} of the surroundings on the system is equal to P, the pressure of the system. If V is the volume of the system, the maximum work W is given by Eq. (1).

$$W = \int_1^2 P\,dV = P\int_1^2 dV = P(V_2 - V_1) \qquad (1)$$

For isobaric processes it is useful to introduce the enthalpy H, which is defined as the total heat content of the system and is given by the sum of the internal energy U and PV. Then Eq. (2) can be

$$Q_p = H_2 - H_1 = m\int_1^2 C_p\,dT \qquad (2)$$

formulated to represent Q_p, the transferred heat at constant pressure, where C_p is the specific heat at constant pressure and m is mass. By the first law of thermodynamics the change of the internal energy in any process is equal to the difference between the heat gained and the work done by the system; thus Eq. (3) holds.

$$Q_p = U_2 - U_1 + W \qquad (3)$$

If the isobaric process is also carried out quasistatically, Eq. (4) is obtained, where S is the en-

$$Q_p = \int_1^2 T\,dS \qquad (4)$$

tropy. *See* ISOMETRIC PROCESS; THERMODYNAMIC PROCESSES. For a comparison of the isobaric process with other processes involving a gas *see* POLYTROPIC PROCESS.

[PHILIP E. BLOOMFIELD/WILLIAM A. STEELE]

Isoelectronic sequence

A term used in spectroscopy to designate the set of spectra produced by different chemical elements ionized in such a way that their atoms or ions contain the same number of electrons. The sequence in the table is an example. Since the neutral atoms

Example of isoelectronic sequence

Designation of spectrum	Emitting atom or ion	Atomic number, Z
CaI	Ca	20
ScII	Sc⁺	21
TiIII	Ti²⁺	22
VIV	V³⁺	23
CrV	Cr⁴⁺	24
MnVI	Mn⁵⁺	25

of these elements each contain Z electrons, removal of one electron from scandium, two from titanium, and so forth, yields a series of ions all of which have 20 electrons. Their spectra are therefore qualitatively similar, but the spectral terms (energy levels) increase approximately in proportion to the square of the core charge, just as they depend on Z^2 in the one-electron sequence H, He⁺, Li⁺⁺, and so forth. As a result, the successive spectra shift progressively toward shorter wavelengths, soon reaching the vacuum ultraviolet region. Isoelectronic sequences are useful in predicting unknown spectra of ions belonging to a sequence in which other spectra are known. *See* ATOMIC STRUCTURE AND SPECTRA.

[F. A. JENKINS/W. W. WATSON]

Isometric process

A constant-volume thermodynamic process in which the system is confined by mechanically rigid boundaries. No direct mechanical work can be done on the surroundings by a system with rigid boundaries; therefore the heat transferred into or out of the system equals the change of internal energy stored in the system. This change in the internal energy, in turn, is a function of the specific heat and the temperature change in the system as in Eq. (1), where Q_V is the heat trans-

$$Q_V = U_2 - U_1 = m \int_1^2 C_V \, dT \qquad (1)$$

ferred at constant volume, U is the internal energy, m is the mass, C_V is the specific heat at constant volume, and T is the absolute temperature. If the process occurs quasistatically (the system going through a continuous sequence of equilibrium states), Eq. (2) holds, where S is the entropy. There

$$Q_V = \int_1^2 T \, dS \qquad (2)$$

is an increase in both the temperature and the pressure of a constant volume of gas as heat is transferred into the system. For a comparison of the isometric process with other processes involving a gas *see* POLYTROPIC PROCESS.

[PHILIP E. BLOOMFIELD]

Isomorphism (crystallography)

A similarity of crystalline form between substances of similar composition. Two substances which are isomorphous have a similar chemical formula, an equal or nearly equal ratio of cation to anion radius, and comparable polarizabilities of their ions. Isomorphism is morphotropism in a narrower, more precise sense. Similarity in the macroscopic characteristics of isomorphous crystals becomes so close that extreme precision is needed to distinguish between them.

Examples of isomorphous substances are $NaNO_3$ and $CaCO_3$, $CaAl_2Si_2O_8$ and $NaAlSi_3O_8$, and $BaSO_4$, $SrSO_4$, and $PbSO_4$. Substances such as ThO_2 and LiO_2 are anti-isomorphous; they both have the calcium fluoride structure, but the positions of the anions and cations are interchanged in the two structures because of the relative sizes of the ions. Isomorphous substances form mixed crystals, while anti-isomorphous substances do not. *See* CRYSTAL STRUCTURE.

[WILLY C. DEKEYSER]

Bibliography: M. J. Buerger, *Crystal Structure Analysis*, 1960, reprint 1979; R. C. Evans, *Introduction to Crystal Chemistry*, 2d ed., 1964.

Isothermal process

A thermodynamic process which occurs with a heat addition or removal rate just adequate to maintain constant temperature. The change in the internal energy per mole U accompanying a change in volume in an isothermal process is given by Eq. (1), where T is the temperature, P is

$$U_2 - U_1 = \int_{V_1}^{V_2} \left[T \left(\frac{\partial P}{\partial T} \right)_V - P \right] dV \qquad (1)$$

the pressure, and V is the volume per mole. The integral in Eq. (1) is zero for an ideal gas, and approximately zero for a condensed phase (solid or liquid) for which the volume changes very little with pressure. Thus, in both these cases, $U_2 = U_1$. For real gases, the integral is nonzero, and the internal energy change is computed using the equation of state of the gas in Eq. (1).

The entropy change in an isothermal process is given by Eq. (2). For relatively incompressible

$$S_2 - S_1 = \int_{V_1}^{V_2} \left(\frac{\partial P}{\partial T} \right)_V dV = -\int_{P_1}^{P_2} \left(\frac{\delta V}{\delta T} \right)_P dP \qquad (2)$$

condensed phases, $S_2 \simeq S_1$, while for the ideal gas Eq. (3) holds, where R is the gas constant. For a

$$S_2 - S_1 = R \ln \frac{V_2}{V_1} = -R \ln \frac{P_2}{P_1} \qquad (3)$$

real (nonideal) gas, the fluid equation of state must be used in Eq. (2) to obtain the numerical value of the entropy change.

In the case of a reversible isothermal process, the work done by the fluid W, and the heat transferred Q, both per mole, are given by Eqs. (4)

$$W = \int_{V_1}^{V_2} P \, dV = Q - (U_2 - U_1) \qquad (4)$$

and (5). For a reversible isothermal process in

$$Q = T(S_2 - S_1) \qquad (5)$$

an ideal gas these equations reduce to Eq. (6).

$$W = Q = RT \ln \frac{V_2}{V_1} = -RT \ln \frac{P_2}{P_1} \qquad (6)$$

See GAS; THERMODYNAMIC PROCESSES.

[STANLEY I. SANDLER]

Isotone

One of two or more atoms which display a constant difference $A - Z$ between their mass number A and their atomic number Z. Thus, despite differences in the total number of nuclear constituents, the numbers of neutrons in the nuclei of isotones are the same. The numbers of naturally occurring isotones provide useful evidence concerning the stability of particular neutron configurations. For example, the relatively large number (six and seven, respectively) of naturally occurring 50- and 82-neutron isotones suggests that these nuclear configurations are especially stable. On the other hand, from the fact that most atoms with odd numbers of neutrons are anisotonic, one may conclude that odd-neutron configurations are relatively unstable. *See* NUCLEAR STRUCTURE.

[HENRY E. DUCKWORTH]

Isotope

One of two or more nuclidic species of an element having identical number of protons (Z) in the nucleus but different number of neutrons (N). Isotopes differ in mass but chemically are the same element. All naturally occurring elements have

Natural isotopic abundances of the elements

Element	Mass no.	Atom %	Element	Mass no.	Atom %	Element	Mass no.	Atom %	Element	Mass no.	Atom %
1 H*	1	99.985	30 Zn	64	48.9	50 Sn	112	1.0	66 Dy	156	0.06
	2	0.015		66	27.8		114	0.7		158	0.1
	3	0.00013		67	4.1		115	0.4		160	2.34
2 He*	4	≈100.		68	18.6		116	14.7		161	18.9
3 Li*	6	7.5		70	0.6		117	7.7		162	25.5
	7	92.5	31 Ga	69	60.0		118	24.3		163	24.9
4 Be	9	100.		71	40.0		119	8.6		164	28.2
5 B*	10	19.8	32 Ge	70	20.7		120	32.4	67 Ho	165	100.
	11	80.2		72	27.5		122	4.6	68 Er	162	0.1
6 C*	12	98.89		73	7.7		124	5.6		164	1.6
	13	1.11		74	36.4	51 Sb	121	57.3		166	33.4
7 N*	14	99.64		76	7.7		123	42.7		167	22.9
	15	0.36	33 As	75	100.	52 Te	120	0.1		168	27.0
8 O*	16	99.756	34 Se	74	0.9		122	2.5		170	15.0
	17	0.039		76	9.0		123	0.9	69 Tm	169	100.
	18	0.205		77	7.6		124	4.6	70 Yb	168	0.1
9 F	19	100.		78	23.5		125	7.0		170	3.1
10 Ne*†	20	90.51		80	49.8		126	18.7		171	14.3
	21	0.27		82	9.2		128	31.7		172	21.9
	22	9.22	35 Br	79	50.69		130	34.5		173	16.2
11 Na	23	100.		81	49.31	53 I	127	100.		174	31.7
12 Mg	24	78.99	36 Kr†	78	0.35	54 Xe†	124	0.1		176	12.7
	25	10.00		80	2.25		126	0.1	71 Lu	175	97.4
	26	11.01		82	11.6		128	1.9		176	2.6
13 Al	27	100.		83	11.5		129	26.4	72 Hf	174	0.2
14 Si	28	92.2		84	57.0		130	4.1		176	5.2
	29	4.7		86	17.3		131	21.2		177	18.5
	30	3.1	37 Rb	85	72.17		132	26.9		178	27.1
15 P	31	100.		87	27.83		134	10.4		179	13.8
16 S*	32	95.00	38 Sr†	84	0.56		136	8.9		180	35.2
	33	0.76		86	9.84	55 Cs	133	100.	73 Ta	180	0.012
	34	4.22		87	7.0	56 Ba	130	0.1		181	99.988
	36	0.02		88	82.6		132	0.1	74 W	180	0.1
17 Cl	35	75.77	39 Y	89	100.		134	2.4		182	26.3
	37	24.23	40 Zr	90	51.4		135	6.6		183	14.3
18 Ar†	36	0.34		91	11.2		136	7.9		184	30.7
	38	0.07		92	17.1		137	11.2		186	28.6
	40	99.59		94	17.5		138	71.7	75 Re	185	37.40
19 K	39	93.26		96	2.8	57 La	138	0.09		187	62.60
	40	0.01	41 Nb	93	100.		139	99.91	76 Os†	184	0.02
	41	6.73	42 Mo	92	14.8	58 Ce	136	0.2		186	1.58
20 Ca	40	96.937		94	9.1		138	0.3		187	1.6
	42	0.65		95	15.9		140	88.4		188	13.3
	43	0.14		96	16.7		142	11.1		189	16.1
	44	2.08		97	9.5	59 Pr	141	100.		190	26.4
	46	0.003		98	24.4	60 Nd	142	27.1		192	41.0
	48	0.19		100	9.6		143	12.2	77 Ir	191	37.4
21 Sc	45	100.	44 Ru	96	5.5		144	23.9		193	62.6
22 Ti	46	8.0		98	1.9		145	8.3	78 Pt	190	0.01
	47	7.5		99	12.7		146	17.2		192	0.79
	48	73.7		100	12.6		148	5.7		194	32.9
	49	5.5		101	17.1		150	5.6		195	33.8
	50	5.3		102	31.6	62 Sm	144	3.1		196	25.3
23 V	50	0.25		104	18.6		147	15.0		198	7.2
	51	99.75	45 Rh	103	100.		148	11.2	79 Au	197	100.
24 Cr	50	4.35	46 Pd	102	1.0		149	13.8	80 Hg	196	0.2
	52	83.79		104	11.0		150	7.4		198	10.1
	53	9.50		105	22.2		152	26.7		199	16.9
	54	2.36		106	27.3		154	22.8		200	23.1
25 Mn	55	100.		108	26.7	63 Eu	151	47.8		201	13.2
26 Fe	54	5.85		110	11.8		153	52.2		202	29.7
	56	91.7	47 Ag	107	51.83	64 Gd	152	0.2		204	6.8
	57	2.14		109	48.17		154	2.2	81 Tl	203	29.5
	58	0.31	48 Cd	106	1.2		155	14.9		205	70.5
27 Co	59	100.		108	0.9		156	20.6	82 Pb†	204	1.4
28 Ni	58	68.3		110	12.4		157	15.7		206	24.1
	60	26.1		111	12.8		158	24.7		207	22.1
	61	1.1		112	24.0		160	21.7		208	52.4
	62	3.6		113	12.3	65 Tb	159	100.	83 Bi	209	100.
	64	0.9		114	28.8				90 Th	232	100.
29 Cu	63	69.2		116	7.6				92 U*	234	0.0054
	65	30.8	49 In	113	4.3					235	0.7200
				115	95.7					238	99.2746

*Isotopic composition of the element may be somewhat variable with specific geological or biological origin of the sample. Commercial chemicals may have, in some cases, quite anomalous composition as the result of processes of isotope separation.

†The element may vary in isotopic composition in some samples because one or more of the isotopes result from radioactive decay, or from nuclear processes in nature, such as spontaneous fission of uranium or α,n reactions on light elements.

radioactive isotopes, and the majority have at least one stable nuclide. Some elements which occur in nature, such as uranium, are radioactive but have isotopes with long half-lives. For a discussion of artificially produced radioisotopes *see* RADIOISOTOPE.

The isotopic composition of an element is generally determined by mass spectrometry. Of the 83 elements present on Earth in significant amounts, 20 possess only a single stable nuclide and are referred to as mononuclidic or anisotopic. The others have 2 to 10 stable isotopes.

Nuclear stability. Of the 287 nuclidic species listed in the table, 168 have even-even structure (even number of protons and even number of neutrons in the nucleus), 57 have even-odd, 53 have odd-even, and only 9 have odd-odd. This indicates the pairing tendency of the nuclear constituents. The extra stability of a 50-proton configuration is indicated by the existence of 10 isotopes of tin. *See* NUCLEAR STRUCTURE.

Isotopic abundance. This refers, unless otherwise specified, to the isotopic composition of the naturally occurring terrestrial element (see table). Some elements are observed to vary in isotopic composition. The variability ranges from a few per mil to a percent or two, although variations greater than this are observed in some samples. This variation occurs for several reasons. In the lighter elements—hydrogen, lithium, and boron, for instance—the isotopes differ enough in mass and to some extent in chemical reactivity that processes of distillation or chemical exchange between different chemical compounds of the element can produce significant differences in isotopic composition. Indeed, exchange reactions are used in the case of hydrogen, lithium, boron, carbon, nitrogen, and oxygen to separate isotopes of these elements on a relatively large scale.

Elements which take part in the life cycle of living organisms will vary somewhat in isotopic composition because of exchange reactions and diffusion through membranes. Slight differences in reaction rates are also important.

The composition of some elements may be variable because one or more of the isotopes are stable products of radioactive decay. Thus three of the four lead isotopes, ^{208}Pb, ^{207}Pb, and ^{206}Pb, are end products of the decay of thorium and of ^{235}U and ^{238}U respectively. The fourth lead isotope, ^{204}Pb, does not come from any known decay chain. The rare potassium isotope, ^{40}K, which is present in only 0.012%, has a half-life of 1.28×10^9 years. It decays by beta-particle emission to stable ^{40}Ar and by electron capture to ^{40}Ca. The argon in the atmosphere, approximately 1.1%, is 99.6% in ^{40}Ar. Argon in potassium-bearing minerals will differ in composition from atmospheric argon just as the composition of lead will depend upon its past association with thorium and uranium. These decays and the decay of rubidium, ^{87}Rb, to ^{87}Sr are the basis of methods used in the determination of geological age.

The three nuclides ^{40}K, ^{40}Ar, and ^{40}Ca are examples of "isobars" in that they have the same mass number, $N + Z$, but differ in the number of protons in the nucleus. The radioactive potassium isotope is an example of an odd-odd nucleus, while the stable ^{40}Ar and ^{40}Ca are both even-even. *See* ISOBAR (ATOMIC PHYSICS).

Anomalous isotopic compositions in some elements occur because of nuclear processes in nature. The discovery, in 1972, of a "fossil reactor" in the Oklo uranium deposit in Gabon (West Africa) is such a case. As a result of a chain reaction, perhaps 1.7×10^9 years ago, much of the uranium in this formation is depleted in the fissionable isotope, ^{235}U. Fission products are present in the composition found in reactor waste, and some elements in the surrounding rocks have been modified in isotopic composition by neutron absorption and subsequent radioactive decay.

The only nonterrestrial materials which are available for comparison of isotopic composition are meteorites and the lunar materials returned to Earth by the several Apollo manned missions and the Soviet unmanned lunar probes. Isotopic compositions in these are generally identical with terrestrial samples within the precision of the measurements. Differences, when they are found, can be identified as caused by radioactive decay, cosmic-ray bombardment and, in the case of lunar surface materials, bombardment by solar "wind" particles. In iron meteorites the spallation of iron nuclei by very energetic particles from cosmic rays produces helium, neon, and some other light elements with anomalous isotopic composition. The helium ^{3}He/^{4}He ratio is used as an indicator of the cosmic ray "bombardment" age. *See* RADIOACTIVITY.

Use of separated isotopes. Isotopes of certain elements possess unique or peculiar properties, and their separation or enrichment is desirable. Deuterium, ^{2}H, which occurs in an abundance of about 0.16% in terrestrial hydrogen, is useful for moderating neutrons in heavy-water reactors and is expected to form the fuel of fusion reactors in the future. Very large quantities of deuterated water have been produced by a variety of distillative, exchange, and electrolysis processes. The desirable fissionable isotope of uranium, ^{235}U, is enriched by gaseous diffusion in uranium hexafluoride gas in very large plants.

Of the biologically important elements, only carbon and hydrogen have radioisotopes of sufficiently long half-lives to be used in tracer studies in living organisms. The use of large amounts of ^{3}H and ^{14}C is not desirable and, in any case, the mass differences are so great that they do not act precisely like the isotopes ^{1}H and ^{12}C. Studies of metabolism, drug utilization, and other reactions in living organisms are best done with stable isotopes like ^{13}C, ^{15}N, ^{18}O, and ^{2}H. Compounds are tagged by introducing a high concentration of the isotope into the molecular structure, and the metabolized products are studied using the mass spectrometer to measure the altered isotopic ratios.

The process of isotope dilution consists of adding a known amount of material containing the tracer isotope, allowing the system to reach chemical equilibrium and then recovering a small sample sufficient in size for a mass spectrometric measurement of the new isotopic composition. This is a method of very wide applicability.

Atomic weight. Atomic weight is the ratio of the average mass per atom of the natural isotopic composition of an element to 1/12 of the mass of an atom of the nuclide ^{12}C. It is a dimensionless number. The atomic weight of a mononuclidic element is simply the atomic mass of that nuclide relative to 1/12 of the mass of $^{12}C = 12$ exactly. These masses can be measured with very high precision. Thus the atomic weight of the mononuclidic element beryllium is 9.01218. Atomic weights of the elements having two or more isotopic species are increasingly based upon calculations from isotopic composition and atomic masses. The mass spectrometer is not an absolute instrument because of certain inherent discriminations. Comparisons must be made with isotopic standards carefully prepared by gravimetric procedures from separated isotopes of high chemical and isotopic purity. This is hardly less exacting than the chemical determination of atomic weights used early in this century by T. W. Richards and coworkers at Harvard University. The "Harvard method" involved the careful precipitation and weighing of silver chloride formed by the stoichiometric reaction of silver nitrate with the chloride of the element being investigated. Thus, one would determine the germanium chloride/silver chloride ratio ($GeCl_4/4AgCl$) and thus the ratio of the atomic weight to that of silver. Even in the 1975 Table of Atomic Weights some of the polynuclidic elements, such as germanium, tin, and mercury, are still based upon chemical ratios measured by the Harvard method. *See* ATOMIC WEIGHT.

[A. E. CAMERON]

Bibliography: J. F. Duncan and G. B. Cook, *Isotopes in Chemistry*, 1968; S. Glasstone, *Sourcebook on Atomic Energy*, 3d ed., 1967; N. E. Holden and F. W. Walker, (eds.), *Chart of the Nuclides*, Educational Relations, General Electric Co., 1972; J. Robos, *Introduction to Mass Spectrometry*, 1968; A. Romer (ed.), *Radiochemistry and the Discovery of Isotopes*, 1970.

Isotope shift

A small difference between the different isotopes of an element in the transition energies corresponding to a given spectral line transition. For a spectral line transition between two energy levels a and b in an atom or ion with atomic number Z, the small difference $\Delta E_{ab} = E_{ab}(A') - E_{ab}(A)$ in the transition energy between isotopes with mass numbers A' and A is the isotope shift. It consists largely of the sum of two contributions, the mass shift (MS) and the field shift (FS), also called the volume shift. The mass shift is customarily divided into a normal mass shift (NMS) and a specific mass shift (SMS); each is proportional to the fractional mass difference $(A' - A)/A'A$. The normal mass shift is a reduced mass correction that is easily calculated for all transitions. The specific mass shift is produced by the correlated motion of different pairs of atomic electrons and is, therefore, absent in one-electron systems.

It is generally difficult to calculate precisely the specific mass shift, which may be 30 times larger than the normal mass shift for some transitions. The field shift is produced by the change in the finite size and shape of the nuclear charge distribution when neutrons are added to the nucleus. Since electrons whose orbits penetrate the nucleus are influenced most, S-P and P-S transitions generally have the largest field shift.

For very light elements, $Z \lesssim 37$, the mass shift dominates the field shift. For $Z = 1$, the 0.13-nm shift in the red Balmer line led to the discovery of deuterium, the $A = 2$ isotope of hydrogen. For me-

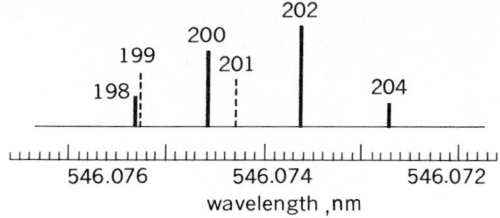

Some isotope shifts in the green line of mercury, $Z = 80$. In this heavy element the contribution of the field shift is much larger than that of the mass shift.

dium-heavy elements, $38 \leq Z \leq 57$, the mass shift and field shift contributions to the isotope shift are comparable. For heavier elements. $Z \gtrsim 58$, the field shift dominates the mass shift. A representative case is shown in the illustration. *See* ATOMIC STRUCTURE AND SPECTRA; RYDBERG CONSTANT.

When isotope shift data have been obtained for at least two pairs of isotopes of a given element, a graphical method introduced by W. H. King in 1963 can be used to evaluate quantitatively the separate contributions of the mass shift and the field shift. Experimentally determined field shifts can be used to test theoretical models of nuclear structure, shape, and multipole moments. Experimentally determined specific mass shifts can be used to test detailed theories of atomic structure and relativistic effects. *See* NUCLEAR MOMENTS; NUCLEAR STRUCTURE.

Experimental techniques that have greatly increased both the amount and the precision of isotope shift data that can be obtained include on-line isotope separators for the study of isotopes with half-lives as short as a few seconds and spectroscopic methods employing high-resolution tunable lasers. Active development of isotope separation schemes based on these laser techniques has been undertaken. Isotope shift data have also been obtained for x-ray transitions of electrons in inner atomic shells and of muons in muonic atoms.

[PETER M. KOCH]

Bibliography: J. Bauche and R. J. Campeau, Recent advances in the theory of atomic isotope shift, *Advan. At. Mol. Phys.*, 12:39–86, 1976.

Isotopic spin

A quantum-mechanical variable or quantum number applied to hadrons, the strongly interacting fundamental particles, and compounds of hadrons (such as nuclear states) to facilitate consideration of the consequences of the charge independence of the strong (nuclear) forces. This variable is rarely labeled isobaric spin or isospin, and most commonly I-spin.

The many strongly interacting elementary particles (hadrons) and the compounds of these particles, such as nuclei, are observed to form sets or multiplets such that the members of the multiplet differ in their electromagnetic charge and electromagnetic properties but are otherwise almost identical. For example, the neutron and proton, with electric charges of zero and plus one fundamental unit (of the magnitude of the electronic charge), form a set of two such states. The pions, one with a unit of positive charge, one with zero charge, and one with a unit of negative charge, form a set of three. It appears that if the effects of electromagnetic forces and the closely related weak nuclear forces (responsible for β-decay) are neglected, leaving only the strong forces effective, the different members of such a multiplet are equivalent and cannot be distinguished in their strong interactions. The strong interactions are thus independent of the different electric charges held by different members of the set; they are charge-independent. *See* ELEMENTARY PARTICLE; FUNDAMENTAL INTERACTIONS; HADRON.

The isotopic spin I of such a set or multiplet of equivalent states is defined such that Eq. (1) is satisfied, where N is the number of states in the set.

$$N = 2I + 1 \qquad (1)$$

isfied, where N is the number of states in the set. Another quantum number, I_3, called the third component of isotopic spin, is used to differentiate the numbers of a multiplet where the values of I_3 vary from $+I$ to $-I$ in units of one. The charge Q of a state and the value of I_3 for this state are connected by the Gell-Mann–Okubo relation, Eq. (2),

$$Q = I_3 + Y/2 \qquad (2)$$

where Y, the charge offset, is called hypercharge. For nuclear states, Y is simply the number of nucleons. Electric charge is conserved in all interactions; Y is observed to be conserved by the strong forces so that I_3 is conserved in the course of interactions mediated by the strong forces. *See* HYPERCHARGE.

Similarity to spin. This description of a multiplet of states with isotopic spin I is similar to the quantum-mechanical description of a particle with total angular momentum or spin of j (in units of $\hbar$, Planck's constant divided by 2π). Such a particle can be considered as a set of states which differ in their orientation or component of spin j_z in a z direction of quantization. There are $2j + 1$ such states, where j_z varies from $-j$ to $+j$. To the extent that the local universe is isotropic (or that there are no external forces on the states which depend upon direction), the components of angular momentum or spin in any direction are conserved, and states with different values of j_z are dynamically equivalent.

There is then a logical or mathematical equivalence between the descriptions of (1) a multiplet of states of definite isotopic spin I and different values of charge and I_3 with respect to charge-independent forces and (2) a multiplet of states of a particle with a definite spin j and different values of j_z with respect to direction-independent forces. In each case, the members of multiplet with different values of the conserved quantity I_3 on the one hand and j_z on the other are dynamically equivalent; that is, they are indistinguishable by any application of the forces in question. *See* ANGULAR MOMENTUM; SPIN.

Importance in reactions and decays. The charge independence of the strong interactions has important consequences defining the intensity ratios of different charge states produced in those particle reactions and decays which are mediated by the strong interactions. As a simple illustration, consider the virtual transitions of a nucleon to a nucleon plus a pion, transitions which are of dominant importance in any consideration of nuclear forces. A proton may undergo a transition to a proton and neutral pion or to a neutron and positive

pion and conserve charge; a neutron can go to a neutron and neutral pion or to a proton and negative pion. The neutron and proton form a nucleon isotopic spin doublet with $I = \frac{1}{2}$; the pions constitute an isotopic spin triplet with $I = 1$. If one starts with an initial set of one proton and one neutron, with no bias in charge state or I_3, and the strong forces responsible for the virtual transitions do not discriminate between states with different charge or different values of I_3, it follows by inspection that the ratio of transition intensities will be defined as shown in the table.

Relative intensities of virtual transitions determined by isotopic spin symmetry

Transition	Relative intensity
$p \rightarrow n + \pi^+$	2/3
$p \rightarrow p + \pi^0$	1/3
$n \rightarrow n + \pi^0$	1/3
$n \rightarrow p + \pi^-$	2/3

If the forces cannot distinguish between charge states and there is no initial charge asymmetry, there can be no charge asymmetry in the final sets of states. If initially there are one neutron and one proton, equal numbers of each charge member of the nucleon isotopic spin doublet, in the final system there must be equal probabilities of finding each charge member of the pion triplet and equal probabilities of finding a neutron or proton. This condition, that the strong interactions cannot differentiate among the members of an isotopic spin multiplet, defines the relative intensities given in the table. These arguments hold equally well for real decays.

The above demonstration considers the decay of an initial (nucleon) doublet with $I = \frac{1}{2}$, to a final state of a (nucleon) doublet and a (pion) triplet with $I = 1$. Using the same kind of argument, it is easy to see that the conditions of equal intensity of each member of a multiplet cannot be fulfilled in a transition from an initial doublet to a final state of a doublet and quartet. Therefore, none of the individual transitions is allowed by charge independence, though charge or I_3 is conserved in the decays. In general, decays are allowed for a transition A $\rightarrow$ B + C only if inequality (3) is satisfied.

$$|I(B) + I(C)| \geq |I(A)| \geq |I(B) - I(C)| \qquad (3)$$

This is analogous to the vector addition rule for spin or angular momentum; the strong interactions conserve isotopic spin in the same manner as angular momentum is conserved. This is an example of the general rule that the whole content of the description of spin or angular momentum can be taken over for isotopic spin. *See* NUCLEAR REACTION; SELECTION RULES.

Classification of states. Isotopic spin considerations provide insight into the total energies or masses of nuclear and particle states. The fundamental constituents of nuclei are the nucleons, the neutron and proton, spin $\frac{1}{2}$ fermions which must obey the Pauli exclusion principle to the effect that the sign of a wave function which describes a set of identical fermions must change sign upon exchange of any two fermions. Similarly, hadrons are now described as compounds of quarks, which are also spin ½ fermions. The two fermions which belong to an isotopic spin doublet can be considered as different charge states of a basic fermion, even as states with spin in the plus and minus direction of quantization are considered as different spin states of the fermion. The extended Pauli exclusion principle then requires that the wave function amplitude change sign upon exchange of spin, charge, and spatial coordinates for two fermions. *See* EXCLUSION PRINCIPLE.

A space state $u(r)$ of two fermions, where r is the vector distance between the two particles, will be even upon exchange of the two particles if $u(r)$ has an even number of nodes, and will be odd under exchange if there is an odd number of nodes. With more nodes, the space wavelength is smaller, and the momentum and energy of the particles are larger. The lowest energy state must then have no spatial nodes and must be even under spatial interchange. From the Pauli principle, the whole wave function must be odd, and then the exchange under spin and isotopic spin coordinates must be odd. Using this kind of argument, Eugene Wigner was able to classify the low-mass (low-energy) states of light nuclei in terms of their isotopic spin symmetries.

An application of the same principle was an important element in the discovery of a new quantum number (labeled color) for quarks, the elementary constituents of the hadrons. The lightest, least energetic baryon states, such as the neutron and proton, were shown to be even under the exchange of quark spin and quark isotopic spin, a result which would seem to require that the space wave function was odd, violating the general energy argument. It is now known that the states are odd under the additional color exchange, allowing the space function to be even, as expected, for the lightest states. *See* COLOR (QUANTUM MECHANICS); QUARKS; SYMMETRY LAWS.

[ROBERT K. ADAIR]

Bibliography: S. deBenedetti, *Nuclear Interactions*, 1964; H. A. Enge, *Introduction to Nuclear Physics*, 1966; M. Gell-Mann and Y. Neeman, *The Eightfold Way*, 1964; D. Lurie, *Particles and Fields*, 1968.

Isotropy

A body is said to be isotropic if its physical properties are not dependent upon the direction in the body along which they are measured. A body displaying isotropy has only one refractive index, one dielectric constant, and so on. Most but not all liquids and aggregates made up of many small crystals randomly oriented in space are isotropic in all their properties. Depending on their symmetry, single crystals may or may not be isotropic with respect to a given property. For example, single crystals with cubic structure are isotropic with respect to electrical resistivity but not with respect to elastic deformability. *See* ANISOTROPY; ELASTICITY.

[DAVID TURNBULL]

J particle

An elementary particle with an unusually long lifetime and large mass which does not fit into any of the schemes for classifying the large number of previously known particles. The discovery of J particles in proton-proton (p,p) and electron-positron (e^-,e^+) collisions created excitement in elementary particle physics in the mid-1970s.

Discovery of J particle. There has been much theoretical speculation on the existence of long-lived neutral (no electric charge) particles with superheavy masses larger than 10 GeV/c^2 (10^{10} electronvolts divided by the speed of light squared, or about ten times the mass of the hydrogen atom). These are thought to play the role in weak interactions that photons play in electromagnetic interactions. There is, however, no theoretical justification, and no predictions exist, for long-lived particles in the mass region $1-10$ GeV/c^2.

The J particles are rarely produced in p-p collisions. Statistically, they occur once after many millions of subnuclear reactions, in which most of the particles are "ordinary" elementary particles, such as kaon (K), pion (π), or proton (p). One searches for the J particle by detecting its e^+e^- decays. A two-particle spectrometer was used by a group from Massachusetts Institute of Technology (MIT) at the Brookhaven National Laboratory (BNL) to discover this particle. A successful experiment must have: (1) A very-high-intensity incident proton beam to produce a sufficient amount of J particles for detection; the Alternating Gradient Synchrotron (AGS) accelerator at BNL provides a beam of 10^{12} 30-GeV protons per second for this experiment. (2) The ability, in a billionth of a second, to pick out the $J{\rightarrow}e^-e^+$ pairs amidst billions of other particles through the detection apparatus.

The detector is called a magnetic spectrometer. A positive particle and a negative particle each traversed one of two 70-ft-long (21-m) arms of the spectrometer. The e^+ and e^- were identified by the fact that a special counter, called a Cerenkov counter, measured their speed as being slightly greater than that of all other charged particles. Precisely measured magnetic fields bent them and measured their energy. Finally, as a redundant check, the particles plowed into high-intensity lead-glass and the e^+ and e^- immediately transformed their energy into light. When collected, this light "tagged" these particles as e^+ and e^-, and not heavier particles such as π, K, or p. The simultaneous arrival of an e^- and e^+ in the two arms indicated the creation of high-energy light quanta from nuclear interactions. The sudden increase in the number of e^+e^- pairs at a given energy (or mass) indicated the existence of a new particle.

The trajectory of electrons was measured by precision devices called multiwire proportional chambers. They consisted of 10,000 very fine gold-plated wires of 2-mm spacing, each with its own amplifier and recording system. The signals from the Cerenkov counters were collected by thin spherical and elliptical mirrors measuring about 3½ ft (1.1 m) in diameter. The counters were filled with gaseous hydrogen so that only energetic e^- (and e^+) would produce the light which is due to the Cerenkov effect. To measure the arrival time of e^+e^- pairs to one-billionth of a second, there were 100 elements of thin plastic scintillation counters, each less than 2 mm thick. *See* CERENKOV RADIATION; SCINTILLATION COUNTER.

By August 1974 the MIT group began to observe abundant numbers of e^+e^- pairs with a total combined mass of 3.112 GeV (see illustration), and line

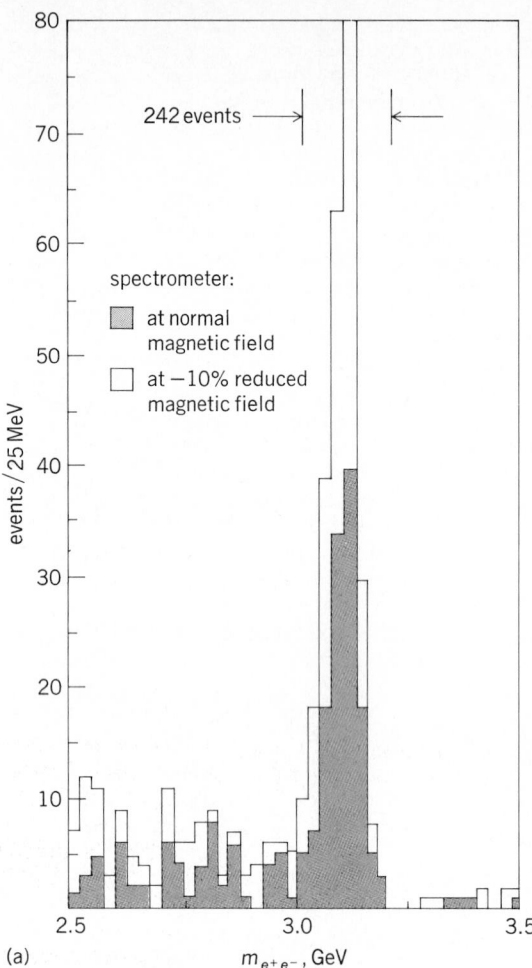

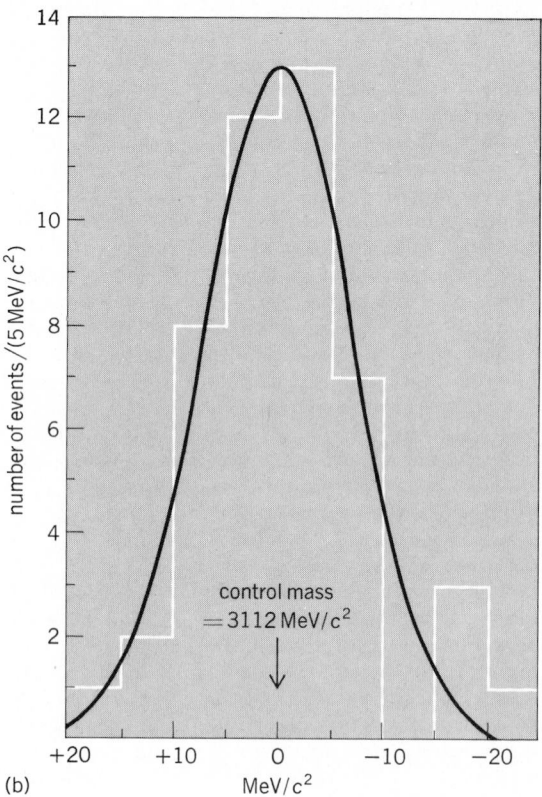

(a) $m_{e^+e^-}$, GeV

(b)   MeV/c^2

Observation of the J particle. (a) On-line data from August and October 1974, showing the existence of the J particle. (b) Measurement of the width of the J particle, showing it has a width less than 5 MeV. (*From J. J. Aubert et al., Discovery of the new particle, J. Nuc. Phys., B89(1):1–18, 1975*)

width Γ much smaller than the resolution of the detector: $\Gamma < 5$ MeV (5×10^6 electronvolts). From August through October, many experimental checks on the detector were made. The most important of these checks consisted of changing the magnetic field of the detecting magnets. This moved the particle trajectories to different regions of the detector. Still the abundance of e^+e^- pairs did not change, indicating a real particle had been discovered. This new particle was called the J particle since J is the symbol used to denote electromagnetic current and spin in elementary particle physics. A joint announcement of the discovery was made together with a team from Stanford Linear Accelerator Center (where it was called the psi particle) in November 1974.

Properties of J particle. The $J \rightarrow e^- e^+$ production rate from proton-proton reactions at an incident energy of 30 GeV is 10^{-34} cm^2, or one part in 10^7 of "ordinary" particle yields. The yield increases by a factor of about 50 with a 300 GeV/c incident neutron beam. This increase of yield with energy is very similar to that of the K meson and antiproton productions. In the Brookhaven experiment, the yield of J decreased by almost a factor of 10 when the incident beam energy was reduced to 20 GeV/c. J particles have also been produced by bombardment of complex nuclear targets with high-energy photons. Production rates seem to be consistent with diffractive production like photoproduction of ordinary vector particles (the ρ, ω, and ϕ). Analyses of photoproduction data indicate the J is not an ordinary intermediate vector boson. Rather, it belongs to a strong-interaction family.

Most of the properties of J particles have been measured at various electron-positron storage rings. The data show that the measured line width of J is less than 2 MeV. By measuring $e^+e^- \rightarrow J \rightarrow e^-e^+$ decay rate, one obtains a total width of less than 100 keV. The observed mass, m_J, varies from 3090 MeV/c^2 to 3112 MeV/c^2 from one laboratory to another. The spin (intrinsic angular momentum) is the same as that of the photon. Modes of decay into $\pi^+\pi^-$ and K^+K^- were not found. From this it has been concluded that the J particle does not belong to families of particles of about equal mass (like the π-mesons, π^+, π^0, π^-, or the two nucleons p, n), but that these particles are "single." Decays of $J \rightarrow \bar{p}p$, $\Lambda\bar{\Lambda}$, $n\pi^0$ have been found. See ISOTOPIC SPIN.

Most of the properties of the J particle have now been measured at various electron-positron storage rings, most notably at Stanford Linear Accelerator, the Deutsches Elektronen-Synchrotron (DESY), and the European Organization for Nuclear Research (CERN). The results have shown that there is a family of particles which, by their emission and absorption of monochromatic gamma rays, transform into each other and into the J particle. The energy spectrum of the J-particle family is very similar to the positronium spectrum. (Positronium is the simplest atom, consisting of an electron bound with a positron.) The similarity between the J-particle spectrum and the positronium spectrum, together with the fact that the lifetime of the J particle is about 1000 to 10,000 times longer than that of other known elementary particles, indicates that the J particle is a bound state of a new type of quark and antiquark. This idea was first proposed by S. Glashow, who called it the charm quark. The J particle therefore is a charmonium state. *See* CHARM; POSITRONIUM.

The existence of the charm quark has stimulated many experiments all over the world to search for more quark states. Indeed, a new quark state (the upsilon) was found at a mass of 9.4 GeV by Leon Lederman and collaborators, again by measuring proton nuclei interactions. *See* UPSILON PARTICLE.

With the construction of a 30-GeV (gigaelectronvolt) electron-positron colliding-beam accelerator (known as PETRA) at DESY, more experiments have shown that the forces between the quarks are transmitted by new kinds of particles, known as gluons, which bind the quarks inside the elementary particles so that they cannot be detected freely. *See* ELEMENTARY PARTICLE; GLUONS; QUARKS.

[SAMUEL C. C. TING]

Bibliography: J. J. Aubert et al., *Phys. Rev. Lett.,* 33:1404 and 1624, 1974; E. J. Augustin et al., *Phys. Rev. Lett.,* 33:1406, 1974; W. Braunschweig et al., *Phys. Lett.,* 53B:393, 1974.

Jahn-Teller effect

A small distortion in the lattice structure of certain crystal defects and, in some cases, of entire crystals, and also in the structure of certain molecules, which reduces the symmetry and removes electronic degeneracy. The Jahn-Teller effect was predicted theoretically in 1937 and was first observed experimentally in 1952, but only since 1965 have its most commonly observable consequences come to be understood. Whereas in 1965 it was regarded as a "mystical effect" to be invoked when all other explanations of anomalous data failed, now it is as well understood as most other phenomena in solid-state physics.

The Jahn-Teller effect is most commonly observed in the optical and microwave spectra of point defects and localized impurity centers in solids. Such a center has a certain point symmetry, that is, there is a group of coordinate transformations which leave the center and its surroundings unaltered. If this symmetry is high enough, some of the electronic states of the center will be orbitally degenerate. The Jahn-Teller theorem states that any such degenerate system is unstable against small distortions of the lattice framework which remove the degeneracy. *See* DEGENERACY (QUANTUM MECHANICS).

Because of the great complexity of the interatomic forces in a real solid, it is customary to think of the center as a molecule consisting of the central impurity or defect and its nearest neighbor atoms. The rest of the crystal is regarded as a fea-tureless heat bath. This "cluster" model preserves the symmetry of the center, which is its most essential attribute for purposes of this discussion, and is appropriate when the electronic wave function is sufficiently localized not to be greatly influenced by motion of atoms outside the nearest neighbor shell. In practice, this is also a condition for electron-lattice coupling to be strong enough for the Jahn-Teller effect to be important. While theoretical treatments have gone beyond the cluster approximation, they have not led to any qualitatively new understanding of the phenomenon.

Simple model for Jahn-Teller distortion. As an illustrative example, consider the center shown in Fig. 1, which is a greatly simplified model of the first excited level of the F-center in an alkali halide. (The F-center is an electron trapped at a negative ion vacancy.) The electron is in a p state and is surrounded by a regular octahedron of positive ions. Because of the cubic symmetry, the three possible p states, $p_x, p_y,$ and p_z (of which only p_z is shown), are obviously degenerate. Now suppose that the positive ions move a small distance in the directions indicated by the arrows. If their mean distance from the center is fixed, the mean energy of the p level is unaltered in first order. The p_z state, in which the electron approaches the positive ions more closely, is lowered in energy by the distortion, while the p_x and p_y states are raised, as shown in Fig. 2. The splitting is initially linear in the displacement Q. The surrounding crystal resists the distortion, and by Hooke's law the additional energy due to this resistance is initially quadratic in Q. Thus the total energy of the center in the p_z state goes through a minimum, and equilibrium is reached at some finite value of Q. It is easy to see that this equilibrium value of Q is A/k, where A is the initial (downward) slope and k the opposing force constant, and that the stabilization energy is $E_{JT} = A^2/2k$. *See* COLOR CENTERS.

Strong electron-lattice coupling or weak interatomic forces favor a strong Jahn-Teller effect. E_{JT} can range from several electron volts (eV) for deep states in diamond and silicon to less than 10^{-3} eV for rare-earth ions in ionic crystals. In transition-metal ions, for which most of the data have been obtained, E_{JT} ranges from 0.01 to 1 eV.

The distortion shown in Fig. 1 could equally well have been along the x or y axes, and the lowest state would have been p_x or p_y, respectively. Thus each electronic state is associated with its own distortion, and if one considers the lowest vibrational state only, the threefold electronic degeneracy is replaced by threefold "vibronic" degeneracy. If no transitions were possible between the different directions of distortion, any one center would remain indefinitely in one state and its corresponding distorted configuration. This distortion might be observable, for instance, in a spin resonance experiment, and in a concentrated crystal, by x-rays. Such an observable reduction in symmetry is called the static Jahn-Teller effect. However, the more common case in isolated centers is the dynamic Jahn-Teller effect, in which transitions between different distorted configurations are rapid compared with the characteristic measurement time. Such transitions can occur through thermal activation, or through quantum mechanical tunneling due to the zero point vibrational motion

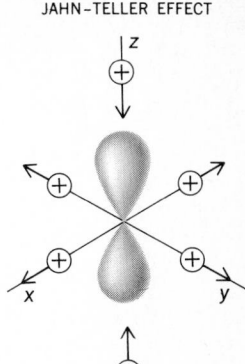

Fig. 1. A p-electron in an octahedral site. The arrows show one possible tetragonal mode of distortion. Only a p_z wave function is shown.

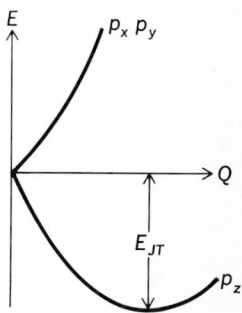

Fig. 2. Energy of the center shown in Fig. 1 as a function of the tetragonal distortion.

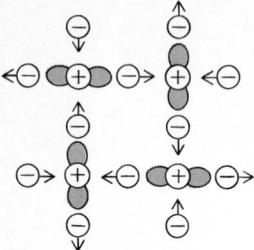

Fig. 3. Part of a two-dimensional model of a cooperative Jahn-Teller system. Arrows indicate the motion in the soft mode.

about the different equilibrium configurations. The observed spectrum now has the same symmetry as if there were no Jahn-Teller effect, and it might be thought that the effect has "disappeared." *See* X-RAY CRYSTALLOGRAPHY.

Observable consequences. In 1965, F. S. Ham pointed out that this is not the case; there are in fact pronounced observable consequences of the dynamic Jahn-Teller effect. These show up in the effects of off-diagonal electronic operators, that is, operators which connect electronic states associated with different distortions. For instance, consider the angular momentum operator $\hbar L$. The p level in the undistorted octahedron of Fig. 1 has $L = 1$. When a magnetic field H is applied parallel to the z axis, the three-fold degenerate level is split into three states with wave functions $\frac{1}{2}\sqrt{2}(p_x + ip_y)$, $\frac{1}{2}\sqrt{2}(p_x - ip_y)$, p_z, corresponding respectively to the $+1$, -1, and 0 eigenvalues of L_z. If one ignores electron spin, the energies are given by $-e\hbar HL_z/2mc$ (in cgs gaussian units). The $(p_x + ip_y)$ state corresponds to clockwise rotation about the field, and the $(p_x - ip_y)$ state to counterclockwise rotation. In the Jahn-Teller distorted case the electron has to "drag" its distortion with it and can no longer rotate freely. Its mass is thus effectively increased and the orbital contribution to the magnetic splitting correspondingly decreased. At 0K, where rotation is only possible at all because of zero point motion, the reduction factor can be shown to be approximately $e^{-3E_{JT}/2\hbar\omega}$, where ω is an effective frequency of vibration for the cluster. Note that the spin operator S is not "quenched" as L is, since the spin direction has no distortion associated with it (except insofar as spin-orbit coupling causes L to follow S).

This reduction in the orbital contribution to magnetic splittings has been observed in the spin resonance and Zeeman spectra of many transition-metal impurity ions in crystals. Other off-diagonal operators, such as the spin-orbit coupling (which can often be written in the form $-\lambda L \cdot S$), are quenched in the same way. A detailed comparison of the Ham theory with experiment has been possible in the optical spectra of $3d$ ions in some cubic crystals. *See* ZEEMAN EFFECT.

If the electronic state is doubly degenerate, it has no orbital momentum or spin-orbit coupling in first order. If the Jahn-Teller interaction is strong, the small local strains which are inevitable in a real crystal are sufficient to stabilize the center in one or other distorted configuration at low temperature. In a crystal containing many centers, three uniaxial spin resonance spectra are seen, corresponding to the three possible directions of distortion. As the temperature is raised, transitions from one distorted configuration to another become possible, and the three anisotropic resonances collapse into one isotropic motionally averaged resonance. This process is exactly analogous to motional narrowing due to diffusion in nuclear magnetic resonance. If the Jahn-Teller effect is sufficiently weak, tunneling will average out the spectrum, even at 0K. The spectrum in this case still retains a cubic anisotropy, qualitatively similar to that expected in the absence of Jahn-Teller interaction. Even if tunneling is too slow to produce this averaging, it can still profoundly affect relaxation processes. For instance, it causes the rapid dephasing of magnetic free induction, and it pro-

duces strong damping of acoustic waves whose period is of the order of the tunneling time.

Cooperative Jahn-Teller effect. In a crystal containing a large concentration of ions with degenerate ground states, a Jahn-Teller distortion of the whole crystal can occur through the strain-mediated interaction between ions. Many rare-earth compounds have been found to undergo second-order phase transitions, in which the crystal symmetry is reduced and electronic degeneracy removed in the manner expected for the Jahn-Teller effect. The electron-lattice interaction is so weak in these ions that, even though their ground states are degenerate, Jahn-Teller effects are not normally detected in the ion as an isolated impurity. However, the ion-ion interaction greatly enhances the effect by involving the cooperation of many ions, and transition temperatures of a few kelvin are typically observed.

Four adjacent lattice cells of a highly simplified two-dimensional model of such a crystal are shown in Fig. 3. Each positive rare-earth ion is surrounded by four negative ions which can distort as shown by the arrows, producing a splitting of its ground state analogous to that in Fig. 2. Because of the shared negative ions, distortions on adjacent cells must be related in phase. It follows that there is an effective coupling between ions analogous to magnetic exchange but of much greater range, since it is mediated by the strain field of the lattice. As an initially symmetric crystal is cooled, a spontaneous distortion begins to appear at a critical temperature T_c, just as spontaneous magnetization appears in a ferromagnet at the Curie temperature. This distortion can be measured directly by x-rays, and by the energy-level splitting that it produces; it also manifests itself in a specific-heat anomaly (see Fig. 4). As T_c is approached from above, the lattice becomes "soft" with respect to the preferred mode of distortion. If this distortion involves a macroscopic strain, the associated elastic (compliance) constant, which is analogous to the magnetic susceptibility, becomes very large at T_c, as illustrated in Fig. 5. Even if the distortion

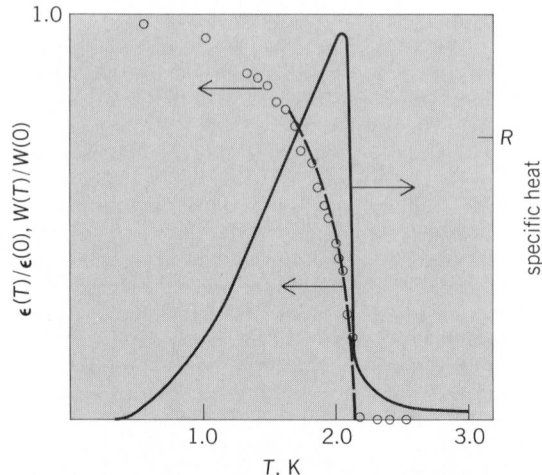

Fig. 4. Normalized lattice distortion (ratio $\epsilon(T)/\epsilon(0)$ of lattice distortion at temperature T to distortion at absolute zero, dashed line); normalized splitting of the lowest energy levels $W(T)/W(0)$ (points); and specific heat (full line) of $TmVO_4$ below T_c.

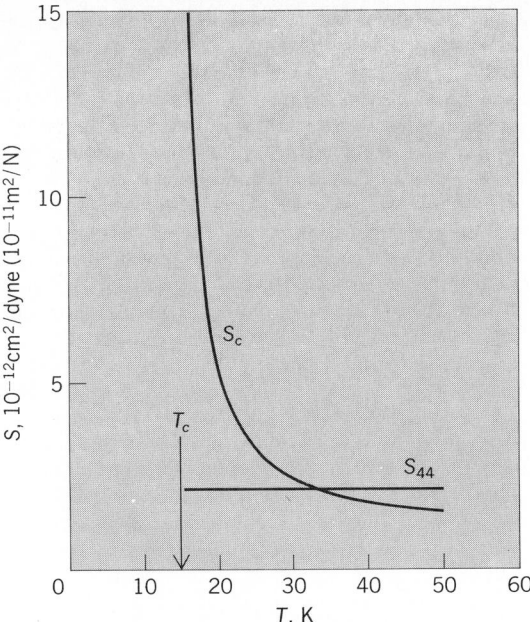

Fig. 5. Elastic constants $S_c = 2(S_{11} - S_{12})$ and S_{44} of DyVO$_4$ just above T_c. S_c couples to the soft mode.

is not macroscopic, it will still show up as a corresponding "softness" in a phonon mode, detectable by Raman effect or neutron diffraction. *See* FERROMAGNETISM; LATTICE VIBRATIONS; NEUTRON DIFFRACTION; PHONON; RAMAN EFFECT.

When the electron-lattice interaction is strong, as in nonmetallic compounds containing Cu^{2+} or Mn^{3+} ions, values of T_c up to 1000 K can be obtained. In metallic compounds, if E_{JT} is less than the bandwidth due to overlap of wave functions, the Jahn-Teller effect is suppressed. However, in some narrow-band metals, such as Nb_3Sn and V_3Si, E_{JT} is so large that Jahn-Teller distortion does occur. The same strong electron-lattice coupling contributes to the high superconducting temperatures of these compounds. *See* SECOND-ORDER TRANSITION; SUPERCONDUCTIVITY.

Vibrational spectra of molecules. The intimate coupling between electronic and nuclear motion which is a consequence of the Jahn-Teller interaction has pronounced effects on the vibrational spectra of molecules. However, it has proved quite difficult to pin down these effects, chiefly because of the difficulty of establishing what the vibrational spectrum would be in the absence of the Jahn-Teller interaction. *See* ATOMIC STRUCTURE AND SPECTRA; CRYSTAL DEFECTS; CRYSTAL STRUCTURE; CRYSTALLOGRAPHY; MOLECULAR STRUCTURE AND SPECTRA; NONRELATIVISTIC QUANTUM THEORY.

[M. D. STURGE]

Bibliography: G. A. Gehring and K. A. Gehring, Cooperative Jahn-Teller effects, *Rep. Progr. Phys.* 38:1–150, 1975; F. S. Ham, The Jahn-Teller effect in EPR spectra, in S. Geschwind (ed.), *Electron Paramagnetic Resonance*, 1971; G. Herzberg, *Electronic Spectra of Polyatomic Molecules*, 1966; B. Reinen and C. Friebel, Local and cooperative Jahn-Teller interactions in model structures, in J. Dunitz (ed.), *Structural Problems*, pp. 1–60, 1979; M. D. Sturge, The Jahn-Teller effect in solids, *Advan. Solid State Phys.*, 20:91–211, 1967.

Josephson effect

The passage of paired electrons (Cooper pairs) through a weak connection (Josephson junction) between superconductors, as in the tunnel passage of paired electrons through a thin dielectric layer separating two superconductors.

Nature of the effect. Quantum-mechanical tunneling of Cooper pairs through a thin insulating barrier (on the order of a few nanometers thick) between two superconductors was theoretically predicted by Brian D. Josephson in 1962. Josephson calculated the currents that could be expected to flow during such superconductive tunneling, and found that a current of paired electrons (supercurrent) would flow in addition to the usual current that results from the tunneling of single electrons (single or unpaired electrons are present in a superconductor along with bound pairs). Josephson specifically predicted that if the current did not exceed a limiting value (the critical current), there would be no voltage drop across the tunnel barrier. This zero-voltage current flow resulting from the tunneling of Cooper pairs is known as the dc Josephson effect. Josephson also predicted that if a constant nonzero voltage V were maintained across the tunnel barrier, an alternating supercurrent would flow through the barrier in addition to the dc current produced by the tunneling of unpaired electrons. The frequency ν of the ac supercurrent is given by Eq. (1), where e is the magni-

$$\nu = 2eV/h \qquad (1)$$

tude of the charge of an electron and h is Planck's constant. The oscillating current of Cooper pairs that flows when a steady voltage is maintained across a tunnel barrier is known as the ac Josephson effect. Josephson further predicted that if an alternating voltage at frequency f were superimposed on the steady voltage applied across the barrier, the ac supercurrent would be frequency-modulated and could have a dc component whenever ν was an integral multiple of f. Depending upon the amplitude and phase of the ac voltage, the dc current-voltage characteristic would display zero-resistance parts (constant-voltage steps) at voltages V given by Eq. (2), where

$$V = nhf/2e \qquad (2)$$

n is any integer. Finally, Josephson predicted that effects similar to the above would also occur for two superconducting metals separated by a thin layer of nonsuperconducting (normal) metal. In 1963 the existence of the dc Josephson effect was experimentally confirmed by P. W. Anderson and J. M. Rowell, and the existence of the ac Josephson effect was experimentally confirmed by S. Shapiro. *See* NONRELATIVISTIC QUANTUM THEORY.

Theory of the effect. The superconducting state has been described as a manifestation of quantum mechanics on a macroscopic scale, and the Josephson effect is best explained in terms of phase, a basic concept in the mathematics of quantum mechanics and wave motion. For example, two sine waves of the same wavelength λ are said to have the same phase if their maxima coincide, and to have a phase difference equal to $2\pi\delta/\lambda$ if their maxima are displaced by a distance δ. An appreciation of the importance that phase can have in physical systems can be gained by considering the radiation from excited atoms in a ruby rod. For a given transition, the atoms emit radiation of the same wavelength; if the atoms also emit the radiation in phase, the result is the ruby laser.

According to the Bardeen-Cooper-Schrieffer (BCS) theory of superconductivity, an electron can be attracted by the deformation of the metal lattice produced by another electron, and thereby be indirectly attracted to the other electron. This indirect attraction tends to unite pairs of electrons having equal and opposite momentum and antiparallel spins into the bound pairs known as Cooper pairs. In the quantum-mechanical description of a superconductor, all Cooper pairs in the superconductor have the same wavelength and phase. It is this phase coherence that is responsible for the remarkable properties of the superconducting state. The common phase of the Cooper pairs in a superconductor is referred to simply as the phase of the

superconductor. *See* COHERENCE; PHASE; QUANTUM MECHANICS.

The phases of two isolated superconductors are totally unrelated, while two superconductors in perfect contact have the same phase. If the superconductors are weakly connected (as they are when separated by a sufficiently thin tunnel barrier), the phases can be different but not independent. If ϕ is the difference in phase between superconductors on opposite sides of a tunnel barrier, the results of Josephson's calculation of the total current I through the junction can be written as Eq. (3), where I_0 is the current due to single elec-

$$I = I_0 + I_1 \sin \phi \qquad (3)$$

tron tunneling, and $I_1 \sin \phi$ is the current due to pair tunneling. The time dependence of the phase is given by Eq. (4). In general, the currents I, I_0,

$$\partial \phi / \partial t = 2\pi (2eV/h) \qquad (4)$$

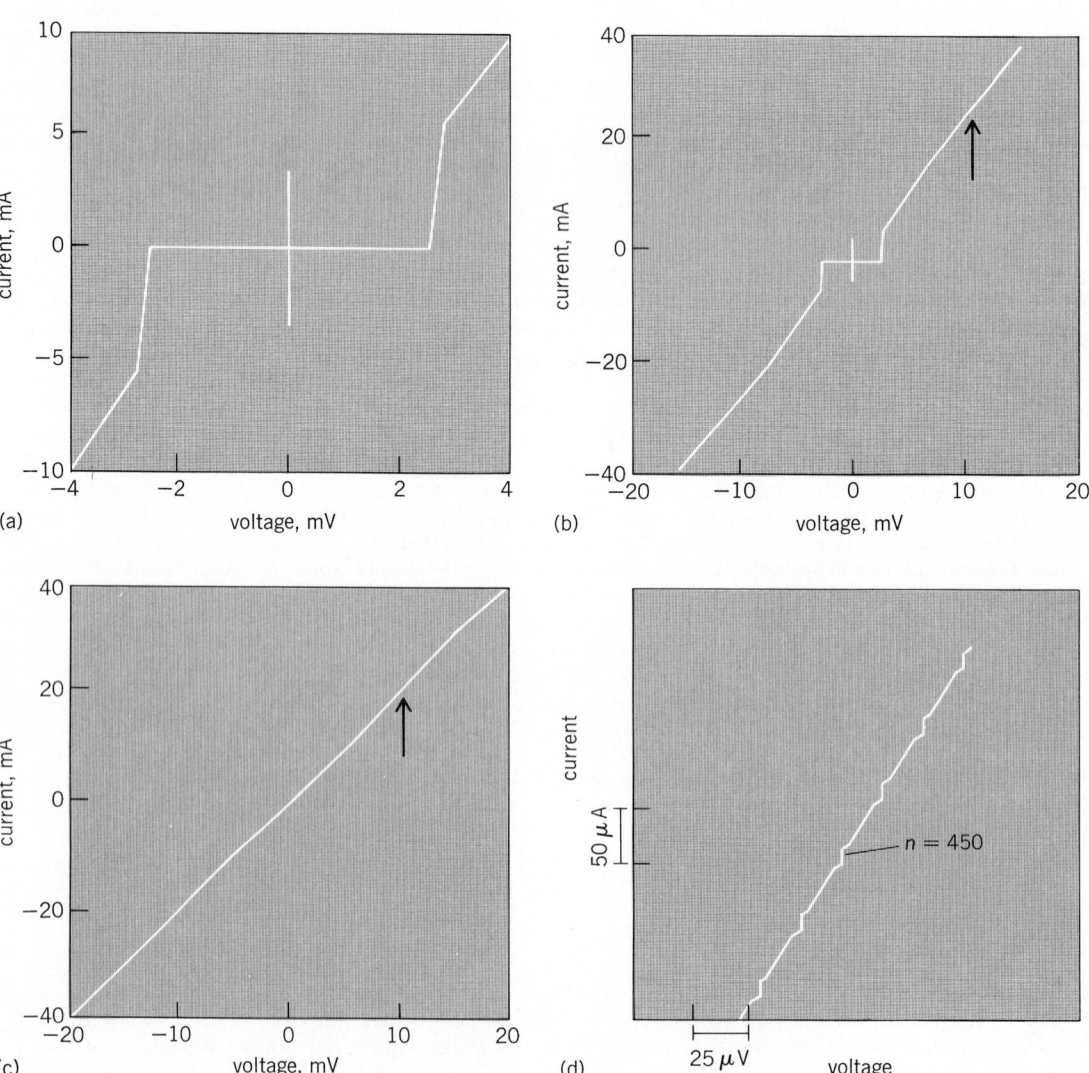

(a)   (b)   (c)   (d)

Fig. 1. DC current-voltage characteristics of lead–lead oxide–lead Josephson tunnel junction at 1.2 K. (a) Without microwave power. (b) Same characteristic with reduced scale. (c) 11-GHz microwave power applied. (d) Expanded portion of c; arrow indicates a constant-voltage step near 10.2 mV corresponding to $n = 450$ in

Eq. (2). This voltage is also indicated by arrows in b and c. (From T. F. Finnegan, A. Denenstein, and D. N. Langenberg, AC-Josephson-effect determination of e/h: A standard of electrochemical potential based on macroscopic quantum phase coherence in superconductors, *Phys. Rev. B*, 4:1487–1522, 1971) and I_1 are all functions of the voltage across the junction. For $V = 0$, I_0 is zero and ϕ is constant. The value of I_1 depends on the properties of the tunnel barrier, and the zero-voltage supercurrent is a sinusoidal function of the phase difference between the two superconductors. However, it is not the phase difference that is under the control of the experimenter, but the current through the junction, and the phase difference adjusts to accommodate the current. The maximum value $\sin \phi$ can assume is 1, and so the zero-voltage value of I_1 is the critical current of the junction.

Integration of Eq. (4) shows the phase changes linearly in time for a constant voltage V maintained across the barrier, and the current through the barrier is given by Eq. (5), where ϕ_0 is a constant.

$$I = I_0 + I_1 \sin [2\pi(2eV/h)t + \phi_0] \qquad (5)$$

The supercurrent is seen to be an ac current with frequency $2eV/h$. The supercurrent time-averages to zero, so the dc current through the barrier is just the single-electron tunneling current I_0.

If the voltage across the junction is $V + v \cos 2\pi ft$, Eqs. (3) and (4) give Eq. (6) for the current.

$$I = I_0 + I_1 \sin [2\pi(2eV/h)t + \phi_0 + (2ev/hf) \sin 2\pi ft] \qquad (6)$$

The expression for the supercurrent is a conventional expression in frequency-modulation theory and can be rewritten as expression (7), where J_n is

$$I_1 \sum_{n=-\infty}^{n=\infty} (-1)^n J_n(2ev/hf) \sin [2\pi(2eV/h)t - 2\pi nft + \phi_0] \qquad (7)$$

an integer-order Bessel function of the first kind. This expression time-averages to zero except when $V = nhf/2e$, in which case the supercurrent has a dc component given by $(-1)^n I_1 J_n(2ev/hf) \sin \phi_0$. As for the zero-voltage dc supercurrent, the phase difference ϕ_0 adjusts to accommodate changes in current at this value of V, and the dc current-voltage characteristic exhibits a constant voltage step. The dc characteristic of a Josephson tunnel junction with and without a microwave-frequency ac voltage is shown in Fig. (1). The straightening of the current-voltage characteristic in the presence of microwave power displayed in Fig. 1c is due to the phenomenon of photon-assisted tunneling, which is essentially identical to classical rectification for the junction and frequency in question. *See* BESSEL FUNCTIONS.

Josephson pointed out that the magnitude of the maximum zero-voltage supercurrent would be reduced by a magnetic field. In fact, the magnetic field dependence of the magnitude of the critical current is one of the more striking features of the Josephson effect. Circulating supercurrents flow through the tunnel barrier to screen an applied magnetic field from the interior of the Josephson junction just as if the tunnel barrier itself were weakly superconducting. The screening effect produces a spatial variation of the transport current, and the critical current goes through a series of maxima and minima as the field is increased. Figure 2 shows the variation of the critical current with magnetic field for a tunnel junction whose length and width are small in comparison with the characteristic screening length of the junction (the Josephson penetration depth, λ_J). The mathematical function which describes the magnetic field dependence of the critical current for this case is

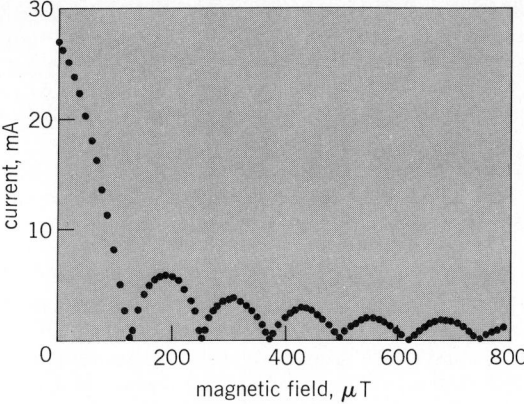

Fig. 2. Magnetic field dependence of the critical current of a Josephson tunnel junction. Data are for a tin–tin oxide–tin junction at 1.2 K, with the magnetic field in the plane of the barrier. (*From D. N. Langenberg, D. J. Scalapino, and B. N. Taylor, The Josephson effects, Sci. Amer., 214(5):30–39, May 1966*)

the same function as that which describes the diffraction pattern produced when light passes through a single narrow slit. *See* DIFFRACTION.

Josephson junctions. The weak connections between superconductors through which the Josephson effects are realized are known as Josephson junctions. Historically, superconductor-insulator-superconductor tunnel junctions have been used to study the Josephson effect, primarily because these are physical situations for which detailed calculations can be made. However, the Josephson effect is not necessarily a tunneling phenomenon, and the Josephson effect is indeed observed in other types of junctions, such as the superconductor–normal metal–superconductor junction. A particularly useful Josephson junction, the point contact, is formed by bringing a sharply pointed superconductor into contact with a blunt superconductor. The critical current of a point contact can be adjusted by changing the pressure of the contact. The low capacitance of the device makes it well suited for high-frequency applications. Thin-film microbridges form another group of Josephson junctions. The simplest microbridge is a short narrow constriction (length and width on the order of a few micrometers or smaller) in a superconducting film known as the Anderson-Dayem bridge. If the microbridge region is also thinner than the rest of the superconducting film, the resulting variable-thickness microbridge has better performance in most device applications. If a narrow strip of superconducting film is overcoated along a few micrometers of its length with a normal metal, superconductivity is weakened beneath the normal metal, and the resulting microbridge is known as a proximity-effect or Notarys-Mercereau microbridge. Among the many other types of Josephson junctions are the superconductor-semiconductor-superconductor and other artificial-barrier tunnel junctions, superconductor–oxide–normal metal–superconductor junctions, and the so-called SLUG junction, which consists of a drop of lead-tin solder solidified around a niobium wire. Some different types of Josephson junctions are illustrated in Fig. 3.

The dc current-voltage characteristics of different types of Josephson junctions may differ, but all show a zero-voltage supercurrent, and constant-voltage steps can be induced in the dc characteristics at voltages given by Eq. (2) when an ac voltage is applied. The dc characteristics of a microbridge and a tunnel junction are compared in Fig. 4.

Applications. The United States legal volt, V_{NBS}, is now defined by Eq. (1) through the assigned value given by Eq. (8), and it is maintained at the Na-

$$2e/h = 483593420 \text{ MHz/V}_{NBS} \qquad (8)$$

tional Bureau of Standards to an accuracy of within a few parts in 10^8 using the ac Josephson effect; the standards of voltage of most other nations as well as the international volt are similarly defined and maintained. This developed as a natural consequence of extremely precise measurements of $2e/h$ via the Josephson effect, and the recognition that a Josephson junction is a precise frequency-to-voltage converter and that atomic frequency standards are inherently more stable than electrochemical voltage standards. *See* ATOMIC CONSTANTS; ELECTRICAL UNITS AND STANDARDS.

The Josephson effect permits measurement of absolute temperature: a voltage drop across a resistor in parallel with a Josephson junction causes

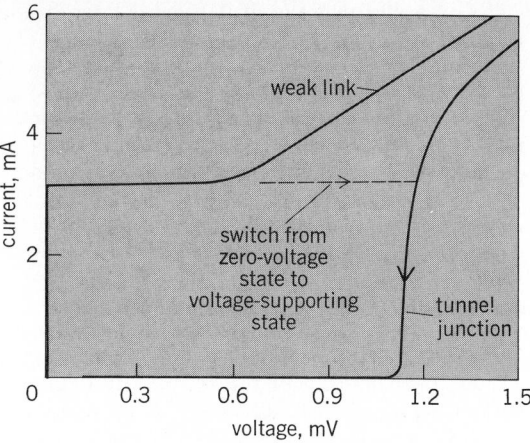

Fig. 4. DC current-voltage characteristics for a weak link and a tunnel junction. (*From D. N. Langenberg, AC Josephson tunneling-experiment, in E. Burstein and S. Lundquist, eds., Tunneling Phenomena in Solids, Plenum, 1969*)

the junction to emit radiation at the frequency given by Eq. (1), but voltage fluctuations resulting from thermal noise produce frequency fluctuations which depend on absolute temperature. The temperature scale below 1 K is maintained at the National Bureau of Standards via this noise thermometry in conjunction with nuclear-orientation thermometry.

Josephson junctions, and instruments incorporating Josephson junctions, are used in other applications for metrology at dc and microwave frequencies, frequency metrology, magnetometry, detection and amplification of electromagnetic signals, and other superconducting electronics such as high-speed analog-to-digital converters and computers. A Josephson junction, like a vacuum tube or a transistor, is capable of switching signals from one circuit to another; a Josephson tunnel junction is capable of switching states in as little as 6 ps and is the fastest switch known. Josephson junction circuits are capable of storing information. Finally, because a Josephson junction is a superconducting device, its power dissipation is extremely small, so that Josephson junction circuits can be packed together as tightly as fabrication techniques will permit. All the basic circuit elements required for a Josephson junction computer have been developed. It has been predicted that the first computer to be made will fill a cube 5 cm on a side and will have a cycle time of 2 nanoseconds, at least 10 times faster than an equivalent high-speed semiconductor-based computer. *See* SUPERCONDUCTIVITY.

[LOUIS B. HOLDEMAN]

Bibliography: E. Burstein and S. Lundquist, *Tunneling Phenomena in Solids*, 1969; D. N. Langenberg, D. J. Scalapino, and B. N. Taylor, The Josephson effects, *Sci. Am.*, 214(5):30–39, May 1966; J. Matisoo, The Superconducting computer, *Sci. Am.*, 242(5):50–65, May 1980; L. Solymar, *Superconductive Tunnelling and Applications*, 1972.

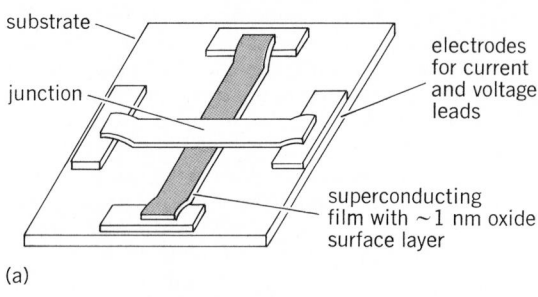

(a)

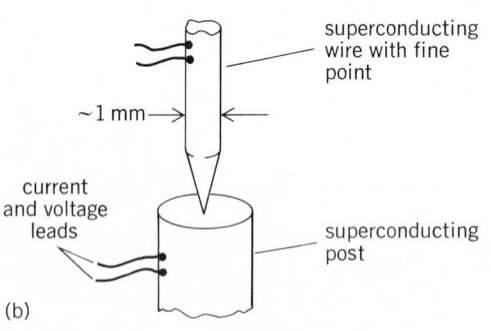

(b)

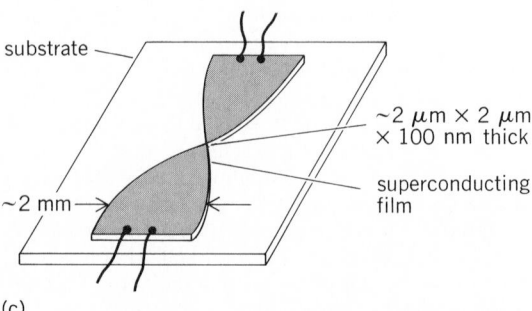

(c)

Fig. 3.   Some types of Josephson junctions. (*a*) Thin-film tunnel junction. (*b*) Point contact. (*c*) Thin-film weak link. (*From D. N. Langenberg, AC Josephson tunneling-experiment, in E. Burstein and S. Lundquist, eds., Tunneling Phenomena in Solids, Plenum, 1969*)

Joule's law

A quantitative relationship between the quantity of heat produced in a conductor and an electric current flowing through it. As experimentally determined and announced by J. P. Joule, the law states that when a current of voltaic electricity is propagated along a metallic conductor, the heat evolved in a given time is proportional to the resistance of the conductor multiplied by the square of the electric intensity. Today the law would be stated as $H = RI^2$, where H is rate of evolution of heat in watts, the unit of heat being the joule; R is resistance in ohms; and I is current in amperes. This statement is more general than the one sometimes given that specifies that R be independent of I. Also, it is now known that the application of the law is not limited to metallic conductors.

Although Joule's discovery of the law was based on experimental work, it can be deduced rather easily for the special case of steady conditions of current and temperature. As a current flows through a conductor, one would expect the observed heat output to be accompanied by a loss in potential energy of the moving charges that constitute the current. This loss would result in a descending potential gradient along the conductor in the direction of the current flow, as usually defined. If E is the total potential drop, this loss, by definition, is equal to E in joules for every coulomb of charge that traverses the conductor. The loss conceivably might appear as heat, as a change in the internal energy of the conductor, as work done on the environment, or as some combination of these. The second is ruled out, however, because the temperature is constant and no physical or chemical change in a conductor as a result of current flow has ever been detected. The third is ruled out by hypothesis, leaving only the generation of heat. Therefore, $H = EI$ in joules per second, or watts. By definition, $R = E/I$, a ratio which has positive varying values. Elimination of E between these two equations gives the equation below,

$$H = RI^2$$

which is Joule's law as stated above. If I changes to a new steady value I', R to R', and H and H', then $H' = R'I'^2$ as before. The simplest case occurs where R is independent of I. If the current is varying, the resulting variations in temperature and internal energy undoubtedly exist and, strictly speaking, should be allowed for in the theory. Yet, in all but the most exceptional cases, any correction would be negligible.

This phenomenon is irreversible in the sense that a reversal of the current will not reverse the outflow of heat, a feature of paramount importance in many problems in physics and engineering. Thus the heat evolved by an alternating current is found by taking the time average of both sides of the equation. Incidentally, the changes in internal energy, if they were included in the theory, would average out. Hence the equation continues to have a similar form, $\overline{H} = \overline{RI^2}$, for ac applications. See OHM'S LAW.

[LLEWELLYN G. HOXTON/JOHN W. STEWART]

Junction detector

A device in which detection of radiation takes place in or near the depletion region of a reverse-biased semiconductor junction. The electrical output pulse is linearly proportional to the energy deposited in the junction depletion layer by the incident ionizing radiation. See CRYSTAL COUNTER; IONIZATION CHAMBER.

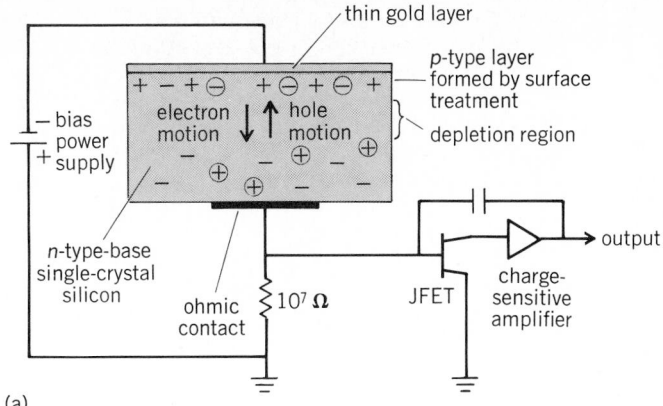

(a)

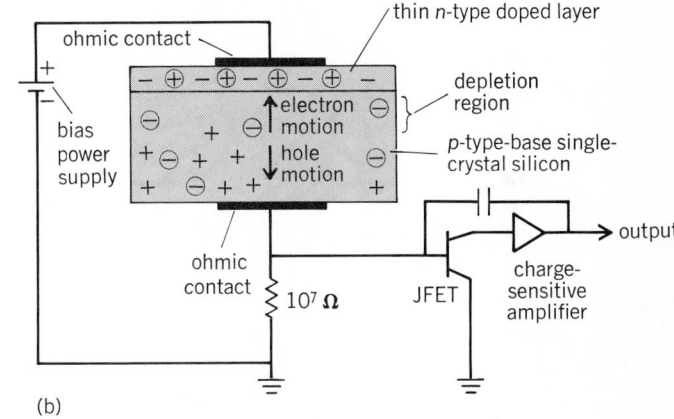

(b)

key:
⊖ p-type dopant ion
⊕ n-type dopant ion
− electron
+ hole

Fig. 1. Silicon junction detectors. (a) Surface barrier detector. (b) pn junction detector. The p-type dopant ions are fixed in the crystal lattice. JFET = junction field-effect transistor.

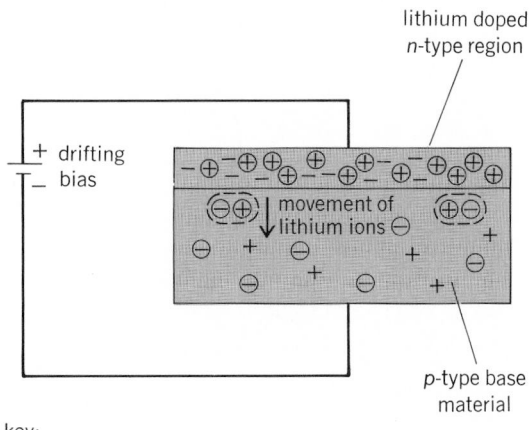

key:
⊖ boron ions   − electrons   ⊕⊖ lithium ion
⊕ lithium ions   + holes   compensating
                                            a boron ion

Fig. 2. Compensation of p-type semiconductor material with lithium (at 100–200°C for silicon, 40–60°C for germanium). Boron ions are fixed in the lattice. Lithium ions are fixed in the lattice, but at elevated temperature can be drifted under an electric field and will compensate boron ions to widen the depletion region.

Introduced into nuclear studies in 1958, the junction detector, or more generally, the nuclear semiconductor detector, revolutionized the field. In the detection of both charged particles and gamma radiation, these devices typically improved

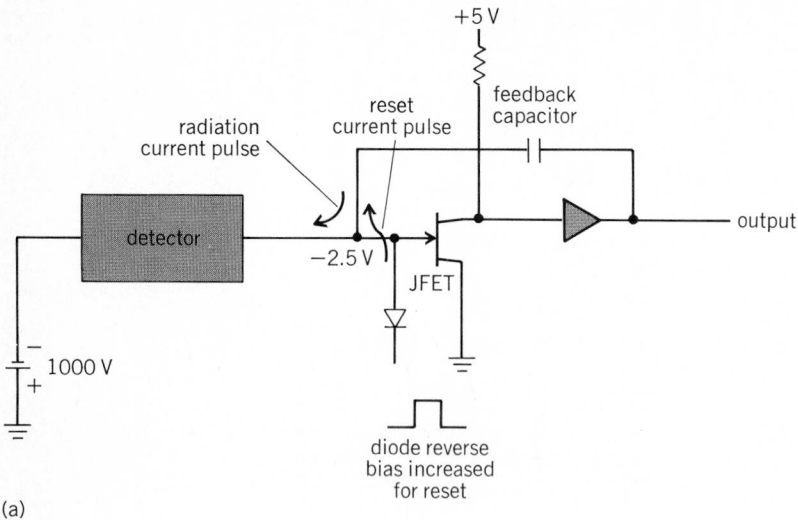

(a)

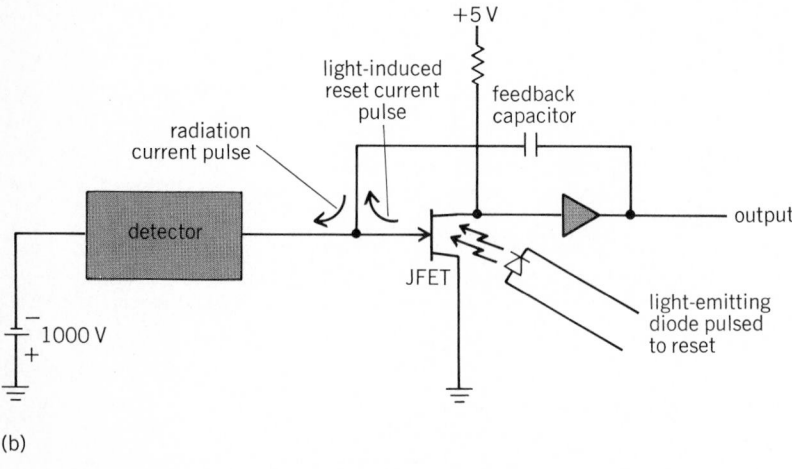

(b)

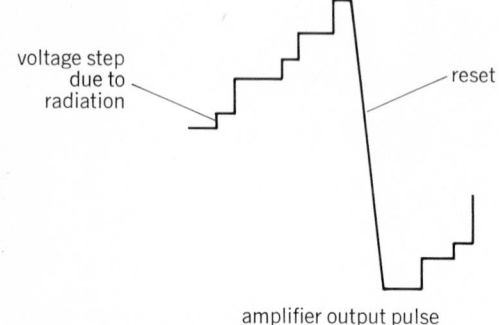

amplifier output pulse

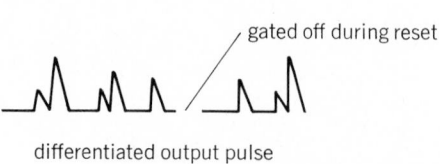

differentiated output pulse

(c)

Fig. 3.   Reset mechanisms for junction detectors. (*a*) Diode reset. (*b*) Optical reset. (*c*) Amplifier output and differentiated output for the pulse height analyzer.

experimentally attainable energy resolutions by about two orders of magnitude over that previously attainable. To this they added unprecedented flexibility of utilization, speed of response, miniaturization, freedom from deleterious effects of extraneous electromagnetic (and often nuclear) radiation fields, low-voltage requirements, and effectively perfect linearity of output response. They are now used for a wide variety of diverse applications, from logging drill holes for uranium to examining the Shroud of Turin. They are used for general analytical applications, giving both qualitative and quantitative analysis in the microprobe and the scanning transmission electron microscopes. They are used in medicine, biology, environmental studies, and the space program. In the last category they continue to play a very fundamental role, ranging from studies of the radiation fields in the solar system to the composition of extraterrestrial surfaces.

Fabrication of diodes. The first practical detectors were prepared by evaporating a very thin gold layer on a polished and etched wafer of *n*-type germanium (Ge). To reduce noise these devices were operated at liquid nitrogen temperature (77 K). Silicon (Si), however, with its larger band gap, 1.107 eV compared to 0.67 eV for germanium, offered the possibility of room-temperature operation. Gold-silicon surface barrier detectors and silicon *p-n* junction detectors were soon developed.

Surface barrier detectors are made from wafers of *n*-type silicon semiconductor crystals. The etching and surface treatments create a thin *p* layer, and the gold contacts this layer (Fig. 1*a*). The *pn* junction silicon detectors are usually made by diffusing phosphorus about 2 μm into the surface of a *p*-type silicon base (Fig. 1*b*). Both techniques give a *pn* junction. When this junction is reverse-biased, a depletion region, or a region devoid of carriers (electrons and holes), forms mainly in the higher-resistivity base material. A high field now exists in this region, and any carriers born or generated in it are rapidly swept from the region. The requirement for detection is that the ionizing radiation must lose its energy by creating electron-hole pairs (2.96 eV/pair in germanium and 3.66 eV/ pair in silicon) in the depletion region or within a carrier diffusion length of this region. Both carriers have to be collected to give an output pulse proportional to the energy of the incident particle. Electrons and holes have similar mobilities in both silicon and germanium, and although carrier trapping occurs it is not as severe as in the II–VI compounds.

Control of depletion region width. The detection of charged particles in the presence of gamma rays or higher-energy particles can be optimized by controlling the width of the depletion region. This width is a function of the reverse bias and of the resistivity of the base material. There is a practical limit to the voltage that can be applied to a junction. Thus detectors for higher-energy or lower-mass particles (electrons) requiring wider depletion regions are made from high-resistivity material. This material occurs, by accident, during the growth of some crystals.

Lithium-drifted silicon detectors. Still wider depletion-width detectors can be made from lithium-drifted silicon. Lithium (Li) is a donor in silicon. In addition, at elevated temperatures (200°C), the

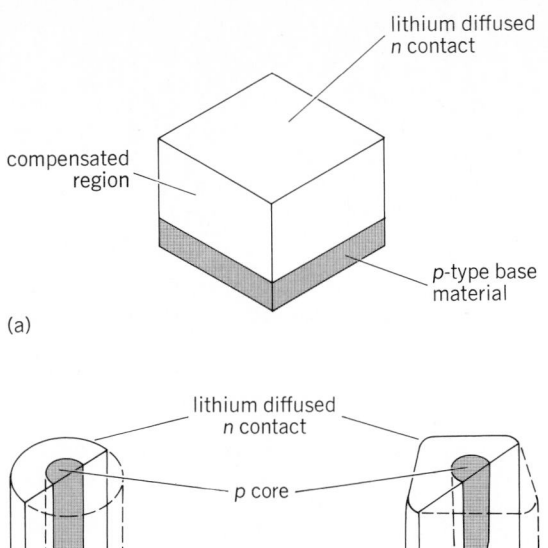

(a)

(b)   (c)

Fig. 4. Lithium-drifted detectors. (a) Planar. (b) Coaxial. (c) Open-one-end coaxial.

lithium ion is itself mobile. Thus when lithium is diffused in p-type silicon, a p-n junction results. Reverse-biasing this junction at elevated temperatures causes the lithium ion, now appearing as a positive charge, to migrate toward the negative side. On the way it encounters an acceptor ion, negatively charged, which is fixed in the crystal lattice. The lithium ion and the acceptor ion compensate each other, and the lithium ion remains in this location. As more lithium ions drift into and across the depletion region, they compensate the

acceptor ions and the region widens (Fig. 2). Depletion regions, or compensated regions, up to 2 cm wide have been achieved with this technique. Lithium-drifted silicon detectors can be operated at room temperature, but the larger volume gives a greater thermally generated leakage current, which degrades the resolution. The best energy resolution is obtained by operating the detectors at low temperature. However, they may be stored at room temperature.

Lithium-drifted silicon detectors are widely used to detect particle- or photon-induced x-rays. The resolution, when operated at 77 K, is sufficient to resolve the K x-rays for all elements higher in atomic number Z than carbon ($Z = 6$). A resolution of 100 eV has been obtained at 2 keV. At the lower x-ray energies the effects of the detector window thickness and the absorption in the window of the mounting are important, and silicon is preferred for these applications. For x-rays the efficiency of a 5-mm depletion-width lithium-drifted silicon detector is about 50% at 30 keV and 5% at 60 keV. Typically these detectors have capacitances of about 2 picofarads and, to minimize noise, are operated with an optical or diode reset mechanism rather than a feedback resistor (Fig. 3). The detector bias is about 1000 V, and the junction field-effect transistor (JFET) gate operates at about −2.5 V. A radiation event causes a pulse of current in the detector. The amplifier drives this current i through the feedback capacitor with capacitance C and in doing so steps a voltage an amount e_{step} proportional to the charge, as given by the equation below.

$$e_{step} = \frac{1}{C} \int i \, dt$$

Each subsequent radiation event causes a voltage step. To keep the amplifier within its dynamic range the feedback capacitor must be discharged. The analyzing circuits are first gated off, and in the diode case (Fig. 3a) the reverse bias on the diode is

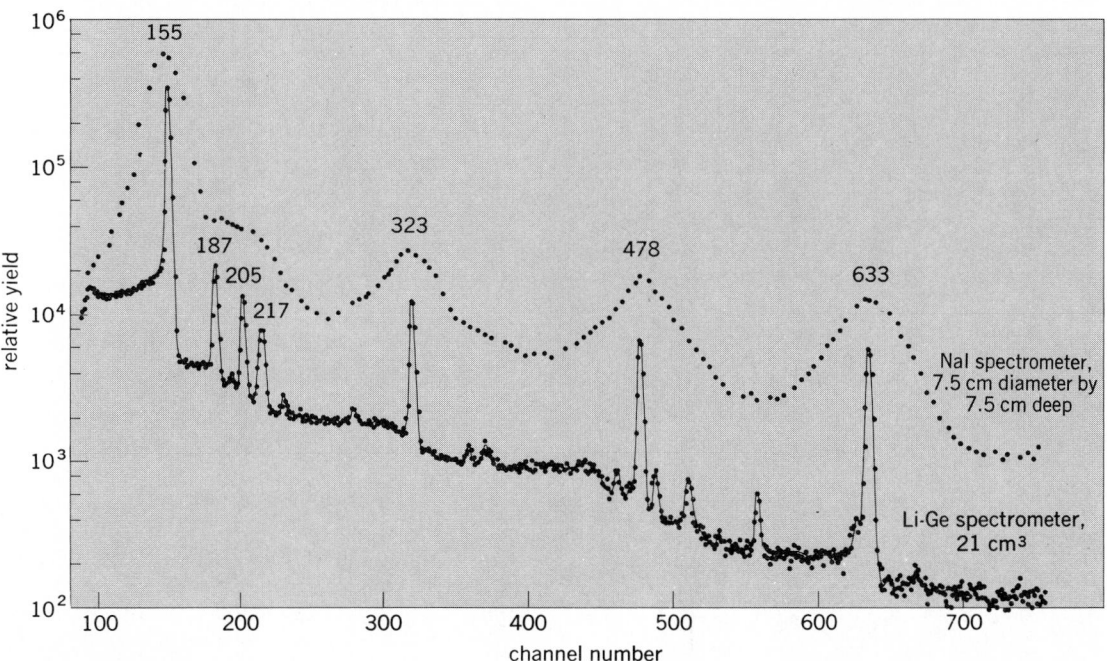

Fig. 5. Gamma-radiation spectra from [188]Os as detected in sodium iodide (NaI) and lithium-germanium (Li-Ge) spectrometers.

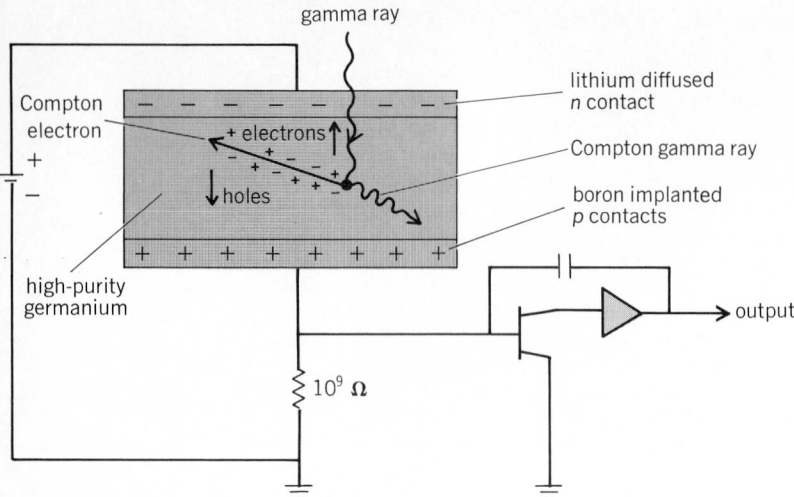

Fig. 6. High-purity germanium gamma-ray detector.

momentarily increased to give a picoampere current pulse. The amplifier output voltage changes to allow this current to flow through the feedback capacitor, discharging it. The analyzing circuit is now gated on, and counting can resume. For the optical reset (Fig. 3b), a light is flashed on the JFET, momentarily increasing the source-to-gate leakage current and discharging the feedback capacitor. The output from the amplifier and the differentiated output for the analyzer are shown in Fig. 3c.

Lithium-drifted germanium detectors. Germanium with its higher atomic number, 32 compared with 14 for silicon, has higher radiation absorption than silicon. Lithium may also be drifted in germanium. But in germanium, lithium is mobile at room temperature and will precipitate or diffuse further if the units, after fabrication, are not kept at liquid nitrogen temperature. Lithium-drifted germanium detectors revolutionized the field of gamma-ray spectroscopy. They may be manufactured in planar, coaxial, or open-one-end coaxial geometry (Fig. 4).

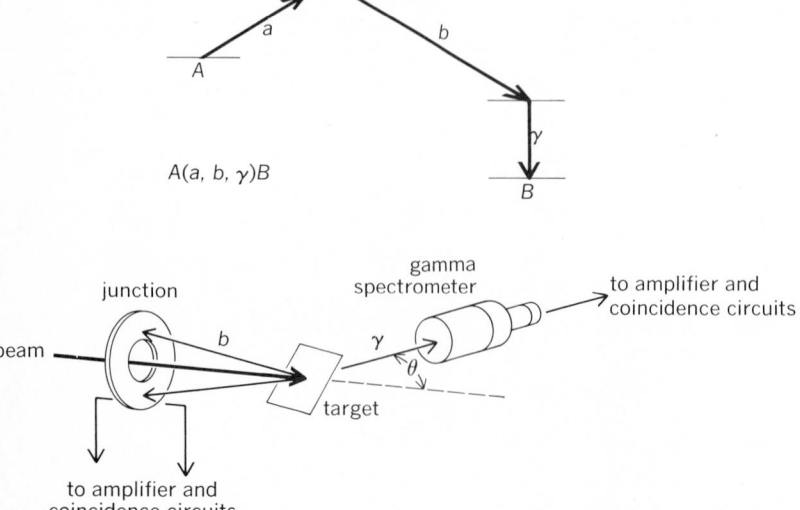

Fig. 7. Schematic of use of annular detector in nuclear reaction studies.

Figure 5 compares the gamma-ray spectrum of ^{188}Os taken with a 21-cm^3 lithium-germanium detector with that from a sodium iodide (NaI) scintillator-type spectrometer which is 7.5 cm in diameter by 7.5 cm deep (330 cm^3). The counting efficiency of the lithium-germanium detector is lower than the scintillator, but the resolution is at least an order of magnitude better. This higher resolution often reduces the actual counting time to adequately identify a particular energy peak even with an order-of-magnitude less sensitive volume. Also, as shown in Figure 5, the lithium-germanium detector is able to resolve more energy groups than the scintillator.

Hyperpure germanium detectors. Intrinsic or hyperpure germanium (Fig. 6) was grown to overcome the low-temperature-storage and the lithium-drifting problems associated with lithium-germanium. Planar detectors with up to a 2-cm-thick depletion region and coaxial detectors with 50 cm^3 volume have been made with the material. Low-temperature processing is used in the fabrication—usually lithium diffused at 280°C for the $n+$ contract and implanted boron for the $p+$ contract. This low-temperature processing is desirable to prevent diffusion of copper, with its subsequent charge trapping, into the germanium. Presently hyperpure germanium detectors cannot be made either as large as, nor with as high a resolution as, lithium-germanium detectors. Both types are operated at liquid nitrogen temperature, 77 K. However, the hyperpure germanium detector is easier to manufacture and can be stored at room temperature when not in use. This is a tremendous practical advantage.

Special detector configurations. Among the many other advantages of semiconductor detectors is the ease with which special detector configurations may be fabricated. One of the simple yet very important examples of this is the annular detector (Fig. 7), which is characteristically used to detect nuclear reaction products from a bombarded target in a tight cone around the incident beam. By examining the decay radiation in coincidence with such products, studies may be carried out only on residual nuclei which have had their spins very highly aligned in the nuclear reaction; this has been shown to provide an extremely powerful nuclear spectroscopic probe, and the annular detector is extensively used in laboratories throughout the world.

Composite detector systems are very readily assembled with the semiconductor devices. For example, it is standard in charged-particle detection to use a very thin detector and a very thick detector (or even two thin and one thick) in series. Multiplication of the resultant signals readily provides a characteristic identification signature for each nuclear particle species in addition to its energy. Three-crystal gamma-ray spectrometers are readily assembled, wherein only the output of the central detector is examined whenever it occurs in time coincidence with two correlated annihilation quanta escaping from the central detector. These systems essentially eliminate background from Compton scattering of other more complex electromagnetic interactions and yield sharp single peaks for each incident photon energy (Fig. 8).

Similarly neutrons may be indirectly detected through examination of recoil protons from a hy-

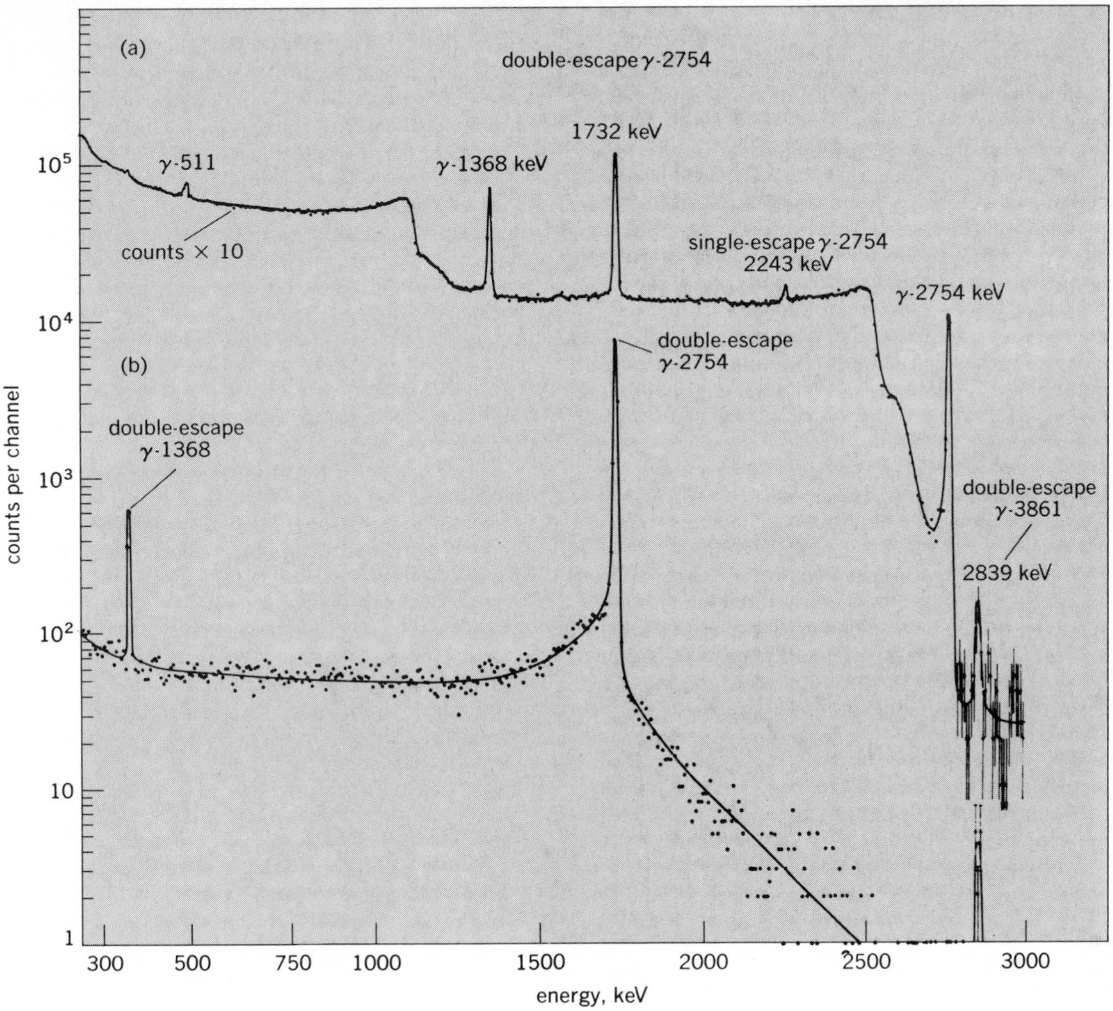

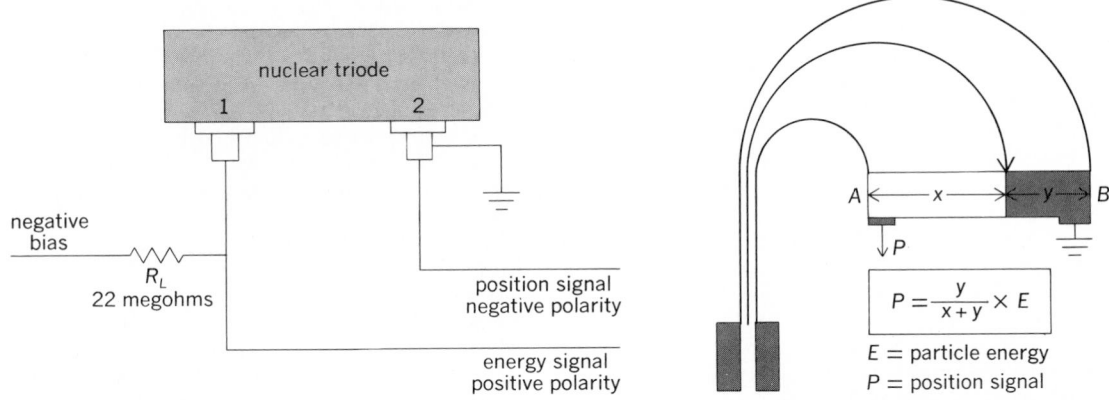

Fig. 8. Comparison of (a) direct single detector and (b) three-crystal spectrometer spectra from ²⁴Na source.

drogenous radiator in the case of high-energy neutrons, or through examination of fission fragments resulting from slow neutrons incident on a fissile converter foil mounted with the semiconductor detectors. (It should be noted that the response of the detectors is essentially perfectly linear all the way from electrons and photons to fission fragments.) Neutrons also may be detected and their energy spectra studied through examination of the charged products of the (nα) reaction (where alpha particles are emitted from incident neutrons) induced in the silicon or germanium base material of the detector itself.

Fabrication of triodes. Whereas the detectors thus far discussed are electrically nothing more than diodes, it has been possible to construct equivalent triodes which have extremely important uses in that they provide not only an output which

Fig. 9. Schematic of position-sensitive (nuclear triode) detector in focal plane of 180° magnetic spectrograph.

is linearly proportional to the energy deposited in them, but also a second output which in combination with the first establishes the precise location on the detector itself where the ionizing radiation was incident. This has very obvious advantages in the construction of simple systems for the measurement of angular distributions, where such position-sensitive detectors are located about a bombarded target. Their most important impact, however, has been in terms of their on-line use in the focal planes of large nuclear magnetic spectrographs. Simultaneous determination of the energy and location of a particle in the focal plane, together with the momentum determination by the magnet itself, establishes unambiguously both the mass and energy of the particle, and does so instantaneously so that additional logical constraints may be imposed through a connected on-line computer—something totally impossible with the earlier photographic plate focal-plane detectors (Figure 9).

A further important utilization of the nuclear triodes has followed their fabrication in an annular geometry similar to that shown in Fig. 7. With radial position sensitivity it becomes possible to correct on-line, and event by event, for the kinematic variation of particle energy with angle over the aperture of the detector. Without this correction possibility all particle group structures in the detector spectrum are smeared beyond recognition. *See* PARTICLE DETECTOR; SEMICONDUCTOR.

[JAMES M. MCKENZIE]

Bibliography: J. M. McKenzie, Development of the semiconductor radiation detector, pp. 49–73, G. T. Ewan, The solid ionization chamber, pp. 75–92, E. Laegsgaard, Position-sensitive semiconductor detectors, pp. 92–111, *Nucl. Instr. Meth.*, vol. 162, 1979.

Kapitza resistance

A resistance to the flow of heat across the interface between liquid helium and a solid. The Kapitza resistance R_K, is given by the equation below,

$$R_K = A\Delta T/\dot{Q} \qquad (\text{cm}^2\,\text{K/W})$$

where $\dot{Q}$ is the heat flow, A is the area of the interface, and ΔT is the temperature discontinuity across the interface. The inverse of this equation is known as the Kapitza conductance. The effect was discovered in 1941 by P. L. Kapitza, who showed that the temperature discontinuity occurred within a few micrometers of the interface, rather than in the bulk material.

The basic explanation treats the heat flow within each material as quantized elastic waves known as acoustic phonons. Because the density and sound velocity of liquid helium are much less than that of a solid, there is an acoustic mismatch at the surface that causes most of the acoustic phonons impinging on the interface to be reflected. This explanation predicts that R_K is proportional to T^{-3}, where T is the absolute temperature. Thus the Kapitza resistance becomes especially large at low temperatures, and usually can be detected only below about 2 K. *See* PHONON.

Experimentally it has been found that the Kapitza resistance is usually less than that predicted by the acoustic-mismatch model. Usually the con-

dition of the surface has a large effect on the value of the Kapitza resistance. The various theories that have been advanced to explain the reason for the lower resistance assume that other conductance mechanisms operate in parallel with the conductance calculated by the acoustic-mismatch model.

Kapitza resistance is an important problem in most experiments below 1 K, since it hampers temperature equilibrium between refrigerator, sample, and thermometer. It is often the dominant thermal resistance at these temperatures. From a theoretical standpoint, Kapitza resistance provides information on the interaction of low-energy phonons and other excitations with each other and with surfaces. *See* CONDUCTION (HEAT); LIQUID HELIUM; LOW-TEMPERATURE PHYSICS; PHONON; QUANTUM SOLIDS.

[R. RADEBAUGH]

Bibliography: L. J. Challis, Kapitza conductance, in J. G. M. Armitage and I. E. Farquhar (eds.), *The Helium Liquids*, 1975; G. L. Pollack, Kapitza resistance, *Rev. Mod. Phys.*, 41:48–81, 1969.

Karman vortex street

A double row of line vortices in a fluid. Under certain conditions a Karman vortex street is shed in the wake of bluff cylindrical bodies when the relative fluid velocity is perpendicular to the generators of the cylinder, as illustrated. This periodic

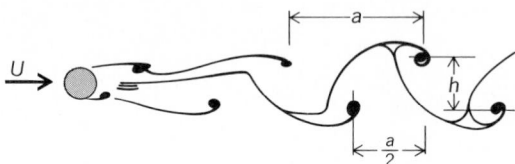

Karman vortex street.

shedding of eddies occurs first from one side of the body and then from the other, an unusual phenomenon because the oncoming flow may be perfectly steady. Vortex streets can often be seen, for example, in rivers downstream of the columns supporting a bridge. The streets have been studied most completely for circular cylinders at low subsonic flow speeds. Regular, perfectly periodic, eddy shedding occurs in the range of Reynolds number (Re) of 50–300, based on cylinder diameter. Above a Re of 300, a degree of randomness begins to occur in the shedding and becomes progressively greater as Re increases, until finally the wake is completely turbulent. The highest Re at which some slight periodicity is still present in the turbulent wake is about 10^6.

Vortex streets can be created by steady winds blowing past smokestacks, transmission lines, bridges, missiles about to be launched vertically, and pipelines aboveground in the desert. The streets give rise to oscillating lateral forces on the shedding body. If the vortex shedding frequency is near a natural vibration frequency of the body, the resonant response may cause structural damage. The aeolian tones, or singing of wires in a wind, is an example of forced oscillation due to a vortex street. T. von Kármán showed that an idealized infinitely long vortex street is stable to small disturbances if the spacing of the vortices is such that $h/a = 0.281$; actual spacings are close to this value. A complete and satisfying explanation of the formation of vortex streets has, however, not yet been given. For $10^3 < \mathrm{Re} < 10^5$ the shedding frequency f for a circular cylinder in low subsonic speed flow is given closely by $fd/U = .21$, where d is the cylinder diameter and U is stream speed; h/a is approximately 1.7. A. Roshko discovered a spanwise periodicity of vortex shedding on a circular cylinder at $\mathrm{Re} = 80$ of about 18 diameters; thus, it appears that the line vortices are not quite parallel to the cylinder axis. *See* FLUID-FLOW PRINCIPLES; VORTEX.

[ARTHUR E. BRYSON, JR.]

Bibliography: V. L. Streeter, *Handbook of Fluid Dynamics*, 1961; C.-S. Yih, *Fluid Mechanics*, rev. ed., 1979.

Kerr effect

Electrically induced birefringence that is proportional to the square of the electric field. When a substance (especially a liquid or a gas) is placed in an electric field, its molecules may become partly oriented. This renders the substance anisotropic and gives it birefringence, that is, the ability to refract light differently in two directions. This effect, which was discovered in 1875 by John Kerr, is called the electrooptical Kerr effect, or simply the Kerr effect.

When a liquid is placed in an electric field, it behaves optically like a uniaxial crystal with the optical axis parallel to the electric lines of force. The Kerr effect is usually observed by passing light between two capacitor plates inserted in a glass cell containing the liquid. Such a device is known as a Kerr cell. There are two principal indices of refraction, n_o and n_e (known as the ordinary and extraordinary indices), and the substance is called a positively or negatively birefringent substance, depending on whether $n_e - n_o$ is positive or negative.

Light passing through the medium normal to the electric lines of force (that is, parallel to the capacitor plates) is split into two linearly polarized waves traveling with the velocities c/n_o and c/n_e, respectively, where c is the velocity of light, and with the electric vector vibrating perpendicular and parallel to the lines of force.

The difference in propagation velocity causes a phase difference δ between the two waves, which, for monochromatic light of wavelength λ_0, is $\delta = (n_e - n_o)x/\lambda_0$, where x is the length of the light path in the medium.

Kerr constant. Kerr found empirically that $(n_e - n_o) = \lambda_0 B E^2$, where E is the electric field strength and B a constant characteristic of the material, called the Kerr constant. Havelock's law states that $B\lambda n/(n-1)^2 = k$, where n is the refractive index of the substance in the absence of the field and k is a constant characteristic of the substance but independent of the wavelength λ. Roughly speaking, the Kerr constant is inversely proportional to the absolute temperature. The phase difference δ is determined experimentally by standard optical techniques. If the wavelength λ_0 is expressed in centimeters, and the field strength E in statvolts/cm (1 statvolt = 300 volts), the Kerr constant for carbon disulfide,

which has been determined most accurately, is $B = 3.226 \times 10^{-7}$. Values of B range from -23.00×10^{-7} for paraldehyde to $+346.0 \times 10^{-7}$ for nitrobenzol.

The theory of the Kerr effect is based on the fact that individual molecules are not electrically isotropic but have permanent or induced electric dipoles. The electric field tends to orient these dipoles, while the normal agitation tends to destroy the orientation. The balance that is stuck depends on the size of the dipole moment, the magnitude of the electric field, and the temperature. This theory accounts well for the observed properties of the Kerr effect.

In certain crystals there may be an electrically induced birefringence that is proportional to the first power of the electric field. This is called the Pockels effect. In these crystals the Pockels effect usually overshadows the Kerr effect, which is nonetheless present. In crystals of cubic symmetry and in isotropic solids (such as glass) only the Kerr effect is present. In these substances the electrically induced birefringence (Kerr effect) must be carefully distinguished from that due to mechanical strains induced by the same field. *See* ELECTROOPTICS.

Kerr shutter. An optical Kerr shutter or Kerr cell consists of a cell containing a liquid (for example, nitrobenzene) placed between crossed polarizers. As such, its construction very much resembles that of a Pockels cell. With a Kerr cell (see illustration), an electric field is applied by means of an electronic driver v and capacitorlike electrodes in contact with the liquid; the field is perpendicular to the axis of light propagation and at 45° to the axis of either polarizer. In the absence of a field, the optical path through the crossed polarizers is opaque. When a field is applied, the liquid becomes birefringent, opening a path through the crossed polarizers. In commercial Kerr cell shutters, the electric field (a typical value is 10 kV/cm) is turned on and off in a matter of several nanoseconds (1 ns = 10^{-9} s). For laser-beam modulation the Pockels cell is preferred because it requires smaller voltage pulses. Kerr cell shutters have the advantage over the Pockels cell of a wider acceptance angle for the incoming light.

A so-called ac Kerr effect or optical Kerr effect has also been observed and put to use in connection with lasers. When a powerful plane-polarized

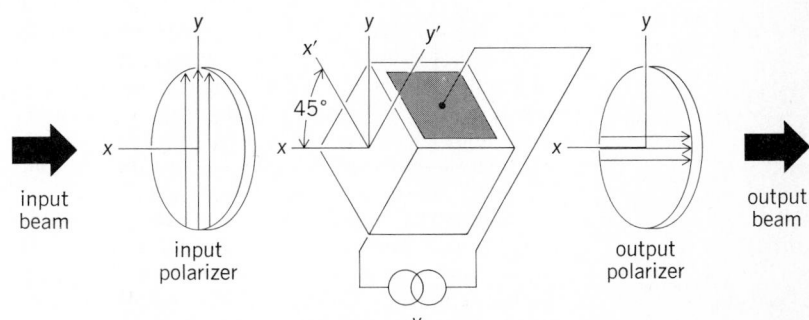

Kerr shutter. The arrows designate a light beam propagating through the cell. Polarizer axes are parallel to x and y axes; electric field of cell is parallel to x' or y' axis. (*Adapted from Amnon Yariv, Quantum Electronics, copyright © 1967 by John Wiley and Sons, Inc.; used with permission*)

laser beam propagates through a liquid, it induces a birefringence through a mechanism that is very similar to that of the ordinary, or dc, Kerr effect. In this case, it is the ac electric field of the laser beam (oscillating at a frequency of several hundred terahertz) which lines up the molecules. By using laser pulses with durations of only a few picoseconds (1 ps $= 10^{-12}$ s), extremely fast optical Kerr shutters have been built in the laboratory. *See* LASER.

[MICHEL A. DUGUAY]

Bibliography: A. Jariv, *Introduction to Optical Electronics*, 2d ed., 1976.

Kinematics

That branch of mechanics which deals with the motion of a system of material particles without reference to the forces which act on the system. Kinematics differs from dynamics in that the latter takes these forces into account. *See* DYNAMICS.

For a single particle moving in a straight line (rectilinear motion), the motion is prescribed when the position x of the particle is known as a function of the time. For uniformly accelerated motion, Eq. (1) holds, where a is the constant acceleration, and hence x is defined as in Eq. (2), where x_0 and

$$\frac{d^2x}{dt^2} = a \qquad (1)$$

$$x = \tfrac{1}{2}at^2 + v_0t + x_0 \qquad (2)$$

v_0 are the initial position and velocity of the particle, respectively. *See* ACCELERATION; RECTILINEAR MOTION; VELOCITY.

Plane kinematics of a particle is concerned with the specification of the position of a particle moving in a plane by means of two independent variables, usually the rectangular cartesian coordinates x and y, but often the polar coordinates $r = (x^2 + y^2)^{1/2}$, $\theta = \tan^{-1}(y/x)$, or even other coordinates that may be especially convenient for a particular problem. Here the x and y components of velocity are dx/dt, dy/dt, and of acceleration d^2x/dt^2, d^2y/dt^2, respectively, but the components of velocity in the directions of increasing r and increasing θ are dr/dt and $r(d\theta/dt)$, and are therefore not so simply given in terms of the time derivatives of the coordinates.

The kinematics of a particle in space is concerned with the ways in which three independent coordinates may be chosen to specify the position of the particle at a given time, and with the relations between the first and second time derivatives of these coordinates and the components of velocity and acceleration of the particle. The motion is specified if three such coordinates are given as functions of the time, and the path of the particle in space is then obtained by eliminating the time between these equations.

For describing the simultaneous position in space of N particles, $3N$ coordinates are necessary, and the configuration of the system may be represented by a point in a $3N$-dimensional space called the configuration space of the system. As the particles move, the corresponding representative point in configuration space traces out a curve called the trajectory of the system. If the particles do not move independently but are subject to m constraints, that is, if m relations between the $3N$ coordinates continue to hold throughout the motion, then it is possible to represent the trajectory in a space of $f = (3N - m)$ dimensions, f being the number of degrees of freedom of the system. For a rigid body, even though N and m are very large, $f = 6$, and the configuration at any time may be represented by six independent variables (usually three to represent the position of the center of mass and three angles to specify the orientation of the body). Kinematics describes the relations between the time derivatives of such angles and the components of the angular velocity of the body. *See* CAYLEY-KLEIN PARAMETERS; CONSTRAINT; DEGREE OF FREEDOM; EULER ANGLES.

Among the coordinate systems studied in kinematics are those used by observers who are in relative motion. In nonrelativistic kinematics the time coordinate for each such observer is assumed to be the same, but in relativistic kinematics proper account must be taken of the fact that lengths and time intervals appear different to observers moving relative to each other. *See* RELATIVITY.

[H. C. CORBEN/BERNARD GOODMAN]

Kinetic theory of matter

A theory which states that the particles of matter in all states of aggregation are in vigorous motion. In computations involving kinetic theory, the methods of statistical mechanics are applied to specific physical systems. The atomistic or molecular structure of the system involved is assumed, and the system is then described in terms of appropriate distribution functions. The main purpose of kinetic theory is to deduce, from the statistical description, results valid for the whole system. The distinction between kinetic theory and statistical mechanics is thus of necessity arbitrary and vague. Historically, kinetic theory is the oldest statistical discipline. Today a kinetic calculation refers to any calculation in which probability methods, models, or distribution functions are involved.

For information which is related to and supplements the present article *see* STATISTICAL MECHANICS. *See also* BOLTZMANN STATISTICS; BOLTZMANN TRANSPORT EQUATION; QUANTUM STATISTICS.

Classes of problems. Kinetic calculations are not restricted to gases, but occur in chemical problems, solid-state problems, and problems in radiation theory. Even though the general procedures in these different areas are similar, there are a sufficient number of important differences to make a general classification useful.

Classical ideal equilibrium problems. In these, there are no interactions between the constituents of the system. The system is in equilibrium, and the mechanical laws governing the system are classical. The basic information is contained in the Boltzmann distribution f (also called Maxwell or Maxwell-Boltzmann distribution) which gives the number of particles in a given momentum and positional range ($d^3x = dx\,dy\,dz$, $d^3v = dv_x\,dv_y\,dv_z$, where x, y, and z are coordinates of position, and v_x, v_y, and v_z are coordinates of velocity). In Eq. (1)

$$f(xyz, v_xv_yv_z) = A\,e^{-\beta\epsilon} \qquad (1)$$

ϵ is the energy, $\beta = 1/kT$ (where k is the Boltzmann constant and T is the absolute temperature), and A

is a constant determined from Eq. (2). The calcula-

$$\int \int d^3x \, d^3v \, f = N \qquad \text{total number of particles} \quad (2)$$

tions of gas pressure, specific heat, and the classical equipartition theorem are all based on these relations.

Classical ideal nonequilibrium problems. Many important physical properties refer not to equilibrium but to nonequilibrium states. Phenomena such as thermal conductivity, viscosity, and electrical conductivity all require a discussion starting from the Boltzmann transport equation, Eq. (3a), where $\mathbf{X}$ is the force per unit mass, ∇ is the gradient on the space-dependence, ∇_v is the gradient on the velocity-dependence, and $(\partial f/\partial t)_{\text{coll}}$ is the change due to collisions. If one deals with states that are near equilibrium, the exact Boltzmann equation need not be solved; then it is sufficient to describe the nonstationary situation as a small perturbation superimposed on an equilibrium state. Even though the rigorous discussion of the nonequilibrium processes is difficult, appeal to simple physical pictures frequently leads to quite tractable expressions in terms of the equilibrium distribution function. Of special importance is the example of the so-called electron gas in a metal. The kinetic treatment of this system forms the basis for the classical (Lorentz) conductivity theory. For states far from equilibrium, no general simple theory exists.

Classical nonideal equilibrium theory. The basic classical procedure for arbitrary systems (systems with interactions taken into account) that allows the calculation of macroscopic entities is that using the partition function, as shown in Eq. (3b), where $\lambda = h/\sqrt{2\pi mkT}$. Here Ψ is the thermo-

$$\frac{\partial f}{\partial t} + \mathbf{v} \cdot \nabla f + \mathbf{X} \cdot \nabla_v f = \left(\frac{\partial f}{\partial t}\right)_{\text{coll}} \quad (3a)$$

$$Z_{cl} = e^{-\psi/kT} = \frac{1}{N! \lambda^{3N}} \int \cdots \int d^3x_1$$
$$\cdots d^3x_N e^{-(1/kT)U(x_1, \ldots, zN)} \quad (3b)$$

dynamic free energy, and h is Planck's constant. Although Eq. (3b) is written so that it may be applied to gases, the partition function may also be written for classical spin systems, such as ferromagnetic and paramagnetic solids. The mathematical problems of evaluating the integrals or the sums are difficult. Equation (3b), with appropriate modifications, is the starting point for all these considerations.

Classical nonideal nonequilibrium theory. This is the most general situation that classical statistics can describe. In general, very little is known about such systems. The Liouville equation applies, and has been used as a starting point for these studies, but no spectacular results have yet been obtained. For studies of the liquid state, however, the results have been quite promising.

Quantum problems. There are quantum counterparts to the classifications just described. In a quantum treatment a distribution function is also used for an ideal system in equilibrium to describe its general properties. For systems of particles which must be described by symmetrical wave functions, such as helium atoms and photons, one has the Bose distribution, Eq. (4), where $\beta = 1/kT$,

$$f(v_x v_y v_z) = \frac{1}{(1/A)e^{\beta\epsilon} - 1} \quad (4)$$

and A is determined by Eq. (2). *See* BOSE-EINSTEIN STATISTICS.

For systems of particles which must be described by antisymmetrical wave functions, such as electrons, protons, and neutrons, one has the Fermi distribution, Eq. (5). Use of these functions

$$f(v_x v_y v_z) = \frac{1}{(1/A)e^{\beta\epsilon} + 1} \quad (5)$$

in calculations is actually not very different from the use of the classical distribution function, but the results are quite different, as are the analytical details. As in the classical case, the treatment of the nonequilibrium state can be reduced to a treatment involving the equilibrium distribution only. The application to electrons as an (ideal) Fermi-Dirac gas in a metal is the basis of the Sommerfeld theory of metals. *See* FERMI-DIRAC STATISTICS; FREE-ELECTRON THEORY OF METALS.

In quantum theory, nonideal systems in equilibrium are described in terms of the quantum partition function, Eq. (6). Here E_n indicates the energy

$$Z_g = \sum_n g_n e^{-E_n/kT} = e^{-\psi/kT} \quad (6)$$

levels of the system, and g_n indicates the weights of these levels. If one defines a Slater sum by Eq. (7), where $U_n(x_1, \ldots, x_N)$ is the wave function of

$$S(x_1, \ldots, x_N) = \sum_n e^{-E_n/kT} |U_n(x_1, \ldots, x_n)|^2 \quad (7)$$

the state n, Z_g may be written as an integral similar to Z, as in Eq. (8). For the applications, the energy

$$Z_g = \int \cdots \int d^3x_1 \cdots d^3x_N S(x_1, \ldots, x_N) \quad (8)$$

levels and the wave functions must be known. In the evaluation of S, given by Eq. (7), the symmetry character of the wave functions must be explicitly introduced. It is sometimes easier to use the grand partition function, Eq. (9). Here μ is the chemical

$$Z_{q \cdot m \cdot gr} = \sum_N \sum_n e^{(\mu N - E_{N,n})/kT} \quad (9)$$

potential, and $E_{N,n}$ is the nth level of a system having N particles. The current theories of quantum statistics, as for example the hard-sphere Bose gas, are for the most part concerned with questions in this area.

The only technique now available is that of the density matrix. It is possible to express certain entities which characterize transport properties, such as the conductivity tensor, in terms of the unperturbed stationary density matrix. A complete discussion of the validity of the approximations is still lacking. In addition, once the conductivity tensor is obtained in terms of the density matrix of the stationary (but still interacting) system, a problem of the same order of difficulty as the evaluation of the quantum mechanical partition function remains, if explicit expressions for these quantities in terms of the forces between atoms are desired.

Classical examples. Kinetic theory gave the first insight into many of the phenomena that take place in gases as well as in metals, where the free (conduction) electrons can be considered as an ideal gas of electrons. The following examples illustrate some of the more fundamental calculations that have been made.

Ideal-gas pressure. A classical ideal gas is described by the Boltzmann distribution of Eq. (1). The constant A is given by Eq. (10), where m is the

$$A = \frac{N}{V}\left(\frac{m\beta}{2\pi}\right)^{3/2} \qquad (10)$$

mass of an individual molecule and V is the total volume of the gas. A gas exerts a force on the wall by virtue of the fact that the molecules are reflected by it. The component of momentum normal to the wall changes its sign as a consequence of this collision. Hence if the normal velocity of the molecule is v_n, the momentum given off to the wall is $2mv_n$. To calculate the total force on a wall, one needs the total momentum transferred to the wall per unit time. Let dS be a small section of the wall, and call θ the angle made by the molecule's velocity vector with the normal to dS (Fig. 1). From Eq. (1), the number of molecules per unit volume that have a speed $c = (v_x{}^2 + v_y{}^2 + v_z{}^2)^{1/2}$ and whose velocity vector makes an angle θ with a given axis (call these c,θ molecules) is given by Eq. (11a). The number of such molecules that in time dt will collide with dS (assuming spatial homogeneity) is given by notation (11b). From this information an interesting side result may be calculated, namely,

$$f(c,\theta) = 2\pi A e^{(-1/2)\beta mc^2} c^2 \sin\theta \, d\theta dc \qquad (11a)$$

$$dS \, c \cos\theta \, dt \, f(c,\theta) \qquad (11b)$$

teresting side result may be calculated, namely, that the number of all collisions with a unit area of the wall per second may be written as expression (12a). If one introduces the average speed $\bar{c}$, defined in Eq. (12b), the total number of collisions

$$2\pi A \int_0^{\pi/2} d\theta \sin\theta \cos\theta \int_0^\infty dc \, c^3 e^{-(1/2)\beta mc^2} \qquad (12a)$$

$$\bar{c} = \frac{1}{N} \int\int d^3x \, d^3v \, c \, f \qquad (12b)$$

may be written as $1/4 \, n\bar{c}$. Here $n = N/V$, the number density. This relation is of importance in calcu-

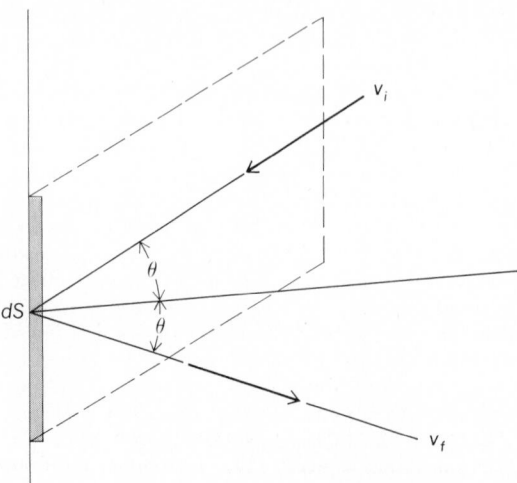

Fig. 1. Change of momentum of gas molecule as it strikes a wall; v_i, initial velocity; v_f, final velocity.

lating the efflux of gases through orifices. Each (c,θ) molecule contributes a momentum of $2mc \cos\theta$ to the wall per collision. Since the pressure p is given by the force per unit area, or the total momentum transferred per unit time per unit area, one obtains for p Eqs. (13) and (14). Since it is

$$p = \int_0^\infty \int_0^{\pi/2} 2mc \cos\theta$$
$$\cdot 2\pi A c \cos\theta \, e^{-(1/2)\beta mc^2} c^2 \sin\theta \, d\theta \, dc \qquad (13)$$

$$p = \tfrac{1}{3} N m \overline{c^2} = (N/V)\beta^{-1} \qquad (14)$$

known experimentally that $pV = NkT$, where k is the Boltzmann constant, $\beta = 1/kT$ is now identified. This pressure calculation is a typical example of a kinetic theory calculation. *See* GAS.

Equipartition theorem and specific heat. In a classical ideal gas, the average kinetic energy associated with any translational degree of freedom is expressed by Eq. (15). It is quite straightforward

$$\tfrac{1}{2}m\overline{v_x{}^2} = \tfrac{1}{2}m\overline{v_y{}^2} = \tfrac{1}{2}m\overline{v_z{}^2} = \frac{1}{2\beta} = \tfrac{1}{2}kT \qquad (15)$$

to show that the average energy associated with any degree of freedom, which occurs as a quadratic term in the expression for the mechanical energy, is given by $1/2 \, kT$. Formally, the result one proves is that, for a Boltzmann distribution written in terms of momenta and generalized coordinates p_i and q_i, one has Eq. (16), which is called the equi-

$$\overline{q_i \frac{\partial\epsilon}{\partial q_i}} = \overline{p_i \frac{\partial\epsilon}{\partial p_i}} = kT \qquad (16)$$

partition theorem. From the equipartition theorem the specific heat C, defined by Eq. (17), may be immediately obtained.

$$C = \frac{\partial\overline{E}}{\partial T} \qquad (17)$$

If one has just the three translational degrees of freedom, the average energy per degree of freedom per molecule is $1/2 \, kT$, hence the average energy $\overline{E} = 3/2 \, NkT$ and $C = 3/2 \, R$. The specific heat is thus constant and independent of T. In the case of a diatomic molecule, one usually has three translational and two rotational degrees of freedom. Therefore $\overline{E} = 5/2 \, NkT$, and $C = 5/2 \, Nk \equiv 5/2 \, R$. Suppose there are N atoms in a solid, each bound by elastic forces to a center. In that case, the mechanical energy will be expressed by Eq. (18), where k_F is the elastic force instant. Accord-

$$\epsilon = \frac{1}{2m}(p_x{}^2 + p_y{}^2 + p_z{}^2) + \frac{1}{2}k_F{}^2(x^2 + y^2 + z^2) \qquad (18)$$

ing to the equipartition theorem, these six terms in ϵ will give a specific heat $C = 3 \, Nk$, the Dulong and Petit value for a monatomic solid. *See* SPECIFIC HEAT OF SOLIDS.

The equipartition theorem is especially useful when a gas is at a temperature so high that it becomes necessary to use relativistic mechanics to describe the system properly. *See* RELATIVISTIC MECHANICS.

This happens at temperatures for which the thermal energy kT is of the same order as mc^2, where c denotes the speed of light. The appropriate relation between energy and momentum is then given by Eq. (19a).

The equipartition relation (16) may then be written as Eq. (19b). From this relation one can obtain the relativistic correction to the average energy of a gas, as shown in Eq. (19c).

$$\epsilon = c[(p_x^2 + p_y^2 + p_z^2) + m^2c^2]^{1/2} \quad (19a)$$

$$\overline{\frac{c^2 p_x^2}{\epsilon}} = kT \quad (19b)$$

$$\overline{E} = {}^3\!/_2 NkT \left(1 + \frac{5}{4}\frac{kT}{mc^2} + \cdots\right) \quad (19c)$$

Electrical conductivity of metals. Experimentally, one observes a proportionality between the applied electric field and the current produced in a metallic conductor. Kinetic procedures give an explanation of this general connection. Consider the electrons in a metal as an ideal gas of electrons of mass m and charge e. In equilibrium, the electrons are described by the Boltzmann distribution, Eq. (1), or the Fermi-Dirac distribution, Eq. (5). Call this distribution f. Because f depends on $v_x^2 + v_y^2 + v_z^2$ only, it is clear that $\bar{v}_x = 0$; no net current can flow in an equilibrium state. The application of an electric field therefore results in the destruction of the spherical symmetry in the velocities. This effect is described by the Boltzmann transport equation, Eq. (20).

$$\frac{\partial f}{\partial t} + \mathbf{v}\cdot\nabla f + \mathbf{X}\cdot\nabla_v f = \left(\frac{\partial f}{\partial t}\right)_{\text{coll}} \quad (20)$$

The collision term causes considerable difficulty. In many problems one is justified in introducing a relaxation time τ which may depend on x and v and which is defined by Eq. (21). Here f_0

$$\left(\frac{\partial f}{\partial t}\right)_{\text{coll}} = -\frac{f - f_0}{\tau} \quad (21)$$

is the distribution function at thermal equilibrium. The introduction of such a relaxation time presupposes that a nonequilibrium state will decay exponentially into an equilibrium state as a consequence of the action of the collisions. This is undoubtedly a good approximation near equilibrium.

Assume the existence of an electric field in the x direction which distorts the initial distribution a small amount. By using Eq. (21), Eq. (20) may be written as Eq. (22). Here E_e is the external electric

$$\frac{eE_e}{m}\frac{\partial f}{\partial v_x} + v_x\frac{\partial f}{\partial x} = -\frac{f - f_0}{\tau} \quad (22)$$

field. If it is now assumed that $(f - f_0)/f_0 \ll 1$, with f very near an equilibrium state, so that quadratic terms may be neglected, and that the parameters A and β in f_0 are independent of x, then an explicit expression for f can be obtained, as shown in Eq. (23).

$$f = f_0 - \tau e E_e v_x \frac{\partial f_0}{\partial \epsilon} \quad (23)$$

The electric current in the x-direction is always given by Eq. (24).

$$j_x = \int e v_x f\, d^3v = -\tau e^2 E_e \int v_x^2 \frac{\partial F_0}{\partial \epsilon} d^3v \quad (24)$$

The first term in Eq. (23) does not contribute to the current, as has already been pointed out. This

is an example of a formal result in transport theory. The result for j_x is in a form in which only a knowledge of the equilibrium distribution is required to obtain an explicit answer.

In the case of Boltzmann statistics, use of Eqs. (1), (2), (10), and (24) gives Eq. (25a). The conductivity $\sigma = Ne^2\tau/m$ cannot be compared with experiment unless the relaxation time τ is known. In the Fermi case the evaluation of Eq. (24) is facilitated by observing that $\partial f_0/\partial\epsilon$ has a δ-function character. Call $A = e^{\beta\mu}$, so that Eq. (5) reads as Eq. (25b).

$$j_x = \frac{Ne^2\tau}{m}E_e \quad (25a)$$

$$f_0 = \frac{1}{e^{\beta(\epsilon - \mu)} + 1} \quad (25b)$$

It is easy to show from Eq. (25b) that expression (26) holds for sufficiently low temperatures. This

$$\frac{\partial}{\partial \epsilon}\frac{1}{e^{\beta(\epsilon - \mu)} + 1} \cong -\delta(\epsilon - \mu) \quad (26)$$

allows the immediate calculation of Eq. (24) also for the case in which the relaxation time depends on the velocity.

Viscosity and mean free path. One of the early successes of kinetic theory was the explanation of the viscosity of a gas. Strictly speaking, this is again a transport property, and as such it should be obtained from the Boltzmann transport equation, Eq. (20). It is possible, however, to give an elementary discussion. Consider a gas that is contained between two walls or plates, the lower one ($y = 0$) at rest and the upper one constrained to move with a given velocity in the x direction (Fig. 2). A force is necessary to maintain the constant velocity of the plate. This force is given by Eq. (27). Here dv_x/dy

$$X_x = \eta A\frac{dv_x}{dy} \quad (27)$$

is the velocity gradient, and $\mathbf{X}$ is the viscous force on the area A, which is perpendicular to the y-axis ($\mathbf{X}_x/A$ is sometimes called the shear stress). Equation (27) defines η, the viscosity coefficient. The physical reason for this force stems from the fact that molecules above the surface S have a greater

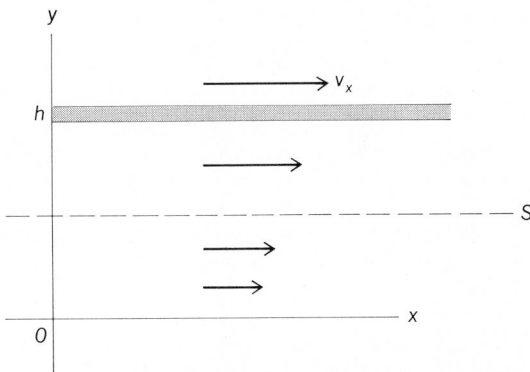

Fig. 2. Explanation of viscosity. The plate at $y = h$ is moving with velocity v_x, and the plate at $y = 0$ is stationary. The horizontal component of velocity of the gas molecules varies from 0 at $y = 0$ to v_x at $y = h$. Momentum therefore must be transferred across surface S by the vertical component of velocity.

flow velocity than those below this surface. (This will certainly be true on the average.) Thus molecules crossing from above to below will carry a larger amount of momentum in the positive x direction than those crossing from below S upwards. Hence the net effect is a transport of momentum in the x direction across the surface. By Newton's second law, this will yield a force. The computation can be carried out in this manner. Consider an area in the xz plane. The number of molecules passing through per second is $fv_y dS$, where f is the distribution function.

The amount of momentum transported in the x direction is (per collision) $fv_y dS \cdot mv_x$.

The force per unit area is the sum of these terms, given by expression (28). For the evaluation

$$fv_y v_x m \qquad (28)$$

of this sum, the notion of mean free path is useful. The mean free path is the average distance traveled by a molecule between collisions, and is usually designated by λ. To investigate this entity, imagine each molecule to be a hard sphere with radius a. If a molecule moves with average speed $\bar{c}$, it sweeps out a volume $4\pi a^2 \bar{c} t$ in time t. If there are $n = N/V$ molecules per unit volume, the number of collisions per second is given by the collision frequency z in Eq. (29).

$$z = n4\pi a^2 \bar{c} \qquad (29)$$

For a typical gas (oxygen) under standard conditions $n = 3 \times 10^{25}$, $\bar{c} = 4.5 \times 10^4$ cm/sec, and $a \cong 1.8 \times 10^{-12}$ cm. Hence, numerically, $z = 5.5 \times 10^9$ collisions/sec. The average distance between collisions, that is, the mean free path, is given by Eq. (30a). Numerically, $\lambda \cong 8 \times 10^{-6}$ cm. This discussion is, of course, exceedingly crude. Making the calculation on the basis of a Boltzmann distribution gives Eq. (30b).

$$\lambda = \frac{\bar{c}t}{n(4\pi a^2)\,\bar{c}t} = \frac{1}{n(4\pi a^2)} \qquad (30a)$$

$$\lambda = \frac{1}{\sqrt{2}\,n(4\pi a^2)} \qquad (30b)$$

Using similar methods, it may be shown that the distribution of the mean free paths (that is, the number of molecules whose mean free path lies between x and $x + dx$) is given by Eq. (31).

$$dN = \frac{N_0}{\lambda} e^{-x/\lambda}\, dx \qquad (31)$$

There is an interesting connection between the mean free path and the relaxation time introduced previously. It should be stressed, however, that this connection follows more from a qualitative discussion than from a rigorous calculation. One would guess that Eq. (32) is valid. This means that

$$\tau = \lambda/\bar{c} \qquad (32)$$

a relaxation time describes the decay from a state so near an equilibrium state that, when the molecules have traveled (on the average) one mean free path, the equilibrium is reestablished. Stated differently, the nonequilibrium state is, on the average, one collision per molecule removed from the equilibrium state.

Expression (28) may now be evaluated in terms of the mean free path as shown in Eq. (33a). From this, the viscosity coefficient follows directly, as shown in Eq. (33b). Introducing the mean free path, as given by Eq. (30a), one sees that Eq. (33c)

$$\text{Force per unit area} = \frac{1}{3}nm\lambda\bar{c}\frac{dv_x}{dy} \qquad (33a)$$

$$\eta = \frac{1}{3}nm\bar{c}\lambda \qquad (33b)$$

$$\eta = \frac{1}{3}\frac{m\bar{c}}{4\pi a^2} \qquad (33c)$$

holds. The remarkable result is that the viscosity is indeed independent of the pressure. Since Eq. (34) holds, the viscosity depends on the tempera-

$$\bar{c} = \sqrt{8kT/\pi m} \qquad (34)$$

ture but not on the pressure. This somewhat unintuitive result was one of the first triumphs of kinetic theory. Both aspects of Eq. (33c) are in good agreement with experiment; for not too low densities η is indeed independent of the pressure; η is also proportional to the square root of the temperature.

[MAX DRESDEN]

Bibliography: R. H. Fowler, *Statistical Mechanics*, 2d ed., 1936; C. Kittel, *Elementary Statistical Physics*, 1958; P. Resibois and M. F. DeLeener, *Classical Kinetic Theory of Fluids*, 1977; F. W. Sears and G. L. Salinger, *Thermodynamics, the Kinetic Theory of Gases and Statistical Mechanics*, 3d ed., 1975.

Kinetics

That part of classical mechanics which deals with the relation between the motions of material bodies and the forces acting upon them. It is synonymous with dynamics of material bodies. *See* DYNAMICS.

Basic concepts. Kinetics proceeds by adopting certain intuitively acceptable concepts which are associated with measurable quantities. These essential concepts and the measurable quantities used for their specification are as follows:

1. Space configuration refers to the positions and orientations of bodies in a reference frame adopted by the observer. It is expressed quantitatively by an arbitrarily chosen set of space coordinates, of which cartesian and polar coordinates are examples. All space coordinates rest on the motion of distance measurement.

2. Duration is expressed quantitatively by time measured by a clock or comparable mechanism.

3. Motion refers to change of configuration with time and is expressed by time rates of coordinate change called velocities and time rates of velocity change called accelerations. The classical assumption that coordinates behave as analytic functions of time permits representation of velocities and accelerations as first and second derivatives, respectively, of the space coordinates with respect to time.

4. Inertia is an attribute of bodies implying their capacity to resist changes of motion. A body's inertia with respect to linear motion is denoted by its mass.

5. Momentum is an attribute proportional to both the mass and velocity of a body. Momentum

of linear motion is expressed as the product of mass and linear velocity.

6. Force serves to designate the influence exercised upon the motion of a particular body by other bodies, not necessarily specified. A quantitative connection between the motion of a body and the force applied to it is expressed by Newton's second law of motion, which is discussed later.

Distance, time, and mass are commonly regarded as fundamental, all other dynamical quantities being definable in terms of them.

Newton's second law. A primary objective of classical kinetics is the prediction of the behavior of bodies which are subject to known forces when only initial values of the coordinates and momenta are available. This is accomplished by use of a principle first recognized by Isaac Newton. Newton's statement of the principle was restricted to the linear motion of an idealized body called a mass particle, having negligible extension in space.

The basic dynamical law set forth by Newton and known as his second law states that the time rate of change of a particle's linear momentum is proportional to and in the direction of the force applied to the particle. This statement, although special in form, serves as a basis for more comprehensive statements of the principle which have since appeared.

Stated analytically, Newton's second law becomes the differential equation, Eq. (1), in which m

$$\frac{d\,(mv)}{dt} = F \tag{1}$$

represents the particle's mass, v its velocity, F the applied force, and t the time. Equation (1) provides a definition of force and of its units if units of mass, distance, and time have previously been adopted. The classical assumption of constancy of mass permits Eq. (1) to be expressed as Eq. (2), where a

$$ma = F \tag{2}$$

represents the linear acceleration. A particle in physical space requires three cartesian coordinates, x, y, and z, to specify its position. Its linear acceleration is a vector with three cartesian components, the second time derivatives of x, y, and z. Equation (2) therefore equates two vectors, requiring equality of their components expressed in detail by Eqs. (3a)–(3c). These are the Newtonian

$$m\,\frac{d^2x}{dt^2} = F_x \tag{3a}$$

$$m\,\frac{d^2y}{dt^2} = F_y \tag{3b}$$

$$m\,\frac{d^2z}{dt^2} = F_z \tag{3c}$$

equations of motion of an unconstrained particle in space. If the three force components F_x, F_y, and F_z are expressed functions of the coordinates and time, the dependence of each coordinate upon the time is implied and can in favorable cases be found as solutions of the equations of motion in the form of Eqs. (4). The primary objective of kinetics is achieved in the discovery of such functions.

$$x = x(t) \qquad y = y(t) \qquad z = z(t) \tag{4}$$

One-dimensional particle motion. The motion of

a particle which remains on the x axis, either because of constraints or initial conditions, is determined by Eq. (3a) alone, whose solution is simplified by the absence of y and z. Such one-dimensional dynamical problems provide an attractively simple introduction to the subject. Examples are the motion of a body falling vertically, subject to gravitational force, and linear harmonic motion.

Two-dimensional particle motion. The motion of a particle remaining in the plane of the x and y axes is determined by Eqs. (3a) and (3b), from which z is absent. Two-dimensional problems are reasonably tractable and include many of physical interest such as the motion of a projectile (exterior ballistics), and of a body attracted toward a central point, as in planetary motion. Solution of a two-dimensional problem is frequently simplified by change of variables which reduces it to a pair of one-dimensional problems.

Three-dimensional particle motion. All three equations of motion, Eqs. (3a)–(3c), apply to an unconstrained particle in space. Complete solutions are possible only when the functions expressing force components are relatively simple in character. Fortunately, many of the solvable cases correspond to important physical examples in which simplicity of the forces allows separation into one- and two-dimensional motions. Three-dimensional projectile motion without friction is an example.

Newton's third law. The behavior of systems composed of two or more interacting particles is treated by Newtonian dynamics augmented by Newton's third law of motion which states that when two bodies interact, the forces they exert on one another are equal and oppositely directed. The important laws of momentum and energy conservation are derivable for such systems (the latter only for forces of special type) and useful in solution of problems. The equations of motion for systems of more than two interacting particles in space are mathematically intractable in the absence of geometrical constraints or special initial conditions, but assumptions approximating the physical situation permit solution of many problems of physical interest. The principles of particle dynamics are transferred to extended bodies by regarding them as systems of particles subject to specified mutual constraints and mutual forces. *See* ACCELERATION; FORCE; GRAVITATION; HARMONIC MOTION; MASS; MOMENTUM; RIGID-BODY DYNAMICS; VELOCITY.

[RUSSELL A. FISHER]

Bibliography: H. Goldstein, *Classical Mechanics*, 2d ed., 1980; C. Kittel, W. D. Knight, and M. A. Ruderman, *Mechanics*, Berkeley Physics Course, vol. 1, 2d ed., 1973; K. R. Symon, *Mechanics*, 3d ed., 1971.

Kirchhoff's laws of electric circuits

Fundamental natural laws dealing with the relation of currents at a junction and the voltages around a loop. These laws are commonly used in the analysis and solution of networks. They may be used directly to solve circuit problems, and they form the basis for network theorems used with more complex networks.

In the solution of circuit problems, it is necessary to identify the specific physical principles involved in the problem and, on the basis of them, to write equations expressing the relations among the unknowns. Physically, the analysis of networks is based on Ohm's law giving the branch equations, Kirchhoff's voltage law giving the loop voltage equations, and Kirchhoff's current law giving the node current equations. Mathematically, a network may be solved when it is possible to set up a number of independent equations equal to the number of unknowns. *See* CIRCUIT.

When writing the independent equations, current directions and voltage polarities may be chosen arbitrarily. If the equations are written with due regard for these arbitrary choices, the algebraic signs of current and voltage will take care of themselves.

Kirchhoff's voltage law. One way of stating Kirchhoff's voltage law is: "At each instant of time, the algebraic sum of the voltage rise is equal to the algebraic sum of the voltage drops, both being taken in the same direction around the closed loop."

The application of this law may be illustrated with the circuit in Fig. 1. First, it is desirable to consider the significance of a voltage rise and a voltage drop, in relation to the current arrow. The following definitions are illustrated by Fig. 1.

A voltage rise is encountered if, in going from 1 to 2 in the direction of the current arrow, the polarity is from minus to plus. Thus, E is a voltage rise from 1 to 2.

A voltage drop is encountered if, in going from 3 to 4 in the direction of the current arrow, the polarity is from plus to minus. Thus, $v_{R1} = R_1 i$ is a voltage drop from 3 to 4. The application of Kirchhoff's voltage law gives the loop voltage, Eq. (1).

$$E = v_{R1} + v_{R2} = R_1 i + R_2 i \qquad (1)$$

In the network of Fig. 2 the voltage sources have the same frequency. The positive senses for the branch currents I_R, I_L, and I_C are chosen arbitrarily, as are the loop currents I_1 and I_2. The voltage equations for loops 1 and 2 can be written using instantaneous branch currents, instantaneous loop currents, phasor branch currents, or phasor loop currents.

The loop voltage equations are obtained by applying Kirchhoff's voltage law to each loop as follows.

By using instantaneous branch currents, Eqs. (2)

$$e_{g1} = Ri_R + L \frac{di_L}{dt} \qquad (2)$$

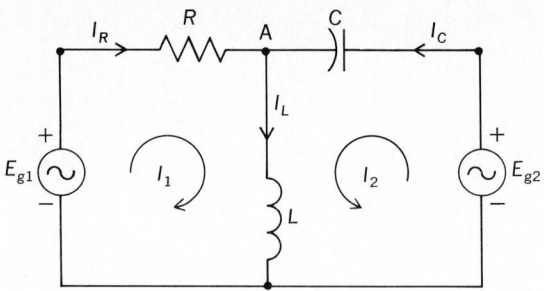

Fig. 2. Two-loop network demonstrating the application of Kirchhoff's voltage law.

and (3) may be obtained. By using instantaneous

$$e_{g2} = \frac{1}{C} \int i_C \, dt + L \frac{di_L}{dt} \qquad (3)$$

loop currents, Eqs. (4) and (5) are obtained. Equa-

$$e_{g1} = Ri_1 + L \frac{d(i_1 + i_2)}{dt} \qquad (4)$$

$$e_{g2} = \frac{1}{C} \int i_2 \, dt + L \frac{d(i_2 + i_1)}{dt} \qquad (5)$$

tions (6) and (7) are obtained by using phasor

$$\mathbf{E}_{g1} = R\mathbf{I}_R + j\omega L\mathbf{I}_L \qquad (6)$$

$$\mathbf{E}_{g2} = -j\frac{1}{\omega C}\mathbf{I}_C + j\omega L\mathbf{I}_L \qquad (7)$$

branch currents. By using phasor loop currents, Eqs. (8) and (9) may be obtained.

$$\mathbf{E}_{g1} = R\mathbf{I}_1 + j\omega L(\mathbf{I}_1 + \mathbf{I}_2) \qquad (8)$$

$$\mathbf{E}_{g2} = -j\frac{1}{\omega C}\mathbf{I}_2 + j\omega L(\mathbf{I}_2 + \mathbf{I}_1) \qquad (9)$$

Kirchhoff's current law. Kirchhoff's current law may be expressed as follows: "At any given instant, the sum of the instantaneous values of all the currents flowing toward a point is equal to the sum of the instantaneous values of all the currents flowing away from the point."

The application of this law may be illustrated with the circuit in Fig. 3. At node A in the circuit in Fig. 3, the current is given by Eq. (10).

$$i_1 + i_2 = i_3 \qquad (10)$$

The current equations at node A in Fig. 2 can be written by using instantaneous branch currents or phasor branch currents.

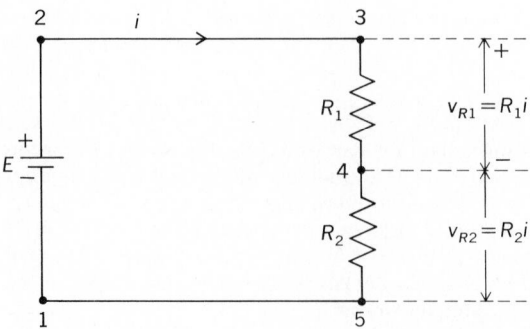

Fig. 1. Simple loop to show Kirchhoff's voltage law.

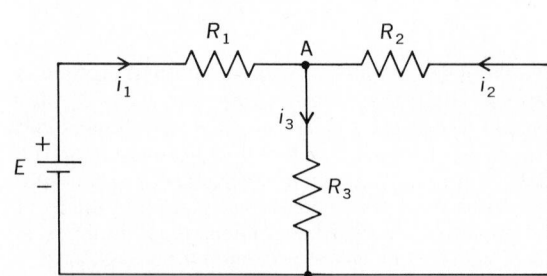

Fig. 3. Circuit demonstrating Kirchhoff's current law.

By using instantaneous branch currents, Eq. (11) is obtained.

$$i_R + i_C = i_L \qquad (11)$$

By using phasor branch currents, Eq. (12) is obtained.

$$\mathbf{I}_R + \mathbf{I}_C = \mathbf{I}_L \qquad (12)$$

[K. Y. TANG/ROBERT T. WEIL]

Knudsen number

In fluid mechanics, the ratio l/L of the mean free path length l of the molecules of the fluid to a characteristic length L of the structure in the fluid stream. When the mean free path of the fluid particles is short relative to the size of the object being considered, the fluid can be treated as a continuum. If the path length between molecular encounters is comparable to or larger than a significant dimension of the flow region, the gas must be treated as consisting of discrete particles. The usual classifications of flow according to Knudsen number are as follows: For $l/L \leqq 0.01$ the flow can be dealt with by the methods of gas dynamics; for $l/L \approx 1$ the behavior is termed slip flow; for $l/L \geqq 10$ the behavior is termed free-molecular flow or rarefied gas dynamics. *See* CLASSICAL FIELD THEORY; GAS DYNAMICS; STATISTICAL MECHANICS; SUPERAERODYNAMICS.

[FRANK H. ROCKETT]

Kondo effect

The large anomalous increase in the resistance of certain dilute alloys of magnetic materials in nonmagnetic hosts as the temperature is lowered. This behavior is contrary to the decrease in electrical resistance with temperature observed in almost all other systems. It is believed that the interaction of the conduction electrons with the spin of the magnetic impurity creates rather long-range correlations in the electrons in the vicinity of the impurity and that this strong interaction tends to inhibit conductivity. This interaction is an exchange interaction similar to that in ferromagnets. *See* FERROMAGNETISM.

For temperatures above a few degrees absolute, however, thermal agitation tends to wash out these fairly delicate correlations, and on heating the sample further the contribution to the resistance diminishes. The temperature below which this effect predominates is referred to as the Kondo temperature, and it depends on the magnetic impurity and the host material.

It has been suggested that Kondo temperatures may range from millidegrees to hundreds of degrees, but for materials studied so far they are below about 30 K. *See* CURIE TEMPERATURE; ELECTRICAL RESISTIVITY.

Importance. The importance of studying the Kondo effect lies in the fact that it is a property of what is, presumably, the simplest possible magnetic system—a single magnetic atom in a nonmagnetic environment. (The alloys used are sufficiently dilute that interaction between different magnetic impurities can be safely ignored. For example, this is true for iron impurities in copper if the iron is less than one part in 10,000 of the alloy.) It is hoped that an understanding of

this system will provide a key to the far more complex and important problem of the electronic structure of magnetic materials themselves, one of the greatest challenges in physics.

Theory. Although the Kondo effect had been experimentally observed many years earlier, the first satisfactory theoretical explanation was not given until J. Kondo's work in 1964. Newer theories predict quite accurately the temperature variation of the resistance, but associated phenomena such as the specific heat and magnetic susceptibility of Kondo-type alloys are not so well understood. Experimentally, these quantities are very hard to measure accurately, and furthermore seem to vary a great deal from one impurity-host system to another. Theoretically, the problem has a deceptive air of simplicity, and all solutions given so far are based on rather broad assumptions and approximations whose importance is hard to assess. Although some understanding of the system has certainly been gained, the picture is far from complete.

Kondo's explanation. Bearing in mind that no similar effect is observed in scattering from nonmagnetic impurities and that the magnetic field associated with the impurity was far too weak to cause the observed effect, Kondo assumed the crucial new feature to be the spin associated with the magnetic impurity. For simplicity he assumed further that the impurity spin had the same magnitude as the electron spin. The impurity can interact in a spin-dependent way with a conduction electron: When the impurity spin flips over, the electron spin flips at the same time, so that the total spin in any direction remains constant. This possible scattering mode causes the total scattering by an impurity, and hence the electrical resistance of the alloy, to depend on the temperature. To understand why this is so, one first considers ordinary (no spins involved) scattering theory and gives an argument showing why this is not temperature-dependent. It can then be demonstrated that the argument breaks down if the concept of "spin-flip" scattering is introduced. *See* ELECTRON SPIN; EXCLUSION PRINCIPLE; FERMI-DIRAC STATISTICS; UNCERTAINTY PRINCIPLE.

Scattering from impurity. In scattering theory the electron is considered to undergo a succession of interactions with the scatterer causing the electron to be in a sequence of so-called intermediate states before finally emerging from the scattering process. In these intermediate states the energy of the electron does not have to be identical with its initial energy. This is a consequence of the Heisenberg uncertainty principle: Since the intermediate states are short-lived (lifetime Δt, say) there is a corresponding arbitrariness in their energy which is given by $\Delta E\,\Delta t \sim h$, Planck's constant. However, there is one important constraint on possible intermediate-state energies—there are many other electrons present, and (from Pauli's exclusion principle) the electron cannot go to an intermediate state already occupied by another electron. Hence the presence of the other electrons modifies the scattering of the electron under consideration. Furthermore, when the temperature changes, the distribution of the other electrons over available energy levels alters, and thus the scattering becomes temperature-dependent. For ordinary (non-

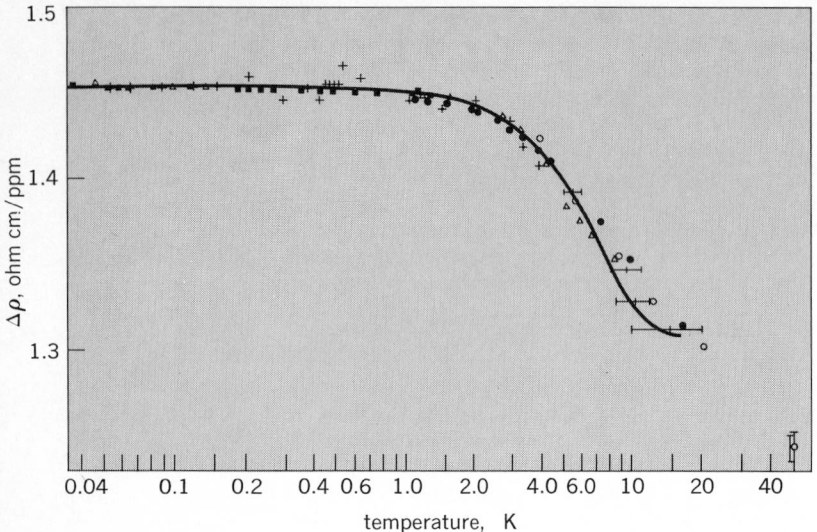

Relationship between impurity contribution to the electrical resistivity ($\Delta\rho$) and the absolute temperature. (After M. Daybell and W. Steyert, Phys. Rev. Lett., 18:398, 1967)

spin-dependent) scattering, this temperature dependence is cancelled by another contribution to the scattering referred to as hole intermediate states. This corresponds to a slightly different sequence of events. For normal scattering (particle intermediate states) an electron with momentum p is scattered to an intermediate state m, then scattered again by the impurity to a state p' (in general, of course, there is a succession of intermediate states). It is also possible for an electron in a state m to be scattered into the state p'; the original electron p subsequently falls into the hole, or vacancy, in state m. Provided the events follow each other closely, that is, the hole in state m is short-lived, the possible energy of the hole is spread by the uncertainty principle; this scattering process will also be temperature-dependent. However, for a given state m the probability of a hole appearing in m and an electron appearing in state p' is just proportional to the probability of the state m having an electron in it in the first place; whereas the probability of the same m being an intermediate state in the normal scattering sequence is proportional to the probability of its being initially vacant (and hence available). Thus when the two types of scattering are taken together, the probability of the state m being initially occupied cancels out and the full scattering becomes temperature-independent. *See* BAND THEORY OF SOLIDS; HOLES IN SOLIDS.

Spin-flip scattering. Kondo demonstrated that if the impurity can flip the spin of the electron, these arguments are no longer valid. The hole intermediate states no longer cancel out the temperature dependence of the particle intermediate states. In fact, he proved that for his model the imbalance was such that the scattering increased as $-\log T$ for T (the temperature) going to zero. It has been shown since that, with further refinement, the scattering levels off below a certain temperature, but at a considerably higher value than the minimum reached at higher temperatures.

In the illustration $\Delta\rho$ is the extra resistance of a piece of copper resulting from the addition of iron impurities. The continuous line is the theoretical

prediction. The different experimental points are for different concentrations. $\Delta\rho$ is measured in ohm-centimeters per part per million of added iron. Since all the points fall on the same curve, the increase in resistance is directly proportional to the amount of added iron. Thus, adding the thousandth iron atom causes exactly the same increase as adding the first iron atom—the iron atoms are not influenced detectably by each other's presence; this suggests that Kondo's idea of considering "one atom" scattering is reasonable.

One example is given here to demonstrate how spin flipping restricts intermediate states differently for particles and holes. Consider an electron in an initial state $p\uparrow$ ($\uparrow$ means spin up) scattered by an impurity initially $I\downarrow$ into a state $p'\uparrow$. Possible particle intermediate states are of the type $m\uparrow\ I\downarrow$ and $m\downarrow\ I\uparrow$. (The total spin remains constant. The interaction in general may or may not flip the spin. A full, rigorous treatment is given by Kondo.)

Now consider the hole intermediate states. Examining the sequence of events in this case, both $p\uparrow$ and $p'\uparrow$ are already present at the same time as the hole m. Hence, since the total initial spin of the system was zero, both the hole m and the impurity I must be in states $m\downarrow$, $I\downarrow$, because at intermediate times also the total spin of the two electrons, one hole, and the impurity must equal zero. Thus the only allowed hole intermediate states are those in which the impurity has its spin pointing down. In contrast, for the particle intermediate states the impurity spin could be up or down. Since this means that there are twice as many possible particle intermediate states as hole intermediate states, the two sets cannot cancel out as in ordinary scattering, and this results in the scattering being temperature-dependent.

[MICHAEL FOWLER]

Kronig-Penney model

An idealized, one-dimensional model of a crystal which exhibits many of the basic features of the electronic structure of real crystals. Consider the potential energy $V(x)$ of an electron illustrated in Fig. 1 with an infinite sequence of potential wells of depth $-V_o$ and width a, arranged with a spacing b. The Schrödinger wave equation can be readily solved for such an arrangement to give electron energy as a function of wave number.

The energy bands thus obtained with the choice of constants given in Eqs. (1) and (2) are shown in

$$(2ma^2V_o/h^2)^{1/2} = 12.0 \tag{1}$$

$$b/a = 0.1 \tag{2}$$

Fig. 2. Notice that the width and the curvatures of

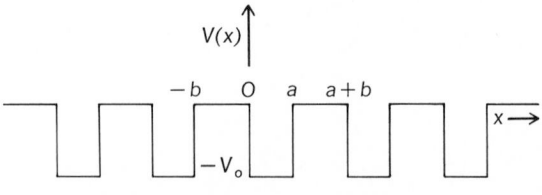

Fig. 1. Potential energy which is assumed for the one-dimensional Kronig-Penney model.

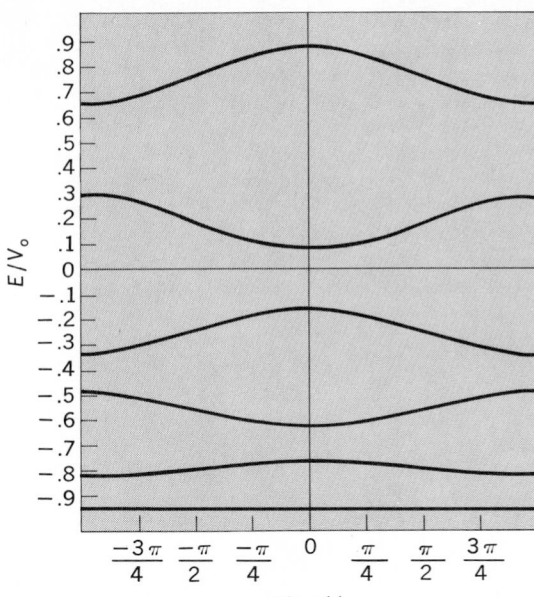

Fig. 2. The ratio of the electron energy E to the magnitude of the well depth V_o shown as a function of electron wave number k times the "lattice constant" $(a + b)$ for the Kronig-Penney model.

the allowed bands increase with energy. The Kronig-Penney model has been extended to include the effects of impurity atoms. A solution can also be derived from the Dirac equation. *See* BAND THEORY OF SOLIDS; NONRELATIVISTIC QUANTUM THEORY. [JOSEPH CALLAWAY]

Bibliography: C. Kittel, *Introduction to Solid State Physics*, 5th ed., 1976; R. de L. Kronig and W. G. Penney, Quantum mechanics of electrons in crystal lattices, *Proc. Roy. Soc. (London) ser. A*, 130:499–513, 1931; G. Wannier, *Elements of Solid State Theory*, 1959.

Lagrange's equations

Equations of motion of a mechanical system for which a classical (non-quantum-mechanical) description is suitable, and which relate the kinetic energy of the system to the generalized coordinates, the generalized forces, and the time. If the configuration of the system is specified by giving the values of f independent quantities $q_1, \ldots, q_f$, there are f such equations of motion.

In their usual form, these equations are equivalent to Newton's second law of motion and are differential equations of the second order for the q's as functions of the time t.

Derivation. Let the system consist of N particles. The masses of the particles are m_ρ and their cartesian coordinates are $x_{\rho i} (\rho = 1, 2, \ldots, N; i = 1, 2, 3)$. These cartesian coordinates are expressible as functions of $f (\leqq 3N)$ generalized coordinates $q_1, \ldots, q_f$ between which there are no constraints. The time does not appear here if the constraints are fixed (as will be assumed), as in Eq. (1).

$$x_{\rho i} = f_{\rho i}(q_1, \ldots, q_f) \tag{1}$$

Then, denoting time differentiation by a dot, Eq. (2) holds. The cartesian velocity components are linear functions of the generalized velocities q_j.

$$\dot{x}_{\rho i} = \sum_{j=1}^{f} \frac{\partial x_{\rho i}}{\partial q_j} \dot{q}_j \tag{2}$$

Let δq_j represent a small displacement of the system. It is automatically consistent with the constraints. In this displacement, the forces of constraint do no work; their only action is to prevent motion contrary to the constraint, so they have no components in the direction of a displacement consistent with the constraints. If W is the work done during the displacement, it is done entirely by the externally applied forces, as in Eq. (3).

$$W = \sum_{\rho,i} F_{\rho i}\, \delta x_{\rho i} = \sum_{\rho,i} m_\rho \ddot{x}_{\rho i}\, \delta x_{\rho i} \tag{3}$$

Each term in the sum of Eq. (3) may receive contributions from forces of constraint, but they cancel when the summation is made. From Eq. (1), Eqs. (4) and (5) are obtained.

$$\delta x_{\rho i} = \sum_j \frac{\partial x_{\rho i}}{\partial q_j} \delta q_j \tag{4}$$

$$\frac{\partial \dot{x}_{\rho i}}{\partial \dot{q}_j} = \frac{\partial x_{\rho i}}{\partial q_j} \tag{5}$$

It is readily verified that Eq. (6) is valid. The quantity in the brackets is thus, for each j, the noncartesian analog of the cartesian $m_\rho \ddot{x}_{\rho i}$.

$$W = \sum_j \left[\frac{d}{dt} \frac{\partial}{\partial \dot{q}_j} \left(\frac{1}{2} m_\rho \dot{x}_{\rho i}{}^2 \right) - \frac{\partial}{\partial q_j} \left(\frac{1}{2} m_\rho \dot{x}_{\rho i}{}^2 \right) \right] \delta q_j \tag{6}$$

The only quantity which appears differentiated in Eq. (6) is the total kinetic energy T of the system. This is easily calculated in generalized coordinates because the connection between the cartesian velocities and the generalized velocities is linear and homogeneous. Usually, the kinetic energy can be written by inspection without using Eq. (2) explicitly. Thus Eq. (7) holds.

$$W = \sum_j \left(\frac{d}{dt} \frac{\partial T}{\partial \dot{q}_j} - \frac{\partial T}{\partial q_j} \right) \delta q_j \tag{7}$$

Transforming the right-hand side of Newton's equation is simpler, as shown in Eqs. (8) and (9),

$$W = \sum_{\rho,i} F_{\rho i}\, \delta x_{\rho i} = \sum_j Q_i\, \delta q_j \tag{8}$$

$$Q_j = \sum_{\rho,i} F_{\rho i} \frac{\partial x_{\rho i}}{\partial q_j} \tag{9}$$

where Eq. (9) is the jth component of the generalized force. By the preceding argument, Q_j depends only on the externally applied forces, the forces of constraint necessarily cancelling in the summation.

The displacement δq_j was entirely arbitrary. Thus, it follows from equating expressions (7) and (8) for W that Eq. (10) is valid. Equation (10) sums

$$\frac{d}{dt} \frac{\partial T}{\partial \dot{q}_j} - \frac{\partial T}{\partial q_j} = Q_j \tag{10}$$

up Lagrange's equations of motion. They are valid also when moving constraints are present. *See* CONSTRAINT.

Examples. Use of this form of Lagrange's equations is shown in the two following examples.

Particle in central force field. Here, the force acting on a particle acts always through a fixed point. Choose this point as origin of a spherical coordinate system with coordinates r, θ, ϕ (Fig. 1). Only radial displacements involve work, so that only Q_r differs from zero, or Eq. (11) holds.

$$Q_\theta = Q_\phi = 0 \qquad (11)$$

The kinetic energy is given by inspection of Fig. 1 and is expressed by Eq. (12).

$$T = \frac{m}{2} (\dot{r}^2 + r^2 \dot{\theta}^2 + r^2 \sin^2 \theta \, \dot{\phi}^2) \qquad (12)$$

Lagrange's equations are Eqs. (13). These may

$$\frac{d}{dt} \frac{\partial T}{\partial \dot{r}} - \frac{\partial T}{\partial r} = m (r - r\dot{\theta}^2 - r \sin^2 \theta \, \dot{\phi}^2) = Q_r$$

$$\frac{d}{dt} \frac{\partial T}{\partial \dot{\theta}} - \frac{\partial T}{\partial \theta} = m \left[\frac{d}{dt} (r^2 \dot{\theta}) - r^2 \sin \theta \cos \theta \, \dot{\phi}^2 \right] = 0$$

$$\frac{d}{dt} \frac{\partial T}{\partial \dot{\phi}} - \frac{\partial T}{\partial \phi} = m \frac{d}{dt} (r^2 \sin^2 \theta \, \dot{\phi}) = 0 \qquad (13)$$

be compared with the cartesian equations, Eqs. (14), in which the force function F appears in all

$$m\ddot{x} = \frac{x}{r} F(x,y,z)$$

$$m\ddot{y} = \frac{y}{r} F(x,y,z) \qquad (14)$$

$$m\ddot{z} = \frac{z}{r} F(x,y,z)$$

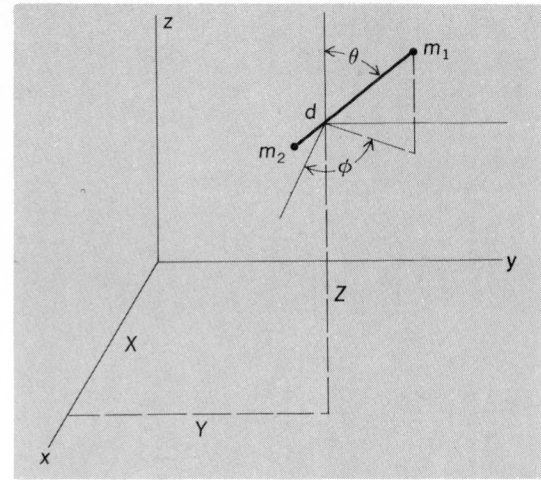

Fig. 2. *X, Y,* and *Z* are cartesian coordinates of the center of mass of dumbbell $m_1 m_2$; θ and ϕ are the polar angles of the dumbbell axis in a set of axes parallel to fixed set of axes and with origin at center of mass.

three of the equations of motion, while two of the three Lagrange equations are independent of the detailed nature of the force.

Two particles, fixed separation. In this system, there is one constraint, the particles being a constant distance d apart, and so there are five degrees of freedom instead of six. Choose as generalized coordinates the cartesian coordinates X, Y, Z of the center of mass and the polar angles θ, ϕ of the line joining the two particles, as in Fig. 2. Then, because the kinetic energy is the sum of the kinetic energy of the center of mass and the kinetic energy relative to the center of mass, Eq. (15) holds.

$$T = \frac{m_1 + m_2}{2} (\dot{X}^2 + \dot{Y}^2 + \dot{Z}^2) + \frac{1}{2} \left(\frac{m_1 m_2}{m_1 + m_2} \right) d^2 (\dot{\theta}^2 + \sin^2 \theta \, \dot{\phi}^2) \qquad (15)$$

The equations of motion may be written down immediately, the generalized forces being evaluated from Eq. (9). There is one equation for each degree of freedom, and the constraint is automatically satisfied.

Conservative systems. In many problems, the forces Q_j are derivable from a potential. Then Eq. (16), and thus Eq. (17), is valid. When this is so, it is

$$W = \sum_{j=1}^{f} Q_j \, dq_j = -dV \qquad (16)$$

$$Q_j = -\frac{\partial V}{\partial q_j} \qquad \frac{\partial V}{\partial \dot{q}_j} = 0 \qquad (17)$$

convenient to define a function, called the Lagrangian, by Eq. (18). Then the equations of motion

$$L(q, \dot{q}t) = T(q, \dot{q}, t) - V(q, t) \qquad (18)$$

become simply Eq. (19), which is the most commonly encountered form of Lagrange's equations.

$$\frac{d}{dt} \frac{\partial L}{\partial \dot{q}_j} - \frac{\partial L}{\partial q_j} = 0 \qquad (19)$$

Example of use of L. A simple vibration problem illustrates the use of generalized coordinates and shows the ease of application of Lagrange's equations when T and V are relatively easy to obtain.

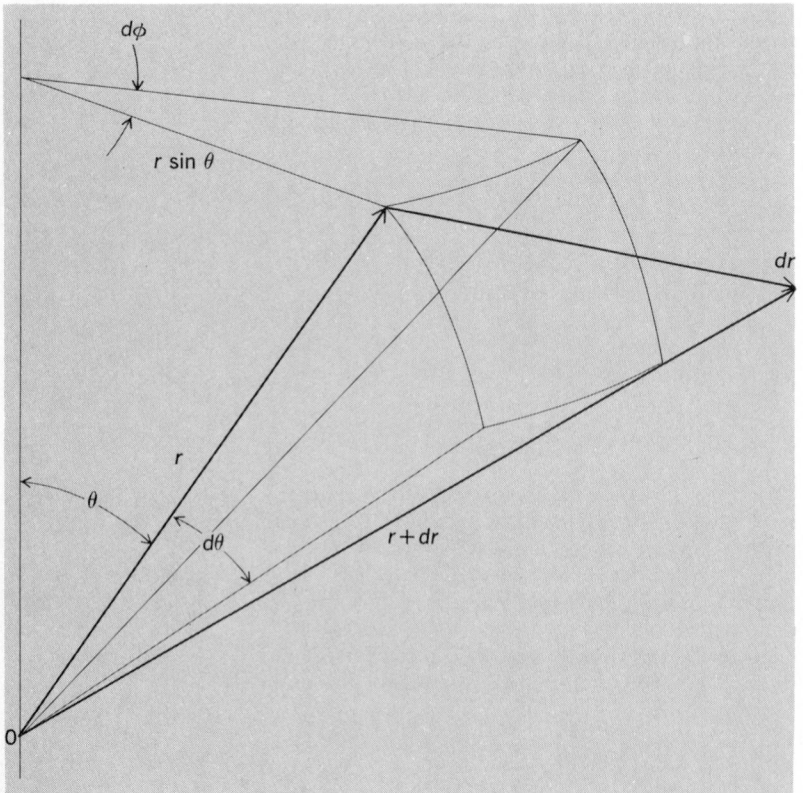

Fig. 1. The vector *dr* is decomposed into three orthogonal vectors of magnitudes $r \sin \theta \, d\phi$, $rd\,\theta$, and *dr*, respectively, r, θ, and ϕ being spherical coordinates of the terminus of the vector r.

Consider the one-dimensional system of two equal masses m connected by a spring of restoring constant k. Let the displacements of the masses from their equilibrium positions be x_1 and x_2, respectively.

Now introduce new coordinates, $q_1 = x_2 - x_1$ and $q_2 = x_2 + x_1$. Then Eqs. (20) hold.

$$V = kq_1^2/2$$
$$T = m\,(\dot{q}_1^2 + \dot{q}_2^2)/4 \tag{20}$$

Using $L = T - V$, Lagrange's equations for the two variables are written as Eqs. (21).

$$\tfrac{1}{2}m\dot{q}_1 - kq_1 = 0$$
$$\tfrac{1}{2}m\dot{q}_2 = 0 \quad \text{or} \quad \dot{q}_2 = 0 \tag{21}$$

The solution to the first of these differential equations is $q_1 = A\sin\sqrt{2k/m}\,t$ (setting $q_1 = 0$ at $t = 0$), which is a simple harmonic vibration of the two masses. The second equation merely states that the acceleration of the center of mass of the system is zero.

Nonconservative systems. Form (19) is also correct even if the system is not conservative, provided one can write Eq. (22), as in the case of a

$$Q_j = \frac{d}{dt}\frac{\partial V}{\partial \dot{q}_j} - \frac{\partial V}{\partial q_j} \tag{22}$$

charged particle in an electromagnetic field. Here V is defined in Eq. (23), where ϕ and $\mathbf{A}$ are the sca-

$$V = e\left(\phi - \frac{\mathbf{v}\cdot\mathbf{A}}{c}\right) \tag{23}$$

lar and vector potentials of the field, respectively, e is the charge of the particle, $\mathbf{v}$ is the particle's velocity, and c is the velocity of light.

Cyclic coordinate. If L does not depend explicitly on a particular coordinate, say q_k, then Eqs. (24)

$$\frac{d}{dt}\frac{\partial L}{\partial \dot{q}_k} = 0 \qquad \frac{\partial L}{\partial \dot{q}_k} = \text{constant} \tag{24}$$

hold. Here q_k is called an ignorable or cyclic coordinate, and $\partial L/\partial \dot{q}_k$ is an integral of motion. In the example of the central force field given previously, ϕ may be ignored.

Total energy. By use of Eq. (19), Eq. (25) is obtained. Hence, if L does not depend explicitly on

$$\frac{d}{dt}\left(\sum_{j=1}^{f}\frac{\partial L}{\partial \dot{q}_j}\dot{q}_j - L\right) = -\frac{\partial L}{\partial t} \tag{25}$$

tained. Hence, if L does not depend explicitly on the time, expression (26) is an integral of motion.

$$\sum_{j=1}^{f}\left(\frac{\partial L}{\partial q_j}q_j - L\right) \tag{26}$$

Usually, it is the total energy of the system. *See* HAMILTON'S EQUATIONS OF MOTION.

Conjugate momentum. The quantity in Eq. (27) is

$$p_k = \frac{\partial L}{\partial \dot{q}_k} \tag{27}$$

defined to be the momentum conjugate to the coordinate q_k. (If a Lagrangian does not exist because of the presence of dissipative forces, the conjugate momentum is sometimes defined as $\partial T/\partial \dot{q}_k$.) This momentum is not necessarily a linear momentum as defined in Newtonian mechanics, since its character depends both on the system and on the nature of the coordinate q_k. If q_k is an ignorable coor-

dinate, its conjugate momentum is a constant of motion.

Kinetic momentum. For a charged particle in an electromagnetic field, the Lagrangian may be written as Eq. (28), and the momentum conjugate to

$$L = \tfrac{1}{2}m\mathbf{v}^2 - e\phi + \frac{e}{c}\mathbf{v}\cdot\mathbf{A} \tag{28}$$

the cartesian coordinate x is given by Eq. (29). The

$$p_x = \frac{\partial L}{\partial \dot{x}} = m\dot{x} + \frac{e}{c}A_x \tag{29}$$

quantity $m\dot{x}$ is called the kinetic momentum. The kinetic momentum is related to p_x much as the kinetic energy is related to the total energy. If ϕ and A are independent of x, it is the momentum p_x and not the kinetic momentum $m\dot{x}$ which is a constant of motion.

Relativistic systems. The equations of motion of a relativistic particle may be written in Lagrangian form. The simplest way to do this is to replace the kinetic energy T by another function τ of the mass and velocity so as to get the desired form, namely, Eq. (30). Here m_0 is the rest

$$\frac{d}{dt}\frac{\partial\tau}{\partial\dot{x}_k} - \frac{\partial V}{\partial x_k} = \frac{d}{dt}\left(\frac{m_0\dot{x}_k}{\sqrt{1 - v^2/c^2}}\right) - \frac{\partial V}{\partial x_k} = 0 \tag{30}$$

mass of the particle. This is accomplished by setting τ as in Eq. (31), which reduces to T in the

$$\tau = (1 - \sqrt{1 - v^2/c^2})m_0 c^2 \tag{31}$$

limit $\mathbf{v}/c \to 0$. The equations of motion in this form are still valid in only one reference frame because the time and the coordinates are treated on different bases. *See* RELATIVISTIC MECHANICS; RELATIVITY. [PHILIP M. STEHLE]

Bibliography: H. C. Corben and P. Stehle, *Classical Mechanics*, 2d ed., 1960, reprint 1974; H. Goldstein, *Classical Mechanics*, 2d ed., 1980; W. Hauser, *Introduction to the Principles of Mechanics*, 1965; L. D. Landau et al., *Mechanics*, 3d ed., 1976; K. R. Symon, *Mechanics*, 3d ed., 1971.

Lagrangian function

A function of the generalized coordinates and velocities of a dynamical system from which the equations of motion in Lagrange's form can be derived. The Lagrangian function is denoted by $L(q_1, \ldots, q_f; \dot{q}_1, \ldots, \dot{q}_f; t)$. *See* LAGRANGE'S EQUATIONS.

For systems in which the forces are derivable from a potential energy V, if the kinetic energy is T, Eq. (1) holds.

$$L = T - V \tag{1}$$

If the system is continuous rather than discrete, the Lagrangian function L is the integral of a Lagrangian density $\mathscr{L}$, as in Eq. (2), where $\eta(x_1, x_2, x_3)$

$$L = \int \mathscr{L}\,(\eta, \text{grad }\eta, x_1, x_2, x_3, t)\,dx_1\,dx_2\,dx_3 \tag{2}$$

describes the displacement of the medium at the point (x_1, x_2, x_3). The equations of motion are in this case written as Eq. (3).

$$\frac{\partial}{\partial t}\frac{\partial\mathscr{L}}{\partial(\partial\eta/\partial t)} + \sum_{j=1}^{3}\frac{\partial}{\partial x_j}\frac{\partial\mathscr{L}}{\partial(\partial\eta/\partial x_j)} - \frac{\partial\mathscr{L}}{\partial\eta} = 0 \tag{3}$$

This formulation of Lagrange's equations applies to the motion of a gas containing sound waves, to a vibrating jelly, or to any medium where discrete masses are replaced by a continuum.

[PHILIP M. STEHLE]

Bibliography: See LAGRANGE'S EQUATIONS.

Laminar flow

Streamline flow of a viscous fluid which satisfies the Navier-Stokes equations of motion. In laminar flow, the fluid moves in layers without large irregular fluctuations. Laminar flow occurs at a low Reynolds number. This corresponds to the conditions of small velocities and dimensions of bodies, to very large viscosity, or to small density of the fluid. Laminar flow plays an important role in several practical problems.

The flow of oil in the bearings for lubrication is laminar. The theory of laminar flow shows that under great normal pressure, the oil in the bearing has only slight frictional resistance. See TURBULENT FLOW.

The motion of a minute particle in a viscous fluid produces laminar flow. The drag coefficient of such a body is inversely proportional to its Reynolds number.

Flow on the surface of modern aircraft and missiles flying at extremely high altitude may be laminar. The Reynolds number for this case is usually moderate, and the viscous effect is confined to the boundary layer region of the body. Laminar boundary-layer flow determines the skin friction and the aerodynamic heating of these bodies. See NAVIER-STOKES EQUATIONS; REYNOLDS NUMBER.

[SHIH I. PAI]

Bibliography: S. I. Pai, *Viscous Flow Theory*, vol. 1, 1956; V. L. Streeter (ed.), *Handbook of Fluid Dynamics*, 1961; V. L. Streeter and E. B. Wylie, *Fluid Mechanics*, 7th ed., 1979.

Langevin function

A mathematical function which is important in the theory of paramagnetism and in the theory of the dielectric properties of insulators. The analytical expression for the Langevin function (see illustration) is shown in the equation below. If $x \ll 1$, $L(x) \cong x/3$. The paramagnetic susceptibil-

$$L(x) = \coth x - 1/x$$

ity of a classical (non-quantum-mechanical) collection of magnetic dipoles is given by the Langevin function, as is the polarizability of molecules hav-

ing a permanent electric dipole moment. In the quantum-mechanical treatment of paramagnetism, the Langevin function is replaced by the Brillouin function. For further discussion see DIELECTRIC CONSTANT; PARAMAGNETISM.

[ELIHU ABRAHAMS; F. KEFFER]

Laplace transform

An integral transform extensively used by P. S. Laplace in the theory of probability. In simplest form it is expressed as Eq. (1). It is thought of as

$$f(s) = \int_0^\infty e^{-st}\phi(t)\, dt \qquad (1)$$

transforming the determining function $\phi(t)$ into the generating function $f(s)$. The variable t is real, the variable s may be real or complex, $s = \sigma + i\tau$. As an example, if $\phi(t) = 1$ the integral converges for $\sigma > 0$, and $f(s) = 1/s$.

The Laplace transform is used for the solution of differential and difference equations, for the evaluation of definite integrals, and in many branches of abstract mathematics (functional analysis, operational calculus, and analytic number theory).

Method. Extensive tables of Laplace transforms exist, and these are used as any table of integrals. To illustrate how a differential equation may be solved, two excerpts (A and B) from such a table can be used.

A. $f(s) = 1/(s - a)$ $\qquad$ $\phi(t) = e^{at}$

B. $f(s) = 1/(s^2 + 1)$ $\qquad$ $\phi(t) = \sin t$

Suppose it is required to find a solution $y(t)$ of Eq. (2) such that $y(0) = 1$, $y'(0) = 2$. Denote the

$$y''(t) + y(t) = 2e^t \qquad \left(y'' = \frac{d^2y}{dt^2}, y' = \frac{dy}{dt}\right) \quad (2)$$

Laplace transform of the unknown function $y(t)$ by $Y(s)$. Integration by parts gives Eq. (3) on the as-

$$\int_0^\infty e^{-st}y''(t)\, dt = -y'(0) - y(0)s + s^2\int_0^\infty e^{-st}y(t)\, dt$$
$$= -2 - s + s^2Y(s) \qquad (3)$$

sumption that the integrated part is zero at $t = \infty$. Applying the Laplace transform to Eq. (2) and using A for the right-hand side, one obtains Eq. (4).

$$-2 - s + s^2Y(s) + Y(s) = \frac{2}{s-1} \qquad (4)$$

The differential equation has become an algebraic one, whose solution is Eq. (5). However, a further

$$Y(s) = \frac{1}{s-1} + \frac{1}{s^2+1} \qquad (5)$$

use of the table shows that the Laplace transform of $y(t) = e^t + \sin t$ is precisely the right-hand side of Eq. (5). Assuming uniqueness, one has thus obtained the required solution. Because its properties can be checked directly, the unproved assumptions need not be verified.

This example illustrates the general method. The unknown function is taken as the determining function and the Laplace transform is applied to the differential (or difference) equation. There results an equation with the generating function as unknown, and this must be solved. Finally the determining function must be determined from the

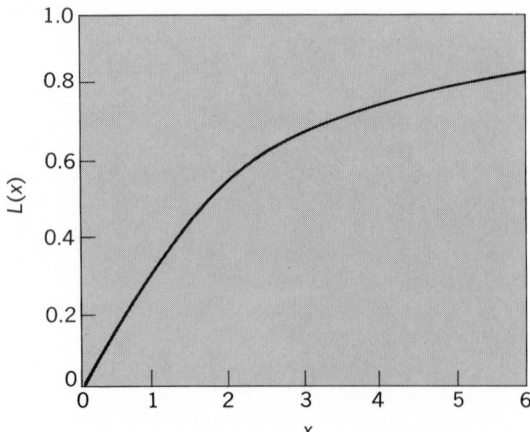

Plot of Langevin function.

generating function, either from tables or by use of an inversion formula. In general, if the original differential equation is partial in any number of independent variables, one application of the Laplace transform reduces the number of these variables by one. If the equation was ordinary (one independent variable), the transformed equation is algebraic, as in the above example.

Properties. Here are the fundamental properties of the Laplace transform:

I. There exists a number σ_c (perhaps $+\infty$ or $-\infty$) called the abscissa of convergence such that the integral in Eq. (1) converges for $\sigma > \sigma_c$, diverges for $\sigma < \sigma_c$. That is, the region of convergence is a half-plane (a half-line if s is real).

II. The generating function is holomorphic for $\sigma > \sigma_c$.

III. The determining function is uniquely determined by the generating function. (Ambiguity is possible only on sets of measure zero.)

IV. The product of two generating functions is in general a generating function. Thus, if Eq. (1) holds for two pairs of functions $f_1(s), \phi_1(t)$ and $f_2(s), \phi_2(t)$, then the product $f_1(s) f_2(s)$ is the transform of the convolution

$$\phi_1(t)*\phi_2(t) = \int_0^t \phi_1(u)\phi_2(t-u)\,du$$

As was evident in the above example, it is very important to be able to derive the determining function $\phi(t)$ from the generating function $f(s)$. This is especially true when tables are unavailable or inadequate. Any expression of $\phi(t)$ in terms of $f(s)$ is called an inversion formula. Many are known. The classical one is Eq. (6). Here the integration is along any line $\sigma = c$ of the complex s-plane on which the integral in Eq. (1) converges absolutely. Another inversion which employs the real variable only is Eq. (7). Here $f^{(k)}(x)$ means the

$$\phi(t) = \frac{1}{2\pi i}\int_{c-i\infty}^{c+i\infty} f(s)e^{st}\,ds \qquad 0 < t < \infty \quad (6)$$

$$\phi(t) = \lim_{k\to\infty} \frac{(-1)^k}{k!} f^{(k)}\left(\frac{k}{t}\right)\left(\frac{k}{t}\right)^{k+1} \qquad 0 < t < \infty \quad (7)$$

kth derivative of $f(x)$. Equation (7) can be illustrated by the example A above. For that pair $f^{(k)}(x)$ is easily computed and Eq. (7) becomes Eq. (8), a familiar result of calculus. *See* INTEGRAL TRANSFORM.

$$e^{at} = \lim_{k\to\infty}\left(1 - \frac{at}{k}\right)^{-k-1} \quad (8)$$

Generalizations. Certain generalizations of Eq. (1) are in frequent use. The transform shown as Eq. (9) is called the bilateral Laplace transform.

$$f(s) = \int_{-\infty}^{\infty} e^{-st}\phi(t)\,dt \quad (9)$$

An inversion is still provided by Eq. (6), which now holds for $-\infty < t < \infty$. If one sets $s = iy$ in Eq. (9), the result is Eq. (10). This equation defines $g(y)$ as

$$g(y) = f(iy) = \int_{-\infty}^{\infty} e^{-iyt}\phi(t)\,dt \quad (10)$$

the Fourier transform of $\phi(t)$. That is, the Laplace transform (9), if considered along a single line, becomes a Fourier transform. By setting $c = 0$ and

$s = iy$ in formula (6) one obtains Eq. (11), the classical inversion of the Fourier transform.

$$\phi(t) = \frac{1}{2\pi i}\int_{-i\infty}^{i\infty} f(s)e^{st}\,ds$$
$$= \frac{1}{2\pi}\int_{-\infty}^{\infty} g(y)e^{iyt}\,dy \quad (11)$$

Another generalization of Eq. (1) is the Laplace-Stieltjes integral, Eq. (12), where now the integral is a Stieltjes integral with respect to the "integrator" function $\alpha(t)$. If $\alpha(t)$ has a derivative $\phi(t)$ the integral (12) becomes the integral (1). On the other hand, if $\alpha(t)$ is a step-function, Eq. (12) reduces to a

$$f(s) = \int_0^\infty e^{-st}\,d\alpha(t) \quad (12)$$

Dirichlet series, Eq. (13), a type of series of great importance in analytic number theory.

$$f(s) = \sum_{k=1}^{\infty} a_k e^{-\lambda_k s} \qu(13)$$

It must not be supposed that one may choose $\phi(t)$ or $f(s)$ arbitrarily in Eq. (1) and expect its mate to exist. For example if $\phi(t) = e^{t^2}$, the integral (1) diverges for all s and $\sigma_c = +\infty$. Again if $f(s) = s$, no corresponding determining function $\phi(t)$ exists, since it is easily seen that every generating function must approach a limit as $s \to +\infty$ along the real axis. Hence it is clearly important to know what functions $\phi(t)$ and $f(s)$ may be used in Eq. (1). So far as $\phi(t)$ is concerned the problem is completely solved by the formula, with $\sigma_c > 0$, shown as Eq. (14). The other problem has been partially

$$\sigma_c = \varlimsup_{t\to\infty} \frac{\log|\alpha(t)|}{t} \quad (14)$$

solved by representation theorems for integral transforms, one striking example of which is presented below.

A function $f(s)$ of the real variable s is said to be completely monotonic on $a < s < \infty$ if and only if relations (15) hold. Examples are $f(s) = 1$, $f(s) =$

$$f(s) \geq 0, f'(s) \leq 0, f''(s) \geq 0, f'''(s) \leq 0, \ldots$$
$$(a < s < \infty) \quad (15)$$

$1/(s-a)$, and $f(s) = e^{-s}$. A theorem of S. Bernstein states that $f(s)$ has a representation (12) converging for $s > a$ and with integrator function $\alpha(t)$ nondecreasing if and only if $f(s)$ is completely monotonic for $a < s < \infty$. For example, if $f(s) = 1$, then $\alpha(t) = 1$, $t > 0$, $\alpha(0) = 0$; if $f(s) = 1/(s-a)$, then $\phi(t) = e^{at}$ and $\alpha(t) = (e^{at}-1)/a$; if $f(s) = e^{-s}$, $\alpha(t) = 0$ for $0 < t < 1$ and $\alpha(t) = 1$ for $1 < t < \infty$. In each case $\alpha(t)$ is nondecreasing, as predicted by Bernstein's theorem. This result is particularly remarkable because the mere signs of the successive derivatives of a function on the real axis determine not only its holomorphic character (property II above) in a half-plane but also its representation in the form of Eq. (12).

[DAVID V. WIDDER]

Bibliography: G. Doetsch, *Introduction to the Theory and Application of the Laplace Transformation*, ed. and trans. by W. Nader, 1974; A. Erdélyi (ed.), *Tables of Integral Transforms*, vol. 1, 1954; P. K. Kuhfittig, *Introduction to the Laplace Transform*, 1978; D. V. Widder, *The Laplace Transform*, 1941; J. Williams, *Laplace Transforms*, 1973.

Laplace's differential equation

Laplace's equation in two independent variables x and y is given as Eq. (1) and is of central impor-

$$\frac{\partial^2 u(x,y)}{\partial x^2} + \frac{\partial^2 u(x,y)}{\partial y^2} = 0 \qquad (1)$$

tance in both pure mathematics and mathematical physics. A function $u(x,y)$ having continuous first and second partial derivatives and satisfying Laplace's equation in a neighborhood of a point is called harmonic at that point. If a plane piece of tinfoil has its edges kept at a temperature which varies from point to point but does not change with time, and if the flow of heat in the tinfoil is steady (that is, independent of the time), the temperature $u(x,y)$ at interior points of the foil is harmonic. Likewise Laplace's equation dominates the flow of electricity (the potential is similarly harmonic) and the flow of any incompressible fluid.

Two-dimensional relations. If $f(z) \equiv u(x,y) + iv(x,y)$ is an analytic function, $u(x,y)$ and $v(x,y)$ are conjugate functions and are harmonic; conversely, if $u(x,y)$ is harmonic in a simply connected region D, one may write Eq. (2), where (x_0, y_0) is fixed in D

$$v(x,y) \equiv \int_{(x_0,y_0)}^{(x,y)} \left(-\frac{\partial u}{\partial y} dx + \frac{\partial u}{\partial x} dy \right) \qquad (2)$$

and (x,y) arbitrary in D. It follows from Green's theorem that the integral over a path in D is independent of the path, so $v(x,y)$ is uniquely defined throughout D; the functions $u(x,y)$ and $v(x,y)$ are conjugate in D, and $f(z) \equiv u + iv$ is analytic there. Under these conditions, let C now be a regular Jordan curve in D; if n denotes the interior normal of C, the equation $\partial u/\partial n = -\partial v/\partial s$ follows from the Cauchy-Riemann equations, whence obtains Eq. (3). The first and last members of this equation

$$\int_C \frac{\partial u}{\partial n} ds = -\int_C \frac{\partial v}{\partial s} ds = -v(x,y) \Big|_C = 0 \qquad (3)$$

form the flux theorem, namely that the total flux (of heat if u is temperature) over C is zero. *See* COMPLEX NUMBERS AND COMPLEX VARIABLES.

If $u(x,y)$ is harmonic in the closed disk bounded by the circumference γ, and $f(z)$ the corresponding analytic function, one can take the real parts of both members of the equations expressing Cauchy's integral formula, as in Eqs. (4).

$$z - z_0 = \rho(\cos\theta + i\sin\theta) \qquad (4a)$$

$$dz = i(z - z_0) d\theta \qquad (4b)$$

$$f(z_0) = u(x_0, y_0) + iv(x_0, y_0) = \frac{1}{2\pi i} \int_\gamma \frac{f(z) dz}{z - z_0}$$

$$= \frac{1}{2\pi} \int_\gamma f(z) d\theta \qquad (4c)$$

$$u(x_0, y_0) = \frac{1}{2\pi} \int_\gamma u(x,y) d\theta \qquad (4d)$$

Equation (4d) expresses Gauss's mean value theorem, that the average of $u(x,y)$ over γ is the value at the center of γ. From this theorem it follows that a function harmonic at a point (x_0, y_0) cannot have a strong local maximum (or minimum) there, and can have a weak local maximum (or minimum) only if identically constant throughout a neighborhood of (x_0, y_0). If $u(x,y)$ is harmonic in a bounded region D, continuous in the correspond-

ing closed region $\overline{D}$, the maximum and minimum of $u(x,y)$ occur on the boundary of D; if a maximum or minimum occurs interior to D, then $u(x,y)$ is identically constant throughout D.

If D is a bounded region with boundary B, and if continuous values $U(x,y)$ are assigned on B, the Dirichlet problem is the problem of determining a function $u(x,y)$ harmonic in D, continuous on $D + B$, equal to $U(x,y)$ on B. If D is a circular region, a Jordan region, or any nonpathological region, the Dirichlet problem has a solution, necessarily (by the absence of nontrivial maxima and minima interior to D) unique. If D is a circular disk of radius a (see illustration), the Dirichlet problem for D is solved by Poisson's integral, Eq. (5), using polar

$$u(r,\theta) = \frac{1}{2\pi} \int_0^{2\pi} \frac{(a^2 - r^2) U(\psi) d\psi}{a^2 - 2ar\cos(\theta - \psi) + r^2} \qquad (5)$$

coordinates (r,θ) with pole the center of D. If D is a less elementary region but with smooth boundary B, the Dirichlet problem is solved by Green's formula, Eq. (6), where n indicates the interior normal in this formula.

$$u(x,y) = \frac{1}{2\pi} \int_B U(\xi,\eta) \frac{\partial g}{\partial n} ds(\xi,\eta) \qquad (6)$$

Green's function $g(x,y;\xi,\eta)$ is harmonic in D except at (x,y), continuous and equal to zero on B, and in the neighborhood of (x,y) has the form $\frac{1}{2}\log[(\xi - x)^2 + (\eta - y)^2] + g_1(\xi,\eta)$, where $g_1(\xi,\eta)$ is harmonic at (x,y). If the boundary B is not smooth, this formula can be expressed in terms of harmonic measure instead of $(\partial g/\partial n) ds$.

Numerous series expansions (for example, Fourier's series) can be used for the solution of the Dirichlet problem for various regions.

n-Dimensional relations. The foregoing remarks apply to Laplace's equation with two independent variables; the facts (but not the methods of proof using analytic functions) apply also in three or more dimensions. Thus, in three dimensions, a point distribution of matter of masses m_k at points (x_k, y_k, z_k) has a potential defined by Eq. (7),

$$u(x,y,z) \equiv \Sigma m_k [(x - x_k)^2 + (y - y_k)^2 + (z - z_k)^2]^{-1/2} \qquad (7)$$

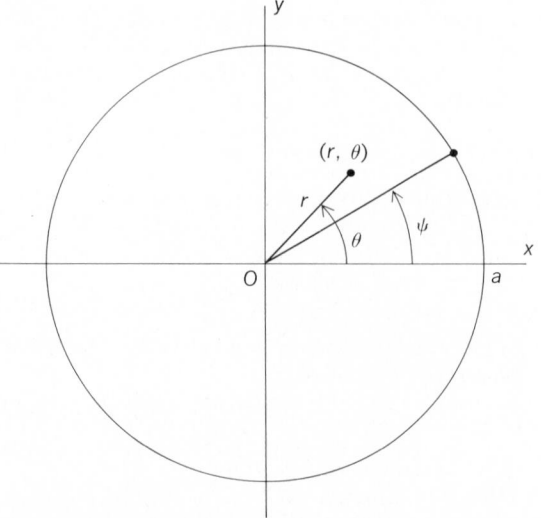

Circular disk D of radius a.

which is harmonic except in the points (x_k,y_k,z_k). Except at such points, the force (Newtonian law of gravitation) exerted by the distribution on a unit exploratory particle at (x,y,z) has the components $(\partial u/\partial x, \partial u/\partial y, \partial u/\partial z)$, and the component of the force in any direction is the directional derivative of $u(x,y)$ in that direction. *See* POTENTIALS; SPHERICAL HARMONICS.

[JOSEPH L. WALSH]

Bibliography: G. F. D. Duff and D. Naylor, *Differential Equations of Applied Mathematics*, 1966; O. D. Kellogg, *Foundations of Potential Theory*, 1929, reprint 1954; T. Masatugu, *Potential Theory in Modern Function Theory*, 1975.

Laplace's irrotational motion

Laplace's equation for irrotational motion of an inviscid, incompressible fluid is partial differential equation (1), where x_1, x_2, x_3 are rectangular carte-

$$\frac{\partial^2\phi}{\partial x_1{}^2}+\frac{\partial^2\phi}{\partial x_2{}^2}+\frac{\partial^2\phi}{\partial x_3{}^2}=0 \tag{1}$$

sian coordinates in an inertial reference frame, and Eq. (2) gives the velocity potential. The fluid

$$\phi=\phi(x_1,x_2,x_3,t) \tag{2}$$

velocity components u_1, u_2, u_3 in the three respective rectangular coordinate directions are given by $u_i=\partial\phi/\partial x_i$, $i=1,2,3$. More generally, in any inertial coordinate system, the equation is div (grad ϕ) $=0$ and the velocity vector is $\mathbf{v}=$ grad ϕ.

Irrotational motion implies that the fluid particles translate without rotation (like the cars on a ferris wheel) and is stated mathematically by saying curl $\mathbf{v}=0$ where $\mathbf{v}=\mathbf{v}(\mathbf{r},t)$ is the velocity vector, $\mathbf{r}$ is the position vector of a particular point in the fluid flow, and t is the time. If the fluid motion is at any time irrotational it will stay irrotational. Thus any motion starting from rest will be irrotational. If curl $\mathbf{v}=0$ then $\mathbf{v}$ may be written as grad ϕ because curl (grad ϕ) is identically zero. For an incompressible fluid, the continuity equation is div $\mathbf{v}=0$; hence, combining this relation with irrotationality gives Laplace's equation, div (grad ϕ) $=0$. *See* FLUID-FLOW PRINCIPLES.

The velocity field $\mathbf{v}(\mathbf{r},t)$ in a certain region is determined by Laplace's equation with a boundary condition given on the entire surface surrounding the region. The two most common boundary conditions are those at a solid surface and at a free surface. At a solid surface the fluid velocity normal to the surface must match the velocity of the surface normal to itself, $\mathbf{v}\cdot\mathbf{n}=\mathbf{v}_s\cdot\mathbf{n}$; that is, $\partial\phi/\partial n=\mathbf{v}\cdot\mathbf{n}$ is given on the boundary. At a free surface, such as one occurring between two fluids of different density, the pressure must be continuous; this boundary condition, in general, involves the use of the nonstationary Bernoulli equation and usually leads to wave motion. *See* BERNOULLI'S THEOREM; WAVE MOTION IN FLUIDS.

[ARTHUR E. BRYSON]

Larmor precession

A precession in a magnetic field of the motion of charged particles or of particles possessing magnetic moments.

Charged particles. The Larmor theorem (J. Larmor, 1897) states that, for electrons moving in a single central field of force, the motion in a uni-

form magnetic field H is, to first order in H, the same as a possible motion in the absence of H except for the superposition of a common precession of angular frequency given by Eq. (1). Here e/c is

$$\omega_L=eH/2mc \tag{1}$$

the magnitude of the electronic charge in electromagnetic units, and m is the electronic mass. The frequency ω_L is called the Larmor frequency and is numerically equal to 2π times 1.40 MHz per oersted or 2π times 111 MHz per SI unit of magnetic field strength (ampere-turn per meter). *See* PRECESSION.

The Larmor theorem is derived in numerous texts. For the special case of an electron moving in a circular orbit of radius r about a fixed nucleus, with H applied normal to the plane of the orbit, the derivation is as follows: The centripetal force holding the electron in orbit must equal $m\omega^2 r$ and is the sum of the Coulomb force Ze^2/r^2 and the Lorentz force, $(e/c)\omega rH$. Therefore Eq. (2) is valid, where

$$\omega=\pm[(eH/2mc)^2+(Ze^2/mr^3)]^{1/2}+(eH/2mc)$$

$$=\pm(\omega_L{}^2+\omega_0{}^2)^{1/2}+\omega_L \tag{2}$$

ω_0 is the angular frequency in the absence of H. If $\omega_0\gg\omega_L$ (bound electron, and first order in H) the approximate angular frequency is given by Eq. (3),

$$\omega=\pm\omega_0+\omega_L \tag{3}$$

which is the Larmor theorem. For a free or unbound electron (no Coulomb force), the approximation breaks down, but direct solution of the equation involving $m\omega^2 r$ and the Lorentz force yields $\omega=eH/mc$. This is twice the Larmor frequency and is called the cyclotron frequency. *See* PARTICLE ACCELERATOR.

In stating the Larmor theorem, use was made of the phrase "a possible motion." If H is applied sufficiently slowly, it can be proved that the motion is the same as in the absence of H, except for the superposition of the Larmor precession. However, a sudden application of H may change, for example, a circular orbit into an elliptical one. For an important application of the Larmor theorem *see* DIAMAGNETISM.

Magnetic moments. According to elementary electromagnetic theory, a current loop of area A and of current I possesses a magnetic moment $\boldsymbol{\mu}$ of magnitude IA and of direction normal to the loop. Thus an electron moving with a velocity v in a circular orbit of radius r, and hence with current $(-e/c)(v/2\pi r)$ in emu, has an orbital magnetic moment of magnitude as given by Eq. (4). *See* MAGNETIC MOMENT.

$$\mu=(-e/c)(v/2\pi r)(\pi r^2)=-(evr/2c) \tag{4}$$

The electron also has orbital angular momentum mvr, which by quantum theory must equal $\hbar J$, where J is an integer and $\hbar$ is Planck's constant h divided by 2π. The ratio of magnetic moment to angular momentum, Eq. (5), is called the magneto-

$$\gamma_J\equiv\frac{\mu}{\hbar J}=\frac{-e}{2mc} \tag{5}$$

gyric (and often the gyromagnetic) factor γ_J. *See* ANGULAR MOMENTUM; GYROMAGNETIC RATIO; NONRELATIVISTIC QUANTUM THEORY.

In terms of the equivalent magnetic moment, Eq. (1) may be written in the form of Eq. (6). In this

$$\omega_L = -\gamma_J H = -(\mu/\hbar J)\, H \qquad (6)$$

form the Larmor precession is exhibited by any magnetic moment **μ**, including magnetic moments associated with spin angular momentum as well as those associated with orbital angular momentum. Equation (6) may also be derived from equating the time rate of change of angular momentum (d/dt) $(\hbar \mathbf{J})$ to the magnetic torque $\boldsymbol{\mu} \times \mathbf{H}$, as in Eq. (7). In

$$d(\hbar \mathbf{J})/dt = \boldsymbol{\mu} \times \mathbf{H} = \gamma_J \hbar \mathbf{J} \times \mathbf{H} \qquad (7)$$

this form the Larmor precession applies to experiments in molecular beams, electron paramagnetic resonance (EPR), and nuclear magnetic resonance (NMR). *See* ELECTRON SPIN; MAGNETIC RESONANCE.

Rotating coordinate system. Let $(\partial/\partial t)$ represent differentiation with respect to a coordinate system rotating with angular velocity **ω**. Then differentiation with respect to a stationary observer (d/dt) is given by Eq. (8). Here **J** is mea-

$$(d\mathbf{J}/dt) = (\partial \mathbf{J}/\partial t) + (\boldsymbol{\omega} \times \mathbf{J}) \qquad (8)$$

sured by the stationary observer. This equation may be combined with Eq. (7) in the form of Eq. (9). Here $\mathbf{H}_r$ is the effective field in the rotating coordinate system as given by Eq. (10). Therefore,

$$\frac{\partial(\hbar\mathbf{J})}{\partial t} = \gamma_J \hbar \mathbf{J} \times \left(\mathbf{H} + \frac{\boldsymbol{\omega}}{\gamma_J}\right) = \gamma_J \hbar \mathbf{J} \times \mathbf{H}_r \qquad (9)$$

$$\mathbf{H}_r = \mathbf{H} + (\boldsymbol{\omega}/\gamma_J) \qquad (10)$$

in a frame which is rotating at the Larmor frequency, the effect of a constant field **H** is reduced to zero.

This result, which is an extension of Larmor's original theorem, also holds in quantum mechanics. It is the basis of simplified analyses of the effects of oscillating magnetic fields on particles with charges and magnetic moments.

[ELIHU ABRAHAMS; FREDERIC KEFFER]

Bibliography: R. P. Feynman, *The Feynman Lectures on Physics,* 1964; H. Goldstein, *Classical Mechanics,* 2d ed., 1980; E. M. Purcell, *Electricity and Magnetism,* Berkeley Physics Course, vol. 2, 1970; K. R. Symon, *Mechanics,* 3d ed., 1971.

Laser

A device that uses the principle of amplification of electromagnetic waves by stimulated emission of radiation and operates in the infrared, visible, or ultraviolet region. The term laser is an acronym for light amplification by stimulated emission of radiation, or a light amplifier. However, just as an electronic amplifier can be made into an oscillator by feeding appropriately phased output back into the input, so the laser light amplifier can be made into a laser oscillator, which is really a light source. Laser oscillators are so much more common than laser amplifiers that the unmodified word "laser" has come to mean the oscillator, while the modifier "amplifier" is generally used when the oscillator is not intended. *See* MASER.

The process of stimulated emission can be described as follows: When atoms, ions, or molecules absorb energy, they can emit light spontaneously (as in an incandescent lamp) or they can be stimulated to emit by a light wave. This stimulated emission is the opposite of (stimulated) absorption, where unexcited matter is stimulated into an excited state by a light wave. If a collection of atoms is prepared (pumped) so that more are initially excited than unexcited, then an incident light wave will stimulate more emission than absorption, and there is net amplification of the incident light beam. This is the way the laser amplifier works.

A laser amplifier can be made into a laser oscillator by arranging suitable mirrors on either end of the amplifier. These are called the resonator. Thus the essential parts of a laser oscillator are an amplifying medium, a source of pump power, and a resonator. Radiation that is directed straight along the axis bounces back and forth between the mirrors and can remain in the resonator long enough to build up a strong oscillation. (Waves oriented in other directions soon pass off the edge of the mirrors and are lost before they are much amplified.) Radiation may be coupled out as shown in Fig. 1 by making one mirror partially transparent so that part of the amplified light can emerge through it. The output wave, like most of the waves being amplified between the mirrors, travels along the axis and is thus very nearly a plane wave. *See* OPTICAL PUMPING.

Comparison with other sources. In contrast to lasers, all conventional light sources are basically hot bodies which radiate by spontaneous emission. The electrons in the tungsten filament of an incandescent lamp are agitated by, and acquire excitation from, the high temperature of the filament. Once excited, they emit light in all directions and revert to a lower energy state. Similarly, in a gas lamp, the electron current excites the atoms to high-energy quantum states, and they soon give up this excitation energy by radiating it as light. In all the above, spontaneous emission from each excited electron or atom takes place independently of emission from the others. Thus the overall wave produced by a conventional light source is a jumble of waves from the numerous individual atoms. The phase of the wave emitted by one atom has no relation to the phase emitted by any other atom, so that the overall phase of the light fluctuates randomly from moment to moment and place to place. The lack of correlation is called incoherence.

Hot bodies emit more or less equally in all directions radiation whose wavelength distribution is dictated by the Planck blackbody radiation curve. For example, the surface of the Sun radiates like a blackbody at a temperature of about 6000 K, and emits a total of 7 kW/cm² , spread out over all wavelengths and directions. Light from gas lamps can be more monochromatic (wavelengths radiat-

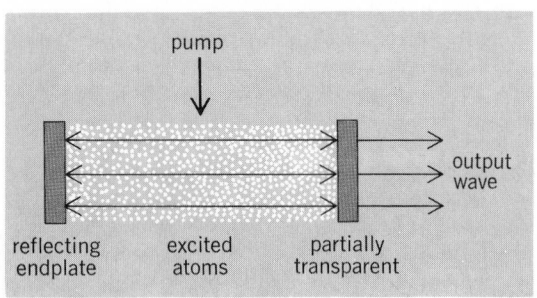

Fig. 1. Structure of a parallel-plate laser.

ed are restricted by the quantized energies allowed in atoms), but radiation still occurs in all directions. In contrast to this, an ideal plane wave would have the same phase all across any wavefront, and the time fluctuations would be highly predictable (coherent). The output of the parallel-plate laser described above is very nearly such a plane wave and is therefore highly directional. This arises because in the laser oscillator atoms are stimulated to emit in phase with the stimulating wave, rather than independently, and the wave that builds up between the mirrors matches very closely the mirrors' surfaces. The output is powerful because atoms can be stimulated to emit much faster than they would spontaneously. It is highly monochromatic largely because stimulated emission is a resonance process that occurs most rapidly at the center of the range of wavelengths that would be emitted spontaneously. Since atoms are stimulated to emit in phase with the existing wave, the phase is preserved over many cycles, resulting in the high degree of time coherence of laser radiation. *See* COHERENCE; HEAT RADIATION.

The various types of lasers are discussed below, classified according to their pumping (or excitation) scheme. The function of the pumping system is to maintain more atoms in the upper than in the lower state, thereby assuring that stimulated emission (gain transitions) will exceed (stimulated) absorption (loss transitions). This so-called population inversion ensures net gain (amplification greater than unity).

Optically pumped lasers. One way to achieve population inversion is by concentrating light (for example, from a flash lamp or the Sun) onto the amplifying medium. Alternatively, lasers may be used to optically pump other lasers. For example, powerful continuous-wave (cw) ion lasers can pump liquid dyes to lase, yielding watts of tunable visible and near-visible radiation. Molecular lasers, like carbon dioxide (CO_2), can pump gases, like deuterium cyanide (DCN), to lase powerfully in the far infrared, operating out to several hundred micrometers wavelength. Optical pumping can be employed to pump gases at very high pressures (for example, 42 atm or 4.3 MPa) to obtain tunability where other excitation methods would be difficult if not impossible to implement.

Many lasers are three-level lasers; that is, ground-state atoms are excited by absorption of light to a broad upper-energy state, from which they quickly relax to the emitting state. Laser action occurs as they are stimulated to emit radiation and so return to the original ground state. Crystalline, glass, liquid, and gaseous systems have been found suitable, but many possible materials remain to be explored. Solid three-level lasers usually make use of ions of a rare-earth element, such as neodymium, or of a transition metal, such as chromium, dispersed in a transparent crystal or glass. For example, ruby, which is crystalline aluminum oxide containing a fraction of a percent of trivalent chromium ions in place of aluminum ions, has been used for lasers to produce red light with wavelengths of 693 to 705 nm. The chromium ions have broad absorption bands in which pumping radiations can be absorbed. Thus broad-band white light can be used to excite the atoms.

Many rare-earth ion lasers use a fourth level above the ground state. This level serves as a ter-

minal level for the laser transition, and is kept empty by rapid nonradiative relaxation to the ground state. This means that, relative to three-level systems, a population inversion is easier to maintain, and therefore such materials require relatively low pumping light intensity for laser action. Neodymium ions can provide laser action in many host materials, producing outputs in the infrared, around 1 μm. In glass, which can be made in large sizes, neodymium ions can generate high-energy pulses or very high peak powers (Fig. 2).

Fig. 2. High-power laser amplifier stage, with a 30-cm aperture, using liquid-cooled slabs of neodynium glass. (*Lawrence Livermore Laboratory*)

Lasers using neodymium ions in crystals such as yttrium aluminum garnet (YAG) can provide continuous output powers up to a kilowatt. The output of either type (in glass or in crystals) can be converted to visible light, near 500 nm, by a harmonic generator crystal, as discussed below. Optically pumped solid-state lasers provide relatively high peak output powers. Tens of kilowatts can easily be obtained in a pulse lasting 100 μs. Much higher peak powers can be obtained by special techniques described below.

The structure of an optically pumped solid-state laser can be as simple as a rod of the light-amplifying material with parallel ends polished and coated to reflect light. Pumping radiation can enter either through the transparent sides or the ends. Other structures can be used. The mirror ends can be spherical rather than plane, with the common focal point of the two mirrors lying halfway between them. Still other structures make use of internal reflection of light rays that strike the surface of a crystal at a high angle.

Liquid lasers have structures generally like those of optically pumped solid-state lasers, except that the liquid is generally contained in a transparent cell. Some liquid lasers make use of rare-earth ions in suitable dissolved molecules, while others make use of organic dye solutions. The dyes can lase over a wide range of wavelengths, depending upon the composition and concentration of the dye or solvent. Thus tunability is obtained throughout the visible, and out to a wavelength of about 0.9 μm. Fine adjustment of the output wavelength can be provided by using a diffraction grating or other dispersive element in place of one of the laser mirrors. The grating acts as a good mirror for only one wavelength, which depends on the angle at which it is set. With further refinements, liquid-dye lasers can be made extremely monochromatic as well as broadly tunable. They may be pumped by various lasers to generate either short, intense pulses or continuous output. Dyes may also be incorporated into solid media such as plastics or gelatin to provide tunable laser action. Then the tuning may be controlled by a regular corrugation in the refractive index of the host medium, which acts as a distributed Bragg reflector. The reflection from any one layer is small, but when the successive alternations of refractive index provide reflections in phase, the effect is that of a strong, sharply tuned reflection. *See* X-RAY DIFFRACTION.

Certain color centers in the alkali halides can be optically pumped to make efficient, tunable lasers for the region 0.8 to 3.3 μm, thus taking over just where the organic dyes fail. Both pulsed (10 kW) and continuous (watts) operation have been achieved. *See* COLOR CENTERS.

In several infrared regions, tunable laser action can be obtained by using an infrared gas laser to pump a semiconductor crystal in a magnetic field, giving amplification by stimulating spin-flip Raman scattering from the electrons in the semiconductor. Tuning is achieved by varying the magnetic field.

Gas-discharge lasers. Another large class of lasers makes use of nonequilibrium processes in a gas discharge. At moderately low pressures (on the order of 1 torr or 10^2 Pa) and fairly high currents, the population of energy levels is far from an equilibrium distribution. Some levels are populated especially rapidly by the fast electrons in the discharge. Other levels empty particularly slowly and so accumulate large numbers of excited atoms. Thus laser action can occur at many wavelengths in any of a large number of gases under suitable discharge conditions. For some gases, a continuous discharge, with the use of either direct or radio-frequency current, gives continuous laser action. Output powers of continuous gas lasers range from less than 1 μW up to about 100 W in the visible region. Wavelengths generated span the ultraviolet and visible regions and extend out beyond 700 μm in the infrared. They thus provide the first intense sources of radiation in much of the far-infrared region of the spectrum.

The earliest, and still most widely used, gas-discharge laser utilizes a mixture of helium and neon. Various infrared and visible wavelengths can be generated, but most commonly they produce red light at a 632.8-nm wavelength, with power outputs of a few milliwatts or less, although they can be as high as about a watt. Helium-neon lasers can be small and inexpensive. Argon and krypton ion gas–discharge lasers provide a number of visible and near-ultraviolet wavelengths with continuous powers typically about 1 to 10 W, but ranging up to more than 100 W. Unfortunately, efficiencies are low.

Many molecular gases, such as hydrogen cyanide, carbon monoxide, and carbon dioxide, can provide infrared laser action. Carbon dioxide lasers can be operated at a number of wavelengths near 10 μm on various vibration-rotation spectral lines of the molecule. They can be relatively efficient, up to about 30%, and have been made large

enough to give continuous power outputs exceeding tens of thousands of watts.

Many gas-discharge lasers, for example, helium-neon, produce only very small optical gain; thus losses must be kept low. Consequently, mirrors with very high reflectivity must be used. Diffraction losses can be kept low by using curved mirrors. One common arrangement, which combines relatively low diffraction losses with good mode selection, uses a flat mirror at one end and a spherically concave mirror at the other. The spacing between mirrors is made equal to the radius of curvature of the spherical mirror. On the other hand, in some of the higher-power lasers even the plane-parallel mirror structure does not give sufficient discrimination against those undesired modes of oscillation which cause the beam to be excessively divergent. It is then helpful to use "unstable" resonators, with at least one of the mirrors convex toward the other.

A smooth, small-bore dielectric tube can guide a light wave in its interior with little loss. Thus a light wave can be amplified by a long, narrow medium without spreading or diffraction. A gas discharge in a hollow optical waveguide can be run at high pressure and benefits from cooling by the nearby walls, so that relatively high-power outputs (watts) can be obtained from a small volume. Waveguide structures are also useful when a medium is pumped optically by another laser, whose narrow beam can be confined within the bore of the tube. For example, pumping of various molecules, such as methyl fluoride, by a carbon dioxide laser has been used to generate coherent light in the very-far-infrared (submillimeter wavelength) region.

Pulsed gas lasers. Pulsed gas discharges permit a further departure from equilibrium. Thus pulsed laser action can be obtained in some additional gases which could not be made to lase continuously. In some of them the length of the laser pulse is limited when the lower state is filled by stimulated transitions from the upper level, and so introduces absorption at the laser wavelength. An example of such a self-terminating laser is the nitrogen laser, which gives pulses of several nanoseconds duration with peak powers from tens of kilowatts up to a few megawatts at a wavelength of 337 nm in the ultraviolet. They are easy to construct and are much used for pumping tunable dye lasers throughout the wavelength range from about 350 to 1000 nm. Very powerful laser radiation in the vacuum ultraviolet region, between 100 and 200 nm, can be obtained from short-pulse discharges in hydrogen, and in rare gases such as xenon at high pressure. (High pressure leads to higher power.) When the gas pressure is too high to permit an electric discharge, excitation may be provided by an intense burst of fast electrons from a small accelerator, the so-called E beam.

Some gases, notably carbon dioxide, which can provide continuous laser action, also can be used to generate intense pulses of microsecond duration. For this purpose, gas pressures of about 1 atm (10^5 Pa) are used, and the electrical discharge takes place across the diameter of the laser column, hence the name transverse-electrical-atmospheric (TEA) laser.

Chemical lasers. It is also possible to obtain laser action from the energy released in some fast chemical reactions. Atoms or molecules produced during the reaction are often in excited states. Under special circumstances there may be enough atoms or molecules excited to some particular state for amplification to occur by stimulated emission. Usually the reacting gases are mixed and then ignited by ultraviolet light or fast electrons. Both continuous infrared output and pulses up to several thousand joules of energy have been obtained in reactions which produce excited hydrogen fluoride molecules.

Pulsed laser action in the ultraviolet (193 to 353 nm) has been obtained from excimer states of rare gas monohalides (for example, KrF, XeF, KrCl, XeCl, XeBr). Such molecules have ground states which are unstable, thereby making a population inversion easy to achieve. Although these lasers require an electrical source for initial gas reaction, laser pumping is dependent on chemical reactions.

Photodissociation lasers. Intense pulses of ultraviolet light can dissociate molecules in such a way as to leave one constituent in an excited state capable of sustaining laser action. The most notable examples are iodine compounds, which have given peak 1.3-μm pulse powers above 10^9 W from the excited iodide atoms.

Nuclear lasers. Laser action in several gases has also been excited by the fast-moving ions produced in nuclear fusion. These fusion products excite and ionize the gas atoms, and make it possible to convert directly from nuclear to optical energy.

Gas-dynamic lasers. When a hot molecular gas is allowed to expand suddenly through a nozzle, it cools quickly, but different excited states lose energy at different rates. It can happen that, just after cooling, some particular upper state has more molecules than some lower one, so that amplification by stimulated emission can occur. Very high continuous power outputs have been generated from carbon dioxide in large gas-dynamic lasers.

Semiconductor lasers. Another method for providing excitation of lasers can be used with certain semiconducting materials. Laser action takes place when free electrons in the conduction band are stimulated to recombine with holes in the valence band. In recombining, the electrons give up energy corresponding nearly to the band gap. This energy is radiated as a light quantum. Suitable materials are the direct-gap semiconductors, such as gallium arsenide. In them, recombination occurs directly without the emission or absorption of a quantum of lattice vibrations. A flat junction between p-type and n-type material may be used. When a current is passed through this junction in the forward direction, a large number of holes and electrons are brought together. This is called recombination, and is accompanied by emission of radiation. A light wave passing along the plane of the junction can be amplified by stimulating such recombination of electrons and holes. The ends of the semiconducting crystal provide the mirrors to complete the laser structure. In indirect-gap semiconductors, such as germanium and silicon, only a small amplification by stimulated emmision is possible because of their requirement for interaction with the lattice vibrations.

Semiconductor lasers can be very small, less than 1 mm in any direction. They can have efficiencies higher than 50% (electricity to light).

Power densities are high, but the thinness of the active layer tends to limit the total output power. Even so, maximum continuous powers are comparable with those of other moderate-size lasers. Since semiconductor lasers are so small, they can be assembled into compact arrays of many units, so as to generate higher peak powers. An alternative excitation method, bombardment of the semiconductor by a high-voltage beam of electrons, may provide laser action in larger crystal volumes, but it is likely to cause damage to the crystal. *See* SEMICONDUCTOR.

Free-electron lasers. Free-electron lasers are of interest because of their potential for efficiently producing very high-power radiation, tunable from the millimeter to the x-ray region. The principle of operation, so far demonstrated only at 3.4 μm and at very low-power levels, involves passage of electrons through a spatially varying magnetic field which causes the electron beam to "wiggle" and hence to radiate. The large Doppler upshift due to relativistic electron velocities can be adjusted, resulting in tunable emission at optical frequencies. *See* DOPPLER EFFECT.

High-power, short-pulse lasers. The output power of pulsed lasers can be greatly increased, with correspondingly shorter pulse durations, by the Q-switch technique. In this method, the optical path between one mirror and the amplification medium is blocked by a shutter. The medium is then excited beyond the degree ordinarily needed, but the shutter prevents laser action. At this time the shutter is abruptly opened and the stored energy is released in a giant pulse (1–100 MW peak power, lasting 1–30 ns for optically pumped solid-state lasers). Still higher peak powers can be obtained by passing this output through a traveling-wave laser amplifier (without mirrors). Peak powers in excess of 100 MW have been obtained in this way.

Still shorter, and higher-power, pulses can be generated by mode-locking techniques. A typical laser without mode locking usually oscillates simultaneously and independently at several closely spaced wavelengths. These modes of oscillation can be synchronized so that the peaks of their waves occur simultaneously at some instant. The result is a very short, intense pulse which quickly ends as the waves of different frequency get out of step. Mode-locked lasers have generated pulses shorter than 1 ps. Since such brief pulses tend to produce somewhat less damage to materials, they can be amplified to very high peak intensities. Power outputs of picosecond pulses as high as 10–100 MW have been obtained, limited as in the Q-switch case by material damage. *See* OPTICAL PULSES.

For the highest peak power, the output may be further intensified by additional stages of laser amplification. The beam diameter is increased by some optical arrangement, such as a telescope, so as to expand (dilute) the beam and thereby prevent damage to the laser material and optics. Sometimes the amplifying medium is divided into flat slabs separated by cooling liquid (Fig. 2). Here the open faces of the light-amplifying slabs present a large area to receive pumping light and liquid cooling.

Development of very large multistage lasers has been undertaken for research on thermonuclear fusion. In a particularly large one, at Lawrence Livermore Laboratory, a single neodymium-glass laser oscillator is designed to drive 20 amplifier chains of glass rods and disks, each delivering a pulse of more than 10^{12} W. Focusing all these pulses onto the surface of a small pellet of heavy hydrogen is designed to heat and compress the pellet by ablation, until it is so hot that the hydrogen nuclei fuse to produce helium and release large amounts of nuclear energy. Ultimately this type of controlled laser fusion may become an important source of thermal, electrical, and chemical energy. *See* NUCLEAR FUSION.

The technology of scaling up lasers to higher and higher powers has been so successful that the limitations are often set by material damage thresholds of the laser medium or associated optics. At the high-power densities attainable by focusing laser beams, the electric fields of the light can be large. Thus when the light intensity is 10^{12} W/cm^2, the corresponding electric field is 10^7 V/cm. To such large fields, many transparent materials have a nonlinear dielectric response. This nonlinearity can be large enough to permit the generation of optical harmonics. It is possible, with careful design and good nonlinear materials, to obtain substantially complete conversion of a laser's output to the second harmonic, at twice the frequency, even for continuous lasers near 1-W output power. Nonlinear dielectrics can be used as mixers to give the sum of difference of two laser frequencies. They also permit the construction of optical parametric oscillators which, when pumped by a laser, can generate coherent light tunable over a wide range of wavelengths. *See* NONLINEAR OPTICS.

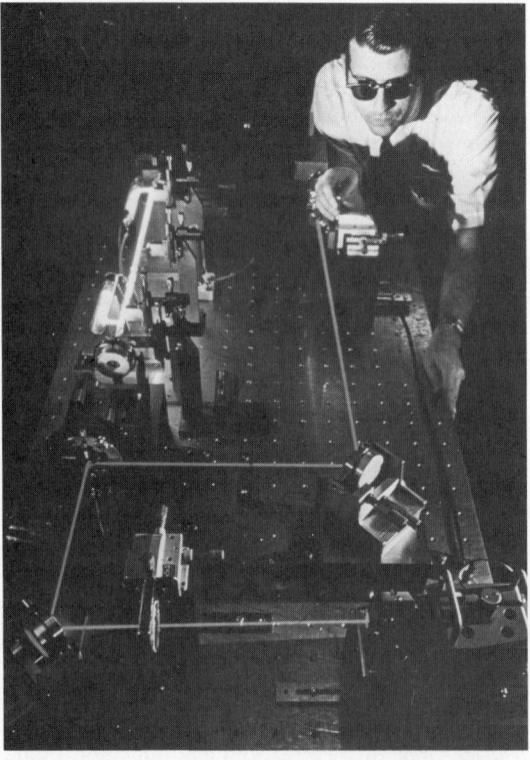

Fig. 3. Experimental arrangement to impress pulse-code modulation on a laser beam for optical communication. (*Bell Laboratories*)

Applications. The variety of technological uses for lasers has increased steadily since their appearance in 1960. Among the noticeable applications are those that utilize high-speed controllability of the tiny focal spot of a laser beam. For example, high-speed automatic scanners identify library cards, ski passes, and supermarket purchases and perform a variety of functions known as optical processing. Other uses for the laser beam's programmable control include information storage and retrieval (including three-dimensional holography and video disk reading), laser printing, micromachining, and automated cutting. Further applications involving high power include weaponry, laser welding, laser surgery (self-cauterizing), laser fusion, and materials processing. Optical communications utilize the laser's high frequency, which makes possible high information capacity (Fig. 3.) Except for space applications, laser light communication is primarily done through glass fibers. Low-loss optical fibers are far more compact and economical than copper wires. Bright laser beams are used by the construction industry to align straight excavations and for surveying. Some of the more specialized laser applications include laser gyros, laser velocity sensing (laser infrared radar or lidar), optical testing, metrology (including the distance to the Moon), laser spectroscopy (including pollution monitoring), and lasers to pump other lasers, and nonlinear processes are used to produce beams of coherent light at other wavelengths from millimeters to x-rays, and for exploration of ultrafast phenomena (by using picosecond laser pulses). *See* HOLOGRAPHY; INTEGRATED OPTICS; OPTICAL FIBERS.

[STEPHEN F. JACOBS; ARTHUR L. SCHAWLOW]

Bibliography: S. F. Jacobs et al. (eds.), *Free Electron Generators of Coherent Radiation*, 1980; S. F. Jacobs et al., *Laser Photochemistry, Tunable Lasers, and Other Topics*, 1976; B. A. Lengyel, *Lasers*, 1969; L. Marton and C. Marton, *Methods of Experimental Physics*, vol. 15, 1979; D. C. O'Shea, W. R. Callen, and W. T. Rhodes, *An Introduction to Lasers and Their Applications*, 1977; A. E. Siegman, *Introduction to Lasers and Masers*, 1971; D. C. Sinclair and W. E. Bell, *Gas Laser Technology*, 1969; A. Yariv, *Introduction to Optical Electronics*, 1976.

Lattice vibrations

Periodic oscillation of the atoms in a crystal lattice about their equilibrium positions. As the crystal is heated, the amplitude of the vibrations increases. If the heating is continued, the temperature of the crystal eventually reaches a value at which the vibrations are so violent that the atoms break away from their lattice sites, and the solid melts. On the other hand, if the crystal is cooled to absolute zero, the amplitude of the vibrations does not subside entirely. A residual vibration of the atoms, which is quantum-mechanical in origin, remains. It is called the zero-point vibration.

Lattice vibrations are involved in many of the temperature-dependent phenomena of a solid. For example, the electrical resistance of a metal at room temperature arises primarily from the scattering of the conduction electrons by the vibrating atoms. The higher the temperature, and hence the more violent the vibrations of the atoms, the more the electrons are scattered, and the higher the electrical resistance becomes. A ferromagnetic solid becomes paramagnetic at temperatures above the Curie temperature because the lattice vibrations are then sufficiently energetic to overcome the interatomic magnetic forces. The Bardeen-Cooper-Schrieffer theory of superconductivity postulates that subtle interactions between lattice vibrations and conduction electrons are mainly responsible for this phenomenon. *See* SUPERCONDUCTIVITY.

Lattice vibrations conduct heat through a crystal; thus a knowledge of the vibration mechanism is essential to an understanding of the heat conductivity of crystalline solids.

The theory of lattice vibrations began as an effort to explain specific heat. Theory shows that the classical treatment of vibrations of atoms in solids leads to the Dulong-Petit law of specific heats, a law which is not obeyed at all temperatures. To account for this failure, Albert Einstein formulated a theory which assumes that each atom vibrates with the same frequency and that the energy of vibration is quantized. Then followed the improvements on the theory made by P. Debye and by M. Born and T. von Kármán. In both of these theories, the single Einstein frequency is replaced by an acoustical spectrum. The theory of Born and von Kármán requires that this spectrum be obtained from the study of the vibrations of a system of point particles distributed in the same manner as the atoms in a crystal lattice. The various models representing a given solid differ from one another essentially in the assumptions made concerning the nature of the forces between the constituent particles. It is usually in the sense of the Born–von Kármán theory and its extensions that one speaks of lattice vibrations. *See* SPECIFIC HEAT OF SOLIDS.

One-dimensional lattice. Some of the main concepts of the Born–von Kármán theory are illustrated by the vibrations of a simple one-dimensional lattice. Figure 1 shows such a lattice composed of equidistant identical particles of mass M. The equilibrium separation of neighboring particles is

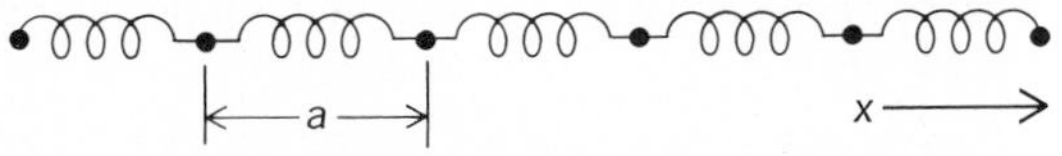

Fig. 1. Schematic diagram of a simple one-dimensional lattice of equidistant identical particles.

a. They are located along an x axis, the location of the nth particle being $x = na$. Assume that the forces between the particles are effective only between nearest neighbors and that these forces obey Hooke's law. In Fig. 1, the forces are represented by springs with Hooke's constant α.

Let u_n denote the displacement along the x axis of the nth particle from its equilibrium position at $x = na$. The force exerted on the nth particle depends upon the distance which the two springs connected to it are stretched. The extensions are $(u_n - u_{n-1})$ for the left-hand spring and $(u_{n+1} - u_n)$ for the right-hand spring. Thus the equation of motion of the nth particle is Eq. (1).

$$M(d^2 u_n/dt^2) = -\alpha(u_n - u_{n-1}) - \alpha(u_n - u_{n+1}) \quad (1)$$

Let the length of the lattice be $L = Na$, so that

there are $(N + 1)$ particles in all. The number N is taken to be large, comparable with Avogadro's number, so that whether the end particles are free to move or are fixed in position is of no practical importance here. To solve Eq. (1), write tentatively the equation of a wave of frequency ν and amplitude A, namely, Eq. (2), where $k = 1/\lambda$, the reciprocal of the wavelength. However, the lattice is dis-

$$u_n = A \exp[2\pi i(\nu t - kx)] \quad (2)$$

crete and the wave has no real existence between the atoms. This leads one to put $x = na$. The trial solution is now given by Eq. (3). Substitution shows

$$u_n = A \exp[2\pi i(\nu t - kna)] \quad (3)$$

that this expression satisfies Eq. (1) provided that Eq. (4) holds, where $\nu_0 = \sqrt{a/\pi^2 M}$. It is evident

$$\nu = \nu_0 \sin \pi ak \quad (4)$$

that ν_0 is the highest permissible frequency. If the end particles are assumed to be immovable, then standing-wave solutions are appropriate and are analogous to those of the continuous string with fixed ends. The restrictions on the wavelengths are $j\lambda/2 = L$, where j is an integer indicating the number of loops in the particular standing wave. Thus Eq. (4) becomes Eq. (5), the range of j being

$$\nu = \nu_0 \sin(\pi j/2N) \quad (5)$$

$1 < j < N$. Each kind of vibration is called a mode of vibration, and there is a mode of vibration for each j. Thus there are N modes. The total number of modes is also equal to the number of particles, because N and $(N + 1)$ are about equal.

The phase velocity U_p and the group velocity U_g can be calculated from Eq. (3) by means of the definitions $U_p = \nu/k$ and $U_g = d\nu/dk$. The results are given by Eqs. (6) and (7). Here U_∞ is the veloc-

$$U_p = U_\infty \frac{\sin \pi ak}{\pi ak} \quad (6)$$

$$U_g = U_\infty \cos \pi ak \quad (7)$$

ity of sound at long wavelengths and has the value $\pi a\nu_0$. By long wavelengths is meant wavelengths which are great compared with the equilibrium distance a.

An important characteristic revealed by this treatment is the dependence of the velocity on wavelength. Such a dependence is called dispersion. It becomes increasingly important as the wavelength approaches the value $2a$. At the cutoff wavelength $\lambda = 2a$, the group velocity is zero. A continuum theory, such as Debye's, gives simply $U_p = U_g = U_\infty = \pi a\nu_0 =$ constant, and the medium is dispersionless.

The number of modes $g(\nu)d\nu$ with frequencies between ν and $\nu + d\nu$ can be calculated from Eq. (5) by noticing that this number is also given by dj. From this, the distribution function $g(\nu)$ is given by Eq. (8). Figure 2 shows a plot of $g(\nu)$ for the lattice

$$g(\nu) = 2N/(\pi \sqrt{\nu_0^2 - \nu^2}) \quad (8)$$

as given by Eq. (8) and for the Debye continuum theory. The latter is simply $g(\nu) = 2N/\pi\nu_0$, with the range $0 < \nu \leqq \pi\nu_0/2$.

The energy E of the lattice is given by Eq. (9),

$$E = \int_0^\infty \frac{g(\nu) h\nu \, d\nu}{e^{(h\nu/kT)} - 1} \quad (9)$$

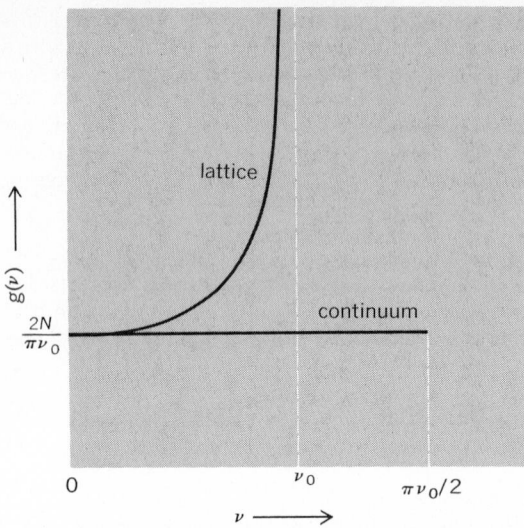

Fig. 2. The function $g(\nu)$ for the one-dimensional lattice and the one-dimensional continuum.

where h is Planck's constant, k is Boltzmann's constant, and T is the absolute temperature.

Three-dimensional lattices. In three-dimensional models, some attempt is usually made to approximate the conditions in actual crystals. At the same time, the considerable mathematical complexity of the problem requires that the model be kept as simple as possible. A decision has to be made concerning both the number of neighbors to be included in the calculation and the types of forces acting between the particles before the equations of motion can be formulated.

The forces are of several kinds. A force which arises when the distance between a pair of neighbors changes is called a central or radial force. This is the type of force assumed for the one-dimensional lattice discussed earlier. If a change in the angle between the pair of lines joining a given particle to two neighbors (bond lines) gives rise to a force, the force is called an angular force. A more general force compounded of radial and angular forces is often referred to as a noncentral force. The most general combination of noncentral forces compatible with symmetry requirements about the particles is commonly called a tensor force.

The types of forces just named are defined for a pair of particles or a pair of bond lines. The motions of the other particles or bond lines in the vicinity of the pair do not affect the force between the pair. In metals, however, the conduction electrons give rise to a type of force which cannot be defined in this manner. To visualize this, imagine that a group of particles (positive ions) move toward each other, thus increasing the local density. The corresponding increase in positive charge density is immediately compensated by a flow of conduction electrons into this region. However, the compressibility of this assembly of electrons, called an electron gas, is small, and the gas resists this increase in its density. The electrons would thus screen the ions less completely, the ions would repeal one another more strongly, and consequently a "stiffening" would occur in the elastic constants involved. To summarize, additional forces arise to oppose any change in particle density. This is clearly a cooperative or collective effect

not covered by the types of forces just listed. The forces produced by this collective effect are sometimes called volume forces.

A common objective of the theory of lattice vibrations is the calculation of the distribution function $g(\nu)$. Analytic methods generally cannot be used and the calculation must be done numerically. R. B. Leighton (1948) calculated $g(\nu)$ for a model of a face-centered cubic lattice with central forces between nearest neighbors and between next-nearest neighbors. Figure 3 shows his result, obtained after much labor, for a selection of force constants approximating copper. The availability of high-speed digital computers has reduced the labor to the extent that calculations of distribution functions are becoming increasingly common.

Figure 3 shows three branches of the distribution function, labeled I, II, and III. In an elastically isotropic solid, branches I and II would correspond to transverse waves and branch III to longitudinal waves, for $\lambda \gg 2a$. In an elastically anisotropic crystal, very few waves are purely transverse or longitudinal, even for $\lambda \gg 2a$. Thus in actual crystals, the waves constituting branches I and II are only predominantly transverse in character and those in branch III only predominantly longitudinal. Lattices in which each atom has exactly the same environment as any other atom are Bravais lattices, and these always have three branches in the distribution function. The distribution functions of non-Bravais lattices such as NaCl and diamond have additional high-frequency branches, called optical branches. The lower-frequency branches, corresponding to the three branches of the Bravais lattices, are acoustical branches.

Experimental studies. The results of the calculations of $g(\nu)$ for a given model are often tested by computing the specific heat and then comparing the result with the measured specific heat. The comparisons which have been made leave little doubt that the Born–von Kármán theory can account satisfactorily for the discrepancies between the Debye theory of specific heats and the experimental measurements. A particular disadvantage of the specific-heat approach is the insensitivity of the specific heat to the detailed shape of the

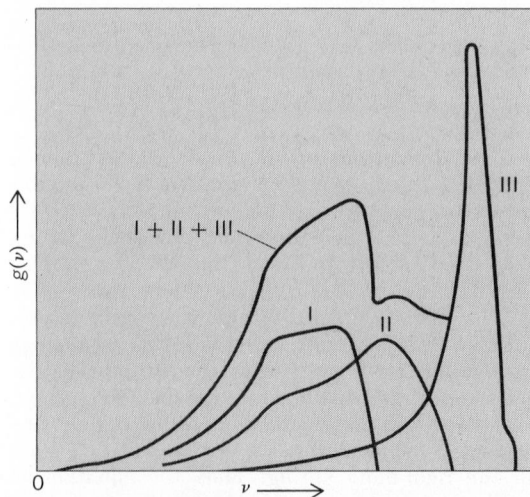

Fig. 3. The function $g(\nu)$ obtained by Leighton for a face-centered cubic lattice. (*After R. B. Leighton, Rev. Mod. Phys.*, 20:165–174, 1948)

distribution function for frequencies beyond the first peak.

Another experimental approach is the study of the scattering of x-rays by lattice vibrations. A static lattice should produce sharp diffraction patterns conforming to the regularity of the lattice. If a wave now passes through such a lattice, a periodic error is produced in the lattice and a pair of spots (ghosts) may occur on either side of the principal diffraction maximum. If the combined effect of all possible waves is considered, the result is a broad diffused spot centered at the diffraction maximum. The theory of this effect was first given in 1918 by H. Faxén and experimentally observed in 1938 by J. Laval. The theory and experimental methods are developed to the extent that dispersion curves for lattice vibrations can be determined from the observations. These are determined for special directions in the crystal, using monochromatic x-rays. The dispersion curves so obtained are then compared with those given by various models. Primary difficulties are the correction for Compton scattering and the determination of the dispersion curve near its ends. Generally, the method seems to be poorest where the specific-heat method is best, so that the two methods can be used advantageously to supplement one another.

A third experimental approach is the study of the inelastic scattering of cold (very slow) neutrons by lattice vibrations. It is possible to use an almost monochromatic beam of neutrons with energies considerably less than those acquired by thermal scattering on passage through the lattice. This is contrary to what happens to x-rays, for which the photon energy is so great as to be scarcely changed by thermal scattering. The theory of neutron scattering becomes simplified if the scattering is also incoherent. *See* NEUTRON DIFFRACTION.

[JULES DE LAUNAY]

Bibliography: N. W. Ashcroft and N. D. Mermin, *Solid State Physics*, 1976; S. Fluegge (ed.), *Handbuch der Physik*, vol. 7, pt. 1, 1955; A. K. Ghatak and L. S. Kothari, *Introduction to Lattice Dynamics*, 1972; C. Kittel, *Introduction to Solid State Physics*, 5th ed., 1976; F. Seitz and D. Turnbull (eds.), *Solid State Physics*, vol. 2, 1956.

Lawson criterion

A necessary but not sufficient condition for the achievement of a net release of energy from nuclear fusion reactions in a fusion reactor. As originally formulated by J. D. Lawson, this condition simply stated that a minimum requirement for net energy release is that the fusion fuel charge must combust for at least enough time for the recovered fusion energy release to equal the sum of energy invested in heating that charge to fusion temperatures, plus other energy losses occurring during combustion.

The result is usually stated in the form of a minimum value of $n\tau$ that must be achieved for energy break-even, where n is the fusion fuel particle density and τ is the confinement time. Lawson considered bremsstrahlung (x-ray) energy losses in his original definition. For many fusion reactor cases, this loss is small enough to be neglected compared to the heating energy. With this simplifying assumption, the basic equation from which the Lawson criterion is derived is obtained by balancing fusion energy release against heat input to the fuel plasma. Assuming hydrogenic isotopes, deuterium

and tritium at densities n_D and n_T respectively, with accompanying electrons at density n_e, all at a maxwellian temperature T, one obtains Eq. (1),

$$n_D n_T \langle \sigma v \rangle Q \tau \eta_r \geq \left[\frac{3}{2}kT\left(n_D + n_T + n_e\right)\right]\frac{1}{\eta_h} \qquad (1)$$

where the recovered fusion energy release is set equal to or greater than the energy input to heat the fuel. Here $\langle \sigma v \rangle$ is the product of reaction cross section and relative ion velocity, as averaged over the velocity distribution of the ions, Q is the fusion energy release, η_r is the efficiency of recovery of the fusion energy, η_h is the heating efficiency, and k is the Boltzmann constant.

For a fixed mixture of deuterium and tritium ions, Eq. (1) can be rearranged in the general form of Eq. (2). For a 50-50 mixture of deuterium and tri-

$$n\tau \geq F(\eta_r, \eta_h, Q)\left[\frac{T}{\langle \sigma v \rangle}\right] \qquad (2)$$

tium (see illustration), the minimum value of $T/\langle \sigma v \rangle$ occurs at about 25 keV ion kinetic temperature (mean ion energies of about 38 keV). Depending on the assumed efficiencies of the heating and recovery processes, the lower limit values of $n\tau$ range typically between about 10^{14} and 10^{15} cm^{-3}s. These values serve as a handy index of progress toward fusion, although their achievement does not alone guarantee success. Under special circumstances (unequal ion and electron temperatures, unequal deuterium and tritium densities, and nonmaxwellian ion distributions), lower $n\tau$ values may be adequate for nominal break-even.

The discussion up to this point has been oriented mainly to situations in which the fusion reactor may be thought of as a driven system, that is, one in which a continuous input of energy from outside the reaction chamber is required to maintain the

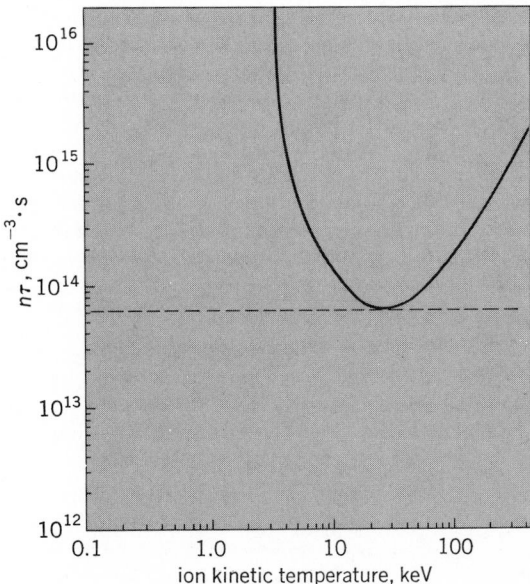

Typical plot of minimum value of $n\tau$ necessary for net release of energy versus ion kinetic temperature in a mixture containing equal amounts of deuterium and tritium. (*From R. F. Post, Nuclear fusion, in J. M. Hollander, ed., Ann. Rev. Energy, 1:213–255, 1976*)

reaction. Provided the efficiencies of the external heating and energy recovery systems are high, a driven reactor generally would require the lowest $n\tau$ values to produce net power. An important alternative operating made for a reactor would be an ignition mode, that is, one in which, once the initial heating of the fuel charge is accomplished, energy directly deposited in the plasma by charged reaction products will thereafter sustain the reaction. For example, in the D-T reaction, approximately 20% of the total energy release is imparted to the alpha particle; in a magnetic confinement system, much of the kinetic energy carried by this charged nucleus may be directly deposited in the plasma, thereby heating it. Thus if the confinement time is adequate, the reaction may become self-sustaining without a further input of energy from external sources. Ignition, however, would generally require $n\tau$ products with a higher range of values, and is thus expected to be more difficult to achieve than the driven type of reaction. However, in all cases the Lawson criterion is to be thought of as only a rule of thumb for measuring fusion progress; detailed evaluation of all energy dissipative and energy recovery processes is required in order properly to evaluate any specific system. *See* Nuclear fusion; Plasma physics.

[RICHARD F. POST]

Least-action principle

Like Hamilton's principle, the principle of least action is a variational statement that forms a basis from which the equations of motion of a classical dynamical system may be deduced. Consider a mechanical system described by coordinates $q_1, \ldots, q_f$ and their canonically conjugate momenta $p_1, \ldots, p_f$. The action S associated with a segment of the trajectory of the system is defined by Eq. (1), where the integral is evaluated along the

$$S = \int_c \sum_j p_j \, dq_j \qquad (1)$$

given segment c of the trajectory. The action is of interest only when the total energy E is conserved. The principle of least action states that the trajectory of the system is that path which makes the value of S stationary relative to nearby paths between the same configurations and for which the energy has the same constant value. The principle is misnamed, as only the stationary property is required. It is a minimum principle for sufficiently short but finite segments of the trajectory. *See* Hamilton's equations of motion; Hamilton's principle; Minimal principles.

Assume that Eq. (2) holds, where $p_j + \delta p_j$ is canonically conjugate to $q_j + \delta q_j$.

$$S + \Delta S = \int_c \sum_j (p_j + \delta p_j) \, d(q_j + \delta q_j) \qquad (2)$$

nonically conjugate to $q_j + \delta q_j$. Neglecting second order term in Eq. (2), one obtains Eq. (3), where an

$$\Delta S = \int_c \sum_j (p_j \, d\,\delta q_j + \delta p_j \, dq_j)$$
$$= \int_c \sum_j (\delta p_j \, dq_j - \delta q_j \, dp_j) \qquad (3)$$

integration by parts has been made, the integrated parts vanishing.

The vanishing of ΔS requires the integrand to be

a perfect differential of a quantity whose end variations vanish. The coefficients of the variations $\delta q_j, \delta p_j$ need not vanish separately because the variations are not independent, the varied qs and ps necessarily being canonically conjugate, as in Eq. (4), with terms defined by Eqs. (5).

$$\sum_j (\delta p_j \, dq_j - \delta q_j \, dp_j) = dU(q,p) \qquad (4)$$

$$\delta p_j = \frac{\partial U}{\partial q_j} \qquad \delta q_j = -\frac{\partial U}{\partial p_j} \qquad (5)$$

Writing $U = -H\,\delta t$ leads to Hamilton's equations of motion, Eqs. (6). The quantity $H(q,p)$, known as

$$\dot{p}_j = -\frac{\partial H}{\partial q_j} \qquad \dot{q}_j = \frac{\partial H}{\partial p_j} \qquad (6)$$

the Hamiltonian function, does not contain the time explicitly because $U(q,p)$ cannot be a function of the time since the end times are not fixed and in general will vary as the path is varied. Thus, the principle is useful only for conservative systems, where H is constant.

If $H(q,p)$ consists of a part H_2 quadratic in the momenta and a part H_0 independent of the momenta, then Eq. (7) holds by Euler's theorem on

$$S = \int_{t_1}^{t_2} \sum_j p_j \dot{q}_j \, dt$$
$$= \int_{t_1}^{t_2} \sum_j p_j \frac{\partial H}{\partial p_j} \, dt$$
$$= 2 \int_{t_1}^{t_2} H_2 \, dt \qquad (7)$$

homogeneous functions. Usually H_2 is the kinetic energy of the system so that the principle of least action may be written as Eq. (8), where v is the potential energy.

$$\Delta \int_{t_1}^{t_2} 2T \, dt = \Delta \int_{t_1}^{t_2} 2(E-V) \, dt = 0 \qquad (8)$$

The principle of least action derives much importance from the fact that it is the action which is quantized in the quantum form of the theory. Planck's constant is the quantum of action. *See* Nonrelativistic quantum theory.

[PHILIP M. STEHLE]

Bibliography: *See* Lagrange's equations.

Legendre functions

Solutions to the differential equation $(1-x^2)y'' - 2xy' + v(v+1)y = 0$.

Legendre polynomials. The most elementary of the Legendre functions, the Legendre polynomial $P_n(x)$ can be defined by the generating function in Eq. (1). More explicit representations

$$(1 - 2xr + r^2)^{-\frac{1}{2}} = \sum_{n=0}^{\infty} P_n(x) r^n \qquad (1)$$

are Eq. (2), and the hypergeometric function, Eq. (3). *See* Hypergeometric functions.

$$P_n(x) = \frac{(-1)^n}{2^n n!} \frac{d^n}{dx^n} (1-x^2)^n \qquad (2)$$

$$P_n(x) = {}_2F_1[-n, n+1; 1; (1-x)/2] \qquad (3)$$

Generating function (1) implies Eq. (4). The

$$(a^2 - 2ar \cos \theta + r^2)^{-\frac{1}{2}} = \frac{1}{a} \sum_{n=0}^{\infty} P_n(\cos \theta)(r/a)^n \quad (4)$$

$$0 < r < a$$

function $(a^2 - 2ar \cos \theta + r^2)^{-\frac{1}{2}}$ represents the potential in an inverse square field at a point P of a source at A, where r and a are the distances from P and A to a fixed point O, and θ is the angle between the segments PO and OA. These functions were extensively studied by A. M. Legendre and P. S. Laplace because they could be used in the study of the celestial mechanics. They had arisen slightly earlier in probabilistic work of J. Lagrange. *See* POTENTIALS.

Since Legendre polynomials are the zonal spherical harmonics on the surface of the unit sphere in three-space, they arise when studying physical phenomena associated with spherical geometry. They arise in many other applications. One application is to the estimation of the smallest eigenvalue of the truncated Hilbert matrix $(a_{i,j})_0^N$, $a_{i,j} = (i+j+1)^{-1}$. This matrix is very hard to invert numerically because the condition number, or the ratio of the largest to the smallest eigenvalue, is very large. One of the essential parts of the proof is an asymptotic formula for Legendre polynomials off the interval $[-1,1]$. *See* MATRIX THEORY; SPHERICAL HARMONICS.

Relation to trigonometric functions. When $x = \cos \theta$, Legendre polynomials have a relation with trigonometric functions, given by Eq. (5), where

$$(\sin \theta)^{\frac{1}{2}} \left(n + \frac{1}{2}\right)^{\frac{1}{2}} P_n(\cos \theta)$$

$$= \sqrt{\frac{2}{\pi}} \cos\left[\left(n + \frac{1}{2}\right)\theta - \frac{\pi}{4}\right] + R(n,\theta) \quad (5)$$

$|R(n,\theta)| \leq A/(n \sin \theta)$, A being a fixed constant, when $c/n \leq \theta \leq \pi - c/n$ for $c > 0$. *See* TRIGONOMETRY.

Properties. A graph of $P_n(x)$ (Fig. 1) illustrates a number of properties. $P_n(x)$ is even or odd as n is even or odd, that is, $P_n(-x) = (-1)^n P_n(x)$. All the zeros of $P_n(x)$ are real and lie between -1 and 1. The zeros of $P_{n+1}(x)$ separate the zeros of $P_n(x)$. $P_n(x)$ satisfies the inequality $|P_n(x)| \leq 1, -1 \leq x \leq 1$. The successive relative maxima of $|P_n(x)|$ increase in size as x increases over the interval $0 \leq x \leq 1$. The closest minimum to $x = 1$ of $P_n(x)$, called $\mu_{1,n}$, satisfies $\mu_{1,n} < \mu_{1,n+1}$. Similar inequalities hold for the first maxima of $P_n(x)$ to the left of $x = 1$, and for the kth relative maxima or minima. All of these results have been used in applications. The last two results about the relative maxima were used to obtain bounds on the phase of scattering amplitudes. From an important limiting relation of F. G. Mehler, Eq. (6), it is easy to show that $\mu_{1,n}$ approaches

$$\lim_{n \to \infty} P_n\left(\cos \frac{\theta}{n}\right) = J_0(\theta) \quad (6)$$

the minimum value of the Bessel function J_0. *See* BESSEL FUNCTIONS.

Differential equation. One of the main reasons for the occurrence of Legendre polynomials is a differential equation they satisfy, Eq. (7). This

$$(1 - x^2)y'' - 2xy' + n(n+1)y = 0 \quad (7)$$
$$y = P_n(x)$$

equation arises from the solution of Laplace's

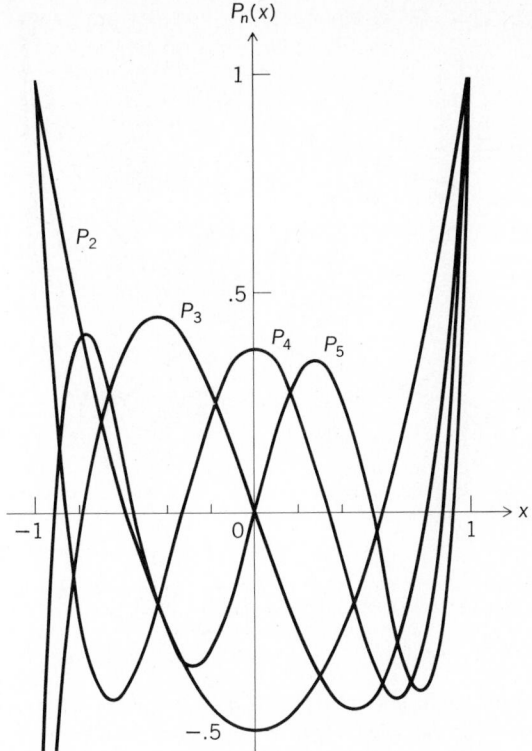

Fig. 1. Legendre polynomials $P_n(x)$, $n = 2, 3, 4, 5$.

equation by separation of variables in spherical coordinates. There is a second solution $Q_n(x)$ given by Eq. (8), when x is not in the interval

$$Q_n(x) = \frac{1}{2} \int_{-1}^{1} \frac{P_n(t)}{x - t} dt \quad (8)$$

$-1 < x < 1$. For $-1 < x < 1$, $Q_n(x)$ is defined by Eq. (9). $Q_n(x)$ has many of the same properties as

$$Q_n(x) = \lim_{\epsilon \to 0^+} \frac{Q_n(x + i\epsilon) + Q_n(x - i\epsilon)}{2} \quad (9)$$

$P_n(x)$, as illustrated by Fig. (2). *See* DIFFERENTIAL EQUATION; LAPLACE'S DIFFERENTIAL EQUATION.

Addition formula. One of the most important formulas satisfied by Legendre polynomials is the addition formula, Eq. (10). The functions $P_n^m(x)$ are

$$P_n(\cos \theta \cos \varphi + \sin \theta \sin \varphi \cos \chi)$$

$$= P_n(\cos \theta) P_n(\cos \varphi)$$

$$+ 2 \sum_{m=1}^{n} \frac{(n-m)!}{(n+m)!} P_n^m(\cos \theta) P_n^m(\cos \varphi) \cos m\psi \quad (10)$$

defined by Eq. (11). These functions, often called

$$P_n^m(x) = (-1)^m (1 - x^2)^{m/2} \frac{d^m}{dx^m} P_n(x) \quad (11)$$

$$-1 < x < 1 \qquad m = 0, 1, \ldots, n$$

associated Legendre functions, satisfy differential equation (12).

$$(1 - x^2)\frac{d^2y}{dx^2} - 2x\frac{dy}{dx} + \left\{n(n+1) - \frac{m^2}{1 - x^2}\right\} y = 0 \quad (12)$$

General Legendre functions. In work of F. G. Mehler on electrical distribution with conical sym-

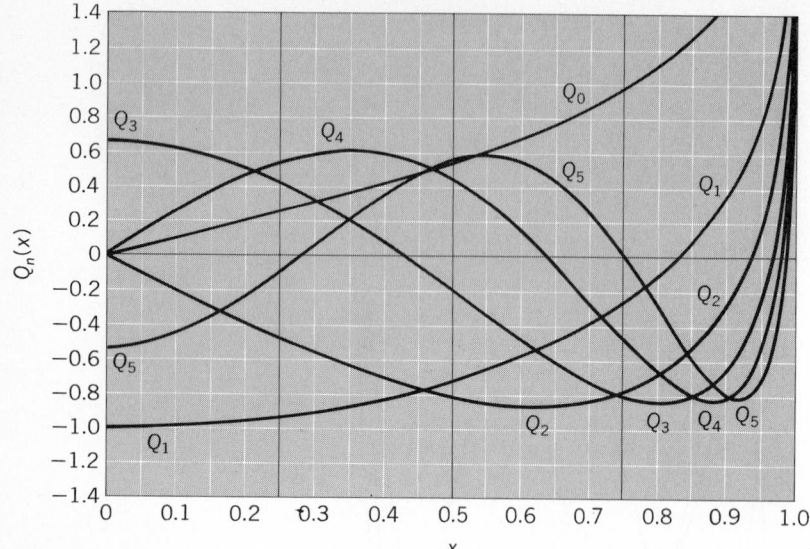

Fig. 2. Legendre functions of the second kind, $Q_n(x)$. (From E. Jahnke and F. Emde, Tables of Functions with Formulae and Curves, 4th ed., Dover Publications, 1945)

$$[P_n(x)]^2 + \frac{4}{\pi}[Q_n(x)]^2$$

$$= \frac{4}{\pi^2}\int_1^\infty Q_n[x^2 + (1-x^2)z](z^2-1)^{-1/2}\,dz \qquad (15)$$

$$-1 < x < 1$$

Gegenbauer functions. There is a related set of functions, called Gegenbauer and associated Gegenbauer functions, which arise as spherical harmonics on higher-dimensional spheres. They are really equivalent to general Legendre functions, since they are algebraic functions times hypergeometric functions which have a quadratic transformation. The only difference is that different algebraic functions are used as multipliers since the functions satisfy different second-order differential equations. A completely analogous theory has been developed for them. There are two extreme cases of Gegenbauer functions which are very important, and in many respects their theory is easier to develop and is used as a model to suggest further developments. One is connected with the unit sphere in two-space (or the circle $x^2 + y^2 = 1$), and the resulting functions are $\cos\lambda\theta$ and $\sin\lambda\theta$. The other comes from an "infinite dimensional sphere" (actually an appropriate limit of finite dimensional spheres), and the spherical functions are Hermite polynomials and their extensions to Hermite functions. See ORTHOGONAL POLYNOMIALS. [RICHARD ASKEY]

Bibliography: A. Erdélyi et al., *Higher Transcendental Functions*, vols. 1–3, 1953–1955; E. W. Hobson, *The Theory of Spherical and Ellipsoidal Harmonics*, 1931, reprint 1975; C. Müller, *Spherical Harmonics*, 1966; I. N. Sneddon, *Special Functions of Mathematical Physics and Chemistry*, 3d ed., 1980; N. Ja. Vilenkin, *Special Functions and the Theory of Group Representations*, 1968.

Length

Extension in space. Length is one of the three fundamental physical quantities (the other two being mass and time), and therefore cannot be defined in terms of simpler quantities. It is measured by comparison with an arbitrary standard called the international meter, defined as 1,650,763.73 times the wavelength of the orange light emitted when a gas of pure krypton-86 is excited in an electrical discharge. Sticks or tapes are calibrated by direct comparison with krypton light by means of an interferometer; multiples and submultiples are also indicated by calibration marks. Lengths are calculated by direct comparison with such sticks or tapes. Decimal multiples and submultiples of the meter are frequently used in specifying length; in English-speaking countries the foot (0.3048 m) is a length unit. [DUDLEY WILLIAMS]

Lens

A curved piece of ground and polished or molded material, usually glass, used for the refraction of light. Its two surfaces have the same axis. Usually this is an axis of rotation symmetry for both surfaces; however, one or both of the surfaces can be toric, cylindrical, or a general surface with double symmetry. The intersection points of the symmetry axis with the two surfaces are called the front and back vertices and their separation is called the thickness of the lens.

metry, solutions arose for Eq. (12) with $m = 0$ and $n = -\frac{1}{2} + it$, where t is a parameter that can assume any real value. When Laplace's equation is solved by separation of variables in toroidal coordinates, this equation occurs with $n = \nu - \frac{1}{2}$ and $m = \mu$, where μ and ν are separation parameters. Thus there are reasons for considering solutions to Eq. (12) when n and m are arbitrary complex numbers. In the general context, n and m are usually called ν and μ. $P_\nu^\mu(z)$ is defined by Eq. (13) for

$$P_\nu^\mu(z) = \frac{1}{\Gamma(1-\mu)}\left(\frac{z+1}{z-1}\right)^{\mu/2} \cdot$$

$$_2F_1[-\nu,\nu+1;1-\mu;(1-z)/2] \qquad (13)$$

$|z-1| < 2$, with the cut $z < 1$ removed. By analytic continuation, this function is extended to the plane cut from 1 to $-\infty$ and then defined for $-1 < z < 1$ by Eq. (14). There is a second solution to dif-

$$P_\nu^\mu(x) = \lim_{\epsilon\to 0^+}\frac{P_\nu^\mu(x+i\epsilon) + P_\nu^\mu(x-i\epsilon)}{2} \qquad (14)$$

ferential equation (12) which is called $Q_\nu^\mu(z)$. The easiest way to carry out the analytic extension of Eq. (13) and to define $Q_\nu^\mu(z)$ is to use the general theory of hypergeometric functions. Legendre functions are just appropriate algebraic functions times a hypergeometric function of the type that has a quadratic transformation. The theory of Legendre functions up to but not including the addition formula can be developed very easily in this fashion.

Addition formulas are best derived by interpreting the functions as spherical harmonics and using the rotation groups which operate on spheres and then analytically continuing the resulting formulas in the appropriate parameters. Addition formula (10) can be extended not only to the general Legendre function of the first kind, $P_\nu^\mu(z)$, but to the general function of the second kind $Q_\nu^\mu(z)$. A number of results can be obtained from this, such as Eq. (15). This is an extension of the formula $\cos^2 x + \sin^2 x = 1$ to Legendre functions.

LENS TYPES

There are three lens types, namely, compound, single, and cemented. These are described in the following sections.

Compound lenses. A compound lens is a combination of two or more lenses in which the second surface of one lens has the same radius as the first surface of the following lens and the two lenses are cemented together. Compound lenses are used instead of single lenses for color correction, or to introduce a surface which has no effect on the aperture rays but large effects on the principal rays, or vice versa. Sometimes the term compound lens is applied to any optical system consisting of more than one element, even when they are not in contact.

A group of lenses used together is a lens system. A symmetrical lens is a lens system consisting of two parts, each of which is the mirror image of the other. If one part is a mirror image of the other magnified m times, the system is called hemisymmetric. When $m = 1$, the system is often said to be holosymmetric.

Single lenses. The lens diameter is called the linear aperture, and the ratio of this aperture to the focal length is called the relative aperture. This latter quantity is more often specified by its reciprocal, called the f-number. Thus, if the focal length is 50 mm and the linear aperture 25 mm, the relative aperture is 0.5 and the f-number is $f/2$. *See* FOCAL LENGTH.

In precalculation formulas, the lens thicknesses (but not the separations of the lenses) can frequently be neglected. This leads to the convenient fiction of a thin lens.

If ρ_1 and ρ_2 are the front and back curvatures of a lens of refractive index n and thickness d, its power is given by Eq. (1). The curvature of the surface is the reciprocal of its radius.

$$\phi = (n-1)(\rho_1 - \rho_2) + \frac{d(n-1)^2}{n}\rho_1\rho_2 \quad (1)$$

The distances from the back vertex to the back nodal point and to the back focal point, respectively, are given by Eq. (2). The last distance is often called the back focus, especially in photographic optics.

$$S'_N\phi = -(d/n)(n-1)\rho_1$$
$$S'_F\phi = 1 + S'_N\phi \quad (2)$$

The bending of a lens is a change in the curvature of the two surfaces by the same amount. It does not change the power of a thin lens, which is $(n-1)(\rho_1 - \rho_2)$. Bending is an important tool of the designer, for it permits the replacing of one lens by another without changing the data of Gaussian optics.

When thick lenses are involved, Gaussian optics remains constant only if both the powers of the thick lenses and the distance between the back nodal point of the first lens and the front nodal point of the second remain unchanged. Thus a bending of a thick lens should be accompanied by such an adjustment.

The optical center of a thick lens is the image of the nodal point produced inside the lens. All finite

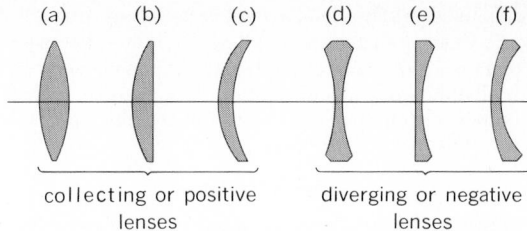

Fig. 1. Common lenses. (*a*) Biconvex. (*b*) Plano-convex. (*c*) Positive meniscus. (*d*) Biconcave. (*e*) Plano-concave. (*f*) Negative meniscus. (*From F. A. Jenkins and H. E. White, Fundamentals of Optics, 3d ed., McGraw-Hill, 1957*)

rays through the optical center emerge parallel to their respective directions at their entrance.

An optical center exists also in a hemisymmetric system. It is the point of symmetry which divides the separation of the two parts in the ratio $1/m$. If negative values of m are permitted, any single lens is a hemisymmetric system and the point dividing the thickness of the lens (the separation of the two vertices) in a ratio equal to the ratio of the two radii is the optical center of the lens.

A lens is said to be a collecting lens if $\phi > 0$ and a diverging lens if $\phi < 0$. When $\phi = 0$, the lens is afocal. Several types of collecting and diverging lenses are shown in Fig. 1.

The surfaces of most lenses are either spherical or planar, but nonspherical surfaces are used on occasion to improve the corrections without changing the power of the lens. *See* OPTICAL SURFACES.

A concentric lens is a lens whose two surfaces have the same center. If the object to be imaged is also at the center, its axis point is sharply imaged upon itself, and since the sine condition is fulfilled, the image is free from asymmetry. Such a lens can be used as an additional system to correct meridional errors.

Another type of lens consists of an aplanatic surface followed by a concentric surface, or vice versa. Such a lens divides the focal length of the original lens to which it is attached by n^2, thus increasing the f-number by a factor of n^2 without destroying the axial correction of the preceding system. It does introduce curvature of field which makes a rebalancing of the whole system desirable. *See* ABERRATION (OPTICS).

Cemented lenses. Consider a compound lens made of two or more simple thin lenses cemented together. Let the power of the κth simple lens be ϕ_κ and its Abbe value ν_κ. The difference between the powers of the combination for wavelengths corresponding to C and F is given by Eq. (3),

$$\Phi_F - \Phi_C = \Phi/N = \Sigma\phi_\kappa/\nu_\kappa \quad (3)$$

where N may be considered to be the effective ν-value of the combination. The ν-values of optical glasses vary between 25 and 70, with the ν-value of fluorite being slightly larger ($\nu = 95.1$). By using compound lenses, effective values of N can be obtained outside this range. Color correction is achieved as N becomes infinite, so that $\Phi_F - \Phi_C = 0$. A lens so corrected is called an achromat. In optical design, it is sometimes desirable to have negative values of N to balance the positive values of the rest of the system containing collect-

LENS

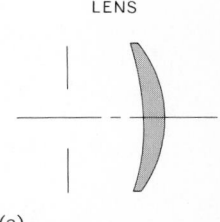

(a)

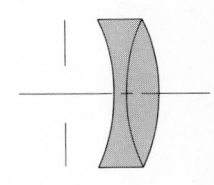

(b)

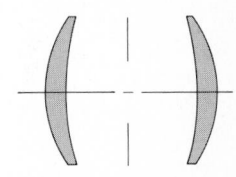

(c)

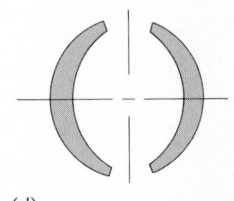

(d)

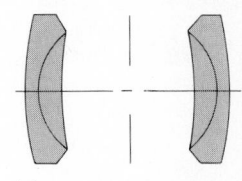

(e)

Fig. 2. Older camera lenses. (*a*) Meniscus. (*b*) Simple achromat. (*c*) Periskop. (*d*) Hypergon wide-angle. (*e*) Symmetrical achromat.

LENS

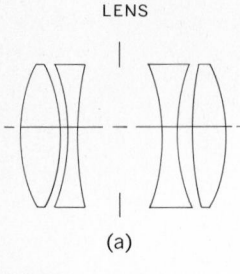

(a)

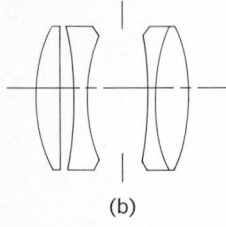

(b)

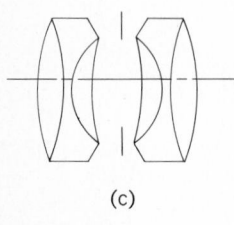

(c)

Fig. 3. Types of anastigmats. (a) Celor. (b) Tessar. (c) Dagor.

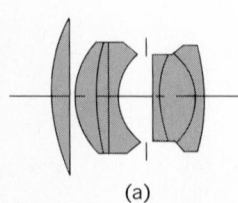

(a)

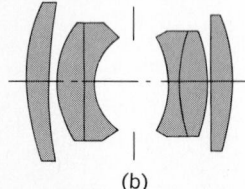

(b)

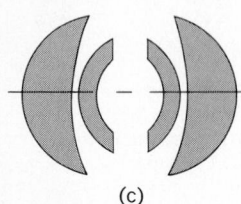

(c)

Fig. 4. Modern camera lenses. (a) Sonnar. (b) Biotar. (c) Topogon.

ing lenses. Such a lens is said to be hyperchromatic. A cemented lens corrected for more than two colors is said to be apochromatic. A lens corrected for all colors of a sizable wavelength range is called a superachromatic lens. *See* CHROMATIC ABERRATION; OPTICAL MATERIALS.

LENS SYSTEMS

Optical systems may be divided into four classes: telescopes, oculars (eyepieces), photographic objectives, and enlarging lenses.

Telescope systems. A lens system consisting of two systems combined so that the back focal point of the first (the objective) coincides with the front focal point of the second (the ocular) is called a telescope. Parallel entering rays leave the system as parallel rays. The magnification is equal to the ratio of the focal length of the first system to that of the second.

If the second lens has a positive power, the telescope is called a terrestrial or Keplerian telescope and the separation of the two parts is equal to the sum of the focal lengths.

If the second lens is negative, the system is called a Galilean telescope and the separation of the two parts is the difference of the absolute focal lengths. The Galilean telescope has the advantage of shortness (a shorter system enables a larger field to be corrected); the Keplerian telescope has a real intermediate image which can be used for introducing a reticle or a scale into the intermediate plane.

Both objective and ocular are in general corrected for certain specific aberrations, while the other aberrations are balanced between the two systems.

Photographic objectives. A photographic objective images a distant object onto a photographic plate or film.

The amount of light reaching the light-sensitive layer depends on the aperture of the optical system, which is equivalent to the ratio of the lens diameter to the focal length. Its reciprocal is called the *f*-number. The smaller the *f*-number, the more light strikes the film. In a well-corrected lens (corrected for aperture and asymmetry errors), the *f*-number cannot be smaller than 0.5.

The larger the aperture (the smaller the *f*-number), the less adequate may be the scene luminance required to expose the film. Therefore, if pictures of objects in dim light are desired, the *f*-number must be small. On the other hand, for a lens of given focal length, the depth of field is inversely proportional to the aperture.

Since the exposure time is the same for the center as for the edge of the field, it is desirable for the same amount of light to get to the edge as gets to the center, that is, the photographic lens should have little vignetting.

The camera lens can be considered as an eye looking at an object (or its image), with the diaphragm corresponding to the eye pupil. The Gaussian image of the diaphragm in the object (image) space is called the entrance (exit) pupil. The angle under which the object (image) is seen from the entrance (exit) pupil is called the object (image) field angle. For most photographic lenses, the entrance and exit pupils are close to the respective nodal points; for such lenses, the object and the image field angles are equal.

In general, photographic objectives with large fields have small apertures; those with large apertures have small fields. The construction of the two types of systems is quite different. One can say in general that the larger the aperture, the more complex the lens system must be.

There exist cameras (so-called pinhole cameras) that do not contain any lenses. The image is then produced by optical projection. The aperture in this case should be limited to *f*/22.

Other types of lenses. A single meniscus lens, with its concave side toward the object and with its stop in front at its optical center, gives good definition at *f*/16 over a total field of 50° (Fig. 2a). The lens can be a cemented doublet for correcting chromatic errors (Fig. 2b). For practical reasons, a reversed meniscus with the stop toward the film is often used.

Combining two meniscus lenses to form a symmetrical lens with central stop makes it possible to correct astigmatic and distortion errors for small apertures as well as large field angles (Fig. 2c).

The basic type of wide-angle objective is the Hypergon, consisting of two meniscus lenses concentric with the regard to stop (Fig. 2d). This type of system can be corrected for astigmatism and field curvature over a total field angle of 180° but it can only be used for a small aperture (*f*/12), since it cannot be corrected for aperture errors. The aperture can be increased to *f*/4 at the expense of field angle by thickening and achromatizing the meniscus lenses and adding symmetrical elements in the center or at the outside of the basic elements.

Two positive achromatic menisci symmetrically arranged around the stop lead to the aplanatic type of lens (Fig. 2e). This type was spherically and chromatically corrected. Since the field could not be corrected, a compromise was achieved by balancing out sagittal and meridional field curvature so that one image surface lies in front and the other in back of the film.

Anastigmatic lenses. The discovery of the Petzval condition for field correction led to the construction of anastigmatic lenses, for which astigmatism and curvature of field are corrected. Such lenses must contain negative components.

The Celor (Gauss) type consists of two airspaced achromatic doublets, one on each side of the stop (Fig. 3a). The Cooke triplet combines a negative lens at the aperture stop with two positive lenses, one in front and the other in back. It is called a Tessar (Fig. 3b) if the last positive lens is a cemented doublet, or a Heliar if both positive lenses are cemented. The Dagor type consists of two lens systems that are nearly symmetrical with respect to the stop, each system containing three or more lenses (Fig. 3c).

Modern lenses. To increase the aperture, the field, or both, it is frequently advantageous to replace one lens by two separated lenses, since the same power is then achieved with larger radii and this means that the single lenses are used with smaller relative apertures. The replacing of a single lens by a cemented lens changes the color balance, and thus the designer may achieve more favorable conditions. Moreover, the introduction of new types of glass (first the glasses containing barium, later the glasses containing rare earths) led to lens elements which for the same power have weaker surfaces and are of great help to the lens designer, since the errors are reduced.

Of modern designs the most successful are the Sonnar, a modified triplet, one form of which is shown in Fig. 4a; the Biotar (Fig. 4b), a modified Gauss objective with a large aperture and a field of about 24°; and the Topogon (Fig. 4c), a periscopic lens with supplementary thick menisci to permit the correction of aperture aberrations for a moderate aperture and a large field. One or two plane-parallel plates are sometimes added to correct distortion.

Special objectives. It is frequently desirable to change the focal length of an objective without changing the focus. This can be done by combining a fixed near component behind the stop with an exchangeable set of components in front of the stop. The designer has to be sure that the errors of the two parts are balanced out regardless of which front component is in use.

The telephoto objective is a specially constructed objective with the rear nodal point in front of the lens, to combine a long focal length with a short back focus.

The Petzval objective is one of the oldest designs (1840) but one of the most ingenious. It consists in general of four lenses ordered in two pairs widely separated from each other. The first pair is cemented and the second usually has a small air space. For a relatively large aperture, it is excellently corrected for aperture and asymmetry errors, as well as for chromatic errors and distortion. It is frequently used as a portrait lens and as a projection lens because of its sharp central definition. Astigmatism can be balanced but not corrected.

Enlarger lenses and magnifiers. The basic type of enlarger lens is a holosymmetric system consisting of two systems of which one is symmetrical with the first system except that all the data are multiplied by the enlarging factor *m*. When the object is in the focus of the first system, the combination is free from all lateral errors even before correction. A magnifier in optics is a lens that enables an object to be viewed so that it appears larger than its natural size.

The magnifying power is usually given as equal to one-quarter of the power of the lens expressed in diopters. *See* MAGNIFICATION.

Magnifying lenses of low power are called reading glasses. A simple planoconvex lens in which the principal rays are corrected for astigmatism for a position of the eye at a distance of 10 in. is well suited for this purpose, although low-power magnifiers are often made commercially with biconvex lenses. A system called a verant consists of two lenses corrected for color, astigmatism, and distortion. It is designed for stereoscopic vision at low magnification.

For higher magnifications, many forms of magnifiers exist. One of the basic designs has the form of a full sphere with a diaphragm at the center, as shown in Fig. 5a. The sphere may be solid or it may be filled with a refracting liquid. When it is solid, the diaphragm may be formed by a deep groove around the equator. Combinations of thin planoconvex lenses as shown in Fig. 5b and c are much used for moderate powers. Better correction can be attained in the aplanatic magnifier of C. A. Steinheil, in which a biconvex crown lens is ce-

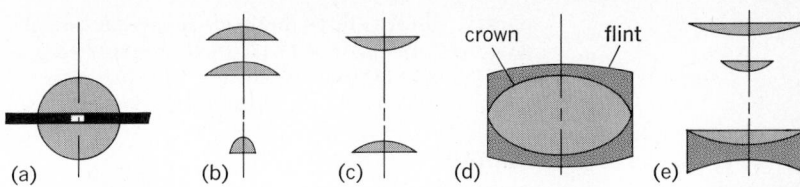

Fig. 5. Typical magnifiers. (a) Sphere with equatorial diaphragm. (b, c) Planoconvex lens combinations. (d) Steinheil triple aplanat. (e) Chevalier type.

mented between a pair of flint lenses (Fig. 5d).

A design by C. Chevalier (Fig. 5e) aims for a large object distance. It consists of an achromatic negative lens combined with a distant collecting front lens. A magnifying power of up to 10× with an object distance up to 3 in. can be attained.

[MAX HERZBERGER]

Bibliography: M. Herzberger, *Modern Geometrical Optics*, 1958, reprint 1973; F. A. Jenkins and M. E. White, *Fundamentals of Optics*, 4th ed., 1976; R. Kingslake et al. (eds.), *Applied Optics and Optical Engineering*, 7 vols., 1965–1979; C. B. Neblette and A. E. Murray, *Photographic Lenses*, 1973.

Lenz's law

A law of electromagnetism which states that, whenever there is an induced electromotive force (emf) in a conductor, it is always in such a direction that the current it would produce would oppose the change which causes the induced emf. If the change is the motion of a conductor through a magnetic field, as in the illustration, the induced

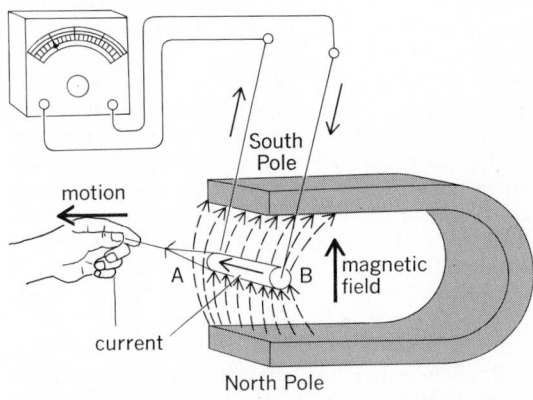

Induced emf in a moving conductor. Direction of current induced in wire AB is indicated by the arrows. (*From M. W. White, K. V. Manning and R. L. Weber, Practical Physics, 2d ed., McGraw-Hill, 1955*)

current must be in such a direction as to produce a force opposing the motion. If the change causing the emf is a change of flux threading a coil, the induced current must produce a flux in such a direction as to oppose the change. That is, if the change is an increase of flux, the flux due to the induced current must be opposite in direction to the increasing flux. If the change is a decrease in flux, the induced current must produce flux in the same direction as the decreasing flux.

Lenz's law is a form of the law of conservation of

energy, since it states that a change cannot propagate itself. *See* CONSERVATION OF ENERGY; ELECTROMAGNETIC INDUCTION.

[KENNETH V. MANNING]

Lepton

An elementary particle having no internal constituents which interacts through the electromagnetic, weak, and gravitational forces, but does not interact through the strong (nuclear) force. Leptons are very small, less than 10^{-18} m in size. This is less than 1/1000 the size of a nucleus and less than 10^{-8} the size of an atom. Indeed, existing measurements are consistent with leptons being point particles.

These properties of the lepton family of particles are to be contrasted with the properties of the hadron family of particles. Hadrons such as the π-meson and the proton are believed to be composed of internal constituents called quarks; and hadrons interact through the strong force as well as through the other forces. Hadrons have sizes about 10^{-15} m, which is at least 1000 times the size of a lepton. Thus leptons are much simpler than hadrons in both their structure and their behavior; hence leptons are considered to be more elementary than hadrons. At present the lepton family and the quark family are thought to lie at the same level of elementariness. *See* HADRON; QUARKS.

The table lists the properties of the six known leptons. There are three known charged leptons: the electron (e), muon (μ), and tau (τ). Associated with each charged lepton is a neutral lepton called a neutrino. *See* ELECTRON; NEUTRINO.

Lepton conservation. The association between charged leptons and their neutrinos comes about through an empirical law called lepton conservation. Take as an example the electron (e^-). There are only two ways in which an e^- can be destroyed: it can be annihilated by combining it with an e^+ (positron); or it can be changed into a ν_e. But an e^- cannot be changed into any other lepton or hadron such as an $e^+, \mu^-, \mu^+, \tau^-, \tau^+, \pi^-$. And an e^- cannot be changed into a muon-associated neutrino ($\nu_\mu, \bar{\nu}_\mu$) or into a tau-associated neutrino ($\nu_\tau, \bar{\nu}_\tau$). Thus there is a unique property of the electron (e^-) which is preserved or conserved in all reactions; the only other particle carrying this unique property is the electron-associated neutrino (ν_e).

Similar lepton conservation laws hold for the μ and τ. The only other particle carrying the unique property of the μ^- is the muon-associated neutrino (ν_μ). The only other particle carrying the unique property of the τ^- is the tau-associated neutrino (ν_τ). There is at present no explanation for these observed lepton conservation laws. *See* SYMMETRY LAWS.

Decay. The electron is a stable particle; that is, it never decays. The muon and tau decay with av-

erage lifetimes listed in the table. The way in which they decay illustrates the lepton conservation laws. The muon decays in only one way, as given in reaction (1), using the μ^- as the example.

$$\mu^- \rightarrow \nu_\mu + e^- + \bar{\nu}_e \qquad (1)$$

The tau, being heavier, decays more rapidly than the muon and can decay in many ways. Some examples are given by (2). In all these examples of

$$\begin{aligned}
\tau^- &\rightarrow \nu_\tau + e^- + \bar{\nu}_e \\
\tau^- &\rightarrow \nu_\tau + \mu^- + \bar{\nu}_\mu \\
\tau^- &\rightarrow \nu_\tau + \pi^- \\
\tau^- &\rightarrow \nu_\tau + \pi^- + \pi^+ + \pi^-
\end{aligned} \qquad (2)$$

τ^- decay, the τ^- is converted to a ν_τ, with other leptons or hadrons also being produced.

Masses. The masses of the leptons (see table) are given in the energy equivalent unit megaelectronvolts (MeV); 1MeV is the energy gained by an electron that moves through a voltage of 10^6 V, and is equal to 1.602×10^{-13} J. The masses of the electron, muon, and their neutrinos are smaller than the masses of any of the hadrons. However, the discovery of the tau, whose mass is larger than that of many hadrons, destroyed the concept that leptons had to be very-small-mass particles.

All measurements of the masses of the neutrinos are upper limits on the masses; that is, it is possible that some or all of the neutrinos have zero mass. Indeed, most theoretical descriptions of lepton behavior assume that all neutrinos are massless.

Production. Leptons can be produced in various ways. One way is through the decay of a hadron. For example, a muon and its associated antineutrino are produced when a pion decays, as in reaction (3). Another way to produce leptons is through the

$$\pi^- \rightarrow \mu^- + \bar{\nu}_\mu \qquad (3)$$

annihilation of an electron with a positron; for example, muons can be produced by reaction (4), and taus can be produced by reaction (5). It was

$$e^+ + e^- \rightarrow \mu^+ + \mu^- \qquad (4)$$

$$e^+ + e^- \rightarrow \tau^+ + \tau^- \qquad (5)$$

through reaction (5) that the tau was discovered. *See* ELEMENTARY PARTICLE. [MARTIN L. PERL]

Bibliography: I. Asimov, *The Neutrino*, 1966; G. Feinberg, *What Is the World Made Of?*, 1977; M. L. Perl, The tau lepton, *Annu. Rev. Nucl. Partic. Sci.*, vol. 30, 1980; M. L. Perl and W. T. Kirk, Heavy leptons, *Sci. Amer.*, 283:50–57, 1978.

Level

Logarithm of the ratio of a given quantity to a reference quantity of the same kind. The base of the logarithm, the reference quantity, and the kind of level must be indicated. Frequently used levels, usually expressed in decibels, are power level,

Properties of leptons

Charged lepton name	Electron	Muon	Tau
Charged lepton symbol	$e^\pm$	$\mu^\pm$	$\tau^\pm$
Dates of discovery	1890s	Late 1930s	1974–1975
Charged lepton mass, MeV	0.51	105.7	~ 1782
Charged lepton lifetime, s	Stable	2.2×10^{-6}	$< 2.3 \times 10^{-12}$
Associated neutrino symbols	$\nu e, \bar{\nu}_e$	ν_μ, ν_μ	$\nu_\tau, \bar{\nu}_\tau$
Associated neutrino mass, MeV	< 0.00006	< 0.57	< 250
	(may be 0)	(may be 0)	(may be 0)

voltage level, and sound pressure level. The decibel is a unit of level when the base of the logarithm is the tenth root of 10, and the quantities concerned, such as voltage squared, are proportional to power. Some other units of level are the neper (Np) and bel (B); 1 Np = 8.686 dB and 1 B = 10 dB. *See* DECIBEL.   [R. W. YOUNG]

Lie group

A topological group with only countably many connected components whose identity component is open and is an analytic group. An analytic group or connected Lie group is a topological group with the additional structure of a smooth manifold such that multiplication and inversion are smooth. Many groups that arise naturally as groups of symmetries of physical or mathematical systems are Lie groups. The study of Lie groups has applications to analytic function theory, differential equations, differential geometry, Fourier analysis, algebraic number theory, algebraic geometry, quantum mechanics, relativity, and elementary particle theory. *See* GROUP THEORY; TOPOLOGY.

Examples of Lie groups. Euclidean n-dimensional space with the standard differentiable structure and with vector addition as the group operation is an analytic group $\mathbf{R}^n$ of dimension n.

The nonsingular n-by-n real matrices form a Lie group $GL(n,\mathbf{R})$ with the differentiable structure obtained as an open set of n^2-dimensional space and with the group operation given by matrix multiplication. Similarly $GL(n,\mathbf{C})$, the group of nonsingular n-by-n complex matrices under matrix multiplication, forms a Lie group. The latter group is connected.

Each closed subgroup of a Lie group, in its relative topology, admits one differentiable structure on its identity component that makes it a Lie group. With $GL(n,\mathbf{R})$ and $GL(n,\mathbf{C})$, the subgroups of upper triangular matrices, matrices of determinant 1, rotation matrices, orthogonal matrices, and unitary matrices are all examples. *See* MATRIX THEORY.

Lie algebra. Let G be a Lie group. The identity component of G is an open subset of G and is a smooth manifold. Thus it makes sense to speak of tangent spaces. Let $\mathfrak{g}$ be the tangent space to G at the identity element. A multiplication $[X,Y]$ within $\mathfrak{g}$ is defined as follows.

By means of the differential of left translation by p, one can associate a tangent vector X_p to the point p of G. If f is a smooth real-valued function on G, let $\widetilde{X}f$ be the function on G with $\widetilde{X}f(p) = X_p f$. Then $\widetilde{X}f$ is smooth. The tangent vector $[X,Y]$ is defined on the function f by Eq. (1).

$$[X,Y]f = X(\widetilde{Y}f) - Y(\widetilde{X}f) \qquad (1)$$

The resulting multiplication satisfies the properties:

(i) $[X,Y]$ is linear in X and in Y.
(ii) $[X,X] = 0$.
(iii) $[X,[Y,Z]] + [Y,[Z,X]] + [Z,[X,Y]] = 0$.

The vector space $\mathfrak{g}$ equipped with this multiplication is called the Lie algebra of the Lie group G.

The Lie algebra $\mathfrak{g}$ determines the multiplication in G near the group identity by a result known as the Campbell-Hausdorff formula. Consequently, many properties of the identity component of G can be deduced from the simpler object $\mathfrak{g}$.

Homomorphisms. If φ is a smooth homomorphism from an analytic group G to an analytic group G', the differential $d\varphi$ of φ at the identity is a Lie algebra homomorphism in the sense that it is linear and satisfies $d\varphi([X,Y]) = [d\varphi(X), d\varphi(Y)]$. If φ is one-to-one or onto, so is $d\varphi$. Conversely, if G is simply-connected, each homomorphism of Lie algebras arises as the differential of some unique smooth group homomorphism.

Abstract Lie algebras. An abstract real Lie algebra is a finite-dimensional real vector space with a multiplication satisfying (i), (ii), and (iii). Each abstract real Lie algebra can be identified with the Lie algebra of some analytic group.

Examples. The Lie algebra of $\mathbf{R}^n$ is an n-dimensional vector space with all $[X,Y]$ equal to 0.

For $GL(n,\mathbf{R})$, the Lie algebra can be identified with the vector space of all n-by-n real matrices with multiplication given in terms of matrix multiplication as $[X,Y] = XY - YX$. The identification of a member L of the Lie algebra with a matrix X is done entry by entry. Namely, the i-jth entry of X is the result of applying L to the i-jth entry function, which is a smooth real-valued function on the group.

For $GL(n,\mathbf{C})$, the Lie algebra can be identified with all n-by-n complex matrices with multiplication $[X,Y] = XY - YX$.

For a closed subgroup G of $GL(n,\mathbf{R})$ or $GL(n,\mathbf{C})$, inclusion is a smooth one-to-one homomorphism and allows one to identify the Lie algebra of G with a Lie subalgebra of matrices. The matrices of the Lie subalgebra can be computed conveniently as all entry-by-entry derivatives $c'(0)$ of smooth curves $c(t)$ that lie completely within the subgroup and have $c(0)$ equal to the identity matrix. For unitary matrices, for example, the result is skew-hermitian matrices. (A skew-hermitian matrix is equal to the negative of the transpose of its complex conjugate.)

Analytic subgroups. Let G be a Lie group. An analytic subgroup H of G is a subgroup of G that admits the structure of an analytic group in such a way that the inclusion mapping of H into G is smooth. The topology of H as a manifold need not be the relative topology inherited from G.

If H is an analytic subgroup of G, the differential of the inclusion mapping identifies the Lie algebra of H with a Lie subalgebra of the Lie algebra of G. Distinct analytic subgroups give rise to distinct Lie subalgebras. Every Lie subalgebra arises from some analytic subgroup by this construction.

Exponential map. Let G be a Lie group with Lie algebra $\mathfrak{g}$. Let $c(t)$ be a curve in G, that is, a smooth mapping from the real line into G, and suppose that $c(0)$ is the group identity. The image of d/dt under the differential dc of $c(t)$ at 0 is a member of $\mathfrak{g}$. For each X in $\mathfrak{g}$ there is one and only one such curve c_X such that $dc_X(d/dt) = X$ and $c_X(s+t) = c_X(s)c_X(t)$ for all s and t. The exponential mapping is the function from $\mathfrak{g}$ to G defined by exp $X = c_X(1)$. It is smooth and, in a small open neighborhood of the group identity, has a smooth inverse function that provides local coordinates for G near the identity.

For $GL(n,\mathbf{R})$ or $GL(n,\mathbf{C})$ the exponential mapping is given by the series expansion for the ordinary exponential function e^X if the Lie algebra is identified with the Lie algebra of all matrices. *See* SERIES.

Fourier analysis. In many applications of Lie groups the notion of group representation is a basic tool. A unitary representation of the Lie group G on a Hilbert space H is a function R from G to the set of unitary operators on H such that $R(g_1 g_2) = R(g_1) R(g_2)$ for all g_1 and g_2 in G and such that the mapping $g \rightarrow R(g)x$ is continuous from G into H for each x in H. [A unitary operator U on H is a linear operator from H onto H that preserves inner products; that is, $(Ux, Uy) = (x,y)$ for each x and y in H.] The representation R is irreducible if there is no proper nonzero closed subspace S such that $R(g)$ maps S into itself for all g in G.

Because of the applications, it is an important mathematical problem to classify the irreducible unitary representations of a given Lie group G and to investigate properties of decompositions of general unitary representations into irreducible ones.

In the analysis of the unitary representations, use of the Lie algebra plays an important role. If X is in the Lie algebra, it is possible to define $R(X)$ as a skew-hermitian operator on a dense subspace by differentiating $R(\exp tX)$ appropriately. Then $iR(X)$ extends to a self-adjoint operator that is easier to deal with than $R(\exp tX)$.

In many applications, H is a space of functions, and some linear operator or set of linear operators is acting in H. Typically, the representation of G is given by a group of transformations that commute with these linear operators. When the representation of G on H is decomposed suitably, one can expect the linear operators to act separately in each of the constituent spaces of H and to be easier to understand on each constituent space than on H.

The prototype for this analysis is the case of classical Fourier series of periodic functions on the line with period 2π. Here G is the circle group (the compact Lie group of real numbers modulo 2π, with addition modulo 2π as group operation), and H is the space $L^2(G)$ of square integrable functions on the interval $-\pi \leq x \leq \pi$. The unitary representation R is given by $[R(x_0)f](x) = f(x_0 + x)$, with addition taken modulo 2π. For this G, the irreducible unitary representations are all one-dimensional and are parametrized by the integers, the nth one being $R_n(x_0) = e^{inx_0}v$ for v in $\mathbf{C}$. The classical Fourier series expansion given in Eq. (2) corresponds to a

$$f(x) \sim \sum_{n=-\infty}^{\infty} c_n e^{inx} \qquad (2)$$

decomposition of R into an infinite orthogonal sum of all R_n, since a single R_n can be realized in the one-dimensional space of multiples of e^{inx} and since Eq. (3) holds. In this context, let L be a

$$R(x_0)f(x) = f(x_0 + x) \sim \sum c_n e^{inx_0}e^{inx}$$
$$= \sum R_n(x_0)(c_n e^{inx}) \qquad (3)$$

bounded linear operator on H that commutes with translations, that is, with all $R(x_0)$. An easy computation shows that L must carry e^{inx} into a multiple of itself. If that multiple is b_n, then L is operating in the space of R_n by multiplication by b_n. Equation (4) is then useful in studying L. *See* OPERATOR THEORY.

$$Lf(x) \sim \sum_{n=-\infty}^{\infty} b_n c_n e^{inx} \qquad (4)$$

Sample applications. Three examples will illustrate the principle of analyzing a linear operator by decomposing it with respect to a group of operators that commute with it.

Rotations and the Fourier transform. Let H be the Hilbert space of square-integrable functions on $\mathbf{R}^n$, let G be the rotation group, and let the representation of G on H be given by rotation of the coordinates. The Fourier transform is a linear operator on H that commutes with each member of G. The space H decomposes under the representation of G into a sequence of simpler spaces, the kth space being all sums of products of a radial function and a harmonic polynomial homogeneous of degree k. The Fourier transform has a simple form on the kth space, given as a one-dimensional integral involving a Bessel function. *See* BESSEL FUNCTIONS; FOURIER SERIES AND INTEGRALS.

Solutions of differential equations. Some information about solutions in $\mathbf{R}^n$ to elliptic partial differential equations with constant coefficients can be obtained by using the Fourier transform. Here the differential operator commutes with the representation of $G = \mathbf{R}^n$ given by translation in the space of square-integrable functions on $\mathbf{R}^n$, and the Fourier transform is the device for decomposing the representation. For an operator with nonconstant coefficients, one can apply a freezing principle, replacing a given equation by one with constant coefficients and obtaining an approximate solution near a point.

For many equations of subelliptic type, which are elliptic except for certain degeneracies, a similar analysis is possible by means of representations of a more complicated Lie group. In the simplest new case, the Lie group is the (nonabelian) group of 3-by-3 real upper-triangular matrices with ones on the diagonal, and the differential equation is Eq. (5) in $\mathbf{R}^3$. Other equations can be analyzed by

$$\left(\frac{\partial^2 f}{\partial x^2} + \frac{\partial^2 f}{\partial y^2} \right) + \left(x \frac{\partial^2 f}{\partial y \partial z} - y \frac{\partial^2 f}{\partial x \partial z} \right)$$
$$+ \frac{1}{4}\left(x^2 + y^2 \right) \frac{\partial^2 f}{\partial z^2} = 0 \qquad (5)$$

a combination of this method and a freezing principle.

The laplacian $\Delta = \Sigma \partial^2 / \partial x_j^2$ in R^n has constant coefficients and can be analyzed by the Fourier transform. But also it commutes with rotations and can be decomposed accordingly. In a ball, for example, the resulting decomposition of solutions of $\Delta u = 0$ corresponds to solving the equation by separation of variables. If the solutions are restricted to a sphere, the analysis amounts to a decomposition of the space of square integrable functions on the sphere into (finite-dimensional) eigenspaces of the spherical laplacian, each of which is the Hilbert space for an irreducible representation of the rotation group. More generally, information about the generalized wave operator given in Eq. (6) can be obtained by doing a Fourier analysis of

$$L = \frac{\partial^2}{\partial x_1^2} + \cdots + \frac{\partial^2}{\partial x_m^2} - \frac{\partial^2}{\partial x_{m+1}^2} - \cdots - \frac{\partial^2}{\partial x_{m+n}^2} \qquad (6)$$

functions on an orbit in R^{m+n} under the action of the noncompact group of linear isometries of a real quadratic form with m plus signs and n minus signs. *See* DIFFERENTIAL EQUATION.

Quantum mechanics. In quantum theory the

things one observes from experiments are eigenvalues (or the spectrum, if there are not discrete eigenvalues) of certain self-adjoint operators on Hilbert spaces of wave functions. Conservation laws come from self-adjoint operators whose associated one-parameter groups of unitary operators commute with the hamiltonian. In this way a system of conservation laws leads to a group of symmetries, namely, a group of unitary operators commuting with the hamiltonian. Frequently this group is a Lie group. Analysis of the representation of the Lie group given by the unitary operators gives information about the physical system. *See* ELEMENTARY PARTICLE; LORENTZ TRANSFORMATIONS; NONRELATIVISTIC QUANTUM THEORY; RELATIVISTIC QUANTUM THEORY; SYMMETRY LAWS. [ANTHONY W. KNAPP]

Bibliography: S. Helgason, *Differential Geometry, Lie Groups, and Symmetric Spaces*, 1978; L. Pukanszky, *Leçons sur les représentations des groupes*, 1967; V. S. Varadarajan, *Lie Groups, Lie Algebras, and Their Representations*, 1974; G. Warner, *Harmonic Analysis on Semi-Simple Lie Groups*, vols. 1 and 2, 1972; H. Weyl, *The Classical Groups, Their Invariants and Representations*, 1939.

Light

The term light, as commonly used, refers to the kind of radiant electromagnetic energy that is associated with vision. In a broader sense, light includes the entire range of radiation known as the electromagnetic spectrum. The branch of science dealing with light, its origin and propagation, its effects, and other phenomena associated with it is called optics. Spectroscopy is the branch of optics that pertains to the production and investigation of spectra.

Any acceptable theory of the nature of light must stem from observations regarding its behavior under various circumstances. Therefore, this article begins with a brief account of the principal facts known about visible light, including the relation of visible light to the electromagnetic spectrum as a whole, and then describes the apparent dual nature of light. The remainder of the article discusses various experimental and theoretical considerations pertinent to the study of this problem.

PRINCIPAL EFFECTS

The electromagnetic spectrum is a broad band of radiant energy which extends over a range of wavelengths running from trillionths of inches to hundreds of miles; wavelengths of visible light are measured in hundreds of thousandths of an inch. Arranged in order of increasing wavelength, the radiation making up the electromagnetic spectrum is termed gamma rays, x-rays, ultraviolet rays, visible light, infrared waves, microwaves, radio and TV waves, and very long electromagnetic waves. Detailed descriptions of these radiations are given separately; *see* ELECTROMAGNETIC RADIATION and the articles listed therein. *See also* OPTICS; POLARIZED LIGHT; RAMAN EFFECT.

Finite velocity. The fact that light travels at a finite speed or velocity is well established, both theoretically and experimentally. This speed is so high, however, and the experimental complexities of measuring it are so great, that this is not a phenomenon the average man can check for himself. The first successful measurement of the speed of light was made in 1676 by the Danish astronomer Ole Roemer, and since that time, numerous increasingly precise experiments have been conducted, leading to the currently accepted value of 299,792.8 km/sec in a vacuum. In round numbers, the speed of light in vacuum or air may be said to be 186,000 mi/sec or 300,000 km/sec.

Diffraction and reflection. One of the most easily observed facts about light is its tendency to travel in straight lines. Careful observation shows, however, that a light ray spreads slightly when passing the edges of an obstacle. This phenomenon is called diffraction. The reflection of light is also well known. The Moon, as well as all other satellites and planets in the solar system, are visible only by reflected light; they are not self-luminous like the Sun. Reflection of light from smooth optical surfaces occurs so that the angle of reflection equals the angle of incidence, a fact that is most readily observed with a plane mirror. When light is reflected irregularly and diffusely, the phenomenon is termed scattering. The scattering of light by gas particles in the atmosphere causes the blue color of the sky. When a change in frequency (or wavelength) of the light occurs during scattering, the scattering is referred to as the Raman effect.

Refraction. The type of bending of light rays called refraction is caused by the fact that light travels at different speeds in different media—faster, for example, in air than in either glass or water. Refraction occurs when light passes from one medium to another in which it moves at a different speed. Familiar examples include the change in direction of light rays in going through a prism, and the bent appearance of a stick partially immersed in water.

Interference and polarization. In the phenomenon called interference, rays of light emerging from two parallel slits combine on a screen to produce alternating light and dark bands. This effect can be obtained quite easily in the laboratory, and is observed in the colors produced by a thin film of oil on the surface of a pool of water. Polarization of light is usually shown with Polaroid disks. Such disks are quite transparent individually. When two of them are placed together, however, the degree of transparency of the combination depends upon the relative orientation of the disks. It can be varied from ready transmission of light to almost total opacity, simply by rotating one disk with respect to the other.

Chemical effects. When light is absorbed by certain substances, chemical changes take place. This fact forms the basis for the science of photochemistry. Rapid progress has been made toward an understanding of photosynthesis, the process by which plants produce relatively complex substances such as sugars in the presence of sunlight. This is but one example of the all-important response of plant and animal life to light.

HISTORICAL APPROACH

Early in the 18th century, light was generally believed to consist of tiny particles. Of the phenomena mentioned in the preceding section, reflection, refraction, and the sharp shadows caused by the straight path of light were well known, and the characteristic of finite velocity was suspected. All of these phenomena except refraction clearly could be expected of streams of particles, and Isaac Newton showed that refraction would occur if the velocity of light increased with the density of the medium through which it traveled.

This theory of the nature of light seemed to be completely upset, however, in the first half of the 19th century. During that time, Thomas Young studied the phenomena of interference, and could see no way to account for them unless light were a wave motion. Diffraction and polarization had also been investigated by that time. Both were easily understandable on the basis of a wave theory of light, and diffraction eliminated the "sharp-shadow" argument for particles. Reflection and finite velocity were consistent with either picture. The final blow to the particle theory seemed to have been struck in 1849, when the speed of light was measured in different media and found to vary in just the opposite manner to that assumed by Newton. Therefore, in 1850, it seemed finally to be settled that light consisted of waves.

Even then, however, there was the problem of the medium in which light waves traveled. All other kinds of waves required a physical medium, but light traveled through a vacuum—faster, in fact, than through air or water. The term ether was proposed by James Clerk Maxwell and his contemporaries as a name for the unknown medium, but this scarcely solved the problem because no ether was ever actually found. Then, near the beginning of the 20th century, came certain work on the emission and absorption of energy that seemed to be understandable only if one assumed light to have a particle or corpuscular nature. The external photoelectric effect, that is, the emission of electrons from the surfaces of solids when light is incident on the surfaces, was one of these. At that time, then, science found itself in the uncomfortable position of knowing a considerable number of experimental facts about light, of which some were understandable regardless of whether light consisted of waves or particles, others appeared to make sense only if light were wavelike, and still others seemed to require it to have a particle nature. *See* ETHER HYPOTHESIS.

THEORY

The study of light deals with some of the most fundamental properties of the physical world and is intimately linked with the study of the properties of submicroscopic particles on the one hand and with the properties of the entire universe on the other. The creation of electromagnetic radiation from matter and the creation of matter from radiation, both of which have been achieved, provide a fascinating insight into the unity of physics. The same is true of the deflection of light beams by strong gravitational fields, such as the bending of starlight passing near the Sun.

A classification of phenomena involving light according to their theoretical interpretation provides the clearest insight into the nature of light. When a detailed accounting of experimental facts is required, two groups of theories appear which, in the majority of cases, account separately for the wave and the corpuscular character of light. The quantum theories seem to obviate questions con-

cerning this dual character of light, and make the classical wave theory and the simple corpuscular theory appear as two very useful limiting theories. It happens that the wave theories of light can cope with a considerable part of the phenomena involving electromagnetic radiation. Geometrical optics, based on the wave theory of light, can solve many of the more common problems of the propagation of light, such as refraction, provided that the limitations of the underlying theory are not disregarded. *See* GEOMETRICAL OPTICS.

Phenomena involving light may be classed into three groups: electromagnetic wave phenomena, corpuscular or quantum phenomena, and relativistic effects. The relativistic effects appear to influence similarly the observation of both corpuscular and wave phenomena. The major developments in the theory of light closely parallel the rise of modern physics. These developments are charted in Fig. 1, and are discussed in the remainder of this article.

Wave phenomena. Interference and diffraction, mentioned earlier, are the most striking manifestations of the wave character of light. Their fundamental similarity can be demonstrated in a number of experiments. The wave aspect of the entire spectrum of electromagnetic radiation is most convincingly shown by the similarity of diffraction pictures produced on a photographic plate, placed at some distance behind a diffraction grating, by radiations of different frequencies, such as x-rays and visible light. The interference phenomena of light are, moreover, very similar to interference of electronically produced microwaves and radio waves.

Polarization demonstrates the transverse character of light waves. Further proof of the electromagnetic character of light is found in the possibility of inducing, in a transparent body that is being traversed by a beam of plane-polarized light, the property of rotating the plane of polarization of the beam when the body is placed in a magnetic field. *See* FARADAY EFFECT.

The fact that the velocity of light had been calculated from electric and magnetic parameters (permittivity and permeability) was at the root of Maxwell's conclusion in 1865 that "light, including heat and other radiations if any, is a disturbance in the form of waves propagated . . . according to electromagnetic laws." Finally, the observation that electrons and neutrons can give rise to diffraction patterns quite similar to those produced by visible light has made it necessary to ascribe a wave character to particles. *See* ELECTRON DIFFRACTION; NEUTRON DIFFRACTION.

Velocity determination. The finite velocity of transmission of signals by means of electromagnetic waves is a fact well established for all wavelengths of radiation in a variety of terrestrial and astronomical experiments. Astronomical observation of so-called eclipsing binaries, both visual and spectroscopic, is taken to indicate that the velocity of light, denoted by c, does not depend on the velocity of the source.

Measurements of c fall basically into two groups, (1) group velocity, or signal velocity, determinations, using the relation $c = $ distance/time; and (2) phase velocity, or wave velocity, determinations, using $c = $ frequency/wavelength.

The group velocity is the average time for a light

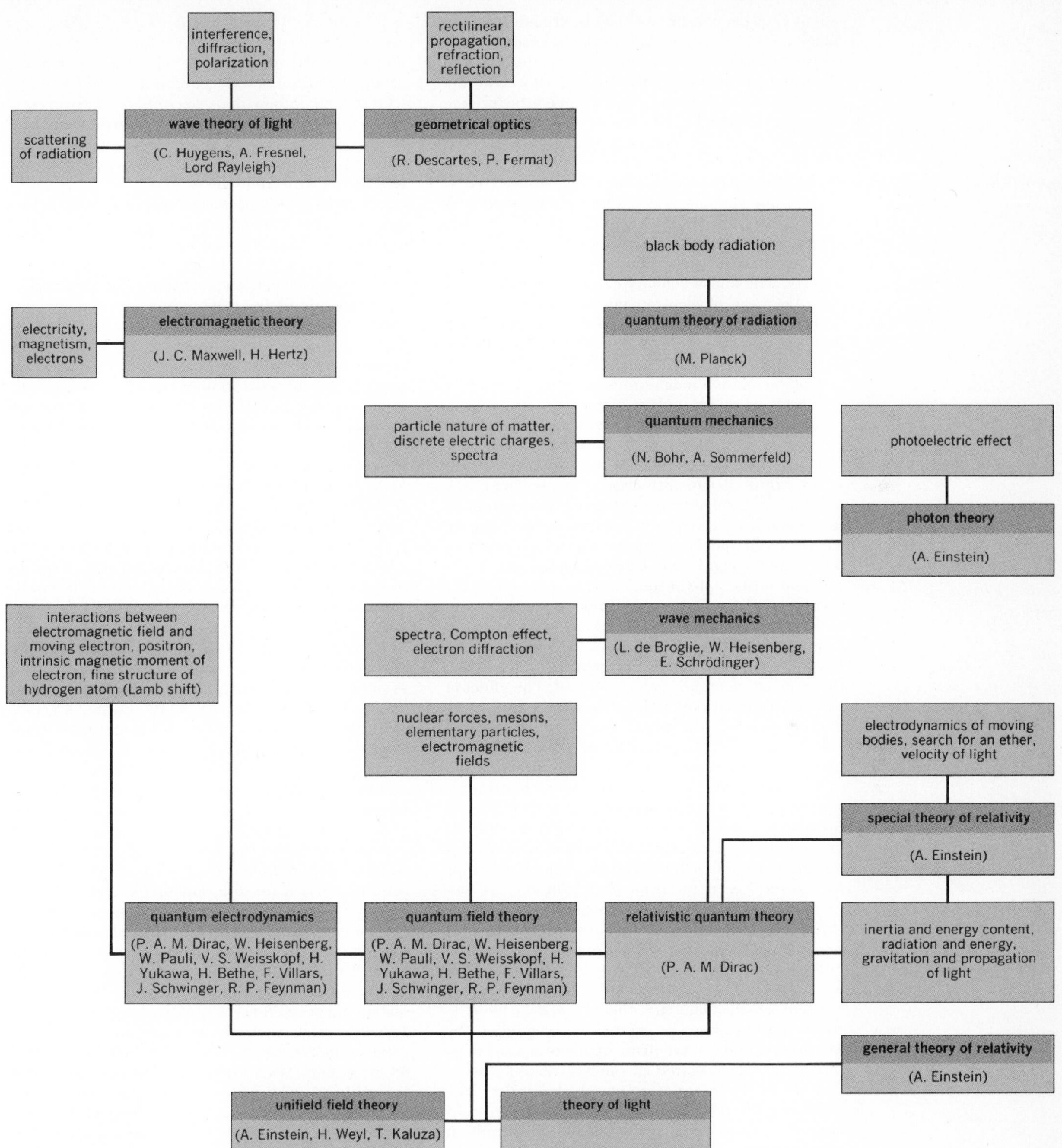

Fig. 1. Developments in the theory of light. (*G. W. Stroke*)

signal, that is, a modulated electromagnetic wave train, to traverse a given distance. Roemer's original determination of the velocity of light was based on careful observations of the time of eclipse of the moons of Jupiter, from various points in the Earth's orbit. Subsequent terrestrial determinations, making use of revolving mirrors (J. L. Foucault and A. A. Michelson), a toothed wheel (H. L. Fizeau), or an electronically modulated light beam (E. Bergstrand), also measured the time required for light to traverse the distance between a source and a reflecting mirror. These are all group velocity measurements, and furnish a value of *c* only if the experiments are carried out in a nondispersive medium, in which the velocity is the same for all wavelengths. Otherwise, the group velocity is smaller than the phase velocity by 0.007% (2.2 km/sec) in air, 1.5% in water, and 2.4% in ordinary crown glass. *See* DISPERSION; GROUP VELOCITY.

Determinations of the phase velocity, which is

simply the speed of the wavefront, are indirect and make the assumption that $c = \lambda f$, where f is the frequency, and λ the free-space wavelength of the electromagnetic radiation. Most of the measurements of this type involve microwave interference in various forms: E. F. Florman with two sources; K. D. Froome in an apparatus similar to a Michelson interferometer; L. Essen, J. P. Gordon, and H. M. Smith in a microwave resonance cavity. D. H. Rank and also E. K. Plyer determined c by calculation from microwave and infrared spectroscopic measurements of the ratio of hertz to waves per centimeter for a certain molecular rotation frequency. The phase velocity can also be calculated from the ratio of electromagnetic to electrostatic units. A detailed description of these experiments is found in the bibliographical references. *See* PHASE VELOCITY.

The measured magnitudes of the velocity of transmission and the phase velocity seem to be in reasonably good agreement among themselves, possibly to a few parts in 10^5. However, the experimental values determined by the two methods are at an admitted and disturbing variance with one another. The velocity of light is not as constant as it is sometimes thought to be. While according to the theory of special relativity, the velocity of light will be the same in any frame of reference, independent of its state of motion, this is true only for frames of the same gravitational potential. Conceivably, local variations of the gravitational potential could lead to variations of the measured velocity of light, although experiments for detecting these variations remain to be devised. The effect of the gravitational potential at the Sun's surface or at other points in the universe is quite appreciable compared with the precision of many measurements of c. *See* POTENTIALS.

Electromagnetic-wave propagation. Electromagnetic waves can be propagated through free space, devoid of matter and fields, and with a constant gravitational potential; through space with a varying gravitational potential; and through more or less absorbing material media which may be solids, liquids, or gases. Radiation can be transmitted through waveguides with cylindrical, rectangular, or other boundaries, the insides of which can be either evacuated or filled with a dielectric medium.

From electromagnetic theory, and especially from the well-known equations formulated by Maxwell, a plane wave disturbance of a single frequency f is propagated in the x direction with a phase velocity $v = \lambda f = \lambda \omega / 2\pi$, where $\omega = 2\pi f$. The wave can be described by the equation $y = A \cos \omega (t - x/v)$. Two disturbances of same amplitude A, of respective angular frequencies ω_1 and ω_2 and of velocities v_1 and v_2, propagated in the same direction, yield the resulting disturbance y', defined in Eq. (1). Here $\Delta \omega = \omega_1 - \omega_2$ and $\omega = 1/2(\omega_1 + \omega_2)$.

$$y' = y_1 + y_2 = 2A \cos \tfrac{1}{2} [(\Delta \omega) t \\ - x\Delta(\omega/v)] \cos (\omega t - \omega x/v) \quad (1)$$

The ratio $u = \Delta \omega / \Delta(\omega/v)$ is defined as the group velocity, just as the ratio $\omega/(\omega/v)$ is identical with the phase velocity. In the limit for small $\Delta \omega$, $u = d\omega/d(\omega/v)$. Noting that $\omega = 2\pi v/\lambda$, $d\omega = 2\pi (\lambda \, dv - v \, d\lambda)/\lambda^2$, and $d(\omega/v) = -2\pi \, d\lambda/\lambda^2$, an important relation, Eq. (2), between group and

$$u = v - \lambda \, dv/d\lambda \quad (2)$$

phase velocity is obtained. This shows that the group velocity u is different from the phase velocity v in a medium with dispersion $dv/d\lambda$. In a vacuum, $u = v = c$. With the help of Fourier theorems, the preceding expression for u can be shown to apply to the propagation of a wave group of infinite length, but with frequencies extending over a finite small domain. Furthermore, even if the wave train were emitted with an infinite length, modulation or "chopping" would result in a degrading of the monochromacy by introduction of new frequencies, and hence in the appearance of a group velocity. Considerations of this nature are not trivial in measurements of the velocity of light, but are quite fundamental to the conversion of instrumental readings to a value of c. Similar considerations apply to the incorporation of the effects of the medium and the boundaries involved in the experiments. Complications arise in the regions of anomalous dispersion (absorption regions), where the phase velocity can exceed c and $dv/d\lambda$ is positive. *See* MAXWELL'S EQUATIONS; WAVE EQUATION; WAVE MOTION.

Refractive index. A plane wavefront, in going from a medium in which its phase velocity is v into a second medium where the velocity is v', changes direction at the interface. By geometry, it can be shown that $\sin i / \sin i' = v/v'$, where i and i' are the angles which the light path forms with a normal to the interface in the two media. It can also be shown that the path between any two points in this system is that which minimizes the time for the light to travel between the points (Fermat's principle). This path would not be a straight line unless $i = 90°$ or $v = v'$. Snell's law states that $n \sin i = n' \sin i'$, where n is the index of refraction of the medium. It follows that the refractive index of a medium is $n = c/v'$, because the refractive index of a vacuum, where $v = c$, has the value 1.

Dispersion. The dispersion $dv/d\lambda$ of a medium can easily be obtained from measurements of the refractive index for different wavelengths of light. The weight of experimental evidence, based on astronomical observations, is against a dispersion in a vacuum. Observation of the light reaching the Earth from the eclipsing binary star Algol, 120 light-years distant (1 light-year $\cong 6 \times 10^{12}$ mi is the distance traversed in vacuum by a beam of light in 1 year), shows that the light for all colors is received simultaneously. The eclipsing occurs every 68 hr 49 min. Were there a difference of velocity for red light and for blue light in interstellar space as great as 1 part in 10^6, this star would show a measurable time difference in the occurrence of the eclipses in these two colors.

Corpuscular phenomena. In its interactions with matter, light exchanges energy only in discrete amounts, called quanta. This fact is difficult to reconcile with the idea that light energy is spread out in a wave, but is easily visualized in terms of corpuscles, or photons, of light.

Blackbody (heat) radiation. The radiation from theoretically perfect heat radiators, called blackbodies, involves the exchange of energy between radiation and matter in an enclosed cavity. The observed frequency distribution of the radiation emitted by the enclosure at a given temperature of the cavity can be correctly described by theory only if one assumes that light of frequency ν is absorbed in integral multiples of a quantum of energy equal to $h\nu$, where h is a fundamental physical constant called Planck's constant. This startling departure from classical physics was made by Max Planck early in the 20th century.

Photoelectric effect. When a monochromatic beam of electromagnetic radiation illuminates the surface of a solid (or less commonly, a liquid), electrons are ejected from the surface in the phenomenon known as photoemission, or the external photoelectric effect. The kinetic energies of the electrons can be measured electronically by means of a collector which is negatively charged with respect to the emitting surface. It is found that the emission of these photoelectrons, as they are called, is immediate, and independent of the intensity of the light beam, even at very low light intensities. This fact excludes the possibility of accumulation of energy from the light beam until an amount corresponding to the kinetic energy of the ejected electron has been reached. The number of electrons is proportional to the intensity of the incident beam. The velocities of the electrons ejected by light at varying frequencies agree with Eq. (3),

$$\tfrac{1}{2}mv^2_{\text{max}} = h(\nu - \nu_0) \qquad (3)$$

where m is the mass of the electron, v_{max} the maximum observed velocity, ν the frequency of the illuminating light beam, and ν_0 a threshold frequency characteristic of the emitting solid.

In 1905 Albert Einstein showed that the photoelectric effect could be accounted for by assuming that, if the incident light is composed of photons of energy $h\nu$, part of this energy, $h\nu_0$, is used to overcome the forces binding the electron to the surface. The rest of the energy appears as kinetic energy of the ejected electron.

Compton effect. The scattering of x-rays of frequency ν_0 by the lighter elements is caused by the collision of x-ray photons with electrons. Under such circumstances, both a scattered x-ray photon and a scattered electron are observed, and the scattered x-ray has a lower frequency than the impinging x-ray. The kinetic energy of the impinging x-ray, the scattered x-ray, and the scattered electron, as well as their relative directions, are in agreement with calculations involving the conservation of energy and momentum. *See* COMPTON EFFECT; HEAT RADIATION; PHOTOELECTRICITY; PHOTON.

Quantum theories. The need for reconciling Maxwell's theory of the electromagnetic field, which describes the electromagnetic wave character of light, with the particle nature of photons, which demonstrates the equally important corpuscular character of light, has resulted in the formulation of several theories which go a long way toward giving a satisfactory unified treatment of the wave and the corpuscular picture. These theories incorporate, on one hand, the theory of quantum electrodynamics, first set forth in articles by P. A. M. Dirac, P. Jordan, W. Heisenberg, and W. Pauli, and on the other, the earlier quantum mechanics of L. de Broglie, Heisenberg, and E. Schrödinger. Unresolved theoretical difficulties persist, however, in the higher-than-first approximations of the interactions between light and elementary particles. The incorporation of a theory of the nucleus into a theory of light is bound to call for additional formulation.

Dirac's synthesis of the wave and corpuscular theories of light is based on rewriting Maxwell's equations in a Hamiltonian form resembling the Hamiltonian equations of classical mechanics. Using the same formalism involved in the transformation of classical into wave-mechanical equations by the introduction of the quantum of action $h\nu$, Dirac obtained a new equation of the electromagnetic field. The solutions of this equation require quantized waves, corresponding to photons. The superposition of these solutions represents the electromagnetic field. The quantized waves are subject to Heisenberg's uncertainty principle. The quantized description of radiation cannot be taken literally in terms of either photons or waves, but rather is a description of the probability of occurrence in a given region of a given interaction or observation. *See* HAMILTON'S EQUATIONS OF MOTION; NONRELATIVISTIC QUANTUM THEORY; QUANTUM ELECTRODYNAMICS; QUANTUM FIELD THEORY; QUANTUM MECHANICS; RELATIVISTIC QUANTUM THEORY; UNCERTAINTY PRINCIPLE.

Relativistic effects. The measured magnitudes of such characteristics as wavelength and frequency, velocity, and the direction of the radiation in a light beam are affected by a relative motion of the source with respect to the observer, occurring during the emission of the signal-carrying electromagnetic wave trains. *See* DOPPLER EFFECT.

A difference in gravitational potential also affects these quantities. Several important observations of this nature are listed in this section, followed by a discussion of several important results of general relativity theory involving light. For extended discussions of both the special theory of relativity and the general theory of relativity *see* RELATIVITY.

Velocity in moving media. In 1818 A. Fresnel suggested that it should be possible to determine the velocity of light in a moving medium, for example, to determine the velocity of a beam of light traversing a column of liquid of length d and of refractive index n, flowing with a velocity v relative to the observer, by measuring the optical thickness nd. The experiment was carried out by Fizeau in a modified Rayleigh interferometer, shown in Fig. 2, by measuring the fringe displacement in O corresponding to the reversing of the direction of flow. If v' is the phase velocity of light in the medium (deduced from the refractive index by the relation $v' = c/n$), it is found that the measured velocity v_m in the moving medium can be expressed as $v_m = v' + v(1 - 1/n^2)$ rather than $v_m = v' + v$, as would be the case with a Newtonian velocity addition.

Aberration. J. Bradley discovered in 1725 a yearly variation in the angular position of stars, the total variation being 41 seconds of arc. This effect is in addition to the well-known parallax effect, and was properly ascribed to the combination of the velocity of the Earth in its orbit and the speed of light. Bradley used the amplitude of the variation to arrive at a value of the velocity of light. George Airy compared the angle of aberration in a telescope before and after filling it with water, and he discovered, contrary to his expectation, that there was no difference in angle.

Michelson-Morley experiment. The famous Michelson-Morley experiment, one of the most

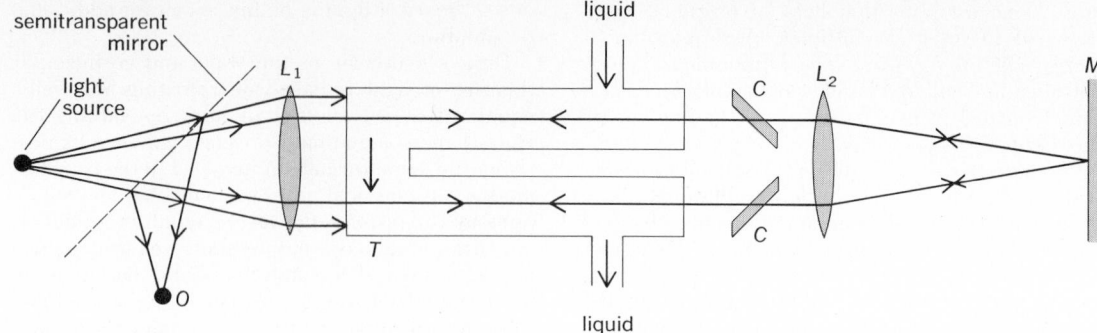

Fig. 2. Fizeau's experiment. *C,* compensator plates; *M,* mirror; *L₁, L₂,* lenses; *T,* tube; *O,* interference fringes.

significant experiments of all time, was performed in 1887 to measure the relative velocity of the Earth through inertial space. Inertial space is space in which Newton's laws of motion hold. Dynamically, an inertial frame of reference is one in which the observed accelerations are zero if no forces act. A point in an orbit is the center of such a frame. *See* FRAME OF REFERENCE.

The rotation of the Earth about its axis, with tangential velocities never exceeding 0.5 km/sec, is easily demonstrated mechanically (Foucault's pendulum, precession of gyroscopes) and optically (Michelson's rectangular interferometer). The surface of the Earth is not an inertial frame. In its orbit around the Sun, on the other hand, the Earth has translational velocities of the order of 30 km/sec, but this motion cannot be detected by mechanical experiments because of its orbital nature. The hope existed, however, that optical experiments would permit the detection (and measurement) of the relative motion of the Earth through inertial space by comparing the times of travel of two light beams, one traveling in the direction of the translation through inertial space, and the other at right angles to it. The hope was based on the now disproven notion that the velocity of a light beam would be equal to the constant c only when measured with respect to the inertial space, but would be measured as smaller $(c-v)$ or greater $(c+v)$ with respect to a reference frame, such as the earth, moving with a velocity v in inertial space if a light beam were projected respectively in the direction and in the opposite direction of translation of this frame. According to classical velocity addition theorems (which, as is now known, do not apply to light), a velocity difference of $2v$ would be detected under such circumstances.

Not only does the Earth move in an orbit around the Sun, but it is carried with the Sun in the galactic rotation toward Cygnus with a velocity of several hundreds of kilometers per second, and the Galaxy itself is moving with a high speed in its local spiral group. Speeds of hundreds and possibly thousands of kilometers per second should be detectable by measurements on Earth in two orthogonal directions, assuming of course that the Earth motion is itself with respect to inertial space, or indeed that such a space has the physical meaning ascribed to it. The unexpected result of the experiment was that no such velocity difference could be detected, that is, no relative motion could be detected by optical means.

The Michelson-Morley apparatus (Fig. 3) consists of a horizontal Michelson interferometer with its two arms at right angles. The mirrors are adjusted so that the central white-light fringe falls on the cross hair of the observing telescope. This indicates equality of optical phase, and therefore an equality of the times taken by the light beams to travel from the beam-splitting surface to each of the two mirrors and back. Rotation of the entire system by 90°, or indeed by any angle, as well as repetition of the experiment at various times of the year all are found to leave the central white-light fringe and associated fringe system undisplaced, indicating no change in the time required by the light to traverse the two arms of the interferometer when their directions relative to the direction of the Earth's motion are varied. Had there been a difference in the velocity of light in the two directions *OM* and *ON*, the two arms would be of unequal length in the initial adjustment. For example, if the light traveled faster in the direction *OM* (on the average, going back and forth), then the corresponding arm would have to be longer so as to make the time of travel equal in both arms. If the apparatus were turned through 90°, the shorter arm would take the place of the longer arm, and the "faster" light would now travel in the shorter arm, and the "slower" light in the longer arm; a

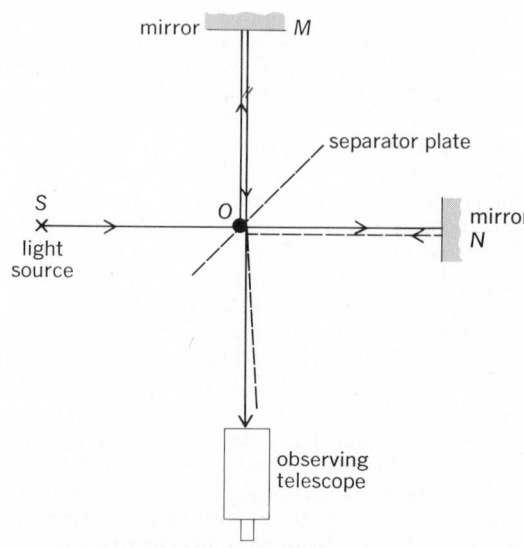

Fig. 3. Michelson-Morley experiment.

noticeable fringe displacement would, but actually does not, take place. *See* INTERFEROMETRY.

Einstein's theory accounts for the null result by the simple explanation that no relative motion between the apparatus and the observer exists in the experiment. No change in measured length has occurred in either direction, and because the propagation of light is isotropic, no velocity difference or detection of relative motion should be expected.

Gravitational and nebular red shifts. Two different kinds of shifts, or displacements of spectral lines toward the red end of the spectrum, are observed in spectrograms taken with starlight or light from nebulae. One is the rare, but extremely significant, gravitational red shift or Einstein shift, which has been measured in some spectra from white dwarf stars. The other, much more widely encountered, is the red shift in spectra from nebulae, usually described as being caused by a radial Doppler effect characterizing an "expansion of the universe."

The most famous example of the gravitational red shift is that observed in the spectrum of the so-called dark companion of Sirius. The existence of this companion was predicted by F. W. Bessel in 1844 on the basis of a perturbation in the motion of Sirius, but because of its weak luminosity (1/360 that of the Sun, 1/11,000 that of Sirius), it was not observed until 1861. This companion is a white dwarf of a mass comparable to that of the Sun (0.95), but having a relatively small radius of 18,000 km and the fantastically high density of 61,000 times that of water. This companion shows a shift in its spectral lines relative to the ones emitted by Sirius itself; the shift of 0.3 A for the Balmer β-line of hydrogen was reliably determined by W. S. Adams.

The nebular red shift is a systematic shift observed in the spectra of all nebulae, best measured with the calcium H and K absorption lines. Distances of nebulae are determined photometrically by measurements of their intensities, and it is found that the wavelength shift towards the red increases with the distance of the nebulae from the Earth. (The Earth does not have a privileged position if expansion of the universe is indeed involved, rather its position is somewhat like that of a man in a crowd which is dispersing—every individual in the crowd will find himself at an ever-increasing distance from every one of his neighbors.) The change of wavelength with distance has been given by E. G. Hubble as shown in Eq. (4), where d

$$\Delta\lambda/\lambda = 1.7 \times 10^{-9}d \qquad (4)$$

is in parsecs (1 parsec = 3×10^{18} cm). Red shifts in nebulae up to 1.1×10^9 light-years distant have been measured in a comprehensive program carried out at the Mount Wilson Observatory by Hubble and M. L. Humason.

Results of general relativity. The propagation of light is influenced by gravitation. This is one of the fundamental results of Einstein's general theory of relativity which has been subjected to experimental tests and found to be verified. Three important results involving light need to be singled out.

1. The velocity of light, measured by the same magnitude c independently of the state of motion of the frame in which the measurement is being carried out, depends on the gravitational potential

Φ of the field in which it is being measured according to Eq. (5). Here $\Phi = -GM/R$, where G is the

$$c = c_0 \left(1 + \frac{\Phi}{c^2}\right) \qquad (5)$$

universal constant of gravitation (6.670×10^{-8} cgs units), M the mass of the celestial body in grams, R the radius of the body in centimeters, and c_0 the velocity of light in a vacuum devoid of fields.

For example, the absolute value of the term Φ/c^2 is about 3000 times greater on the Sun than on Earth, making the measurements of c smaller by two parts in 10^6 on the Sun as compared to measurements on Earth.

2. The frequency ν of light emitted from a source in a gravitational field with the gravitational potential Φ is different from the frequency ν_0 emitted by an identical source (atomic, nuclear, molecular) in a field-free region, according to Eq. (6).

$$\nu = \nu_0 \left(1 + \frac{\Phi}{c^2}\right) \qquad (6)$$

Spectral lines in sunlight should be displaced toward the red by two parts in 10^6 when compared to light from terrestrial sources.

3. Light rays are deflected when passing near a heavenly body according to Eq. (7), where α is the

$$\alpha = \frac{4GM}{c^2R} \qquad (7)$$

angular deflection in radians, and R the distance of the beam from the center of the heavenly body of mass M. The deflection is directed so as to increase the apparent angular distance of a star from the center of the Sun when starlight is passing near the edge of the Sun. The deflection according to this equation should be 1.75 seconds of arc, a value which compares favorably with eclipse measurements of the star field around the Sun in 1931. These measurements indicated values up to 2.2 seconds of arc when compared with photographs of the same field 6 months earlier. This most sensational prediction of Einstein's theory might seem less surprising today when the corpuscular-photon character of light is widely known, and when a Newtonian M/R^2 attraction might be considered to be involved in the motion of a corpuscle with the velocity c past the Sun. However, application of Newton's law predicts a deviation only half as great as the reasonably well verified relativistic prediction.

Matter and radiation. The possibility of creating a pair of electrons—a positively charged one (positron) and a negatively charged one (negatron) —by a rapidly varying electromagnetic field (γ-rays of high frequency) was predicted as a consequence of Dirac's wave equation for a free electron and has been experimentally verified. I. Curie and F. Joliot, as well as J. Chadwick, P. M. S. Blackett, G. P. Occhialini, and others have compared the number of positrons and negatrons ejected by γ-rays passing through a thin sheet of lead (and other materials) and have found them to be the same, after accounting for two other groups of electrons also appearing in the experiment (photoelectrons and recoil electrons). Other examples of negatron-positron pair production include the collision of two heavy particles, a fast electron passing through the field of a nucleus, the direct collision of two electrons, the collision of two light quanta in a vacuum, and the action of a nuclear field on a γ-ray emitted by the nucleus involved in the action.

Evidence of the creation of matter from radiation, as well as that of radiation from matter, substantiates Einstein's equation, Eq. (8), which

$$E = mc^2 \qquad (8)$$

was first expressed in the following words: "If a body [of mass m] gives off the energy E in the form of radiation, its mass diminishes by E/c^2."

In regard to exchanges of energy and momentum, electromagnetic waves behave like a group of particles with energy as in Eq. (9) and momentum as in Eq. (10).

$$E = mc^2 = h\nu \qquad (9)$$

$$p = h\nu/c = h/\lambda \qquad (10)$$

Finally, many experiments with photons show that they also possess an intrinsic angular momentum, as do particles. Circularly polarized light, for example, carries an experimentally observable angular momentum, and it can be shown that, under certain circumstances, an angular momentum can be imparted to unpolarized or plane-polarized light (plane wave passing through a finite circular aperture). In any case, the angular momentum will be quantized in units of $h/2\pi$.

The inverse process to the creation of electron pairs is the annihilation of a positron and a negatron, resulting in the production of two γ-ray quanta (two-quantum annihilation). Nuclear chain reactions are known to involve similar processes. *See* CHAIN REACTION; ELECTRON-POSITRON PAIR PRODUCTION; ELEMENTARY PARTICLE; INERTIA OF ENERGY.

Unified field theories. In conclusion, one might mention the so-called unified field theories. These are classical theories that attempt to represent electromagnetism and gravitation as aspects of space-time geometry. Although there is some experimental evidence for a connection between electromagnetism and gravitation, the subject remains speculative. *See* SPACE-TIME.

[GEORGE W. STROKE]

Bibliography: P. G. Bergmann, *Introduction to the Theory of Relativity*, 1976; M. Born and E. Wolf, *Principles of Optics*, 5th ed., 1975; S. Fluegge (ed.), *Handbuch der Physik*, vol. 24, 1956; F. A. Jenkins and H. E. White, *Fundamentals of Optics*, 4th ed., 1976; R. Loudon, *The Quantum Theory of Light*, 1973; H. H. Skilling, *Fundamentals of Electric Waves*, 2d ed., 1948, reprint 1974; A. J. W. Sommerfeld, *Lectures on Theoretical Physics*, vol. 4, 1954; J. Strong, *Concepts of Classical Optics*, 1958.

Light amplifier

In the broadest sense, a device which produces an enhanced light output when actuated by incident light. A simple photocell relay–light source combination would satisfy this definition. To make the term more meaningful, common usage has introduced two restrictions: (1) a light amplifier must be a device which, when actuated by a light image, reproduces a similar image of enhanced brightness; and (2) the device must be capable of operating at very low light levels without introducing spurious brightness variations (noise) into the reproduced image. The term is used synonymously with image intensifier. The light amplifier increases the brightness of an image which is below the visual threshold to a level where it can be readily seen with the unaided eye. It is, of course, impossible to see under conditions of complete darkness. Indeed, there is a fundamental lower limit of illumination under which an image of a given quality can be recognized. This limitation arises because of the corpuscular nature of light. *See* PHOTON.

Photons arriving through a lens, or other optical system, onto an image area are random in time. If the number of photons per unit area arriving during the time allotted for image formation (for example, the period of the persistence of vision) is too low, the statistical fluctuation will be greater than the variation in number due to true differences in image brightness. Under these circumstances, image recognition is impossible.

Image-intensifier tubes. Intensifier tubes may consist of a semitransparent photocathode which emits electrons with a density distribution proportional to the distribution of light intensity incident on it. Thus, when a light image falls on one side of the cathode, an electron-current image is emitted from the other side. An electron optical system, which acts on electrons in much the same way as does a glass lens system on light, focuses the electron image onto an intensifier element. This electron optical system may be purely electrostatic or may utilize magnetic focusing. The intensified electron image from the other side of the intensifier element is again focused, by a second electron lens, onto a second intensifier element (if additional amplification is required) or onto a fluorescent viewing screen, where the electron energy is converted to visible light.

Proximity tubes are a class of intensifiers in which the photocathode and the fluorescent viewing screen are parallel and separated by only a short distance. No focusing is required for a proximity tube, since the electrons generated at the photocathode are accelerated along the tube axis by the high electric fields.

The spectral response of the image-intensifier tube depends on the type of photocathode employed, and the tube may serve as a wavelength converter as well as an intensifier.

Intensifier elements. Two types of intensifier elements are ordinarily employed. One consists of a transparent support (either a thin film or a fiber optics plate) with a phosphor screen on the side on which the electron image is incident, and a photocathode on the other. An electron striking the phosphor produces a flash of light which causes a release of 50 or more electrons from the photoemitter directly opposite the point of impact. Thus, the intensifier element increases the electron-image current density by a factor of 50. Two such intensifier elements in cascade give a gain of 2500. The illustration shows diagrammatically a three-stage electrostatic image-intensifier tube.

A second type of high-gain intensifier element is a thin secondary-emission current amplifier called a microchannel plate, which is placed between the photocathode and screen. It consists of a parallel bundle of small, hollow glass cylinders, where the

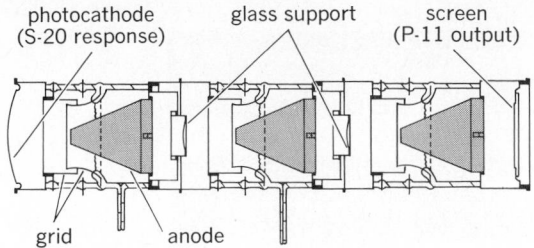

photocathode (S-20 response)   glass support   screen (P-11 output)

grid   anode

Three-stage electrostatically focused image tube.

inside walls of the cylinders are coated with a secondary emitting material. Electrons emitted from the photocathode strike the inside walls of the cylinders, causing secondary electron generation. These secondary electrons in turn continue to cascade down the inside walls of the cylinders, resulting in a high total current gain.

Television camera tube. The intensifier image device can be combined with television pickup tubes to make a television camera whose sensitivity is very close to the threshold determined by the photon "shot noise" from the scene being televised. A SIT (silicon intensifier target) camera tube has a wide dynamic range and operates at very low light levels. In the SIT tube, the photoelectrons accelerated from the photocathode are focused onto a thin silicon target consisting of an array of *pn* junctions. The high-energy photoelectrons generate multiple hole-electron pairs in the silicon target, resulting in gain. The signal is read out by scanning the reverse side of the silicon target with an electron beam, similar to the method used in vidicon camera tubes. In other types of EBS (electron-bombarded silicon) tubes, the silicon *pn* junction target is replaced with scanned photodiode arrays or with CCD (charge-coupled device) imagers. In these cases, the electronic readout of the silicon targets is accomplished without the use of a scanned electron beam, and this results in even lower light-level performance.

Solid-state image intensifiers. A great deal of exploratory work has been done on solid-state light amplifiers. The form which has been extensively investigated consists of a photoconductive film in contact with an electroluminescent screen. Current flows through the photoconductor at illuminated areas, causing light to be generated in the electroluminescent layer. This type of intensifier gives considerable image enhancement at intermediate light levels, but fails at very low light levels because of extreme time lag.

Another type of solid-state light amplifier which works at intermediate light levels is the liquid-crystal light valve. The light valve consists of an electrically biased multilayer sandwich which has a liquid-crystal layer on the front face and a photoconductive layer on the rear face. In this system, the incident light is focused on the photoconductive side, while a bright, polarized light is projected onto the liquid-crystal side. The incident light from the photoconductive side causes an increase in voltage drop across the liquid crystal. This increased voltage causes a rotation in the orientation of the liquid-crystal molecules, which in turn modulates the polarization of the reflected projection light. Only the modulated reflected light is allowed to be viewed.

Applications. In addition to their application for night vision, light amplifiers have been useful in many fields of science, such as astronomy, nuclear physics, and microbiology.

[DEAN R. COLLINS]

Bibliography: D. R. Collins et al., Development of a CCD for ultraviolet imaging using a CCD photocathode combination, *Proceedings of the Symposium on Charge Coupled Device Technology for Scientific Imaging Applications*, Jet Propulsion Laboratory, pp. 163–174, 1975; A. D. Jacobson et al., A new color-TV projector, *SID77 Digest*, Society for Information Displays, pp. 106–107, 1977; *RCA Electro-Optics Handbook*, 1974; Symposium of Photoelectronic Image Devices, *Advances in Electronics and Electron Physics*, 1960.

Line integral

The line integral of a vector function $\mathbf{F}$ of position over a path C is represented by Eq. (1), where F_x,

$$\int_C \mathbf{F} \cdot d\mathbf{r} = \int_C F_x(x,y,z)\, dx$$
$$+ \int_C F_y(x,y,z)\, dy + \int_C F_z(x,y,z)\, dz \quad (1)$$

F_y, F_z are the scalar components of $\mathbf{F}$ along the coordinate axes. The path C is supposed to be a curve, smooth at least in part, defined parametrically by equations of form (2) for each smooth por-

$$x = x(p) \qquad y = y(p) \qquad z = z(p) \quad (2)$$

tion. The functions $F_x(x,y,z)$, etc., must be defined at all points of C. When this is so, the line integral can be evaluated by writing Eq. (3), where the

$$\int_C \mathbf{F} \cdot d\mathbf{r} = \int_{p_1}^{p_2} F_x(p) x'(p)\, dp$$
$$+ \int_{p_1}^{p_2} F_y(p) y'(p)\, dp + \int_{p_1}^{p_2} F_z(p) z'(p)\, dp \quad (3)$$

prime means differentiation with respect to the parameter, and p_1, p_2 are values of the parameter at the end points of the path C or of a smooth piece. The integral has been converted into an ordinary definite integral.

When C is a closed curve, the line integral is called a circuit integral, and is written as notation (4).

$$\oint \mathbf{F} \cdot d\mathbf{r} \quad (4)$$

Some physical applications of the line integral follow. If $\mathbf{F}$ is a force, the line integral is the work done in moving a mass along the curve C. If $\mathbf{F}$ is the velocity of flow of a fluid, the line integral is the circulation of the fluid along the curve. If $\mathbf{F}$ is the electrostatic field strength, the integral is the electric potential difference between the end points of the curve. If $\mathbf{F}$ is the electric field strength of an electromagnetic field, the circuit integral is the electromotive force of the circuit. In each example $d\mathbf{r}$ is physically a length. *See* INTEGRATION.

[MC ALLISTER H. HULL, JR.]

Line spectrum

A discontinuous spectrum characteristic of excited atoms, ions, and certain molecules in the gaseous phase at low pressures, to be distinguished from band spectra, emitted by most free molecules, and continuous spectra, emitted by matter in the solid, liquid, and sometimes gaseous phase.

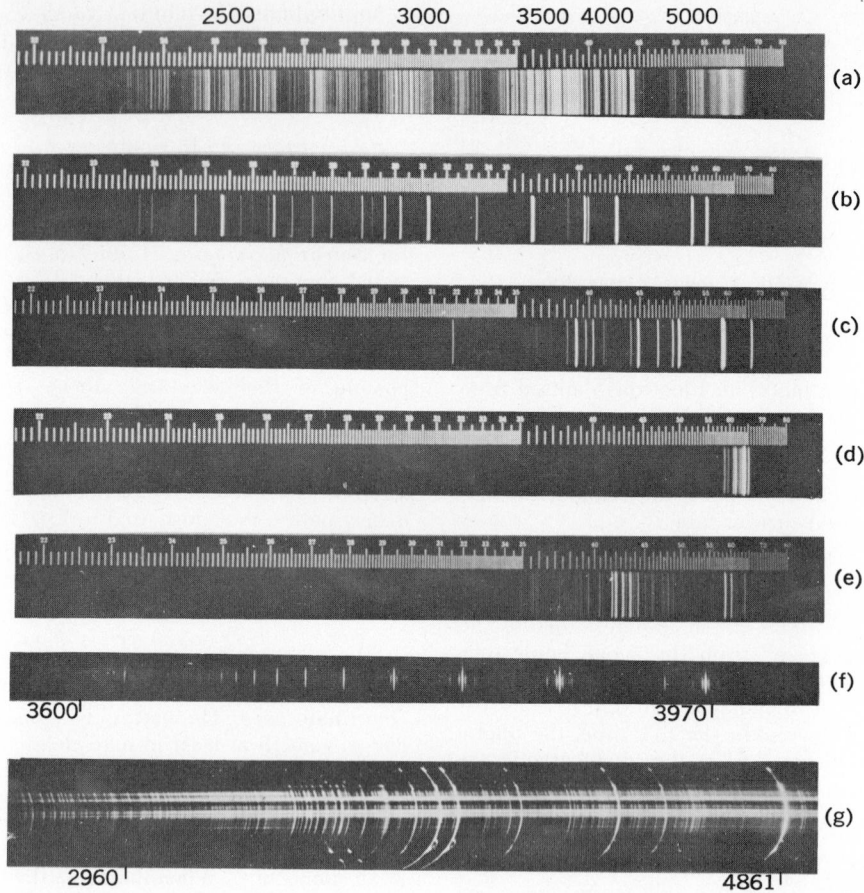

Photographs of common line spectra, wavelengths given in angstroms; emission spectra (a–e) all taken with the same quartz spectrograph. (a) Spectrum of iron arc. (b) Mercury spectrum from an arc enclosed in quartz. (c) Helium in a glass discharge tube. (d) Neon in a glass discharge tube. (e) Argon in a glass discharge tube. (f) Balmer series of hydrogen in the ultraviolet, photographed with a grating spectrograph. (g) Emission spectrum from gaseous chromosphere of the Sun, a grating spectrum taken without a slit at the instant immediately preceding a total eclipse, when the rest of the Sun is covered by the Moon's disk. Two strongest images, H and K lines of calcium, show marked prominences, or clouds, of calcium vapor. Other strong lines are caused by hydrogen and helium. (*From F. A. Jenkins and H. E. White, Fundamentals of Optics, 3d ed., McGraw-Hill, 1957*)

If an electric arc or spark between metallic electrodes, or an electric discharge through a low-pressure gas, is viewed through a spectroscope, images of the spectroscope slit are seen in the characteristic colors emitted by the atoms or ions present.

To avoid the overlapping of close spectral images, the slit illuminated by the light source is made very narrow. The spectrum then appears as an array of bright line slit images on a dark background (see illustration). Under certain conditions spectra show dark absorption lines against a bright background. *See* ATOMIC STRUCTURE AND SPECTRA.

[GEORGE R. HARRISON]

Linear algebra

That branch of mathematics which deals with solutions of systems of linear equations and the related geometric notions of vector spaces and linear transformations. It is fundamental in the theory of the calculus of functions of several variables and hence is of great importance in the application of mathematics to physical and biological sciences, economics, and so on.

The word linear is derived from the fact that the equation of a line in two-dimensional analytic geometry has the form shown in Eq. (1) and a

$$ax + by = c \qquad (1)$$

system of linear equations has the corresponding form shown in Eq. (2), where $i = 1, 2, \ldots, m$.

$$a_{i1}x_1 + a_{i2}x_2 + \cdots + a_{in}x_n = b_i \qquad (2)$$

The a_{ij} and b_i are fixed quantities belonging to a specified field, for example, the field of real numbers, and solutions $(x_1, x_2, \ldots, x_n)$ are sought in the same field. If $m = n = 2$, the problem of solving the equations is equivalent to the geometric one of finding the points common to two lines (which may be coincident).

Basic concepts. Let R denote the field of real numbers and let R^2 be the set of pairs (x, y) of real numbers x, y. Such a pair can be regarded as the coordinates of a point P in a two-dimensional euclidean space. The point P determines a vector, or directed segment **OP** with initial point the origin O and terminal point P, as shown in the figure.

Such vectors are basic in representing certain physical entities, for example, acceleration, velocity, force. Addition of these physical entities corresponds to the addition of vectors. If $\mathbf{v} = \mathbf{OP}$ and $\mathbf{v'} = \mathbf{OP'}$ where P' is the point (x',y'), then $\mathbf{v} + \mathbf{v'} = \mathbf{OQ}$, where $Q = (x+x', y+y')$. If t is a real number, a stretching of the vector $\mathbf{v}$ in the ratio $t{:}1$ gives rise to the vector $t\mathbf{v} = \mathbf{OM}$, where $M = (tx,ty)$.

On the algebraic side these considerations generalize from pairs of real numbers (x,y) to n-tuples of real numbers $(x_1,x_2, \ldots ,x_n)$, and one can replace the field R of real numbers by any field F, for example, the field of rational numbers.

Accordingly, let F^n denote the set of n-tuples $\mathbf{x} = (x_1,x_2, \ldots ,x_n)$, where the x_i are in a given field F. Such an n-tuple is called a vector. If $\mathbf{y} = (y_1,y_2, \ldots ,y_n)$, then one defines the sum of x and y by Eq. (3). *See* ABSTRACT ALGEBRA.

$$\mathbf{x} + \mathbf{y} = (x_1 + y_1, x_2 + y_2, \ldots , x_n + y_n) \quad (3)$$

The elements of the field F are called scalars. If t is such an element, one defines the multiplication of the vector $\mathbf{x}$ by the scalar t by means of Eq. (4).

$$t\mathbf{x} = (tx_1, tx_2, \ldots , tx_n) \quad (4)$$

The basic properties of addition of vectors and multiplication of vectors by scalars are listed below.

$A_1.$ $(\mathbf{x} + \mathbf{y}) + \mathbf{z} = \mathbf{x} + (\mathbf{y} + \mathbf{z})$
$A_2.$ $\mathbf{x} + \mathbf{y} = \mathbf{y} + \mathbf{x}$
$A_3.$ There exists a vector $\mathbf{O}$ such that $\mathbf{x} + \mathbf{O} = \mathbf{x} = \mathbf{O} + \mathbf{x}$ for all $\mathbf{x}$
$A_4.$ For any vector $\mathbf{x}$ there exists a vector $-\mathbf{x}$ such that $\mathbf{x} + (-\mathbf{x}) = \mathbf{O} = (-\mathbf{x}) + \mathbf{x}$
$S_1.$ $t(\mathbf{x} + \mathbf{y}) = t\mathbf{x} + t\mathbf{y}$
$S_2.$ $(s + t)\mathbf{x} = s\mathbf{x} + t\mathbf{x}$
$S_3.$ $(st)\mathbf{x} = s(t\mathbf{x})$
$S_4.$ $1\mathbf{x} = \mathbf{x}$

Here $\mathbf{x}$, $\mathbf{y}$, $\mathbf{z}$ are vectors, s, t are scalars, and 1 is the unit element of F. These properties are immediate consequences of the definitions and the properties of the field F. If one defines unit vectors $\mathbf{e}$ by means of Eq. (5), where the 1 occurs

$$\mathbf{e}_i = (0, \ldots ,0,1,0, \ldots ,0) \quad (5)$$

in the ith place, then a consequence of the definition of vector addition is that any vector $\mathbf{x} = (x_1,x_2, \ldots ,x_n)$ can be expressed as the sum shown in Eq. (6). In addition, this way of present-

$$\mathbf{x} = x_1\mathbf{e}_1 + x_2\mathbf{e}_2 + \cdots + x_n\mathbf{e}_n \quad (6)$$

ing the vector $\mathbf{x}$ as a linear combination of the vectors $\mathbf{e}_i$ is unique.

Axiomatic formulation. It is possible to obtain better insight and power in dealing with vectors if one axiomatizes the situation described above. Accordingly, one defines the notion of a vector space V over a field F as a set V with an addition composition $\mathbf{x} + \mathbf{y}$, and a multiplication $t\mathbf{x}$ of elements $\mathbf{x}$ of V by scalars t(elements of F). These are required to satisfy the axioms $A_1 - A_4$ and $S_1 - S_4$. It is worth noting that conditions $A_1 - A_4$ state that V and its addition composition constitute a commutative group. A subset B of V is called a base if every vector $\mathbf{x}$ can be written in one and only one way as a linear combination $\mathbf{x} = x_1\mathbf{b}_1 + x_2\mathbf{b}_2 + \cdots + x_m\mathbf{b}_m$, where the $\mathbf{b}_i$ are in B and the x_i are in F. Every vector space has a base and any two bases have the same cardinal

number which is called the dimension of V. The vector space F^n of n-tuples of elements of the field F has dimension n with base $B = \{\mathbf{e}_1,\mathbf{e}_2, \ldots ,\mathbf{e}_n\}$ defined above. *See* GROUP THEORY.

A subset U of V is called a subspace if the element $s\mathbf{u} + t\mathbf{v}$ is in U whenever $\mathbf{u}$ and $\mathbf{v}$ are in U and s and t are in F. The factor space V/U is the set of cosets $\mathbf{x} + U$, that is, the set of vectors of the form $\mathbf{x} + \mathbf{u}$ where $\mathbf{x}$ is fixed and $\mathbf{u}$ ranges over U. Addition and scalar multiplication in V/U are defined by $(\mathbf{x} + U) + (\mathbf{y} + U) = \mathbf{x} + \mathbf{y} + U$, $t(\mathbf{x} + U) = t\mathbf{x} + U$, respectively.

Linear transformations. If U and V are vector spaces over the same field F, then one defines a linear transformation $\mathbf{A}$ of U into V as a correspondence $x \rightarrow \mathbf{A}x$ (also written as $x\mathbf{A}$) of U into V such that Eq. (7) is satisfied for any $\mathbf{x}$, $\mathbf{y}$ in U and

$$\begin{aligned} \mathbf{A}(\mathbf{x} + \mathbf{y}) &= \mathbf{A}\mathbf{x} + \mathbf{A}\mathbf{y} \\ \mathbf{A}(t\mathbf{x}) &= t(\mathbf{A}\mathbf{x}) \end{aligned} \quad (7)$$

any t in F. The image $\mathbf{A}U$ of U under the linear transformation $\mathbf{A}$ is the set of all vectors $\mathbf{A}\mathbf{x}$ where $\mathbf{x}$ ranges over all of U. $\mathbf{A}U$ is a subspace of V. The kernel of $\mathbf{A}$ is the set of vectors $\mathbf{z}$ in U such that $\mathbf{A}\mathbf{z} = 0$. This is also a subspace of U. If U and V are finite dimensional and $\mathbf{b}$ is vector in V, then the determination of the vectors $\mathbf{x}$ of U such that $\mathbf{A}\mathbf{x} = \mathbf{b}$ is equivalent to the determination of the set of solutions of a system of linear equations.

If $\mathbf{A}$ and $\mathbf{B}$ are linear transformations of a vector space U into another vector space V and if t is in F, then one defines $\mathbf{A} + \mathbf{B}$ and $t\mathbf{A}$ by $(\mathbf{A} + \mathbf{B})\mathbf{x} = \mathbf{A}\mathbf{x} + \mathbf{B}\mathbf{x}$, $(t\mathbf{A})\mathbf{x} = t(\mathbf{A}\mathbf{x})$, respectively. If $\mathbf{C}$ is a linear transformation of V into a third vector space W, then one defines $\mathbf{A}\mathbf{C}$ by $(\mathbf{A}\mathbf{C})\mathbf{x} = \mathbf{A}(\mathbf{C}\mathbf{x})$. The mappings $\mathbf{A} + \mathbf{B}$, $t\mathbf{A}$, $\mathbf{A}\mathbf{C}$ are all linear transformations. One denotes the set of linear transformations of U into V by hom (U,V). This is itself a vector space relative to the addition $\mathbf{A} + \mathbf{B}$ and the multiplication $t\mathbf{A}$ defined above. In the special case in which $V = F$, the vector space $V^* = $ hom (V,F) is called the conjugate space of V or the space of the linear functions on V. If $\mathbf{A}$, $\mathbf{B}$, $\mathbf{C}$ are in hom(U,U), then $(\mathbf{A}\mathbf{B})\mathbf{C} = \mathbf{A}(\mathbf{B}\mathbf{C})$, $\mathbf{A}(\mathbf{B} + \mathbf{C}) = \mathbf{A}\mathbf{B} + \mathbf{A}\mathbf{C}$, $(\mathbf{B} + \mathbf{C})\mathbf{A} = \mathbf{B}\mathbf{A} + \mathbf{C}\mathbf{A}$, $t(\mathbf{A}\mathbf{B}) = (t\mathbf{A})\mathbf{B} = \mathbf{A}(t\mathbf{B})$. These properties imply that hom (U,U) is an algebraic system of a type called an associative algebra.

If U and V are vector spaces over F with bases $\{\mathbf{u}_1,\mathbf{u}_2, \ldots ,\mathbf{u}_m\}$ and $\{\mathbf{v}_1,\mathbf{v}_2, \ldots ,\mathbf{v}_n\}$, respectively, and $\mathbf{A}$ is a linear transformation of U into V, then according to the definition of a base, one can write $\mathbf{A}\mathbf{u}_i = \sum_{j=1}^{n} a_{ji}\mathbf{v}_j$. In this way one can associate with $\mathbf{A}$ the n by m matrix (a_{ji}). The correspondence between hom (U,V) and the set of n by m matrices with entries in F is 1-1 and preserves addition and multiplication by scalars. That is, to each transformation in hom (U,V) there corresponds one and only one matrix, and the matrix which corresponds to the sum of two transformations is the sum of the matrices corresponding to the separate transformations. Also if $V = U$ then the correspondence preserves multiplication. In this way one sees that the theory of linear transformations between finite-dimensional vector spaces is equivalent to matrix theory. *See* MATRIX THEORY.

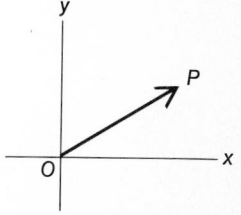

LINEAR ALGEBRA

Geometric representation of a vector **OP**.

Multilinear algebra and tensors. Linear algebra is generally regarded as including also multilinear algebra, the basic notion of which is that of a bilinear mapping of the pair of vector spaces U, V into a third vector space W (over the same field F). This is a mapping B of the set of pairs (x,y) with x in U and y in V into W such that Eq. (8) is satisfied

$$B(x+x', y) = B(x,y) + B(x',y)$$
$$B(x, y+y') = B(x,y) + B(x,y') \qquad (8)$$
$$B(tx,y) = tB(x,y) = B(x,ty)$$

for x, x' in U; y, y' in V; and t in F. This leads to the notion of the tensor product $U \otimes V$, which is a vector space together with a bilinear mapping of U and V into this space which is maximal in a certain sense. One can iterate the construction of $U \otimes V$ to obtain tensor spaces of higher degree. Such spaces are fundamental for applications of algebra to differential geometry and relativity theory.

Quadratic forms. An important part of linear algebra is concerned with the study of a vector space relative to a given quadratic form. This is defined to be a correspondence $Q: x \rightarrow Q(x)$ of V into F such that $Q(tx) = t^2 Q(x)$ and such that $B(x,y) = Q(x+y) - Q(x) - Q(y)$ is bilinear from the pair V, V to F. If $F = R$, the field of real numbers, and V is complete in the topology defined by Q as a norm (an absolute value). Replacement of Q by a hermitian form defines (complex) Hilbert space. The theory of finite-dimensional real Hilbert spaces is equivalent to the study of n-dimensional euclidean geometry. Infinite-dimensional Hilbert spaces and their generalization called Banach spaces play an important role in modern analysis. [NATHAN JACOBSON]

Bibliography: H. Anton, *Elementary Linear Algebra*, 3d ed., 1981; J. W. Daniels, *Elementary Linear Algebra and Its Applications*, 1981; P. R. Halmos, *Finite Dimensional Vector Spaces*, 2d ed., 1958; N. Jacobson, *Lectures in Abstract Algebra*, vol. 2, 1953, reprint 1975; W. G. Strang, *Linear Algebra and Its Applications*, 2d ed., 1980.

Lines of force

Imaginary lines in fields of force whose tangents at any point give the direction of the field at that point and whose number through unit area perpendicular to the field represents the intensity of the field. The concept of lines of force is perhaps most common when dealing with electric or magnetic fields.

Electric lines of force. These are lines drawn to represent, or map, an electric field graphically in the space around a charged body. They are of great help in visualizing an electric field and in quantitative thinking about such a field. The direction of an electric line of force at any point is drawn parallel to the direction of the electric field intensity E at that point. The quantitative convention is that the number of electric lines of force drawn through an imaginary unit area of surface perpendicular to the field shall be numerically equal to E at that area. From this quantitative convention, and using the rationalized mks system of units, q/ϵ_0 lines of force originate on a positive charge q (in coulombs) and a like number terminate on a negative charge q of the same magnitude. Here $\epsilon_0 = 8.85 \times 10^{-12}$ coulomb^2/newton-m^2 is the permittivity of empty space. *See* ELECTRIC FIELD.

Magnetic lines of force. A magnetic field of strength H may also be represented by lines called lines of force. The number of lines of force per unit area of a surface perpendicular to H is numerically equal to the value of H. The lines of magnetic force due to currents are closed curves, as are the lines of magnetic induction. Magnetic lines of force due to magnets originate on north poles and terminate on south poles, both inside and outside the magnet. *See* MAGNETIC FIELD.

[RALPH P. WINCH]

Bibliography: A. F. Kip, *Fundamentals of Electricity and Magnetism*, 2d ed., 1968; E. M. Purcell, *Electricity and Magnetism*, Berkeley Physics Course, vol. 2, 1970; R. Resnick and D. Halliday, *Physics*, 3d ed., 1977.

Liquid

A state of matter intermediate between that of crystalline solids and gases. Macroscopically, liquids are distinguished from crystalline solids in their capacity to flow under the action of extremely small shear stresses and to conform to the shape of a confining vessel. Liquids differ from gases in possessing a free surface and in lacking the capacity to expand without limit. On the scale of molecular dimensions liquids lack the long-range order that characterizes the crystalline state, but nevertheless they possess a degree of structural regularity that extends over distances of a few molecular diameters. In this respect, liquids are wholly unlike gases, whose molecular organization is completely random.

Thermodynamic relations. The thermodynamic conditions under which a substance may exist indefinitely in the liquid state are described by its phase diagram, shown schematically in the figure. The area designated by L depicts those pressures and temperatures for which the liquid state is energetically the lowest and therefore the stable state. The areas denoted by S and V similarly indicate those pressures and temperatures for which only the solid or vapor phase may exist. The connecting lines OC, OB, and OA define pressures and temperatures for which the liquid and its vapor, the solid and its liquid, and the solid and its vapor, respectively, may coexist in equilibrium. They are usually termed phase boundary or phase coexistence lines. The intersection of the three lines at O defines a triple point which, for the three states of matter under discussion, is the unique pressure and temperature at which they may coexist at equilibrium. Other triple points exist in the phase diagram of a substance that possesses two or more crystalline modifications, but the one depicted in the figure is the only triple point for the coexistence of the vapor, liquid, and solid. Line OA has its origin at the absolute zero of temperature and OB, the melting line, has no upper limit. The liquid-vapor pressure line OC is different from OB, however, in that it terminates at a precisely reproducible point C, called the critical point. Above the critical temperature no pressure, however large, will liquefy a gas. *See* TRIPLE POINT.

Along any of the coexistence curves the relationship between pressure and temperature is given by the Clausius-Clapeyron equation: $dP/dT = \Delta H / T \Delta V$, where ΔV is the difference in molar volume of the corresponding phases (gas-liquid, gas-solid, or liquid-solid) and ΔH is the molar heat of

LIQUID

pressure

B

S

C

L

V

O

A

temperature

Phase diagram of a pure substance.

transition at the temperature in question. By means of this equation the change in the melting point of the solid or the boiling point of the liquid as a function of pressure may be calculated. When a liquid in equilibrium with its vapor is heated in a closed vessel, its vapor pressure and temperature increase along the line OC. ΔH and ΔV both decrease and become zero at the critical point, where all distinction between the two phases will vanish.

Transport properties. Liquids possess important transport properties, notably their capacity to transmit heat (thermal conductivity), to transfer momentum under shear stresses (viscosity), and to attain a state of homogeneous composition when mixed with other miscible liquids (diffusion). These nonequilibrium properties of liquids are well understood in macroscopic terms and are exploited in large-scale engineering and chemical-process operations. Thus, the rate of flow of heat across a layer of liquid is given by $Q = \kappa \, dT/dx$, where $\dot{Q}$ is the heat flux, dT/dx is the thermal gradient, and κ is the coefficient of thermal conductivity. Similarly, the shearing of one liquid layer against another is resisted by a force equal to the momentum transfer: $F = \dot{p} = \eta \, dv/dx$, where dv/dx is the velocity gradient and η is the coefficient of viscosity. Likewise, the rate of transport of matter under nonconvective conditions is governed by the gradient of concentration of the diffusing species: $J = -D \, dC/dx$, where J is the matter flux and D is the coefficient of diffusion. Each of these transport coefficients depends upon temperature, pressure, and composition and may be determined experimentally. An a priori calculation of κ, η, and D is a very difficult problem, however, and only approximate theories exist.

Theoretical explanations. In fact, although a great deal of effort has been expended, there still exists no satisfactory theory of the liquid state. Even so commonplace a phenomenon as the melting of ice has no adequate theoretical explanation. The reason for this state of affairs lies in the tremendous structural and dynamical complexity of the liquid state. To understand this, it is useful to compare the structural and kinetic properties of liquids with those of crystalline solids on the one hand and with those of gases on the other.

In crystals, atoms or molecules occupy well-defined positions on a three-dimensional lattice, oscillating about them with small amplitudes; their kinetic energy is entirely distributed among these quantized vibrational states up to the melting point. This nearly perfect spatial order is revealed by diffraction techniques, which utilize the coherent scattering of x-rays or particles having wavelengths comparable with interatomic spacings. The structural and dynamical properties are sufficiently tractable mathematically so that the theory of solids is quite well understood.

The theory of gases is also simple, but for quite a different reason. No vestige of positional regularity of atoms remains in gases, and their energy resides entirely in high-speed translational motion. Except for collisions, which deflect their motions into new straight-line trajectories, atoms in gases do not interact with one another; vibrational modes in monatomic gases are absent.

Liquids, by contrast, lie intermediate between gases and crystals from both a structural and dynamic point of view. Kinetic energy is partitioned among translational and vibrational modes, and diffraction studies reveal a degree of short-range order that extends over several molecular diameters. Moreover, this "structure" is continually changing under the influence of vibrational and translational displacements of atoms. Physical reality may be attributed to this short-range structure, nevertheless, in the sense that a time average over the huge number of possible configurations of atoms may show that a fairly definite number of neighboring atoms lie close to any arbitrary atom. At a somewhat greater distance from this reference atom, the density of neighbors oscillates above and below the average density of atoms in the liquid as a whole.

Information about the degree of local order is contained in the radial distribution function, a mathematical property which may be deduced from diffraction measurements. This is the starting point for a theory of the liquid state, and although efforts by J. G. Kirkwood and his students, by H. Eyring and his collaborators, and by J. E. Lennard-Jones and A. F. Devonshire have yielded partial successes, prodigious mathematical difficulties lie in the path of an entirely satisfactory solution. *See* KINETIC THEORY OF MATTER; VISCOSITY; X-RAY DIFFRACTION.

[NORMAN H. NACHTRIEB]

Bibliography: P. A. Egelstaff (ed.), *An Introduction to the Liquid State*, 1967; J. P. Hansen and I. R. McDonald, *The Theory of Simple Liquids*, 1976; P. Kruus, *Liquids and Solutions: Structure and Dynamics*, 1977; H. N. Temperley and D. H. Trevena, *Liquids and Their Properties: A Molecular and Microscopic Treatise*, 1978.

Liquid helium

Helium liquefies at a substantially lower temperature, 4.2 K ($-269°C$), than any other substance; and below 2.2 K the liquid exhibits the extraordinary properties of superfluidity, most notably the ability to flow with complete absence of friction. Helium remains in a liquid state at absolute zero. All of these characteristics are due to the weakness of the attractive force between two helium atoms and to the small atomic mass, which according to the laws of quantum mechanics makes the atoms difficult to localize.

Liquid helium was first produced in 1908 by H. Kamerlingh Onnes. Although extensive thermodynamic studies followed, the superfluid properties were not discovered until 1938. In addition to the common isotope of atomic weight 4, helium has a rare isotope of atomic weight 3 with quite different properties. In 1972 a superfluid transition in liquid ^{3}He was discovered at a temperature near 0.002 K.

The use of helium is almost indispensable for maintaining temperatures below about 14 K (the solidification temperature of hydrogen). A bath of liquid helium may be cooled by evaporation to a little below 1 K by pumping off the vapor. The number of laboratories with access to this temperature range increased rapidly after the commercial Collins liquefier appeared in 1946, and since about 1962 commercial liquid helium has become widely available. There has been a corresponding expansion of work on properties of helium itself as well as in other areas requiring these temperatures, notably superconductivity. *See* CRYOGENICS.

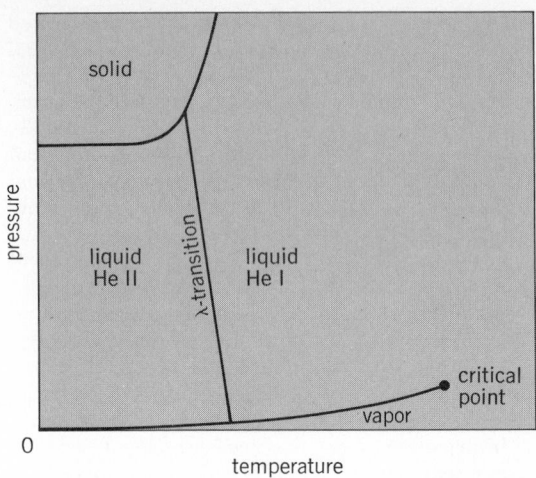

Fig. 1. Phase diagram for ⁴He. The critical point is at $T_c = 5.20$ K, $P_c = 2.29$ bars.

At 4.2 K liquid helium is colorless and of low refractive index ($n = 1.024$), with a density of 0.125 g/cm³. The latent heat of vaporization, 5 cal/g (1 cal = 4.19 J), is very small, so care must be taken to reduce the heat input by conduction and radiation into the storage container or research apparatus. The classical container consists of two vacuum-insulated vessels of silvered glass (Dewar flask or Thermos bottle) or metal, the inner vessel containing the liquid helium being immersed in a larger outer vessel filled with liquid nitrogen. It is possible to dispense with the liquid nitrogen if the evolving helium vapor is used efficiently to cool radiation shields in the vacuum space surrounding the liquid helium.

Phase diagram. The phase diagram of ⁴He (Fig. 1) shows several remarkable characteristics. Helium remains a liquid down to absolute zero unless a pressure greater than 25.3 bars (1 bar = 0.987 atm = 10⁵ Pa) is applied. This is true despite the requirement of thermodynamics that the system have the same degree of order as a perfect crystal. A more subtle feature is a transition at the lambda line between two different liquid phases. This λ-transition is so named because the specific

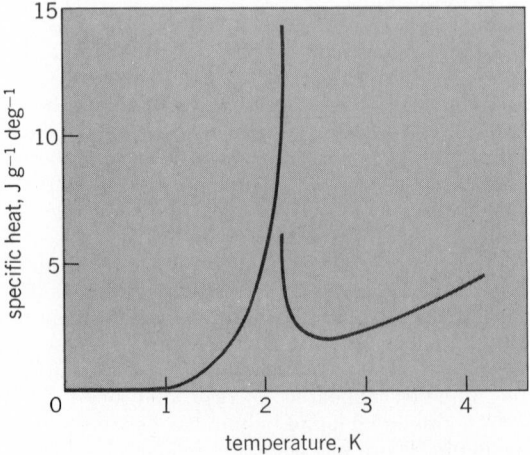

Fig. 2. Specific heat of liquid helium at saturated vapor pressure as a function of temperature. (From K. Atkins, Liquid Helium, Cambridge University Press, 1959)

heat (Fig. 2) has a singularity resembling a backwards Greek lambda. There is no latent heat; such a transition is called second-order. The high-temperature liquid phase, called helium I, is a rather ordinary liquid. The λ-transition at 2.172 K (at vapor pressure) marks the onset of superfluidity, which is the characteristic property of the low temperature phase, helium II.

LIQUID HELIUM II

When liquid helium is cooled by pumping, it boils vigorously until the λ-temperature is reached at a vapor pressure of 0.050 bar. The λ-transition is manifest by a sudden cessation of bubbling, and the liquid becomes completely quiescent. In 1937, 29 years after this phenomenon was first observed, it was discovered that He II has an enormous effective thermal conductivity, hundreds of times larger than that of the best metallic conductors, such as pure copper. This explains the absence of bubbling, as evaporation need occur only at the surface; but it raises the greater problem of discovering the heat transport mechanism.

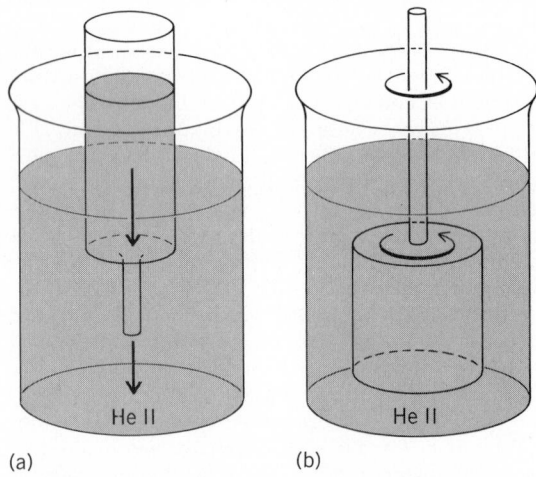

Fig. 3. Apparatus for viscosity measurement. (a) Flow through a narrow channel; (b) rotating cylinder.

Superfluidity. In 1938 P. Kapitza attempted to measure the viscosity of liquid helium by studying its rate of flow through a very narrow slit between two glass plates (Fig. 3a). He discovered that He II flowed thousands of times faster than He I at the same (small) pressure head, and the flow appeared to be turbulent. From his data he inferred an upper limit of 10⁻⁹ cgs units (1 cgs unit = 0.1 Pa · s) for the viscosity (the viscosity of air is about 2×10^{-4} cgs units), and he suggested that He II should be considered a "superfluid" state analogous to the superconducting state which appears in certain metals at low temperatures. Kapitza's experiments were concurrent with similar work by J. F. Allen and A. D. Misener. *See* SUPERFLUIDITY; VISCOSITY.

Among many subsequent superfluid flow experiments, perhaps the most remarkable are studies of persistent currents by J. D. Reppy and coworkers and by I. Rudnick and associates. They set up currents of helium circulating around annular containers packed with fine powder or other porous material. Provided that the initial speed of the helium is not too great, the helium will continue to circulate for hours with no measurable loss of kinetic energy. From these measurements one can place an upper limit of the order of 10^{-19} cgs unit on the viscosity of the circulating superfluid at low speeds.

Kapitza's flow measurements created a paradox, because a different type of viscosity measurement, based on the drag torque on a cylinder rotating in the liquid (Fig. 3b), had been done in 1935 and had shown that the viscosity of He II was only about three times smaller than that of He I.

Thermomechanical effect. If two vessels partially filled with He II are connected by narrow channels and heat is supplied to one vessel, helium will flow into the warmer vessel and cause the level to rise against gravity. If the top of the heated vessel is shaped into a nozzle, it can shoot a jet of liquid to a considerable height. This thermomechanical or "fountain" effect was discovered by Allen and coworkers in 1938. There is a converse "mechanocaloric" effect: If He II is forced by pressure through narrow channels, such as a porous plug, the upstream liquid becomes warmer and the downstream liquid becomes colder.

THEORY: THE TWO-FLUID MODEL

The remarkable and paradoxical experimental results described above can be understood on the basis of a theoretical model proposed in 1941 by L. Landau and independently at about the same time by L. Tisza. In this model, He II is thought of as consisting of two interpenetrating components, the "superfluid" (with density ρ_s) and the "normal fluid" (with density ρ_n). These densities depend on temperature as shown in Fig. 4. The superfluid density equals the total density at absolute zero and decreases to zero at the λ-temperature. The total density ρ of the liquid is given by the sum $\rho = \rho_s + \rho_n$. Each component can move independently with its own velocity, v_s and v_n, respectively. The superfluid flows without friction and carries no entropy or thermal energy. The normal fluid flow is viscous and carries all of the liquid's thermal energy.

Now the viscosity paradox can be understood as follows: In flow through narrow channels the superfluid moves, while the normal flow is kept nearly immobile by its viscosity; thus the flow is fric-

tionless. But in a rotating cylinder experiment the surface of the cylinder moves past the normal fluid, which therefore exerts a drag force on it. The thermomechanical effect is a consequence of thermodynamics. Any process which transports mass without thermal energy, such as pure superfluid flow, must be driven by the difference in chemical potential. This quantity is proportional to the sum of the pressure difference and a term proportional to the negative of the temperature difference (this term is analogous to an osmotic pressure due to concentration of thermal energy).

The mechanocalorical effect also arises because pure superfluid flow carries no thermal energy. Thus the superfluid dilutes the normal fluid in the downstream reservoir, lowering its temperature, and leaves the normal fluid more concentrated in the upstream reservoir, raising its temperature.

If heat is added to He II, some superfluid is converted into normal fluid. This produces a "counterflow" of superfluid toward the heat source and of normal fluid away from the heat source. The counterflow is driven by the thermomechanical effect and is limited only by the normal fluid viscosity. This is a much more effective heat transport process than ordinary diffusion or convection, and it accounts for the extraordinary thermal conductivity of He II.

Foundations of two-fluid model. The central idea on which the model is based is that all the helium atoms in liquid helium at absolute zero occupy a single quantum state, analogous to an electron orbital in an atom, but extending throughout the volume of the liquid. (In the case of electrons and ^{3}He atoms, which have spin 1/2, the exclusion principle of quantum mechanics allows only one particle in each quantum state, but this restriction does not apply to ^{4}He atoms, which have zero spin.) The motion of atoms in this macroscopic quantum state is highly correlated, so that there is no entropy or disorder in a thermodynamic sense. Some vestige of this quantum state remains at finite temperatures up to the λ-transition and constitutes the superfluid component. The superfluid responds in a reversible, frictionless way to deformations of the container, giving rise to superfluid flow.

At any nonzero temperature there are thermal vibrations superimposed on the superfluid state, just as in a crystal there are thermal vibrations distorting the regular crystal lattice. These vibrational excitations are waves which travel freely through the superfluid (or crystal) until they scatter off other excitation waves or the walls of the container. Thus the excitation waves behave like a gas superimposed on the superfluid. This excitation gas, which constitutes the normal fluid, can have an average drift velocity v_n (the normal fluid velocity) which is entirely independent of the superfluid velocity v_s.

According to quantum mechanics the excitation waves are created in discrete increments, with energy E proportional to vibration frequency and momentum p inversely proportional to wavelength. The energy-momentum relation for excitations in liquid He II is shown in Fig. 5, as determined experimentally from neutron-scattering measurements. For long wavelengths (small momentum and energy) the excitations are simply sound waves and are called phonons. The ratio of energy

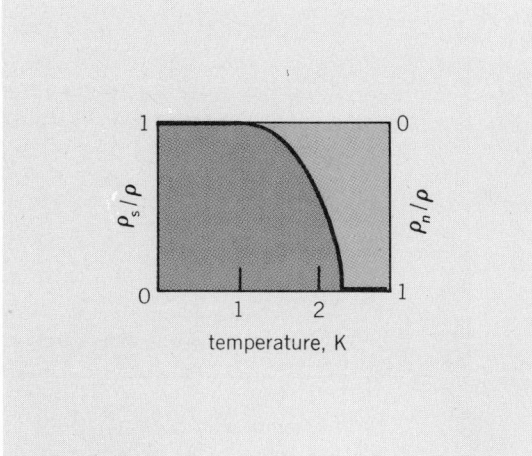

Fig. 4. Fractional densities of superfluid (left scale) and normal fluid (right scale).

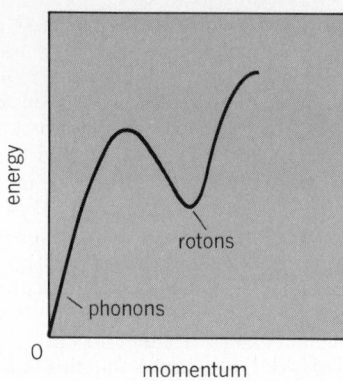

Fig. 5. Energy versus momentum relation for excitations in liquid helium II. (*From R. A. Cowley and A. D. B. Woods, Inelastic scattering of thermal neutrons from liquid helium, Can. J. Phys., 49:177–200, 1971*)

to momentum is the speed of sound. When the wavelength becomes comparable to the average separation of atoms, the curve is distorted. Excitations near the bottom of the dip are referred to as rotons. *See* NEUTRON DIFFRACTION; PHONON.

The excitation energy-momentum relation (Fig. 5) determines the thermodynamic properties of liquid He II, according to the rules of statistical mechanics. For instance, the specific heat as a function of temperature rises rapidly above about 0.7 K, as rotons begin to be excited. The normal fluid density also increases rapidly, as was confirmed in measurements with a torsion pedulum.

Microscopic theory. The two-fluid model is not a complete theory, as it is based on an assumed or experimentally determined form for the excitation energy-momentum relation (Fig. 5). A microscopic theory would be based only on the interaction forces between helium atoms and the laws of quantum mechanics. Among the things one would like to calculate are: (1) properties of the ground state of liquid helium, especially the nature of the long-range coherence among the motions of the helium atoms; (2) the excitation energy-momentum relation; and (3) residual interactions among these excitations, and the effect of these interactions on the normal transport properties, on the thermodynamic properties at high excitation densities (approaching the λ-transition), and on the excitation curve itself. These tasks are difficult because the repulsive interactions between helium atoms are strong, and it is not easy to find approximations which are both realistic and allow calculations to be carried out explicitly.

FURTHER PROPERTIES OF HELIUM II

Some further predictions of the two-fluid model are discussed in the following sections.

Sound modes. The two-fluid model predicted that two types of longitudinal vibrational wave modes should propagate in bulk He II with different speeds, and this prediction was subsequently confirmed. First sound is a compressional wave in which the superfluid and normal fluid components oscillate together. It travels at a speed which decreases from 237 m/s near 0 K to 217 m/s at the λ-transition, where it joins continuously (but with discontinuous slope) to the speed of ordinary sound in He I. Second sound is a unique mode in

which the superfluid and normal fluid move in opposite directions, leaving the total density constant but producing a temperature oscillation. This temperature wave propagates at a speed near 20 m/s between 1 and 2 K, above which the speed decreases, going to zero at the λ-transition, as shown in Fig. 6. *See* SECOND SOUND.

A propagating oscillation in thickness of a helium film on a solid surface is called third sound, while fourth sound designates a compressional wave of superfluid in a porous medium, in which the normal fluid is immobilized by its viscosity. All four modes have proved valuable as a means for studying the properties of He II. *See* LOW-TEMPERATURE ACOUSTICS.

Rotation and quantized vortex lines. Landau predicted that the flow of superfluid must be irrotational, but subsequent experiments with He II in rotating containers showed that the superfluid as well as the normal fluid apparently comes into uniform rotation at the angular velocity of the container, just like any ordinary fluid. The experiments included observations of the curvature of the surface and direct measurements of angular momentum. L. Onsager, and in more detail, R. P. Feynman, then reexamined the theoretical basis for a constraint against rotation, which arises because the superfluid is a macroscopic quantum state. They concluded that the flow of superfluid is constrained by a condition analogous to the Bohr orbit condition for electrons in an atom. Thus the circulation of superfluid around any closed path in the liquid is quantized in integer muliples of $h/m = 1.0 \times 10^{-3}$ cm^2/s (h is Planck's constant and m is the mass of a helium atom). It is this quantization which accounts for the metastability of persistent currents.

In a simply connected volume of He II, the possibility of integers different from zero (which was missed by Landau) implies the existence of singular lines in the form of quantized vortex lines. The circulation quantum number for a closed path is simply the net number of vortex lines which the

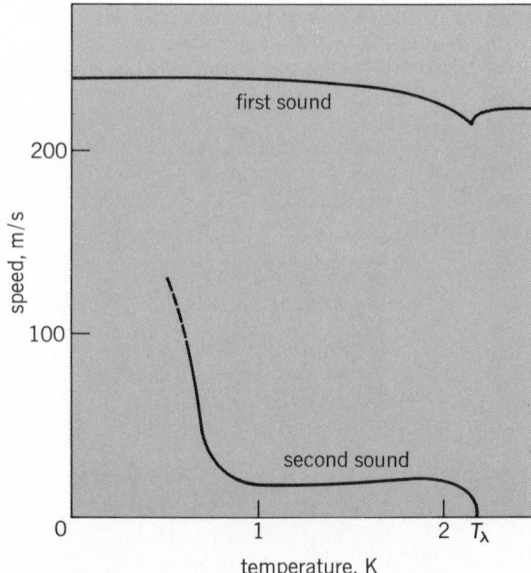

Fig. 6. Propagation speed of sound modes in bulk liquid helium. (*From W. E. Keller, Helium-3 and Helium-4, Plenum Press, 1969*)

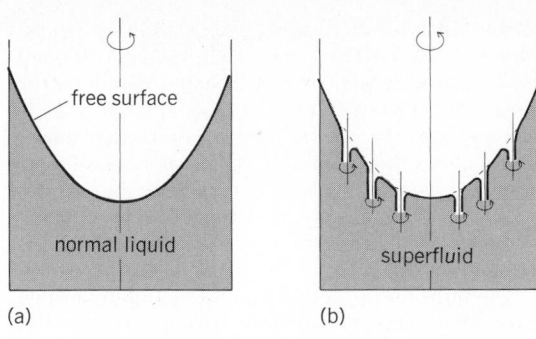

(a)   (b)

Fig. 7. Cross section of the surface of a rotating liquid. (a) Uniform rotation; (b) rotation due to quantizied vortices. (*From K. Atkins, Liquid Helium, Cambridge University Press, 1959*)

path encircles. Around a single vortex line the superfluid moves in circles, with speed increasing inversely as the distance decreases, like a miniature bathtub vortex. The velocity cannot become infinite at the line, so there must be a small but finite core empty of superfluid. Now the experiments can be explained: The quantum of circulation h/m is sufficiently small that extraordinary sensitivity is required to distinguish rotation due to an array of parallel vortex lines from uniform rotation (see Fig. 7). The dimples on the surface at the ends of the vortices are so small that no one has succeeded in detecting them. *See* QUANTIZED VORTICES.

Among the subsequent experiments which lent credence to the existence of quantized vortices was a study by H. E. Hall and W. F. Vinen of the extra attenuation of second sound when He II is rotating or in turbulent flow. Vinen also devised a sensitive technique for measuring the superfluid circulation around a wire and found that it tended to stabilize at the predicted quantum values. But the most convincing data come from studies of ions and their interactions with vortices, discussed in the next section. In 1974 R. E. Packard produced visual images of the ends of vortices on the surface of rotating He II. He was able to photograph the vortex ends and to observe their motion on a television monitor.

Ions in liquid helium. Electrons or positive ions injected into liquid helium form rather large complexes. An electron, which interacts repulsively with helium atoms, reduces its zero point energy by forming a cavity or bubble of 1.7-nm radius, within which it is confined. Thus the negative ion bubble displaces about 480 helium atoms. A He$^+$ ion attracts neutral helium atoms to form a "snowball" with an effective radius of 0.6 nm and an effective mass of about 44 helium atoms.

If quantized vortices are present, ions are attracted to them by the Bernoulli force and can become trapped on the vortex core. The binding energy is due to shortening of the vortex line by one ion diameter and is great enough to hold an ion against thermal fluctuations only below about 1.7 K for the negative ion and 0.7 K for the smaller positive ion. This interaction has been exploited to study properties of both ions and vortices. It is this possibility of attaching electric charge to a vortex array which allowed its direct visualization, as described above. After the vortices are charged with electron bubbles, the electrons are extracted

through the surface by an electric field parallel to the rotation axis and accelerated into a fluorescent screen.

In an applied electric field, ions move with a drift velocity limited by their collisions with the thermal excitations, particularly rotons. Working at temperatures well below 1 K where few rotons remain, it was discovered that an ion accelerated by an electric field can create a quantized vortex ring (similar to a smoke ring) and become trapped on it. The more energy this charged vortex ring is given, the slower it moves—a unique characteristic of a vortex ring.

Critical velocities. Superfluid flow is frictionless only if the velocity is not too large. Above some "critical" velocity a pressure drop appears along the flow channel (or more precisely, a chemical potential difference). Figure 8 shows the typical form of the pressure head versus superfluid velocity relation. The magnitude of the critical velocity may be less than 1 mm/s in a wide channel (near 1-mm diameter) or close to 10 m/s in a very narrow or short channel. The value may also depend on temperature; on approaching the λ-temperature the critical velocity decreases to zero.

Landau suggested that a superfluid flow could lose kinetic energy by creating phonons if the velocity exceeded the speed of sound, or by creating rotons at a somewhat lower velocity (about 60 m/s), but velocities this large are never observed. Feynman first suggested that the energy dissipation mechanism involves creation of quantized vortices, perhaps in the form of vortex rings. This idea predicts the correct order of magnitude and channel-size dependence for the critical velocity, as a consequence of the velocity versus diameter relation of a quantized vortex ring. Experiments using ions and second sound have confirmed the presence of vortices. It now seems probable that critical velocities which are dependent on channel size and independent of temperature result from pinning of vortices at the channel walls, whereas temperature-dependent critical velocities observed in other situations, especially near the λ-temperature, are determined by nucleation of vortex rings by thermal fluctuations of the liquid.

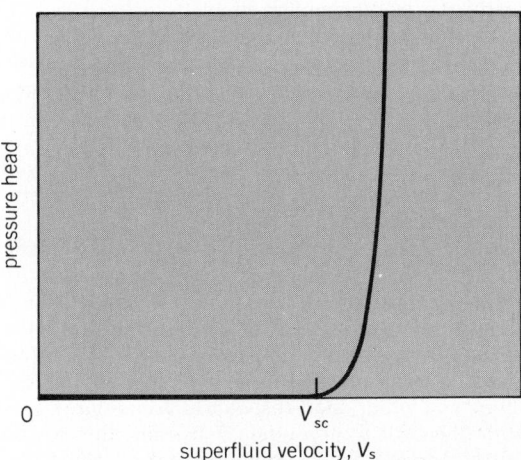

Fig. 8. Pressure head (chemical potential difference) as a function of superfluid velocity, for superfluid flow through a narrow channel. The critical velocity is V_{sc}.

When vortices are present in the superfluid, they interact with the rotons and phonons which constitute the normal fluid and with the walls. These interactions are sometimes allowed for by adding empirical mutual friction and superfluid friction terms to the hydrodynamic equations of motion.

Lambda-transition singularities. At the λ-line in liquid helium, as at other second-order phase transitions or critical points, many of the thermodynamic properties are singular, that is, not smooth functions of temperature. For instance, the specific heat (Fig. 2) has a large cusp with infinite slope at the λ-temperature, and the superfluid density approaches zero with infinite slope. The measured superfluid density near the λ-temperature (T_λ) is accurately represented by Eq. (1). The "critical" exponent ζ is very close to 2/3 and A is a constant.

$$\rho_s = A(T_\lambda - T)^\zeta \qquad (T < T_\lambda) \qquad (1)$$

There has been much theoretical interest in second-order phase transitions because the nature of the singularities—for instance, critical exponents such as ζ in Eq. (1)—appear to depend only on very general symmetry properties of the system in question and not at all on details like the force between atoms. If this is true, then certain "scaling laws" must hold which relate the critical exponents for singularities in different thermodynamic properties. The λ-transition in helium is one experimentally attractive case for testing scaling laws, because it is relatively easy to attain high purity, homogeneity, and temperature stability.

Film flow. Liquid helium wets all solid surfaces, because a helium atom is more strongly attracted to other substances, which have molecules that are more polarizable than helium atoms. Consequently the surface of any solid in contact with liquid helium or its saturated vapor is covered, in thermodynamic equilibrium, by a helium film a few tens of nanometers thick. What is unique in the case of He II is that the film, like the bulk liquid, is superfluid and can flow as fast as about 30 cm/s without friction. Thus the equilibrium film is established rapidly. Helium in a dewar flask flows up the inside walls of the dewar, at a rate of about 10^{-4} cm³/s per centimeter of perimeter, until it approaches a warmer region and evaporates (this flow is driven by the thermomechanical effect).

Thinner helium films are formed on solid surfaces exposed to helium vapor at less than the saturated vapor pressure. As the film becomes thinner, the onset of superfluidity is depressed progressively farther below the λ-temperature of the bulk liquid. Superfluid flow has been observed in films as thin as two atomic layers, at temperatures well below 1 K.

At the onset temperature T_c in these unsaturated films, the superfluid mass per unit area increases abruptly from zero to a finite value, which is proportional to T_c. This is inferred from the speed of third sound, and more directly from torsional pendulum measurements. The size of the jump and other features of the transition agree with theoretical predictions. In this theory the onset temperature is characterized by the exponentially rapid disappearance of free quantized vortices, perpendicular to the film, which are present in thermal equilibrium at higher temperatures.

In the presence of vortices, superfluid flow is possible, but it is not frictionless. The concentration of vortices can be determined in experiments on the decay of persistent currents and, at slightly higher temperatures, on the resistance to thermomechanically driven film flow. This has been of wide interest because the same theory, with minor modifications, should apply to certain other two-dimensional phase transitions, including melting. *See* PHASE TRANSITIONS.

A helium film of less than one complete atomic layer on a "smooth" substrate (exfoliated graphite) exhibits a number of remarkable phenomena. These include the occurrence of two-dimensional gas, solid, and perhaps liquid phases and, at a certain density, a λ-like singularity in specific heat due to long-range ordering in conformity with the substrate lattice.

LIQUID HELIUM-3

The stable isotope ³He occurs naturally as about 1 part per million (ppm) of atmospheric helium and 0.1 ppm of well helium. Commercial ³He is obtained from beta decay of tritium.

The vapor pressure of ³He is considerably greater than that of ⁴He at the same temperature, so that it is possible to pump a liquid ³He bath to lower temperature by about a factor of 3—a little below 0.3 K.

Liquid ³He-⁴He mixtures. The addition of ³He to liquid ⁴He depresses the λ-transition to lower temperatures, as shown in the phase diagram (Fig. 9). There is a region of the phase diagram below 0.87 K in which the homogeneous solution is unstable and separates into a nonsuperfluid ³He-rich phase floating on top of a superfluid ⁴He-rich phase. In the superfluid phase, ³He acts as part of the normal fluid.

In general when a two-component system exhibits a liquid-liquid phase separation, the top of the phase-separation curve is rounded and the highest point is a critical point. By contrast in the helium isotope case, the two branches of the phase-sepa-

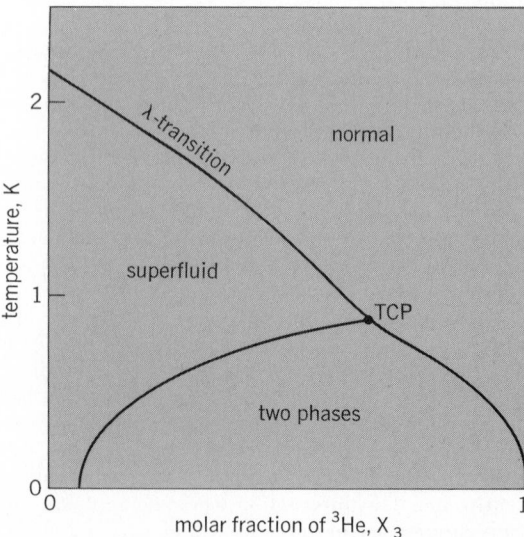

Fig. 9. Phase diagram for ³He-⁴He mixtures at saturated vapor pressure in the temperature-³He concentration plane. The tricritical point (TCP) is at $X_3 = 0.67$, $T = 0.87$ K.

ration curve meet in an angular contact with the λ-line. The junction, called a tricritical point, marks the change of the second-order λ-transition into a first-order phase-separation transition. Singularities in various thermodynamic functions at the tricritical point (in addition to those associated with the λ-line) have drawn considerable theoretical interest.

As the temperature is lowered toward absolute zero, the ^{3}He-rich phase rapidly becomes pure ^{3}He, but ^{3}He remains soluble up to a concentration of about 6% in the ^{4}He-rich phase. An insignificant number of phonons and rotons remain in the ^{4}He at temperatures well below 0.5 K, and it is essentially in its ground state. This pure superfluid ^{4}He acts as an inert background for a gaslike solute of ^{3}He atoms, except that it increases the effective mass of the ^{3}He atoms and weakens their interaction. This has a very important practical consequence: It is possible to "evaporate" ^{3}He into superfluid ^{4}He with absorption of latent heat far below the temperature at which the true vapor pressure of ^{3}He becomes insignificant. Helium-3 dilution refrigerators based on this principle can attain temperatures of about 10 mK in continuous, recirculating operation and 4 mK in a one-time evaporation.

Pure liquid helium-3. Liquid ^{3}He solidifies at absolute zero only at a pressure of 34 bars, as shown in the phase diagram (Fig. 10). There is no λ-transition, because the exclusion principle prevents condensation of particles with half-integer spin such as ^{3}He atoms into a macroscopic quantum state, as occurs in superfluid ^{4}He. Instead, below about 0.5 K the liquid gradually becomes "degenerate" in the same sense as the electrons in a normal metal are degenerate. That is, one ^{3}He atom with each of the two allowed nuclear spin directions occupies each of the lowest momentum quantum states. At low temperatures practically all the quantum states with energies up to the so-called Fermi level are doubly occupied, and practically all above this energy are empty. More precisely, the system is described in terms of the coordinates of weakly interacting quasiparticles, each moving in the mean field of the rest of the system. This Fermi liquid theory, originated by

Landau, works well below about 100 mK. In the same range, high-frequency compressional waves no longer propagate via particle collisions, but go over to collisionless zero sound. The mean field for nuclear spin interaction is such that the liquid is nearly ferromagnetic.

One consequence of the Fermi degeneracy is that the liquid has lower entropy than the solid, in which the nuclear spins are randomly oriented down to about 1 mK, where nuclear antiferromagnetic ordering occurs. In accord with the Clausius-Clapeyron equation of thermodynamics, the melting curve then has negative slope (see Fig. 10). This has a practical application known as Pomeranchuk cooling: if liquid ^{3}He at the solidification pressure in the region of negative slope is further compressed adiabatically, it begins to solidify, and this is accompanied by cooling. The amount of cooling can be very substantial if the initial temperature is sufficiently low. *See* FREE-ELECTRON THEORY OF METALS.

Superfluid ^{3}He. In 1957 a theory developed by J. Bardeen, L. N. Cooper, and J. R. Schrieffer (BSC theory) succeeded in explaining how superconductivity occurs in certain metals, by a mechanism of electron pairing. Electrons of opposite spin and momentum become bound in pairs, of zero net spin, which condense into a macroscopic quantum state responsible for superconductivity. It became apparent that a similar pairing mechanism might result in a superfluid transition in liquid ^{3}He. Although it was difficult to predict the transition temperature, this possibility motivated the development of techniques for studying ^{3}He at progressively lower temperatures. No transition was found in the temperature region accessible with the ^{3}He-dilution refrigerator, and further efforts required an additional cooling stage at the bottom of a dilution refrigerator. This last stage can be a Pomeranchuk compression cell, an adiabatic demagnetization stage using powdered cerium magnesium nitrate (CMN) or a related paramagnetic salt immersed in liquid ^{3}He, or adiabatic demagnetization of nuclear spins in a suitable metal. *See* ADIABATIC DEMAGNETIZATION; SUPERCONDUCTIVITY.

In 1972, in a Pomeranchuk cell, two phase transitions near 2.7 and 2.2 mK were discovered, which proved to be transitions to two distinct superfluid phases of liquid ^{3}He, now designated liquid A and liquid B (see Fig. 11). Measurements on these phases were extended to pressures less than the solidification pressure with magnetic cooling.

The strong short-range repulsion between ^{3}He atoms prevents their pairing in a state of zero relative angular momentum like electron pairs in a superconductor. Instead the ^{3}He pair has one quantum unit of orbital angular momentum. Quantum statistics then require that the nuclear spins occupy a symmetric state, for which the three possibilities are ↑ ↑, ↓ ↓, and triplet ↑ ↓ (which are analogous, in the notation used for atomic spins, to $S = 1$; $m_S = +1, -1$, and 0 respectively). In the B phase, pairs of all three spins states are coupled into a superfluid which is isotropic, except in more subtle aspects of the spin configuration. The A phase involves only ↑ ↑ and ↓ ↓ pairs, which couple coherently to give a macroscopic orbital angular momentum **L**, and a spin angular momentum **S**, which is oriented in the plane perpendicular to **L**. The superfluid proper-

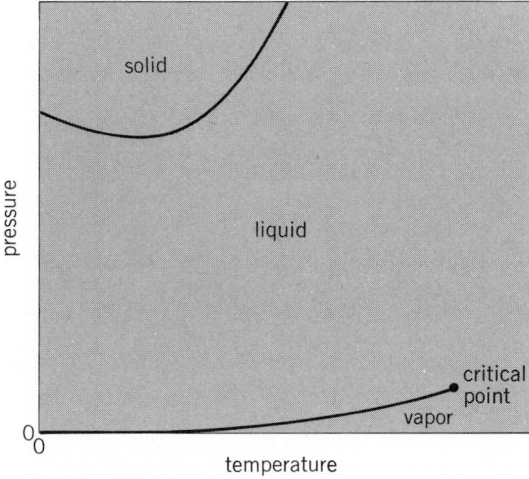

Fig. 10. Phase diagram for ^{3}He. The critical point is at $T_c = 3.32$ K, $P_c = 1.17$ bars. The minimum of the melting curve is at $T = 0.32$ K, P = 29.3 bars.

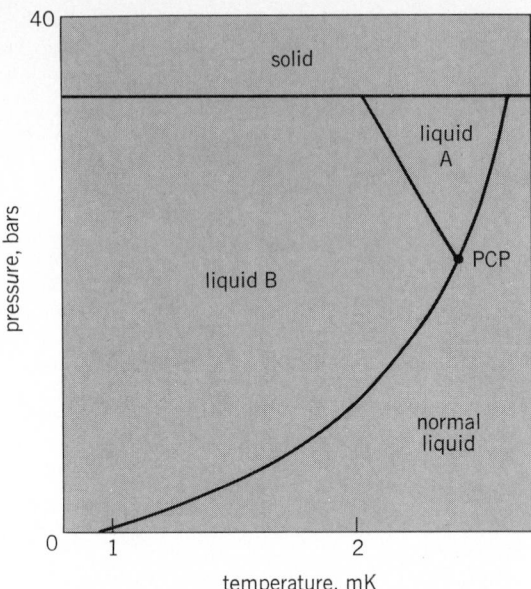

Fig. 11. Phase diagram for ^{3}He in the millikelvin temperature range, with no magnetic field, showing the superfluid phases A and B. PCP is the polycritical point. (*From O. V. Lounasmaa, The superfluid phases of liquid ^{3}He, Contemp. Phys., 15(4):353–374, 1974*)

ties are anisotropic, with axial symmetry about the direction of **L**. *See* Angular momentum; Atomic structure and spectra; Spin.

The phase diagram in Fig. 11 shows that liquid B is stable over a much wider range than liquid A. In fact, the A phase is stable in zero magnetic field only by virtue of the nearly ferromagnetic character of liquid ^{3}He. The A-B transition is first-order, with a small latent heat, whereas the A-normal or B-normal transition is second-order, with a jump in the specific heat. These transitions join at the polycritical point. In the presence of even a small magnetic field (approximately 10^{-2} tesla), the A-B transition is depressed in temperature, the polycritical point disappears, and a narrow A-phase region is believed to extend down to zero pressure. In magnetic fields of order 1 tesla, the B phase is completely suppressed. The A-normal transition becomes split, with the appearance of a narrow intermediate phase A_1, in which only ↑ ↑ pairs (or perhaps only ↓ ↓ pairs) are superfluid.

The superfluid ^{3}He phases, particularly the A phase, exhibit complex "textures" due to the variation in space of the macroscopic orbital and spin angular momentum directions under a variety of influences. The spin axis is aligned by a magnetic field, and couples via the nuclear dipole field to **L**, with the result that **L** tends to align perpendicular to an applied magnetic field. **L** also tends to align perpendicular to walls, but parallel to the direction of superfluid flow. These features make nuclear magnetic resonance (NMR) a very powerful technique for studying this superfluid. Ultrasonic attenuation, which depends on the direction of propagation with respect to **L**, can be used to study the orientations of **L** in the absence of a magnetic field.

Evidence for frictionless flow, which is the fundamental characteristic of a superfluid, is available from several types of flow experiments. The most exacting are studies of the propagation of fourth sound in powder-packed cells.

[George B. Hess]

Bibliography: W. E. Keller, *Helium-3 and Helium-4*, 1969; A. J. Leggett, A theoretical description of the new phases of liquid ^{3}He, *Rev. Mod. Phys.*, 47:331–414, 1975; K. Mendelssohn, *The Quest for Absolute Zero*, 1966; J. C. Wheatley, Helium three, *Phys. Today*, 29(2):32–42, 1976; J. Wilks, *Properties of Liquid and Solid Helium*, 1967.

Liquid scintillation detector

A particle detector in which the sensitive medium is a liquid scintillator. For a general discussion of scintillators as particle detectors *see* Scintillation counter.

Liquid and plastic scintillators are the most extensively used forms of organic scintillator. They are noted for their fast response, for the ease with which they may be formed in arbitrary shapes, large or small, and for the economy they offer in achieving large, sensitive detecting volumes. The timing precision achieved by the faster liquids and plastics is typically in the range 10^{-9} to 10^{-10} s. Liquid scintillators are used in volumes ranging from a few milliliters or less through several tens of cubic meters. The largest scintillators are used in high-energy physics to study neutrino interactions. Liquid scintillators also have the capacity to assimilate other liquids or substances and form homogeneous media, thereby providing an efficient and simple means for measuring the products of nuclear reaction or radioactive decay in those substances. *See* Neutrino.

Solvent and solute. A simple liquid scintillator might consist of 5 g of solute, such as *p*-terphenyl, PPO (2,5-diphenyloxazole), or PBD [2-phenyl,5-(4-biphenyl)-1,3,4,oxadiazole], dissolved in 1 liter of solvent, such as xylene, toluene, or benzene. The solvent, being the bulk constituent, absorbs virtually all of the energy deposited by the ionizing particle being detected in the scintillator. For efficient operation, the excitation energy imparted to the solvent molecules must transfer rapidly through the solvent to the solute molecules, which then deexcite and produce the scintillation emission. Large liquid scintillators usually incorporate an additional small quantity (0.1 g/liter) of secondary solute or "wavelength shifter," such as POPOP [1,4-bis-{2-(5-phenyloxazolyl)}-benzene], which captures excitation energy from the primary solute. The resulting scintillation output from the secondary solute is emitted at longer wavelengths than that of the primary solute, and is therefore more efficiently transmitted through the scintillator to the surrounding photomultiplier tubes.

Pulse shape discrimination. The responses of most liquid scintillators can be improved significantly if dissolved oxygen is either removed from the solution or displaced by nitrogen or an inert gas. This procedure also endows many liquid scintillators with the ability to pulse-shape-discriminate, that is, to exhibit scintillation decay properties which depend on the type of ionizing particle causing the scintillation. Most liquid scintillators contain a high proportion of hydrogen, and therefore detect fast neutrons (energy greater than 0.1 MeV) via the proton recoils generated by internal neutron-proton scattering. Pulse shape discrimination may be used to select these proton scintillations while rejecting others, especially those caused by gamma rays. Liquid scintillators are therefore very effective in detecting and identifying neutrons, even when the background of gamma radiation is high. The response of liquid scintillators to electrons is linear, that is, proportional to electron energy. However, their response to heavier particles is nonlinear; and for different particles of the same energy, the response is smaller, the heavier the particle.

Additions of compounds and samples. Appreciable quantities of compounds of a number of elements can be incorporated in liquid scintillators (in appropriate chemical form) without degrading the scintillation performance excessively. Some examples are: boron, gadolinium, or cadmium compounds, which facilitate the detection of low-energy neutrons (less than about 0.1 MeV) through neutron capture in these elements; and tin or lead compounds, which enhance the scintillator sensitivity to low energy gamma rays or x-rays (less than about 0.1 MeV). Hexafluorobenzene may be used as a solvent, instead of benzene, to obtain a hydrogen-free, neutron-insensitive liquid detector for gamma rays. Organic liquid scintillators are extensively used in the medical, biological, and environmental fields to detect radioactive samples and tracer compounds labeled with the beta emitters ^{14}C or ^3H, or with other radionuclides. A carrier solution containing the sample or tracer compound is assimilated or dissolved in the liquid scintillator to form a uniform counting sample or scintillating medium with a volume of a few milliliters. Such counting samples are often processed in large batches by using sophisticated instrumentation which incorporates automatic sample changing and the capacity to carry out all the necessary data reduction automatically.

Manufacture and handling. The handling of liquid scintillators requires extreme care, because many are highly inflammable and because small concentrations of common impurities can drastically impair the scintillation performance by quenching the energy-transfer mechanism of the scintillation process. Commercial manufacturers offer a wide variety of liquid scintillators packaged in a variety of forms, for example, bottled in bulk form or encapsulated in glass or metal-and-glass containers. Encapsulated scintillators may have reflector paint applied to all surfaces other than that which is to be coupled to the photomultiplier cathode. *See* PARTICLE DETECTOR.

[FRANK D. BROOKS]

Bibliography: J. B. Birks, *The Theory and Practice of Scintillation Counting,* 2d ed., 1967; F. D. Brooks, Development of organic scintillators, *Nucl. Instr. Meth.,* 162:477–505, 1979; A. A. Noujaim, C. Ediss, and L. I. Weibe (eds.), *Liquid Scintillation Science and Technology,* 1976.

Logarithm

An exponent of a suitably chosen positive number (base) larger than unity. Logarithms are of value in mathematical computation and in the equations and formulas used in expressing natural phenomena.

The invention of natural logarithms (base $e = 2.718$ . . .) is usually attributed to John Napier, who published a table of logarithms in Edinburgh in 1614 under the title *Mirifici Logarithmorum Canonis Descriptio,* even though the functions tabulated by Napier are merely related to, and not identical with, natural logarithms.

The invention of common logarithms (base 10) is generally attributed to Henri Briggs; Napier's share is acknowledged, however, by Briggs himself in his *Arithmetica logarithmica* in 1624.

Theory. If $b^l = n$, where b is a positive number larger than unity, then l is called the logarithm of n to the base b and is written $l = \log_b n$. From this definition it follows at once that all positive numbers larger than unity have positive logarithms, all positive numbers smaller than unity have negative logarithms, and the logarithm of unity is equal to 0. Since $b^0 = 1$ irrespective of the value of b, it follows that the logarithm of unity is equal to 0. From the known properties of exponentials expressed by relations (1), it follows immediately that (1) the log-

$$b^{l_1} \cdot b^{l_2} = b^{l_1 + l_2} \qquad b^{l_2} \div b^{l_2} = b^{l_1 - l_2}$$
$$(b^l)^m = b^{lm} \qquad \sqrt[m]{b^l} = b^{l/m} \qquad (1)$$

arithm of a product of two or more factors is equal to the sum of the logarithms of the factors; (2) the logarithm of the ratio of two numbers is equal to the difference between the logarithm of the numerator and the logarithm of the denominator; (3) the logarithm of the mth power of a number is the product of m and the logarithm of the number; and (4) the logarithm of the mth root of a number is equal to the logarithm of the number divided by m. These properties are responsible for the great simplification in the task of carrying out numerical computations involving multiplication, division, raising numbers to certain powers, or extracting the roots of a certain order of given numbers. It also follows from these properties that in constructing a table of logarithms to a certain base from the beginning, the computer need only compute the logarithms of prime numbers, because the logarithm of any number which is not prime can be obtained from the logarithms of its prime factors by simple operations of addition and multiplication.

Although any positive number b larger than unity might have been chosen as the base of a system of logarithms, actually two numbers have been chosen in the construction of tables of logarithms, namely, the number $b = 10$ and the number e defined as the sum of infinite series (2). The system

$$1 + \frac{1}{1} + \frac{1}{1 \cdot 2} + \frac{1}{1 \cdot 2 \cdot 3} + \cdots$$
$$+ \frac{1}{1 \cdot 2 \cdot 3 \cdots n} + \cdots \quad (2)$$

of logarithms to the base 10 is usually referred to as common logarithms; the system of logarithms to the base e is called natural logarithms.

Common logarithms have certain obvious advantages not shared by natural logarithms. All numbers between 1 and 10 have logarithms between 0 and 1, all numbers between 10 and 100 have logarithms between 1 and 2, and so on. Thus the integral part of the common logarithm of a number larger than unity is one unit less than the number of digits before the decimal point. This integral part is called the characteristic; the decimal part is called the mantissa. Similarly, because the common logarithms of 0.1, 0.01, and 0.001 are −1, −2, and −3, it follows that the logarithm of a number smaller than unity having p zeros after the decimal point may be expressed as the sum of a negative characteristic equal to $−(p+1)$ and a positive mantissa.

Natural logarithms. Because the number $e = 2.718 \ldots$, the base of the system of natural logarithms is irrational; that is, it cannot be expressed as the ratio of two integers (and therefore when expressed as a decimal number, it involves an infinite number of decimals with no repeating groups of decimals). It might seem odd, therefore, that it has been chosen as the base of a system of logarithms. The primary motivation for this choice lies in the fact that the solutions of numerous problems in applied mathematics are most naturally expressed in terms of powers of e. Thus, for instance, the solutions of the problems of equilibrium of a flexible cable, the transient flow of electric current in a circuit, and the disintegration of radioactive elements are expressed in terms of e^x, where x is either positive or negative and depends on the physical parameters of the problem in question. Thus, the tabulation of the function e^x for both positive and negative values of x was an indispensable aid in obtaining the solutions of many physical problems. Because the logarithmic function is the inverse of the exponential function, a table of natural logarithms can be constructed with relative ease from a table of e^x by the process of inverse interpolation. Another motivation for constructing tables of natural logarithms is the fact that natural logarithms arise in the integration of quotients of polynomials and of functions which may be approximated by quotients of polynomials. Indeed, it is a well-known fact that if the n roots of a polynomial $Pn(x)$ are known so that $Pn(x) = a_0(x−x_1)(x−x_3) \cdots (x−x_n)$ and if $Q(x)$ is a polynomial of degree smaller than n, then the quotient $Q(x)/Pn(x)$ may be expressed in the form of expression (3) when the A_ks may be easily determined.

$$\sum_{k=1}^{n} \frac{A_k}{x−x_k} \quad (3)$$

Accordingly Eq. (4) may be written.

$$\int \{Q(x)/Pn(x)\}\, dx = \sum_{k=1}^{n} A_k \int \frac{dx}{x−x_k}$$
$$= \sum_{k=1}^{n} A_k \log_e (x−x_k) \quad (4)$$

Relation between logarithm bases. Let l_1 be the known natural logarithm of a number N. What is the common logarithm of N? Clearly $N = e^{l_1}$, and therefore $\log_{10} N = l_l \log_{10} e$. Since $\log_{10} e = 0.43429$ (to five decimals) it can be concluded that natural logarithms are converted into common logarithms

by multiplying the natural logarithm by the factor 0.43429, which is called the modulus. Similarly the common logarithm is converted into a natural logarithm by division by the modulus 0.43429 or multiplication by 2.303. The table shows natural and common logarithms for some numbers from 0.01 to 1000.

Logarithms of complex numbers. Since b^l is positive when b is positive, negative numbers have no real logarithms. Nevertheless, Euler's famous formula $e^{i\theta} = \cos \theta + i \sin \theta$, where $i = \sqrt{−1}$, makes it possible to define not only the logarithm of a negative number but also the logarithm of a complex number. Indeed, any complex number $a + ib$ may be written in the form $\rho e^{i\theta}$, where $\rho = \sqrt{a^2 + b^2}$ and $\theta = \arctan (b/a)$. It follows that the logarithm of a complex number $a + ib$ is a complex number whose real part is the logarithm of its modulus ρ and whose imaginary part is its phase (or argument) θ. Because $e^{i\theta} = e^{i(\theta + 2n\pi)}$, it follows that a complex number has an infinite number of logarithms given by Eq. (5), where n is a positive or negative integer.

$$\log (a+ib) = \log \sqrt{a^2+b^2} + i(\theta + 2n\pi) \quad (5)$$

As previously mentioned, the evaluation of natural logarithms may be based on inverse interpolation in a table of e^x. For small values of x, positive or negative, it is possible to use expansion (6). This

$$\log_e (1+x) = x − \frac{x^2}{2} + \frac{x^3}{3} − \cdots$$
$$+ (−1)^{n+1} \frac{x^n}{n} + \cdots \quad (6)$$

expansion may be derived from Taylor's series, notation (7), by writing $f(x) = \log_e x$. After a

$$f(x+a) = f(a) + \frac{x}{1} f'(a) + \frac{x^2}{1 \cdot 2} f''(a)$$
$$+ \cdots \frac{x^n}{1 \cdot 2 \cdot 3 \cdot m} f^{(n)}(a) + \cdots \quad (7)$$

sufficiently accurate table of $\log (1 \pm x)$, for $x = k10^{−n}$, $k = 1,2,3 \ldots 9$, and $n = 1,2,3 \ldots N$, where N is sufficiently large (perhaps 15) has been computed, this table may be used to compute the logarithm of any number. From the expansion (6) for $\log_e (1+x)$ it is easy to derive expansion (8),

$$\log_e p = \frac{1}{2}\{\log_e (p−1) + \log_e (p+1)\} + S(p) \quad (8)$$

where Eq. (9) holds; this has the virtue that $S(p)$

$$S(p) = \frac{1}{2p^2−1} + \frac{1}{3(2p^2−1)^3}$$
$$+ \cdots \frac{1}{(2n+1)(2p^2−1)^{2n+1}} + \cdots \quad (9)$$

converges quite rapidly. Thus for $p > 10,000$ the error resulting from neglecting all but the first term in $S(p)$ is considerably less than $10^{−20}$. See NUMERICAL ANALYSIS.

[ARNOLD N. LOWAN/SALOMON BOCHNER]

Lorentz transformations

The family of mathematical transformations used in the special theory of relativity to relate the space and time variables of uniformly moving (inertial) reference systems. They were discovered in part by H. A. Lorentz during his studies of matter moving in an electromagnetic field and were formulated more completely by H. Poincaré and

Table of logarithms

Number	Natural logarithm	Common logarithm
0.01	5.586 − 10	−2 (8 − 10)
1.0	0	0
$e = 2.718 \ldots$	1	0.4343
10	2.303	1
1000	6.909	3

by Albert Einstein. If S and S' are two inertial reference frames with space-time coordinates (x,y,z,t) and (x',y',z',t'), respectively, the equations connecting these variables have the form $(x^0 = ct, x^1 = x, x^2 = y, x^3 = z)$ of Eq. (1), where a^0, a^1, a^2, a^3 are

$$x'^{\alpha} = \sum_{\beta=0}^{3} L_{\beta}{}^{\alpha}(x^{\beta} - a^{\beta}) \qquad (\alpha = 0, 1, 2, 3) \quad (1)$$

arbitrary real constants. Forming the two real matrices (2) and (3), the matrix L is subject to the condition shown as Eq. (4), where L^{-1} and L_t are the

$$L = \begin{bmatrix} L_0{}^0 & L_1{}^0 & L_2{}^0 & L_3{}^0 \\ L_0{}^1 & L_1{}^1 & L_2{}^1 & L_3{}^1 \\ L_0{}^2 & L_1{}^2 & L_2{}^2 & L_3{}^2 \\ L_0{}^3 & L_1{}^3 & L_2{}^3 & L_3{}^3 \end{bmatrix} \qquad (2)$$

$$\eta = \begin{bmatrix} 1 & 0 & 0 & 0 \\ 0 & -1 & 0 & 0 \\ 0 & 0 & -1 & 0 \\ 0 & 0 & 0 & -1 \end{bmatrix} \qquad (3)$$

dition shown as Eq. (4), where L^{-1} and L_t are the inverse and transposed matrices of L.

$$L^{-1} = \eta L_t \eta \qquad (4)$$

The Lorentz transformation group is composed of four classes of transformations, which can be organized as indicated in the table, depending on the algebraic signs of the pivotal element $L_0{}^0$ of the matrix L, and the determinant of L, which is designated by $D(L)$. Positive signs are indicated by + and negative signs by − in the table.

Classes of Lorentz transformations

	$L_0{}^0$	$D(L)$
Proper	+	+
Improper	+	−
	−	+
	−	−

The improper Lorentz transformations involve changes of sign (reflections) of the space coordinates, the time coordinate, or both. For additional discussion of improper Lorentz transformations *see* SPACE-TIME.

The proper Lorentz transformations form a 10-parameter group which can be generated from translations of the space-time coordinates, rotations of the space reference frame, and transformations to uniformly moving systems. *See* RELATIVITY.

[E. L. HILL]

Bibliography: W. J. Kaufmann, *Relativity and Cosmology*, 2d ed., 1977; H. M. Schwartz, *Introduction to Special Relativity*, 1968, reprint 1977.

Low-temperature acoustics

The application of acoustics to research on the properties of condensed matter at low temperatures. Acoustic techniques are readily adaptable to the cryogenic environment and make possible many measurements of the structural and thermodynamic properties of materials at temperatures approaching absolute zero (0 K, which is −273°C). The study of sound propagation has also yielded major insights into the low-temperature phenomena of superconductivity in metals and superfluidity in liquid helium.

LOW-TEMPERATURE PROPERTIES OF SOLID MATERIALS

Acoustic measurements have been used to characterize the properties of a wide variety of solid-state materials, such as metals, dielectric crystals, amorphous solids, and magnetic materials. A measurement of the velocity of sound in a substance gives information on its elastic properties, while the attenuation of the sound characterizes the interaction of the lattice vibrations with the electronic and structural properties of the material.

Experimental methods. Ultrasonic frequencies, in the range from 20 kHz to 100 MHz and above, are commonly employed in these measurements because of the ease of generating and detecting the sound with piezoelectric quartz crystals. In a typical experimental arrangement (Fig. 1), the opposite sides of the material being investigated are cut flat and the surfaces polished. Two quartz transducers are cemented to the flat sides. The sound is generated by exciting the quartz with a short voltage pulse at its resonant frequency. The pulse of sound propagates through the sample and is detected by the receiver-transducer. By measuring the time of flight across the sample, the sound velocity can be obtained. As the sound then reflects back and forth across the sample, the attenuation of the wave is found by measuring the rate of decay of the echo pulses. The temperature of the sample can be varied by bringing it into contact with cryogenic fluids such as liquid nitrogen (77 K) or liquid helium (1–4.2 K). Various refrigeration techniques using these liquids allow measurements over the entire range from room tem-

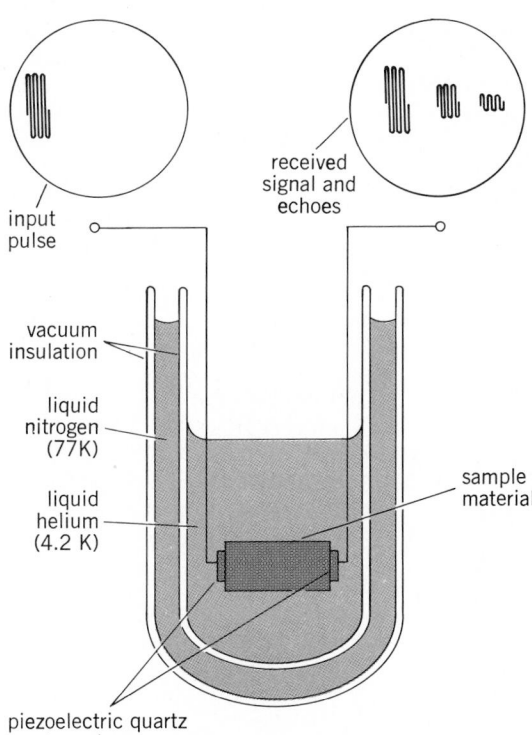

Fig. 1. Experimental arrangement to measure the velocity and attenuation of sound in a sample cooled to 4.2 K using liquid helium.

perature (300 K) down to temperatures well below 1 K. *See* ULTRASONICS.

Metals and crystals. Because the sound velocity effectively measures elastic constants, such measurements are used to characterize phase transitions in crystals where the structure of the lattice changes. For example, in solid metallic sodium there is a transition near the temperature of 36 K, where the crystal lattice changes from body-centered cubic to hexagonal close-packed. Both the sound velocity and the attenuation change abruptly when the transition occurs, and provide a means of monitoring the nature of the transition. *See* CRYSTAL STRUCTURE; PHASE TRANSITIONS.

The attenuation of sound in many crystals is due to defects and impurities in the crystal lattice and provides information on such structures. If there is a line of missing atoms (a line dislocation), the stress from the sound field causes it to move back and forth. However, because of pinning forces, the dislocation cannot freely follow the sound oscillation and hence absorbs energy from it. In a metal at very low temperatures, the dominant source of attenuation is the interaction of the sound with the conduction electrons. Oscillations of the positively charged ions making up the crystal lattice couple to the negatively charged electrons. When the mean free path of the electrons becomes longer than the acoustic wavelength (which occurs at low temperatures), the sound wave loses energy to the electrons. Studies of this process are important in determining the strength of the electron-phonon coupling in a material (phonons are thermally excited quantized lattice vibrations). The coupling between the electrons and phonons is the mechanism responsible for superconductivity. *See* CRYSTAL DEFECTS; LATTICE VIBRATIONS; PHONON; SOUND ABSORPTION.

A large variety of magnetoacoustic effects are observed in metals and crystals. In these measurements, changes in the sound attenuation occur as the strength of a magnetic field applied to the sample is increased. One example is the phenomenon of nuclear acoustic resonance. The nuclear spins in a crystal interact with vibrations of the lattice. When the frequency of the sound matches the Larmor frequency of the spins (which is determined by the magnetic field), the sound can induce transitions in the spin states to higher energies. This results in absorption of the sound, and a resonance is observed in the graph of attenuation versus field. There are also a number of other magnetoacoustic effects in metals which are useful in determining the orbits followed by the conduction electrons in the metal. *See* DE HAAS—VAN ALPHEN EFFECT; LARMOR PRECESSION.

Amorphous materials. Sound propagation is becoming increasingly useful for studying amorphous materials, in which there is no regular crystal structure. In materials such as silica glass (amorphous silicon dioxide, SiO_2), it has been found that at low temperatures there are only two quantum energy levels that are important. These levels correspond to two nearly equivalent arrangements of the atoms, with one arrangement having slightly higher energy. The stress from an imposed sound field can cause a transition from one arrangement to the other. If the relaxation rate back to the original configuration is comparable to the sound frequency, there will be a net absorption of energy from the sound wave. A peak in the attenuation in silica glass near 50 K has been identified as being due to this process, and measurements as a function of frequency allow a determination of the relaxation rate. *See* AMORPHOUS SOLID.

Superconductors. When a metal is cooled below its superconducting transition temperature, there are striking changes in the attenuation of sound. At the transition some of the electrons near the Fermi surface begin to pair together, due to the attractive electron-phonon coupling. Once this occurs, the electrons can no longer exchange momentum with the lattice, and hence have zero resistance. This also means that the paired electrons no longer absorb energy from the sound wave, and the attenuation is from the remaining unpaired normal electrons. As the temperature is lowered well below the transition, the density of the unpaired electrons drops rapidly, and the attenuation becomes very small. In the Bardeen-Cooper-Schrieffer (BCS) theory of superconductivity, the ratio of the sound attenuation α_s in the superconducting state to that in the normal state α_n is given by Eq. (1), where T is the temperature, Δ

$$\frac{\alpha_s}{\alpha_n} = \frac{2}{e^{\Delta/kT} + 1} \qquad (1)$$

is the energy gap of the electrons, and k is Boltzmann's constant. Measurements of this ratio in metallic tin as a function of temperature are shown by data points in Fig. 2. The drop in attenuation is well predicted by the BCS theory, shown by the curve. By inverting Eq. (1), values of the energy gap Δ can be obtained, and this is the fundamental parameter characterizing the superconducting state.

The sound attenuation is also used to study the characteristics of type II superconductors. In these superconductors the magnetic flux applied to a sample is expelled until a critical field H_{c1} is reached. Above that field the flux enters the sample by threading the materials with an array of flux

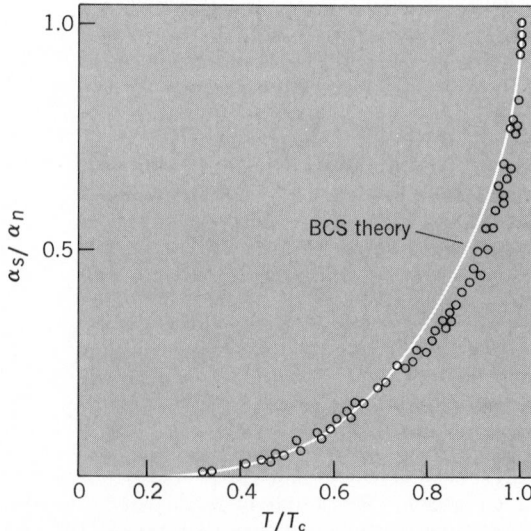

Fig. 2. Ratio of the sound attenuation α_s in the superconducting state to that in the normal state α_n for metallic tin, as a function of temperature T. The superconducting transition temperature T_c is 3.71 K for tin.

tubes known as quantized vortex lines. The motion of these pinned vortex lines contributes to the sound attenuation. As the magnetic field is increased, there is a further sharp increase in the attenuation at the second critical field H_{c2} where the superconductivity is destroyed. *See* SUPERCONDUCTIVITY.

SUPERFLUID HELIUM

Sound propagation has been extensively used to probe many of the unusual properties of superfluid helium. When liquid ^{4}He is cooled below 2.17 K, it undergoes a phase transition into the superfluid state. A successful theoretical model for describing the superfluid properties, known as the two-fluid model, was developed independently by L. D. Landau and L. Tisza in the 1940s. In this model the liquid helium below $T_\lambda = 2.17$ K is treated as being a mixture of two fluids, a superfluid component with density ρ_s and a normal fluid component with density ρ_n. These densities add to give the actual density ($\rho = 0.14$ g/cm^3) of the liquid, $\rho_s + \rho_n = \rho$. Both ρ_s and ρ_n depend on temperature. Just below the transition at 2.17 K, the superfluid density ρ_s is nearly zero, but it increases rapidly as the liquid is cooled, and below 1 K ρ_s is nearly equal to ρ.

The superfluid component has the unusual property that its viscosity is identically zero, and it is able to flow through the tiniest of capillaries. The normal fluid component, on the other hand, has the properties of an ordinary viscous liquid and is completely immobilized in a small capillary. A second characteristic of the superfluid is that it has zero entropy because it is an ordered quantum system, and all of the liquid's entropy is carried by the normal fluid component. This leads to an unusual flow property of the superfluid, namely, that it flows in response to both pressure gradients and temperature gradients in the liquid. An ordinary fluid is accelerated only by pressure gradients. These novel features of the superfluid (zero viscosity and entropy) give rise to a rich variety of different types of sound which can propagate in the superfluid helium. Five distinct sound modes have been identified and observed experimentally. The sound velocities of a number of these modes are shown in Fig. 3 as a function of temperature.

First sound. First sound is a pressure wave which propagates in the bulk liquid. It is quite similar to sound in ordinary fluids. The velocity C_1 is given by Eq. (2), where B is the adiabatic bulk

$$C_1^2 = \frac{B}{\rho} \tag{2}$$

modulus of the helium. The velocity is relatively temperature-independent at a value of about 237 m/s, except for the region near the superfluid transition at 2.17 K. At that point the velocity dips in a sharp cusp. There is no jump in the velocity, but the slope is discontinuous. High-resolution measurements of the sound velocity and attenuation in this region yield valuable information on the critical properties of the superfluid transition.

Second sound. Second sound is an unusual type of wave: it is a temperature wave in the bulk superfluid. In this mode the normal fluid and superfluid move in opposite directions. This keeps the density constant, and hence there are no pressure oscillations in the wave (as in first sound); but because

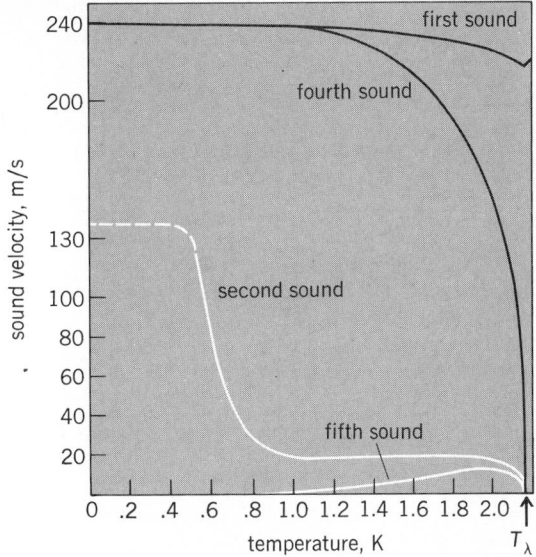

Fig. 3. Velocity of the various types of sound in superfluid ^{4}He as a function of temperature.

only the normal fluid carries entropy, there are oscillations in the entropy and thus in the temperature of the liquid. The second sound can be generated with a heater wire and detected with a carbon resistance thermometer. Landau first predicted the existence of second sound, and its observation experimentally (in Moscow in 1944) confirmed many aspects of the two-fluid model of liquid helium. According to the theory, the sound velocity is given by Eq. (3), where T is the temperature, S the

$$C_2^2 = \frac{\rho_s}{\rho_n} \frac{TS^2}{C} \tag{3}$$

specific entropy, and C the specific heat. The velocity is zero at T_λ (since $\rho_s/\rho = 0$ there), and then increases at lower temperatures to a value of about 20 m/s. Below 1 K there is a further rapid increase to a value $C_2 = C_1/\sqrt{3} = 140$ m/s, but in this region the mean free path of the normal fluid phonon excitations becomes longer than the sound wavelength, and the second sound can no longer be observed below about 0.6 K. *See* SECOND SOUND.

Third sound. Third sound is a wave which propagates in very thin films of helium. If helium gas is introduced into a chamber at low temperatures, some of the helium atoms are attracted to solid surfaces within the chamber by van der Waals forces, forming a thin layer of liquid on all of the surfaces. The thickness of the film can be varied between about 0.5 and 30 nm by adjusting the pressure of the gas in the chamber. The third sound is a wave in which the thickness of the film varies, somewhat like waves in a tank of water. Because the films are so thin, only the superfluid can move, the normal fluid being immobilized by its viscosity. The van der Waals force is the restoring force for the wave, and the velocity is given by Eq. (4), where α is the van der Waals coefficient

$$C_3^2 = \frac{\rho_s}{\rho} \frac{3\alpha}{d^3} \tag{4}$$

and d is the film thickness. Experiments have verified this formula.

Third sound has been used to investigate the

superfluid transition in thin films, which occurs at temperatures lower than T_λ and is found to depend on the exact thickness of the helium film. When the superflow vanishes at the transition, the third sound can no longer propagate. However, the sound velocity remains finite as the transition is approached, and from Eq. (4) this means that ρ_s/ρ is finite just before the transition. These observations have served to verify a general theory of two-dimensional matter, which predicts that ρ_s should drop discontinuously from a finite value to zero at the transition in the thin films.

Fourth sound. Fourth sound is a pressure wave which propagates in superfluid helium when it is confined in a porous material such as a tightly packed powder. In such a situation the normal fluid is immobilized, and only the superfluid can flow freely because of its zero viscosity (the porous materials are often called superleaks for this reason). The fourth sound is analogous to first sound because it involves density and pressure oscillations and can be generated and detected by pressure transducers. Because only the superfluid component can move in the superleak, the velocity is modified from that of first sound and is given by Eq. (5). As seen in Fig. 3 this velocity is zero at T_λ

$$C_4^2 = \frac{\rho_s}{\rho} C_1^2 \qquad (5)$$

where $\rho_s/\rho = 0$, while at lower temperatures it approaches C_1 as ρ_s/ρ approaches 1. Because C_1 is known quite accurately, measurements of C_4 can be used with Eq. (5) to obtain highly precise values of ρ_s/ρ as a function of temperature.

Fifth sound. Fifth sound is a temperature wave which can propagate in helium confined in a superleak. It is analogous to second sound, except that again only the superfluid component can flow. This modifies the velocity from that of second sound to Eq. (6). This velocity is zero both at T_λ (where C_2 is

$$C_5^2 = \frac{\rho_n}{\rho} C_2^2 \qquad (6)$$

zero) and $T = 0$ (where $\rho_n/\rho = 0$). The fifth sound is quite a low-velocity mode, reaching a maximum value of about 12 m/s at 1.9 K. The mode can be observed experimentally using a heater wire drive and a thermometer receiver on opposite sides of a superleak partially filled with helium. Measurements of the sound velocity are in excellent agreement with Eq. (6).

Superfluid ^{3}He. Acoustic methods have also been very important in determining the superfluid properties of the isotope ^{3}He. A phase transition was discovered in liquid ^{3}He in 1973 at ultralow temperatures, below 0.003 K. The first conclusive evidence that this was a transition to a superfluid state came when it was discovered that fourth sound can propagate in a superleak filled with the ^{3}He. The fourth sound exists only if there is a frictionless superfluid component, and the measurements of ρ_s derived from C_4 were in general agreement with theoretical predictions.

The transition has also been studied using very high-frequency first sound, which is known as zero sound. When the sound frequency is increased to a value greater than the rate of collisions between the ^{3}He atoms, the characteristics of the wave change from first sound to zero sound (as first pre-

dicted by Landau), where the velocity is slightly higher and the attenuation is reduced. At the superfluid transition there is a sharp peak in the attenuation as a function of temperature, and further variation is found at lower temperatures. In the magnetic A phase, the zero sound attenuation is found to depend on the angle between the direction of the sound beam and the direction of a magnetic field applied on the liquid. This anisotropy is an unusual feature, and the zero sound measurements are able to probe the microscopic structure of the paired ^{3}He atoms. *See* LIQUID HELIUM.

[GARY A. WILLIAMS]

Bibliography: *Physical Acoustics*, vols. 1–5 ed. by W. P. Mason, vols. 6–14 ed. by W. P. Mason and R. N. Thurston, 1964–1979; S. J. Putterman, *Superfluid Hydrodynamics*, 1974.

Low-temperature physics

A study of the properties of gross matter at such low temperature that the quantum character of the substance becomes observable. In a general way, this implies temperatures of 4 K and below. Quantum-mechanical coherence effects extend over readily measurable distances and give rise to superconductivity and superfluidity. Zero-point energy "blows up" light atoms and greatly influences the bulk properties of the hydrogens and the heliums in the solid state. The existence of quantized vortices plays a large role in the basic behavior of superconductors and the liquid heliums. The existence of discrete atomic energy levels shows up via heat capacity anomalies (the Schottky effect) and various magnetic and nuclear phenomena. The "ortho-para conversion" is a striking thermal phenomenon in hydrogen compounds. Minima in electrical resistance (the Kondo effect) and thermal boundary resistances (Kapitza resistance) are other peculiarities under active study. The topic of cryogenics, most strictly interpreted, is the method of producing low temperatures. In addition to the considerable technology involved in refrigeration, there is a very important section of low-temperature physics concerned with temperature measurement and related instrumentation. *See* CRYOGENICS; KAPITZA RESISTANCE; KONDO EFFECT; LIQUID HELIUM; QUANTIZED VORTICES; QUANTUM SOLIDS; SCHOTTKY ANOMALY; SUPERCONDUCTIVITY; SUPERFLUIDITY.

Production of low temperature. Low temperatures such as those of liquid hydrogen (normal boiling point 20.4 K) and liquid helium (normal boiling point 4.2 K) are produced from the gas phase by the use of three mechanical devices: (1) the gas compressor or pump, (2) heat exchangers at several places along the path of flow of the gas, and (3) devices for expanding the gas such as a piston expander and a throttle valve. These are shown schematically in Fig. 1, where the flow of helium gas may be traced from the intake of the compressor, past the water-cooled heat exchanger, and into the liquefier.

A basic concept to grasp in understanding why the temperature drops when the gas is allowed to expand against the piston is that the process has occurred at constant entropy. Entropy S is a thermodynamic property of the system of gaseous molecules, and it is useful to think of its physical meaning as being that of the degree of disorder of

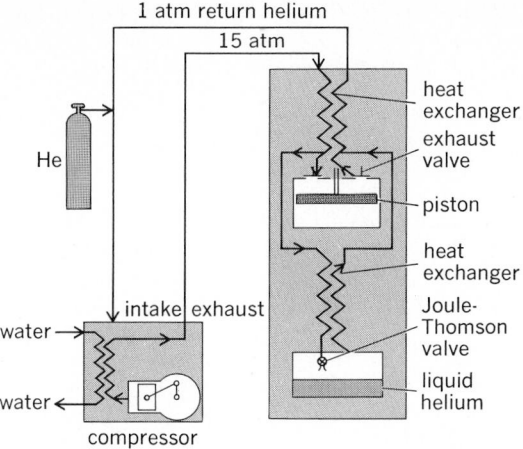

Fig. 1. Equipment for producing liquid helium. 1 atm= 101 kPa.

the system. The entropy-versus-temperature plot in Fig. 2 shows, for the gas system, a series of constant-pressure curves. The thermodynamic process at the compressor (with water cooling to keep the temperature constant) is shown as the path from point α to the point β. The entropy decrease $\Delta S = \Delta Q/T$, where ΔQ is the heat removed at temperature T during this reversible process that is assumed to be ideal. The diagram also shows an expansion of the gas against the piston as the path from point β to the point γ. This is done at constant entropy so that one may think of the pressure and temperature variables as having shifted suitably to preserve the degree of disorder. Energy was removed from the gaseous system by the work it did on the piston. Figure 1 shows still another expansion at the throttle valve or Joule-Thomson valve at which a further temperature drop and some liquefaction may occur. The throttle valve expansion is at constant enthalpy and is somewhat more subtle.

Thus, thermodynamic variables may be manipulated in a straightforward manner to produce a low temperature and liquefy gases such as helium. By vigorous pumping on the liquid helium, the vapor pressure may be greatly reduced, and the temperature of the liquid becomes of the order 1 K. Liquid ³He has a higher vapor pressure, for a given temperature, than ⁴He and so may be cooled by this same means to about 0.3 K without great difficulty. *See* ENTROPY.

Adiabatic demagnetization. To produce temperatures below 0.3 K, it is necessary to use another system which contains disorder (entropy) and to operate in such a way as to bring about order and so lower the entropy. Such a system is the set of electrons within a paramagnetic salt whose magnetic dipole moments may be oriented in a powerful magnetic field. The process (termed magnetic cooling or adiabatic demagnetization) is entirely analogous to that discussed above for the expansive cooling of compressed gases, and may be illustrated equally well by reference to an S-T diagram such as Fig. 2. The individual curves, however, will be characterized by values of magnetic induction B rather than pressure. Upon "demagnetizing," the order created by the powerful external field at 1 K is maintained by very weak magnetic fields within

the salt because of the very substantial drop in temperature.

A discussion of the electron spin system and magnetic cooling naturally leads to the matter of nuclear orientation and further reduction in temperature. But because typical nuclear magnetic moments are 1000 times smaller than electronic moments, nuclear cooling requires a matching increase in the initial value of the quotient B/T. A factor of 500, at least, is practicable by raising B fivefold and decreasing T from 1 K to 10 mK, either by means of an electronic first stage in a two-stage process or by using a ³He-⁴He dilution refrigerator. It is also possible to use a modest external field to magnetize the nuclear spin system via the nuclear hyperfine interaction. *See* ADIABATIC DEMAGNETIZATION.

Pomeranchuk cooling. Another system which retains substantial entropy even below 1 K is solid ³He. This entropy, S_s, remains at the value $R \ln 2$, where R is the gas constant, due to the nuclear spin degeneracy, until—taking the case along the melting curve—the temperature falls below 10 mK and the exchange interaction between spins becomes strong enough, relative to kT, where k is Boltzmann's constant, to cause the onset of nuclear ordering. Liquid ³He, on the other hand, behaves as a Fermi liquid with entropy S_l proportional to T, and S_l along the melting curve crosses the value $R \ln 2$ around 0.32 mK. This brings about the unusual situation that the melting curve, that is, the graph of melting pressure P_m versus temperature T, goes through a minimum at 0.32 K, $T_{\min}$, where, in accord with the Clausius-Clapeyron equation, the slope of the curve dP/dT falls to zero. Moreover, below 0.32 mK, where S_s is larger than S_l and thus dP/dT is negative, an increase in pressure will move one along the melting curve toward lower temperatures. By this method, one can reach temperatures as low as 1 mK. At the minimum, the pressure $P_{\min}$ is 2.93 MPa, and in the millikelvin region the melting curve is flat at a pressure value of 3.44 MPa. The method, conceived by I. Pomeranchuk in 1950, has been used successfully by numerous workers. Care must be taken to minimize frictional heating in the cell containing the liquid-solid mixture during compression. It is also necessary for the ³He cell to be in bellows form so that the enclosed fluid may be compressed mechanically; the simple direct approach would merely result in the pressurizing tube becoming plugged with solid ³He at the point where $T = T_{\min}$ as soon as P reached the value $P_{\min}$.

³He-⁴He dilution refrigerator. A method of producing very low temperatures by dissolving liquid ³He into liquid ⁴He was proposed by H. London, G. R. Clarke, and E. Mendoza in 1962, though it can be traced to earlier ideas by London. Experiments in 1965 and 1966 led to dilution refrigerators capable of continuously achieving a temperature of 10 mK. Subsequent improvements have pushed the limit to about 2.5 mK. The ³He-⁴He dilution refrigerator has opened up exciting research possibilities, especially for nuclear orientation and nuclear cooling. The operation of the dilution refrigeration depends on the basic properties of the superfluid formed by the isotope ⁴He in which the ³He is dissolved.

Figure 3 indicates the important components of

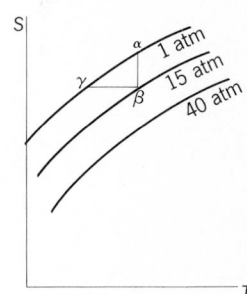

Fig. 2. Entropy-temperature plot for helium gas. 1 atm= 101 kPa.

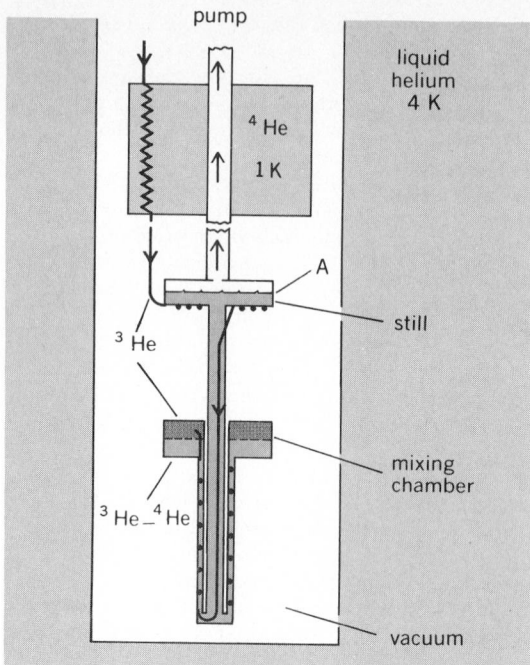

Fig. 3. Schematic drawing of a ³He–⁴He dilution refrigerator.

a ³He-⁴He dilution refrigerator. The refrigerator itself is surrounded by liquid helium of the normal and abundant isotope ⁴He at 4.2 K. There is also usually a shield of liquid nitrogen at 78 K. A special chamber (at the top of the drawing) contains liquid ⁴He at reduced vapor pressure and acts as a thermal dam holding a temperature of about 1 K. The dilution refrigerator suspended in the high-vacuum region below the 1 K thermal dam is to be viewed as a device into which liquid ³He is continuously injected and out of which an equivalent molar volume of ³He gas is extracted. The circulating ³He is contaminated by some ⁴He isotope, and attention must be given to the heat load caused by the contamination.

The two chambers marked "still" and "mixing chamber" are connected by a tube which also has the function of a heat exchanger. This tube doubles back on itself for efficient and essential operation of the device. When the refrigerator is put into operation, the still and mixing chambers are filled to the A level in the still with a mixture of 40% ³He and 60% liquid ⁴He. The gas phase above the liquid in the still is very rich in isotope ³He because of its high vapor pressure relative to ⁴He. When these vapors are pumped vigorously, the still begins to drop in temperature because of ³He evaporation. Because of the very high heat conductivity of superfluid liquid ⁴He along the connecting tube, the mixing chamber also drops in temperature. After a short time this cooling produces a temperature of about 0.6 K in the mixing chamber, and a solution very rich in ³He begins to separate out of the original mixture. Being less dense, the solution rich in ³He floats on top in the mixing chamber. The dashed line shows the separation, with nearly pure ³He on top of the now depleted ³He-⁴He mixture at the bottom of the mixing chamber. Figure 4 shows the phase separation diagram for ³He-⁴He mixtures. Remarkably, as $T \to 0$ K, the lower phase,

shown at the left side of Fig. 4, continues to contain about 6% of ³He in the mixture. Thus even at a very low temperature of, say, 0.015 K, the following may happen in the mixing chamber: If ³He atoms are removed from the lower region via the heat exchanger and the still, then more ³He will dissolve into the ³He-⁴He mixture and the latent heat of dilution will produce refrigeration sufficient to balance heat leaks or other sources of heat and so continuously maintain a temperature of 0.015 K. Thus the refrigeration produced in the mixing chamber by dissolving liquid ³He into liquid ³He-⁴He mixture is the reason that the temperature of the mixing chamber steadily drops from the initial value of about 0.6 K established at the still.

The upper layer of ³He in the mixing chamber is continuously replenished with nearly pure ³He. Figure 3 shows the supply of ³He coming through temperature cooling stages until it is so thoroughly precooled that it is delivered to the mixing chamber at very nearly that chamber's final steady low temperature—say about 0.015 K. Of course, the incoming ³He is just the recirculated ³He atoms (contaminated by ⁴He) which have been removed moments before from the still. Pumps do the external work of sucking on the still and compressing the ³He for the incoming supply line. Baths of liquid nitrogen and liquid helium are heat sinks for removing energy from the incoming ³He. The entropy of a small part of a total system has been lowered in order to produce low temperatures. The entropy of the total system, of course, has increased. *See* THERMODYNAMIC PRINCIPLES.

Low-temperature phenomena. Consider a solid specimen of matter having a collection of nuclei each with magnetic moment μ, which is taken as parallel to the spin angular momentum characterized by I (for example, the ⁷Li nuclei in a crystal of LiF). A considerable polarization will occur only if the magnetic energy μB is large compared to the thermal energy kT. Magnetic fields of 10 teslas must be combined with temperatures of 0.01 K before significant polarization of the nuclei will be

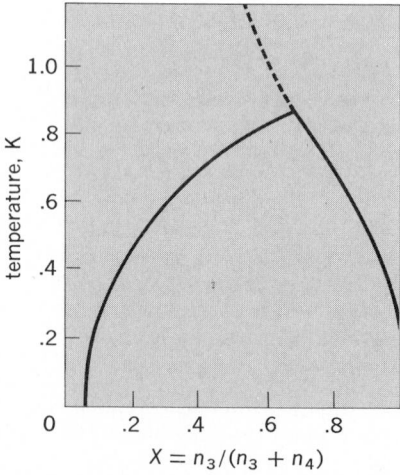

Fig. 4. Phase separation diagram for ³He–⁴He mixtures given by solid line such that at 0.6 K the heavy liquid will be at the bottom of the mixing chamber with about 3 atoms of ³He to 7 atoms of ⁴He. The dotted line represents superfluid properties to the left and normal liquid properties to the right; n_3 and n_4 in the equation refer to the concentrations of ³He and ⁴He, respectively.

obtained. An alternative approach to nuclear polarization is to obtain a larger magnetic field at the nucleus by using the intrinsic magnetic field of the atomic electrons (of the order 100 teslas at such close range). The interaction of these fields with μ gives rise to hyperfine structure in optical spectra. Thus at 0.01 K, almost all the nuclei are in the lowest energy state, which entails a parallel alignment of the nuclear and electronic moments in each individual atom. The application of a small external field of only 0.01 tesla will orient the electron spin system nearly to saturation value, and the coupled nuclear spins will be simultaneously spatially ordered; that is, a significant nuclear polarization appears. *See* HYPERFINE STRUCTURE; NUCLEAR ORIENTATION.

Nuclear orientation becomes particularly interesting if the oriented nuclei are radioactive. Anisotropy of the intensity of α-, β-, and γ-radiation from oriented nuclei has been observed experimentally. These experiments have led to important conclusions in the field of theoretical physics of nuclear interactions. *See* PARITY.

The properties of matter at low temperatures can be most surprising. There are a large number of metals which become superconductors at low temperatures such that the electric field **E** within the metal is zero. The magnetic field described by the induction vector **B** is also zero within the metal superconductor. These properties have been utilized to produce numerous measuring devices of unprecedented sensitivity. It is possible, too, to generate magnetic fields as intense as 15 teslas, with no dissipation of electric energy, using superconductors. Many metals exhibit at low temperatures an unusual diamagnetism with a peculiar periodic dependence on the applied magnetic field. The effect was first noted by W. de Haas and P. van Alphen and has been extensively studied by D. Shoenberg and others. Magnetic nuclear resonance studies on a number of systems have been useful in understanding the properties of matter at low temperature. *See* DE HAAS–VAN ALPHEN EFFECT; SQUID.

Liquid ^{4}He undergoes a change of state at the "lambda point," 2.17 K, below which it exhibits superfluid properties. In this new state, the liquid manifests enormous heat conductivity, flow unimpeded by viscosity, and numerous unusual modes for the transmission of thermal and acoustic waves. *See* LIQUID HELIUM; LOW-TEMPERATURE ACOUSTICS; SECOND SOUND; SUPERFLUIDITY.

D. D. Osheroff, R. C. Richardson, and D. M. Lee discovered the superfluid phases of liquid ^{3}He at temperatures below 2.6 mK and at pressures just below the approximate 30 bar needed to push the liquid into the solid state. Under zero pressure the superfluid transition occurs at 0.93 mK. To obtain these low temperatures below 3 mK, new methods had to be developed, such as adiabatic compression of the liquid-solid mixture of ^{3}He along the melting curve (Pomeranchuk's method) and adiabatic demagnetization of both electronic and nuclear spins, as discussed above.

[RALPH P. HUDSON]

Bibliography: C. J. Gorter (ed.), *Progress in Low Temperature Physics*, vols. 1–6, 1956–1970; O. V. Lounasmaa, Solid helium-three, *Contemp. Phys.*, 15:373–422, 1974; O. V. Lounasmaa, *Experimental Principles and Methods below 1 K*, 1974; K. Mendelssohn, *The Quest for Absolute Zero*, 1977; J. C. Wheatley, Experimental properties of superfluid ^{3}He, *Rev. Mod. Phys.*, 47:415–470, 1975.

Luminance

The luminous intensity of any surface in a given direction per unit of projected area of the surface viewed from that direction. The International Commission on Illumination defines it as the quotient of the luminous intensity in the given direction of an infinitesimal element of the surface containing the point under consideration, by the orthogonally projected area of the element on a plane perpendicular to the given direction. Simply, it is the luminous intensity per unit area. Luminance is also called photometric brightness.

Since the candela is the unit of luminous intensity, the luminance, or photometric brightness, of a surface may be expressed in candelas/cm^2, candelas/in.^{2}, and so forth.

Mathematically, luminance L may be found from the equation below, where θ is the angle be-

$$L = dI/(dA \cos \theta)$$

tween the line of sight and the normal to the surface area A considered and I is the luminous intensity. *See* LUMINOUS INTENSITY.

The stilb is a unit of luminance (photometric brightness) equal to 1 candela/cm^2. It is often used in Europe, but the practice in America is to use the term candela/cm^2 in its place.

The apostilb is another unit of luminance sometimes used in Europe. It is equal to the luminance of a perfectly diffusing surface emitting or diffusing light at the rate of 1 lumen/m^2. *See* PHOTOMETRY.

[RUSSELL C. PUTNAM]

Luminescence

Light emission that cannot be attributed merely to the temperature of the emitting body. Various types of luminescence are often distinguished according to the source of the energy which excites the emission. When the light energy emitted results from a chemical reaction, such as in the slow oxidation of phosphorus at ordinary temperatures, the emission is called chemiluminescence. When the luminescent chemical reaction occurs in a living system, such as in the glow of the firefly, the emission is called bioluminescence. In the foregoing two examples part of the energy of a chemical reaction is converted into light. There are also types of luminescence that are initiated by the flow of some form of energy into the body from the outside. According to the source of the exciting energy, these luminescences are designated as cathodoluminescence if the energy comes from electron bombardment; radioluminescence or roentgenoluminescence if the energy comes from x-rays or from γ-rays; photoluminescence if the energy comes from ultraviolet, visible, or infrared radiation; and electroluminescence if the energy comes from the application of an electric field. By attaching a suitable prefix to the word luminescence, similar designations may be coined to characterize luminescence excited by other agents. Since a given substance can frequently be made to luminesce by a number of different external exciting agents, and since the atomic and electronic phenomena that cause luminescence are basically the same regardless of the mode of excitation, the classification of luminescence phenomena into the foregoing categories is essentially only a matter of convenience, not of fundamental distinction.

When a luminescent system provided with a special configuration is excited, or "pumped," with sufficient intensity of excitation to cause an excess of excited atoms over unexcited atoms (a so-called population inversion), it can produce laser action. (Laser is an acronym for light amplification by stimulated emission of radiation.) This laser emission is a coherent stimulated luminescence, in contrast to the incoherent spontaneous emission from most luminescent systems as they are ordinarily excited and used. *See* LASER; OPTICAL PUMPING.

Fluorescence and phosphorescence. A second basis frequently used for characterizing luminescence is its persistence after the source of exciting energy is removed. Many substances continue to luminesce for extended periods after the exciting energy is shut off. The delayed light emission (afterglow) is generally called phosphorescence; the light emitted during the period of excitation is generally called fluorescence. In an exact sense, this classification, based on persistence of the afterglow, is not meaningful because it depends on the properties of the detector used to observe the luminescence. With appropriate instruments one can detect afterglows lasting on the order of a few thousandths of a microsecond, which would be imperceptible to the human eye. The characterization of such a luminescence, based on its persistence, as either fluorescence or phosphorescence would therefore depend upon whether the observation was made by eye or by instrumental means. These terms are nevertheless commonly used in the approximate sense defined here, and are convenient for many practical purposes. However, they can be given a more precise meaning. For example, fluorescence may be defined as a luminescence emission having an afterglow duration which is temperature-independent, while phosphorescence may be defined as a luminescence with an afterglow duration which becomes shorter with increasing temperature. For details on various types of luminescence *see* FLUORESCENCE; PHOSPHORESCENCE; THERMOLUMINESCENCE.

Type of radiation emitted. Because of their many practical applications, materials that give a visible luminescence have been studied and developed more intensively than those which emit in other spectral regions. Luminescence, however, may consist of radiation in any region of the electromagnetic spectrum. The production of x-radiation by the bombardment of a metal target by a fast electron beam is an example of luminescence. Certain fluorescent lamps, called black-light lamps, are coated with a luminescent powder chosen for its ability to emit ultraviolet light of approximately 360 nm rather than visible light. A number of luminescent solids have been developed which luminesce in the near infrared under excitation by visible light, by a cathode-ray beam, or by electric fields.

Luminescent substances. The ability to luminesce is not confined to any particular state of matter. Luminescence is observed in gases, liquids, and amorphous and crystalline solids. The passage of an electrical discharge through a gas will excite the gaseous atoms or molecules to luminesce under certain conditions. An example of such a gaseous luminescence is the mercury-vapor lamp, in which an electrical discharge excites the mercury vapor to emit both visible and ultraviolet light. Many liquids, such as oil or solutions of certain dyestuffs in various solvents, luminesce very strongly under ultraviolet light. A large number of solids, including many natural minerals as well as thousands of synthetic inorganic and organic compounds, luminesce under various types of excitation. Applications of gas luminescence, formerly confined to advertising signs (neon signs) and fluorescent lamps, have multiplied greatly with the invention of the laser, many atomic and molecular gases providing the laser-active materials. The same is true of dye solutions and inorganic gases. The major nonlaser applications of luminescence involve solid luminescent materials.

The term phosphor, originally applied to certain solids that exhibit long afterglows, has been extended to include any luminescent solid regardless of its afterglow properties. Other terms sometimes used synonymously with phosphor are luminophor, fluor, or fluorphor. The term luminophor is preferable, since it carries no connotation that afterglow times are long or short. In conformity with current usage, however, the term phosphor is used in the succeeding discussion of solid materials.

Comparatively few pure solids luminesce efficiently, at least at normal temperatures. In the category of organic solids, the pure aromatic hydrocarbons, such as naphthalene and anthracene, which consist exclusively of condensed phenyl rings, are luminescent, as are many heterocyclic closed-ring compounds. However, the closed phenyl ring structure is not in itself a guarantee of efficient luminescence, since certain substituents for hydrogen in the structure, particularly halogens, tend to reduce luminescence efficiency. This quenching of efficiency is called internal conversion. Among the pure inorganic solids that luminesce efficiently at room temperature are the tungstates, uranyl salts, platinocyanides, and a number of salts of the rare-earth elements.

Activators and poisons. The development of a large number of inorganic phosphors is due to the discovery that certain impurities, called activators, when present in amounts ranging from a few parts per million to several percent, can confer luminescent properties on the compounds (host or matrix compounds) in which they are incorporated. The activator and its nearby atoms are often referred to as the luminescent center.

By the same token, small amounts of other impurities or imperfections, called poisons or quenchers, can inhibit or destroy the luminescence, evidently by providing alternative mechanisms for the radiationless dissipation of the energy imparted to the phosphor or by emitting radiation characteristic of the poison itself which may not be in a spectral region of interest for the purpose at hand.

Manganese is a particularly effective activator in a wide variety of matrices when incorporated in amounts ranging from a small trace up to the order of several percent. The emission spectrum of these manganese-activated phosphors generally lies in the green, yellow, or orange spectral regions.

Other frequently used activators are copper, silver, thallium, lead, cerium, chromium, titanium, antimony, tin, and the rare-earth elements. Poisons in inorganic phosphors generally are from the iron-nickel-cobalt group. The most important host materials are silicates, phosphates, aluminates, sulfides, selenides, the alkali halides, and oxides of calcium, magnesium, barium, and zinc. Preparation of phosphors and the incorporation of the activators is generally done by high-temperature reaction of well-mixed, finely ground powders of the components. A number of semiconducting compounds such as gallium phosphide (GaP) and gallium arsenide (GaAs) can be made to luminesce and display laser action when prepared so that charges can be injected into the semiconductor. *See* SEMICONDUCTOR.

In order for ultraviolet light to provoke luminescence in a substance, the substance must be able to absorb it. Because of variations in its ability to absorb ultraviolet light of different wavelengths, a material may be nonluminescent under ultraviolet light of one wavelength and strongly luminescent under a different wavelength in the ultraviolet region. For a similar reason a substance may not luminesce under ultraviolet light at all and yet be strongly luminescent under x-ray irradiation.

LUMINESCENCE IN ATOMIC GASES

The processes that occur in luminescence may be most simply illustrated in the case of an atom in a gas. The atom can exist only in certain specific states of energy, some of which are shown schematically in Fig. 1. The lowest energy level of the atom, E_1, corresponds to the atom in its unexcited, or ground, state, and the higher energy levels, E_2 and E_3, represent electronically excited states of the atom. The excitation of the atom from state E_1 to state E_2 requires the absorption of an amount of energy $\Delta E = E_2 - E_1$. If this excitation is to be produced by the absorption of light, the energy of the acting light photon, E_{absorbed}, must equal ΔE. The frequency of the exciting light must therefore be $\nu = E_{\text{absorbed}}/h = (E_2 - E_1)/h$, and the wavelength of the exciting light must be $\lambda_{\text{absorbed}} = hc/(E_2 - E_1)$, where h is Planck's constant and c is the velocity of light. In an isolated atom this extra energy cannot be dissipated and is emitted as radiation when the atom eventually returns to its ground state. The emitted light will therefore again correspond in energy to ΔE and it will have a wavelength $\lambda_{\text{emitted}} = \lambda_{\text{absorbed}}$. When $\lambda_{\text{emitted}} = \lambda_{\text{absorbed}}$, the emitted light is sometimes referred to as resonance luminescence or resonance radiation. *See* ATOMIC STRUCTURE AND SPECTRA.

If a large number of atoms are excited and the excitation then removed, the luminescence intensity will decrease with time exponentially according to the equation $I_t = I_0 e^{-t/\tau_l}$, where I_t is the intensity of luminescence at a time t after removal of the excitation, I_0 is the intensity at $t = 0$, and τ_l is the average time required by an atom to make a spontaneous luminescent transition. The quantity τ_l, called the radiative lifetime, is independent of temperature, and it is this temperature independence that is emphasized in the more precise definition of fluorescence given earlier. If the transition between the energy levels E_1 and E_2 is highly probable (a so-called permitted or allowed transition), τ_l is very small, of the order of 10^{-8} sec for transitions involving visible light.

The excited atom can also lose a certain amount of energy and fall to an energy state E_3, of intermediate energy between E_1 and E_2. This can happen, for example, if the excited atom collides with another atom. If the transition from state E_3 to the ground state E_1 can occur with a reasonably high probability, fluorescence will occur starting from this intermediate excited state. In this case the fluorescent wavelength $\lambda_{\text{emitted}} = hc/(E_3 - E_1)$. Since $(E_3 - E_1)$ is smaller than $(E_2 - E_1)$, the fluorescence in this case will be of longer wavelength than the resonance radiation. Although the quantum efficiency of luminescence is unity (one photon of emitted light per photon of absorbed light), the energy efficiency is less than unity.

If, however, a transition between state E_3 and state E_1 is highly improbable (a so-called forbidden transition), state E_3 is known as a metastable state. The atom can remain in this state for long periods of time and cannot return to the ground state with the emission of radiation. Luminescence can occur under these circumstances only if the atom regains the energy $(E_2 - E_3)$ by a collision with another atom or by some other process. Once the atom has been brought back to state E_2, a transition to the ground state is again allowed, and luminescence corresponding to $(E_2 - E_1)$ will be emitted. The existence of metastable states like E_3 explains the delayed emission termed phosphorescence. The atom may spend a considerable amount of time in such a state before some external influence causes it to return to an emitting state such as E_2, in which case the luminescence is correspondingly delayed and appears as an afterglow. In order for the atom to get from E_3 to E_2 it must absorb energy somehow. The rate of return to the emitting state, and hence the duration of the afterglow, will therefore depend to a very large extent on temperature. At high temperatures the atoms will be excited back to the emitting state rapidly, and there will be a bright afterglow of short duration. At lower temperatures the atoms will be raised back to the emitting state very slowly, and the afterglow will be of long duration but of low intensity. This temperature dependence is the basis for the more precise definition of phosphorescence that was given

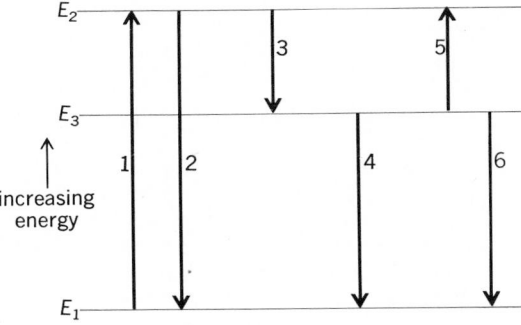

Fig. 1. Schematic representation of energy levels and electronic transitions in an atomic gas. E_1, ground state; E_2 and E_3, excited states; 1, excitation; 2, emission of resonance luminescence; 3, radiationless transition to lower excited state; 4, luminescence emission, if transition $E_3 \rightarrow E_1$ is allowed (if it is not allowed, 4 does not occur and E_3 is called a metastable state); 5, stimulation of atom back to emitting state; 6, radiationless transition of atom back to ground state (quenching).

earlier. As an alternative to regaining the energy $(E_2 - E_3)$, the atom may lose the energy $(E_3 - E_1)$ by a competing radiationless process, for example, by collision with another atom. The energy of excitation is thus dissipated without luminescence, and the quantum yield of luminescence is zero (quenching).

The principles illustrated in the foregoing discussion may be extended with little modification to the case where the primary absorption or excitation act completely removes an electron from an atom (that is, ionizes the atom) instead of merely raising the atom to an excited state. Under these conditions the electron can be trapped temporarily by other atoms, and its return to the parent atom can also be delayed by this mechanism.

The same principles are also operative in the case of complex configurations of atoms, such as in organic molecules or solids. However, the forbiddenness of radiative transitions can be modified in these cases, and efficient luminescent emissions can consequently be observed due to radiative transitions from metastable states, albeit with longer afterglows.

When nonlinear optics conditions exist, luminescence can be excited in certain systems by multiphoton absorption, the resulting luminescence being of higher frequency than the exciting light because the energy of two or more photons of the latter are combined to give one photon of emitted light. There are also cases where a single photon is absorbed by a pair of activator atoms or ions, each of which then emits a luminescence photon, leading to a quantum efficiency of 2. However, the preceding discussion and the exposition of those principles which follow deal primarily with the more usual case of single-photon absorption and emission. *See* NONLINEAR OPTICS.

CONFIGURATION COORDINATE CURVE MODEL

Luminescence in atomic gases is adequately described by the concepts of atomic spectroscopy, but luminescence in molecular gases, in liquids, and in solids introduces two major new effects which need special explanation. One is that the emission band appears on the long wavelength (low-energy) side of the absorption band; the other is that emission and absorption often show as bands tens of nanometers wide instead of as the lines found in atomic gases.

Both of these effects may be explained by using the concept of configuration coordinate curves illustrated in Fig. 2. As in the case of atomic gases, the ground and excited states represent different electronic states of the luminescent center, that is, the region containing the atoms, or electrons, or both, involved in the luminescent transition. On these curves the energy of the ground and excited state is shown to vary parabolically as some configuration coordinate, usually the distance from the activator to its nearest neighbors, is changed. There is a value of the coordinate for which the energy is a minimum, but this value is different for the ground and excited states because of the different interactions of the activator with its neighbors. Absorption of light gives rise to the transition from A to B. This transition occurs so rapidly that the ions in the luminescent center do not have time to rearrange. Once the system is at B it gives up heat energy to its surroundings by

means of lattice vibrations and reaches the new equilibrium position at C. Emission occurs when the system makes the transition from C to D, and once again heat energy is given up when the system goes from D back down to A. This loss of energy in the form of heat causes the energy associated with the emission $C \rightarrow D$ to be less than that associated with the absorption $A \rightarrow B$. *See* FRANCK-CONDON PRINCIPLE; LATTICE VIBRATIONS.

When the system is at an equilibrium position, such as C of the excited-state curve, it is not at rest but migrates over a small region around C because of the thermal energy of the system. At higher temperatures these fluctuations cover a wider range of the configuration coordinate. As a result, the emission transition is not just to point D on the ground state curve but covers a region around D. In the vicinity of D the ground state curve shows a rapid change of energy, so that even a small range of values for the configuration coordinate leads to a large range of energies in the optical transition. This explains the broad emission and absorption bands that are observed. An analysis of this sort predicts that the widths of the band (usually measured in energy units between the points at which the emission or absorption is half its maximum value) should vary as the square root of the temperature. For many systems this relationship is valid for temperatures near and above room temperature. At low temperatures, quantum-mechanical effects, described below, become dominant.

Two other phenomena which can be explained on the basis of the model described in Fig. 2 are temperature quenching of luminescence and the variation of the decay time of luminescence with temperature. In Fig. 3 a curve of temperature quenching for the emission in thallium-activated potassium chloride is shown. At low temperatures

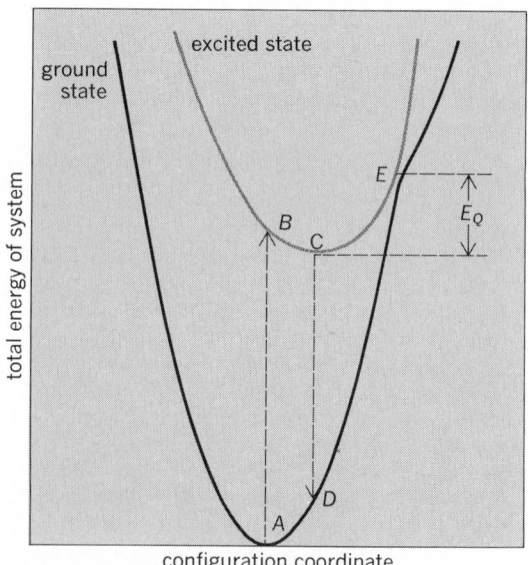

Fig. 2. Configuration coordinate curves for a simple luminescence center. Transition $A \rightarrow B$ shows absorption, and transition $C \rightarrow D$ shows emission. Energy E_Q is that necessary for the excited state to reach point E, from which a transition to the ground state can be made without luminescence.

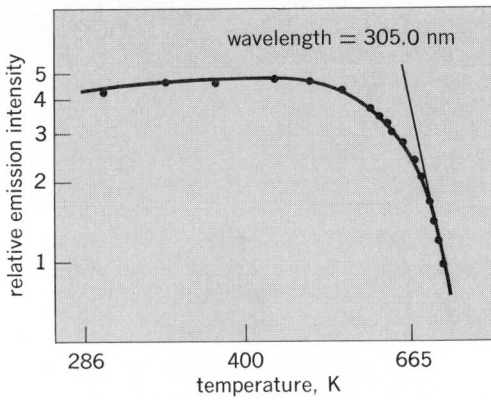

Fig. 3. Variation of the brightness of a thallium-activated potassium chloride phosphor with temperature.

the temperature is reduced to absolute zero (0 K). This is not the case, as is shown in Fig. 4, which gives the width of the absorption band of the *F*-center in potassium chloride as a function of the square root of the temperature. (An *F*-center is the simplest type of color center, a color center being a lattice defect which can absorb light.) At high temperatures the previously quoted results are valid, but at low temperatures the width of the band is constant. *See* COLOR CENTERS.

This phenomenon may be explained by treating the configuration coordinate curves quantum mechanically. The curves have the energy versus displacement characteristics of the simple harmonic oscillator. For this case the quantum-mechanical analysis shows that there is a series of equally spaced energy levels separated by an energy of $h\nu$, where h is Planck's constant and ν is the frequency of vibration. The lowest of these levels occurs at a value of $1/2\ h\nu$ above the minimum of the classical curve, and this energy is called the zero point energy. Its importance is that

there is very little change in brightness with temperature, but at elevated temperatures the luminescence efficiency decreases rapidly (so-called thermal quenching). On the scheme of Fig. 2 this is interpreted as meaning that the thermal vibrations become sufficiently intense to raise the system to point *E*. From point *E* the system can fall to the ground state by emitting a small amount of heat, or infrared radiation. If point *E* is at an energy E_Q above the minimum of the excited state curve, it may be shown that the quantum efficiency, η, of luminescence is given by Eq. (1), where *C* is a

$$\eta = 1 + C \exp\left(-E_Q/kT\right)^{-1} \quad (1)$$

constant, k is Boltzmann's constant, and T is the temperature on the Kelvin scale. By fitting an expression of this form to the data of Fig. 3, a value of 0.60 eV is obtained for E_Q.

Another result of the onset of thermal quenching is that the luminescence decays faster, since the excited state is now depopulated by two processes simultaneously—a dissipative process in parallel with the luminescent process.

Quantum-mechanical corrections. Although the configuration coordinate curve model of Fig. 2 is successful in describing many aspects of luminescence in solids, it predicts that emission and absorption bands should become narrow lines as

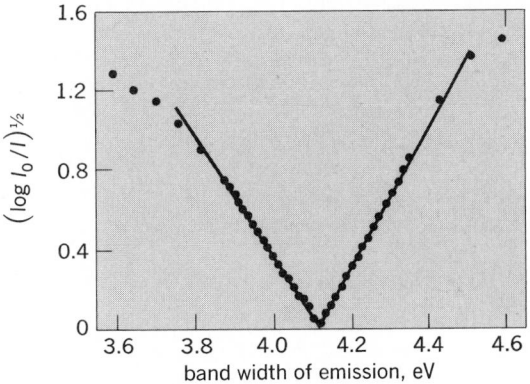

Fig. 5. Emission of thallium-activated potassium chloride at 4 K. In a plot of this particular form, a Gaussian curve would consist of two straight lines which make equal angles with the abscissa.

even at absolute zero the system is not at rest but varies over a range of configuration coordinates characteristic of this lowest vibrational level. Analysis shows that under these conditions the widths of the bands, ΔE, vary as in Eq. (2), where A is a

$$\Delta E = A[\coth\left(h\nu/2kT\right)]^{1/2} \quad (2)$$

constant. A curve of this form is drawn through the experimentally obtained points of Fig. 4 and shows satisfactory agreement.

One other result of the introduction of quantum mechanics is that this simple model predicts that both absorption and emission bands should be Gaussian in shape at all temperatures; that is, they should be of the form of Eq. (3), where I is the

$$I = I_0 \exp\left[-A(E - E_0)^2\right] \quad (3)$$

emission intensity or absorption strength for light of energy E, and I_0 and E_0 refer to corresponding quantities at the maximum of the curve. Figure 5 shows the emission spectrum of thallium-activated potassium chloride plotted in such a way that the expression of Eq. (3) would give two straight lines making equal angles with the horizontal. Although there is some disagreement in the wings of the

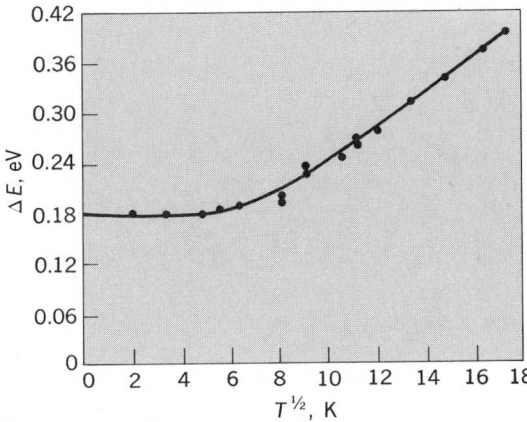

Fig. 4. Variation of the width of the *F*-center absorption band in potassium chloride at its half-maximum points, ΔE, as a function of the square root of the temperature.

emission spectrum, and although the lines are not quite at the same angle, there still is fairly good agreement with the predictions of Eq. (3).

High dielectric constant materials. The use of configuration coordinate curves is justifiable only when the electron taking part in an optical transition is tightly bound to a luminescent center and interacts primarily with its nearest neighbors. This appears to be generally the case in materials with low dielectric constants such as the alkali halides. For high dielectric constant materials the situation is very different. It has been estimated that for boron in silicon the electron is spread over about 500 atoms so that its interaction with the nearest neighbors is small. In these cases the center interacts with the lattice during an optical transition through the creation of many vibrational phonons (sound quanta) at relatively large distances from the center. A case of this sort is illustrated in Fig. 6, which shows edge emission in cadmium sulfide near the absorption edge of the material. The major peaks correspond to an optical transition with simultaneous emission of 0, 1, 2, and 3 phonons, respectively, starting with the highest peak. The peaks are equidistant in energy, and the spacing is just that to be expected for the creation of phonons. Although both short- and long-range interactions of a center with its environment occur in all cases, the short-range interaction dominates for tightly bound electrons in low dielectric constant materials, while the long-range interaction dominates for loosely bound electrons in high dielectric constant materials. *See* PHONON.

SENSITIZED LUMINESCENCE

A process of considerable interest occurs in systems where one type of activator absorbs the exciting light and transfers its energy to a second type of activator which then emits. The transfer does not involve motion of electrons. This process is often called sensitized luminescence, and it is widespread in both inorganic and organic luminescent systems. The principal phosphors used in fluorescent lamps are of this type. The results on calcium carbonate ($CaCO_3$) shown in Fig. 7 illustrate this kind of system. If divalent manganese (Mn^{++}) alone is incorporated as an activator into $CaCO_3$, the system gives only a very weak Mn^{++}

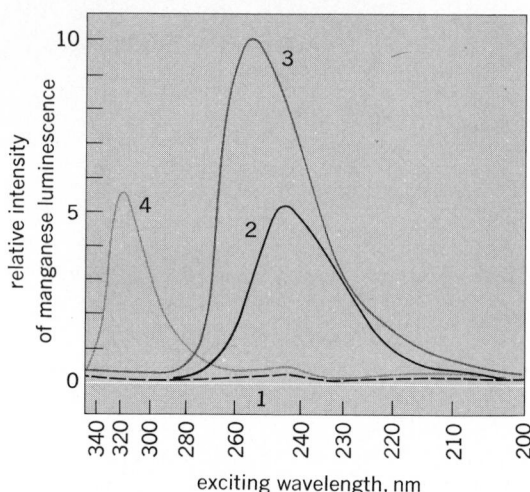

Fig. 7. Excitation spectra for emission of calcium carbonate phosphors, using manganese as an activator, with different impurity sensitizers: 1, no sensitizer; 2, thallium; 3, lead; and 4, cerium.

emission for all wavelengths of exciting ultraviolet light. This is because the Mn^{++} ion has only forbidden transitions in this spectral region, as shown by its very low absorption strength and slow decay of luminescence. In other words, the Mn^{++} does not absorb very much ultraviolet light. Bright phosphors may be prepared by incorporating divalent lead (Pb^{++}), monovalent thallium (Tl^+), or trivalent cerium (Ce^{+++}) along with Mn^{++} in $CaCO_3$. In each case the Mn^{++} emission is the same, but the wavelengths which excite these phosphors vary as the other activator (called the sensitizer) is changed. All these absorptions occur for energies well below the energy at which free electrons are formed in the solid. The ultraviolet light is absorbed by the Pb^{++}, Tl^+, or Ce^{+++} sensitizers, which have allowed transitions leading to strong absorption of the light; this absorbed energy is then passed on to nearby Mn^{++} activators and excite the Mn^{++} emission. It is important to know over what distances the energy may be transferred from absorber to emitter. Experiments on calcium silicate with Pb^{++} and Mn^{++} as added impurities have shown that if a Mn^{++} ion is in any one of the 28 nearest lattice sites around the Pb^{++} ion, the energy transfer may take place.

Resonant transfer. The transfer of energy has been treated theoretically using quantum mechanics and assuming a model of resonance between coupled systems. A number of cases have been examined; in each case the sensitizer has an allowed transition since only in this case will it absorb the exciting light appreciably.

The first case is one in which the activator undergoes an allowed transition. The probability of energy transfer, P (defined as the reciprocal of the time required for a transition to occur) varies as in expression (4). Here R is the distance between

$$P \propto \frac{1}{R^6 n^4} \int f_s F_A \, dE \qquad (4)$$

sensitizer and activator; n is the index of refraction of the host material, which may be liquid or solid; f_s is a function describing the emission band of the sensitizer as a function of energy, and is normal-

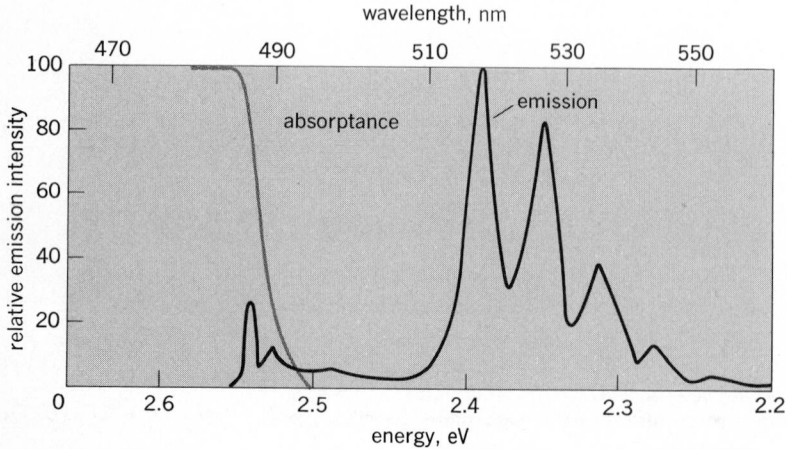

Fig. 6. Graph showing the absorptance (100 minus percent of transmission) and edge emission of cadmium sulfide at 4 K.

ized so that the area under the curve is unity; and F_A is a function describing the absorption band of the activator, also normalized to unity. The integral of expression (4) measures the overlap of these bands and determines the resonance transfer. In typical cases the sensitizer will transfer energy to an activator if the activator occupies any one of the 1000–10,000 nearest available sites around the sensitizer. Another case is that of a forbidden "quadrupole" transition in the activator; in this case the number of sites for transfer would be about 100. If the transition in the activator is even more strongly forbidden, quantum-mechanical "exchange" effects predominate and the number of available sites for transfer should be reduced to 40 or less. In both of the cases just mentioned the integral measuring the overlap enters in the same way as it does in expression (4). From this theoretical treatment it appears that phosphors with Mn^{++} as the activator probably receive their energy by exchange interactions.

Concentration quenching. Another phenomenon related to the resonant transfer of energy is that of concentration quenching. If phosphors are prepared with increasing concentrations of activators, the brightness will first increase but eventually will be quenched at high concentrations. It is believed that at high concentrations the absorbed energy is able to move from one activator to a nearby one by resonant transfer and thus migrate through the solid or liquid. If there are "poisons," or quenching sites, distributed in the material, the migrating energy may reach one of these, be transferred to it, and dissipate without luminescence. Impurity atoms, vacancies, jogs at dislocations, normal lattice ions near dislocations, and even a small fraction of the activator ions themselves, when associated in pairs or higher aggregates, can act as poisons. As the concentration is increased, the speed of migration is increased, and the quenching process becomes increasingly important.

LUMINESCENCE INVOLVING ELECTRON MOTION

In an important group of luminescent materials the transfer of energy to the luminescent center is brought about by the motion of electrons. Many oxides, sulfides, selenides, tellurides, arsenides, and phosphides are of this type, and of these, zinc sulfide has been most widely studied. It is frequently used as the luminescent material in cathode-ray tubes and electroluminescent lamps.

Electronic processes in insulating materials are described by a band model such as that illustrated in Fig. 8. The electron energy increases vertically and the horizontal dimension shows position in a crystal. In the shaded area, called the filled band, or the valence band, all the energy levels are filled with electrons. No electron can be accelerated or moved to higher energies within this band since the higher levels are already filled. Thus the material is an insulator. Above the valence band is an energy region called the forbidden band, which has no energy levels in it for pure materials. However, imperfections may introduce a local energy level in this region, as illustrated by the short line above 3 in Fig. 8. Above the forbidden band is an energy region called the conduction band. Here there are energy levels but, in an insulator, no electrons. If an electron in the filled band absorbs light

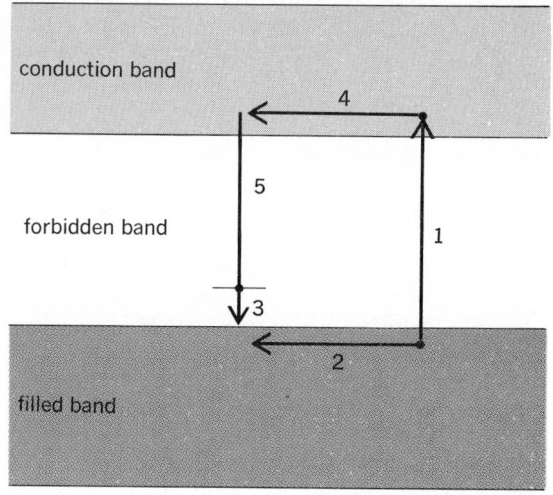

Fig. 8. Energy-band model for luminescence processes in zinc sulfide. Electron energy increases vertically.

of sufficiently high energy, it may jump up into the conduction band, as shown in 1 of Fig. 8. The empty position left behind in the filled band, called a hole, has properties which allows it to be described as an electron except that it has a positive charge. Both the electron in the conduction band and the hole in the filled band are free to move and gain energy. As a result, current can flow when a voltage is applied externally. The electron and hole can also diffuse far from their origin and can thus transport energy to a distant luminescent center. *See* HOLES IN SOLIDS.

A simple luminescent transition is illustrated in Fig. 8. Assume that the impurity level (black dot in the forbidden band) is due to a luminescent center and that there is an electron in the level at the beginning of the process. Transition (1) shows the creation of a free electron and hole due to the absorption of light. The hole migrates to the center (2), and the electron in the impurity level falls into the hole (3), thus destroying it. The free electron now migrates toward the center (4) and falls into it (5), giving off luminescence. The cycle is complete and the center once again has an electron. It is important to note that in this process both electrons and holes must be assumed to move.

In zinc sulfide the normal activators are monovalent metals such as Ag^+ or Cu^+. They replace Zn^{++} ions. To maintain electrical neutrality it is necessary also to incorporate into the lattice ions such as Cl^- for S^{--} or Al^{+++} for Zn^{++}. These additional ions are called coactivators. The situation is now quite complex, and is illustrated for ZnS with Cu and Al as added impurities in Fig. 9. The various arrows in this figure show some of the electronic transitions that give rise to observable effects. Arrows 1 and 4 show two different luminescent transitions. Arrow 2 shows the excitation of an electron into the luminescent center, which may occur by absorption of infrared light or by thermal fluctuations. In either case the center is prevented from luminescing and thus is quenched. Transition 3 is the excitation of an electron into the conduction band. The electron may be excited by thermal energy and its appearance may be detected by the appearance of transition 1. This process is called thermoluminescence.

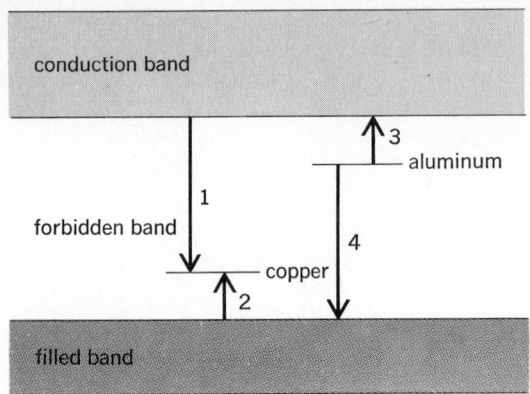

Fig. 9. Energy-band model for luminescence processes in zinc sulfide which uses copper as the activator and aluminum as the coactivator.

The wide variety of luminescence phenomena, the insight they give into the constitution of very different classes of materials, and their many important applications to light production, imaging devices, and radiation detectors make luminescence a perennially challenging field of study.

[CLIFFORD C. KLICK; JAMES H. SCHULMAN]

Bibliography: R. S. Becher, *Theory and Interpretations of Fluorescence and Phosphorescence*, 1969; J. H. Crawford, Jr., and L. M. Slifkin (eds), *Point Defects in Solids*, vol. 1, 1972; D. Curie, *Luminescence in Crystals*, 1960; R. J. Elliott and A. F. Gibson, *An Introduction to Solid State Physics*, 1974; P. G. Goldberg (ed.), *Luminescence of Inorganic Solids*, 1966; H. W. Leverenz, *An Introduction to the Luminescence of Solids, Fluorescence and Phosphorescence*, 1949; F. Seitz and D. Turnbull (eds), *Solid State Physics*, vol. 5, 1957; S. Shionoya et al. (eds), *Proceedings of the 1975 International Conference on Luminescence, Tokyo*, 1976.

Luminous efficacy

There are three ways this term can be used: (1) The luminous efficacy of a source of light is the quotient of the total luminous flux emitted divided by the total lamp power input. Light is visually evaluated radiant energy. Luminous flux is the time rate of flow of light. Luminous efficacy is expressed in lumens per watt. (2) The luminous efficacy of radiant power is the quotient of the total luminous flux emitted divided by the total radiant power emitted. This is always somewhat larger for a particular lamp than the previous measure, since not all the input power is transformed into radiant power. (3) The spectral luminous efficacy of radiant power is the quotient of the luminous flux at a given wavelength of light divided by the radiant power at that wavelength. A plot of this quotient versus wavelength displays the spectral response of the human visual system. It is, of course, zero for all wavelengths outside the range from 380 to 760 nanometers. It rises to a maximum near the center of this range. Both the value and the wavelength of this maximum depend on the degree of dark adaptation present. However, an accepted value of 683 lumens per watt maximum at 555 nanometers represents a standard observer in a light-adapted condition. The reciprocal of this maximum spectral luminous efficacy of radiant power is sometimes referred to as the mechanical equivalent of light, with a probable value of 0.00147 watt per lumen.

If the spectral luminous efficacy values at all wavelengths are each divided by the value at the maximum (683 lumens per watt), the spectral luminous efficiency of radiant power is obtained. It is dimensionless.

For purposes of illuminating engineering, light is radiant energy which is evaluated in terms of what is now simply and descriptively the spectral luminous efficiency curve, V_λ. In this case it is the intent that an illuminating engineer have a measure for the capacity or capability for the production of luminous flux from radiant flux, or that radiant flux has efficacy for the production of luminous flux.

Since the human eye is sensitive only to radiations of wavelengths between about 400 and 700 nanometers, it follows that the sensation of sight evoked by a radiator is due only to the radiation within this limited wavelength band. If all the wavelengths within this band were equally effective for the purpose of vision, then the area under the graph of radiant power distribution would be a measure of the light output.

The illustration shows the spectral distribution curve for an electric tungsten filament lamp. The total area under the curve between the wavelength limits $\lambda = 0$ and $\lambda = \infty$ is proportional to the power of the total radiation. In illustration *a* the shaded area between the limits of $\lambda = 400$ and $\lambda = 700$ nanometers is a measure of the power of visible radiation. However, this is not the same thing as luminous efficacy because of the variable sensitivi-

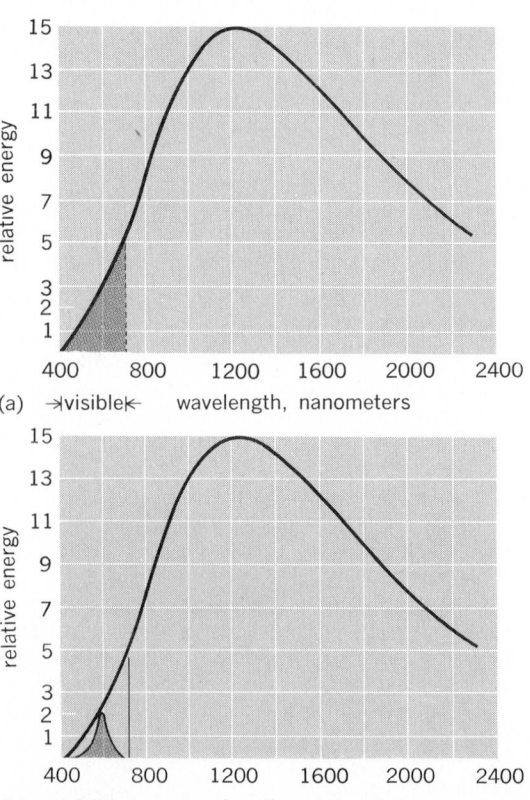

Spectral energy distribution of electric filament lamp. (*a*) Colored area is measure of power of visible radiation. (*b*) Colored area is measure of light output.

ty of the eye with respect to wavelength. If each ordinate, say at 10-nanometer wavelength intervals, is multiplied by the appropriate value of V_λ, a new distribution is obtained, as shown in the shaded area of illustration *b*. This area is a measure of the light output of the source. The ratio of the shaded area of illustration *b* to the area under the larger curve is the luminous efficacy in terms of lumens per watt, and this is the method of determination that is generally used. The name luminous efficacy expresses exactly what is meant by the effectiveness of 1 watt of radiant power in producing luminous flux. *See* LUMINOUS EFFICIENCY; LUMINOUS FLUX; PHOTOMETRY.

[G. A. HORTON]

Luminous efficiency

Visual efficacy of visible radiation, a function of the spectral distribution of the source radiation in accordance with the "spectral luminous efficiency curve," usually for the light-adapted eye or photopic vision, or in some instances for the dark-adapted eye or scotopic vision. Account is taken of this fact in photometry, or the measurement of light. Actually, instead of indicating directly the physical radiant power of light, photometers are designed to indicate the corresponding luminous flux of the visually effective value of the radiation.

The spectral luminous efficiency of radiant flux is the ratio of luminous efficacy for a given wavelength to the value of maximum luminous efficacy. It is a dimensionless ratio. Values of spectral luminous efficiency for photopic vision V_λ at 10-nanometer-wavelength intervals were provisionally adopted by the International Commission on Illumination in 1924 and by the International Committee on Weights and Measures in 1933, as a basis for the establishment of photometric standards of types of sources differing from the primary standard in spectral distribution of radiant flux. These standard values of spectral luminous efficiency were determined by visual photometric observations with a 2° photometric field having a moderately high luminance (photometric brightness); photometric evaluations based upon them, consequently, do not agree exactly to other conditions of observations, such as in physical radiometry. Watts weighted in accordance with these standard values are often referred to as light-watts.

Values of spectral luminous efficiency for scotopic vision V'_λ at 10-nanometer intervals were provisionally adopted by the International Commission on Illumination in 1951. These values of spectral luminous efficiency were determined by visual observations by young, dark-adapted observers, using extra-foveal vision at near-threshold luminance.

The main difference between the terms luminous efficacy and luminous efficiency is that the latter came into use a century or so before the former. It should also be noted that in engineering, general values of efficiency never exceed 1.0, and illuminating engineers were the only ones who had values of efficiency up to 683, but this has now been changed by the usage of the term luminous efficacy. In view of the above discussion, the term luminous efficiency is evidently a misnomer. *See* LUMINOUS EFFICACY; PHOTOMETRY.

[G. A. HORTON]

Bibliography: Illuminating Engineering Society, *American National Standard Nomenclature and Definitions for Illuminating Engineering*, ANSI/IES R16–1980, 1980.

Luminous energy

The radiant energy in the visible region or quantity of light. It is in the form of electromagnetic waves, and since the visible region is commonly taken as extending 380–760 nanometers (millimicrons) in wavelength, the luminous energy is contained within that region. It is equal to the time integral of the production of the luminous flux. *See* PHOTOMETRY.

[RUSSELL C. PUTNAM]

Luminous flux

The time rate of flow of light. It is radiant flux in the form of electromagnetic waves which affects the eye or, more strictly, the time rate of flow of radiant energy evaluated according to its capacity to produce visual sensation. The visible spectrum is ordinarily considered to extend from 380 to 760 nanometers (millimicrons) in wavelength; therefore, luminous flux is radiant flux in that region of the electromagnetic spectrum. The unit of measure of luminous flux is the lumen. *See* PHOTOMETRY.

[RUSSELL C. PUTNAM]

Luminous intensity

The solid angular luminous flux density in a given direction from a light source. It may be considered as the luminous flux on a small surface normal to the given direction, divided by the solid angle (in steradians) which the surface subtends at the source of light. Since the apex of a solid angle is a point, this concept applies exactly only to a point source. The size of the source, however, is often extremely small when compared with the distance from which it is observed, so in practice the luminous flux coming from such a source may be taken as coming from a point. For accuracy, the ratio of the diameter of the light source to the measuring distance should be about 1:10, although in practice ratios as large as 1:5 have been used without excessive error.

Mathematically, luminous intensity I is given by the equation below, where ω is the solid angle

$$I = dF/d\omega$$

through which the flux from the point source is radiated, and F is the luminous flux. The luminous intensity is often expressed as candlepower. *See* CANDLEPOWER; PHOTOMETRY.

[RUSSELL C. PUTNAM]

Mach number

In fluid mechanics, the ratio v/c of the free stream velocity v to the velocity of sound c in the fluid at the same condition, such as temperature and pressure. Mach number is also the ratio of the inertia force of the fluid to the force of compressibility or the elastic force. In most fluid systems compressibility effects become important for values of the Mach number greater than about 0.3. A body moving through a fluid at a velocity less than sonic is preceded by a region of gradually varying density and pressure that controls the flow around the body. At Mach numbers equal to or greater than unity the gradual transition of pressure cannot exist, and shock waves, or regions of abruptly altered pressure and density, form at critical sections on or near the surface of the body and extend outward. As a result, the fluid force pattern on the moving body is markedly different at supersonic velocities from the pattern at subsonic velocities. The theory for calculating the pressure patterns is well established for ideal fluids. *See* SHOCK WAVE.

When compressibility effects alone are significant, geometrically similar bodies will develop identical flow and shock wave patterns when operated at equal Mach numbers. However, in the hypersonic region, generally considered to mean a Mach number of 5 or more, there is appreciable interaction between the boundary layer and the shock pattern. When this occurs, the Reynolds number as well as the Mach number is significant and the similitude requirements become more difficult to satisfy. *See* DYNAMIC SIMILARITY.

Techniques of transformation of variables have been devised whereby boundary conditions may be altered to accommodate nonconformance to the Mach requirement. That is, data from incompressible flow may be used to give accurate predictions under compressible-flow conditions using distortion of geometry to balance distortion of velocity.

[GLENN MURPHY]

Madelung constant

A numerical constant α_M in terms of which the electrostatic energy U of a three-dimensional periodic crystal lattice of positive and negative point charges q_+, $-q_-$, N in number, is given by Eq. (1),

$$U = -\frac{1}{2}\frac{Nq_+q_-}{d}\alpha_M \qquad (1)$$

where d is the nearest-neighbor distance between positive and negative charges and N is large. Knowledge of such electrostatic energies as given by the Madelung constant is of importance in the calculation of the cohesive energies of ionic crystals and in many other problems in the physics of solids.

Designate the lattice sites by indices i or j, and let the distance between sites i and j be given by r_{ij}. Then U is given by the sum of the Coulomb interaction energies of all pairs of charges as shown in Eq. (2). In the summation, i and j range over all

$$U = \frac{1}{2}\sum_{i,j}{}' \frac{q_iq_j}{r_{ij}} \qquad (2)$$

sites in the lattice, and the prime on the summation sign indicates the exclusion of terms for which $i=j$. The factor $1/2$ avoids counting pairs of charges twice. In Eq. (2) $q_i = q_+$ or q_-, depending upon whether the lattice site i is occupied by a positive or a negative ion. From this expression for U it can be seen by comparison of Eq. (2) with Eq. (1) that Eq. (3) holds.

$$\alpha_M = -\frac{1}{2}\sum{}'\left(\frac{q_i}{q_+}\right)\left(\frac{q_j}{q_-}\right)\left(\frac{d}{r_{ij}}\right) \qquad (3)$$

This number is called the Madelung constant after E. Madelung, who first calculated the sum in connection with the cohesive energies of ionic crystals. It is characteristic of the lattice structure

Madelung constants for some common ionic crystals

Crystal structure	Madelung constant, α_M
Sodium chloride, NaCl	1.7476
Cesium chloride, CsCl	1.7627
Zinc blende, α-ZnS	1.6381
Wurtzite, β-ZnS	1.641
Fluorite, CaF_2	5.0388
Cuprite, CuO_2	4.1155
Rutile, TiO_2	4.816
Anatase, TiO_2	4.800
Corundum, Al_2O_3	25.0312

but independent of the dimensions of the lattice. Its numerical value does, however, depend on the choice of the unit distance within the crystal, d. In Eq. (1), d has been chosen to be the nearest-neighbor positive-negative charge separation. Occasionally other choices of d are used instead, such as the cube root of the molecular volume or the unit cube edge in cubic crystals.

The calculation of Madelung constants requires some care since they are given by slowly and conditionally convergent series. Attempts at direct summation can involve thousands of terms without impressive accuracy. Various ingenious schemes for calculation have been used, but the most general and powerful is Ewald's method, also called the theta function method.

The Madelung constants for a number of common ionic crystal structures are given in the table. For these cases d is chosen as the nearest-neighbor distance. *See* CRYSTAL STRUCTURE.

[B. GALE DICK]

Bibliography: *See* IONIC CRYSTALS.

Magic numbers

Numbers of neutrons or protons in nuclei which correspond to particularly stable structures and closed shells. In nature the magic numbers for both neutrons and protons are 2, 8, 14, 20, 50, and 82. The next neutron magic numbers are 126 and 184; the next proton magic numbers are expected to be 114 and perhaps 164. Nuclei having magic numbers of both neutrons and protons have been found to have spherical equilibrium shapes and special stability; they have anomalously low capture probabilities for additional neutrons and protons respectively. *See* NUCLEAR STRUCTURE.

Magic number effects were first observed as systematic trends in neutron-capture cross sections (see illustration), in nuclear electric quadrupole moments, in isotopic and isotonic abundances, and in nuclear magnetic dipole moments. The illustration shows neutron-capture cross sections for a wide range of nuclei; these data provide clear evidence for the magic character of neutron numbers 50, 82, and 126. These data led in turn to the establishment of the nuclear shell model in analogy to the electronic shell structure of atoms; in both cases the magic numbers of nucleons (neutrons and protons) and of electrons are those required to close the shells, that is, to complete particularly stable orbital structures. In the nuclear case, however, the magic numbers are quite different from those in the atomic case because of the strong nuclear spin-orbit effect; the energy of a nuclear configuration depends sensitively upon the relative orientations of the orbital and the spin angular

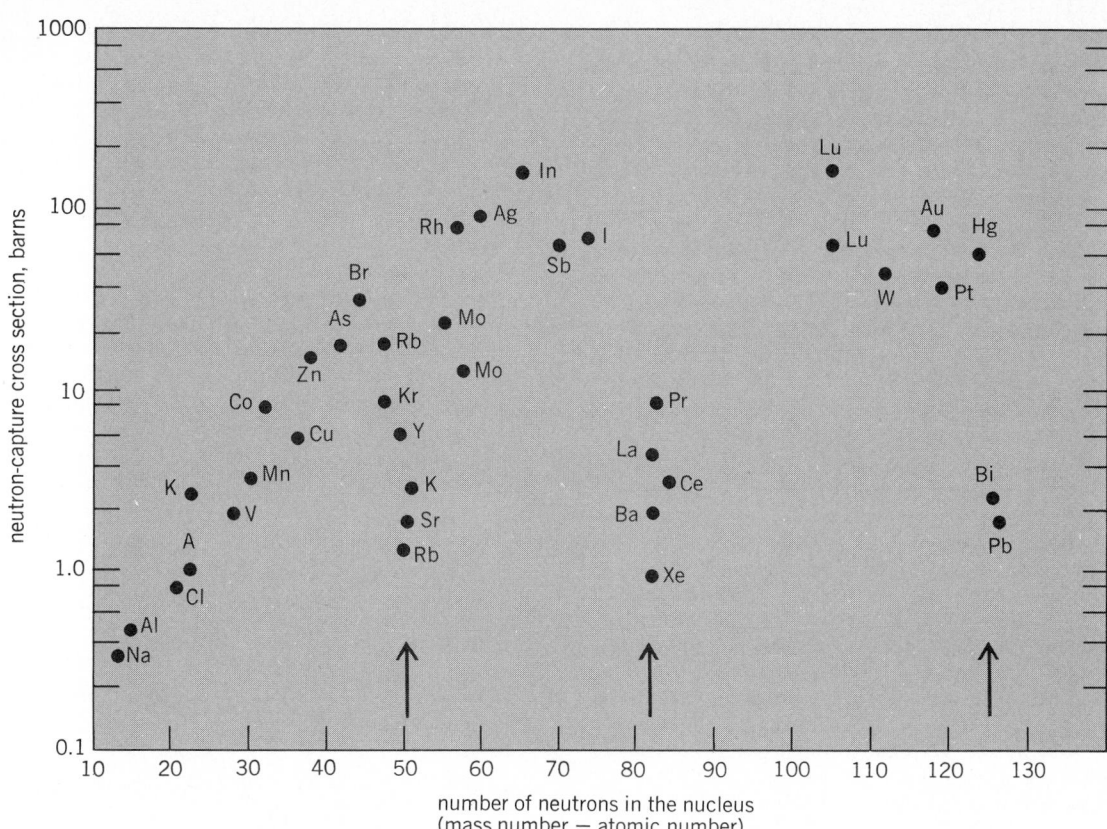

Measured thermal neutron-capture cross sections for various nuclei having the indicated number of neutrons already present. Arrows indicate magic neutron numbers 50, 82, and 126.

momenta. *See* ATOMIC STRUCTURE AND SPECTRA; NEUTRON SPECTROMETRY; NUCLEAR MOMENTS.

The abundant isotopes of helium, oxygen, silicon, calcium, tin, and lead owe their special nuclear stability to magic numbers of protons and (in all cases except tin) to magic numbers of neutrons as well. *See* ISOTOPE.

The hope of finding stable or quasistable superheavy nuclear species well beyond any occurring naturally, and beyond the transuranic species (between elements number 92 and 106) thus far studied with heavy-ion accelerators, depends critically upon the assumed stability of the doubly magic species having 114 protons and 184 neutrons. *See* SUPERTRANSURANICS.

[D. ALLAN BROMLEY]

Magnet

A piece of ferromagnetic material whose domains have been aligned sufficiently so that it produces a net magnetic field outside itself and can experience a net torque when placed in a magnetic field produced by some other source. For a discussion of ferromagnetic domains *see* FERROMAGNETISM.

If a bar of magnetically hard ferromagnetic material, such as Alnico 5, is magnetized to saturation in the magnetic field of a current flowing in a coil of wire, and the current is then turned off, the magnetic flux density B in the bar drops down the hysteresis curve to a point below the retentivity point. The flux density B goes below this point because the field H becomes negative due to the demagnetizing field of the magnet's poles. The important point here is that the bar magnet does retain an appreciable magnetization and, if the material is well chosen, this magnetization will remain more or less permanently, provided the magnet is not heated, dropped, exposed to too strong a magnetic field, or in general, treated too roughly. Such a magnet is called a permanent magnet. *See* MAGNETIC HYSTERESIS.

Keeper. This is a piece of soft (easily magnetized) ferromagnetic material (such as soft iron) which (when the magnet is not in use) extends from one pole to the other, and through which the magnetic flux lines between the poles are concentrated. A keeper is used especially with U-shaped or horseshoe magnets. Its presence decreases the demagnetizing effect of the poles and thus helps to keep the magnet strong.

Atomic theory. Modern theory indicates that the magnetic properties of bulk matter can be attributed to the orbital motions and spins of the electrons of the atoms which make up the matter. Orbital motions of electrons and electron spins are the equivalent of tiny electric current loops, and whenever there is a current, a magnetic field is always produced. A current loop has a magnetic moment M which, by the Sommerfeld proposal, is given in mks units by Eq. (1), where I is the current in am-

$$M = IA \qquad (1)$$

peres flowing around the periphery of the area A, the latter being expressed in square meters. The direction of the vector for A (and thus for M) is the direction of the extended thumb of the right hand when the fingers encircle A in the direction of the flow of I. (By the Kennelly proposal in the mks sys-

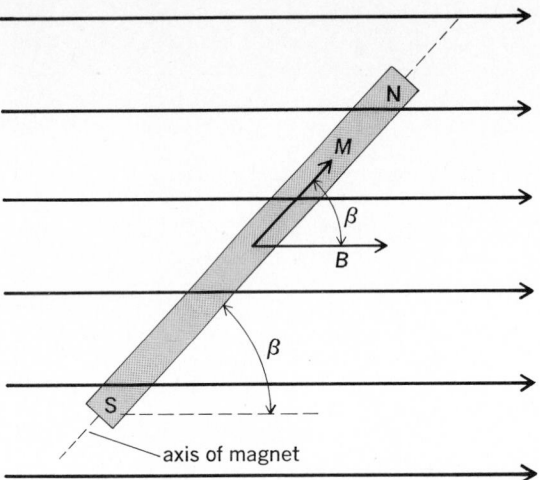

Fig. 1. Permanent bar magnet located in a uniform magnetic field, which exerts a torque on the magnet.

tem $M = \mu_0 IA$, where $\mu_0 = 4\pi \times 10^{-7}$ weber/amp-meter is the permeability of free space.) The magnetic moment of an atom is the vector sum of all the magnetic moments of the orbital motions and spins of all the electrons in the atom. Each atom of a ferromagnetic material has an appreciable net magnetic moment and, in addition, all the magnetic moments of the atoms within a domain are in alignment in the same direction, so the domain is magnetized to saturation. Thus, each domain has a sizable net magnetic moment. The magnetic dipole moment of a bar magnet is the vector sum of the dipole moments of all the domains in the magnet. The preceding discussion gives the fundamental reason why a magnet produces a net magnetic field external to itself, and why it can experience a net

torque when properly placed in a magnetic field. *See* Dipole moment; Electrical units and standards; Electron spin.

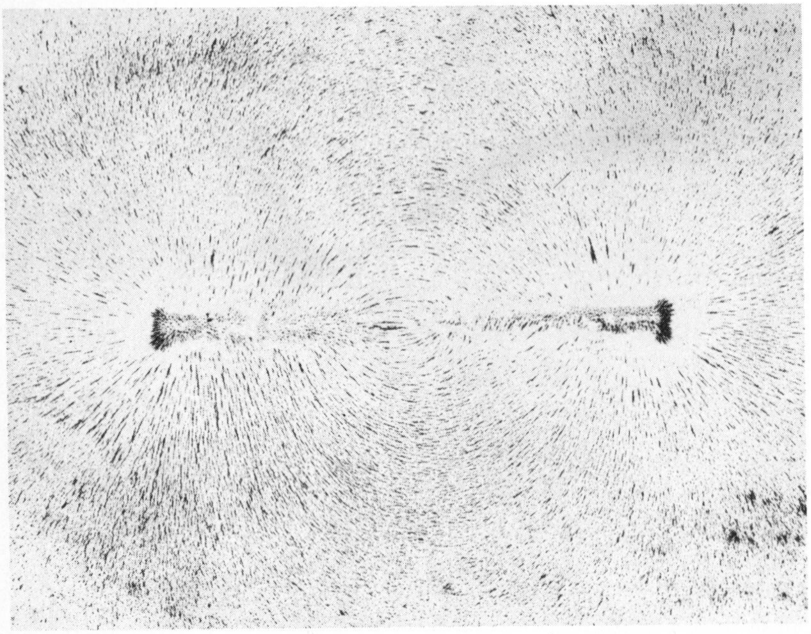

North-south orientation. If a bar magnet is free to turn in the Earth's magnetic field, it will set itself in a generally north-south direction. The north-seeking end is called the north pole of the magnet and the other end, the south pole. *See* Geomagnetism.

Magnetic dipole moment. Consider the torque experienced by a bar magnet in any uniform magnetic field of known magnetic flux density B and, as shown in Fig. 1, with its axis at an angle β with respect to the direction of the applied field. The results of a systematic experiment show that (1) the torque τ tries to align the axis of the magnet parallel with the magnetic flux lines of the field; (2) τ is independent of the axis of rotation selected, therefore τ is due to a couple (equivalent to the action of two equal and opposite forces); (3) τ is

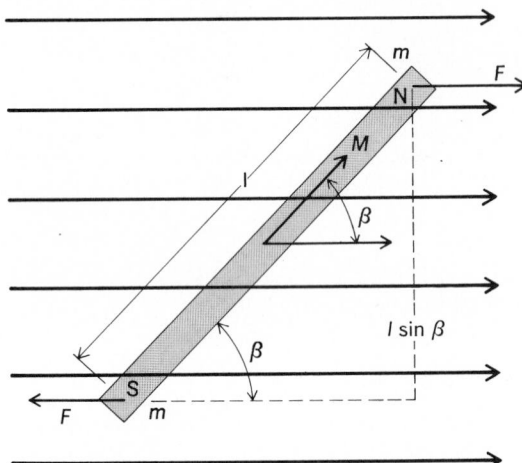

Fig. 3. Forces F on magnet poles in a magnetic field.

directly proportional to B; and (4) τ is directly proportional to $\sin \beta$. Hence, using M as a constant of proportionality, Eq. (2) holds. The value of M is

$$\tau = MB \sin \beta \qquad (2)$$

found to be constant for a given magnet, but different for different magnets, so M belongs to the magnet and is called the magnetic dipole moment of the magnet. If $\beta = 90°$, $M = \tau/B$, and the magnetic dipole moment of a magnet is defined as equal to the moment of force that the magnet experiences when it is perpendicular to a magnetic field of unit magnetic flux density. In the mks system, as Eq. (2) shows, the units of M are newton-meter^3/weber or, the equivalent, ampere-meter^2.

The definition of M given by this empirical method is the same as the one given earlier, which says that M is the vector sum of the magnetic moments of the domains of which the magnet is composed. This follows from the fact that the magnet experiences a torque in Fig. 1 only because the applied B field exerts a torque on the individual atoms of the bar as a result of the spins and orbital motions of their electrons. It is known that electron spin is much more important than electron

Fig. 2. Photograph of the iron-filing map of the magnetic field of a permanent bar magnet. Note that the magnetic flux lines can be traced by the lines of iron filings, which act like tiny compass needles.

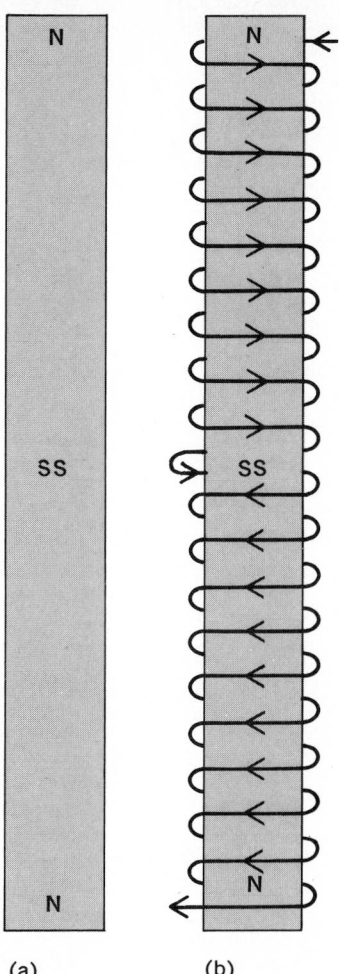

Fig. 4. Consequent poles. (*a*) Magnet with this polarity. (*b*) Diagram to show windings of coils which could produce the polarity shown in *a*.

orbital motion in the magnetic behavior of the atoms of ferromagnetic materials.

Pole strength. Figure 2 is a photograph of an iron-filing map of the magnetic field of a permanent bar magnet. The iron filings arrange themselves like tiny compass needles along the magnetic flux lines. The regions called the poles are visible because the iron filings cluster in these regions and stand on end directly over the poles. The poles are simply regions where magnetic flux lines from the internal domains enter and leave the magnet. (The lines leave at the north pole and enter at the south pole.) However, one may look upon these poles as fictitious seats of magnetic effects and assign a pole strength m to each of them. When this pole point of view is used, the fictitious poles replace the uncompensated electron spins and orbital motions as the source of the magnetic behavior. Many complicated problems are considerably simplified by the use of the pole approach to behavior of magnets.

From the iron-filing map, one can more or less locate the centers of the poles. The effective magnetic length l of the magnet is the distance between the poles, as shown in Fig. 3. With a long slender magnet, l can be determined with fair accuracy. The pole strength m of each pole is defined

as the quotient obtained when the magnitude M of the magnetic moment of the magnet is divided by the magnetic length l, as in Eq. (3). The units of m

$$m = M/l \qquad (3)$$

in the mks system, using the Sommerfeld proposal, are ampere-meters. (By the mks Kennelly proposal $M = \mu_0 \tau/B$ when $\beta = 90°$, and M has the units of weber-meter; $m = M/l$ and thus the unit of M is the weber.)

When the pole point of view is taken, an external applied magnetic field, as in Fig. 3, exerts a fictitious force F on each fictitious pole as shown. The two forces are to be of such magnitude that they will produce the actual torque experienced by the magnet. The two forces must be equal and opposite because, as already pointed out, the torque on a magnet in a uniform magnetic field is produced by a couple, and these two equal forces constitute such a couple. From Fig. 3, $\tau = Fl \sin \beta$, and from Eqs. (2) and (3), $\tau = mlB \sin \beta$; thus Eq. (4) expresses the force on each magnet pole in a

$$F = mB \qquad (4)$$

magnetic field of flux density B. It follows from Eq. (4), and the fact that F is the same on both poles, that the pole strength must be the same for both poles of a magnet. Experiment shows that unlike poles attract each other and like poles repel.

Consequent poles. These may be additional poles in a magnet beyond the normal two. Figure 4 shows a magnet with a north pole at each end and the corresponding south poles together in the center and a diagram of the coil winding which could produce this type of polarity. *See* ELECTRET; MAGNETIC FIELD; MAGNETIC FLUX; MAGNETISM; MAGNETIZATION; MAGNETOSTATICS.

[RALPH P. WINCH]

Bibliography: B. I. Bleaney and B. Bleaney, *Electricity and Magnetism*, 3d ed., 1976; E. M. Pugh and E. W. Pugh, *Principles of Electricity and Magnetism*, 2d ed., 1970.

Magnetic circuits

Closed paths of magnetic flux; also a design method using such paths to compute the magnetic field of a core geometry that is often encountered, for instance, high-permeability flux-path segments (and their associated air gaps), each segment having reasonably definite length and area. Examples of magnetic circuits are transformer cores, relay frames, and iron parts of electrical machinery. The magnetic-field equations for these devices look so similar to dc circuit equations that they are called magnetic circuits.

Reluctance. If NI ampere-turns link one closed core, then Eqs. (1) hold, where l is length, A is

$$\Phi \times \frac{l}{\mu A} = NI$$

(Flux) × (reluctance)
  = magnetomotive force (mmf) $\qquad (1)$

$$\Phi \times R = \text{mmf}$$

cross section, and μ is permeability large enough so that the flux Φ is nearly completely confined to the core (Fig. 1).

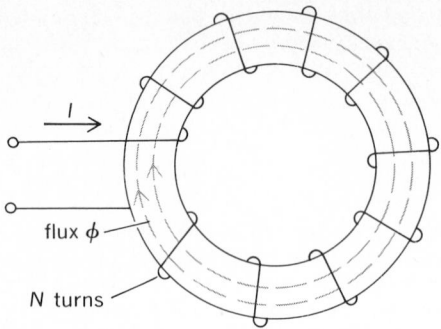

Fig. 1. Diagram of a toroidal magnetic circuit. (*A. E. Fitzgerald, D. E. Higginbotham, and A. Grabel, Basic Electrical Engineering, McGraw-Hill, 1967*)

If the flux path is a sequence of dissimilar segments, its total reluctance is the sum of segment reluctances as given by Eq. (2).

$$R = \sum \frac{l_n}{\mu_n A_n} \qquad (2)$$

If the (constant) flux divides among several parallel flux paths, it does so in inverse proportion to their reluctances.

These series and parallel reluctances have the same algebra as do series and parallel dc resistances; hence the fundamental equation is often called Ohm's law for magnetic circuits, even though mmf is not a force, Φ is not a flow, and the equation is not Ohm's but Hopkinson's.

Specific use of magnetic-circuit reluctance is nearly always qualitative; it is handy for explanation and discussion. How much flux change occurs when reluctance is altered is usually determined by a nonlinear calculation which does not compute a numerical value for reluctance itself. Accordingly, there is no common name for a reluctance unit.

Quantitative calculations. Three principles are applied:

1. Magnetic-flux lines are endless: Φ is the same at every point in the flux bundle, even though the lines are specially crowded at some places.

2. The relation between flux density B and field vector H at any point is a property of the matter at that point. From (experimental) charts for the material, either B or H can be found from the other (Fig. 2). *See* MAGNETIZATION.

3. Ampère's line integral $\oint H \cdot dl = \Sigma I_{\text{linked}}$ is taken along the path followed by the flux bundle being analyzed.

Given total flux Φ, flux-density in the nth segment is $B_n = \Phi/A_n$. Each H_n is then found from its B_n by chart, and the Ampère integral is evaluated as shown in Eq. (3).

$$\oint H \cdot dl = \Sigma H_n l_n = NI \qquad (3)$$

Units. Numerical evaluation of the Ampère integral requires consistent units. Modern practice uses the ones listed in the table.

Many published curves still use the older cgs units: Φ in maxwells, B in gauss, and H in oersteds, with $\mu_{\text{air}} = 1$. Conversion to one of the systems in the table is recommended; it avoids the absolute amperes and factors of 4π that occur in the

Commonly employed units for magnetic circuits

Quantity	mks units	Engineering units
Flux density B	Webers/m^2	Kilolines/in.^{2}
Flux Φ	Webers	Kilolines
Field vector H	Amp-turns/m	Amp-turns/in.
$\mu_{\text{air}} = B/H$ in air	$4\pi \times 10^{-7} = \dfrac{1}{7.95 \times 10^5}$	$3.18 \times 10^{-3} = \dfrac{1}{313}$
Length	Meters	Inches

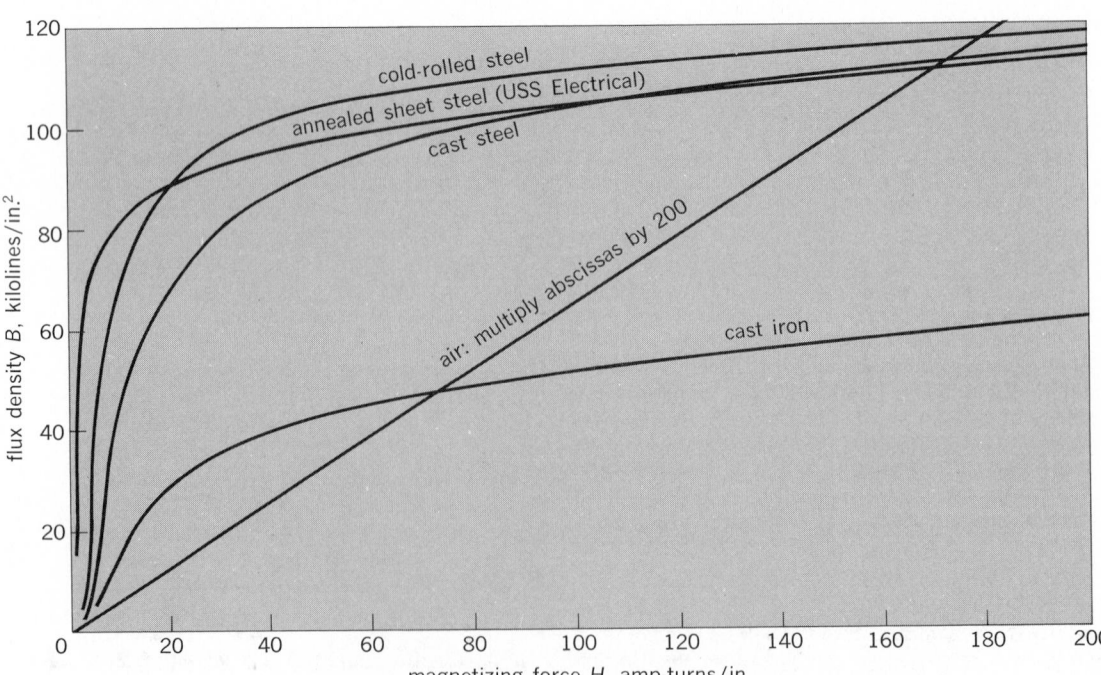

Fig. 2. Graph of the normal magnetization curves for some magnetic materials. (*A. E. Fitzgerald, D. E. Higginbotham, and A. Grabel, Basic Electrical Engineering, McGraw-Hill, 1967*)

cgs equations. The conversion below is straight-forward because the flux lines keep their identity.

One weber is 10^5 kilolines (10^8 lines or 10^8 max-wells).

One weber/m² is 64.5 kilolines/in.² (10^4 lines/cm² or 10^4 gauss).

One amp-turn/m is 2.51×10^{-2} amp-turns/in. and corresponds to $4\pi \times 10^{-7}$ weber/m² in air (or to $4\pi \times 10^{-3}$ gauss in air or $4\pi \times 10^{-3}$ oersteds).

Special situations. Some rules of thumb allow for leakage flux and for reduction of flux density when the lines spread in air gaps. Magnetic circuits that include permanent magnets or superconductors or plasmas need care in their analysis. For example, when a slice of permanent-magnet material in inserted as one of the flux-path segments, it supplies a magnetomotive force that requires calculation. Instead of puzzling over what H is in a permanent magnet, it is better to replace with a slice of normal material (that is, of the same dimensions and same incremental permeability) plus ampere-turns around its edge to make the same magnetic moment as the original magnet slice.

In each piece of superconducting material, two things happen: (1) Unless the superconductor is very thin, all the flux lines detour around the outside of it, and (2) any attempt to change Φ near a superconductor starts nondecaying eddy currents that hold Φ to (almost) its initial value. More accurate (quantum) descriptions of these phenomena have been published. The flux rejection of a superconductor can be used to channel the flux when this is desired; several devices employ very thin superconductors into which some flux can be driven to modify the onset of superconductivity. *See* SUPERCONDUCTIVITY.

The added ampere-turns from a few ions in the flux path can be found by eddy-current methods if time variation is not too violent. However, when there are enough of these ions to be called a plasma they interact by collision and the resulting currents and magnetomotive force are not easily found. *See* EDDY CURRENT; MAGNETOHYDRO-DYNAMICS.   [MARK G. FOSTER]

Bibliography: A. E. Fitzgerald et al., *Basic Electrical Engineering*, 5th ed., 1981; G. Schmidt, *Physics of High Temperature Plasmas*, 2d ed., 1979; R. J. Smith, *Electronics: Circuits and Devices*, 1973; R. Stein and W. T. Hunt, *Electrical Power System Components: Transformers and Rotating Machines*, 1979; M. Tinkham, *Introduction to Superconductivity*, 1975.

Magnetic field

A condition existing in the vicinity of a magnetic body (or a current-carrying medium) whereby the magnetic forces due to the body (or current) are detectable. In a magnetic field, there is a force on a moving charge in addition to the electrostatic (Coulomb) forces between charges, or there is a force on a magnetic pole. The description of the field may be in terms of magnetic induction (flux density) or in terms of magnetic field strength. *See* MAGNETIC INDUCTION.

The magnetic induction B of the field is defined from the force F on a moving charge q or current element of length l carrying a current I by Eq. (1).

$$B = \frac{F}{qv \sin \theta} = \frac{F}{Il \sin \theta} \qquad (1)$$

The direction of the magnetic induction is that direction in which the force on the moving charge is zero. The factor $v \sin \theta$ is then the component of the velocity of the charge in a direction perpendicular to B. The mks unit of B is the newton/ampere-meter or weber/meter².

The magnetic induction at a given point due to a current may be found from Ampère's law, Eq. (2), which may also be expressed as Eq. (2b). Equation (2a) gives the contribution of a current element Idl

$$dB = \frac{\mu_0}{4\pi} \frac{Idl \sin \theta}{r^2} \qquad (2a)$$

$$B = \frac{\mu_0}{4\pi} \int \frac{Idl \sin \theta}{r^2} \quad \text{(vector sum)} \qquad (2b)$$

to the magnetic induction, where μ_0 is the permeability of empty space, r the distance of the point to the current element, and θ the angle between the current element and the line joining the element to the point.

The direction of B, from Ampère's law, is perpendicular to the plane determined by a line tangent to dl and the line joining the current element to the point at which B is determined. *See* AMPERE'S LAW.

A second magnetic vector quantity, the magnetic field strength or intensity H, may be defined in part from Ampère's law. The part of the field strength that is due to currents is given by the vector sum from Eq. (3). Thus, H is computed in the

$$H = \frac{1}{4\pi} \int \frac{Idl \sin \theta}{r^2} \qquad (3)$$

same manner as the flux density B, but without the factor μ_0. The direction is specified in exactly the same manner as B. Thus, H depends only on the current present, and not upon the properties of the surrounding medium. This definition of H is only partial because it does not include contributions by magnetic poles if they are present in the neighborhood.

The mks unit of magnetic field strength appears from the defining equation, when current is in amperes and dl and r are in meters, as ampere per meter. Because many equations derived from the defining equation involve the number of turns N of a coil times the current, the ampere-turn per meter (amp-turn/m) is also used as an equivalent unit.

Magnetic poles. A body can be magnetized by bringing it into a magnetic field due to currents or magnets. Except in the case of a ring magnetized along its circumference, the field associated with a magnetized body extends to the region surrounding the body. The external effect usually appears in limited regions of the body called poles. A magnetized bar of iron has two poles, one at either end; and from the fact that the bar will set itself in an approximate north-south direction in the Earth's field, it appears that there are two kinds of poles. The pole that is at the north end of the bar is called a north-seeking pole; that at the south end is called a south-seeking pole. The two poles at the ends are merely indications of the continuous magnetization within the body. An indication of the validity of this statement is the fact that when a bar magnet is broken into two parts, two new poles appear at the break, and the orientation of the poles in each fragment is the same as it was in the original magnet. *See* MAGNETIZATION.

It is observed that magnetic poles exert forces on each other and upon moving charges in the region near the poles. There is a field near the pole, and the pole may be considered as the cause of that field.

If the poles of a magnetized body are small enough that they may be considered point poles, the force that one pole exerts upon another is found to be proportional to the product of the pole strengths m and m' and inversely proportional to the square of the distance r between them. This statement is called Coulomb's law of magnetostatics and is written as Eq. (4). The proportionality

$$F = k' \frac{mm'}{r^2} \qquad (4)$$

factor k' depends upon the units used and upon the medium between the poles. For empty space in the mks system, k' is assigned the value $\frac{1}{4}\pi\mu_0$. The unit of pole strength associated with this choice is the weber. *See* MAGNET; MAGNETIC PERMEABILITY; MAGNETOSTATICS.

The magnetic field strength or magnetic intensity H may be expressed as the force per unit north-seeking magnetic pole as in Eq. (5). Then,

$$H = \frac{F}{m'} \qquad (5)$$

from Coulomb's law, the contribution of a point pole of strength m to the magnetic field strength near the pole is given by Eq. (6). The direction of H

$$H = \frac{F}{m'} = \frac{1}{4\pi\mu_0} \frac{m}{r^2} \qquad (6)$$

is away from north-seeking poles and toward south-seeking poles. The contribution of several poles is the vector sum of the contributions of the individual poles as shown in Eq. (7). If the H due to

$$H = \frac{1}{4\pi\mu_0} \sum \frac{m}{r^2} \quad \text{(vector sum)} \qquad (7)$$

poles is to be in the same units as the H due to currents, the unit of pole strength must be chosen properly. If H is to be in amperes per meter when μ_0 is in webers per ampere-meter and r is in meters, then m must be in webers.

If the poles are distributed over surfaces or throughout volumes, the summation becomes an integral as shown in Eq. (8).

$$H = \frac{1}{4\pi\mu_0} \int \frac{dm}{r^2} \quad \text{(vector sum)} \qquad (8)$$

The general expression for the field strength due to both currents and poles is given by Eq. (9), where the integrals represent vector sums.

$$H = \frac{1}{4\pi} \int \frac{Idl \sin\theta}{r^2} + \frac{1}{4\pi\mu_0} \int \frac{dm}{r^2} \qquad (9)$$

If a toroidal coil has an iron core, the iron is magnetized by the current of the coil. The magnetic field strength within the core is given entirely by the first term of the equation for H, since there are no poles. For a long straight solenoid with an air core (illustration *a*), H is entirely due to the current and is found by integration to be $H_I = NI/l$, where N is the number of turns of the coil, and l is the length of the solenoid. When an iron core is inserted into the solenoid (illustration *b*), the iron becomes magnetized, and poles appear at each end.

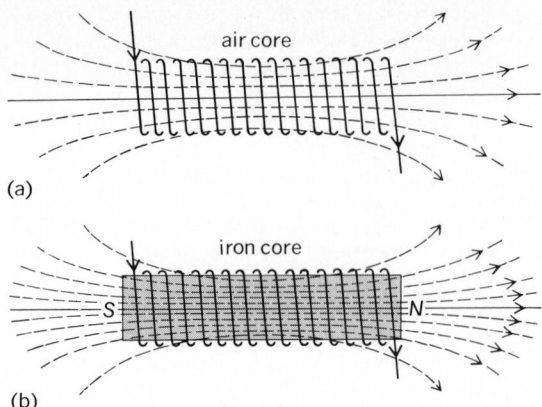

(a)

(b)

Flux in a solenoid (*a*) containing only air (essentially vacuum) and (*b*) containing an iron core.

The contribution of the current to H remains the same as before, but the second term now contributes to H components that are opposite in direction to H_I and that vary along the bar because of the variation in distances from the poles and because of the variation of the permeability of the iron. The effect of the poles on H is greatest at the ends near the poles. The magnetization of the iron is not uniform. The effect of the poles is essentially a demagnetizing effect; it is large for short magnets, and becomes negligible at the center of a long magnet.

Lines of force. As in the case of magnetic induction B, which can be represented by lines called magnetic flux, the vector quantity H may be represented by lines called lines of force. The number of lines of force per unit area of a surface perpendicular to H is made equal to the value of H. The direction of the lines of force is the direction of the field. The lines of force are closed curves, as are the lines of induction. *See* LINES OF FORCE; MAGNETIC FLUX.

Energy of a magnetic field. Consider a Rowland ring (a ring-shaped sample of magnetic material) surrounded by a coil of N turns in which there is a current I. The field strength H within the ring is given from Ampère's law by Eq. (10), where l is the mean circumference of the ring.

$$H = \frac{NI}{l} \qquad (10)$$

In building up the current in the coil, energy must be supplied that becomes energy of the magnetic field. This energy is given by Eq. (11), where

$$W = \tfrac{1}{2} L I^2 \qquad (11)$$

L is the self-inductance of the coil. *See* INDUCTANCE.

But L is defined by Eq. (12), where Φ is the flux,

$$L = \frac{N\Phi}{I} \qquad (12)$$

and W by Eq. (13), where V is the volume of the

$$W = \frac{1}{2} \frac{N\Phi}{I} I^2 = \tfrac{1}{2} N\Phi I = \tfrac{1}{2} NBAI$$

$$= \frac{1}{2} \frac{NI}{l} BV = \tfrac{1}{2} HBV \qquad (13)$$

core, and A is the mean cross-sectional area of

the ring. The energy per unit volume of the field is then given by Eq. (14). If μ is the permeability of

$$\frac{W}{V} = \tfrac{1}{2}HB \qquad (14)$$

the core, $B = \mu H$ and the energy density of the field may be written down as in Eq. (15).

$$\frac{W}{V} = \tfrac{1}{2}HB = \tfrac{1}{2}\mu H^2 = \frac{1}{2}\frac{B^2}{\mu} \qquad (15)$$

Magnetic potential. When a magnetic pole is in a magnetic field, there will be a force F acting on it. If it is moved a distance ds in the field, work dW done against the field is given by Eq. (16), where θ

$$dW = -F \cos \theta \, ds = -Hm \cos \theta \, ds \qquad (16)$$

is the angle between the positive sense of H and the positive direction of s. The magnetic potential difference V may be defined as the work done per unit pole in taking the pole from one point to the other as in Eq. (17a) or (17b).

$$dV = \frac{dW}{m} = -H \cos \theta \, ds \qquad (17a)$$

$$V = -\int_{s_1}^{s_2} H \cos \theta \, ds \qquad (17b)$$

The resulting equation for the magnetic potential does not include the concept of the magnetic pole, and can be used as a defining equation for the magnetic potential. The integral represents the sum along a path of the products of ds and the component of H in the direction of ds. Such an integral is called a line integral and is represented by the symbol $\oint$. The defining equation for magnetic potential may be written as Eq. (18). From this

$$V = -\oint H \cos \theta \, ds \qquad (18)$$

equation the relationship between magnetic potential and H may be written as Eq. (19). Thus the

$$H \cos \theta = -\frac{dV}{ds} \qquad (19)$$

component of field strength in any direction is the negative of the magnetic potential gradient in that direction. This statement is similar to the relation between electric field intensity and electric potential gradient.

The rise in magnetic potential around any closed path in a magnetic field may be deduced from consideration of the field about a long straight conductor. *See* BIOT-SAVART LAW.

For this special case, the lines of force are concentric circles about the conductor and at a distance a from the conductor and Eq. (20) holds. For

$$H = \frac{I}{2\pi a} \qquad (20)$$

a path that follows a circular line of force in a sense opposite to that of H, $\theta = 180°$ and $\cos \theta = -1$. Then the rise in magnetic potential around the closed path is given by Eq. (21). This result is inde-

$$V = -\oint H \cos \theta \, ds = \int_0^{2\pi a} \frac{I}{2\pi a} \, ds = I \qquad (21)$$

pendent of the radius of the circle followed and, in fact, is independent of the shape of the path, since any path may be resolved into components that are along circles and along radii. The contributions of the radial parts are zero, since they are perpendicular to H.

The result deduced from this special case may be generalized for a closed loop of any shape. The rise in magnetic potential around the path is equal to the line integral of $H \cos \theta$ around the path, and this in turn is equal to the current through the surface bounded by the path as shown in Eq. (22). If

$$\oint H \cos \theta \, ds = I \qquad (22)$$

the path taken includes no current, the integral is zero. If there are N equal currents inside the path, the integral becomes NI.

By analogy to an electric circuit, in which the line integral of the electric intensity around the circuit is the electromotive force of the circuit, the rise in magnetic potential around the closed path, that is, the line integral of $H \cos \theta$ around the path, is called the magnetomotive force (mmf). Thus, the mmf of a coil of N turns in which there is a current I is NI. *See* ELECTRIC FIELD; MAGNETOMOTIVE FORCE. [KENNETH V. MANNING]

Bibliography: S. S. Attwood, *Electric and Magnetic Fields*, 3d ed., 1949, reprint 1966; *Berkeley Physics Course*, vol. 2: *Electricity and Magnetism*, 1970; B. I. Bleaney and B. Bleaney, *Electricity and Magnetism*, 3d ed., 1976; E. M. Pugh and E. W. Pugh, *Principles of Electricity and Magnetism*, 2d ed., 1970; R. Resnick and D. Halliday, *Physics*, 3d ed., 1977; F. W. Sears et al., *University Physics*, 5th ed., 1976.

Magnetic flux

Lines used to represent the magnetic induction **B** in a magnetic field. *See* MAGNETIC FIELD; MAGNETIC INDUCTION.

The vector quantity **B** is defined from the equation of force **F** on a charge q moving with speed v in a magentic field at an angle θ with the direction of the field. The magnitude of **B** is given by Eq. (1).

$$B = \frac{F}{qv \sin \theta} \qquad (1)$$

Lines of flux used to represent the field in magnitude and direction are selected so that they are parallel to **B** at each point, and the number of lines of flux per unit area of a surface perpendicular to the field is equal to B. The total number of lines of induction through a surface is the magnetic flux **Φ**. Magnetic induction is the flux per unit area or flux density. The flux $d\Phi$ through an element of area $d\mathbf{A}$ perpendicular to **B** is given by Eq. (2). If the

$$d\Phi = \mathbf{B} \, d\mathbf{A} \qquad (2)$$

surface is not perpendicular to **B**, Eq. (3) holds,

$$d\Phi = B \cos \theta \, dA \qquad (3)$$

where θ is the angle between **B** and the normal to the surface. Then $B \cos \theta$ is the normal component of **B**. For a surface of any orientation and for varying **B**, the flux through the surface is given by Eq. (4), where the integral is taken over the whole surface.

$$\Phi = \int B \cos \theta \, dA \qquad (4)$$

Since lines of induction are always closed curves, it follows that for any closed surface in the magnetic field, every line that enters the closed surface must leave it. Therefore, the integral of the normal component of **B** over any closed surface in the magnetic field must be zero as shown in Eq. (5).

$$\int B \cos \theta \, dA = 0 \qquad (5)$$

In the mks system, a line of induction is called a weber, and the magnetic induction or flux density is in webers per square meter. Since, from the defining equation of magnetic induction, the unit of **B** is the newton/ampere-meter, that unit is equivalent to the weber per square meter.

[KENNETH V. MANNING]

Magnetic hysteresis

The lagging of magnetization behind magnetizing force as the magnetic condition of a ferromagnetic material is changed.

When a ferromagnetic sample that is initially demagnetized is subjected to a continuously increasing magnetizing force H, the relation between H and flux density B is shown by the normal magnetization curve Oab of the figure. The point a indicates the magnetic condition as the increasing magnetic intensity has reached H_1. If H is increased to a maximum value H_2 and then decreased again to H_1, the decreasing flux density does not follow the path of increase, but decreases at a rate less than that at which it rose. This lag in the change of B behind the change of H is called hysteresis. If the value of H is further reduced from H_1 to zero, B is not reduced to zero but to a value B_r. The specimen has retained a permanent magnetism. This ordinate B_r is called the retentivity or remanence. The value of B may be reduced to zero at e by reversing the direction of H and increasing its value to H_c. This value H_c is called the coercive force or coercivity.

Hysteresis loop. As H is increased in the negative direction, the magnetization proceeds along the curve of the figure until at f the values of B and H are the same as those at b, but opposite in direction. When the reverse changes in H are made, the magnetization changes along the curve $fghb$. This

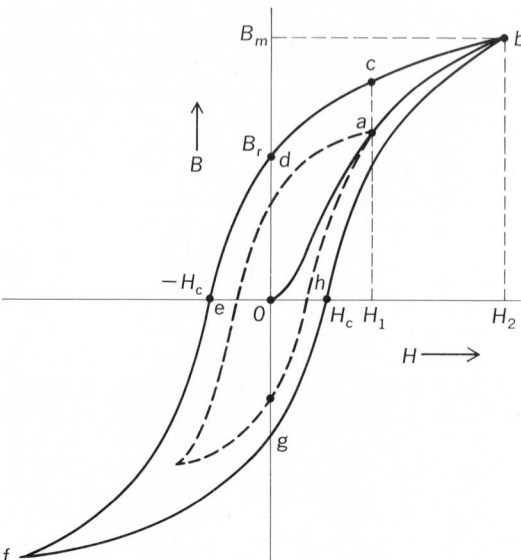

Hysteresis loop; symbols explained in text.

entire loop $bdefghb$ is called a hysteresis loop. If the hysteresis loop starts from another point on the normal magnetization curve, such as a, there will be a smaller hysteresis loop entirely within the larger, such as the dotted loop of the figure. For additional information on magnetization curves *see* MAGNETIZATION.

Energy. In magnetizing the core, energy must be supplied. As H and B increase along $fghb$, the energy gain is proportional to the area under that portion of the curve. Along the path $bcdef$, there is a loss in energy proportional to the area under $bcdef$. The net loss in energy per cycle per unit volume is given by Eq. (1) where the integral is taken around

$$W = \oint dW = \oint H \, dB \qquad (1)$$

the closed loop. But $\oint H \, dB$ is the area of the hysteresis loop, and the energy loss per unit volume per cycle is equal to the area of the hysteresis loop. This energy is converted into heat. If B is expressed in webers/m² and H in ampere-turns/m. the energy loss is in joules/(m³) (cycle).

C. P. Steinmetz found an empirical relation between the energy loss per unit volume per cycle W and the maximum value B_m of the flux density during the cycle given in Eq. (2), where η is called

$$W = \eta \, (B_m)^n \qquad (2)$$

the Steinmetz coefficient. Steinmetz found a value of about 1.6 for the exponent n for many materials, but it varies from about 1.5 to 2.5 for others. Values of η have been tabulated.

In alternating-current machinery, masses of iron are in fields that are constantly reversing. Therefore, the iron is constantly being carried around hysteresis paths, and there is an energy loss per cycle that depends upon the hysteresis loop for the particular iron that is used. This hysteresis loss results in undesired heating of the iron as well as waste of energy. For more information on hysteresis loss *see* CORE LOSS; HYSTERESIS MOTOR; THERMAL HYSTERESIS.

[KENNETH V. MANNING]

Bibliography: B. Bleaney and B. I. Bleaney, *Electricity and Magnetism*, 3d ed., 1976; E. M. Pugh and E. W. Pugh, *Principles of Electricity and Magnetism*, 2d ed., 1970; E. M. Purcell, *Electricity and Magnetism*, Berkeley Physics Course, vol. 2, 1965; R. Resnick and D. Halliday, *Physics*, 3d ed., 1977; F. W. Sears et al., *University Physics*, 5th ed., 1976.

Magnetic induction

A vector quantity that is used as a quantitative measure of a magnetic field. It is defined in terms of the force on a charge moving in the field by Eq. (1), where B is the magnitude of the magnetic in-

$$B = \frac{F}{qv \sin \theta} \qquad (1)$$

duction and F is the force on the charge q, which is moving with speed v in a direction making an angle θ with the direction of the field. The direction of the vector quantity B is the direction in which the force on the moving charge is zero.

The magnetic induction may also be expressed in terms of the force F on a current element of length l and current I, as in Eq. (2). The meter-

$$B = \frac{F}{Il \sin \theta} \qquad (2)$$

kilogram-second (mks) unit of magnetic induction is derived from this equation by expressing the force in newtons, the current in amperes, and the length in meters. The unit of B is thus the newton/ampere-meter.

Magnetic flux density is the magnetic flux per unit area through a surface perpendicular to the magnetic induction. Magnetic flux density and magnetic induction are equivalent terms. *See* MAGNETIC FIELD; MAGNETIC FLUX.

The magnetic induction may be represented by lines that are drawn so that at every point in the field the tangent to the line is in the direction of the magnetic induction. To represent the magnetic induction qualitatively, as many such lines may be drawn as are necessary to portray the field. If, however, the lines are to represent the magnetic induction quantitatively, an arbitrary choice must be made for the number of lines to represent a given condition. One such choice is that in which the number of lines per square meter of a surface perpendicular to B is set equal to the value of B. These lines are called magnetic flux. One line of induction as here selected is called a flux of 1 weber. The corresponding unit of flux density is then the weber per square meter. From the manner of defining flux used here, it follows that the weber per square meter is equivalent to the newton/ampere-meter.

Another unit of flux density is defined by using centimeter-gram-second (cgs) units in both the defining equation for magnetic induction and in the area in which there is unit flux. This cgs unit of flux density is called the gauss. The relationship between the gauss and the weber is given by 1 weber/m² = 10^4 gauss. *See* GAUSS.

For discussion of an important device known as a betatron, which utilizes the principles of magnetic induction *see* PARTICLE ACCELERATOR.

[KENNETH V. MANNING]

Bibliography: See MAGNETIC HYSTERESIS.

Magnetic moment

The relationship between a magnetic field and the torque exerted on a magnet, a current loop, or a charge that is moving in the field.

When a magnet is placed in a magnetic field of strength H, there is a torque L exerted on the magnet by the field. The torque is a maximum when the axis of the magnet is perpendicular to the field. The ratio of the torque for this position to the strength of the field is called the magnetic moment M of the magnet, as defined in Eq. (1). *See* MAGNET.

$$M = \frac{L}{H} \qquad (1)$$

If a flat coil of wire of N turns and area A, in which there is a current I, is placed in a magnetic field of flux density B, the coil experiences a torque L given by Eq. (2), where θ is the angle between the

$$L = NIAB \sin \theta \qquad (2)$$

field and the normal to the plane of the coil. The torque is maximum when $\theta = 90°$, that is, when the plane of the coil is parallel to the field. The ratio of the maximum torque to the flux density B is the

magnetic moment of the coil, as shown in Eq. (3).

$$M = \frac{L}{B} = NIA \qquad (3)$$

Alternatively, the magnetic moment of the coil may be defined as the ratio of L to H. For this definition Eq. (4) holds, since $B = \mu_0 H$ in empty

$$M = \frac{L}{H} = \mu_0 NIA \qquad (4)$$

space, μ_0 being the permeability of empty space.

An electron in its orbit about an atomic nucleus has an orbital magnetic moment. If I represents the equivalent current as the electron moves in its orbit, and T is the time of one revolution, Eq. (5)

$$I = \frac{q}{T} = \frac{q\omega}{2\pi} \qquad (5)$$

holds, where ω is the angular velocity and q is the charge of the electron. Then the orbital magnetic moment is given by Eq. (6), where A is the orbital area.

$$M_o = IA = \frac{q\omega}{2\pi} A \qquad (6)$$

If a charge is spinning, there is a charge in motion and thus an electric current. The spin is equivalent to a tiny current loop which has a magnetic moment. *See* ELECTRON MAGNETIC MOMENT; ELECTRON SPIN.

Atomic nuclei also possess magnetic moments. For a discussion of these *see* NUCLEAR MOMENTS. *See also* MAGNETON.

[KENNETH V. MANNING]

Bibliography: B. I. Bleaney and B. Bleaney, *Electricity and Magnetism*, 3d ed., 1976; F. W. Sears et al., *University Physics*, 5th ed., 1976; G. Shortley and D. Williams, *Elements of Physics*, 5th ed., 1971.

Magnetic monopoles

Magnetic charges; hypothetical entities designed to put magnetism on an equal footing with electricity. In this article, gaussian units are used throughout.

Electric fields are produced by electric charges according to the formula $\mathbf{E}(\mathbf{r}) = q\,\mathbf{r}/r^3$, where q is a point charge at the origin of coordinates, $\mathbf{r}$ is the position of observation at distance r from q, and $\mathbf{E}$ is the resulting field at $\mathbf{r}$. They exert forces on point charges $\mathbf{F} = q\mathbf{E}$, where $\mathbf{F}$ is the force, q is the charge, and $\mathbf{E}$ is the electric field from all other charges. A magnetic field is produced by moving electric charges. The magnetic field about a wire carrying electric current is a well-known example. In turn, magnetic fields exert forces on charges according to $\mathbf{F} = q(\mathbf{v}/c) \times \mathbf{B}$, where $\mathbf{v}$ is the velocity, $\mathbf{B}$ is the magnetic field, and c is the speed of light. This force only acts on a moving charge and is perpendicular to the direction of motion and to the magnetic field. It is natural to speculate that there may be magnetic charges which act as sources of magnetic fields in the same way that electric charges produce electric fields. Two such magnetic charges would obey the famous rules that like charges repel and unlike charges attract each other. *See* ELECTRIC FIELD; ELECTROMAGNETIC FIELD.

The name monopole or one-pole for a magnetic

charge is meant as a contrast to dipole or two-pole. A magnetic dipole is a system producing a magnetic field which looks like that of two equal but opposite magnetic charges close together, when the field is measured at a large distance from the dipole. Such a field can be produced by a loop of wire carrying an electric current. In this case, it is obviously impossible to tear the dipole apart and get two magnetic monopoles.

Parity violation. If there were only magnetic charges in the world, and no electric charges, one could simply interchange the names electric and magnetic and proceed to describe all electromagnetic phenomena in the same way one does now. Clearly, the interesting new results must come from the interaction of magnetic monopoles with electric charges. Consider an electric charge q in the presence of a magnetic monopole m. If both particles are at rest, there is no force between them. If the electric particle moves, it suffers a force perpendicular to its velocity and to the line of centers from q to m. Imagine that m is situated at the center of a clock face, and q, which is moving upward, is located at 3 o'clock. Then the resultant force will be into the clock if the product qm is positive, and out of the clock if qm is negative. If q were located at 9 o'clock, but still moving upward, the resultant force would be reversed from the previous case. Thus, the simultaneous presence of electric and magnetic charges would permit an absolute distinction between left and right. In other words, the laws of motion of charges would no longer be symmetric under reflection; they would violate parity. However, the laws would still be symmetric under a "generalized reflection," in which all magnetic charges present are reversed in sign when spatial coordinates are inverted. This is significant because a magnetically neutral system, even if it contained monopoles, could still obey reflection symmetry. Similar considerations apply to symmetry with respect to the direction of flow of time. Again, a generalized time-reversal symmetry is obeyed even if monopoles exist, provided there are also particles with exactly the same properties except for opposite magnetic charge. Monopoles cannot explain the well-known parity violation effects in weak interactions, but they might be connected with the much feebler violation of time-reversal symmetry found in the decays of neutral K-mesons. *See* PARITY; SYMMETRY LAWS.

Charge quantization. The quantity $S = qm/c$ has the dimensions of angular momentum. In fact, S is the magnitude of the angular momentum stored in the electromagnetic field when a point charge and a monopole are simultaneously present. The axis of this angular momentum is the line from charge to monopole. During the motion of a charge past a stationary monopole, the angular momentum $\mathbf{L} = \mathbf{r} \times \mu \mathbf{v}$ of the charge is not conserved (here $\mathbf{r}$ is the position and μ the mass of the charged particle). This is clear because $\mathbf{L}$ is perpendicular to the plane defined by the position and velocity, but the magnetic force tilts $\mathbf{v}$ out of the plane it is in at a given moment, so that $\mathbf{L}$ is also tilted. There does exist a total angular momentum which is conserved, $\mathbf{J} = \mathbf{L} + \mathbf{S}$, the sum of the angular momentum of the electric charge and that stored in the electromagnetic field. In quantum theory, angular momentum around a given axis is measured in multiples of $\hbar/2$, where $\hbar$ is Planck's constant divided by 2π. If this principle is applied to the angular momentum S, one has the Dirac quantization condition $qm/c = n\hbar/2$, where n is any integer. This result was obtained in 1931 by P. A. M. Dirac on the basis of quantum-mechanical analysis. As Dirac pointed out, the condition implies that the charge of any monopole must be a multiple of $m_0 = \hbar c/2e$, where e is the smallest known charge, that of an electron. Reciprocally, the existence of even one monopole m in the universe implies that electric charge must come in multiples of $q_0 = \hbar c/2m$. Thus, the existence of monopoles might be said to explain the quantization of electric charge. This argument is somewhat less attractive now than when Dirac first proposed it, since many properties of elementary particles besides charge are known to be quantized. Furthermore, in theories where the electromagnetic field belongs to a set of fields related by symmetry transformations in some abstract space (as happens in so-called non-Abelian gauge field theories), it is often possible to deduce the quantization rules from the symmetry, analogous to the deduction of angular momentum quantization from rotation symmetry. Some such theories actually predict "composite" magnetic monopoles, made by coupling electromagnetic and other fields together.

Properties of Dirac monopoles. A monopole obeying the condition $m = nm_0$ is called a Dirac monopole. The minimum magnetic charge is $m_0 = \hbar c/2e = e/2\alpha \approx 69e$, where α is the fine-structure constant $e^2/\hbar c$. Several characteristics would make a Dirac monopole easy to identify. Because of its great charge, a rapidly moving monopole would ionize atoms far more effectively than an electron, thus producing a unique, very dense track. The end of the track would show a decrease in ionization density, in contrast to that of an electric particle. This happens because the electric force on a monopole is proportional to its velocity. One would expect monopoles to be bound in ferromagnetic materials, with binding to the bulk material that could approach the kilo-electron-volt level, and even stronger binding to small clusters of atoms. In other solids, electron-volt binding might occur. If these estimates were correct, then monopoles incident on the Earth (or the Moon!) could be trapped in surface material for geologic times. Even such trapped monopoles could be accelerated to very high energies by applied magnetic fields, making them easily liberated and detected. Alternatively, material containing a monopole and passed through a wire coil would cause a transient rise in voltage across the coil, by Faraday's law of magnetic induction. No other entity could duplicate this effect.

The special features of monopoles have all been used in one or more of the many searches in magnetic or other materials, with particle accelerators, and in cosmic radiation. In 1975, one track, which could have been produced by a monopole of charge 137 times the electron charge, was reported from a balloon-borne cosmic-ray experiment. The track might have been made by an electrically charged particle of about 70 electron charges but with a mass more than 10^3 times greater than the proton mass. However, some more mundane effects might have combined in an unusual way to imitate the track of such a bizarre object. Taken at face value, this event corresponds to a monopole flux of about 10^{-13} monopole/cm²s incident on the upper atmosphere (the rate is determined using all balloon-borne cosmic-ray experiments). If confirmed such a large flux would be hard to explain in view of previous null results from surface-based experiments as well as Moon rock studies. A possible explanation would be that monopoles are so exceedingly massive that even at nonrelativistic velocities they could easily penetrate the Earth, and only a rare, super-slow one could be trapped.

[A. S. GOLDHABER]

Bibliography: P. A. M. Dirac, Quantised singularities in the electromagnetic field, *Proc. Roy. Soc. (London), Ser. A,* 133:60−72, 1931; K. W. Ford, Magnetic monopoles, *Sci. Amer.,* 209:122−131, December 1963; A. S. Goldhaber and J. Smith, Hypothetical particles, *Rep. Progr. Phys.,* 38:731−770, 1975; G. 't Hooft, Magnetic monopoles in unified gauge theories, *Nucl. Phys.,* B79:276−284, 1974; P. B. Price et al., Evidence for detection of a moving magnetic monopole, *Phys. Rev. Lett.,* 35:487−490, 1975.

Magnetic permeability

A factor, characteristic of a material, that is proportional to the magnetic flux density (magnetic induction) B produced in the material by a magnetic field divided by the intensity of the field H. Permeability is usually represented by the Greek letter μ.

Absolute permeability. Consider a solenoid that has been bent into a circular form so that the ends are joined together. This winding is called a toroid (Fig. 1). For a closely wound toroid, almost all the flux is in the interior. The intensity of the magnetic field inside the toroid is given by Eq. (1), where l is

$$H = \frac{NI}{l} \qquad (1)$$

the mean circumference of the toroid and I is the current in the toroid coil. *See* MAGNETIC FIELD.

The flux density B within the toroid is given, from Ampère's law, by Eq. (2) if the toroid is in empty space. Then Eq. (3) holds. If a medium

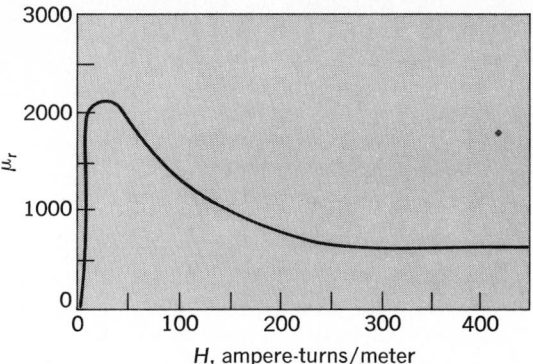

Fig. 2. Permeability of wrought iron as a function of magnetic field strength.

takes the place of the empty space within the toroid, the value of B changes for the same value of H. The ratio of B to H, given in Eq. (4), is called

$$B_0 = \mu_0 \frac{NI}{l} \qquad (2)$$

$$\mu_0 = \frac{B_0}{H} \qquad (3)$$

$$\mu = \frac{B}{H} \qquad (4)$$

the absolute permeability of the medium. *See* AMPÈRE'S LAW.

Since the mks unit of B is the weber per square meter, and the corresponding unit of H is the ampere per meter, the mks unit of permeability is the weber per ampere-meter. A second unit is the henry per meter. That these two units are equivalent is seen from the relationship $L = N\Phi/I$. Since the inductance L is in henrys when the flux Φ is in webers and I is in amperes, the henry is a weber per ampere. Thus, a henry per meter is the same as a weber per ampere-meter. *See* INDUCTANCE.

Relative permeability. It is convenient to define another quantity, called the relative permeability, μ_r, as the ratio of the permeability μ of the material to the permeability μ_0 of empty space, as in Eq. (5). Relative permeability is a pure number,

$$\mu_r = \frac{\mu}{\mu_0} \qquad (5)$$

and independent of the system of units used. The permeability of free space has the numerical value of $4\pi \times 10^{-7}$ henry per meter. *See* ELECTRICAL UNITS AND STANDARDS.

Materials may be classified in terms of their relative permeabilities. Diamagnetic materials have values of μ_r a little less than unity; paramagnetic materials have μ_r a little greater than unity. Ferromagnetic materials are those that have relative permeabilities considerably greater than unity, which are variable, depending upon the value of H and the previous magnetic history. For some ferromagnetic materials, the maximum value of μ_r may be in the thousands (Fig. 2). *See* DIAMAGNETISM; FERROMAGNETISM; PARAMAGNETISM.

Permeability measurement. The toroidal coil may be used to measure the permeability of a material if the material is used as the core of the ring.

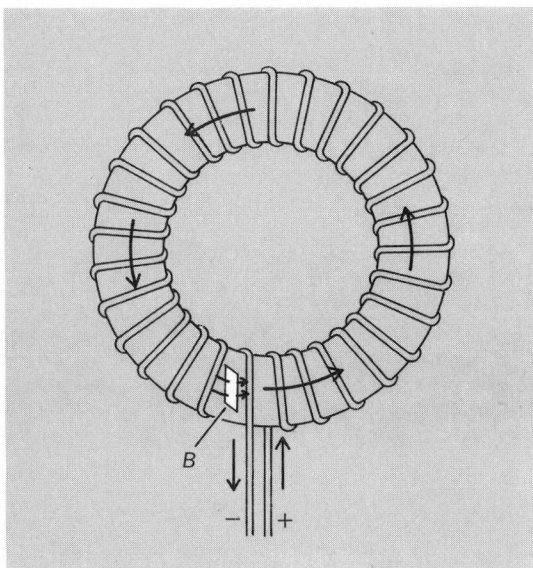

Fig. 1. A solenoid wound in the form of a toroid.

Such a ring is called a Rowland ring. A second small coil, called a search coil, is wound around part of the ring and connected to a ballistic galvanometer. When the current in the ring is reversed, there is a change in flux of 2Φ. The deflection of the galvanometer gives a measure of Φ. The value of B is then found by dividing Φ by the cross-sectional area of the ring. The value of H is found by the use of the relation $H = NI/l$, and μ is computed. The Rowland ring is used for this measurement because in the ring, the flux is all in one material that is being tested; thus, there are no poles, and the magnetic intensity H is that due to the current alone. Other devices for measurement of permeability are called permeameters. They are based on the same principle as the Rowland ring, but allow more rapid and convenient operation. *See* GALVANOMETER; MAGNETIC INDUCTION; SEARCH COIL.

[KENNETH V. MANNING]

Bibliography: B. I. Bleaney and B. Bleaney, *Electricity and Magnetism*, 3d ed., 1976; W. C. Michels, *Electrical Measurements*, 3d ed., 1957; F. W. Sears et al., *University Physics*, 5th ed., 1976.

Magnetic relaxation

The relaxation or approach of a magnetic system to an equilibrium or steady-state condition as the magnetic field is changed. This relaxation is not instantaneous but requires time. The characteristic times involved in magnetic relaxation are known as relaxation times.

Magnetism is associated with angular momentum called spin, because it usually arises from spin of nuclei or electrons. The spins may interact with applied magnetic fields, the so-called Zeeman energy; with electric fields, usually atomic in origin; and with one another through magnetic dipole or exchange coupling, the so-called spin-spin energy. Relaxation which changes the total energy of these interactions is called spin-lattice relaxation; that which does not is called spin-spin relaxation. (As used here, the term lattice does not refer to an ordered crystal but rather signifies degrees of freedom other than spin orientation, for example, translational motion of molecules in a liquid.) Spin-lattice relaxation is associated with the approach of the spin system to thermal equilibrium with the host material; spin-spin relaxation is associated with an internal equilibrium of the spins among themselves. *See* MAGNETISM; SPIN.

The measurement of relaxation times is of interest in the study of molecular motions and interactions. An important practical application of magnetic relaxation is the attainment of very low temperatures. For example, a sample of a paramagnetic salt can be cooled by liquid helium to about 1 K, then subjected to a magnetic field. Heat, evolved in the magnetization, will be dissipated into the liquid helium. If thermal contact to the helium is broken, and the magnetic field is then removed, the relaxation of the sample to more random orientation will be accompanied by a further decrease in temperature, in some cases to 0.01 K. A similar adiabatic demagnetization of copper, in this case involving nuclear spin, has cooled the sample to 0.00002 K. *See* ADIABATIC DEMAGNETIZATION.

Relaxation process. In an unmagnetized sample (ferromagnets are excluded for the moment) the spin orientations are random. In the presence of a static magnetic field and at thermal equilibrium, an excess of the spin vectors point in the lower energy orientation along the field. The build-up of magnetization following application of the field requires that some of the spins turn from antiparallel to parallel orientations. The energy given up in this turning process may go to either or both of two reservoirs. If the applied field is weak, the spin-spin energy constitutes an effective reservoir. If the applied field is strong, the spin-spin (free) energy is negligible and magnetization requires an interchange with the lattice. At low fields the transfer of energy into the spin-spin reservoir may be thought of as corresponding to a heating of the spin system. The spin-lattice coupling will eventually bring the spins into thermal equilibrium with the lattice (by changing either the spin "temperature" or the lattice temperature, or both). Thus for low fields the process of magnetization would exhibit two stages: (1) the redistribution of energy in the spin system, or spin-spin relaxation, and (2) equilibration of the spin system with the lattice, or spin-lattice relaxation. For high fields, only spin-lattice relaxation plays a role in establishing the magnetization.

Although relaxation exhibits itself in a macroscopic manner, an explanation of the rates of relaxation requires consideration of atomic processes. Two requirements must be met for spin-lattice relaxation: (1) There must be a coupling of the spin to the lattice, and (2) the coupling must be time-dependent, its frequency spectrum containing natural frequencies of the spin system (for example, the precession frequency of the spins in the applied field).

Types of relaxation. Relaxation has been studied for nuclear magnetism, electron paramagnetism, and ferromagnetism.

Nuclear relaxation. For nuclei, the Zeeman and spin-spin interactions are always important, and in some cases, the nonspherical charge distribution (electric quadrupole moment) of a nucleus affects the relaxation time. When the quadrupolar energy may be neglected, the magnetization usually approaches thermal equilibrium exponentially after application of a strong static field, the time constant being the spin-lattice relaxation time. Nuclear relaxation is almost invariably studied by magnetic resonance; the weak nuclear paramagnetism cannot be detected by other methods. *See* MAGNETIC RESONANCE.

If bulk nuclear magnetization $\mathbf{M}$ is tilted away from the direction of the static field $\mathbf{H}$, the $\mathbf{M}$ vector precesses about $\mathbf{H}$. The component of $\mathbf{M}$ parallel to $\mathbf{H}$ returns to its thermal equilibrium value by means of spin-lattice relaxation. The spin-spin coupling gives a spread in precession frequencies, causing the nuclei to get out of step with one another, as manifested by the decay of the components of $\mathbf{M}$ perpendicular to $\mathbf{H}$ without energy exchange with the lattice (spin-spin relaxation).

Rapid motion of a nucleus relative to its neighbors lengthens the spin-spin relaxation. The times range from tens of microseconds, in almost all solids, to several seconds, in cases of rapid nuclear motion. Measurement of spin-spin relaxation has been extensively used to study atomic motion.

Spin-lattice relaxation arises from magnetic dipole coupling with electrons in the case of metals, paramagnetic substances, or diamagnetic insulators possessing paramagnetic impurities; with other nuclear moments when the relative nuclear positions undergo large changes due to sufficiently rapid self-diffusion or molecular rotation in liquids, gases, or even some solids; or from the nuclear electric quadrupole coupling to the electric fields of the host material. Spin-lattice relaxation times range from less than 1 msec to several hours.

Paramagnetic relaxation. Resonance methods can be used for the study of paramagnetic relaxation. However, the observation of the ability of the magnetization to follow a changing field and even the observation of the change in the lattice temperature brought about by spin-lattice relaxation have also been used. The first interest in magnetic relaxation, in fact, stemmed from the desire to cool the lattice by adiabatic demagnetization of the spins of paramagnetic ions. Large electrostatic interactions between the ions and their surroundings make major contributions to the total spin energy, and it is rarely possible to neglect this effect. Other important contributions to the spin energy arise from dipolar coupling to applied fields, spin-spin coupling by both magnetic dipole and exchange mechanisms, and magnetic coupling to nuclei, particularly the nuclei of the paramagnetic ions.

The principal spin-lattice relaxation mechanisms involve modulation of the spin-orbit coupling by lattice motion, modulation of the magnetic coupling to nuclei by lattice motion, and coupling with paramagnetic impurity atoms. Typical spin-lattice relaxation times vary from less than 10^{-8} sec at room temperature to tens of seconds at liquid helium temperature. Although the study of paramagnetic relaxation was begun in the 1930s, it is not as thoroughly understood as the more recently studied nuclear relaxation, primarily because of the complexity of the electronic energy levels of the paramagnetic ions.

Ferromagnetic relaxation. In ferromagnetic substances, strong exchange coupling between electron spins, causing them to prefer an alignment parallel to one another, and demagnetizing effects cause ferromagnetic relaxation to differ from paramagnetic relaxation.

The strong tendency of ferromagnetic spins to remain parallel makes it convenient to distinguish between two types of relaxation: (1) that in which the total magnetization changes direction but not magnitude, thereby keeping the exchange energy constant, and (2) that in which the magnitude of the magnetization changes, producing a change in the exchange and also in the demagnetizing energies. The second mechanism depends on the sample shape and may arise from irregularities in the ferromagnetic lattice (impurities and nonstochiometric composition, as in ferrites) as well as from thermal vibrations. Ferromagnetic relaxation has been studied primarily by magnetic resonance.

[CHARLES P. SLICHTER]

Magnetic resonance

A phenomenon exhibited by the magnetic spin systems of certain atoms whereby the spin systems absorb energy at specific (resonant) frequencies when subjected to alternating magnetic fields. The magnetic fields must alternate in synchronism with natural frequencies of the magnetic system.

In most cases the natural frequency is that of precession of the bulk magnetic moment **M** of constituent atoms or nuclei about some magnetic field **H**. Because the natural frequencies are highly specific as to their origin (nuclear magnetism, electron spin magnetism, and so on), the resonant method makes possible the selective study of particular features of interest. For example, it is possible to study weak nuclear magnetism unmasked by the much larger electronic paramagnetism or diamagnetism which usually accompanies it.

Nuclear magnetic resonance (that is, resonance exhibited by nuclei) reveals not only the presence of a nucleus such as hydrogen, which possesses a magnetic moment, but also its interaction with nearby nuclei. It has therefore become a most powerful method of determining molecular structure. The detection of resonance displayed by unpaired electrons, called electron paramagnetic resonance, is also an important application. These two phenomena, as well as other related resonance phenomena, are discussed in this article. *See* MAGNETISM.

Origin. Because **M** has its origin in circulating currents or intrinsic spins, there is always an angular momentum **J** associated with it. The vector quantities **M** and **J** are related by Eq. (1), where γ is called the gyromagnetic ratio.

$$\mathbf{M} = \gamma \mathbf{J} \qquad (1)$$

For a system to exhibit a magnetic resonance it must possess a magnetic moment, possess angular momentum, and experience torques. The magnetic moment may arise from nuclei of atoms, from orbital electronic motion, from electronic spins, or from moving nuclear charges during molecular rotation. The angular momentum arises from the same sources. The torques may arise from externally applied magnetic fields; from magnetic dipole fields exerted by neighboring nuclei, atoms, or molecules; from electric fields acting on, for example, a nuclear electric quadrupole moment or the nonspherical electron cloud of an atom; or from electron exchange coupling. In any given case it is necessary to decide which interactions are large and which are small. Thus for a paramagnetic atom possessing a nuclear magnetic moment a distinction might be made between small applied static fields, in which the coupled nucleus and electron angular momenta act as a unit, and large magnetic fields, which decouple them so that they act independently. An effective angular momentum and magnetic moment can often be defined, giving an effective γ (in analogy to the Landé *g* factor of optical spectroscopy).

Several types of resonances have been observed; these differ in one or more of the three basic requirements. However, the principal features can be understood by assuming that the torques arise from an effective static applied magnetic field **H**. The torque **M** × **H** causes the angular momentum to change with time according to Eq. (2). The resultant motion of **M** is a precession

$$\frac{d\mathbf{J}}{dt} = \mathbf{M} \times \mathbf{H} \quad \text{or} \quad \frac{d\mathbf{M}}{dt} = \gamma \mathbf{M} \times \mathbf{H} \qquad (2)$$

at angular frequency γH about the direction of **H**. At thermal equilibrium **M** is parallel to **H**, and no precession occurs. Application of an alternating

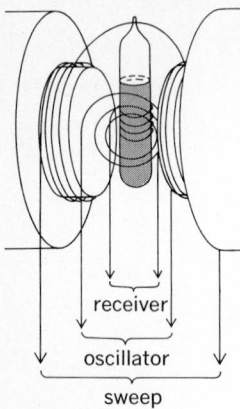

receiver

oscillator

sweep

Fig. 1. Arrangement of a sample and coils in a nuclear induction apparatus. (*From J. D. Roberts, Nuclear Magnetic Resonance, McGraw-Hill, 1959*)

magnetic field $H_x \cos \omega t$ perpendicular to **H** causes **M** to tilt away from **H** with a consequent absorption of energy, provided the resonant condition $\omega = \gamma H$ is satisfied. In practice the absorption takes place over a narrow range of frequency on both sides of γH. The magnetization M_x parallel to H_x obeys Eq. (3), where $\chi'(\omega)$ and $\chi''(\omega)$ are the real

$$M_x = H_x \chi'(\omega) \cos \omega t + H_x \chi''(\omega) \sin \omega t \qquad (3)$$

and imaginary parts of the complex magnetic susceptibility $\chi = \chi'(\omega) - j\chi''(\omega)(j = \sqrt{-1})$, and characterize dispersion and absorption respectively. For a typical resonance, $\chi''(\omega)$ attains maximum value for a region of frequencies near $\omega = \gamma H$.

For many cases it is necessary to analyze magnetic resonance by quantum theory. Consider a system composed of many (weakly interacting) identical parts (atoms or nuclei), each with angular momentum quantum number F (total angular momentum $= \sqrt{F(F+1)}\hbar$, where $\hbar$ is Planck's constant h divided by 2π). The spatial quantization in the magnetic field H gives $2F+1$ equally spaced energy levels labeled by the quantum number $M_F = (F, F-1, \ldots, -F)$ and energy spacing between adjacent levels of $\gamma\hbar H$. The field $H_x \cos \omega t$ produces transitions with the selection rule $\Delta M_F = \pm 1$. To satisfy the Bohr frequency condition, $\hbar\omega = \gamma\hbar H$, in agreement with the classical result, $\omega = \gamma H$.

According to quantum theory, the probability of transition from any energy level A to any other B is the same as that from B to A; thus a net absorption of energy requires that the population of the lower energy states be greater than that of the upper. The reverse situation in which the upper states are more populated leads to an induced emission and is the basis for the operation of the solid-state maser. At thermal equilibrium, as a result of spin-lattice relaxation, the lower energy states are more heavily populated in accordance with the classical Maxwell-Boltzmann statistics (ordinarily it is unnecessary to use either Fermi-Dirac or Bose-Einstein statistics). When H_x becomes sufficiently large, the level populations become disturbed from thermal equilibrium. The population difference between states joined by H_x decreases, a phenomenon known as saturation, because it causes χ' and χ'' to diminish with increasing H_x. Population differences between pairs of states other than A and B may simultaneously be increased, or even inverted, as in the three-level maser. The intensity of the alternating field necessary to produce saturation depends on the width of the absorption line and on the spin-lattice relaxation time (wider lines or shorter times require larger H_x). *See* MAGNETIC RELAXATION; MASER; NONRELATIVISTIC QUANTUM THEORY.

Observation. Experimentally it is possible to detect magnetic resonance by measuring the absorption of magnetic energy of a circuit containing the magnetic material or by measuring the change in inductance or resonant frequency of the circuit. The two methods measure $\chi''(\omega)$ or $\chi'(\omega)$ respectively. The resonant condition $\omega = \gamma H$ may be produced by varying ω, or more customarily, by changing H. In some experiments one tilts **M** away from the direction of **H** by alternating fields of short duration and then observes voltages induced

by the subsequent free precession of **M**. This method is particularly useful for studying relaxation times.

Nuclear magnetic resonance (NMR). The nuclei of many atoms possess angular momentum (spin) and nonvanishing magnetic moments. The former may be characterized by an angular momentum quantum number I (integer or half integer) of the nuclear particles. As far as is known, stable nuclei with an even number of neutrons and even number of protons have zero spin and magnetic moment, hence are incapable of exhibiting magnetic resonance. *See* NUCLEAR MOMENTS.

Nuclear resonances have been observed in insulators, metals, paramagnetic salts, antiferromagnetic substances, and other solids, and in gases and liquids. Often, to observe NMR a sample is placed between the poles of an electromagnet (Fig. 1) which in addition to the main winding carries a small auxiliary winding or sweep. A coil connected to an oscillator surrounds the sample, as does a second coil at right angles both to the oscillator coil and the sweep winding, to avoid direct coupling. The oscillator frequency is fixed and the sweep circuit is used to vary the magnetic field strength continuously. When a resonance frequency of the sample is reached, a signal induced in the second coil is detected and amplified. Typical resonance frequencies in a field of 10,000 gauss lie in the radio-frequency region (1–45 MHz). For example, the H^1 nucleus shows a resonance frequency of 42.6 MHz at this field strength. C^{13} nuclei give a much weaker signal, further decreased in a sample containing this isotope in its natural abundance of 1.1%.

Nuclei with quadrupole moments. If a nucleus has a spin ≥ 1, it generally has a nonvanishing electric quadrupole moment (there are good grounds for believing that all nuclei have zero electric dipole moments). The electrical interaction between the nucleus and electric potentials $V(x,y,z)$ from other charges depends on the nuclear orientation (specified by the direction of nuclear spin) and on the spatial second derivatives of the potential $\partial^2 V/\partial x^2$, $\partial^2 V/\partial x\,\partial y$, and so on, at the position of the nucleus. For potentials of spherical, tetrahedral, or cubic symmetry, the interaction energy is independent of orientation and may be disregarded. When the electric quadrupole interaction is nonzero but nevertheless much weaker than the static magnetic interaction, the unique resonance condition $\omega = \gamma H$ is changed, and the resonance line splits into $2I$ components centered about γH. A convenient method of determining the nuclear spin is thereby provided.

The name pure quadrupole resonance is used when one dispenses with the static field H and observes the reorientation of the nucleus among its various quantized orientations with respect to the electric potential alone. Classically, the nonspherical nuclear charge experiences a torque which causes precession.

Applications. Nuclear magnetic resonance has been used widely to measure nuclear magnetic moments, electric quadrupole moments, and spins. Because the resonance lines may be on occasion very sharp (1 cycle wide at 40 MHz), nuclear resonance is frequently employed to measure magnetic fields with great precision.

The extensive use of NMR in molecular structure determinations arises from the slight shift in the resonance frequency of an atom—commonly that of a proton—due to the environment of neighboring atoms. Because the magnitude of this shift depends on the type of environment, it is called the chemical shift. Figure 2 shows the NMR spectrum of ethyl alcohol, CH_3CH_2OH. The three main resonance frequencies are due to protons in the OH, CH_2, and CH_3 groups, respectively, and the spacing between them (which varies with the field strength) shows the chemical shift characteristic of the protons in these three typical structural groups.

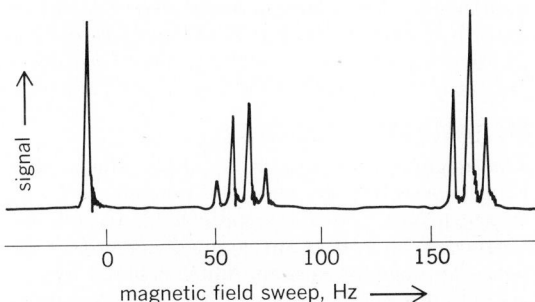

Fig. 2. Proton resonance spectra of ethyl alcohol at 40 MHz. (*From J. D. Roberts, Nuclear Magnetic Resonance, McGraw-Hill, 1959*)

The separate peaks at each frequency are due to spin-spin splitting. Often, the presence of n protons will split the frequency of a given, structurally different proton into $n + 1$ peaks, in direct analogy to ordinary spectral lines. The triplet in the NMR spectrum of ethyl alcohol results from the splitting of the frequency of the CH_3 protons by the two adjacent CH_2 protons; the quadruplet is at the typical CH_2 frequency and is split into four peaks by the three protons of the CH_3 group. *See* MOLECULAR STRUCTURE AND SPECTRA.

Because the time required for nuclear transitions is relatively long, substances undergoing fast reactions show altered NMR spectra, and rates of such rapid processes as ionization or intramolecular rotation can be measured.

Electron-nuclear double resonance (ENDOR). In this technique the magnetic resonance of a nucleus is detected by observing that of a nearby electron. The magnetic coupling between the nuclear and electronic magnetic moments gives rise to a back reaction on the electron resonance when the nuclei are brought into resonance. The sample under study is placed in a conventional electron resonance apparatus. In addition, an oscillator drives a coil to produce alternating magnetic fields at the sample under study, the frequency of alternation being in an appropriate range to produce nuclear transitions. Typically, the static magnetic field is adjusted to produce electron resonance. With the static magnetic field and the frequency of the electron resonance apparatus held fixed, the frequency of the nuclear resonance oscillator is then swept. As it passes through the resonant frequency of any nucleus that is coupled to the electron, a change in the electron absorption occurs.

The electron resonance and the nuclear resonance both represent transitions between energy levels of the combined system of electron and nucleus, each resonance being between a pair of levels. The strength of the electron resonance depends on the difference in population between its two levels. Thus, if one of these levels takes part in both resonances, the nuclear resonance may influence the electron resonance by changing the population of the common energy level.

Since the individual quanta absorbed in an electron resonance are much larger in energy than those absorbed in a nuclear resonance, the double resonance often permits the detection of the resonant absorption of a smaller number of nuclei than it would be possible to detect by a direct observation of the nuclear resonance. Since ENDOR requires both electron and nuclear resonances, it is applicable only to systems possessing both. It has been used, for example, to study the nuclear resonance associated with impurity atoms in semiconductors and that associated with F-centers in alkali halides, as well as to measure nuclear moments of rare elements. Important results have included the determination of the electronic structure of point imperfections and the measurement of nuclear magnetic moments and hyperfine anomalies.

Paramagnetic resonance. Magnetic resonance arising from electrons in paramagnetic substances or from electrons in paramagnetic centers in diamagnetic substances is called paramagnetic resonance. For applied fields of several thousand gauss the electron paramagnetic resonance (EPR) experiments are done at microwave frequencies, commonly at 3-cm or at 1-cm wavelengths. In some instances, nuclear resonance apparatus has been used with correspondingly lower applied fields. The most sensitive apparatus detects approximately 10^{12} electron spins for a line 1 gauss broad, a sensitivity far greater than obtained by nonresonant methods, such as those utilizing paramagnetic susceptibility.

Resonances have been observed in atoms of the iron group, rare earths, and other transition elements; in paramagnetic gases; in organic free radicals; in color centers in crystals (such as F- and V-centers); in metals (conduction electron spin); and in semiconductors (both conduction electron and impurity center spins).

When two paramagnetic ions or molecules approach one another, the spins become coupled via the exchange interaction. Much weaker couplings than the usual exchange interactions within atoms, in chemical bonds, or in ferromagnets produce pronounced effects.

Ferromagnetic resonance. In the case of both nuclear and paramagnetic resonance, the spins of neighboring atoms are nearly randomly oriented with respect to one another. In contrast, the electron spins in one domain of a ferromagnet are nearly all parallel for temperatures sufficiently below the Curie point. The alignment may be described in terms of the exchange coupling between neighboring spins, or equivalently in terms of the Weiss molecular magnetic field $\mathbf{H}_w$.

It is simplest to consider the case where magnetization is uniform throughout the sample. Neglecting relaxation effects, the equation of motion is still Eq. (2), but an effective field is substituted for $\mathbf{H}$, consisting of the applied static and alternating fields, the demagnetizing corrections (from the electron magnetic dipolar fields), and the effects of crystalline anisotropy. Because the Weiss molecular field is always parallel to **M**, **H**$_w$ exerts no torque and plays no role as long as the magnetization is uniform throughout the sample. (The exchange energy between spins does not change as long as their relative orientation does not change.)

The crystalline anisotropy can be shown to be equivalent to a magnetic field **H**$_A$ along the direction of easy magnetization as long as **M** points nearly in that direction.

Because of the demagnetizing effects, the resonant frequency depends on sample geometry. For an infinite plane perpendicular to the applied field, the resonant angular frequency ω is given by $\omega = \gamma\sqrt{BH}$, but for a sphere, $\omega = \gamma H$.

The large demagnetizing and exchange fields are the principal difference between ferromagnetic and paramagnetic resonance. The demagnetizing field has components $-N_x M_x$, $-N_y M_y$, and $-N_z M_z$, where N_x, N_y, and N_z are the demagnetizing coefficients. Thus, suppose the static field and **M** lie along the z direction. Application of the alternating field tilts **M** away from z, changing the effective z field. If the change brings the spins closer to resonance, **M** may tilt out more. It is possible for such a nonlinear effect to be unstable for sufficiently large alternating fields. This instability is utilized in the Suhl ferromagnetic amplifier.

Antiferromagnetic resonance. The two sublattices of spins in an antiferromagnet are strongly coupled together by exchange forces. If both magnetizations (**M**$_1$ and **M**$_2$) are tilted together, away from the normal direction of magnetization in the crystal (call this the z direction), the only change in energy results from the anisotropy fields. However, because the anisotropy fields are reversed in direction for the two lattices, the magnetizations **M**$_1$ and **M**$_2$ tend to precess in opposite directions, bringing about a change in the exchange energy. An external field along the z direction aids one anisotropy field but opposes the other. The resonant angular frequency ω for a sphere is given by Eq. (4),

$$\omega = \gamma(H \pm \sqrt{H_A(H_A + 2H_E)}) \qquad (4)$$

where H is the applied field, H_A the equivalent anisotropy field, and H_E the equivalent exchange field. The plus and minus signs refer to two opposite directions of rotating magnetic fields which may be used to observe the resonance. If H_E is 10^6 oersted, and H_A is 10^4 oersted, the corresponding frequency is 3×10^{11} Hz.

Ferrimagnetic resonance. Magnetic resonance in ferrites is called ferrimagnetic resonance. Ferrites are the natural generalization of antiferromagnets, containing two or more sublattices which may differ in magnetization. The basic coupling terms are still anisotropy fields, exchange fields, and the applied fields. The resonant angular frequency ω for the case of two sublattices is given by Eq. (5), where η is a parameter measuring the

$$\omega = \gamma\left[H - \frac{\eta H_E}{2}\right.$$
$$\left. \pm \sqrt{\left(\frac{\eta H_E}{2}\right)^2 + H_E H_A(2 - \eta) + H_A{}^2}\,\right] \quad (5)$$

relative sizes of the two magnetization vectors. Taking **M**$_1$ to be the smaller magnetization, η is defined by Eq. (6). This equation assumes the

$$\mathbf{M}_1 = (1 - \eta)\,\mathbf{M}_2$$

magnetizations to be at saturation and the two sublattices to have the same γ (deviations might differ from spin-orbit coupling). *See* MOLECULAR BEAMS.

[CHARLES P. SLICHTER]

Bibliography: A. Carington and A. D. McLachlan, *Introduction to Magnetic Resonance with Applications to Chemistry and Chemical Physics*, 1979; E. Kundla et al., *Magnetic Resonance and Related Phenomena*, 1980; K. A. McLauchlan, *Magnetic Resonance*, 1972; G. E. Pake, Magnetic resonance, *Sci. Amer.*, 199(2):58–66, 1958; R. T. Schumacher, *Introduction to Magnetic Resonance*, 1970; F. Seitz and D. Turnbull (eds.), *Solid State Physics*, vol. 2, 1956; C. P. Slichter, *Principles of Magnetic Resonance*, 2d ed., 1978.

Magnetic susceptibility

The magnetization of a material per unit applied field. It describes the magnetic response of a substance to an applied magnetic field. If M is the magnetization and H the applied magnetic field, then the magnetic susceptibility, denoted by χ, is given by Eq. (1). In the case that M is not parallel

$$\chi = M/H \qquad (1)$$

to H, χ is a tensor. Otherwise, it is a simple number. For a crystalline material, χ may depend upon the direction of H with respect to the axes of the crystal because of anisotropic effects. For an elementary discussion of M and H *see* MAGNETISM. *See also* MAGNETIZATION.

The magnetic susceptibility is expressed in a variety of ways: per gram, per atom, per unit volume, and per mole. In this article, electromagnetic units are used. In Eq. (1), the units of χ are ergs per oersted per unit volume. Figure 1 shows the atomic susceptibilities χ_A (ergs per oersted per atom) of the elements. The static susceptibility is measured in constant applied magnetic fields. The frequency-dependent susceptibility is measured in alternating magnetic fields. It is usually a complex quantity in which both the real and imaginary parts are functions of frequency.

Ferromagnetic susceptibility. The general behavior of the susceptibility of ferromagnetic materials above the Curie temperature follows the Curie-Weiss law, Eq. (2). This behavior is followed

$$\chi = \frac{C}{T - \theta} \qquad (2)$$

in the region well above the ferromagnetic Curie temperature T_c. The paramagnetic Curie temperature θ is usually slightly greater than the temperature of transition T_c. Comparison of θ and T_c for three ferromagnetic metals is given in the table.

All ferromagnetic materials exhibit paramagnetic behavior above their ferromagnetic Curie points. The magnitude of the paramagnetic susceptibility is determined by the Curie constant C which appears in Eq. (2). A typical value of C is 0.2 K/cm³ for iron. For the theory of ferromagnetic susceptibility *see* FERROMAGNETISM. *See also* CURIE TEMPERATURE; CURIE-WEISS LAW.

In the region just a fraction of a degree above the "critical point," or Curie temperature T_c, the susceptibility is found to approximate Eq. (3), with

$$\chi = C'/(T - T_c)^\gamma \qquad (3)$$

γ generally very close to 1.33. The theory is extremely complicated and not entirely satisfactory.

Below the Curie point, the static susceptibility is not usually defined for a ferromagnetic substance, since the ferromagnet may have a finite magnetization in zero applied field.

The initial permeability is the slope of the magnetization curve (B plotted against H) for magnetic field strength $H = 0$. Initial permeabilities vary from almost 300 (platinum-cobalt) to 100,000 (supermalloy).

The frequency-dependent permeability is that measured in an alternating magnetic field. The experiments are usually carried out in small magnetic fields so that it is the frequency dependence of the initial permeability which is measured. Such experiments give information on the structure of ferromagnetic domains and the motion of domain walls.

Paramagnetic susceptibility. Most paramagnetic substances at room temperature have a static susceptibility which follows a Langevin-Debye law, Eq. (4), where N is the number of magnetic

$$\chi = Np^2\mu_B^2/3kT + N\alpha \qquad (4)$$

dipoles per unit volume, p is the effective magneton number, μ_B is the Bohr magneton, k is Boltzmann's constant, T is the absolute temperature, and α is the temperature-independent contribution of Van Vleck paramagnetism. The first term of Eq. (4) is referred to as the Curie law because of its $1/T$ dependence. Experimental results are often expressed in terms of what the effective magneton numbers must be in order to account for the variation of χ with $1/T$. The interpretation of the experimental data reveals the nature of the energy levels of the paramagnetic ions, the symmetry of paramagnetic crystals, the effects of crystalline electric fields on the energy levels, and the influence of the paramagnetic ions on one another. *See* PARAMAGNETISM.

Saturation of the paramagnetic susceptibility occurs when a further increase of the applied magnetic field fails to increase the magnetization, because practically all the magnetic dipoles are already oriented parallel to the field. Saturation will occur either at very strong fields or at very low temperatures; that is, in cases when the approximation leading to the Langevin-Debye formula fails. Some paramagnetic solids have susceptibilities which follow a Curie-Weiss law rather than a Curie law.

Diamagnetic susceptibility. The susceptibility of diamagnetic materials is negative, since a diamagnetic substance is magnetized in a direction opposite to that of the applied magnetic field. The diamagnetic susceptibility is independent of temperature.

Diamagnetic susceptibility depends upon the distribution of electronic charge in an atom and upon the energy levels. Interpretation of the experimental data reveals the nature of the atomic wave functions of the atom in question. For the theory of diamagnetic susceptibility and a listing of values of the molar diamagnetic susceptibilities of several rare gases and ions in crystals *see* DIAMAGNETISM.

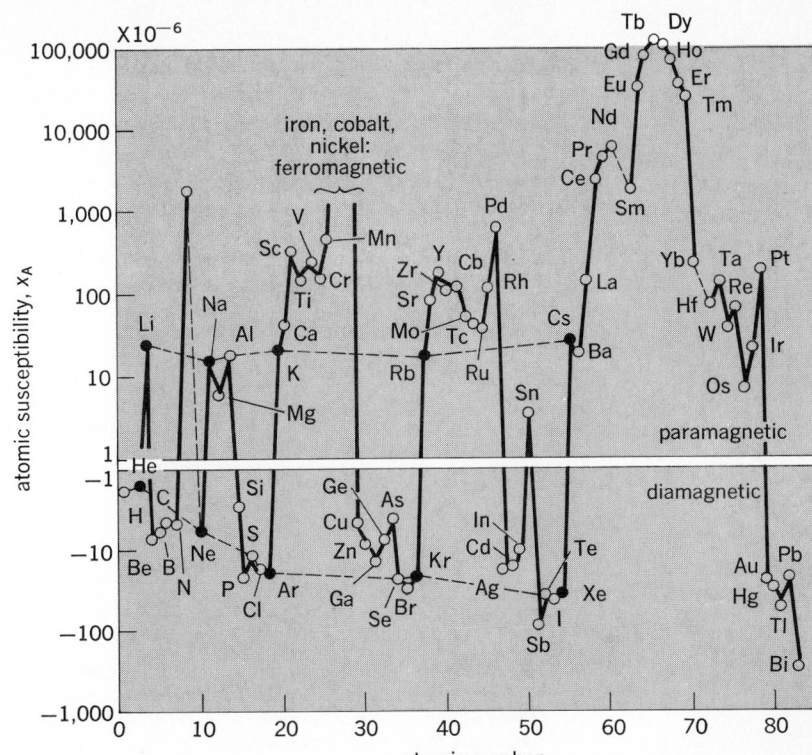

Fig. 1. Atomic susceptibilities of the elements at room temperature. The broken lines connect the alkali metals (paramagnetic) and the rare gases (diamagnetic). (*After R. M. Bozorth*)

Antiferromagnetic susceptibility. The susceptibility of antiferromagnetic materials above the Néel point, which marks the transition from antiferromagnetic to paramagnetic behavior, follows a Curie-Weiss law with a negative paramagnetic Curie temperature $-\theta$, as in Eq. (5). The Néel temperature is always somewhat less than θ.

$$\chi = C/(T + \theta) \qquad (5)$$

The antiferromagnetic susceptibility is a maximum at the Néel temperature. Below the Néel temperature, it behaves in a way determined by the angle between the field direction and the crystal axes — it usually decreases as the temperature lowers. For the theory of antiferromagnetic susceptibility *see* ANTIFERROMAGNETISM.

Measurement of susceptibility. Magnetic susceptibilities may be measured by several methods, depending upon the quantity sought. Ferromagnetic static permeabilities may be measured by the Rowland ballistic method. In this technique, the specimen is cut in the shape of a ring, and a magnetic field is applied by means of a primary winding on the ring. A secondary winding is connected to a ballistic galvanometer. The current through the primary is changed suddenly. The resulting abrupt change in magnetic field H causes a change in magnetic induction B ($B = H + 4\pi M$) in the specimen which in turn induces a voltage in the secondary. Thus, the galvanometer suffers a deflection proportional to the change in B. One starts with a known value of B (usually zero) and plots the magnetization curve in this way.

Paramagnetic and diamagnetic susceptibilities may be measured by the balance method. A magnetized substance in an inhomogeneous magnetic

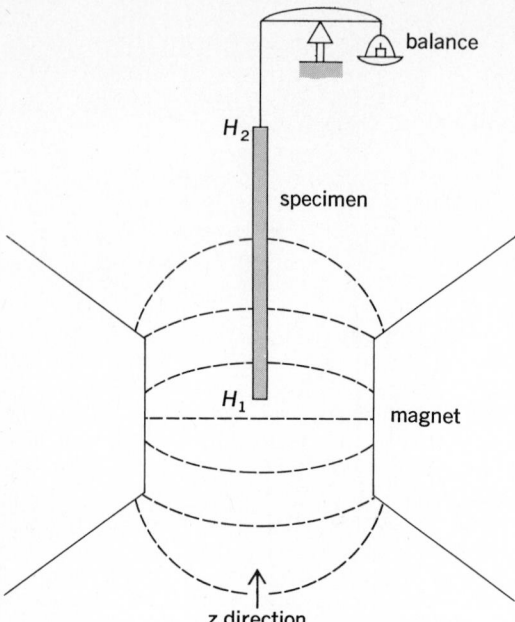

Fig. 2. Gouy balance method for measurement of magnetic susceptibilities. (*After C. Kittel, Introduction to Solid State Physics, 3d ed., copyright © 1966 by John Wiley and Sons, Inc.; used with permission*)

field experiences a force given by Eq. (6).

$$F = \frac{1}{2}\ \text{grad}\ \int M \cdot H\, dV \qquad (6)$$

The integral is over the volume of the specimen. The susceptibility is $\chi = M/H$. Therefore Eq. (7) holds.

$$F = \frac{1}{2}\chi\ \text{grad}\ \int H^2 dV \qquad (7)$$

If the magnetic field now varies only in one direction, say the z-direction, then Eq. (8) is valid,

$$F = \frac{1}{2}\chi A \int \frac{d}{dz}(H^2)\, dz = \frac{1}{2}\chi A(H_1{}^2 - H_2{}^2) \qquad (8)$$

where the specimen is in the shape of a rod of cross-sectional area A and is suspended in the z-direction in the inhomogeneous field (Fig. 2). In Fig. 2, H_1 and H_2 are the values of the magnetic field strength at the two ends of the rod. The force is measured and, when H_1 and H_2 (or H and dH/dz) are known, the susceptibility can be determined. This is known as the Gouy balance method. Note that the total magnetic susceptibility is measured, both para- and diamagnetic contributions.

The paramagnetic susceptibility may sometimes be measured separately by spin-resonance techniques. This is especially true of the Pauli paramagnetic susceptibility of conduction electrons in alkali metals. *See* MAGNETIC RESONANCE.

[ELIHU ABRAHAMS; FREDERIC KEFFER]

Bibliography: N. W. Ashcroft and N. D. Mermin, *Solid State Physics*, 1976; C. Kittel, *Introduction to Solid State Physics*, 5th ed., 1976; J. H. Van Vleck, *The Theory of Electric and Magnetic Susceptibilities*, 1932.

Magnetism

Magnetism comprises those physical phenomena involving magnetic fields and their effects upon materials. Magnetic fields may be set up on a macroscopic scale by electric currents or by magnets.

On an atomic scale, individual atoms cause magnetic fields when their electrons have a net magnetic moment as a result of their angular momentum. A magnetic moment arises whenever a charged particle has an angular momentum. It is the cooperative effect of the atomic magnetic moments which causes the macroscopic magnetic field of a permanent magnet. *See* ELECTRON SPIN; MAGNET; MAGNETIC MOMENT.

History. The name magnet was used by the Greeks for a stone which was capable of attracting other pieces of the same material and iron as well. This original magnet (lodestone) is the naturally occurring magnetic iron oxide, magnetite. Some of the properties of magnetite were discovered earlier than 600 B.C., although it is only in the 20th century that physicists have begun to understand why substances behave magnetically. Magnetism is one of the earliest known physical phenomena of solid materials.

William Gilbert (1540–1603) was the first to apply scientific methods to a systematic exploration of magnetic phenomena. His greatest contribution was the discovery that the Earth is itself a magnet. He distinguished clearly between electricity and magnetism. Quantitative studies of magnetism began in the early 18th century. Charles Coulomb (1736–1806) established the inverse-square law of force between magnetic poles and also between electric charges. S. D. Poisson (1781–1840) set up the basis of magnetostatics through application of potential theory to the problem of the forces between magnetized bodies. In the 19th century James Clerk Maxwell (1831–1879) established the relationship between magnetism and electricity by developing the science of electromagnetism. *See* COULOMB'S LAW; ELECTRICITY; ELECTROMAGNETISM; MAGNETOSTATICS.

Magnetic field. A magnetic field is said to occupy a region when the magnetic effect of an electric current or of a magnet upon a small test magnet which is brought in the vicinity is detectable. Because of the magnetic effect, a torque will be exerted on the test magnet until it becomes oriented in a particular direction. The magnitude of this torque is a measure of the strength of the magnetic field, and the preferred direction of orientation is the direction of the field. In electromagnetic units, the magnitude of the torque is given by Eq. (1),

$$L = -\,|\boldsymbol{\mu}|\,|\mathbf{H}|\,\sin\theta \qquad (1)$$

where $\boldsymbol{\mu}$ is the magnetic moment of the test magnet, $\mathbf{H}$ is the magnetic field strength, and θ is the angle between the direction of $\boldsymbol{\mu}$ and the direction of $\mathbf{H}$.

When a magnetic material is placed in a magnetic field $\mathbf{H}$, it becomes magnetized; that is, it becomes itself a magnet. The intensity of this induced magnetism is called the magnetization $\mathbf{M}$. More precisely, $\mathbf{M}$ is the magnetic moment per unit volume of the material. A vector field, the magnetic induction $\mathbf{B}$, is often defined to describe the magnetic forces anywhere in space. In electromagnetic units, the definition of $\mathbf{B}$ is given by Eq. (2). *See* MAGNETIC FIELD; MAGNETIC FLUX; MAGNETIC INDUCTION; MAGNETIZATION.

$$\mathbf{B} = \mathbf{H} + 4\pi\mathbf{M} \qquad (2)$$

Classification of substances. Materials can be grouped according to their magnetic behavior, although there is some overlap among groups.

1. Diamagnetic substances have a negative magnetic susceptibility; they are magnetized in a direction opposite to that of an applied magnetic field. *See* DIAMAGNETISM; MAGNETIC SUSCEPTIBILITY.

2. Paramagnetic substances have a positive magnetic susceptibility such that their magnetization is parallel and usually proportional to the applied magnetic field. *See* PARAMAGNETISM.

3. A ferromagnetic substance is one with net atomic moments which, within a certain temperature range, tend to line up in such a way that there can exist a net magnetization in the absence of an applied field. This category includes all ferromagnets and ferrimagnets and some helimagnets. At sufficiently high temperatures all these substances become paramagnetic. *See* CURIE TEMPERATURE; FERRIMAGNETISM; FERROMAGNETISM; HELIMAGNETISM.

4. An antiferromagnetic substance is one with net atomic moments which, within a certain temperature range, tend to line up in such a way that there can exist no net magnetization in the absence of an applied field. This category includes all antiferromagnets and some helimagnets. *See* ANTIFERROMAGNETISM; LOW-TEMPERATURE PHYSICS; MAGNETIC PERMEABILITY; MAGNETIC RESONANCE; MAGNON.

[ELIHU ABRAHAMS; FREDERIC KEFFER]

Bibliography: A. H. Morrish, *The Physical Principles of Magnetism*, 1965, reprint 1976; D. Wagner, *Introduction to the Theory of Magnetism*, 1972.

Magnetization

The process of becoming magnetized; also the property and in particular the extent of being magnetized. Magnetization has an effect on many of the physical properties of a substance. Among these are electrical resistance, specific heat, and elastic strain. *See* MAGNETOCALORIC EFFECT; MAGNETORESISTANCE; MAGNETOSTRICTION.

The magnetization **M** of a body is caused by circulating electric currents or by elementary atomic magnetic moments, and is defined as the magnetic moment per unit volume of such currents or moments. In the electromagnetic system of units (emu), **M** is measured in gauss; in the mks system, **M** is measured in webers per square meter (webers/m^2). For **M**, 1 weber/m$^2 = 10^4/4\pi$ gauss.

The magnetic induction or magnetic flux density **B** is given by Eq. (1), where **B** and **M** are in gauss,

$$\mathbf{B} = \mathbf{H} + 4\pi\mathbf{M} \quad \text{(emu)} \tag{1}$$

and **H**, the applied magnetic field, is in oersteds (equivalent to gauss); or by Eq. (2), where **B** and **M**

$$\mathbf{B} = \mu_0\mathbf{H} + \mathbf{M} \quad \text{(mks)} \tag{2}$$

are in weber/m^2, **H** is in amp-turns/m, and μ_0, the permeability of free space, is defined as $4\pi \times 10^{-7}$ henries/m, that is, webers/(amp-turn)(m). *See* ELECTRICAL UNITS AND STANDARDS; MAGNETIC INDUCTION.

The permeability μ of a substance is defined as the ratio **B/H** (in emu) or **B**/μ_0**H** (in mks). The magnetic susceptibility χ is defined as the ratio **M/H** (in emu) or **M**/μ_0**H** (in mks). From Eqs. (1) and (2) Eq. (3) is obtained, where the 4π is to be

$$\mu = 1 + (4\pi)\chi \tag{3}$$

used only in the emu system. It is to be noted that the magnitude of μ is the same in both systems of units, but the magnitude of χ differs by a factor of 4π. Both μ and χ may be tensors, although usually they are simple numbers. *See* MAGNETIC PERMEABILITY; MAGNETIC SUSCEPTIBILITY.

The topic of magnetization is generally restricted to materials exhibiting spontaneous magnetization, that is, magnetization in the absence of **H**. All such materials will be referred to as ferromagnets, including the special category of ferrimagnets. A ferromagnet is composed of an assemblage of spontaneously magnetized regions called domains. Within each domain, the elementary atomic magnetic moments are essentially aligned, that is, each domain may be envisioned as a small magnet. An unmagnetized ferromagnet is composed of numerous domains, oriented in some fashion as shown in Fig. 1, so that the total magnetization is zero.

The process of magnetization in an applied field **H** consists of growth of those domains oriented most nearly in the direction of **H** at the expense of others (Fig. 2a), followed by rotation of the direction of magnetization against anisotropy forces (Fig. 2b). For discussions of anisotropy and of domains *see* FERROMAGNETISM.

On removal of the field **H**, some magnetization will remain, called the remanence $\mathbf{M}_r$.

Magnetization curves. These curves, sometimes called *B-H* curves, are used to describe magnetic materials. They are plotted with **H** as abscissa and with either **M** or **B** as ordinate. In Fig. 3, $\mathbf{B}_r$ is the remanent induction ($\mathbf{B}_r = 4\pi\mathbf{M}_r$); $\mathbf{H}_c$ is the coercive force, or reverse field required to bring the induction **B** back to zero; and $\mathbf{M}_s$ is the saturation magnetization, or magnetization when all domains are aligned. The saturation magnetization is equal to the spontaneous magnetization of a single domain, except that it is possible to increase this magnetization slightly by application of an extremely large field. Saturation magnetization is temperature dependent, and disappears completely above the Curie temperature T_c where a ferromagnet changes into a paramagnet. The table lists $\mathbf{M}_s$ and T_c for a few ferromagnetic materials.

The initial permeability μ_i of a substance is the slope of the magnetization curve at **H** = 0. There is a definite correlation between initial permeability

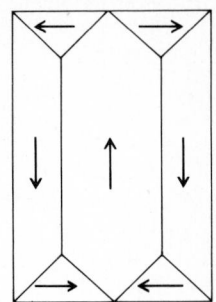

Fig. 1. Possible arrangement of domains in an unmagnetized single crystal. The arrangement is much more haphazard in polycrystals.

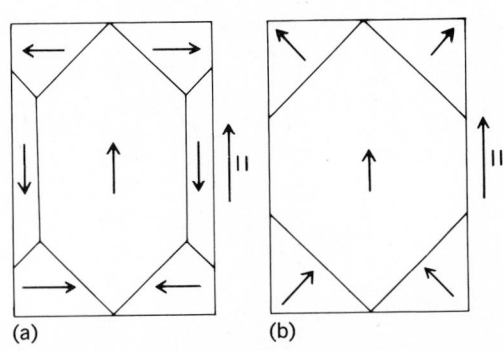

Fig. 2. The process of magnetization. (a) Domain growth. (b) Rotation.

Values of saturation magnetization M_s and ferromagnetic Curie temperature T_c

Substance	M_s at 293 K, gauss	M_s at 0 K, gauss	T_c, K
Fe	1707	1752	1043
Co	1400	1446	1394
Ni	485	510	631
Gd	0	1980	293
MnBi	600	675	670
Fe_3O_4	480	510	858
EuO	0	1910	77
$Y_3Fe_5O_{12}$	135	195	560

and coercive force; that is, materials of large μ_i have small H_c and vice versa.

The relationship between **B** and **H** may be studied by means of a Rowland ring or toroid in which the core is the material to be studied. For this arrangement, the numerical value of **H** can be computed from the relation in Eq. (4), where N is

$$H = \frac{NI}{l} \qquad (4)$$

the number of turns of the coil, I is the current in the coil, and l is the length of the mean circumference of the coil. If the radius of the ring is large compared to the radial distance across the core, the flux density will be nearly uniform.

The flux density **B** is not a linear function of **H**, and furthermore, its value depends not only upon **H** but upon the previous magnetic history of the sample used.

The ferromagnetic core of the Rowland ring may be demagnetized by successive reversals of the current as the current is gradually reduced from a sufficiently high initial current to a current that will produce a magnetizing force less than any value to be used subsequently. *See* DEMAGNETIZATION.

Changes in **B**, but not **B** itself, can be measured by wrapping a search coil on the toroid and observing the throw of a ballistic galvanometer.

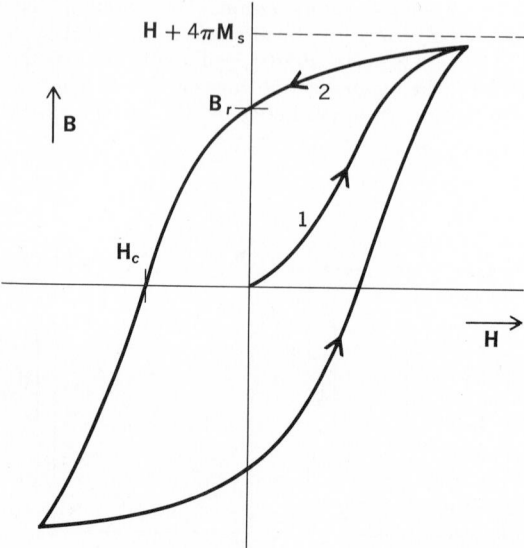

Fig. 3.   Magnetization curves.

Normal magnetization curve. When the core of the Rowland ring is initially demagnetized, $\mathbf{B} = 0$ and $\mathbf{H} = 0$. If the current is quickly increased to give a predetermined value $\mathbf{H_1}$, a flux $\Phi_1 - 0$ will be measured by the throw of the ballistic galvanometer. From the flux and the area A of cross section of the core, the value of $\mathbf{B} = \Phi/A$ corresponding to the selected **H** is computed. Then, without reducing the current, a second increase in current is made to reach $\mathbf{H_2}$. Again, the throw of the galvanometer measures the change in flux $\Phi_2 - \Phi_1$, and hence Φ_2 is found and $\mathbf{B_2}$ is computed corresponding to $\mathbf{H_2}$. By continuing this process for as many pairs of **B** and **H** as are necessary, points on the magnetization curve are obtained. The curve obtained starting with the core demagnetized is called the normal magnetization curve (Fig. 4). If the specimen were not initially demagnetized, the process here described would yield a magnetiza-

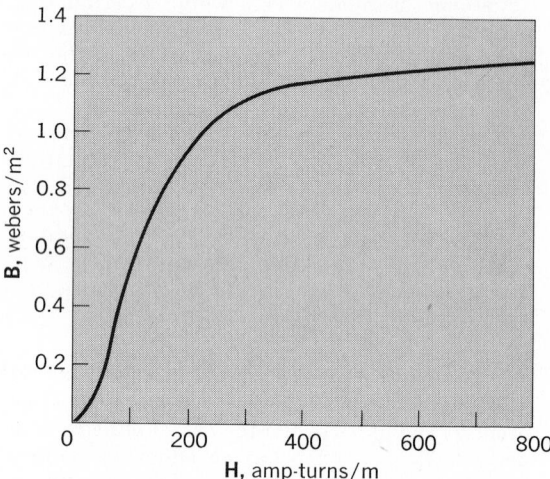

Fig. 4.   Normal magnetization curve.

tion curve that would differ in shape from the normal magnetization curve and, in general, would not pass through the origin.

In the curve of Fig. 4, **B** increases slowly at first as **H** rises, then increases rapidly until the "knee" of the curve is reached. Here, the rate of change of **B** with respect to **H** decreases and becomes small as saturation is approached.

Single crystals. Magnetization curves of single crystals depend upon the direction of **H** with respect to the crystallographic axes. If a weak field is applied to a crystal, those domains whose magnetic moments are most nearly parallel to **H** increase in size at the expense of those in other directions. There is then a small net contribution of the domains to the flux. As the field increases, the boundaries of the domains continue to change, and **B** increases rapidly along the steep part of the curve. Near the end of this change, all the magnetic moments in the domains rotate in a direction parallel to those that have been increasing. The rotation within the domains requires greater change in **H**, and thus the slope changes at the knee. As **H** is further increased, the domains rotate until their magnetic moments are all parallel to the field and

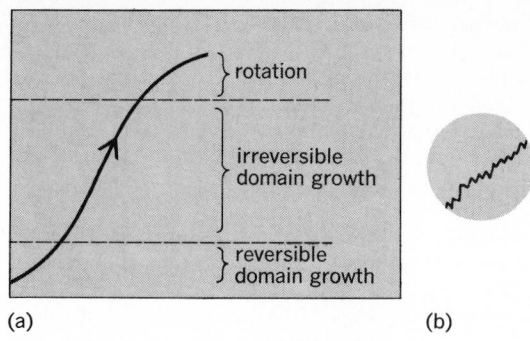

(a)   (b)

Fig. 5. Domain growth. (a) Portion of a magnetization curve, showing magnetization processes. (b) Enlarged section of path, showing Barkhausen effect.

saturation has been reached.

The ferromagnetic materials in common use are polycrystalline; that is, a piece of the material consists of a tremendous number of single crystals of random orientation. In each single crystal, there are many domains. The magnetization of the whole body consists of the magnetization of the various single crystals within that body.

Domain growth. The process of magnetization is shown in Fig. 5a. Domain growth proceeds by movement of the so-called Bloch wall between domains. This takes place reversibly at first, then irreversibly. Irreversible motion causes sudden almost discontinuous changes in magnetization, called the Barkhausen effect (Fig. 5b).

Hysteresis. The irreversible nature of magnetization is shown most strikingly by the fact that the path of demagnetization does not retrace the path of magnetization—the path 2 of Fig. 3 does not retrace path 1. There is a tendency for the magnetization to show hysteresis, that is, to lag behind the applied field, and the loop of Fig. 3 is called a hysteresis loop. The area of the loop is a measure of the energy lost to heat, per cycle, in going around the loop, and therefore is related to the energy lost by a magnet in an alternating field. An additional energy loss is that from eddy currents; this is absent in the nonconducting ferrites and may be minimized in metals by lamination. *See* MAGNETIC HYSTERESIS.

Demagnetizing fields. The effect of an applied field **H** on a sample is always reduced by a demagnetizing field $-N\mathbf{M}$ coming from the surface "poles," where N, the demagnetizing factor, depends upon the sample shape and orientation with respect to **H**. It varies from nearly 0 in a sample long and thin in the direction of **H**, to nearly 4π (in emu) in a sample short and fat, like a disk of revolution about **H**. Magnetization curves are generally obtained from long, thin samples to avoid demagnetization. The demagnetization effects are extremely important in ferromagnetic resonance. *See* MAGNETIC RESONANCE.   [ELIHU ABRAHAMS; FREDERIC KEFFER; KENNETH V. MANNING]

Bibliography: B. I. Bleaney and B. Bleaney, *Electricity and Magnetism*, 3d ed., 1976; S. Chikazumi and S. H. Charap, *Physics of Magnetism*, 1964, 2d ed., 1978; E. M. Pugh and E. W. Pugh, *Principles of Electricity and Magnetism*, 2d ed., 1970; H. Zijlstra, *Experimental Methods in Magnetism*, vol. 2, 1967.

Magnetocaloric effect

The reversible change of temperature accompanying the change of magnetization of a ferromagnetic or paramagnetic material. Thermodynamics theory shows that, for an adiabatic change of field ΔH, the change of temperature ΔT is given by the equation below, where c_H is the specific heat per unit

$$\frac{\Delta T}{\Delta H} = -\frac{T}{c_H}\left(\frac{\partial M}{\partial T}\right)_H$$

volume at constant H, and M is the magnetization.

This change in temperature may be of the order of 1°C, and is not to be confused with the much smaller hysteresis heating effect, which is irreversible. *See* THERMAL HYSTERESIS.

Except in antiferromagnets, $(\partial M/\partial T)_H$ is negative and an adiabatic decrease in H causes T to drop. This is the basis of the Giauque-Debye adiabatic demagnetization of paramagnetic salts, a technique which has achieved extremely low temperatures. *See* ADIABATIC DEMAGNETIZATION.

[ELIHU ABRAHAMS; FREDERIC KEFFER]

Magnetohydrodynamics

The science that deals with the dynamics or motion of a fluid interacting with a magnetic field. The fluid must be a good conductor of electricity and hence can be a liquid metal or, more usually, an ionized gas or plasma. Magnetohydrodynamics is important in the development of controlled thermonuclear reactors. In these devices, the fusion reaction takes place in a high-temperature plasma composed of heavy hydrogen isotopes; the plasma is surrounded by a magnetic field which serves to confine the plasma and isolate it from the walls of the reaction chamber. Magnetohydrodynamics is employed to study the usefulness of different plasma and magnetic field configurations for this purpose. *See* PLASMA PHYSICS.

Other applications include simulation of hypersonic flight conditions, ionic thrust for outer-space propulsion, space-vehicle braking upon reentry to the atmosphere, high-energy particle accelerators, microwave generators, thermionic energy-conversion devices, application of thin metallic coatings, and the study of cosmic and upper atmospheric phenomena.

Magnetohydrodynamics is alternately called hydromagnetics or magnetogas dynamics.

The conducting fluid and magnetic field interact through electric currents that flow in the fluid. The currents are induced as the conducting fluid and the magnetic field lines move across each other. In turn, the currents influence both the magnetic field and the motion of the fluid. Qualitatively, the magnetohydrodynamic interactions tend to link the fluid and the field lines so as to make them move together.

The generation of the currents and their subsequent effects are governed by the familiar laws of electricity and magnetism. The motion of a conductor across magnetic lines of force causes a voltage drop or electric field at right angles to the direction of the motion and the field lines; the induced voltage drop causes a current to flow as in the armature of a generator.

The currents surround themselves with magnetic field lines, heat the conductor, and give rise to mechanical ponderomotive forces when flowing across a magnetic field. (These are the forces which cause the armature of an electric motor to turn.) In a fluid, the ponderomotive forces combine with the pressure forces to determine the fluid motion. *See* ELECTRICITY; GENERATOR; MAGNETISM; MOTOR.

It has been suggested that magnetic disturbances which develop in the very hot and turbulent center of the Sun whip up to the surface along magnetic lines of force in this way to produce sunspots. Other theories relate the heating of the solar corona and the acceleration of cosmic rays to Alfvén waves. It has been possible to generate these waves in the laboratory by twisting a column of liquid sodium which is placed in a strong magnetic field.

Although some early work in magnetohydrodynamics has been concerned with liquid metals, a much wider interest has developed in phenomena which involve ionized gases or plasmas. A large electrical conductivity in a plasma requires a high density of high-energy electrons. This can occur for a plasma either in thermal equilibrium at relatively high temperatures, from a few electron volts upward (1 electronvolt is equivalent to a temperature of 11,400 K), or in a nonequilibrium situation where the ions and molecules remain at a low temperature and the electrons are supplied with energy by an external source such as a microwave generator or ultraviolet radiation.

Plasmas are encountered in interstellar space, in hot stars, and in the upper atmosphere, as well as in man-made devices into which energy is fed from electrical, chemical, or nuclear sources. Strong shock waves forming ahead of a blunt object traveling with hypersonic velocities through low-density air may heat the air sufficiently to ionize it.

FUNDAMENTAL LAWS

Magnetohydrodynamic phenomena involve two well-known branches of physics, electrodynamics and hydrodynamics, with some modifications to account for their interplay. *See* ELECTRODYNAMICS; HYDRODYNAMICS.

The basic laws of electrodynamics as formulated by J. C. Maxwell apply without any change. However, Ohm's law, which relates the current flow to the induced voltage, has to be modified.

It is useful to consider first the extreme case of a fluid with a very large electrical conductivity σ. Maxwell's equations predict, according to H. Alfvén, that for a fluid of this kind the lines of the magnetic field **B** move with the material. The picture of moving lines of force is convenient but must be used with care because such a motion is not observable. It may be defined, however, in terms of observable consequences by either of the following statements: (1) a line moving with the fluid, which is initially a line of force, will remain one; or (2) the magnetic flux through a closed loop moving with the fluid remains unchanged.

If the conductivity is low, this is not true and the fluid and the field lines slip across each other. This is similar to a diffusion of two gases across one another and is governed by similar mathematical laws. Numerically, the distance the magnetic field will slip through the fluid in a time t is $\delta = \sqrt{t/\mu\sigma}$,

where μ is the magnetic permeability (a constant depending upon the magnetic properties of the fluid). The condition that the conductivity be very large can now be stated more precisely: σ should be large enough so that the distance δ for the time of interest t is small compared to the dimension of the system L.

As in ordinary hydrodynamics, the dynamics of the fluid obeys theorems expressing the conservation of mass, momentum, and energy. These theorems treat the fluid as a continuum. This is justified if the mean free path λ of the individual particles is much shorter than the distances that characterize the structure of the flow. Although this assumption does not generally hold for plasmas, one can gain much insight into magnetohydrodynamics from the continuum approximation. The ordinary laws of hydrodynamics can then easily be extended to cover the effect of magnetic and electric fields on the fluid by adding a magnetic force to the momentum-conservation equation and electric heating and work to the energy-conservation equation.

The mathematical descriptions of electrodynamic and hydrodynamic phenomena—Maxwell's equations for the electromagnetic field and the equations of ordinary fluid dynamics—both involve a set of partial differential equations.

Maxwell's equations. Maxwell's equations, written in the rationalized mks system of units, are Eqs. (1)–(4). Equations (3) and (4) lead to the con-

$$\nabla \times \mathbf{E} + \frac{\partial \mathbf{B}}{\partial t} = 0 \tag{1}$$

$$\nabla \cdot \mathbf{B} = 0 \tag{2}$$

$$\nabla \times \mathbf{H} - \frac{\partial \mathbf{D}}{\partial t} = \mathbf{j} \tag{3}$$

$$\nabla \cdot \mathbf{D} = \rho_e \tag{4}$$

servation of charge density ρ_e defined in Eq. (5), where **j** is the current density.

$$\frac{\partial \rho_e}{\partial t} + \nabla \cdot \mathbf{j} = 0 \tag{5}$$

For definitions of the symbols and operations used for vectors and tensors in these and subsequent equations *see* CALCULUS OF TENSORS; CALCULUS OF VECTORS.

The electric and magnetic fields **E** and **B** are related to the electric displacement **D** and the magnetic induction **H** by the equations $\mathbf{B} = \mu\mathbf{H}$ and $\mathbf{D} = \epsilon\mathbf{E}$, where μ is the magnetic permeability and ϵ is the dielectric constant for the medium. For good electrical conductors, a very small local excess or deficiency of electrons compared to the positive charge carriers would be removed almost instantly by the resulting electric field. The charge equalization takes place in a time which is roughly the larger one of the two characteristic times $t_1 = \epsilon/\sigma$ and $t_2 = \sqrt{m\epsilon/ne^2}$, where n, m, and e are the number density, mass, and charge of the electrons. In a metallic conductor, t_1 is roughly 10^{-18} sec; in the ionosphere it drops from 10^{-9} sec at a height of 100 km (E layer) to 10^{-12} sec at 250 km (F layer). Usually t_2 is larger, and the corresponding values are 6×10^{-17} sec, 6×10^{-8} sec, and 4×10^{-8} sec.

With these high rates of charge neutralization,

it is not practical to calculate the electric field from the charge density through Eq. (4). The electric field is related much more effectively to the current distribution through Ohm's law, Eq. (6), which

$$\mathbf{E} = -\mathbf{v} \times \mathbf{B} + \frac{1}{\sigma}\mathbf{j} + \frac{m}{ne^2}\frac{\partial \mathbf{j}}{\partial t} \qquad (6)$$

has been reformulated to include an electric field induced by the velocity $\mathbf{v}$ of the fluid across a magnetic field and the effect of electron inertia. The latter, however, need be retained only for very rapid oscillation; otherwise the last term in Eq. (6) can be dropped.

The displacement current $\partial\mathbf{D}/\partial t$ in Eq. (3) is important only when currents can pile up electrical charges. Because of the high rate with which charges are neutralized in a good conductor, it is usually possible to drop the term. This brings about a considerable simplification of Maxwell's equations, because one can now, with the help of Ohm's law, eliminate the electric field altogether and arrive at the relation in Eq. (7).

$$\frac{\partial \mathbf{B}}{\partial t} = \nabla \times (\mathbf{v} \times \mathbf{B}) + \frac{1}{\sigma\mu}\nabla^2\mathbf{B} \qquad (7)$$

In this equation, the first term on the right makes the fluid and the field lines move together; the second term makes them slip across each other. The magnetic Reynolds number $R_m = vL\sigma\mu$ is a measure of the effect the motion of the fluid has on the magnetic field. When this number is very large, the second term can be dropped, producing a still simpler equation which is the mathematical basis for Alfvén's description.

Hydrodynamic equations. The conservation equations of hydrodynamics need additional terms, to take into account the interaction of the fluid with the electromagnetic field. The mass equation can remain unchanged. In the momentum equation one must add the force density $\mathbf{j} \times \mathbf{B}$. Just as the hydrodynamic force is expressed (when viscosity effects are included) as the divergence of the pressure tensor, so the magnetic force $\mathbf{j} \times \mathbf{B}$ can be expressed as the divergence of the magnetic part of Maxwell's stress tensor. This obvious analogy has led to regarding certain components of Maxwell's tensor as a magnetic pressure.

The energy equation requires the addition of a term, $\mathbf{j} \cdot \mathbf{E}$, to account for the transfer of energy from the electromagnetic field to the fluid. Thus, the three hydrodynamic conservation equations modified for magnetohydrodynamics are Eqs. (8)–(10). Gravitational forces can be added readily if necessary.

$$\frac{\partial \rho}{\partial t} + \nabla \cdot (\rho\mathbf{v}) = 0 \qquad (8)$$

$$\rho\frac{d\mathbf{v}}{dt} = -\nabla \cdot \mathbf{P} + \mathbf{j} \times \mathbf{B} \qquad (9)$$

$$\rho\frac{d}{dt}\left(E + \frac{v^2}{2}\right) = -\nabla \cdot (\mathbf{Pv} + \mathbf{Q}) + (\mathbf{j} \cdot \mathbf{E}) \qquad (10)$$

If the mean free path $\lambda \ll L$, the fluid is everywhere nearly in a state of thermal equilibrium, and one can express all state variables in terms of the density ρ, the temperature T, and the velocity $\mathbf{v}$. The components of the pressure tensor $\mathbf{P}$ and of the heat flow vector $\mathbf{Q}$ are given in Eqs. (11) and

(12), where $\delta_{ij} = 1$, if $i = j$; and $\delta_{ij} = 0$, if $i \neq j$. The

$$P_{ij} = \left[p + \tfrac{2}{3}\eta\left(\nabla \cdot \mathbf{v}\right)\right]\delta_{ij} - \eta\left(\frac{\partial v_i}{\partial x_j} + \frac{\partial v_j}{\partial x_i}\right) \qquad (11)$$

$$Q_i = -K\frac{\partial T}{\partial x_i} \qquad (12)$$

scalar pressure p, the viscosity η, the heat conductivity K, and the internal energy E per unit mass are functions of ρ and T which depend upon the fluid and which can be found experimentally or from kinetic theory.

Using Ohm's law in the form $\mathbf{j} = \sigma(\mathbf{E} + \mathbf{v} \times \mathbf{B})$, one can split $(\mathbf{j} \cdot \mathbf{E}) = \mathbf{v} \cdot (\mathbf{j} \times \mathbf{B}) + \mathbf{j}^2/\sigma$ into work done by the force $\mathbf{j} \times \mathbf{B}$ and the joule heating $\mathbf{j}^2/\sigma$. The three conservation equations lead to the entropy equation, Eq. (13). Here the rate of viscous

$$\rho T\frac{dS}{dt} = \rho\left(\frac{dE}{dt} + p\frac{d}{dt}\frac{1}{\rho}\right)$$
$$= \nabla \cdot (K\nabla T) + \eta\phi + \mathbf{j}^2/\sigma \qquad (13)$$

dissipation $\eta\phi$ is positive. In many problems the three dissipation terms on the right can be ignored and the entropy S then is conserved along a streamline.

Ohm's law. The formulation of Ohm's law given in Eq. (6) is still incomplete. Yet to be added are two effects due to the force $\mathbf{j} \times \mathbf{B}$ and the different rate of diffusion of electrons and ions in the presence of pressure and temperature gradients. The nature of these effects depends on the size of λ in relation to the radii $r_i = \sqrt{3m_iKT}/e\mathbf{B}$ of the spiral paths followed by particles with charge e and speed $\sqrt{3KT/m_i}$ in a field $\mathbf{B}$.

For the first time the detailed motion of the individual charged particles in the plasma must be considered. In a magnetic field, both positively and negatively charged particles describe helixes about the field lines (Fig. 1).

If a magnetic field $\mathbf{B}$ combines with an electric field $\mathbf{E}$, the center of the spiral drifts sideways in the direction of $\mathbf{E} \times \mathbf{B}$ rather than staying on a line of force. This drift is reduced roughly by a factor $1/[1 + (r_i/\lambda)^2]$ because of collisions. If $\mathbf{B}$ is small, the ion drift practically vanishes compared to the electron drift and the latter constitutes the Hall effect. If $\mathbf{B}$ is large, the two components drift nearly alike; this causes a mass flow but nearly cancels the Hall current. In a strong magnetic field the effective mean free path of a charged particle in a direction perpendicular to $\mathbf{B}$ is its radius of gyra-

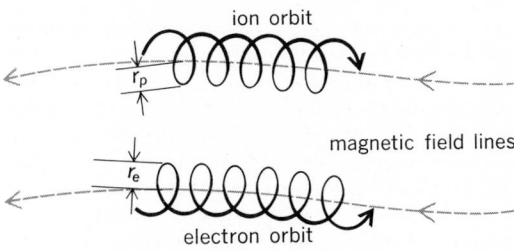

Fig. 1. Charged particles moving in a magnetic field describe helixes about field lines. The sense of rotation for positively charged ions is opposite that for negatively charged electrons. (*From Fusion/1958, Brochure for U.S. Fusion Research Exhibit, 2d International Conference on Peaceful Uses of Atomic Energy, Geneva, 1958*)

tion. Thus, the diffusion current is also reduced significantly by a strong magnetic field. Two forms of a generalized Ohm's law can be used:

For small **B** Eq. (14) holds.

$$\mathbf{E} = -\mathbf{v} \times \mathbf{B} + \mathbf{j}/\sigma + (\mathbf{j} \times \mathbf{B} - \nabla p_e)/ne \quad (14)$$

For large **B** Eqs. (15) hold.

$$\mathbf{E}_\| = \mathbf{j}_\| /\sigma$$
$$\mathbf{E}_\perp = -\mathbf{v} \times \mathbf{B} + \left(\mathbf{j}_\perp + \frac{3n}{4\mathbf{B}^2} \nabla kT \times \mathbf{B}\right) \Big/ \sigma_\perp \quad (15)$$

The symbols $\|$ and $\perp$ indicate the direction relative to **B**, p_e stands for the partial pressure of the electron gas, and the conductivity $\sigma_\perp$ is approximately $\sigma/2$. For some applications it is desirable to solve Eq. (14) for **j** and this leads to Eq. (16),

$$\mathbf{j} = \sigma \mathbf{E}_\|^1 + \sigma_1 \mathbf{E}_\perp^1 + \sigma_2 \frac{\mathbf{B} \times \mathbf{E}^1}{B} \quad (16)$$

where $\mathbf{E}^1 = \mathbf{E} + \mathbf{v} \times \mathbf{B} + \nabla p_e/ne$, $\sigma_1 = \sigma/1 + \alpha^2$, $\sigma_2 = \alpha\sigma$, and $\alpha = \sigma B/ne$.

The parameter α is of the same order as λ/r_e, so that Eq. (16) does not apply in the limit of large α.

The electrical conductivity σ can be found experimentally or from kinetic theory. The latter leads to the formula $\sigma = ne^2\tau/m$, where τ is the effective collision time of an electron, that is, the time in which collisions alone would bring the velocity of an electron into equilibrium with that of the surrounding ions.

A partially ionized gas can be regarded as a mixture of a completely ionized plasma and a neutral gas. One can consider separate densities ρ_p, ρ_n and velocities $\mathbf{v}_p, \mathbf{v}_n$ for these two components. The density and velocity of the mixtures are $\rho = \rho_p + \rho_n$ and $\mathbf{v} = (\rho_p\mathbf{v}_p + \rho_n\mathbf{v}_n)/\rho$. Ohm's law retains the form given by Eq. (14) if one replaces **v** by $\mathbf{v}_p$. If one writes τ_{rs} for the effective collision time of a particle of type r with all particles of type s, the effective collision time in the formula for σ can be expressed by the relation in Eq. (17). The indices e,

$$\frac{1}{\tau} = \frac{1}{\tau_{ei}} + \frac{1}{\tau_{en} + \tau_{in}} \quad (17)$$

i, and n refer to electrons, ions, and neutral atoms.

The motion of the plasma relative to the neutral component is called ambipolar diffusion. It gives rise to drag forces which dissipate energy in the form of heat. The part of the dissipation resulting from the drag between ions and neutrals is most often not included in (although it is often larger than) the joule heating.

MAGNETOHYDRODYNAMIC PHENOMENA

The combination of fluid motion and electromagnetic effects can lead to much more varied phenomena than either one alone. The analysis, however, can often be simplified by observing that the equations permit similar solutions which differ only by scaling factors. The results of a particular calculation or experiment can then be applied to an entire class of arrangements. Scaling laws can be most conveniently used by specifying the value of certain dimensionless parameters such as the magnetic Reynolds number R_m mentioned before, the ratio of magnetic to kinetic energy $S = B^2/\mu\rho v^2$, and others yet to be introduced.

Equilibrium. Before turning to the peculiar interplay of hydrodynamic and electromagnetic forces, the conditions for equilibrium in the absence of motion can be considered. The formulation of equilibrium conditions can be simplified by expressing the force density as the divergence of Maxwell's stress tensor, as in Eq. (18). The stress-

$$(\mathbf{j} \times \mathbf{B})_i = \frac{\partial}{\partial x_k} (B_i B_k/\mu - B^2 \,\delta_{ik}/2\mu) \quad (18)$$

es represented by this tensor are two-fold: a pressure $B^2/2\mu$ at right angles to the field and a tension $-B^2/2\mu$ along the field lines. The field lines thus tend to repel each other, and they also tend to contract like elastic strings. When the magnetic field lines are straight and parallel to each other, equilibrium is obtained for $p + B^2/2\mu =$ constant. When the magnetic field lines form circles around the axis of a cylinder (Fig. 2a), the equilibrium condition is given by Eq. (19), which

$$\frac{dp}{dr} = \frac{B}{\mu r}\frac{d}{dr}(rB) = 0 \quad (19)$$

requires an additional relation for its integration. For example, B can be assumed to be proportional to r, and this leads to $p + B^2/\mu =$ constant. The magnetic field in this geometry can balance a pressure twice as large as in the first example. Intuitively this can be understood by considering that the tension along the field lines causes an inward force adding to the magnetic pressure gradient. The particular geometry of the second example arises in the theories of the filaments in the solar atmosphere and of the static pinch in controlled fusion.

In discussing equilibrium configurations, it is quite common to assume simple mathematical expressions for the field. Although one usually does not consider how to generate or maintain the electric currents which give rise to the magnetic field assumed in the mathematical model, one tries to use only plausible current distributions. For instance, the magnetic field in the cylindrical geometry above could be generated by a constant current density along the axis.

It is possible to maintain appropriate combinations of toroidal and polar fields in equilibrium

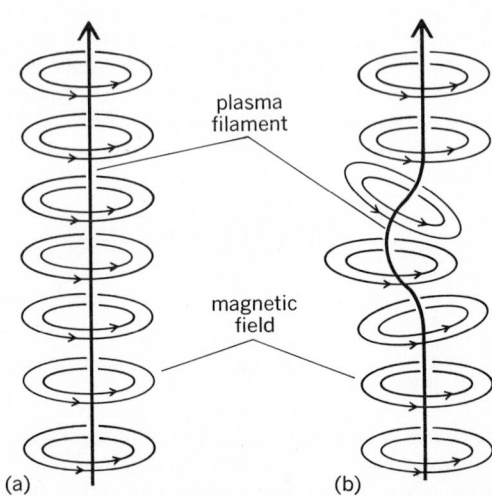

Fig. 2. Plasma-magnetic-field configuration for cylindrical pinch. (a) Equilibrium configuration of plasma filament and magnetic field generated by axial current flow through plasma. (b) Onset of kink instability.

"Magnetohydrodynamics 613"

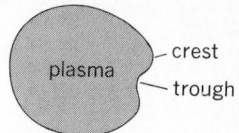

Fig. 3. Cross section of plasma cylinder with flute instability. Magnetic field is into the paper.

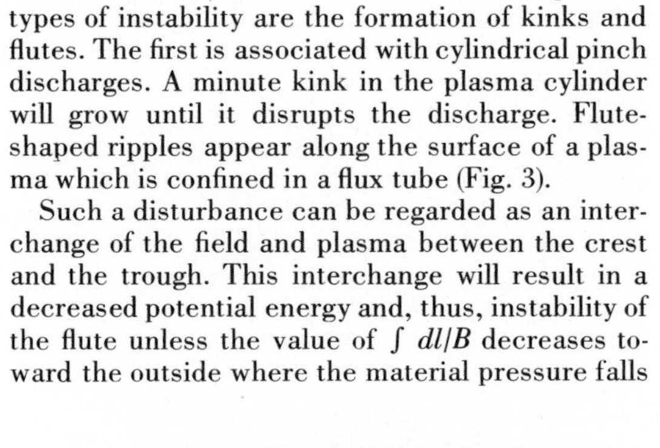

with both pressure and gravitational forces in axially symmetric configurations, where the fields are confined to an interior region. Specific solutions are known for the sphere and the infinite cylinder. The latter is a possible model for explaining the relatively strong magnetic fields in the spiral arms of the Galaxy.

If **B** and **j** are parallel, no force is exerted on the fluid. Such force-free fields are in equilibrium in the absence of other forces. A case of special interest is encountered if $\nabla \times \mathbf{B} = \alpha \mathbf{B}$ with constant α. Whereas normally the decay of fields due to finite conductivity creates forces which are not balanced, this special type remains force-free. An example is the twisted field in a cylinder in which the axial and tangential components of the field are proportional to the Bessel functions $J_0(\alpha r)$ and $J_1(\alpha r)$.

Stability. It is important to establish whether an equilibrium configuration is stable. To determine this, one can use an energy principle. One defines a potential energy as the sum of internal energy E, magnetic energy as in Eq. (20), and, if necessary,

$$E_m = \int (\mathbf{B}^2/2\mu) \, dv \qquad (20)$$

gravitational energy E_g. If an arbitrary deformation of the plasma always leads to an increased potential energy, the equilibrium is stable. In applying this principle, the magnetic field lines are considered to be attached to the plasma.

Another method of ascertaining stability considers small amplitude disturbances by linearizing the equations of motion. One carries out an analysis for the normal modes of oscillation and the equilibrium is unstable if any of the modes are. This analysis furnishes the rate of growth of an instability.

Generally, plasma contained on the concave side of curved field lines is unstable. Important types of instability are the formation of kinks and flutes. The first is associated with cylindrical pinch discharges. A minute kink in the plasma cylinder will grow until it disrupts the discharge. Flute-shaped ripples appear along the surface of a plasma which is confined in a flux tube (Fig. 3).

Such a disturbance can be regarded as an interchange of the field and plasma between the crest and the trough. This interchange will result in a decreased potential energy and, thus, instability of the flute unless the value of $\int dl/B$ decreases toward the outside where the material pressure falls

off. In the stellarator the confining flux tube is bent to form a torus (Fig. 4). The coils producing the field can be arranged so that the lines (except for one called the magnetic axis) do not close on themselves after a single turn around the torus. By this means it is possible to prevent interchange and thus interchange instability.

In a pinch discharge a magnetic field superimposed along the axis of the pinch stabilizes some but not all modes of deviation from equilibrium.

An attempt to confine the plasma on the convex side of the field has led to the cusp geometry (Fig. 5).

A plasma configuration in equilibrium with magnetic and gravitational forces will be unstable if the magnetic field is too large. The virial theorem leads to an upper limit $E_m < |E_g|$ for the magnetic field energy. Actually, no definitely stable configurations are known. However, in some configurations the buildup rate of instabilities is so slow they are practically stable. This is true for the cylinders which are proposed as models for galactic spiral arms.

Some interesting results appear if one enlarges the class of gravitational equilibrium configurations to include internal motions. In a rotating equilibrium configuration with axial symmetry, all points along a line of force rotate with the same angular velocity. This is known as the law of isorotation. The inclusion of internal motions has made possible the proof of the stability of all axis-symmetric solutions whose motion is $\mathbf{v} = \mathbf{B}/\sqrt{\rho\mu}$ and whose pressure p is given by $p/\rho + v^2/2 + \phi = $ constant, where ϕ is the gravitational potential.

Steady flow. A perfectly conducting fluid tends to push the field lines out of the way. With a finite conductivity, however, the field lines slip through the fluid. Generally, a large magnetic Reynolds number indicates a strong effect of the flow on the magnetic fields in its path. This criterion does not depend upon the size of the magnetic field. The extent to which a magnetic field influences the motion of the fluid, on the other hand, must clearly increase with the magnitude of the field. If viscous forces are negligible in the momentum balance, the ratio of ponderomotive to inertial force $N_p = B^2 L\sigma/\rho v$ is a measure of this influence.

As an example consider the one-dimensional flow of a fluid at supersonic speed coming from a field-free region and passing through a region of width L with a strong magnetic field at right angles to the direction of the flow. The magnetic field exerts a drag on the fluid and an increase in N_p causes a reduction in the rate of flow. When a critical value is passed, the flow will become subsonic. However, the fluid cannot pass continuously from supersonic to subsonic conditions. Instead, a standing shock wave will form in the magnetic region effecting the transition to subsonic flow.

In the previous example the magnetic field region was bounded by parallel planes perpendicular to the flow. Consider next an axial magnetic field confined to a cylindrical region perpendicular to the flow. In this case, a two-dimensional flow pattern will develop. The flow will be deflected sideways as it passes through the field because of the Hall effect. A reaction force will push the magnetic field region in the opposite direction. It has been suggested that this lift may have aerodynamic ap-

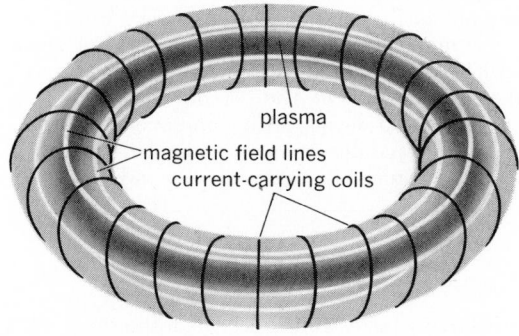

Fig. 4. Stellarator geometry. Plasma is confined in magnetic-field torus generated by external windings. Additional windings (not shown) cause field lines to rotate about chamber axis to improve stability.

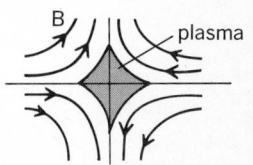

Fig. 5. Cusp geometry of plasma confined on convex side of magnetic field.

plications. For strong fields the flow pattern goes around the cylinder as if it were a solid obstacle.

When a fluid goes around an object which has a magnetic field at right angles to its surface, it has to cross magnetic field lines. This slows the flow down and increases the size of the stagnation region (the region of zero velocity in front of the object) so that velocity and temperature gradients near the stagnation point are reduced. The magnitude of these effects also depends upon the parameter N_p where L is a typical linear dimension of the object.

Near a wall, viscous forces influence the flow and the viscosity also enters the scaling laws. The profile of a pressure-induced incompressible flow between parallel insulating walls (Poiseuille flow) with a magnetic field at right angles depends in the laminar region upon the Hartmann number $N_H = BL\sqrt{\sigma/\rho v}$, where $2L$ is the distance between the walls (Fig. 6). For $N_H = 0$ one has the parabolic

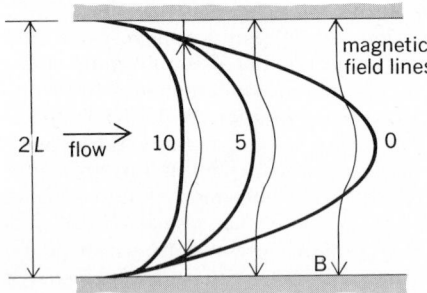

Fig. 6. Approximate profiles of pressure-induced incompressible flow between insulating walls for three values of N_H keeping the pressure gradient unchanged.

profile of classical Poiseuille flow. For large N_H the velocity is nearly constant across the channel, decreasing only in the layers adjacent to the walls. The pressure gradient necessary to maintain the same average flow velocity increases by the factor in notation (21). The lines of force remain perpendicular at the walls but bulge in direction of the

$$N_H{}^2/3(N_H \coth N_H - 1) \approx 1 + N_H{}^2/15 \quad (21)$$

dicular at the walls but bulge in direction of the flow at the center line. The size of the bulge is proportional to R_m.

Frequently it is good approximation to assume that R_m is so small that the flow does not distort the magnetic field lines, and this approximation greatly simplifies the solution of otherwise difficult flow problems. One can, for example, determine the flow around a sphere in a field which is parallel to the axis of the flow by assuming that the field is constant throughout the region of the flow. The drag on such a sphere is increased by the factor in notation (22), over the classical Stokes value.

$$1 + \frac{3}{8} N_H + \frac{7}{960} N_H{}^2 + \cdots \quad (22)$$

In general, laminar flow through channels or past objects shows an increased drag with application of a magnetic field which has a component normal to the walls. The heat transfer to the wall, on the other hand, is generally reduced.

In the presence of a magnetic field a steady-state solution for the boundary layer flow along a

semi-infinite flat plate can exist only when the Alfvén speed $v_a = B/(\mu\rho)^{1/2}$ is less than the speed of the flow or equivalently, $S < 1$. This conclusion is based upon a mathematical approximation procedure carried out for the case of a longitudinal magnetic field. Either $S = 0$ or $R_m/R = \mu v \sigma = 0$ reduces this theory to the classical case treated by H. Blasius. If $\mu v \sigma$ is small but not zero, the Blasius solution is not significantly modified unless S is very close to one. In the limit as $S \to 1$ the thickness of the boundary layer increases to infinity and the skin friction decreases to zero.

The case of a transverse field has been treated by an expansion in powers of $N_{px} = \sigma B^2 x/\rho v$, where x is the variable distance along the plate measured from the leading edge. This procedure does not, however, answer the question of whether there is a limit to the existence of steady flow. One can distinguish the cases in which the field is tied either to the motion of the plate or of the fluid. In the first case one finds a reduction of the skin friction coefficient at the wall of $(1 - 2.7N_{px} + 1.1N_{px}{}^2)$ and in the second case an increase of $(1 + 3.4N_{px} - 4.2N_{px}{}^2)$ compared to the Blasius solution.

Small-amplitude waves. One difficulty of the magnetohydrodynamic equations is their nonlinear character. By adopting a restriction to small-amplitude motions it is possible to linearize these equations. For Fourier analysis one can show the existence of a large variety of waves. Some of these are predominantly electromagnetic (with no fluid motion involved). In the presence of a dc magnetic field $\mathbf{B}_0$ there are, however, low-frequency wave modes in which there is a strong interaction between fluid motion and fields. These waves are similar to ordinary sound waves, but in contrast to the ordinary fluid situation there are three sound speeds. These speeds depend strongly upon the direction of propagation relative to $\mathbf{B}_0$.

The tendency of the fluid and the field lines to move together causes characteristic restoring forces which combine with the ordinary pressure forces in a number of different ways. This gives rise to the larger variety of wave types. In particular, it is possible to have shear waves with the fluid moving at right angles to the wave motion. The most simple shear wave, called the Alfvén wave, moves along the magnetic field lines with the Alfvén speed $v_a = B_0/\sqrt{\rho\mu}$. The speed of compression waves at right angles to the magnetic field is $\sqrt{v_a{}^2 + v_s{}^2}$, where v_s is the ordinary sound speed.

For an arbitrary angle θ between the direction of propagation and the direction of the magnetic field, one of the modes is a shear wave where the fluid is moving at right angles to the plane formed by the field $\mathbf{B}_0$ and the wave vector $\mathbf{k}$. The speed of this wave is $v_a \cos \theta$. The two other modes are in general hybrid forms with components of fluid motion both parallel and perpendicular to $\mathbf{k}$ and in the $(\mathbf{B}_0, \mathbf{k})$ plane. The two velocities in Eq. (23) are

$$v = \sqrt{\frac{1}{2}[v_a{}^2 + v_s{}^2 \pm \sqrt{(v_a{}^2 + v_s{}^2)^2 - 4v_a{}^2 v_s{}^2 \cos^2 \theta}]}$$

$$(23)$$

respectively faster and slower than the shear wave velocity.

In the two extreme cases where $\mathbf{k}$ is parallel or perpendicular to $\mathbf{B}_0$, the two hybrid modes are unscrambled into pure shear and compression modes. In the parallel case, k and $\mathbf{B}_0$ do not define a plane so that the two shear modes become undistinguishable. In the perpendicular case, the velocity of both shear modes formally goes to zero and only the compression mode exists.

The strong coupling between fluid motion and field disappears as the frequency of the wave approaches $\omega_i = eB/m_i$, the angular velocity of ions spiraling in the field B. Another limit is the frequency $V_a^2 \mu \sigma$, at which the slipping of the fluid across the field lines destroys the wave structure.

Alfvén waves have been generated in the laboratory by twisting a column of liquid sodium in a strong magnetic field. In the design of this experiment, scaling laws have been very useful. The Lundquist number $N_L = R_m \sqrt{S} = BL\sigma \sqrt{\mu/\rho}$ seems to be natural to the description of this phenomenon. This number (where L stands for the radius of the column) must be large compared to 1, to permit the experiment. For mercury under normal laboratory conditions, say $L = 10$ cm and $B = 1000$ gauss, N_L is 10^{-1}. By using liquid sodium it is possible to increase this to $N_L = 5$. It is of interest to compare this to the interior of the Earth and the Sun, where $N_L \sim 10^3$ and $N_L \sim 10^7$ respectively.

Shock waves. The threefold structure of magnetohydrodynamic waves that follows from the linearized theory also shows up in the nonlinear case. One is led to three speeds depending on p, ρ, and B in the same manner as before. These speeds define the typical motion of certain disturbances.

Just as do ordinary hydrodynamic waves, magnetohydrodynamic waves tend to get steeper in front and develop into shocks, that is, surfaces across which the physical state changes discontinuously. These changes are restricted by the conservation laws, which relate the state variables on one side of the shock front with those on the other side and with the speed u of the front. If $\Delta F = F_{\text{ahead}} - F_{\text{behind}}$ is the difference between the values of a quantity F on the two sides of the shock front and $\mathbf{n}$ is the unit vector normal to the shock front pointing ahead, the shock conditions for a fluid with infinite conductivity are given by Eqs. (24)–(28).

$$\Delta \mathbf{B} \cdot \mathbf{n} = 0 \qquad (24)$$

$$\Delta[(\mathbf{v} \cdot \mathbf{n} - u)\mathbf{B} - (\mathbf{B} \cdot \mathbf{n})\mathbf{v}] = 0 \qquad (25)$$

$$\Delta[(\mathbf{v} \cdot \mathbf{n} - u)\rho\mathbf{v} + (p + \mathbf{B}^2/2\mu)\mathbf{n} \\ - (\mathbf{B} \cdot \mathbf{n})\mathbf{B}/\mu] = 0 \qquad (26)$$

$$\Delta[(\mathbf{v} \cdot \mathbf{n} - u)\rho] = 0 \qquad (27)$$

$$\Delta\Big[(\mathbf{v} \cdot \mathbf{n} - u)\left(\frac{1}{2}\rho v^2 + \rho E + \mathbf{B}^2/2\mu\right) \\ + (\mathbf{v} \cdot \mathbf{n})(p + \mathbf{B}^2/2\mu) - (\mathbf{B} \cdot \mathbf{n})(\mathbf{B} \cdot \mathbf{v})/\mu\Big] = 0 \qquad (28)$$

These equations can be supplemented by the statement that a discontinuity of the tangential component of $\mathbf{B}$ implies the flow of a sheet current $I = (\mathbf{n} \times \mathbf{B})/\mu$ along the front.

One can again distinguish fast, slow, and intermediate shocks. Special cases of interest are parallel and perpendicular shocks where the names indicate the direction of $\mathbf{B}$ relative to $\mathbf{n}$. In the first of these the hydrodynamic motion is not coupled to the magnetic field and the shock proceeds as in ordinary hydrodynamics, so that the slow and intermediate types do not exist for this special case. Parallel as well as slow and fast perpendicular shocks can be considered in a frame in which the flow velocity is normal on both sides.

Shocks which are neither parallel nor perpendicular are called oblique. For these a frame of reference can be introduced in which the stream lines are parallel to $\mathbf{B}$ on both sides of the shock front but change their direction in passing through. In this frame the electric field vanishes on both sides.

In the three shock modes the change of the tangential component $\mathbf{B}_t$ of the magnetic field is distinctly different. Across a fast shock, $\mathbf{B}_t$ retains its direction and increases in magnitude; across a slow shock it retains or reverses its direction and decreases in magnitude; and across an intermediate shock it changes its direction and retains its magnitude.

In a fast shock it is possible for $\mathbf{B}_t$ to change from zero ahead of the shock front to a nonzero value behind; this is called a switch-on shock. In a slow shock it is similarly possible for $\mathbf{B}_t$ to change to zero behind the front; this is called a switch-off shock. Switch-on shocks exist only if $v_a > v_s$ ahead of the shock front and if the shock strength lies below a critical valve. Switch-off shocks always exist if behind the shock front $v_a < v_s$, and they exist for $v_a > v_s$ provided the shock is strong enough.

Transient flow. Transient flows can be started either by setting an object into motion or by switching on electrical circuits which create fields.

The impulsive motion of an infinite flat plate (Rayleigh problem) in a transverse magnetic field starts a transient flow which approaches the steady state in a time of the order $\rho/\sigma B^2$.

The flow which develops along the semi-infinite flat plate in a parallel field ($S < 1$) has a different character in different regions. Between the leading edge and a point $x = (1 - \sqrt{S})vt$ the flow approaches the steady-state solution discussed earlier. Beyond $x = (1 + \sqrt{S})vt$ the flow approaches the infinite plate solution. In between there is a transition region.

The sudden release of the energy stored in a large condenser and its conversion to mechanical and thermal energy of a plasma give rise to problems whose theoretical treatment requires fast computers. One can simplify the theory of a collapsing pinch discharge considerably by assuming that all material which has been swept up by the contracting magnetic field is piled up in a very thin layer which is snow plowed toward the axis. In this manner one can set up an ordinary differential equation for the radius of this layer whose numerical integration can be carried out with more modest equipment. The time of collapse of such a pinch is given by $Et \approx r(\rho\mu)^{1/4}$, where ρ and r are the initial density and radius of the plasma, and E is the electric field causing the discharge. *See* Thermonuclear reaction.

Flow instability. Laminar flow breaks down when R rises beyond a critical number. It has been demonstrated both theoretically and experimentally that a magnet field improves flow stability; the critical R increases in proportion to N_H. For flow between parallel flat plates with a transverse magnetic field the predicted onset of instability takes place at $R = 50,000 N_H$. The measured suppression of turbulence takes place at the much lower value $R = 225 N_H$. Between coaxial cylinders, where the inner one rotates faster than the outer one, an axial magnetic field raises the critical angular velocity by a factor which approaches $N_H/2$ for $N_H \geqq 20$ (the difference in radii is used to define N_H).

Turbulence. The mathematical structure of Eq. (7) for the field vector **B** is identical with that of Eq. (29) for the vorticity vector $\boldsymbol{\omega}$ in ordinary hydrodynamics.

$$\frac{\partial \boldsymbol{\omega}}{\partial t} = \nabla \times (\mathbf{v} \times \boldsymbol{\omega}) + \nu \nabla^2 \boldsymbol{\omega} \qquad (29)$$

namics. The first term on the right-hand side of either equation tends to increase the mean square value of the respective field vector; the second term causes a decrease due to resistive and viscous losses. In ordinary turbulence the increase and decrease of $\overline{\omega^2}$ roughly balance each other. It has been suggested that small magnetic disturbances in a turbulent flow field will increase if resistive losses are less than viscous ones, that is, if $\sigma \mu \nu > 1$. In a Fourier expansion of the magnetic field energy, the spectrum within an uncertain range of wave numbers k tends toward a distribution $\sim k^{-5/3}$ (Kolmogoroff spectrum). Within that range the ratio of magnetic to kinetic energy is about 1.6.

Ordinary turbulence requires large values of R and magnetohydrodynamic turbulence requires even larger values of R_m. Such conditions are encountered only in geophysical and astrophysical situations.

TRANSPORT-EQUATION DESCRIPTION

The continuum approach to magnetohydrodynamics ceases to be valid when the mean free paths of the particles are of the order of or larger than the lengths characterizing the structure of the flow. A description of more general validity can be based on the functions $f_i(r,v,t)$ that give the densities of particles with velocity v at position r for the various particle species identified by the index i. These functions obey "transport equations," Eq. (30), where **E** and **B** are macroscopic fields that

$$\frac{\partial f_i}{\partial t} + v \cdot \nabla f_i + \frac{e_i}{m_i} (\mathbf{E} + v \times \mathbf{B}) \, \nabla_v f_i = \left(\frac{df_i}{dt}\right)_{\text{coll}} \qquad (30)$$

satisfy Maxwell's equations. The rate of change due to collisions $(df_i/dt)_{\text{coll}}$ can be brought into manageable form by certain assumptions. One of these is that only two particles participate in any collision. This assumption yields the Boltzmann equation. Another assumption is that collisions produce predominantly small deflections. In this case one can make certain expansions leading to the Fokker-Planck equation. Both these forms can be used to derive continuum-type equations. *See* BOLTZMANN TRANSPORT EQUATION; KINETIC THEORY OF MATTER.

Although a charged particle in a plasma collides simultaneously with very many other particles, each collision has only a minute effect. For this situation, the mean free path can be defined as the distance a particle travels until its momentum or energy is appreciably changed by the random addition of minute changes. This effective mean free path increases as the square of the temperature and can become very large. In this case one can set $(df_i/dt)_{\text{coll}} = 0$.

In a magnetic field, charged particles spiral around the lines of force; this restricts their motion at right angles to the field. If the radii of the spirals are small compared to L, these gyrations will cause some degree of randomness of the particle motion even in the absence of collisions. Gyrations are not as effective in doing this as collisions, and in particular exclude any energy transfer between the components perpendicular and parallel to the magnetic field. Nevertheless, they may produce a nonthermal equilibrium in which a modified continuum approach is still possible.

One such modification replaces the scalar pressure by a tensor with two pressures $p_\parallel$ and $p_\perp$ parallel and perpendicular to **B**. With no collisions, viscous and resistive losses are absent, and therefore the assumption of no heat flow in the energy equation of the ordinary continuum theory (Eq. 13) leads to constant entropy along stream lines. In the modified theory, no heat flow leads to the constancy of $p_\perp/\rho \mathbf{B}$ and of $p_\parallel \mathbf{B}^2/\rho^3$ along stream lines. Such constants of motion are used mainly in investigating the stability of equilibria by means of a variational principle. However, it is not obvious why the no-heat-flow assumption should hold.

Small changes from an equilibrium distribution as they occur in applying a variational principle can also be handled by a direct use of the transport equation. This theory requires the Debye length $\sqrt{\epsilon KT/ne^2}$ to be small compared to L, which is true for quite general conditions in the plasma. The stability obtained in this manner is stronger than that obtained using the constant-entropy approximation and weaker than that obtained from the two-pressure modification. *See* PINCH EFFECT.

[ROLF LANDSHOFF]

Bibliography: T. G. Cowling, *Magnetohydrodynamics*, 2d ed., 1976; L. Spitzer, Jr., *Physics of Fully Ionized Gases*, 2d ed., 1962.

Magnetomotive force

The magnetomotive force (mmf) around a magnetic circuit is the work per unit magnetic pole required to carry the pole once around the circuit. It is the analog of electromotive force. *See* ELECTROMOTIVE FORCE (EMF); MAGNETIC CIRCUITS.

It is expressed mathematically in Eq. (1), where

$$\text{mmf} = \oint H \cos \theta \, ds \qquad (1)$$

$H \cos \theta$ is the component of magnetic field strength in the direction of a length of path ds. The line integral is taken around any closed path in the field.

The magnetomotive force is the rise in magnetic potential around the path. For a discussion of magnetic potential *see* MAGNETIC FIELD.

For a path that encloses a current I, Eq. (2)

$$\oint H \cos \theta \, ds = I \qquad (2)$$

holds, and for a path that encloses N equal currents, for example, a path that loops through a coil of N turns, Eq. (3) is valid. If no current is enclosed by the path, the line integral is zero.

$$\oint H \cos \theta = NI \qquad (3)$$

The meter-kilogram-second (mks) unit of magnetomotive force is the ampere-turn.

[KENNETH V. MANNING]

Magneton

A unit of magnetic moment used for atomic, molecular, or nuclear magnets. *See* MAGNETIC MOMENT.

The Bohr magneton μ_B has the value of the classical magnetic moment of the electron, which can theoretically be calculated as shown in Eq. (1),

$$\mu_B = \mu_0 = \frac{e\hbar}{2mc}$$

$$= (0.92733 \pm .00002) \times 10^{-20} \text{ erg/oersted} \quad (1)$$

where e and m are the electronic charge and mass, $\hbar$ is Planck's constant divided by 2π, and c is the velocity of light. A consistent relativistic treatment of the magnetic moment of the free electron shows that corrections to the classical calculation are necessary, so that the electron moment is about 0.1% larger than μ_0. For an extended discussion of the electron magnetic moment *see* ELECTRON SPIN.

The magnetic moment of an atom or molecule results from contributions from both the orbital angular momentum of the atomic electrons and the electronic moments themselves (attributed to the electron spin). When certain groups of atoms are compared, their moments show simple ratios. Observation of this fact led to the definition of the Weiss magneton (before the Bohr magneton) on a purely experimental basis as the unit for these moments. Its value is given by Eq. (2).

$$\mu_W = 0.1853 \times 10^{-20} \text{ erg/oersted} \quad (2)$$

The nuclear magneton is obtained from the Bohr magneton by replacing m by the proton mass; it is thus 1836.31 times smaller than the Bohr magneton. The nuclear magneton is hardly a first approximation to the value of the nuclear magnetic moment, which is nearly three nuclear magnetons in the case of the proton. Meson current effects are held responsible for this deviation, as well as for the fact that the neutron, which has no net charge, nevertheless has a magnetic moment of the same order (but opposite sign) as the proton. For a discussion of the neutron magnetic moment *see* NEUTRON. *See also* NUCLEAR MOMENTS.

[MC ALLISTER H. HULL, JR.]

Magnetooptics

That branch of physics which deals with the influence of a magnetic field on optical phenomena. Considering the fact that light is electromagnetic radiation, an interaction between light and a magnetic field would seem quite plausible. It is, however, not the direct interaction of the magnetic field and light that produces the known magnetooptic effects, but the influence of the magnetic field upon matter which is in the process of emitting or absorbing light.

Zeeman effect. This produces a splitting of spectrum lines when the emitting light source is placed in a magnetic field. The inverse Zeeman effect refers to a similar splitting of absorption lines when the absorbing substance is in a magnetic field. The Zeeman effect for spectrum lines originating from closely spaced levels is called the Paschen-Back effect. The explanation of most other magnetooptical phenomena is based on the Zeeman effect, which therefore may be regarded as the basic magnetooptic effect. *See* PASCHEN-BACK EFFECT; ZEEMAN EFFECT.

Faraday effect. This is the rotation of the plane of polarization of light when light traverses certain substances in a magnetic field. *See* FARADAY EFFECT.

Voigt effect. An anisotropic substance placed in a magnetic field becomes birefringent (doubly refracting), and its optical properties are similar to those of a uniaxial crystal. The Faraday effect is the result of this birefringence when observations are made parallel to the magnetic field. The analogous observations perpendicular to the magnetic lines of force are more difficult and were not successfully carried out until 1898 because of the smallness of the effect. The transverse magnetooptic birefringence is called the Voigt effect after its discoverer, W. Voigt.

The Voigt effect (also called magnetic double refraction) can easily be calculated for substances having a normal Zeeman effect. For more complicated Zeeman effects the results can also be theoretically predicted but are less simple quantitatively, though not essentially different from those in the simpler cases.

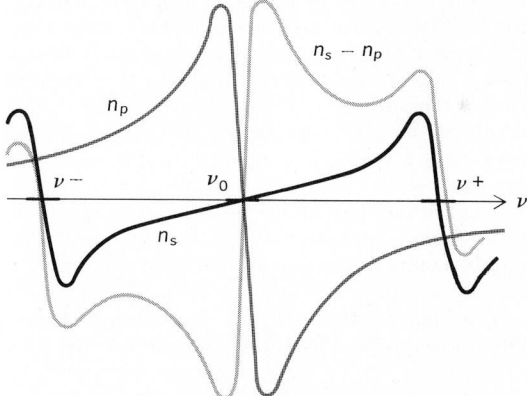

Index of refraction of light polarized parallel (n_p) and perpendicular (n_s) to the magnetic field in the vicinity of a Zeeman triplet. The Voigt effect is proportional to $n_s - n_p$ (shown as the curve labeled as such).

The Voigt effect depends on the fact that the indices n_s and n_p for light polarized perpendicular or parallel to the magnetic lines of force respectively are different from one another in a magnetic field where the absorption line shows a Zeeman effect. The value of n_p is independent of the magnetic field because the central component does not change while $n_s = \frac{1}{2}(n^+ + n^-)$. The appropriately labeled curve in the figure gives $n_s - n_p$, on which the observed effects of the transverse magnetic double refraction depend.

The theoretical formulas which represent the double refraction, though readily derived, are quite complicated. When the wavelength is considerably removed from the Zeeman triplet, the phase difference is as given by Eq. (1), where ν_0 is absorp-

$$\delta = \frac{2\pi x}{c}(n_p - n_s) = \frac{e^4 f x}{32\pi^2 c^3 n_0 (\nu - \nu_0)^3} N H^2 \quad (1)$$

tion line frequency, ν is frequency of transmitted light, e is charge of electron, x is path length, c is velocity of light, n_0 is index of refraction without field, N is number of absorbing atoms per unit volume, H is magnetic field strength, and f is so-called oscillator strength, measure of strength of absorption.

Contrary to the situation existing in the Faraday effect, first-order effects of magnetic double refraction are canceled out because of the presence of two perpendicular, symmetrically placed Zeeman components of combined strength equal to that of the parallel component. Because of this the Voigt effect can be observed only in the vicinity of sharply defined absorption lines, that is, in gases and in certain crystals having sharp lines, such as rare-earth salts. In the case of the rare-earth salts a linear Voigt effect is also possible at extremely low temperatures.

Since the formula for phase difference contains the oscillator strength f, the Voigt effect may be used to measure this important quantity.

Cotton-Mouton effect. This effect is concerned with the double refraction of light in a liquid when the liquid is placed in a transverse magnetic field. It is analogous to the electrooptical Kerr effect and is observed in liquids with complicated molecular structure. If the molecule has a magnetic moment, the field tries to orient the molecule, but the thermal motion tends to oppose this action. There is thus a degree of orientation, which is dependent on the temperature. If the molecule itself is optically anisotropic, the liquid also will be anisotropic and will exhibit double refraction. *See* KERR EFFECT.

The Cotton-Mouton effect is observed chiefly in nitrobenzene and aromatic organic liquids. Aliphatic compounds have a considerably smaller effect.

The phase difference of the Cotton-Mouton effect is expressed by Eq. (2), where x is the path

$$\delta = \lambda C_m x H^2 \quad (2)$$

length and C_m is called the Cotton-Mouton constant. For nitrobenzene at a temperature of 16.3°C and a wavelength λ of 5780 A, $C_m = 2.53 \times 10^{-12}$ in absolute cgs units. With large magnets ($H = 46,500$ oersteds) under the most favorable circumstances, rotations of the plane of polarization up to 27° have been observed.

The dispersion of the Cotton-Mouton effect is given by Havelock's law as in the Kerr effect.

Magnetooptic Kerr effect. This deals with the changes that are produced in the optical properties of a reflecting surface of a ferromagnetic substance when the substance is magnetized. In a typical case this will result in elliptically polarized light appearing in reflection, when the ordinary rules of metallic reflection would give only plane-polarized light. The component produced by the magnetic field when the magnetization is close to saturation is only of the order 10^{-3} of that normally present. The explanation must be sought in the fact that the conduction electrons made to vibrate by the incident light will have a curved path in the magnetic field.

Majorana effect. This deals with optical anisotropy of colloidal solutions. The effect probably is caused by the orientation of the particles in the magnetic field.

Magnetooptic effects have played an increasingly important role in microwave spectroscopy, where transitions between the Zeeman components of a single level can be observed directly.

[G. H. DIEKE/WILLIAM W. WATSON]

Magnetoresistance

The change of electrical resistance produced in a current carrying conductor or semiconductor on application of a magnetic field H. Magnetoresistance is one of the galvanomagnetic effects. It is observed both with H parallel to and transverse to the current flow. The change of resistance usually is proportional to H^2, except in very large fields, where it becomes proportional to H. *See* GALVANOMAGNETIC EFFECTS.

In most metals, the change of resistance is positive; however, it is generally negative in alloys of a noble metal and a transition metal and in ferromagnets above saturation.

In semiconductors, the magnetoresistance is unusually large (especially in indium antimonide) and is highly anisotropic with respect to the direction of current flow in single crystals. The latter property is of great value in determining the band structure. Magnetoresistance measurements also yield information about current carrier mobilities.

[ELIHU ABRAHAMS; FREDERIC KEFFER]
Bibliography: *See* GALVANOMAGNETIC EFFECTS.

Magnetostatics

The study of magnets and the fields produced by magnets. It should not be confused with the science of electromagnetism, which is concerned with the study of magnetic fields produced by currents. When considering magnetostatics, one looks upon magnet poles as seats of magnetism, and Coulomb's law of force between point poles in empty space is of fundamental importance. This law, expressed in rationalized meter-kilogram-second (mks) units, and using the Sommerfeld proposal, is written as Eq. (1), where $\mu_0 = 4\pi \times 10^{-7}$ we-

$$F = \frac{\mu_0}{4\pi} \frac{m_1 m_2}{r^2} \quad (1)$$

ber/amp-m is the permeability of empty space; m_1 is the pole strength of one point pole and m_2 of the other, with pole strengths expressed in ampere-meters; F is the force in newtons; and r is the distance in meters between the point poles. If rationalized mks units are used but the Kennelly proposal followed, Eq. (1) becomes $F = m_1 m_2 / 4\pi \mu_0 r^2$ where μ_0 has the same value and meaning as before, m_1 and m_2 are pole strengths expressed in webers, F is in newtons, and r in meters. The force F is one of attraction if the poles are unlike and one of repulsion if they are like poles. *See* ELECTRICAL UNITS AND STANDARDS.

The magnetic field intensity H at a distance r meters in empty space from a point magnet pole of pole strength m ampere-meters is given by $H = m/4\pi r^2$ where H is in amperes per meter.

For an assembly of point poles, the principle of superposition shows that the resultant H at a point P is the *vector sum*, as shown in Eq. (2), where the

$$H = \frac{1}{4\pi} \sum_{i=1}^{n} \frac{m_i}{r_i^2} \qquad (2)$$

south poles have minus signs and the north poles have plus signs. The direction of H at P is the direction that a north pole at P would tend to move. *See* MAGNET; MAGNETIC FIELD; MAGNETISM.

[RALPH P. WINCH]

Bibliography: B. I. Bleaney and B. Bleaney, *Electricity and Magnetism*, 3d ed., 1976; E. M. Pugh and E. W. Pugh, *Principles of Electricity and Magnetism*, 2d ed., 1970.

Magnetostriction

The change of length of a ferromagnetic substance when it is magnetized. More generally, magnetostriction is the phenomenon that the state of strain of a ferromagnetic sample depends on the direction and extent of magnetization. The phenomenon has an important application in devices known as magnetostriction transducers.

Physical cause. Magnetostriction results from the dependence of the crystalline anisotropy energy upon the state of strain of the crystalline lattice. If the crystal deforms (for example, suffers a change in length) the anisotropy energy may be lowered more than the elastic energy is raised. Thus, a strained state will be favored. For a discussion of crystalline anisotropy energy *see* FERROMAGNETISM.

The total energy of a ferromagnetic substance depends upon the state of strain and the direction of magnetization through three contributions. The first two consist of the crystalline anisotropy energy of the unstrained lattice plus a correction which takes into account the dependence of the anisotropy energy on the state of strain. The third contribution is that of the elastic energy, which is independent of magnetization direction and is a minimum in the unstrained state. The state of strain of the crystal will be that which makes the sum of the three contributions to the energy a minimum. The result is that, when magnetized, the lattice is always distorted from the unstrained state, unless there is no anisotropy.

Since spontaneous magnetization occurs below the Curie temperature, there will always be a spontaneous lattice distortion which depends on magnetization direction in the ferromagnetic state. In nickel, the lattice spacing parallel to the magnetization is always smaller than the lattice spacing perpendicular to the magnetization.

Magnetoelastic coupling constants. These determine the magnitude of the correction arising from strains to the anisotropy energy. In cubic crystals, there are two magnetoelastic coupling constants B_1 and B_2. The constant B_1 determines the change in anisotropy due to a diagonal component of strain, and B_2 that due to a mixed component. The values of the strain components which lead to a minimum in the total magnetoelastic energy are given in terms of the magnetoelastic coupling constants, the elastic constants, and the direction cosines of the magnetization with respect to the crystal axes, α_1, α_2, α_3. In the strain state of lowest magnetoelastic energy, that part of the energy which depends upon magnetization direction is given by the equation below, where ΔK is a

$$U = (K_1 + \Delta K)(\alpha_1^2\alpha_2^2 + \alpha_2^2\alpha_3^2 + \alpha_3^2\alpha_1^2) + \cdots$$

correction to the first order anisotropy constant K_1. The quantity ΔK depends only upon the magnetoelastic constants and the elastic constants.

When a high permeability (soft) magnetic material is required, the magnetostriction should be small in order that anisotropy not be induced by lattice distortions. Magnetostriction appears to be a major source of transformer hum. As the silicon content of soft magnetic steel is increased toward 6.5%, the magnetostriction disappears. Unfortunately, the metal becomes excessively brittle.

Applications. The magnetostrictive effect is exploited in transducers used for the reception and transmission of high-frequency sound vibrations. Nickel is often used for this application. *See* ELASTICITY; ULTRASONICS.

[ELIHU ABRAHAMS; FREDERIC KEFFER]

Bibliography: S. Chikazumi and S. H. Charap, *Physics of Magnetism*, 1964, reprint 1978; F. Seitz and D. Turnbull (eds.), *Solid State Physics*, vol. 3, 1956.

Magnification

A measure of the effectiveness of an optical system in enlarging or reducing an image. For an optical system that forms a real image, such a measure is the lateral magnification m, which is the ratio of the size of the image to the size of the object. If the magnification is greater than unity, it is an enlargement; if less than unity, it is a reduction.

The ratio of the longitudinal (with respect to the optical axis) dimensions of the image to the corresponding dimensions of the object is known as longitudinal magnification, which in first order equals the square of the lateral magnification.

The angular magnification γ is the ratio of the angles formed by the image and the object at the eye. The relation $n'\gamma m = n$ relates angular to lateral magnification. Here n and n' are the refractive indices of the media containing the object and image, respectively. In telescopes the angular magnification (or, better, the ratio of the tangents of the angles under which the object is seen with and without the lens, respectively), can be taken as a measure of the effectiveness of the instrument.

A small off-axis element is imaged with different magnification in both the meridional and sagittal directions. This may be called differential magnification.

Magnifying power is the measure of the effectiveness of an optical system used in connection with the eye. The magnifying power of a spectacle lens is the ratio of the tangents of the angles under which the object is seen with and without the lens, respectively. The magnifying power of a magnifier or an ocular is the ratio of the size under which an object would appear seen through the instrument at a distance of 10 in. (the distance of distinct vision) divided by the object size.

E. Abbe suggested defining the magnifying power of an optical system as the ratio of the tangent of the visual angle under which the object appears to the object size. This quantity is approximately equal to the power of the system, which is the reciprocal of the focal length, $1/f'$. *See* LENS; OPTICAL IMAGE.

[MAX HERZBERGER]

Bibliography: M. Born and E. Wolf, *Principles of Optics*, 5th ed., 1975; M. Herzberger, *Modern*

Geometrical Optics, 1958, reprint 1978; F. A. Jenkins and H. E. White, *Fundamentals of Optics*, 4th ed., 1976; J. Strong, *Concepts of Classical Optics*, 1958.

Magnon

A quasi-particle which is introduced to describe small departures from complete magnetic ordering in ferro-, ferri-, antiferro-, and helimagnetic arrays. *See* ANTIFERROMAGNETISM; FERRIMAGNETISM; FERROMAGNETISM; HELIMAGNETISM.

A sinusoidal variation of that angular momentum which is associated with magnetism (mostly spin momentum of the electrons) is called a spin wave. A single spin wave represents a decrease by one unit of $\hbar$ of the total sample spin momentum; it may be thought of as a ripple of amplitude $\hbar$ propagating across a sea of magnetism. A magnon is a quantized spin wave, much as a phonon is a quantized lattice-vibration wave or a photon is a quantized electromagnetic wave.

The low-temperature behavior of ordered magnetic materials is very accurately described in terms of the statistical excitation of magnons. In particular, the departure of ferromagnetic magnetization from complete saturation is found to fall off proportionally to $T^{3/2}$ as the temperature T increases from 0 K to about half the Curie temperature. At higher temperatures the interactions between magnons cause the theory to become extremely complex, and an exact treatment has yet (1968) to be worked out. The gross experimental behavior, however, can roughly be described by the phenomenological Weiss molecular field approximation.

The energy of a spin wave, or magnon, increases with increasing wave vector $\mathbf{k}$, that is, with decreasing wavelength of the ripple. The reason is that the so-called exchange forces which produce magnetic ordering are very short-range, causing only neighboring spins strongly to resist alignment. The result in ferromagnets is that the magnon energy at large $\mathbf{k}$ is roughly proportional to $\mathbf{k}^2$.

As $\mathbf{k}$ approaches zero, the ripple wavelength becomes of the order of the sample dimensions. In a ferromagnet there is no relative magnetic misalignment of neighboring spins; instead the magnetization of the entire sample moves as a single unit. The energy to excite the $\mathbf{k}=0$ ferromagnetic magnon is thus independent of exchange forces; the disturbance is equivalent to a precession without change in length of the total sample magnetization about the resultant of any existing applied and anisotropy fields. The eigenfrequency ω_0 is equal to the Larmor frequency in the resultant field, provided the sample is spherical (otherwise ω_0 is altered by the torques arising from differences in surface demagnetizing fields). The magnon energy is $\hbar\omega_0$.

Volume dipolar fields cause magnon energies to vary with the direction of wave number $\mathbf{k}$, the energies being greatest for $\mathbf{k}$ normal to the bulk magnetization. There results a degenerate manifold appropriate to wavelengths less than about one-tenth of sample dimensions. For longer wavelengths the dipolar fields are position-dependent, that is, inhomogeneous across the sample, and spin waves are not normal modes of excitation. Instead, the sample can oscillate in certain characteristic shape-dependent magnetostatic modes,

one of which is the uniform mode expressible as a $\mathbf{k}=0$ spin wave (because in this special case the dipolar fields are homogeneous).

Resonance experiments. In a typical ferromagnetic resonance experiment a sample of millimeter dimensions is placed in the homogeneous oscillating magnetic field region of a microwave cavity, and an applied steady field is slowly varied until resonant absorption is observed. With microwaves of 3-cm or 1-cm wavelength, the microwave magnetic field is essentially uniform across the sample and thus sets the sample magnetization into a homogeneous precession, that is, excites $\mathbf{k}=0$ magnons. *See* MAGNETIC RESONANCE.

The inhomogeneous modes can be resonantly excited if the sample is placed in the appropriately inhomogeneous magnetic field region of a microwave cavity.

Ferromagnetic resonance experiments yield information concerning magnon interaction processes. Theoretical studies indicate that ordinary magnon-magnon, magnon-phonon, and magnon-electron interactions lead to lifetimes of about 10^{-9} sec for short-wavelength magnons (thermal magnons) but to infinitely long lifetimes for the $\mathbf{k}=0$ magnons excited in resonance. The key to their relaxation, that is, to the observed resonance linewidth, is their energy degeneracy with $\mathbf{k}\neq 0$ magnons along the dashed line of the illustration. Energy can thus be conserved in the process of transforming a $\mathbf{k}=0$ magnon into one of shorter wavelength. Momentum, however, can be conserved only if the transformation occurs at a nonperiodic sample imperfection. In very pure single crystals of yttrium iron garnet, ferromagnetic resonance linewidth is controlled by sample surface pits and can be narrowed by surface polishing.

Observation with neutrons. A magnon interacts directly with the magnetic moment of a neutron. Measurement of the inelastic scattering of a beam of neutrons by a magnetic material yields magnon energy versus momentum relations, that is, dispersion curves such as shown in the illustration. In this way the general magnon theory has been confirmed in detail.

Other modes. The type of spin-wave behavior discussed above is restricted to those modes in which all magnetic ions in a unit cell precess in the same fashion. In antiferromagnets and ferrimagnets, resonant modes are possible in which the magnetic moments within a cell undergo antiphase oscillations (analogous to optical-type lattice vibrations). These usually have characteristic frequencies in that largely unexplored borderline region between infrared and microwaves (0.1- to 1-mm wavelength).

Application. The precessions of interacting spins are very complicated and, in general, the equations of motion contain nonlinear terms. These become important for the large-amplitude precessions induced by intense microwave fields. There results an instability of certain $\mathbf{k}\neq 0$ modes which drain off the microwave power when the uniform $\mathbf{k}=0$ precession reaches a critical angle (usually a few degrees). It is thus possible to "pump" $\mathbf{k}\neq 0$ magnons. H. Suhl has invented a parametric microwave amplifier based on these nonlinear phenomena.

Large-amplitude precessions can also break up into acoustic vibrations, provided the acoustic eigenfrequencies equal the difference between the microwave (pump) frequency and that of some magnetostatic mode. This is called magnetoacoustic resonance.

[FREDERIC KEFFER]

Bibliography: S. Fluegge (ed.), Encyclopedia of Physics, vol. 18, 1966; A. H. Morrish, The Physical Principles of Magnetism, 1965, reprint 1979.

Many-body theory

The theoretical study of interacting systems of identical particles. The implied conceptual limit is that of an infinite number of particles, and the properties of interest, such as energy per particle and particle density, have their natural setting in this limit. Mixtures, that is, systems with several classes of particles, provide an extension that leads to consideration of structured particles whose internal state, for example, spin for electrons, electronic state for atoms, and vibrational state for molecules, must be specified as well.

Scope. Many-body theory encompasses in principle a range of systems from nucleons in a small nucleus at one extreme to the bulk matter of statistical thermodynamics at the other, including perhaps even idealizations, such as lattice gases, which do not exist in continuous physical space. Included systems might have energies ranging from the minimum energy of the quantum-mechanical ground state to high energy as in a partially ionized plasma with its bewildering array of transitory components. The state of the system may correspond to a single-phase normal fluid or ideal solid system, the intricate correlations in a two-phase system, the dramatic fluctuational structure of a system at the critical point, or an entire temperature region of anomalous properties as in superfluid helium-4. See LIQUID HELIUM; PLASMA PHYSICS.

Quantities of interest. The desired information may refer to static zero-temperature bulk properties such as the spectrum of excitation energies, momentum distribution, correlations between density fluctuations, and to the corresponding data at finite temperature, bounded volume, or both. Time dependence may occur incidentally, as in the transport coefficients that determine the relationship between stationary currents and the forces needed to maintain them. It can also occur overtly when the rate of decay of an initially nonequilibrium state is required or in the time correlations between various fluctuations.

This information may be qualitative or quantitative, but the latter type will be emphasized. It should of course be compared with that obtained from physical measurements. Energy spectra are available for intrinsically finite many-body systems such as a nucleus, an atom, or a molecule. Analysis of extended systems, such as fluids, typically involves observation of the scattering of incident beams. X-rays used in this way scatter essentially elastically and probe mean electron density as a function of location, but if this is uniform, they pick up the structure factor—the spatial correlation between density fluctuations. At higher energy, Compton scattering samples the momentum distribution. Low-energy neutrons scatter inelastically from nuclei, and the consequent changes of momentum and energy permit the evaluation of the Van Hove function, or density-density correlation with respect to both space and time separation. Combinations of the same quantity, at long wavelength and low frequency, are provided (via the Green-Kubo relation) by frequency- and wave-vector-dependent linear transport coefficients. See COMPTON EFFECT; NEUTRON DIFFRACTION; X-RAY DIFFRACTION.

Increasingly, computer simulation is being used to provide the numerical information that must be reproduced by a successful theory. When feasible, this is perhaps more appropriate than physical experiments, since one can guarantee that the theoretical treatment uses precisely the same particle properties and interactions between particles that occur in the simulation. Most simulations refer to classical equilibrium statistical mechanics, either by following the mechanical motion of the system for a long time (molecular dynamics) or by constructing a diffusion process that reproduces the configurations (states of the system at given times) of the equilibrium system, but not necessarily in the same temporal order (Monte Carlo simulation). The molecular dynamics approach has been extended in many cases to compute time-dependent quantities. However, analysis of systems of particles with many internal degrees of freedom, such as water molecules, is still in a primitive stage.

Quantum-mechanical systems at zero temperature can now be treated as well: boson fluids or solids by a diffusion analog of the Schrödinger equation, and fermion systems similarly but with much greater technical difficulty. Finite temperature or time-dependent simulations are virtually nonexistent. See MONTE CARLO METHOD; STATISTICAL MECHANICS.

Model systems. Theoretical techniques for analyzing many-body systems range from general methods which are readily applied but whose quantitative accuracy may be poor, to more incisive methods which take advantage of the special properties of the system in question. Model systems, those which include only selected aspects of reality, can often be solved in great detail and play a crucial role in theoretical developments. The ability to analyze a model system depends, however, upon the nature of the information to be extracted. The basic model system is that of independent particles, moving in some external field but not interacting with each other. In classical mechanics or statistical mechanics, this is equivalent to many copies of a one-body problem, and is eminently solvable. In quantum mechanics, even the one-body problem is not trivial, and the statistical coupling of the particles due to symmetry requirements imposed by Bose or Fermi statistics must also be taken into account. In the absence of an external field, both systems are indeed solvable, and concepts such as Fermi sphere and Bose condensation take on explicit meaning. A simple yet accurate description of a fermion system in the presence of an arbitrary external field has yet to be found. But for special external potentials, such as the harmonic oscillator, the solution is easy and has played an important role in the shell model of the nucleus. In fact, special interactions can also be accommodated here, for example, the Elliott model with its implications for nuclear rotational structure. See BOSE-EINSTEIN STATISTICS; FERMI-DIRAC STATISTICS;

NUCLEAR STRUCTURE.

The independent entities do not have to be real particles. In the ideal crystal lattice, or any system with a deep potential minimum, they are vibrational degrees of freedom, or phonons. Alternatively, they may correspond to quasiparticles: particles surrounded by clouds of correlated particles. These quasiparticles form the basis of the Fermi liquid model of L.D. Landau, which is very useful, and in extended form, of the BCS model of superconductivity, the Bogolubov model of a Bose superfluid, and the Sawada model of the electron plasma. One realization of this picture is through the mean field model, in which fluctuations in particle density are neglected, so that a given particle feels a fixed mean external field at each point in space because of the full interactions of its neighbors—the Hartree-Fock model for fermions, or the Debye-Hückel model for classical electrolytes. The latter also results from the mean spherical model, in which the particle densities at different spatial points are assumed to have independent gaussian distributions. The continuum density field viewpoint appears as well in the local thermodynamic description of an inhomogeneous system, which is regarded as a connected succession of homogeneous regions, constrained, for example, by thermodynamic equilibrium conditions. *See* SUPERCONDUCTIVITY; SUPERFLUIDITY.

There are also a number of very specialized models, primarily in one-dimensional space, that can be solved: the full classical statistical dynamics of hard rods, the quantum thermodynamics of the same system, the Luttinger fermion model with linear kinetic energy but arbitrary interaction potential, statistical classical thermodynamics for any interaction restricted to nearest-neighbor particles—even the longer-range Coulomb or Yukawa interaction. The two-dimensional classical electron plasma has been solved at a special temperature, but there are very few nontrivial two- or three-dimensional models in real space.

Techniques. Discussion will be restricted to systems far from the critical point, where renormalization group techniques and field representations are particularly appropriate, and far from the relativistic region, in which the dynamics of the interaction itself cannot be neglected. The input information consists of the nature of the particles involved, their interaction (generally in terms of pairs), and the parameters that control the state of the system. If any technique can be said to be routine, it is that of perturbation expansion of the interaction about that of a reference or model system, with the plethora of expansion terms organized by a diagrammatic representation. Convergence is accelerated by a self-consistent choice of reference systems (Hartree-Fock, optical model, coherent component, and so forth) and by compression of the diagrammatic structure by selected explicit summation of sequences of diagrams (binary collision expansion, ring cluster, and so forth). Compression can be done in order to render convergent a series which was not initially so, and thus, for example, to treat systems which are qualitatively different from the independent particle models that they formally perturb, as in the cases of the electron plasma, superconductor, and superfluid systems. The Mayer-Ursell expansion in classical statistical mechanics, the Brueckner expansion in quantum many-body theory, and their resummations are standard examples, the latter most elegantly in the form of an expansion of the resolvent operator. A number of physically significant quantities arise in the resummation process—the direct correlation function, the effective propagator, and others. *See* PERTURBATION (QUANTUM MECHANICS); RENORMALIZATION.

Perturbation expansions are also available from various integral relations among several-particle distribution functions. These are infinite hierarchies, corresponding to the infinite series so generated, but they can be truncated by an additional approximate relation among distribution functions. Depending upon the information desired, the distributions involved may be classical few-particle coordinate distributions or phase-space distributions, quantum-mechanical reduced-density matrices, or Green functions. Truncations include the random phase approximation, explicit neglect of higher correlations, and the probability superposition approximation, as well as more specialized closures leading, for example, to the BGY, CHNC, and PY equations of classical statistical mechanics. Hybrid few-body distributions arise through projection techniques and are similarly used in time-dependent problems, for example, via the memory function.

Variational principles are commonly employed for equilibrium free energies, which serve as generating functions for equilibrium expectations. Here model n-body distributions with free parameters can be used—for example, Jastrow functions or intrinsic state models—or a large class of functions can be simply inserted, as in the configuration interaction technique of quantum chemistry. It is also possible to write down variational principles for few-body distributions, but these must be coupled with an extensive unmanageable class of inequalities and equalities. However, the latter, for example, those representing linear response relationships, also serve as brief hierarchies.

There are of course numerous techniques in which advantage can be taken of the special properties, known and surmised, of the system under study. One of these is sufficiently general to deserve mention: the inclusion of collective variables—center of mass, axes of inertia, acoustic modes—as an additional set of variables. This results in an overcomplete description, and the problem of applying subsidiary restrictions must be handled. At the other extreme, there are extrapolation methods, for example, the Pade expansion, in which an approximation is carried well past the parameter range that is valid from first principles by imposing general mathematical conditions on the nature of the parametric variation.

[J. K. PERCUS]

Bibliography: J. P. Hansen and I. R. McDonald, *Theory of Simple Liquids*, 1976; G. D. Mahan, *Many Particle Physics*, 1981; N. H. March, W. H. Young, and S. Sampanthar, *Many Body Problem in Quantum Mechanics*, 1967; J. K. Percus, *The Many-Body Problem*, 1963.

Maser

A device for coherent amplification or generation of electromagnetic waves by use of excitation energy in resonant atomic or molecular systems. The word is an acronym for *m*icrowave *a*mplification by *s*timulated *e*mission of *r*adiation. The device uses an unstable ensemble of atomic or molecular particles which may be stimulated by an electromagnetic wave to radiate excess energy at the same frequency and phase as a stimulating wave, thus providing coherent amplification. Masers, however, are not limited to the microwave region; this type of amplification has been extended to include a frequency range from audio to infrared or optical frequencies. Maser-type amplifiers and oscillators are also sometimes referred to as molecular, or quantum-mechanical, since they involve processes on a molecular scale, and since some types cannot be adequately described by classical mechanics, but show characteristic quantum-mechanical phenomena.

Maser amplifiers can have exceptionally low noise, and come close to effectively amplifying a single quantum of radiation in the microwave region; that is, they approach the limits, set by the uncertainty principle, on the precision with which phase and energy of a wave may be amplified. Their inherently low noise makes maser oscillators that use very narrow atomic or molecular resonances extremely monochromatic, providing a basis for frequency standards. Since atoms or molecules may have resonances and effective amplification over a wide frequency range and to very short wavelengths, masers are useful as coherent amplifiers of millimeter, infrared, optical, and perhaps also ultraviolet wavelengths, where older types of circuit elements are not effective.

Because of their low noise and high sensitivity, maser amplifiers are particularly useful for reception and detection of very weak signals in radio astronomy, microwave radiometry, long-distance radar, and long-distance microwave communications. They also provide research tools for very sensitive amplification or detection of electromagnetic radiation.

Thermodynamic equilibrium of an ensemble of particles—such as atoms, molecules, electrons, or nuclei—which have discrete energy levels and which may radiate electromagnetic energy, requires that the number n_1 of particles in a lower level 1 be related to the number n_2 in an upper level 2 by the Boltzmann distribution condition $n_1/n_2 = e^{(E_1 - E_2)/kT}$, where E_1 and E_2 are the respective energies of the two levels; k is Boltzmann's constant; and T is the absolute temperature, a positive number. Thermodynamic equilibrium also requires that phases of oscillation of the particles, or relative phases of quantum-mechanical wave functions for the various states, be random. A violation of either condition can result in instabilities which may release electromagnetic radiation. The frequency ν of radiation released is characteristically given by $h\nu = E_2 - E_1$, where h is Planck's constant. *See* BOLTZMANN STATISTICS.

The particles may be stimulated by an electromagnetic wave to make transitions from the lower to the upper level, absorbing energy from the wave, or from the upper to the lower level, imparting energy to the wave, and thereby increasing the wave amplitude coherently. Stimulated transitions from the upper to the lower state and those from the lower to the upper state are equally probable. For equilibrium at any positive temperature, the Boltzmann distribution requires that n_1 be greater

than n_2. Therefore, there is a net absorption of energy from the wave, because particles which absorb are more numerous than those which emit. If the condition $n_2 > n_1$ occurs, the system may be said to have a negative absolute temperature, because the Boltzmann condition is fitted only by a negative value of T. If there are not too many counterbalancing losses from other sources, this condition allows a net amplification, because particles which emit energy are more numerous than those which absorb.

Gas masers. An amplifier where $n_2 > n_1$ is the beam-type maser (Fig. 1). Operated in 1954, this was the first type of maser to be suggested. Ammo-

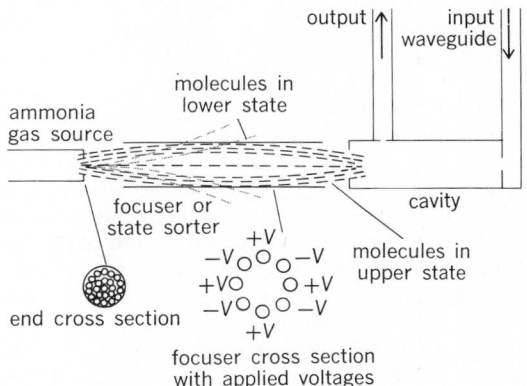

Fig. 1. Schematic of beam-type maser.

nia gas issues from a small orifice into a vacuum system to form a molecular beam. Molecules in the lower of the two states are deflected away from the axis of the state sorter or focuser by inhomogeneous electric fields which act on their dipole moments. Those molecules in the upper state are deflected toward the axis and sent into the microwave-resonant cavity. If losses in the cavity walls and coupling holes are sufficiently small, or if the number of molecules is sufficiently large, amplification or oscillation will occur. This maser is particularly useful as a frequency or time standard because of the relative sharpness and invariance of resonances of the ammonia beam.

The condition for oscillations to occur is given by Eq. (1), where h is Planck's constant, V the cavi-

$$n_2 - n_1 \geqq hV\,\Delta\nu/8\pi\,\mu^2 Q \qquad (1)$$

ty volume, $\Delta\nu$ the width at half maximum of resonant response of the molecules, μ the molecular dipole moment (matrix element), and Q the quality factor of the loaded cavity. The maximum power output is approximately $h\nu$ multiplied by the rate at which molecules enter the cavity, and is very small. A wave impinging on the cavity will be amplified on reflection if it occurs near the resonant response and if $n_2 - n_1 \geqq hV\Delta\nu/8\pi\mu^2 Q_0$, where Q_0 is the quality factor of the unloaded cavity.

Other masers using gaseous molecules have been proposed which involve production of a nonequilibrium distribution by excitation of the gas by means of externally applied radiation of shorter wavelength. Normally, such a system requires molecules with at least three energy levels, two used for amplifying a wave and a third higher level to which molecules are excited from the lowest level, as indicated in Fig. 2. In decaying from the higher level, molecules return, at least in part, to fill up the intermediate level and to satisfy the condition $n_2 > n_1$. If it is light radiation which excites molecules to higher levels, the system is said to use optical pumping.

Solid-state masers. Solid-state masers usually involve the electrons of paramagnetic atoms or molecules in a static or slowly varying magnetic field. In the simplest case, the two-level solid-state maser, only one electron on each molecule is affected. The energy of the electron is quantized into two levels, according to whether the magnetic moment, associated with the electron spin, is parallel or antiparallel to the magnetic field. At thermal equilibrium there are more magnetic moments parallel than antiparallel to the field, corresponding to $n_1 > n_2$. This situation may be reversed, so that $n_2 > n_1$, by interchanging the two populations n_1 and n_2. The interchange is accomplished by rapid variation of the frequency of an intense electromagnetic field through resonance, by application of a pulse of resonant electromagnetic radiation, or in principle by sudden reversal of the magnetic field. *See* ELECTRON SPIN.

Electron-spin moments are more weakly coupled to the electromagnetic field than are molecular electric-dipole moments (by a factor of about 10^4). A much larger preponderance in the upper state is required than for the maser of Fig. 1. If requirements for amplification are met, however, electron-spin moments give correspondingly greater power output. Furthermore, their resonant frequencies are easily tunable by variation of the magnetic field, because their energies involve interaction between electronic magnetic moments and the field. In the simplest cases, the resonant frequency in megacycles is approximately 2.8 times the magnetic field strength in oersteds. Electron-spin resonances in paramagnetic materials allow amplification over broader bandwidths (one to a few hundred megahertz) than do gas systems. Favorable conditions are usually obtainable only at very low temperatures, such as occur in a liquid-helium cryostat; hence, cryogenic problems are often involved and materials used are normally solids rather than liquids. The two-level solid-state maser is most easily operated in pulses, between which the populations of the two levels are readjusted.

The popular three-level solid-state maser also uses paramagnetic material containing electronic magnetic moments in a magnetic field. It has many of the characteristics of the two-level solid-state maser, but can be operated continuously with much more convenience. In Fig. 2, the spacing between the three levels is shown to correspond to microwave frequencies ν_1, ν_2, and ν_3. Usually a few milliwatts of microwave power at frequency ν_3, called the pumping frequency, are sufficient and amplification occurs at a lower frequency with a maximum power output of a few microwatts. Under the simplest assumptions, the number of systems in levels 1 and 3 is equal and the number in level 2 is greater if $\nu_1 < \nu_2$, giving amplification at frequency ν_1. If $\nu_1 > \nu_2$, simple assumptions predict amplification for frequency ν_2.

For three suitable levels to occur in a paramagnetic material, each paramagnetic center must involve the magnetic moment of more than one electron, and must interact with a surrounding array of atoms which are not cubically arranged. The energy levels and frequencies still respond to an externally applied magnetic field. They are, however, no longer simply related to it but may vary widely in accordance with fields internal to the crystalline material. This allows responses at high frequencies with relatively low applied fields.

Three-level solid-state maser amplifiers have been made which have a noise temperature less than about 5 K (noise figure $\leqq 1.02$). Although a wide variety of paramagnetic materials may be used, synthetic ruby, containing paramagnetic chromic (Cr^{3+}) ions, is favored. It has provided amplification both in resonant cavities and in traveling-wave structures.

Optical and infrared masers. Optical and infrared masers utilize a variety of principles for excitation of the nonequilibrium distribution. All involve multimode cavities, usually consisting of two optically flat plates between which the radiation is reflected many times through the excited medium. Such masers are particularly valuable as oscillators, producing coherent light that is extremely monochromatic and can be focused either to a very intense small spot or radiated in a remarkably parallel beam. *See* LASER.

Maser oscillators in the megahertz and audio range have been proposed; they use nuclear moments in an applied magnetic field or, with pumping at a higher frequency, in internal crystalline fields. The magnetic moments of protons in liquid water have provided a successful maser of this type. Small impurities of a paramagnetic ion furnish the higher energy level needed and transfer their excitation to the protons. The proton resonances in a magnetic field must be extremely narrow. Such a maser may be used as a very monochromatic oscillator with frequency proportional to the magnetic field strength. Hence it serves as a very precise magnetometer.

Deviation from thermodynamic equilibrium of the second type, involving phase coherence, also allows maser-type amplification and is present in many masers. In the beam-type maser oscillator, molecules decay toward the lower state. They continue to amplify after the probability of their being found in the lower state is greater than that of being found in the upper, because they oscillate coherently and in such a phase that they transfer energy to the electromagnetic field.

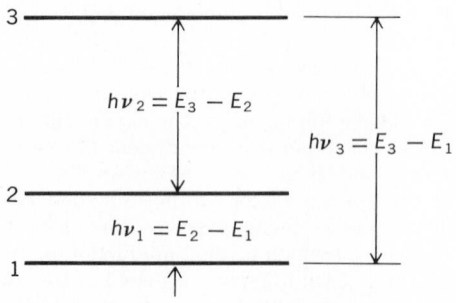

Fig. 2.   Energy levels for a three-level maser.

Raman-type masers rely on a type of phase coherence. Molecules with two levels, separated by energy $h\nu_1$, may be strongly driven by an electromagnetic field of frequency ν_2. If the majority of systems is in the lower state, and if $\nu_2 > \nu_1$, the Raman effect can allow amplification at frequency $\nu_2 - \nu_1$. This requires an intense driving field or a very strong coupling of the systems to the field, such as occurs in ferromagnetic electron resonances. In ferromagnetic materials, large numbers of electrons act in unison, thus providing coupling to an oscillating field which is strong enough to make Raman effects prominent. This can also be discussed in terms of classical theory if nonlinearities are allowed for, and is closely related to ferromagnetic amplifiers and to other parametric amplifiers. *See* MAGNETIC RESONANCE; MOLECULAR BEAMS; MOLECULAR STRUCTURE AND SPECTRA; NUCLEAR MOMENTS; RAMAN EFFECT.

Circuits. Maser circuits characteristically involve atoms or molecules which provide resonant reactances, positive or negative resistances, and coupling between two or more frequencies or circuit components. They also use certain elements of spectroscopic systems and a wide variety of components found in other radio-frequency and microwave devices.

In the simplest cases the molecular resonances behave like a resonant LC circuit with a positive or negative series resistor, or like a large number of such circuits in parallel and tuned over a distribution of frequencies.

The more classical circuit elements normally involved in masers supply the following functions:

1. Means of ensuring sufficiently strong interaction of an electromagnetic wave with material which amplifies the wave by stimulated emission.

2. Input and output coupling for the wave.

3. Auxiliary circuits which take appropriate advantage of maser characteristics in an overall system.

4. Where electromagnetic excitation is used, circuits which supply energy to the material and produce an unstable state which can radiate.

5. Magnetic field or other components for controlling frequencies of resonance of the material. The schematic of a three-level solid-state maser amplifier, shown in Fig. 3, illustrates each of these functional parts.

The resonant cavity, fulfilling function 1, must have sufficiently low internal losses and produce a sufficiently intense oscillating field (in this case a magnetic field) in the region of the amplifying material. If the cavity is uniformly filled with this material, the condition that amplification be obtainable is given by Eq. (2), where Q_0 is the unload-

$$Q_0 \geqq -Q_M = \frac{h\Delta\nu V}{8\pi\mu^2(n_1 - n_2)} \qquad (2)$$

ed quality factor of the cavity, h is Planck's constant, $\Delta\nu$ is the width of the molecular or atomic resonance, V is the cavity volume, n_1 is the number of particles in the upper state, n_2 is the number in the lower state, and μ is the effective dipole moment (matrix element) of the atoms or molecules.

A maser of this type is very similar to any other

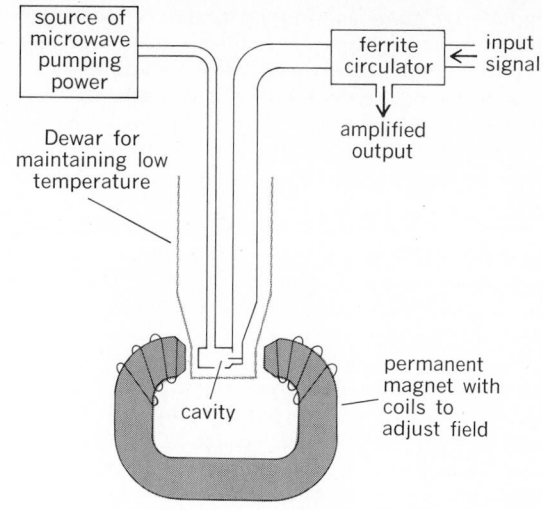

Fig. 3. Maser amplifier system for 9000 MHz.

amplifier with positive feedback. There is an effective negative resistance with which may be associated the negative quantity Q_M, giving a fractional gain per cycle in energy stored of $2\pi/|Q_M|$. As the losses, characterized by the loaded Q of the cavity, are decreased, the amplifier gain increases until it becomes unstable and oscillates when $1/Q + 1/Q_M \leqq 0$. Here $1/Q = 1/Q_0 + 1/Q_E$, where Q_E applies to the external coupling. The power gain is $G = [(2Q/Q_E) - 1]^2$. The bandwidth B decreases with gain in such a way that the bandwidth-voltage gain product is nearly constant under typical conditions, as in Eq. (3), where ν is

$$(\sqrt{G} + 1)\,B = 2\nu/Q_E \qquad (3)$$

the frequency for maximum gain. Thus, for given characteristics of the maser material specified by Q_M, the gain and bandwidth can be adjusted over certain limits by variation of the cavity losses, or Q_0, and the coupling, or Q_E.

The noise temperature T for the amplifier alone (Fig. 3) is given by Eq. (4), where k is Boltzmann's

$$T = \frac{h\nu}{k \ln\left[\dfrac{n_2}{n_1} + \dfrac{n_1 - n_2}{n_1}\dfrac{Q_M}{Q_0}\right]} \qquad (4)$$

constant. Minimum noise temperatures require a Q_0 appreciably larger than Q_M and a small ratio n_2/n_1. A noise temperature of $h\nu/k$, or near 1K for microwave frequencies, is the minimum needed, since this allows effective amplification of approximately one quantum, the limit set by quantum mechanics.

The input and output use the same coupling hole in Fig. 3, and require a directional coupler, or preferably a circulator, for their separation. Separate input and output coupling holes may be used, but this tends to decrease the gain-bandwidth product and to increase noise.

To take full advantage of the low noise in the maser amplifier, the input and output circuits must be prevented from radiating excess noise into the amplifier. For example, attenuation in parts of the input wave guide and in the circulators, which are not at low temperature, results in noise radiation into the amplifier. If there is 0.1 dB loss in some part which is at room temperature, noise radiation into the amplifier corresponding to about 7K will result. Since most solid-state maser amplifiers operate at low temperatures, some components of the input or output circuits may conveniently be cooled to minimize their noise radiation. If input and output coupling holes are used, the output wave may be passed through a cooled isolator to avoid noise radiation from warm components of the output circuit.

If energy is supplied to the amplifying material by an electromagnetic drive, the circuits used for this purpose must provide a sufficiently strong controllable drive (usually constant in amplitude and frequency). If the interaction between the driving or pumping radiation and the material is strong, a relatively small amount of power may be needed. In Fig. 3 the cavity must be resonant at both the pumping and the amplifying frequencies. In some cases the orientation of fields at both frequencies must also be carefully controlled. Usually excess pumping power is available so that careful design for coupling it to the amplifying material is not essential.

Masers using a resonant cavity may also be operated as superregenerative amplifiers, for which various components such as the driving field, the cavity losses, the couplings, or the "static" magnetic field may be modulated.

Traveling-wave masers normally use slow-wave structures, which need only be effective over the range of response of the maser material. Amplification of a reflected wave may be controlled in the usual ways by matching, by attenuators or isolators, or by arrangement of the maser material itself to absorb a reflected wave.

The overloading of a maser amplifier is qualitatively like that of any other amplifier, the overload occurring when the molecular energy becomes exhausted. Recovery times vary between microseconds and seconds, depending especially on the maser material used; *TR* (transmit-receive) devices are sometimes necessary.

Maser oscillators used as frequency standards require especially stable cavities, in which the field is strongly coupled to the excited molecules. The cavities should be decoupled from external circuits, as by an attenuator or isolator.

Amplification or generation of electromagnetic waves at infrared or optical frequencies by maser techniques requires circuits which superficially appear quite different from those at lower frequencies but which serve the same functions. The resonant circuit is usually provided by two surfaces between which the radiation is reflected. The short wavelengths in this region imply that all man-made circuit elements are large compared with a wavelength. Partially transparent surfaces, multilayers of dielectrics, lenses, and other normal infrared and optical components fill the role of circuit elements.   [CHARLES H. TOWNES; J. P. GORDON]

Bibliography: T. K. Ishii, *Maser and Laser Engineering*, 1980; A. E. Siegman, *Introduction to Lasers and Masers*, 1971; H. G. Unger, *Introduction to Quantum Electronics*, 1970; A. Yariv, *Quantum Electronics*, 1975.

Mass

The quantitative or numerical measure of a body's inertia, that is, of its resistance to being accelerated.

Before this rather abstract definition is developed, it is useful to consider a description of mass that, although less general, is more easily grasped intuitively. Isaac Newton said that the mass of a body is the measure of the quantity of matter the body contains. This description is useful for comparing the masses of samples of a particular type of matter, say, sugar, because it correctly suggests that the mass of a sample is the basic factor in determining the extent to which the sample possesses the fundamental unchangeable properties peculiar to that type of matter. Thus a given mass of sugar, that is, a quantity of sugar that exhibits a given measure of inertia, contains a definite number of molecules and will therefore sweeten a definite number of cups of coffee or supply a definite number of calories if it is burned or eaten. Twice the mass of sugar will contain twice as many molecules.

These properties are inherent or inalienable. Their extent can be changed only by adding more sugar to the sample or taking some away, that is, only by changing mass; it will not be changed by taking the sample of sugar into a spaceship or to the Moon. Other quantities associated with a given sample of sugar are nonpermanent. Thus its volume can be changed by pressure and its weight changed by taking it to a different altitude. Conversely, equal volumes of sugar may contain different masses and thus different numbers of molecules. In contrast to volume or weight of a body, mass is a measure of something that is fundamental and permanent. (However, see the subsequent discussion of mass and energy.)

Gravitation and inertia. The preceding discussion shows that the mass of a body could be defined in terms of one of the properties of the material of the body, for example, in the case of sugar, in terms of the amount of energy released when the sugar is burned in a specified way. Because, however, it is often necessary to compare masses of such dissimilar bodies as a sample of sugar, a sample of air, an electron, and the Moon, it is necessary to define mass in terms of a property that not only is inherent and permanent but is also universal in that it is possessed by all known forms of matter. All matter possesses two properties, gravitation and inertia. The property of gravitation is that every material body attracts every other material body. The property of inertia is that every material body resists any attempt to change its motion. A body's motion is said to change if the body is accelerated, that is, if it increases or decreases its speed or changes the direction of its motion. Because of its inertia a body cannot be accelerated unless a force is exerted on it. The greater the inertia of a body, the less will be the acceleration produced by a given force. *See* GRAVITATION.

As indicated in the beginning of this article, the present definition of mass is in terms of inertia. The masses of two bodies are compared by applying equal forces to the bodies and measuring their accelerations. For example, the two bodies may be allowed to collide. According to Newton's third law, each body will then experience an equally strong force. If there are no external forces, and if a_1 and a_2 are the measured accelerations of the two bodies, the ratio of the masses of the two bodies is by definition given by Eq. (1).

$$\frac{m_1}{m_2} = \frac{a_2}{a_1} \qquad (1)$$

This equation gives only ratios of masses; it is therefore necessary to designate the mass of some one body as the standard mass to which the masses of all other bodies can be compared. The body that has been chosen for this purpose is a cylinder of platinum-iridium alloy. It is known as the international standard of mass; its mass is called 1 kilogram (kg), and it is kept at the International Bureau of Weights and Measures near Paris, France. Replicas of the standard mass, kept at various national laboratories, are periodically compared with this standard.

It would be possible to define mass in terms of the property of gravitation. This gravitational mass of a body—to distinguish it from the inertial mass already defined—could be determined from the force of attraction exerted by the body on some standard body at a specified distance. It could also be determined (by Newton's third law, which must give the same result) from the force of attraction exerted on the body by the standard body. For example, to compare the gravitational masses of two bodies, the forces with which the Earth attracts them at any one point on the Earth's surface could be compared. The force of the Earth's attraction on a body is called the weight of the body. Thus the gravitational masses of bodies can be compared by simply weighing the bodies. *See* WEIGHT.

The ease and precision of weighing contrasts strongly and favorably with the difficult and necessarily imprecise experimental determination of the inertial mass of a body that requires measurement of acceleration. It is therefore extremely fortunate that R. von Eötvös in the 19th century proved experimentally with great precision that the inertial masses of such bodies as could be tested in the laboratory are equal to their gravitational masses. Astronomical observations support this finding. Finally, Albert Einstein, in his general theory of relativity, presented strong theoretical arguments that the two definitions of mass—in terms of inertia and in terms of gravitation—are identical. *See* RELATIVITY.

At present, mass is defined in terms of inertia but is measured by weighing. The result is a combination of the more useful of two definitions and of the simpler of two experimental determinations. The relation between the weight W and the mass m of a body is very simple. In a locality where the acceleration of gravity is g, $W = mg$ (in appropriate units).

Other properties. Besides inertia and gravitational attraction, mass has two other properties that point up the genius of Newton in abstracting, from the infinity of possible observations and descriptions, a concept that is as simple as it is basic. The first of these properties is that mass is linearly additive; for example, the total mass of a 1-kg body and a 3-kg body is 4 kg. (For exceptions, see the subsequent discussion of mass and en-

ergy.) As a consequence of this experimental fact, it is easy to define masses that are multiples or fractions of the standard kilogram mass. Thus 1 gram is defined as a mass, a thousand of which add up to 1 kg; 1 pound, as a mass of 1/2.204622 kg. This awkward conversion factor is used in order to make the current definition of a 1-lb mass agree closely with the old definition, which was made in terms of a standard pound, given by the mass of a block of platinum. Another useful unit of mass is the slug, which is defined as the mass of a body whose inertia is such that a force of 1 lb, which is the weight of a 1-lb mass at a certain specified locality, gives the body an acceleration of 1 ft/sec². One slug is equal to 32.174 lb.

The second property of mass is that it is conserved; that is, it can neither be created nor destroyed; a loss of mass by a system of bodies is always accompanied by an equal gain by some other system. *See* CONSERVATION OF MASS.

Mass and energy. Einstein's special theory of relativity predicts that the inertia of a body should increase if the energy of the body increases. This prediction has been conclusively verified experimentally. It follows that the mass of a body will increase if, for example, the body gains speed (addition of kinetic energy), or its temperature rises (addition of heat energy), or the body is compressed (addition of elastic energy). *See* INERTIA OF ENERGY.

The increase Δm of a body's mass (in grams), if its energy is increased by ΔE (in ergs), is given by $\Delta m = \Delta E/c^2$, where $c = 3 \times 10^{10}$ cm/sec is the velocity of light in vacuum. Thus the mass of a body increases by about 1.1×10^{-21} g for each erg of energy added to it. It follows that changes in mass are observable only if energy changes are extremely large. Nevertheless, experimental confirmation of Einstein's relation between mass and energy is excellent. For example, the mass m of an electron moving with a speed v that is near the speed of light has been experimentally found to be greater than its mass m_0 when at rest, by observing that a given force produces smaller acceleration of the rapidly moving electron than of an electron at rest. The results agree with Einstein's prediction defined in Eq. (2).

$$m = \frac{m_0}{\sqrt{1 - (v^2/c^2)}} \qquad (2)$$

Another example is given by neutrons and protons which when combined to form a stable atomic nucleus have, because of tremendous forces of mutual attraction, much lower energy than when they are free and separated. The mass of such a nucleus is measurably smaller than the sum of the masses of its constituent particles. On the other hand, in the explosion of dynamite, the mass of the products of explosion after they have come to rest is smaller than the original mass of the dynamite by less than one part in 10^{10}. Even in the explosion of a plutonium bomb, the "loss" of mass of plutonium is a small fraction of 1%. *See* MASS DEFECT.

The mass of a body can be changed not only by cutting a part of the body off, for example, but by merely changing the body's energy; thus Newton's description of mass as the measure of the quantity of matter is not useful when extremely high energies are involved, and may even be misleading. Indeed, mass must be assigned to all forms of energy and therefore to such nonmaterial entities as light. Although so-called particles of light, called photons, have zero rest mass, they have energy when they are—as they always are—in motion, and thus must have mass, equal to their energy divided by c^2. The mass of one of the more massive known photons, the photon of a γ-ray of wavelength 2×10^{-11} cm, is approximately the same as the mass of an electron. This means that such a photon will have just as much inertia as an electron and will weigh as much in the same location. *See* PHOTON.

Even with this broader, ultra-material concept of mass, mass is conserved. That is, if a system loses some of its mass either by losing some matter or some energy, another system must gain just as much mass. Matter can be destroyed or created by converting it into the energy of light, or vice versa, but the total mass remains unchanged. Mass, however, is no longer simply additive. Two 1-kg blocks, when brought close together, will have less gravitational potential energy than when far apart, and the mass of the combination will be less than 2 kg. The difference, however, is too small to be detectable now or in the foreseeable future. *See* LIGHT.

[LEO NEDELSKY]

Bibliography: F. Bueche, *Principles of Physics*, 3d ed., 1977; A. Einstein and L. Infeld, *The Evolution of Physics*, 1938, reprint 1960; L. B. Macurdy, Standards of mass, *Phys. Today*, 4(4):7–11, 1951.

Mass defect

The difference between the mass of an atom and the sum of the masses of its individual components in the free (unbound) state. The mass of an atom is always less than the total mass of its constituent particles; this means, according to Albert Einstein's well-known formula, that an energy of $E = mc^2$ has been released in the process of combination, where m is the difference between the total mass of the constituent particles and the mass of the atom, and c is the velocity of light. *See* INERTIA OF ENERGY.

The mass defect, when expressed in energy units, is called the binding energy, a term which is perhaps more commonly used. *See* NUCLEAR BINDING ENERGY.

[W. W. WATSON]

Mass number

The mass number A of an atom is the total number of its nuclear constituents, or nucleons, as the protons and neutrons are collectively called. The mass number is placed (by North American practice) following and above the elemental symbol, thus U^{238}, or (by international agreement) before and above it, thus ^{238}U. Because of the approximate equality of the proton and neutron masses, and the relative insignificance of that of the electron, the mass number gives a useful rough figure for the atomic mass; for example, $H^1 = 1.00814$ atomic mass units (amu), $U^{238} = 238.124$ amu, and so on. The mass number is reduced by four during α-emission, but it is not altered during β-decay or electron capture. *See* ATOMIC NUMBER.

[HENRY E. DUCKWORTH]

Mathematical physics

The area of physics aimed at deducing the consequences of the more established physical theories by relying mainly on the method of mathematical solution, presuming that the basic laws of physics are known. A fruitful approach is possible largely because there is close analogy between the mathematical problems arising in different fields of theoretical physics The same partial differential equations are encountered in many different contexts.

Some examples of problems in mathematical physics are given below.

1. The theory of the motion of planets, particularly the classical three-body problem; for example, the motion of an asteroid under the combined influence of the Sun and Jupiter. Gyroscopic motion of rigid bodies.

2. Potential theory, applicable primarily in electrostatics and hydrodynamics of nonviscous flow. Many important special functions such as Bessel functions and Legendre polynomials have been developed in connection with potential theory. Functions of a complex variable are useful for two-dimensional problems.

3. Theory of vibrations, determining the normal modes of electromagnetic or elastic vibration of regions of a given shape, or of systems of bodies interacting in various ways. Among other fields this is important for the theory of microwave cavities, for acoustics, and for seismology. Again special mathematical functions are important.

4. Wave propagation, including the exact solution of diffraction problems such as for electromagnetic or acoustic waves.

5. Solution of problems in wave mechanics, such as are encountered in the helium atom or the hydrogen molecule or in scattering problems, which are too complicated for direct analytical solution and yet simple enough to be solved accurately. The variational method is most useful here.

6. Diffusion problems, such as diffusion of neutrons in matter, conduction of heat, and transport phenomena in statistical mechanics.

7. Dispersion theory, in which the reactions of a system to external forces of different frequencies are related. Optical properties of matter, plasma physics, and high-energy physics are examples.

8. Nonlinear problems in hydrodynamics, elasticity theory, and so on.

9. Problems in probability theory related to statistical mechanics.

Through World War II, the main technique of mathematical physics was the analytical mathematical solution of problems. Since World War II, high-speed computing machines have become increasingly important and have made numerical solutions possible for many problems where the analytical technique did not work.

The term mathematical physics is sometimes used synonymously with theoretical physics. *See* THEORETICAL PHYSICS.

[HANS A. BETHE]

Matrix calculus

That branch of mathematics which deals with matrices whose elements are functions of one or more independent variables.

The derivative of a matrix $A(t)$ whose elements

$a_{ij}(t)$ are functions of a variable t is defined in Eq. (1), where da_{ij}/dt represents the matrix whose

$$\frac{dA}{dt} = \lim_{\Delta t \to 0} \frac{A(t+\Delta t) - A(t)}{\Delta t} = \left(\frac{da_{ij}}{dt}\right) \qquad (1)$$

elements are da_{ij}/dt. Thus dA/dt is formed by replacing the elements of $A(t)$ by their derivatives. *See* MATRIX THEORY.

If the matrices A and B are functions of t, then Eqs. (2) are satisfied by the operation of differen-

$$\begin{aligned}
\frac{d}{dt}(A+B) &= \frac{dA}{dt} + \frac{dB}{dt} \\
\frac{d}{dt}(AB) &= \frac{dA}{dt}B + A\frac{dB}{dt}
\end{aligned} \qquad (2)$$

tiation. In differentiating a product the order of the factors must be preserved. Thus Eq. (3) is

$$\frac{d}{dt}A^2 = \frac{dA}{dt}A + A\frac{dA}{dt} \qquad (3)$$

obtained for $(d/dt)A^2$ and not $2A(dA/dt)$. From $A^{-1}A = I$ Eq. (4) is found.

$$\frac{dA^{-1}}{dt} = -A^{-1}\frac{dA}{dt}A^{-1} \qquad (4)$$

The integral of $A(t)$ is defined as the matrix whose elements are integrals of $a_{ij}(t)$.

Every square n by n matrix satisfies a polynomial equation of lowest degree, called its minimum equation, which is shown in Eq. (5). Equa-

$$A^m + a_1 A^{m-1} + \ldots + a_m I = 0 \quad (m \leq n) \qquad (5)$$

tion (5) may be used to express all powers of $A > m-1$ in terms of $I, A, A^2, \ldots, A^{m-1}$. Therefore any matrix polynomial $f(A)$ of degree $k > m-1$ may be replaced by a polynomial $F(A)$ of degree $< m$.

If $f(z) = \sum_{k=0}^{\infty} c_k z^k$ is a power series whose radius of convergence is r, and if all eigenvalues of A are less than r in absolute value, then $f(A)$ is defined as the matrix series shown in Eq. (6).

$$f(A) = c_0 I + \sum_{k=0}^{\infty} c_k A^k \quad |\lambda_A| < r \qquad (6)$$

Thus the binomial series yields Eq. (7) for ration-

$$(I+A)^n = I + \sum_{k=1}^{\infty} \binom{n}{k}A^k \quad |\lambda_A| < 1 \qquad (7)$$

al n; and since the exponential series for e^z converges for all z, Eq. (8) can be set up for arbitrary A.

$$e^A = I + \sum_{k=1}^{\infty} \frac{1}{k!}A^k \qquad (8)$$

The matrix exponential has the properties $e^0 = I$, $e^A e^{-A} = I$; also, $e^A e^B = e^{A+B}$ when $AB = BA$. From the series for e^{tA}, $de^{tA}/dt = Ae^{tA}$.

The minimum equation labeled (5) may also be used to replace a power series for $f(A)$ by a polynomial $F(A)$. If Eq. (5) has distinct roots $\lambda_1, \lambda_2, \ldots, \lambda_m$ (eigenvalues of A), $F(A)$ has the eigenvalues $F(\lambda_i) = f(\lambda_i)$. The polynomial $F(\lambda)$ is completely determined by the m values $f(\lambda_i)$ it assumes when $\lambda = \lambda_i$. Hence $F(\lambda)$, and also $F(A) = f(A)$, may be determined by Lagrange's interpolation formulas given in Eqs. (9) and (10), where each

product has $m-1$ factors.

$$f(A) = f(\lambda_1)C_1(A) + \ldots + f(\lambda_m)C_m(A) \qquad (9)$$

$$C_j(A) = \prod_{i \neq j}(A - \lambda_i I) \bigg/ \prod_{i \neq j}(\lambda_j - \lambda_i) \qquad (10)$$

When Eq. (5) has repeated roots, say $\lambda_1 = \lambda_2 = \lambda_3$, the corresponding terms in the modified equation, numbered (9), are given by formula (11), and Eq.

$$f(\lambda_1)C_1(A) + f'(\lambda_1)C_2(A) + f''(\lambda_1)C_3(A) \qquad (11)$$

(10) does not apply. But since the $C_j(A)$ do not depend on f, they can be found by choosing m functions $f_i(\lambda)$ so that the m linear equations numbered (9) and (10) in C_j have a nonzero determinant, and then solving them for the $C_j(A)$.

Consider, for example, a 4 by 4 matrix A with $\lambda_1 = \lambda_2 = \lambda_3$, $\lambda_4 \neq \lambda_1$. If $m = 4$, the basic equation (9) is now Eq. (12). By choosing $f(\lambda) = 1$, $\lambda - \lambda_1$,

$$f(A) = f(\lambda_1)C_1 + f'(\lambda_1)C_2 \\ + f''(\lambda_1)C_3 + f(\lambda_4)C_4 \qquad (12)$$

$(\lambda - \lambda_1)^2$, and $(\lambda - \lambda_1)^3$, respectively, Eqs. (13) are obtained, which give C_4, C_3, C_2, and C_1.

$$\begin{aligned}
I &= C_1 + 0 + 0 + C_4 \\
A - \lambda_1 I &= 0 + C_2 + 0 + (\lambda_4 - \lambda_1)C_4 \\
(A - \lambda_1 I)^2 &= 0 + 0 + 2C_3 + (\lambda_4 - \lambda_1)^2 C_4 \\
(A - \lambda_1 I)^3 &= 0 + 0 + 0 + (\lambda_4 - \lambda_1)^3 C_4
\end{aligned} \qquad (13)$$

Eigenmatrices. Any n by n matrix A is reduced to its Jordan form J by finding its eigenmatrix E whose columns are the eigenvectors $e_1, \ldots, e_n$ belonging to the eigenvalues $\lambda_1, \lambda_2, \ldots, \lambda_n$. If λ_1 is simple, A has a proper eigenvector e_1 such that $(A - \lambda_1 I)e_1 = 0$. If λ_1 is k-tuple, with one proper eigenvector e_1 and $k-1$ generalized eigenvectors $e_2, e_3, \ldots, e_k$ which satisfy Eqs. (14),

$$(A - \lambda_1 I)e_j = e_{j-1} \quad j = 2, 3, \ldots, k \qquad (14)$$

then $E = (e_1 | e_2 | \ldots | e_n)$ and Eqs. (15) are valid.

$$E^{-1}AE = J \qquad A = EJE^{-1} \qquad (15)$$

If $f(z)$ is an analytic function, then Eq. (16)

$$f(A) = f(EJE^{-1}) = Ef(J)E^{-1} \qquad (16)$$

holds, which gives $f(A)$ when the simpler $f(J)$ is known. Thus, when A has distinct eigenvalues $\lambda_1, \ldots, \lambda_n$, J and $f(J)$ have the form shown in Eqs. (17). Equation (16) is an explicit formula for

$$\begin{aligned}
J &= \mathrm{diag}(\lambda_1, \ldots, \lambda_n) \\
f(J) &= \mathrm{diag}[f(\lambda_1), \ldots, f(\lambda_n)]
\end{aligned} \qquad (17)$$

$f(A)$ and agrees with the former interpolation method; it requires, however, the eigenmatrix E and its reciprocal. Both methods are shown in the following example.

The eigenvalues of $A = \begin{pmatrix} 0 & 1 \\ -2 & 3 \end{pmatrix}$ are $\lambda_1 = 1$, $\lambda_2 = 2$ and $e_1 = (1,1)^T$, $e_2 = (1,2)^T$. Then Eq. (16) gives Eq. (18).

$$\begin{aligned}
f(A) &= \begin{pmatrix} 1 & 1 \\ 1 & 2 \end{pmatrix}\begin{pmatrix} f(1) & 0 \\ 0 & f(2) \end{pmatrix}\begin{pmatrix} 2 & -1 \\ -1 & 1 \end{pmatrix} \\
&= \begin{pmatrix} 2 & -1 \\ 2 & -1 \end{pmatrix}f(1) + \begin{pmatrix} -1 & 1 \\ -2 & 2 \end{pmatrix}f(2) \qquad (18)
\end{aligned}$$

The interpolation method uses Eq. (9), which in

this case reduces to Eq. (19). Choose $f(\lambda)$ as $\lambda - 1$,

$$f(A) = f(1)C_1 + f(2)C_2 \qquad (19)$$

$\lambda - 2$, respectively; then C_2 and C_1 are given by Eqs. (20) in agreement with Eq. (18).

$$C_2 = A - I = \begin{pmatrix} -1 & 1 \\ -2 & 2 \end{pmatrix}$$
$$-C_1 = A - 2I = \begin{pmatrix} -2 & 1 \\ -2 & 1 \end{pmatrix} \qquad (20)$$

As an example with multiple eigenvalues, consider A given by Eq. (21). The modified Eq. (19)

$$A = \begin{pmatrix} 1 & 1 & 0 \\ 0 & 1 & 1 \\ 0 & 0 & 1 \end{pmatrix} \qquad \lambda_1 = \lambda_2 = \lambda_3 = 1 \quad (21)$$

is now Eq. (22). Choose $f(\lambda)$ as 1, $\lambda - 1$, $(\lambda - 1)^2$,

$$f(A) = f(1)C_1 + f'(1)C_2 + f''(1)C_3 \qquad (22)$$

respectively; then C_1, C_2 and C_3 are given by Eqs. (23).

$$I = C_1 \qquad A - I = C_2 \qquad (A - I)^2 = 2C_3 \qquad (23)$$

Let $f(\lambda) = e^{\lambda t}$; then $f'(\lambda) = te^{\lambda t}$, $f''(\lambda) = t^2 e^{\lambda t}$, and so e^{tA} is given by Eq. (24), which can be expressed as Eq. (25).

$$e^{tA} = \begin{pmatrix} 1 & 0 & 0 \\ 0 & 1 & 0 \\ 0 & 0 & 1 \end{pmatrix} e^t + \begin{pmatrix} 0 & 1 & 0 \\ 0 & 0 & 1 \\ 0 & 0 & 0 \end{pmatrix} te^t$$
$$+ \begin{pmatrix} 0 & 0 & 1/2 \\ 0 & 0 & 0 \\ 0 & 0 & 0 \end{pmatrix} t^2 e^t \qquad (24)$$

$$e^{tA} = \begin{pmatrix} 1 & t & \frac{1}{2}t^2 \\ 0 & 1 & t \\ 0 & 0 & 1 \end{pmatrix} e^t \qquad (25)$$

Differential equations. Let $x_1(t), \ldots, x_n(t)$ be n unknown functions and $X(t)$ their column vector. Then if $A(t)$ is an n by n matrix whose elements are continuous functions of t, the system of n linear differential equations shown in Eqs. (26) may

$$dX/dt = AX \qquad X(0) = X_0 \qquad (26)$$

be solved by successive approximations. The nth approximation is obtained from the $(n-1)$th by integrating Eqs. (27) from $t = 0$ to t. Then a solution

$$dX_n/dt = AX_{n-1} \qquad n = 1, 2, \ldots \quad (27)$$

is obtained in the form of Eq. (28), where Ω is

$$X = \lim_{n \to \infty} X_n = \Omega X_0 \qquad (28)$$

given by Eq. (29) and is called the matricant of A.

$$\Omega = I + \int_0^t A\, dt + \int_0^t A\, dt \int_0^t A\, dt + \ldots \quad (29)$$

When A has constant elements, Eq. (29) reduces to Eq. (30).

$$\Omega = I + At + \frac{1}{2!}A^2 t^2 + \frac{1}{3!}A^3 t^3 + \ldots = e^{tA} \quad (30)$$

Consider, for example, the system given by Eqs. (31), or in matrix form by Eq. (32). Then from Eq. (25) one obtains Eq. (33).

$$\frac{dx_1}{dt} = x_1 + x_2$$
$$\frac{dx_2}{dt} = x_2 + x_3 \qquad (31)$$
$$\frac{dx_3}{dt} = x_3$$

$$\frac{dX}{dt} = \begin{pmatrix} 1 & 1 & 0 \\ 0 & 1 & 1 \\ 0 & 0 & 1 \end{pmatrix} X = AX \qquad (32)$$

$$X = e^{tA}X_0 = e^t \begin{pmatrix} 1 & t & \frac{1}{2}t^2 \\ 0 & 1 & t \\ 0 & 0 & 1 \end{pmatrix} X_0 \qquad (33)$$

In a linear system with variable coefficients $p_{ij}(t)$, the differential Eq. (26) becomes Eqs. (34).

$$\frac{dX}{dt} = P(t)X \qquad P(t) = [p_{ij}(t)] \qquad (34)$$

If X_1, X_2, X_3 are three linearly independent solutions of Eqs. (34), let $Y = (X_1 | X_2 | X_3)$ denote the 3 by 3 matrix whose columns are X_1, X_2, X_3. Then Eqs. (34) are equivalent to the matrix equation shown in Eq. (35). If one differentiates $\det Y$ by

$$\frac{dY}{dt} = P(t)Y \qquad (35)$$

rows and adds the resulting determinants, one obtains Eq. (36), where $\operatorname{tr} P$ denotes the trace of P

$$\frac{d}{dt}(\det Y) = (p_{11} + p_{22} + p_{33})\det Y$$
$$= (\operatorname{tr} P)\det Y \qquad (36)$$

and $\det Y$ is the determinant of Y. On integration this yields the Jacobi identity given in Eq. (37).

$$\det Y = c \exp \int_{t_0}^t \operatorname{tr} P\, dt \qquad (37)$$

Since $\det Y$ is not identically zero, $c \neq 0$; hence the matrix integral $Y(t)$ is nonsingular; that is, $\det Y \neq 0$.

If Y_1 is a particular solution of Eq. (35) and C is an arbitrary 3 by 3 constant matrix, then $Y = Y_1 C$ is also a solution. Moreover, all solutions of Eq. (35) are contained in $Y = Y_1 C$. For if Y is a solution of Eq. (35), then Eq. (4) gives Eq. (38), hence $Y_1^{-1}Y = C$ and $Y = Y_1 C$.

$$\frac{d}{dt}(Y_1^{-1}Y) = \left(-Y_1^{-1}\frac{dY_1}{dt}Y_1^{-1}\right)Y + Y_1^{-1}\frac{dY}{dt}$$
$$= -Y_1^{-1}PY + Y_1^{-1}PY = 0 \qquad (38)$$

If the coefficients $p_{ij}(t)$ are periodic with the same period ω, that is, $P(t + \omega) = P(t)$, then the system shown in Eq. (35) can be reduced to one with constant coefficients by a linear transformation $Y = L(t)Z$, where $L(t)$ is a nonsingular matrix of period ω. Let $Y_1(t)$ be the solution of Eq. (35) for which $Y_1(0) = I$; then $Y_1(t + \omega)$ is also a solution, and hence $Y_1(t + \omega) = Y_1(t)C$, where C is a nonsingular matrix of constants. If a matrix B is determined so that $e^{B\omega} = C$, then Eqs. (39) and (40)

$$Y = L(t)Z \qquad (39)$$

$$L(t) = Y_1(t)\,e^{-Bt} = L(t+\omega) \qquad (40)$$

result, which reduce Eq. (35) to $dZ/dt = BZ$, whose solution $Z_1(t)\,e^{Bt}$ corresponds to $Y_1(t)$. Note that $Y_1(t) = Y_1(t)e^{-Bt}Z_1(t)$ requires $Z_1(t) = e^{Bt}$, $dZ_1/dt = BZ_1$; and since all solutions of this equation have the form $Z(t) = CZ_1(t)$, $dZ/dt = BZ$.

[LOUIS BRAND]

Bibliography: R. A. Frazer et al., *Elementary Matrices and Some Applications to Dynamics and Differential Equations,* 1946, repr. 1977; F. R. Gantmacher, *The Theory of Matrices,* 2 vols., 1959; A. Halanay, *Differential Equations,* 1966.

Matrix mechanics

A formulation of quantum theory in which the operators are represented by time-dependent matrices. For a discussion of quantum-mechanical operators and other information essential to an understanding of this article *see* NONRELATIVISTIC QUANTUM THEORY.

Matrix mechanics is not useful for obtaining quantitative solutions to actual problems; on the other hand, because it is concisely expressed in a form independent of special coordinate systems, matrix mechanics is useful for proving general theorems. For the purposes of the following brief discussion it is sufficient to consider a one-dimensional spinless system, described by a wave function $\psi(x,t)$.

A matrix A is a rectangular array of numbers; the number of rows or columns of the array may be finite or infinite; A_{mn} is the element (number) in the mth row and the nth column; m, n may be discrete or continuous indices.

Symbolic addition $C = A + B$ is defined as meaning that $C_{mn} = A_{mn} + B_{mn}$ for each m and n, with the implication that the rows of A, B have the same indexing m and that the columns also have the same indexing m. Symbolic multiplication $C = AB$ is defined to mean $C_{mn} = \Sigma_j A_{mj} B_{jn}$, summed over all values of j, again with the implication that the columns of A and the rows of B have the same indexing j; when j is continuous, the sum in the definition of C_{mn} is replaced by an integral. The adjoint $A^\dagger$ of A is obtained by interchanging rows and columns of A and then taking the complex conjugate of each element, that is, $A_{mn}^\dagger = (A_{nm})^* \equiv A_{nm}^*$; $(AB)^\dagger = B^\dagger A^\dagger$. *See* MATRIX THEORY.

Suppose $\psi(x,t)$ is expanded in terms of any complete set of orthonormal functions $u_n(x)$, as shown in Eq. (1). The matrix elements (in this u

$$\psi(x,t) = \sum_n \psi_n(t) u_n(x) \qquad (1)$$

representation) of any quantum-mechanical operator A are defined by Eq. (2). For instance, when

$$A_{mn} = \int_{-\infty}^{\infty} dx\, u_m^*(x) A u_n(x) \qquad (2)$$

$A \equiv p_x$, A is replaced by $(\hbar/i)\,\partial/\partial x$ in Eq. (2). From Eqs. (1) and (2) and the orthonormality of $u_n(x)$, if Eq. (3) holds, then Eq. (4) can be obtained. This

$$A\psi \equiv \phi = \sum_n \phi_n u_n \qquad (3)$$

$$\phi_m = \sum_n A_{mn} \psi_n \qquad (4)$$

result shows that the equation $\phi = A\psi$, which states that ϕ is the function resulting from the operation A on ψ, can be interpreted equally well as a matrix equation, provided ψ, ϕ are represented by single-column matrices whose elements are respectively ψ_n, ϕ_n. Similarly, the expectation value of the operator A in the state ψ is given by Eq. (5), that is, in matrix notation $\langle A \rangle = \psi^\dagger A\psi$; the adjoint of a column matrix ψ_m is the row ψ_m^*.

$$\langle A \rangle = \int \psi^* A\psi = \int \psi^* \phi = \sum_{m,n} \psi_m^* A_{mn} \psi_n \qquad (5)$$

The time-dependent Schrödinger equation implies $\psi(t) = \exp\,(-iHt/\hbar)\,\psi(0)$. Thus, noting that, because the operator H is hermitian, the matrix H is self-adjoint, $H^\dagger = H$, and the expectation value of A at time t is as shown in Eq. (6). Equation (6)

$$\langle A \rangle_t = \psi^\dagger(0)\,\exp\,(iHt/\hbar) A \exp-(iHt/\hbar)\psi(0) \qquad (6)$$

shows that $\langle A \rangle_t$ can be computed in two equivalent ways: (i) the conventional Schrödinger representation, in which ψ is time-dependent and the A_{mn}s are time-independent; (ii) the Heisenberg representation, in which ψ is time-independent and equal to $\psi(0)$ but A is represented by time-dependent matrix elements given in Eq. (7). Differentiating

$$A_{mn}(t) = [\exp\,(iHt/\hbar) A(0) \exp-(iHt/\hbar)]_{mn} \qquad (7)$$

Eq. (7), one obtains Eq. (8); that is, $dA/dt = (i\hbar)^{-1}(A,H)$, where $(A,B) \equiv AB - AB$.

$$\frac{d}{dt} A(t) = \frac{iH}{\hbar} \exp\,(iHt/\hbar) A(0) \exp-(iHt/\hbar)$$

$$- \exp\,(iHt/\hbar) A(0) \exp-(iHt/\hbar)\frac{iH}{\hbar}$$

$$= \frac{1}{i\hbar}\,[A(t)H = HA(t)] \qquad (8)$$

It can be proved that (i) operators can consistently be replaced by equivalent matrices, defined by Eq. (2); and (ii) the Heisenberg time-dependent matrix formulation of quantum theory is completely equivalent to the Schrödinger time-dependent wave function formulation. [EDWARD GERJUOY]

Matrix theory

A matrix is a rectangular array of numbers, or other elements, of form (1). The array A is an m by

$$A = \begin{pmatrix} a_{11} & a_{12} & \cdots & a_{1n} \\ a_{21} & a_{22} & \cdots & a_{2n} \\ \cdots\cdots\cdots\cdots\cdots\cdots\cdots \\ a_{m1} & a_{m2} & \cdots & a_{mn} \end{pmatrix} \qquad (1)$$

n matrix with m rows and n columns, and the size of A is said to be m by n. The rows of a matrix are always numbered from the top down and the columns from left to right. The position of each element in the array is given by its subscripts; that is, a_{ij} is the element in the ith row and jth column. Since every element of A is represented by a_{ij} as i takes on the values $1, 2, \ldots, m$ and j the values $1, 2, \ldots, n$, a_{ij} is called the typical element of A, and the compact notation $A = (a_{ij})$ is used when the size of A is given.

Matrices have application as computational devices in such widely diversified fields as economics, psychology, statistics, engineering, physics, and mathematics. In mathematics, matrices are useful tools in the study of linear systems of algebraic equations, linear differential equations, linear mappings and transformations, and bilinear and quadratic forms.

For example, in the linear system of equations

$$-3x + 2y - 6z = 10$$
$$7x - y + 3z = 0$$
$$x + y - 5z = 1$$

the letters x, y, and z are merely symbols which stand for possible numerical solutions, and the only significant features of this system are the numbers which appear in the equations and their relative positions. Therefore, these equations are completely described by the 3 by 4 matrix which is labeled (2).

$$\begin{pmatrix} -3 & 2 & -6 & 10 \\ 7 & -1 & 3 & 0 \\ 1 & 1 & -5 & 1 \end{pmatrix} \qquad (2)$$

Similarly the properties of the linear substitution

$$x = x' \cos \theta - y' \sin \theta$$
$$y = x' \sin \theta + y' \cos \theta$$

which describes a rotation of axes of a cartesian coordinate system, are completely determined by the square 2 by 2 matrix

$$\begin{pmatrix} \cos \theta & -\sin \theta \\ \sin \theta & \cos \theta \end{pmatrix}$$

If $m = n$ in (1), A is called a square matrix of order n. If $m = 1$, A is a row matrix, and if $n = 1$, A is a column matrix. The elements a_{ij} of A, for which $i = j$, are the principal diagonal elements. A diagonal matrix is a matrix such that $a_{ij} = 0$ if $i \neq j$, and a scalar matrix is a square diagonal matrix with equal diagonal elements. An identity matrix is a scalar matrix in which the common diagonal element is the number 1. An n by n identity matrix is denoted by I_n.

Matrices can be regarded as generalized numbers, and their utility in applications depends on the possibility of combining them in certain definite ways. The matrix operations of addition, subtraction, and multiplication are defined in terms of these same operations for the elements, and they satisfy some, but not all, of the rules of ordinary algebra. In discussing these matrix operations it is assumed that the elements of the matrices are numbers.

Two matrices $A = (a_{ij})$ and $B = (b_{ij})$ are equal if they have the same size m by n and $a_{ij} = b_{ij}$ for all i, j. Matrices $A = (a_{ij})$ and $B = (b_{ij})$ of the same size m by n are added by adding correspondingly placed elements; that is, $A + B = C = (c_{ij})$ is an m by n matrix where $c_{ij} = a_{ij} + b_{ij}$ for all i, j. It follows that matrix addition is associative and commutative, that is, $(A + B) + C = A + (B + C)$ and $A + B = B + A$, as is the case for numbers. The m by n matrix with 0 in every position is denoted by 0, and is called a null matrix. Then $A + 0 = 0 + A = A$. The matrix $-A = (-a_{ij})$ is the negative of the matrix $A = (a_{ij})$, and $A + (-A) = 0$. Subtraction of m by n matrices is defined by $B - A = B + (-A) = (b_{ij} - a_{ij})$.

A matrix $B = (b_{ij})$ is conformable with respect to a matrix $A = (a_{ij})$ if B has size n by q and A has size m by n; that is, B has the same number of rows as A has columns. The matrix product AB is defined only when B is conformable with respect to A. The product $C = AB$ is an m by q matrix and the element in the i, j position of C is obtained by multiplying the n elements in the ith row of A into the n elements in the jth column of B, term by term, and adding these products. Thus if

$$A = \begin{pmatrix} 2 & 0 & -1 \\ 4 & 1 & \frac{1}{2} \end{pmatrix} \quad \text{and} \quad B = \begin{pmatrix} 5 \\ 1 \\ -2 \end{pmatrix}$$

then

$$AB = \begin{pmatrix} 12 \\ 20 \end{pmatrix}$$

If A and B are square matrices of the same size, then both AB and BA are defined. Matrix multiplication is associative. If A is m by n, B is n by q and C is q by r, then $(AB)C$ and $A(BC)$ are equal m by r matrices. Also, when the matrices A, B, and C have the proper sizes for the operations to be defined, $A(B + C) = AB + AC$ and $(A + B)C = AC + BC$. If A is m by n, then for identity matrices of the proper sizes, $AI_n = I_m A = A$. Unlike the case for numbers, it may happen for matrices that $AB \neq BA$ and $AB = 0$ with $A \neq 0$ and $B \neq 0$.

The product of a matrix A and a number a is called a scalar product and is obtained by multiplying every element a_{ij} of A by a. Thus,

$$\frac{1}{2}\begin{pmatrix} 4 & -3 & \frac{1}{3} \\ 0 & 2 & 1 \end{pmatrix} = \begin{pmatrix} 2 & -\frac{3}{2} & \frac{1}{6} \\ 0 & 1 & \frac{1}{2} \end{pmatrix}$$

The transpose of an m by n matrix A is the n by m matrix B which has as its ith row the ith column of A and as its jth column the jth row of A for all i, j. If the transpose of a matrix A is denoted by A', and B is conformable with respect to A, then $(AB)' = B'A'$. A matrix A is symmetric if $A = A'$. A symmetric matrix is necessarily square.

A square n by n matrix is nonsingular if the determinant of A is not zero. Otherwise A is singular. A nonsingular matrix A has a unique inverse, that is, a matrix A^{-1} such that $AA^{-1} = A^{-1}A = I_n$. The inverse of a nonsingular matrix is easily described but is difficult to compute for matrices of large size. Many important applications require the calculation of the inverse, and numerical methods are used to approximate the elements of the inverse.

Two m by n matrices A and B are equivalent if there exist nonsingular matrices P and Q such that $B = PAQ$.

The matrices A and B are equivalent if, and only if, B can be obtained from A by a sequence of elementary transformations which consists of the following operations: interchanging two rows (or columns) of A; multiplying the elements of a row (or column) of A by a fixed number and adding to the corresponding elements of another row (or column) of A; multiplying the elements of a row (or column) of A by a nonzero number. A matrix A can be carried into a matrix with r ones on the principal diagonal and zeros elsewhere by a sequence of elementary transformations. The number r is called the rank of A, and the process of reducing A to this diagonal form is called the reduction of A to canonical form. The rank of A can be defined intrinsically as the largest order r of a nonvanishing minor of A. For example, matrix (2) has rank 3 since the three-rowed minor

$$\begin{vmatrix} -3 & 2 & -6 \\ 7 & -1 & 3 \\ 1 & 1 & -5 \end{vmatrix} = 22 \neq 0$$

and the canonical form of this matrix is

$$\begin{pmatrix} 1 & 0 & 0 & 0 \\ 0 & 1 & 0 & 0 \\ 0 & 0 & 1 & 0 \end{pmatrix}$$

Two square n by n matrices A and B are similar if there exists a nonsingular matrix P such that $B = PAP^{-1}$. A linear transformation, or substitution, labeled (3), can be written $y = Ax$, where y and x is

$$
\begin{aligned}
y_1 &= a_{11}x_1 + a_{12}x_2 + \cdots + a_{1n}x_n \\
y_2 &= a_{21}x_1 + a_{22}x_2 + \cdots + a_{2n}x_n \\
&\cdots\cdots\cdots\cdots\cdots\cdots\cdots \\
y_n &= a_{n1}x_1 + a_{n2}x_2 + \cdots + a_{nn}x_n
\end{aligned}
\qquad (3)
$$

are n by 1 column matrices and $A = (a_{ij})$ is the matrix of the transformation. If new variables $z = Py$ and $w = Px$, where P is nonsingular, are substituted, $P^{-1}z = AP^{-1}w$ or $z = PAP^{-1}w$ is obtained. Thus the given linear transformation in terms of the new variables has a matrix which is similar to A. The theory of a single linear transformation is given by finding canonical forms for the matrix of the transformation under similarity.

If $A = (a_{ij})$ is an n by n square matrix and x is a variable, the matrix

$$Ix - A = \begin{pmatrix} x - a_{11} & a_{12} & \cdots & a_{1n} \\ a_{21} & x - a_{22} & \cdots & a_{2n} \\ \cdots\cdots\cdots\cdots\cdots\cdots\cdots\cdots \\ a_{n1} & a_{n2} & \cdots & x - a_{nn} \end{pmatrix}$$

is called the characteristic matrix of A. The determinant $|Ix - A|$ of $Ix - A$ is a polynomial in x of degree n, and the equation $|Ix - A| = 0$ is called the characteristic equation of A. The roots of the characteristic equation are the characteristic values, or eigenvalues, of A. Many applications in mathematics and physics require information about the characteristic values of a matrix.

A quadratic form in n variables $x_1, x_2, \ldots, x_n$ is a polynomial of degree 2 which can be written as the matrix product xAx', where $x = (x_1, x_2, \ldots, x_n)$ is a row matrix and A is an n by n symmetric matrix. If a change of variable $x = yP$ is made, the result is $xAx' = yPAP'y'$. If P is nonsingular, the new matrix PAP' is said to be congruent to A. The simplification of a quadratic form consists of simplifying a symmetric matrix A by a congruence transformation. A matrix A can be reduced to various diagonal forms by congruence transformations, the particular form depending on the number system (that is, rational, real, or complex numbers) which contains the elements of A and the elements of the transforming matrix P. When A has been reduced to diagonal form, the corresponding quadratic form is a sum of squares of the new variables $y_1, y_2, \ldots, y_n$.

A square matrix A with complex number elements is hermitian if $A = \bar{A}'$, where $\bar{A}$ is the matrix obtained from A by replacing each element of A by its complex conjugate. The reduction of a hermitian matrix to diagonal form by a matrix transformation of the form $P A \bar{P}'$ is called a conjunctive reduction.

A nonsingular matrix P with real number elements is called orthogonal if $P' = P^{-1}$. The congruence transformation which replaces a square matrix A with real number elements by PAP', where P is an orthogonal matrix, is called an orthogonal transformation. If A is a real symmetric matrix,

then A can be reduced by an orthogonal transformation to a diagonal matrix which has the n characteristic values of A on the diagonal. The characteristic values of a hermitian matrix and a real symmetric matrix are real numbers.

A nonsingular matrix P with complex number elements is called unitary if $\bar{P}' = P^{-1}$. The similarity transformation PAP^{-1} where P is unitary is a unitary transformation. The diagonal form obtained for a hermitian matrix A by a unitary transformation has the characteristic values of A on the diagonal.   [ROSS A. BEAUMONT]

Bibliography: H. G. Campbell, *Introduction to Matrices and Linear Programming*, 2d ed., 1977; D. T. Finkbeiner, *Introduction to Matrices and Linear Transformations*, 3d ed., 1978; A. Graham, *Matrix Theory and Applications for Engineers and Mathematicians*. 1979.

Matthiessen's rule

An empirical rule which states that the total resistivity of a crystalline metallic specimen is the sum of the resistivity due to thermal agitation of the metal ions of the lattice and the resistivity due to the presence of imperfections in the crystal. This rule is a basis for understanding the resistivity behavior of metals and alloys at low temperatures.

The resistivity of a metal results from the scattering of conduction electrons. Lattice vibrations

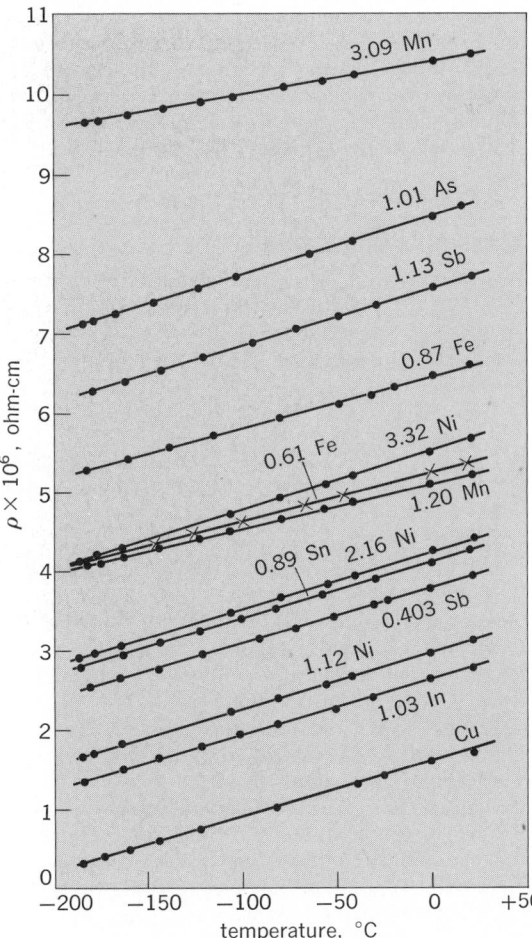

Resistivity of copper and copper alloys. Solute concentrations are given in atomic percent. (*After J. O. Linde*)

scatter electrons because the vibrations distort the crystal. Imperfections such as impurity atoms, interstitials, dislocations, and grain boundaries scatter conduction electrons because in their immediate vicinity the electrostatic potential differs from that of the perfect crystal. If lattice scattering and imperfection scattering are independent processes and are isotropic, it can be shown that Eq. (1)

$$\rho(T) = \rho_L(T) + \rho_I \tag{1}$$

holds, where $\rho(T)$ is the resistivity at temperature T, $\rho_L(T)$ is the resistivity due to lattice scattering (the ideal resistivity, which is temperature-dependent), and ρ_I is the so-called residual resistivity due to imperfections, which is presumably independent of temperature. An equivalent statement is given in Eq. (2).

$$\frac{\partial \rho(T)}{\partial T} = \frac{\partial \rho_L(T)}{\partial T} \tag{2}$$

As exemplified by the illustration, Matthiessen's rule is generally obeyed, although some deviations do occur. Such deviations arise for the following reasons: (1) The introduction of impurities generally alters the elastic properties, influences the lattice vibration spectrum, and thereby changes $\rho_L(T)$; (2) some imperfections, such as dislocations, do not scatter isotropically; (3) thermal expansion lowers the Fermi energy (since impurity scattering depends on the velocity of the electrons at the Fermi energy, the resistivity due to impurities depends also on the temperature through the thermal expansion); and (4) the impurity ions also participate in the thermal vibrations of the crystal lattice. However, since the scattering potential of an impurity differs from that of a solvent ion, lattice scattering from the former also differs from that due to a solvent ion.   [FRANK J. BLATT]

Bibliography: See ELECTRICAL RESISTIVITY.

Maxwell's equations

Four differential equations proposed by James Clerk Maxwell in 1864 as the basis of the theory of electromagnetic waves. They may be written, in vector notation, as Eqs. (1)–(4), where **D** is the

$$\nabla \cdot \mathbf{D} = \rho \tag{1}$$

$$\nabla \cdot \mathbf{B} = 0 \tag{2}$$

$$\nabla \times \mathbf{E} = -\frac{\partial \mathbf{B}}{\partial t} \tag{3}$$

$$\nabla \times \mathbf{H} = \mathbf{i} + \frac{\partial \mathbf{D}}{\partial t} \tag{4}$$

electric displacement, **B** the magnetic flux density, **E** the electric field strength or intensity, **H** the magnetic field strength or intensity, ρ the charge density, and **i** the current density.

The first equation states that electric flux lines, if they end at all, will do so on electric charges. The second states that magnetic flux lines never terminate. The third is a form of Faraday's law of induction, which states that the rate of change of the magnetic flux threading a circuit equals the electromotive force or line integral of **E** around the circuit so that Eq. (5) holds, where **n** is a unit vec-

$$-\frac{\partial}{\partial t}\int_S \mathbf{B} \cdot \mathbf{n} \, dS = \oint \mathbf{E} \cdot d\mathbf{s} = \int_S \nabla \times \mathbf{E} \cdot \mathbf{n} \, dS \tag{5}$$

tor normal to the surface S. The third integral comes from the second by Stokes' theorem. If this holds for any surface S, then the integrands of the surface integrals are equal and Eq. (3) follows. The fourth integral is based partially on A. M. Ampère's experiments on steady currents which show that the line integral of the magnetic intensity **H** (or $\mathbf{B}/\mu$, where μ is the permeability) around a closed curve equals the current encircled, which is found by integration of the normal current density over the enclosed area. Thus Eq. (6) holds. Stokes'

$$\int_S \mathbf{i} \cdot \mathbf{n} \, dS = \oint \mathbf{H} \cdot d\mathbf{s} = \int_S \nabla \times \mathbf{H} \cdot \mathbf{n} \, dS \tag{6}$$

theorem transforms the second integral into the third. The surface S is arbitrary so the integrands of the surface integrals are equal, and this gives Eq. (4) without the term $\partial \mathbf{D}/\partial t$. Maxwell realized that for fluctuating currents, where charges may accumulate, the $\partial \mathbf{D}/\partial t$ term was needed to satisfy the equation of continuity which requires that, if charge is to be conserved, its rate of increase in any region must equal the flow of current into that region so that Eq. (7) holds. When the divergence of Eq. (4) is taken and **D** is eliminated

$$\nabla \cdot \mathbf{i} + \frac{\partial \rho}{\partial t} = 0 \tag{7}$$

vergence of Eq. (4) is taken and **D** is eliminated by Eq. (1), Eq. (4) results. The quantity $\partial \mathbf{D}/\partial t$ is called the displacement current. *See* DISPLACEMENT CURRENT; EQUATION OF CONTINUITY.

Rectangular coordinates. When the components of Eqs. (1)–(4) are written in rectangular coordinates, the result is given by Eqs. (8)–(11). In

$$\frac{\partial D_x}{\partial x} + \frac{\partial D_y}{\partial y} + \frac{\partial D_z}{\partial z} = \rho \tag{8}$$

$$\frac{\partial B_x}{\partial x} + \frac{\partial B_y}{\partial y} + \frac{\partial B_z}{\partial z} = 0 \tag{9}$$

$$\frac{\partial E_{z,x,y}}{\partial u_{y,z,x}} - \frac{\partial E_{y,z,x}}{\partial u_{z,x,y}} = -\frac{\partial B_{x,y,z}}{\partial t} \tag{10}$$

$$\frac{\partial H_{z,x,y}}{\partial u_{y,z,x}} - \frac{\partial H_{y,z,x}}{\partial u_{z,x,y}} = i_{x,y,z} + \frac{\partial D_{x,y,z}}{\partial t} \tag{11}$$

Eqs. (10) and (11), the first, second, or third subscript must be used throughout, depending on the component desired, and u_x, u_y, and u_z are identical with x, y, z.

Ratio of units. In Maxwell's time, **E** and **B** were measured in a different system of units from **E** and **D**, the ratio of the units of charge being c. In a vacuum, where **i** is zero, this introduces a factor $1/c$ on the right sides of Eqs. (3) and (4), so that when Eq. (4) is used to eliminate the curl of **B** from the curl of Eq. (3), there results Eq. (12). The first

$$\nabla \times \nabla \times \mathbf{E} = \nabla^2 \mathbf{E} = \frac{1}{c} \frac{\partial^2 \mathbf{E}}{\partial t^2} \tag{12}$$

and second terms are equal because $\nabla \cdot \mathbf{E}$ is zero in a vacuum. This is the differential equation for a wave of velocity c. Thus, the experimental verification of Maxwell's prediction that the velocity of light equals the ratio of the electromagnetic unit to the electrostatic unit of charge proves the electromagnetic nature of light. *See* ELECTRICAL

UNITS AND STANDARDS; ELECTROMAGNETIC RADIATION; LIGHT; WAVE EQUATION.

Conducting media. If γ is the conductivity of a substance, Ohm's law states that the current density $\mathbf{i}$ in Eq. (4) may be replaced by $\gamma\mathbf{E}$. If this is done and $\mathbf{H}$ and $\mathbf{D}$ are replaced by $\mathbf{B}/\mu$ and $\epsilon\mathbf{E}$, then the use of Eq. (3) to eliminate the curl of $\mathbf{E}$ from the curl of Eq. (4) gives Eq. (13), which is the

$$\nabla^2\mathbf{B} = \mu\gamma\frac{\partial\mathbf{B}}{\partial t} + \mu\epsilon\frac{\partial^2\mathbf{B}}{\partial t^2} \qquad (13)$$

differential equation for a damped wave. This equation also holds if $\mathbf{E}$ replaces $\mathbf{B}$. When γ is much larger than ϵ, the second term on the right may be dropped so that Eq. (13) becomes the differential equation for eddy currents used in skin-effect and induction-heating calculations.

Index of refraction. When γ is zero, the solutions of Eq. (13) represent undamped waves of velocity $(\mu\epsilon)^{-1/2}$. In optics, the index of refraction of a medium is defined as the ratio of the velocity of light in a vacuum to that in the medium. Thus, by Maxwell's theory, the index of refraction n may be expressed in terms of the permeability and the dielectric constant of the medium by Eq. (14), where

$$n = \frac{c}{v} = \left(\frac{\mu\epsilon}{\mu_0\epsilon_0}\right)^{1/2} \qquad (14)$$

μ_0 and ϵ_0 are the vacuum values of the permeability and the dielectric constant, and v is the velocity of light in the medium. For everything except ferromagnetic materials, μ is approximately equal to μ_0, so these symbols cancel. The dielectric constant ϵ is a function of frequency, so static values should be used with caution, although they often hold in the radio-frequency range. *See* REFRACTION OF WAVES.

Integral form. In some ways, the integral forms of Eqs. (1)–(4) give a clearer physical concept than the differential forms. With the aid of Eqs. (5) and (6), Maxwell's equations may be written as Eqs. (15)–(18).

$$\int\mathbf{D}\cdot\mathbf{n}\,dS = \int\rho\,dv \qquad (15)$$

$$\int\mathbf{B}\cdot\mathbf{n}\,dS = 0 \qquad (16)$$

$$\oint\mathbf{E}\cdot d\mathbf{s} = -\frac{\partial}{\partial t}\int_S\mathbf{B}\cdot\mathbf{n}\,dS \qquad (17)$$

$$\oint\mathbf{H}\cdot d\mathbf{s} = \int_S\left(\mathbf{i} + \frac{\partial\mathbf{D}}{\partial t}\cdot\mathbf{n}\right)dS \qquad (18)$$

Lorentz invariance. The form of Maxwell's equations is the same for all observers whose coordinate systems move with a uniform translational velocity relative to each other. The values of the observed fields, as well as ρ, $\mathbf{i}$, μ, ϵ, and γ, may be quite different. This Lorentz transformation can sometimes be used to remove one of the fields entirely, thus greatly simplifying the calculation. *See* CALCULUS OF VECTORS; LORENTZ TRANSFORMATIONS.

[WILLIAM R. SMYTHE]

Bibliography: B. I. Bleaney and B. Bleaney, *Electricity and Magnetism*, 3d ed., 1976; J. D. Jackson, *Classical Electrodynamics*, 2d ed., 1975; E. M. Pugh and E. W. Pugh, *Principles of Electricity and Magnetism*, 2d ed., 1970; A. Shadowitz, *The Electromagnetic Field*, 1974.

Mean free path

The average distance traveled between two similar events. The concept of mean free path is met in all fields of science and is classified by the events which take place. The concept is most useful in systems which can be treated statistically, and is most frequently used in the theoretical interpretation of transport phenomena in gases and solids, such as diffusion, viscosity, heat conduction, and electrical conduction. The types of mean free paths which are used most frequently are for elastic collisions of molecules in a gas, of electrons in a crystal, of phonons in a crystal, and of neutrons in a moderator.

An elementary formula for the mean free path for elastic collision of a molecule in a gas can be derived in the following way. In a gas at a pressure P, let n be the average number of molecules per unit volume and a the average radius of a molecule. The distance between centers of the molecules during a collision will be $2a$. Assume that one molecule has a radius $2a$ and travels with a velocity c, and that all other molecules are mass-points at rest. In a time t the moving molecule will "sweep out" a volume $4\pi a^2ct$, and since there are n molecules per unit volume, there will be $4\pi a^2ctn$ collisions. The distance traveled is ct; therefore the average distance between collisions is $1/(4\pi a^2 n)$. Other more exact methods of taking averages give values which are closer to those determined experimentally. In hydrogen at normal temperature and pressure the mean free path is 1.7×10^{-5} cm. *See* KINETIC THEORY OF MATTER.

[W. DEXTER WHITEHEAD]

Mechanical impedance

For a system executing simple harmonic motion, the mechanical impedance is the ratio of force to particle velocity. If the force is that which drives the system and the velocity is that of the point of application of the force, the ratio is the input or driving-point impedance. If the velocity is that at some other point, the ratio is the transfer impedance corresponding to the two points.

As in the case of electrical impedance, to which it is analogous, mechanical impedance is a complex quantity. The real part, the mechanical resistance, is independent of frequency if the dissipative forces are proportional to velocity; the imaginary part, the mechanical reactance, varies with frequency, becoming zero at the resonant and infinite at the antiresonant frequencies of the system. *See* ACOUSTIC IMPEDANCE; ELECTRICAL IMPEDANCE; FORCED OSCILLATION; HARMONIC MOTION.

[MARTIN GREENSPAN]

Bibliography: R. E. D. Bishop, *Vibration*, 2d ed., 1979; C. M. Harris and C. E. Crede, *Shock and Vibration Control Handbook*, 3d ed., 1976.

Mechanical vibration

The continuing motion, repetitive and often periodic, of a solid or liquid body within certain spatial limits. Vibration occurs frequently in a variety of natural phenomena such as the tidal motion of the oceans, in rotating and stationary machinery, in structures as varied in nature as buildings and ships, in vehicles, and in combinations of these various elements in larger systems. The sources of vibration and the types of vibratory motion and their propagation are subjects that are complicated and depend a great deal on the particular characteristics of the systems being examined. Further, there is strong coupling between the notions of mechanical vibration and the propagation of vibration and acoustic signals through both the ground and the air so as to create possible sources of discomfort, annoyance, and even physical damage to people and structures adjacent to a source of vibration.

Mass-spring-damper system. Although vibrational phenomena are complex, some basic principles can be recognized in a very simple linear model of a mass-spring-damper system (Fig. 1). Such a system contains a mass M, a spring with spring constant k that serves to restore the mass to a neutral position, and a damping element which opposes the motion of the vibratory response with a force proportional to the velocity of the system, the constant of proportionality being the damping constant c. This damping force is dissipative in nature, and without its presence a response of this mass-spring system would be completely periodic.

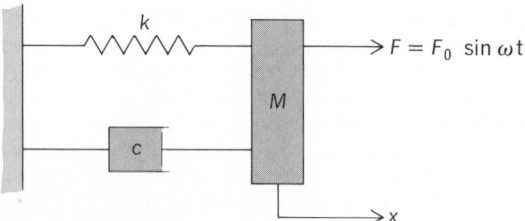

Fig. 1. Vibrating linear system (mass-spring-damper) with one degree of freedom.

Free vibrations. In the absence of an exciting force and of the damping component, the spring-mass system can execute free, periodic vibration at a natural frequency ω_n given by Eq. (1). This is

$$\omega_n = \sqrt{k/M} \qquad (1)$$

the so-called angular frequency and must be divided by 2π to get the actual frequency f_n in hertz (with dimensions time^{-1}). Equation (1) demonstrates that the natural frequency, also called the fundamental or resonant frequency, increases with the stiffness of the system and decreases as the mass of the system is increased. Thus, very stiff systems have high natural frequencies and very massive systems could have very low natural frequencies. In some of the more complex structures referred to above, there are in fact multiple (and even infinite) resonant frequencies, but they all have in common the basic dependence upon the stiffness of the system and the system's mass. *See* HARMONIC MOTION.

Effect of damping. The inclusion of the damping element with the constant of proportionality c results in a change of the natural frequency of the system and of its response characteristics. If the damping is relatively small in magnitude (the scale against which smallness is to be measured will be given below), then the natural frequency of the system does not change substantially, and the free response of a perturbed mass-spring-damper system is periodic within the confines of a slow expo-

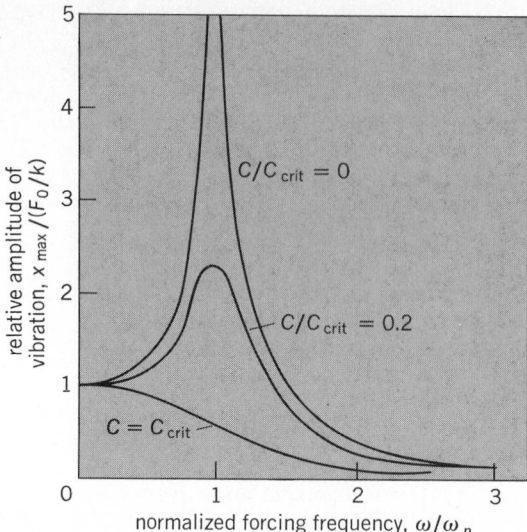

Fig. 2. Variation of amplitude of vibration with forcing frequency and damping. x_{max} = maximum displacement.

nential decay. However, if the damping constant c becomes large, the motion of the system is not vibratory but is in fact one where the mass creeps back, along an exponential curve, to its initial position without any oscillation. The lowest value of damping for which this loss of vibratory response occurs is called critical damping, and it is defined by a damping constant, Eq. (2). Damping is considered

$$c_{crit} = 2\sqrt{kM} \qquad (2)$$

ered to be small when the ratio c/c_{crit} is small compared to unity. *See* DAMPING.

Impedance and resonance. When the mass-spring-damper system is excited by a periodic force, for example, $F = F_0 \sin \omega t$, where the angular frequency $\omega \ (= 2\pi f)$ can be chosen freely, the response of the system is such that it can be displayed in one of two commonly used forms. One of these (Fig. 2) is a plot of the relative amplitude of vibration (the dimensionless ratio of the actual motion of the perturbed system divided by the corresponding elastic response of the spring alone to the same force) against the normalized (dimensionless) frequency ratio of the forcing frequency divided by the undamped natural frequency. The curves shown in Fig. 2 indicate the response for no damping ($c/c_{crit} = 0$), small damping, and critical damping. It can be seen that in the undamped case ($c/c_{crit} = 0$), the amplitude of the response is infinite when the forcing frequency is identically equal to the natural frequency. This is the condition of resonance. It can also be seen that when there is damping, even though it may be small, the system responds so as to peak very near the undamped resonant frequency, but the response is not infinite in amplitude. Finally, if the damping is greater than or equal to c_{crit}, the response is simply the decaying curve that exhibits no peak at the resonant frequency.

Another way of presenting these results is to look at the quantity called the impedance Z which may be defined here as the ratio of the magnitude of the input force to the magnitude of the velocity of the mass-spring system (Fig. 3). For the case where there is no damping, the impedance can be

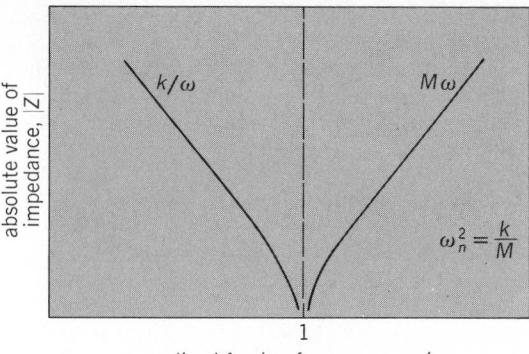

absolute value of impedance, $|Z|$

k/ω

$M\omega$

$$\omega_n^2 = \frac{k}{M}$$

1

normalized forcing frequency, ω/ω_n

Fig. 3. Impedance of a simple oscillator.

shown to be in the form of Eq. (3). It follows from

$$Z = k\left(\frac{1}{\omega} - \frac{M\omega}{k}\right) \qquad (3)$$

Eq. (1), which relates the stiffness k and the mass M to the resonant frequency ω_n, and from the impedance formula that when the forcing frequency is well below the resonant frequency, the impedance can be approximated by Eq. (4). This is considered the stiffness-controlled region wherein the

$$Z_k \cong \frac{k}{\omega} \qquad \omega \ll \omega_n \qquad (4)$$

sidered the stiffness-controlled region wherein the spring-mass system responds largely as a static spring. For frequencies that are significantly above the resonant frequency, the impedance can be written as Eq. (5).

$$Z_m \cong -M\omega \qquad \omega \gg \omega_n \qquad (5)$$

In this regime the response of the system is said to be mass-controlled, and at high frequencies, well above the resonant frequency, the dynamics of the system are governed by the mass of the system. It follows from Eq. (3) that the impedance vanishes at resonance in the absence of damping. Thus resonance may also be defined as the condition where a very small input force produces an infinite response in the absence of damping. When damping is incorporated in the calculation of impedance, it then turns out that at resonance the impedance is exactly equal to the damping constant; that is, $Z_{res} = c$. See FORCED OSCILLATION; MECHANICAL IMPEDANCE; RESONANCE (ACOUSTICS AND MECHANICS).

Linear response. One other feature in the simple model is that the response is linear; that is, there is a direct proportionality of the magnitude of the input to the magnitude of the output, and signals of response for different inputs can be superposed to give the total response to a combined signal.

Relation to real systems. This simple model of the response of the single-degree-of-freedom system, as it is commonly called, serves as a model or paradigm for almost all mechanical vibration problems. However, real systems are often extremely complex and involve multiple resonances and multiple sources of excitation. Further, the types of vibrational response that can be seen in different solid bodies are very much dependent not only on the nature of the excitation in terms of its time his-

tory, but also on the spatial distribution of the exciting force and the geometry of the system that is responding. There are many different ways in which a solid body can vibrate, and these different ways couple into vibratory and acoustic motions of surrounding media such as air or water in very different ways. Some of the sources of vibration, the types of vibrational response exhibited by solid bodies, and their interactions that result in perceptible, audible, and even dangerous vibration and acoustic signals will be discussed. First, however, some further comments about the general complexity of vibration problems are appropriate in order that the context of the above simple model is more clearly and sharply defined.

Analysis of complex systems. The foregoing model of the linear spring-mass-damper system contains within it a number of simplifications that do not reflect conditions of the real world in any obvious way. These simplifications include the periodicity of both the input and, to some extent, the response; the discrete nature of the input, that is, the assumption that it is temporal in nature with no reference to spatial distribution; and the assumption that only a single resonant frequency and a single set of parameters are required to describe the mass, the stiffness, and the damping. The real world is far more complex. Many sources of vibration are not periodic. These include impulsive forces and shock loading, wherein a force is suddenly applied for a very short time to a system; random excitations, wherein the signal fluctuates in time in such a way that its amplitude at any given instant can be expressed only in terms of a probabilistic expectation; and aperiodic notions, wherein the fluctuation in time may be some prescribed nonperiodic function or some other function that is not readily seen to be periodic.

However, in most cases the notions of periodic response to periodic excitation can be used because it is generally possible to describe the variation in time of a nonperiodic forcing function as a sum of periodic components using the tools of Fourier analysis and Fourier series. In such an approach, the nonperiodic signal is expressed as a sum of periodic signals, each of whose amplitudes and individual periods (or resonant frequencies) is tailored to represent the total nonperiodic signal being described. In such a Fourier analysis of a vibrating system, it is customary to look at the response of the complex system as a sum of responses of linear spring-mass-damper systems to each of the given periodic inputs, and then the simple model outlined earlier can be readily applied.

In any real system there are multiple resonances, each reflecting a different balance of stiffness and mass within a complex elastic system. Again, it is possible in general to describe the response of a complicated system in terms of individual spring-mass oscillators, the stiffness and mass of each of which represents some average of the overall stiffness and mass distributions of the complicated system. Thus, the analysis of complex systems can again be reduced to the superposition of systems of the elementary type. Very often this reduction of the complex to the simple not only involves Fourier analysis in time, but it also represents a Fourier analysis in space, wherein spatial averages are used to reduce spatial variations of both excitation and response to the discrete forms in which they appear in the simple model. Thus, for a continuous system where the variables depend on space as well as time, a Fourier analysis can be used to generate a system that involves multiple degrees of freedom, each degree of freedom being a single discrete spring-mass-damper oscillator. *See* FOURIER SERIES AND INTEGRALS.

Sources of vibration. There are many sources of mechanical and structural vibration that the engineer must contend with in both the analysis and the design of engineering systems. The most common form of mechanical vibration problem is motion induced by machinery of varying types, often but not always of the rotating variety; the machinery vibration thus induced can be a problem both for the machinery itself and in terms of vibration and noise propagated to adjacent systems. For example, the vibrations of a compressor in an air-conditioning system in a building can create fatigue and stress problems for the compressor and adjacent piping itself, and can set into motion the floor and walls of the room wherein the compressor is housed. This in turn can set into vibrational motion other walls and components of the building and thus produce vibration and acoustic signals that are both perceptible and audible throughout the building. This simple example could thus run a gamut of problems from structural failure to human annoyance.

Such machinery-induced vibration can also be propagated into adjacent buildings through the coupling of a building foundation with the ground around it, and it is not uncommon for heavy machinery in a factory to produce perceptible vibration at considerable distances from the factory itself. This is more likely to be the case with extremely heavy pieces of equipment, for example a large-scale shredder used in a solid-waste disposal system.

Another source of vibration wherein ground-borne propagation is important is vibration due to construction. Construction vibration is a major source of environmental concern, particularly in urban environments, and especially in the construction of large projects within such environments such as urban transportation systems.

Another source of vibration in urban environments is transportation systems themselves and may include vibration from heavy vehicles on conventional pavement as well as vibratory signals from the rail systems common in many metropolitan areas. For such rapid-transit system vibration, the sources include tunnels, at-grade rails, and elevated guideway systems. The propagation characteristics for each of these important vibration sources are different and have been the subject of extensive research.

Other vibrations are induced by natural phenomena, including earthquakes and wind forces. Wave motion is a source of vibration in mechanical and structural systems associated with offshore structures. In the analysis and design of systems that are used in environments where these natural phenomena are important, the basic principles of vibration analysis, such as the elementary model outlined above, remain the same. Thus, although such motion is not in a very strict sense due to ordinary mechanical excitation, it is in terms of practice and design a strongly related phenomenon.

Types of mechanical signals and waves. Due to the varying geometry of both the structural elements involved in a machine or structure and of the loading itself, several different types of vibratory motion may be induced in a solid body. Unlike the acoustic signals that are generated in air and water, wherein all the motion is in the form of waves that are termed body or compressional waves in which the fundamental mechanism is volumetric contraction or expansion, solid bodies support waves of several different varieties and thus produce several different mechanical vibratory signals. These include torsional vibration, longitudinal vibration, and the lateral vibrations of beams, plates, and membranes. The fundamental mechanisms in each case are different in that the elastic restoring forces that are analogous to the spring constant k in the simple model are very different and are produced by different physical effects in these bodies. Each type of vibration will be discussed briefly, although the major emphasis will be placed on vibrations of and in machinery.

Torsional vibration. An engine-driven system (usually diesel or internal combustion powering a generator, a ship's propeller, or other load) has many degrees of freedom and, hence, many natural frequencies of which only the lowest two or three are of practical importance. The installation may, if excited by alternating components in its torque, execute torsional vibrations between its various parts of a magnitude of a fraction of a degree. This relatively small alternating motion is superposed on, and independent of, the continuous rotation of the engine shaft. Whereas the continuous rotation causes no extreme stress in the shaft, the small (quarter degree) angles of vibration wind up the shaft and relax it, causing alternating stresses that on many occasions have caused failure.

The torque on a crankshaft caused by a single cylinder and piston having one explosion for each one or two revolutions has a highly irregular time history containing many harmonics. (Harmonics are integral multiples of the fundamental frequency.) The torques developed by the various cylinders of a multicylinder engine combine in accordance with their times of occurrences. Thus, the exciting torque contains components of the firing frequency (one cycle per one or two revolutions) and multiples of that firing frequency, which are of practical importance up to the 16th multiple or harmonic. Because these 16 exciting or forced frequencies are proportional to the engine speed and the two or three natural frequencies are independent of it, and each exciting frequency can resonate with each natural one, there are as many as 48 resonances or critical speeds to be considered. Many of these will lie outside the habitual running range of the engine.

The designer may shift the severity of the resonances by using flexible couplings or extra flywheels in the engine, or by changing the arrangement of the cylinders, the V-angle of the engine, and the firing order. These suffice to make an installation satisfactory for any one running speed, but it is usually not possible to avoid all dangerous resonances in a wide range of running speeds. The designer must then resort to dampers to keep the amplitude of motion and the stress down. Various types of dampers exist; the most familiar is the pulley in the front of an automobile engine which drives the fan belt. This pulley is usually a small flywheel coupled to the engine shaft through a rubber insert. The assembly serves simultaneously as a torsional spring and as a dashpot damper. The flywheel is so tuned that it holds the principal torsional critical speed of the engine below dangerous torsional stress levels. Fatigue failures of crankshafts and of other shafting due to torsional vibration were common in the past, but such failures are avoidable by proper design and are now rare.

By the principle of action equals reaction, the gas torque acting on the piston-crankshaft assembly is equal and opposite to the gas torque acting on the engine frame. Hence, when the engine frame is rigidly attached to a foundation, that foundation experiences an alternating torque. To protect the foundation and to prevent the vibration from spreading through the structure, the engine is often mounted on metal springs or on rubber bushings. This is now universal practice with automobile and aircraft piston engines.

Longitudinal vibration. A large ship's shafting excited by the propeller blades may come to resonance. The shafting and propeller must be designed to allow for such resonance. This is one cause of vibration in passenger liners. It occurs at a frequency of the shaft revolutions multiplied by the number of propeller blades.

The air or gas column in the suction or discharge lines of internal combustion engines acted upon by the piston motions can come to resonance at certain speeds. This principle has been used to increase the power of the engine by designing those lines so that during the intake more air than usual enters the cylinder, and during exhaust more air than usual goes out, thereby decreasing the backpressure.

In internal combustion engines the gas pressure exerted on the piston and thence on a crank throw tends to lengthen the crankshaft. Because this effect is periodic, longitudinal vibrations in internal combustion engine shafts have been observed; however, they are of less importance than the torsional ones and become serious only if a longitudinal and a torsional natural frequency fall close together causing coupling oscillations.

Lateral vibrations. The most important of all machinery vibrations are the lateral or bending vibrations of shafts and other rotors caused by the centrifugal force of unbalance. The unbalance consists of a very small deviation of the center of gravity of the rotor from the geometric axis connecting the bearing centers, or of a small angular deviation between that bearing center line and a principal axis of inertia of the rotor. Thus the excitation always has the frequency of the shaft rotation, and when that coincides with one of the natural bending frequencies of the rotor, there is resonance or a critical speed where the bending stresses in the shaft can be a hundred times higher than those caused by the centrifugal force of the unbalance directly.

Slow-speed machines are always designed to run well below the lowest or first critical speed, but for high-speed ones this often is not possible. Steam turbines and electric generators of large power stations have for many years been designed to run between the first and second critical speeds; the newest and largest units are between their second and third criticals. The flexibility of the supporting (nonrotating) bearings is an important factor in the calculation of critical speeds or natural frequencies of rotors, because the flexible bearing decreases the natural frequencies by some 10% on the average below those that would be present if rigid bearings were used. For rotors of a high diameter-length ratio such as in steam turbines with large diameter disks mounted on a comparatively thin shaft, the effect of rotating inertia, sometimes called the gyroscopic effect of the disks, is an important consideration.

Besides the classical unbalance critical speed, which always has a vibration frequency equal to that of the rpm, a number of secondary critical speeds have been observed and explained, in which the frequency of vibration is a multiple of, usually twice, the rpm. The practical importance of these is secondary with respect to the ordinary critical speed.

Beam vibration. Whenever a part of a structure or the entire structure itself has a natural frequency resonating with an alternating excitation, nearby (or sometimes far removed) severe vibrations may result. A typical example is an unbalanced piston machine, such as an air compressor, which frequently causes objectionable vibration in some locations in the building where it is installed. Another example is the vibration of an entire ship in the mode of a free-free (totally unsupported) beam, excited by the propeller blade frequency.

With the advent of jet engines, the effect of high intensity airborne noise on the very light structures of airplanes and missiles has become important. It is characteristic of jet noise, or indeed of all cases of turbulent flow, that the excitation is distributed continuously over a wide band of frequencies. This type of excitation is random and is playing an increasingly important role in the design of aircraft and missiles.

Membrane vibration. A tightly stretched skin, which has negligible bending stiffness like a drumhead, is a membrane. (The diaphragm in a telephone receiver possesses considerable bending stiffness, and is not stretched; hence technically it is a plate, although sometimes it is also called a membrane.) The theory of vibration of a membrane of circular shape has been known for a century and is one of the more beautiful illustrations of the mathematics of Bessel functions. The lowest frequency of vibration corresponds to a shape where the entire membrane bulges in and out, without nodal lines, the periphery remaining fixed. The higher modes of motion possess nodal lines, which are concentric circles or angularly equidistant diameters. The frequency formula for the circular membrane vibration is given in Eq. (6). Here

$$f = \alpha \sqrt{Tg/\gamma r^2} \qquad (6)$$

T is the tension in the membrane, g is the acceleration due to gravity, γ is the surface weight density of the membrane, r is the radius, and a is a numerical factor having the values shown in the table.

Numerical factor for membrane frequency

Number of nodal circles	Number of nodal diameters			
	0	1	2	3
0	0.383	0.610	0.819	1.02
1	0.880	1.12	1.34	1.55
2	1.38	1.62	1.85	2.08

Plate vibration. Plate vibrations are very important for a variety of reasons. They occur in large, flat, solid surfaces that are excited in planes normal to that surface. The reason that such vibrations are important in mechanical systems is that they occur frequently in practice and in addition they are very good acoustical radiators; that is, they couple extremely well with the air around them and produce significant quantities of vibratory energy that can be both felt and heard as acoustic signals. The basic mechanism by which a plate operates is similar to a beam in that the action consists of the bending of the plate about its middle surface, except that whereas a beam is a one-dimensional element with bending in only one direction, a plate is a surface that has bending in two orthogonal directions as well as a torsional or twisting type of bending at angles between these two orthogonal directions. Thus, more complicated curvature and surface interaction effects are involved.

Self-excited vibrations. In the cases discussed so far, the existence of an alternating exciting force (or torque) has been assumed which would continue to exist by itself even after the vibratory motion was prevented or stopped. These are called forced vibrations. There is another class of motions, called self-excited, in which the exciting force is generated by the vibrating motion itself and hence disappears with that motion.

A familiar mechanical example of self-excited vibration is the piston of an engine. The back and forth motion (vibration) is maintained by an alternating gas or steam pressure steered by the valve mechanism. The initial source of energy is without alternating properties (the gasoline supply or the boiler steam) but is made alternating by the valves, which are operated by the vibratory piston itself.

A familiar electrical example is the self-oscillating electronic circuit which has a steady source of energy (for example, a battery), and a transistor that plays the role of the valve.

The oldest practical examples of useful self-excited vibrations are musical instruments. The violin operates on the behavior of the friction between the rosined bow and the string, which has aptly been described as stick-slip friction. Vibrations of this type have appeared repeatedly and still are often met within machinery with insufficient lubrication; the most vexing and difficult case is that of the chattering machine tool cutter which leaves a wavy cut instead of a smooth one.

The clarinet operates on the passage of air (from the mouth to the instrument) though a narrow leakage opening whose width varies periodically with the vibration of the reed. Serious vibrations of this character occur in steam, gas, or hydraulic turbines, heat exchangers, and other apparatus in which a fluid or gas passes through narrow passages. Other self-excited vibrations are the pulsating flow sometimes observed in fans and blowers, and the shimmying motion of wheels, which has been a serious problem in automobiles and in the landing gear of aircraft. Self-excited vibrations appear in autopilots of aircraft and missiles, and in general are apt to occur in servomechanisms with high gain.

A class of dangerous vibrations is provided by the various phenomena of flutter, whereby an elastic system becomes self-excited in the stream of air, gas, or fluid of sufficient speed. This aviation problem first arose in airplane wings, but as speeds increased, other parts of airplanes and other machinery, such as turbine blades, displayed flutter. Flutter theory has grown into a subject by itself.

Serious vibrations have occurred in rocket engines and other types of combustion chambers, whereby the combustion becomes unstable and the burning gas mass enters a state of self-excited vibration. These phenomena have been known for a century and still are not completely understood.

Effect of vibrations. The most serious effect of vibration, especially in the case of machinery is that sufficiently high alternating stresses can produce fatigue failure in machine and structural parts. Less serious effects include increased wear of parts, general malfunctioning of apparatus, and the propagation of vibration through foundations and buildings to locations where the vibration or

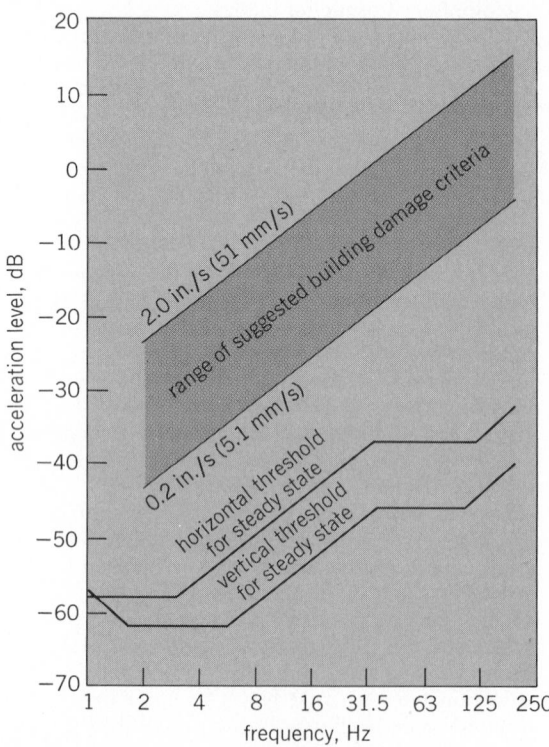

Fig. 4. Comparison of suggested building damage criteria and human threshold of perception curves for whole-body vibration. Acceleration level in dB = 20 log acceleration level/g. Because they are negative logarithms, the decibel values on the vertical axis indicate increasing acceleration levels from bottom to top. (From T. G. Gutowski, L. E. Wittig, and C. L. Dym, Some aspects of the ground vibration problem, Noise Control Eng., 10(3): 94–100, 1978)

its acoustic realization is intolerable either for human comfort or for the successful operation of sensitive measuring equipment.

Whole-body vibration. Figure 4 shows a comparison of suggested building damage criteria and human threshold-of-perception curves for whole-body vibration. The threshold curves for human perception are based upon tests wherein people are seated on shakers, so the perception is of motion that affects the entire body. There are a variety of ranges of vibration, and these ranges are strongly dependent upon the frequency (expressed here in hertz, equal to $\omega/2\pi$). The ordinate of this graph is expressed as an acceleration level (in decibels) with reference to g (the acceleration of gravity) as the base acceleration level. The decibel notation indicates a logarithmic dependence such that an acceleration level of -20 dB represents an acceleration level that is 0.1 g in magnitude, an acceleration of -40 dB represents a level that is 0.01 g in level, and so on. The magnitudes of vibration that are felt by humans are extremely small; that is, on the order of -60 dB or 0.001 g in magnitude. While identifying the human threshold of perception from whole-body laboratory experiments is very useful, other factors must be considered in practice. Human annoyance may be related to secondary vibratory effects such as vibration-induced noise of the rattling of dishes, visual cues such as the vibration-induced motion of household items such as plants and mirrors, and so on. Thus, human annoyance could depend on the observer's location and activity as well as other suggested factors (Fig. 4). The values for which damage to a building might be realized are significantly higher than those for which vibration is perceptible by humans. In fact, the difference tends to be on the order of two orders of magnitude (a multiplicative factor of 100 in amplitude).

Generation of acoustic noise. Figure 5 relates vibration magnitudes both to perceptible vibration level and to audible acoustical levels. These data demonstrate that an acoustic signal can be generated by a mechanical vibration and produce auditory noise that can be very uncomfortable. For example, as discussed above, noise can propagate in a building due to an air-conditioning compressor. The curves displayed in Fig. 5 also show so-called NC-equivalent curves used to describe architectural levels of quietness in a design environment. Thus an NC-20 equivalent curve is the design standard for a first-class concert hall, while an NC-40 criterion would be adequate for a grade-school classroom. The acoustic signals are normally expressed in logarithmic amplitudes of acoustic pressure, but are expressed here in terms of amplitudes of vibration, that is, acceleration levels. An equivalence between the two is based on the recognition that when a large surface vibrates, it simultaneously sets into motion the acoustic medium around it, whether it be air or water. The bending vibrations of plates, for example, couple extremely well with the compressional waves of a surrounding acoustic medium. Thus when a building wall or floor is set into motion by a mechanical source of vibration, it in turn causes a very effective propagation of an acoustical signal, or noise, in the surrounding air. Another significant example

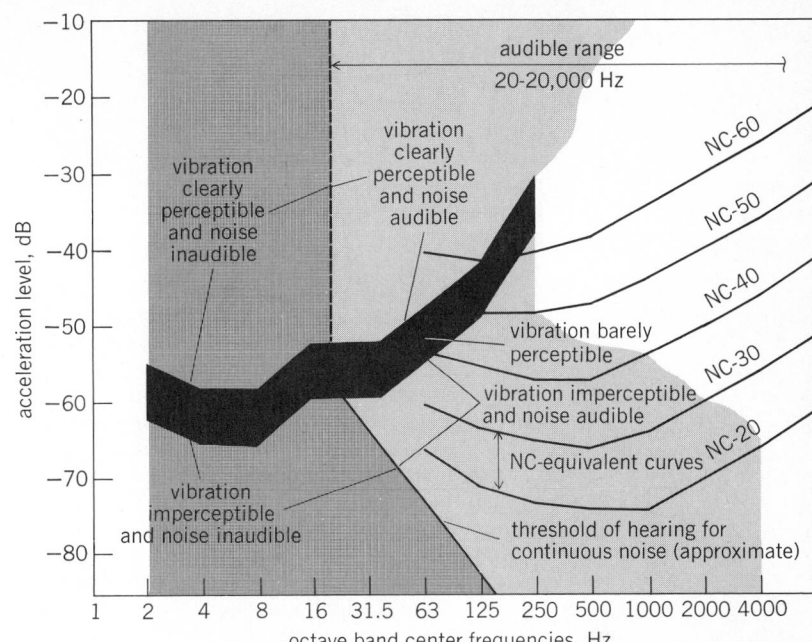

Fig. 5. Levels of perceptibility of vibration and audibility of noise. Acceleration level in dB = 20 log acceleration level/*g*. Because they are negative logarithms, the decibel values on the vertical axis indicate increasing acceleration levels from bottom to top. *(From C. L. Dym and D. Klabin, Architectural implications of structural vibration, Architect. Rec., 157(9):125–127, 1975)*

of good coupling is the mechanical excitation of a drumhead by a drumstick. The lateral vibrations of the stretched membrane (the drumhead) also couple well with surrounding air to produce the corresponding tympanic sound of the drum which can be heard at great distances from the drum itself.

There are many factors that enter into the coupling of a vibrating structural element and the surrounding acoustical medium. The principal factors are, however, the surface area of the mechanical source that is exposed to the medium and the nature of the mechanical vibration, that is, whether it is perpendicular to the surface (as with bending vibrations of beams and plates or lateral vibrations of membranes) or along the direction of the surface (as in torsional and longitudinal vibration). It is the perpendicular motion of large surfaces that produces the best acoustical coupling and, hence, the best chance of radiation of noise from vibration. *See* Sound; Vibration.

[CLIVE L. DYM; J. P. DEN HARTOG]

Bibliography: R. E. D. Bishop, *Vibration*, 2d ed., 1979; J. P. Den Hartog, *Mechanical Vibrations*, 4th ed., 1956; C. L. Dym and D. Klabin, Architectural implications of structural vibration, *Architect. Rec.*, 157(9):125–127, September 1975; C. M. Harris and C. E. Crede, *Shock and Vibration Handbook*, 2d ed., 1976.

Mechanics

In its original sense, mechanics refers to the study of the behavior of systems under the action of forces. Statics deals with cases where the forces either produce no motion or the motion is not of interest. Dynamics deals properly with motions under forces. Mechanics is subdivided according to the types of systems and phenomena involved.

An important distinction is based on the size of the system. Those systems that are large enough can be adequately described by the Newtonian laws of classical mechanics. In this category, for example, are celestial mechanics, the study of the motions of planets, stars, and other heavenly bodies, and fluid mechanics, which treats liquids and gases on a macroscopic scale. Fluid mechanics is a part of a larger field called continuum mechanics or (by some physicists) classical field theory, involving any essentially continuous distribution of matter, whether rigid, elastic, plastic, or fluid. On the other hand, the behavior of microscopic systems such as molecules, atoms, and nuclei can be interpreted only by the concepts and mathematical methods of quantum mechanics.

From its inception, quantum mechanics had two apparently different mathematical forms—the wave mechanics of E. Schrödinger, which emphasizes the spatial probability distributions in the quantum states, and the matrix mechanics of W. Heisenberg, which emphasizes the transitions between states. These are now known to be equivalent.

Mechanics may also be classified as nonrelativistic or relativistic mechanics, the latter applying to systems with material velocities comparable to the velocity of light. This distinction pertains to both classical and quantum mechanics.

Finally, statistical mechanics uses the methods of statistics for both classical and quantum systems containing very large numbers of similar subsystems to obtain their large-scale properties. *See* CLASSICAL FIELD THEORY; CLASSICAL MECHANICS; DYNAMICS; FLUID MECHANICS; QUANTUM MECHANICS; RELATIVISTIC MECHANICS; STATICS; STATISTICAL MECHANICS.

[BERNARD GOODMAN]

Meissner effect

The expulsion of magnetic flux from the interior of a superconducting metal when it is cooled in a magnetic field to below the critical temperature, near absolute zero, at which the transition to superconductivity takes place. It was discovered by Walther Meissner in 1933, when he measured the magnetic field surrounding two adjacent long cylindrical single crystals of tin and observed that at 3.72 K the Earth's magnetic field was expelled from their interior. This indicated that at the onset of superconductivity they became perfect diamagnets. This discovery showed that the transition to superconductivity is reversible, and that the laws of thermodynamics apply to it. The Meissner effect forms one of the cornerstones in the understanding of superconductivity, and its discovery led F. London and H. London to develop their phenomenological electrodynamics of superconductivity. *See* DIAMAGNETISM; THERMODYNAMIC PRINCIPLES.

The magnetic field is actually not completely expelled, but penetrates a very thin surface layer where currents flow, screening the interior from the magnetic field.

Discovery. Before the discovery of the effect, it had been pointed out that simple relations exist between the specific heat, a thermal property, and the magnetic properties of superconductors. These relations imply that the change from the normal to the superconducting state of a metal is a thermodynamic transition. The conclusion that a magnetic field would be expelled in the transition to the superconducting state, however, was not drawn.

Shortly before the discovery of the Meissner effect, it was pointed out that because of the inertia of the electrons, screening currents flow in a thin surface layer when a superconductor is suddenly exposed to a magnetic field. Similarly, it was expected that a current passed through a superconducting rod would also be confined to a thin surface layer. This conjecture formed the starting point for Meissner's investigation. By passing the current down one rod and up an identical, adjacent rod, and by measuring the magnetic field surrounding the two rods, it is possible to decide whether the currents flow in a thin surface layer or uniformly over the whole cross section. Since Meissner initially had not canceled the Earth's magnetic field, he was able to observe that the magnetic field was expelled at the onset of superconductivity. Subsequently Meissner also found that indeed the current flows in a very thin surface layer.

Limitations. Meissner prepared two single crystal rods of tin for his experiment. If he had used polycrystalline samples, especially of a noncubic metal, he would not have observed the full diamagnetism. Even when a polycrystalline sample is well annealed and strain-free at room temperature, severe strains are set up in the process of cooling it down, because the thermal contraction is different in different crystal directions. Since the critical temperature shifts slightly with strain, such samples fail to become uniformly superconducting, allowing persistent current to flow in the regions which become superconducting first. These persistent currents then lock in the magnetic flux. *See* MAGNETISM.

In impure samples having a short electronic mean free path, the effect is not observed because the energy associated with a boundary between a superconducting and a normal domain is negative. This allows the formation of a stable array of superconducting flux tubes carrying one quantum of magnetic flux (approximately 2×10^{-7} gauss cm^2 = 2×10^{-15} weber) each. *See* QUANTIZED VORTICES.

The effect is also not observed in a round flat disk with the magnetic field parallel to its axis of rotation, since the magnetic field also would have to be excluded from roughly hemispherical regions above and below the disk. For a disk, however, the difference between the free energies in the superconducting and normal states is too small to balance the change in the magnetic field energy in these two regions. The disk therefore breaks up into normal and superconducting domains or, if extremely thin, exhibits a flux tube structure similar to impure superconductors. *See* SUPERCONDUCTIVITY.

[HANS W. MEISSNER]

Meson

The generic name for any hadronic particle with baryon number zero. Such particles were first envisaged in 1935 by H. Yukawa, who pointed out that the main features of nuclear forces would be explained if these forces were transmitted between nucleons through an intermediate field coupled with nucleons, provided that its quanta

Table 1. Semistable pseudoscalar mesons now known

	$\pi^+(\pi^-)$	π^0	$K^+(K^-)$	K^0_S	K^0_L	η	$D^+(\bar{D}^-)$	$D^0(\bar{D}^0)$	$F^+(\bar{F}^-)$	$B^+(\bar{B}^-)$
Mass, MeV	139.567	134.962	493.67	497.69		548.8	1868.3	1863.1	2018	5270
σ_M	±0.001	±0.004	±0.015	±0.13		±0.6	±0.9	±0.9	±0.9	±10
Lifetime, s	2.603×10^{-8}	0.83×10^{-16}	1.237×10^{-8}	0.892×10^{-10}	5.18×10^{-8}	7.74×10^{-19}	$\sim2.5\times10^{-13}$	$\sim3.5\times10^{-13}$	?	$<10^{-12}$
σ_τ	±0.002	±0.06	±0.003	±0.002	±0.04	±0.11				
Major decay modes	$\mu^+\nu_\mu$ 100%	$\gamma\gamma$ 99%	$\mu^+\nu_\mu$ 63.5%	$\pi^+\pi^-$ 69%	$3\pi^0$ 22%	$3\pi^0$ 30%	$K^-\pi^+\pi^+$ 4%	$K^\pm X$ $\approx35\%$	$\eta\pi^+$	$e^+\nu_e X$ } 25%
	$e^+\nu_e$ 1.27×10^{-4}	γe^+e^- 1.15%	$\pi^0\mu^+\nu_\mu$ 3.2%	$\pi^0\pi^0$ 31%	$\pi^+\pi^-\pi^0$ 12%	$\pi^+\pi^-\pi^0$ 24%	$K^0_S X$ $\approx20\%$	$K^0_S X$ $\approx30\%$	$2\pi^+\pi^-\pi^0$	$\mu^+\nu_\mu X$
		$2e^+2e^-$ 3.3×10^{-5}	$\pi^0 e^+\nu_e$ 4.8%		$\pi^\pm e^\mp$ 39%	$\pi^+\pi^-\gamma$ 5%	$e^+\nu_e X$ 8%	$e^+\nu_e X$ 8%	$K^+K^0_S\pi^+\pi^-$	
			$\pi^+\pi^0$ 21.2%		$\pi^\pm\mu^\mp$ 27%	$\gamma\gamma$ 38%			observed	
			$\pi^+\pi^+\pi^-$ 5.6%							
			$\pi^+\pi^0\pi^0$ 1.7%							
Nonzero flavor numbers	$I_3=+1(-1)$	$I_3=0$	$I_3=+\frac{1}{2}(-\frac{1}{2})$ $s=+1(-1)$	Mixture of two states with $(I_3,s)=\pm(-\frac{1}{2},1)$		$I_3=0$	$I_3=+\frac{1}{2}(-\frac{1}{2})$ $C=+1(-1)$	$I_3=-\frac{1}{2}(+\frac{1}{2})$ $C=+1(-1)$	$s=-1(+1)$ $C=+1(-1)$	$I_3=+\frac{1}{2}(-\frac{1}{2})$ $b=+1(-1)$

(nuclear force mesons) were massive [200 to 300 electron masses (m_e)] and could carry electric charge between the nucleons. *See* BARYON; HADRON; NUCLEAR STRUCTURE; QUANTUM FIELD THEORY.

Discovery. In 1937 C. Anderson established the existence of positively and negatively charged μ-mesons (now known as muons) of mass 105.6 MeV ($\simeq207\ m_e$) in cosmic radiation, but these were soon shown to have very weak coupling with nucleons, that is, to be nonhadronic and therefore not "mesons," as this name is used today. They are members of the lepton family, together with the electron and the neutrinos. The nuclear force mesons turned out to be the π-mesons (pions), of mean mass about 138 MeV ($\simeq270\ m_e$), first identified in cosmic radiation by C. F. Powell in 1947. Pions with positive, negative, or zero charge are produced copiously in nuclear collisions of sufficiently high energy. Their properties have been studied intensively in the laboratory, using the powerful beams of charged pions now available from many types of high-energy particle accelerators. *See* LEPTON; PARTICLE ACCELERATOR.

Already in 1948, studies of the cosmic radiation indicated the existence of heavier mesons (masses about 495 MeV), now known as K mesons (kaons). Their detailed study developed much later, partly because their laboratory production required the development of multi-GeV proton accelerators and partly because their production is less copious than that for pions at all proton energies available to date (typically, several percent of pion production). As discussed below, there are four K mesons, K^+, K^-, and two neutral particles, K^0_S and K^0_L, and strong K^+, K^-, and K^0_L, beams are obtainable from higher-energy (≥30 GeV) proton accelerators. Since the $\pi^\pm$, $K^\pm$, and $K^0_{L,S}$ mesons have relatively long lifetimes (about 10^{-8} to 10^{-10}s), they will be referred to as semistable; their properties are listed in Table 1.

In 1961 the ω-meson of mass 783 MeV was discovered through its rapid decay to three pions in a bubble chamber investigation at Berkeley. It is highly unstable, with lifetime 6.6×10^{-23} s, as corresponds to the natural width 10.1 MeV observed in its mass value. [From the uncertainty relation for energy and time, the lifetime τ and the natural width Γ of any state are related by $\tau(\text{s})\times\Gamma(\text{MeV})=6.582\times10^{-22}$.] Such a short lifetime implies that the ω-meson decays through hadronic interactions. As discussed below, many such highly unstable heavy mesons are now established, with lifetimes shorter than 10^{-22} s, decaying hadronically to lighter mesons, and more continue to be discovered, as higher mass ranges are explored and as the statistics of earlier experiments are increased. *See* BUBBLE CHAMBER.

In 1974 the J/ψ meson of mass 3097 MeV was discovered at Brookhaven National Laboratory in proton experiments and at the electron-positron storage ring SPEAR at Stanford. Although it decays to mesons, its natural width is only 0.063 MeV (lifetime 10^{-20} s). Also, it decays electromagnetically, yielding an e^+e^- or $\mu^+\mu^-$ pairs, at a rate comparable with those for mesonic final states. This new phenomenon gave the first indication of the existence of heavy new quarks. In consequence, its study has often been termed the "new physics." *See* J PARTICLE; QUARKS.

Quark structure. Hadrons are now considered to be composite, consisting of spin-$\frac{1}{2}$ quarks (q), corresponding antiquarks ($\bar{q}$), and some number of gluons (g), the last being the quanta of the intermediate field which binds the quarks and antiquarks to form hadrons. Baryon number $B=+\frac{1}{3}$ holds for a quark q, $B=-\frac{1}{3}$ for antiquark $\bar{q}$, while $B=0$ holds for a gluon. In this view, the simplest possibility is that each meson is a quark-antiquark ($q-\bar{q}$) pair bound together by the gluon field, and this model does account quite well for most of the known mesons and their properties. However, more complicated systems (for example, consisting of two quarks with two antiquarks) can be considered and may even be required by some of the present data. The known quarks are listed in Table 2. They must be assigned fractional charge values, relative to the proton charge. *See* GLUONS.

Table 2. Quarks now established†

Quark type:	u	d	s	c	b	$t?\ldots$
Flavor name:	(charge, isospin I_3)		(strange)	(charm)	(bottom)	(top)
Charge Q/e_p	2/3	$-1/3$	$-1/3$	2/3	$-1/3$	(2/3)
Mass, GeV	$\simeq0.01$	$\simeq0.01$	$\simeq0.5$	$\simeq1.5$	$\simeq4.7$	—
Flavor‡	$I_3=\frac{1}{2}$	$I_3=-\frac{1}{2}$	$s=-1$	$C=+1$	$b=+1$	$t=+1$

†To each quark, there exists an antiquark with the opposite flavor values and with opposite intrinsic parity.
‡Only nonzero flavor values are entered on this table.

Color and quantum chromodynamics. An exactly conserved color variable may be attributed to the quarks. The facts about the baryons require the color space to be three-dimensional. This law of conservation of the color current is then equivalent to an invariance of all hadronic interactions through all the transformations of the special unitary symmetry group $SU(3)_C$ acting on this color space. Quantum chromodynamics is the gauge theory associated with this exact symmetry, just as quantum electrodynamics is the gauge theory associated with the law of conservation of electric charge. The quanta of its gauge field are known as the gluons. These are neutral vector particles, eight in number—a color octet—coupling directly with the color current, just as the photon, the quantum of the electromagnetic field, couples directly with the electric current. *See* COLOR (QUANTUM MECHANICS); SYMMETRY LAWS.

Asymptotic freedom. The coupling constant for the gluon interaction with the color current depends on the momentum transfer. Although it is large for low-momentum transfer, since it is responsible for the strong interactions which bind quarks and antiquarks to form hadrons, it approaches zero as the momentum transfer increases asymptotically to infinity.

Color current. The net color current receives separate contributions from each quark type, as well as from the gluon field. Since each quark type is color triplet, each of these contributions has the same form, except for the label α (known as the quark flavor) which characterizes the quark type. Since the gluon couples with the total color current, the theory of quantum chromodynamics is well defined, apart from the universal coupling constant. Further, the color currents do not change quark flavor, so that, for any hadronic reaction, Eq. (1) holds, where $n(q_\alpha)$ and $\bar{n}(\bar{q}_\alpha)$ denote the numbers

$$n(q_\alpha) - \bar{n}(\bar{q}_\alpha) = \text{constant} \qquad (1)$$

bers of quarks and of antiquarks, respectively, for each quark flavor α.

Confinement. The confinement dogma holds that only color singlet states are physically realizable. With it, neither a quark nor a gluon can exist in isolation. Thus, for any color singlet system of quarks and antiquarks, quantum chromodynamics must generate an appropriate confining potential. This has not yet been proved to be the case, although some arguments have been given which suggest that it may be so.

A mesonic system therefore consists of an equal number of quarks and antiquarks, with perhaps some gluons, all in a color singlet configuration. *See* QUANTUM CHROMODYNAMICS.

Mesons as quark-antiquark systems. Quarks q (and antiquarks $\bar{q}$) have spin $1/2$, so that a $q - \bar{q}$ system may have net spin $S = 0$ or 1 (unit $\hbar$). The relative angular momentum within the $q - \bar{q}$ system must be quantized, with value $L\hbar$, where $L = 0, 1, 2 \ldots$ is a positive integer. The total angular momentum J is the vector sum of L and S, and so takes the value $J = L$ for $S = 0$, and $J = L, L \pm 1$ for $S = 1$. The notation used for these configurations is $^{2S+1}L_J$, the letters S, P, D, F, . . . being used for $L = 0, 1, 2, 3$ . . ., respectively. There will be a series of radial excitations for each configuration, so that it is also necessary to specify a principal

quantum number n. *See* ANGULAR MOMENTUM; SPIN.

Dirac's theory of spin-$1/2$ particles indicates that an antiquark has intrinsic parity opposite that for the corresponding quark. Since orbital angular momentum L contributes parity $(-1)^L$, the net parity P is $(-1)^{L+1}$. Another operation of importance, known as charge conjugation, is the replacement of quark by antiquark, and vice versa, in the system. This is of interest only when the system contains an antiquark $\bar{q}_\alpha$ for every quark q_α it contains, so that the operation can reproduce the original system, albeit with coefficient ± 1, known as the charge conjugation parity C. This operation is clearly of interest only to systems with value zero for charge, baryon number, and all other flavors. For the $q - \bar{q}$ system with configuration $^{2S+1}L_J$, C has the value $(-1)^{L+S}$ and C is conserved through all hadronic interaction processes. The phenomenological notation JPC will be used to characterize a multiplet which includes one state having zero values for all flavor quantum numbers, C being the value for this state. The photon has $C = -1$, which leads to the forbiddenness of some $q - \bar{q}$ radiative transitions, such as $^3S_1 \rightarrow \gamma^1P_1$ and $^1S_0 \rightarrow 3\gamma$. *See* PARITY; RELATIVISTIC QUANTUM MECHANICS.

Unitary flavor symmetries. It appears plausible that all the flavor dependence of mesonic states is due to the mass differences between different quark flavors. The quark mass values known are given in Table 2; these are effective mass values for quarks as constituents within hadrons.

The u and d quarks have essentially the same mass $\bar{m}$. The color current contributed by them is proportional to expression (2), as is also the contri-

$$\sim (u^*u + d^*d) \qquad (2)$$

bution of their masses to the total energy, in the limit where $m_u = m_d = \bar{m}$. Form (2) remains the same under all transformations of the $SU(2)$ group, acting in the (u,d) flavor space, this being the group of all the complex linear transformations in a two-dimensional space which do not change the magnitude of vectors in that space. This is well known, as the isospin group, isomorphic with the group of real rotations in three-dimensional space. It is this invariance which is responsible for the fact that hadronic particle states are observed to occur in isospin multiplets, characterized by a total isospin I, there being $(2I + 1)$ states with the same mass but different charge values, running from $(Q_0 + I_3)$ in unit steps to $(Q_0 - I_3)$, where the value of Q_0 depends on the system being considered. *See* ISOTOPIC SPIN.

Similarly, the color current for (u,d,s) quarks is proportional to expression (3), which remains in-

$$(u^*u + d^*d + s^*s) \qquad (3)$$

variant under all transformations of the $SU(3)_f$ group, acting in the (u,d,s) flavor space, as indicated by the suffix f. However, the energy due to the quark masses takes form (4), neglecting the

$$\bar{m}(u^*u + d^*d + s^*s) + (m_s - \bar{m})(s^*s) \qquad (4)$$

small difference between m_u and m_d, and this is invariant under $SU(3)_f$ only to the extent that the physical quantities considered may be insensitive to this quark mass difference $(m_s - \bar{m}) \simeq 500$ MeV. This is the $SU(3)$ symmetry proposed for the hadronic interactions by M. Gell-Mann and by Y.

Ne'eman in 1961. It works remarkably well in many contexts, but its approximate nature is evident from the above discussion.

This discussion can be extended to the space of (u,d,s,c) quarks, with a corresponding unitary flavor symmetry group SU(4)$_f$. In this case, there are mass differences of order $(m_c - \bar{m}) \simeq 1500$ MeV, so that the symmetry is very badly broken. Nevertheless, some remnants of this symmetry remain, and it provides a useful basis for the classification of states and for discussing relationships between processes related by SU(4)$_f$ transformations, if used with caution.

The situation for vector mesons is illustrated on Fig. 1, where the mesonic states have been plotted in a symmetrical three-dimensional pattern. The charm C has values -1, 0, and $+1$ on the three planes shaded, going from bottom to top. The plane containing the four D-mesons holds all the mesons with zero strangeness; those to the right of it have $s = -1$, those to the left have $s = +1$. On each shaded plane there is a definite SU(3)$_f$ pattern; on the bottom and top planes, the patterns correspond to the 3 and 3* representations of SU(3)$_f$ symmetry, respectively. The isospin multiplets are marked by prominent heavy lines. On the central plane, the pattern consists of an SU(3)$_f$ octet together with two singlets. The latter mix strongly with the $I = 0$, $s = 0$ member of the SU(3)$_f$ octet, owing to the strong breaking of the SU(3)$_f$ and SU(4)$_f$ symmetries. The physical states ω, ϕ, and ψ at the center of the pattern are quite well approximated by $(\bar{u}u + \bar{d}d)/\sqrt{2}$, $(\bar{s}s)$, and $(\bar{c}c)$, respectively, rather than by the $C = 0$, $s = 0$, $I = 0$ states characteristic of an SU(4)$_f$ symmetry, and their masses range from 782 to 3097 MeV. *See* UNITARY SYMMETRY.

Quark and antiquark interactions. These are of several classes.

Hadronic interactions. At the fundamental level, these are due to the quark-gluon and gluon-gluon couplings discussed above. They are flavor-independent, apart from their dependence on the quark masses m_α. The quark-antiquark interactions have some close parallels with the electromagnetic interactions, although the former are much stronger; the quark-gluon coupling constant α_s has a value $\simeq 0.5$ for (momentum transfer)$^2 = 5(\text{GeV/c})^2$, compared with $\alpha \simeq 1/137$ for the electromagnetic coupling constant.

At the phenomenological level, prominent contributions may still arise from more specific mechanisms. In this respect the pion plays an outstanding role, since it is the lightest meson, the next being the kaon, about 3.5 times heavier. For example, the long-range part of the nucleon-nucleon interaction remains dominated by the one-pion-exchange mechanism. *See* STRONG NUCLEAR INTERACTIONS.

Electromagnetic interactions. These are carried by the electromagnetic field which couples with the electric current of the quarks. As can be seen from the quark charge values listed in Table 2, these interactions are flavor-dependent, violating even the SU(2) isospin symmetry.

Weak interactions. These have been unified with the electromagnetic interactions, forming the electroweak interaction theory developed by S. Glashow, A. Salam, and S. Weinberg. This unification requires the existence of some very massive weak

bosons, the $W^\pm$ boson at about 80 GeV and the Z^0 boson at about 90 GeV, which (together with the massless photon) carry the electroweak interactions between the currents with which these bosons and the photon interact. However, when the energies of interest are so low relative to the weak boson masses, the weak interactions separate out from the electromagnetic interactions, being effectively pointlike and many orders of magnitude weaker than the latter. In this low-energy limit, the structure of the weak interactions is a sum of direct current × current terms, each being of the general form proposed by E. Fermi in 1933.

The electroweak interaction is universal, and the $W^\pm$-current interaction therefore has the same coupling constant e as holds for the electromagnetic interaction. The observed decay interactions are weak because of the large $W^\pm$ mass. For a purely leptonic decay mediated by the W^+ boson, such as $\mu^+ \rightarrow \bar{\nu}_\mu e^+ \nu_e$, the interaction amplitude is given in the current × current form as expression (5), where

$$\left(\frac{e^2}{M_W^2}\right) J(\mu^+ \rightarrow \bar{\nu}_\mu) \cdot J(\text{Vac} \rightarrow e^+ \nu_e) \qquad (5)$$

J represents the current giving the transition indicated. Its coefficient defines the weak coupling amplitude G, whose value is well known empirically. Thus, Eq. (6) is valid, and it is essentially this

$$G = 4\pi\alpha/(M_W^2\sqrt{2}) = (1.02678(2)) \times 10^{-5}/M_P^2 \qquad (6)$$

equation which gives the value for M_W quoted above. This small value causes the weak decay processes relevant for semistable mesons to have decay rates of order 10^{-12} relative to those for unstable mesons, whereas meson decay processes involving electromagnetic effects are suppressed typically by only a factor of about 10^{-3} per photon interaction involved, as in the case of $\pi^0 \rightarrow \gamma\gamma$ decay, even though the electromagnetic and weak interactions are now part of the one electroweak interaction.

The study of the weak decays of the semistable mesons is an important means of learning the precise constitution of the hadronic weak currents. Little is known of those involving the b quark, for example, and this is an area of very active research. *See* FUNDAMENTAL INTERACTIONS; WEAK NUCLEAR INTERACTIONS.

Pseudosclalar mesons. These have q-$\bar{q}$ configuration $^1S_0$, corresponding to total spin zero and odd parity. In most cases, they are the lightest meson for their set of flavor quantum numbers, presumably because the spin dependence of the flavor-independent q-$\bar{q}$ potential has most attraction when $S = 0$. These mesons (Table 1) are semistable because the only decay processes energetically available to them involve violations of flavor conservation, which can occur only through the weak interactions. The exceptions to these remarks are the states with zero for all flavor quantum numbers, the π^0, η, η', and η_c; π^0 and η undergo electromagnetic decay, whereas η' and η_c have strong hadronic decay modes.

Pions. The pions π^-, π^0, and π^+ form a charge triplet (isospin $I = 1$). The $\pi^\pm$ mesons decay dominantly by reactions (7). The muons in cosmic radiation or in beams from accelerators originate mainly from these π-μ decay processes, following

$$\pi^+ \rightarrow \mu^+ + \nu_\mu \qquad \pi^- \rightarrow \mu^- + \bar{\nu}_\mu \qquad (7)$$

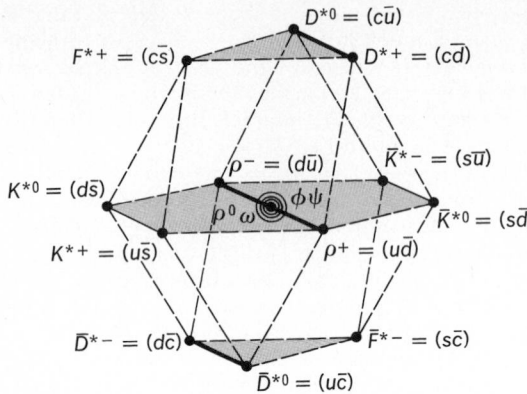

Fig. 1.   The 16 vector mesons arrayed on a flavor SU(4), plot.

pion production in high-energy nuclear collisions. The muon emitted in $\pi^+(\pi^-)$ decay spins in a right-handed (left-handed) sense relative to its direction of emission, which illustrates directly the nonconservation of both parity and charge-conjugation parity in the π-μ decay process. Since its flavor quantum numbers are all zero, and it has charge-conjugation parity $C = +1$, the π^0 meson can decay to $\gamma\gamma$ electromagnetically. It therefore decays rather rapidly, with lifetime $\sim 10^{-16} \times$ s.

Kaons. The kaons form two charge doublets (isospin $I = 1/2$), (K^+, K^0) and $(\bar{K}^0, K^-)$, each doublet being antiparticle to the other. The K^+ meson has many decay modes, the dominant being to $(\mu^+ \nu_\mu)$ with branching ratio 63.5%. The K^- meson decays to the corresponding antiparticle systems, with the same branching ratios. It was recognized quite early that K^+ decay led to two systems, $\pi^+\pi^0$ and $\pi^+\pi^+\pi^-$, which have opposite parity. This discrepancy was resolved in 1957, when it was shown experimentally that parity conservation does not generally hold for the weak interactions, so that the existence of both pionic modes for K^+ decay was simply an illustration of this fact. Parity-nonconservation is demonstrated explicitly by the observation of longitudinal polarization for the μ^+ meson emitted in K^+ semileptonic decay modes.

Two neutral kaons are known empirically, both semistable (Table 1). The K^0_S meson is short-lived and decays dominantly to $\pi\pi$ states. The K^0_L meson has lifetime several times longer than $\pi^\pm$ and $K^\pm$, so that K^0_L beams can also be prepared at accelerators and used for experiments on K^0_L properties. A series of ingenious experiments has established that (because of weak interaction effects) the K^0_L meson is just a little heavier than the K^0_S meson, the masses differing by $3.52 \pm 0.015 \times 10^{-12}$ MeV (the K^0_S lifetime width is 7.4×10^{-12} MeV). To a good first approximation, these K^0_S and K^0_L states are the coherent superpositions (8) of the K^0 and

$$\psi(K^0_S) = \psi_+ = \{\psi(K^0) - \psi(\bar{K}^0)\}/\sqrt{2} \qquad (8a)$$

$$\psi(K^0_L) = \psi_- = \{\psi(K^0) + \psi(\bar{K}^0)\}/\sqrt{2} \qquad (8b)$$

$\bar{K}^0$ states, which have definite values under the combined operation CP, $+1$ for K^0_S and -1 for K^0_L. Charge conjugation C induces $K^0 \leftrightarrow \bar{K}^0$, and the parity P has the value -1 for both K^0 and $\bar{K}^0$. For $J = 0$, CP has the value $+1$ for the final nonleptonic state $\pi\pi$, and -1 for the final state 3π, assuming that the latter has only S-wave internal motions, as is compatible with the data on this decay mode. This suggested for a time that, although both charge-conjugation and space-reflection symmetries are separately violated in the weak interactions, the weak interactions might still be invariant under the combined operation CP.

In 1964 observation of the decay mode $K^0_L \rightarrow \pi^+ \pi^-$ was reported but with a low branching ratio, now known to be $2.03 \pm 0.05 \times 10^{-3}$, by J. Cronin, V. Fitch, and coworkers. This observation demonstrates directly the failure of CP invariance in the weak interactions, at the level of 10^{-3}. This failure of CP invariance was soon confirmed by the observation of the decay mode $K^0_L \rightarrow \pi^0\pi^0$, with branching ratio $0.94 \pm 0.18 \times 10^{-3}$. It follows that the physical K^0_L meson does not correspond to ψ_- in Eqs. (8), but to ψ_- with a small admixture of ψ_+, the latter with amplitude $\simeq 10^{-3}$; and vice versa for K^0_S.

Despite considerable exploration, no other evidence has been found for the failure of CP invariance, nor for the failure of time-reversal invariance T, which is considered equivalent, in view of a rather general theorem that CPT invariance must always hold, in any other weak interaction process. These $K^0_L \rightarrow \pi\pi$ phenomena enable an experiment to be prescribed for a laboratory far out of contact with the solar system, whose results will determine absolutely whether that laboratory is made of matter or antimatter. Also, the observation of $K^0_S - K^0_L$ interference effects in high-momentum neutral kaon beams has demonstrated the property of quantum-mechanical coherence over macroscopic distances, measured in tens of meters. *See* CO-HERENCE.

Eta mesons. The neutral $\eta(549)$ decays all require the intervention of the electromagnetic interaction. Even the mode $\eta \rightarrow \pi^+\pi^-\pi^0$ violates isospin conservation, and virtual electromagnetic effects are the most plausible cause of this. The neutral $\eta'(958)$ is the ninth member of this SU(3) nonet, lying far above all the other members in mass. Its dominant decay mode $\eta' \rightarrow \eta\pi\pi$ is hadronic but slow, the η' full width being only (0.24 ± 0.1) MeV; photon-emitting modes, $\rho^0\gamma$ and even $\gamma\gamma$, compete quite strongly with it, their branching fractions being 30% and 2%, respectively.

Vector mesons. These mesons have $JPC = (1--)$, consistent with the parameters of the lowest $(\bar{q}q)$ state with the configuration $^3S_1$. The 16 states made from (u,d,s,c) quarks and $(\bar{u},\bar{d},\bar{s},\bar{c})$ antiquarks are arrayed on Fig. 1, and their masses, widths, and dominant decay processes are given in Table 3.

The "old" vector mesons form the nonet depicted on the central $C = 0$ plane of Fig. 1, excluding the $(c\bar{c})$ state J/ψ. They were all discovered in production reactions, and identified by the analysis of their decay processes.

The existence of some of these mesons had already been conjectured before 1961, on the basis of knowledge of the electromagnetic structure of

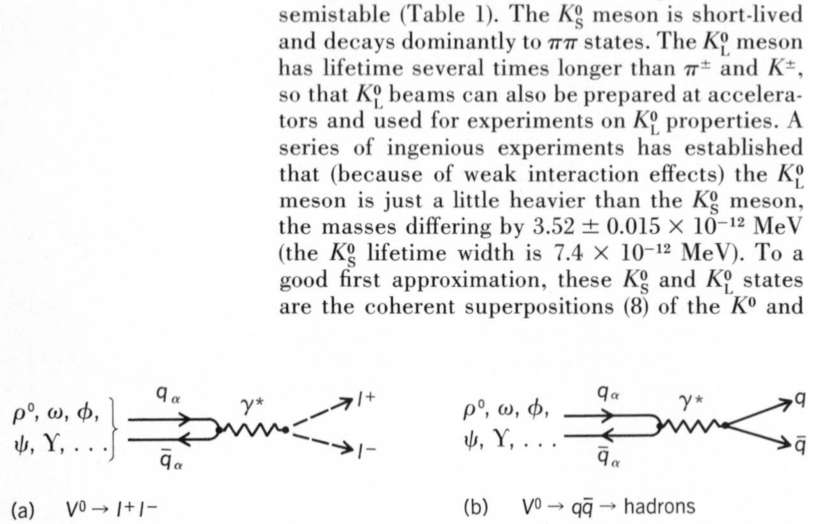

(a) $V^0 \rightarrow l^+ l^-$

(b) $V^0 \rightarrow q\bar{q} \rightarrow$ hadrons

Fig. 2.   Quark and lepton line figures showing annihilation of quark q_α with antiquark $\bar{q}_\alpha$ in a vector meson whose flavor quantum numbers are all zero, to give rise to (a) to a lepton pair or (b) to a quark pair through an intermediate virtual photon γ^*.

the proton and neutron. The neutral mesons ρ^0, ω, ϕ, and ψ all have the same quantum numbers as the photon and therefore couple directly with it; it is therefore natural that the ρ^0 and ω mesons, which are made of the same quark flavors as are the nucleons, should play an important role in the photon-nucleon interaction. A free photon cannot transform in vacuum to become a free vector meson V, of course, but a photon with sufficiently high energy can so transform if there is some particle present which can absorb the (small) momentum transfer then involved. Reaction (9) provides such

$$\gamma + \text{nucleus} \rightarrow \text{nucleus} + V \qquad (9)$$

a situation, one where the nucleons can even act coherently, taking up this momentum transfer while leaving the nucleus unexcited. Process (9) is then diffractive in nature, and its investigation has been important in the study of vector mesons.

However, the major part of present knowledge of the vector mesons comes from formation experiments, through the annihilation processes between head-on electron and positron beams in a storage ring, process (10). (Here and on the figures, γ^* de-

$$e^+ + e^- \rightarrow \gamma^* \rightarrow \text{vector meson } V \rightarrow$$
$$V\text{-meson decay products} \qquad (10)$$

notes an intermediate electromagnetic field such that its energy is not equal to its momentum, as is required for a free photon. The technical term used to describe this field is virtual photon.) The inverse process, the decay of a vector meson, all of whose flavor quantum numbers are zero, into a lepton pair, is shown in Fig. 2a, while the decay of such a meson to a quark pair (seen subsequently as two jets of outgoing hadrons) through an intermediate virtual photon is shown in Fig. 2b. (However, the major contribution to the latter decay mode arises from transitions due to intermediate gluons, discussed below.) The rates observed for e^+e^- production of hadrons are shown on Fig. 3, where the ratio $R = \sigma(e^+e^- \rightarrow \text{hadrons})/\sigma(e^+e^- \rightarrow \mu^+\mu^-)$, where σ stands for the indicated cross section, is plotted against the center-of-mass energy $E_{cm} = 2E_e$, the electron and positron having the same energy E_e, but opposite directions, in the storage ring. The excitation of the ρ^0 and ω mesons is a

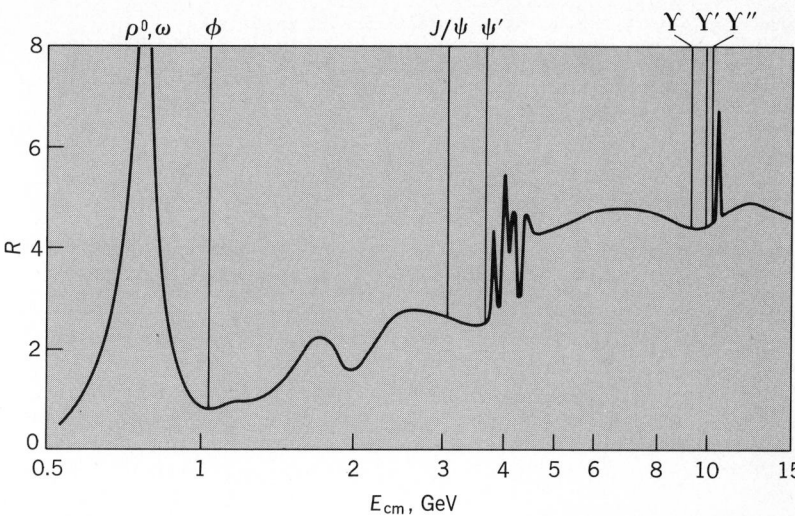

Fig. 3.   Observed ratio $R = \sigma(e^+e^- \text{ hadrons})/\sigma(e^+e^- \rightarrow \mu^+\mu^-)$ plotted as a function of center-of-mass energy E_{cm}.

strong and broad resonance peak. The excitation of the ϕ meson is strong, but this state has a width of only 4 MeV, too narrow to be shown.

The ρ^0 decay is depicted in terms of quarks on Fig. 4a. From the $u\bar{u}$ part of the ρ^0 state, this decay requires the spontaneous creation of a $d\bar{d}$ pair, and their interaction with $\bar{u}$ and u, respectively, leading to the configuration $(\pi^+ + \pi^-)$; from the $d\bar{d}$ part, it requires the spontaneous creation of a $u\bar{u}$ pair, leading to the same final state. The kinetic energy released is about 500 MeV, and the transition $\rho^0 \rightarrow \pi^+\pi^-$ is rapid, as the observed width $\Gamma(\rho^0) = 158$ MeV attests.

The ϕ decays are illustrated on Figs. 4b and c. Without the gluon coupling in Fig. 4b, the initial ϕ and final $\pi^+\rho^-$ systems would have no quarks in common. Such disconnected graphs are to be disregarded, as noted by S. Okubo, G. Zweig, and J. Iizuka (OZI), since the creation point of the final quarks is then not causally connected with the initial state. However, the gluons have a strong interaction with quarks, and they can provide the necessary connection between the initial and final

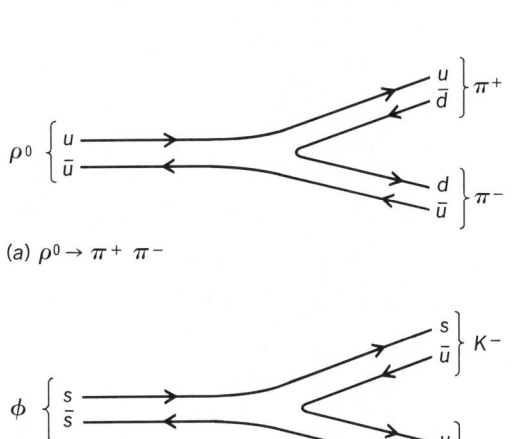

(a) $\rho^0 \rightarrow \pi^+\pi^-$

(c) $\phi \rightarrow K^+K^-$

Fig. 4.   Quark line figures to illustrate the decay processes of vector mesons. (a) Decay $\rho^0 \rightarrow \pi^+\pi^-$. (b) Decay $\phi = (s\bar{s}) \rightarrow \pi^+\rho^-$. (c) Decay $\phi \rightarrow K^+K^-$. (d) Decay $(c\bar{c}) =$

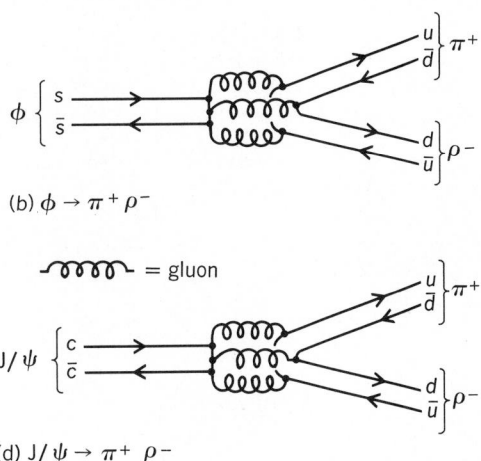

(b) $\phi \rightarrow \pi^+\rho^-$

$\sim\!\!\sim\!\!\sim$ = gluon

(d) $J/\psi \rightarrow \pi^+\rho^-$

$J/\psi \rightarrow \pi\rho$, which is observed to have branching fraction $1.2 \pm 0.1\%$.

Table 3. Lightest vector mesons established for the u, d, s, c, and b quarks

	ρ	$K^*(\bar{K}^*)$	ω	ϕ	J/ψ	$D^*(\bar{D}^*)$	$F^*(\bar{F}^*)$	T	$B^*(\bar{B}^*)$
Mass, MeV	776 ± 3	891.8 ± 0.4	782.4 ± 0.2	1019.6 ± 0.1	3097 ± 1	2007 ± 1	2140 ± 60	9448 ± 10	$\approx 5320 \pm 10$
Width, MeV	158 ± 5	50 ± 1	10.1 ± 0.3	4.1 ± 0.2	0.063 ± 0.009	<2	?	0.045 ± 0.015	?
Nonzero flavor Numbers	$I=1$	$s=+1(-1)$ $I=\frac{1}{2}$				$C=+1(-1)$ $I=\frac{1}{2}$	$C=+1(-1)$ $s=+1$		$b=+1(-1)$ $I=\frac{1}{2}$
Dominant decay Modes	$\pi\pi$	$K\pi$	$\pi^+\pi^-\pi^0$ 90% $\pi^0\gamma$ 9%	K^+K^- 49% $K^0_L K^0_S$ 35% $\pi^+\pi^-\pi^0$ 15%	e^+e^- 7% $\mu^+\mu^-$ 7% $2\pi^+2\pi^-\pi^0$ 4% + many hadronic modes	$D\pi$ ~80% $D\gamma$ ~20%	$F\pi$? $F\gamma$ seen	$\mu^+\mu^-$ 3% e^+e^- ? + very many hadronic modes	$B\gamma$?

quarks. But they do not make this decay transition rapidly on account of the property of asymptotic freedom for the gluon interactions. Annihilation between s and $\bar{s}$ quarks requires their very close approach. By the uncertainty principle, small spatial distance implies large momentum transfer p, and asymptotic freedom requires that the hadronic coupling constant $\alpha_s(p^2)$ approach zero logarithmically as p^2 increases to infinity. This $s\bar{s}$ annihilation also requires the formation of at least three gluons, so that the amplitude $(s\bar{s}) \rightarrow \pi^{\pm}\rho^{\pm}$ is proportional to α_s^3 and the asymptotic freedom of quantum chromodynamics is easily sufficient to make it very small. In fact, even the transitions $\phi \rightarrow \pi^+\pi^-\pi^0$ observed (Table 3) are attributable to a small admixture of ω state with $(s\bar{s})$ in the physical ϕ state. This conclusion about the role of asymptotic freedom in heavy vector meson decay

does not affect the amplitude for $\rho^0 \rightarrow \pi^+\pi^-$ of Fig. 4a, nor that for $(s\bar{s}) \rightarrow K^+ + K^-$ in Fig. 4c. See UNCERTAINTY PRINCIPLE.

The decay mode $\phi \rightarrow K\bar{K}$ (Fig. 4c) is the natural one for a state $(s\bar{s})$, involving the pair creation of only the lightest quarks. However, the K^+K^- threshold mass is at 987.3 MeV, whereas $M(\phi) = 1019.6$ MeV, so that the kinetic energy release is only 32 MeV. Since the decay $\phi \rightarrow K\bar{K}$ requires orbital angular momentum $l_{K\bar{K}} = 1$, there is a centrifugal barrier against this decay, and the small energy release means that its rate is very much suppressed relative to that for the otherwise parallel decay mode $\rho^0 \rightarrow \pi^+\pi^-$.

With this background, the immediate interpretation of the narrow resonant states shown on Fig. 3 at around 3 to 4 GeV and around 10 GeV is that each cluster of states signals the threshold for a new quark flavor. The $J/\psi(3097)$ is interpreted as the $^3S_1$ state for the charmed system $(c\bar{c})$ and ψ' (3685) as its first excited $^3S_1$ state, that is, the first radial excitation of the J/ψ meson. The lowest state of the system $(c\bar{u})$ is the pseudoscalar $(^1S_0)$ meson D^0 (1863), so the threshold $D\bar{D}$ analogous to the $K\bar{K}$ threshold in the case of the ϕ meson lies at 3727 MeV. Since the mass values of the J/ψ and ψ' mesons are below this $D\bar{D}$ threshold, they can decay only through the OZI-forbidden processes mediated by intermediate gluons, such as given in Fig. 4d or through the electromagnetic processes shown in Fig. 2a. The strong suppression of J/ψ decay to hadrons by the effect of asymptotic freedom is underlined by the fact that the decays given by Fig. 2a for the leptons $l = e$ and μ amount to 14% of the total J/ψ decay rate. The peaks occurring just above the ψ' mass are due to the excitation of further states in the $(^3S_1, ^3D_1)$ $c\bar{c}$ system. These states have widths characteristic of normal hadronic processes, since they lie above the $D\bar{D}$ threshold.

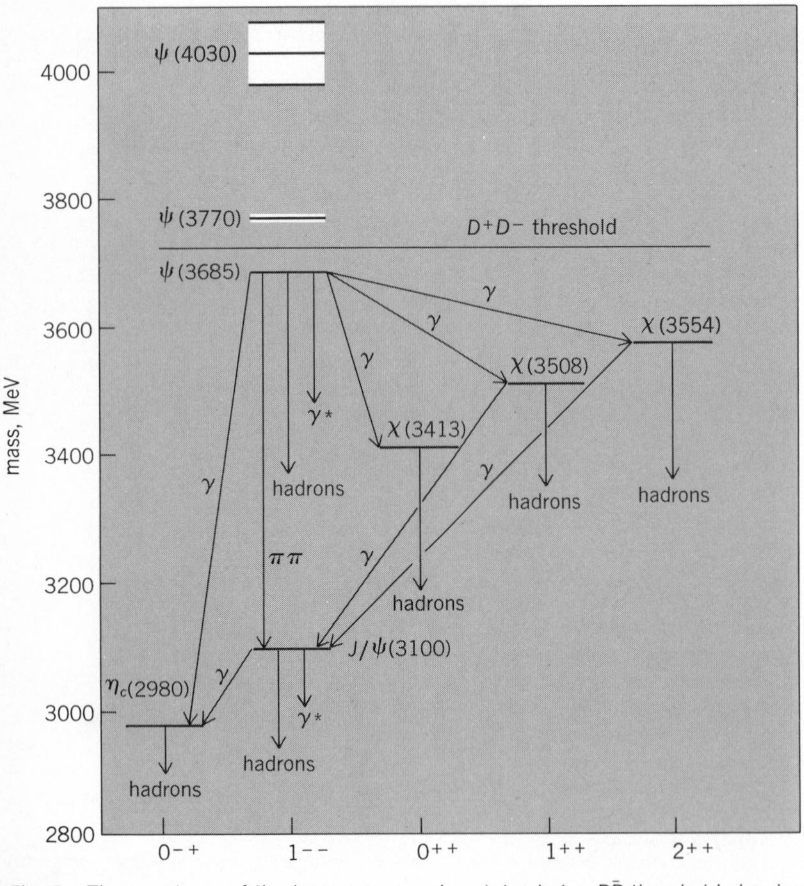

All 16 of the vector mesons appropriate to the (u,d,s,c) system are now known empirically. Their known properties are given in Table 3, and they are displayed symmetrically on Fig. 1, although their masses range from 776 to 3097 MeV. All of the corresponding pseudoscalar mesons are now known and may be arrayed similarly. Those which are semistable are entered in Table 2; the D-meson doublet are the lightest mesons with $(s,C) = (0, +1)$, and the F-meson singlet the lightest meson with $(s,C) = (+1,+1)$. The others are the isosinglet mesons η' (958) and η_c (2980), both of which can decay hadronically, through OZI-forbidden processes; their decays involve only two gluons, so that their widths are less suppressed than are the J/ψ hadronic decay modes.

The situation for the T mesons around 10 GeV is

Fig. 5. The spectrum of the known ψ, χ, and η_c states below $D\bar{D}$ threshold showing their γ and $\pi\pi$ transitions, together with the nature (that is, hadronic or mediated by a virtual photon γ^*) of their absorptive decay processes.

quite similar to the above, and corresponds to the onset of processes involving the fifth quark b. There are three narrow vector states, T(9448), T′(10005), and T‴(10340), and a fourth state at about 10565 MeV, which is clearly in the continuum, having width 13 ± 6 MeV. This implies that the B-meson, with structure $(b\bar{u})$, has mass between 5170 and 5283 MeV. Some of its decay modes have been identified, and indications are that its mass is about 5270 MeV.

There is great confidence in this $(c\bar{c})$ model for the ψ states, in consequence of the understanding it has given of their detailed γ-spectroscopy. The known states (Fig. 5) are those expected from this model, and simple calculations based on the flavor-independent $q\bar{q}$ potential of quantum chromodynamics, consisting of a linear confining potential and an attractive one-gluon-exchange potential, provide an acceptable account of the various branching fractions measured, including the net hadronic rate, for each of the ψ and χ states.

Meson multiplets with positive parity. The $q\bar{q}$ model for meson multiplets suggests that the first excited multiplets should correspond to $L = 1$, the net parity then being $+$. Four such multiplets should exist: ^3L_J with $J = 0$, 1, and 2, for $L = 1$, and ^1L_J with $J = L = 1$. The former three have $C = +1$, and there is quite detailed evidence for their existence as χ states in the ψ spectroscopy. The $^1P_1$ state has $C = -1$; for the $(c\bar{c})$ system, it is difficult to reach from the initial ψ' state, so that the present experiments would not be expected to find it. Positive parity states for the systems $(c\bar{u})$, $(c\bar{d})$, and $(c\bar{s})$ are also difficult to produce and there is no knowledge of them yet.

In the narrower realm of (u,d,s) quarks, many resonances with positive parity have become established, requiring the existence of a nonet for each of these four P configurations (Table 4). The $(2++)$ nonet is well established, and the $(1++)$ nonet has also become established, after long controversy, while the $1(+-)$ nonet is not yet complete.

The $(0++)$ states offer serious problems. The states $\epsilon(1300)$ and $K^*(1500)$ can plausibly belong to a common nonet; similarly the states $\delta(981)$ and $S^*(980)$ can plausibly belong to a common nonet.

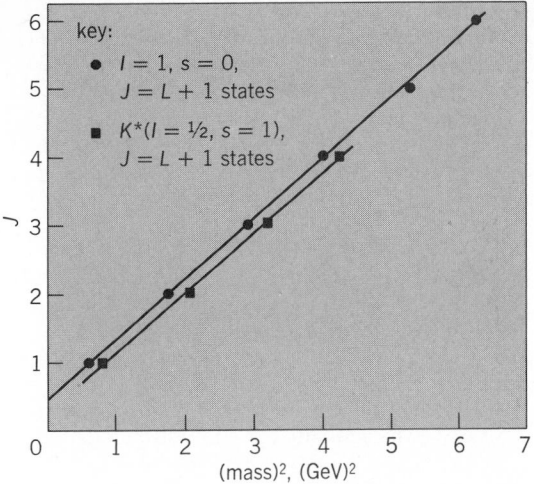

Fig. 6.   Spin J plotted against (mass)^2 for the $I=1$, $s=0$ mesons and the $I=1/2$, $s=1$ mesons with $J=L+1$, fitted by straight lines.

However, all four of these states can scarcely belong to the same nonet. For example, $K\bar{K}$ is the dominant decay mode for $S^*(980)$ while $\pi\pi$ is dominant for $\epsilon(1300)$, just opposite to expectation for a nonet.

Higher excited mesonic states. Rotationally excited nonets are expected to exist, with mass increasing with increasing L. No nonets with $L \geq 2$ are completely established yet, but the $(I=1, s=0)$ state with $J=L+1$ is known for all $L \leq 6$. The value (mass)^2 has been plotted against J on Fig. 6, and is well fitted by a linear dependence, passing through 0.5 at (mass)$^2 = 0$. This line is known as a Regge trajectory—a trajectory of rotationally excited states based on the ρ-meson as its lowest member—hence the ρ-trajectory.

The K^* member of the $J = L + 1$ nonets is known for $L \leq 3$, and their (mass)^2 values are also plotted against J on Fig. 6. They are well fitted by a linear K^* trajectory, parallel with the ρ trajectory. *See* REGGE POLE.

Radial excitations of these nonets also occur, denoted by $n^{(2S+1)}L_J$, where $n = 1$ denotes the lowest state. These may be sought most easily for the $JPC = (1--)$ nonets, since the e^+e^- annihilations excite selectively those states with the same quantum numbers as the photon. The states involving only (u,d,s) quarks will be relatively broad, since the decay channels available for the $n = 1$ states will necessarily be equally available for the radially excited $n \geq 2$ states. Three candidates for the $n = 2$ vector meson nonet, $\rho'(1600)$, $K^{*\prime}(1650)$ and $\phi'(1670)$, have been identified.

Other excitations. Higher-mass mesonic states can also result from the excitation of further $q\bar{q}$ pairs, or of the gluonic degrees of freedom. The simplest excitation is that offered by the four-quark configuration $(\bar{q}^2 q^2)$. Even with all internal orbital angular momenta limited to zero values, and considering only (u,d,s) quarks, this configuration would predict the existence of a very large number of positive parity mesonic multiplets, some of which are predicted to lie quite low in mass. It has been suggested, for example, that the $\delta(980)$ and $S^*(980)$ $(0++)$ states which appeared so anomalous above may be states of this type. Con-

Table 4. "Old" mesonic states below about 1800 MeV assigned to $(q\bar{q})$ messon nonets on the basis of their (JPC) values†

	(JPC)	(1,0)	(0,0)	(0,0)	(½,1)
			(I,s)		
$L=0$	(0−+)	$\pi(137)$	$\eta(549)$	$\eta'(958)$	$K(495)$
	(1−−)	$\rho(776)$	$\omega(782)$	$\phi(1020)$	$K^*(892)$
$L=1$	(0++)	$\delta(981)?$	$\epsilon(1300)$	$S^*(980)?$	$K^*(1500)$
	(1++)	$A_1(1260)$	$D(1285)$	$E(1425)$	$Q_1(1270)$
	(1+−)	$B(1231)$	$H(1190)$	?	$Q_2(1400)$
	(2++)	$A_2(1317)$	$f(1273)$	$f'(1516)$	$K^*(1434)$
$L=0^*$	(1−−)	$\rho'(1600)$	?	$\phi'(1670)$	$K^{*\prime}(1650)$
$L=2$	(3−−)	$g(1700)$	$\omega(1666)$	?	$K^*(1785)$
	(2−+)	$A_3(1670)$	?	?	$K^*(1820)?$

†A question mark indicates either that the assignment given is controversial or that no meson has been found to have the required quantum numbers. $L=0^*$ means the first radially excited $L=0$ state. The lightest charmed meson $D(1866)$ lies just above this mass range.

figurations of the type $\bar{q}^2 q^2$, or more complicated configurations, have also been suggested for certain resonances that appear to favor baryon-antibaryon final states in their decay. *See* ELEMENTARY PARTICLE.   [RICHARD H. DALITZ]

Bibliography: K. Berkelman, New flavour spectroscopy, in L. Durand and L. G. Pondrom (eds.), *Proceedings of the 20th International Conference on High Energy Physics*, Amer. Inst. Phys. Conf. Publ. no. 68, pp. 1499–1529, 1981; S. L. Glashow, Quarks with color and flavor, *Sci. Amer.*, 233(4): 38–50, 1975; A. Hendry and D. B. Lichtenberg, The quark model, *Rep. Progr. Phys.*, 41: 1707–1780, 1980; R. L. Kelly et al., Review of particle properties, *Rev. Mod. Phys.*, 52:S1–286, 1980; L. Montanet, Light quark-hadron spectroscopy, in L. Durand and L. G. Pondrom (eds.), *Proceedings of the 20th International Conference on High Energy Physics*, Amer. Inst. Phys. Conf. Publ. no. 68, pp. 1196–1233, 1981; G. t'Hooft, Gauge theories of the forces between elementary particles, *Sci. Amer.*, 242(6):90–116, 1980; S. Weinberg, Unified theories of elementary-particle interaction, *Sci. Amer.* 231(1):50–59, 1974.

Metastable state

In quantum mechanics, an excited stationary energy state whose lifetime is unusually long. *See* EXCITED STATE; STATIONARY STATE.

The stationary states of atoms and nuclei, or indeed of any material system, ordinarily are computed ignoring the interactions between matter and the electromagnetic radiation field. When these interactions are included, a system in an excited state always has some probability of emitting radiation in the form of one or more photons, and thereby decaying to a lower or less excited state. For a metastable state, the probability of making such a transition is unusually small. Most atomic excited states decay with the emission of a single photon, in a time of the order of 10^{-8} sec. However, the necessity for angular momentum and parity conservation forces the second excited state $(2S_{1/2})$ of atomic hydrogen to decay by simultaneous emission of two photons; consequently, the lifetime is increased to an estimated value of 0.15 sec. Similarly, emission of a γ-ray photon by an excited nucleus usually occurs in 10^{-13} sec or less; however, the lifetime of one excited state of the In^{113} nucleus is about 100 min. Because the transition probabilities decrease rapidly with decreasing frequency, a low-lying excited state may have a lifetime longer than usual, and yet not be truly metastable; that is, emission from the state may not be hindered by any general requirement or selection rule, such as is invoked in the case of hydrogen. *See* NONRELATIVISTIC QUANTUM THEORY; NUCLEAR ISOMERISM.
    [EDWARD GERJUOY]

Microscopic quantization effects

The discrete values of the properties of matter and radiation observed at small distances; the prediction of such properties by quantum mechanics. Besides those properties well described by quantum mechanics (such as the discrete energy states and angular momenta of electromagnetic radiation, molecules, atoms, and nuclei, and the quantized flow patterns of superfluids) there are other particular features of matter in the small that are often regarded as microscopic quantization effects

in expectation of their future explanation in terms of a more advanced quantum mechanics. These include the existence, masses, charges, and intrinsic angular momenta of the elementary particles. The quantized units of properties predicted by quantum mechanics are all multiplied by Planck's constant $\hbar$, which is so small ($6.63 \cdot 10^{-27}$ erg-sec) that microscopic quantization effects cannot be observed without the aid of instruments. Nevertheless, cooperative behavior of a very large number of particles may occur in such effects as in the case of quantized superfluid vortices. *See* ELEMENTARY PARTICLE; QUANTIZATION; QUANTIZED VORTICES; QUANTUM MECHANICS; QUANTUM NUMBERS.

    [LAURENCE J. CAMPBELL]

Bibliography: G. Gamow, *Mr. Tompkins in Paperback*, 1967; D. A. Park, *Introduction to the Quantum Theory*, 2d ed., 1974.

Minimal principles

In the treatment of physical phenomena, it can sometimes be shown that, of all the processes or conditions which might occur, the ones actually occurring are those for which some characteristic physical quantity assumes a minimum value. These processes or conditions are known as minimal principles. The application of minimal principles provides a powerful method of attacking certain problems that would otherwise prove formidable if approached directly from first principles.

One simple minimal principle asserts that the state of stable equilibrium of any mechanical system is the state for which the potential energy is a minimum. Other general theorems of classical dynamics that are related to minimal principles are Hamilton's principle and the principle of least action. *See* HAMILTON'S PRINCIPLE; LEAST-ACTION PRINCIPLE.

Minimal principles are important in branches of physics other than mechanics. Fermat's principle in optics, for example, states that, of all possible paths of light transmission between two points, the actual path is the path for which the transmission time is a minimum. Minimal principles also find wide application in thermodynamics.

    [DUDLEY WILLIAMS]

Bibliography: E. V. Condon and H. Odishaw, *Handbook of Physics*, 1967; H. Goldstein, *Classical Mechanics*, 2d ed., 1980; W. A. Thirring, *A Course in Mathematical Physics*, 2 vols., 1978, 1980.

Mirror optics

The use of plane or curved reflecting surfaces for the purpose of reverting, directing, or forming images. The most familiar use of reflecting optical surfaces is for the examination of one's own reflected image in a flat or plane mirror. A single reflection in a flat mirror produces a virtual image which is reverted or reversed in appearance. The use of one or more reflecting surfaces permits light or images to be directed around obstacles, with each successive reflection producing a reversal of the image. A curved mirror, either spherical or conic in form, will produce a real or virtual image in much the same manner as a lens, but generally with reduced aberrations. There will be no chromatic aberrations since the law of reflection is independent of the color or wavelength of the incident light. *See* ABERRATION (OPTICS); OPTICAL IMAGE.

An optical surface which specularly reflects the largest fraction of the incident light is called a reflecting surface. Such surfaces are commonly fabricated by polishing of glass, metal, or plastic substrates, and then coating the surface of the substrate with a thin layer of metal, which may be covered in addition by a single or multiple layers of thin dielectric films. The law of reflection states that the incident and reflected rays will lie in the plane containing the local normal to the reflecting surface and that the angle of the reflected ray from the normal will be equal to the angle of the incident ray from the normal. This law is a special case of the law of refraction in that the angles rather than the sines of the angles of incidence and reflection are equal. Formally, this relation is commonly used in calculations by setting the effective index of refraction after reflection equal to the negative of the index of refraction prior to incidence on the surface. When this concept is introduced, all of the formulas relating to lenses are applicable to reflective optics. In this article, however, the imaging relations will be described in the most appropriate form for reflecting surfaces. *See* GEOMETRICAL OPTICS.

Plane mirrors. The formation of images in the plane mirrors is easily understood by applying the law of reflection. Figure 1 illustrates the formation of the image of a point formed by a plane mirror. Each of the reflected rays appears to come from a point image located a distance behind the mirror equal to the distance of the object point in front of the mirror. In Fig. 1, the face of the observer can be considered as a set of points, each of which is imaged by the plane mirror. Since the observer is viewing the facial image from the object side of the mirror, the face will appear to be reversed left for right in the virtual image formed by the mirror. Such a virtual image cannot, of course, be projected on a screen, but can be viewed by a lens, in this case the eyes of the observer. Figure 1 also indicates the redirection of light by a plane mirror, in that a viewer who cannot observe the object point directly can observe the virtual image of the point formed by the mirror. A simple optical de-

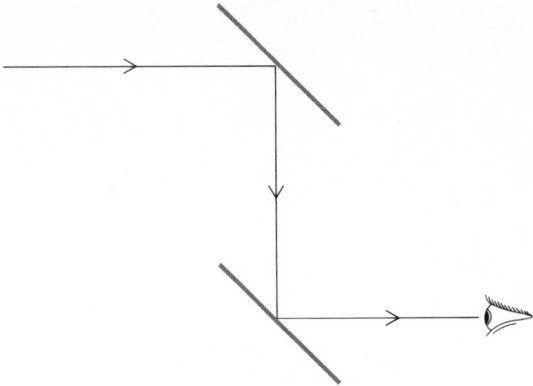

Fig. 2. Simple mirror periscope.

vice which is based on this principle is the simple mirror periscope (Fig. 2), which uses two mirrors to permit viewing of scenes around an obstacle. In this case two reflections are present and provide an image which is correctly oriented, and not reverted, to the observer. The property of reversion in a complicated mirror system depends upon the location and view direction of the observer, as well as the number of reflections that take place and the orientation of the planes through which the light is directed.

Prisms. These are solid-glass optical components that use reflection at the faces to provide redirection of the optical pencils passing through them. The advantage of the use of a prism is that the reflecting surfaces are maintained in accurate location with respect to each other by the integrity of the glass material making up the body of the prism. Difficulties with prisms are that very homogeneous glass is required since the light may make many passes through the prism and that a prism is optically equivalent to insertion of a long block of glass into the imaging system. The insertion of such a glass block often results in a system which is mechanically shorter in space, but the aberration balance of the imaging system is changed, frequently requiring a redesign of the associated optical components in order to accommodate the increased glass path. It is obvious that the use of solid glass prisms may introduce more weight into the optical design than the equivalent metal mounts required for a similar arrangement of mirrors in air. In certain cases, the angle of incidence on a reflecting surface within the prism may exceed the critical angle of incidence, and no reflective coating may be required on such a surface. Figure 3 shows some common types of optical prisms with reflecting surfaces. The applications of such prisms range from simple redirection of light to variable angle of rotation of the image passed through the prism and binocular combination of images. A special case of reflection is the use of a beam splitter which permits the splitting of light, or the combining of two beams by the use of a surface which is partially reflecting and partially transmitting. *See* OPTICAL PRISM.

Spherical mirrors. These are reflecting components that are used in forming images. The optics of such mirrors are almost identical to the properties of lenses, with their ability to form real and virtual images. In the case of spherical mirrors, there is a reversal of the direction of light at the mirror

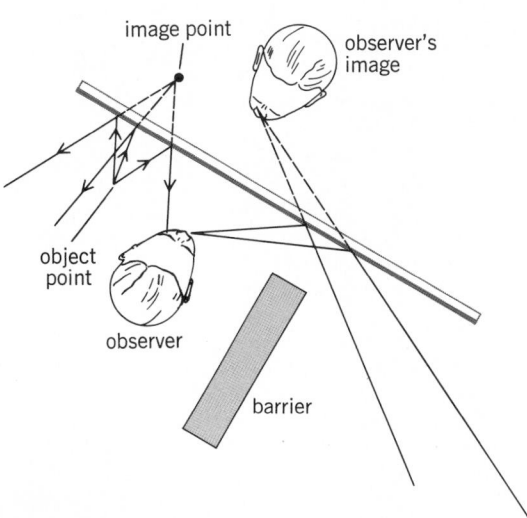

Fig. 1. Formation of images by a plane mirror.

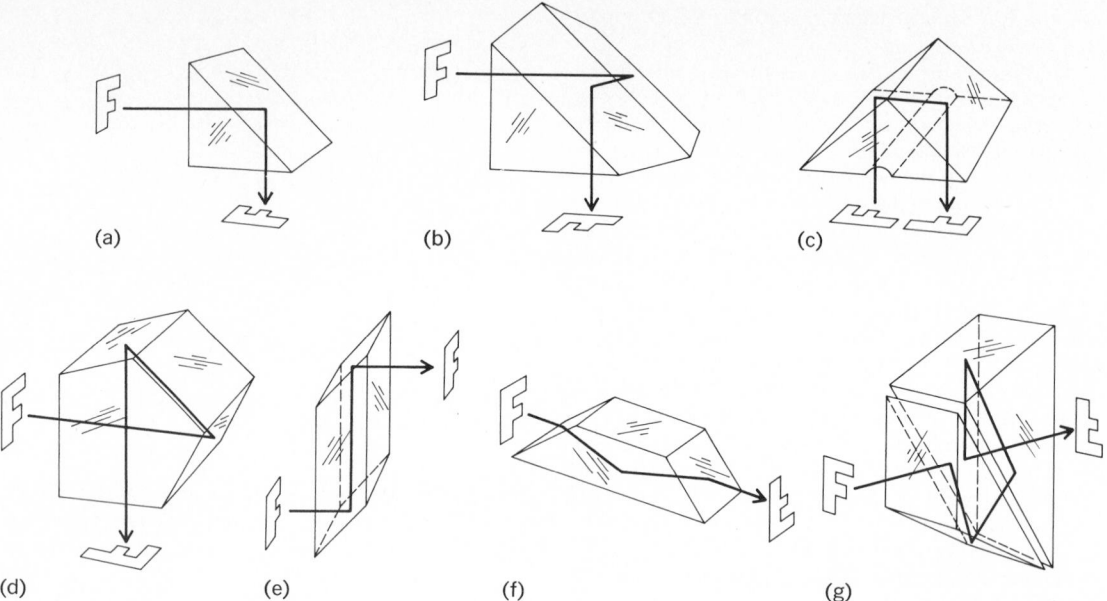

Fig. 3.   Prism types. (*a*) Right-angle. (*b*) Amici roof. (*c*) Porro. (*d*) Pentaprism. (*e*) Rhomboid. (*f*) Dove. (*g*) Pechan.

so that real images are formed on the same side of the mirror as the object, while virtual images are viewed from the object side but appear to exist on the opposite side of the mirror. Both concave and convex spherical mirrors are commonly encountered. Only a virtual erect image of a real object will be formed by a convex mirror. Such mirrors are commonly used as wide-angle rearview mirrors in automobiles or on trucks. The image formed appears behind the mirror and is greatly compressed in space, with a demagnification dependent on the curvature of the mirror. A concave spherical mirror can form either real or virtual images. The virtual image will appear to the observer as erect and magnified. A common application is the magnifying shaving mirror frequently found in bathrooms. A real image will be inverted, as is the real image formed by a lens, and will actually appear in space between the observer and the mirror.

Figure 4 shows the formation of real and virtual images by a spherical mirror. The equation which applies to all of the image relations is given below.

$$\frac{1}{S'}+\frac{1}{S}=\frac{2}{R}$$

The distances S and S' are measured from the surface of the spherical mirror; when either is negative, a virtual image is formed behind the mirror. The constant R is the radius of curvature of the mirror. The magnification of the image is the ratio of the image distance S' to the object distance S.

Conic mirrors. These are a special case of the spherical mirror with improved image quality. A spherical mirror will form an image which is not perfect, except for particular conjugate distances. The use of a mirror which has the shape of a rotated conic section, such as a parabola, ellipsoid, or hyperboloid, will form a perfect image for a particular set of object-image conjugate distances and will have reduced aberrations for some range of conjugate relations. Two of the most familiar applications for conic mirrors are shown in Fig. 5. Figure 5*a* shows the use of a paraboloid of revolution about the optical axis to form the image of an object at an infinite distance. In this drawing the image to be viewed by the observer at the eyepiece is relayed to the side of the telescope tube by a flat folding mirror in what is called a newtonian form of a telescope. This demonstrates one of the difficulties that is found with the use of reflecting optical components to form real images; namely, that the image must often be relayed out of the incident path on the image-forming mirror, otherwise the

Fig. 4.   Formation of images by a spherical mirror. (*a*) Real image. (*b*) Virtual image.

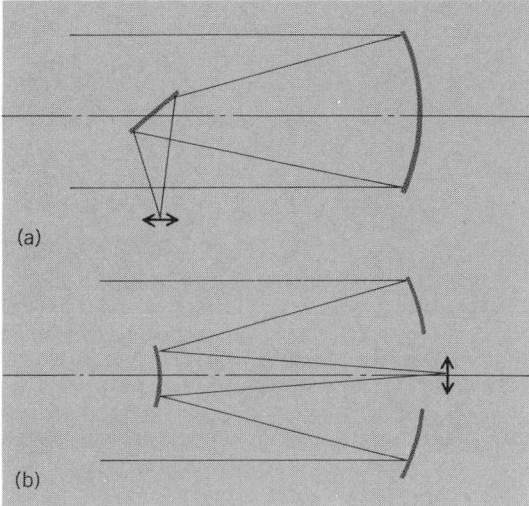

Fig. 5. Applications of conic mirrors. (*a*) Newtonian telescope. (*b*) Cassegrain telescope.

observer will block some of the light from the object. Not all reflecting systems carry out this relaying in the same manner. The Cassegrain system uses a curved secondary mirror to achieve magnification of the final image while allowing the image to fall outside the telescope barrel.

Figure 5*b* shows the use of a paraboloid as the primary mirror with the image relayed to the final image location by a hyperbolic secondary mirror. Such a use of two mirrors permits the construction of a long-focal-length telescope within a relatively short space. This latter form, usually referred to as a Cassegrain telescope, serves as the principal type of modern reflecting astronomical telescope. One of the advantages of this type of telescope design is the freedom from chromatic aberration that would be present in a refracting telescope. *See* OPTICAL SURFACES.

Mirror coatings. The reflectivity of a mirror depends on the material used for coating the reflecting surface. The conventional coatings for glass mirror surfaces are silver or aluminum, which are vacuum-deposited or sputtered onto the surface. In some cases, chemical deposition will be used. Most mirrors intended for noncritical uses, such as looking glasses or wall mirrors, will have the reflecting metallic coating placed on the back side of the glass, thus using the glass to protect the coating from oxidation by the atmosphere. Mirrors for most critical or scientific uses require the use of front-surface reflectors, with the reflecting coating on the exposed front surface of the glass. In this case, a hard overcoat of a thin layer of silicon dioxide is frequently deposited over the metal to protect the delicate thin metal surface. The reflectivity of the mirror with respect to wavelength depends on the choice of the metal for the reflector and the material and thickness of material layers in the overcoat. In some cases, a fully dielectric stack will be used as a reflecting coating with special spectral selective properties to form a dichroic beam splitter, as in a color television camera, or as an infrared-transmitting "cold mirror" for a movie projector illumination system. *See* REFLECTION OF ELECTROMAGNETIC RADIATION.

[R. R. SHANNON]

Mode of vibration

A characteristic manner in which vibration occurs. In a freely vibrating system, oscillation is restricted to certain characteristic patterns of motion at certain characteristic frequencies; these motions are called normal modes of vibration.

An ideal string, for example, can vibrate as a whole with a characteristic frequency defined in the equation below, where L is the length of string

$$f = (1/2L) \sqrt{T/m}$$

between rigid supports, T the tension, and m the mass per unit length of the string. The displacements of different parts of the string are governed by a characteristic shape function. More specifically, the motion of any part of the string is proportional to $\sin (\pi x/L) \sin (2\pi ft)$, where x is the distance of the part of the string from a fixed end and t is the time. This simplest kind of vibration is the first, or fundamental, mode of vibration of the string; its frequency is the fundamental frequency. All parts of the string vibrate with the same frequency, and move outward from equilibrium at the same time.

The string is also capable of vibrating in two segments, one of which goes outward from equilibrium in a positive direction at the same time as the other part goes outward in a negative direction, and conversely. Again, the motion of any part of the string can be described by the product of a space function by a sinusoidal function of time: $\sin (2\pi x/L) \sin (2 \times 2\pi ft)$. All parts of the string move together as a sinusoidal function of the time and at the same frequency; the space function governs the motion in opposite directions. The frequency of the second mode of vibration is twice that of the first mode. Similarly, modes of higher order have frequencies that are integral multiples of the fundamental frequency.

Because the frequencies are in the ratios 1:2:3 . . . , the modes of vibration of an ideal string are properly called harmonics. Not all vibrating bodies have harmonic modes of vibration, however. The ideal drumhead, for example, vibrates freely with frequencies in the ratios 1:1.59:2.14:2.30 . . . . In fact, most real systems vibrating freely have modes of vibration whose frequencies are not exactly in the ratios of integers. *See* VIBRATION.   [ROBERT W. YOUNG]

Mole

One mole is the mass numerically equal (in grams) to the relative molecular mass of a substance. It is the amount of a substance that contains the same number of molecules as there are atoms in 0.012 kilogram of carbon-12. The mole is an individual unit of mass, that is, it relates only to a given substance. If the relative molecular mass is μ, 1 mole $= \mu$ grams, and the M of one mole is μ g/mol.

Moles of different elements or compounds contain the same number of molecules. This number, called the Avogadro number N_0, is 6.023×10^{23}.

Since they contain the same number of molecules, moles of ideal gases occupy the same volume at the same temperature and pressure. At 0°C and 1 atmosphere (101,325 Pa) pressure, this volume, called the molar volume, is 22.4 liters. *See* AVOGADRO NUMBER; GAS; RELATIVE MOLECULAR MASS.   [THOMAS C. WADDINGTON]

Molecular beams

Utilization of well-directed streams of atoms or molecules in vacuum. This is a cornerstone technique in the investigation of molecular structure and interactions. Molecular beams are usually formed at sufficiently low particle density for the interaction of one beam molecule with another to be negligible. This ensemble of truly isolated molecules is available for the spectroscopic study of molecular energy levels using photon probes from the radio-frequency to optical portions of the electromagnetic spectrum. Some of the best-determined fundamental knowledge of physics comes from spectroscopic molecular-beam experiments. Beyond this, beams can be applied as probes of the multifaceted nature of gases, plasmas, surfaces, and even the structure of solids. An application intermediate in complexity is the study of molecular interactions by means of two colliding beams, where one might be a beam of charged particles such as ions or electrons. This subfield of physics, a part of atomic collisions research, is increasingly applied to basic problems underlying chemical and biological processes. It is also of value in understanding the component molecular reactions determining the properties of plasma and electric discharge devices, the nature of the upper atmosphere, and some aspects of the cooler astrophysical regions. *See* SCATTERING EXPERIMENTS (ATOMS AND MOLECULES).

Production and detection. One simple means of forming a beam is to permit gas from an enclosed chamber to escape through a small orifice into a second chamber maintained at high vacuum by means of large pumps. Figure 1*a* shows such effusion into a collimating chamber from an oven chamber, generally heated to control the vapor pressure of the gas. The molecules coming from the orifice are distributed in angle according to a cosine law, illustrated by the circle downstream of the orifice which represents the envelope of relative beam flux vectors. A useful number of molecules passes forward along the horizontal axis of the apparatus. A well-collimated beam is then formed by requiring that those molecules entering the test chamber where an experiment is to be performed pass not only through the orifice but also through a second small hole separating the collimating and test chambers. A property of beams formed this way is that the velocities of the individual molecules have a large thermodynamic spread in values centered on a mean value of order 10^3 m/s determined by the oven temperature.

Charge-exchange system. If higher velocities are desired, then a charge-exchange beam system can be used. In this scheme, ions are produced by some ionizing process such as electron impact on atoms within a gas discharge. Since the ions are electrically charged, they can be accelerated to the desired velocity and focused into a beam using electric or magnetic fields. The last step in neutrally charged beam formation is to pass the ions through a neutralizing gas where electrons from the gas molecules are transferred to the beam ions in charge-exchange molecular collisions. If the acceleration voltage is relatively high, then the ion-beam velocity will be 10^5 m/s or greater upon entering the neutralizer. In many cases, such energetic charge-exchange collisions produce beam atoms or molecules in internally excited energy levels rather than just in the ground-state level of lowest energy occupied by atoms in low-temperature gases. On the other hand, useful beams of internally excited molecules can be formed from ground-state molecules which have been excited by energetic electron- or ion-beam impact or by photon absorption from a laser or other light beam. *See* EXCITED STATE; GROUND STATE.

Secondary electron detection. The faster types of neutral molecular-beam particles are easy to detect by secondary electron ejection from a solid. A collision of a beam molecule with a surface is sufficiently violent for one or more electrons to be ejected. For intense beams, the rate of electron production is so high that the electrons can be collected and measured electronically as an electric current. At lower rates, the effect of each electron can be multiplied by means of an alternating sequence of electron acceleration and surface ejection steps, to produce a burst of 10^6 or more electrons from a single beam molecule. This pulse of current is adequate for electronic pulse counting, leading to a very sensitive overall beam-detection technique useful at beam intensities as low as five molecules per minute.

Other detection techniques. For molecular beams effusing from an oven, a variety of detection techniques has been devised. Alkali atoms have such small ionization potentials that their valence electrons can be transferred to a heated metal surface having a large work function, such as tungsten. The resultant ions can then be detected by current measurement or particle multiplication techniques, depending again upon the beam intensity. A second special technique useful for some beams of reactive or excited molecules is electron ejection from a surface, powered by the internal energy of these special beam molecules. A universal detector for any kind of slow molecule employs initial conversion into ions by electron- or light-beam impact; the ions are then accelerated into a fast beam for easy detection.

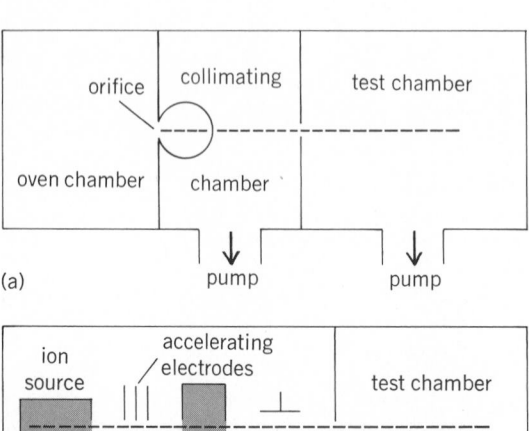

Fig. 1. Schematic diagrams of systems for producing molecular beams. (*a*) Conventional oven-beam system. (*b*) Charge-exchange beam system.

Special beams. Occasionally a beam containing atoms or molecules in a specific quantum-mechanical state is needed. Energy-resonant transitions between the ground state and the specific excited state can be utilized, often induced by single-frequency laser radiation. Beams of slow molecules possessing magnetic (or electric) dipole moments can be selected according to the direction of orientation of their moments; spatial separation into component beams is achieved through the orientation-dependent interaction between the dipole moment and an externally applied strong nonuniform magnetic (or electric) field. Some precision spectroscopic molecular-beam experiments use such beams, and form the basis for atomic-clock precision time standards.

Molecular-beam spectroscopy. Much of molecular spectroscopy involves the absorption or emission of light by molecules in a gas sample. The frequency of the light photon is proportional to the separation of molecular energy levels involved in the spectroscopic transition. However, the molecule density in typical gas samples is so high that the energy levels are slightly altered by collisions between molecules, with the transition frequency no longer characteristic of the free molecule. The use of low-density molecular beams with their sensitive detection techniques can reduce this collision alteration problem, with the result that atomic properties can be measured to accuracies of parts per million or even better. If the very simplest atoms or molecules are employed, then the basic electromagnetic interactions holding the component electrons and nuclei together can be precisely studied. This is of great importance to fundamental physics, since theoretical understanding of electromagnetic interactions through quantum electrodynamics represents the most successful application of quantum field theory to elementary particle physics problems. *See* QUANTUM ELECTRODYNAMICS; QUANTUM FIELD THEORY.

Properties measured. Among the basic quantities of physics measured by molecular beams are: the fine-structure constant $\alpha = e^2/\hbar c = 1/137.0361$ (e is the electron charge in electrostatic units, $\hbar$ is Planck's constant divided by 2π, and c is the speed of light), obtained from microwave spectroscopy studies of the fine and hyperfine energy-level splittings in one- and two-electron atoms; the value 1.5210326×10^{-3} of the magnetic moment of the proton in Bohr magnetons, obtained with the hydrogen maser; the purely quantum-electrody-

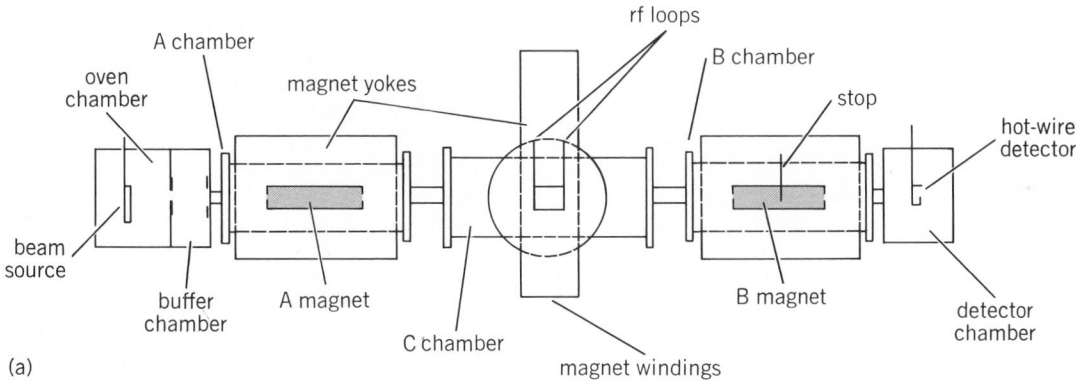

(a)

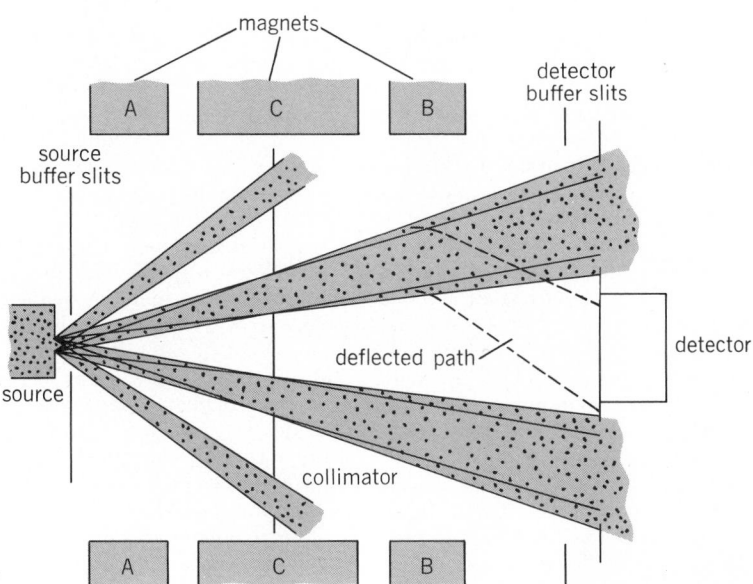

(b)

Fig. 2. Molecular-beam magnetic resonance spectroscopy experiment. (*a*) Apparatus used in the experiment. (*b*) Spatially resolved multiple-beam configuration showing typical molecular trajectories.

namic shifts of atomic energy levels called Lamb shifts; the nuclear magnetic dipole moments of several hundred isotopes; the equality in magnitude of the unit electron and nuclear charges to an accuracy of better than parts in 10^{18}; and the absence of intrinsic electric dipole moments for the electron and proton, which is a test of parity and time reversal as symmetry properties obeyed by the electromagnetic interaction. The isolated, unperturbed nature of atoms in a beam is needed for the device used as the length standard, based upon the 605.7-nm wavelength of krypton atoms being reproducible to parts in 10^9. The time standard is the cesium-beam atomic clock operating on the ground-state hyperfine splitting microwave transition with a reproducibility over hours of parts in 10^{12}. *See* ATOMIC CONSTANTS.

Apparatus. The molecular-beam magnetic resonance apparatus used in a number of the spectroscopy experiments is shown in Fig. 2, along with a diagram showing typical molecular trajectories. On the left end of Fig. 2a is an oven beam source, here of alkali atoms; on the right end is a hot-wire surface ionization detector of these atoms. Between the ends are three separate regions where external fields are applied to the beam. The A region is a state selector, and the B region a state analyzer; that is, some property of the beam atoms is well defined in the A region, and then this situation is checked in the B region. In the example of Fig. 2 the selected property is the direction of the atoms' magnetic moment. The center of the apparatus is the transition region marked C, where a resonant radio-frequency transition can be induced between atomic energy levels associated with different directions for the magnetic moment. This alteration of the atoms entering the analyzing region results in a change in atom trajectories, with a corresponding change in the intensity of the beam striking the hot-wire detector. For example, in Fig. 2b, molecules undergoing a transition within the C region follow the deflected path and are detected. When the frequency of the radio waves in the C region is not resonant, such a change does not occur. Determining the frequency for maximum effect results in a measure of the atomic energy level splitting, which for the example of Fig. 2 is directly related by theory to the atom's nuclear magnetic moment.

The power of molecular-beam techniques derives, in large measure, from the ability, in principle, of the A and B regions to contain any kind of state-selective device utilizing combinations of static or time-varying electromagnetic fields. Also, the interaction region C can involve resonant interactions with light or other photons or even nonresonant-state destructive interactions such as collisions with atoms in an introduced gas. If the A and B regions do not control beam atom trajectories, but instead some other property such as the state of internal energy, then correspondingly the nature of the beam detector might need to be sensitive to the controlled property.

Use of fast beams. The fast molecular beams produced by charge exchange increasingly have been employed in spectroscopy experiments and offer some advantages in addition to easy detection. Among these is the fractionally well-defined and controllable molecule velocity determined by the ion acceleration voltage, a feature useful in the study of time-dependent quantum-mechanical interference effects on transition rates. The ability to transport rapidly decaying excited atoms through an apparatus is also enhanced using fast beams. A highly accurate atomic fine-structure measurement employed a fast hydrogen atom beam, as did an experiment on multiphoton microwave transitions between highly excited atomic states.

Scattering experiments. The molecular-beam study of interactions occurring during individual collisions of two species of electrons, ions, atoms, or molecules can be divided into two categories, depending upon whether the targets for the particles in one beam are those in a gas or those in another beam. In addition, the nature of the collisional interaction depends strongly on the velocity of impact of the two particles; if this is very low, all that can happen is that the two particles elastically scatter off each other, with no change in the internal nature of either particle being energetically possible. At higher collision velocities, inelastic processes can occur in which a part of the energy of collision is converted into a change in internal energy of one or both particles. Particle rearrangement can also occur, such as the electron transfer used previously in charge-exchange atomic-beam sources, or proton transfer in chemical reactions between molecules. Collisions involving fast electrons or fast bare nuclei are primarily probes of the structure of the target molecule. The slower collision processes are more concerned with the composite molecular system transiently formed during the collision time. An example is the negative ion state formed when an electron joins a neutral atom or molecule. Such compound states cannot live indefinitely since the total energy of the system is positive; in a rearrangement collision, however, the compound state breaks up into molecules different from the originally colliding ones.

Quantities measured. The simplest quantity to measure in a molecular-beam collision experiment is the probability for a particular elastic or inelastic event to result from a single collision of one beam particle with one target particle. This probability is usually expressed in terms of an apparent size of the target particle for the process, called a total cross section. This quantity is a sum over contributions from all possible distances of closest approach of the two particles, since molecules are so microscopic that individual ones cannot be aimed at other individual ones in any controlled manner. One must work with collections of beam particles and target particles, and divide out their numbers to obtain a cross section. However, distances of approach are often uniquely correlated with definite angles of scattering of beam particles relative to the incident beam direction. Thus, the necessary averaging of the distances of approach contained in a total cross section can be avoided if one instead measures angular distributions of scattered beam particles. Different scattering angles are associated with beam particles probing the strength of the collisional interaction at different particle separations, and so angular distribution studies investigate the detailed nature of molecular interactions in a way similar to angular scattering experiments in nuclear and high-energy physics.

Electron-beam scattering. The apparatus for a comparatively sophisticated electron-beam angu-

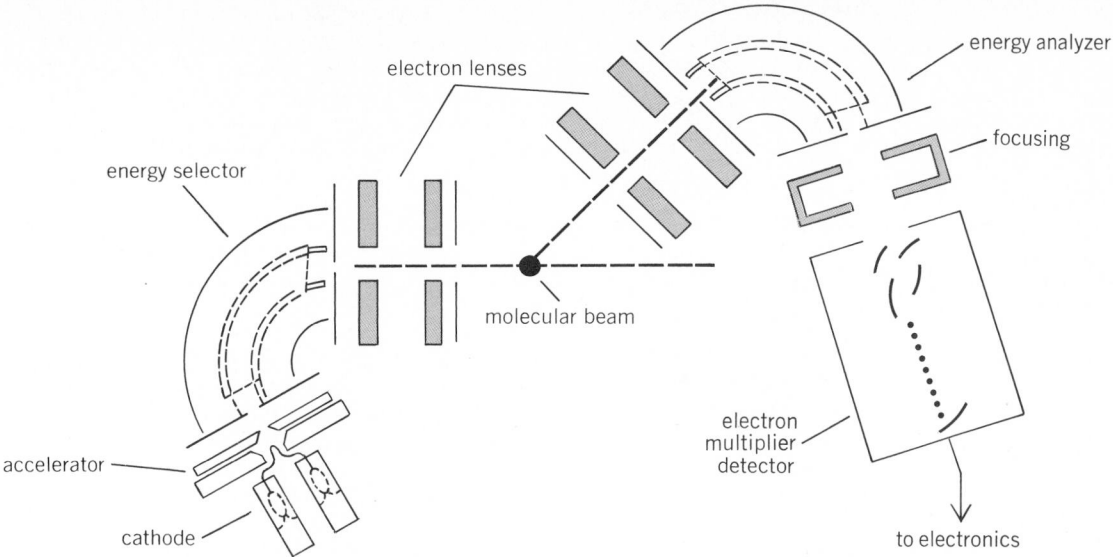

Fig. 3. Apparatus for electron-beam angular scattering study.

lar-scattering study is indicated schematically in Fig. 3. The target for the electrons is a molecular beam such as previously discussed, and is shown here as a dot in the center of the experiment. The molecular beam is coming out of the paper. No collision-induced changes in the properties of that beam are detected, although this actually can be done in coincidence with observation of electrons scattered from the electron beam.

The electrons are produced by thermionic surface emission from a hot cathode, and have a Maxwellian spread in energies of order 0.1 eV. They can be electrostatically accelerated if desired, and then energy-selected with a cylindrical electrostatic analyzer consisting of two 60° electric field plates on arcs of circles about a common origin. The resolution of such analyzers can be as good as 0.01 eV, a value sometimes needed to observe sharp resonance peaks in the energy dependence of cross sections that are associated with the excitation of long-lived compound negative-ion states. After focusing, the energy-selected

electron beam is scattered by the molecular beam. Electrons scattered at various angles are selected by rotation of the remainder of the apparatus about the molecular-beam axis. Scattered electron focusing is followed by energy analysis in a second cylindrical analyzer and then by acceleration into an electron multiplier for detection by pulse counting. Thus, the experiment is divided into source, state-selector, interaction, state-analyzer, and beam-detector regions just as in the molecular-beam magnetic resonance experiment.

Figure 4 shows the collision energy dependence observed at various scattering angles for elastic electron scattering by beams of He and N_2 molecules. The change with angle in the shape of the He^- resonance near 19.3 eV makes possible a positive identification of the orbital angular momentum of the compound state. The resonances in the N_2 scattering curves are associated with the excitation of different amounts of internuclear vibration in the N_2^- ion. Searches for new states of molecules can be made by using electron scattering by ion beams. Detailed studies of resonances in slow electron scattering are used to investigate with considerable success the nature of energy exchange between the various bonds of molecules as large as benzene.

Double-molecular-beam scattering. Figure 5 schematically shows a typical double-molecular-beam scattering apparatus used in studies of chemical reactions at the molecular level. Elastic scattering can also be studied to obtain data sensitive to the molecule-molecule interaction potential. The beams from the primary and secondary ovens collide in an interaction region. The intensity of the primary beam is monitored by an on-axis detector, and the scattered secondary-beam molecules or chemical reaction products are measured with a second detector. Shown in the primary beam line before the interaction region are two devices commonly used in such experiments: a beam chopper for phase-sensitive beam detection and a velocity selector. The latter is a set of disks rotating together, each with beam-transmission holes placed at different angles about the axis of

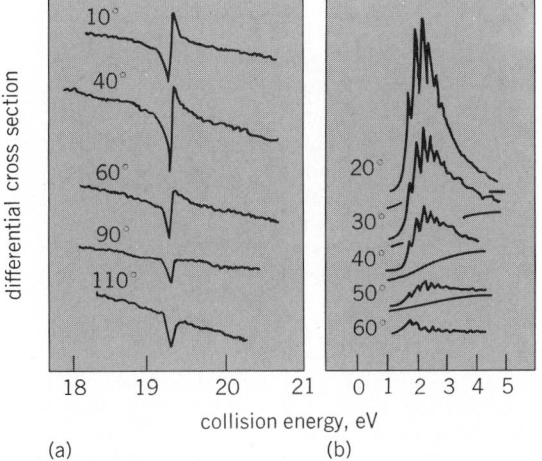

Fig. 4. Resonances observed at various scattering angles for elastic electron scattering by beams of (a) He and (b) N_2 molecules.

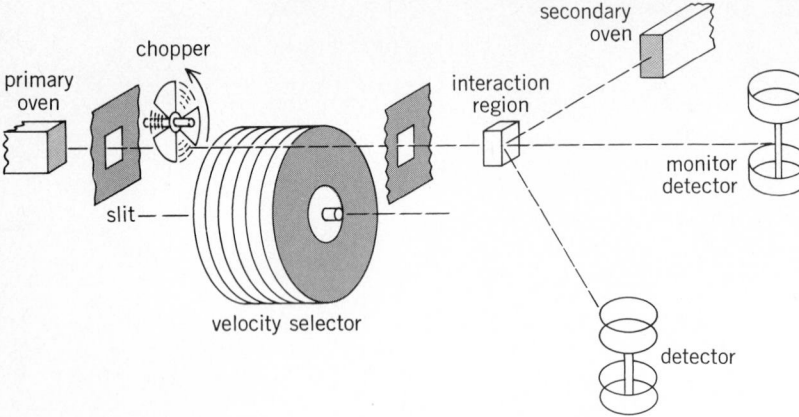

Fig. 5. A double-molecular-beam experiment on chemical reactions.

rotation. Only those beam molecules having a certain velocity pass through all the holes as the disks rotate. The beam chopper breaks the beam into pulses with known timing, and detected scattered particles are electronically checked for the same timing pattern. Thus beam-beam scattering can be detected in the presence of a poor vacuum producing very large amounts of beam scattering in the background gas.

The results of such chemical molecular-beam experiments can tell much about the mechanisms underlying chemical reactions. For instance, the alkali halide product molecule formed in $K + Br_2 \rightarrow KBr + Br$ reactions is produced primarily in small-angle or forward scattering (called a stripping reaction), whereas in $K + CH_3I$ collisions this product is seen largely in backward scattering (a rebound reaction). Large-reaction total cross sections are correlated with stripping reactions, small ones with rebound reactions; this is observed even though the sums of reaction and elastic scattering cross sections are equal in the two cases. One concludes that the total effective sizes of the targets are the same in the two cases, and that the reactive process steals scattering probability away from the elastic process where no particle rearrangement occurs.

A detailed description of the reaction mechanism can be simplified into a sequence of events occurring as the collision progresses. The sequence involves the so-called harpoon model based upon a long-range sudden jumping of the valence electron on the primary alkali atom over to the electronegative halide molecule. The jump occurs at a molecular separation R_c, where the interaction potential energies for the composite systems $K + Br_2$ and $Kr^+ + Br_2^-$ happen to become almost equal, a point where little collision kinetic energy needs to be converted into a change in potential energy. The electron's jump is sufficiently violent for the Br_2^- to be vibrationally excited, and in addition the strong attraction of the two oppositely charged ions polarizes the Br_2^-, two factors leading to the second step in the model: a dissociation of Br_2^- into $Br^- + Br$. The reaction product KBr at last is formed by the attractive interaction of the K^+ and Br^- ions while the neutral Br atom is still nearby.

The difference between the rebound and stripping cases now can be seen as associated respec-

tively with small and large values of the electron jump distance R_c. When reached, small values of molecular separation correspond to strong molecular interaction and thus to large scattering angles. If R_c itself is small, then reaction occurs only for large scattering angles; at the smaller angles the molecules never become close enough for the electron to jump. The total reaction cross section πR_c^2 will be small in this case since R_c is small. On the other hand, if R_c is large, then the reaction occurs at small angles and the cross section is large as well.

Although the harpoon model is not always applicable, it does characterize the ideas pursued as a result of molecular-beam scattering experiments. Development of apparatus for the scattering of molecules polarized with their electric dipole orientations fixed by external fields has been undertaken, and this should permit further progress in understanding chemical reaction mechanisms, including those of interest to biology.

Laser excitation. The development of tunable, strong laser sources of single-frequency light beams has added another dimension to molecular-beam experiments. With laser radiation resonantly tuned to excite a molecule from its normal ground state to one of its infinite number of vibrationally, rotationally, and electronically excited states, the number of possible studies and applications of excited molecular beams becomes enormous. Many experiments using laser-excited fast or slow beams have been undertaken. Of basic importance is the fact that excited molecules can be either highly reactive or good carriers of stored potential energy. Excited-state atomic- and molecular-beam experiments will certainly contribute to understanding in physics, chemistry, and biology. *See* MOLECULAR STRUCTURE AND SPECTRA; NUCLEAR STRUCTURE. [JAMES E. BAYFIELD]

Bibliography: Faraday Society, *Molecular Beam Scattering*, vol. 55, 1973; P. Kusch and V. W. Hughes, *Atomic and Molecular Beam Spectroscopy*, vol. 37 of *Handbuch der Physik*, 1958; G. zu Putlitz, E. W. Weber, and A. Winnacker, *Atomic Physics 4*, 1975; N. F. Ramsey, *Molecular Beams*, 1956; C. Schlier, *Molecular Beams and Reaction Kinetics*, 1970.

Molecular physics

The study of the physical properties of molecules. Molecules possess a far richer variety of physical and chemical properties than do isolated atoms. This is attributable primarily to the greater complexity of molecular structure, as compared to that of the constituent atoms. Molecules also possess additional energy modes because they can vibrate; that is, the constituent nuclei oscillate about their equilibrium positions and rotate when unhindered. These modes give rise to additional spectroscopic properties, as compared to those of an atom; molecular spectroscopy in the optical, infrared, and microwave regions is one of the physical chemist's most powerful means of identifying and understanding molecular structure. Molecular spectroscopy has also given rise to the rapidly growing field of molecular astronomy.

Molecular physics is primarily concerned with the study of properties of isolated molecules, as contrasted to the more general study of molecular reactions, which is the domain of physical chemistry. Such properties, in addition to the broad field of spectroscopy, include electron affinities (for the formation of molecular negative ions); polarizabilities (the "distortability" of the molecule along its various symmetry axes by external electric fields); magnetic and electric multipole moments, attributable to the distributions of electric charge; currents and spins of the molecule; and the (nonreactive) interactions of molecules with other molecules, atoms, and ions.

One of the most important areas of molecular physics is the rapidly developing application of computational techniques to the calculation of molecular wave functions from basic principles and, consequently, of molecular charge distributions and of all the physical properties noted above.

Intermolecular forces are responsible for such varied phenomena as condensation, surface tension, gaseous diffusion, and, most importantly, for the formation of crystalline and noncrystalline solids. Thus, the field is basic to achieving an understanding of the physical world. It would be difficult to list all the articles in the present volumes which could be classed under the domain of molecular physics. For a representative sampling *see* INTERMOLECULAR FORCES; MOLECULAR BEAMS; MOLECULAR STRUCTURE AND SPECTRA.

[BENJAMIN BEDERSON]

Molecular structure and spectra

Until the advent of quantum theory, ideas about the structure of molecules evolved gradually from analysis and interpretation of the facts of chemistry. Chemists developed the concept of molecules as built from atoms in definite proportions, and identified and constructed (synthesized) a great variety of molecules. Later, when the structure of atoms as built from nuclei and electrons began to be understood with the help of quantum theory, a beginning was made in seeing why atoms can combine in definite ways to form molecules; also, infrared spectra began to be used to obtain information about the dimensions and the nuclear motions (vibrations) in molecules. However, a fundamental understanding of chemical binding and molecular structure became possible only by application of the present form of quantum theory, called quantum mechanics. This theory makes it possible to obtain from the spectra of molecules a great deal of information about the nature of molecules in their normal as well as excited states, and about dissociation energies and other characteristics of molecules.

Molecular sizes. The size of a molecule varies approximately in proportion to the numbers and sizes of the atoms in the molecule. Simplest are diatomic molecules. These may be thought of as built of two spherical atoms of radii r and r', flattened where they are joined. The equilibrium value R_e of the distance R between their nuclei is then smaller than the sum of the atomic radii (Fig. 1). However, the nuclei of atoms in two different molecules cannot normally approach more closely than a distance $r + r'$; r and r' are called the van der Waals radii of the atoms. The smallest molecule is hydrogen (H_2), with two electrons whose negative charges equal the positive charges of the two nuclei. Here r is about 1.2 A (1 A $= 10^{-8}$ cm) giving $r + r' = 2.4$ A, but R_e is only 0.74 A. In

HCl $r = 1.2$ for H and 1.8 A for Cl, but R_e is only 1.27 A.

To describe a polyatomic molecule, one must specify not merely its size but also its shape or configuration. For example, carbon dioxide (CO_2) is a linear symmetrical molecule, the O—C—O angle being 180°. The H—O—H angle in the nonlinear water (H_2O) molecule is 105°. Many molecules which are essential for life contain thousands or even millions of atoms. Proteins are often coiled or twisted and cross-linked in curious ways which are important for their biological functioning.

Dipole moments. Most molecules have an electric dipole moment. In atoms, the electron cloud surrounds the nucleus so symmetrically that its electrical center coincides with the nucleus, giving zero dipole moment; in a molecule, however, these coincidences are disturbed, and a dipole moment usually results.

Thus, when the atoms of HCl come together, there is some shifting of the H-atom electron toward the Cl. A complete shift would give H^+Cl^-, which would constitute an electric dipole of magnitude eR_e, where e is the electronic charge. But in fact the dipole moment is only 0.17 eR_e. This is because the actual electronic shift is fractional.

Although in molecules such as H_2, N_2, and CO_2 partial shifts of electronic charge from the original atoms do take place, these necessarily occur so symmetrically that no dipole moment results. Many larger molecules also have zero dipole moments by virtue of high symmetry. Examples are methane (CH_4), uranium hexafluoride (UF_6), and benzene (C_6H_6).

In general, the dipole moment of a neutral molecule is defined as the vector sum of quantities $+Q\mathbf{S}$ for the nuclei and $-e\mathbf{s}$ for the electrons. Here Q is the charge on any nucleus and $\mathbf{S}$ its vector distance from any fixed point in the molecule; $\mathbf{s}$ is the average vector distance of any electron from the same point. To calculate a dipole moment with these definitions, quantum mechanics must be used.

However, a study of what is known experimentally about molecular dipole moments has led to useful semiempirical generalizations. Bond moments and group moments have been obtained for various types of chemical bonds and of chemical groups or radicals. By adding these vectorially, the actual dipole moment of a large molecule can often be reproduced fairly accurately. In CH_4 or CO_2, one can assume a moment for each C—H or C=O

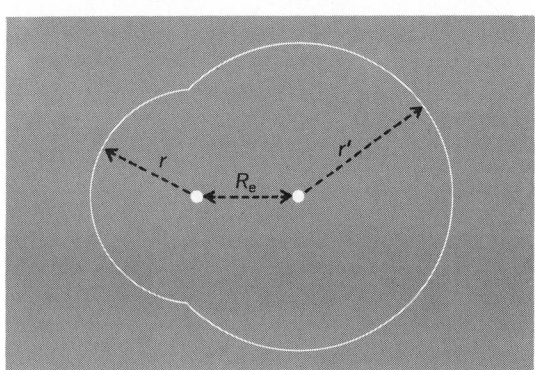

Fig. 1. Diatomic molecule with nuclei at distance R_e apart, built from atoms of radii r and r'.

bond, even though these cancel out vectorially to give a zero resultant. In the linear molecule OCS, the unequal moments of the C=O and C=S bonds give a nonzero resultant. Because of the zero moment of CH_4, the CH bond moment and the CH_3 group moment must be equal and opposite. In CH_3Cl, the total moment can be thought of as the vector sum of the H_3C group moment and the C—Cl bond moment. For additional information *see* DIELECTRIC CONSTANT.

When molecules vibrate, their dipole moments usually vary. Figure 2 shows how the dipole moment μ in a diatomic molecule may vary with R; the quantity previously discussed is μ_e, the value of μ at R_e. When μ_e is zero because of symmetry, it remains zero for symmetrical vibrations but, in polyatomic molecules, varies during unsymmetrical vibrations. Molecules may possess magnetic as well as electric dipole moments.

Molecular polarizability. In the preceding consideration of dipole moments, the discussion has been in terms of atoms and molecules free from external forces. An electric field pulls the electrons of an atom or molecule toward it and pushes the nuclei away, or vice versa. This action creates a small induced dipole moment, whose magnitude per unit strength of the field is called the polarizability.

Molecular polarizabilities can be expressed as sums of atomic polarizabilities, plus corrections depending on the types of bonds present. Polarizabilities increase rather rapidly in such series as F, Cl, Br, I, and also from HF to HI, or F_2 to I_2.

Molecular polarizabilities can also be expressed approximately as sums of bond polarizabilities. These polarizabilities are anisotropic, being greater along bonds than perpendicular to bonds.

Molecular energy levels. The states of motion of nuclei and electrons in a molecule, or of electrons in an atom, are restricted by quantum mechanics to special forms with definite energies. The state of lowest energy is called the ground state; all others are excited states. In analogy to water levels, one speaks of energy levels. Excited states exist only momentarily, following an electrical or other stimulus. *See* QUANTUM MECHANICS.

Energy levels are either discrete or continuous. The levels of a self-contained atom or molecule are restricted to special, sharply defined values

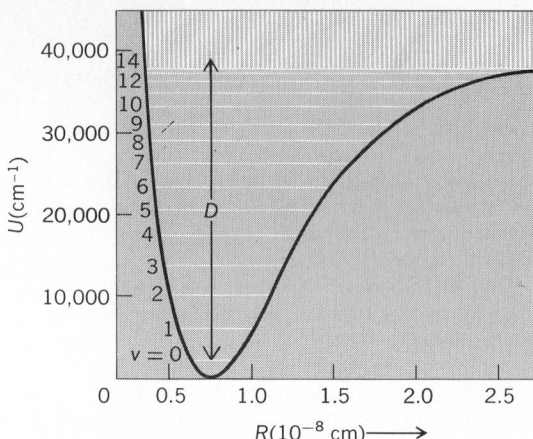

Fig. 3. $U(R)$ curve of ground electronic state of H_2 with vibrational levels and dissociation continuum. D indicates the dissociation energy. Maximum v here is 14. (*After G. Herzberg, Molecular Spectra and Molecular Structure, vol. 1, 2d ed., Van Nostrand, 1950*)

(discrete levels). When an atom or molecule is ionized, that is, when one of its electrons has enough energy to escape completely, the energy can take on any value exceeding the minimum escape energy. One then speaks of continuous levels or of an ionization continuum. Molecules also have dissociation continua, which are discussed later in this article.

Excitation of an atom consists of a change in the state of motion of its electrons. Electronic excitation of molecules can also occur, but alternatively or additionally, molecules can be excited to discrete states of vibration and rotation.

In a diatomic vibration, R varies periodically above and below R_e. The possible vibration energies E_v are given by Eq. (1), where $c\omega_e$ is just

$$E_v = hc\omega_e[(v + \frac{1}{2}) - x_e(v + \frac{1}{2})^2] + \cdots \quad (1)$$

the small-amplitude vibration frequency, and h is Planck's constant (6.62×10^{-27} erg-sec); x_e is a small quantity which is nearly always positive. The vibrational quantum number v can take whole-number values 0, 1, 2, etc. The $+ \cdots$ in Eq. (1) indicates small correction terms. The zero-point vibration energy $\frac{1}{2}hc\omega_e(1 - \frac{1}{2}x_e)$ present in the ground vibration state ($v = 0$) is a characteristic manifestation of quantum theory.

The value of $c\omega_e$ depends on the masses m_1 and m_2 of the atoms and the force constant k, as shown in Eq. (2). The frequency $c\omega_e$ (c = speed of light) is

$$c\omega_e = \sqrt{k[(1/m_1) + (1/m_2)]} \quad (2)$$

written in this manner for reasons of convenience in spectroscopic work, where the factor c is usually dropped.

The quantities R_e, k, and the dissociation energy D are the most important properties of a potential curve, which shows how the energy of attraction $U(R)$ of the atoms varies with R; k is d^2U/dR^2 taken at R_e. The $U(R)$ curve and vibrational levels for the ground electronic state of H_2 are shown in Fig. 3. Similar curves, but with other R_e, k, and D values, exist for other electronic states and other molecules. Molecules have also repulsive electronic states, whose $U(R)$ curves rise steadily with de-

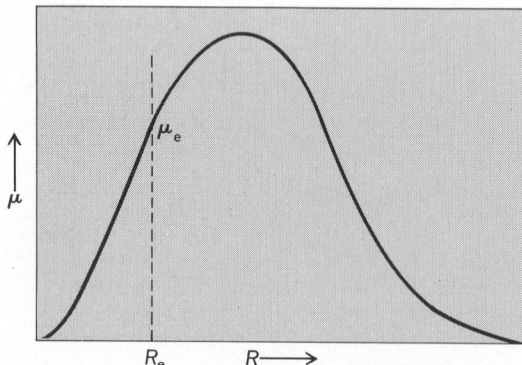

Fig. 2. Electric dipole moment μ of typical diatomic molecule as function of internuclear distance R; μ_e is the dipole moment at the equilibrium distance R_e.

creasing R. These are often important for spectroscopy and in atomic collisions. For stable (attractive) $U(R)$ curves, the vibrational levels decrease in spacing as v increases, until finally, as the spacing approaches zero, a maximum v is reached; in Fig. 3 this is 14. After a small gap, a dissociation continuum of energy levels then sets in. Here the atoms have enough mutual kinetic energy to fly apart. For repulsive states, there is only a dissociation continuum, with no vibrational levels. Figure 4 illustrates how strongly vibration level spacings can vary: both k and $1/m$, and therefore $c\omega_e$, decrease from H_2 to O_2 to I_2. Figure 4 likewise illustrates the effect of mass in isotopic molecules..

The total energy of any molecule can be written as Eq. (3). Both the electronic energy E_{el} and vibra-

$$E = E_{el} + E_v + (E_r + E_{fs} + E_{hfs} + E_{ext}) \qquad (3)$$

tion energy E_v can be discrete or continuous. The quantities E_r, E_{fs}, and E_{hfs} denote rotational, fine-structure, and hyperfine-structure energies. The last two appear as small or minute splittings of the rotation levels. The spacings ΔE of adjacent discrete levels of each type are usually in the order given in notation (4).

$$\Delta E_{el} \gg \Delta E_v \gg \Delta E_r \gg \Delta E_{fs} \gg \Delta E_{hfs} \qquad (4)$$

The fine structures of rotational levels differ strongly for different types of electronic states. The simplest diatomic electronic states are called $^1\Sigma$ states, and include $^1\Sigma^+$ and $^1\Sigma^-$ types for heteropolar and $^1\Sigma^+_g$, $^1\Sigma^+_u$, $^1\Sigma^-_g$, and $^1\Sigma^-_u$ for homopolar molecules. Most even-electron diatomic and linear polyatomic molecule ground states are $^1\Sigma^+$ states ($^1\Sigma^+_g$ if homopolar). The rotational levels of $^1\Sigma$ states have no fine structure; hyperfine structure, because of interaction with nuclear spins, is usually on too small a scale to detect by optical spectroscopy, to which the present article is limited. The E_{ext} term in Eq. (3) refers to additional fine structure which appears on subjecting molecules to external magnetic fields (Zeeman effect) or electric fields (Stark effect). *See* FINE STRUCTURE; HYPERFINE STRUCTURE; STARK EFFECT; ZEEMAN EFFECT.

The rotational levels of any $^1\Sigma$ state are given by Eq. (5). The quantity B_v is related to the moment of inertia I $[I = m_1 m_2 R^2/(m_1 + m_2)]$, and to v, by Eq. (6).

$$E_r = hcB_v J(J+1) + \cdots \qquad (5)$$

$$B_v = (h/8\pi^2 c)\,\overline{(1/I)}_v = B_e - \alpha_e(v + 1/2)$$
$$+ \cdots = B_0 - \alpha_e v + \cdots \qquad (6)$$

The rotational quantum number J can have any whole number value from 0 up, and corresponds to an angular momentum $(h/2\pi\sqrt{J(J+1)}$. The averaging of $1/I$ in Eq. (6) normally results in a slow decrease of B with increasing v (α_e is usually a small positive quantity). The quantity B_e refers to a hypothetical nonvibrating molecule ($R = R_e$).

Figure 5 illustrates how enormously rotational level spacings can vary because of differences in m and R_e (both are much greater for I_2 than H_2). The effect of mass for isotopic molecules is illustrated for O_2. Comparison with Fig. 4 illustrates the relation $\Delta E_v \gg \Delta E_r$ mentioned earlier.

Polyatomic molecules have much more compli-

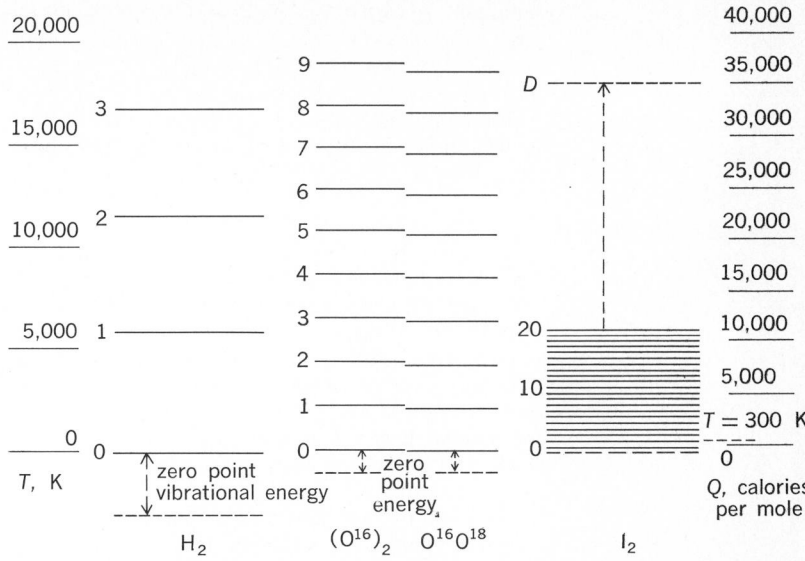

Fig. 4. Lowest vibrational levels of H_2, O_2, and I_2, numbered by vibrational quantum number v. Vibration level spacings decrease with increasing v. Where spacing reaches zero, the molecule dissociates; dissociation level D is indicated for I_2. Energies are given by the scale at right. The scale at left shows the average energy of vibration at various temperatures.

cated patterns of vibrational and (usually) of rotational energy levels than diatomic molecules. The number of normal modes (independent forms) of vibration for a molecule with n atoms is $3n - 6$ for nonlinear molecules, and $3n - 5$ for linear molecules. Each normal mode is a cooperative vibration of some or all the atoms moving with the same frequency, characteristic of the mode. Sometimes two or even three modes are so related in form that their frequencies are identical. These are called degenerate vibrations.

Figure 6 depicts the normal modes of H_2O and CO_2. They are labeled by symbols which also denote their frequencies. The arrows indicate the directions of motion of the atoms during one phase of vibration. The CO_2 frequency ν_2 is twofold de-

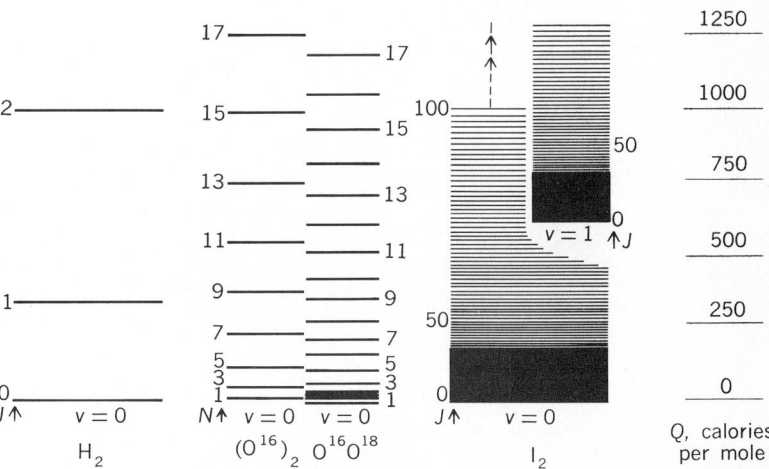

Fig. 5. Lowest rotational levels of H_2, O_2, and I_2. For H_2 and I_2, J is the rotational quantum number, according to Eq. (5) in the text O_2 is in a Hund's case b triplet state, and the rotational levels are designated by N, where the total angular momentum $J = N + 1, N$, and $N - 1$. This narrow spin tripling is indicated for the $N = 1$ level of $O^{16}O^{18}$ only. Energies are given by the scale shown at right.

generate: there are two independent modes corresponding to motion in either of two planes at right angles. The other two CO_2 modes, and all three H_2O modes, are nondegenerate.

Molecular spectra. The frequencies $c\nu$ of electromagnetic spectra obey the Einstein-Bohr equation, Eq. (7). The quantities ν, in waves per centi-

$$hc\nu = E' - E'' \qquad (7)$$

meter, or wave numbers (cm^{-1}), will hereafter be called frequencies, as is usual in spectroscopy, although properly only the $c\nu$ are frequencies. Molecular emission spectra accompany jumps in energy from higher to lower levels; absorption spectra accompany jumps from lower to higher levels. Both E' and E'' can be either discrete or continuous levels. If both are discrete, they give a spectrum of discrete frequencies; otherwise, they give a continuous spectrum. Discrete spectra are the main type considered here. Discrete frequencies are usually called spectrum lines because of their appearance when recorded by an optical spectrograph.

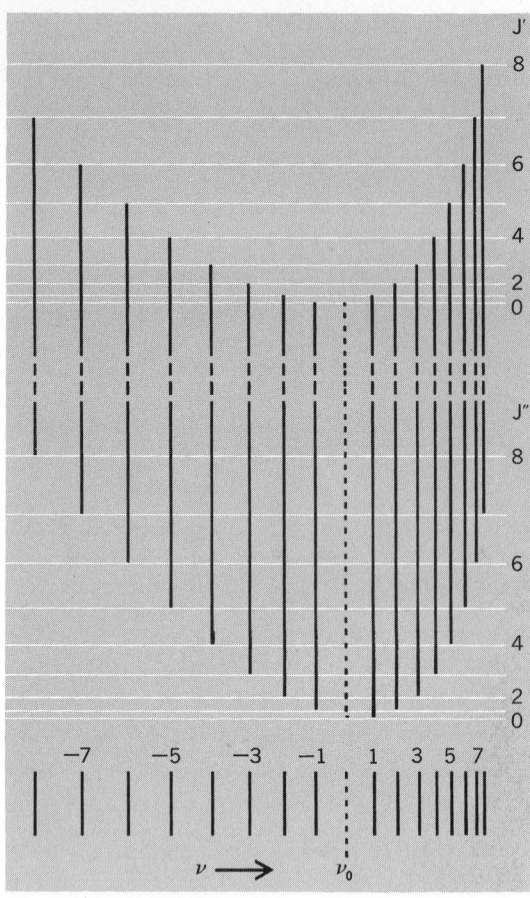

Fig. 7. Relation of band lines (lower part), [see Eqs. (8) and (9)] to rotational levels [see Eq. (6)] for a vibration-rotation band or an electronic band. In the former case, the upper and lower sets of rotational levels belong to two vibrational levels of a $^1\Sigma$ electronic state. In the latter case, they belong to two different $^1\Sigma$ states. Positive M values, R branch; negative M values, P branch.

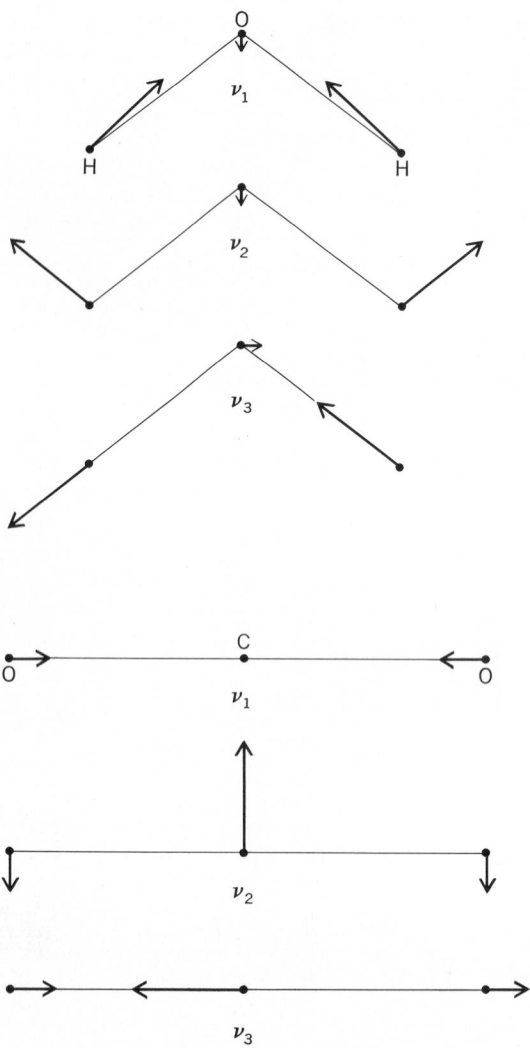

Fig. 6. Normal vibration modes of H_2O and CO_2. Synchronized displacements of atoms occur in proportion to lengths of the arrows. Diagram corresponds to snapshot taken at one phase of vibration.

Besides its frequency, the intensity and width of a spectrum line are important. Intensities vary over wide ranges. In the extreme case of nearly zero intensity for a spectroscopic transition, the transition is called forbidden. Only a small minority of all pairs of levels yield allowed transitions. These are governed by selection rules derivable from quantum theory. *See* SELECTION RULES.

Under disturbing influences, however, some lines are seen, weakly, which violate these rules. Further, the usual selection rules are electric dipole rules, and additional transitions become very weakly allowed if magnetic dipole, electric quadrupole, and other selection rules are also considered. The following discussion is confined to spectra which obey the electric dipole rules.

Molecular spectra can be classified as fine-structure or low-frequency spectra, rotation spectra, vibration-rotation spectra, and electronic spectra. Low-frequency spectra are discussed elsewhere. *See* MAGNETIC RESONANCE; MOLECULAR BEAMS.

Pure rotation spectra. Transitions between energy levels differing only in rotational state give rise to pure rotation spectra. For diatomic molecules in $^1\Sigma$ states, Eq. (5), the relation is given by Eq. (8). The transitions obey the selection rule

$$hc\nu = E'_r - E''_r = hcB_v[J'(J'+1) \\ -J''(J''+1)] + \cdots \quad (8)$$

$\Delta J = 1$ (ΔJ means $J' - J''$). Putting $J' = J'' + 1$, Eq. (9) is obtained. Equation (9) represents a sequence

$$\nu = 2B_v(J''+1) + \cdots \quad (9)$$

of lines spaced almost equidistantly ($2B_v$, $4B_v$, $6B_v$, . . .), and lying in the far infrared or (for small B or low J'') the microwave region. Their intensities are proportional to μ_e^2, where μ_e is the electric moment at R_e (Fig. 2); hence homopolar molecules (H_2, N_2, and so on) show no pure rotation spectra. The intensities are proportional also to the lower-state (v'', J'') level population and to ν (for absorption) or ν^4 (for emission).

Pure rotation spectra of linear polyatomic molecules are like those of diatomic molecules. Polyatomic molecules having $\mu_e = 0$, whatever their shape (examples are CO_2, CH_4, C_6H_6), have no pure rotation spectra. In other cases, the spectra can be obtained using $hc\nu = E'_r - E''_r$ with appropriate E_r expressions and selection rules.

Vibration-rotation bands. Spectra involving only vibrational and rotational state changes lie mainly in the infrared. For a $^1\Sigma$ diatomic state, using Eqs. (1), (5), and (7), Eq. (10) is obtained, with ν_0 defined in Eq. (11). In Eq. (10) B' and B'' mean B_v for v' and

$$\nu = \nu_0(v',v'') + [B'J'(J'+1) \\ -B''J''(J''+1)] + \cdots \quad (10)$$

$$\nu_0 = \omega_e(1-x_e)(v'-v'') - x_e\omega_e(v'^2 - v''^2) \quad (11)$$

v'', respectively. Each band consists of two sets of rotational lines, one on each side of its ν_0. Each line corresponds to a particular rotational transition conforming to $\Delta J = \pm 1$. The two series (branches) have frequencies defined in Eq. (12) for R or positive branch ($J' = J'' + 1$), and in Eq. (13), for P or negative branch ($J' = J'' - 1$). Both can be represented by a single equation, Eq. (14), by let-

$$\nu = \nu_0 + 2B''(J''+1) \\ + (B'-B'')(J''+1)(J''+2) \quad (12)$$

$$\nu = \nu_0 - 2B''J + (B'-B'')J''(J''-1) \quad (13)$$

$$\nu = \nu_0 + (B'+B'')M \\ + (B'-B'')M^2 + \cdots \quad (14)$$

ting $M = J'' + 1$ for the R and $M = -J''$ for the P branch. Neglecting the term in M^2, Eq. (14) represents a series of equidistant lines with one missing ($M = 0$) at ν_0. Figure 7 shows how the line positions are related to the upper (v') and lower (v'') sets of rotational levels.

Since $B' - B''$ is a small negative quantity [see Eq. (6), noting that $v' > v''$], the M^2 term makes the P line spacing increase and the R line spacing decrease slowly as M increases. This is shown, exaggerated, in Fig. 7. At some large M value, the R branch turns back on itself, but usually the lines have become weak before this value is reached.

The relative intensities of band lines depend primarily on the initial rotational distribution of molecules. More precisely, Eq. (15) holds. Here

Intensity
$$= C(v',v'')\nu^n(J'+J''+1)e^{-B_{in}J_{in}(J_{in}+1)hc/kT} \quad (15)$$

B_{in}, J_{in}, and n in ν^n are B', J', and 4, respectively,

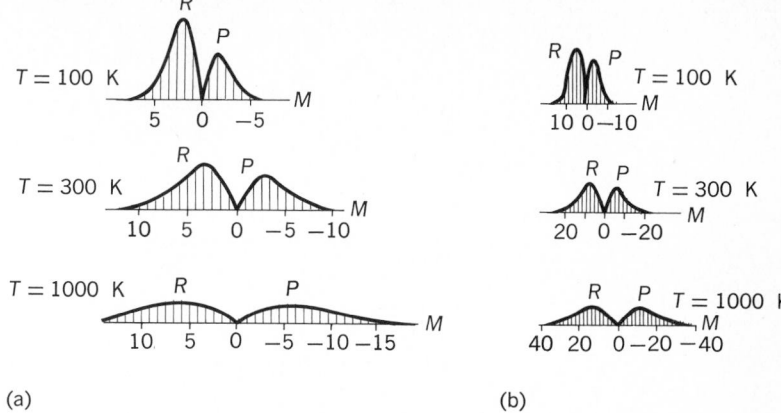

Fig. 8. Intensity distribution at several temperatures for a diatomic absorption band. Line positions are based on Eq. (9) assuming $B' = B''$ for simplicity; frequency increases toward the left (opposite to Fig. 7). (*a*) and (*b*) correspond respectively to B values of HCl ($B = 10.44$ cm^{-1}) and of 2 cm^{-1} (approximately the value for CO, for which $B = 1.93$ cm^{-1}). (*After G. Herzberg, Molecular Spectra and Molecular Structure, vol. 1, 2d ed., Van Nostrand, 1950*)

for an emission band, and B'', J'', and 1, respectively, for an absorption band. Figure 8 shows diagrammatically how the values of B and T affect the appearance of a typical absorption band ($B' = B''$ has been assumed for simplicity in Fig. 8). Figure 9 shows the appearance of an actual HCl band. The weaker HCl[37] lines are at slightly lower frequencies than the HCl[35] lines, mainly because ω_e is smaller [see Eqs. (2) and (11)].

The factor $C(v',v'')$ is largest by far for fundamental bands ($\Delta v = 1$), and falls rapidly with increasing Δv in the overtone bands or harmonics (Δv is $v' - v''$). For fundamental bands, C depends on the slope of the $\mu(R)$ curve (Fig. 2), being approximately proportional to $(d\mu/dR)^2$ taken at R_e. For overtone bands, C depends on the detailed shapes of both the $\mu(R)$ and $U(R)$ curves. Fundamental or overtone bands arising from $v'' > 0$ are called hot bands.

Vibration-rotation absorption bands of liquids and solutions are widely used in chemical analysis. Here the rotational structure is blurred out, and

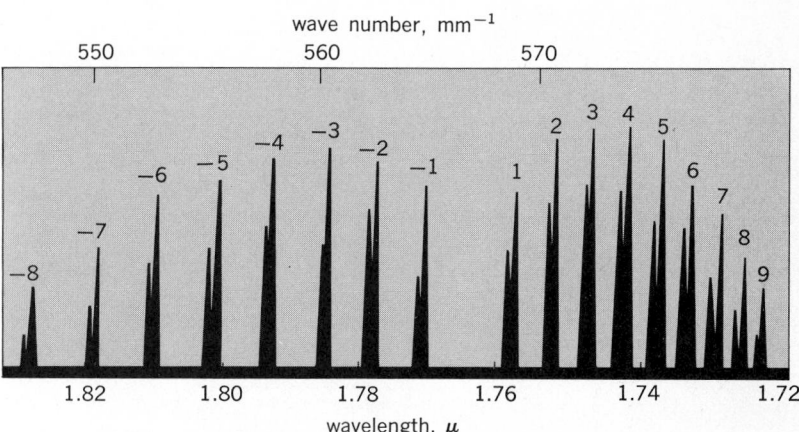

Fig. 9. First harmonic (2,0) vibration-rotation band of HCl in absorption. R branch to right, P branch to left, showing intensity distribution. The stronger lines are HCl[35]; the weaker companions, at lower frequencies, are HCl[37]. (*After C. F. Meyer and A. A. Levin, Phys. Rev., 34:44, 1929*)

only an "envelope" is seen. For many purposes, it is sufficient to know empirically the spectrum of each molecule which may be present. Also, groups of atoms which recur in many molecules often have nearly constant frequencies, of use for identification and in determining molecular structures.

Electronic band spectra. These are the most general type of molecular spectra. The characteristic feature is a change of electronic state. From Eqs. (3) and (7), neglecting fine structure, Eq. (16)

$$\nu = \frac{(E'_{el} - E''_{el}) + (E'_v - E''_v) + (E'_r - E''_r)}{hc} \quad (16)$$

and (17) are obtained. Diatomic electronic spectra

$$\nu = \nu_{el} + \nu_v + \nu_r = \nu_0 + \nu_r \quad (17)$$

are often observed in emission, while the electronic spectra of polyatomic molecules are usually absorption spectra. Depending mainly on the magnitude of ν_{el}, electronic spectra occur in the infrared, visible, ultraviolet, or vacuum ultraviolet.

For any one electronic transition, the spectrum consists typically of many bands. These lie in general at frequencies both above and below ν_{el}, since ν_v can be positive or negative. They constitute a band system. Each band consists of numerous rotational lines arranged in two or more branches and lying on both sides of a central position ν_0.

For diatomic molecules, ν_0 depends on a single v' and v'' and, using Eq. (1) for each electronic state, is given by Eq. (18). Since ω_e and $x_e\omega_e$ are

$$\nu_0(v', v'') =$$
$$\nu_{el} + [\omega'_e(v' + \tfrac{1}{2}) - x'_e\omega'_e(v' + \tfrac{1}{2})^2 + \cdots]$$
$$- [\omega''_e(v'' + \tfrac{1}{2}) - x''_e\omega''_e(v'' + \tfrac{1}{2})^2 + \cdots] \quad (18)$$

now different (often strongly) in the upper and lower states, $\nu_0(v', v'')$ cannot be reduced to as simple an expression as the corresponding Eq. (11) for vibration-rotation bands. Eq. (18) is more convenient when rewritten as Eq. (19), where Eqs. (20)

$$\nu_0(v', v'') = \nu_{00} + (\omega'_0 v' - a' v'^2)$$
$$- (\omega''_0 v'' - a'' v''^2) + \cdots \quad (19)$$

$$\nu_{00} = \nu_{el} + \tfrac{1}{2}(\omega'_e - \omega''_e) - \tfrac{1}{4}(x'_e\omega'_e - x''_e\omega''_e)$$
$$\omega'_0 = \omega'_e(1 - x'_e) \qquad a' = x'_e\omega'_e, \text{ etc.} \quad (20)$$

apply. The relative intensities of the bands depend on (1) the initial distribution of molecules among vibrational levels, and (2) the relative transition probabilities from any initial to various final vibrational levels.

The simplest example is the absorption spectrum of a cool gas of low molecular weight, for which all molecules initially have $v'' = 0$. The spectrum then consists of one "v' progression," a single series of bands with various values of v'; the frequencies are given by $\nu = \nu_{00} + \omega_0 v' - a' v'^2$. For a hot or a heavy gas, additional weaker v' progressions with $v'' > 0$ also appear.

In emission spectra, the initial population usually ranges over a number of v' values, from each of which transitions occur to a number of v'' values, so that the system contains many bands on both

sides of ν_{00}. In the special case of fluorescence spectra, the molecule is excited to various v' values by absorbing light; it then emits light belonging to the same (or sometimes another) electronic transition. From each v', it can descend not only to the original v'' but also to various other, mainly larger, values. Hence fluorescence bands lie mainly at lower frequencies than the absorption bands used to excite them.

Relative transition probabilities are governed by the Franck-Condon principle. This takes note of the very great rapidity of electronic motions as compared with those of the far more massive nuclei, and concludes that during the extremely brief time for an electronic transition, the nuclei tend to remain unchanged in their positions and momenta. It is applicable to both polyatomic and diatomic spectra. Consider first a diatomic molecule starting from the $v'' = 0$ level of a ground state $U(R)$ curve like the lower curves in Fig. 10. A vertical line drawn from the bottom point A ($v'' = 0$ if zero-point vibration is neglected) to point B on any one of the upper curves corresponds to an electronic absorption transition in which the nuclei have not moved.

In the case of Fig. 10a, point B corresponds to $v' = 0$, and the conclusion is that this is the most probable v' for $v'' = 0$. In the case of Fig. 10b, point B corresponds to an excited molecule at the inner turning point of a vibration with a v' of possibly 3 or 4, in a typical case. One then concludes (with J. Franck) that the strongest absorption bands for $v'' = 0$ have $v' = 3.0$ and 4.0. To obtain more exact information, a quantum-mechanical calculation (first carried out by E. U. Condon) is necessary.

In the case of Fig. 10c, point B corresponds to an energy level in the dissociation continuum above the asymptote of the upper $U(R)$ curve. According to the Franck-Condon principle, the absorption spectrum will have maximum intensity in a continuous range of frequencies, with $hc\nu$ about equal to the energy difference AB. The quantum-mechanical calculation shows that the actual spectrum will

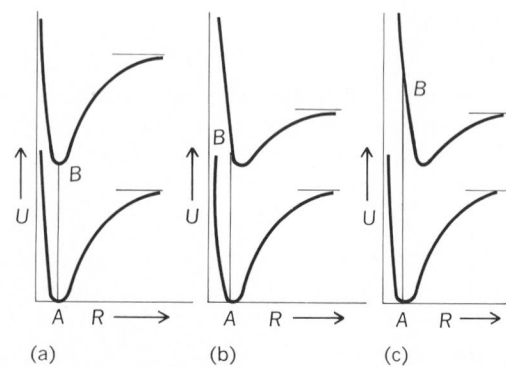

Fig. 10.   Diatomic $U(R)$ curves for three cases to explain the vibrational intensity distribution according to the Franck-Condon principle. The asymptote of each curve for large R corresponds to dissociation into atoms, with one or both atoms excited in the case of the upper curves. Starting in each case from the bottom of the lower curve (essentially $v'' = 0$), the most probable transition in absorption is (a) to $v' = 0$, (b) to $v' = 3$ or 4, and (c) to the dissociation continuum, as shown by vertical lines. (*After G. Herzberg, Molecular Spectra and Molecular Structure, vol. 1, 2d ed., Van Nostrand, 1950*)

extend with appreciable intensity over a range of both higher and lower frequencies than this, including, on the lower-frequency side, a number of high-v' bands. Actual examples of such spectra (a long v' progression followed by a strong continuum) are the far-ultraviolet Schumann-Runge bands of oxygen and the visible bands of iodine. By measuring the frequency at which the continuum begins, one obtains an exact value of the dissociation energy of each of these molecules. In so doing, any excitation energy of the atomic dissociation products to which the upper $U(R)$ curve leads is subtracted.

The Franck-Condon method is useful in understanding intensity distributions and structure in emission as well as absorption band systems. For diatomic spectra, various patterns of intensity as functions of v' and v'' occur, depending largely on the R_e values of the two $U(R)$ curves and, of course, also on the initial distribution among v' levels. Sometimes the upper-state $U(R)$ curve is stable (has a minimum) but the lower state is repulsive. Continuous emission spectra then occur, with the atoms flying apart on reaching the lower state. The H_2 molecule shows such a spectrum, as do rare gas molecules such as He_2 and Kr_2, which are stable only in excited or ionized states.

Molecular electronic states. Before discussing the structures of electronic bands, one must consider the nature of molecular electronic states. Each electronic state has orbital and spin characteristics. The spin quantum number S has a whole-number value if the number of electrons is even, a half-integral value if it is odd. Electronic states with $S = 0$ are called singlet states, all others multiplet states. The orbital characteristics differ sharply for linear (including diatomic) and nonlinear molecules.

For linear molecules only, there is a quantum number Λ such that $\pm \Lambda h/2\pi$ is the component of angular momentum around the line of nuclear centers. Linear-molecule electronic states can be discussed under three headings: (1) singlet states; (2) multiplet states with strong spin coupling (Hund's case a); and (3) multiplet states with weak spin coupling (Hund's case b). Strictly speaking, actual multiplet states are intermediate between cases a and b, or between these and certain other cases called c and d. The discussion to follow is largely restricted to singlet electronic states.

Singlet states with $\Lambda = 0$ include $^1\Sigma^+$ and $^1\Sigma^-$ states: states with $\Lambda = 1, 2, \ldots$ are called $^1\Pi, ^1\Delta,$ and so on. In linear molecules with a center of symmetry (H_2, CO_2 and so on), one must further distinguish even and odd (g and u) states: $^1\Sigma^+_g$, $^1\Sigma^+_u$, $^1\Sigma^-_g$, $^1\Sigma^-_u$, $^1\Pi_u$, $^1\Pi_g$, $^1\Delta_g$, $^1\Delta_u$, etc. The rotational levels of singlet states obey the symmetric rotor formula, Eq. (21). Here J is restricted to integral values equal to or greater than Λ.

$$E_r = hc[BJ(J+1) - \Lambda^2] + \cdots \qquad (21)$$

For $\Lambda > 0$, each rotational level is a narrow doublet (Λ-doubling). Corresponding fine structure [see E_{fs} in Eq. (3)] can usually be detected in electronic bands, but (for ground states only) it can be much more accurately studied in low-frequency spectra. Hyperfine structure [see E_{hfs} in Eq. (3)] is usually on too small a scale to be detected in electronic band lines, but has been found in a few

cases. Hyperfine structure is best studied in low-frequency spectra.

Electronic band structures. The simplest electronic bands occur for transitions between singlet electronic states. The possible types of electronic transitions are limited by the selection rule $\Delta\Lambda = 0, \pm 1$. The structures of $^1\Sigma - ^1\Sigma$ bands are essentially the same as for the $^1\Sigma$-state vibration-rotation bands described earlier. Equations (12) to (15) and Fig. 7, also Fig. 8, for the intensities in absorption are still applicable if Eq. (18) instead of Eq. (11) is used for ν_0, and it is recognized that B' and B'' now belong to two different electronic states.

The quantity $B' - B''$ in Eq. (14), instead of always being a small negative quantity, may now be either positive or negative, and $(B' - B'')/(B' + B'')$ is often fairly large (although it can also be nearly zero). As a result, it is usual in electronic bands to find so-called heads. A head is a position of maximum or minimum frequency; by using Eq. (14) to obtain $d\nu/dM = 0$, one finds $M_{head} = (B' + B'')/2(B' - B'')$. Then, on inserting M_{head} into Eq. (14), one obtains $\nu_{head} = \nu_0 - (B' + B'')^2/4(B' - B'')$. [Since $(B' + B'')/2(B' - B'')$ is not usually a whole number, the actual M_{head} is the nearest whole-number M to that calculated.] According to whether $B' - B''$ is negative or positive, the positive (R) or the negative (P) branch forms the head. Figure 7, if continued to somewhat larger M values, illustrates the formation of an R-branch head at a calculated M of 10.5; the actual head is formed by the two coincident lines $M = 10$ and 11.

Although homopolar molecules (H_2, N_2, and so on) have no pure rotation or vibration-rotation spectra, they do have electronic spectra. For homonuclear homopolar molecules, the band lines show alternating intensities. The lines in each branch are alternately stronger and weaker as M increases, this effect being superposed on the otherwise smoothly varying intensity distribution. The alternation ratio depends on the nuclear spin I and has been, in several cases, the means of determining I. When $I = 0$, alternate lines are completely missing. Heteronuclear molecules, even if homopolar (for example, HD or $O^{16}O^{18}$) do not show alternating intensities.

Polyatomic electronic spectra. These differ from diatomic electronic spectra because several initial and final vibration quantum numbers are involved, and because the rotational structure (except for linear molecules) is usually much more complicated. However, the detailed structures of the electronic spectra of a number of simple molecules and radicals in the vapor state in emission and in absorption have been studied. Nevertheless, for the most part, the spectra of polyatomic molecules are examined as absorption spectra in solution. The rotational structure is then completely blurred out, but the vibrational structure can be seen.

The Franck-Condon principle is here a useful guide. One of its corollaries, which amounts almost to a selection rule, is that only totally symmetric vibrations (vibrations during which the equilibrium symmetry of the molecule is preserved) undergo quantum number changes. This greatly simplifies the vibrational structure, especially of absorption spectra where most molecules are initially mainly in the $v'' = 0$ state of all vibrations. One finds then mostly v' progressions of one or a very few totally symmetric vibrations, and combinations of these.

Rather often, polyatomic band systems do not even show obvious vibrational structure. This can happen for any of several reasons: The upper state may involve dissociation; in CH_3I, for example, the first ultraviolet absorption region yields $CH_3 + I$; there may be so many low-frequency, upper-state vibrations that the spectrum looks continuous; or there may be a combination of these and other reasons. Such continuous or pseudocontinuous band systems are often loosely referred to as bands. For complicated molecules, the spectra of several different electronic transitions often overlap strongly so that it is difficult even to separate one system from another. *See* ATOMIC STRUCTURE AND SPECTRA; ELECTRON SPIN; INTERMOLECULAR FORCES; NEUTRON SPECTROMETRY; RAMAN EFFECT; SCATTERING EXPERIMENTS (ATOMS AND MOLECULES).

[ROBERT S. MULLIKEN]

Bibliography: C. N. Barswell, *Fundamentals of Molecular Spectroscopy*, 2d ed., 1973; P. R. Bunker, *Molecular Symmetry and Spectroscopy*, 1979; W. H. Flygare, *Molecular Structure and Dynamics*, 1978; I. N. Levine, *Molecular Spectroscopy*, 1975; J. Steinfeld, *Molecules and Radiation: An Introduction to Modern Molecular Spectroscopy*, 1978.

Molecule

A molecule may be thought of either as a structure built of atoms bound together by chemical forces or as a structure in which two or more nuclei are maintained in some definite geometrical configuration by attractive forces from a surrounding swarm of negative electrons. Besides chemically stable molecules, short-lived molecular fragments called free radicals can be observed under special circumstances, for example, at high temperatures, in electrical discharges, in chemical reactions, and even (but in small quantities) frozen tions. Free radicals are really just highly active molecules. *See* MOLECULAR STRUCTURE AND SPECTRA.

[ROBERT S. MULLIKEN]

Moment of inertia

A relation between the area of a surface or the mass of a body to the position of a line. The analogous positive number quantities, moment of inertia of area and moment of inertia of mass, are involved in the analysis of problems of statics and dynamics respectively.

The moment of inertia of a figure (area or mass) about a line is the sum of the products formed by multiplying the magnitude of each element (of area or of mass) by the square of its distance from the line. The moment of inertia of a figure is the sum of moments of inertia of its parts.

Moment of inertia of area. In practice, only moments of inertia of a plane area about mutually perpendicular axes (lines) in or normal to its plane are useful (Fig. 1).

The moment of inertia of plane area A about the X and Y axes in its plane are respectively $I_x = \int y^2 \, dA$ and $I_y = \int x^2 \, dA$. In these, x and y are the coordinate locations of area element dA.

The polar moment of inertia of a plane area is its moment of inertia about an axis normal to the plane of area. The polar moment of area A about

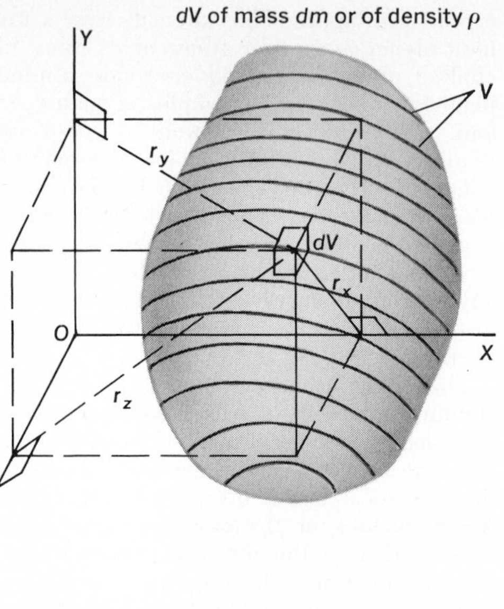

MOMENT OF INERTIA

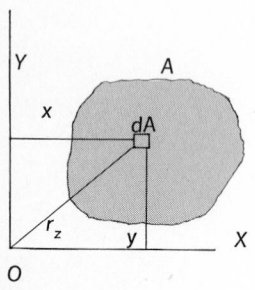

Fig. 1. Moment of inertia of an area.

dV of mass dm or of density ρ

Fig. 2.   Moment of inertia of a volume.

the Z axis is $J_z = \int r_z^2 \, dA$. As referred to a common origin of axes $J_z = I_x + I_y$.

Moment of inertia of area is measured in quartic length units, such as ft^4.

Moment of inertia of mass. For a body of mass distributed continuously within volume V, the moment of inertia of the mass about the X axis is given by either $I_x = \int r_x^2 \, dm$ or $I_x = \int r_x^2 \rho \, dV$, where dm is the mass included in volume element dV at whose position the mass per unit volume is ρ (Fig. 2). Similarly $I_y = \int r_y^2 \rho \, dV$ and $I_z = \int r_z^2 \rho \, dV$. Mass moment of inertia is measured in units of mass times length units squared, such as $g\text{-}cm^2$.

Parallel-axis theorem. The moment of inertia of a figure about any axis is the sum of its moment of inertia about a parallel axis containing the centroid of the figure and the product formed by multiplying the magnitude of the figure (its area or mass) by the distance squared between the parallel axes; for area, $I = \bar{I} + AD^2$; for mass, $I = \bar{I} + MD^2$. Accordingly the moment of inertia about a centroidal axis is less than its moment about any parallel axis.

Principal axes of inertia. The moments of inertia of a figure about lines which intersect at a common point are generally unequal. The moment is greatest about one line and least about another line perpendicular to the first one. A set of three orthogonal lines consisting of these two and a line perpendicular to both are the principal axes of inertia of the figure relative to that point. If the point is the figure's centroid, the axes are the central principal axes of inertia. The moments of inertia about principal axes are principal moments of inertia. *See* CENTROIDS (MATHEMATICS); PRODUCT OF INERTIA; RADIUS OF GYRATION.   [NELSON S. FISK]

Momentum

Linear momentum is the product of the mass and the linear velocity of a body. It is defined by Eq. (1),

$$\mathbf{P} = m\mathbf{v} \qquad (1)$$

where m is the mass and $\mathbf{v}$ is the linear velocity. Since linear momentum is the product of a scalar and a vector quantity, it is a vector and hence has both magnitude and direction.

The angular momentum of a body is defined as the product of its moment of inertia and its angular velocity. *See* ANGULAR MOMENTUM.

No special names are given to the units of linear momentum. The units are gram-centimeters per second, kilogram-meters per second, and slug-feet per second in the centimeter-gram-second, meter-kilogram-second, and British engineering systems of units, respectively.

According to the general statement of Newton's second law, for a force **F**, a momentum **P**, and a time t, Eq. (2) holds. Thus Newton's second law

$$\mathbf{F} = d\mathbf{P}/dt \qquad (2)$$

involves the time rate of change of momentum. Usually the mass of a body is constant, and the time rate of change of momentum of a body equals the product of its mass and acceleration. However, under certain conditions the mass can change, as when a rocket moves through space by consuming part of its mass as fuel. Whenever a change in mass occurs, the total time rate of change of momentum must be considered in describing the motion. Changes of momentum are important in collision processes. *See* COLLISION.

When a group of bodies is subject only to forces that members of the group exert on one another, the total momentum of the group remains constant. *See* CONSERVATION OF MOMENTUM; IMPULSE (MECHANICS).

[PAUL W. SCHMIDT]

Monte Carlo method

A technique for estimating the solution x of a numerical mathematical problem by means of an artificial sampling experiment. The estimate is usually given as the average value, in a sample, of some statistic whose mathematical expectation is equal to x. In most of the useful applications, the mathematical problem itself arises in a problem of probability, either in physics or in operational research.

The importance of the method arises primarily from two sources: the practical need to solve equations that are too complicated to solve by analytic methods alone, and the increased importance of all numerical methods because of the advent of the electronic computer.

The method described above is identical to the earlier method known as artificial or model sampling, or simulation. In fact the name simulation is most appropriate when the mathematical problem arises out of a mathematical model of a real-world situation. The main justification for the name Monte Carlo is that during the 1950s several tricks were introduced for improving the efficiency of the method, so that the subject has assumed a new flavor.

One of the earliest examples of the use of artificial sampling was the experimental estimation of π by the French naturalist G. L. L. Buffon in 1773. The method is to throw a needle on a striped tablecloth and see how often it falls touching more than one stripe. If the width of each stripe is equal to the length of the needle, then the proportion of "successes" will be close to $2/\pi$ for a long series of trials.

Classification of Monte Carlo methods. The usual method of applied mathematics is to replace a physical problem P by a mathematical problem M, by assuming an adequate mathematical model, to solve the mathematical problem, and thus to solve P (perhaps only approximately). However, sometimes the mathematical problem, when replaced by a numerical problem, is solved with the aid of a calculating machine, so that M is replaced by a new physical problem, P'. One may think of the matter in this way especially if the calculating machine is a so-called analog machine, that is, a machine using nonradix arithmetic. The method then becomes $P \to M \to P'$, where the arrow means "is replaced by." If $P = P'$, it can be said that P is solved by the crude method, or direct experimental method, in which M is inessential.

A special type of experiment with physical apparatus is a statistical experiment S, which makes use of games of chance or their equivalent, such as throwing dice, coins, or even needles. In practice, random sampling numbers are usually used, that is, a sequence of digits generated in such a manner that each selection has an independent chance of 0.1 of giving each of the digits 0, 1, 2, . . . , 9. If a distinction is made between physical and other statistical experiments, there are, among others, the following methods of solving problems: $S \to S$, $S \to S'$, $S \to M$, $M \to S$, $S \to M \to S'$. The method $S \to S$ is the method of estimating the solution of a real-life statistical problem by direct sampling. It may be regarded as a crude form of the Monte Carlo method. In the method $S \to S'$, a real-life statistical problem is replaced by a simpler model from which estimation is made by sampling. This method is perhaps best called simulation. The method $S \to M$ is the ordinary form for the application of mathematical statistics. The method $M \to S$ is exemplified by Buffon's needle problem and was suggested for more serious applications by John von Neumann and S. Ulam in 1948. At that time it might have been reasonably called the true Monte Carlo method, but this name is no longer appropriate, because of change of emphasis. It is the method $S \to M \to S'$ which has been of most interest recently. This method can be of varying degrees of sophistication, depending on the amount of mathematical ingenuity required in order to transform S into S' via M.

Advantages and disadvantages. The main advantage of Monte Carlo is that many numerical problems are in practice too complicated to solve in any other method. A familiar example is the estimation of the probability of winning a game of pure chance: Sometimes the only reasonably simple method of estimation is to play the game several times. There are also numerical problems that can be solved by deterministic methods but are logically simpler to solve approximately by the Monte Carlo method. The work can then be handed over to unskilled workers with some economy. Sometimes poor approximations are satisfactory because the aim is merely to determine the strategic variables of a problem. This is likely to be a fruitful technique in mathematical economics. To find out how to simplify a complicated economic model without losing touch with reality, one could carry out an approximate Monte Carlo solution for a complicated model, find out which of the variables are important, and then perhaps attempt a mathematical solution using only the important variables.

Another situation where a poor approximation is

satisfactory occurs when there is available an iterative method of calculation, that is, a method of successive approximation, which converges to the right answer in a reasonable time provided that the first trial solution is not too far from the truth. The Monte Carlo method may then perhaps be used for obtaining a first trial solution.

The main disadvantage of the Monte Carlo method is that for each extra decimal place required it is necessary to multiply the sample size by 100. To calculate π to five decimal places by throwing a needle would require about 10^{10} throws, by which time the tablecloth would be worn out.

An advantage claimed for Monte Carlo methods is that they do not become much more complicated when the physical dimensionality of a problem is increased. They therefore offer some hope for obtaining qualitative information about the solution of partial differential equations in three or more dimensions and about the values of multidimensional definite integrals.

Applications. The method has been applied to the following problems among others:

1. Size of cosmic-ray showers.
2. Critical size of nuclear reactors.
3. Other neutron transport problems, concerning, for example, the shielding properties of water or graphite. The probability that a neutron will cause a branching process that penetrates a shield may be as low as 10^{-10}. In a sample of random walks of reasonable size there will be no "successes" unless some tricks are used.
4. Enumeration of high-polymer molecules or number of self-avoiding walks on a diamond lattice.
5. Percolation of a liquid through a solid.
6. Brownian motion and diffusion.
7. The birth-and-death branching stochastic process.
8. Autoregressive time series.
9. Theory of queues and other problems of commercial importance, such as storage, equipment replacement and maintenance, and insurance problems.
10. Laplace's partial differential equation.
11. Schrödinger's partial differential equation.
12. Integral equations.
13. Inversion of matrices.
14. Evaluation of definite integrals.
15. Random rounding off.
16. Discovery of the t-distribution by Student.

Techniques. When a Monte Carlo method is performed on an electronic computer, the random sampling numbers may be replaced conveniently by deterministic pseudorandom sampling numbers. To do so has the advantage that every job can be checked precisely. Having generated pseudorandom or random sampling numbers, it is trivial to produce random variables that are uniformly distributed in an interval. There are also methods for producing random variables having other probability distributions. Other techniques are concerned with the reduction of the variance of estimates. Example 3 above shows the need for such techniques. If x is a random variable with probability density function $p(x)$, the mathematical expectation of a function $f(x)$ may be estimated by replacing f and p by new functions, f^* and p^*, in such a

way that $f^*(x)p^*(x) = f(x)p(x)$, where the integral of $p^*(x)$ is unity (so that p^* can be regarded as a probability point function of a new random variable). It is now adequate to estimate the expectation of f^* for the new random variable. This technique is useful if $f(x)$ is negligible (or zero) except when $p(x)$ is very small. Some other techniques are familiar in sampling survey work, such as the use of statistical regression and stratified sampling. *See* PROBABILITY; PROBABILITY (PHYSICS); STATISTICS; STOCHASTIC PROCESS.

[IRVING J. GOOD]

Bibliography: K. Binder (ed.), *Monte Carlo Methods in Statistical Physics,* 1979; J. M. Hammersley and D. C. Handscomb, *Monte Carlo Methods,* 1964; I. M. Sobol, *The Monte Carlo Method,* 2d ed., transl. by P. Fortini et al., 1975; J. Spanier and E. M. Gelbard, *Monte Carlo Principles and Neutron Transport Problems,* 1969.

Mössbauer effect

Recoil-free gamma-ray resonance absorption. The Mössbauer effect, also called nuclear gamma resonance fluorescence, has become the basis for a type of spectroscopy which has found wide application in nuclear physics, structural and inorganic chemistry, biological sciences, the study of the solid state, and many related areas of science.

Theory of effect. The fundamental physics of this effect involves the transition (decay) of a nucleus from an excited state of energy E_e to a ground state of energy E_g with the emission of a gamma ray of energy E_γ. If the emitting nucleus is free to recoil, so as to conserve momentum, the emitted gamma ray energy is $E_\gamma = (E_e - E_g) - E_r$, where E_r is the recoil energy of the nucleus. The magnitude of E_r is given classically by the relationship $E_r = E_\gamma^2/2mc^2$, where m is the mass of the recoiling atom. Since E_r is a positive number, the E_γ will always be less than the difference $E_e - E_g$, and if the gamma ray is now absorbed by another nucleus, its energy is insufficient to promote the transition from the nuclear ground state E_g to the excited state E_e.

In 1957 R. L. Mössbauer discovered that if the emitting nucleus is held by strong bonding forces in the lattice of a solid, the whole lattice takes up the recoil energy, and the mass in the recoil energy equation given above becomes the mass of the whole lattice. Since this mass typically corresponds to that of 10^{10} to 10^{20} atoms, the recoil energy is reduced by a factor of 10^{-10} to 10^{-20}, with the important result that $E_r \approx 0$ so that $E_\gamma = E_e - E_g$; that is, the emitted gamma-ray energy is exactly equal to the difference between the nuclear ground-state energy and the excited-state energy. Consequently, absorption of this gamma ray by a nucleus which is also firmly bound to a solid lattice can result in the "pumping" of the absorber nucleus from the ground state to the excited state. The newly excited nucleus remains, on the average, in its upper energy state for a time given by its mean lifetime τ (a quantity dependent on energy, spin, and parity of the nuclear states involved in the deexcitation process) and then falls back to the ground state by reëmission of the gamma ray. An important feature of this reemission process is the fact that it is essentially isotropic; that is, it occurs with equal probability

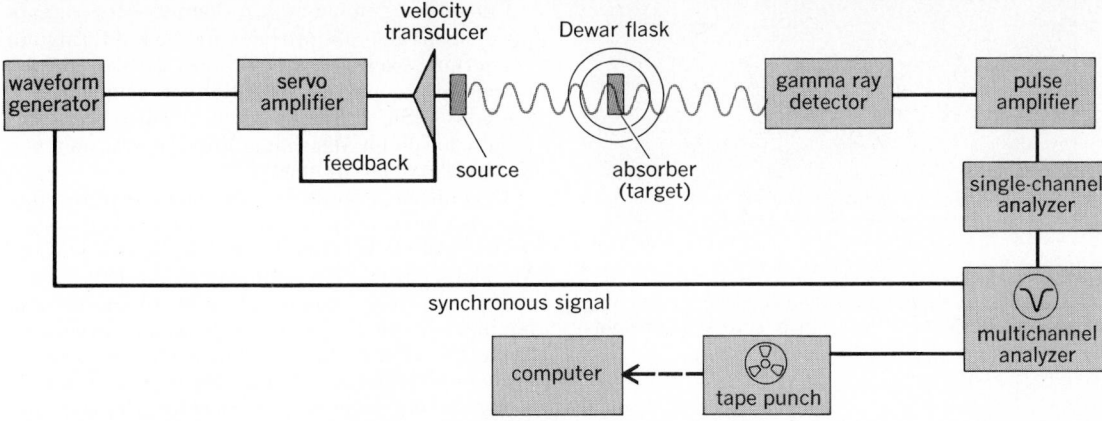

Fig. 1. Experimental arrangement for performing Möss-bauer effect spectroscopy. This typical Mössbauer experiment is with ^{57}Fe or ^{119}Sn. (*From R. H. Herber, Mössbauer spectroscopy, Sci. Amer., 225(4):86–95, October 1971*)

in all directions. *See* ENERGY LEVEL; EXCITED STATE; GROUND STATE.

Energy modulation. Before this phenomenon of resonance fluorescence can be turned into a spectroscopic technique, it is necessary to provide an appropriate energy modulation of the gamma ray emitted in the initial decay process. An estimate of the energy needed to accomplish this can be calculated from a knowledge of the inherent width or sharpness of the excited-state nuclear level. This is given by the Heisenberg uncertainty principle as $\Gamma = h/2\pi\tau$ (h is Planck's constant and τ is the mean lifetime of the excited state). In the case of ^{57}Fe, a nucleus for which resonance fluorescence is especially easy to observe experimentally, $\Gamma = 4.6 \times 10^{-12}$ keV. In order to modulate the emitted gamma-ray energy, which in this case corresponds to 14.4 keV, one can take advantage of the Doppler phenomenon which states that if a radiation source has a velocity relative to an observer of v, its energy will be shifted by an amount equal to $E = (v/c)E_\gamma$. Setting the required Doppler energy equal to the width of the nuclear level and E_γ equal to the nuclear transition energy leads to $v = c(\Gamma/E_\gamma) = 3 \times 10^{10}$ cm/s $\times (4.6 \times 10^{-12}/14.4) = 0.0096$ cm/s, and relative velocities of this order of magnitude can be used to modulate the gamma ray emitted in a typical Mössbauer transition, that is, to "sweep through" the energy width of the nuclear transition. For the experimental demonstration of this effect and its interpretation in terms of the fundamental physical principles involved, Mössbauer was awarded the Nobel Prize in Physics for 1961. *See* DOPPLER EFFECT; RADIO-ACTIVITY; UNCERTAINTY PRINCIPLE.

Experimental realization. The experimental realization of gamma-ray resonance fluorescence can be achieved with the arrangement illustrated schematically in Fig. 1. In a typical Mössbauer experiment the radioactive source is mounted on a velocity transducer which imparts a smoothly varying motion (relative to the absorber, which is held stationary), up to a maximum of several centimeters per second, to the source of the gamma rays. These gamma rays are incident on the material to be examined (the absorber). Some of the gamma rays are absorbed and reemitted in all directions, while the remainder of the gamma rays traverse the absorber and are registered in an ap-

propriate detector which causes one or more pulses to be stored in a multichannel analyzer. The electronics are so arranged that the location (address) in the multichannel analyzer, where the transmitted pulses are stored, is synchronized with the magnitude of the relative motion of source and absorber.

A typical display of a Mössbauer spectrum, which is the result of many repetitive scans through the velocity range of the transducer, is shown in Fig. 2. Such a Mössbauer spectrum is characterized by a position δ of the resonance maximum (corresponding to a minimum in the intensity of the transmitted radiation), a line width Γ, and a resonance effect magnitude ϵ corresponding to the total area A under the resonance curve.

In the case of the Mössbauer active nuclides ^{57}Fe and ^{119}Sn, among others, two additional features which are of great interest to chemists and physicists may be experimentally elucidated. One of these is the quadrupole coupling which is observed if the Mössbauer nuclide is located in an environment where the electric charge distribution does not have cubic (that is, tetrahedral or octahedral) symmetry. Such a spectrum is shown in Fig. 3, in which the magnitude of the quadrupole interaction Δ is equal to $e^2qQ/2$, where e is the electron charge, q is the gradient of the electrostatic field at the nucleus, and Q is the nuclear quadrupole moment. Finally, a Mössbauer spectrum can also give information on the magnitude of

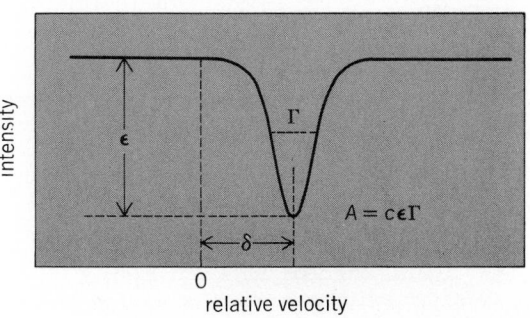

Fig. 2. Mössbauer spectrum of an absorber which gives an unsplit resonance line. The spectrum is characterized by a position δ, a line width Γ, and an area A related to the effect magnitude ϵ.

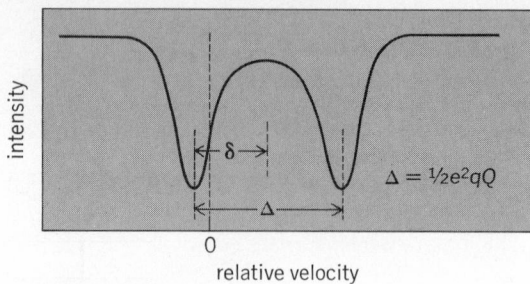

Fig. 3. Mössbauer spectrum of an absorber (containing for example ^{57}Fe or ^{119}Sn) which shows quadrupole splitting Δ.

the magnetic field H_0 acting on the nucleus through the magnetic hyperfine interaction. This is illustrated in Fig. 4, where only a single resonance line would be observed in the absence of a magnetic interaction. *See* HYPERFINE STRUCTURE; NUCLEAR MOMENTS.

Moreover, all of these parameters—δ, Δ, Γ, A, and H_0—are temperature-dependent quantities, and their study over a range of temperatures and conditions can shed a great deal of light on the nature of the environment in which the Mössbauer nuclide is located in the sample under investigation. One hundred Mössbauer transitions, involving 43 different elements, have been experimentally observed and reported.

Application. Mössbauer effect experiments have been used to elucidate problems in a very wide range of scientific disciplines, and only a few examples can be cited as representative of the information extracted from such studies.

Nuclear physics and chemistry. One of the narrowest resonance lines which has been observed is that from the 6.8-μs, 6.2-keV gamma transition in ^{181}Ta, and detailed Mössbauer effect measurements using a source of 140-day ^{181}W have shown that the magnetic moment of the spin 9/2 excited state in ^{181}Ta is $+5.35 \pm 0.09$ nm and the nuclear quadrupole moment of this state is $(+4.4 \pm 0.05) \times 10^{-24}$ cm^{-2}. Such data are of considerable use to nuclear physicists in refining models which describe the fundamental interaction forces in the nucleus. Similarly, the 93.26-keV resonance in ^{67}Zn has been used to determine the magnetic moment of the 1/2$^-$ state (spin = 1/2, negative parity) in this nuclide and leads to the conclusion that 1/2$^-$ and 5/2$^-$ states can be considered minus-quasiparticle states. Such nuclear information is difficult to obtain by non-Mössbauer effect methods. *See* NUCLEAR STRUCTURE.

Recoilless gamma-ray resonance experiments have been able to provide detailed information concerning excited-state lifetimes involved in the nuclear decay process. The lifetime values for the nuclides ^{119}Sn, ^{197}Au, and ^{73}Ge, among others, are largely based on Mössbauer effect measurements. *See* NUCLEAR ISOMERISM.

Recoilless gamma fluorescence spectroscopy has also been used to study the chemical consequences of nuclear decay, and the lifetimes of the Mössbauer transition (typically about 10^{-8} s) provide a convenient time scale with which one can distinguish rapid electronic relaxation processes (typically 10^{-12} to 10^{-14} s) from atomic translation processes (typically slower than 10^{-6} s), and thus study the chemical fate of an atom which results from the decay of a radioactive parent nuclide.

Solid-state physics. Mössbauer effect spectroscopy has made significant contributions to the study of problems in solid-state physics, especially of the nature of the magnetic interactions in iron-containing alloys and the dependence of the magnetic field on composition, temperature, pressure, and other parameters which are of importance in metallurgical processes, solid-state-device fabrication, the structural use of metals and alloys, and numerous related problems of great practical importance.

Combining Mössbauer effect spectroscopy with vibrational spectroscopic studies has led to a clearer understanding of the nature of inter- and intramolecular forces and the relationship of these forces to the properties of polymeric materials.

It has also been possible, using this technique, to study the effect of high pressure and isotropic compressibility on the chemical properties of materials, especially in the case of experiments with ^{57}Fe, ^{181}Ta, and ^{119}Sn. Such studies have led to the design of high-pressure processes in preparative metallurgy and materials science.

At the extremely low end of the temperature scale, Mössbauer effect spectroscopy has been useful in examining the nature of those materials which become superconductive at sufficiently low temperatures and the relationship between chemical composition and structure on the one hand and the superconductive transition on the other, as in the dichalcogen layer compounds TaS$_2 \cdot$ Sn and TaS$_2 \cdot$ Sn$_{1/3}$ (both studied using the nuclide ^{119}Sn). Nb$_3$Sn, a material widely used in the construction of superconducting magnets, has been subjected to detailed Mössbauer effect investigations. *See* SUPERCONDUCTIVITY.

Structural chemistry. The two Mössbauer nuclides most widely exploited by chemists are ^{57}Fe and ^{119}Sn, although a growing body of data resulting from experiments with ^{129}I, ^{99}Ru, ^{121}Sb, and others has been reported. The position of the resonance maximum δ, also called the isomer or chemical shift, can be related to the systematics of the electron configuration of the atom, and extensive isomer shift correlations for iron- and tin-containing compounds have been tabulated. In particular, the isomer shift of tin compounds is readily related to the oxidation state, since it has been observed that all stannous (Sn^{+2}) isomer shifts are larger than that observed for metallic tin (β-Sn), while those for stannic compounds (Sn^{+4}) are smaller than this value. This observation allows an assignment of oxidation state to be made on the basis of

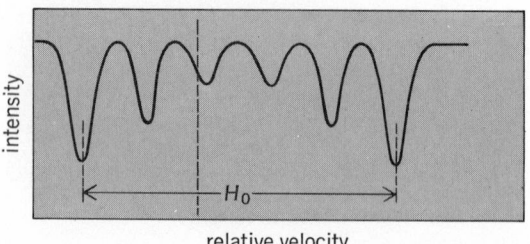

Fig. 4. Mössbauer spectrum of metallic iron showing the splitting of the resonance line by the internal magnetic field ($H_0 = 330$ kG at room temperature).

the isomer shift parameter, as in the two-dimensional layer compound $SnTa_3S_6$ in which the tin atom is clearly identified as a stannous ion, contrary to expectations based on theory.

Similarly, the isomer shifts reported for a number of ruthenium compounds, which have been studied using the 89.36-keV resonance in ^{99}Ru, can be correlated systematically with the number of $4d$ electrons involved in the bonding of the metal atom to its nearest-neighbor ligands. Such Mössbauer effect studies have led to a clearer understanding of the nature of "ruthenium red," a trimeric ammonia ruthenium oxide, and a number of other compounds of this relatively rare transition-metal homolog of iron.

In the field of organometallic chemistry, Mössbauer effect spectroscopy has served to clarify the structure of a number of compounds of iron and tin which are of considerable synthetic and industrial importance, including $Fe_3(CO)_{12}$, $[(\pi C_5H_5)_2Fe(CO)_2]_2$, $[(C_4H_9)_3Sn]_2SO_4$, and the organotin thioglycolates which are used as stabilizers in the plastics industry.

Biological science. Many molecules of biological importance, including hemoproteins, iron-sulfur proteins, and iron storage and transport proteins, offer an ideal system in which Mössbauer effect spectroscopy can be used to elucidate the structure and bonding properties of the metal atom in complex systems. The first measurements on such molecules were reported in 1961, and a very large number of iron-containing systems have been studied since then. Paramagnetic iron compounds can be studied at temperatures below the magnetic ordering point (Néel temperature), and it is thus possible, by means of Mössbauer effect spectroscopy, to determine the sign and magnitude of the magnetic field acting on an iron atom in a complex biological material with molecular weights ranging up to 50,000 or more.

It is also possible to study antiferromagnetically coupled iron atoms in biological molecules by carrying out the Mössbauer effect measurements in an external magnetic field over a range of temperatures. Typical of such a study is that of oxidized and reduced putidaredoxin, an iron-sulfur protein (molecular weight = 12,500) which acts as a one-electron transfer enzyme. The Mössbauer experiments on this material clearly showed that in the oxidized material the two irons atoms in the molecule occupy chemically equivalent sites. On one electron reduction, one iron atom remains ferric (Fe^{3+}), while the other becomes ferrous (Fe^{2+}), and the two atoms couple antiferromagnetically to give an electronic ground state of $S = 1/2$. Such detailed knowledge of the chemical behavior of the iron atoms in this molecule can elucidate the action of biological catalysts (enzymes) on the molecular level. *See* ANTIFERROMAGNETISM.

Related fields. Mössbauer effect studies have also played a role in studies in many related fields of science, including archeology, geology, engineering studies, theoretical (relativity) physics, chemical kinetics, and biology. The samples of surface material returned from the Moon by the United States Apollo program have been carefully scrutinized by Mössbauer techniques, as have core samples extracted from deep-drilling experiments on the Earth's outer layer. The geographical distribution of ancient Greek pottery has been traced by

making use of characteristic Mössbauer effect data, and the pigments used in painting and decorating have been similarly investigated using this technique.

[ROLFE H. HERBER]

Bibliography: G. M. Bancroft, *Mössbauer Spectroscopy*, 1973; S. G. Cohen and M. Pasternak (eds.), *Perspectives in Mössbauer Spectroscopy*, 1973; H. Frauenfelder, *The Mössbauer Effect*, 1962; V. I. Goldanskii and R. H. Herber (eds.), *Chemical Applications of Mössbauer Spectroscopy*, 1968; N. N. Greenwood and T. C. Gibb, *Mössbauer Spectroscopy*, 1971; I. J. Gruverman (ed.), *Mössbauer Effect Methodology*, vols. 1–9, 1965–1974; International Atomic Energy Agency, *Mössbauer Spectroscopy and its Applications*, 1972; J. G. Stevens and V. Stevens, *Mössbauer Effect Data Index*, 1969–1973; G. K. Wertheim, *Mössbauer Effect; Principles and Applications*, 1964.

Motion

If the position of a material system as measured by a particular observer changes with respect to time, that system is said to be in motion with respect to the observer. Absolute motion, then, has no significance, and only relative motion may be defined; what one observer measures to be at rest, another observer in a different frame of reference may regard as being in motion. *See* FRAME OF REFERENCE; RELATIVE MOTION.

The time derivatives of the various coordinates used to specify the system may be used to prescribe the motion at any instant of time. How the motion develops in subsequent instants is then determined by the laws of motion. In classical dynamics it is supposed that in principle the motion and configuration of the system may be specified to an arbitrary precision, although in quantum mechanics it is recognized that the measurement of the one disturbs the other.

For a system of f degrees of freedom, the motion may be represented by a point in an f-dimensional velocity space, the coordinates of which are the time rates of change of the coordinates that describe the configuration of the system. For a system under no forces that is described by rectangular cartesian coordinates, these time derivatives are constants. For a single particle, this result is the first of Newton's laws of motion, namely, that a particle remains at rest or in a state of uniform motion unless acted upon by an external force.

The most general theory of motion that has yet been developed is quantum field theory, which combines both quantum mechanics and relativity theory, as well as the experimentally observed fact that elementary particles can be created and annihilated. *See* DEGREE OF FREEDOM; DYNAMICS; EULER'S EQUATIONS OF MOTION; HAMILTON'S EQUATIONS OF MOTION; HARMONIC MOTION; KINEMATICS; KINETICS; LAGRANGE'S EQUATIONS; NEWTON'S LAWS OF MOTION; OSCILLATION; PERIODIC MOTION; QUANTUM FIELD THEORY; QUANTUM MECHANICS; RECTILINEAR MOTION; RELATIVITY; ROTATIONAL MOTION.

[HERBERT C. CORBEN/BERNARD GOODMAN]

Multipole radiation

Gamma rays, internal conversion electrons, or positron-electron pairs of defined characteristics emitted from an atom when the nucleus makes a

Spin and parity changes for various multipoles

ΔI	0	1	2	3	4	5
Multipole	Monopole	Dipole	Quadrupole	Octupole	2^4-pole	2^5-pole
Electric	E0 no	E1 yes	E2 no	E3 yes	E4 no	E5 yes
Magnetic		M1 no	M2 yes	M3 no	M4 yes	M5 no

transition between two energy states. The multipole order is the number of units of angular momentum removed by the radiation. This number is not necessarily equal to the difference between the spins of the nucleus in its initial and final states because the nuclear spin direction may change. Thus a quadrupole radiation will result when a state of spin 2 makes a transition to a state of spin 0, but a transition from a state of spin 2 to one of spin 1 may also result in quadrupole radiation if there is an appropriate change in nuclear spin direction. If I_i and I_f are the spins of the nucleus in its initial and final states, then the multipole order ΔI must be as in the equation shown below in order to conserve angular momentum.

$$|I_i - I_f| \leq \Delta I \leq |I_i + I_f|$$

From multipole radiation measurements, the static and dynamic properties of nuclear energy states may be determined, and this information may be used in theories of nuclear structure.

In addition to the energy and angular momentum values, a third characteristic of a nuclear state is the parity. The parity of a given state is odd ($-$) or even ($+$), depending on whether the quantum-mechanical wave function describing the nucleus in that state changes sign upon transposing the function to a reflected coordinate system. There are two classes of multipole radiation, the electric and the magnetic, and the designation of a given radiation depends upon both the angular momentum change and whether the parities of the initial table illustrates the spin and parity changes for the various multipoles. *See* PARITY; SPIN.

The characteristics of multipole radiations may be determined experimentally by measurements of one or more of the following: the angular correlation of the γ-ray with another coincident or cascade γ-ray; the relative intensities of internal conversion electrons and γ-rays; the relative intensities of internal conversion electrons from various electron shells of the atom; the characteristics of internal positron-electron pair formation; and the half-life of the transition. Nuclear transitions generally take place with emission of the lowest possible order of multipole radiation, because the higher the order the longer will be the half-life. *See* RADIOACTIVITY; SELECTION RULES.

[DAVID E. ALBURGER]

Bibliography: J. M. Blatt and V. F. Weisskopf, *Theoretical Nuclear Physics*, 1952, reprint 1979; W. E. Burcham, *Elements of Nuclear Physics*, 1979; W. E. Meyerhof, *Elements of Nuclear Physics*, 1967; M. A. Preston and R. K. Bhaduri, *Structure of the Nucleus*, 1975.

Muonium

An exotic atom, Mu or (μ^+e^-), formed when a positively charged muon (μ^+) and an electron are bound by their mutual electrical attraction. It is a light, unstable isotope of hydrogen, with a muon replacing the proton. Muonium has a mass 0.11 times that of a hydrogen atom due to the lighter mass of the muon, and a mean lifetime of 2.2 μs, determined by the spontaneous decay of the muon ($\mu^+ \rightarrow e^+ \nu_e \bar{\nu}_\mu$).

Muonium is formed when beams of μ^+ produced in particle accelerators are stopped in certain nonmetallic targets. The μ^+ beams are generally spin-polarized; that is, the average spin angular momentum of the muons in the beam points in a definite spatial direction. The muon spin retains this spatial orientation after picking up an electron to form muonium in the reaction $\mu^+ + X \rightarrow Mu + X^+$, where X represents an atom of the target material. Since the positron in μ^+ decay is emitted preferentially along the muon spin direction, the muonium polarization can be monitored very simply by measuring the spatial distribution of the decay positron. This technique of polarization measurement allows extremely small amounts of muonium to be detected and studied. *See* SPIN.

Muonium was first observed by V. W. Hughes and coworkers in 1960. The characteristic Larmor precession of the muonium polarization in a magnetic field when positive muons were stopped in a target of argon gas indicated the formation of muonium. The observed Larmor frequency of 14 GHz/tesla is a unique signature of the triplet ($F = 1$) bound state of an electron and a muon. *See* LARMOR PRECESSION.

Since muonium is a system consisting only of leptons, it serves as an ideal testing ground for the theory of quantum electrodynamics (QED), which describes the electromagnetic interaction between particles. Indeed, experimental measurement of the hyperfine structure interval of the ground-state muonium levels have shown that the quantum electrodynamic theory of this system is accurate to the level of one part per million. Such measurements also provide the best available values for the muon mass and magnetic moment. *See* LEPTON; QUANTUM ELECTRODYNAMICS.

Muonium chemistry and muonium spin rotation (MSR) are two rapidly developing subfields which concentrate on the study of muonium in matter. Chemical reaction rates for a wide variety of muonium reactions have been measured and compared with hydrogen rates. The formation, spin precession, and depolarization of muonium in gases, semiconductors, and insulators have been studied in some detail. Such experiments seek to understand the chemical and physical behavior of a light hydrogen isotope in matter, and to use the extremely sensitive muon polarization measurement technique to probe the structure of materials. *See* POSITRONIUM.

[PATRICK O. EGAN]

Bibliography: V. W. Hughes, Muonium, *Phys. Today*, 20(12):29, 1967.

Navier-Stokes equations

Three scalar partial differential equations that describe conservation of momentum for the motion of a viscous, incompressible fluid. They may be expressed vectorially as one equation, Eq. (1),

$$\rho \frac{\partial \mathbf{v}}{\partial t} + \rho (\mathbf{v} \cdot \boldsymbol{\nabla}) \mathbf{v} = -\nabla p + \rho \mathbf{f} + \mu \, \nabla^2 \mathbf{v} \qquad (1)$$

where ρ is fluid density, $\mathbf{v}$ is fluid velocity vector, p is fluid pressure, $\mathbf{f}$ is body force (such as gravity) per unit mass, μ is fluid viscosity coefficient, and t is time. These equations, together with the continuity relation, $\boldsymbol{\nabla} \cdot \mathbf{v} = 0$, and suitable boundary conditions determine the flow field; for example, $\mathbf{v}$ and p are determined as functions of position in space and of time. One of these boundary conditions is that of no slip at the surface of a body; that is, the fluid immediately at the body surface "sticks" to it and thus has the same velocity as the surface itself. *See* NEWTONIAN FLUID.

Few mathematical solutions to this complicated set of nonlinear partial differential equations are known, except for simple geometries. The importance of viscosity in determining the flow depends on the relative size of the body. Approximations to the Navier-Stokes equations for small Reynolds number Re give good results. For $Re < 1$, the acceleration forces, those on the left-hand side of the equation, are negligible, leaving only linear terms on the right. Such an approximation is called a Stokes-flow approximation, and one of the most famous applications, made by R. A. Millikan, is to the slow motion of a tiny spherical oil droplet in air. Lubrication theory makes use of the Stokes-flow approximation as well as even further approximations. For $Re \gg 1$, the effects of viscosity are confined to a thin layer near the surface of bodies in the fluid. Outside this layer the fluid acts essentially as an inviscid fluid, and this behavior is the reason that inviscid fluid theory is of any use at all. *See* BOUNDARY-LAYER FLOW; REYNOLDS NUMBER.

For a compressible, viscous fluid, the viscous term $\mu \, \nabla^2 \mathbf{v}$ must be replaced by the divergence of the viscous stress tensor in which the bulk viscosity coefficient λ occurs. Using rectangular cartesian coordinates x_1, x_2, x_3, the force in the x_i direction ($i = 1,2,3$) can be written as notation (2), where

$$\sum_{j=1,2,3} \frac{\partial \tau_{ij}}{\partial x_j} \qquad (2)$$

Eqs. (3) and (4) apply. Here μ and λ are functions

$$\tau_{ij} = (\lambda - {}^2\!/_3\mu)\, \delta_{ij} (\boldsymbol{\nabla} \cdot \mathbf{v}) + \mu \left(\frac{\partial u_i}{\partial x_j} + \frac{\partial u_j}{\partial x_i} \right) \qquad (3)$$

$$\delta_{ij} = \begin{cases} 1 & i=j \\ 0 & i \neq j \end{cases} \qquad (4)$$

of temperature and, in liquids, of pressure. *See* FLUID-FLOW PRINCIPLES.

[ARTHUR E. BRYSON JR.]

Negative temperature

The property of a thermodynamical system which satisfies certain conditions and whose thermodynamically defined absolute temperature is negative. The essential requirements for a thermodynamical system to be capable of negative temperature are: (1) the elements of the thermodynamical system must be in thermodynamical equilibrium among themselves in order for the system to be described by a temperature at all; (2) there must be an upper limit to the possible energy of the allowed states of the system; and (3) the system must be thermally isolated from all systems which do not satisfy both of the first two requirements, that is, the internal thermal equilibrium time among the elements of the system must be short compared to the time during which appreciable energy is lost to or gained from other systems.

The second condition must be satisfied if negative temperatures are to be achieved with a finite energy. If W_m is the energy of the mth state for one element of the system, then in thermal equilibrium the number of elements in the mth state is proportional to the Boltzmann factor exp $(-W_m/kT)$, where k is Boltzmann's constant and T is the absolute temperature. For negative temperatures, the Boltzmann factor increases exponentially with increasing W_m, and the high-energy states are therefore occupied more than the low-energy ones, which is the reverse of the positive-temperature case. Consequently, with no upper limit to the energy, negative temperatures could not be achieved with a finite energy. Most systems do not satisfy this condition; for example, there is no upper limit to the possible kinetic energy of a gas molecule. It is for this reason that systems of negative temperatures occur only rarely. *See* KINETIC THEORY OF MATTER; STATISTICAL MECHANICS.

Systems of interacting nuclear spins, however, have the characteristic that under suitable circumstances they can satisfy all three of the above conditions, in which case the nuclear spin system can be at negative absolute temperature, as first demonstrated by R. V. Pound, E. M. Purcell, and N. F. Ramsey and discussed theoretically by Ramsey and others. *See* NUCLEAR ORIENTATION.

From the point of view of thermodynamics, the absolute temperature T is given by the equation below, where S is the entropy, U the internal en-

$$T = \left(\frac{\partial S}{\partial U} \right)_B^{-1}$$

ergy, and B the externally applied magnetic field. A system at negative absolute temperature must therefore be in a condition such that the slope of S as a function of U is negative. This is clearly achievable for a spin system in an external field B, since the entropy S falls to a low value when all the nuclear spins are in their highest energy state and thereby all pointing in the same direction. *See* ENTROPY; THERMODYNAMIC PRINCIPLES.

It is apparent either from the Boltzmann factor or from the above equation applied to a spin system that the transition between positive and negative temperatures is through infinite temperature, not absolute zero; negative absolute temperatures should therefore not be thought of as colder than absolute zero, but as hotter than infinite temperature. *See* ABSOLUTE ZERO; TEMPERATURE.

[NORMAN F. RAMSEY]

Bibliography: E. M. Purcell and R. V. Pound, A nuclear spin system at negative temperature, *Phys. Rev.*, 81:279, 1951; N. F. Ramsey, Thermodynamics and statistical mechanics at negative absolute temperatures, *Phys. Rev.*, 103:20−28, 1956; N. F. Ramsey and R. V. Pound, Nuclear audiofrequency spectroscopy by resonant heating of nuclear spin system, *Phys. Rev.*, 81:278, 1951.

Neutrino

An elementary particle designated by the Greek symbol ν, with zero rest mass and zero electric charge. It is classified as a lepton (the other classes of elementary particles being the photon, meson, and baryon). The elementary particles are known to interact with each other by four basic types of forces: strong nuclear, electromagnetic, weak, and gravitational, in order of decreasing strength. All particles with nonzero electric charge, as well as the photon, have electromagnetic interactions, and the mesons and baryons have strong interactions. The neutrino is the only known particle that has only weak interactions. Thus, the neutrino is a unique tool in the study of weak forces, since the interactions are free of the effects of the strong interactions and the electromagnetic interactions, which are many orders of magnitude stronger than the weak interactions. For this reason, ν interactions have been the subject of active study at large particle accelerators. *See* ELEMENTARY PARTICLE; FUNDAMENTAL INTERACTIONS; LEPTON; WEAK NUCLEAR INTERACTIONS.

Basic properties. The existence of the neutrino was postulated in 1930 to explain the apparent nonconservation of energy in beta-decay process $n \rightarrow p + e^- + \bar{\nu}$. It was not until 1953 that the existence of the neutrino was experimentally verified by observation of the interactions caused by free neutrinos. Quantitative studies of ν interactions, undertaken in 1961 by utilizing neutrinos produced by the Brookhaven National Laboratory alternating gradient synchrotron resulted in the discovery that there are two distinct neutrinos—the electron neutrino ν_e, associated with beta decay, and the muon neutrino, associated with pion decay, $\pi^+ \rightarrow \mu^+ + \nu_\mu$. From the absence of neutrinoless double beta decay, it can be inferred that ν_e is not identical to its antiparticle $\bar{\nu}_e$. It is thus believed that there are four distinct neutrinos, namely, ν_e, ν_μ, $\bar{\nu}_e$, and $\bar{\nu}_\mu$.

In the late 1970s a new heavy charged lepton, the tau ($\tau^\pm$), was discovered to exist in addition to the previously known charged leptons, the electron ($e^\pm$) and the muon ($\mu^\pm$). There is some theoretical speculation that two new additional types of neutrinos, the tau-neutrino (ν_τ) and the anti-tau-neutrino ($\bar{\nu}_\tau$), which are associated with this new heavy lepton, exist in addition to the four types of neutrinos mentioned above, and experimental searches for these new types of neutrinos have been undertaken.

Considerations of angular momentum conservation in pion decay establish the spin of the neutrino to be ½ in units of $h/2\pi$, where h is Planck's constant. The masses of the neutrinos are experimentally consistent with zero; the best upper limits are $M_{\nu_e} \leq 60$ eV and $M_{\nu_\mu} \leq 1.2$ MeV. The experimental upper limit on the neutrino charge is less than 10^5 times the charge of the electron. It is therefore generally assumed that the rest mass and electric charge of the neutrinos are identically zero. However, there has been some experimental indication and theoretical speculation that the mass of at least some of the neutrinos is not identically zero but has some finite, although very small, value. This possibility has fundamental cosmological implications since astrophysicists believe that there are a very large number of neutrinos in the universe, and if the neutrinos had a nonzero mass they would constitute a significant fraction of the total mass of the universe. Nonzero-mass neutrinos might also lead to some interesting experimental consequences, generally referred to as neutrino oscillations, in which the different types of neutrinos could transmute into each other. *See* SPIN.

According to the rules of quantum mechanics, a massless spin-1/2 particle such as the neutrino can have its spin lined up along its direction of motion or opposite to its direction of motion; these are called the helicity states, right-handed and left-handed respectively. Thus, in general, such a particle has four components—the particle and its antiparticle, right-handed and left-handed. It was, however, predicted by the two-component theory of the neutrino, and shown by experiment, that only two of the four components exist for the neutrinos—left-handed neutrinos and right-handed antineutrinos. *See* HELICITY; RELATIVISTIC QUANTUM THEORY.

Study of interactions. The interactions of neutrinos with electrons, protons, and neutrons have been studied in a number of experiments at large proton accelerator laboratories: the 12-GeV zero-gradient synchrotron at the Argonne National Laboratory (Argonne, IL); the 28-GeV proton synchrotron (the PS) and the 400-GeV proton synchrotron (the SPS) at CERN (Geneva, Switzerland); the 30-GeV alternating-gradient synchrotron at Brookhaven National Laboratory (Upton, NY); the 70-GeV proton synchrotron at Serpukhov (Soviet Union); and the 400-GeV proton synchrotron at the Fermi National Accelerator Laboratory or Fermilab (Batavia, IL). At all of these laboratories, the synchrotrons are used to accelerate on the order of 10^{12} protons/sec to their peak energy. The proton beam then is focused onto a small metal target in which, among other things, π and K mesons are produced. Magnetic focusing lenses are used to form a beam of these mesons, which are then allowed to drift in a decay path, where their decays ($\pi \rightarrow \mu + \nu_\mu$, or $K \rightarrow \mu + \nu_\mu$, for example) are the source of the desired neutrinos. The decay path is followed by a thick shield (made out of iron or earth) that will absorb the μs, and πs and Ks that did not decay, but allow the neutrinos to pass through. *See* MESON.

The neutrino detector, in which the neutrino interactions occur, is located beyond the shield. Typically, 10^9 neutrinos traverse the detector per second. The probability of interactions for the neutrinos is so small that typically less than 1 of these 10^9 neutrinos interact in the detector. Two kinds of neutrino detectors have been used: large bubble chambers, containing $10,000-40,000$ liters of liquid hydrogen, deuterium, or some other liquid; and spark chamber detectors, with aluminum, iron, or lead plates, with total masses in the range $10-100$ tons (9–90 metric tons). The interactions of the neutrinos in these detectors are studied by the observation of tracks left by the charged particles produced in the interactions.

The weak interactions are theoretically understood to be a current-current type of interaction, where the interaction is mediated by an exchange current or particle (the intermediate boson), much as the electromagnetic interaction is mediated by the exchange of a photon (see illustration). Since the photon γ carries no electric charge, there is no change in charge between the incoming and the outgoing electron (the upper vertex in illustration *a*). The weak interactions can be classified in two types: the charged current interactions, in which the exchanged current or particle (the $W^\pm$) carries one unit of electric charge, and thus, for example, an incoming neutral lepton such as the ν_μ in illustration *b* is changed into a charged lepton, the μ^-; and the neutral current interactions, where the exchanged current or particle (the Z^0) carries no electric charge, and thus an incident neutral ν_μ remains an outgoing ν_μ (illustration *c*).

Charged current interactions. A large number of experiments have been carried out studying charged current neutrino interactions such as the quasielastic process, $\nu_\mu + n \rightarrow \mu^- + p$, single-pion production $\nu_\mu + p \rightarrow \mu^- + p + \pi^+$, and the inclusive inelastic process, $\nu_\mu + N \rightarrow \mu^- + $ hadrons, as well as similar reactions with incident $\bar{\nu}_\mu$. The most striking result is related to the study of the internal structure of the proton and the neutron. According to the currently favored quark-parton model, the proton and the neutron are not the simple fundamental particles they were originally believed to be but are composites made up of more fundamental constituents called quarks or partons. The scattering of neutrinos by neutrons or protons can be used to study this internal structure, just as the scattering of alpha particles (Rutherford scattering) was used to investigate the internal structure of the atom. The neutrino scattering data were thus very important in establishing the basic correctness of the quark-parton model. A more complete theory of the forces between the quarks that bind them together in the proton and the neutron is called quantum chromodynamics (QCD). A detailed analysis of the neutrino scattering data has both provided a test of the validity of the QCD theory and yielded measurements of the distributions of the different kinds of quarks inside the proton and the neutron. *See* QUANTUM CHROMODYNAMICS; QUARKS.

Neutral current interactions. Since their discovery in 1973, the neutral current interactions of neutrinos have been studied extensively in a large variety of processes such as neutrino-electron scattering, $\nu + e^- \rightarrow \nu + e^-$, neutrino-proton scattering, $\nu + p \rightarrow \nu + p$, single-pion production, $\nu + p \rightarrow \nu + p + \pi^0$, and inclusive neutrino scattering, $\nu + p \rightarrow \nu + $ any number of hadrons. The extensive amount of data collected on these processes can all be understood in terms of the Weinberg-Salam model of the weak and the electromagnetic interactions. In this model there are two families of intermediate bosons that serve as the exchanged particles mediating these interactions: a triplet, the W^+, W^0, and W^-, and a singlet, the B^0 (the superscripts indicate the sign of the electric charge carried by the particles). The neutral members of these families form two quantum-mechanical mixtures which are the physically observable particles, the γ and the Z^0, according to the equations below. In this model the $W^\pm$ mediate the

$$\gamma = -\sin\theta\, W^0 + \cos\theta\, B^0$$

$$Z^0 = \cos\theta\, W^0 + \sin\theta\, B^0$$

charged current weak interactions, the Z^0 mediates the neutral current weak interactions, and

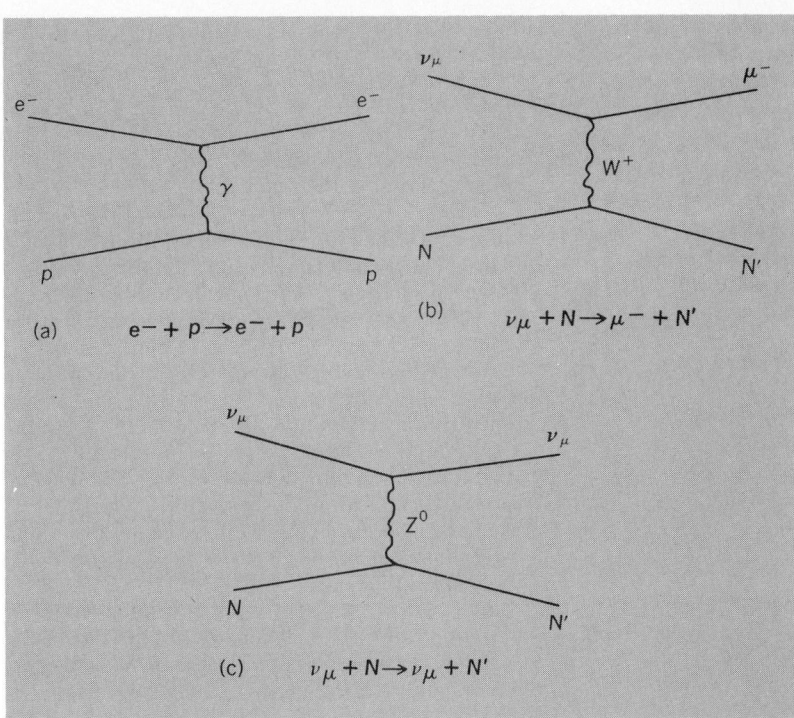

(a) $e^- + p \rightarrow e^- + p$

(b) $\nu_\mu + N \rightarrow \mu^- + N'$

(c) $\nu_\mu + N \rightarrow \nu_\mu + N'$

Diagrams for various interactions. (*a*) Electromagnetic interaction. (*b*) Charged-current weak interaction. (*c*) Neutral-current weak interaction.

the γ is the well-known photon that mediates the electromagnetic interactions (see illustration). Thus this model unifies the electromagnetic and the weak forces into a single force, the electroweak force, reducing the number of basic forces or interactions of nature from four (strong, electromagnetic, weak, and gravitational) to three.

The experimental data determine the value of the mixing angle to be $\sin^2\theta = 0.23 \pm 0.02$. With this value the model predicts the masses of the intermediate bosons $W^\pm$ and Z^0 to be 78 and 88 GeV, respectively (the mass of the proton is approximately 1 GeV). Presently existing particle accelerators do not have sufficient energy to produce such massive particles. However, extensive experimental searches for these very fundamental particles are planned for the next generation of facilities now under construction at the CERN laboratories in Switzerland and at the Fermi National Accelerator Laboratory and Brookhaven National Laboratory in the United States.

Charm production by neutrinos. One difficulty encountered in the early days of the Weinberg-Salam model was that the neutral current interactions predicted by the model and observed in neutrino interactions, as discussed above, also predicted certain K-meson decays such as $K^0 \rightarrow \mu^+ + \mu^-$, which were not observed experimentally. This difficulty was resolved by S. L. Glashow, J. Iliopoulos, and L. Maiani, who postulated a new quantum number (or property) of hadrons that they called charm, such that the effects of this new quantum number canceled out the unwanted neutral current contributions to K decays. This new quantum number also implied the existence of a new and until then experimentally unobserved class of particles, the so-called charmed particles. The first experimental evidence for the existence of this new class of particles came from neutrino interactions. Experiments at the Fermi National Accelerator lab observed events of the type $\nu_\mu + N \rightarrow \mu^+ + \mu^- +$ hadrons and $\nu_\mu + N \rightarrow \mu^- + e^+ + K^0 +$ hadrons, where the μ^+ or the e^+ and the K^0 were the decay products of the new charmed particles. An experiment at Brookhaven National Laboratory found an event of the reaction $\nu_\mu + p \rightarrow \mu^- + \Lambda^0 + \pi^+ + \pi^+ + \pi^+ + \pi^-$, where the Λ^0 and the four πs were interpreted to be the decay products of a charmed particle. Conclusive evidence for the existence of charmed particles was obtained in $e^+ + e^-$ collisions at the Stanford Linear Accelerator Center in California and in photoproduction experiments at Fermilab. Subsequently the production of four different charmed particles has been seen in neutrino reactions: the $D(1.86)$, the $D^*(2.01)$, the $\Lambda_c(2.26)$ and the $\Sigma_c(2.43)$, where the numbers in parentheses are the masses of the particles in units of GeV. The properties of these particles as produced in neutrino interactions have been found to be consistent with the predictions of the model of Glashow, Iliopoulus, and Maiani. *See* CHARM.

[CHARLES BALTAY]

Bibliography: J. S. Allen, *The Neutrino*, 1958; C. Baltay, Interactions of muon neutrinos, in V. W. Hughes and C. S. Wu (eds.), *Muon Physics*, ch. 4, 1975; C. Baltay and K. Tittel, *Proceedings of the 19th International Conference on High Energy Physics*, Tokyo, July 1978; D. C. Cundy, Report on neutrino physics, in *Proceedings of the 17th International Conference on High Energy Physics, London*, pp. 1–131, 1974; L. M. Lederman, Neutrino physics, in E. H. S. Burhop (ed.), *Elementary Particle Physics*, vol. 2, 1967; D. H. Perkins, Neutrino interactions, in J. D. Jackson and A. Roberts (eds.), *Proceedings of the 16th International Conference on High Energy Physics*, vol. 4, 1972.

Neutron

An elementary particle having approximately the same mass as the proton, but lacking a net electric charge. It is indispensable in the structure of the elements, and in the free state it is an important reactant in nuclear research and the propagating agent of fission chain reactions. Neutrons, in the form of highly condensed matter, constitute the substance of neutron stars.

Neutrons in nuclei. Neutrons and protons are the constituents of atomic nuclei. The number of protons in the nucleus determines the chemical nature of an atom, but without neutrons it would be impossible for two or more protons to exist stably together within nuclear dimensions, which are of the order of 10^{-13} cm. The protons, being positively charged, repel one another by virtue of their electrostatic interactions. The presence of neutrons weakens the electrostatic repulsion, without weakening the nuclear forces of cohesion. In light nuclei the resulting balanced, stable configurations contain protons and neutrons in almost equal numbers, but in heavier elements the neutrons outnumber the protons; in ^{238}U, for example, 146 neutrons are joined with 92 protons. Only one nucleus, ^{1}H, contains no neutrons. For a given number of protons, neutrons in several different numbers within a restricted range often yield nuclear stability—and hence the isotopes of an element. *See* ISOTOPE; NUCLEAR STRUCTURE; PROTON.

Sources of free neutrons. Free neutrons have to be generated from nuclei, and since they are bound therein by cohesive forces, an amount of energy equal to the binding energy must be expended to get them out. Usually the binding energy for each neutron amounts to 1–1.4 picojoules (6–8 MeV). Nuclear machines, such as cyclotrons and electrostatic generators, induce many nuclear reactions when their ion beams strike target material. Some of these reactions release neutrons, and these machines are sources of high neutron flux. If the accelerator is sharply pulsed, the time-of-flight method can be used for accurate energy resolution of the neutrons up to energies of about 0.3 pJ (2 MeV), the flight paths being up to 200 m long. *See* NEUTRON SPECTROMETRY; NUCLEAR BINDING ENERGY; NUCLEAR REACTION.

There are several kinds of portable neutron sources. Some consist of an intimate mixture of an α-emitting radionuclide with beryllium powder. Neutrons are released from the nuclear reaction ^{9}Be$(\alpha,n)^{12}$C, which is the reaction by which the neutron was discovered in 1932. Intense sources that emit about 5×10^{10} n/s are now made, using mixtures of beryllium with, for example, ^{238}Pu (half-life 89 years) or ^{241}Am (half-life 458 years). Such a source generates several hundred watts of heat. Pure ^{252}Cf (half-life 2.65 years) needs no admixture of beryllium, because neutrons are emitted in its spontaneous fission; such a source is especially compact. The neutrons emitted by the foregoing sources have energies that extend up to 0.8–1 pJ (5–6 MeV). A source that gives neutrons of lower energy is the Sb-Be photoneutron source. Here the 1.70-MeV γ-rays of ^{124}Sb (half-life 60 days) slightly exceed the binding energy of neutrons in beryllium (1.67 MeV), so the neutrons have an energy of $1.70 - 1.67 = 0.03$ MeV (4.6 femtojoules).

Neutrons are released in the act of fission, and nuclear reactors are unexcelled as intense neutron sources. The absorption of one neutron by a ^{235}U nucleus is required to induce fission, but 2.5 neutrons are on the average released; this regeneration makes possible the nuclear chain reaction. A powerful research reactor may generate neutrons in such abundance that 1 cm^2 of a sample placed therein would be traversed by 10^{15} neutrons per second. A hole through the surrounding shield can yield a collimated beam having a unidirectional flux of 10^9 neutrons/(cm^2)(s). The explosion of a 10-kiloton (4×10^{13} J) nuclear bomb releases about 10^{30} neutrons in about 1 μs. *See* CHAIN REACTION; NUCLEAR FISSION.

Neutrons occur in cosmic rays, being liberated from atomic nuclei in the atmosphere by collisions of the high-energy primary or secondary charged particles. They do not themselves come from outer space. For information on neutrons of another origin *see* DELAYED NEUTRON. *See also* COSMIC RAYS.

Penetrating power. Neutrons resulting from nuclear reactions usually possess kinetic energies of the order of 0.2 pJ (1 MeV). Having no electric charge, they interact so slightly with atomic electrons in matter that energy loss by ionization and atomic excitation is essentially absent. Consequently they are vastly more penetrating than charged particles of the same energy. The main energy-loss mechanism occurs when they strike nuclei. As with billiard balls, the most efficient slowing-down occurs when the bodies that are struck in an elastic collision have the same mass as the moving bodies; hence the most efficient neutron moderator is hydrogen, followed by other light elements: deuterium, beryllium, and carbon.

The great penetrating power of neutrons imposes severe shielding problems for reactors and other nuclear machines, and it is necessary to provide walls, usually of concrete, several feet in thickness to protect personnel. The currently accepted health tolerance levels for an 8-h day correspond for fast neutrons to a flux of 20 neutrons/$(cm^2)(s)$; for slow neutrons, $700/(cm^2)(s)$. On the other hand, fast neutrons are useful in some kinds of cancer therapy.

Detection of neutrons. In pulse counting, neutrons are allowed to produce exothermic (energy-releasing) nuclear reactions, the ionizing products of which are made to generate electrical impulses, in a proportional counter, ionization chamber, or scintillation counter, that can be amplified for individual counting. A proportional counter containing boron, either as a coating on the inner walls or as a filling gas (boron trifluoride), counts neutrons by virtue of the reaction $^{10}B(n,\alpha)^7Li+0.451$ pJ (2.78 MeV). An ionization chamber coated internally with ^{235}U gives ionization pulses from the energy of fission fragments as they travel through the gas. A lithium iodide crystal (europium-activated) scintillates because of the energy released by the reaction $^6Li(n,\alpha)^3H + 0.765$ pJ (4.78 MeV). The light pulses (scintillations) are reflected onto a photomultiplier, which transforms them to electrical pulses. Capture γ-rays emitted from strong neutron absorbers, such as cadmium, can similarly be registered by scintillation counting. Large and sensitive neutron detectors have been made by dissolving cadmium or boron salts in tanks containing scintillating liquids. *See* Ionization chamber; Liquid scintillation detector; Particle detector; Scintillation counter.

In detection by activation, advantage is taken of the fact that many elements become radioactive under neutron irradiation. A sample is exposed, and its radioactive strength is subsequently measured by conventional counting equipment. Gold and indium foils are convenient and sensitive detectors of this kind. Their applications can be further specialized by taking advantage of resonance absorption. If, for example, gold foil is enclosed in cadmium, the cadmium will exclude thermal neutrons, and the gold will be activated mainly by neutrons with an energy of 0.78 attojoule (4.9 eV) because gold has a neutron capture resonance at that energy. Other elements can be similarly used for other selected energies. The converse also occurs; for example, if a thick plug of ^{57}Fe is placed in a beam of fast neutrons, it will preferentially transmit neutrons with an energy of 4 pJ (25 keV) because at that energy the neutrons interact only weakly with the ^{57}Fe nuclei.

Detection by recoil is particularly applicable to the counting of fast neutrons. A counter with hydrogenous walls or filling gas (for example, methane) gives pulses because the protons produce ionization when they recoil after being struck by the fast neutrons.

Intrinsic properties. Free neutrons are themselves radioactive, each transforming spontaneously into a proton, an electron (β^--particle), and an antineutrino. The energy release is 0.125 pJ (0.782 MeV) per event, and the half-life is 10.61 ± 0.16 min. This instability is a reflection of the fact that neutrons are slightly heavier than hydrogen atoms. The neutron's rest mass is 1.0086652 atomic mass units on the unified mass scale (1.67495 $\times 10^{-24}$ g), as compared with 1.0078252 atomic mass units for the hydrogen atom.

Neutrons are, individually, small magnets. This property permits the production of beams of polarized neutrons, that is, beams of neutrons whose magnetic dipoles are aligned predominantly parallel to one direction in space. The magnetic moment is -1.913042 nuclear magnetons. The magnetic structure has a finite size, being roughly exponential in intensity, with a root-mean-square radius of 0.9×10^{-13} cm. Neutrons spin with an angular momentum of $^1/_2$ in units of $h/2\pi$, where h is Planck's constant. The negative sign which is attached to the magnetic moment indicates that the magnetic moment vector and the angular momentum vector are oppositely directed. *See* Magneton; Nuclear moments; Nuclear orientation; Spin.

Despite its overall neutrality, the neutron does have an internal distribution of electric charge, as has been revealed by means of scattering experiments. On a still finer scale, the neutron can also be presumed to have a quark structure like that of the proton. *See* Quantum chromodynamics; Quarks.

If the centers of the $+$ and $-$ charge distributions in the neutron should be slightly displaced from each other, the neutron would have an electric dipole moment. This possiblity has a fundamental importance because it is linked through various interaction theories with the conservation of parity and with the symmetry of time reversal. (If time-reversal invariance holds, the neutron should have no electric dipole moment.) So far it has been found that if the separated charges are equal to $\pm e$ (the electronic charge), the distance between their centers must be less than 10^{-24} cm. This limit is not yet sufficiently small to give a conclusive choice between the various forms of theoretically possible interactions, but it is likely that the sensitivity of the experiments can be increased through the use of ultracold neutrons. *See* Parity; Symmetry laws.

Ultracold neutrons. When neutrons are completely slowed down in matter, they have a maxwellian distribution in energy that corresponds to the temperature of the moderator with which they are in equilibrium. At room temperature their mean energy is about 0.004 aJ (0.025 eV), their mean velocity is about 2200 m/s, and their de Broglie wavelength is about 0.2 nm. (The approximate coincidence of this wavelength with the interatomic distances in solids is the basis for the science of neutron diffraction.) The maxwellian distribution has a tail extending to very low energies, and a few neutrons (about 10^{-11} of the main neutron flux) at this extreme have energies less than 5×10^{-8} aJ (3×10^{-7} eV), and hence velocities of less than about 7 m/s. The de Broglie wavelength of these ultracold neutrons is greater than 50 nm, which is so much larger than interatomic distances in solids that they interact with regions of a surface rather than with individual atoms, and as a result they are reflected from polished surfaces at all angles of incidence. *See* NEUTRON DIFFRACTION; THERMAL NEUTRONS.

A typical source of ultracold neutrons consists of a "converter" in the reflector of a neutron reactor, together with an internally smooth, evacuated tube several centimeters in diameter that leads the neutrons out through the shield. The lead-out duct has three or four bends; the ultracold neutrons are preferentially reflected at these bends and are thus selected from the numerous faster neutrons. The neutrons can be further slowed either by sending them upward against gravity (they can rise only 2–3 m, and the lead-out duct can be vertical if desired), or by means of a "neutron turbine," which is a paddle wheel whose curved blades move in the same direction as the neutrons, but with lower velocity. The neutrons can be polarized by passage through or reflection from a sheet of magnetized material.

The ultracold neutrons can be stored in "neutron bottles," of which there are two kinds. One is simply a vacuum vessel with a door that can be closed after a batch of neutrons has entered. Populations of about 100 neutrons have been retained in such vessels, but the storage times are considerably shorter than the half-life of the neutrons against their natural radioactive decay, and the nature of the extra loss mechanisms is not yet fully understood. The other kind of bottle is again a vacuum vessel, but it uses a multipolar magnetic field that contains the neutrons, because the field gradients act upon the neutrons' magnetic dipole moments so as to keep the neutrons away from the walls. Such a configuration is realized in a torus with hexapole windings around its major circumference, or a sphere with polar and equatorial windings carrying opposing currents.

Ultracold neutrons can be expected to be important in basic physics and to have applications in studies of surfaces and of the structure of inhomogeneities and magnetic domains in solids. *See* ANTINEUTRON; ELEMENTARY PARTICLE.

[ARTHUR H. SNELL]

Bibliography: T. von Egidy (ed.), *Fundamental Physics with Reactor Neutrons and Neutrinos*, Conf. Ser. no. 42, Institute of Physics, London, 1978; R. Golub and J. M. Pendlebury, Ultra-cold neutrons, *Rep. Progr. Phys.*, 42:439, 1979; E. Sheldon (ed.), *Proceedings of International Conference on the Interactions of Neutrons with Nuclei*, vols. 1 and 2, Technical Information Center, Department of Energy, Oak Ridge, 1976.

Neutron diffraction

The phenomenon associated with the interference processes which occur when neutrons are scattered by the atoms within solids, liquids, and gases. The use of neutron diffraction as an experimental technique is relatively new compared to electron and x-ray diffraction, since successful application requires high thermal-neutron fluxes, which can be obtained only from nuclear reactors. (A thermal neutron is defined as a neutron possessing a kinetic energy of about 0.025 eV.) These diffraction investigations are possible because thermal neutrons have energies with equivalent wavelengths near 1 A and are therefore ideally suited for interatomic interference studies.

The scattering of low-energy neutrons is generally considered a tool for the study of solid-state phenomena, but many significant investigations have also been performed to obtain information necessary for the understanding of nuclear processes. Diffraction techniques have been employed to measure numerous coherent neutron-scattering amplitudes under special conditions, and these determinations have provided details on the interaction between nuclear forces, potential scattering, and resonance effects. Experiments have also helped to establish upper limits for values of a possible small neutron electric charge and a possible small neutron electric dipole moment. The most numerous and important investigations by neutron diffraction have been concentrated on solid-state problems, because these experiments offer unique methods to obtain information on crystallographic properties, magnetic phenomena, and the dynamics of crystal lattices. In its applications to solid-state problems, neutron diffraction is very similar in both theory and experiment to x-ray diffraction, but its importance arises from the significant differences in the scattering of the two types of radiation.

The scattering of x-rays by atoms results from a scattering interaction with the atomic electrons, and the scattering amplitudes are approximately proportional to the atomic number of the scatterer. Since the electrons are distributed within the atom at distances comparable to the x-ray wavelength, interference effects occur which produce an angular distribution of the scattering, usually referred to as a form factor, that is descriptive of the spatial distribution of the electrons. In the scattering of neutrons by atoms, there are two important interactions. One is the short-range, nuclear interaction of the neutron with the atomic nucleus. This interaction produces isotropic scattering because the atomic nucleus is essentially a point scatterer relative to the wavelengths of thermal neutrons. Strong resonances associated with the scattering process prevent any regular variation of the nuclear scattering amplitudes with atomic number. These resonances can cause the scattering amplitudes to have imaginary components of appreciable size, and they can affect the phase changes between the incident and scattered neutron waves so that the scattering amplitudes can be either positive or negative. The other important process for the scattering of neutrons by atoms is the interaction of the magnetic moment of the neutron with the spin and orbital magnetic moments of the atom. The amplitude of the interaction varies with the size and orientation of the atomic magnetic moment, and the intensity of scattering has a form-factor angular dependence that is representative of the magnetic electrons within the atom. *See* SCATTERING EXPERIMENTS (ATOMS AND MOLECULES); SCATTERING EXPERIMENTS (NUCLEI).

Techniques. Although thermal neutron beams from nuclear reactors have intensities that are lower than those obtained from efficient x-ray tubes, most of the methods developed for x-ray diffraction can be used with neutrons. Furthermore, since the neutron absorption cross section for many materials is very small, diffraction effects can also be investigated by observations of the neutrons transmitted through a sample. In most experiments the sample is irradiated with monochromatic neutrons, and the scattered radiation is measured with a neutron detector, such as a proportional counter filled with boron trifluoride gas, BF_3. In structure determinations and other experiments where there is no energy change in the scattering process, only the angular distribution of the scattered neutrons is required. In inelastic scattering experiments, that is, experiments involving an increase or decrease in neutron energy, the energy distribution of the scattered neutrons must also be measured. The neutron energies can be determined with a crystal spectrometer or by analyzing the time of flight of the scattered neutrons. Both polycrystalline and single-crystal specimens can be examined, and the type of specimen is usually determined by the conditions of the experiment. However, since single-crystal techniques provide much better resolution of the diffraction peaks, this method is required for the study of complicated structures.

For those experiments requiring monochromatic neutron beams, the neutrons must be obtained by isolating a narrow slice of the neutron spectrum from the reactor, because these neutrons have a continuous energy distribution with no pronounced peaks. Monochromatization is usually accomplished by diffraction of the reactor neutrons from large single crystals, but filters and mechanical neutron velocity selectors also can be used. In investigations of certain magnetic properties, it is frequently necessary to use neutron beams that are both monochromatic and polarized. Such beams can be obtained in the monochromating process by using single crystals, because the neutrons scattered under specific conditions from

particular ferromagnetic crystals are almost completely polarized. Diffraction experiments with polarized neutrons have been particularly important for precise determinations of magnetic form factors, and techniques using polarization analysis provide a unique method for separating neutrons scattered by magnetic and nuclear interactions.

Another technique for neutron diffraction utilizes pulses containing the entire spectrum of reactor neutrons and employs time-of-flight energy analysis of the neutrons scattered at a fixed angle. This technique is particularly useful with pulsed neutron sources, and it offers definite advantages in certain types of experiments where the range of scattering angles is limited. However, in most investigations with continuous neutron sources, it is not competitive with the conventional diffraction methods. Photographic techniques can also be used in neutron experiments, but they have been restricted almost completely to quick qualitative examinations.

Auxiliary apparatus for controlling the conditions of the samples can be constructed easily because of the relatively low neutron cross sections of most materials. Consequently, many diffractometers and spectrometers have low-temperature cryostats, furnaces, and electromagnets as integral parts of the instruments. Furthermore, the methods of investigation are readily adaptable to automation, and some of the newer instruments at high-flux research reactors are controlled directly by on-line computers.

Chemical crystallography. Since the nuclear scattering amplitudes for neutrons do not vary uniformly with atomic number, there are certain

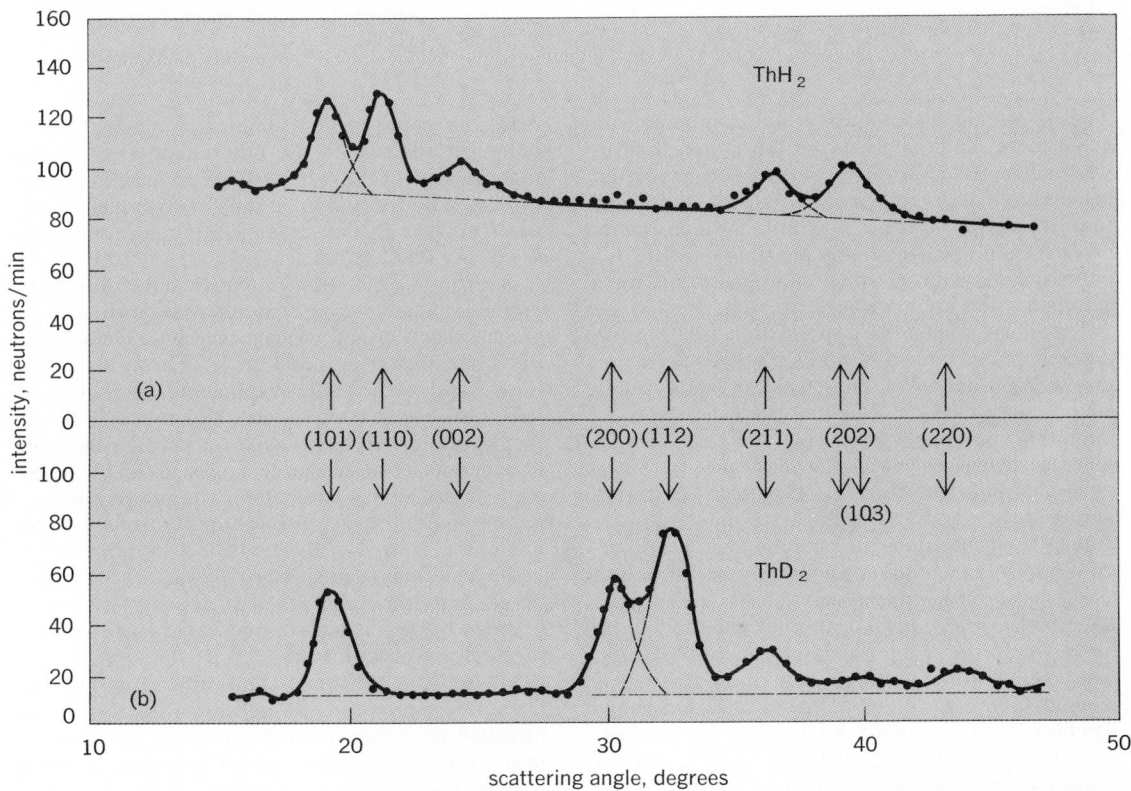

Fig. 1.   Neutron diffraction patterns from (a) polycrystalline thorium hydride, ThH_2, and (b) polycrystalline thorium deuteride, ThD_2. Differences in the patterns are caused primarily by differences in the nuclear scattering from hydrogen and deuterium atoms.

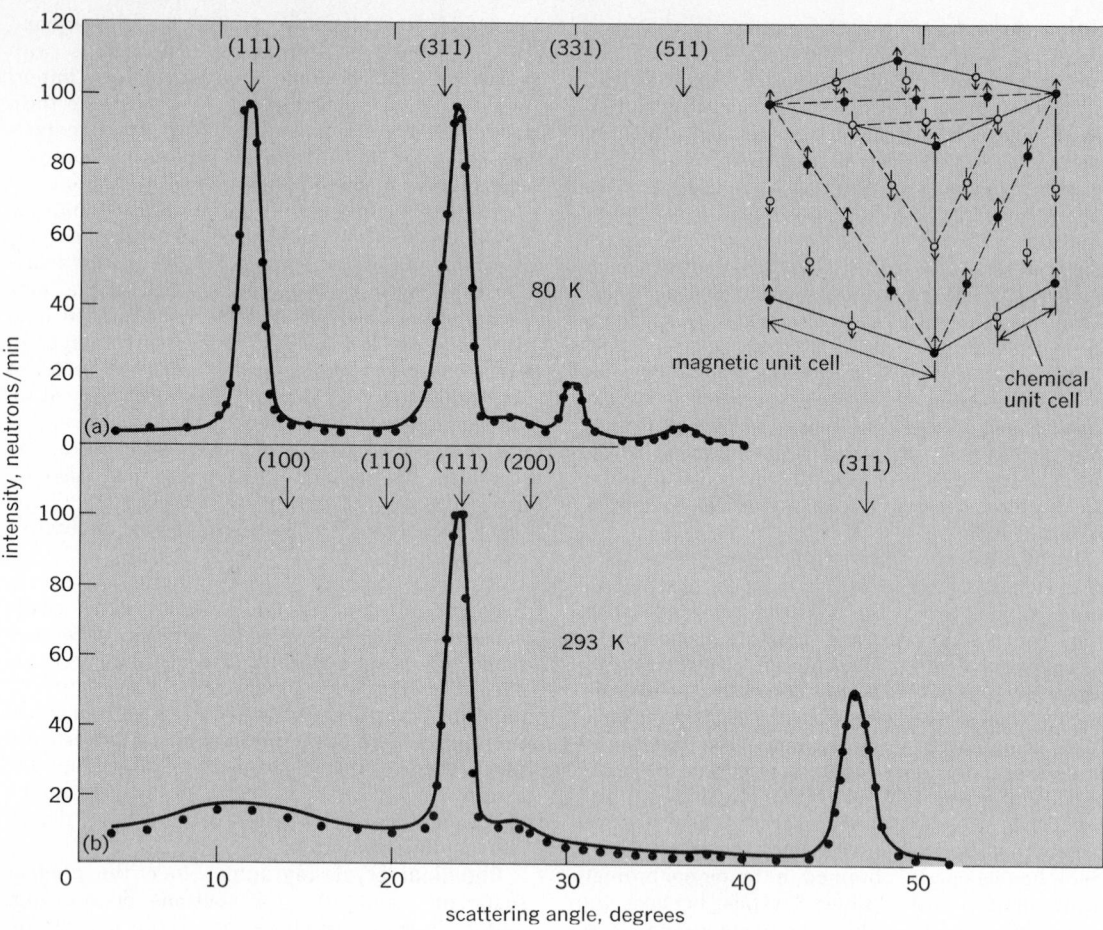

Fig. 2. Neutron diffraction patterns from polycrystalline manganese oxide, MnO, at temperatures (a) below and (b) above the antiferromagnetic transition at 122 K. At 293 K, only nuclear reflections are observed, while at 80 K, additional reflections are obtained from the indicated antiferromagnetic structure. The atomic magnetic moments in this structure are directed along a magnetic axis within the (111) planes.

types of chemical structures which can be investigated more readily by neutron diffraction than by x-ray diffraction. Moreover, since neutron scattering is a nuclear process, when the scattering amplitude of an element is not favorable for a particular investigation, it is frequently possible to substitute an enriched isotope which has scattering characteristics that are markedly different.

The most significant application of neutron diffraction in chemical crystallography is the structure determination of composite crystals which contain both heavy and light atoms, and the most important compounds in this general classification are the hydrogen-containing substances. Since hydrogen and deuterium have neutron scattering amplitudes that are comparable to those of other atoms, their positions in crystals can be determined by this technique, whereas x-ray diffraction usually gives little information about them (Fig. 1). Most of the early investigations concerned relatively simple inorganic compounds, but with the construction of higher-flux reactors and the use of computers in data collection and processing, more complex crystal structures have been examined. The crystal structure of sucrose, $C_{12}H_{22}O_{11}$, which required the measurement and analysis of 2800 independent Bragg reflections, was determined by neutron diffraction, and studies have also been made on some of the simplest biological molecules. In addition to the inherent interest in the materials, all of these investigations have helped to provide a better understanding of hydrogen bonds in crystals. For a discussion of Bragg reflections see X-RAY DIFFRACTION.

Neutron diffraction techniques have also been applied to many other compounds with special chemical or physical properties and with unfavorable x-ray scattering amplitudes. These investigations include the ionic displacements associated with ferroelectric transitions, the rotation of molecular groups in compounds, and order-disorder phenomena in alloy systems composed of atoms with almost the same atomic number. Furthermore, since the scattering of neutrons by the nucleus is isotropic, the neutron technique is advantageous in investigations of liquids, gases, amorphous materials, and other structures where the features of the diffraction pattern at large scattering angles are significant.

Magnetic scattering. The interaction of the magnetic moment of the neutron with the orbital and spin moments in magnetic atoms makes neutron scattering a unique tool for the study of a wide variety of magnetic phenomena, because information is obtained on the magnetic properties of the individual atoms in a material. This interaction depends on the size of the atomic magnetic moment and also on the relative orientation of the neutron spin and of the atomic magnetic moment with respect to the scattering vector and with respect to each other. Consequently, detailed information can be obtained on both the magnitude and orientation of magnetic moments in any substance which displays magnetic properties.

Each type of magnetic lattice has a characteristic neutron diffraction pattern. For paramagnetic materials, where the atomic moments are uncoupled and randomly oriented in direction, the magnetic scattering is diffuse. For ordered magnetic lattices the magnetic scattering is found in Bragg reflections. Magnetic reflections from ferromagnetic materials occur superimposed on the nuclear reflections, but for antiferromagnetic materials, in which the atomic moments are oriented with no net magnetization per unit volume, superlattice reflections are observed at other scattering angles, as shown in Fig. 2. Since ferrimagnetic materials have atomic moments with antiparallel components but also possess a net ferromagnetic moment, magnetic reflections are observed at both nuclear and other positions. Thus neutron diffraction experiments can determine the magnetic transition temperature, type of magnetic order, temperature variation of the magnetic order, and detailed magnetic configuration in the ordered lattice. This information is basic to understanding the magnetic exchange interactions that are responsible for producing an ordered magnetic structure. *See* ANTIFERROMAGNETISM; FERRIMAGNETISM; FERROMAGNETISM; PARAMAGNETISM.

The investigation of antiferromagnetic and ferrimagnetic substances is one of the most important applications of the neutron diffraction technique, because detailed information on the magnetic configuration in these systems cannot be obtained by other methods. Several hundred antiferromagnetic structures have been investigated, and various types of systems that have been determined are shown schematically in Fig. 3. In most antiferromagnetic substances the magnetic moments are found in truly antiparallel arrays, but more complicated systems have been encountered. Structures have been determined in which the magnetic moments are canted with respect to each other, and a number of systems have been found to have a long-range modulation of the moment distribution. The latter configurations require long-range magnetic interactions for stability, and in the heavy rare-earth metals and alloys which have such configurations, a long-range interaction through the conduction electrons can explain many of their unusual magnetic properties. Similar complex structures have also been observed in certain types of ferrimagnetic materials.

The magnetic moments of a simple ferromagnet can usually be obtained from saturation magnetization experiments, but in substances such as ferromagnetic alloys, although the moments are arranged in parallel alignment, different types of atoms have different moment values. Since magnetic measurements can give only the average moment of the alloy, the determination of the individual magnetic moments of the constituent atoms has been another important aspect of neutron diffraction. Alloys with both ordered and disordered arrangements of the atoms can be studied, and experiments on very dilute concentrations provide information on the effect of magnetic and nonmagnetic impurities.

The form factor for the magnetic scattering of neutrons can be interpreted in terms of the spatial distribution and angular momentum characteristics of the magnetic electrons within the atoms. Determinations of these form factors can be made from either measurements of magnetic intensities in coherent reflections or from measurements of paramagnetic scattering. However, the most precise measurements of this type have been made on reflections from ferromagnetic and ferrimagnetic materials, utilizing polarized neutrons. This technique takes advantage of cross terms in the combined nuclear and magnetic scattering to provide an accuracy not readily obtainable with an unpolarized beam. Measurements on the ferromagnetic iron-group metals have provided detailed maps showing the distribution of magnetic electrons throughout the unit cells.

A variety of changes can be produced in neutron diffraction patterns by application of magnetic fields sufficiently strong to change the orientation of atomic magnetic moments within the sample. The use of magnetic fields in these experiments can therefore provide information on ferromagnetic and antiferromagnetic domains, on the magnetic anisotropy within magnetic structures, and on the nature and strength of magnetic exchange interactions.

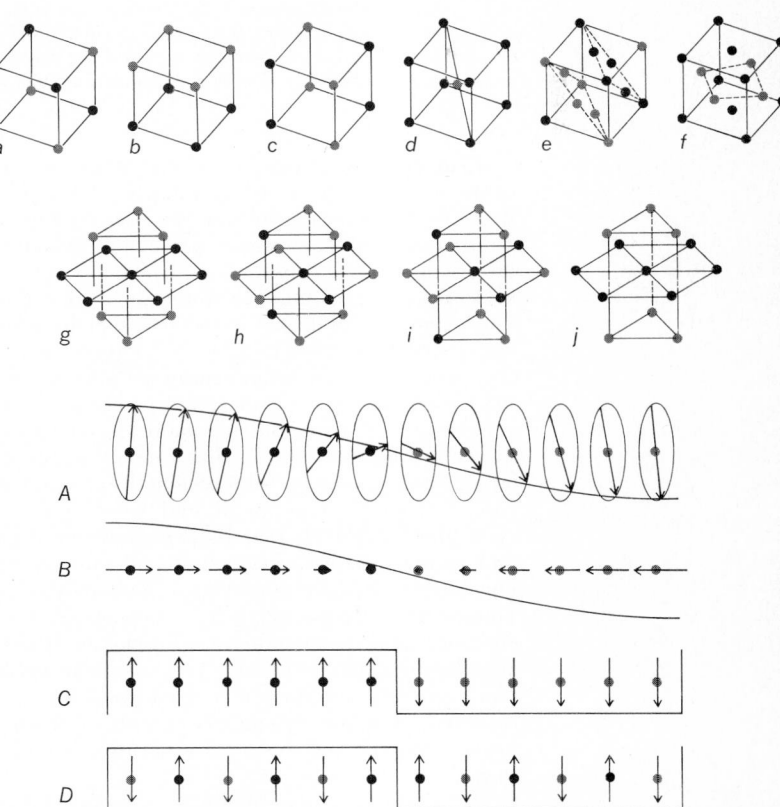

Fig. 3. Schematic representation of various antiferromagnetic systems studied by neutron diffraction. In structures *a* through *j* there is a single magnetic axis and the atomic moments at the darker circles are antiparallel to those at the lighter circles. Figures *A* through *D* indicate types of antiferromagnetism with a long-range modulation of the moment distribution.

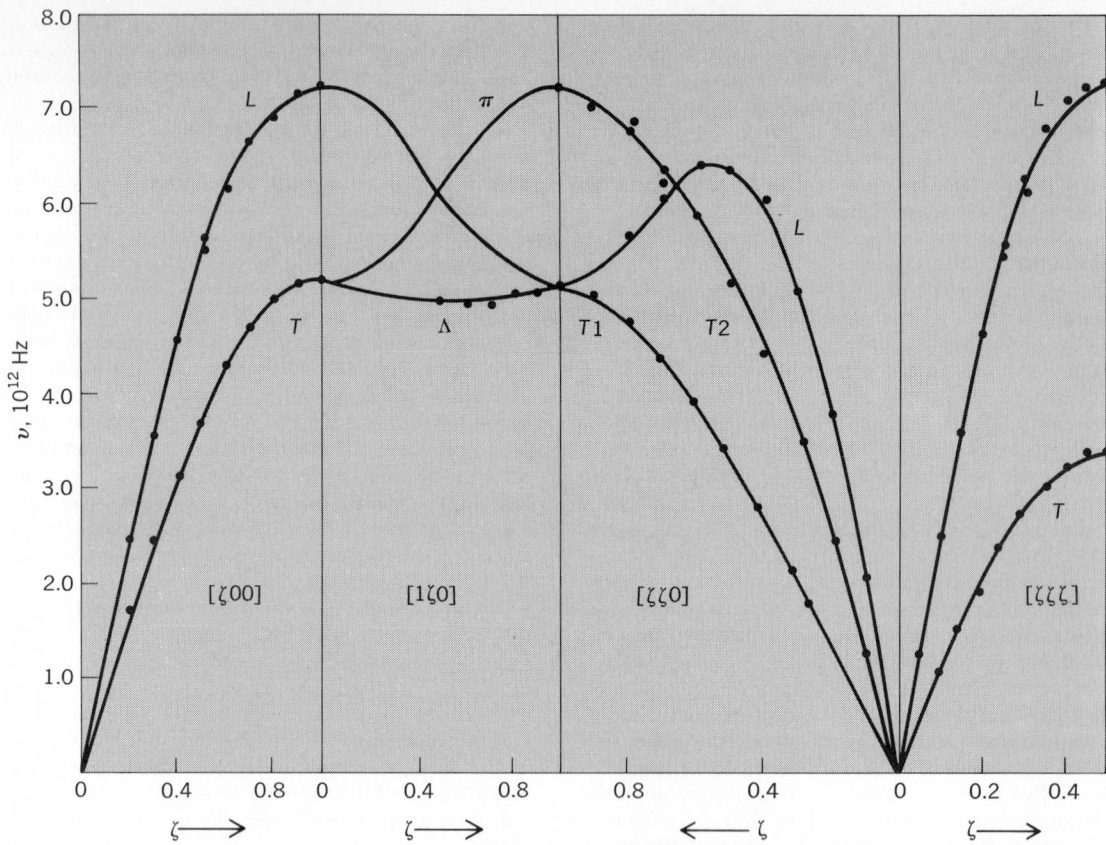

Fig. 4. Phonon dispersion curves for copper at 49°K, which relate phonon frequency ν to phonon wave vector ζ along major symmetry directions indicated in brackets. Solid circles are results from inelastic neutron scattering experiments and smooth curves are calculations based on an axially symmetric interatomic force model extended to six nearest neighbors. (L and T correspond to longitudinal and transverse modes of vibration, respectively, while π and Λ represent modes of vibration with both longitudinal and transverse components.)

Inelastic scattering. Scattering investigations of thermal neutrons, in which the neutrons undergo an energy change, fall into the broad scope of neutron diffraction. However, since both the angular distribution and the energy distribution of the scattered neutrons must be determined, these investigations are different from those usually associated with diffraction experiments. Because of the favorable values of energies and wavelengths associated with thermal neutrons, these measurements provide a method for studying many physical properties of solids and liquids that cannot be studied by any other method. The wavelengths are comparable with atomic separations, and the energies are of the order of the characteristic energies of solids and liquids, so that the energy and momentum changes resulting from many interactions can be measured easily by diffraction techniques. Furthermore, analogous to the case for elastic scattering, these inelastic scattering processes can result from a nuclear interaction or from a magnetic interaction, and the dynamical properties of both atomic systems and magnetic systems can be investigated.

One of the most important uses of inelastic neutron scattering is the study of thermal vibrations of atoms about their equilibrium positions, because lattice vibration quanta, or phonons, can be excited or annihilated in their interactions with low-energy neutrons. The measurements provide a direct determination of the dispersion relations for the normal vibrational modes of the crystal and do not require the large corrections necessary in similar x-ray investigations. These measured dispersion relations furnish the best experimental information available on interatomic forces that exist in crystals (Fig. 4). Similar measurements can be made on the quantized motion of magnetic moments about the equilibrium direction in an ordered magnetic lattice, and these magnon dispersion curves can be interpreted in terms of the magnetic forces between atoms. Furthermore, neutron scattering techniques are not restricted to solids but can be used to investigate details of atomic motion in liquids.

With the availability of higher neutron fluxes and more sophisticated techniques, it has been possible to extend these investigations to more difficult problems, such as the effect of impurities on interatomic forces. Localized vibrational modes associated with impurities can be observed, and in certain types of experiments information can be obtained on the degee of spatial localization of the modes in the crystal. Dispersion curves have been measured with sufficient precision to permit observation of additional effects, including the interaction between phonons and magnons and the interaction between phonons and conduction electrons. The latter observations may prove particularly useful for determining the Fermi surface in metals as a function of crystallographic direction. *See* ELECTRON DIFFRACTION; MAGNON; NEUTRON SPECTROMETRY.   [MICHAEL K. WILKINSON]

Bibliography: G. E. Bacon, *Neutron Diffraction*, 3d ed., 1975; H. Dachs (ed.), *Neutron Diffraction*, 1978; P. A. Egelstaff (ed.), *Thermal Neutron Scattering*, 1965; Y. A. Izyumov and R. P. Ozerov, *Magnetic Neutron Diffraction*, 1970.

Neutron optics

A title by analogy of certain phases of neutron physics in which the wave character of neutrons dominates and leads to behavior similar to that of light. Neutrons can be reflected at small glancing angles from plane surfaces; they show various scattering phenomena with similarity to light, and they can be diffracted by crystals. Although they can also be polarized, the analogy with light is in this case invalid because the polarization of neutrons depends upon their possession of a constant magnetic moment, which light waves lack. *See* NEUTRON; NEUTRON DIFFRACTION.

[ARTHUR H. SNELL]

Bibliography: L. Koester and A. Steyerl, *Neutron Physics*, 1977.

Neutron spectrometry

A generic term applied to experiments in which neutrons are used as the probe for measuring excited states of nuclides and for determining the properties of these states. The term neutron spectroscopy is also used. The strength of the interaction between a neutron and a target nuclide can vary rapidly as a function of the energy of the incident neutron, and it is different for every nuclide. At particular neutron energies the interaction strength for a specific nuclide can be very strong; these narrow energy regions of strong interactions are called resonances. The strength of the interaction, expressing the probability that an interaction of a given kind will take place, can be considered as the effective cross-sectional area presented by a nucleus to an incident neutron. This cross-sectional area is expressed in barns (1 barn = 10^{-28} m²) and is represented by the symbol σ. The neutron total cross section of the nuclide ^{231}Pa from 0.01 to 10 eV is shown in Fig. 1. Even though the neutron has zero charge, neutron energies are measured in electronvolts (1 eV = 1.60×10^{-19} joule). Neutron spectroscopy covers the vast energy range from 10^{-3} eV to 10^3 MeV.

Unbound and bound states of nuclides. Each resonance corresponds to an unbound excited state of the compound nucleus (Fig. 2) at an excitation energy that is the sum of the energy of the neutron and the binding energy of the neutron (4–11 MeV) which has been added to the target nuclide. The compound nucleus has a mass number which is one more than that of the target nuclide. Near the ground state of the compound nuclide the spacing of energy levels may be 10^4 to 10^6 eV. However, for a heavy nuclide, such as the compound nuclide ^{232}Pa, the excited states at an excitation energy just above the binding energy of approximately 5.5 MeV are less than 1 eV apart. To observe the individual states, the neutron energy resolution must be smaller than the level spacing. This can be achieved with low-energy neutrons, because they provide the requisite reso-

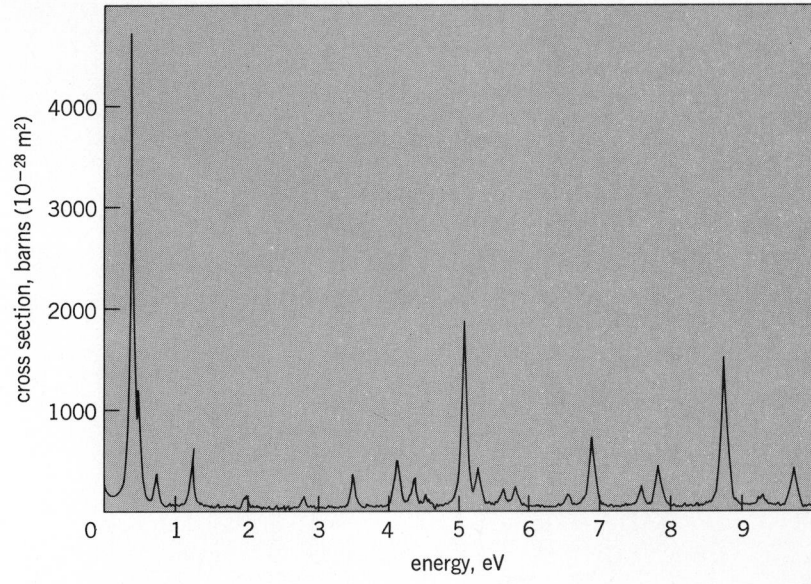

Fig. 1. Neutron total cross section of ^{231}Pa + neutron. The variation in sizes of the resonances and the nonuniform spacing of resonances are apparent.

lution, and there is no Coulomb repulsion to prevent them from entering the target nucleus. Neutron spectroscopy is presently the only technique which can provide this detailed information. For light nuclei (atomic weights ≤ 40) the spacing between the excited states can be many keV, and

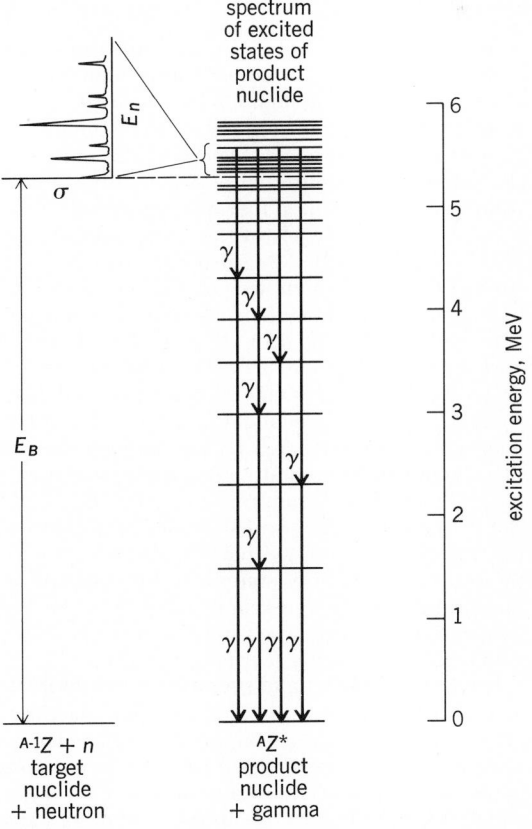

Fig. 2. Energy-level diagram for the product nucleus $^A Z^*$. The asterisk emphasizes that the product nucleus is in an excited state.

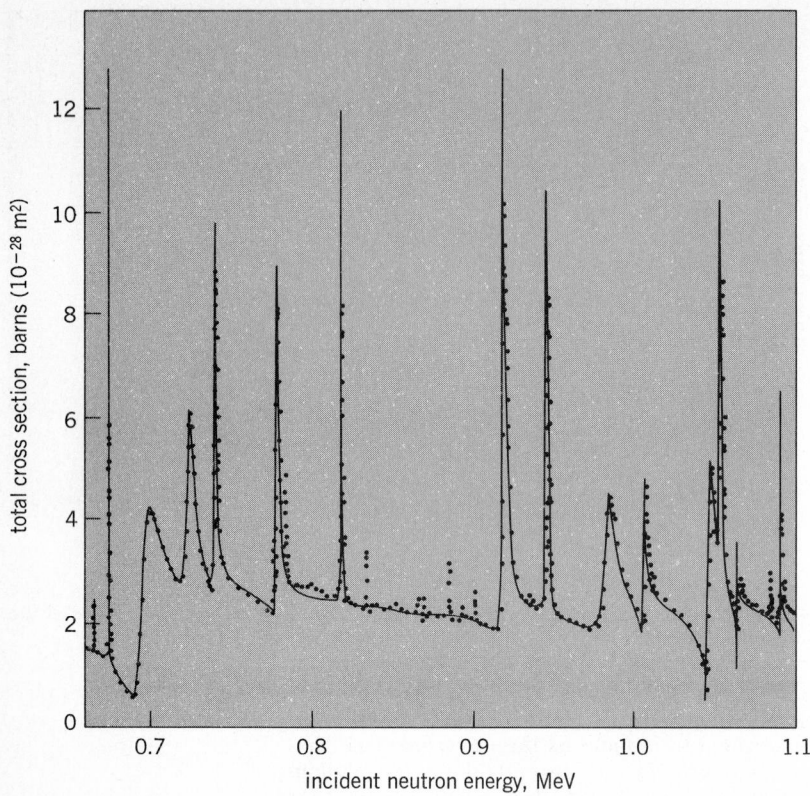

Fig. 3. Experimental neutron total cross section of sulfur compared to a theoretical fit. The fit does not include contributions from small resonances and minor isotopes of sulfur. The asymmetry of some resonances arises from the interference of resonance and potential scattering.

resonances can be resolved up to neutron energies of several MeV (Fig. 3). Lower-energy bound excited states of the compound nucleus below the neutron binding energy can also be studied by gamma-ray spectroscopy, observing the energy of the gamma rays emitted after the capture of neutrons at resonances or at thermal energy (Fig. 2).

Nuclear energy levels of the target nuclide can also be determined by measuring the energy spectrum of neutrons which are inelastically scattered by the target under bombardment from monoenergetic MeV incident neutrons (Fig. 4). The energy of an excited state is equal to the difference in energies between the incident and scattered neutrons, $E_n - E_{n'}$. If the incident neutron in Fig. 4 has energy E_{n1}, it has enough energy to excite any of the six lowest levels and emit a neutron of lesser energy than E_{n1}. A neutron of energy E_{n2} could excite only the two lowest levels and emit a neutron of lesser energy than E_{n2}. Information on these same low-energy states can be obtained by measuring the energies and intensities of the gamma rays from the deexcitation of these states excited by inelastically scattered neutrons.

Neutron reactions and resonance parameters.
The abbreviated notation for neutron reactions, (n,n) and so on, lists the bombarding particle before the comma, and the emitted particle or particles after the comma. The standard symbols are: n (neutron), p (proton), d (deuteron), α (alpha particle), γ (gamma ray), f (fission), and T (total). A more complete description of the reaction lists also the target and product nuclides, for example, $^A Z(n,\gamma)$ ^{A+1}Z. The reactions most useful for neutron spec-

troscopy are: the total interaction; elastic scattering (n,n); radiative capture (n,γ); fission (n,f); inelastic scattering (n,n'); charged particle emission (n,p), (n,α), and (n,d); and three-body breakup or sequential decay $(n,2n)$ and (n,np). See NUCLEAR REACTION.

The resonances observed in these various reactions can be fitted by a theoretical formula to give parameters of the resonances (E_0, Γ, Γ_f, Γ_γ, Γ_n, and so forth) which correspond to detailed properties of the excited states in the compound nucleus. For example, E_0 is the resonance energy; the fission width, Γ_f, is obtained from the fission cross section; the radiation width, Γ_γ, from the capture cross section, and so forth. The neutron width, Γ_n, can be obtained from the scattering cross section or the total cross section. The total width, Γ, can be obtained if the energy resolution is less than or equal to Γ. In addition, two other properties, the angular momentum of the neutron forming the resonance and the spin, J, of the state can often be determined. For narrow resonances where Γ (in eV) $\leq 0.05 \sqrt{E_0}$ (in eV), it is necessary to consider the Doppler broadening of resonances due to the thermal motion of the target nuclides.

Neutron cross sections. The measurement of a cross section for a particular reaction consists of measuring the number of such reactions produced by a known number of neutrons incident on a known number of target nuclides. When the probability of all neutron interactions with the target nucleus is small, the number of reactions of a particular process i, per unit area and unit time using a beam of neutrons equals $(nv)(Nx)\sigma_i$. The quantity nv is the number of incident neutrons per unit area normal to their direction per unit time, N is the number of target nuclei per unit volume, x is the thickness of the target in the direction of

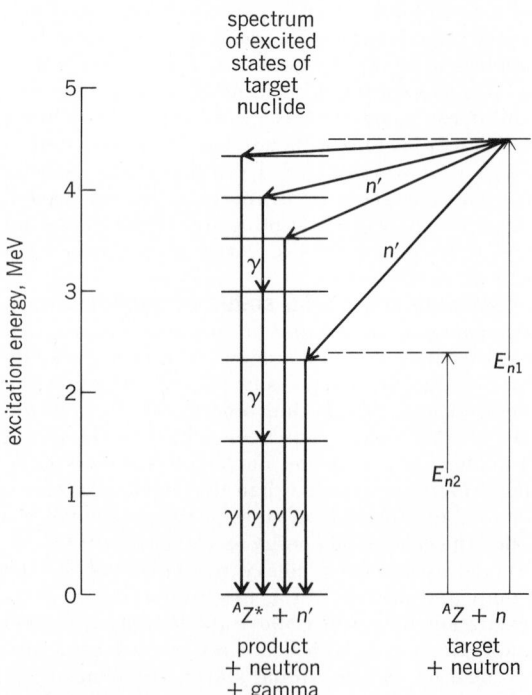

Fig. 4. Energy diagram of the target nuclide showing excitation of levels by the inelastic scattering of MeV neutrons.

the incident neutrons, and σ_i is the cross section per target nucleus for a particular reaction expressed in units of area. If the probability of all interactions is not small, the incident beam will be attenuated exponentially, as $\exp(-Nx\sigma_T)$, in passing through the sample, where Nx is the number of target nuclei per unit area normal to the beam, and σ_T is the neutron total cross section.

The most common type of neutron cross-section measurement, which can usually be made with the highest neutron energy resolution and usually with the most accuracy, is that of the total cross section. This measurement consists of measuring the transmission of a well-collimated beam of neutrons through a sample of known thickness; the transmission through the sample is simply the ratio of the intensity of the beam passing through the sample to that incident on the sample. The intensity of the incident beam is reduced in passing through the sample because the incident neutrons are absorbed or scattered by the target particles. The total cross section, σ_T, is determined from the equation $\sigma_T = -[\log_e (\text{transmission})]/Nx$.

In order to measure partial cross sections, more elaborate equipment is needed, in general, than for total cross-section measurements, and the measurements are considerably more difficult. For example, to measure the differential elastic scattering cross section, it is necessary to measure the number of elastically scattered neutrons as a function of angle of the scattered neutron relative to that of the incident neutron. MeV-energy neutrons, in addition to being elastically scattered, can also lose energy when scattered from a target nucleus [inelastic scattering, $(n,n'\,\gamma)$]. Several techniques have been developed for determining inelastic cross sections, both by measuring the energy spectrum of the inelastically scattered neutrons and by measuring the energies and intensities of the gamma rays emitted from the excited nuclei. These cross sections can also be measured as a function of the angle relative to the direction of the incident neutron.

Techniques for neutron spectroscopy. Neutron spectroscopy can be carried out by two different techniques (or a combination): (1) by the use of a time-pulsed neutron source which emits neutrons of many energies simultaneously, combined with the time-of-flight technique to measure the velocities of the neutrons; this time-of-flight technique can be used for neutron measurements from 10^{-3} eV to about 200 MeV; (2) by the use of a beam of nearly monoenergetic neutrons whose energy can be varied in small steps approximately equal to the energy spread of the neutron beam; however, useful "monoenergetic" neutron sources are not available from about 10 eV to about 10 keV.

Time-of-flight neutron spectrometers. Time-of-flight neutron spectrometers are the most widely used spectrometers for most neutron cross-section measurements. The time-of-flight technique requires an intense pulse of neutrons which contains neutrons of many energies and a flight path to measure the velocities of the neutrons. Various detectors are placed at the end of the flight path depending on the type of cross-section measurement. Burst widths from 10^{-5} to 10^{-9} s and flight paths from 1 to 1000 m have been used. The resolution of a time-of-flight spectrometer is often quoted in microseconds or nanoseconds per meter. The

energy resolution $(\Delta E/E)$ is equal to $2\Delta t/t$, where Δt is the time width of the neutron burst plus the time spread in the detector, and t is the time of flight of the neutron [t (in μs) $= 72.3 \times$ path length (in meters)$/\sqrt{E}$ (in eV)]. The time between pulses, the flight path length, and filters in the beam must be selected so that low-energy neutrons from previous pulses do not interfere with the high-energy neutrons from following pulses. By the use of multichannel storage, usually a computer, the complete neutron spectrum (or cross section) can be obtained in one measurement with good energy resolution over a broad energy range.

The most valuable neutron source for neutron time-of-flight spectroscopy is an electron or other charged-particle accelerator capable of producing intense pulses of neutrons of short duration (on the order of 10^{-9} s). Excellent neutron cross-section measurements can be made with these spectrometers from 0.01 eV to 200 MeV. For example, a beam of 140-MeV electrons incident on a tantalum target produces neutrons with an energy distribution that has a peak at about 1 MeV and extends up to about 80 MeV. The peak of the neutron distribution for protons or deuterons incident on a heavy target occurs at a higher neutron energy than for electrons; for deuterons a broad peak occurs at about half the energy of the deuterons. Lower-energy neutrons from these accelerators are obtained by placing a moderator (about 2 cm thick) around or near the target. The duration of these moderated neutron pulses in nanoseconds is approximately $2/\sqrt{E}$ (in MeV). The flux distribution of these moderated neutrons approximately follows the relation $E^{-0.8}$ down to thermal neutron energies. With a moderated neutron source the pulse repetition must be sufficiently low, depending on the flight path length, beam filter, and energy range, to prevent overlap of neutrons from successive pulses. Typical high-resolution results obtained using a moderated source are shown in Figs. 1 and 3. *See* PARTICLE ACCELERATOR.

Before the development of short-pulse accelerators, a mechanical chopper rotating at high speed in a well-collimated beam from a moderated fission reactor was used to produce bursts of electronvolt-energy neutrons. The neutron pulses were sufficiently short (about 10^{-6} s) and of sufficient intensity for measurements to be made up to a neutron energy of about 10^4 eV using neutron flight paths up to about 100 m in length. In order to produce pulses of only 10^{-6} s duration, the neutron beam had to be collimated to narrow slits (0.05 cm) to match the narrow slits through the chopper. Only when the rotation of the chopper was such that the slits in the rotor lined up with the slits in the collimator was a neutron beam with a broad energy spread passed. A fast-chopper time-of-flight spectrometer is particularly useful for transmission measurements on samples which are available in small amounts, since the sample only needs to be large enough to cover the beams passing through the narrow slits in the collimator.

For time-of-flight measurements in the energy region from about 10 keV to 1 MeV, a pulsed electrostatic accelerator using the ^{7}Li (p,n) reaction is capable of producing neutron pulses of short duration (10^{-9} s) with sufficient intensity for measurements with flight paths of a few meters. By selecting the proton-bombarding energy and a suitable target thickness, neutrons produced in the reaction at a given angle can have well-defined upper and lower energy limits. With no low-energy neutrons and short flight paths, rather high repetition rates of 10^6 Hz can be used and an energy resolution of about 1% can be realized.

The most intense pulsed neutron source used for neutron time-of-flight spectroscopy is that achieved from an underground nuclear explosion. The burst duration is about 80 ns, and the neutron distribution extends down to about 20 eV. Fission cross-section measurements have been made on very small samples of many radioactive heavy nuclides using such a source and an approximately 300-m flight path length. The availability of this source is obviously rather restricted, but it is unique for measurements on highly radioactive samples.

Neutron time-of-flight measurements have also been made using a pulsed fission reactor where the duration of burst is about 40 μs. Finally, subcritical boosters have been used to multiply the intensity of the neutron pulses from electron accelerators by factors of $10-200$, which results in pulse durations of $0.08-4$ μs.

Monoenergetic neutron spectrometers. The best technique for obtaining an intense beam of low-energy neutrons ($\leqslant 10$ eV) with an energy spread of only about 1% is to use a crystal monochromator placed in a well-collimated beam of neutrons from a high-flux moderated fission reactor. If a single crystal (such as beryllium, copper, or lead) is properly oriented in a collimated neutron beam, neutrons of a discrete energy, E, will be elastically scattered from a particular set of planes of atoms in the crystal through an angle 2θ given by Bragg's law $n\lambda = 2d \sin \theta$. In this equation, the integer n is the order of the reflection, λ is the neutron wavelength, d is the spacing between the planes of atoms of the particular set in the crystal, and θ is the angle of incidence between the direction of the neutron beam and the set of planes of atoms being considered. The neutron wavelength λ in centimeters equals $0.286 \times 10^{-8}/\sqrt{E}$, where E is in eV. The energy of the diffracted beam can be continuously varied by changing the angle of the crystal. Measurements of many rare-earth nuclides and heavy nuclides have been made with crystal spectrometers up to 10 eV neutron energy. Capture gamma-ray spectra have also been studied as a function of neutron energy, specifically from different neutron resonances. *See* NEUTRON DIFFRACTION.

"Monoenergetic" neutrons in the energy range from a few keV to 20 MeV can be obtained by bombarding various thin targets with protons or deuterons from a variable-energy accelerator such as an electrostatic accelerator. The most useful (p,n) reactions to cover the energy range from a few keV to a few MeV are those on lithium and tritium targets. The (d,n) reaction on deuterium is useful from about 1 to 10 MeV, and the (d,n) reaction on tritium from 10 to 20 MeV. In the energy range up to 1 MeV an energy resolution of about 1 keV is possible, but this resolution is usually not adequate for neutron spectroscopy for neutrons with energies less than 10^4 eV. The measurement of a complete cross-section spectrum up to 1 MeV may require 1000 sequential measurements at slightly different neutron energies. Monoenergetic neutron sources are also useful for measurements such as

activation, which cannot be done with the time-of-flight technique.

Applications. Neutron spectroscopy has yielded a mass of valuable information on nuclear systematics for almost all nuclides. The distribution of the spacings between nuclear levels and the average of these spacings have provided valuable tests for various nuclear theories. The properties of these levels, that is, the probabilities that they decay by neutron or gamma-ray emission, or by fission, and the averages and distribution of these probabilities have stimulated much theoretical effort.

In addition, knowledge of neutron cross sections is fundamental for the optimum design of thermal fission power reactors and fast neutron breeder reactors, as well as fusion power reactors now in the conceptual stage. Cross sections are needed for nuclear fuel materials such as ^{235}U or ^{239}Pu, for fertile materials such as ^{238}U, for structural materials such as iron and chromium, for coolants such as sodium, for moderators such as beryllium, for shielding materials such as concrete. The optimum choice of materials for the energy region under consideration is critical to the success of the project and is of great economic significance. *See* NUCLEAR STRUCTURE.

[JOHN A. HARVEY]

Bibliography: J. A. Harvey (ed.), *Experimental Neutron Resonance Spectroscopy*, 1970; D. J. Hughes, *Neutron Cross Sections* 1957; J. E. Lynn, *The Theory of Neutron Resonance Reactions*, 1968: J. B. Marion and J. L. Fowler (eds.), *Fast Neutron Physics*, vol. 1, 1959, and vol. 2, 1963; S. F. Mughabghab and D. I. Garber (compilers), *Neutron Cross Sections*, 3d ed., vol. 1: *Parameters*, 1973; and D. I. Garber and R. R. Kinsey, 3d ed., vol. 2: *Curves*, 1976.

Newtonian fluid

A fluid in which the state of stress at any point is a linear function of the time rate of strain at that point. The fluid thus bears a direct analogy to a hookean solid, for which the state of stress is a linear function of the strain. Many gases and liquids are closely newtonian over a wide range of pressures and temperatures.

The simplest example of newtonian fluid flow is Couette flow, the low-speed steady motion of a viscous fluid between two infinite plates moving parallel to each other with relative velocity U, as illustrated. The shear stress τ_{21} in the fluid is constant and equals $\mu(\partial u_1/\partial x_2) = \mu(U/\delta)$. The time rate of strain at a point is a tensor quantity given by Eq. (1), where u_i are velocity components of fluid and x_i

$$\varepsilon_{ij} = \frac{1}{2}\left(\frac{\partial u_i}{\partial x_j} + \frac{\partial u_j}{\partial x_i}\right) \qquad (1)$$

are rectangular cartesian coordinates with $i = 1,2,3$. Fluids are inherently isotropic so that the most general linear relationship between stress τ_{ij} and ϵ_{ij} is given by Eq. (2). Here μ is the ordinary

$$\tau_{ij} = -P\,\delta_{ij} + (\lambda - {}^2\!/_3\mu)\,\delta_{ij}\epsilon_{mm} + 2\mu\epsilon_{ij} \qquad (2)$$

viscosity coefficient, λ is the bulk or volume viscosity coefficient, and P is the pressure. In gases, $\lambda = 0$ if the molecules have no internal degrees of freedom or if the internal motions are not excited. For low Mach numbers, $\epsilon_{mm} = 0$, so λ does not appear in the stress relationship. For most liquids, μ de-

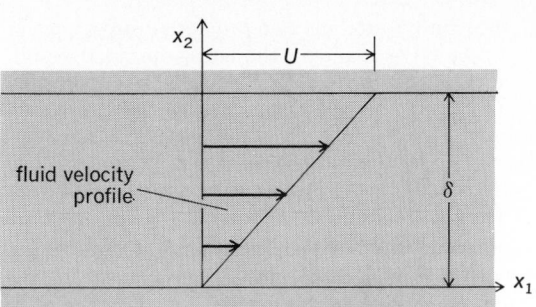

Newtonian fluid. Top plate moves relative to bottom one to produce Couette flow of intervening viscous fluid.

creases with temperature and increases with pressure; for gases it increases with temperature and is almost independent of pressure. *See* FLUID-FLOW PRINCIPLES; FLUIDS.   [ARTHUR E. BRYSON, JR.]

Newton's laws of motion

Three fundamental principles which form the basis of classical, or newtonian, mechanics. They are stated as follows:

First law: A particle not subjected to external forces remains at rest or moves with constant speed in a straight line.

Second law: The acceleration of a particle is directly proportional to the resultant external force acting on the particle and is inversely proportional to the mass of the particle.

Third law: If two particles interact, the force exerted by the first particle on the second particle (called the action force) is equal in magnitude and opposite in direction to the force exerted by the second particle on the first particle (called the reaction force).

The first law, sometimes called Galileo's law of inertia, can now be regarded as contained in the second. At the time of its enunciation, however, it was important as a negation of the Aristotelian doctrines of natural placement and continuing force.

The third law, sometimes called the law of action and reaction, was also to some extent established prior to Newton's statement of it. However, Newton's formulation of the three laws as a mutually consistent set, with the nature of force clearly defined in the second law, provided the basis for classical dynamics.

The newtonian laws have proved valid for all mechanical problems not involving speeds comparable with the speed of light (approximately 300,-000 km/sec) and not involving atomic or subatomic particles. The more general classical methods of Lagrange and of Hamilton are elaborations of the newtonian principles. *See* DYNAMICS; FORCE; HAMILTON'S EQUATIONS OF MOTION; KINETICS; LAGRANGE'S EQUATIONS; MOTION.

[DUDLEY WILLIAMS]

Noneuclidean geometry

A system of geometry based upon a set of axioms different from those of the euclidean geometry based on the three-dimensional space of experience. Noneuclidean geometries, especially Riemannian geometry, are useful in mathematical physics. This article will describe some of the basic concepts and axioms of noneuclidean geome-

tries and then draw some comparisons with euclidean concepts. *See* RIEMANNIAN GEOMETRY.

The famous names related to hyperbolic (noneuclidean) geometry are J. Bolyai and N. I. Lobachevski. Also A. Cayley, E. Beltrami, K. F. Gauss, and G. F. B. Riemann have given many outstanding contributions to the subject of noneuclidean geometry.

Projective space. Let K denote a given field of elements, called the ground field K. In actual applications, K will be simply the field of real numbers. Consider the class Σ of ordered sets of $(n+1)$-tuples $x = (x^0, x^1, \ldots, x^n)$, where each x^i is an element of the ground field K. The index i is not an exponent but is merely a superscript. In this article, both superscripts and subscripts will be used. If $x = (x^0, x^1, \ldots, x^n)$ and $y = (y^0, y^1, \ldots, y^n)$ are two elements in the class Σ, then the sum $(x+y)$ is defined to be the element $x+y = (x^0 + y^0, x^1 + y^1, \ldots, x^n + y^n)$ in Σ. Also, if $x = (x^0, x^1, \ldots, x^n)$ is an element of Σ and if ρ is an element of the ground field K, then the scalar product ρx is defined to be the element $\rho x = (\rho x^0, \rho x^1, \ldots, \rho x^n)$ in Σ.

If $x_1, x_2, \ldots, x_m$ are m elements in Σ and if $\rho_1, \rho_2, \ldots, \rho_m$ is a set of m elements from the ground field K, then the element $x = \rho_1 x_1 + \rho_2 x_2 + \cdots + \rho_m x_m$ of Σ is called a linear combination of the elements $x_1, x_2, \ldots, x_m$ of Σ.

The class Σ^1 is a proper subset of Σ which is composed of all ordered sets of $(n+1)$-tuples $x = (x^0, x^1, \ldots, x^n)$ of Σ such that at least one x^i of the $(n+1)$ elements $x^0, x^1, \ldots, x^n$ of the ground field K is not zero. Two elements $x = (x^0, x^1, \ldots, x^n)$ and $y = (y^0, y^1, \ldots, y^n)$, of the class Σ^1, are said to be equivalent if, and only if, there exists an element $\rho \neq 0$, of the ground field K, such that $y = \rho x$. This establishes an equivalence relation among the elements $x = (x^0, x^1, \ldots, x^n)$ of the class Σ^1. The equivalence classes in Σ^1, formed relative to this equivalence relation, are called points P. The collection of all these points P forms a projective space S_n of n dimensions over the ground field K.

A point P in S_n is a class in Σ^1 of all ordered sets of $(n+1)$-tuples equivalent to a given set of $(n+1)$-tuples, namely, $x = (x^0, x^1, \ldots, x^n)$. Each of these ordered sets of $(n+1)$-tuples is in the equivalence class called the point P and is said to be a set of homogeneous point coordinates for the point P. Consequently, if $x = (x^0, x^1, \ldots, x^n)$ is a set of homogeneous point coordinates of a point P in S_n, any other set of homogeneous point coordinates for the same point P is $\rho x = (\rho x^0, \rho x^1, \ldots, \rho x^n)$, where $\rho \neq 0$ is an arbitrary element of the ground field K.

A collection of $(p+1)$ points $x_0, x_1, \ldots, x_p$ in S_n is said to be linearly dependent if, and only if, at least one of them is a linear combination of the remaining ones such that at least one of the scalar coefficients is not zero. Otherwise they are said to be linearly independent. Clearly, if $(p+1)$ points $x_0, x_1, \ldots, x_p$ in S_n are linearly independent, then $0 \leq p \leq n$.

Let a collection of $(p+1)$ linearly independent points $x_0, x_1, \ldots, x_p$ in S_n be given. Then $0 \leq p \leq n$. The S_p determined by them is composed of all points x in S_n such that Eq. (1) holds,

$$x = y^0 x_0 + y^1 x_1 + \cdots + y^p x_p \qquad (1)$$

where at least one of the scalar multiples y^0,

$y^1, \ldots, y^p$ is not zero. This S_p is a projective space of p dimensions over the ground field K, and is a projective subspace of S_n. A set of homogeneous point coordinates for an arbitrary point P in this S_p is $(y^0, y^1, \ldots, y^p)$.

Define S_{-1} to be a vacuous collection of points P in S_n. Such an S_{-1} is called a projective subspace of S_n of dimension -1. Evidently every S_0 is a point P and conversely. A line is an S_1, a plane is an S_2, a three-space is an S_3, and a p-space is an S_p. In particular, a hyperplane is an S_{n-1}.

Dual coordinates. An S_p is a locus of points x whose homogeneous point coordinates $x = (x^0, x^1, \ldots, x^n)$ satisfy a system of $(n-p)$ linear homogeneous equations, Eq. (2), such that the rank of the coefficient matrix (u_{ij}) is $(n-p)$.

$$\sum_{j=0}^{n} u_{ij} x^j = 0 \quad (i = 1, 2, \ldots, n-p) \quad (2)$$

In particular, a hyperplane S_{n-1} is a locus of points x whose homogeneous point coordinates $x = (x^0, x^1, \ldots, x^n)$ satisfy a single linear homogeneous equation, Eq. (3), where at least one of the $(n+$

$$ux = \sum_{i=0}^{n} u_i x^i = u_0 x^0 + u_1 x^1 + \cdots + u_n x^n = 0 \quad (3)$$

1) elements $u_0, u_1, \ldots, u_n$, of the ground field K, is not zero.

Because an ordered set of $(n+1)$-tuples $u = (u_0, u_1, \ldots, u_n)$ of this type uniquely defines an S_{n-1}, it is called a set of homogeneous hyperplane coordinates, for the particular S_{n-1}. If $u = (u_0, u_1, \ldots, u_n)$ is a set of homogeneous hyperplane coordinates of an S_{n-1}, any other set of homogeneous hyperplane coordinates for the same S_{n-1} is $\sigma u = (\sigma u_0, \sigma u_1, \ldots, \sigma u_n)$, where $\sigma \neq 0$ is an arbitrary element of the ground field K.

In the projective space S_n of n dimensions, a point P and a hyperplane S_{n-1} are termed dual objects. Thus, homogeneous point coordinates $x = (x^0, x^1, \ldots, x^n)$ and homogeneous hyperplane coordinates $u = (u_0, u_1, \ldots, u_n)$ are said to be dual systems of coordinates.

A point P with the homogeneous point coordinates $x = (x^0, x^1, \ldots, x^n)$ is on the hyperplane S_{n-1} with the homogeneous hyperplane coordinates $u = (u_0, u_1, \ldots, u_n)$ if, and only if, the condition of Eq. (3) is satisfied.

By mathematical induction, it follows that the dual of a projective subspace S_p of p dimensions is a projective subspace S_{n-p-1} of $(n-p-1)$ dimensions. For example, the dual of S_{-1} is S_n, the dual of an S_0 is an S_{n-1}, and the dual of an S_1 is an S_{n-2}.

The projective subspace S_p is said to be contained in the projective subspace S_q, or the projective subspace S_q contains the projective subspace S_p, if, and only if, every point of S_p is a point of S_q. In particular, S_{-1} is contained in every S_q, and S_n contains every S_p. This relation is written as $S_p \subset S_q$, or $S_q \supset S_p$.

The dual of the relation $S_p \subset S_q$ is the dual relation $S_{n-p-1} \supset S_{n-q-1}$.

Principle of duality. For every theorem concerning the S_ps and the relations $S_p \subset S_q$, there is obtained a dual theorem wherein each S_p is replaced by the dual object S_{n-p-1}, and each relation $S_p \subset S_q$ is replaced by the dual relation $S_{n-p-1} \supset S_{n-q-1}$.

If S_p and S_q are any two projective subspaces of the projective space S_n, the largest projective subspace contained in both S_p and S_q is denoted by $S_p \cap S_q$, and the smallest projective subspace containing both S_p and S_q is denoted by $S_p \cup S_q$. Clearly, $S_p \cap S_q$ is the intersection or meet, and $S_p \cup S_q$ is the union or join of S_p and S_q. See SET THEORY.

The two relations $S_p \cap S_q$ and $S_p \cup S_q$ are dual. Concerning the inclusion relation $S_p \subset S_q$, the projective subspaces of the projective space S_n form a complemented modular lattice.

Finally, let $\dim(S_p)$ denote the dimension p of the projective subspace S_p of the projective space S_n. The dimension theorem may be stated in the form shown in Eq. (4).

$$\dim(S_p \cap S_q) + \dim(S_p \cup S_q)$$
$$= \dim(S_p) + \dim(S_q) \quad (4)$$

Projective group. A collineation T of the projective space S_n is a one-to-one correspondence between the points of S_n, of the form shown in Eq. (5),

$$\rho \bar{x}^i = \sum_{j=0}^{n} a_j^i x^j \quad (i = 0, 1, \ldots, n) \quad (5)$$

where the rank of the coefficient matrix (a_j^i) is $(n+1)$, and $\rho \neq 0$ is an arbitrary constant of proportionality. Of course, all elements are from the original ground field K.

A collineation T is not only a one-to-one correspondence between the points S_0 of S_n, but also a one-to-one correspondence between the S_ps of S_n; that is, under T, any S_p is carried into one, and only one, $\bar{S}_p$, and conversely.

The set of collineations T of S_n forms the collineation group G of S_n. It is composed of $n(n+2)$ essential parameters.

The fundamental theorem of projective geometry may be stated in the following form: There is a collineation T which carries a given set of $(p+1)$ linearly independent points into a prescribed set of $(p+1)$ linearly independent points. Moreover, if $p = n$, then T is uniquely determined.

A correlation Γ of the projective space S_n is a one-to-one correspondence between the points S_0 and the hyperplanes S_{n-1} of S_n, of the form shown in Eq. (6), where the rank of the coefficient matrix (b_{ij}) is $(n+1)$, and $\sigma \neq 0$ is an arbitrary constant of proportionality.

$$\sigma \bar{u}_i = \sum_{j=0}^{n} b_{ij} x^j \quad (i = 0, 1, \ldots, n) \quad (6)$$

The set C of correlations Γ of S_n is composed of $n(n+2)$ essential parameters. In general, C is not a group.

The total projective group G^* of the projective space S_n is composed of collineations T and correlations Γ. It is a mixed group G^* of $n(n+2)$ essential parameters. Obviously, the collineation group G is a subgroup of the total projective group G^*.

Projective geometry consists of the qualitative and quantitative invariants of the total projective group G^*.

Cross ratio. Let S_{r-1} and S_{r+1} be two fixed projective subspaces of S_n of dimensions $(r-1)$ and $(r+1)$, respectively, such that S_{r-1} is contained in S_{r+1}. Evidently $0 \leq r \leq n-1$. A pencil is composed of all the S_rs that contain the given S_{r-1} and that are contained in the given S_{r+1}.

For example, where $r = 0$, one obtains a pencil of points, all of which are in a fixed line S_1. Similarly when $r = n-1$, there is defined a pencil of hyperplanes S_{n-1}, all of which contain a fixed projective subspace S_{n-2} of $(n-2)$ dimensions.

A pencil of elements P is a projective space of one dimension. Therefore, the elements P of a pencil are defined by a system of homogeneous coordinates (ρ, τ), where at least one of the elements ρ and τ of the ground field K is not zero. If it is assumed that τ is always one, then ρ is said to be the nonhomogeneous coordinate of the element P which is not the element at infinity $(1,0)$. The element of infinity $(1,0)$ in this system of coordinates is denoted by the symbol ∞.

If (ρ_1, τ_1), (ρ_2, τ_2), (ρ_3, τ_3), (ρ_4, τ_4) are the homogeneous coordinates of four distinct elements P_1, P_2, P_3, P_4, their cross ratio R is defined to be the numerical invariant given in Eq. (7). This is the fundamental projective invariant. In nonhomogeneous coordinates, this is Eq. (8).

$$R(P_1 P_2, P_3 P_4) = \frac{(\rho_3 \tau_1 - \rho_1 \tau_3)(\rho_2 \tau_4 - \rho_4 \tau_2)}{(\rho_2 \tau_3 - \rho_3 \tau_2)(\rho_4 \tau_1 - \rho_1 \tau_4)} \quad (7)$$

$$R(P_1 P_2, P_3 P_4) = \frac{(\rho_3 - \rho_1)(\rho_2 - \rho_4)}{(\rho_2 - \rho_3)(\rho_4 - \rho_1)} \quad (8)$$

In particular, Eqs. (9) and (10) hold.

$$R(\infty P_3, P_1 P_2) = \frac{\rho_3 - \rho_2}{\rho_3 - \rho_1} \quad (9)$$

$$R(\infty 0, P_1 P_2) = \frac{\rho_2}{\rho_1} \quad (10)$$

Of importance in projective geometry is the concept of a harmonic set of elements of a pencil provided that the ground field K is not of characteristic two. Four elements P_1, P_2, P_3, P_4 of a pencil are said to form a harmonic set of elements if, and only if, $R(P_1 P_2, P_3 P_4) = -1$.

Quadrics. Consider a correspondence Γ of S_n in which every point P is converted into a single hyperplane S_{n-1}, or S_n. It is assumed that Γ carries every line S_1 into a single S_{n-2}, or S_{n-1}, or S_n. Finally, it is supposed that if Γ carries a point P into an S_p and if $\overline{P}$ is any point of this S_p, then Γ converts the point $\overline{P}$ into an S_q which passes through the original point P. Such a correspondence Γ is called a polarity Γ.

In homogeneous coordinates, a polarity Γ is given by Eq. (11), where the matrix (g_{ij}) is symmetric,

$$\sigma \bar{u}_i = \sum_{j=0}^{n} g_{ij} x^j \quad (i = 0, 1, \ldots, n) \quad (11)$$

that is, $g_{ij} = g_{ji}$, and where its rank is $(r+1)$ for which $0 \leqq r \leqq n$. If $0 \leqq r < n$, the polarity Γ is said to be singular. Otherwise Γ is said to be nonsingular.

In the homogeneous point coordinates, the polarity Γ is given by Eq. (12).

$$\sum_{i,j=0}^{n} g_{ij} x^i \bar{x}^j = 0 \quad (12)$$

The dual of a polarity Γ is also a polarity Γ^*. Such a polarity Γ^* is given in homogeneous coordinates by Eq. (13), where the matrix (g^{ij}) is symme-

$$\rho \bar{x}^i = \sum_{j=0}^{n} g^{ij} u_j \quad (i = 0, 1, \ldots, n) \quad (13)$$

tric, that is, $g^{ij} = g^{ji}$, and where its rank is $(r+1)$ for which $0 \leqq r \leqq n$.

In homogeneous hyperplane coordinates, the polarity Γ^* is given by Eq. (14).

$$\sum_{i,j=0}^{n} g^{ij} \bar{u}_i \bar{u}_j = 0 \quad (14)$$

If the polarity Γ is nonsingular, then the dual Γ^* is the original polarity Γ. In that event, the two matrices (g_{ij}) and (g^{ij}) can be considered to be inverse matrices. Thus a nondegenerate polarity can be given by Eqs. (11), (12), (13), or (14).

A hyperplane element (P, π) is composed of a point P and a hyperplane π of dimension $(n-1)$ which passes through the point P. A quadric Q is a locus of hyperplane elements (P, π) such that under a given polarity Γ or Γ^*, the point P is transformed into the hyperplane π, or the hyperplane π is carried into the point P.

The polarity Γ or Γ^* is said to be a polarity relative to the corresponding quadric Q.

In homogeneous point coordinates, the equation of a quadric Q is Eq. (15).

$$\sum_{i,j=0}^{n} g_{ij} x^i x^j = 0 \quad (15)$$

In homogeneous hyperplane coordinates, the equation of a quadric Q is Eq. (16).

$$\sum_{i,j=0}^{n} g^{ij} u_i u_j = 0 \quad (16)$$

Euclidean and noneuclidean geometries. Consider a real projective space S_n. That is, S_n is defined over the real number system K. Sometimes it is convenient to consider that S_n is immersed in a complex projective space $S_n{}^*$, which is defined over the complex number system K^*.

Let $k \neq 0$ be either a positive real number or infinite (that is, $1/k = 0$), or else a pure imaginary number of the form $k = il$, where $i^2 = -1$ and l is a positive real number. Consider the fundamental quadric Σ whose homogeneous point equation is Eq. (17), where the superscripts exterior to the

$$f(x,x) = (kx^0)^2 + (x^1)^2 + (x^2)^2 + \cdots + (x^n)^2 = 0 \quad (17)$$

parentheses denote exponents. The homogeneous hyperplane equation of this fundamental quadratic Σ is Eq. (18).

$$F(u,u) = \left(\frac{u_0}{k}\right)^2 + u_1{}^2 + u_2{}^2 + \cdots + u_n{}^2 = 0 \quad (18)$$

The set of all collineations T of S_n which carry this fundamental quadric Σ into itself is a group $G(k)$ of $n(n+1)/2$ essential parameters. When k is a positive real number, this is the elliptic group G_E of elliptic geometry. When $1/k = 0$, this is the euclidean group G_P of euclidean geometry. Finally when $k = il$ where $i^2 = -1$ and l is a positive real number, this is the hyperbolic group G_H of hyperbolic geometry.

The study of the qualitative and quantitative invariants of these three groups G_E, G_P, and G_H

constitutes the three subjects of elliptic, euclidean, and hyperbolic geometries. The elliptic and hyperbolic geometries are usually referred to as the noneuclidean geometries.

In elliptic geometry, two distinct lines contained in a single plane always meet in a single point. Therefore, if L is a fixed line and if P is a given point not in L, there cannot be a line M passing through this point P which is parallel to the given line L.

In euclidean geometry, the ideal hyperplane π_∞ or the hyperplane π_∞ at infinity is the one whose point equation is $x^0 = 0$, or whose hyperplane equations are $u_1 = 0,\ u_2 = 0,\ \ldots,\ u_n = 0$. The proper S_ps for $p = -1, 0, 1, \ldots, n$ are those which are not contained in the ideal hyperplane π_∞. The improper S_ps for $p = 0, 1, \ldots, n-1$ are those which are contained in the ideal hyperplane π_∞. In euclidean geometry, only proper S_ps are studied.

If a proper S_p and a distinct proper S_q intersect in an improper S_r, then S_p and S_q are said to be parallel. Thus, two distinct lines in euclidean space are said to be parallel if, and only if, they intersect in an ideal point.

From the preceding discussion, Euclid's fifth parallel postulate is an easy consequence. That is, if in euclidean space L is a fixed line and if P is a fixed point not on L, there is one, and only one, line M parallel to L and passing through P.

In hyperbolic geometry, the points and the tangent S_ps for $p = 0, 1, 2, \ldots, n-1$ of the fundamental quadric Σ given by Eqs. (17) or (18) are said to be ordinary improper or ordinary infinite. The S_ps for $p = 0, 1, 2, \ldots, n-1$, which are in the exterior of this quadric Σ, are said to be ultraimproper or ultrainfinite. The proper points in hyperbolic geometry are those which are in the interior of this quadric Σ. The proper S_ps for $p = 1, 2, \ldots, n-1, n$ are those which contain proper points and are considered to be sets of these proper points.

If a proper S_p and a distinct proper S_q intersect in an ordinary improper S_r or in an ultraimproper S_r, then S_p and S_q are said to be ordinary parallel or ultraparallel.

In hyperbolic geometry, if L is a fixed proper line and if P is a given proper point not on this line L, then there are two distinct proper lines M_1 and M_2 passing through P which are ordinary lines parallel to L.

Also passing through P, there is an infinite number of proper lines M which are ultraparallel to L. These lines M belong to the flat pencil with vertex at P and determined by the lines M_1 and M_2.

Distance. Let P and Q be two distinct points given by the homogeneous point coordinates $x = (x^0, x^1, \ldots, x^n)$ and $y = (y^0, y^1, \ldots, y^n)$. In euclidean and hyperbolic geometries, it is understood that P and Q are proper points. The line L determined by the two points P and Q intersects the fundamental quadric Σ in two distinct points P_∞ and Q_∞. The distance $s = s(P,Q)$ between these two points P and Q is defined by formula (19). It is

$$s = s(P,Q) = \frac{k}{2i} \log \mathrm{R}(PQ, P_\infty Q_\infty) \qquad (19)$$

understood that $s = s(P,Q)$ is a real nonnegative number. Also it is assumed in elliptic geometry that $0 \leq s/k \leq \pi$.

Let $f(x,y)$ denote the expression in Eq. (20). The

$$f(x,y) = k^2 x^0 y^0 + x^1 y^1 + x^2 y^2 + \cdots + x^n y^n \qquad (20)$$

two points P and Q are polar reciprocal or orthogonal relative to the fundamental quadric Σ if and only if $f(x,y) = 0$. The point P is on Σ if, and only if, $f(x,x) = 0$.

A point R whose homogeneous point coordinates are $z = (z^0, z^1, \ldots, z^n)$ is on the line L determined by the two points P and Q if, and only if, a number ρ exists such that Eqs. (21) hold.

$$z^0 = x^0 + \rho y^0$$
$$z^1 = x^1 + \rho y^1, \ldots, z^n = x^n + \rho y^n \qquad (21)$$

This point R is on the fundamental quadric Σ whose point equation is Eq. (17) if, and only if, ρ satisfies the quadratic equation, Eq. (22). Because

$$f(x,x) + 2\rho f(x,y) + \rho^2 f(y,y) = 0 \qquad (22)$$

the two points P and Q are distinct, this will have two distinct roots ρ_1 and ρ_2. The two points P_∞ and Q_∞ are on the line L whose homogeneous point equations are Eqs. (21), corresponding to the two distinct roots ρ_1 and ρ_2.

In elliptic and euclidean geometries, these two points P_∞ and Q_∞ are conjugate-imaginary. In hyperbolic geometry, they are real.

From Eqs. (21) and (22), it is seen that Eq. (23) holds.

$$\mathrm{R}(PQ, P_\infty Q_\infty) = \frac{\rho_1}{\rho_2}$$
$$= \frac{-f(x,y) - \sqrt{f^2(x,y) - f(x,x) f(y,y)}}{-f(x,y) + \sqrt{f^2(x,y) - f(x,x) f(y,y)}} \qquad (23)$$

Then, from Eq. (23), Eq. (24) applies.

$$\mathrm{R}(PQ, P_\infty Q_\infty)$$
$$= \frac{[f(x,y) + \sqrt{f^2(x,y) - f(x,x) f(y,y)}\,]^2}{f(x,x) f(y,y)} \qquad (24)$$

Consequently the distance $s(P,Q)$ is given by formula (25).

$$s(P,Q)$$
$$= \frac{k}{i} \log \frac{f(x,y) + \sqrt{f^2(x,y) - f(x,x) f(y,y)}}{\sqrt{f(x,x)}\ \sqrt{f(y,y)}} \qquad (25)$$

Set $s = s(P,Q)$. Then s is given by Eqs. (26).

$$\cos \frac{s}{k} = \frac{f(x,y)}{\sqrt{f(x,x)}\ \sqrt{f(y,y)}}$$
$$\sin \frac{s}{k} = \frac{\sqrt{f^2(x,y) - f(x,x) f(y,y)}}{i \sqrt{f(x,x)}\ \sqrt{f(y,y)}} \qquad (26)$$

Of course, from Eq. (19), this distance $s = s(P,Q)$ is invariant under each of the groups G_E, G_P, G_H of elliptic, euclidean, and hyperbolic geometries.

By Eq. (20), Eqs. (26) may be written in the forms shown in Eqs. (27), in which $\epsilon = +1$ or $\epsilon = -1$, according to whether the geometry is elliptic (including the euclidean case) or hyperbolic.

In hyperbolic geometry, $k = il$ where $i^2 = -1$ and l is a positive real number. Equations (27) can be written in the forms shown in Eqs. (28).

$$\cos\frac{s}{k}=\frac{\epsilon(k^2x^0y^0+x^1y^1+x^2y^2+\cdots+x^ny^n)}{\sqrt{[(kx^0)^2+(x^1)^2+\cdots+(x^n)^2][(ky^0)^2+(y^1)^2+\cdots+(y^n)^2]}}$$

(27)

$$\sin\frac{s}{k}=\sqrt{\frac{\epsilon\left[k^2\sum_{i=1}^n\begin{vmatrix}x^0&x^i\\y^0&y^i\end{vmatrix}^2+\frac{1}{2}\sum_{i,j=1}^n\begin{vmatrix}x^i&x^j\\y^i&y^j\end{vmatrix}^2\right]}{[(kx^0)^2+(x^1)^2+\cdots+(x^n)^2][(ky^0)^2+(y^1)^2+\cdots+(y^n)^2]}}$$

$$\cosh\frac{s}{l}=\frac{l^2x^0y^0-x^1y^1-x^2y^2-\cdots-x^ny^n}{\sqrt{[(lx^0)^2-(x^1)^2-\cdots-(x^n)^2][(ly^0)^2-(y^1)^2-\cdots-(y^n)^2]}}$$

(28)

$$\sinh\frac{s}{l}=\sqrt{\frac{l^2\sum_{i=1}^n\begin{vmatrix}x^0&x^i\\y^0&y^i\end{vmatrix}^2-\frac{1}{2}\sum_{i,j=1}^n\begin{vmatrix}x^i&x^j\\y^i&y^j\end{vmatrix}^2}{[(lx^0)^2-(x^1)^2-\cdots-(x^n)^2][(ly^0)^2-(y^1)^2-\cdots-(y^n)^2]}}$$

Return to the general case of Eqs. (27). Let $x=(x^0,x^1,\ldots,x^n)$ denote affine coordinates in a euclidean space of $(n+1)$ dimensions. Define the special quadric Σ^* by Eq. (29). This is a central

$$f(x,x)=(kx^0)^2+(x^1)^2+\cdots+(x^n)^2=k^2 \quad (29)$$

quadric Σ^*. Each of the two noneuclidean geometries can be visualized as the geometry on this quadric Σ^* in which diametrically opposite points are identified. In particular, for elliptic geometry, Σ^* can be considered a sphere.

On this quadric Σ^*, Eqs. (27) can be written in the forms labeled Eqs. (30). Here k is a positive real

$$\cos\frac{s}{k}=x^0y^0+\frac{1}{k^2}(x^1y^1+x^2y^2+\cdots+x^ny^n)$$

$$k\sin\frac{s}{k}=\sqrt{\sum_{i=1}^n\begin{vmatrix}x^0&x^i\\y^0&y^i\end{vmatrix}^2+\frac{1}{2k^2}\sum_{i,j=1}^n\begin{vmatrix}x^i&x^j\\y^i&y^j\end{vmatrix}^2}$$

(30)

number and the distance s, such that $0\leqq s/k\leqq\pi$, is given by the preceding equations.

Euclidean geometry can be considered to be the limiting case of either elliptic or hyperbolic geometry as k becomes infinite. In this case, one can always regard x^0 as unity. Then from Eqs. (30), the distance formula $s=s(P,Q)$ for euclidean geometry is given by Eq. (31). In this case, coordinates $x=$

$$s=s(P,Q)$$

$$=\sqrt{\begin{array}{c}(x^1-y^1)^2+(x^2-y^2)^2\\+\cdots+(x^n-y^n)^2\end{array}} \quad (31)$$

$(x^1,x^2,\ldots,x^n)$ of a point P are said to be rectangular or cartesian.

The final case is that of hyperbolic geometry. Here $k=il$, where $i^2=-1$ and l is a positive real number. From Eqs. (30), the distance formula $s=s(P,Q)$ is given by Eqs. (32).

$$\cosh\frac{s}{l}=x^0y^0-\frac{1}{l^2}(x^1y^1+x^2y^2+\cdots+x^ny^n)$$

$$l\sinh\frac{s}{l}=\sqrt{\sum_{i=1}^n\begin{vmatrix}x^0&x^i\\y^0&y^i\end{vmatrix}^2-\frac{1}{2l^2}\sum_{i,j=1}^n\begin{vmatrix}x^i&x^j\\y^i&y^j\end{vmatrix}^2}$$

(32)

In each of the three geometries it is assumed that the relationship $s=s(P,Q)$ is zero if, and only if, the two points P and Q are identical.

Each of the three geometries is an abstract metric space. That is,

(i) $s(P,Q)\geqq 0$, and $s(P,Q)=0$ if, and only if, $P=Q$

(ii) $s(P,Q)=s(Q,P)$

(iii) $s(P,Q)+s(Q,R)\geqq s(P,R)$

The condition (i) is that of positive definiteness; the condition (ii) is that of symmetry; the condition (iii) is that of the well-known triangular inequality.

Angle. By dualizing the concept of distance $s=s(P,Q)$, the concept of angle $\theta=\theta(\pi,\sigma)$ between two hyperplanes π and σ is obtained. Let π and σ be two distinct hyperplanes which are given by the homogeneous hyperplane coordinates $u=(u_0,u_1,\ldots,u_n)$ and $v=(v_0,v_1,\ldots,v_n)$. In euclidean and hyperbolic geometries, it is understood that π and σ are proper hyperplanes which determine a pencil λ. In this pencil λ there are two distinct hyperplanes π_∞ and σ_∞ which are tangent to the fundamental quadric Σ. The angle $\theta=\theta(\pi,\sigma)$, where $0\leqq\theta\leqq\pi$, between these two hyperplanes π and σ is defined by formula (33). This angle $\theta=\theta(\pi,\sigma)$ is given by Eqs. (34), in which $\epsilon=+1$ or

$$\theta=\theta(\pi,\sigma)=\frac{l}{2i}\log R(\pi\sigma,\pi_\infty\sigma_\infty) \quad (33)$$

$\epsilon=-1$ according to whether the geometry is elliptic (including the euclidean case) or hyperbolic.

$$\cos\theta=\frac{\epsilon\left(\dfrac{u_0v_0}{k^2}+u_1v_1+u_2v_2+\cdots+u_nv_n\right)}{\sqrt{\left(\dfrac{u_0^2}{k^2}+u_1^2+\cdots+u_n^2\right)\left(\dfrac{v_0^2}{k^2}+v_1^2+\cdots+v_n^2\right)}}$$

(34)

$$\sin\theta=\sqrt{\frac{\epsilon\left[\dfrac{1}{k^2}\sum_{i=1}^n\begin{vmatrix}u_0&u_i\\v_0&v_i\end{vmatrix}^2+\dfrac{1}{2}\sum_{i,j=1}^n\begin{vmatrix}u_i&u_j\\v_i&v_j\end{vmatrix}^2\right]}{\left(\dfrac{u_0^2}{k^2}+u_1^2+\cdots+u_n^2\right)\left(\dfrac{v_0^2}{k^2}+v_1^2+\cdots+v_n^2\right)}}$$

It is evident that when $\epsilon=+1$ and k becomes infinite, the preceding formulas become those for the angle θ between the two hyperplanes π and σ in a euclidean space of n dimensions.

The two hyperplanes π and σ are said to be orthogonal if, and only if, $\mathbf{R}(\pi\sigma,\pi_\infty\sigma_\infty) = -1$. Let σ_1 and σ_2 be two distinct hyperplanes which intersect in a line L. Then L is said to be orthogonal to a hyperplane π if σ_1 and σ_2 are each orthogonal to π. If L and M are two distinct lines which pass through a point P and if M is contained in a hyperplane π orthogonal to L, then L and M are said to be orthogonal.

If π is a fixed hyperplane and if P is a point not in π, then there is one and only one line L passing through the point P orthogonal to the hyperplane π.

Differential of arc length. If P and Q are two nearby points on the quadric Σ^* defined by the two sets of coordinates $(\bar{x}^0, \bar{x}^1, \ldots, \bar{x}^n)$ and $(\bar{x}^0 + d\bar{x}^0, \bar{x}^1 + d\bar{x}^1, \ldots, \bar{x}^n + d\bar{x}^n)$, then by Eqs. (30), the square of the differential ds of arc length s between the points P and Q is Eq. (35).

$$ds^2 = \sum_{i=1}^n (\bar{x}^0 \, d\bar{x}^i - \bar{x}^i \, d\bar{x}_0)^2 + \frac{1}{2k^2} \sum_{i,j=1}^n (\bar{x}^i \, d\bar{x}^j - \bar{x}^j \, d\bar{x}^i)^2 \quad (35)$$

For the point P on this quadric Σ^*, introduce a new set of coordinates $(x^1, x^2, \ldots, x^n)$ defined by Eq. (36), where it is understood that $\bar{x}^0 \neq -1$.

$$x^i = \frac{2\bar{x}^i}{1 + \bar{x}^0} \quad (i = 1, 2, \ldots, n) \quad (36)$$

In this new set of coordinates, the quadric Σ^* is given by Eq. (37).

$$(x^1)^2 + (x^2)^2 + \cdots + (x^n)^2 + 4k^2 = \frac{8k^2}{1 + \bar{x}^0} \quad (37)$$

In the euclidean and noneuclidean geometries, the square of the differential ds of arc length s is given by Eq. (38).

$$ds^2 = \frac{(dx^1)^2 + (dx^2)^2 + \cdots + (dx^n)^2}{\left[1 + \dfrac{(x^1)^2 + (x^2)^2 + \cdots + (x^n)^2}{4k^2}\right]^2} \quad (38)$$

As k approaches infinity, the differential $d\sigma$ of arc length σ of euclidean geometry is obtained. If ds represents the differential of arc length s in elliptic or hyperbolic geometry, then Eq. (39) holds,

$$ds = \rho \, d\sigma \quad (39)$$

in which the scale ρ is given by Eq. (40).

$$\rho = \frac{1}{1 + \dfrac{(x^1)^2 + (x^2)^2 + \cdots + (x^n)^2}{4k^2}} \quad (40)$$

Thus, each of the noneuclidean geometries may be visualized as a conformal image of euclidean geometry.

Each of these three geometries is a special case of Riemannian geometry. Let the $g_{ij}(x)$ be $n(n+1)/2$ real functions which are continuous and possess continuous partial derivatives of as high an order as is necessary in a certain region of real n-dimensional space for which the coordinates of a point P are $x = (x^1, x^2, \ldots, x^n)$. The g_{ij}s are symmetric; that is, $g_{ij} = g_{ji}$, for all $i, j = 1, 2, \ldots, n$. The quadratic form, expression (41), is assumed to

$$\sum_{i,j=1}^n g_{ij} \lambda^i \lambda^j \quad (41)$$

be positive definite, that is, expression (41) is nonnegative; it is zero if and only if $\lambda^i = 0$, for all $i = 1, 2, \ldots, n$. Then the Riemannian space V_n is one for which the square of the differential ds of arc length s between two nearby points P and Q is given by Eq. (42).

$$ds^2 = \sum_{i,j=1}^n g_{ij} \, dx^i \, dx^j \quad (42)$$

For this Riemannian space V_n, the Christoffel symbols of the first kind are shown as Eq. (43), for

$$\Gamma_{ij;k} = \frac{1}{2}\left(\frac{\partial g_{ik}}{\partial x^j} + \frac{\partial g_{jk}}{\partial x^i} - \frac{\partial g_{ij}}{\partial x^k}\right) \quad (43)$$

$i, j, k = 1, 2, \ldots, n$. The Christoffel symbols of the second kind are shown as Eq. (44), for $i, j, k = $

$$\Gamma_{ji.}{}^i = \sum_{l=1}^n g^{il}\Gamma_{jk;l} \quad (44)$$

$1, 2, \ldots, n$. Of course, (g^{il}) is the inverse matrix of (g_{ij}). The Christoffel symbols of the second kind are also called the affine connections of V_n.

The vector geodesic curvature κ^i of a curve C in V_n is Eq. (45). A curve C of V_n is said to be a geodes-

$$\kappa^i = \frac{d^2 x^i}{ds^2} + \sum_{j,k=1}^n \Gamma_{jk}{}^i \frac{dx^j}{ds} \frac{dx^k}{ds} \quad (45)$$

ic if, and only if, $\kappa^i = 0$ for $i = 1, 2, \ldots, n$ at every point P of C.

Each of the three spaces already discussed is a Riemannian space. The geodesics of any one such space are the lines of the space.

A Riemannian space of constant curvature is applicable to (that is, can be mapped isometrically into) elliptic, euclidean, or hyperbolic space. In this case, this constant curvature is equal to the Gaussian curvature G. From Eq. (35) or (38), this constant Gaussian curvature G is given by Eq. (46).

$$G = 1/k^2 \quad (46)$$

For elliptic space, G is of positive constant curvature. For euclidean space, G is identically zero. Finally for hyperbolic space, G is of negative constant curvature.

In each of these three geometries, consider a geodesic triangle. This is formed by three points P, Q, R, not all on one geodesic, and the three geodesics passing through every two of the three points P, Q, R. If A, B, C are the angles of this geodesic triangle, and if T denotes its area, then Eq. (47) holds.

$$A + B + C - \pi = T/k^2 \quad (47)$$

Thus, according to whether the geometry is elliptic, euclidean, or hyperbolic, the sum of the angles of a geodesic triangle is greater than π, equal to π, or less than π.

Elliptic geometry of two dimensions may be represented upon a sphere in euclidean space of three dimensions. On the other hand, hyperbolic geometry of two dimensions can be depicted on a pseudosphere in euclidean space of three dimensions. The pseudosphere is obtained by revolving the tractrix about its asymptote.   [JOHN DE CICCO]

Bibliography: H. S. M. Coxeter, *Introduction to Geometry*, 2d ed., 1969; H. S. M. Coxeter, *Non-Euclidean Geometry*, 5th ed., 1965; L. P. Eisenhart, *Non-Riemannian Geometry*, 1922, reprint 1972; L. P. Eisenhart, *Riemannian Geometry*, 1949; E. Kasner and J. Newman, *Mathematics and the Imagination*, 1940, reprint 1962.

Nonlinear optics

A field of study concerned with the interaction of electromagnetic radiation and matter in which the matter responds in a nonlinear manner to the incident radiation fields. The nonlinear response can result in intensity-dependent variation of the propagation characteristics of the radiation fields or in the creation of radiation fields at new frequencies. Nonlinear effects can take place in solids, liquids, gases, and plasmas, and may involve one or more electromagnetic fields as well as internal excitations of the medium. The wavelength range of interest generally coincides with the spectrum covered by lasers, extending from the far infrared to the vacuum ultraviolet, but some nonlinear interactions have been observed at wavelengths extending into the x-ray range. Historically, nonlinear optics precedes the laser, but most of the work done in the field has made use of the high powers available from lasers. *See* LASER.

Nonlinear materials. Nonlinear effects of various types are observed in all materials. It is convenient to characterize the response of the medium mathematically by expanding it in a power series in the electric and magnetic fields of the incident optical waves. The linear terms in such an expansion give rise to the linear index of refraction, linear absorption, and the magnetic permeability of the medium, while the higher-order terms give rise to nonlinear effects. Direct calculations of some nonlinear susceptibilities, the coefficients in the expansion that describe the strength of the nonlinear interactions, have been made in some materials, notably atomic and molecular gases and some crystals, using various theories. More generally, however, the nonlinear coefficients are determined by measurement.

In general, nonlinear effects associated with the electric field of the incident radiation dominate over magnetic interactions. Effects involving the electric dipole response of the medium, termed the polarization, are generally the most important of the electric field interactions. The even-order dipole susceptibilities are zero except in media which lack a center of symmetry, such as certain classes of crystals, certain symmetric media to which external forces have been applied, or at boundaries between certain dissimilar materials. Odd-order terms can be nonzero in all materials regardless of symmetry. Generally the magnitudes of the nonlinear susceptibilities decrease rapidly as the order of the interaction increases. Second- and third-order effects have been the most extensively studied of the nonlinear interactions, although effects up to order 29 have been observed. In some situations such as dielectric breakdown or saturation of absorption, effects of different order cannot be separated, and all orders must be included in the response. *See* ELECTRIC SUSCEPTIBILITY; POLARIZATION OF DIELECTRICS.

Second-order effects. Second-order effects involve a polarization with the dependence $P_{nl}^{(2)} =$ dE^2, where E is the electric field of the optical waves and d is a nonlinear susceptibility. The second-order polarization has components that oscillate at sum and difference combinations of the incident frequencies, and also a component that does not oscillate. The oscillating components produce a propagating polarization wave in the medium with a propagation vector that is the appropriate sum or difference of the propagation vectors of the incident waves. The nonlinear polarization wave serves as a source for an optical wave at the corresponding frequency in a process that is termed three-wave parametric mixing.

Phase matching. The strongest interaction occurs when the phase velocity of the polarization wave is the same as that of a freely propagating wave of the same frequency. The process is then said to be phase-matched. Dispersion in the refractive indices, which occurs in all materials, usually prevents phase matching from occurring unless special steps are taken. Phase matching in crystals may be achieved by using noncollinear beams, materials with periodic structures, anomalous dispersion near an absorption edge, or compensation using free carriers in a magnetic field, or by using the birefringence possessed by some crystals. In the birefringence technique, the one used most commonly for second-order interactions, one or two of the interacting waves propagate as an extraordinary wave in the crystal. The phase-matching conditions are achieved by choosing the proper temperature and propagation direction. Usually these conditions depend on the wavelengths of the individual waves involved. The conditions for phase-matched three-wave parametric mixing can be summarized by Eqs. (1) and (2), where the ν's

$$\nu_3 = \nu_1 \pm \nu_2 \tag{1}$$

$$\mathbf{k}_3 = \mathbf{k}_1 \pm \mathbf{k}_2 \tag{2}$$

are the frequencies of the waves and the $\mathbf{k}$'s are the propagation constants. When phase matching is accomplished, the power in the generated wave is many orders of magnitude greater than that generated in non-phase-matched interactions. *See* ABSORPTION OF ELECTROMAGNETIC RADIATION; CRYSTAL OPTICS; REFRACTION OF WAVES.

Frequency mixing. In three-wave sum- and difference-frequency mixing, two incident waves at ν_1 and ν_2 are converted to a third wave at ν_3 according to Eq. (1). The simplest interaction of this type is second-harmonic generation in which $\nu_3 = 2\nu_1$. For this interaction the phase-matching condition reduces to $n(\nu_3) = n_1(\nu_1)/2 + n_2(\nu_1)/2$, where $n(\nu_3)$ is the refractive index at the harmonic wavelength and $n_1(\nu_1)$ and $n_2(\nu_1)$ are the refractive indices of the incident waves at the fundamental frequency. By using radiation from pulsed lasers, conversion efficiencies of over 90% have been achieved in second-harmonic generation. With continuous lasers, it is more efficient if the nonlinear crystal is placed in the laser cavity. Conversion efficiencies of over 30% of the internal laser power have been obtained in this way for continuous lasers.

Second-harmonic generation and second-order frequency mixing have been demonstrated at wavelengths ranging from the infrared to the ultraviolet, generally coinciding with the transparency range of the nonlinear crystals. Sum-frequency mixing has been used at wavelengths as short as 185 nm. Difference-frequency mixing can be used to generate both visible and infrared radiation. Radiation has been generated in this way to wavelengths of about 2 mm. If one or more of the incident wavelengths is tunable, the generated wavelength will also be tunable, providing a source of radiation that is useful for high-resolution spectroscopy. *See* LASER SPECTROSCOPY.

Parametric generation. Parametric generation is the reverse process of sum frequency mixing. In parametric generation a single input wave at ν_3 is converted to two lower-frequency waves according to the relation $\nu_3 = \nu_1 + \nu_2$. The individual values of ν_1 and ν_2 are determined by the simultaneous satisfaction of the phase-matching condition in Eq. 2. Generally this condition can be satisfied for only one pair of frequencies at a time for a given propagation direction. By changing the phase-matching conditions, the individual longer wavelengths can be tuned. Parametric generation can be used to amplify waves at lower frequencies than the pump wave or, in an oscillator, as a source of tunable radiation in the visible or infrared over a wavelength range similar to that covered in difference-frequency mixing.

Optical rectification. The component of the second-order polarization that does not oscillate produces an electrical voltage. The effect is called optical rectification and is a direct analog of the rectification that occurs in electrical circuits at much lower frequencies. It has been used in conjunction with ultrashort mode-locked laser pulses to produce electrical pulses that, at a duration of several picoseconds, are among the shortest electrical pulses that have been generated. *See* OPTICAL PULSES.

Third-order interactions. Third-order interactions give rise to several types of nonlinear effects. Four-wave parametric mixing involves interactions which generate waves at sum- and difference-frequency combinations of the form $\nu_4 = \nu_1 \pm \nu_2 \pm \nu_3$ with the corresponding phase-matching condition $\mathbf{k}_4 = \mathbf{k}_1 \pm \mathbf{k}_2 \pm \mathbf{k}_3$. Phase matching in liquids is usually accomplished by using noncollinear pump beams. Phase matching in gases can also be accomplished by using anomalous dispersion near absorption resonances or by using mixtures of gases, one of which exhibits anomalous dispersion. The use of gases, which are usually transparent to longer and shorter wavelengths than are the solids used for second-order mixing, allows the range of wavelengths covered by parametric mixing interactions to be extended considerably. Four-wave mixing processes of the type $\nu_4 = \nu_1 - \nu_2 - \nu_3$ have been used to generate far-infrared radiation out to wavelengths of the order of 25 μm. Sum-frequency mixing and third-harmonic generation have been used to generate radiation extending into the vacuum ultraviolet. Radiation down to 38 nm has been obtained with still higher-order processes.

The nonlinear susceptibility is greatly increased, sometimes by four to eight orders of magnitude, through resonant enhancement that occurs when the input frequencies or their multiples or sum or difference combinations coincide with appropriate energy levels in the nonlinear medium. Two-photon resonances are of particular importance since they do not involve strong absorption of the incident or generated waves. Resonant enhancement in gases has allowed tunable dye lasers to be used for nonlinear interactions, providing a source of tunable radiation in the vacuum ultraviolet and in the far infrared. Such radiation is useful in spectroscopic studies of atoms and molecules.

Four-wave sum-frequency generation can be used to convert infrared radiation to the visible where it is more easily detected. These interactions have been used for infrared spectroscopic studies and for infrared image conversion. Similar infrared up-conversion can also be done with three-wave mixing.

Intensity-dependent effects. Nonlinear polarization components at the same frequencies as those in the incident waves can result in effects that change the index of refraction or the absorption coefficient, quantities that are constants in linear optical theory.

Multiphoton absorption and ionization. Materials that are transparent at low optical intensities can have their absorption increase at high intensities. This effect involves simultaneous absorption of multiple numbers of photons from one or more incident waves. When the process involves transitions to discrete upper levels in gases or to conduction bands in solids that obey certain quantum-mechanical selection rules, it is usually termed multiphoton absorption. When transitions to the continuum of gases are involved, the process is termed multiphoton ionization.

Saturable absorption. In materials which have a

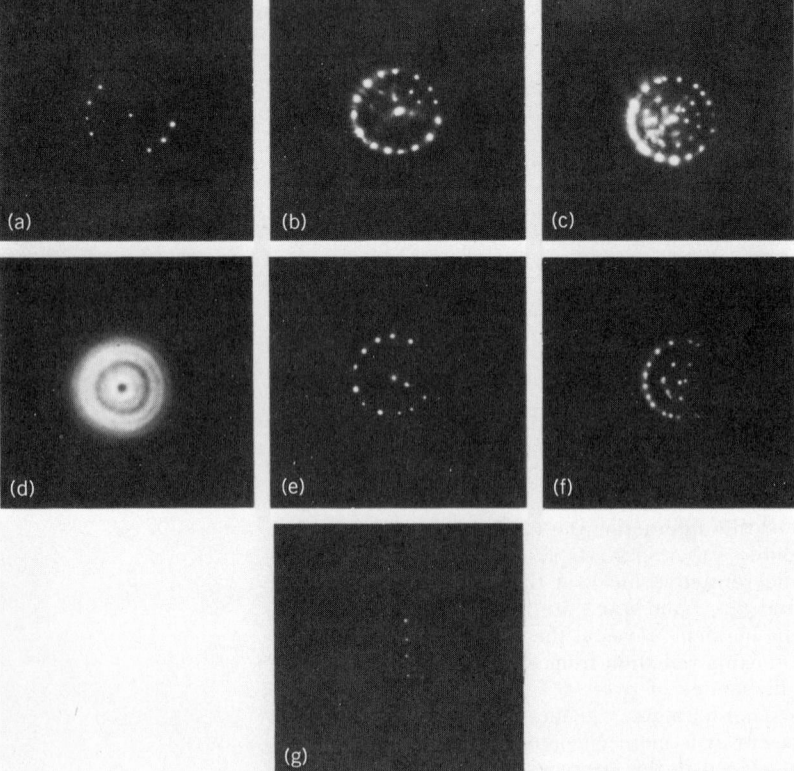

Fig. 1. Breakup of an optical beam with a circular ring structure due to self-focusing. (*a–c*) Patterns obtained from a beam of Fresnel number 7 and increasing intensity. (*d*) Near-field pattern of beam with Fresnel number 4. (*e, f*) Corresponding far-field patterns (focal spot patterns). (*g*) Focal-spot pattern for a beam that has been diffracted from a straight edge. (*From A. J. Campillo, S. L. Shapiro, and B. R. Suydam, Periodic breakup of optical beams due to self focusing, Appl. Phys. Lett., 23:628–630, 1973*)

strong linear absorption at the incident frequency, the absorption can decrease at high intensities. This effect, termed saturable absorption, was observed long before lasers were available. Saturable absorbers are useful in operating Q-switched and mode-locked lasers.

Self-focusing and -defocusing. Intensity-dependent changes in the refractive index can affect the propagation characteristics of a laser beam. For many materials, for example, many solids, Kerr active liquids, and some gases, the index of refraction increases with increasing optical intensity. If the laser beam has a profile that is more intense in the center than at the edges, the profile of the refractive index corresponds to that of a positive lens, causing the beam to focus. This effect, termed self-focusing, can cause an initially uniform laser beam to break up into many smaller spots with diameters of the order of a few micrometers and intensity levels that are high enough to damage many solids. Such a beam breakup is illustrated in the patterns shown in Fig. 1. This is a primary mechanism that limits the maximum intensities that can be obtained from some high-power pulsed solid-state lasers. *See* KERR EFFECT.

In other materials the refractive index decreases as the optical intensity increases. The resulting nonlinear lens causes the beam to defocus, an effect termed self-defocusing. When encountered in media that are weakly absorbing, the effect is termed thermal blooming. It is prominent, for example, in the propagation of high-power infrared laser beams through the atmosphere.

Broadening of spectrum. When pulsed laser fields are involved, the nonlinear refractive index can lead to a broadening of the laser spectrum.

Phase conjugation. The same interactions that give rise to the intensity-dependent effects can also cause nonlinear interactions between waves that are at the same frequency but are otherwise distinguishable, for example, by their direction of polarization or propagation. Termed degenerate four-wave mixing, this interaction gives rise to a number of effects such as amplification of a weak probe wave and generation of a fourth wave that propagates in a direction opposite to that of a probe wave. In one of the most interesting aspects of this interaction, the backward wave has the same distribution of phase variations as the probe wave but with their sense reversed. This effect, termed phase conjugation or time reversal, can allow the correction of phase distortions of light beams caused by propagation through inhomogeneous media, by aberrated optics, or by mode dispersion in optical waveguides. It has important applications in the area of image transmission through optical fibers and in various forms of optical processing and information transfer. The correction of phase aberrations is illustrated in Fig. 2.

Low-frequency electric, magnetic, or acoustic fields can be used to control the polarization or propagation of an optical wave. These effects, termed electro-, magneto-, or acoustooptic effects, are useful in the modulation and deflection of light waves and are used in information-handling systems. *See* ACOUSTOOPTICS; ELECTROOPTICS; MAGNETOOPTICS.

Nonlinear spectroscopy. The variation of the nonlinear susceptibility near the resonances that

correspond to sum- and difference-frequency combinations of the input frequencies forms the basis for various types of nonlinear spectroscopy which allow study of energy levels that are not

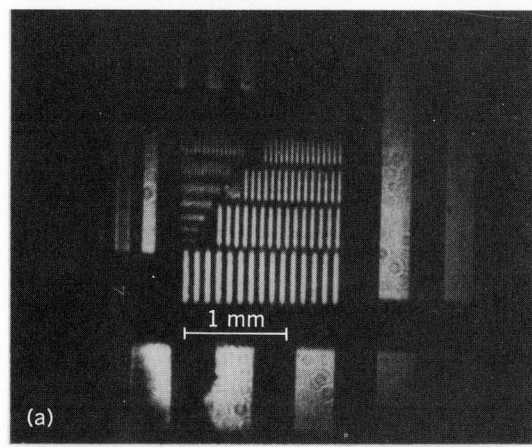

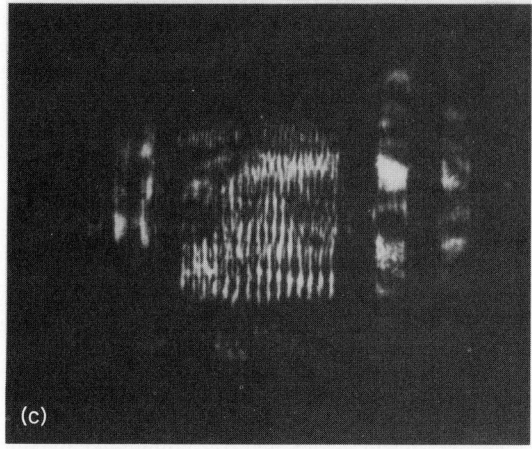

Fig. 2. Photographs showing correction of phase aberrations by degenerate four-wave mixing. (*a*) Image of a resolution chart taken without a phase-aberrating plate (a distorting piece of glass) in place. (*b*) Same image with a phase-aberrating plate but with corrections due to degenerate four-wave mixing. (*c*) Distorted image that results when the corrections of the four-wave mixing system are incomplete. The aberrating plate used in these pictures was a piece from a glass bottle. (*From D. M. Bloom and G. C. Bjorklund, Conjugate wave-front generation and image reconstruction by four-wave mixing, Appl. Phys. Lett., 31:592–594, 1977*)

normally accessible with linear optical spectroscopy.

Nonlinear spectroscopy can be performed with many of the interactions discussed earlier. Multiphoton absorption spectroscopy can be performed by using two strong laser beams, or a strong laser beam and a weak broadband light source. If two counterpropagating laser beams are used, spectroscopic studies can be made of energy levels in gases with spectral resolutions much smaller than the Doppler limit. The use of multiphoton resonances associated with multiphoton ionization has allowed the identification of many new energy levels with principal quantum numbers as high as 60 in several elements. *See* RYDBERG ATOM.

Many types of four-wave mixing interactions can also be used in nonlinear spectroscopy. The most widespread of these processes is a four-wave interaction of the form $\nu_4 = 2\nu_1 - \nu_2$, and utilizes Raman resonances which coincide with the energy difference $\nu_1 - \nu_2$. Termed coherent anti-Stokes Raman spectroscopy (CARS), it offers the advantage of greatly increased signal levels over linear Raman spectroscopy for the study of certain classes of materials. The spectroscopy of energy levels near the generated frequency in the vacuum ultraviolet has been studied by using four-wave mixing interactions of the type $\nu_4 = 2\nu_1 + \nu_2$. Other types of nonlinear spectroscopy involve polarization changes or higher-order processes. *See* RAMAN EFFECT.

Time measurements. Various nonlinear interactions, such as second harmonic generation, multiphoton absorption followed by fluorescence, or coherent Raman scattering, have provided a means for measurement of pulse duration and excited-state lifetimes in the picosecond and subpicosecond time regimes.

Inelastic scattering. Light can scatter inelastically from fundamental excitations in the medium, resulting in the production of radiation at new frequencies that differ from the incident frequencies by an amount that is equal to the frequency of the fundamental excitation. The difference in photon energy between the incident and scattered fields is taken up or given up by the medium. Inelastic scattering can take place from any of the possible excitations in the medium, and include: Brillouin scattering from acoustic vibrations; various forms of Raman scattering involving internal vibrations or rotational modes of molecules, optical phonons, electronic states in atoms or molecules, polaritons, plasmons, spin waves, Landau levels in semiconductors, and transitions that involve spin flips in semiconductors; Rayleigh scattering involving entropy fluctuations; and scattering from density variations or concentration fluctuations in gases. Inelastic scattering may occur in liquids, solids, gases, or plasmas and may be influenced by the application of external fields or other externally applied forces.

At power levels that are available from pulsed lasers, the scattered radiation can become stimulated, resulting in exponential growth of the scattered power. In stimulated scattering the incident light can be almost completely scattered. Stimulated scattering has been observed for most of the interactions that were mentioned above. The stimulated scattered light has many of the properties of the incident laser light, including collimation and narrow line width. By using different materials, stimulated scattering can provide laserlike sources in spectral regions that are not directly covered by lasers. In some interactions the use of external fields can allow the production of tunable stimulated scattered radiation. Under certain conditions the scattered field is phase-conjugate to the incident field and can be used to correct phase aberrations. *See* SCATTERING OF ELECTROMAGNETIC RADIATION.

Coherent effects. Another class of third-order effects involves excitations of the medium in which the phase of the atomic wave function is preserved. These effects are generally observed only for short light pulses, of the order of several nanoseconds or less. In one interaction, termed self-induced transparency, a pulse of light of the proper shape, magnitude, and duration can propagate unattenuated in a medium which is otherwise absorbing.

Other coherent effects involve changes of the propagation speed of a light pulse or production of a coherent pulse of light at a characteristic time after two pulses of light spaced by a time τ have entered the medium. The generated pulse in this last effect is termed a photon echo. Still other coherent interactions involve oscillations of the atomic polarization, giving rise to effects known as optical nutation and free induction decay. Two-photon coherent effects are also possible.

[JOHN F. REINTJES]

Bibliography: D. H. Auston, in S. L. Shapiro (ed.), *Ultra Short Light Pulses*, 1976; D. C. Hanna, M. A. Yuratich, and D. Cotter, *Nonlinear Optics of Free Atoms and Molecules*, 1979; H. Rabin and C. L. Tang (eds.), *Nonlinear Optics, Quantum Electronics*, vol. 1, 1975; A. Yariv, *Quantum Electronics*, 2d ed., 1975.

Non-newtonian fluid

A fluid whose flow behavior departs from that of an ideal newtonian fluid. In a newtonian fluid the rate of shear in the fluid under isothermal conditions is proportional to the corresponding stress at the point under consideration. Consider, for example, two flat plates of area A containing a layer of fluid of thickness y and caused to move parallel to each other at a relative velocity u as in Fig. 1, where F is shearing force. The shear stress is then F/A, and the rate of shear is u/y. For a newtonian fluid, Eq. (1) can be written, where μ, the constant

$$\frac{F}{A} \propto \frac{u}{y} \qquad \text{or} \qquad \frac{F}{A} = \mu \frac{u}{y} \qquad (1)$$

of proportionality, is called the newtonian viscosity and completely characterizes the fluid; that is,

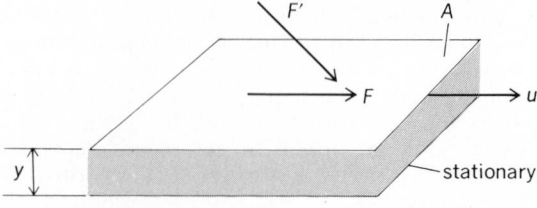

Fig. 1. Illustration of shear stress.

the relationship between the shear stress and the rate of shear is linear, as shown in Fig. 2. *See* NEWTONIAN FLUID.

A non-newtonian fluid is one for which this rather strict requirement is not met. There are several ways in which real fluids can depart from the relatively simple newtonian condition, but it is better to think in terms of non-newtonian regions rather than non-newtonian fluids because many fluids exhibit both newtonian and non-newtonian behavior, depending on the conditions of flow, and cannot be strictly classified as being exclusively one type or the other.

Non-newtonian fluids can be broadly classified into three types as follows:

1. Time-independent fluids for which the rate of shear at any point in the fluid is some function of the shear stress at that point and depends on nothing else. (A newtonian fluid is a special case of a fluid in this category.)

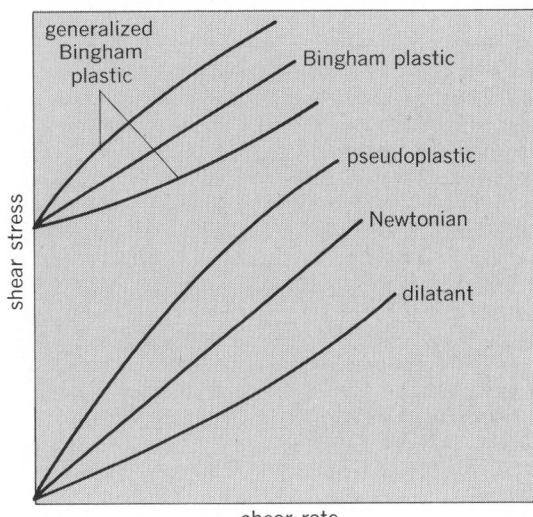

Fig. 3. Non-newtonian flow curves.

2. More complex time-dependent fluids for which the relationship between shear stress and shear rate depends on the time the fluid has been sheared, that is, on its previous history.

3. Fluids which have the characteristics of both viscous liquids and elastic solids and exhibit partial elastic recovery after deformation, the so-called viscoelastic fluids.

Time-independent fluids. The rheological equation for these fluids is Eq. (2), where $\dot{\gamma}$, the rate of

$$\dot{\gamma} = f(\tau) \qquad (2)$$

shear, is a function only of τ, the corresponding shear stress. These fluids may be conveniently divided into three distinct types, depending on the nature of the function in Eq. (2). These types, the flow curves for which are shown in Fig. 3, are as follows.

Bingham plastics. These exhibit a yield stress τ_y which is the stress that must be exceeded before flow starts. Thereafter the rate-of-shear curve is linear. This behavior is found in such materials as toothpaste, oil paints, and drilling muds. There are other materials which also exhibit a yield stress,

but the flow curve is thereafter not linear. These are usually called generalized Bingham plastics.

Pseudoplastic fluids. These show no yield value, but the ratio of shear stress to the rate of shear, which may be termed the "apparent viscosity," falls progressively with shear rate. This phenomenon of shear thinning is characteristic of suspensions of asymmetric particles or solutions of polymers such as cellulose derivatives.

Dilatant fluids. These are similar to pseudoplastics in that they show no yield stress, but the apparent viscosity for dilatant materials increases with increasing rates of shear. Such "shear thickening" is observed with suspensions of solids at high solids content (approaching point of tightest packing). Corn-flour pastes can be dilatant.

Time-dependent fluids. The apparent viscosity of some fluids depends not only on the rate of shear but also on the time the shear has been applied. These fluids may be subdivided into two classes—thixotropic fluids and rheopectic fluids—according to whether the shear stress decreases or increases with time when the fluid is sheared at a constant rate.

Thixotropic materials. The structure breakdown of these materials is a function of the duration of shear as well as of the rate of shear. Consider a thixotropic material which is sheared at a constant rate after a period of rest. The structure will be progressively broken down, and the apparent viscosity will decrease with time. The rate of breakdown of structure during shearing at a given rate will depend on the number of linkages available for breaking and therefore must decrease with time. (This could be compared with the rate of a first-order chemical reaction.) Also, the simultaneous rate of reformation of structure will increase with time as the number of possible new structural linkages increases. Eventually a state of dynamic equilibrium is reached when the rate of buildup of structure equals the rate of breakdown. This equilibrium position depends on the rate of shear and moves toward greater breakdown at increasing rates of shear.

As an example, consider the material confined in the annular gap of a cylindrical viscometer. After the material has been resting for a long time, one of the cylinders is rotated at a constant speed. The torque on the other cylinder then decreases with time, as shown in Fig. 4. The rate of decrease and the final torque both depend on the speed, that is, on the rate of shear. Thixotropy is a reversible process, and after resting, the structure of the material builds up again gradually. This type of

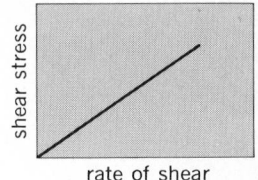

Fig. 2. Newtonian shear stress–shear rate curve.

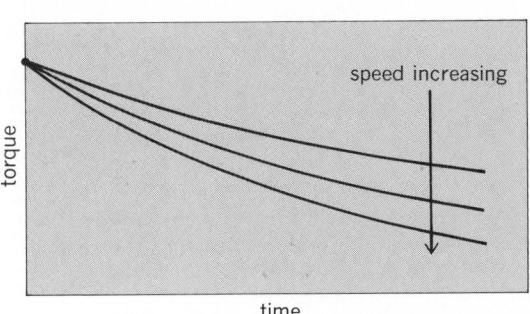

Fig. 4. Cylinder torque decrease due to thixotropy.

behavior leads to a kind of hysteresis loop on the shear stress–shear rate curve if the curve is plotted for the rate of shear increasing at a constant rate followed by the curve for the rate of shear decreasing at a constant rate.

Rheopectic fluids. These are fluids for which the structure builds up on shearing, and this phenomenon can be regarded as the reverse of thixotropy—in fact, rheopectic behavior is often referred to as antithixotropy. Such materials are comparatively rare, and few if any are of industrial significance.

Pseudoplastic fluids. These are analogous to thixotropic fluids but with the difference that in the case of pseudoplastic fluids the time required for structure breakdown is negligible; that is, the time effect is not observable in the apparatus being used for the testing of the so-called pseudoplastic fluids. The difference, then, is only a matter of degree.

In the same way rheopectic fluids are superficially similar to their time-independent counterparts, dilatant fluids, in which the time for structure buildup is insignificantly small. Here, however, the analogy is not so close because rheopexy is a case where buildup is often brought about by small shearing rates only. In some cases there is an upper limit to the shear rate beyond which the analogy breaks down.

Viscoelastic fluids. A viscoelastic material is one which possesses both viscous and elastic properties; that is, although the material might be viscous, it exhibits a certain elasticity of shape and is capable of storing energy of deformation. This is easily demonstrated in the laboratory by making the shearing force F in Fig. 1 a function of time, as in Eq. (3), where F_0 is the force amplitude, ω is the

$$F = F_0 \cos \omega t \qquad (3)$$

circular frequency of oscillation, and t is the elapsed time. The corresponding time-dependent velocity is then given by Eq. (4). The relation between F_0 and U_0 is illustrated on the phase dia-

$$U = U_0 \cos (\omega t + \phi) \qquad (4)$$

gram in Fig. 5.

If the substance is purely viscous, no energy of deformation can be stored and the directions of F_0 and U_0 coincide. If the response of the material is purely elastic, no energy can be dissipated and F_0 will lag U_0 by 90°. In the case of viscoelastic materials, some energy is stored and $0 < \phi < 90°$; that is, the material possesses both viscous and elastic properties.

The constitutive equation for a viscoelastic material is a relationship between stress and strain and their time derivatives. One simple equation which describes linear viscoelasticity and approximates the behavior of some real fluids is that proposed by Maxwell, Eq. (5), where τ is the stress and

$$\tau + \lambda \frac{d\tau}{dt} = \mu \dot{\gamma} \qquad (5)$$

$\dot{\gamma}$ is the rate of shear. The constant λ, which has dimensions of time, is referred to as the relaxation time of the fluid. It is the time constant of the exponential decay of stress at constant strain; that is, if the motion is suddenly stopped, the strain will decay as $\exp(-t/\lambda)$. In general, after a suddenly

imposed strain, which is subsequently held constant, the stress in viscoelastic materials decays in time according to an equation of the form of Eq. (6), where $G(t - t')$ is a time-dependent rigidity

$$d\tau(t) = G(t - t') \frac{d\gamma}{dt'}(t') \, dt' \qquad (6)$$

modulus, $\tau(t)$ is the current stress, and $d\gamma(t')$ is the strain increment at some previous time t'.

For an arbitrary strain history, integration over the entire strain history is required and may extend to $-\infty$, as in Eq. (7).

$$\tau(t) = \int_{-\infty}^{t} G(t - t') \frac{d\gamma(t')}{dt'} \, dt' \qquad (7)$$

This type of behavior, which is typical of solutions of macromolecules and molten polymers, is analogous to mechanical springs and dashpots in series.

Real materials behave as though they comprised a large number of such spring-dashpot combinations, each with its own relaxation time defined as the ratio of viscosity to the corresponding rigidity modulus in the combination. Much experimental work is devoted to finding the viscosity contribution of each flow unit with a given relaxation time to the total bulk viscosity of the material from steady-flow measurements.

Nonlinear viscoelasticity. In the steady flow of solutions of macromolecules and polymer melts, nonlinear elastic effects manifest themselves quite readily. The first of them to occur is that of the forces F in Fig. 1 moving out of the planes of shear and assuming directions inclined to these planes, as F', even though the planes themselves continue to move parallel to each other. This phenomenon arises from the state of anisotropic stress which exists at each point within the flow. The stress state on an element of fluid is illustrated in Fig. 6. If the fluid is newtonian, then Eq. (8) holds, where

$$\sigma'_x = \sigma'_y = \sigma'_z = 0 \qquad (8)$$

$\sigma_x = \sigma'_x - p$, $\sigma_y = \sigma'_y - p$, and $\sigma_z = \sigma'_z - p$. Generally, σ'_x, σ'_y, σ'_z, the so-called extra stresses, are generated by the shearing motion while p is an isotropic stress whose value is determined by the boundary conditions in specific problems. Note that on one pair of faces of the cube in Fig. 6 no shear stresses are present.

A further nonlinear effect appears if the shear rate is increased, namely, that of variable viscosity. A simple proportionality between shear stress and shear rate no longer applies; instead, the viscosity varies as shown in Fig. 7. In the region of the low-shear-rate newtonian viscosity, the extra

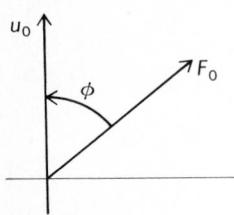

NON-NEWTONIAN FLUID

Fig. 5. The phase relationship between the applied force and motion for viscoelastic materials.

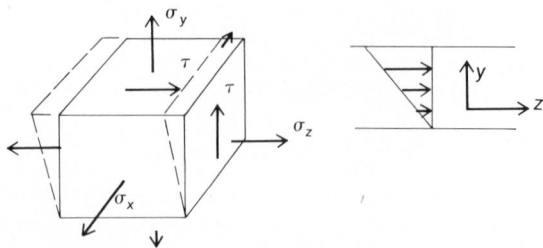

Fig. 6. The state of stress on a fluid element.

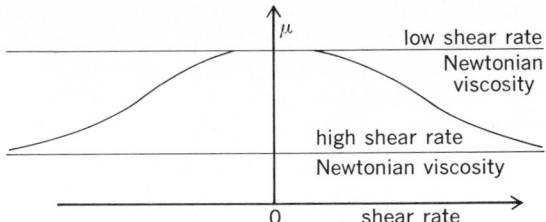

Fig. 7. The variation of viscosity with shear rate for solutions of macromolecules.

stresses $\sigma'_x, \sigma'_y, \sigma'_z$ are a function of the square of the shear rate and vanish at a faster rate than the shear stress which is an odd function in shear rate. However, in the high-shear-rate newtonian viscosity zone, normal stress would be expected to be present always.

There are many theories available in the literature today which accommodate the above nonlinear effects; none have achieved more than limited acceptance.

Experimental characterization. There are two main viscometric methods for the determination of the rheological properties of time-independent fluids, as follows:

1. Determination of the shear stress–shear rate relationship directly by subjecting the entire sample to a uniform rate of shear in a suitably designed instrument and measuring the corresponding shear stress. These viscometers are usually rotational instruments of the coaxial cylinders or cone and plate type.

2. To infer the shear stress – shear rate relationship indirectly from observations on the pressure gradient and volumetric flow rate in a straight pipe or capillary tube viscometer. In these instruments the rate of shear is not constant but rises from zero at the center of the pipe to a maximum at the wall, and consequently the interpretation of the results is not easy.

Viscoelastic fluids are difficult to characterize quantitatively. Much useful information can be obtained by subjecting the material to sinusoidal shearing motions and following the amplitude and phase shift of the induced stress, that is, a frequency-response analysis. Normal stresses can also be measured, for example, in a cone and plate viscometer.

The characterization of time-dependent fluids of the thixotropic type is also difficult, and only qualitative results can be obtained by testing in steady shear in capillary or rotational viscometers. Oscillatory testing offers more promise, but this is a newer technique attempted on this type of fluid. *See* FLUIDS.

[J. HARRIS; W. L. WILKINSON]

Bibliography: F. R. Eirich, *Rheology*, vols. 1–5, 1958–1970; J. Harris, *Rheology and Non-Newtonian Flow*, 1978; W. R. Schowalter, *Mechanics of Non-Newtonian Fluids*, 1977.

Nonrelativistic quantum theory

The modern theory of matter and its interaction with electromagnetic radiation, applicable to systems of material particles which move slowly compared to the velocity of light and which are neither created nor destroyed.

This article details the formal structure of quantum theory, which can be summarized in the form of postulates as unequivocal as are Newton's laws of classical mechanics; a less logically rigorous presentation is adopted here, however. Even so, the reader unfamiliar with quantum theory is advised to read first another article (*see* QUANTUM MECHANICS) which more qualitatively discusses the novel (from the standpoint of classical physics) features of nonrelativistic quantum theory.

That article and this one attempt to make convincing the thesis that the formalism of nonrelativistic quantum theory is an unarbitrary and physically reasonable extension of the older classical theories. Belief in quantum theory stems as much from acceptance of this thesis as from the broad range, barely hinted at in these articles, of successful application of the theory. For added details concerning special formal topics *see* MATRIX MECHANICS; PERTURBATION (QUANTUM MECHANICS); SPINOR. For generalizations of nonrelativistic quantum theory to relativistic particles (particles with speeds close to the velocity of light), or to systems in which particle creation and destruction can occur, *see* QUANTUM ELECTRODYNAMICS; QUANTUM FIELD THEORY; RELATIVISTIC QUANTUM THEORY.

WAVE FUNCTION AND PROBABILITY DENSITY

Basic to quantum mechanics is the belief that the wave properties of matter are associated with a function, the wave function, obeying an equation called a wave equation. The simplest possible wave in three-dimensional space is a so-called scalar wave, in which the wave disturbance is wholly characterized by a single function $\psi(x,y,z,t) \equiv \psi(\mathbf{r},t)$. It is natural, therefore, to postulate that a wave function $\psi(x,yz,z,t)$ provides a complete description of the simplest possible physical system, namely, a single particle moving in a force field specified by a potential $V(\mathbf{r})$. It is further postulated that $|\psi(\mathbf{r},t)|^2$, which classically would be proportional to the wave intensity, is the probability density, that is, $|\psi(\mathbf{r},t)|^2 \, d\mathbf{r}$ is the probability, at time t, of finding the particle in the volume $dxdydz \equiv d\mathbf{r}$ of space lying between x and $x + dx$, y and $y + dy$, z and $z + dz$.

There is the obvious generalization that a wave function $\psi(\mathbf{r}_1, \ldots, \mathbf{r}_g, t)$ will completely describe a system of g particles, with $|\psi(\mathbf{r}_1, \ldots, \mathbf{r}_g, t)|^2 \, d\mathbf{r}_1 \cdots d\mathbf{r}_g$ the probability at time t of simultaneously finding particle 1 in the volume element $d\mathbf{r}_1 = dx_1 dy_1 dz_1, \ldots$, particle g in $d\mathbf{r}_g$. Moreover, for a system of g distinguishable particles, the probability $P_j(\mathbf{r})d\mathbf{r}$ of finding particle j in the volume element $d\mathbf{r}$ at $\mathbf{r} \equiv x, y, z$ of physical space is given by Eq. (1),

$$P_j(\mathbf{r})d\mathbf{r} = d\mathbf{r} \int \{ d\mathbf{r}_1 \cdots d\mathbf{r}_{j-1} \, d\mathbf{r}_{j+1} \cdots d\mathbf{r}_g$$
$$\times |\psi(\mathbf{r}_1, \ldots, \mathbf{r}_{j-1}, \mathbf{r}, \mathbf{r}_{j+1}, \ldots, \mathbf{r}_g)|^2 \} \quad (1)$$

where $|\psi(\mathbf{r}_1, \ldots, \mathbf{r}_g)|^2$ is integrated over all positions of particles 1 to $j-1$ and $j+1$ to g, with $\mathbf{r}_j$ put equal to $\mathbf{r}$.

Normalization. Because each of the particles $1, \ldots, g$ must be somewhere in physical space, Eq. (1) demands that Eq. (2a) be integrated over all positions of all g particles. When Eq. (2a) is satisfied, ψ is said to be normalized. If $\psi'(\mathbf{r}_1, \ldots, \mathbf{r}_g, t)$ is a proposed wave function which satisfies Eq. (2b), the probabilities specified by ψ'

$$\int d\mathbf{r}\, P_j(\mathbf{r})$$

$$= \int d\mathbf{r}_1 \cdots d\mathbf{r}_g |\psi(\mathbf{r}_1, \ldots, \mathbf{r}_g)|^2 = 1 \quad (2a)$$

$$\int d\mathbf{r}_1 \cdots d\mathbf{r}_g |\psi'|^2 = C \neq 1 \quad (2b)$$

are found from Eq. (1) using the normalized $\psi = C^{-1/2}\psi' \exp(i\eta)$, provided C is not infinite, that is, provided ψ' is quadratically integrable; the phase factor $\exp(i\eta)$, η being real, can be chosen arbitrarily. Though the absolute probabilities $P_j(\mathbf{r})$ are not defined when $\psi'(\mathbf{r}_1, \ldots, \mathbf{r}_g)$ is not quadratically integrable, $|\psi'|^2$ may remain a useful measure of the relative probability of finding particles $1, \ldots, g$ at $\mathbf{r}_1, \ldots, \mathbf{r}_g$.

One need be concerned only with wave functions which can represent actual physical systems. It is postulated that an admissible (physically possible) wave function $\psi'(\mathbf{r}_1, \ldots, \mathbf{r}_g)$ is quadratically integrable (type 1), or fails to be quadratically integrable (type 2) only because ψ' vanishes too slowly or at worst remains finite as infinity is approached in the $3g$-dimensional space of $\mathbf{r}_1, \ldots, \mathbf{r}_g$. Convincing physical and mathematical reasons can be found for excluding wave functions other than these types. One-particle wave functions $\psi'(\mathbf{r})$ of type 2 correspond to classically unconfined systems, for example, an electron ionized from a hydrogen atom; for these systems $\psi(\mathbf{r}) = \infty^{-1/2}\psi' = 0$ has the obvious interpretation that an unconfined particle is sure to be found outside any given finite volume. Such a $\psi'(\mathbf{r})$ also can represent a very large (effectively infinite) number of independently moving identical particles in a very large (effectively infinite) volume, for example, a beam of free electrons issuing from an electron gun; in this event $|\psi'(\mathbf{r})|^2$, although it continues to describe the likelihood of observing an electron, can be thought to equal the actual number density of electrons at any point $\mathbf{r}$, with $C = \infty$ in Eq. (2b) indicating that the number of particles in all of space is infinite. These considerations can be extended to quadratically nonintegrable many-particle wave functions $\psi'(\mathbf{r}_1, \ldots, \mathbf{r}_g)$.

Spin. The preceding formalism accepts the presumption of classical physics that a particle is a structureless entity, about which "everything" is known when its position $\mathbf{r}$ is known. This presumption is inaccurate (and therefore the formalism is not wholly adequate) for systems of electrons, protons, and neutrons, these being the fundamental particles which compose what usually is termed stable matter. In particular an electron, proton, or neutron cannot be described completely by a single wave function $\psi(\mathbf{r},t)$. For each of these particles two wave functions $\psi_1(\mathbf{r},t)$ and $\psi_2(\mathbf{r},t)$ are required, which may be regarded as components of an overall two-component wave function $\psi(\mathbf{r},t)$. The need for a multicomponent wave function has the immediate classical interpretation that the particle has internal degrees of freedom, that is, that knowledge of the position of the particle is not "everything." It can be shown that these internal degrees of freedom are associated with so-called intrinsic angular momentum or spin. See SPIN.

For electrons, protons, or neutrons the spin is $1/2\hbar$, where $\hbar = h/2\pi$, and h is Planck's constant; the only allowed values of s_z, the z component of the spin, are $\pm\frac{1}{2}$ (in units of $\hbar$). Thus when the system contains a single particle of spin $\frac{1}{2}$, an electron, say, $|\psi_1(\mathbf{r},t)^2\, d\mathbf{r}$ can be interpreted as the probability of finding, in the volume $d\mathbf{r}$, an electron with $s_z = +\frac{1}{2}$; $|\psi_2(\mathbf{r},t)|^2$ is the probability density for finding an electron with $s_z = -\frac{1}{2}$. The normalization condition replacing Eq. (2a) is given by Eq. (3).

$$\int d\mathbf{r}\, [|\psi_1(\mathbf{r},t)|^2 + |\psi_2(\mathbf{r},t)|^2] = 1 \quad (3)$$

This formalism is readily extended to many-particle systems. For example, when the system contains g particles of spin $\frac{1}{2}$, the overall wave function ψ has 2^g components ψ_j; $|\psi_1(\mathbf{r}_1, \ldots, \mathbf{r}_g)|^2$ is the probability density for finding each of particles 1 to g with spin oriented along $+z$; and the normalization condition is given by Eq. (4), summed from

$$\sum_j \int d\mathbf{r}_1 \cdots d\mathbf{r}_g |\psi_j(\mathbf{r}_1, \ldots, \mathbf{r}_g)|^2 = 1 \quad (4)$$

$j = 1$ to 2^g. The appropriate reinterpretations when the wave function is not quadratically integrable are obvious. Complications which arise from particle indistinguishability are discussed later in the article.

OPERATORS

Whereas in classical mechanics particle coordinates $\mathbf{r}$ and momenta $\mathbf{p}$ are numbers which can be specified independently at any instant of time, in quantum mechanics the components of $\mathbf{r}$ and $\mathbf{p}$ are linear operators, as also are functions $f(\mathbf{r},\mathbf{p})$ of the coordinates and momenta. It is postulated that (i) the operator $\mathbf{x}$ (here distinguished by boldface from the x coordinate to which $\mathbf{x}$ corresponds) simply multiplies a wave function $\psi(x)$ by x, that is, $\mathbf{x}\psi = x\psi$; (ii) the operator corresponding to the canonically conjugate x component of momentum of that particle is $p_x = (\hbar/i)\partial/\partial x$, that is, $p_x\psi = (\hbar/i) \partial/\psi\partial/x$. Thus $A\psi$ denotes the new wave function $\psi' = A\psi$, resulting from the linear operation A on a given wave function ψ. When A, B, C are linear operators, and ψ, ξ are any two functions, Eqs. (5)

$$A(\psi + \xi) = A\psi + A\xi$$
$$(A + B)\psi = A\psi + B\psi \quad (5)$$
$$AB\psi = A(B\psi)$$

hold. Moreover, if ξ and ψ can be added or equated, then ξ and ψ must have the same number of components and depend upon the same space and spin coordinates; if $\xi = \psi$, corresponding components of ξ and ψ are equal. A spin-independent operator performs the same operation on each component of a many-component wave function, for example, $(p_x\psi)_j = p_x\psi_j = (\hbar/i)\partial\psi_j/\partial x$. Spin-dependent operators are more complicated; for instance, in a one-particle system of spin $\frac{1}{2}$ the components of $\psi' = s_z\psi$, where s_z denotes the z component of the spin operator, are given by Eqs. (6), using the notation adopted previously in the discussion of spin.

$$\psi'_1 = {}^{1/2}\hbar\psi_1 \qquad \psi'_2 = -{}^{1/2}\hbar\psi_2 \quad (6)$$

The operators A and B are said to commute when, for any ψ, $A(B\psi) = B(A\psi)$, implying that Eq. (7) is valid. The operator $(AB - BA)$ is termed the

$$AB - BA = 0 \quad (7)$$

commutator of A and B. Any operator $f(A)$ expressible as a power series in the operator A commutes with A. By performing the indicated operations, Eq. (8) is obtained, which shows that pairs of oper-

$$(xp_x - p_x x)\psi = i\hbar\psi \qquad (8)$$

ators need not commute. In a g-particle system, all particle coordinates $x_1, y_1, z_1, \ldots, x_g, y_g, z_g$ commute with each other; all momentum coordinates $\mathbf{p}_1, \ldots, \mathbf{p}_g$ commute with each other; any component of $\mathbf{r}_1$ or $\mathbf{p}_1$ commutes with all components of $\mathbf{r}_2, \ldots, \mathbf{r}_g$ and of $\mathbf{p}_2, \ldots, \mathbf{p}_g$; the x coordinate of any particle commutes with p_y and p_z of that particle, and so on.

Hermitian operators. An operator A relevant to a given system of g particles is termed Hermitian if Eq. (9) holds for all pairs of sufficiently well-be-

$$\sum_j \int d\mathbf{r}_1 \cdots d\mathbf{r}_g \xi_j^* (A\psi)_j$$
$$= \sum_j \int d\mathbf{r}_1 \cdots d\mathbf{r}_g (A\xi)_j^* \psi_j \qquad (9)$$

haved quadratically integrable wave functions, $\xi(\mathbf{r}_1, \ldots, \mathbf{r}_g), \psi(\mathbf{r}_1, \ldots, \mathbf{r}_g)$.

In Eq. (9) the asterisk denotes the complex conjugate, and the sum is over all components j of ξ, $A\psi, A\xi, \psi$. Evidently every particle coordinate $x_1, y_1, z_1, \ldots, x_g, y_g, z_g$ is a Hermitian operator, as is any reasonably well-behaved $f(\mathbf{r}_1, \ldots, \mathbf{r}_g)$. Recalling that quadratically integrable functions vanish at infinity, integration by parts shows that every component of $\mathbf{p}_1, \ldots, \mathbf{p}_g$ is Hermitian, as is any $f(\mathbf{p}_1, \ldots, \mathbf{p}_g)$ expressible as a power series in components of $\mathbf{p}_1, \ldots, \mathbf{p}_g$; for example, in the simple one-dimensional spinless case, Eq. (10)

$$\int_{-\infty}^{\infty} dx\, \xi^* \frac{\hbar}{i} \frac{\partial\psi}{\partial x} = -\frac{\hbar}{i} \int_{-\infty}^{\infty} dx \frac{\partial\xi^*}{\partial x}\psi$$
$$= \int_{-\infty}^{\infty} dx \left(\frac{\hbar}{i}\frac{\partial\xi}{\partial x}\right)^* \psi \qquad (10)$$

holds. It is implied that ξ and ψ are continuous; otherwise the integration by parts yields extra terms on the right side of Eq. (10). Similarly, p_x^2 has the desired Hermitian property, defined by Eq. (9), only when ξ and ψ are continuous and have continuous first derivatives at all points.

When $A, B, C, \ldots$, are individually Hermitian Eq. (11) holds. For simplicity the integration varia-

$$\int \xi^* [(ABC \cdots)\psi]$$
$$= \int [(\cdots CBA)\xi]^* \psi \qquad (11)$$

bles and the summation over components are not indicated explicitly in Eq. (11). When A and B are Hermitian, Eq. (11) implies (i) $1/2\,(AB + BA)$ is Hermitian; (ii) AB and BA are not Hermitian unless A and B commute. For example, xp_x and $p_x x$ are not Hermitian, but $1/2\,(xp_x + p_x x)$ is; classically, of course, there is no distinction between $xp_x, p_x x$, or $1/2(xp_x + p_x x)$. By appropriately symmetrizing, taking note of Eq. (11), one can construct the quantum mechanical Hermitian operator corresponding to any classical $f(\mathbf{r}_1, \ldots, \mathbf{r}_g; \mathbf{p}_1, \ldots, \mathbf{p}_g)$ expressible as a power series in components of coordinates and momenta.

If one supposes that the forces acting on a quantum-mechanical system of g particles are precisely the same as in the classical case, the energy operator is the classical Hamiltonian in Eq.

(12), where the kinetic energy $T = p_1^2/2m_1 + $

$$H(\mathbf{r}_1, \ldots, \mathbf{r}_g; \mathbf{p}_1, \ldots, \mathbf{p}_g) = T + V \qquad (12)$$

$\cdots + p_g^2/2m_g$; $V(\mathbf{r}_1, \ldots, \mathbf{r}_g)$ is the potential energy; and m_i is the mass of particle i. *See* HAMILTON'S EQUATIONS OF MOTION.

Quantum-mechanical nonclassical forces, with associated potential energy operators more complicated than $V(\mathbf{r}_1, \ldots, \mathbf{r}_g)$, are not uncommon, however. For example, the interaction between two neutrons is believed to include space exchange as in Eq. (13), where $J(r_{12})$ is an ordinary

$$V\psi_j(\mathbf{r}_1, \mathbf{r}_2) \equiv J(r_{12})\, P_{12}\psi_j(\mathbf{r}_1, \mathbf{r}_2)$$
$$= J(r_{12})\psi(\mathbf{r}_2, \mathbf{r}_1) \qquad (13)$$

function of the distance r_{12} between the particles, and the space exchange operator P_{12} interchanges the space coordinates of particles 1 and 2 in each component of ψ. *See* NUCLEAR STRUCTURE.

Real eigenvalues. The eigenvalue equation for an operator A is given by Eq. (14), where the num-

$$Au(\alpha) = \alpha u(\alpha) \qquad (14)$$

ber α is the eigenvalue, and the corresponding eigenfunction $u(\alpha)$ is a not identically zero wave function solving Eq. (14) for that value of α. The eigenvalue equation $Hu = Eu$ for the energy operator H has special importance and is known as the time-independent Schrödinger equation. Since the eigenvalues α are identified with the results of measurement, it is desirable that (i) the eigenvalues α all be real numbers and not complex numbers; (ii) the corresponding eigenfunctions form a complete set, the meaning and importance of which is explained subsequently. Property (i) is important because actual measurements yield real numbers only, for example, lengths, meter readings, and the like. If the eigenvalue of A were complex, it could not be maintained that each value of α represented a possible result of exact measurement of A. This assertion is not negated by the fact that it is formally possible to combine real quantities into complex expressions; for example, the coordinates of $\mathbf{r}$ in the xy plane form the complex vector $x + iy$. *See* EIGENFUNCTION; EIGENVALUE.

The eigenvalues α_n belonging to the quadratically integrable eigenfunctions $u(\alpha_n) \equiv u_n$ of a Hermitian operator A are necessarily real. In the notation of Eq. (11), letting $\xi = \psi = u_n$ in Eq. (9) and employing Eq. (14), one obtains Eq. (15).

$$\alpha_n \int u_n^* u_n = \int u_n^* (\alpha_n u_n) = \int u_n^* (Au_n) = \int (Au_n)^* u_n$$
$$= \int (\alpha_n u_n)^* u_n = \alpha_n^* \int u_n^* u_n \qquad (15)$$

The equality of the first and last terms of Eq. (15) demonstrates that $\alpha_n = \alpha_n^*$. For this reason it is postulated that all "observable" operators are Hermitian operators, and conversely that all Hermitian operators represent observable quantities; henceforth all operators are supposed Hermitian. In addition, it is necessary to require that the allowed eigenfunctions preserve the hermiticity property of Eq. (9); otherwise Eq. (15) would not hold. For the important class of Hamiltonian operators $H = T + V$, except for highly singular (discontinuous) potentials, the boundary conditions that u and $\partial u/\partial x$ must be continuous guarantee reality of the eigenvalue corresponding to a quadratically integrable u solving Eq. (14). Admissible wave functions, defined following Eq. (2b), may be quadratically nonintegrable, however. It always is assumed that the physically desirable properties (i) and (ii) that were discussed above follow from the equally physically desirable simple requirement that the eigenfunctions $u(\alpha)$ of A must be admissible, provided that u, $\partial u/\partial x$, etc., satisfy the continuity conditions which make Eq. (15) correct for quadratically integrable u_n. In systems containing a single spinless particle this assumption has been justified rigorously for operators of interest; the widespread quantitative successes of quantum theory support the belief that the assumption is equally valid in more complicated systems.

Orthogonality. The eigenvalues α corresponding to quadratically integrable eigenfunctions typically form a denumerable (countable), though possibly infinite, set and compose the discrete spectrum of A. The admissible nonquadratically integrable eigenfunctions typically correspond to a continuous set of real eigenvalues composing the continuous spectrum of A. An eigenvalue α is degenerate, with order of degeneracy $d \geqq 2$, if there exist d independent eigenfunctions $u_1, \ldots, u_d$ corresponding to the same value of α, whereas every $d+1$ such eigenfunctions are dependent. The d functions $\psi_1, \ldots, \psi_d$ are (linearly) dependent if the relation in Eq. (16) can be true with not all the con-

$$c_1\psi_1 + \ldots + c_d\psi_d = 0 \qquad (16)$$

stants $c_1, \ldots, c_d$ equal to zero. Eigenvalues which are either discrete or continuous or both may be degenerate. While so-called accidental degeneracy can occur, degeneracy of the eigenvalues α of A ordinarily is associated with the existence of one or more operators which commute with A. In the absence of degeneracy the eigenfunctions $u(\alpha)$ are uniquely indexed by α, and can be chosen to satisfy the orthonormal (orthogonality and normalizing) relations [compare Eq. (4)], as shown in Eq. (17).

$$\sum_j \int d\mathbf{r}_1 \cdots d\mathbf{r}_g u_j^*(\alpha)\, u_j(\alpha')$$
$$= \delta_{aa'} \text{ or } \delta\,(\alpha - \alpha') \quad (17)$$

In Eq. (17) the Kronecker symbol $\delta_{aa'}$ is employed when α lies in the discrete spectrum; $\delta_{aa'} = 0$ for $\alpha \neq \alpha'$, $\delta_{aa'} = 1$ for $\alpha = \alpha'$. The Dirac delta function $\delta\,(\alpha - \alpha')$ is employed when α lies in the continuous spectrum; $\delta(\alpha - \alpha') = 0$ when $\alpha \neq \alpha'$, but has a finite integral defined by Eq. (18a). For a wide variety of functions $f(\alpha)$ these properties of $\delta(\alpha - \alpha')$ imply that Eq. (18b) is valid.

$$\int_{-\infty}^{\infty} d\alpha\, \delta(\alpha - \alpha') = \int_{-\infty}^{\infty} d\alpha'\, \delta(\alpha - \alpha') = 1 \quad (18a)$$

$$\int_{-\infty}^{\infty} d\alpha f(\alpha)\, \delta(\alpha - \alpha')$$
$$= \int_{-\infty}^{\infty} d\alpha f(\alpha)\, \delta(\alpha' - \alpha) = f(\alpha') \quad (18b)$$

When Eq. (17) holds, $u(\alpha)$ are said to be normalized on the α-scale. Evidently $\delta(x)$ is highly singular at $x = 0$, and is an even function of x.

Equation (17) asserts that eigenfunctions corresponding to different eigenvalues always are orthogonal, that is, that the integral in Eq. (17) equals zero whenever $\alpha \neq \alpha'$. For discrete α (or α') this

assertion is readily proved by an argument similar to Eq. (17); in fact, the orthogonality for $\alpha \neq \alpha'$ holds whether or not the eigenvalues are degenerate. When α and α' both lie in the continuous spectrum, however, the integral of Eq. (17) does not converge and therefore actually is not defined. Thus the delta function $\delta\,(\alpha - \alpha')$ primarily is a useful formal device; here and elsewhere in the theory, delta functions always are eventually integrated over their arguments, as in Eqs. (18); such integration makes expressions like the left side of Eq. (17) effectively convergent. A mathematically rigorous justification of the use of the delta function in quantum theory, or a rigorous justification of Eq. (17), encounters difficulties. These are related to the difficulties in establishing rigorously the aforementioned properties (i) and (ii). For operators and eigenfunctions of physical interest, experience suggests no reason to doubt the basic correctness of the mathematical procedures of nonrelativistic quantum theory. *See* OPERATOR THEORY.

ILLUSTRATIVE APPLICATIONS

The immediately following subheadings apply the preceding formalism to some representative problems involving a one-dimensional spinless particle free to move in the x direction only; in every case, one begins by seeking the appropriate solution to Eq. (14). As subsequent discussion makes clear, results for this simplest of systems are pertinent to more complicated systems.

Momentum. Equation (14) is written as Eq. (19),

$$p_x u = \frac{\hbar}{i}\frac{\partial u}{\partial x} = \hbar k u \qquad (19)$$

where $\hbar k$ is the eigenvalue; conventionally the eigenfunctions are indexed by the wave number k. Solutions to Eq. (19) have the form $u(x,k) = C(k) \exp[ikx]$, where $C(k)$ is a normalizing constant. When k is real, $u(x,k)$ is finite for all x, $-\infty \leqq x \leqq \infty$. Consequently, the spectrum is continuous and includes all real k, $-\infty \leqq k \leqq \infty$. If k has an imaginary part, that is, if $k = k_1 + ik_2$, $k_2 \neq 0$, then $u(x, k_1 + ik_2)$ is not admissible since it becomes infinite at either $x = +\infty$ or $x = -\infty$. If the eigenfunction could vanish identically for $|x| \geqq a$, it would be quadratically integrable for complex k; in this event the eigenfunction would be discontinuous at $x = a$. However, Eq. (20) holds unless $C(k) = 0$. Thus the require-

$$|C(k)\exp[ikx]| \neq 0 \qquad (20)$$

ments that u be admissible and continuous ensure that the eigenvalues of p_x are real, as asserted previously in the discussion of real eigenvalues. The $u(x,k)$ are quadratically nonintegrable, as expected for the continuous spectrum; each $u(x,k)$ can be regarded as representing a beam of particles, all moving with the same velocity. Because $\exp[ikx]$ is periodic with wavelength $\lambda = 2\pi/k$, such a beam will demonstrate wavelike properties; in fact, $p_x = \hbar k = h/\lambda$, in agreement with the de Broglie relations. This agreement is not trivial; although the conclusion that the quantum-mechanical momentum operator is $p_x = (\hbar/i)\,\partial/\partial x$ can be argued starting from the de Broglie relations, the form of p_x also can be inferred directly from Eq. (8), which in turn can be argued from the formal analogy between the properties of the commutator and the Poisson bracket, without any reference to the de

Broglie relations. *See* CANONICAL TRANSFORMATIONS.

Normalized on the k scale, the eigenfunctions are given by Eq. (21a), for which corresponding to Eq. (17) Eq. (21b) may be written. Equation (21b)

$$u(x,k) = \frac{1}{\sqrt{2\pi}} e^{ikx} \qquad (21a)$$

$$\int_{-\infty}^{\infty} dx\, u^*(k)u(k')$$

$$= \frac{1}{2\pi} \int_{-\infty}^{\infty} dx\, e^{i(k'-k)x} = \delta(k-k') \quad (21b)$$

amounts to a nonrigorous statement of the Fourier integral transform theorem, as shown by Eqs. (22).

$$\psi(x) = \frac{1}{\sqrt{2\pi}} \int_{-\infty}^{\infty} dk\, e^{ikx} c(k) \qquad (22a)$$

$$c(k) = \frac{1}{\sqrt{2\pi}} \int_{-\infty}^{\infty} dx\, e^{-ikx} \psi(x) \qquad (22b)$$

See FOURIER SERIES AND INTEGRALS.

Equation (22b) can be derived by mathematically rigorous procedures. The quantum-theoretic derivation multiplies Eq. (22a) by $(2\pi)^{-1/2} \exp(-ik'x)$ and integrates over all x. There is obtained, after interchanging the orders of x and k integration, Eq. (23).

$$\frac{1}{\sqrt{2\pi}} \int_{-\infty}^{\infty} dx\, e^{-ik'x} \psi(x)$$

$$= \frac{1}{2\pi} \int_{-\infty}^{\infty} dk\, c(k) \int_{-\infty}^{\infty} dx\, e^{i(k-k')x}$$

$$= \int_{-\infty}^{\infty} dk\, c(k)\, \delta(k-k') = c(k') \qquad (23)$$

Kinetic energy. For the kinetic energy $T_x = p_x^2/2m$, Eq. (14) is written as Eq. (24). The eigen-

$$T_x u = -\frac{h^2}{2m} \frac{\partial^2 u}{\partial x^2} = Eu \qquad (24)$$

value is E. Admissible solutions are given by Eq. (25), where $\sqrt{E}$ must be real. Thus the spec-

$$u(x,E) = C(E) \exp\left(\frac{i}{\hbar} x \sqrt{2mE}\right) \qquad (25)$$

trum is continuous and runs from $E = 0$ to $E = +\infty$. The square root may be either positive or negative in Eq. (25) so that there are two independent eigenfunctions at each value of E; that is, the spectrum is degenerate. The eigenfunctions may be labeled $u_+(x,E)$ and $u_-(x,E)$; normalized on the E scale, Eq. (26) obtains, in which the square root always is pos-

$$u_\pm(x,E) = \left(\frac{m}{2h^2E}\right)^{1/4} \exp\left(\pm\frac{i}{\hbar} x\sqrt{2mE}\right) \quad (26)$$

itive, u_+ and u_- individually satisfy Eq. (17), but the sets u_+ and u_- are orthogonal to each other. Introducing $k = \sqrt{2mE}/h$, the eigenfunctions can be labeled by the single parameter k, $-\infty \le k \le \infty$, instead of by E, $0 \le E \le \infty$, and the subscripts $+$ or $-$. Evidently the $u(x,k)$ of Eq. (21a) are eigenfunctions not only of p_x but also of T_x; each $u(x,k)$ corresponds to a different eigenvalue of p_x, but $u(x,k)$ and $u(x,-k)$ correspond to the same eigenvalue of T_x. Interpreted physically, these results mean that the energy E of a free particle is known exactly when its momentum p_x is known exactly, and that

$E = p_x^2/2m = (-p_x)^2/2m$ just as in classical mechanics. The normalizations in Eqs. (21a) and (26) are different because $dE = (\hbar^2 k/m)\, dk$. The eigenfunctions in Eqs. (27) also are normalized on the

$$u'_+(x,E) = \frac{1}{\sqrt{2}}[u_+(x,E) + u_-(x,E)]$$

$$= \left(\frac{2m}{h^2E}\right)^{1/4} \cos\frac{x}{\hbar}\sqrt{2mE} \quad (27a)$$

$$u'_-(x,E) = \frac{1}{\sqrt{2}}[u_+(x,E) - u_-(x,E)]$$

$$= \left(\frac{2m}{h^2E}\right)^{1/4} \sin\frac{x}{\hbar}\sqrt{2mE} \quad (27b)$$

E scale, and are an alternative set to $u_\pm(x,E)$ of Eq. (25).

Typical energy operator. Equation (14) is written as Eq. (28), where the operators in the brackets

$$Hu = (T_x + V)u$$
$$= \left[\frac{-\hbar^2}{2m}\frac{\partial^2}{\partial x^2} + V(x)\right]u = Eu \quad (28)$$

operate on u. Typically, but not always (see the two subheadings immediately following), $V(x)$ is an everywhere smooth finite function which approaches zero at $x = \pm\infty$. Figure 1 is a plot of such a $V(x)$, having attractive force centers (negative potential energy) near $x = \pm a$, and a repulsive region (positive potential energy) near $x = 0$; in this case the minimum potential at $x = \pm a$ is $V_0 < 0$.

Equation (28) is rewritten in the form of Eqs. (29).

$$\frac{\partial^2 u}{\partial x^2} = Ku \qquad (29a)$$

$$K(x,E) = \frac{2m}{\hbar^2}[V(x) - E] \qquad (29b)$$

When $K > 0$, Eq. (29b) implies that $\partial/\partial x(\partial u/\partial x)$ has the same sign as u; for example, $\partial u/\partial x$ increases with increasing x if $u > 0$. In other words, at points x where $K(x,E) > 0$, solutions of Eq. (29a), are convex toward the x axis; where $K(x,E) < 0$, $u(x,E)$ is concave toward the x axis. Provided $V(x)$ decreases to zero sufficiently rapidly, $K(\pm\infty,E) < 0$ when $E > 0$, so that at $x = \pm\infty$ solutions to Eq. (28) are approximated by linear combinations of the oscillatory functions of Eqs. (27): when $E < 0$, however, $K(\pm\infty,E) > 0$, so that at $x = \pm\infty$ solutions to Eq. (28) are approximated by linear combinations of the convex functions [compare Eq. (25)] in Eqs. (30).

$$u_1(x,E) = \exp[x(2m|E|)^{1/2}/\hbar]$$
$$u_2(x,E) = \exp[-x(2m|E|)^{1/2}\ \hbar] \qquad (30)$$

Function u_1 is infinite at $x = +\infty$; u_2 is infinite at $x = -\infty$. Thus, for $E < 0$, every admissible solution of Eq. (28) must behave like u_1 at $x = -\infty$ and like u_2 at $x = +\infty$, that is, must be asymptotic to the x axis at both $x = +\infty$ and $x = -\infty$. When $V_0 < E < 0$ there are values of x where the solution is concave, so that a smooth (satisfying the boundary condition that u and $\partial u/\partial x$ must be continuous) transition from a curve like G or L (Fig. 1) on the left to F on the right may be possible. When $E < V_0$ solutions to Eq. (28) are everywhere convex and, like F, G, L of Fig. 1, never are asymptotic to the x axis at both $x = \pm\infty$.

It is concluded that (i) the continuous spectrum

NONRELATIVISTIC
QUANTUM THEORY

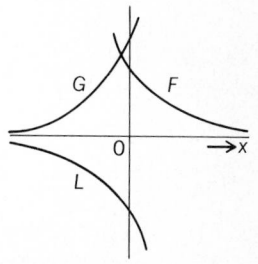

Fig. 1. Everywhere convex solutions asymptotic to x axis.

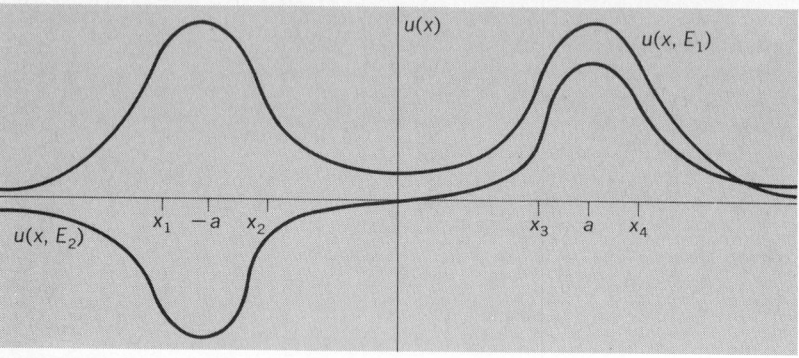

Fig. 2. Eigenfunctions indicated for the two lowest eigenvalues of the Hamiltonian with the potential of Fig. 1.

includes all positive values of E, $0 \leqq E \leqq \infty$, and at each such E there are two independent, nonquadratically integrable (oscillatory at $x = \pm\infty$) eigenfunctions, as in the previously discussed example of kinetic energy; (ii) there are no eigenvalues $E < V_0$; (iii) there may but need not be discrete eigenvalues $V_0 < E < 0$ corresponding to quadratically integrable eigenfunctions. The lowest eigenfunction, for $E = E_1$, has the least region of concavity and therefore joins F of Fig. 1 to G without crossing the x axis, as in Fig. 2. The eigenfunction $u(x,E_2)$ corresponding to the next higher eigenvalue $E = E_2 > E_1$ has one node (one zero), decreases less rapidly at $x = \pm\infty$ than $u(x,E_1)$, and links F to L; the next higher eigenfunction corresponding to $E_3 > E_2$ has two nodes, again links F to G, and so on.

The horizontal lines in the neighborhood of $x = \pm a$ in Fig. 3 show the energy levels of the two lowest eigenvalues E_1, E_2. The points x_1, x_2, where $E = V_1$ and where the curvature of $u(x,E_1)$ changes sign (Fig. 2), are the turning points (velocity equals zero) of a classical particle oscillating with total energy E_1 in the attractive potential well at $x = -a$; x_3, x_4 are similar turning points near $x = a$. For $E_1 < E < E_2$ a solution $u(x,E)$ starting out as does F at $x = +\infty$ crosses the x axis but becomes negatively infinite at $x = -\infty$, because it is insufficiently concave to join smoothly with L; this makes under-responding to quadratically integrable eigenfunctions are discrete.

Potential barrier. Consider Eq. (28) with $V(x) = 0$ for $x < 0$; $V(x) = V_1 > 0$ for $x > 0$, and $0 < E < V_1$, as shown in Fig. 4a. As previously explained, u and $\partial u / \partial x$ must be everywhere continuous, even though $V(x)$ is discontinuous at $x = 0$. Recollecting

Eqs. (26), (29), and (30) for $0 < E < V_1$, there must be one and only one independent eigenfunction, having the form shown in Eqs. (31), where R and S

$$u(x,E) = \exp\left[ix(2mE)^{1/2}/\hbar\right] + R\exp\left[-ix(2mE)^{1/2}/\hbar\right] \quad x < 0 \quad (31a)$$

$$u(x,E) = S\exp\left[-x(2m|E - V_1|)^{1/2}/\hbar\right] \quad x > 0 \quad (31b)$$

are constants. It is reasonable and customary to interpret the first exponential in Eq. (31a) as a beam of particles moving with momentum $p_x = (2mE)^{1/2}$ toward the classical barrier or turning point $x = 0$; the second exponential in Eq. (31a) represents a reflected beam. Since the amplitude of the incident beam has been normalized to unity [not a normalization consistent with Eq. (17)], $|R|^2$ must be the reflection coefficient of the barrier. The continuity requirements at $x = 0$ yield Eqs.

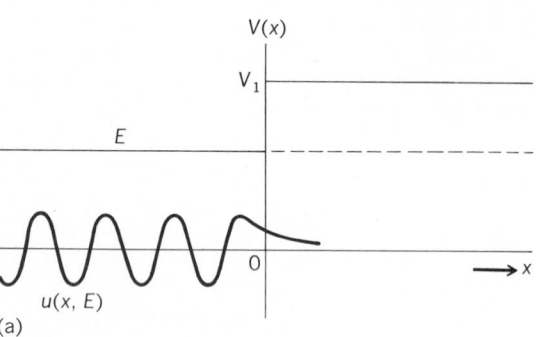

(a)

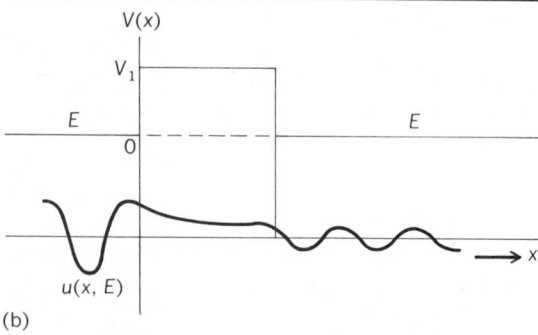

(b)

Fig. 4. Reflections (a) from a barrier of infinite thickness, and (b) from one of finite thickness.

(32), showing that $|R|^2 = 1$ in agreement with class-

$$R = \frac{iE^{1/2} + (V_1 - E)^{1/2}}{iE^{1/2} - (V_1 - E)^{1/2}}$$

$$S = \frac{2iE^{1/2}}{iE^{1/2} - (V_1 - E)^{1/2}} \quad (32)$$

ical expectation for $E < V_1$. *See* REFLECTION AND TRANSMISSION COEFFICIENTS.

The eigenfunction $u(x,E)$ is sketched in Fig. 4a. Since $|u|^2 \neq 0$ at $x > 0$, particles penetrate the classically inaccessible region to the right of the barrier; because $|u|^2 = 0$ at $x = +\infty$, all particles eventually are turned back, however, consistent with $|R|^2 = 1$. This penetration, and Eqs. (31) and (32), closely resembles the penetration (and corresponding classical optics expressions) of a totally reflected light wave into a medium with smaller

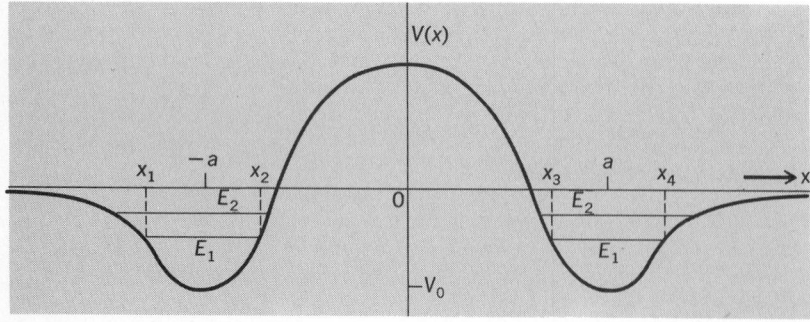

Fig. 3. A potential $V(x)$, capable of trapping particles near $x = \pm a$.

index of refraction. Hence Eqs. (31) and (32), which are unforced results of solving Eq. (28) for $V(x)$ of Fig. 4a, manifest the wave-particle duality inherent in the quantum theoretic formalism.

For the barrier of finite thickness a (Fig. 4b), there are two solutions at every $0 < E < V_1$. The solution representing a beam incident from the left, transmitted with coefficient $|T|^2$ and reflected with coefficient $|R|^2$, is found from Eq. (31a) together with Eqs. (33). Equations (33) mean that, except for the incident $\exp[ix(2mE)^{1/2}]$ at $x = -\infty$,

$$u(x,E) = C \exp[-x(2m|E-V_1|)^{1/2}/\hbar]$$
$$+ D \exp[x(2m|E-V_1|)^{1/2}/\hbar]$$
$$0 < x < a \quad (33a)$$

$$u(x,E) = T \exp[ix(2mE)^{1/2}/\hbar] \quad x > a \quad (33b)$$

the solution at $x = \pm\infty$ must consist of waves traveling out toward infinity; a similar outgoing boundary condition specifies the continuum solution $E > 0$ representing more complicated collisions, for example, the scattering of a beam of particles by target particles in a foil. Outgoing boundary conditions are employed in classical wave theories as well. *See* SCATTERING EXPERIMENTS (ATOMS AND MOLECULES); SCATTERING EXPERIMENTS (NUCLEI); SCATTERING OF ELECTROMAGNETIC RADIATION.

The continuity requirements lead to Eq. (34a), and $|R|^2 = 1 - |T|^2$, as is necessary if $|R|^2$ and $|T|^2$ are to represent, respectively, reflection and transmission coefficients. Equation (34a), and the corresponding expression when $E > V_1$, resemble the equations for transmission of light through thin films. When $(a/\hbar)\sqrt{2m(V_1-E)}$ is $\gg 1$, Eq. (34a) is closely approximated by Eq. (34b), where the expo-

$$|T|^2 = \left\{ 1 + \frac{V_1^2 \sinh^2[(a/\hbar)\sqrt{2m(V_1-E)}]}{4E(V_1-E)} \right\}^{-1} \quad (34a)$$

$$|T|^2 = \frac{16E(V_1-E)}{V_1^2} \exp[-2a(2m|E-V_1|)^{1/2}/\hbar] \quad (34b)$$

nential factor, sometimes termed the barrier penetrability, is $\ll 1$. The transmission through a less simple barrier $V(x)$ than Fig. 4b is measured by the penetrability factor defined by Eq. (35), where x_1,

$$P = \exp \frac{-2}{\hbar} \int_{x_1}^{x_2} dx\,(2m|E-V(x)|)^{1/2} \quad (35)$$

x_2 are the turning points $E = V(x_1) = V(x_2)$. The barrier penetrability governs the rates at which (i) an incident proton, whose kinetic energy is less than the height of the repulsive Coulomb potential barrier surrounding a nucleus, is nonetheless able to penetrate the nucleus to produce nuclear reactions; (ii) α-particles can escape from a radioactive nucleus; (iii) electrons can be pulled out of a metal by a strong electric field. *See* FIELD EMISSION; NUCLEAR FUSION; RADIOACTIVITY.

Harmonic oscillator. The potential $V(x = 1/2Kx^2$ (Fig. 5) describes a classical harmonic oscillator whose equilibrium position is $x = 0$; the classical oscillator frequency is $\nu = (2\pi)^{-1}\sqrt{K/m}$. Since $V(x)$ becomes infinite at $x = \pm\infty$, there is no continuous spectrum, but there must be an infinite number of discrete eigenvalues. Various sophisticated

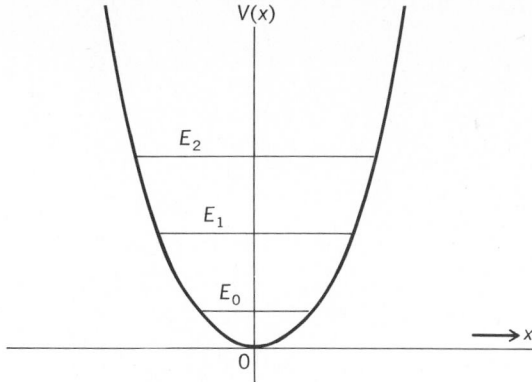

Fig. 5. Harmonic oscillator potential.

methods exist for finding the eigenfunctions and eigenvalues. *See* HARMONIC OSCILLATOR.

The energy levels turn out to be as in Eq. (36),

$$E_n = (n + \tfrac{1}{2})h\nu \quad (36)$$

where $n = 0, 1, 2, \ldots,$ (Fig. 5). The corresponding first three eigenfunctions $u_0(\xi)$, u_1, u_2 are sketched in Fig. 6, with $\xi = x(mK)^{1/4}/\hbar^{1/2}$ a convenient dimensionless variable; the dashed vertical lines indicate the turning points at $\xi = \pm\sqrt{2n+1}$. Figure 7 plots the probability density $|u(\xi)|^2$ for $n = 10$. The classical probability of finding a particle in the x interval dx is proportional to the time spent in dx. The curved dashed line in Fig. 7 plots the classical probability density for a classical oscillator whose energy is $E_{10} = 2\tfrac{1}{2}h\nu$, Eq. (36). *See* ENERGY LEVEL.

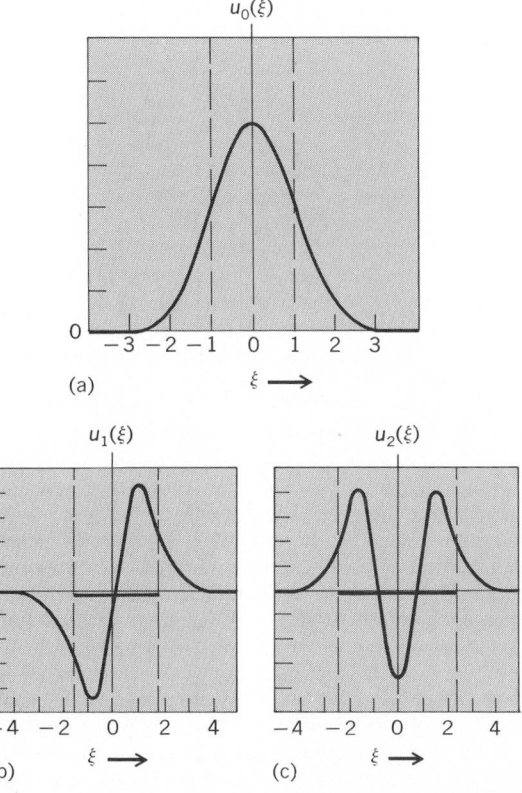

Fig. 6. Harmonic oscillator eigenfunctions. (a) $n=0$. (b) $n=1$. (c) $n=2$. (L. Pauling and E. B. Wilson, Jr., *Introduction to Quantum Mechanics*, McGraw-Hill, 1935)

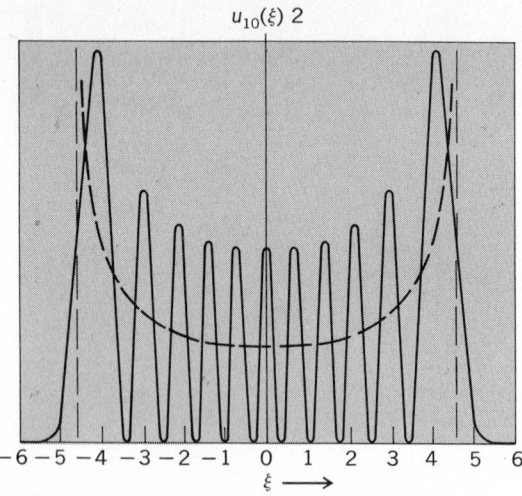

$u_{10}(\xi)\,2$

$$-6\;-5\;-4\;-3\;-2\;-1\;\;0\;\;1\;\;2\;\;3\;\;4\;\;5\;\;6$$

$\xi \longrightarrow$

Fig. 7. Probability density for harmonic oscillator in state $n=10$. (L. Pauling and E. B. Wilson, Jr., Introduction to Quantum Mechanics, McGraw-Hill, 1935)

The agreement between the classical probability density and the average over oscillations of $|u_{10}(\xi)|^2$ illustrates the connection between classical particle mechanics and the more fundamental dual wave-particle interpretation of quantum theory. With increasing n the oscillations of $|u_n(\xi)|^2$ become more rapid, and the agreement with the classical probability density improves, in accordance with the correspondence principle. These harmonic oscillator results are a good first approximation to the energy levels and eigenfunctions of vibrating atoms in isolated molecules and in solids. See LATTICE VIBRATIONS; MOLECULAR STRUCTURE AND SPECTRA; SPECIFIC HEAT OF SOLIDS.

EXPECTATION VALUES

Suppose B is an operator that commutes with A. If $Au_n = \alpha_n u_n$, $BAu_n = B(\alpha_n u_n)$, so that, using Eq. (7), one obtains Eq. (37), which means that Bu_n also

$$A(Bu_n) = \alpha_n(Bu_n) \qquad (37)$$

is an eigenfunction of A corresponding to the eigenvalue α_n. When α_n is not degenerate, Eq. (16) means Bu_n must be a multiple of u_n, that is, $Bu_n = \beta u_n$, β being a constant; thus u_n is simultaneously an eigenfunction of the pair of commuting operators A, B. When the order of degeneracy of α_n is d, Bu_n need not be a multiple of u_n, but d independent eigenfunctions $u_{n1}(\alpha_n,\beta_1), \ldots, u_{nd}(\alpha_n,\beta_d)$ always can be found, such that $u_{ns}(\alpha_n,\beta_s)$ is an eigenfunction of B corresponding to the eigenvalue β_s, as well as an eigenfunction of A corresponding to α_n. If all the β_s are different, $u_{n1}, \ldots, u_{nd}$, being eigenfunctions corresponding to different eigenvalues of B, are mutually orthogonal (see the earlier discussion on orthogonality). If all the β_s are not different, a third operator C, simultaneously commuting with A and B, is sought, and so on. For the system of g particles of spin 1/2, one determines in this fashion a complete set of simultaneously commuting observables, A, B, C, . . . , whose corresponding eigenvalues α, β, γ, . . . (except for accidental degeneracy) uniquely index an orthonormal set of 2^g-compo-

nent eigenfunctions as in Eq. (38), which satisfies Eq. (39).

$$u_j(\mathbf{r}_1, \ldots, \mathbf{r}_g; \alpha,\beta,\gamma, \ldots), j=1 \text{ to } 2^g \qquad (38)$$

$$\int u^*(\alpha,\beta,\gamma, \ldots) u(\alpha',\beta',\gamma', \ldots)$$
$$= \delta(\alpha-\alpha')\,\delta(\beta-\beta')\,\delta(\gamma-\gamma') \cdots \qquad (39)$$

In Eq. (39), and henceforth unless specifically indicated otherwise, the integral is in the simplified notation of Eq. (11). Equation (39) generalizes Eq. (17); $\delta(\alpha-\alpha')$ is replaced by a Kronecker symbol $\delta_{\alpha\alpha'}$ when α, α' are discrete, and so on.

The meaning of property (ii) mentioned in the earlier discussion of real eigenvalues can now be explained. The eigenfunctions of an operator A having been indexed as in Eq. (39), any sufficiently smooth quadratically integrable 2^g-component wave function $\psi(\mathbf{r}_1, \ldots, \mathbf{r}_g)$ can be expressed in the form of Eq. (40), where each eigenvalue is inte-

$$\psi(\mathbf{r}_1, \ldots, \mathbf{r}_g) = \int d\alpha \int d\beta \int d\gamma \cdots$$
$$\times c(\alpha,\beta,\gamma, \ldots) u(\mathbf{r}_1, \ldots, \mathbf{r}_g; \alpha,\beta,\gamma, \ldots)$$
$$(40)$$

grated over its entire spectrum, with the understanding (further to simplify the notation) that in the discrete spectrum integration is replaced by summation. Employing Eq. (39), the constants $c(\alpha,\beta,\gamma, \ldots)$ are found to be as shown in Eq. (41).

$$c(\alpha,\beta,\gamma, \ldots) = \int u_j^*(\alpha,\beta,\gamma, \ldots)\psi \qquad (41)$$

Equations (40) and (41) are consistent with the often instructive interpretation that $c(\alpha,\beta,\gamma, \ldots)$ are the projections of the vector ψ on a complete set of orthogonal unit vectors $u(\alpha,\beta,\gamma, \ldots)$ in an infinite-dimensional vector space.

Probability considerations. When ψ is normalized, Eqs. (4), (18), (39), and (40) imply Eq. (42), in

$$\int d\alpha \int d\beta \int d\gamma \cdots |c(\alpha,\beta,\gamma, \ldots)|^2 = 1 \qquad (42)$$

the notation of Eq. (40). Equations (40) and (42) make it reasonable to postulate that whenever a system is instantaneously described by the normalized quadratically integrable $\psi(\mathbf{r}_1, \ldots, \mathbf{r}_g)$, the instantaneous probability of finding the system in the state corresponding to $A=\alpha$ is given by Eq. (43a).

Similarly, the simultaneous probability of finding the system in the state corresponding to $A=\alpha$ and $B=\beta$ is given, by Eq. (43b), and so on.

$$P(\alpha) = \int d\beta \int d\gamma \cdots |c(\alpha,\beta,\gamma, \ldots)|^2 \qquad (43a)$$

$$P(\alpha,\beta) = \int d\gamma \cdots |c(\alpha,\beta,\gamma, \ldots)|^2 \qquad (43b)$$

Clearly Eqs. (1) and (2) are a special case of Eqs. (42) and (43), in which the coordinate operators, $\mathbf{x}_1$, $\mathbf{y}_1$, $\mathbf{z}_1$, . . . , $\mathbf{x}_g$, $\mathbf{y}_g$, $\mathbf{z}_g$ of the g spinless particles form the complete set of simultaneously commuting observables, and $\psi(\mathbf{x}_1,\mathbf{y}_1,\mathbf{z}_1, \ldots, \mathbf{x}_g,\mathbf{y}_g,\mathbf{z}_g)$ is the projection of ψ on the eigenfunction simultaneously corresponding to $\mathbf{x}_1=x_1$, $\mathbf{y}_1=y_1$, . . . , $\mathbf{z}_g=z_g$.

The total probability $\int d\alpha\, P(\alpha)$ is unity, Eq. (42), and therefore any measurement of the observable A must yield a number equal to one of its eigenvalues. Moreover, by the very meaning of probability, the average or expectation value of the observable A must be as given in Eq. (44a), which shows that the same operator may have different forms in different representations. For instance, in the momentum representation discussed earlier where $\alpha \equiv k$, $p_x c(k)$ must equal simply $\hbar k c(k)$; consequently, to satisfy Eq. (8), $\mathbf{x}c(k)$ must equal $i(\partial c/\partial k)$. When ψ is a wave function for which Eq. (9) holds with $\xi \equiv \psi$, Eqs. (5), (14), (39), and (40) imply that Eq. (44b) is equivalent to Eq. (44a). Equation (40b),

$$\langle A \rangle = \int d\alpha \, \alpha \, P(\alpha) \qquad (44a)$$

$$\langle A \rangle = \int \psi^*(A\psi) \qquad (44b)$$

usually more convenient than Eq. (44a), predicts the expectation value of any given observable in an arbitrary physical situation described by a resonably well-behaved quadratically integrable wave function; expectation values are not defined for nonquadratically integrable ψ. Equation (9) guarantees that $\langle A \rangle$ computed from Eq. (44b) is a real number, necessary if $\langle A \rangle$ is to represent the results of measurement (see the earlier discussion on real eigenvalues). *See* PROBABILITY.

Uncertainty principle. A precise measure of the spread or uncertainty in the value of A is ΔA defined by Eq. (45). Here $(\Delta A)^2$ is the average

$$(\Delta A)^2 = \langle (A - \langle A \rangle)^2 \rangle = \langle A^2 \rangle - (\langle A \rangle)^2 \qquad (45)$$

square deviation of A from its average $\langle A \rangle$. The quantity $\Delta A = 0$ when and only when ψ is an eigenfunction of A, that is, $A\psi = \alpha\psi$, in which event $\langle A \rangle = \alpha$, $\langle A^2 \rangle = \alpha^2$. The discussion following Eq. (37) implies ΔA and ΔB simultaneously can be zero; that is, A and B are simultaneously exactly measurable, whenever the commutator $AB - BA$ is zero. If $AB - BA \neq 0$, introduce $A' = A - \langle A \rangle$, $B' = B - \langle B \rangle$; use Eqs. (11) and (44b); and employ the so-called Schwartz inequality. Then Eq. (46) holds.

$$(\Delta A)^2 (\Delta B)^2 = \int (A'\psi)^*(A'\psi) \int (B'\psi)^*(B'\psi)$$
$$\geq |\int (A'\psi)^*(B'\psi)|^2 = |\int \psi^*(A'B'\psi)|^2 \qquad (46)$$

But since Eq. (47) is valid, it leads, after further

$$A'B' = \frac{1}{2}(A'B' - B'A') + \frac{1}{2}(A'B' + B'A') \qquad (47)$$

manipulation, to Eq. (48), which is the rigorous

$$(\Delta A)^2 (\Delta B)^2 \geq \frac{1}{4}|\langle AB - BA \rangle|^2 \qquad (48)$$

quantum theoretic formulation of the uncertainty principle. When $A = x$, $B = p_x$, Eq. (8) yields Eq. (49). This simple derivation of the uncertainty

$$(\Delta x)(\Delta p_x) \geq \hbar/2 \qquad (49)$$

principle demonstrates anew the necessity for a dual wave-particle interpretation of the operator formalism.

TIME DEPENDENCE

The procedures developed thus far predict the results of measurement at any given instant of time t_0 that the wave function $\psi(x,t_0)$ is known. Given ψ at t_0, to predict the results of measurement at some future instant t, it is necessary to know the time-evolution of ψ from t_0 to t. It is postulated that this evolution is determined by the time-dependent Schrödinger equation, Eq. (50),

$$H\psi = i\hbar \frac{\partial \psi}{\partial t} \qquad (50)$$

where H is the Hamiltonian or energy operator; ψ is supposed to be a continuous function of t. If H were a number, the solution to Eq. (50) would be as shown in Eq. (51). Equation (51) remains valid

$$\psi(t) = \exp\left[\frac{-iH(t - t_0)}{\hbar}\right]\psi(t_0) \qquad (51)$$

when H is an operator, provided Eq. (52) holds,

$$\exp\left[\frac{-iH(t - t_0)}{\hbar}\right] \equiv I - \frac{i(t - t_0)}{\hbar}H$$
$$+ \frac{[-i(t - t_0)]^2}{2\hbar^2}H^2 + \cdots \qquad (52)$$

with I the unit operator, $I\psi = \psi$; the right side of Eq. (52) is the usual series expansion of the exponential on the left side. These so-called operational methods, which manipulate operators like numbers, are widely employed in quantum theory; they must be used cautiously, but usually lead rapidly to the same results as more conventional mathematical techniques.

When $H\psi(t_0) = E\psi(t_0)$, that is, when $\psi(t_0)$ is known to be an eigenfunction $u(E,t_0)$ of the energy operator, Eq. (51) shows that $\psi(t) \equiv u(E,t) = u(E,t_0) \exp[-iE(t - t_0)/\hbar]$, so that $|u(E,t)|^2 = |u(E,t_0)|^2$; similarly, the expectation values $\langle A \rangle$ of all operators A are time-independent when the system is in a stationary state $u(E,t)$. In an arbitrary state $\psi(t)$, Eqs. (5), (11), (44b), and (50) imply [assuming that A is not explicitly time-dependent, for example, $A = A(\mathbf{r},\mathbf{p})$ but not $A(\mathbf{r},\mathbf{p},t)$] Eq. (53), which uses the

$$\frac{d}{dt}\langle A \rangle = \int \left[\psi^*\left(A\frac{\partial \psi}{\partial t}\right) + \frac{\partial \psi^*}{\partial t}(A\psi)\right]$$
$$= \frac{1}{i\hbar}\int [\psi^*(AH\psi) - (H\psi)^*A\psi] \qquad (53)$$
$$= \frac{1}{i\hbar}\int [\psi^*(AH - HA)\psi] = \frac{1}{i\hbar}\langle (AH - HA) \rangle$$

notation of Eq. (11). Of course $\langle (AH - HA) \rangle = 0$ whenever $\psi(t)$ is a stationary state $u(E,t)$. Equation (53) shows, however, that if A commutes with the Hamiltonian, then $\langle A \rangle$ is independent of time whether or not the system is in a stationary state. Consequently, operators commuting with the Hamiltonian are termed constants of the motion; a system initially described by an eigenfunction $u(E,\beta,\gamma,\ldots)$ of the simultaneously commuting observables $H, B, C, \ldots$, remains in an eigenstate of $H, B, C, \ldots$, as the wave function evolves in time.

Equation (53) is closely analogous to the classical mechanics expression for $dA(\mathbf{r},\mathbf{p})/dt$, where $A(\mathbf{r},\mathbf{p})$ is the classical quantity corresponding to the quantum mechanical operator A. If A is put equal to I in Eq. (53), one obtains Eq. (54).

$$\frac{d}{dt}\int \psi^*\psi = \frac{1}{i\hbar}\int [\psi^*(H\psi) - (H\psi)^*\psi] \qquad (54)$$

Therefore the requirement that H be Hermitian, Eq. (9), which has been justified on the grounds that the eigenvalues of H must be real, has the further important consequence that the right side of Eq. (54) is zero, that is, that Eq. (2a) is obeyed at all times $t > t_0$ if it is obeyed at $t = t_0$. This result is necessary for the consistency of the formalism; otherwise it could not be claimed that $|\psi(t)|^2$ from

Eq. (50) is the probability density at $t > t_0$. For a single spinless three-dimensional particle, with $H = p^2/2m + V(x,y,z)$, it follows directly from Eq. (50) that one obtains Eq. (55a), where Eq. (55b)

$$\frac{\partial}{\partial t}(\psi^*\psi) + \frac{\partial}{\partial x}S_x + \frac{\partial}{\partial y}S_y + \frac{\partial}{\partial z}S_z = 0 \quad (55a)$$

$$S_x = \frac{\hbar}{2mi}\left(\psi^*\frac{\partial\psi}{\partial x} - \psi\frac{\partial\psi^*}{\partial x}\right) \quad (55b)$$

holds, and so on. In Eq. (55a) S_x, S_y, S_z can be interpreted as the components of a probability current vector **S**, whose flow across any surface enclosing a volume τ accounts for the change in the probability of finding the particle inside τ. See EQUATION OF CONTINUITY; MAXWELL'S EQUATIONS.

For a nonquadratically integrable ψ in the one-particle case where Eq. (55b) is applicable, the probability current at infinity generally has the value $|\psi|^2 \mathbf{v}$, where **v** is the classical particle velocity at infinity; the one-dimensional plane waves of Eq. (21a) trivially illustrate this assertion. Consequently (see the preceding discussion of normalization), **S** of Eq. (55b) is interpretable as particle current density when $|\psi|^2$ is nonvanishing at infinity. These considerations may be generalized to more complicated systems and are important in collision problems, where the incoming and outgoing currents at infinity determine the cross section.

Invariance. Extremely general arguments support the view that the form of the Schrödinger equation (50) for any g-particle system isolated from the rest of the universe must be (i) translation invariant, that is, independent of the origin of coordinates; (ii) rotation invariant, that is, independent of orientation of the coordinate axes; and (iii) reflection invariant, that is, independent of whether one chooses to use a left-handed or right-handed coordinate system.

The only known failures of these general requirements occur for reflections, in a domain outside the scope of nonrelativistic quantum theory, namely, in phenomena, such as beta decay, that are connected with the so-called weak interactions. Correspondingly, it can be inferred that the Hamiltonian operator H for any such isolated system must commute with (i) the total linear momentum operator $\mathbf{p}_R = \mathbf{p}_1 + \cdots + \mathbf{p}_g$; (ii) the total angular momentum operator **J**; (iii) the parity operator P, which reflects every particle through the origin, that is, changes $\mathbf{r}_1$ to $-\mathbf{r}_1, \ldots, \mathbf{r}_g$ to $-\mathbf{r}_g$. For additional information *see* PARITY; SYMMETRY LAWS.

In quantum mechanics as in classical mechanics, therefore, linear momentum and total angular momentum are conserved, that is, are constants of the motion. Since for an infinitesimal displacement ϵ in the x direction Eq. (56) holds, the connec-

$$\psi(x_1 + \epsilon, y_1, z_1, x_2 + \epsilon, y_2, z_2, \ldots, x_g + \epsilon, y_g, z_g)$$

$$= \psi(x_1, y_1, z_1, \ldots, x_g, y_g, z_g)$$

$$+ \epsilon\left(\frac{\partial\psi}{\partial x_1} + \frac{\partial\psi}{\partial x_2} + \cdots + \frac{\partial\psi}{\partial x_g}\right)$$

$$= \psi + \epsilon p_{Rx}\psi \quad (56)$$

tion between $p_{1x} + \cdots + p_{gx}$ and translation in the x direction can be understood; the connection between **J** and rotation is understood similarly. Because a discontinuous change in position, from **r**

to $-\mathbf{r}$, is inconceivable classically, the conservation of parity concept has no relevance to classical mechanics.

Transition probability. Frequently the Hamiltonian H of Eq. (50) has the form $H_0 + V'(t)$, where the time-dependent potential energy $V'(t)$ represents an externally imposed interaction, for example, with measuring equipment; it is supposed that $V'(t) = 0$ for $t < 0$ and $t > t_1$. Usually the system is in a stationary state $u(E_i)$ of H_0 at time $t < 0$, and one wishes to compute the probability of finding the system in some other stationary state $u(E_f)$ of H_0 at times $t > t_1$. From Eq. 40) and the discussion preceding Eq. (53), Eq. (57) is obtained,

$$\psi(t) = \int dE\, d\beta\, c(E,\beta,t)\exp(-iEt/\hbar)u(E,\beta) \quad (57)$$

which, when substituted in Eq. (50) for times $0 \leq t \leq t_1$, yields Eq. (58).

$$i\hbar\int dE\, d\beta\,\frac{dc(E,\beta,t)}{dt}\exp(-iEt/\hbar)\,u(E,\beta)$$

$$= \int dE\, d\beta\, c(E,\beta,t)\exp(-iEt/\hbar)\,V'(t)u(E,\beta) \quad (58)$$

In Eqs. (57) and (58), E is α of Eq. (40) and β stands for all other indices $\beta, \gamma, \ldots$, necessary to make $u(E)$ a complete orthonormal set; if $V'(t)$ were zero, the projections $c(E,\beta)$ would equal $\delta(E - E_i)\delta(\beta - \beta_i)$ independent of time.

Most problems in quantum theory cannot be solved exactly; therefore some approximate treatment using perturbations must be devised. In the present case it is assumed that $c(E,\beta,t)$ do not change appreciably from their initial values at $t = 0$, so that it is a reasonable approximation to replace $c(E,\beta,t)$ by $c(E,\beta,0)$ on the right side of Eq. (58). With the further approximation that V' is constant during the interval $0 \leq t \leq t_1$, one finds, using the notation of Eq. (11), that Eq. (59) obtains, where $\hbar\omega = E_f - E_i$.

$$|c(E_f,\beta_f,t_1)|^2 \equiv \frac{4}{\hbar^2}\left|\int u^*(E_f,\beta_f)V'u(E_i,\beta_i)\right|^2\frac{\sin^2\frac{1}{2}\omega t_1}{\omega^2}$$

$$\equiv \frac{4}{\hbar^2}|V'_{fi}|^2\frac{\sin^2\frac{1}{2}\omega t_1}{\omega^2} \quad (59)$$

Equation (59) shows that the probability of finding the system in some new stationary state $u(E_f,\beta_f)$ after the system is no longer perturbed is proportional to (i) $|V'_{fi}|^2$, the square of the matrix element of the perturbation between initial and final states; (ii) an oscillating factor which for given t_1 has a peak $t_1^2/4$ at $\omega = 0$.

$\sin^2(1/2\omega t_1)/\omega^2$ is plotted as a function of ω in Fig. 8; evidently $|c(E_f,\beta_f)|^2$ is relatively small for energies such that $|\omega| > \sim\pi/t_1$. The most likely occupied states after the perturbation conserve energy $(E_f = E_i)$, and the spread in energy of the final states is $\Delta E = \hbar\Delta\omega = \sim 2\pi\hbar/t_1 \equiv h/\Delta t$, where $t_1 \equiv \Delta t$ is, for example, the duration of the measurement. Thus Eq. (59) provides a version of the uncertainty principle between energy and time. As t_1 approaches infinity, the area under the curve in Fig. 8 becomes proportional to t_1, since the main peak has a height proportional to t_1^2 and a width proportional to $1/t_1$. Therewith one obtains a widely employed formula giving the approximate transition probability w per unit time for making transitions, under the influence of a steady perturbation V', from an initial stationary $u(E_i,\beta_i)$ to a set of

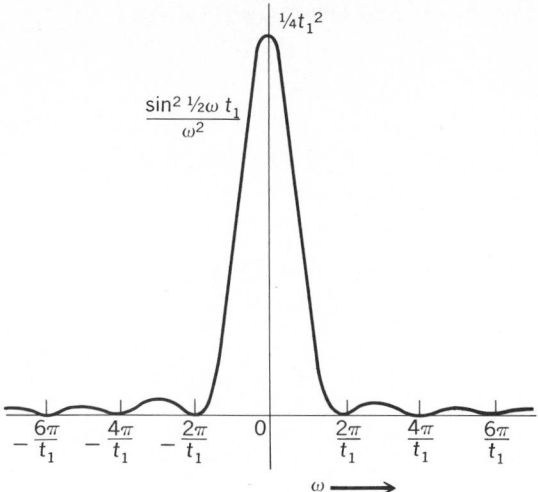

Fig. 8. Plot of $\sin^2 \frac{1}{2}\omega t_1 / \omega^2$ versus $\omega = (E_i - E_l)/n$. (L. I. Schiff, Quantum Mechanics, 2d ed., McGraw-Hill, 1955)

final states of the same energy, namely, Eq. (60),

$$w = \frac{2\pi}{\hbar} \rho(E_f) |V'_{fi}|^2 \qquad (60)$$

where $\rho(E_f) = d\beta \, d\gamma \, . \, . \, .$ is the density of independent final states in the neighborhood of $E = E_f = E_i$, $\beta = \beta_f$, $\gamma = \gamma_f$, and so on. For instance, with u_i a plane wave e^{ikz} moving in the z direction, Eq. (21a), and u_f a plane wave in some other direction, Eq. (60) yields the Born approximation to the cross section for elastic scattering by a potential. Equation (60) is also applicable to problems outside the domain of nonrelativistic quantum theory, for example, to the theory of β-decay wherein new particles are created.

The preceding considerations are important for understanding how a measurement of an operator A not commuting with the Hamiltonian H can cause an initially stationary state $u(E)$ to evolve into an eigenstate $u(\alpha)$ of A. Equations (50)–(53), with $H = T + V = H_0$, the unperturbed Hamiltonian, hold only in the intervals between measurements; during the measurement $u(E)$, though an eigenfunction of H_0, is not a stationary state of the complete Hamiltonian. This paragraph does not do justice, however, to the subtle questions involved in the quantum theory of measurements.

FURTHER ILLUSTRATIVE APPLICATIONS

When g particles are noninteracting, the Hamiltonian has the trivially separable form $H \equiv H_1 + H_2 \cdots + H_g$, and Eq. (61) holds, where $u_1(\mathbf{r}_1,$

$$(H_1 + \cdots + H_g)[u_1(\mathbf{r}_1,E_1,\alpha_1) \cdots u_g(\mathbf{r}_g,E_g,\alpha_g)]$$

$$= (E_1 + \cdots + E_g)$$

$$[u_1(\mathbf{r}_1,E_1,\alpha_1) \cdots u_g(\mathbf{r}_g,E_g,\alpha_g)] \quad (61)$$

$E_1,\alpha_1)$ are a complete orthonormal set of eigenfunctions of H_1 for particle 1, and so on. Thus, because an operation performed solely on particle 1 always commutes with an operation performed solely on particle 2, the products $u_1(\mathbf{r}_1,E_1,\alpha_1) \cdots u_g$ $(\mathbf{r}_g,E_g,\alpha_g)$ are a complete orthonormal set of eigenfunctions of $H = H_1 + \cdots + H_g$; also the energy levels of H are the set of possible sums of individual particle energies, $E = E_1 + \cdots + E_g$. Similarly, if $H(x) = T_x + V(x)$, with eigenfunctions $u(x,E_x)$, is a Hamiltonian for a one-dimensional spinless particle, then any quadratically integrable $\psi(x,y,z)$ describing a three-dimensional spinless particle can be expanded in a series of products $u(x,E_x)u(y,E_y)u(z,E_z)$. This paragraph explains the relevance, to three-dimensional many-particle systems of the results obtained for the illustrative applications previously discussed. The immediately following subheadings continue to illustrate the general formalism. The reader is cautioned that the remaining contents of this article, although very important, especially for applications to atomic and nuclear structure, are for the most part admittedly more condensed than the material presented heretofore.

Parity. Because $P^2\psi(\mathbf{r}) = P\psi(-\mathbf{r}) = \psi(\mathbf{r})$, the parity operator has but two eigenvalues, namely, +1 and −1; the corresponding eigenfunctions are said to have even or odd parity. Evidently P commutes with the harmonic oscillator Hamiltonian $p_x^2/2m + (1/2)Kx^2$. The harmonic oscillator eigenvalues are nondegenerate, and therefore every harmonic oscillator eigenfunction (Fig. 6) has either even or odd parity. Similarly the eigenfunctions u'_+, u'_- of T_x, Eqs. (27), have even and odd parity, respectively; eigenfunctions of T_x which do not have definite parity also exist, however, Eq. (26), because the eigenvalues E are degenerate. The eigenfunctions of p_x do not have definite parity, Eq. (21a), because $p_x P + P p_x = 0$, that is, p_x does not commute, but instead anticommutes, with P.

Time evolution of packet. A wave function ψ representing a single (spinless) particle localized in the neighborhood of a point is termed a wave packet. Assuming $H = p^2/2m + V(\mathbf{r})$, Eq. (53) yields Eq. (62a) by employing Eq. (8); similarly, one obtains Eq. (62b). Equations (62) mean (i) the average position of the particle

$$\frac{d}{dt}\langle \mathbf{x} \rangle = \frac{1}{i\hbar}\langle \mathbf{x}H - H\mathbf{x} \rangle = \left\langle \frac{p_x}{m} \right\rangle \qquad (62a)$$

$$\frac{d^2}{dt^2}\langle \mathbf{x} \rangle = \frac{d}{dt}\left\langle \frac{p_x}{m} \right\rangle = \frac{-1}{m}\left\langle \frac{\partial V}{\partial x} \right\rangle \qquad (62b)$$

tion of the particle, that is, the center of the packet, moves with a velocity given by the expectation value of the momentum; (ii) the acceleration of the center of the packet is found from the expectation value of the classical force $-\partial V/\partial x$. Equations (62) illustrate the correspondence between quantum and classical mechanics and show that the classical description of particle motion is valid when the spread of the packet about its mean position can be ignored. When the particle is free, $H = p^2/2m$ Eqs. (8), (45), and (53) lead to Eq. (63), where the

$$(\Delta x)_t^2 = (\Delta x)_0^2 + \left\{ \frac{t}{m}\langle xp + px \rangle_0 - 2\langle x \rangle_0\langle p \rangle_0 \right\}$$

$$+ \frac{t^2}{m^2}(\Delta p_x)_0^2 \quad (63)$$

subscripts t, 0 refer, respectively, to expectation values at t and at initial time zero. Equation (63) shows that, although an unconfined free wave packet may contract for awhile, ultimately it will spread over all space; when the minimum spread happens to occur at $t = 0$, the term linear in t vanishes in Eq. (63).

Orbital angular momentum. The quantum mechanical operators representing the components of orbital angular momentum **L** have the same form as in classical mechanics, namely, for each particle $L_x = yp_z - zp_y$, and so on. By using Eq. (8), one obtains Eqs. (64), and so on, where $L^2 \equiv L_x^2 +$

$$L^2 L_z - L_z L^2 = 0 \qquad (64a)$$

$$L_x L_y - L_y L_x = i\hbar L_z \qquad (64b)$$

$L_y^2 + L_z^2$. According to Eq. (48), therefore (i) L^2 and L_z are simultaneously exactly measurable; (ii) once the values of L^2 and L_z are specified, the values of L_x and L_y must be uncertain. In spherical coordinates $z = r \cos\theta$, $x = r \sin\theta \cos\phi$, $y = r \sin\theta \sin\phi$, L_z and L^2 becomes Eqs. (65).

$$L_z = \frac{\hbar}{i} \frac{\partial}{\partial \phi} \qquad (65a)$$

$$L^2 = -\hbar^2 \left(\frac{1}{\sin\theta} \frac{\partial}{\partial \theta} \sin\theta \frac{\partial}{\partial \theta} + \frac{1}{\sin^2\theta} \frac{\partial^2}{\partial \phi^2} \right) \qquad (65b)$$

Equation (14) for L_z is solved by $u(\phi, m) = \exp(im\phi)$, where $m\hbar$ is the eigenvalue. It can be argued that $u(\phi, m)$ must have a unique value at any point x, y, z, meaning $u(\phi + 2\pi, m) = u(\phi, m)$, so that m must be a positive or negative integer, or zero. With $\partial^2/\partial\phi^2 = -m^2$ in Eq. (65b), the eigenvalues of L^2 turn out to be $l(l + 1)\hbar^2$, where $l = 0, 1, \ldots$, independent of m, except that for each l the allowed values of the magnetic quantum number m are $m = -l, -l + 1, \ldots, l - 1, l$; thus each l has order of degeneracy $2l + 1$. Because L^2 and L_z commute with P, the eigenfunctions $u(l, m)$ have definite parity; in fact, Eq. (66) holds.

$$P_u(l, m) = (-1)^l u(l, m) \qquad (66)$$

In a two-particle system, the components of the total orbital angular momentum $\mathbf{L} = \mathbf{L}_1 + \mathbf{L}_2$ obey the same commutation, Eqs. (64); as a result the eigenvalues of L^2 and L_z (but not the eigenfunctions) are the same as in the one-particle case. L^2 and L_z commute with L_1^2 and L_2^2, but L^2 does not commute with L_{1z} or L_{2z}. Consequently the total orbital angular momentum eigenfunctions are labeled by l, m, l_1, l_2. For given l_1, l_2, the possible values of l are the positive integers from $l = l_1 + l_2$ down to $l = |l_1 - l_2|$; the corresponding eigenfunctions have parity $(-)^{l_1 + l_2}$ independent of l, m. These rules for combining angular momenta are readily generalized to more complicated systems including spin, are well established, and form the basis for the vector model of the atom. *See* ATOMIC STRUCTURE AND SPECTRA.

Coulomb potential. The Hamiltonian for an (assumed spinless) electron of mass m_e in the field of a fixed charge Ze is $H = \mathbf{p}^2/2m_e + V(r)$, where $V(r) = -Ze^2/r$. In spherical coordinates, Eq. (14) for the eigenfunctions $u(r, \theta, \phi)$ is written as Eq. (67),

$$Hu = \left[\frac{-\hbar^2}{2m_e} \frac{1}{r^2} \frac{\partial}{\partial r} \left(r^2 \frac{\partial}{\partial r} \right) + \frac{1}{2m_e r^2} L^2 + V(r) \right] u$$
$$= Eu \qquad (67)$$

with L^2 defined by Eq. (65b). Now H commutes with L^2 and L_z and $r^2 H$ is separable; that is, Eq. (68)

$$r^2 H(r, \theta, \phi) = H_1(r) + (2m_e)^{-1} L^2(\theta, \phi) \qquad (68)$$

holds [compare Eq. (61)]. Thus $u(r)u(l, m)$ are a complete set of eigenfunctions; $L^2 u(r)u(l, m) = l(l + 1)\hbar^2 u(r)u(l, m)$, and therefore the radial eigenfunc-

tions $u(r) \equiv u(E, l)$ must satisfy Eq. (69).

$$\left[\frac{-\hbar^2}{2m_e} \frac{1}{r^2} \frac{\partial}{\partial r} \left(r^2 \frac{\partial}{\partial r} \right) + \frac{l(l+1)\hbar^2}{2m_e r^2} + V(r) \right] u(r)$$
$$= Eu(r) \qquad (69)$$

The positive term $l(l + 1)\hbar^2/2m_e r^2$ acts as an added repulsive potential; it can be understood in terms of the classical centripetal force needed to maintain the angular momentum. For $E < 0$, admissible solutions to Eq. (69) must be exponentially decreasing at $r = \infty$; moreover, because of the r^{-1} and r^{-2} terms in Eq. (69), an eigenfunction $u(E, l)$ which behaves properly at $r = \infty$ becomes infinite at $r = 0$ unless E is specially chosen. Thus (as always) the quadratically integrable eigenfunctions form a discrete set. The corresponding bound state energies $E < 0$ are given by Eq. (70).

$$E = -\frac{m_e Z^2 e^4}{2\hbar^2 n^2} \qquad (70)$$

In Eq. (67) the principal quantum number $n = 1$, $2, 3, \ldots$; for given l and n the number $n_r \geq 0$ of zeros (between $r = 0$ and $r = \infty$) of the corresponding $u(E, l)$ is $n_r = n - l - 1$. Because $dx\, dy\, dz = r^2 \sin\theta\, dr\, d\theta\, d\phi$, the radial probability density is $r^2 |u(r, E, l)|^2$. Figure 9 shows the radial probability density plotted versus r (in units of the Bohr radius $a_0 = \hbar^2/m_e e^2 \approx 5 \times 10^{-9}$ cm) for several low-lying stationary states of atomic hydrogen, $Z = 1$. The notation for the eigenfunctions is standard in atomic physics: the principal quantum number is supplemented by a lower case letter s, p, d, $\ldots$ corresponding to $l = 0, 1, 2, \ldots$; for example, a $3d$ state has $l = 2$ and therefore $n_r = 0$ or no radial nodes, as in Fig. 9. The eigenfunction $u(l = 0, m = 0)$ is a constant; that is, an s state is spherically symmetric; $|u(r, \theta, \phi)|^2$ is proportional to $\cos^2\theta$ or to $\sin^2\theta$ in p states, and so on. The eigenfunc-

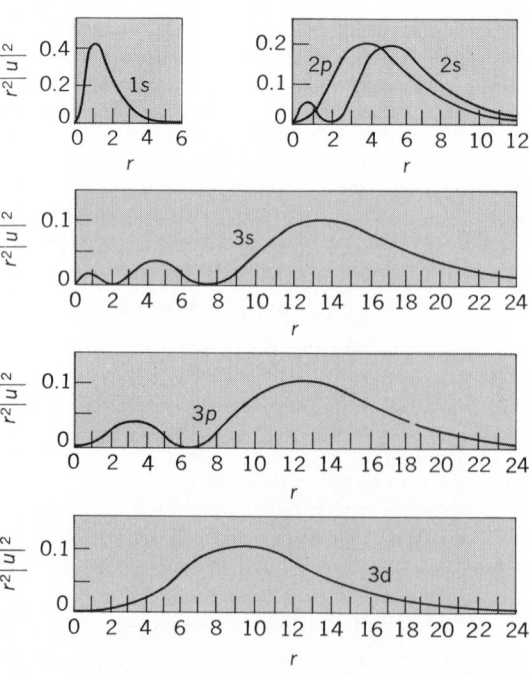

Fig. 9. Radial probability density in atomic hydrogen. (F. K. Richtmyer, E. H. Kennard, and T. Lauritsen, *Introduction to Modern Physics*, 5th ed., McGraw-Hill, 1955)

tions, although they are spread over all space, have their maxima at about the radii expected on the older Bohr theory of Eq. (70).

In the actual hydrogen atom the nucleus, of mass M, is not fixed. The Hamiltonian is Eq. (71a), where the subscripts 1 and 2 refer to the electron and the nucleus, respectively. Introducing the center of mass $(X,Y,Z) \equiv \mathbf{R} = (M + m_e)^{-1}[m_e\mathbf{r}_1 + M\mathbf{r}_2]$, $(x,y,z) \equiv \mathbf{r} = \mathbf{r}_1 - \mathbf{r}_2$, Eq. (71$a$) takes the separable form of Eq. (71b), where $(\mathbf{p}_R)_x = p_X = (\hbar/i)\partial/\partial X$,

$$H = \frac{p_1^2}{2m_e} + \frac{p_2^2}{2M} - \frac{Ze^2}{|r_1 - r_2|} \qquad (71a)$$

$$H = \frac{p_R^2}{2(M + m_e)} + \frac{p^2}{2\mu} - \frac{Ze^2}{r} \qquad (71b)$$

etc., are the components of the total momentum $\mathbf{p}_R = \mathbf{p}_1 + \mathbf{p}_2$; $\mathbf{p}_x = (\hbar/i)\partial/\partial x$, etc.; and the reduced mass $\mu = m_e M (M + m_e)^{-1}$. Therefore $\mathbf{p}_R$ is a constant of the motion, as asserted in connection with Eq. (56); moreover, Eq. (71b) is separable in $\mathbf{R}$ and $\mathbf{r}$. In other words, the center of mass moves like a free particle, completely independent of the internal $\mathbf{r}$ motion. Comparing Eqs. (67) and (71b), and recalling Eq. (61), the eigenvalues of Eq. (71a), after the kinetic energy of the center of mass is subtracted, are given by Eq. (70) with μ (depending on m_e/M) replacing m_e. The independence of internal and center-of-mass motion means that the temperature broadening of spectral lines can be explained quantum mechanically in terms of the classical Doppler effect for a moving fixed-frequency source. This paragraph illustrates the correspondence between the classical and quantum theories. *See* CENTER OF MASS.

The eigenvalues E of Eq. (70) have order of degeneracy n^2; this degeneracy stems from (i) the fact that $V(r)$ is spherically symmetric, which permits H to commute with $\mathbf{L}$; (ii) a special symmetry of Eq. (67) for the specially simple Coulomb potential, causing solutions of Eq. (69) for different l to have the same energy. For an arbitrary spherically symmetric $V(r)$, the bound-state eigenvalues $E(n_r, l)$ of Eq. (69) do not coincide for different l, and each bound state has degeneracy $2l + 1$, corresponding to the $2l + 1$ possible values $m = -l$ to l of the magnetic quantum number m; an energy level associated with orbital angular momentum l has parity $(-)^l$. For any such $V(r)$ the bound-state energies (i) increase with increasing n_r for a constant value of l, for the same reasons that were discussed in connection with the eigenvalues of Eq. (28); (ii) increase with increasing l for constant n_r because the rotational kinetic energy $(2m_e r^2)^{-1} l(l+1)\hbar^2$ increases. Except for these regularities, the order and spacing of the levels depends on the details of $V(r)$. For potentials $V(r)$ decreasing more rapidly than $1/r$, the total number of bound states generally is finite, whereas this number is infinite for the Coulomb $V(r) = -Ze^2/r$.

Removal of degeneracy. The Hamiltonian of Eq. (67) is spin-independent, which is why Eq. (67) has been treated as if the wave function had only one component; compare the remarks following Eqs. (5). Corresponding to any one-component solution u of Eq. (67), there are two independent two-component eigenfunctions; (i) $\psi_1 = u$, $\psi_2 = 0$ and (ii) $\psi_1 = 0$, $\psi_2 = u$; compare the earlier discussion of spin. Thus for an electron the degeneracy of the

energy levels in a Coulomb field is $2n^2$; similarly the degeneracy for neutrons, protons, or electrons in an arbitrary spherically symmetric potential is $2(2l + 1)$. The energy operator for an electron in an actual atom, for example, hydrogen, is not spin-independent, however. Relativistic effects add, to the central $V(r)$ of Eq. (67), noncentral spin-orbit potentials $V'(r)[L_x s_x + L_y s_y + L_z s_z] \equiv V'(r)\mathbf{L}\cdot\mathbf{s}$; here $\mathbf{s}$ is the spin operator and obeys the same commutation Eqs. (64) as $\mathbf{L}$.

Equations (64) show that $V'3r)\mathbf{L}\cdot\mathbf{s}$ commutes with L^2, s^2, $(\mathbf{L} + \mathbf{s})^2$, and $L_z + s_z$, but not with L_z or s_z, illustrating the principle of conservation of total angular momentum $\mathbf{J} = \mathbf{L} + \mathbf{s}$; compare the remarks preceding Eq. (56). Consequently, referring to the final part of the discussion of orbital angular momentum (i) $J^2 = j(j+1)\hbar^2$; (ii) for given $l \neq 0$ (and $s = 1/2$); j has but two possible values $l \pm 1/2$; (iii) the energy levels are labeled by j and have a $(2j + 1)$-fold degeneracy corresponding to the $2j + 1$ possible orientations of $J_z = -j$ to $+j$; (iv) because $2\mathbf{L}\cdot\mathbf{s} = (\mathbf{L} + \mathbf{s}^2 - L^2 - s^2)$, levels of different j have different energies, and the splitting of the energies depends predictably on l; (v) L_z (and s_z) no longer are constants of the motion, although L^2 and $s^2 = (1/2)\cdot(3/2)\hbar^2$ still are. Moreover, because $2j + 1 = 2l + 2$ for $j = l + 1/2$, and $= 2l$ for $j = l - 1/2$, the total number of independent eigenfunctions associated with given l (and n_r) remains $2(2l + 1) = (2l + 2) + (2l)$.

In the independent particle model of atoms and nuclei, one assumes that to a first approximation each particle, particle i, say, moves in a potential $V(r_i)$ which has been averaged over the coordinates of all the other particles, so that $V(r_i)$ depends only on the distance of i from the atomic or nuclear center. To a first approximation, therefore, the energy levels are associated with configurations of one-particle eigenfunctions; for example, the ground state of atomic Be is $1s^2 2s^2$. In higher approximation one introduces two-body interactions $V(r_i, r_j)$, which may be said to mix different configurations. The considerations of this and the paragraphs just preceding, together with the exclusion principle discussed subsequently, account for the periodic system of the elements and are the basis for the highly successful nuclear shell model.

The observation that splitting the levels does not change the number of independent eigenfunctions illustrates a general principle and justifies the postulate that the statistical weight of a discrete level equals its order of degeneracy. This principle can be understood on the basis that the number of bound-state eigenfunctions should be a continuous function of the parameters in a reasonably well-behaved Hamiltonian; because this number by definition is an integer, it must change discontinuously and, therefore, except under unusual mathematical circumstances, cannot change at all. *See* BOLTZMANN STATISTICS; STATISTICAL MECHANICS.

In an external magnetic field B the Hamiltonian of a many-electron atom (i) no longer is independent of the orientation of the coordinate axes, so that the degeneracy associated with this symmetry is removed; (ii) retains symmetry with respect to rotation about the magnetic field, so that (with the z axis along B) J_z commutes with the Hamiltonian. Thus in a magnetic field a level associated with the total angular momentum quantum number j should split into $2j + 1$ levels, each of which is associated with one of the magnetic quantum numbers $-j$ to $+j$. This prediction is thoroughly confirmed in the Zeeman effect and in the Stern-Gerlach experiment. *See* ZEEMAN EFFECT.

Radiation. The classical Hamiltonian for a charged particle in an electromagnetic field has the form given by Eq. (72), where $\mathbf{A}$ and ϕ are the

$$H = \frac{1}{2m}\left(\mathbf{p} - \frac{e\mathbf{A}}{c}\right)^2 + e\phi \qquad (72)$$

scalar and vector potentials, respectively.

It is postulated that when properly symmetrized, that is, when $\mathbf{A} \cdot \mathbf{p} + \mathbf{p} \cdot \mathbf{A}$ replaces the classical $2\mathbf{A} \cdot \mathbf{p}$ (see the earlier discussion on Hermitian operators), H of Eq. (72) is the quantum mechanical energy operator. The presence of terms linear in $\mathbf{p}$ modifies some of the formulas which have been given, for example, Eq. (55a). When plane waves of light (frequency f, wavelength λ, and moving in the z direction) fall on a hydrogen atom, $\mathbf{A}$ is proportional to $\cos[2\pi(z/\lambda - ft)]$, and $e\phi$ is the Coulomb potential $V(r)$ of Eq. (67).

Proceeding as in Eqs. (57)–(60), noting that $\mathbf{A}$ contains terms proportional to $\exp(2\pi i f t)$ and $\exp(-2\pi i f t)$, and neglecting the small (as can be shown) A^2 terms, one obtains an expression similar to Eq. (59), except that $\omega \pm 2\pi f$ replaces $\omega \equiv \hbar^{-1}(E_f - E_i)$. In other words, after a long time there are appreciable transition probabilities only to final states f whose energies satisfy $E_f - E_i \pm hf = 0$, in agreement with the notion that a quantum of energy hf has been emitted or absorbed; the $+$ sign corresponds to emission, the $-$ sign to absorption. The corresponding transition probabilities are given by Eq. (60) with V' proportional to $\exp[2\pi i z/\lambda]$, and u_i, u_f stationary-state atomic wave functions satisfying the radiation-unperturbed Eq. (67). The final expressions are analogous to the classical formulas for emission or absorption of radiation; for instance, when the wave-length λ is large compared to atomic dimensions, expanding $\exp[2\pi i z/\lambda]$ in powers of z/λ shows that the leading term in the transition probability is the matrix element, between initial and final states, of the dipole moment ez corresponding to classical electric dipole emission or absorption. Because z changes sign on reflection through the origin, the dipole matrix element vanishes unless u_i and u_f have opposite parities. This is one of the selection rules for electric dipole radiation; other selection rules, connected with angular momentum conservation, are obtained similarly. *See* ELECTROMAGNETIC RADIATION; SELECTION RULES.

The theory starting with Eq. (72) is termed semiclassical, because it does not replace the classical $\mathbf{A}$, ϕ by a quantum mechanical operator description of the electromagnetic field. This semiclassical theory has led to the induced emission and absorption probabilities in the presence of external radiation, but not to the spontaneous probability of emission of a photon in the absence of external radiation. The spontaneous transition probabilities can be inferred from the induced probabilities by thermodynamic arguments. The spontaneous transition probability is deduced directly, however, without appeal to the arguments of thermodynamics, when the radiation field is quantized.

PARTICLE INDISTINGUISHABILITY

For systems of g identical, and therefore indistinguishable, particles, the formalism is further complicated because the probability P of finding a given particle in a specified volume $dx\,dy\,dz$ of space must be $P_1 + P_2 + \cdots + P_g$, where $P_1, \ldots, P_g$ are the probabilities given previously for distinguishable particles; expectation values and normalizations must be reinterpreted accordingly. Moreover, the Pauli exclusion principle asserts that the only physically permissible wave functions must change sign when the space and spin coordinates of any pair of indistinguishable particles of spin $\frac{1}{2}$ are interchanged. *See* EXCLUSION PRINCIPLE.

To amplify this assertion, consider the four-component wave function of the two electrons in atomic helium in Eq. (73), where $|\psi_{++}(\mathbf{r}_1,\mathbf{r}_2)|^2$ is the

$$\psi = \psi_{++}(\mathbf{r}_1,\mathbf{r}_2),\ \psi_{+-}(\mathbf{r}_1,\mathbf{r}_2),$$
$$\psi_{-+}(\mathbf{r}_1,\mathbf{r}_2),\ \psi_{--}(\mathbf{r}_1,\mathbf{r}_2) \qquad (73)$$

probability density for finding both electrons with spin along $+z$, and so forth. The exclusion principle requires validity of Eqs. (74).

$$\psi_{++}(\mathbf{r}_1,\mathbf{r}_2) = -\psi_{++}(\mathbf{r}_2,\mathbf{r}_1)$$
$$\psi_{--}(\mathbf{r}_1,\mathbf{r}_2) = -\psi_{--}(\mathbf{r}_2,\mathbf{r}_1) \qquad (74)$$
$$\psi_{+-}(\mathbf{r}_1,\mathbf{r}_2) = -\psi_{-+}(\mathbf{r}_2,\mathbf{r}_1)$$

In the independent particle approximation, all components of the ground-state eigenfunctions of atomic He are composed of products $u(\mathbf{r}_1)u(\mathbf{r}_2)$, where $u(\mathbf{r})$ is the lowest $1s$ eigenfunction solving Eq. (67) for the spherically symmetric $V(r)$ appropriate to He. In this approximation, therefore, $\psi_{++} = \psi_{--} = 0$; if $\psi_{+-} = u(\mathbf{r}_1)u(\mathbf{r}_2)$, ψ_{-+} is necessarily equal to $-u(\mathbf{r}_1)u(\mathbf{r}_2)$. Of the four independent $1s^2$ eigenfunctions originally possible, only one remains, which can be shown to be a total spin zero eigenfunction; the exclusion principle literally has excluded the three eigenfunctions corresponding to total spin one.

These results are summarized by the rule that at most two electrons, $s_z = \pm\frac{1}{2}$, can occupy one-particle states with the same quantum numbers n_r, l, and $m = -l$ to l. By the general principle explained previously in the discussion of removal of degeneracy, the introduction of two-particle interactions $V(\mathbf{r}_1,\mathbf{r}_2)$ does not change the number of independent eigenfunctions consistent with the exclusion principle. In the next higher $1s2s$ configuration of He, ψ_{++} will equal $u_{1s}(\mathbf{r}_1)u_{2s}(\mathbf{r}_2) - u_{1s}(\mathbf{r}_2)u_{2s}(\mathbf{r}_1)$. This antisymmetrized wave function makes nonclassical exchange energy contributions, of the form given by expression (75) to the expectation value

$$\int u_{1s}*(\mathbf{r}_1)\,u_{2s}*(\mathbf{r}_2)V(\mathbf{r}_1,\mathbf{r}_2)u_{1s}(\mathbf{r}_2)u_{2s}(\mathbf{r}_1) \qquad (75)$$

$\langle V(\mathbf{r}_1,\mathbf{r}_2)\rangle$ of the interaction energy. Exchange energies are important for understanding chemical binding.

Except for the complication of spin, this article presupposes that electrons, neutrons, and protons are immutable structureless mass points. Actually, modern theories of nuclear forces and high-energy scattering experiments indicate that this assumption is untrue. When creation, destruction, or other alterations of fundamental particle structure are improbable, however, the general separability of internal and center-of-mass motion [see the discussion following Eq. (71b)] implies that each fundamental particle is sufficiently described by the position of its center of mass and by its spin orientation, that is, by the many-component wave functions used here.

In circumstances wherein more obviously composite systems undergo no changes in internal structure, they too can be treated as particles. For instance, in the slow collisions of two atoms the slowly changing potentials acting on the electrons do not induce transitions to new configurations, and the collision can be described by solving an equation of the form (67) for the relative motion; in rapid collisions electron transitions occur, and the many-electron Schrödinger equation must be employed. Similarly, since a deuteron is a neutron-proton bound state, with total angular momentum unity, in a deuterium molecule (i) each deuteron can be treated as if it were a fundamental particle of spin 1; (ii) the wave function of a deuterium molecule must be symmetric under interchange of the space and spin coordinates of the two deuterons, which interchange involves successive antisymmetric interchanges of the two neutrons and of the two protons. In other words (when they can be treated as particles) deuterons and other composite systems of integral spin obey Bose-Einstein statistics; composite systems of half-integral spin obey Fermi-Dirac statistics.

When the particles composing a many-particle system can be represented by nonoverlapping wave packets which spread an amount $\ll \Delta x$ as the center packet moves a distance equal to its width Δx, individual classical particle trajectories can be distinguished; under these circumstances, the particles, whether or not identical, are in effect distinguishable classical particles, and one expects Bose-Einstein and Fermi-Dirac statistics to reduce to the classical Maxwell-Boltzmann statistics. The well-known condition for the validity of classical statistics in an electron gas, $Nh^3 (2\pi mkT)^{-3/2} \ll 1$, implies that such packets can be constructed; here N is the electron density, k is Boltzmann's constant, and T is the absolute temperature. In the lower vibrational states of a molecule such packets cannot be constructed for the vibrating nuclei, however (compare the earlier discussion of the harmonic oscillator), so that quantum statistics cannot be ignored, for example, in the specific heat of H_2 at low temperatures. *See* QUANTUM STATISTICS.

[EDWARD GERJUOY]

Bibliography: C. Cohen-Tanmoudji et al., *Quantum Mechanics*, 2 vols., 1978; P. A. M. Dirac, *Principles of Quantum Mechanics*, 4th ed., 1958; L. D. Landau and E. M. Lifshitz, *Quantum Mechanics: Non-relativistic Theory*, 3d ed., 1977; D. A. Park, *Introduction to the Quantum Theory*, 1974; L. I. Shiff, *Quantum Mechanics*, 3d ed., 1968; R. Winter, *Quantum Physics*, 1979.

Nuclear binding energy

The amount by which the mass of an atom is less than the sum of the masses of its constituent protons, neutrons, and electrons expressed in units of energy. This energy difference accounts for the stability of the atom. In principle, the binding energy is the amount of energy which was released when the several atomic constituents came together to form the atom. Most of the binding energy is associated with the nuclear constituents (protons and neutrons), or nucleons, and it is customary to regard this quantity as a measure of the stability of the nucleus alone. *See* NUCLEAR STRUCTURE.

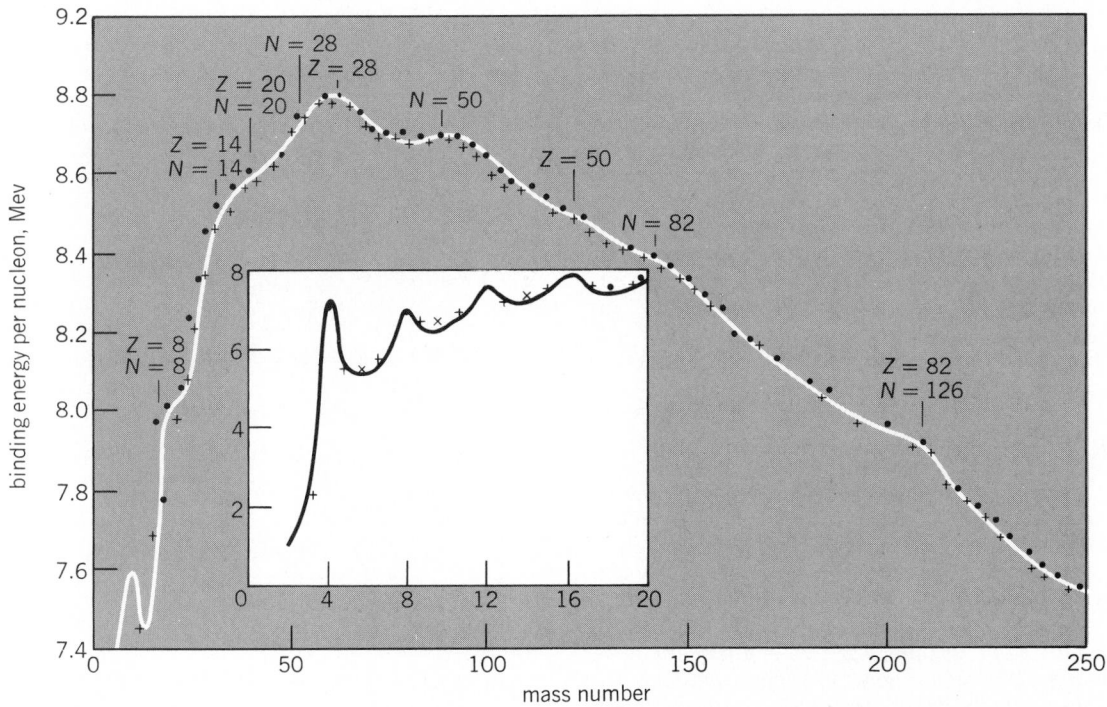

Binding energy per nucleon (in MeV) versus mass number. N = number of neutrons; Z = atomic number. (*From A. H. Wapstra, Isotopic measure, part 1, where A is less than 34, Physica, 21:367–384, 1955*)

A widely used term, the binding energy (BE) per nucleon, is defined by the equation below, where

$$BE/nucleon = \frac{[ZH + (A-Z)n - _ZM^A]c^2}{A}$$

$_ZM^A$ represents the mass of an atom of mass number A and atomic number Z, H and n are the masses of the hydrogen atom and neutron, respectively, and c is the velocity of light. The binding energies of the orbital electrons, here practically neglected, are not only small, but increase with Z in a gradual manner; thus the BE/nucleon gives an accurate picture of the variations and trends in nuclear stability. The figure shows the BE/nucleon (in million electronvolts) plotted against mass number for $A > 40$.

The BE/nucleon curve at certain values of A suddenly changes slope in such a direction as to indicate that the nuclear stability has abruptly deteriorated. These turning points coincide with particularly stable configurations, or nuclear shells, to which additional nucleons are rather loosely bound. Thus there is a sudden turning of the curve over $A = 52$ (28 neutrons); the maximum occurs in the nickel region (28 protons, $\sim A = 60$); the stability rapidly deteriorates beyond $A = 90$ (50 neutrons); there is a slightly greater than normal stability in the tin region (50 protons, $\sim A = 118$); the stability deteriorates beyond $A = 140$ (82 neutrons) and beyond $A = 208$ (82 protons plus 126 neutrons).

The BE/nucleon is remarkably uniform, lying for most atoms in the range 5 – 9 Mev. This near constancy is evidence that nucleons interact only with near neighbors; that is, nuclear forces are saturated.

The binding energy, when expressed in mass units, is known as the mass defect, a term sometimes incorrectly applied to quantity $M - A$, where M is the mass of the atom. [H. E. DUCKWORTH]

The term binding energy is sometimes also used to describe the energy which must be supplied to a nucleus in order to remove a specified particle to infinity, for example, a neutron, proton, or α-particle. A more appropriate term for this energy is the separation energy. This quantity varies greatly from nucleus to nucleus and from particle to particle. For example, the binding energies for a neutron, a proton, and a deuteron in O^{16} are 15.67, 12.13, and 20.74 Mev, respectively, while the corresponding energies in O^{17} are 4.14, 13.78, and 14.04 Mev, respectively. The usual order of neutron or proton separation energy is 7 – 9 Mev for most of the periodic table.

[D. H. WILKINSON]

Nuclear fission

An extremely complex nuclear reaction representing a cataclysmic division of an atomic nucleus into two nuclei of comparable mass. This rearrangement or division of a heavy nucleus may take place naturally (spontaneous fission) or under bombardment with neutrons, charged particles, gamma rays, or other carriers of energy (induced fission). Although nuclei with mass number A of approximately 100 or greater are energetically unstable against division into two lighter nuclei, the fission process has a small probability of occurring, except with the very heavy elements. Even for

these elements, in which the energy release is of the order of 200,000,000 electronvolts (eV), the lifetimes against spontaneous fission are reasonably long. *See* NUCLEAR REACTION.

Liquid-drop model. The stability of a nucleus against fission is most readily interpreted when the nucleus is viewed as being analogous to an incompressible and charged liquid drop with a surface tension. Such a droplet is stable against small deformations when the dimensionless fissility parameter X in Eq. (1) is less than unity, where the

$$X = \frac{(charge)^2}{10 \times volume \times surface\ tension} \quad (1)$$

charge is in esu, the volume is in cm³, and the surface tension is in ergs/cm². The fissility parameter if given approximately, in terms of the charge number Z and mass number A, by the relation $X = Z^2/50\ A$.

Long-range Coulomb forces between the protons act to disrupt the nucleus, whereas short-range nuclear forces, idealized as a surface tension, act to stabilize it. The degree of stability is then the result of a delicate balance between the relatively weak electromagnetic forces and the strong nuclear forces. Although each of these forces results in potentials of several hundred million electron volts, the height of a typical barrier against fission for a heavy nucleus, because they are of opposite sign but do not quite cancel, is only 5,000,000 or 6,000,000 eV. Investigators have used this charged liquid-drop model with great success in describing the general features of nuclear fission and also in reproducing the total nuclear binding energies. *See* NUCLEAR BINDING ENERGY; NUCLEAR STRUCTURE; SURFACE TENSION.

Shell corrections. The general dependence of the potential energy on the fission coordinate representing nuclear elongation or deformation for a heavy nucleus such as ^{240}Pu is shown in Fig. 1. The expanded scale used in this figure shows the large decrease in energy of about 200 MeV as the

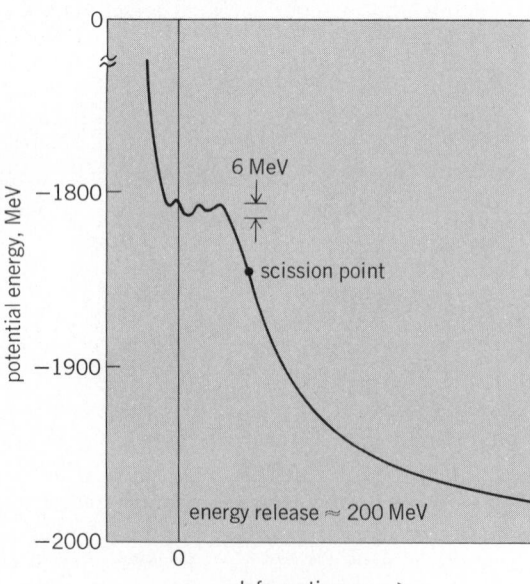

Fig. 1. Plot of the potential energy in MeV as a function of deformation for the nucleus ^{240}Pu. (*From M. Bolsteli et al., New calculations of fission barriers for heavy and superheavy nuclei, Phys. Rev., 5C:1050–1077, 1972*)

fragments separate to infinity. It is known that ^{240}Pu is deformed in its ground state, which is represented by the lowest minimum of −1813 MeV near zero deformation. This energy represents the total nuclear binding energy when the zero of potential energy is the energy of the individual nucleons at a separation of infinity. The second minimum to the right of zero deformation illustrates structure introduced in the fission barrier by shell corrections, that is, corrections dependent upon microscopic behavior of the individual nucleons, to the liquid-drop mass. Although shell corrections introduce small wiggles in the potential-energy surface as a function of deformation, the gross features of the surface are reproduced by the liquid-drop model. Since the typical fission barrier is only a few million electron volts, the magnitude of the shell correction need only be small for irregularities to be introduced into the barrier. This structure is schematically illustrated for a heavy nucleus by the double-humped fission barrier in Fig. 2, which represents the region to the right of zero deformation in Fig. 1 on an expanded scale. The fission barrier has two maxima and a rather deep minimum in between. For comparison, the single-humped liquid-drop barrier is also schematically illustrated. The transition in the shape of the nucleus as a function of deformation is schematically represented in the upper part of the figure.

Double-humped barrier. The developments which led to the proposal of a double-humped fission barrier were triggered by the experimental discovery of spontaneously fissionable isomers by S. M. Polikanov and colleagues in the Soviet Union and by V. M. Strutinsky's pioneering theoretical work on the binding energy of nuclei as a function of both nucleon number and nuclear shape. The double-humped character of the nuclear potential energy as a function of deformation arises, within the framework of the Strutinsky shell-correction method, from the superposition of a macroscopic smooth liquid-drop energy and a shell-correction energy obtained from a microscopic single-particle model. Oscillations occurring in this shell correction as a function of deformation lead to two minima in the potential energy, shown in Fig. 2, the normal ground-state minimum at a deformation of β_1 and a second minimum at a deformation of β_2. States in these wells are designated class I and class II states, respectively. Spontaneous fission of the ground state and isomeric state arises from the lowest-energy class I and class II states, respectively. *See* NUCLEAR ISOMERISM.

The calculation of the potential-energy curve illustrated in Fig. 1 may be summarized as follows. The smooth potential energy obtained from a macroscopic (liquid-drop) model is added to a fluctuating potential energy representing the shell corrections, and to the energy associated with the pairing of like nucleons (pairing energy), derived from a non-self-consistent microscopic model. The calculation of these corrections requires several steps, namely, (1) specification of the geometrical shape of the nucleus, (2) generation of a single-particle potential related to its shape, (3) solution of the Schrödinger equation, and (4) calculation from these single-particle energies of the shell and pairing energies.

The oscillatory character of the shell corrections as a function of deformation is caused by varia-

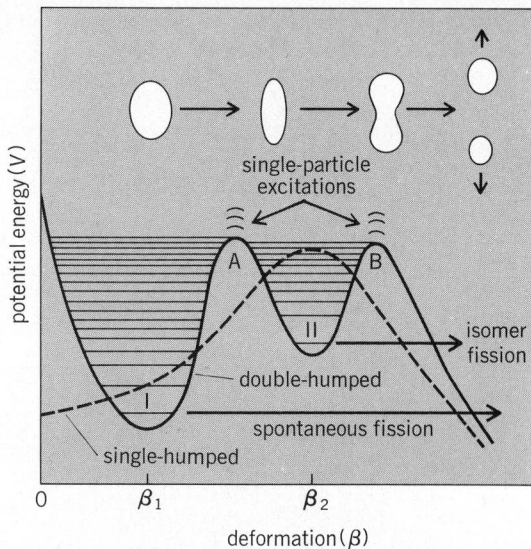

Fig. 2. Schematic plots of single-humped fission barrier of liquid-drop model and double-humped barrier introduced by shell corrections. (*From J. R. Huizenga, Nuclear fission revisited, Science, 168:1405–1413, 1970*)

tions in the single-particle level density in the vicinity of the Fermi energy. For example, the single-particle levels of a pure harmonic oscillator potential arrange themselves in bunches of highly degenerate shells at any deformation for which the ratio of the major and minor axes of the spheroidal equipotential surfaces is equal to the ratio of two small integers. Nuclei with a filled shell, that is, with a level density at the Fermi energy that is smaller than the average, will then have an increased binding energy compared to the average, because the nucleons occupy deeper and more bound states; conversely, a large level density is associated with a decreased binding energy. It is precisely this oscillatory behavior in the shell correction that is responsible for spherical or deformed ground states and for the secondary minima in fission barriers, as illustrated in Fig. 2. *See* NONRELATIVISTIC QUANTUM THEORY.

More detailed theoretical calculations based on this macroscopic-microscopic method have revealed additional features of the fission barrier. In these calculations the potential energy is regarded as a function of several different modes of deformation. The outer barrier B (Fig. 2) is reduced in energy for shapes with pronounced left-right asymmetry (pear shapes), whereas the inner barrier A and deformations in the vicinity of the second minimum are stable against such mass asymmetric degrees of freedom. Similar calculations of potential-energy landscapes reveal the stability of the second minimum against gamma deformations, in which the two small axes of the spheroidal nucleus become unequal, that is, the spheroid becomes an ellipsoid.

Experimental consequences: The observable consequences of the double-humped barrier have been reported in numerous experimental studies. In the actinide region more than 30 spontaneously fissionable isomers have been discovered between uranium and berkelium, with half-lives ranging from 10^{-11} to 10^{-2} s. These decay rates are faster by 20 to 30 orders of magnitude than the fission

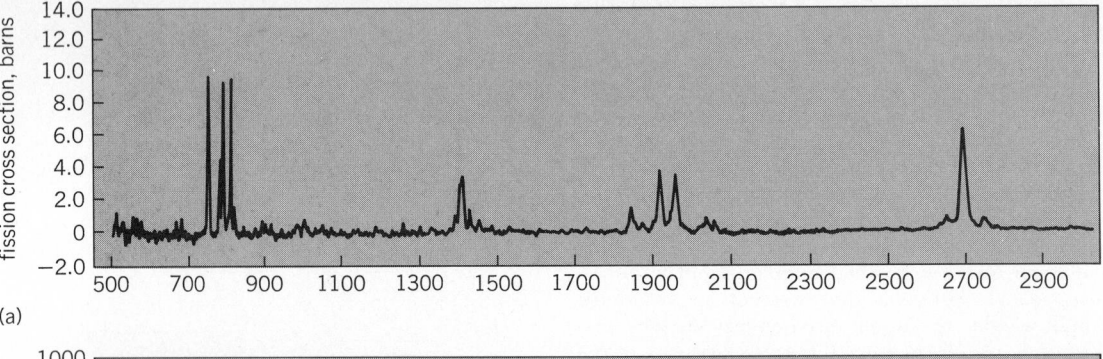

(a)

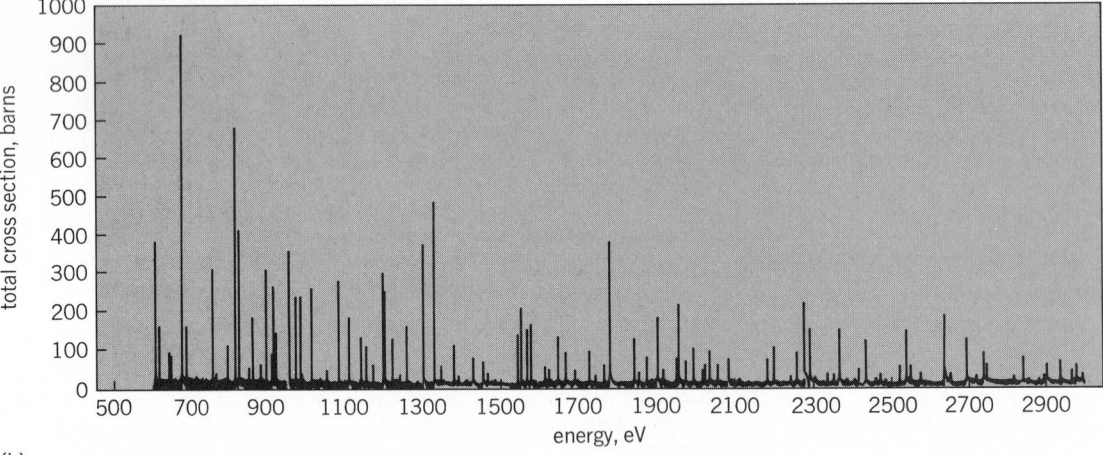

(b)

Fig. 3. Grouping of fission resonances demonstrated by (a) the neutron fission cross section of ^{240}Pu and (b) the total neutron cross section. (*From V. M. Strutinsky and*

H. C. Pauli, Shell-structure effects in the fissioning nucleus, Proceedings of the 2d IAEA Symposium on Physics and Chemistry of Fission, Vienna, pp. 155–177, 1969)

half-lives of the ground states, because of the increased barrier tunneling probability (see Fig. 2). Several cases in which excited states in the second minimum decay by fission are also known. Normally these states decay within the well by gamma decay; however, if there is a hindrance in gamma decay due to spin, the state (known as a spin isomer) may undergo fission instead.

Qualitatively, the fission isomers are most stable in the vicinity of neutron numbers 146 to 148, a value in good agreement with macroscopic-microscopic theory. For elements above berkelium the half-lives become too short to be observable with available techniques; and for elements below uranium, the prominent decay is through barrier A into the first well, followed by gamma decay. It is difficult to detect this competing gamma decay of the ground state in the second well (called a shape isomeric state), but identification of the gamma branch of the 200-ns ^{238}U shape isomer has been reported. *See* RADIOACTIVITY.

Direct evidence of the second minimum in the potential-energy surface of the even-even nucleus ^{240}Pu has been obtained through observations of the E2 transitions within the rotational band built on the isomeric 0+ level. The rotational constant (which characterizes the spacing of the levels and is expected to be inversely proportional to the effective moment of inertia of the nucleus) found for this band is less than one-half that for the ground state and confirms that the shape isomers have a deformation β_2 much larger than the equilibrium ground-state deformation β_1. From yields and an-

gular distributions of fission fragments from the isomeric ground state and low-lying excited states some information has been derived on the quantum numbers of specific single-particle states of the deformed nucleus (Nilsson single-particle states) in the region of the second minimum.

At excitation energies in the vicinity of the two barrier tops, measurements of the subthreshold neutron fission cross sections of several nuclei have revealed groups of fissioning resonance states with wide energy intervals between each group where no fission occurs. Such a spectrum is illustrated in Fig. 3a, where the subthreshold fission cross section of ^{240}Pu is shown for neutron energies between 500 and 3000 eV. As shown in Fig. 3b, between the fissioning resonance states there are many other resonance states, known from data on the total neutron cross sections, which have negligible fission cross sections. Such structure is explainable in terms of the double-humped fission barrier and is ascribed to the coupling between the compound states of normal density in the first well to the much less dense states in the second well. This picture requires resonances of only one spin to appear within each intermediate structure group illustrated in Fig. 3a. In an experiment using polarized neutrons on a polarized ^{237}Np target, it was found that all nine fine-structure resonances of the 40-eV group have the same spin and parity: $I = 3+$. Evidence has also been obtained for vibrational states in the second well from neutron (n,f) and deuteron stripping (d,pf) reactions at energies below the barrier tops (f indicates fission of the nucleus). *See* NEUTRON SPECTROMETRY.

A. Bohr suggested that the angular distributions of the fission fragments are explainable in terms of the transition-state theory, which describes a process in terms of the states present at the barrier deformation. The theory predicts that the cross section will have a steplike behavior for energies near the fission barrier, and that the angular distribution will be determined by the quantum numbers associated with each of the specific fission channels. The theoretical angular distribution of fission fragments is based on two assumptions. First, the two fission fragments are assumed to separate along the direction of the nuclear symmetry axis so that the angle θ between the direction of motion of the fission fragments and the direction of motion of the incident bombarding particle represents the angle between the body-fixed axis (the long axis of the spheroidal nucleus) and the space-fixed axis (some specified direction in the laboratory, in this case the direction of motion of the incident particle). Second, it is assumed that the transition from the saddle point (corresponding to the top of the barrier) to scission (the division of the nucleus into two fragments) is so fast that Coriolis forces do not change the value of K (where K is the projection of the total angular momentum I on the nuclear symmetry axis) established at the saddle point.

In several cases, low-energy photofission and neutron fission experiments have shown evidence of a double-humped barrier. In the case of two barriers, the question arises as to which of the two barriers A or B is responsible for the structure in the angular distributions. For light actinide nuclei like thorium, the indication is that barrier B is the higher one, whereas for the heavier actinide nuclei, the inner barrier A is the higher one. The heights of the two barriers themselves are most reliably determined by investigating the probability of induced fission over a range of several mega-electron volts in the threshold region. Many direct reactions have been used for this purpose, for example, (d,pf), (t,pf), and $(^3\text{He}, df)$. There is reasonably good agreement between the experimental and theoretical barriers. The theoretical barriers are calculated with realistic single-particle potentials and include the shell corrections.

Fission probability. The cross section for particle-induced fission $\sigma(y,f)$ represents the cross section for a projectile y to react with a nucleus and produce fission, as shown by Eq. (2). The quanti-

$$\sigma(y,f) = \sigma_R(y)\,(\Gamma_f/\Gamma_t) \qquad (2)$$

ties $\sigma_R(y)$, Γ_f, and Γ_t are the total reaction cross sections for the incident particle y, the fission width, and the total level width, respectively where $\Gamma_t = \Gamma_f + \Gamma_n + \Gamma_y + \cdots$ is the sum of all partial-level widths. All the quantities in Eq. (2) are energy-dependent. Each of the partial widths for fission, neutron emission, radiation, and so on, is defined in terms of a mean lifetime τ for that particular process, for example, $\Gamma_f = \hbar/\tau_f$. Here $\hbar$, the action quantum, is Planck's constant divided by 2π and is numerically equal to 1.0546×10^{-34} J s $= 0.66 \times 10^{-15}$ eV s. The fission width can also be defined in terms of the energy separation D of successive levels in the compound nucleus and the number of open channels in the fission transition nucleus (paths whereby the nucleus can cross the barrier on the way to fission), as given by expression (3), where I is the angular momentum and i is

$$\Gamma_f(I) = \frac{D(I)}{2\pi} \sum_i N_{fi} \qquad (3)$$

an index labeling the open channels N_{fi}. The contribution of each fission channel to the fission width depends upon the barrier transmission coefficient, which, for a two-humped barrier (see Fig. 2), is strongly energy-dependent. This results in an energy-dependent fission cross section which is very different from the total cross section shown in Fig. 3 for ^{240}Pu.

When the incoming neutron has low energy, the likelihood of reaction is substantial only when the energy of the neutron is such as to form the compound nucleus in one or another of its resonance levels (see Fig. 3b). The requisite sharpness of the "tuning" of the energy is specified by the total level width Γ. The nuclei ^{233}U, ^{235}U, and ^{239}Pu have a very large cross section to take up a slow neutron and undergo fission (see table) because both their absorption cross section and their probability for decay by fission are large. The probability for fission decay is high because the binding energy of the incident neutron is sufficient to raise the energy of the compound nucleus above the fission barrier. The very large, slow neutron fission cross sections of these isotopes make them important fissile materials in a chain reactor. *See* CHAIN REACTION.

Cross sections for neutrons of thermal energy to produce fission or undergo capture in the principal nuclear species, and neutron yields from these nuclei*

Nucleus	Cross section for fission, σ_f, 10^{-24} cm^2	σ_f plus cross section for radiative capture, σ_r	Ratio, $1+\alpha$	Number of neutrons released per fission, ν	Number of neutrons released per slow neutron captured, $\eta = \nu/(1+\alpha)$
^{233}U	525 ± 2	573 ± 2	1.093 ± 0.003	2.50 ± 0.01	2.29 ± 0.01
^{235}U	577 ± 1	678 ± 2	1.175 ± 0.002	2.43 ± 0.01	2.08 ± 0.01
^{239}Pu	741 ± 4	1015 ± 4	1.370 ± 0.006	2.89 ± 0.01	2.12 ± 0.01
^{238}U	0	2.73 ± 0.04			0
Natural uranium	4.2	7.6	1.83	2.43 ± 0.01	1.33

*Data from *Brookhaven National Laboratory 325*, 2d ed., suppl. no. 2, vol. 3, 1965. The data presented are the recommended or least-squares values published in this reference for 0.0253-eV neutrons. All cross sections are in units of barns (1 barn $= 10^{-24}$ cm$^2 = 10^{-28}$ m^2).

Scission. The scission configuration is defined in terms of the properties of the intermediate nucleus just prior to division into two fragments. In heavy nuclei the scission deformation is much larger than the saddle deformation at the barrier, and it is important to consider the dynamics of the descent from saddle to scission. One of the important questions in the passage from saddle to scission is the extent to which this process is adiabatic with respect to the particle degrees of freedom. As the nuclear shape changes, it is of interest to investigators to know the probability for the nucleons to remain in the lowest-energy orbitals. If the collective motion toward scission is very slow, the single-particle degrees of freedom continually readjust to each new deformation as the distortion proceeds. In this case, the adiabatic model is a good approximation, and the decrease in potential energy from saddle to scission appears in collective degrees of freedom at scission, primarily as kinetic energy associated with the relative motion of the nascent fragments.

On the other hand, if the collective motion between saddle and scission is so rapid that equilibrium is not attained, there will be a transfer of collective energy into nucleonic excitation energy. Such a nonadiabatic model, in which collective energy is transferred to single-particle degrees of freedom during the descent from saddle to scission, is usually referred to as the statistical theory of fission. *See* PERTURBATION (QUANTUM MECHANICS).

The experimental evidence indicates that the saddle to scission time is somewhat intermediate between these two extreme models. The dynamic descent of a heavy nucleus from saddle to scission depends upon the nuclear viscosity. A viscous nucleus is expected to have a smaller translational kinetic energy at scission and a more elongated scission configuration. Experimentally, the final translational kinetic energy of the fragments at infinity, which is related to the scission shape, is measured. Hence, in principle, it is possible to es-

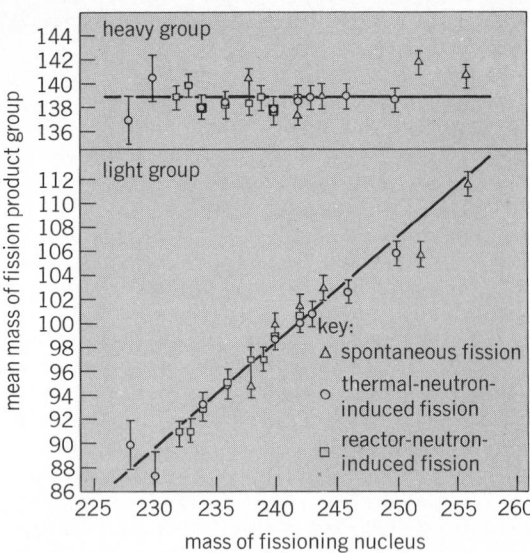

Fig. 5. Average masses of the light- and heavy-fission product groups as a function of the masses of the fissioning nucleus. Energy spectrum of reactor neutrons is that associated with fission. (*From K. F. Flynn et al., Distribution of mass in the spontaneous fission of ^{256}Fm, Phys. Rev., 5C:1725–1729, 1972*)

timate the nuclear viscosity coefficient by comparing the calculated dependence upon viscosity of fission-fragment kinetic energies with experimental values. The viscosity of nuclei is an important nuclear parameter which also plays an important role in collisions of very heavy ions.

The mass distribution from the fission of heavy nuclei is predominantly asymmetric. For example, division into two fragments of equal mass is about 600 times less probable than division into the most probable choice of fragments when ^{235}U is irradiated with thermal neutrons. When the energy of the neutrons is increased, symmetric fission (Fig. 4) becomes more probable. In general, heavy nuclei fission asymmetrically to give a heavy fragment of approximately constant mean mass number 139 and a corresponding variable-mass light fragment (see Fig. 5). These experimental results have been difficult to explain theoretically. Calculations of potential-energy surfaces show that the second barrier (B in Fig. 2) is reduced in energy by up to 2 or 3 MeV, if octuple deformations (pear shapes) are included. Hence, the theoretical calculations show that mass asymmetry is favored at the outer barrier, although direct experimental evidence supporting the asymmetric shape of the second barrier is very limited. It is not known whether the mass asymmetric energy valley extends from the saddle to scission; and the effect of dynamics on mass asymmetry in the descent from saddle to scission has not been determined. Experimentally, as the mass of the fissioning nucleus approaches $A \approx 260$, the mass distribution approaches symmetry. This result is qualitatively in agreement with theory.

A nucleus at the scission configuration is highly elongated and has considerable deformation energy. The influence of nuclear shells on the scission shape introduces structure into the kinetic energy and neutron-emission yield as a function of fragment mass. The experimental kinetic energies for

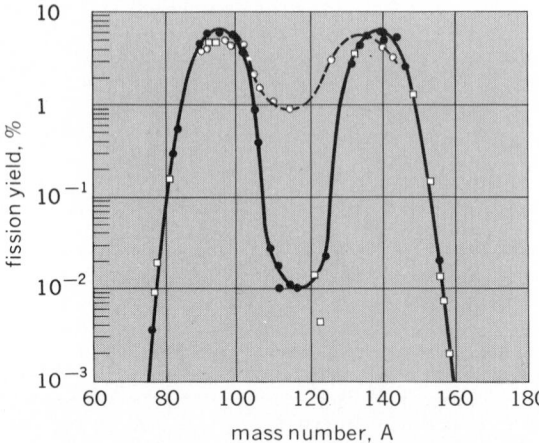

Fig. 4. Mass distribution of fission fragments formed by neutron-induced fission of $^{235}U + n = {}^{236}U$ when neutrons have thermal energy, smooth curve (*Plutonium Project Report, Rev. Mod. Phys., 18:539, 1964*), and 14-MeV energy, dashed curve (*based on R. W. Spence, Brookhaven National Laboratory, AEC-BNL (C-9), 1949*). Quantity plotted is 100 × (number of fission decay chains formed with given mass)/(number of fissions).

the neutron-induced fission of ^{233}U, ^{235}U, and ^{239}Pu have a pronounced dip as symmetry is approached, as shown in Fig. 6. (This dip is slightly exaggerated in the figure because the data have not been corrected for fission fragment scattering.) The variation in the neutron yield as a function of fragment mass for these same nuclei (Fig. 7) has a "saw-toothed" shape which is asymmetric about the mass of the symmetric fission fragment. Both these phenomena are reasonably well accounted for by the inclusion of closed-shell structure into the scission configuration.

A number of light charged particles (for example, isotopes of hydrogen, helium, and lithium) have been observed to occur, with low probability, in fission. These particles are believed to be emitted very near the time of scission. Available evidence also indicates that neutrons are emitted at or near scission with considerable frequency.

Postscission phenomena. After the fragments are separated at scission, they are further accelerated as the result of the large Coulomb repulsion. The initially deformed fragments collapse to their equilibrium shapes, and the excited primary fragments lose energy by evaporating neutrons. After neutron emission, the fragments lose the remainder of their energy by gamma radiation, with a lifetime of about 10^{-11} s. The kinetic energy and neutron yield as a function of mass are shown in Figs. 6 and 7. The variation of neutron yield with fragment mass is directly related to the fragment excitation energy. Minimum neutron yields are observed for nuclei near closed shells because of the resistance to deformation of nuclei with closed shells. Maximum neutron yields occur for fragments that are "soft" toward nuclear deformation. Hence, at the scission configuration, the fraction of the deformation energy stored in each fragment depends on the shell structure of the individual fragments. After scission, this deformation energy is converted to excitation energy, and, hence, the neutron yield is directly correlated with the frag-

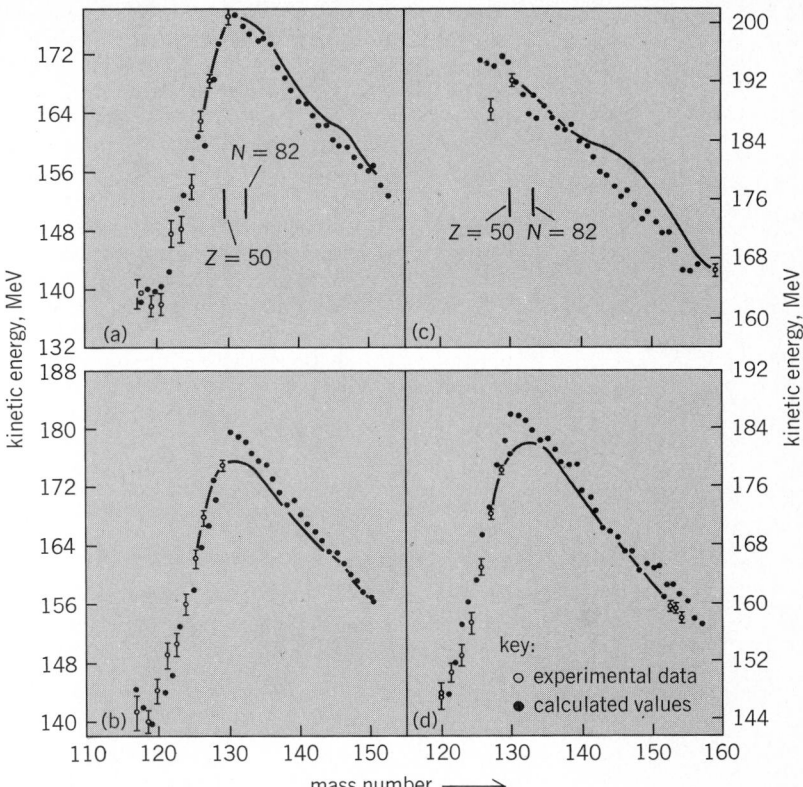

Fig. 6. Average total kinetic energy of fission fragments as a function of heavy fragment mass for fission of (a) ^{235}U, (b) ^{233}U, (c) ^{252}Cf, and (d) ^{239}Pu. Curves indicate experimental data. (*From J. C. D. Milton and J. S. Fraser, Time-of-flight fission studies on ^{233}U, ^{235}U and ^{239}Pu, Can. J. Phys., 40:1626–1663, 1962*)

ment shell structure. This conclusion is further supported by the correlation between the neutron yield and the final kinetic energy. Closed shells result in a larger Coulomb energy at scission for fragments that have a smaller deformation energy and a smaller number of evaporated neutrons.

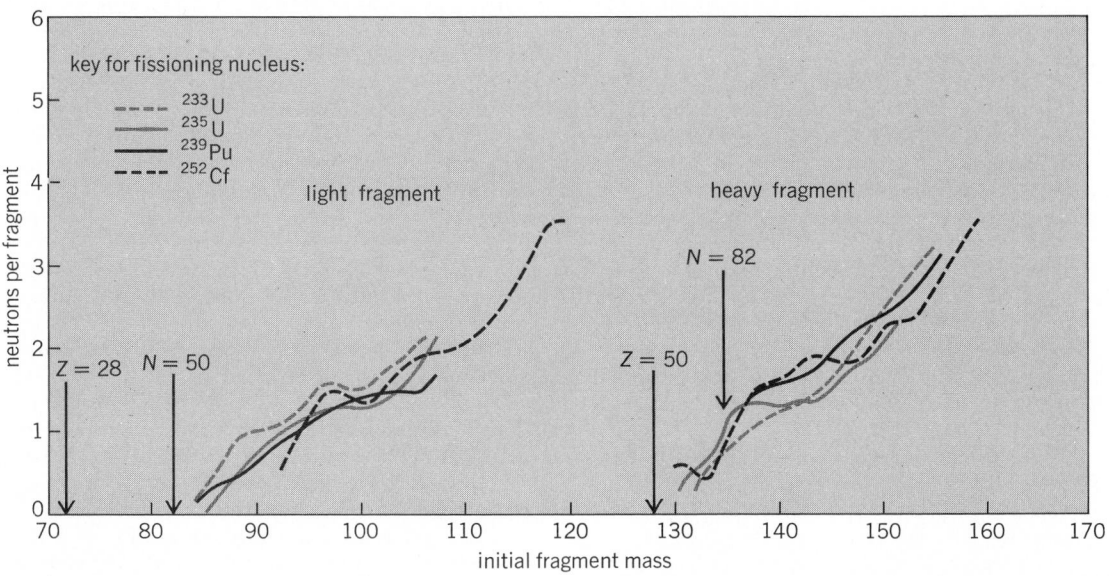

Fig. 7. Neutron yields as a function of fragment mass for four types of fission as determined from mass-yield data. Approximate initial fragment masses corresponding to various neutron and proton "magic numbers" N and Z are indicated. (*From J. Terrell, Neutron yields from individual fission fragments, Phys. Rev., 127:880–904, 1962*)

After the emission of the prompt neutrons and gamma rays, the resulting fission products are unstable against β-decay. For example, in the case of thermal neutron fission of ^{235}U, each fragment undergoes on the average about three β-decays before it settles down to a stable nucleus. For selected fission products (for example, ^{87}Br and ^{137}I) β-decay leaves the daughter nucleus with excitation energy exceeding its neutron binding energy. The resulting delayed neutrons amount, for thermal neutron fission of ^{235}U, to about 0.7% of all the neutrons given off in fission. Though small in number, they are quite important in stabilizing nuclear chain reactions against sudden minor fluctuations in reactivity. *See* NEUTRON.

[JOHN R. HUIZENGA]

Bibliography: *Proceedings of the 3d IAEA Symposium on Physics and Chemistry of Fission*, Rochester, NY, 1973; R. Vandenbosch and J. R. Huizenga, *Nuclear Fission*, 1973.

Nuclear fusion

One of the primary nuclear reactions, the name usually designating an energy-releasing rearrangement collision which can occur between various isotopes of low atomic number. *See* NUCLEAR REACTION.

Interest in the nuclear fusion reaction arises from the expectation that it may someday be used to produce useful power, from its role in energy generation in stars, and from its use in the fusion bomb. Since a primary fusion fuel, deuterium, occurs naturally and is therefore obtainable in virtually inexhaustible supply (by separation of heavy hydrogen from water, 1 atom of deuterium occurring per 6000 atoms of hydrogen), solution of the fusion power problem would permanently solve the problem of the present rapid depletion of chemically valuable fossil fuels. As a power source, the lack of radioactive waste products from the fusion reaction is another argument in its favor as opposed to the fission of uranium.

In a nuclear fusion reaction the close collision of two energy-rich nuclei results in a mutual rearrangement of their nucleons (protons and neutrons) to produce two or more reaction products, together with a release of energy. The energy usually appears in the form of kinetic energy of the reaction products, although when energetically allowed, part may be taken up as energy of an excited state of a product nucleus. In contrast to neutron-produced nuclear reactions, colliding nuclei, because they are positively charged, require a substantial initial relative kinetic energy to overcome their mutual electrostatic repulsion so that reaction can occur. This required relative energy increases with the nuclear charge Z, so that reactions between low-Z nuclei are the easiest to produce. The best known of these are the reactions between the heavy isotopes of hydrogen, deuterium and tritium.

Fusion reactions were discovered in the 1920s when low-Z elements were used as targets and bombarded by beams of energetic protons or deuterons. But the nuclear energy released in such bombardments is always microscopic compared with the energy of the impinging beam. This is because most of the energy of the beam particle is dissipated uselessly by ionization and single-particle collisions in the target; only a small fraction of the impinging particles actually produce reactions.

Nuclear fusion reactions can be self-sustaining, however, if they are carried out at a very high temperature. That is to say, if the fusion fuel exists in the form of a very hot ionized gas of stripped nuclei and free electrons termed a plasma, the agitation energy of the nuclei can overcome their mutual repulsion, causing reactions to occur. This is the mechanism of energy generation in the stars and in the fusion bomb. It is also the method envisaged for the controlled generation of fusion energy. *See* PLASMA PHYSICS.

PROPERTIES OF FUSION REACTIONS

The cross sections (effective collisional areas) for many of the simple nuclear fusion reactions have been measured with high precision. It is found that the cross sections generally show broad maxima as a function of energy and have peak values in the general range of 0.01 barn (1 barn = 10^{-24} cm^2) to a maximum value of 5 barns, for the deuterium-tritium (D-T) reaction. The energy releases of these reactions can be readily calculated from the mass differences between the initial and final nuclei or determined by direct measurement.

Simple reactions. Some of the important simple fusion reactions, their reaction products, and their energy releases in millions of electronvolts (MeV) are given by reactions (1).

$$\begin{array}{llll}
D & +D \rightarrow & He^3 + n + & 3.25 \text{ MeV} \\
D & +D \rightarrow & T + p + & 4.0 \text{ MeV} \\
T & +D \rightarrow & He^4 + n + & 17.6 \text{ MeV} \\
He^3 & +D \rightarrow & He^4 + p + & 18.3 \text{ MeV} \\
Li^6 & +D \rightarrow & 2He^4 & +22.4 \text{ MeV} \\
Li^7 & +p \rightarrow & 2He^4 & +17.3 \text{ MeV}
\end{array} \quad (1)$$

If it is remembered that the energy release in the chemical reaction in which hydrogen and oxygen combine to produce a water molecule is about 1 eV per reaction, it will be seen that, gram for gram, fusion fuel releases more than 1,000,000 times as much energy as typical chemical fuels.

The two alternative D-D reactions listed occur with about equal probability for the same relative particle energies. Note that the heavy reaction products, tritium and helium-3, may also react, with the release of a large amount of energy. Thus it is possible to visualize a reaction chain in which six deuterons are converted to two helium-4 nuclei, two protons, and two neutrons, with an overall energy release of 43 MeV—about 10^5 kilowatt-hours (kWh) of energy per gram of deuterium. This energy release is several times that released per gram in the fission of uranium, and several million times that released per gram by the combustion of gasoline.

Cross sections. Figure 1 shows the measured values of cross sections as a function of bombarding energy up to 100 keV for the total D-D reaction (both D-D,n and D-D,p), the D-T reaction, and the D-He^3 reaction. The most striking feature of these curves is their extremely rapid falloff with energy as bombarding energies drop to a few kilovolts. This effect arises from the mutual electrostatic repulsion of the nuclei, which prevents them from approaching closely if their relative energy is small. *See* NUCLEAR STRUCTURE.

The fact that reactions can occur at all at these energies is attributable to the finite range of nucle-

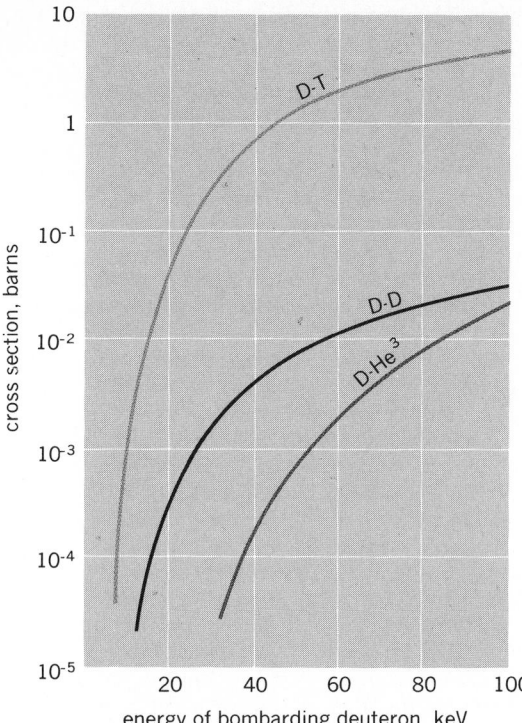

Fig. 1. Cross sections versus bombarding energy for three simple fusion reactions. *(From R. F. Post, Fusion power, Sci. Amer., 197(6):73–84, December 1957, copyright © 1957 by Scientific American, Inc.; all rights reserved)*

ar interaction forces. In effect, the boundary of the nucleus is not precisely defined by its classical diameter. The role of quantum mechanical effects in nuclear fusion reactions has been treated by G. Gamow and others. It is predicted that the cross sections should obey an exponential law at low energies. This is well borne out in energy regions reasonably far removed from resonances (for example, below about 30 keV for the D-T reaction). Over a wide energy range at low energies, the data for the D-D reaction can be accurately fitted by a Gamow curve, the result for the cross section being given by Eq. (2), where the bombarding energy W is in kiloelectronvolts.

$$\sigma_{\text{D-D}} = \frac{288}{W} e^{-45.8W^{-1/2}} \times 10^{-24} \text{ cm}^2 \quad (2)$$

The extreme energy dependence of this expression can be appreciated by the fact that, between 1 and 10 keV, the predicted cross section varies by about 13 powers of 10, that is, from 3×10^{-42} to 1.5×10^{-29} cm^2.

Energy division. The kinematics of the fusion reaction stipulates that the reaction can occur only if two or more reaction products result. This is because both mass energy and momentum balance must be preserved. When there are only two reaction products (which is the case in all of the important reactions), the division of energy between the reaction products is uniquely determined, the lion's share always going to the lighter particle. The energy division (disregarding the initial bombarding energy) is as in reaction (3). If reaction (3)

$$A_1 + A_2 \rightarrow A'_1 + A'_2 + Q \quad (3)$$

holds, with the As representing the atomic masses of the particles and Q the total energy released, then Eqs. (4) are valid, where $W(A'_1)$ and $W(A'_2)$ are the kinetic energies of the reaction products.

$$W(A'_1) + W(A'_2) = Q$$

$$W(A'_1) = Q\left(\frac{A'_2}{A'_1 + A'_2}\right) \quad (4)$$

$$W(A'_2) = Q\left(\frac{A'_1}{A'_1 + A'_2}\right)$$

Thus in the D-T reaction, for example, A'_1, the mass of the α-particle, is four times A'_2, the mass of the neutron, so that the neutron carries off four-fifths of the reaction energy, or 14 MeV.

Reaction rates. When nuclear fusion reactions occur in a high-temperature plasma, the reaction rate per unit volume depends on the particle density n of the reacting fuel particles and on an average of their mutual reaction cross sections σ and relative velocity v over the particle velocity distributions. *See* THERMONUCLEAR REACTION.

For dissimilar reacting nuclei (such as D and T), the reaction rate is given by Eq. (5).

$$R_{12} = n_1 n_2 \langle \sigma v \rangle_{12} \quad \text{reactions}/(\text{cm}^3)(\text{s}) \quad (5)$$

For similar reacting nuclei (for example, D and D), the reaction rate is given by Eq. (6).

$$R_{11} = \frac{1}{2} n^2 \langle \sigma v \rangle \quad (6)$$

Note that both expressions vary as the square of the total particle density (for a given fuel composition).

If the particle velocity distributions are known, $\langle \sigma v \rangle$ can be determined as a function of energy by numerical integration, using the known reaction cross sections. It is customary to assume a maxwellian particle velocity distribution, toward which all others tend in equilibrium. The values of $\langle \sigma v \rangle$ for the D-D and D-T reactions are shown in Fig. 2. In this plot the kinetic temperature is given in units of kiloelectronvolts; 1 keV kinetic temperature $= 1.16 \times 10^7$ K. Just as in the case of the

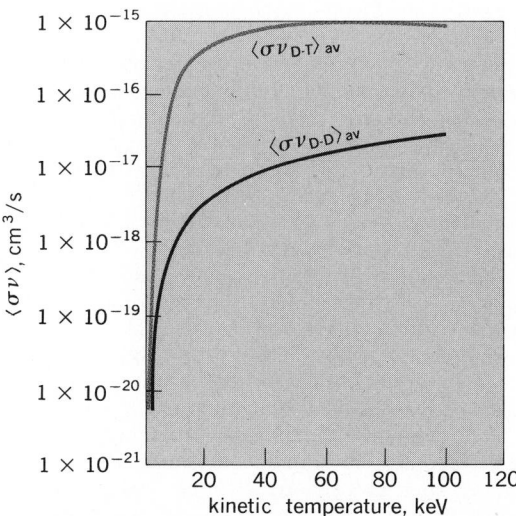

Fig. 2. Plot of the values of $\langle \sigma v \rangle$ versus kinetic temperature for the D-D and D-T reactions.

cross sections themselves, the most striking feature of these curves is their extremely rapid falloff with temperature at low temperatures. For example, although at 100 keV for all reactions $\langle \sigma v \rangle$ is only weakly dependent on temperature, at 1 keV it varies as $T^{6.3}$ and at 0.1 keV as T^{133}! Also, at the lowest temperatures it can be shown that only the particles in the "tail" of the distribution, which have energies large compared with the average, will make appreciable contributions to the reaction rate, the energy dependence of σ being so extreme.

Critical temperatures. The nuclear fusion reaction can obviously be self-sustaining only if the rate of loss of energy from the reacting fuel is not greater than the rate of energy generation by fusion reactions. The simplest consequence of this fact is that there will exist critical or ideal ignition temperatures below which a reaction could not sustain itself, even under idealized conditions. In a fusion reactor, ideal or minimum critical temperatures are determined by the unavoidable escape of radiation from the plasma. A minimum value for the radiation emitted from any plasma is that emitted by a pure hydrogenic plasma in the form of x-rays or bremsstrahlung. Thus plasmas composed only of isotopes of hydrogen and their one-for-one accompanying electrons might be expected to possess the lowest ideal ignition temperatures. This is indeed the case: It can be shown by comparison of the nuclear energy release rates with the radiation losses that the critical temperature for the D-T reaction is about 4×10^7 K. For the D-D reaction it is about 10 times higher. Since both radiation rate and nuclear power vary with the square of the particle density, these critical temperatures are independent of density over the density ranges of interest. The concept of the critical temperature is a highly idealized one, however, since in any real cases additional losses must be expected to occur which will modify the situation, increasing the required temperature.

FUSION REACTOR

Intense interest in nuclear fusion arises from its promise as a safe and inexhaustible source of energy for the future. Fusion reactors do not yet exist, but studies of the physics and technology that will be needed to construct such reactors have been underway since the 1950s.

The two key problems in achieving net power from a fusion reactor are, first, to heat the fusion fuel charge to its required high temperature, and second, to confine the heated fuel for a long enough time for the fusion energy released to exceed the energy required to heat the fuel to its conbustion temperature, including all relevant losses. *See* LAWSON CRITERION.

The problem of achieving fusion power is in fact dominated by the quantitative requirements associated with the fusion process. The plasma heating technique employed must be capable of raising the fusion fuel charge to kinetic temperatures of order 100,000,000° or higher. The confinement system must be capable of satisfying stringent requirements on confinement time (which could be as long as seconds in some cases). At the same time it must be capable of sustaining the strong outward gas pressure exerted by the fuel charge. Furthermore, since the rate at which

fusion power is generated varies as the square of the fuel density, for any continuously operating fusion reactor engineering limits on heat transfer must be taken into account, thus limiting the fuel density for such systems to a small fraction of the particle density of atmospheric air. At higher density it is only possible to conceive of pulsed operation—basically microexplosions. The quantitative requirements of fusion therefore strongly limit the possible approaches to fusion.

Two generically different approaches have emerged as constituting the most promising avenues to the eventual achievement of net fusion power, namely, magnetic confinement and pellet fusion.

Magnetic confinement relies on the fact that at fusion temperatures the fusion fuel charge will be completely ionized, that is, it will exist solely in the plasma state. Charged particles can be held trapped by a properly shaped magnetic field, and are thereby isolated from the reactor chamber walls, physical contact with which would instantly cool the plasma.

Pellet fusion aims at the same objective, but by an entirely different route. Here the idea to rapidly heat and compress a tiny fuel pellet, carrying out the entire operation so quickly that fusion can take place before the pellet flies apart—that is, the confinement is properly called "inertial." The major technical effort on pellet fusion is centered on the use of high-powered lasers to accomplish the heating and compression; substantial activity is also being devoted to pellet fusion induced by bombardment of the pellet with very-high-intensity electron beams; and use of heavy ions as the ignition probe is now receiving serious study.

Magnetic confinement. Magnetic confinement of a fusion plasma depends on the nature of the plasma state. The plasma may be viewed as an electrically conducting gas that exerts an outward pressure, or as a collection of free positive and negative charges. The pressure exerted by the plasma can be resisted by the electromagnetic stresses associated with a strong magnetic field;

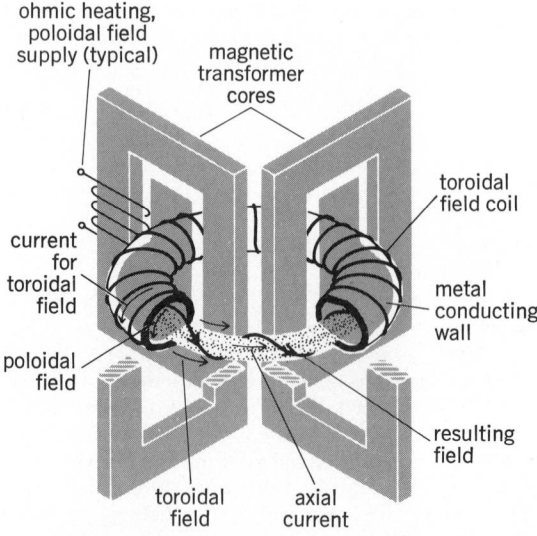

Fig. 3. Schematic illustration of a tokamak device. *(From R. F. Post and F. L. Ribe, Fusion reactors as future energy sources, Science, 186:397–407, 1974)*

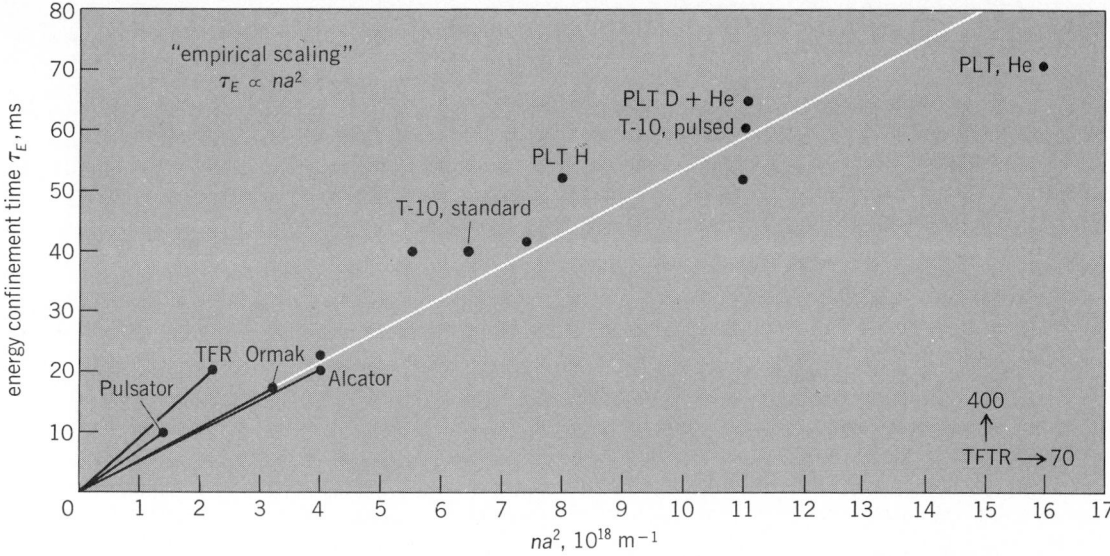

Fig. 4. Plot of "empirical scaling law" results for several tokamak devices, including results projected for the large tokamak fusion test reactor (TFTR).

the individual charged particles can at the same time be guided by a properly shaped magnetic field that forces these particles to execute orbits that remain within the vacuum chamber surrounding the plasma without contact with the walls.

Adequate stability of the confined plasma is a prime requirement for effective magnetic confinement; otherwise, particles can escape prematurely, before having a sufficient probability to fuse. Thus, finding means for suppressing the inherent tendency for confined plasma to become unstable has been one of the central goals of nuclear fusion research since its inception.

There have been many types of magnetic confinement systems proposed since the inception of fusion research. Three generic types appear to have the most promise: the tokamak, the mirror and tandem mirror systems, and field-reversed systems. *See* WAVES AND INSTABILITIES IN PLASMAS.

Tokamak. The tokamak (Fig. 3) is a closed or toroidal (doughnut-shaped) confinement system. It uses confining fields that represent a combination of a strong toroidal field (that is, field lines directed the long way around the toroid), with a weaker poloidal field (field lines circling the short way around the torus). The toroidal field is generated by external coils that encircle the chamber; the poloidal field is generated by a strong toroidal electric current induced to flow in the plasma by transformer action. The field line pattern that results is helical. In the tokamak, the circulating current not only provides the main confining force through the generation of the poloidal magnetic field but also performs the important function of initial heating (ohmic heating) of the plasma.

The strong toroidal field has the main function of stabilizing the plasma against "kinking" magnetohydrodynamic instability modes. The necessity of having a strong toroidal external field has the disadvantage of limiting the beta value of the tokamak to a few percent. Beta, in magnetic fusion, is the ratio of the energy density of the plasma to that of the applied field. High beta is desirable from an economic standpoint to maximize the utilization of

the externally generated field, thus minimizing the capital cost of the magnet system relative to the fusion power output.

As a closed device the tokamak has the important advantage that its plasma confinement time increases as the square of the radius, a, of the plasma column (actually, empirically, as na^2, where n is the plasma density; Fig. 4). This property is a consequence of the fact that as long as gross stability is maintained, plasma particles can be lost only by diffusion across the field, and such losses proceed on a time scale which increases with the square of the characteristic distances involved. Thus, adequate confinement times can always be achieved by scaling up the dimensions. Though large (meters), the plasma diameters for tokamak plasmas as projected from present experiments appear to be acceptable for practical fusion power plants.

Mirror machine and tandem mirror. The mirror machine (Fig. 5) is an open-ended system in which a hot plasma is held trapped by repeatedly reflecting its particles between magnetic mirrors (regions of intensified magnetic field at each end of the confinement chamber).

An important property of magnetic mirror systems is that their fields can be shaped so as to create a magnetic well, that is, a confining field that has a nonzero minimum surrounded by closed contours of increasing magnetic intensity. One type of magnetic well system is the baseball coil (Fig. 6). When confined in a magnetic well, a plasma is restrained from exhibiting any form of gross instability, up to plasma pressures comparable to those of the confining field, that is, up to beta values of order unity.

Mirror systems are in contrast with the tokamak or other closed systems. In the latter there is no need to confine particles as far as their motion along the field lines is concerned. In a mirror machine the required longitudinal confinement is provided by the repelling force exerted on charged particles as they spiral along field lines that are converging, that is, as they move toward regions of

NUCLEAR FUSION

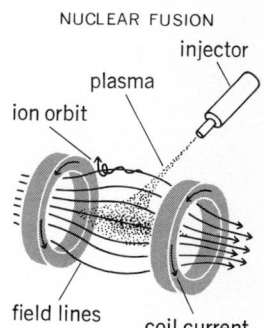

Fig. 5. Schematic illustration of mirror machine using simple mirrors. (*From R. F. Post and F. L. Ribe, Fusion reactors as future energy sources, Science, 186: 397–407, 1974*)

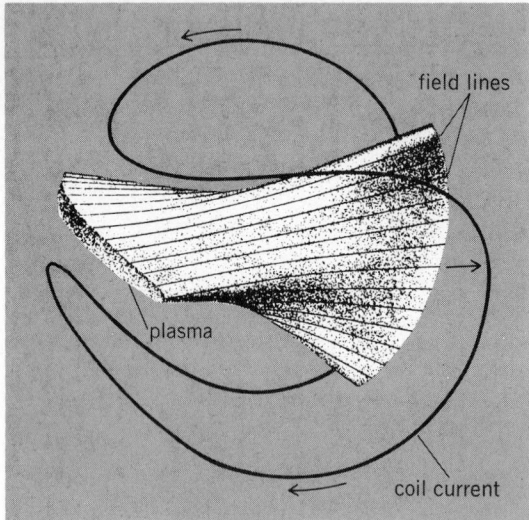

Fig. 6. "Baseball" coil configuration, producing a magnetic well mirror field.

increasing field strength. Particles which spiral with sufficiently steep helical pitch angles will be repelled strongly enough to be reflected, that is, trapped, by the mirrors. It follows that this type of mirror system cannot confine an isotropic plasma; only particles whose pitch angles are sufficiently large to lie between the loss cones defined by the strength of the mirror fields relative to the field intensity between the mirrors will be contained.

These systems must therefore rely on the injection of intense beams of energetic neutral atoms to maintain the plasma temperature and density in competition with particle leakage through the mirrors.

An important disadvantage of conventional mirror systems for fusion purposes is that the leakage of particles through the mirrors, arising as it does from collisions that deflect the ions into pitch angles lying within the mirror loss cone, occurs at a rate comparable to the rate of fusion reactions. In this circumstance the energy gain factor Q — the ratio of fusion power to plasma heating power — is at best not much larger than 1, implying an economically unacceptably large fraction of recirculated power needed to maintain the plasma. This deficiency in the mirror concept has stimulated the development of the tandem mirror and the field-reversed mirror concepts, discussed below.

An important aspect of mirror confinement is the positive ambipolar potential of the plasma that arises naturally from its operation. Since the collision frequency for electrons is higher than that for ions (because of the electrons' higher velocities), other factors being equal, they would tend to diffuse into the loss cone and be lost much more rapidly than would the ions. But any such differential loss rate would result in the buildup of a net positive charge in the confined plasma, thus driving it to a positive potential with respect to its surroundings, a potential sufficient to bring the electron loss rate to equality with that of the ions.

In the tandem mirror (Figs. 7 and 8) the ambipolar potentials are used to confine a fusion plasma, resulting in greatly improved confinement relative to that in a single mirror cell. A large-volume central chamber, in which the fusion plasma is to be

confined, is stoppered at each end by two small-volume mirror cells having a high-ion-temperature plasma. The ambipolar potential of each of these plugs, being relatively more positive than that of the central cell, serves to confine (axially) the ions of the central cell plasma; its electrons are confined by the overall positive potential of the system.

Field-reversal systems. Magnetically confined plasmas exhibit the property of diamagnetism. That is, their confinement necessarily involves the existence of internal electric currents that act to reduce the strength of the confining magnetic field within the plasma relative to the vacuum value it would have in the absence of the plasma. It is in fact these internal diamagnetic currents that, by interacting with the externally generated field, produce a body force balancing the outward-acting pressure of the plasma. If these diamagnetic currents could be employed to provide the major part of the confining force, the confining effect of the external field would be greatly enhanced, with consequent economic and other practical benefits.

The earliest attempted embodiment of this concept was the toroidal pinch, the progenitor of the tokamak, which was unsuccessful because its self-constricted plasma column was subject to rapidly growing kinking instabilities of hydromagnetic origin. However, with increased understanding of magnetic confinement, ways to circumvent such problems in pinch-confined plasmas have been proposed. These ideas, generically describable as field-reversed systems, have taken four distinguishable forms: field-reversing particle rings; the

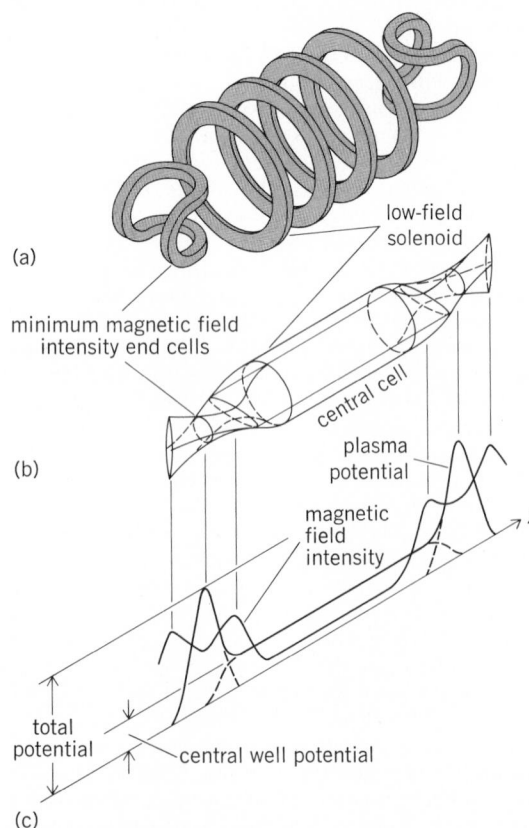

Fig. 7. Tandem mirror system. (*a*) Coils. (*b*) Configuration. (*c*) Variations of magnetic field intensity and plasma potential.

Fig. 8. Fusion power plant based on tandem mirror concept. (*Lawrence Livermore Laboratory*)

field-reversed mirror; the field-reversed pinch; and self-field tokamaks. *See* PINCH EFFECT.

1. Field-reversing particle rings. These forms employ a ring of high-energy charged particles. The particles circle in large orbits in an external mirror field and generate a diamagnetic current that is sufficiently strong to reverse the direction of the field within the ring, thereby producing a poloidal field with closed field lines that traps the fusion plasma.

2. Field-reversed mirror. This is a mirror system where tangentially injected neutral beams maintain ion diamagnetic currents sufficient to create a field-reversed state, thus inhibiting the rate of escape of plasma ions, which must now diffuse across the field-reversed region before they reach the open field lines of the mirror field (Fig. 9).

3. Field-reversed pinch. This is a toroidal pinch in which the pinch (poloidal) field has superposed on it a toroidal field, the direction of which is reversed in the deep interior of the plasma, relative to its direction in the exterior parts.

4. Self-field tokamaks. These are tokamaklike configurations in which the main confining fields, both poloidal and toroidal, are the result of diamagnetic currents flowing within the plasma. An example is the spheromak configuration, in which the plasma and the conducting chamber around it (needed to preserve stability) are roughly spherical.

All of the field-reversed magnetic confinement systems described above have as their objective the more efficient use of magnetic field for fusion confinement purposes. Particularly as exemplified by the field-reversed mirror, the plasma beta value, as determined relative to the externally generated magnetic field, is very high. This could result in compact fusion power systems with improved

economics, that is, lowered capital costs for the magnet and its support structures. For example, calculations for the field-reversed mirror indicate a fusion power density with the D-T reaction of order 500 times that for conventional tokamak. Alternatively, the increased magnetic efficiency could in principle allow the use of so-called advanced fusion fuel cycles, involving the D-^{3}He or other reactions, having special advantages in terms of reduced neutron fluxes or suitability for direct energy conversion. The basic physics understanding is, however, at present too limited to assure the practical success of any of the above systems, although there are indications that all of them can exhibit stable confinement.

Progress. In theory, magnetic confinement has been long perceived as an almost ideal solution to the fusion problem. In practice, over the earlier years of fusion research it was found very difficult to achieve confinement that approached the optimistic theoretical predictions, owing to the ubiquitous problem of plasma instabilities. However, refinement of the theory, coupled with experiments performed with technologically sophisticated, scaled-up apparatus, has led to major advances toward the quantitative goals of fusion power. Plasma temperatures and density−confinement time products (discussed below) have both been brought within a factor of 10 or less of breakeven values, and the next generation of tokamaks will probably come close to, or actually achieve, an energy breakeven situation (fusion energy yield equaling heat energy content of the confined plasma).

Increasingly, the emphasis in fusion research is on problems of practical or economic origin. Scientifically, this means a search for magnetic config-

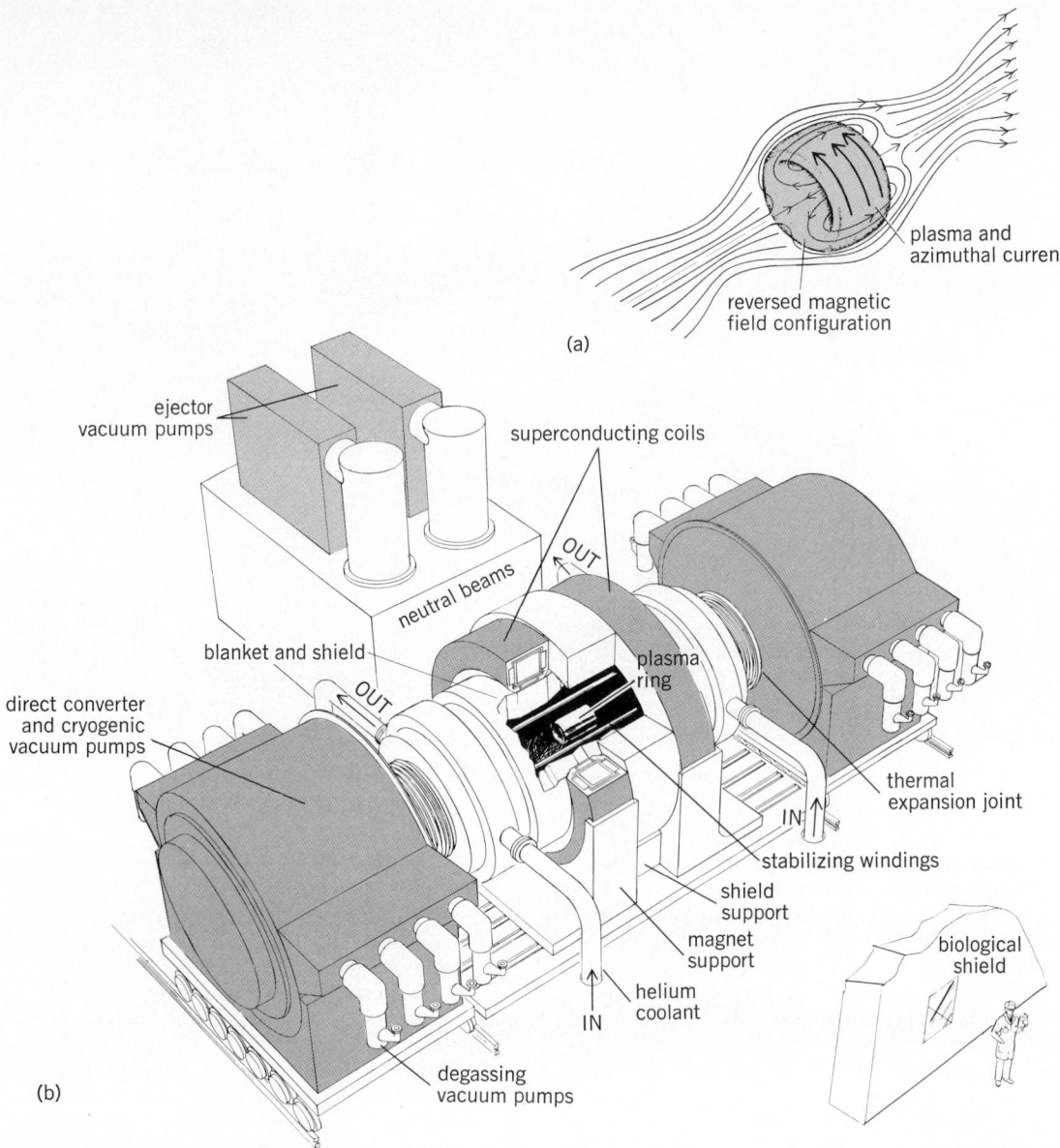

Fig. 9. Field-reversed mirror system. (*a*) Configuration, showing field lines and diamagnetic currents. (*b*) 20- MW demonstration power plant. (*Lawrence Livermore Laboratory*)

Table 1. Plasma parameters attained by magnetic confinement experiments compared with values believed needed for a fusion power plant

Approach	Device	Location	Plasma diameter (d), cm	Particle density (n), cm^{-3}	Ion temperature (T_i), keV	Confinement time (τ), ms	Lawson product $(n\tau)$, s/cm^3	β_{max}
Tokamak	T-4 (1970)	Kurchatov Institute, Moscow	30	3×10^{13}	0.4	10	6×10^{11}	.0006
	TFR (1974)	Fontenay-Aux-rose, Paris	40	4×10^{13}	0.8	15	6×10^{11}	.0016
	Alcator A (1975)	Massachusetts Institute of Technology	18	6×10^{14}	0.8	20	1.2×10^{13}	.006
	PLT (1976)	Princeton University	90	6×10^{13}	1.0	50	3×10^{12}	.005
	PLT (neutral beams, 1978)	Princeton University	90	4×10^{13}	5.0	25	$\sim 10^{12}$	.013
Conventional mirror	PR-6 (1971)	Kurchatov Institute, Moscow	10	2×10^{12}	0.5	0.15	3×10^{8}	.0016
	Baseball II (1972)	Lawrence	20	2×10^{9}	2.0	1000	2×10^{9}	—
	2XIIB (1977)	Livermore	15	$\sim 10^{14}$	13.0	1.	$\sim 10^{11}$	>1.0
Tandem mirror	TMX (1979)	Laboratory	50	2×10^{13}	0.2	$\sim 5.$	$\sim 10^{11}$	$\sim .2$

	Ion temperature (T_i), keV	Lawson product $(n\tau)$, s/cm^3	β	Remarks
		Power plant requirements		
Tokamak	15	$>10^{14}$	0.1	~ 1000 MWe, ignited mode
Conventional mirror	150	$>10^{13}$	0.8	High recirculated power required
Tandem mirror	30	$>10^{14}$	0.5	~ 500 MWe, ignited mode

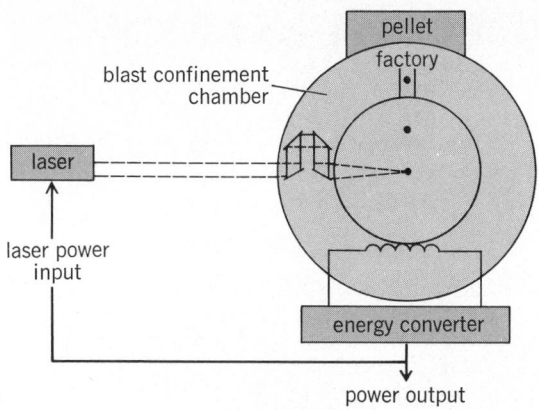

Fig. 10.   Conceptual laser pellet fusion power plant.

high-energy neutral atoms and charged particles (hydrogen and helium ions) escaping from the plasma.

A rough (and sometimes misleading) index of progress toward fusion requirements, mentioned above, is the $n\tau$ confinement factor used in the Lawson criterion, where n is the particle density and τ is the average confinement time of the plasma. This parameter and the plasma ion kinetic temperature T_i at which the confinement is achieved, together provide a useful index of progress toward the goal of fusion power. Table 1 compares values $n\tau$ and T_i achieved in experiments in magnetic confinement with nominal values estimated to be required for a fusion power plant.

Pellet fusion. The basic idea behind pellet fusion (Fig. 10) is the rapid implosion of a high-density fusion fuel pellet to produce a heated core that will fuse before it can fly apart. As usually conceived, the implosion would result from the rapid heating and subsequent ablation of the surface of the pellet, giving rise to an inward-acting reaction force that compresses the core. But for this process to yield a net energy, that is, for it to achieve the required $n\tau$ value, very large compression factors, of order 10,000, are required. That this should be the case can be seen from simple considerations: As matter is compressed spherically, its density (n in the $n\tau$ product) increases as the tube of the

urations that will result in smaller, more practical, or more efficient confinement geometries. Technologically, this means that more emphasis is being placed on the problems of high-field superconducting magnet coils, high-power neutral beams, vacuum technique, and materials problems associated with the inner wall of the confinement chamber, which must withstand high fluxes of energetic neutrons as well as relatively

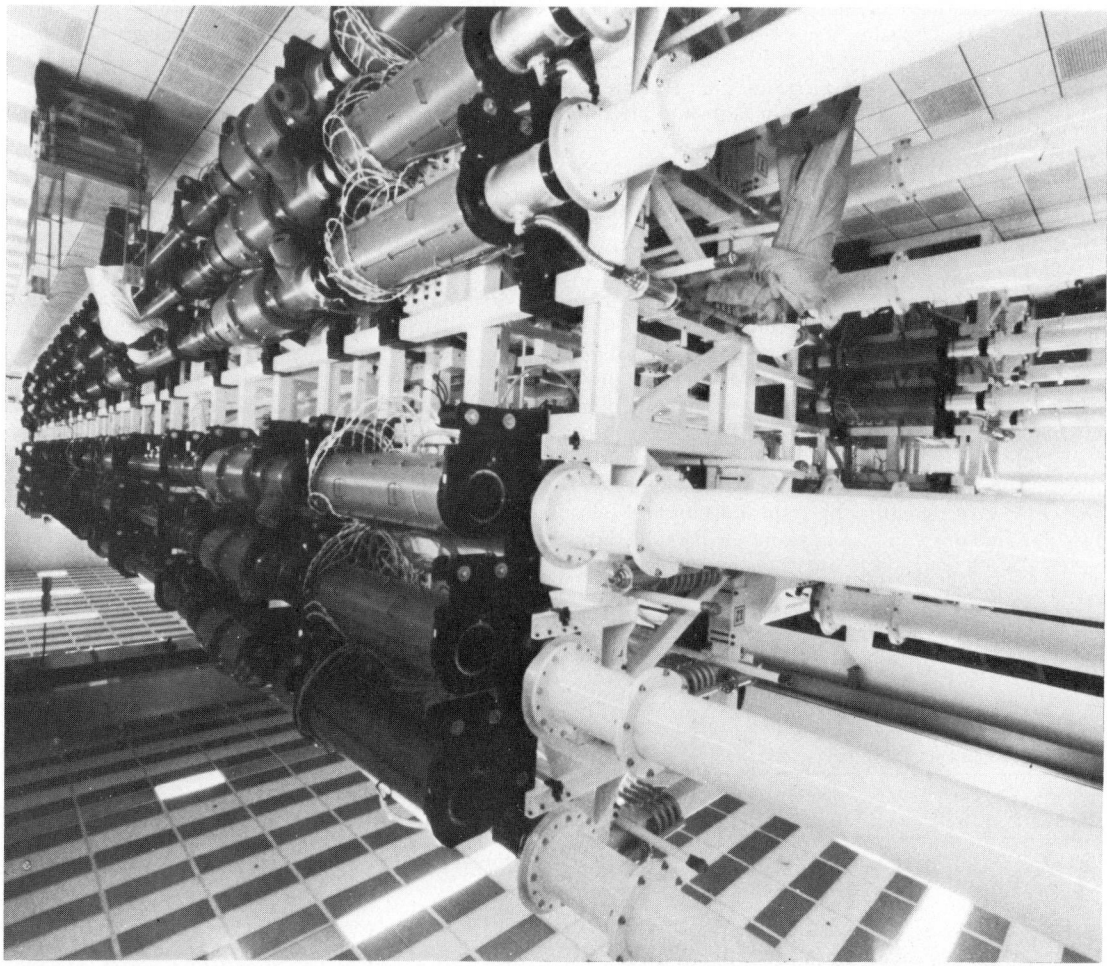

Fig. 11.   Portion of the Shiva laser pellet fusion facility at Lawrence Livermore Laboratory, showing 6 of the 20 laser amplifier trains, looking at the output end. System is designed to deliver more than 30 TW of optical power in less than 10^{-9} s. (*Lawrence Livermore Laboratory*)

Fig. 12. Design of Shiva Nova laser pellet fusion experimental facility at Lawrence Livermore Laboratory. System is designed to provide enough power (200–300 TW) and energy (200 kJ) to obtain significant gain from laser fusion targets. (*Lawrence Livermore Laboratory*)

radial compression factor. But the confinement time τ, here measured by the time of flight of an average particle out of the compressed core, decreased only as the first power of the radius. Thus $n\tau$ increases with the square of the radial compression factor. However, it is necessary to use tiny pellets to make this approach to fusion technically accessible from the standpoint of engineering limits on the amount of energy deliverable from the pellet drivers (lasers or particle accelerators), and on the amount of fusion energy released from the pellet that can be absorbed in surrounding structures. These limitations, taken together with the Lawson requirement, dictate the need for very large density compression factors, in turn leading to a requirement for very high compression forces [of order 10^{12} atmospheres (10^{17} Pa), ten times greater than those existing at the center of the Sun].

The problems thereby posed are twofold. First, very high pulsed powers of hundreds of terawatts must be focused down to millimeter dimensions and delivered in times of nanoseconds or less. The implied peak power densities are thus of order 10^{16} W/cm^2, made necessary by the requirement that an almost instantaneous ablation of the outer surface of the pellet should occur before heat can flow into the interior of the pellet; premature heat flow would prevent reaching the required compression factors. This problem can in principle be solved by using a sufficiently large array of high-power lasers, focusing their beams so as to uniformly illuminate the surface of the pellet (Figs. 11 and 12). Alternatively, converging beams of electrons or ions from particle accelerators would be employed.

Second, if the compression process is not carried out with high uniformity, it will go askew; small errors in uniformity can lead to a major reduction in the achievable compression. Sophisticated aiming and timing techniques must be used in the driver, and any instabilities must not be allowed to spoil the symmetry of the compression or lead to undue mixing or preheating.

High temperatures and substantial $n\tau$ values, comparable to those achieved in magnetic fusion, have been attained in pellet fusion experiments. Table 2 lists some of the parameters achieved in laser pellet experiments. Continued progress can be expected, as new and larger facilities come on line, but increase in performance by some orders of magnitude is needed before breakeven can be attained.

POWER PLANTS

Preliminary studies have been made of the forms that fusion power plants might take, following some of the approaches outlined above. These studies cannot of course be definitive, but they have helped to indicate the sizes, capital costs, and special engineering problems that are likely to characterize fusion power plants, insofar as they can now be visualized.

The types of power plants that have been studied encompass both pulsed and steady-state fusion systems, operating in either a driven mode or one in which plasma ignition would be achieved. A driven fusion power system is one in which the plasma temperature is maintained primarily by a continuous input of energy—for example, by the

Table 2. Examples of laser fusion facilities

Name	Location	Energy pulse, kJ	Peak power, TW	Compression achieved (× liquid density)
Zeta	University of Rochester	1.2	3–4	7–20
Chrome I	KMS Fusion, Inc., Ann Arbor, MI	1.0	2	7–35
Helios	Los Alamos Scientific Laboratory	5–10	10–20	8–30
Shiva	Lawrence Livermore Laboratory	10	26	30–160

injection of high-intensity beams of energetic neutral fuel atoms—thereby maintaining the required kinetic temperature and density of the plasma. Electrical power needed to produce the beams would be obtained by recirculating a portion of the electrical output of the plant. A positive power balance, that is, net output power ($Q > 1$), would be possible here only when the recirculated power is less than the electrical output as recovered from the plasma, including not only the energy content of the reaction products but that of the unreacted part (electrons and heated fuel ions); economical power would likely be possible only if the recirculated power fraction were to be made much smaller than the power produced (that is, Q much greater than 1). By contrast, a fusion system operated in an ignition mode is one where the energy deposited directly within the plasma by charged reaction products (3.5-MeV alpha particles in the case of the D-T reaction, that is, only 20% of the total fusion energy release) is sufficient to maintain the plasma temperature, including the requirement for heating up new cold fuel particles introduced to maintain the fuel plasma density as fusion "combustion" proceeds. The ignition mode is therefore more demanding with respect to confinement time than is a driven mode, but in principle could be technically simpler, since it does not impose as strict requirements on the efficiencies of the energy recovery and plasma heating systems.

By exploiting the increase in confinement time associated with an increase in plasma radius and extrapolating from the performance of present devices, it appears that large tokamaks could operate in an ignition mode. Conventional and tandem mirror systems, requiring as they do the maintenance of plasma in the mirror cells, would need to be operated in a driven mode. However, in the case of the tandem, it seems that it will be possible to achieve ignition in the central cell plasma, with attendant simplifications and other advantages. Laser pellet systems, although the pellet itself would be expected to ignite, necessarily require recirculated power to initiate the burn. To achieve high net power relative to recirculated power would seem to imply the need for pellet energy gains (Q) of 100 or more.

Considering the magnetic confinement approach, systems studies have led to some important conclusions: Fusion power plants based on the tokamak principle in its conventional form will be relatively large both in size and in power output; electrical power outputs of 500 to some thousands of megawatts are likely to be typical. Power plants utilizing the tandem idea might be somewhat smaller in physical size than conventional tokamaks, and possibly also capable of somewhat lower plant electrical outputs than tokamaks, still satisfying economic requirements. Power plants based on field reversal, such as the field-reversed mirror, might be the smallest of all, exhibiting the highest fusion power density while being both compact in size and permitting (at least in the demonstration phase) electrical power outputs as low as tens of megawatts.

Another general result of the design studies is to show the importance of choice of materials and heat transfer characteristics of the inner wall of the containment chamber. The flux of 14-MeV neutrons through this wall coming from D-T reactions in the plasma will cause localized heating, radiation damage, and induced radioactivity. Thus the design of this portion of any D-T fusion power plant can be expected to be of critical importance. The materials chosen need to be picked not only for their resistance to radiation damage but also for minimum activation (that is, minimal yield and short half-life for the neutron-induced radioactivity). It does appear that first-wall materials having the desired characteristics can be developed. Another critical factor is that of the generation of the confining magnetic field. Here the criterion is to achieve the required field (which may be very high for some approaches) at the least capital cost and for the least expenditure of energy. Fortunately, the development of practical high-current-density, high-field superconductors appears to provide an almost ideal solution to this problem.

[RICHARD F. POST]

Bibliography: F. Chen, *Introduction to Plasma Physics*, 1974; N. Krall and A. Trivelpiece, *Principles of Plasma Physics*, 1973; R. F. Post, Controlled fusion research and high temperature plasmas, *Annu. Rev. Nucl. Sci.*, 20:509–558, 1970; G. Schmidt, *Physics of High Temperature Plasmas*, 2d ed., 1979; L. Spitzer, *Physics of Fully Ionized Gases*, 1962.

Nuclear isomerism

The existence of excited states of atomic nuclei with unusually long lifetimes. A nucleus may exist in an excited quantum state with well-defined excitation energy (E_x), spin (J), and parity (π). Such states are unstable and decay, usually by the emission of electromagnetic radiation (γ-rays), to lower excited states or to the ground state of the nucleus. The rate at which this decay takes place is characterized by a half-life $(\tau_{1/2})$, the time in which one-half of a large number of nuclei, each in the same excited state, will have decayed. If the lifetime of a specific excited state is unusually long, compared with the lifetimes of other excited states in the same nucleus, the state is said to be isomeric. The definition of the boundary between isomeric and normal decays is arbitrary, and the term is therefore used loosely. *See* EXCITED STATE; PARITY; SPIN.

Spin isomerism. The predominant decay mode of excited nuclear states is by γ-ray emission. The rate at which this process occurs is determined largely by the spins, parities, and excitation energies of the decaying state and of those to which it is decaying. In particular, the rate is extremely sensitive to the difference in the spins of initial and final states and to the difference in excitation energies. Both extremely large spin differences and extremely small energy differences can result in a slowing of the γ-ray emission by many orders of magnitude, resulting in some excited states having unusually long lifetimes and therefore being termed isomeric.

Occurrence. Isomeric states have been observed to occur in almost all known nuclei. However, they occur predominantly in regions of nuclei with neutron numbers N and proton numbers Z close to the so-called magic numbers at which shell closures occur. This observation is taken as important evidence for the correctness of the nuclear shell model which predicts that high-spin, and therefore isomeric, states should occur at quite low excitation energies in such nuclei. *See* MAGIC NUMBERS.

Examples. Three examples of isomeric states are shown in the illustration. In the case of ^{90}Zr (illustration *a*), the 2.319-MeV, $J^\pi = 5^-$ state has a half-life of 809 milliseconds, compared to the much shorter lifetimes of 93 femtoseconds and 61 nanoseconds for the 2.186-MeV and 1.761-MeV states, respectively. The spin difference of 5 between the 2.319-MeV state and the ground and 1.761 $J^\pi = 0^+$ states, together with the spin difference of 3 and small energy difference between the 2.319-MeV state and the 2.186-MeV, $J^\pi = 2^+$ state, produce this long lifetime.

For ^{42}Sc (illustration *b*), the γ-decay of the $J^\pi = 7^+$ state at 0.617 MeV is so retarded by the large spin changes involved that the intrinsically much slower β^+ decay process can take place, resulting in a half-life of 62 s.

Yet another process takes place in the case of the high-spin isomer in ^{212}Po (illustration *c*), which decays by α-particle emission with a half-life of 45 s rather than γ-decays.

The common feature of all these examples is the slowing of the γ-ray emission process due to the high spin of the isomeric state.

Other mechanisms. Not all isomers are the spin isomers described above. Two other types of isomers have been identified. The first of these arises from the fact that some excited nuclear states represent a drastic change in shape of the nucleus from the shape of the ground state. In many cases this extremely deformed shape displays unusual stability, and states with this shape are therefore isomeric. A particularly important class of these shape isomers is observed in the decay of heavy nuclei by fission, and the study of such fission isomers has been the subject of intensive effort. The possibility that nuclei may undergo sudden changes of shape at high rotational velocities has spurred searches for isomers with extremely high spin which may also be termed shape isomers. *See* NUCLEAR FISSION.

A more esoteric form of isomer has also been observed, the so-called pairing isomer which results from differences in the microscopic motions of the constituent nucleons in the nucleus. A state of this type has a quite different character from the ground state of the nucleus, and is therefore also termed isomeric. *See* NUCLEAR STRUCTURE.

[RUSSELL BETTS]

Bibliography: B. L. Cohen, *Concepts of Nuclear Physics*, 1971; C. M. Lederer and V. S. Shirley, *Table of Isotopes*, 7th ed., 1978; K. Siegbahn (ed.), *Alpha, Beta and Gamma-Ray Spectroscopy*, vols. 1 and 2, 1965.

Nuclear molecule

A quasistable entity of nuclear dimensions formed in nuclear collisions and comprising two or more discrete nuclei that retain their identities and are bound together by strong nuclear forces. Whereas the stable molecules of chemistry and biology consist of atoms bound through various electronic mechanisms, nuclear molecules do not form in nature except possibly in the hearts of giant stars; this simply reflects the fact that all nuclei carry positive electrical charges, and that under all natural conditions the long-range electrostatic repulsion prevents nuclear components from coming within the grasp of the short-range attractive nuclear force which could provide molecular binding. But in energetic collisions this electrostatic repulsion can be overcome.

Observation. Nuclear molecules were first suggested by D. A. Bromley, J. A. Kuehner, and E. Almqvist to explain very surprising experimental results obtained in the first studies on collisions between carbon nuclei carried out under conditions of high precision. The yields of various radiations in this experiment are shown in Fig. 1. In each case the yield has been summed over a large number of transitions of the type in question. The molecules appear as resonances in all reaction channels at energies just below the Coulomb barrier. These data have stimulated a great amount of work, both experimental and theoretical, since their discovery; but although additional examples have been found in other than the carbon-carbon system, no completely adequate explanation for the molecular binding is yet available.

Formation mechanism. As higher-energy heavy-ion beams have become available, entirely new classes of nuclear molecular phenomena have emerged. Much evidence has been accumulated

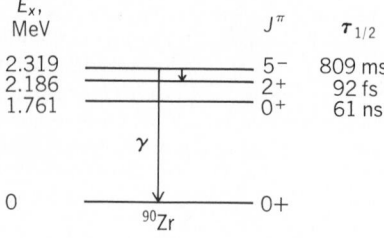

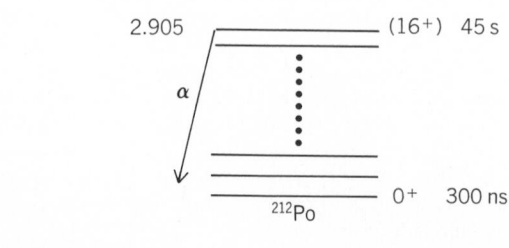

Examples of nuclear isomerism in (*a*) ^{90}Zr, (*b*) ^{42}Sc, and (*c*) ^{212}Po. (*From C. M. Lederer and V. S. Shirley, Table of Isotopes, 7th ed., copyright © 1978 by John Wiley and Sons, Inc.; used with permission*)

for a double-resonance mechanism first proposed by B. Imanishi and extended by W. Greiner and colleagues. In this mechanism the incident ion orbits the target and, while doing so, temporarily loses kinetic energy to inelastic excitation of its own, or the target's, excited quantum states. Under the appropriate conditions this temporary decrease in kinetic energy drops the ion into a quasibound state in the interaction potential; this resonant excitation of the quasibound state then couples with the orbiting resonance to yield the observed phenomena.

Figure 2 illustrates the Imanishi proposal in the case of a carbon-carbon collision. The incident ion, with energy E (shown *above* the Coulomb barrier for clarity), interacts with the potential shown to represent the interaction. The typical molecular shape of this potential results from the summation of a short-range, specifically nuclear part, a longer-range centrifugal part, and an even-longer-range Coulomb part. E^* is, in the carbon case, equal to 4.43 MeV, the excitation energy of the first quantum state; if E is chosen appropriately, then $E - E^* = E_B$, the energy of a quasibound state in the molecular potential. In the decay of the molecular state the excitation energy can be returned to the elastic channel, and the incident ion reemerges with its original energy or, as is illustrated in Fig. 2, the molecular state can decay into formation of a magnesium-24 compound nucleus, which subsequently emits decay radiations, including those shown in Fig. 1. In the Greiner double-resonance model, in addition, E coincides with an orbiting resonance for this potential.

Highly excited nuclei. It has long been a question as to what happened as more and more energy is added to a nucleus. If it were simply a matter of thermodynamic increase in the mean kinetic energy of the component nucleons, the nucleus would be a very dull entity indeed. Direct evidence has been obtained demonstrating that at very high energies of 35 to 45 MeV—many times that required to enable decay by particle emission—special new long-lived nuclear configurations of a molecular character exist, for example, $^{12}C + ^{12}C$ dinuclear molecules at high excitation in ^{24}Mg and $^{12}C + ^4He + ^{12}C$ trinuclear molecules in ^{28}Si. Here the available energy is largely utilized in the internal binding of the component nuclear clusters, with only relatively little appearing as kinetic energy of relative motion.

In the case of ^{24}Mg, for example, it is known that normally the nucleus is strongly football-shaped and that this intrinsic shape can be rotated ever faster to yield a characteristic quantum-mechanical rotational band spectrum of excited states. The new molecular states correspond to a dumbbell structure, and the corresponding new observed rotational band at high excitation shows the much larger moment of inertia which would be expected of the molecular dumbbell.

Prospects. The discovery of this class of molecular states opens up a new area of nuclear spectroscopy—both experimental and theoretical. It is already clear that molecular complexes play an important role in the collisions of massive nuclei, they appear to be a high-excitation feature of nuclear structure, and they may play a significant role in the burning of heavier nuclear fuels in the

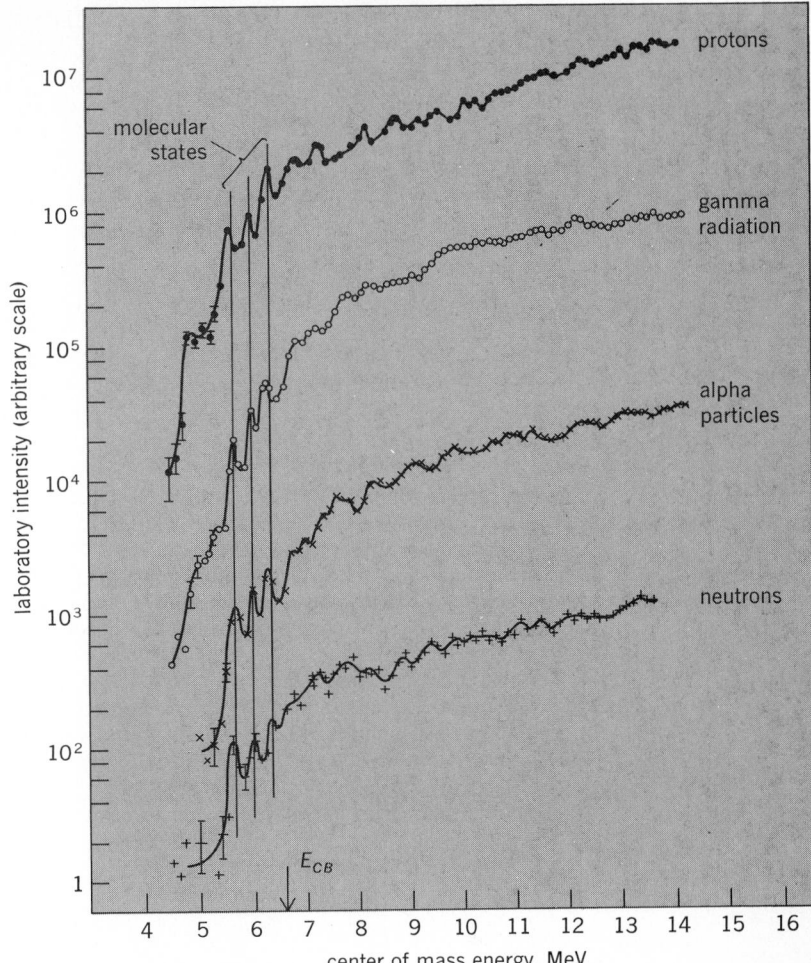

Fig. 1. Yields of various radiations from the bombardment of a carbon-12 target with a carbon-12 projectile beam as functions of the center-of-mass energy. E_{CB} indicates the energy of the Coulomb barrier for the carbon plus carbon system.

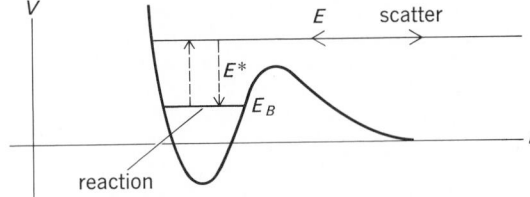

Fig. 2. Diagram illustrating mechanism proposed for formation of a nuclear molecule in collision of two carbon-12 nuclei. Curve shows interaction potential V as a function of separation of centers of nuclei r.

twilight phases of stellar evolution. *See* NUCLEAR STRUCTURE; SCATTERING EXPERIMENTS (NUCLEI).

[D. ALLAN BROMLEY]

Bibliography: D. A. Bromley, Nuclear molecules, *Sci Amer.*, 239(6):58–69, 1978; K. A. Erb and D. A. Bromley, Nuclear molecular resonances in heavy-ion collisions, *Phys. Today*, 32(1):34–42, 1979.

Nuclear moments

Intrinsic properties of atomic nuclei: electric moments result from deviations of the nuclear charge distribution from spherical symmetry;

magnetic moments are a consequence of the intrinsic spin and the rotational motion of nucleons within the nucleus. The classical definitions of the magnetic and electric multipole moments are written in general in terms of multipole expansions. *See* NUCLEAR STRUCTURE; SPIN.

Parity conservation allows only even-rank electric moments and odd-rank magnetic moments to be nonzero. The most important terms are the magnetic dipole, given by Eq. (1), and the electric

$$\vec{\mu} = \int \vec{M}(\vec{r}) \, dv \qquad (1)$$

monopole, quadrupole, and hexadecapole, given by Eq. (2), for $l = 0, 2, 4$. Here m is the projection of

$$Q = \frac{1}{e} \int r^l Y_{lm}(\theta, \phi) \rho(\vec{r}) \, dv \qquad (2)$$

the orbital angular momentum l on a z-axis appropriately chosen in space, $\vec{M}(\vec{r})$ is the magnetization density of the nucleus and depends on the space coordinates $\vec{r}$, e is the electronic charge, $\rho(\vec{r})$ is the charge density in the nucleus, and Y_{lm} are normalized spherical harmonics that depend on the angular coordinates θ and ϕ. *See* MAGNETIZATION; PARITY; SPHERICAL HARMONICS.

Quantum mechanically only the z components of the effective operators have nonvanishing values. Magnetic dipole and electric quadrupole moments are usually expressed in terms of the nuclear spin I through Eqs. (3) and (4), where g, the nuclear gy-

$$\mu/\mu_N = gI \qquad (3)$$

$$Q(m_I) = \frac{[3m_I^2 - I(I+1)]Q}{I(2I-1)} \qquad (4)$$

romagnetic factor, is a measure of the coupling of nuclear spins and orbital angular momenta, $\mu_N = eh/4\pi M_p c = 5.0508 \times 10^{-24}$ erg/gauss $= 5.0508 \times 10^{-27}$ joule/tesla, is the nuclear magneton, M_p is the proton mass, h is Planck's constant, e is the electron charge, and c is the speed of light. $Q(m_I)$ is the effective quadrupole moment in the state m_I, and Q is the quadrupole moment for the state $m_I = I$. All angular momenta are expressed in units of $h/2\pi$. The magnitude of g varies between 0 and 1.8, and Q is of the order of 10^{-25} cm^2. *See* ANGULAR MOMENTUM; MAGNETON.

In special cases nuclear moments can be measured by direct methods involving the interaction of the nucleus with an external magnetic field or with an electric field gradient produced by the scattering of high-energy charged particles. In general, however, nuclear moments manifest themselves through the hyperfine interaction between the nuclear moments and the fields or field gradients produced by either the atomic electrons' currents and spins, or the molecular or crystalline electronic and lattice structures. *See* HYPERFINE STRUCTURE.

Effects of nuclear moments. In a free atom the magnetic hyperfine interaction between the nuclear spin $\vec{I}$ and the effective magnetic field $\vec{H}_e$ associated with electronic angular momentum $\vec{J}$ results in an energy $W = -\vec{\mu} \cdot \vec{H}_e = ha\vec{I} \cdot \vec{J}$, which appears as a splitting of the energy levels of the atom. The magnetic field at the nucleus due to atomic electrons can be as large as $10-100$ T for a neutral atom. The constant a is of the order of 1000 MHz. *See* ATOMIC STRUCTURE AND SPECTRA.

The electric monopole moment is a measure of the nuclear charge and does not give rise to hyperfine interactions. The quadrupole moment Q reflects the deviation of the nuclear charge distribution from a spherical charge distribution. It is responsible for a quadrupole hyperfine interaction energy W_Q, which is proportional to the quadrupole moment Q and to the spatial derivative of the electric field at the nucleus due to the electronic charges, and is given by Eq. (5), where q is the average of expression (6). Here, r_i is the radius

$$W_Q = \frac{e^2 Q q}{2I(2I-1)J(2J-1)}$$
$$\cdot [3(\vec{I} \cdot \vec{J})^2 + \tfrac{3}{2}(\vec{I} \cdot \vec{J}) - I(I+1)] \qquad (5)$$

$$\sum_i (3\cos^2 \theta - 1)/r_i^{-3} \qquad (6)$$

vector from the nucleus to the ith electron, and θ_i is the angle between r_i and the z axis. *See* MÖSSBAUER EFFECT.

In free molecules the hyperfine couplings are similar to those encountered in free atoms, but as the charge distributions and the spin coupling of valence electrons vary widely, depending on the nature of the molecular bonding, a greater diversity of magnetic dipole and quadrupole interactions is met. *See* MOLECULAR STRUCTURE AND SPECTRA.

In crystals the hyperfine interaction patterns become extremely complex, because the crystalline electric field is usually strong enough to compete with the spin orbit coupling of the electrons in the ion. Nevertheless, the energy-level structure can often be resolved by selective experiments at low temperatures on dilute concentrations of the ion of interest. *See* MAGNETIC RELAXATION; MAGNETIC RESONANCE.

Measurement. The hyperfine interactions affect the energy levels of either the nuclei or the atoms, molecules, or ions, and therefore can be observed either in nuclear parameters or in the atomic, molecular, or ionic structure. The many different techniques that have been developed to measure nuclear moments can be grossly grouped in three categories: the conventional techniques based mostly on spectroscopy of energy levels, the methods based on the detection of nuclear radiation from aligned excited nuclei, and techniques involving the interactions of fast ions with matter or of fast ions with laser beams.

Hyperfine structure of spectral lines. The hyperfine interaction causes a splitting of the electronic energy levels which is proportional to the magnitude of the nuclear moments, to the angular momenta I and J of the nuclei and their electronic environment, and to the magnetic field or electric field gradient at the nucleus. The magnitude of the splitting is determined by the nuclear moments, and the multiplicity of levels is given by the relevant angular momenta I or J involved in the interaction. The energy levels are identified either by optical or microwave spectroscopy.

Optical spectroscopy (in the visible and ultraviolet) has the advantage of allowing the study of atomic excited states and of atoms in different states of ionization. Furthermore, optical spectra provide a direct measure of the monopole moments, which are manifested as shifts in the energy levels of atoms of different nuclear isotopes exhibiting different nuclear radii and charge distributions. Optical spectroscopy has a special advantage over other methods in that the intensity of the lines often yields the sign of the interaction constant a. *See* ISOTOPE SHIFT.

Microwave spectroscopy is a high-resolution technique involving the attenuation of a signal in a waveguide containing the absorber in the form of a low-pressure gas. The states are identified by the observation of electric dipole transitions of the order of 20,000 MHz. The levels are split by quadrupole interactions of the order of 100 MHz. Very precise quadrupole couplings are obtained, as well as vibrational and rotational constants of molecules, and nuclear spins.

Atomic and molecular beams and nuclear resonance. Atomic and molecular beams passing through inhomogeneous magnetic fields are deflected by an amount depending on the nuclear moment. However, because of the small size of the nuclear moment, the observable effect is very small. The addition of a radio-frequency magnetic field at the frequency corresponding to the energy difference between hyperfine electronic states has vastly extended the scope of the technique. For nuclei in solids, liquids, or gases, the internal magnetic fields and gradients of the electric fields may be quenched if the pairing of electrons and the interaction between the nuclear magnetic moment and the external field dominate. The molecular beam apparatus is designed to detect the change in orientation of the nuclei, while the nuclear magnetic resonance system is designed to detect absorbed power (resonance absorption) or a signal induced at resonance in a pick-up coil around the sample (nuclear induction). The required frequencies for fields of about 0.5 T are of the order of 1–5 MHz. The principal calibration of the field is accomplished in relation to the resonant frequency for the proton whose g-factor is accurately known. Sensitivities of 1 part in 10^8 are possible under optimum experimental conditions. The constant a for ^{133}Cs has been measured to 1 part in 10^{10}, and this isotope is used as a time standard. *See* MOLECULAR BEAMS.

The existence of quadrupole interactions produces a broadening of the resonance line above the natural width and a definite structure determined by the value of the nuclear spin.

In some crystals the electric field gradient at the nucleus is large enough to split the energy levels without the need for an external field, and pure quadrupole resonance spectra are observed. This technique allows very accurate comparison of quadrupole moments of isotopes.

Atomic and molecular beams with radioactive nuclei. The conventional atomic and molecular beam investigations can be applied to radioactive nuclei if the beam current measurement is replaced by the much more sensitive detectors of radiations emitted in a radioactive decay. Moments of nuclei with half-lives down to the order of minutes have been determined. *See* RADIOACTIVITY.

Perturbed angular correlations. The angular distribution and the polarization of radiation emitted by nuclei depend on the angle between the nuclear spin axis and the direction of emission. In radioactive sources in which the nuclei have been somewhat oriented either by a nuclear reaction or by static or dynamic polarization techniques at low temperatures, the ensuing nonisotropic angular correlation of the decay radiation can be perturbed by the application of external magnetic fields, or by the hyperfine interaction between the nuclear moment and the electronic or crystalline fields acting at the nuclear site. Magnetic dipole and electric quadrupole moments of ground and excited nuclear states with half-lives as short as 10^{-9} s have been measured by these techniques. *See* DYNAMIC NUCLEAR POLARIZATION.

Techniques involving interactions of fast ions. Techniques involving the interaction of intense light beams from tuned lasers with fast ion beams have extended the realm of resonance spectroscopy to the study of exotic species, such as nuclei far from stability, fission isomers, and ground-state nuclei with half-lives shorter than minutes. The hyperfine interactions in beams of fast ions traversing magnetic materials result from the coupling between the nuclear moments and unpaired polarized s-electrons, and are strong enough (H_e is of the order of 2×10^3 T) to extend the moment measurements to excited states with lifetimes as short as 10^{-12} s. Progress in atomic and nuclear technology has contributed to the production of hyperfine interactions of increasing strength, thus allowing for the observation of nuclear moments of nuclei and nuclear states of increasing rarity.

[NOÉMIE KOLLER]

Bibliography: P. Averbuch (ed.), *Magnetic Resonance and Radiofrequency Spectroscopy*, 1969; H. Kopferman, *Nuclear Moments*, 1958; *Proceedings of the International Conference on Nuclear Moments and Nuclear Structure*, vol. 43, Physics Society, Japan, 1973.

Nuclear orientation

The directional ordering of an assembly of nuclear spins I with respect to some axis in space. Under normal conditions nuclei are not oriented; that is, all directions in space are equally probable. For a system of nuclear spins with rotational symmetry about an axis, the degree of orientation is completely characterized by the relative populations a_m of the $2I+1$ magnetic sublevels $m(= I, I - 1, \ldots, -I)$. There are just $2I$ independent values of a_m, since they are normalized to unity, namely,

$$\sum_m a_m = 1$$

Rather than specify these populations directly, it turns out to be more useful to form the moments

$$\sum_m m^\nu a_m$$

since these occur in the theoretical calculations. There are $2I$ independent linear combinations of these moments which are called orientation parameters, $f_k(I)$, and are defined by the equation below. Here $f_0(I) = 1$ and all $f_k(I)$ with $k \geq 2I + 1$ are zero.

$$f_k(I) = \binom{2k}{k}^{-1} I^{-k}$$

$$\sum_m \sum_{\nu=0}^{k} (-1)^\nu \frac{(I-m)!(I+m)!}{(I-m-\nu)!(I+m-k+\nu)!} \binom{k}{\nu}^2 a_m$$

Nuclear polarization and alignment. Nuclear polarization is said to be present when one or more $f_k(I)$ with k-odd is not zero, regardless of the even $f_k(I)$ values. In this case the nuclear spin system is polarized. If all the $f_k(I)$ for k-odd are zero and at least one $f_k(I)$ for k-even is not zero, nuclear alignment is said to be present; that is, the nuclear spin system is aligned. Simply stated, if the z axis is the axis of quantization of the nuclear spin system, polarization represents a net moment along the z axis, whereas alignment does not. Unfortunately, the term nuclear polarization is usually associated with $f_1(I)$, and nuclear alignment with $f_2(I)$, although their meanings are in fact much more general. There are other definitions of nuclear orientation parameters; they are mathematically equivalent to the one above. If the nuclear spin system does not have cylindrical symmetry, a more general definition of nuclear orientation is needed leading to the statistical tensors. *See* SPIN.

Production. Nuclear orientation can be achieved in various ways. The most obvious way is to modify the energies of the $2I + 1$ magnetic sublevels so as to remove their degeneracy and thereby change the populations of these sublevels. The spin degeneracy can be removed by a magnetic field interacting with the nuclear magnetic dipole moment, or by an inhomogeneous electric field interacting with the nuclear electric quadrupole moment. Significant differences in the populations of the sublevels can be established by cooling the nuclear sample to low temperatures T such that T is in the region around $\Delta E/k$, where ΔE is the energy separation of adjacent magnetic sublevels of energy E_m, and k is the Boltzmann constant. If the nuclear spin system is in thermal equilibrium, the populations a_m are given by the normalized Boltzmann factor

$$\exp\left(-E_m/kT\right)\Big/ \sum_m \exp\left(-E_m/kT\right)$$

This means of producing nuclear orientation is called the static method. In contrast, there is the dynamic method, which is related to optical pumping in gases. There are other ways to produce oriented nuclei; for example, in a nuclear reaction such as the capture of polarized neutrons (produced by magnetic scattering) by unoriented nuclei, the resulting compound nuclei could be polarized. In addition to polarized neutron beams, polarized beams of protons, deuterons, tritons, helium-3, lithium-6, and other nuclei have been produced. *See* BOLTZMANN STATISTICS; DYNAMIC NUCLEAR POLARIZATION; OPTICAL PUMPING.

Applications. Oriented nuclei have proved to be very useful in various fields of physics. They have been used to measure nuclear properties, for example, magnetic dipole and electric quadrupole moments, spins, parities, and mixing ratios of nuclear states. Oriented nuclei have been used to examine some of the fundamental properties of nuclear forces, for example, nonconservation of parity in the weak interaction. Measurement of hyperfine fields, electric-field gradients, and other properties relating to the environment of the nucleus have been made by using oriented nuclei. Nuclear orientation thermometry is one of the few sources of a primary temperature scale at low temperatures. Oriented nuclear targets used in conjunction with beams of polarized and unpolarized particles have proved very useful in examin-

ing certain aspects of the nuclear force. *See* NUCLEAR MOMENTS; NUCLEAR STRUCTURE; PARITY.

With the advent of helium-3/helium-4 dilution refrigerators, large superconducting magnets, and new methods of producing oriented nuclei and polarized beams, the field of nuclear orientation physics is expected to grow substantially. *See* CRYOGENICS.

[HARVEY MARSHAK]

Bibliography: S. R. De Groot, H. A. Tolhoek, and W. J. Huiskamp, Orientation of nuclei at low temperatures, in K. Siegbahn (ed.), *Alpha-, Beta- and Gamma Ray Spectroscopy*, 1968; R. P. Hudson et al., Recent advances in thermometry below 300 mK, *J. Low Temp. Phys.*, vol. 20, no. 1, 1975; M. L. Marshak (ed.), *High Energy Physics with Polarized Beams and Targets*, 1976; W. J. Thompson and T. B. Clegg, Physics with polarized nuclei, *Phys. Today*, pp. 32–39, February 1979.

Nuclear physics

The discipline involving the structure of atomic nuclei and their interactions with each other, with their constituent particles, and with the whole spectrum of elementary particles that is provided by very large accelerators. The nuclear domain occupies a central position between the atomic range of forces and sizes and those of elementary-particle physics, characteristically within the nucleons themselves. As the only system in which all the known natural forces can be studied simultaneously, it provides a natural laboratory for the testing and extending of many fundamental symmetries and laws of nature. *See* ATOMIC NUCLEUS; ATOMIC STRUCTURE AND SPECTRA; ELEMENTARY PARTICLE; SYMMETRY LAWS.

Containing a reasonably large, yet manageable number of strongly interacting components, the nucleus also occupies a central position in the universal many-body problem of physics, falling between the few-body problems, characteristic of elementary-particle interactions, and the extreme many-body situations of plasma physics and condensed matter, in which statistical approaches dominate; it provides the scientist with a rich range of phenomena to investigate—with the hope of understanding these phenomena at a microscopic level. *See* PLASMA PHYSICS; STATISTICAL MECHANICS.

Activity in the field centers on three broad and interdependent subareas. The first is referred to as classical nuclear physics, wherein the structural and dynamic aspects of nuclear behavior are probed in numerous laboratories, and in many nuclear systems, with the use of a broad range of experimental and theoretical techniques. Second is higher-energy nuclear physics (referred to as medium-energy physics in the United States), which emphasizes the nuclear interior and nuclear interactions with mesonic probes. Third is heavy-ion physics, internationally the most rapidly growing subfield, wherein accelerated beams of nuclei spanning the periodic table are used to study previously inaccessible nuclear phenomena.

Nuclear physics is unique in the extent to which it merges the most fundamental and the most applied topics. Its instrumentation has found broad applicability throughout science, technology, and medicine; nuclear engineering and nuclear medicine are two very important areas of applied specialization. *See* NUCLEAR RADIATION.

Nuclear chemistry, certain aspects of condensed matter and materials science, and nuclear physics together constitute the broad field of nuclear science; outside the United States and Canada elementary particle physics is frequently included in this more general classification. *See* ANALOG STATES; FUNDAMENTAL INTERACTIONS; ISOTOPE; NUCLEAR FISSION; NUCLEAR FUSION; NUCLEAR ISOMERISM; NUCLEAR MOMENTS: NUCLEAR REACTION; NUCLEAR SPECTRA; NUCLEAR STRUCTURE; PARTICLE ACCELERATOR; PARTICLE DETECTOR; RADIOACTIVITY; SCATTERING EXPERIMENTS (NUCLEI); WEAK NUCLEAR INTERACTIONS.

[D. ALLAN BROMLEY]

Nuclear quadrupole resonance

A selective absorption phenomenon observable in a wide variety of polycrystalline compounds containing nonspherical atomic nuclei when placed in a magnetic radio-frequency field. Nuclear quadrupole resonance (NQR) is very similar to nuclear magnetic resonance (NMR), and was originated in the late 1940s by H. G. Dehmelt and H. Krüger as an inexpensive (no stable homogenous large magnetic field is required) alternative way to study nuclear moments. It has since gained a modest popularity, especially in developing countries. *See* MAGNETIC RESONANCE.

Principles. In the simplest case, for example, ^{35}Cl in solid Cl_2, NQR is associated with the precession of the angular momentum of the nucleus, depicted in the illustration as a flat ellipsoid of rotation, around the symmetry axis (taken as the z axis) of the Cl_2 molecule fixed in the crystalline solid. (The direction of the nuclear angular momentum coincides with those of the symmetry axis of the ellipsoid and of the nuclear magnetic dipole moment μ.) The precession, with constant angle θ between the nuclear axis and symmetry axis of the molecule, is due to the torque which the inhomogeneous molecular electric field exerts on the nucleus of electric quadrupole moment eQ. This torque corresponds to the fact that the electrostatic interaction energy of the nucleus with the molecular electric field depends on the angle θ. The interaction energy is given by Eq. (1), where ρ is the nu-

$$E = \int \phi \rho \, dV \qquad (1)$$

clear charge density distribution and ϕ is the potential of the molecular electric field. Its dependence on θ is given by Eq. (2), where ϕ_{zz} is the axial

$$E = eQ\phi_{zz}(3\cos^2\theta - 1)/8 \qquad (2)$$

gradient of the (approximately) axially symmetric molecular electric field. The quantum-mechanical analog of this expression is Eq. (3), where I and m

$$E_m = eQ\phi_{zz}[3m^2 - I(I+1)]/4I(2I-1) \qquad (3)$$

denote the quantum numbers of the nuclear angular momentum and its z component I_z. The absorption occurs classically when the frequency of the rf field ν and that of the precessing motion of the angular momentum coincide, or quantum-mechanically when Eq. (4) is satisfied, where m and m'

$$h\nu = |E_{m'} - E_m| \qquad (4)$$

are given by Eqs. (5) and $m' - m = \pm 1$, correspond-

$$\begin{aligned} m,m' &= 0 \pm 1, \pm 2 \ldots \pm I \quad \text{for integer } I \geq 1 \\ m,m' &= \pm\tfrac{1}{2}, \pm\tfrac{3}{2} \ldots \pm I \quad \text{for half-integer } I \geq \tfrac{3}{2} \end{aligned} \qquad (5)$$

ing to magnetic dipole transitions. *See* NUCLEAR MOMENTS.

It is not necessary that the rf field direction is perpendicular to z; a nonvanishing perpendicular component suffices. This eliminates the necessity of using single crystals and makes it practical, unlike in the NMR of solids, to use polycrystalline samples of unlimited mass and volume. In fact, a polycrystalline natural sulfur sample of 3 liters volume was used in work on the rare (0.74% abundance) ^{33}S isotope.

Techniques. The original NQR work was done on approximately 50 cm³ of frozen *trans*-dichloroethylene submerged in liquid air using a superregenerative detector. The oscillator incorporated a vibrating capacitor driven by the power line to sweep a frequency band about 50 kHz wide over the approximately 10-kHz-wide ^{35}Cl and ^{37}Cl resonances near 30 MHz, and an oscilloscopic signal display was used. This work demonstrated the good sensitivity and rugged character of these simple, easily tunable circuits which are capable of combining high rf power levels with low noise. Their chief disadvantage is the occurrence of side bands spaced by the quench frequency which may confuse the line shape. For nuclear species of low abundance it becomes important to use nuclear modulation. In the ^{33}S work zero-based magnetic-field pulses periodically smearing out the absorption line proved satisfactory.

Application. NQR spectra have been observed in the approximate range 1–1000 MHz. Such a range clearly requires more than one spectrometer. Most of the NQR work has been on molecular crystals. While halogen-containing (Cl, Br, I) organic compounds have been in the forefront since the inception of the field, NQR spectra have also been observed for K, Rb, Cs, Cu, Au, Ba, Hg, B, Al, Ga, In, La, N, As, Sb, Bi, S, Mn, Re, and Co isotopes. For molecular crystals the coupling constants $eQ\phi_{zz}$ found do not differ very much from those measured for the isolated molecules in microwave spectroscopy. The most precise nuclear information which may be extracted from NQR $eQ\phi_{zz}$ data are quadrupole moment ratios of isotopes of the same element, since one may assume that ϕ_{zz} is practically independent of the nuclear mass. As far as ϕ_{zz} values may be estimated from atomic fine structure data, for example, for Cl_2 where a pure p-bond is expected and the molecular nature of the solid is suggested by a low boiling point and so forth, fair Q values may be obtained. However, it has also proved very productive to use the quadrupole nucleus as a probe of bond character and orientation and crystalline electric fields and lattice sites, and a large body of data has been accumulated in this area.

[HANS DEHMELT]

Bibliography: I. P. Biryukov, M. G. Voronkov, and I. A. Safin, *Tables of Nuclear Quadrupole Resonance Frequencies*, 1969; T. P. Das and E. L. Hahn, *Nuclear Quadrupole Resonance Spectrosco-*

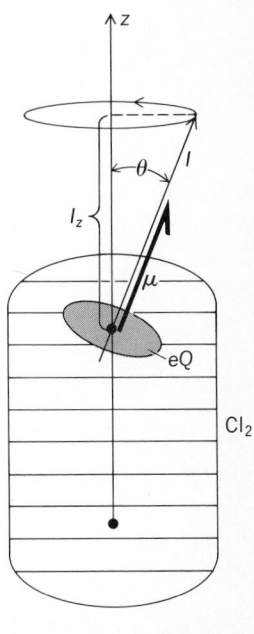

NUCLEAR QUADRUPOLE RESONANCE

Interaction of ^{35}Cl nucleus with the electric field of a Cl_2 molecule.

py, 1958; H. G. Dehmelt, Nuclear quadrupole resonance (in solids), *Amer. J. Phys.*, 22:110–120, 1954, and *Faraday Soc. Discuss.*, 19:263–274, 1955; G. K. Semin, T. A. Babushkina, and G. G. Yakobson, *Nuclear Quadrupole Resonance in Chemistry*, 1975.

Nuclear radiation

All particles and radiations emanating from an atomic nucleus due to radioactive decay and nuclear reactions. Thus the criterion for nuclear radiations is that a nuclear process is involved in their production. The term was originally used to denote the ionizing radiations observed from naturally occurring radioactive materials. These radiations were alpha rays (energetic helium nuclei), beta rays (negative electrons), and gamma rays (electromagnetic radiation with wavelength much shorter than visible light).

Nuclear radiations have traditionally been considered to be of three types based on the manner in which they interact with matter as they pass through it. These are the charged heavy particles with masses comparable to that of the nuclear mass (for example, protons, alpha particles, and heavier nuclei), electrons (both negatively and positively charged), and electromagnetic radiation. For all of these, the interactions with matter are considered to be primarily electromagnetic. (The neutron, which is also a nuclear radiation, behaves quite differently.) The behavior of mesons and other particles is intermediate between that of the electron and heavy charged particles. *See* CHARGED PARTICLE BEAMS.

A striking difference in the absorption of the three types of radiations is that only heavy charged particles have a range. That is, a monoenergetic beam of heavy charged particles, in passing through a certain amount of matter, will lose energy without changing the number of particles in the beam. Ultimately, they will be stopped after crossing practically the same thickness of absorber. The minimum amount of absorber that stops a particle is its range. The greatest part of the energy loss results from collisions with atomic electrons, causing the electrons to be excited or freed. The energy loss per unit path length is the specific energy loss, and its average value is the stopping power of the absorbing substance.

For electromagnetic radiation (gamma rays) and neutrons, on the other hand, the absorption is exponential; that is, the intensity decreases in such a way that the equation below is valid, where I is

$$-\frac{dI}{I} = \mu \, dx$$

the intensity of the primary radiation, μ is the absorption coefficient, and dx is the thickness traversed. The difference in behavior reflects the fact that charged particles are not removed from the beam by individual interactions, whereas gamma radiation photons (and neutrons) are. Three main types of phenomena involved in the interaction of electromagnetic radiation with matter (namely, photoelectric absorption, Compton scattering, and electron-positron production) are responsible for this behavior.

Electrons exhibit a more complex behavior. They radiate electromagnetic energy easily because they have a large charge-to-mass ratio and hence are subject to violent acceleration under the action of the electric forces. Moreover, they undergo scattering to such an extent that they follow irregular paths. *See* ELECTRON.

Whereas in the case of the heavy charged particles, electrons, or gamma rays the energy loss is mostly due to electromagnetic effects, neutrons are slowed down by nuclear collisions. These may be inelastic collisions, in which a nucleus is left in an excited state, or elastic collisions, in which the colliding nucleus acquires part of the energy of the nucleus as kinetic energy. In the first instance, the neutron must have enough kinetic energy (of the order of 1 MeV) to excite the collision partner. With less kinetic energy, only elastic scattering can slow down the neutron, a process which is effective down to thermal energies (about 1/40 keV). At this stage the collision, on the average, has no further effect on the neutron's energy. *See* NEUTRON.

As noted previously, the other nuclear radiations such as mesons have behaviors which are intermediate between that of heavy charged particles and electrons. Another radioactive decay product is the neutrino; because of its small interaction with matter, it is not ordinarily considered to be a nuclear radiation. *See* MESON; NEUTRINO; NUCLEAR REACTION

[DENNIS G. KOVAR]

Nuclear reaction

A process that occurs as a result of interactions between atomic nuclei when the interacting particles approach each other to within distances of the order of nuclear dimensions ($\approx 10^{-12}$ cm). While nuclear reactions occur in nature, understanding of them and use of them as tools have taken place primarily in the controlled laboratory environment. In the usual experimental situation, nuclear reactions are initiated by bombarding one of the interacting particles, the stationary target nucleus, with nuclear projectiles of some type, and the reaction products and their behaviors are studied. The study of nuclear reactions is the largest area of nuclear and subnuclear (or particle) physics; the threshold for producing pions has historically been taken to be the energy boundary between the two fields.

Types of nuclear interaction. As a generalized nuclear process, consider a collision in which an incident particle strikes a previously stationary particle, to produce an unspecified number of final products. If the final products are the same as the two initial particles, the process is called scattering. The scattering is said to be elastic or inelastic, depending on whether some of the kinetic energy of the incident particle is used to raise either of the particles to an excited state. If the product particles are different from the initial pair, the process is referred to as a reaction.

The most common type of nuclear reaction, and the one which has been most extensively studied, involves the production of two final products. Such reactions can be observed, for example, when deuterons with a kinetic energy of a few million electronvolts (MeV) are allowed to strike a carbon nucleus of mass 12. Protons, neutrons, deuterons, and alpha particles are observed to be emitted, and reactions (1)–(4) are responsible. In these

$$_1^2H + _6^{12}C \rightarrow _1^2H + _6^{12}C \tag{1}$$

$$_1^2H + _6^{12}C \rightarrow _1^1H + _6^{13}C \tag{2}$$

$$_1^2H + _6^{12}C \rightarrow _0^1n + _7^{13}N \tag{3}$$

$$_1^2H + _6^{12}C \rightarrow _2^4He + _5^{10}B \tag{4}$$

equations the nuclei are indicated by the usual chemical symbols; the subscripts indicate the atomic number (nuclear charge) of the nucleus, and the superscripts the mass number of the particular isotope. These reactions are conventionally written in the compact notation $^{12}C(d,d)^{12}C$, $^{12}C(d,p)^{13}C$, $^{12}C(d,n)^{13}N$, and $^{12}C(d,\alpha)^{10}B$, where d represents deuteron, p proton, n neutron, and α alpha particle. In each of these cases the reaction results in the production of an emitted light particle and a heavy residual nucleus. The (d,d) process denotes the elastic scattering as well as the inelastic scattering processes that raise the ^{12}C nucleus to one of its excited states. The other three reactions are examples of nuclear transmutation or disintegration where the residual nuclei may also be formed in their ground states or one of their many excited states. The processes producing the residual nucleus in different excited states are considered to be the different reaction channels of the particular reaction. If the residual nucleus is formed in an excited state, it will subsequently emit this excitation energy in the form of gamma rays or, in special cases, electrons. The residual nucleus may also be a radioactive species, as in the case of ^{13}N formed in the $^{12}C(d,n)$ reaction. In this case the residual nucleus will undergo further transformation in accordance with its characteristic radioactive decay scheme. *See* RADIOACTIVITY.

Nuclear cross section. In general one is interested in the probability of occurrence of the various reactions as a function of the bombarding energy of the incident particle. The measure of probability for a nuclear reaction is its cross section. Consider a reaction initiated by a beam of particles incident on a region which contains N atoms per unit area (uniformly distributed), and where I particles per second striking the area result in R reactions of a particular type per second. The fraction of the area bombarded which is effective in producing the reaction products is R/I. If this is divided by the number of nuclei per unit area, the effective area or cross section $\sigma = R/IN$. This is referred to as the total cross section for the specific reaction, since it involves all the occurrences of the reaction. The dimensions are those of an area, and total cross sections are expressed in either square centimeters or barns (1 barn = 10^{-24} cm^2). The differential cross section refers to the probability that a particular reaction product will be observed at a given angle with respect to the beam direction. Its dimensions are those of an area per unit solid angle (for example, barns per steradian).

Requirements for a reaction. Whether a specific reaction occurs and with what cross section it is observed depend upon a number of factors, some of which are not always completely understood. However, there are some necessary conditions which must be fulfilled if a reaction is to proceed.

Coulomb barrier. For a reaction to occur, the two interacting particles must approach each other to within the order of nuclear dimensions ($\approx 10^{-12}$ cm). With the exception of the uncharged neutron, all incident particles must therefore have sufficient kinetic energy to overcome the electrostatic (Coulomb) repulsion produced by the intense electrostatic field of the nuclear charge. The kinetic energy must be comparable to or greater than the so-called Coulomb barrier, whose magnitude is approximately given by the expression $E_{Coul} \approx Z_1Z_2/(A_1^{1/3} + A_2^{1/3})$ MeV, where Z and A respectively refer to the nuclear charge and mass number of the interacting particles 1 and 2. It can be seen that while, for the lightest targets, protons with kinetic energies of a few hundred thousand electronvolts (keV) are sufficient to initiate reactions, energies of many hundred millions of electronvolts are required to initiate reactions between heavier nuclei. In order to provide energetic charged particles, to be used as projectiles in reaction studies, particle accelerators of various kinds (such as Van de Graaff generators, cyclotrons, and linear accelerators) have been developed, making possible studies of nuclear reactions induced by projectiles as light as protons and as heavy as ^{208}Pb. *See* ELECTROSTATICS; PARTICLE ACCELERATOR.

Since neutrons are uncharged, they are not repelled by the electrostatic field of the target nucleus, and neutron energies of only a fraction of an electronvolt are sufficient to initiate some reactions. Neutrons for reaction studies can be obtained from nuclear reactors or from various nuclear reactions which produce neutrons as reaction products. There are two other means of producing nuclear reactions which do not fall into the general definition given above. Both electromagnetic radiation and high-energy electrons are capable of disintegrating nuclei under special conditions. However, both interact much less strongly with nuclei than with nucleons or other nuclei, through the electromagnetic and weak nuclear forces, respectively, rather than the strong nuclear force responsible for nuclear interactions. *See* NEUTRON.

Q value. For a nuclear reaction to occur, there must be sufficient kinetic energy available to bring about the transmutation of the original nuclear species into the final reaction products. The sum of the kinetic energies of the reaction products may be greater than, equal to, or less than the sum of the kinetic energies before the reaction. The difference in the sums is the Q value for that particular reaction. It can be shown that the Q value is also equal to the difference in the masses (rest energies) of the reaction products and the masses of the initial nuclei. Reactions with a positive Q value are called exoergic or exothermic reactions, while those with a negative Q value are called endoergic or endothermic reactions.

In reactions (1)–(4), where the residual nuclei are formed in their ground states, the Q values are $^{12}C(d,d)^{12}C$, $Q = 0.0$ MeV; $^{12}C(d,p)^{12}C$, $Q = 2.72$ MeV; $^{12}C(d,n)^{13}N$, $Q = -0.28$ MeV; and $^{12}C(d,\alpha)^{10}B$, $Q = -1.34$ MeV. For reactions with a negative Q value, a definite minimum kinetic energy is necessary for the reaction to take place. While there is no threshold energy for reactions with positive Q values, the cross section for the reactions induced by charged particles is very small unless the energies are sufficient to overcome the Coulomb barrier. A nuclear reaction and its inverse are reversible in the sense that the Q values are the same but have the opposite sign (for example, the Q value for the $^{10}B(\alpha,d)^{12}C$ reaction is $+1.39$ MeV).

Conservation laws. It has been found experimentally that certain physical quantities must be the same both before and after the reaction. The quantities conserved are electric charge, number of nucleons, energy, linear momentum, angular momentum, and in most cases parity. Except for high-energy reactions involving the production of mesons, the conservation of charge and number of nucleons allow one to infer that the numbers of protons and neutrons are always conserved. The conservation of the number of nucleons indicates that the statistics governing the system are the same before, during, and after the reaction. Fermi-Dirac statistics are obeyed if the total number is odd, and Bose-Einstein if the number is even. The conservation laws taken together serve to strongly restrict the reactions that can take place, and the conservation of angular momentum and parity in particular allow one to establish spins and parities of states excited in various reactions. *See* Angular momentum; Parity; Quantum statistics; Symmetry laws.

Reaction mechanism. What happens when a projectile collides with a target nucleus is a complicated many-body problem which is still not completely understood. Progress made in the last decades has been in the development of various reaction models which have been extremely successful in describing certain classes or types of nuclear reaction processes. In general, all reactions can be classified according to the time scale on which they occur, and the degree to which the kinetic energy of the incident particle is converted into internal excitation of the final products. A large fraction of the reactions observed has properties consistent with those predicted by two reaction mechanisms which represent the extremes in this general classification. These are the mechanisms of compound nucleus formation and direct interaction.

Compound nucleus formation. As originally proposed by N. Bohr, the process is envisioned to take place in two distinct steps. In the first step the incident particle is captured by (or fuses with) the target nucleus, forming an intermediate or compound nucleus which lives a long time ($\simeq 10^{-16}$ s) compared to the approximately 10^{-22} s it takes the incident particle to travel past the target. During this time the kinetic energy of the incident particle is shared among all the nucleons, and all memory of the incident particle and target is lost. The compound nucleus is always formed in a highly excited unstable state, is assumed to approach thermodynamic equilibrium involving all or most of the available degrees of freedom, and will decay, as the second step, into different reaction products, or through so-called exit channels. In most cases the decay can be understood as a statistical evaporation of nucleons or light particles. In the examples of reactions (1)–(4), the compound nucleus formed is ^{14}N, and four possible exit channels are indicated (Fig. 1). In reactions involving heavier targets (for example, $A \simeq 200$), one of the exit channels may be the fission channel where the compound nucleus splits into two large fragments. *See* Nuclear fission.

The essential feature of the compound nucleus formation or fusion reaction is that the probability for a specific reaction depends on two independent probabilites: the probability for forming the compound nucleus, and the probability for decaying into that specific exit channel. While certain features of various interactions cannot be completely explained within the framework of the compound nucleus hypothesis, it appears that the mechanism is responsible for a large fraction of reactions occurring in almost all projectile-target interactions. Fusion reactions have been extremely useful in several kinds of spectroscopic studies. Particularly notable have been the resonance studies performed with light particles, such as neutrons, protons, deuterons, and alpha particles, on light target nuclei, and the gamma-ray studies of reactions induced by heavy projectiles, such as ^{16}O and ^{32}S, on target nuclei spanning the periodic table. These studies have provided an enormous amount of information regarding the excitation energies and spins of levels in nuclei. *See* Nuclear spectra.

Direct interactions. Some reactions have properties which are in striking conflict with the predictions of the compound nucleus hypothesis. Many of these are consistent with the picture of a mechanism where no long-lived intermediate system is formed, but rather a fast mechanism where the incident particle, or some portion of it, interacts with the surface, or some nucleons on the surface, of the target nucleus. Models for direct processes make use of a concept of a homogeneous lump of nuclear matter with specific modes of excitation, which acts to scatter the incident particle through forces described, in the simplest cases, by an ordinary spherically symmetric potential. In the process of scattering, some of the kinetic energy may be used to excite the target, giving rise to an inelastic process, and nucleons may be exchanged, giving rise to a transfer process. In general, however, direct reactions are assumed to involve only a very small number of the available degrees of freedom.

Most direct reactions are of the transfer type where one or more nucleons are transferred to or from the incident particle as it passes the target, leaving the two final partners either in their ground states or in one of their many excited states. Such

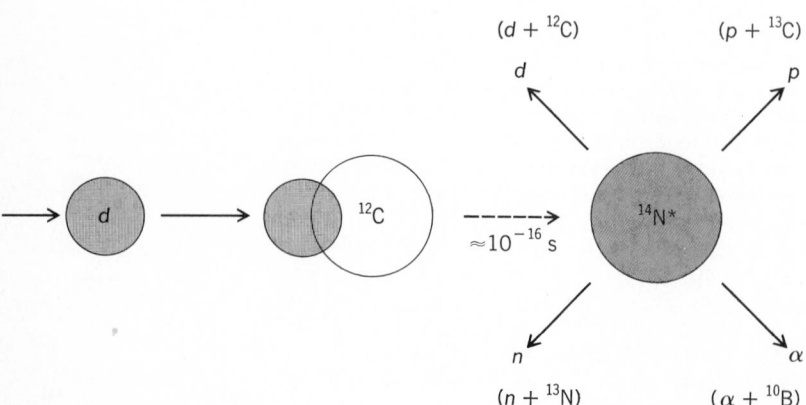

$(d + {}^{12}C)$ $(p + {}^{13}C)$

$(n + {}^{13}N)$ $(\alpha + {}^{10}B)$

Fig. 1. Formation of the compound nucleus ^{14}N after capture of the deuteron by ^{12}C. Four exit channels are indicated.

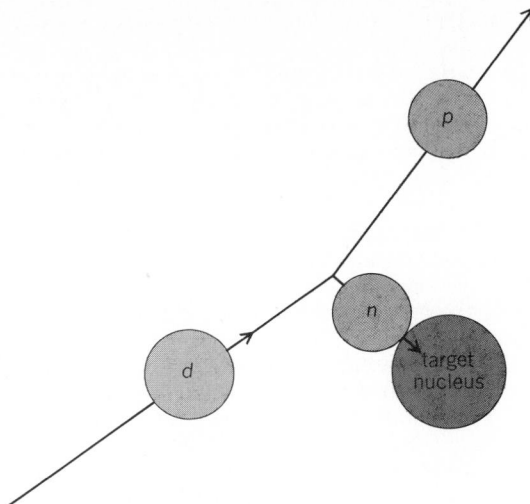

Fig. 2. A (*d,p*) transfer reaction.

transfer reactions are generally referred to as stripping or pick-up reactions, depending on whether the incident particle has lost or acquired nucleons in the reaction. The *(d,p)* reaction is an example of a stripping reaction, where the incident deuteron is envisioned as being stripped of its neutron as it passes the target nucleus, and the proton continues along its way (Fig. 2).

The properties of the target nucleus determine the details of the reaction, fixing the energy and angular momentum with which the neutron must enter it. The energy of the outgoing proton is determined by how much of the deuteron's energy is taken into the target nucleus by the neutron, and indeed serves to identify the final state populated by the *Q* value of the reaction as a whole. The angular distribution of the differential cross sections will, at appropriate bombarding energies, not be smooth but rather will show a distinct pattern of maxima and minima which are indicative of the spin and parity of the final state. The cross section for populating a specific final state in the target nucleus depends on the nuclear structure of the nuclei involved. This sensitivity has been used in studies of single-nucleon, two-nucleon, and four-nucleon transfer reactions, such as *(d,p)*, *(p,d)*, (³He, *d*), *(t,p)*, and (⁶Li, *d*), to establish the validity and usefulness of the shell-model description of nuclei. Multinucleon transfer reactions with heavier projectiles have been powerful tools for reaching nuclei inaccessible by other means and in producing new isotopes.

Inelastic scattering is also a direct reaction whose angular distribution can provide information about the spin and parity of the excited state. Whereas the states preferentially populated in transfer reactions are those of specific single-particle or shell-model structure, the states preferentially excited in inelastic scattering are collective in nature. The states are most easily understood in the framework of macroscopic descriptions in which they are considered to be oscillations in shape about a spherical mean (vibrations) or to be the rotations of a statically deformed shape. The cross section for inelastic scattering is related to the shape or deformation of the target nucleus in its various collective excitations. Inelastic excita-

tion can be caused by both a nuclear interaction and an electromagnetic interaction known as Coulomb excitation, where the target nucleus interacts with the rapidly changing electric field caused by the passage of the charged particle. Coulomb excitation is an important process at low bombarding energies in interactions involving heavy projectiles. Studies in inelastic scattering to low-lying states of nuclei across the periodic table have provided graphic demonstrations of the collective aspects of nuclei.

Elastic scattering is the direct interaction which leaves the interacting particles unchanged. For charged particles at low bombarding energies, the elastic scattering is well described in terms of the inverse-square force law between two electrically charged bodies. In this case the process is known as Rutherford scattering. At higher energies the particles come into the range of the nuclear force, and the elastic scattering deviates from the inverse-square behavior. *See* NUCLEAR STRUCTURE; SCATTERING EXPERIMENTS (NUCLEI).

More complex reaction mechanisms. Processes intermediate between direct and compound nucleus formation do occur. The best example of such a process is the so-called preequilibrium emission, where light particles are emitted before the kinetic energy has been shared among all the nucleons in the compound nucleus. Another example is seen in the interaction of two massive nuclei, such as ⁸⁴Kr + ²⁰⁹Bi, where the probability for formation of the compound nucleus is found to be very small. The experimental observations indicate that the nuclei interact for a short time and then separate in what appears to be a direct reaction. Although the interaction times for these so-called strongly damped or deep inelastic collisions are very small compared to that for compound nucleus formations, they are sufficiently long for considerable mass transfer and loss of relative kinetic energy to occur.

Nuclear reaction studies. In most instances the study of nuclear reactions is directed toward the long-range goal of obtaining information about the properties and structure of nuclei. Such studies usually proceed in two stages. In the first stage the attention focuses on the mechanism of the reaction and on establishing the dependence on the nuclei involved. The specific models proposed continue to be modified and improved, as they are confronted with more experimental results, until their predictions become reliable. At that point the second stage is entered, in which the focus of the effort is on extraction of information about nuclei.

There are other studies which are focused on reaction cross-section behaviors for other purposes. Examples of this are the neutron-capture reactions on heavy target nuclei that fission, and the ³H(*d,n*)⁴He reaction, important in thermonuclear processes, which continue to be studied because of their application as energy sources. Studies of the interactions of light nuclei at low energies are intensely studied because of their implications for astrophysics and cosmology. *See* NUCLEAR FUSION; THERMONUCLEAR REACTION.

[DENNIS G. KOVAR]

Bibliography: J. Cerny (ed.), *Nuclear Spectroscopy and Reactions*, 1974; W. M. Gibson, *The Physics of Nuclear Reactions*, 1980; P. E. Hodgson, *Nuclear Reactions and Nuclear Spectroscopy*, 1971.

Nuclear spectra

The distribution in energy or momentum of radiations emitted by radioactive nuclei or in a nuclear reaction; also, the graphical display of data from instruments used to measure such radiations (for example, magnetic spectrometers and scintillation detectors).

Radiations can occur when the total energy of a nuclear system (or state) is higher than that of a different configuration (or state) to which a transition can take place. The specific type of radiation which is emitted in such a process is determined by the characteristics of the initial and final nuclear systems as well as the properties of the interaction mechanism. The radiations not only remove energy from the initial system but can also lead to changes in mass, charge, angular momentum (spin), and parity (or symmetry characteristic) of the system.

Nuclear spectra are widely used in both basic and applied research. For the former, it is usually the information which can be inferred from studies of such spectra rather than the radiations themselves which are of primary interest. From experimental determinations of the type of radiation, its energy, and the changes that it causes in angular momentum and parity, one is able to deduce information pertaining to the static and dynamic properties of the initial and final nuclear configurations. This, in turn, enables one to gain insight into the structure of nuclei as well as the forces between nucleons. In applied research, it is the radiations themselves or the effects which they produce that are of most importance. *See* NUCLEAR STRUCTURE.

The distribution in energy (momentum) of radiations emitted during transitions between nuclear configurations can be discrete or continuous. Transitions which give rise to discrete spectra are those in which a single type of radiation is emitted. The emitted particle then has a specific energy determined by the two nuclear configurations and type of radiation involved. On the other hand, if two or more radiations are emitted during a transition, the energy will be shared among them. In this case, the energy of each particle emitted can take on a continuum of values between well-defined limits. This then gives rise to a continuous spectrum.

Mathematically a spectrum is described as the number of particles with a given energy (that is, relative intensity) as a function of energy. Graphically, the relative intensity is usually plotted along the ordinate and the energy along the abscissa. A spectrum can be composed of a series of discrete lines (or peaks), a smooth curve, or a combination of these, depending upon the specifics of the transitions involved. In practice, the spectra of particles emitted during nuclear transitions are distorted by the detection instruments used to measure them. For discrete spectra, the instruments are usually calibrated so that the central position of the peak along the abscissa defines the energy of the detected radiation, and the area under the peak is proportional to its relative intensity. The full width at half maximum of the peak is a measure of the effective resolution. A continuous spectrum has an end point corresponding to the energy change of the transition, and it often has a definite shape that is related to the characteristics of the radiation.

In early years, the study of nuclear spectra was mainly confined to naturally occurring radioisotopes (unstable nuclei). The development of nuclear particle accelerators has greatly broadened the scope of such research so that nuclear spectra characteristics of over 1600 nuclei have been measured. Some types and characteristics of nuclear spectra are described below, with those associated with the decay of radioisotopes being discussed first. *See* ISOTOPE; PARTICLE ACCELERATOR; RADIOISOTOPE.

Beta-ray spectrum. Beta rays are electrons emitted by a nucleus of atomic number Z which, in its ground state, is unstable with respect to one of its neighboring isobars of charge $Z+1$ or $Z-1$. Beta decay can also take place from an excited state (isomer) if radiative decay to the ground state is greatly hindered. The charged-particle emission consists of a negative electron to the $Z+1$ nucleus or a positive electron (positron) to the $Z-1$ nucleus, and the transition can take place to either the ground or to an excited state of the daughter nucleus.

A transition involving the emission of a beta ray is accompanied by the simultaneous emission of a neutrino, a particle that has neither mass nor charge and is exceedingly difficult to detect. The total energy of such a transition is a fixed quantity, and since this energy is shared between the beta ray and the neutrino, the beta ray can emerge with any energy from zero up to the maximum transition energy. Thus the spectrum of beta rays emitted from an ensemble of nuclei which undergo such a transition is continuous. As a result of the statistical manner in which the energy is shared between the beta ray and neutrino, the spectrum of beta-ray intensity plotted versus energy has a bell shape with a broad maximum located at somewhat less than half the maximum energy.

Beta-ray spectra are usually analyzed by a so-called Fermi-Kurie plot. Such a plot converts the experimental intensity-versus-energy spectrum (corrected for source thickness and instrumental effects) into a straight-line distribution which intersects the abscissa at the maximum energy (end-point energy). For negative beta-ray emitters, the end-point energy is equal to the transition energy, except for a nuclear recoil correction. In most instances, this correction is negligible due to the small mass of the beta ray relatives to the nucleus. For a positron emitter, the transition energy is equal to the end-point energy plus 1.02 MeV. If the Fermi-Kurie plot of the experimental data, under the assumption of a purely statistical intensity-versus-energy distrubution, is not a straight line, the beta-ray transition is said to have a nonstatistical shape and to be of a forbidden type. The degree of forbiddenness is determined by the correction factors needed to straighten out the Fermi-Kurie plot. Whether a beta-ray transition has the allowed or forbidden shape depends upon the characteristics of the initial and final nuclear states involved.

Alpha-particle spectra. The emission of an alpha particle (equivalent to the nucleus of a mass-4 helium atom) can occur when the state of a nucleus with charge Z, mass A, is unstable with respect to a state of a nucleus with charge $Z-2$, mass $A-4$. In order for the alpha particle to emerge, it must overcome the Coulomb-charge-potential barrier of the nucleus and also a centrifugal barrier which depends upon the angular momentum removed. Alpha-particle spectra of radioisotopes are discrete, and the peak energies are less than the corresponding transition energies by an amount equal to the nuclear recoil energy.

Spontaneous fission. Some radioisotopes decay by spontaneous fission; that is, they break up into two fragments (occasionally three). The primary fission fragments and a number of their daughters are themselves radioactive and give rise to nuclear spectra. *See* NUCLEAR FISSION.

Gamma-ray spectra. Gamma rays are emitted when a transition takes place from an excited state to a lower state in the same nucleus. Of the various types of nuclear radiation, gamma rays produce the least amount of nuclear recoil, although in certain kinds of experiments the recoil energy shift is observable. Thus, the gamma-ray energy is almost exactly equal to the energy difference between the states. A process which sometimes competes with gamma-ray emission is internal electron conversion. This process creates holes in the atomic shell structure which, when filled, are accompanied by the emission of x-rays. Hence, it is not unusual to find x-rays in the low-energy regions of gamma-ray spectra.

When gamma rays interact with matter, they produce secondary radiations by means of the photoelectric and Compton effects as well as by pair production (the emission of a positron-electron pair). Instruments (such as ionization chambers, scintillators, and magnetic spectrometers) that measure gamma rays are based upon the detection of these secondary radiations. Of the various types of gamma-ray detectors, it is the scintillator (especially with LiGe crystals) which has been responsible for the gathering of enormous quantities of gamma-ray data. From a determination of gamma-ray energies and intensities, it has been possible to construct nuclear level schemes for many nuclei. Since there is little probability that the level structure of any two nuclei will be identical, precise measurements of gamma-ray transitions can serve as a means to uniquely identify a particular isotope. This feature of gamma-ray spectra is used extensively in applied research. *See* COMPTON EFFECT; ELECTRON-POSITRON PAIR PRODUCTION; PHOTOELECTRICITY.

Nuclear spectra from reactions. There is a myriad of nuclear reactions by which one can produce nuclear spectra of various types (for example, gamma rays, neutrons, protons, and many nuclei). Nuclear reactions can be categorized roughly into three types: (1) elastic and inelastic scattering; (2) transfer reactions in which one or more nucleons (that is, protons or neutrons) are transferred between the projectile and target nucleus; and (3) reactions in which the projectile and target nucleus coalesce to form a compound system which then decays in some manner. In past years, the particles that could be used as projectiles were limited to the lighter elements. However, the development of heavy-ion accelerators has been undertaken to make possible the use of even the heaviest elements as projectiles.

The spectra of particles produced in a nuclear reaction depend upon the target nucleus, the reaction energy release, and the kinematics involved (including geometric and recoil effects). Usually, more than one type of radiation is emitted, and, in practice, the researcher chooses which particles to measure on the basis of the nuclear structure or reaction properties under study. Whether the spectrum of particles of any given type is discrete or continuous depends upon the specifics of the reaction used.

As an example, consider the case of inelastic scattering. Inelastic scattering takes place when the projectile gives up some of its energy to a target nucleus and is scattered. If the energy transferred to the target is sufficient only to excite discrete levels, the spectrum of the scattered particles will consist of a series of peaks corresponding to each level excited. However, if the energy transferred were sufficient to excite the nucleus up to the point where the density of states is continuous, then the corresponding portion of the spectrum would also be continuous. In the same reaction, nuclear radiations of one type or another (such as gamma rays and neutrons) will also be emitted following the decay of the states which have been excited. In most cases (but not always), the decay of discrete states excited in such reactions takes place by the emission of gamma rays and the spectra involved are also discrete. The decay of excited states in the continuum region usually takes place by neutron or proton emission, and these spectra are usually continuous. *See* NUCLEAR REACTION; RADIOACTIVITY.

[D. J. HOREN]

Bibliography: J. Cerny (ed.), *Nuclear Spectroscopy and Reactions*, pts. A, B, C, and D, 1974; J. H. Hamilton and J. C. Manthuruthil (eds.), *Radioactivity in Nuclear Spectroscopy*, vols. 1 and 2, 1972; K. Siegbahn (ed.), *Alpha-, Beta-, and Gamma-Ray Spectroscopy*, vols. 1 and 2, 1965.

Nuclear structure

The atomic nucleus is at the center of the atom and contains 99.975% of the total mass of the atom. Its average density is about 3×10^{11} kg/cm^3; its diameter is about 10^{-12} cm, and thus much smaller than the diameter of the atom, which is about 10^{-8} cm.

The nucleus is composed of protons and neutrons. The number of protons is usually denoted by Z, while that of neutrons is denoted by N. The number of protons is equal to the number of electrons in the atom. Since the proton is positively charged, the electron is negatively charged, and the neutron is neutral, the atom as a whole is neutral under normal conditions. The total number of protons and neutrons in a nucleus is called the mass number and denoted by $A = N + Z$. Nuclei having the same proton number but different neutron number are called isotopes. Nuclei having the same neutron number but different proton number are called isotones. Finally, nuclei with the same mass number are called isobars. *See* ATOMIC STRUCTURE AND SPECTRA; ELECTRON; NEUTRON; PROTON.

Bulk properties. The bulk properties of nuclei include their sizes, density distributions, and masses.

Nuclear densities and sizes. The average radius of the proton distribution of the nucleus can be easily measured by using the scattering of fast electrons and the energies of x-rays emitted by μ-

mesic atoms. When a high-velocity electron approaches a nucleus, it is deflected from its path because of the electromagnetic interaction with the protons in the nucleus. A measurement of this deflection permits the determination of the charge distribution of the nucleus. When a negative μ-meson (of mass $m_\mu \simeq 207\, m_e$, where m_e is the mass of the electron) is brought to rest in matter, it is attracted by the positive charge of the nucleus and captured into an orbit around it, thus forming a μ-mesic atom. The μ-meson can then jump from one orbit to another, emitting x-rays. The lowest orbits of the μ-meson have diameters comparable to the size of the nucleus, and thus the x-ray energies are greatly affected by the actual charge distribution of the nucleus. A measurement of these energies provides information on the proton distribution in nuclei.

From both methods, it has been possible to determine that the charge distribution in nuclei can be quite accurately described by Eq. (1), where

$$\rho(r) = \rho(0)[1 + e^{(r-R)/a}]^{-1} \qquad (1)$$

$\rho(r)$ is the proton density at a distance r from the center, and $\rho(0)$ is the density at the center. This bell-shaped distribution (called a Woods-Saxon distribution) is shown in Fig. 1 for nuclei of cobalt and bismuth. The quantity R is the point at which the nuclear density has fallen to one-half of its central value, and is referred to as the nuclear radius. It is usually measured in fermis (1 F $= 10^{-13}$ cm $= 10^{-15}$ m $=$ 1 fm). The dependence of R on the mass number is approximately given by Eq. (2). The

$$R = (1.07 \pm 0.02)\, A^{1/3}\ \text{F} \qquad (2)$$

quantity a is called the surface thickness and is given by Eq. (3). The average radius R is propor-

$$a = (0.55 \pm 0.07)\ \text{F} \qquad (3)$$

tional to $A^{1/3}$, which implies that the volume is proportional to A and thus that the mean density is independent of the size of the nucleus. The phenomenon is usually referred to as saturation.

It is much more difficult to measure the neutron distribution. Since the neutrons have no electric charge, a measurement of their distribution must be done with strongly interacting particles, such as α-particles. However, unlike electrons, these particles are not pointlike, but have a finite extent of their own. Moreover, the strong interaction is not so well known as the electromagnetic interaction,

and this introduces further uncertainties in the measurement of the neutron distribution. Although most theories predict a neutron radius in heavy nuclei larger than the proton radius, measurements are consistent with a neutron radius approximately equal to the proton radius ($R_n - R_p \cong$ 0.1 – 0.2 F, where R_n is the neutron radius and R_p the proton radius). A method of measurement based on the scattering of positively and negatively charged π-mesons may provide more definitive information on the difference $R_n - R_p$. *See* SCATTERING EXPERIMENTS (NUCLEI).

Nuclear masses. Nuclear masses are usually measured in unified mass units (symbol u). One mass unit equals one-twelfth of the mass of the carbon atom, which has $A = 12$. The observed masses of the proton and neutron are 1.007277 and 1.008665 u, respectively, while that of the hydrogen atom is 1.007825 u. By convention, whenever one speaks of a nuclear mass, one includes the mass of the Z electrons, thus quoting the mass of the corresponding neutral atom and not of the nucleus alone.

Atomic masses are, to a good approximation, described by a semiempirical mass formula. The following terms contribute to this formula. First, one has the contribution of the Z protons, N neutrons, and Z electrons, altogether given by 1.007825 Z + 1.008665 N. There is then a term which is roughly proportional to the volume of the nucleus. Since from Eq. (2), R is proportional to $A^{1/3}$, this gives a contribution proportional to A, $-c_v A$. Next, there is a term proportional to the area of the surface of the nucleus, $c_a A^{2/3}$. Another term comes from the Coulomb repulsion between protons. This term can be written in the form $c_c Z^2/A^{1/3}$. The Coulomb repulsion makes it more favorable to have in the nucleus more neutrons than protons. However, this repulsion is counterbalanced by the strong nuclear interaction, that which holds nuclei together. This interaction favors the situation in which the number of protons and neutrons is equal, and contributes $c_s(N-Z)^2/A$. Finally there is a term which takes into account the fact that nucleons in nuclei tend to pair together, gaining energy when doing so. This last term is of the form $\pm c_p A^{-3/4}$, where the plus sign is used for odd-odd nuclei and the minus sign is used for even-even nuclei. The term is assumed to be zero in even-odd nuclei. Inserting the appropriate numerical constants, the mass of an atom, $M(A,Z,N)$, is given by Eq. (4).

$$\begin{aligned}
M(A,Z,N) = {} & 1.007825\, Z + 1.008665\, N - 0.015\, A \\
& + 0.014\, A^{2/3} + 0.021\, \frac{(N-Z)^2}{A} \\
& + 0.000627\, \frac{Z^2}{A^{1/3}} \pm 0.036\, A^{-3/4} \qquad (4)
\end{aligned}$$

A convenient way to present miscellaneous data concerning nuclei is offered by the chart of nuclei, in which each nucleus is represented by a unit square in a plot of Z versus N. The general layout of this chart is shown in Fig. 2. Stable nuclei are found along a stability valley. This valley departs more and more from the line $Z = N$ the heavier the nucleus. Nuclei away from the stability line decay either by converting protons into neutrons (or vice versa), creating in the process an electron and a neutrino (β-decay), or by boiling off particles, such as neutrons (n-decay), protons (p-decay), and He

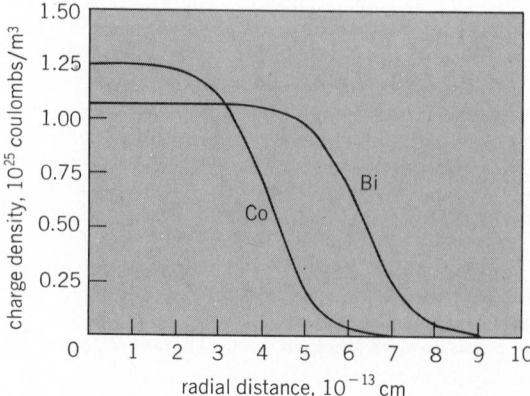

Fig. 1. Charge density in nuclei of cobalt and bismuth.

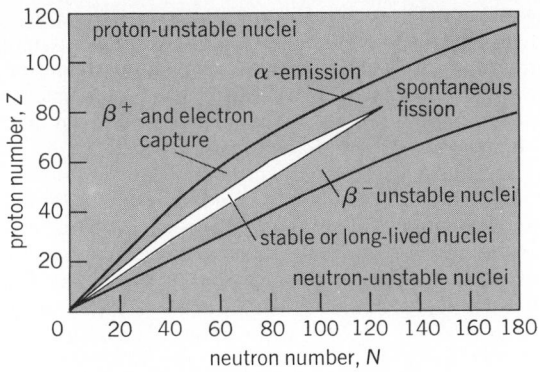

120
100
80
60
40
20

proton-unstable nuclei

α-emission

spontaneous fission

β^+ and electron capture

β^- unstable nuclei

stable or long-lived nuclei

neutron-unstable nuclei

proton number, Z

0   20   40   60   80   100 120 140 160 180

neutron number, N

Fig. 2. General arrangement of the chart of nuclei, with lines of stability against various break-up modes.

nuclei (α-decay), or finally by dividing into two or more pieces (fission). The number of nuclei which have been investigated experimentally has increased considerably since 1970. For each isotopic chain (constant Z) there are, at present, on the average about 15 nuclei known. *See* NUCLEAR FISSION; RADIOACTIVITY.

Detailed properties. The detailed properties of a nucleus include the energies, angular momenta, and parities of its quantum states, and its magnetic and electric moments.

Energy-level diagram. The nucleus is a quantum-mechanical system. Its properties are best described by a diagram, called the energy-level diagram, in which its quantum states are listed, together with the expectation values of all measurable quantities. Three of these are particularly important: the energy, the angular momentum, and the parity. Energies are usually measured from the lowest state (called the ground state) and are given in millions of electronvolts (1 MeV = 1.60219×10^{-13} J). Angular momenta are labeled by a quantity J which is half-integer for odd-even nuclei and integer for even-even and odd-odd nuclei. In terms of this quantity the square of the angular momentum is given by $P_j^2 = J(J+1)h^2$, where $h = 1.05459 \times 10^{-34}$ J·s. With no exception, the ground states of even-even nuclei have been found to have $J = 0$. Parity, denoted by π, is a purely quantum-mechanical concept, which describes the transformation of the wave function of the system under reflection $\vec{x} \to -\vec{x}$. It is either plus or minus. *See* ANGULAR MOMENTUM; ENERGY LEVEL; PARITY.

Magnetic and electric moments. In addition to energies, angular momenta, and parities, other properties are often measured. Among these, especially important are the intensities of the electromagnetic transitions between two energy levels. Nuclei, like atoms, in a state of higher energy can decay to states of lower energy by emitting photons. The corresponding transitions can be either electric *(E)* or magnetic *(M)*, and have multipolarity $l = 0, 1, 2, \ldots$. Electromagnetic transitions satisfy certain selection rules which are related to the angular momenta and parities of the energy levels between which the transition is taking place. For electric transitions of multipolarity l, the parity of the initial, π_i, and final, π_f, states are related by $\pi_i \pi_f = (-)^l$. For magnetic transitions, they are related by $\pi_i \pi_f = (-)^{l+1}$. The multipolarity of the transition l is limited by $|J_i - J_f| \leq l \leq |J_i + J_f|$.

Thus an electric-dipole, $l = 1$, transition (*E*1) between a state with $J_i^{\pi_i} = 2^+$ and $J_f^{\pi_f} = 0^+$ is not allowed. Conversely, the same transition is allowed between $J_i^{\pi_i} = 1^-$ and $J_f^{\pi_f} = 0^+$. Other measured properties of nuclear states are their static magnetic and electric moments. Magnetic dipole moments are measured in nuclear magnetons (1 nuclear magneton = 1/1836.15 Bohr magneton = 5.0508×10^{-27} J/T). In the scale of nuclear magnetons, the proton has a magnetic moment of 2.79285 nuclear magnetons, and the neutron of −1.91304 nuclear magnetons. Electric quadrupole moments are usually measured in electron-barns (1 eb = 10^{-24} e cm^2, where e is the magnitude of the charge of the electron). *See* MAGNETON; NUCLEAR MOMENTS.

Nuclear models. A number of models have been developed to account for the properties of nuclei.

Shell model. The basic model for the description of nuclear properties is the shell model. The strong nuclear force binds together protons and neutrons, called by the single name nucleons. Each individual nucleon moves in the average potential generated by all the others. This potential has the same shape as the nuclear matter distribution of Fig. 1, but upside down. The states of a single nucleon in the average potential cluster together into layers or shells, much like the single-particle states in atoms. Attempts to predict the location of the shells were not very successful until 1949, when M. G. Mayer and J. H. Jensen introduced a new term in the average potential field. The term, spin-orbit interaction, describes an interaction of the intrinsic spin of the nucleon with its orbital angular momentum, and it can be written in the form of Eq. (5), where $f(r)$ is a function of the radial dis-

$$V_{so} = f(r) (\vec{s} \cdot \vec{l}) \qquad (5)$$

tance r, and $\vec{s}$ and $\vec{l}$ are the intrinsic and orbital angular momenta of the nucleon. The introduction of this term gives rise to the single-particle structure shown in Fig. 3. In this figure the closed shells appear at nucleon numbers 2, 8, 20, 28, 50, 82, 126, in agreement with experiment. The numbers at which closed shells occur are called magic numbers. *See* MAGIC NUMBERS; SPIN.

Nuclei with few valence particles. Closed-shell nuclei behave, in many respects, as inert. Most properties of nuclei can be described by considering as active only nucleons outside the closed shells. These are called valence nucleons. In addition to the average potential in which they move, it is assumed that there is a residual interaction between valence nucleons. When the number of valence nucleons is small (up to about four), the effects of the residual interaction can be calculated easily. Figure 4 shows a comparison between the calculated energy-level diagram, called a spectrum, of an even-even nucleus with two valence nucleons, and the experimental scheme. The agreement between theory and experiment is usually found to be good.

Nuclei with many valence particles. When the number of valence nucleons is large (greater than 5–6), some striking regularities develop in the level scheme. These regularities, first recognized by J. Rainwater, A. Bohr, and B. R. Mottelson, indicate the presence of collective features in nuclei. The collective features arise from two special properties of the residual interaction between val-

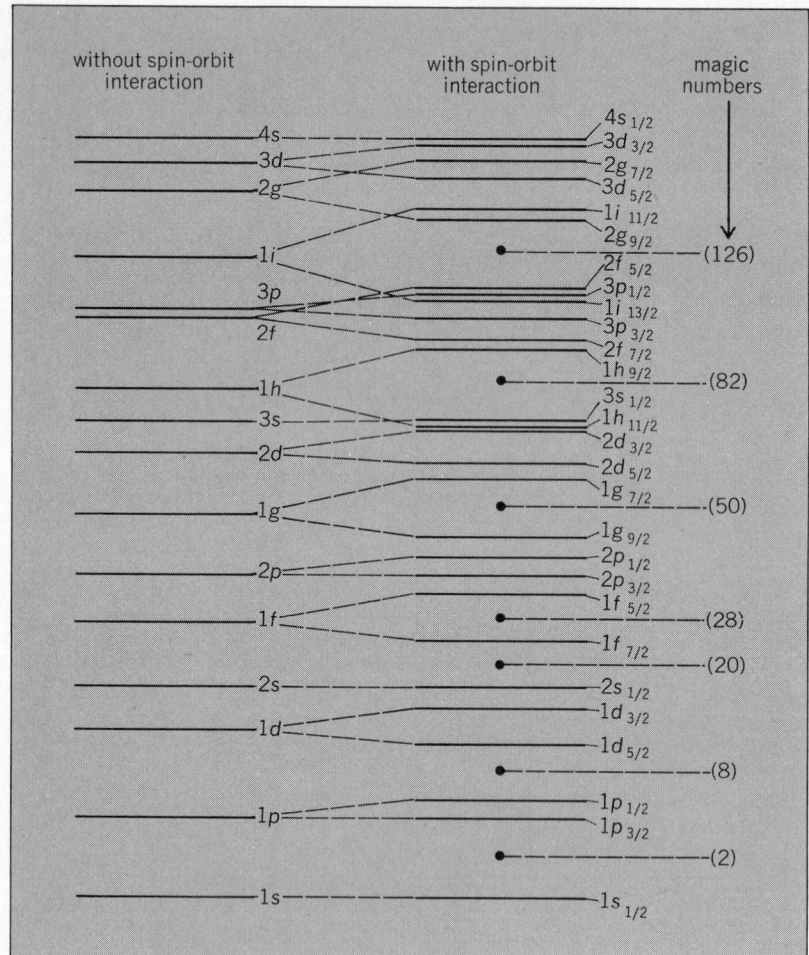

Fig. 3. Single-particle levels according to the shell model. Symbols at right indicate their spectroscopic notation. The letter indicates orbital angular momentum (*s, p, d, f, . . .*, for *l* = 0, 1, 2, 3, . . .), the integer indicates the order of levels with the same *l*, and the subscript indicates the *j* (total angular momentum) value.

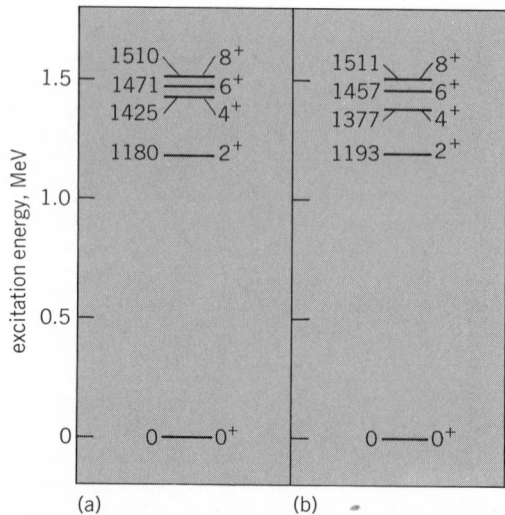

Fig. 4. Comparison of (a) experimental level scheme of $^{210}_{84}Po_{126}$ (A = 210, Z = 84, N = 126) and (b) that calculated using the shell model. The numbers on the left of each level are the excitation energies in kiloelectronvolts, and the symbols on the right are the angular momenta and parities.

ence nucleons: first, its pairing property which tends to pair off nucleons, one with angular momentum up and one with angular momentum down, to give zero resultant; and second, a quadrupole property which favors a distortion of the nuclear surface in the shape of an ellipsoid.

A large variety of collective spectra of quadrupole type have been observed in nuclei. These spectra can be discussed in terms of the geometry of the nuclear surface. The nuclear surface can be characterized by giving its radius R. For nuclei with quadrupole distortions, R can be written as in Eq. (6), where $Y_{2\mu}(\theta,\phi)$ is the spherical harmonic of

$$R = R_0 \left(1 + \sum_{\mu=-2}^{+2} \alpha_\mu Y_{2\mu}(\theta,\phi)\right) \quad (6)$$

order two and the quantities α_μ ($\mu = 0, \pm 1, \pm 2$) are called deformation parameters. When $\alpha_\mu = 0$, the nucleus is said to be spherical; when $\alpha_\mu \neq 0$, the nucleus is said to be deformed. For spherical nuclei, the radius R is given by Eq. (2). Instead of α_μ, it has become customary to introduce another set of variables, related to the α_μ's by Eqs. (7). These

$$\alpha_0 = \beta \cos \gamma$$
$$\alpha_2 = \alpha_{-2} = \frac{1}{\sqrt{2}} \beta \sin \gamma$$
$$\alpha_1 = \alpha_{-1} = 0 \quad (7)$$

are called Bohr variables. Collective spectra can then be classified according to the values of β and γ which describe the shape of a nucleus. The observed spectra fall into three major categories:

1. Rotational spectra of deformed nuclei with axial symmetry. These correspond to shapes characterized by a value of $\beta = \beta_{equ} \neq 0$, called equilibrium deformation, and $\gamma = 0^0$. The spectra consist of a series of energy levels, connected by large electric quadrupole (E2) transitions, called bands. Within each band, the excitation energies are approximately given by Eq. (8), where κ is a constant

$$E = \kappa J(J + 1) \quad (8)$$

typical of a given nucleus and J is the angular momentum of the level. In even-even nuclei, the band built on the ground state is composed of levels with $J = 0, 2, 4, . . .$; levels with odd angular momenta, $J = 1, 3, . . .$, are missing. Other bands, with higher excitation energy, may or may not begin with angular momentum $J = 0$. If they do, the angular momentum sequence is the same as in the ground-state band. Otherwise, both even and odd angular momenta appear. An example is shown in Fig. 5.

2. Rotational spectra of deformed nuclei with γ instability. These correspond to shapes characterized by a value $\beta = \beta_{equ} \neq 0$, but no fixed value of γ. Here again, the spectrum consists of a series of bands, but the band structure is very different from the previous case, and energies within a band are given by Eq. (9), where A is a constant typical

$$E = A\tau(\tau + 3) \quad (9)$$

of a given nucleus and τ is a quantum number which labels the energy levels ($\tau = 0, 1, 2, . . .$). For each value of τ, there can be one or more levels with different J. The values of J belonging to each τ are given, for the lowest levels, by $\tau = 0, J = 0$; $\tau = 1, J = 2$; $\tau = 2, J = 4, 2$; and $\tau = 3, J = 6, 4, 3, 0$.

3. Vibrational spectra of spherical nuclei. These correspond to shapes characterized by a value of $\beta = \beta_{equ} = 0$ and no fixed value of γ. The spectrum consists of a series of vibrational multiplets, whose energy is approximately given by Eq. (10),

$$E = \epsilon n \tag{10}$$

where ϵ is the vibrational energy, typical of a given nucleus, and n is a vibrational quantum number ($n = 0, 1, 2, \ldots$). Here again, for each value of n, there can be one or more levels with different J. The values of J belonging to each n are given, for the lowest levels, by $n = 0, J = 0$; $n = 1, J = 2$; $n = 2$, $J = 4, 2, 0$; $n = 3, J = 6, 4, 3, 2, 0$. See QUANTUM NUMBERS.

Instead of discussing collective spectra in terms of the geometry of the nuclear surface, it is also possible to describe them in terms of representations of symmetry groups. The group structure then expresses the symmetry of the nuclear surface. The three major categories given above correspond then to invariance of the hamiltonian describing the nucleus, under the appropriate symmetry group. The three symmetry groups of the collective quadrupole motion in nuclei are called SU(3), SO(6), and SU(5), respectively. See NONRELATIVISTIC QUANTUM THEORY; SYMMETRY LAWS.

Even-even nuclei with several valence particles show spectra which are either of these three types or intermediate between them. A similar, but more complex, classification scheme of collective spectra has also been developed for odd-even and odd-odd nuclei.

Statistical model. The number of energy levels of a nucleus below an excitation energy of 2 MeV is usually small. As shown above, these levels can be described by using the shell model and allowing only for excitations of the valence nucleons from some single-particle levels to others nearby in energy. At higher excitation energies, the number of observed states increases considerably. These states arise from the excitation of the valence nucleons to higher single-particle levels and from excitations of nucleons from the closed shells to the valence shell. Because of their large number, it is no longer possible to describe properties of individual states. Thus, statistical methods are employed to describe average properties of states of this sort. An important statistical property is the average density of states ρ as a function of the excitation energy E. This is given by Eq. (11), where

$$\rho(E) = b \frac{1}{E^2} \exp\{2(aE)^{1/2}\} \tag{11}$$

a is a coefficient which varies from nucleus to nucleus. For a nucleus with mass number $A = 150$, $a \cong 16$ MeV^{-1}. The constant b depends on the spin J of the level, but not on the parity π, since it is assumed that at a given excitation energy E there are an equal number of states with parity $+$ as with parity $-$. The principal source of information on the quantities b and a comes from the study of the resonances observed in the interaction of slow neutrons with nuclei. See NEUTRON SPECTROMETRY.

Simple modes. It is an interesting property of nuclei that simple collective modes of excitation occur even at higher excitation energy. Unlike the

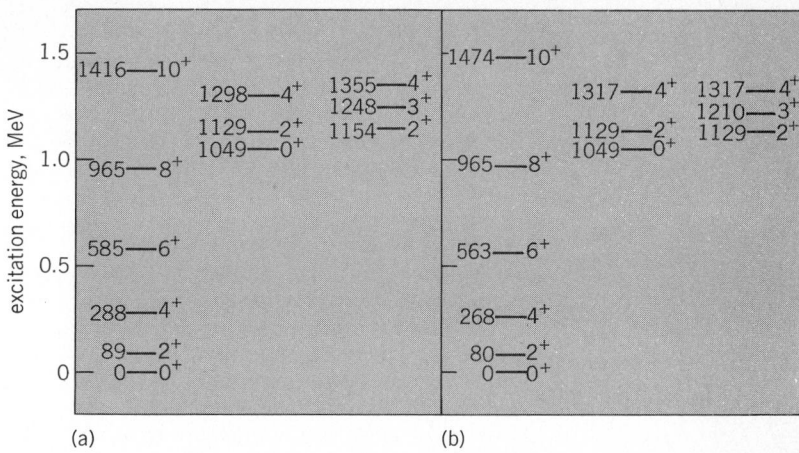

Fig. 5.   Comparison of (a) experimental level scheme of $^{156}_{64}$Gd$_{92}$ and (b) that calculated using the collective model. The numbers on the left of each level are the excitation energies in kiloelectronvolts, and the symbols on the right are the angular momenta and parities.

low-lying collective modes, these are no longer necessarily associated with pairing and quadrupole correlations. They arise from some coherent excitation of nucleons to shells other than the valence shell. These modes are called giant resonances, and are observed in inelastic scattering of strongly interacting particles off nuclei (α-scattering, for example), in inelastic scattering of electrons off nuclei, and in photoabsorption. The resonances appear as bumps in the cross section, indicating that the special properties of these collective states are spread over many states in a certain energy region. The spreading occurs because, in the same energy region, there exist many states of the type described above (random excitations). The collective state then mixes with these underlying states, sharing its properties with them. In addition to its energy, angular momentum, and parity, a giant resonance is characterized by its spreading width, Γ, that is, the interval in energy over which its properties are spread (Fig. 6). The first giant mode to be observed in nuclei was the $J^\pi = 1^-$ mode, called giant dipole resonance. The energy of this mode follows approximately Eq. (12).

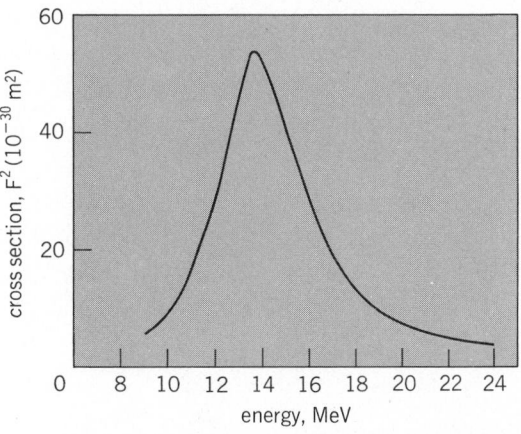

Fig. 6.   Total photoabsorption cross section for $^{197}_{79}$Au$_{11}$. The resonance has energy $E = 13.9$ MeV, width $\Gamma = 4.2$ MeV, and spin and parity $J^\pi = 1^-$.

$$E = 79A^{-1/3} \text{ MeV} \qquad (12)$$

Other modes observed are the giant quadrupole resonance, $J^\pi = 2^+$, and the giant monopole resonance, $J^\pi = 0^+$. *See* GIANT NUCLEAR RESONANCES.

Nuclear forces. The nucleons which form the nucleus are held together by the strong nuclear forces.

Free interaction. The forces which act between two nucleons when they are free (not inside the nucleus) have been studied by performing scattering experiments of one nucleon on the other, and by analyzing the properties of the deuteron, the only bound state of two nucleons. Because the proton and the neutron share many properties, it is convenient to consider them as two different states of the same particle, the nucleon. In order to distinguish protons from neutrons, the nucleon is given a sort of intrinsic angular momentum, called isospin. Since there are only two states, the isospin of the nucleon is taken to be $t = 1/2$, and the proton and the neutron correspond to the two components $t_z = \pm 1/2$ of t. Nuclear forces depend on the total isospin T of the nucleons. This isospin can be $T = 0$ or $T = 1$. A study of low-energy scattering of nucleons on nucleons has revealed that the proton-proton, neutron-neutron, and proton-neutron $T = 1$ forces are practically identical. This property is called charge independence. The proton-neutron $T = 0$ force is rather different. The free nucleon-nucleon force strongly depends also on the total intrinsic angular momentum S of the two particles. Since each nucleon has spin $s = 1/2$, the total spin can be $S = 0$ or $S = 1$, called singlet and triplet, respectively. *See* DEUTERON; ISOTOPIC SPIN.

To a good approximation, the free nucleon-nucleon force can be written as a sum of a spin-independent interaction, $V_c(r)$, sometimes called a central interaction, an interaction which depends on the total intrinsic, S, and orbital, L, angular momentum, $V_{LS}(r)(\vec{S} \cdot \vec{L})$, called a spin-orbit interaction, and a more complex interaction which in-

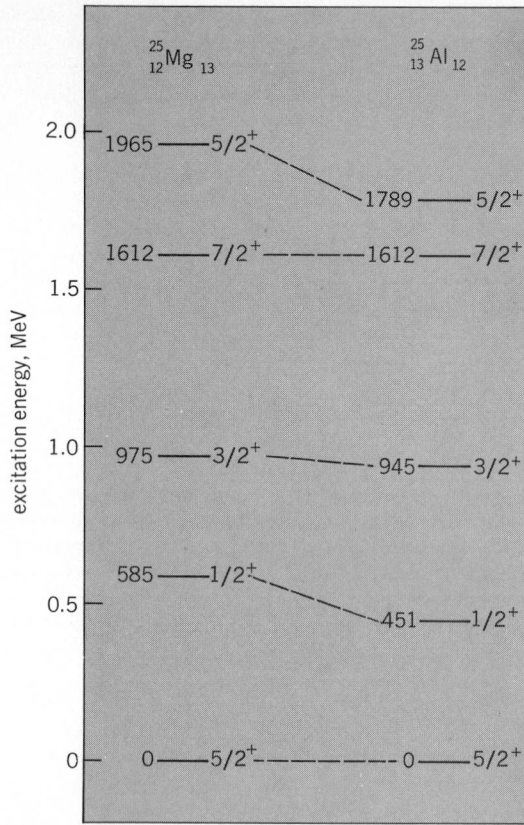

Fig. 8. Comparison of the experimental level schemes of the mirror nuclei $^{25}_{12}\text{Mg}_{13}$ and $^{25}_{13}\text{Al}_{12}$. The Coulomb energy difference has been removed by placing the two ground states at the same energy. The numbers on the left of each level are the excitation energies in kiloelectronvolts, and the symbols on the right are the spins and parities.

volves both spins simultaneously, $V_T(r)[3(\vec{s}_1 \cdot \hat{r}) \cdot (\vec{s}_2 \cdot \hat{r}) - (\vec{s}_1 \cdot \vec{s}_2)]$, called a tensor interaction. Here r is the relative distance between the two nucleons. The form of $V_c(r)$ is shown in Fig. 7. It consists of a short-range repulsion ($r \lesssim 0.6$ F) and a long-range attraction ($r \gtrsim 0.6$ F). It is rather similar to the interaction between two neutral atoms but, on the contrary, very different from the Coulomb interaction between two pointlike charges $V_{\text{Coul}}(r) \propto 1/r$. The nucleon-nucleon interaction decreases as $e^{-\mu r}/r$ for large distances, $r \to \infty$. The inverse length μ is of the order of 0.7 F^{-1}.

Forces inside nuclei. Nuclear structure calculations require the knowledge of the residual interaction between nucleons. This interaction is different from the free nucleon-nucleon interaction because of the modifications introduced by the presence of the other nucleons in the nucleus. Most calculations are done by assuming a strong, attractive interaction of zero range, usually written as $-V_0\delta(\vec{r} - \vec{r}')$. The function $\delta(\vec{r} - \vec{r}')$, called the Dirac delta function, states that the interaction is different from zero only when the relative distance between the two nucleons is zero. In order to take into account the dependence on intrinsic angular momenta, $\vec{s}_1$, $\vec{s}_2$, and isospins, $\vec{t}_1$, $\vec{t}_2$, the full interaction is written as in Eq. (13), where W, B, H, and

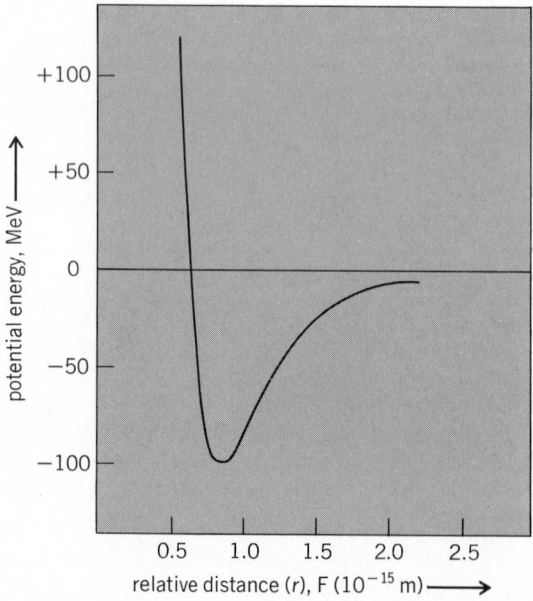

Fig. 7. Shape of the nucleon-nucleon potential for two nucleons in a singlet $S = 0$ state and isotopic spin $T = 1$ as a function of the relative distance r.

$$V(\vec{r} - \vec{r}') = -V_0\delta(\vec{r} - \vec{r}')[W + B(\vec{s}_1 \cdot \vec{s}_2) \\ + H(\vec{t}_1 \cdot \vec{t}_2) + M(\vec{s}_1 \cdot \vec{s}_2)(\vec{t}_1 \cdot \vec{t}_2)] \quad (13)$$

M are constants which describe the spin-isospin dependence of the interaction and V_0 is an overall strength. The interaction inside nuclei is called the effective interaction, and its relationship to the free interaction is not fully understood.

One property of the effective interaction is its charge independence. This property arises from the charge independence of the free nucleon-nucleon interaction. If charge independence were exactly true, the spectra of the two nuclei $(Z, N + 1)$ and $(Z + 1, N)$, called mirror nuclei, would be exactly identical. This cannot be actually correct since, in addition to the nuclear interaction, there is the electrostatic repulsion between nucleons. However, if this Coulomb piece is removed, the two spectra should be identical. In the example of Fig. 8, the two spectra are, to a large extent, identical. The observed differences are due to high-order effects of the electrostatic interaction between the protons which are not removed by subtracting the Coulomb energy. *See* ANALOG STATES; NUCLEAR REACTION; NUCLEAR SPECTRA; STRONG NUCLEAR INTERACTIONS.   [F. IACHELLO]

Bibliography: A. Bohr and B. R. Mottelson, *Nuclear Structure*, vols. 1–2, 1969–1975; A. deShalit and H. Feshbach, *Theoretical Nuclear Physics*, vol. 1: *Nuclear Structure*, 1974; H. A. Enge, *Introduction to Nuclear Physics*, 1966.

Nuclide

A species of atom that is characterized by the constitution of its nucleus, in particular by its atomic number Z and its neutron number $A - Z$, where A is the mass number. Whereas the terms isotope, isotone, and isobar refer to families of atomic species possessing common atomic number, neutron number, and mass number, respectively, the term nuclide refers to a particular atomic species. The total number of stable nuclides is approximately 275. About a dozen radioactive nuclides are found in nature, and in addition, hundreds of other have been created artificially.

[HENRY E. DUCKWORTH]

Numerical analysis

The development and analysis of computational methods (and ultimately of program packages) for the minimization and the approximation of functions, and for the approximate solution of equations, such as linear or nonlinear (systems of) equations and differential or integral equations. Originally part of every mathematician's work, the subject is now often taught in computer science departments because of the tremendous impact which computers have had on its development. Research focuses mainly on the numerical solution of (nonlinear) partial differential equations and the minimization of functions.

Numerical analysis is needed because answers provided by mathematical analysis are usually symbolic and not numeric; they are often given implicitly only, as the solution of some equation, or they are given by some limit process. A further complication is provided by the rounding error which usually contaminates every step in a calculation (because of the fixed finite number of digits carried).

Even in the absence of rounding error, few numerical answers can be obtained exactly. Among these are (1) the value of a piecewise rational function at a point and (2) the solution of a (solvable) linear system of equations, both of which can be produced in a finite number of arithmetic steps. Approximate answers to all other problems are obtained by solving the first few in a sequence of such finitely solvable problems. A typical example is provided by Newton's method: A solution c to a nonlinear equation $f(c) = 0$ is found as the limit $c = \lim_{n \to \infty} x_n$, with x_{n+1} a solution to the linear equation $f(x_n) + f'(x_n)(x_{n+1} - x_n) = 0$, that is, $x_{n+1} = x_n - f(x_n)/f'(x_n)$, $n = 0, 1, 2, \ldots$. Of course, only the first few terms in this sequence $x_0, x_1, x_2, \ldots$ can ever be calculated, and thus one must face the question of when to break off such a solution process and how to gauge the accuracy of the current approximation. The difficulty in the mathematical treatment of these questions is exemplified by the fact that the limit of a sequence is completely independent of its first few terms.

In the presence of rounding error, an otherwise satisfactory computational process may become useless, because of the amplification of rounding errors. A computational process is called stable to the extent that its results are not spoiled by rounding errors. The extended calculations involving millions of arithmetic steps now possible on computers have made the stability of a computational process a prime consideration.

Interpolation and approximation. Polynomial interpolation provides a polynomial p of degree n or less which uniquely matches given function values $f(x_0), \ldots, f(x_n)$ at corresponding distinct points $x_0, \ldots, x_n$. The interpolating polynomial p is used in place of f, for example in evaluation, integration, differentiation, and zero finding. Accuracy of the interpolating polynomial depends strongly on the placement of the interpolation points, and usually degrades drastically as one moves away from the interval containing these points (that is, in case of extrapolation). *See* EXTRAPOLATION.

When many interpolation points (more than 5 or 10) are to be used, it is often much more efficient to use instead a piecewise polynomial interpolant or spline. Suppose the interpolation points above are ordered, $x_0 < x_1 < \ldots < x_n$. Then the cubic spline interpolant to the above data, for example, consists of cubic polynomial pieces, with the ith piece defining the interpolant on the interval $[x_{i-1}, x_i]$ and so matched with its neighboring piece or pieces that the resulting function not only matches the given function values (hence is continuous) but also has a continuous first and second derivative.

Interpolation is but one way to determine an approximant. In full generality, approximation involves several choices: (1) a set P of possible approximants, (2) a criterion for selecting from P a particular approximant, and (3) a way to measure the approximation error, that is, the difference between the function f to be approximated and the approximant p, in order to judge the quality of approximation. Much studied examples for P are the polynomials of degree n or less, piecewise polynomials of a given degree with prescribed breakpoints, and rational functions of given numerator and denominator degrees. The distance between f and p is usually measured by a norm, such as the L_2 norm $(\int |f(x) - p(x)|^2 dx)^{1/2}$ or the uniform norm $\sup_x |f(x) - p(x)|$. Once choices 1 and 3 are made, one often settles 2 by asking for a best approximation to f from P, that is, for an element of P whose distance from f is as small as possible. Questions of existence, uniqueness, characterization, and numerical construction of such best approximants have been studied extensively for various choices of P and the distance measure. If P is linear, that is, if P consists of all linear combinations

$$\sum_{i=1}^{n} a_i p_i$$

of certain fixed functions $p_1, \ldots, p_n$, then determination of a best approximation in the L_2 norm is particularly easy, since it involves nothing more than the solution of n simultaneous linear equations.

Solution of linear systems. Solving a linear system of equations is probably the most frequently confronted computational task. It is handled either by a direct method, that is, a method which obtains the exact answer in a finite number of steps, or by an iterative method, or by a judicious combination of both. Analysis of the effectiveness of possible methods has led to a workable basis for selecting the one which best fits a particular situation.

Direct methods. Cramer's rule is a well-known direct method for solving a system of n linear equations in n unknowns, but it is much less efficient than the method of choice, elimination. In this procedure the first unknown is eliminated from each equation but the first by subtracting from that equation an appropriate multiple of the first equation. The resulting system of $n-1$ equations in the remaining $n-1$ unknowns is similarly reduced, and the process is repeated until one equation in one unknown remains. The solution for the entire system is then found by back-substitution, that is, by solving for that one unknown in that last equation, then returning to the next-to-last equation which at the next-to-last step of the elimination involved the final unknown (now known) and one other, and solving for that second-to-last unknown, and so on.

This process may break down for two reasons: (1) when it comes time to eliminate the kth unknown, its coefficient in the kth equation may be zero, and hence the equation cannot be used to eliminate the kth unknown from equations $k+1$, $\ldots, n$; and (2) the process may be very unstable. Both difficulties can be overcome by pivoting, in which one selects, at the beginning of the kth step, a suitable equation from among equations $k, \ldots,$ n, interchanges it with the kth equation, and then proceeds as before. In this way the first difficulty may be avoided provided that the system has one and only one solution. Further, with an appropriate pivoting strategy, the second difficulty may be avoided provided that the linear system is stable. Explicitly, it can be shown that, with the appropriate pivoting strategy, the solution computed in the presence of rounding errors is the exact solution of a linear system whose coefficients usually differ by not much more than roundoff from the given ones. The computed solution is therefore close to the exact solution provided that such small changes in the given system do not change its solution by much. A rough but common measure of the stability of a linear system is the condition of its coefficient matrix. This number is computed as the product of the norm of the matrix and of its inverse. The reciprocal of the condition therefore provides an indication of how close the matrix is to being noninvertible or singular.

Iterative methods. The direct methods described above require a number of operations which increases with the cube of the number of unknowns. Some types of problems arise wherein the matrix of coefficients is sparse, but the unknowns may number several thousand; for these, direct methods are prohibitive in computer time required. One frequent source of such problems is the finite difference treatment of partial differential equations (discussed below). A significant literature of iterative methods exploiting the special properties of such equations is available. For certain restricted classes of difference equations, the error in an initial iterate can be guaranteed to be reduced by a fixed factor, using a number of computations that is proportional to $n \log n$, where n is the number of unknowns. Since direct methods require work proportional to n^3, it is not surprising that as n becomes large, iterative methods are studied rather closely as practical alternatives.

The most straightforward iterative procedure is the method of substitution, sometimes called the method of simultaneous displacements. If the equations for $i = 1, \ldots, n$ are as shown in Eq. (1),

$$\sum_{j=1}^{n} a_{ij} x_j = b_i \tag{1}$$

then the rth iterate is computed from the $(r-1)$st by solving the trivial equations for $x_i^{(r)}$ shown in Eq. (2) for $i = 1, \ldots, n$, where the elements $x_i^{(0)}$

$$\sum_{j \neq i} a_{ij} x_j^{(r-1)} + a_{ii} x_i^{(r)} = b_i \tag{2}$$

are chosen arbitrarily. If for $i = 1, \ldots, n$, the inequality

$$\sum_{j \neq i} |a_{ij}| \leq |a_{ii}|$$

holds for some i, and the matrix is irreducible, then $x_i^{(r)} \underset{r}{\rightarrow} x_i$ is the solution. For a matrix to be irreducible, the underlying simultaneous system must not have any subset of unknowns which can be solved for independently of the others. For practical problems for which convergence occurs, analysis shows the expected number of iterations required to guarantee a fixed error reduction to be proportional to the number of unknowns. Thus the total work is proportional to n^2.

The foregoing procedure may be improved several ways. The Gauss-Seidel method, sometimes called the method of successive displacements, represents the same idea but uses the latest available values. Equation (3) is solved for $i = 1, \ldots,$

$$\sum_{j<i} a_{ij} x_j^{(r)} + a_{ii} x_i^{(r)} + \sum_{j>i} a_{ij} x_j^{(r-1)} = b_i \tag{3}$$

n. The Gauss-Seidel method converges for the conditions given above for the substitution method and is readily shown to converge more rapidly.

Further improvements in this idea lead to the method of successive overrelaxation. This can be thought of as calculating the correction associated with the Gauss-Seidel method and overcorrecting by a factor ω. Equation (4) is first solved for y.

$$\sum_{j<i} a_{ij} x_j^{(r)} + a_{ii} y + \sum_{j>i} a_{ij} x_j^{(r-1)} = b_i \tag{4}$$

Then $x_i^{(r)} = x_i^{(r-1)} + \omega(y - x_i^{(r-1)})$. Clearly, choosing $\omega = 1$ yields the Gauss-Seidel method. For problems of interest arising from elliptic difference equations, there exists an optimum ω which guarantees a fixed error reduction in a number of iterations proportional to $n^{1/2}$, and thus in total work proportional to $n^{3/2}$.

A number of other iterative techniques for systems with sparse matrices have been studied. Primarily they depend upon approximating the given matrix with one such that the resulting equations can be solved directly with an amount of work proportional to n. For a quite large class of finite difference equations of interest, the computing work to guarantee a fixed error reduction is proportional to $n^{5/4}$. The work requirement proportional only to $n \log n$ quoted earlier applies to a moderately restricted subset.

Overdetermined linear systems. Often an overdetermined linear system has to be solved. This happens, for example, if one wishes to fit the model

$$p(x) = \sum_{j=1}^{n} a_j p_j$$

to observations $(x_i, y_i)_{i=1}^{m}$ with $n < m$. Here one would like to determine the coefficient vector $\mathbf{a} = (a_1, \ldots, a_n)^T$ so that $p(x_i) = y_i$, $i = 1, \ldots, m$. In matrix notation, one wants $A\mathbf{a} = \mathbf{y}$, with A the m-by-n matrix $[p_j(x_i)]$. If $n < m$, one cannot expect a solution, and it is then quite common to determine $\mathbf{a}$ instead by least squares, that is, so as to minimize the "distance" $(\mathbf{y} - A\mathbf{a})^T(\mathbf{y} - A\mathbf{a})$ between the vectors $\mathbf{y}$ and $A\mathbf{a}$. This leads to the so-called normal equations $A^TA\mathbf{a} = A^T\mathbf{y}$ for the coefficient vector $\mathbf{a}$. But unless the "basis functions" $p_1, \ldots, p_n$ are chosen very carefully, the condition of the matrix A^TA may be very bad, making the elimination process outlined above overly sensitive to rounding errors. It is much better to make use of a so-called orthogonal decomposition for A. *See* LEAST-SQUARES METHOD.

Assume first that A has full rank (which is the same thing as assuming that the only linear combination p of the functions $p_1, \ldots, p_n$ which vanishes at all the points $x_1, \ldots, x_m$ is the trivial one, the one with all coefficients zero). Then A has a QR decomposition, that is, $A = QR$, with Q an orthogonal matrix (that is, $Q^T = Q^{-1}$), and R an m-by-n matrix whose first n rows contain an invertible upper triangular matrix R_1, while its remaining m-n rows are identically zero. Then $(\mathbf{y} - A\mathbf{a})^T \cdot (\mathbf{y} - A\mathbf{a}) = (Q^T\mathbf{y} - R\mathbf{a})^T(Q^T\mathbf{y} - R\mathbf{a})$ and, since the last m-n entries of $R\mathbf{a}$ are zero, this is minimized when the first n entries of $R\mathbf{a}$ agree with those of $Q^T\mathbf{y}$, that is, $R_1\mathbf{a} = [(Q^T\mathbf{y})(i)]_1^n$. Since R_1 is upper triangular, this system is easily solved by back-substitution, as outlined above. The QR decomposition for A can be obtained stably with the aid of Householder transformations, that is, matrices of the simple form $H = I - (2/\mathbf{u}^T\mathbf{u})\mathbf{u}\mathbf{u}^T$, which are easily seen to be orthogonal and even self-inverse, that is, $H^{-1} = H$. In the first step of the process, A is premultiplied by a Householder matrix H_1 with $\mathbf{u}$ so chosen that the first column of H_1A has zeros in rows $2, \ldots, m$. In the next step, one premultiplies H_1A by H_2 with $\mathbf{u}$ so chosen that H_2H_1A retains its zeros in column 1 and has also zeros in column 2 in rows $3, \ldots, m$. After $n-1$ such steps, the matrix $R := H_{n-1} \ldots H_1A$ is reached with zeros below its main diagonal, and so $A = QR$ with $Q := H_1 \ldots H_{n-1}$.

The situation is somewhat more complicated when A fails to have full rank or when its rank cannot be easily determined. In that case, one may want to make use of a singular value decomposition for A, which means that one writes A as the product USV, where both U and V are orthogonal matrices and $S = (s_{ij})$ is an m-by-n matrix that may be loosely termed "diagonal," that is, $s_{ij} = 0$ for $i \neq j$. Calculation of such a decomposition is more expensive than that of a QR decomposition, but the singular value decomposition provides much more information about A. For example, the diagonal elements of S, the so-called singular values of A, give precise information about how close A is to a matrix of given rank, and hence make it possible to gauge the effect of errors in the entries of A on the rank of A. *See* INTERPOLATION; MATRIX THEORY.

Differential equations. Classical methods yield practical results only for a moderately restricted class of ordinary differential equations, a somewhat more restricted class of systems of ordinary differential equations, and a very small number of partial differential equations. The power of numerical methods is enormous here, for in quite broad classes of practical problems relatively straightforward procedures are guaranteed to yield numerical results, whose quality is predictable.

Ordinary differential equations. The simplest system is the initial value problem in a single unknown, $y' = f(x,y)$, and $y(a) = \eta$, where y' means dy/dx, and f is continuous in x and satisfies a Lipschitz condition in y; that is, there exists a constant K such that for all x and y of interest, $|f(x,y) - f(x,z)| \leq K|y - z|$. The problem is well posed and has a unique solution.

The Euler method is as follows: $y_0 = \eta$, Eq. (5)

$$y_{i+1} = y_i + hf(x_i, y_i) \tag{5}$$

holds, and $i = 0, 1, 2, \ldots, (b - a)/h$. Here h is a small positive constant, and $x_i = a + ih$. Analysis shows that as $h \to 0$, there exists a constant C such that $|y_k - y(x_k)| \leq Ch$, where $y(x_k)$ is the value of the unique solution at x_k, and $a \leq x_k \leq b$. This almost trivial formulation of a numerical procedure thus guarantees an approximation to the exact solution to the problem that is arbitrarily good if h is sufficiently small, and it is certainly easy to implement. A trivial extension of this idea is given by the method of Heun, Eq. (6).

$$y_{i+1} = y_i + \tfrac{1}{2}h[f(x_i, y_i) + f(x_i + h, y_i + hf(x_i, y_i))] \tag{6}$$

The method of Heun similarly is guaranteed to approximate the desired solution arbitrarily well since there exists another constant C_1 such that the $\{y_i\}$ satisfy relation (7). This is clearly asymp-

$$|y_k - y(x_k)| \leq C_1 h^2 \tag{7}$$

totically better than the method of Euler. It is readily found to be practically superior for most problems. Further improvement of this type is offered by the classical Runge-Kutta method, Eq. (8), where $\phi(x,y,h) = \tfrac{1}{6}[k_1 + 2k_2 + 2k_3 + k_4]$,

$$y_{i+1} = y_i + h\phi(x_i, y_i, h) \tag{8}$$

and $k_1 = f(x,y)$, $k_2 = f(x + h/2, y + hk_1/2)$, $k_3 = $

$f(x+h/2, y+hk_2/2)$, and $k_4 = f(x+h, y+hk_3)$. For this method there exists a constant C_2 such that $|y_k - y(x_k)| \leq C_2 h^4$.

The foregoing methods are called single-step since only y_i is involved in the computation of y_{i+1}. The single-step methods yielding the better results typically require several evaluations of the function f per step. By contrast, multistep methods typically achieve high exponents on h in the error bounds without more than one evaluation of f per step. Multistep methods require use of $y_{i-\alpha}$ or $f(x_{i-\alpha}, y_{i-\alpha})$ or both for $\alpha = 0, 1, \ldots, j$ to compute y_{i+1}. Typical is the Adams-Bashforth method for $j=5$, Eq. (9).

$$y_{i+1} = y_i + \frac{h}{1440}[4277\,f(x_i,y_i) - 7923\,f(x_{i-1},y_{i-1}) + 9982\,f(x_{i-2},y_{i-2}) - 7298\,f(x_{i-3},y_{i-3}) + 2277\,f(x_{i-4},y_{i-4}) - 475\,f(x_{i-5},y_{i-5})] \quad (9)$$

Analysis shows the solution of this Adams-Bashforth procedure to satisfy relation (10) for

$$|y_k - y(x_k)| \leq C_3 h^6 \quad (10)$$

some constant C_3. A large number of valuable multistep methods have been studied. If f is nontrivial to evaluate, the multistep methods are less work to compute than the single-step methods for comparable accuracy. The chief difficulty of multistep methods is that they require j starting values and, therefore, cannot be used from the outset in a computation.

Partial differential equations. Methods used for partial differential equations differ significantly, depending on the type of equation. Typically, parabolic equations are considered for which the prototype is the heat flow equation, Eq. (11), with $u(x,0)$

$$\frac{\partial^2 u}{\partial x^2} = \frac{\partial u}{\partial t} \quad (11)$$

given on $x\epsilon[0,1]$, say, and $u(0,t)$ and $u(1,t)$ given for $t > 0$. A typical finite difference scheme is shown in Eq. (12), where $i - 1, \ldots, 1/h - 1$, $w_{0,n} =$

$$(w_{i,n})_{x\bar{x}} = \frac{w_{i,n+1} - w_{in}}{k} \quad (12)$$

$u(0,t_n)$, and $w_{1/h,n} = u(1,t_n)$, with $(w_i)_x = (w_{i+1} - w_i)/h$, $(w_i)_{\bar{x}} = (w_{i-1})_x$, and $w_{i,n}$ is the function defined at $x_i = ih$, $t_n = nk$. Analysis shows that as $h,k \to 0$, the solution $w_{i,n}$ satisfies $|w_{i,n} - u(x_i,t_n)| < C(h^2 + k)$ for some constant C if $k/h^2 \leq 1/2$, but for k/h^2 somewhat larger than $1/2$, $w_{i,n}$ bears no relation at all to $u(x_i,t_n)$.

The restriction $k/h^2 \leq 1/2$ can be removed by using the implicit difference equation, Eq. (13), but

$$(w_{i,n+1})_{x\bar{x}} = \frac{w_{i,n+1} - w_{in}}{k} \quad (13)$$

now simultaneous equations must be solved for $w_{i,n+1}$ each step. The inequality, $|w_{i,n} - u(x_i,t_n)| < C(h^2 + k)$, still holds for some constant C. An improved implicit formulation is the Crank-Nicolson equation, Eq. (14).

$$\tfrac{1}{2}(w_{i,n} + w_{i,n+1})_{x\bar{x}} = \frac{w_{i,n+1} - w_{i,n}}{k} \quad (14)$$

As $h,k \to 0$, solutions satisfy relation (15) for

$$|w_{i,n} - u(x_i,t_n)| < C(h^2 + k^2) \quad (15)$$

some constant C; again h/k is unrestricted. Such techniques can readily extend to several space variables and to much more general equations. The work estimates given above for iterative solution of simultaneous equations, such as Eq. (14), apply for two-space variables.

Work using variational techniques to approximate the solution of parabolic and elliptic equations has been most fruitful for broad classes of nonlinear problems. The technique reduces partial differential equations to systems of ordinary differential equations. Analysis shows that solutions obtained approximate the desired solution, as within a constant multiple of the best that can be achieved within the subspace of the basis functions used. Practical utilization suggests this to be the direction for most likely future developments in the treatment of partial differential systems. *See* DIFFERENTIAL EQUATION. [CARL DE BOOR]

Bibliography: P. G. Ciarlet, *The Finite Element Method for Elliptic Problems*, 1978; S. Conte and C. de Boor, *Elementary Numerical Analysis*, 3d ed., 1980; J. M. Ortega and W. C. Rheinboldt, *Iterative Solution of Nonlinear Equations in Several Variables*, 1970; R. D. Richtmyer and K. W. Morton, *Difference Methods for Initial-Value Problems*, 2d ed., 1967; J. H. Wilkinson, *The Algebraic Eigenvalue Problem*, 1965.

Nutation

In mechanics, the term nutation refers to a bobbing or nodding up-and-down motion of a spinning rigid body, such as a top, as it precesses about its vertical axis. Astronomical nutation refers to irregularities in the precessional motion of the equinoxes caused by the varying torque applied to the Earth by the Sun and Moon. Astronomical nutation, which is sometimes called nutational wandering of the terrestrial poles, should not be confused with nutation as defined in mechanics; the latter is present even if the source of the torques is unvarying.

Nutation of tops. The general motion of a spinning top, easily observed at low spin rates, consists of both precession and nutation. Figure 1*a* shows a symmetrical top spinning about a fixed point with its axis tracing out this general motion. Figure 1*b* shows the motion for the case when the axis of the spinning top is released with an initial angular velocity in the direction of precession; Fig. 1*c*, that with an initial velocity opposite to the precession; Fig. 1*d*, that with zero initial angular velocity (axis of spin released from rest). *See* PRECESSION.

The angular frequency of the nutation of a top axis at a high spin rate is given by Eq. (1), where I_z

$$\omega_n = \frac{I_z}{I_x} S \quad (1)$$

and I_x are moments of inertia about the z and x axes, respectively, and S is the angular velocity of spin. Furthermore, the rate of precession ω_p for the general motion is not uniform but varies harmonically with time with the same frequency as does the nutation, as shown in Eq. (2). The aver-

$$\omega_p = \frac{Wl}{I_z S}(1 - \cos \omega_n t) \quad (2)$$

age precessional frequency is given by Eq. (3).

$$(\omega_p)_{av} = \frac{Wl}{I_z S} \qquad (3)$$

As the spin rate S is increased the frequency of nutation increases, as shown by Eq. (1), and the nutational displacement $(\theta_2 - \theta_1)$ decreases very rapidly. Furthermore, as S is increased in Eq. (2), the frequency of the precessional variation increases, but from Eq. (3) the average rate of precession decreases. Therefore, in practice, for a sufficiently fast top, the nutation is so small and fast that it is damped out by the friction at the pivot and is unobservable. The top appears to precess uniformly about the vertical axis for this common case.

Nutation of projectiles. The motion relative to the centers of mass of bullets and shells stabilized by high rates of spin is identical to that of a spinning top relative to the fixed point of contact. Torques about the centers of mass due to aerodynamic forces acting on such bodies during flight cause precession and nutation to occur (Fig. 2). Furthermore, finned missiles, which usually rotate or spin rather slowly during flight unless prohibited from doing so by a suitable control system, also develop precessional and nutational angular velocities, similar to those of a spinning pendulum.

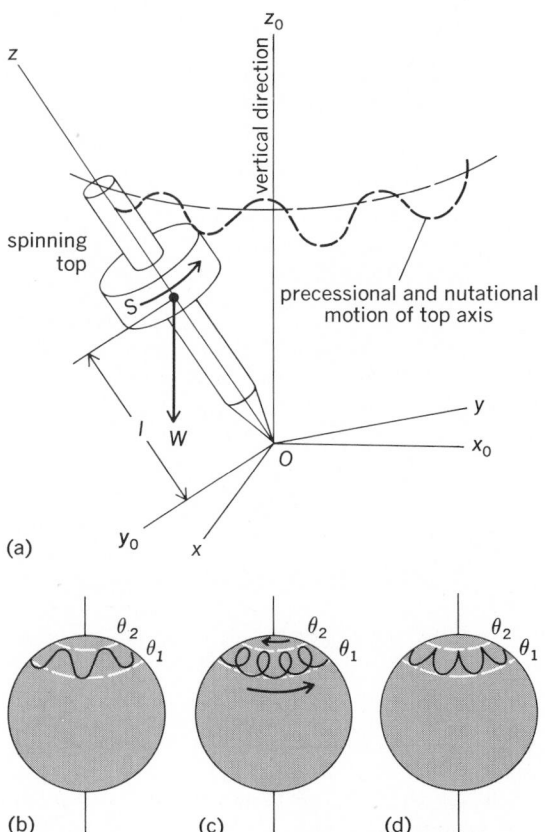

Fig. 1.   Motion of spinning top showing typical traces of the top spin axis on unit spheres for different initial conditions. (a) General motion. (b) Top released with initial angular velocity in direction of precession. (c) Top released with initial angular velocity opposite to direction of precession. (d) Axis of spin released from rest.

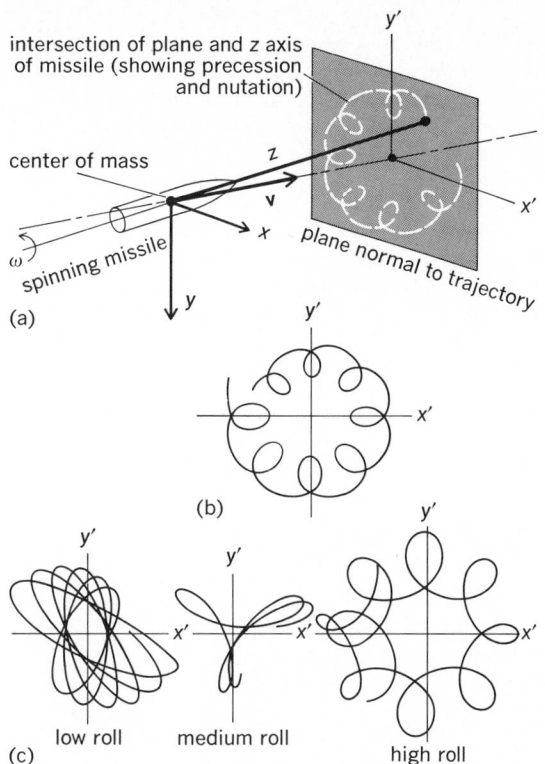

Fig. 2.   Precessional and nutational displacements of missiles in flight. (a) Intersection of missile axis on a plane. The velocity of the center of mass of the missiles is indicated by **v**. (b) Typical motion of spin-stabilized missiles (bullets, shells, and the like). (c) Typical motion of rolling fin-stabilized missiles for low, medium, and large rates of roll, respectively.

Astronomical nutation. The rotating Earth can be regarded as a spinning symmetrical top with small angular speed but large angular momentum, the latter due to its large mass. The gravitational attractions of the Sun and Moon cause the Earth's axis to describe a cone about the normal to the plane of its orbit. However, the magnitude of these gravitational attractions is continually varying, due to the changing positions in space of Sun, Moon, and Earth. The Moon's orbit is continually changing its position in such a way that the celestial pole undergoes a nodding (nutation) as well as a periodic variation in the rate of advance. The largest nutation is about 9″.2, and occurs in a period of a little less than 19 years; that is, the celestial pole completes a small ellipse of semimajor axis 9″.2 in about 19 years.

There are lesser nutation effects which are due to the motion of the Moon's nodes, the changing declination of the Sun, and the changing declination of the Moon.

Nutation of gyroscopes. Still another example of the nutation of a spinning symmetrical body is given by the general motion of a gyroscope.

[RAY E. BOLZ]

Bibliography: E. P. Federov and M. L. Smith (eds.), *Nutation and the Earth's Rotation: Proceedings*, I.A.U. Symposium no. 78, Kiev, May 23–28, 1980; H. Goldstein, *Classical Mechanics*, 2d ed., 1980.

Ohm's law

The direct current flowing in an electrical circuit is directly proportional to the voltage applied to the circuit. The constant of proportionality R, called the electrical resistance, is given by the equation below, in which V is the applied voltage and I is the current.

$$V = RI$$

This relationship was first described by Georg Simon Ohm in 1827 and was based on his experiments with metallic conductors. Since that time numerous deviations from this simple, linear relationship have been discovered. *See* CONDUCTIVITY; ELECTRICAL RESISTANCE; ELECTRICAL RESISTIVITY; SEMICONDUCTOR; SKIN EFFECT; THERMAL CONDUCTION IN SOLIDS.

[CHARLES E. APPLEGATE]

Opaque medium

One which is impervious to rays of light, that is, not transparent to the human eye. By extension, a medium may be described as opaque if it does not transmit infrared waves or other regions of the electromagnetic spectrum, such as the x-ray, ultraviolet, and microwave regions. The property of zero transmittance does not necessarily imply total reflectance; that is, opacity can result both from reflection and from absorption of incident rays. *See* ABSORPTION OF ELECTROMAGNETIC RADIATION.

[M. G. MELLON]

Operator theory

At one level of abstraction an operator is simply a function whose arguments and values are real- (or complex-) valued functions of one or more real variables; in more naive terms an operator is a rule for converting such real- (or complex-) valued functions into others. The following are simple examples: (i) the operator which takes each differentiable real-valued function of one variable into its derivative; (ii) the operator which takes each twice-differentiable function f of one variable into expression (1); (iii) the operator which takes

$$\left(\frac{df}{dx}\right)^2 + x^2 \frac{d^2f}{dx^2} \tag{1}$$

each twice-differentiable function f of three variables into expression (2); and (iv) the operator which

$$\frac{\partial^2 f}{\partial x^2} + \frac{\partial^2 f}{\partial y^2} + \frac{\partial^2 f}{\partial z^2} \tag{2}$$

takes the continuous function f of one real variable into the function g where relation (3) holds.

$$g(x) \equiv \int_0^1 \sqrt{x+y}\, f(y)\, dy \tag{3}$$

Since an operator is a function, the usual functional notation is applicable. $L(f)$ may be used to denote the result of operating on f with the operator L. The set of all functions f for which $L(f)$ is defined is called the domain of L, and the set of all functions g such that $L(f) = g$ for some f in the domain of L is called the range of L. It is obvious that solving a differential or integral equation is equivalent (in many ways) to solving an operator equation $L(f) = g$, where g and L are given and it is required to find f. Moreover the operator concept can be very useful both in theory and practice, producing a great variety of illuminating insights.

In large part the fruitfulness of the operator concept can be traced to two sources. One of these is the possibility of adding and multiplying operators in such a way that many, though not all, of the laws of ordinary algebra hold. The other is the fact that the ranges and domains of operators behave in many respects like ordinary space and, indeed, may be regarded as contained in infinite dimensional generalizations of the familiar three-dimensional space of solid geometry. This makes it possible to think of an operator as a geometrical transformation and to exploit one's spatial intuition.

Let D be a family of real-valued functions such that $\lambda f + \mu g$ is in D whenever λ and μ are real numbers and f and g are in D. Let L and M be operators with domain D and range included in D. Then the operator which takes each f in D into $L[M(f)]$ is called the product of L and M and is denoted by LM. Moreover the operator which takes each f in D into $L(f) + M(f)$ is called the sum of L and M and denoted by $L + M$. In particular one may form powers L^2, L^3, ..., polynomials $a_0 + a_1 L + a_2 L^2 + \cdots + a_r L^r$, and in suitably restricted contexts, power series. It is important to note that, while it is always true that $L + M = M + L$, it is not always true that $LM = ML$. On the other hand, it is possible to show that $(LM)N = L(MN)$ and that $(L + M) + N = L + (M + N)$ for all L, M, and N so that parentheses may be omitted just as in ordinary algebra.

In many but not all cases the sets of functions with which one deals derive their spacelike properties from the possibility of assigning a distance $\rho(f,g)$ to each pair f and g of members of the set D under consideration. This is done in such a manner that $\rho(f,g) = \rho(g,f)$, $\rho(f,g) > 0$ if $f \neq g$; $\rho(f,f) = 0$ and $\rho(f,g) \leqq \rho(f,h) + \rho(h,g)$ for all f, g, and h in D. When D is as described in the preceding paragraph, ρ is often chosen so that $\rho(f,g) = \|f - g\|$ where $\|f\| = \rho(f,0)$. If $\|\lambda f\| = |\lambda|\ \|f\|$ for all real numbers λ, then $\|f\|$ is said to be a norm for D. There will usually be more than one way of norming a given D. For example, if D is the set of all real-valued continuous functions defined on the interval $0 \leqq x \leqq 1$, Eq. (4) gives one value for D.

$$\|f\| = \max_{0 \leqq x \leqq 1} |f(x)| \tag{4}$$

and Eq. (5) gives another value for D. The analogy

$$\|f\|_1 = \sqrt{\int_0^1 |f(x)|^2\, dx} \tag{5}$$

with the familiar space of experience is closest when the second norm is used, but the first is useful also.

The operator L is said to be linear if $L(\lambda f + \mu g)$ is defined and equal to $\lambda L(f) + \mu L(g)$ whenever f and g are in the domain of L, and λ and μ are numbers. Insofar as there is a general theory of operators, it is largely concerned with linear operators, and this article discusses linear operators exclusively. It is useful to develop this theory from axioms.

Axioms. Let F denote either the field of all real numbers or the field of all complex numbers. A vector space over F is a set or collection X whose members are of an unspecified character except that they may be added together and multiplied by the members of F in such a way that the following formal laws are satisfied:

1. $(f+g)+h=f+(g+h)$ and $f+g=g+f$ for all f, g, and h in X.

2. There is a unique zero vector 0 in X such that $f+0=f$ for all f in X.

3. $\lambda(\mu f)=(\lambda\mu)f$, $(\lambda+\mu)f=\lambda f+\mu f$, and $\lambda(f+g)=\lambda f+\lambda g$ for all f and g in X and all λ and μ in F.

4. $1f=f$ for all f in X.

By generalizing the more special and concrete definition in the obvious fashion, a linear operator is defined to be a function L whose domain is a vector space X, whose range is in a vector space Y, and for which it is true that $L(\lambda f+\mu g)=\lambda L(f)+\mu L(g)$ whenever f and g are in X and λ and μ are in F.

Finite dimensional case. A vector space X is said to be finite dimensional if it contains a finite subset $v_1, v_2, \ldots, v_n$ spanning the space in the sense that every element in the space may be written in the form $\lambda v_1+\lambda_2 v_2+\cdots+\lambda_n v_n$ where the λ_j are in F. The representation $f=\lambda_1 v_1+\lambda_2 v_2+\cdots+\lambda_n v_n$ is unique if, and only if, no v_j is in the span of the rest. In this case $v_1, v_2, \ldots, v_n$ is said to form a basis for X, and the λ_j are said to be the coordinates of f with respect to this basis. It is not hard to show that any two bases for the same space have the same number of elements. This number is called the dimension of the space.

Let L be a linear operator whose domain X is finite dimensional. It follows immediately that the range is also finite dimensional and that the dimension d_R of the range is less than, or equal to, the dimension d_X of the domain. Let Y be the vector space containing the range of L, and let d_Y denote the dimension of Y. This gives $d_R \leqq d_Y$ and $d_R \leqq d_X$. The differences d_Y-d_R and d_X-d_R measure the extent to which the operator equation $L(f)=g$ fails to have a unique solution for all g. In fact if X_1 and X_2 are finite-dimensional subvector spaces of a vector space and $X_1 \supseteqq X_2$, then $X_1=X_2$ if, and only if, X_1 and X_2 have the same dimension. Thus $L(f)=g$ is always solvable if, and only if, $d_Y=d_R$ and d_Y-d_R is a measure of the size of the set of gs for which no solution exists. On the other hand, it is easily seen that if f_0 is a particular solution of $L(f)=g$, then the general solution is f_0+h where h is any element of X such that $L(h)=0$. The set of all such h is a vector space N, called the null space of L, whose dimension d_N measures the extent to which the equation $L(f)=g$ has multiple solutions. It is not hard to show that $d_N+d_R=d_X$ so that $d_N=d_X-d_R$. In the special but important case in which $X=Y$, $d_X-d_R=d_Y-d_R$. Thus either $L(f)=g$ has a unique solution for all g (nonsingular case) or else for many values of g, $L(f)=g$ has no solutions, and whenever it has any nonzero solutions, it has many (singular case).

There is a certain sense in which most operators are nonsingular. Let I denote the identity operator which takes every vector into itself. Then it can be shown that $L-\lambda I$ is nonsingular for all but a finite number of values of λ. Indeed $L-\lambda I$ will be singular if, and only if, there exists a nonzero vector f such that $L(f)-\lambda f=0$; that is, $L(f)=\lambda f$. Such an f is called a proper vector (or eigenvector) belonging to the proper value (or eigenvalue) λ. It is easy

to show that proper vectors belonging to distinct proper values are linearly independent in the sense that no one is in the span of the rest. Thus the number of distinct proper values cannot exceed the dimension of the space. *See* DIFFERENTIAL EQUATION.

Knowledge of the proper values of an operator L yields a great deal of information about the nature of L, especially in the important case in which the domain X admits a basis made up of proper vectors. Let $v_1, v_2, \ldots, v_n$ be a basis for X, and let $L(v_j)=\lambda_j v_j$ where the λ_j are (not necessarily distinct) members of F. Very simple computations lead to the following observations.

1. Every proper value of L is equal to some λ_j.

2. L is nonsingular if, and only if, no λ_j is zero.

3. If L is nonsingular, the inverse operator carries $\mu_1 v_1+\mu_2 v_2+\cdots+\mu_n v_n$ into $(\mu_1/\lambda_1)v_1+(\mu_2/\lambda_2)v_2+\cdots+(\mu_n/\lambda_n)v_n$.

4. If P is any polynomial with coefficients in F, then $P(L)$ carries $\mu_1 v_1+\mu_2 v_2+\cdots+\mu_n v_n$ into $P(\mu_1)v_1+P(\mu_2)v_2+\cdots+P(\mu_n)v_n$.

The structure of $P(L)$ revealed by observation (4) suggests a definition of $F(L)$ where F, instead of being a polynomial, is an arbitrary function with domain and range in F. To wit: $F(L)(\mu_1 v_1+\mu_2 v_2+\cdots+\mu_n v_n)=F(\mu_1)v_1+F(\mu_2)v_2+\cdots+F(\mu_n)v_n$. A similar, but of course more subtle, definition in certain infinite-dimensional cases is the source of the modern rigorization of the celebrated operational calculus of O. Heaviside.

Returning to the case in which X need not equal Y, let $v_1, v_2 \ldots, v_n$ and $w_1, w_2, \ldots, w_n$ be bases for X and Y, respectively. Let $L(v_j)=\alpha_{j1}w_1+\alpha_{j2}w_2+\cdots+\alpha_{jn}w_n$ where each α_{ji} is in F. Then $L(x_1 v_1+\cdots+x_n v_n)=x_1 L(v_1)+x_2 L(v_2)+\cdots+x_n L(v_n)=y_1 w_1+y_2 w_2+\cdots+y_n w_n$ where $y_i=a_{1i}x_1+\alpha_{2i}x_2\cdots\alpha_{ni}x_n$. The rectangular array in which α_{ji} is in the ith row and jth column is called the matrix of L with respect to the basis in question. It is clear that solving the operator equation $L(f)=g$ in the finite-dimensional case is equivalent to solving m linear algebraic equations in n unknowns. The theory sketched above is the theory of such equations couched in the language of operator theory. When $X=Y$ and $v_j=w_j$ for all j, then the condition $\alpha_{ji}=\bar{\alpha}_{ij}$ (where the overbar denotes complex conjugate) implies that there exists a basis for X made up of proper vectors of L. However, this condition is by no means a necessary one.

Infinite dimensional case. When X is not assumed to be finite dimensional, such simple and general theorems as those described above are no longer available. In certain contexts, however, more complicated and less complete analogs of them may be proved. It is with these that the general theory of linear operators is mainly concerned.

Let X be a vector space which is not necessarily finite dimensional but instead is equipped with a norm—defined in the abstract case as suggested by the definition given above for real functions spaces. Such a normed vector space is said to be complete if each sequence $f_1, f_2, \ldots$ of members of X which is convergent in the sense defined by Eq. (6) is also convergent in the sense defined by

$$\lim_{\substack{n\to\infty \\ m\to\infty}} \|f_n-f_m\|=0 \qquad (6)$$

Eq. (7) for some f in X. This f is easily seen to be

$$\lim_{n \to \infty} \|f_n - f\| = 0 \qquad (7)$$

unique and is called the limit of the sequence $\{f_n\}$. A complete normed vector space is called a Banach space. A normed vector space X is said to be separable if there exists a sequence $f_1, f_2, \ldots, f_n$ of elements of X such that every element of X is the limit of some subsequence. The space of continuous functions defined earlier is a separable Banach space. Now let $X = Y$ and let X be a Banach space. Let L be completely continuous in the sense that whenever $f_1, f_2, \ldots$ is a sequence of elements such that $\|f_n\| \le 1$ for all n, there is a subsequence $f_{n1}, f_{n2}, \ldots$ such that $L(f_{n1}), L(f_{n2})$, $\cdots$ is convergent in the sense defined in Eq. (6). Then the following theorem can be proved. For each λ in F with $\lambda \ne 0$ there is a pair of vector subspaces M_λ and N_λ of X such that: (1) $L(f)$ is in M_λ for all f in M_λ, and $L(f)$ is in N_λ for all f in N_λ; (2) every f in X can be written uniquely in the form $f_1 + f_2$ where $f_1 \in M_\lambda$ and $f_2 \in N_\lambda$; (3) for each $g_1 \in M_\lambda$ there is one and only one element $f_1 \in M_\lambda$ such that $L(f_1) - \lambda f_1 = g_1$; and (4) N_λ is finite dimensional. It follows easily that existence and uniqueness questions for the operator equation $L(f) - \lambda f = g$ reduce to the corresponding questions for the restriction of L to the finite-dimensional space N_λ and, hence, that the simple analysis given earlier applies. It may be proved further that M_λ coincides with X for all values of λ except those in a sequence λ_1, $\lambda_2, \ldots$ such that $\lim_{n \to \infty} \lambda_n = 0$.

Let K be a continuous real-valued function defined on the unit square $0 \le x \le 1$, $0 \le y \le 1$. Let X be the Banach space of all continuous real-valued functions defined on the interval $0 \le x \le 1$ with $\|f\| = \max_{0 \le x \le 1} |f(x)|$. Then it can be proved that the operator L_K which takes f into g, where Eq. (8)

$$g(x) = \int_0^1 K(x,y) f(y) \, dy \qquad (8)$$

holds, is a completely continuous linear operator. Application of the theorems quoted above yields most of the results of the Fredholm theory of integral equations. Such integral operators occur in inverting the members of a large class of linear differential operators. There are similar results for integral operators in $2n$ variables, it being possible to establish complete continuity whenever the region of integration is bounded.

Infinite-dimensional versions of theorems about bases of proper vectors and functions of operators take their simplest and most complete form when the underlying Banach space is a Hilbert space; that is, when there is defined an F-valued "inner product" $f \cdot g$ for each f and g in X which satisfies the following conditions: (1) $(\lambda f + \mu g) \cdot h = \lambda(f \cdot h) + \mu(g \cdot h)$; (2) $(f \cdot g) = \overline{(g \cdot f)}$; and (3) $(f \cdot f) = \|f\|^2$ for all f, g, and h in X and all λ and μ in F. The simplest (and original) example of a Hilbert space is the vector space of all sequences $c_1, c_2, \ldots$ of complex numbers such that $|c_1|^2 + |c_2|^2 + \cdots < \infty$, based on the definition $(c_1, c_2, \ldots) \cdot (c_1', c_2', \ldots) = c_1 \bar{c}_1' + c_2 \bar{c}_2' + \cdots$.

Now let L be a completely continuous linear operator, the domain of which is a separable Hilbert space H and whose range is contained in H.

Let L be self adjoint in the sense that $L(f) \cdot g = f \cdot L(g)$ for all f and g in H. Then it is a theorem that there exists a sequence v_1, v_2, of members of H which has the following properties: (1) Each v_j is a proper vector for L; (2) if f is any element in H and $c_j = f \cdot v_j$, then $f = c_1 v_1 + c_2 v_2 + \cdots$ in the sense that the partial sums of this series have f as a limit; and (3) $v_i \cdot v_j = 0$ if $i \ne j$ and $v_i \cdot v_i = 1$. Such a sequence is said to be an orthonormal basis for H. Just as in the finite-dimensional case, one can form more or less arbitrary functions of the operator L.

In rough terms the celebrated spectral theorem is a generalization of the preceding theorem in which the operator L, though self-adjoint, is not required to be completely continuous. Instead of a discrete basis of proper vectors, one finds a sort of continuous basis. More precisely it is possible to map H onto a Hilbert space H' whose elements are complex-valued functions in such a manner that the norms and the vector space operations are preserved and so that L becomes the operator of multiplying the elements of H' by a fixed real-valued function.

In addition to the abstract theory there are many detailed studies of particular operators of importance, such as the Laplace and Fourier transforms, and numerous applications to differential and integral equations. Moreover, in addition to the theory of single operators there are extensive theories of certain kinds of collections of operators. A ring of operators contains the sum, product, and difference of any two of its members, and a group of operators contains the inverse of every operator in it and the product of each two operators in it. The theory of groups and rings of operators is related to the algebraic theory of groups and rings much as the theory of a single operator is related to systems of linear algebraic equations. It has applications to harmonic analysis and to the conceptual foundations of quantum mechanics. *See* COMPLEX NUMBERS AND COMPLEX VARIABLES; GROUP THEORY; INTEGRAL TRANSFORM; SET THEORY; TOPOLOGY.   [GEORGE W. MACKEY]

Bibliography: S. K. Berberian, *Lectures in Functional Analysis and Operator Theory*, 1972; A. Brown and C. Pearcy, *Introduction to Operator Theory*, 1977; N. Dunford and J. T. Schwartz, *Linear Operators*, 3 vols., 1958–1971; E. Kreyszig, *Introductory Functional Analysis with Applications*, 1978; A. Papoulis, *The Fourier Integral and Its Applications*, 1962; I. E. Segal and R. A. Kunze, *Integrals and Operators*, 2d ed., 1978.

Optical fibers

Flexible transparent fiber devices, used for either image or data transmission, in which light propagates by total internal reflection. The optical fiber, or light guide, has a core of material with a refractive index higher than that of the surrounding cladding material. Fiber properties and requirements for image transfer, in which information is continuously transmitted over relatively short distances, are quite different than for data transfer where on-off pulses of light are used to transmit information over much longer distances. *See* REFLECTION OF ELECTROMAGNETIC RADIATION; REFRACTION OF WAVES.

Types of fibers. There are three basic types of light guides (see illustration). Multimode, stepped

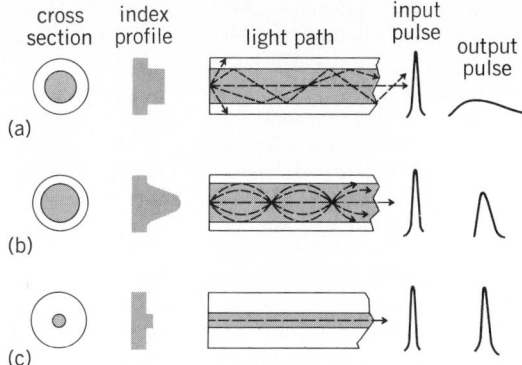

Types of optical fiber. (*a*) Multimode, stepped refractive index profile. (*b*) Multimode, graded index. (*c*) Single-mode, stepped index.

refractive index profile fibers (illustration *a*) are typically used for conventional image transfer as well as for short-distance data transmission. The number of rays or "modes" of light which are guided is determined by the core size and core/clad refractive index difference. In such a fiber, an initially sharp pulse made up of many modes broadens as it travels long distances through the fiber, since high-angle modes have a longer distance to travel than low-order modes. This limits the data transmission rate and distance because it determines how closely input pulses can be spaced without overlap at the output end.

A graded index multimode fiber (illustration *b*), where the core refractive index decreases with increased radial distance, can be used to minimize pulse broadening due to mode dispersion. Light rays travel more slowly near the core center than near the edge, resulting in the speed of the high-order modes approximating the speed of the low-order modes. This type of fiber is suitable for intermediate-distance, intermediate-data-rate transmission systems.

In the single-mode fiber type (illustration *c*), with low refractive index difference and small core size, the effect of mode dispersion on pulse dispersion is eliminated since only one mode is guided. This fiber is useful for long-distance, high-data-rate applications.

Attenuation. For transmission over any distance, the attenuation or loss of the light intensity is also an important property, and is caused principally by light absorption and scattering. Every material has some fundamental absorption due to its constituents. In addition, the presence of impurities can cause the absorption of light at specific wavelengths. Fluctuations in a material on a molecular scale cause intrinsic scattering of light, while fiber core diameter variations or the presence of defects such as bubbles can also cause scattering light loss. Because high-silica-content glasses as materials have very low intrinsic absorption and scattering at the near-infrared wavelengths where light sources (lasers and light-emitting diodes) operate, fibers made from such glasses are most suitable for transmission applications requiring very low attenuation. Very specialized glassmaking techniques, however, are necessary in order to eliminate light loss due to defects or to impurities such as iron and water. Plastics, whose

inherent attenuation is higher, can be used for many image transfer appplications. *See* ABSORPTION OF ELECTROMAGNETIC RADIATION; OPTICAL MATERIALS; SCATTERING OF ELECTROMAGNETIC RADIATION.

Many other fiber properties, such as strength, dimensional control, and chemical durability, are also important and dictated by the particular application.

[SUZANNE R. NAGEL]

Bibliography: N. S. Kapany, *Fiber Optics Principles and Applications*, 1967; S. E. Miller and A. G. Chynoweth (eds.), *Optical Fiber Telecommunications*, 1979.

Optical image

The image formed by the light rays from a self-luminous or an illuminated object that traverse an optical system. The image is said to be real if the light rays converge to a focus on the image side and virtual if the rays seem to come from a point within the instrument (see illustration).

The optical image of an object is given by the light distribution coming from each point of the object at the image plane of an optical system. The ideal image of a point according to geometrical optics is obtained when all rays from an object point unite in a single image point. However, diffraction theory teaches that even in this case the image is not a point but a minute disk. The diameter of this disk is about $1.22 \lambda/A$, where λ is the wavelength of the light considered and A is the numerical aperture, the sine of the largest cone angle on the image side multiplied by its refractive index (which is usually equal to unity).

Aberrations. From the standpoint of geometrical optics, if this most desirable type of image formation cannot be achieved, the next best objective is to have the image free from all but aperture errors (spherical aberration). In this case the light distribution in the image plane is still circular, resem-

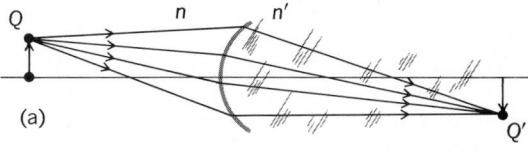

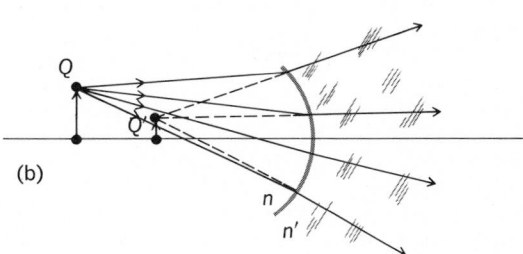

Optical images. (*a*) Real image. Rays leaving object point *Q* and passing through the refracting surface separating media *n* and *n'* are brought to a focus at the image point *Q'*. (*b*) Virtual image. Rays leaving *Q* and refracted by the concave surface separating *n* and *n'* appear to be coming from the virtual image point *Q'*. As the rays are diverging, they cannot be focused at any point. (*Modified from F. A. Jenkins and H. E. White, Fundamentals of Optics, 3d ed., McGraw-Hill, 1957*)

bling the point image; there is a true coordination of object point and image, although the image may be slightly unsharp. If the aperture errors are small, or if the image is viewed from a distance, such an image formation may be very satisfactory.

Asymmetry and deformation errors may be very disturbing if not held in check, because the light distribution of the image of a point in this case has a decidedly undesirable shape.

When the image of an axis point is considered, the rays through a fixed aperture circle converge to an axis point. For this type of imagery, the term half-sharp image will be used. A small object at the object point is then imaged by a circular stop at the focus of the image bundle with a magnification as given by Eq. (1) where u and u' are the angles of

$$m = (n \sin u)/(n' \sin u') \tag{1}$$

the imaging cone in object and image space, respectively, and n and n' are the corresponding refractive indices.

If the axis point is sharply imaged, an object of finite extent is sharply imaged if, and only if, $m = m_o$ (the Gaussian magnification) for all values of u (sine condition).

In the case of aperture errors, the most desirable image formation for an axis point is attained when the different images appear under the same angle from the exit pupil. If k' is the distance of the image point from the exit pupil, $\Delta s'$ is the aperture aberration, and if Eq. (2) gives the magnification

$$\Delta m = \frac{n \sin u}{n' \sin u'} - m_o \tag{2}$$

error compared with the magnification m_o on the axis, the condition is given by Eq. (3). The

$$\Delta s'/k' - \Delta m/m_o = \text{constant} \tag{3}$$

fulfillment of this condition gives equal quality for an object near the axis of a system with rotation symmetry.

Corresponding conditions can be ascertained for the image of an off-axis element if all the asymmetry errors and deformation errors are balanced. *See* ABERRATION (OPTICS).

Resolution. Two points are resolved by an optical system if the two images lie apart. Photometric analysis of an image may indicate the existence of two object points even if their images overlap, but in such an analysis the illumination of the object, as well as the imagery, plays a role.

In interference experiments, it is found that the image of two self-luminous points (that is, two light sources that are sufficiently separated) is incoherent; that is, the intensities of the two beams simply add. If the two object points are illuminated by the same light source, however, the phase relation of the light at the two points has to be taken into consideration. This is of the greatest importance for microscopes and telescopes, which image very small or distant objects. In this case an artificial change of phase by phase plates and apodization may improve resolution. *See* DIFFRACTION; RESOLVING POWER.

Resolving power is not the only consideration in image formation. The eye recognizes only contrast differences, and therefore objects may not be discerned if the contrast difference is too small. Again, for the image of a point, or an object illuminated by a point light source, means can be found

to change the apparent contrast, making it possible to discern biological objects, for example, having small differences of refractive index.

Image analysis. Methods have been suggested for obtaining information about optical images by sine-wave analysis. A sinusoidal test object is imaged by an optical system as a sinusoidal image, but altered in phase and amplitude. A large number of sinusoidal test objects with different frequencies (number of maxima per millimeter) are imaged, and the amplitude and phase are measured.

The curve of amplitude versus frequency gives a measure for resolving power and contrast as a function of the frequency of the test object, whereas the adjusted curve of phase versus frequency describes the lack of symmetry in the image. These amplitude-frequency curves can be measured as well as calculated from the spot diagrams, onto which the effects of diffraction can be superimposed if necessary. [MAX HERZBERGER]

Bibliography: E. U. Condon and H. Odishaw, *Handbook of Physics*, 1967; M. Herzberger, *Modern Geometrical Optics*, 1958, reprint 1978; F. A. Jenkins and H. E. White, *Fundamentals of Optics*, 4th ed., 1976; Optical Society of America, *Handbook of Optics*, 1978.

Optical materials

Generally, all substances used to reflect, refract, filter, polarize, modulate, detect, or disperse infrared, visible, or ultraviolet radiation or light; more specifically, the transparent materials usually used for windows and lenses. These materials are often some formulation of glass, a plastic in either bulk or thin-film form, or crystals either single or in aggregates. The most important optical properties are the degree and spectral region of transparency, the value of the refractive index n over the same spectral region, and the uniformity of the sample. Desirable materials are usually hard, strong, and relatively insensitive to temperature variations. *See* ABSORPTION OF ELECTROMAGNETIC RADIATION; INFRARED RADIATION; ULTRAVIOLET RADIATION.

Transmission loss is caused by reflection at the surface, absorption in the bulk, and scattering in both places. Surface reflection is given by Fresnel's equation, which for normal incidence is $[(n-1)/(n+1)]^2$ for each surface, and varies from about 4% per surface for a material such as glass ($n = 1.5$) to 36% per surface for germanium ($n = 4$). The change with wavelength is quite gradual between regions of strong absorption, being about $10^{-5}/°C$ for many materials. In general, denser materials have higher refractive indices. Absorption is the process by which incident radiation is converted to some other form of energy in the material. Absorption may be due to increased motion of electrons in the solid or the vibration of atoms or ions in the lattice or partially ordered network. The absorption by electrons is a smooth function, increasing gradually in proportion to the square of the wavelength. Absorption by vibration is characterized by relatively narrow bands at distinct wavelengths. Scattering of radiation occurs when a local disturbance, such as a fleck of dirt, impurity ion, grain boundary, or imperfection of the surface, is encountered by the incoming wave. If the wavelength is large compared to the disturbance, then

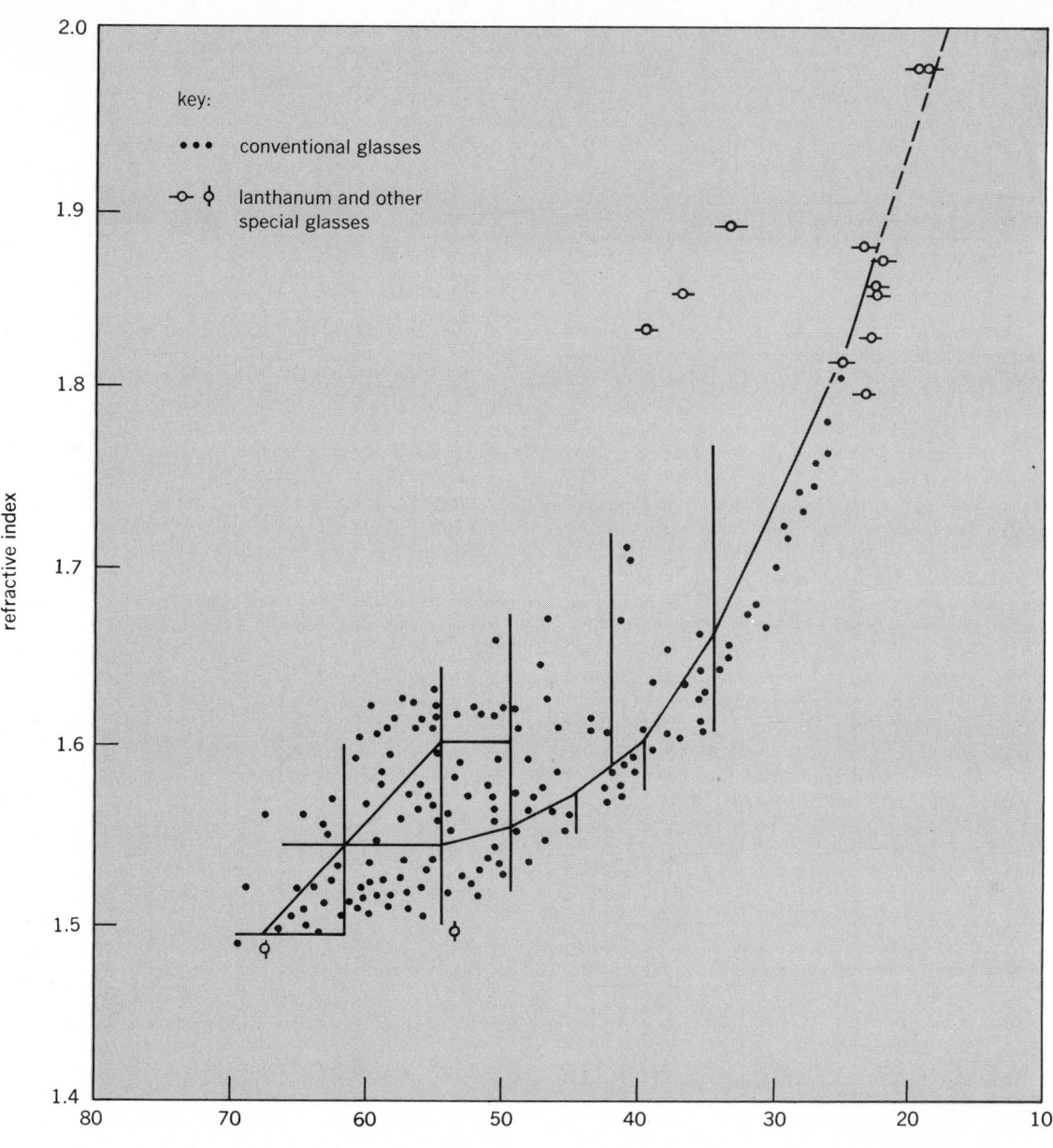

Fig. 1. Refractive index and nu values of some liquids, crystals, and conventional, lanthanum, and other special glasses. Glasses are grouped into types: Below the curve (right to left) are heavy flints, flints, light flints, extra-light flints, short crowns, crowns, borosilicate crowns (triangular area), and fluor crowns; above the curve (left to right) are phosphate crowns, heavy phosphate crowns, and barium crowns (triangular area); above them are heavy barium crowns and light barium flints; above them are extra-heavy barium crowns and similar glasses, barium flints, and heavy barium flints; field above and to right of heavy barium flints is occupied by lanthanum and similar glasses. (*From G. W. Morey, The Properties of Glass, 2d ed., Reinhold, 1954*)

the transmission varies as λ^{-4}; if small, a more complicated oscillating dependence is observed. *See* REFRACTION OF WAVES; SCATTERING OF ELECTROMAGNETIC RADIATION.

Optical glass. The optical material most widely used is optical glass, which is available in a wide range of refractive indices and dispersions and differs from ordinary glass in its freedom from imperfections. It must be free from unmelted particles or "stones," from bubbles, and from chemical inhomogeneity, which gives rise to regions of variable refractivity known as cords or striae.

The refractive index is sometimes followed by a subscript indicating the wavelength of the light in vacuum; n_{D} indicates the refractive index for the mean D line of sodium, $\lambda = 589.3$ nanometers (nm). In optical glass catalogs, it is customary to give the refractive indices for a group of spectral lines chosen by E. Abbe for the convenience with which they could be obtained for spectrometric work. Two more wavelengths, one in the near-ultraviolet and one in the near-infrared, have been added chiefly to accommodate calculations for photographic systems sensitive in these areas. The source, designation, and wavelength in vacuum of these lines are given in Table 1, together with some other lines which are being used increasingly because of their greater convenience.

Table 1. Designation, source, and wavelength of spectral lines used in spectrometric measurements*

Source	Hg	Hg	Hg	H	H	He	Hg
Designation	—	h	g	G'	F	—	e
Wavelength, nm	365.0	404.7	435.8	434.1	486.1	492.2	546.1

Source	He	Na (mean)		H	He	K (mean)	Hg
Designation	d	D		C	b	A	—
Wavelength, nm	587.6	589.3		656.3	706.5	768.2	1014.0

*From G. W. Morey, *Properties of Glass*, 2d ed., Reinhold, 1954, supplemented, 1968.

Optical glass catalogs also give the mean dispersion, commonly designated by $(n_C - n_F)$ and other partial dispersions, as well as the dispersion ratios $(n_D - n_C)/(n_F - n_C)$, and so forth, in which the symbols C, D, and F refer to the refractive indices for the spectral lines as designated in Table 1. Another number given under various names is the nu value, $\nu = (n_D - 1)/(n_F - n_C)$. This expression occurs in the calculation of the color correction of lenses.

Glass types. The early optical glasses were crowns and flints. The crown glasses were essentially of the same type as window glass, with low index and dispersion; the flint glasses contained lead oxide and had higher index and dispersion. These glasses, roughly indicated by solid lines in Fig. 1, did not have sufficient range in optical properties to enable desirable corrections to be made in optical systems. A great step forward was made by O. Schott in the introduction of the borosilicate, fluoride, and barium crown and flint types, and a second advance was made by G. W. Morey in the rare-earth glasses, now usually called lanthanum crowns and flints, also indicated by Fig. 1.

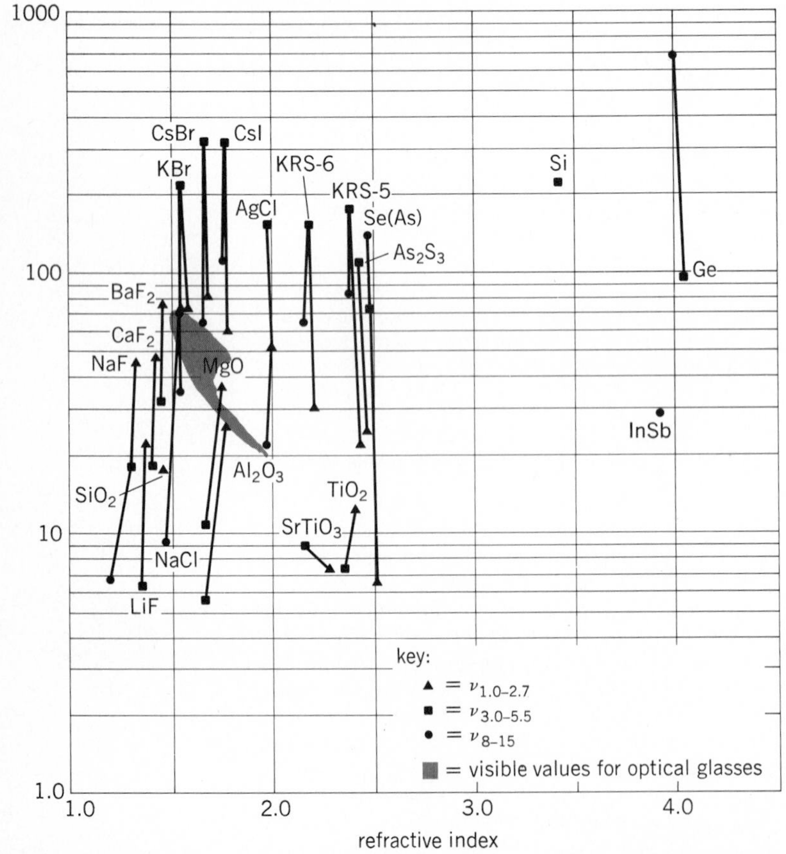

Effect of absorption. When the absorption of light in glass is fairly uniformly distributed throughout the visible spectrum and the amount of absorption is small, the glass appears colorless and limpid when viewed in white light; when the amount of uniform absorption increases, the glass takes on a grayish hue. If the absorption is significantly greater for light of any particular color, the transmitted light will appear of a complementary color; since the absorption is never monochromatic, the color will depend on the thickness of the sample. Comparisons are best made by means of curves showing the change of absorption with wavelength for samples of standard thickness.

Silica glass, in thicknesses such as are used in cameras and optical instruments, transmits all the radiation to which ordinary photographic plates are sensitive, that is, down to a wavelength of 220 nm, but wavelengths of 193 nm and less are almost completely absorbed. The transmission of the usual colorless glass is limited chiefly by absorptive bands, as is indicated by the dispersion curve. The limit of transmission in ultraviolet light is determined largely by the content of Fe_2O_3, which shows strong absorption in the near-ultraviolet. Special glasses should contain, and have been made to contain, as little as less than 5 ppm of iron oxide. The limit of transmission in the very-near-infrared is determined largely by the content of FeO, which shows strong absorption at about 1 μm. The best optical glasses have a transparency of over 99% throughout the visible spectrum (380–780 nm), a result achieved only by using the greatest care in excluding impurities, especially iron. Ordinary window glass removes the 313-nm line of mercury vapor and shorter wavelengths. Typical transmission curves of glasses are shown in Fig. 2.

Colored glass. The coloring agent or colorant in glass is usually produced by (1) substances dissolved in the glass which absorb characteristic frequencies, (2) particles of submicroscopic dimensions, such as gold or copper in ruby glass, or (3) particles of microscopic or larger dimensions, either themselves colored, as in the aventurine glasses, or colorless, as in opal glass. The coloring agents which act by virtue of characteristic absorption spectra are all elements belonging to the transition rows of the periodic system, and especially to the first of these rows, to which belong titanium, vanadium, chromium, manganese, iron, cobalt, nickel, and copper, the commonest and most effective colorants of glass.

Available glasses and limitations. Most companies that sell glass for use in lenses for optical instruments also produce glass charts, like that shown in Fig. 1, of their products. Designers often

Fig. 2. Reciprocal dispersion of some infrared optical materials. (*From W. Wolfe, ed., Handbook of Military Infrared Technology, Government Printing Office, 1966*)

Table 2. NATO optical glass

Six-digit code	Schott name	N_d	V_d	P_{Fd}	Stain	Bubble	Density	Transmission
511604	K-7	1.51112	60.41	0.6950	0	0–1	2.53	0.986
517642	BK-7	1.51680	64.17	0.6928	0	0	2.51	0.972
522595	K-5	1.52249	59.48	0.6956	0	2	2.63	0.98
523602	K-50	1.52257	60.18	0.6956	0	0	2.62	0.986
527511	KzF-6	1.52682	51.13	0.6969	2	1	2.54	0.88
529517	KzF-2	1.52944	51.68	0.6970	2	0–1	2.55	0.91
548458	LLF-1	1.54814	45.75	0.7020	0	0–1	2.94	0.985
573575	BaK-1	1.57250	57.55	0.6966	1	1	3.19	0.97
581409	LF-5	1.58144	40.85	0.7040	0	0–1	3.22	0.988
613443	KzFS-N4	1.61340	44.30	0.6997	2	0–1	3.20	0.65
613574	SK-19	1.61342	57.37	0.6967	1	0–1	3.56	0.91
613586	SK-4	1.61272	58.63	0.6966	2	0–1	3.57	0.92
618551	SSK-4	1.61765	55.14	0.6982	1	0–1	3.63	0.94
620364	F-2	1.62004	36.37	0.7062	0	0	3.61	0.90
620603	SK-16	1.62041	60.33	0.6953	4	0–1	3.58	0.88
648339	SF-2	1.64769	33.85	0.7076	0	0–1	3.86	0.84
650392	BaSF-10	1.65016	39.15	0.7057	2–3	2	3.91	0.80
652585	LaK-N7	1.65160	58.52	0.6961	2	1	3.84	0.845
653397	KzFS-5	1.65332	39.71	0.7013	4–5	0–1	3.44	0.14
670471	BaF-N10	1.67003	47.11	0.7017	3	0–1	3.76	0.29
691547	LaK-N9	1.69100	54.71	0.6959	1	1	3.55	0.745
699301	SF-15	1.69895	30.07	0.7102	0	1	4.06	0.01
702410	BaSF-52	1.70181	41.01	0.7030	5	1	3.96	0.43
717480	LaF-N3	1.71700	47.98	0.7007	5	1	4.34	0.62
720504	LaK-10	1.72000	50.41	0.6974	2	1	3.81	0.755
744448	LaF-N2	1.74400	44.77	0.7023	4–5	1	4.46	0.47
755276	SF-4	1.75520	27.58	0.7114	2	1	4.78	0.31
805254	SF-6	1.80518	25.43	0.7129	3–4	0–1	5.18	0.03

*1 lb = 0.45 kg.

search for very special substances for extra color correction of lenses, but these materials are usually scarce and expensive. Agreement has been reached by many users to utilize only the NATO glasses listed in Table 2 for all but the most unusual requirements. NATO glasses are a set of glass formulations approved by a NATO group as those for which almost all optical designs will be done. Economies should come from this.

Most glasses are uniformly transparent over the entire visible spectrum, suffering only reflection losses. The limits of transmission (and therefore usefulness) in the neighboring infrared and ultraviolet portions of the spectrum are usually set by the content of Fe_2O_3, which absorbs strongly in the near-ultraviolet, and by FeO, which absorbs strongly at 1 μm. However, even for completely pure glass, the vibrations between Si and O in the silicate structure limit transmission to about 2.5 μm, and freeing electrons from the lattice determines the short-wavelength limit. Research on other matrices such as Ge-O and $CaAl_2O_3$ has produced somewhat softer substances that transmit to about 5 μm, and nonoxide compositions of combinations Si, Ge, Se, As, Te, S. These are softer still, melt at low temperatures (~500°C), but transmit to about 14 μm.

Crystalline materials. Because of the limitations of glasses, single-crystal or polycrystalline aggregates are used for special applications. The most useful of these are of cubic symmetry, so that the refractive index and other physical properties are isotropic. The properties of some of the more useful of these materials are listed in Table 3. The short-wavelength edge of the transmission region is determined by the width of the energy gap; the long-wavelength edge is determined by the onset of molecular vibrations or rotations. The long wavelength is proportional to the square-root of the mass of the constituent molecules and inversely proportional to the square root of the binding energy. Many single-crystal materials are available in cylinders about 30 cm in diameter by 30 cm long. Some polycrystalline samples are available in even larger diameters (1 m) in thinner sections.

Plastics. The most significant properties of plastics (compared to glass) are their softness and mo-

Table 3. Optical materials for the ultraviolet and infrared and selected properties

Material Name	Symbol	λ_{min}, μm	λ_{max}, μm	T_m, K*	H_K, mm²/kg†
Quartz	SiO_2	0.12	5	174	740–460
Sapphire	Al_2O_3	0.14	6	2300	1370
Rutile	MgO	0.45	6	307	690
Diamond	C	0.25	80+		
Salt	NaCl	0.21	2	1075	15
Potassium iodide	KI	0.25	45	1000	
Cesium bromide	CsBr	0.25	80	900	20
KRS-5	KI-BnI	0.6	40	690	
Magnesium fluoride	MgF_2	0.11	8		
Calcium fluoride	CaF_2	0.13	12	163	160
Cadmium telluride	CdTe	0.9	16	1310	
Zinc sulfide	ZnS	2	20	1293	
Calcium aluminate (glass)	$CaAl_2O_3$	0.4	6		
Lithium fluoride	LiF	0.12	9	1140	100
Calcite	$CaCO_3$	0.2	6	1170	
Arsenic trisulfide glass	As_2S_3	0.6	13	48	100
Silicon	Si	1.2	15	1690	1150
Germanium	Ge	1.8	20	1210	
Barium fluoride	BaF_2	0.25	15	1550	82
Potassium chloride	KCl	0.21	30	1060	

*T_m = melting temperature.   †H_K = Knoop hardness.

lecular complexity. In the visible spectrum they can be used for relatively cheap lenses but must be protected from abrasion. They can be used in the infrared at relatively short and very long wavelengths. The resonance of the molecular bond between carbon and hydrogen prevents the use of most plastics beyond about 3 μm in all but the thinnest membranes. Polyethylene and polystyrene are particularly good choices for thin window coverings throughout the infrared.

Materials for infrared and ultraviolet. The lack of a good transparent glass for these spectral regions caused the long and continuing search for appropriate substitutes, which included single crystals, polycrystalline compacts, and plastics. These are listed in Table 3 along with some of their more important properties. The nu value or Abbe number is meaningless for materials that are used for infrared or ultraviolet work, but a generalized coefficient can be used $(\Delta n/\Delta\lambda)/(n-1)$, where $\Delta n/\Delta\lambda$ is the slope of the refractive index versus wavelength curve in the useful region well between the absorption bands. Some representative materials are shown in Fig. 2 for the main infrared regions, and the data of Fig. 1 are contained within the shaded area.

For the ultraviolet region of the spectrum, the main refractors are LiF, CaF_2, and SiO_2; the shortest wavelength attainable is with LiF and is about 0.12 μm. The region short of this is known as the vacuum ultraviolet, where the atmosphere and all known materials are opaque. In the very far infrared (which usually means beyond about 100 μm) most of the alkali halides "open up" and transmit reasonably well. Many plastics are useful if not thicker than several thousandths of an inch. These are used primarily as windows in the form of membranes to maintain a prescribed interior atmosphere.

Cautions. The required quality of an optical material depends, of course, on its use. The properties that should be considered in general include the transmission, refractive index, melting or softening temperature, strength, hardness, possibility of good surface finish, stain, scratch, internal scatter, homogeneity, and strength. The influence of each of these properties can be complex, and the degree of importance varies with the application.

[WILLIAM L. WOLFE]

Bibliography: N. J. Kreidl and J. L. Rood, Optical materials, in R. Kingslake (ed.), *Applied Optics and Optical Engineering*, vol. 1, 1965; W. A. Weyl, Coloured glasses, in D. D. Gray (ed.), *American Institute of Physics Handbook*, 3d ed., 1972; W. L. Wolfe, Properties of optical materials, sec. 7, in Optical Society of America, *Handbook of Optics*, 1978.

Optical phase conjugation

A process that involves the use of nonlinear optical effects to precisely reverse the direction of propagation of each plane wave in an arbitrary beam of light, thereby causing the return beam to exactly retrace the path of the incident beam. The process is also known as wavefront reversal or time-reversal reflection. The unique features of this phenomenon suggest widespread application to the problems of optical beam transport through distorting or inhomogeneous media. Although

closely related, the field of adaptive optics will not be discussed here.

Fundamental properties. Optical phase conjugation is a process by which a light beam interacting in a nonlinear material is reflected in such a manner as to retrace its optical path. As Fig. 1 shows, the image-transformation properties of this reflection are radically different from those of a conventional mirror. The incoming rays and those reflected by a conventional mirror (Fig. 1a) are related by inversion of the component of the k-vector or wave vector k normal to the mirror surface. Thus a light beam can be arbitrarily redirected by adjusting the orientation of a conventional mirror. In contrast, a phase-conjugate reflector (Fig. 1b) inverts the vector quantity $\vec{k}$ so that, regardless of the orientation of the device, the reflected conjugate light beam exactly retraces the path of the incident beam. This retracing occurs even though an aberrator (such as a piece of broken glass) may be in the path of the incident beam. Looking into a conventional mirror, one would see one's own face, whereas looking into a phase-conjugate mirror, one would see only the pupil of the eye. This is because any light emanating from, say, one's chin would be reversed by the phase conjugator and return to the chin, thereby missing the viewer's eye. A simple extension of the arrangement in Fig. 1b indicates that the phase conjugator will reflect a diverging beam as a converging one, and vice versa. These new and remarkable image-transformation properties (even in the presence of a distorting optical element) open the door to many potential applications in areas such as laser fusion, atmospheric propagation, fiber-optic propagation, image restoration, real-time holography, optical data processing, nonlinear microscopy, laser resonator design, and high-resolution nonlinear spectroscopy.

Optical conjugation techniques. Optical phase conjugation can be obtained in many materials whose properties are affected by strong applied optical fields. The response of the material may permit many beams to combine in such a way as to generate a new beam that is the phase-conjugate of one of the input beams. Processes associated with degenerate four-wave mixing, scattering from saturated resonances, stimulated Brillouin scattering, stimulated Raman scattering, photon echoes, and three-wave mixing have all been utilized to generate optical phase-conjugate reflections.

Degenerate four-wave mixing. As shown in Fig. 2, two strong counterpropagating (pump) beams with k-vectors $\vec{k}_1$ and $\vec{k}_2$ (at frequency ω) set up a standing wave in a clear material whose index of refraction varies linearly with intensity. This arrangement provides the conditions in which a third (probe) beam with k-vector $\vec{k}_3$, also at frequency ω, incident upon the material from any direction would result in a fourth beam with k-vector $\vec{k}_4$ being emitted in the sample precisely retracing the third one. (The term degenerate indicates that all beams have exactly the same frequency.) In this case, phase matching (even in birefringent materials) is obtained independent of the angle between $\vec{k}_3$ and $\vec{k}_1$. The electric field of the conjugate wave E_4 is given by the equation below, where

$$E_4 = E_3{}^* \tan\left(\frac{2\pi}{\lambda_0}\delta n\ell\right)$$

OPTICAL PHASE CONJUGATION

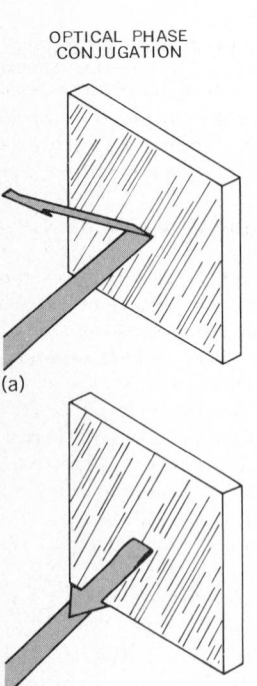

(a)

(b)

Fig. 1. Comparison of reflections (a) from a conventional mirror and (b) from an optical phase conjugator. (From I. J. Bigio et al., High efficiency phase-conjugate reflection in germanium and in inverted CO_2, in V. J. Corcoran, ed., Proceedings for the International Conference on Laser '78 for Optical and Quantum Electronics, STS Press, McLean, VA, 1979)

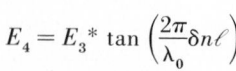

δn is the index change induced by one strong counterpropagating wave, λ_0 is the free-space optical wavelength, and ℓ is the length over which the probe beam overlaps the conjugation region. The conjugate reflectivity is defined as the ratio of reflected and incident intensities, which is the square of the above tangent function. The essential feature of phase conjugation is that E_4 is proportional to the complex conjugate of E_3. Although degenerate four-wave mixing is a nonlinear optical effect, it is linear in the field one wishes to conjugate. This means that a superposition of E_3's will generate a corresponding superposition of E_4's; therefore faithful image reconstruction is possible.

To visualize the degenerate four-wave mixing effect, consider first the interaction of the weak probe wave with pump wave number two. The amount by which the index of refraction changes is proportional to the intensity $(E_3 + E_2)^2$, and the cross term corresponds to a phase grating (periodic phase disturbance) appropriately oriented to scatter the pump wave number one into the $\vec{k}_4$ direction. Similarly, the very same scattering process occurs with the roles of pump waves reversed. Creating the phase gratings can be thought of as "writing" of a hologram, and the subsequent scattering can be thought of as "reading" the hologram. Thus the four-wave mixing process is equivalent to volume holography in which the writing and reading are done simultaneously.

Conjugation using saturated resonances. Instead of using a clear material, as outlined above, the same beam geometry can be set up in an absorbing or amplifying medium partially (or totally) saturated by the pump waves. When the frequency of the light is equal to the resonance frequency of the transition, the induced disturbance corresponds to amplitude gratings which couple the four waves. Because of the complex nature of the resonant saturation process, one does not obtain the simple $\tan^2 [(2\pi/\lambda_0)(\delta n\ell)]$ expression for the conjugate reflectivity. Instead, this effect is maximized when the intensities of the pump waves are about equal to the intensity which saturates the transition.

Stimulated Brillouin and Raman scattering. Earliest demonstrations of optical phase conjugation were performed by focusing an intense optical beam into a waveguide containing materials that exhibit backward stimulated Brillouin scattering. More recently, this technique has been extended to include the backward stimulated Raman effect. In both cases, the conjugate wave is downshifted by the frequencies characteristic of the effect.

Practical applications. Many practical applications of optical conjugators utilize their unusual image-transformation properties. Because the conjugation effect is not impaired by interposition of an aberrating material in the beam, the effect can be used to repair the damage done to the beam by otherwise unavoidable aberrations. This technique can be applied to improving the output beam quality of laser systems which contain optical phase inhomogeneities or imperfect optical components. In a laser, one of the two mirrors could be replaced by a phase-conjugating mirror, or in laser amplifier systems, a phase conjugate reflector could be used to reflect the beam back through the amplifier in a double-pass configuration. In both cases, the optical-beam quality would not be degraded by inhomogeneities in the amplifying medium, by deformations or imperfections in optical elements, windows, mirrors, and so forth, or by accidental misalignment of optical elements.

Aiming a laser beam through an imperfect medium to strike a distant target may be another application. The imperfect medium may be turbulent air, the air-water interface, or the focusing mirror in a laser-fusion experiment. Instead of conventional approaches, one could envision a phase-conjugation approach in which the target would be irradiated first with a weak diffuse probe beam. The glint returning from the target would pass through the imperfect medium, and through the laser system, and would then strike a conjugator. The conjugate beam would essentially contain all the information needed to strike the target after passing through both the amplifier and the imperfect medium a second time. Just as imperfections between the laser and the target would not impair the results, neither would problems associated with imperfections in the elements that constitute the laser amplifier.

Other applications are based upon the fact that four-wave mixing conjugation is a narrow-band mirror that is tunable by varying the frequency of the pump waves. There are also applications to fiber-optic communications. For example, a spatial image could be reconstructed by the conjugation process after having been "scrambled" during passage through a multimode fiber. Also, a fiber-optic communication network is limited in bandwidth by the available pulse rate; this rate is determined to a large extent by the dispersive temporal spreading of each pulse as it propagates down the fiber. The time-reversal aspect of phase conjugation could undo the spreading associated with linear dispersion and could therefore increase the possible data rate. *See* HOLOGRAPHY; LASER; NONLINEAR OPTICS; RAMAN EFFECT.

[ROBERT A. FISHER; BARRY J. FELDMAN]

Bibliography: A. Yariv, *IEEE J. Quantum Electron.*, QE(14):650–660, 1978; B. Ya. Zel'dovich et al., *Kvant. Elektron.*, 5:1800–1803, 1978 (transl.. *Sov. J. Quantum Electron.*, 8:1021–1023, 1978).

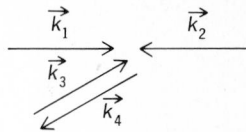

OPTICAL PHASE
CONJUGATION

Fig. 2. Geometry of *k*-vectors for optical phase conjugation using degenerate four-wave mixing.

Optical prism

An optical system consisting of two or more usually plane surfaces of a transparent solid or embedded liquid at an angle with each other. Prisms are used for deviating light. Since the amount of deviation depends on the refractive index of the prism, which varies with wavelength, prisms can also be used for dispersing light. *See* DISPERSION; REFRACTION OF WAVES.

Reflecting prisms. Prisms can be used instead of mirrors for deviating light, with the added advantage that the reflecting surfaces are protected against corrosion. In this case there is at least one internal reflection. When the angles of incidence and emergence are zero, there is no dispersion. The overall dispersion is also zero when the geometry of the prism is such that the dispersion at the entering surface is compensated by dispersion in the opposite sense at the emergent surface. For a detailed discussion of important types of reflecting prisms *see* MIRROR OPTICS.

Dispersing prisms. Dispersing prisms deviate light of different wavelengths by different amounts, and they can therefore be used to separate white light into its monochromatic parts. A parallel

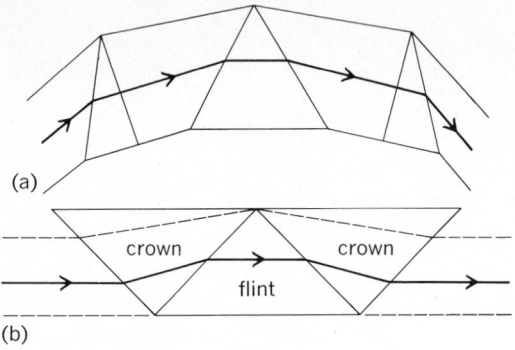

Fig. 1. Some types of dispersing prisms. (*a*) Rayleigh prism system. (*b*) Amici direct-vision system consisting of a flint-glass prism and two crown-glass prisms.

beam of light entering the prism leaves the prism as a parallel beam of light but its diameter may be changed. The ratio of its diameter after refraction to its diameter before refraction can be considered as the magnification of the prism.

The prism magnification for a bundle parallel to the prism edge is always equal to unity, whereas it varies in the meridional plane (normal to the edge) as the angle of incidence is varied. It is equal to unity in this plane only if the prism is traversed at minimum deviation. If α is the prism angle, n the refractive index, and δ the deviation, it can be shown that, for minimum deviation, the equation below holds.

$$\sin (\delta + \alpha)/2 = n \sin \alpha/2$$

To increase the dispersion, several prisms with their refracting edges parallel can be used. The Rayleigh prism, shown in Fig. 1*a*, is an example of such a system. By using a prism made of a material, such as flint glass, that has a high dispersion, and adding one or more prisms made of a material having a low dispersion, such as crown glass (Fig. 1*b*), the deviation can be neutralized without neutralizing the dispersion to give a direct-vision prism system. The arrangement shown in Fig. 1*b* is known as the Amici prism system. By using a similar arrangement but adjusting the angle so that the dispersion, but not the deviation, is neutralized, it is possible to make a prism system that is achromatic over a small part of the spectrum, like an achromatic lens. *See* LENS.

An achromatic prism in front of an optical system with its refracting edge normal to the meridional plane can be used to change the magnification of the optical system in that plane. This amount can be varied by rotating prism A in Fig. 2*a* about an axis normal to the meridional plane. A second achromatic prism, B, with its edge parallel to the meridional plane, can be used to adjust the sagittal magnification of the optical system. Therefore, an arrangement of two such prisms with their motions linked together can be used for the purpose of forming a variable-focal-length lens system, which is commonly called a zoom lens.

A thin prism is one whose angle is so small that the prism angle expressed in radians is practically equal to the tangent. Such prisms are used in ophthalmology, and their powers are usually expressed in prism diopters. The Risley prism sys-

OPTICAL PRISM

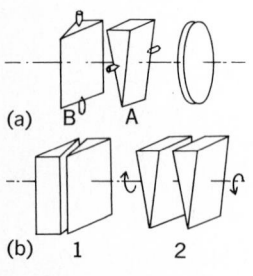

Fig. 2. Prisms. (*a*) Pair used for varying magnification, as in zoom system. (*b*) Pair (Risley prism system) used for varying deviation.

tem, used for testing ocular convergence, consists of two thin prisms mounted so that they can be rotated simultaneously in opposite directions, as shown in Fig. 2*b*. When they are in the orientation sketched at 1, their combined deviation is zero; when both have been rotated by 90° in opposite directions, as shown at 2, their combined deviation is a maximum; at intermediate positions, their combined deviation lies between zero and the maximum, but the plane of deviation is constant. A similar pair of rotating wedges is used in certain types of rangefinders. *See* BINOCULARS; GEOMETRICAL OPTICS; OPTICAL MATERIALS; RESOLVING POWER.

[MAX HERZBERGER]

Bibliography: G. A. Boutry, *Instrumental Optics*, 1962; E. U. Condon and H. Odishaw, *Handbook of Physics*, 1967; D. F. Horne, *Optical Instruments and Their Applications*, 1980; F. A. Jenkins and H. E. White, *Fundamentals of Optics*, 4th ed., 1976; Optical Society of America, *Handbook of Optics*, 1978.

Optical pulses

Short flashes of light, used to isolate moments of time. As the technology of making short light pulses has advanced, new types of processes have become open to investigation. Spark or flash photography can freeze the most rapid movement of macroscopic objects with flashes of 10^{-7} s in duration. High-speed flash lamps and electronics play an important role in the study of fast photophysical and photochemical processes with a resolution of 10^{-10} s.

In 1966 a laser technique was used to generate the first optical pulses of less than 10 picoseconds in duration, a time too short to measure by conventional electronic means. This work provided the stimulus for the rapid development of a variety of new methods for pulse measurements and diagnostics. Laser techniques have been refined and new pulse sources invented to provide even shorter pulses. Pulses as short as a tenth of a picosecond have been produced with the continuous dye laser system. *See* LASER.

Principle of generation. The more frequencies or modes oscillating in a laser, the shorter the optical pulse that can be generated. Organic dyes are ideal for generating pulses, because dyes are capable of oscillating over a broad range of frequencies or colors. To make a pulse, each of these oscillating frequencies must maintain a fixed phase relationship with one another to be coherently added to produce a short optical pulse. This process is called mode locking.

Applications. The primary importance of these ultrashort flashes of light is to provide a new tool for experimental investigations on a very short time scale. Clearly, picosecond flashes of light are not necessary for stopping the action of large objects encountered in daily life. Rather, it is the movement of molecules and atoms which takes place in this very short time period and which is revealed on this time scale. Some of the most important applications of these short optical pulses have been to the study of ultrafast processes in molecules, biological systems, and solids. The primary photoprocesses in vision and photosynthesis have been unraveled with picosecond pulse studies. The fundamental nature of fast chemical reactions and vibrations in molecules and solids has been investigated with ultrashort optical pulse techniques.

The prospect of high-data-rate communication with picosecond flashes of light lies in the future. The propagation of picosecond pulses for kilometers in optical-fiber light guides has already been demonstrated. See OPTICAL FIBERS.

[C. V. SHANK]

Bibliography: S. L. Shapiro (ed.), *Ultrashort Light Pulses, Picosecond Techniques and Applications*, 1977.

Optical pumping

The process of causing strong deviations from thermal equilibrium populations of selected quantized states of different energy in atomic or molecular systems by the use of optical radiation (that is, light of wavelengths in or near the visible spectrum), called the pumping radiation.

At thermal equilibrium at temperature $T°K$, the relative numbers of atoms, N_2/N_1, in quantized levels E_2 and E_1, respectively, where E_2 is the higher, are given by $N_2/N_1 = e^{-(E_2-E_1)/kT}$, where k is Boltzmann's constant. The number of atoms in the higher level is, at equilibrium, always less than that in the lower, and as the energy difference between the two levels increases the number in the higher level becomes very small indeed. By exposing a suitable system to optical radiation, one can, so to speak, pump atoms from a lower state to an upper state so as greatly to increase the number of atoms in the upper state above the equilibrium value. See ENERGY LEVEL.

In an early application of the principle, the levels E_2 and E_1 were not far apart, so that the equilibrium populations of the atoms in the two levels were not greatly different. The system was chosen to possess a third level E_3, accessible from E_1 but not from E_2 by the absorption of monochromatic visible light. (The states involved were the paramagnetic Zeeman components of atomic states, and the necessary selectivity of the transitions excitable by the visible light was secured by appropriate choice of the state of polarization of this light). The visible light excites atoms from E_1 to E_3, from which they return, with spontaneous emission, with about equal probability, to the lower states E_2 and E_1. After a period of time, provided there has been sufficiently intense excitation by the visible light, most of the atoms are in state E_2 and few are in the lower state E_1—atoms have been pumped from E_1 to E_2 by way of the highly excited state E_3. See ATOMIC STRUCTURE AND SPECTRA; ZEEMAN EFFECT.

Optical pumping is vital for light amplification by stimulated emission in an important class of lasers. For example, the action of the ruby laser involves the fluorescent emission of red light by a transition from an excited level E_2 to the ground level E_1. In this case E_2 is relatively high above E_1 and the equilibrium population of E_2 is practically zero. Amplification of the red light by laser action requires that N_2 exceed N_1 (population inversion). The inversion is accomplished by intense green and violet light from an external source which excites the chromium ion in the ruby to a band of levels, E_3, above E_2. From E_3 the ion rapidly drops without radiation to E_2, in which its lifetime is relatively long for an excited state.

Sufficiently intense pumping forces more luminescent ions into E_2 by way of the E_3 levels than remain in the ground state E_1, and amplification of the red emission of the ruby by stimulated emission can then occur. See LASER. [WILLIAM WEST]

Bibliography: D. C. O'Shea, W. R. Callen, and W. T. Rhodes, *An Introduction to Lasers and Their Applications*, 1977; O. Svelto, *Principles of Lasers*, 1976; A. Yariv, *Introduction to Optical Electronics*, 1976.

Optical surfaces

Many common optical instruments contain optical materials bounded by spherical surfaces with a wide range of curvatures. A surface that is not a sphere is called an aspherical surface. Examples of such aspherical surfaces with symmetry of rotation that are occasionally used in optical instruments are conic sections (ellipsoids, hyperboloids, and paraboloids). Cylindrical and toroidal lenses are used in anamorphotic systems.

Since it is much easier to grind and polish spherical surfaces than aspherical ones, most optical systems consist of lenses that are bounded by two spherical surfaces. These surfaces are arranged so that their centers lie on a line known as the axis of the system. The point of intersection of the surface with the axis is called the vertex. A plane surface is generally considered as a special case of a spherical surface with its center at infinity and which can therefore be said to have an infinite radius.

Aspherical surfaces. Rays normal to a given wave surface can be refracted or reflected by a single aspherical surface so that they are normal to a desired wave surface. This is most frequently done to improve the aperture errors of a system and is successful when the field to be imaged is very small. For example, large telescopes use aspherical mirrors. If the basic system is well corrected, the asphericity may not destroy the other corrections. This is the secret of the Schmidt camera, the best-known application of an aspherical optical element. Aspherical surfaces can also be used to correct the errors of the principal rays (the rays through the center of the diaphragm).

Aspherical surfaces are difficult to manufacture. When many identical aspherical elements are to be made in series, special methods of grinding and polishing must be used. For instance, templates are used for spectacle lenses and for paraboloidal and ellipsoidal mirrors. Where extreme correction is desired, for instance, in astronomical telescopes, retouching by hand is the proper procedure. For condensers and for other lenses where extreme accuracy is not needed, a molding process has been developed. Large paraboloidal mirrors for searchlights and ellipsoidal mirrors for arc lamps used to project motion pictures are made by a "dropping" process. A sheet of plate glass is laid on a suitable concave mold and heated until it softens and can be sucked into the mold.

Lenses with more than one aspherical surface have been proposed and designed by M. Linnemann to correct aperture and first-order asymmetry errors.

There are a few patents in which two or more aspherical surfaces are successfully used to balance errors, but a complete theory for doing this has not yet been developed.

Ray-tracing formulas. To obtain ray-tracing formulas that are valid for all surfaces, it is convenient to place the coordinate origin at the vertex of the refracting (reflecting) surface, though for a refracting sphere alone, formulas with the origin at the center are somewhat simpler. If the z axis is in the direction of the system axis and the x and y axes are normal to it, the equations of the surface are then Eqs. (1), where $\bar{u} = \frac{1}{2}(\bar{x}^2 + \bar{y}^2 + \bar{z}^2)$.

$$\begin{aligned}
\bar{z} &= 0 \quad \text{(for the plane)} \\
\bar{z} &= \rho\bar{u} \quad \text{(for the sphere of curvature } \rho) \\
\bar{z} &= \rho\bar{u} + \frac{1}{2}A_2\bar{u}^2 + \frac{1}{6}A_3\bar{u}^3 + \cdots \\
&\quad \text{(for an aspherical surface)}
\end{aligned} \quad (1)$$

The object point and the object ray may be given by the coordinates x, y of the intersection point with the plane at the vertex and the optical direction cosines ξ, η, ζ of the ray. The procedure is to find the value of a parameter λ such that Eqs. (2)

$$\begin{aligned}
u &= \frac{1}{2}[(x+\lambda\xi)^2 + (y+\lambda\eta)^2 + (\lambda\zeta)^2] \\
\bar{z} &= \lambda\zeta
\end{aligned} \quad (2)$$

fulfill Eqs. (1). This can be achieved by an iteration process starting from $\lambda = 0$. With λ found, the coordinates of the point of intersection are computed as in Eqs. (3).

$$\begin{aligned}
\bar{x} &= x + \lambda\zeta \\
\bar{y} &= y + \lambda\eta \\
\bar{z} &= \lambda\zeta
\end{aligned} \quad (3)$$

Equations (1) give Eqs. (4) for the direction O of

$$\begin{aligned}
O_1 &= \bar{z}_u\bar{x}/[1 + \bar{z}_u^2(2\bar{u} - \bar{z})]^{1/2} \\
O_2 &= \bar{z}_u\bar{y}/[1 + \bar{z}_u^2(2\bar{u} - \bar{z})]^{1/2} \\
O_3 &= [1 + \bar{z}_u^2(2\bar{u} - \bar{z})]^{-1/2}
\end{aligned} \quad (4)$$

the unit vector along the surface normal (coordinates O_1, O_2, O_3).

Here $\bar{z}_u = \frac{1}{2}\rho$ (for the sphere), $\bar{z}_u = \frac{1}{2}\rho + A_2\bar{u} + \frac{1}{2}A_3\bar{u}^2$ (for an aspherical surface), and $\bar{z}_u = 0$ (for the plane). The directions of the normal and of the entering ray being known, the refraction law enables the refracted ray to be computed. The corresponding formulas are Eqs. (5), with $\Gamma =$

$$\begin{aligned}
\xi' - \xi &= \Gamma O_1 \\
\eta' - \eta &= \Gamma O_2 \\
\zeta' - \zeta &= \Gamma O_3
\end{aligned} \quad (5)$$

$n' \cos i' - n \cos i$. Here n and n' are the refractive indices of the media, separated by the refracting surface, i and i' are the angles formed by the incident and refracted rays with the surface normal, and $\cos i$ and $\cos i'$ are given by Eqs. (6).

$$\begin{aligned}
\cos i &= \xi O_1 + \eta O_2 + \zeta O_3 \\
\cos i' &= +[n'^2 - n^2 + \cos^2 i]^{1/2}
\end{aligned} \quad (6)$$

Refracting sphere. Rays through the center of a sphere are unrefracted. The center of a sphere is sharply imaged, and since the sine condition is fulfilled, a surface element through the center is imaged without errors of asymmetry.

The points having a center distance $c = -n'r/n$ are sharply imaged upon points with center distance $c' = -nr/n'$. The spheres on which these points respectively lie are called aplanatic spheres. Either the object or the image is virtual. The magnification with which these two spheres are imaged upon each other is $m = n^2/n'^2$. Again the

sine condition is fulfilled, and thus first-order asymmetry errors are corrected. Lenses consisting of a refracting centered sphere or a refracting aplanatic sphere, or both, are often added to a given system to achieve a desired effect without destroying corrections previously achieved.

Cartesian surfaces. The centered sphere and the aplanatic sphere are special cases of surfaces which image the rays coming from an object point so that they all converge to another point. The general surface of this kind is determined by the fact that on all rays the light path from the object point to the surface and thence to the image point is constant. Such a surface is in general of the fourth order. For an infinitely distant object, it is a hyperboloid. For zero light path, it becomes the aplanatic sphere.

The conic sections are reflecting cartesian surfaces. The reflecting ellipsoid images the rays from one geometrical focus to the other; the paraboloid images the geometrical focus sharply at infinity, or conversely, a set of parallel rays sharply at the focus; the hyperboloid images the rays from a real (or virtual) point sharply at a virtual (or real) point. Both points are situated at the geometrical foci of the hyperboloid. The only cartesian surface free from asymmetry errors is the aplanatic sphere. *See* GEOMETRICAL OPTICS.

[MAX HERZBERGER]

Bibliography: M. Herzberger, *Modern Geometrical Optics*, 1958, reprint 1978; F. A. Jenkins and H. E. White, *Fundamentals of Optics*, 4th ed., 1976; B. Jurek, *Optical Surfaces*, 1976; Optical Society of America, *Handbook of Optics*, 1978.

Optics

Narrowly, the science of light and vision; broadly, the study of the phenomena associated with the generation, transmission, and detection of electromagnetic radiation in the spectral range extending from the long-wave edge of the x-ray region to the short-wave edge of the radio region. This range, often called the optical region or the optical spectrum, extends in wavelength from about 10 A to about 1 mm. For information on the various branches of optics *see* GEOMETRICAL OPTICS; PHYSICAL OPTICS.

In ancient times there was some isolated elementary knowledge of optics, but it was the discoveries of the experimentalists of the early 17th century which formed the basis of the science of optics. The statement of the law of refraction by W. Snell, Galileo Galilei's development of the astronomical telescope and his discoveries with it, F. M. Grimaldi's observations of diffraction, and the principles of the propagation of light enunciated by C. Huygens and P. de Fermat all came in this relatively short period. The publication of Isaac Newton's *Opticks* in 1704, with its comprehensive and original studies of refraction, dispersion, interference, diffraction, and polarization, established the science.

So great were the contributions of Newton to optics that a hundred years went by before further outstanding discoveries were made. In the early 19th century many productive investigators, foremost among them Thomas Young and A. J. Fresnel, established the transverse-wave nature of light. The relationship between optical and magnetic phenomena, discovered by M. Faraday in the 1840s, led to the crowning achievement of classical optics—the electromagnetic theory of J. C. Maxwell. Maxwell's theory, which holds that light consists of electric and magnetic fields propagated together through space as transverse waves, provided a general basis for the treatment of optical phenomena. In particular, it served as the basis for understanding the interaction of light with matter and, hence, as the basis for treatment of the phenomena of physical optics. In the hands of H. A. Lorentz, this treatment led at the end of the 19th century and the beginning of the 20th to an explanation of many optical phenomena, such as the Zeeman effect, in terms of atomic and molecular structure. The theories of Maxwell and Lorentz are regarded as the culmination of classical optics. *See* ELECTROMAGNETIC RADIATION; LIGHT; MAXWELL'S EQUATIONS.

In the 20th century optics has been in the forefront of the revolution in physical thinking caused by the theory of relativity and especially by the quantum theory. To explain the wavelength dependence of heat radiation, the photoelectric effect, the spectra of monatomic gases, and many other phenomena of physical optics, radical departure from the ideas of Lorentz and Maxwell about the mechanism of the interaction of radiation and matter and about the nature of radiation itself has been found necessary. The chief early quantum theorists were M. Planck, A. Einstein, and N. Bohr; later came L. de Broglie, W. Heisenberg, P. A. M. Dirac, E. Schrödinger, and others.

The science of optics finds itself in a position that is satisfactory for practical purposes but less so from a theoretical standpoint. The theory of Maxwell is sufficiently valid for treating the interaction of high-intensity radiation with systems considerably larger than those of atomic dimensions. The modern quantum theory is adequate for an understanding of the spectra of atoms and molecules and for the interpretation of phenomena involving low-intensity radiation, provided one does not insist on a very detailed description of the process of emission or absorption of radiation. However, a general theory of relativistic quantum electrodynamics valid for all conditions and systems has not been worked out.

The development of the laser has been an outstanding event in the history of optics. The theory of electromagnetic radiation from its beginnings was able to comprehend and treat the properties of coherent radiation, but the controlled generation of coherent monochromatic radiation of high power was not achieved in the optical region until the work of C. H. Townes and A. L. Schawlow in 1958 pointed the way. Many achievements in optics, such as holography and interferometry over long paths, have resulted from the laser. *See* HOLOGRAPHY; INTERFEROMETRY; LASER.

[RICHARD C. LORD]

Bibliography: M. Born and E. Wolf, *Principles of Optics*, 5th ed., 1975; F. A. Jenkins and H. E. White, *Fundamentals of Optics*, 4th ed., 1976; D. C. O'Shea, W. R. Callen, and W. T. Rhodes, *An Introduction to Lasers and Their Applications*, 1977; J. A. Stratton *Electromagnetic Theory*, 1941; J. Strong, *Concepts of Classical Optics*, 1958; R. W. Wood, *Physical Optics*, 3d ed., 1934.

Orthogonal polynomials

A special case of orthogonal functions that arise in many physical problems (often as the solutions of differential equations), in the study of distribution functions, and in certain other situations where one approximates fairly general functions by polynomials. *See* PROBABILITY.

Each set of orthogonal polynomials is defined with respect to a particular averaging procedure. The average value of a suitable function f is denoted by $E\{f\}$. Examples are shown in Eqs. (1)–(4).

$$E\{f\} = \frac{1}{2} \int_{-1}^{1} f(x)\,dx \qquad (1)$$

$$E\{f\} = \frac{\int_{-1}^{1} f(x)(1-x)^\alpha(1+x)^\beta\,dx}{\int_{-1}^{1} (1-x)^\alpha(1+x)^\beta\,dx} \qquad (2)$$

$$(\alpha, \beta > -1)$$

$$E\{f\} = \int_{0}^{\infty} f(x)e^{-x}\,dx \qquad (3)$$

$$E\{f\} = (2\pi)^{-1/2} \int_{-\infty}^{\infty} f(x)e^{-x^2}\,dx \qquad (4)$$

In general an averaging procedure has the form shown in Eq. (5), a Stieltjes integral, where σ is a

$$E\{f\} = \int_{-\infty}^{\infty} f(x)\,d\sigma(x) \qquad (5)$$

distribution function, that is, an increasing function with $\sigma(-\infty) = 0$ and $\sigma(+\infty) = 1$. In the above examples σ has the form $\sigma(x) = b^{-1} \int_{-\infty}^{x} \omega(y)\,dy$, where ω is a nonnegative weight function and $b = \int_{-\infty}^{\infty} \omega(y)\,dy$. Consideration will be given only to averaging procedures for which all the moments $\mu_n = E\{x^n\} = \int_{-\infty}^{\infty} x^n\,d\sigma(x)$ exist and for which $E\{|P|\} > 0$ for every polynomial P.

Orthogonal functions. Two functions f and g are said to be orthogonal with respect to a given averaging procedure if $E\{f\bar{g}\} = 0$ where the bar denotes complex conjugation. By the system of orthogonal polynomials associated with the averaging procedure is meant a sequence P_0, P_1, P_2, ... of polynomials P_n having exact degree n, which are mutually orthogonal, that is, $E\{P_m\bar{P}_n\} = 0$ for $m \neq n$. This last condition is equivalent to the statement that each P_n is orthogonal to all polynomials of degree less than n. Thus P_n has the form $P_n(x) = a_0 + a_1x + a_2x^2 + \cdots + a_nx^n$ where $a_n \neq 0$ and is subject to the n conditions $E\{x^kP_n\} = 0$ for $k = 0, 1, \ldots, n-1$. This gives n linear equations in the $n+1$ coefficients of P_n, leaving one more condition, called a normalization, to be imposed. The method of normalization differs in different references. Orthogonal polynomials arising from the average of Eq. (1), Legendre polynomials, satisfy Legendre's differential equation. With the normalization $P_n(1) = 1$ the first few Legendre polynomials are $P_0(x) = 1$, $P_1(x) = x$, $P_2(x) = 3/2x^2 - 1/2$, $P_3(x) = 5/2x^3 - 3/2x$. The average in Eq. (1) is the special case of Eq. (2) with $\alpha = \beta = 0$; the orthogonal polynomials corresponding to averages of Eq. (2) are called Jacobi polynomials; those associated with Eq. (3), Laguerre polynomials; with Eq. (4), Hermite polynomials.

The proper setting for the study of expansions in terms of orthogonal polynomials is the Hilbert space H of functions f such that $E\{|f|^2\}$ exists and is finite. The inner product is $(f,\bar{g}) = E\{f\bar{g}\}$. In analogy with the procedure for Fourier series one can write down a formal expansion, relation (6), where

$$f(x) \sim \sum_{n=0}^{\infty} c_n P_n(x) \tag{6}$$

the coefficients are given by Eq. (7). The Nth partial sum of the series shown in Eq. (8) has the prop-

$$c_n = E\{f\bar{P}_n\}/E\{|P_n|^2\} \tag{7}$$

tial sum of the series shown in Eq. (8) has the property that among all polynomials p of degree not exceeding N, the minimum of the quadratic deviation $E\{|f-p|^2\}$ is achieved uniquely by $p = s_N$. If

$$s_N(x) = \sum_0^N c_n P_n(x) \tag{8}$$

erty that among all polynomials p of degree not exceeding N, the minimum of the quadratic deviation $E\{|f-p|^2\}$ is achieved uniquely by $p = s_N$. If the only function f in H with the property that $E\{x^k f\} = 0$ for every k is the zero function, one says that the polynomials are "complete" in H. In this case the coefficients in Eq. (7) uniquely determine the function f, and the properties of the series in Eq. (6) are quite analogous to the properties of Fourier series of functions in L^2. The polynomials are always complete when the average is taken over a finite interval, but in general some extra assumption is required. The divergence of the series $\sum \mu_{2n}^{-1/2n}$ is a sufficient condition for the completeness of the polynomials. (It is fulfilled in each of the examples cited.) *See* FOURIER SERIES AND INTEGRALS.

The orthogonality property entails certain algebraic properties for the polynomials. For example, the zeros of P_n are all distinct, they lie in the interior of the interval over which the average is taken, and they separate the zeroes of P_{n-1}. Let $X_1^{(n)}, \ldots, X_n^{(n)}$ be the zeros of P_n. One can find constants $b_1^{(n)}, \ldots, b_n^{(n)}$ such that $Q_n\{1\} = 1, Q_n\{P_k\} = 0$ for $0 < k < n$, where $Q_n\{f\} = \sum_{j=1}^n b_j^{(n)} f(X_j^{(n)})$. In the case of an average over a finite interval, $\lim_{n \to \infty} Q_n\{f\} = E\{f\}$ for every continuous f. This is of interest in approximate integration, because the integral $E\{f\}$ is approximated by an expression $Q_n\{f\}$ which depends only on the values of f at n points and, what is remarkable, $Q_n\{f\} = E\{f\}$ whenever f is a polynomial of degree $\leq 2n-1$ whereas one would ordinarily expect an n-point approximation to be exact only for polynomials of degree $\leq n$.

Ultraspherical polynomials. There exists a theory of orthogonal polynomials in several variables. The most important applications involve averages over spheres in m dimensions. Complete sets of orthogonal polynomials may be chosen among the homogeneous, harmonic polynomials. A polynomial $P(x_1, \ldots, x_m)$ is homogeneous of degree n if $P(\lambda x_1, \ldots, \lambda x_m) = \lambda^n P(x_1, \ldots, x_m)$ for each λ; it is harmonic if it satisfies Laplace's differential equation. Let P be such a polynomial with the property that $P(1, 0, \ldots, 0) \neq 0$. Consider $P(x, y_1, \ldots, y_{m-1})$ as a polynomial in the $m-1$ variables $y_1, \ldots, y_{m-1}$ and take the average over a sphere centered at the origin in $m-1$ dimensions. The result is a polynomial $P_n(x)$ of degree n.

The orthogonality over the sphere in m dimensions translates itself into orthogonality on the interval $[-1,1]$ with the weight function $\omega(x) = (1-x^2)^{(n-3)/2}$. For fixed m, the polynomials obtained this way are the ultraspherical polynomials, special cases of the Jacobi polynomials with $\alpha = \beta = (m-3)/2$. Where $m=3$ corresponds to three-dimensional space, these are Legendre polynomials. *See* DIFFERENTIAL EQUATION; LAPLACE'S DIFFERENTIAL EQUATION; POLYNOMIAL SYSTEMS OF EQUATIONS; RIEMANNIAN GEOMETRY.   [CARL S. HERZ]

Bibliography: H. F. Davis, *Fourier Series and Orthogonal Functions*, 1963; J. A. Shohat and J. D. Tamarkin, *The Problem of Moments*, Math. Surv. no. 1, 1943; G. Szegö, *Orthogonal Polynomials*, AMS Colloq. Publ. no. 23, 1939.

Oscillation

Any effect that varies in a back-and-forth or reciprocating manner. Examples of oscillation include the variations of pressure in a sound wave and the fluctuations in a mathematical function whose value repeatedly alternates above and below some mean value.

The term oscillation is for most purposes synonymous with vibration, although the latter sometimes implies primarily a mechanical motion. A device designed to reduce a person's weight by shaking him is likely to be called a vibrator, whereas an electronic device that produces an electric current which reverses its direction periodically is usually called an oscillator. The alternating current and the associated electric and magnetic fields are referred to as electric (or electromagnetic) oscillations.

If a system is set into oscillation by some initial disturbance and then left alone, the effect is called a free oscillation. A forced oscillation is one in which the oscillation is in response to a steadily applied periodic disturbance.

Any oscillation that continually decreases in amplitude, usually because the oscillating system is sending out energy, is spoken of as a damped oscillation. An oscillation that maintains a steady amplitude, usually because of an outside source of energy, is undamped. *See* ANHARMONIC OSCILLATOR; DAMPING; FORCED OSCILLATION; HARMONIC OSCILLATOR; MECHANICAL VIBRATION; VIBRATION.   [JOSEPH M. KELLER]

Paramagnetism

A property exhibited by substances which, when placed in a magnetic field, are magnetized parallel to the field to an extent proportional to the field (except at very low temperatures or in extremely large magnetic fields). Paramagnetic materials always have permeabilities greater than 1, but the values are in general not nearly so great as those of ferromagnetic materials. Paramagnetism is of two types, electronic and nuclear.

Paramagnetic substances. The following types of substances are paramagnetic:

1. All atoms and molecules which have an odd number of electrons. According to quantum mechanics, such a system cannot have a total spin equal to zero; therefore, each atom or molecule has a net magnetic moment which arises from the electron spin angular momentum. Examples are organic free radicals and gaseous nitric oxide.

2. All free atoms and ions with unfilled inner

electron shells and many of these ions when in solids or in solution. Examples are transition, rare-earth, and actinide elements and many of their salts. This includes ferromagnetic and antiferromagnetic materials above their transition temperatures. For a discussion of these materials *see* ANTIFERROMAGNETISM; FERRIMAGNETISM; FERROMAGNETISM.

3. Several miscellaneous compounds including molecular oxygen and organic biradicals.

4. Metals. In this case, the paramagnetism arises from the magnetic moments associated with the spins of the conduction electrons and is called Pauli paramagnetism.

Relatively few substances are paramagnetic. Aside from the Pauli paramagnetism found in metals, the most important paramagnetic effects are found in the compounds of the transition and rare-earth elements which have partially filled $3d$ and $4f$ electron shells respectively.

Electronic paramagnetism. This arises in a substance if its atoms or molecules possess a net electronic magnetic moment. The magnetization arises because of the tendency of a magnetic field to orient the electronic magnetic moments parallel to itself. The magnitudes of electronic magnetic moments are of the order of a Bohr magneton, which is equal to 9.27×10^{-21} electromagnetic unit, or emu (erg/gauss). See ELECTRON SPIN.

Nuclear paramagnetism. This arises when there is a net magnetic moment due to the magnetic moments of the nuclei in a substance. An example is solid sodium, in which each sodium atom has a nuclear magnetic moment of 2.217 nuclear magnetons. One nuclear magneton is equal to 5.05×10^{-24} emu. Nuclear magnetic moments are about 10^3 times smaller than electron magnetic moments. As a result, nuclear paramagnetism produces effects 10^6 times smaller than electron paramagnetic or diamagnetic effects. Therefore, it is usually impossible to detect nuclear paramagnetism by static methods since it will be masked by electronic effects. (An exception is the case of nuclear paramagnetism arising from the protons in solid hydrogen.) However, paramagnetic effects of nuclei are directly observable in resonance experiments. *See* MAGNETIC RESONANCE; NUCLEAR MOMENTS.

Langevin theory. The Langevin theory of paramagnetism (P. Langevin, 1905) treats the paramagnetic substance as a classical (non-quantum-mechanical) collection of permanent magnetic dipoles with no interactions between them. The dipoles are the magnetic moments of the paramagnetic atoms or ions in the substance. The first task of a theory of paramagnetism is to account for the experimentally observed susceptibility (ratio of magnetization to applied field). *See* MAGNETIC SUSCEPTIBILITY.

If an external magnetic field is applied to the paramagnet, each magnetic dipole experiences a torque. Associated with the force which produces this torque is a potential energy given by Eq. (1),

$$V = -\mu H \cos \theta \qquad (1)$$

where μ is the magnetic moment of the dipole, H is the applied magnetic field intensity, and θ is the angle between the dipole and the direction of H. Now, in the absence of thermal agitation each permanent magnetic dipole will become oriented

in such a way that this potential energy is minimized, that is, oriented parallel to the magnetic field. With all the dipoles lined up, the magnetization (magnetic moment per unit volume), if there are N dipoles per unit volume, would be given by Eq. (2), where the direction of the magnetization

$$M = N\mu \qquad (2)$$

would be that of the applied field. Note that in this case an arbitrarily small magnetic field causes all the dipoles to line up so that the susceptibility $\chi = M/H$ would be infinite. In the actual case there is thermal agitation which in part offsets the aligning tendency of the magnetic field. The Langevin theory takes this into account and predicts the paramagnetic susceptibility as a function of temperature.

In the presence of thermal agitation, the magnetic dipoles are not all lined up in the direction of the magnetic field, but there is some distribution of angles made with the field. In this case the magnetization is given by Eq. (3), where $\overline{\cos \theta}$ is the aver-

$$M = N\mu \overline{\cos \theta} \qquad (3)$$

age of the cosine of the angle between dipole and field. The average is taken over the distribution of dipoles in thermal equilibrium. According to statistical mechanics, this average is given by Eq. (4),

$$\overline{\cos \theta} = \int e^{(-V/kT)} \cos \theta \, d\Omega / \int e^{(-V/kT)} \, d\Omega \qquad (4)$$

where $d\Omega$ is the element of solid angle and $e^{(-V/kT)}$ is the Boltzmann distribution in energy $V = -\mu H \cos \theta$ [Eq. (1)] of a dipole at angle θ with respect to the applied field at absolute temperature T. The integrations may be performed, and the result is $L(a)$, the Langevin function of $a = \mu H/kT$. The result may be combined with Eq. (3) to give Eq. (5).

$$M = N\mu L(a) \qquad (5)$$

If $a \ll 1$, then $L(a) \cong a/3$ so that Eq. (6) holds. This

$$M \cong N\mu^2 H/3kT \qquad (6)$$

is a good approximation except at low temperatures or extremely high fields. The susceptibility is given by Eq. (7). The $1/T$ dependence of the susceptibili-

$$\chi = M/H = N\mu^2/3kT = C/T \qquad (7)$$

ty is known as the Curie law. The Curie law was established empirically by P. Curie in 1895 and is obeyed by many gases, liquids, and solids. There are some paramagnetic solids which obey the Curie-Weiss law $\chi = C/(T - \Theta)$ in a certain temperature range. Here Θ is the Curie temperature. The modification often arises because of effective interactions between the dipoles which are neglected in the preceding development. It may also be due to distortion effects. *See* CURIE-WEISS LAW; LANGEVIN FUNCTION.

Experimental data for the paramagnetic susceptibility are often expressed in terms of the effective magnetic moment which must be used for μ in the Curie law [Eq. (7)] in order to give the observed slope of the curve of χ plotted against $1/T$.

Quantum theory. The quantum-mechanical theory of paramagnetism was worked out in detail by J. H. Van Vleck in 1928. This theory is based on the fact that the magnetic moment of the permanent magnetic dipole arises from the total angular momentum of the electrons in the paramagnetic atom, ion, or molecule. Thus an atom with total angular momentum quantum number J has $(2J + 1)$ energy levels in a magnetic field. A collection of such atoms will be distributed among these levels according to a Boltzmann distribution. The magnetization of such a system may be computed by finding the average component of angular momentum parallel to the field. The result is Eq. (8),

$$M = NgJ\mu_B B_J(a^*) \qquad (8)$$

where g is the spectroscopic splitting factor (the measure of the energy level splittings of the system), μ_B is the Bohr magneton, $a^* = gJ\mu_B H/kT$, and $B_J(a^*)$ is the Brillouin function of a^* expressed in Eq. (9). The Brillouin function also enters the

$$B_J(a^*) = \frac{2J+1}{2J} \coth \frac{(2J+1)a^*}{2J} - \frac{1}{2J} \coth \frac{a^*}{2J} \qquad (9)$$

theory of ferromagnetism. If a^* is much less than unity, which is a good approximation except at very low temperatures or in large fields, then Eq. (10) holds. In this case a Curie law again prevails

$$B_J(a^*) \cong g(J+1)\mu_B H/3kT \qquad (10)$$

as in Eq. (11). The effective magneton number is

$$\chi = M/H = NJ(J+1)g^2\mu_B{}^2/3kT \qquad (11)$$

defined by $g\sqrt{J(J+1)}$ and is the quantity usually given in experimental results. *See* NONRELATIVISTIC QUANTUM THEORY.

If only the electron spin contributes to the total angular momentum, $J = 1/2$ and $B_{1/2}(a^*) = \tanh(a^*)$ so that, except at low temperatures or high fields, Eq. (12) holds, which agrees with the classical result. This case is referred to as the spin-only case.

$$\chi = N\mu_B{}^2/3kT \qquad (12)$$

Rare-earth ions. The paramagnetism of rare-earth ions at room temperature is summarized by some representative examples in Table 1.

Table 1. Paramagnetism of some trivalent rare-earth ions

Ion	Electron configuration	Effective magneton number	
		Calculated	Experimental
Ce^{3+}	$4f^1 5s^2 p^6$	2.54	2.4
Nd^{3+}	$4f^3 5s^2 p^6$	3.62	3.5
Sm^{3+}	$4f^5 5s^2 p^6$	0.84	1.5
Eu^{3+}	$4f^6 5s^2 p^6$	0.00	3.4
Gd^{3+}	$4f^7 5s^2 p^6$	7.94	8.0
Yb^{3+}	$4f^{13} 5s^2 p^6$	4.54	4.5

The calculated effective magneton numbers in Table 1 are the theoretical values for isolated ions. The experimental values are derived from Eq. (11), using experimental values of the paramagnetic susceptibility χ. There is good agreement for all rare-earth ions with the exception of europium and samarium. The experimental results of Table 1 refer to the paramagnetic behavior of rare-earth ions in crystals; different salts of the same ion give the same results.

The experimental result is therefore that at room temperature a crystal containing a number of trivalent rare-earth ions has the paramagnetic susceptibility of that number of free trivalent ions. The reason that there is little influence of the crys-

talline electric fields on the magnetic behavior is that the electrons responsible for the magnetic moments are in the $4f$ state and therefore occupy an electronic shell lying well inside the ion, a shell that is shielded from outside influence by the $5s$ and $5p$ electrons. This is in contrast to the behavior of iron-group ions discussed later.

At lower temperatures the influence of the crystalline electric fields on the electrons becomes more important and the behavior of the susceptibility can become quite complex. In this case, the susceptibility depends upon the orientation of the magnetic field with respect to the crystal axes.

The behavior of europium and samarium at room temperature is still explainable on the basis of a theory of free ions if the effect of Van Vleck paramagnetism is included.

Van Vleck paramagnetism. This arises when the energy states of an atom or ion divide into two groups, those within an energy kT of the ground (lowest energy) state and those which are separated from the ground state by an energy greater than kT. Here k is Boltzmann's constant and T is the absolute temperature. The situation is shown in the illustration. The low-lying states give rise to a susceptibility which follows a Curie law. If these low-lying states arise from a single value of the total angular momentum J, as in the figure, then the quantum-mechanical derivation applies [Eq. (11)]. The high-lying states give rise to a small temperature-independent susceptibility, an effect which is known as Van Vleck paramagnetism. In intermediate cases, such as in the trivalent europium and samarium ions, the upper states are only a little more than kT away from the ground state so that the temperature dependence is still more complicated.

Iron-group ions. The paramagnetism of iron-group ions in crystals is summarized in Table 2.

Quenching of orbital angular momentum is exhibited in crystals containing ions of the iron group. The last three columns of Table 2 indicate

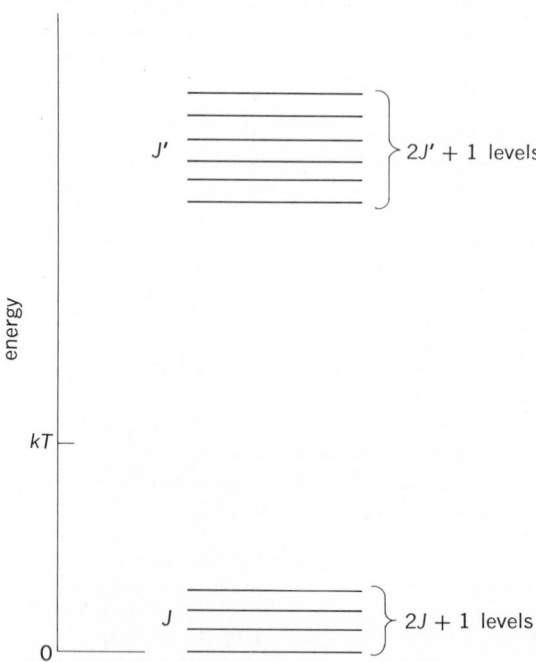

Energy levels in Van Vleck paramagnetism.

Table 2. Paramagnetism of iron-group ions

Ion	Electron config-uration	Effective magneton number		
		Calcu-lated with J	Calcu-lated with S only	Experi-mental
Ti^{3+}, V^{4+}	$3d^1$	1.55	1.73	1.8
V^{3+}	$3d^2$	1.63	2.83	2.8
Cr^{3+}, V^{2+}	$3d^3$	0.77	3.87	3.8
Mn^{3+}, Cr^{2+}	$3d^4$	0.00	4.90	4.9
Fe^{3+}, Mn^{2+}	$3d^5$	5.92	5.92	5.9
Fe^{2+}	$3d^6$	6.70	4.90	5.4
Co^{2+}	$3d^7$	6.54	3.87	4.8
Ni^{2+}	$3d^8$	5.59	2.83	3.2
Cu^{2+}	$3d^9$	3.55	1.73	1.9

that the orbital angular momentum makes no contribution to the magnetic moment but the iron-group ions behave in crystals as free ions with only the spin S contributing to the magnetic effects. This is evidenced by the fact that the spin-only values of the effective magneton numbers agree well with the experimental results. The orbital angular momentum is quenched because the $3d$ electronic shell, which gives rise to the paramagnetism, is outermost for the iron group; it is therefore exposed to the strong crystalline electric fields arising from neighboring ions. These asymmetric electric fields decouple the orbital angular momentum from the spin angular momentum. This means that the energy levels are no longer specified by the total angular momentum quantum number J; S alone may determine the levels. More precisely, the $(2L + 1)$ degenerate orbital angular momentum states of orbital angular momentum quantum number L may be split by the crystal fields so that the lowest orbital state is nondegenerate (singlet). Then there is no possibility of orienting the orbital angular momentum by a magnetic field so that only the spin contributes to the magnetic moment. It is often said that the crystal field "locks" the orbital angular momentum so that it cannot be oriented by a magnetic field. Partial quenching occurs when the orbital degeneracy is only partially removed by the crystal field. Partial quenching and anisotropic effects can also be caused by spin-orbit coupling. This influence may be observed in spin-resonance and specific-heat experiments. *See* ADIABATIC DEMAGNETIZATION; MAGNETIC RESONANCE.

Pauli paramagnetism. This is the paramagnetism associated with the conduction electrons of a metal. A metal is usually described in terms of a collection of positive ions with closed shells which are arranged on a crystal lattice plus electrons which are essentially free to move about the crystal. Each electron has an intrinsic spin angular momentum, and these momenta give rise to a paramagnetic magnetic moment. At first sight it would seem to be correct to apply the Langevin formula to this "gas" of electrons, but the experimental facts are that the paramagnetic susceptibility of conduction electrons is about one-hundredth of that predicted by the Langevin formula [Eq. (7)]. Furthermore, the susceptibility is temperature independent rather than varying as $1/T$ (Curie law). The explanation was given by W. Pauli in 1927:

The electrons obey the quantum statistics of E. Fermi and P. A. M. Dirac rather than the classical statistics which are used in the derivation of the Langevin formula. This means that a given energy state can be occupied at most by two electrons, and their spin angular momenta must be in opposite directions. As a result the net angular momentum is zero, even on application of a magnetic field. Thus most of the electrons in a metal contribute in sum no magnetic moment. That is to say, an electron's spin angular momentum may not orient parallel to an applied magnetic field because there is already an electron in that energy state with its spin parallel to the field. There are, however, a few electrons which are not "paired off," and the spins of these can be oriented by the field. These electrons contribute to the susceptibility according to a Curie law, but the number of them is proportional to the temperature. The combination of the two temperature dependences leads to a temperature-independent susceptibility, smaller than the prediction of the Langevin formula for N electrons per unit volume because only a fraction of these may contribute. The Pauli susceptibility may be written (for a free electron gas) as Eq. (13),

$$\chi = 3N\mu_B{}^2/2kT_F \qquad (13)$$

where N is the number of electrons per unit volume and kT_F is the Fermi energy characteristic of the metal. The fraction of electrons contributing to the susceptibility at temperature T is of the order T/T_F. For sodium, for example, $kT_F = 3.12$ electron volts, $T_F = 37,000$ K. The Pauli paramagnetism of metals has been observed in spin-resonance experiments. The total susceptibility arises from Pauli paramagnetism and diamagnetic contributions from conduction electrons and ion cores. *See* EXCLUSION PRINCIPLE; FREE-ELECTRON THEORY OF METALS.

[ELIHU ABRAHAMS; FREDERIC KEFFER]

Bibliography: E. A. Boudreaux and L. N. Mulay, *Theory and Applications of Molecular Paramagnetism*, 1976; C. Kittel, *Introduction to Solid State Physics*, 5th ed., 1976; A. H. Morrish, *The Physical Principles of Magnetism*, 1965, reprint 1979; J. H. Van Vleck, *The Theory of Electric and Magnetic Susceptibilities*, 1932.

Parametric arrays

Arrays of sources (or receivers) of sound formed by variation of appropriate parameters of the propagation medium. Normally, these parameters are the local sound speed c and the particle velocity u, which vary because of the presence of large-amplitude pump, or primary, sound waves.

Parametric acoustic sources. The usual parametric source configuration simply consists of a directional transducer (often a plane piston or planar array) driven at two frequencies near the transducer resonance, forming a dual-frequency sound beam called the primary beam. Because sound-wave propagation is not a completely linear process, signals at new frequencies are formed effectively through the interaction of sound with sound as the beam progresses and are generated along the length of the primary beam. The lowest of these new frequencies is the difference of the two primary frequencies, and so the primary beam acts as an end-fire array of sources at the difference frequency. The effective length of the array will be

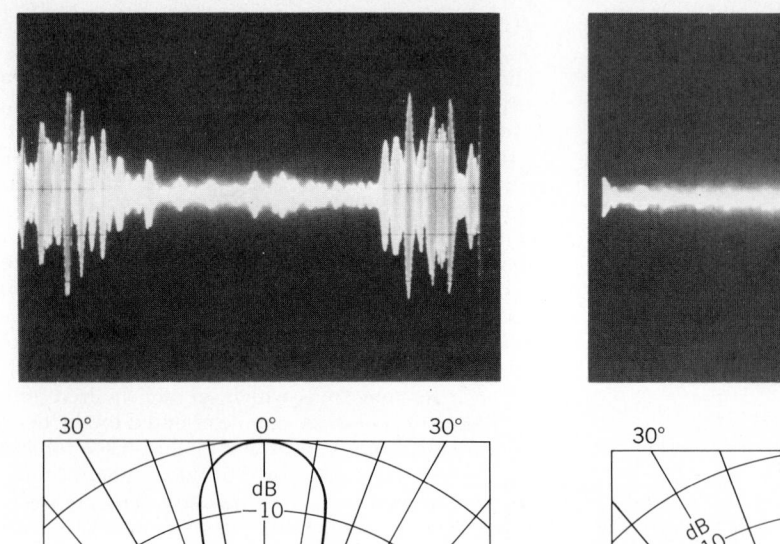

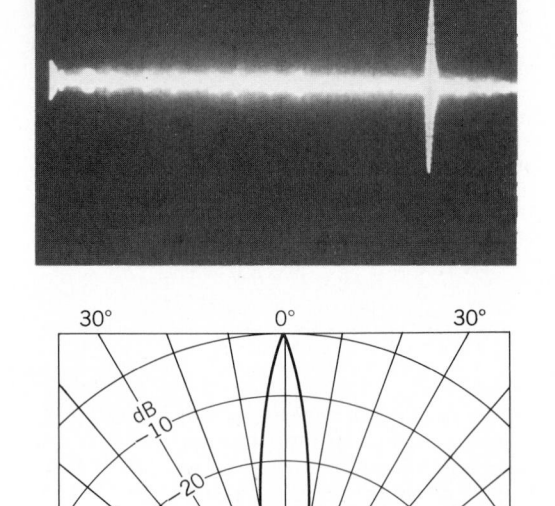

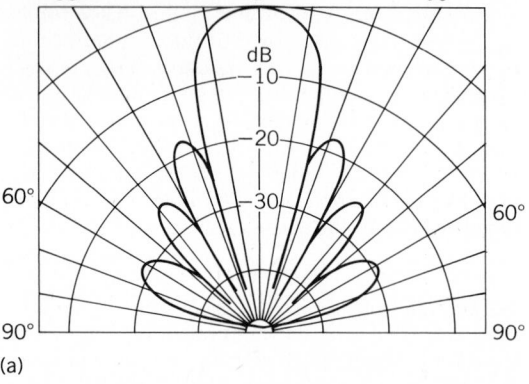

(a)

(b)

Fig. 1. Oscillograms of echoes received from a cylindrical target (top), and the respective directivity patterns (bottom). (a) Conventional acoustic source. (b) Parametric acoustic source. (*Naval Underwater Systems Center*)

determined by the attenuation of the primary beam, which occurs either as a result of small-signal absorption or, for sufficiently high primary amplitudes, as a result of nonlinear losses due to the generation of harmonics of the primary frequencies and other intermodulation components, such as the sum-frequency component.

Applications. Most applications of parametric sources have been to underwater acoustics, but their use in air, as well as in other media, may be expected in the future. Because the effective length of a parametric source can be made quite long in practice, it is possible to generate highly directional difference-frequency beams, and because the primary amplitude is shaded very gradually along the length of the array, these beams can be made practically side-lobe-free. Figure 1 illustrates the difference between a parametric source and a conventional source of the same size and frequency. In each case, the projector was a plane piston transducer 0.25 m in diameter. The conventional beam (Fig 1*a*) was obtained by simply driving the projector directly at 25 kHz. For parametric operation the projector was driven at primary frequencies of 250 ± 12.5 kHz in order to generate the 25-kHz beam depicted in Fig. 1*b*. Figure 1 also illustrates oscillograms of echoes received from a cylindrical target 1.3 cm in diameter and about 7.6 cm in length at a range of 36 m. The conventional echo is obscured by reverberation from the surface and bottom of the quarry in which the experiment was performed, whereas the parametric source exhibits no reverberation. Thus, the parametric source may be expected to be useful in

reverberation-limited situations where one desires a narrow beam from a small projector. Such applications include precision fathometry, subbottom profiling, echo ranging, communications, and Doppler navigation logs. In order to obtain the advantages of a parametric source, however, one must be willing to tolerate low efficiency and low search rate.

Theory. The basic theoretical approach is that outlined by P. J. Westervelt, in which the secondary pressure wave is generated by a volume distribution of acoustic sources whose strength depends on the primary pressure. The acoustic source density $q(\mathbf{r}',t)$, due to a finite-amplitude (primary) acoustic pressure field $p_o(\mathbf{r}',t)$, is given by Eq. (1), where ρ is the static density, c is the

$$q = \beta \rho^{-2} c^{-4} \, \partial p_o^2 / \partial t \qquad (1)$$

ambient sound speed, and β involves the parameter of nonlinearity of the fluid ($\beta \approx 1.2$ for air and 3.5 for water). Then the secondary pressure field is given by Eq. (2), where $\mathbf{r}$ is the position vector of

$$p(\mathbf{r}, t) = \frac{\rho}{4\pi} \iiint \frac{d^3\mathbf{r}'}{|\mathbf{r} - \mathbf{r}'|} \frac{\partial}{\partial t} \left[q\left(\mathbf{r}', t - \frac{|\mathbf{r} - \mathbf{r}'|}{c}\right) \right] \qquad (2)$$

the observation point and $\mathbf{r}'$ is that of the source point. The primary pressure p_0 must be evaluated at the retarded time, $t - |\mathbf{r} - \mathbf{r}'|/c$, for each source point and then integrated over all regions having appreciable primary-wave intensity.

In the most common parametric-source configuration, the primary waveform consists of two discrete frequency components of equal amplitude which generate a difference-frequency wave, that is, the secondary signal of interest. The primary beam is usually produced by a directional piston transducer. One can express the (averaged) acoustic pressure at the face of the transducer by Eq. (3), where f_0 and f are the average and dif-

$$P_0[\cos 2\pi(f_0 - \tfrac{1}{2}f)t - \cos 2\pi(f_0 + \tfrac{1}{2}f)t]$$
$$= 2P_o \sin(2\pi f_0 t)\sin(\pi f t) \quad (3)$$

ference of the primary frequencies and P_0 is a constant. In the absence of linear and nonlinear absorption, the amplitude of each primary component in the far field would be approximately $P_0 R_0 D_0(\theta', \varphi')/r'$, where r' is the radial distance from the center of the transducer, θ' and φ' are angular coordinates, $D_0(\theta',\varphi')$ is the directivity function for the primary beam (assumed to be similar at the two primary frequencies), and R_0 is the Rayleigh length given by Eq. (4), that is, the ratio

$$R_0 = A_0/\lambda_0 = A_0 f_0/c \quad (4)$$

of transducer area A_0 to the primary wavelength λ_0.

The root-mean-square source level SL_0 for each primary component is then given by Eq. (5), where

$$SL_0 = 20\log(P_0 R_0/\sqrt{2}) \quad (5)$$

SL_0 is given in decibels above 1 micropascal-meter, P_0 is given in micropascals, and R_0 is given in meters. The difference-frequency source level SL may be defined in terms of the parametric gain G by Eq. (6). (G is always a negative quantity.) Figure

$$G = SL - SL_0 \quad (6)$$

2 shows computed values of the parametric gain for $f_0/f = 10$ in water. The parameter for the curves is αR_0, the amount of primary absorption loss occurring within the Rayleigh length. The abscissa

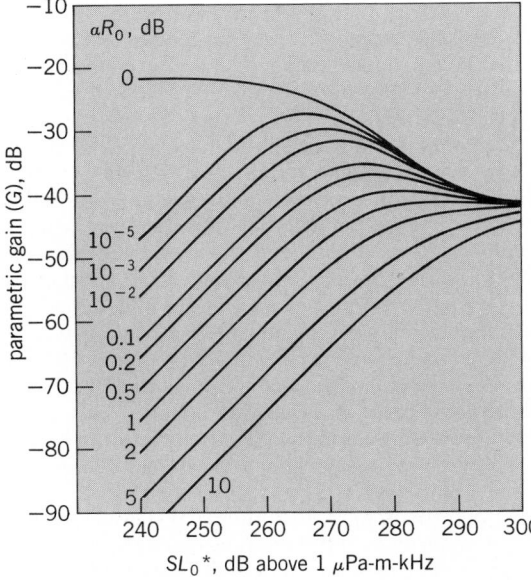

Fig. 2. Parametric-gain curves for $f_0/f = 10$ in water. (From M. B. Moffett and R. H. Mellen, Model for parametric acoustic sources, J. Acoust. Soc. Amer., 61:325–337, 1977)

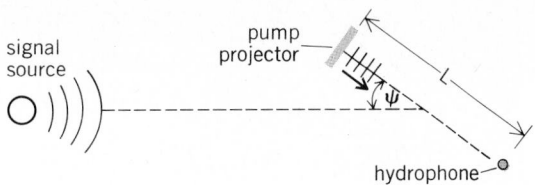

Fig. 3. Parametric acoustic receiver. (Naval Underwater Systems Center)

is a scaled root-mean-square primary source level given by Eq. (7), where f_0 is given in kilohertz. At

$$SL_0^* = SL_0 + 20\log f_0 \quad (7)$$

low levels the gain is proportional to SL_0^*, but at high levels, when the primary waveform undergoes nonlinear distortion and consequent loss of energy to harmonics, saturation, illustrated by the leveling-off of the gain curves at the right of Fig. 2, occurs.

Parametric acoustic receivers. The parametric acoustic receiving array consists of a large-amplitude pump source directed at a receiving hydrophone, such as in Fig. 3. When an acoustic signal arrives at the array, it interacts with the pump wave to form sum- and difference-frequency components which are received at the hydrophone. After filtering, these side bands can be detected, and their level will be proportional to the signal amplitude. The beam pattern (for rotation about the angle ψ) is identical to that of a continuous end-fire line array whose length is equal to the pump-hydrophone separation distance L.

The potential advantages of parametric receivers over their conventional counterparts are not as obvious as for parametric sources. Since the beam pattern is identical to that of a continuous end-fire line array, one would expect that the simplicity of the latter would render it more suitable than a parametric receiver, which requires a power source and sophisticated circuitry to filter the side bands from the strong pump component. Furthermore, the parametric receiver cannot be electrically steered away from the end-fire direction, whereas electrical beam steering is standard practice with conventional line arrays. Still, the fact that no hardware is required between the pump source and the receiving hydrophone may eventually prove to be a useful feature. See SOUND.

[MARK B. MOFFETT]

Bibliography: R. T. Beyer, Nonlinear Acoustics, 1974; M. B. Moffett and R. H. Mellen, Model for parametric acoustic sources, J. Acoust. Soc. Amer., 61:325–337, 1977; P. H. Rogers et al., Parametric detection of low-frequency acoustic waves in the near-field of an arbitrary directional pump transducer, J. Acoust. Soc. Amer., 55:528–534, 1974; P. J. Westervelt, Parametric acoustic array, J. Acoust. Soc. Amer., 35:535–537, 1963.

Parity

A physical property of a wave function which specifies the wave function's behavior under simultaneous reflection of all spatial coordinates of the wave function through the origin, that is, when x is replaced by $-x$, y by $-y$, and z by $-z$. If the wave function ψ satisfies Eq. (1), it is said to have

$$\psi(x,y,z) = \psi(-x,-y,-z) \qquad (1)$$

even parity. If, on the other hand, Eq. (2) holds, the

$$\psi(x,y,z) = -\psi(-x,-y,-z) \qquad (2)$$

wave function is said to have odd parity. These two expressions can be combined in Eq. (3), where $P =$

$$\psi(x,y,z) = P\psi(-x,-y,-z) \qquad (3)$$

± 1 is a quantum number having only the two values $+1$ (designated as even parity) and -1 (odd parity). The physical property defined by P is quantized and is called parity. More precisely, parity is defined as the eigenvalue of the operation of space inversion. Parity is a concept that has meaning only for waves and therefore has no meaning in classical particle physics.

Parity does have a meaning for the Schrödinger wave function of a particle in quantum mechanics. It likewise has meaning for the wave function of any system. Corresponding to the fact that the wave function of a complex system is the product of the wave function of the coordinates of the subsystems into which the system may be subdivided times the internal wave functions of those subsystems, the parity of the system is the product of the parity of the wave function of the coordinates of the subsystems multiplied by the intrinsic parities of these subsystems. *See* QUANTUM MECHANICS.

Conservation. The conservation of parity is a consequence of the inversion symmetry of space. To show this formally, let $\mathcal{P}$ be the parity operator which inverts space; that is, $\mathcal{P}$ acting on a wave function yields the wave function at the inverse point of space, $\mathcal{P}\psi(\mathbf{r}) = \psi(-\mathbf{r})$. Similarly, for an operator A, $\mathcal{P}A(\mathbf{r})\mathcal{P}^{-1} = A(-\mathbf{r})$. The statement that the world is symmetrical to inversion means that the Hamiltonian H after inversion is the same as before, that is, $\mathcal{P}H\mathcal{P}^{-1} = H$; and thus $\mathcal{P}H - H\mathcal{P} = [\mathcal{P},H] = 0$. Since $\mathcal{P}$ commutes with H, it is a constant of the motion. Further, H and $\mathcal{P}$ can be simultaneously diagonal, that is, the eigenfunctions of H can be simultaneously eigenfunctions of $\mathcal{P}$. In fact, if for an eigenvalue E of H there is only a single eigenfunction (nondegenerate level), this eigenfunction must be an eigenfunction of $\mathcal{P}$. As for the eigenvalues of $\mathcal{P}$, note that $\mathcal{P}^2 = 1$, from which it follows that the possible eigenvalues of $\mathcal{P}$ are $+1$ or -1. That is, an eigenfunction of $\mathcal{P}$ satisfies $\mathcal{P}\psi_\pm(\mathbf{r}) = \psi_\pm(-\mathbf{r}) = \pm\psi_\pm(\mathbf{r})$, where the upper (lower) sign indicates an eigenfunction of positive (negative) parity, also known as even (odd) parity, as stated above. *See* EIGENFUNCTION; EIGENVALUE; NONRELATIVISTIC QUANTUM THEORY.

Thus, parity would be conserved if the statement of physical laws were independent of the handedness of the coordinate system used. Of course, the fact that most people are right-handed is not a physical law but an accident of evolution; there is nothing in the laws of physics which favors a right-handed to a left-handed human. The same holds for optically active organic compounds, such as the amino acids. However, the statement that the neutrino is left-handed *is* a physical law. *See* NEUTRINO.

All the strong interactions between elementary particles (for example, nuclear forces) and the electromagnetic interactions are symmetrical to inversion, so that parity is conserved by these interactions. As far as is known, only the β-interac-

tions (which involve neutrinos) and the other weak interactions are not symmetrical to inversion and do not conserve parity. The weak interactions contribute little to all processes except the decays of elementary particles (including β-decay of nuclei), so that in all other processes parity is very nearly conserved.

Atomic and nuclear energy states are characterized by a definite parity (which may be different for different energy states of the same nucleus), and the conservation of parity has an important bearing on atomic and nuclear reactions. Operators representing dynamical variables may also be classified in terms of the parity concept, depending upon how they are affected by an inversion of their spatial coordinates.

Orbital parity. Since parity is conserved in strong and electromagnetic interactions, it is termed a good quantum number, and an energy eigenstate (unless it is degenerate) must be an eigenstate of parity. The parity of a one-particle state of orbital angular momentum l is given by $P = (-)^l$, that is, even ($+1$) for s, d, . . . , states, and odd (-1) for p, f, . . . , states. Thus the deuteron, whose state is a linear combination of $^3S_1$ and $^3D_1$, has even parity; there cannot be any admixture of $^3P_1$. The orbital parity of an n-particle system is the product of the parities of the $n-1$ relative orbital angular momentum states: $P_{orb} = (-)^{l + \cdots + l_{n-1}}$. Thus the parity of an atom is the product of the parities of the one-electron orbital wave function; all configurations which mix must have the same parity. The Laporte rule of atomic spectroscopy, which states that an electric dipole transition can occur only between states of opposite parity, depends on the fact that the electric dipole radiation field has odd parity. *See* ANGULAR MOMENTUM.

Intrinsic parity. The intrinsic parities of the particles composing a system must be multiplied by the orbital parity to yield the total parity. But the intrinsic parity of a conserved particle is irrelevant and can be omitted. For if the particle is conserved in a reaction, so is the contribution of its intrinsic parity to the total parity, so that its intrinsic parity is irrelevant to the balance of parity in the reaction. In fact, if a particle is conserved in all reactions, its intrinsic parity can never be determined. The photon is an unconserved particle; its intrinsic parity is odd. The parity of the π^0-meson (a pseudoscalar) is odd, so that to conserve parity it must be emitted by a nucleon into a P state. By charge independence, the charged π-meson must also be emitted in a P state; it is natural to call the parity of the charged π-meson odd also, which amounts to defining the parity of the neutron and proton to be the same. An electron by itself is conserved, but an electron plus a positron can annihilate. Thus the product of the parities of an electron and a positron must be well defined. According to the Dirac equation of relativistic quantum theory, the product of their parities is -1. The same result holds for any fermion particle-antiparticle pair. Thus the parity of positronium is -1 times its orbital parity, that is, $-(-)^l$. *See* ELEMENTARY PARTICLE; RELATIVISTIC QUANTUM THEORY.

Spin and momentum correlations. The symmetry of the strong and electromagnetic interactions with respect to inversion implies statements about possible correlations of momenta and spins of the particles emitted as a result of such reactions. The principle is that the probability of a configuration of momenta and spins must be a scalar, in order that it not change under inversion of the coordinate system. Thus in a reaction yielding three particles with momenta $\mathbf{p}_1, \mathbf{p}_2, \mathbf{p}_3$, the angular distribution might be of the form $a + b\mathbf{p}_1 \cdot \mathbf{p}_2$ but not $a + b\mathbf{p}_1 \cdot \mathbf{p}_2 \times \mathbf{p}_3$, for under inversion the last term changes sign as in Eq. (4). This triple product is a

$$\mathbf{p}_1 \cdot \mathbf{p}_2 \times \mathbf{p}_3 \rightarrow$$
$$(-\mathbf{p}_1) \cdot (-\mathbf{p}_2) \times (-\mathbf{p}_3) = -\mathbf{p}_1 \cdot \mathbf{p}_2 \times \mathbf{p}_3 \quad (4)$$

pseudoscalar, and the description of the angular distribution would not be independent of the handedness of the coordinate system, because the coefficient b would appear to change sign. Orbital angular momentum $\mathbf{L} = \mathbf{r} \times \mathbf{p}$ is a pseudovector, since under inversion $\mathbf{L} \rightarrow +\mathbf{L}$; the same must hold for spin angular momentum $\mathbf{S}$. Thus $\mathbf{S} \cdot \mathbf{p}$ is a pseudoscalar, and so such a term cannot occur in the angular distribution of a parity conserving process. This term, $\mathbf{S} \cdot \mathbf{p}$, in an angular distribution, would correlate a particle's spin with its momentum; that is, it would imply a polarization in the momentum direction, or longitudinal polarization, which is accordingly absent in strong and electromagnetic reactions. Transverse polarizations, indicated by terms such as $\mathbf{S}_1 \cdot \mathbf{p}_1 \times \mathbf{p}_2$, are of course always possible.

Nonconservation. One of the selection rules which follows from parity conservation is the following: The same spin zero boson cannot decay both into two π-mesons and three π-mesons, because these final states have opposite parities, even and odd respectively. But the positive K-meson is observed to do just this: It has both the $K_{\pi 2}$ and $K_{\pi 3}$ decay modes, and its spin is zero as deduced from the distribution of momenta in the $K_{\pi 3}$ mode. Thus the conclusion is that parity is not conserved in this decay. In 1956, T. D. Lee and C. N. Yang made the bold hypothesis that parity also is not conserved in β-decay. They reasoned that the magnitude of the β-decay coupling is about the same as the coupling which leads to decay of the K-meson, so these decay processes may be manifestations of a single kind of coupling. Also, there is a very natural way to introduce parity nonconservation in β-decay, namely by assuming a restriction on the possible states of the neutrino (two-component theory). They pointed out that no β-decay experiment had ever looked for the spin-momentum correlations, which would indicate parity nonconservation; they urged that these correlations be sought.

The first experiment to show parity nonconservation in β-decay was done by C. S. Wu in collaboration with physicists at the National Bureau of Standards. Here the spins $\mathbf{S}_{Co}$ of the β-active nuclei cobalt-60 were polarized with a magnetic field $\mathbf{H}$ at low temperature; the decay electrons were observed to be emitted preferentially in directions opposite to the direction of the ^{60}Co spin (see illustration). Thus they found a $\mathbf{S}_{Co} \cdot \mathbf{p}_e$ correlation or, in terms of macroscopic quantities, an $\mathbf{H} \cdot \mathbf{p}_e$ correlation. The magnitude of this correlation shows that the parity-nonconserving and parity-conserving parts of the β-interaction are of equal size, substantiating the two-component neutrino theory.

It is believed that parity conservation fails in all

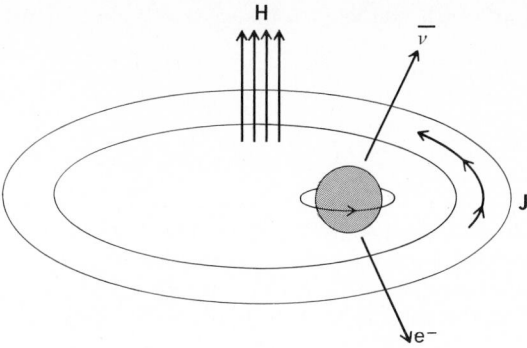

Beta decay from polarized cobalt-60 nuclei. When the spin axes of the cobalt nuclei are not polarized, the preferential emission of the electrons and antineutrinos in the directions shown is not detectable.

the weak decays, which includes all decays of the quasi-stable elementary particles (except the electromagnetic decays $\pi^0 \rightarrow 2\gamma$ and $\Sigma^0 \rightarrow \Lambda + \gamma$).

It was at first somewhat disconcerting to find parity not conserved, for that seemed to imply a handedness of space which would then not be the empty thing which (since the demise of the ether hypothesis) most physicists think it to be. That is, an ether would be needed to provide a standard of handedness at each point of space, to tell ^{60}Co which direction to decay into. But this is not really the situation; the saving thing is that anti-^{60}Co decays in the opposite direction. Thus, after all, there is **nothing intrinsically left-handed** about the world, just as there is nothing intrinsically positively charged about nuclei. What really exists here is a correlation between handedness and sign of charge. *See* SELECTION RULES.

[CHARLES J. GOEBEL]

Bibliography: M. Gardner, *The Ambidextrous Universe*, 2d ed., 1979; T. D. Lee, Weak interactions and nonconservation of parity, *Science*, 127(3298): 569–573, 1958; P. Morrison, The overthrow of parity, *Sci. Amer.*, 196(4): 45–53, 1957; C. N. Yang, Law of parity conservation and other symmetry laws, *Science*, 127(3298): 565–569, 1958.

Partial differentiation

A mathematical operation performed on functions of more than one variable. In this article only two or three variables are considered; however, the principles apply to functions of n variables, for any positive integer $n > 1$. If $z = f(x,y)$, the partial derivative $\partial z/\partial x$ is defined as the derivative of $f(x,y)$ with respect to x, y being regarded as fixed; that is,

$$\frac{\partial z}{\partial x} = \lim_{h \to 0} \frac{f(x+h,y) - f(x,y)}{h}$$

Another notation for $\partial z/\partial x$ is $f_1(x,y)$. The other first partial derivative is $\partial z/\partial y$, also written $f_2(x,y)$. For values at particular points the notation is

$$\left(\frac{\partial z}{\partial x}\right)_{(a,b)} = f_1(a,b)$$

In the case of a function of three variables, $f(x,y,z)$, the expression is

$$\frac{\partial f}{\partial z} = f_3(x,y,z)$$

The second derivatives of $f(x,y)$ are given by

Eqs. (1). It can happen that $f_{12}(x,y) \neq f_{21}(x,y)$, but

$$f_{11}(x,y) = \frac{\partial}{\partial x}\left(\frac{\partial f}{\partial x}\right) \quad f_{12}(x,y) = \frac{\partial}{\partial y}\left(\frac{\partial f}{\partial x}\right)$$

$$f_{21}(x,y) = \frac{\partial}{\partial y}\left(\frac{\partial f}{\partial y}\right) \quad f_{22}(x,y) = \frac{\partial}{\partial y}\left(\frac{\partial f}{\partial y}\right) \tag{1}$$

this will not happen in common practice, especially with elementary functions. If f_1, f_2, f_{12}, f_{21} are defined in a neighborhood of (a,b), and if f_{12}, f_{21} are continuous at (a,b), then $f_{12}(a,b) = f_{21}(a,b)$. In addition, there are more delicate theorems relating to this matter.

Differentials. The notions of a differential, and of the differentiability of a function, are fundamental in the theory of partial differentiation. The requirement that $f(x,y)$ be differentiable is not the same as the requirement that $f_1(x,y)$ and $f_2(x,y)$ both exist; it is a more inclusive requirement. The geometric meaning of f being differentiable at (a,b) is that the surface defined by $z = f(x,y)$ has a tangent plane not parallel to the z axis when $x = a$, $y = b$. In analytic terms the condition is that if

$$\epsilon = f(a+h, b+k) - f(a,b) - f_1(a,b)h - f_2(a,b)k$$

then
$$\lim_{(h,k) \to (0,0)} \frac{\epsilon}{|h| + |k|} = 0$$

When f is differentiable at (a,b), the expression $f_1(a,b)\,dx + f_2(a,b)\,dy$ is called the differential of f at (a,b) with independent increments dx and dy. It is a linear function of dx and dy, and among all linear functions $A\,dx + B\,dy$, it is the best approximation (in a definite sense of the word) to the expression

$$f(a+dx, b+dy) - f(a,b)$$

In the usual notation $z = f(x,y)$, the differential, evaluated at (x,y), is written as Eq. (2).

$$dz = \frac{\partial z}{\partial x}\,dx + \frac{\partial z}{\partial y}\,dy \tag{2}$$

Here dx and dy are independent variables and dz is a dependent variable.

A sufficient condition that f be differentiable at (a,b) is that the partial derivatives f_1, f_2 be defined at all points near (a,b), and continuous at (a,b).

The chain rule. The prime importance of the differentiability concept is that the differentiability property is needed in proving the chain rule for functions of several variables. This rule asserts that a differentiable function of a differentiable function is differentiable, and the rule tells how to compute partial derivatives of the composite function. For example, if $x = f(s,t)$, $y = g(s,t)$, where f and g are differentiable, and if $z = F(x,y)$, where F is differentiable, then the composite function is $G(s,t) = F[f(s,t), g(s,t)]$, and its differential is

$$\frac{\partial F}{\partial x}\,dx + \frac{\partial F}{\partial y}\,dy$$

where dx and dy, instead of being independent, are given by Eqs. (3), where ds and dt are independent.

$$dx = \frac{\partial f}{\partial s}\,ds + \frac{\partial f}{\partial t}\,dt \quad dy = \frac{\partial g}{\partial s}\,ds + \frac{\partial g}{\partial t}\,dt \tag{3}$$

Then $z = G(s,t)$ is differentiable as a function of s and t, and

$$\frac{\partial G}{\partial s} = \frac{\partial F}{\partial x}\frac{\partial f}{\partial s} + \frac{\partial F}{\partial y}\frac{\partial g}{\partial s} \quad \frac{\partial G}{\partial t} = \frac{\partial F}{\partial x}\frac{\partial f}{\partial t} + \frac{\partial F}{\partial y}\frac{\partial g}{\partial t}$$

These equations, expressing the formal part of the chain rule, are often written in the form

$$\frac{\partial z}{\partial s} = \frac{\partial z}{\partial x}\frac{\partial x}{\partial s} + \frac{\partial z}{\partial y}\frac{\partial y}{\partial s} \quad \frac{\partial z}{\partial t} = \frac{\partial z}{\partial x}\frac{\partial x}{\partial t} + \frac{\partial z}{\partial y}\frac{\partial y}{\partial t}$$

At one occurrence, the status of x and y is that of independent variables, as in $z = F(x,y)$, where they are called variables of the first class. But x, y also occur as dependent variables, depending on the independent variables s, t, which are called variables of the second class.

The chain rule is valid for situations in which there are any number of variables of the first class and, quite independently, any number (the same or different) of variables of the second class.

A typical use of the chain rule occurs when transformations are made on the variables in a problem. For example, one may switch from rectangular to polar coordinates. Then derivatives with respect to x, y, or both, must be converted into expressions involving derivatives with respect to r and θ. Transformations of variables are quite extensively used in studying partial differential equations.

Another interesting instance of the chain rule occurs in the so-called particle differentiation in the flow of fluids. If $\rho = F(x,y,z,t)$ is the density at (x,y,z) in the fluid at time t, and if in a given motion one follows a certain selected particle, denoting the density of the fluid at this particle by $\rho = G(t)$, then

$$G'(t) = \frac{d\rho}{dt} = \frac{\partial \rho}{\partial x}\frac{dx}{dt} + \frac{\partial \rho}{\partial y}\frac{dy}{dt} + \frac{\partial \rho}{\partial z}\frac{dz}{dt} + \frac{\partial \rho}{\partial t}$$

Here ρ has two different meanings: On the left $\rho = G(t)$, and on the right $\rho = F(x,y,z,t)$. In dx/dt, dy/dt, dz/dt, the point (x,y,z) is the position of the particle being followed. Here the variables of the first class are x, y, z, t, and there is just one variable of the second class, namely, t. The role of t also is different on the left and on the right in the equation.

Taylor developments. There is a Taylor's formula with remainder and a Taylor's series of functions of several variables. The easiest way to deal with these things is to think of them as being reduced back to the case of one variable by a device. If one wants to express $f(a+h, b+k)$ by a formula proceeding by terms of various degrees in h and k, consider $g(t) = f(a+th, b+tk)$, develop $g(t)$ in powers of t, and then set $t = 1$. The chain rule is needed to compute the derivatives of g. The general formula is Eq. (4). Here a symbolic nota-

$$g^{(n)}(t) = \left[\left(h\frac{\partial}{\partial x} + k\frac{\partial}{\partial y}\right)^n f(x,y)\right]_{x=a+th, y=b+tk} \tag{4}$$

tion with rather evident meaning is used on the right.

Implicit functions. Suppose $F(x,y,z)$ is a function of three variables whose domain of definition is a certain collection of points (x,y,z) in space of three dimensions. As a general rule, it will not be the case that the locus of points for which $F(x,y,z) = 0$ is the graph of an equation $z = f(x,y)$, where f is a single-valued function of two variables. But it may happen that, if (x_0, y_0, z_0) is a point of the locus $F(x,y,z) = 0$, there is a neighborhood of (x_0, y_0, z_0), consisting of all points inside a certain rectangular box centered at (x_0, y_0, z_0), such that the part of the locus $F(x,y,z) = 0$ inside this box is the graph of a function $z = f(x,y)$. There is a standard "implicit function theorem" which covers this situation. It states: Suppose F and its first partial derivatives F_1, F_2, F_3 are continuous throughout some specified neighborhood N of (x_0, y_0, z_0). Suppose also that $F(x_0, y_0, z_0) = 0$ and $F_3(x_0, y_0, z_0) \neq 0$. Then there exist certain positive constants a, b, c and a function f of x and y meeting all the following conditions. Let B denote the boxlike region composed of all (x,y,z) such that $|x - x_0| < a$, $|y - y_0| < b$, $|z - z_0| < c$, and let R denote the rectangle in the xy plane composed of all (x,y) such that $|x - x_0| < a$, $|y - y_0| < b$. The region B is contained in N; the function f is defined in R, and the graph of $z = f(x,y)$ is composed of precisely all the points in B at which $F(x,y,z) = 0$; f is continuous and has continuous first partial derivatives in R, given by

$$f_1(x,y) = -\frac{F_1(x,y,z)}{F_3(x,y,z)} \qquad f_2(x,y) = -\frac{F_2(x,y,z)}{F_3(x,y,z)}$$

where $z = f(x,y)$.

This theorem has two kinds of generalizations: One of the type in which F is a function of n variables and f is a function of n-1 variables, and the other of the type in which the locus $F(x,y,z) = 0$ is replaced by a locus defined by k equations in n variables $(n > k)$, while the equation $z = (x,y)$ is replaced by k equations involving k functions of n-k variables. Sample: $F(x,y,z,u,v) = 0$, $G(x,y,z,u,v) = 0$, $u = f(x,y,z)$, $v = g(x,y,z)$. Implicit function theorems of this second type are proved by mathematical induction with respect to k. The conditions in these theorems involve what are called jacobian determinants.

Jacobians. If $F_1, \ldots, F_k$ are k functions of $z_1, \ldots, z_k$, determinant (5) is called the jacobian

$$J = \begin{vmatrix} \dfrac{\partial F_1}{\partial z_1} & \dfrac{\partial F_1}{\partial z_2} & \cdots & \dfrac{\partial F_1}{\partial z_k} \\ \dfrac{\partial F_2}{\partial z_1} & \cdots & & \dfrac{\partial F_2}{\partial z_k} \\ \cdots & \cdots & \cdots & \cdots \\ \dfrac{\partial F_k}{\partial z_1} & \cdots & & \dfrac{\partial F_k}{\partial z_k} \end{vmatrix} \qquad (5)$$

of $F_1, \ldots, F_k$ with respect to $z_1, \ldots, z_k$, and is denoted by

$$J = \frac{\partial(F_1, \ldots, F_k)}{\partial(z_1, \ldots, z_k)}$$

Notice that the subscripts on the Fs are for distinguishing different functions, and do not indicate partial derivatives.

The general implicit function theorem for a system of equations

$$F_1(x_1, \ldots, x_r, z_1, \ldots, z_k) = 0, \ldots,$$
$$F_k(x_1, \ldots, x_r, z_1, \ldots, z_k) = 0$$

guarantees a local solution of the form

$$z_1 = f_1(x_1, \ldots, x_r) \cdots z_k = f_k(x_1, \ldots, x_r)$$

near a set of values $x_i = a_i$, $z_j = b_j$ for which $F_1 = \cdots = F_k = 0$ and $J \neq 0$. This is on the assumption that the Fs have continuous first partial derivatives. The derivatives of the fs are given by the formulas

$$\frac{\partial f_i}{\partial x_p} = -\frac{J_{ip}}{J}$$

where J_{ip} is what J becomes when its ith column is replaced by $\partial F_1/\partial x_p, \ldots, \partial F_k/\partial x_p$.

If $u = f(x,y)$, $v = g(x,y)$ defines a one-to-one mapping of a region R_1 of the xy plane onto a region R_2 of the uv plane, and if it is known that f and g have continuous partial derivatives and the jacobian $J = [\partial(f,g)/\partial(x,y)]$ is never zero in R_1, then a simple closed curve C_1 in R_1 maps onto a simple closed curve C_2 in R_2 and, as a point P_1 goes counterclockwise around C_1, its image P_2 goes counterclockwise or clockwise around C_2 according to whether $J > 0$ or $J < 0$. Also, if A_1 and A_2 are the areas enclosed by C_1 and C_2 respectively, there is some point inside C_1 such that A_2/A_1 is the value of $|J|$ at that point. If a double integral with respect to u and v, over the region R_2, is converted into a double integral with respect to x and y, over the region R_1, $du\,dv$ is replaced by $|J|dx\,dy$. These results generalize to the case of mappings in space of more than two dimensions. For example, in the passage from rectangular coordinates x, y, z to spherical polar coordinates r, θ, ϕ, by the equations

$$x = r \sin \phi \cos \theta$$
$$y = r \sin \phi \sin \theta$$
$$z = r \cos \phi$$

the jacobian $\partial(x,y,z)/\partial(r,\theta,\phi)$ has the value $-r^2 \sin \phi$, and $dx\,dy\,dz$ is replaced by $r^2 \sin \phi\,dr\,d\theta\,d\phi$ in triple integrals.

If the equations $u = f(x,y)$, $v = g(x,y)$ define a one-to-one mapping from the xy plane to the uv plane (in restricted regions), then

$$\frac{\partial(u,v)}{\partial(x,y)} = \left[\frac{\partial(x,y)}{\partial(u,v)} \right]^{-1}$$

Functional dependence. If $f(x,y)$ and $g(x,y)$ are functionally dependent in a region R of the xy plane, then $[\partial(f,g)/\partial(x,y)] = 0$ in that region. An example of functional dependence would be: $g(x,y) = [f(x,y)]^2 + \sin[f(x,y)]$. In general, f and g are called functionally dependent in R if there is some function F of u and v such that $F[f(x,y),g(x,y)] = 0$ at all points of R, and yet $F(u,v)$ is not zero throughout any two-dimensional portion of the uv plane. Conversely, if $[\partial(f,g)/\partial(x,y)] = 0$ at all points (x,y) in a neighborhood of (x_0, y_0), then usually f and g are functionally dependent in some (perhaps smaller) neighborhood of the point.

Homogeneous functions. One calls a function $F(x_1, \ldots, x_k)$ positively homogeneous of degree n if $F(tx_1, \ldots, tx_k) = t^n F(x_1, \ldots, x_k)$ for all $t > 0$ and for all $(x_1, \ldots, x_k)$ in the domain of definition of F. The index n need not be an integer. If F is differentiable and positively homogeneous of degree n, the Euler relation

$$x_1 \frac{\partial F}{\partial x_1} + \cdots + x_k \frac{\partial F}{\partial x_k} = nF(x_1, \ldots, x_k)$$

holds. Conversely, if F is differentiable in an open region which contains $(tx_1, \ldots, tx_k)$ for all $t > 0$, provided it contains $(x_1, \ldots, x_k)$, then the validity of Euler's relation in the region implies that F is positively homogeneous in degree n.

Lagrange's method in extremal problems. If $F(x,y,z)$ is a differentiable function of three independent variables in an open region R of (x,y,z) space, and if F reaches a relative maximum or minimum value at a point of R, then necessarily $(\partial F/\partial x) = (\partial F/\partial y) = (\partial F/\partial z) = 0$ there. Sufficient conditions, and tests for discrimination between maximum and minimum values, are sometimes stated in terms of second partial derivatives. Lagrange's method is concerned with the situation in which x, y, z are not independent, but are restricted by a side condition $G(x,y,z) = 0$, where G is a specified function. Example: What is the maximum value of $x^2 y^2 z^2$ subject to the restriction $(x^2/25) + (y^2/16) + (z^2/9) - 1 = 0$? On the assumption that F and G have continuous first partial derivatives and that one never has $G = G_1 = G_2 = G_3 = 0$ at one point, the Lagrange procedure is to set $u = F + \lambda G$, where λ is a parameter. Then, among all the values of F attained for (x,y,z) such that $G(x,y,z) = 0$, if there is a maximum or minimum value, it will occur for an (x,y,z) point which satisfies the equations $F_i + \lambda G_i = 0$ $(i = 1, 2, 3)$ and $G = 0$, for a certain value of λ. These four equations can in theory be solved for x, y, z, λ, and the extreme value can be located. The method extends to other numbers of variables and to more than one side condition. *See* CALCULUS; DIFFERENTIATION; OPERATOR THEORY.

[ANGUS E. TAYLOR]

Bibliography: T. M. Apostol, *Mathematical Analysis*, 2d ed., 1974; R. Courant, *Differential and Integral Calculus*, 2 vols., 1936–1937; W. Fulks, *Advanced Calculus*, 1961; A. E. Taylor and R. W. Mann, *Advanced Calculus*, 2d ed., 1972.

Particle

In classical mechanics, a body having finite mass but negligible extension. A particle has inertia and possesses gravitational properties. Because of its negligible extension, forces acting on a particle cannot cause rotational acceleration; therefore, the motion of a particle is regarded as one of pure translation. An extended body is composed of particles. The translational motion of an extended body is equivalent to that of a single particle located at the center of mass of the body and having a mass equal to that of the entire body. *See* RIGID BODY; RIGID-BODY DYNAMICS.

The term particle is also used in physics as a synonym for "elementary particle." *See* ELEMENTARY PARTICLE.

[DUDLEY WILLIAMS]

Particle accelerator

An electrical device which accelerates charged atomic or subatomic particles to high energies. The particles may be charged either positively or negatively. If subatomic, the particles are usually electrons or protons and, if atomic, are charged ions of various elements and their isotopes throughout the entire periodic table of the elements. Before the advent of accelerators the only source of energetic particles for research was the naturally occurring radioactive atoms that emit subatomic particles such as electrons and alpha particles at various energies ranging from kilovolts to over 8 MeV.

The energy of an accelerated atomic or subatomic particle is usually expressed in units of electron-volts (eV). An electronvolt is the amount of energy that a particle with unit charge, such as an electron or proton, receives when accelerated through a potential difference of 1 volt. Therefore, when a proton is accelerated with an electrostatic accelerator operating at 1,000,000 volts, its energy will be 1,000,000 electron volts; equivalently, if the particle has q units of charge, its resultant energy in this example will be q million electronvolts. Commonly used multiple units are kiloelectronvolts (keV), million electronvolts (MeV), or billion (giga) electronvolts (GeV). *See* ELECTRONVOLT.

Very-high-energy atomic and subatomic particles are accelerated in outer space, and constantly bombard the Earth in the form of cosmic rays ranging in energy from hundreds of keV to hundreds of thousands of MeV. These natural sources of particles are uncontrollable and must be used for research in whatever energy, intensity, and kind of natural radiation is available. Particle accelerator devices, on the other hand, allow these same particles to be accelerated to precise energies with complete control of the intensity and energy over wide ranges. The particles may also be directed to collide with specific target materials in whatever way desired so as to greatly expand knowledge of the fundamental interactions of charged particles with each other and with other materials. Moreover, with accelerators it is possible to produce a wide variety of secondary beams of exotic particles with such short half-lives that they would otherwise be unavailable for controlled experimentation. *See* ELEMENTARY PARTICLE.

Accelerators that produce various subatomic particles at high intensity have many practical applications in industry and medicine as well as in basic research. Electrostatic generators, pulse transformer sets, cyclotrons, and electron linear accelerators are used to produce high levels of various kinds of radiation that in turn can be used to polymerize plastics, provide bacterial sterilization without heating, and manufacture radioisotopes which are utilized in industry and medicine for direct treatment of some illnesses as well as research. They can also be used to provide high-intensity beams of x-rays with extreme penetrating power that can be used for cancer therapy, as well as for x-ray radiographic determination of flaws and structural problems in heavy industrial steel castings and other types of structures.

Particle accelerators fall into two general classes — electrostatic accelerators that provide a steady dc potential, and varieties of accelerators that employ various combinations of time-varying electric and magnetic fields.

Electrostatic accelerators. Electrostatic accelerators in the simplest form either accelerate the charged particle from the source of high voltage to ground potential or from ground potential to the source of high voltage. The high-voltage dc potential may be either positive or negative, and consequently positive or negative particles will be accelerated by being either attracted to or repelled from the high voltage.

All particle accelerations are carried out inside an evacuated tube so that the accelerated particles do not collide with air molecules or atoms and may follow trajectories characterized specifically by the electric fields utilized for the acceleration. Usually the evacuated tube is provided with a series of electrodes arranged to have gradually increasing potentials from ground to the maximum high-voltage potential. In this way, the high voltage is distributed uniformly along the acceleration tube, and the acceleration process is thereby simplified and better control permitted. The maximum energy available from this kind of accelerator is limited by the ability of the voltage generator to provide some maximum high voltage. This limitation in energy was a severe problem in the early days of nuclear and atomic research, when electrostatic accelerators were used, and led to the development of the second kind of accelerator, which uses time-varying electric or magnetic fields, or both, and is not restricted by any particular limit of potential that can be maintained.

Time-varying field accelerators. In contrast to the high-voltage-type accelerator which accelerates particles in a continuous stream through a continuously maintained potential, the time-varying accelerators must necessarily accelerate particles in small discrete groups or bunches. Since the voltage on any given electrode is varying in time, at certain times the voltage will be suitable for acceleration, while at other times it would actually decelerate the particles. For this reason the electrodes must be arranged so that the particle bunches appear in their vicinity only when the voltage is correct for acceleration.

Linear accelerators. An accelerator that varies only in electric field and does not use any magnetic guide or turning field is customarily referred to as a linear accelerator or linac. In the simplest version of this kind of accelerator, the electrodes that are used to attract and accelerate the particles are connected to a radio-frequency (rf) power supply or oscillator in such a way that alternate electrodes are of opposite polarity. In this way each successive gap between adjacent electrodes is alternately accelerating and decelerating. If these acceleration gaps are appropriately spaced to accommodate the increasing velocity of the accelerated particle, the frequency can be adjusted so that the particle bunches are always experiencing an accelerating electric field as they cross each successive gap. In this way modest voltages can be used to accelerate bunches of particles indefinitely, limited only by the physical length of the accelerator construction. Some of the larger linacs are over a mile in length.

Circular accelerators. As accelerators are carried to a higher and higher energy, a linac eventually reaches some practical construction limit because of length. This problem of extreme length can be circumvented conveniently by accelerating the particles in a circular path maintained by either static or time-varying magnetic fields. Accelerators utilizing steady magnetic fields as guide paths are usually referred to as cyclotrons or synchrocyclotrons, and are arranged to provide a steady magnetic field over relatively large areas that allow the particles to travel in a circular orbit of gradually increasing diameter as they increase in energy. After many accelerations through various electrode configurations, the particles eventually achieve an orbit as large as the maximum diameter of the magnetic field and are then extracted for utilization in research and other kinds of applications. *See* CYCLOTRON; MAGNETIC FIELD.

A circular accelerator utilizing a time-varying magnetic field that in turn produces an electric field for an acceleration through the induction principle is called a betatron and has been used only for the acceleration of electrons. Betatrons are limited to the acceleration of electrons to energies not much in excess of 300 MeV and have largely been replaced by synchrotron accelerators, which use a time-varying electric field for acceleration rather than a field produced by induction. A few small betatrons designed for 20 to 30 MeV are still used for producing extremely hard x-rays, which are employed in radiographic testing of thick steel castings in industry and for other similar applications. *See* ELECTROMAGNETIC INDUCTION.

Early electron synchrotrons used betatron acceleration in the initial phase of acceleration of each bunch of particles and subsequently continued the acceleration process through energizing of one or more rf cavities through which the particles passed in each orbit. Renewed interest in such electron synchrotron devices has arisen from the fact that they constitute a powerful source of ultraviolet or x-radiation which results from the electrons constrained to move in circular orbits. This synchrotron radiation has important applications in solid-state physics, chemistry, and biology. *See* SYNCHROTRON RADIATION.

Practical limitations of magnet construction have kept the size of circular proton accelerators with static magnetic fields to the vicinity of 100 to 1000 MeV. For even higher energies, up to 400 GeV per nucleon in the maximum-size proton accelerator in operation, it is necessary to vary the magnetic field as well as the electric field in time. In this way the magnetic field can be of a minimal practical size, which is still quite extensive for a 400-GeV accelerator. This circular magnetic containment region or "race track" is injected with relatively low-energy particles that can coast around the magnetic ring when it is at minimum field strength. The magnetic field is then gradually increased to stay in step with the higher magnetic rigidity of the particles as they are gradually accelerated with a time-varying electric field. Again, when the particles achieve an energy corresponding to the maximum magnetic field possible in the circular guide field, they are extracted for utilization in research programs, and the magnetic field is cycled back down to its low or near-zero value and the acceleration process is repeated.

Focusing. One of the chief problems in any accelerator system is that of maintaining spatial control over the beam; this requires some form of focusing. In addition, the presence of focusing elements makes it possible to trade off beam cross-sectional area and angular divergence as the specific utilization may require.

A wide variety of focusing principles and devices are used in different accelerator types. The natural bowing-out of magnetic field lines in a cyclotron magnet results in a net focusing action as will be discussed below; so also does the electrostatic field line configuration between drift tubes of many simple linear accelerators. But these systems are classed generally as "weak-focusing" approaches. The particle beams under their control can make relatively large excursions from the desired equilibrium orbits with the consequence that large vacuum envelopes are essential if the beams are not to be lost through collision with the envelope walls.

Fortunately, an alternate "strong-focusing" approach has made possible very substantial improvements in this area. This approach was described in 1952 by E. Courant, M. S. Livingston, and H. S. Snyder, who reported the invention of electrostatic and magnetic lens systems arranged alternately positive and negative (in the sense of convex and concave optical elements) so that, as in the optical analog, they have a net- and strong-focusing action on charged particle beams. This alternation of positive and negative elements has been implemented in many ways, with many devices and for a variety of geometries. Perhaps its most frequent appearance is in quadrupole (four-pole) or hexapole (six-pole) magnets used as variable-focal-length, variable-astigmatism elements in beam transport both within large accelerators themselves and in extensive systems external to the accelerator.

Superconducting magnets. The study of the fundamental structure of nature and all associated basic research require an ever increasing energy in order to allow finer and finer measurements on the basic structure of matter. Since the voltage-varying and magnetic-field-varying accelerators also have limits to their maximum size in terms of cost and practical construction problems, the only way to increase particle energies even further is to provide higher-varying magnetic fields through superconducting magnet technology, which can extend electromagnetic capability by a factor of 4 to 5. Construction of large superconducting cyclotrons for heavy-ion acceleration has been undertaken in the United States and Canada.

Storage rings. Beyond this limit the only other possibility is to accelerate particles in opposite directions and arrange for them to collide at certain selected intersection regions around the accelerator. The main technical problem is to provide adequate numbers of particles in the two colliding beams so that the probability of a collision is moderately high. Such storage ring facilities are in operation for both electrons and protons, and design and construction of much larger ones have been undertaken. Currently, the colliding-beam storage ring facilities provide the most fundamental measurements on the basic structure of the proton and its constituent particles and components.

Collective accelerators. Development of a completely different kind of accelerator, the collective-effect accelerator, has been undertaken is several countries. The idea is based on the fact that electrons, because of their extremely low mass, can be accelerated close to the velocity of light with very modest voltages compared with those required for a particle approximately 2000 times heavier, such as the proton, or for much heavier particles, such as atomic ions. An acceleration voltage of only 0.5 million volts (MV) will accelerate an electron to 7/10 the velocity of light, while an effective voltage of over 900 MV is required to achieve the same velocity for protons.

The idea of a collective accelerator is to accelerate a cloud or bunch of electrons containing within the electrostatic well established by its cloud structure one or a few protons or heavier atoms in such a way that the electrical forces providing the containment are sufficiently strong to literally drag the heavy particle along. An electron bunch containing a number of protons or other atoms, when accelerated to 7/10 the velocity of light by 0.5 MV, would carry along the heavier particles at the same velocity equivalent to many hundreds of MeV. The idea is very appealing because such an accelerator would be much simpler than conventional acceleration techniques; however, in 1980 the system had not yet been proved practical although some devices have accelerated various heavy ions to a few tens of MeV.

One such device that has been used extensively in studies in both the United States and other countries is the so-called electron ring accelerator, wherein the cloud of electrons is in the form of a kind of "smoke ring" in which the heavier particles are contained. Primary acceleration is obtained by magnetically compressing the smoke ring, and then the entire compressed ring is accelerated as described above to achieve higher energies.

Performance characteristics. This introduction has shown that there are many kinds of accelerators, varying in size and performance characteristics and capable of accelerating many kinds of particles at varying intensities and energies, in the form of bunched or pulsed beams of particles or as a steady dc current. Table 1 lists a few of the more common types of accelerators and their basic performance characteristics. This table is not intended to be complete. The machine characteristics listed are for the largest machine in each class, and many accelerators of a given type can have widely different performance characteristics depending on their use. Accelerators used principally for industrial applications are not listed.

ELECTROSTATIC ACCELERATORS

Electrostatic accelerators are used to accelerate atomic or subatomic particles to high energies by means of a high voltage potential that is maintained through some electrical or mechanical means of transporting charge from ground to the high voltage potential. Modern machines of larger size all use mechanical charge transport, but electrical transport systems are widely used in small Cockcroft-Walton or dynamitron-type electrostatic accelerator units. The most commonly used electrostatic generator was invented in the 1930s by R. J. Van de Graaff and utilizes a high-speed rubberized-fabric belt to transport the charge. Many machines utilizing this principle have been home-made or built by various research groups throughout the world.

Most electrostatic accelerators are housed inside a large high-pressure vessel that is filled to a pressure as high as 15 atmospheres (1.5×10^6 Pa) with very dry insulating gas. The insulating gas may be pure sulfur hexafluoride (SF_6) or a mixture with carbon dioxide (CO_2) and nitrogen (N_2) at a dew point of $-60°F$ ($-51°C$) or less. This high-pressure insulating gas allows the machine to be housed in a much smaller space than would be possible in air at atmospheric pressure (for example, 6-ft or 1.8-m clearance from the high-voltage terminal to ground has been necessary to insulate successfully up to 17 MV).

The Pelletron electrostatic accelerator, invented

Table 1. Operating characteristics of particle accelerators

Accelerator type	Particle accelerated	Energy range†	Beam current (average; peak)	Duty cycle	Energy spectrum*	Beam geometry	Development status (1975)
ELECTROSTATIC ACCELERATORS							
Cockcroft-Walton	p, d, α, e, heavy ions	To 4 MV	1 mA; 10 mA	Continuous	~0.01%	Small focal spot	4-MeV operating
Dynamitron	p, d, α, e, heavy ions	To 4 MV	1 μA; 50 mA	Continuous	~0.01%	Small focal spot	4-MV operating
Tandem Van de Graaff	p, d, α, e, heavy ions	To 25 MV	1 μA; 50 μA	Continuous	~0.01%	Small focal spot	14-MV operating; 16-MV under construction
Tandem pelletron	p, d, α, e, heavy ions	To 25 MV	1 μA; 50μA	Continuous	~0.01%	Small focal spot	20-MV operating; 25-MV under construction
TIME-VARYING FIELD ACCELERATORS							
Circular magnetic types (radio-frequency resonance accelerators)							
Sector or isochronous cyclotron	p, d, α, heavy ions	To 590 MeV (p)	20 μA; 2 mA	Continuous	~0.01%	Internal target or external beam with small focal spot	590-MeV (p) operating
Synchrocyclotron	p, d, α	To 1 GeV (p)	1 μA; depends on duty cycle	$< 10^{-2}$	0.1%	Internal target or external beam of fair collimation at lower intensity; external neutrons; external meson beams	1-GeV (p), mostly closed down
Synchrotron (weak focusing)	p, e, heavy ions	1 – 12.5 GeV	0.1 μA; depends on duty cycle	30%	0.1%	Internal targets: external beam of fair collimation at lower intensity; external secondary-particle beams	12.7-GeV (p), 6.3-GeV (e) closed down; 2-GeV/nucleon heavy ion operating (that is, 28-GeV ^{14}N)
Alternating-gradient synchrotron	p, e	10 – 400 GeV	0.1 μA (p)	30%	0.1%	(p) internal targets; external beam of fair collimation; external secondary-particle beams	400-GeV (p) operating; 1000 GeV under construction
Linear accelerators							
Heavy-ion linear accelerator	p, d, α, heavy ions	To 15 MeV/nucleon (that is, 300 MeV ^{20}Ne)	10 μA; 130 mA	~10%	0.5%	Well-collimated and well-focused external beam	15-MeV/nucleon (heavy ion) operating
Linear accelerator	p	50 – 800 MeV	1 mA; 100 mA	~10%	0.1%	Well-collimated and well-focused external beam	800-MeV (p) operating
Electron linear accelerator	e	6 MeV to 22 GeV	60 μA; 400 μA	~6%	~0.2%	Well-collimated and well-focused external beam	22-GeV operating
Colliding-beam storage rings							
Electron storage ring	e^+, e^-	0.3 – 22 GeV	To 300 mA	Continuous	0.1%	Small-diameter internal beam	22-GeV operating
Proton storage ring	p	10 – 28 GeV	To 20A	Continuous	0.1%	Small-diameter internal beam	28-GeV operating; 400 GeV under construction

*Spread in energy of beam expressed as a percentage of total energy of beam; that is, 1% for the cyclotron means for a 1-MeV beam a spread in energy of 0.01 MeV.

†Voltage range is given for electrostatic accelerators.

by R. G. Herb and manufactured by the National Electrostatic Corporation, utilizes a chain-charging system instead of the rubberized-fabric belt. The continuous-looped charging chain is fabricated of alternate metal and insulated links and travels at high speed on special pulleys from ground to high potential, an arrangement mechanically similar to the charging belt system. Instead of the charge being sprayed onto the charging belt system from sharp corona points, each metal link is charged and discharged by induction methods so that no electrical sparking or erosion is involved. A large Pelletron accelerator at the Oak Ridge National Laboratory designed to operate at 25 MV

has been raised to test voltages as high as 32 MV without acceleration tubes installed.

Tandem electrostatic accelerator. These large electrostatic accelerators are constructed in the form of tandem electrostatic accelerators that utilize the high voltage potential for acceleration more than once; the first of these was installed in Chalk River Nuclear Laboratories of Atomic Energy of Canada in the late 1950s, although the concept in its present form was presented by Willard Bennett before World War II and independently by Luis Alvarez after World War II.

The tandem electrostatic accelerator gets its name from the fact that it is effectively two electrostatic accelerators back to back, arranged so that the high voltage potential is first used to accelerate negative ions from ground potential up to the high positive potential. The charge of the particle is then switched from negative to positive, and the positive ions are additionally accelerated from the high voltage potential back to ground, continuing in the same direction as the original acceleration. The charge-changing inside the high-voltage terminal allows the high voltage to be used to accelerate the particles twice and consequently produce much higher energies than would be possible in the conventional single-stage electrostatic accelerator. A third stage of acceleration can be accomplished by first accelerating negative particles from high voltage to ground and then injecting them into a conventional tandem instead of utilizing negative particles injected directly from an ion source at low energy as is more customary.

The charge-changing in the high-voltage terminal is accomplished by either a gas or foil stripper. The negative ions which are accelerated from ground potential to the high-voltage terminal have enough energy to penetrate either thin foils or a high-pressure gas region within an open-ended tube. In so doing, the collisions with other atoms violently strip away many electrons from the negative ion, leaving the ion with a net positive charge. In cases of the lighter heavy ions such as carbon or oxygen, it is possible to strip away all the electrons from the atom, leaving the bare nucleus of the atom, and thereby accomplish the maximum possi-

ble acceleration, represented by the high charge state of the accelerated ion. Each charge provides a particle with an energy in MeV equivalent to the voltage of the high-voltage terminal in MV; therefore an oxygen ion with charge 8$^+$, $^{16}O^{8+}$, receives an energy of 80 MeV from its acceleration through a potential of 10 MV. Secondary strippers can be placed farther down the acceleration tube from the high-voltage terminal to produce even higher charge states for heavier ions and consequently higher energies. Iodine atoms, for example, have been accelerated to energies in excess of 350 MeV by such multiple stripping techniques.

One of the larger electrostatic accelerator facilities in the world is at Brookhaven National Laboratory and utilizes the three-stage principle with two large tandem electrostatic accelerators arranged in line, the first of which has a negative ion source inside its high-voltage terminal in order to provide the first stage of the three-stage acceleration. Figure 1 shows the second of the two large tandem accelerators in the foreground with the injector tandem in the distant background. The accelerator room is 300 ft (90 m) long, and the two tandem accelerators are each 80 ft (24 m) long. Construction of much larger two-stage tandem electrostatic accelerators has been undertaken, such as the 30-MV vertical tandem electrostatic accelerator at the Daresbury Nuclear Physics Laboratory in Daresbury, England, and the 25-MV folded tandem at the Holifield National Laboratory in Oak Ridge, TN. Figure 2 is an artist's conception of this

Fig. 1. Three-stage MP (Emperor) tandem facility at Brookhaven National Laboratory. The accelerated ions come out through the large evacuated tube on the right and are directed to experimental stations. (*From Physics in Perspective, vol. 2, pt. A: The Core Subfields of Physics, Physics Survey Committee, National Academy of Sciences, 1972*)

large machine, which has been constructed by the National Electrostatic Corporation.

Folded tandem acceleration. The folded tandem design has been developed for large electrostatic accelerators. They are constructed like a large single-stage electrostatic accelerator; however, the column which supports the high-voltage terminal contains two acceleration tubes—one for the acceleration of negative ions from ground potential to the high-voltage terminal, and the other for acceleration of positive ions from the high-voltage terminal down to ground potential. In Fig. 2 negative ions produced in the ion source on the left are directed vertically by the magnet and then accelerated up to the high-voltage terminal. The large magnet inside the terminal bends the ions 180° to a downward direction after they have been stripped to high positive-charge states. They are then accelerated in the second acceleration tube on the right, back down to ground potential to achieve their final energy, and then directed horizontally by a final magnet to experimental stations. The advantage of the folded design is that both the accelerator and building enclosure are much shorter and the complex negative-ion source is at ground level instead of several stories above the top of a very tall accelerator. However, the diameter does have to be increased somewhat, at least in the vicinity of the high-voltage terminal. The added problem of providing a reliably operating large bending magnet inside the high-voltage terminal is a complexity not found in the conventional tandem, in which the particles pass from ground-to-ground potential in a straight line.

Control and adjustment. These large electrostatic accelerators can be controlled very precisely in terms of the absolute voltage of the high-voltage terminal, so that particles can be accelerated with an accuracy of better than 1 part in 10^4 on a routine basis and better than 1 part in 10^5 when additional control measures are invoked.

Since singly charged negative ions can be made of most elements or some compound of the elements, it is a straightforward procedure to accelerate almost any atom throughout the periodic table from hydrogen to uranium. The large electrostatic accelerator facility at Brookhaven National Laboratory has accelerated as many as 35 different elements throughout the periodic table, and more than 25 have been used in different kinds of research applications. Since the electrostatic accelerator identically accelerates different elements or mass particles having the same charge, only minimal adjustments are required in order to change from the acceleration of one mass particle to another. In contrast, the accelerators that utilize varying magnetic and electric fields require considerably more adjustment because of the different velocities encountered as the heavy ions are changed. Although the adjustments are more complex, the field-varying accelerators are not limited in the maximum energy obtainable by the high voltage potential, unlike the electrostatic accelerators.

Plans have been undertaken at several electrostatic accelerator facilities for various kinds of field-varying accelerators as energy booster accelerators, so that their heavy-ion energy range can be greatly extended. In this dual accelerator arrangement, the tandem accelerator is arranged to inject

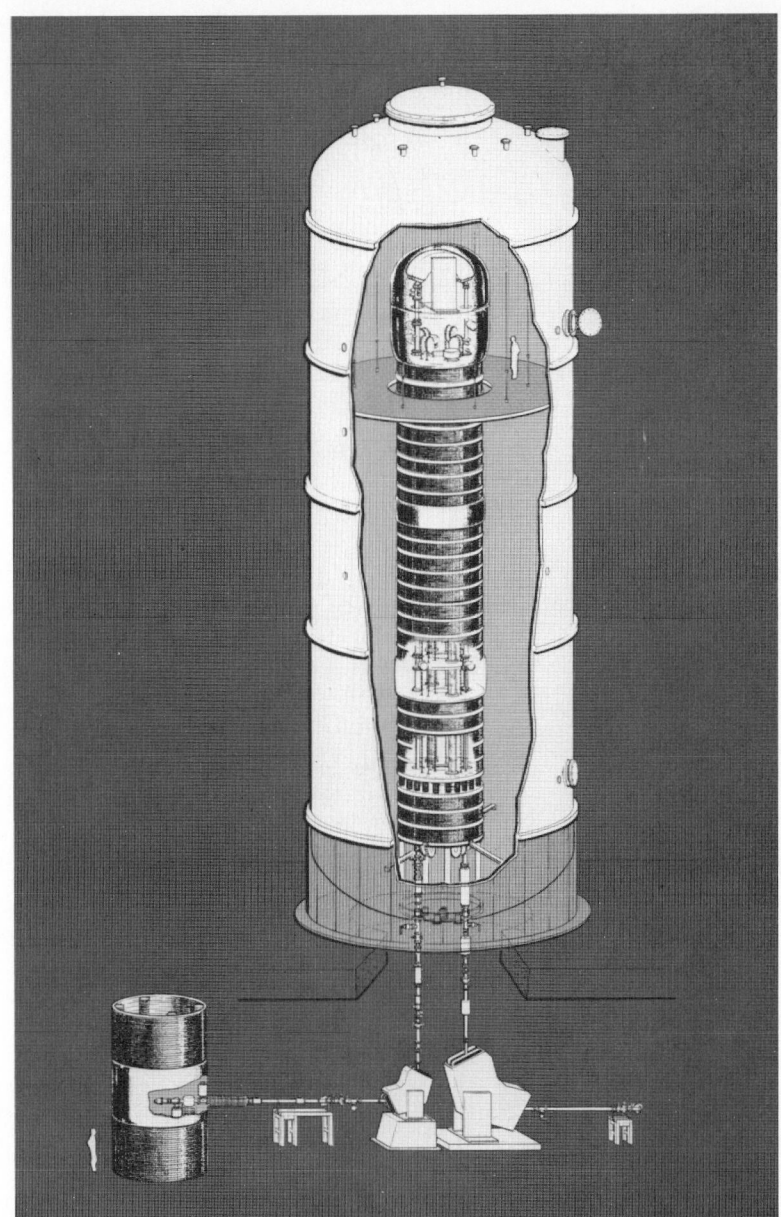

Fig. 2.   The 25-MV pelletron (up-down) tandem accelerator designed for the Oak Ridge National Laboratory. The human figure on the left of the ion source and another on the elevator platform near the high-voltage terminal indicate the large size of this machine. (*National Electrostatics Corporation*)

a pulsed beam matched to the timing requirements of the field-varying accelerator. At present, a small, single-stage electrostatic accelerator is used to inject an open sector cyclotron at the Hahn-Meitner Institut in Berlin. Tandem accelerators are also used to inject a superconducting linac at the Argonne National Laboratory in Illinois and a room-temperature linac at the Max Planck Institut für Physik in Heidelberg. Construction of facilities for tandem injection of superconducting cyclotrons and other superconducting linacs has been undertaken.

Applications. The larger electrostatic accelerators are used primarily for basic research in the study of nuclear structure and heavy-ion reactions, including fusion and fission processes. The smaller machines are used throughout industry for many applications, including neutron and x-ray produc-

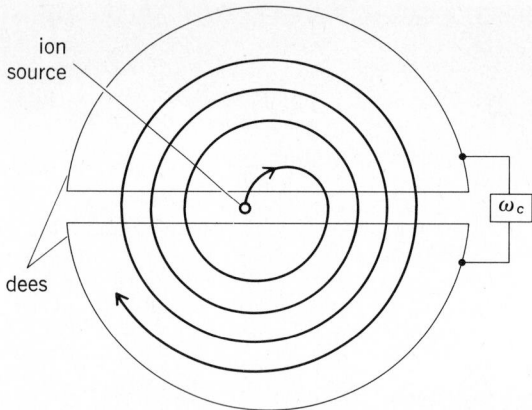

Fig. 3. Principle of the cyclotron. The line shows the path of an ion in the accelerator. The magnetic field is perpendicular to the page.

tion for diagnostic measurements, polymerization and other treatments of plastics and various materials, ion implantation, thin-film analysis, and many other related applications.

[HARVEY E. WEGNER]

CIRCULAR ACCELERATORS

Circular accelerators utilize a magnetic field to bend charged-particle orbits and confine the extent of particle motion.

Cyclotrons. The cyclotron is a circular accelerator in which the particle orbits start at the center and spiral outward in a guide magnetic field which is constant in time. The cyclotron concept

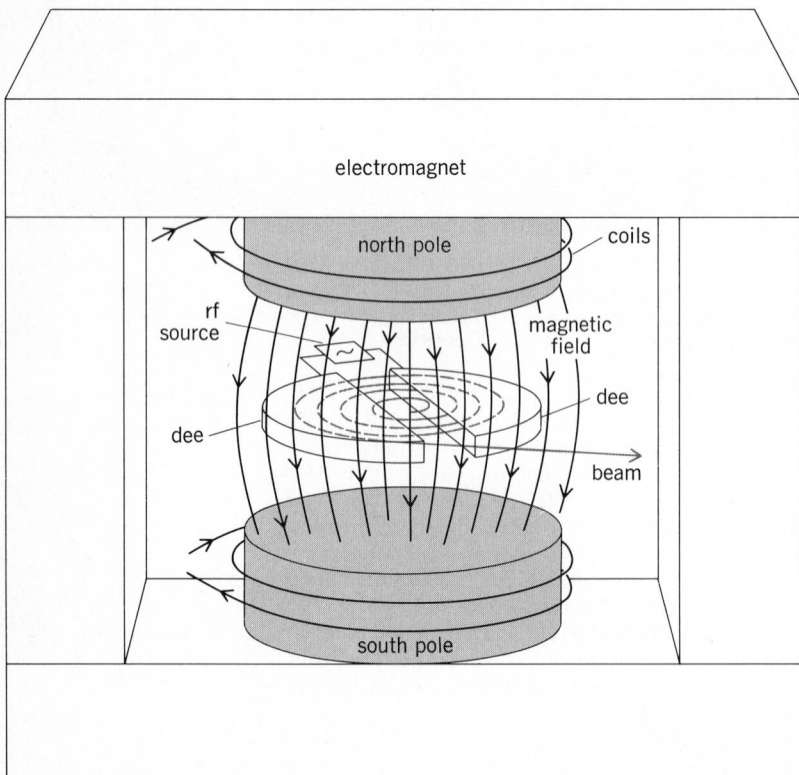

Fig. 4. A cyclotron, showing a particle beam accelerated from the center by dees. A vacuum chamber (not shown) encloses the dee system and beam. (*Lawrence Berkeley Laboratory*)

was put forward by E. O. Lawrence in 1930; the exploitation of this concept is still the basis for major developments in accelerator science. The survival of the cyclotron idea has resulted from successive blending of the basic concept with a series of later ideas (phase stability, strong focusing, and superconductivity). In this process the classic Lawrence cyclotron of the 1930s evolved into the synchrocyclotron of the 1950s, then into the isochronous cyclotron of the 1960s, and finally into the superconducting cyclotron of the late 1970s.

Basic principles. The basic cyclotron concept is given by Eq. (1), the Lawrence equation for the

$$\omega_c = qB/m \qquad (1)$$

angular frequency of rotation of a particle in a perpendicular magnetic field, where q and m are the charge and mass of the particle and B is the magnetic induction of the perpendicular magnetic field. (SI or mksa units are used throughout this article.) One can derive this equation by noting that (1) $\omega = v/r$, where v is the linear velocity and r is the orbit radius; (2) the magnetic force is perpendicular to the velocity and is given by qvB; (3) this perpendicular force produces a centripetal acceleration and, from Newton's second law, gives $qvB = mv^2/r$. Solving the last equation for v/r leads directly to Eq. (1). The striking statement of Eq. (1) is that if field, charge, and mass are all constant, then angular velocity is likewise constant and, in particular, does not depend on the linear velocity. Faster-moving particles travel on circles of larger radius—the increase in the length of a revolution due to the larger radius just matches the increase in speed, and fast or slow particles thus require the same amount of time to make a 360° revolution. Thus, an accelerating electrode operating at constant frequency can be used. This is the key idea of the Lawrence cyclotron.

The essential features of such a cyclotron are diagrammed in Fig. 3. A pair of D-shaped accelerating electrodes, called dees or D's, are electrically attached to an rf voltage source whose frequency is matched to the rotation frequency of the particle in a surrounding magnetic field. Ions are formed in the ion source at the center and are drawn out by the high voltage on the facing dee. Particles which cross the gap between the two dees at a time when the rf voltage between the dees is in such a direction that the particle is speeded up by the resulting electric field are accepted for acceleration; the remainder are lost. Beyond the accelerating gap the particle passes inside the dee, which is an electrically shielded region; at the same time it is pulled into a circular path by the perpendicular magnetic field, and after 180° it arrives again at the gap between the dees. Since the frequency of the rf source which drives the dees was selected to match the rotational frequency of the particle, the voltage between the dees will have reversed while the particle was making its 180° turn, and therefore the particle is again speeded up as it crosses from dee to dee. This process can clearly be repeated as often as desired. The particle will thus be repetitively accelerated and, as it speeds up, will gradually move on circles of larger and larger radius. Finally the particle will come to the limit of the magnetic field. At this point an additional electrode—referred to as a deflector—can be inserted

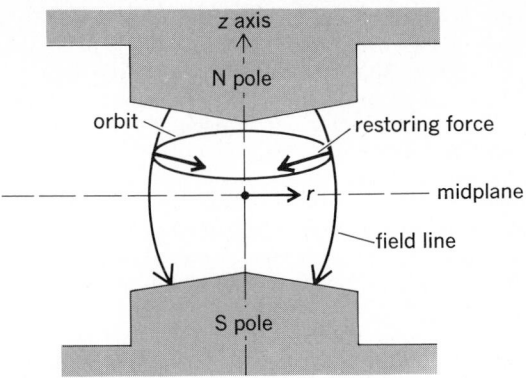

Fig. 5. Focusing forces in an axially symmetric magnetic field. If the field lines curve as shown, toward the axis, the magnetic force has a component parallel to the midplane (which holds the particle in its circular orbit) and perpendicular to the midplane (which pulls the particle back toward that plane).

to direct all particles out along a single path. Figure 4 shows the dees in the magnetic field provided by an electromagnet.

Limitations. This theoretical picture of exactly synchronized orbits and voltage is, of course, an oversimplification, and more detailed investigation by H. A. Bethe and M. E. Rose in 1937 revealed two fundamental problems with the cyclotron concept. First of all, as any particle speeds up, its mass increases according to Albert Einstein's equation, Eq. (2), where m_0 is the particle's rest

$$m = m_0 / \sqrt{1 - v^2/c^2} \qquad (2)$$

mass. When this accurate expression is inserted into Eq. (1), the rotation frequency then depends on velocity through the v^2/c^2 term, although for low velocities the dependence is weak because the factor c, the velocity of light, is large. Nevertheless, in a constant magnetic field, the gradual increase in mass causes the rotation frequency to steadily decrease as the particle speeds up, and the particle then gradually lags behind a constant rf frequency. Ultimately the particle will reach an accelerating gap so late that the voltage will have already shifted to the reverse direction, and thereafter the particle will steadily slow down at successive gaps. The mass increase thus sets a limit on the acceleration process. *See* RELATIVISTIC ELECTRODYNAMICS; RELATIVISTIC MECHANICS; RELATIVITY.

A second limitation comes from the magnetic field factor in Eq. (1). Thus far, motion in a plane has been assumed; however, to prevent particles from drifting away from the plane due to their initial velocities there must be a restoring force which pushes particles back toward the plane. This is called axial focusing. A way to obtain the needed restoring force is to make the magnetic field lines bend as shown in Fig. 5. The magnetic force (always perpendicular to the field lines) slants downward for an orbit above the midplane and upward for an orbit below the midplane, so that a particle displaced from the desired plane always feels a component of the magnetic force pushing it back toward the plane. From Fig. 5 it can be seen that such a restoring force requires the magnetic field lines to bend toward the axis of rotation of the particle. However, allowed magnet-

ic field shapes are limited by Maxwell's equations; applying these equations, one finds that when lines of force bend toward the axis as in Fig. 5, the field strength perpendicular to the plane of motion must decrease as the radius becomes larger, producing a slowing-down according to Eq. (1). This effect adds to that of the mass increase. In their 1937 paper, Bethe and Rose calculated both of these effects for the Lawrence cyclotron and concluded that such a cyclotron could not achieve an energy greater than 10 MeV per nucleon, corresponding to a mass increase of about 1%. A Lawrence cyclotron, constructed at Oak Ridge, in fact achieved an energy of 23 MeV per nucleon (23 MeV protons) by raising the voltage on each accelerating electrode to 200 kV; this energy was much higher than Bethe and Rose thought possible, but actually in precise accord with the Bethe-Rose prediction when corrected for the higher dee voltage. *See* MAXWELL'S EQUATIONS.

Synchrocyclotron. A way to avoid the Bethe-Rose energy limit was proposed in 1946 in independent papers by E. M. McMillan in the United States and V. Veksler in the Soviet Union: namely, to slow down the frequency of the rf power source at the same rate as the slowing down of the rotational frequency of the particle. The usefulness of this process depends on the question of stability. Particles which are not exactly at the design values of energy and phase will either oscillate about the design values (stability) and accelerate, or diverge from the design values (instability) and not accelerate. Figure 6 sketches the ideas involved in establishing that the accelerating process is stable. The frequency of the driving oscillator is adjusted to just match the rotation frequency of a reference particle, shown in Fig. 6. Its phase is called the synchronous phase ϕ_s.

Particles which arrive at the accelerating gap before the synchronous particle receive a larger accelerating voltage and energy gain than the reference particle; this increases their mass, which causes them to rotate slower according to Eq. (1). They thus wait for the reference particle to catch up and are gradually restored to the synchronous phase. Similarly, particles which reach the accelerating gap late receive a smaller energy increase

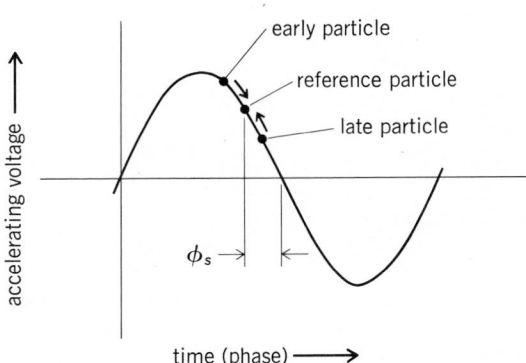

Fig. 6. Principle of phase stability in the synchrocyclotron. The reference particle has resonance energy and synchronous phase. The early particle accelerates faster to larger radius, where it has a lower rotation frequency which moves it later in time toward the central particle. The late particle is correspondingly moved earlier in time toward the central particle.

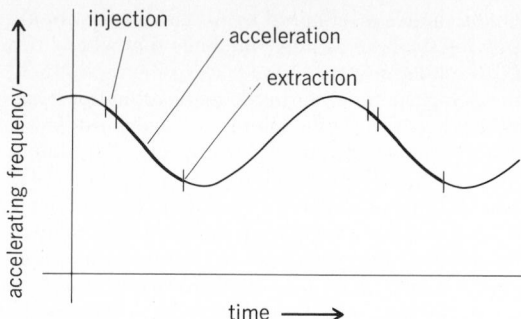

Fig. 7. Accelerating frequency as a function of time in the synchrocyclotron. The frequency decreases during acceleration to match the particle's frequency change due both to mass increase and to magnetic field reduction with increasing radius.

than the reference particle. Their mass increase is therefore less, which causes them to rotate faster, enabling them to catch up. This restoring action relative to the phase of the synchronous particle is referred to as phase focusing, and the idea is referred to as the principle of phase stability. This is the basic concept in the frequency-modulated cyclotron or synchrocyclotron. Figure 7 shows the cyclic nature of the acceleration process in a synchrocyclotron. This concept provided the leading advances in energy in the post–World War II years. The size of such cyclotrons is limited only by cost and construction difficulty. The largest which have been built achieve an energy of about 1000 MeV protons.

In changing from the Lawrence cyclotron to the synchrocyclotron, however, a very valuable beam property is given up, namely, beam intensity. In the Lawrence cyclotron, particles can leave the ion source on every rf cycle and start their acceleration journey, because every rf cycle is exactly identical. The synchrocyclotron is, in contrast, a batch system, as shown in Fig. 7. A group of particles leave the source, and the frequency of the accelerating system is then steadily lowered to stay matched with this reference particle while it is being accelerated. During this time other particles "wanting" to leave the ion source "see" a frequency which no longer matches their rotation frequency and must therefore "wait" until the reference group has been accelerated all the way to full en-

ergy and the rf returned to the frequency corresponding to the rotation of particles just leaving the source. This results in a large intensity loss. The beam coming from a synchrocyclotron typically has about 1/100 as many particles as the beam from a Lawrence cyclotron under comparable conditions. This low beam intensity became the limiting factor in a major class of precise physics experiments which require stringent filtering of beams or which use primary proton beams for production of secondary beams such as pions or muons.

Isochronous cyclotron. Responding to this experimental need, the isochronous or sector cyclotron was introduced in the late 1950s. Although the basic ideas for the isochronous cyclotron had actually been put forward in 1938 by L. H. Thomas (immediately after the Bethe-Rose paper), its practical application was delayed because of the complicated equations involved. Computers were not then available to solve the equations numerically and, as a result, Thomas's idea was generally overlooked for nearly 2 decades. Thomas's idea

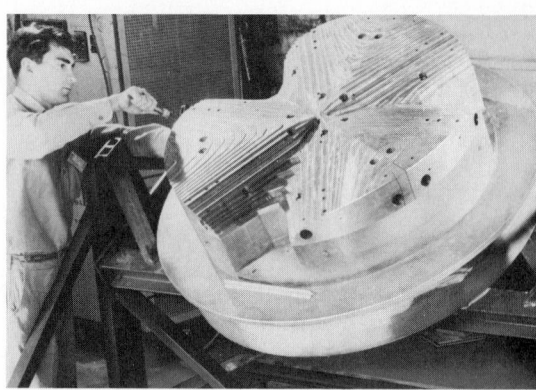

Fig. 9. The magnet pole for the first isochronous cyclotron, which operated at the Lawrence Berkeley Laboratory in 1950. (*Lawrence Berkeley Laboratory*)

can be understood by starting with Eq. (1). As the particle speeds up, its mass increases, as discussed earlier, and this contributes a slowing-down effect which cannot be avoided; however, if magnetic field B is increased in a compensating way, then the frequency is again constant. Since faster (heavier) particles move in circles of larger radius, a magnetic field is required whose strength increases with the radius. Such a field can readily be built, but the discussion of Fig. 5 applies: if the strength of the magnetic field increases with radius, then, according to Maxwell's equations, the field lines must curve away from the axis and therefore produce force components pushing displaced particles further away from the central plane. In the absence of other restoring forces, the number of particles surviving the acceleration process would then be extremely small.

The discussion in Fig. 5, however, is specifically applicable only to fields which are axially symmetrical. If azimuthal variations are included in the magnetic field, additional terms in Maxwell's equations come into play, and upon analysis one finds additional focusing forces. The origin of these forces is shown in Fig. 8, which shows a particular-

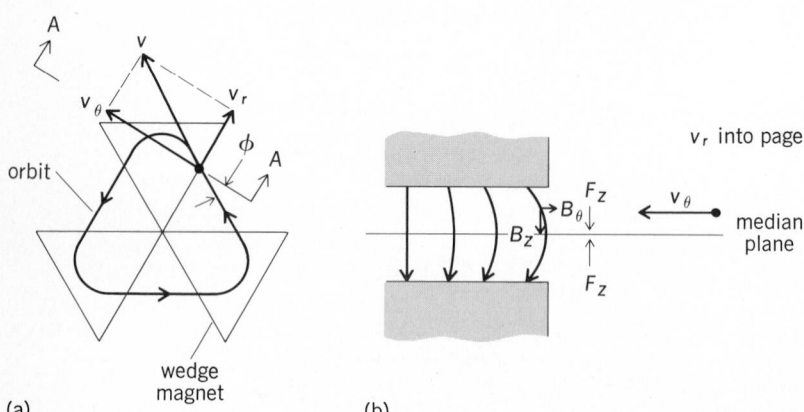

Fig. 8. Additional axial focusing in an azimuthally varying field. (*a*) View along axis of cyclotron. (*b*) Section *A-A* through median plane of *a*.

Fig. 10. Workers assembling the yoke for a typical isochronous cyclotron, the 50-MeV variable-energy cyclotron at Michigan State University. (*Michigan State University*)

ly simple version of the azimuthally varying field, or isochronous, cyclotron. In this example the magnetic field comes from three wedge-shaped magnets which meet at a point in the center. In the region between the magnets, the orbits are an approximate straight line (since very little magnetic field is present in this region). They then bend sharply through 120° in the region within the magnets and continue to repeat this pattern. The orbit

crosses the edge of each sector at a nonperpendicular angle denoted by ϕ in Fig. 8a. As a result, the orbit velocity at the field edge has a component $v_r = |v| \sin \phi$ in the radial direction. As indicated in Fig. 8b, the magnetic field at this point has an azimuthal component B_θ due to the inevitable bowing of the field lines at the edge of any magnet. This component is pointed away from the magnet above the median plane and toward the magnet below the median plane. Taking the vector product of the r component of velocity times the θ component of the field gives an axial force F_z toward the median plane which pushes particles which are out of the median plane back into the plane. It is easy to verify that the z force is likewise focusing at the point where the orbit leaves the wedge since both the r component of velocity and the θ component of the field have reversed direction. This "edge" focusing is then an additional force introduced by the sector structure; it can be used to override defocusing which comes from having the average field increase with radius in order to maintain constant angular frequency in Eq. (1).

Figure 9, 10, and 11 give an idea of the broad spectrum of sizes and types of isochronous cyclotrons. Figure 9 shows the pole tip of the first isochronous cyclotron, which was put into operation in 1950 in a classified program at the Lawrence Berkeley Laboratory, but not generally known until declassified and published in 1956. The elaborate shaping of the pole gave an approximately

Fig. 11. Partially assembled magnet for the TRIUMF cyclotron, showing the spiral iron sectors of the lower pole. (*TRIUMF, University of British Columbia*)

sinusoidal variation of field strength with angle, the specific shape studied by Thomas. (Later, computer studies indicated that almost any azimuthal form for the field variation would produce focusing, and elaborate pole tip contours were thereafter unnecessary.) Figure 10 shows the basic magnet core of a typical isochronous cyclotron for nuclear physics research, the 50-MeV machine at Michigan State University. This cyclotron has achieved an overall energy resolution of 1 part in 20,000. Figure 11 shows the magnet for the largest isochronous cyclotron (17 m in diameter), that of the TRIUMF Laboratory, Vancouver, B.C., Canada. The sectors in this design have a spiral shape to increase the edge axial focusing. Many isochronous cyclotrons use the spiral concept. This cyclotron accelerates negative hydrogen ions to an energy of 500 MeV. The cyclotrons at TRIUMF and at Zurich, Switzerland, have produced over 100 microamperes of proton beams, and are known as meson factories, since large quantities of mesons are produced when these beams hit a target. Development plans for isochronous cyclotrons include construction of two coupled 400-MeV cyclotrons injected by a small cyclotron at the GANIL national laboratory at Caen, France. *See* MESON; NUCLEAR PHYSICS.

Superconducting cyclotron. Work has begun on the superconducting cyclotron, another major evolution of the cyclotron concept. In contrast to earlier cyclotron concepts, the superconducting cyclotron is mainly a technical change rather than a

Fig. 13. Magnet of $K = 500$ superconducting cyclotron at Michigan State University with upper pole cap raised. (*From H. G. Blosser, The Michigan State University superconducting cyclotron program, IEEE Trans. Nucl. Sci., NS-26(2), pt. 1:2040–2047, 1979*)

conceptual one. The only superconducting element in the superconducting cyclotron is in fact the main coil, which is typically housed in an annular cryostat as illustrated in Fig. 12. Conventional room-temperature components, including pole tips, accelerating system, vacuum system, and ion source, are inserted in the warm bore of the cryostat from top and bottom. The superconducting coil allows the strength of the magnetic field to be greatly increased up to the level of approximately 50,000 gauss (5 teslas), or three times higher than was previously typical of cyclotrons. This increase in field reduces the linear size of the cyclotron to approximately one-third, areas to approximately one-ninth, and so on, compared to a normal cyclotron of the same energy. The result is a large reduction in the cost of many cyclotron components. Construction of the first magnets for cyclotrons of this type, designed to operate at approximately 500 MeV, has been undertaken at Chalk River Nuclear Laboratories, Canada, and at Michigan State University in the United States (Figs. 12 and 13). Construction of coupled 500-MeV and 800-MeV superconducting cyclotrons has been undertaken at Michigan State University. The superconducting cyclotron has a variety of potential applications, particularly for accelerators designed to accelerate very heavy nuclei such as lead or uranium to energies needed for nuclear reactions. *See* NUCLEAR REACTION.

Ion sources. An important property of an accelerated ion is its charge state, q. The maximum

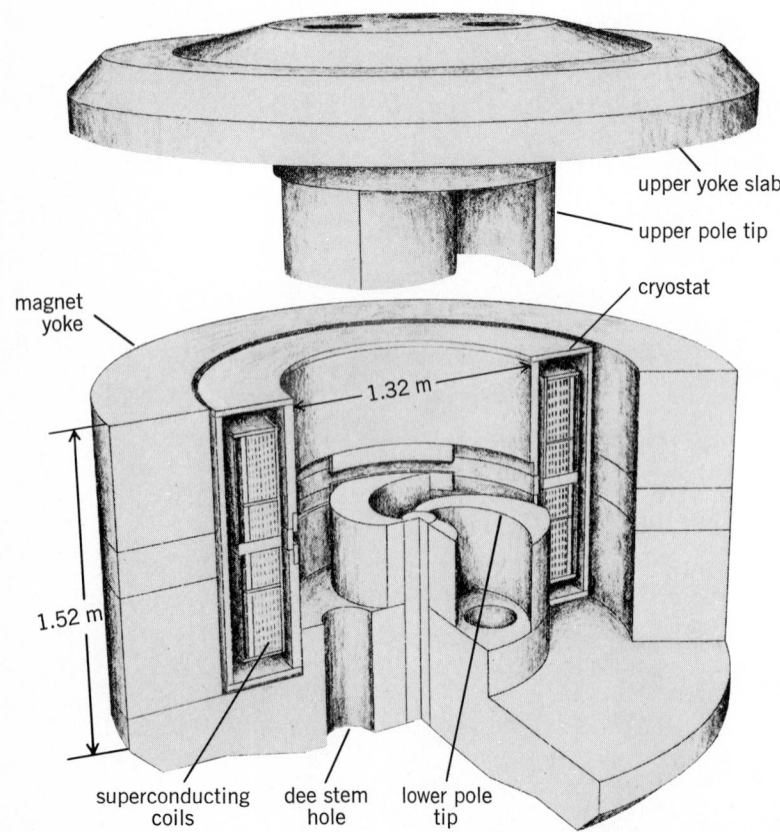

Fig. 12. Conceptual drawing of $K = 500$ superconducting cyclotron at Michigan State University. The upper pole cap is shown in the raised position used for maintenance. (*From H. G. Blosser, The Michigan State University superconducting cyclotron program, IEEE Trans. Nucl. Sci., NS-26(2), pt. 1:2040–2047, 1979*)

energy E of an ion in a cyclotron is given by Eq. (3),

$$E = Kq^2/m \qquad (3)$$

where K is the energy constant (usually in MeV) and q and m are in units of proton charge and mass. This equation comes from the relations $E = \frac{1}{2} mv^2$, $v = \omega r$, and Eq. (1), setting B equal to the maximum field available, and r equal to the maximum radius. For protons, $q = m = 1$ and thus $E = K$. For heavier ions there is a strong incentive to produce high q ions with the ion source, since $E \sim q^2$.

The standard ion source is based on the principle of the Penning ion gage (PIG). The highest-charge-state ions produced by a PIG source in useful intensities are, for example, $q = 5$ for nitrogen (N^{5+}). New concepts in ion sources have been developed which will produce N^{7+} (fully stripped) beams, doubling the nitrogen energy, Eq. (3). These advanced sources are called the electron cyclotron resonance (ECR) source and the electron-beam ion source (EBIS). The ECR source uses microwave-heated electrons in a magnetically confined plasma to produce high-charge-state ions. The EBIS uses an electron beam to ionize the feed material to very high charge states (Xe^{44+}). These advanced sources offer important opportunities to increase the heavy-ion energies of existing cyclotrons, and are also of interest for use at other types of accelerators such as synchrotrons. *See* ION SOURCES.

Applications. There are approximately 90 cyclotrons in operation. They are used principally for basic research in nuclear physics and chemistry, aimed at understanding the structure of nuclei, nuclear forces, and nuclear reactions. Practical applications being pursued in many laboratories include isotope production for medical use, secondary neutron beams for cancer therapy, and analysis of environmental material. Twenty-eight cyclotrons are used primarily for medical applications. *See* NUCLEAR STRUCTURE.

[DAVID J. CLARK]

Electron synchrotron. In the electron synchrotron the electrons being accelerated follow a roughly circular path of fixed radius in a magnetic guide field which varies cyclically and is perpendicular to the plane of the orbit. As the electrons pass through rf accelerating cavities placed along the orbit, their kinetic energy is increased. In this way the accelerating action of a rather modest cavity system can be applied repetitively to the same electrons, thus multiplying their initial energy by factors of hundreds or thousands. Electron synchrotrons are used primarily for research in elementary particle physics, but have found increasing use as intense sources of ultraviolet and x-radiation in solid-state and materials science.

For normal injection energies, as a consequence of their low mass, the velocity of the electrons injected into a synchrotron is usually so close to the velocity of light that the change in velocity during acceleration is insignificant. This is in contrast to the situation in proton synchrotrons or cyclotrons where the velocity of the particles being accelerated may vary by a factor of more than 10 during acceleration. As a result, the frequency of revolution of electrons in the synchrotron is a constant, as is the frequency of the power applied to the accelerating cavities. Injection into electron synchrotrons is usually done from a preaccelerator such as a linac or microtron having an output energy of tens or hundreds of MeV.

The orbit radii of machines in use range from a few meters for a synchrotron which achieves a few hundred MeV peak energy, to 100 m for the largest electron synchrotron, the 12-GeV instrument at Cornell University. Most modern synchrotrons employ the alternating-gradient or strong-focusing principle which minimizes the volume of the magnetic guide field, so that while the circumference of the orbit may be almost a kilometer, the cross-sectional dimensions of the evacuated volume surrounding the beam need be only a few centimeters.

Accelerating cycle. In a typical accelerating cycle, electron currents of several milliamperes are injected into the synchrotron in a burst lasting up to a few microseconds. During this time the value of the guiding magnetic field is rather low, being in the range 10^{-3} to 10^{-2} tesla, and almost constant in time. Upon completion of the injection phase, the strength of the guide field is increased, and the kinetic energy of the electrons increases proportionally as energy is received from the accelerating cavities by electromagnetic induction. This proportionality is ensured by the principle of phase stability and maintains the orbit radius constant throughout the acceleration process. After executing a few thousand revolutions with each revolution providing an energy increment as great as a few MeV, the beam becomes accelerated to the desired energy. At this point the beam is utilized either by steering it onto a metal target within the vacuum chamber, thus converting its kinetic energy into x-radiation, or by extracting it from the synchrotron to bombard an external target material directly. The strength of the guide field is held approximately constant during the time the beam is being used. This period may last a few milliseconds. After the beam is spent, the guide field strength is lowered again to its injection value, and the cycle repeats. Repetition rates of 60 times per second are common. Table 2 lists the parameters of two large electron synchrotrons in operation. Figure 14 shows the Cornell synchrotron.

Beam extraction. The x-ray beam that is created when electrons strike an internal metal target emerges in a very narrow cone tangent to the orbit. At energies of a few GeV, this beam will have a

Table 2. Some parameters describing the two largest electron synchrotrons in use

	Cornell	DESY*
Energy, GeV	12	7.5
Radius, meters	100	50
Cycling rate, cycles per second	60	50
Beam intensity, particles per pulse	3×10^{10}	10^{11}
Maximum magnetic field, tesla	0.40	0.79
Injection energy, MeV	150	300
Total weight of iron in magnet, tons	200	570
Number of magnet units	192	48

*Deutsches Elektronen Synchrotron, located at Hamburg, Germany.

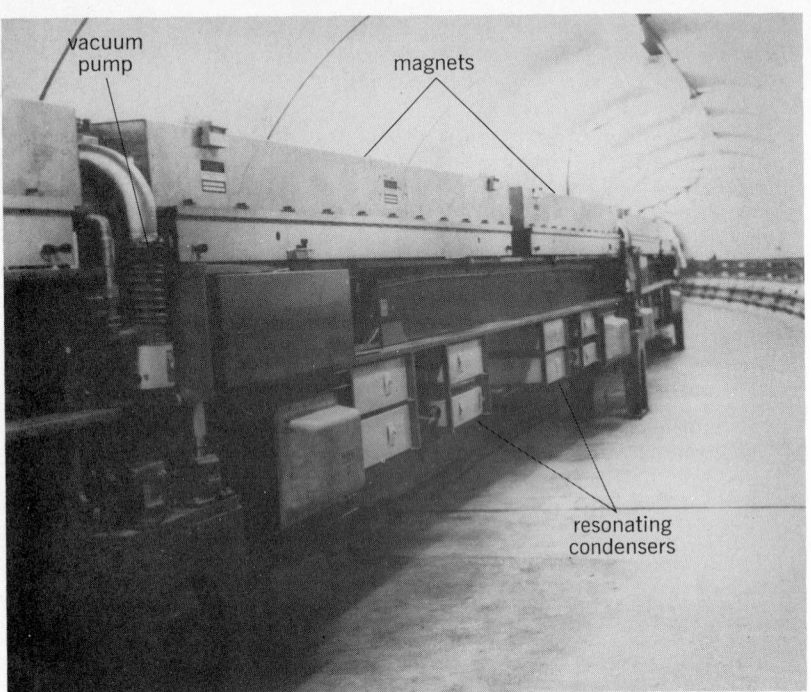

Fig. 14. The 12-GeV electron synchrotron guide field assembly installed in an underground tunnel at Cornell University.

diameter of a few centimeters at a distance of 100 m from the target. When the electron beam itself is to be extracted from the synchrotron, the method of resonant extraction is frequently employed. This method employs a small perturbation of the focusing characteristics of the guide field in order to force individual particles of the beam to oscillate about their normal orbit with ever growing amplitude. The rapidly increasing amplitude of oscillation causes the electrons to jump to the outside of a thin, current-carrying septum where the magnetic guide field has been drastically weakened by the septum. Thus particles whose oscillation amplitude has carried them to the outside of the septum find themselves suddenly in a weak guide field, and they fly off nearly tangent to their initial orbit.

Magnets. Because of the relatively low electrical losses in the electromagnet which provides the guide field, it is common practice to operate the magnet as part of a resonant circuit consisting of the magnet itself and a capacitor bank, with the combination being resonant at the power line frequency. To minimize eddy current and hysteresis losses, the magnet is built up of thin laminations of electrical steel of the kind used in making common transformers. *See* EDDY CURRENT; MAGNETIC HYSTERESIS.

Synchrotron radiation. In addition to the energy which must be supplied to the beam by the cavities in order to match the rising strength of the guide field, energy must also be supplied to replenish that which is emitted by the beam as "synchrotron radiation." This electromagnetic radiation is given off by all charged particles which follow a curved path, and is directed in a narrow beam tangent to the path. The energy lost by this mechanism is inversely proportional to the fourth power of the rest mass of the particle. For energies now achiev-

able in the laboratory this loss is of practical consequence only for electrons. This kinetic energy loss is also proportional to the fourth power of the kinetic energy. At 12 GeV in the Cornell machine, for example, the loss per revolution is about 19 MeV. This radiation is given off in a continuous spectrum that extends from radio wavelengths, through visible light, and into the hard x-ray region. It is so intense that, for many experiments in atomic and solid-state physics and biology, synchrotron radiation is superior to that produced by ultraviolet lamps and x-ray tubes and is being exploited extensively for these purposes. Serious planning has been undertaken for synchrotrons totally devoted to such studies. Besides being an energy drain on the acceleration system, the synchrotron radiation is quantized and emits individual photons which tend to make the beam diffuse radially into a flat ribbon at very high energies. In the Cornell machine the beam is about a centimeter wide at 12 GeV, whereas it is about a millimeter wide at 8 GeV.   [M. TIGNER]

Proton synchrotron. The proton synchrotron developed from the cyclotron by way of the synchrocyclotron. This progression has resulted in proton accelerators with higher and higher beam energies. A 400-GeV accelerator is in operation in the United States, while several European countries have constructed a similar facility near Geneva, Switzerland.

The cyclotron is limited in its ultimate energy because of the relative increase of mass during acceleration. This causes the beam to get out of step with the frequency of the accelerating system and to be lost. The idea of slowly varying the frequency of the accelerating voltage to maintain synchronism dramatically extended the energy regime of the circular proton accelerator. Then the limitation became the size and cost of the large magnet within which all the beam acceleration took place. The incorporation of time-varying magnetic fields made it possible to overcome this limitation. The use of numerous but small magnets whose fields could track the proton beam energy led to the major breakthrough that made possible higher proton beam energies at reasonable cost.

The essential feature of a proton synchrotron is that the average radius of the accelerated beam is a constant. This is accomplished by a careful synchronization of the magnetic field and the rf accelerating frequency. The magnets are arranged in an approximate circle and are powered in such a way that the excitation current can be varied from low to high values during acceleration. Since the momentum of the protons at injection is small compared to the final momentum, the magnetic field must be proportionately small at the beginning of acceleration. As the momentum of the proton increases during acceleration, the magnetic field is increased in a manner that keeps the guide path of the protons nearly constant. This permits the bundle of protons to be contained within a relatively small vacuum chamber located within the magnets.

In the proton synchrotron the ion source and initial accelerator system are located away from the accelerator. As seen in Fig. 15, the proton beam is injected into the ring of magnets at a low energy and accelerated to higher energies as the magnetic field is increased. At the peak energy it can be

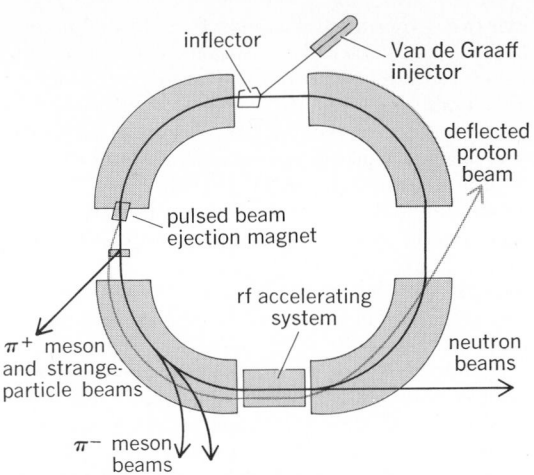

inflector

Van de Graaff injector

deflected proton beam

pulsed beam ejection magnet

rf accelerating system

neutron beams

π^+ meson and strange-particle beams

π^- meson beams

Fig. 15. Schematic diagram of the principal components of a proton synchrotron. Note the various possibilities of external neutral beams, charged beams, and extracted primary beams.

used on internal targets or it can be deflected (extracted from the accelerator to be used externally).

Accelerating cycle. The magnetic field B must be changed together with the momentum in order to keep the radius r constant. As the magnetic field increases, the beam of protons passes repeatedly through the rf electric fields, and gains energy steadily. The proper coordination of these fields is essential to the success of the proton synchrotron. In order to minimize the frequency variations, it is customary for protons to be injected at a relatively

high energy. For example, at Fermi National Accelerator Laboratory (Fermilab, Fig. 16), where the ultimate proton-beam energy is 400 GeV, the protons are injected into the main ring at 8 GeV. At 8 GeV the particles are already traveling at the speed of light, and therefore the frequency variation during the course of acceleration is less than 1%. At injection the magnetic guide field strength is 0.36 kG (36 milliteslas) and increases to 18 kG (1.8 teslas) during the 3 sec of acceleration. A typical curve showing the increase of the magnetic field (or the energy) is shown in Fig. 17. Here the beam at 8 GeV circles the magnetic guide field until the accelerating voltage is applied. As the accelerating voltage increases the beam momentum, the beam tends to move toward the outside of the vacuum chamber. In order to bring the beam back and contain it within the circular vacuum chamber, the magnetic guide field is proportionally increased. Steadily the beam gains energy as the guide field increases in strength, always containing the beam within the vacuum chamber. The entire acceleration process brings the circular plasma of protons from the 8 GeV of energy at injection to 400 GeV in 3 sec.

Beam focusing. In order to hold the plasma of protons together, it is necessary to have some restoration force that keeps the protons within the vacuum chamber. On older "weak-focusing" accelerators, focusing of the beam and its maintenance in a circular path are accomplished simultaneously by the gentle shaping of the iron pole faces of the guide field magnets. The largest such machine in the United States was the zero-gradient synchrotron (ZGS) at Argonne National Laboratory in Illinois, now closed, which accelerated protons

experimental areas

meson area   neutrino area   proton area

main accelerator ring

central laboratory building

200-MeV linear accelerator

8-GeV synchrotron

injector   accelerator

Fig. 16. A portion of the Fermilab site and the 400-GeV accelerator. The accelerator is located below ground in a circle with a diameter of 2 km. (*Fermi National Accelerator Laboratory*)

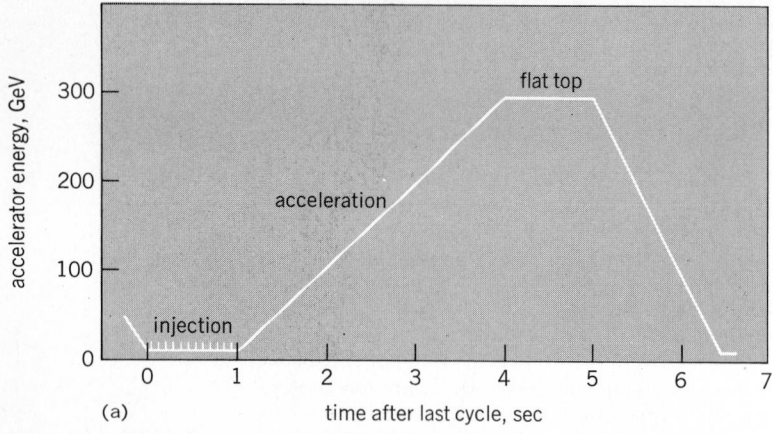

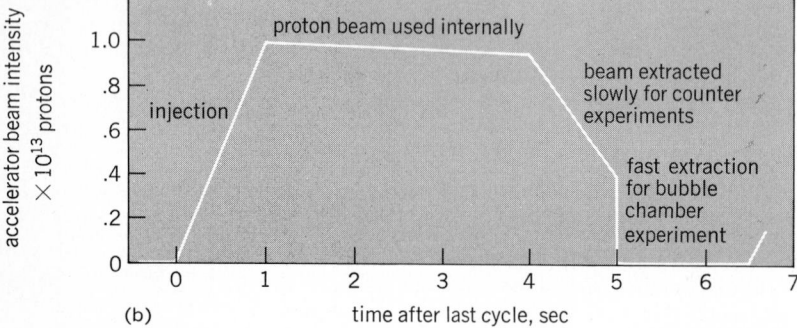

Fig. 17. Curves of (a) the energy and (b) the intensity of the circulating protons during a cycle of the main accelerator at Fermilab. (*Fermi National Accelerator Laboratory*)

external experimental areas. The extraction process typically takes approximately 1 sec, during which time the beam is slowly removed from the accelerator in a steady stream of particles. The protons are directed onto targets external to the accelerator, where they are used to produce beams of other particles desirable for high-energy physics experiments. An innovation in the targeting of protons results in the ability to divide the stream of protons into several separate bunches. The protons can thus be directed simultaneously onto several targets. The intensity is divided among the proton branches in a manner that is consistent with the needs of the experimental program. With electrostatic septa and magnetic deflecting magnets, less than 1% of the beam is lost in the splitting process.

The use of one extracted beam coupled with splitting stations has considerably added to the efficiency of operation of proton synchrotrons. This makes it possible for a number of experiments to run simultaneously so that eight to ten experiments can be in operation at any given time. [JAMES R. SANFORD]

LINEAR ACCELERATORS

A linac accelerates particles in a straight line by means of rf electric fields. Since a given field location is traversed only once, such fields must be produced along the entire particle orbit; hence very high field strengths are necessary. This implies megawatt rf power levels for room-temperature structures, but in the more recently developed superconducting linacs a few watts suffice.

Principles of operation. Possible synchronism between particles and field is attained by either of two methods: traveling-wave acceleration, wherein a wave with an accelerating field component is produced whose phase velocity is equal to the particle velocity; or standing-wave acceleration, wherein a standing-wave pattern is produced by the superposition of a forward wave and a back-

to 12.5 GeV. In order to keep the volume of the magnetic field as small as possible, stronger focusing is needed to reach higher energies. At Fermilab this is accomplished by using quadrupole (four-pole) magnets that act like lenses in keeping particles together. These focusing units, which control the beam geometry and do not provide any net bending to the particles, constitute about 10% of the magnets in the circular accelerator. Since their effect is not combined with that of the guide field dipoles which determine the mean orbit shape, the structure is called a separated-function strong-focusing accelerator (Fig. 18). The ability to independently adjust the excitation of the two sets of magnets led to an accelerator that can be used very flexibly. For example, the separate excitation of the quadrupole fields aids in the process of beam extraction.

Magnet current control. The magnets are powered by excitation currents that must be programmed to provide current pulses with just the right properties. The electrical power is available as alternating high voltage which is rectified to quasidirect current for magnet excitation. The current must be controlled so that the magnetic field is pulsed in synchronism with the accelerating particles. The current is controlled by the application of low voltage pulses to the rectifiers. These voltages are in turn provided by minicomputers. With modern computers it is possible to accelerate the protons in pulses that lead to the efficient capture of the injected beam and low-loss acceleration of the protons.

Beam extraction. At the peak energy the beam is extracted from the accelerator to be used in the

Fig. 18. Interior of the main accelerator tunnel at Fermilab, showing the guide field and focusing magnets. (*Fermi National Accelerator Laboratory*)

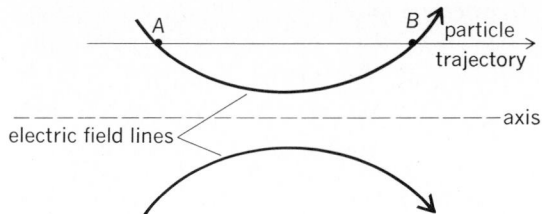

Fig. 19. Phase focusing in the linear accelerator in the absence of grids. If the field increases during particle transit across an accelerating gap, defocusing momentum imparted at *B* is greater than focusing momentum at *A*. Action is thus defocusing but phase-stable.

ward wave in the structure. In the latter method, either the phase velocity of the forward wave is made equal to the particle velocity, as in the traveling-wave accelerator just described, or a set of drift tubes is introduced into the structure to shield the particles from the fields when they are passing through regions where the fields would otherwise be decelerating.

Particle energies. The energy *T* to which a particle of charge *q* can be accelerated in a structure of length *L* fed by rf power sources operating at a wavelength λ and dissipating power *P* is given by Eq. (4), where *K* is a constant depending somewhat

$$T = qK(PL)^{1/2}\lambda^{-1/4} \qquad (4)$$

on the structure chosen and also on whether standing-wave or traveling-wave acceleration is used; hence length and power are equally instrumental in attaining high energies. Short wavelengths are advantageous from the point of view of power economy, but also have disadvantages: (1) In principle, higher power levels are available from microwave tubes at longer wavelengths, and higher powers can be dissipated in the accelerator. (2) The absolute tolerances of the mechanical structure are larger at longer wavelengths. (3) Higher currents can be accelerated at longer wavelengths. In practice, the wavelengths used are 3−30 cm for electron accelerators, 30−200 cm for proton accelerators, and 500−1000 cm for heavy-ion accelerators. The high power levels require that room-temperature linear accelerators be operated as pulsed rather than as continuously operated machines, but this requirement is imposed by practical and economic considerations and not by fundamental principles of operation, as in the case of the phase-stable circular machines.

Orbit dynamics. The focusing action and phase stability of linear accelerators in the absence of external lenses that could be placed around the beam will now be discussed. Consider a region in space where the accelerating field increases in time as the particle crosses the region (Fig. 19); assume that no field lines end inside the beam. It is clear that there will be a net defocusing action since, in an increasing field, the transverse momentum imparted at *B* is greater than that imparted at *A*. However, the action is phase-stable, since a particle arriving late will receive a larger acceleration and will catch up in the next accelerating region. If a particle crosses the region during a period when the field is decreasing, the conclusions are reversed; that is, the transverse momentum imparted at *A* will be greater than that imparted at *B*, but a late-arriving particle will receive a smaller

acceleration and will fall farther behind. Thus focusing and phase stability are incompatible in the simple linear accelerator that operates at a fixed phase difference between the field and the particle. This incompatibility of focusing and phase stability appeared to be a serious obstacle to the design of linear accelerators, but it has been circumvented in various ways as follows: (1) A very short accelerator can be operated even though unstable. (2) Charge can be included in the beam to terminate the lines in Fig. 19 inside the beam, thus producing a convergent action even at phase-stable transit phases; this charge can be induced on grids placed in the field. This was the earliest solution but suffers from grid interception and has been abandoned. (3) As the particle velocity approaches *c*, the apparent incompatibility becomes irrelevant because, for relativistic velocities, the action of the radial electric time-varying field is almost canceled by the accompanying time-varying magnetic field; also, because the velocity is almost invariable, the particles are in neutral equilibrium longitudinally. (4) External magnetic (either solenoidal or strong-focusing) or electrostatic lenses have been used in all modern designs. (5) By alternating the phase difference, a limited region of complete stability appears for certain field shapes. This has often been proposed but not yet adopted in practice.

Electron accelerators. Almost all electron linear accelerators are of the traveling-wave type. The wave is produced in a loaded waveguide, which is a waveguide excited in a TM mode (having a longitudinal electric field component) and loaded by disks or other means to produce the correct phase velocity to match the particle. An unloaded or uniform waveguide is not suitable since the phase velocity in such a guide exceeds the velocity of light. *See* PHASE VELOCITY.

A typical field configuration is shown in Fig. 20; the field pattern should be visualized as translating along the axis with a velocity equal to that of the electron shown. The electron velocities approach *c* very rapidly; hence, except for the first few feet of the machine, all radial forces can be neglected. This means that the momentum component transverse to the axis remains essentially constant, and since the longitudinal component increases continuously, the angle of divergence of the beam decreases continuously; for uniform energy gain per unit length, this corresponds to a beam angle varying as the inverse of the distance along the machine and to a beam radius increasing logarithmically. A set of weak external magnetic lenses can cause the beam diameter to decrease instead of increase.

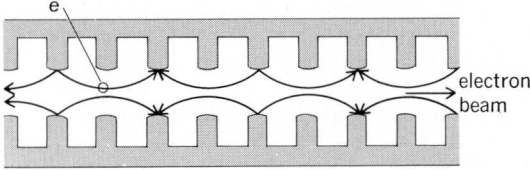

Fig. 20. Electric field configuration in a traveling-wave electron linear accelerator. Waveguide shown is loaded by four disks per wavelength to obtain a phase velocity matching the speed of particle.

Fig. 21. Stanford Linear Accelerator Center. (a) Aerial view, showing the 2-mi (3.2 km) accelerator and asso- ciated laboratories, shops, and research facilities. (b) The accelerator installed in underground housing.

The accelerator waveguide must be constructed to very close tolerances to control the phase velocity to the required accuracy, since there is no phase stability once the energy has grown to several times the rest energy of the electron. However, the required machining accuracy is relieved somewhat by two considerations: (1) The structure can be tuned after final assembly by appropriate distortion of its outer walls. (2) Each independently fed section of the accelerator can be individually phased during operation.

Particles are injected from an electron gun which can easily inject electrons at a velocity of approximately $c/2$, corresponding to about 80 keV. The first accelerator section can be a special bunching device to accentrate the particles near the crest of the traveling wave of the succeeding sections.

The energy possibilities of the electron linear accelerator are not limited by fundamental considerations since electrons accelerated in a straight line do not lose energy by radiation by an appreciable amount. Since the operating frequency is in the microwave region, the performance of the accelerator is closely tied to the available power sources. Some of the earlier machines were powered by magnetron oscillators; newer machines are powered by klystron amplifiers driven from a common master oscillator. The largest electron linac is the 2-mi-long (3.2 km) 20-GeV machine located at Stanford University (Fig. 21).

The acceleration of electrons to useful energies by means of accelerators operating in the usual temperature ranges requires a supply of peak rf power measured in megawatts. The costs of power and power sources limit the beam duty cycle to values that are usually a few tenths of a percent. These low duty cycles restrict the kinds of experiments which can be done, and several methods have been developed to achieve much higher duty cycles. In the case of the 400-MeV linac at MIT, the duty cycle was raised to several percent using conventional traveling-wave structures at lower than maximum fields but with long pulses and fast repetition rates. This and a similar 500-MeV linac at Saclay, France, are limited by the average power of the klystrons. The other development is the superconducting electron linac which can operate continuously with dissipations of about a watt per meter. The development of standing-wave structures of niobium metal operated at 1 to 4 K has been pioneered at the Stanford University High Energy Physics Laboratory. There have been severe metallurgical and electron loading problems, and the initial hopes of very high accelerating gradients have not been realized. However, these machines are practical for continuous operation at 2–3 MeV/m and are being used with recirculating beams (microtron) at the University of Illinois and elsewhere.

Proton accelerators. Existing proton accelerators are of the standing-wave type since protons, unlike electrons, do not reach relativistic velocities at injection voltages and traveling-wave structures are not feasible. The earliest and still popular design is the Alvarez structure shown in Fig. 22. The drift tubes, supported by transverse stems, are metallic cylinders and no rf field is present inside. The distance between tube centers is called a cell and is one guide wavelength long ($v\lambda/c$, where λ is the free-space exciting wavelength and v the particle velocity at the cell). The particles cross each gap at the same phase, usually 20–30° before the time maximum, and the beam is automatically phase-stable. In early machines, focusing was performed by grids at the entrance to each drift tube; in later designs, focusing is done by magnetic quadrupoles inside the tubes. An assembly of cells with end plates is a resonant cavity if each cell is resonant at exactly the same frequency, usually about 200 MHz. Manufacturing tolerances, ohmic losses, and beam loading perturb this exact resonance and set severe limits on the cavity length.

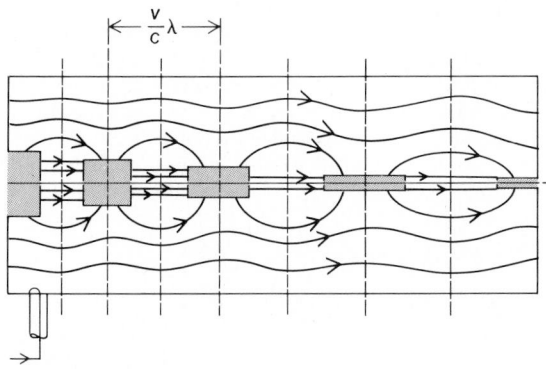

Fig. 22. Electric field configuration in the standing-wave proton accelerator employing a cavity resonant in the TM_{01} mode. Drift tubes are shown placed to shield the particles when the phase is decelerating. The length of the drift tubes is designed so that the particles cross each gap at approximately the same phase.

For a large accelerator several cavities are placed in tandem, each fed by a power source at the same frequency. Protons are injected from a 0.5–2 MeV electrostatic accelerator; the beam emerges from the output end and is directed to the experiment or injected into another accelerator.

The trend has been toward much higher beam currents, more precise energy, and much higher energies. These have been achieved largely by advances in structure design, in particular by the use of multiple resonances or cells of double periodicity which greatly relax the mechanical and beam-loading tolerances and allow efficient acceleration and precise control of the fields together with good power transfer along the structure (high group velocity; the Alvarez structure has zero group velocity). In modern designs the Alvarez structure with resonant posts or stems is preferred up to about 200 MeV. The newest Brookhaven National Laboratory injector for the alternating-gradient synchrotron and the similar 200-MeV injector for the 400-GeV Fermi National Accelerator Laboratory synchrotron are such machines. Above 200 MeV, coupled-cavity structures of great precision and efficiency have been developed. The 200–800 MeV section of the proton accelerator at the Los Alamos Meson Physics Facility (LAMPF) uses these structures. They have beam currents of tens of milliamperes and energy precisions of better than 0.1%. *See* GROUP VELOCITY.

Heavy-ion accelerators. Heavy ion linacs are similar to low-energy proton accelerators, with differences dictated by the relatively low charge-to-mass ratios available from ion sources. For a given electric field and structure length, the ion velocities are much lower than for protons, and cell lengths at 200 MHz would be impractically short; operating frequencies of 30 to 70 MHz have been used in various designs. After passing through one or more cavities and reaching velocities of about 4% of c (energies of 1 MeV per atomic mass unit), the ions are passed through a stripper—a very thin foil, gas-filled tube, or gas jet—and electrons are stripped from the ions; typically the ion charge is doubled, and about half the ions are in a single-charge state. Subsequent accelerato 15 MeV/AMU (v/c of 0.12 to 0.18) and is chosen to allow nuclear reactions between any of the accelerated ions and the heaviest target nuclei. tion in later cavities is then twice as efficient. The final energy of the accelerators is in the range 7

Earlier linacs were limited to ions lighter than argon, but later ones are designed to accelerate any ion up to uranium. Superhilac (Fig. 23) at the University of California Lawrence Berkeley Laboratory uses a 0.75-MV dc injector for the lighter ions or a 2-MV injector for heavier ions, followed by an Alvarez cavity with quadrupoles in each drift tube, a stripper, several Alvarez cavities, and finally several single-gap cavities which can be used to change the beam energy. The maximum output energy is 7.4 MeV/AMU for all ions. Unilac at Darmstadt, Germany, is a very elaborate series of linacs, strippers, and independent cavities at the output end. Its technical innovations include Wideröe structures (cells of length $v\lambda/2c$, based on alternately tapping a resonant two-wire transmission line) for the four prestripper cavities, which operate at 27 MHz, followed by a stripper and two Alvarez cavities which operate at the fourth harmon-

Fig. 23.   Inside view of the prestripper cavity of the Superhilac at the University of California Lawrence Berkeley Laboratory. Note the drift tube structure and the radio-frequency coupling loops which are seen on the edge of the tank. (*From A. Ghiorso, Progress with the Superhilac, IEEE Trans. Nucl. Sci., NS-20(3)151–154, 1973*)

ic, 108 MHz, and up to 20 independent cavities at the end. It is designed to accelerate all ions up to uranium to energies of 1.5 to 7 MeV/AMU or higher for lighter ions.

Considerable development work has been undertaken to perfect a practical superconducting structure for heavy ions. To be attractive, it should have a small diameter, high accelerating field, and low frequency. Various problems remain to be solved with all of the proposed designs, but the principle is very attractive, particularly because of the very large power consumption of room-temperature heavy-ion linacs. Superconducting linacs will probably be used as postaccelerators injected by a large tandem Van de Graaff accelerator.

[ROBERT BERINGER]

STORAGE RINGS

Storage rings consist of annular vacuum chambers in which beams of high-energy charged particles can be stored and caused to collide in near head-on collisions.

Principles of colliding-beam systems. The motivation for using this technique is found in the following kinematic considerations. Unlike fixed-target systems (cyclotrons, synchrotrons, linear accelerators) in which a beam of accelerated particles traverses a fixed target, colliding-beam systems use the full energy of each particle to produce reactions. When two particles of mass m_1 and m_2 and energy E_1 and E_2 interact ($E = T + mc^2$, where T is the kinetic energy and mc^2 the rest-mass energy), the center-of-mass energy $E_{\rm cm}$ measures the energy available for reactions. For a fixed-target accelerator which produces a beam of high-energy relativistic particles 1 ($E_1 > m_1c^2$) hitting stationary target particles 2, this energy is given approximately by Eq. (5). The useful energy is on-

$$E_{\rm cm} = [2(m_2c^2)E_1]^{1/2} \qquad (5)$$

ly a fraction of the available beam energy and increases slowly as its square root. By contrast, when two particles of the same mass and energy E_1 collide head-on in a storage ring, $E_{\rm cm}$ is given by Eq. (6). All of the available energy is useful. For

$$E_{\rm cm} = 2E_1 \qquad (6)$$

instance, the Stanford University e^+e^- storage ring PEP (Positron-Electron Project) is designed to operate with beams of 18 GeV yielding $E_{\rm cm} = 36$ GeV. With a fixed e^- target, an e^+ beam of 1,300,000 GeV would be required to reach the same value of $E_{\rm cm}$.

Low reaction rate. The high value of $E_{\rm cm}$ of colliding-beam systems is obtained at the expense of reaction rate, which is proportional to the density of target particles and to the beam intensity. It is helpful to think of one of the stored beams as the target for the other. In a typical e^+e^- storage ring, the beam density is about 10^{14} particles/cm³; a fixed target contains about 10^{24} electrons/cm³. This huge difference is partially compensated by the larger beam intensities in storage rings which result from the accumulation of particles injected from a fixed-target accelerator. It is because of the importance of accumulation for obtaining useful reaction rates that beam storage is required in high-energy colliding-beam systems. However, there are limits to the density and intensity of beams that can be stored and made to collide, and thus colliding-beam reaction rates are always much lower than fixed-target reaction rates. Storage ring experiments are designed to take full advantage of the high center-of-mass energy within the limitations imposed by the relatively low reaction rates.

Luminosity. The performance of colliding-beam systems is measured by the luminosity L defined by Eq. (7), where R = reaction rate or number of

$$L \equiv \frac{R}{\sigma} = \frac{N_1 N_2 f}{Ab} \qquad (7)$$

interactions per second; σ = interaction cross section; N_1, N_2 = number of particles in each circulating beam; A = beam cross-sectional area at the interaction point; f = number of stored particle revolutions per second; b = number of bunches in each beam. Thus, the reaction rate is given by $R = L\sigma$. Typical values are in the $L = 10^{29} - 10^{32}$ cm⁻² sec⁻¹ range.

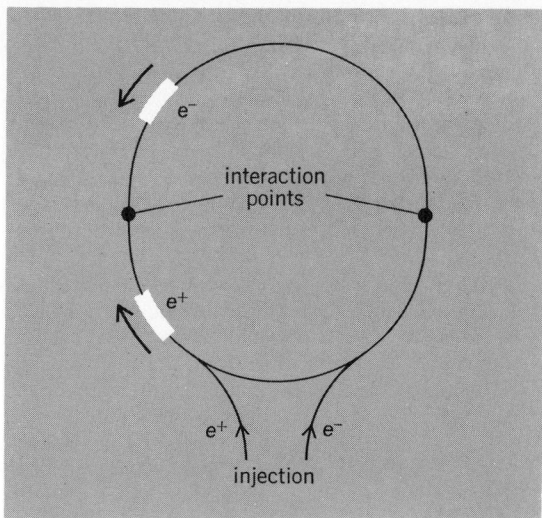

Fig. 24. Schematic diagram of e^+e^- storage ring.

Positron-electron storage rings. A positron-electron (e^+e^-) storage ring consists of a ring of bending and focusing magnets enclosing a doughnut-shaped vacuum chamber in which counterrotating beams of e^+ and e^- are stored for periods of several hours. The two beams are made to collide with each other about 10^6 times per second in straight interaction sections of the vacuum chamber (Fig. 24). These are surrounded with detectors for the observation of collision products. An alternate design consists of two separate but intersecting rings, one to store e^-, the other to store e^- or e^+, with collisions taking place at the intersection points.

Collisions between antiparticle and particle such as e^+e^- are of particular interest since most of the resulting processes proceed through annihilation which forms a particularly simple final state with the quantum numbers of the photon. This is one of the main reasons why e^+e^- collisions have proved to be particularly fruitful in deepening understanding of the fundamental structure of matter.

Much has been learned about the charm-anticharm quark system (the J/ψ meson family) and other mesons containing charmed quarks with the e^+e^- storage rings SPEAR (Stanford Positron-Electron Accelerating Ring) at Stanford Linear Accelerator Center (SLAC), in California, and DORIS at DESY (Deutsches Elektronen Synchrotron), Hamburg, Germany, where beams are stored up to 5 GeV. As a result, new, higher-energy storage rings have been constructed. PETRA (Positron Electron Tandem Ring Accelerator), at DESY, started operation in 1978 and is designed to operate with beams of up to 18 GeV. New results on hadron jets produced by quark and gluon decays have been observed. Cornell Electron Storage Ring (CESR) at Cornell University, Ithaca, New York, started operation in 1979 and is designed to store beams of up to 10 GeV. This storage ring has carried out a very important study of the bottom-antibottom quark system (the Υ meson family). PEP, at SLAC, covering an energy region similar to PETRA, started operation in 1980. *See* CHARM; J PARTICLE; UPSILON PARTICLES.

Beam instabilities. When particles from beam 1 (e^+ or e^-) pass through a more intense beam 2 (e^- or e^+) in the interaction region, they feel a strong force due to the electromagnetic field set up by beam 2. Thus, beam 2 acts as a very nonlinear lens on beam 1 and tends to cause a diffusive growth in the size of the latter. The force increases rapidly with the density of particles in beam 2 until it becomes so strong that the beam 1 area becomes suddenly very large and the luminosity decreases drastically; this is beam-beam instability. There also exist single-beam instabilities which are caused by electromagnetic forces between particles in the same beam or by fields set up by induced currents in the walls of the vacuum chamber. Maximum luminosity is achieved by operating with two beams of equal intensity just below the beam-beam instability limit, which in a well-designed storage ring is reached before single-beam instabilities become a problem.

Synchrotron radiation. The limitations discussed so far apply equally well in principle to e^+e^- as to p-p colliding-beam systems. There are important differences, however. Because they have a small mass, e^+ and e^- in circular orbits radiate a substantial amount of energy in the form of synchrotron radiation, just as they do in electron synchrotrons. The power $P_{\rm rad}$ radiated by N_1 positrons and N_2 electrons of energy E revolving in circular orbits of radius r is given by Eq. (8),

$$P_{\rm rad} = \left(\frac{1}{4\pi\epsilon_0}\right)\frac{2}{3}\left(N_1 + N_2\right)e^2c\left(\frac{E}{mc^2}\right)^4\frac{1}{r^2} \qquad (8)$$

where e and mc^2 are the electron charge and rest-mass energy and $\epsilon_0 = 8.854 \times 10^{-12}$ fermis/m. This power, which has to be supplied by the rf system to keep particles on the design orbit, is appreciable and costly; at the peak operating energy it amounts typically to several megawatts in multi-GeV storage rings. In the design of a storage ring for a given energy E and luminosity L, the most economic solution requires a careful balance between the cost of magnets and buildings (which increases with increasing r) and the cost of the rf system (which decreases with increasing r since less power is needed). At the highest operating energies the amount of available rf power limits the luminosity, since for fixed values of $P_{\rm rad}$ and r only a steep decrease in $(N_1 + N_2)$ can compensate for an increase of E. Figure 25 shows the variation with E of the maximum luminosity L that is theoretically possible in an e^+e storage ring through adjustment of the focusing magnets. E_0 is the energy at which L is designed to have its maximum value L_0. L is limited by beam-beam instability for $E < E_0$ and by available rf power for $E > E_0$. At $E = E'$ the focusing magnets become saturated and can no longer be adjusted for maximum luminosity.

There are also very beneficial consequences from the emission of synchrotron radiation. This emission dampens the amplitude of e^- and e^+ oscillations around the equilibrium orbit, and produces a gradual transverse polarization of the stored beams. As in electron synchrotrons, the radiation itself constitutes a very intense source of x-rays which has proved very useful in various branches of biology, chemistry, and physics.

Ring. The ring consists of circular sectors of bending and focusing magnets and of straight sec-

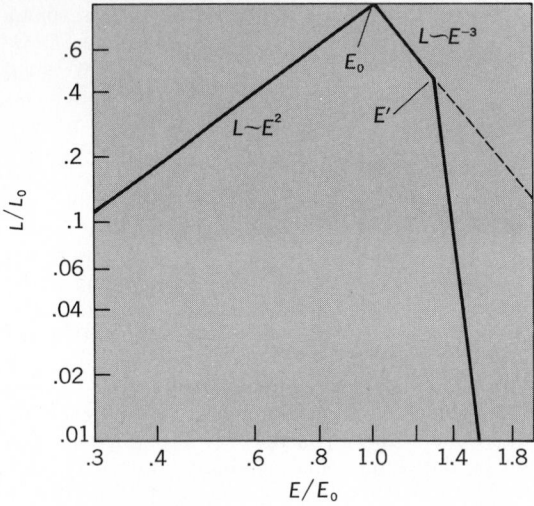

Fig. 25. Variation of luminosity with energy in an e^+e^- storage ring (log-log scale).

tions for rf cavities and interaction regions. The bending magnets guide the beam particles around the ring, and the focusing magnets drive them toward the equilibrium orbit around which they execute radial and vertical oscillations. The use of separate magnets for guiding and for focusing permits tight control of beam size to maximize the luminosity as the energy of operation is changed.

Injector. To achieve high luminosity, it is desirable to inject and accumulate e^- and e^+ with energies close to the desired collision energy. Linear accelerators and synchrotrons are used as injectors. The e^- are accelerated directly; e^+ are collected from a target bombarded with high-energy e^- and then accelerated. Accumulation occurs through the addition of particles first to the circulating e^+ beam, then to the circulating e^- beam until desired intensities have been reached. Typical accumulation times are seconds for e^- and minutes for e^+.

Vacuum chamber and system. Very low pressures of 10^{-9} and 10^{-10} torr (1 torr = 133 Pa) are required to prevent the outscattering of beam particles by collisions with gas molecules, which reduces storage times below useful values. Special techniques are used to achieve such high vacuum in the presence of high-power synchrotron radiation hitting the chamber walls.

The rf power system. The power from the rf system replaces the energy emitted as synchrotron radiation and accelerates the stored beams to the desired energy. In a multi-GeV storage ring, several megawatts of power at 400–500 MHz are typically delivered to several rf cavities located in the ring.

Beam structure. The stored beams consist of bunches (typically 5 cm long), since rf power is being supplied continuously. For a given number of stored particles, maximum luminosity is achieved with the smallest number of bunches in each beam according to Eq. (7). This number equals half the number of interaction regions. (For example, in Fig. 24 there are two interaction regions and one bunch circulating in each direction.)

Interaction region. Magnet-free straight sections at the center of which e^+e^- collisions occur are provided for the installation of detectors. Special magnets are usually installed on each side of the interaction region to focus the beams to a very small area at the interaction point so as to maximize the luminosity; this is called low beta insertion.

[KARL STRAUCH]

Proton storage rings. Proton storage rings consist of a pair of annular vacuum chambers in which high-energy protons can be stored and caused to collide nearly head-on. The motivation for using this technique is the same as for electron-positron storage rings. Namely, the colliding-beam storage ring permits studying reactions at energies that cannot be achieved with economically feasible conventional accelerators.

In 1980 there was only one proton storage ring device in operation—the Intersecting Storage Rings (ISR) at CERN near Geneva, Switzerland. Construction of a larger device, Project Isabelle, has been undertaken at Brookhaven National Laboratory, Upton, NY.

The ISR consists of two nearly circular rings arranged as in Fig. 26. Each ring consists of 132 magnets of a type similar to those used in a large proton synchrotron. The total circumference of each ring is nearly a kilometer. Protons of 26 GeV from the CERN proton synchrotron (PS) are injected into the rings. A "stacking" procedure permits accumulating protons from many PS pulses into the rings until currents of up to 50 amperes are stored. The entire stack of protons can then be slowly accelerated to an energy of 31 GeV. Collisions occur in all eight intersection regions (Fig. 27), permitting an extensive and varied research program. Experiments at this facility have made important contributions to high-energy physics, particularly in the study of total cross sections, the production of particles which have large transverse momenta, direct lepton production, and scaling laws in weak and electromagnetic interactions.

Limitations. The foremost technical problem facing the builders of proton storage rings is achieving currents of sufficient intensity and density to provide adequate reaction rates for the experiments. The ISR has realized a design luminosity of 5×10^{31} (cm^2 s)$^{-1}$, which exceeds the design luminosity by a factor of 5. From Eq. (7), it is clear that the most direct way to increase luminosity is to increase the current in the rings. As with positron-electron storage rings, there are rather fundamental considerations which limit the current that can be stacked. First, the current density of any accelerator is limited by fundamental thermodynamic arguments and, given economic limits on size of magnets and vacuum chambers, there is a limit on the total current that can be stacked in the storage ring. The second limit is imposed by self-fields from the beam. For the large charge density of the beam of a proton storage ring, there are electrostatic potentials of a few kilovolts acting on the beam caused by the beam itself (space charge forces). Similarly there are magnetic fields of a few oersteds from the beam current. These electromagnetic fields are sufficient to cause deterioration of the beam quality or even sudden loss of the beam through collective oscillations of the particles forced by the self-fields. The third limit, not anticipated during design and construction of the ISR, is set by a beam-induced pressure rise in the vacuum system. Ions formed by ionization of the residual gas by the beam are driven into the vacuum chamber walls by the electrostatic potential of the beam. The ion impact causes desorption of molecules from the wall surface. At a sufficiently large beam current, the result is an avalanchelike increase in pressure, causing loss of the beam. The fourth limit is imposed by the large stored energy of the beam. The stored energy in each of the beams in a proton storage ring can be many megajoules and can cause serious damage to components of the facility if dumped in an uncontrolled fashion.

The luminosity of a storge ring can also be increased by decreasing the beam size or reducing the crossing angle. These changes are done with special bending and focusing magnets in the intersection region. Such features will be designed into future storage rings, permitting considerably higher luminosity than can be achieved in the ISR at CERN. There is, however, a limit to how far this technique can be pushed. Just as the small size and small crossing angle increase the luminosity, they also enhance the electromagnetic interaction of one beam on another. The proton storage ring is much more sensitive to the resulting beam-beam instability than is the electron storage ring because in the latter device the diffusive growth of the beam is overcome by radiation damping, at least for slow rates of the diffusion.

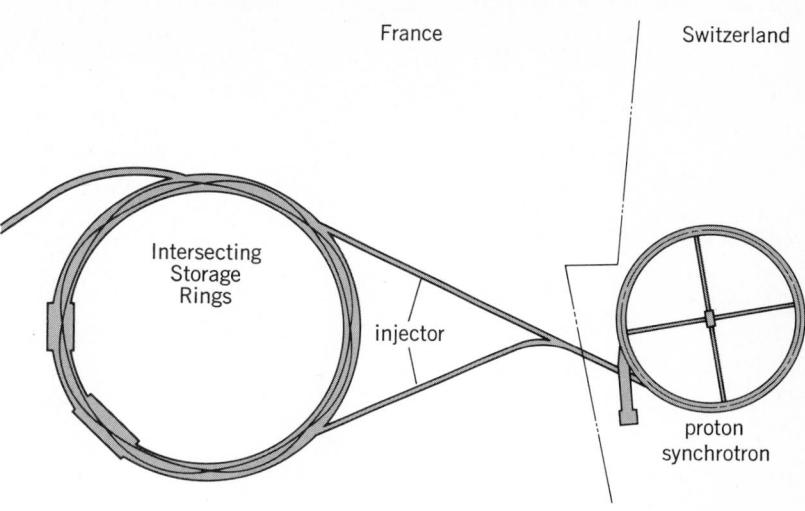

Fig. 26. A schematic layout of the intersecting storage rings at CERN and the injector.

Technology. The technology of constructing a proton storage ring is very similar to that for a proton synchrotron but with three important differences. The magnets must be of greater precision and stability and have an elaborate set of corrections to optimize stability of the proton beam against perturbations by space charge forces and

Fig. 27. Inside of the ISR tunnel, with most equipment installed during a late stage of construction. One of the intersection points is clearly visible. (*From K. Johnsen, Nucl. Instr. Meth., 108(2):205–223, 1973; Photo CERN*)

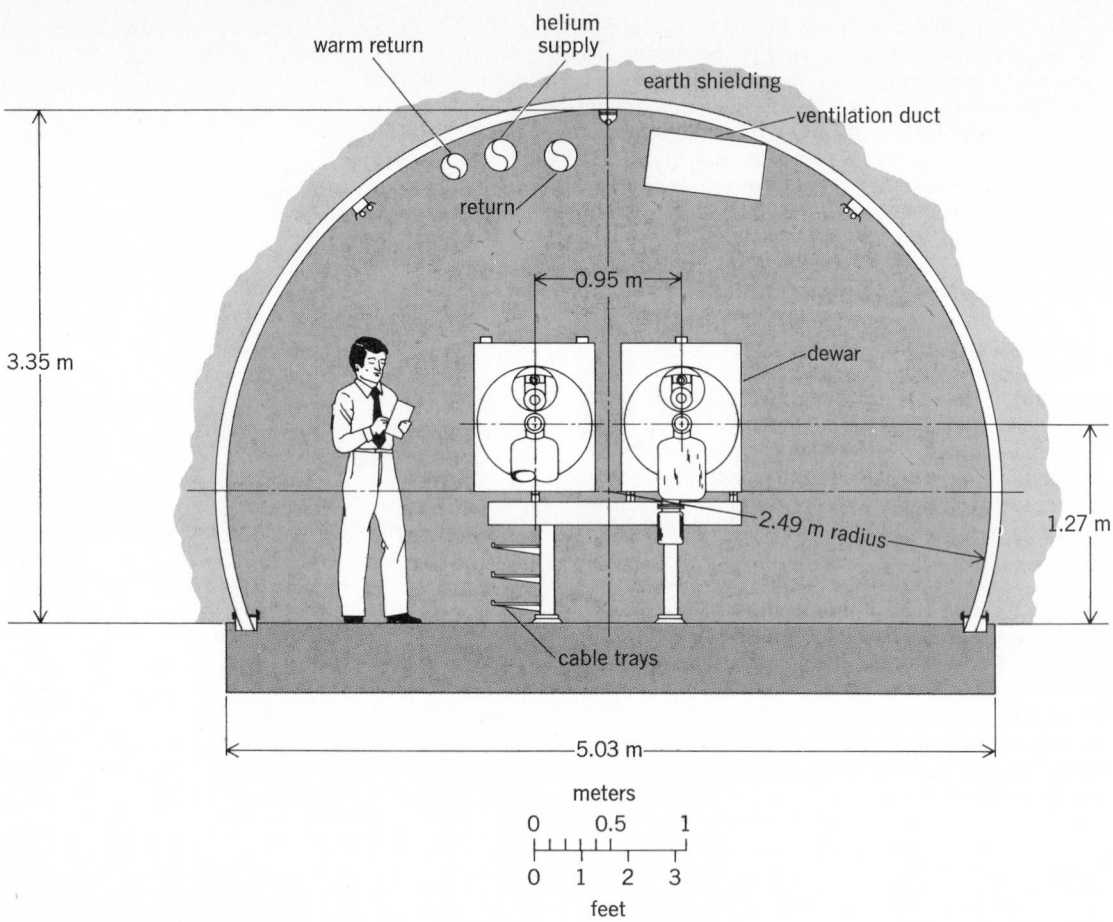

Fig. 28. Sketch of the cross section of the Isabelle project magnet enclosure.

imperfections in the magnets. Second, the vacuum requirements are very severe. A pressure of 10^{-10} torr or less is required to prevent excessive beam loss and to minimize the background in the experiments caused by beam particles colliding with residual gas molecules. The vacuum chamber also must be designed and constructed to minimize the pressure bump phenomenon described earlier. Finally, in a storage ring there is a much more intimate connection between the machine and the experiments than is the case for conventional accelerators. The experiments must be designed

Fig. 29. Isabelle magnet enclosure during an early stage of construction. (*Brookhaven National Laboratory*)

about the specific properties of the beams in the interaction region, and the machine builders must build the storage rings as if they are part of the experiments. In particular, the machine builders must pay particular attention to the problem of maintaining good access to the intersecting regions and keep these areas free of magnets, pumps, and other equipment incompatible with the experiments. The beam losses must be kept very small from all causes because the experiments, which are typically designed to study very low event rates, are extremely sensitive to background events associated with the lost particles. Loss rates of less than a part per million per minute have been achieved at the ISR. At this rate nearly half of the loss is associated with real beam-beam collisions. These loss rates are so low that it is possible to have continuous experimental physics runs of several days without refilling the rings.

The Isabelle project (Fig. 28) should extend the ISR technology to energies of 800 GeV in the center of mass (400 GeV in each ring). To achieve such an energy with a conventional accelerator, a ring of about 1000 mi (1600 km) diameter would be required. Isabelle is to consist of two rings of 3.8 km circumference, which intersect in six regions. Thirty-GeV protons from the Brookhaven AGS proton synchrotron will be injected and stacked into the Isabelle rings until currents of 8 amperes are achieved. The stacks will then be slowly accelerated to 400 GeV and made to collide for high-energy physics experiments. Luminosities in excess of 2×10^{32} (cm² s)$^{-1}$ are expected.

Superconducting magnets are used in Isabelle because they consume much less electrical power. Since superconducting magnets operate at much higher magnetic fields (5 teslas) than conventional magnets, the circumference is correspondingly reduced, which is beneficial in limiting construction costs and improving performance. More than 1000 superconducting magnets are required. The largest helium refrigerator ever built is required to keep the magnets at a temperature below 4 K. Construction (Fig. 29) is to be completed in 1985–1986.　　　　　　　　　[MARK BARTON]

ELECTRON RING ACCELERATOR

An electron ring accelerator (ERA) is a device for imparting to protons or other positive ions kinetic energies of several to several hundred MeV per nucleon by the electron drag force of stable ringlike configuration of electrons that is accelerated to the requisite velocity.

Collective acceleration. Conventional particle accelerators are subject to limitations in the strengths of the fields that can feasibly be employed to guide and accelerate the desired particles, and accordingly may be massive and expensive. This situation has motivated the development of techniques for accelerating a small group of particles by means of their interaction with another group of charges or with some form of plasma wave. This "collective acceleration" employs fields developed by charged particles within the accelerator itself. Such fields can be quite large and are not subject to the technological limitations that restrict externally applied fields. *See* WAVES AND INSTABILITIES IN PLASMAS.

Electron ring collective field. A direct and attractive method of collective acceleration is employed in the electron ring accelerator. The fields of significance in the ERA are produced by a slender ring of electrons circulating in a magnetic field. The electron ring is formed by injecting an electron beam of several hundred amperes into a magnetic field that is rapidly deformed to trap one or more injected turns into closely spaced circular orbits and that then is further pulsed to shrink the ring dimensions to suitable size. The pulsed magnetic field serves both to increase the energy of the circulating electrons (through the transformer action of induced electric fields) and to reduce the major and minor dimensions of the ring (compression). In alternative experimental devices (at the University of Maryland) these rings are formed by injecting a hollow relativistic electron beam into a cusp-shaped magnetic field, wherein the forward motion of the electrons is converted into circulatory motion.

At the edge of the ring of major and minor radii R and a (R much greater than a), formed of N circulating electrons of charge e, electrostatic fields as great as those given by Eq. (9) arise. Thus a field

$$E = \frac{Ne}{4\pi^2 \epsilon_0 R a} \tag{9}$$

of 7.3×10^7 volts/m is obtained at the edge of a ring of 10^{13} electrons, with $R = 0.025$ m and $R/a = 10$. This electrostatic field binds the ions to the ring, since they are formed in the ring volume by electron collisional ionization of neutral gas introduced into the vacuum chamber. The trapped ions can be accelerated with the electron ring, if this is accelerated in a direction along the magnetic field lines. The ions obtain substantial kinetic energies, which are in the initial stages and for small ion loading larger by a factor $M/\gamma m_0$ than the energy associated with the transverse velocity imparted to the individual electrons, where M/m_0 denotes the ratio of ion to electron rest mass and γ is the relativistic factor for the electrons. Hence the effectiveness of the externally applied acceleration field can be amplified by this factor—for example, by 60.4 for proton acceleration in a ring of 15-MeV electrons ($\gamma = 30.4$). The acceleration mechanism, moreover, is applicable to ions of any desired type, and indeed ions of differing charge-to-mass ratio can be simultaneously accelerated.

Figure 30 shows the cross section of an experimental electron ring compressor. The compression of the electron rings, which are formed by injecting electrons perpendicular to the plane of the figure into the median plane of the compressor, is obtained by the sequentially pulsed coil sets 1–3. The magnetic field thus forms a "magnetic bottle." The equilibrium and the single-particle stability of the ring are ensured by the focusing force that is achieved by the field index n, defined by Eq. (10),

$$n = -\frac{r}{B_z} \frac{\partial B_z}{\partial r} \tag{10}$$

where r is the distance from the center of the ring and B_z is the magnetic induction perpendicular to the ring. The focusing is necessary to compensate for the difference between electrostatic repulsion and magnetic attraction of the ring electrons. The field index n must lie between 0 and 1. The electron ring is released from the magnetic bottle by creating an unbalance between the currents in the left-

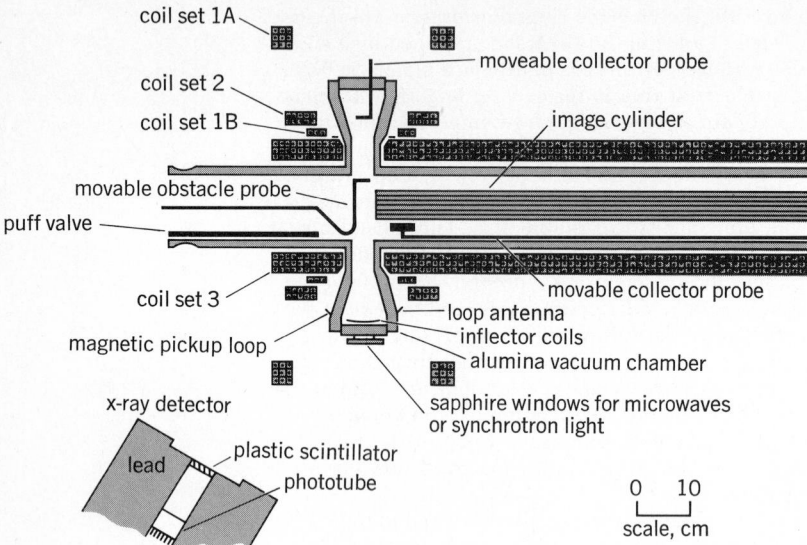

coil set 1A
moveable collector probe
coil set 2
coil set 1B
image cylinder
movable obstacle probe
puff valve
coil set 3
movable collector probe
loop antenna
magnetic pickup loop
inflector coils
alumina vacuum chamber
sapphire windows for microwaves or synchrotron light
x-ray detector
lead
plastic scintillator
phototube

0   10

scale, cm

Fig. 30. Cross section of an experimental electron ring compressor employed at the Lawrence Berkeley Laboratory. (*Lawrence Berkeley Laboratory Report 2090*)

and right-hand portions of the coils, and gets into the long solenoid to permit study of collective acceleration. The "puff valve" introduces a short gas burst for ion loading of the electron rings. The probes, loops, and x-ray detector provide information concerning the performance of the entire device. *See* PLASMA PHYSICS.

Ring stability. Potentially destructive instabilities can arise in the motion of the individual electrons and in collective motion of the ring as a whole. The resonant excitation of single-particle transverse oscillations can be avoided during compression by selecting suitable values for the field index n, by controlling other characteristics of the field, and by passing rapidly during compression through the most dangerous values of the field index. When the ring is released into the acceleration column, the axial magnetic field no longer provides a focusing action, and the beam must then be preserved by the agency of image fields arising from nearby conducting (or dielectric) ma-

terial or through the imposition of a supplemental azimuthally directed magnetic field B_ϕ.

Collective instabilities have proven to be more troublesome in achieving high-quality electron rings. An azimuthal instability, which can lead to bunching of the electron beam, is of particular importance. Effective means for the suppression of this instability include the provision of conducting surfaces close to the ring and the presence of an appreciable spread in the energies of the circulating electrons (Landau damping). An energy spread acts to increase the radial minor dimension of the ring (thus diminishing its maximum electric field), however, and may preclude the placement of conducting surfaces close to the ring. Additional instabilities may result from oscillations of the ring with respect to the trapped ions, but are subject to some control through adjustment of the ion abundance. A theoretical analysis of instabilities suggests that the dominant azimuthal instability should not preclude the attainment of peak fields of approximately $80B$ MV/m (effective acceleration fields of about $40B$ MV/m) in a ring situated in a guide field of B tesla. It is noteworthy that this limit increases in direct proportion to B.

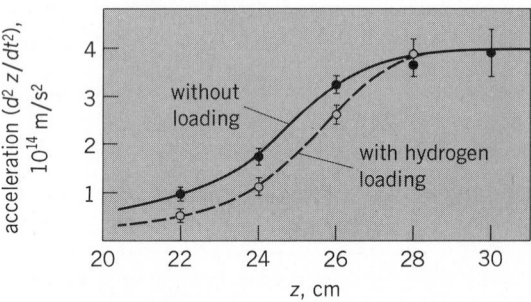

Fig. 32. Acceleration of the electron rings with and without hydrogen loading measured along the distance z from the compression plane. (*From U. Schumacher et al., Collective acceleration of protons and helium ions in the Garching ERA, IEEE Trans. Nucl. Sci., NS-22(3):989–991, 1975*)

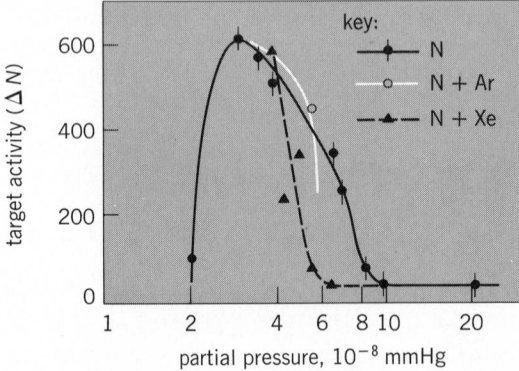

Fig. 31. Dependence of the target activity of deuterium on the partial pressures of the gases nitrogen, argon, and xenon.10^{-8} mmHg$= 1.33 \times 10^6$ Pa. (*From G. V. Dolbilov et al., Experiments on nitrogen ion acceleration in prototype of collective heavy ion accelerator, in N. Rostoker and M. Reiser, eds., Collective Methods of Acceleration, Harwood Academic Publishers, pp. 83–95, 1979*)

Developmental work with electron rings has employed techniques that act to suppress instabilities in an electron ring compressor. The effect of single-particle resonances has been greatly reduced by the use of a fast compression cycle (compression time less than 10 μs, and all dangerous resonances traversed in less than 100 ns). Collective radial ion-electron oscillations also were suppressed by use of shaped conducting surfaces designed to modify the field index n and the derivative $\partial n/\partial r$ at the end of compression. An image cylinder (Fig. 30), a "squirrel-cage" structure formed of longitudinal conducting strips near the ring, has been used to provide image fields designed to preserve the integrity of the beam during the initial portion of the acceleration period. The longitudinal bunching instability has been greatly reduced by carefully designed conducting surfaces close to the ring.

Methods of acceleration. The acceleration of the electron ring and ions attached to it requires the release of the ring from the magnetic bottle in

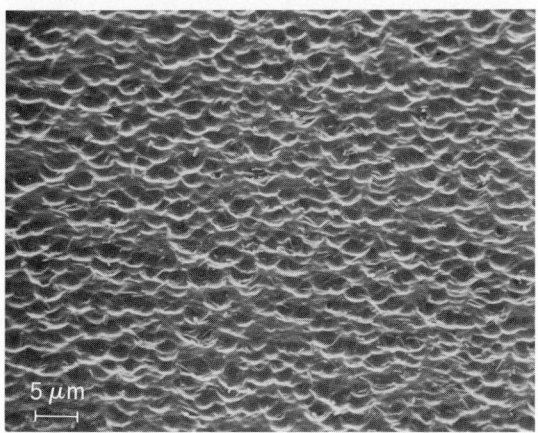

Fig. 33. Electron scanning microscope picture of a cellulose nitrate foil area bombarded with collectively accelerated helium ions. *(From U. Schumacher et al., Collective acceleration of ions by relativistic electron rings, Phys. Lett., 51A:367–369, 1975)*

which it is held and the provision of fields that will accelerate the ring axially down an acceleration column. An axial magnetic field, together with suitable provisions for maintenance of ring stability, must be present throughout the length of this column in order to preserve the electron ring configuration. For acceleration to high energies the use of a sequence of rf or pulsed cavities appears necessary.

Initial experiments have focused, however, on a simpler method of acceleration wherein the force that accelerates the ring arises from a small radial component B_r of the static magnetic field in the acceleration column (magnetic expansion acceleration). With this method the axial velocity imparted to the electrons arises specifically from the Lorentz force $ev_\phi B_r$ (where v_ϕ is the azimuthal component of the velocity) acting on them, and the axial motion of the electrons and ions arises entirely at the expense of the energy initially associated with the azimuthal motion of the electrons.

Results of experiments. Two groups have reported evidence for ion acceleration in short magnetic-expansion acceleration columns. A group at Dubna, in the Soviet Union, reported in 1971 the production of He^{2+} ions of 30 MeV energy by magnetic-expansion acceleration through a distance estimated as 0.4 m. In 1978 this group succeeded in accelerating nitrogen ions with heavily loaded electron rings. An ion energy gain of about 4 MeV/nucleon-m over an acceleration length of about half a meter for ion numbers of as much as $(5 \pm 2) \times 10^{11}$ ions per cycle was achieved. The acceleration parameters were measured by the method of activation analysis. Figure 31 shows the dependence of the target activity ΔN of a deuterium target on the partial pressure of nitrogen, argon, and xenon mixtures, indicating that a certain pressure and hence ion loading range resulted in sufficient focusing and collective acceleration conditions.

The well-monitored experiments at the Max Planck Institute (West Germany) with rings of relatively low holding power produced both hydrogen and helium ions of approximately 400 keV kinetic energy by acceleration through a few centimeters.

The acceleration of the electron rings with and without hydrogen or helium loading was measured with the aid of magnetic probes. Figure 32 shows measurements of the acceleration of electron rings with and without hydrogen loading over the axial distance z from the compression plane. The increase of the acceleration with z can be used to determine the holding power of the electron rings from the maximum acceleration up to which the difference in the ring inertia persists. The holding power of this experiment (performed in 1975) corresponds to 3 to 4 MV/m, a value which was subsequently slightly improved. The collective acceleration of helium ions was confirmed by nuclear track registration in thin cellulose nitrate foils. Figure 33 shows an electron scanning microscope picture of a foil area bombarded with collectively accelerated helium ions, the tracks made visible by etching after exposure.

In addition to its potential for achieving collective-field acceleration of ions, a compressed electron ring offers a unique means for the production and spectroscopic study of highly stripped ions as well as for an intense synchrotron radiation source.

[UWE SCHUMACHER]

Bibliography: D. A. Bromley (ed.), *Large Electrostatic Accelerators*, 1974; CERN, *The 300 GeV Programme*, 1972; G. E. Fischer and R. T. Nelson, *Catalogue of High-Energy Accelerators*, Stanford Linear Accelerator Center, 1974; H. Fraunfelder and E. M. Henley, *Subatomic Physics*, 1974; R. Gourian, *Particles and Accelerators*, 1967; H. Hahn, M. Month, and R. R. Rau, Proton-proton intersecting storage accelerator facility Isabelle at the Brookhaven National Laboratory, *Rev. Mod. Phys.*, 49:625–679, 1977; K. Johnson, The CERN intersecting storage rings, *Nucl. Instr. Met.*, 108(2): 205–223, 1973; D. Keefe, Collective-effect accelerators, *Sci. Amer.*, 226(4):22–33, 1972; P. M. Lapostolle and A. L. Septier, *Linear Accelerators*, 1970; M. S. Livingston and J. P. Blewett, *Particle Accelerators*, 1966; 1971 Particle Accelerator Conference, *IEEE Trans. Nucl. Sci.*, NS-18, no. 3, 1971, and similar conferences every 2 years following; G. K. O'Neill, Particle storage rings, *Sci. Amer.*, 215(5):107–116, 1966; *Physics in Perspective*, vol. 11, pt. A: *The Core Subfields of Physics*, Physics Survey Committee, National Academy of Sciences, 1972; *Proceedings of the 9th International Conference on High Energy Accelerators*, Stanford Linear Accelerator Center, 1974; *Proceedings of the 7th International Conference on Cyclotrons and Their Applications*, Aug. 19–22, 1975; F. K. Richtmyer, E. H. Kennard, and J. N. Cooper, *Introduction to Modern Physics*, 1969; N. Rostoker and M. Reiser (eds.), *Collective Methods of Acceleration*, 1979; M. Sands, *The Physics of Electron Storage Rings: An Introduction*, Stanford Linear Accelerator Center, Rep. no. 121, 1970; W. Schnell, Report on the ISR, *IEEE Trans. Nucl. Sci.*, NS-22(5):1358–1362, 1975; U. Schumacher, Collective ion acceleration with electron rings, in G. Höhler (ed.), *Springer Tracts in Modern Physics*, vol. 84, pp. 145–232, 1979; H. Semat and J. Albright, *Introduction to Atomic and Nuclear Physics*, 1972; V. I. Veksler et al., Collective linear acceleration of ions, *Proceedings of the 6th International Conference on High Energy Accelerators*, Cambridge Electron Accelerator CEAL-2000, pp.

289–304, 1967; R. R. Wilson and R. M. Littauer, *Accelerators, Machines of Nuclear Physics*, 1960; R. R. Wilson, The Batavia accelerator, *Sci. Amer.*, 230(2):72–83, 1974; R. R. Wilson, Fantasies of future Fermilab facilities, *Rev. Mod. Phys.*, 51:259–273, 1979; R. R. Wilson, The next generation of particle accelerators, *Sci. Amer.*, 242(1):42–57, January 1980.

Particle detector

A device used to detect and measure radiations characteristically emitted in nuclear processes, including γ- or x-rays, lightweight charged particles (electrons or positrons), nuclear constituents (neutrons, protons, and heavier ions), and subnuclear constituents such as mesons. The device is also known as a radiation detector. Since human senses do not respond to these types of radiation, detectors are essential tools for the discovery of radioactive minerals, for all studies of the structure of matter at the atomic, nuclear, and subnuclear levels, and for protection from the effects of radiation. They have also become important practical tools in the analysis of materials using the techniques of neutron activation and x-ray fluorescence analysis.

Classification by use. A convenient way to classify radiation detectors is according to their mode of use: (1) For detailed observation of individual photons or particles, a pulse detector is used to convert each such event (that is, photon or particle) into an electrical signal. (2) To measure the average rate of events, a mean current detector, such as an ion chamber, is often used. Radiation monitoring and neutron flux measurements in reactors generally fall in this category. Sometimes, when the total number of events in a known time is to be determined, an integrating version of this detector is used. (3) Position-sensitive detectors are used to provide information on the location of particles or photons in the plane of the detector. (4) Track-imaging detectors image the whole three-dimensional structure of a particle's track. The output may be recorded by immediate electrical readout or by photographing tracks as in the bubble chamber. (5) The time when a particle passes through a detector or a photon interacts in it is measured by a timing detector. Such information is used to determine the velocity of particles and when observing the time relationship between events in more than one detector.

Role of ionization. Any radiation-induced effect in a solid, liquid, or gas can be used in a detector. To be useful, however, the effect must be directly or indirectly interpretable in terms of either the quantity or quality (that is, type, energy, and so on) of the incident radiation or both. The ionization produced by a charged particle is the effect most commonly employed.

Gas ionization detectors. In the basic type of gas ionization detector, an electric field applied between two electrodes separates and collects the electrons and positive ions produced in the gas by the radiation to be measured. Depending on the intensity of the electric field, the charge signal in the external circuit may be equal to the charge produced by the radiation, or it may be much larger. In a proportional counter, the output charge is larger than the initial charge by a factor called the gas amplification. A Geiger-Müller counter provides still larger signals, but each signal is independent of the original amount of ionization. All three types of gas ionization detectors can be used as pulse detectors, mean current detectors, or, with an indicator such as a quartz-fiber electroscope, as integrating detectors.

Position-sensitive detectors and track-imaging detectors are nearly all based on the ionization process. Multiwire proportional chambers and spark chambers are position-sensitive adaptations of gas detectors. The signal division or time delay that occurs between the ends of an electrode made of resistive material is sometimes used to provide position sensitivity in gas and semiconductor detectors. Track-imaging detectors rely on a secondary effect of the ionization along a particle's track to reveal its structure. In Wilson cloud chambers, ionization triggers condensation along particle tracks in a supercooled vapor; bubble formation in liquids is the basis for operation of bubble chambers. A secondary effect of ionization is also employed in photographic emulsions used as radiation detectors where ionization triggers the formation of an image. *See* BUBBLE CHAMBER.

Semiconductor detectors. In this type of detector, a solid replaces the gas of the previous example. The "insulating" region (depletion layer) of a reverse-biased *pn* junction in a semiconductor is employed. Choice of materials is very limited; very pure single crystals of silicon or germanium are presently the only fully suitable materials, although other semiconductors can be used in noncritical applications. Collection of the primary ionization is normally used, but an avalanche mode is sometimes employed in which an intense electric field causes charge multiplication. Although semiconductor devices are mostly used as pulse detectors, mean-current and integrating modes are possible when the significant leakage currents of *pn* junctions can be tolerated.

Since solids are approximately 1000 times denser than gases, absorption of radiation can be accomplished in relatively small volumes. A less obvious but fundamental advantage of semiconductor detectors is the fact that much less energy is required (~ 3 eV) to produce a hole-electron pair than that required (~ 30 eV) to produce an ion-electron pair in gases. This results in better statistical accuracy in determining radiation energies. For this reason, semiconductor detectors have become the main tools for nuclear spectroscopy, and they have also made neutron activation analysis and x-ray fluorescence analysis of materials practical tools of great value.

Scintillation detector. In addition to producing free electrons and ions, the passage of a charged particle through matter temporarily raises electrons in the material into excited states. When these electrons fall back into their normal state, light may be emitted and detected as in the scintillation detector. The early scintillation detectors consisted of a layer of powder (zinc sulfide, for example) that, when struck by charged particles, produced light flashes, which were observed by eye and counted. The meaning of the term "scintillation detector" has changed with time to refer to the combination of a scintillator and a photomultiplier tube that converts light scintillations into signals that can be processed electronically. Various organic and inorganic crystals, plastics, liquids, and glasses are used as scintillators, each having particular virtues in regard to radiation absorption, speed, and light output. *See* LIQUID SCINTILLATION DETECTOR.

Neutral particles. Neutral particles, such as neutrons, cannot be detected directly by ionization. Consequently, they must be converted into charged particles by a suitable process and then observed by detecting the ionization caused by these particles. For example, high-energy neutrons produce "knock-on" protons in collisions with light nuclei, and the protons can be detected. Slow neutrons are usually detected by using a nuclear reaction in which the neutron is captured and a charged particle is emitted. For example, in boron trifluoride (BF_3) detectors, neutrons react with boron to produce alpha particles which are detected.

Other detector types. Although ionization detectors dominate the field, a number of detector types based on other radiation-induced effects are used. Notable examples are: (1) transition radiation detectors, which depend on the x-rays and light emitted when a particle passes through the interface between two media of different refractive indices; (2) track detectors, in which the damage caused by charged particles in plastic films and in minerals is revealed by etching procedures; (3) thermo- and radiophoto-luminescent detectors, which rely on the latent effects of radiation in creating traps in a material or in creating trapped charge; and (4) Cerenkov detectors, which depend on measurement of the light produced by passage of a particle whose velocity is greater than the velocity of light in the detector medium. *See* CERENKOV RADIATION.   [FRED S. GOULDING]

Bibliography: G. Bertolini and A. Coche, *Semiconductor Detectors*, 1968; J. B. Birks, *The Theory and Practice of Scintillation Counting*, 1964; J. Litt and R. Meunier, Cerenkov counter technique in high-energy physics, *Annu. Rev. Nucl. Sci.*, Vol. 23:1, 1973; P. B. Price and R. L. Fleischer, Solid state track detector applications, *Annu. Rev. Nucl. Sci.*, 21:296, 1971; R. P. Shutt (ed.), *Bubble and Spark Chambers*, vols. 1 and 2, 1967; A. H. Snell (ed.), *Nuclear Instruments and Their Uses*, vol. 1, 1962.

Particle track etching

A technique of selective chemical etching to reveal tracks of heavy nuclear particles in a wide variety of solid substances. Developed in order to see fossil particle tracks in extraterrestrial materials, the technique finds application in many fields of science and technology.

Identification of nuclear particles. An etchable track is produced if the charged particle has a sufficiently high radiation-damage rate and if the damaged region in the solid is permanently localized. Thus only highly ionizing particles are detectable; only nonconductors record tracks; and radiation-sensitive plastics can detect lighter particles than can radiation-insensitive minerals and glasses. The conical shape of the etched track depends on the ratio of the rate of etching along the track to the bulk etching rate of the solid. Careful measurements with accelerated ions of known atomic number Z and velocity $v = \beta c$ (where $c =$ velocity of light) have shown that this ratio is an increasing function of Z/β. Measurements of the shapes of etched tracks thus serve to identify particles (Fig.

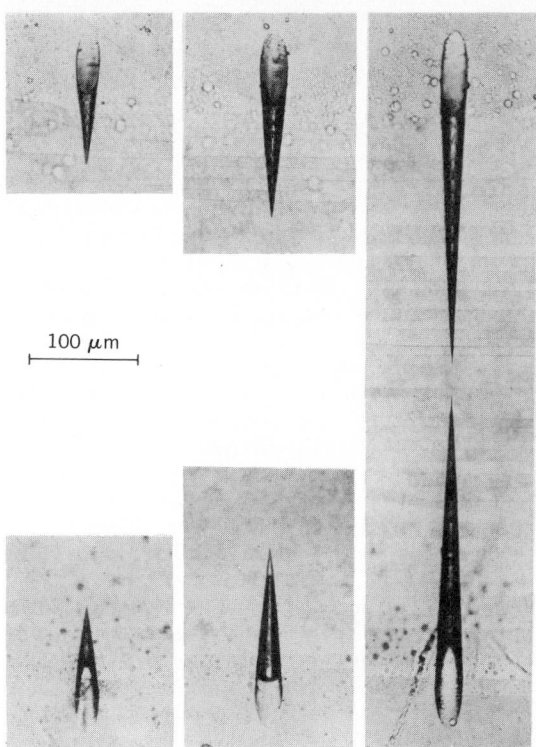

Fig. 1. Increasing radiation damage produced by a high-energy uranium nucleus in the cosmic radiation as it penetrates a thick stack of Lexan plastic sheets and slows down. The lengths of the etched cones in these three sheets, taken at intervals of several millimeters from within the stack, permit the atomic number and energy of the nucleus to be determined. Lower images show the exit points of the tracks.

1). The most sensitive detector, a plastic known as CR-39, detects particles with Z as low as 1 (Fig. 2) provided $Z/\beta \gtrsim 10$. The clarity and contrast of the images and its high sensitivity make CR-39 a very attractive track-etch detector for nuclear physics, cosmic-ray research, element mapping, personnel neutron dosimetry, and many other applications. Minerals, on the other hand, are insensitive to par-

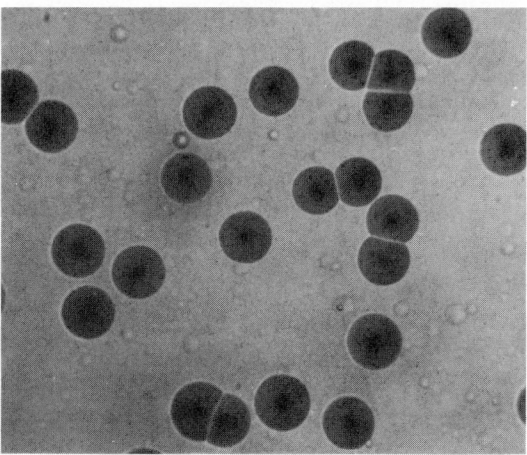

Fig. 2. Etched conical pits in a sheet of allyl diglycol carbonate (CR-39 plastic) irradiated at normal incidence with 60-MeV alpha particles. (*From P. B. Price et al., Do energetic heavy nuclei penetrate deeply into the Earth's atmosphere?, Proc. Nat. Acad. Sci., 77:44–48, 1980*)

ticles with $Z/\beta \lesssim 150$ and are therefore useful in recording rare, very heavily ionizing, relatively low-energy nuclei with $Z \gtrsim 25$. See CHARGED PARTICLE BEAMS.

Solar and galactic irradiation history. The lunar surface, meteorites, and other objects exposed in space have been irradiated by charged particles from a variety of sources in the Sun and the Galaxy. The particles of lowest energy are produced by the expanding corona of the Sun—the solar wind. Arriving in prodigious numbers, but penetrating only some millionths of a centimeter, the solar wind particles quickly produce an amorphous radiation-damaged layer on crystalline grains. At depths from about a thousandth of a millimeter to about 1 mm, individual tracks produced in solar flares (sporadic energetic outbursts on the Sun) can be resolved by electron microscopy or sometimes by optical microscopy. At depths greater than 1 mm, most of the particle tracks are produced by heavy nuclei in the galactic cosmic radiation.

Comparison of fossil particle tracks in lunar rocks and meteorites with spacecraft measurements of present-day radiations has established that solar flares and galactic cosmic rays have not changed over the last 2×10^7 years—the typical time a lunar rock exists before being shattered by impacting interplanetary debris. Observations of grains in stratified lunar cores and lunar and meteoritic breccias (Fig. 3) enable the particle track record to be extended back more than 4×10^9 years in time. Breccias, which are complex grain assemblages, often contain grains that have high solar flare track densities on their edges, indicating exposure to free space prior to breccia formation. Dating work, using spontaneous-fission tracks, indicates that some of the breccias were assembled soon after the beginning of the solar system some 4.6×10^9 years ago. The study of the time history of energetic radiations in space, using lunar samples and meteorites, elucidates various dynamic processes such as rock survival lifetimes, microerosion of rocks, and the formation and turnover rates of planetary regoliths.

Nucleosynthesis. In terrestrial crystals, which are well shielded from external radiations, the dominant source of tracks is the spontaneous fission of ^{238}U. Certain meteorites and lunar rocks (Fig. 3a) contain additional fission tracks due to the presence of ^{244}Pu (half-life, 8×10^7 years) when the crystal was formed. After corrections are made for chemical fractionation of Pu and U, the data indicate that a large spike of newly synthesized elements was produced at the time of formation of the Galaxy, followed by a period of continuous synthesis and relatively rapid mixing.

Current solar and galactic irradiation. Studies of tracks in a piece of glass from the *Surveyor 3* spacecraft after a 2.6-year exposure on the lunar surface, and of tracks in plastic detectors exposed briefly above the Earth's atmosphere in rockets, have led to the surprising discovery that the Sun preferentially ejects heavy elements in its flares rather than an unbiased sample of its atmosphere. These data for a nearby, well-known cosmic accelerator provide constraints on the interpretation of data on the relative abundances of elements in galactic cosmic rays.

The existence of galactic cosmic rays with atomic number greater than 30 was discovered in 1966 when fossil particle tracks were first studied in meteorites. Since then many stacks of various types of plastics and nuclear emulsions up to 20 m² in area have been exposed in high-altitude balloons and in Skylab in order to map out the composition of the heaviest, rarest cosmic rays. Several particles heavier than uranium have been detected, indicating that cosmic rays originate in sources where synthesis has proceeded explosively beyond uranium. Exposures of giant detectors about 100 m² in area for a year in space are planned. Hybrid detectors using stacks of track-recording plastics to measure range, and photomultiplier tubes to measure light from Cerenkov radiation and from scintillation detectors, have made it possible to determine the relative abundances of the isotopes of very heavy elements in the cosmic rays. These data bear both on the history of cosmic rays and on nucleosynthesis in their sources.

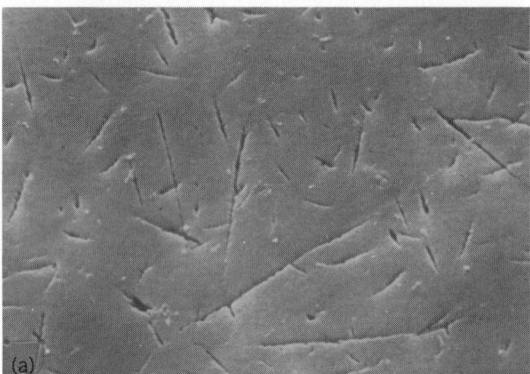

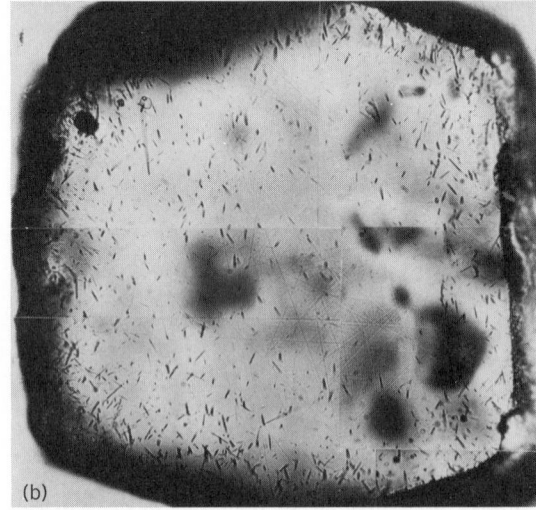

Fig. 3. Etched fossil particle tracks. (*a*) Tracks from spontaneous fission of ^{238}U and ^{244}Pu in a zircon crystal from a lunar breccia, seen by scanning electron microscopy (*from D. Braddy et al., Crystal chemistry of Pu and U and concordant fission track ages of lunar zircons and whitlockites, Proceedings of the 6th Lunar Science Conference, pp. 3587–3600, 1975*). (*b*) Tracks of energetic iron nuclei from solar flares, showing a decreasing concentration from edges to center of a 150-μm-diameter olivine crystal from a carbonaceous chondritic meteorite. The irradiation occurred about 4×10^9 years ago, before the individual grains were compacted into a meteorite (*from P. B. Price et al., Track studies bearing on solar-system regoliths, Proceedings of the 6th Lunar Science Conference, pp. 3449–3469, 1975*).

Nuclear and elementary particle physics.
Unique advantages of etched-track detectors are
their ability to distinguish heavy-particle events in
a large background of lightly ionizing radiation and
their ability to detect individual rare events by a
specialized technique such as electric-spark scanning or ammonia penetration through etched
holes. These advantages have permitted such advances as the measurement of very long fission
half-lives; the discovery of ternary fission; the determination of fission barriers; the production of
numerous isotopes of several far-transuranic elements; the discovery of several light, neutron-rich
nuclides such as ^{20}C at the limit of particle stability; and highly sensitive searches for magnetic
monopoles, superheavy elements, and anomalously dense nuclear matter in nature and in accelerators. *See* MAGNETIC MONOPOLES; NUCLEAR FIS-
SION; SUPERTRANSURANICS.

Geochronology. The spontaneous fission of ^{238}U,
present as a trace-element purity, gives tracks that
can be used to date terrestrial samples ranging
from rocks to human artifacts. Because fission
tracks are erased in a particular mineral at a well-
defined temperature (for example, 100°C in apa-
tite), one can use the apparent fission-track ages as
a function of distance from the heat source to measure the thermal (tectonic) history of regions.
Examples are the rate of sea-floor spreading (about
1 cm per year) and the surprisingly rapid rate of
uplift of the Alps (0.3–1.4 mm per year) and of the
Wasatch range in Utah (0.01–0.4 mm per year).
Because of its very high uranium concentration
(typically a few hundred parts per million), a few
tiny zircon crystals less than 0.1 mm in diameter
suffice to give a fission track age of sedimentary
volcanic ash and thus to determine absolute ages
of stratigraphic boundaries. Occasional, anomalously low fission-track ages can sometimes be
used to locate valuable ore bodies and even petroleum.

Geophysics and element mapping. In these
applications a track detector placed next to the
material being studied records the spatial distribution of certain nuclides that either spontaneously
decay by charged particle emission or are induced
to emit a charged particle by a suitable bombardment. A resolution of a few micrometers is easily
attained. Such micromaps make it possible to
identify radionuclides in atmospheric aerosols, to
measure the distribution and transport of radon,
thorium, and uranium, and to measure sedimenta-
tion rates at ocean floors. Thermal neutrons, readily available in a reactor, can be used to map ^{235}U
(via fission), ^{10}B (via alpha particles), ^{6}Li (via
tritons), ^{14}N (via protons), and several other nuclides. Deuterium, a tracer in biological studies,
and lead and bismuth can also be mapped by using
different irradiations.

Practical applications. Filters are produced by
irradiating thin plastic sheets with fission fragments and then etching holes to the desired size.
Uses include biological research, wine filtration,
and virus sizing. In virus sizing a single hole is
formed that separates two halves of a conducting
solution. When a virus or other tiny object passes
through the hole, the resistance increases drastically, and the size, shape, and speed of the object
can be determined by analyzing the electric signal.

There are numerous other uses. A uranium ex-

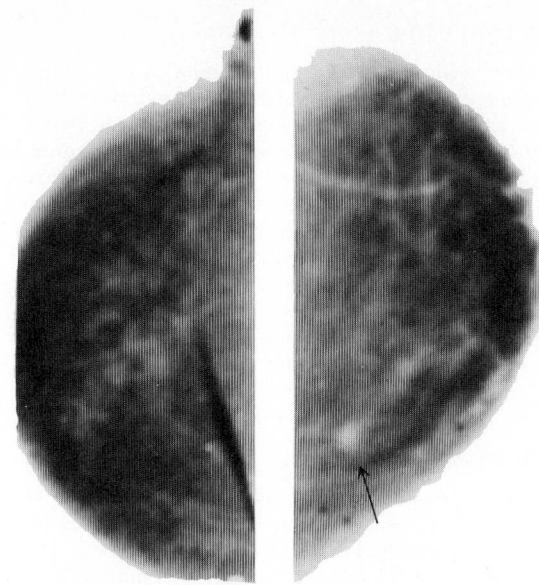

Fig. 4. Reconstructed images of right and left breasts
of a patient taken by using a 3600-MeV carbon-ion beam
at the Lawrence Berkeley Laboratory Bevalac accelerator. (*From C. A. Tobias et al., Lawrence Berkeley Laboratory*)

ploration method relies on a survey of radon ema-
nation, as measured by alpha-particle tracks in
plastic detectors, to locate promising locations in
which to drill. CR-39 plastic detectors are used in a
Fresnel zone-plate imaging technique to make
high-resolution images of the thermonuclear burn
region in laser fusion experiments. Plastic detectors are also used in conjunction with a beam of
high-energy heavy ions to take radiographs of cancer patients that reveal details not detectable in x-
rays. (Figure 4 shows reconstructed images of right
and left breasts of a patient taken by using a 3600-
MeV carbon-ion beam at the Lawrence Berkeley
Laboratory Bevalac accelerator. A single beam
pulse is passed through each breast as it is immersed in water, and the stopping points of the
ions are recorded in a stack of 30 cellulose nitrate
sheets, each 0.025 cm thick, located behind the
water bath. After the sheets are etched, the information on each sheet is converted to digital data
by a scanning system, and two images capable of
revealing slight differences in density are displayed on a television screen. The white spot indicated by an arrow in the right radiograph is a carci-
noma. The patient dose for such images is about 20
to 100 millirads (200 to 1000 micrograys), considerably lower than the usual dose from x-ray mammo-
graphic imaging.

The journal *Nuclear Track Detection* is devoted
to techniques and applications of particle track
etching.

[P. BUFORD PRICE]

Bibliography: B. G. Cartwright, E. K. Shirk, and
P. B. Price, A nuclear-track-recording polymer of
unique sensitivity and resolution, *Nucl. Instr.
Meth.*, 153:457–460, 1978; R. L. Fleischer, Where
do nuclear tracks lead?, *Amer. Sci.*, 67:194–203,
1979; R. L. Fleischer, P. B. Price, and R. M. Walker, *Nuclear Tracks in Solids*, 1975; R. M. Walker,
Interaction of energetic nuclear particles in space
with the lunar surface, *Annu. Rev. Earth Planet.
Sci.*, 3:99–128, 1975.

Pascal's law

A law of physics which states that a confined fluid transmits externally applied pressure uniformly in all directions. Blaise Pascal, using the mercury-column barometer of Evangelista Torricelli, demonstrated the decrease in atmospheric pressure with increasing height and determined that atmospheric force at a point exerted equal pressure in all directions. More exactly, in a static fluid, force is transmitted at the velocity of sound throughout the fluid. The force acts normal to any surface. This natural phenomenon is the basis of the pneumatic tire, balloon, hydraulic jack, and related devices. *See* HYDROSTATICS.

[KARL ARNSTEIN; ROBERT S. ROSS]

Paschen-Back effect

An effect on spectral lines obtained when the light source is placed in a very strong magnetic field, first explained by F. Paschen and E. Back in 1921. In such a field the anomalous Zeeman effect, which is obtained with weaker fields, changes over to what is, in a first approximation, the normal Zeeman effect. The term "very strong field" is a relative one, since the field strength required depends on the particular lines being investigated. It must be strong enough to produce a magnetic splitting that is large compared to the separation of the components of the spin-orbit multiplet. *See* ATOMIC STRUCTURE AND SPECTRA; ZEEMAN EFFECT.

The illustration shows, as an example, the Paschen-Back effect of the red line $2s^2S - 2p^2P$ of lithium. The natural separation of this doublet is very small, only 0.175 A, so that in the field of 44,200 oersteds for which the diagram is drawn the normal Zeeman splitting of 0.929 A greatly exceeds it. The Paschen-Back effect is therefore practically complete, and the resulting pattern is nearly a normal triplet. There is still, however, a residual splitting, which theory gives as two-thirds of the field-free separation. The weak field patterns shown in the figure require a field of only about 1800 oersteds and correspond to the anomalous patterns obtained in the Zeeman effect.

Theory explains the transformation from the Zeeman to the Paschen-Back effect as due to the uncoupling of the orbital and spin vectors L and S by the magnetic field. Whereas in a weak field these vectors are coupled magnetically to form a

resultant J, a sufficiently strong field causes them to precess independently about the field direction. The correlation of the energy levels between weak and strong fields follows from the principle that the magnetic field is incapable of changing the angular momentum component along the field; that is, the magnetic quantum number M of a level remains constant for all field strengths. A further rule is that two levels having the same value of M (which in the strong field equals $M_L + M_S$) do not cross each other. These rules lead to the correlation of lines shown in the figure. Certain lines fade out during the transition, as is indicated by the change from a solid to a dotted line, and these are the ones for which the direction of polarization would be altered.

[F. A. JENKINS/W. W. WATSON]

Peltier effect

A phenomenon discovered in 1834 by J. C. A. Peltier, who found that at the junction of two dissimilar metals carrying a small current the temperature rises or falls, depending upon the direction of the current. Many different pairs of metals were investigated; bismuth and copper were among the first. The temperature rises at a junction where the flow of positive charge is from Cu to Bi and falls where the flow is from Bi to Cu. A reversible output of heat occurs at the first-named junction and a reversible intake at the second. In view of the experiments of Quintus Icilius (1853), which established that the rate of intake or output of heat is proportional to the magnitude of the current, it can be shown that an electromotive force resides at a Cu-Bi junction, directed from Bi to Cu. Electromotive forces of this type are called Peltier emfs. *See* SEEBECK EFFECT; THERMOELECTRICITY; THOMSON EFFECT.

[JOHN W. STEWART]

Pendulum

A rigid body mounted on a fixed horizontal axis, about which it is free to rotate under the influence of gravity. The period of the motion of a pendulum is virtually independent of its amplitude and depends primarily on the geometry of the pendulum and on the local value of g, the acceleration of gravity. Pendulums have therefore been used as the control elements in clocks, or inversely as instruments to measure g.

Pendulum motion. In the schematic representation of a pendulum shown in the figure, O represents the axis and C the center of mass. The line OC makes an instantaneous angle θ with the vertical. In rotary motion of any rigid body about a fixed axis, the angular acceleration is equal to the torque about the axis divided by the moment of inertia I about the axis. If m represents the mass of the pendulum, the force of gravity can be considered as the weight mg acting at the center of mass C. Therefore, the angular acceleration α is determined by the relation in Eq. (1) where h is the distance OC, and t represents time.

$$-mgh \sin \theta = I\alpha = I \, d^2\theta/dt^2 \qquad (1)$$

If the amplitude of motion is small, $\sin \theta \approx \theta$ in radian measure. In this approximation the motion is simple harmonic. Equation (2) has for its solution Eq. (3) where the amplitude A and the phase δ

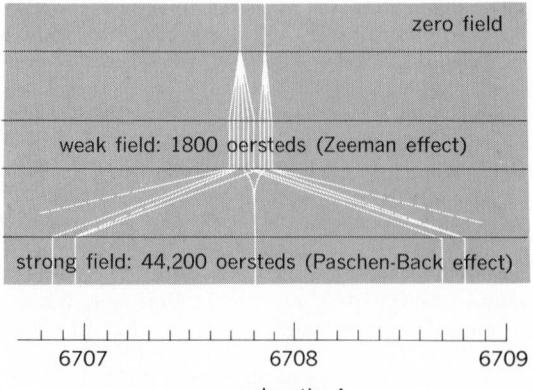

Zeeman and Paschen-Back effects of red lithium doublet, whose natural separation is 0.175 A.

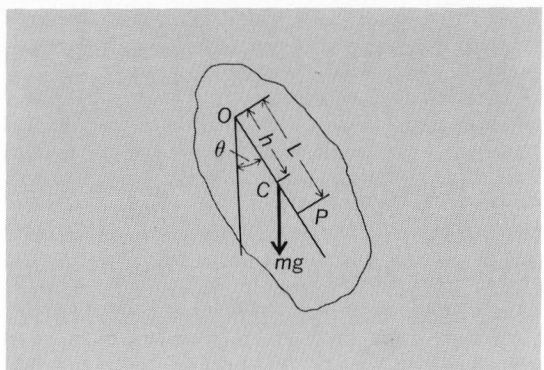

Schematic diagram of a pendulum. O represents the axis, C is the center of mass and P the center of oscillation.

$$-mgh\theta = I\, d^2\theta/dt^2 \tag{2}$$

$$\theta = A\sin(\omega t - \delta) \tag{3}$$

are arbitrary constants. The angular frequency ω is given by Eq. (4). The period T, time for a com-

$$\omega^2 = mgh/I \tag{4}$$

plete vibration (for example, from the extreme displacement right to the next extreme displacement right), and frequency f, number of vibrations per unit time, are given by Eq. (5). *See* HARMONIC MOTION.

$$T = 1/f = 2\pi/\omega = 2\pi\sqrt{I/mgh} \tag{5}$$

The actual form of a pendulum often consists of a long, light bar or a cord that serves as a support for a small, massive bob. The idealization of this form into a point mass on the end of a weightless rod of length L is known as a simple pendulum. An actual pendulum is sometimes called a physical or compound pendulum. In a simple pendulum the lengths h and L become identical, and the moment of inertia I equals mL^2. Equation (5) for the period becomes Eq. (6). Because the value of g in metric

$$T = 2\pi\sqrt{L/g} \tag{6}$$

units (about 9.8 m/sec^2) is very nearly equal to π^2, a simple pendulum 1 m in length has a period very close to 2 sec; the time for a single swing from right to left or left to right is approximately 1 sec.

Center of oscillation. Equation (6) can be used to define the equivalent length of a physical pendulum. Comparison with Eq. (5) shows that Eq. (7)

$$L = I/mh \tag{7}$$

holds. The point P on line OC of the figure, whose distance from the axis O equals L, is called the center of oscillation. Points O and P are reciprocally related to each other in the sense that if the pendulum were suspended at P, O would be the center of oscillation.

The proof of this relation follows from the parallel axis property of moments of inertia. If the moment of inertia of the pendulum about its center of mass is defined by Eq. (8), then Eq. (9) holds; and if defined by Eq. (7), then Eq. (10) holds.

$$I_0 = mb^2 \tag{8}$$

$$I = m(h^2 + b^2) \tag{9}$$

$$hL = h^2 + b^2 \tag{10}$$

For a given value of b (the radius of gyration about the center of mass) and L, h is determined by Eq. (10) to be either of the quantities given in Eq. (11). The sum of these two values is equal to L;

$$h = L/2 \pm \sqrt{(L^2/4) - b^2} \tag{11}$$

therefore, if one value of h is the distance OC, then CP must represent the other value of h that will give the same equivalent length.

If some particular body with a definite value for b is to be mounted about an arbitrary axis to make a pendulum, Eq. (10) shows that L can never be less than $2b$, and that L will have this minimum value (and the period T will be a maximum) if h is made equal to b.

Center of percussion. The points O and P share another reciprocal property. If the body is free to move in the plane of the figure, instead of fixed on an axis, and an impulsive force is applied to the body at O, the initial motion of the body will be a rotation about P. For this reason P is sometimes called the center of percussion about O.

If the motion of a pendulum is not limited to small amplitudes, Eq. (2) is not an adequate substitute for the correct Eq. (1). The angular velocity, $d\theta/dt$, can be derived as a function of displacement by multiplying both sides of Eq. (1) by $d\theta/dt$ and then integrating. The result, which can also be obtained directly from the principle of conservation of energy, is given in Eq. (12), where θ_0 is the

$$(d\theta/dt)^2 = 2\omega^2(\cos\theta - \cos\theta_0) \tag{12}$$

maximum displacement or amplitude of the motion. Here ω is still the characteristic constant of the pendulum defined by $\omega^2 = mgh/I = g/L$, although the relation between ω and frequency is no longer as simple as in the approximate Eq. (5).

To obtain θ as a function of time, introduce an angle ψ by the relation in Eq. (13), where $k \equiv \sin(\theta_0/2)$.

$$\sin\psi \equiv (1/k)\sin(\theta/2) \tag{13}$$

Equation (12) becomes Eq. (14). If time is chosen

$$\begin{aligned}\omega\, dt &= \pm\frac{d\theta}{\sqrt{2(\cos\theta - \cos\theta_0)}}\\ &= \pm\frac{d\theta}{2\sqrt{\sin^2(\theta_0/2) - \sin^2(\theta/2)}}\\ &= \pm\frac{d\psi}{\sqrt{1 - k^2\sin^2\psi}}\end{aligned} \tag{14}$$

zero when θ is zero, then Eq. (15) holds, where F

$$\omega t = F(k,\psi) \tag{15}$$

(k,ψ) is the standard elliptic integral of the first kind, as defined by Eq. (16). Conversely, the angle

$$F(k,\psi) = \int_0^\psi \frac{dz}{\sqrt{1 - k^2\sin^2 z}} \tag{16}$$

θ can be expressed as an elliptic function of time, as in Eq. (17).

$$\sin(\theta/2) = k\,\mathrm{sn}(\omega t) \tag{17}$$

The accurate expression for the period T, obtained from Eq. (15), can be written in terms of the complete elliptic integral of the first kind, $K(k)$, as Eq. (18). Numerical values for the ratio of the peri-

Ratio of the period for amplitude θ_0 to the period for infinitesimal amplitude

θ_0	$T(\theta_0)/T(O)$	θ_0	$T(\theta_0)/T(O)$
0	1.0000	100°	1.2322
20°	1.0077	120°	1.3729
40°	1.0313	140°	1.5944
60°	1.0732	160°	2.0075
80°	1.1375	180°	∞

$$\omega T = 4F(k, \pi/2) = 4K(k) \tag{18}$$

od for amplitude θ_0 to the period for infinitesimal amplitude are listed in the table.

Pendulum types. The following paragraphs describe important types of gravity pendulums.

Kater's reversible pendulum. This type is designed to measure g, the acceleration of gravity. It consists of a body with two knife-edge supports on opposite sides of the center of mass as at O and P (and with at least one adjustable knife-edge). If the pendulum has the same period when suspended from either knife-edge, then each is located at the center of oscillation of the other, and the distance between them must be L, the length of the equivalent simple pendulum. The value for g follows from Eq. (6) or Eq. (18).

Ballistic pendulum. This is a device to measure the momentum of a bullet. The pendulum bob is a block of wood into which the bullet is fired. The bullet is stopped within the block and its momentum transferred to the pendulum. This momentum is determined from the amplitude of the pendulum swing.

Spherical pendulum. This is a simple pendulum mounted on a pivot so that its motion is not confined to a plane. The bob then moves over a spherical surface. A Foucault pendulum is a spherical pendulum suspended so that its plane of oscillation is free to rotate. Its purpose is to demonstrate the rotation of the Earth. If such a pendulum were mounted at the North Pole, the rotation of the Earth under the pendulum would make it appear to a terrestrial observer that the plane of the pendulum's motion rotated 360° once every day. The plane of motion of a Foucault pendulum set up at a lower latitude rotates at a reduced rate, proportional to the sine of the latitude. *See* FOUCAULT PENDULUM.

Torsional pendulum. Despite its name, a torsional pendulum is not a pendulum. It is an example of a torsional harmonic oscillator, consisting of a disk or other body of large moment of inertia mounted on one end of a torsionally flexible rod. The other end of the rod is held fixed. If the disk is twisted and released, the torsional pendulum oscillates harmonically. Gravitation plays no part in a motion. *See* DIMENSIONAL ANALYSIS.

[JOSEPH M. KELLER]

Bibliography: R. A. Becker, *Introduction to Theoretical Mechanics*, 1954; W. W. Hagerty and H. J. Plass, Jr., *Engineering Mechanics*, 1967, reprint 1975; K. R. Symon, *Mechanics*, 3d ed., 1971.

Period

The time interval of a single repetition of a varying quantity of a motion or phenomenon which repeats itself regularly. The period is the reciprocal of the frequency. *See* FREQUENCY.

Waves which have regularly repeated time-varying quantities are termed periodic. In general, any complex periodic wave can be described by a Fourier analysis as the sum of a series of sine and cosine partial waves whose periods are integral multiples of a single period known as the fundamental period of the complex wave. *See* PERIODIC MOTION; WAVE MOTION.

[WILLIAM J. GALLOWAY]

Periodic motion

Any motion that repeats itself identically at regular intervals. If $x(t)$ represents the displacement of any coordinate of the system at time t, a periodic motion has the property defined by Eq. (1) for every

$$x(t + T) = x(t) \tag{1}$$

value of the variable time t. The fixed time interval T between repetitions, or the duration of a cycle, is known as the period of the motion. Frequency is the number of repeating cycles per unit time, and is numerically equal to the reciprocal of the period T.

The motion of the escapement mechanism of a watch, the motion of the Earth about the Sun, and the more complicated motion of the crankshaft, piston rods, and pistons in an engine running at uniform speed are all examples of periodic motion.

The vibration of a piano string after it is struck is a damped periodic motion, not strictly periodic according to the definition. Although the motion very nearly repeats itself, and with a fixed repetition time, each successive cycle has a slightly smaller amplitude. *See* DAMPING.

Any periodic motion can be expressed as a Fourier series — a sum of sine and cosine terms whose frequencies are integral multiples of the frequency f of the periodic motion. Thus Eq. (2) holds, where

$$x(t) = A_0 + \Sigma_n A_n \cos(2\pi nft) + \Sigma_n B_n \sin(2\pi nft) \tag{2}$$

the As and Bs are constant coefficients, and the sums may be taken over all positive integer values of n. For the special case in which the coefficients all vanish for $n > 1$ *see* HARMONIC MOTION. *See also* FOURIER SERIES AND INTEGRALS.

Many systems with more than one degree of freedom, whose motion is not simply periodic, are multiply periodic. The motion may be resolved into parts (for example, horizontal and vertical components, radial and tangential components), each of which is periodic, but with periods that are not commensurate. One example is the vibration of a bell, whose overtone frequencies are not simply related to the fundamental frequency. The motion of the solar system is multiply periodic because it never exactly repeats itself, even though each planet moves periodically. *See* VIBRATION; WAVE MOTION.

[JOSEPH M. KELLER]

Permittivity

The permittivity ϵ of a material medium is related to the permittivity ϵ_0 of empty space by the equation $\epsilon = k\epsilon_0$, where k is the relative dielectric constant of the medium. The permittivity of empty space ϵ_0 has the value 8.85×10^{-12} coulomb^2/newton-meter^2. *See* COULOMB'S LAW; DIELECTRIC CONSTANT; ELECTRICAL UNITS AND STANDARDS.

[RALPH P. WINCH]

Perturbation (mathematics)

A modification in the mathematical structure of a problem changing the problem from one that can be solved exactly, the unperturbed problem, to one, the perturbed problem, for which it is usually possible to obtain only an approximate solution. The methods employed for this purpose form perturbation theory. These methods attempt to express the solution of the perturbed problem in terms of the properties of the solutions of the unperturbed problem.

Examples. Examples of perturbation problems can be found in nearly every branch of mathematics and physics, and in astronomy. The simplest case occurs in ordinary algebra. Suppose that the roots of the equation $f(x) = 0$ are known (the unperturbed problem), and that the roots of the equation $f(x) + \epsilon g(x) = 0$ are to be found (the perturbed problem). The parameter ϵ measures the size of the perturbation. Another set of examples occurs in linear differential equations and in particle dynamics. Possible perturbations include changes in the forces considered to be acting on the particle as well as changes in initial conditions.

Several examples occur in partial differential equations. One physical realization occurs in the theory of wave propagation where the perturbations can be changes in the index of refraction, changes in initial conditions, or changes in the nature or shape of the surfaces encountered by the waves. All of these changes can occur separately or concurrently. The first of these changes is called a volume perturbation, the second a perturbation of initial conditions, and the third a perturbation of boundary conditions. Similar examples can be taken from quantum mechanics, where the volume perturbation corresponds to a change in the Hamiltonian, and perturbation of initial conditions to quantum mechanical time-dependent perturbation theory. Other partial differential equations of physics, such as the Laplace equation, the diffusion equation, and the equations of hydrodynamics, furnish further examples. *See* PERTURBATION (QUANTUM MECHANICS).

As a final illustration of these various types of perturbation, consider possible modifications in an equation (as well as boundary and initial conditions) describing the motion of particles such as neutrons or electrons moving through a medium which can scatter and absorb them. The equation is known as the Lorentz-Boltzmann equation and changes in it occur as a consequence of modifications of the laws of scattering and absorption, that is, because of changes in the medium.

All of these problems are linear and can therefore be cast into an equation of the form $A\psi = \lambda\psi$, where ψ is the unknown quantity, λ is a constant, and A is an operator involving among other possibilities differentiation and integration. The quantity ψ may be a scalar, a vector, or more generally a matrix quantity. When solutions can be obtained for only special values of λ, the eigenvalues, the equation is called the eigenvalue equation, and the associated problem is called the eigenvalue problem. The operator A contains the perturbation; that is, A equals $A_0 + \epsilon A_1$, where A_0 is the unperturbed operator and ϵA_1, the perturbing term.

Iteration method. The method generally employed to obtain an approximate solution is called the iteration method. Rewrite the equation $A\psi = \lambda\psi$ as $(A_0 - \lambda)\psi = -\epsilon A_1\psi$. Let the unperturbed solution be ϕ_0, where $(A_0 - \lambda_0)\phi_0 = 0$. Then ψ_1, a first approximation to ψ, is obtained as a solution of $(A_0 - \lambda)\psi_1 = -\epsilon A_1\phi_0$. A second approximation is the solution of $(A_0 - \lambda)\psi_2 = -\epsilon A_1\psi_1$. The nth approximation is obtained in terms of the $(n-1)$ approximation, ψ_{n-1}, from the equation $(A_0 - \lambda)\psi_n = -\epsilon A_1\psi_{n-1}$. It is assumed that the properties of the unperturbed operator, A_0, are completely known so that the solution of these equations can be obtained. If the sequence $\phi_0, \psi_1, \ldots, \psi_n \ldots$, converges, it will converge to a solution of the problem. For an eigenvalue problem, the procedure must be modified. The first approximation to λ is λ_0; the nth approximation is λ_n. Then the equation determining ψ_n in terms of ψ_{n-1} is $(A_0 - \lambda_{n-1})\psi_n = \epsilon A_1\psi_n$. It is important for the practicality of this procedure that the approximation λ_n can be expressed in terms of the approximation ψ_{n-1} and the operators A_0 and ϵA_1.

In a related and more familiar formulation both ψ and λ are expanded in a power series in ϵ; that is, $\psi = \phi_0 + \epsilon\phi_1 + \epsilon^2\phi_2 + \cdots$ and $\lambda = \lambda_0 + \epsilon\lambda_1 + \epsilon_2\lambda_2 + \cdots$. Then the equation $A\psi = \lambda\psi$ reduces to a set of equations for ϕ_n. For example, the equation for ϕ_1 is $(A_0 - \lambda_0)\phi_1 = -(A_1 - \lambda_1)\phi_0$ and the equation for ϕ_2 is $(A_0 - \lambda_0)\phi_2 = -(A_1 - \lambda_1)\phi_1 + \lambda_2\phi_0$, and so on. This formulation is more complex, and often yields slower rates of convergence than the method just outlined.

The iteration method can be generalized in two respects. First, it is not necessary to use ϕ_0 as the zeroth approximation to ψ. If by reason of other information a better approximation, say ψ_0, is known, the iteration sequence starts with the equation $(A_0 - \lambda)\psi_1 = -\epsilon A_1\phi_0$. Second, the iteration method can be employed in the treatment of nonlinear as well as the linear problems discussed in detail here.

For the iteration method to be at all possible, it is necessary for the sequence $\psi_0, \psi_1, \ldots$ to exist and to converge. The usefulness of the method increases with increasing rate of convergence. The sequence exists only if the singularities of the perturbation are not too strong, or if the initial zero approximation is properly chosen, or both. When the sequence exists, it will converge for a range in values of the parameter ϵ. The largest value of ϵ is the radius of convergence. This is found to be that value of ϵ for which the equation $A\psi = \lambda\psi$ has at least two degenerate solutions, that is, solutions with identical values of λ. There are various methods of increasing the radius of convergence. For example, the general techniques of analytic continuation, such as the Euler transformation, can often be employed. A clever choice of ψ_0, the zeroth approximation, will often produce the desired effect. The variational method can generate the appropriate choice for ψ_0. A more general method was developed by I. Fredholm in which the solution of $A\psi = \lambda\psi$ is given as the ratio of two functions, each of which can be expressed as a series of ϵ. For a wide class of operators A, each of these series will have an infinite radius of convergence.

The eigenvalue problem can be reduced to the problem of the solution of a set of homogeneous linear simultaneous equations which is generally but not always infinite. A nontrivial solution of these equations is possible only if the determinant of the coefficients is zero. Because the coefficients involve the eigenvalue λ, this condition yields an equation, the secular equation, which determines the possible values of λ. The determinant is known as the secular determinant. An example follows. Let the solutions of the unperturbed problem be $\phi^{(p)}$ with eigenvalues $\lambda^{(p)}$; that is, $A_0\phi^{(p)} = \lambda^{(p)}\phi^{(p)}$. Moreover, suppose that the set $\phi^{(p)}$ is complete, which roughly means that an arbitrary function can be represented as a linear combination of $\phi^{(p)}$. Therefore let ψ, the solution of the perturbed problem, be $C_0\phi^{(0)} + C_1\phi^{(1)} + C_2\phi^{(2)} + \cdots$, where C_p are constants. By substituting this expression for ψ in the equation $A\psi = \lambda\psi$ and employing the properties of the set $\phi^{(p)}$ which follow from the nature of the operator A_0, it is possible to obtain a set of equations for C_p. In a typical case these equations have the form:

$$C_0(\lambda - \lambda^{(0)}) + C_1(\epsilon A_1)_{10} + C_2(\epsilon A_2)_{20} + \cdots = 0$$
$$C_0(\epsilon A_1)_{01} + C_1(\lambda - \lambda^{(1)}) + C_2(\epsilon A_1)_{21} + \cdots = 0$$
$$C_0(\epsilon A_1)_{02} + C_1(\epsilon A_1)_{12} + C_2(\lambda - \lambda^{(2)}) = 0$$

and so on. The elements $(\epsilon A_1)_{pq}$ are numbers which depend upon $\phi^{(p)}$, $\phi^{(q)}$ and the operator ϵA_1. The consequent secular equation is obtained by setting the determinant of the coefficients of C_p in this sequence of equations equal to zero. The solution of these simultaneous equations for the coefficients C_p can be obtained by the iteration method, which yields a particular representation of each of the approximations ψ_n. If there are only a finite number of C_p, the secular determinant is of finite order and reduces to a finite polynomial in λ so that in the finite case solutions of the secular equation can always be obtained without using perturbation methods. Once allowed values of λ are known, corresponding values of C_p can be determined. *See* EIGENVALUE.

Degenerate perturbation theory. A special technique is required when the unperturbed problem is degenerate, that is, when there are several solutions of the equation $A_0\phi = \lambda\phi$ for a single value of the eigenvalue λ. The number of such independent solutions is the order of the degeneracy. The corresponding method is designated degenerate perturbation theory. The objective of the special method adopted for this case is the determination of the appropriate linear combinations of these degenerate solutions for use as the initial approximation, ψ_0, in the iterative method. To this end, all terms of the equations for C_p are dropped when p refers to an unperturbed solution which is not one of the degenerate solutions under consideration, and only those C_p which do refer to these degenerate solutions are retained. The resulting secular equation for λ has a number of roots equal to the order of the degeneracy. For each root there is a corresponding set of values for C_p which determine a particular linear combination of the degenerate unperturbed solutions. Each of these combinations can be employed as the initial approximation, ψ_0, in the iterative method. It is often the case that the determination of the possible ψ_0 is sufficient for the evaluation of the major effects of the perturbation.

[HERMAN FESHBACH]

Bibliography: P. M. Morse and H. Feshbach, *Methods of Theoretical Physics*, 1953; A. H. Neyfeh, *Introduction to Perturbation Techniques*, 1980.

Perturbation (quantum mechanics)

An expansion technique useful for solving complicated quantum-mechanical problems in terms of solutions for simple problems. Perturbation theory in quantum mechanics provides an approximation scheme whereby the physical properties of a system, modeled mathematically by a quantum-mechanical description, can be estimated to a required degree of accuracy. Such a scheme is useful because very few problems occurring in quantum mechanics can be solved analytically. Consequently an approximation technique must be employed in order to give an approximate analytic solution or to provide suitable algorithms for a numerical solution. Even for problems which admit an exact analytic solution, the exact solution may be of such mathematical complexity that its physical interpretation is not apparent. For these situations, perturbation techniques are also desirable. Here the discussion of the application of perturbation techniques to quantum mechanics will be limited to the domain of nonrelativistic quantum theory. Applications, of a similar but mathematically more intricate nature, have also been made in quantum electrodynamics and quantum field theory. *See* NONRELATIVISTIC QUANTUM THEORY; QUANTUM ELECTRODYNAMICS; QUANTUM FIELD THEORY; QUANTUM MECHANICS.

Perturbation theory is applied to the Schrödinger equation, $H\Psi = (H_0 + \lambda V)\Psi = i\hbar(\partial/\partial t)\Psi$ [where $\hbar$ is Planck's constant h divided by 2π, and $(\partial/\partial t)$ represents partial differentiation with respect to the time variable t], for which the exact hamiltonian H is split into two parts: the approximate (unperturbed) time-independent hamiltonian H_0 whose solutions of the corresponding Schrödinger equation are known analytically, and the perturbing potential λV. The basic idea is to expand the exact solution Ψ in terms of the solution set of the unperturbed hamiltonian H_0 by means of a power series in the coupling constant λ. Such a procedure is expected to be successful if the system characterized by the unperturbed hamiltonian closely resembles that characterized by the exact hamiltonian. Supposedly the differences are not singular in character, but change as a continuous function of the parameter λ.

Perturbation theory is used in two contexts to provide information about the state of the system, which in quantum mechanics is determined by the wave function Ψ. If λV is time-independent, an objective may be to find the stationary states of the system Ψ_n whose time dependence is given by exp $(-iE_nt/\hbar)$, where $i = \sqrt{-1}$ and E_n represents the energy of the stationary state labeled by n. If λV is either time-independent or time-dependent, an objective may be to find the time evolution of a state which at some specified time was a stationary state of the unperturbed hamiltonian. The perturbing potential is then considered as causing transitions from the original state to other states of the unperturbed hamiltonian, and application of time-dependent perturbation theory provides the probability of such transitions.

Time-independent perturbation theory. In time-independent perturbation theory, the stationary state $\Psi_n = \psi_n \exp(-iE_n t/\hbar)$, which satisfies the time-independent Schrödinger equation, Eq. (1), is solved by a perturbation technique

$$(H_0 + \lambda V)\psi_n = E_n \psi_n \qquad (1)$$

employing the normalized set of eigenfunctions of H_0, Eq. (2). The approach used in Rayleigh-

$$H_0 \phi_m = \epsilon_m \phi_m \qquad (2)$$

Schrödinger perturbation theory, which was first utilized by Lord Rayleigh in his study of acoustic modes of vibration, is to expand both the eigenfunctions ψ_n and the eigenvalues E_n in a power series in λ, as in Eqs. (3) and (4), where s is a

$$\psi_n = \sum_{m=0}^{\infty} \lambda^m \psi_n^{(m)} = \sum_{m=0}^{\infty} \lambda^m \sum_s a_{ns}^{(m)} \phi_s \qquad (3)$$

$$E_n = \sum_{m=0}^{\infty} \lambda^m E_n^{(m)} \qquad (4)$$

label that ranges over the complete set of the unperturbed set ϕ_s [the symbolic sum over s representing a sum over discrete (bound-state) unperturbed eigenfunctions, as well as an integral over continuum unperturbed eigenfunctions], and $a_{ns}^{(m)}$ are numerical coefficients to be determined. The labels s are arranged so that in the limit $\lambda \to 0$, $E_n \to \epsilon_n$, and $\psi_n \to \phi_n$. If only one eigenfunction ϕ_n has the value ϵ_n as its associated eigenvalue (that is, the eigenvalue is nondegenerate), then equating the coefficients of like powers of λ in the time-independent Schrödinger equation leads to Eqs. (5)–(7), where generically $V_{mn} = $

$$a_{ns}^{(0)} = \delta_{ns} \qquad E_n^{(0)} = \epsilon_n \qquad (5)$$

$$a_{nn}^{(1)} = 0 \qquad a_{ns}^{(1)} = -V_{sn}/(\epsilon_s - \epsilon_n) \qquad \text{for } n \neq s \qquad (6)$$

$$E_n^{(1)} = V_{nn} \qquad E_n^{(2)} = -\sum_m{}' V_{nm}V_{mn}/(\epsilon_m - \epsilon_n) \qquad (7)$$

$\int \phi_m^* V \phi_n$ (integration taking place over all configuration space), and the prime indicates that the $m = n$ term is excluded. The expansion also provides recursion relations from which higher-order contributions may be obtained. The criteria for validity of the expansion is usually taken to be $|a_{ns}^{(1)}| \ll 1$ for $n \neq s$, although formal proofs of convergence exist only in a limited number of cases.

Degenerate perturbation theory. If $N > 1$ zero-order (unperturbed) eigenfunctions have the same value of the eigenvalue ϵ_n, the zeroes in the denominators invalidate the preceding expressions, and one must employ degenerate perturbation theory. Essentially this consists of taking appropriate linear combinations of the zero-order degenerate wave functions for the unperturbed basis. The degenerate set $\{\phi_s\}$ is replaced by the j linear combinations of Eq. (8), and matching

$${}^j\phi_s = \sum_{s'=1}^{N} {}^j b_{ss'} \phi_{s'} \qquad j = 1, 2, \ldots, N \qquad (8)$$

like powers of λ then leads to the condition of Eq. (9). The first-order energy corrections for the

$$\sum_{s'=1}^{N} {}^j b_{ss'} \left[V_{ks'} - \delta_{ks'} {}^j E_s^{(1)} \right] = 0 \qquad (9)$$

jth combination, ${}^j E_s^{(1)}$, are found by setting the determinant of the coefficients of ${}^j b_{ss'}$ equal to zero. This technique, which may be considered as

a first-order diagonalization process, together with the normalization condition on the coefficients, gives the coefficients ${}^j b_{ss'}$ for the jth combination of unperturbed wave functions. In general, the first-order shifts ${}^j E_s^{(1)}$ split the original unperturbed energy levels ϵ_s, and higher-order correction formulas contain denominators involving differences of the first-order shifts. If the shifts are the same for two or more different values of j, a linear combination of the corresponding ${}^j\phi_s$ must then be taken for the unperturbed starting basis for the second-order treatment. The appropriate combination is determined by a second-order diagonalization process. Higher-order degeneracies are removed in a similar fashion.

Brillouin-Wigner expansion. An alternative approach to time-independent perturbation theory is provided by the Brillouin-Wigner expansion. In this approach the exact time-independent wave function ψ_n is expanded in terms of unperturbed eigenfunctions, $\psi_n = \sum_s a_{ns} \phi_s$, which leads to relation (10). The basic idea here is to use this ex-

$$a_{nm}(E_n - \epsilon_m) = \lambda \sum_s a_{ns} V_{ms} \qquad (10)$$

pression to generate an iterative expansion for the off-diagonal coefficients a_{ns} (that is, for $n \neq s$). It follows that Eq. (11) holds, where the prime

$$E_m - \epsilon_m = \lambda V_{mm} + \lambda(a_{mm})^{-1} \sum_s{}' a_{ms} V_{ms} \qquad (11)$$

indicates only a sum over off-diagonal a's, and the expansion is provided by repeated iteration of Eq. (12). The convergence of the Brillouin-Wigner

$$\frac{a_{ms}}{a_{mm}} = \frac{\lambda V_{sm}}{E_m - \epsilon_s} + \sum_k{}' \frac{\lambda a_{mk} V_{sk}}{E_m - \epsilon_s} \qquad (12)$$

series is usually much faster than the convergence of the corresponding Rayleigh-Schrödinger series, but since the energy levels E_m are given only implicitly (to a given order in terms of a solution of a polynomial), the Brillouin-Wigner series expansion may be less convenient, although formally the expansion is much simpler.

Applications. Among the problems for which time-independent perturbation theory is employed are estimation of level splitting in the Zeeman effect, calculation of electric and magnetic susceptibilities, and a host of other problems concerning energy-level determination in atomic and molecular physics. Special techniques are employed in many-body perturbation theory in which a product basis of single-particle unperturbed eigenfunctions is utilized.

WKB method. An approximation technique for solving the time-independent Schrödinger equation by exploiting the classical limit of quantum mechanics is the WKB (Wentzel-Kramers-Brillouin) method. Here one substitutes $\psi = A \exp[iW/\hbar]$ into Schrödinger's equation, and expands the amplitude A and phase W in powers of h. In the zeroth approximation as $h \to 0$, the phase W satisfies the classical Hamilton-Jacobi equation; higher-order contributions give the peculiarly quantum-mechanical effects. The applicability of the WKB approximation is usually confined to one-dimensional potential well or barrier problems, or three-dimensional problems for which the well or barrier depends only on the radial coordinate.

The WKB approximation is expected to give reliable results when $d|\lambda(x)|/dx \ll 1$, $\lambda(x)$ being the local de Broglie wavelength $h/p = h[2m(E - \lambda V)]^{-1/2}$. The WKB method has been used to give an approximate treatment of alpha-particle emission from a nucleus, as well as a number of other problems where penetration of a barrier is involved. *See* HAMILTONIAN-JACOBI THEORY; WENTZEL-KRAMERS-BRILLOUIN METHOD.

Time-dependent perturbation theory. In time-dependent perturbation theory, where the perturbing potential may or may not be time-dependent, the time-dependent wave function Ψ is expanded in terms of the set of time-dependent unperturbed eigenfunctions by means of expansion coefficients which are themselves time-dependent. This method is sometimes known as the method of variation of constants. Specifically one expresses a particular state, labeled by n, by means of Eq. (13). The coefficients c then obey the

$$\Psi_n = \sum_s c_{ns}(t)\phi_s \exp(-i\epsilon_s t/\hbar) \exp\left[\frac{-i\lambda}{\hbar}\int_{t_0}^t V_{ss'} dt'\right] \quad (13)$$

rate relation, Eq. (14), where $\gamma_{ms} = (\epsilon_m + \lambda V_{mm}) -$

$$i\hbar \frac{\partial c_{nm}}{\partial t} = \lambda \sum_{s \neq m} c_{ns} V_{ms} \exp\left[\frac{i}{\hbar}\int_{t_0}^t \gamma_{ms} dt'\right] \quad (14)$$

$(\epsilon_s + \lambda V_{ss})$ is the difference of the perturbed energies for the states m and s. The coefficient c_{ns} is defined in Eq. (13) so as not to include the phase involving the integral over V_{ss}, in order to prevent so-called secular terms (those which have a non-oscillatory behavior at large times) from appearing in an iterative expansion of the coefficients. The appearance of such terms would inhibit convergence of the resulting expansion. Usually, in time-dependent problems, one assumes that at time $t = t_0$ the system is in state n so that $c_{ns}(t_0) = \delta_{ns}$. The problem then is to determine the probability of transition to other states at a later time.

Transient perturbations. For transient perturbations, that is, those for which $V \to 0$ for $t \to \pm\infty$, the initial time t_0 may be taken as $-\infty$, and it is clear that as $t \to \infty$, the expansion coefficients of Eq. (15) become constant since the ϕ's are eigen-

$$a_{ns}(t) = c_{ns}(t) \exp\left[\frac{-i\lambda}{\hbar}\int_{t_0}^t V_{ss} dt'\right] \quad (15)$$

functions of the final hamiltonian operator. Calculations of $a_{ns}(\infty)$ to successive orders in λ, leading to what is called successive Born approximations, give the long-term excitation probability $P_{nm}(\infty) = |a_{nm}(\infty)|^2$ for excitation from initial state n to final state m. Problems such as scattering of particles in a continuum state from atomic systems, and inelastic collisions of atomic systems, are amenable to this type of treatment. *See* SCATTERING EXPERIMENTS (ATOMS AND MOLECULES).

Persistent perturbations. When a persistent perturbation is present, the perturbing potential which was originally zero ultimately reaches a nonvanishing time-independent limit. A first-order solution for the coefficients c gives Eqs. (16), where

$$c_{nn}(t) = 1 \qquad c_{nm}(t) = \alpha_{nm}(t) + \beta_{nm}(t) \qquad n \neq m \quad (16)$$

the notation of Eqs. (17) and (18) is used. The coef-

$$\alpha_{nm}(t) = \int_{t_0}^t \left[\exp\frac{i}{\hbar}\int_{t_0}^{t'} \gamma_{mn} dt'' \right] \frac{d}{dt'}\left[\frac{V_{mn}(t')}{\gamma_{mn}(t')}\right] dt' \quad (17)$$

$$\beta_{nm}(t) = \frac{-\lambda V_{mn}(t)}{\gamma_{mn}(t)} \exp\left[\frac{i}{\hbar}\int_{t_0}^t \gamma_{mn} dt'\right] \quad (18)$$

ficients $\beta_{nm}(t)$ merely provide for the first-order propagation of the initial state in time, whereas the coefficients $\alpha_{nm}(t)$ give the excitation probability $P_{nm}(t) = |\alpha_{nm}(t)|^2$. If the perturbation increases from zero to its asymptotic value in such a short time that the exponential in the definition of α_{nm} is essentially constant, that is, in times shorter than $\hbar/(\epsilon_m - \epsilon_n)$, then $|\alpha_{nm}(\infty)|$ may be approximated by $|\lambda V_{mn}/\gamma_{mn}|$ where the relevant quantities are evaluated at their asymptotic values. This approximation, known as the sudden approximation, leads to useful results for transition probabilities arising from short-time perturbations, such as occur in the radioactive decay of a ^{3}H atom into a doubly ionized ^{3}He ion.

Adiabatic approximation. Yet another approach, the adiabatic approximation, is useful when the perturbing potential changes very slowly. This technique employs a set of basis wave functions χ_s (with a time-independent phase) and associated eigenvalues $E_s(t)$ satisfying Eq. (19), which would

$$[H_0 + \lambda V(t)]\chi_s = E_s(t)\chi_s \quad (19)$$

follow for the exact solution of the time-dependent Schrödinger equation if the time variation of the potential were neglected. The basic idea is to expand the exact time-dependent wavefunction Ψ_n in terms of this basis, as in Eq. (20), obtain the

$$\Psi_n = \sum_s c_{ns}(t)\chi_s \exp\left[\frac{-i}{\hbar}\int^t E_s(t') dt'\right] \quad (20)$$

rate equation giving the time-development of the coefficients, and solve for these coefficients to the desired order in λ by an iterative technique, assuming that the coefficients c_{ns} differ little from their initial values. The use of the basis χ rather than an unperturbed basis hopefully provides for a more satisfactory lower-order treatment. This approach has been successfully used in atomic-collision problems for which the interaction distance changes slowly with time. *See* PERTURBATION (MATHEMATICS).

[DAVID M. FRADKIN]

Phase

The fractional part of a period through which the time variable of a periodic quantity (alternating electric current, vibration) has moved, as measured at any point in time from an arbitrary time origin. In the case of a sinusoidally varying quantity, the time origin is usually assumed to be the last point at which the quantity passed through a zero position from a negative to a positive direction. It is customary to choose the origin so that the fractional part of the period is less than unity.

In comparing the phase relationships at a given instant between two time-varying quantities, the phase of one is usually assumed to be zero, and the phase of the other is described, with respect to the first, as the fractional part of a period through which the second quantity must vary to achieve a zero of its own (see illustration). In this case, the

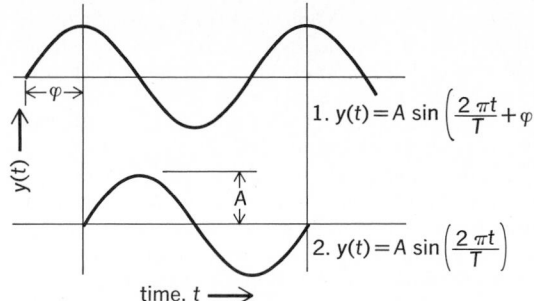

1. $y(t) = A \sin\left(\dfrac{2\pi t}{T} + \varphi\right)$

2. $y(t) = A \sin\left(\dfrac{2\pi t}{T}\right)$

An illustration of the meaning of phase for a sinusoidal wave. The difference in phase between waves 1 and 2 is φ and is called the phase angle. For each wave, A is the amplitude and T is the period.

fractional part of the period is usually expressed in terms of angular measure, with one period being equal to 360° or 2π radians. Thus two sine waves of a given frequency are said to be 90°, or $\pi/2$, out of phase when the second must be displaced in time, with respect to the first, by 1/4 period in order for it to achieve a zero value. *See* SINE WAVE.

[WILLIAM J. GALLOWAY]

Phase rule

A relationship used to determine the number of state variables F, usually chosen from among temperature, pressure, and species compositions in each phase, which must be specified to fix the thermodynamic state of a system in equilibrium. It was derived by J. Willard Gibbs between 1875 and 1878. The phase rule (in the absence of electric, magnetic, and gravitational phenomena) is given by Eq. (1), where C is the number of chemical spe-

$$F = C - P - M + 2 \tag{1}$$

cies present at equilibrium, P is the number of phases, and M is the number of independent chemical reactions. Here phase is used to indicate a homogeneous, mechanically separable portion of the system, and the term independent reactions refers to the smallest number of chemical reactions which, upon forming various linear combinations, includes all reactions which occur among the species present. The number of independent state variables F is referred to as the degrees of freedom or variance of the system.

Examples. The system constituted by liquid water and water vapor contains two phases ($P = 2$) and one component ($C = 1$), and there are no chemical reactions ($M = 0$). Therefore this system has $F = 1 - 2 - 0 + 2 = 1$ degree of freedom. This is in accord with the observation that the vapor and liquid forms of water exist in equilibrium only for values of temperature and pressure along the coexistance curve, so that specifying either of these variables fixes the other.

At high temperatures the three reactions shown in (2) occur between sulfur and oxygen in the gas

$$S + O_2 = SO_2 \tag{2a}$$

$$SO_2 + \tfrac{1}{2} O_2 = SO_3 \tag{2b}$$

$$S + \tfrac{3}{2} O_2 = SO_3 \tag{2c}$$

phase. There are only two independent chemical reactions in this single-phase, four-component

system, since the last reaction is the sum of the first two. Therefore this system has $F = 4 - 1 - 2 + 2 = -3$ degrees of freedom.

Derivation. The derivation of the phase rule starts with the experimental observation that a single phase of C nonreacting components has $C + 1$ degrees of freedom (for example, once temperature, pressure, and $C - 1$ mole fractions are specified, no other intensive variables of the system can vary). Now consider chemical and phase equilibrium in a more general system in which there are C components and P phases, and in which M independent chemical reactions occur. It might appear, based on the experimental observation above, that such a system should have $P(C + 1)$ degrees of freedom. However, since the system is in equilibrium, the degrees of freedom are reduced as follows:

1. At equilibrium the temperature of each phase must be the same. Thus it is not possible to set the temperature of each of the P phases separately; once the temperature of one phase is specified, the temperature of all phases is fixed. Consequently the requirement that the temperature must be the same in all phases at equilibrium eliminates $P - 1$ degrees of freedom.

2. At equilibrium the pressure must be the same in each phase. This eliminates another $P - 1$ degrees of freedom.

3. At equilibrium the chemical potential (partial molar Gibbs free energy) of each species must be the same in each phase. This restriction, which holds for each of the C components, eliminates an additional $C(P - 1)$ degrees of freedom.

4. For each chemical reaction (3) to be in equilibrium, Eq. (4) must be satisfied, where μ_i is the

$$aA + bB + \ldots = rR + sS + \ldots \tag{3}$$

$$a\mu_A + b\mu_B + \ldots = r\mu_R + s\mu_S + \ldots \tag{4}$$

chemical potential of species i. (If this equation is satisfied in one phase, it is, by the equality of the chemical potentials of a given species in its various phases, satisfied in all phases.) This requirement eliminates another M degrees of freedom.

Therefore, the actual number of degrees of freedom in a multicomponent, multiphase, chemically reacting system is given by Eq. (5), which is the

$$\begin{aligned} F &= P(C + 1) - 2(P - 1) - C(P - 1) - M \\ &= C - P - M + 2 \end{aligned} \tag{5}$$

same as Eq. (1). *See* THERMODYNAMIC PROCESSES.

[STANLEY I. SANDLER]

Phase transitions

Changes of state brought about by a change in an intensive variable (for example, temperature or pressure) of a system. Some familiar examples of phase transitions are the gas-liquid transition (condensation), the liquid-solid transition (freezing), the normal-to-superconducting transition in electrical conductors, the paramagnet-to-ferromagnet transition in magnetic materials, and the superfluid transition in liquid helium.

CHARACTERISTICS

Typically the phase transition is brought about by a change in the temperature of the system. The temperature at which the change of state occurs is called transition temperature (usually denoted by T_c). For example, the liquid-solid transition occurs at the freezing point.

Order and entropy. The two phases above and below the phase transition can be distinguished from each other in terms of some ordering that takes place in the phase below the transition temperature. For example, in the liquid-solid transition, the molecules of the liquid get "ordered" in space when they form the solid phase. In a paramagnet, the magnetic moments on the individual atoms can point in any direction (in the absence of an external magnetic field), but in the ferromagnetic phase the moments are lined up along a particular direction, which is then the direction of ordering. Thus in the phase above the transition, the degree of ordering is smaller than in the phase below the transition. One measure of the amount of disorder in a system is its entropy, which is the negative of the first derivative of the thermodynamic free energy with respect to temperature. When a system possesses more order, the entropy is lower. Thus at the transition temperature the entropy of the system changes from a higher value above the transition to some lower value below the transition. *See* ENTROPY; FERROMAGNETISM; PARAMAGNETISM.

Continuous and discontinuous transitions. This change in entropy can be continuous or discontinuous at the transition temperature. In other words, the development of order in the system at the transition temperature can be gradual or abrupt. This leads to a convenient classification of phase transitions into two types, namely, discontinuous and continuous.

Discontinuous transitions involve a discontinuous change in the entropy at the transition temperature. A familiar example of this type of transition is the freezing of water into ice. As water reaches the freezing point, order develops without any change in temperature. Thus there is a discontinuous decrease in the entropy at the freezing point. This is characterized by the amount of latent heat that must be extracted from the water for it to be "ordered" into the solid phase (ice). Discontinuous transitions are also called first-order transitions.

In a continuous transition, entropy changes continuously, and hence the growth of order below T_c

is also continuous. There is no latent heat involved in a continuous transition. Continuous transitions are also called second-order transitions. The paramagnet-to-ferromagnet transition in magnetic materials is an example of such a transition. *See* SECOND-ORDER TRANSITION.

Order parameter. The degree of ordering in a system undergoing a phase transition can be made quantitative in terms of an order parameter. At temperatures above the transition temperature the order parameter has a value zero, and below the transition it acquires some nonzero value. For example, in a ferromagnet the order parameter is the magnetic moment per unit volume (in the absence of an externally applied magnetic field). It is zero in the paramagnetic state since the individual magnetic moments in the solid may point in any random direction. Below the transition temperature, however, there exists a preferred direction of ordering, and as the temperature is lowered below T_c, more and more individual magnetic moments start to align along the preferred direction of ordering, leading to a continuous growth of the magnetization or the macroscopic magnetic moment per unit volume in the ferromagnetic state. Thus the order parameter changes continuously from zero above to some nonzero value below the transition temperature. In a first-order transition, the order parameter would change discontinuously at the transition temperature.

The behavior of systems very near the onset of phase transitions is an important field of study in itself. For a detailed description of this behavior *see* CRITICAL PHENOMENA.

EXAMPLES

Some specific examples of phase transitions will be discussed. A number of well-known transitions are excluded from the discussion here; for these transitions *see* FERROMAGNETISM; SECOND-ORDER TRANSITION; SUPERCONDUCTIVITY; SUPERFLUIDITY.

Glass transitions. When certain substances are taken very rapidly from the liquid state to the solid state, or condensed from the vapor phase to the solid phase, they are found to have an amorphous or glassy structure, whose atomic arrangement differs from those of both crystals and gases. In the crystal, the atoms are placed on a regular, periodic

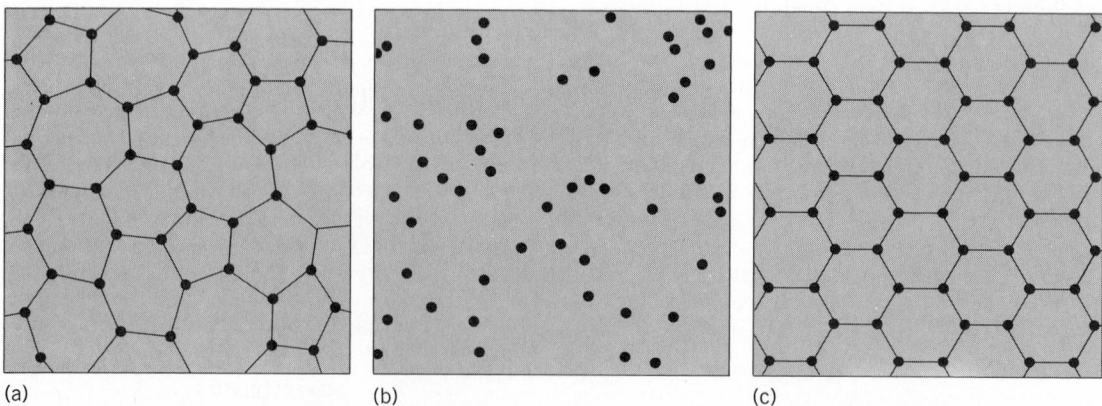

(a)   (b)   (c)

Fig. 1. Schematic in two dimensions of atomic arrangements in three phases. Each dot represents position of an atom at a given instant of time. (*a*) Crystal. (*b*) Gas. (*c*) Glass. (*From E. W. Montroll and J. L. Leibowitz, Fluctuation Phenomena, North-Holland Publishing Co., 1979*)

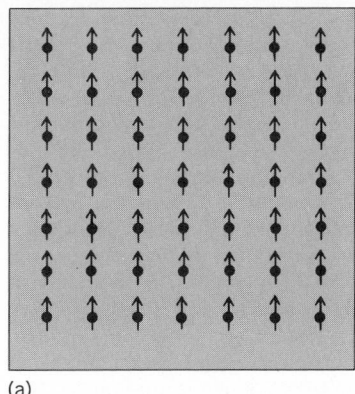 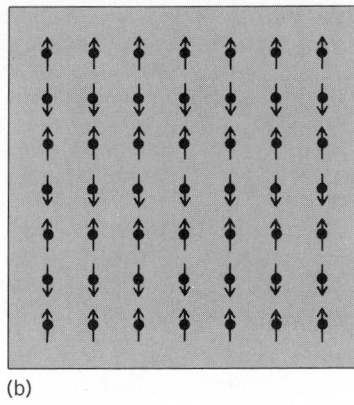 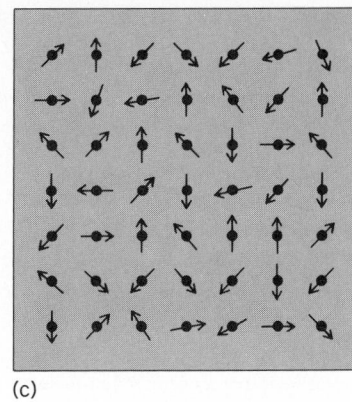

(a)   (b)   (c)

Fig. 2. Ordering of spins in a crystal lattice. Arrows show the direction in which the spin (or the magnetic mo-ment) at a particular lattice site is pointing. (*a*) Ferromagnet. (*b*) Antiferromagnet. (*c*) Paramagnet.

lattice, and only discrete interatomic spacings between the atoms are possible (Fig. 1*a*); this describes a state of long-range order. In the gas, the atoms are constantly colliding with each other and the walls of the container (Fig. 1*b*)—there is no long-range order. The glass phase has a structure intermediate between that of the crystal and gas (Fig. 1*c*). The local environment or short-range order of each atom in the glass shown is similar to that of the crystal, but there is no long-range order. Some glasses have atomic arrangements, at any instant of time, very similar to those of a liquid of the same composition. The structure of certain amorphous metallic alloys is similar to that which would be obtained by packing hard spheres into a container to obtain the largest density.

The glass transition temperature T_g is defined to be the temperature below which the atoms are frozen into relatively stable positions, even though the glassy phase is a metastable one whose energy is higher than the crystalline phase. When a glass is heated above T_g, the material can transform into the crystalline state by the formation of nucleation centers in the solid state. In metallic glasses this crystallization process can take place with only a little heating above T_g. In other glasses, such as fused silica, crystallization does not take place above T_g, no matter how long one waits; the next transition in such a material as the temperature is increased is that to the liquid state, that is, melting. Some examples of insulating or semiconducting glasses are: silicon dioxide (SiO_2), arsenic selenide (As_2Se_3), silicon, (Si), and germanium (Ge). Examples of metallic glasses are $Pd_{80}Si_{20}$, $Fe_{80}B_{20}$, $Gd_{65}Co_{35}$, $Zr_{40}Cu_{60}$, and $Mg_{70}Zn_{30}$. Both classes of materials have been intensively studied with a view toward significant technological applications. *See* AMORPHOUS SOLID.

Ferromagnetism and antiferromagnetism. As a preliminary to discussing several more complicated magnetic phase transitions, the fundamentals of simple ferromagnetic and antiferromagnetic transitions will be outlined. The fundamental interaction energy E_{ij} between two spins S_i and S_j can be represented by Eq. (1), where $\mathscr{J}_{ij}$ is called the

$$E_{ij} = -\mathscr{J}_{ij} \vec{S}_i \cdot \vec{S}_j \qquad (1)$$

exchange interaction because it originates in the quantum-mechanical requirement that the wave function of a system of electrons must be antisym-metric (change sign) under exchange of the coordinates of any two electrons. Although Eq. (1) is a quantum-mechanical effect, it will be discussed from a classical viewpoint since much insight can be gained in this way. Classically, Eq. (1) can be written as Eq. (2), where θ_{ij} is the angle between

$$E_{ij} = -\mathscr{J}_{ij} S_i S_j \cos \theta_{ij} \qquad (2)$$

the two spins. When $\mathscr{J}_{ij}$ is positive, E_{ij} is minimized for $\vec{S}_i$ and $\vec{S}_j$ parallel; when $\mathscr{J}_{ij}$ is negative, E_{ij} is minimized for $\vec{S}_i$ and $\vec{S}_j$ antiparallel. These two cases are shown schematically in Fig. 2*a* and *b* as ferromagnetic and antiferromagnetic order, respectively. The magnetic moment vector associated with each spin is proportional to the magnitude of the spin and points along the same line. In the absence of thermal fluctuations, that is, at absolute temperature $T = 0$, the spins of a ferromagnet are all parallel in a single domain so that a large magnetization results (Fig. 2*a*). For an antiferromagnet, again at $T = 0$, there are two sublattices whose spins are oppositely directed so that there is no net magnetization (Fig. 2*b*). As the temperature is raised from near zero the spins in either case begin to deviate from their original positions until, at the magnetic ordering temperature T_0, a phase transition to the paramagnetic phase takes place (Fig. 2*c*). In the paramagnetic phase there is a complete lack of long-range order in the spin directions. *See* ANTIFERROMAGNETISM.

Magnetism in crystalline alloys. A disordered crystalline alloy is a mixture of two elements in which the atoms of the mixture are found at more or less random positions on a crystal lattice. If the concentration of one element is x (and the other 1-x), then some typical examples of disordered binary alloys are $Cu_{1-x}Mn_x$, $Au_{1-x}Fe_x$, and $Ag_{1-x}Pd_x$. If one of the constituents of the alloy has a magnetic moment and therefore spin associated with it, there will be disorder in the atomic positions of the spins, but yet it is possible for certain types of magnetic order to exist. In the three examples mentioned above, the manganese (Mn) and iron (Fe) atoms do have localized magnetic moments associated with them.

Spin-glass state. Consider a relatively dilute alloy such as $Cu_{0.95}Mn_{0.05}$. In such a case the conduction electrons will scatter from the manganese impurity atoms and create a disturbance in the net spin density in the vicinity of each manganese

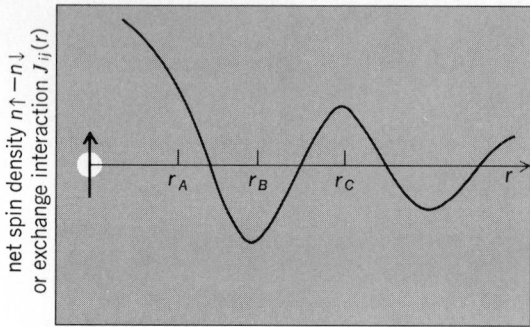

Fig. 3. Net spin density of conduction electrons (number of electrons per unit volume with spin up $n\uparrow$ minus number with spin down $n\downarrow$) as function of distance r from magnetic moment. Graph also shows exchange interaction $\mathscr{J}_{ij}(r)$ between magnetic moments i and j as function of distance r between them. Magnetic moment at the origin will couple ferromagnetically to magnetic moments at positions r_A and r_C, but antiferromagnetically to a magnetic moment at r_B.

atom. This disturbance (net spin up minus spin down) is shown in Fig. 3. If another manganese atom is found in the vicinity of the atom at the origin, it will have a spin that is coupled to the latter according to Eqs. (1) and (2). However, in this case $\mathscr{J}_{ij}$ is a function of the distance r between the two spins and has the behavior shown by Fig. 3. Thus $\mathscr{J}_{ij}(r)$ oscillates in sign and decreases in magnitude as r increases. A magnetic moment at the origin of Fig. 3 will couple ferromagnetically to magnetic moments at positions r_A and r_C, but antiferromagnetically to a magnetic moment at r_B. Since there will be a considerable number of different Mn-Mn interatomic spacings, there will be a mixture of ferromagnetic and antiferromagnetic exchange interactions in such an alloy. This mixture leads to a new magnetic phase called a spin glass. In analogy with the ordinary glass shown in Fig. 1(c), a spin glass has short-range magnetic order but no long-range magnetic order, even below the spin-glass ordering temperature T_{SG} (Fig. 4a). Above T_{SG}, thermal fluctuations will destroy the ordered spin-glass state, and the magnetic system will become paramagnetic, with no short- or long-range magnetic order (Fig. 4b). See SPIN GLASS.

Ferromagnetic state. In certain disordered alloys such as $Au_{1-x}Fe_x$, for x values larger than about 0.16, the spin-glass phase is replaced by the ferromagnetic state (Fig. 4c). At this iron concentration, called the percolation limit, it is possible for nearest-neighbor ferromagnetic couplings of iron spins to propagate completely throughout the crystal, thus leading to the state in which all iron spins are parallel. This situation depends on $\mathscr{J}_{ij}(r)$ having positive values for nearest neighbors, as depicted schematically in Fig. 3 for $r = r_A$. If the nearest-neighbor magnetic couplings are negative (antiferromagnetic), as is thought to be the case in $Cu_{1-x}Mn_x$, then the ordered state at higher manganese concentrations becomes a random antiferromagnetic state in which, to the extent possible, nearest-neighbor spins are antiparallel.

The question of the existence of a true phase transition in the spin-glass case, and the spin-glass to ferromagnetic transition, have been intensively studied.

Speromagnetic and asperomagnetic states. Speromagnetic and asperomagnetic transitions, which occur in substances having amorphous atomic structures, were characterized only in the 1970s. In Fig. 5a a schematic diagram is shown of a glass in which each atom possesses a magnetic moment which is assumed to be coupled to its neighbors by the Heisenberg interaction of Eq. (1). Even though the interatomic spacings are smeared out compared to the case of a crystal, if all the exchange interactions $\mathscr{J}_{ij}$ are positive the state of magnetic order will be as shown in Fig. 5a. An example of a ferromagnetic metallic glass is $Gd_{65}Cu_{35}$, in which all the gadolinium (Gd) moments are aligned while the copper (Cu) atoms possess no moments.

A new state of order can exist in rare-earth glasses such as $Dy_{75}Au_{25}$, $Tb_{75}Ga_{25}$, and $Er_{65}Co_{35}$. In these cases the rare-earth atoms have magnetic moments which are due to both electron spin and orbital-angular momentum. The electric fields which exist in such a glass in combination with the spin-orbit interaction make it difficult for the total angular momentum vector $\vec{J}$ and hence the magnetic moment of each rare-earth atom to be aligned in any arbitrary direction. There can be an "easy" axis $\hat{k}_i$ for each $\vec{J}_i$ such that $\vec{J}_i$ tries to align itself

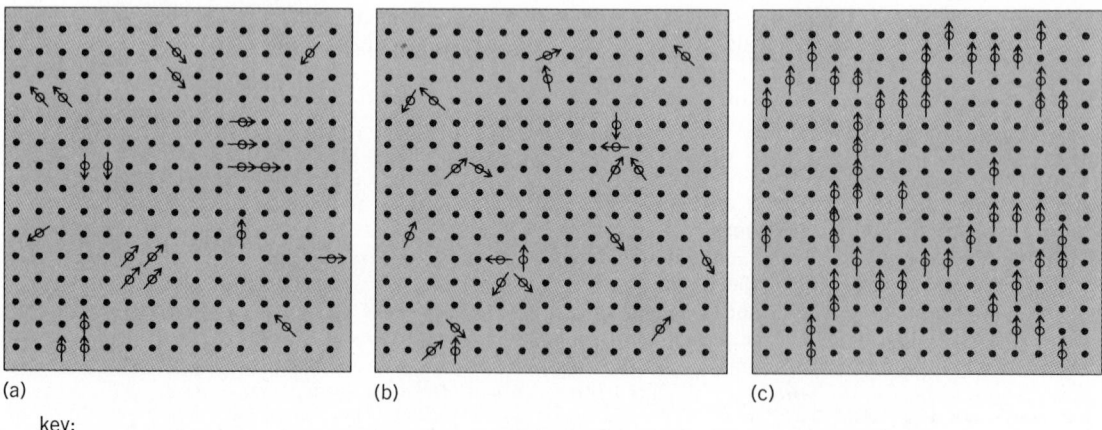

(a)   (b)   (c)

key:
• nonmagnetic atom $\updownarrow$ atom with magnetic moment

Fig. 4. Arrangement of magnetic moments in crystal lattice of a disordered crystalline alloy. (a) Spin-glass state. (b) Paramagnetic state. (c) Ferromagnetic state.

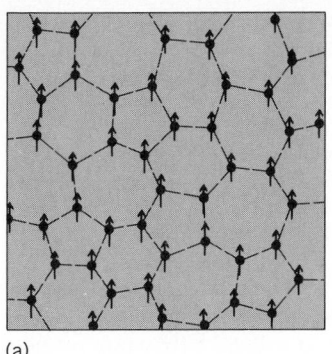 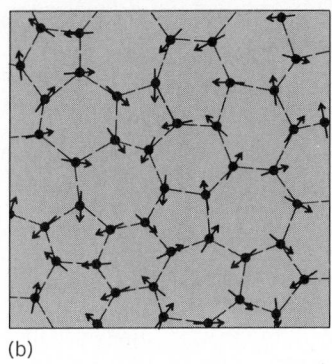 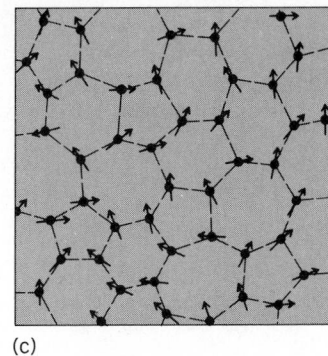

(a)   (b)   (c)

Fig. 5. Arrangement of magnetic moments in (a) ferromagnetic glass, (b) speromagnetic glass, and (c) asperomagnetic glass.

parallel or antiparallel to $\hat{k}_i$. A model that describes this situation of the energy of the ith ion given by Eq. (3), where $\hat{k}_i$ is a unit vector and

$$E_i = -D(\hat{k}_i \cdot \vec{J}_i)^2 \tag{3}$$

D the strength of the local anisotropy. The random anisotropy model of magnetism in glasses assumes that the amorphous structure leads to completely random anisotropy axes $\hat{k}_i$ for each magnetic moment. Then the total energy is given by Eq. (4).

$$E = -\mathscr{J}\sum_{i,j} \vec{J}_i \cdot \vec{J}_j - D\sum_i (\hat{k}_i \cdot \vec{J}_i)^2 \tag{4}$$

If $\mathscr{J}$ is positive, the first term tends to cause parallel alignment of the $\vec{J}_i$ vectors while the second term causes a scattering of the $\vec{J}_i$ vectors. This leads in the ordered state to either a speromagnetic structure (Fig. 5b) or an asymmetric speromagnetic or asperomagnetic structure (Fig. 5c). The origin of the term speromagnetism is the Greek word for scattered, *spero*. The speromagnetic state has a more or less random distribution of moment directions which are frozen in below the magnetic ordering temperature (Fig. 6a). This state of magnetic order has some common features with the spin-glass state, but the origin of the scatter in the frozen moment directions is different. In the asperomagnetic state, the positive exchange interaction is assumed to lead to mostly parallel or nearly parallel near-neighbor moment directions so that the moment distribution would be in the upper hemisphere (Fig. 6b). The true ground state of the system, even with all exchange interactions positive, has been the subject of considerable controversy, but research suggests that the ground state is not the asperomagnetic state but rather has some fraction of the moments in the lower hemisphere.

Metallic glasses of the form $Fe_{80}G_{20}$, where G represents a glass-forming element such as boron, phosphorus, or silicon, may have possible uses as soft magnetic materials in transformer cores. Metallic glasses containing rare-earth elements and transition metals may find applications as magnetic bubble computer memory devices.

Charge-density waves. These transitions have been observed in low-dimensional conductors, that is, materials that conduct electricity in only one or two dimensions. For example, if a material consists of linear chains of molecules such that the interaction between neighboring molecules on different chains is much smaller than that between neighboring molecules on the same chain, the elec-

trons can move relatively easily along the chain, but motion of electrons from one chain to another is highly unlikely. In this case the material behaves as a quasi one-dimensional conductor. Some examples are $NbSe_3$ and $TaSe_3$ (Fig. 7a). There are also materials whose molecules are arranged in sheets, with the interaction between neighboring molecules on different sheets being much smaller than that between neighboring molecules in the same sheet. In this case the electrons can move relatively easily within the sheet, but motion of electrons between sheets is highly restricted. Such materials behave as quasi two-dimensional conductors. Some examples of two dimensional conductors are $NbSe_2$ and $TaSe_2$ (Fig. 7b).

Below the transition temperature the electronic charge-density is modulated, the periodicity depending upon the topology of the Fermi surface of the electrons. The electronic charge density as a function of position $\vec{r}$ can be represented by Eq. (5), where $P_0(\vec{r})$ is the electronic charge density

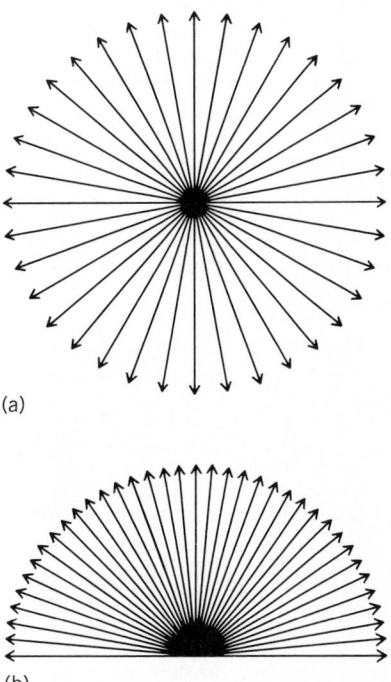

(a)

(b)

Fig. 6. Distribution of magnetic moments in (a) speromagnetic glass and (b) asperomagnetic glass.

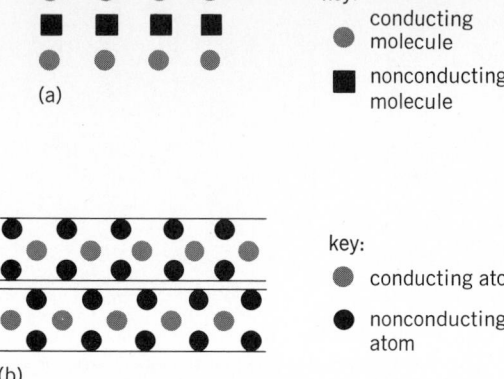

(a)

key:

● conducting
molecule

■ nonconducting
molecule

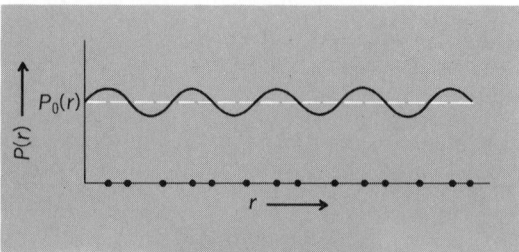

(b)

key:

● conducting atom

● nonconducting
atom

Fig. 7. Schematic representations of low-dimensional conductors. (a) One-dimensional conductor. Conducting molecules are separated by nonconducting molecules. (b) Two-dimensional conductor. Conducting atoms, for example, niobium, form a sheet. Nonconducting atoms, for example, sulfur, form sheets above and below the sheet of conducting atoms and thus help to isolate it from the next layer or sheet of conducting atoms.

$$P(\vec{r}) = P_0(\vec{r})[1 + \alpha \cos(\vec{q}_0 \cdot \vec{r} + \phi)] \quad (5)$$

above the transition (that is, in the normal state), α is the amplitude of the charge-density wave (which is zero above the transition and grows as the temperature is lowered below the transition temperature), $\vec{q}_0$ is the wave vector of the charge-density wave which determines the periodicity, and ϕ is the phase. This modulation of the negative electronic charge alone is energetically unfavorable due to the Coulomb interaction. But the modulation of the electronic charge density produces a distortion in the lattice of the ions, such that the attractive Coulomb interaction between electronic charge and ionic charge helps to stabilize the charge-density wave (Fig. 8). It is this accompanying lattice distortion that can be observed by x-ray, neutron, and electron diffraction techniques and thus provides evidence for the charge-density-wave transition.

The lattice distortion causes a gap in the electronic energy spectrum, which leads to a reduction in the number of electrons available for conduction and hence causes an increase in the resistivity of the material. The size of the gap grows from zero at the transition temperature to a maximum at the

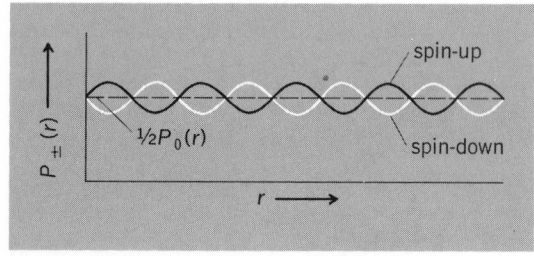

Fig. 8. Charge-density wave. Horizontal straight line shows electronic charge density in absence of wave. Curve shows sinusoidal variation of charge density below transition temperature. Dots at bottom show distorted positions of atoms in lattice.

absolute zero of temperature. In some cases the gap is so large that the material becomes an insulator at the onset of the lattice distortion, leading to a metal-insulator transition. *See* CHARGE-DENSITY WAVE.

Spin-density waves. In a charge-density-wave state below the transition, the modulation in the electronic density has the same phase for electrons of both spins. If the modulation in the density of spin-up electrons is 180° out of phase with the modulation in density of spin-down electrons, then the charge density would remain unmodulated, but there would be a modulation of the spin density. Such a state is called a spin-density-wave state. The electronic density for spin-up (+) and spin-down (−) electrons can then be written in a fashion similar to that for the charge-density-wave case as

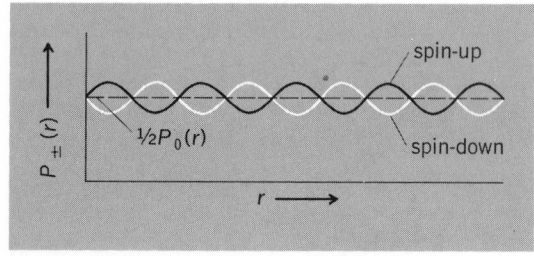

Fig. 9. Spin-density wave. The two curves show the variations of the densities of spin-up and spin-down electrons.

Eq. (6) [Fig. 9]. Spin-density-wave transitions are

$$P_\pm(\vec{r}) = \tfrac{1}{2}P_0(\vec{r})[1 \pm \alpha \cos(\vec{q}_0 \cdot \vec{r} + \phi)] \quad (6)$$

much less common than charge-density waves, but are known to exist in metallic chromium below 312 K. *See* SPIN-DENSITY WAVE.

Valence transitions. In certain rare-earth and actinide materials it has been observed that the valence, or in other words the electronic occupation of the 4f or 5f orbital, of the rare-earth or the actinide atom may change as a function of temperature, pressure, or composition, or a combination of these. If the external pressure and the composition of the material remain constant, the valence may undergo either a first-order or a second-order transition as the temperature is varied. In most cases the transition is discontinuous (and is thus first-order).

This transition is attributed to the delocalization of a 4f (or 5f) electron. If an electron is delocalized from a 4f (or 5f) shell, it contributes to the conduction band of the solid, and the valence of the atom from which this electron came increases by one. Valence transitions are accompanied by changes in physical properties of the material, such as resistivity and the lattice constant.

[D. J. SELLMYER; S. JAFAREY]

Bibliography: P. Chaudhari, B. C. Giessen, and D. Turnbull, Metallic glasses, *Sci. Amer.*, 242(4): 98–117, April 1980; F. J. DiSalvo, Jr., and T. M. Rice, Charge-density waves in transition-metal compounds, *Phys. Today*, 32(4):32–38, April 1979; L. D. Landau and E. M. Lifshitz, *Statistical Physics*, 1958; R. A. Levy and R. Hasegawa (eds.), *Amorphous Magnetism II*, 1977.

Phase velocity

The velocity of propagation of a pure sine wave of infinite extent. In one dimension, for example, the form of the disturbance for such a wave is $y(x,t) = A \sin[2\pi(x/\lambda - t/T)]$. Here x is the position at which the disturbance $y(x,t)$ exists at time t, λ is the wavelength, T is the period which is related to the wave frequency by $T = 1/f$, and A is the disturbance amplitude. The argument of the sine function is called the phase. The phase velocity is the speed with which a point of constant phase can be said to move. Thus $x/\lambda - ft = $ constant, so the phase velocity v_p is given by $dx/dt = v_p = \lambda f$. This is the basic relationship connecting phase velocity, wavelength, and frequency. *See* PHASE; SINE WAVE; WAVE MOTION.

For a simple sine wave on a string, one can perceive this phase velocity (provided the wave does not move too quickly) by merely focusing attention on any particular wave crest and observing its apparent motion along the string. A fairly slack, heavy cord has a slow easily observable phase velocity.

The phase velocity for waves in a medium is determined in part by intrinsic properties of the medium. For all mechanical waves in elastic media, the square of the phase velocity is proportional to the ratio of the appropriate elastic property of the medium to the appropriate inertia property. For example, the square of v_p for transverse waves of small amplitude in a stretched string is $v_p^2 = T/\mu$, where T is the tension in the string and μ is the mass per unit length. The tension is proportional to Young's modulus of elasticity Y for the material from which the string is made, so $v_p^2 \alpha Y/\mu$. The phase velocity of electromagnetic waves depends upon the medium as well. In vacuum or (usually to good enough approximation) air, the phase velocity c is given by $c^2 = 1/\epsilon_0 \mu_0 \approx 9 \times 10^{16}$ m²/s², where ϵ_0 and μ_0 are respectively the permitivity and permeability of the vacuum. *See* ELECTROMAGNETIC RADIATION.

Phase velocity may also depend upon the mode of wave propagation. For example, transverse waves in a bar travel with a phase velocity which is different from that for longitudinal waves of the same frequency traveling in the same bar.

Phase velocity will also depend, in general, upon the frequency of the wave. Waves of different frequencies will travel at different speeds, resulting in a phenomenon called dispersion. A beautiful example of this is the dispersion of white light into the colors of the visible spectrum by a prism. There the phase velocity is given by c/n, where n is the index of refraction of the glass from which the prism is made, which depends quite markedly upon the frequency (hence wavelength, hence color) of the light. The equation for determining the disturbance at any place and time, given suitable starting values, is called the wave equation. If this wave equation involves only second-order rates of change of the disturbance with respect to both space coordinates and time, then the phase velocity is frequency-independent. Otherwise, the phase velocity will be frequency-dependent. *See* GROUP VELOCITY; LIGHT; REFRACTION OF WAVES; WAVE EQUATION.

[S. A. WILLIAMS]

Bibliography: D. Halliday and R. Resnick, *Fundamentals of Physics*, rev. printing, 1974; F. Lokkowicz and A. C. Melissinos, *Physics for Scientists and Engineers*, 1975.

Phonon

A sound quantum. The energy of a phonon is $h\nu$, where h is Planck's constant and ν the frequency of vibration of the sound wave. The phonon is thus analogous to the photon, a light quantum.

In treatments of the scattering of electrons and other particles by thermal waves (short sound waves) in matter, the selection rules which arise bear a formal resemblance to the laws of conservation of energy and momentum holding for collisions between particles. This leads to the concept of a phonon as a packet of sound waves, the wave packet having particlelike aspects. The concept is particularly convenient in the theory of the thermal conductivity of insulators, where one may speak of a phonon gas, collisions between phonons, and a phonon mean free path. In the theory of the properties of superfluid helium, the quanta of longitudinal sound waves in the liquid helium are called phonons. *See* CONDUCTION (HEAT); THERMAL CONDUCTION IN SOLIDS. [JULES DE LAUNAY]

Phosphorescence

A delayed luminescence, that is, a luminescence that persists after removal of the exciting source. It is sometimes called afterglow.

This original definition is rather imprecise, because the properties of the detector used will determine whether or not there is an observable persistence. There is no generally accepted rigorous definition or uniform usage of the term phosphorescence. In the literature of inorganic luminescent systems, some authors define phosphorescence as delayed luminescence whose persistence time decreases with increasing temperature. According to this usage, luminescence whose persistence time is independent of temperature is called fluorescence regardless of the length of the afterglow; a temperature-independent afterglow of long duration is called simply a slow fluorescence, which implies that the atomic or molecular transition involved is forbidden to a greater or lesser degree by the spectroscopic selection rules. In nonphotoconductive inorganic systems, phosphorescence arises when some excitation process has placed an atom (or ion or molecule) in a metastable energy state M (from which transitions to the state of lowest energy, or ground state G, are highly improbable or forbidden), and energy from the thermal vibrations of the system subsequently raises the atom to a higher energy state E from which luminescent transitions are highly probable or allowed (see illustration). The most common mechanism of phosphorescence in photoconductive inorganic systems, however, occurs when electrons or holes, set free by the excitation process and trapped at lattice defects, are expelled from their traps by the thermal energy in the system and recombine with oppositely charged carriers with the emission of light. In these cases, the level M represents the state of the system with the electron or hole trapped. *See* HOLES IN SOLIDS; SELECTION RULES.

In the study of organic systems much attention

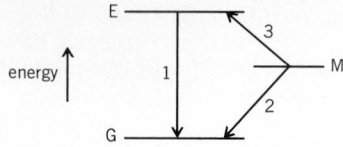

1 = allowed transition
2 = forbidden transition
3 = thermally excited (radiationless) transition

process	usage		
	1	2	3 followed by 1
inorganic	fluorescence	fluorescence	phosphorescence
organic	fluorescence	phosphorescence	delayed fluorescence

Atomic and molecular transitions involved in luminescence, and corresponding terminology.

has been given to the G→E→M process, which is the excitation of a molecule from its ground state to an excited state, both usually spectroscopic singlet states, followed by a radiationless transition (called an intersystem crossing) from the singlet excited state to the metastable triplet state M. In the organic literature the term phosphorescence is reserved for the forbidden luminescent transition M → G, while the afterglow corresponding to the M→E→G process is called delayed fluorescence. The spectrum (color) of organic "phosphorescence," defined in this way, is necessarily different from the spectrum of the ordinary fluorescence, because the emitting states (E and M) in the two cases are different and the final (ground) state G is the same. Conversely, the spectra of ordinary fluorescence and delayed fluorescence in organic systems are the same, because the luminescent transition takes place between the same emitting state E and the ground state G in both cases.

The temperature-dependent (M→E→G process) luminescence of a given system can exhibit a wide range of persistence times. At very low temperature where there is insufficient energy available to raise atoms from metastable to emitting states, or to expel electrons from traps, little or no afterglow is observed. At some higher temperature a low-intensity, long-lived afterglow will be observed; at a still higher temperature the afterglow will be brighter but of shorter duration. Finally, at some high temperature where rate of expulsion of atoms from metastable states or rate of expulsion of electrons from traps is very rapid, afterglow can become immeasurably short.

The time dependence of the luminescence intensity (the decay law) can be extremely complex, depending on the number and energies of the metastable states or electron traps involved. Phosphors which give a phosphorescent emission visible to the eye for about half a day at normal temperatures have been synthesized. See ABSORPTION OF ELECTROMAGNETIC RADIATION; FLUORESCENCE; LIGHT; LUMINESCENCE.

[CLIFFORD C. KLICK; JAMES H. SCHULMAN]

Photoconductivity

The increase in electrical conductivity caused by the excitation of additional free charge carriers by light of sufficiently high energy in semiconductors and insulators. Effectively a radiation-controlled electrical resistance, a photoconductor can be used for a variety of light- and particle-detection applications, as well as a light-controlled switch. Other major applications in which photoconductivity plays a central role are television cameras (vidicons), normal silver halide emulsion photography, and the very large field of electrophotographic reproduction. The phenomena related to photoconductivity have also played a large part in the understanding of electronic behavior and crystalline imperfections in a variety of different materials. See PARTICLE DETECTOR.

Since the electrical conductivity σ of a material is given by the product of the carrier density n, its charge q, and its mobility μ [Eq. (1)], an increase in

$$\sigma = nq\mu \tag{1}$$

the conductivity $\Delta\sigma$ can be formally due to either an increase in n, Δn, or an increase in μ, $\Delta\mu$. Although cases are found in which both types of effects are observable, photoconductivity ($\Delta\sigma$) in single-crystal materials is due primarily to Δn, with only small effects at low temperatures due to $\Delta\mu$ if photoexcitation decreases the density of charged impurities that scatter charge carriers. In polycrystalline materials, on the other hand, where transport may be limited by potential barriers between the crystalline grains, an increase in mobility $\Delta\mu$ due to photoexcitation effects on these intergrain barriers may dominate the photoconductivity.

The increase in carrier density Δn can be conveniently related to the photoexcitation density f (excitations per unit volume per second) by the simple relation (2), where τ is the lifetime of the

$$\Delta n = f\tau \tag{2}$$

photoexcited carrier, that is, the length of time that this carrier stays free and able to contribute to the conductivity before it loses energy and returns to its initial state via recombination with another carrier of opposite type (that is, electrons with holes, or holes with electrons). In Eq. (2) the photoexcitation term f includes all the processes of optical absorption (excitation across the bandgap of the material, excitation from or to imperfection states in the material, generation of excitons that are thermally dissociated to form the free carriers), and the lifetime τ includes all the processes of recombination (free electron with free hole, free electron with trapped hole, free hole with trapped electron). An understanding of the detailed processes of photoconductivity therefore requires a comprehensive understanding of the variation of optical absorption with photon energy, and of the dependence of recombination on imperfection density, capture cross section, photoexcitation intensity, and temperature. See ABSORPTION OF ELECTROMAGNETIC RADIATION; BAND THEORY OF SOLIDS; EXCITON; HOLES IN SOLIDS; TRAPS IN SOLIDS.

Photosensitivity. Although all insulators and semiconductors may be said to be photoconductive, that is, they show some increase in electrical conductivity when illuminated by light of sufficiently high energy to create free carriers, only a few materials show a large enough change, that is, show a large enough photosensitivity, to be practically useful in applications of photoconductors. There are several ways that the magnitude of the photosensitivity can be defined, depending on the application in mind.

Lifetime-mobility product. Comparison of Eqs. (1) and (2) shows that the basic measure of material photosensitivity is given by the $\tau\mu$ product in the common case where $\Delta\sigma$ results primarily from Δn. The mobility does vary from material to material, but in most practical photoconductors μ has values between 10^2 and 10^4 cm^2/V-s at room temperature, and of course the choice of a particular material is usually dominated by its desirable optical absorption characteristics. The free carrier lifetime τ, on the other hand, can take on a wide range of values from 10^{-9} to 10^{-2} s, depending on the particular density and properties of imperfections present in the material.

Detectivity. One of the major applications for photoconductors has been in the detection of small signals in the infrared portion of the spectrum, where the principal objective is to be able to detect the smallest signal possible with the detecting system. Since in this case the photoconductivity is usually much smaller than the dark conductivity, an ac technique is used in which the light signal is chopped and ac amplification stages are used. The limit to detectability is reached when the light-generated signal is comparable to the electrical noise in the photoconductor. Thus the photosensitivity in this particular case is often defined as a detectivity, which is a normalized radiation power required to give a signal equal to the noise.

Gain. A third device-oriented definition of photosensitivity is that of photoconductivity gain. The gain is defined as the number of charge carriers that circulate through the circuit involving the photoconductor for each charge carrier generated by the light. The time required for a charge carrier to pass through the photoconductor from one electrode to the other, called the transit time, t_r, is given by Eq. (3), where L is the distance between electrodes and V is the applied voltage. The gain is

$$t_r = L^2/\mu V \qquad (3)$$

trodes and V is the applied voltage. The gain is given then by τ/t_r for each type of possible charge carrier, giving Eq. (4) for the total gain if both elec-

$$\text{Gain} = (\tau_e\mu_e + \tau_h\mu_h)V/L^2 \qquad (4)$$

trons and holes contribute. Gains of hundreds or thousands can be readily achieved if the lifetimes are sufficiently long. Gains greater than unity require electrical contacts to the photoconductor that are able to replenish charge carriers that pass out of the opposite contact in order to maintain charge neutrality; such contacts are called ohmic contacts. If nonohmic contacts are used, so that charge carriers cannot be replenished, the maximum gain is simply unity, since only the initially created charge carrier contributes to the current flow. Historically, unity-gain currents of this latter type have been called primary photocurrents, whereas high-gain currents described by Eq. (4) have been called secondary photocurrents.

Spectral response. The variation of photoconductivity with photon energy is called the spectral resonse of the photoconductor. The spectral response of 10 typical photoconductors is shown in the illustration. These curves typically show a fairly well-defined maximum at a photon energy close to that of the bandgap of the material, that is, the minimum energy required to excite an electron from a bond in the material into a higher-lying conduction band where it is free to contribute to

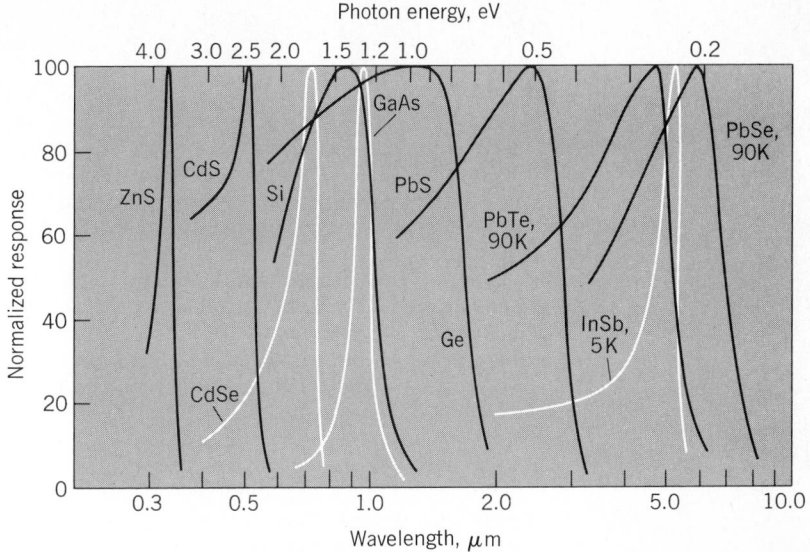

Spectral response of photoconductivity for 10 common photoconducting materials (*From R. H. Bube, Photoconductivity of Solids, John Wiley and Sons, Inc., 1960; reprint, Krieger, 1978*)

the conductivity. This energy ranges from 3.7 eV, in the ultraviolet, for zinc sulfide (ZnS) to 0.2 eV, in the infrared, for cooled lead selenide (PbSe). Photoconductivity associated with excitation across the bandgap of the material is called intrinsic photoconductivity. For photon energies smaller than the bandgap, the light is not strongly absorbed by the material, and the photoconductivity decreases. For photon energies larger than the bandgap, the optical absorption is large, and absorption takes place close to the surface of the material; since the surface has in general more imperfections than the bulk, the carrier lifetime at the surface is generally smaller, and hence the photoconductivity decreases. If the bulk of the material contains a sufficiently high density of imperfections contributing localized levels within the bandgap of the material, it is often possible to detect photoconductivity corresponding to optical excitation from an occupied imperfection level to the conduction band or to an unoccupied imperfection level from the valence band of the material. This photoconductivity occurs for photon energies smaller than the bandgap, and is called extrinsic photoconductivity.

Speed of response. A third major characteristic of a photoconductor of practical concern is the rate at which the conductivity changes with changes in photoexcitation intensity. If a steady photoexcitation is turned off at some time, for example, the length of time required for the current to decrease to $1/e$ of its initial value is called the decay time of photoconductivity, t_d. The magnitude of the decay time is determined by the lifetime τ and by the density of carriers trapped in imperfections as a result of the previous photoexcitation, which must now also be released in order to return to the thermal equilibrium situation. If the photoexcitation intensity is high, or the density of imperfections is small, the decay time t_d approaches the lifetime τ as a minimum limiting value. For low light intensities or high imperfection densities, where the density of trapped carriers is much larger than the density of free carriers, the decay of photoconductivity is controlled not by the free carrier recombination rate but by the rate of thermal freeing of trapped carriers, and can be many orders of magnitude larger than the lifetime.

Device forms. Photoconductive detectors are made as single-crystal or polycrystalline film homogeneous materials, or as *pn* or *npn* semiconductor junctions. The behavior of a *pn* junction is similar to that of a homogeneous material with nonohmic contacts; the maximum gain is unity. The behavior of an *npn* junction is similar to that of a homogeneous material with ohmic contacts; a photoexcited hole remaining in the *p*-type base causes injection of electrons which traverse the device until the hole diffuses out of the base. Thus the gain for an *npn* junction can be much larger than unity. The advantage of the junctions is that they can be made with standard silicon semiconductor technology. *See* JUNCTION DIODE.

Polycrystalline film photodetectors can be made from a variety of methods involving vacuum evaporation, powder sintering, and chemical solution deposition. The photoconductive behavior can be dominated by quite different effects in different materials systems. For example, the photoconductivity in cadmium sulfide (CdS) films deposited by spray pyrolysis is usually controlled by the modulation of intergrain barriers by photoexcitation, so that $\Delta\mu \gg \Delta n$. On the other hand, standard infrared detecting films of lead sulfide (PbS), deposited from chemical solution, exhibit a photoconductivity for which $\Delta n \gg \Delta\mu$, even though the effects of intergrain barriers are clearly measurable in the mobility.

The television camera vidicon and electrophotography are two applications of photoconductivity in which the device form is dictated by the specific nature of the information processing system involved. In both cases an electrical charge is deposited on one side of a high-resistivity photoconducting material; subsequent illumination of the material increases the conductivity locally and allows the charge to leak off through the material. The local absences of charge are then detected and used to produce or reproduce the original light pattern. The material involved must have special characteristics: it must have a high enough dark resistivity so that the deposited charge does not leak off by ordinary dark conduction, and a high enough photosensitivity so that the charge will leak off as quickly as desired. In the vidicon the charge is deposited by scanning by an electron beam; absence of charge is detected by current flow in a subsequent scanning by the beam. In electrophotography the charge is deposited by a corona discharge; absence or presence of charge is fixed by a subsequent printing process.

[RICHARD H. BUBE]

Bibliography: R. H. Bube, Photoconductivity of semiconductors, in H. Eyring, D. Henderson, and W. Jost (eds.), *Physical Chemistry,* vol. 10: *Solid State,* pp. 515–578, 1970; R. H. Bube, *Photoconductivity of Solids,* 1960, reprint 1978; J. Mort and D. M. Pai (eds.), *Photoconductivity and Related Phenomena,* 1976; A. Rose, *Concepts in Photoconductivity and Allied Problems,* 1963.

Photoelectricity

The process by which electromagnetic radiation incident on a solid, liquid, or gas liberates electrical charge, which is detectable in an electric field.

The process is strictly quantum in nature.

Historically, this phenomenon was first explained by Albert Einstein, who invoked the quantum character of electromagnetic radiation, namely the photon, with characteristic energy $h\nu$, where h is Planck's constant and ν the frequency. The liberation of electrons from matter is governed by Einstein's photoelectric equation $E = h\nu - \phi$. where E is the kinetic energy of the emitted electron and ϕ represents the binding energy of the electron.

Early photoelectric experiments involved the photoemission of electrons from the surface of metals into a vacuum, or the liberation of electrons and positively charged ions in a gas by the process of photoionization. Values of ϕ for these processes typically are a few electronvolts, and as such these experiments were limited primarily to the visible and ultraviolet regions of the spectrum. The common photomultiplier tube is an example of a photoemissive device.

The quantum nature of electromagnetic radiation also manifests itself in the liberation of electrons and positive holes within the interior of a solid, giving rise to photoconductive and photovoltaic effects. These phenomena are readily observable in the class of solids known as semiconductors. The binding energy ϕ of electrons and holes in semiconductors can be as little as a few millielectronvolts, and as such the region of the spectrum involved transcends both the visible and the infrared, out into the far infrared. Devices based upon these effects are in widespread use in the fields of thermal imaging, solar energy conversion, security monitoring, data readout, and product control. *See* PHOTOCONDUCTIVITY; PHOTOVOLTAIC EFFECT.   [MICHAEL A. KINCH]

Photometry

That branch of science which deals with the calculation and measurement of light or of its time rate of flow. The term light is usually restricted to electromagnetic radiations of wavelengths that are capable of affecting the human eye. However, in some cases photometry has come to mean also the measurement of radiations in the nearby ultraviolet and infrared regions.

Photometry is usually concerned with measurements of luminous intensity, luminous flux, luminous flux density, luminance, light distribution, color, and the reflectance and transmittance of light; by extension it may even include visibility measurements. *See* ILLUMINANCE; LUMINANCE; LUMINOUS FLUX; LUMINOUS INTENSITY.

Since light is defined as radiant energy that is capable of producing visual sensation, photometric measurements either are made by the human eye or are based upon its visual responses. Because the spectral response of the human eye differs with the individual, the International Commission on Illumination (ICI) has adopted a standard spectral luminous efficiency curve which has been accepted as being that of photopic vision of the normal eye.

Photometric laws. Many photometric measurements are based upon the inverse-square law and upon Lambert's cosine laws of incidence and emission.

Inverse-square law and Lambert's cosine law of incidence. This combined law states that for a point source of light the illumination E on a surface varies directly with the luminous intensity I of the source and inversely as the square of the distance D between the source and the surface and with the cosine of the angle of incidence, θ, measured between the normal to the surface and the direction of D. The relationship is given by Eqs. (1). In prac-

$$E = \frac{I}{D^2}\cos\theta \qquad I = ED^2/\cos\theta \qquad (1)$$

tical photometry, where the source is of finite size, near the source the values of ED^2 vary in a marked manner, but as D increases the values all approach a constant and sooner or later the value of ED^2 becomes essentially independent of D. This limiting value of ED^2 divided by $\cos\theta$ is a measure of the luminous intensity I of the source. The distance from the source at which the inverse-square law therefore begins to apply depends upon the size and shape of the source and the precision of measurement required. Several measurements of E and D for a particular source will reveal the minimum value of D that should be used.

Lambert's cosine law of emission. This law is concerned with light sources and states that the luminous intensity in a given direction radiated or reflected by a perfectly diffusing plane surface varies as the cosine of the angle between that direction and the normal to the surface.

Photometric measurements. The accuracy of photometric measurements when made visually is lower than that of most physical measurements since the quantity measured is not of a physical nature but is a sensation which is never exactly the same for different observers.

Modernization of photometric measurements using photoelectric solid-state cells rather than visual observers has removed the problem of the visual individuality of the observer. Such cells generate electric currents when irradiated and cover the range of the visual spectrum. Although their spectral response is not the same as the standard ICI spectral luminous efficiency curve, their response can be modified by filters to match the ICI curve. Also, geometric corrections can be accomplished by properly shaped translucent shields to correct for deviations from Lambert's cosine law of incidence caused by shadows or reflections.

Although visual photometry has been largely supplanted by physical photometry, photometric measurements in both cases are eventually based on the primary standard of luminous intensity, the candela, cd.

Standards based on lamps. The earliest standards of luminous intensity were wax candles constructed of specific substances and dimensions. In 1884 a device known as the Hefner lamp was adopted in Germany as a reference. After the invention of the incandescent lamp, carbon-filament lamps were used in the English-speaking countries and in France. In 1909, following comparisons among the standard lamps being used in various countries, compromises were reached to establish an International candle recognized by the United States, Great Britain, and France and related quantitatively to the Hefner unit of Germany. This standard was adopted by ICI in 1921.

Differences caused by color variations were theoretically solved by the adoption of the ICI spectral luminous efficiency curve of 1924. However,

changes caused by evaporation of the filament were recognized as potential problems. *See* LUMINOUS EFFICIENCY.

Standard based on blackbody radiator. Radiation properties as determined by the temperature of a substance are a well-developed science. A standard luminous intensity source determined through such properties was the next step. In 1937 the International Committee on Weights and Measures adopted a system based on the luminance of a blackbody, or full radiator at the temperature of the solidification of platinum. In 1948 an agreement upon a luminance of 60 cd/cm² for the standard was reached. In 1967 a specification was added of the atmospheric pressure at which the observation was to be made.

A good approximation of a blackbody radiator can be constructed as shown in the illustration. Centrally contained in a crucible C of thorium oxide is a tube B of the same substance within which is a small quantity of powdered fused thorium oxide A. The space between B and C is filled with platinum P. The platinum is melted by a high-frequency induction furnace, and the whole of the tube (B and A) acquires the temperature of the platinum. The remainder of the system (D, E, F, and G) provides thermal insulation to retard the flow of heat outward from P, B, and A.

When the furnace is switched off, the platinum P and the inner core (B and A) slowly cool. As the change of state of the platinum from liquid to solid occurs, a plateau of temperature is reached and remains constant for about 20 min; during this period the luminance of the radiator is viewed and measured from above. If the loss of heat radiation through the view aperture is negligible, the incandescent body of thorium oxide at the temperature of solidification of platinum obeys the planckian radiation formula for a complete radiator. *See* HEAT RADIATION.

Planck's radiation formula contains two constants: $c_1 = 2\pi hc^2$ (h is Planck's constant from his quantum hypothesis, and c is the velocity of electromagnetic radiation in free space); and $c_2 = hc/k_B$ (k_B is Boltzmann's distribution law constant). Planck's radiation formula also contains the temperature T. The luminance L (established by ICI at 60 cd/cm² in 1948) is related to Planck's radiation spectral power density R_λ and the spectral luminous efficiency curve v_λ by Eq. (2), where k is a

$$\pi L = k \int_0^\infty v_\lambda R_\lambda \, d\lambda \qquad (2)$$

constant converting radiant flux to luminous flux in lumens per watt. The π, derivable from Lambert's cosine law of emission from a perfectly diffuse source, converts luminance to luminous flux density.

While the 1948 ICI definition of the candela fixed the luminous units of the system, the radiated power values depend on the values of h, c, k_B, and T_{Pt} (the solidification temperature of platinum.) Improvements in experimental techniques have resulted in increasingly accurate determinations of the values of these quantities. A 1970 report of the ICI gave best estimated data with probable deviations as $c_1 = (3.74150 \pm 0.00009) \times 10^{-16}$ W·m² and $c_2 = (1.43880 \pm 0.00006) \times 10^{-2}$ m·K. The report listed 2045 K as the best estimate of T_{Pt}. A calculation for k using these values gives $k = 672.7$ lm/W. The report concluded that it is advisable to temporarily continue the use of the previously adopted value of $k = 680$ lm/W, which was based on earlier data, although the range of the measured values is $k = 673 \pm 8$ lm/W.

No statements appear in the ICI reports of the relative dimensions of the viewing aperture with respect to the diameter and length of the crucible tube of the blackbody radiator. A perfect radiator should be completely closed, but an aperture is necessary for the observation. Appreciable loss of heat radiation from the aperture degrades the performance of the radiator. A standardization of the geometry of the approximation of the blackbody radiator should improve the uniformity of results of measurements among those laboratories building and using such radiators.

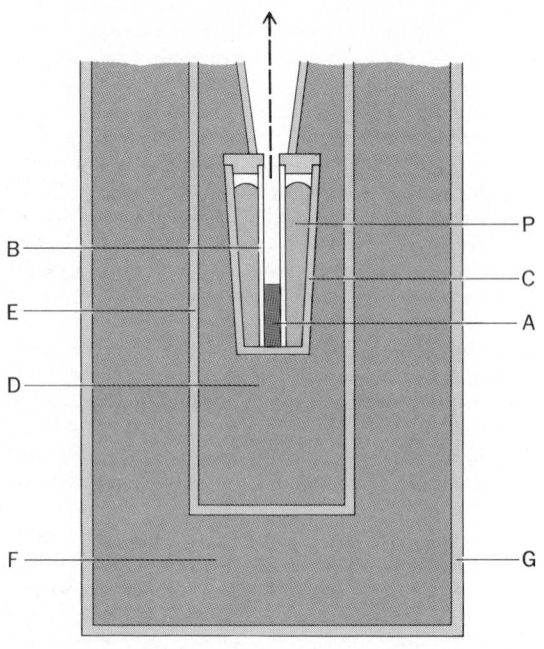

Approximation of a blackbody radiator, which serves as a standard of luminous intensity.

The blackbody radiator shown in the illustration is difficult to construct and to use and is generally only found in the national physical laboratories; elsewhere, lamps, known as secondary standards, are carefully calibrated and used in its place. Working standard lamps are then calibrated with reference to the secondary standards and used as standards of luminous intensity and luminous flux by industrial and educational laboratories.

Standard based on monochromatic radiation. In October 1979 the General Conference on Weights and Measures redefined the base SI unit candela as the luminous intensity, in a given direction, of a source that emits monochromatic radiation of frequency 540×10^{12} hertz and of which the radiant intensity in that direction is 1/683 watt per steradian. The new definition fixes k at 683 lm/W.

If a physically realizable monochromatic radiation at 540×10^{12} Hz can be accomplished, perhaps by the use of a tunable dye laser with a sufficiently narrow frequency bandwidth, then this source could become the primary standard as used in the national and other physical laboratories. The specified frequency corresponds to a wavelength of 555 nm in free space, which is at the maximum visual response of the human eye.

If such a monochromatic radiation cannot be physically accomplished, the ICI may have plans, presently unannounced, for alleviating the problem for any future variations of both watts and lumens that could occur if the blackbody radiator is to be continued as the laboratory standard with the new definition.

Photometric surveys. Photometric tests are made of light sources, lighting fixtures (luminaires), lighting materials, and lighting installations. The measurements made of light sources and luminaires are of luminous intensity, luminous flux, efficiency, luminance, and light distribution curves in one plane.

The Illuminating Engineering Society of North America has standardized various techniques of photometry computation and methods of presenting test data for different types of lighting installations.   [WARREN B. BOAST]

Bibliography: W. B. Boast, *Illumination Engineering*, 2d ed., 1953; Illuminating Engineering Society, *IES Lighting Handbook*, 5th ed., 2d printing, 1978; IES Nomenclature Committee, Proposed American national standard nomenclature and definitions for illuminating engineering, *J. IES*, 9(1):2–46, October 1979; International Commission on Illumination, *Principles of Light Measurements*, CIE no. 18 (E-1.2), Paris, 1970; H. A. E. Keitz, *Light Calculations and Measurements*, 2d rev. ed., 1971; National Bureau of Standards, Guideline for use of the modernized metric system *Dimensions*, 63(12):13–19, 1979.

Photon

A quantum of a single mode (that is, single wavelength, direction, and polarization) of the electromagnetic field. There are also two other definitions of photon in use, not entirely consistent with the first definition or each other: an elementary light particle or "fuzzy ball," and an informal unit of light energy. The fuzzy-ball definition emphasizes a particle character of light suggested, for example, by momentum exhibited in the Compton effect and light levitation phenomena. Although this definition is often justified by the random arrivals of counts in photoelectron detection, light waves incident on a quantum-mechanical detector yield the same behavior. More critically, the fuzzy-ball picture lacks a rigorous foundation and is not required for the explanation of any fundamental phenomenon. As an informal unit of energy, the photon equals $h\nu$, where h is Planck's constant ($= 6.626 \times 10^{-34}$ joule-second), and ν is the frequency of the light in hertz.

The definition as a single-mode light quantum has rigorous foundation in quantum electrodynamics, and contradicts the fuzzy-ball definition in that, according to Fourier analysis, light of a single wavelength must be spread out. Other theories, typified by "neoclassical" theory, attempt to explain the interaction between light and matter by quantizing only the matter's response, that is, without using the photon. However, quantum electrodynamics remains the only theory capable of quantitatively explaining spontaneous emission, the Lamb shift, and the anomalous magnetic moment of the electron. *See* COMPTON EFFECT; QUANTUM; QUANTUM MECHANICS.

[MURRAY SARGENT III]

Bibliography: C. Cohen-Tannoudji, B. Diu, and F. Laloë, *Quantum Mechanics*, 1976; M. O. Scully and M. Sargent III, The concept of the photon, *Phys. Today*, 25(3):38–47, March 1972.

Photovoltaic effect

A term most commonly used to mean the production of a voltage in a nonhomogenous semiconductor, such as silicon, by the absorption of light or other electromagnetic radiation. In its simplest form, the photovoltaic effect occurs in the common photovoltaic cell, used, for example, in solar batteries and exposure meters. The photovoltaic cell consists of an *np* junction between two different semiconductors, an *n*-type material in which conduction is due to electrons, and a *p*-type material in which conduction is due to positive holes. When light is absorbed near such a junction, new mobile electrons and holes are released, as in photoconduction. An additional feature of a photovoltaic cell, however, is that there is an electric field in the junction region between the two semiconductor types. The released charge moves in this field. This current flows in an external circuit without the need for a battery as required in photoconduction. If the external circuit is broken, an "open-circuit photovoltage" appears at the break.

In certain rather complex electrolytic systems, illumination of the electrodes may give rise to a voltage classed as photovoltaic. *See* PHOTOCONDUCTIVITY; SEMICONDUCTOR.

[L. APKER]

Bibliography: L. Azaroff and J. J. Brophy, *Electronic Processes in Materials*, 1963; A. Van der Ziel, *Solid State Physical Electronics*, 3d ed., 1976.

Physical measurement

Quantitative information on physical conditions, properties, or relations essential for coordination of activities, efficiency of communication, and understanding of the nature of things in science and engineering and in much of everyday life. Time, distance, mass, temperature, force, power, and all other physical quantities (or parameters or variables), as well as the properties of matter, materials, and devices, must be described and measured in terms which have the same meaning for everyone. The measuring device or instrument is calibrated (that is, the functional relationship between its indication and the magnitude of the measured quantity is determined) by direct or indirect comparison with a standard which embodies, possesses, or generates a fixed or reproducible magnitude of the physical quantity which is taken as the unit or some multiple or fraction of the unit. Any measured quantity may thus be expressed by a number (the magnitude ratio) and the name of the unit, for example, a length of 1.54 meters. The general area of scientific activity relating to standards and units and the accuracy of measurement is called metrology. *See* UNITS OF MEASUREMENT.

UNITS AND STANDARDS OF MEASUREMENT

From earliest history, nations have had standards for length, volume, and mass. These have differed from country to country and from time to time, so that a large number of units for mass, length, volume, and area came to be in widespread use by the 18th century, some by the same name being of different size in different areas.

In 1793 the French government adopted the decimal metric system wherein the basic unit of length was defined as one ten-millionth of the Earth's polar quadrant (as determined from latitude surveys), to be called the meter. The basic unit for mass was defined as the mass of a cubic decimeter of water, to be called the kilogram. For working standards, a platinum bar was marked with fine lines a meter apart, and a platinum iridium cylinder was constructed equal in mass to a cubic decimeter of water. When later refinements in measurement showed that neither of these standards exactly realized the units as originally defined, the discrepancies were eliminated by redefining the units in terms of the material standards which had been constructed.

In 1866 the United States legalized the use of the metric system, without making its use mandatory, and in 1875 signed a treaty with 18 other countries providing for international cooperation in maintaining and refining international standards of measurement through the General Conference of Weights and Measures (abbreviated CGPM, from the French Conférence Générale des Poids et Mésures), to be convened periodically; a continuing International Committee on Weights and Measures; and a laboratory, established near Paris, known as the International Bureau of Weights and Measures (BIPM).

In 1893 the United States received prototype no. 27 of the international standard meter bar and prototype no. 20 of the international standard kilogram. In that year it was announced that these would be recognized as the fundamental United States standards of length and mass. Later legislation adopted the international metric definitions for electrical and photometric units and provided that the legal standards for these quantities would be maintained by the National Bureau of Standards.

Because the electrical quantities became important only after the metric system had been widely adopted in scientific and industrial work, the metric units for them became the customary units in all countries. Also, metric units for all quantities were increasingly used in the English-speaking countries in scientific education, research, and publication. Thus, to a considerable extent, the whole world has been using metric units for many years. In 1959 the English-speaking countries adopted common definitions for the inch, as 2.54 centimeters exactly, or 1 yard = 0.9144 meter, and for the pound as 0.453 592 37 kilogram. However, the United States gallon is only about 5/6 of the imperial gallon, and other differences remain between United States and imperial measures of capacity.

In 1965 the British government announced a policy of moving toward full metric usage in industry and trade, hoping that this could be generally accomplished within 10 years. By 1980, most major industries had successfully converted and encouraging progress was being made in remaining areas, but total conversion had not been attained.

In 1968 the U.S. Congress authorized the Department of Commerce to undertake a 3-year study of the advantages and disadvantages of wider usage of metric units in United States industry and commerce, the costs involved in making such wider usage, and the optimum rate of change if a concerted change were to be made, and to recommend policies, programs and implementing of legislation as appropriate.

In 1975, the United States adopted the Metric Conversion Act, declaring that "the policy of the U.S. shall be to coordinate and plan the increasing use of the metric system in the United States," and established the U.S. Metric Board "to coordinate the voluntary conversion to the metric system." This board has actively promoted metric education, not only in schools but in industry and every day life. In accordance with the voluntary nature of the programs, the United States may expect many years of mixed units; for example, football fields will probably remain 100 yards in length; track and field events will adapt to international custom. Many technical and some nontechnical publications now give measurement data in both customary and metric units. As familiarity and practice justify, the customary units will be phased out.

However, English units have become almost universal in some worldwide industries—for example, dimensions of oil-drilling equipment, or altitude measurement in aviation. Product sizes can of course remain the same if economic considerations make it desirable; they can be expressed easily in metric units. The widespread practice in aviation flight control is to separate aircraft flying in various directions by 500, 1000, or 2000 feet in altitude. Altimeter pointers, making one revolution for each 1000 feet, provide a conveniently readable index. To use a metric separation basis (300 meters, or even 500 meters) would result in reading inconvenience and some loss of flight levels or of flight safety. Also, by international agreement, through the International Civil Aviation Organization, the unit for speed in air navigation (as in marine navigation) is the knot, defined as 1 nautical mile (1 minute of arc on the Earth's surface) per hour. In view of existing practice, the CIPM has accepted the nautical mile and the knot with a number of other units to be used temporarily with the International System. Thus it is likely that there will always be exceptions to uniformity, requiring special knowledge of special units for at least some people even as the whole world "goes metric" in principle.

The General Conference of Weights and Measures has adopted numerous changes in the definitions of some of the international units and the standards by which they may be realized. There have been concerted and largely successful efforts by scientists and standards laboratories in all nations to devise standards which would not only have greater precision and reliability, but which would be based on fundamental properties of matter or physical phenomena. Such standards permit the units to be realized independently in various laboratories, free from reliance on comparisons with perishable material standards.

At present the International System of Units (abbreviated SI, from the French Système International d'Unités) is constructed from seven base units for independent quantities plus two supplementary units for plane angle and solid angle (Table 1). Units for all other quantities are derived from these nine units. In Table 2 are listed 19 SI derived units with special names. These units are derived from the base and supplementary units in a coherent manner, which means they are expressed as products and quotients of the nine base and supplementary units without numerical factors. All other SI derived units, such as those in Tables 3 and 4, are similarly derived in a coherent manner from the 28 base, supplementary, and special-name SI units. For use with the SI units, there is a set of 16 prefixes (Table 5) to form multiples and submultiples of these units. For mass, the prefixes are to be applied to the gram instead of to the SI unit, the kilogram. *See* DIMENSIONAL ANALYSIS.

The SI units together with the SI prefixes provide a logical and interconnected framework for measurements in science, industry, and commerce.

Natural units. In some cases, quantities are commonly expressed in terms of fundamental constants of nature, and use of these constants or "natural units" is acceptable.

Many other physical quantities, less commonly used are not included in the SI tables, for example, further time and space derivatives. Likewise, many properties of matter or materials are not listed. Units for all of them can, of course, be expressed as a function of the base units or other derived units.

Typical examples of natural units, with their symbols, are:

elementary charge	e
electron mass	m_e
proton mass	m_p
Bohr radius	a_o
electron radius	r_e
Compton wavelength of electron	λ_c
Bohr magneton	μ_B
nuclear magneton	μ_N
speed of light	c
Planck's constant	h

Table 1. SI base and supplementary units

Quantity*	Unit name	Unit symbol
SI base units		
Length	meter	m
Mass	kilogram	kg
Time	second	s
Electric current	ampere	A
Thermodynamic temperature	kelvin	K
Amount of substance	mole	mol
Luminous intensity	candela	cd
SI supplementary units		
Plane angle	radian	rad
Solid angle	steradian	sr

*Quantity here and in Tables 2, 3, 4, and 7 means a measurable attribute.

Table 2. SI derived units with special names

Quantity	SI unit			
	Name	Symbol	Expression in terms of other units	Expression in terms of SI base units
Frequency	hertz	Hz		s^{-1}
Force	newton	N		$m \cdot kg \cdot s^{-2}$
Pressure, stress	pascal	Pa	N/m^2	$m^{-1} \cdot kg \cdot s^{-2}$
Energy, work, quantity of heat	joule	J	$N \cdot m$	$m^2 \cdot kg \cdot s^{-2}$
Power, radiant flux	watt	W	J/s	$m^2 \cdot kg \cdot s^{-3}$
Quantity of electricity, electric charge	coulomb	C	$A \cdot s$	$s \cdot A$
Electric potential, potential difference, electromotive force	volt	V	W/A	$m^2 \cdot kg \cdot s^{-3} \cdot A^{-1}$
Capacitance	farad	F	C/V	$m^{-2} \cdot kg^{-1} \cdot s^4 \cdot A^2$
Electric resistance	ohm	Ω	V/A	$m^2 \cdot kg \cdot s^{-3} \cdot A^{-2}$
Conductance	siemens	S	A/V	$m^{-2} \cdot kg^{-1} \cdot s^3 \cdot A^2$
Magnetic flux	weber	Wb	$V \cdot s$	$m^2 \cdot kg \cdot s^{-2} \cdot A^{-1}$
Magnetic flux density	tesla	T	Wb/m^2	$kg \cdot s^{-2} \cdot A^{-1}$
Inductance	henry	H	Wb/A	$m^2 \cdot kg \cdot s^{-2} \cdot A^{-2}$
Celsius temperature	degree Celsius	°C		K
Luminous flux	lumen	lm		$cd \cdot sr^*$
Illuminance	lux	lx	lm/m^2	$m^{-2} \cdot cd \cdot sr^*$
Activity (of a radionuclide)	becquerel	Bq		s^{-1}
Absorbed dose, specific energy imparted, kerma, absorbed dose index	gray	Gy	J/kg	$m^2 \cdot s^{-2}$
Dose equivalent, dose equivalent index	sievert	Sv	J/kg	$m^2 \cdot s^{-2}$

*In this expression the steradian (sr) is treated as a base unit.

Units acceptable for use with SI. Certain units which are not part of the SI are used so widely that it is impractical to abandon them. The units that are accepted for continued use with the International System are listed in Table 6. It is likewise necessary to recognize, outside the International System, the following units which are used in specialized fields:

electronvolt	eV
unified atomic mass unit	u
astronomical unit	AU
parsec	pc

Logarithmic measures such as pH, dB (decibel), and Np (neper) are acceptable.

The units shown with an asterisk in Table 7 are

Table 3. Some SI derived units expressed in terms of base units

Quantity	SI unit	Unit symbol
Area	square meter	m^2
Volume	cubic meter	m^3
Speed, velocity	meter per second	m/s
Acceleration	meter per second squared	m/s^2
Wave number	1 per meter	m^{-1}
Density, mass density	kilogram per cubic meter	kg/m^3
Current density	ampere per square meter	A/m^2
Magnetic field strength	ampere per meter	A/m
Concentration (of amount of substance)	mole per cubic meter	mol/m^3
Specific volume	cubic meter per kilogram	m^3/kg
Luminance	candela per square meter	cd/m^2

Table 4. Some SI derived units expressed by means of special names

Quantity	SI unit		
	Name	Symbol	Expression in terms of SI base units
Dynamic viscosity	pascal second	Pa·s	$m^{-1} \cdot kg \cdot s^{-1}$
Moment of force	newton meter	N·m	$m^2 \cdot kg \cdot s^{-2}$
Surface tension	newton per meter	N/m	$kg \cdot s^{-2}$
Power density, heat flux density, irradiance	watt per square meter	W/m²	$kg \cdot s^{-3}$
Heat capacity, entropy	joule per kelvin	J/K	$m^2 \cdot kg \cdot s^{-2} \cdot K^{-1}$
Specific heat capacity, specific entropy	joule per kilogram kelvin	J/(kg·K)	$m^2 \cdot s^{-2} \cdot K^{-1}$
Specific energy	joule per kilogram	J/kg	$m^2 \cdot s^{-2}$
Thermal conductivity	watt per meter kelvin	W/(m·K)	$m \cdot kg \cdot s^{-3} \cdot K^{-1}$
Energy density	joule per cubic meter	J/m³	$m^{-1} \cdot kg \cdot s^{-2}$
Electric field strangth	volt per meter	V/m	$m \cdot kg \cdot s^{-3} \cdot A^{-1}$
Electric charge density	coulomb per cubic meter	C/m³	$m^{-3} \cdot s \cdot A$
Electric flux density	coulomb per square meter	C/m²	$m^{-2} \cdot s \cdot A$
Permittivity	farad per meter	F/m	$m^{-3} \cdot kg^{-1} \cdot s^4 \cdot A^2$
Permeability	henry per meter	H/m	$m \cdot kg \cdot s^{-2} \cdot A^{-2}$
Molar energy	joule per mole	J/mol	$m^2 \cdot kg \cdot s^{-2} \cdot mol^{-1}$
Molar entropy, molar heat capacity	joule per mole kelvin	J/(mol·K)	$m^2 \cdot kg \cdot s^{-2} \cdot K^{-1} \cdot mol^{-1}$
Exposure (x- and γ-rays)	coulomb per kilogram	C/kg	$kg^{-1} \cdot s \cdot A$
Absorbed dose rate	gray per second	Gy/s	$m^2 \cdot s^{-3}$

used in limited fields and have been authorized by the International Committee for Weights and Measures (CIPM). It is recommended that the term "weight" should be avoided in technical publications except under circumstances in which its meaning is completely clear. It is also recommended that the terms atomic weight and molecular weight be replaced by relative atomic mass and relative molecular mass in accordance with established international practice.

The internationally accepted definitions for the seven base units are given below, with brief descriptions of the standards in use for the most precise measurements and calibrations in terms of the units so defined.

Table 5. SI prefixes

Factor	Prefix	Symbol	Factor	Prefix	Symbol
10^{18}	exa	E	10^{-1}	deci	d
10^{15}	peta	P	10^{-2}	centi	c
10^{12}	tera	T	10^{-3}	milli	m
10^{9}	giga	G	10^{-6}	micro	μ
10^{6}	mega	M	10^{-9}	nano	n
10^{3}	kilo	k	10^{-12}	pico	p
10^{2}	hecto	h	10^{-15}	femto	f
10^{1}	deka	da	10^{-18}	atto	a

Mass. The kilogram (kg) is equal to the mass of the International Prototype Kilogram. The International Prototype is a platinum-iridium cylinder preserved at the International Bureau of Weights and Measures at Sèvres, France.

Prototype no. 20 is kept at the U.S. National Bureau of Standards; equivalent prototypes are kept by other countries. Mass is the only one of the base quantities for which the standard is an arbitrarily defined object. No basic property of matter involving mass can be measured with more precision than is possible in comparing kilogram masses by weighing, about 1 part in 10^8.

The standard prototype kilogram embodies the unit of mass; since masses may be compared by weighing, other mass standards are easily adjusted for equality with the unit, within the uncertainty set by the reproducibility of the weighting equipment and the weighing procedure. Sets of standard masses (or weights) are obtained by adding or subdividing unit masses. Within practical ranges (say, 10^{-9} to 10^4 kilograms) any given mass can be measured by weighing it against combinations of standard masses, with an uncertainty equal to the smallest standard used in the weighing. Most weighing balances have additional means for indicating the mass required to remove the remaining imbalance.

Table 6. Units in use with the International System

Name	Symbol	Value in SI unit
Minute	min	1 min = 60 s
Hour	h	1 h = 60 min = 3,600 s
Day	d	1 d = 24 h = 86,400 s
Degree	°	$1° = (\pi/180)$rad
Minute	'	$1' = (1/60)° = (\pi/10,800)$ rad
Second	"	$1'' = (1/60)' = (\pi/648,000)$ rad
Liter	L*	$1 L = 1 dm^3 = 10^{-3} m^3$
Metric ton	t	$1 t = 10^3 kg$
Hectare	ha	$1 ha = 10^4 m^2$

*An alternative symbol for liter is "l". Since "l" can be easily confused with the numeral 1, the symbol "L" is recommended for United States use.

Length. The meter (m) is the length equal to 1 650 763.73 wavelengths in vacuum of the radiation corresponding to the transition between the levels $2p_{10}$ and $5d_5$ of the krypton-86 atom (an orange-red line).

The standard is based on the krypton atom; it is an apparatus for generating the selected radiations and relating its wavelength to other lengths. The generator—a krypton lamp—is a quartz bulb containing krypton gas at low pressure and electrodes for exciting the atoms sufficiently to radiate the chosen spectral line. An optical interferometer illuminated by the krypton radiation exhibits a cycle of variations in the position or intensity of interference bands or spots as the length of the light path is varied by one wavelength.

Other lines of krypton 86 and several lines of mercury 198 and of cadmium 114 are recommended as secondary standards.

Two monochromatic radiations, one in the visible, the other in the infrared spectral region, produced by helium-neon lasers stabilized on a saturated absorption line of iodine or of methane, are recommended as wavelength standards: iodine 127, R(127), band 11−5, component i [wave-length in a vacuum, 632 991.399 × 10^{-12} m]; methane, P(7), band v_3 [wavelength in a vacuum 3 393 231.40 × 10^{-12} m]. These lines are reproducible with an uncertainty of the order of 1 in 10^{10}; the value of their wavelength in meters is subject to the uncertainty of the standard (the wavelength of the ^{86}Kr line) estimated to be 4 in 10^9. By measuring the beat frequencies of neighboring lines (for example, various components of the hyperfine multiplet of iodine) very exact values of the wavelength differences are obtained.

The wavelength of the methane line mentioned above multiplied by its frequency (measured by comparison with the ^{133}Cs transition of the definition of the second) yields the speed of propagation of electromagnetic waves in vacuum c = 299 792 458 m/s. This value of c will probably be kept unaltered in the future and will permit the meter to be defined as the distance electromagnetic radiation travels in a vacuum in 1/299 792 458 s. *See* ELECTROMAGNETIC RADIATION.

Using a krypton light source in an interferometer, lengths of a meter or more can be measured with an uncertainty of about 4 parts in 10^9 in a vacuum. Some carefully stabilized lasers can extend the range of direct measurement to kilometers with equal precision in a vacuum, or to $1 : 10^7$ in air. Powerful ground-built radar transmitters and receivers (radio telescopes) used in conjunction with transponders on spacecraft can extend direct measurement, in terms of radio wavelengths, to interplanetary distances (10^{11} meters) with comparable or greater relative precision. The relative precision increases with range, as long as the count of successive wavelengths is not lost.

On the atomic scale, the size of nuclei (10^{-15} meter) can be measured to ±5%. (They are not all spherical!) *See* LENGTH; NUCLEAR PHYSICS.

Time interval. The second (s) is the duration of 9 192 631 770 periods of the radiation corresponding to the transition between the two hyperfine levels of the ground state of the cesium-133 atom.

Table 7. Examples of conversion factors from non-SI units to SI*

Quantity	Name of unit	Symbol for unit	Definition in SI units
Length	inch	in.	2.54×10^{-2} m
Length	nautical mile*	nmi	1852 m
Length	angstrom	Å	10^{-10} m
Velocity	knot*	kn	(1852/3600) m/s
Cross section	barn*	b	10^{-28} m²
Acceleration	gal	gal	10^{-2} m/s²
Mass	pound (avoirdupois)	lb	0.453,592,37 kg
Force	Kilogram-force	kgf	9.806,65 N
Pressure	millimeter of mercury at 0°C	mmHg	133.322 Pa†
Pressure	atmosphere	atm	101,325 Pa
Pressure	torr	torr	(101,325/760) Pa
Pressure	bar*	bar	10^5 Pa
Stress	pound-force per square inch	lbf/in²	6,894.757 Pa†
Energy	British thermal unit (International Table)	Btu	1055.056 J†
Energy	kilowatt-hour	kWh	3.6×10^6 J
Energy	calorie (thermochemical)	cal	4.184 J
Activity (of a radionuclide)	curie*	Ci	3.7×10^{10} Bq
Exposure (x- or γ-rays)	roentgen*	R	2.58×10^{-4} C·kg⁻¹
Absorbed dose	rad*	rd	1×10^{-2} Gy
Dose equivalent	rem*	rem	1×10^{-2} Sv

*The Committee for Weights and Measures has sanctioned the temporary use of these units.
†Approximate; all other conversion factors are exact.

Frequency (s^{-1}) is the reciprocal of the period for regularly repetitive events; the unit of frequency has been named the hertz (Hz).

As these definitions imply, a time standard (clock) is the combination of a constant frequency generator and a period (or cycle) counter.

Oversimplified, the cesium frequency generator involves a beam of cesium atoms (evaporating off a heated bit of cesium metal through collimating holes) passing through a nonhomogeneous magnetic field which separates the atoms as to their energy states, a variable frequency electromagnetic field stimulating the transition, and a detector of the degree of stimulated transition. An automatic feedback control system varies the frequency of the applied radio-frequency field to achieve maximum effect (resonance), which occurs when the applied frequency is equal to the natural frequency of the transition. In the best equipments the stability and accuracy correspond to an uncertainty of 1 in 10^{12} or even 1 in 10^{13}.

Using refinements of conventional techniques, the cesium-stabilized frequency is divided down to provide other standard frequencies, some of which are made widely available by radio broadcast.

Time intervals are obtained by counting periods of any of the signals of known frequency.

There are other standards besides the cesium beam, among them the hydrogen maser, rubidium clocks, and quartz frequency standards and clocks. Their frequency is controlled by comparison with a cesium standard, either directly, or by means of radio transmissions.

The second was long defined, for physical measurements as well as for civil affairs, as 1/86,400 of the time required for an average complete rotation of the Earth on its axis with respect to the Sun. Because of the slight slowing of the Earth's rotation rate, now averaging about 1 second per year (that is, 3 parts in 10^8) but with erratic and unexplained fluctuations, the universal second thus defined is not a constant. A time scale called Coordinated Universal Time (UTC) recommended by CGPM in 1975 is defined in such a manner that it differs from international atomic time (TAI) by an exact whole number of seconds. This difference is adjusted occasionally by the use of a positive or negative leap second at the end of certain months to keep UTC in agreement with the time defined by the rotation of the Earth with an approximation better than 9/10 second. The legal times of most countries are further offset by a whole number of hours (time zones and "summer time").

Temperature. The kelvin (K), the unit of thermodynamic temperature, is the fraction 1/273.16 of the thermodynamic temperature of the triple point of water.

The unit kelvin and its symbol K should also be used to express an interval or differences of temperature.

Absolute zero is defined as the condition in which all kinetic energy of random motion has been abstracted from the atoms or molecules. While this condition can be approached very closely, it cannot be fully attained. Its attainment would, in fact, be counter to the laws of thermodynamics, since heat can be abstracted only in an environment of lower temperature. The problems of attaining or measuring extremely low tempera-

ture are both complicated and facilitated by the quantum nature of the energy of internal motions of the atoms or molecules (spins, vibrations, or rotations), as well as the inherent zero-point energy of particles in close proximity to others. *See* ABSOLUTE ZERO; TEMPERATURE.

There are a number of physical phenomena that in theory relate thermodynamic temperature to other quantities which can be measured quite accurately, even if subject to many corrections, for example, pressure and volume of gases, specific heats of elements, electrical noise (voltage) in resistors, radiation intensity and spectral distribution, and speed of sound waves in gases. These and others are used in various experiments to determine the numerical values of temperatures below and above the triple point of water, that is, to establish the thermodynamic temperature scale. None of these phenomena, however, provides convenient methods for practical measurements over the entire range of achievable temperatures from 10^{-6} to 10^6 K.

To provide convenient and adequately accurate means for practical realization and measurement of temperature, the International Practical Temperature Scale is used, based on the assigned values of the temperatures of a number of reproducible equilibrium states (defining fixed points) and on standard instruments calibrated at those temperatures. Interpolation between the fixed-point temperatures is provided by formulas used to establish the relation between indications of the standard instruments and values of International Practical Temperature. An extensive revision, which was made in 1968, is called the ITPS-68.

The defining fixed points are established by realizing specified equilibrium states between phases of pure substances: triple points or boiling points (under specified pressures) of hydrogen, neon, and oxygen; the triple point and boiling point of water; and freezing points of tin, zinc, silver, and gold. These fixed points are distributed over the range from 13.81 to 1337.58 K.

The standard instrument used from 13.81 K to 630.74°C is the platinum resistance thermometer.

The standard instrument used from 630.74°C to 1064.43°C is the platinum-10% rhodium/platinum thermocouple, the electromotive force-temperature relation of which is represented by a quadratic equation.

Above 1337.58 K (1064.43°C) the International Practical Temperature of 1968 is defined by Planck's law of radiation, with 1337.58 K as the reference temperature and the value of constant c_2 taken as 0.014388 meter kelvin.

Above the gold point, the standard techniques and the standard instruments of optical pyrometry are employed in conjunction with a blackbody at the temperature that is to be measured.

While the temperature of the triple point is defined exactly (273.16 K), the present uncertainty in independently reproducing it in different laboratories and with different types of apparatus is $0.0002-3$ K, that is, about 1 part in 10^6. The uncertainties in the thermodynamic temperatures of the defining fixed points of the IPTS-68 are not more than 0.01 K up to the boiling point of water, and not more than 0.2 K at the freezing point of gold (1337.58 K), but increase to about 4 K at the melting point of tungsten (3600 K). *See* PLANCK'S RADIATION LAW; RADIATION; THERMODYNAMIC PRINCIPLES.

The degree Celsius (°C), earlier known as the degree Centigrade, is the same magnitude as the kelvin. The Celsius scale assigns 0°C to the freezing temperature of water (273.15 K) and 100°C to the boiling point of water (373.15 K), so that temperature K = temperature C + 273.15.

On the Fahrenheit scale, the freezing point of water is 32°F and the boiling point is 212°F. The degree Fahrenheit, °F, is therefore 5/9 of a kelvin, so that temperature F = 1.8 temperature C + 32, or temperature F + 40 = 1.8 (temperature C + 40).

Electric current. The ampere (A) is that constant current which, if maintained in two straight parallel conductors of infinite length and of negligible circular sections, and placed 1 meter apart in a vacuum, would produce between these conductors a force equal to 2×10^{-7} newton per meter of length.

The experimental realization with highest precision is difficult. It is impractical to measure the force in the idealized geometry; coils of many turns are used. The force is measured by a balance in terms of the local acceleration of gravity. The local acceleration of gravity may be calculated for a given latitude, longitude, and elevation by interpolation from geodetic and gravity surveys; measured by local pendulum experiments (seldom better than 1 part in 10^6); or measured by very elaborate time-of-fall experiments using standard atomic frequencies and wavelength (with uncertainties less than 1 in 10^7).

Constant values of other related electrical quantities — voltage, resistance, capacitance, and inductance — can be maintained more easily than the ampere, some with much better precision. Capacitance of a tubular capacitor of variable length can be calculated quite accurately. The step from capacitance to resistance uses special bridges at various frequencies, and thus the ohm can be determined to about 1 part in 10^7.

The known current going through a standard resistor provides a known voltage which can be used to calibrate voltage standards such as the electrochemical cells on which the United States legal volt is based.

A dc voltage can be maintained with much better precision and stability in terms of a measured frequency by means of the ac Josephson effect. The uncertainties in the ampere and the volt are believed to be about 5 parts in 10^6, but the volt is maintained (thanks to the Josephson apparatus) with a precision and stability better than 5 parts in 10^8. *See* CAPACITANCE; ELECTRIC CURRENT; ELECTRICAL RESISTANCE; ELECTRICAL UNITS AND STANDARDS; ELECTRICITY; INDUCTANCE.

Luminous intensity. The CGPM, in 1979, redefined the base SI unit candela as the luminous intensity, in a given direction, of a source that emits monochromatic radiation of frequency 540×10^{12} hertz and of which the radiant intensity in that direction is 1/683 watt per steradian.

The new definition for the candela is based on monochromatic radiation rather than, as previously, on white light, and provides a definite numerical relationship between the photometric quantities and the watt. It thus links the fields of photometry and radiometry.

The particular frequency, 540×10^{12} hertz (or 555-nanometer wavelength in a vacuum) was chosen because the "spectral luminous efficiency" curves for light- and darkadapted vision have very closely the same value at this frequency. The CIPM had previously ratified the spectral luminous efficiency curves — or weighting functions — as adopted by the International Commission on Illumination. *See* LIGHT; LUMINOUS EFFICACY; LUMINOUS EFFICIENCY; LUMINOUS INTENSITY; PHOTOMETRY; RADIOMETRY.

Amount of substance. The mole is the amount of substance of a system which contains as many elementary entities as there are atoms in 0.012 kilogram of carbon-12. When the mole is used, the elementary entities must be specified, and may be atoms, molecules, ions, electrons, other particles, or specified groups of such particles.

Since the discovery of the fundamental laws of chemistry, units of amount of substance, called, for instance, gram-atom and gram-molecule, have been used to specify amounts of chemical elements or compounds. These units had a direct connection with "atomic weights" and "molecular weights," which were originally referred to the atomic weight of oxygen (by general agreement, taken as 16). But whereas physicists separated isotopes in the mass spectrograph and attributed the value 16 to one of the isotopes of oxygen, chemists attributed that same value to the (slightly variable) mixture of isotopes 16, 17, 18, which was for them the naturally occurring element oxygen. Finally an agreement between the International Union of Pure and Applied Physics (IUPAP) and the International Union of Pure and Applied Chemistry (IUPAC) brought this duality to an end in 1959–1960. Physicists and chemists have ever since agreed to assign value 12 to the isotope 12 of carbon. The unified scale thus obtained gives values of "relative atomic mass." *See* ATOMIC MASS UNIT; ATOMIC WEIGHT; RELATIVE ATOMIC MASS; RELATIVE MOLECULAR MASS.

UNCERTAINTY IN PRACTICAL MEASUREMENTS

Although the seven base units, and others derived from them, are thus exactly defined, their practical availability requires the development and refinement of standard devices or apparatus to realize each of them with high precision. Extensive theoretical studies and laboratory experiments are involved in the selection and refinement of operating principles and in design, construction, and operation of these standards. Once compared with the base standard, other subordinate standards and reference instruments, specimen objects, signal sources (generators or modifiers of the quantity), and so on can be used for further calibrations or measurement.

As noted above, a kilogram mass standard can be calibrated only through a series of comparisons, starting from the International Prototype. The units for the other five base quantities, and all quantities derived solely from them, are in principle independently realizable; that is, the standard apparatus may be constructed equally well in many laboratories. In practice, however, inevitable minor differences between standards constructed independently, even with equal care, and among the instruments, environments, and operators individually are bound to introduce small discrepancies. Periodic comparison of standards and the resolution of these discrepancies is required for compatibility among domestic standards laboratories, as well as internationally. Within the United States, the National Bureau of Standards provides calibration services for industrial, educational, and other governmental standards laboratories and cooperates with them in conducting measurement agreement comparisons. Periodic intercomparisons of NBS standards with those of other countries are made through the International Bureau of Weights and Measures, through international scientific organizations, or by direct arrangement.

Frequency and time comparisons within the United States are made between the NBS, the U.S. Naval Observatory, and other users of high-precision frequency standards. The data from worldwide astronomical observations and from standards laboratories in many countries are coordinated by the International Bureau of the Hour, which coordinates differences between the International Atomic Time scale (a running count of atomic seconds) and the Coordinated Universal Time scale and announces the time when the UTC offset should be changed by ± 1 second to keep UTC in phase with the solar year. Most radio broadcast signals are based on UTC.

In general, any measurement has less than perfect accuracy; there is some error or uncertainty as to the true, or exact, numerical ratio between the magnitude of the measured quantity and the unit. Associated with the measuring instrument itself are the errors made in its calibration, the uncertainty in the constancy or reproducibility of the standard by which it was calibrated, and the possible changes of its response after its calibration. Other uncertainties, somewhat more under the control of the observer making a measurement, include those in taking readings, in correcting for environmental effects, and in allowing for characteristics of the instrument as well as of the system or object undergoing measurement.

Inherent characteristics of materials and structures cause the phenomena of drift, lag, hysteresis, damping, and resonance in measuring instruments or systems, as well as in the systems undergoing measurement.

These general phenomena may affect the relationship of any quantity or condition (mechanical, electrical, thermal, and so on,) to any parameter. In terms of instrument reading and measured quantity, they may be defined as follows:

Drift. This is the gradual continued change of instrument reading after a change to a different but constant value of the measured quantity.

Lag. This is the failure of the instrument reading to follow changes in the measured quantity instantly. The time constant of an instrument is the time required for the indication to change by $1/e$ (0.37) of a sudden change in the measured quantity in response to this change.

Hysteresis. This results from lag or drift, and is the difference between readings of the measured quantity for corresponding actual magnitudes of that quantity when the quantity is increasing and when it is decreasing. *See* MAGNETIC HYSTERESIS; MAGNETISM; THERMAL HYSTERESIS.

Damping. This is the dissipation of energy (electrical, magnetic, or mechanical) caused by a change in the measured quantity. With critical damping the indication changes to its new value with minimum lag without overshoot, following a sudden change in the measured quantity. With less damping, the indication overshoots the new reading and may oscillate about it with decreasing amplitude; with greater damping, the indication changes to its new value more slowly.

Resonance. This is the condition of enhanced response or oscillation which results when the rapidity of change of the measured quantity is close to the natural rapidity of response of the instrument. *See* CIRCUIT; RESONANCE (ACOUSTICS AND MECHANICS); RESONANCE (ALTERNATING-CURRENT CIRCUITS).

The change of properties of materials or structures with temperature, pressure, humidity, radiation, vibration, or other environmental conditions may cause further uncertainties in the response of the instrument and in the characterization of the quantity or property being measured.

The interaction of the measuring instrument or process with the quantity being measured is often not negligible, and uncertainties remain even after corrections; for example, measuring a voltage usually requires some power, which generally tends to lower the measured voltage. On the atomic scale, Heisenberg's uncertainty principle states that it is impossible to determine simultaneously both the position and the momentum of a particle with the product of the uncertainties less than about 10^{-34} joule second. See UNCERTAINTY PRINCIPLE.

Thus, the analysis of experiments to detect all possible sources of error, the design of experimental procedures to minimize them, and the development of mathematical techniques for estimating their probable magnitudes are all important parts of any measurement in which highest accuracy is essential. The effect of random disturbances and reading errors may be minimized by averaging repeated measurements. Averages with various observers may reduce the effects of systematic reading errors of a given observer.

From the dispersion of repeated observations a statistical estimate of the imprecision (often called precision or repeatability) of the measurement may be obtained. *See* STATISTICS.

The inaccuracy (often called accuracy) of a measurement (or calibration) is the total uncertainty, including not only the imprecision of observation, but also the systematic uncertainties associated with the measuring instrument and with the measuring process.

The reduction of uncertainty in measurement is one of the continuing objectives in science; it is also one of the most fruitful contributions to science. Scientific theories live or die as they are tested by more precise measurements. New theories are evolved to fit phenomena revealed by more precise observations, and new advances in many fields follow on improved measurement capability for any one quantity.

Spectacular reductions in uncertainty—by a factor of 10^6 in 20 years—occurred in the determination of time interval and frequency after the introduction and refinement of the atomic standards. For all quantities, the average improvement has been about a factor of 10 in the same time. While further progress becomes continually more difficult and expensive, there appears to be no fundamental reason why the same average rate of improvement could not be maintained, or exceeded, for at least the rest of the 20th century.

MEASUREMENT TECHNIQUES

The comparison of quantities as to equality, the counting of units, and the determination of the coincidence of events—all involved in physical measurement—may be done by an observer (usually using his visual, aural, or tactile faculties) or by instruments which display or record the results, or apply them to automatic computation or control.

Direct measurement involves comparison with a standard (such as a meter bar) or measurement by a calibrated instrument (such as a voltmeter). Indirect measurements are those derived from measurements of related quantities; for example, the mass of the electron can be derived from measurements on the bending of the path of electrons of a known velocity in a known magnetic field and from separate measurements of the electron charge.

Measuring instruments or systems may respond to the physical quantity to be measured by generating or modifying another quantity or series of quantities. The final (output) quantity, having a functional relationship to the quantity measured, may be compared with a standard of its own kind, appropriately scaled and labeled to represent units of the quantity being measured. For example, in a voltmeter, the voltage across a resistor induces a current which reacts with a magnetic field to generate a force or torque which deforms a mechanical spring and moves a pointer along a scale. Thus the final quantity generated is a length, the displacement of the pointer, which is compared with a scale graduated in intervals of length (or angle) appropriately proportioned to indicate numerically the applied voltage. The proportionality factor is the product of the successive transformation factors from volts to amperes to force to displacement along the scale. Devices that effect this conversion or transformation of one quantity to another are called transducers.

The hundreds of physical phenomena relating one quantity to another supply a rich reservoir of alternatives for the designer of measurement experiments or measuring instruments. For example, the determination of concentration or composition is widely important in both science and industry. An increasing variety of spectroscopies—radiofrequency, infrared, ultraviolet, x-ray, gamma-ray, and so on—utilize atomic and molecular characteristics for absorption, emission, scattering, or reemission of electromagnetic radiation or nuclear particles to determine not only composition but even molecular and crystalline structure.

Many atomic phenomena involve several quantities and may serve as a basis for extending the range or accuracy of measurement of each of them, and others. *See* JOSEPHSON EFFECT; LASER; MASER; MÖSSBAUER EFFECT.

Some techniques for reducing uncertainty, extending range, or providing flexibility and convenience in measurements in general are briefly mentioned below.

1. Many natural phenomena and the fundamental properties of matter, such as the speed of light in a vacuum or the charge of the electron, provide constant values of certain quantities, or combinations of quantities which may be used in physical measurement or instrument design. For example, the precession frequency of the spin of the proton in the hydrogen atom, or hydrogen compounds, is proportional to the magnetic field, or to a current generating it, which may be determined from this frequency. It in turn is indicated by varying the frequency of a high-frequency electric field, applied at right angles to the magnetic field, to achieve maximum coupling, through synchronized spins, with a similar passive circuit detecting an induced electric field at right angles to both the magnetic and applied electric fields.

2. The quantity to be measured or one related to it in a known manner may be allowed to modulate or attenuate the value of another quantity—as when the thickness of metal or paper is determined from its attenuation of an x-ray beam, which is converted by any of several types of radiation meters (transducers) to a change in voltage or current and then to a visual indication or record. As is common in spectroscopy, uncertainty due to variation of the source may be reduced by taking the ratio of signals with and without the sample in the beam.

3. The measured quantity itself, or its signal, may be time-modulated to permit better discrimination between the measured quantity and extraneous effects (noise) in the measuring instrument.

4. In weighing on a balance, known weights are substituted for unknown weights previously balanced by counterweights, so that such uncertainties as knife-edge placement and beam lengths do not affect the comparison.

5. The effect of a quantity to be measured may be nearly offset by a similar known and fixed quantity, and the differences measured by an instrument of lesser range and higher sensitivity.

6. If the standard offsetting quantity can be divided into sufficiently small or continuously variable fractions, it can be adjusted for equality to within the smallest limits detectable. This is the so-called null, or balancing, method of measurement.

7. A number of small and equal magnitudes of the same quantity may be combined so that the cumulative magnitude is appropriate for measurement with available methods.

8. The undesired effects of environmental factors, such as temperature and external magnetic fields, may be compensated for in the instrument system by elements which are responsive to the disturbing factor and interact with the measuring or indicating means to offset such effects.

9. Based on careful design of experiments, observations may be made under a wide variety of conditions, with controlled variation in the factors considered as possible sources of error, permitting statistical estimation of the magnitude of the various errors and appropriate correction for them.

10. In measuring properties of materials, uncertainties may be reduced by comparing measurements on the sample with those made on a sample having closely similar, and accurately known, properties. Hundreds of such reference materials or standard samples are available from government or industrial sources with an indication of the

value (and the uncertainty) of the characteristic property, be it composition, purity, size, radioactivity, viscosity, or some other. Measurements on such reference materials obviously provide a calibration check of measuring instruments; if their properties are close to those of the unknowns, the differences may be measured by more sensitive methods or devices.

11. Published values for carefully measured properties of generally available materials may obviate the need for repetitive measurements, and may also provide a basis for checking calibration.

12. Data obtained by the "round-robin" circulation of two test pieces, of somewhat different (and undisclosed) values, for measurement by each of a number of individuals or laboratories, provide a simple and practical method for self—as well as group—evaluation.

[WILLIAM A. WILDHACK]

Bibliography: C. W. Churchman and P. Tatoosh (eds.), *Measurement: Definitions and Theories*, 1959; *Dimensions*, December 1979; B. Ellis, *Basic Concepts of Measurement*, 1959; *Flow: Its Measurement and Control in Science and Industry, 1974*, Symposium Vols., 1981; (also symposia on *Humidity*, 1965, and *Temperature*, 1972); R. Garnap, *Philosophical Foundations of Physics*, 1960; J. S. Hunter, The national system of scientific measurement, *Science*, 210:869–874, Nov. 21, 1980; *The International System of Units (SI)*, NBS Spec. Publ. 330, 1977; *NBS Standard Reference Material Catalog*, NBS Spec. Publ. 260, 1979–1980; R. S. Sangster, *Structure and Functions of the National Measurement System*, NBS IR 75-949, 1977. See also *Analytical Chemistry*; *Instruments and Measurements* (Soviet Union); *Journal of Physical and Chemical Reference Data*; *Journal of Scientific Instruments* (British); *Metrologia*; *NBS Journal of Research*; *Review of Scientific Instruments*.

Physical optics

The study of the interaction of electromagnetic waves in the optical range with material systems. The optical range of wavelengths may be taken as the range from about 10 A (10^{-6} mm) to about 1 mm. More narrowly, physical optics deals with the relationship between the atomic structure of a system and the manner in which the system affects light sent into it. The chief founder of this branch of science was Michael Faraday, who in 1845 provided the first clue to the electromagnetic nature of light by showing that the optical properties of glass could be altered by a magnetic field. *See* FARADAY EFFECT.

The explanation of the absorption, reflection, scattering, polarization, and dispersion of light by a material medium in terms of the properties of the atoms and molecules making up the medium is the objective of physical optics. In the course of seeking this objective, physicists have found that optical investigations are powerful methods of determining the structures of atoms and molecules and of larger systems composed thereof. *See* ABSORPTION OF ELECTROMAGNETIC RADIATION; ATOMIC STRUCTURE AND SPECTRA; CRYSTAL OPTICS; DIFFRACTION; DISPERSION; ELECTROMAGNETIC RADIATION; ELECTROOPTICS; FLUORESCENCE; INTERFERENCE OF WAVES; LASER; LIGHT; MAG-NETOOPTICS; MOLECULAR STRUCTURE AND SPECTRA; POLARIZED LIGHT; REFLECTION OF ELECTROMAGNETIC RADIATION; REFRACTION OF WAVES; SCATTERING OF ELECTROMAGNETIC RADIATION.

[RICHARD C. LORD]

Bibliography: M. Born and E. Wolf, *Principles of Optics*, 6th ed., 1980; F. A. Jenkins and H. E. White, *Fundamentals of Optics*, 4th ed., 1976; R. S. Longhurst, *Geometrical and Physical Optics*, 2d ed., 1974.

Physics

Formerly called natural philosophy, physics is concerned with those aspects of nature which can be understood in a fundamental way in terms of elementary principles and laws. In the course of time, various specialized sciences broke away from physics to form autonomous fields of investigation. In this process physics retained its original aim of understanding the structure of the natural world and explaining natural phenomena.

Basic parts. The most basic parts of physics are mechanics and field theory. Mechanics is concerned with the motion of particles or bodies under the action of given forces. The physics of fields is concerned with the origin, nature, and properties of gravitational, electromagnetic, nuclear, and other force fields. Taken together, mechanics and field theory constitute the most fundamental approach to an understanding of natural phenomena which science offers. The ultimate aim is to understand all natural phenomena in these terms. *See* CLASSICAL FIELD THEORY; MECHANICS; QUANTUM FIELD THEORY.

The older, or classical, divisions of physics were based on certain general classes of natural phenomena to which the methods of physics had been found particularly applicable. These consisted of classical mechanics with branches in celestial mechanics, hydrodynamics, and ballistics; heat and thermodynamics; kinetic theory of gases and statistical mechanics; optics; acoustics; and electricity and electromagnetism. These divisions are all still current, but many of them tend more and more to designate branches of applied physics or technology, and less and less inherent divisions in physics itself.

Branches. The divisions or branches, of modern physics are made in accordance with particular types of structures in nature with which each branch is concerned. Thus particle physics, or high-energy physics, is the most recent branch and is concerned with understanding the properties and behavior of elementary particles, and more particularly of the heavy particles—mesons, baryons, and their antiparticles—which are produced in collisions involving energies in a range measured in billions of electron volts. The next branch in this classification is nuclear physics, which is concerned with associations of neutrons and protons forming the nuclei of atoms; their structure, properties, and energy states; reactions between nuclei, including scattering processes and radioactivity; and related phenomena, such as the interaction of high-speed nuclear particles with matter. Atomic physics is concerned with the structure and properties of atoms as determined by the electrons outside the nucleus; the states of motion of these electrons, including such topics as energy levels, angular momentum properties, and magnetic moments; and the absorption and emission of radiation by atoms.

Continuing with this classification in ascending complexity there is molecular physics, which is concerned with systems of atoms formed into molecules, the nature of intermolecular forces, chemical binding, vibration and rotation spectra of molecules, and the like. Next in order are solid-state physics; physics of liquids; physics of gases; and plasma physics, which deals with properties of highly ionized atoms forming a mixture of bare nuclei and electrons called an ion plasma.

In this same classification could also be included biophysics, which deals with the application of physical methods and types of explanation to biological systems and structures.

Other more specialized classifications may be made in accordance with particular instruments or techniques, such as x-ray diffraction, neutron diffraction, mass spectrometry, infrared spectroscopy, and seismology. The special field of low-temperature physics is characterized not only by special instruments involved in the production and measurement of low temperatures in the range of liquid helium but also by the phenomena of superconductivity and superfluidity which occur only in this temperatue range. Other fields, such as astrophysics and geophysics, are concerned with aspects of other sciences to which physics is applicable.

Mathematical physics is the study of physical phenomena by means of mathematics, and includes the more mathematical parts of all branches of physics, as well as most of the content of statistical mechanics, quantum mechanics, relativity, and field theory. A distinction is often made between mathematical physics and theoretical physics, in which the latter, although still entirely mathematical in form, is thought of as being more closely related to experimental physics. Neither mathematical nor theoretical physics can really be separated from experimental physics, since a complete understanding of nature can only be obtained by the application of both theory and experiment.

Aim. In every area physics is characterized not so much by its subject-matter content as by the precision and depth of understanding which it seeks. The aim of physics is the construction of a unified theoretical scheme in mathematical terms whose structure and behavior duplicates that of the whole natural world in the most comprehensive manner possible. Where other sciences are content to describe and relate phenomena in terms of restricted concepts peculiar to their own disciplines, physics always seeks to understand the same phenomena as a special manifestation of the underlying uniform structure of nature as a whole. In line with this objective, physics is characterized by accurate instrumentation, precision of measurement, and the expression of its results in mathematical terms.

For the major areas of physics and for additional listings of articles in physics *see* ACOUSTICS; ATOMIC PHYSICS; CLASSICAL MECHANICS; ELECTRICITY; ELECTROMAGNETISM; HEAT; LOW-TEMPERATURE PHYSICS; MOLECULAR PHYSICS; NUCLEAR PHYSICS; OPTICS; SOLID-STATE PHYSICS; THEORETICAL PHYSICS.

[WILLIAM G. POLLARD]

Piezoelectricity

Electricity, or electric polarity, resulting from the application of mechanical pressure on a dielectric crystal. The application of a mechanical stress produces in certain dielectric (electrically nonconducting) crystals an electric polarization (electric dipole moment per cubic meter) which is proportional to this stress. If the crystal is isolated, this polarization manifests itself as a voltage across the crystal, and if the crystal is short-circuited, a flow of charge can be observed during loading. Conversely, application of a voltage between certain faces of the crystal produces a mechanical distortion of the material. This reciprocal relationship is referred to as the piezoelectric effect. The phenomenon of generation of a voltage under mechanical stress is referred to as the direct piezoelectric effect, and the mechanical strain produced in the crystal under electric stress is called the converse piezoelectric effect. *See* POLARIZATION OF DIELECTRICS.

Piezoelectric materials are used extensively in transducers for converting a mechanical strain into an electrical signal. Such devices include microphones, phonograph pickups, vibration-sensing elements, and the like. The converse effect, in which a mechanical output is derived from an electrical signal input, is also widely used in such devices as sonic and ultrasonic transducers, headphones, loudspeakers, and cutting heads for disk recording. Both the direct and converse effects are employed in devices in which the mechanical resonance frequency of the crystal is of importance. Such devices include electric wave filters and frequency-control elements in electronic oscillator circuits. *See* ULTRASONICS.

Necessary condition. The necessary condition for the piezoelectric effect is the absence of a center of symmetry in the crystal structure. Of the 32 crystal classes, 21 lack a center of symmetry, and with the exception of one class, all of these are piezoelectric. In the crystal class of lowest symmetry, any type of stress generates an electric polarization, whereas in crystals of higher symmetry, only particular types of stress can produce a piezoelectric polarization. For a given crystal, the axis of polarization depends upon the type of the stress. There is no crystal class in which the piezoelectric polarization is confined to a single axis. In several crystal classes, however, it is confined to a plane. Hydrostatic pressure produces a piezoelectric polarization in the crystals of those 10 classes that show pyroelectricity in addition to piezoelectricity. The pyroelectric axis is then the axis of polarization. *See* CRYSTALLOGRAPHY; PYROELECTRICITY.

The converse piezoelectric effect is a thermodynamic consequence of the direct piezoelectric effect. When a polarization P is induced in a piezoelectric crystal by an externally applied electric field E, the crystal suffers a small strain S which is proportional to the polarization P. In crystals with a normal dielectric behavior, the polarization P is proportional to the electric field E, and hence the strain is proportional to this field E. Superposed upon the piezoelectric strain S is a much smaller

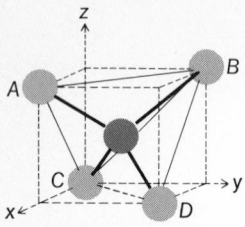

z

A B

C D

x→ →y

● zinc (or sulfur)

● sulfur (or zinc)

Fig. 1. Tetrahedral structure of zincblende, ZnS. Only part of unit cell is shown. Size of circles has no relation to size of ions.

strain which is proportional to P^2 (or E^2). This strain is called the electrostrictive strain. It is present in any dielectric. *See* ELECTROSTRICTION.

Matrix formulation. The relation of the six components T_j of the stress tensor (three compressional components and three shear components) to the three components P_i of the polarization vector can be described by a scheme (matrix) of 18 piezoelectric moduli d_{ij}. The same scheme (d_{ij}) also relates the three components E_i of the electric field to the six components S_j of the strain:

		Compression			Shear		
		S_1 T_1	S_2 T_2	S_3 T_3	S_4 T_4	S_5 T_5	S_6 T_6
E_1	P_1	d_{11}	d_{12}	d_{13}	d_{14}	d_{15}	d_{16}
E_2	P_2	d_{21}	d_{22}	d_{23}	d_{24}	d_{25}	d_{26}
E_3	P_3	d_{31}	d_{32}	d_{33}	d_{34}	d_{35}	d_{36}

The direct effect is obtained by reading this scheme in rows, as in Eq. (1). The converse effect is obtained by reading it in columns, as in Eq. (2).

$$P_i = -\sum_{j=1}^{6} d_{ij} T_j \qquad i = 1, 2, 3 \qquad (1)$$

$$S_j = \sum_{i=1}^{3} d_{ij} E_i \qquad j = 1, 2, \ldots, 6 \qquad (2)$$

An analogous matrix (e_{ij}) relates the strain to the polarization and the electric field to the stress, as in Eqs. (3).

$$P_i = \sum_{j=1}^{6} e_{ij} S_j \qquad i = 1, 2, 3$$

$$T_j = -\sum_{i=1}^{3} e_{ij} E_j \qquad j = 1, 2, \ldots, 6 \qquad (3)$$

The matrices (d_{ij}) and (e_{ij}) are not independent, but are related by expressions involving the elasticity tensor c_{jh}^E (for constant electric field E), as in Eq. (4).

$$e_{mh} = \sum_{j=1}^{6} d_{mj} c_{jh}^E \qquad \begin{matrix} m = 1, 2, 3 \\ h = 1, 2, \ldots, 6 \end{matrix} \qquad (4)$$

Alternative formulations can be made by introducing the dielectric displacement D or visualizing the simultaneous action of electrical and mechanical stresses. *See* ELASTICITY.

The number of independent matrix elements d_{ij} or e_{ij} depends upon the symmetry elements of the crystal. For the lowest symmetry, all 18 matrix elements are independent, whereas piezoelectric classes of higher symmetry can have as few as one independent element in the matrix (d_{ij}). The matrix takes its simplest form if the natural symmetry axes of the crystal are chosen for the coordinate system.

Electromechanical coupling. The direct piezoelectric effect makes a crystal a generator, and the converse effect makes it a motor. Consequently, a piezoelectric crystal has many properties in common with a motor-generator. For example, the electrical properties, such as the dielectric constant, depend upon the mechanical load; conversely, the mechanical properties, such as the elastic constants, depend upon the electric boundary conditions. The electromechanical coupling factor

k can be defined as follows. Suppose electrodes are attached to a piezoelectric crystal and connected to a battery. Then the ratio of the energy stored in mechanical form to the electrical energy delivered by the battery is equal to k^2. In general, k ranges from below 1 to about 30%. In quartz, for example, the coupling is roughly 10%. In ferroelectric crystals, k can approach unity in certain circumstances. *See* FERROELECTRICS.

In quartz, a stress of 1 newton/m applied along the diad axis produces a polarization of about 2×10^{-12} coulomb/m^2 along the same axis. Conversely, an electric field of 10^4 volts/m produces a strain of about 2×10^{-8}. In ferroelectric crystals, such as rochelle salt and KH_2PO_4, and in certain antiferroelectrics, such as $NH_4H_2PO_4$ (ADP), these effects can be several orders of magnitude larger.

Molecular theory. Quantitative theories based on the detailed crystal structure are very involved. Qualitatively, however, the piezoelectric effect is readily understood for simple crystal structures. Figure 1 illustrates this for a particular cubic crystal, zincblende (ZnS). Every Zn ion is positively charged and is located in the center of a regular tetrahedron $ABCD$, the corners of which are the centers of sulfur ions, which are negatively charged. When this system is subjected to a shear stress in the xy plane, the edge AB, for example, is elongated, and the edge CD of the tetrahedron becomes shorter. Consequently, these edges are no longer equivalent, and the Zn ion will be displaced along the z axis, thus giving rise to an electric dipole moment. The dipole moments arising from different octahedrons sum up because they all have the same orientation with respect to the axes x, y, and z.

Another simple type of piezoelectric structure is encountered in barium titanate, $BaTiO_3$, as shown in Fig. 2. The positive Ti ions are surrounded by an almost regular octahedron of negative oxygen ions. The Ti ions are not in the center of the octahedron, but somewhat displaced along the z axis. This structure already has a dipole moment

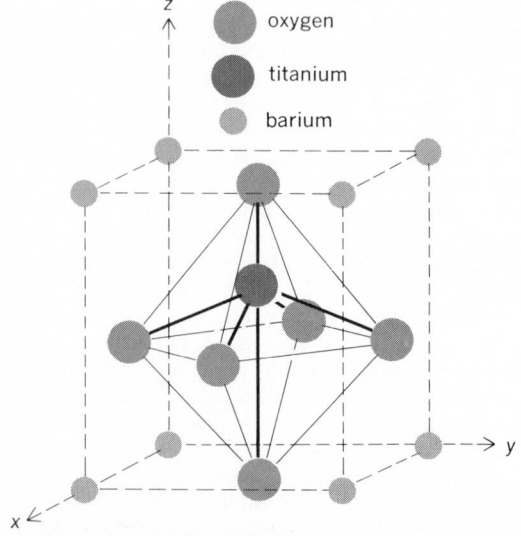

z

● oxygen

● titanium

● barium

x

→ y

Fig. 2. Unit cell of tetragonal barium titanate, $BaTiO_3$. Deviation from cubic symmetry is exaggerated. Size of the circles has no relation to the size of the ions.

or spontaneous polarization in the absence of externally applied stresses. It is clear from Fig. 2 that the Ti ion is pushed more off center when the crystal is mechanically compressed in the xy plane or elongated along z. The additional polarization associated with this deformation is the piezoelectric polarization.

Piezoelectric ceramics. Barium titanate and a few related compounds have the remarkable property that, by means of a sufficiently strong electric field, the direction of the spontaneous polarization can be switched to any one of the x, y, or z axes. This makes it possible to produce polycrystalline samples (ceramics) which are piezoelectric. The electromechanical coupling factors of such ceramics can reach about 50%.

Piezoelectric resonator. The piezoelectric strains that can be induced by a static electric field are very small, except in certain ferroelectrics. Larger strains can be obtained when a piezoelectric crystal is driven by an alternating voltage, the frequency of which is equal to a mechanical resonance frequency of the crystal. The vibrating crystal reacts back on the circuit through the direct piezoelectric effect. In the range of a mechanical resonance, this reaction is equivalent to the response of the network shown in Fig. 3, provided

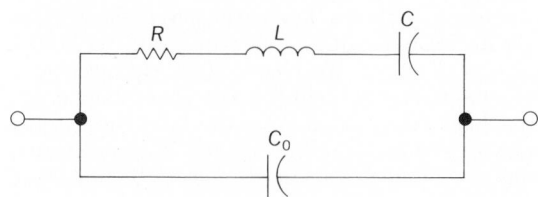

Fig. 3. Network equivalent to a piezoelectric resonator near and at a resonance frequency.

that the series resonance frequency of the network is equal to a mechanical resonance frequency of the crystal, as in Eq. (5). An important difference

$$f_R = 1/(2\pi\sqrt{LC}) \tag{5}$$

between the network of Fig. 3 and the piezoelectric resonator is that the latter has many discrete modes of vibration, whereas the network has only one resonance frequency.

Network elements. The elements L, C, and C_0 of the equivalent network can be calculated from the physical constants of the crystal. Consider, for example, the simple resonator shown in Fig. 4. A rectangular crystal bar with the dimensions $l_1 \gg l_2 \gg l_3$ is excited to compressional lengthwise vibrations. The xy faces have adherent electrodes, and the bar is oriented with respect to the natural crystal axes so that an electric field E_3 along z causes a strain S_1 along the bar according to the equation $S_{1(\text{piezoel})} = d_{31}E_3$. A mechanical stress T_1 along the bar causes a strain $S_{1(\text{mech})} = s_{11}{}^E T$, where $s_{11}{}^E$ is the elastic compliance measured at constant electric field E_3. The resonance frequency for the fundamental lengthwise compressional mode is then given by Eq. (6), where ρ is

$$f_R = 1/(2l_1\sqrt{\rho s_{11}{}^E}) \text{ Hz} \tag{6}$$

the density of the crystal. The parallel capacitance C_0 is the static capacitance of the crystal, as in Eq.

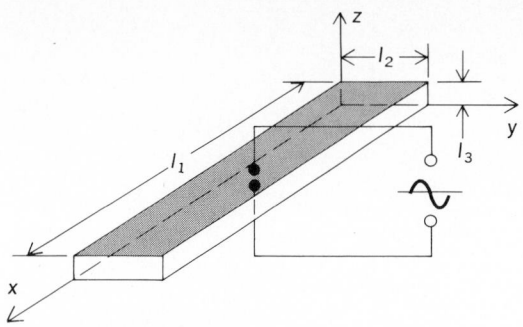

Fig. 4. Simple piezoelectric resonator. A voltage applied to the electrodes shortens or lengthens the bar, thus exciting longitudinal vibrations.

(7). Here ϵ is the relative dielectric constant along

$$C_0 = 8.85\epsilon l_1 l_2 / l_3 \text{ picofarad} \tag{7}$$

z. For C and L, the analysis yields Eqs. (8) and (9).

$$C = \frac{70.8 d_{31}{}^2 \int_1 \int_2}{\pi^2 S_{11} E \int_3} \text{ picofarad} \tag{8}$$

$$L = \frac{\rho (S_{11}{}^E)^2 \int_1 \int_3}{8 d_{31}{}^2 \int_2} \text{ henry} \tag{9}$$

(All physical constants are in mks units.) For the nth overtone, C_0 and L remain the same, whereas C must be divided by n^2. The losses (damping) represented by the resistance R in Fig. 3 arise, for example, from ultrasonic radiation, friction in the crystal mount, internal friction in the crystal originating in various imperfections, and dielectric relaxation.

At the mechanical resonance frequency f_R, the alternating current is maximum and is determined by R. At the antiresonant frequency, given by Eq. (10), the current is minimum. The difference $\Delta f =$

$$f_A = \sqrt{(C_0 + C)/LCC_0} \tag{10}$$

$f_A - f_R$ increases with increasing electromechanical coupling according to relation (11).

$$\Delta f \approx 4k^2/\pi^2 \tag{11}$$

The reactance depends upon frequency, as shown in Fig. 5. For a typical piezoelectric crystal such as quartz, resonating at about 10^5 Hz, the

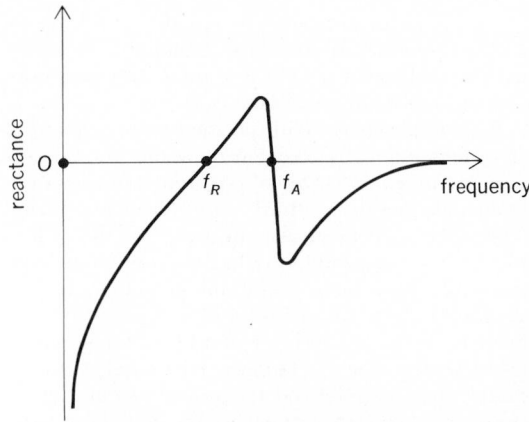

Fig. 5. Diagram showing reactance versus frequency for a piezoelectric resonator.

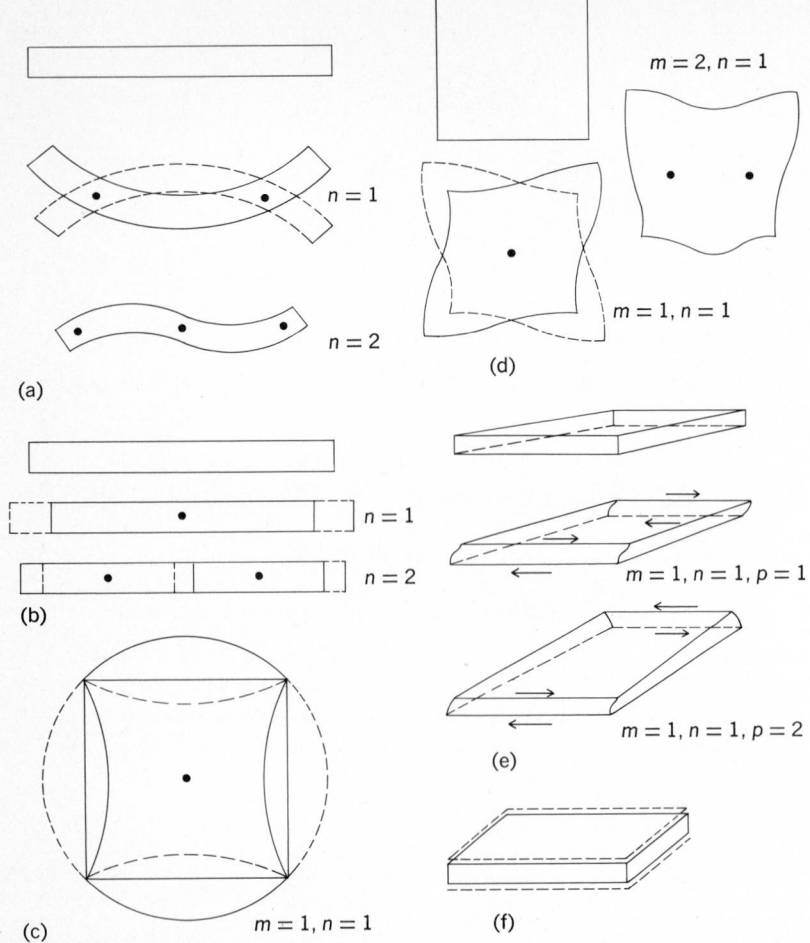

(a)

(b)

(c) $m = 1, n = 1$

(d) $m = 2, n = 1$ $m = 1, n = 1$

(e) $m = 1, n = 1, p = 1$ $m = 1, n = 1, p = 2$

(f)

$n = 1$

$n = 2$

Fig. 6. Diagrammatic representation of examples of vibration modes of bars and plates. (a) Flexural vibrations of a bar. (b) Longitudinal vibrations of a bar. (c) Longitudinal vibration of a plate. (d) Face shear vibrations of a plate. (e) Thickness shear vibrations of a plate. (f) Thickness vibration of a plate.

orders of magnitude given by relations (12) are typical for the elements of the equivalent network.

$$L \approx 10^2 \text{ henry}$$
$$C \approx 0.02 \text{ picofarad} \quad (12)$$
$$C_0 \approx 5 \text{ picofarad}$$

The damping resistance R varies from about 10^2 to 10^4 ohms; that is, the Q factors, given by Eq. (13),

$$Q = \frac{1}{R}\sqrt{\frac{L}{C}} \quad (13)$$

are in the range between 10^6 and 10^4, and the resonances are very sharp. These characteristics cannot be achieved with conventional coils and condensors as circuit elements.

Vibration modes. With piezoelectric resonators of various types, the range from audio frequencies to many megahertz can be covered. The vibration modes frequently used are, in order of increasing frequency: (1) flexural vibrations of bars and plates, (2) longitudinal vibrations of bars and plates, (3) face shear vibrations of plates, and (4) thickness shear vibrations and compressional vibrations of plates. Figure 6 illustrates some of these modes. The excitation of particular vibration modes can be achieved by proper orientation of the resonator with respect to the natural crystal axes, by proper positioning of the electrodes, and by proper mounting. A simple example is illustrat-

ed by Fig. 7. A bar is oriented so that an electric field along x causes an expansion or contraction along y. The electrodes are split and cross-connected so that the bar flexes in the yz plane when a voltage is applied. The fundamental flexure mode is easily excited with this arrangement; however, excitation of higher even-numbered flexural modes is also possible. Interesting resonators are possible with piezoelectric ceramics (BaTiO$_3$ type) because different parts of the resonator can be polarized in different directions. *See* VIBRATION.

Common applications. The sharp resonance curve of a piezoelectric resonator makes it useful in the stabilization of the frequency of radio oscillators. Quartz crystals are used almost exclusively in this application. The main advantages of quartz are high Q factor, stability with respect to aging, and the possibility of orienting the resonator with respect to the natural crystal axes so that the temperature coefficient of the resonance frequency vanishes near the operating temperature. Figure 8 illustrates the orientation of commonly used cuts.

In vacuum-tube oscillators, the crystal generally is part of the feedback circuit. In the circuit proposed by G. W. Pierce, the conditions for oscillation are not satisfied unless the crystal reactance is positive. Hence, the oscillation frequency is between the resonant and antiresonant frequency of the crystal (Fig. 5). Circuits of this type hold the frequency within a few parts per million. Much greater stability can be achieved with the bridge circuit of L. A. Meacham. Here the oscillation conditions are fulfilled by zero phase shift in the feedback circuit, that is, at the exact series resonance frequency of the crystal. Long-term frequency stability of about one part in 10^8 and short-term stability of one part in 10^9 can be achieved with such oscillators.

Selective band-pass filters with low losses can be built by using piezoelectric resonators as circuit elements. With a simple network consisting of resonating crystals only, a passband of twice the difference between resonant and antiresonant frequency can be obtained. For quartz resonators, this passband is about 0.8%. At relatively low operating frequencies, this band is too narrow, and combinations of crystal resonators with coils and condensors are generally used. A synthetic piezoelectric crystal which is often substituted for quartz in this application is ethylene diamine tartrate.

Piezoelectric crystals provide the most convenient means for generation and detection of vibrations in gases, liquids, and solids at frequencies

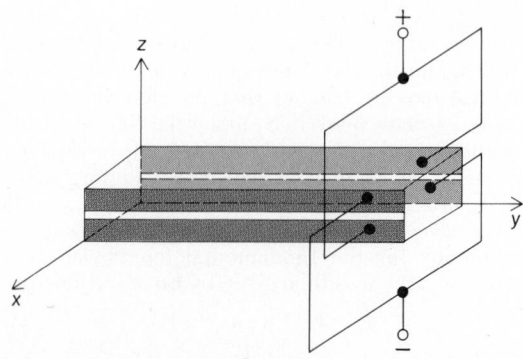

Fig. 7. Excitation of flexure mode by split electrodes.

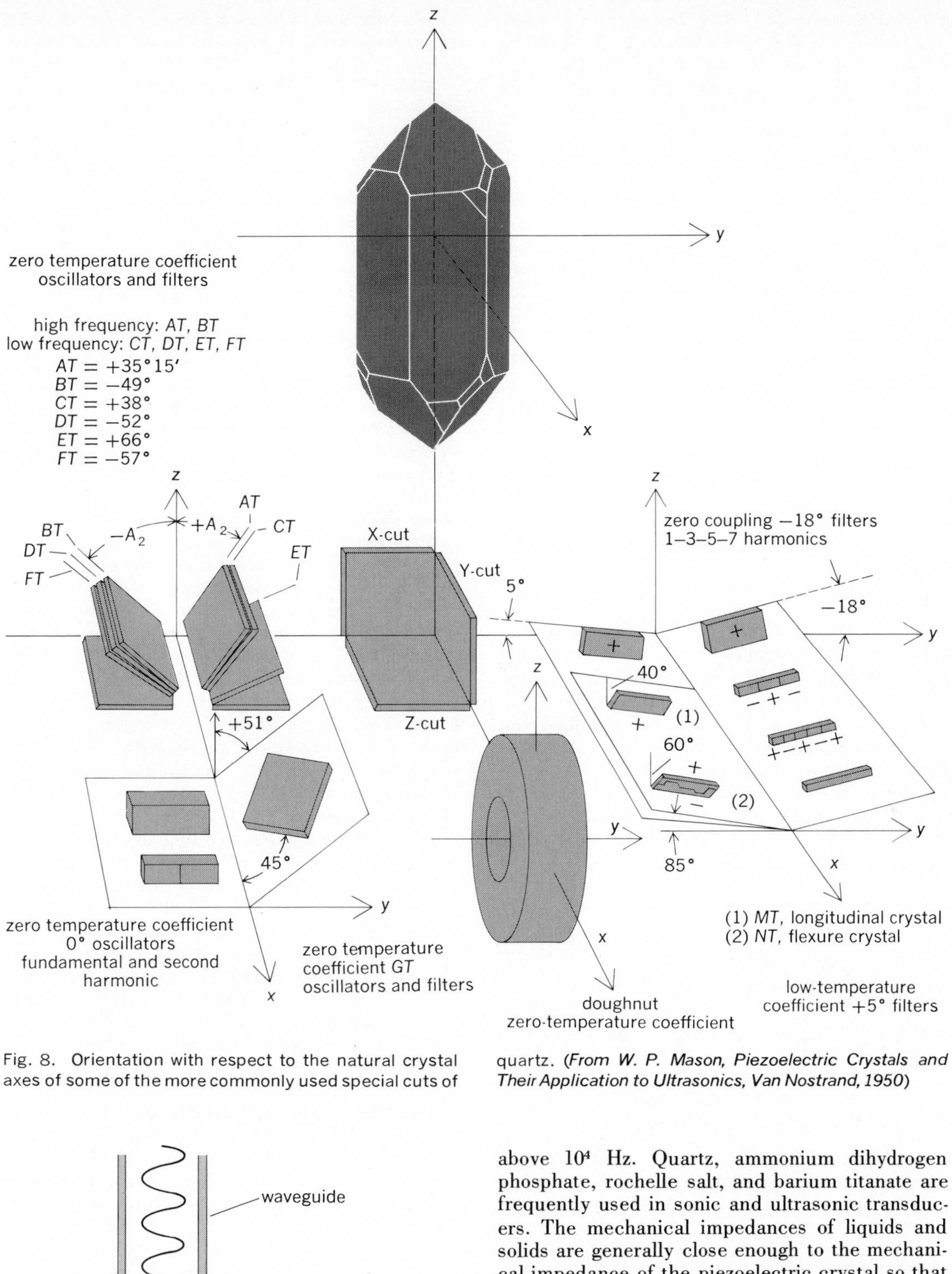

zero temperature coefficient
oscillators and filters

high frequency: *AT, BT*
low frequency: *CT, DT, ET, FT*
AT = +35°15′
BT = −49°
CT = +38°
DT = −52°
ET = +66°
FT = −57°

BT −A₂ +A₂ AT CT
DT   ET
FT

X-cut

Y-cut
5°

zero coupling −18° filters
1–3–5–7 harmonics

−18°

+51°

Z-cut

40°
(1)
60°
(2)

45°

zero temperature coefficient
0° oscillators
fundamental and second
harmonic

zero temperature
coefficient GT
oscillators and filters

doughnut
zero-temperature coefficient

85°

(1) *MT*, longitudinal crystal
(2) *NT*, flexure crystal

low-temperature
coefficient +5° filters

Fig. 8. Orientation with respect to the natural crystal axes of some of the more commonly used special cuts of quartz. (*From W. P. Mason, Piezoelectric Crystals and Their Application to Ultrasonics, Van Nostrand, 1950*)

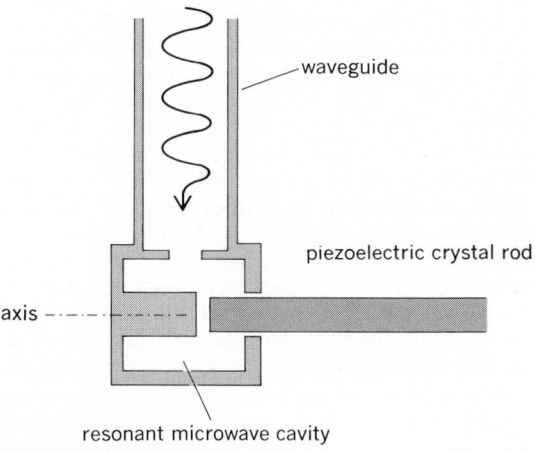

waveguide

piezoelectric crystal rod

axis

resonant microwave cavity

Fig. 9. Diagram showing experimental arrangement for generation of ultrasound at microwave frequencies by means of a piezoelectric crystal.

above 10^4 Hz. Quartz, ammonium dihydrogen phosphate, rochelle salt, and barium titanate are frequently used in sonic and ultrasonic transducers. The mechanical impedances of liquids and solids are generally close enough to the mechanical impedance of the piezoelectric crystal so that efficient energy transfer is possible. The intensity of ultrasonic radiation that can be achieved is mainly limited by the mechanical strength of the piezoelectric crystal. The maximum ultrasonic intensity theoretically obtainable in water by means of quartz or ammonium dihydrogen phosphate is of the order 2000 watts/cm² and 200 watts/cm², respectively. For gases, the mechanical impedance match is so poor that the corresponding values are about 4000 times smaller. However, the mechanical impedance match can be greatly improved by using piezoelectric devices consisting of two differently oriented crystal cuts cemented together in such a way that a voltage applied to the electrodes causes the elements to deform in opposite directions, and a twisting or bending action results. Assembles of this type (bimorphs) with $BaTiO_3$ ceramics or rochelle salt are widely used in such devices as microphones, earphones, and phonograph pickup cartridges.

Ultrasonic waves at microwave frequencies up to 2.4×10^{10} Hz have been generated by means of the piezoelectric effect. The arrangement is shown in Fig. 9. The end surface of a piezoelectric crystal rod is exposed to a strong microwave electric field in a resonant reentrant cavity. The ultrasonic waves travel through the crystal rod in a guided wave mode. The attenuation is low only at very low temperatures.

[H. GRANICHER]

Bibliography: D. A. Berlincourt, D. R. Curran, and H. Jaffe, Piezoelectric and piezomagnetic materials and their function in transducers, in W. P. Mason (ed.), *Physical Acoustics*, vol. 1, pt. A, 1964; W. G. Cady, *Piezoelectricity*, 1964; W. P. Mason, *Crystal Physics of Interaction Processes*, 1966; J. F. Nye, *Physical Properties of Crystals: Their Representation by Tensors and Matrices*, 1957.

Pinch effect

A name given to manifestations of the magnetic self-attraction of parallel electric currents having the same direction. The effect at modest current levels of a few amperes can usually be neglected, but when current levels approach a million amperes such as occur in electrochemistry, the effect can be damaging and must be taken into account by electrical engineers. Since the late 1940s the pinch effect in a gas discharge has become the subject of intensive study in laboratories throughout the world, since it presents a possible way of achieving the magnetic confinement of a hot plasma (a highly ionized gas) necessary for the successful operation of a thermonuclear or fusion reactor.

Ampere's law. The law of attraction which describes the interaction between parallel electric currents was discovered by A. M. Ampère in 1820 and can be stated as follows: The force of attraction in newtons per meter of length between two thin straight wires r meters apart carrying currents I_1 and I_2 amperes, respectively, is $2 \times 10^{-7} I_1 I_2 / r$. The law applies equally to the attraction between the individual components of a current in a single wire, in which case, for a cylindrical wire of radius r meters carrying a total surface current of I amperes, it manifests itself as an inward pressure on the surface (Fig. 1) given by $I^2 / 2 \times 10^7 \pi r^2$ Pa.

For the electric currents of normal experience, this force is small and passes unnoticed, but it is significant that the pressure increases with the square of the current, I^2. For example, at 25,000 A the pressure amounts to about 1 atm (100 kPa) for a wire of 1-cm radius, but at 10^6 A the pressure is about 1600 atm or about 12 tons in.$^{-2}$ (160 MPa).

Manifestations. The pinch effect first showed up practically in certain early types of induction electric furnaces in which large low-frequency alternating currents of the order of 100,000 A were induced at low voltage in a horizontal ring-shaped fused-metal load (Fig. 2). At these currents the pinch pressure can be larger than the hydrostatic pressure exerted by the fused metal, and as indicated above ($I^2 / 2 \times 10^7 \pi r^2$), the pinch pressure increases as the radius of the conductor decreases. Consequently, once the pinch process starts, the pressure at a narrow neck in the ring of fused metal can squeeze out the fluid metal until the neck pinches off completely, cutting off the current. This led to very uneven heating of the load. The term pinch effect was given to this process by C. Hering in 1907. The technical difficulty was eventually overcome by making the plane of the ring vertical and submerging it deeply below the surface of the fused metal. The force of the pinch effect also manifests itself by the crushing of tubular conductors exposed to large impulsive currents, such as occur in lightning strokes or high-power short circuits, and is used in certain metal-forming techniques.

Thermonuclear applications. One of the conditions for the attainment of a profitable balance between energy expended in heating and energy released in fusion from a thermonuclear reaction in a plasma composed of deuterium and tritium (DT, the most favorable case) is that the temperature shall be not less than about 10 keV (1.16×10^8 K). This enormous temperature can be reached and maintained only if the hot plasma is effectively isolated from the material walls of the container by vacuum. The isolation has the function of preventing cooling by contact with materials at normal temperatures and allowing the plasma to be heated. For the plasma to remain confined under these conditions, its outward pressure must be balanced by the inward pressure of nonmaterial origin, for example, a magnetic field. A profitable energy balance also depends on the density n of the confined plasma and on τ, the time it is confined. The product of $n\tau$ must exceed a certain minimum, which is 10^{14} ions cm^{-3} s for DT. *See* LAWSON CRITERION;

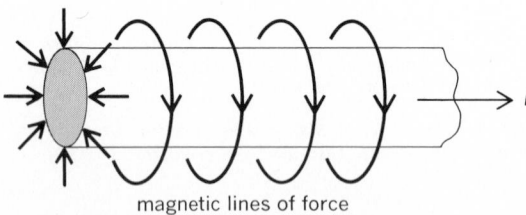

magnetic lines of force

Fig. 1. Pinch pressure on a current-carrying conductor. Arrows at left show direction of pinch pressure.

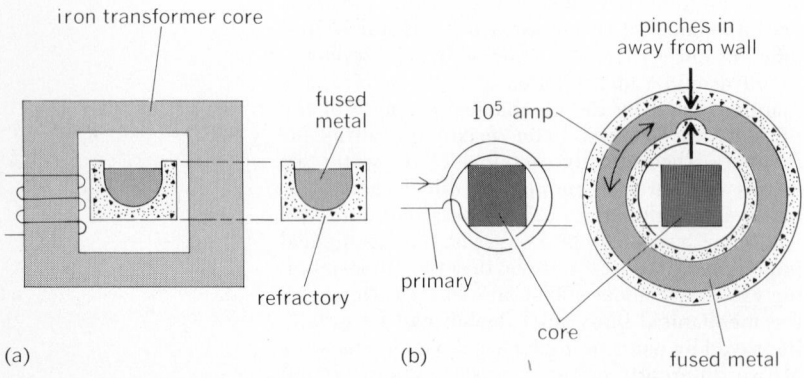

iron transformer core

fused metal

refractory

(a)

10^5 amp

pinches in away from wall

primary

core

fused metal

(b)

Fig. 2. Early type of ring induction electric furnace. (a) Side view. (b) Plan view.

MAGNETOHYDRODYNAMICS; PLASMA PHYSICS; THERMONUCLEAR REACTION.

There are a number of ways in which a magnetic field can be arranged around the plasma to hold it together, and one of these methods is the pinch effect. A fusion reactor using this type of confinement would ideally be a toroidal tube in which the confined plasma would carry a large electric current induced in it by magnetic induction from a transformer core passing through the major axis of the torus. The current would have the double function of ohmically heating the plasma and compressing the plasma toward the center of the tube.

The fundamental equation for the pinch effect in a gas, derived theoretically by W. Bennett in 1934, gives the current required for the inward pinch pressure to balance the outward gas pressure, as shown in the equation below, where I is the total

$$I^2/(2 \times 10^7) = Nk(T_i + T_e)$$

current in amperes, N is the number of electrons (also the number of ions) per meter of length of the pinch, $k = 1.4 \times 10^{-23}$ J/K (Boltzmann's constant), and T_i and T_e are the temperatures in kelvins of the ions and electrons, respectively.

Experimental studies. In general, two types of apparatus were used in early studies of the pinch effect: (1) straight discharge tubes of quartz or porcelain, with metal electrodes at each end, intended for short-duration studies, in which the cooling of the plasma by the relatively cold electrodes was slight during the time of the experiment, and (2) toroidal discharge tubes, also composed of quartz or porcelain, in which the pinch was endless and consequently was more effectively confined than in the first type of apparatus, and the current was induced into the discharge by magnetic coupling to a primary winding. In both cases, currents of 50,000–500,000 A were obtained by electric fields of 10–100 V/cm along the pinch. The primary power sources in early experiments used charged capacitors with capacitances of 4–50 microfarads, charged to 10–100 kV.

Instability. Characteristically, as can be shown by high-speed photography, the pinch forms at the inner surface of the discharge tube wall and contracts radially inward, forming an intense line, the pinch, on the axis (Fig. 3); the pinch rebounds slightly; the contracted discharge rapidly develops necks and kinks; and in a few microseconds all structure is lost in an apparently turbulent glowing gas which fills the tube. Thus, the pinch turns out to be unstable, and plasma confinement is soon lost by contact with the wall. The cause of the instability is easily seen qualitatively: The pinch confinement can be described as being caused by the magnetic field lines encircling the pinch which are stretched longitudinally but which are in compression transversely (Fig. 4). For a uniform cylindrical pinch, the magnetic pinch pressure is everywhere equal to the outward plasma pressure, but at a neck or on the inward side of a kink, the magnetic field lines crowd together, creating a higher magnetic pressure than the outward gas pressure. Consequently, the neck contracts still further, the kink cuts in on the concave side and bulges out on the convex side, and both perturbations grow. The instability has a disastrous effect on τ, limiting it to 10^{-6} s or less in light atom plasmas such as DT.

Neutrons have been produced by deuterium

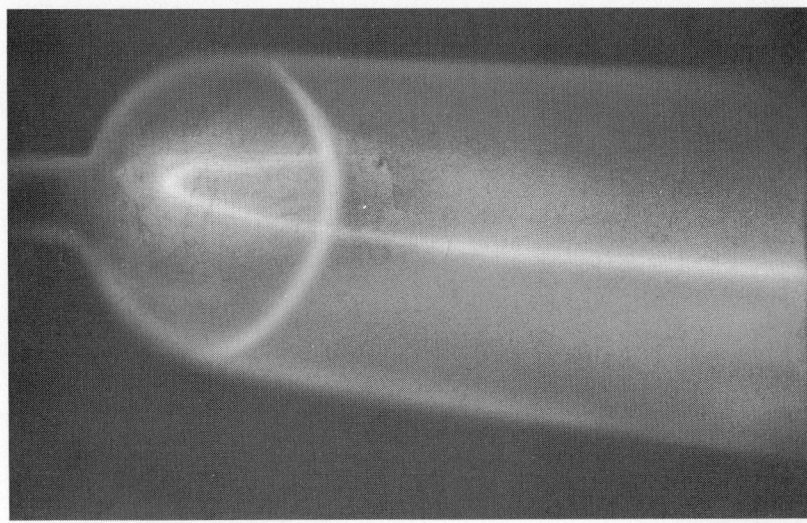

Fig. 3. Xenon pinched discharge in Perhapsatron torus.

pinches in large numbers. For a time (1952–1953), they were thought to be evidence of thermonuclear reaction, but it has since been shown that they are emitted preferentially in certain directions, and are associated with the instability of the pinch and the violent accelerations that are produced. Such neutrons are then not a product of thermal collisions and are not thermonuclear.

Great efforts have been devoted to overcoming the basic instability of the simple pinch. One such measure was to add an axial magnetic field by means of an external winding around the pinch tube. This might be expected to resist the neck and kink deformation by stiffening the discharge. Also a conducting wall located close to the discharge tube has the effect of trapping the magnetic field between the pinch and the wall, cushioning and reflecting the moving pinch back to the center.

Studies which involved such modifications were in progress on a worldwide scale during the period 1955–1963, notably including Zeta (Harwell, United Kingdom), Alpha (Leningrad, Soviet Union), and the Perhapsatron (Los Alamos, United States). In general these measures were disappointing, and work on pinches of this type declined.

Reversed-field Z pinch. In the 1970s, encouraged by analysis of Zeta results and improved theoretical understanding of pinch instability, there was renewed interest in the pinch formed by adding

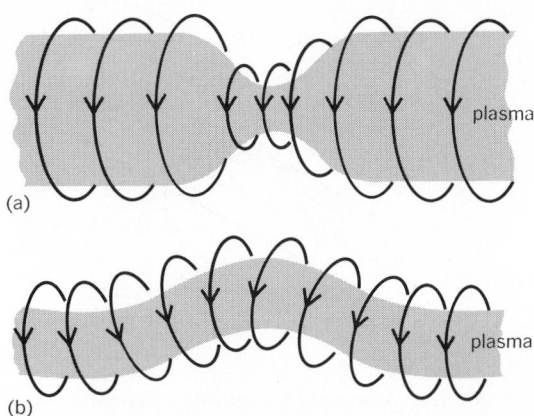

Fig. 4. Instability. (*a*) Sausage type. (*b*) Kink type.

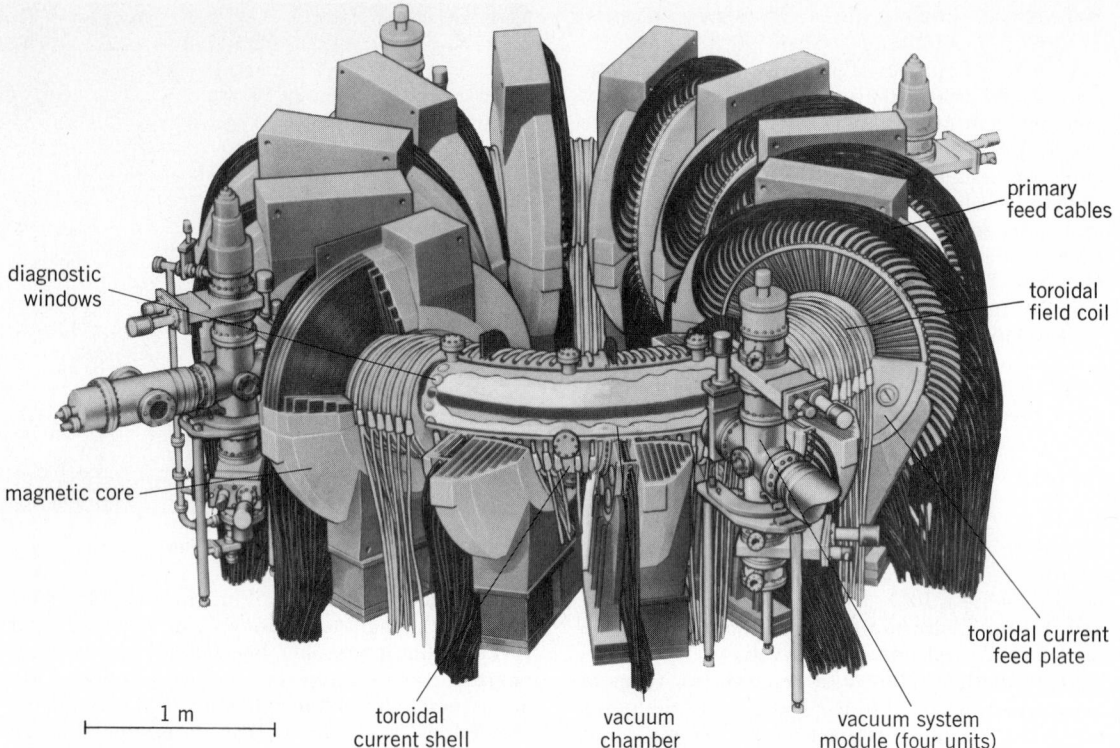

primary
feed cables

diagnostic
windows

toroidal
field coil

magnetic core

toroidal current
feed plate

1 m

toroidal
current shell

vacuum
chamber

vacuum system
module (four units)

Fig. 5.   ZT-40 reversed-field Z pinch experiment. (*Los Alamos Scientific Laboratory*)

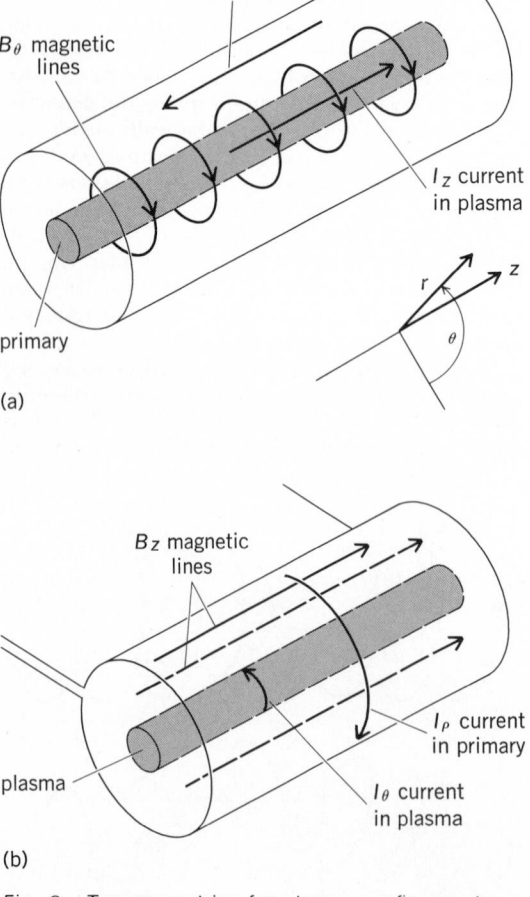

I_z current
in primary

B_θ magnetic
lines

I_z current
in plasma

r   z

θ

primary

(a)

B_z magnetic
lines

plasma

I_ρ current
in primary

I_θ current
in plasma

(b)

Fig. 6.   Two geometries for plasma confinement systems. (*a*) Z pinch. (*b*) θ pinch.

another longitudinal field outside the pinch in the opposite direction to that inside the pinch described above. This geometry is known as the reversed-field Z pinch. Confirmation of improved Z-pinch stability was undertaken in ZT-40 (Fig. 5; Los Alamos, United States), ETA Beta II (Padua, Italy), HBTX (Culham, United Kingdom), and elsewhere.

High-power pulsed systems. The simple pinch is still of considerable interest because the reciprocal functional relation between n and τ for power production means that there is always the possibility of achieving a net power output no matter how much τ is reduced by instabilities, by the use of very high n. Such plasmas will require extreme high-pulsed electric power to heat and confine them, with perhaps the only limitation being that the resultant output burst of thermonuclear power may destroy the machine. For such pulsed systems, the pinch must always rank highly—it is uniquely the most efficient magnetic confinement system expressed in terms of magnetic energy expended per unit of plasma energy confined, and furthermore the pressure it confines can exceed the strength of any known material.

Theta pinch. The term theta pinch has come into wide usage to denote an important plasma confinement system which relies on the repulsion of oppositely directed currents and which is thus not in accord with the original definition of the pinch effect (self-attraction of currents in the same direction). Plasma confinement systems based on the original pinch effect are known as Z pinches. Figure 6 defines the two geometries. The first laboratory thermonuclear reaction was in the small θ-pinch experiment, Scylla, at Los Alamos in 1957.

Tokamak. The stabilized Z pinch became the

Tokamak performance

Parameter	Tokamak	Target values
T_i (keV)	6	10
$n\tau$ (ions cm^{-3} s)	3×10^{13}	10^{14}

subject of intense interest when L. Artsimovich of the Soviet Union announced in 1969 the achievement of a 20-ms confinement and an ion temperature of 0.5 keV (5×10^6 K) in the T-3 tokamak experiment. Tokamak is essentially a low-density, slow Z pinch in a torus with a very strong longitudinal field such that the pinch current is below the Kruskal-Shafranov limit. This means that the helical magnetic field lines, resultant from the externally applied field and that of the pinch, do not close, that is, complete one revolution of the minor axis in going around the major axis of the torus once. This is known theoretically to prevent the growth of certain helical distortions of the plasma. The achievements of tokamak experiments are listed in the table, together with the target values for achieving a positive power balance.

The temperatures achieved are higher than the 0.5 keV that could be reached by resistive heating of the plasma by the pinch current, such as was used in the early 1970s. Auxiliary heating is necessary, and the most successful has been the injection of energetic neutral beams.

The performance of tokamak experiments has raised the possibility of achieving a net power balance in the mid-1980s. Construction of several large tokamak installations has been undertaken, including the TFTR (Princeton, United States), JET (European Community at Culham, United Kingdom), JT-60 (Tokyo), and T-15 (Moscow). *See* NUCLEAR FUSION. [JAMES A. PHILLIPS]

Bibliography: International Fusion Research Council, Status report on controlled thermonuclear fusion, *Nucl. Fusion*, 18:137–149, 1978; C. L. Longmire, J. L. Tuck, and W. B. Thompson (eds.), Plasma physics and thermonuclear research, *Progr. Nucl. Energ.*, Ser. 11, 1963; Theoretical and experimental aspects of controlled nuclear fusion, in *Proceedings of the 2d International Conference on Peaceful Uses of Atomic Energy*, vols. 31–32, 1958; J. L. Tuck, Artsimovich talks about controlled-fusion research, *Phys. Today*, 22(6):54–57, 1969.

Pionium

An exotic atom, also called the pi-mu atom, which is similar in structure to the hydrogen atom but with the proton replaced by a pion and the electron replaced by a muon. Pionium is unique among atoms that have been observed in the laboratory in that all of its constituents are unstable particles not found in ordinary matter. The pion, a particle involved in nuclear forces, lives only 26×10^{-9} s, and the muon, a particle like the electron except that it is 207 times heavier, lives 2.2×10^{-6} s. The lifetime of this atom is thus determined by the pion lifetime. Due to the large mass of the muon, the atom is about 120 times smaller in radius than is hydrogen. *See* ELEMENTARY PARTICLE; LEPTON; MESON.

Most exotic atoms, such as muonium or positronium, contain only one short-lived particle and are formed by causing that particle to interact with an ordinary atom. This method is impossible for pionium since neither constituent is found in normal material. Instead, pionium is formed during the decay of a certain heavier particle called the neutral kaon. A kaon has many modes of decay, one of which results in the formation of a pion, a muon, and a neutrino ($K_L^0 \rightarrow \pi^+ \mu^- \nu$). For this decay, both the muon and pion occasionally have almost exactly the same speed and direction, resulting in the formation of a bound atomic system. Although this process is extremely rare, less than 1 per 1,000,000 kaon decays, pionium has been observed in the laboratory. This decay of a particle of radius 0.5×10^{-13} cm into an atom of radius 0.5×10^{-10} cm is in striking contrast to the usual scheme of things where atoms are broken into smaller constituents. *See* MUONIUM; POSITRONIUM.

Pionium is of interest to physicists for two reasons. First, the rate of formation provides information about the details of kaon decay. The observed formation rate is indeed compatible with present understanding of such decays. Another potentially interesting application of pionium involves atomic spectroscopy. Although pionium is usually formed in the ground state, it is presumably also produced in the excited *2S* state. A measurement of the Lamb shift, which is the energy difference between the *2S* and the *2P* state, would serve to measure the pion radius. *See* ATOMIC STRUCTURE AND SPECTRA; WEAK NUCLEAR INTERACTIONS.

The nomenclature of exotic atoms is not well established, and the name pionium may also refer to the pion-electron atom. Although this atom is probably formed when pions are stopped in gases, there has been no interest in studying it in the laboratory. [PAUL A. SOUDER]

Bibliography: R. Coombes et al., Detection of π-μ Coulomb bound states, *Phys. Rev. Lett.*, 37: 249–252, 1976; I. L. Nemenov, Atomic decays of K_L^0 mesons, *Sov. J. Nucl. Phys.*, 16:67–69, 1973 (*Yad. Fiz.*, 16:125, 1973); Pionium, another new quasi-atom, *Sci. News*, 109:356–357, 1976.

Planck's constant

A fundamental physical constant which represents the elementary quantum of action, action being defined as energy multiplied by time. Introduced by Max Planck in 1900, it has the value $h = 6.6256 \times 10^{-27}$ erg-sec or 6.6256×10^{-34} joule-sec. The symbol $\hbar$, sometimes called the Dirac h, is often used for convenience in physics to denote the quantity $h/2\pi$, where $\pi = 3.1416$. . . .

The unique feature of Planck's constant is this: As used by Planck in deriving his radiation law, h multiplied by the frequency of radiation represented a bundle of energy, that is, a quantum of energy. Radiant energy at any wavelength can occur only as multiples of this energy; thus energy is quantized. This was a fundamental departure from the beliefs of physics up to Planck's time and was indeed quite startling to Planck himself. Until Planck deduced his law to satisfy experimental data, the general belief was that energy could be divided indefinitely. The quantization of energy implied by Planck's constant laid the foundations of thought upon which much of modern physics has flourished. The frequency ν of emitted radiation is related to the quantized energy, ΔE, by the relation $\Delta E = h\nu$.

The concept of energy quantization first began to win general acceptance after 1905 when Albert Einstein showed that it gave a good account of some of the then puzzling features of the photoelectric effect. Later A. H. Compton showed that the electromagnetic quanta also carry momentum p which is related to the wavelength λ by $p = h/\lambda$.

For an extended discussion of Planck's radiation law *see* HEAT RADIATION. For a discussion of the role of Planck's constant in theoretical physics *see* QUANTUM MECHANICS. *See also* ATOMIC CONSTANTS.

[HEINZ G. SELL; PETER J. WALSH]

Planck's radiation law

A law of physics which gives the spectral energy distribution of the heat radiation emitted from a so-called blackbody at any temperature. Discovered by Max Planck early in the 20th century, this law laid the foundation for the advent of the quantum theory because it was the first physical law to postulate that electromagnetic energy exists in discrete bundles, or quanta. *See* HEAT RADIATION; QUANTUM MECHANICS.

[HEINZ G. SELL; PETER J. WALSH]

Plasma physics

That field of physics which relates to the study of highly ionized gases. A gas which is composed of a nearly equal number of positive and negative free charges (positive ions and electrons) is called a plasma after the original definition by I. Langmuir. Because it is composed of charged particles, a plasma exhibits many phenomena not encountered in ordinary gases. In addition to their importance in many new areas of applied science, these effects are evident in astrophysical phenomena (most of the matter in this universe exists in the plasma state), both in stellar atmospheres and in interstellar space. Plasma phenomena have also been observed in the tenuous ionized gases of the Earth's outer atmosphere.

Practical interest in plasma physics arises from various applications of gas discharges, from the study of electron beams in electron tubes, and from the new research fields of ultra-high-temperature processes and controlled fusion. These latter experiments require millions or even hundreds of millions of degrees kinetic temperature (to be defined later). To obtain a comprehensive physical picture of plasma it is necessary to consider its behavior from two different aspects:

1. The microscopic picture, relating to its particlelike properties, such as the effects of interparticle collisions in producing diffusion and other transport phenomena, ionization, x-radiation, and other particulate processes. In this picture a plasma exhibits properties some of which are much like those of any gas.

2. The macroscopic picture, where the collective or fluidlike properties are most evident. These properties include conduction of electricity, propagation of various kinds of waves, and ability to support classes of unstable and turbulent behavior peculiar to conducting fluids. Many of the macroscopic behavioral properties of plasma are related to the general field of magnetohydrodynamics. *See* MAGNETOHYDRODYNAMICS.

The charged particles of a plasma interact with each other through the electrostatic or Coulomb field with which each is surrounded. On the microscopic scale, these electrostatic fields give rise to localized attractive or repulsive forces between the particles as they pass near to each other, resulting in mutual deflection. On the macroscopic scale, the summation of the many infinitesimal electrostatic and magnetic fields produced by the moving plasma particles results in a smeared-out or averaged electromagnetic field. The plasma then reacts collectively, that is, as a conducting fluid, to the total electromagnetic field in which is is immersed. This field consists of the combination of the plasma electromagnetic field and any externally imposed fields. The coupled nature of the plasma motion and the electromagnetic field in which it moves causes most of the complexity of plasma behavior. *See* NUCLEAR FUSION; PINCH EFFECT.

CRITERION FOR PLASMA PHENOMENA

A length scale which approximately divides the microscopic domain from the macroscopic domain in a plasma is the so-called Debye screening distance λ_D. It can be shown that as long as the distance between two passing particles is appreciably less than λ_D, normal Coulomb attraction or repulsion will exist and one can define the encounter as a simple collision, to which the ordinary laws of particle dynamics apply. However, if the minimum distance of approach of two particles is greater than λ_D, the collective motions of the surrounding plasma electrons induced by the passage of the particle will be such as to screen the test particle from feeling the influence of the other particle (or any others beyond the distance λ_D). *See* COLLISION.

The length λ_D depends on the density n_e and the kinetic temperature T_e of the plasma electrons. It is usually defined through the relationship in Eq. (1), where e is the charge on the electron and k is Boltzmann's constant.

$$\lambda_D = \sqrt{\epsilon_0 k T_e / n_e e^2} \ \text{m} \qquad \text{(SI units)} \qquad (1)$$

Values of λ_D for typical plasma densities and electron kinetic temperatures of interest in laboratory experiments are listed in the table. Kinetic temperature refers to a measure of temperature in terms of the kinetic energy of random motion of the particles of gas. In a Maxwellian gas the mean kinetic energy $\overline{W} = 3/2kT$, where T is the absolute temperature in kelvins. A convenient measure of W is the electron volt, so that kinetic temperature is often measured in electron volts: 1 eV kinetic temperature $= kT = 11,600$ K $= (2/3)\overline{W}$. *See* KINETIC THEORY OF MATTER.

In the last column of the table, the approximate mean particle separation $d = (1/n_e)^{1/3}$ is given for comparison. Except at the highest densities, λ_D is seen to be substantially larger than d, corresponding to the fact that many particles are contained within a Debye sphere, so that each particle lies within collision range of many other particles at any given moment in time. This is of importance to the understanding of certain collision effects in a plasma.

The numerical value of λ_D provides an important criterion by which to decide whether in a plasma of given size collective phenomena are to be expected. Certainly, if the overall dimensions of a region containing plasma are small compared to λ_D, only

Values of λ_D for typical densities and temperatures

n_e, electrons/m³	$T_e(\mathrm{K})\begin{cases} \end{cases}$	10^5 (8.6 eV)	10^6 (86 eV)	10^7 (860 eV)	10^8 (8.6 keV)	d, m $(1/n_e)^{1/3}$
		λ_D, m				
10^{14}		2.2×10^{-3}	6.9×10^{-3}	0.022	0.069	2.1×10^{-5}
10^{16}		2.2×10^{-4}	6.9×10^{-4}	2.2×10^{-3}	6.9×10^{-3}	4.8×10^{-6}
10^{18}		2.2×10^{-5}	6.9×10^{-5}	2.2×10^{-4}	6.9×10^{-4}	1.0×10^{-6}
10^{20}		2.2×10^{-6}	6.9×10^{-6}	2.2×10^{-5}	6.9×10^{-5}	2.1×10^{-7}
10^{22}		2.2×10^{-7}	6.9×10^{-7}	2.2×10^{-6}	6.9×10^{-6}	4.8×10^{-8}
10^{24}		2.2×10^{-8}	6.9×10^{-8}	2.2×10^{-7}	6.9×10^{-7}	1.0×10^{-8}

simple collisional or single-particle behavior is to be expected, the plasma will behave as an ordinary low-density gas, and collective processes will not be important. Conversely, if the dimensions of a plasma region are very much larger than λ_D, the possibility exists for collective plasma phenomena. Thus, as seen from the table, in the laboratory, where dimensions are measured in centimeters, plasmas with electron densities less than about 10^{12} to 10^{14} m^{-3} would not be expected to exhibit collective behavior. These are particle densities which are typically encountered in conventional particle accelerators. On the other hand, in the Earth's upper atmosphere, λ_D would be much smaller than other typical dimensions for all electron densities higher than a few electrons per cubic centimeter. In such cases collective effects could be expected to be possible even at the lowest particle densities encountered.

The condition just given admits of another simple and useful physical interpretation. Consider a sphere of plasma with radius λ_D, and assume that nearly all of the electrons of the plasma are removed to infinity from this region. The removal of these electrons will leave an uncanceled positive charge and a resulting radial electric field. From the definition of λ_D it is then easily shown that in this case the energy necessary to remove the additional electrons from the plasma is (within a factor of 2) equal to their mean kinetic energy. It follows that if $\lambda_D \ll$ the plasma dimensions, any influence which tends to separate the bulk of the plasma changes by a distance greater than λ_D will give rise to strong electrostatic restoring forces which will prevent any further separation. Conversely, if $\lambda_D \gg$ the plasma dimensions, the electrostatic forces arising from even a complete separation of charge will have little influence on the motions of the individual charges so that collective effects will be unimportant.

The strong tendency for plasmas where $\lambda_D \ll$ the plasma dimensions to maintain near charge neutrality leads to some of the most important collective phenomena in plasma. Among these are the development of so-called ambipolar potentials of the plasma and the existence of sheathlike regions in plasmas in contact with bounding surfaces. Whenever a difference in the intrinsic loss rates between electrons and ions exists, a plasma will automatically assume an electric potential of such a size and magnitude as to bring these loss rates to equality (as required for a steady state to exist). In the case of ordinary gas discharges where metallic conductors are present, the result is to develop a sheath region, a few Debye lengths in extent, adjacent to the metallic surface. Such sheaths arise from the following considerations: Being much more mobile than the ions, the electrons in a discharge plasma would try to escape much more rapidly than the ions. In so doing, they would charge the plasma positive with respect to its boundary. Since deep within such a plasma dc electric fields are suppressed, owing to the high electrical conductivity of plasma, the ambipolar potential drop is constrained to appear only in sheaths at the boundary, where electron and ion density fall self-consistently to zero. Many other examples of ambipolar phenomena are found to occur in plasmas, where they may strongly influence the propagation of waves, transport across magnetic fields, and other processes.

CREATION OF A PLASMA

To create a plasma in the laboratory it is usually necessary to start with an ordinary gas, at a small fraction of atmospheric pressure, and then to heat it by electrical or other means until the mean kinetic energy of the gas particles becomes comparable to the ionization potential of the gas. Mutual collisions of the gas particles will then result in a cascading ionization of the gas. Since ionization potentials are always several volts, such effects are only important at kinetic temperatures of several electron volts, so that the threshold temperature for most plasma experimentation is $50,000-100,000$ K, and ranges up to tens or hundreds of millions of degrees. *See* IONIZATION POTENTIAL.

Returning to the question of producing the plasma by ionization of a gas, one recalls that such processes occur in ordinary gaseous discharges, for example, in fluorescent lamps. However, in such discharges, the ions and electrons of the plasma are continually and rapidly being cooled and recombined by contact with the chamber walls, so that the temperature is low and the state of ionization is only partial and can be maintained only by a large continuous input of energy. A necessary alternative is to find some means of electromagnetic confinement of the plasma, once created, so that its particles cannot touch the chamber walls. Effective confinement is the prime objective of high-temperature plasma research and is the basis for hopes of achieving controlled nuclear power from nuclear fusion reactions in a hot plasma. Without such means a high-temperature plasma, even if it could be created, would have only a fleeting existence, typically less than a millionth of a second. The most commonly employed method of confining plasma involves the use of magnetic fields. To employ magnetic confinement effectively, however, requires the solution of problems of plasma instability.

Instabilities. A plasma immersed in a magnetic field can exhibit unstable motions that can exert a major influence on the plasma behavior, dominating all other phenomena. A hot plasma localized in a magnetic field represents a system that is far from being in a state of thermodynamic equilibrium with its surroundings. In time, interparticle collisions would bring the entire system to such a state, that is, to a state of uniform particle density, in temperature equilibrium with its surroundings. Instabilities in the plasma, driven by the free-energy sources created by temperature and density gradients, represent another way that the plasma can reach equilibrium, usually by a much faster route than collisions. The most violent plasma instabilities can destroy confinement in about the transit time of an ion across the system, a short time, of the order of a few microseconds, for hot plasmas contained in systems of the order of a meter in dimension.

There are two general classes of instabilities: (1) gross or hydromagnetic or magnetohydrodynamic (MHD) instabilities, driven generally by gross electric currents in the plasma or by the free-energy reservoir that is tapped when the plasma can expand toward a region of weakening magnetic field; and (2) microinstabilities, arising from streaming or relative drift motions of the plasma particles with respect to each other. Therefore the particulate nature of the plasma must be explicitly considered in determining its susceptibility to microinstabilities, while it is unimportant in MHD modes, where only gross fluidlike phenomena are involved.

One of the simplest of the hydromagnetic instabilities is the kink instability. This may occur when a plasma column carries a strong longitudinal current. Here a straight column of plasma typically rapidly develops a helical shape, thereby

converting magnetic field energy into transverse energy of motion of the column.

A simple example of a microinstability is the so-called two-stream instability, which can arise when counterstreaming motion of special groups of particles within the plasma exists (one group of electrons counterstreaming against another, for example). For sufficiently strong counterstreaming, unstable waves build up within the plasma, converting particle stream energy into wave energy.

In fact, a general property of microinstabilities is that they act to convert ordered particle motion (departures from a Maxwellian distribution) of the plasma into wave motion. In this respect, they closely resemble laser action, through which an ordered (nonequilibrium) distribution of atomic states is converted into a coherent light wave. *See* LASER.

The intimate relationship between waves and instabilities in a plasma is also brought out by the above example. Plasma waves are the carriers of the plasma instabilities. Whenever an adequately strong driving source (for example, internal particle streaming motion) can couple constructively with a plasma wave, that wave may grow in amplitude, starting from normal fluctuation levels, until it reaches such large amplitudes that it can cause major disruptions in the plasma behavior, such as premature loss of confinement. *See* WAVES AND INSTABILITIES IN PLASMAS.

Magnetic confinement. The use of specially shaped magnetic fields to confine high temperature plasma is one of the promising approaches to the achievement of power from nuclear fusion reactions. Magnetic confinement becomes possible because a plasma is composed of charged particles whose motions can be influenced by a magnetic field. Confining fields can be externally imposed, generated by currents flowing in the plasma itself, or formed by combinations of the two. There have emerged many approaches to magnetic confinement. Generally speaking, these approaches have been selected for their ability to avoid, or minimize, the deleterious effects of both MHD and microinstability modes of plasma instability. These approaches can generally be divided into one of two categories:

1. In closed systems, the magnetic lines of force of the confining field remain within the confinement chamber, an endless tube of toroidal (doughnut) shape. Particles are lost from confinement only when they are able to cross the magnetic lines of force of the confining field. In such a situation the average particle confinement time will necessarily increase with the size of the systems. If governed by normal diffusion processes, confinement time will vary as the square of the dimension scaling factor. Two examples of closed systems are the stellarator and the tokamak. In the stellarator (invented by L. Spitzer, Jr.) the confining fields are generated solely by external field windings. To avoid various instabilities of the plasma which would be present if the stellarator were constructed as a simple torus, the field lines are given a helical twist. In the first stellarators (Fig. 1) this was done by making the tube in the form of a figure eight; in modern stellarators it is accomplished by means of auxiliary helical coils. In the tokamak, developed first in the Soviet

Fig. 1. Early stellarator, constructed in the 1950s. The figure-eight shape prevented certain transverse drifts of the plasma.

Union, the confining field is composed of the combination of a strong externally generated toroidal field plus the magnetic field originating from a heavy circulating current induced in the plasma by transformer action. The resultant field again forms a helical pattern of field lines. The tokamak is the most widely studied method of plasma confinement.

2. In open systems, the configuration is such that the field lines leave the confinement region. The best-known example of an open system is the mirror machine. In the mirror machine the strength of the magnetic field at each end of the confinement region is intensified. These regions are the mirrors. The charged particles of the plasma may be trapped between these mirrors, provided that they spiral with a sufficiently steep pitch angle. The Earth's magnetic field forms a natural mirror machine, whereby particles are trapped in the Van Allen belts lying between the north and south magnetic poles. In open systems the confinement time is determined by the mean time it takes for a trapped particle to be deflected, by interparticle collisions or any other process, into the loss cone of the mirror machine, consisting of pitch angles such that the particle's trajectory becomes too nearly parallel to the field-line direction for the mirroring action to be effective. Since once a particle is deflected into the loss cone it is immediately lost upon its next encounter with a mirror, confinement time in mirror machines does not generally depend on the size of the system but only on the rates of diffusion of particles into the loss cone. These rates decrease with increasing plasma temperature, if the diffusion is due to interparticle collision rates rather than to collective effects. Therefore, good plasma confinement in mirror systems generally requires both a highly quiescent plasma state and high plasma-ion temperatures.

MHD instabilities can be readily suppressed in open-ended systems by shaping the confining field into a magnetic-well configuration, one in which the plasma is confined by fields which increase outward in every direction. The presence of the loss cone, however, creates gradients in velocity-space which tend to excite microinstability modes unless care is taken to minimize such tendencies.

In closed systems it is not possible to create a field which everywhere possesses the magnetic-well property, so that MHD instabilities are not as easily suppressed as in open systems. Consequently, a residual level of MHD instability is often found to be present. On the other hand, a properly designed closed system can avoid most of the loss-cone effects associated with open systems, so that microinstabilities generally represent much less of a problem than they do for mirror systems.

Variant systems. In the pursuit of improved magnetic confinement, some important variants on the existing closed and open geometry systems have been studied. These variants share a common property: In one way or another they utilize effects arising from the presence of the plasma to improve the plasma's own confinement. Three such approaches will be discussed here: the tandem mirror; "compact torus" configurations (field-reversed mirror and "spheromak"); and the electron-ring bumpy torus.

Tandem mirror. The tandem mirror takes advantage of the fact that a high-temperature plasma confined in a mirror cell automatically develops a high positive "ambipolar" potential, of order five times the electron temperature. This potential arises from the intrinsic disparity in the scattering rates of electrons and ions in an open-ended mirror system, leading to a tendency for the electrons to escape through the mirrors more rapidly than the ions, thereby leaving behind a net positive charge. However, in steady state, electron and ion loss rates must balance on the average; otherwise the potential would climb indefinitely. The ambipolar potential thus builds up to the point where the electron loss rate (occurring now only from the high-energy "tail" of the electron distribution) equals the ions loss rate.

In the tandem mirror, two extra mirror cells within which a high-temperature plasma is maintained are placed at opposite ends of a central mirror cell. Under the proper conditions of relative plasma density and electron temperature, potential peaks produced by the end cells act as electrostatic "stoppers" for the ions of the central plasma, its electrons being confined by the net positive potential of the end cells (with respect to their surroundings). Electrostatic confinement can be made to be much more effective than simple mirror confinement, at the expense only of maintaining small volumes of hot plasma in the end plugs (that is, small compared to the volume of the plasma in the central cell).

Compact torus. Compact torus systems, examples of which are the field-reversed mirror and the spheromak, utilize diamagnetic currents flowing in the plasma to drastically reshape an externally applied confining field, for example, a mirror field. By enhancing these currents, through plasma heating techniques, the use of special plasma guns, or other means, a pattern of field lines that close on themselves within the plasma volume can be created. Resembling somewhat a smoke ring, among all possible configurations this type of magnetic configuration is potentially the most compact and magnetically efficient (highest plasma pressure for a given externally applied magnetic field strength). The issue is how to maintain adequate plasma stability under the demanding conditions represented by field reversal.

Electron-ring bumpy torus. The electron-ring bumpy torus consists of an externally generated dc toroidal confining field that is favorably perturbed by exciting within it (using incident microwave power at the electron cyclotron frequency) a periodic series of hollow "rings" populated by high-energy electrons. These electron rings locally depress the toroidal field, reshaping it so that plasma can be held in equilibrium inside the region occupied by the rings. An advantage of such a system over the tokamak is that it permits steady-state confinement.

Pressure balance and diamagnetism. The possibility of magnetic confinement of a plasma, considered as a high-pressure gas of charged particles, can be identified with a diamagnetic effect associated with the plasma; that is, the introduction of a plasma into a region containing a magnetic field tends to weaken the magnetic field in that region. This effect is readily seen in the pressure balance relationship for a confined plasma. If a (usually) small effect arising from the curvature of magnetic lines is neglected, this relationship is given by Eq. (2), where $p_\perp$ is the

$$\nabla\left(p_\perp + \frac{B^2}{2\mu_0}\right) = 0 \qquad (2)$$

local value of the component of the plasma pressure perpendicular to the lines of force, and B is the local field strength. This expression, which shows the constancy of the sum of the plasma pressure and the magnetic pressure, can be integrated to yield Eq. (3), where B_0 is the strength of the

$$p_\perp + \frac{B^2}{2\mu_0} = \text{const} = \frac{B_0{}^2}{2\mu_0} \qquad (3)$$

magnetic field just outside the plasma. There exist an infinite number of allowed solutions to this equation, so that other circumstances, such as diffusion, will dictate the actual equilibrium solution in any given physical case. *See* DIAMAGNETISM.

It is clear from Eq. (3) that the plasma pressure can never exceed $B_0{}^2/2\mu_0$. This value corresponds to a complete diamagnetic exclusion of the magnetic field from the plasma. Other circumstances, such as plasma instabilities, may impose additional limitations on the plasma pressure, leading to values less than $B_0{}^2/2\mu_0$. A convenient representation of Eq. (3) can be given in terms of a parameter β, which is defined as the ratio of plasma pressure to externally applied magnetic pressure by Eq. (4).

$$\beta = \frac{p_\perp}{B_0{}^2/2\mu_0} \qquad (4)$$

In terms of β, Eq. (3) becomes Eq. (5). The para-

$$\beta = \left[1 - \left(\frac{B}{B_0}\right)^2\right] < 1 \qquad (5)$$

meter β measures the effect which the plasma can have on the applied field. If $\beta \ll 1$, the applied field will be only slightly affected by the presence of the plasma.

The diamagnetic effect of a plasma arises from persistent electric currents flowing throughout its volume. In fact, the magnetic confining force is just the body force $\mathbf{F} = \mathbf{J} \times \mathbf{B}$. But the existence of currents in the plasma must imply a dissipation of energy. Thus, unless maintained, the confining currents will decay with time, so that plasma and magnetic field will gradually intermingle, leading to eventual escape of the plasma. Magnetic confinement of a plasma is therefore necessarily a transient process, with a time scale set by the electrical conductivity of the plasma. This situation is analogous to the slow penetration of a suddenly applied magnetic field into a large metallic conductor. The application of the field induces eddy currents in the surface of the conductor which decay with time, leading to the eventual penetration of the field.

The effectiveness of a magnetic field in confining a hot plasma for a long time depends on the electrical resistivity of the plasma. This is given theoretically by the approximate expression in Eq. (6),

$$\rho \cong \frac{7.6 \times 10^2 Z}{T_e{}^{3/2}} \quad \text{ohm-m} \qquad (6)$$

where Z is the mean ionic charge. Theoretically ρ is independent of density, a result which cannot be expected to hold at very low plasma densities. Note also that plasma has a negative temperature coefficient of resistivity. At 10^7 K, a hydrogenic plasma has a theoretical resistivity about as low as pure copper at room temperature, and at temperatures of 10^8 K or higher it is much less.

PARTICLE DYNAMICS IN PLASMAS

Magnetic confinement, diffusion of a plasma across a magnetic field, and other characteristics of plasma can be better understood by returning to the microscopic picture. Each charged particle of the plasma moves in the smoothed-out electromagnetic field arising from the combination of the external applied fields and the electric and magnetic fields produced by the plasma itself. In many cases collisions are infrequent and the fields are well enough known to allow prediction of the particle motions by relatively simple techniques.

The equation of motion of a charged particle of mass M, velocity $\mathbf{v}$, the charge e in an arbitrary electric and magnetic field is Eq. (7).

$$\mathbf{F} = M\frac{d\mathbf{v}}{dt} = e(\mathbf{E} + \mathbf{v} \times \mathbf{B}) \qquad \text{(SI units)} \quad (7)$$

In many cases of practical interest $\mathbf{E} \ll \mathbf{B}$, and $\mathbf{B}$ will be relatively slowly varying in time and space. In such cases the solutions to Eq. (8) represent orbits which are approximate helices (coil-spring–shaped), corresponding to a rotation of the particle around a line of force superposed on a translational motion along the lines of force (parallel or antiparallel to the direction of B). For the component of motion perpendicular to the direction of B, Eq. (7) reduces simply to a centrifugal force equation, Eq. (8), that is, a rotation at the

$$F = \frac{Mv_\perp^2}{r_c} = ev_\perp B \qquad \text{(SI units)} \quad (8)$$

cyclotron angular frequency given by Eq. (9). This

$$\omega_c = \frac{v_\perp}{r_c} = \frac{eB}{M} \qquad (9)$$

frequency is clearly much higher for electrons than for positive ions.

Corresponding to these frequencies of rotation, the radius of curvature r_c of the orbit is also given by solving Eq. (8) for r_c as in Eq. (10), where $W_\perp = (1/2)Mv_\perp^2$.

$$r_c = \frac{M}{eB} = \left(\frac{2W_\perp}{M}\right)^{1/2} = \frac{1}{\omega_c}\left(\frac{2W_\perp}{M}\right)^{1/2} \qquad (10)$$

For the same $W_\perp$, r_c is much smaller for electrons of mass m than for ions, corresponding to a value given by Eq. (11). For ions Eq. (12) holds, where A is the mass of the ion in atomic mass units.

$$r_{ce} = 3.4 \times 10^{-6} W_\perp{}^{1/2}/B \quad \text{m} \qquad (11)$$

$$(W_\perp \ll mc^2 = 511 \text{ keV}, W_\perp \text{ in eV})$$

$$r_{ci} = 1.45 \times 10^{-4}(A^{1/2}/Z)(W_\perp{}^{1/2}/B) \quad \text{m} \qquad (12)$$

To take an example, if $A = Z = 1$ (protons), $W_\perp = 10^4$ eV, and $B = 1$ tesla, then $r_{ci} = 1.45$ cm. At the same energy and field strength, r_{ce} is only 0.34 mm. In this case $\omega_{ci} = 9.5 \times 10^7$ rad/sec and $\omega_{ce} = 1.75 \times 10^{11}$ rad/sec.

Returning to the equation of motion, Eq. (7), it is possible now to precisely define the conditions under which simple solutions are obtained. It is only required that B should vary slowly in time compared with $(\omega_{ci})^{-1}$ and by a small amount, percentagewise, over an orbit radius r_c, as in Eqs. (13) and (14). These are the so-called conditions of adiabaticity for the particle motion.

$$\tau = \left(\frac{1}{B}\frac{\partial B}{\partial t}\right)^{-1} \gg 1/\omega_c \quad \text{sec} \qquad (13)$$

$$\lambda = \left(\frac{1}{B}\frac{\partial B}{\partial r}\right)^{-1} \gg r_c \quad \text{m} \qquad (14)$$

If these conditions are satisfied, some very important consequences follow. These are (i) that the motion of any charged particle can be well represented by following the motion of its instantaneous center of rotation or guiding center, (ii) that the guiding centers will move about with respect to the magnetic lines with slow drift velocities predictable from simple laws, and (iii) that many of the salient features of the motion can be prescribed in terms of nearly constant quantities known as adiabatic invariants.

Drift velocities. If conditions in Eqs. (13) and (14) are satisfied, it is relatively easy to calculate the guiding center drift velocities. The simplest and most important of these velocities follows by inspection of Eq. (7). If the plasma particles move in an electric field with a component transverse to B, then the guiding center of each particle will drift in a direction perpendicular both to E and to B, with a velocity v_o which is independent of the charge, mass, or energy of the particle. This velocity is given by Eq. (15).

$$v_o = \frac{E \times B}{B^2} \quad \text{m/sec} \qquad (15)$$

The derivation of Eq. (15) from Eq. (7) follows immediately if it is noted that in a frame of reference moving at velocity v_o, the motional electric field just cancels the applied electric field and the particle motions are again helices. It is concluded that an electric field component perpendicular to B causes the plasma to move and distort locally so as always to make the electric field vanish in the plasma's own frame of reference. This is just the behavior which would be expected for a compressible, highly conducting gas. Further analysis shows that this property is also identifiable with a strong tendency for the plasma to preserve constant magnetic flux through each of its volume elements as it moves from place to place or is subjected to slowly varying magnetic fields.

The drift velocity v_o is also related to the paradoxical situation whereby a jet of plasma can cross through an evacuated region containing a strong transverse magnetic field. If a plasma jet is impelled at velocity v into such a region, a very slight separation of charges (Fig. 2) is all that is required to generate an internal polarization electric field in the plasma which maintains the velocity. However, if the plasma should pass into a strongly conducting region (another plasma), the polarization fields would not persist, and the motion would stop, the momentum being taken up by the magnetic field or converted into rotation.

In another case, if a plasma is immersed in a magnetic field which is then increased slowly, in-

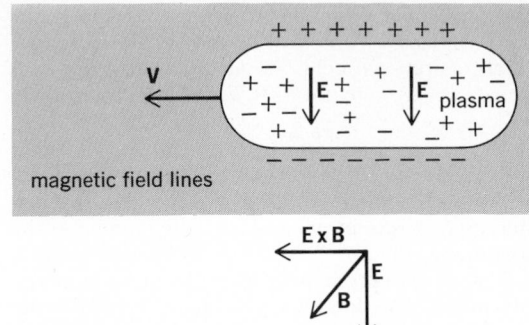

Fig. 2. Schematic illustration of drift of a plasma across a magnetic field in vacuum.

duction electric fields will appear which will cause the plasma to be compressed toward the magnetic symmetry axis of the system (where $E = 0$). In a uniform magnetic field of circular cross section this would result in a uniform radial compression of the plasma. Throughout the process, the flux through the plasma would remain constant (ignoring diffusion effects).

These illustrations point up the importance of the $E \times B$ drift velocity. Any situation in which electric fields appear in the plasma because of induction effects or charge separation will produce this drift. In some cases drifts of this kind can be self-perpetuating (charge separation leading to a drift, leading to more charge separation, and so on), so that a plasma instability results. In other cases the $E \times B$ drift is accompanied by plasma heating.

If the magnetic field in which the plasma particles move is inhomogeneous, other drift motions occur. If there is a gradient of the magnetic field strength perpendicular to the direction of the field, the radius of curvature of each particle will clearly be smaller on the high field side of its orbit than on the low field side, and a slow transverse drift perpendicular to the directions of both ∇B and B will occur. Since oppositely charged particles spiral in opposite directions, this drift will be oppositely directed for electrons and positive ions, so that it produces a tendency for charge separation to occur. The magnitude of the drift is given by Eq. (16).

$$v_b = \frac{r_c v_\perp}{2}\left(\frac{\nabla_\perp B}{B}\right) \qquad (16)$$

As the charged particles move in helical paths along the magnetic lines of force they may encounter regions where the flux tubes (bundles of magnetic lines) are curved. While the particles are being guided around these curves by the magnetic field, centrifugal forces will arise which will produce a drift. The centrifugal drift is also oppositely directed for ions and electrons and has the magnitude given by Eq. (17), where $v_{\parallel}$ is the velocity of

$$v_c = v_{\parallel}^2/R\omega_c \qquad (17)$$

motion of the particle along the lines of force and R is the local radius of curvature of the magnetic lines.

The centrifugal drift v_c is an example of a more general gyroscopic kind of drift that can be expected to arise in situations where the plasma particles are subjected to a force which is perpendicular to the local field direction. Another example is the drift velocity v_g which will occur in the presence of a gravitational field. The magnitude of this drift velocity is given by Eq. (18), where $g_\perp$ is the compo-

$$v_g = g_\perp/\omega_c \qquad (18)$$

nent of gravity perpendicular to the magnetic field. In strong magnetic fields v_g is very small, but it may play a role in geophysical phenomena in the upper atmosphere.

It is essentially impossible to study a confined plasma in the laboratory without encountering magnetic field gradients and therefore stimulating the drifts v_b or v_c. Since either of these drifts can give rise to charge separation effects, unless care is taken in choice of the field configuration electrostatic fields within the plasma can be set up which will cause the $\mathbf{E} \times \mathbf{B}$ drift v_o to occur, leading to a rapid escape of the plasma. A classic example is the simple torus with magnetic field lines parallel to the torus walls, as shown in Fig. 3. This

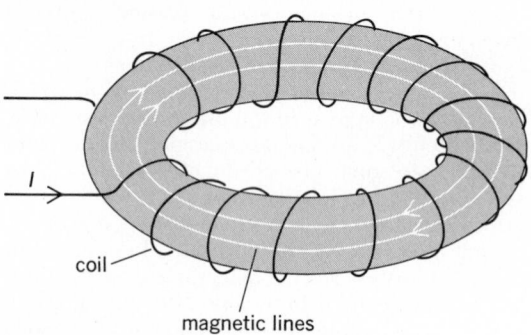

coil

magnetic lines

Fig. 3. Magnetic field lines in simple torus.

geometry, the forerunner of the stellarator, is characterized by a magnetic gradient which is everywhere inwardly directed (the field at the inner wall of a simple torus is always stronger than at the outer wall, since the line integral $\oint \mathbf{B} \cdot d\mathbf{l}$ is the same for all paths taken around the inside of the torus). Thus, if a plasma is confined in a simple torus, upward and downward drift motions will tend to occur, as predicted by Eqs. (16) and (17), producing a vertically directed electric field and a subsequent common outward drift of the plasma toward the outer wall, that is, toward the region of weakest magnetic field. In the first stellarators, this drift was cancelled, on the average, by twisting the torus into a figure eight. Thus, like the famous Möbius strip of mathematics, the inside wall on one curve of the stellarator becomes the outside wall at the opposite curve; consequently no net drift results.

Adiabatic invariants. Under the adiabatic conditions defined by Eqs. (13) and (14) there are some quantities which are approximate constants of the motion, the so-called adiabatic invariants alluded to earlier. These invariants are useful in predicting many of the details of plasma behavior and have been used as a starting point for erudite theoretical analyses. The simplest of these invariants is μ, the magnetic moment associated with the rotation of any charged particle in a magnetic field. A moving charge represents a current, and a circling charge therefore represents a circular current element,

with which is associated a dipolelike magnetic field similar to that of a simple loop current. By analogy to Lenz's law, this current loop is always diamagnetic; that is, it tends to depress infinitesimally the field strength inside the loop, so that an assembly of charged particles, a plasma, usually exhibits a bulk diamagnetic effect arising from the vector summation of the individual particle diamagnetic effects.

The magnitude of μ is given by Eq. (19). The

$$\mu = W_\perp/B \quad \text{J/tesla} \qquad (19)$$

magnetic moment μ may be expected to be a constant provided the adiabatic conditions in Eqs. (13) and (14) are fulfilled. More specifically, examination of some theoretical work suggests that, except for the effect of collisions, the maximum fluctuations in μ occurring as a particle moves back and forth in a varying magnetic field should be roughly representable by Eq. (20), where a and

$$\left| \frac{\delta\mu}{\mu} \right| = ae^{-b/\epsilon} \qquad (20)$$

b are constants of order 1, and ϵ is either $2r_c/\lambda$ for spatially varying fields or $1/\omega_c\tau$ for time-varying fields. It is clear that if ϵ is 0.1 or less, the fluctuations in μ should be very small. It is concluded that there are many plasma situations where the constancy of μ between particle collisions should be a valid assumption.

The importance of μ in plasma calculations can be illustrated by using it to calculate the condition for reflection of a charged particle by a magnetic mirror. Suppose that a spiraling particle approaches a magnetic mirror from a region where the field is B_O, and is to be reflected when it reaches the point M where the field intensity is B_M. From Eq. (19), it is seen that Eq. (21) holds. The last part of

$$\mu = \frac{W(O)}{B_o} = \frac{W_\perp(M)}{B_M} = \frac{W}{B_M} \qquad (21)$$

the equation follows since the entire energy resides in rotational motion at the point of reflection. It follows that the condition for reflection at a point where the field is B_M is given by Eq. (22), where R

$$\frac{W_\perp(O)}{W} = \frac{B_o}{B_M} = \frac{1}{R} \qquad (22)$$

is called the mirror ratio. Thus, since Eq. (23)

$$\frac{W_\perp(O)}{W} = \frac{v_\perp^2}{v^2} = \sin^2\theta \qquad (23)$$

holds, θ being the pitch angle of the helix at $B = B_O$, the condition that a particle be reflected at or before it penetrates to M is just that the pitch angle should be greater than a critical angle θ_c, where Eq. (24) applies. This conditions is independent of

$$\sin\theta_c = \frac{1}{R^{1/2}} \qquad (24)$$

charge, mass, or total energy of the reflected particle, except as limited by the requirements of the adiabatic assumptions. It is equally easy to show that in the general case, the pitch angle θ transforms in accordance with the relationship in Eq. (25), where $R(u) = B(u)/B(O)$. Equation (24) follows

$$\sin\theta(u) = [R(u)]^{1/2}\sin\theta(O) \qquad (25)$$

by setting $\theta(u) = \pi/2$ (reflection). Equation (25) resembles Snell's law of optics, with $R^{1/2}$ playing the role of the index of refraction.

Returning to the question of magnetic mirror reflection, it is apparent that if two magnetic mirrors are used, one at each end of a weaker central field, the charged particles of a plasma can be trapped in the "magnetic bottle" between the two mirrors, provided the particles satisfy the pitch angle requirements. This is the mirror machine mentioned earlier. If a plasma containing particles with randomly oriented pitch angles were to be suddenly thrust into such a magnetic bottle, all particles with pitch angles less than θ_c would immediately escape. The remainder of the particles would be trapped, however, independent of their charge, mass, or energy, and could escape only as mutual collisions deflected them into unfavorable pitch angles. At high temperatures this process is predicted to be quite slow.

Another consequence of the constancy of μ relates to plasma heating by magnetic compression. It has already been pointed out that an increasing magnetic field will compress a plasma, which tends to maintain constant flux through its volume. This property is also true for the flux threading each orbit. Since $\mu = W_\perp / B = \text{constant}$, and $r_c \sim (1/B) W^{1/2}$, this implies that $B r_c^2 \sim W_\perp / B = \text{constant}$; that is, the flux through the orbit circle area $\pi r_c^2 B$ is a constant. The heating effect follows simply by noting that since μ is constant, $W_\perp$ must increase in direct proportion to B. This is the process of adiabatic compression heating often alluded to in the literature.

To complete the list of adiabatic invariants of plasma motion, two more will be mentioned. Suppose that the plasma particles are trapped in a magnetic bottle and move in a periodic way back and forth between limits (as in the mirror machine). It can then be shown that the so-called action integral is an adiabatic invariant. The integral is just the line integral of the momentum component parallel to the direction of the field, taken along the path of the guiding center as in Eq. (26).

$$\oint p_\parallel \, du = A = \text{const} \qquad (26)$$

The action integral has the dimensions of an area in $p_\parallel,u$ (phase) space. Its constancy implies that this area is a constant. Physically, it simply means that as a trapped particle moves about in the confinement volume, its guiding center motion must always be such as to keep the phase space area constant (Liouville's theorem). If axial compression occurs, this implies that the axial momentum must increase correspondingly (Fig. 4). One of the important results of adiabatic theory is to show that the constancy of A implies that, in the absence of collisions, and for slowly varying fields, the drift motion of trapped particles in a magnetic bottle (such as the Earth's magnetic field) is such as to generate an imaginary fixed closed surface for each trapped particle, to which it would be forever bound. The now-famous Argus experiment provided a substantial degree of confirmation of these ideas. *See* STATISTICAL MECHANICS.

The last adiabatic invariant to be mentioned has already been hinted at. It has been demonstrated that the magnetic flux enclosed by the aforementioned surfaces which are generated as a consequence of the constancy of A is also an adiabatic invariant. This fact is very useful in analyzing novel or complicated magnetic confinement geometries.

In summary of the particle dynamics of plasmas, it has been shown that plasma particles in a strong magnetic field are generally constrained to move in helical orbits. In the presence of field gradients and other perturbations they tend to drift from line of force to line of force, tracing out closed surfaces as they move about. Throughout the motion of the plasma particles, the collective motion will be constrained to be such as to preserve near equality of total positive and negative charge, since any departure from neutrality gives rise to strong electric restoring forces.

WAVES IN PLASMAS

Because of the coupled nature of the motion of plasma and its electromagnetic field environment, it can support unusual oscillatory or wave motions, both stable and unstable. Instabilities have already been discussed.

A rough subdivision of the stable wave motions is possible in terms of four characteristic frequencies, or periods of a plasma, assumed immersed in a magnetic field. In many typical cases these characteristic frequences occur in the following order: (i) interparticle collision frequencies (lowest); (ii) ion cyclotron frequency; (iii) electron cyclotron frequency; and (iv) the plasma frequency, $\omega_p = \sqrt{n_e e^2 / \epsilon_0 m}$ (highest), where m is the electron mass. Only a few of the features of wave propagations will be sketched, the details of plasma wave propagation being remarkably complicated.

At frequencies below (i) a plasma would behave as an ordinary gas and propagate a simple elastic wave. Such waves would be important only for high density, low temperature, or large plasmas.

At frequencies between (i) and (ii) a characteristic plasma wave, the hydromagnetic or Alfvén wave, is possible. This wave resembles that which would be propagated by a loaded elastic string. Here the "strings" are the magnetic lines of force, loaded by the plasma mass (that is, primarily the ions). Satisfying the adiabatic criteria, particles and field distort together (field lines "stick" to the plasma and vice versa), providing a loaded elastic medium for the waves. The Alfvén wave velocity is given by Eq. (27), where ρ is the plasma density.

$$v_A = \sqrt{B^2 / \mu_0 \rho} \quad \text{m/sec} \qquad (27)$$

The expression is valid only for frequencies $\ll \omega_{ci}$ and for values of B and ρ such that $v_A \ll c$; that is, $B^2/2\mu_0 \ll (1/2)\rho c^2$ (magnetic energy density small compared with one-half the mass energy density).

At frequencies in the vicinity of (ii), resonance of the wave motion with the ion cyclotron frequency occurs and the waves become highly dispersive.

At frequencies between (ii) and (iv), wave propagation is allowed only for special directions of propagation with respect to the magnetic field, and types of waves are propagated which resemble the ordinary and extraordinary light wave in a birefringent crystal. *See* CRYSTAL OPTICS.

Above frequency (iv), the plasma propagates simple electromagnetic waves with a phase velocity greater than c; that is, it behaves like a medium with an index of refraction less than unity. The index of refraction ν is given by Eq. (28).

$$\nu = \sqrt{1 - (\omega_p/\omega)^2} \quad \omega > \omega_p \qquad (28)$$

PLASMA PHYSICS

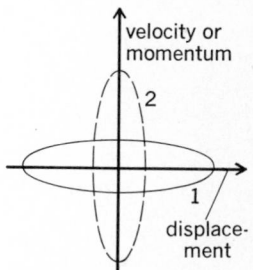

Fig. 4. Schematic illustration of conservation-of-action integral, Eq. (26).

This property of a plasma can be used to determine its electron density. By measuring the phase shift in the transmission of a microwave beam propagated through the plasma the mean index of refraction can be determined, and from this the density, through use of Eq. (28). The numerical value of the plasma frequency is given by Eq. (29).

$$f_p = \frac{\omega_p}{2\pi} = 9(n_e)^{1/2} \quad (29)$$

For a density of 10^{20} electrons/m³, $f_p = 9 \times 10^{10}$ Hz, corresponding to a free-space wavelength of 3.3 mm.

The plasma frequency itself represents a classic example of cooperative effects in a plasma. It represents a natural frequency of oscillation of the electrons of the plasma, occurring even in the absence of any magnetic field. It arises from the fact that any net displacement of the plasma electrons with respect to the ions produces a polarization electric field which acts as a restoring force. Thus, if such a displacement occurs, volume oscillations of the electrons about the position of charge neutrality will result. The ions, being much more massive, will remain essentially stationary, like plums in a pudding. If a more detailed analysis is performed, it can be shown that the effect of the thermal velocity of the electrons is to modify slightly the allowed frequency of these electrostatic plasma oscillations, so that a dispersion relation results, that is, a frequency-wavelength relationship, showing that such oscillations can occur over about an octave, bounded on the low frequency side by ω_p. *See* WAVES AND INSTABILITIES IN PLASMAS.

COLLISION PROCESSES IN PLASMAS

In the absence of instability mechanisms which could destroy a state of order in a plasma, the drive toward a state of higher disorder comes about through interparticle collisions. The dominant interparticle collision process in a high temperature plasma is Rutherford scattering, elastic scattering arising from the mutual Coulomb electrostatic fields of the charged particles of the plasma. Such processes lead to the deflection of and energy exchange between the plasma particles. These collision processes are important, since they determine the basic rate of all collisional transport processes in the plasma. The cross section or effective collisional area σ_c for "close" collisions between plasma particles of equal mass and charge can be simply estimated from the classical minimum distance of approach of two equal charges (where mutual potential energy becomes just equal to initial kinetic energy; that is $Z^2 e^2/4\pi\epsilon_0 r_{min} = W$) as in Eq. (30).

$$\sigma_c = \pi Z^4 e^4/(4\pi\epsilon_0)^2 W^2 \quad m^2 \quad (30)$$

However, if the classical minimum distance of approach is compared with the value of the Debye length λ_D given in the table, it will be readily seen that at the usual particle energies λ_D is typically some 10^9 times larger. Thus the probability of a particle "colliding" with some particle that is distant yet still within the range λ_D is some 10^{18} times larger than the probability that it should make a simple "hard" close collision. Thus, even though the effect of each individual distant encounter

(within λ_D) is infinitesimal, the statistical sum of their effects is not, and in fact is about 10 times more important than that of close collisions. In the calculation of this effect, the logarithm of the ratio $(\lambda_D/r_{min}) = \Lambda$ appears; $\ln \Lambda \approx 20$ in most cases. For an approximate value of the scattering cross section owing to distant collisions the value given by Eq. (31) is found.

$$\sigma_d \cong (\pi Z^4 e^4/16\pi^2\epsilon_0^2 W^2)(\ln \Lambda/2) \quad (31)$$

Written in terms of the particle energy in eV, this becomes Eq. (32) where W is in eV.

$$\sigma_d \cong 10\sigma_c \cong 6 \times 10^{-17}(Z^4/W^2) \quad m^2 \quad (32)$$

Some quantitative values are of interest. If $Z = 1$ (hydrogen ions, or singly charged heavier ions), and if $W = 1$ eV, then $\sigma_d \cong 6 \times 10^{-17}$ m², or some 10^3 to 10^4 times normal atomic or gas kinetic cross sections. However, if $W = 1$ keV, σ_d is 10^6 times smaller (about 6×10^{-23} m²) and is already much much smaller than atomic cross sections. It can be seen that even disregarding the fourth power dependence on Z, the range of collision cross sections encountered in hot plasmas may be enormous. It follows that the collision mean free path d can vary over an even more extreme range, considering the range of densities which is of interest. For example, for a hydrogenic plasma density of 10^{23} m⁻³ (about the upper limit for most high-temperature plasma studies) and a mean ion energy of 10 ev, $d = 1/n\sigma_d \cong 1.6 \times 10^{-6}$ m, whereas at $n = 10^{18}$ m⁻³ and 1 keV, $d \cong 16$ km. Clearly the physical behavior of the plasma can be expected to be quite different over the ranges of temperature and density which may be encountered in even a single experiment. *See* SCATTERING EXPERIMENTS (ATOMS AND MOLECULES).

Relaxation time. The collision processes in a plasma serve to define the rate of randomization of the energy and direction of motion of the plasma particles. One reason that these processes are important is that they act in opposition to magnetic confinement (especially in the mirror machine) and set an upper limit on its duration. From fundamental work by S. Chandrasekhar and Spitzer, the so-called relaxation time for a large angular deflection or for energy exchange comparable with the original energy of a given ion (or electron) colliding with particles of the same kind can be calculated. To a sufficient approximation, this time is given by Eq. (33), where W is in eV.

$$\tau = 5.7 \times 10^{11}(A^{1/2}/Z^4)(W^{3/2}/n) \quad \text{sec} \quad (33)$$

Returning to the numerical examples given, in this case there are, for protons ($A = Z = 1$), at a density of 10^{23} m⁻³ and an energy of 10 eV, $\tau = 1.8 \times 10^{-10}$ sec; while for electrons ($A = 1/1836$), $\tau = 4 \times 10^{-12}$ sec, which is a very short time indeed. On the other hand, at $n = 10^{18}$ m⁻³ and energy of 1 keV, these times become equal to 18 msec and 0.42 msec respectively. It is quite clear that in the first example the times are small even compared with the cyclotron frequencies of the plasma particles (normally the shortest characteristic times in a magnetically confined plasma) so that heating by magnetic compression would be slight and magnetic confinement would be essentially inoperative, the plasma diffusing like an ordinary gas. On the other hand, in the second example, the times are very long compared with cyclotron periods for reasonable magnetic field strengths, and in slowly varying fields the adiabatic assumptions would be well satisfied. Further increase in the temperature or magnetic field strength would make the conditions even more accurately satisfied.

Degrees of freedom. Another interesting consequence of plasma situations where the collision times are long is that for operations performed on the plasma in times short compared with the collision times, the degrees of freedom of the plasma motion are essentially uncoupled. That is to say, compressions perpendicular to the field direction involve only the rotational degrees of freedom, so that the plasma acts as a two-dimensional gas. Similarly, compression along the field lines results in the plasma acting as a one-dimensional gas. The laws of thermodynamics relate the value of the adiabatic gas constant γ to the number of degrees of freedom f; $\gamma = (2 + f)f$. Thus for compression transverse to B, $f = 2$, so that $\gamma = 2$. For longitudinal compression $\gamma = 3$. The same general thermodynamics laws show that the temperature-versus-particle density law for an adiabatic compression of a gas is given by relation (34).

$$T \propto n^{\gamma - 1} \tag{34}$$

Thus for transverse compression $T_\perp \propto n$, but for longitudinal compression $T_\parallel \propto n^2$; that is, $T_\parallel$ varies as the square of the compression ratio. However if the compression were carried out in a time which was long compared with collision times, $f = 3$, $\gamma = 5/3$ (all degrees of freedom coupled), and $T \propto n^{2/3}$, a substantially less pronounced variation with density.

Dynamical friction. While collisions between an energetic ion and the electrons of a plasma lead to little deflection of the ion's path, owing to the small mass of the electrons, another important effect, dynamical friction, can arise. Provided only that the mean energy of the ion is greater than that of the electrons, the statistically averaged effect of collisions with electrons within the Debye length leads to a frictional drag on the ion which will eventually reduce its mean energy to that of the electrons. This can be an important effect, leading to a damping of the energy of energetic ions immersed in a cold plasma. On the other hand, if the mean ion energy is lower than that of the electrons, the electron collisions will lead to an increase of the ion energy by heating. This latter effect is important in cases where plasma heating is accomplished via heating the electrons first, as in joule or resistive heating.

A single expression can be derived which represents both these effects. This is given by Eq. (35).

For ion energies small compared with the mean electron energy, the rate of ion heating (T_e in eV) is given by Eq. (36).

$$\left(\frac{dW}{dt}\right)_{ei} = 4\pi \sqrt{2} \ln \Lambda \, \frac{n_e Z^4 e^4}{(4\pi\epsilon_0)^2 (\pi m k T_e)^{1/2}}$$

$$\cdot \frac{m}{M}\left(1 - \frac{W}{3kT_e/2}\right) \tag{35}$$

$$\left(\frac{dW}{dt}\right)_{ei} = 8.8 \times 10^{-14} \frac{Z^2}{A} \frac{n_e}{T_e^{1/2}} \quad \text{eV/sec} \tag{36}$$

$$W \ll 3kT_e/2$$

When the ion energy is substantially greater than the mean electron energy, the general expression takes the form of Eq. (37).

$$\frac{1}{W}\frac{dW}{dt} = -5.7 \times 10^{-8} \frac{n_e Z^2/A}{T_e^{3/2}} \quad \text{sec}^{-1} \tag{37}$$

$$W \gg T_e$$

This can be integrated to yield an expression, Eq. (38), for the exponential decay of ion energy as a function of time where Eq. (39) applies.

$$W = W_0 e^{-t/t_e} \tag{38}$$

$$t_e = 1.8 \times 10^{13} (T_e^{3/2}/n_e)(A/Z^2) \tag{39}$$

If the electron temperature is low, t_e may be quite small compared with the mean ion-ion collision time, and therefore will dominate the energy exchange times for high-energy ions. For example, if $n_e = 10^{20}$ m^{-3} and $T_e = 10$ eV, then for energetic protons, $t_e = 5.5$ μsec. By contrast, if the proton energy is, for example, 10 keV, the corresponding ion-ion scattering time τ calculated from Eq. (33) is roughly 6 msec or 1000 times larger. At an electron temperature of 1 keV, however, the two times become about equal.

The expression for t_e holds as long as the ion velocity is less than the mean velocity of the electrons, that is, as long as Eq. (40) is valid.

$$W < (M/m)(3/2)kT_e \tag{40}$$

When this is not satisfied, the dynamical friction rate is somewhat smaller than that just predicted. For protons, this occurs when $W = 2760 \, kT_e$, that is, only at very high energies.

The entire question of energy transfer rates between ions and electrons is one of great importance in plasma research, since in many cases these rates are critical in determining the relative importance of other processes, such as the rate of loss of energy from the plasma by radiation (which arises from the electrons).

RADIATION FROM PLASMAS

Radiation provides a direct cooling mechanism for a high-temperature plasma. Fortunately indeed for the future of plasma research, theory shows that at even the highest particle densities used in the laboratory, the radiation rates from a plasma are much less than the Planck or blackbody value. For example, at a radiation temperature of 10^8 K, the Planck radiation, being proportional to T^4, would amount to the almost inconceivable value of 6×10^{24} W/m^2. But the fortunate fact that a tenuous plasma is optically very "thin" over almost all of its emission spectrum means that, as might be expected from Kirchhoff's law, radiation is greatly reduced compared with the Planck value, so that under the proper circumstances a plasma with a kinetic temperature of 10^8 K might radiate at a rate equivalent to the radiation rate from a blackbody at radiation temperatures of only a few hundred kelvins. Nevertheless, in many experiments great care must be taken to avoid certain impurity problems so as to keep the radiation from rising to a value where it would overwhelm the means for heating the plasma and keeping it hot. *See* HEAT RADIATION.

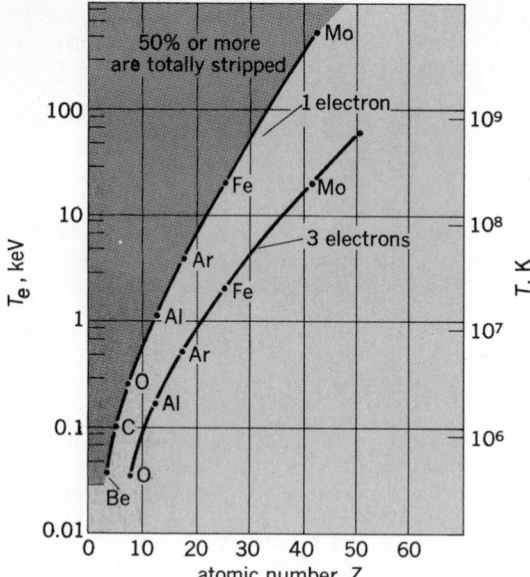

Fig. 5. Stripping curve.

Common mechanisms.

Common mechanisms. Considering collisional processes, there exist three important mechanisms for radiation from a plasma.

The first is the generation of x-rays (bremsstrahlung), which occurs when the plasma electrons are deflected by encounters with the ions. The second mechanism is a similar radiation which occurs when electron-electron collisions occur. This process is important only at very high electron temperature, where the electron motion becomes relativistic (T_e of order $mc^2 = 511$ keV). The third mechanism which can occur, and one which under some circumstances may overwhelmingly dominate the radiation losses, is that process which might be called excitation radiation, radiation resulting from the collision of electrons with partially stripped ions (ions with remaining bound electrons) with the production of excited states of the bound electrons and subsequent radiation.

Electron-ion bremsstrahlung can be calculated by the methods of quantum mechanics, from which is found, for a Maxwellian distribution, the approximate expression for the radiation per unit volume, Eq. (41).

$$p_{ei} = 1.4 \times 10^{-40} n_e n_i Z^2 T_e^{1/2} \quad \text{W/m}^3 \quad (41)$$

Here Z is the ionic charge, and T_e is the kinetic temperature measured in kelvins. Except at high densities, the radiation rate is nominal, being only some 14 kW/m³ for a hydrogenic plasma at 10^8 K and a density of 10^{20}/m³ ($Z=1$, $n_e = n_i$). At $n = 10^{23}$/m³ this would be 10^6 times larger, reaching the respectable figure of 14,000 MW/m³. Also for a plasma composed entirely of higher Z ions, since electrical neutrality of the plasma requires that $n_e = Z n_i$, the bremsstrahlung radiation rate varies as Z^3 and would therefore be much larger. *See* BREMSSTRAHLUNG.

Excitation radiation rates can be calculated if the degree of stripping of the ions in the plasma is known. In the past it has been customary to assume that by the time kinetic temperatures of $10^5 - 10^6$ K or higher have been reached, all atoms in a plasma will have been completely stripped of their bound electrons. In this case, there would, of course, be no excitation radiation emitted. The assumption of complete stripping would be valid for a hydrogenic plasma, whose atoms have only a single, easily stripped electron. Unfortunately, it is not possible to create in the laboratory an absolutely pure hydrogen plasma; there will always exist some appreciable number of higher Z containment atoms, such as oxygen.

By means of calculations similar to those employed by astrophysicists in calculating the radiation from the Sun's corona, it is possible to compute the degree of stripping of impurity atoms immersed in a low Z plasma. It is found that the degree of stripping is much less than would have been intuitively assumed. As a result, the role of excitation radiation can be very important, even at quite high kinetic temperatures.

In the calculations it is found that the most interesting and hardy ions are those which are stripped down to one to three or four remaining electrons. Using approximate ionization cross sections, a stripping curve has been calculated for the expected relative abundances of one-electron (hydrogenlike) and three-electron (lithiumlike) ions as a function of atomic number and kinetic temperature (Fig. 5). The figure shows loci for the curve dividing the region of temperature below which a given atom has a probability greater than 50% of having one or more bound electrons, and for a similar curve for three or more bound electrons. These curves represent steady-state values of the stripping. It will be seen that high Z impurity atoms become completely stripped only at very high temperatures.

Using calculations of this kind, it is possible to determine the expected rates of excitation radiation, per impurity atom, as a function of kinetic temperature. This should be done for all existing states to obtain the total radiation rate. However, it is sufficiently informative to present the results for the two states (one electron and three electrons) just given. The results of these calculations, using approximate excitation cross sections, are presented in Fig. 6. The results are normalized on an atom-for-atom basis, against the ordinary hydrogenic bremsstrahlung rate. The feature that is immediately apparent from the curves is that at temperatures of the order 10^6 K, the excitation radiation rate is enormous compared with ordinary bremsstrahlung. This radiation is all emitted in the

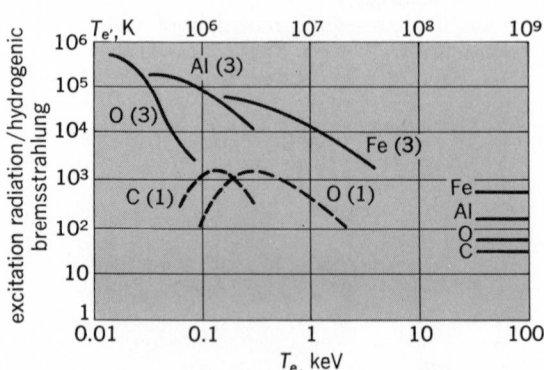

Fig. 6. Ratio of two plasma radiation mechanisms to temperature calculated for iron, aluminum, oxygen, and carbon.

vacuum ultraviolet region of the spectrum: $\lambda \cong$ 10–100 nm. Only as the temperature reaches 10^7 K or more does it drop to more reasonable values. This shows that at the lower temperatures excitation radiation may provide a very rapid cooling process for a hydrogenic plasma, even with relatively small percentages of impurities. Secondly, it is apparent that once very high temperatures are reached, the rate falls to about the ordinary bremsstrahlung values (indicated by the marked lines along the right edge of the plot in Fig. 6).

The preceding calculations represent steady-state rates. However, in experimental plasmas the duration of the experiment may be too short for the plasma ions to reach the steady-state degree of stripping. In this case, calculations show that the radiation rate may be substantially higher than that just indicated.

At relatively low plasma temperatures and very high impurity densities, self-absorption of the radiation may become important and reduce the calculated loss somewhat. This usually occurs only at a very high absolute radiation flux, however, since the onset of self-absorption signals the approach to equilibrium or Planck radiation levels.

Other mechanisms. The mechanisms described thus far produce their main radiation flux in the x-ray and the vacuum ultraviolet part of the spectrum. Going toward longer wavelengths, the next significant radiation region is not reached until the long-wave infrared or the short-wavelength microwave region is reached. In this region, the plasma again possesses mechanisms which can produce appreciable radiation fluxes. Being limited by the Rayleigh-Jeans value (which varies as the square of frequency), this long-wavelength radiation does not constitute an appreciable energy loss mechanism except at the highest electron temperatures. The radiation in this region is important only for a plasma immersed in a strong magnetic field. It arises simply from the centrifugal acceleration of the electrons as they move in helical orbits in the magnetic field. Measurement of it provides another possible way to determine the electron temperature of the plasma, by applying techniques similar to those employed in radio astronomy.

[RICHARD F. POST]

Bibliography: F. Chen, *Introduction to Plasma Physics*, 1974; N. Krall and A. Trivelpiece, *Principles of Plasma Physics*, 1973; R. F. Post, Controlled fusion research and high temperature plasmas, *Annu. Rev. Nuc. Sci.* 20:509–558, 1970; G. Schmidt, *Physics of High-Temperature Plasmas*, 2d ed., 1979; L. Spitzer, *Physics of Fully Ionized Gases*, 1962; B. Tanenbaum, *Plasma Physics*, 1967.

Pleochroism

In some colored transparent crystals, the effect wherein the color is quite different in different directions through the crystals. In such a crystal the absorption of light is different for different polarization directions. Tourmaline offers one of the best-known examples of this phenomenon. In colored transparent tourmaline the effect may be so strong that one polarized component of a light beam is wholly absorbed, and the crystal can be used as a polarizer. For a fuller discussion of the effect *see* DICHROISM; TRICHROISM.

[BRUCE H. BILLINGS]

Poinsot's method

A method of describing, by means of geometrical construction, the motion of a rigid body with a point fixed in space and with zero torque or moment acting on the body about the fixed point. If a rigid body is constrained to rotate about a smooth fixed axis, under no moments except those due to the axis reactions, the motion is simply one of constant angular velocity. If, however, the body is constrained to move with only one point fixed in space, the motion, even with no moment acting about that point, is much more complicated. Furthermore, the motion in this latter case is identical to that of a rigid body relative to its own center of mass, and it is, therefore, a more general type of motion for those cases where zero or negligible moments act about the center of mass. Such a body might be a top spinning on a frictionless table in a gravityless system, a body mounted within a Cardan suspension, or a spinning rocket flying in space outside of the atmosphere but in a gravity field. *See* CENTER OF MASS.

Cardan's suspension. Consider a heavy body mounted with only one point fixed, constructed from light rings in an arrangement known as Cardan's suspension. In Fig. 1, point O is the fixed point, and it is assumed that the frictional torques can be made negligible and that the mass of the suspension system compared with the heavy body is negligible. Let O be the center of a coordinate system composed of the principal axes of the body x, y, and z, with unit directional vectors $\mathbf{i}$, $\mathbf{j}$, and $\mathbf{k}$.

The vector angular velocity $\boldsymbol{\omega}$ and vector angular momentum $\mathbf{H}$ of the body are given by Eqs. (1),

$$\boldsymbol{\omega} = \omega_x \mathbf{i} + \omega_y \mathbf{j} + \omega_z \mathbf{k}$$
$$\mathbf{H} = I_x \omega_x \mathbf{i} + I_y \omega_y \mathbf{j} + I_z \omega_z \mathbf{k} \tag{1}$$

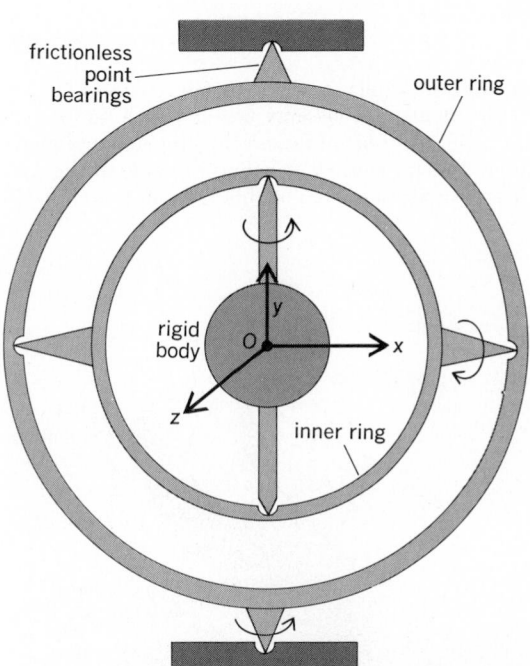

Fig. 1. Cardan's suspension. Point O of the rigid body is fixed in space while the body is free to rotate about any axis in space under no external moments.

where I_x, I_y, and I_z are moments of inertia about the x, y, and z axes respectively, and the products of inertia are zero about the principal axis. The kinetic energy of the body is constant and is given by Eq. (2). *See* RIGID-BODY DYNAMICS.

$$T = \frac{I_x\omega_x^2 + I_y\omega_y^2 + I_z\omega_z^2}{2} \qquad (2)$$

The angular momentum **H** is constant in magnitude and direction. Its magnitude is given by Eq. (3). From Eqs. (1) and (2), Eq. (4) is obtained.

$$H^2 = (I_x\omega_x)^2 + (I_y\omega_y)^2 + (I_z\omega_z)^2 \qquad (3)$$

$$\boldsymbol{\omega} \cdot \mathbf{H} = 2T \qquad (4)$$

Ellipsoid equations. If now a line OA, called the invariable line, is drawn in the fixed direction of **H** (Fig. 2), and OB is the vector angular velocity $\boldsymbol{\omega}$ at any instant, then the line BC, drawn perpendicular to OA, determines line OC such that Eq. (5) holds.

$$OC = \frac{\boldsymbol{\omega} \cdot \mathbf{H}}{H} \qquad (5)$$

From Eqs. (4) and (5), Eq. (6) is obtained. Therefore,

$$OC = \frac{2T}{H} \qquad (6)$$

fore, C is a fixed point and the plane through C normal to OA is a fixed plane, called the invariable plane. The terminus of $\boldsymbol{\omega}$ (point B, Fig. 2) moves on the invariable plane during motion of the rigid body. If point B is given coordinates x_1, y_1, and z_1, then Eq. (7) holds and Eq. (2) becomes Eq. (8).

$$\boldsymbol{\omega} = \mathbf{i}x + \mathbf{j}y + \mathbf{k}z \qquad (7)$$

$$I_x x^2 + I_y y^2 + I_z z^2 = 2T \qquad (8)$$

Equation (8) is the equation of the Poinsot ellipsoid which is fixed in the rigid body tangent to the invariable plane at B, and with center at O. The semiaxes are given by Eqs. (9). As the body moves,

$$a = \sqrt{\frac{2T}{I_x}} \qquad b = \sqrt{\frac{2T}{I_y}} \qquad c = \sqrt{\frac{2T}{I_z}} \qquad (9)$$

the ellipsoid rolls on the invariable plane because it has an angular velocity vector which terminates at B, the point of contact of the ellipsoid and plane (Fig. 2). The point B is called the pole of the axis of rotation. The locus of this pole on the ellipsoid

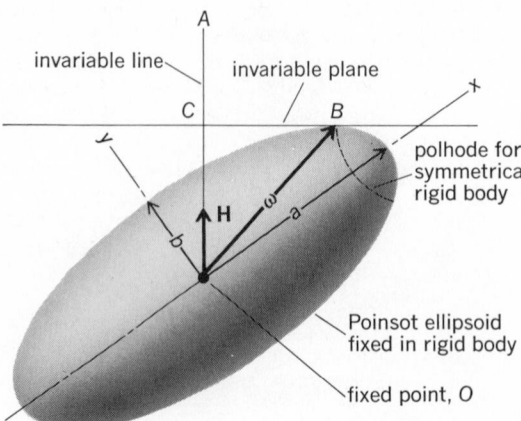

Fig. 2.   Poinsot ellipsoid.

was called by L. Poinsot the polhode and the locus on the tangent plane the herpolhode.

The resulting motion is, in general, quite complicated except when the Poinsot ellipsoid is a surface of revolution. If the invariable plane were represented by a sheet of paper and if an ellipsoid given by Eq. (8) were constructed, coated with ink, and rolled on the paper holding center O fixed, the ink would trace out a herpolhode curve. On joining the fixed point O to all points on this curve, the space cone of the $\boldsymbol{\omega}$ vector would result. In this way the motion of the rigid body under no moments can be visualized and represented.

[RAY E. BOLZ]

Bibliography: H. Goldstein, *Classical Mechanics*, 2d ed., 1980; K. R. Symon, *Mechanics*, 3d ed., 1971.

Point source

A source having definite position but no extension in space. In discussing radiation, it is convenient to define the concept of point source. If the radiation propagates in radially straight lines (or, which is the same thing, in spherical waves) from the point source, conservation of energy demands that the intensity of the radiation decrease in any direction inversely as the square of the distance from the source. No physical source is actually a mathematical point, but for distances sufficiently large compared to dimensions of the source, the inverse-square law may be a good approximation. *See* INVERSE-SQUARE LAW.

[MC ALLISTER HULL, JR.]

Polar molecule

A molecule possessing a permanent electric dipole moment. Molecules containing atoms of more than one element are polar except where forbidden by symmetry; molecules formed from atoms of a single element are nonpolar (except ozone). The dipole moments of polar molecules result in stronger intermolecular attraction, increased viscosities, higher melting and boiling points, and greater solubility in polar solvents than in nonpolar molecules. *See* DIPOLE MOMENT.

The electrical response of polar molecules depends in part on their partial alignment in an electric field, the alignment being opposed by thermal agitation forces. This orientation polarization is strongly temperature-dependent, in contrast to the induced polarization of nonpolar molecules. *See* DIELECTRIC CONSTANT; FERROELECTRICS; MOLECULAR STRUCTURE AND SPECTRA; POLARIZATION OF DIELECTRICS.   [ROBERT D. WALDRON]

Polarity

A term which refers to the designation of positive and negative terminals in an electric circuit carrying a current. The choice of positive charge, and thereby of positive potential, is a purely arbitrary one which had been made long before the discovery of the electron. It is now known that the primary current carriers in most circuit elements are electrons, which possess negative charge.

For a seat of electromotive force, such as a battery or a generator, the positive terminal is by convention the one at which electrons enter from the external circuit; the negative terminal is the one from which they leave. For a "passive" element, such as an ammeter or a resistor, the positive terminal is the one through which the electrons leave the element; the negative terminal is the one through which they enter from the circuit. The potential of the positive terminal of any element is by definition greater in value than that of the negative terminal. In all cases the conventional current flows in the opposite direction to the electrons.

In the case of alternating current the term polarity may be used in a different sense. Usually in commercial circuits one leg is maintained at zero potential, while the other is alternately above and below ground. In some electrical appliances, it makes a difference which side is connected to the "hot" lead. Turning the plug around in the wall socket to obtain the best operation of a particular device is often referred to as "reversing the polarity." *See* POTENTIALS.

[JOHN W. STEWART]

Polarization of dielectrics

A vector quantity representing the electric dipole moment per unit volume of a dielectric material. *See* DIELECTRICS; DIPOLE MOMENT.

The polarization **P** is related to the macroscopic electric parameters by the equation shown below,

$$\mathbf{P} = \frac{\mathbf{D} - \epsilon_0 \mathbf{E}}{\gamma} = \frac{\epsilon_0 (\kappa' - 1) \mathbf{E}}{\gamma} = \chi \epsilon_0 \mathbf{E}$$

where **D** is the electric displacement, **E** is the electric field strength, ϵ_0 is the permittivity of vacuum, κ' is the dielectric constant, χ is the electric susceptibility, and γ is a geometrical factor. In cgs electrostatic units $\epsilon_0 = 1$ and $\gamma = 4\pi$; in rationalized mks units $\epsilon_0 = 8.854 \times 10^{-12}$ farad/m and $\gamma = 1$. The dimensions of polarization are statcoulomb per square centimeter in the cgs system and coulomb per square meter in the rationalized mks system. *See* ELECTRICAL UNITS AND STANDARDS.

Dielectric polarization arises from the electrical response of individual molecules of a medium and may be classified as electronic, atomic, orientation, and space-charge or interfacial polarization, according to the mechanism involved.

Electronic polarization represents the distortion of the electron distribution or motion about the nuclei in an electric field. This polarization occurs for all materials and is nearly independent of temperature and frequency up to about 10^{14} Hz for insulators.

Atomic polarization arises from the change in dipole moment accompanying the stretching of chemical bonds between unlike atoms in molecules. This mechanism contributes to polarization at frequencies below those of the vibrational modes of molecules (about 10^{12} to 10^{14} Hz). For a discussion of molecular vibrations *see* MOLECULAR STRUCTURE AND SPECTRA.

Orientation polarization is caused by the partial alignment of polar molecules, that is, molecules possessing permanent dipole moments, in an electric field. This mechanism leads to a temperature-dependent component of polarization at lower frequencies.

Space-charge or interfacial polarization occurs when charge carriers are present which can migrate an appreciable distance through a dielectric but which become trapped or cannot discharge at an electrode. This process always results in a distortion of the macroscopic field and is important only at low frequencies. *See* DIELECTRIC CONSTANT; ELECTRIC FIELD; ELECTRIC SUSCEPTIBILITY.   [ROBERT D. WALDRON]

Polarization of waves

Polarization is the phenomenon which is exhibited when a transverse wave is polarized. The term polarization is also used to describe the process of polarizing a wave.

In an unpolarized wave, the vibrations in a plane perpendicular to the ray appear to be oriented in all directions with equal probability. In a polarized wave the displacement direction of the vibrations is completely predictable. For certain disturbances, such as the transverse acoustic wave produced when a steel bar is struck, the polarization is complete. Electromagnetic radiation is normally unpolarized if it is generated by atomic processes. Thus ultraviolet, visible, and infrared radiations produced by heated bodies or electrical discharges are generally unpolarized. Radiation generated by vacuum-tube oscillators or transistor oscillators is always polarized. The probability waves (matter waves) associated with atomic or nuclear particles are generally unpolarized. *See* ELECTROMAGNETIC RADIATION; QUANTUM MECHANICS.

Some of the different types of polarization, as well as the technique of producing polarization in an unpolarized wave, are described in another article. The electric vector can lie in a plane or it can follow a path whose projection at right angles to the direction of propagation is a circle or an ellipse. The same types of polarization can be produced in any transverse wave. *See* POLARIZED LIGHT.

Electromagnetic radiation is difficult to polarize in certain spectral regions, and few techniques exist for analysis. This is true in the ultraviolet below 190 nm. No dichroic polarizers have been found for this region, and transparent birefringent materials from which Nicol or Wollaston polarizing prisms could be made do not seem to exist. Polarization by reflection is possible, but very little work has been done with this technique. In the infrared region from the end of the visible spectrum to approximately 2 μm, sheet polarizers exist. To around 4 μm, polarizing prisms can be made. From 4 μm to 80 μm, reflection from a single plate or transmission through a pile of transparent plates is the common procedure. All these techniques produce linear polarization. Elliptical or circular polarization is more difficult to achieve.

X-ray photons, electrons, neutrons, and other particles can be polarized most easily by scattering.

[BRUCE H. BILLINGS]

Polarized light

Light which has its electric vector oriented in a predictable fashion with respect to the propagation direction. In unpolarized light, the vector is oriented in a random, unpredictable fashion. Even in short time intervals, it appears to be oriented in all directions with equal probability. Most light sources seem to be partially polarized so that some fraction of the light is polarized and the remainder unpolarized. It is actually more difficult to produce a completely unpolarized beam of light than one which is completely polarized.

The polarization of light differs from its other properties in that human sense organs are essentially unable to detect the presence of polarization. The Polaroid Corp. with its polarizing sunglasses and camera filters has made millions of people conscious of phenomena associated with polarization. Light from a rainbow is completely linearly polarized; that is, the electric vector lies in a plane. The possessor of polarizing sunglasses discovers that with such glasses, the light from a section of the rainbow is extinguished.

According to all available theoretical and experimental evidence, it is the electric vector rather than the magnetic vector of a light wave that is responsible for all the effects of polarization and other observed phenomena associated with light. Therefore, the electric vector of a light wave, for all practical purposes, can be identified as the light vector. *See* CRYSTAL OPTICS; ELECTROMAGNETIC RADIATION; LIGHT; POLARIZATION OF WAVES.

One of the simplest ways of producing linearly polarized light is by reflection from a dielectric surface. At a particular angle of incidence, the reflectivity for light whose electric vector is in the plane of incidence becomes zero. The reflected light is thus linearly polarized at right angles to the plane of incidence. This fact was discovered by E. Malus in 1808. Brewster's law shows that at the polarizing angle the refracted ray makes an angle of 90° with the reflected ray. By combining this relationship with Snell's law of refraction, it is found that Eq. (1) holds, where i is the angle of incidence

$$\tan i = n \tag{1}$$

and n is the refractive index. This provides a simple way of measuring refractive indices. *See* REFRACTION OF WAVES.

Law of Malus. If linearly polarized light is incident on a dielectric surface at Brewster's angle (the polarizing angle), then the reflectivity of the surface will depend on the angle between the incident electric vector and the plane of incidence. When the vector is in the plane of incidence, the reflectivity will be zero. When it is at right angles, the reflectivity will be at a maximum. To compute the complete relationship, the incident light vector **A** is broken into components, one vibrating in the plane of incidence and one at right angles to the plane, as in Eqs. (2) and (3), where θ is the angle

$$A_{\parallel} = A \sin \theta \tag{2}$$

$$A_{\perp} = A \cos \theta \tag{3}$$

between the light vector and a plane perpendicular to the plane of incidence. Since the component in the plane of incidence is not reflected, the reflected ray can be written as Eq. (4), where r is

$$B = Ar \cos \theta \tag{4}$$

the reflectivity at Brewster's angle. The intensity is given by Eq. (5). This is the mathematical statement of the law of Malus.

$$I = B^2 = A^2 r^2 \cos^2 \theta \tag{5}$$

Linear polarizing devices. The angle θ can be considered as the angle between the transmitting axes of a pair of linear polarizers. When the polarizers are parallel, they are transparent. When they

are crossed, the combination is opaque. The first polarizers were glass plates inclined so that the incident light was at Brewster's angle. Such polarizers are quite inefficient since only a small percentage of the incident light is reflected as polarized light. More efficient polarizers can be constructed.

Dichroic crystals. Certain natural materials absorb linearly polarized light of one vibration direction much more strongly than light vibrating at right angles. Such materials are termed dichroic. For a description of them *see* DICHROISM.

Tourmaline is one of the best-known dichroic crystals, and tourmaline plates were used as polarizers for many years. A pair was usually mounted in what were known as tourmaline tongs.

Birefringent crystals. Other natural materials exist in which the velocity of light depends on the vibration direction. These materials are called birefringent. The simplest of these structures are crystals in which there is one direction of propagation for which the light velocity is independent of its state of polarization. These are termed uniaxial crystals, and the propagation direction mentioned is called the optic axis. For all other propagation directions, the electric vector can be split into two components, one lying in a plane containing the optic axis and the other at right angles. The light velocity or refractive index for these two waves is different. *See* BIREFRINGENCE.

One of the best-known of these birefringent crystals is transparent calcite (Iceland spar), and a series of polarizers have been made from this substance. W. Nicol (1829) invented the Nicol prism, which is made of two pieces of calcite cemented together as in Fig. 1. The cement is Canada bal-

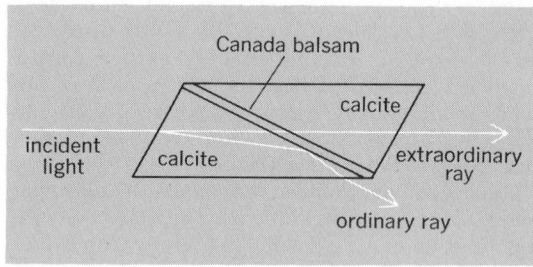

Fig. 1. Nicol prism. The ray for which Snell's law holds is called the ordinary ray.

sam, in which the wave velocity is intermediate between the velocity in calcite for the fast and the slow ray. The angle at which the light strikes the boundary is such that for one ray the angle of incidence is greater than the critical angle for total reflection. Thus the rhomb is transparent for only one polarization direction.

Canada balsam is not completely transparent in the ultraviolet at wavelengths shorter than 400 nm. Furthermore, large pieces of calcite material are exceedingly rare. A series of polarizers has been made using quartz, which is transparent in the ultraviolet and which is more commonly available in large pieces. Because of the small difference between the refractive indices of quartz and Canada balsam, a Nicol prism of quartz would be tremendously long for a given linear aperture.

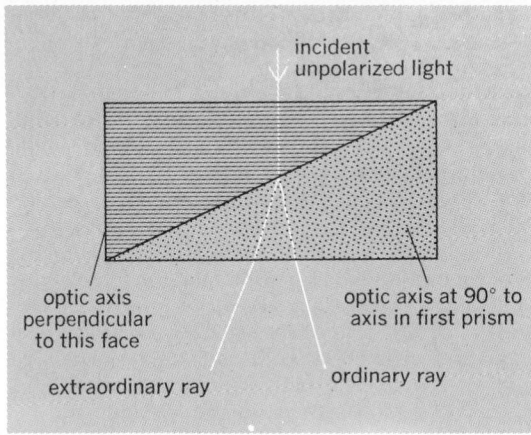

Fig. 2. Wollaston prism.

A different type of polarizer, made of quartz, was invented by W. H. Wollaston and is shown in Fig. 2. Here the vibration directions are different in the two pieces so that the two rays are deviated as they pass through the material. The incoming light beam is thus separated into two oppositely linearly polarized beams which have an angular separation between them. By using appropriate optical stops (obstacles that restrict light rays) in the system, it is possible to select either beam.

In the Wollaston prism, both beams are deviated; since the quartz produces dispersion, each beam is spread into a spectrum. This is not the case of a prism which was invented by A. Rochon. Here the two pieces are arranged as in Fig. 3. One beam proceeds undeviated through the device and is thus achromatic.

Sheet polarizers. A third mechanism for obtaining polarized light is the Polaroid sheet polarizer invented by E. H. Land. Sheet polarizers fall into three types. The first is a microcrystalline polarizer in which small crystals of a dichroic material are oriented parallel to each other in a plastic medium. Typical microcrystals, such as needle-shaped quinine iodosulfate, are embedded in a viscous plastic and are oriented by extruding the material through a slit.

The second type depends for its dichroism on a property of an iodine-in-water solution. The iodine appears to form a linear high polymer. If the iodine is put on a transparent oriented sheet of material such as polyvinyl alcohol (PVA), the iodine chains apparently line themselves parallel to the PVA molecules and the resulting dyed sheet is strongly dichroic. A third type of sheet polarizer depends for its dichroism directly on the molecules of the plastic itself. This plastic consists of oriented polyvinylene. Because these polarizers are commercially available and can be obtained in large sheets, many experiments involving polarized light have been performed which would have been quite difficult with the reflection polarizers or the birefringent crystal polarizer.

Characteristics. There are several characteristics of linear polarizers which are of interest to the experimenter. First is the transmission for light polarized parallel and perpendicular to the axis of the polarizer; second is the angular field; and third is the linear aperture. A typical sheet polarizer has a transmittance of 48% for light parallel to the axis

and $2 \times 10^{-4}\%$ for light perpendicular to the axis at a wavelength of 550 nm. The angular field is 60°, and sheets can be many feet in diameter. The transmittance perpendicular to the axis varies over the angular field.

The Nicol prism has transmittance similar to that of the Polaroid sheeting but a much reduced linear and angular aperture.

Polarization by scattering. When an unpolarized light beam is scattered by molecules or small particles, the light observed at right angles to the original beam is polarized. The light vector in the original beam can be considered as driving the equivalent oscillators (nuclei and electrons) in the molecules. There is no longitudinal component in the original light beam. Accordingly, the scattered light observed at right angles to the beam can only be polarized with the electric vector at right angles to the propagation direction of the original beam. In most situations, the scattered light is only partially polarized because of multiple scattering. The best-known example of polarization by scattering is the light of the north sky. The percentage polarization can be quite high in clean country air. The late A. H. Pfund invented a technique for using measurements of sky polarization to determine the position of the sun when it is below the horizon. *See* SCATTERING OF ELECTROMAGNETIC RADIATION.

Types of polarized light. Polarized light is classified according to the orientation of the electric vector. In linearly polarized light, the electric vector remains in a plane containing the propagation direction. For monochromatic light, the amplitude of the vector changes sinusoidally with time. In circularly polarized light, the tip of the electric vector describes a circular helix about the propagation direction. The amplitude of the vector is constant. The frequency of rotation is equal to the frequency of the light. In elliptically polarized light, the vector also rotates about the propagation direction, but the amplitude of the vector changes so that the projection of the vector on a lane at right angles to the propagation direction describes an ellipse.

These different types of polarized light can all be broken down into two linear components at right angles to each other. These are defined by Eqs. (6)

$$E_x = A_x \sin(\omega t + \varphi_x) \qquad (6)$$

and (7), where A_x and A_y are the amplitudes, φ_x

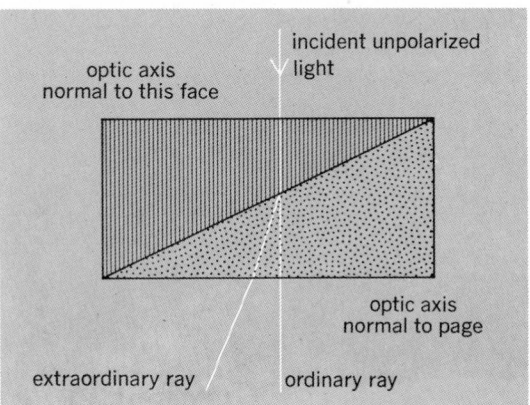

Fig. 3. Rochon prism.

$$E_y = A_y \sin(\omega t + \varphi_y) \tag{7}$$

and φ_y are the phases, ω is 2π times the frequency, and t is the time. For linearly polarized light Eqs. (8) hold. For circularly polarized light Eqs. (9) hold.

$$\varphi_x = \varphi_y \quad A_x \neq A_y \tag{8}$$

$$\varphi_x = \varphi_y \pm \frac{\pi}{2} \quad A_x = A_y \tag{9}$$

For elliptically polarized light Eqs. (10) hold.

$$\varphi_x \neq \varphi_y \quad A_x \neq A_y \tag{10}$$

In the last case, it is always possible to find a set of orthogonal axes inclined at an angle α to x and y along which the components will be E'_x and E'_y, such that Eqs. (11) hold.

$$\varphi'_x = \varphi'_y \pm \frac{\pi}{2} \quad A'_x \neq A'_y \tag{11}$$

In this new system, the x' and y' amplitudes will be the major and minor axes a and b of the ellipse described by the light vector and α will be the angle of orientation of the ellipse axes with respect to the original coordinate system. The relationships between the different quantities can be written as in Eqs. (12) and (13). The terms are defined by Eqs.

$$\tan 2\alpha = \tan 2\gamma \cos \varphi \tag{12}$$

$$\sin 2\beta = \sin 2\gamma \sin \varphi \tag{13}$$

(14)–(17). These same types of polarized light can

$$\tan \gamma = A_y / A_x \tag{14}$$

$$\varphi = \varphi_x - \varphi_y \tag{15}$$

$$\tan \beta = \pm b/a \tag{16}$$

$$A_x{}^2 + A_y{}^2 = a^2 + b^2 \tag{17}$$

also be broken down into right and left circular

components or into two orthogonal elliptical components. These different vector bases are useful in different physical situations.

Production of polarized light. Linear polarizers have already been discussed. Circularly and elliptically polarized light are normally produced by combining a linear polarizer with a wave plate. A Fresnel rhomb can be used to produce circularly polarized light.

Wave plate. A plate of material which is linearly birefringent is called a wave plate or retardation sheet. Wave plates have a pair of orthogonal axes which are designated fast and slow. Polarized light with its electric vector parallel to the fast axis travels faster than light polarized parallel to the slow axis. The thickness of the material can be chosen so that for light traversing the plate, there is a definite phase shift between the fast component and the slow component. A plate with a 90° phase shift is termed a quarter-wave plate. The retardation in waves is given by Eq. (18), where $n_s - n_f$ is

$$\delta = \frac{(n_s - n_f)d}{\lambda} \tag{18}$$

the birefringence; n_s is the slow index at wavelength λ; n_f is the fast index; and d is the plate thickness.

Wave plates can be made by preparing X-cut sections of quartz, calcite, or other birefringent crystals. For retardations of less than a few waves, it is easiest to use sheets of oriented plastics or of split mica. A quarter-wave plate for the visible or infrared is easy to fabricate from mica. The plastic wrappers from many American cigarette packages seem to have almost exactly a half-wave retardation from green light. Since mica is not transparent in the ultraviolet, a small retardation in this region is most easily achieved by crossing two quartz plates which differ by the requisite thickness.

Linearly polarized light incident normally on a quarter-wave plate and oriented at 45° to the fast axis can be split into two equal components parallel to the fast and slow axes. These can be represented, before passing through the plate, by Eqs. (19) and (20), where x and y are parallel to the

$$E_x = A_x \sin(\omega t + \varphi_x) \tag{19}$$

$$E_y = A_x \sin(\omega t + \varphi_x) \tag{20}$$

wave-plate axes. After passing through the plate, the two components can be written as Eqs. (21) and (22), where E_x is now advanced one quarter-wave with respect to E_y.

$$E_x = A_x \sin\left(\omega t + \varphi_x + \frac{\pi}{2}\right) \tag{21}$$

$$E_y = A_x \sin(\omega t + \varphi_x) \tag{22}$$

It is possible to visualize the behavior of the light by studying the sketches in Fig. 4, which show the projection on a plane $z=0$ at various times. It is apparent that the light vector is of constant amplitude, and that the projection on a plane normal to the propagation direction is a circle. If the linearly polarized light is oriented at −45° to the fast axis, the light vector will revolve in the opposite direction. Thus it is possible with a quarter-wave plate and a linear polarizer to make either right or left circularly polarized light. If the linearly polarized light is at an angle other than 45° to the fast axis, the transmitted radiation will be elliptically polar-

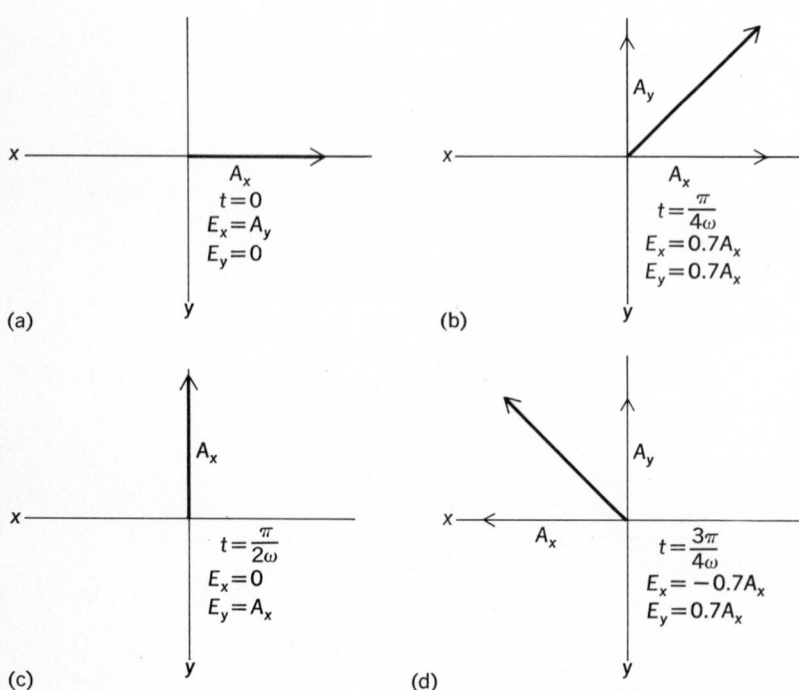

Fig. 4. Projection of light vector of constant amplitude on plane $z=0$ for circularly polarized light. (a) $t=0$. (b) $t=\pi/4\omega$. (c) $t=\pi/2\omega$. (d) $t=3\pi/4\omega$.

ized. When circularly polarized light is incident on a quarter-wave plate, the transmitted light is linearly polarized at an angle of 45° to the wave-plate axes. This polarization is independent of the orientation of the wave-plate axis. For elliptically polarized light, the behavior of the quarter-wave plate is much more complicated. However, as was mentioned earlier, the elliptically polarized light can be considered as composed of two linear components parallel to the major and minor axes of the ellipse and with a quarter-wave phase difference between them. If the quarter-wave plate is oriented parallel to the axes of the ellipse, the two transmitted components will either have zero phase difference or a 180° phase difference and will be linearly polarized. At other angles, the transmitted light will still be elliptically polarized, but with different major and minor axes. Similar treatment for a half-wave plate shows that linearly polarized light oriented at an angle θ to the fast axis is transmitted as linearly polarized light oriented at an angle $-\theta$ to the fast axis.

Wave plates all possess a different retardation at each wavelength. This appears immediately from Eq. (18). It is conceivable that a substance could have dispersion of birefringence, such as to make the retardation of a plate independent of wavelength. However, no material having such a characteristic has as yet been found.

Fresnel rhomb. A quarter-wave retardation can be provided achromatically by the Fresnel rhomb. This device depends on the phase shift which occurs at total internal reflection. When linearly polarized light is totally internally reflected, it experiences a phase shift which depends on the angle of reflection, the refractive index of the material, and the orientation of the plane of polarization. Light polarized in the plane of incidence experiences a phase shift which is different from that of light polarized at right angles to the plane of incidence. Light polarized at an intermediate angle can be split into two components, parallel and at right angles to the plane of incidence, and the two components mathematically combined after reflection.

The phase shifts can be written as Eqs. (23) and (24), where $\varphi_{\parallel}$ is the phase shift parallel to the

$$\tan\frac{\varphi_{\parallel}}{2}=\frac{n\sqrt{n^2\sin^2 i-1}}{\cos i} \qquad (23)$$

$$\tan\frac{\varphi_{\perp}}{2}=\frac{\sqrt{n^2\sin^2 i-1}}{n\cos i} \qquad (24)$$

plane of incidence; $\varphi_{\perp}$ is the phase shift at right angles to the plane of incidence; i is the angle of incidence on the totally reflecting internal surface; and n is the refractive index. The difference $\varphi_{\parallel}-\varphi_{\perp}$ reaches a value of about $\pi/4$ at an angle of 52° for $n=1.50$. Two such reflections give a retardation of $\pi/2$. The Fresnel rhomb shown in Fig. 5 is cut so that the incident light is reflected twice at 52°. Accordingly, light polarized at 45° to the principal plane will be split into two equal components which will be shifted a quarter-wave with respect to each other, and the transmitted light will be circularly polarized. Nearly achromatic wave plates can be made by using a series of wave plates in series with their axes oriented at different specific angles with respect to a coordinate system.

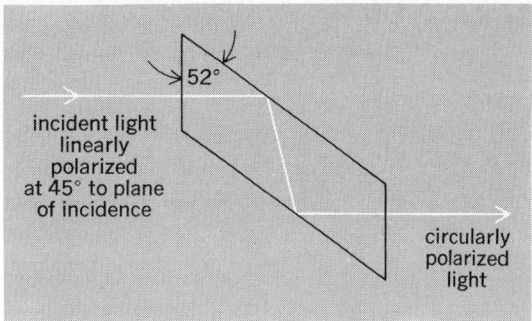

Fig. 5. Fresnel rhomb.

Analyzing devices. Polarized light is one of the most useful tools for studying the characteristics of materials. The absorption constant and refractive index of a metal can be calculated by measuring the effect of the metal on polarized light reflected from its surface. *See* REFLECTION OF ELECTROMAGNETIC RADIATION.

The analysis of polarized light can be performed with a variety of different devices. If the light is linearly polarized, it can be extinguished by a linear polarizer and the direction of polarization of the light determined directly from the orientation of the polarizer. If the light is elliptically polarized, it can be analyzed with the combination of a quarter-wave plate and a linear polarizer. Any such combination of polarizer and analyzer is called a polariscope. As explained previously, a quarter-wave plate oriented parallel to one of the axes of the ellipse will transform elliptically polarized light to linearly polarized light. Accordingly, the quarter-wave plate is rotated until the beam is extinguished by the linear polarizer. At this point, the orientation of the quarter-wave plate gives the orientation of the ellipse and the orientation of the polarizer gives the ratio of the major to the minor axis. Knowledge of the origin of the elliptically polarized light usually gives the orientation of the components which produced it, and from these various items, the phase shifts and attenuations produced by the experiment can be deduced.

One of the best-known tools for working with polarized light is the Babinet compensator. This device is normally made of two quartz prisms put together in a rhomb. One prism is cut with the optic axis in the plane of incidence of the prism, and the other with the optic axis perpendicular to the plane of incidence. The retardation is a function of distance along the rhomb; it will be zero at the center, varying to positive and negative values in opposite directions along the rhomb. It can be used to cancel the known or unknown retardation of any wave plate.

Retardation theory. It is difficult to see intuitively the effect of a series of retardation plates or even of a single plate of general retardation δ on light which is normally incident on the plate and which is polarized in a general fashion. This problem is most easily solved algebraically. The single-wave plate is assumed to be oriented normal to the direction of propagation of the light, which is taken to be the z direction of a set of cartesian coordinates. Its fast axis is at an angle α to the x axis. The incident light can be represented by Eqs. (25)

and (26). A first step is to break the light up into

$$E_x = A_x \sin (\omega t + \varphi_x) \qquad (25)$$

$$E_y = A_y \sin (\omega t + \varphi_y) \qquad (26)$$

components $E_{x'}$ and $E_{y'}$ parallel to the axes of the plate. It is possible to write Eqs. (27) and (28).

$$E_{x'} = E_x \cos \alpha - E_y \sin \alpha \qquad (27)$$

$$E_{y'} = E_x \sin \alpha + E_y \cos \alpha \qquad (28)$$

These components can also be written as Eqs. (29) and (30). After passing through the plate, the components become Eqs. (31) and (32).

$$E_{x'} = A_{x'} \sin (\omega t + \varphi_{x'}) \qquad (29)$$

$$E_{y'} = A_{y'} \sin (\omega t + \varphi_{y'}) \qquad (30)$$

$$E_{x''} = A_{x'} \sin (\omega t + \varphi_{x'} + \delta) \qquad (31)$$

$$E_{y''} = A_{y'} \sin (\omega t + \varphi_{y'}) \qquad (32)$$

In general, it is of interest to compare the output with the input. The transmitted light is thus broken down into components along the original axes. This results in Eqs. (33) and (34).

$$E_{x'''} = E_{x''} \cos \alpha + E_{y''} \sin \alpha \qquad (33)$$

$$E_{y'''} = -E_{x''} \sin \alpha + E_{y''} \cos \alpha \qquad (34)$$

With this set of equations, it is possible to compute the effect of a wave plate on any form of polarized light.

Jones calculus. Equations (33) and (34) still become overwhelmingly complicated in any system involving several optical elements. Various methods have been developed to simplify the problem and to make possible some generalizations about systems of elements. One of the most straightforward, proposed by R. C. Jones, involves reducing Eqs. (33) and (34) to matrix form. The Jones calculus for optical systems involves the polarized electric components of the light vector and is distinguished from other methods in that it takes cognizance of the absolute phase of the light wave.

The Jones calculus writes the light vector in complex form as in Eqs. (35) and (36). Matrix oper-

$$\begin{aligned} E_x &= A_x e^{i(\omega t + \varphi_x)} \\ E_y &= A_y e^{i(\varphi t + \varphi_y)} \end{aligned} \qquad (35)$$

$$E = \begin{vmatrix} A_x e^{i\varphi_x} \\ A_y e^{i\varphi_y} \end{vmatrix} e^{i\omega t} \qquad (36)$$

ators are developed for different optical elements. From Eqs. (25)–(34), the operator for a wave plate can be derived directly. *See* MATRIX THEORY.

The Jones calculus is ordinarily used in a normalized form which simplifies the matrices to a considerable extent. In this form, the terms involving the actual amplitude and absolute phase of the vectors and operators are factored out of the expressions. The intensity of the light beam is reduced to unity in the normalized vector so that Eq. (37) holds. Under this arrangement, the

$$A_x^2 + A_y^2 = 1 \qquad (37)$$

matrices for various types of operations can be written as Eq. (38). This is the operator for a wave

$$G(\delta) = \begin{vmatrix} e^{i(\delta/2)} & 0 \\ 0 & e^{-i(\delta/2)} \end{vmatrix} \qquad (38)$$

plate of retardation δ and with axes along x and y. Equation (39) gives the operator for a rotator which

$$S(\alpha) = \begin{vmatrix} \cos \alpha & -\sin \alpha \\ \sin \alpha & \cos \alpha \end{vmatrix} \qquad (39)$$

rotates linearly polarized light through an angle α. Equation (40) gives the operator for a perfect linear

$$P_h = \begin{vmatrix} 1 & 0 \\ 0 & 0 \end{vmatrix} \qquad (40)$$

polarizer parallel to the x axis. A wave plate at an angle α can be represented by Eq. (41). A series

$$G(\delta,\alpha) = S(\alpha)G(\delta)S(-\alpha) \qquad (41)$$

of optical elements can be represented by the product of a series of matrices. This simplifies enormously the task of computing the effect of many elements. It is also possible with the Jones calculus to derive a series of general theorems concerning combinations of optical elements. Jones has described three of these, all of which apply only for monochromatic light.

1. An optical system consisting of any number of retardation plates and rotators is optically equivalent to a system containing only two elements, a retardation plate and a rotator.

2. An optical system containing any number of partial polarizers and rotators is optically equivalent to a system containing only two elements – one a partial polarizer and the other a rotator.

3. An optical system containing any number of retardation plates, partial polarizers, and rotators is optically equivalent to a system containing four elements – two retardation plates, one partial polarizer, and one rotator.

As an example of the power of the calculus, a rather specific theorem can be proved. A rotator of any given angle α can be formed by a sequence of three retardation plates, a quarter-wave plate, a retardation plate at 45° to the quarter-wave plate, and a second quarter-wave plate crossed with the first, as in Eq. (42), where β is the angle between

$$S(\alpha) = S(\beta)G\left(-\frac{\pi}{2}\right)S(-\beta)S\left(\beta+\frac{\pi}{4}\right)$$

$$\cdot G(\delta)S\left(-\beta-\frac{\pi}{4}\right)S(\beta)G\left(\frac{\pi}{2}\right)S(-\beta) \qquad (42)$$

the axis of the first quarter-wave plate and the x axis, and δ is the retardation of the plate in the middle of the sandwich.

The first simplification arises from the fact that the axis rotations can be done in any order. This reduces Eq. (43) to Eq. (44). Now Eqs. (45) and (46)

$$S\left(\beta+\frac{\pi}{4}\right) = S(\beta)S\left(\frac{\pi}{4}\right) = S\left(\frac{\pi}{4}\right)S(\beta) \qquad (43)$$

$$S(\alpha) = S(\beta)G\left(-\frac{\pi}{2}\right)S\left(\frac{\pi}{4}\right)$$

$$\cdot G(\delta)S\left(-\frac{\pi}{4}\right)G\left(\frac{\pi}{2}\right) \qquad (44)$$

$$S\left(-\frac{\pi}{4}\right)G\left(\frac{\pi}{2}\right) = \frac{1}{\sqrt{2}}\begin{vmatrix} 1 & -i \\ -i & 1 \end{vmatrix} \qquad (45)$$

$$G\left(-\frac{\pi}{2}\right)S\left(\frac{\pi}{4}\right) = -\frac{1}{\sqrt{2}}\begin{vmatrix} 1 & i \\ i & 1 \end{vmatrix} \qquad (46)$$

hold. When the multiplication is carried through, Eq. (47) is obtained.

$$S(\alpha) = S(\beta)$$

$$-\frac{1}{2} \begin{vmatrix} e^{i\delta/2} + e^{-i\delta/2} & -ie^{i\delta/2} + ie^{-i\delta/2} \\ ie^{i\delta/2} - ie^{-i\delta/2} & e^{i\delta/2} + e^{-i\delta/2} \end{vmatrix} S(-\beta)$$

$$= S(\beta)\, S\!\left(\frac{\delta}{2}\right) S(-\beta) = S\!\left(\frac{\delta}{2}\right) \quad (47)$$

The rotation angle is therefore equal to one-half the phase angle of the retardation. This combination is a true rotator in that the rotation is independent of the azimuth angle of the incident polarized light.

A variable rotator can be made by using a Soleil compensator for the central element. This consists of two quartz wedges joined to form a plane parallel plate. The lower wedge is cemented to a plane parallel quartz plate.

Mueller matrices. In the Jones calculus, the intensity of the light passing through the system must be obtained by calculation from the components of the light vector. A second calculus is frequently used in which the light vector is split into four components. This also uses matrix operators which are termed Mueller matrices. In this calculus, the intensity I of the light is one component of the vector and thus is automatically calculated. The other components of the vector are given by Eqs. (48)–(50). The matrix of a perfect polarizer

$$M = A_x^2 - A_y^2 \quad (48)$$

$$C = 2A_x A_y \cos(\varphi_x - \varphi_y) \quad (49)$$

$$S = 2A_x A_y \sin(\varphi_x - \varphi_y) \quad (50)$$

parallel to the x axis can be written as Eq. (51).

$$P = \tfrac{1}{2} \begin{vmatrix} 1 & 1 & 0 & 0 \\ 1 & 1 & 0 & 0 \\ 0 & 0 & 0 & 0 \\ 0 & 0 & 0 & 0 \end{vmatrix} \quad (51)$$

This calculus can treat unpolarized light directly. Such a light vector is given by Eq. (52). The vec-

$$\begin{vmatrix} I \\ M \\ C \\ S \end{vmatrix} = \begin{vmatrix} 1 \\ 0 \\ 0 \\ 0 \end{vmatrix} \quad (52)$$

tor for light polarized parallel to the x axis is written as expression (53). In the same manner as in the Jones calculus, matrices can be derived for retardation plates, rotators, and partial polarizers.

$$\begin{vmatrix} 1 \\ 1 \\ 0 \\ 0 \end{vmatrix} \quad (53)$$

This calculus can also be used to derive various general theorems about various optical systems. *See* FARADAY EFFECT; INTERFERENCE OF WAVES.

[BRUCE H. BILLINGS]

Bibliography: M. Born and E. Wolf, *Principles of Optics*, 6th ed., 1980; D. Clarke and J. F. Grainger, *Polarized Light and Optical Measurements*, 1971; F. A. Jenkins and H. E. White, *Fundamentals of Optics*, 4th ed., 1976; J. Strong, *Concepts of Classical Optics*, 1958; W. Swidell (ed.), *Polarized Light*, 1975.

Polaron

The object that results when an electron in the conduction band of a crystalline insulator or semiconductor polarizes or otherwise deforms the lattice in its vicinity. The polaron comprises the electron plus its surrounding lattice deformation. (Polarons can also be formed from holes in the valence band.) If the deformation extends over many lattice sites, the polaron is "large," and the lattice can be treated as a continuum. Charge carriers inducing strongly localized lattice distortions form "small" polarons. *See* BAND THEORY OF SOLIDS; HOLES IN SOLIDS; SEMICONDUCTOR.

Large polaron. The concept of the large polaron is most useful when the carrier mobility is high and the carrier density and temperature are both low.

Properties of free polarons. In the standard large polaron model of H. Fröhlich, a slowly moving conduction-band electron in a polar crystal is assumed to interact, via its Coulomb field, with longitudinal optical (LO) phonons, which are the quanta of the lattice polarization waves of the crystal. The strength of the coupling between electrons and LO phonons is measured by the dimensionless Fröhlich coupling constant α, defined by Eq. (1),

$$\alpha = e^2/(2r_0 \bar{\epsilon} \hbar \omega_{LO}) \quad (1)$$

where, in a given crystal, $\hbar \omega_{LO}$ is the energy of the long-wavelength LO phonons; $\bar{\epsilon}$, an effective dielectric constant, is that part of the static dielectric constant due to lattice polarizability; e is the charge of the electron; r_0 is the "polaron radius," the radius of the induced lattice polarization charge surrounding the electron, defined by Eq. (2), with m the band mass of the electron. Typi-

$$r_0 = (\hbar/2m\omega_{LO})^{1/2} \quad (2)$$

cally $0.01 < \alpha < 1$ for electrons in direct-gap compound semiconductors and $1 \lesssim \alpha \lesssim 4$ in polar insulators. *See* LATTICE VIBRATIONS; PHONON.

In the absence of external electric or magnetic fields, a polaron of low kinetic energy moves with fixed quasimomentum, p. The energy of such a polaron can be written as expression (3) if

$$E_0 + p^2/2m^*(p) \quad (3)$$

$p^2/2m^*(p) < \hbar\omega_{LO}$, with the polaron ground-state energy, E_0, and effective mass, $m^*(p=0)$, quite accurately given, for $\alpha \lesssim 3$, by weak-coupling perturbation theory; the perturbation results are very well approximated by Eqs. (4) and (5). Considerable

$$E_0 = (-\alpha - 0.01592\alpha^2)\hbar\omega_{LO} \quad (4)$$

$$m^*(p=0) = (1 + \alpha/6 + 0.02363\alpha^2)m \quad (5)$$

theoretical attention has been devoted to calculating E_0 and $m^*(p=0)$ in the strong-coupling regime ($\alpha \gg 6$), where it is found that E_0 is much lower than the result of Eq. (4) and m* much greater than values inferred from Eq. (5). The Feynman path integral method has provided useful approximation formulas for E_0 and $m^*(p=0)$ for all values of α. *See* FEYNMAN INTEGRAL; PERTURBATION (QUANTUM MECHANICS).

Observing polarons. Energy separations between quantized energy levels of polarons in magnetic or Coulomb fields are shifted from the theoretically predicted positions of the corresponding band electron levels; measurements of optical transition frequencies between quantized levels of carriers have given strong evidence for large polarons in a number of polar materials. For example, the effective mass, m_c, of carriers as determined by a low-temperature cyclotron resonance experiment is usually defined by Eq. (6), where ω is the

$$m_c = eH/c\omega \qquad (6)$$

measured cyclotron resonance frequency, H is the magnitude of the applied magnetic field, and c is the speed of light. If the carrier is an electron, one expects, in a simple band, $m_c = m$, independent of ω. However, for polarons, m_c varies with ω, approaching m^* ($p = 0$) only when $\omega \ll \omega_{LO}$, and increasing significantly as ω approaches ω_{LO} from below. Theory and experiment agree well where tested. See CYCLOTRON RESONANCE EXPERIMENTS.

If two discrete electronic levels are separated by an energy close to $\hbar\omega_{LO}$, the polaron level corresponding to the upper electronic level is split into two states, one lying above, the other below, the upper electronic level. This resonant polaron effect has been observed even in materials where α is very small.

Polaron effects are undoubtedly important in producing the large observed binding energies of excitons in insulators and the more polar semiconductors; in most of these materials the exciton radius is comparable to r_0, making the binding sensitive to the polarization charge distribution around the electron and hole. *See* EXCITON.

The scattering and emission of LO phonons by carriers is another important manifestation of the carrier–LO phonon interaction. Such processes tend to limit the mobility of carriers (especially of hot carriers) in polar crystals; they also give rise, in these crystals, to characteristic mobility oscillations as a function of magnetic field (the magnetophonon effect) and to the appearance of LO-phonon-assisted optical transitions.

[DAVID M. LARSEN]

Small polaron. A small polaron is a quasiparticle comprising a self-trapped electronic charge localized within a small region of a solid (of spatial extent comparable to an interatomic dimension) and the atomic displacement pattern which produces the potential well within which the charge is bound (Fig. 1). Small-polaron formation is typically associated with a short-range interaction between the displaced atoms and the charge carrier. In contrast to the large polaron, there need not be a long-range dipolar field. Thus the term polaron is a misnomer in this case. Electron and hole small polarons are found in both polar and nonpolar, crystalline and amorphous materials, semiconductors, and insulators. These include numerous oxides and molecular and rare-gas solids.

Energetics. When small-polaron formation occurs, the equilibrium positions of the atoms surrounding the self-trapped carrier are displaced from their carrier-free equilibrium locations. The energy of a small polaron is the sum of three terms: the strain energy required to displace the atoms to new equilibrium positions, the energy of the electronic carrier bound in the potential well created by the atomic displacements, and the banding energy associated with transferring the small polaron, as a unit, to other sites of the solid. Because the intersite transfer of a small polaron

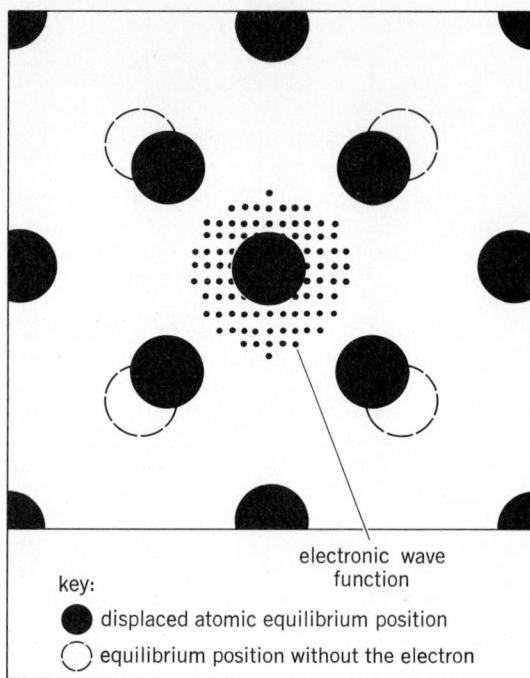

Fig. 1.   An electron self-trapped by the equilibrium atomic displacement pattern around it to form a small polaron.

requires the alteration of the pattern of atomic displacements, it is very much impeded. As a result, small-polaron bandwidths are typically sufficiently small ($\widetilde{<}\ 10^{-3}$ eV) so that, to first approximation, the small polaron is usually viewed as being localized. When energetically stable, the electron small-polaron band lies lower in energy than the conduction band (Fig. 2). Depending on the physical parameters, one or both of those bands may exist as dynamically stable.

Formation. With a short-range interaction between a charge carrier and the atoms of a three-dimensional solid, a charge carrier, if it self-traps at all, will self-trap to form a small polaron. That is, unlike the situation of a Fröhlich polaron, the polaron cannot be of intermediate size. In instances in which small-polaron formation occurs, there is, with a short-range interaction, an energy barrier for self-trapping. Namely, as the atoms of such a solid are progressively moved from their carrier-free equilibrium positions to those appropriate to small-polaron formation, the net energy of the system rises before it ultimately falls. As a result, upon insertion of an excess charge into a solid, the carrier will wander without being self-trapped until the atoms surrounding it undergo sufficiently large and suitable (energetically unfavorable) displace-

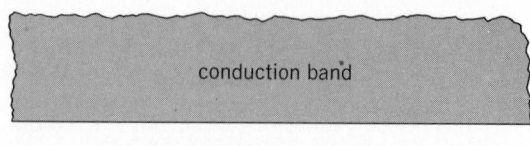

Fig. 2.   Band structure of material with energetically stable small polaron.

ments from their equilibrium positions so as to correspond to surmounting the energy barrier. In this circumstance the carrier is self-trapped, and the system can relax to form a small polaron. The time, in excess of that required for a forced atomic displacement (a vibrational period), associated with small-polaron formation is referred to as the time delay for self-trapping.

Transport properties. The motion of a small polaron is qualitatively different from that of a quasifree nonpolaronic carrier. Namely, the self-trapped carrier can move only when the atoms surrounding it undergo an appropriate change of configuration. Thus, small-polaron motion typically proceeds relatively slowly via a succession of phonon-assisted hopping events. As a result, the small-polaron drift mobility increases as the thermal agitation of the solid increases. Above a temperature comparable to the solid's phonon temperature, the diffusivity increases in an Arrhenius manner with reciprocal temperature. While the Hall mobility of a quasifree carrier is comparable to its drift mobility, for a small polaron the Hall and drift mobilities differ in magnitude, temperature dependence, and even, in some instances, in sign. That is, a small polaron in a magnetic field may be deflected in a sense opposite to that of a free carrier of the same electronic charge. *See* HALL EFFECT.

[DAVID EMIN]

Bibliography: J. T. Devreese (ed.), *Polarons in Ionic Crystals and Polar Semiconductors*, 1972; D. Emin, The formation and motion of small polarons, in J. T. Devreese and V. E. van Doren (eds.), *Linear and Nonlinear Transport in Solids*, pp. 409–433, 1976; D. Emin, The Hall effect in hopping conduction, in C. L. Chien and C. R. Westgate (eds.), *The Hall Effect and Its Applications*, pp. 281–298, 1980; S. P. Keller, J. C. Hensel, and F. Stern (eds.), *Proceedings of the 10th International Conference on the Physics of Semiconductors*, pp. 145–153, 1970; C. G. Kuper and G. D. Whitfield (eds.), *Polarons and Excitons*, 1962.

Polymorphism

The property of crystallizing in two or more forms. The term is applied to crystals of the same substance having a different structure. Substances such as $CaCO_3$ which exist in two crystal forms are said to be dimorphous, while substances such as TiO_2 which appear in three forms are termed trimorphous. The polymorphic modifications of $CaCO_3$ are aragonite, which is orthorhombic, and calcite, which is trigonal. The modifications of TiO_2 are rutile (tetrogonal), anatase (tetragonal), and brookite (orthorhombic). These modifications are stable at normal temperature and in a comparable temperature range. Other polymorphic forms, such as those of SiO_2 (quartz, cristoballite, and trydimite), have a specific nonoverlapping stability range. P. W. Bridgman discovered many polymorphic modifications which are only stable under high pressure. The polymorphic forms of the elements are called allotropic modifications. *See* CRYSTAL STRUCTURE.

From a structural standpoint, molecular polarization and its change with temperature are important factors in bringing about changes from one structure to another. For substances having polymorphic forms stable at the same temperature, the atomic or ionic ratios are such that they are at the limit of the stability of a structure. Therefore the structure is sensitive to external secondary conditions, such as temperature at which the crystals are formed, pressure, and impurities.

Polytypism is polymorphism in a narrow and specific sense. This term is applied to substances having structures like that of zinc blende. The zinc blende structure can be described as a close packing of sulfur layers, zinc layers in a similar arrangement being sandwiched in between. Calling *A, B, C* the three possible positions of sulfur layers and *A', B', C'* the zinc layers, the structure of zinc blende can be written symbolically as *AA'BB' CC'AA'BB'CC'*. A polytype is a structure in which longer sequences such as *ABC* are periodically repeated. Many polytypes of carborundum are known. The occurrence of such polytypes has been fully explained by the spiral growth of crystals. [WILLY C. DEKEYSER]

Bibliography: A. C. Bishop, *Outline of Crystal Morphology*, 1970; P. J. Brown and J. B. Forsyth, *The Crystal Structure of Solids*, 1978; R. C. Evans, *Introduction to Crystal Chemistry*, 2d ed., 1964.

Polynomial systems of equations

Systems of mathematical equations of the form of system (1). Each $f_i(x_1, x_2, \ldots, x_n)$, $i = 1, 2, \ldots$,

$$
\begin{aligned}
f_1(x_1, x_2, \ldots, x_n) &= 0 \\
f_2(x_1, x_2, \ldots, x_n) &= 0 \\
&\cdots\cdots\cdots\cdots\cdots \\
f_m(x_1, x_2, \ldots, x_n) &= 0
\end{aligned}
\tag{1}
$$

m, is a sum of terms of the form shown as expression (2), where the coefficient $a_{i_1 i_2 \cdots i_n}$ is a con-

$$
a_{i_1 i_2 \cdots i_n} x_1^{i_1} x_2^{i_2} \cdots x_n^{i_n}
\tag{2}
$$

stant, or fixed number, and the exponent i_j of the variable x_j is a nonnegative whole number. An example of such a system in two variables is system (3). The expressions $f_i(x_1, x_2, \ldots, x_n)$ are

$$
\begin{aligned}
x^2 - xy + y^2 - 1 &= 0 \\
x^2 + xy - 3y^2 - 2x + 2y + 1 &= 0
\end{aligned}
\tag{3}
$$

called polynomials in several variables. The problem posed by system (1) is to find necessary and sufficient conditions that there exist values of the variables $x_1 = a_1$, $x_2 = a_2$, $\ldots$, $x_n = a_n$ which simultaneously satisfy each equation of the system, and to find all such sets of values, which are called solutions of the system. In example (3), a complete set of solutions is given by $x = 1$, $y = 0$; $x = 0$, $y = 1$; $x = 1$; $y = 1$; and $x = -1$, $y = -1$.

The equations of system (1) can be written as polynomials in one of the variables, for example, x_1, with coefficients which are polynomials in the remaining variables $x_2, x_3, \ldots, x_n$. If system (1) has a solution, then for certain values of the variables $x_2 = a_2$, $x_3 = a_3$, $\ldots$, $x_n = a_n$, the equations of the system, as polynomials in x_1, have a common root $x_1 = a_1$. The process of finding a condition involving the variables $x_2, x_3, \ldots, x_n$ which is both necessary and sufficient for the equations to have a common root $x_1 = a_1$ is called eliminating x_1 from the equations. In example (3), it can be shown that if x is eliminated from the equations, the condition $12y(y-1)^2(y+1)=0$ is obtained. Corresponding to the values $y=0, 1, -1$ which satisfy this condition, the four solutions of the system given earlier are obtained.

Example (3) illustrates a system of two polynomial equations, which can be written in the form of system (4), where the coefficients are either constants or polynomials in the variables $y, z, \ldots$.

$$f(x) = a_n x^n + a_{n-1} x^{n-1} + \cdots + a_1 x + a_0 = 0$$
$$g(x) = b_m x^m + b_{m-1} x^{m-1} + \cdots + b_1 x + b_0 = 0 \qquad (4)$$

The resultant of the polynomials $f(x)$ and $g(x)$ of system (4) is the following determinant, with elements which are the coefficients of the given polynomials:

$$R_x(f,g)$$

$$= \left. \begin{vmatrix} a_n & a_{n-1} & \cdots & a_1 & a_0 & & & \\ & a_n & a_{n-1} & \cdots & a_1 & a_0 & & \\ & & & \cdots & & & & \\ & & & a_n & a_{n-1} & \cdots & a_1 & a_0 \\ b_m & b_{m-1} & \cdots & b_1 & b_0 & & & \\ & b_m & b_{m-1} & \cdots & b_1 & b_0 & & \\ & & & \cdots & & & & \\ & & & b_m & b_{m-1} & \cdots & b_1 & b_0 \end{vmatrix} \right\} \begin{matrix} m \text{ rows} \\ \\ n \text{ rows} \end{matrix}$$

where it is understood that the blank spaces should be filled with zeros. The resultant in this form is called Sylvester's determinant.

It can be proved that the condition $R_x(f,g)=0$ is both necessary and sufficient for $f(x)=0$ and $g(x)=0$ to have a common root $x=a$, except when both $a_n=0$ and $b_m=0$. In example (3),

$$R_x(f,g)$$

$$= \begin{vmatrix} 1 & -y & y^2-1 & 0 \\ 0 & 1 & -y & y^2-1 \\ 1 & y-2 & -3y^2+2y+1 & 0 \\ 0 & 1 & y-2 & -3y^2+2y+1 \end{vmatrix}$$

$$= 12y(y-1)^2(y+1)$$

It does not matter in which order the polynomials $f(x)$ and $g(x)$ are taken, as it can be shown that $R_x(g,f) = (-1)^{mn} R_x(f,g)$. Let $r_1, r_2, \ldots, r_n$ be the roots of $f(x)=0$ and $s_1, s_2, \ldots, s_m$ be the roots of $g(x)=0$. The resultant can be written in the following factored forms:

$$R_x(f,g) = a_n^m b_m^n (r_1 - s_1)(r_1 - s_2) \cdots (r_1 - s_m)$$
$$(r_2 - s_1)(r_2 - s_2) \cdots (r_2 - s_m)$$
$$\cdots\cdots\cdots\cdots\cdots\cdots\cdots\cdots\cdots$$
$$(r_n - s_1)(r_n - s_2) \cdots (r_n - s_m)$$
$$= a_n^m g(r_1) g(r_2) \cdots g(r_n)$$
$$= (-1)^{nm} b_m^n f(s_1) f(s_2) \cdots f(s_m)$$

This form of the resultant is often useful if all of the roots of one of the polynomials are known.

The method of solving systems consisting of more than two equations will be illustrated by considering system (5). As before, f, g, and h are considered to be polynomials in x with coefficients which are polynomials in y and z. If the system has a solution, $x=a$, $y=b$, $z=c$, then for $y=b$, $z=c$ each pair of equations has a common root, $x=a$. Hence $R_x(f,g)=0$, $R_x(g,h)=0$, and $R_x(f,h)=0$.

$$\begin{aligned} f(x,y,z) &= 0 \\ g(x,y,z) &= 0 \\ h(x,y,z) &= 0 \end{aligned} \qquad (5)$$

The last equations are polynomials in y and z, and for $z=c$, they have a common root $y=b$. Hence $R_y[R_x(f,g), R_x(g,h)]=0$, $R_y[R_x(g,h), R_x(f,h)]=0$, and $R_y[R_x(f,g), R_x(f,h)]=0$. These three equations are polynomials in z, and give necessary conditions for system (5) to have a solution. However, these conditions are not, in general, sufficient. That is, it is possible for a value of z to satisfy all three equations without system (5) having a solution. Nevertheless, in many cases these equations make it possible to tell whether system (5) has solutions and to find them. Example (6) illustrates the

$$\begin{aligned} f(x,y,z) &= x^2 - y + z = 0 \\ g(x,y,z) &= x + z^2 - y = 0 \\ h(x,y,z) &= x + z + 1 = 0 \end{aligned} \qquad (6)$$

method. This gives $R_x(f,g) = y^2 - 2z^2 y - y + z^4 + z$, $R_x(g,h) = y - z^2 + z + 1$, and $R_x(f,h) = -y + z^2 + 3z + 1$. The necessary conditions are shown as system (7). Now $z = -1/2$ is the only value of z satis-

$$\begin{aligned} R_y[R_x(f,g), R_x(g,h)] &= 2(2z+1) = 0 \\ R_y[R_x(g,h), R_x(f,h)] &= 2(2z+1) = 0 \\ R_y[R_x(f,g), R_x(f,h)] &= 4z(2z+1) = 0 \end{aligned} \qquad (7)$$

fying the necessary conditions. In system (6), $z = -1/2$ gives $x = -1/2$ and $y = -1/2$. Since these values satisfy all three equations, this is the unique solution of system (6).

Other methods of elimination are applicable when the equations of system (1) have a special form. If one of the equations is linear, as in system (6), such an equation can be solved for one of the variables and this variable can be eliminated. Systems that are linear in powers of the variables may be solved by the methods applicable to linear systems. *See* EQUATIONS, THEORY OF.

[ROSS A. BEAUMONT]

Bibliography: R. A. Barnett, *College Algebra*, 2d ed., 1979; P. K. Rees, F. W. Sparks, and C. S. Rees, *Intermediate Algebra*, 5th ed., 1978.

Polytropic process

A process which occurs with an interchange of both heat and work between the system and its surroundings. The nonadiabatic expansion or compression of a fluid is an example of a polytropic process. The interrelationships between the pressure (P) and volume (V) and pressure and temperature (T) for a gas undergoing a polytropic process are given by Eqs. (1) and (2), where a and b are the

$$PV^a = \text{constant} \qquad (1)$$

$$P^b / T = \text{constant} \qquad (2)$$

polytropic constants for the process of interest. These constants, which are usually determined from experiment, depend upon the equation of state of the gas, the amount of heat transferred, and the extent of irreversibility in the process. Once these constants are known, Eqs. (1) and (2) can be used with the initial-state conditions (P_1 and T_1 or V_1) and one final-state condition (for example, P_2) to determine the temperature or specific volume of the final state. From this information, and other equations of thermodynamics, the entropy change and the enthalpy or internal energy change for the process can be evaluated, so that the power produced by a turbine or necessary to drive a compressor can be computed.

Equations (1) and (2) for a polytropic process may be compared with Eq. (3) for the isothermal expansion or compression of an ideal gas, and Eq. (4) for the reversible adiabatic (isentropic) expansion or compression of an ideal gas. In Eqs. (3) and (4), R is the gas constant, and C_p and C_v the ideal

$$PV = \text{constant} \qquad (3)$$

$$PV^{-(C_p/C_v)} = \text{constant} \qquad (4)$$

$$P^{(R/C_p)}/T = \text{constant}$$

gas heat capacity at constant pressure and volume, respectively. *See* GAS; ISENTROPIC PROCESS; ISOTHERMAL PROCESS; THERMODYNAMIC PROCESSES.   [STANLEY I. SANDLER]

Positron

An elementary particle with mass equal to that of the electron, and positive charge equal in magnitude to the electron's negative charge. The positron is thus the antiparticle (charge-conjugate particle) to the electron. Its existence was predicted by P. A. M. Dirac. It was first observed by C. D. Anderson in 1932. The positron has the same spin and statistics as the electron. Positrons, like electrons, appear as decay products of many heavier particles; electron-positron pairs are produced by high-energy photons in matter. *See* ELECTRON; ELECTRON-POSITRON PAIR PRODUCTION; ELEMENTARY PARTICLE.

A positron is, in itself, stable, but cannot exist indefinitely in the presence of matter, for it will ultimately collide with an electron. The two particles will be annihilated as a result of this collision, and photons will be created. However, a positron can first become bound to an electron to form a short-lived "atom" termed positronium. *See* POSITRONIUM.

The virtual production of electron-positron pairs by an electromagnetic field produces a polarization of the vacuum. This results in effects such as the scattering of light by light and modification of the electrostatic Coulomb field at short distances. *See* QUANTUM ELECTRODYNAMICS. [CHARLES GOEBEL]

Bibliography: S. Gasiorowicz, *The Structure of Matter: A Survey of Modern Physics*, 1979; J. M. Jauch and F. Rohrlich, *The Theory of Photons and Electrons*, rev. ed., 1975; M. R. Wehr, J. A. Richards, and T. W. Adair, *Physics of the Atom*, 3d ed., 1978.

Positronium

The bound state of an electron and a positron. Positronium was discovered by studies of the so-called annihilation radiation from positrons stopped in gases. It is formed in a collision between a positron and a gas atom which results in the capture of an atomic electron by the positron. The positron is the antiparticle to the electron and hence has an inertial mass equal to that of the electron, a positive charge equal in magnitude to the charge of the electron, and a spin of $\hbar/2$, where $\hbar$ is Planck's constant h divided by 2π. *See* POSITRON.

Positronium is of particular interest because it is the two-body system to which quantum electrodynamics is applicable, and its study has served as an important confirmation of the theory of quantum electrodynamics. *See* QUANTUM ELECTRODYNAMICS.

No states of positronium other than the ground

$n = 1$ state ($n = 1, 2, 3, \ldots$, being the principal quantum number) have been found. Studies of positron annihilation in solids and liquids indicate that a perturbed form of positronium exists under certain conditions.

Energy levels. The approximate energy levels of positronium can be calculated from the Schrödinger equation with the nonrelativistic Hamiltonian, H_0, as shown in Eq. (1), where $p_1(p_2)$ is the electron

$$H_0 = \frac{p_1^2}{2m} + \frac{p_2^2}{2m} - \frac{e^2}{r} \qquad (1)$$

(positron) linear momentum, m is the mass of the electron or positron, $-e$ is the charge of the electron, and r is the distance between the positron and the electron. *See* NONRELATIVISTIC QUANTUM THEORY.

The energy levels of the bound states are given by Eq. (2), where the quantity r_{yp} is defined as the

$$W_n = -\frac{\pi^2 m e^4}{h^2 n^2} = -\frac{r_{yp}}{n^2} \qquad (2)$$

Rydberg constant for positronium. The binding energies W_n of positronium are one-half the corresponding binding energies of the hydrogen atom (if the proton-to-electron mass ratio is considered infinite). In particular, the ionization energy of positronium (the binding energy of the ground $n = 1$ state) is 6.8 ev. *See* RYDBERG CONSTANT.

Fine structure to the energy levels of positronium as given by Eq. (2) of the order $\alpha^2 r_{yp}$ (here $\alpha = e^2/\hbar c \cong \frac{1}{137}$ is called the fine-structure constant) arises from relativistic effects, including the electron and positron spin magnetic moments, and from the interaction with the electromagnetic field, which causes electron-positron pair annihilation. Since the electron and positron intrinsic spin angular momenta are $\frac{1}{2}$ in units of $\hbar$, the total spin angular momentum quantum number S of positronium can be either 0 (singlet state, parapositronium) or 1 (triplet state, orthopositronium). For each n value, positronium can exist in either a singlet or a triplet state. The orbital angular momentum quantum number L can assume the values $L = 0, 1, \ldots, n-1$. In particular, the ground $n = 1$ state of positronium is split into two levels, $^1S_0$ and $^3S_1$, which are separated in energy by the amount given by Eq. (3). The term of order $\alpha^3 r_{yp}$ arises from virtual

$$W(^3S_1) - W(^1S_0)$$
$$= \alpha^2 r_{yp} \left[\frac{7}{3} - \frac{2\alpha}{\pi} \left(\frac{16}{9} + \ln 2 \right) \right] \qquad (3)$$

quantum electrodynamic processes. This energy separation, often called the hyperfine structure of the ground state of positronium, corresponds to a frequency difference $\Delta \nu$ of 2.0337×10^5 MHz. *See* FINE STRUCTURE; HYPERFINE STRUCTURE.

The dependence of the energy levels of positronium on an external magnetic field (Zeeman effect) can be determined from the Hamiltonian term given in Eq. (4), in which subscript 1 refers to the

$$H_H = \mu_0 g_{s_1} \mathbf{s}_1 \cdot \mathbf{H} + \mu_0 g_{l_1} \mathbf{l}_1 \cdot \mathbf{H}$$
$$+ \mu_0 g_{s_2} \mathbf{s}_2 \cdot \mathbf{H} + \mu_0 g_{l_2} \mathbf{l}_2 \cdot \mathbf{H} \qquad (4)$$

electron and subscript 2 to the positron, μ_0 is the Bohr magneton ($= e\hbar/2mc$), g_l is the orbital g value ($=1$), g_{s_1} is the electron spin g value [$=2(1+$

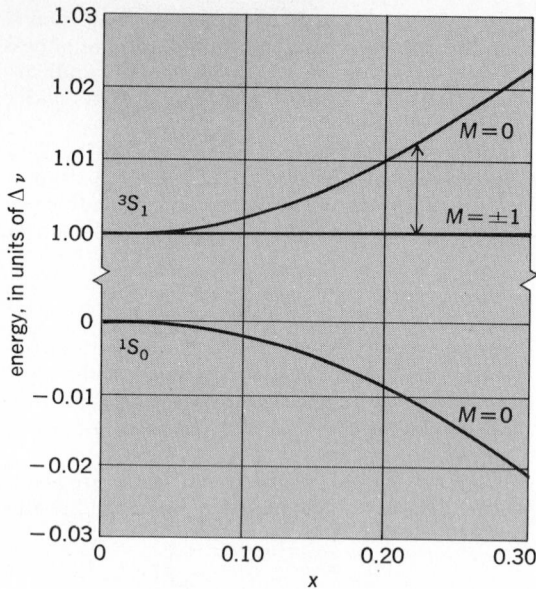

Zeeman energy levels of positronium in its ground $n=1$ state. The quantity $\Delta\nu$ is the hyperfine structure separation between the $^3S_1$ and the $^1S_0$ states of positronium at zero static magnetic field. The M values designate the magnetic substates, and $x = 2g_{s_1}\mu_0 H/(h\Delta\nu)$.

$\alpha/2\pi - 0.328\ \alpha^2/\pi^2)]$, $g_{s_2} = -g_{s_1}$, $l_{1(2)}$ is the electron (positron) orbital angular momentum, $s_{1(2)}$ is the electron (positron) spin angular momentum, and **H** is the external magnetic field intensity. Positronium has no permanent magnetic moment, but there can be a magnetic moment induced by the external magnetic field, and hence an energy level can depend on H^2 or higher powers of H. The energy level diagram for the ground state of positronium in a magnetic field is shown in the figure. From measurements of the frequency of the Zeeman transition $\Delta M = \pm 1$ between the magnetic sublevels of the $^3S_1$ state, the hyperfine structure interval $\Delta\nu$ has been determined and is given by Eq. (5). This experimental value agrees with the

$$\Delta\nu = (2.0333 \pm 0.0004) \times 10^5\ \text{MHz} \qquad (5)$$

theoretical value, and the agreement constitutes the principal test of the quantum electrodynamics of the two-body problem. *See* ZEEMAN EFFECT.

Decay. Positronium is an unstable atom and annihilates with the emission of photons. From its ground $^1S_0$ state a positronium atom at rest decays into two γ-rays, each having an energy of mc^2 (~ 510 kev) with a decay rate of 8.03×10^9 sec^{-1}; from its ground $^3S_1$ state a positronium atom at rest decays into three γ-rays whose energies total $2\ mc^2$ with a decay rate of 7.21×10^6 sec^{-1}.

[VERNON HUGHES]

Bibliography: S. DeBenedetti and H. C. Corben, Positronium, *Annu. Rev. Nucl. Sci.*, 4:191–218, 1954; M. Deutsch, Annihilation of positrons, *Progr. Nucl. Phys.*, 3:131–138, 1953; S. Fluegge (ed.), *Handbuch der Physik*, vol. 35, 1956; J. Green and J. Lee, *Positronium Chemistry*, 1964.

Potential barrier

A region including a maximum of potential energy which prevents a particle on one side of the region from passing to the other side. According to classical physics, a particle must possess an energy exceeding the height of the potential barrier to surmount it. However, quantum mechanics shows that a particle with less energy has a finite probability for penetrating the barrier. The probability decreases rapidly as the particle energy decreases. Such barriers are exemplified by the negative field of the electrons surrounding an atom, by the positive charge in the nucleus which repels a positively charged particle, or by barriers to electrons or holes created in solid-state devices as a result of fixed charge distributions.

In the case of a nucleus, if the bombarding particle is positively charged, $+Z'e$, it will feel the Coulomb electrostatic repulsion to which corresponds a potential energy which varies as $1/r$, where r is the distance from the center of the nucleus, Z' is the atomic number of the bombarding particle, and e is the charge of the proton. The potential barrier increases to the edge of the nucleus and is then overcome by the attractive nuclear forces. The maximum height of this Coulomb barrier is at the nuclear surface, given by the expression $ZZ'e^2/R$, where R is the radius of the nucleus and Z its charge. For protons, the barrier is about 4 MeV for neon and 17 MeV for uranium. In the case of solids, potential barriers arise at the interface between semiconductors and metals or between regions of semiconductors that have different types of dopants (acceptors or donors). The existence of these potential barriers in the case of solids causes the rectifying behavior for metal semiconductor contacts and transister behavior for semiconductors. Barrier heights are measured in electronvolts rather than millions of electronvolts, typical for atomic scattering. *See* NUCLEAR STRUCTURE; SEMICONDUCTOR

[ROBERT STRATTON]

Potential flow

Fluid flow which can be specified by a velocity potential. In contrast to creeping flow, it represents a condition where inertia is controlling and viscous forces are negligible. When the effect of viscosity of a fluid is negligible, in most cases the fluid flow will be irrotational at all times if it starts from rest. For irrotational flow, there exists a velocity potential such that the velocity vector is the gradient of the velocity potential. For an incompressible fluid, the velocity potential satisifies the Laplace equation, which has been investigated very thoroughly from a mathematical point of view. For compressible fluid flow, potential flow exists for subsonic flow. In transonic flow over a thin body, potential flow may be considered as a good approximation. However, in supersonic flow, behind a strong shock wave, potential flow does not exist. *See* FLUID FLOW; FLUID-FLOW PRINCIPLES; HYDRODYNAMICS; LAPLACE'S IRROTATIONAL MOTION; POTENTIALS.

The potential flow satisfies the Navier-Stokes equations of a viscous and incompressible fluid, but not the boundary conditions at a solid wall for viscous flow. Hence the potential flow applies to most flow problems of a fluid of small viscosity, such as air and water, far away from a solid wall but not in the boundary-layer flow near the wall. *See* BOUNDARY-LAYER FLOW; NAVIER-STOKES EQUATIONS.

[S. I. PAI]

Bibliography: H. Lamb, *Hydrodynamics*, 1932; S. I. Pai, *Introduction to the Theory of Compressible Flow*, 1959; H. Schlichting, *Boundary Layer Theory*, 1968.

Potentials

Functions or sets of functions from whose first derivatives a vector can be formed. A vector is a quantity which has a magnitude and a direction, such as force.

A single function, the scalar potential, is used in gravitation theory, electricity and magnetism, fluid mechanics, and other areas. The vectors obtained from it by partial differentiation are in these cases the gravitational, electric and magnetic field strengths, and the velocity, respectively. The vector potential is a set of three functions whose first derivatives give the magnetic induction.

Potential theory. The mathematical theory of the potential is basically the study of differential equations (1) and (2), called Laplace's equation

$$\frac{\partial^2 \phi}{\partial x^2} + \frac{\partial^2 \phi}{\partial y^2} + \frac{\partial^2 \phi}{\partial z^2} = 0 \tag{1}$$

$$\frac{\partial^2 \phi}{\partial x^2} + \frac{\partial^2 \phi}{\partial y^2} + \frac{\partial^2 \phi}{\partial z^2} = -C\rho \tag{2}$$

and Poisson's equation, respectively. Here ϕ is a function defined in part or all of space, C is some constant, and ρ is a function defined wherever ϕ is defined. The theory is best expressed in terms of vector calculus. The three vectors of unit magnitude which point in the positive directions of the x, y, and z axes of a right-handed cartesian coordinate system are denoted by $\mathbf{i}$, $\mathbf{j}$, and $\mathbf{k}$. The radius vector which points from the origin to an arbitrary point with coordinates x, y, z is then $\mathbf{r} = x\mathbf{i} + y\mathbf{j} + z\mathbf{k}$. Suppose that to each point in a region of space (or all of space) there is associated a number ϕ such that the function $\phi(\mathbf{r})$ has partial derivatives with respect to x, y, and z, of first and second order everywhere in that region; and suppose further that the second derivatives satisfy $\partial^2 \phi / \partial x \partial y = \partial^2 \phi / \partial y \partial x$, and corresponding equations for x and z, and for y and z. Then $\phi(\mathbf{r})$ is called a scalar potential. The vector $\mathbf{V}$ given by Eq. (3) is defined by this

$$\mathbf{V}(\mathbf{r}) = \mathbf{i} V_x + \mathbf{j} V_y + \mathbf{k} V_z = -\mathbf{i}\frac{\partial \phi}{\partial x} - \mathbf{j}\frac{\partial \phi}{\partial y} - \mathbf{k}\frac{\partial \phi}{\partial z} \tag{3}$$

equation for every $\mathbf{r}$ in the region and is called the gradient vector field associated with this potential. Equation (3) is abbreviated as $\mathbf{V}(\mathbf{r}) = -\nabla \phi(\mathbf{r})$, in which ∇ is called the gradient operator. It is a vector and a differentiation operator. The minus sign is a matter of convention. The curl of $\mathbf{V}$, a vector denoted by $\nabla \times \mathbf{V}$ and defined by Eq. (4), satisfies

$$\nabla \times \mathbf{V} = \mathbf{i}\left(\frac{\partial V_z}{\partial y} - \frac{\partial V_y}{\partial z}\right) + \mathbf{j}\left(\frac{\partial V_x}{\partial z} - \frac{\partial V_z}{\partial x}\right) + \mathbf{k}\left(\frac{\partial V_y}{\partial x} - \frac{\partial V_x}{\partial y}\right) \tag{4}$$

$\nabla \times \mathbf{V} = 0$ if $\mathbf{V}$ is defined by Eq. (3). Conversely, whenever a vector $\mathbf{V}$ satisfies $\nabla \times \mathbf{V} = 0$ in a certain region, there exists a scalar potential ϕ in that region such that $\mathbf{V} = -\nabla \phi$. If, furthermore, the divergence of $\mathbf{V}$ vanishes, that is, if $\mathbf{V}$ satisfies Eq. (5), which can be written as $\nabla \cdot \mathbf{V} = 0$, then ϕ

$$\frac{\partial V_x}{\partial x} + \frac{\partial V_y}{\partial y} + \frac{\partial V_z}{\partial z} = 0 \tag{5}$$

satisfies Laplace's equation, Eq. (1), that is, $\nabla^2 \phi = 0$. In that case ϕ is called a harmonic function.

If ϕ satisfies Laplace's equation $\nabla^2 \phi = 0$, in a finite region of three-dimensional space enclosed by the closed surface S, then the solution ϕ inside S is uniquely determined by specifying either the values of ϕ on S or the values of the derivatives of ϕ normal to S, $\partial \phi / \partial n$, on S. In the former case ϕ is given by Eq. (6), in the latter case by Eq. (7).

$$\phi(\mathbf{r}) = -\frac{1}{4\pi} \int\!\!\int \phi(\mathbf{r}') \frac{\partial}{\partial n}\left(\frac{1}{R}\right) dS \tag{6}$$

$$\phi(\mathbf{r}) = \frac{1}{4\pi} \int\!\!\int \frac{\partial \phi(\mathbf{r}')}{\partial n} \frac{1}{R} dS \tag{7}$$

Here the normal derivative is taken with the normal pointing out of the enclosed volume; the quantity R is the distance between the end point of $\mathbf{r}$ which is inside the surface, and the end point of $\mathbf{r}'$ which is on the surface, $R = \sqrt{(x-x')^2 + (y-y')^2 + (z-z')^2}$, and the integration is extended over the whole closed surface S. The two cases are called the Dirichlet and Neumann boundary conditions, respectively. One can also specify mixed boundary conditions such as ϕ on part of S and $\partial \phi / \partial n$ on the remaining part of S.

If the vector field satisfies $\nabla \times \mathbf{V} = 0$, and $\nabla \cdot \mathbf{V} = C\rho$, where ρ is some function defined in the region and whose integral over the region exists, then ϕ satisfies Poisson's equation, $\nabla^2 \phi = -C\rho$. A solution of this equation is given by Eq. (8). The integral

$$\phi(\mathbf{r}) = \frac{C}{4\pi} \int\!\!\int\!\!\int \frac{\rho(\mathbf{r}')}{R} dx' dy' dz' \tag{8}$$

extends over the whole three-dimensional region and R is again the distance from $\mathbf{r}$ to $\mathbf{r}'$, $R = |\mathbf{r} - \mathbf{r}'|$. The most general solution of Poisson's equation is obtained by adding to the particular solution given by Eq. (8) the general solution of Laplace's equation.

Laplace's equation also has important applications in two dimensions for problems in a plane. *See* LAPLACE'S DIFFERENTIAL EQUATION; LAPLACE'S IRROTATIONAL MOTION.

Gravitational potential. According to Isaac Newton's theory of gravitation, every spherical body of mass M produces a force field $\mathbf{F}$. Any other body, of mass m, will experience an attractive force, directed toward M, and of the magnitude given in Eq. (9), where distance between the cen-

$$F = G\frac{Mm}{r^2} \tag{9}$$

ters of gravity of the masses M and m is denoted by r; if M is at the origin, $\mathbf{r}$ is the position vector of m and r is its magnitude; G is Newton's constant of gravitation. Such a force is present at every point in the space surrounding the mass M at which one chooses to put the mass m. This gravitational force $\mathbf{F}$ can be expressed in terms of the gravitational field strength $\mathbf{f}$ defined by $\mathbf{f} = \mathbf{F}/m$, that is, the force per unit mass. The gravitational potential is that function ϕ whose negative gradient equals $\mathbf{f}$, $\mathbf{f} = -\nabla \phi$, or $f_x = -\partial \phi / \partial x$, etc. This definition of ϕ is made unique by requiring ϕ to vanish as the distance from M goes to infinity. Then $\phi = -GM/r$ for the gravitational potential of the mass M, for all points outside the mass M. The gravitational force exerted by the mass M on the mass m can be written as Eqs. (10): the differentiation is with respect

$$\mathbf{F} = -m\nabla\phi, \quad \phi = -GM/r \tag{10}$$

to the position of the mass m; ϕ is negative because $\mathbf{F}$ is attractive. The quantity $V = m\phi$ is called the gravitational potential energy. The negative gradient ϕ gives f; the negative gradient of V gives $\mathbf{F}$. The potential energy of m in the field of M increases (becomes less negative) as the distance increases.

If the body is not spherically symmetric, $\mathbf{F}$ in Eq. (9) and ϕ in Eqs. (10) will hold only at large distances. But for any distance and arbitrary shape, if the mass M is spread out over a certain volume with mass density $\rho(\mathbf{r})$, then ϕ satisfies Poisson's equation (2) in the form $\nabla^2\phi = +4\pi G\rho$. Its solution is given by Eq. (8) with $C = -4\pi G$.

When there are several masses present, M_1, $M_2, \ldots, M_n$, each of them will produce a force on a given mass m. The total force on m can be obtained by adding all the forces $\mathbf{F}_1, \mathbf{F}_2, \ldots, \mathbf{F}_n$ by vector addition. However, it is obtained more easily by adding the potentials $\phi_1, \phi_2, \ldots, \phi_n$, produced by these masses and then finding the total force $\mathbf{F}$ from $\mathbf{F} = -m\nabla\phi$, where ϕ is the sum of the potentials. This is the principle of superposition. It is one reason why potentials are so useful.

Those sets of points at which ϕ has a fixed value form two-dimensional surfaces in space. These surfaces are called equipotential surfaces. For a spherically symmetric mass distribution these surfaces are concentric spheres whose centers are at the center of mass. Since the vector ϕ is parallel to the normal of the surface $\phi = $ constant, the lines of force are perpendicular to the equipotential surfaces.

Electrostatic potential. An electrically charged object with total electric charge Q will, when at rest, produce an electrostatic force field $\mathbf{F}$. If another charge q is placed at a distance r from it which is large compared to the size of the two charges, this force is given by Coulomb's law. Its magnitude is as given in Eq. (11), and it is repulsive

$$\mathbf{F} = \frac{1}{4\pi\epsilon}\frac{Qq}{r^2} \tag{11}$$

or attractive depending on whether the two charges Q and q have the same signs (both positive or both negative) or have opposite signs. The constant ϵ depends on the dielectric properties of the medium and on the units in which $\mathbf{F}$, r, and the charges are measured. In vacuum, ϵ is denoted by ϵ_0. In Gaussian units $(4\pi\epsilon_0)^{-1} = 1$, in rationalized mks units $(4\pi\epsilon_0)^{-1} = c^2/10^7$, where c is the velocity of light. The analogy between Eqs. (9) and (11) is obvious; an important difference is that gravitational forces are always attractive.

If the charge Q is spherically symmetric, of radius a, and centered at the origin, the electrostatic potential ϕ at a distance $r > a$ from the origin is given by Eq. (12). This formula also holds when Q is not spherically symmetric, but of characteristic

$$\phi = \frac{1}{4\pi\epsilon}\frac{Q}{r} \tag{12}$$

size a, provided r is much larger than a. The electrostatic field strength is then $\mathbf{E} = -\nabla\phi$. It is the analog of f in the gravitational case. In general, the electrostatic force on a charge q is $\mathbf{F} = -q\nabla\phi$. In the special case when ϕ is given by Eq. (12), this force can be expressed by Eq. (13). The magnitude

$$\mathbf{F} = -q\frac{\mathbf{r}}{r}\frac{d\phi}{dr} = \frac{1}{4\pi\epsilon}\frac{Qq}{r^3}\mathbf{r} \tag{13}$$

of this vector is given by Eq. (11). *See* COULOMB'S LAW; ELECTRIC FIELD.

If the charge Q is spread out over some volume with charge density $\rho(\mathbf{r})$ so that $\int\int\int\rho dxdydz = Q$, then ϕ satisfies Poisson's equation (2) in the form $\nabla^2\phi = -\rho/\epsilon$. Its solution is given by Eq. (8) with $C = 1/\epsilon$. The Coulomb potential given by Eq. (12) is a special case of this result. If the charge is not distributed over a volume, but over a surface such as the surface of a sphere [surface charge density $\sigma(\mathbf{r})$], then the triple integral over $\rho dx' dy' dz'$ in Eq. (8) must be replaced by the double integral over σdS, where dS is the differential surface element.

As in the gravitational case, the total force acting on a charge q due to the presence of a set of charges $Q_1, \ldots, Q_n$ can be obtained by adding the potentials of these charges and finding $\mathbf{F} = -q\nabla\phi$; the principle of superposition holds. Equipotential surfaces are defined as in gravitation theory.

The electrostatic potential energy V of a charge q in a potential ϕ is given by the relation $V = q\phi$.

Magnetic scalar potential. If the magnetic induction vector $\mathbf{B}$ satisfies $\nabla\times\mathbf{B} = 0$, for example, in the absence of current densities and time-dependent electric fields, then $\mathbf{B}$ can be derived from a scalar potential ψ by the relation $\mathbf{B} = -\nabla\psi$. Maxwell's equations require $\nabla\cdot\mathbf{B} = 0$, so that the magnetic scalar potential must be a solution of Laplace's equation $\nabla^2\psi = 0$ which has the form of Eq. (1). *See* MAXWELL'S EQUATIONS.

Logarithmic potential. A straight electrically charged cylinder of circular cross section (typically a wire) and of effectively infinite length poses an electrostatic problem in two dimensions, in the plane perpendicular to the cylinder axis. If λ is the charge per unit length on the cylinder, and ϵ is the dielectric constant of the surrounding medium, then the electrostatic field at a distance r greater than the radius of the cylinder is given by the relation $\mathbf{E} = \lambda/(2\pi\epsilon\mathbf{r})$ and is directed radially outward if λ is positive. The potential ϕ whose negative gradient gives this field $\mathbf{E}$, is called logarithmic potential, and is given by Eq. (14). Here ln

$$\phi(r) = -\frac{\lambda}{2\pi\epsilon}\ln r + \text{constant} \tag{14}$$

is the natural logarithm. The additive constant can be so chosen that $\phi = 0$ for $r = a$, where a is the radius of the cylinder. Then Eq. (14) can be written as Eq. (15).

$$\phi(r) = -\frac{\lambda}{2\pi\epsilon}\ln\frac{r}{a} \tag{15}$$

Work and potential difference. When a charge q is moved in an electrostatic potential field, work is being done. This work W amounts to a loss of energy and is therefore negative. If q is moved from point 1 to point 2, W will be given by Eq. (16).

$$W = -\int_1^2 \mathbf{F} \cdot d\mathbf{r} = q \int_1^2 \nabla \phi \cdot d\mathbf{r} = q(\phi_2 - \phi_1)$$
$$= V_2 - V_1 \tag{16}$$

The work done equals the difference in potential energy between the final and initial positions. Thus, W does not depend on the path along which q is moved, but only on the initial and the final position. This is a direct consequence of the fact that $\mathbf{F}$ is a gradient vector field. It follows, in particular, that no work is being done if the charge q is moved in an arbitrary path which returns to the initial position, for example, a path which is closed. Mathematically this means $\oint \mathbf{F} \cdot d\mathbf{r} = 0$ whenever $\mathbf{F}$ is a gradient field. Physical systems in which all forces can be derived from potentials are therefore called conservative systems.

The potential difference between two points 1 and 2 in a potential field is the work per unit charge that needs to be done to move a charge from 1 to 2. The electrostatic potential at $\mathbf{r}$ given by Eq. (12), in which ϕ is chosen to vanish at infinite distance, is the work per unit charge necessary to move a positive charge from infinity to the position $\mathbf{r}$. In an electric network the potential difference between two points is the work per unit charge that must be done to move a positive charge through the network from one point to the other. It is usually measured in volts.

Neither potential nor potential energy has absolute physical meaning. They must be measured relative to some arbitrary reference potential or potential energy. For example, one chooses it to be zero at infinite distance from a charged body; or in electric networks one connects one point to the ground and defines this point to have zero potential. Only voltage difference is a physically meaningful quantity, not absolute voltage.

In gravitation theory the potential difference between a body located at a distance h above the Earth's surface and its location on the Earth's surface is, according to Eq. (10), expressed as $\phi(h) - \phi(0) = -GM/(R+h) + GM/R = GMh/R^2$, where the Earth's radius and mass are R and M, respectively, and h is much smaller than R. This potential difference is written gh where $g = GM/R^2$ is the gravitational acceleration on the Earth's surface. If the body has a mass m, the two locations differ in potential energy by mgh.

Vector potential. The electromagnetic vector potential $\mathbf{A}$ is a vector, that is, a set of three functions A_x, A_y, A_z, whose partial derivatives give the magnetic induction vector $\mathbf{B}$ through the relation $\mathbf{B} = \nabla \times \mathbf{A}$. The curl operation is defined in Eq. (4). This definition ensures that one of the Maxwell equations, $\nabla \cdot \mathbf{B} = 0$, is identically satisfied. $\mathbf{A}$ is not uniquely defined by $\nabla \times \mathbf{A}$. Since the curl of a gradient vanishes identically, a gradient added to $\mathbf{A}$ will not change $\mathbf{B}$. This nonuniqueness is called the gauge invariance of $\mathbf{B}$.

If the magnetic field strength $\mathbf{H}$ and the induction $\mathbf{B}$ are linked by a constant permeability μ such that $\mathbf{B} = \mu \mathbf{H}$, then the vector potential $\mathbf{A}$ satisfies Eq. (17), where $\mathbf{j}$ is the current density

$$\nabla^2 \mathbf{A} = -\mu \mathbf{j} \tag{17}$$

which produces the magnetic field. Each component of this vector equation is a Poisson equation

of the type of Eq. (2). Its solution is given by Eq. (18), where R is the magnitude of the vector $\mathbf{r} - \mathbf{r}'$.

$$\mathbf{A}(\mathbf{r}) = \frac{\mu}{4\pi} \iiint \frac{\mathbf{j}(\mathbf{r}')}{R} dx' dy' dz' \tag{18}$$

If the current I does not have a density $\mathbf{j}$ but is a line current, Eq. (16) is replaced by Eq. (19), the

$$\mathbf{A}(\mathbf{r}) = \frac{\mu}{4\pi} \oint \frac{I ds}{R} \tag{19}$$

integral being a line integral along the whole current loop. When this loop is small compared to R, Eq. (17) can be transformed by Stoke's integral theorem into the potential of a magnetic dipole, which is given by Eq. (20). Here R is the distance

$$\mathbf{A}(\mathbf{r}) = \frac{\mu}{4\pi} \frac{\mathbf{M} \times \mathbf{R}}{R^3} \tag{20}$$

of the end point of $\mathbf{r}$ from the dipole, and $\mathbf{M}$ is the magnetic dipole moment. When $\mathbf{M}$ is produced by a current I forming a circle of radius a, $M = \pi a^2 I$.

Electromagnetic potentials. In the general time-dependent case, the electric field strength $\mathbf{E}$ and the magnetic induction $\mathbf{B}$ can be expressed in terms of the scalar potential ϕ and the vector potential $\mathbf{A}$ as in Eqs. (21). This ensures that the

$$\mathbf{E} = -\phi - \frac{1}{c} \frac{\partial \mathbf{A}}{\partial t} \qquad \mathbf{B} = \nabla \times \mathbf{A} \tag{21}$$

homogeneous Maxwell equations are identically satisfied. In free space, outside a source distribution of charge density ρ and current density $\mathbf{j}$, the potentials satisfy inhomogeneous wave equations (22), where ϵ_0 and μ_0 are constants which depend

$$\nabla^2 \phi - \frac{1}{c^2} \frac{\partial^2 \phi}{\partial t^2} = -\rho/\epsilon_0$$
$$\nabla^2 \mathbf{A} - \frac{1}{c^2} \frac{\partial^2 \mathbf{A}}{\partial t^2} = -\mu_0 \mathbf{j} \tag{22}$$

on the units chosen. When the potentials are time independent, Eqs. (22) reduce to the Poisson equations. *See* WAVE EQUATION.

The potentials ϕ and $\mathbf{A}$ are not uniquely determined by Eqs. (21) if the fields are known. One can always add a gradient $\nabla \Lambda$ to $\mathbf{A}$ and a term $-\partial \Lambda / c \partial t$ to ϕ without affecting $\mathbf{E}$ and $\mathbf{B}$. Such a transformation of the potentials is called a gauge transformation. The invariance of the fields under this transformation is gauge invariance. This freedom of ϕ and $\mathbf{A}$ can be restricted in a relativistically invariant way by the Lorentz condition on the potentials, which is $\nabla \cdot \mathbf{A} + \frac{1}{c} \frac{\partial \phi}{\partial t} = 0$.

In the special theory of relativity the four functions ϕ, A_x, A_y, A_z transform under Lorentz transformations like four components of a four-vector. They are sometimes collectively called four-potential. *See* RELATIVITY.

Retarded and advanced potentials. Electromagnetic fields propagate through empty space with a finite velocity c, the velocity of light, rather than with infinite velocity. Consequently, it takes a certain time for an electric or magnetic field produced by a charge or current at time t_Q and at position $\mathbf{r}_Q$ to reach another position $\mathbf{r}_P$. This time is $|\mathbf{r}_p - \mathbf{r}_Q|/c$. The time t_Q is the retarded time of the fields which are observed at time $t_p = t_Q + |\mathbf{r}_p - \mathbf{r}_Q|/c$. The fields present at t_p and produced at $t_Q < t_p$ are the retarded fields. The potentials from which they are derived are the retarded potentials. They are important in problems of electromagnetic radiation.

If time would run backwards, the fields and potentials observed at time t_p at point $\mathbf{r}_p$ would have to be produced at some future time $t_{Q'}$ and at a position $\mathbf{r}_{Q'}$ of the source such that $t_p = t_{Q'} - |r_{Q'} - r_p|/c$. The potentials at time t_p which are produced at the later time $t_{Q'} > t_p$ are the advanced potentials. While they have no direct physical meaning, they play an important role in the mathematical analysis of radiation theory.

Liénard-Wiechert potentials. The retarded and advanced electromagnetic scalar and vector potentials produced by a moving point charge can be expressed in terms of the (retarded or advanced) position and velocity of that charge. This was first done by A. Liénard in 1898, and E. Wiechert in 1900.

Polarization potentials. One can define two vectors $\mathbf{\Pi}_e$ and $\mathbf{\Pi}_m$ by $\mathbf{A} = 1/c\ \partial\mathbf{\Pi}_e/\partial t + \nabla \times \mathbf{\Pi}_m$ and $\phi = -\nabla \cdot \mathbf{\Pi}_e$. The electromagnetic fields $\mathbf{E}$ and $\mathbf{B}$ can then be expressed in terms of $\mathbf{\Pi}_e$ and $\mathbf{\Pi}_m$. These are the polarization potentials. The Lorentz condition is identically fulfilled. The fields $\mathbf{E}$ and $\mathbf{B}$ are now given by second derivatives of $\mathbf{\Pi}_e$ and $\mathbf{\Pi}_m$, an exception to the usual definition of potentials, which involves only first derivatives. Heinrich Hertz in 1889 was the first to use such potentials.

Debye potential. Radiation or scattering of electromagnetic waves by a distribution of localized sources in a homogeneous isotropic medium leads to fields $\mathbf{E}$ and $\mathbf{B}$ which can be expressed in terms of only two scalar potentials Π_e and Π_m. They were first introduced by Peter Debye in 1909 and are related to the polarization potentials.

Velocity potential. In the theory of fluid dynamics the absence of vortices is expressed by $\nabla \times \mathbf{v} = 0$, where $\mathbf{v}$ is the velocity vector which is a function of position inside the fluid. If that equation holds, there always exists a velocity potential, that is, a scalar function ϕ with the property $\mathbf{v} = -\nabla\phi$. This potential satisfies Laplace's equation (1) if the fluid is incompressible and in steady motion, so that the continuity (conservation of mass) equation requires that $\nabla \cdot \mathbf{v} = 0$.

Thermodynamic potential. In thermodynamics the Helmholtz function of free energy A (or F) and the Gibbs function G are both referred to as thermodynamic potentials. The reason for this name is the analogy to mechanical potential energy rather than to potential. When an isolated system is in equilibrium under various conditions, these quantities are a minimum. Thus thermodynamic equilibrium can be attained at constant temperature and volume, in which case A is a minimum; alternatively, it can be attained at constant temperature and pressure, in which case G is a minimum.

Chemical potential. A thermodynamic system in general consists of several chemical constituents. Let n_k be the number of moles of constituent k. Then the chemical potential of this constituent is the rate of change of the Gibbs function G with n_k, $\partial G/\partial n_k = \mu_k$. Even when constituent k is not initially present, μ_k can be different from zero, so that its

addition to the system will affect the Gibbs function. Chemical potentials play an essential role in phase transitions and chemical equilibrium.

[F. ROHRLICH]

Bibliography: M. Born and E. Wolf, *Principles of Optics*, 5th ed., 1975; H. L. Helms, *Introduction to Potential Theory*, 1969, reprint 1975; O. D. Kellogg, *Foundations of Potential Theory*, 1929, reprint 1953; F. Rohrlich, *Classical Charged Particles*, 1965; W. R. Smythe, *Static and Dynamic Electricity*, 3d ed., 1968; J. Wermer, *Potential Theory*, 1974; M. W. Zemansky and R. Dittman, *Heat and Thermodynamics*, 6th ed., 1981.

Power

The time rate of doing work. Like work, power is a scalar quantity, that is, a quantity which has magnitude but no direction. Some units often used for the measurement of power are the watt (1 joule of work per second) and the horsepower (550 ft-lb of work per second). *See* WORK.

Usefulness of the concept. Power is a concept which can be used to describe the operation of any system or device in which a flow of energy occurs. In many problems of apparatus design, the power, rather than the total work to be done, determines the size of the component used. Any device can do a large amount of work by performing for a long time at a low rate of power, that is, by doing work slowly. However, if a large amount of work must be done rapidly, a high-power device is needed. High-power machines are usually larger, more complicated, and more expensive than equipment which need operate only at low power. A motor which must lift a certain weight will have to be larger and more powerful if it lifts the weight rapidly than if it raises it slowly. An electrical resistor must be large in size if it is to convert electrical energy into heat at a high rate without being damaged.

Electrical power. The power P developed in a direct-current electric circuit is $P = VI$, where V is the applied potential difference and I is the current. The power is given in watts if V is in volts and I in amperes. In an alternating-current circuit, $P = VI \cos \phi$, where V and I are the effective values of the voltage and current and ϕ is the phase angle between the current and the voltage. *See* ALTERNATING CURRENT.

Power in mechanics. Consider a force F which does work W on a particle. Let the motion be restricted to one dimension, with the displacement in this dimension given by x. Then by definition the power at time t will be given by Eq. (1). In this

$$P = dW/dt \qquad (1)$$

equation W can be considered as a function of either t or x. Treating W as a function of x gives Eq. (2). Now dx/dt represents the velocity v of the

$$P = \frac{dW}{dt} = \frac{dW}{dx}\frac{dx}{dt} \qquad (2)$$

particle, and dW/dx is equal to the force F, according to the definition of work. Thus Eq. (3) holds.

$$P = Fv \qquad (3)$$

This often convenient expression for power can be generalized to three-dimensional motion. In this case, if ϕ is the angle between the force $\mathbf{F}$ and the velocity $\mathbf{v}$, which have magnitudes F and v, respectively, Eq. (4) expresses quantitatively the observa-

$$P = \mathbf{F} \cdot \mathbf{v} = Fv \cos \phi \qquad (4)$$

tion that if a machine is to be powerful, it must run fast, exert a large force, or do both.

[PAUL W. SCHMIDT]

Power factor

The ratio of watts average (or active) power to the apparent power of an alternating-current circuit. By definition, and of general application, the equation below holds, which is the ratio of instrument

$$\text{Power factor (pf)} = \frac{\text{watts average power}}{\text{rms volts} \times \text{rms amperes}}$$

readings. A watt-meter indicates average power and electrodynamometer or iron-vane instruments show rms voltage and current. For the steady-state ac circuit under sinusoidal voltage and current, pf $= \cos \theta$, where θ is the phase angle between the voltage and current. This definition is restricted to sine waves of the same frequency.

[BURTIS L. ROBERTSON]

Poynting's vector

A vector, the outward normal component of which, when integrated over a closed surface in an electromagnetic field, represents the outward flow of energy through that surface. It is given by Eq. (1),

$$\mathbf{\Pi} = \mathbf{E} \times \mathbf{H} = \mu^{-1} \mathbf{E} \times \mathbf{B} \qquad (1)$$

where $\mathbf{E}$ is the electric field strength, $\mathbf{H}$ the magnetic field strength, $\mathbf{B}$ the magnetic flux density, and μ the permeability. This can be shown with the aid of Maxwell's equations, Eqs. (2) where $\mathbf{D}$ is

$$\mathbf{H} \cdot (\nabla \times \mathbf{E}) - \mathbf{E} \cdot (\nabla \times \mathbf{H}) = \nabla \cdot (\mathbf{E} \times \mathbf{H})$$
$$= -\mathbf{i} \cdot \mathbf{E} - \mathbf{E} \cdot \frac{\partial \mathbf{D}}{\partial t} - \mathbf{H} \cdot \frac{\partial \mathbf{B}}{\partial t} \qquad (2)$$

the electric displacement and $\mathbf{i}$ the current density. Integration over any volume v and use of the divergence theorem to replace one volume integral by a surface integral give Eq. (3), where $\mathbf{n}$ is a unit vec-

$$-\int (\mathbf{E} \times \mathbf{H}) \cdot \mathbf{n} \, dS$$
$$= \int_v \left[\frac{\partial}{\partial t} \left(\tfrac{1}{2} \mathbf{B} \cdot \mathbf{H} \right) + \frac{\partial}{\partial t} \left(\tfrac{1}{2} \mathbf{D} \cdot \mathbf{E} \right) + \mathbf{E} \cdot \mathbf{i} \right] dv \qquad (3)$$

tor normal to dS. In the volume integral, $\tfrac{1}{2} \mathbf{B} \cdot \mathbf{H}$ is the magnetostatic energy density, and $\tfrac{1}{2} \mathbf{D} \cdot \mathbf{E}$ is the electrostatic energy density, so the integral of the first two terms represents the rate of increase of energy stored in the magnetic and electric fields in v. The product of $\mathbf{E} \cdot \mathbf{i}$ is the rate of energy dissipation per unit volume as heat; or, if there is a motion of free charges so that $\mathbf{i}$ is replaced by $\rho \mathbf{v}$, ρ being the charge density, it is the energy per unit volume used in accelerating these charges. The net energy change must be supplied through the surface, which explains the interpretation of Poynting's vector.

It should be noted that this proof permits an interpretation of Poynting's vector only when it is integrated over a closed surface. In quantum theory, where the photons are localized, it could be interpreted as representing the statistical distribution of photons over the surface. Perhaps this

justifies the common practice of using Poynting's vector to calculate the energy flow through a portion of a surface.

When an electromagnetic wave is incident on a conducting or absorbing surface, theory predicts that it should exert a force on the surface in the direction of the difference between the incident and the reflected Poynting's vector. *See* ELECTROMAGNETIC RADIATION; MAXWELL'S EQUATIONS; RADIATION PRESSURE; WAVE EQUATION.

[WILLIAM R. SMYTHE]

Precession

An angular velocity of the axis of spin of a spinning rigid body, which arises as a result of external torques acting on the body. Examples are the precession of a spinning top, the Earth (precession of the equinoxes), an airplane propeller, or a gyroscope. Of these, the most familiar is undoubtedly the precession of the equinoxes, a slow change in the direction of orientation of the Earth's axis of rotation which results in a gradual westward motion of the equinoxes.

The uniform precession of a charged spinning body in a uniform magnetic field is called Larmor precession. This motion, similar to that of a rapidly spinning top, is of great importance in atomic physics. *See* LARMOR PRECESSION.

Motion of a spinning top. Precession is best explained by a discussion of the behavior of a symmetrical spinning top or of any rigid body spinning about an axis of symmetry. Consider a top spinning rapidly about its z axis of symmetry and placed horizontally on a point support at O (Fig. 1). The top does not fall as a result of the pull of gravity (its weight, W), as might be supposed; rather, the z axis of the top rotates slowly about the vertical y axis while maintaining its position in the horizontal xz plane. This rotation about the y axis is called precession and the motion can be predicted from Eq. (1), where $\mathbf{M}_O$ is the moment, or torque, about the point O; H_x, H_y, and H_z are the x, y, and

$$\mathbf{M}_O = \mathbf{i}(\dot{H}_x + \omega_y H_z - \omega_z H_y)$$
$$+ \mathbf{j}(\dot{H}_y + \omega_z H_x - \omega_x H_z)$$
$$+ \mathbf{k}(\dot{H}_z + \omega_x H_y - \omega_y H_x) \qquad (1)$$

z components of the angular momentum $\mathbf{H}$ of the top; and $\mathbf{i}$, $\mathbf{j}$, and $\mathbf{k}$ are unit vectors along x, y, and z respectively. For the derivation of Eq. (1) *see* RIGID-BODY DYNAMICS.

If $\mathbf{k}S$ is the spin velocity of the top about the z axis relative to the x, y, z coordinate system, then ω_x, ω_y, and ω_z are the angular velocity components of the x, y, and z axes which are attached to the top at O and move with it except for this relative spin.

The angular momentum of the top is given by Eq. (2), where $I_x = I_y$, I_z are the moments of inertia of the top about the x, y, z axes through O.

$$\mathbf{H} = \mathbf{i} I_x \omega_x + \mathbf{j} I_y \omega_y + \mathbf{k} I_z (\omega_z + S) \qquad (2)$$

Substitution of Eq. (2) into Eq. (1) (for steady-state motion $\omega_x = \omega_y = \omega_z = S = 0$) gives Eq. (3).

$$\mathbf{M}_O = \mathbf{i}[\omega_y(\omega_z + S)I_z - \omega_z \omega_y I_x]$$
$$+ \mathbf{j}[\omega_z \omega_x I_x - \omega_x(\omega_z + S)I_z]$$
$$+ \mathbf{k}[\omega_x \omega_y I_x - \omega_y \omega_x I_x] \qquad (3)$$

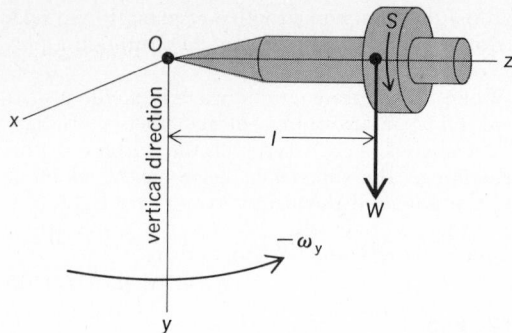

Fig. 1.   A fast-spinning top supported at point O and released in a horizontal plane precesses about the vertical y axis.

For the case where ω_x, ω_y, and ω_z are negligible compared to the spin S, Eq. (3) becomes Eqs. (4). For the case illustrated in Fig. 1, $M_x = -Wl$; $M_y = 0$ (no moment due to W about y); and $M_z = 0$ (no frictional torque about z, $S = 0$).

$$M_x \cong S\omega_y I_z \qquad (4a)$$

$$M_y \cong -S\omega_x I_z \qquad (4b)$$

$$M_z = 0 \qquad (4c)$$

From Eq. (4a), verified by experiment, the top slowly precesses about the y axis with a precessional velocity that is counterclockwise when observed from above and given by Eq. (5).

$$\omega_y = \frac{-Wl}{I_z S} \quad \text{(a constant)} \qquad (5)$$

From Eq. (4b), $\omega_x = 0$, and the top maintains its position in the horizontal plane. From Eqs. (4), it is

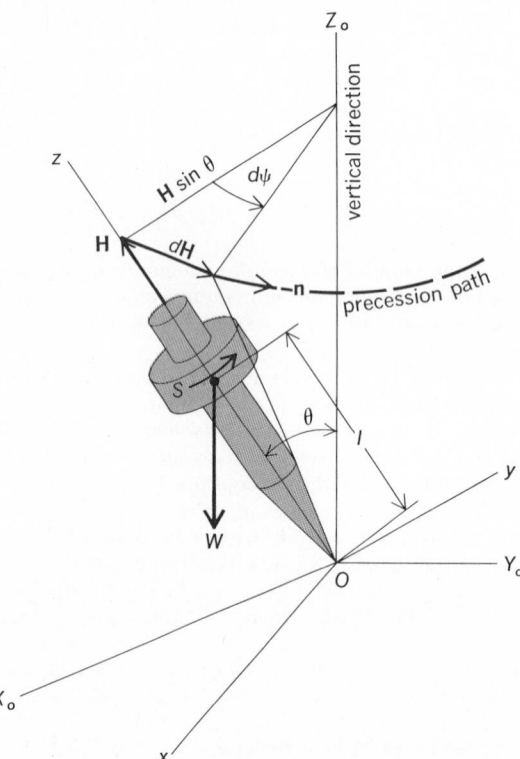

Fig. 2.   A fast-spinning top supported at point O precesses about the vertical Z_o axis.

seen that any moment exerted on the spinning body about the x axis produces an angular velocity of precession about the y axis, and any moment exerted about the y axis produces a negative precession about the x axis. Such a spinning body is the heart of the gyroscope and such moments M_x and M_y are called gyroscopic moments.

If the top of Fig. 1 is now placed in the more general position shown in Fig. 2, a similar precession about a fixed vertical axis (Z_o) due to the gravity moment will result. This is most easily seen by considering the angular momentum vector **H**. Because S is very large as compared to the angular velocity components of the coordinate system x, y,

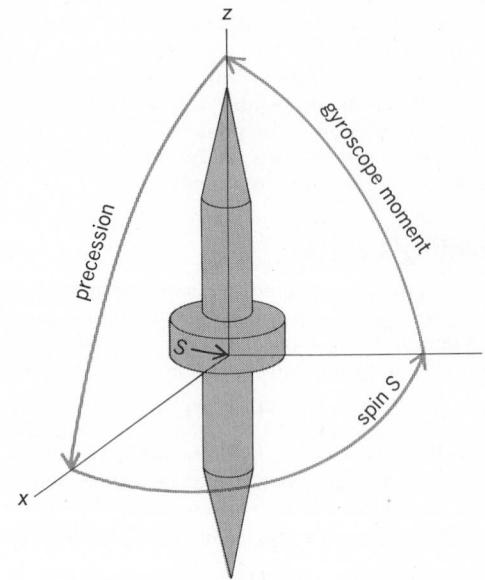

Fig. 3.   Schematic drawing shows the relation between spin, moment, and precession of a spinning body.

z attached to the top, the vector **H** is directed very nearly along the spin axis z, and is given by $\mathbf{H} \cong kI_z S$. From Fig. 2, the moment vector due to gravity is directed normal to the plane of the z and Z_o axes at all times and this produces a change $d\mathbf{H}$, in time dt, of the angular momentum vector. From Fig. 2, $d\mathbf{H} = H (\sin \theta)(d\psi)\mathbf{n}$ ($\mathbf{n}$ = unit vector normal to plane zZ_o), $d\mathbf{H}/dt = H \sin \theta (d\psi/dt)\mathbf{n}$, and $\mathbf{M} = Wl (\sin \theta)\mathbf{n}$. The equation of angular motion $\mathbf{M} = d\mathbf{H}/dt$ becomes $Wl \sin \theta = H \sin \theta (d\psi/dt)$. Therefore Eq. (6) is the precessional velocity.

$$\frac{d\psi}{dt} \cong \frac{Wl}{I_z S} \qquad (6)$$

The equation of motion is satisfied if the axis of the top swings or precesses around Z_o with a velocity $Wl/I_z S$ and with θ, the inclination of the top to the vertical, unchanged. Note that Eq. (6) is identical to Eq. (5) and for large spin the precession is small, satisfying the assumptions of the analysis, that the components of $\boldsymbol{\omega}$ are negligible compared to S.

When the value of S becomes smaller and no longer fulfills this inequality, the general motion of a spinning top must be considered. For the general case, the resultant motion consists of precession and a second angular velocity of the axis of spin called nutation.

Gyroscope motion. A gyroscope consists of a rapidly spinning rigid body with an axis of symmetry. Generally such a body is mounted in a Cardan's suspension so that its mass center is fixed and set spinning with a large angular velocity. Equations (4) are valid and (4a) and (4b) can be expressed as a single vector equation, Eq. (7),

$$\boldsymbol{\omega} = \frac{1}{I_z S^2}\,(\mathbf{S} \times \mathbf{M}) \qquad (7)$$

where $\mathbf{S}$ has a direction along z, the axis of spin. The precessional angular velocity $\boldsymbol{\omega}$ of the axis of spin is always at right angles to z and $\mathbf{M}$. Therefore, from Eq. (7), the axis of spin always turns toward the resultant moment acting on the gyroscope. *See* POINSOT'S METHOD.

This simple analysis and result shows the stability which a large rotation imparts to a body, because under such rotation the body refuses to change the direction of its axis unless large moments are applied. When such moments are applied, the resulting displacements due to the precession are obtained according to Eq. (7) and are shown in Fig. 3. As a result a gyroscope is applicable for stabilizing ships in rolling seas or a monorail car, for inertial navigation instruments, and as the heart of the gyrocompass. Furthermore, the dynamical analysis of the gyroscope explains how bullets and shells that are spun obtain aerodynamic stability in flight.

Airplane maneuvers. Rotating masses on aircraft, such as propellers, gas turbine rotors, jet engine compressors, and the like, can exert gyroscopic moments on the airplane during maneuvers.

As an example, consider an airplane which is making a sharp turn in a horizontal plane (Fig. 4). As the airplane follows the flight path curve C, it is

forcing the rotating parts with spin $\mathbf{S}$ to precess with velocity $\omega_y = V/R$. From Eq. (4), then, the airplane must exert a moment $M_x = S\omega_y I_z$, where I_z is the moment of inertia about z of the rotating parts. If the elevators are not set to produce the required moment M_x, the nose of the airplane will rise.

All flight paths which involve sharp and rapid turns and loops will be accompanied by gyroscope moments exerted by the rotating parts, although such moments are usually small as compared with the aerodynamic moments acting. [RAY E. BOLZ]

Bibliography: V. D. Barger and M. G. Olsson, *Classical Mechanics: A Modern Perspective*, 1973; H. Goldstein, *Classical Mechanics*, 2d ed., 1980; L. Meirovitch, *Methods of Analytical Dynamics*, 1970; K. R. Symon, *Mechanics*, 3d ed., 1971.

Pressure

The ratio of force to area. The force per unit area at the interior of the Sun is estimated to be 3×10^{17} dynes/cm². In interstellar space, pressure approaches zero. Atmospheric pressure at the surface of Earth is in the vicinity of 14 lb/in². Pressures in enclosed containers less than this value are spoken of as vacuum pressures; for example, the vacuum pressure inside a cathode-ray tube is 10^{-8} mm of Hg, meaning that the pressure is equal to the pressure that would be produced by a column of mercury, with no force acting above it, that is 10^{-8} mm high. This is absolute pressure measured above zero pressure as a reference level. Inside a steam boiler, the pressure may be 800 lb/in² or higher. Such pressure, measured above atmospheric pressure as a reference level, is gage pressure, designated psig.

[FRANK H. ROCKETT]

Probability

Although probability theory derives its notion and terminology from intuition, a vague statement such as "John will probably come" is as remote from it as the statement "John is forceful and energetic" is remote from mechanics. Probability theory constructs abstract models, mostly of a qualitative nature, and only experience can show whether these reasonably describe laws of nature or life. As always in mathematics, only logical relations and implications enter the theory, and the notion of probability is just as undefinable (and as intuitive) as are the notions of point, line, or mass. An actual assignment of numerical probabilities is frequently unnecessary or impossible. For example, telephone exchanges are based on a theoretical comparison of several possible systems; only the optimal ones are built and the others discarded. Thus a huge industry depends on theoretical models of exchanges which will never exist.

An uncomplicated illustration of the nature of probability models is found in Lord Rutherford's experiment.

Example (a). To measure radioactive intensity, Lord Rutherford proceeded as follows. Observers A_1 and A_2 counted scintillations on a screen and observed, respectively, N_1 and N_2 scintillations; of these, N_{12} were common to both observers. To estimate the unknown true number X, Rutherford assumed that each scintillation has fixed probabilities p_1 and p_2 to be observed by A_1 and A_2, and furthermore that the observations are independent in the sense that a scintillation observed by A_1 has

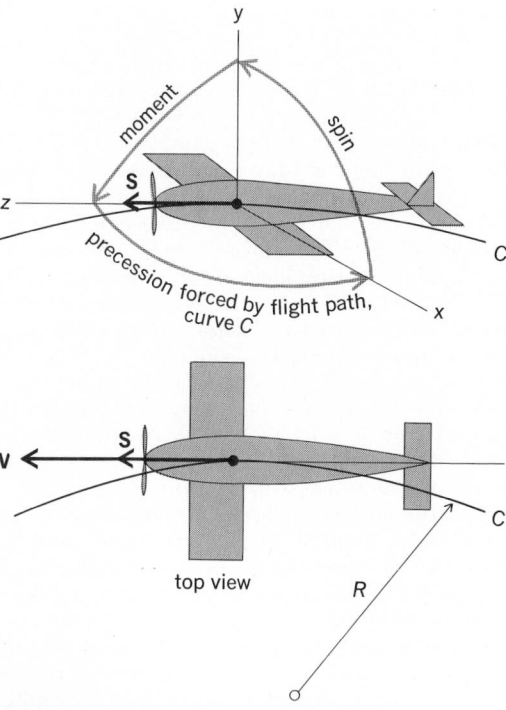

Fig. 4. Rotating elements such as propeller on airplane on curved flight path exert gyroscopic moments.

still probability p_2 to be observed by A_2. In reality the likelihood of observing a scintillation varies with growing fatigue and the proximity of the preceding scintillation; also, the observers are affected by common causes and are therefore not independent. Equating probabilities with observed frequencies (another approximation), Rutherford set $N_1 = Xp_1$, $N_2 = Xp_2$, $N_{12} = Xp_1p_2$, whence $X = N_1N_2/N_{12}$. The three equations may be solved for p_1 and p_2, but these "probabilities" are purely fictitious and, as experience shows, inaccessible to experimental verification. The model is justified by plausibility and success.

The sample space. One speaks of probabilities only in connection with conceptual (not necessarily performable) experiments and must first define the possible outcomes. Thus, by convention, tossing coins results in heads H or tails T; regardless of experimental or philosophical difficulties the age of a person is taken as an exact number and each positive number is taken as a possible age. Throwing two dice results in one of the 36 combinations (1.1), (1.2), . . . , (6,6). An outcome such as "sum 4" is a compound event which can be further decomposed by enumeration: Sum 4 occurs if the outcome is (1.3), (2.2), or (3.1). Thus it is necessary to distinguish between elementary (indivisible) and compound outcomes or events. Each elementary outcome is called sample point; their aggregate is the sample space. The conceptional experiment is defined by the sample space, and it must be introduced and established at the outset.

Example (b). The experiment "distributing 3 balls in 3 cells" has 27 possible outcomes (sample points) listed in tabulation (1).

1. $\{abc|-|-\}$
2. $\{-|abc|-\}$
3. $\{-|-|abc\}$
4. $\{ab|c|-\}$
5. $\{ac|b|-\}$
6. $\{bc|a|-\}$
7. $\{ab|-|c\}$
8. $\{ac|-|b\}$
9. $\{bc|-|a\}$
10. $\{a|bc|-\}$
11. $\{b|ac|-\}$
12. $\{c|ab|-\}$
13. $\{a|-|bc\}$
14. $\{b|-|ac\}$
15. $\{c|-|ab\}$
16. $\{-|ab|c\}$
17. $\{-|ac|b\}$
18. $\{-|bc|a\}$
19. $\{-|a|bc\}$
20. $\{-|b|ac\}$
21. $\{-|c|ab\}$
22. $\{a|b|c\}$
23. $\{a|c|b\}$
24. $\{b|a|c\}$
25. $\{b|c|a\}$
26. $\{c|a|b\}$
27. $\{c|b|a\}$

$$(1)$$

Note that "n balls in 7 cells" may represent the distribution of n hits among 7 targets, or of n accidents in 7 weekdays, and so on.

Consider next the experiment of placing 3 indistinguishable balls into 3 cells. Whether or not actual balls are indistinguishable is irrelevant; they are treated as such and, by convention, there is now a space of only 10 sample points. It is listed in tabulation (2).

1. $\{***|-|-\}$
2. $\{-|***|-\}$
3. $\{-|-|***\}$
4. $\{**|*|-\}$
5. $\{**|-|*\}$
6. $\{*|**|-\}$
7. $\{*|-|**\}$
8. $\{-|**|*\}$
9. $\{-|*|**\}$
10. $\{*|*|*\}$

$$(2)$$

In playing roulette, each point on a circle represents a possible outcome and the sample space is the interval $0 \leqq \vartheta < 2\pi$. When one observes the motion of a particle under diffusion, every function $x(t)$ represents a conceivable outcome and the sample space is a complicated function space.

Events. In examining a bridge hand, one may ask whether it contains an ace or satisfies some other condition. In principle each such event may be described by specifying the sample points which do satisfy the stipulated condition. Thus every compound event is represented by an aggregate of sample points, and in probability theory these terms are synonymous. The standard notations of set theory are used to describe relations among events. *See* SET THEORY.

Given an event A one may consider the case that A does not occur. This is the negation or complement of A, denoted by A'; it consists of those sample points that do not belong to A. Given two events A and B, the event C that either A or B or both occur is the union of A and B and denoted by $C = A \cup B$. In particular $A \cup A'$ is the whole sample space $\mathfrak{S}$ which therefore represents certainty. The event D, both A and B occur, is the intersection of A and B and written $D = A \cap B$. It consists of the points common to A and B. If there are no such common points (as in the case of A and A'), A and B cannot occur simultaneously and they are called mutually exclusive, written $A \cap B = 0$. The event "A but not B" is simply $A \cap B'$.

Example (c). In tabulation (1), the event A "one cell multiply occupied" is the aggregate of the points numbered $1 - 21$. The event B "first cell not empty" is the aggregate of the points 1, $4 - 15$, and $22 - 27$. Because every point belongs either to A or to B (or both), $A \cup B = \mathfrak{S}$ is the certain event. Next, $D = A \cap B$ consists of the points 1, $4 - 14$. Finally, A' may be described as "no cell empty."

Probabilities in finite spaces. If the sample space $\mathfrak{S}$ contains only N points $E_1, \ldots, E_N$ their probabilities may be any numbers such that $P\{E_j\} \geqq 0$ and $P\{E_1\} + \cdots + P\{E_N\} = 1$. The probability $P\{A\}$ of an event A is the sum of the probabilities of all points contained in A; thus $P\{\mathfrak{S}\} = 1$. To find $P\{A \cup B\}$ one considers all points belonging to either A or B, but those belonging to both A and B are counted only once. Therefore $P\{A \cup B\} = P\{A\} + P\{B\} - P\{A \cap B\}$. In particular, for mutually exclusive events, there is the addition rule $P\{A \cup B\} = P\{A\} + P\{B\}$.

Frequently considerations of symmetry lead one to consider all E_j as equally likely; that is, to set $P(E_j) = 1/N$. In this case $P(A) = n/N$ where n is the number of points in A; for a gambler betting on A, these represent the "favorable cases." For example, in throwing a pair of "perfect" dice, one naturally assumes that the 36 possible outcomes are equally likely. This model does not lose its justification or usefulness by the fact that actual dice do not live up to it. The assumption of perfect randomness in games, card shuffling, industrial quality control, or sampling is rarely realized, and the true usefulness of the model stems from the experience that noticeable departures from the ideal scheme lead to the detection of assignable causes and thus to theoretical or experimental improvements.

How the success of probability theory depends on the disregard of preconceived philosophical ideas and on the readiness to adapt models to unexpected circumstances is illustrated by Bose-Einstein statistics.

Example (d): Bose-Einstein statistics. In the example of tabulation (1), the notion of perfect randomness leads to the assignment of probability $1/27$ to each point. In the case of indistinguishable balls, tabulation (2), it has been argued that an experiment is unaffected by failure to distinguish between balls; physically there remain 27 possibilities grouped in 10 distinguishable forms. This argument leads to assigning probability 1/27 to each of the points 1–3, probability 1/9 to each of the points 4–9, and 2/9 to point 10. This reasoning (sound in certain situations) has been accepted as evident in statistical mechanics for the distribution of r particles in n cells (Maxwell-Boltzmann statistics). Surprisingly, it turned out that no physical particles behave this way and it was revolutionary when Bose and Einstein showed that for one type of particle all distinguishable arrangements are equally likely. This model assigns probability 1/10 to each point of tabulation (2). *See* BOLTZMANN STATISTICS; BOSE-EINSTEIN STATISTICS.

A useful, although vague, intuitive description of probability describes $P\{A\}$ as the relative frequency of the event A if the experiment is repeated many times under identical circumstances. The laws of large numbers render this more precise, but the description often lacks operational meaning. Experiments in agriculture and human sampling cannot be repeated under remotely similar conditions, and in the case of telephone exchanges, useful probability models refer to situations which will never materialize.

Probabilities in infinite spaces. Two examples may illustrate the novel features of this topic.

Example (e): unending coin tossing. In the study of limit laws, one must consider potentially infinite sequences of coin tossings. The possible outcome of this experiment is an infinite sequence of heads and tails, and every sequence such as *HTHTTH* . . . represents a simple point. Finitely many tosses are the beginning of an infinite sequence, and the event "first four trials resulted in *HTTH*" is the aggregate of the infinitely many sequences with the prescribed beginning. Such an event is called an interval of length 4. There are 2^n intervals of length n, and they are mutually exclusive. For reasons of symmetry, one attributes the probability 2^{-n} to each interval of length n. Thus the assignment of basic probabilities refers to intervals rather than to points. A point such as *HTHT* . . . is the limit of an infinite sequence of contracting intervals $H, HT, HTH, . . . ,$ and therefore probability zero must be attributed to each individual point.

The probabilities of other events are similarly defined by limiting procedures. For example, consider the event A that an infinitely prolonged sequence of trials never produces a run of at least two consecutive heads, or two consecutive tails. It is more convenient to enumerate the points of the complementary event A' that two equal symbols do occur in succession. Clearly A' is the union of the infinitely many mutually exclusive intervals *HH, TT*; *HTT, THH*; *HTHH, THTT*, and so on. Here there are 2 intervals of length $n \geq 2$, and therefore $P\{A'\} = 2(2^{-2} + 2^{-3} + 2^{-4} + \cdots) = 1$, whence $P\{A'\} = 0$. The indicated result of the experiment is thinkable, but probability zero is attributed to it. A similar, although more complicated, limiting procedure leads to the law of large numbers according to which the event "the frequencies of H and T in the first n trials tend to 1/2 as $n \to \infty$" has probability one.

Example (f): roulette. Here the sample space consists of the angles $0 \leq \vartheta < 2\pi$, and the notion of a perfect roulette assumes equal probabilities for intervals of equal length; thus an interval of length a carries probability $a/2\pi$. If the roulette is divided into 32 equal numbered intervals, the event "even number" consists of 16 intervals and has probability 1/2.

The situation encountered here is not peculiar to probability but is common in measure theory. One starts with a collection of basic events, called intervals, and attributes probabilities to them. By simple and natural limiting procedures, probabilities can then be defined for a much wider class $\mathfrak{F}$ of events which are obtainable by applying the operations of set theory to intervals (in finite or infinite numbers). $\mathfrak{F}$ is the Borel field generated by the intervals. Probability is simply a measure on $\mathfrak{F}$; that is, to each event A in $\mathfrak{F}$, there corresponds a probability $P\{A\} \geq 0$ which is completely additive. If A is the union of the mutually exclusive events $A_1, A_2, . . . ,$ then $P\{A\} = \Sigma P\{A_i\}$. The probability of the whole space is, of course, unity.

The extension of the addition rule from finitely to infinitely many summands may be defended by considerations of continuity, but ultimately it is justified by its simplicity and its success.

Conditional probability—independence. Suppose that a population of N people includes N_A color-blind persons and N_H females. To the event A "a randomly chosen person is color-blind" can be ascribed probability $P\{A\} = N_A/N$, and similarly for the event H that a person be female one has $P\{B\} = N_H/N$. If N_{AH} is the number of color-blind females, the ratio N_{AH}/N_H may be interpreted as probability that a randomly chosen female be color-blind; here the experiment "random choice in the population" is replaced by a selection from the female subpopulation. In the original experiment, N_{AH}/N is the probability of the simultaneous occurrence of both A and H, so that $N_{AH}/N_H = P\{A \cap H\}/P\{H\}$. Similar situations occur so frequently that it is convenient to introduce the equation

$$P\{A|H\} = \frac{P\{A \cap H\}}{P\{H\}} \qquad (3)$$

and to call this the conditional probability of the event A relative to H. This concept is useful whenever it is desired to restrict the consideration to those cases where the event H occurs (or where the hypothesis H is fulfilled). Thus, in betting on an event A the knowledge that H occurred would induce one to replace $P\{A\}$ by $P\{A|H\}$. If all sample points are equally likely, $P\{A|H\}$ still represents the ratio of favorable cases to the total of cases possible when it is known that H has occurred.

Despite its simplicity the notion of conditional probability is exceedingly important, and frequently the probabilities in sample space are defined only in terms of conditional probabilities.

Example (g). In a bolt factory three machines manufacture, respectively, 25, 35, and 40% of the total. Of their output 5, 4, and 2% are defective bolts. Classification of the bolts according to the number of the machine and the quality (d for defective, c for conforming) gives the six categories $c_1, c_2, c_3, d_1, d_2,$ and d_3. A random choice of a bolt results in one of these six outcomes, but their probabilities are not given directly. Instead, the data relating to the first machine are given by

$$P\{c_1 \cup d_1\} = 0.25$$
$$P\{d_1 | c_1 \cup d_1\} = 0.25$$

It follows that $P\{d_1\} = 0.0125$ and similarly for the other points. This example may also serve to illustrate the reasoning following Bayes concerning the probability of causes. Supposing a bolt was found to be defective (hypothesis H), what is the probability that it came from the first machine (cause A)? Here

$$P\{H\} = P\{d_1\} + P\{d_2\} + P\{d_3\}$$
$$= 0.0125 + 0.0140 + 0.0080 = 0.0345$$

$$P\{A \cap H\} = P\{D_1\} = 0.0125$$

and thus the required answer is given by

$$P\{A | H\} = 0.0125/0.0345 = 25/69$$

In example (*b*) the probability of H "ball a is in the first cell" equals 1/3, and the probability of A "first cell is multiply occupied" = 7/27. Now given that the ball a is in the first cell, the conditional probability that this cell is multiply occupied becomes 5/9. The knowledge that H has occurred should increase one's readiness to bet on A. By contrast, for the event B "ball b is in the second cell," $P\{B|H\} = 1/3 = P\{B\}$, and so the knowledge that H has occurred gives no clue as to B. Therefore, B is said to be independent of H if $P\{B|H\} = P\{B\}$, that is, if

$$P\{B \cap H\} = P\{B\}P\{H\}$$

Clearly, in this case $P\{H|B\} = P\{H\}$ so that H is also independent of B. Accordingly, two events B and H are independent of each other if the probability of their simultaneous occurrence follows the multiplication rule $P\{B \cap H\} = P\{B\}P\{H\}$. This notion carries over to systems of more than two events.

Independent trials. The intuitive frequency interpretation of probability is based on the concept of experiments repeated under identical conditions; a theoretical model for this concept can be developed.

Consider an experiment described by a sample space $\mathfrak{S}$; for simplicity of language it can be assumed that $\mathfrak{S}$ consists of finitely many sample points $E_1, \ldots, E_N$. When the same experiment is performed twice in succession, the thinkable outcomes are the N^2 pairs of sample points (E_1, E_1), $(E_1, E_2), \ldots, (E_N, E_N)$, and these now constitute the new sample space. It is called the combinatorial product of $\mathfrak{S}$ by itself and denoted by $\mathfrak{S} \times \mathfrak{S}$; with reference to analytic geometry, one speaks of the first and second coordinate of the point (E_i, E_j). These notions apply equally to infinite sample spaces and to products $\mathfrak{S} \times \mathfrak{S} \times \mathfrak{S} \cdots$ of more than two factors. For example, the cartesian plane of points (x,y) is the product of the real line by itself. In tossing a coin once, $\mathfrak{S}$ contains only the points H and T; tossing the coin n times leads to the n-tuple product $\mathfrak{S} \times \cdots \times \mathfrak{S}$ whose points have n coordinates and are of the form $(HT \cdots T)$.

Probabilities must be assigned to the events in $\mathfrak{S} \times \mathfrak{S}$. The case of dependent trials will be treated in the next section; if the second trial is independent of the first, the probabilities in $\mathfrak{S} \times \mathfrak{S}$

follow the productive rule $P\{E_i, E_j\} = P\{E_i\}P\{E_j\}$.

In the case of n tossings of a coin, this rule leads to the probability 2^{-n} for each sample point in agreement with the requirement of equally likely cases. The present approach is more flexible and more general as shown by the Bernoulli trials.

Example (h): Bernoulli trials. Suppose each trial results in success S or failure F, and $P\{S\} = p$, $P\{F\} = q$ where $p + q = 1$. (This may be considered as the model of a skew coin.) A succession of n independent trials of this kind leads to the sample space of n-tuples ($SFFS \cdots FS$), and the probability of such a point is the product ($pqqp \cdots qp$) obtained on replacing each S by p and each F by q.

This model has obvious applications to repeated observations and to gambling. Independence is an assumption to be verified experimentally. Conceivably a coin could be endowed with memory and avoid runs of more than 17 successive heads. That the sex distribution within families resembles Bernoulli trials is purely a matter of experience. Many gamblers fully accept the independence and yet believe that they can influence fate by using "systems," for example, by skipping the game after each failure, or waiting for a run of 3 successes, and so on. The theorem on systems shows this to be a fallacy; a gambler not endowed with foresight may use any system or random choice of the times when he plays or skips the game; he remains confronted with Bernoulli trials and is exactly in the same situation as if he played at each trial. *See* DISTRIBUTION (PROBABILITY).

Example (i): geometric probabilities. In the interval $0 < x < 1$ a point is chosen at random. This interval is the sample space $\mathfrak{S}$ and the probability of each subinterval equals its length. The sample space $\mathfrak{S} \times \mathfrak{S}$ is the unit square of the x,y plane, and the probability of any figure equals its area. The event "the two successive choices result in a sum <1" is represented by the triangle below the main diagonal and has probability 1/2. The event "the greater of the two choices is $<t$" is represented by the square $0 < x < t$, $0 < y < t$ and has probability t^2.

Dependent trials; Markov chains. Many phenomena can be analyzed in terms of dependent trials. In their description the convenient and picturesque terminology of urn models shall be adopted, which should not detract from the general nature of the schemes being presented.

Consider an urn containing N balls, of which r are red R and $b = N - r$ black B. Assuming perfect randomness, the probability that a randomly drawn ball be red equals r/N. If the ball is replaced and the procedure repeated, the result is Bernoulli trials with $p = r/N$. Without replacement, the sample space corresponding to two drawings contains four points RR, RB, BR, and BB, to which probabilities are assigned as follows: If the first ball drawn is red (probability r/N), the conditional probabilities of R and B at the second trial become $(r-1)/(N-1)$ and $b/(N-1)$. By Eq. (3), therefore,

$$P[RR] = r(r-1)/N(N-1)$$

$$P[RB] = P[BR] = rb/N(N-1)$$

$$P[BB] = b(b-1)/N(N-1)$$

A more general urn model is obtained by letting the composition of the urn vary from trial to trial.

For definiteness consider the following scheme: Each time a ball is drawn, it is replaced, and c balls of the color drawn and d balls of the opposite color are added to the urn. Here c and d are fixed numbers which may be negative. This scheme contains interesting special cases such as the following:

1. When $c=d=0$, drawing with replacement occurs, and for $c=-1$, $d=0$, drawing without replacement occurs. In the latter case, the process terminates after N drawings.

2. The Polya model of contagion is the special case when $c>0$ is fixed and $d=0$. Here the drawing of either color increases the probability of the same color at subsequent trials, just as in the contagious disease each occurrence increases the probability of further occurrences. This model represents only a crude first approximation to phenomena of contagion, but it leads to comparatively simple formulas and has been applied with astonishing success to a variety of experiences from sickness insurance to baseball scores.

3. The Ehrenfest model for heat exchange considers two containers, I and II, and N particles distributed in them. A particle is chosen at random and removed from its container into the other. This scheme differs only linguistically from the urn scheme. If the particles in I are called red and those in II black, then each trial changes the color of one ball and gives the special case $c=-1$ and $d=1$.

The probabilities of the various possible outcomes in the general scheme are obtained as above. For example, $P\{RBR\}=r(b+d)(r+c+d)/N(N+c+d)(N+2c+2d)$, and so on.

Markov chains represent another important scheme for dependent trials. Suppose that at each trial the possible outcomes are $E_1, \ldots, E_N$ and that whenever E_i occurs the conditional probabilty of E_j at the next trial is p_{ij}, independently of what happened at the preceding trials. Here, of course, $p_{ij} \geq 0$ and $p_{i1}+p_{i2}+\cdots+p_{iN}=1$ for each i. The p_{ij} are called transition probabilities. The whole process is now determined if the initial probabilities, π_i, at the first trial are known. For example, $P\{E_a E_b E_c\}=\pi_a p_{ab} p_{bc}$. The probability of the event "E_c at the third trial" is obtained by summation over all a and b, and so on. Markov chains, and their analog with continuous time, represent the simplest type of stochastic process. The Ehrenfest model considered above may be treated as a Markov chain by letting E_i represent the event that container I contains i particles. Then $p_{i,i-1}=i/N$, $p_{i,i+1}=(N-i)/N$, and $p_{ij}=0$ for all other combinations of ij. Other examples of Markov chains are the gambler's accumulated fortune, the composition of a deck of cards under random shuffling, and random walks. Important applications are to queueing theory where one encounters also processes with more complicated aftereffects. See STOCHASTIC PROCESS.

Random variables and their distributions. The theory of probability traces its origin to gambling, and the gambler's gain may still serve as the simplest example of a random variable. With every possible outcome (sample point) there is associated a number, namely, the corresponding gain. In other words, the gain is a function on the sample space, and such functions are called random variables. (In infinite spaces the idea is the same, but a somewhat more cautious definition is in order.)

With the same experiment, one may associate many random variables. As an example, consider the sample space of tabulation (1) with probability 1/27 for each point. A typical random variable is the number N of occupied cells; it assumes the value 1 at the three points numbered $1-3$; the value 2 at the eighteen points $4-21$; and the value 3 at the six points $22-27$. One says, therefore, that the probability distribution of N is given by $P\{N=1\}=1/9$, $P\{N=2\}=2/3$, $P\{N=3\}=2/9$. Another variable is the number X of balls in the first cell. An inspection of tabulation (1) shows that its probability distribution is given by $P\{X=0\}=8/27$, $P\{X=1\}=12/27$, $P\{X=2\}=6/27$, $P\{X=3\}=1/27$. One may also consider the two variables simultaneously and find, for example, that the combination $N=1$, $X=0$ occurs at two points, whence $P\{N=1, X=0\}=2/27$. The probabilities of all pairs are given by the joint probability distribution of N and X exhibited in tabulation (4). Adding the entries in the rows and columns gives the distribution of N and X, respectively, and they are there-

N \ X	0	1	2	3	Distribution of N	
1	2/27	0	0	1/27	3/27	
2	6/27	6/27	6/27	0	18/27	(4)
3	0	6/27	0	0	6/27	
Distribution of X	8/27	12/27	6/27	1/27		

fore occasionally called marginal distributions. See COMBINATORIAL THEORY.

Example *(i)* may be used to illustrate the case of continuous random variables. This example considers two consecutive selections of a point in the interval $0<x<1$. Let S be the random variable denoting the sum of the two choices, and L the larger of the two. One sees that for $0<t<1$ the event $L \leq t$ has probability t^2; thus, setting $P\{L<t\}=F(t)$ gives $F(t)=t^2$ when $0 \leq t \leq 1$; for $t<0$ and for $t>1$ one has trivially $F(t)=0$ and $F(t)=1$, respectively. This is the distribution function of L. From it can be calculated all probabilities relating to L. Similarly the event $S \leq u$ is represented by the region in the unit square below the line $x+y=u$; therefore, the distribution function of S, namely, $P\{S \leq u\}=G(u)$, is given by $G(u)=0$ for $u \leq 0$, $G(u)=(1/2)u^2$ for $0 \leq u \leq 1$, $G(u)=1-(1/2)(2-u)^2$ for $1 \leq u \leq 2$, and $G(u)=1$ for $u \geq 2$. In like manner, the joint distribution function $P\{L \leq t, S \leq u\}=H(t,u)$ of the pair L,S can be calculated.

Every random variable X has a distribution function $F(t)=P\{X \leq t\}$. If X assumes only finitely many values, then $F(t)$ is a step function. Thus, in tabulation (4), $F(t)$ assumes the values 0, 8/27, 20/27, 26/27, and 1, respectively, in intervals $t<0$, $0 \leq t < 1$, $1 \leq t < 2$, $2 \leq t < 3$, and $t \geq 3$. In such cases the notion of distribution function is used mainly for uniformity of language. The notion is really convenient when $F(t)$ is not only continuous but also has a derivative $f(t)=F'(t)$; then $f(t)$ is called the probability density of X. In the above example the variable L has a density defined by $2t$ for $0<t<1$ and 0 elsewhere; the density of S is 0 for $u<0$ and $u>2$; it equals u for $0<u<1$ and equals $2-u$ for $1<u<2$.

The notion of independence carries over: Two random variables X and Y are independent if $P\{X \leq x, Y \leq t\} = P\{X \leq s\} \cdot P\{Y \leq t\}$. It is easily seen that for independent variables with distribution functions $F(t)$ and $G(t)$ the distribution function of the sum $S = X + Y$ is given by the convolution

$$P\{S \leq u\} = \int_{-\infty}^{+\infty} F(u-s) \, dG(s)$$

$$= \int_{-\infty}^{+\infty} G(u-s) \, dF(s) \qquad (5)$$

In terms of densities, Eq. (5) reads

$$h(u) = \int_{-\infty}^{+\infty} f(u-s)g(s) \, ds$$

$$= \int_{-\infty}^{+\infty} g(u-s)f(s) \, ds \qquad (6)$$

In the random choice example, the coordinates of the points chosen are independent variables with the rectangular density $f(s) = g(s) = 1$ for $0 < s < 1$. The distribution of their sum S which has been calculated above can be found also using Eq. (6).

Expectations. Given a random variable X one may interpret its distribution function $F(t)$ as describing the distribution of a unit mass along the real axis such that the interval $a < x \leq b$ carries mass $F(b) - F(a)$. In the case of a discrete variable assuming the values $x_1, x_2, \ldots$ with probabilities $p_1, p_2, \ldots$ the entire mass is concentrated at the points x_i; if $F'(x) = f(x)$ exists, it represents the ordinary mass density as defined in mechanics. The center of gravity of this mass distribution is called the expectation of X; the usual symbol for it is $E(X)$, but physicists and engineers use notations such as $\langle X \rangle$, $\langle X \rangle_{\text{Av}}$, or $\overline{X}$. In the cases mentioned,

$$E(X) = \Sigma p_i x_i$$

$$E(X) = \int_{-\infty}^{+\infty} x f(x) \, dx$$

In all cases $E(X)$ is given by the Stieltjes integral over $x \, dF(x)$. (To be precise, one speaks of expectations only when the integral converges absolutely.)

Before discussing the signficance of the new concept, a few frequently used definitions are appropriate. Put $m = E(X)$. Then $(X - m)^2$ is, of course, a random variable. In mechanics, its expectation represents the moment of inertia of the mass distribution. In probability, it is called variance of X; its positive root is the standard deviation. Clearly,

$$\text{Var}(X) = E(X - m)^2 = E(X^2) - m^2$$

The variance is a measure of spread: It is zero only if the entire mass is concentrated at the point m, and it increases as the mass is moved away from m. In the case of two variables X_1 and X_2 with expectations m_1 and m_2 it is necessary to consider not only the two variances $s_i^2 = E[(X_i - m_i)^2]$ but also the covariance $\text{Cov}(X_1, X_2) = E[(X - m_1)(X_2 - m_2)] = E(X_1 X_2) = m_1 m_2$. The covariance divided by $s_1 s_2$ is called the correlation coefficient of X_1 and X_2. If it vanishes, X_1 and X_2 are called uncorrelated. Every pair of independent variables is uncorrelated, but the converse is not true.

If $X_1, X_2, \ldots, X_n$ are random variables with expectations $m_1, \ldots, m_n$ and variances $s_1^2,$ $\ldots, s_n^2$, the expectation of their sum $S_n = X_1 + \cdots + X_n$ is always given by $E(S_n) = m_1 + \cdots + m_n$; if all the covariances of X_i and X_j vanish, then clearly $\text{Var}(S_n) = s_1^2 + \cdots + s_n^2$.

When X represents a physical quantity, then $X^* = (X - m)s^{-1}$ represents the same quantity measured from a different origin and in new units. In the physicist's terminology, X^* is the quantity X referred to dimensionless units. In probability, X^* is called the reduced or standardized variable.

It was once assumed that every reasonable random variable has finite expectation and variance. Modern theory refutes this assumption. Many recurrence times in important physical processes have no finite expectations. Even in the simple coin-tossing game, the number of trials up to the time when the gambler's accumulated gain first reaches a positive level has infinite expectation.

Laws of large numbers. To explain the meaning of the expectation and, at the same time, to justify the intuitive frequency interpretation of probability, consider a gambler who at each trial may gain the amounts $x_1, x_2, \ldots, x_n$ with probabilities $p_1, p_2, \ldots, p_n$. The gains at the first and second trials are independent random variables X_1, X_2 with the indicated distribution and the common expectation $m = \Sigma p_i x_i$. The event that an individual gain equals x_i has probability p_i, and the frequency interpretation of probability leads one to expect that in a large number n of trials this event should happen approximately np_i times. If this is true, the total gain $S_n = X_1 + X_2 + \cdots + X_n$ should be approximately nm; that is, the average gain $(1/n)S_n$ should be close to m. The law of large numbers in its simplest form asserts this to be true. More precisely, for each $\varepsilon > 0$ it assures one that

$$P\left\{ \left| \frac{1}{n} S_n - m \right| > \varepsilon \right\} \to 0 \quad \text{as } n \to \infty$$

This law holds also when the distribution function is not discrete.

As a special case, one can obtain a frequency interpretation of probability. In fact, consider an event A with $P\{A\} = p$ and suppose that in a sequence of independent trials a gambler receives a unit amount each time when A occurs. Then the expectation of the individual gain equals p, and S_n is the number of times the event A has occurred in n trials. It follows that

$$P\left\{ \left| \frac{1}{n} S_n - p \right| > \varepsilon \right\} \to 0$$

That is, the relative frequency of the occurrence of A is likely to be close to p.

Without this theorem, probability theory would lose its intuitive foundation, but its practical value is minimal because it tells one nothing concerning the manner in which the averages $n^{-1}S_n$ are likely to approach their limit m. In the regular case where the X_j have finite variances, the central limit theorem gives much more precise and more useful information; for example, it tells one that for large n the difference $S - np$ is about as likely to be positive as negative and is likely to be of the magnitude $n^{1/2}$. When the X_k have no finite variances, the central limit theorem fails and the sums S_n may behave oddly in spite of the law of large numbers. For example, it is possible that $E(X_k) = 0$ but

$P\{S_n < 0\} \to 1$. In gambling language this game is "fair," and yet the gambler is practically certain to sustain an ever-increasing loss.

There exist many generalizations of the law of large numbers, and they cover also the case of variables without finite expectation, which play an increasingly important role in modern theory. *See* PROBABILITY (PHYSICS); STATISTICS.

[WILLIAM FELLER]

Bibliography: W. Feller, *An Introduction to Probability Theory and Its Applications*, vol. 1, 3d ed., 1968, vol. 2, 2d ed., 1971.

Probability (physics)

To the physicist the concept of probability is like an iceberg. The part of it which he uses and which is, therefore, in full view for him is but a small fraction of what is hidden in other and larger disciplines. The philosophy and mathematics of probability have become increasingly interesting and important. *See* PROBABILITY.

Bernoulli's problem. One of the most basic problems encountered in the application of probability to physics was solved by J. Bernoulli. It concerns the probability of achieving a specified number (x) of successes in n independent trials when the probability of success in a single trial is known. Denote by p the probability of success; let $q = 1 - p$ be the probability of failure. Equation (1) represents the well-known binomial coefficient. The

$$C_x{}^n \equiv \frac{n!}{x!(n-x!)} \tag{1}$$

sents the well-known binomial coefficient. The probability in question is given by Eq. (2). For

$$w_n(x) = C_x{}^n p^x q^{n-x} \tag{2}$$

example, one may consider an urn containing a black balls and b white balls; n drawings are to be made from this urn, with replacement of the drawn ball each time. The probability that x white balls shall turn up in these n drawings is required. In this example, Eqs. (3) hold.

$$p = \frac{b}{a+b}$$
$$w_n(x) = C_x{}^n \left(\frac{b}{a+b}\right)^x \left(\frac{a}{a+b}\right)^{n-x} \tag{3}$$

Equation (2) defines Bernoulli's distribution; because of the Newtonian binomial coefficients which it involves, it is sometimes called Newton's formula. It satisfies relations (4)–(6). Hence, the

$$\sum_{x=0}^{n} w_n(x) = 1 \tag{4}$$

$$\bar{x} = \sum_x x w_n(x) = np \tag{5}$$

$$\sigma^2 = \sum_x (x - \bar{x})^2 w_n(x) = \bar{x^2} - \bar{x}^2 = npq \tag{6}$$

σ^2 is called the dispersion of the distribution $w_n(x)$. The so-called standard deviation σ equals $\sqrt{npq}$.

The physicist regards the emission of particles from nuclei, of photons from hot bodies, and of electrons from hot filaments as random phenomena controlled by probability laws. One of the simplest methods for testing this assumption is based upon these formulas. Suppose that a radioactive specimen emits particles for a finite period T, and

that their number is x. By repeated observations on the emission of particles in a period of T seconds, one obtains a series of numbers, x_1, x_2, x_3, etc. Now imagine the period T to be subdivided into a very large number, n, of intervals, each of length τ, so that $n\tau = T$. Indeed, τ will be assumed to be so small that, at most, one particle is emitted within the interval τ. A single observation made during that infinitesimal interval would therefore yield the result: either no emission or emission, say, with probabilities p or q. By hypothesis, n such observations yield x emissions. It is seen, therefore, that the problem involves an application of Bernoulli's distribution. While it is difficult to specify the values of p, q, and τ, the result given by Eq. (6) must nevertheless be true. Since q is extremely small, p is very nearly 1. Hence, in view of Eq. (5), $\sigma^2 = npq = nq = \bar{x}$. This relation has been tested experimentally and found to be true in all instances.

Approximations. In many practical applications, the values of x and n are very large, and a direct use of Newton's formula becomes impossible because the binomial coefficients are difficult to evaluate for large x and n. The example of particle emission just discussed is a case in point. Under these circumstances, two approximations to Newton's formula are available, one derived by C. F. Gauss, the other one by S. D. Poisson and bearing his name. Gauss's law is the limiting form of Eq. (2), when both x and n are large, so large in fact that $1/\bar{x} = 1/np$ and $1/npq$ are both negligible. This implies that p and q are numbers not greatly different from unity. In that case, Eq. (7) holds. This is Gauss's law.

$$\lim_{n \to \infty} w_n(x) = \frac{1}{\sqrt{2\pi p \bar{x}}} \exp\left[\frac{-(x-\bar{x})^2}{2px}\right]$$
$$= \frac{1}{\sqrt{2\pi}\sigma} \exp\left[-(x-\bar{x})^2/2\sigma^2\right] \tag{7}$$

If p is so small that the mean np is of the order of unity in any given application, Bernoulli's distribution is approximated by Poisson's law, Eq. (8).

$$\lim_{n \to \infty} w_n(x) = \frac{(np)^x e^{-np}}{x!} \tag{8}$$

The validity of this formula has also been confirmed in many instances of particle emission.

The density fluctuations which occur in a finite volume of gas present an interesting application of Gauss's law. Suppose that a vessel of volume V contains but a single molecule. One may then select a small element of volume v and imagine successive observations to be made in order to determine whether the molecule is in this element or not. The probability of its being there, which will be denoted by p, is clearly the ratio of the small volume element to the volume of the total container, $p = v/V$. Now let n observations be made upon the volume v. The probability that, in x of them, the one molecule is found in v is given by Newton's formula. If the vessel contains not one but n molecules, then the probability of finding x of them in v simultaneously is the same as that of finding the one molecule x times in v on n successive occasions. Therefore, the probability of observing x of the n molecules in v is given by Eq. (9).

$$w_n(x) = C_x{}^n p^x q^{n-x} \tag{9}$$

Equation (9) may be approximated by Gauss's law. Thus, on substituting p and q, as shown in Eqs. (10) and (11), Eq. (12) results. The expected

$$p = \frac{v}{V} = \frac{1}{k} \qquad (10)$$

$$q = 1 - p \qquad (11)$$

$$w(x)\,dx =$$

$$\frac{k}{\sqrt{2\pi n(k-1)}}\,\exp\left[-\frac{(kx-n)^2}{2n(k-1)}\right]dx \quad (12)$$

mean and the dispersion of this distribution must be the same as the corresponding quantities for Bernoulli's distribution. Hence Eqs. (13) and 14) hold. If k is sufficiently large, $\sigma^2 = \bar{x}$ approximately.

$$\bar{x} = np = \frac{n}{k} \qquad (13)$$

$$\sigma^2 = npq = \frac{n}{k^2}(k-1) = \bar{x}\left(1 - \frac{\bar{x}}{n}\right) \qquad (14)$$

It is often convenient to introduce a quantity δ, known as the relative fluctuation and defined by Eq. (15).

$$\delta = \frac{x - \bar{x}}{\bar{x}} \qquad (15)$$

In view of Eq. (13), Eq. (16) holds.

$$\delta = \frac{kx - n}{n} \qquad (16)$$

In terms of δ, the law of fluctuations takes on the simple form given by Eq. (17). Here $\bar{x}$ is the mean

$$w(\delta)\,d\delta = \sqrt{\frac{\bar{x}}{2\pi q}}\,\exp\left[\frac{-x}{2q}\delta^2\right]d\delta \qquad (17)$$

number of molecules within v, and q is very nearly equal to 1 if $v \ll V$. Equation (17) is well supported by experiment.

Theory of errors. Perhaps the most important application of probability theory to exact science occurs in the treatment of errors. Measurements are accompanied by errors of two kinds: determinate and random. The former arise from actual mistakes, either on the part of the observer or from faulty instruments; they are not susceptible of mathematical treatment. Random errors, however, because they are numerous, small, and likely to combine in linear fashion, are subject to the laws of probability analysis.

Gauss error law. Let there be n measurements of some physical quantity resulting in the numbers $X_1, X_2, X_3, \ldots, X_n$. If the true value of the quantity (usually unknown) is denoted by X, then the errors are given by Eqs. (18). It can be shown mathe-

$$x_1 = X_1 - X, \quad x_2 = X_2 - X, \ldots, \\ x_n = X_n - X \qquad (18)$$

matically and has been confirmed by numerous observations that the relative frequency of occurrence of an error x is represented by Bernoulli's distribution which, under the present conditions, takes on the form of Gauss's law, Eq. (19). In this

$$N(x)\,dx = \frac{h}{\sqrt{\pi}}e^{-h^2x^2}\,dx \qquad (19)$$

formula the parameter h^2 equals $1/(2\sigma^2)$ and is

called the index of precision. Clearly, the greater the value of h the narrower the distribution given by Eq. (19). The latter is often called the Gauss error law or the normal distribution of errors. In writing it, the assumption has of course been made that the numbers x_1, x_2, etc., may be replaced by a continuous distribution.

The probability that a single measurement will contain an error between the limits $-a$ and $+a$ is given by Eq. (20). The integral occurring here is

$$\frac{h}{\sqrt{\pi}}\int_{-a}^{a}e^{-h^2x^2}\,dx = \frac{2h}{\sqrt{\pi}}\int_0^a e^{-h^2x^2}\,dx \qquad (20)$$

called the error function, and is denoted by erf (a); it is tabulated in most textbooks that discuss the theory of errors. The factor in front of the integral has been chosen so that Eq. (21) holds.

$$\int_{-\infty}^{\infty}N(x)\,dx = 1 \qquad (21)$$

Returning to the set of measured values $X_1, X_2, \ldots, X_n$, one may ask: Which is the most probable value (the one most likely to be true) consistent with this set of numbers? The answer is given by the principle of least squares, which affirms that the most probable value is the one for which the sum of the squared errors $x_1^2 + x_2^2 + \ldots + x_n^2$ is a minimum. *See* LEAST-SQUARES METHOD.

From this one may prove directly that the most probable value of X is its arithmetical mean, given by Eq. (22).

$$\overline{X} = \frac{X_1 + X_2 + \cdots + X_n}{n} \qquad (22)$$

Kinds of errors. The reliability of a set of measurements, such as the sequence $X_1, X_2, \ldots, X_n$, is specified by certain measures called errors, but in a slightly different sense from that previously employed. Three kinds of "error" will be described: the average error a, the root mean square error m, and the probable error r. All three refer, not to a single measurement as did the quantity x_i, but to the entire distribution in Eq. (19).

The average error a is the arithmetical mean of all individual errors without regard to sign, as shown in Eq. (23).

$$a = \frac{\Sigma|x_i|}{n} \qquad (23)$$

When the averaging process is carried out with the use of Eq. (19), Eq. (24) is obtained.

$$a = \int_{-\infty}^{\infty}|x|N\,dx = \frac{2h}{\sqrt{\pi}}\int_0^{\infty}xe^{-h^2x^2}\,dx = \frac{1}{h\sqrt{\pi}} \qquad (24)$$

The measure of precision h has already been defined. It is an inverse measure of the width of the Gaussian curve, but it cannot be established without ambiguity from a finite set of measurements X_i. If the most probable value of h is calculated, it turns out to be the root-mean-square error given by Eq. (25).

$$m = \sqrt{\Sigma x_i^2/n} \qquad (25)$$

Comparison with Eq. (24) shows that Eq. (26) is

$$a = m\sqrt{2/\pi} \qquad (26)$$

valid. The quantity m is identical with what was previously called the standard deviation.

The probable error r is defined as follows: It

marks that value of x which divides the area under the curve $N(x)$ between zero and infinity into two equal parts. Hence, r is an error such that a given error x_i has an equal chance of being greater or smaller than r. Mathematically, the value of r is found from Eq. (27).

$$\text{erf } (hr) = 1/2 \qquad (27)$$

The numerical relations between r, m, and a are given by Eqs. (28).

$$r = 0.6745m = 0.8453a$$
$$m = 1.4826r = 1.2533a \qquad (28)$$
$$a = 0.7979m = 1.1829r$$

Probability in statistical mechanics. A complex physical system made up of many constituents, for example, molecules, obeys the empirical laws of thermodynamics. Statistical mechanics is that science which attempts to explain these laws by an appeal to the laws of ordinary mechanics. In doing so, it encounters problems of the following sort.

Suppose one wishes to explain why a gas exerts a pressure upon the walls of a container. Pressure is the momentum lost by the molecules per unit time as they strike a unit area of wall. The molecules are very numerous, and their momentum losses vary erratically from instant to instant of collision, and from point to point on the surface. It is necessary, therefore, that some average be taken. Should this average be over the different collisions which a single molecule experiences in time? Or should one take the average at a given time over the whole area under consideration? Clearly, there are many ways of computing the average, each involving a particular collective in the aforementioned sense and each requiring a specification of elementary probabilities.

The science of mechanics does not clearly dictate which collective is the proper one to be used, and there is a considerable latitude of choice. One of these collectives which has proved most successful, the ensemble of J. W. Gibbs, will be discussed briefly.

Phase space. The motion of a single molecule is described in terms of two variables—its position and its momentum at any given time. In the simplest case of motion along a single axis, for example, the x axis, two numbers suffice to describe the motion; they are the position x and the momentum p_x. If a plane is constructed with x and p_x laid off on two perpendicular axes within that plane, the motion of the molecule is represented by a curve in the plane, and the plane is called the phase space of the moving molecule. A molecule whose motion is in a plane requires x, y, p_x, and p_y, that is, four numbers, for its complete description, and these four numbers define a point in a phase space of four dimensions. Similarly, a molecule moving in three-dimensional space has a phase space of six dimensions (axes x, y, z, p_x, p_y and p_z) and its dynamic behavior is depicted by a curve in this six-dimensional space.

This idea can be generalized and applied to a gas containing N molecules. The phase space of the gas will have $6N$ dimensions, and the motion of the entire gas corresponds to the trajectory of a single point in this $6N$-dimensional space. A single point within it describes the physical condition of the entire gas.

Ensembles. Here Gibbs introduced the notion of an ensemble. He imagined a great number of thermodynamic systems, all similar to the given one. If the latter be a vessel filled with gas, he imagined a very large number of similar vessels, all filled with the same quantity of the same gas. This collection of imaginary vessels is called an ensemble. Each member of the ensemble will have its fate represented by a point moving in $6N$-dimensional phase space, and the whole ensemble, when viewed in that space, will appear like a cloud of dust, with each individual dust particle following its own path. The density of this cloud of dust will differ from place to place and will change in time at any given place. From the laws of mechanics it may be shown that this imaginary cloud of dust behaves like an incompressible fluid.

There is a set of conditions, however, under which the cloud will not change its density in time, even though its individual points are in motion. One such condition amounts to the existence of a special density distribution known as the canonical distribution. It has the simple form given by Eq. (29), where H is the energy, T the temperature of the gas, and k is Boltzmann's constant.

$$D(x_1 \cdots p_n) =$$
$$\text{constant} \times \exp \left[\frac{-H(x_1 \cdots p_n)}{kT} \right] \qquad (29)$$

At this point, contact is made with the earlier considerations. All dynamical variables, such as the pressure of the preceding example, which need to be averaged in order to correspond to the observables of thermodynamics, are to be averaged over the probability distribution D. When this is done, the laws of thermodynamics follow; in that sense, Gibbs' probability distribution provides an explanation of thermodynamics.

The success of Gibbs' theory in classical mechanics is remarkable. In order for it to be applicable, however, to systems which follow the laws of quantum mechanics, certain modifications are necessary. *See* BOLTZMANN STATISTICS; QUANTUM STATISTICS; STATISTICAL MECHANICS; STATISTICS.

[HENRY MARGENAU]

Bibliography: A. B. Clarke and R. L. Disney, *Probability and Random Processes for Engineers and Scientists*, 1970; A. G. Frodesen et al., *Probability and Statistics in Particle Physics*, 1979; F. Mandl, *Statistical Physics*, 1971; A. Papoulis, *Probability, Random Variables and Stochastic Processes*, 1965; J. B. Thomas, *An Introduction to Applied Probability and Random Processes*, 1971, reprint 1980.

Product of inertia

The product of inertia of area A relative to the indicated XY rectangular axes is $I_{XY} = \int xy\, dA$ (Fig. 1). The product of inertia of the mass contained in volume V relative to the XY axes is $I_{XY} = \int xy\, \rho\, dV$ (Fig. 2). Similarly for I_{YZ} and I_{ZX}.

The product of inertia of area, like moment of inertia, is measured in quartic length units such as ft^4; for mass, it is measured in mass multiplying length squared units as g-cm^2. Unlike moment of inertia, the product of inertia may be positive or negative.

Relative to principal axes of inertia, the product

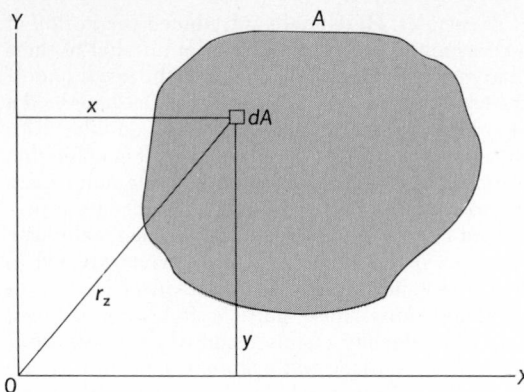

Fig. 1. Product of inertia of an area.

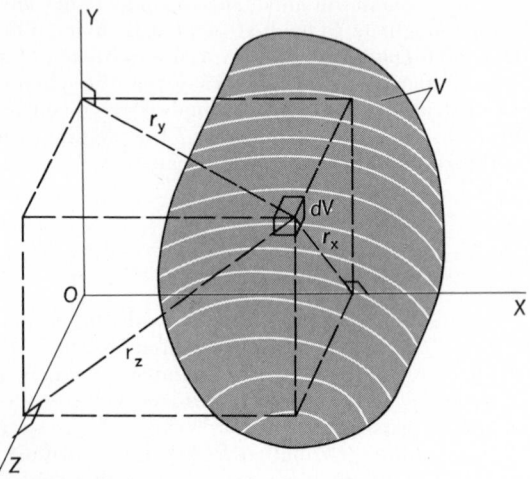

Fig. 2. Product of inertia of a volume.

of inertia of a figure is zero. If a figure is mirror symmetrical about a YZ plane, $I_{ZX} = I_{XY} = 0$. *See* MOMENT OF INERTIA.   [NELSON S. FISK]

Progression

Ordered, countable sets of numbers, x_1, x_2, x_3, . . . , not necessarily all different. In general such sets are called sequences, whereas the term progression is usually confined to the special types: the arithmetic, in which the difference $x_k - x_{k-1}$ between successive terms is constant; the geometric, in which the ratio x_k / x_{k-1} is constant; and the harmonic, in which the reciprocals of the terms are in arithmetic progression.

Arithmetic progressions. If the first term is a and the common difference b,

$$x_1 = a, x_2 = a + b, x_3 = a + 2b, \ldots, \qquad (1)$$
$$x_n = a + (n-1)\, b, \ldots$$

In the sum of n terms S_n, two terms equidistant from the ends always have the same sum $x_1 + x_n$; hence $2S_n = n(x_1 + x_n)$ and

$$S_n = n \frac{x_1 + x_n}{2} = n\left(a + \frac{n-1}{2}\, b\right)$$

If x_1, x_2, x_3 are in arithmetic progression, $x_2 = (x_1 + x_3)/2$ is called the arithmetic mean of x_1 and x_3.

Geometric progressions. If the first term is a and the common ratio r,

$$x_1 = a, x_2 = ar, x_3 = ar^2, \ldots, \qquad (2)$$
$$x_n = ar^{n-1}, \ldots$$

Excluding the case $r = 1$ (when all terms are the same), the sum S_n of n terms satisfies $S_n - rS_n = a - ar^n$; hence,

$$S_n = a \frac{1 - r^n}{1 - r}$$

If $|r| < 1$, $r^n \to 0$ as $n \to \infty$; hence the sum of the infinite geometric series

$$\sum_{n=1}^{\infty} ar^{n-1} = \frac{a}{1-r} \qquad |r| < 1$$

If x_1, x_2, x_3 are in geometric progression, $x_2 = \sqrt{x_1 x_3}$ is called the geometric mean of x_1 and x_3. Since

$$(\sqrt{x_1} - \sqrt{x_3})^2 = x_1 + x_3 - 2\sqrt{x_1 x_3} \geqq 0$$

and

$$\frac{x_1 + x_3}{2} \geqq \sqrt{x_1 x_3}$$

the arithmetic mean of two unequal positive numbers exceeds their geometric mean.

The arithmetic mean A and the geometric mean G of n positive numbers are defined as

$$A = \frac{x_1 + x_2 + \cdots + x_n}{n} \quad \text{and} \quad G = \sqrt[n]{x_1 x_2 \cdots x_n}$$

Also, $A \geqq G$.

Harmonic progression. The reciprocals of sequence (1) form a harmonic progression. There is no compact expression for the sum of n terms. If x_1, x_2, x_3 are in harmonic progression,

$$x_2 = \frac{2x_1 x_3}{x_1 + x_3}$$

is called their harmonic mean.

Sum sequence. A general method of summing a sequence of n terms depends upon a theorem in the difference calculus which is the analog of the fundamental theorem of the differential calculus.

If $x_n = f(n)$ is defined for $n = 0, 1, 2, \ldots,$ the difference $f(n)$ is defined as

$$\Delta f(n) = f(n+1) - f(n)$$

Δ is a linear operator; that is,

$$\Delta[af(n) + bg(n)] = a\,\Delta f(n) + b\,\Delta g(n)$$

In the difference calculus the factorial powers $n^{(k)}$ for integral k are defined by

$$n^{(k)} = \begin{cases} n!/(n-k)! & k < n \\ n! & k = n \\ 0 & k > n \end{cases}$$

Thus

$$n^{(k)} = n(n-1)(n-2) \cdots (n-k+1) \\ \qquad\qquad\qquad 0 < k \leqq n$$

$$n^{(-k)} = \frac{1}{(n+1)(n+2) \cdots (n+k)} \qquad k > 0$$

In particular

$$n^{(0)} = 1, n^{(1)} = n, n^{(-1)} = 1/(n+1)$$

Moreover, $n^{(k)}$ satisfies the functional equation

$$n^{(k)} \cdot (n-k)^{(h)} = n^{(h)} \cdot (n-h)^{(k)} = n^{(h+k)}$$

This leads to the general definition of the factorial power

$$x^{(k)} = \frac{\Gamma(x+1)}{\Gamma(x+1-k)}$$

for all real values of x and k for which the gamma functions exist. When both gammas are infinite (for arguments $0, -1, -2, \ldots$), define $x^{(k)} = \lim (x+\boldsymbol{\epsilon})^{(k)}$ as $\varepsilon \to 0$; thus

$$(-1)^{(2)} = \lim (\varepsilon - 1)^{(2)} = \lim \Gamma(\varepsilon)/\Gamma(\varepsilon - 2)$$
$$= \lim (\varepsilon - 1)(\varepsilon - 2) = 2!$$

The last two entries of Table 1 follow from

$$\Delta e^{in\alpha} = (e^{i\alpha} - 1)e^{in\alpha} = 2i \sin \alpha/2 e^{i(n+1/2)\alpha}$$

on taking real and imaginary parts.

Table 1. Differences

$f(n)$	$\Delta f(n)$
constant	0
r^n	$(r-1)r^n$
$n^{(k)}$	$kn^{(k-1)}$
$\cos n\alpha$	$-2 \sin \tfrac{1}{2}\alpha \sin (n+\tfrac{1}{2})\alpha$
$\sin n\alpha$	$2 \sin \tfrac{1}{2}\alpha \cos (n+\tfrac{1}{2})\alpha$

If $\Delta F(n) = f(n)$, $F(n)$ is called the antidifference of $f(n)$ and written $\Delta^{-1}f(n)$. Two antidifferences of $f(n)$ differ at most by a constant (or by a periodic function of period 1 for functions $f(x)$ of a continuous variable). Table 1 implies the table of antidifferences (Table 2).

Table 2. Antidifferences

$f(n)$	$\Delta^{-1}f(n)$	
0	constant	
r^n	$\dfrac{r^n}{r-1}$	$(r \neq 1)$
$\begin{cases} n^{(k)} \\ \\ 1 \end{cases}$	$\dfrac{n^{(k+1)}}{k+1}$ n	$(k \neq -1)$ $(k=0)$
$\cos n\alpha$	$\dfrac{\sin (n-\tfrac{1}{2})\alpha}{2 \sin \alpha/2}$	
$\sin n\alpha$	$\dfrac{\cos (n-\tfrac{1}{2})\alpha}{2 \sin \alpha/2}$	

Just as antiderivatives are used to compute definite integrals, antidifferences are used to compute definite sums:

$$\sum_{n=p}^{q} f(n) = \Delta^{-1}f(n) \Big|_{p}^{q+1} = F(q+1) - F(p)$$

The proof is immediate on replacing $f(n)$ by $\Delta F(n) = F(n+1) - F(n)$ and performing the indicated summation.

Example 1. The arithmetic progression, as in notation (1):

$$\sum_{n=1}^{N} a + b(n-1) = (a-b)n + \tfrac{1}{2}bn^{(2)} \Big|_{1}^{N+1}$$
$$= (a-b)N + \tfrac{1}{2}b(N+1)N$$

Example 2. The geometric progression, as in notation (2):

$$\sum_{n=1}^{N} ar^{n-1} = \frac{ar^{n-1}}{r-1} \Big|_{1}^{N+1} = a \frac{r^N - 1}{r-1}$$

Example 3. The cosine sequence:

$$\sum_{n=1}^{N} \cos n\alpha = \frac{\sin (n-\tfrac{1}{2})\alpha}{2 \sin \tfrac{1}{2}\alpha} \Big|_{1}^{N+1}$$
$$= \frac{\sin (N+\tfrac{1}{2})\alpha - \sin \tfrac{1}{2}\alpha}{2 \sin \tfrac{1}{2}\alpha}$$

In order to sum a polynomial $P(n)$ of degree k, $P(n)$ may be expressed in terms of factorial powers $n^{(1)}, n^{(2)}, \ldots, n^{(k)}$ by Newton's theorem:

$$P(n) = P(0) + \frac{\Delta P(0)}{1!} n^{(1)} + \frac{\Delta^2 P(0)}{2!} n^{(2)}$$
$$+ \cdots + \frac{\Delta^k P(0)}{k!} n^{(k)}$$

The differences of $P(n)$ when $n=0$ are computed as in Table 3.

Table 3. Sequence of cubes

n	0	1	2	3	
$P(n)$	0	1	8	27	$P(0) = 0$
$\Delta P(n)$	1	7	19		$\Delta P(0) = 1$
$\Delta^2 P(n)$	6	12			$\Delta^2 P(0)/2! = 3$
$\Delta^3 P(n)$	6				$\Delta^3 P(0)/3! = 1$

In Table 3, $P(n) = n^3$ and hence

$$n^3 = n^{(1)} + 3n^{(2)} + n^{(3)}$$

$$\sum_{n=1}^{N} n^3 = \tfrac{1}{2}n^{(2)} + n^{(3)} + \tfrac{1}{4}n^{(4)} \Big|_{1}^{N+1}$$

$$= \tfrac{1}{4}n^2(n-1)^2 \Big|_{1}^{N+1} = \tfrac{1}{4}(N+1)^2 N^2$$

Summation by parts,

$$\Delta^{-1}[f(n) \Delta g(n)]$$
$$= f(n) g(n) - \Delta^{-1}[g(n+1) \Delta f(n)]$$

is frequently useful in finding antidifferences. With $g(n) = (-1)^n$, $\Delta g(n) = 2(-1)^{n-1}$, this gives

$$2\Delta^{-1}[(-1)^{n-1}f(n)]$$
$$= (-1)^n f(n) - \Delta^{-1}[(-1)^{n-1} \Delta f(n)]$$

When $P(n)$ is a polynomial of degree k, $\Delta^{k+1}P(n) = 0$, and this formula repeatedly applied gives

$$\Delta^{-1}[(-1)^{n-1}P(n)]$$

and hence the sum $\sum_{n=1}^{N} (-1)^{n-1}P(n)$. The following tabulation permits the summation of alternating powers $(-1)^{n-1}n^k$:

k	$\Delta^{-1}[(-1)^{n-1}n^k]$
1	$\tfrac{1}{2}(-1)^n(n-\tfrac{1}{2})$
2	$\tfrac{1}{2}(-1)^n n(n-1)$
3	$\tfrac{1}{2}(-1)^n(n-\tfrac{1}{2})(n^2-n-\tfrac{1}{2})$
4	$\tfrac{1}{2}(-1)^n n(n-1)(n^2-n-1)$
5	$\tfrac{1}{2}(-1)^n(n-\tfrac{1}{2})(n^2-n-1)^2$

Summation by parts also gives antidifferences

such as $\Delta^{-1}[r^n P(n)]$. Thus $\Delta^{-1}(nr^n) = r^{n-1}p(n-p)$, where $p = r/(r-1)$. *See* SERIES.

[LOUIS BRAND]

Bibliography: L. Brand, *Differential and Difference Equations*, 1966; C. Jordan, *Calculus of Finite Differences*, 1965.

Proton

A particle that is the positively charged constituent of ordinary matter. Together with the neutron, the proton is the building stone of all atomic nuclei; a single proton constitutes the nucleus of the hydrogen atom. The most important properties that characterize the proton are its charge, which is identical in magnitude but of opposite sign to that of the electron (the negatively charged constituent of ordinary matter) and has the value of 4.8033×10^{-10} esu $= 1.6022 \times 10^{-19}$ coulomb; its mass, 1.6726×10^{-24} g $= 1836.1 m_e$ (m_e is the mass of the electron); its spin, $(1/2)\hbar = (1/2)$ 1.0546×10^{-27} erg-sec ($\hbar$ is Planck's constant h divided by 2π); its magnetic moment, 1.4106×10^{-23} erg/gauss; its lifetime, which, according to all available evidence, is infinite; and the fact that it obeys the Pauli exclusion principle, that is, it is a fermion (obeys Fermi-Dirac statistics). *See* ELEMENTARY PARTICLE.

Mass and charge measurements. The determination of the mass and charge of the proton can be made, with different degrees of precision, in many ways. Deflection of proton beams in electric and magnetic fields gives the ratio of the charge to the mass of the proton. The fact that the hydrogen atom is neutral guarantees the equality is absolute value of the charge of the electron and of the proton, and thus every method for measuring the charge of the electron, such as the celebrated oil-drop experiment of R. A. Millikan (1909), also gives the charge of the proton. Spectroscopic observations also contribute to this determination, mostly giving the ratio between the charge and mass of the electron. The spin of the proton may be ascertained by many diverse effects, ranging from the intensity in the rotational lines of the band spectra of H_2 to the specific heat of liquid hydrogen. The magnetic moment may be measured directly by molecular beam methods, by detailed spectroscopic studies of the hydrogen atom spectrum, or by nuclear induction methods. Moreover, all methods which give Avogadro's number indirectly give in addition the mass of the proton. Indeed, the strict connection and interdependence of many lines of approach to such quantities as the charge and mass of the proton are among the strongest and most striking supports of the modern theories of atomic and nuclear physics. *See* ATOMIC CONSTANTS.

Charge distribution in proton. For many purposes it is sufficient to approximate a proton with a point charge, but it is now possible to analyze its structure more accurately. Scattering of high-energy particles, mainly electrons or muons, shows that the protonic electric charge is extended over a region of a radius of about 10^{-13} cm. If one assumes that the charge density decreases as exp $(-r/r_0)$, r_0 is 0.23×10^{-13} cm. The magnetic moment also may be interpreted as due to a density of magnetization extended over a sphere of similar size. This analysis may be refined by considering the Fourier transform of the charge density called the form factor. This is a function of the momentum transfer of the impinging particle.

Range of protons in matter. Protons of high velocity lose their energy in matter by several mechanisms. Occasionally they strike another nucleus and then may be elastically scattered or may produce a nuclear reaction. These events drastically alter the proton energy and if the proton is in a beam, it is removed from the beam. In addition to this type of event, protons (as well as all other charged particles of mass considerably larger than that of the electron) lose energy by imparting it to the electrons of the medium in which they move, without being appreciably deflected from their trajectory; the energy loss simply slows down the heavy particle. Ultimately the heavy particle (proton) comes to rest and the distance it travels between the point where it has an energy E and the point where it comes to rest is called the range of the proton. The range is a function of the energy of the proton and of the medium in which it moves. A semiempirical formula connecting the energy and range in air (normal temperature and pressure) for protons is given below, where R is in meters and E

$$R = \left(\frac{E}{9 \cdot 3} \right)^{1.8}$$

is in MeV. This relation is valid between a few MeV and about 200 MeV. For other substances it is necessary to consider the relative stopping power, which is a number giving the reciprocal of the ratio of the thickness in g/cm² of two layers producing the same energy loss. The stopping power decreases with Z, the atomic number. *See* SCATTERING EXPERIMENTS (NUCLEI).

[EMILIO G. SEGRÉ]

Pyroelectricity

The property of certain crystals to produce a state of electric polarity by a change of temperature. Certain dielectric (electrically nonconducting) crystals develop an electric polarization (dipole moment per unit volume) ΔP when they are subjected to a uniform temperature change ΔT. For a small change ΔT, the components ΔP_i of the polarization vector are given by Eq. (1). This pyroelec-

$$\Delta P_i = p_i \Delta T \qquad i = x, y, z \qquad (1)$$

tric effect occurs only in crystals which lack a center of symmetry and also have polar directions (that is, a polar axis). These conditions are fulfilled for 10 of the 32 crystal classes. Typical examples of pyroelectric crystals are tourmaline, lithium sulfate monohydrate, cane sugar, and ferroelectric barium titanate.

Pyroelectric crystals can be regarded as having a built-in or permanent electric polarization. When the crystal is held at constant temperature, this polarization does not manifest itself because it is compensated by free charge carriers that have reached the surface of the crystal by conduction through the crystal and from the surroundings. However, when the temperature of the crystal is raised or lowered, the permanent polarization changes, and this change manifests itself as pyroelectricity. There is no easy way to determine the magnitude of the total permanent polarization, except for these special pyroelectric crystals, called ferroelectrics, in which the polarization can be reversed by an electric field. *See* FERROELECTRICS.

The magnitude of the pyroelectric effect depends upon whether the thermal expansion of the crystal is prevented by clamping or whether the crystal is mechanically unconstrained. In the clamped crystal, the primary pyroelectric effect is observed, whereas in the free crystal, a secondary pyroelectric effect is superposed upon the primary effect. The secondary effect may be regarded as the piezoelectric polarization arising from thermal expansion. The secondary effect is generally much larger than the primary effect. When heated nonuniformly, even piezoelectric crystals produce a polarization due to the temperature gradients and corresponding nonuniform stresses and strains. This phenomenon is called tertiary, or false, pyroelectricity, because it may be found in nonpyroelectric crystals, that is, crystals which have no polar directions. *See* PIEZOELECTRICITY.

In a typical pyroelectric crystal such as tourmaline, a temperature change of 1°C produces at room temperature a polarization of about 10^{-5} coulomb/m². The same amount of polarization can be produced in a crystal at constant temperature by applying an external electric field of about 70 kv/m. The pyroelectric effect is much larger in ferroelectrics, in particular at temperatures close to the Curie point. Applications of the pyroelectric effect, such as pyroelectric laser calorimeters and sensitive infrared detectors, have been proposed. The figure of merit, defined as the pyroelectric coefficient divided by the dielectric constant, is high for ferroelectrics such as triglycine sulfate. Thermal response times of 2 μsec and less have been reported. *See* DIELECTRIC CONSTANT.

As can be shown in thermodynamics, the pyroelectric effect has an inverse, the linear electrocaloric effect. A temperature change ΔT results when the permanent polarization is altered by an externally applied electric field ΔE, as in Eq. (2). In

$$\Delta T = \frac{T}{\rho C_P} \sum_i p_i \Delta E_i \qquad i = x, y, z \qquad (2)$$

this expression, ρ is the density, C_P the specific heat at constant pressure, p_i are the pyroelectric coefficients, and T is the absolute temperature. The temperature changes that can be realized in typical pyroelectrics are of the order of magnitude 0.01°C.

[H. GRANICHER]

Bibliography: S. B. Lang, *Sourcebook of Pyroelectricity*, 1974; J. F. Nye, *Physical Properties of Crystals: Their Representation by Tensors and Matrices*, 1960.

Pyrometer

A temperature-measuring device, originally an instrument that measures temperatures beyond the range of thermometers, but now in addition a device that measures thermal radiation in any temperature range. This article discusses radiation pyrometers.

Technique. Figure 1 shows a very simple type of radiation pyrometer. Part of the thermal radiation emitted by a hot object is intercepted by a lens and focused onto a thermopile. The resultant heating of the thermopile causes it to generate an electrical

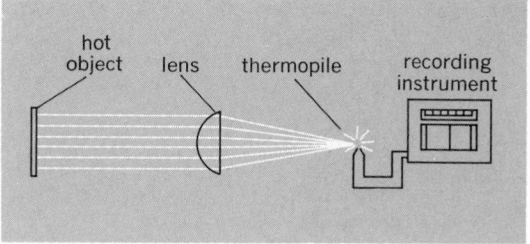

Fig. 1. Elementary radiation pyrometer. (*From D. M. Considine, ed., Process Instruments and Controls Handbook, McGraw-Hill, 1957*)

signal (proportional to the thermal radiation) which can be displayed on a recorder.

Unfortunately, the thermal radiation emitted by the object depends not only on its temperature but also on its surface characteristics. The radiation existing inside hot, opaque objects is so-called blackbody radiation, which is a unique function of temperature and wavelength and is the same for all opaque materials. However, such radiation, when it attempts to escape from the object, is partly reflected at the surface. The fraction reflected back into the interior (and hence also the fraction emitted) depends on the type of material, surface roughness, surface films, and microstructure.

The pyrometer readings, then, are ambiguous, since a measured rate of thermal radiation may be due to a temperature of 400°C and an internal reflectance of 0.80 (emittance of 0.20) on the one hand, or to a temperature of 500°C and an internal reflectance of 0.90 (emittance of 0.10) on the other hand. In order to use the output of the pyrometer as a measure of target temperature, the effect of the surface characteristics must be eliminated from the measurements. *See* BLACKBODY.

This elimination is accomplished as follows. A cavity is formed in an opaque material and the pyrometer is sighted on a small opening extending from the cavity to the surface. The opening has no surface reflection, since the surface has been eliminated. Such a source is called a blackbody source, and is said to have an emittance of 1.00. Any other object, when its surface temperature is measured, has some actual surface reflectance and therefore has an emittance less than 1.00.

By attaching thermocouples to the blackbody source, a curve of pyrometer output voltage versus blackbody temperature can be constructed. Similarly, for a given material (without the cavity) thermocouples can be attached to the surface and the radiation output to the pyrometer output voltage determined at known material temperatures. The emittance factor is taken to be this voltage divided by the voltage output observed for the blackbody at the same temperature. Then, in making measurements on this material remotely and without contact (as on a moving process line), the voltage readings are divided by the proper emittance factor to obtain the true temperature from the blackbody curve.

The effects of target size and target distance on the pyrometer readings must also be eliminated. One solution is to design the field of view to increase as the square of the target distance. Thus, as long as the field of view is filled by the target,

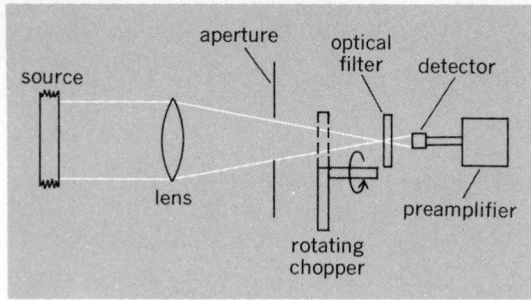

Fig. 2. Narrow-band pyrometer.

the increasing target size just compensates for the decreased signal per unit target area as the target distance increases.

Types. Pyrometers can be classified generally into types requiring that the field of view be filled, such as narrow-band and total-radiation pyrometers; and types not requiring that the field of view be filled, such as optical and ratio pyrometers. The latter depend upon making some sort of comparison between two or more signals.

Narrow-band (spectral) pyrometers. These pyrometers (Fig. 2) respond to a limited band of wavelengths. (For example, a typical lead sulfide pyrometer will have a filter that limits its spectral response to the $1.8-2.7$ μm region.) The output may vary as the 10th to 15th power of the temperature. Time constants are of the order of milliseconds (as in lead sulfide) or microseconds (as in silicon cells). Fields of view are of the order of 1-cm diameter at 1 m. The detectors are usually of the photoconductive or photovoltaic type. *See* RADIOMETRY.

Total-radiation pyrometers. The lens, or window and mirror, because of limited spectral transmission, allows only broad-band response, rather than "total" radiation response. Hence the output usually varies as about the 4.5 or 5th power. Detectors are usually of the thermal type (that is, the detector generates an output because of a rise in its temperature), such as a thermopile. Time constants are usually of the order of 1 sec or longer. Fields of view of about 5 cm diameter at 1 m are typical (Fig. 3).

Optical pyrometer. This instrument should more strictly be called the disappearing-filament pyrometer. In operation, an image of the target is focused in the plane of a wire that can be heated electrically. A rheostat is used to adjust the current through the wire until the wire blends into the

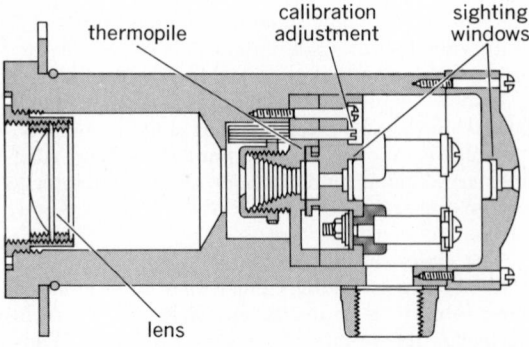

Fig. 3. Radiaton pyrometer. (*Honeywell Inc.*)

image of the target (equal brightness condition), and the temperature is then read from a calibrated dial on the rheostat (Fig. 4).

Ratio, or "two-color," pyrometer. This pyrometer makes measurements in two wavelength regions and electronically takes the ratio of these measurements. If the emittance is the same for both wavelengths, the emittance cancels out of the result, and the true temperature of the target is obtained. This so-called graybody assumption is sufficiently valid in some cases so that the "color temperature" measured by a ratio pyrometer is close to the true temperature.

A given target may be said to have a certain "brightness temperature," "radiation temperature," "color temperature" and, of course, true temperature, depending on how the temperature is measured. These may all differ before the readings are corrected for emittance. The terms luminance temperature and distribution temperature have been adopted to replace brightness temperature and color temperature, respectively.

Radiation pyrometers are particularly useful for (1) temperatures above the practical operating

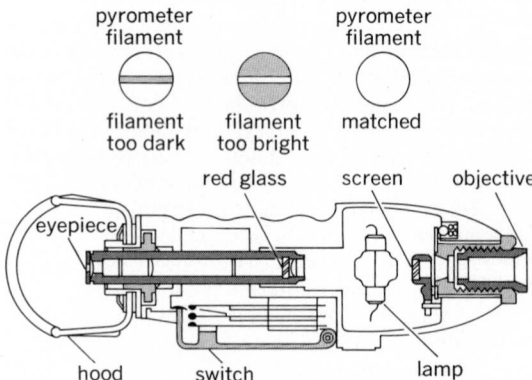

Fig. 4. Optical pyrometer. (*Leeds and Northrup Co.*)

ranges of thermocouples; (2) environments that contaminate and limit the useful life of thermocouples; (3) moving targets; (4) visible targets not easily accessible; (5) targets that would be damaged by contact or insertion; (6) average temperatures of large areas; and (7) rapidly changing temperatures (using fast response detectors).

Errors in temperatures measured by radiation pyrometers can be caused by (1) uncertainty in the emittance value of the target; (2) dust, steam, or smoke in the radiation path; (3) dirt on lens or window; (4) water or dirt on the target surface; and (5) reflection of spurious radiation from the target.
[THOMAS P. MURRAY]

Bibliography: H. H. Plumb et al. (eds.), *Temperature: Its Measurement and Control in Science and Industry*, 1972; A. Stimson, *Photometry and Radiometry for Engineers*, 1974.

Q (electricity)

Often called the quality factor of a circuit, Q is defined in various ways, depending upon the particular application. In the simple RL and RC series circuits, Q is the ratio of reactance to resistance, as in Eqs. (1), where X_L is the inductive reactance,

$$Q = X_L/R \quad Q = X_C/R \quad \text{(a numerical value)} \quad (1)$$

X_C is the capacitive reactance, and R is the resistance. An important application lies in the dissipation factor or loss angle when the constants of a coil or capacitor are measured by means of the alternating-current bridge.

Q has greater practical significance with respect to the resonant circuit, and a basic definition is given by Eq. (2), where Q_0 means evaluation at res-

$$Q_0 = 2\pi \frac{\text{max stored energy per cycle}}{\text{energy lost per cycle}} \quad (2)$$

onance. For certain circuits, such as cavity resonators, this is the only meaning Q can have.

For the RLC series resonant circuit with resonant frequency f_0, Eq. (3) holds, where R is the total

$$Q_0 = 2\pi f_0 L/R = 1/2\pi f_0 CR \quad (3)$$

circuit resistance, L is the inductance, and C is the capacitance. Q_0 is the Q of the coil if it contains practically the total resistance R. The greater the value of Q_0, the sharper will be the resonance peak.

The practical case of a coil of high Q_0 in parallel with a capacitor also leads to $Q_0 = 2\pi f_0 L/R$. R is the total series resistance of the loop, although the capacitor branch usually has negligible resistance.

In terms of the resonance curve, Eq. (4) holds,

$$Q_0 = f_0/(f_2 - f_1) \quad (4)$$

where f_0 is the frequency at resonance, and f_1 and f_2 are the frequencies at the half-power points. *See* RESONANCE (ALTERNATING-CURRENT CIRCUITS).

[BURTIS L. ROBERTSON]

Quantization

A term applied to the transition from a description of a system of particles or fields in the classical approximation where canonically conjugate variables commute to a description where these variables are treated as noncommuting operators. In this transition, Poisson bracket relations among dynamical variables are replaced by commutation relations. For a point particle these variables may be taken as the particle coordinates and corresponding components of the momentum. Under quantization they become linear operators satisfying, for example, for the x components, $xp_x - p_x x = i\hbar$. So-called second quantization is the analogous treatment of a field, such as the electromagnetic field solution of Maxwell's equations, the wave-function solution of Schrödinger's equation, or the solution of Dirac's equation, with commutation relations imposed upon the field as the canonical coordinate and with the generalized momentum defined through the Lagrangian. The wave equation, especially in the latter two cases, resulted from the first quantization. *See* CANONICAL TRANSFORMATIONS; LAGRANGE'S EQUATIONS; NONRELATIVISTIC QUANTUM THEORY; QUANTUM ELECTRODYNAMICS; QUANTUM FIELD THEORY.

A result of treating a system through quantum mechanics is that some observable quantities become quantized, or assume discrete values. Independently of the above operator discussion, the process leading to discrete values for an observable is also called quantization. For example, en-

ergy and angular momentum historically were first treated this way through quantization postulates hypothesizing discrete eigenvalue spectra. *See* ATOMIC STRUCTURE AND SPECTRA; QUANTUM MECHANICS.

[KENNETH E. LASSILA]

Quantized vortices

A type of flow pattern exhibited by superfluids, such as liquid ^{4}He below 2.17 K. The term vortex designates the familiar whirlpool pattern where the fluid moves circularly around a central line and the velocity diminishes inversely proportionally to the distance from the center. The strength of a vortex is determined by the circulation κ, which is the line integral of the velocity around any path enclosing the central line. For an ordinary vortex, κ can possess any value; for a superfluid vortex, κ is restricted to a quantized multiple of Planck's constant h divided by m, the mass of the helium atom. Hence the expression quantized vortex line. Although h and m are microscopic (that is, atomic) quantities, their ratio h/m is rather large, being equal to 10^{-3} cm^2 s^{-1}. *See* VORTEX.

Theory. The possible existence of quantized vortices was suggested on theoretical grounds as early as 1949. A superfluid is believed to be characterized by a macroscopic (that is, large-scale) quantum-mechanical wave function ψ. This wave function locks the superfluid into a coherent state. Considerations on how a uniform flow might affect ψ lead to realization that the superfluid velocity v_s is intimately connected to the manner in which ψ might vary in space. From elementary considerations, one then deduces that quantized vortices should exist.

Since the velocity around the vortex increases without limit as the center is approached, the superfluid density and thus ψ must vanish at the center in order to avoid an infinite energy. Thus the central core of the vortex marks the zeros, or nodal lines, in the macroscopic wave function. *See* NONRELATIVISTIC QUANTUM THEORY; QUANTUM MECHANICS.

Production. Quantized vortex lines are usually produced by rotating a vessel containing superfluid helium. At very low rotation speeds, no vortices exist: the superfluid remains at rest while the vessel rotates. At a certain speed the first vortex appears and corresponds to the first excited rotational state of the system. If the container continues to accelerate, additional quantized vortices will appear. At any given speed the vortices form a regular array which rotates with the vessel.

Experimental investigation. Quantized vortex lines were first detected in the mid-1950s by their influences on superfluid thermal waves traveling across the lines. Shortly thereafter the quantum of circulation was directly measured by studying the precessional motion of a fine vibrating wire immersed in rotating superfluid helium. If no circulation is present, the plane of vibration of a circular-cross-section wire would remain stationary. In the presence of circulation, the vibration plane precesses at a rate determined by the circulation. It has been found that the stable values of circulation are close to h/m.

Mechanical properties. The mechanical properties of vortices have been studied by detecting various types of bending waves in the vortices. In one early experiment, well-defined normal modes were observed, which were consistent with predictions. Since the effective tension of a single vortex is very small, these mechanical measurements have usually been concerned with systems containing many lines. However, some progress has been made in measuring the very small angular momentum of a single line.

Use of ions to probe vortices. In the late 1950s it was discovered that electrons in liquid helium form tiny charged bubbles which can become trapped on the vortex core but can move quite freely along the line. These electron bubbles (often referred to as ions) have been one of the most useful probes of quantized vortices. The most accurate quantitative proof of the quantized nature of the vortices emerged from the discovery that at low temperatures (below 0.5 K) a rapidly moving ion can create a vortex ring. These rings were found to move precisely like classical smoke rings except that the circulation κ was equal to h/m to within 1%.

Detection of single vortex lines. Researchers have been able to use ions to detect single quantized vortex lines. In one experiment the vortices in a rotating vessel were charged with ions, and subsequently the amount of trapped charge was measured. As the vessel slowly accelerates from rest, the trapped charge is seen to increase as a stepwise function of rotation speed, thus proving that the superfluid comes into rotation in a sequence of quantum steps.

In an extension of the ion technique, the trapped ions are pulled out at the top of the vortex lines, accelerated, and focused onto a phosphor screen. The pattern of light thus produced on the phosphor is a map of the position of the vortices where they contact the liquid meniscus. The illustration shows some of the stationary vortex states photographed from the phosphor signal. The arrays of vortices agree well with theoretical predictions.

Vortices in helium-3 and neutron stars. Although quantized vortices have been studied only in superfluid ^{4}He, it is expected that they will be found to exist in superfluid ^{3}He. Because the ^{3}He superfluid exists only at temperatures a thousand times colder than ^{4}He superfluid, the necessary experiments are more difficult to perform. In addition to the helium isotopes, there may exist other quantum liquids. For example, it is believed that the interiors of neutron stars are filled with a very dense superfluid. Since these distant objects are rotating, their interiors should be filled with quantized vortices. Although it is unlikely that scientists can perform laboratory experiments on neutron superfluid, the star's motion itself is to some extent a probe of the interior. Indeed, scientists have been able to explain some features of the neutron stars' motion as being due to the quantized vortices in the interior.

Since quantized vortices should be common to all superfluid systems, more research on this topic can be expected as more exotic superfluid systems are discovered. *See* LIQUID HELIUM; SUPERFLUIDITY. [RICHARD E. PACKARD]

Bibliography: G. A. Williams and R. E. Packard, A technique for photographing vortex positions, *J. Low Temp. Phys.*, 39:553, 1980.

Quantum

A term characterizing an excitation in a wave or field, connoting fundamental particlelike properties such as energy or mass, momentum, and angular momentum for this excitation. In general, any field or wave equation that is quantized, including systems already treated in quantum mechanics that are second-quantized, leads to a particle interpretation for the excitations which are called quanta of the field. This term historically was first applied to indivisible amounts of electromagnetic, or light, energy usually referred to as photons. The photon, or quantum of the electromagnetic field, is a massless particle, best interpreted as such by quantizing Maxwell's equations. Analogously, the electron can be said to be the quantum of the Dirac field through second quantization of the Dirac equation, which also leads to the prediction of the existence of the positron as another quantum of this field with the same mass but with a charge opposite to that of the electron. In similar fashion, quantization of the gravitational field equations suggests the existence of the graviton, a particle which is the quantum of gravitational field. The pi meson or pion was theoretically predicted as the quantum of the nuclear force field by H. Yukawa; other particles associated with nuclear forces have since been found. Another quantum is the quantized lattice vibration, or phonon, which can be interpreted as a quantized sound wave since it travels through a quantum solid or fluid, or through nuclear matter, in the same manner as sound goes through air. The energy of a phonon is $h\nu$, h being Planck's constant (called the quantum of action in "old" quantum mechanics) and ν the frequency of

Stationary configurations of vortices which appear when a cylindrical container of superfluid ^{4}He is rotated about its axis. As the rotation speed increases from *a* to *l*, more vortices appear and the patterns become more complex. (*From E. J. Yarmchuk, M. J. V. Gordon, and R. E. Packard, Observation of stationary vortex arrays in rotating superfluid helium, Phys. Rev. Lett., 43:214–217, 1979*)

oscillation of the matter through which the wave propagates. The energy of a photon is also $h\nu$, with h standing for Planck's constant and ν standing for the frequency of the light wave.

The use of the word, quantum, as an adjective (quantum mechanics, quantum electrodynamics) implies that the particular subject is to be treated according to the modern rules that have evolved for quantized systems. *See* ELEMENTARY PARTICLE; GRAVITATION; GRAVITON; MAXWELL'S EQUATIONS; MESON; NONRELATIVISTIC QUANTUM THEORY; PHONON; PHOTON; QUANTIZATION; QUANTUM ELECTRODYNAMICS; QUANTUM FIELD THEORY; QUANTUM MECHANICS.

[KENNETH E. LASSILA]

Quantum chromodynamics

A theory of the strong ("nuclear") interactions among quarks, which are regarded as fundamental constituents of matter. Quantum chromodynamics (QCD) seeks to explain why quarks combine in certain configurations to form the observed patterns of subnuclear (or "elementary") particles, such as the proton and pi meson. According to this picture, the strong interactions among quarks are mediated by a set of force particles known as gluons. Strong interactions among gluons may lead to new structures that correspond to as yet undiscovered particles. The long-studied nuclear force which binds protons and neutrons together in atomic nuclei is regarded as a collective effect of the elementary interactions among constituents of the composite protons and neutrons. *See* NUCLEAR STRUCTURE.

It is believed that free quarks and gluons cannot be isolated, and it is hoped that quantum chromodynamics will provide a justification for this belief. By construction, quantum chromodynamics embodies many features abstracted from empirical observations. However, new mathematical inventions appear required before all the consequences of the theory can be reliably calculated. In part because of the shortcomings of known techniques for extracting predictions from the theory, quantum chromodynamics has not yet been subjected to rigorous experimental tests. Several qualitative predictions of quantum chromodynamics do seem to have been borne out. Part of the esthetic appeal of the theory is due to the fact that quantum chromodynamics is nearly identical in mathematical structure to quantum electrodynamics (QED) and to the unified theory of weak and electromagnetic interactions put forward by S. Weinberg and A. Salam. This resemblance encourages the hope that a unified description of the strong, weak, and electromagnetic interactions may be at hand. *See* QUANTUM ELECTRODYNAMICS; WEAK NUCLEAR INTERACTIONS.

Gauge theories. At the heart of current theories of the fundamental interactions is the idea of gauge invariance, which draws its name from some early investigations by H. Weyl into a possible connection between scale changes and the equations of electrodynamics. Weyl attempted to deduce electromagnetism from a symmetry principle, the conjectured invariance of physical laws under a change of length scale chosen independently at every position of space and time. This specific undertaking ran afoul of quantum mechanics, but the general strategy and the name

have survived. It is widely believed that gauge theories constructed to embody various symmetry principles represent the correct quantum-mechanical descriptions of the strong, weak, and electromagnetic interactions. *See* SYMMETRY LAWS.

Electromagnetism. The simplest example of a gauge theory is electromagnetism. It may be derived from a symmetry principle as follows. Quantum-mechanical observables do not depend upon the phase of the complex wave function which describes the state of a system. Therefore, one has the freedom to rotate the phase of a wave function by an amount which is the same at all times and all places without affecting the physical consequences of the theory. The choice of phase is thus conventional, as opposed to observable. This is known as a global symmetry principle. It is natural to ask whether it should not be possible to choose this arbitrary convention independently at each point of space-time, again without affecting the physical consequences of the theory. It turns out to be possible to construct a quantum theory which is invariant under local (that is, position- and time-dependent) phase rotations that are proportional to the electric charge of the particles, but only if the theory contains an electromagnetic field with precisely the observed properties as summarized by Maxwell's equations. In the quantum theory, a massless spin-1 particle identified as the photon mediates the electromagnetic interaction. The interactions of matter with electromagnetism thus are essentially prescribed by the requirement of local phase invariance.

Local gauge invariance. Local phase rotations of the kind described above are the simplest examples of local gauge transformations. For a continuous symmetry, global gauge invariance implies the existence of a set of conserved currents. In the case of electromagnetism, it is the electric current which is conserved. A local gauge invariance requires in addition the existence of a massless gauge field corresponding to each conserved current. The photon is the gauge field of electromagnetism, corresponding to the set of phase transformations which forms the one-parameter unitary group U(1). The theory of electromagnetism was codified by James Clerk Maxwell more than 60 years before the local gauge invariance of its equations was discovered. However, it frequently happens in physics that the symmetries respected by a phenomenon are recognized before a complete theory has been developed. The question therefore arises as to whether the notion of local gauge invariance can be used to deduce the theory of nuclear forces.

Yang-Mills theory. This question was addressed in 1954 by C. N. Yang and R. L. Mills, and independently by R. Shaw. Early in the study of nuclear forces it was established that the nuclear interaction is charge-independent; it acts with the same strength between proton and proton, or proton and neutron, or neutron and neutron. This may be understood by saying that the proton and neutron represent two states of the same particle, called the nucleon. Just as an electron can be in a state with spin-up or spin-down, a nucleon can be in a state with the internal quantum number isospin-up (defined as the proton) or isospin-down (defined as the neutron). Charge independence then would reflect the invariance of the strong interactions under isospin rotations, characterized by the group SU(2). If isospin is regarded as a gauge group, local gauge invariance requires the existence of three massless spin-1 gauge particles, corresponding to the three generators of SU(2). The interactions of the gauge particles with nucleons are prescribed by the gauge principle. All of this is entirely parallel to the theory of electromagnetism. What distinguishes this SU(2) gauge theory from its U(1) counterpart is that the SU(2) gauge fields carry isospin and thus couple among themselves, whereas the photon is electrically neutral and does not interact with itself. Interacting gauge fields are an attribute of any theory based upon a nonabelian gauge group. *See* ISOTOPIC SPIN.

Spontaneous symmetry breaking. Its mathematical properties notwithstanding, the Yang-Mills theory was unacceptable as a description of nuclear forces because, as H. Yukawa had shown, they are mediated by massive particles, whereas the gauge particles are required to be massless. Attempts in the early 1960s to generalize the Yang-Mills theory to the newly discovered SU(3) symmetry of the strong interactions encountered similar experimental objections. Gauge theories nevertheless continued to hold considerable appeal for theorists. Beginning in the late 1950s, a succession of gauge theories of the weak interactions appeared. At first these too foundered on the prediction of massless gauge bosons, but it was ultimately learned from the work of P. Higgs and others that spontaneous breakdown of the gauge symmetry would endow the gauge bosons with masses. All the elements were successfully combined in 1967 in the theory of weak and electromagnetic interactions proposed separately by Weinberg and Salam. When in 1971 G. 't Hooft and others demonstrated that spontaneously broken gauge theories were renormalizable, and hence calculable in the same sense as quantum electrodynamics, it stimulated experimental interest in the predictions of the Weinberg-Salam model, and renewed theoretical enthusiasm for gauge theories in general. Spontaneous symmetry breaking was not a cure for the shortcomings of gauge theories of the strong interactions. Instead, thanks to parallel developments described below, a new candidate emerged for the strong gauge group, and with it arose the idea of quantum chromodynamics.

Color. It had been shown in 1963 by M. Gell-Mann and G. Zweig, working independently, that the observed pattern of strongly interacting particles, or hadrons, could be explained if the hadrons were composed of fundamental constituents called quarks. According to the quark model, a baryon such as the proton, which has half-integral spin in units of $\hbar$ (Planck's constant h divided by 2π) is made up of three quarks. An integral-spin meson, such as the pi meson, is made up of one quark and one antiquark. Three varieties (or flavors) of quarks, denoted up, down, and strange, could be combined to make all of the known hadrons, in precisely the families identified according to the eightfold way.

Although the quark model reproduced the properties of the observed hadron states, a theoretical inconsistency arose from the fact that the quarks must be spin-$\frac{1}{2}$ particles. According to the Pauli exclusion principle (first applied to the electrons in

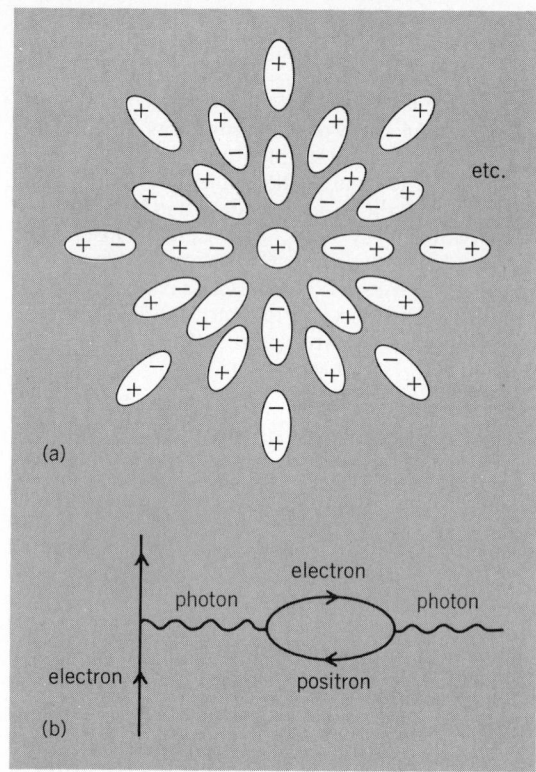

Fig. 1. Screening in electrodynamics. (*a*) Polarization of a dielectric medium by a test charge. (*b*) Feynman diagram contributing to vacuum polarization.

atoms), identical spin-$\frac{1}{2}$ particles cannot occupy the same quantum state. The properties of several of the baryon resonances can be reproduced only by having three quarks of the same flavor in a configuration which is symmetric under the interchange of any pair of quarks, contrary to the exclusion principle. It was shown by the work of O. W. Greenberg and others that the quark model could be made to conform with the Pauli principle if a new distinguishing property was imputed to the quarks. Each quark flavor is thus said to come in three colors, arbitrarily labeled red, blue, and green. (This nomenclature is purely fanciful, and has no connection with the color of visible light.) A baryon composed of a red quark, a blue quark, and a green quark can then have the desired properties and be consistent with the laws of quantum mechanics. In this picture, the mesons are described as colorless quark-antiquark pairs. The color hypothesis is supported by a number of reaction rates which are sensitive to the number of distinct quark species, and hence to the number of colors. *See* COLOR (QUANTUM MECHANICS).

The quarks are the constituents of strongly interacting particles. The leptons, of which the electron and neutrino are the most common examples, are the fundamental particles which do not interact strongly. Each lepton flavor appears in only a single species. In other words, the leptons are colorless. In other respects, leptons resemble quarks: they are spin-$\frac{1}{2}$ particles which have no internal structure, at the current limits of resolution. Color may therefore be regarded as the strong-interaction analog of the electric charge. Color cannot be created or destroyed by any of the known interactions. Like electric charge, it is said to be conserved. *See* LEPTON.

In the face of evidence that color could be regarded as the conserved charge of the strong interactions, it was natural to seek a gauge symmetry which would have color conservation as its consequence. An obvious candidate for the gauge symmetry group is the unitary group SU(3), now to be applied to color rather than flavor. The theory of strong interactions among quarks which is prescribed by local color gauge symmetry is known as quantum chromodynamics. The mediators of the strong interactions are eight massless spin-1 bosons, one for each generator of the symmetry group. These strong-force particles are named gluons because they make up the "glue" which binds quarks together into hadrons. Gluons also carry color, and hence have strong interactions among themselves.

Asymptotic freedom. The theoretical description of the strong interactions has historically been inhibited by the very strength of the interaction, which renders low-order perturbative calculations untrustworthy. In 1973 a remarkable observation was reported by H. D. Politzer and by D. J. Gross and F. Wilczek. While studying the properties of Yang-Mills theories, they found that in many circumstances the effective strength of the interaction becomes increasingly feeble at short distances. For quantum chromodynamics, this implies that the interaction between quarks becomes weak at small separations. This discovery potentially has numerous important consequences, some of which will be described below. It raises the hope that some aspects of the strong interactions might be treated by using familiar computational techniques that are predicated upon the smallness of the interaction strength.

The physical basis for the change in the strength of the strong interaction as a function of the distance may be understood by examining first the corresponding question in electrodynamics, and analyzing the behavior of a dielectric medium containing an electric charge (Fig. 1a). The test charge polarizes the medium. Charges of opposite sign are oriented toward the test charge, while those with like charge are repelled, as illustrated. This familiar screening effect means that the influence of the charge is diminished by the surrounding medium. Viewed from afar (but within the medium), the charge appears smaller in magnitude than its true or unscreened value. Only by inspecting the test charge at short range is it possible to feel its full effects. A closely related phenomenon in quantum electrodynamics is known as vacuum polarization (Fig. 1b). Here the virtual photons surrounding a point charge may dissociate into virtual pairs of charged particles and antiparticles, such as electrons and positrons, which serve to screen the charge at long distances. The polarizability of the vacuum is related to the number of species of charged particles which take part in the screening.

The situation in quantum chromodynamics is complicated by the fact that gluons carry the strong (color) charge, whereas photons are electrically neutral. This means that in addition to a color polarization phenomenon like the related charge screening of quantum electrodynamics (Fig. 2a

and b), it is possible for the color charge of a quark to be shared with the gluon cloud (Fig. 2c and d). Because the color charge is spread out rather than localized, the effective color charge will tend to appear larger at long distances and smaller at short distances. The outcome of the competition between these opposing tendencies depends on the number of gluon species that can share the color charge and on the number of quark types that can screen the color charge. If the color gauge group is SU(3), the net effect is one of antiscreening, that is, of a smaller effective charge at short distances, provided the number of quark flavors is less than 17. Only five quark flavors are now

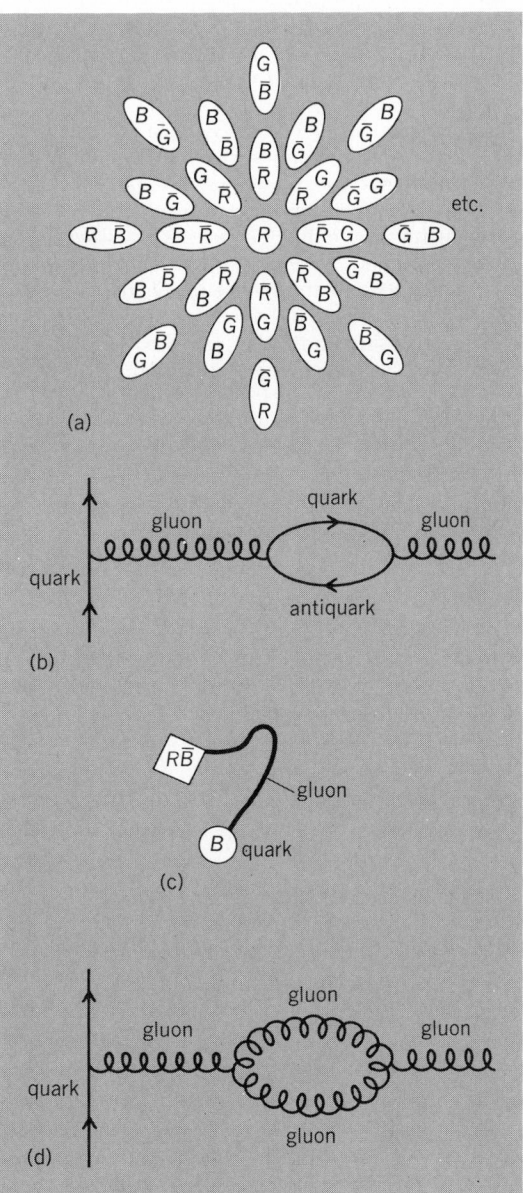

Fig. 2. Screening and antiscreening in quantum chromodynamics. (a) Screening of a colored quark by vacuum polarization. Color charges are denoted by R = red, $\bar{R}$ = anti-red, B = blue, $\bar{B}$ = anti-blue, G = green, $\bar{G}$ = antigreen. (b) Feynman diagram contributing to color polarization of the vacuum. (c) Delocalization of the color charge by gluon radiation. (d) Feynman diagram contributing to color antiscreening.

known. Extremely close to a quark, the effective color charge becomes vanishingly small, so that nearby quarks behave as if they are noninteracting free particles. This is the origin of the term asymptotic freedom.

Experimental consequences. Asymptotic freedom has the immediate consequence, within the context of quantum chromodynamics, of providing a partial justification of the parton model description of violent scattering processes. The parton model was invented to explain features of electron-proton collisions in which many particles are created. Experiments first carried out at the Stanford Linear Accelerator Center in 1968 indicated that in these deeply inelastic collisions the electron "sees" the proton, not as an amorphous whole, but as a collection of structureless entities that have been identified as quarks. According to this picture, an electron scatters from a single parton, free from the influence of any other parton. This is reminiscent of electron scattering from nuclei, in which an electron may scatter from an individual proton or neutron as if it were a free particle. There is an important difference. Protons and neutrons are lightly bound in nuclei, and may be liberated in violent collisions. In contrast, free quarks have not been observed, so they must be regarded as very deeply bound within hadrons. Asymptotic freedom offers a resolution to the paradox of quasifree quarks that are permanently confined. At the short distances probed in deep inelastic scattering, the effective color charge is weak, so the strong interactions between quarks can largely be neglected. As quarks are separated, the effective color charge grows, so the strong interaction becomes more formidable. This characteristic may provide a mechanism for quark confinement, but calculational difficulties have prevented a verification of this conjecture.

Although quantum chromodynamics provides support for the spirit of the parton model, it also exposes the incompleteness of the parton model description. Because quantum chromodynamics is an interacting field theory of quarks and gluons, probes of different wavelengths, which are analogous to microscopes of different resolving power, may map out different structures within the proton. This is indicated in Fig. 3, which shows, under increasing magnification, the virtual dissociation of quarks into quarks and gluons. There is evidence from high-energy muon-nucleon and neutrino-nucleon scattering of changes in the perceived structure of the proton of the kind suggested by quantum chromodynamics. Whether these changes are quantitatively described by the theory remains to be seen. Some indirect evidence for the radiation of gluons from quarks has been reported.

Quarkonium. The light hadrons are sufficiently large that the forces between the quarks within them are quite formidable. A fundamental description of the spectroscopy of light hadrons therefore awaits solution of the confinement problem. It is, however, possible to imagine special situations in which hadron spectroscopy is completely tractable by using available theoretical techniques. T. Appelquist and H. D. Politzer suggested late in 1974 that the bound system of an extremely massive quark with its antiquark would be so small that the strong force would be extremely feeble. In this case, the binding between quark and antiquark is mediated by the exchange of a single massless gluon, and the spectrum of bound states resembles that of an exotic atom composed of an electron and an antielectron (positron) bound electromagnetically in a Coulomb potential generated by the exchange of a massless photon (Fig. 4a). Since the electron-positron atom is known as positronium, the heavy quark-antiquark atom has been called quarkonium. Two families of heavy quark-antiquark bound states, the ψ/J system composed of charmed quarks (Fig. 4b) and the Y system made up of b quarks, have been discovered. Both have level schemes characteristic of atomic spectra, which have been analyzed by using tools of nonrelativistic quantum mechanics developed for ordinary atoms. Neither system is so massive that the one-gluon-exchange description originally envisaged is entirely adequate, but the atomic analogy has proved extremely fruitful for studying the strong interaction. *See* CHARM; J PARTICLE; POSITRONIUM.

Future prospects. Quantum chromodynamics is an extremely promising candidate to be the correct theory of the strong interactions. In the regime in which asymptotic freedom renders some aspects of the strong interactions calculable, sharp predictions and incisive experiments are on the horizon. It is extremely important to refine the evidence that gluons (with the required properties) are radiated by quarks, and to seek verification that gluons interact among themselves. At the opposite extreme of very formidable strong interactions, the issue of permanent confinement of quarks and gluons remains to be resolved. A full understanding will necessarily bring with it solutions of two corollary problems: how quarks and gluons created in violent collisions materialize into the observed hadrons, and a complete description of the spectrum of hadrons. Finally, because quantum chromodynamics and the unified theory of weak and electromagnetic interactions have the same gauge theory structure, it is interesting to contemplate the possi-

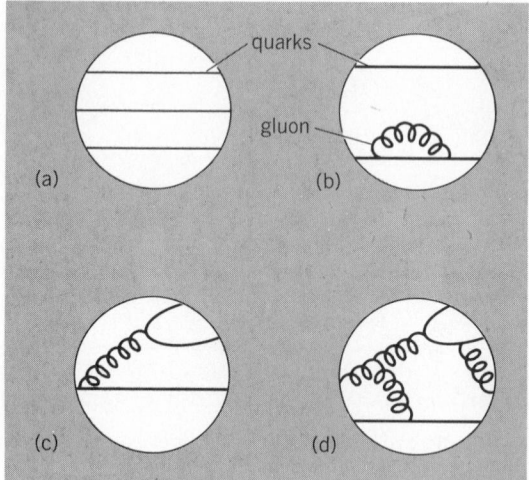

Fig. 3. Structure of the proton as observed with probes of increasing resolution. (*a*) Quarks appear as simple and noninteracting objects under "low magnification." (*b–d*) Quarks reveal increasing complexity because of their interactions with gluons as resolving power is improved.

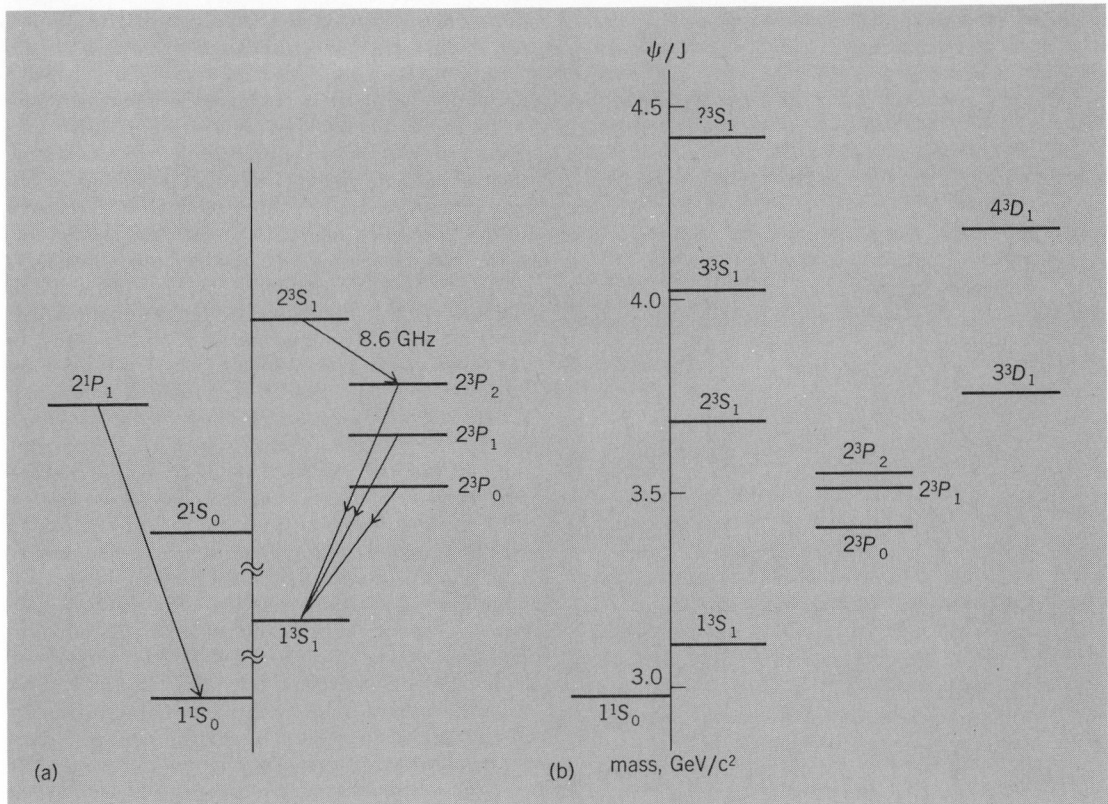

Fig. 4. Spectra of bound particle-antiparticle systems. (a) Positronium. (b) ψ/J family, known as charmonium.

bility that all three interactions have a common origin in a single gauge symmetry. The construction of such "grand unified theories" is a very active, though highly speculative, area of theoretical research. *See* ELEMENTARY PARTICLE; FUNDAMENTAL INTERACTIONS; GLUONS; QUANTUM FIELD THEORY; QUARKS; STRONG NUCLEAR INTERACTIONS.

[C. QUIGG]

Bibliography: N. Calder, *The Key to the Universe*, 1977; W. J. Marciano and H. Pagels, Quantum chromodynamics, *Phys. Rep.*, 36C:137–276, 1978; Y. Nambu, The confinement of quarks, *Sci. Amer.*, 235(5):48–60, November 1976; H. D. Politzer, Asymptotic freedom: An approach to strong interactions, *Phys. Rep.*, 14C:129–180, 1978; C. Rebbi, Solitons, *Sci. Amer.* 242(2):92–116, February 1979.

Quantum electrodynamics

The study of the properties of electromagnetic radiation and its interaction with electrically charged matter; in particular, with atoms and their constituent electrons. The fundamental equations governing quantum electrodynamics actually are believed to encompass all of atomic physics, chemistry, properties of bulk matter, and classical electromagnetic theory. Almost all the phenomena readily perceived by the senses are believed to be ultimately understandable in terms of the laws of quantum electrodynamics. In the scope of its application and implications, and in the simplicity of the underlying assumptions, quantum electrodynamics is rivaled only by gravitational theory. In practice, the term quantum electrodynamics refers to those phenomena specific to the quantum nature of electromagnetic radiation. This includes the study of emission and absorption of light by atoms and the basic interactions of light with electrons and other fundamental particles. *See* ELECTROMAGNETIC RADIATION; GRAVITATION; QUANTUM FIELD THEORY.

Quantum electrodynamics was formulated by P. A. M. Dirac, W. Heisenberg, and W. Pauli shortly after the foundations of quantum mechanics were laid down in 1925. This formulation gave a satisfactory interpretation of the wave-particle duality of light, in which it is possible for light to manifest itself both in ways appropriate to a particle (photon) description or a wavelike description, in accordance with the experimental foundations of quantum mechanics. About the same time, Dirac discovered an equation describing the motion of electrons which incorporated both the requirements of quantum theory and of relativity. However, introduction of electromagnetic interaction into Dirac's equation, while successfully describing the magnetic properties of the electron and the existence of its oppositely charged counterpart, or antiparticle, the positron, led to mathematical disasters which inhibited the further development of the theory until after World War II. Stimulated by experiments using microwave techniques developed during World War II, S. Tomonaga, R. P. Feynman, J. S. Schwinger, F. J. Dyson, and many others found a way to bypass, but not solve, these difficulties. Since that time the theory has been confronted by very sophisticated experimental tests and has passed them satisfactorily. The accuracy of the comparisons of theory and experiment is well illustrated by the measurements of the magnetic moment (the strength of the elementary magnet associated with an electron) of the electron, which gives Eqs. (1) and (2), where $e =$

$$\text{Experiment:} \quad \mu = \frac{eh}{4\pi m_e c}\left(\substack{1.001\ 159\ 652\ 41 \\ \pm 20}\right) \quad (1)$$

$$\text{Theory:} \quad \mu = \frac{eh}{4\pi m_e c}\left(\substack{1.001\ 159\ 652\ 21 \\ \pm 60}\right) \quad (2)$$

electron charge, $h =$ Planck's constant, $m_e =$ electron mass, and $c =$ velocity of light. *See* MAGNETIC MOMENT; QUANTUM MECHANICS.

General applications. The equations of quantum electrodynamics have found many applications in the study of the delicate details of the structure of atoms (the so-called fine-structure and hyperfine structure of atoms). In particular, in the hydrogen atom the frequency of a certain spectral line (for the $2S_{1/2} \to 2P_{1/2}$ transition) is predicted, in the absence of quantum electrodynamic effects, to be almost zero. The effects of the radiation field modify this prediction. This shift in frequency was accurately measured by W. E. Lamb and computed by H. A. Bethe and others, and the theory and experiment agree to within 200 parts per million. Several similar modifications in the frequency of spectral lines in hydrogen have been accurately computed and compared with experiment. *See* ATOMIC STRUCTURE AND SPECTRA.

As mentioned in the introduction, the magnitude of the magnetic moment of the electron and μ-meson (an elementary particle with properties similar to the electron, except for a larger mass) is also predicted by quantum electrodynamics. In all cases, measurements agree with theory to very high accuracy.

Quantum electrodynamics also provides an account of collision processes of electrons, positrons, and photons in matter. For example, an electron in passing through matter is accelerated by the Coulomb fields of the atoms and radiates light. The light (photons) in turn can materialize into matter, provided its frequency is high enough; this effect in fact was predicted by the equations of quantum electrodynamics. The matter produced is almost always an electron and a positron. These in turn, in passing through matter, are accelerated and radiate more light, which produces more electrons and positrons. In this way an avalanche of electrons, positrons, and photons is created; this is called a cascade shower. The details of the evolution of such showers can be computed from the equations of quantum electrodynamics and they agree with experiment, even for enormous showers containing 10,000,000 electrons created by very energetic cosmic rays entering the Earth's atmosphere. The collision processes involved in the cascade showers can also be studied under more controlled conditions by using particle accelerators. In this way, detailed tests probing the structure of the theory have been made employing very energetic beams of photons, electrons, positrons, and μ-mesons from the accelerators. Again, the agreement with the theoretical predictions has been consistently excellent. Examples of these collision process are given below. *See* PARTICLE ACCELERATOR.

Free electromagnetic field. The simplest electrodynamic system is that of radiation in free space, described classically by the Maxwell equations. To construct a quantum theory, one may first visual-

ize the classical radiation confined in a box (such as a microwave cavity) and consider separately the normal modes, that is, components oscillating with a well-defined frequency. If for each normal mode (in musical terminology, fundamental, first harmonic, and so on) the amplitude of the oscillating electromagnetic field is given, the complete field may be constructed by adding together the contribution from each mode. In this way the dynamics of the electromagnetic field is reduced to the dynamics of each normal-mode amplitude, which in turn undergoes sinusoidal oscillation in time, just like a mass hung from a spring. In quantum theory one finds that any oscillator (such as the mass on a spring) can be found only in a discrete set of quantum states, with energy $E_n = h\nu(n + \frac{1}{2})$, where h is Planck's constant and ν is the oscillator frequency. For the radiation field, the same picture is taken over. The state above represents, in this case, n photons each of energy $h\nu$ and associated with the normal mode of frequency ν. The energy in the field is changed in discrete amounts corresponding to transitions between different oscillator states. Such changes in the state of the radiation field are associated physically with the addition or removal of photons from the box. The state of no photons present (the vacuum) is characterized by all the radiation oscillators in their lowest state $n = 0$. However, the energy of such a state $E = \sum_\nu \frac{1}{2}h\nu$, where the sum is taken over all normal modes and is not zero, because of the existence of quantum fluctuations associated with the uncertainty principle. These quantum fluctuations of the radiation field have observable consequences. As an example, one may consider the energy contained between two neutral parallel conducting plates of area A, separated by a distance d. As d is decreased, fewer normal modes can fit inside the plates, and therefore the vacuum energy is decreased. This leads to an attractive force between the plates $F = hc\,A/360\,d^4$, which, although quite small, has been measured and agrees with theory. Another example of the existence of such quantum fluctuations is the Lamb shift of the spectral lines in hydrogen, mentioned in the previous section. *See* UNCERTAINTY PRINCIPLE.

Free electron field. Because the electron is the lightest of all charged particles, it is most easily accelerated, and it most easily emits radiation. Most experimental studies of quantum electrodynamics have dealt with the electromagnetic properties of electrons. The first requirement in a theory of the electron is to have an equation of motion which incorporates both the theory of special relativity and quantum theory. Such an equation was found by Dirac. The wave function, or electron field, replacing the two-component Schrödinger-Pauli field $\psi(x)$ for a nonrelativistic electron, now has four components. The two new components are associated with negative energy states. Although this appears at first to be unacceptable physically, Dirac found a way around this problem, and the negative energy states were eventually associated with the electron's antiparticle, the positron. This question is dealt with in the next section. Aside from the negative-energy problem, which in time has become its most striking success, the Dirac equation immediately accounted for many of the observed properties of the electron. It predicts correctly an electron spin angular

QUANTUM ELECTRODYNAMICS

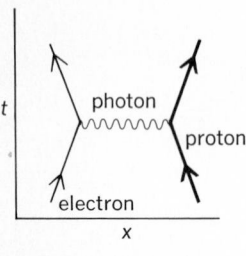

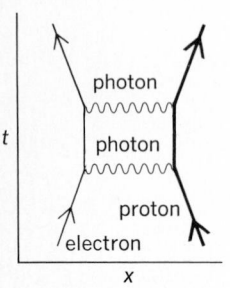

Fig. 1. Space-time picture of electron-proton collision.

momentum of $h/4\pi$, and when coupling to external electromagnetic potentials are introduced, the correct interaction of electron spin with magnetic field (magnetic moment) is predicted. The relativistic corrections to the frequencies of hydrogen spectral lines are also correctly given from the solutions of the Dirac equation for an electron moving in the proton Coulomb field.

Interaction between radiation and electrons. The classical Maxwell equations contain in a well-defined way the current of charged particles as the source of radiation. This current is constructed from the electron field and in turn determines the nature of coupling of electrons to the radiation field. The resulting equations, just as their classical counterparts, have resisted exact solutions. However, the dynamics of the coupled electron-photon system can be determined in terms of a power series expansion in a dimensionless parameter $\alpha = 2\pi e^2/hc \sim 1/137$, which is fortunately small (for a reason not totally understood). An intuitive physical picture of the meaning of this expansion was given by Feynman. The collision of a moving electron with a proton is visualized in space-time as occurring by means of exchange of photons, as in Fig. 1. Plotted there is the trajectory of the particles in space-time. For each such picture representing the collision process, there is a corresponding quantum-mechanical probability amplitude, given by well-defined rules having to do with the nature of the lines and vertices (points where the photon lines end). One of these rules is that the amplitude is proportional to e^n, with n the number of vertices in the corresponding diagram. The full probability amplitude is given by the sum over all diagrams of the amplitude for each diagram. However, the complicated diagrams generally give a very small contribution because they are proportional to a large power of α.

One of the most interesting uses of the diagrams by Feynman is his interpretation of the negative-energy particles predicted by Dirac's equation for the electron. Feynman tolerates the existence of such particles by giving them acausal behavior; that is, negative-energy objects always propagate backward in time, not forward. Consider the process in Fig. 2a. A negative-energy electron, "produced" in the future, propagates backward in time to the present, interacts with radiation, absorbing energy, and then propagates into the future with positive energy. The creation of negative energy and negative charge in the future, which then propagates backward in time, is physically equivalent to absorption of positive energy and

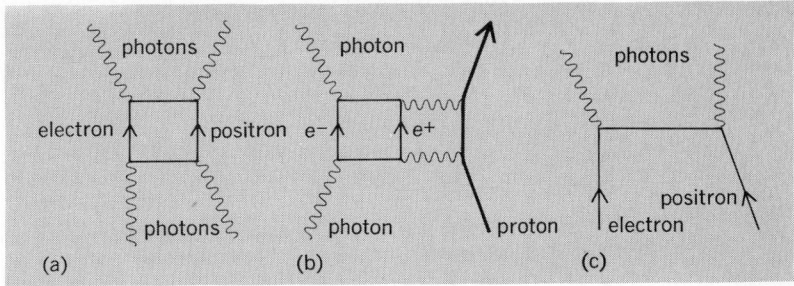

Fig. 3. Higher-order electromagnetic processes (*a*) Collision of light with light. (*b*) Indirect observation of the collision. (*c*) Fate of positrons in matter.

positive charge propagating forward in time. Therefore, one can interpret the negative-energy acausal electrons as causal particles with opposite charge, but the same mass. These are the positrons, the electron antiparticles. The process of Fig. 2a is physically the materialization of a photon, in the presence of a proton, into a pair of electron and positron, as shown in Fig. 2b. Illustrated in Fig. 3 are other collision processes predicted by quantum electrodynamics. The first is the scattering (or collision) of light with light, which has been observed indirectly through the process in Fig. 3b. Figure 3c shows the fate of positrons in matter: Each annihilates with an electron into electromagnetic radiation. *See* ELECTRON-POSITRON PAIR PRODUCTION.

Theory deficiencies. Despite the complete success of the comparison of the predictions of quantum electrodynamics with experiment, there is great uneasiness about the foundations of the theory. The basic equation of Dirac contains parameters, the "bare mass" and "bare charge" of the electron, which appear to be infinite. These physical constants, which represent the hypothetical mass and charge of the electron in the absence of electromagnetic corrections, are modified by effects represented in the diagrams in Fig. 4 (among others), which turn out to be infinite. The great achievement of the postwar theorists was to find a way to avoid these infinities, but without resolving the basic difficulties and to get numbers out of the equations to compare with measurements. However, the infinities are believed to be an indication of error in the theory, and a continuing series of experiments probing the theory in its most sensitive places is being carried out.

[J. D. BJORKEN]

Bibliography: J. Bialyniccy-Birula, *Quantum Electrodynamics*, 1975; J. Bjorken and S. Drell, *Relativistic Quantum Mechanics*, 1964; S. N. Gupta, *Quantum Electrodynamics*, 1977; J. M. Jauch and F. Rohrlich, *The Theory of Photons and Electrons*, 2d ed., 1975; G. Kaellen, *Quantum Electrodynamics*, 1972; J. Schwinger, *Selected Papers on Quantum Electrodynamics*, 1958.

Quantum field theory

Quantum theory of physical systems possessing an infinite number of degrees of freedom, such as the electromagnetic field, gravitational field, or wave fields in a medium. Its major applications lie in the attempted description of fundamental particles and their associated wave fields under circumstances in which both the effects of quantum mechanics and of special relativity are important.

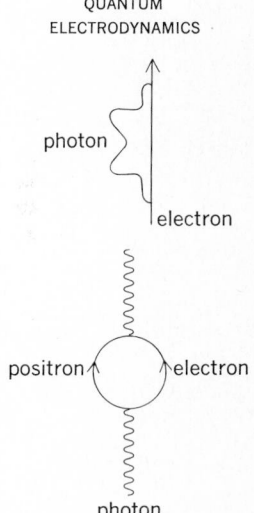

QUANTUM ELECTRODYNAMICS

photon

electron

positron   electron

photon

Fig. 4. Self-interaction of electrons and photons.

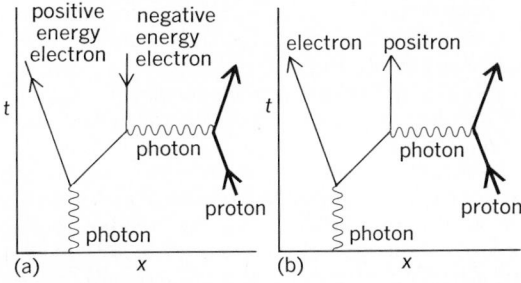

Fig. 2. Electron-positron pair production by radiation. (*a*) Materialization of a photon, in the presence of a proton, into (*b*) a pair of electron and positron.

Analogy between
(a) electromagnetic
force via photon
exchange and
(b) nuclear force
via meson exchange.

Quantum field theory is also used in nonrelativistic quantum theory for systems of many particles such as electrons in a metal, or for sound waves in liquids and solids. Here the discussion is mainly confined to the use of quantum field theory as a description of the properties of fundamental particles. *See* NONRELATIVISTIC QUANTUM THEORY; QUANTUM MECHANICS; RELATIVISTIC QUANTUM THEORY; RELATIVITY.

Yukawa force. The classical electromagnetic field is a dynamical system which must be subjected to the rules of quantum mechanics. The consequence of this requirement is the emergence of a particle (photon) interpretation from the classical Maxwell equations, manifesting the particle-wave duality of quantum mechanics in its most perfect form. The experimental success of this theory for electrodynamics has led to its imitation for the more complex interactions of other fundamental particles. For example, in 1935 H. Yukawa tried to interpret the strong force between protons and neutrons responsible for the existence of nuclei in the same way. This force is only effective when the protons and neutrons are within a distance, 10^{-13} cm, of each other. Yukawa supposed the nuclear force was due to the exchange of a particle, called a meson, between the nucleons, just as the electrostatic force between charged particles is interpreted in terms of exchange of a photon, as illustrated. *See* ELEMENTARY PARTICLE; MESON; QUANTUM ELECTRODYNAMICS.

The short range of the force can be understood as a consequence of a finite rest mass m for the meson. When the nucleon emits the meson, the energy cost is approximately mc^2. According to the uncertainty principle, if the energy of the system is observed to the accuracy ΔE, the observation must be made over an interval of time Δt longer than $h/\Delta E$. Thus, an energy fluctuation of a magnitude of $\sim mc^2$ can only take place over a time interval $\Delta t < h/mc^2$, consistent with the uncertainty principle. In this time interval, the meson can travel no further than $c\Delta t < h/mc^2$, a distance identified with the range of the force. In this way Yukawa was led to predict the mass of the meson ($m \simeq 270\, m_{\text{electron}}$) intermediate between the electron and proton mass. To describe this Yukawa meson, now known as π-meson, one writes an analog of the Maxwell field equations, described below, and then passes to the quantum description in a way similar to that described for the electromagnetic field. A satisfactory particle description is thereby obtained, again embodying the characteristic wave-particle duality of the quantum theory. *See* UNCERTAINTY PRINCIPLE.

However, although the theory of isolated mesons is satisfactory, when interactions such as the couplings to the nucleons are introduced, chaos emerges. The field equations have not been solved. There are grave doubts as to whether solutions exist. There is, aside from the electromagnetic interactions, no small parameter in terms of which there might exist a series expansion. Finally, it is not even known what equation to write. For by now there are dozens of strongly interacting particles, none of which appear any more fundamental than any other. If field variables of all these particles appear in the basic dynamical equations, they are complicated indeed. *See* CLASSICAL FIELD THEORY; ELECTROMAGNETIC RADIATION.

Faced with this dilemma, the theoretical physicists have concentrated efforts on finding general properties of any relativistic quantum field theory, rather than detailed consequences of a specific set of equations. These efforts have been fruitful in three different areas: (1) the quantum theory of symmetries and their connection with conservation laws such as conservation of energy, momentum, and angular momentum; (2) the study of certain exact properties; in particular the mathematical property of analyticity of the functions used to describe collision processes; (3) the use of the diagrammatic method of R. P. Feynman, developed for quantum electrodynamics and appropriately modified for the study of more general interactions, to interpret qualitatively many of the features of collision processes between relativistic particles.

Axiomatic quantum field theory. For the study of general and presumably exact properties of a quantum field theory, one begins by assuming that it makes sense and that it satisfies the following axioms: (i) The solutions are consistent with the theory of special relativity and with basic quantum mechanical principles. (ii) The vacuum state (empty space) exists and is the state of lowest energy. (iii) The field variables are local; if the space-time coordinates of two field variables cannot be connected by a signal traveling with a velocity less than or equal to the velocity of light, the fields are dynamically independent. This is the principle of microscopic causality. One also assumes that the dynamical equations do not depend explicitly upon absolute position in space or upon the orientation of the coordinates. These requirements lead, via the field theory formalism, to the existence of conservation laws of energy, momentum, and angular momentum, and are a deep and important result of the fundamentals of quantum field theory. Other more subtle symmetry properties are found as a consequence of the third, locality assumption. It is found, for example, that to every particle there must exist an antiparticle with opposite charge but precisely the same mass (in some cases, the particle and antiparticle can be identical). This prediction has been well verified experimentally, and the equality of mass of particle and antiparticle has been verified in one case to 1 part in 10^{17}. Furthermore, aggregates of identical particles of integer spin angular momentum 0, h, $2h$. . . must have wave functions symmetric under interchange of any two particles, while aggregates of identical half-integer spin ($h/2$, $3h/2$, . . .) particles must have antisymmetric wave functions. Finally, the simultaneous operation (TCP) of space-inversion (P), time reversal (T), and replacement of particle by antiparticle (C) should be a symmetry of nature; that is, all dynamical equations are left unchanged in form by the TCP operation.

All these results are a consequence of "axioms" (i), (ii), and (iii), and form a cornerstone of axiomatic quantum field theory. In addition to these results, study of the analyticity properties of the functions describing collision processes has led to "dispersion relations," relations similar to the Kramers-Kronig relation in optics which relate the real part of the index of refraction at one frequency to an integral over the imaginary part at all frequencies. In the case of quantum field theory, the corresponding index of refraction is that associated with an elementary particle such as a π-meson propagating through matter and interacting with it by means of the strong nuclear force. The main input required to obtain these relations, both in the case of optics and the present case, is axiom (iii) of microscopic causality, that no signal is propagated with a velocity greater than that of light. *See* DISPERSION RELATIONS.

Feynman diagrams. For the special field theory of quantum electrodynamics, Feynman diagrams are extremely useful because the parameter e coupling the radiation field to matter is small. This means only the simplest diagrams are in practice important. For the strong interactions of elementary particles responsible for the nuclear force, there are no known small parameters, and complicated diagrams are just as important as simple ones, if indeed the concept may even be used under these circumstances. However, in practice it is found that for qualitative interpretation of many phenomena simple diagrams may be used to provide a partial understanding of the underlying dynamics; they remain a tool for the "theoretical laboratory," but their ultimate role in a correct theory is not known.

Mathematical structure. As an illustration of the mathematical structure, the description of the Yukawa π-meson, which has zero spin angular momentum, is given below.

The first step in building a field theory for this meson is to construct a classical field equation analogous to the Maxwell equations. This is done by writing the relativistic energy-momentum relation $E^2 = \mathbf{p}^2 c^2 + m^2 c^4$, making the quantum-mechanical replacement $E \rightarrow i\hbar\,(\partial/\partial t)$, $\mathbf{p} \rightarrow -i\hbar\nabla$. One obtains in Eq. (1) the relativistic version of the

$$-\hbar^2 \frac{\partial^2}{\partial t^2} \phi(\mathbf{x},t) = (-\hbar^2 c^2 \nabla^2 + m^2 c^4)\,\phi(\mathbf{x},t) \quad (1)$$

Schrödinger equation, but treated as a classical equation. Upon Fourier transformation, one finds that Eq. (2) holds. For each Fourier component,

$$-\hbar^2 \frac{\partial^2}{\partial t^2} \widetilde{\phi}(\mathbf{p},t) = (\mathbf{p}^2 c^2 + m^2 c^4)\,\widetilde{\phi}(\mathbf{p},t) \quad (2)$$

this is the equation for a classical harmonic oscillator. *See* HARMONIC OSCILLATOR.

In passing to quantum mechanics, each $\widetilde{\phi}(\mathbf{p},t)$ becomes an operator, just as coordinate and momentum in the nonrelativistic quantum mechanics, which acts on an abstract wave function, or "state vector." Because the dynamics is that of an infinite assembly of independent harmonic oscillators, one for each momentum $\mathbf{p}$, the stationary states are labeled by the oscillator quantum numbers $n(\mathbf{p})$ for each and every $\mathbf{p}$, with the energy of such a state given by Eq. (3). For example, the

$$E = \sum_{\mathbf{p}} E\,n(\mathbf{p}) = \sum_{\mathbf{p}} \left(n(\mathbf{p}) + \frac{1}{2}\right) \sqrt{\mathbf{p}^2 c^2 + m^2 c^4} \quad (3)$$

state with $n(\mathbf{p}) \neq 0$ and $n(\mathbf{p}') = 0$ for $\mathbf{p}' \neq \mathbf{p}$ represents a state of n mesons of momentum $|\mathbf{p}|$. The field operator $\widetilde{\phi}(\mathbf{p},t)$, the oscillator "coordinate," acting on a state containing $n(\mathbf{p})$ mesons changes $n(\mathbf{p})$ by ± 1, as in the nonrelativistic oscillator. It therefore "creates" or "destroys" single mesons of momentum $\mathbf{p}$. In this way the particle interpretation emerges from the quantum field formalism.

For the electron, as described by the wave equation of P. A. M. Dirac, a similar procedure may be carried out. The equation is considered a classical field equation, a Fourier transformation performed, and quantum conditions imposed upon the Fourier coefficients $\widetilde{\psi}(\mathbf{p},t)$, again treated as operators acting upon an abstract state vector or wave function. The only difference, aside from more complicated algebra, is that while meson operators for different momenta are taken to commute, as in Eq. (4) they must for electrons be taken to anticom-

$$\widetilde{\phi}(\mathbf{p},t)\widetilde{\phi}(\mathbf{p}'t) = \widetilde{\phi}(\mathbf{p}'t)\widetilde{\phi}(\mathbf{p},t) \quad (4)$$

mute as in Eq. (5), to obtain a satisfactory physical

$$\widetilde{\psi}(\mathbf{p},t)\widetilde{\psi}(\mathbf{p}'t) = -\widetilde{\psi}(\mathbf{p}'t)\widetilde{\psi}(\mathbf{p},t) \quad (5)$$

interpretation. This change of sign is related to the antisymmetry of electron many-particle wave functions, as compared with the symmetry of meson wave functions alluded to previously.

When interactions are introduced such as coupling of mesons to nucleons, the meson theory is equivalent to an infinite assembly of coupled nonlinear oscillators, treated quantum-mechanically. It is no wonder that progress has been slow in studying such a complex dynamical system.

Nonrelativistic applications. The Schrödinger equation, as well as the Dirac equation, can also be treated as a classical field equation and subjected to quantization. When this is done, a many-particle nonrelativistic theory of electrons is obtained. This theory, along with the Feynman diagram techniques, has been applied successfully in the study of properties of electrons in solids, notably in the theory of collective oscillations (plasma oscillations) of electrons in metals and in the presently accepted theory of superconductivity. Along with quantum electrodynamics, these studies are thus far the most successful applications of quantum field theory to physical phenomena. *See* FREE-ELECTRON THEORY OF METALS; SUPERCONDUCTIVITY; SYMMETRY LAWS.

[J. D. BJORKEN]

Bibliography: J. Bjorken and S. Drell, *Relativistic Quantum Fields*, 1965; N. Bogoliubov and D. Shirkov, *Introduction to the Theory of Quantized Fields*, 3d ed., 1980; L. D. Fadeev and A. A. Slavonov, *Quantum Theory of Gauge Fields*, 1980; E. G. Harris, *Pedestrian Approach to Quantum Field Theory*, 1972; C. Itzykson and J. B. Zuber, *Quantum Field Theory*, 1980; D. Thouless, *The Quantum Mechanics of Many-Body Systems*, 1961.

Quantum mechanics

The modern theory of matter, of electromagnetic radiation, and of the interaction between matter and radiation; also, the mechanics of phenomena to which this theory may be applied. Quantum mechanics, also termed wave mechanics, generalizes and supersedes the older classical mechanics and Maxwell's electromagnetic theory. Atomic and subatomic phenomena provide the most striking evidence for the correctness of quantum mechanics and best illustrate the differences between quantum mechanics and the older classical physical theories. Quantum mechanics is needed to explain many properties of bulk matter, for instance, the temperature dependence of the specific heats of solids. These, along with numerous other applications, are more fully discussed in the articles given as cross-references and in the bibliography.

The formalism of quantum mechanics is not the same in all domains of applicability. In approximate order of increasing conceptual difficulty, mathematical complexity, and likelihood of future fundamental revision, these domains are the following: (i) Nonrelativistic quantum mechanics, applicable to systems in which particles are neither created nor destroyed, and in which the particles are moving slowly compared to the velocity of light, as shown by relation (1). Here a particle is

$$c \cong 3 \times 10^{10} \text{ cm/sec} \qquad (1)$$

defined as a material entity having mass, whose internal structure either does not change or is irrelevant to the description of the system. (ii) Relativistic quantum mechanics, applicable in practice to a single relativistic particle (one whose speed equals or nearly equals c); here the particle may have zero rest mass, in which event, its speed must equal c. (iii) Quantum field theory, applicable to systems in which particle creation and destruction can occur; the particles may have zero or nonzero rest mass.

This article is concerned mainly with nonrelativistic quantum mechanics, which apparently applies to all atomic and molecular phenomena, with the exception of the finer details of atomic spectra. Nonrelativistic quantum mechanics also is well established in the realm of low-energy nuclear physics, meaning nuclear phenomena wherein the particles have kinetic energies less than about 10^8 eV (1 eV = 1 electronvolt = 1.6×10^{-12} erg, is the energy gained by an electron in traversing a potential difference of 1 volt). Many quantum-mechanical predictions are not as quantitatively accurate for nuclei as for atomic and molecular systems, however, because nuclear forces are not accurately known. *See* ATOMIC STRUCTURE AND SPECTRA.

For the formal mathematical structure of nonrelativistic quantum mechanics *see* NONRELATIVISTIC QUANTUM THEORY. That article provides justification for many of the assertions made under the subheadings immediately following, wherein are described the novel (from the standpoint of classical physics) features of nonrelativistic quantum mechanics. Some of these features are retained, others modified, in the more complicated domains of relativistic quantum mechanics and quantum field theory. *See* QUANTUM ELECTRODYNAMICS; QUANTUM FIELD THEORY; RELATIVISTIC QUANTUM THEORY; SYMMETRY LAWS.

Planck's constant. The quantity 6.61×10^{-27} erg-sec, first introduced into physical theory by Max Planck in 1901, is a basic ingredient of the formalism of quantum mechanics. Most of the fundamental quantum-mechanical relations, for example, Schrödinger's equation and Heisenberg's uncertainty principle, explicitly involve Planck's constant, as do many of the well-verified consequences of quantum mechanics; for example, the formula for the energy levels of atomic hydrogen. Planck's constant plays no role in the classical theories. Planck's constant commonly is denoted by the letter h; the notation $\hbar = h/2\pi$ also is standard.

Uncertainty principle. In classical physics the observables characterizing a given system are assumed to be simultaneously measurable (in principle) with arbitrarily small error. For instance, it is thought possible to observe the initial position and velocity of a particle and therewith, using Newton's laws, to predict exactly its future path in any assigned force field. According to the uncertainty principle (W. Heisenberg, 1927), accurate measurement of an observable quantity necessarily produces uncertainties in one's knowledge of the values of other observables. In particular, for a single particle relation (2a) holds, where Δx repre-

$$\Delta x \, \Delta p_x > \hbar \qquad (2a)$$

sents the uncertainty (error) in the location of the x coordinate of the particle at any instant, and Δp_x is the simultaneous uncertainty in the x component of the particle momentum. Relation (2a) asserts that under the best circumstances, the produce $\Delta x \Delta p_x$ of the uncertainties cannot be less than about 10^{-27} erg-sec; of course, with poor measurements, the product can be much greater than $\hbar$. On the other hand, there is no restriction on the simultaneous determination of position along x and momentum along y; that is, the product $\Delta x \, \Delta p_y$ may equal zero. Other typical uncertainty inequalities for a single particle are given by relations (2b) and (2c).

$$\Delta \phi \, \Delta l_z \gtrsim \hbar \qquad (2b)$$

$$\Delta x \, \Delta E \gtrsim \frac{\hbar}{m} p_x \qquad (2c)$$

In relation (2b) the particle location is specified in spherical coordinates, with polar axis along z; $\Delta \phi$ is the uncertainty in azimuth angle; and Δl_z is the uncertainty in the z component of the orbital angular momentum. In relation (2c) ΔE is the uncertainty in energy and m is the particle mass. The uncertainty relation (2d) is derived and interpreted

$$\Delta t \, \Delta E \gtrsim \hbar \qquad (2d)$$

somewhat differently than relations (2a)–(2c); it asserts that for any system, an energy measurement with error ΔE must be performed in a time not less than $\Delta t \sim \hbar/\Delta E$. If a system endures for only Δt sec, any measurement of its energy must be uncertain by at least $\Delta E \sim \hbar/\Delta t$ ergs.

Because the numerical value of $\hbar$ is so small, and since $\Delta p_x = m \Delta v_x$, v denoting velocity, the restrictions implied by relations (2a)–(2c) are utterly inconsequential for macroscopic systems, wherein masses are of the order of grams. For an electron, however, whose mass is 9.1×10^{-28} g, $\hbar/m \sim 1$ cm²/sec, and the uncertainty principle cannot be ignored. Similarly, relation (2d) is unimportant for macroscopic systems, wherein energies are of the order of ergs, but is significant for atomic systems where $\Delta E = \hbar/\Delta t$ need not be negligible compared to the actual energy E. *See* UNCERTAINTY PRINCIPLE.

Wave-particle duality. It is natural to identify such fundamental constituents of matter as protons and electrons with the mass points or particles of classical mechanics. According to quantum mechanics, however, these particles, in fact all material systems, necessarily have wavelike properties. Conversely, the propagation of light, which, by Maxwell's electromagnetic theory, is understood to be a wave phenomenon, is associated in quantum mechanics with massless energetic and momentum-transporting particles called photons. The quantum-mechanical synthesis of wave and particle concepts is embodied in the de Broglie realtions, given by Eqs. (3a) and (3b). These give

$$\lambda = h/p \qquad (3a)$$

$$f = E/h \qquad (3b)$$

the wavelength λ and wave frequency f associated with a free particle (a particle moving freely under no forces) whose momentum is p and energy is E; the same relations give the photon momentum p and energy E associated with an electromagnetic wave in free space (that is, in a vacuum) whose wavelength is λ and frequency is f. See PHOTON.

The wave properties of matter have been demonstrated conclusively for beams of electrons, neutrons, atoms (hydrogen, H, and helium, He), and molecules (H_2). When incident upon crystals, these beams are reflected into certain directions, forming diffraction patterns. Diffraction patterns are difficult to explain on a particle picture; they are readily understood on a wave picture, in which wavelets scattered from regularly spaced atoms in the crystal lattice interfere constructively along certain directions only. Moreover, the wavelengths of these "matter waves," as inferred from the diffraction patterns, agree with the values computed from Eq. (3a), as first demonstrated by C. J. Davisson and L. H. Germer in 1927. See ELECTRON DIFFRACTION; NEUTRON DIFFRACTION.

Photoelectric effect; Compton effect. The particle properties of light waves are observed in the photoelectric effect. When light of frequency f causes electrons to be emitted from a surface, all the electrons have very nearly the same maximum kinetic energy; the maximum kinetic energy is independent of the light intensity; the number of electrons emitted in unit time is proportional to the light intensity; as f is varied, the maximum electron kinetic energy W varies linearly with f, in fact, $W = hf - C$, C being a constant characteristic of the emitting material. These observations are difficult to understand on the wave picture, wherein the magnitude of the electric field vector (which presumably exerts the force which ejects the electron) is proportional to the square root of the incident light intensity, and is not directly related to nor limited by, the incident light frequency f. The photoelectric effect is interpreted readily on the assumptions that the energy in the light beam is carried in quanta of energy $E = hf$; that emission of an electron results from absorption of a single quantum (a single photon); and that absorption of half or any fraction of a quantum is not possible, because the photons act as discrete indivisible entities.

The particle properties of electromagnetic waves also are demonstrated in the Compton effect, wherein the wavelengths of x-rays are lengthened by scattering from free electrons. The change in wavelength is predicted quantitatively, assuming the scattering results from elastic collisions between photons and electrons, and using Eqs. (3a) and (3b) for the photon momentum and energy. The diffraction of x-rays by crystals was regarded, in the prequantum era, as conclusive proof that x-rays are waves and not "corpuscles." *See* COMPTON EFFECT; ELECTROMAGNETIC RADIATION; MAXWELL'S EQUATIONS; X-RAY DIFFRACTION.

Interference and diffraction. Wave propagation is distinguished from particle propagation by the phenomena of interference and diffraction. It is a general result of wave theories that interference and diffraction effects largely are confined to an angle (relative to the incident beam) which in radians equals about λ/d, where d is a characteristic dimension of the system causing the diffraction or interference, for example, the width of the slit diffracting the wave, or the distance between two interfering scattering centers. This fact and the magnitudes of λ inferred from Eq. (3a) are sufficient to explain why wave effects are not observed in the propagation of ordinary macroscopic bodies, but can be observed in the propagation of electrons, neutrons, and light atoms or molecules. For example, for a mass of 1 g moving at 1 cm/sec, $\lambda = 6.6 \times 10^{-27}$ cm. But for a neutron or hydrogen atom (using $p = Mv = \sqrt{2ME}$, where $M = 1.66 \times 10^{-24}$ g) moving at velocity corresponding to room temperature (300 K), $E = 3/2\ kT$ ($k =$ Boltzmann's constant $= 1.38 \times 10^{-16}$ erg deg^{-1}), and λ turns out to equal 1.45×10^{-8} cm. For an electron with an energy of 100 eV, $\lambda = 1.22 \times 10^{-8}$ cm. For a proton with an energy 10^6 eV $= 1$ MeV, $\lambda = 2.88 \times 10^{-12}$ cm. These numerical results and the discussion of this paragraph also explain the ability of crystals, wherein interatomic spacings are about 10^{-8} cm, to give a good demonstration of electron and molecular diffraction; suggest the need for quantum mechanics to "understand" atomic systems, wherein atomic dimensions are $\sim 10^{-8}$ cm and electron energies are ~ 10 eV; suggest the need for quantum mechanics to "understand" atomic nuclei, wherein nuclear dimensions are $\sim 10^{-13}$ cm and neutron or proton energies are ~ 10 MeV; and explain why quantum effects are more readily observed in H_2 and He than in heavier gases, and at low temperatures rather than high. *See* DIFFRACTION; INTERFERENCE OF WAVES.

Relationship to uncertainty principle. Wave-particle duality is intimately connected with the uncertainty principle in that the uncertainty inequalities can be derived from analyses of specific experiments. For a nonrelativistic particle the connection can be seen from the following argument, which contains the basic elements of the rigorous formal treatment. As explained later, the probability that the x coordinate of the particle will lie in the interval x to $x + dx$ is $|\psi(x)|^2\ dx$, where $\psi(x)$ is called the wave function. Suppose measurement has ascertained that the particle lies in an interval of width Δx centered at $x = 0$; that is, measurement has determined that the dependence of $\psi(x)$ on x is approximately as shown in Fig. 1. Because of wave-particle duality, the wave packet of Fig. 1 can be looked upon as a superposition of waves. Since $\psi(x)$ rises from and falls to a very small value in an interval Δx, the packet must contain waves whose half wavelengths are as small as Δx; furthermore, it can be proved that because $\psi(x)$ does not change sign, the packet must contain waves of very long wavelength. Thus, in the packet, the wavelengths λ run from about $2\Delta x$ to ∞; the reciprocal wavelengths λ^{-1} run from about 0 to $1/(2\Delta x)$. But from Eq. (3a), $\Delta p_x = h\Delta\lambda^{-1}$. Hence

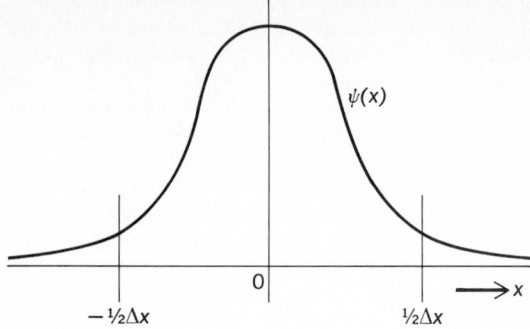

Fig. 1.   Plot of $\psi(x)$ versus x.

$\Delta p_x \, \Delta x$ cannot be less than about $h/2$. Considering the simplicity of the argument, this result is close enough to relation (2a). Similarly, relation (2d) can be understood from Eq. (3b) and from the fact that to transmit information in a time Δt, that is, to turn a measuring instrument on and off in an interval Δt, it is necessary to use frequencies higher than about $(2\Delta t)^{-1}$. Using relation (2a), the discussion of this paragraph suggests that in atoms, whose dimensions Δx are about 10^{-8} cm, one must expect to find electrons with energies as given by Eq. (4),

$$E = p^2/2m = \hbar^2/2m(\Delta x)^2 = 3.8 \text{ eV} \qquad (4)$$

which is of the order of magnitude observed. Similarly, in atomic nuclei, whose dimensions Δx are about 10^{-13} cm, one must expect to find neutrons or protons with energies about 20 MeV, again of the order of magnitude observed. On the other hand, if atomic nuclei contained electrons, the electron energies would be as high as 200 MeV. In computing this value, it is necessary to use the relativistic relation $E = \sqrt{m^2c^4 + c^2p^2}$ connecting E and p. These energies are much too high to be explained by electrostatic forces between electrons and protons at separations of about 10^{-13} cm. Thus, the uncertainty principle leads to the inference that electrons are not contained in atomic nuclei, which inference is in accord with current theories of nuclear structure and β-decay. *See* NUCLEAR STRUCTURE; RADIOACTIVITY; RELATIVISTIC MECHANICS.

Complementarity. Wave-particle duality and the uncertainty principle are thought to be examples of the more profound principle of complementarity, first enunciated by Niels Bohr (1928). According to the principle of complementarity, nature has "complementary" aspects; an experiment which illuminates one of these aspects necessarily simultaneously obscures the complementary aspect. To put it differently, each experiment or sequence of experiments yields only a limited amount of information about the system under investigation; as this information is gained, other

equally interesting information (which could have been obtained from another sequence of experiments) is lost. Of course, the experimenter does not forget the results of previous experiments, but at any instant, only a limited amount of information is usable for predicting the future course of the system.

The well-known double-slit experiment provides a good illustration of the principle of complementarity. Light from a monochromatic point source P (Fig. 2) is diffracted by the two slits S_1 and S_2 in the screen Q_1. On the screen Q_2, an interference pattern of alternating bright and dark bands is formed in the region $D_1 D_2$, where the two diffraction patterns (from the slits S_1 and S_2) overlap. Assuming that P is equidistant from the slits, and also that the slits are very narrow compared to the distances PS_1 or $S_1 O$, it follows that interference maxima (bright bands) are observed whenever $S_2 O - S_1 O$ equals $n\lambda$, λ being the wavelength of the light and n an integer; when $S_2 O - S_1 O = (n + 1/2)\lambda$, interference minima (dark bands) are observed. In moving from any maximum to an adjacent maximum, the path difference $S_2 O - S_1 O$ changes by precisely one wavelength. Consequently, measurement of the distance Y between successive maxima, and knowledge of $S_1 S_2$ and of the distance X between Q_1 and Q_2, yields λ via the formula $\lambda = Yd/X$ (valid when $S_1 S_2 = d$ is much smaller than X). Evidently, the double-slit experiment is understandable in terms of, and provides information concerning, the wave properties of light.

The double-slit experiment yields no information concerning the particle properties of light; in fact, introducing the particle picture leads only to conceptual difficulties. These difficulties appear with the recognition that reducing the source intensity does not modify the interference pattern; after a sufficiently long exposure, a photographic film at Q_2 will show exactly the same interference pattern $D_1 D_2$ as is observed by the eye using a more intense source. Since it is possible to make the source intensity so low that two photons almost never will be emitted during the very small time required for light to travel from P to Q_2 via either of the slits, it is necessary to conclude that the interference pattern is produced by independent individual photons, and not by interference between two or more different photons. On the other hand, the interference pattern is destroyed when either of the slits is closed. Thus, the question arises: How can a stream of independent photons, each of which presumably passes through only one of the slits, and half of which on the average pass through S_1, produce an interference pattern that is destroyed by closing one of the slits? Or to put it differently, how can closing or opening a slit through which a photon does not pass affect the likelihood of that photon reaching any particular point on Q_2?

The principle of complementarity meets these difficulties with the assertion that the possibility of demonstrating that the photons have well-defined trajectories through one or the other slit (a particle-like property) is complementary to the possibility of demonstrating the wavelike property of interference. In the double-slit experimental setup which has been described, until the photon is localized at Q_2 (by the visible evidence that a chemical effect has occurred in a photographic film), it is not possi-

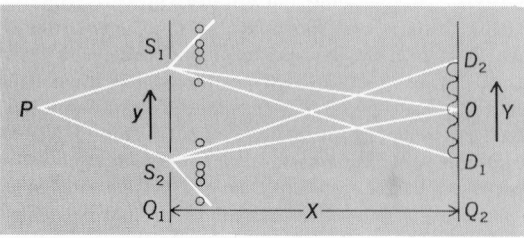

Fig. 2.   Double-slit experiment.

ble to locate the photon at any particular point in space, nor is it legitimate to insist that the photon must have passed through only one of the slits. Moreover, according to the principle of complementarity, modifying the experimental setup so as to localize the photon at one of the slits, and thereby to determine through which slit the photon passes, necessarily destroys the interference pattern. This last assertion is supported by analysis of various photon-detection schemes, recognizing that the proposed experiments perforce are entirely *Gedanken* (in the mind); an actual measurement of the slit through which the photon passes demands extreme precision, and this has not been attempted. It is concluded that quantum mechanics involves no inconsistencies of paradoxes. From the standpoint of the complementarity principle, the questions presented in the preceding paragraph, and other similar difficulties, rest always on the specious assumption of more information than actually is obtainable. *See* SUPERPOSITION PRINCIPLE.

The limitations on the experiments are imposed by the requirements of the uncertainty principle, relations (2), and wave-particle duality, Eqs. (3). To illustrate the analysis, suppose indicators (symbolized by circles), free to move in the vertical y direction, are placed immediately behind the slits S_1 and S_2 in Fig. 2; the recoil of an indicator signifies a collision with a photon and places the photon at the indicator. In order that the recoil establish the slit through which the photon has passed, the uncertainty in the vertical location of each indicator must be much less than $d/2$. According to relation (2a), this means that the vertical momentum p_y of each indicator will be uncertain by amount $\Delta p_y \gg 2\hbar/d$. There can be no assurance that the indicator has recoiled unless δp_y, the momentum transferred from photon to indicator, equals or exceeds Δp_y. When the slits are narrow and $X \gg d$, a momentum transfer δp_y deflects the photon through an angle $\theta \sim \delta p_y/p$, and changes the point Y where the photon strikes the screen by an amount $\Delta Y = \theta X \cong X \delta p_y/p$, with $p = h/\lambda$, Eq. (3a). It follows that when there are observable recoils localizing the photon at one of the slits, then $\Delta Y \gg \lambda X/\pi d$; that is, $\Delta Y \gg Y/\pi$, where $Y/2 = \lambda X/2d$ is the distance between adjacent maxima and minima of the interference pattern. Thus, determining the slit through which the photon passes necessarily gives the photon an uncontrollable random vertical deflection (on the screen Q_2), which is very much larger than the distance between adjacent maxima and minima. The photons now spread with uniform intensity over vertical distances ΔY, which are large compared to the original widths of the light and dark interference bands; in other words, the interference pattern is destroyed.

Quantization. In classical physics the possible numerical values of each observable, meaning the possible results of exact measurement of the observable, generally form a continuous set. For example, the x coordinate of the position of a particle may have any value between $-\infty$ and $+\infty$; similarly p_x, the x component of momentum, may have any value between $-\infty$ and $+\infty$; the kinetic energy T of a particle may have any value between 0 and $+\infty$; the total energy (kinetic plus potential) of an electron in the field of a proton may have any value between $-\infty$ and $+\infty$; the orbital angular momen-

tum vector $\mathbf{l}$ of a particle moving in a central field, for example, an electron in a hydrogen atom, may have any magnitude between 0 and ∞; if the magnitude of $\mathbf{l}$ is known to be l, then, since the plane of the orbit may be arbitrarily oriented, the z component of $\mathbf{l}$ may have any value between $-l$ and $+l$. In quantum mechanics the possible numerical values of an observable need not form a continuous set, however. For some observables, the possible results of exact measurement form a discrete set; for other observables, the possible numerical values are partly discrete, partly continuous; for example, the total energy of an electron in the field of a proton may have any positive value between 0 and $+\infty$, but may have only a discrete set of negative values, namely, -13.6, $-13.6/4$, $-13.6/9$, $-13.6/16$ eV, ... Such observables are said to be quantized; often there are simple quantization rules determining the quantum numbers which specify the allowable discrete values. For example $\mathbf{L}^2$, the square of the orbital angular momentum of a particle, must equal $l(l+1)\hbar^2$, where l is zero or a positive integer; the only allowed values of $\mathbf{J}^2$, the square of the total angular momentum (orbital angular momentum plus intrinsic angular momentum or spin, are $\mathbf{J}^2 = j(j+1)\hbar^2$ where $j = 0, 1/2, 1, 3/2, 2, . . . $; that is, $j\hbar$ is an integral or half-integral multiple of $h/2\pi$. If the magnitude of the total angular momentum is given by some one of these values, $j\hbar = 3\hbar/2$, for example, then the only allowed values of the z component J_z of $\mathbf{J}$ are: $-3\hbar/2$, $-\hbar/2$, $\hbar/2$, $3\hbar/2$; that is, for a given j, the allowed values of $J_z = -j, -j+1, -j+2, . . . , j-1, j$, in units of $\hbar$. On the other hand, the observables x, p_x, and T for a relativistic particle are not quantized, and these observables have precisely the same allowed values in quantum mechanics as they do classically. In the formal theory each observable is a linear operator, whose eigenvalues (characteristic values) are the allowed values of that observable. The set of eigenvalues is termed the spectrum of the operator, which spectrum may be discrete, continuous, or mixed. *See* QUANTUM NUMBERS.

The fact that nature is quantized has been amply verified experimentally. For instance, the quantization of energy and momentum in light waves is demonstrated in the photoelectric and Compton effects, described earlier.

Stern-Gerlach experiment. The quantization of angular momentum is strikingly exhibited in the Stern-Gerlach experiment, wherein a beam of, for example, hydrogen atoms moving through a region of space containing an inhomogeneous magnetic field breaks up into two separate beams. Classically, the force on a hydrogen atom in an inhomogeneous magnetic field depends on the angle between the magnetic field and the plane of the electron's orbit. Thus, classically, the original beam, in which all orientations of the plane of the orbit are possible, is expected to spread out or defocus in the inhomogeneous magnetic field, but is not expected to form two distinct focused beams. Formation of two beams agrees, however, with the quantum mechanical prediction that the square of the total angular momentum of atomic hydrogen is $\mathbf{J}^2 = 3\hbar^2/4 = 1/2(1/2+1)\hbar^2$, and therefore that the vector $\mathbf{J}$ (which classically always is perpendicular to the orbit plane) must be either parallel or antiparallel to the applied magnetic field (along z); for these two directions of **J**, and only for these two, the value of J_z equals one of the permitted values $J_z = -\hbar/2$ or $J_z = \hbar/2$. The Stern-Gerlach experiment also can be performed with beams of other atoms and molecules, as well as with neutron beams. In all cases the observations agree with quantum mechanical expectation. Refinements of the experiment yield accurate measurements of the parameters characterizing the magnetic interactions of atoms, molecules, and atomic nuclei. *See* Magnetic resonance; Molecular beams; Nuclear moments.

Atomic spectra. Spectroscopy, especially the study of atomic spectra, probably provides the most detailed quantitative confirmation of quantization. Granting that the energy in a light wave is quantized, it follows from conservation of energy and Eq. (3b) that when an atom emits a photon of frequency f, its initial energy E_i and final energy E_f after emission are related by Eq. (5a).

Similarly, when a photon of frequency f is absorbed, the initial energy E'_i, and final energy E'_f after absorption, satisfy Eq. (5b).

$$E_i - E_f = hf \qquad (5a)$$

$$E'_f - E'_i = hf \qquad (5b)$$

In quantum mechanics, as in classical mechanics, an electron remains in the vicinity of a proton, that is, it is "bound" to the proton, when and only when the magnitude of its kinetic energy T is less than the magnitude of its negative potential energy V (V is set equal to zero at infinite separation). Thus, the total energy of a stable hydrogen atom necessarily is negative; in other words, the only allowed energies of atomic hydrogen are $-R/n^2$, where n is an integer and $R = -13.6$ ev. Consequently, the radiation emitted by atomic hydrogen must consist of a discrete set of frequencies, or lines, obeying the relation, from Eq. (5a), given by Eq. (6), where m and n are integers, $m > n$. This

$$f = \frac{R}{h}\left(\frac{1}{n^2} - \frac{1}{m^2}\right) \qquad (6)$$

simple argument provides a convincing explanation of the observation that atomic hydrogen has a line spectrum and not a continuous spectrum; in other words, it radiates discrete frequencies rather than a continuous band of frequencies. The observed lines very accurately satisfy Eq. (6); moreover, except for small relativistic and quantum field theory effects, the nonrelativistic Schrödinger equation accurately predicts not merely the frequencies of the lines, but also their relative intensities and widths. Because on the average the levels E_i and E_f of Eq. (5a) endure only for a limited time, their energies are uncertain by an amount ΔE, as explained in the earlier discussion of the uncertainty principle; therefore, the observed lines have a width not less than $\Delta f \cong h^{-1}(\Delta E_i + \Delta E_f)$, according to Eq. (5a) Heavier atoms have more complicated line spectra than hydrogen, and the frequencies they emit cannot be described by formulas as simple as Eq. (6). For these atoms, the agreement between experiment and the predictions of the nonrelativistic Schrödinger equation, though always very good, is not as precise as for hydrogen. All the evidence indicates, however, that the discrepancies arise because the Schrödinger equa-

tion cannot be solved exactly in many-electron atoms, so that the theoretical predictions necessarily are approximate. If approximations were not necessary (for example, if perfect computing machines were available), there is every reason to think that the predictions of the Schrödinger equation would be exactly correct, except for the aforementioned small relativistic and field theory effects.

Probability considerations. The uncertainty and complementarity principles, which limit the experimenter's ability to describe a physical system, must limit equally the experimenter's ability to predict the results of measurement on that system. Suppose, for instance, that a very careful measurement determines that the x coordinate of a particle is precisely $x = x_0$. This is permissible in nonrelativistic quantum mechanics. Then formally, the particle is known to be in the eigenstate corresponding to the eigenvalue $x = x_0$ of the x operator. Under these circumstances, an immediate repetition of the position measurement again will indicate that the particle lies at $x = x_0$; if the particle is moving in a one-dimensional force field, described by the potential $V(x)$, the particle's potential energy will be exactly $V(x_0)$. Knowing that the particle lies at $x = x_0$ makes the momentum p_x of the particle completely uncertain, however, according to relation (2a). A measurement of p_x immediately after the particle is located at $x = x_0$ could yield any value of p_x from $-\infty$ to $+\infty$; a measurement of $T = p_x^2/2m$ could yield any value from 0 to $+\infty$, and in fact the average or expectation value of T in these circumstances would be infinite.

More generally, suppose the system is known to be in the eigenstate corresponding to the eigenvalue α of the observable A. Then for any observable B, which is to some extent complementary to A, that is, for which an uncertainty relation of the form of relations (2) limits the accuracy with which A and B can simultaneously be measured, it is not possible to predict which of the many possible values $B = \beta$ will be observed. However, it is possible to predict the relative probabilities $P_\alpha(\beta)$ of immediately thereafter finding the observable B equal to β, that is, of finding the system in the eigenstate corresponding to the eigenvalue $B = \beta$. If the system is prepared in the eigenstate α of A a great many times, and each time the observable B is measured immediately thereafter, the average of these observed values of B will equal the expectation value of B, defined by Eq. (7), summed over all

$$\langle B \rangle = \Sigma \beta P_\alpha(\beta) \qquad (7)$$

eigenvalues of B; when the spectrum is continuous, the summation sign is replaced by an integral. To the eigenvalues correspond eigenfunctions, in terms of which $P_\alpha(\beta)$ can be computed. In particular, when α is a discrete eigenvalue of A, and the operators depend only on x and p_x, the probability $P_\alpha(\beta)$ is postulated as in Eq. (8), where $u(x,\alpha)$ is the

$$P_\alpha(\beta) = \left| \int_{-\infty}^{\infty} dx \, v^*(x,\beta) u(x,\alpha) \right|^2 \qquad (8)$$

eigenfunction corresponding to $A = \alpha$; $v(x,\beta)$ is the eigenfunction corresponding to $B = \beta$; and the asterisk denotes the complex conjugate. Since measurement of A in the state $A = \alpha$ must yield the result $A = \alpha$, it is necessary that the states $u(x,\alpha)$

satisfy the normalizing and orthogonalizing relation in Eq. (9), where $\delta_{\alpha\alpha'} = 0$ when $\alpha \neq \alpha'$ (eigenfunctions normalized); the eigenfunctions $v(x,\beta)$

$$\left| \int_{-\infty}^{\infty} dx\, u^*(x,\alpha')\, u(x,\alpha) \right|^2 = \delta_{\alpha\alpha'} \qquad (9)$$

functions normalized); the eigenfunctions $v(x,\beta)$ are similarly orthonormal. The integral in Eq. (8) is called the projection of $u(x,\alpha)$ on $v(x,\beta)$. The projection of the eigenfunction corresponding to $A = \alpha$ on the eigenfunctions of the x operator is $u(x,\alpha)$, and $|u(x,\alpha)|^2\, dx$ is the probability that the system, known to be in the eigenstate $A = \alpha$, will be found in the interval x to $x + dx$. *See* EIGENFUNCTION; EIGENVALUE.

The formalism just described embodies the essential feature that each measurement on an individual system, as it develops new information, necessarily loses or makes untrue some information gained in the past; in fact, this formalism leads to a rigorous derivation of uncertainty relations (2). For example, suppose B is measured with the system in the state $A = \alpha$, and it is found that $B = \beta$ exactly. The act of measurement necessarily and unavoidably disturbs the system, with the result that after the measurement, the system is in the eigenstate $v(\beta,x)$. After the measurement, therefore, it no longer is certain that $A = \alpha$; rather the probability of finding $A = \alpha$ is $P_\beta(\alpha)$, which by Eq. (8) equals $P_\alpha(\beta)$. Thus, after starting with $A = \alpha$, and then measuring $B = \beta$, it is possible to find that A equals $\alpha' \neq \alpha$. Of course, as stressed previously, these considerations are unimportant for macroscopic systems, where the limitations imposed by the uncertainty principle are inconsequential. Even if all observations were extremely accurate by usual standards, when the momentum p_x of a 1-g mass at x is measured, the position x' immediately thereafter should be indistinguishable from x. Were x' and x distinguishable, the particle position seemingly would have changed discontinuously from x to x', contrary to all classical (macroscopic) experience.

This formalism yields predictions in excellent agreement with observation; furthermore, it can be seen that the formalism is internally consistent. Consequently, the following doctrine, embodied in a formalism, appears well established: Although it is possible to predict the average of a large number of observations on identical systems, the result of a measurement on a single (microscopic) system generally is unpredictable and largely a matter of chance. Nonetheless, some physicists have refused to accept this inherent indeterminancy of nature and believe that this doctrine is a serious deficiency of present physical theory. To put the problem in simplest terms, consider a gram of radium, containing approximately 10^{21} atoms. According to generally accepted theory, it is not possible to predict when any one atom will decay, but it is possible to predict very accurately the average number of atoms decaying every second. The objectors to this doctrine feel that it must be possible to predict the subsequent history of every individual atom; failure to do so represents, not an inherent indeterminism in nature, but rather a lack of obtainable information—and therefore a lack of understanding—concerning the mechanism of radioactive decay. To mention but one possible alternative, nonrelativistic quantum theory can be reinterpreted in terms of hidden variables, which in principle determine the precise behavior of an individual system but whose values are not ascertained in measurements of the type which now can be carried out. This alternative has not led to new predictions, however, and contains some unappealingly ad hoc features. *See* CAUSALITY; PROBABILITY (PHYSICS).

Wave function. When the system is known to be in the eigenstate corresponding to $A = \alpha$, the eigenfunction $u(x,\alpha)$ is the wave function; that is, it is the function whose projection on an eigenfunction $v(x,\beta)$ of any observable B gives the probability of measuring $B = \beta$. The wave function $\psi(x)$ may be known exactly; in other words, the state of the system may be known as exactly as possible (within the limitations of uncertainty and complementarity), even though $\psi(x)$ is not the eigenfunction of a known operator. This circumstance arises because the wave function obeys Schrödinger's wave equation. Knowing the value of $\psi(x)$ at time $t = 0$, the wave equation completely determines $\psi(x)$ at all future times. In general, however, if $\psi(x,0) = u(x,\alpha)$, that is, if $\psi(x,t)$ is an eigenfunction of A at $t = 0$, then $\psi(x,t)$ will not be an eigenfunction of A at later times $t > 0$. For example, suppose at $t = 0$ a free particle (a particle moving under no forces) is known to be in an eigenstate for which the uncertainty in x is $(\Delta x)_0$; $(\Delta x)_0$ is approximately the x interval within which $|\psi(x,0)|^2$ is not negligibly small. Suppose further that, at $t = 0$, the product of the uncertainties in position and momentum is as small as possible: $(\Delta x)_0 (\Delta p_x)_0 \cong \hbar$, compare relation (2a). Then it can be proved that $(\Delta x)_t$, the uncertainty in x at time t, satisfies Eq. (10), where

$$(\Delta x)_t = \left[(\Delta x)_0{}^2 + \frac{t^2}{m^2} (\Delta p_x)_0{}^2 \right]^{1/2} \qquad (10)$$

m is the particle mass. Equation (10) is readily interpreted. The root-mean-square spread at time t results from $(\Delta x)_0$ and from an uncertainty in the distance the particle has traveled; the latter uncertainty is $t(\Delta v_x)_0 = t(\Delta p_x)_0/m$. If $(\Delta x)_0 = 0$, meaning $\psi(x,0)$ is an eigenfunction of the x operator, $(\Delta x)_t$ is infinite, showing that $\psi(x,t)$ cannot be an eigenfunction of the x operator. When the particle is free, the projections of the wave function on the eigenfunctions of the momentum operator do not change with time, corresponding to the classical result that the momentum of a free particle does not change. Thus, the probability of measuring any value of the momentum does not change, $(\Delta p_x)_t = (\Delta p_x)_0$, and Eq. (10) shows that $(\Delta x)_t (\Delta p_x)_t$ grows with time for a free particle, whatever the value of $(\Delta x)_0$, provided $(\Delta x)_0 (\Delta p_x)_0 \cong \hbar$. Nonetheless, $\psi(x,t)$ is known no less exactly than the initial wave function $\psi(x,0)$. The magnitude of $\Delta x\, \Delta p_x$ at any instant is no measure of the exactness with which the state of the system is known; increased uncertainties in position or momentum may be the price for increased certainty in the value of some other observable.

A system described by a wave function is said to be in a pure state. Not all systems are described by wave functions, however. Consider, for example, a beam of hydrogen atoms streaming in the x direction out of a small hole in a hydrogen discharge tube. According to the formal theory, if the beam were described by a wave function $\psi(x)$, then Eq. (11) holds, where $u^+(z)$ is the eigenfunction corresponding to finding a hydrogen atom with its z

$$\psi(x) = C_+(x)\,u^+(z) + C_-(x)\,u^-(z) \qquad (11)$$

component J_z of total angular momentum equal to $\hbar/2$; $u^-(z)$ is the corresponding eigenfunction for finding $J_z = -\hbar/2$; $|C_+(x)|^2$ is the probability of finding $J_z = \hbar/2$ at any point x along the beam; $|C_-(x)|^2$ is the corresponding probability of finding $J_z = -\hbar/2$. Since there are only two possibilities, $J_z = \pm\hbar/2$, $|C_+(x)|^2 + |C_-(x)|^2 = 1$, and since there seems no reason to favor either of these possibilities, it is reasonable to suppose that $|C_+(x)|^2 = |C_-(x)|^2 = 1/2$. As Eq. (11) shows, however, to specify $\psi(x)$ it is necessary to know not merely the relative magnitudes of the complex numbers $C_+(x)$ and $C_-(x)$ but also their relative phase. It can be shown that each choice of relative magnitude and phase of C_+ and C_- corresponds to a direction γ for which there is probability 1 of finding $J_\gamma = \hbar/2$, and probability zero of finding $J_\gamma = -\hbar/2$. Thus each choice of relative magnitude and phase of C_+ and C_- puts the system in an eigenfunction $u^+(\gamma)$ or $u^-(\gamma)$, that is, in a pure state. On the other hand, the discharge tube singles out no particular direction in space, so that in a Stern-Gerlach experiment the original beam must break up into two beams of equal intensity, whatever the direction of the external magnetic field. Consequently, the original beam is not in a pure state, but can be regarded as a statistical ensemble or mixture of pure states oriented with equal probability in all directions. Equivalently, Eq. (11) can be used for the original beam, provided calculations are averaged over all relative phases of C_+ and C_-. The distinction between mixtures and pure states is strongly analogous to the distinction between polarized and unpolarized light beams; consequently, beams of particles in pure spin states are termed polarized.

Schrödinger equation. Equation (12) describes

$$\psi(x,t) = A(\lambda)\,\exp\!\left[2\pi i\!\left(\frac{x}{\lambda} - ft\right)\right] \qquad (12)$$

a plane wave of frequency f, wavelength λ, and amplitude $A(\lambda)$, propagating in the positive x direction. The previous discussion concerning wave-particle duality suggests that this is the form of the wave function for a beam of free particles moving in the x direction with momentum $p = p_x$, with Eq. (3) specifying the connections between f, λ and E, p. Differentiating Eq. (12), it is seen that Eqs. (13)

$$p_x\psi = \frac{h}{\lambda}\psi = \frac{\hbar}{i}\frac{\partial\psi}{\partial x} \qquad (13a)$$

$$E\psi = hf\psi = -\frac{\hbar}{i}\frac{\partial\psi}{\partial t} \qquad (13b)$$

hold. Since for a free particle $E = p^2/2m$, it follows also that Eq. (14) is valid. *See* WAVE MOTION.

$$\frac{-\hbar^2}{2m}\frac{\partial^2\psi}{\partial x^2} = -\frac{\hbar}{i}\frac{\partial\psi}{\partial t} \qquad (14)$$

Equation (14) holds for a plane wave of arbitrary λ, and therefore for any superposition of waves of arbitrary λ, that is, arbitrary p_x. Consequently, Eq. (14) should be the wave equation obeyed by the wave function of any particle moving under no forces, whatever the projections of the wave function on the eigenfunctions of p_x. Equations (13) and (14) further suggest that for a particle

whose potential energy $V(x)$ changes, in other words, for a particle in a conservative force field, $\psi(x,t)$ obeys Eq. (15).

$$\frac{-\hbar^2}{2m}\frac{\partial^2\psi}{\partial x^2} + V(x)\psi = -\frac{\hbar}{i}\frac{\partial\psi}{\partial t} \qquad (15)$$

Equation (15) is the time-dependent Schrödinger equation for a one-dimensional (along x), spinless particle. Noting Eq. (13b), and observing that Eq. (15) has a solution of the form of Eq. (16), it is inferred that $\psi(x)$ of Eq. (16) obeys the time-independent

$$\psi(x,t) = \psi(x)\,\exp\,(-iEt/\hbar) \qquad (16)$$

ferred that $\psi(x)$ of Eq. (16) obeys the time-independent Schrödinger equation, Eq. (17). *See* FORCE.

$$\frac{-\hbar^2}{2m}\frac{\partial^2\psi}{\partial x^2} + V(x)\psi = E\psi \qquad (17)$$

Equation (17) is solved subject to reasonable boundary conditions, for example, that ψ must be continuous and must not become infinite as x approaches $\pm\infty$. These boundary conditions restrict the values of E for which there exist acceptable solutions $\psi(x)$ to Eq. (17), the allowed values of E depending on $V(x)$. In this manner, the allowed energies of atomic hydrogen listed in the earlier discussion of quantization are obtained. When a $\psi(x)$ solving Eq. (17) exists, all probabilities inferred from the corresponding $\psi(x,t)$ are independent of time; see Eq. (8). Thus the allowed energies E of Eq. (17) are the energies of the stationary states (states not changing with time) of the system. *See* STATIONARY STATE.

The forms of Eqs. (13a), (15), and (17) suggest that the classical observable p_x must be replaced by the operator $(\hbar/i)\,(\partial/\partial x)$. With this replacement, Eq. (18) holds. In other words, whereas the classi-

$$(xp_x - p_x x)\psi = i\hbar\psi \qquad (18)$$

cal canonically conjugate variables x and p_x are numbers, obeying the commutative law in Eq. (19a), the quantum-mechanical quantities x and p_x are noncommuting operators, obeying Eq. (19b).

$$xp_x - p_x x = 0 \qquad (19a)$$

$$xp_x - p_x x = i\hbar \qquad (19b)$$

In the formal theory the noncommutativity of x, p_x, leads directly to uncertainty relation (2a). For a derivation of relation (2a) from Eq. (19b), as well as for generalizations and more sophisticated derivations of Eqs. (13), (15), and (19b), *see* MATRIX MECHANICS; NONRELATIVISTIC QUANTUM THEORY. *See also* HAMILTON'S EQUATIONS OF MOTION.

Correspondence principle. Since classical mechanics and Maxwell's electromagnetic theory accurately describe macroscopic phenomena, quantum mechanics must have a classical limit in which it is equivalent to the older classical theories. Although there is no rigorous proof of this principle for artibrarily complicated quantum-mechanical systems, its validity is well established by numerous illustrations, such as those mentioned in the preceding discussions of the uncertainty principle and wave-particle duality. In general, the classical limit is approached when (i) $h \to 0$; (ii) the mass becomes large; (iii) wavelengths become small; (iv) dimensions become large; (v) quantum numbers become large. (The notation $h \to 0$ refers to a mathematical operation in which one adheres to a fixed set of values of the other quantities involved and considers the effect of making h smaller and smaller.) These simple criteria must be employed cautiously, since they have not been stated in terms of dimensionless parameters; obviously the classical limit, in (ii) for instance, cannot depend on the unit of mass. Nonetheless, these criteria are useful guides; for example, Eq. (19a) is the limit of Eq. (19b) as $h \to 0$.

Before the introduction of the Schrödinger equation (1926) made possible exact determination of energy levels and related quantum numbers, the correspondence principle was very effectively employed as a heuristic means of arriving at the quantization rules. In particular, for periodic orbits there was evolved the rule in Eq. (20), where n is

$$\oint p \, dq = nh \tag{20}$$

an integer, q and p are canonically conjugate position and momentum variables, and the integration is performed along the orbit for one complete cycle. *See* HAMILTON-JACOBI THEORY; PERTURBATION (QUANTUM MECHANICS).

Equation (20) is not always exact, but nonetheless is useful, especially in the limit of high quantum numbers. In the case of a one-dimensional harmonic oscillator, for example, whose classical frequency is f and whose potential energy is $V(x) = 2\pi^2 m f^2 x^2$, Eq. (20) implies that the allowed energies are as defined by Eq. (21a), whereas the correct result, deduced from the Schrödinger equation, is given by Eq. (21b).

$$E = nhf \tag{21a}$$

$$E = (n + 1/2)hf \tag{21b}$$

See HARMONIC OSCILLATOR.

The existence of the zero-point energy $\frac{1}{2}hf$ in the ground state $n = 0$ has been confirmed in analyses of molecular spectra and has significance for many phenomena.

Quantum statistics; indistinguishability. Equation (20) means that the allowed orbits, plotted as functions of q and p, have area nh; the area between two allowed orbits equals h. Since in classical statistical mechanics (i) all orbits are allowed, and (ii) the statistical weight of a volume $dq \, dp$ of q, p (phase) space is proportional to $dq \, dp$, the correspondence principle suggests that in quantum statistics each allowed orbit replaces a set of classical orbits which cover an area h when plotted in the q, p plane. It follows that in quantum statistics: (i) each allowed quantized orbit should have the same statistical weight, which may be set equal to unity; and (ii) with unit weight for a quantized orbit, the weight of an unquantized volume $dp \, dq$ of phase space must be $(dp \, dq)/h$. Because the average number of atoms in a state of energy E is proportional to $\exp(-E/kT)$, where k is Boltzmann's constant and T the absolute temperature, it is reasonable that quantum statistics yields different predictions than classical (Boltzmann) statistics when the energy-level spacing is large compared to kT. However, in the limit that the level spacing becomes small compared to kT, which corresponds to the limit $h \to 0$, quantum statistics must be equivalent to Boltzmann statistics. The quantum theory of specific heat and the theory of blackbody radiation illustrate and confirm the considerations of this paragraph. *See* BOLTZMANN STATISTICS; HEAT RADIATION; QUANTUM STATISTICS; SPECIFIC HEAT OF SOLIDS; STATISTICAL MECHANICS.

Quantum statistics is further complicated by the fact that identical particles are indistinguishable. Classically, if two He atoms in their ground states are placed in a box, the statistical weights of states are computed as if the atoms can be distinguished, that is, as if each atom had an identifying mark. Quantum mechanics rejects the possibility of this identification and simultaneously modifies the statistical weights which otherwise would be used in computing statistical averages. Formally, this modification is accomplished by insisting that for a system of identical particles, the wave function must be symmetric in the coordinates of all the particles, including spin coordinates specifying spin orientations, if the particles have integral spin, namely, $0, \hbar, 2\hbar, \ldots$; and antisymmetric in the coordinates of all the particles, including spin coordinates, if the particles have half-integral spin, namely, $\hbar/2, 3\hbar/2, 5\hbar/2, \ldots$. *See* BOSE-EINSTEIN STATISTICS; FERMI-DIRAC STATISTICS.

A symmetric wave function is unchanged under interchange of coordinates, for example, for two particles $\psi(x_1, x_2) = \cos(x_1 - x_2)$ is symmetric; an antisymmetric wave function changes sign under interchange of coordinates, for example, $\psi(x_1, x_2) = \sin(x_1 - x_2)$ is antisymmetric. It is postulated that wave functions having other symmetry properties than these specified here never occur, and therefore must not be included in any enumeration of available states, even though without these symmetry restrictions, they might be acceptable solutions of the Schrödinger equation (for the many-particle system). These symmetry restrictions have profound consequences for macroscopic as well as microscopic systems, and are extremely well established for nuclei as well as atoms. The reason for the connection between spin and statistics is beyond the scope of this article, but appears to be a consequence of requirements imposed by the special theory of relativity. *See* EXCLUSION PRINCIPLE.   [EDWARD GERJUOY]

Bibliography: D. Bohm, *Quantum Theory*, 1951; R. Eisberg and R. Resnick, *Quantum Physics of Atoms, Molecules, Solids, Nuclei, and Particles*, 1974; S. G. Gasiorowicz, *Quantum Physics*, 1974; A. Messiah, *Quantum Mechanics*, vol. 1, 1961, vol. 2, 1962; F. K. Richtmyer, E. H. Kennard, and J. N. Cooper, *Introduction to Modern Physics*, 6th ed., 1969; M. R. Wehr, J. A. Richards, and T. W. Adair, *Physics of the Atom*, 3d ed., 1978.

Quantum numbers

The quantities, usually discrete with integer or half-integer values, which are needed to characterize a physical system of one or more atomic or subatomic particles. Specification of the set of quantum numbers serves to define such a system or, in other words, to label the possible states the system may have. In general, quantum numbers are obtained from conserved quantities determinable by performing symmetry transformations consisting of arbitrary variations of the system which leave the system unchanged. For example, since the behavior of a set of particles should be independent of the location of the origin in space and time (that is, the symmetry operation is translation in space-time), it follows that momentum and energy are rigorously conserved. In similar fashion, the invariance of the system to arbitrary rotations of the three axes defining three-dimensional space leads to conservation of total angular momentum. Inversion of these three axes through their origin is likewise a symmetry operation and leads to conservation of parity. (Parity is not conserved in weak interactions and cannot be used as a quantum number to specify the state of a set of particles undergoing such interactions. For such a case, parity is not a "good" quantum number.) For a list of invariance or symmetry operations and the corresponding conserved quantities *see* SYMMETRY LAWS.

Of these conserved quantities, the ones that can be used to label a physical state are momentum, energy, angular momentum, charge, baryon number, lepton number, parity, charge parity, product of parity and charge parity, isotopic spin and third component thereof, hypercharge, isotopic parity (or G parity), and unitary spin. In general, each physical system must be studied individually to find the symmetry transformations, and thus the conserved quantities and possible quantum numbers. The quantum numbers themselves, that is, the actual state labels, are usually the eigenvalues of the physical operators corresponding to the conserved quantities for the system in question. *See* EIGENVALUE; ELEMENTARY PARTICLE; NONRELATIVISTIC QUANTUM THEORY; PARITY.

It is not necessary that the conserved quantity be "quantized" in order to be regarded as a quantum number; for example, a free particle possesses energy and momentum, both of which can have values from a continuum but which are used to specify the state of the particle. For a particle at rest the momentum is zero, and the energy is given by the rest mass; the additional specification of values for the remaining (discrete) quantities in the list mentioned above completely identifies the particle. The determination of these quantum numbers for the numerous strongly interacting particles being discovered in experiments with high-energy accelerators constitutes the developing field of meson and baryon (particles with baryon number zero and one, respectively) spectroscopy.

Some important numbers. Probably the most familiar quantum numbers are those associated with the restricted (or quantized) energy values which appear naturally in quantum mechanics whenever one or more particles are confined to a small region of space. The wave equation (Schrödinger equation in the nonrelativistic case) for the confined particle can be solved to express these allowed energy values in terms of one or more quantum numbers. The number of nodes, or zeros, that the solution of the wave equation possesses always enters such an expression and is usually called the principal quantum number n. In the case of a charged particle confined to the vicinity of an oppositely charged particle according to Coulomb's law (for example, the electron to the proton in a hydrogen atom), the Schrödinger equation yields possible energies, $-E_B/n^2$, where $n = 1, 2, 3, \ldots$ and for hydrogen $E_B = 13.6$ ev. The magnitude of n is also directly related to the most probable distance from the center of the confining region where the electron may be found. *See* ATOMIC STRUCTURE AND SPECTRA; COULOMB'S LAW.

The wave equation for the confined particle also leads to quantized angular momentum, with possible values or quantum numbers $L = 0, 1, 2, 3, \ldots$ in units of $\hbar$ (Planck's constant divided by 2π). In addition, each particle, whether confined or not, possesses an intrinsic spin angular momentum S with possible values $0, \frac{1}{2}, 1, \frac{3}{2}, \ldots$, in units of $\hbar$. The electron, for example, has a spin of $\frac{1}{2}$. *See* SPIN.

Since angular momentum is a quantized vector quantity (a direction in space is associated with it), it might be expected that the three space components would also be quantized observables. However, only one, the Z-axis projection, is observable and the associated quantum number m_L has possible values $-L, -L+1, \ldots, L-1, L$; correspondingly, the spin angular momentum M_S has possible values $-S, -S+1, \ldots, S-1, S$. These are often called magnetic quantum numbers because they determine how the energy changes when a system with given n, L, S is placed in a magnetic field.

Determination of numbers. Thus the properties or state of a confined electron (for example, in the hydrogen atom) are known when the quantum numbers n, L, M_L, and M_S ($S = \frac{1}{2}\hbar$) are specified. Actually, when two or more angular momenta enter, as for the confined electron with an intrinsic $S = \frac{1}{2}\hbar$ in a state with orbital angular momentum L or especially for a system with more than one particle, the overall rotational symmetry of the system is best expressed by a total angular momentum quantum number J and a total magnetic quantum number M. The resultant J is obtained by vectorially adding all individual angular momenta, and M is obtained by summing all the separate M_S's and M_L's. For a many-electron atom or many-nucleon nucleus, the preferred way to perform the vector addition is often indicated by the dynamics, that is, how the individual particle L and S values enter into interparticle interactions. In such a case an intermediate result in the determination of J may be a good quantum number; for example, the seniority quantum number is related to the number of particles whose spins are paired. The objective of atomic and nuclear spectroscopy is the determination of the quantum numbers for the states of many-electron and many-nucleon systems, respectively. Nucleon systems have the additional quantum numbers of isotopic spin and third-axis projection. *See* ELECTRON CONFIGURATION; NUCLEAR STRUCTURE. [KENNETH E. LASSILA]

Bibliography: B. Lichtenberg, *Unitary Symmetry and Elementary Particles*, 2d ed., 1978; Particle Data Group, Review of particle properties, *Rev. Mod. Phys.*, vol. 52, no. 2, pt. 11, 1980; M. A. Preston and R. K. Bhaduri, *Structure of the Nucleus*, 1975; P. Roman, *Advanced Quantum Theory*, 1965; J. J. Sakurai, *Invariance Principles and Elementary Particles*, 1964; H. F. Schaefer, *Modern Theoretical Chemistry*, vols. 3 and 4, 1971.

Quantum solids

A class of solids whose atoms or molecules undergo large zero-point motion even in the quantum ground state (at temperature $T = 0$ K) as a result of their small mass and the weak attractive part of their interaction potential. The most striking examples are the isotopes of helium, ^{3}He and ^{4}He, which have a root-mean-square displacement from their lattice sites of approximately 25%. Further examples are the molecular hydrogens, H_2, D_2, and HD, as well as some heavier molecular solids. *See* INTERMOLECULAR FORCES; NONRELATIVISTIC QUANTUM THEORY; QUANTUM MECHANICS.

Phase diagrams of ^{3}He and ^{4}He. These materials display quantum effects in their bulk properties when cooled to temperatures near absolute zero so that the chaotic thermal motion is reduced. Some of these effects can be seen in the phase diagrams of ^{3}He and ^{4}He, as shown in Figs. 1 and 2. Both of the isotopes remain liquid all the way to absolute zero, unless external pressure ($\sim$3 MPa $\approx$ 30 atm) is applied. This is because the atoms are not at rest at 0 K; the zero-point motion acts as an internal pressure which must be overcome in order to bring the atoms close enough together for solidification. All other substances, including the hydrogens, freeze under their own vapor pressure above 10 K.

The two melting curves are quite different in detail because of the different types of quantum statistics which the particles obey. Helium-4 has zero spin (angular momentum) and therefore obeys Bose statistics, while ^{3}He has a spin of $\frac{1}{2}$ and obeys Fermi statistics. In ^{4}He the melting curve becomes very flat shortly after the intersection with the λ-line marking the superfluid transition. This is because both the liquid and the solid have very small amounts of entropy remaining. The freezing process is then largely a mechanical one of decreasing the volume, with very little latent heat involved. *See* BOSE-EINSTEIN STATISTICS; FERMI-DIRAC STATISTICS; LIQUID HELIUM.

The pronounced minimum in the ^{3}He melting pressure is unique, and can be understood by considering the entropies of the liquid S_l and the solid S_s. Ordinarily a liquid has more entropy than the solid in equilibrium with it. However, in solid ^{3}He the spin system retains its orientational entropy of $R \ln 2$ down to a few millikelvin (mK). (Here R is the gas constant, and the factor of 2 comes from the two possible spin orientations.) Consequently, below 0.32 K, the temperature of the pressure minimum, the solid has the greater entropy. The slope of the melting curve $(dP/dT)_m$ is related to the entropy difference $S_l - S_s$ through the Clausius-Clapeyron equation given here, in which V_l and V_s

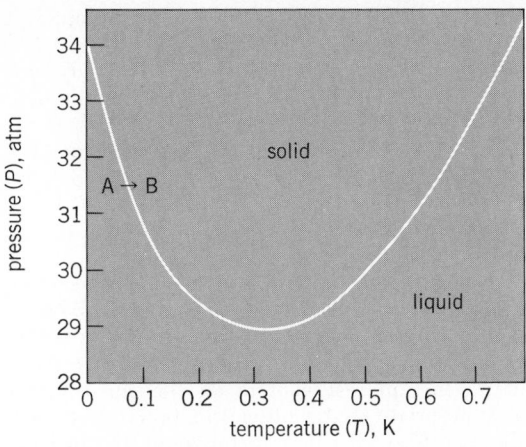

Fig. 2. Phase diagram of ^{3}He (the origin of the vertical scale is suppressed). The minimum melting pressure at $T = 0.32$ K occurs when the liquid and solid entropies are equal; 1 atm = 101.325 kPa.

$$\left(\frac{dP}{dT}\right)_m = \frac{S_l - S_s}{V_l - V_s}$$

are the volumes of the liquid and solid; this slope is therefore negative below 0.32 K. *See* ENTROPY.

The pressure minimum leads to the bizarre situation of the addition of heat causing freezing, as illustrated by the process A → B in Fig. 2. The inverse of this process, the adiabatic formation of the solid by compression, is an important process which has been used extensively to cool liquid and solid ^{3}He to temperatures of approximately 1 mK. *See* CRYOGENICS.

Nuclear spin alignment in solid ^{3}He. In solids with heavy particles which behave classically, the weak dipolar interactions between nuclear spins lead to their alignment (ordering) only at temperatures on the order of 10^{-7} K. (It is this feature which allows temperatures of this order to be attained by nuclear demagnetization of materials such as copper.) However, in solid ^{3}He there is a quantum effect which produces alignment at 10^{-3} K. This results from the combination of an antisymmetric wave function required by the Fermi statistics and a sizeable overlap of wave functions of neighboring atoms because of their large zero-point motion. Since these circumstances make it possible for particles to exchange position in the lattice, the effect is called an exchange interaction. This interaction in solid ^{3}He turns out to be 10^4 times larger than the dipolar interaction, and leads to ordering of the spins in a magnetic phase transition at 10^{-3} K. *See* NUCLEAR ORIENTATION.

Molecular quantum solids. In the molecular quantum solids such as H_2, the main interest is in the orientational degrees of freedom. Hydrogen molecules consist of two types: para molecules with zero angular momentum, J, and ortho molecules with $J = 1$. In the region below about 1 K, cooperative orientational effects in the nonspherical ortho molecules produce a phase with long-range order. Orientational transitions are also of interest in heavy quantum solids such as CH_4 and N_2.

[E. DWIGHT ADAMS]

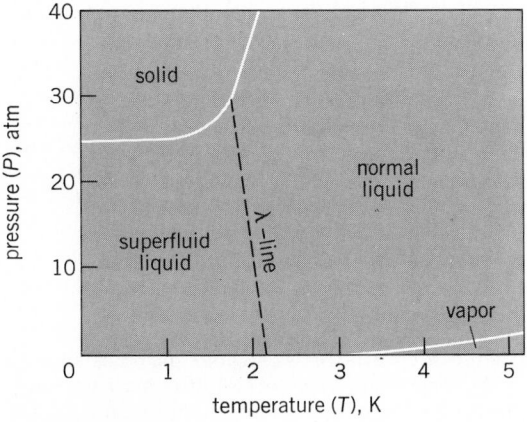

Fig. 1. Phase diagram of ^{4}He in the pressure-temperature plane; 1 atm = 101.325 kPa.

Bibliography: H. R. Glyde, Solid helium, in M. L. Klein and J. A. Venables (eds.), *Rare Gas Solids*, vol. 1, pp. 382–504, 1976; R. A. Guyer, Solid ^{3}He: A magnet in search of a hamiltonian, *J. Low Temp. Phys.*, 30:1–50, 1978; L. H. Nosanow, Theory of quantum crystal, *Phys. Rev.*, 146:120–133, 1966; S. B. Trickey, E. D. Adams, and J. W. Dufty (eds.), *Quantum Fluids and Solids*, 1977.

Quantum statistics

The statistical description of particles or systems of particles whose behavior must be described by quantum mechanics rather than by classical mechanics. As in classical, that is, Boltzmann statistics, the interest centers on the construction of appropriate distribution functions. However, whereas these distribution functions in classical statistical mechanics describe the number of particles in given (in fact, finite) momentum and positional ranges, in quantum statistics the distribution functions give the number of particles in a group of discrete energy levels. In an individual energy level there may be, according to quantum mechanics, either a single particle or any number of particles. This is determined by the symmetry character of the wave functions. For antisymmetric wave functions only one particle (without spin) may occupy a state; for symmetric wave functions, any number is possible. Based on this distinction, there are two separate distributions, the Fermi-Dirac distribution for systems described by antisymmetric wave functions and the Bose-Einstein distribution for systems described by symmetric wave functions.

In relativistic quantum theory it is shown that particles having integer spin necessarily obey Bose-Einstein statistics, while those having half-integer spin necessarily obey Fermi-Dirac statistics. (Particles obeying Bose-Einstein statistics are often called bosons; particles obeying Fermi-Dirac statistics, fermions.) For sufficiently high temperatures, both forms of distribution functions go over into the familiar Boltzmann distribution, although strictly speaking no system is correctly described by this distribution. In practice, of course, the Boltzmann distribution gives an exceedingly good description of the experiments, but there are situations, such as those involving the behavior of electrons in metals and liquid helium, where the quantum description is essential. *See* BOLTZMANN STATISTICS; BOSE-EINSTEIN STATISTICS; EXCLUSION PRINCIPLE; FERMI-DIRAC STATISTICS; KINETIC THEORY OF MATTER; NONRELATIVISTIC QUANTUM THEORY; QUANTUM MECHANICS; RELATIVISTIC QUANTUM THEORY; SPIN; STATISTICAL MECHANICS.

[MAX DRESDEN]

Quarks

The basic constituent particles, of which "elementary" particles are now believed to be composed. Theoretical models built on the quark concept have been very successful in understanding the predicting many phenomena in particle physics. However, the experimental observation of free quarks remains ambiguous.

Search for fundamental constituents. Physics research for almost 2 centuries has been probing progressively deeper into the structure of matter in order to seek at every stage the constituents of each previously "fundamental" entity. Thus a sequence proceeding from crystals through molecules, atoms, and nuclei to nucleons and mesons has been revealed. The energies required to dissociate each entity increase from thermal energies to gigaelectronvolts in proceeding from crystals to mesons. It is therefore only a logical extrapolation of past patterns to expect that hadrons—mesons and baryons—might be dissociated into more fundamental constituents if subjected to a sufficiently high energy. *See* ATOMIC PHYSICS; HADRON; NUCLEAR PHYSICS.

Evidence supporting the quark model. There were at least three factors which made a quark model plausible in the 1960s. First, the classic electron-proton elastic scattering experiments demonstrated that a proton has a finite form factor. Nonrelativistically, this is equivalent to a finite radial extent of the electric charge and magnetic moment distributions. It was plausible that the charge cloud which constitutes a proton is a probability distribution of some smaller, perhaps pointlike constituents, just as the charge cloud of an atom was learned to be a probability distribution of point electrons.

Second, the evolution in the late 1950s and early 1960s of hadron spectroscopy revealed an order and symmetry among the states of hadronic matter that could be interpreted in terms of representations of the SU(3) symmetry group. This in turn was interpreted by M. Gell-Mann, and independently by G. Zweig, as a consequence of the grouping of elementary constituents of fractional electric charge, (christened quarks by Gell-Mann) in pairs and triplets to form the observed hadrons. The general features of the quark model of hadrons have withstood the tests of time, and many of the static properties of hadrons are consistent with predictions of this model. *See* SYMMETRY LAWS; UNITARY SYMMETRY.

Third, the deep inelastic scattering of electrons on protons revealed form factors corresponding to pointlike constituents of the proton. This is altogether consistent with the interpretation of the finite proton elastic form factor suggested above, in analogy to atomic electrons. J. D. Bjorken referred to these proton constituents as partons, although from the beginning it was recognized the partons and quarks might be merely different manifestations of the same entities. R. P. Feynman also invoked the parton notion of proton structure to explain the nature of secondary particle (pion) distributions in high-energy proton-proton collisions.

Physicists now believe that the proton and neutron are not fundamental constituents of matter, but that they are made of quarks, very much as the nuclei of ^{3}H and ^{3}He are made of protons and neutrons and as the molecules of NO_2 and N_2O are made of oxygen and nitrogen atoms.

Kinds of quarks. Until 1974 only three flavors of quarks were known; two of very nearly equal mass, of which the proton, neutron, and pi mesons are composed, and a third, more massive quark which is a constituent of strange particles such as the K mesons and hyperons such as $\Lambda°$. The names attached to these quarks are the up quark *(u)*, the down quark *(d)*, and the strange quark *(s)*. Baryons are presumed to be composed of three quarks, such as the proton *(uud)*, neutron *(udd)*, $\Lambda°$ *(uds)*,

Properties of quarks

Flavor	u	d	c	s	t^*	b
Mass, $(GeV/c^2)^\dagger$	0.39	0.39	1.55	0.51	(>15)	4.72
Electric charge	+2/3	−1/3	+2/3	−1/3	(+2/3)	−1/3
Baryon number	1/3	1/3	1/3	1/3	(1/3)	1/3
Spin in units of ($\hbar$)	1/2	1/2	1/2	1/2	(1/2)	1/2
Isotopic spin	+1/2	−1/2	0	0	(0)	0
Strangeness	0	0	0	−1	(0)	0
Charm	0	0	+1	0	(0)	0

*The t quark has not yet been found. Listed are its predicted properties.

†Masses are uncertain. Values listed here are half the mass of the lowest-lying quark-antiquark vector meson.

and $\Xi^-(dss)$. Mesons are composed of a quark-antiquark pair, such as the $\pi^+(u\bar{d})$, $\pi^-(\bar{u}d)$, $K^+(u\bar{s})$, and $K^-(\bar{u}s)$. Antiparticles such as the antiprotons are formed by the antiquarks of those forming the particle, for example, the antiproton $\bar{p}$ $(\overline{uud})$. See BARYON; MESON.

The quantum numbers of quarks are simply added to give the quantum numbers of the elementary particle which they form on combination. The natural unit of electric charge of a quark is $+\frac{2}{3}$ or $-\frac{1}{3}$ of the charge on a proton (1.6×10^{-19} coulomb), and the baryon number of each quark is $+\frac{1}{3}$; the charge, baryon number, and so forth, of each antiquark is just the negative of that for each quark.

The properties of quarks, to the extent that they are now understood, are presented in the table, where additional quarks discussed below are also included.

The manner in which the u, d, and s quarks and their antiquarks may combine to form families of mesons is indicated in Fig. 1 for one such multiplet, a meson pseudoscalar nonet. On this plot, "hypercharge" is a quantum number related to the quark strangeness and baryon number, whereas "isospin" (here the third component of isotopic

spin) is a quantum number related to the u-d quark difference. See HYPERCHARGE; ISOTOPIC SPIN.

In the strong interactions of elementary particles, quark-antiquark pairs may be created if sufficient energy is present, but a quark does not transform into another quark. The weak interaction, however, does permit quark transformations to occur, so that a u quark becomes a d quark in the radioactive decay of a neutron, and elementary particles containing an s quark (strange particles) decay to nonstrange particles only through the weak interaction. See STRANGE PARTICLES; WEAK NUCLEAR INTERACTIONS.

A particularly interesting set of particles is the group of vector mesons composed of a quark-antiquark pair. An example is the ϕ meson, composed of an $s\bar{s}$ quark pair. It may decay to a K^+K^- meson pair, but is inhibited from decaying to a $\pi^+\pi^-$, as the latter contain no s or $\bar{s}$ quarks. The meson may also decay electromagnetically to a $\mu^+\mu^-$ or e^+e^- lepton pair; conversely it may be produced by a photon (which coincidentally has the same quantum numbers) or by the collision of an electron and positron.

Such quark-antiquark systems are very analogous to the positronium atom, composed of a positron, e^+, and an electron, e^-. The force binding this atom together is the electromagnetic Coulomb attraction, whereas the force binding the quark-antiquark system is the strong interaction. In both cases, the energy levels of excited states of the system are directly related to the details of the force holding the particles together. See POSITRONIUM.

Charm. In 1974 two very different experiments, at the Brookhaven National Laboratory and the Stanford Linear Accelerator Center, found evidence for unusual elementary particles with lifetimes about 1000 times longer than typical mesons which decay through the strong interaction. These particles (labeled the J or ψ) were the first evidence for the existence of a new quark, the c (charm) quark. In the Brookhaven experiment of S. C. C. Ting and his collaborators, a high-intensity proton beam of 30 GeV was directed onto a beryllium target, and the invariant mass of produced e^+e^- pairs studied for evidence of the new particle. B. Richter with a Stanford–University of California group found a sharp resonance in the cross section for e^+ e^- interactions in an electron-positron colliding-beam storage ring. Ting labeled his particle the J; Richter independently called it the ψ.

The J/ψ meson is understood to be the vector meson composed of the $c\bar{c}$ quark pair, and its decay into mesons is strongly inhibited, so that its lifetime is about 1000 times typical heavy, unstable mesons. This new quark had been predicted earlier by S. Glashow. Subsequently mesons and baryons containing the c quark were found and studied. The two-dimensional diagram of Fig. 1 may be elaborated by adding a third dimension corresponding to the charm quantum number, with integral values of charge, and so forth, defining a polyhedron whose vertices correspond to identified and to predicted charmed mesons. Such a diagram is reproduced in Fig. 2 for the pseudoscalar mesons, including those of Fig. 1. See CHARM; J PARTICLE.

Upsilon particle. At the 400-GeV proton synchrotron of the Fermi National Accelerator Labo-

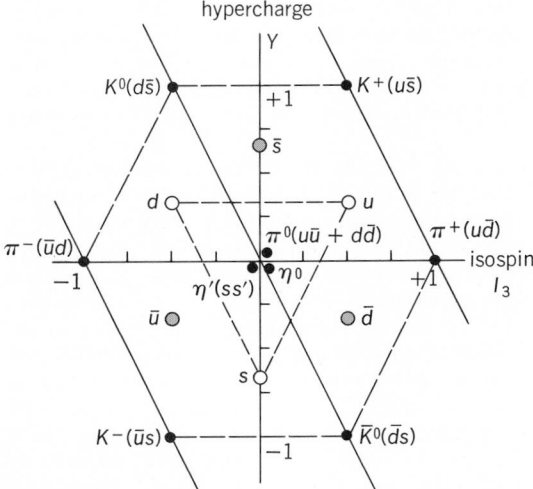

Fig. 1. The pseudoscalar meson nonet and the u, d and s quarks and antiquarks as represented in isospin (I_3)–hypercharge space. The diagonal lines sloping downward to the right are lines of constant electric charge.

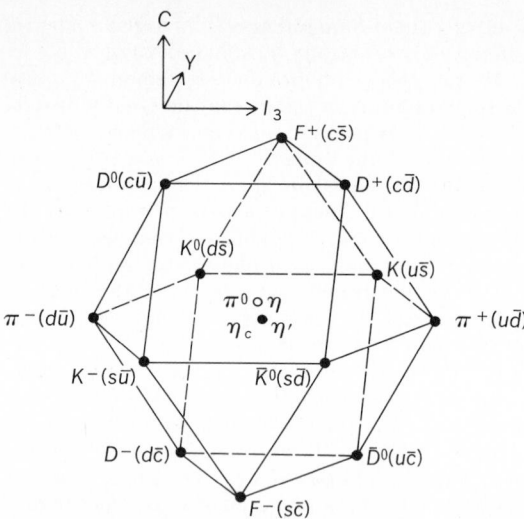

Fig. 2. Pseudoscalar mesons composed of $u, d, s,$ and c quarks. The three orthogonal axes are the z component of isospin I_3, hypercharge Y, and charm C. (From C. Quigg, *Lectures on charmed particles*, Fermi National Accelerator Laboratory, Fermilab-Conf-78/37-THY, April 1978)

ratory near Chicago, L. M. Lederman and his colleagues studied the production of $\mu^+\mu^-$ meson pairs in an experiment which in many ways resembled a scaled-up version of Ting's Brookhaven experiment. In 1977 the group reported a new resonance at about 9.4 GeV; subsequent detailed study by this and other groups confirmed that this is the lowest-lying state of a new quark system. Experiments at the electron-positron storage ring at the

Deutsches Elektronen Synchrotron (DESY) laboratory near Hamburg strongly indicated that the new quark has a charge of $-\frac{1}{3}$. The quarks which make up Lederman's Y (upsilon) particle have been called b for "bottom" (or "beauty"). There is almost surely a new quantum number associated with the b, and there is some evidence for a meson containing a b quark and an ordinary u or d antiquark. One could then imagine a four-dimensional extension of Fig. 2 to include another hierarchy of mesons, such as $b\bar{u}$, $b\bar{s}$, and $c\bar{b}$.

Theoretical physicists expect that there is a heavier quark, as yet undiscovered, which may have the same relationship to b that the charm quark does to the strange quark. This quark, together with its presumed quantum numbers, is indicated in the table as the t quark, where t stands for "top" or "truth." However, careful searches for the $t\bar{t}$ particle have been negative, up to total energies of 31 GeV. *See* UPSILON PARTICLE.

Color. The most natural spin assignment for quarks is $\frac{1}{2}$, such that their intrinsic angular momentum is $\hbar/2$ (where $\hbar$ is Planck's constant h divided by 2π), exactly as for the electron and muon. This results in spins of 0,1, or higher integers (for mesons composed of quark-antiquark pairs with orbital angular momentum about each other), and in spins for three-quark baryons (protons and so forth) of $\frac{1}{2}$, $\frac{3}{2}$, and so forth. A problem arose when the structure of observed baryons required two or, in some cases, three quarks of the same flavor in the same quantum state, a situation which is forbidden for spin-$\frac{1}{2}$ particles by the Pauli exclusion principle. In order to accommodate this contradiction, the concept of color was introduced, and it was proposed that each quark could be red, green, or blue. This color quantum number then breaks the degeneracy and allows up to three quarks of the same flavor to occupy a single quantum state. Since the original proposal, confirmation of the color concept has been obtained from experiments with electron-positron storage rings, and theory of quantum chromodynamics (analogous to quantum electrodynamics) has been developed. *See* COLOR (QUANTUM MECHANICS); EXCLUSION PRINCIPLE; QUANTUM CHROMODYNAMICS.

Searches for free quarks. One outstanding mystery remains: why are quarks not observed as free objects in high-energy collisions? Their fractional electric charge should render them easily detectable. The reason is that the ionization, and hence the signal, produced by a charged particle of very high energy passing through a detector is proportional to the square of its electric charge. A quark would then give a signal of $\frac{1}{9}$ or $\frac{4}{9}$ that from an electron, proton, meson, or any other known charged particle.

One possible explanation for the failure of experiments to detect quarks would be that their rest mass is so great that searches at particle accelerators have failed to reach the threshold for their production. Alternatively (or in addition) they might be produced so rarely that the quark production cross section might be very low.

Sensitive searches have been made at all high-energy particle accelerators and with cosmic rays. In 1969 positive evidence for quarks was reported from cosmic rays by C. B. A. McCusker of Sydney,

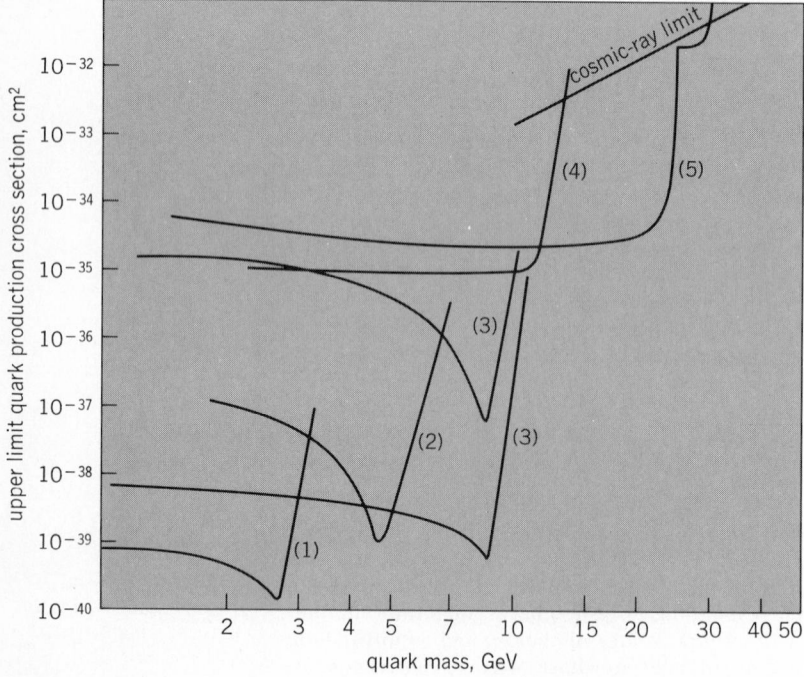

Fig. 3. Upper-limit cross section for production of quarks of charge-$\frac{1}{3}$ e as a function of quark mass from experiments at various particle accelerators and cosmic rays. The different experiments are indicated by numbers: (1) CERN 28-GeV proton synchrotron; (2) Serpukov 70-GeV proton synchrotron; (3, 4) Fermi National Accelerator Laboratory synchrotron; (5) CERN intersecting storage rings. (From L. W. Jones, *A review of modern quark search experiments*, Rev. Mod. Phys., 49:717–752, 1977)

Australia, and by some other cosmic-ray groups. Subsequent, more sensitive cosmic-ray experiments have, however, failed to confirm those observations.

In an experiment at the intersecting storage rings (ISR) of the European Organization for Nuclear Research (CERN), several "telescopes" of scintillation counters, Cerenkov detectors, and multiwire proportional chambers were assembled to search for particles of fractional charge emerging from the proton-proton (p-p) intersection region. These p-p collisions at 52 GeV in the center of mass (equivalent to a proton of 1500 GeV striking a proton at rest) would be energetically above threshold for the production of quarks in pairs of up to 23 GeV rest mass. No quarks were seen among the products of 1.2×10^{10} proton-proton interactions, corresponding to an upper-limit quark production cross section of 4×10^{-35} cm^2, or less than one quark per billion proton-proton interactions.

Two other experiments were carried out at the synchrotron of the Fermi National Accelerator Laboratory (FNAL) near Chicago. Here protons of up to 300 GeV struck a beryllium target, and beams of secondary particles were detected emerging from it. In the first experiment, a telescope of scintillation counters was used to search for particles of fractional electric charge. No quarks were found among 1.4×10^9 negative particles (mesons) corresponding to an upper-limit production cross section of about 10^{-35} cm^2. With 300-GeV protons incident, this search was sensitive to quarks of up to 12 GeV rest mass. The second search was made with similar beams over a somewhat longer period of time and with additional detectors. No quarks were detected with an upper-limit production cross section of 10^{-39} cm^2 for 9 GeV/c^2 rest mass quarks of $\frac{1}{3}e$ charge and 10^{-38} cm^2 for 11 GeV/c^2 quarks of $-\frac{2}{3}e$ charge. See PARTICLE ACCELERATOR.

These searches, together with accelerator and cosmic-ray searches reported earlier, are summarized in Fig. 3 for quarks of charge $-\frac{1}{3}$. The area of production cross section–quark mass to the lower right of the curves represents still possible quark production. As quarks are not observed even at very high energy, physicists are inclined to believe that they may not be heavy objects after all but may have a mass of only $100-500$ MeV/c^2 (as in the table), and that they cannot be observed as free particles. It has also been suggested that they interact through a long-range force that increases with the separation between two quarks.

Search for quarks in stable matter. It is also possible to seek quarks in stable matter by exploring the net electric charge on small samples, such as oil droplets or metal pellets. The charge of the electron e was first determined in 1911 by R. A. Millikan in his oil drop experiments, and variations of that technique have been used to search for net electric charges of $\pm\frac{1}{3}e$. Whereas a number of experimenters have reported negative results of searches for fractional charges, W. M. Fairbank and colleagues have reported positive evidence for quarks. In their experiment, Fairbank's group suspended superconducting niobium pellets of about 100 μg in a magnetic potential well, and then caused them to oscillate under an applied electric field. Several of the nine pellets that they have studied have given evidence for fractional electric charge. A peculiar aspect of their evidence for free quarks is that the pellets seem to gain or lose their quarks somewhat capriciously. It is also disturbing that this is the only group to claim evidence for quarks, in spite of considerable efforts and comparable sensitivity elsewhere. The Fairbank result is difficult but not impossible to reconcile with the otherwise negative evidence. Most physicists would prefer to await an independent confirmation of this result before accepting the idea of free quarks.

Understanding of quarks. Quarks seem firmly established as fundamental constituents of matter. Theoretical physicists have evolved a theoretical structure, quantum chromodynamics, to understand and predict the behavior of elementary particles, based on the interactions of their constituent quarks. With the understanding of quarks, it may be possible to evolve a grand unified theory combining the forces of electricity and magnetism, the strong and the weak interactions, all into one coherent theoretical framework, and a major theoretical effort in this direction has been undertaken. See FUNDAMENTAL INTERACTIONS.

Unresolved questions. In spite of the dramatic progress in the understanding of elementary particles which has accompanied the quark concept and the corresponding revolution in the concept of matter at its most fundamental level, at least three major questions remain.

First, it has not been determined whether or not there are free quarks, and whether Fairbank's observations are correct, or there is an error in his experiment. Second, the number of flavors (kinds) of quarks has not been determined. The predicted t quark has not been found, and it is not known whether there are an infinite number of quarks, or whether the flavor spectrum stops at five or six. Third, it is not known whether quarks are truly fundamental entities, or whether quarks in turn have their own internal substructure. See ELEMENTARY PARTICLE.   [LAWRENCE W. JONES]

Bibliography: R. P. Feynman, Structure of the proton, *Science,* 183:601–610, 1974; S. L. Glashow, Quarks with color and flavor. *Sci. Amer.,* 233(4): 38–50, 1975; L. W. Jones, A review of quark search experiments, *Rev. Mod. Phys.,* 49(4): 717–752, 1977; L. M. Lederman, The upsilon particle, *Sci. Amer.,* 239(4):72, 1978; R. F. Schwitters, Fundamental particles with charm, *Sci. Amer.,* 237(4):56–70, 1977.

Quasiatom

A transient electronic structure, formed in an atom-atom or ion-atom collision, which closely approximates the characteristics usually identified with a stable atom whose atomic number equals the combined charge carried by the colliding nuclei. Such short-lived atomic states can be formed in energetic atomic collisions, when the distance of closest approach between nuclear charge centers shrinks beyond the mean classical orbiting radius of the most-bound electrons, so that even these innermost electrons see a nearly monopolar electric field generated by the single charge center momentarily formed by the two nuclear collision partners. Thus, for the short time of this close nuclear promixity, all the surrounding electrons behave as if their motion were governed by this single charge center, leading to the formation of an electronic energy-level structure and electronic shells associated with a united atom having atomic number $Z = Z_1 + Z_2$, where Z_1 and Z_2 are the individual atomic numbers of the colliding system. *See* ATOMIC STRUCTURE AND SPECTRA.

Quasiatom formation. The underlying theoretical basis for the validity of the quasiatomic concept derives from the condition that the collision time is sufficiently long compared with the classical orbiting time of the electrons that, to a good approximation, the electrons can adjust continuously to their changing environment at each stage of the collision as the internuclear separation changes with time. This condition is closely realized for collisions of heavy systems at bombarding energies which bring the two nuclear surfaces to a near-touching configuration against the Coulomb repulsion of the positive charge centers; the electron velocities are then about 10 times larger than the velocities associated with the nuclear motion. Figure 1 illustrates schematically how some of the

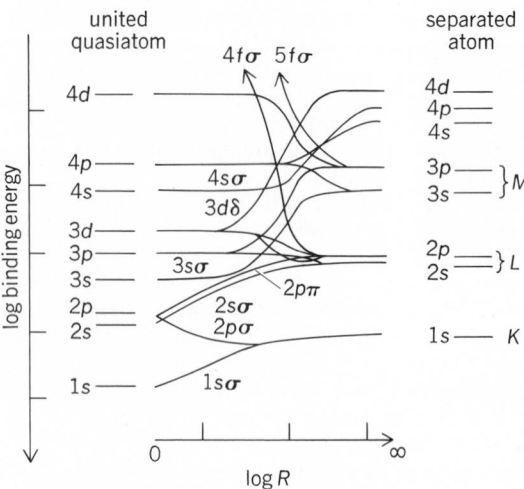

Fig. 1. Schematic representation of evolving quasimolecular states as a function of internuclear separation R. Both the separated atom and united quasiatomic limits are also shown.

tightest-bound states are expected to evolve quasiadiabatically as a function of time and internuclear separation under such bombarding conditions. The continuously changing electronic states during the collision include (1) the atomic states of the separated atom at very large internuclear separation, where the individual atoms are free from any interactions with each other; (2) the quasimolecular orbital states at intermediate internuclear separations, where the motion of the electrons of both atoms is governed principally by both nuclear centers and the potential binding of each electron varies with the distance between nuclei; and (3) the quasiatomic states of the combined atom in the united-atom limit at the distance of closest approach. The quasiatomic states are only the limit at the united atom of the quasimolecular potential binding as a function of the internuclear separation. The independent particle states shown in Fig. 1, calculated at each internuclear separation by freezing the nuclear motion at a chosen separation,

form the basis for understanding a number of observed collision phenomena in atom-atom scattering. *See* MOLECULAR STRUCTURE AND SPECTRA.

Electron promotion. For example, an important feature of this model of quasimolecular orbitals, first introduced by F. Hund and R. Mulliken in 1927, is electron promotion, whereby the principal quantum number of some molecular orbitals is elevated at the united atom limit relative to its value for the separated atoms. A particularly striking example of this behavior is illustrated in Fig. 1, where the $2p$ state of the separated atom evolves into the $4f\sigma$ orbital which climbs steeply in binding energy with decreasing internuclear distance, and, in the independent particle motion diabatic approximation, goes through many crossings with other states. Upon separation after the collision, the electrons revert to their atomic states, but not necessarily their lowest atomic orbitals, since the promoted electrons can be trapped at the crossings of the promoted-inner-shell molecular orbitals with the outer-shell orbitals. Thus ensuing from the collision are highly excited atoms with inner-shell vacancies which then may deexcite by emitting fast electrons in an Auger process or x-rays in a radiative decay. The formation of quasimolecules in atomic collisions and electron promotion had been suggested in 1932 to account for the observed thresholds for the ionization of noble gases by impact with alkali-metal ions, and subsequently to account for similar thresholds in inner-shell vacancy production. This mechanism particularly explains the very large inner-shell ionization cross sections observed at low relative velocities, where the inner-shell vacancy production sometimes exceeds by 14 orders of magnitude the predicted ionization probability from the direct Coulomb interaction between the electrons and the impinging nuclear Coulomb field of the projectile. *See* AUGER EFFECT; X-RAYS.

Superheavy atoms. An especially important potential utilization of quasiatom formation was suggested in 1969 as a possible vehicle for investigating superheavy atoms, considerably beyond the heaviest stable systems available. Such atomic systems are of intrinsic interest in quantum field theory, and particularly in the quantum electrodynamics of the very strong fields which can be generated in atoms when an electron is bound by a charge center whose value Z exceeds $1/\alpha \cong 137$, where α is the fine-structure constant. Under the latter conditions the usual perturbative expansion of all quantum-electrodynamic effects in terms of the coupling constant $Z\alpha$ is of doubtful validity, and any new theoretical approaches require experimental verification. Thus far, the discrete set of solutions of the Dirac equation with total energy between $+m_0c^2$ and $-m_0c^2$ ($m_0c^2 = 511$ keV is the energy equivalent of the rest mass of the electron) have constituted the manifold of states investigated in atomic and molecular physics, with the heaviest atom studied being fermium with a binding energy of approximately 142 keV.

A particularly interesting aspect of the many questions to be explored with superheavy atomic systems occurs when the Coulombic binding increases beyond a critical value so that the lowest bound state (the $1s$ state) becomes degenerate with the negative energy continuum states, below $-m_0c^2$, of the Dirac equation. For stable atoms, taking into account the finite size of the nuclei involved, the threshold critical charge Z_{cr}, for such a supercritical system where the bound states are expected to dive into the negative energy continuum is predicted to be about 173. This critical charge defines a boundary of fundamental significance and marks the threshold of the onset of a new phenomena. Viewed in the context of the Dirac hole theory, as a vacant $1s$ state joins the negative-energy continuum, it can be filled spontaneously (no external energy required) with an electron from the Dirac sea with the simultaneous emission of a free positron (the ensuing hold in the Dirac sea) of a well-defined kinetic energy. This phenomenon is equivalent to pair creation by photons, except that in this case the electron now fills the vacancy in the supercritical diving $1s$ state and no electron appears. Thus the critical charge defines the boundary between endoergic and exoergic pair production in atoms. As a result of the spontaneous pair production in supercritical fields, a supercritically charge nucleus is inherently unstable. The lowest atomic state of the system surrounding the bare supercritically charged nucleus is no longer free of particles and neutral, but instead contains a bound electron cloud which is negatively charged. This new form of instability in supercritical fields can be viewed as a breakdown of the neutral vacuum. The resulting charged lowest-energy state, the so-called charged vacuum, differs inherently from the usual charged atomic state in subcritical systems. In the latter case, removing all the electrons from the nuclear charge leads to a stable neutral state. On the other hand, the neutral vacuum of a supercritical system always spontaneously decays to a charged vacuum and positrons, and therefore the space surrounding a supercritically charged nucleus can never be free of charges. Thus the charged vacuum is a necessary new concept for supercritical fields.

The decisive experiment to verify these new aspects of quantum electrodynamics is to observe the spontaneously emitted positions. For this purpose atoms with $Z \geq 173$ are required with the supercritically bound states unoccupied. With the region of supercritically charged nuclei being far beyond the present capability of producing superheavy nuclei, the formation of quasiatoms seems to offer the only opportunity to examine these very strong field phenomena. For quasiatoms, the conditions for achieving supercritical fields are established at small internuclear separations when $(Z_1 + Z_2) \geq Z_{cr}$. For example, Fig. 2 illustrates schematically that for uranium + uranium (U + U) collisions the two nuclei act as a combined source of a supercritical field for internuclear separations smaller than 34×10^{-15} m. The experimental verification of the spontaneous positron emission process would represent a major contribution to the understanding of the quantum electrodynamics of strong fields. *See* POSITRON; QUANTUM ELECTRODYNAMICS; QUANTUM FIELD THEORY; RELATIVISTIC QUANTUM THEORY.

Positron creation in quasiatoms. In contrast to the stable atom situation referred to previously, in dynamical systems such as quasiatoms, other positron creation mechanisms, not involving supercritically bound states, complicate the observation of spontaneous positron emission. In addition to the spontaneous process which proceeds without

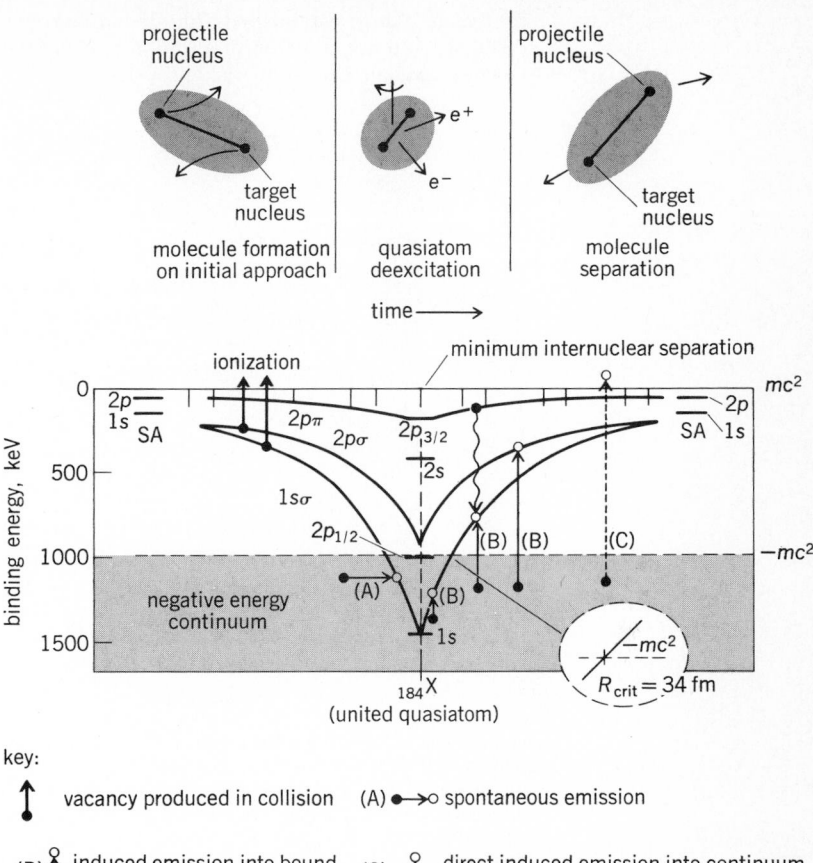

Fig. 2. Positron production mechanisms in heavy-ion collisions, shown on plot of energies of bound states as a function of time for U + U system.

an external energy source, the time dependence of the electric field produced by the nuclear charges can also induce transitions of electrons from the negative-energy continuum both to unbound states and to unoccupied bound states even above the critical binding energy. These two subcritical processes, induced emission into bound states and direct induced emission into continuum states, are shown in Fig. 2, along with spontaneous emission, in a plot of the evolution of the bound states with time during the collision. The energy required for each of these subcritical pair creation processes is supplied by the nuclear motion. Although they are undesirable backgrounds in the detection of spontaneous positron emission, the two effects are of considerable interest in themselves since they also reflect new aspects of the theory when strong electromagnetic fields are involved and when perturbation theory is not applicable.

In 1976 a series of measurements were initiated to investigate positron creation in heavy-ion collisions. To selectively study both the subcritical and supercritical positron production processes discussed above, the studies included lead + lead, lead + uranium, uranium + uranium, and uranium + curium (Pb + Pb, Pb + U, U + U, and U + Cm) collision systems. The first generation of experiments have developed information on the dynamic positron creation mechanisms (B) and (C) in Fig. 2. They established that in superheavy collision systems, positron production is observed considerably in excess of the intensity expected

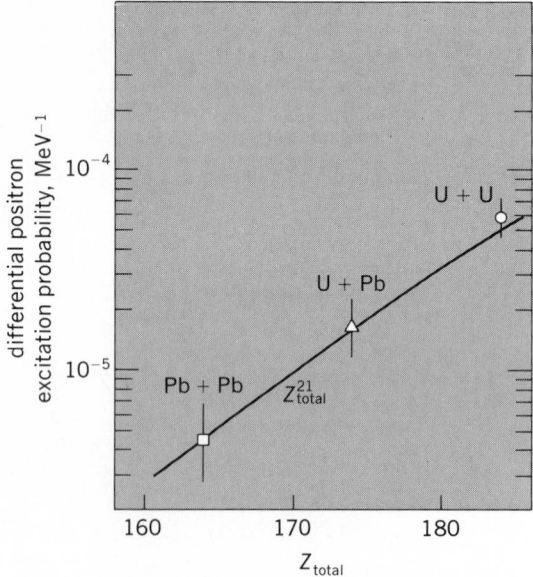

Fig. 3. Measured differential positron excitation probability as a function of the combined atomic number Z_{total} of the target and projectile. The distance of minimum approach and the relative ion velocity have been kept constant at 30 fm and 3.3×10^7 m/s respectively, and a positron energy, $E_{e+} = 478^{+54}_{-53}$ keV, has been selected near the peak of the positron spectrum.

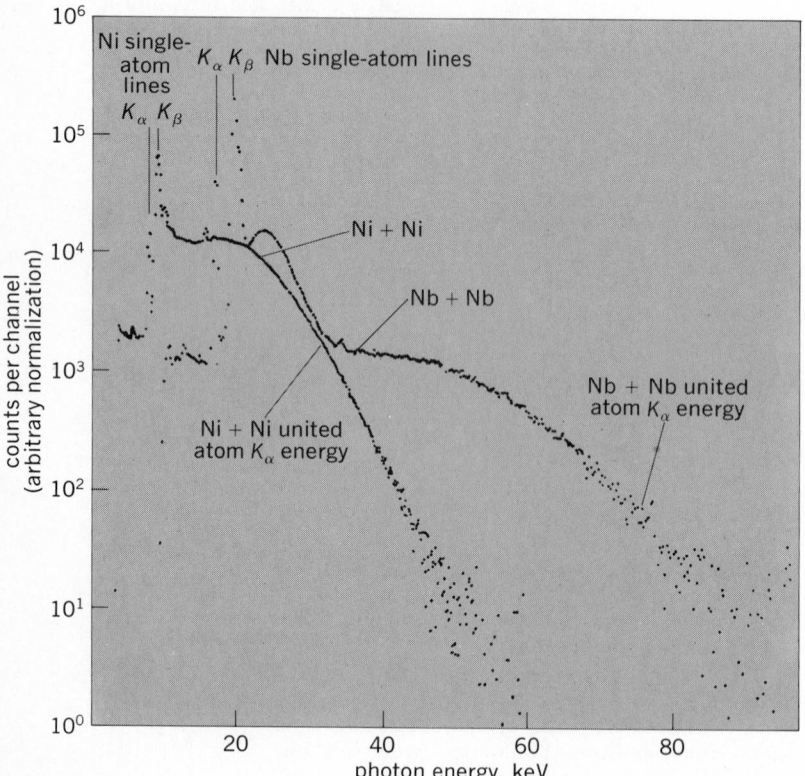

Fig. 4. Quasimolecular K x-ray spectra from Ni + Ni and Nb + Nb collisions. Incoming ion energies are 64.8 MeV for Ni + Ni and 75.9 MeV for Nb + Nb, and x-rays are observed at angle of 90° to beam direction. Extension of spectra beyond the united atom energies is produced by collision broadening. The characteristic K_α and K_β single-atom lines are also shown.

from nuclear background processes, and that the magnitude of the cross sections, their dependence on the minimum internuclear separation and particularly the almost exponential increase of the positron creation probability with increasing total nuclear charge $(Z_1 + Z_2)$ cannot be accounted for without invoking the quasiatomic picture of $(Z_1 + Z_2)$ acting in unison to produce very strong electric fields.

The general features predicted by the theory developed by W. Greiner and coworkers have been verified, although some discrepancies remain to be explained. The most striking of these first observations is the rapid increase of positron production with increasing Z_{total} $(Z_1 + Z_2)$. Expressed as a power law proportional to $(Z_1 + Z_2)^n$, n takes on a value of about 20 (Fig. 3). This rapid increase with Z_{total} finds no other known analog in nature and singularly reflects the relativistic phenomena associated with very strong electric fields and their strongly bound atomic states.

The identification of a definitive signature for spontaneous positron emission from the decay of the vacuum has met with less success. Experiments designed to explore details of the positron production probability, in an effort to find such a signature, have revealed a peaked-structure in the positron spectra from U + U and U + Cm collisions. These two systems are expected to exhibit supercritical binding under the bombarding conditions employed. Further experiments have been undertaken to associate these structure features with the spontaneous positron emission process.

Quasimolecular x-ray spectra. The observation of x-radiation from radiative fillings of inner-shell vacancies has provided one of the more important sources of information on quasimolecules and quasiatoms. The principal features of these x-ray spectra are determined by the time evolution of the molecular orbital energy levels, by the finite lifetime of the vacancy, by the collision time, and most significantly, by the small ratio of collision time to radiative lifetime typical of heavy systems. Since for a given transition, such as that shown in Fig. 2, the transition energy is a continuous function of internuclear separation, and since radiations can occur at all internuclear separations because of the small collision-time to radiative-lifetime ratio, the quasimolecular x-ray spectrum is a continuum extending to approximately the transition energy of the united quasiatom. The spectrum reflects the quasimolecular energy level structure, but may be modified significantly by collision broadening.

Pioneered for light systems and L x-rays in 1972, the study of quasimolecular x-rays has been extended to M, L, and K radiations in a variety of collision systems ranging from carbon + carbon (C + C) to lead + lead. Examples of the continuum K radiations from nickel + nickel (Ni + Ni) and niobium + niobium (Nb + Nb) collisions are shown in Fig. 4. The demonstration that these continuum radiations exhibited a characteristic photon-frequency-dependent anistropic emission pattern, which was shown to be associated only with molecular systems, provided an important identification feature for the quasimolecular x-rays, and opened up new possibilities for the development of a spectroscopy of quasimolecular and quasiatomic states. Techniques have been developed to selectively identify L- and K-type quasimolecular transitions in the continuum spectra by exploring cascade relationships between these radiations and characteristic x-rays. Due to gamma-ray background radiations from nuclear excitations during the collision, the exceptionally nuclear-stable Pb + Pb collision system ($Z_{total} = 164$) has been the heaviest quasiatom explored via quasimolecular x-rays. Information on the quasimolecular energy level structure beyond this combined nuclear charge has been extracted from studies of the K vacancy excitation probability as a function of internuclear separation in the Pb + U ($Z_{total} = 174$) and Pb + Cm ($Z_{total} = 178$) collision systems. With these and other experimental approaches, including the study of positron production in superheavy collision systems discussed above, information on quasiatoms with subcritical and supercritical binding is being accumulated.

[JACK S. GREENBERG]

Bibliography: H. Backe et al., Observation of positron creation in superheavy collision systems, *Phys. Rev. Lett.*, 40:1443–1446, 1978; B. Grasemann, *Atomic Inner Shell Processes*, 1975; J. S. Greenberg, Radiative processes in superheavy quasiatoms: A search for positron emission, in N. Oda and K. Takayanagi (eds.), *Electronic and Atomic Collisions*, pp. 351–368, 1980; C. Kozhuharov et al., Positrons from 1.4 GeV uranium atom collisions, *Phys. Rev. Lett.*, 42:376–379, 1979; W. Lichten, The quasimolecular model of atomic collisions, in *Proceedings of the 4th International Conference on Atomic Physics*, Heidelberg, pp. 249–285, 1975; J. Rafelski, L. P. Fulcher, and A. Klein, Fermions and bosons interacting with arbitrarily strong external fields, *Phys. Rep.*, 38C: 227–361, 1978; J. Reinhardt and W. Greiner, Quantum electrodynamics of strong fields, *Rep. Prog. Phys.*, 40:219–295, 1977.

Quaternions

An associative, noncommutative algebra based on four linearly independent units or basal elements. Quaternions were originated in Dublin, Ireland, on Oct. 16, 1843, by W. R. Hamilton (1805–1865), who is famous because of his canonical functions and equations of motion which are important in both classical and quantum dynamics.

The four linearly independent units in quaternion algebra are commonly denoted by 1, i, j, k, where l commutes with i, j, k and is called the principal unit or modulus. These four units are assumed to have the following multiplication table:

$$1^2 = 1 \quad i^2 = j^2 = k^2 = ijk = -1$$
$$i(jk) = (ij)k = ijk$$
$$li = il \quad lj = jl \quad lk = kl$$

The i, j, k do not commute with each other in multiplication, that is, $ij \neq ji, jk \neq kj, ik \neq ki$, etc. But all real and complex numbers do commute with i, j, k; thus if c is a real number, then $ic = ci, jc = cj$, and $kc = ck$. On multiplying $ijk = -1$ on the left by i, so that $iijk = i(-1) = -i$, it is found, since $i^2 = -1$, that $jk = i$. Similarly $jjk = ji = -k$; when exhausted, this process leads to all the simple noncummutative relations for i, j, k, namely,

$$ij = -ji = k \quad jk = -kj = i \quad ki = -ik = j$$

More complicated products, for example, $jikjk =$

$-kki = i$, are evaluated by substituting for any adjoined pair the value given in the preceding series of relations and then proceeding similarly to any other adjoined pair in the new product, and so on until the product is reduced to $\pm 1, \pm i, \pm j,$ or $\pm k$. Multiplication on the right is also permissible; thus from $ij = k$, one has $ijj = kj$, or $-i = kj$. Products such as jj and jjj may be written j^2 and j^3.

All the laws and operations of ordinary algebra are assumed to be valid in the definition of quaternion algebra, except the commutative law of multiplication for the units i, j, k. Thus the associative and distributive laws of addition and multiplication apply without restriction throughout. Addition is also commutative, for example, $i + j = j + i$.

Now if s, a, b, c are real numbers, rational or irrational, then a real quaternion q and its conjugate q' are defined by

$$q = s + ia + jb + kc \quad q' = s - (ia + jb + kc)$$

In this case $qq' = q'q = s^2 + a^2 + b^2 + c^2 = N$, and N is called the norm of q; the real quantity $T = \sqrt{N}$ is called the tensor of q, and s, a, b, c are components (or coordinates) of q. The part $\gamma = ia + jb + kc$ is the vector of q, and it may be represented by a stroke or vector in a frame of cartesian coordinates, a, b, c being its components. Let now $p = w + ix + jy + kz$ be another real quaternion; if $pq = 0$, either p or q or both are zero, which is called the product law. If, for example, $p = 0$, then all of its components w, x, y, z are zero. When $p = q$, that is, $p - q = 0$, then one must have $w = s, x = a, y = b, z = c$; otherwise

$$(w - s) + i(x - a) + j(y - b) + k(z - c) = 0$$

would constitute a linear relation between l, i, j, k, which would be in conflict with their original definitions as linearly independent.

Multiplication. The product of two quaternions may be found by a straightforward process, and in full is

$$pq = ws - (ax + by + cz) + i(aw + sx + cy - bz)$$
$$+ j(bw + sy + az - cx) + k(cw + sz + bx - ay)$$
$$qp = ws - (ax + by + cz) + i(aw + sx - cy + bz)$$
$$+ j(bw + sy - az + cx) + k(cw + sz - bx + ay)$$

and hence

$$pq - qp =$$
$$2 [i(cy - bz) + j(az - cx) + k(bx - ay)]$$

which is zero only when $cy - bz = az - cx = bx - ay = 0$. This shows that $pq \neq qp$ except under special conditions; quaternion multiplication is not, in general, commutative.

In q, if any two of a, b, c are zero, one has, in effect, an ordinary complex number; if all of a, b, c are zero, then $q = s$ and is an ordinary real number of scalar. Hence real quaternions include the real and ordinary complex numbers as special cases. It will be evident that real quaternions are a kind of extension of the ordinary complex numbers $z = x + \sqrt{-1} \, y$.

So far the case in which s, a, b, c are complex quantities has not been included, thus making q a complex quaternion, and for present purposes complex quaternions will be put aside.

It may be noted that the invention of vector analysis was inspired by Hamilton's quaternions; as early as 1846–1852 Rev. M. O'Brien published papers in which he assumed $i^2 = j^2 = k^2 = 1$, and thus paved the way to the dot or inner product of vector analysis. Fundamentally, quaternion algebra provides much deeper concepts and consequences, and in some practical problems it presents clear advantages over vector analysis.

Division. In the division of quaternions, reciprocals of the quaternion units i, j, k are easily found, thus

$$i^{-1} = \frac{1}{i} = \frac{i}{ii} = -i \qquad j^{-1} = -j$$

$$k^{-1} = -k \qquad kk^{-1} = -k^2 = 1$$

More complicated quotients may be evaluated if care is taken to observe the defined conventions. Thus

$$ij/j = ijj^{-1} = i$$

but

$$ji/j = jij^{-1} = -kj^{-1} = kj = -i$$

It is best to write denominators with negative exponents and place them properly in the numerator to avoid errors. Similarly, the reciprocals and quotients of real quaternions yield unique results. If in $q = s + ia + jb + kc$, all of s, a, b, c are not zero, then

$$q^{-1} = 1/q = q'/qq' = q'/N, N = s^2 + a^2 + b^2 + c^2$$

Further, if $p = w + ix + jy + kz$ is a second real quaternion then

$$p/q = pq'/qq' = pq'/N$$

Accordingly, real quaternions admit of division. If $rq = p$, then $r = p/q = pq'/N$; and if $qr = p$, $r = q^{-1}p = q'p/N$. Hence both right division and left division yield unique quotients. If, for example, $s = a = 1$, $b = \sqrt{-2}$, $c = 0$, then $N = 1 + 1 - 2 + 0 = 0$; this result is one reason why the discussion has been limited to real quaternions.

Hamilton adopted the name vectors for directed lines in space. If vectors are denoted with Greek letters, then $\alpha = ix + jy + kz$ is a vector whose components along a conventional right-handed rectangular cartesian coordinate system are x, y, z, respectively. If $\beta = ix' + jy' + kz'$ is another vector in the same coordinate system, their products are

$$\alpha\beta = -(xx' + yy' + zz') + i(yz' - y'z)$$
$$+ j(x'z - xz') + k(xy' - x'y) = -u + \gamma$$
$$\beta\alpha = -(xx' + yy' + zz') - [i(yz' - y'z)$$
$$+ j(x'z - xz') + k(xy' - x'y)] = -u - \gamma$$

Both products yield a scalar u and a vector γ as a sum, and such a sum is by definition a quaternion. Further,

$$\alpha\beta + \beta\alpha = -2(xx' + yy' + zz')$$

That is, in general, multiplication of nonparallel vectors is not commutative, which is a special case of the noncommutation multiplication of quaternions. If one sets $x = x'$, $y = y'$, $z = z'$, then

$$\alpha^2 = -(x^2 + y^2 + z^2)$$

which, with negative sign, is the square of the length of the vector α whose components are x, y, z. The quotient α/β emerges easily since

$$\alpha/\beta = -\alpha\beta/(x'^2 + y'^2 + z'^2)$$

Hence α/β has a unique value if x', y', z' are all real and not all zero. The quotient of two vectors is a quaternion, and previously it was stated that a quaternion can be defined as the ratio of two vectors.

By multiplication a real q and its conjugate q' are found to commute

$$qq' = q'q = s^2 + a^2 + b^2 + c^2 = T^2$$

to give a real positive scalar. Further, $q + q' = 2s$, and hence q and q' are the roots of a quadratic equation with real coefficients

$$t^2 - 2st + T^2 = 0$$

When this equation is solved in the field of ordinary complex numbers, one finds

$$t = s \pm \sqrt{-(a^2 + b^2 + c^2)}$$

This simple but important result emphasizes the fact that, in asking for a solution of a given algebraic equation, the field of the quantities for which a solution is desired must be specified.

Applications. Next, let $r = s' + ia' + jb' + kc'$ be another real quaternion, then

$$qr = ss' - (aa' + bb' + cc')$$
$$+ i(as' + a's + bc' - b'c)$$
$$+ j(bs' + b's + a'c - ac')$$
$$+ k(cs' + c's + ab' - a'b)$$
$$= S + iA + jB + kC$$

where $S = ss' - (aa' + bb' + cc')$, $A = as' + a's - bc' - b'c$, etc. for B and C. Also,

$$r'q' = S - (iA + jB + kC)$$
$$= (qr)'qr(qr)' = S^2 + A^2 + B^2 + C^2$$

But $qr(qr)' = qrr'q' = qq'rr'$, since rr' is a scalar and commutes with q and q'. Therefore, one has Euler's famous result that

$$S^2 + A^2 + B^2 + C^2$$
$$= (s^2 + a^2 + b^2 + c^2)(s'^2 + a'^2 + b'^2 + c'^2)$$

which is important in number theory since it is used in the proof that every positive rational integer can be represented as a sum of four squares.

The quotient of real vectors α/β is the sum of a scalar and vector and hence is a quaternion, for example, q. If l_1, m_1, n_1 are the direction cosines of α in a rectangular cartesian coordinate frame, then

$$\alpha = a_0(il_1 + jm_1 + kn_1) \quad a_0 = \sqrt{x^2 + y^2 + z^2}$$
$$\beta = b_0(il_2 + jm_2 + kn_2) \quad b_0 = \sqrt{x'^2 + y'^2 + z'^2}$$

and

$$q = \frac{\alpha}{\beta} = \frac{a_0}{b_0}[l_1l_2 + m_1m_2 + n_1n_2 + i(m_2n_1 - m_1n_2)$$
$$+ j(l_1n_2 - l_2n_1) + k(l_2m_1 - l_1m_2)]$$
$$= s + ia + jb + kc = s + \gamma$$

Then $\alpha = q\beta$; that is, q is an operator which turns and stretches β to coincide with α. From analytical geometry $l_1l_2 + m_1m_2 + n_1n_2 = \cos\omega$, where ω is the angle $(<\pi)$ between the vectors α and β. It can be noted that

$$l_1(m_2n_1 - m_1n_2) + m_1(l_1n_2 - l_2n_1)$$
$$+ n_1(l_2m_1 - l_1m_2) = 0$$

and hence the vector part γ of q is perpendicular to α, and the same is true for γ and β. This suggests the relationship

$$q = \frac{a_0}{b_0}(\cos\omega + \epsilon\sin\omega) = TUq$$

where $a_0{}^2/b_0{}^2 = s^2 + a^2 + b^2 + c^2 = T^2$, $\cos\omega = s/T$, and ϵ is the unit vector $\epsilon = (ia + jb + kc)/\sqrt{a^2 + b^2 + c^2}$, which is perpendicular to both α and β. Because $\epsilon^2 = -1$, the square of any real unit vector is -1. The factor $Uq = \cos\omega + \epsilon\sin\omega$ turns β through the angle ω to coincide with the direction of α and is called the versor of q. U is one of several symbols encountered in the grammar of earlier quaternion theory. If $U'q = \cos\omega - \epsilon\sin\omega$, then $UqU'q = 1$.

There is an important result in quaternion algebra which seems to be due to A. Cayley. With q as before, then its reciprocal is $q^{-1} = q'/T^2$, as before. If some other real quaternion is $p = w + ix + jy + kz$, then the expression qpq^{-1} is called the conical rotation (or transform) of p, and it finds important application in the specification of motions of rigid bodies. One has

$$qpq^{-1} = \{wT^2 + i[(s^2 + a^2 - b^2 - c^2)x + 2(ab - cs)y + 2(ac + bs)z] + j[2(ab + cs)x + (s^2 - a^2 + b^2 - c^2)y + 2(bc - as)z] + k[2(ac - bs)x + 2(as + bc)y + (s^2 - a^2 - b^2 + c^2)z]\}/T^2$$
$$= w + iX + jY + kZ = w + \tau$$

where $T^2 = s^2 + a^2 + b^2 + c^2$. If $T^2 = 1$, $w = w$ and

$$X = (s^2 + a^2 - b^2 - c^2)x + 2(ab - cs)y + 2(ac + bs)z$$
$$Y = 2(ab + cs)x + (s^2 - a^2 + b^2 - c^2)y + 2(bc - as)z$$
$$Z = 2(ac - bs)x + 2(as + bc)y + (s^2 - a^2 - b^2 + c^2)z$$

which is a linear transformation of X, Y, Z to x, y, z.

It is easy to show that, for example,

$$(s^2 + a^2 - b^2 - c^2)^2 + 4(ab - cs)^2 + 4(ac + bx)^2$$
$$= T^4 = 1, 2(ab + cs)(s^2 + a^2 - b^2 - c^2)$$
$$+ 2(s^2 - a^2 + b^2 - c^2)(ab - cs)$$
$$+ 4(bc - as)(ac + bs) = 0$$

Hence the coefficients of x, y, z are direction cosines of the frame X, Y, Z in the x, y, z frame of coordinates, and the transformation is orthogonal (unitary). The a, b, c, s are Euler's parameters (ξ, η, ζ, χ). Clearly the angle ψ between $\alpha = ix + jy + kz$ and $\gamma = ia + jb + kc$ is given by $g(ax + by + kz) = \cos\psi$, with g a known quantity, and some easy algebra shows that the angle between the vector part τ of qpq^{-1} and γ of $q = s + \gamma$ is given by the same expression. Hence qpq^{-1} has rotated α conically about γ, and the magnitude of the rotation comes out, remarkably enough, to be just 2ω, where again $\cos\omega = s/T = s$. As an example, let

$$a = b = w = 0 \qquad s = \cos\omega \qquad c = \sin\omega$$

so $q = \cos\omega + k\sin\omega$; then the formulas above yield easily

$$X = x\cos 2\omega - y\sin 2\omega$$
$$Y = x\sin 2\omega + y\cos 2\omega$$
$$Z = z$$

which shows that p has been rotated conically about z and through the angle 2ω.

If λ, μ, ν are the angles which the vector $\gamma = ia + jb + kc$ makes with x, y, z, respectively, and if $T^2 = 1$, then the relations $a = \cos\lambda\sin\omega$, $b = \cos\mu\sin\omega$, $c = \cos\nu\sin\omega$, $s = \cos\omega$, and $a^2 + b^2 + c^2 = \sin^2\omega$ are consistent with the above analysis. An-

other formulation for s, a, b, c is

$$a = \sin\frac{\theta}{2}\sin\frac{1}{2}(\psi - \phi) \qquad b = \sin\frac{\theta}{2}\cos\frac{1}{2}(\psi - \phi)$$
$$c = \cos\frac{\theta}{2}\sin\frac{1}{2}(\psi + \phi) \qquad s = \cos\frac{\theta}{2}\cos\frac{1}{2}(\psi + \phi)$$

The angles θ, ϕ, ψ are the familiar Eulerian angles relating a fixed cartesian frame to another with the same origin, the second being assumed, in kinematics, to be fixed in a rigid body moving about a fixed point, namely, the common origin of the two frames.

If $\alpha, \beta, \gamma, \delta$ are defined by $\alpha = s + ic$, $\beta = a + ib$, $\gamma = -a + ib$, $\delta = s - ic$, $i = \sqrt{-1}$, then $\alpha\delta - \beta\gamma = s^2 + a^2 + b^2 + c^2 = 1$,

$$\alpha = \cos\frac{\theta}{2}e^{i(\phi + \psi)/2} \qquad \beta = i\sin\frac{\theta}{2}e^{i(\phi - \psi)/2}$$
$$\gamma = i\sin\frac{\theta}{2}e^{i(\psi - \phi)/2} \qquad \delta = \cos\frac{\theta}{2}e^{i(-\psi - \phi)/2}$$
$$i = \sqrt{-1}$$

These are the Cayley-Klein parameters which lend themselves to homographic (bilinear) transformation, $z' = \alpha z + \beta/\gamma z + \delta$, in the complex z plane so that the motion of a solid body in space can be represented on a plane. The quantities $u = -i\gamma$, $v = i\delta$ are also components of a unit spinor, a function occurring in higher algebra and in the quantum theory (similarly for α, β). If $a_x = (1/2)(u^2 - v^2)$, $a_y = (-i/2)(u^2 + v^2)$, $a_z = uv$ are the components of a complex vector, then $a_x{}^2 + a_y{}^2 + a_z{}^2 = 0$ and the vector is of length zero. Such a vector is the starting point for some treatments of spinor theory.

Returning now to the quaternion units $1, i, j, k$, they may, as appears to have been first discovered by Sylvester, be represented by various matrices, one 2×2 set being

$$1 \leftrightharpoons \begin{bmatrix} 1 & 0 \\ 0 & 1 \end{bmatrix} \qquad i \leftrightharpoons \begin{bmatrix} \sqrt{-1} & 0 \\ 0 & -\sqrt{-1} \end{bmatrix}$$
$$j \leftrightharpoons \begin{bmatrix} 0 & 1 \\ -1 & 0 \end{bmatrix} \qquad k \leftrightharpoons \begin{bmatrix} 0 & \sqrt{-1} \\ \sqrt{-1} & 0 \end{bmatrix}$$

and the last three, when multiplied by $\sqrt{-1}$, are the Pauli spin matrices occurring in the quantum theories of electron spin (spin $= 1/2$). It has been shown by S. Bochner that no set of 3×3 matrices have the multiplication table corresponding to $1, i, j, k$. A set of 4×4 matrices, due to A. S. Eddington, which represents $1, i, j, k$ is

$$1 \leftrightharpoons \begin{bmatrix} 1 & 0 & 0 & 0 \\ 0 & 1 & 0 & 0 \\ 0 & 0 & 1 & 0 \\ 0 & 0 & 0 & 1 \end{bmatrix} \qquad i \leftrightharpoons \begin{bmatrix} 0 & 1 & 0 & 0 \\ -1 & 0 & 0 & 0 \\ 0 & 0 & 0 & 0 \\ 0 & 0 & -1 & 0 \end{bmatrix}$$
$$j \leftrightharpoons \begin{bmatrix} 0 & 0 & 0 & -1 \\ 0 & 0 & -1 & 0 \\ 0 & 1 & 0 & 0 \\ 1 & 0 & 0 & 0 \end{bmatrix} \qquad k \leftrightharpoons \begin{bmatrix} 0 & 0 & -1 & 0 \\ 0 & 0 & 0 & 1 \\ 1 & 0 & 0 & 0 \\ 0 & -1 & 0 & 0 \end{bmatrix}$$

These are four out of a group of sixteen 4×4 matrices used by Eddington in his fundamental theory; five of the sixteen, not including the above, are, when multiplied by $\sqrt{-1}$, the matrices occurring in Dirac's theory of the relativistic wave equation for the electron. Another set of four matrices for $1, i, j, k$ is

$$1 \leftrightharpoons \begin{bmatrix} 1 & 0 & 0 & 0 \\ 0 & 1 & 0 & 0 \\ 0 & 0 & 1 & 0 \\ 0 & 0 & 0 & 1 \end{bmatrix} \qquad i \leftrightharpoons \begin{bmatrix} 0 & 1 & 0 & 0 \\ -1 & 0 & 0 & 0 \\ 0 & 0 & 0 & -1 \\ 0 & 0 & 1 & 0 \end{bmatrix}$$

$$j \leftrightharpoons \begin{bmatrix} 0 & 0 & 1 & 0 \\ 0 & 0 & 0 & 1 \\ -1 & 0 & 0 & 0 \\ 0 & -1 & 0 & 0 \end{bmatrix} \qquad k \leftrightharpoons \begin{bmatrix} 0 & 0 & 0 & 1 \\ 0 & 0 & -1 & 0 \\ 0 & 1 & 0 & 0 \\ -1 & 0 & 0 & 0 \end{bmatrix}$$

and if, by ordinary matrix addition rules, $x + ix_1 + jx_2 + kx_3 = q$ are formed, then

$$q = \begin{bmatrix} x & x_1 & x_2 & x_3 \\ -x_1 & x & -x_3 & x_2 \\ -x_2 & x_3 & x & -x_1 \\ -x_3 & -x_2 & x_1 & x \end{bmatrix}$$

and the determinant of q, $|q|$, has the form

$$|q| = (x^2 + x_1{}^2 + x_2{}^2 + x_3{}^2)^2$$

which is an intriguing result.

The basal quaternion elements 1, i, j, k, and their negatives, are the elements of a non-Abelian group of order eight, which is called the quaternion group. The group contains proper, self-conjugate (invariant) cyclical subgroups, one of the order 1 and, significantly, three of order 4; and there are five conjugate classes, three of which are of order 2 and the others each of order 1.

Hamilton's interest in analytical geometry led to many applications of quaternions in this field, especially to conics and quadric surfaces. In his hands and especially in those of P. G. Tait, P. Kelland, J. McAulay, C. J. Joly, and A. S. Hardy there have been many applications of quaternions to classical geometry and mathematical physics. These include the use of Hamilton's partial differential operator

$$\nabla = i \frac{\partial}{\partial x} + j \frac{\partial}{\partial y} + k \frac{\partial}{\partial z}$$

The great advances made in quantum theory, relativity, number theory, algebra, and group theory are associated with scholars who had an easy familiarity with quaternions. *See* CALCULUS OF VECTORS; GROUP THEORY; MATRIX THEORY.

[DON M. YOST]

Bibliography: H. F. Baker, *Abel's Theorem and the Allied Theory*, 1897, reprint 1976; H. C. Brinkman, *Spinor Invariants*, 1956; H. Goldstein, *Classical Mechanics*, 2d ed., 1980; W. R. Hamilton, *Elements of Quaternions*, 3d ed., C. J. Joly (ed.), 1901, reprint 1969; G. H. Hardy and E. M. Wright, *Introduction to the Theory of Numbers*, 5th ed., 1980; R. Hermann, *Spinors, Clifford and Cayley Algebras*, 1974; H. Jeffreys and B. Jeffreys, *Mathematical Physics*, 3d ed., 1956; L. D. Landau and E. M. Lifshitz, *Quantum Mechanics–Non-Relativistic Theory*, 3d ed., 1977; L. W. Shapiro, *Introduction to Abstract Algebra*, 1975; J. C. Slater, *Quantum Theory of Molecules and Solids*, 2 vols., 1963, 1965, 1967, 1974.

Radiance

The physical quantity that corresponds closely to the visual brightness of a surface. A simple radiometer for measuring the (average) radiance of an incident beam of optical radiation (light, including invisible infrared and ultraviolet radiation) con-

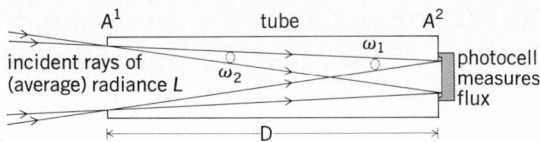

A simple radiometer.

sists of a cylindrical tube, with a hole in each end cap to define the beam cross section there, and with a photocell against one end to measure the total radiated power in the beam of all rays that reach it through both holes (see illustration). If A_1 and A_2 are the respective areas of the two holes, D is the length of the tube (distance between holes), and Φ is the radiant flux or power measured by the photocell, then the (average) radiance is approximately given by the equation below. In the alter-

$$L = \Phi/(A_1 \cdot A_2/D^2) \text{ W} \cdot \text{m}^{-2} \cdot \text{sr}^{-1}$$
$$= \Phi/(A_1 \cdot \omega_2) = \Phi/(A_2 \cdot \omega_1)$$

nate forms on the second line, $\omega_2 = A_2/D^2$ is the solid angle subtended at A_1 by A_2 (at a distance D), and vice versa for ω_1. These approximate relations are good to about 1% or better when D is at least 10 times the widest part of either hole, but the accuracy degenerates rapidly for larger holes or shorter distances between them. A more exact formula, in terms of the ray radiance in each direction at every point across an area through which a beam flows, requires calculus. Then the radiance of a ray in a given direction through a given point is defined as the radiant flux per unit projected area perpendicular to the ray at the point and unit solid angle in the direction of the ray at the point. *See* LIGHT.

Power (flux) is given in SI units of watts (W), areas in square meters (m²), and solid angles in steradians (sr), and so the units of radiance are, as shown, watts per square meter and steradian [W · m⁻² · sr⁻¹ or W/(m² · sr)]. If the flux of visible radiation is given in units weighted for standardized eye response, called lumens (lm), instead of watts, with everything else exactly the same, the corresponding photometric quantity of luminance is obtained in lm · m⁻² · sr⁻¹. When concerned with interactions between radiation and matter, flux can also be measured in numbers of photons or quanta per second (q · s⁻¹) rather than watts.

A complete specification of a beam of optical radiation requires the distribution of radiance, not only as a function of ray position and direction, but also as a function of wavelength, time, and polarization. A distribution in space and wavelength λ is spectral radiance L_λ in watts per square meter, steradian, and nanometer of wavelength (W · m⁻² · sr⁻¹ · nm⁻¹). Spectral radiance may also be given in terms of wave number σ in reciprocal centimeters (cm⁻¹) or of frequency ν in terahertz (THz). No special names are given to the distributions with respect to time (or frequency of fluctuation or modulation) and polarization, and they are often overlooked even though their effects may be significant.

All of the foregoing assumes that radiant energy is propagated along the rays of geometrical optics that can form sharp shadows and images. The results may be in error when there are significant diffraction or interference effects, as is often the case with coherent radiation such as that from a laser. *See* DIFFRACTION; INTERFERENCE OF WAVES; LASER.

[FRED E. NICODEMUS]

Bibliography: American Association of Physics Teachers, *Radiometry: Selected Reprints*, 1971; National Bureau of Standards, *A Self-study Manual on Optical Radiation Measurements*, NBS Tech. Note, 1976; F. E. Nicodemus, Radiance, *Amer. J. Phys.*, 31:368–377, 1963.

Radiation

The emission and propagation of energy; also, the emitted energy itself. The etymology of the word implies that the energy propagates rectilinearly, and in a limited sense, this holds for the many different types of radiation encountered.

The major types of radiation may be described as electromagnetic, acoustic, and particle, and within these major divisions there are many subdivisions.

For example, electromagnetic radiation, which in the most familiar energy ranges behaves in a manner usually characteristic of waves rather than of particles, is classified roughly in order of decreasing wavelength as radio, microwave, visible, ultraviolet, x-rays, and γ-rays. In the last three subdivisions, and frequently in the visible, the behavior of the radiation is more particlelike than wavelike.

Since the energy of a photon (light quantum) is inversely proportional to the wavelength, this classification is also on the basis of increasing photon energy. *See* ELECTROMAGNETIC RADIATION.

Acoustic or sound radiation may be classified by frequency as infrasonic, sonic, or ultrasonic in order of increasing frequency, with sonic being between about 16 and 20,000 Hz. Infrasonic sound can result, for example, from explosions or other sources so loud that exceptional waves are set up because the large amplitudes of the source vibrations exceed the elastic limit of the transmitting medium. Ultrasonic sound can be produced by means of crystals which vibrate rapidly in response to alternating electric voltages applied to them. There is a nearly infinite variety of sources in the sonic range. *See* SOUND.

The traditional examples of particle radiation are the α- and β-rays of radioactivity. Cosmic rays also consist largely of particles—protons, neutrons, and heavier nuclei, along with β-rays, mesons, and the so-called strange particles. *See* ELEMENTARY PARTICLE.

[MC ALLISTER H. HULL, JR.]

Radiation pressure

Pressure exerted by electromagnetic radiation on objects on which it impinges. This pressure is caused by the fact that electromagnetic radiation transmits energy and possesses momentum. In the case of a plane electromagnetic wave incident normally on a plane absorbing sheet, the mean pressure is $\epsilon E_0{}^2$, where E_0 is the amplitude of the electric field and ϵ is the dielectric constant of the medium. If the wave impinges normally on a perfectly reflecting, plane conducting sheet, then standing waves are formed, and the average pressure is twice that on the absorbing sheet. These pressures are very small (about 10^{-9} newton/m² if

E is a few volts per meter), but were measured successfully by E. F. Nichols and G. F. Hull in 1903. The effect is conspicuous in the case of a comet near the Sun, where the radiation pressure from the Sun forces the lighter cometary constituents away from the Sun. *See* ELECTROMAGNETIC RADIATION; MAXWELL'S EQUATIONS; WAVE EQUATION.

[WILLIAM R. SMYTHE]

Radioactivity

A phenomenon resulting from an instability of the atomic nucleus in certain atoms whereby the nucleus experiences a spontaneous but measurably delayed nuclear transition or transformation with the resulting emission of radiation. The discovery of radioactivity by H. Becquerel in 1896 was an indirect consequence of the discovery of x-rays a few months earlier by W. Röntgen, and marked the birth of nuclear physics. Studies of the radioactive decays of new isotopes continue as one of the major frontiers in nuclear research.

In 1934 I. Curie and F. Joliot demonstrated that radioactive nuclei can be made in the laboratory. All chemical elements may be rendered radioactive by adding or by subtracting (except for hydrogen and helium) neutrons from the nucleus of the stable ones. The availability of this wide variety of radioactive isotopes has stimulated their use in science and technology in an enormous number of applications.

A particular radioactive transition may be delayed by less than a microsecond or by more than a billion years, but the existence of a measurable delay or lifetime distinguishes a radioactive nuclear transition from a so-called prompt nuclear transition, such as is involved in the emission of most gamma rays. The delay is expressed quantitatively by the radioactive decay constant, or by the mean life, or by the half-period for each type of radioactive atom, discussed below.

There are five types of radioactivity commonly found in nearly all elements, each characterized by the particular type of nuclear radiation which is emitted by the transforming parent nucleus (the

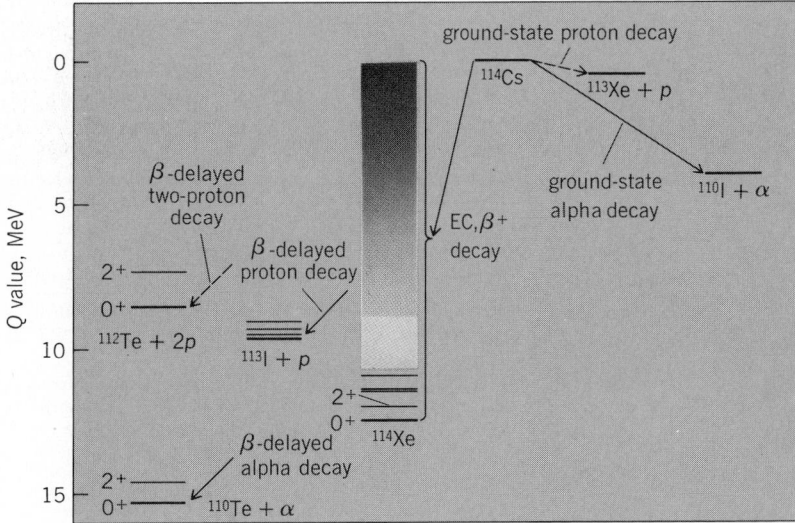

Fig. 1. Decay modes of ¹¹⁴Cs based on *Q* values from the droplet-model formula for nuclear masses. (*From E. Roeckl, Recent experiments at the GSI on-line separation, in J. H. Hamilton et al., eds., Future Directions in Studies of Nuclei Far from Stability, pp. 397–404, 1980*)

Table 1. Types of radioactivity

Type	Symbol	Particles emitted	Change in atomic number, ΔZ	Change in atomic mass number, ΔZ	Example†
Alpha	α	Helium nucleus	-2	-4	$^{226}_{86}\text{Ra} \rightarrow {}^{222}_{84}\text{Rn} + \alpha$
Beta negatron	β^-	Negative electron and antineutrino	$+1$	0	$^{24}_{11}\text{Na} \rightarrow {}^{24}_{12}\text{Mg} + e^- + \bar{\nu}$
Beta positron	β^+	Positive electron and neutrino	-1	0	$^{22}_{11}\text{Na} \rightarrow {}^{22}_{10}\text{Ne} + e^+ + \nu$
Electron capture	EC	Neutrino	-1	0	$^{7}_{4}\text{Be} + e^- \rightarrow {}^{7}_{3}\text{Li} + \nu$
Isomeric transition	IT	Gamma rays or conversion electrons or both (and positive-negative electron pair)‡	0	0	$^{137m}_{56}\text{Ba} \rightarrow {}^{137}_{56}\text{Ba} + \gamma$ or c.e.
Proton	p	Proton	-1	-1	$^{53m}_{27}\text{Co} \rightarrow {}^{52}_{26}\text{Fe} + p$ §$^{114}_{55}\text{Cs} \rightarrow {}^{113}_{54}\text{Xe} + p$
Spontaneous fission	SF	Heavy fragments and neutrons	Various	Various	$^{238}_{92}\text{U} \rightarrow {}^{133}_{50}\text{Sn} + {}^{103}_{42}\text{Mo} + 2n$
Isomeric spontaneous fission	ISF	Heavy fragments and neutrons	Various	Various	$^{244f}_{95}\text{Am} \rightarrow {}^{134}_{53}\text{I} + {}^{107}_{42}\text{Mo} + 3n$
Beta-delayed spontaneous fission	$(\text{EC} + \beta^+)\text{SF}$	Positive electron, neutrino, heavy fragments, and neutrons	Various	Various	$^{246}_{99}\text{Es} \rightarrow \beta^+ + \nu + {}^{246f}_{98}\text{Cf} \rightarrow {}^{138}_{54}\text{Xe} + {}^{107}_{44}\text{Ru} + n$
	$\beta^-\text{SF}$	Negative electron, antineutrino, heavy fragments, and neutrons	Various	Various	$^{236}_{91}\text{Pa} \rightarrow \beta^- + \bar{\nu} + {}^{236f}_{92}\text{U} \rightarrow {}^{139}_{53}\text{I} + {}^{94}_{39}\text{Y} + 3n$
Beta-delayed neutron	$\beta^- n$	Negative electron, neutrino, and neutron	$+1$	-1	$^{11}_{3}\text{Li} \rightarrow \beta^- + \bar{\nu} + {}^{11}_{4}\text{Be}^* \rightarrow {}^{10}_{4}\text{Be} + n$
Beta-delayed two-neutron (three-neutron)	$\beta^- 2n\,(3n)$	Negative electron, antineutrino, and two (three) neutrons	$+1$	-3	$^{11}_{3}\text{Li} \rightarrow \beta^- + \bar{\nu} + {}^{11}_{4}\text{Be}^* \rightarrow {}^{9(8)}_{4}\text{Be} + 2n\,(3n)$
Beta-delayed proton	$\beta^+ p$ or $(\beta^+ + \text{EC})p$	Positive electron, neutrino, and proton	-2	-1	$^{114}_{55}\text{Cs} \rightarrow \beta^+ + \nu + {}^{114}_{54}\text{Xe}^* \rightarrow {}^{113}_{53}\text{I} + p$
Beta-delayed alpha	$\beta^+\alpha$	Positive electron, neutrino, and alpha	-3	-4	$^{114}_{55}\text{Cs} \rightarrow \beta^+ + \nu + {}^{114}_{54}\text{Xe}^* \rightarrow {}^{110}_{52}\text{Te} + \alpha$
	$\beta^-\alpha$	Negative electron, antineutrino, and alpha	-1	-4	$^{214}_{83}\text{Bi} \rightarrow \beta^- + \bar{\nu} + {}^{114}_{84}\text{Po}^* \rightarrow {}^{210}_{82}\text{Pb} + \alpha$
Double beta decay§	$\beta^-\beta^-$	Two negative electrons and two antineutrinos	$+2$	0	¶$^{130}_{52}\text{Te} \rightarrow {}^{130}_{54}\text{Xe} + 2\beta^- + 2\bar{\nu}$
	$\beta^+\beta^+$	Two positive electrons and two neutrinos	-2	0	§$^{130}_{56}\text{Ba} \rightarrow {}^{130}_{54}\text{Xe} + 2\beta^+ + 2\nu$
Double electron capture§	EC EC	Two neutrinos	-2	0	§$^{130}_{56}\text{Ba} + 2e^- \rightarrow {}^{130}_{54}\text{Xe} + 2\nu$
Two-proton§	$2p$	Two protons	-2	-2	§$^{114}_{55}\text{Cs} \rightarrow {}^{112}_{53}\text{I} + 2p$
Beta-delayed two-proton§	$\beta^+ 2p$	Positive electron, neutrino, and two protons	-3	-2	§$^{114}_{55}\text{Cs} \rightarrow \beta^+ + \nu + {}^{114}_{54}\text{Xe}^* \rightarrow {}^{112}_{52}\text{Te} + 2p$
Neutron§	n	Neutron	0	-1	
Two-neutron§	$2n$	Two neutrons	0	-2	

†Excited states with relatively long measured half-lives are called isomeric and are identified by placing the symbol m for metastable after the mass number, as in ^{137m}Ba. Excited states with essentially prompt decay are identified by asterisks, as in ^{11}Be*.

‡Occurs as an additional decay mode when the decay energy exceeds 1.022 MeV.

§Theoretically predicted but not established experimentally.

¶Some indirect evidence for this particular decay has been reported, but one cannot say double beta decay is established.

first five types in Table 1). In addition, there are several other decay modes that are observed more rarely in specific regions of the periodic table, as shown in the lower half of Table 1. Several of these rarer processes are in fact two-step processes, as shown in Figs. 1 and 2. In addition, there are several other processes predicted theoretically that remain to be verified. There is some indirect evidence for double beta decay of ^{130}Te.

In the first type in Table 1, in alpha radioactivity

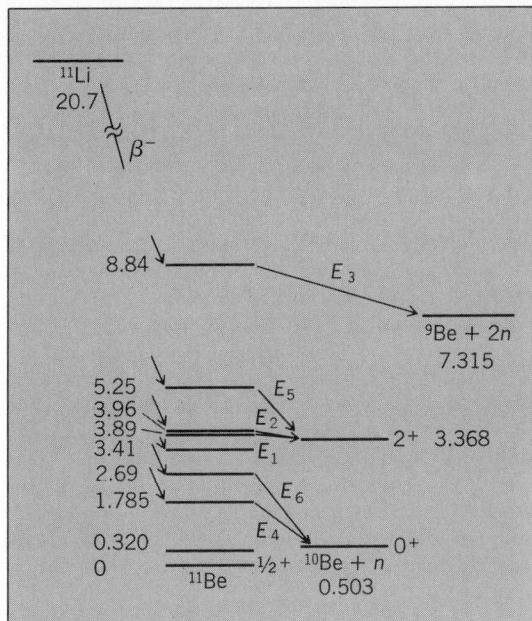

Fig. 2. Observed decay modes of ^{11}Li. Energies in MeV. (*From P. G. Hansen, On-line experiments with high-energy protons: Recent results and possible future directions, in J. H. Hamilton et al., eds., Future Directions in Studies of Nuclei Far from Stability, pp. 405–417, 1980*)

the parent nucleus spontaneously emits an alpha ray; then the atomic number, or nuclear charge, of the decay product is 2 units less than that of the parent, and the nuclear mass of the product is 4 atomic mass units less than that of the parent, because the emitted alpha particle carries away this amount of nuclear charge and mass. This decrease of 2 units of atomic number or nuclear charge between parent and product means that the decay product will be a different chemical element, displaced by 2 units to the left in a periodic table of the elements. For example, radium has atomic number 88 and is found in column II of the periodic table. Its decay product after the emission of an alpha ray is a different chemical element, radon, whose atomic number is 86 and whose position is in column 0 of the periodic table.

TRANSITION RATES AND DECAY LAWS

This section covers radioactive decay constant, dual decay, exponential decay law, mean life, and half-period.

Radioactive decay constant. The rate of radioactive transformation, or the activity, of a source equals the number A of identical radioactive atoms present in the source, multiplied by their characteristic radioactive decay constant λ. Thus Eq. (1) holds, where the decay constant λ has dimensions of second^{-1}. The numerical value of λ

$$\text{Activity} = A\lambda \text{ disintegrations per second} \quad (1)$$

expresses the statistical probability of decay of each radioactive atom in a group of identical atoms, per unit time. For example, if $\lambda = 0.01$ s^{-1} for a particular radioactive species, then each atom has a chance of 0.01 (1%) of decaying in 1 s, and a chance of 0.99 (99%) of not decaying in any given 1-s interval. The constant λ is one of the most im-

portant characteristics of each radioactive nuclide: λ is essentially independent of all physical and chemical conditions such as temperature, pressure, concentration, chemical combination, or age of the radioactive atoms. There are a few cases where measureable effects are observed for different chemical combinations. One of the largest observed is a 3.2% change in λ for the 24-s isomer in ^{90}Nb. The half-period is inversely proportional to λ.

The identification of some radioactive samples can be made simply by measuring λ, which then serves as an equivalent of qualitative chemical analysis. For the most common radioactive nuclides, the range of λ extends from 3×10^6 s^{-1} (for thorium C') to 1.6×10^{-18} s^{-1} (for thorium).

Dual decay. Many radioactive nuclides have two or more independent and alternative modes of decay. For example, ^{238}U can decay either by alpha-ray emission or by spontaneous fission. A single atom of ^{64}Cu can decay in any of three competing independent ways: negatron beta-ray emission, positron beta-ray emission, or electron capture. When two or more independent modes of decay are possible, the nuclide is said to exhibit dual decay.

The competing modes of decay of any nuclide have independent partial decay constants given by the probabilities $\lambda_1, \lambda_2, \lambda_3, \ldots$, per second, and the total probability of decay is represented by the total decay constant λ, defined by Eq. (2). If there

$$\lambda = \lambda_1 + \lambda_2 + \lambda_3 + \cdots \quad (2)$$

are A identical atoms present, the partial activities, as measured by the different modes of decay, are $A\lambda_1, A\lambda_2, A\lambda_3, \ldots$, and the total activity $A\lambda$ is given by Eq. (3). The partial activities, $A\lambda_1, \ldots$,

$$A\lambda = A\lambda_1 + A\lambda_2 + A\lambda_3 + \cdots \quad (3)$$

such as positron beta-rays from ^{64}Cu, are proportional to the total activity, $A\lambda$, at all times.

The branching ratio is the fraction of the decaying atoms which follow a particular mode of decay, and equals $A\lambda_1/A\lambda$ or λ_1/λ. For example, in the case of ^{64}Cu the measured branching ratios are $\lambda_1/\lambda = 0.40$ for negatron beta-decay, $\lambda_2/\lambda = 0.20$ for positron beta-decay, and $\lambda_3/\lambda = 0.40$ for electron capture. The sum of all the branching ratios for a particular nuclide is unity.

Exponential decay law. The total activity, $A\lambda$, equals the rate of decrease $-dA/dt$ in the number of radioactive atoms A present. Because λ is independent of the age t of an atom, integration of the differential equation of radioactive decay, $-dA/dt = A\lambda$, gives Eq. (4), where ln represents the natu-

$$\ln \frac{A}{A_0} = -\lambda(t - t_0) \quad (4)$$

ral logarithm to the base e, and A atoms remain at time t if there were A_0 atoms initially present at time t_0. If $t_0 = 0$, then Eq. (4) can be rewritten as the exponential law of radioactive decay in its most common form, Eq. (5). The initial activity at $t = 0$

$$A = A_0 e^{-\lambda t} \quad (5)$$

was $A_0\lambda$, and the activity at t, when only A atoms remain untransformed, is $A\lambda$. Because λ is a constant, the fractional activity $A\lambda/A_0\lambda$ at time t and the fractional amount of radioactive atoms A/A_0 are given by Eq. (6). In cases of dual decay, the

$$\frac{A\lambda}{A_0\lambda} = \frac{A}{A_0} = e^{-\lambda t} \tag{6}$$

partial activities $A\lambda_1$, $A\lambda_2$, . . . , also decrease with time as $e^{-\lambda t}$, not as $e^{-\lambda_1 t}$, . . . , because $A\lambda_1/A_0\lambda_1 = A/A_0 = e^{-\lambda t}$ where λ is the total decay constant. This is because the decrease of each partial activity with time is due to the depletion of the total stock of atoms A, and this depletion is accomplished by the combined action of all the competing modes of decay.

Mean life. The actual life of any particular atom can have any value between zero and infinity. The average or mean life of a large number of identical radioactive atoms is, however, a definite and important quantity.

If there are A_0 atoms present initially at $t = 0$, then the number remaining undecayed at a later time t is $A = A_0 e^{-\lambda t}$, by Eq. (5). Each of these A atoms has a life longer than t. In an additional infinitesimally short time interval dt, between time t and $t + dt$, the absolute number of atoms which will decay on the average is $A\lambda dt$, and these atoms had a life-span t. The total L of the life-spans of all the A_0 atoms is the sum or integral of $tA\lambda\,dt$ from $t = 0$ to $t = \infty$, which is given by Eq. (7). Then the

$$L = \int_0^\infty tA\lambda\,dt = \int_0^\infty tA_0\lambda e^{-\lambda t}\,dt = \frac{A_0}{\lambda} \tag{7}$$

average lifetime L/A_0, which is called the mean life τ, is given by Eq. (8), where λ is the total radioactive decay constant of Eq. (2). Substitution of $t =$

$$\tau = 1/\lambda \tag{8}$$

$\tau = 1/\lambda$ into Eq. (6) shows that the mean life is the time required for the number of atoms, or their activity, to fall to $e^{-1} = 0.368$ of any initial value.

Half-period. The time interval over which the chance of survival of a particular radioactive atom is exactly one-half is called tle half-period T. From Eq. (4), Eq. (9) is obtained. Then the half-period T

$$-\ln(A/A_0) = \ln(A_0/A) = \ln 2 = 0.693 = \lambda T \tag{9}$$

is related to the total radioactive decay constant λ, and to the mean life τ, by Eq. (10). For mnemonic

$$T = 0.693/\lambda = 0.693\tau \tag{10}$$

reasons, the half-period T is much more frequently employed than the total decay constant λ or the mean life τ. For example, it is more common to speak of ^{232}Th as having a half-period of 1.4×10^{10} years than to speak of its mean life of 2.0×10^{10} years or its total decay constant of 1.6×10^{-18} s^{-1}, although all three are equivalent statements of the average longevity of ^{232}Th atoms.

The relationships between T, τ, and λ are summarized graphically in Fig. 3. Any initial activity $A_0\lambda$ is reduced to $1/2$ in 1 half-period T, to $1/e$ in 1 mean life τ, to $1/4$ in 2 half-periods $2T$, and so on. The slope of the activity curve, or rate of decrease of activity, is $d(A\lambda)/dt = -\lambda dA/dt = -\lambda(A\lambda)$. Thus the initial slope is $-\lambda(A_0\lambda) = -(A_0\lambda)\tau$. The area under the activity curve, if integrated to $t = \infty$, is simply A_0, the total initial number of radioactive atoms. Also, the initial activity $A_0\lambda$, if it could continue at a constant value for one mean life τ, would exactly destroy all the atoms because $(A_0\lambda)\tau = A_0$.

RADIOACTIVE SERIES DECAY

In a number of cases a radioactive nuclide A decays into a nuclide B which is also radioactive; the nuclide B decays into C which is also radioactive, and so on. For example, $^{232}_{90}$Th decays into a series of 10 successive radioactive nuclides. Substantially all the primary products of nuclear fission are negatron beta-ray emitters which decay through a chain or series of two to six successive beta-ray emitters before a stable nuclide is reached as an end product. *See* Nuclear fission.

Let the initial part of such a series be represented by reaction (11), where radioactive atoms of

$$A \xrightarrow{\lambda_A} B \xrightarrow{\lambda_B} C \xrightarrow{\lambda_C} D \xrightarrow{\lambda_D} \cdots \tag{11}$$

types A, B, C, D, . . . , have radioactive decay constants given by λ_A, λ_B, λ_C, λ_D, ... Then if there are initially present, at time $t = 0$, A_0 atoms of type A, the numbers A, B, C, . . . , of atoms of types A, B, C, . . . , which will be present at a later time t are given by Eqs. (12)–(14), and the

$$A = A_0 e^{-\lambda_A t} \tag{12}$$

$$B = A_0 \frac{\lambda_A}{\lambda_B - \lambda_A}(e^{-\lambda_A t} - e^{-\lambda_B t}) \tag{13}$$

$$C = A_0 \left(\frac{\lambda_A}{\lambda_C - \lambda_A} \frac{\lambda_B}{\lambda_B - \lambda_A} e^{-\lambda_A t} \right.$$
$$+ \frac{\lambda_A}{\lambda_A - \lambda_B} \frac{\lambda_B}{\lambda_C - \lambda_B} e^{-\lambda_B t}$$
$$\left. + \frac{\lambda_A}{\lambda_A - \lambda_C} \frac{\lambda_B}{\lambda_B - \lambda_C} e^{-\lambda_C t} \right) \tag{14}$$

activities of A, B, C, . . . , are $A\lambda_A$, $B\lambda_B$, $C\lambda_C$, ... General equations describing the amounts and activities of any number of radioactive decay products are more complicated and are given in standard texts.

Figure 4 illustrates the growth and decay of the activity of a short series of radioactive decay products in accord with Eqs. (12)–(14).

Radioactive equilibrium. In Fig. 4 the ratio $B\lambda_B/A\lambda_A$ of the activities of the parent A and the daughter product B change with time. The activity $B\lambda_B$ is zero initially and also after a very long time, when all the atoms have decayed. Thus $B\lambda_B$ passes through a maximum value, and it can be shown

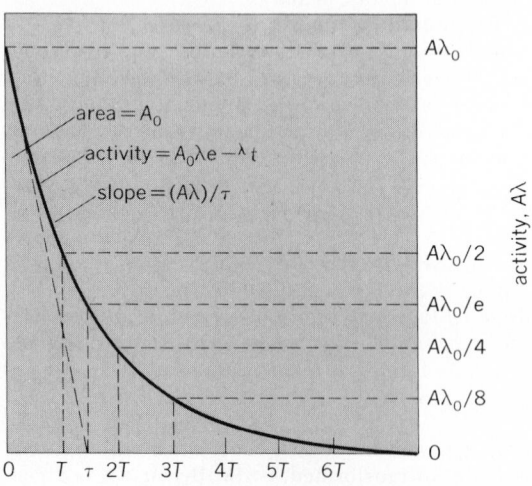

Fig. 3.   Graphical representation of relationships in decay of a single radioactive nuclide.

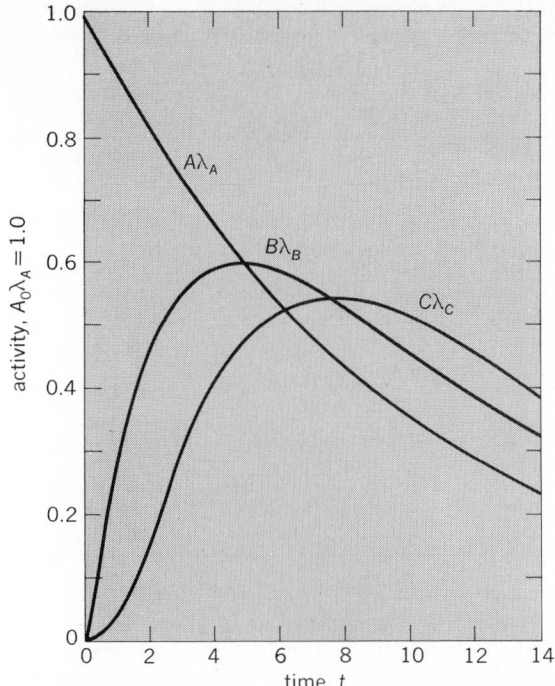

Fig. 4. Growth and decay of the activity $B\lambda_B$ of the daughter product, and $C\lambda_C$ of the granddaughter product, in an initially pure source of a radioactive parent whose activity at $t = 0$ is $A_0\lambda_A$.

that this occurs at a time t_m given by Eq. (15). The

$$t_m = \frac{\ln(\lambda_B/\lambda_A)}{(\lambda_B - \lambda_A)} \qquad (15)$$

situation in which the activities $A\lambda_A$ and $B\lambda_B$ are exactly equal to each other is called ideal equilibrium, and exists only at the moment t_m.

If the parent A is longer-lived than the daughter B, as occurs in many cases, then at a time which is long compared with the mean life τ_B of B, the activity ratio approaches a constant value given by Eq. (16), where T_A and T_B are the half-periods of A and

$$\frac{B\lambda_B}{A\lambda_A} = \frac{\lambda_B}{\lambda_B - \lambda_A} = \frac{T_A}{T_A - T_B} \qquad (16)$$

of B. When the activity ratio $B\lambda_B/A\lambda_A$ is constant, a particular type of radioactive equilibrium exists. This is spoken of as secular equilibrium if the activity ratio is experimentally indistinguishable from unity, as occurs when T_A is very much greater than T_B.

Equilibrium concepts are applied also between a long-lived parent and any of its decay products in a long series. For example, in a sufficiently old uranium ore, radium ($T = 1620$ years) is in secular equilibrium with its ultimate parent uranium ($T = 4.5 \times 10^9$ years) although there are four intermediate radioactive substances intervening in the series between uranium and radium. Here, secular equilibrium expresses the fact that the activities of radium and uranium continue to be equal to each other even though the activity of the parent uranium is decreasing with time.

When T_B is comparable with T_A, Eq. (16) shows that the equilibrium ratio will clearly exceed unity; this situation is spoken of as transient equilibrium.

For example, in fission-product decay series (17)

$$^{140}\text{Ba} \rightarrow {}^{140}\text{La} \rightarrow {}^{140}\text{Ce} \qquad (17)$$

the half-period of ^{140}Ba is 307 hours and that of ^{140}La is 40 h. In an initially pure source of ^{140}Ba the activity of ^{140}La starts at zero, rises to a maximum at $t_m = 135$ h [Eq. (15)], then decreases, and after a few hundred hours is in transient equilibrium with its parent, when the ^{140}La activity [by Eq. (16)] is $307/(307 - 40) = 1.15$ times the activity of its parent ^{140}Ba.

Radioactivity in the Earth. A number of isotopes of elements found in the Earth are radioactive. All known or theoretically predicted isotopes of elements above bismuth are radioactive. Because the Earth is composed of atoms which were believed to have been created more than 3×10^9 years ago, the naturally occurring parent radioactive isotopes are those which have such long half-periods that detectable residual activity is still observable today. As a general rule, one can detect the presence of a radioactive substance for about 10 half-lives. Therefore activities with $T \leq 0.3 \times 10^9$ years should not be found in the Earth. For example, present-day uranium is an isotopic mixture containing 99.3% ^{238}U, whose half-period is 4.5×10^9

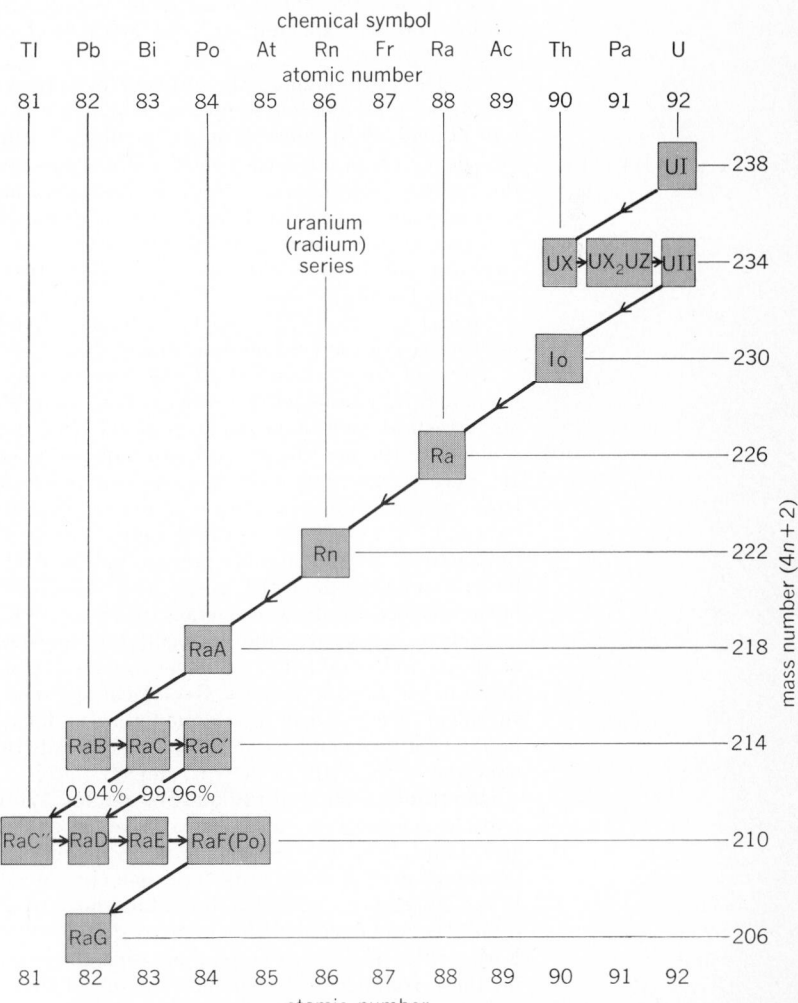

Fig. 5. Main line of decay of uranium series, or $4n + 2$ series, of heavy radioactive nuclides, headed in nature by uranium-238. Each member has a mass number given by $4n + 2$, where n is an integer.

Table 2. Parent radioactive nuclides found in nature

Nuclide		Percent abundance in nature	Half-period, years	Radioactive transitions observed	Disintegration energy, MeV
Atomic number, Z	Mass number, Z				
19 K	40	0.0117	1.3×10^9	β^-, EC	β^- 1.3 – EC 1.5
37 Rb	87	27.83	4.8×10^{10}	β^-	0.3
48 Cd	113	12.2	9×10^{15}	β^-	0.3
49 In	115	95.77	5.1×10^{14}	β^-	0.5
52 Te	130	34.49	2×10^{21}	Growth of $^{130}_{54}$Xe†	1.6
57 La	138	0.089	1.1×10^{11}	β^-, EC	β^- 1.0 – EC 1.75
60 Nd	144	23.8	2.1×10^{15}	α	1.9
62 Sm	147	15.07	1.1×10^{11}	α	2.3
62 Sm	148	11.3	8×10^{15}	α	1.99
64 Gd	152	0.20	1.1×10^{14}	α	2.2
71 Lu	176	2.6	3.6×10^{10}	β^-, γ	0.6
72 Hf	174	0.16	2×10^{15}	α	2.5
75 Re	187	62.6	4×10^{10}	β^-	0.003
78 Pt	190	0.013	6×10^{11}	α	3.24
90 Th	232	100.	1.4×10^{10}	α	4.08
92 U	235	0.715	7.0×10^8	α	4.68
92 U	238	99.28	4.5×10^9	α	4.27

†Indirect evidence for $\beta^-\beta^-$ decay.

years, and only 0.7% of the shorter-lived uranium isotope ^{235}U, whose half-period is 0.7×10^9 years, whereas these isotopes were produced in roughly equal amounts in the Earth a few billion years ago. Geophysical evidence indicates that originally some ^{236}U was present also, but none is found in nature now as expected with its half-period of 0.02×10^9 years.

^{238}U decays through a long series of 14 radioactive decay products before ending as a stable isotope of lead, ^{206}Pb. Some of these members of the ^{238}U decay chain have very short half-periods, so their existence in nature is entirely dependent on the presence of their long-lived parent, and thus is a genealogical accident. For example, radium occurs in nature only in the minerals of its parent, uranium. The decay series of ^{235}U supports 14, and the decay series of ^{232}Th supports 10, short-lived radioactive substances found in nature.

A few of the common elements contain long-lived, naturally radioactive isotopes. For example, all terrestrial potassium contains 0.012% of the radioactive isotope ^{40}K, which has a half-period of 1.3×10^9 years, and emits negatron or positron beta-rays and gamma rays in a dual decay to stable ^{40}Ca and ^{40}A. This isotope is the principal source of radioactivity in the normal human being; each human contains about 0.1 microcurie (3.7×10^3 becquerel) of the radioactive potassium isotope ^{40}K.

Table 2 summarizes the radioactive properties of all the well-established cases of radioactivities found in the Earth's surface. Geological age measurements are based on the accumulation of decay products of these long-lived isotopes, especially in the cases of ^{40}K, ^{87}Rb, ^{232}Th, ^{235}U, and ^{238}U.

Laboratory produced radioactive nuclei. With particle accelerators and nuclear reactors, over 1650 radioactive isotopes not found in detectable quantities in the Earth's crust have been produced in the laboratory since 1935, including those of at least 14 new chemical elements up to element 106. Earlier titles of induced or artificial radioactivities for these isotopes are misnomers. Many of these now have been identified in meteorites and in stars, and others are produced in the atmosphere by cosmic rays.

For example, carbon-14 is a negatron beta-ray emitter, with a half-period of about 5600 years, which can be produced in the laboratory as the product of a variety of different nuclear transmutation experiments. Nuclear bombardment of ^{11}B nuclei by alpha rays (helium nuclei) can produce excited compound nuclei of ^{15}N which promptly emit a proton (hydrogen nucleus), leaving ^{14}C as the end product of the transmutation. The same end-product ^{14}C can be produced by bombarding ^{14}N with neutrons, resulting in nuclear reaction (18). This reaction is easily carried out by using

$$^{14}\text{N} + \text{neutron} \rightarrow {}^{15}\text{N}^* \rightarrow {}^{14}\text{C} + \text{proton} \qquad (18)$$

neutrons from nuclear accelerators or a nuclear reactor. This particular transmutation reaction is one which occurs in nature also, because the nitrogen in the Earth's atmosphere is continually bombarded by neutrons which are produced by cosmic rays, thus producing radioactive ^{14}C. Mixing of ^{14}C with stable carbon provides the basis for radiocarbon dating of systems that absorb carbon for times up to about 50,000 years ago (10 half-lives). *See* NUCLEAR REACTION; PARTICLE ACCELERATOR.

Radioactive hydrogen, ^{3}H, is also formed in the atmosphere from the $^{14}\text{N} + \text{neutron} \rightarrow {}^{12}\text{C} + {}^3\text{H}$ reaction. Also, ^{3}H is produced in the Sun, and the Earth's water as well as satellites shows an additional concentration of ^{3}H from the Sun. Over two dozen radioactive products, ranging in half-life from a few days to millions of years, have been identified in meteorites that have fallen to Earth. The carbon and hydrogen burning cycles that produce energy for stars produce radioactive ^{13}N, ^{15}O, ^{3}H. At higher temperatures the radioactivities ^{7}Be and even ^{8}Be ($T \approx 10^{-16}$ s) help burn hydrogen and helium. In addition to the production of radioactive as well as stable isotopes prior to the formation of the solar system, nucleosynthesis continues to go on in stars with the production of many short-lived radioactive atoms by different processes.

The yield of any radioactivity produced in the laboratory is the initial rate of the activity under the particular conditions of nuclear bombardment. When a target material A is bombarded to produce a radioactive product B whose radioactive decay constant is λ_B, the number of atoms B which are present after a bombardment of duration t, and

Table 3. Names, symbols, and radioactive properties of members of the three naturally occurring radioactive transformation series

Conventional name	Conventional symbol	Atomic number	Mass number	Isotopic symbol	Half-period	Type of decay
Uranium ($4n+2$) series						
Uranium I	UI	92	238	^{238}U	4.5×10^9 y	α
Uranium X_1	UX_1	90	234	^{234}Th	24 d	β^-
Uranium X_2	UX_2	91	234	^{234m}Pa	1.2 m	It,β^-
Uranium Z	UZ	91	234	^{234}Pa	6.7 h	β^-
Uranium II	UII	92	234	^{234}U	2.5×10^5 y	α
Ionium	Io	90	230	^{230}Th	8×10^4 y	α
Radium	Ra	88	226	^{226}Ra	1600 y	α
Radon	Rn	86	222	^{222}Rn	3.8 d	α
Radium A	RaA	84	218	^{218}Po	3.0 m	α
Radium B	RaB	82	214	^{214}Pb	27 m	β^-
Radium C	RaC	83	214	^{214}Bi	20 m	β^-,α
Radium C'	RaC'	84	214	^{214}Po	1.6×10^{-4} s	α
Radium C''	RaC''	81	210	^{210}Tl	1.3 m	β^-
Radium D	RaD	82	210	^{210}Pb	22 y	β^-
Radium E	RaE	83	210	^{210}Bi	5.0 d	β^-
Radium F	RaF	84	210	^{210}Po	138 d	α
Polonium	Po	84	210	^{210}Po	138 d	α
Radium G	RaG	82	206	^{206}Pb	Stable	Stable
Thorium ($4n$) series						
Thorium	Th	90	232	^{232}Th	1.4×10^{10} y	α
Mesothorium$_1$	MsTh$_1$	88	228	^{228}Ra	5.8 y	β^-
Mesothorium$_2$	MsTh$_2$	89	228	^{228}Ac	6.1 h	β^-
Radiothorium	RdTh	90	228	^{228}Th	1.9 y	α
Thorium X	ThX	88	224	^{224}Ra	3.7 d	α
Thoron	Tn	86	220	^{220}Rn	56 s	α
Thorium A	ThA	84	216	^{216}Po	0.15 s	α
Thorium B	ThB	82	212	^{212}Pb	10.6 h	β^-
Thorium C	ThC	83	212	^{212}Bi	1.0 h	β^-,α
Thorium C'	ThC'	84	212	^{212}Po	3×10^{-7} s	α
Thorium C''	ThC''	81	208	^{208}Tl	3.1 m	β^-
Thorium D	ThD	82	208	^{208}Pb	Stable	Stable
Actinium ($4n+3$) series						
Actinouranium	AcU	92	235	^{235}U	7.0×10^8 y	α
Uranium Y	UY	90	231	^{231}Th	26 h	β^-
Protactinium	Pa	91	231	^{231}Pa	3.3×10^4 y	α
Actinium	Ac	89	227	^{227}Ac	22 y	β^-,α
Radioactinium	RdAc	90	227	^{227}Th	19 d	α
Actinium K	AcK	87	223	^{223}Fr	22 m	β^-,α
Actinium X	AcX	88	223	^{223}Ra	11 d	α
Astatine	At	85	219	^{219}At	0.9 m	α,β^-
Actinon	An	86	219	^{219}Rn	4.0 s	α
Actinium A	AcA	84	215	^{215}Po	1.8×10^{-3} s	α
Actinium B	AcB	82	211	^{211}Pb	36 m	β^-
Actinium C	AcC	83	211	^{211}Bi	2.2 m	α,β^-
Actinium C'	AcC'	84	211	^{211}Po	0.5 s	α
Actinium C''	AcC''	81	207	^{207}Tl	4.8 m	β
Actinium D	AcD	82	207	^{207}Pb	Stable	Stable

their activity $B\lambda_B$, are given by Eq. (19), where the

$$B\lambda_B = \frac{Y}{\lambda_B}(1 - e^{-\lambda_B t}) \qquad (19)$$

yield Y has dimensions equivalent to curies of activity produced per second of bombardment. The yield Y depends on the number of atoms A present in the target, the intensity of beam of bombarding particles, and the cross section, or probability of the reaction per bombarding particle under the conditions of bombardment.

Radioactive transformation series. As noted in Eqs. (12)–(14), many radioactive substances have decay products which are also radioactive. Thus many long chains or series of radioactive transformations are known.

The three naturally occurring transformation series are headed by ^{232}Th, ^{235}U, and ^{238}U. Their genealogical relationships are summarized in Fig. 5 and Table 3.

Each of the naturally occurring radioactive isotopes in these transformation series has two synonymous names. For example, the commercially important radioisotope whose classical name is mesothorium-1 is now known to be an isotope of radium with mass number of 228 and is designated as radium-228 (^{228}Ra). Table 3 summarizes the names, symbols, and some radioactive properties of these three transformation series. However, these chains are not complete, and their uniqueness or importance as chains is an accident of the very long half-lives of ^{232}Th, ^{235}U, and ^{238}U. For example, element 105 of mass 260 has a succession of seven alpha decays and one electron capture and positron decay to ^{232}Th. The special importance of the chains in Table 3 is related to the fact that they were essentially the only early sources of radioactive materials, and more recently to their role in nuclear power.

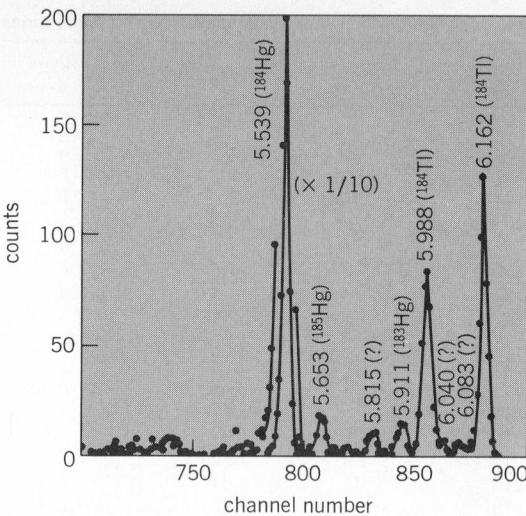

Fig. 6. Alpha groups in the decay of ^{184}Tl ($T = 11$s) and ^{184}Hg and weak groups from ^{183}Hg and ^{185}Hg, very far off stability (17 neutrons less than the lightest stable thallium isotope). Energies in MeV. (*From K. S. Toth et al., Observation of α-decay in thallium nuclei, including the new isotopes ^{184}Tl and ^{185}Tl, Phys. Lett., 63B:150–153, 1976*)

Transformation series are now known for every element in the periodic table except hydrogen. Chains of neutron-rich isotopes have been produced and studied among the products of nuclear fission. First, heavy-ion-induced reactions and subsequently high-flux reactors have been used to extend knowledge of the elements beyond uranium. Both proton- and heavy-ion-induced reactions have extended knowledge of chains of neutron-deficient isotopes of the stable elements.

ALPHA-RAY DECAY

Alpha-ray decay is that type of radioactivity in which the parent nucleus expels an alpha ray (a helium nucleus). The alpha ray is emitted with a speed of the order of 1 to 2×10^7 m/s, that is, about 1/20 of the velocity of light.

In the simplest case of alpha decay, every alpha ray would be emitted with exactly the same velocity and hence the same kinetic energy. However, in most cases there are two or more discrete energy groups called lines, as shown in Fig. 6 in the spectrum of alpha rays from ^{184}Tl and ^{184}Hg decays. For example, in the alpha decay of a large group of ^{238}U atoms, 77% of the alpha decays will be by emission of alpha rays whose kinetic energy is 4.20 MeV, while 23% will be by emission of 4.15-MeV alpha rays. When the 4.20-MeV alpha ray is emitted, the decay product nucleus is formed in its ground (lowest energy) level. When a 4.15-MeV alpha ray is emitted, the decay product is produced in an excited level, 0.05 MeV above the ground level. This nucleus promptly transforms to its ground level by the emission of a 0.05-MeV gamma ray or alternatively by the emission of the same amount of energy in the form of a conversion electron and the associated spectrum of characteristic x-rays. Thus in all alpha-ray spectra, the alpha rays are emitted in one or more discrete and homogeneous energy groups, and alpha-ray spectra are accompanied by gamma-ray and conversion elec-

tron spectra whenever there are two or more alpha-ray groups in the spectrum.

Geiger-Nuttall rule. Among all the known alpha-ray emitters, most alpha-ray energy spectra lie in the domain of 4−6 MeV, although a few extend as low as 2 MeV ($^{147}_{62}$Sm) and as high as 10 MeV (ThC$'$). There is a systematic relationship between the kinetic energy of the emitted alpha rays and the half-period of the alpha emitter. The highest-energy alpha rays are emitted by short-lived nuclides, and the lowest-energy alpha rays are emitted by the very-long-lived alpha-ray emitters. H. Geiger and J. M. Nuttall showed that there is a linear relationship between log λ and the energy of the alpha ray.

The Geiger-Nuttall rule is inexplicable by classical physics, but emerges clearly from quantum, or wave, mechanics. In 1928 the hypothesis of transmission through nuclear potential barriers, as introduced by G. Gamow and independently by R. W. Gurney and E. U. Condon, was shown to give a satisfactory account of the alpha-decay data, and it has been altered subsequently only in details. The form of the barrier-penetration equations is such that correlation plots of log λ against $1/\sqrt{E}$ give nearly straight lines.

Nuclear potential barrier. At distances r which are large compared with the nuclear radius, the potential energy of an alpha ray, whose charge is $2e$, in the field of a residual nucleus, whose charge is $(Z - 2)e$, is $2(Z - 2)e^2/r$. At very close distances this electrostatic repulsion is opposed and overcome by short-range, specifically nuclear, attractive forces. The net potential energy U as a function of the separation r between the alpha ray and its residual nucleus is called the nuclear potential barrier.

One of several operating definitions of the nuclear radius R is the distance $r = R$ at which the attractive nuclear forces just balance the repulsive electrostatic forces. At this distance, called the top of the nuclear barrier, the potential energy is about 25−30 MeV for typical cases of heavy, alpha-emitting nuclei, as indicated in Fig. 7. *See* NUCLEAR STRUCTURE; POTENTIAL BARRIER; QUANTUM MECHANICS.

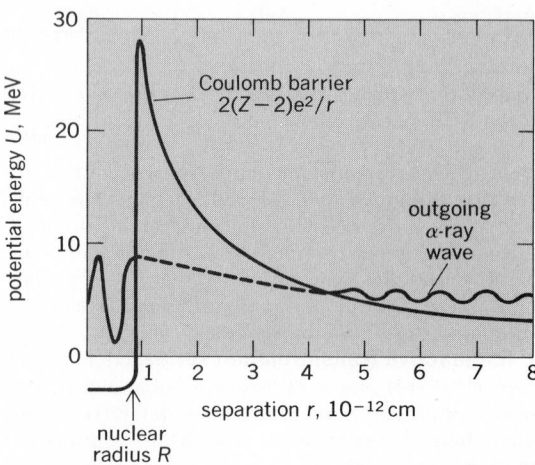

Fig. 7. Schematic of nuclear potential barrier, illustrating emission of an α-ray as a wave which can be transmitted through the barrier.

Inside the nucleus the alpha particle is represented as a de Broglie matter wave. According to wave mechanics, this wave has a very small but finite probability of being transmitted through the nuclear potential energy barrier and thus of emerging as an alpha ray emitted from the nucleus. The transmission of a particle through such an energy barrier is completely forbidden in classical electrodynamics but is possible according to wave mechanics. This transmission of a matter wave through an energy barrier is analogous to the familiar case of the transmission of ordinary visible light through an opaque metal such as gold: if the gold is thin enough, some light does get through, as in the case of the thin gold leaf which is sometimes used for lettering signs on store windows.

The wave-mechanical probability of the transmission of an alpha particle through the nuclear potential barrier is very strongly dependent upon the energy of the emitted alpha ray. Analytically the probability of transmission T depends exponentially upon a barrier transmission exponent γ according to Eq. (20). To a good approximation, Eq. (21) holds, where $h = 6.626 \times 10^{-34}$ joule-second

$$T = e^{-\gamma} \qquad (20)$$

$$\gamma = \left(\frac{4\pi^2}{h}\right)\frac{(Z-2)2e^2}{V} - \left(\frac{8\pi}{h}\right)[2(Z-2)2e^2MR]^{1/2} \quad (21)$$

is Planck's constant, and M is the so-called reduced mass of the alpha particle. For the alpha decay of ^{226}Ra, the numerical value of γ is about 71: hence $T = e^{-71} = 10^{-31}$. The first term on the right side of Eq. (21) is about 154 and is therefore the dominant term. When this term is taken alone, $e^{-(4\pi^2/h)(Z-2)2e^2/V}$ is called the Gamow factor for barrier penetration.

Inspection of Eq. (21) shows that the barrier transmission decreases with increasing nuclear charge $(Z-2)e$, increases with increasing velocity V of emission of the alpha ray, and increases with increasing radius R of the nucleus. When the experimentally known values of alpha-decay energy are substituted into Eq. (21), with R about 10^{-12} cm and Z about 90, the transmission coefficient $T = e^{-\gamma}$ is found to extend over a domain of about 10^{-20} to 10^{-40}. This range of about 10^{20} is just what is needed to relate the alpha-disintegration energy to the broad domain of known alpha-decay half-periods. Equation (21) thus explains the Geiger-Nuttall rule very successfully. Figure 8 presents a modern form of the Geiger-Nuttall relationship. The individual points show the measured half-periods and alpha-disintegration energies (alpha-ray energy plus recoil energy) for a number of high-Z emitters of alpha rays. The smooth curves are drawn using the wave-mechanical theory of transmission through nuclear barriers, with a nuclear radius of $R = 1.48 \times 10^{-15}\,A^{1/3}$ m, where A is the mass number of the alpha-ray decay product. The agreement between experiment and theory is good.

Since 1970, knowledge of alpha-emitting isotopes has been greatly enlarged through the identification of many isotopes far off stability in the region just above tin and in the broad region from neodymium all the way to uranium. For example, fusion reactions between 290-MeV ^{58}Ni ions and ^{58}Ni and ^{63}Cu targets have been used to produce and study very-neutron-deficient radioactive iso-

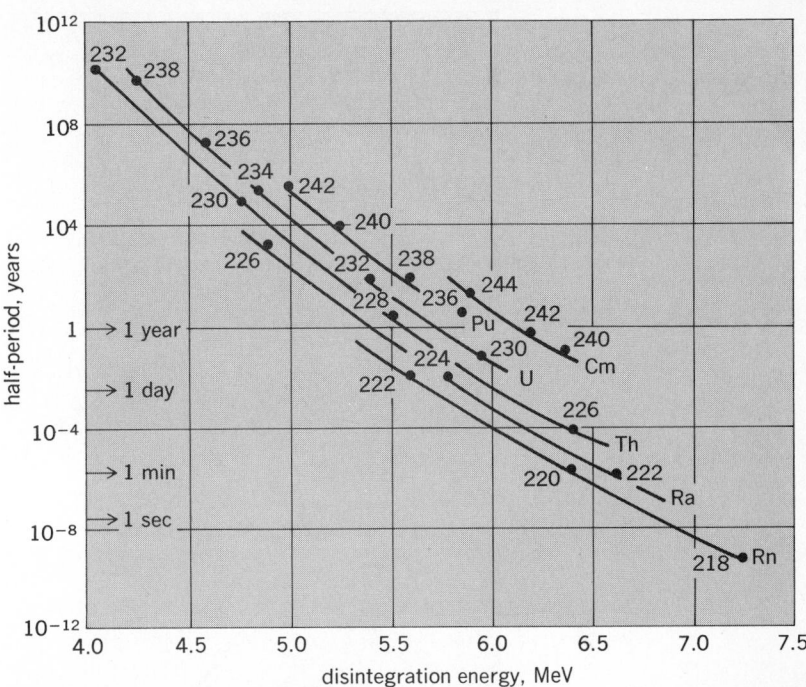

Fig. 8. Systematics of the broad range of half-periods for α-ray decay and their strong dependence on α-decay energy and weaker dependence on nuclear charge. Numbers beside experimental points are mass numbers of parent α-ray emitters. Lines connect parent isotopes and are drawn using wave-mechanical theory of α-ray transmission through nuclear potential barriers.

topes, including 12 alpha emitters between tin and cesium. These results provide important data on the atomic masses of nuclei far from the stable ones in nature. These data test understanding of nuclear mass formulas and their validity in new regions of the periodic table.

BETA-RAY DECAY

Beta-ray decay is a type of radioactivity in which the parent nucleus emits a beta ray. There are two types of beta decay established: in negatron beta decay (β^-) the emitted beta ray is a negatively charged electron (negatron); in positron beta decay (β^+) the emitted beta ray is a positively charged electron (positron). In beta decay the atomic number shifts by one unit of charge, while the mass number remains unchanged (Table 1). In contrast to alpha decay, when beta decay takes place between two nuclei which have a definite energy difference, the beta rays from a large number of atoms will have a continuous distribution of energy.

The continuous number-versus-energy distribution of emitted beta rays is illustrated in Fig. 9. For each beta-ray emitter, there is a definite maximum or upper limit to the energy spectrum of beta rays. This maximum energy, E_{max}, corresponds to the change in nuclear energy in the beta decay. Thus $E_{max} = 0.57$ MeV for β^- decay of ^{64}Cu, and $E_{max} = 0.66$ MeV for β^+ decay of ^{64}Cu. As in the case of alpha decay, most beta-ray spectra are not this simple, but include additional continuous spectra which have less maximum energy and which leave the product nucleus in an excited level from which gamma rays are then emitted.

For nuclei very far from stability, the energies of these excited states populated in beta decay are so

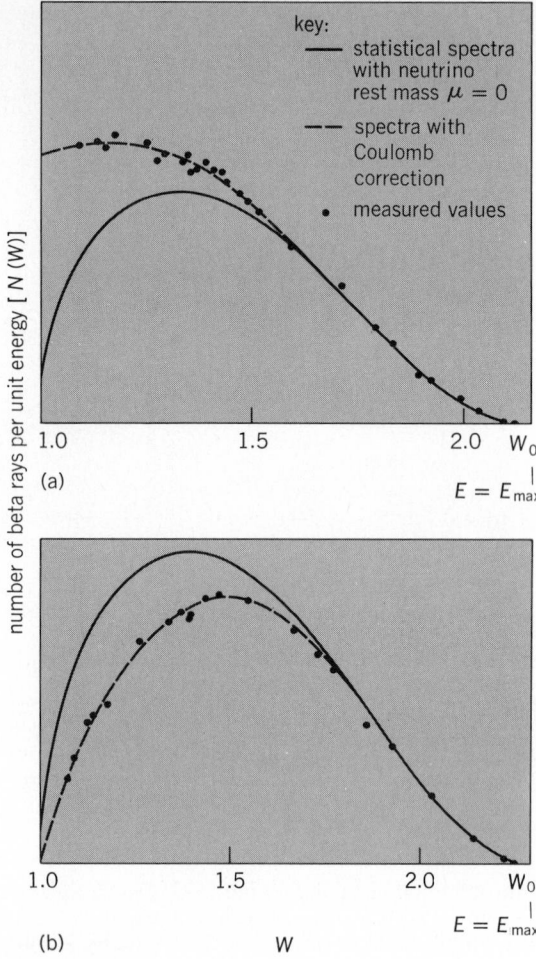

Fig. 9. Spectra of beta rays. (a) β^- decay of ^{64}Cu. (b) β^+ decay of ^{64}Cu. (From L. M. Langer, R. D. Moffat, and H. C. Price, The beta-spectrum of Cu^{64}, Phys. Rev., 76:1725–1726, 1949)

large that the excited states may decay by proton, neutron, two-neutron or alpha emission, or spontaneous fission and, it is predicted theoretically, by two-proton emission. In some cases, the energies are so great that the number of excited states to which beta decay can occur is so large that only the gross strength of the beta decays to many states can be studied.

Neutrinos. The continuous spectrum of beta-ray energies shown in Fig. 9 implies the simultaneous emission of a second particle besides the beta ray, in order to conserve energy and angular momentum for each decaying nucleus. This particle is the neutrino. The sum of the kinetic energy of the neutrino and the beta ray equals E_{max} for the particular transition involved except in the rare cases where internal bremsstrahlung or shake-off electrons are emitted. The neutrino has zero charge and presumably zero rest mass, travels at the same speed as light (3×10^{10} cm/s), and is emitted as a companion particle with each beta ray.

Earlier careful measurements of the beta spectra of ^{3}H established an upper limit for the neutrino rest mass as less than 0.0005 times the rest energy of the electron (Fig. 10). In 1980, however, some indirect evidence and a new ^{3}H beta spectra measurement yielded evidence for a rest mass much smaller than this limit, but finite. If the neutrino

does have a nonzero rest mass, this will have many consequences, although they will not radically change the general features of the beta decay as presented here.

Two forms of neutrinos are distinguished. In positron beta decay, a proton p in the nucleus transforms into a neutron n in the nucleus, thus reducing the nuclear charge by 1 unit. At the time of this transition, two particles, the positron β^+ and the neutrino ν, are created and emitted. The emitted β^+ and ν together carry away the energy E_{max} of the transition and provide for conservation of energy, momentum, angular momentum, charge, and statistics. Thus positron beta decay is represented by reaction (22). Negatron beta decay is a

$$p \rightarrow n + \beta^+ + \nu \qquad (22)$$

closely related process, except that a neutron n changes to a proton p in the nucleus, and a negatron beta-ray β^- and its characteristic companion particle, the antineutrino $\bar{\nu}$, are emitted. Thus reaction (23) is written. The antineutrino is the anti-

$$n \rightarrow p + \beta^- + \bar{\nu} \qquad (23)$$

particle of the neutrino as the β^+ is the antiparticle of the β^-. They have the same properties of zero charge and zero rest mass, and differ only with respect to the direction of alignment of their intrinsic spin along their direction of motion. In most beta-decay contexts, the term neutrino includes both its forms, neutrino and antineutrino. Because of the fact that the neutron rest mass is greater than the proton rest mass, free neutrons can undergo beta decay (23), but protons must use part of the nuclear energy available to make up the rest mass difference.

The interaction of neutrinos with matter is exceedingly feeble. A neutrino can pass all the way through the Sun with little chance of collision. The thickness of lead required to attenuate neutrinos by the factor $\frac{1}{2}$ is about 10^{18} m, or 100 light-years of lead! See ANTIMATTER; NEUTRINO; POSITRON.

Average beta energy. Charged particles, such as beta rays or alpha rays, are easily absorbed in matter, and their kinetic energy is thereby converted into heat. In beta decay the average energy E_{av} of the beta rays is far less than the maximum energy E_{max} of the particular beta-ray spectrum. The detailed shape of beta-ray spectra and hence the exact value of the ratio E_{av}/E_{max} varies somewhat with Z, E_{max}, the degree of forbiddenness of the transition, and the sign of charge of the emitted beta ray. A rough rule of thumb which covers many practical cases is $E_{av} = (0.40 \pm 0.05) \, E_{max}$, with slightly higher values for positron beta-ray spectra than for negatron beta-ray spectra. The remaining disintegration energy is emitted as kinetic energy of neutrinos and is not recoverable in finite absorbers.

There are other processes that carry off part of the energy of beta decay, including internal bremsstrahlung (gamma rays) and shake-off electrons (atomic electrons). The total probabilities for these additional two processes are the order of 1% or much less per beta decay, and the probability of their emission decreases rapidly with increasing energy so they are mainly low-energy (less than about 50 keV) radiations. In internal bremsstrahlung, through an interaction of the beta ray and the emitting nucleus, part of the decay energy is emitted as a gamma ray. In the shake-off process, part of the beta-decay energy is given to one of the atomic electrons. The gamma rays are not absorbed in matter as easily as the beta rays. In addition, if one tries to absorb the beta rays in matter, the beta rays can interact with the atoms and give off external bremsstrahlung (gamma rays). The number of these gamma rays again is a strongly decreasing function of energy, but their emission extends up to the maximum energies of the beta rays.

Fermi theory. By postulating the simultaneous emission of a beta ray and a neutrino, as in reaction (22). E. Fermi developed in 1934 a quantum-mechanical theory which satisfactorily gives the shape of beta-ray spectra (Fig. 9), and the relative half-periods of beta-ray emitters for allowed beta decays. The energy distribution of beta rays in allowed transitions is then given by Eq. (24).

$$N(W)\,dW$$
$$= \frac{|P|^2}{\tau_0}\,F(Z,W)\,(W^2-1)^{1/2}(W_0-W)^2 W\,dW \quad (24)$$

$N(W)\,dW =$ number of beta rays in energy range W to $W+dW$

$W = 1 + E/(m_0 c^2) =$ total energy of beta ray in units of rest energy $m_0 c^2 = 0.51$ MeV for an electron ($m_0 =$ electron mass, $c =$ velocity of light)

$W_0 = 1 + E_{max}/(m_0 c^2) =$ maximum energy of the beta-ray spectrum

$|P|^2 =$ squared matrix element for the transition, and is of the order of unity for allowed transitions

$\tau_0 =$ time constant $\cong 7000$ s

$F(Z,W) =$ complex, dimensionless function involving the nuclear radius, nuclear charge, beta-ray energy, and whether the decay is β^- or β^+

Physically this distribution function involves the product of the energy W and momentum $(W^2-1)^{1/2}$ of the beta ray times the energy (W_0-W) and the momentum $(W_0-W)/c$ of the neutrino. The product of these factors gives a "statistical" distribution for the number of beta rays as a function of energy as shown in Fig. 9. The observed spectra show an excess of low-energy β^- and a deficiency of low energy β^+ particles. This arises because of the Coulomb attraction and repulsion of the nucleus for β^- and β^+. The statistical spectrum is corrected by the Fermi function, $F(Z,W)$, and the new distribution agrees with experiments as shown in Fig. 9.

Equation (24) essentially matches the energy spectra of allowed beta-ray transitions and therefore furnishes one type of experimental verification of the properties of neutrinos. Its counterpart in terms of the beta-ray momentum spectrum is often used for the analysis of spectra, and is given by Eq. (25). The momentum distribution is much more

$$N(\eta)\,d\eta = \frac{|P|^2}{\tau_0}\,F(Z,\eta)\,(W_0-W)^2\eta^2\,d\eta \quad (25)$$

$N(\eta)\,d\eta =$ number of beta rays in the momentum interval from η to $\eta+d\eta$

$\eta = (W^2-1)^{1/2} =$ momentum of the beta ray in units of $m_0 c$

$F(Z,\eta) = F(Z,W)$ of Eq. (24)

nearly symmetric than its corresponding energy spectrum.

Konopinski-Uhlenbeck theory. After the work of Fermi which explained allowed decay, E. J. Konopinski and G. E. Uhlenbeck in 1941 developed the theory of forbidden beta decay. Allowed decays occur between nuclear states which differ in spin by 0 or 1 unit and which have the same parity. Konopinski and Uhlenbeck developed a theory to describe beta decays where energy is available for decay but the allowed selection rules on spin or parity or both are violated. These beta transitions occur at a slower rate and are called forbidden transitions. In 1949 the theory of forbidden beta decay was confirmed by L. M. Langer and H. C. Price. The orders of forbiddenness, which retard the rate of decay, are: once-forbidden decay when the change in nuclear spin ΔI is again 0 or 1 as in allowed decay, but a parity charge $\Delta\pi$ occurs; once-forbidden unique decay when $\Delta\pi$ changes and $\Delta J = 2$; n-times forbidden decay when $\Delta J = n$, $\Delta\pi = (-)^n$, where $\Delta\pi = -$ indicates a parity change; and n-times forbidden unique decay when $\Delta J = n+1$, $\Delta\pi = (-)^n$. These are illustrated in Table 4. In forbidden decays the first-order allowed matrix elements of the Fermi theory in Eq. (24) vanish because of the selection rules on angular momentum and spin. Then the much smaller higher-order matrix elements that can be neglected compared to the large allowed matrix elements come into play. *See* PARITY; SELECTION RULES.

Comparative half-lives, fT. The half-period T of beta decay can be derived from Eq. (24) because the radioactive decay constant $\lambda = 0.693/T$ is simply the total probability of decay, or $N(W)dW$ integrated over all possible values of the beta-ray energy from $W = 1$ to $W = W_0$.

For allowed decays, the matrix elements are not functions of the beta energy and can be factored out of Eq. (24), so Eq. (26) is valid, where f is given by Eq. (27), and the constants include $|P|^2$ pf Eq. (24). Equation (26) can be rearranged as Eq. (28).

$$\lambda = 0.693/T = \text{constants} \times f \quad (26)$$

$$f = \int_1^W F(Z,W)\,(W^2-1)^{1/2}(W_0-W)^2 W\,dW \quad (27)$$

$$fT = \frac{0.693}{\text{constants}} = \text{comparative half-life} \quad (28)$$

For different beta decays, T varies over a range greater than 10^{18} and inversely depends on the

Table 4. Selection rules for beta decay and log _fT_ values

Type	ΔJ	$\Delta\pi$	Log fT	Examples
Allowed (favored)	0 or 1	No	3	n, ^{3}H
Allowed (normal)	0 or 1	No	4 to 7	^{35}S, ^{30}P
Allowed (l-forbidden)	1	No	6 to 9	^{32}P, ^{65}Ni
Once-forbidden	0 or 1	Yes	6 to 8	^{111}Ag, ^{143}Pr
Once-forbidden (unique)	2	Yes	8 to 9	^{42}K, ^{91}Y
Twice-forbidden	2	No	11 to 14	^{36}Cl, ^{59}Fe
Twice-forbidden (unique)	3	No	12 to 14	^{22}Na, ^{60}Co
Third-forbidden	3	Yes	17 to 19	^{87}Rb, ^{138}La
Third-forbidden (unique)	4	Yes	(~18)	^{40}K
Fourth-forbidden	4	No	~24	^{115}In
Fourth-forbidden (unique)	5	No		

beta-decay energy in analogy to the Geiger-Nuttall rule for alpha decay. However, Eq. (28) says that the comparative half-life should be a constant. Indeed it is found experimentally that different classes of beta decay do have very similar fT values. It is generally easier to give the $\log_{10} fT$ for comparison. The groups are illustrated in Table 4 and include, in addition to the forbidden decays, three classes of allowed decays: the favored or superallowed decays of nuclei whose structures are very similar so that the matrix element in the denominator of Eq. (28) is large and $\log fT$ is small; normal allowed; and allowed l-forbidden where the total angular momentum selection rule holds, but the individual particle that is undergoing beta decay has a change of 2 units of orbital angular momentum. The matrix elements for each degree of forbiddenness get progressively smaller and so $\log fT$ values increase sharply with each degree of forbiddenness. The ranges of these fT values for each degree of forbiddenness are in general so well established that measurements of fT values can be used to establish changes in spins and parities between nuclear states in beta decay.

Kurie plots. For allowed transitions, the transition matrix element $|P|^2$ is independent of the momentum η. Then Eq. (25) can be put in the form of Eq. (29). Therefore a straight line results when

$$\left(\frac{N(\eta)}{\eta^2 F(Z,\eta)}\right)^{1/2} = \text{const}\,(W_0 - W) \qquad (29)$$

the quantity $\sqrt{N/\eta^2 F}$ is plotted against beta-ray

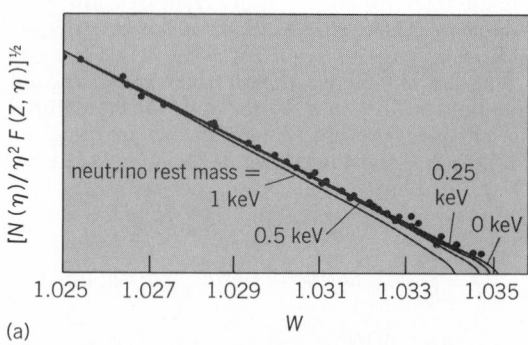

(a)

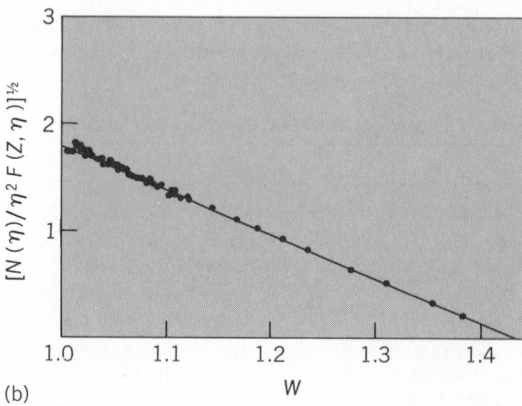

(b)

Fig. 10. Kurie plots. (a) Allowed decay of ^{3}H (from L. M. Langer, and R. J. D. Moffat, The beta-spectrum of tritium and the mass of the neutrino, Phys. Rev., 88:689–694, 1952). (b) Once-forbidden decay of ^{147}Pm (from J. H. Hamilton, L. M. Langer, and W. G. Smith, The shape of the ^{143}Pr spectrum, Phys. Rev., 112:2010–2019, 1958).

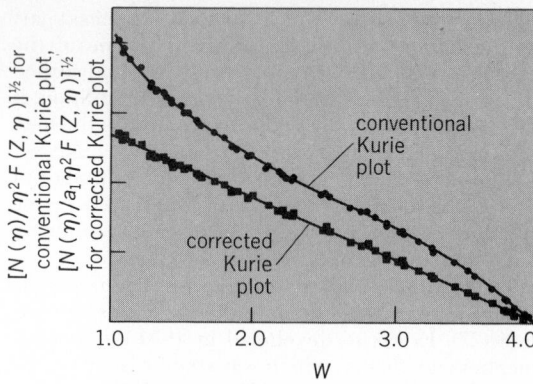

Fig. 11. Once-forbidden spectrum of ^{91}Y: conventional Kurie plot and Kurie plot corrected by the unique shape factor, $a_1 = W^2 - 1 + (W_0 - W)^2$, given by Konopinski and Uhlenbeck, which linearizes the data. (From L. M. Langer and H. C. Price, Shape of the beta spectrum of the forbidden transition of yttrium 91, Phys. Rev., 75:1109, 76:641, 1949)

energy, either as W or as E, on a linear scale. Such graphs are called Kurie plots, Fermi plots, or Fermi-Kurie plots. These are especially useful for revealing deviations from the allowed theory and for obtaining the upper energy limit E_{max} as the extrapolated intercept of $\sqrt{N/\eta^2 F}$ on the energy axis. Practically all results on the shapes of beta-ray spectra are published as Kurie plots, rather than as actual momentum or energy spectra.

Figure 10 shows representative Kurie plots for ^{3}H and ^{147}Pm. When spectral data give a straight line, such as these, then $N(\eta)$ is in agreement with the Fermi momentum distribution, Eq. (25); and the intercept of this straight line, on the energy axis, gives the disintegration energy E_{max}. In Fig. 10a, theoretical curves are given for various values of the neutrino rest mass, and the data points, which are experimental values, lie on the curve corresponding to zero mass.

In addition to allowed decays, all but one known once-forbidden decays have Kurie plots that are essentially linear in energy (Fig. 10b). The once-forbidden unique decays have a pronounced characteristic energy dependence for their matrix elements, and thus the conventional Kurie plot has a characteristic shape that differs from a straight line (Fig. 11). When the data are corrected by the unique shape factor, given by Konopinski and Uhlenbeck, a linear Kurie plot is again obtained. This unique shape was the key to the discovery of forbidden beta decay by Langer and Price, and Fig. 11 is, in fact, from the data with which they made this discovery. The higher-order forbidden spectra each show different strong energy dependences in their Kurie plots, each characteristic of their degree of forbiddenness.

Double beta decay. When the ground state of a nucleus differing by 2 units of charge from nucleus A has lower energy than A, then it is theoretically possible for A to emit two beta particles, either $\beta^+\beta^+$ or $\beta^-\beta^-$ as the case may be, and two neutrinos or antineutrinos, and go from Z to $Z \pm 2$. Here two protons decay into two neutrons, or vice versa. This is a second-order process and so should go much slower than beta decay. There are a number of cases where such decays should occur, but their half-lives are of the order of 10^{20} years or greater.

These are obviously very difficult to detect and
have not been seen directly. There is indirect evidence for double beta decay in one case, that of
$^{130}_{52}$Te, from the observed buildup of $^{130}_{54}$Xe in samples. However, it cannot be said with certainty that
this process has been shown to occur in nature.

Electron-capture transitions. Whenever it is
energetically allowed by the mass difference between neighboring isobars, a nucleus Z may capture one of its own atomic electrons and transform
to the isobar of atomic number $Z - 1$ (Table 1).
Usually the electron-capture (EC) transition involves an electron from the K shell of atomic electrons, because these innermost electrons have the
greatest probability density of being in or near the
nucleus. *See* ELECTRON CAPTURE.

In EC transitions, a proton p bound in the parent
nucleus absorbs an electron e^- and changes to a
bound neutron n. The disintegration energy is carried away by an emitted neutrino ν as in Eq. (30).

$$p + e^- \rightarrow n + \nu \qquad (30)$$

The residual nucleus may be left either in its
ground level or in an excited level from which
gamma-ray emission follows.

EC transitions compete with all cases of positron beta-ray decay. EC has an energetic advantage over β^+ decay equivalent to the mass of two
electrons, or 1.02 MeV, because in Eq. (30) one
electron mass e^- enters the reaction and is available, whereas in Eq. (24) one electron mass β^+
must be produced as a product of the positron
beta-ray decay. For example, in the radioactive
decay of $^{64}_{29}$Cu, twice as many transitions go by
EC to $^{64}_{28}$Ni as go by positron beta decay to the
same decay product. In the heavy, high-Z elements, EC is greatly favored over the competing
β^+ decay, and examples of measurable β^+ decay
are practically unknown for Z greater than 80, although there are a large number of examples of
electron capture. As the energy for decay increases beyond 1.02 MeV, the probability of β^+
decay increases relative to EC and dominates at
several MeV of energy.

Several examples are known of completely pure
EC radioactivity in which there is insufficient nuclear energy to allow any positron beta-ray decay.
For example, $^{55}_{26}$Fe emits no positron beta rays,
but transforms with a half-period of 2.6 years entirely by EC to the ground level of $^{55}_{25}$Mn. This
radioactivity is detectable through the K-series
x-rays which are emitted from ^{55}Mn when the
atomic electron vacancy, produced by nuclear
capture of a K electron, refills from the L shell of
atomic electrons. Also, the process of double electron capture, analogous to double beta decay, is
theoretically predicted to exist. Here two atomic
electrons are captured and two neutrinos emitted.
See X-RAYS.

GAMMA-RAY DECAY

Gamma-ray is a transition between two excited
levels of a nucleus, or between an excited level and
the ground level. A nucleus in its ground level
cannot emit any gamma radiation. Therefore
gamma-ray decay occurs only as a sequel of one of
the processes in Table 1 or of some other process
whereby the product nucleus is left in an excited
state. Such additional processes include gamma

rays observed following the fusion of two nuclei, as
occurs in bombarding ^{58}Ni with ^{16}O to form an excited compound nucleus of ^{74}Kr. This compound
nucleus first promptly gives off a few particles like
two neutrons to leave ^{72}Kr* or two protons to leave
^{72}Se*, both of which will be in excited states which
will emit gamma rays. Or one may excite states in
a nucleus by the Coulomb force between two nuclei when they pass close to each other but do not
touch (their separation is greater than the sum of
the radii of the two nuclei). There are also other
nuclear reactions such as induced nuclear fission
that leave nuclei in excited states to undergo
gamma decay.

A gamma ray is high-frequency electromagnetic
radiation (a photon) in the same family with radio-
waves, visible light, and x-rays. The energy of a
gamma ray is given by $h\nu$, where h is Planck's constant and ν is the frequency of oscillation of the
wave in hertz. The gamma-ray or photon energy $h\nu$
lies between 0.05 and 3 MeV for the majority of
known nuclear transitions. Higher-energy gamma
rays are seen in neutron capture and some reactions. *See* ELECTROMAGNETIC RADIATION.

Gamma rays carry away energy, linear momentum, and angular momentum, and account for
changes of angular momentum, parity, and energy between excited levels in a given nucleus.
This leads to a set of gamma-ray selection rules for
nuclear decay and a classification of gamma-ray
transitions as "electric" or as "magnetic" multipole radiation of multipole order 2^l, where $l = 1$ is
called dipole radiation, $l = 2$ is quadrupole radiation, and $l = 3$ is octupole, l being the vector
change in nuclear angular momentum. Ths most
common type of gamma-ray transition in nuclei is
the electric quadrupole (E2). There are cases
where several hundred gamma rays with different
energies are emitted in the decays of atoms of
only one isotope.

Mean life for transitions. A reasonably successful approximate theory of the mean life for gamma-
ray decay was developed by V. F. Weisskopf in
1951, using the single-particle shell model of nuclei. Figure 12 summarizes the numerical consequences of this theory. An E2 transition of about 1
MeV is expected to take place with a mean life, τ_{el},
or mean delay in the upper level, of about 10^{-11} s.
Thus most gamma-ray transitions are prompt transitions, in which the mean life of the excited level
is too short to be measured easily. Figure 12 is
for electric multipole transitions. The mean life
τ_{mag} for magnetic multipoles is of the order of
30 (for $A = 20$) to 150 (for $A = 200$) times longer
than τ_{el}.

At low energies or high Z, or both, the internal
conversion process becomes a very important additional mode of decay that markedly shortens the
mean lives of the nuclear levels. In addition, in
many cases the structure of the nucleus comes
into play and alters the observed mean lives considerably compared to those in Fig. 12. Electric
dipole (E1) transitions are generally retarded
(longer mean lives) by factors of 10^6 over the Weisskopf estimates of Fig. 12. On the other hand, A.
Bohr and B. Mottelson developed a model of
collective nuclear motions where E2 transitions
are enhanced by factors of 100 or more (shorter τ)
over the Weisskopf single-particle estimates, and
these predictions are confirmed by experiments.

The magnetic dipole (M1) transitions are also often hindered by factors of 100 or more. Measurements of the mean lives for gamma-ray decay provide important tests of nuclear models. *See* MULTIPOLE RADIATION.

Internal conversion. An alternative type of deexcitation which always competes with gamma-ray emission is known as internal conversion. Instead of the emission of a gamma ray, the nuclear excitation energy can be transferred directly to a bound electron of the same atom. Then the nuclear energy difference is converted to energy of an atomic electron, which is ejected from the atom with a kinetic energy E_i given by Eq. (31). Here B_i

$$E_i = W - B_i \qquad (31)$$

is the original atomic binding energy of the particular electron, which is ejected, and W is the nuclear transition energy which would otherwise have been emitted as a gamma-ray photon having energy $h\nu = W$.

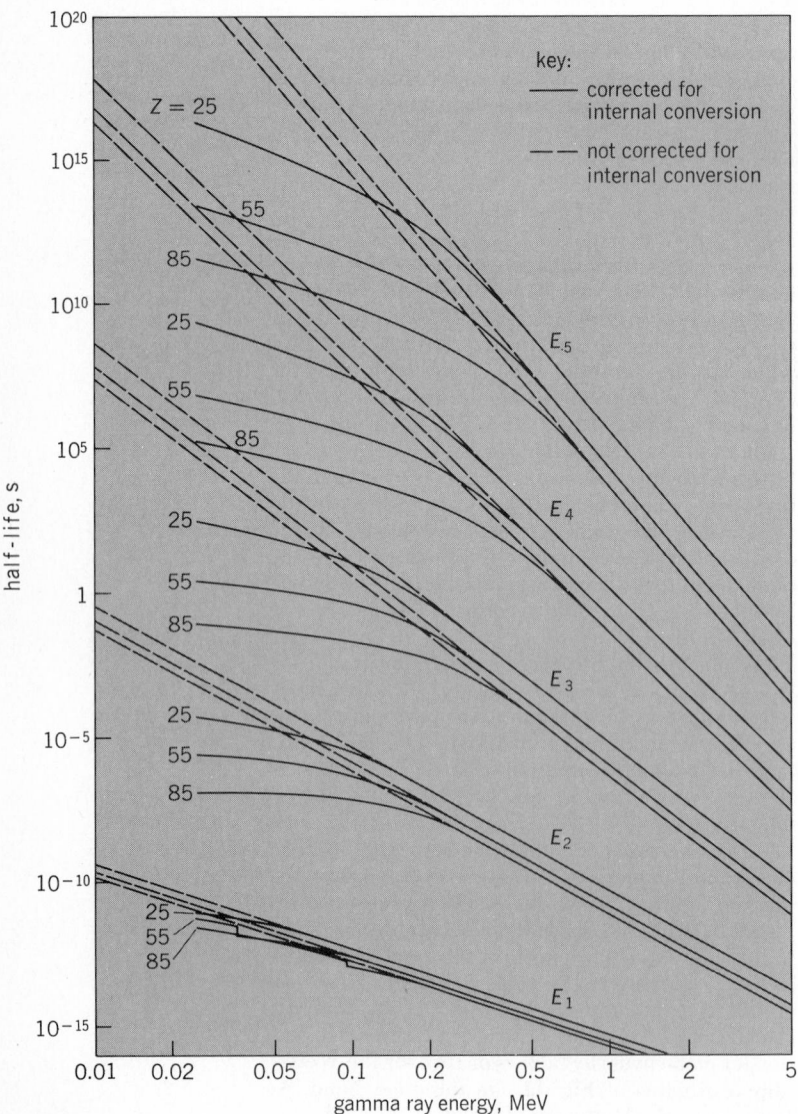

Fig. 12.   Theoretical values of half-lives for decay of nuclear levels by emission of gamma rays and conversion electrons, for electric multipoles. (*From A. H. Wapstra, G. J. Nijgh, and R. van Lieshout, Nuclear Spectroscopy Tables, North-Holland, 1959*)

The spectrum of internal conversion electrons is then a series of discrete energies, or "lines," each corresponding to an individual value of B_i, for the K, L (L_1, L_2, L_3), M, . . . , electrons in each shell and subshell of the atom. Thus conversion electron spectra are much more complex than gamma spectra. From the spacing of the E_i values in this conversion electron spectrum, it is possible to assign definitely the atomic number Z of the atom in which the nuclear transition W took place. In this way it is known that the conversion electron and the competing gamma-ray emission are sequels and not antecedents of alpha decay, beta decay, and electron-capture transitions. Partial electron spectra, showing K, L, and M shell conversion, and gamma-ray spectra are shown in Fig. 13 for the decay of ^{186}Tl. By comparing the K and L electron intensities of the $402 + 405$ and 522 keV transitions with the gamma-ray spectrum, it can be seen that the strong 522-keV electron transition has no gamma ray associated with it. The strong 511-keV gamma ray is from the annihilation of positrons and is not a nuclear transition, and so has no conversion electrons of this energy. One can improve the energy resolution by factors of $100 - 1000$ over that in Fig. 13 with magnetic spectrometers so that, for example, one can separate the lines with different energies from even the five M subshells.

The internal conversion coefficient α is the ratio of the number of transitions proceeding by internal conversion to the number going by gamma-ray emission, for any particular nuclear transformation from an excited level to a lower-lying level. In general, this probability of internal conversion relative to gamma-ray emission increases with increasing atomic number Z, with increasing multipole order 2^l, and with decreasing nuclear deexcitation energy W. In middle-weight elements, for $W = 1$ MeV, α is of the order of 10^{-2} to 10^{-4}; while for $W = 0.2$ MeV, α is of the order of 0.1 for electric $l = 2$ transitions, and 10 or larger for electric $l = 5$ transitions.

Radiationless transitions. There are cases where gamma-ray emission is strictly forbidden and conversion electron emission allowed. This occurs when both nuclear states have zero spin and the same parity. The conversion electrons are called electric monopole radiations, E0. These transitions occur because of the penetration of the atomic electrons into the nuclear volume where they interact directly with the nucleus. An example of the E0 decay is shown in Fig. 13. E0 radiation can occur in principle whenever two states have the same spin and same parity, but in practice, E0 decays are found to be very, very small in these cases. There are some exceptions in well-deformed nuclei, as was first shown for ^{154}Gd, where they totally dominate the electron emission for certain transitions that involve large shape changes.

The E0 decays which arise because of the penetration of the atomic electrons into the nuclear volume are thus sensitive measures of changes in shape between two nuclear states, and have played important roles in establishing vibrations of the nuclear shape and the coexistence of states with quite different deformation in the same nucleus. There also are other circumstances where the penetration of the atomic electron into the nuclear

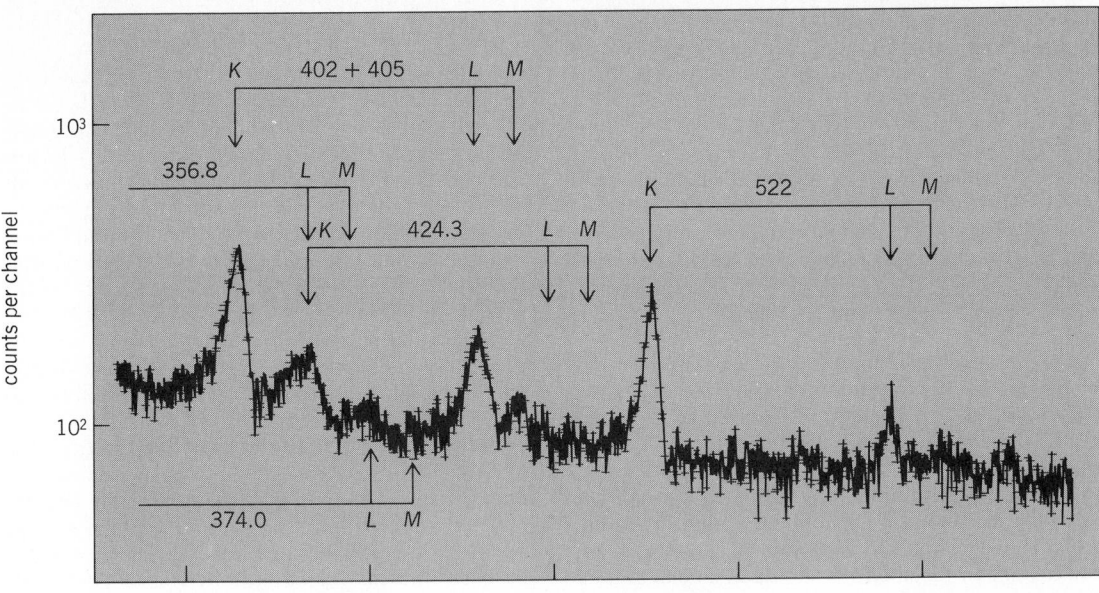

(a)

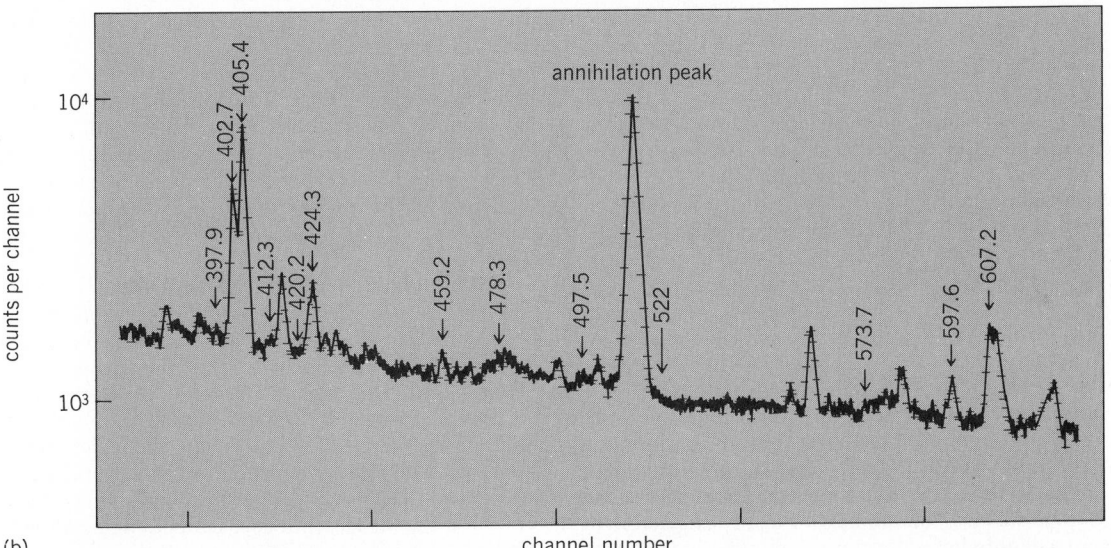

(b)

Fig. 13. Spectra from the decay of 30-s half-life ^{186}Tl far off stability (15 neutrons less than the lightest stable thallium isotope). (*a*) Internal conversion electrons. (*b*) Gamma rays. Nuclear transition energies are given in keV. (*From J. H. Hamilton et al., Shape coexistence in ^{186}Hg and the decay of ^{186}Tl, Phys. Rev., C16:2010–2018, 1977*)

volume give rise to additional contributions to the conversion-electron decay. Again these penetration effects probe details of the structure of the nucleus.

Internal pair formation. When the energy of a nuclear gamma-ray transition exceeds 1.022 MeV, twice the rest mass energy of an electron, it is possible for a nucleus to give up its excess energy to an electron-positron pair—a pair creation process. This is a third alternate mode to gamma decay and conversion electron decay. This process becomes more important as the gamma-ray energy increases. It is relatively unimportant below 2–3 MeV of decay energy.

Isomeric transitions. Measurably delayed radioactive transitions from an excited level of a nucleus are known as isomeric transitions. The measurably long-lived excited level is called an isomer-

ic or metastable level or an isomer of the ground level. What constitutes an isomer is not well defined. The terminology grew up when it was difficult to measure mean lives shorter than 10^{-7} s. States with longer mean lives were isomers. Now mean lives down to 10^{-13} s can be measured for many transitions in different nuclei, but these are not generally called isomers. The break point is simply not defined.

Figure 12 shows that if the excitation energy is small (say, 0.5 MeV or less) and the angular momentum difference l is large (say, $l = 3$ or more) then the mean life of an excited level for gamma ray or conversion-electron emission can be of the order of 1 s up to several years.

Most of the long-lived isomers occur in nuclei which have odd mass number A. Then either the number of protons Z in the nucleus is odd, or the number of neutrons N in the nucleus is odd. The frequency distribution of odd-A isomeric pairs, excited level and ground level, displays so-called islands of isomerism in which the odd-proton or odd-neutron number is less than 50, or less than 82. The distribution is one of several lines of evidence for closed shells of identical nucleons at N or $Z = 50$ or 82 in nuclei, and it plays an important role in the so-called shell model of nuclei. *See* MAGIC NUMBERS; NUCLEAR ISOMERISM.

SPONTANEOUS FISSION

This involves the spontaneous breakup of a nucleus into two heavy fragments and neutrons, as shown in Table 1. After the discovery of fission in 1939, it was subsequently discovered that isotopes like ^{238}U had very weak decay branches for spontaneous fission, with branching ratios on the order of 10^{-6}. New isotopes subsequently identified like ^{252}Cf have large (3.1%) spontaneous fission branching. In these cases, the nucleus can go to a lower energy state by spontaneously splitting apart into two heavy fragments of rather similar mass plus a few neutrons. This process liberates a large amount of energy compared to any other decay mode. Thus, ^{252}Cf is becoming important in many applications in medicine and industry as a compact energy source or as a source of nuclear radiation, since the fragments themselves are left in excited states and so emit gamma rays.

An important isomeric decay mode was discovered in the early 1960s in the very heavy elements, spontaneous fission isomers. Here the nucleus in an excited state, rather than emit a gamma ray or conversion electron, spontaneously breaks apart into two heavy fragments plus neutrons exactly as in spontaneous fission. To identify these isomers, the symbol f is often placed after their atomic mass, for example, $^{244f}_{95}$Am. Their half-lives are generally short, 10^{-3} to 10^{-9} s. It is now understood that these fission isomers are states with much larger deformation than the ground states of these isotopes. The Coulomb barrier against fission is in fact a double-hump barrier with the fission isomers in the valley at large deformation (Fig. 14). The study of these fission isomers has provided important tests of understanding of the behavior and structure of nuclei with very large deformation.

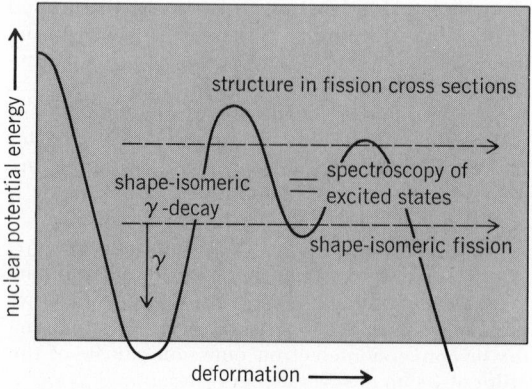

Fig. 14. Observable consequences of a double-humped nuclear potential barrier against fission. The potential well at the larger deformation gives rise to isomeric spontaneous fission.

DELAYED PARTICLE EMISSIONS

As shown in Table 1, there are six types of beta-delayed particle emissions which have been reported, and a seventh is being sought. One might also expect to find beta-delayed deuteron (^{2}H) and triton (^{3}H) emissions which are not shown there. There are now over 100 beta-delayed particle radioactivities known. Theoretically, the number of isotopes which can undergo beta-delayed particle emission can exceed 1000. Thus, this mode, which was observed in only a few cases prior to 1965, is among the important ones in nuclei very far from the stable ones in nature. Studies of these decays can provide insights into the nucleus which can be gained in no other way.

Beta-delayed alpha radioactivity. The β^- decay of ^{214}Bi to ^{214}Po leaves the nucleus in such a high-energy excited state that it can emit an alpha particle and go to ^{210}Pb as an alternative to gamma-ray decay to lower levels in ^{214}Po. This is a two-step process with beta decay the first step. After beta decay the nucleus is in such a highly excited state that it can emit either an alpha particle or gamma ray. Several different beta-delayed particle emissions are now known.

β^--delayed alpha emission has been found relatively rarely, but the additional process of β^+-delayed alpha emission has been discovered. In proton-rich nuclei far from stability, the conditions are more favorable for beta-delayed alpha emission because of the excess of nuclear charge, and a number of such β^+-delayed alpha emitters are now known.

Beta-delayed neutron radioactivity. In 1939, shortly after the discovery of nuclear fission, it was proposed that the delayed neutrons observed following fission were in fact beta-delayed neutrons. That is, after the nucleus fissioned, the beta decay of the neutron-rich fission fragments populated high-energy excited states that could promptly undergo dual decay, emitting either a gamma ray or neutron. Beta-delayed neutron emission is illustrated in Fig. 2. The processes of beta-delayed two- and three-neutron emission were discovered in 1979 and 1980 in the decay of ^{11}Li. The former is shown in Fig. 2, and β^-2n decays were subsequently observed in other nuclei.

Beta-delayed proton radioactivities. Proton radioactivity itself is a mode of radioactive decay, generally expected to arise in proton-rich nuclei far from the stable isotopes, in which the parent nucleus changes its chemical identity by emission of a proton in a single-step process. Its physical interpretation parallels almost exactly the quantum-mechanical treatment of alpha-ray decay. It is also theoretically predicted that one can have the simultaneous emission of two protons—two-proton radioactivity (Fig. 1). In addition, one can have β-delayed proton and β-delayed two-proton radioactivities which again ultimately result in emission of protons from the nucleus. These latter processes also occur in quite proton-rich nuclei with very high decay energies; however, they are complex two-step decay modes whose fundamental first step is β-ray decay.

Over 40 nuclei ranging from $^9_6$C to $^{183}_{80}$Hg have been identified to decay by the two-step mode of β^+-delayed proton radioactivity. Figure 15 presents the observed proton energy spectrum arising in the decay of $_{18}^{33}Ar$, with a half-life of 173 ms; it was produced by the $_{16}^{32}S + _2^3He \rightarrow _{18}^{33}Ar + 2n$ reaction. This isotope decays by superallowed and allowed β^+ decay to a number of levels in its daughter nucleus $_{17}^{33}Cl$, which immediately (in less than 10^{-17} s) breaks up into $_{16}^{32}S$ and a proton. More than 30 proton groups arising from the decay of $_{18}^{33}Ar$ are observed, ranging in energy from 1 to approximately 6 MeV and varying in intensity over four orders of magnitude. Although it is normally very difficult to study many β-decay branches in the decay of a particular nuclide—because of the continuous nature of the energy spectrum of the emitted beta particles—it is possible to do so when investigating β^+-delayed proton emitters. The observed proton group energies and intensities can be correlated with the levels fed in the preceding beta decay and their transition rates, thereby permitting sensitive tests via beta decay of nuclear wave functions arising from different models of the nucleus. β^+-delayed two-proton decay is being sought (Fig. 1).

Proton radioactivity. Although proton radioactivity has been of considerable theoretical interest since 1951 and is expected to be a general phenomenon, so far only one example of this decay mode has been observed, because of the experimental difficulties associated with producing extremely proton-rich nuclei. Figure 16 presents the decay scheme of the first nuclide found, in 1970, to decay by proton radioactivity. It is $_{27}^{53m}Co$, where the m (metastate) denotes a (relatively) long-lived isomeric state. Because of its very high angular momentum of 19/2 and odd parity, gamma decay is highly forbidden. This mode of decay is essentially the same as that of β-delayed proton emission, except that now the energy of the excited nuclear level is low, and angular momentum selection rules highly forbid gamma-ray decay so the state lives a relatively long time in comparison to those states populated in beta decay. It was produced in the laboratory by the compound nucleus reactions $_8^{16}O + _{20}^{40}Ca \rightarrow _{27}^{53m}Co + p + 2n$ and $_{26}^{54}F + p \rightarrow _{27}^{53m}Co + 2n$. This 247-ms isomer exhibits two different decay modes: though it predominantly decays by positron (β^+) emission to a similar 19/2$^-$ level in $_{26}^{53}Fe$, a 1.5% branch in its decay occurs via direct emission of a 1.59-MeV proton to the $_{26}^{52}Fe$ ground state. The calculated half-life that $_{27}^{53m}Co$ would possess if proton radioactivity were the only decay mode (its partial half-life for this decay branch) is the surprisingly long time of 17 s.

Because of the lower charge on the proton compared to the alpha particle, the Coulomb barrier indicated in Fig. 7 is of less importance in proton radioactivity than in alpha-particle radioactivity; however, the effect of the angular momentum, or centrifugal, barrier is much more significant because of the lower mass of the proton. This can be seen from the mathematical form of the centrifugal barrier, which is $\hbar^2(l + 1)/2Mr^2$; here M, the so-called reduced mass of the emitted particle, appears in the denominator; additionally, $\hbar$ is Planck's constant divided by 2π, l is the angular momentum of the emitted proton (or alpha particle), and r is its radial separation. A high centrifugal barrier arises in the decay of $_{27}^{53m}Co$ because of the large change of angular momentum of 9 units required for the proton (of intrinsic angular momentum $1/2$) to be emitted from this angular momentum 19/2 isomer (of odd parity), leaving the

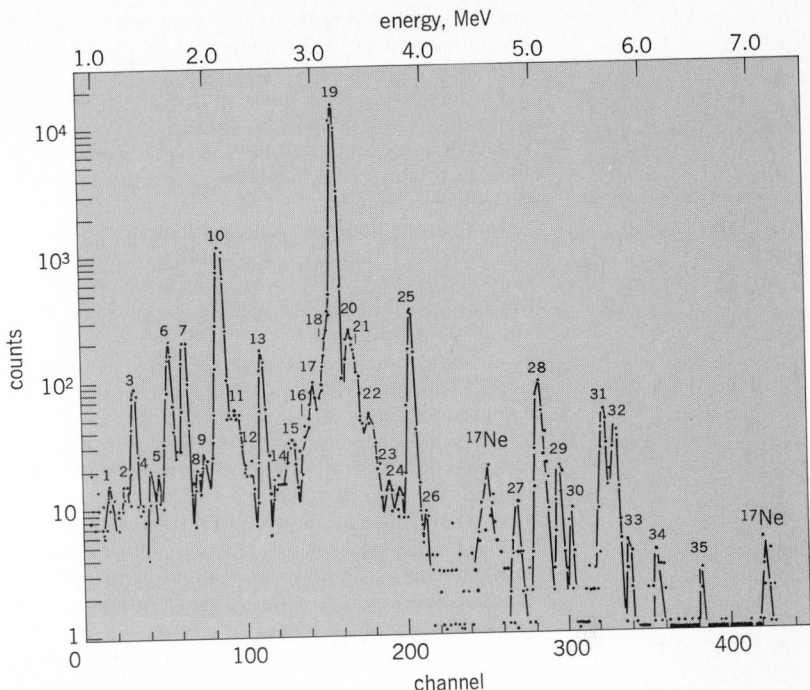

Fig. 15.   Spectrum of β^+-delayed protons from the decay of $_{18}^{33}Ar$ as observed in a counter telescope; the proton laboratory energy is indicated at the top. Proton groups are numbered 1 through 35. (*From J. C. Hardy et al., Isospin purity and delayed-proton decay: ^{17}Ne and ^{33}Ar, Phys. Rev., C3:700–718, 1971*)

daughter nucleus $_{26}^{52}Fe$ in its angular-momentum zero ground state (of even parity).

Searches for proton and two-proton radioactivities from ground states of nuclei are being carried out. Figure 1 illustrates one of the prime candidates for such decays.

Beta-delayed spontaneous fission. There are also observed beta-decay processes where the excited nucleus following beta decay has a probability of undergoing spontaneous fission rather than gamma-ray decay. This is the same process as in

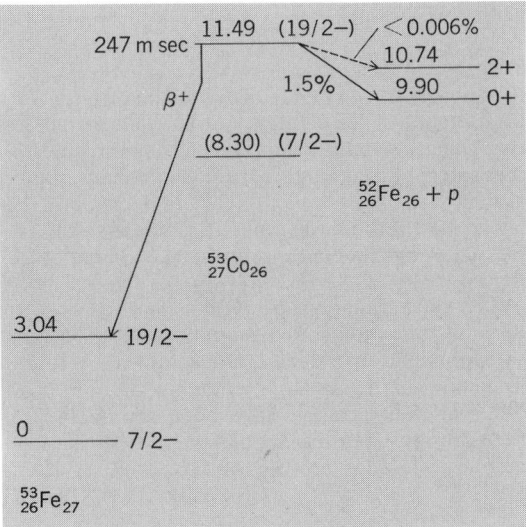

Fig. 16.   The decay scheme of $_{27}Co^{53*}$. Numbers to left of levels represent energies in MeV, relative to ground state of $_{26}Fe_{27}^{53}$. Symbols to right of levels are spin and parity. (*From J. Cerny et al., Further results on the proton radioactivity of Co^{53*}, Nucl. Phys., A188:666–672, 1972*)

spontaneous or isomeric spontaneous fission. The excitation energy of the nuclear level provides the extra energy to make fission possible. The nucleus splits into two nearly equal fragments plus some neutrons. This process is like isomeric spontaneous fission except that the lifetime of the nuclear level is so short that the level would not normally be called an isomer.

Neutron radioactivity. In very neutron-rich nuclei near the boundary line of nucleus stability, one may find nuclei with ground states which are unstable to the emission of one or two neutrons. Here there is no Coulomb barrier, but one can have a centrifugal barrier that may give rise to one- or even two-neutron radioactivity. These processes for ground states would be very near the limits where nuclei become totally unstable to the addition of a neutron, the neutron drip line, and very difficult to even make much less measure. However, there may be neutron-rich nuclei with high-spin isomeric states where the high spin analogous to the one in ^{53m}Co gives rise to a large centrifugal barrier. Such isomeric states may undergo one-or two-neutron radioactivity.

[JOSEPH H. HAMILTON]

Bibliography: J. M. Eisenberg and W. Greiner, *Nuclear Theory*, 3 vols., 1975; R. D. Evans, *The Atomic Nucleus*, 1955; J. H. Hamilton (ed.), *Internal Conversion Processes*, 1965; J. H. Hamilton et al. (eds.), *Future Directions in Studies of Nuclei Far from Stability*, 1980; J. H. Hamilton and J. C. Manthurathil (eds.) *Radioactivity in Nuclear Spectroscopy*, 1972; W. D. Hamilton (ed.), *Electromagnetic Interaction in Nuclear Spectroscopy*, 1975; J. C. Hardy, Nuclear spectroscopy from delayed particle emission, in J. Cerny (ed.), *Nuclear Spectroscopy and Reactions*, Part C, 417, 1974; I. Kaplan, *Nuclear Physics*, 2d ed., 1963; E. J. Konopinski, *The Theory of Beta Radioactivity*, 1966; C. M. Lederer et al. (eds.), *Table of Isotopes*, 7th ed., 1978; National Bureau of Standards, Tables for the analysis of beta spectra, *Applied Mathematics*, ser. 13, 1952; S. C. Pancholi (ed.), *Gamma-Ray Transition Probabilities*, 1977; F. Rösel et al., *Atomic and Nuclear Data Tables*, 21:91, 1978.

Radioisotope

A radioactive isotope (as distinguished from a stable isotope) of an element. Atomic nuclei are of two types, unstable and stable. Those in the former category are said to be radioactive and eventually are transformed, by radioactive decay, into the latter. One of the three types of radioactive ray (α-, β-, and γ-rays) is emitted during each stage of the decay.

The term radioisotope is also loosely used to refer to any radioactive atomic species. Whereas approximately a dozen radioisotopes are found in nature in appreciable amounts, hundreds of different radioisotopes have been artifically produced by bombarding stable nuclei with various atomic projectiles. *See* ISOTOPE; RADIOACTIVITY.

[HENRY E. DUCKWORTH]

Radiometry

The detection and measurement of radiant electromagnetic energy. Conventionally, radiometry is concerned with infrared radiation. Generally, the devices used in radiometry can be, and are, used with visible light. However, the use of devices applicable only to visible light is commonly excluded from the term radiometry. For a summary of the methods used for detection and measurement of the electromagnetic spectrum as a whole *see* ELECTROMAGNETIC RADIATION. *See also* INFRARED RADIATION; PHOTOMETRY.

In 1800 Sir William Herschel studied the ability of sunlight to heat a sensitive mercury-in-glass thermometer. He detected radiation beyond the red end of the visible spectrum, hence the name infrared. He also noted that radiation could be detected from moderately hot objects showing no visible light. Detectors such as the thermometer which respond to the increase in temperature resulting from the absorption or radiant energy are termed thermal detectors.

As early as 1843, E. Becquerel secured a photographic effect with near-infrared radiation. The effect does not depend upon a rise in temperature but upon the freeing of a bound electron by the absorption of a single quantum of radiation. Detectors utilizing this principle are termed quantum detectors, or photodetectors.

Thermal detectors. The mercury-in-glass thermometer is sluggish and relatively insensitive. So also is the Crookes radiometer, which consists of two blackened vanes at the ends of a horizontal rod suspended from a fine quartz fiber in an atmosphere of air or other gas at about 0.1 mm Hg pressure. Radiation absorbed by one of the vanes heats it, and gas molecules, recoiling with added momentum from the warm vane, tend to turn it about the suspension. The Crookes radiometer survives in jewelers' windows as a "perpetual motion" device.

Improvement in thermal detectors has been concerned with securing a large and rapid rise in temperature and high sensitivity in the detection of changes in temperature. The temperature of any thermal detector will increase until the rate of loss of heat to its surroundings is equal to the rate at which radiant energy is absorbed. To secure the largest rise for a given radiation, the detector should absorb it as completely as possible; the loss of heat, by reradiation to its surroundings, conduction through its supports, and gas conduction and convection, must be as small as possible. The heat capacity of a thermal detector should be small, so that a rapid rise in temperature will occur. The area of the absorbing surface should be small to

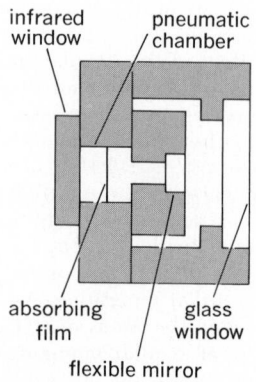

infrared window
pneumatic chamber
absorbing film
glass window
flexible mirror

Schematic diagram of the detecting part of a Golay pneumatic radiometer.

permit the measurement of narrow beams of radiation. These criteria lead to detectors which are very thin, with thin black coatings, supported commonly by two wires. Enclosing the detector in a vacuum decreases the heat loss, but requires a window capable of transmitting the radiation.

The thermocouple, one of the oldest and still one of the best radiation detectors, produces an electromotive force (emf) when heated. The sensitive portion of a bolometer undergoes a change in resistance when heated.

The essential part of a Golay pneumatic radiometer, a type of thermal detector, is shown in the figure. At the center of an air- or gas-filled cell is a radiation-absorbing film. The gas, heated by the absorption of radiation, expands and changes the curvature of the mirror. This mirror consists of a thin plastic film coated by evaporation with a layer of antimony. A beam of light reflected from this mirror to a photocell permits slight changes of curvature in the mirror to be measured by means of changes in reflected light. The Golay cell is intended for use in a chopped beam of radiation. (To secure the alternating potential desired for amplification, it is general practice to "chop" the radiation with a rotating sector disk or by some other means.)

Quantum detectors. Photoemissive cells and photomultipliers are of interest for visible radiation, but have limited application in the near infrared to about 1 μm. Photoconductive cells using materials such as lead sulfide, lead selenide, and lead telluride are useful to 5 μm. Photoconductive cells of single crystals of gold-doped germanium, cooled with liquid helium, have sensitivity extending to much longer wavelengths.

Such quantum detectors have short time constants, permitting high chopping rates, and are superior to thermal detectors for following rapidly changing radiation flux. Their sensitivity is more dependent on wavelength, and they are not operable as far into the infrared as are thermal detectors.

Performance of detectors. Interest in detection and measurement of radiation arises in a great variety of circumstances for which a variety of detectors are suitable or adequate. For full sunlight, a receiver of large area, consisting of a number of thermocouples in series and known as a thermopile, together with a simple galvanometer, may be adequate. For the radiation of a star, a single thermocouple and an amplifier might be used. For rapid scanning of an infrared molecular spectrum, a maximum of sensitivity and speed of response are desirable.

The response of thermal detectors is generally independent of wavelength over a range from the ultraviolet to wavelengths of the order of magnitude of the dimensions of the receiver. Quantum detectors generally have maximum sensitivity in the visible or near-infrared regions, and they are unresponsive above a particular cutoff wavelength.

It might appear that amplification could be increased as much as desired by the use of an electronic amplifier. However, as amplification of any signal by any type of amplifier is increased, random fluctuations in the signal (noise) are also increasingly amplified. For radiation detection, the limiting noise is not in the amplifier but in the detector itself. Thermocouples and bolometers are generally limited by so-called Johnson noise (also called thermal noise), which is caused by thermal agitation of electrons and which has a magnitude given by the equation below.

$$e^2 = 4kTR\,(f_1 - f_2)$$

$e =$ root-mean-square (rms) noise voltage
$k =$ Boltzmann constant
$T =$ absolute temperature
$R =$ electrical resistance of detector
$(f_1 - f_2) =$ frequency limits between which rms voltage is measured

A commonly used measure of the sensitivity of a detector is the noise-equivalent power, that is, the watts of radiation which produce a response equal to the noise for an amplifier bandwidth $(f_1 - f_2)$ of 1 Hz.

This and some other characteristics of three kinds of thermal detectors are given in the table.

[H. W. RUSSELL/GEORGE R. HARRISON]

Bibliography: D. E. Gray (ed.), *American Institute of Physics Handbook*, 3d ed., 1972; R. J. Keyes (ed.), *Optical and Infrared Detectors*, 2d ed., 1981; R. H. Kingston, *Detection of Optical and Infrared Radiation*, 1978; W. L. Wolfe and G. J. Zissis, *The Infrared Handbook*, 1978.

Characteristics of thermal detectors

Type	Thermo-couple	Bolometer	Golay pneumatic
Material	Bi-Sb vs. Bi-Sn	Platinum	Gas-filled
Time constant, sec	0.036	0.016	0.015
Area, mm²	0.5	1.6	8.0
Frequency of measurement, Hz	0.5	10	10
Resistance, ohms	5	40	
Noise-equivalent power, watts	0.5×10^{-10}	1.7×10^{-10}	1.5×10^{-10}

Radius of gyration

A relation of the area or mass of a figure to its moment of inertia. If I is the moment of inertia about a line of a figure whose area is A, the figure's radius of gyration with respect to that line is $k = +\sqrt{I/A}$. Accordingly, $I = k^2 A$. For a figure of mass M, $k = +\sqrt{I/M}$; $I = k^2 M$. In these equations, k is measured in length units such as feet. Geometrically similar figures have equal radii of gyration about corresponding centroidal axes. If the radius of gyration of a figure with respect to an axis is k and with respect to a parallel centroidal axis is $\overline{k}$, $k^2 = \overline{k^2} + D^2$, where D is the distance between the parallel axes. *See* MOMENT OF INERTIA.

[NELSON S. FISK]

Raman effect

A phenomenon observed in the scattering of light as it passes through a material medium, whereby the light suffers a change in frequency and a random alteration in phase. Raman scattering differs in both these respects from Rayleigh and Tyndall scattering, in which the scattered light has the same frequency as the unscattered and bears a definite phase relation to it. The intensity of normal Raman scattering is roughly one-thousandth

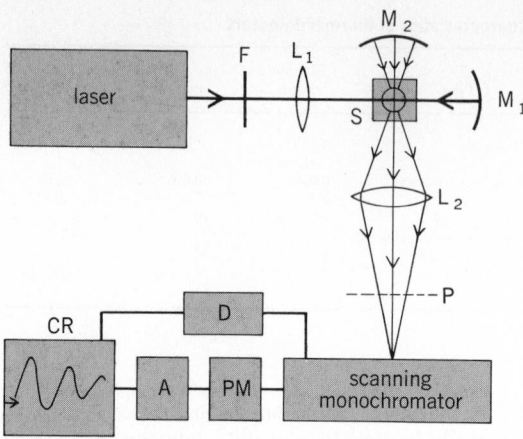

Fig. 1. Laser-Raman spectroscopic system.

light. They then showed, by using light of a single frequency from a mercury arc, that the new frequencies in the scattered radiation were characteristic of the scattering medium. Within a few months of Raman and Krishnan's first announcement of their discovery, the Soviet physicists G. Landsberg and L. Mandelstam communicated their independent discovery of the existence of the effect in crystals. In Soviet literature the phenomenon is referred to as combination scattering, and not Raman effect. *See* COMPTON EFFECT.

The development of the laser has led to a resurgence of interest in the Raman effect and to the discovery of a number of related phenomena. A beam of laser radiation is intense, polarized, and coherent; it can be made monochromatic, small in diameter, and highly collimated. The laser is therefore nearly ideal for the production of the Raman effect, and other kinds of sources are seldom employed. Many different wavelengths in the visible spectrum and adjacent regions are available. The argon-ion and krypton-ion lasers are most commonly used, since they have high continuous-wave power (1 to 10 W), but tunable dye lasers are also often employed in excitation of resonance Raman scattering. *See* LASER.

Raman spectroscopy. Raman scattering is analyzed by spectroscopic means. The collection of new frequencies in the spectrum of monochromatic radiation scattered by a substance is characteristic of the substance and is called its Raman spectrum. Although the Raman effect can be made to occur in the scattering of radiation by atoms, it is of greatest interest in the spectroscopy of molecules and crystals.

Because of the laser beam's small diameter and

that of Rayleigh scattering in liquids and smaller still in gases. For an extended discussion of Rayleigh scattering *see* SCATTERING OF ELECTROMAGNETIC RADIATION. *See also* TYNDALL EFFECT.

Discovery. Because of its low intensity, the Raman effect was not discovered until 1928, although the scattering of light by transparent solids, liquids, and gases had been investigated for many years before. Prompted by A. H. Compton's observation of frequency changes in x-rays scattered by electrons (Compton effect), the Indian physicists C. V. Raman and K. S. Krishnan examined sunlight scattered by a number of liquids. With the help of complementary filters, they found that there were frequencies in the scattered light that were lower than the frequencies in the filtered sun-

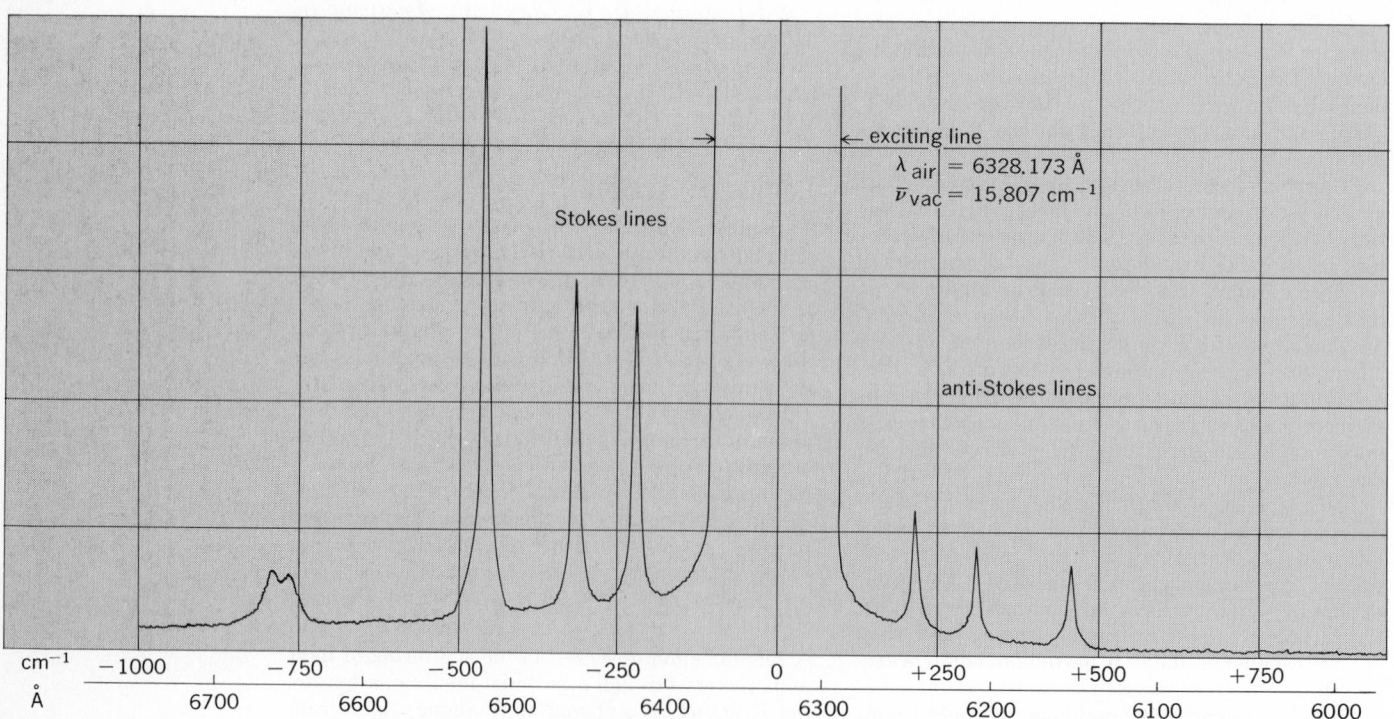

Fig. 2. Photoelectric recording of the Raman spectrum of carbon tetrachloride excited by He-Ne laser line at 6328 A. Intensity of the radiation is recorded vertically against the horizontal wave-number scale (cm⁻¹) measured from the exciting line as zero. The Rayleigh-scattered exciting line is three orders of magnitude more intense than the Raman lines, and its maximum is therefore far off-scale. The Stokes lines appear at lower frequencies, and the less intense anti-Stokes lines at higher frequencies, than those of the exciting line. The lower scale shows wavelengths in angstroms.

high collimation, it can easily be used to excite the Raman effect. A typical optical arrangement is shown in Fig. 1. Monochromatic radiation from the laser impinges on the sample S in an appropriate transparent cell. It may be desirable to condense or expand the laser beam by means of a lens system L_1 and to remove unwanted radiation from the beam by a narrow-band optical filter F. A concave mirror M_1 can return unscattered radiation for a second passage through the sample.

Raman scattering is approximately uniform in all directions and is usually studied at right angles, as shown in Fig. 1. In this way the intense radiation of the laser beam interferes least with the observation of the weak scattered light. This light is collected by a lens system L_2 and focused on the slit of a scanning monochromator, which analyzes it spectroscopically. As the spectrum is scanned, the dispersed radiation from the monochromator is detected by a photomultiplier PM, further amplified and processed electronically at A, and then recorded by a strip-chart recorder CR. The recorder is driven in synchronism with the monochromator by a suitable mechanism D. The concave mirror M_2 may be used to augment the amount of scattered radiation by collecting light scattered at $-90°$ and returning it to the $+90°$ direction. The polarization characteristics of the scattered radiation are frequently of interest, especially since the laser radiation itself is linearly polarized. An analyzing device for evaluating the degree of polarization of the scattered radiation may be inserted at point P.

The appearance of a photoelectrically recorded Raman spectrum of liquid carbon tetrachloride as excited by the red line of the helium-neon laser at 632.8 nm (6328 A, power incident on the sample of about 50 mW) is shown in Fig. 2. Intensity of the scattered light on an arbitrary scale is plotted vertically against the wave number in cm^{-1} measured with respect to the wave number of the exciting line taken as zero. For convenience, it is usual to express the data of Raman spectroscopy in cm^{-1} rather than frequency units (s^{-1}). (Frequency ν in $s^{-1} = c\bar{\nu}$ in cm^{-1}, where c is the velocity of light in vacuum in cm/s.)

A spectrum of the most intense line in Fig. 2 is shown at higher resolution (smaller spectral slit width $\Delta\bar{\nu}$) in Fig. 3. The line is seen to consist of several closely spaced components. These result from the presence of the two isotopes of chlorine, ^{35}Cl and ^{37}Cl, which produce five isotopic species $C^{35}Cl_n{}^{37}Cl_{4-n}$, $n = 0,1,2,3,4$. The line due to the least abundant species, $n = 0$, is not visible, but the other four are readily identified.

Theory. The mechanism of the Raman effect can be envisaged either by the corpuscular picture of light or from the point of view of the wave theory. Both pictures merge in the basic quantum theory of radiation. The corpuscular model of light scattering envisages light quanta or photons as particles which have linear and angular momenta. On passing through a material medium, these particles collide with atoms or molecules. If the collision is elastic, the photons bounce off the molecules with unchanged energy E and momentum, and hence with unchanged frequency ν. Such a process gives rise to Rayleigh scattering. If the collision is inelastic, the photons may gain energy from, or lose it to, the molecules. A change ΔE in

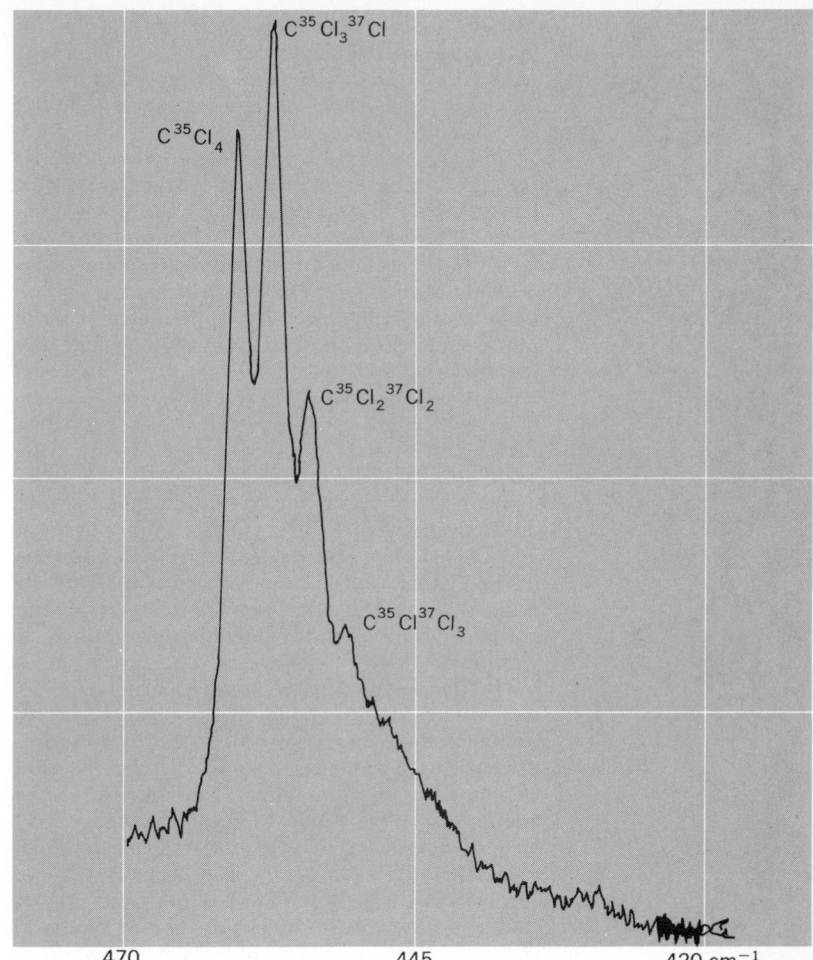

Fig. 3. The Stokes line at -460 cm^{-1} in the Raman spectrum of carbon tetrachloride. This spectrum, taken at about 10 times better resolution than that of Fig. 2, shows lines due to four of five isotopic species resulting from the 3:1 ratio of the chlorine isotopes ^{35}Cl and ^{37}Cl. The instrumental wave-number scale requires a calibration correction to increase the wave numbers by 1.4 cm^{-1}.

the photon energy by Planck's relationship: $E = h\nu$, must produce a change in the frequency $\Delta\nu = \Delta E/h$. Such inelastic collisions are rare compared to the elastic ones, and the Raman effect is correspondingly much weaker than Rayleigh scattering. *See* LIGHT; QUANTUM MECHANICS; SCATTERING EXPERIMENTS (ATOMS AND MOLECULES).

In the wave picture of the effect, the electromagnetic waves which constitute the incoming monochromatic radiation sweep through the material medium. Since the atoms and molecules composing the medium are made up of negatively charged electrons and positively charged nuclei, the electric field of the light waves sets the electrons to oscillating, chiefly with the frequency of the incoming radiation. The oscillating electrons re-create the alternating electric field of the incoming light, thus passing the light wave along through the medium. This process is analogous to the elastic collisions given by the corpuscular picture.

The ability of the electrons and nuclei in a molecule to be displaced by an electric field is called the molecular polarizability α. It is not a simple property of the molecule, but depends in a complicated way on the frequency of the electric field, on the orientation of the molecule, and on the internal motions of the nuclei and electrons. Thus the molecular polarizability α varies periodically with molecular rotation and vibration, and thereby the effect of a light wave on the electrons and nuclei of a molecule can be changed.

When a monochromatic light wave sweeps through a transparent medium containing rotating and vibrating molecules, most of the wave is re-created unchanged by the oscillating electrons, but because of the periodic changes produced in α by rotation and vibration, new frequencies are added to the light wave. The appearance of these new frequencies, whose values are determined by the rotational and vibrational energies of the molecules, is analogous to the result of the inelastic collisions of the corpuscular model. For the wave picture of the Raman effect, the quantity α is the basic quantity. The intensity of the Raman effect depends on the magnitude of the changes produced in α by molecular rotation and vibration, and the number and values of new frequencies (usually expressed as frequency shifts $\Delta\nu$ from the original monochromatic frequency) depend on the variation of α with the frequencies of rotation and vibration.

The temperature of the scattering molecules is an additional factor which affects the intensity of Raman frequencies higher than the exciting frequency (the anti-Stokes lines of Fig. 2). The anti-Stokes lines, having higher frequencies, correspond to photons which have higher energy than that of the exciting light, and this energy must come from the molecules. If the molecules do not have any available vibrational or rotational energy, that is, if they are at the absolute zero of temperature, there is no possibility of inelastic collisions in which energy is transferred from a molecule to a photon. So, anti-Stokes lines vanish at absolute zero. At nonzero temperatures the intensity ratio of an anti-Stokes line to a Stokes line is approximated by the ratio of the number of molecules which can give up the corresponding energy to the number which can accept it from the light wave.

Special forms. The development of lasers resulted in the discovery of a number of kinds of Raman scattering.

Resonance Raman effect. When the exciting radiation falls within the frequency range of a molecule's absorption band in the visible or ultraviolet spectrum, the radiation may be scattered by two different processes, resonance fluorescence or the resonance Raman effect. Both these processes give much more intense scattering than the normal nonresonant Raman effect. Resonance fluorescence differs from the resonance Raman effect in that the absolute frequencies of the fluorescent spectrum do not shift when the exciting radiation's frequency is changed, so long as the latter does not move outside the absorption band. The absolute frequencies of the resonance Raman effect, on the contrary, shift by exactly the amount of any shift in the exciting frequency, just as do those of the normal Raman effect. Thus the main characteristic of the resonance as compared to the normal Raman effect is its intensity, which may be greater by two or three orders of magnitude. *See* FLUORESCENCE.

The resonance Raman effect was anticipated by G. Placzek in 1934 in his pioneering development of the polarizability theory of Raman scattering. It was actually observed before the discovery of lasers, but tunable lasers are the most effective sources for the study of its various aspects. A typical resonance Raman spectrum is shown in Fig. 4, in which oxyhemoglobin is excited by the 568.2-nm (5682 A) wavelength of singly ionized krypton. The top spectrum $I_{||}$ is taken with the polarizer P of Fig. 1 set to pass the components parallel to the direction of laser polarization; the bottom spectrum $I_\perp$ is taken with P set to pass perpendicular components. Lines in which $I_\perp$ is much greater than $I_{||}$ are said to have inverse polarization and are seen only in the resonance Raman effect; these include the lines at 1305, 1342, and 1589 cm⁻¹ and numerous others.

Hyper-Raman effect. The nature of this effect is most easily described in terms of the corpuscular picture of the Raman effect. With an intense laser source, the number of monochromatic photons impinging on the molecules of a medium per unit volume and unit time may be extremely large. If so, the probability that two photons will collide simultaneously with the same molecule is very much larger than in normal scattering, and there is considerable chance that the two photons will unite and be scattered as a single photon of approximately twice the frequency. The rules governing the scattering in such three-photon processes (two incoming and one outgoing photons) are quite different from those for normal (two-photon) Rayleigh and Raman scattering. For example, in molecules that are centrosymmetric, the collision must be inelastic, that is, the molecule must absorb or give up an amount of energy ΔE during the process. The frequency of the scattered photon will therefore not be exactly twice the frequency of the incident photons but will differ from it by $\Delta\bar\nu = \Delta E/hc$. Such scattered radiation is called the hyper-Raman effect. Even in molecules that are not centrosymmetric, the likelihood of elastic collisions is much smaller than in the normal case, so that the intensity of hyper-Rayleigh scattering may

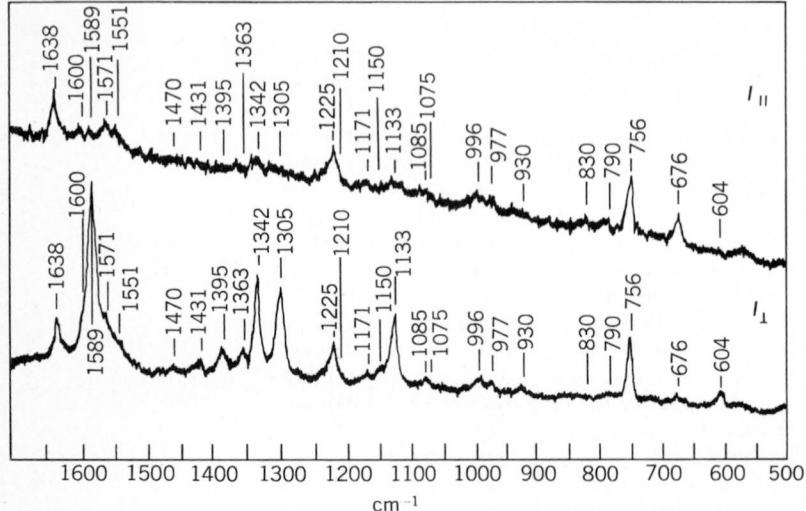

Fig. 4. Resonance Raman spectrum of oxyhemoglobin. (*From T. G. Spiro and T. C. Strekas, Resonance Raman spectra of hemoglobin and cytochrome c. Proc. Nat. Acad. Sci. USA, 69:2622–2626, 1972*)

be substantially weaker than hyper-Raman scattering.

As implied above, the selection rules for the vibrational and rotational transitions in the hyper-Raman effect are different from those of the normal Raman effect. Thus certain transitions are observable in the hyper-Raman effect that are normally forbidden. This is one virtue of the hyper-Raman effect; the other is that it is observed in a spectral region whose frequency is far removed from that of the incoming radiation (and is, in fact, twice that of the latter). The effect is therefore observable without interference from the normal Rayleigh line.

Stimulated Raman effect. The mechanism of the stimulated Raman effect depends on the coherent pumping of the molecules of the sample into an excited vibrational state by the powerful electric field of the laser beam. In view of the large discrepancy of one or two orders of magnitude between the frequency of the vibration and the frequency of the laser, this can be accomplished only if the field of the light wave has a very high value (the threshold power) and if the mismatch in frequency is compensated by the generation of coherent radiation with a frequency equal to that of the laser minus the vibrational frequency. The coherent radiation so produced is called stimulated Raman scattering. It was first observed by R. Woodbury and A. Ng in 1962. They found the effect in liquid nitrobenzene, which they were using as an electrooptical shutter within a laser system. *See* OPTICAL PUMPING.

In addition to its high intensity and its coherence, there are other new features of stimulated Raman scattering. Since the pumping power of the incident laser beam must exceed a certain threshold for the scattering to take place, when the laser power is used up in exciting one vibrational mode, there is insufficient power available to excite other modes. Therefore the stimulated Raman effect usually contains only one frequency, though in rare cases the power may be divided between two vibrational modes of roughly the same threshold. However, the power in the scattered radiation may itself produce further stimulated Raman emission by a repetition of the initial process. This results in a new frequency, which is the laser frequency minus exactly twice the frequency of vibrational mode that is being scattered. This fact shows that the mechanism does not involve a double jump in the vibrational levels; such a double jump would give a frequency shift that is not exactly twice that of the vibrational fundamental because of vibrational anharmonicity.

Another striking and unusual effect in stimulated Raman scattering is the excitation of intense anti-Stokes radiation. This radiation may be even stronger than the Stokes radiation in certain circumstances. Moreover, it can be observed at such low temperatures that the initial populations of the excited vibrational levels needed for normal anti-Stokes Raman scattering are zero. It arises from the above-mentioned pumping of molecules from the ground vibrational state into upper excited states by the initial laser power. These excited molecules can then be pumped by further radiation back into the ground state, with a simultaneous stimulated emission of coherent radiation at a frequency that equals the laser plus the molecular vibrational frequency.

The development of tunable lasers has led to a special technique for stimulated Raman scattering called coherent anti-Stokes Raman spectroscopy (CARS). In this technique, two lasers are used, one of fixed and the other of tunable frequency. The two beams enter the sample at angles differing only by some appropriate small amount (approximately 2°) and simultaneously impinge on the sample molecules. Whenever the frequency difference between the two lasers coincides with the frequency of a Raman-active vibration of the molecules, emission of coherent radiation (both Stokes and anti-Stokes) is stimulated. Thus the total Raman spectrum can be scanned in stimulated emission by varying the frequency of the tunable laser. An advantage of CARS, in addition to the high intensity of the scattering, is that its elevated frequency avoids interference from sample fluorescence, which always has frequencies below that of the exciting radiation.

Applications. Raman spectroscopy is of considerable value in determining molecular structure and in chemical analysis. Molecular rotational and vibrational frequencies can be determined directly, and from these frequencies it is sometimes possible to evaluate the molecular geometry, or at least to find the molecular symmetry. *See* MOLECULAR STRUCTURE AND SPECTRA.

Even when a precise determination of structure is not possible, much can often be said about the arrangement of atoms in a molecule from empirical information about the characteristic Raman frequencies of groups of atoms. This kind of information is closely similar to that provided by infrared spectroscopy; in fact, Raman and infrared spectra often provide complementary data about molecular structure. The complex structures of biologically important molecules, for example, are the subjects of current spectroscopic research. Both normal and resonance Raman spectroscopy are valuable techniques in molecular biology (see Fig. 4). Raman spectra also provide information for solid-state physicists, particularly with respect to lattice dynamics but also concerning the electronic structures of solids. *See* LATTICE VIBRATIONS.   [RICHARD C. LORD]

Bibliography: N. Bloembergen, *Non-Linear Optics*, 1965; A. J. Clark and R. E. Hester (eds.), *Advances in Infrared and Raman Spectroscopy*, vols. 1–7, 1975–1980; G. Herzberg, *Infrared and Raman Spectra of Polyatomic Molecules*, vol. 2, 2d ed., 1945; D. A. Long, *Raman Spectroscopy*, 1977; M. C. Tobin, *Laser Raman Spectroscopy*, 1971, reprint 1980; E. B. Wilson, J. C. Decius, and P. C. Cross, *Molecular Vibrations*, 1955.

Reactance

The opposition that inductance and capacitance offer to alternating current through the effect of frequency. Reactance alters the magnitude of current and also changes the circuit phase angle.

Inductive reactance X_L equals $2\pi f L$, where f is the frequency in hertz and L is the self-inductance in henrys. The voltage E across an inductance reaches its peak 90° before the current I reaches its peak, and $I = E/X_L$ amperes. Capacitive reactance X_C equals $1/(2\pi f C)$, where C is the capacitance in farads. The voltage E across a capacitance reaches its peak 90° after the current reaches its peak, and $I = E/X_C$ amperes.

Reactances are components of impedance which, in general, includes resistance R and reactance. Impedance is given by Eq. (1) for the series

$$Z = \sqrt{R^2 + (X_L - X_C)^2} \text{ ohms} \qquad (1)$$

RLC circuit. In terms of complex quantities, Eq. (2) holds. Both reactances have magnitude and

$$Z = R + jX_L - jX_C = R + j(X_L - X_C) \quad \text{ohms} \qquad (2)$$

angle: $+j$ means +90° for X_L, and $-j$ means −90° for X_C, the angles by which the voltages across them lead, or lag, the current. The phase angle between voltage and current is given by Eq. (3), and current lags, or leads, the voltage depending upon

$$\theta = \arctan\left[(X_L - X_C)/R\right] \qquad (3)$$

whether $X_L - X_C$ is positive or negative.

[BURTIS L. ROBERTSON]

Reciprocity principle

In the scientific sense, a theory that expresses various reciprocal relations for the behavior of some physical systems. Reciprocity applies to a physical system whose input and output can be interchanged without altering the response of the system to a given excitation. Optical, acoustical, electrical, and mechanical devices that operate equally well in either direction are reciprocal systems, whereas unidirectional devices violate reciprocity. The theory of reciprocity facilitates the evaluation of the performance of a physical system. If a system must operate equally well in two directions, there is no need to consider any nonreciprocal components when designing it.

Examples of reciprocal systems. Some systems that obey the reciprocity principle are any electrical network composed of resistances, inductances, capacitances, and ideal transformers; systems of antennas, with restrictions given according to Eq. (2); mechanical gear systems; and light sources, lenses, and reflectors.

Devices that violate the theory of reciprocity are transistors, vacuum tubes, gyrators, and gyroscopic couplers. Any system that contains the above devices as components must also violate the reciprocity theory. The gyrator differs from the transistor and vacuum tube in that it is linear and passive, as opposed to the active and nonlinear character of the other two devices.

Rayleigh's theorem of reciprocity. Reciprocity is concisely expressed by a theorem originally proposed by Lord Rayleigh for acoustic systems and later generalized by J. R. Carson to include electromagnetic systems. Both mathematical expressions of the theory of reciprocity are closely related to the mathematical theorem known as Green's theorem. The acoustical reciprocity theorem of Lord Rayleigh is as follows: In an acoustic system consisting of a fluid medium having boundary surfaces $s_1, s_2, \ldots, s_k$ and subject to no impressed body forces, surface integral (1) holds. Here p_1 and

$$\int_s (p_1 v_{2n} - p_2 v_{1n}) ds = 0 \qquad (1)$$

p_2 are the pressure fields produced respectively by the components of the fluid velocities v_{1n} and v_{2n}

normal to the boundary surfaces $s_1, s_2, \ldots, s_k$. The integral is evaluated over all boundary surfaces.

For a region containing only one simple source H. L. F. Helmholtz has shown that the theorem can be expressed as follows: A simple source at A produces the same sound pressure at B as would have been produced at A had the source been located at B. In other words, the response of a human ear at B due to a vibrating tuning fork at A is the same as the response at A due to the same tuning fork when located at B. The human ear, tuning fork, and intervening acoustical media constitute a physical system that obeys the theory of reciprocity.

Electromagnetic systems. The generalization of Lord Rayleigh's theorem to electromagnetic systems can be mathematically expressed by volume integral (2), where $\mathbf{E}_1$ and $\mathbf{H}_1$ are the electric

$$\int_v \nabla \cdot (\mathbf{E}_1 \times \mathbf{H}_2 - \mathbf{E}_2 \times \mathbf{H}_1) \, dv = 0 \qquad (2)$$

and magnetic field vectors describing a state due to one electromagnetic sound and $\mathbf{E}_2$ and $\mathbf{H}_2$ describe another state due to a second source. The above relation is valid as long as the medium is isotropic and the field vectors are finite and continuous, and vary according to a linear law (thus excluding ferromagnetic materials, electronic space charges, and ionized gas phenomena).

By means of Maxwell's equations, relation (2) can be expressed in another form when restricted to systems of conduction current only where $\mathbf{J}_1$ and $\mathbf{J}_2$ are the conduction current densities in an electromagnetic system due to the action of the external electric fields $\mathbf{E}_1$ and $\mathbf{E}_2$, respectively.

Equation (3) is readily applied to antennas and

$$\int_v (\mathbf{E}_1 \cdot \mathbf{J}_2 - \mathbf{E}_2 \cdot \mathbf{J}_1) \, dv = 0 \qquad (3)$$

radiation. If, in Fig. 1, $\mathbf{J}_1$ is the resulting current density in antenna B due to an electric field $\mathbf{E}_1$ established by antenna A, and $\mathbf{J}_2$ is the current density in antenna A due to electric field $\mathbf{E}_2$ established by antenna B, then $\mathbf{J}_1 = \mathbf{J}_2$, provided $\mathbf{E}_1 = \mathbf{E}_2$. The two emfs need not be applied at the same instant of time. The integral in Eq. (3) over all space reduces to an integral over the two antennas since $\mathbf{J}_1$ and $\mathbf{J}_2$ are zero elsewhere. From this particular

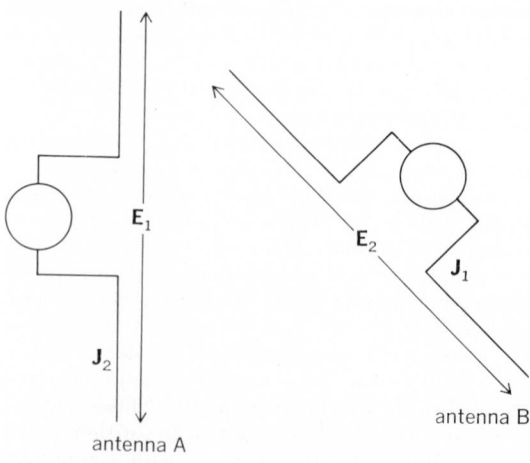

Fig. 1. Antenna system.

application of the reciprocity theorem it is seen that the transmitting and receiving patterns of an antenna are the same.

Equation (3), when evaluated over an N-mesh electrical network, reduces to Eq. (4), where a and

$$\sum_{j=1}^{N} V_{aj}i_{bj} = \sum_{j=1}^{N} V_{bj}i_{aj} \qquad (4)$$

b are two different states of the network and the j subscript denotes in which of the N meshes the voltage and current are measured. For the two-mesh network in Fig. 2, Eq. (4) gives Eq. (5). Expressed in words: If an emf source of magnitude V and zero internal impedance, when applied to terminals 1–1, produces a current I at terminals 2–2, then the same current I will be measured at terminals 1–1 when the emf V is applied to terminals 2–2. This statement, that is, Eq. (5), is probably the most familiar form of the theorem of reciprocity.

$$V_{a1}i_{b1} = V_{b2}i_{a2} \qquad (5)$$

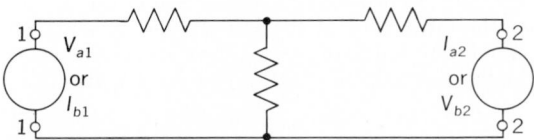

Fig. 2. Two-mesh network.

Electrostatic systems. The statement of reciprocity for electrostatics is given by Eq. (6), where V_1 and V_2 are the electric potentials produced at some arbitrary point due to the volume charge distributions ρ_1 and ρ_2, respectively. When the integral expression in Eq. (6) is applied to the

$$\int_v \rho_1 V_2 \, dv = \int_v \rho_2 V_1 \, dv \qquad (6)$$

electrostatic system of two charged conductors in Fig. 3, it becomes Eq. (7). Here V_a is the poten-

$$V_a q_a = V_b q_b \qquad (7)$$

tial on conductor a due to charge q_b on conductor b; the remaining quantities are similarly defined. In other words, if a charge q_b on conductor b raises the potential of conductor a to V, then the same charge on conductor a raises the potential of conductor b to V.

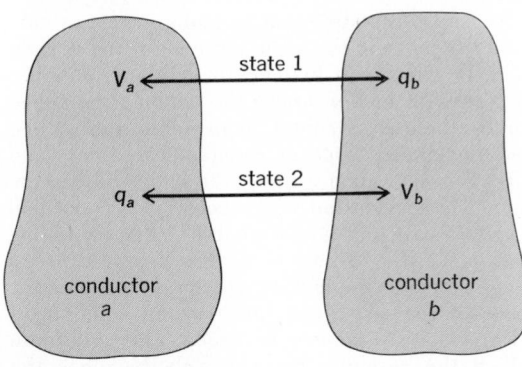

Fig. 3. Charged conducting bodies.

Electrical networks. A somewhat different approach to reciprocity is the so-called black box, or two-terminal pair, method illustrated in Fig. 4. The box might contain a mechanical, acoustical, optical, or electrical system. The applied excitation or cause is E and the response or effect is E'. The ratio of E/E' (or E'/E) is the transfer function G for the system within the black box. Using the subscript notation of G_{12} when E is impressed at terminals 1–1 and E' is measured at terminals 2–2, then G_{21} represents a response measured at 1–1 for an excitation at 2–2. Mathematically the general behavior of the box to excitations at both sets of terminals can be expressed by Eqs. (8), as long as the response bears a linear relation to the excitation.

$$E_1 = G_{11}E'_1 + G_{12}E'_2$$
$$E_2 = G_{21}E'_1 + G_{22}E'_2 \qquad (8)$$

If, in addition to its linear characteristic, the system satisfies Eq. (9), the principle of reciprocity is

$$G_{12} = G_{21} \qquad (9)$$

obeyed, and the device will operate equally in either direction. Whenever $G_{12} \neq G_{21}$, the system violates the theory of reciprocity, with the result that the response in one direction is different from that obtained in the other direction.

[HUGH S. LANDES]

Bibliography: B. Bleaney and B. I. Bleaney, *Electricity and Magnetism*, 3d ed., 1976; D. E. Gray (ed.), *American Institute of Physics Handbook*, 3d ed., 1972; Howard W. Sams Engineering Staff, *Reference Data for Radio Engineers*, 6th ed., 1975; J. D. Kraus, *Antennas*, 1950; J. D. Kraus and K. R. Carver, *Electromagnetics*, 2d ed., 1973; J. A. Stratton, *Electromagnetic Theory*, 1941.

Rectilinear motion

Motion is defined as continuous change of position of a body. If the body moves so that every particle of the body follows a straight-line path, then the motion of the body is said to be rectilinear. *See* MOTION.

When a body moves from one position to another, the effect may be described in terms of motion of the center of mass of the body from a point A to a point B (Fig. 1). If the center of mass of the body moves along a straight line connecting the points A and B, then the motion of the center of mass of the body is rectilinear. If the body as a whole does not rotate while it is moving, then the path of every particle of which the body is composed is a straight line parallel to or coinciding with the path of the center of mass, and the body as a whole executes rectilinear motion. This is shown by the straight line connecting points P_1 and P_2 in Fig. 1. *See* CENTER OF MASS.

Rectilinear motion is an idealized form of motion which rarely, if ever, occurs in actual experience, but it is the simplest imaginable type of motion and thus forms the basis for the analysis of more complicated motions. However, many actual motions are approximately rectilinear and may be treated as such without appreciable error. For example, a ball thrown directly upward may follow, for all practical purposes, a straight-line path. The motion of a high-speed rifle bullet fired horizontally

RECIPROCITY PRINCIPLE

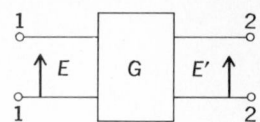

Fig. 4. Four-terminal, black-box network.

RECTILINEAR MOTION

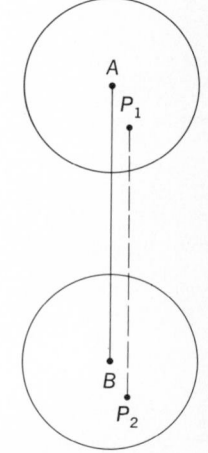

Fig. 1. Rectilinear motion. All points move parallel to the center of mass.

may be essentially rectilinear for a short length of path, even though in its larger aspects the ideal path is a parabola. The motion of an automobile traveling over a straight section of roadway is essentially rectilinear if minor variations of path are neglected. The motion of a single wheel of the car is not rectilinear, although the motion of the center of mass of the wheel may be essentially so.

Curvilinear motion. Rectilinear motion must be distinguished from curvilinear motion. In the former the direction of motion is constant, while in the latter the direction of motion is continuously changing. If a body is moved from A to B along a curved path (Fig. 2), the resulting position of the object is the same as if the body had undergone rectilinear motion from A to B (dotted line) followed by a rotation. Thus any displacement of a body may always be described in terms of rectilinear motion plus a rotation. *See* ROTATIONAL MOTION.

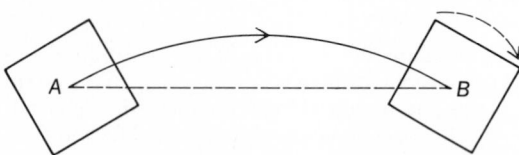

Fig. 2. Curvilinear motion, which is equivalent to rectilinear motion plus a rotation.

Motion with constant velocity. When a body moves from one location to another it is said to be displaced. The linear displacement is the distance from the first position to the second position. Motion cannot be instantaneous; it must involve a time interval during which the displacement takes place. This interval is called the elapsed time. When the ratio of the displacement to the elapsed time is constant regardless of how small an interval is chosen, the motion is uniform and the body is said to move with a constant velocity since neither the magnitude of the motion (its speed) nor the direction of the motion is changing. Motion with constant velocity is the simplest form of rectilinear motion. *See* DISPLACEMENT (MECHANICS).

Average velocity and acceleration. Rectilinear motion in general may be motion with an increasing velocity (positive acceleration), with a decreasing velocity (negative acceleration or deceleration), or with a variable velocity. When the velocity is not constant, the ratio of the displacement to the elapsed time is the average velocity for the interval. The limiting value of the ratio of the displacement to elapsed time as the interval is made smaller and smaller is the instantaneous velocity, sometimes called the velocity at a point. *See* ACCELERATION; VELOCITY.

The displacement s of a moving body in a given time interval t is the average velocity v_{av} multiplied by the time interval, as shown in Eq. (1). If the velocity is uniformly increasing or decreasing, the average velocity is half the sum of the initial velocity v_0 at the beginning of the interval and the final velocity v_f at the end of the interval, as shown in Eq. (2).

$$s = v_{av}t \tag{1}$$

$$v_{av} = \frac{v_0 + v_f}{2} \tag{2}$$

The gain (or loss) of velocity during the interval is $v_f - v_0$ and the gain or loss per unit time is called the acceleration a and is given by Eq. (3). From

$$a = \frac{v_f - v_0}{t} \tag{3}$$

this, the gain (or loss) of velocity $v_f - v_0$ is equal to the acceleration times the elapsed time, as in Eq. (4). For a body uniformly accelerated from rest, the

$$v_f - v_0 = at \tag{4}$$

initial velocity v_0 is zero and the average velocity is $v_f/2$; this equals $at/2$ since the final velocity of a uniformly accelerated body starting from rest is $v_f = at$. Consequently, the displacement or space traversed by a uniformly accelerated body starting from rest is given by Eq. (5). If the initial velocity of

$$s = v_{av}t = \left(\frac{v_f}{2}\right)t = \left(\frac{at}{2}\right)t = \tfrac{1}{2}at^2 \tag{5}$$

the body is not zero, the displacement due to the continuing action of the initial velocity must be added to that due to the acceleration, as shown in Eq. (6). The units in which velocity is usually meas-

$$s = v_0 t + \tfrac{1}{2}at^2 \tag{6}$$

ured are meters per second (m/sec), feet per second (ft/sec), or miles per hour (mph). For acceleration, the most common units are meters per second per second (m/sec²) or feet per second per second (ft/sec²).

Formula (6) is the solution to the basic problem in the kinematics of rectilinear motion. The problem is as follows: Given the initial conditions, that is, the initial velocity v_0 and the acceleration a, find the position of the body after any elapsed time t. This requires that both the velocity and acceleration be vector quantities denoting direction as well as magnitude and that the initial position of the object be known. If this initial position is denoted by s_0, then in general the position s after any elapsed time t is given by Eq. (7). For rising and

$$s = s_0 + v_0 t + \tfrac{1}{2}at^2 \tag{7}$$

falling bodies, it is necessary only to substitute for a the known acceleration of gravity g, which is approximately 9.8 m/sec² or 32.2 ft/sec² for bodies near the surface of the Earth.

Newton's laws. When all the forces acting on a body are balanced, there is no net unbalanced force and the body either remains at rest or continues to move in the same direction with constant velocity (Newton's first law of motion). Whether at rest or in motion under the action of balanced forces, the body is said to be in equilibrium. When an unbalanced force of constant magnitude and direction acts upon a body, the body will be continuously accelerated with a constant acceleration (Newton's second law of motion). A freely falling body near the surface of the Earth would be an almost ideal example if it were not for the resisting force of the air. Since this resistance increases with velocity, a point is reached sooner or later at which the resisting force exactly balances the downward pull of gravity. The body is then in equilibrium, the acceleration is zero, and the body continues to fall with a constant velocity called the terminal velocity. Other, more complicated, motions can be described in terms of more complicated force functions, such as one in which the force (1) varies in direct proportion to the distance the body has traveled, (2) varies as the square of the distance, or (3) varies inversely as the distance or (4) inversely as the square of the distance. *See* FREE FALL; NEWTON'S LAWS OF MOTION.

Force and motion. The basic formula relating force and motion and expressing both of Newton's first two laws of motion is $F = ma$ or $a = F/m$; it indicates that the acceleration produced in any body of mass m by an unbalanced force F is directly proportional to the force and inversely proportional to the mass. If the force is zero, the acceleration is zero and the motion (if any) is nonaccelerated; that is, it has constant velocity. In the symbols of calculus this is defined by Eq. (8). If the appropriate force function is put into

$$F = m\frac{d^2s}{dt^2} \qquad (8)$$

this equation as a function of the displacement s, the equation expresses the fundamental relations of the problem of linear motion in Newtonian mechanics. *See* FORCE.   [ROGERS D. RUSK]

Reflection and transmission coefficients

When an electromagnetic wave passes from a medium of permeability μ_1 and dielectric constant ϵ_1 to one with values μ_2 and ϵ_2, part of the wave is reflected at the boundary and part transmitted. The ratios of the amplitudes in the reflected wave and the transmitted wave to that in the incident wave are called the reflection and transmission coefficients, respectively. For oblique incidence, the reflection and refraction formulas of optics are most convenient, but for normal incidence of plane waves on plane boundaries, such as occur with transmission lines, waveguides, and some free waves, the concept of wave impedance and characteristic impedance is useful.

For a z-directed wave with electric intensity $\mathbf{E}$ in the x direction and magnetic intensity $\mathbf{H}$ in the y direction, the total phasor fields on the incident side are given by Eqs. (1) and (2), where primes are

$$\check{E}_x = E_0 e^{-jkz} + \check{E}'_0 e^{jkz} \qquad (1)$$

$$\check{H}_y = (\eta)^{-1}(E_0 e^{-jkz} - E'_0 e^{jkz}) \qquad (2)$$

used for reflected quantities and η is the wave impedance. The sign difference in Eqs. (1) and (2) is due to the fact that Poynting's vector, $\frac{1}{2}\mathbf{E} \times \mathbf{H}$, is positive for the incident and negative for the reflected wave. For the transmitted wave, Eqs. (3)

$$\check{E}''_x = E''_0 e^{-jkz} \qquad \check{H}''_y = (\eta'')^{-1} E''_0 e^{-jkz} \qquad (3)$$

hold. Since the tangential components of $\mathbf{E}$ and $\mathbf{H}$ are continuous across the boundary at $z = 0$, $\check{E}_x = \check{E}''_x$ and $\check{H}_y = \check{H}''_y$, so that Eqs. (4) hold. The ratios

$$E_0 + E'_0 = E''_0 \qquad \eta''(E_0 - E'_0) = \eta E''_0 \qquad (4)$$

for the reflected and transmitted fields obtained by

solving these equations are given by Eqs. (5),

$$\frac{E'_0}{E_0} = \frac{\eta'' - \eta}{\eta'' + \eta} \qquad \frac{E''_0}{E_0} = \frac{2\eta''}{\eta'' + \eta} \qquad (5)$$

which are the reflection and transmission coefficients, respectively. *See* ELECTROMAGNETIC RADIATION; POYNTING'S VECTOR.

Coefficients for optics. Equations (5) hold for normal incidence in optics, if the velocities v and v'' are written for η and η''. For a plane wave whose electric vector is normal to the plane of incidence and whose direction makes an acute angle θ with the normal to the interface, the reflection and transmission coefficients are given by Eqs. (6),

$$\frac{E'_0}{E_0} = -\frac{\sin(\theta - \theta'')}{\sin(\theta + \theta'')} \qquad \frac{E''_0}{E_0} = \frac{2\sin\theta''\cos\theta}{\sin(\theta + \theta'')} \qquad (6)$$

where $v'' \sin\theta = v \sin\theta''$. When the electric vector lies in the plane of incidence, the coefficients are given by Eqs. (7). The ratio v/v'' is the index of refraction. *See* REFRACTION OF WAVES.

$$\begin{aligned}\frac{E'_0}{E_0} &= \frac{\tan(\theta - \theta'')}{\tan(\theta + \theta'')} \\[2mm] \frac{E''_0}{E_0} &= \frac{2\sin\theta''\cos\theta}{\sin(\theta + \theta'')\sin(\theta - \theta'')}\end{aligned} \qquad (7)$$

Waveguides. In waveguides, as in free space, the characteristic impedance is defined as the ratio of the transverse electric field E_t to the transverse magnetic field H_t. For waveguides, this radio depends on the frequency and the dimensions of the waveguide, as well as on the permeabilities and dielectric constants. For a transverse interface, the boundary conditions used for Eqs. (4) on the tangential fields still hold. Thus, Eqs. (5) for the reflection and transmission coefficients are valid if η and η'' are replaced by the characteristic impedances on the incident and emergent sides, respectively.

Transmission lines and networks. Let $\check{Z}$ and $\check{Z}''$ be the characteristic impedances on the incident and emergent sides of a discontinuity in a transmission line or on the two sides of a junction between two networks. Then the relations between the potentials and currents of the incident, reflected, and transmitted waves are given by Eqs. (8a)–(8c), respectively. At the discontinuity, po-

$$\check{V} = \check{Z}\check{I} \qquad (8a)$$

$$\check{V}' = -\check{Z}\check{I}' \qquad (8b)$$

$$\check{V}'' = \check{Z}''\check{I}'' \qquad (8c)$$

tential and current must be continuous so that Eqs. (9) hold. Solution for the ratios gives Eqs. (10),

$$\check{V} + \check{V}' = \check{V}'' \qquad \check{I} + \check{I}' = \check{I}'' \qquad (9)$$

$$\frac{\check{V}'}{\check{V}} = \frac{\check{Z}'' - \check{Z}}{\check{Z}'' + \check{Z}} \qquad \frac{\check{V}''}{\check{V}} = \frac{2\check{Z}''}{\check{Z}'' + \check{Z}} \qquad (10)$$

which are the reflection and transmission coefficients.

Coefficients for acoustics. Equations (10) hold in acoustics, provided acoustic impedance is sub-

stituted for electrical impedance. Acoustic impedance is defined as the product of the density of a
medium by the speed of sound in it. There are two
types of waves in solid mediums, longitudinal
waves and shear waves, and thus there are two
impedances. *See* ACOUSTIC IMPEDANCE.

[WILLIAM R. SMYTHE]

Bibliography: D. Dearholt and W. McSpadden,
Electromagnetic Wave Propagation, 1973; D. E.
Gray (ed.), *American Institute of Physics Hand-
book*, 3d ed., 1972; J. D. Kraus and K. R. Carver,
Electromagnetics, 2d ed., 1973; P. Lorrain and
D. R. Corson, *Electromagnetic Fields and Waves*,
2d ed., 1970; S. Ramo, J. R. Whinnery, and T.
Van Duser, *Fields and Waves in Communication
Electronics*, 1965; W. R. Smythe, *Static and Dy-
namic Electricity*, 3d ed., 1968.

Reflection of electromagnetic radiation

The returning or throwing back of electromagnetic
radiation such as light, ultraviolet rays, radio
waves, or microwaves by a surface upon which the
radiation is incident. In general, a reflecting surface is the boundary between two materials of different electromagnetic properties, such as the
boundary between air and glass, air and water, or
air and metal. Devices designed to reflect radiation
are called reflectors or mirrors.

Reflection angle. The simplest reflection laws
are those that govern plane waves of radiation. The
law of reflection concerns the incident and reflected rays (as in the case of a beam from a flash-
light striking a mirror) or, more precisely, the wave
normals of the incident and reflected waves. The
law states that the incident and reflected rays and
the normal to the reflecting surface all lie in one
plane, called the plane of incidence, and that the
reflection angle θ_{refl} equals the angle of incidence
θ_{inc} as in Eq. (1) (see Fig. 1). The angles θ_{inc} and

$$\theta_{refl} = \theta_{inc} \qquad (1)$$

θ_{refl} are measured between the surface normal and
the incident and reflected rays, respectively. The
surface (in the above example, that of the mirror) is
assumed to be smooth, with surface irregularities
small compared to the wavelength of the radiation.
This results in so-called specular reflection. In
contrast, when the surface is rough, the reflection

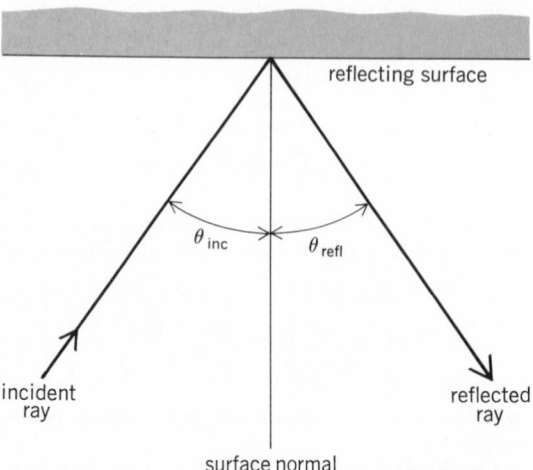

Fig. 1. Reflection of electromagnetic radiation from a
smooth surface.

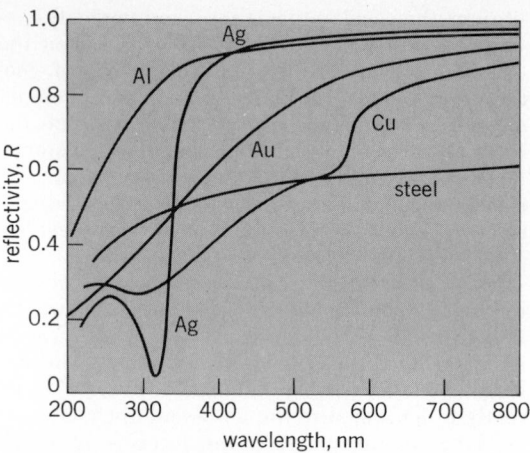

Fig. 2. Reflectivity of some common metals for normal
incidence as a function of wavelength. (*From F. A. Jenkins
and E. White, Fundamentals of Optics, 4th ed., McGraw-
Hill, 1976*)

is diffuse. An example of this is the diffuse scattering of light from a screen or from a white wall
where light is returned through a whole range of
different angles.

Reflectivity. The reflectivity of a surface is a
measure of the amount of reflected radiation. It is
defined as the ratio of the intensities of the reflected and incident radiation. The reflectivity
depends on the angle of incidence, the polarization
of the radiation, and the electromagnetic properties of the materials forming the boundary surface.
These properties usually change with the wavelength of the radiation. Reflecting materials are
divided into two groups: transparent materials and
opaque conducting materials. Radiation penetrating a transparent material propagates essentially
unattenuated, while radiation penetrating a conducting material is heavily attenuated. Transparent materials are also called dielectrics. In the
wavelength range of visible light, typical dielectrics are glass, quartz, and water. Conducting materials are usually metals such as gold, silver, or
aluminum, which are good reflectors at almost all
wavelengths. *See* ABSORPTION OF ELECTROMAG-
NETIC RADIATION.

Reflection from metals. The reflectivity of polished metal surfaces is usually quite high. Silver
and aluminum, for example, reflect more than 90%
of visible light. The reflectivity of some common
metals as a function of wavelength is given in Fig.
2 for normal incidence ($\theta_{inc} = 0$). The reflectivities
vary considerably with wavelength, generally falling off toward the shorter wavelengths, with silver
exhibiting a reflection "window" at 320 nm. The
reflectivity values depend somewhat on the way
the metal surface was prepared; for example,
whether or not it was polished or was produced by
evaporation. The presence of an oxidation layer is
also a factor influencing (and usually decreasing)
the reflectivity. In ordinary mirrors the reflecting
surface is the interface between metal and glass,
which is thus protected from oxidation, dirt, and
other forms of deterioration. When it is not permissible to use this protection for technical reasons,
one uses "front-surface" mirrors, which are usually coated with evaporated aluminum. The aluminum has a high reflectivity and deteriorates

relatively little in the atmosphere. Front-surface mirrors are used in scientific applications such as interferometry and in large reflecting telescopes such as that on Mount Palomar.

Reflection from dielectrics. The material property that determines the amount of radiation reflected from an interface between two dielectric media is the phase velocity v of the electromagnetic radiation in the two materials. In optics one uses as a measure for this velocity the refractive index n of the material, which is defined by Eq. (2) as the

$$n = c/v \qquad (2)$$

ratio of the velocity of light c in vacuum and the phase velocity in the material. For visible light, for example, the refractive index of air is about $n = 1$, the index of water is about $n = 1.33$, and the index of glass is about $n = 1.5$. *See* PHASE VELOCITY; REFRACTION OF WAVES.

For normal incidence ($\theta_{inc} = 0$) the reflectivity R of the interface is given by Eq. (3), in which the

$$R = \left(\frac{v_1 - v_2}{v_1 + v_2}\right)^2 = \left(\frac{n_2 - n_1}{n_2 + n_1}\right)^2 \qquad (3)$$

material constants are labeled 1 and 2 as shown in Fig. 3, where the radiation is incident in material 1. The reflectivity of an air-water interface is about 2% ($R = 0.02$) and that of an air-glass interface about 4% ($R = 0.04$); the other 98% or 96% are transmitted through the water or glass, respectively.

A ray incident upon the interface at an oblique, nonnormal angle θ_1 is deviated as it penetrates material 2 as shown in Fig. 3. This is called refraction, and the refracted angle θ_2 follows from Snell's law of refraction, Eq. (4). Again, there is partial

$$n_1 \sin \theta_1 = n_2 \sin \theta_2 \qquad (4)$$

reflection, with the reflectivity depending on the angle of incidence θ_1 and the polarization of the radiation. When the electric field is polarized parallel to the plane of incidence, the reflectivity $R_{\parallel}$ is given by Eq. (5), and when the electric field is polarized perpendicular to the plane of incidence, the

$$R_{\parallel} = \frac{\tan^2(\theta_1 - \theta_2)}{\tan^2(\theta_1 + \theta_2)} \qquad (5)$$

reflectivity $R_{\perp}$ is given by Eq. (6). These formulas

$$R_{\perp} = \frac{\sin^2(\theta_1 - \theta_2)}{\sin^2(\theta_1 + \theta_2)} \qquad (6)$$

are known as the "Fresnel Formulas." For unpolarized radiation, in which the electric field varies rapidly in a random, unpredictable manner, the reflectivity $\bar{R}$ is the average of $R_{\parallel}$ and $R_{\perp}$. Figure 4 shows $R_{\parallel}$, $R_{\perp}$, and $\bar{R}$ for an air-glass interface as a function of angle. The reflectivity approaches 100% at grazing incidence ($\theta_1 \approx 90°$) for both polarizations. *See* REFLECTION AND TRANSMISSION COEFFICIENTS.

At an angle of about $\theta_1 = \theta_B = 56°$ the reflectivity $R_{\parallel}$ assumes zero value. This angle is called Brewster's angle, which is, in general, obtained from formula (7). At this angle, which is also called

$$\tan \theta_B = n_2/n_1 \qquad (7)$$

the polarizing angle, only radiation polarized per-

pendicular to the plane of incidence is reflected.

Total internal reflection. Total reflection, that is, reflection of 100% of the incident radiation, occurs at the interface of two dielectrics when the radiation is incident in the denser medium, that is, when $n_1 > n_2$ and when the angle of incidence θ_1 is larger than the critical angle θ_0 given by Eq. (8).

$$\sin \theta_0 = n_2/n_1 \qquad (8)$$

Total internal reflection can be observed, for example, by a submerged diver looking up at the water-air interface for which the critical angle is about $\theta_0 = 49°$. For a glass-air interface the critical angle is approximately $\theta_0 = 42°$.

Selective reflection from crystals. The discussion of reflection from dielectrics to this point has been concerned with reflectivity of nonabsorbing media far removed from absorption bands. These bands are located in the spectral regions where the frequency of the radiation corresponds to a resonance frequency of the atoms, molecules, or crystal lattice of the medium. Since this radiation is strongly absorbed, it is also strongly reflected. The metallic sheen of dye crystals, which have very strong absorption bands in the visible spectrum, is caused by selective reflection. Crystalline solids such as rocksalt or quartz, the lattices of which are built up of atoms bearing net electric charges, show strong selective reflection in the infrared region at wavelengths near those of the strong absorption bands associated with lattice vibrations in the crystal. By reflecting an infrared beam several times from such a material, highly monochromatic radiation can be obtained at the specific wavelengths. These monochromatic beams are referred to as residual rays or reststrahlen. Figure 5 shows residual rays for some crystals.

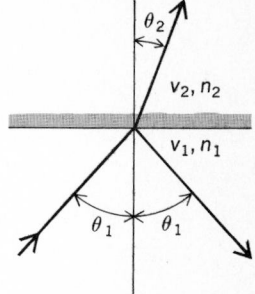

Fig. 3. Reflection and refraction of electromagnetic radiation at an interface between two dielectric media.

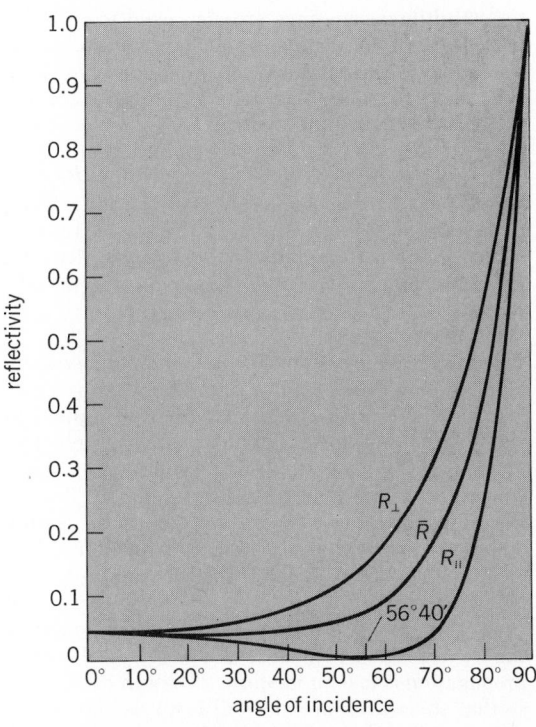

Fig. 4. Reflectivity as a function of angle of incidence for an air-glass interface. (*From M. Born and E. Wolf, Principles of Optics, 5th ed., Pergamon Press, 1975*)

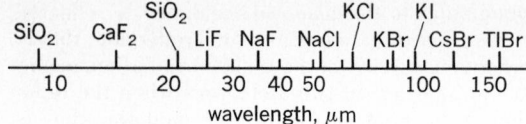

SiO₂   CaF₂   SiO₂   LiF NaF NaCl   KCl KI CsBr TlBr
                                              KBr

| 10 | 20 | 30 40 50 | 100 | 150 |

wavelength, μm

Fig. 5. Residual rays for various crystals.

Antireflection coatings. In order to reduce undesired reflections from surfaces of optical components such as photographic lenses, one can coat the surface with a thin film. To minimize reflection, the refractive index n_f of this film should be the geometric mean of the indices n_1 and n_2 of the incident and refracting media, Eq. (9), and the film thickness d_f should be equal to a quarter of the wavelength of light in the film material, that is, it should satisfy Eq. (10), where λ is the vacuum

$$n_f^2 = n_1 n_2 \qquad (9)$$

$$n_f d_f = \lambda/4 \qquad (10)$$

wavelength of the light for which the reflectivity of the surface is minimized. Such a film is called a quarter-wave layer. Materials such as magnesium fluoride with an index of $n = 1.38$ are used for antireflection coatings of air-glass interfaces. Antireflection coatings have been made which reduce the reflectivity to less than 0.1% on certain materials; on glass surfaces less than 1/2% has been achieved.

High-reflectivity multilayers. A series of dielectric quarter-wave layers of alternating high and low refractive index can be used to produce reflectors of high reflectivity at a specified wavelength. This is called a quarter-wave stack. The high reflectivity is due to an interference effect. Stacks with about 20 layers are used to make reflectors with reflectivities higher than 99%. Mirrors such as these are used for the construction of laser resonators. Multilayer coatings are also used to enhance the reflectivity of metals such as aluminum. About four quarter-wave layers can enhance the metal reflectivity to values better than 99%. *See* GEOMETRICAL OPTICS; MIRROR OPTICS.

[HERWIG KOGELNIK]

Bibliography: American Institute of Physics Handbook, 3d ed., 1972; M. Born and E. Wolf, *Principles of Optics*, 5th ed., 1975; F. A. Jenkins and E. White, *Fundamentals of Optics*, 4th ed., 1976; J. Strong, *Concepts of Classical Optics*, 1958.

Refraction of waves

The change of direction of propagation of any wave phenomenon which occurs when the wave velocity changes. The term is most frequently applied to visible light, but it also applies to all other electromagnetic waves, as well as to sound and water waves.

The physical basis for refraction can be readily understood with the aid of Fig. 1. Consider a succession of equally spaced wavefronts approaching a boundary surface obliquely. The direction of propagation is in ordinary cases perpendicular to the wavefronts. In the case shown, the velocity of propagation is less in medium 2 than in medium 1, so that the waves are slowed down as they enter the second medium. Thus the direction of travel is bent toward the perpendicular to the boundary

surface (that is, $\theta_2 < \theta_1$). If the waves enter a medium in which the velocity of propagation is faster than in their original medium, they are refracted away from the normal.

Snell's law. The simple mathematical relation governing refraction is known as Snell's law. If waves traveling through a medium at speed v_1 are incident on a boundary surface at angle θ_1 (with respect to the normal), and after refraction enter the second medium at angle θ_2 (with the normal) while traveling at speed v_2, then Eq. (1) holds. The

$$\frac{v_1}{v_2} = \frac{\sin \theta_1}{\sin \theta_2} \qquad (1)$$

index of refraction n of a medium is defined as the ratio of the speed of waves in vacuum c to their speed in the medium. Thus $c = n_1 v_1 = n_2 v_2$, and therefore Eq. (2) holds. The refracted ray, the

$$n_1 \sin \theta_1 = n_2 \sin \theta_2 \qquad (2)$$

normal to the surface, and the incident ray always lie in the same plane.

The relative index of refraction of medium 2 with respect to that of medium 1 may be defined as $n = n_1/n_2$. Snell's law then becomes Eq. (3). For

$$\sin \theta_1 = n \sin \theta_2 \qquad (3)$$

sound and other elastic waves which require a medium in which to propagate, only this last form has meaning. Equation (3) is frequently used for light when one medium is air, whose index of refraction is very nearly unity.

When the wave travels from a region of low velocity (high index) to one of high velocity (low index), refraction occurs only if $(n_1/n_2) \sin \theta_1 \leqq 1$. If θ_1 is too large for this relation to hold, then $\sin \theta_2 > 1$, which is meaningless. In this case the waves are totally reflected from the surface back into the first medium. The largest value that θ_1 can have without total internal reflection taking place is known as the critical angle θ_c. Thus $\sin \theta_c = n_2/n_1$. When the angle of incidence $\theta_1 < \theta_c$, refraction occurs, as in Fig. 2a. When $\theta_1 = \theta_c$, the emergent ray just grazes the surface (Fig. 2b). Total internal reflection (Fig. 2c) represents the only practical case for which 100% of the incident energy is reflected and none is absorbed. When it is desired

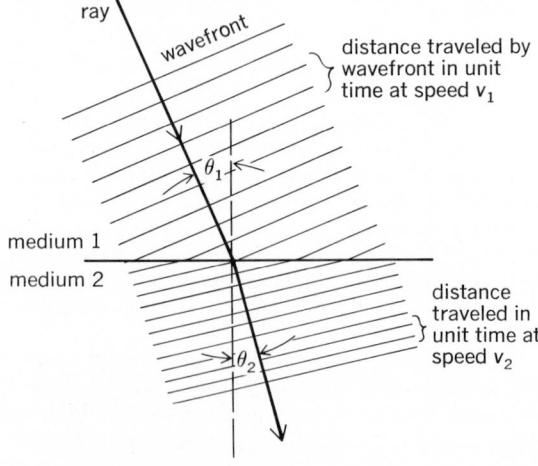

Fig. 1. Physical basis for Snell's law.

to change the direction of a beam of light without loss of energy, totally reflecting prisms are often used, as in prism binoculars.

If waves travel through a medium having a continuously varying index of refraction, the rays follow smooth curves with no abrupt changes of direction. Suppose (Fig. 3) that $n = n(y)$, and that the incident ray lies in the xy plane. If θ is the angle between the direction of the ray and the y axis, then Snell's law can be written in the differential form given by Eq. (4). In a particular case Eq. (4) can be integrated to give the path of the ray.

$$\frac{d\theta}{dn} = -\frac{1}{n}\tan\theta \tag{4}$$

Visible light. Many interesting cases of refraction occur for visible light. The refraction of light by a prism in air affords a particularly simple and useful example.

As a ray passes through the prism of Fig. 4, its total deflection or deviation $D = \theta_1 + \theta_4 - A$, where A is the vertex angle of the prism. Also, by Snell's law, Eq. (5) holds. It is found that the deviation is a

$$n = \frac{\sin\theta_1}{\sin\theta_2} = \frac{\sin\theta_4}{\sin\theta_3} \tag{5}$$

minimum when the ray passes through the prism symmetrically (that is, when $\theta_1 = \theta_4$). For minimum deviation, Eq. (6) holds. For a given prism, the dis-

$$n = \frac{\sin\frac{1}{2}(A+D)}{\sin\frac{1}{2}A} \tag{6}$$

persion, or lateral spread of the spectrum formed, is maximum for that wavelength of light which passes through the prism at minimum deviation. *See* OPTICAL PRISM.

For most optical materials, the dispersion $dn/d\lambda$ (λ is the wavelength) is negative; therefore red light is bent less than blue light. Typical values of n for optical materials range from 1.5 for ordinary crown glass, 1.7 or 1.8 for dense flint glass, and up to 2.42 for diamond. For water, n is 1.33. Some special substances have even higher values. Many substances show anisotropy in the refraction of light, with different indices of refraction in different directions. *See* OPTICAL MATERIALS.

For a lens (Fig. 5), refraction occurs at both surfaces. If the lens is thin and the rays all make small angles with the axis of the system, application of Snell's law to the two spherical surfaces yields the well-known lens formula, Eq. (7), where s is the

$$\frac{1}{s} + \frac{1}{s'} = \frac{1}{f} \tag{7}$$

object distance from the lens, s' the image distance, and f the focal length of the lens. Magnifying instruments such as binoculars, telescopes, microscopes, and projectors make use of refraction by lenses or prisms in their operation. *See* LENS.

Double refraction. Some anisotropic single crystals such as those of calcite and quartz are birefringent, or doubly refracting. If one looks through such a crystal at a dot on a piece of paper, he sees two images. As the crystal is rotated in the plane of the paper, one image remains stationary while the

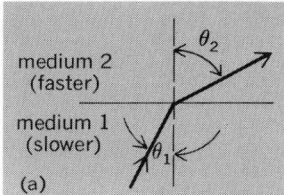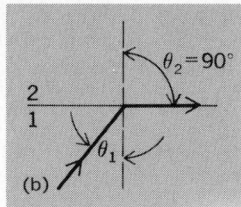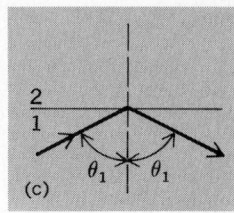

Fig. 2. Behavior of ray traveling from medium of high refractive index to medium of low refractive index. (*a*) When $\theta_1 < \theta_c$, ray is refracted. (*b*) When $\theta_1 = \theta_c$, ray grazes surface. (*c*) When $\theta_1 > \theta_c$, ray is reflected.

other appears to rotate about it.

Two separate rays propagate through the crystal; they are called the ordinary ray and the extraordinary ray. These rays are linearly polarized at right angles to each other. The ordinary ray obeys Snell's law; the extraordinary ray in general does not. The extraordinary ray does not propagate perpendicularly to its wavefronts. The separation between the two rays depends upon the direction in which the light travels through the crystal relative to that of the optic axis of the crystal. Light traveling parallel to the optic axis is only singly refracted.

Birefringent crystals are either uniaxial or biaxial, depending upon whether they have one optic axis or two. They are said to be positive or negative, depending upon whether the velocity of propagation (within the crystal) of the extraordinary wave is greater or less than that of the ordinary wave. Calcite is a uniaxial negative crystal, quartz a uniaxial positive crystal. The most commonly used biaxial crystal is mica. Doubly refracting crystals are frequently employed as polarizers, such as the nicol prism. *See* CRYSTAL OPTICS; POLARIZED LIGHT.

Refractometry. The measurement of indices of refraction, called refractometry, can be made in several ways. A very accurate technique is to determine in a prism spectrometer the minimum deviation D for a prism made from the material in question. The value of n is then calculated from Eq. (6). Hollow prisms can be used in this manner to determine the values of indices of refraction of various liquids. Alternatively, the critical angle for total internal reflection may be measured. Another method is to observe visually the apparent thickness of a slab of material by looking straight through it, and to compare the apparent thickness with the real thickness as measured with a micrometer. Then Eq. (8) holds.

REFRACTION OF WAVES

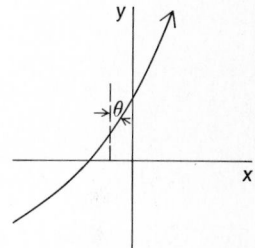

Fig. 3. Path of light in medium having continuously varying index of refraction $n = n(y)$.

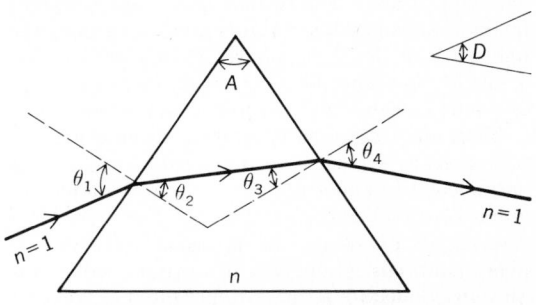

Fig. 4. Refraction of light by a prism.

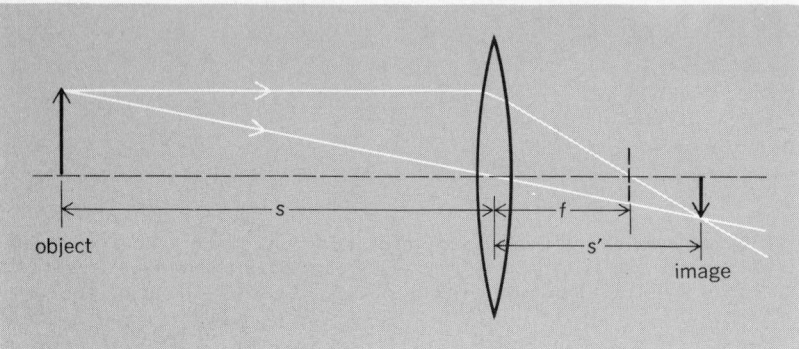

Fig. 5.   Refraction of light by a lens.

$$n = \frac{\text{real thickness}}{\text{apparent thickness}} \qquad (8)$$

Interferometric methods are particularly convenient for gases. In the Jamin refractometer, for example, a simple count of fringes as the gas is slowly admitted to an initially evacuated tube in the optical path yields n. These techniques can also be used for solids, particularly when the material is available in the form of thin films. *See* INTER-FEROMETRY.

Refractometry is an important tool in analytical chemistry. For example, information about the composition of an unknown solution can frequently be obtained by measurement of its index of refraction.

Atmospheric refraction. Gases have indices of refraction only slightly greater than unity. In general, $n - 1$ is proportional to the density of the gas, or to the ratio of pressure to absolute temperature. The index of refraction of the Earth's atmosphere increases continuously from 1.000000 at the edge of space to 1.000293 (yellow light) at 0°C and 760 mm Hg pressure. Thus celestial bodies as seen in the sky are actually nearer to the horizon than they appear to be. The effect decreases from a maximum of about 35 minutes of arc for an object on the horizon to zero at the zenith, where the light enters the atmosphere at perpendicular incidence. Thus the Sun (and all other bodies) appears to rise 2 or more minutes earlier (depending upon latitude) and to set 2 or more minutes later than would be the case without refraction. This must be taken into account when the altitude of a celestial body is observed for navigational purposes.

Other manifestations of atmospheric refraction are the mirages and "looming" of distant objects which occur over oceans or deserts, where the vertical density gradient of the air is quite uniform over a large area. The twinkling of stars is caused by the rapid small fluctuations in density along the light path in the atmosphere. Rainbows are produced by the multiple reflections, refraction, and dispersion of sunlight by spherical raindrops.

Other electromagnetic waves. Although refraction is most frequently encountered for the visible portion of the spectrum, it is of importance for other electromagnetic radiation. For very-long-wavelength radiation, the index of refraction of many materials is equal to the square root of the dielectric constant k. In general, $dn/d\lambda$ is negative except in the regions of so-called anomalous dispersion near absorption bands. On the short-wave-length side of an absorption band, n can be less than 1.00. Since it is the phase velocity of the wave rather than the group velocity which is involved in the definition of the index of refraction, this does not represent a violation of the principle of relativity, that is, that energy cannot be propagated at a velocity faster than the velocity of light in vacuum. At very high frequencies, the index of refraction of all materials is also slightly less than unity. *See* ABSORPTION OF ELECTROMAGNETIC RADIATION; GROUP VELOCITY; PHASE VELOCITY; RELATIVITY.

Refraction plays a role in the propagation beyond the line of sight of radio waves in the Earth's atmosphere.

The interaction of electromagnetic radiation with more or less opaque substances is often described in terms of a complex index of refraction. The real part of this quantity has the usual meaning for the small amount of light which penetrates into the material before it is absorbed. The imaginary part is a measure of the absorption.

Sound waves. The velocity of sound in a gas is proportional to the square root of the absolute temperature. Because of the vertical temperature gradients in the atmosphere, refraction of sound can be quite pronounced. As in mirage formation, to allow large-scale refraction the temperature at a given height must be uniform over a rather large horizontal area. If the temperature decreases with altitude (the usual situation), sound waves initially traveling at a small angle with the horizontal are refracted upward. A sound out of doors is thus not normally audible at a great distance. However, if there is a temperature inversion (as over a body of water on a calm sunny day), the waves would be refracted downward. This is the main reason that sound carries long distances across water on a calm day. On a windy day the horizontal temperature strata are broken up and the sound is dissipated.

Refraction accompanied by reflection accounts for the fact that large explosions are sometimes heard in several distinct regions at surprisingly large distances, with zones of silence in between. A temperature inversion at high levels can refract the waves downward into a zone of audibility. The sound is then reflected from the ground, and must again be refracted downward to give the next zone of audibility. *See* SOUND; WAVE MOTION IN FLUIDS.

Seismic waves. The velocity of elastic waves in a solid depends upon the modulus of elasticity and upon the density of the material. Waves propagating through solid earth are refracted by changes of material or changes of density. Worldwide observations of earthquake waves enable scientists to draw conclusions on the distribution of density within the Earth. These waves may be totally internally reflected at the boundary of the core. It was through such observations that the existence of the much denser core of the Earth was first postulated.

Refraction of compressional waves from explosions set off on the ground is (combined with reflection) used in prospecting for oil, natural gas, and minerals which have large differences in density and elastic constants from the surrounding rocks.

Water waves. The speed of water waves in shallow water is proportional to the square root of the depth. As the waves enter shallower water they travel more slowly. As a train of waves approaches a coastline obliquely, its direction of travel becomes more nearly perpendicular to the shore because of refraction. *See* GEOMETRICAL OPTICS; WAVE MOTION IN LIQUIDS.

[JOHN W. STEWART]

Bibliography: M. Born and E. Wolf, *Principles of Optics*, 6th ed., 1980; D. Halliday and R. Resnick, *Physics*, 3d ed., 1978; F. A. Jenkins and H. E. White, *Fundamentals of Optics*, 4th ed., 1976.

Regge pole

Pole singularity of the scattering amplitude in the complex angular momentum plane. The concept of Regge pole arises in the theory of nonrelativistic two-body quantum dynamics. To understand the rather complex ideas which follow, it is necessary to begin by reviewing a few elements of potential scattering. *See* NONRELATIVISTIC QUANTUM THEORY.

Theory. A system composed of two elementary spinless particles is described by the wave function $\psi(\vec{x})$ ($\vec{x}$ is the relative distance between the two particles), which is the solution, with appropriate asymptotic conditions, of the Schrödinger equation, Eq. (1), where $\hbar$ is the Planck constant, M the reduced mass, and U the potential.

$$-\frac{\hbar^2}{2M} \Delta \psi(\vec{x}) + U(\vec{x}) \psi(\vec{x}) = k^2 \psi(\vec{x}) \quad (1)$$

For the sake of simplicity assume $U(\vec{x}) = U(x)$.

The dynamics of the system is fully specified by the interaction potential U.

The solution ψ, which describes the collision of a plane wave with the scattering center, satisfies the asymptotic conditions given by Eq. (2).

$$\psi(\vec{x}) \underset{x^{\text{large}}}{\longrightarrow} e^{i\vec{k} \cdot \vec{x}} + \frac{e^{ikx}}{x} F(k, \cos \theta) \quad (2)$$

$$\cos \theta = \frac{\vec{k} \cdot \vec{x}}{k\,x}$$

This formula, in conjunction with Eq. (1), defines the scattering amplitude F. The knowledge of F is equivalent to the knowledge of the Heisenberg S matrix. In turn the S matrix gives a complete description of any scattering process associated with Eq. (1). In order to compute F one needs to find an explicit solution of the Schrödinger equation. F is directly connected with the quantum flux of scattered particles according to well-known formula (3). This flux (differential cross section) is directly measurable in a scattering experiment. *See* SCATTERING MATRIX.

$$\frac{d\sigma}{d\Omega} = |F|^2 \quad (3)$$

The modern quantum theory of scattering has recognized the importance of analyzing the mathematical properties of the scattering amplitude for wide classes of potentials. The general goal is the study of the properties (analytic properties) of the amplitude $F(k, \cos \theta)$ defined in the complex plane of the variables k and $\cos \theta$. The extension to the complex domain provides the physicist with a comprehensive algorithm; general properties of classes of $U(x)$ reflect themselves in general analytic properties of the function F. In particular, singularities in the k plane represent definite fea-

tures of the physical system and they are directly related to the properties of $U(x)$. According to W. Heisenberg, a pole of F in the k plane can be associated with either a resonant state or a bound state of the system. Other analytic properties also admit related interpretation, but for the sake of simplicity they are not discussed here.

The theory is considerably simplified if the partial wave amplitude defined by Eq. (4) is used.

$$a_l(k) = \frac{1}{2} \int_{-1}^{1} F(k, \cos \theta) P_l(\cos \theta)\, d \cos \theta \quad (4)$$

Here the $P_l(\cos \theta)$ are the standard spherical harmonics.

The $a_l(k)$ depend on the angular momentum l instead of the angle θ. In this way the symmetry properties of the scattering amplitude under rotations are explicitly exhibited. *See* SYMMETRY LAWS.

The study of $a_l(k)$ is related to the solution of an ordinary differential equation, the radial Schrödinger equation, instead of the three-dimensional one.

The connections between the radial Schrödinger equation and the partial wave amplitude are expressed by Eqs. (5)–(7), where $\delta_l(k)$ is the well-

$$\phi''(x) + k^2 \phi(x) - \frac{l(l+1)}{x^2} \phi(x)$$
$$- U(x) \phi(x) = 0 \quad (5)$$

$$\lim_{x \to 0} \phi(x) = 0 \qquad \phi(x) \underset{x \to \infty}{=} \sin\left[kx - \frac{\pi l}{2} + \delta_l(k)\right] \quad (6)$$

$$a_l(k) = \frac{1}{k} e^{i\delta_l(k)} \sin \delta_l(k) \quad (7)$$

known phase shift. Given $a_l(k)$, the full $F(k, \cos \theta)$ can be reconstructed through the partial wave expansion in Eq. (8). It is obvious that this expan-

$$F(k, \cos \theta) = \sum_0^\infty (2l+1)\, a_l(k)\, P_l(\cos \theta) \quad (8)$$

sion is particularly useful when only a few partial waves are stimulated. This situation certainly arises when the energy $E = k^2$ is close to a pole of $a_l(k)$. If this is the case, the system is resonating in a state of angular momentum l, and the pole in the lth component in the expansion is the predominant contribution to the amplitude $F(k, \cos \theta)$.

T. Regge extends the study of the partial wave amplitude a_l by generalizing it to be a function of l as a complex variable. This extension is made by considering Eqs. (5)–(7) for complex values of l.

The study of the analytic properties of $a(l,k)$ for complex l is particularly interesting for the restricted class of Yukawian potentials given by Eq. (9). Indeed, these potentials yield amplitudes $a(l,k)$

$$U(x) = \int_m^\infty \frac{e^{-\mu x}}{x} \sigma(\mu)\, d\mu \quad (9)$$

with remarkably simple properties: In the right-hand plane Re $l \geq -1/2$, $a(l,k)$ is meromorphic; that is, it has only simple poles. These are called Regge poles. The position and residues of Regge poles are functions of the variable k^2. In the formula $a(l,k) \simeq \beta(k^2)/l - \alpha$ the function $\alpha(k^2)$ is called the Regge trajectory and $\beta(k^2)$ is called the Regge residue.

In the region Re $l < -1/2$ the analytic properties of $a(l,k)$ are complicated and depend on the interaction U in a somewhat unstable way.

Note that the values of $a(l,k)$ for noninteger l do

not enter directly into the Rayleigh-Faxen formula, Eq. (8).

By the Watson-Sommerfeld transformation a convenient substitute can be obtained for Eq. (8) which at the same time provides a physical interpretation of the theory.

Suppose that the amplitude $a(l,k)$ in Re $l > -1/2$ has only a simple pole for $l = \alpha(k^2)$, the Watson transformation implies that Eq. (10) is valid. The

$$
\begin{aligned}
F(k, &\cos\theta) \\
&= -i \int_{-1/2-i\infty}^{-1/2+i\infty} (2l+1)\,\frac{a(l,k)}{\sin \pi l}\,P_l(-\cos\theta) \\
&\quad + \frac{\beta(k^2)}{\sin \pi \alpha(k^2)}\,P_{\alpha(k^2)}(-\cos\theta) \quad (10)
\end{aligned}
$$

term $\beta P_\alpha/\sin \pi\alpha$ is the contribution to F of the pole of $a(l,k)$ at $l = \alpha$. The integral between $-1/2 - i\infty$ and $-1/2 + i\infty$ is called the background integral. It takes into account the contribution to $F(k,\cos\theta)$ of all the singularities of $a(l,k)$ on the left plane Re $l < -1/2$.

Consider the behavior of $F(k,\cos\theta)$ for $E = k^2$ real and increasing. The simple pole of $a(l,k)$ moves along the trajectory $\alpha(E)$. Suppose, moreover, that for E close to a value E_0 Eq. (11) holds,

$$
\alpha = n + \epsilon(E) + i\gamma(E) \quad (11)
$$

where n is integer and ϵ, γ are small real numbers. If ϵ, γ were zero at $E = E_0$, F would have a pole for $E = E_0$ of residue $\beta(E_0)\,P_n(\cos\theta)$. This would correspond to a bound state of physical angular momentum n. However, it should be noted that this phenomenon occurs only when $E < 0$.

In general, for $E > 0$, γ never vanishes, and whenever the ϵ, γ are small the scattering can be described rather as occurring via an intermediate resonance. Near the point $E = E_0$, where ϵ, $\gamma \ll 1$, Eq. (12) holds. Supposing $\epsilon(E_0) = 0$ one may keep

$$
a(l,E) \simeq \frac{\beta(E_0)}{l - \alpha(E)} \simeq \frac{\beta(E_0)}{l - n - \epsilon(E) - i\gamma(E)} \quad (12)
$$

the linear terms in $E - E_0$ only, in the denominator. From the general theory, γ and β can be taken as constant. It follows approximately that Eqs. (13) and (14) are valid. If $l = n$ it becomes obvious that

$$
l - n - \epsilon - i\gamma = l - n - (E - E_0)\frac{d\epsilon}{dE} - i\gamma \quad (13)
$$

$$
a(l,E) \simeq \frac{\beta}{l - n - (E - E_0)\dfrac{d\epsilon}{dE} - i\gamma} \quad (14)
$$

$a(l,E)$ has a complex pole in E, as shown in Eq. (15),

$$
E = E_0 - \frac{i\Gamma}{2} \quad (15)
$$

Γ being defined by Eq. (16). This corresponds to

$$
\Gamma = \frac{2\gamma}{\dfrac{d\epsilon}{dE}} \quad (16)
$$

the classical picture of a resonance of width Γ and life $\hbar/\Gamma$ at the energy E_0. If instead E is real, $a(l,E)$ is singular in l at a value given by Eq. (17), and for

$$
l = n + (E - E_0)\frac{d\epsilon}{dE} + i\gamma = n + \epsilon(E) + i\gamma \quad (17)
$$

$E = E_0$ the angular momentum has integer real part plus an imaginary part γ.

The angular life of the system can be considered to be $1/\gamma$. As E varies, $\alpha(E)$ may pass close to different integers, and therefore it may originate several different resonances. In this way resonances and bound states are grouped into families given by the same trajectory $\alpha(E)$.

As clearly seen by the Watson transformation in Eq. (10), knowledge of the family of resonances is not enough to reconstruct the full $F(k,\cos\theta)$ because of the contribution of the background integral. Nevertheless, the properties of the background integral are sufficiently known to allow one to extract from the inversion formula (10) the important asymptotic result in Eq. (18), which gives a

$$
F(k,\ \cos\theta)\underset{\cos\theta\to\infty}{\simeq} (\cos\theta)^{\alpha(E)} + O(\cos\theta^{-1/2}) \quad (18)
$$

precise prediction for the asymptotic behavior of $F(k,\cos\theta)$ in the large $\cos\theta$ limit (unphysical region) and which is important in the theory of potential scattering as a step in the proof of the Mandelstam representation.

Hypothesis. The Regge pole hypothesis can be stated as follows: The properties of the scattering amplitude which have been proved by Regge in potential theory represent a deep property of nature and are indeed true also for physical relativistic amplitudes. The consequences of the hypothesis are far-reaching.

The relativistic theory of strong interacting particles shares with potential theory the general concept of S matrix but does not have the counterpart of the Schrödinger equation.

It is generally agreed that elementary particles, bound states, and resonances are described by poles of the S-matrix elements as functions of the energy. Researchers have obtained experimental evidence for the existence of hundreds of new particles. As it is extremely difficult to establish a clear-cut distinction between elementary and compound particles, the very idea of elementary particles has become the target of mounting criticism. Heisenberg, and later G. F. Chew, proposed theories in which the idea of elementarity is altogether abolished. This is not the place to discuss the merit of such theories. The conjecture that the mechanism of Regge poles holds for relativistic amplitudes has provided physicists with a workable criterion by which it will be possible to reach a decision on whether elementary of a given particle exists. Moreover, the Regge pole hypothesis leads to extremely interesting predictions on the high-energy behavior of scattering processes. This statement is illustrated in the figure. Consider the scattering process $A + B \to C + D$ symbolized by the diagram. Here p_a, p_b, $-p_c$, $-p_d$ are the energy-momentum four vectors associated with A, B, C, D respectively. For simplicity assume $A, \ldots, D$ to have equal mass M and to be spinless. The invariants are defined by Eqs. (19), where $p^2 = -(\vec{p})^2 + E_p{}^2$.

$$
(p_a + p_b)^2 = s \qquad (p_a + p_c)^2 = \mathrm{t} \quad (19)
$$

Also, $s = (p_c + p_d)^2$ because of the conservation of energy-momentum as in Eq. (20). The scatter-

$$
p_a + p_b + p_c + p_d = 0 \quad (20)
$$

ing amplitude will be a function of s, t, only, as in Eqs. (21).

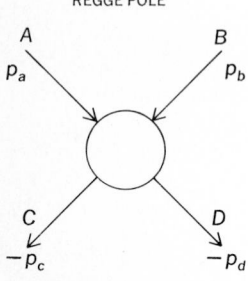

REGGE POLE

Scattering process.

$$A = A(s, t)$$

$$s = 4M^2 + k_s{}^2 \qquad (21)$$

$$t = -2k_s{}^2(1 - \cos\theta_s)$$

Here k_s, $\sqrt{s}$ are respectively the relative momentum and total energy in the center of mass system of A and B, and θ_s is the scattering angle. For a physical scattering process one has $s > 4M^2$, $t < 0$. Consider now the scattering process in relation (22), where $\overline{C}, \overline{B}$ are the antiparticles of C, B. Let

$$A + \overline{C} \rightarrow \overline{B} + D \qquad (22)$$

p'_a, $-p'_b$, p'_c, $-p'_d$ be the energy momentum vectors of the particles $A\overline{BC}D$ respectively, and let t' and s' be defined by Eqs. (23). The corresponding amplitude is written as $A'(s', t')$.

$$t' = (p'_a + p'_c)^2 \qquad (23)$$
$$s' = (p'_a + p'_b)^2$$

Notice that here t' plays the role of (square) energy in the CM system. It was long conjectured and subsequently proved by V. Glaser and coworkers that $A(s, t)$ and $A'(s', t')$ are actually the same analytic function in different domains, for the $A + B \rightarrow C + D$ process $s > 4M^2$, $t < 0$ and for $A + \overline{C} \rightarrow \overline{B} + D$, $t' \geq 4M^2$, $s' < 0$. Accordingly the distinction between A', s', t' and A, s, t is dropped. Suppose now that in the channel $A + \overline{C} \rightarrow \overline{B} + D$, a scattering angle θ_t is defined by Eq. (24). If the

$$s = -2k_t{}^2(1 - \cos\theta_t) \qquad (24)$$

Regge pole hypothesis holds, then for $\cos\theta_t \rightarrow \infty$ one should have $s \rightarrow \infty$ and $A(s,t) \cong \beta(t)s^{\alpha(t)}$, where $\alpha(t)$ is a pole in the partial scattering amplitude $a_l(t)$ as a function of t, l defined by Eq. (25). Although the region $s > 0$ is unphysical for the

$$A(s,t) = \int_{-\infty}^{\infty} \frac{dl(2l+1)}{\sin\pi l} a_l(t) P_l(-\cos\theta_t) \qquad (25)$$

though the region $s > 0$ is unphysical for the process $A + \overline{C} \rightarrow \overline{B} + D$, it is not so for $A + B \rightarrow C + D$, and this yields a definite prediction for the asymptotic behavior of $A(s,t)$ for large s and constant t of the kind given by Eq. (26).

$$A(s,t) \simeq \beta(t)\, s^{\alpha(t)} \qquad (26)$$

The Regge pole hypothesis essentially assumes that this behavior is valid for $t < 0$ and not only for $t > 4M^2$, as implied analogously by the $A + \overline{C} \rightarrow \overline{B} + D$ reaction.

In this sense the conjecture is a definite statement on the behavior of the scattering amplitude for constant transmitted momentum t and high energies. The exponent $\alpha(t)$ is tied to an angular momentum of an exchanged Regge pole of (mass)$^2 = t (t < 0)$. Since α is not integer in this region, the pole does not correspond to any physical particle. The trajectory $\alpha(t)$ may, however, assume integer values for $t > 0$, and here it may correspond to physical particles of angular momentum $\alpha(t)$ and (mass)$^2 = t$. In this way the outcome of high-energy scattering experiments may be related to $\alpha(t)$ and therefore to a family of particles. As the existence of a trajectory is inferred from the particular potential scattering mechanism, it is clear that a Regge pole ought always to be interpreted as a compound state.

If now a particle lies on a trajectory, there is a strong indication that the particle is not elemen-

tary. It has been conjectured by Chew and others that all particles lie on Regge trajectories (or that at least the strongly interacting particles do so) and that the concept of elementarity should be abandoned. Certainly, if the Regge pole hypothesis is correct, physicists will have at least one objective criterion for elementarity. Elementary particles will correspond in general to fixed poles, that is, poles of $a_l(t)$ whose position does not depend on t. *See* ELEMENTARY PARTICLE.

Considerable theoretical and experimental research has been carried out on basis of this conjecture, and although the basic elements for an axiomatic foundation of such a dynamics have not been formulated, from a pragmatic point of view much has been achieved. Much mathematical work has been done to clarify the concept of Regge pole contribution to high-energy scattering for any reaction involving spinning particles, and an equally great effort has been made in the statistical analysis of the experimental data in this scheme by using the most powerful computers available.

It is hoped that this great amount of work will be the basis for a future theory. *See* DISPERSION RELATIONS; SCATTERING EXPERIMENTS (ATOMS AND MOLECULES); SCATTERING EXPERIMENTS (NUCLEI). [LUIGI SERTORIO]

Bibliography: P. D. Collins, *An Introduction to Regge Theory and High Energy Physics*, 1977; R. J. Eden et al., *The Analytic S-matrix*, 1966; R. Omnes and M. Froissart, *Mandlestam Theory and Regge Poles: An Introduction for Experimentalists*, 1963.

Relative atomic mass

The ratio of the average mass per atom of the natural nuclidic composition of an element to 1/12 of the mass of an atom of nuclide ^{12}C. For example, $\mu(Cl) = 35.453$. Relative atomic mass replaces the concept of atomic weight. It is also known as relative nuclidic mass. *See* NUCLIDE.

[THOMAS C. WADDINGTON]

Relative molecular mass

The ratio of the average mass per formula unit of the natural nuclidic composition of a substance to 1/12 of the mass of an atom of nuclide ^{12}C. For example, $\mu(KCl) = 74.555$. Relative molecular mass replaces the concept of molecular weight. *See* NUCLIDE. [THOMAS C. WADDINGTON]

Relative motion

All motion is relative to some frame of reference. The simplest laboratory frame of reference is three mutually perpendicular axes at rest with respect to an observer. Such a system is commonly used in the laboratory when various types of motion are being studied. The general effects of other motions to which the system as a whole is subjected are then neglected and the system is said to be isolated. In terms of the frame of reference of an observer some distance from Earth, the laboratory frame of reference would be moving with Earth as it rotates on its axis and as it revolves about the Sun. What would be a simple form of motion in the laboratory frame of reference would appear to be a much more complicated motion in the frame of reference of the distant observer. *See* FRAME OF REFERENCE.

Motion means continuous change of position of

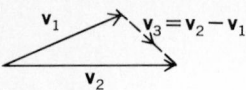

Fig. 1. Relative velocity of two bodies moving at an angle with respect to one another.

an object with respect to an observer. To another observer in a different frame of reference the object may not be moving at all, or it may be moving in an entirely different manner. The motions of the planets were found in ancient times to appear quite complicated in the laboratory frame of reference of an observer on Earth. By transferring to the frame of reference of an imaginary observer on the Sun, Johannes Kepler showed that the relative motion of the planets could be simply described in terms of elliptical orbits. The validity of one description is no greater than the other, but the latter description is far more convenient.

Relative velocity. That motion is relative to an observer must have been implicit in the earliest ideas of motion. In the mechanics of Galileo and Isaac Newton these ideas became clarified, and methods were developed for finding the relative velocity of two bodies, each moving with a different velocity. If the velocity of one body is v_1, represented in Fig. 1 by the magnitude and direction of the vector v_1, and if a second body has a velocity v_2 represented by the vector v_2, then v_3 is the vector to be added to v_1 to make the sum equal to v_2, and, consequently, the vector v_3 is equal to the vector difference $v_2 - v_1$. Therefore, the relative velocity of the second body with respect to the first is that velocity represented in magnitude and direction by the vector v_3, all vectors being drawn to a suitable scale.

In the simplest case the velocities v_1 and v_2 are parallel (Fig. 2). The relative velocity of the second body with respect to the first is again v_3, and its

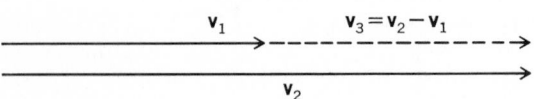

Fig. 2. Relative velocity of two bodies moving in the same direction; that is, velocities are parallel.

magnitude is the difference in the numerical values of v_2 and v_1. When the two velocities are antiparallel (in opposite directions, Fig. 3), one is negative with respect to the other, and the magnitude of the relative velocity of v_2 with respect to v_1 is again the vector difference v_3, which is now the numerical sum of the two. To obtain the relative velocity of v_1 with respect to v_2, the arrowhead on v_3 should be reversed. For example, when two automobiles move with velocities of 20 and 30 m/s respectively in the frame of reference of an observer stationed by the roadside, the relative velocity of the two is 10 m/s. If they were moving in opposite directions, the relative velocity would be 50 m/s.

The flight of an airplane illustrates the principles of relative motion. The common statement that the speed of an airplane is, for instance, 150 m/s is essentially a meaningless statement since the speed is not designated with any particular

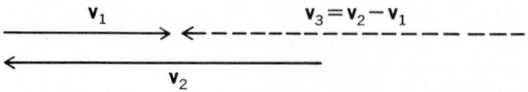

Fig. 3. Relative velocity of two bodies moving in opposite directions; that is, velocities are antiparallel.

frame of reference in mind and the listener must make an assumption that it is perhaps the ground speed in still air. For the airplane to sustain itself with normal lift, it is the airspeed or speed with respect to the air that is important. In a head wind of 50 m/s the airplane would have a ground speed of 100 m/s. In a similar tail wind it would have a ground speed of 200 m/s. If the wind velocity is at an angle to the direction of flight, the relative velocities must be considered. The velocity with respect to the ground would be the vector sum of the velocity of the plane with respect to the air and the velocity of the wind with respect to the ground.

Relative acceleration. Acceleration, like velocity, is relative to the observer's frame of reference. An automobile starting from rest is accelerated with respect to the Earth, but the driver of the car does not see himself or the car accelerated forward. He sees objects at rest with respect to the roadway accelerated backward with respect to himself and the car. Persons by the roadside see him and the car accelerated forward. In the driver's frame of reference, the car is at rest.

Acceleration is a vector quantity involving both magnitude and direction, just as velocity is. Just as the velocities of two objects may be represented by vectors and their relative velocity obtained by subtracting one vector from the other, so may the accelerations of two bodies be represented by vectors and the relative acceleration of one with respect to the other obtained by taking the vector difference.

Relativity. Since, according to Einstein's theory of relativity, the velocity of light is the limiting velocity that any physical object can attain, the addition or subtraction of very large velocities cannot be accomplished by Galilean-Newtonian methods, and the rules to be followed are derived from relativistic theory. *See* RELATIVITY.

[ROGERS D. RUSK]

Relativistic electrodynamics

The study of the interaction between charged particles and electric and magnetic fields when the velocities of the particles approach that of light. Relativistic electrodynamics, which can be considered as an extension of the everyday laws of electricity and magnetism, is an important consideration in high-energy particle accelerators, in high-current, high-voltage vacuum tubes, and in electromagnetic radiation.

The laws which relate the electric and magnetic fields to the charges and currents which produce them are known as Maxwell's equations. Charged particles or current elements in such fields experience a force which is called a Lorentz force. The motion of these charged particles and current elements in such fields is then determined by Newton's laws, appropriately generalized for relativistic velocities. *See* MAXWELL'S EQUATIONS; RELATIVISTIC MECHANICS; RELATIVITY.

J. C. Maxwell's contribution to relativistic electrodynamics was to formulate his four equations and to introduce the concept of displacement current. Although each equation was originally deduced for static or other restrictive conditions, Maxwell implied the validity of each for fields which varied in arbitrary ways with time. *See* DISPLACEMENT CURRENT.

The Lorentz force describing the influence of

the electric field $\mathbf{E}$ and magnetic field $\mathbf{B}$ on a moving charge q of velocity $\mathbf{v}$ is given by Eqs. (1),

$$\mathbf{F} = q(\mathbf{E} + \mathbf{v} \times \mathbf{B})$$

$$F_x = q(E_x + v_y B_z - v_z B_y), \text{ etc.}$$

(1)

which, together with the equation of continuity, Eqs. (2), and Newton's second law, Eqs. (3), supply

$$\text{div } \mathbf{J} + \frac{\partial \rho}{\partial t} = \text{div } (\rho \mathbf{v}) + \frac{\partial \rho}{\partial t} = 0$$

(2)

or

$$\frac{\partial J_x}{\partial x} + \frac{\partial J_y}{\partial y} + \frac{\partial J_z}{\partial z} + \frac{\partial \rho}{\partial t} = 0$$

$$\mathbf{F} = \frac{d\mathbf{p}}{dt} = \frac{d}{dt}\left(\frac{m_0 \mathbf{v}}{\sqrt{1 - \dfrac{v^2}{c^2}}}\right)$$

(3)

or

$$F_x = \frac{dp_x}{dt} = \frac{d}{dt}\left(\frac{m_0 v_x}{\sqrt{1 - \dfrac{v^2}{c^2}}}\right), \text{ etc.}$$

a self-consistent set of equations to determine the fields and motions of charges under various conditions. In the preceding equations, ρ is the electric charge density, $\mathbf{J}$ is the current density, $\mathbf{p}$ is the momentum of the charge, m_0 is the rest mass of the charge, and c is the velocity of light. *See* EQUATION OF CONTINUITY.

One direct consequence of Maxwell's equations comes from the fact that the fields $\mathbf{E}$ and $\mathbf{B}$, in the absence of charges and currents, satisfy a wave equation with wave velocity given by Eq. (4). The

$$c = \frac{1}{\sqrt{\epsilon_0 \mu_0}} \approx 3 \times 10^8 \text{ m/sec}$$

(4)

fact that this is the same as the velocity of light led Maxwell to infer that light, like electricity, magnetism, and optics, is an electromagnetic wave phenomenon.

Invariance of Maxwell equations. The property of the Maxwell equations that makes them applicable to problems in relativistic electrodynamics is their relativistic invariance. Specifically, this means that Maxwell's equations will seem correct to an observer traveling with a constant velocity as well as to an observer at rest. One must realize, however, that a magnetic field in the rest frame will appear to be both an electric and a magnetic field in the moving frame. *See* FRAME OF REFERENCE.

As an example, consider a charge moving through an externally applied static electric and magnetic field with a velocity such that the Lorentz force vanishes. In this case Eqs. (5) hold. In a

$$\mathbf{F} = q(\mathbf{E} + \mathbf{v} \times \mathbf{B}) = 0 \qquad \mathbf{E} = -\mathbf{v} \times \mathbf{B}$$

(5)

frame of reference moving with velocity $\mathbf{v}$, the charge will seem to be at rest and will therefore experience a force $\mathbf{F}' = q\mathbf{E}'$, responding only to the electric field in the new system. Since the electron now remains at rest, the force $\mathbf{F}'$, and therefore $\mathbf{E}'$, must vanish. (A magnetic field $\mathbf{B}'$ will still be present.) It is therefore apparent that the electric and magnetic fields must change when a Lorentz transformation is made. The fields mix in such a way as to make Maxwell's equations appear to be the same to all observers. *See* LORENTZ TRANSFORMATIONS.

Relativistic beams. One interesting consequence of the way in which the fields transform is the diminution of the repulsive force between charges moving with high velocity in parallel paths. In the reference system moving in the z direction with velocity v, in which the charges are at rest, the magnetic field vanishes and the electric field of one at the location of the other is given by Eq. (6),

$$E'_y = \frac{q}{y'^2}$$

(6)

where y' is the separation of the charges. In the rest system, the fields can be shown to transform to the form of Eqs. (7), so that the force between

$$E_y = \frac{E'_y}{\sqrt{1 - \dfrac{v^2}{c^2}}} \qquad B_x = -\frac{v}{c^2}\frac{E'_y}{\sqrt{1 - \dfrac{v^2}{c^2}}} \qquad y' = y \quad (7)$$

the charges is given by Eq. (8). The force is thus

$$F_y = q(E_y + vB_x)$$

$$= qE_y\left(1 - \frac{v^2}{c^2}\right) = qE'_y \sqrt{1 - \frac{v^2}{c^2}} \quad (8)$$

reduced by a factor $1 - (v^2/c^2)$ compared with the electric force alone, and by a factor $[1 - (v^2/c^2)]^{1/2}$ compared with the electric force which exists when the charges are at rest.

Another way of illustrating this cancellation is to consider the force on the outer charges in a cylindrical beam of current I, consisting of charges moving with velocity v in the z direction. The tangential magnetic field at the surface ($r = a$) is, by Ampère's law, given by Eq. (9), where μ_0 is the

$$B_\theta = \frac{2\mu_0 I}{a}$$

(9)

magnetic inductive capacity of free space $\approx 1.257 \times 10^{-6}$ henry/m. Continuity of charge requires the electric charge density per unit length to satisfy Eq. (10), in which case the radial electric field

$$\tau = \frac{I}{v}$$

(10)

at the surface is, by Gauss' law, given by Eq. (11),

$$E_r = \frac{2I}{\epsilon_0 va}$$

(11)

where ϵ_0 is the electric inductive capacity of free space $\approx 8.85 \times 10^{-12}$ farad/m. The force on a charged particle at the surface is therefore given by Eq. (12), again confirming the almost complete

$$F_r = q(E_r - vB_\theta) = \frac{2Iq}{\epsilon_0 av}\left(1 - \frac{v^2}{c^2}\right) \quad (12)$$

cancellation of the electric field by the magnetic field for velocities near that of light. A further significant effect has to do with the fact that the transverse motion (considered to be slow compared to the longitudinal or z motion) of this charge configuration is determined from Eq. (13a), where

$$\frac{d^2r}{dt^2} = \frac{F_r}{m_t} = \frac{F_r}{m_0}\left(1 - \frac{v^2}{c^2}\right)^{1/2} = \frac{2Iq}{\epsilon_0 m_0 av}\left(1 - \frac{v^2}{c^2}\right)^{3/2} \quad (13a)$$

$$m_t = m_0\left(1 - \frac{v^2}{c^2}\right)^{-1/2} \quad (13b)$$

Relativistic form of common equations

Effect	Nonrelativistic form	Relativistic form
1. Acceleration of charge through potential difference V	$E_{kin} = \dfrac{m_0 v^2}{2} = qV$	$E_{kin} = m_0 c^2 \left[\dfrac{1}{\sqrt{1 - \dfrac{v^2}{c^2}}} - 1 \right] = qV$
	$v = \sqrt{\dfrac{2E_{kin}}{m_0}} = \sqrt{\dfrac{2qV}{m_0}}$	$v = \sqrt{\dfrac{2E_{kin}\left(1 + \dfrac{E_{kin}}{2m_0 c^2}\right)}{m_0\left(1 + \dfrac{E_{kin}}{m_0 c^2}\right)}} = \sqrt{\dfrac{2qV}{m_0}} \sqrt{\dfrac{1 + \dfrac{qV}{2m_0 c^2}}{1 + \dfrac{qV}{m_0 c^2}}}$
	$p = m_0 v = \sqrt{2m_0 E_{kin}}$ $= \sqrt{2m_0 qV}$	$p = \dfrac{m_0 v}{\sqrt{1 - \dfrac{v^2}{c^2}}} = \sqrt{2m_0 E_{kin}\left(1 + \dfrac{E_{kin}}{2m_0 c^2}\right)}$ $= \sqrt{2m_0 qV}\,\sqrt{1 + \dfrac{qV}{2m_0 c^2}}$
2. Circular motion in a uniform magnetic field	$\dfrac{m_0 v^2}{r} = qvB$	$\dfrac{m_0 v^2}{r\sqrt{1 - \dfrac{v^2}{c^2}}} = qvB$
	$p = m_0 v \times qBr$	$p = \dfrac{m_0 v}{\sqrt{1 - \dfrac{v^2}{c^2}}} = qBr$
	$\omega = \dfrac{v}{r} = \dfrac{qB}{m_0}$	$\omega = \dfrac{v}{r} = \dfrac{qB}{m_0}\sqrt{1 - \dfrac{v^2}{c^2}} = \dfrac{qBc}{\sqrt{q^2 B^2 r^2 + m_0^2 c^2}}$
	$E_{kin} = \dfrac{q^2 B^2 r^2}{2m_0}$	$E_{kin} = \sqrt{q^2 B^2 r^2 c^2 + m_0^2 c^4} - m_0 c^2$
3. Langmuir-Child law	$J = \dfrac{4\epsilon_0}{9}\sqrt{\dfrac{2q}{m_0}}\dfrac{V^{3/2}}{L^2}$	$J \approx \dfrac{4\epsilon_0}{9}\sqrt{\dfrac{2q}{m_0}}\dfrac{V^{3/2}}{L^2}\left(1 - \dfrac{3}{28}\dfrac{qV}{m_0 c^2}\right)$

notation (13b) gives the relativistic mass of the charged particle. The radial motion is thus reduced even further from the value it would have if the charges were at rest.

An important application of this effect is to relativistic beams of particles which might be expected to disperse as a result of space-charge forces. In linear electron accelerators, for example, this transverse divergence can be neglected because of the small value of $[1 - (v^2/c^2)]^{3/2}$ occurring in Eq. (13). Another way of looking at the phenomenon is that in the rest system of the electrons, the accelerator appears extremely short because of the Lorentz contraction, and the space-charge forces have little time to spread the beam apart. (The argument is actually more complicated since one must deal with an accelerated frame of reference, but the conclusion is the same.)

The tendency for relativistically charged beams not to diverge is of course not restricted to accelerators. It is important in consideration of electron optics in high-current, high-voltage vacuum tubes, and may even be important in controlled fusion devices. Applications in which the transverse motion of ion beams is made convergent by introducing an opposing beam of electrons have even been considered. Partial charge neutralization results in a decrease of the electric defocusing force but

leaves the magnetic focusing force of almost the same magnitude unchanged, thus providing a considerable net focusing action. In addition, it is often possible to diminish the space-charge defocusing of positive ion beams by space-charge neutralization using electrons.

Other relativistic phenomena. As a result primarily of relativistic dynamics, many of the formulas for interaction of particles and fields have been altered. Some of these altered expressions are listed in the table.

The significance of the phenomena included in the table lies mainly in the fact that the relativistic treatment gives a correction, small but not negligible, to the usual nonrelativistic situations in which these phenomena are encountered. There are many other applications in which the relativistic aspects are not small corrections but are indeed the main considerations involved. Included in this group are most of the particle accelerators with energies in the relativistic range (for example, electrons with energy above 1 Mev, protons with energy above 1 Bev).

Linear particle accelerators. Most electron linear accelerators produce extremely relativistic electrons (more than 500 Mev in the traveling-wave accelerator at Stanford University, compared with the rest energy of 1/2 Mev). The considerations regarding the transverse motion of the electrons have already been described: The magnetic interaction between the moving electrons reduces the repulsive force by several orders of magnitude, and no other means of controlling the transverse motion is necessary. The longitudinal motion is also inhibited by the large mass of the electrons. Indeed, the phase oscillations which are responsible for longitudinal stability become extremely slow; since all electrons are effectively traveling at the same velocity c, they neither fall behind nor overtake one another and are therefore less inclined to get out of phase. In fact, the phase motions are inversely proportional to the longitudinal mass, so called because the acceleration is, in this case, in the direction of the initial velocity. The relationship is given by Eq. (14), giving a value which is extremely large in the relativistic range.

$$m_l = \frac{m_0}{\left(1 - \dfrac{v^2}{c^2}\right)^{3/2}} \qquad (14)$$

Similar consideration applies to relativistic linear ion accelerators (above 1 Bev), although these of necessity have a large section in which nonrelativistic phenomena predominate since conventional ion sources yield particles in the Mev range.

Circular particle accelerators. The cyclotron is basically a nonrelativistic device with a uniform field in which a fixed radio frequency accelerates particles according to the relation in Eq. (15). As

$$\omega = \frac{qB}{m} = \frac{qB}{m_0}\sqrt{1 - \frac{v^2}{c^2}} \qquad (15)$$

long as m is nonrelativistic ($\cong m_0$), the rotational frequency $\omega_{\text{rot}}/2\pi$ of the particles is the same as the applied radio frequency, and they are accelerated. Once the particles become relativistic, however, ω_{rot} and ω_{rf} are no longer the same; to accelerate the particles beyond this region, either the frequency must be made to decrease with time (synchrocyclotron) or the fields must be made to increase with radius (one of the features of the fixed field alternating gradient, or FFAG, cyclotron).

The synchrotron (both electron and proton) is a device which has a magnetic field varying in time in such a way as to keep the orbit radius constant. Specifically, Eqs. (16) give the necessary relation

$$B = \frac{mv}{qr} = \frac{m_0\omega}{q}\left(1 - \frac{\omega^2 r^2}{c^2}\right)^{-1/2} \qquad \omega = \frac{v}{r} \quad (16)$$

between the magnetic field and the frequency for a given radius. In the case of relativistic electrons, $v \approx c$, and only the magnetic field need be varied to accelerate the particle.

A betatron is an electron accelerator in which orbits of constant radius are maintained without radio-frequency fields. The electrons are accelerated by an increase of magnetic flux linking the orbit; the relation between the linked flux and the field at the orbit may be derived from Faraday's law of induction and from the relativistic relation for circular motion. Faraday's law is given by Eq. (17), which indicates that the particle gains a kinetic energy, given by Eq. (18), during each turn. The

$$\text{emf} = -\frac{d\Phi}{dt} \qquad (17)$$

$$\Delta E_{\text{kin}} = q\frac{\Delta\Phi}{\Delta t} = \frac{qv}{2\pi r}\Delta\Phi \qquad (18)$$

symbol Δ refers to the change per turn, and Φ is the flux linked by the orbit. The particle will remain in the same orbit provided the magnetic field and momentum increase in such a way that Eq. (19) holds. It is also possible to derive Eq. (20). This

$$\Delta p = qr\,\Delta B \qquad (19)$$

$$\Delta E_{\text{kin}} = F\,\Delta x = \frac{\Delta p}{\Delta t}\Delta x = \frac{\Delta x}{\Delta t}\Delta p = v\,\Delta p \quad (20)$$

condition between the increase of energy and momentum per turn can therefore be met only if Eq. (21) holds, that is, if the change in the average field enclosed by the orbit is twice the change of the field at the orbit. Equation (21) is known as the betatron condition. For additional

$$\frac{\Delta\Phi}{\pi r^2} = 2\,\Delta B \qquad (21)$$

information on the betatron condition and detailed information on particle motion in all the important types of particle accelerators *see* PARTICLE ACCELERATOR.

Electromagnetic radiation. Another important area of application for relativistic phenomena is in the field of electromagnetic radiation. In fact, the phenomenon of radiation itself involves a complete relativistic treatment, as can be safely assumed from the appearance of c, the velocity of light, in all relevant formulas.

When one considers a configuration of current sources varying sinusoidally in time, the low-frequency solution corresponds to the coherent superposition of the radiation fields from each element of exciting current. As the frequency is increased, an interesting relativistic aspect of the problem appears: Disturbances from different current elements take different times to reach the point of observation, and the relative phase of these contributions to the total signal is changed. One can take this retardation into account by adding to the phase an amount $kr_{c,o}$ where $r_{c,o}$ is the distance between the element of current under consideration and the observation point. The wave number k is given, in terms of the frequency, by Eq. (22), where λ is the wavelength of the radia-

$$k = 2\pi f/c = 2\pi/\lambda \qquad (22)$$

tion. Clearly, this effect is important if the phase difference between opposite extremes of the current distribution is comparable to or greater than 2π, that is, if $kD \gtrsim 2\pi$ or $D \gtrsim \lambda$. Here D is a typical linear dimension of the current distribution. In general, retardation effects must be taken into account if the dimensions of the radiating system are comparable to or greater than the wavelength.

Synchrotron radiation. As a further example of a purely relativistic phenomenon, consider the radiation of a charged particle which undergoes acceleration during circular motion at constant speed. The classical Larmor formula for slowly moving charges gives Eq. (23) for the rate of energy

$$R = \frac{2}{3} \frac{q^2}{4\pi\epsilon_0} \frac{a^2}{c^3} \qquad (23)$$

radiation per second. Here a is the particle acceleration. The appropriate relativistic generalization in the case of circular motion is given by Eq. (24), where $a = v^2/r$, r being the orbit radius.

$$R = \frac{2}{3} \frac{q^2}{4\pi\epsilon_0} \frac{a^2}{c^3} \left(1 - \frac{v^2}{c^2}\right)^{-4} \qquad (24)$$

One can see directly the radical change for relativistic velocities. This radiation, called synchrotron radiation, represents a serious limitation on the magnitude of energies attainable in circular electron accelerators.   [ROBERT L. GLUCKSTERN]

Bibliography: J. D. Jackson, *Classical Electrodynamics*, 2d ed., 1975; E. Konopinski, *Electromagnetic Fields and Relativistic Particles*, 1981; W. K. H. Panofsky and M. Phillips, *Classical Electricity and Magnetism*, 2d ed., 1962; J. C. Slater and N. H. Frank, *Electromagnetism*, reprint 1969; A. Sommerfeld, *Electrodynamics*, vol. 3, 1952; J. A. Stratton, *Electromagnetic Theory*, 1941.

Relativistic mechanics

An extension of Newtonian mechanics conforming to the principles of special relativity. Suitably generalized, its results are also incorporated in the general theory. The energy-momentum conservation laws of relativistic mechanics enter in the development of relativistic quantum mechanics, and they find important application in high-energy physics.

Of Newton's laws of motion, it is the third law whose relativistic formulation presents particular challenge. Its application is simplest in the idealized context of strictly contiguous action, whether in continuous media or in discrete particle systems—the only case discussed here. Without this assumption, one must cope with the problem of retarded action at a distance, on which much exploratory work has been done. *See* NEWTON'S LAWS OF MOTION; RELATIVITY.

Equations of motion of a particle. The first correct relativistic equation of motion was that of a charged particle in an electromagnetic field developed by Max Planck in 1906. The restriction on the applied force was later removed by Hermann Minkowski, who was also the first to use the four-dimensional space-time formalism systematically. This formalism greatly facilitates a concise discussion of the subject, and will be employed here from the start. *See* RELATIVISTIC ELECTRODYNAMICS; SPACE-TIME.

The relativistic generalization of Newton's second law of motion for a particle, Eq. (1), must satisfy

$$m\mathbf{a} = \mathbf{f} \qquad (1)$$

fy two conditions: it must include the Newtonian equation as a limiting case for small particle speeds; it must be covariant (that is, retain its form) under Lorentz transformations, which is automatically assured when it is expressed in four-vector form. As to Newton's first law, its relativistic extension is immediate, because it serves only to define the family of inertial frames relative to which the second law is enunciated, and that family plays an identical role in the special theory of relativity. *See* LORENTZ TRANSFORMATIONS.

To arrive at the relativistic generalization of Eq. (1), it is first rewritten, using the particle momentum, $\mathbf{p} = m \, d\mathbf{x}/dt$, as $d\mathbf{p}/dt = \mathbf{f}$ (which is actually a closer representation of the law as stated in Newton's *Principia*). One then replaces dx, dy, dz by dx^α ($\alpha = 0,1,2,3$; $x^0 \equiv ct$), m by the rest mass m_0 (that is, the mass of the particle as measured in the reference frame in which it is instantaneously at rest), and dt by its four-scalar extension $d\tau = dt\sqrt{1 - v^2/c^2}$. Retaining, as is customary, the symbol p for the generalization of $\mathbf{p}$, so that $p^\alpha = m_0 \, dx^\alpha/d\tau$ ($\alpha = 0,1,2,3$), the resulting four-vector equation assumes the form of Eq. (2), which may be called Minkowski's equation of motion.

$$dp^\alpha/d\tau = F^\alpha \qquad (\alpha = 0,1,2,3) \qquad (2)$$

The significance of Eq. (2) becomes apparent when the time t is reintroduced, resulting in Eqs. (3) and (4), where v is the instantaneous speed of

$$dp^\alpha/dt = f^\alpha$$
$$p^\alpha = m \, dx^\alpha/dt \equiv mv^\alpha \qquad (\alpha = 0,1,2,3) \qquad (3)$$

$$m = \gamma m_0$$
$$f^\alpha = F^\alpha/\gamma \qquad (\alpha = 0,1,2,3) \qquad (4)$$
$$\gamma \equiv 1/\sqrt{1 - v^2/c^2}$$

the particle. Since $\gamma = 1 + v^2/2c^2 + \ldots$, it follows that when v^2/c^2 can be neglected, Eqs. (3) for $\alpha = 1,2,3$ coincide indeed with the corresponding Newtonian equations. Moreover, it is F^α and not f^α that are the components of a four-vector. This four-vector generalization of the Newtonian force is known as the Minkowski force.

The meaning of Eqs. (3) for $\alpha = 0$ can be established with the aid of the four-scalar equation (5)

$$(p^0)^2 - \mathbf{p}^2 = m_0^2 c^2 \qquad (5)$$

derivable from Eqs. (4) and (3), where $\mathbf{p} \equiv (p^1, p^2, p^3)$ (not to be confused with the previous $\mathbf{p}$). Differentiating this equation with respect to t and using Eqs. 3, one finds: $d(mc^2)/dt = cf^0 = \mathbf{f} \cdot \mathbf{v}$. Hence the quantity defined by Eq. (6) is an energy, and the four-vector p, the four-momentum, is thus also commonly called the energy-momentum vector. By Eqs. (5) and (6), it follows that E can also be expressed in terms of the "relativistic" $\mathbf{p}$ as in Eq. (7). In addition to E, which is usually called the

$$E = cp^0 = mc^2 \qquad (6)$$
$$E = \sqrt{m_0^2 c^4 + c^2 \mathbf{p}^2} \qquad (7)$$

total energy, it is useful to introduce a relativistic kinetic energy (vanishing with v): $m_0 c^2 (\gamma - 1)$, where γ is defined in Eqs. (4). The expansion of γ shows that when v^2/c^2 can be neglected, this expression reduces indeed to the Newtonian kinetic energy $mv^2/2$.

When $m_0 \neq 0$, Eqs. (4) and (6) show that for $v = 0$ the energy E has the nonvanishing value $E_0 = m_0 c^2$. This rest energy is thus an intrinsic energy of a particle associated entirely with its rest mass. The universality of this inertia-of-energy result, as established principally by Albert Einstein, and extensively confirmed by experiment, constitutes one of the most far-reaching conclusions of the theory of relativity. *See* INERTIA OF ENERGY.

Another important conclusion from Eqs. (4),

which accords with other deductions from the relativistic principles, is that a particle of nonvanishing rest mass may approach but can never attain the speed c. On the other hand, the speed of a particle of vanishing rest mass, such as a photon, can only be c; otherwise, contrary to theory and experience, the particle's energy and momentum would be zero, as is seen, for instance, by taking the limit $m_0 \rightarrow 0$ in Eqs. (3). Equation (7) shows, moreover, that the energy and momentum of such a particle are related by the simple equation $E = c|\mathbf{p}|$; and also, that when $m_0 \neq 0$ this relation holds as an ultra-high-energy approximation ($m_0 c \ll |\mathbf{p}|$).

Special systems of particles. Under the restrictive assumption that the particles of the system are involved only in spatially and temporally localized interactions, it follows easily that Newton's third law can be formulated in a relativistically covariant fashion. Moreover, if the system is closed, that is, subject to no external action, then as in the case of Newtonian mechanics the total energy and momentum of the system are conserved quantities: $\Sigma E =$ constant, $\Sigma \mathbf{p} =$ constant. Or, in four-vector notation Eq. (8) is satisfied.

$$P^\alpha \equiv \Sigma p^\alpha = \text{constant} \qquad (8)$$
$(\alpha = 0,1,2,3;$ summation over the particles$)$

It is possible to extend this result to allow for the (quantum) processes of the creation and annihilation of particles, so that the number of particles of the system need not be conserved, and the summation symbol in Eq. (8) refers to the particles existing at the instant under consideration. In particular, for a particle decay process or a localized binary collision of any type, Eq. (8) can be written more explicitly as Eq. (9), with $m = 1$ or 2, respec-

$$\sum_{i=1}^{m} p_i^\alpha \text{ (initial)} = \sum_{j=1}^{n} p_j^\alpha \text{ (final)} \qquad (9)$$

tively. An early and historically very important application of this formula was made by Arthur Compton in 1922 to the collision of an x-ray photon and a practically free electron. The extended formula has been applied to numerous high-energy particle reactions, and it appears to represent one of the most firmly established laws in physics. *See* COMPTON EFFECT.

As in the treatment of low-energy collisions, so also in the treatment of high-energy particle reactions it is helpful to deal not only with the laboratory frame but also with the center-of-mass (or center-of-momentum) frame of the mechanical system under consideration. The latter is defined by the condition that the spatial part of the total four-momentum of the system vanishes. In addition to the usual advantages of the center-of-mass frame, it is, in a practical sense, especially useful in the case of very-high-energy collisions. With the aid of the Lorentz transformations connecting the components of the total four-momentum P defined by Eq. (8) in the laboratory and center-of-mass frames, it can be shown that if instead of bombarding some type of particles, which are effectively stationary in the laboratory, by a beam of similar particles, one directs two such beams against each other so that for the latter system the laboratory serves as a center-of-mass frame, the energy released in the collision is substantially greater in the second case, the advantage increasing with the

rise in particle energy. This has been utilized in positron-electron and proton-proton colliding-beam devices called storage rings. *See* PARTICLE ACCELERATOR.

The center-of-mass frame has also independent interest. If it is denoted by S_0, and S is an arbitrary inertial frame, it can be shown from Lorentz transformations of the total four-momentum that $\mathbf{P} = \gamma P^0_{(0)}\mathbf{v}/c = P^0\mathbf{v}/c$, where $\mathbf{v}$ represents the velocity of S_0 relative to S, and P is defined in Eq. (8), P^α_0 being the components of this four-vector in S_0. It follows that one can consider the quantity $M_0 \equiv P^0_{(0)}/c$ as the rest-mass of the system, and that $M \equiv P^0/c = \gamma M_0$, $\mathbf{P} = M\mathbf{v}$. The latter relations are identical with the appropriate relations in Eqs. (3) and (4) for a single particle of rest mass M_0 and velocity $\mathbf{v}$ relative to S. One can show, further, that $M_0 = \Sigma m_0 + \Sigma K_0/c^2$, the sums being taken over the particles of the system, with m_0 and K_0 representing, respectively, the rest mass and the relativistic kinetic energy of a particle relative to S_0. The last term is an interesting manifestation of the inertia of energy. [H. M. SCHWARTZ]

Bibliography: A. Einstein, *The Meaning of Relativity*, 5th ed., 1956; C. Kittel, W. D. Knight, and M. A. Ruderman, *Mechanics*, 1965; H. M. Schwartz, *Introduction to Special Relativity*, 1968; R. D. Sard, *Relativistic Mechanics*, 1970; C. Moller, *The Theory of Relativity*, 2d. ed., 1972.

Relativistic quantum theory

The quantum theory of particles which is consistent with the special theory of relativity, and thus can describe particles moving arbitrarily close to the speed of light. It is now realized that the only satisfactory relativistic quantum theory is quantum field theory; the attempt to relativize the Schrödinger equation for the wave function of a single particle fails, as shown below. However, with a change of interpretation, relativistic wave equations do correctly describe some aspects of the motions of particles in an electromagnetic field. *See* QUANTUM FIELD THEORY; QUANTUM MECHANICS; RELATIVITY.

Invariance. The Schrödinger equation for the wave function $\psi(\mathbf{r},t)$ of a particle [$|\psi(\mathbf{r},t)|^2$ is the density of probability of finding the particle at point $\mathbf{r}$ at time t] is Eq. (1), where E is the energy

$$E\psi = H(\mathbf{p},\mathbf{r})\psi \qquad (1)$$

operator $i\hbar(\partial/\partial t)$, $\mathbf{p}$ is the momentum operator $-i\hbar\nabla$, and $H(\mathbf{p},\mathbf{r})$ is the classical Hamiltonian. For a nonrelativistic particle in free space, $H = \mathbf{p}^2/2m$. The naive way to relativize Eq. (1) would be to use the relativistic Hamiltonian, Eq. (2).

$$H = \sqrt{(mc^2)^2 + \mathbf{p}^2 c^2} \qquad (2)$$

Although Eq. (1) with (2) gives the correct relativistic relation between frequency and wave number, and hence energy and momentum, the equation itself is not relativistically invariant, essentially because E and $\mathbf{p}$ do not occur in it in a similar manner. A further concrete difficulty is this: Suppose that at $t = 0$ the particle is localized at $\mathbf{r} = 0$, that is, $\psi(\mathbf{r},0) = 0$ for $\mathbf{r} \neq 0$. Then it can be shown that, according to Eq. (1) with (2), at any later time $\psi(\mathbf{r},t) \neq 0$ for all $\mathbf{r}$. But, according to relativity, the particle could not have traveled faster than light, speed c, so that $\psi(\mathbf{r},t)$ should be zero for $\mathbf{r} > ct$. *See* SYMMETRY LAWS.

Klein-Gordon equation. An equation without the above defects is the so-called Klein-Gordon equation, Eq. (3). But $\varphi(\mathbf{r},t)$, which satisfies this equa-

$$E^2\varphi = [(mc^2)^2 + \mathbf{p}^2c^2]\varphi \qquad (3)$$

tion, cannot be a wave function, for two related reasons: (i) Equation (3) is second-order in $\partial/\partial t$, and so not only $\varphi(\mathbf{r},0)$ but also $\partial\varphi/\partial t$ is needed to determine the future values of φ; (ii) the only possible density of a conserved quantity formed from φ is of the form shown in (4). But this cannot be a

$$\rho \propto \varphi^*E\varphi - \varphi E\varphi^* \qquad (4)$$

probability density, because it is not positive definite (it changes sign when φ is replaced by φ^*).

But ρ, in (4), can be interpreted as a charge density (when multiplied by a unit charge e); φ is then to be interpreted as a matrix element of a field operator Φ of a quantized field whose quanta are particles with mass m and charge e and no spin. [Equation (3) is obeyed by $\Phi(\mathbf{r},t)$, and hence by any of its matrix elements.] The same is true in an electromagnetic field, where (3) and (4) are to be modified by notation (5), where $\phi(\mathbf{r},t)$ and $\mathbf{A}(\mathbf{r},t)$ are the

$$E \rightarrow E - e\phi \qquad \mathbf{p} \rightarrow \mathbf{p} - e\mathbf{A} \qquad (5)$$

scalar and vector potentials, respectively, of the electromagnetic field. Eigenstates of φ in a static field can be found in the usual way, which yields the energy levels of a charged spinless particle, say a π^--meson in the electric field of a nucleus, correctly except for the effects of radiative corrections (the electromagnetic interactions of the particle with itself and other virtually present particles) and of any internal structure of the particle. *See* EIGENFUNCTION; ENERGY LEVEL.

Dirac equation. P. A. M. Dirac found a relativized form of Eq. (1) which is both linear in $\mathbf{E}$ and has a positive definite density form given by Eq. (6),

$$E\psi = [\beta mc^2 + \boldsymbol{\alpha} \cdot \mathbf{p}c]\psi \quad (\rho \propto \psi^*\psi) \qquad (6)$$

where β and $\boldsymbol{\alpha}$ are constants which obey Eqs. (7)

$$\alpha_i\alpha_j + \alpha_j\alpha_i = 0, \quad i \neq j \qquad (7)$$

$$\alpha\beta + \beta\alpha = 0, \quad \alpha_i^2 = 1, \beta^2 = 1, i, j = 1, 2, 3$$

Applying the equation a second time, Eq. (8) is

$$E^2\psi = [\beta mc^2 + \boldsymbol{\alpha} \cdot \mathbf{p}c]^2\psi = [(mc^2)^2 + p^2c^2]\psi \qquad (8)$$

obtained, that is, the Klein-Gordon equation, which assures that the energy and momentum of the particle are correctly related. Obviously the four constants β and α_i cannot be numbers; however, they can be 4×4 matrices, and ψ is then a four-component object called a Dirac spinor. Consider plane wave solutions of Dirac's equation (6); $\mathbf{p}$ has the eigenvalue $\mathbf{p}$. Taking Eq. (6) as an eigenequation for E, one finds four eigenstates (because H is a 4×4 matrix), two with $E = +\sqrt{(mc^2)^2 + p^2c^2}$ and two with $E = -\sqrt{(mc^2)^2 + p^2c^2}$. The interpretation of the two positive energy states is that they are the two spin states of a particle with spin $\frac{1}{2}\hbar$; in fact an operator representing the spin angular momentum can be constructed out of $\boldsymbol{\alpha}$. But the two negative energy states are an embarrassment; even if one started with a particle in a positive energy state, it would quickly make radiative transitions down through the negative energy states. Dirac's solution was to observe that if the particle described by ψ obeyed the Pauli principle (as does

the electron; in fact any spin $\frac{1}{2}\hbar$ particle must), then one can suppose that all the negative energy states are already filled with particles, thus excluding any more. *See* SPINOR.

There are still four single-particle states for a given momentum $\mathbf{p}$: the two spin states of a particle with positive energy, and the two states obtained by subtracting a negative energy particle (of momentum $-\mathbf{p}$). These last states ("hole states") have positive energy and a charge opposite the charge of the particle. The hole is in fact the antiparticle; if the particle is an electron, the hole is a positron. However, with the filling up of the negative energy states, one no longer has a single-particle system, and ψ, just as in the Klein-Gordon case, no longer can be interpreted as a wave function but as a matrix element of a field operator Ψ. Again, Eq. (6), with the modifications in Eq. (5), gives correct results for the eigenstates of a charged spin $\frac{1}{2}\hbar$ particle in an electromagnetic field, except for the effects of radiative corrections and internal structure; the latter is present only in hadrons such as the proton, but absent for the electron and muon. Particularly significant is the fact that the correct magnetic moment (Dirac moment) is given for the electron and muon (up to radiative corrections). *See* ELECTRON-POSITRON PAIR PRODUCTION; ELEMENTARY PARTICLE; HADRON; POSITRON; POSITRONIUM.

For a massless spin $\frac{1}{2}\hbar$ particle, β does not appear in Dirac equation (6), and the conditions in Eqs. (7), with β omitted, can be satisfied with the α_i being only 2×2 matrices. The particle, and likewise the antiparticle, then have only one spin state. But it can be shown that the particle has spin $\frac{1}{2}\hbar$; the particle (antiparticle) can only have its spin parallel (antiparallel) to its momentum. In an inverted coordinate system these statements would reverse, so that the "two-component Dirac equation" is not invariant to space inversion (parity). This "economical" version of the Dirac equation is satisfied by the only two known massless spin $\frac{1}{2}\hbar$ particles, the ν_e and ν_μ.

Wave equations for higher spin particles also exist. A well-known example is Maxwell equations for a (massless) spin $1\hbar$ particle. The equations become more complicated as the spin becomes higher, and they will not be written here. It should be remarked that when the spin is $\geq 1\hbar$, the interaction of a "structureless" particle with the electromagnetic field seems to be not unique, and so the magnetic moment is not determined, unlike the Dirac case of spin $\frac{1}{2}\hbar$. *See* MAGNETIC MOMENT; MAXWELL'S EQUATIONS; QUANTUM ELECTRODYNAMICS; RELATIVISTIC MECHANICS.

[CHARLES J. GOEBEL]

Bibliography: See NONRELATIVISTIC QUANTUM THEORY.

Relativity

A general theory of physics, primarily conceived by Albert Einstein, which involves a profound analysis of time and space, leading to a generalization of physical laws, with far-reaching implications in important branches of physics and in cosmology. Historically, the theory developed in two stages. Einstein's initial formulation in 1905 (now known as the special, or restricted, theory of relativity) does not treat gravitation, and one of the two principles on which it is based, the principle of relativity (the other being the principle of the constancy of the speed of light), stipulates the form invariance of physical laws only for inertial reference systems. Both restrictions were removed by Einstein in his general theory of relativity developed in 1915, which exploits a deep-seated equivalence between inertial and gravitational effects, and leads to a successful "relativistic" generalization of Isaac Newton's theory of gravitation. Because of the extreme weakness of gravitation compared with all other known forces, the non-newtonian consequences of the general theory have been verified only in a few instances, and their applications are largely confined to cosmology, the structure of neutron stars and black holes, and the motion of bodies in the solar system. For the same reason, where it can be applied, the special theory suffices in the treatment of relativistic effects in atomic, nuclear, and high-energy physics, and its confirmation in numerous instances of this kind renders it among the most securely established of modern scientific theories.

SPECIAL (RESTRICTED) THEORY

The scientific developments that led to the birth of special relativity arose from a dilemma confronting physicists in the latter part of the 19th century in their attempts to reconcile Maxwell's electromagnetic equations with the principles of newtonian mechanics. The latter were then widely held to have universal validity, but Maxwell's equations do not preserve their form under galilean transformations, such as that given by Eqs. (1) [see Fig. 1], that is, they do not satisfy the gali-

$$x' = x - vt \qquad y' = y \qquad z' = z \qquad t' = t \qquad (1)$$

lean principle of relativity. Hence, it appeared necessary to assume that these equations apply only in one special inertial frame, one in which the electromagnetic, or luminiferous, ether is at rest, and which, it was also natural to suppose, served to define Newton's absolute space and absolute velocity. In particular, the absolute velocity of the Earth in its orbital motion around the Sun could be expected to reveal itself in appropriate electromagnetic or optical experiments. However, all such experiments yielded negative results. *See* FRAME OF REFERENCE; MAXWELL'S EQUATIONS.

The most famous of these, the interferometer experiment of A. A. Michelson and E. W. Morley in 1881, presented the greatest challenge. It was

the first and, until the beginning of the 20th century, the only experiment whose negative result involved an effect of order v^2/c^2 (second-order effect), v being the average orbital speed of the Earth and c the speed of light (so that $v^2/c^2 \sim 10^{-8}$). The simplest explanation—that this negative result arises from the dragging of the ether by the Earth, a hypothesis proposed in another connection earlier (1845) by G. G. Stokes—was shown by H. A. Lorentz to be untenable because of difficulties in explaining stellar aberration. Lorentz was therefore led in 1892 to postulate that a body whose absolute speed of translational motion is v is contracted in the direction of motion by the factor $\sqrt{1 - v^2/c^2}$. This hypothesis was independently suggested by G. F. FitzGerald, and is thus known as the Lorentz-FitzGerald contraction hypothesis. With its aid, Lorentz succeeded in accounting for the Michelson-Morley experiment, and he also explained it on the basis of ideas from his theory of electrons. However, other second-order experiments performed at the turn of the 20th century, as well as critical methodological considerations, led Lorentz to arrive gradually, in 1899 and 1904, at a general solution of the problem posed by all these unexpected negative ether-drift results. This solution involved replacing galilean transformations (1) by space-time transformation equations (2), and introducing associated transformation

$$\begin{aligned} x' &= \frac{x - vt}{\sqrt{1 - v^2/c^2}} \\ y' &= y \\ z' &= z \\ t' &= \frac{t - vx/c^2}{\sqrt{1 - v^2/c^2}} \end{aligned} \qquad (2)$$

equations for the electromagnetic field quantities. However, the theory was partly logically flawed, and it was Henri Poincaré who corrected and completed it in 1905 and 1906. Poincaré named Eqs. (2) Lorentz transformations, and he was the first to discuss fully the complete group of Lorentz transformations, including rotations of the spatial coordinate axes. *See* LIGHT; LORENTZ TRANSFORMATIONS.

Although the mathematical structure of the Lorentz-Poincaré theory is the same as that of the theory advanced by Einstein in 1905, independently of Lorentz's work of 1904 and of Poincaré's work, there is nevertheless a vast difference in the underlying conceptual framework. Lorentz retained the prevailing notions of absolute time and space, and he clung to the belief that an absolutely stationary ether is necessary for a consistent explanation of electromagnetic phenomena. Einstein, on the contrary, started with the idea that the absoluteness of time and space is imposed not by nature but by custom, and that therefore, in particular, the notion of simultaneity must be reexamined critically. In addition, he did not consider the existence of an electromagnetic ether essential to electrodynamics. This intellectual independence enabled him to develop an exceptionally viable formulation of what is in consequence commonly referred to as Einstein's theory of relativity.

Einstein's formulation. The theory is erected on two fundamental postulates and two definitions. The postulates, which Einstein called the principle

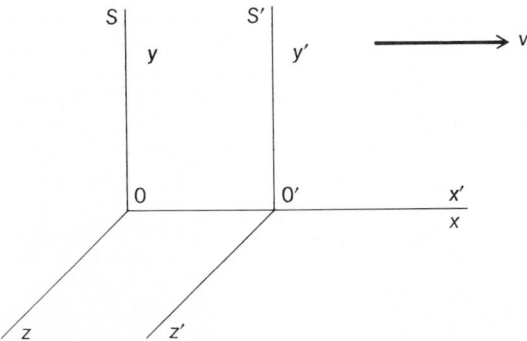

Fig. 1. Inertial reference frames S, S' with parallel coordinate axes; $\mathbf{v}$ is the velocity of S' relative to S.

of relativity and the principle of the constancy of the velocity of light, can be stated respectively as follows: (I) The analytical form of physical laws is the same in all inertial reference systems. (II) The speed of light in vacuum is a universal constant.

Postulate I is a generalization of the Galilean principle of relativity to embrace all phenomena rather than only those of mechanics (excepting, however, gravitational phenomena). It is II that, taken together with I, clearly represents a radical breaking away from traditional thinking. The new formulation of simultaneity is also intimately connected with II. Two spatially separated localized occurrences (or events) are simultaneous when the readings of two identical clocks adjacent to the events are the same, and it is known that the clocks are synchronized. However, when the clocks are not near each other, their synchronism must be defined.

The definition given by Einstein is tied to II, and can be stated as follows: (A) Two identical clocks C, C', situated at two distant points P, P' fixed in a given inertial frame S, synchronize in S, when the respective C and C' times t_1, t_1' of the sending of a light signal at P and its arrival at P' are connected by the formula $t_1' - t_1 = t_2 - t_1'$ with the C time t_2 of its return to P after reflection at P' back to P.

As will be shown below, two events that are simultaneous in terms of clocks synchronized in S according to definition A are not simultaneous when referred to identical clocks synchronized according to A in a reference frame that is moving relative to S. In other words, simultaneity is a relative concept: it depends on the reference system under consideration. This relativity is at the heart of many relativistic phenomena.

Einstein's second definition clarifies the operational meaning of the length of a moving rod, and can be stated as follows: (B) The length, as measured in a given inertial frame S, of a rigid rod that is in uniform motion relative to S is the length between the instantaneous positions of the ends of the rod, where the instantaneousness and the length are determined respectively by clocks synchronized in S and by meter sticks at rest in S.

With the aid of postulates I and II and definition A, Einstein succeeded for the first time in deriving the Lorentz transformations. Then, using B and I he deduced the Lorentz-FitzGerald contraction: Suppose a rigid rod is at rest in the system S' (of Fig. 1) along the x' axis, its ends having the coordinates x_1', x_2' ($x_1' < x_2'$), then by Eq. (2) for a constant value t_0 of t, $x_k' = \gamma(x_k - vt_0)$, $k = 1, 2$, where $\gamma \equiv 1/\sqrt{1 - v^2/c^2}$, so that $x_2' - x_1' \equiv \Delta x' = \gamma \Delta x$, $\Delta x = \Delta x'/\gamma \equiv \sqrt{1 - v^2/c^2}\, \Delta x'$, which is the required result, since $\Delta x'$ is by I the same as the length of the rod when at rest in S and measured in S. By transformation equations (3), inverse to Eqs. (2), a similar result is obtained when the roles

$$
\begin{aligned}
x &= \gamma(x' + vt') \\
y &= y' \\
z &= z' \\
t &= \gamma(t' + vx'/c^2)
\end{aligned}
\tag{3}
$$

of S and S' are interchanged, as is in any case required by I. It can also be concluded from Eqs. (2) that the relativistic contraction of a body takes place only in the direction parallel to that of the relative motion of S and S'.

Time dilation. Another remarkable result concerns the relativity of time. Setting $x' = $ constant in Eqs. (3), one finds that $\Delta t = \gamma \Delta t'$, and hence that Eq. (4) is true. Since Eqs. (2) involve of course the

$$
\Delta t' = \sqrt{1 - v^2/c^2}\, \Delta t
\tag{4}
$$

assumption that the clocks in S and S' (and the meter sticks) are identical, Eq. (4) states that as measured in S', the time interval between the given pair of events is smaller by the factor $\sqrt{1 - v^2/c^2}$ than it is as measured in S. In this respect, too, there is complete reciprocity, as there must be, between S and S': the time interval between two events is a relative concept. Einstein pointed out also the following corollary of Eq. (4): If one of two identical clocks, initially synchronized and adjacently at rest in an inertial frame, makes a round trip, it will lag behind the other upon its return. This is proved by integrating Eq. (4) over the closed path of the moving clock, on the assumption that in case of accelerated motion relative to an inertial frame, the kinematic relativistic results hold during each infinitesimal interval—an assumption, usually tacit, which is involved in many relativistic topics. If the clocks are replaced by the beating hearts of identical twins, it follows that the traveling one will upon return be younger than the stay-at-home. Convincing verification of the effect indicated by Eq. (4) is provided by the measured lifetimes of fast mesons, whose increase with the speed of the particles conforms to Eq. (4), and by more precise experiments employing the Mössbauer effect. The Michelson-Morley-type experiment, as well, has been repeated with greater precision using laser light. *See* CLOCK PARADOX; MÖSSBAUER EFFECT.

Composition of velocities. If $\mathbf{u}'$ is the velocity of a projectile relative to S' as measured in S', and $\mathbf{u}$ is its velocity relative to S as measured in S, one finds by Eqs. (3) that the components of the two velocities are related by Eqs. (5). If $\mathbf{u}'$ is taken to

$$
\begin{aligned}
u_x &= (u_x' + v)/(1 + vu_x'/c^2) \\
u_y &= u_y'/\gamma(1 + vu_x'/c^2) \\
u_z &= u_z'/\gamma(1 + vu_x'/c^2)
\end{aligned}
\tag{5}
$$

represent, instead, the velocity of the projectile relative to S' as measured in S, then $u_x = u_x' + v, u_y = u_y', u_z = u_z'$, as in newtonian physics. In any case, Eqs. (5) reduce to the latter equations when vu_x'/c^2 and v^2/c^2 are negligible. On the other hand, if $u_x' = c$ (and hence, $u_y' = u_z' = 0$), then $u_x = c$, $u_y = u_z = 0$, in agreement with postulate II.

The preceding discussion involves the tacit assumption that particle speeds cannot exceed c. That c does indeed represent the maximum speed of energy propagation is indicated by an argument of Einstein (1907) employing Eqs. (5), which shows that in the contrary case it would be possible to transmit information into the past. Such a situation, while perhaps not entailing a formal logical contradiction, is unacceptable to the great majority of physicists, although the possibility of the existence of particles with speed greater than c, dubbed tachyons, has been suggested. However, such particles would represent a category quite distinct from known particles, since the latter cannot be accelerated to speeds greater than c. *See* RELATIVISTIC MECHANICS; TACHYON.

Minkowski's formulation. Hermann Minkowski's geometric approach, formulated in 1908, which was an important link in the development of general relativity, centers on absolute quantities in the special theory, and provides it with the efficient formalism of tensors in space-time.

Minkowski's space-time, his "world," is the continuum of all events (Minkowski's "world points"), the primitive absolute quantities of relativistic kinematics. It is a four-dimensional, quasi-euclidean space, with the line element ds^2 (the distance squared between two neighboring events) given by Eq. (6), defining the metric of the space the "Minkowski metric." *See* SPACE-TIME.

$$ds^2 = \sum_{\alpha,\beta=0}^{3} \eta_{\alpha\beta}\, dx^\alpha\, dx^\beta \equiv (dx^0)^2 - \sum_{i=1}^{3} (dx^i)^2 \qquad (6)$$
$$x^0 = ct \qquad x^1 = x \qquad x^2 = y \qquad x^3 = z$$
$$\eta_{\alpha\beta} = 0 \text{ (if } \alpha \neq \beta)$$
$$\eta_{00} = -\eta_{11} = -\eta_{22} = -\eta_{33} = 1$$

The tensors of special relativity represent absolute quantities, which are determined by their components, relative quantities, obeying the appropriate transformation rules associated with the Lorentz group. Because the metric of Eq. (6) is not definite ($\eta_{\alpha\beta} \neq \delta_{\alpha\beta}$, since $\eta_{ii} = -1$ for $i = 1, 2, 3$) one must distinguish between contravariant, covariant, and mixed components of a tensor, which differ, however, at most in sign; for example, $\eta_{\alpha\beta}$ and dx^α are covariant and contravariant components of the respective tensors, and $\eta_{\alpha\beta} = \eta^{\alpha\beta}$ while $dx_i = -dx^i$ ($i = 1, 2, 3$). One usually attaches the prefix "four" to relativistic tensors. Thus, the scalar of Eq. (6) is a four-scalar, and the dx^α represent a four-vector. *See* CALCULUS OF TENSORS.

One may replace dx^α in Eqs. (6) by finite Δx^α, and according as Δs^2 is greater than 0, less than 0, or equal to 0, one has a timelike, spacelike, or null interval (displacement four-vector). The first and last cases correspond to possible displacements of a particle with $m_0 \neq 0$ or $m_0 = 0$. Considering the motion of a particle ($m_0 \neq 0$), one finds from Eq. (6) that $ds^2/c^2 \equiv d\tau^2 = dt^2(1 - v^2/c^2)$, $\mathbf{v} = d\mathbf{x}/dt$. The integral $\int d\tau$ along the particle's space-time path (or "world line") gives the time as measured by a clock accompanying the particle: the "proper time" on the world line.

Applications. The branches of classical physics can usually be relativistically generalized by applying postulate I in Minkowski's formulation of expressing the physical laws in tensor form.

Mechanics. Minkowski's method is successful in the case of a particle moving under the action of an external field of force, and important but limited results exist for systems of interacting particles. A dramatic consequence is Einstein's mass-energy equivalence, which has implications ranging from nuclear reactor energy production to atomic bombs.

Continuum mechanics and fields. The space-time formulation is here especially effective. Maxwell's equations of the free electromagnetic field are already Lorentz-invariant, and their four-tensor form is immediate. Using relativistic-mechanics results, the four-tensor reformulation of the Maxwell-Lorentz equations for the electromagnetic field generated by a system of elementary charges and for the motion of the charges in the electromagnetic field, and of the phenomenological equations of polarizable and magnetizable media, first developed by Minkowski in 1909, present few difficulties. Of central importance in this area is the energy-momentum tensor, whose components T^{00}, T^{ij}, T^{0i}, and T^{i0} ($i = 1, 2, 3$) represent the respective densities of energy, momentum flux (stress), (energy flux)$/c$, and $c \cdot$ (momentum). The symmetry of $T(T^{\alpha\beta} = T^{\beta\alpha})$ implies, among other things, the inertia of energy. For closed systems, the equations of motion assume the compact form

$$\sum_{\beta=0}^{3} \partial T^{\alpha\beta}/\partial x^\beta = 0 \qquad (\alpha = 0, 1, 2, 3)$$

and the integrals $\int T^{00} dV$ over space for any fixed t give the conserved total energy and momentum of the system in energy units. *See* INERTIA OF ENERGY; RELATIVISTIC ELECTRODYNAMICS.

Optics. From the four-scalar property of the phase function $2\pi\nu$ $(t - \mathbf{n} \cdot \mathbf{x}/c)$ of a plane light wave, it follows that if a light ray is emitted in S' in a direction making in S' the angle θ' with the x' axis, then in S its direction with the x axis is given by the angle θ such that $\cos \theta = (\cos \theta' + v/c)/1 + \nu \cos \theta'/c)$; and if the frequency of the emitted light is ν' in S', then in S it is $\nu = \nu'/\gamma(1 - \nu \cos \theta/c)$. The two formulas are respectively the relativistic generalizations of the laws governing the aberration of light and the Coppler effect. *See* DOPPLER EFFECT.

Statistical physics. The relativistic extension of classical thermodynamics presents some conceptual difficulties. Unlike Einstein's definition of the length of a moving body, the natural operational definition of, say, the temperature of a moving body is neither obvious nor always independent of the thermodynamic state of the body. This circumstance explains in part the paradoxical situation which arose regarding the conceptual framework of relativistic thermodynamics, when three different sets of Lorentz transformation formulas were being advocated. No such difficulties attach to statistical mechanics and kinetic theory, and the simplest topics of these subjects were treated relativistically not long after the rise of relativity. Considerable progress has been achieved in studies of relativistic kinetic theory and statistical mechanics of both reversible and irreversible processes, involving proper choices of the energy-momentum tensor of the thermodynamic systems. These studies have been mostly motivated by astrophysical problems presented by superdense stars such as pulsars and certain white dwarfs and by some speculative cosmological theories. In most of these applications intense gravitational fields are involved that require general-relativistic treatment. However, the special-relativistic formulations represent an essential step.

[H. M. SCHWARTZ]

GENERAL THEORY

General relativity is the geometric theory of gravitation developed by Einstein in 1915. It is a generalization of special relativity, and includes the classical gravitational theory of Newton as the limiting case when the gravitational fields involved are weak and the velocities of all the bodies involved are small compared to the speed of light c. The most important applications of the theory are to the structure of neutron stars and black holes, the large-scale cosmological description of the universe, and the motion of bodies in the solar system. *See* GRAVITATION.

Need for relativistic theory. The special theory of relativity proposed in 1905 gained acceptance rapidly among physicists by virtue of its theoretical elegance and experimental success, and was rather well established by 1915. One of the basic tenets of special relativity is that no physical effect can propagate with a velocity greater than the speed of light, and so c represents a universal speed limit.

On the other hand, classical gravitational theory describes the gravitational field of a body throughout space as a function of its instantaneous position, which is equivalent to the assumption that gravitational effects propagate with an infinite velocity; that is, classical gravitational theory is an action-at-a-distance theory. Thus, special relativity and classical gravitational theory are inconsistent, and a modified theory of gravity is necessary. This is the theory searched for and found by Einstein from 1905 to 1915, following his discovery of special relativity.

Principle of equivalence. It had long been considered a fundamental and puzzling question why bodies of different mass fall with the same acceleration in a gravitational field, or equivalently why the trajectory of a test body is independent of its mass. This situation was explained by Newton with the statement that both the gravitational force on a body and its inertial resistance to acceleration are proportional to its mass. Thus the mass cancels out of the mathematical description of the motion. In laboratory experiments early in the 20th century by L. von Eötvös, it was found that this cancellation is true to a few parts in 10^8. Later work by R. Dicke improved the accuracy to a few parts in 10^{11}, and V. Braginsky obtained an accuracy of a few parts in 10^{12}. Thus the independence of the motion of a test body on its mass is one of the most accurately tested experimental facts in physics.

The explanation by Newton is not very profound and is more in the nature of an ad hoc description.

A deeper and more natural explanation occurred to Einstein. In physics there are numerous examples of forces other than gravitation which are mass-proportional; these generally arise due to the use of accelerated coordinate systems to describe the motion. One well-known example is the centrifugal force encountered in a rotating coordinate system. Consider one observer in the gravitational field of the Earth and another in an accelerating elevator or rocket in free space (Fig. 2). If both drop a test body, they will observe it to accelerate relative to the floor. According to classical theory, the Earth-based observer would attribute this to a gravitational force and the elevator-based observer would attribute it to the accelerated floor overtaking the uniformly moving body. However, Einstein reasoned that the effects are identical and the theory of gravity should provide an equivalent description of the two systems. This is the famous principle of equivalence that Einstein made the physical cornerstone of general relativity; it states that on a local scale the physical effects of a gravitational field are indistinguishable from the physical effects of an accelerated coordinate system. From the point of view of the principle of equivalence, it is evident why the motion of a test body in a gravitational field is independent of its mass. *See* CENTRIFUGAL FORCE.

The principle of equivalence is strictly local and applicable only to a region of space and time sufficiently small that inhomogeneities in the gravitational field can be ignored. There is an intrinsic difference between gravitational and accelerative effects on a finite scale. This is well illustrated by considering in a nonuniform gravitational field two nearby test bodies, which, being in slightly different parts of the field, follow slightly different trajectories. The relative deviations of the trajectories characterize the inhomogeneities of the field. They intrinsically distinguish the effects of gravity and acceleration, which is impossible on a strictly local scale according to the principle of equivalence.

The principle of equivalence is heuristic and somewhat imprecise; despite its logical imprecision, it has played a very important historical role.

Tensor field equations. The close connection between gravity and accelerating coordinate systems suggests that the equations describing the gravitational field be cast in a form that is manifestly independent of the coordinate system in order to achieve the maximum simplicity and elegance. The study of systems of equations that are independent of coordinate systems was begun by K. Gauss in connection with his study of geometry on curved two-dimensional surfaces, and was carried to a high state of development by B. Riemann and T. Levi-Civita in the tensor calculus. Tensors are a basically simple generalization of vectors. *See* CALCULUS OF TENSORS.

Only tensors in the four-dimensional space-time of relativity need be considered. Suppose that the points in four-dimensional space-time are labeled by two essentially arbitrary systems of coordinates x^μ and x'^ν (the indices μ and ν range from 0 to 3) and that the two coordinate systems are related to each other by an infinitely differentiable set of four functions, $x'^\mu = x'^\mu(x^\alpha)$ and $x^\beta = x^\beta(x'^\nu)$. Then contravariant tensors of rank 0, rank 1, rank 2, etc. are defined as sets of 1, 4, 16, etc. numbers or func-

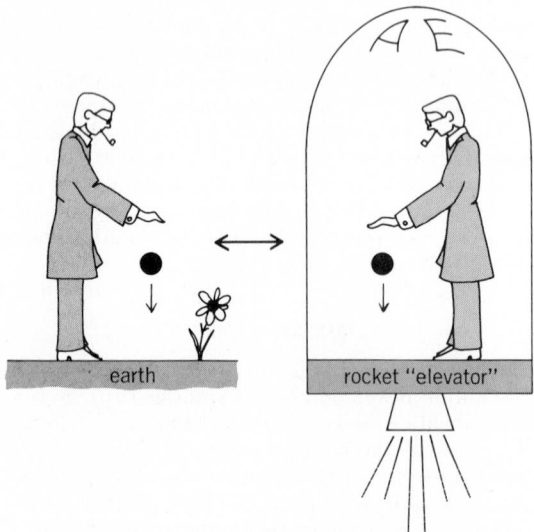

Fig. 2.   Einstein's elevator, illustrating the principle of equivalence. The elevator acceleration is equal to g, the acceleration of gravity.

tions whose values at the same point in the two systems of coordinates are related by transformation equations (7). Covariant tensors are similarly defined by Eqs. (8). In these equations the indices

$$\phi' = \phi$$
$$\text{rank } 0$$

$$\eta'^\mu = \frac{\partial x'^\mu}{\partial x^\alpha} \eta^\alpha \tag{7}$$
$$\text{rank } 1$$

$$T'^{\mu\nu} = \frac{\partial x'^\mu}{\partial x^\alpha} \frac{\partial x'^\nu}{\partial x^\beta} T^{\alpha\beta}, \text{ etc.}$$
$$\text{rank } 2$$

$$\sigma'_\mu = \frac{\partial x^\alpha}{\partial x'^\mu} \sigma_\alpha$$
$$\text{rank } 1$$
$$\tag{8}$$
$$\omega'_{\mu\nu} = \frac{\partial x^\alpha}{\partial x'^\mu} \frac{\partial x^\beta}{\partial x'^\nu} \omega_{\alpha\beta}$$
$$\text{rank } 2$$

that appear twice in one expression are to be summed over; this is the Einstein summation convention. Its use allows a great saving of effort and space by avoiding numerous explicit summation signs Σ. Tensors of rank 0 are also called scalars or invariants since they have the same value in all coordinate systems and are thus of great importance in the description of physical quantities. Tensors of rank 1 are also called four-vectors. One example of a scalar is the inner product of a contravariant four-vector ξ^α and covariant four-vector η_α defined as $\xi^\alpha\eta_\alpha$.

The space-time of relativity contains one covariant second-rank tensor of particularly great importance, called the metric tensor $g_{\mu\nu}$, which is a generalization of the Lorentz metric of special relativity $\eta_{\alpha\beta}$, introduced in Eqs. (6). Nearby points in space-time, known as events, which are separated by coordinate distances dx^μ have an invariant physical separation whose square, the line element, is given by Eq. (9). This quantity is a generalization of the line element in special relativity and has the same relation to the concept of proper time; that is, when ds^2 is positive, it represents the square of the time interval between nearby events (multiplied by c^2) measured by an observer who moves at approximately constant velocity in such a way as to be present at both nearby events. (A space with a quadratic metric form as in Eq. (9) is called a Riemann space by mathematicians.) *See* RIEMANNIAN GEOMETRY.

Tensor equations are equations in which one tensor of a given rank and type (contravariant or covariant) is set equal to another of the same type, for example, $T^\mu = S^\mu$. In a different coordinate system, denoted by a prime, the definition of a tensor implies that both sides transform in the same way so that the partial derivatives of the transformation cancel from the equation and it has the same form as in the original system, for example, $T'^\mu = S'^\mu$. Tensor equations are thus called form-invariant, or covariant.

The field equations of general relativity are ten-

sor equations for the metric tensor, which completely describes the geometry of the space. To present the field equations, a Cristoffel symbol (not a tensor) is first defined by Eq. (10). Here $g^{\mu\tau}$ is the

$$\begin{Bmatrix} \mu \\ \alpha\beta \end{Bmatrix} = \frac{1}{2} g^{\mu\sigma} \left(\frac{\partial g_{\sigma\alpha}}{\partial x^\beta} + \frac{\partial g_{\sigma\beta}}{\partial x^\alpha} - \frac{\partial g_{\alpha\beta}}{\partial x^\sigma} \right) \tag{10}$$

inverse of $g_{\mu\tau}$, considered as a matrix. The Riemann tensor (or curvature tensor) is defined by Eq. (11). This tensor plays a central role in the geo-

$$R^\alpha_{\mu\beta\nu} = \frac{\partial}{\partial x^\nu} \begin{Bmatrix} \alpha \\ \beta\mu \end{Bmatrix} - \frac{\partial}{\partial x^\mu} \begin{Bmatrix} \alpha \\ \beta\nu \end{Bmatrix}$$
$$+ \begin{Bmatrix} \alpha \\ \tau\nu \end{Bmatrix} \begin{Bmatrix} \tau \\ \beta\mu \end{Bmatrix} - \begin{Bmatrix} \alpha \\ \tau\beta \end{Bmatrix} \begin{Bmatrix} \tau \\ \mu\nu \end{Bmatrix} \tag{11}$$

metric structure of a space; if it is zero, the space is termed flat and has no gravitational field; if nonzero, the space is termed curved. In terms of the contracted Riemann tensor, a Riemann tensor summed over $\alpha=\beta$, the Einstein field equations for free space are given by Eq. (12). In a region of

$$R^\alpha_{\mu\alpha\nu} \equiv R_{\mu\nu} = 0 \tag{12}$$

space containing matter or energy, the zero on the right side of this equation is replaced by a tensor representing the energy content of space, usually written $-(8\pi G/c^2)(T_{\mu\nu} - 1/2 g_{\mu\nu} T^\alpha_a)$, where G is the gravitational constant (equal to 6.670×10^{-11} N · m² · kg⁻²), and $T_{\mu\nu}$ is called the energy-momentum tensor. *See* ENERGY; MOMENTUM.

The field equations are a set of 10 second-order partial differential equations since the four-by-four symmetric tensor $R_{\mu\nu}$ has 10 independent components; they are to be solved for the metric tensor. A solution in a given coordinate system defines an Einstein space-time. The curvature of this space corresponds to the intrinsic presence of a gravitational field. That is, the concept of a field of mechanical force in classical gravitational theory is replaced by the geometric concept of curved space in relativity theory. *See* DIFFERENTIAL EQUATION.

Motion of test bodies. Within the context of tensor notation and the description of gravity by means of curved space, the equations of motion are almost obvious. The path of a test body is a generalization of a straight line in euclidean space; it is the shortest "distance" (in terms of intervals ds) between points in space-time, known as a geodesic. The differential equation of a geodesic, Eq. (13), involves derivatives of the coordinates along the path and the Christoffel symbols. In

$$\frac{d^2x^\mu}{ds^2} + \begin{Bmatrix} \mu \\ \alpha\beta \end{Bmatrix} \frac{dx^\alpha}{ds} \frac{dx^\beta}{ds} = 0 \tag{13}$$

the special case of no gravitational field, with the metric equal to the constant Lorentz metric, the Christoffel symbols vanish and this is indeed the equation of a straight line. Basically the scenario for gravitational influence in the general theory of relativity is that matter curves space in its vicinity, in accord with the Einstein field equations, and in this curved space test bodies move on geodesics, or generalized straight lines. The motion is thereby clearly independent of the mass of the test body, consistent with the principle of equivalence.

The basic equations of relativity were thought at first to be the Einstein field equations and the separately postulated geodesic equations of motion. However, general relativity theory possesses an extraordinary property; because the field equations are nonlinear, unlike newtonian theory, the motion of a test body in a gravitational field is not arbitrary since the body itself has mass and contributes to the field. Indeed, it turns out that the field equations are so restrictive that the geodesic equation of motion is a necessary consequence and need not be treated as a separate postulate. This has been shown by Einstein, L. Infeld, and B. Hoffman, by Levi-Civita, and by numerous others. This remarkable property of the field equations is unique to general relativity among the accepted theories in physics.

Test bodies are defined as bodies without significant extent or structure, and do not materially affect the field. Bodies with nonnegligible size should not be considered as test bodies, and indeed have more complicated equations of motion.

Schwarzschild solution. A very important solution of the field equations was obtained by K. Schwarzschild in 1916, surprisingly soon after the inception of general relativity. This solution represents the field in free space around a spherically symmetric body such as the Sun. It is the basis for a relativistic description of the solar system and all of the experimental tests of general relativity which have been carried out. In spherical coordinates r, θ, ϕ and time coordinate $x^0 = ct$ the solution is represented by a line element (valid only outside the body) given by Eq. (14). Here M is the

$$ds^2 = \left(1 - \frac{2GM}{c^2 r}\right) c^2 dt^2 - \left(1 - \frac{2GM}{c^2 r}\right)^{-1} dr^2$$
$$- r^2 (d\theta^2 + \sin^2\theta \, d\phi^2) \quad (14)$$

mass of the body. In the limit of $M = 0$ this represents the line element of special relativity written in spherical coordinates. This solution is the relativistic analog of the classical gravitational potential field $\Phi = -GM/r$. Indeed, there is an important approximate relation between Φ and g_{00}, $g_{00} = 1 + 2\Phi/c^2$, which is necessary and sufficient to show that general relativity has classical gravitational theory as its limit for very weak fields and low velocities of test bodies.

Gravitational red shift. There are a number of ways to show that electromagnetic radiation of a given frequency emitted in a gravitational field will appear to an outside observer to have a lower frequency; that is, it will be red-shifted. For the Schwarzschild solution the fractional decrease in frequency is equal to the difference in $GM/c^2 r$ between emission point and observation point. The red shift was first verified for radiation of optical frequency emitted by atoms on the Sun. However, nongravitational effects on such radiation make the measurement of the 1-part-in-10^6 effect difficult and somewhat uncertain. More accurate and dependable measurements involve the use of the Mössbauer effect in terrestrial experiments. Certain radioactive nuclei, such as ^{57}Fe, in crystals emit and absorb gamma radiation in extremely narrow frequency bands of fractional width about 10^{-12}. A gamma ray emitted upward by such a crystal will be red-shifted about 1 part in 10^{15} in

100 ft (30 m) and will not be resonantly absorbed by a receiving crystal, unless the receiving crystal is given a small downward velocity to compensate for the gravitational red shift with a Doppler shift; the velocity necessary to reestablish absorption provides a measurement of the red shift. With this technique the red shift has been measured by R. Pound and G. Rebka to be within about 1% of the value predicted by general relativity theory.

The most accurate test of the red shift to date was performed using a hydrogen maser on a rocket that reached an altitude of about 10^4 km. Comparison of the maser frequency with Earth-based masers by R. Vessot gave a measured red shift in agreement with theory to about 1 part in 10^4.

The red shift can be derived from the principle of equivalence without the use of the Schwarzschild solution, so these experiments do not test the Schwarzschild solution or the Einstein field equations.

Perihelion shift. The equations of motion can be solved for a planet considered as a test body in the Schwarzschild field of the Sun. As should be expected, the orbits obtained are very similar to the ellipses of classical theory. Small differences occur, however, the most interesting of which is that the ellipse rotates very slowly in the plane of the orbit so that the perihelion, the point of closest approach of the planet to the Sun, is at a slightly different angular position on each orbit. This shift is extremely small. It is greatest in the case of the planet Mercury, whose perihelion advance is predicted to be only 43 seconds of arc in a century. This is in excellent agreement with the value for the discrepancy between classical theory and observation, which was well known and unaccountable for many years before the discovery of general relativity.

Modern tests of the perihelic motion involve the planet Mars in addition to Mercury and also the asteroid Icarus in a comprehensive model of the solar system. Very precise planetary and asteroid position measurements using radar have been made by I. Shapiro, and the results are consistent with general relativity theory. In particular the perihelion shift of Mercury has been determined to an accuracy of about half a percent and agrees with theory.

Some question still exists about the interpretation of these experimental results. The reason is that the Sun's quadrupole moment is not known precisely, and such a quadrupole moment would produce a perihelion advance analogous to the relativistic effect.

Deflection of star light. The principle of equivalence suggests an extraordinary phenomenon of gravity. Light or other electromagnetic radiation crossing the Einstein elevator horizontally will appear to be deflected downward in a parabolic arc because of the upward acceleration of the elevator. The same phenomenon must occur for light in the gravitational field of the Sun; it must be deflected toward the Sun. A calculation of this deflection gives 1.75 seconds for the net deflection of starlight grazing the edge of the Sun. A star near the edge of the Sun viewed from the Earth will appear to be artificially displaced away from the Sun. Early measurements of this effect, notably by A. S. Eddington, were made by photographing stars near the Sun during total solar eclipses and comparing their positions with those when the Sun is in a different part of the sky. They verified the qualitative correctness of the deflection phenomenon. Modern measurements are made by tracking quasars as they pass near or behind the Sun. With these techniques, free of the limitations imposed by infrequent eclipses and photographic problems, the deflection has been measured to be within 1% of the value predicted by general relativity.

Radio time delay. In the curved space around the Sun the distance between points in space, for example between two planets, is not the same as it would be in flat space. In particular, the round-trip travel time of a radar signal sent between the Earth and Venus will be measurably increased by the curvature effect when the Earth, the Sun, and Venus are approximately lined up. Although the maximum time delay is only a few hundred microseconds, it has been accurately measured and found to agree with the predictions of general relativity to within 4%. Subsequent experiments used the Viking spacecraft, which contained a transponder, instead of the planet Venus, and were able to achieve an accuracy of about one-half of 1%. These measurements provide the first qualitatively new test of general-relativity theory in 50 years.

Precession of a gyroscope. A particularly interesting test of general-relativity theory involves the motion of a highly accurate and stable gyroscope in orbit around the Earth. Relativity theory predicts that such a gyroscope will precess at a rate of about 7 seconds of arc per year.

Most of this relativistic precession can be understood as due to the curvature of space described by the Schwarzschild metric in Eq. (14); this is called the geodetic precession. However, since the Earth spins, the Schwarzschild metric does not describe its field completely, and small additional terms occur in the metric. Loosely speaking, these terms arise because in relativity theory a spinning body partially drags space around with it. This is known as the dragging of inertial frames. It is sometimes also referred to as the Lense-Thirring effect after its discoverers J. Lense and H. Thirring. There is no classical analog; the classical gravitational field of a spinning spherical body is identical to that of a nonspinning spherical body.

A small part, about 1%, of the precession of the gyroscope is due to the dragging of the inertial frame. Moreover, for a satellite in a polar orbit the precession due to the Lense-Thirring effect is at right angles to the dominant geodetic precession. This fact makes it possible to separate and measure the geodetic effect and the Lense-Thirring effect in the same experiment. Thus, if sufficiently accurate, the gyroscope experiment can provide a test of this novel effect of general-relativity theory. A gyroscope and associated experimental apparatus have been developed that should allow the precession due to the Lense-Thirring effect to be measured to an accuracy of about 2%, and the geodetic precession to about 1 part in 10^4.

Neutron stars and black holes. The astronomical sources of regularly spaced pulses of electromagnetic radiation, known as pulsars, are believed to be small, rapidly spinning stars of extremely high density known as neutron stars.

For ordinary stars, such as the Sun, the gravitational field is sufficiently weak that classical gravitational theory is an adequate approximation for studying the internal structure. This is not true for neutron stars, which contain a core composed largely of neutrons and other elementary particles at nearly nuclear density. The gravitational field in these stars can be quite large, and it is necessary to describe them using hydrodynamic equations derived from the Einstein field equations in the presence of matter.

One of the most remarkable theoretical properties of neutron stars is that they have an upper mass limit of about 2 solar masses. A nonrotating, spherically symmetric neutron star with a larger mass is not stable and undergoes a process known as gravitational collapse, one of the most exotic concepts of relativity theory. In such an unstable neutron star the pressure produced by compression of the star is not sufficient to balance the inward force of gravity, and the star shrinks in size; the remarkable fact is that it shrinks indefinitely, the outward pressure never being able to balance the force of gravity, no matter how large the pressure becomes. Viewed by an observer far outside the very strong gravitational field region, such a star appears to asymptotically approach a sphere of constant radius $r = 2GM/c^2$, which is known as the Schwarzschild radius. This is the radius at which the Schwarzschild solution becomes singular, and is equal to about 1.5 km for a star with the mass of the Sun. Light emitted by a source at this radius suffers a red shift to zero frequency and thus cannot be observed. As a consequence, the collapsing star will asymptotically approach a radius where it is invisible, and becomes what is known as a black hole. In fact, not only is light unable to escape from the black hole, but no physical effect is able to reach the external world from the surface or interior of the black hole. Matter and radiation can fall into the black hole, but can never emerge. The surface acts as a one-way membrane that separates space into two disjoint parts; it is impossible to communicate with the outside world from the surface or interior of a black hole.

Among the end products of the evolution of stars it is expected that occasionally a configuration should occur that leads to gravitational collapse and a black hole. Such stellar black holes may have already been observed. It is believed by many astronomers that the x-ray source known as Cygnus X-1 is a binary system of an ordinary star and a black hole in orbit, and that gas approaching the black hole from the ordinary star is strongly heated by gravitational compression, emits x-rays in very short random bursts, and ultimately falls into the black hole (Fig. 3).

If a collapsing star is rotating, as is most likely to be the case, the asymptotic state is believed to be a generalization of the Schwarzschild solution to rotating systems discovered by R. Kerr. Like the Schwarzschild solution, the Kerr solution has a spherical singular surface and describes a black hole, but the black hole is surrounded by an annular region known as the ergosphere. The ergosphere can contain an enormous concentration of energy, equal to a significant fraction of the total rest energy of the star. It is possible that this phenomenon is of great astrophysical importance if the energy can be liberated by naturally occurring astronomical processes.

Studies by S. Hawking indicate that black holes are not totally stable but essentially evaporate

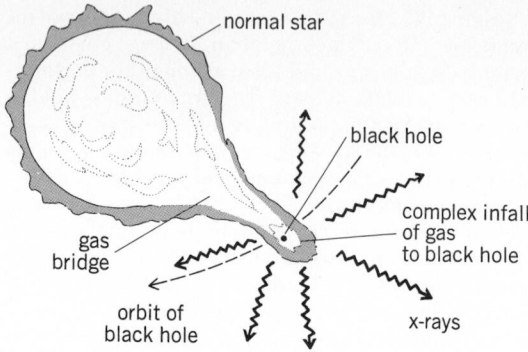

Fig. 3. Gas falling into a black hole and producing x-rays. (*From R. J. Adler, M. Bazin, and M. M. Schiffer, Introduction to General Relativity, McGraw-Hill, 1975*)

away by emitting radiant energy. For stellar-mass black holes the rate of evaporation is completely negligible, but if very-low-mass black holes exist, possibly formed at the birth of the universe, they would evaporate explosively with large energy release. Such bursts of radiation have not yet been detected. *See* BLACK HOLE.

Cosmology. The geometric viewpoint of relativity provides an elegant description of the shape, origin, and evolution of the universe on the cosmological scale; this scale is so large that entire galaxies and even clusters of galaxies are treated in theory like atoms in a gas. On this scale the universe appears observationally to be spatially isotropic and homogeneous, that is, it is substantially the same in all directions and at all points in space. It is thus assumed that the large-scale three-dimensional geometry of the universe is isotropic and homogeneous. This assumption leads to a type of three-dimensional metric known as the Robertson-Walker metric, after its discoverers. There are three subtypes of Robertson-Walker metric: (1) hyperspherical, (2) hyperplane, (3) hyperpseudospherical. These are precisely the three-dimensional analogs of (1) the surface of a sphere, (2) a plane surface, (3) a negative-curvature two-dimensional surface studied in the last century by N. Lobachevski; it unfortunately cannot be simply visualized.

The "galactic atoms" in this picture are at fixed three-dimensional coordinate positions, but the three-space has a single arbitrary metrical scale factor, generally time-dependent, which relates the constant coordinate distance between galaxies to the physical distance. This description is based on observationally motivated geometrical assumptions, and is independent of the Einstein field equations. The hyperspherical universe is of particular philosophical interest since it represents a three-dimensional model of the universe which has no boundaries but is of finite volume, analogous to the surface of a sphere, which has no boundaries but is of finite area. It is thus a closed universe.

The Einstein field equations enter the cosmological problem by providing a differential equation for the metrical scale factor. They cannot determine the geometrical type of the metric, which is an observational problem. A. Friedman in 1922 obtained the solutions of the field equations representing the special case in which the pressure of the "galactic gas" is negligible compared to its

density. This is realistic at the present epoch, but not for very early epochs. For the three subtypes of Robertson-Walker metric the solutions have the following properties: (1) For the hyperspherical case, the scale factor increases from zero, reaches a maximum, and then contracts again to zero. That is, the average intergalactic distance expands from zero to a maximum and then contracts to zero. (2) For the hyperplane case, the scale factor increases from zero and becomes indefinitely large. (3) For the hyperpseudospherical case, the scale factor increases from zero, becoming indefinitely large, more rapidly than in the hyperplane case. In all cases the universe begins with an explosive "big bang" (here ignoring details) and then expands. For sufficiently early epochs the behavior of the scale factor is in fact nearly the same for the three cases, and at the present epoch it is difficult to distinguish between these three models.

It is remarkable that Einstein's first attempts at cosmological solutions of the field equations yielded time-evolutionary solutions, some years before the observational discovery of the expansion of the universe as evidenced by the Doppler shift of distant galaxies. He was forced to introduce a new cosmological term in the field equations to obtain static solutions which he thought were necessitated by observation; the subsequent discovery of the universal expansion removed the need for static solutions and made the cosmological term unnecessary.

Observational cosmology mainly involves the measurement of two important numbers. The first is Hubble's constant H, with the units of inverse time; the fractional change in wavelength of light from a galaxy at distance L is given by HL/c. This number H can be related to the scale factor of the Friedmann model universes. The observational value for $1/H$ of about 2×10^{10} years is a rough measure of the age of the universe. The second number is the deceleration parameter q_0, which determines whether the expansion of the universe is decelerated $q_0 > 1/2$, uniform $q = 1/2$, or accelerated $q < 1/2$. These three cases correspond to the three subtypes of Robertson-Walker metric, hyperspherical, hyperplane, or hyperpseudospherical. Measurements of q_0 involve very distant galaxies and are quite difficult and subject to large uncertainties. The difficulty stems from lack of understanding of the evolution of galaxies and their intrinsic luminosity. As a result, there is considerable controversy over the observational value of the deceleration parameter and thus over the question of whether the universe is closed (hyperspherical) or open (hyperplane or hyperpseudospherical). *See* HUBBLE CONSTANT.

There is another method of determining whether the universe is open or closed. It is a remarkable consequence of the Einstein equations that, given the assumption of a Robertson-Walker metric, if the average density of mass and energy in the universe exceeds a critical value of about 2×10^{-29} g/cm³, then the universe must be closed. The density of visible material, such as that contained in stars and luminous gas, is observed to be only about 2×10^{-31} g/cm³. However, it does not follow from this that the universe is open, since there could well be a great deal of mass and energy that is not visible. The density of material in the form of very dim stars, dark gas and dust in galactic halos, black holes, neutrinos, gravitational radiation, and so forth cannot easily be determined. Studies of the gravitational dynamics of galaxies do in fact indicate the presence of large amounts of invisible material, often called the missing mass. This is an active area of research, and it is not yet clear if sufficient density is present to indicate a closed universe.

Gravitational radiation. Gravitational radiation is closely analogous to electromagnetic radiation; whereas electromagnetic radiation is emitted by charges in accelerated motion, gravitational radiation is emitted by masses in accelerated motion. Such motion produces small ripples or waves in the gravitational field that propagate at the velocity of light. The usual method of studying these ripples is with linearized general-relativity theory; in this theory the metric tensor is expressed as the constant Lorentz metric of special relativity plus a small perturbation term representing small gravitational ripples in an almost flat space. Quantities of higher than first-order in this perturbation are discarded in the field equations. In this manner the nonlinear field equations become linear approximate equations. This simplifies the mathematics enormously and, in view of the small amplitude of gravitational radiation, it is an excellent approximation. The forces exerted by plane gravitational waves are at right angles to the direction of propagation of the wave; that is, they are transverse. The directions of these forces are such as to produce the distortions shown in Fig. 4 in a circular ring of particles for a periodically varying wave moving perpendicular to the paper. As with electromagnetic waves, there are two polarizations possible for gravitational waves. Both polarizations exert forces as illustrated in Fig. 4, but the force field of one polarization is rotated 45° to the other. *See* ELECTROMAGNETIC RADIATION; PERTURBATION (MATHEMATICS); POLARIZATION OF WAVES.

The only known sources of gravitational radiation strong enough to be directly detectable are violent astrophysical events such as supernova explosions or the gravitational collapse of stars. The number of such catastrophic events that occur in the Milky Way Galaxy can only be roughly estimated. Supernovae are seen only about once every 100 years in the Milky Way, but others may be obscured by dust or gas, or may involve too little visual display to be seen on Earth. Thus the frequency of about 1 per 100 years should be taken as a rough lower limit. This number is consistent with the observed frequency of supernovae seen in other galaxies. For example, in the Virgo cluster of about 2000 galaxies there is about one supernova seen per month.

The number of collapsing stars in the Galaxy can be estimated from the number of pulsars observed, which is about 450, and the average lifetime of a pulsar. Unfortunately, characteristic lifetimes of pulsars vary greatly from about 10^3 to 10^8 years. If a lifetime of 5000 years is adopted, somewhat arbitrarily, a rate of neutron star formation of about 1 per 10 years is obtained. The frequency of catastrophic events in the Galaxy is thus very roughly 1 every 10 or 100 years, a disappointingly small number.

On the other hand, it is possible that intense gravitational radiation is produced in extragalactic sources such as the collapse of entire galactic nuclei, the quasars, or the collisions of black holes. Thus, the amount of gravitational radiation that impinges on the Earth cannot be accurately predicted, and the only reliable procedure is to make observational searches.

When gravitational radiation passes over a solid body, it exerts forces that slightly distort the body. These distortions can be utilized to make massive bodies act as antennas. For example, a large metallic cylinder will ring like a bell when struck with a pulse of gravitational radiation, and the small displacements can in principle be measured. A number of experiments have already been performed with such cylinders, but no reproducible and definitive evidence for gravitational radiation has been obtained. To improve the sensitivity of such antennas, it is necessary to use low-temperature techniques, with cylinders of very high Q cooled to only a few kelvins. The sensitivity should soon approach the limits set by the uncertainty principle of quantum theory, and it is expected that by 1990 many groups of experimentalists will possess antennas capable of detecting most gravitational-radiation-producing events occurring in the Milky Way. If the frequency of such events is in the optimistic end of the range discussed above (about one every 10 years), there could well be an unambiguous detection of gravitational radiation. *See* UNCERTAINTY PRINCIPLE.

The prospects for detecting extragalactic sources, such as supernovae in the Virgo cluster, are much less promising since the radiation from such distant sources would be about 10^6 times weaker than from the Milky Way. Very sophisticated means for obtaining great sensitivity would be necessary, but the high frequency of events makes it an intriguing prospect.

Whereas the direct detection of gravitational radiation is probably many years in the future, there exists very convincing indirect evidence of such radiation. The pulsar PSR 1913+16, discovered by J. Taylor and R. Hulse in 1975, appears to be one component of a close binary system with an orbital period of only 8 h. By using the short pulses from the pulsar as calibrations, observers have been able to obtain very precise information about its orbit, in particular the decrease in orbital period. This decrease is interpreted as due to orbit decay via the emission of energy in the form of gravitational radiation. The numerical value agrees with the predictions of the theory within an accuracy of about 10%.

Relation to electromagnetism. General relativity is the only fundamentally geometric theory in physics and has stood apart from other theories since its inception. Its geometric nature is intimately connected with the principle of equivalence and the independence of the trajectory of a test body on the properties of the body such as its mass. There is no analog of the principle of equivalence in electromagnetism: the motion of a charged test body in an electromagnetic field depends directly on its charge and mass. The lack of such a principle has hindered the development of a true geometric theory of electromagnetism. Many attempts at a classical unified field theory of gravitation and electromagnetism were made by Einstein, H. Weyl, and others, but the results were not very convincing. These attempts seem less interesting since the discovery of other forces in nature

RELATIVITY

key:
— location of undistorted ring of particles
-- distorted ring
forces

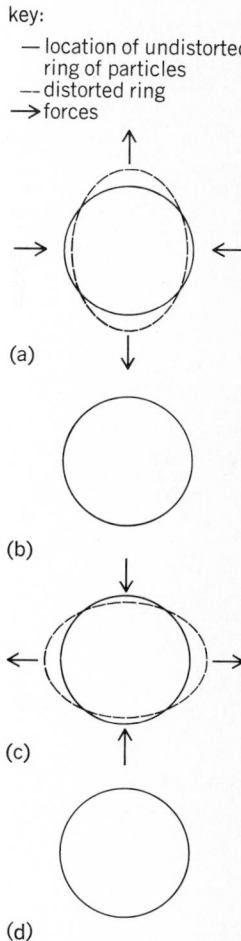

Fig. 4. The forces and distortions produced by a gravitational wave. (*a*) Begin cycle. (*b*) 1/4 cycle (forces equal zero). (*c*) 1/2 cycle (forces reverse). (*d*) 3/4 cycle (forces equal zero).

besides gravitation and electromagnetism. It would be desirable ultimately to unify the strong and weak nuclear forces, electromagnetism, and gravity, and such a grand unification would necessarily involve quantum theory.

Although there is no convincing classical unified field theory, the Maxwell equations of electromagnetism, when expressed in covariant form, are completely consistent with the ideas and equations of general relativity, and no geometric interpretation of the electromagnetic field is logically necessary. In this view the electromagnetic field operates conventionally in the curved space described by the gravitational field equations. In turn, the electromagnetic field contains energy and is thus the source of some of the curvature of the space. Much work has gone into the elucidation of the properties of the coupled Einstein-Maxwell equations in vacuum, sometimes referred to as already-unified field theory, and interesting formal results and interpretations have emerged. *See* Unified field theory.

Constancy of gravity. The ratio of the electrostatic to the gravitational force between the electron and the proton in a hydrogen atom is about 10^{39}. Such an enormous number presents a manifest challenge for theoreticians to explain on some fundamental level; moreover, it is hard to see how a theoretical unification of gravity with the other forces in nature could be accomplished without some understanding of the extreme weakness of gravity.

P. Dirac suggested an interesting approach to this problem. A natural unit for the measurement of time is the time interval required for light to cross an elementary particle diameter, about 10^{-23} s. In terms of this unit the lifetime of the universe is about 10^{40}, roughly the same as the ratio of electrostatic to gravitational forces in the hydrogen atom. Dirac argued that the rough equality of such enormous dimensionless numbers must be more than a coincidence, and suggested that the numbers may be actually equal at all times. This is referred to as the large-number hypothesis. It implies that the inverse of the gravitational constant G may be proportional to the lifetime of the universe, and not be a constant at all. G should decrease by roughly 1 part in 10^{10} per year. That is, gravity weakens with time.

Many approaches to searching for a variation in G have been made using geology, paleontology, celestial mechanics, and so forth, but the only positive result is somewhat uncertain and controversial. The upper limit on the variation of G is now about 1 part in 10^{10} per year, and work is continuing in order to improve the accuracy of the observations.

Relation to quantum theory. Fusing the ideas of general relativity and those of the quantum theory is a difficult problem that is the subject of extensive research. The most naive approach to this problem is to assume that some invariant generalizations of the conventional quantum equations are valid in the curved space specified by the Einstein equations. This approach has led to many important results and insights such as the evaporation of black holes, discussed above. However, it does not provide a natural way to quantize the gravitational field itself and thus is not a consistent approach. For example, the uncertainty principle of quantum theory could be subverted, in principle, by gravitational measurements, which presents a manifest inconsistency.

A more fruitful approach is to deemphasize the geometric interpretation of gravity and develop the theory in analogy with successful quantum field theories, in particular quantum electrodynamics. Quantum electrodynamics accurately describes the interactions of charged particles, such as electrons, with photons, which are the quanta of the electromagnetic field. This theory has already served as a model for the development of quantum field theories of both the weak and strong nuclear interactions, and the two theories of quantum electrodynamics and the weak interactions have been combined by S. Weinberg and A. Salam in a very successful unified theory. The relatively successful theory of the strong interactions, known as quantum chromodynamics, describes the interactions of quarks with the quanta of the strong nuclear force, called gluons. Much attention has been focused on grand unified theories, which attempt to describe in a coherent way the strong, electromagnetic, and weak interactions—that is, all the fundamental forces of nature except gravity. *See* Fundamental interactions; Gluons; Quantum chromodynamics; Quantum electrodynamics; Quarks; Strong nuclear interactions; Weak nuclear interactions.

The basis of much of the success of these quantum field theories has been their renormalizability. This means that certain divergences or infinite results that occur during a calculation do not appear in the final answer. Thus the predictions for physically measurable effects are finite. The divergences disappear by rather subtle cancellations.

When a quantum theory of gravity is constructed by analogy with quantum electrodynamics, it is convenient to abandon or at least deemphasize the geometric interpretation of the gravitational field as necessary to the basic physics. Then the formal construction of the theory is in fact not difficult, and the quanta of the field, called gravitons, can be studied. However, quantum general relativity is not renormalizable (as yet) and can predict infinite results for real experiments, which is manifestly unacceptable.

Some progress has been made in this problem with a generalization of the theory known as supergravity. In supergravity the quanta of the gravitational field are grouped into a family with the quanta of another field, and some of the divergences of the theory do indeed vanish. However, supergravity and variations are still in a preliminary stage of development and as yet offer only hopes and hints as to how theoretical study might proceed in the future.

Quantum field theories at present are based on the assumption that a space-time continuum exists down to indefinitely small distances and short times. Many physicists have noted that this assumption cannot in fact be justified by experience and that difficulties with the concept probably occur for distance scales of about 10^{-31} cm and times of 10^{-42} s. The geometric view of gravity suggests that the energy density of quantum fluctuations at such a scale should distort space-time out of all recognition and usefulness. Field theories other than gravity may avoid such difficulties by the process of renormalization, but it may be that a successful quantum theory of gravity will require a much deeper understanding of space and time at very small scales. *See* NONRELATIVISTIC QUANTUM THEORY; RELATIVISTIC QUANTUM THEORY.

[RONALD J. ADLER]

Bibliography: R. J. Adler, M. Bazin, and M. M. Schiffer, *Introduction to General Relativity*, 1975; A. Einstein, *The Meaning of Relativity*, 1956; N. D. Mermin, *Space and Time in Special Relativity*, 1968; C. Misner, K. Thorne, and J. Wheeler, *Gravitation*, 1973; W. Pauli, *Theory of Relativity*, 1958; W. Rindler, *Essential Relativity*, 1969; H. M. Schwartz, *Introduction to Special Relativity*, 1968, corrected reprint, 1977; S. Weinberg, *Gravitation and Cosmology*, 1972.

Relaxation time of electrons

The characteristic time for a distribution of electrons in a solid to approach or "relax" to equilibrium after a disturbance is removed. The most familiar example is a transport property such as electrical resistivity, in which the disturbance is the electric field, equilibrium is the state of zero current, and relaxation time is inversely proportional to the resistivity. Closely related is the concept of a lifetime, which is related to equilibrium properties rather than transport properties. The lifetime is the mean time that an electron will remain in a given quantum state before changing its state as a result of a collision. A related concept is the electron mean free path, the average distance traveled before a collision. Although characteristic collision times are quite short (of the order of 10^{-14} s at room temperature), mean free paths are surprisingly long (ranging from about 100 atom distances at room temperature to 1,000,000 atom distances in a pure metal near absolute zero), due to the large velocity (of the order of 10^6 m/s) with which electrons at the Fermi level travel. This is surprising because atoms in a solid are very densely packed. It corresponds classically to the unlikely event that a rifle bullet might travel for miles through a dense forest without hitting a tree. From a practical point of view, that is what makes metals useful as electrical conductors. The explanation of the great length of the mean free path is one of the major successes of the theory of solids. *See* FREE-ELECTRON THEORY OF METALS.

Definition. A relaxation time appears in the simplest approximate expression for the electrical conductivity σ, given by Eq. (1), where n is the

$$\sigma = ne^2\tau/m \qquad (1)$$

number density of electrons able to participate in conduction, m is the electron's effective mass in the solid, and τ is the relaxation time. The term "relaxation" is used because the conduction process may be viewed as the steady-state balance between the accelerating effect of an electric field and the decelerating effect of collisions. This process is best viewed in terms of the probability distribution function for the electron gas, $f(\mathbf{k},\mathbf{r},t)$, where $\mathbf{k}$ specifies the momentum, $\mathbf{r}$ the position, and t the time ($\mathbf{k}$ also labels the quantum number of the electrons). Viewed in $\mathbf{k}$-space, the entire distribution shifts from its equilibrium distribution f_0 under the influence of a perturbation such as an electric field. For example, in f_0 the collection of occupied states in $\mathbf{k}$-space is bounded by the Fermi

surface centered at the origin, while in f this region is shifted (Fig. 1). A steady state is reached due to collisions which tend to restore the distribution to f_0; these collisions are due to lattice vibrations, impurities or lattice imperfections, or other electrons. The relaxation time is most simply defined in terms of the rate at which the perturbed distribution approaches its equilibrium value once the perturbation is removed, as given in Eq. (2).

$$\left(\frac{\partial f}{\partial t}\right)_{\text{coll.}} = \frac{f - f_0}{\tau} \qquad (2)$$

Constructive interference. Electron mean free paths in solids are extremely long because the electron wave (viewed quantum-mechanically) readjusts in a perfectly periodic crystal lattice to avoid the atom cores and spend most of its time in the spaces between. In the analogy of the rifle in the forest, the bullet will not travel very far, but the sound of the gun can, because sound waves bend around the trees. In a perfectly periodic lattice, the scattering vanishes because the electron waves scattered from the ion cores interfere in a coherent and constructive way. Any disturbance to the lattice periodicity tends to destroy this wave pattern, resulting in less transmission of the electron wave. The conductivity of solids at room temperature, for example, is limited by the scattering produced by atomic vibrations. At lower temperatures, when the vibrations are reduced, the conductivity is limited by scattering from impurities and lattice imperfections. *See* CRYSTAL DEFECTS; ELECTRICAL RESISTIVITY; LATTICE VIBRATIONS.

Local measurements. The direct measurement of electron lifetimes for particular electron states in $\mathbf{k}$-space has become possible. The techniques generally involve some kind of resonance in space (for example, radio-frequency size effect), time (cyclotron resonance), or energy (de Haas–van Alphen effect), and are generally observed at low temperatures where such resonances are strong. The richness of detail made possible by such "$\mathbf{k}$-space microscopes" has greatly aided the understanding of the interactions between conduction electrons and impurities or lattice imperfections. *See* CYCLOTRON RESONANCE EXPERIMENTS; MAGNETIC RESONANCE.

For example, measurements of conduction electron scattering from lattice vibrations (often called

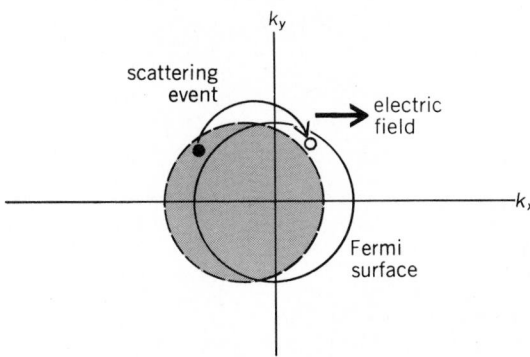

Fig. 1. Effect of electric field on electron distribution in a solid, viewed in $\mathbf{k}$-space. Shaded area indicates occupied states in distribution f which result when field is applied.

phonons) reveal that the strongest interaction occurs when the electron's wave vector is nearly commensurate with the lattice. This is shown for copper in Fig. 2. The scattering rate is equal to γT^3, where T is the temperature and the coefficient γ is determined by the experiment. The wave vectors $\mathbf{k}$ of conduction electrons lie near the Fermi surface shown in Fig. 2a, and as shown in Fig. 2b,

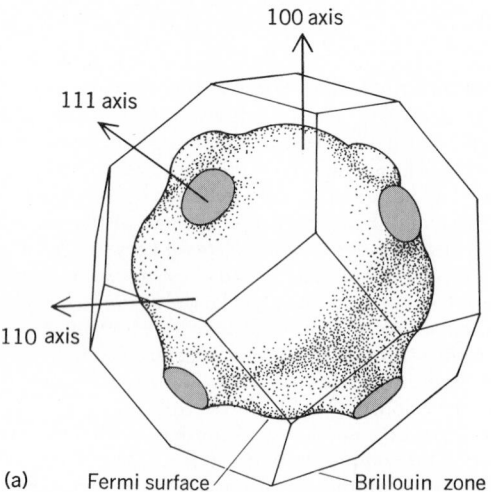

(a) Fermi surface   Brillouin zone

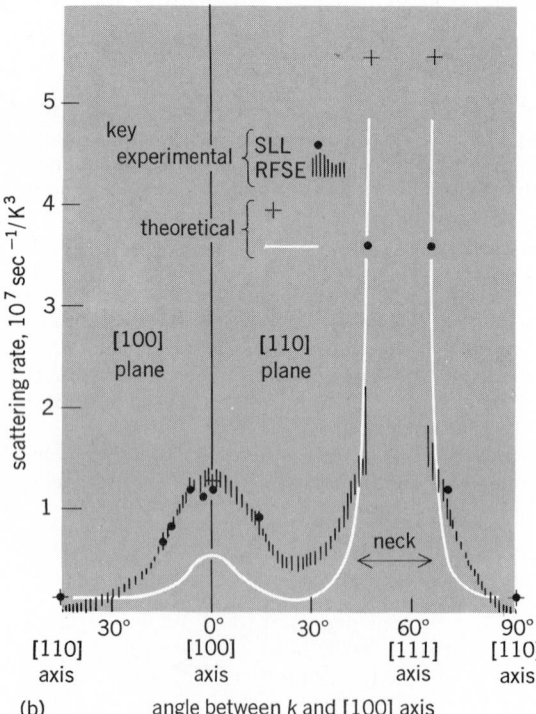

(b)   angle between k and [100] axis

Fig. 2. Scattering of conduction electrons in copper from lattice vibrations. (a) Fermi surface and Brillouin zone of copper. (b) Dependence of scattering coefficient γ on the direction of the wave vector of the conduction electron, according to theoretical calculations and experimental measurements using the radio-frequency size effect (RFSE) and the cyclotron resonance of surface Landau levels (SLL). (From J. F. Koch and R. E. Dozema, The anisotropy of thermal scattering on the Fermi surface of metals, in Proceedings of the 14th International Conference on Low Temperature Physics, pp. 314–338, 1975)

the scattering is strongest when $\mathbf{k}$ is near the [111] axis, where it is commensurate with the lattice periodicity.

Measurements of impurity scattering have shown that the symmetry character of both the impurity and host are important in the scattering. The scattering is strongest where the symmetry is most nearly similar and where a "resonant" scattering is most nearly achieved. Thus, studies of impurity scattering rates allow the symmetry of the scattering to be deduced. For example, for nickel in copper (Fig. 3a), the scattering of conduction electrons is largest near the "belly" region of the Fermi surface at the [110] axis where the host is very d-like, establishing that the scattering is mostly from the nickel d-band resonance. For germanium in copper (Fig. 3b), by contrast, the scattering is low at [110] and large at the [100] "belly" and [111] "neck" regions, establishing that the scattering due to the extra charge on the germanium involves mostly s- and p-like components.

Measurements of conduction electron scattering from lattice dislocations, the disturbance of lattice periodicity which makes metals deformable, show that the lifetime differs by more than three orders of magnitude in different kinds of measurement, depending upon the extent to which the measurement is sensitive to small-angle scattering. The long-range elastic strain of the dislocation gives small-angle scattering which is seen, for example, in the de Haas–van Alphen effect but not in resistivity. This is shown graphically in Fig. 4. The differential scattering cross sections are the same for the three techniques shown, but the scattering rates, corresponding to the total shaded areas, differ drastically because of the minimum angle θ to which the measurements are sensitive. For a dislocation "forest" in which 1 out of every 10,000 atoms in a row is out of place, the rate is 10^{11} s^{-1} in the de Haas–van Alphen effect, but only 10^8 s^{-1} in resistivity, and is intermediate in the radio-frequency size effect, which is moderately sensitive to small-angle scattering. These differences arise because in electrical conductivity small-angle scattering leaves the electron still contributing to current flow, whereas for a quantum effect even a tiny scattering event destroys phase coherence. See BAND THEORY OF SOLIDS.

Transport measurements. A measure of τ may be extracted from Eq. (1) by combining the conductivity with a measurement of the Hall coefficient R_H, obtained from the voltage V_H appearing at right angles to the transport current density J and an applied magnetic field B by Eq. (3), where w

$$R_H = V_H/(wBJ) = 1/nec \qquad (3)$$

is the width of the specimen, e is the electronic charge, and c is the speed of light. The carrier density n drops out in the combination of Eq. (4).

$$cR_H\sigma = e\,\tau/m = \mu \qquad (4)$$

The mobility μ is a measure of the speed of response and also of the power dissipation in semiconductor devices, and is thus an important figure of merit in integrated circuits. See HALL EFFECT.

The behavior of semiconductor devices is dominated by another lifetime, the minority carrier recombination time, the mean time an electron can survive before recombining with a hole in p-type

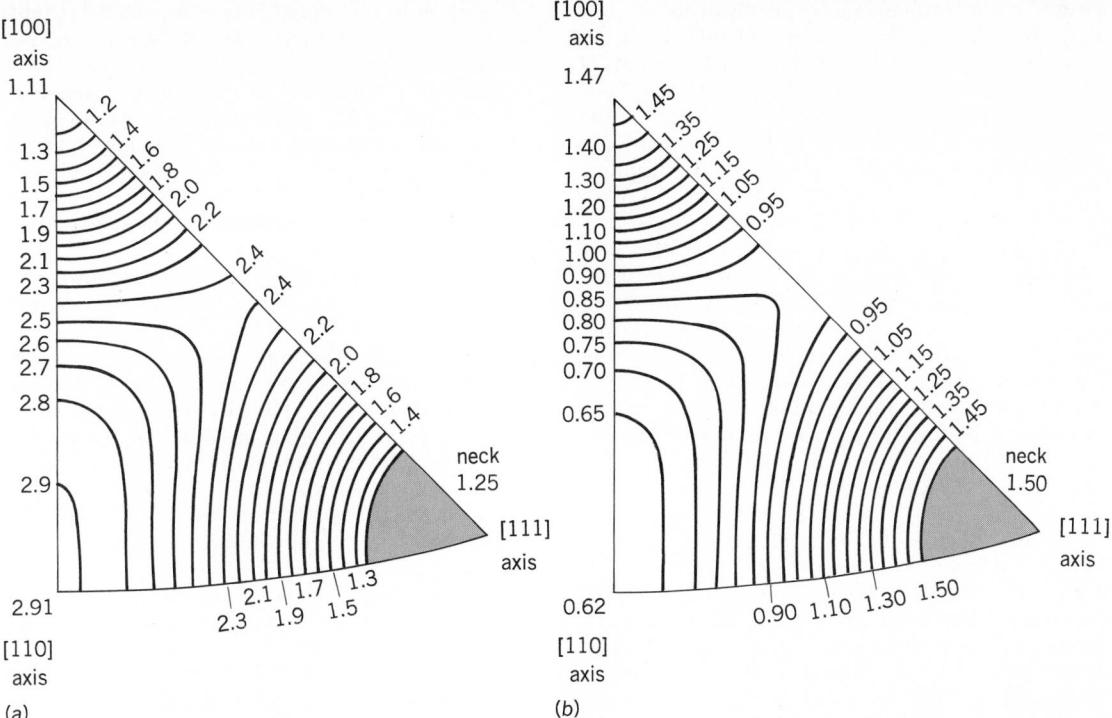

(a)

(b)

Fig. 3. Topographic maps of impurity scattering rate of conduction electrons on the Fermi surface of copper. Contours connect wave vectors with equal scattering rates. Numbers indicate value of scattering rate, in units of 10^{13} s^{-1}. (a) Nickel in copper, (b) Germanium in cop-per. (*From R. G. Poulson, D. L. Randles, and M. Springford, Studies of conduction electron scattering in copper using the de Haas–van Alphen effect, J. Phys., F4:981–998, 1974*)

material (or vice versa). The key feature in the invention of the junction or bipolar transistor was fabricating a base region thin enough so a minority carrier injected from the adjacent emitter could survive recombination long enough to reach the collector. Recombination also limits the efficiency of solar cells, since the electron-hole pairs created by light must survive long enough to contribute to the current at the cell's terminals. *See* HOLES IN SOLIDS; TRANSISTOR.

Recombination times are in the range 10^{-4} to 10^{-9} s, many orders of magnitude longer than the transport relaxation time above, and may be measured by methods such as photocurrent decay, the characteristic time for the excess current created by a light beam to decay away after the light is turned off. The dynamics of minority carrier transport may be studied in more detail by injecting excess carriers into a material in an electrical field and observing the decay of the pulse "downstream" in the current flow.

Recombination centers or traps are most effective if located near the center of the forbidden bandgap in the material, in contrast to normal doping centers which are located near band edges. The recombination process which begins by capture of an electron must end with capture of a hole to complete the process, rather than by reemission of the electron which is more feasible toward band edges. Recombination centers may be impurities, lattice imperfections, or surface states. As few as 10^{16} gold atoms/cm^3 in silicon (less than 1 part in 10^6) will shorten the recombination time to 10^{-9} s. *See* TRAPS IN SOLIDS.

Intrinsic surface states delayed the perfection of the metal oxide semiconductor (MOS) transistor. A clean silicon surface contains one dangling bond per atom, which acts as a trap. MOS transistor and integrated circuit fabrication includes passivation of these surface states during oxide formation, by processes often poorly understood. An internal version of surface states is the dangling bond on

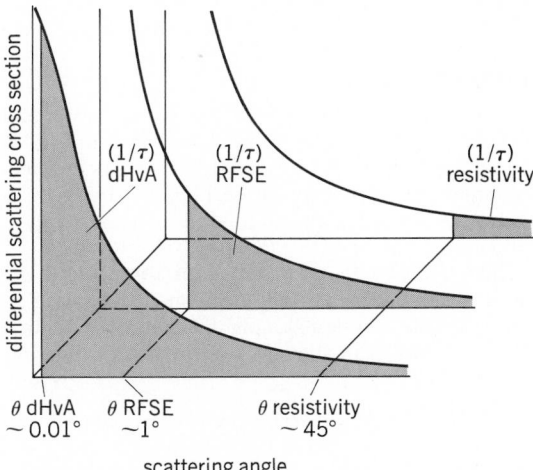

Fig. 4. Scattering of conduction electrons in copper from lattice dislocations. Graphs of differential scattering cross section for de Haas–van Alphen effect (dHvA), radio-frequency size effect (RFSE), and resistivity are shown in perspective. Magnitudes of shaded areas indicate corresponding scattering rates $(1/\tau)$. (*Y. K. Chang and R. J. Higgins, DeHaas–van Alphen effect study of dislocations in copper, Phys. Rev., B12:4261–4281, 1975*)

lattice imperfections such as dislocations and grain boundaries. This is why semiconductor devices are ordinarily fabricated on single crystals of high perfection. Advances in the use of amorphous or polycrystalline materials, particularly in the fabrication of large-area solar cells, have centered on discovering methods to neutralize the trapping action of dangling bonds and thereby increase minority carrier recombination time.

The energy given up in the recombination process may be taken away either as heat or as light. The second process is enhanced in so-called direct gap materials such as gallium arsenide (GaAs), where the conduction band is located vertically above the valence bond in momentum space, and leads to the light-emitting diode used in pocket calculators and watches, and to the solid-state laser important in telecommunications. *See* LASER; SEMICONDUCTOR.

[RICHARD J. HIGGINS]

Bibliography: N. W. Ashcroft and N. D. Merman, *Solid State Physics*, 1976; M. Ya Azbel', M. I. Kaganov, and I. M. Lifshitz, Conduction electrons in metals, *Sci. Amer.*, 228(1):88–98, 1973; F. J. Blatt, *Physics of Electronic Conduction in Solids*, 1968; A. S. Grove, *Physics and Technology of Semiconductor Devices*, 1967; D. H. Lowndes and F. M. Mueller (eds.), Electron lifetimes in metals, *Physics of Condensed Matter*, vol. 19, pp. 1–434, 1975; M. Springford, The anisotropy of conduction electron scattering in the noble metals, *Advances in Physics*, vol. 20, pp. 483–550, 1971; S. M. Sze, *Physics of Semiconductor Devices*, 1969.

Reluctance

A property of a magnetic circuit analogous to resistance in an electric circuit.

Every line of magnetic flux is a closed path. Whenever the flux is largely confined to a well-defined closed path, there is a magnetic circuit. That part of the flux that departs from the path is called flux leakage. *See* MAGNETIC CIRCUITS; MAGNETIC FLUX.

For any closed path of length l in a magnetic field H, the line integral of $H \cos \alpha \, dl$ around the path is the magnetomotive force (mmf) of the path, as in Eq. (1), where α is the angle between H and the path. If the path encloses N conductors, each

$$\text{mmf} = \oint H \cos \alpha \, dl \qquad (1)$$

with current I, Eq. (2) holds. *See* MAGNETOMOTIVE FORCE.

$$\text{mmf} = \oint H \cos \alpha \, dl = NI \qquad (2)$$

Consider the closely wound toroid shown in the figure. For this arrangement of currents, the magnetic field is almost entirely within the toroidal coil, and there the flux density or magnetic induction B is given by Eq. (3), where l is the mean circumference of the toroid and μ is the permeability.

$$B = \mu \frac{NI}{l} \qquad (3)$$

The flux Φ within the toroid of cross-sectional area A is given by either form of Eqs. (4), which is simi-

$$\Phi = BA = \frac{\mu A}{l} NI \qquad (4)$$

$$\Phi = \frac{NI}{l/\mu A} = \frac{\text{mmf}}{l/\mu A} = \frac{\text{mmf}}{\mathscr{R}}$$

RELUCTANCE

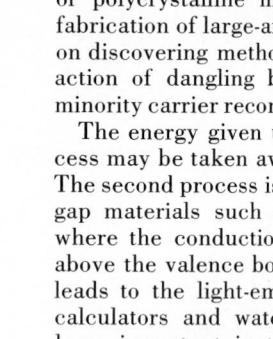

A toroidal coil.

lar in form to the equation for the electric circuit, although nothing actually flows in the magnetic circuit. The factor $l/\mu A$ is called the reluctance $\mathscr{R}$ of the magnetic circuit. The reluctance is not constant because the permeability μ varies with changing flux density. From the defining equation for reluctance, it is seen that when the mmf is in ampere-turns and the flux is in webers, the unit of reluctance is the ampere-turn/weber. *See* MAGNETIC PERMEABILITY.

Reluctances in series. For the simple toroid, all parts of the magnetic circuit have the same μ and the same A. More complicated circuits may include parts that differ in permeability, in cross section, or in both. Suppose a small gap were cut in the core of the toroid. The flux would fringe out at the gap, but as a rough approximation, the area of the gap may be considered the same as that of the core.

The magnetic path then has two parts, the core of length l_1 and reluctance $l_1/\mu_1 A$, and the air gap of length l_2 and reluctance $l_2/\mu_2 A$. Since the same flux is in both core and gap, this is considered a series circuit and Eq. (5) holds. Since the

$$\mathscr{R} = \mathscr{R}_1 + \mathscr{R}_2 = \frac{l_1}{\mu_1 A} + \frac{l_2}{\mu_2 A} \qquad (5)$$

relative permeability of the ferromagnetic core is several hundred or even several thousand times that of air, the reluctance of the short gap may be much greater than that of the much longer core. For any combination of paths in series, $\mathscr{R} = \Sigma l/\mu A$. Then Eq. (6) holds.

$$\Phi = \frac{\text{mmf}}{\Sigma \mathscr{R}} = \frac{\text{mmf}}{\Sigma l/\mu A} \qquad (6)$$

Reluctances in parallel. If the flux divides in part of the circuit, there is a parallel magnetic circuit and the reluctance of the circuit has the same relation to the reluctances of the parts as has the analogous electric resistance. For the parallel circuit Eq. (7) is valid.

$$\frac{1}{\mathscr{R}} = \frac{1}{\mathscr{R}_1} + \frac{1}{\mathscr{R}_2} + \cdots \qquad (7)$$

[KENNETH V. MANNING]

Renormalization

A program in quantum field theory consisting of a set of rules for calculating S-matrix amplitudes which are free of ultraviolet (or short-distance) divergences, order by order in perturbative calculations in an expansion with respect to coupling constants. *See* SCATTERING MATRIX.

Divergences in quantum field theory. To describe the nature of the problem, it is useful to consider the simple example of a ϕ^4 theory defined by the lagrangian density $L(x)$ in Eq. (1) in four-

$$L(x) = \frac{1}{2} \partial_\mu \phi(x) \partial^\mu \phi(x) - \frac{1}{2} m^2 \phi(x)^2 - \frac{1}{4!} \lambda \phi(x)^4 \qquad (1)$$

dimensional Minkowski space-time. Here, $\phi(x)$ is a quantum field operator depending on the four-vector x, ∂_μ and ∂^μ represent differentiation with respect to space-time coordinates, and m and λ are parameters. *See* LAGRANGIAN FUNCTION; RELATIVITY.

If one attempts to calculate any physical process in a perturbative expansion with respect to the coupling constant λ, in terms of Feynman diagrams, one has to confront divergent integrals. An example is the one-loop contribution to two-body scattering which is of order λ^2 (the second diagram in the illustration), and is proportional to the logarithmically divergent momentum integral given by notation (2), where p_1 and p_2 are the initial momenta of the two particles.

$$\lambda^2 \int d^4k[(p_1+k)^2 - m^2 + i\epsilon]^{-1} \\ [(p_2-k)^2 - m^2 + i\epsilon]^{-1} \quad (2)$$

These ultraviolet (large-momentum) divergences have their origin in short-distance singularities occurring in the product of the quantum field operators (or their matrix elements) such as $\phi(x_1)\phi(x_2)$ as x_1 approaches x_2. Therefore, the quantum theory is not well defined since the lagrangian (or hamiltonian) contains these fields multiplied at the same point $x_1 = x_2 = x$. This suggests a procedure for constructing a well-defined finite quantum field theory, and provides the beginning of the resolution of the problem by renormalization.

Regularization procedure. First the nature and kind of singularities must be identified, and then they should be removed to define the physical theory. The theory is first rendered finite by introducing a regularization parameter so that as it approaches a limiting value the divergences appear as definite singularities in this parameter. This process is called regularization. There are many methods of regularization. For example, in the cutoff method the integral in Eq. (2) will be rendered finite if it is cut off at $k^2 = \Lambda^2$, where Λ is finite. The result of integration, which is a function of Λ, will be seen to diverge as log Λ when $\Lambda \to \infty$. Another, more modern method is dimensional regularization: The integral is first performed in n dimensions rather than four. It will converge for $n < 4$ and will have a definite dependence on n. An example of this dependence is given by Eq. (3), where $\Gamma(\alpha)$ is

$$\int d^nk(-k^2+c)^{-\alpha} = i\pi^{n/2}c^{n/2-\alpha}\Gamma(\alpha-n/2)/\Gamma(\alpha) \quad (3)$$

the Euler gamma function. The answer is then analytically continued to complex values of n and the four-dimensional theory is defined as the limit of n approaching 4. In this way, as long as $n \neq 4$ the gamma functions in Eq. (3) are finite, and only finite well-defined quantities are manipulated. In this method the singularities reappear as poles of the form $(n-4)^{-k}$ as n approaches 4. The behavior of the integral of notation (2) is given by notation (4). The remainder of this article is restricted to

$$-\frac{2i\pi^2}{n-4} + (\text{finite}) + 0(n-4) \quad (4)$$

dimensional regularization.

Renormalization of parameters. In a regularized quantum field theory, one can, in principle, calculate finite expressions for all the Feynman diagrams that contribute to any physical process, to any desired order in perturbation theory. The answer will be a well-defined function of the momenta of the external legs, the parameters of the theory, such as m and λ in Eq. (1), and the number of dimensions n, which is taken as a complex number. As n approaches 4, the singularities can be studied and the method for removing them can be given. The degree of divergence of a graph is

determined by Weinberg's theorem. In a field theory such as ϕ^4 of Eq. (1), one finds only two types of Feynman graphs that diverge as $n \to 4$. Those are the graphs with two external legs and four external legs and any other graph that contains these two types of graphs as subgraphs. Therefore, to further study the singularities it is sufficient to concentrate only on the two- and four-point functions of notation (5). The singularities oc-

$$<o|T(\phi(x_1)\phi(x_2))|o> \\ <o|T(\phi(x_1)\phi(x_2)\phi(x_3)\phi(x_4))|o> \quad (5)$$

cur at short distances in the product of the field operators as points approach each other. This suggests that they can be removed from all physical processes by properly defining the local product of operators that appear in the lagrangian of Eq. (1). It turns out that indeed it is possible to absorb and remove all the infinities by a simple redefinition or renormalization of the field $\phi(x)$, the mass parameter m, and the coupling constant λ, as will be outlined below. Because this is possible in the ϕ^4 theory, it is called a renormalizable theory. In fact, in all renormalizable theories it is possible to eliminate all infinities by a simple renormalization of the finite number of constants that define the theory. A theory is called nonrenormalizable when this procedure fails, and it can be rendered finite only by introducing an infinite number of renormalization parameters.

Nonphysical nature of parameters. The possibility of renormalization by a redefinition of parameters such as masses and coupling constants that appear in the lagrangian hinges on the fact that these are not the physical quantities that would be observed as the prediction of the theory. For example, the physical finite coupling constant is defined as the strength of the interaction that an experimentalist will observe in the process shown in the illustration. The strength of the interaction is defined by the measured value of the physical probability amplitude in the illustration, $A(p_1,p_2,p_3,p_4,\lambda,m)$, at an agreed value of the external momenta. This clearly is a complicated function of the parameter λ when the perturbation series is summed. The strength of the interaction is equal to λ only in lowest order and in general differs from it in the full theory. Similar remarks apply to the parameter m. The true mass of the particle in the theory differs from m. It is in general a complicated function of m and λ and is defined as the location of the pole in the full propagator. Similarly, the field $\phi(x)$ does not create the correctly

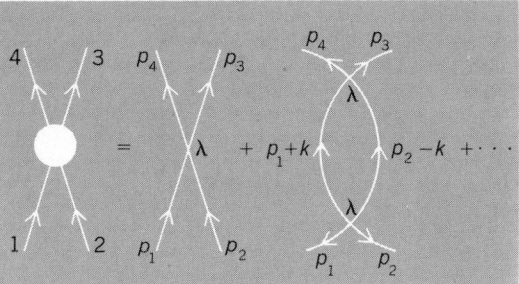

Two-body scattering in lowest-order perturbative expansion.

normalized interacting particle when applied on the vacuum. It differs from the correctly normalized field ϕ_R by the multiplicative wave function renormalization constant $\sqrt{Z}$. Renormalization theory is the process of rewriting systematically the field theory in terms of the physical, finite, renormalized coupling constant λ_R, mass m_R, and field ϕ_R. If the theory is renormalizable and rewritten in terms of renormalized quantities, then all infinities cancel and all physical quantities, such as S-matrix elements, are finite and meaningful, as the regularization parameter is removed.

Power series in the renormalized parameters. An outline of this procedure can be given as follows. Since it has been agreed that λ, m and the renormalization of the field are not the observed physical quantities, they need not be finite. They are reexpressed as power series in terms of finite renormalized parameters λ_R and m_R and the regularization parameter. In the dimensional regularization scheme they take the form of Eqs. (6),

$$\lambda = \mu^{4-n}\left(\lambda_R + \frac{a_1(\lambda_R)}{n-4} + \frac{a_2(\lambda_R)}{(n-4)^2} + \cdots\right)$$

$$m^2 = m_R^2\left(1 + \frac{b_1(\lambda_R)}{n-4} + \frac{b_2(\lambda_R)}{(n-4)^2} + \cdots\right) \quad (6)$$

$$Z = 1 + \frac{c_1(\lambda_R)}{n-4} + \frac{c_2(\lambda_R)}{(n-4)^2} + \cdots$$

where the renormalized field ϕ_R is defined as $\phi = \sqrt{Z}\phi_R$. Here $a_i(\lambda_R)$, $b_i(\lambda_R)$, $c_i(\lambda_R)$ are power series in the finite parameter λ_R; that is, $a_i(\lambda_R) = \Sigma a_{ik}\lambda_R^k$, and so forth, where the coefficients a_{ik}, b_{ik}, c_{ik} are to be determined so as to render the theory finite as described below. Here μ is a mass parameter called the renormalization point. All Green's functions or probability amplitudes such as the one in the illustration will be renormalized by setting the momenta at $p_i^2 = \mu^2$, and so forth.

The perturbation series is now arranged as an expansion in the finite parameter λ_R and not in λ. This can be done by rewriting the lagrangian in terms of renormalized field ϕ_R and the parameters λ_R and m_R by substituting directly from Eqs. (6) into Eq. (1). Thus, the lagrangian takes the form of Eq. (7), where $\Delta L_{\text{counterterm}}$, the "counterterm"

$$L = \frac{1}{2}\partial_\mu\phi_R\partial^\mu\phi_R - \frac{1}{2}m_R^2\phi_R^2$$
$$- \mu^{4-n}\lambda_R\phi_R^4 + \Delta L_{\text{counterterm}} \quad (7)$$

piece, has the same functional dependence on ϕ_R as above. This piece would diverge as $n \to 4$ and is a power series in λ_R involving the coefficients a_{ik}, b_{ik}, c_{ik} mentioned above.

Minimal subtraction. These coefficients are determined in order to cancel all infinities by the following procedure: The amplitude $A(p_i;\lambda_R)$ in the illustration and the propagator $\Delta_F(p,\lambda_R)$ are calculated to any desired order in λ_R in terms of Feynman graphs [these are related to the Green's functions in Eq. (5)]. To these there are contributions from the explicit part of the lagrangian in Eq. (7) and also from the counterterms which involve the coefficients a_{ik}, b_{ik}, and c_{ik} up to the desired order in λ_R. Then, order by order, one requires that the poles $(n - 4)^{-i}$ must completely cancel in the three measurable quantities given in Eq. (8). This is called the minimal subtraction

$$A(p_i,\lambda_R)\big|_{p_2=\mu_2} = \text{finite as } n \to 4$$
$$\Delta_F^{-1}(p,\lambda_R)\big|_{p_2=\mu_2} = \text{finite as } n \to 4$$
$$\frac{\partial^2}{\partial p^2}\Delta_F^{-1}(p,\lambda_R)\big|_{p_2=\mu_2} = \text{finite as } n \to 4 \quad (8)$$

scheme, and it completely fixes the coefficients a_{ik}, b_{ik}, and c_{ik}. There are other subtraction schemes which require the above quantities to be equal to a fixed observed value rather than being simply finite as $n \to 4$. The observable quantities such as the S-matrix elements are independent of the particular subtraction scheme adopted.

Finite S-matrix. As mentioned above, the only "primitive" infinities in the ϕ^4 theory appear in the Green's functions of Eq. (5) containing two or four external legs and in those graphs containing them as subgraphs. With the above renormalization procedure (choosing a_{ik}, b_{ik}, c_{ik}) all such infinities in these two- and four-point functions have been arranged to cancel. It remains to be shown that as $n \to 4$ (removal of cutoff) the S-matrix is finite. This is done by mathematical induction: Assume that it is true to order m in perturbation theory, then prove it to order $m + 1$. This is carried out successfully if the theory is renormalizable. Therefore, all infinities can be removed by a proper definition of the products of the field operators appearing in the lagrangian. This amounts to properly identifying the observable finite quantities which led to a renormalization of the field, the coupling constant and the mass as in Eq. (6). Any other physical quantity can now be calculated and will be a function of only the renormalized finite parameters λ_R and m_R, in addition to momenta.

Examples of renormalizable fields. So far the only field theories known to be renormalizable in four dimensions are those which include spin-0, spin-$\frac{1}{2}$ and spin-1 fields such that no term in the lagrangian exceeds operator dimension 4. The operator dimension of any term is calculated by assigning dimension 1 to bosons and derivatives ∂_μ, and dimension 3/2 to fermions. Spin-1 fields are allowed only if they correspond to the massless gauge potentials of a locally gauge-invariant Yang-Mills-type theory associated with any compact Lie group. The gauge invariance can remain exact or can be allowed to break via spontaneous breakdown without spoiling the renormalizability of the theory. In the latter case the spin-1 field develops a mass. The successful quantum chromodynamics theory describing the strong forces and the $\text{SU}(2) \times \text{U}(1)$ S. Weinberg–A. Salam–S. Glashow gauge model of unified electroweak particle interactions are such renormalizable gauge models containing spin-0, $-\frac{1}{2}$, and -1 fields. The renormalization procedure in a gauge theory is much more complicated than in the simple ϕ^4 theory because of gauge fixing and lack of either manifest unitarity or Lorentz invariance, but gauge theories have been shown to be renormalizable. *See* FUNDAMENTAL INTERACTIONS; QUANTUM CHROMODYNAMICS; WEAK NUCLEAR INTERACTIONS.

Renormalization group. An important topic in renormalization theory is the renormalization group. This is the study of the dependence of the theory on the renormalization point μ that appeared in the subtraction procedure of Eqs. (6) and (8). Observable quantities such as the S-matrix elements, the measured strength of the interaction in the illustration, and so forth, do not depend on μ, but the finite expansion parameter λ_R or the mass parameter m_R do (the measurable quantities are functions of these). Changing the value of the renormalization point induces new values of λ_R and m_R in such a way as to keep the measurable S-matrix elements unchanged. Since changing the mass parameter μ can be viewed as a change of scales, the renormalization group is intimately connected to scale transformations in the theory. It makes it possible to study the high- or low-energy behavior of the field theory by applying scale transformations to the momenta of the particles involved in a particular reaction. It is then found that the value of the effective expansion parameter $\lambda_R(\mu)$ of the theory depends on the energy scales of the reaction under study. For certain gauge theories the effective coupling constant, which is a measure of the interaction, decreases as the energy scale increases. This behavior is called asymptotic freedom, and it explains why quarks act as free particles (small effective coupling constant) when they collide at very high energies and come to within very short distances of each other. *See* QUANTUM FIELD THEORY; QUARKS; SYMMETRY LAWS.   [ITZHAK BARS]

Bibliography: C. Itzykson and J. B. Zuber, *Quantum Field Theory*, 1980; W. Marciano and H. Pagels, *Phys. Rep.*, C36:137, 1978.

Resolving power

A quantitative measure of the ability of an optical instrument to produce separable images. The images to be resolved may differ in position because they represent (1) different points on the object, as in telescopes and microscopes, or (2) images of the same object in light of two different wavelengths, as in prism and grating spectroscopes. For the former class of instruments, the resolving limit is usually quoted as the smallest angular or linear separation of two object points, and for the latter class, as the smallest difference in wavelength or wave number that will produce separate images. Since these quantities are inversely proportional to the power of the instrument to resolve, the term resolving power has generally fallen into disfavor. It is still commonly applied to spectroscopes, however, for which the term chromatic resolving power is used, signifying the ratio of the wavelength itself to the smallest wavelength interval resolved. The figure quoted as the resolving power or resolving limit of an instrument may be the theoretical value that would be obtained if all optical parts were perfect, or it may be the actual value found experimentally. Aberrations of lenses or defects in the ruling of gratings usually cause the actual resolution to fall below the theoretical value, which therefore represents the maximum that could be obtained with the given dimensions of the instrument. This maximum is fixed by the wave nature of light and may be calculated for given conditions by diffraction theory. *See* DIFFRACTION.

Chromatic resolving power. The chromatic resolving power R of any spectroscopic instrument, including prisms, gratings, and interferometers, is defined by Eq. (1), where $\delta\lambda$ represents the

$$R = \frac{\lambda}{\delta\lambda} \qquad (1)$$

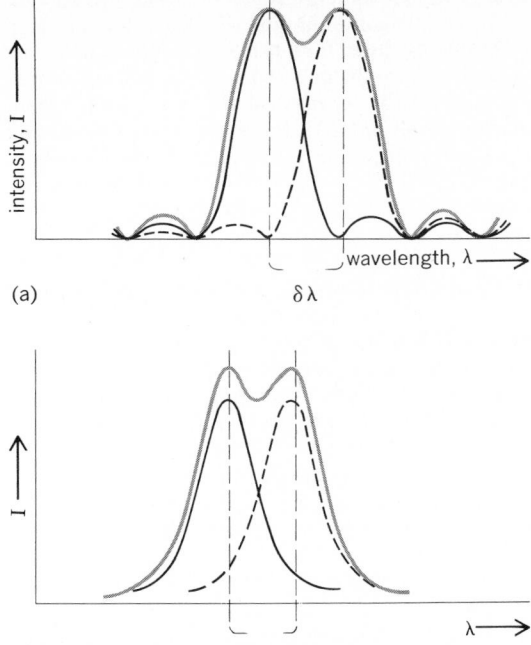

Resolution of two spectrum lines (*a*) when the shape is determined by diffraction (Rayleigh criterion), and (*b*) when the shape follows the Airy formula. The latter is applicable to multiple-beam interferometers.

difference in wavelength of two equally strong spectrum lines that can barely be separated by the instrument, and λ the average wavelength of these two lines. It is necessary to specify more precisely the term "barely separated," and for prisms and gratings, in which the width of the lines is determined by diffraction, this is done by use of Rayleigh's criterion. Illustration *a* shows the contours of two similar spectrum lines which are at the limit of resolution according to this criterion. The lighter curve represents the line shape due to Fraunhofer diffraction for the wavelength λ, the dashed curve that for $\lambda + \delta\lambda$, and the heavy curve the sum of the two. Rayleigh's criterion specifies that the lines are resolved when the principal maximum of one falls exactly on the first minimum (zero intensity) of the other. Diffraction theory shows that the intensity I of either pattern at the central crossing point is $4/\pi^2$ of that at the maximum, so that the curve representing the sum dips to 81% at the center. The theory also shows that the angular separation $\delta\theta$ of the rays forming the two maxima is λ/a, where a is the linear width of the beam of light emerging from the prism or grating. Hence, quite generally for such an instrument, the resolving power may be defined by Eq. (2).

$$R = \frac{\lambda}{\delta\lambda} = \frac{\lambda}{\delta\theta}\frac{d\theta}{d\lambda} = a\frac{d\theta}{d\lambda} \qquad (2)$$

Expressed in words, Eq. (2) means that

$$\begin{pmatrix} \text{Chromatic} \\ \text{resolving power} \end{pmatrix} =$$

$$\begin{pmatrix} \text{width of} \\ \text{emergent beam} \end{pmatrix} \times (\text{angular dispersion})$$

In a given instrument, the calculation of resolving power thus involves finding the last two quantities.

Resolving power of prisms. When a prism is used at minimum deviation, the resolving power depends on the length b of the base of the prism and the slope $dn/d\lambda$ of the dispersion curve giving the wavelength variation of the refractive index n. Thus Eq. (3) holds. Here the assumption is made

$$R = b\frac{dn}{d\lambda} \tag{3}$$

that the prism is completely filled by the beam of light. If it is not, b must represent the difference in path length between the longest and shortest rays through the prism. See OPTICAL PRISM.

Resolving power of gratings. This equals the product of the order of interference m and the total number of rulings N. The order m may be expressed in terms of the grating space s and the angles α and β of incidence and diffraction. Thus Eq. (4) holds. Here w is the width of the ruled area of

$$R = mN = \frac{Ns(\sin\alpha + \sin\beta)}{\lambda}$$
$$= \frac{w(\sin\alpha + \sin\beta)}{\lambda} \tag{4}$$

the grating. For the limiting case of grazing angles of incidence and diffraction, the maximum possible R is seen to be $2w/\lambda$, or the number of wavelengths in twice the width of the grating. See DIFFRACTION GRATING.

Resolving power of interferometers. For the type of interferometer most commonly used, the Fabry-Perot interferometer, the resolving power may be expressed as the product of the order of interference $m = 2t/\lambda$, where t is the separation of the interferometer mirrors, and an effective number N_{eff} of interfering beams. For interferometers the line contour of the spectrum lines is not that of Fraunhofer diffraction, but is given by a relation called the Airy formula. This contour has no points of zero intensity but has the general shape shown in illustration b. Therefore, the Rayleigh criterion cannot be applied in the usual way. If, however, the two curves are made to cross at the half-intensity point of each, it is found that there is a dip of approximately 20% in the resultant curve. The value of N_{eff} is thereby specified, and the resolving power R is given by Eq. (5), where ρ designates the

$$R = mN_{eff} = m\left(\frac{\pi\sqrt{\rho}}{1-\rho}\right) \tag{5}$$

reflectance of the interferometer plates. See INTERFEROMETRY.

Resolving power of telescopes. This depends on the size of the diffraction maximum produced when light from a distant point source passes through a circular aperture of size equal to that of the objective lens or mirror. A graph of the intensity in the diffraction pattern plotted against radial distance closely resembles one of the curves of illustration a, and hence the pattern consists of a central spot surrounded by faint rings. The angular radius of the first dark ring corresponds, by the Rayleigh criterion, to the angular separation of two point sources that are barely resolved. Theory gives this angle, which represents the resolving limit, as defined by Eq. (6) for $\lambda = 5600$ A and d, the

$$\alpha = \frac{1.220\lambda}{d} \text{ radians} = \frac{14.1}{d} \text{ seconds of arc} \tag{6}$$

diameter of the objective lens, in centimeters.

Resolving power of microscopes. This is determined by diffraction of a circular aperture representing the exit pupil of the microscope objective. There are two important differences between the resolving power of microscopes and that of telescopes. First, the resolving limit of microscopes is expressed in terms of the smallest distance l between two points on the object that are just resolved. Second, this limit depends on the mode of illumination of the object. If the illumination is incoherent, so that there is no constant phase relation between light from adjacent points, the resolving limit is given by Eq. (7), where n is the refrac-

$$l = \frac{0.61\lambda}{n\sin\alpha} \tag{7}$$

tive index of the material (for example, oil) in the object space, and α the angle that the extreme ray entering the objective makes with the axis of the instrument. The quantity $n\sin\alpha$ is called the numerical aperture of the objective. With coherent illumination the resolving limit is given by this formula, with 1.0 in place of 0.61, provided the illumination is central. When the object is illuminated from a point slightly to one side, the factor may be reduced to 0.5.

[FRANCIS A. JENKINS/GEORGE R. HARRISON]

Bibliography: G. A. Boutry, *Instrumental Optics*, 1962; D. F. Horne, *Optical Instruments and Their Applications*, 1980; F. A. Jenkins and H. E. White, *Fundamentals of Optics*, 4th ed., 1976.

Resonance (acoustics and mechanics)

When a mechanical or acoustical system is acted upon by an external periodic driving force whose frequency equals a natural free oscillation frequency of the system, the amplitude of oscillation becomes large and the system is said to be in a state of resonance.

When a simple oscillator of mass m, stiffness constant s, and mechanical damping constant R is driven by a periodic driving force $F\cos 2\pi ft$, it vibrates with a velocity amplitude given by the equation below, which implies that (1) the amplitude

$$V = \frac{F}{[R^2 + (2\pi fm - s/2\pi f)^2]^{1/2}}$$

becomes a maximum when the driving frequency is $f = (1/2\pi)\sqrt{s/m}$, that is, at the natural free oscillation frequency of the oscillator, (2) small damping constants R are associated with large amplitudes of vibration at resonance, and (3) the smaller R, the more rapidly the amplitude decreases as the driving frequency departs from the resonance frequency. In addition, driving any vibrating system at its resonance frequency is characterized by a maximum dissipation of power.

A knowledge of both the resonance frequency and the sharpness of resonance is essential to any discussion of driven vibrating systems. When a vibrating system is sharply resonant, careful tuning is required to obtain the resonance condition. Mechanical standards of frequency must be sharply resonant so that their peak response can easily be determined. In other circumstances, resonance is undesirable. For example, in the faithful recording and reproduction of musical sounds, it is necessary either to have all vibrational resonances of the system outside the band of frequencies being reproduced or to employ heavily damped systems. *See* SYMPATHETIC VIBRATION; VIBRATION.

[LAWRENCE E. KINSLER]

Resonance (alternating-current circuits)

A condition in a circuit characterized by relatively unimpeded oscillation of energy from a potential to a kinetic form. In an electrical network there is oscillation between the potential energy of charge on capacitance and the kinetic energy of current in inductance. This is analogous to the mechanical resonance seen in a pendulum.

Three kinds of resonant frequency in circuits are officially defined. Phase resonance is the frequency at which the phase angle between sinusoidal current entering a circuit and sinusoidal voltage applied to the terminals of the circuit is zero. Amplitude resonance is the frequency at which a given sinusoidal excitation (voltage or current) produces the maximum oscillation of electric charge in the resonant circuit. Natural resonance is the natural frequency of oscillation of the resonant circuit in the absence of any forcing excitation. These three frequencies are so nearly equal in low-loss circuits that they do not often have to be distinguished.

Phase resonance is perhaps the most useful in many practical situations, as well as being slightly simpler mathematically. The following discussion considers phase resonance in passive, linear, two-terminal networks.

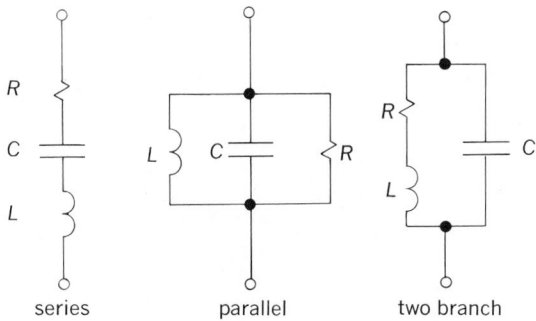

Fig. 1. Resonant circuits.

Resonance can appear in two-terminal networks of any degree of complication, but the three circuits shown in Fig. 1 are simple and typical. The first illustrates series resonance and the second, parallel resonance; the third is a series-parallel resonant circuit of two branches (sometimes referred to as antiresonance). Series resonance is highly practical for providing low impedance at the resonant frequency. Parallel resonance is the dual of series resonance, but it is not practical because it assumes an inductive element with no resistance. The third example, however, shows an eminently practical means of providing the typical characteristic of parallel resonance, which is high impedance at the resonant frequency.

Use. Resonance is of great importance in communications, permitting certain frequencies to be

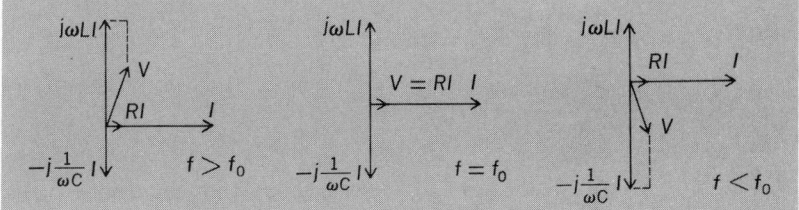

Fig. 2. Phasor diagrams at frequencies near resonance ($Q = 5$).

passed and others to be rejected. Thus a pair of telephone wires can carry many messages at the same time, each modulating a different carrier frequency, and each being separated from the others at the receiving end of the line by an appropriate arrangement of resonant filters. A radio or television receiver uses much the same principle to accept a desired signal and to reject all the undesired signals that arrive concurrently at its antenna; tuning a receiver means adjusting a circuit to be resonant at a desired frequency.

Many frequency-sensitive circuits are not truly resonant, and oscillations of a certain frequency can be produced or enhanced by networks that do not involve inductance. It is difficult and expensive to provide inductance with integrated circuits, but frequency selection can be provided by the use of capacitance and resistance, a large amount of amplification being obtained from the semiconductor material employed.

Series resonance. Figure 2 shows a phasor diagram of the voltage, resulting from a given current (steady alternating current) in a series-resonant circuit, such as shown in Fig. 1. The component voltages across the three circuit elements add to give the total applied voltage V, as shown for a frequency slightly above resonance, for the resonant frequency, and for a frequency below resonance. It is of course possible in low-loss (high Q) circuits for the voltage across the capacitance and the voltage across the inductance each to be many times greater than the applied voltage.

Analytically, the impedance of the series-resonant circuit is given by Eq. (1). The resonant fre-

$$Z = R + j\omega L + \frac{1}{j\omega C} = R + j\left(\omega L - \frac{1}{\omega C}\right) \quad (1)$$

quency f_0 is the frequency at which Z is purely real (phase resonance), so $\omega_0 L = 1/\omega_0 C$, or $2\pi f_0 L = 1/2\pi f_0 C$, from which Eq. (2) obtains.

A more convenient notation is expressed by Eq. (3).

$$f_0 = \frac{1}{2\pi\sqrt{LC}} \quad (2)$$

$$Z = R_0\left(\frac{R}{R_0} + jQ_0\delta\frac{2+\delta}{1+\delta}\right) \quad (3)$$

Z = impedance at the terminals of the series-resonant circuit
R, L, C = the three circuit parameters
R_0 = resistance (effective) at resonant frequency
Q_0 = $\omega_0 L/R_0$
δ = $(\omega - \omega_0)/\omega_0$
ω = $2\pi f$, where f is frequency (hertz or cps)
ω_0 = $2\pi f_0$, where f_0 is resonant frequency

Equation (3) is true for all series-resonant circuits, but interest is mainly in circuits for which Q_0, the quality factor at the resonant frequency, is high (20 or more) and for which δ, the fractional detuning, is low (perhaps less than 0.1). Assuming high Q_0 and low δ, which means a low-loss circuit and a frequency near resonance, Eq. (4) is very nearly the relative admittance of the series-resonant circuit.

$$\frac{Y}{Y_0} = \frac{Z_0}{Z} = \frac{1}{1 + j2Q_0\delta} \qquad (4)$$

Universal resonance curve. The magnitude and the real and imaginary components of Eq. (4) are usefully plotted in the universal resonance curve of Fig. 3. Since Y/Y_0 is plotted as a function of $Q_0\delta$, this curve can be applied to all series-resonant circuits. (If $Q_0 = 20$, the error in Y barely exceeds 1% of Y_0 for any δ, and is less for small δ.)

Moreover, because of the duality of the network, the curve can also be applied to any parallel-resonant circuit (Fig. 1) provided Q_0 is now interpret-

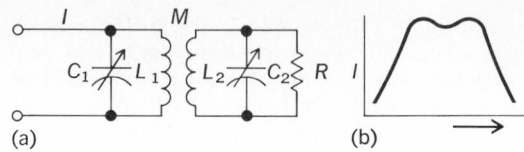

Fig. 4. (a) Resonance in a double-tuned network. (b) Current in R as function of frequency.

ed as $Q_0 = R_0/\omega_0L$. When used for a parallel-resonant circuit, the curve of Fig. 3 gives not Y/Y_0 but the relative input impedance Z/Z_0.

Finally, the universal resonance curve of Fig. 3 can also be applied (with the same slight approximations) to the two-branch resonant circuit of Fig. 1. For this purpose the curve shows Z/Z_0 (as for the three-branch parallel-resonant circuit), but the value of Q to be used is $Q_0 = \omega_0L/R_0$, exactly as with the series-resonant circuit. Note that Z_0 for this circuit is given by Eq. (5) (instead of being

$$Z_0 = (\omega_0L)Q_0 = R_0Q_0^2 \qquad (5)$$

equal to R_0 as it is in the other two circuits of Fig. 1).

Multiple resonance. If two or more coupled circuits are resonant at slightly different frequencies, many valuable characteristics can be obtained. Figure 4 shows a double-tuned network and a typical curve of current in R, the load, as a function of frequency.

[HUGH HILDRETH SKILLING]

Bibliography: D. Bell, *Fundamentals of Electric Circuits*, 2d ed., 1981; J. R. Duff and M. Kauffman, *Alternating Current Fundamentals*, 1980; Institute of Electrical and Electronics Engineers, *IEEE Standard Dictionary of Electrical and Electronics Terms*, 2d ed., 1977.

Resonance ionization spectroscopy

A form of atomic and molecular spectroscopy in which wavelength tunable light sources are used to remove electrons from (that is, ionize) a given kind of atom or molecule. Laser-based resonance ionization spectroscopy (RIS) techniques have been developed and used with ionization detectors such as the proportional counter to show that single atoms can be detected. Both RIS and one-atom detectors find a wide range of applications in physics, chemistry, and oceanography, and in the environmental sciences.

Theory. When an atom (or molecule) is subjected to a light source that provides photons of an angular frequency ω, these photons can be absorbed by the atom if the photon energy $\hbar\omega$ ($\hbar$ is Planck's constant divided by 2π) is almost exactly the difference in energy between an atom in its normal or ground state and some excited state. Suppose, as in scheme 1 of Fig. 1, that a light source is tuned to a frequency which excites a given kind of atom, A. If the light source is a tunable pulsed laser of very narrow bandwidth, it is highly unlikely that any other kind of atom will be excited. But the atoms which are in an excited state can be further excited to the ionization continuum where electrons are set free, provided that the ionization potential of the atoms is less than $2\hbar\omega_1$. While the final ionization step can occur with photons of any energy

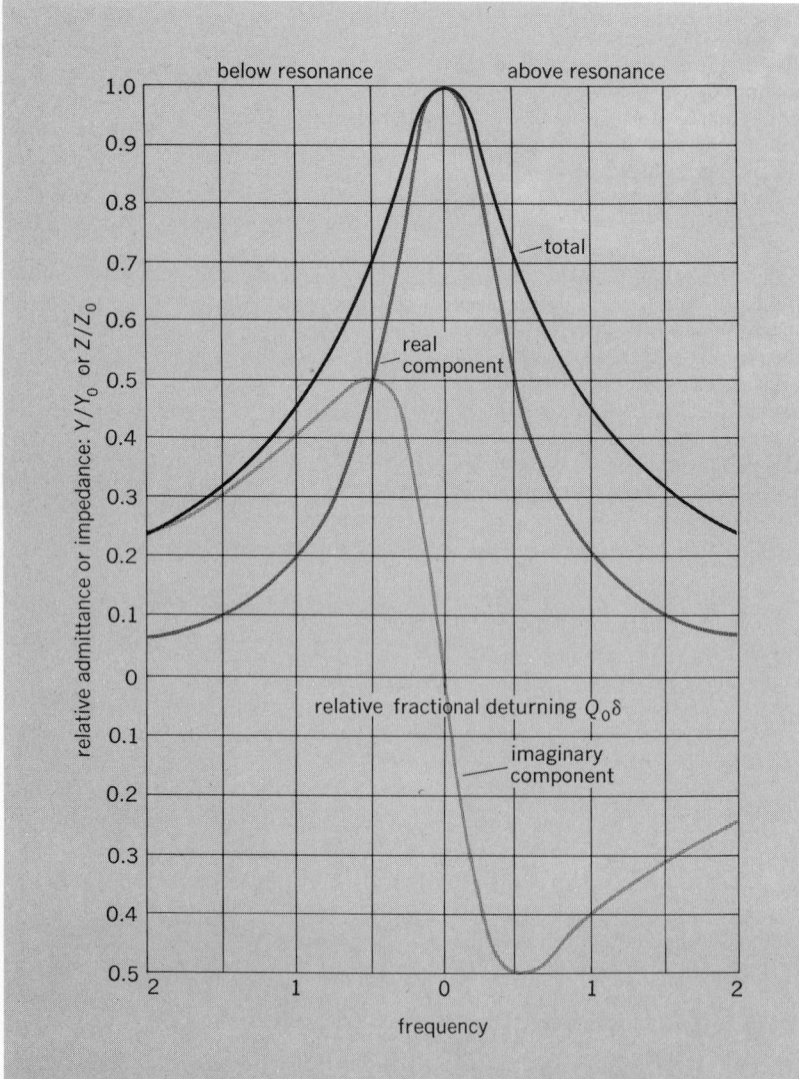

Fig. 3. Universal resonance curve. (*From H. H. Skilling, Electrical Engineering Circuits, 2d ed., copyright © 1965 by John Wiley and Sons, Inc.; used with permission*)

above a threshold, the entire process is a resonance process. In sharp contrast with other ionization means—for example, x-rays or radioactive sources—resonance ionization spectroscopy is a selective process in which only those atoms that are in resonance with the light source are ionized. Modern pulsed lasers are excellent tunable sources for resonance ionization spectroscopy; furthermore, they provide enough light in a single pulse to remove one electron from each atom of the selected type. A laser that provides 100 millijoules of photons in a single pulse of 10^{-6} s duration can be tuned to ionize nearly all of the atoms of a given type that may happen to be gas contained in a virtual test tube whose diameter is 1 cm and which can be very long (even meters), consistent with the divergence of the laser beam. *See* ATOMIC STRUCTURE AND SPECTRA; LASER.

Laser schemes. Many laser schemes can be used, as shown in Fig. 1. The notation $A[\omega_1,\omega_1 e^-]A^+$, taken from the standard notation for describing nuclear reactions, is used for the two-step process described above. On the other hand, the frequency of a laser can be doubled to $2\omega_1$, so that scheme 2 requires only one laser, while schemes 3, 4, and 5 involve two lasers to generate photons at frequencies of ω_1 and ω_2. With these five schemes only, it is possible to selectively ionize every known element in the periodic table except two of the noble gases, helium and neon (Fig. 2). *See* NUCLEAR REACTION.

One-atom detection. Both the high selectivity and the extraordinary sensitivity of resonance ionization spectroscopy were demonstrated by pulsing a laser directly through a proportional counter (Fig. 3). It was shown by S. C. Curran, J. Angus, and A. L. Cockcroft in 1949 that an improved version of the 1908 Rutherford-Geiger electrical counter (now known as a proportional counter) can be used to count single electrons at thermal energy. Therefore, if lasers are used to remove one electron from all of the atoms of a selected type, one-atom detection is possible. Proportional counters are normally filled with gases like argon (90%) and methane (10%). A pulsed laser tuned, for example, to 455.5 nm can detect even one atom of cesium without producing background ionization of the counting gas. In the original demonstration of one-atom detection, it was proved that one atom of cesium could be selected out of 10^{19} atoms of the counting gas (argon and methane). *See* IONIZATION CHAMBER.

Another important form of one-atom detection involves the time-resolved detection of a single daughter atom in flight following the decay of a parent atom. Thus, it was shown that an atom of cesium could be detected from the fission decay of an individual atom of the isotope ^{252}Cf. The energy released in the fission process generated a signal that triggered the laser used to accomplish the resonance ionization spectroscopy process $Cs[\omega_1,\omega_1 e^-]Cs^+$. The success of that experiment proved that daughter atoms can be detected in coincidence with the decay of parent atoms. Such techniques could eventually work for most of the daughter atoms associated with radioactive decay, and could possibly be used to greatly reduce backgrounds in low-level counting facilities. *See* RADIOACTIVITY.

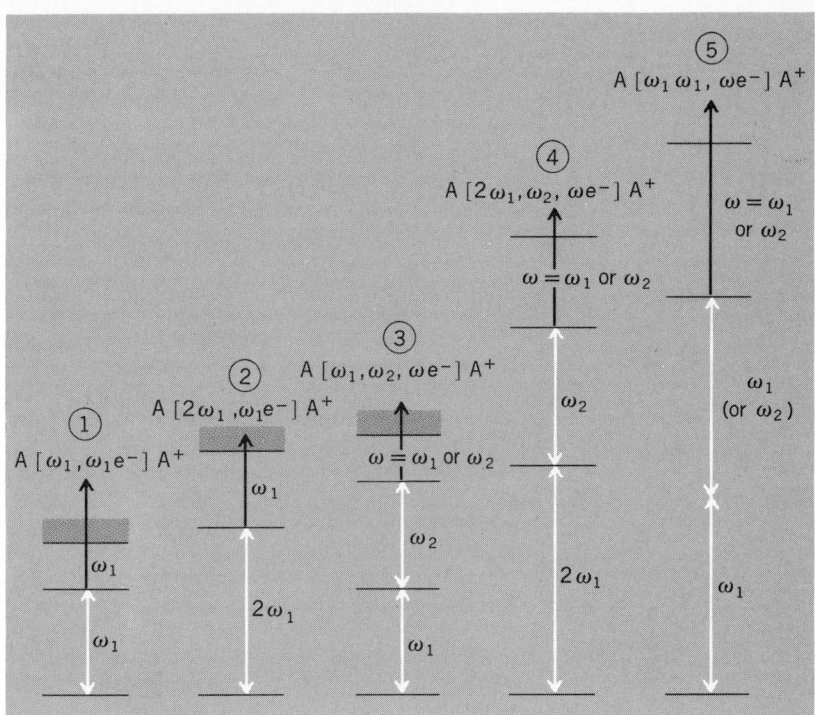

Fig. 1. Various laser schemes used in resonance ionization spectroscopy. (*Office of Health and Environmental Research, U.S. Dept. of Energy under contract W-7405-eng-26 with the Union Carbide Corp.*)

Applications. Resonance ionization spectroscopy and one-atom detection find a variety of interesting applications.

Classical chemical physics. The capability for detecting a population of just a few atoms has made it possible to investigate some problems in classical chemical physics which previously were difficult or impossible. Precision measurements of the diffusion of free atoms among other atoms and molecules have been made in sufficient detail to test the basic diffusion equation in both time and space domains. The determination of rates of reactions of extremely reactive substances (such as alkali atoms) with other atoms or molecules is now possible. Since only a few of these reactive atoms need to be produced, several problems concerning corrosion of the apparatus and the production of complicated chemical by-products are avoided. Study of a population of a few atoms (for example, 10) to observe their statistical behavior has been made.

Modern physics. One-atom detection makes it feasible to detect extremely rare events. Measurements of neutrinos from the Sun are crucial to testing both solar models and neutrino physics. After prolonged exposure to the Sun, the neutrinos may produce on the order of 100 atoms of a particular type in a very large tank. Previously, targets have been rich in ^{37}Cl so that neutrino capture would produce ^{37}Ar, a radioactive atom which can be counted by the standard methods of radioactivity. Resonance ionization spectroscopy and one-atom detection are making possible a much wider variety of neutrino targets. For example, in a lithium-rich target neutrinos produce ^{7}Be, which can be detected by observing daughter (lithium) atoms in time coincidence with the decay of the parent atoms.

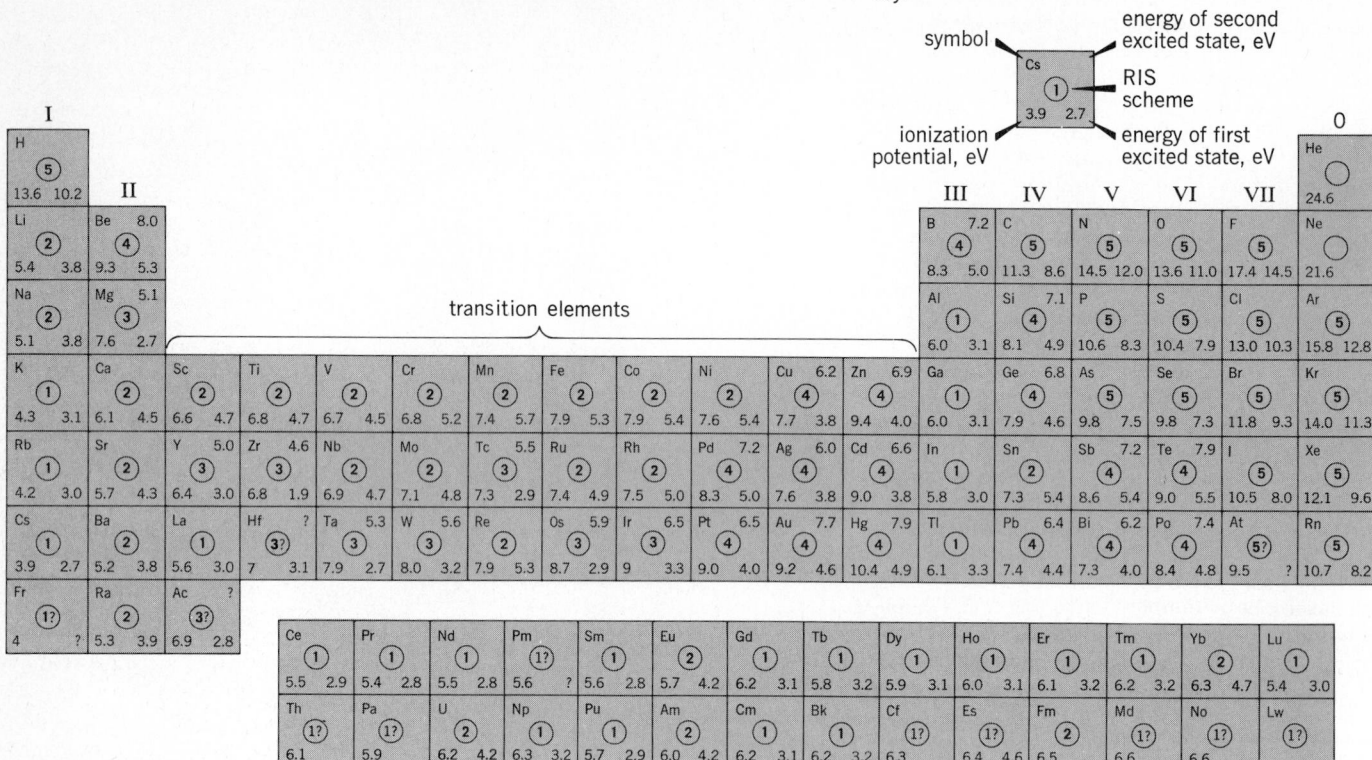

Fig. 2. Resonance ionization spectroscopy schemes for the elements of the periodic table. Numbers identifying schemes are those given in Fig. 1. Subgroups of elements are not designated by the letters a and b, in agreement with modern usage. (*Office of Health and Environmental Research, U.S. Dept. of Energy under contract W-7405-eng-26 with the Union Carbide Corp.*)

Another experiment involves bromine-rich targets, where the neutrino capture produces ^{81}Kr. Radioactive ^{81}Kr can be counted directly (before it decays) by another technique made possible by resonance ionization spectroscopy. Other problems in weak interaction physics are also amenable to the one-atom detection techniques. Some meson interactions with nuclei have extremely low cross sections and thus produce only a few product atoms which, however, can be detected with resonance ionization spectroscopy techniques. *See* MESON; NEUTRINO; WEAK NUCLEAR INTERACTIONS.

Environmental. Oceanographers have considered the use of ^{39}Ar as a tracer for ocean water circulation. Measurements of ^{81}Kr in the natural environment to obtain the ages of polar ice caps and old groundwater deposits have been suggested. Several techniques made possible by the development of resonance ionization spectroscopy and one-atom detection have been developed for these applications.

[G. S. HURST]

Bibliography: S. C. Curran, J. Angus, and A. L. Cockroft, Investigation of soft radiations by proportional counters—I, *Phil. Mag.*, 40:36–52, 1949; R. Davis, Jr., D. S. Harmer, and K. C. Hoffman, Search for neutrinos from the Sun, *Phys. Rev. Lett.*, 20:1205–1209, 1968; G. S. Hurst et al., Resonance ionization spectroscopy and one-atom detection, *Rev. Mod. Phys.*, 51:767–819, 1979.

Rest mass

A constant associated with a material body which determines its inertial properties and its internal energy content. It is sometimes called the inertial mass.

For a particle of rest mass m_o, Newton's second law of motion is given by Eq. (1), where $\mathbf{a} = d\mathbf{v}/dt$ is

$$m_o \, d\mathbf{v}/dt = \mathbf{F} \qquad (1)$$

the acceleration of the particle and $\mathbf{F}$ is the force

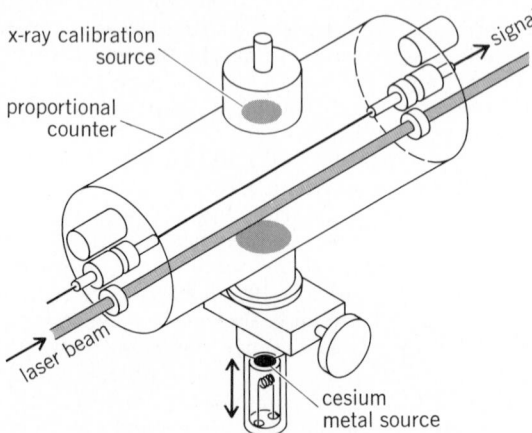

Fig. 3. Apparatus for experiment conducted to prove that resonance ionization spectroscopy can be used to detect a single atom. (*Office of Health and Environmental Research, U.S. Dept. of Energy under contract W-7405-eng-26 with the Union Carbide Corp.*)

applied to it. For a particle moving with a speed near that of light Eq. (1) must be replaced by the relativistic equation of motion, Eq. (2), where c is

$$\frac{d}{dt}\left(\frac{m_o\mathbf{v}}{\sqrt{1-(v^2/c^2)}}\right)=\mathbf{F} \qquad (2)$$

the speed of light. These equations are used for the measurement of the rest masses of particles by deflection in suitable force fields. The masses of macroscopic bodies are normally measured by weighing in the Earth's gravitational field.

The formula connecting internal energy E_o and rest mass (Einstein's mass-energy relation) is written as Eq. (3).

$$E_o = m_o c^2 \qquad (3)$$

See INERTIA OF ENERGY; RELATIVISTIC MECHANICS; RELATIVITY. [EDWARD L. HILL]

Reverberation

After sound has been produced in, or enters, an enclosed space, it is reflected repeatedly by the boundaries of the enclosure, even after the source ceases to emit sound. This prolongation of sound after the original source has stopped is called reverberation. A certain amount of reverberation adds a pleasing characteristic to the acoustical

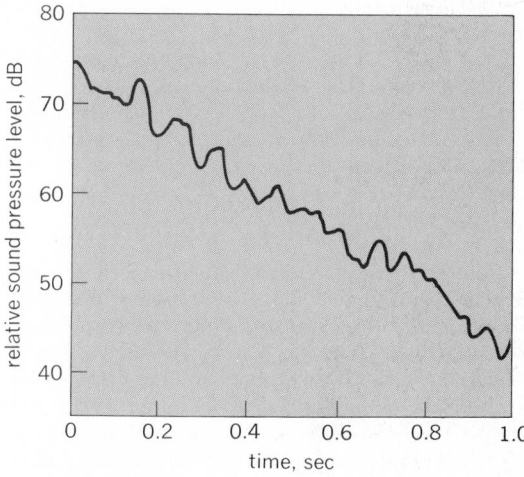

Fig. 1. Typical decay curve illustrating reverberation.

qualities of a room. However, excessive reverberation can ruin the acoustical properties of an otherwise well-designed room. A typical record representing the sound-pressure level at a given point in a room plotted against time, after a sound source has been turned off, is given in the decay curve shown in Fig. 1. The rate of sound decay is not uniform but fluctuates about an average slope.

Reverberation time. Because of the importance of the proper control of reverberation in rooms, a standard of measure called reverberation time (abbreviated t_{60}) has been established. Reverberation time is the time required for sound to die away to one-thousandth of its initial pressure, that is, to drop 60 decibels (dB) in sound-pressure level.

Optimum reverberation time is a matter of individual preference. A critical study of empirical

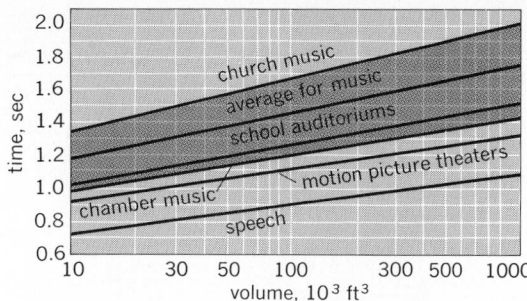

Fig. 2. Optimum reverberation time at 500 Hz for different types of rooms as a function of room volume. This figure should be used in conjunction with Fig. 3 to obtain optimum reverberation time as a function of frequency. 10^3 ft = 28.3 m³. (*After V. O. Knudsen and C. M. Harris, Acoustical Designing in Architecture, copyright © 1950 by John Wiley and Sons, Inc.; used with permission*)

data based upon preference evaluations in the United States and abroad has been made by V. O. Knudsen and C. M. Harris (Fig. 2). Since the optimum reverberation time for music depends on the type of music, it is represented in the figure by a broad band. The optimum reverberation time for a room used primarily for speech is considerably shorter; reverberation times longer than those shown for speech result in a decrease in speech intelligibility. The optimum reverberation time at frequencies other than 500 hertz (Hz) is obtained by multiplying the 500-Hz value by the ratio R, which is given in Fig. 3. Note that R is unity for frequencies above 500 Hz and is given by a band for frequencies below 500 Hz. For large rooms R may have any value within the indicated band; for small rooms preferred ratios are in the lower part of the band.

Mean free path. According to the principles of geometrical acoustics, sound radiated from a source in an enclosure is successively reflected by its boundaries. The average distance between reflections is defined as the mean free path. The mean free path of a sound ray in a room depends on the shape and size of the room, and to some extent on the distribution and nature of the absorptive material. However, in most cases, it is approxi-

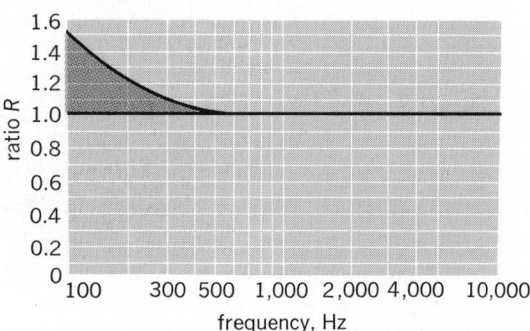

Fig. 3. Chart for computing optimum reverberation time. The time at any frequency is given in terms of the ratio R, which should be multiplied by the optimum time at 500 Hz (from Fig. 2) to obtain the optimum time at that frequency. (*After V. O. Knudsen and C. M. Harris, Acoustical Designing in Architecture, copyright © 1950 by John Wiley and Sons, Inc.; used with permission*)

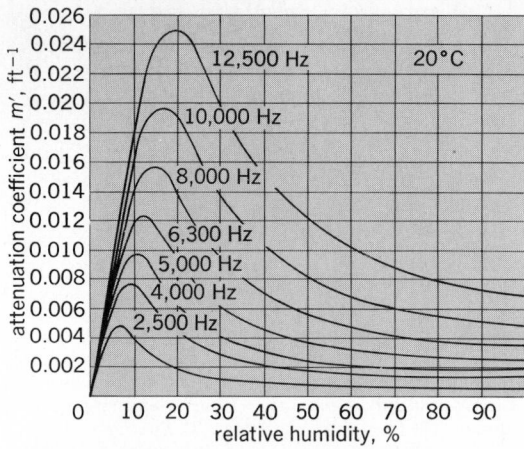

Fig. 4. Values of the attenuation coefficient m as a function of relative humidity for different frequencies. 1 ft^{-1} = 3.28 m^{-1}. (*C. M. Harris, Absorption of sound in air versus humidity and temperature, J. Acoust. Soc. Amer., 40:148–159, 1966*)

mately $4V/S$, where V is the volume of the room and S the total surface area.

Decay rate. The number of reflections per second of a decaying sound wave is numerically equal to the distance sound will travel in 1 sec, that is, the velocity of sound c, which is about 343 m/sec (1130 ft/sec) in air at 20°C, divided by the average distance between reflections, or the mean free path. Hence the number of reflections per second is $cS/4V$. Each time a wave strikes one of the boundaries, on the average, a fraction ($\bar{\alpha}$) of the energy is absorbed, and a fraction ($1 - \bar{\alpha}$) is reflected. Here $\bar{\alpha}$ is the average absorption coefficient given by Eq. (1), where α_1 is the coefficient of ab-

$$\bar{\alpha} = \frac{\alpha_1 S_1 + \alpha_2 S_2 + \alpha_3 S_3 + \cdots}{S_1 + S_2 + S_3 + \cdots} \quad (1)$$

sorption of surface S_1, and so forth. Because sound pressure is proportional to the square root of sound intensity, the ratio of the average reflected pressure to incident pressure is given by $(1 - \bar{\alpha})^{1/2}$, and the average decrease in the sound pressure level is therefore given by notation (2).

$$10 \log_{10} \frac{1}{1 - \bar{\alpha}} \text{ dB/reflection} \quad (2)$$

Since there are $cS/4V$ reflections per second, the average decay rate is given by Eq. (3a), where S is in square meters and V is in cubic meters, or (3b), where S' is in square feet and V' is in cubic feet.

Decay = $373(S/V)[-2.30 \log_{10}(1 - \bar{\alpha})]$ dB/sec (3a)

Decay = $1230(S'/V')[-2.30 \log_{10}(1 - \bar{\alpha})]$ dB/sec (3b)

Reverberation-time formulas. From Eqs. (3) for decay rate, it follows that the time it takes for the sound-pressure level to decay 60 dB, that is, the reverberation time, is given by Eqs. (4).

$$t_{60} = \frac{0.161V}{S[-2.30 \log_{10}(1 - \bar{\alpha})]} \text{ sec} \quad (4a)$$

$$t_{60} = \frac{0.049V'}{S'[-2.30 \log_{10}(1 - \bar{\alpha})]} \text{ sec} \quad (4b)$$

When $\bar{\alpha} \ll 1$, Eqs. (4) becomes Eqs. (5). For fre-

$$t_{60} = \frac{0.161V}{S\bar{\alpha}} \text{ sec} \quad (5a)$$

$$t_{60} = \frac{0.049V'}{S'\bar{\alpha}} \text{ sec} \quad (5b)$$

quencies above 2000 Hz, especially in large auditoriums, the effects of air absorption must be included in the reverberation time formulas. The corresponding equations are then Eqs. (6) and (7), where m is the attenuation coefficient in inverse meters and m' is the attenuation coefficient in inverse feet, given in Fig. 4. It can be shown that air absorption is molecular in origin.

$$t_{60} = \frac{0.161V}{S[-2.30 \log_{10}(1 - \bar{\alpha}) + 4mV]} \text{ sec} \quad (6a)$$

$$t_{60} = \frac{0.049V'}{S'[-2.30 \log_{10}(1 - \bar{\alpha}) + 4m'V']} \text{ sec} \quad (6b)$$

$$t_{60} = \frac{0.161V}{S\bar{\alpha} + 4mV} \text{ sec} \quad (7a)$$

$$t_{60} = \frac{0.049V'}{S'\bar{\alpha} + 4m'V'} \text{ sec} \quad (7b)$$

The above reverberation time formulas apply only in rooms in which sound is diffuse, a condition that was assumed in their derivation. Thus, large discrepancies may be noted between the calculated and observed values of reverberation time in a room whose shape or absorptive treatment (or both) do not promote diffuse conditions of sound. *See* SOUND.

[CYRIL M. HARRIS]

Bibliography: C. M. Harris (ed.), *Handbook of Noise Control*, 2d ed., 1979; V. O. Knudsen and C. M. Harris, *Acoustical Design in Architecture*, 1950; H. F. Olson, *Acoustical Engineering*, 3d ed., 1957.

Reynolds number

In fluid mechanics, the ratio $\rho vd/\mu$ of the inertia force ρvd to the viscous force μ, where ρ is fluid density, v is velocity, d is a characteristic length, and μ is fluid viscosity. The Reynolds number is significant in the design of a model of any system in which the effect of viscosity is important in controlling the velocities or the flow pattern. In the evaluation of drag on a body submerged in a fluid and moving with respect to the fluids, the Reynolds number is important. If the model is operated in the same fluid as is the prototype, the similarity requirement based on the Reynolds number yields the equation below.

$$v_m = nv$$

It is evident that if both gravity and viscosity are significant and if the same gravitational field and same fluid are used in both the model and prototype, similarity must be achieved for Froude and for Reynolds numbers. The only possible solution is that n be equal to unity, or a model be equal in size to the prototype. If the length scale is to be greater than 1 (model smaller than the prototype), either the model must be tested in a different gravitational field or a different fluid must be used to satisfy similitude requirements. *See* DYNAMIC SIMILARITY.

The Reynolds number also serves as a criterion of type of fluid motion. In a pipe, for example, laminar flow normally exists at Reynolds numbers less than 2000, and turbulent flow at Reynolds numbers above about 3000.

[GLENN MURPHY]

Rheology

The study of the deformation and flow of matter. The states of matter differ strikingly in their density and in the ease with which they can be deformed. The less dense the state of matter, the more easily deformable it ordinarily is. The viscosity of a gas arises from the crossing-over of molecules from a fast-moving layer into a neighboring more slowly moving layer, and vice versa. Because this crossing-over increases with temperature rise, the viscosity of a gas increases with temperature. Solids and liquids, however, become more fluid with temperature rise. *See* VISCOSITY.

Viscosity and structure. A perfect crystal, according to theoretical calculations, should be orders of magnitude stronger than crystals customarily are found to be. This is because of the presence in actual crystals of various types of imperfections which greatly facilitate their deformation. The thermodynamic properties of crystals are readily calculated by assuming that the atoms form a perfectly ordered lattice. The rheological properties of crystals cannot be calculated from the perfect lattice model, however, but only by considering the number and nature of the lattice imperfections. Besides various types of vacant lattice sites, extra interstitial atoms are frequently present. For details on the important types of crystal imperfections *see* CRYSTAL DEFECTS.

The extra fluidity of liquids, as compared with solids, arises from the great increase in the number of imperfections introduced with the 10% expansion in melting shown by normal liquids. Ice is an exception since it contracts by about 10% on melting. This contraction results from a change from the tetrahedral hydrogen-bonded structure displayed by one crystalline form of ice into some denser tetrahedral forms, such as ice III. As a result, water has a viscosity of about 17 millipoises at the melting point, which is close to the normal value for ordinary liquids. (A poise is a unit of viscosity equal to 1 dyne sec/cm^2.) A solid melts at that temperature at which the entropy from imperfections multiplied by the temperature equals the heat of introducing these imperfections.

Metals melt with only 3% expansion instead of the usual 10% expansion of normal un-ionized molecules because the positive ion, which is only about one-third as large as the atom, requires only one-third the space for extra equilibrium positions.

Molten salts, on the other hand, expand about 22% on melting. In this case, new equilibrium positions require about double the usual space. This is probably because a sodium chloride molecule must be accommodated in the added equilibrium position instead of a Na$^+$ or a Cl$^-$ ion separately.

That the viscosity of a system closely reflects structure is exemplified by the fact that for nearly all simple substances the viscosity of the liquid at the melting point is approximately 2 centipoises. Also, the reciprocal of the viscosity, the fluidity, is a linear function of the molecular volume.

Because holes are necessary to permit viscous flow and since pressure tends to decrease the number of holes in a system, it follows that pressure normally increases viscosity. The pressure coefficient of viscosity indicates that the empty space a molecule requires in order to flow viscously is about one-seventh its normal volume. To provide such vacant sites, an activation energy of about one-third the heat of vaporization is required, as shown by the temperature coefficient of viscosity. The normal effect of pressure rise and temperature drop in increasing viscosity is modified in the rare cases where such changes promote significant modifications in structure. Thus, increasing the pressure on water just above the melting point increases the fluidity of water because the liquid structure is shifted away from the hydrogen-bonded tetrahedral state toward the more fluid ice-III-like structure. At a temperature of about 160°C, sulfur changes from a comparatively fluid, straw-colored, eight-membered ring to a dark, viscous, high polymer, having thousands of atoms in a linear chain. Lubricants likewise are made relatively temperature independent by adding high polymers which are soluble only with difficulty and which unfold with temperature rise, thus counteracting the usual decrease of viscosity of liquids with temperature rise.

When linear high polymers change their state, they introduce holes only in directions normal to the length of the chain. Thus a linear polymer melts in two directions but retains its solidlike properties along the chain length. The result is that high polymers progress by wriggling a segment at a time, as shown by the fact that they exhibit the expected pressure and temperature coefficients of a molecule the size of a segment. The viscosity is much higher for a high polymer because of a negative entropy of activation corresponding to correlation of segment motion. Extensive quantitative correlations of rheological properties are available.

Significant structures in liquids. To obtain quantitative expressions for the viscosity, a model of the liquid state is important. Upon melting, liquids ordinarily expand and become orders of magnitude more fluid. Since x-rays indicate that melting does not materially change nearest-neighbor distances, the expansion must be due to holes in the liquid. These holes should approximate molecular size, since smaller holes would not be as effective in facilitating molecular motion, that is, are wasteful of entropy, and larger holes are unduly wasteful of energy. Melting occurs because a thermodynamically more stable state appears at the melting temperature where the kinetic energy can support the fluidized vacancies, characteristic of the liquid state, against the potential energy of melting, which tends to collapse the holes. Such considerations lead to a model of the liquid in which fluidized vacancies in the liquid closely mirror the vapor molecules in their size, motion, and concentration. The partition function f_l embodies this significant structure theory.

The partition function for argon takes the form of Eq. (1).

With appropriate modifications the partition function can be written for any liquid. Here Eq. (2) is the Einstein partition function for a solid, and Eq. (3) is the partition function of a gas. The signifi-

$$f_l = \left[\left(f_s\left\{1+10.7\frac{V-V_s}{V_s}\right\}\right.\right.$$
$$\left.\left.\exp\left(-\frac{.0052\,E_s V_s}{(V-V_s)\,RT}\right)\right)^{Vs/V} f_g^{(V-Vs)/V}\right]^N \quad (1)$$

$$f_s = \frac{e^{Es/RT}}{(1-e^{-\theta/T})^3} \quad (2)$$

$$f_g = \frac{(2\pi mkT)^{3/2}}{h^3}\frac{eV}{N} \quad (3)$$

cant structure theory yields satisfactory values for all thermodynamic properties of liquids. In Eqs. (1) and (2), E_s, V_s, and θ are the energy of sublimation, the solid molal volume at melting, and the Einstein characteristic temperature, respectively; V, T, m, h, k, E and N are the liquid molal volume, absolute temperature, molecular mass, Planck's constant, Boltzmann's constant, internal energy, and Avogadro's number, respectively. The two numerical constants, 10.7 and .0052, are dimensionless and are calculated from theoretical considerations.

The viscosity η is defined by Eq. (4). Here f is

$$\eta = \frac{f}{\dot{S}} = \frac{f\lambda}{u} \quad (4)$$

the shear stress, that is, the force per unit area along the plane of shear. The equation for rate of shear is $\dot{S} = u/\lambda$, where u is the relative velocity of contiguous planes of molecules parallel to the shear plane, and λ is the distance between these neighboring planes. If one has a mixture of structures subject to laminar flow in which the fractional area occupied along the shear plane of the ith structure is χ_i, the viscosity is represented by Eq. (5).

$$\eta = \frac{f}{\dot{S}} = \frac{\Sigma \chi_i f_i}{\dot{S}} = \sum_i \chi_i \eta_i \quad (5)$$

In a liquid the gaslike structures occupy a fraction of the surface $\chi_i = (V-V_s)/V$, and the solidlike structures occupy the remaining fraction V_s/V of the planar surface. Thus Eq. (6) for the viscosity of a liquid is formed.

$$\eta = \frac{V_s}{V}\eta_s + \frac{V-V_s}{V}\eta_g \quad (6)$$

The viscosity of a gas is usually represented by Eq. (7). The significant structure theory of the viscosity of solidlike molecules in the liquid is given by Eq. (8). In the preceding equations N is Avogad-

$$\eta_g = \frac{2}{3d^2}\left(\frac{mkT}{\pi^3}\right)^{1/2} \quad (7)$$

cosity of solidlike molecules in the liquid is given by Eq. (8). In the preceding equations N is Avogad-

$$\eta_s = \frac{Nh}{Z\kappa}\frac{V}{V_s}\frac{6}{\sqrt{2}}\frac{\psi}{V-V_s}\exp\left(\frac{aE_s V_s}{(V-V_s)RT}\right)$$
$$\exp\left(-\frac{P(V-V_s)}{RT}\right) \quad (8)$$

ro's number; $Z=12$ is the number of nearest neighbor positions, both full and empty; λ is the distance a molecule jumps in viscous flow; $\lambda_2\lambda_3$ is the area parallel to the shear plane occupied by a

molecule; d is the molecule diameter and m its mass; and ψ is the vibrational partition function for the degree of freedom which becomes the reaction coordinate in the activated state. The significant structure theory of liquids leads to the rate equation, Eq. (9), for molecules to jump into a neighboring position at a distance λ under zero stress.

$$k' = \kappa\frac{kT}{h\psi}Z\frac{(V-V_s)}{V}\exp\left(-\frac{aE_s V_s}{(V-V_s)RT}\right)$$
$$\exp\left(\frac{P(V-V_s)}{RT}\right) \quad (9)$$

This result used in Eq. (4) leads to Eq. (8). For argon the term a has the value .0052 appearing in Eq. (1) and a dimensionless constant calculated from the model. For energy levels of the partition function ψ higher than $E_s/3$, the oscillator is replaced by a translator with a solid volume calculated by using van der Waals's b. The term exp$(P(V-V_s)/RT)$ in Eq. (8) gives only a small change in the fluidity, with an increase in the pressure and temperature at constant volume. Whether it is an increase or a decrease depends on whether P or T increases most rapidly as both increase. The transmission function $\kappa = .375$ is a nonthermodynamic quantity that is chosen to fit the data and is of the expected magnitude. Figure 1 shows agreement with the experimental findings of fluidity, $\theta = \eta^{-1}$, of N. F. Zhadanova over the entire liquid and gaseous range of argon.

Table 1 shows the agreement between calculated and observed viscosities of argon under its vapor pressure.

High polymer viscosity. Liquid high polymers shear as a result of individual kinetic units jumping in a direction to reduce the stress. If the high polymer is long enough, there will be entanglements and a given molecule will be obligated to "slalom" around a neighbor with which it is entangled. If the small contribution of gaslike molecules to the viscosity in Eq. (6) is neglected, Eq. (10) results.

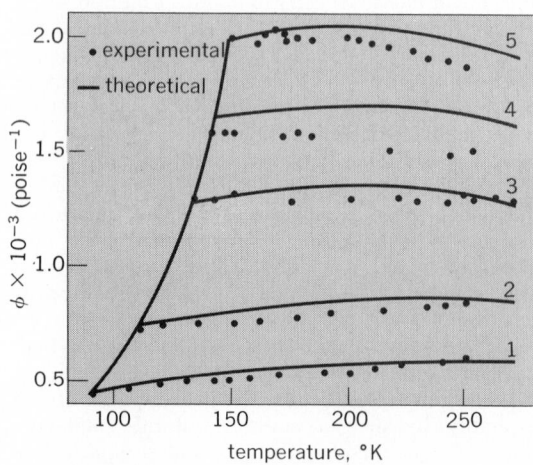

Fig. 1. Fluidities (poise⁻¹) of argon at constant volume versus temperatures (°K). Curve 1 at $V=29.660$ cm³; curve 2 at $V=33.416$ cm³; curve 3 at $V=40.59$ cm³; curve 4 at $V=46.48$ cm³; curve 5 at $V=52.78$ cm³. (*Experimental data by N. F. Zhadanova, in T. Ree, T. S. Ree, and H. Eyring, Proc. Nat. Acad. Sci., 48:501, 1962*)

Table 1. Viscosities (millipoise) of argon under its vapor pressure*

T, K	η, calculated	η, observed	Δ, %
94.25	2.91	2.82	3.2
86.90	2.60	2.56	1.6
90	2.32	2.32	0.0
111	1.35	1.37	−2.2
133.5	0.81	0.77	2.6
143	0.65	0.63	3.2
149	0.55	0.50	2.0

*N. F. Zhadanova, Soviet physics, *JETP*, 4:749, 1957.

$$\eta_1 = \frac{\lambda/kT}{2\lambda^2(\lambda_2\lambda_3)k'}\frac{V}{(V-V_s)} \tag{10}$$

Here k' is the rate for a segment to jump into an empty site given in Eq. (9), and should not change with molecular weight after the molecule becomes long enough to flow by segments; $\lambda_2\lambda_3$ is the area of a segment upon which the shear stress acts to make the segment flow and does not change as additional segments are added to the molecule, and λ is the distance between successive layers of molecules normal to the shear plane.

A molecule consists then of $n_1 \equiv M/m_1$ kinetic segments and $n_2 = M/m_2$ tangling segments; that is, there are $(n_2 - 1)$ points about which the molecule has to slalom. Here m_1 is the mass of a kinetic segment, m_2 is the mass of a tangling segment, and M is the molecular weight. The subscript p indicates a property for the polymer and the subscript s the property for the kinetic segment. Accordingly, Eqs. (11), (12), and (13) are expected.

$$(\lambda_1)_p = (\lambda_1)_s n_s^{1/3} \tag{11}$$

$$(\lambda)_p = (\lambda)_s/(n_1 n_2) \tag{12}$$

$$(k')_p = (k')_s n_1 \tag{13}$$

Substituting these results into Eq. (10) leads to the following two cases: For $M < m_2$ Eq. (14) is expected to hold, and for $M > m_2$, Eq. (15) is expected to hold.

$$\eta_p = \eta_s n_1^{4/3} = (\eta_s/m_1^{4/3})M^{4/3} \tag{14}$$

$$\eta_p = \eta_s n_1^{4/3} n_2^2 = (\eta_s/(m_1^{4/3} m_2^2))M^{3.33} \tag{15}$$

Here η_p is the viscosity of the high polymer, and η_s is the viscosity of the kinetic segment. Equation (11) expresses the idea that the effective distance between the centers of gravity of successive layers of molecules normal to the shear plane should increase in proportion to their molecular volume to the one-third power, and therefore to the one-third power of the molecular weight. Equation (12) follows because the motion of a kinetic segment advances the center of gravity of the high polymer containing n_1 segments by the fraction, $1/n_1$, of the amount the kinetic segment itself is advanced. Further, when there is tangling, only one out of n_2 tangling segments is free to choose a course which advances the molecule, since the remaining n_2^{-1} segments must slalom and so do not advance the center of gravity.

F. Beuche, proceeding differently, arrives at a power of the molecular weight equal to unity in Eq. (14) and equal to 3.5 in Eq. (15). The experimental results, as may be seen in Fig. 2 and Table 2, are in reasonable argeement with theory. Equations (11), (12), and (13), and the neglect of any effect of molecular weight on other factors in Eq. (10), thus seem justified, but more important, the model provides a simple method of interpretation of other cases that may arise. For example, simple branching of a high polymer should not greatly affect the dependence of viscosity on molecular weight when the situation is analyzed in terms of kinetic and tangling segments.

Non-newtonian viscosity. Frequently, systems are found which do not obey Newton's equation for viscosity; instead, a small increase in stress is found to disproportionately increase the rate of flow. For such a system one has Eq. (16) for the rate of shear.

$$\dot{S} = \frac{Z\lambda k'}{\lambda_1} <\cos\theta_i>$$

$$\sinh(f\lambda_2\lambda_3\lambda <\cos\theta_1>/2kT) \tag{16}$$

By introducing Eqs. (17) and (18), Eq. (16) becomes Eq. (19) or Eq. (20).

$$\beta^{-1} = Z\lambda k' <\cos\theta_1> \tag{17}$$

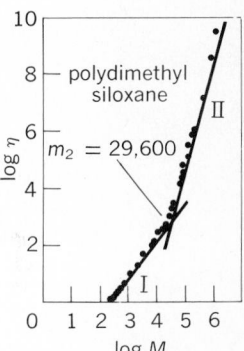

RHEOLOGY

Fig. 2. The curve of $\log\eta$ versus $\log M$ for polydimethyl siloxane. (*From H. Eyring, T. Ree, and M. Hirai, Proc. Nat. Acad. Sci., 44:1213, 1958*)

Table 2. Effect of molecular weight on viscosity of high polymers*

Polymer	s_1	s_2	$-A$*	$-B$†	Log m_2 Calculated	Log m_2 Observed	Temp., °C
Polyisobutylene	1.75	3.40	13.7	6.30	3.70	4.29	217
Polystyrene	1.58	3.44	13.6	5.03	4.29	4.68	217
Polydimethyl siloxane	1.34	3.70	15.7	5.14	5.28	4.71	25
Polymethyl methacrylate†	1.40	3.40	14.2	5.50	4.35	4.20	60
Polydecamethylene adipate	1.34	3.40	12.1	4.47	3.82	3.68	109
Polydecamethylene sepacate	1.23	3.23	11.4	4.17	3.62	3.67	109
Polydiethylene adipate	1.26	2.92	10.0	4.11	2.95	3.56	109
Poly (ε-capcolactam)							
Linear chain	1.66	3.52	12.8	5.79	3.52	3.78	253
Dichain	1.70	3.50	12.5	5.80	3.35	3.72	253
Tetrachain	1.31	3.24	11.8	4.29	3.76	3.79	253
Octachain	1.31	3.24	12.7	4.75	3.93	4.09	253

*Here s_1 and s_2 are the observed slopes in the regions I and II of Fig. 2 for the substances listed, and are explained theoretically by Eqs. (14) and (15). $A \equiv -\log(\eta_1/m_1^{4/3} m_2^2)$. $B \equiv -\log(\eta_1/M_1^{4/3})$.
†25% solution in diethyl phthalate.

$$\alpha = (\lambda_2\lambda_3\lambda < \cos\theta_i > /2rT) \qquad (18)$$

$$\dot{S} = \beta^{-1}\sinh(\alpha f) \qquad (19)$$

$$f = \frac{1}{\alpha}\sinh^{-1}(\beta\dot{S}) \qquad (20)$$

Here β^{-1} is the mean rate of shear per second in any direction under zero stress, and may be called the intrinsic rate of shear. Thus β, aside from a shear factor near unity, is a relaxation time. The term α^{-1} has the dimensions of stress and may be thought of as an intrinsic stress characterizing the flow unit.

When several different types of flow units occur in a shear plane, with the type i occupying the fraction χ_1 of the area along the shear plane, the viscosity is given by Eq. (5). Combining Eqs. (20) and (5) yields Eq. (21).

$$\eta = \sum_i \frac{\chi_i\beta_i}{\alpha_i}\frac{\sinh^{-1}(\beta_i\dot{S})}{(\beta_i\dot{S})}$$

$$= \sum_i \chi_i\eta_i\frac{\sinh^{-1}(\beta_i\dot{S})}{(\beta_i\dot{S})} \qquad (21)$$

The factor $\sinh^{-1}(\beta_i\dot{S})/(\beta_i\dot{S})$ approaches unity for low values of the argument $(\beta_i\dot{S})$ and falls off toward zero as the argument increases. For example, two types of flow units, with the following assigned values $\chi_1\beta_1/\alpha_1 = 3\times10^{-2}$ poise, $\chi_2/\alpha_2 = 179.3$ (dynes/cm^2), and $\beta_2 = 7.25$ sec, suffice to fit the experimental data of W. Philippoff for 10% nitrocellulose in 99% butylacetate. Philippoff's data for 11% ethylcellulose in cyclohexanone was also fitted by using $\chi_1/\alpha_1 = 5\times10^4$ dynes, $\beta_1 = 1\times10^{-3}$

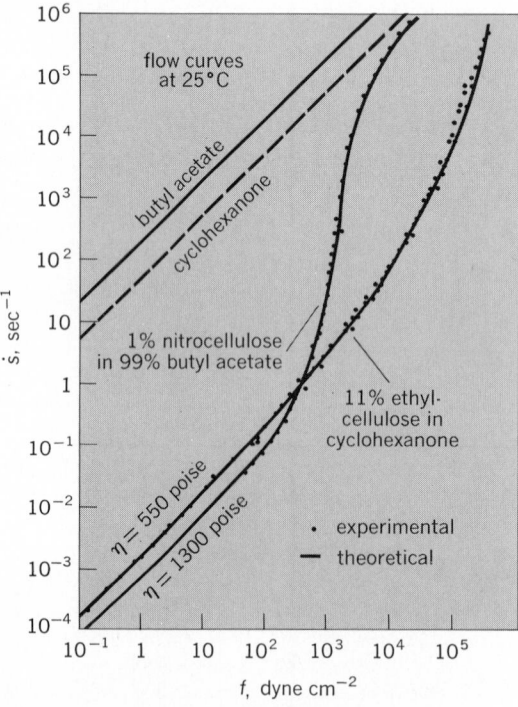

Fig. 3. Flow curves for 1% nitrocellulose in 99% butyl acetate and 11% ethylcellulose in cyclohexanone. Flow curves for solvents are shown by the straight lines of 45° slope. (*From T. Ree, T. S. Ree, and H. Eyring, Ind. Eng. Chem., 50:1036, 1958*)

sec, $\chi_2/\alpha_2 = 1000$ (dynes/cm^2), and $\beta_2 = .50$ sec. The results are plotted in Fig. 3. Clearly, viscosity is a simple concept only for low rates of shear, where a system is newtonian. In the non-newtonian range the intrinsic rates of shear β_i^{-1}, the intrinsic stress α^{-1}, and the fractional area χ_i, covered by units of the kind i, are the properties which characterize a system rather than the viscosity. Equation (21) accordingly applies over the entire newtonian and non-newtonian range.

[HENRY EYRING]

Bibliography: H. J. Cantow et al., *Advances in Polymer Science*, vol. 5, 1968; F. R. Eirich (ed.), *Rheology: Theory and Applications*, 5 vols., 1956–1970; H. Eyring et al., *Statistical Mechanics and Dynamics*, 2d ed., 1979.

Riemannian geometry

The geometry of an N-dimensional space which bears a coordinate system (x) and has associated with it and the given coordinate system a set of N^2 functions $g_{ab}(x)$ which are involved in the determination of certain fundamental geometric magnitudes, including lengths of arcs in the space, lengths of vectors, angles between vectors, the measures of parts of the space (areas, volumes, and hypervolumes), and curvatures. The arc length s, for example, of a parameterized arc $x^a = x^a(t)$ of class C' is given by formula (1). Here, the

$$\Delta s = s(t_2) - s(t_1) = \int_{t_1}^{t_2}\sqrt{g_{ab}x'^a x'^b}\,dt \qquad (1)$$

prime ($'$) indicates differentiation with respect to t so that $x'^a = dx^a/dt$ and Einstein's summation convention is in force. The quantities g_{ab} are, as indicated, functions of the N coordinate variables x^1, x^2, . . . , x^N denoted briefly by x. The general subject may be classified into subspecies by means of the value of N and the conditions imposed on the fundamental quantities g_{ab} either directly or through derived quantities.

Riemannian geometry was initiated in 1854 by G. F. B. Riemann (1826–1866), with the quadratic forms $g_{ab}x'^a x'^b$ of course limited to the positive definite case. In later developments this restriction is removed.

The importance of this subject is due to several facts. First, it includes both euclidean geometry and the geometry of surfaces in euclidean space and, in addition, provides their most natural extension, a generalization of great richness and scope. Second, it provides a geometrical realization or application for certain of the major abstractions of tensor analysis and together with that discipline, with which it merges, forms the pattern and much of the motivation for general relativity. Third, most of the work on the incorporation of electricity and magnetism into the mold of general relativity (unified field theory) almost necessarily involves a Riemannian geometry of one type or another (nonsymmetric g_{ab}'s, symmetric g_{ab} with $N=5$, g_{ab}'s in projective coordinates). Fourth, Riemannian geometry provides the steppingstone to a variety of generalized differential geometries.

Finally, in the absence of contrary evidence, the success of the general theory of relativity suggests that actual physical space may correspond more closely to a noneuclidean Riemannian geometry than to the euclidean variety. *See* RELATIVITY; UNIFIED FIELD THEORY.

Riemannian geometry presents an impressive array of special devices and procedures. Among these are coordinate systems (geodesic and Riemannian) having the property that for at least one point $\{^a_{bc}\} = 0$, systems of N vector fields (orthogonal ennuples), operations based on mappings, and the use of anticommutative algebra (exterior differential forms).

Basic concepts. In the case of a Riemann space immersed in a euclidean space, the formulation of certain fundamental entities is greatly facilitated by the introduction of Gibbs' vectors (directed line segments). A second advantage is the revelation of some otherwise obscure vectorial characteristics. A relatively simple Riemannian geometry is furnished by a two-dimensional surface embedded in ordinary euclidean space E_3. A vector formulation of the basic concepts for this space may be obtained as follows. Let y^1, y^2, y^3 be a rectangular cartesian coordinate system in E_3 and e_1, e_2, e_3 be the corresponding base vectors. The radius (or position) vector r for E_3 is then given by $r = y^i e_i$, i: 1 to 3. Next let x^1, x^2 be rectangular cartesian coordinates in an auxiliary plane E_2 and D a domain in E_2. If each of the variables y^i (i: 1 to 3) is made a single-valued function of x^1, x^2 for x^1, x^2 in D, then r becomes a vector function $r(x^1,x^2)$ and each point in D determines r (and its endpoint in E_3). Further, let $r(x^1,x^2)$ be of class C^M, $M \geqq 1$, and meet the requirement $r(a,b) \neq r(c,d)$ for each pair of distinct points in D. (Class C^M implies the existence and continuity of Mth-order derivatives.) This procedure defines a one-to-one reciprocal and continuous correspondence between the points in the two-dimensional domain D and the set R_2 of end points of $r(x^1,x^2)$. Thus, each admissible ordered number pair a, $b(x^1 = a, x^2 = b)$ determines a point in D and simultaneously a point in R_2. Thus, the space R_2 bears a coordinate system, that is, a scheme for associating ordered number sets (pairs in the present case) to points in the space. The equations $x^2 = b$, with x^1 variable, and $x^1 = a$, with x^2 variable, determine a pair of coordinate lines in D and at the same time the parameterized arcs $r = r(x^1,b)$ and $r = r(a,x^2)$ (coordinate curves) in R_2. The derivatives $r_a(r_1 = \partial r/\partial x^1, r_2 = \partial r/\partial x^2)$ at a point (a,b) in R_2 are tangent vectors to the two coordinate curves through (a,b). More generally, the equations $x^1 = x^1(t)$ and $x^2 = x^2(t)$ determine an arc C in D and also the mate arc $r = r[x^1(t), x^2(t)]$ in R_2. The vector $r'(r' = D_t r = r_1 x'^1 + r_2 x'^2 = r_a x'^a)$ is a tangent vector to C. The magnitude of r' expressed in terms of surface quantities is given by Eq. (2). Equation (2) is valid

$$|r'|^2 = [(r_a x'^a) \cdot (r_b x'^b)]$$

$$= (g_{ab} x'^a x'^b), \quad g_{ab} \overset{\text{def}}{=\!=} r_a \cdot r_b \quad (2)$$

for C considered as a curve in E_3 except that the range of indices is now 1 to 3 and x^1, x^2, x^3 must denote the variables of a coordinate system in E_3. For example, in terms of cartesian coordinates x, y, z, with associated unit orthogonal base vectors i, j, f, C is given by $r = x(t)i + y(t)j + z(t)f$, $r_1 = i$, $r_2 = j$, $r_3 = f$, $g_{ab} = r_a \cdot r_b = 0$ or 1 according to whether $a \neq b$ or $a = b$, and $|r'|^2 = (x')^2 + (y')^2 + (z')^2$. Furthermore, according to a formula of calculus,

Δs (arc length) is expressed by Eq. (3), and so for

$$\Delta s = \int_{t_1}^{t_2} [(x')^2 + (y')^2 + (z')^2]^{1/2} dt \quad (3)$$

C regarded either as a curve in R_2 or E_3 it follows that Eq. (4) holds. Thus, in summary R_2 is a space

$$\Delta s = \int_{t_1}^{t_2} |r'| dt = \int_{t_1}^{t_2} (g_{ab} x'^a x'^b)^{1/2} dt \quad (4)$$

which bears a coordinate system (x), and there is associated with R_2 and (x) a set of functions g_{ab} which are involved in the determination of arc length. Essentially the same may be said of E_3.

The quantities g_{ab} are involved in other metric relationships. For example, if C_1 and C_2 are two arcs in R_2 which intersect at a point P and if the values of x'^a for C_1 and C_2 are denoted respectively by U^a and V^a, then the angle A between the tangents at P is given by Eq. (5). Also, it develops that

$$|r'(C_1)| \, |r'(C_2)| \cos A = r'(C_1) \cdot r'(C_2)$$

$$= (r_a U^a) \cdot (r_b V^b) = g_{ab} U^a V^b \quad (5)$$

a suitable extension of the concept area A from plane configurations to curved surface patches is furnished by the definition expressed in Eqs. (6).

$$A = \int \int G^{1/2} dx^1 dx^2 \qquad G = g_{11} g_{22} - g_{12} g_{21} \quad (6)$$

For most of the developments in Riemannian geometry, it is necessary or convenient to impose the restriction that for a, b in D, the determinant $G(a,b) \neq 0$.

Because the quantities g_{ab} are functions of x^1 and x^2, they may be associated with the domain D and if other surfaces R_2 are introduced with the same domain of definition D, then there becomes associated with D more than one set of functions $g_{ab}(x)$, assuming that the same letters x^1, x^2 are used. Each set of g_{ab}'s applied to mathematical entities in D determines measurements of related entities in the associated R. Thus, there arises a variety of problems concerning correspondences and mappings.

Generalizations to higher dimensions. The step from two to N dimensions can be made easily by making the coordinate variables $y^i(i$: 1 to $L)$ of an L-dimensional euclidean space E_L functions of N variables x^a with $N < L$, thus $y^i = y^i(x^1, \ldots, x^N)$. If N of these equations (for example, the first N) determine the x's as functions of $y^1, \ldots, y^N$, $x^a = x^a(y^1, \ldots, y^N)$, then substitution into the remaining $L - N$ equations will relate the $y^i(i$: $N+1$ to $L)$ to $y^1, \ldots, y^N$. Thus, this procedure separates out a subset R_N of the points of the original E_L. If the coordinate variables y are cartesian and the symbols i_i denote the corresponding base vectors in E_L, then $r = \mathbf{OP} = y^i(x)i_i = r(x^1, \ldots, x^N)$ is the vector equation of the subspace R_N. Examination of the development of Eqs. (2), (4), and (5) will show that these relationships may be taken over provided the range of indices is changed from 1, 2 to 1, \ldots, N. Thus, the space R_N is Riemannian.

Transformation theory. Substitution from a set of transformation equations $x^a = x^a(\bar{x}^1, \ldots, \bar{x}^N) = x^a(\bar{x})$ of the general type of tensor analysis, into the vector function $r(x)$ yields the mate function $\bar{r}(\bar{x})$; thus, $\bar{r}(\bar{x}) = r[x(\bar{x})]$, and the equation of the subspace R_N in terms of the variables $\bar{x}$ is $r = \bar{r}(\bar{x})$. Differentiation of the identities $\bar{r}(\bar{x}) = r[x(\bar{x})]$, $r(x) =$

r̄[x̄(x)] yields the transformation equations, Eqs. (7a). It follows from the definitions in Eq. (7b)

$$\frac{\partial \bar{\mathfrak{r}}}{\partial \bar{x}^r} \equiv \bar{\mathfrak{r}}_r = \mathfrak{r}_a X_r{}^a \tag{7a}$$

$$\mathfrak{r}_a = \bar{\mathfrak{r}}_r X_a{}^r \left(X_r{}^a = \frac{\partial x^a}{\partial \bar{x}^r}, X_a{}^r = \frac{\partial \bar{x}^r}{\partial x^a} \right)$$

$$\bar{g}_{rs} = \bar{\mathfrak{r}}_r \cdot \bar{\mathfrak{r}}_s, g_{ab} = \mathfrak{r}_a \cdot \mathfrak{r}_b \tag{7b}$$

and the relations in (8) that Eq. (9) holds. That is,

$$\bar{\mathfrak{r}}_r \cdot \bar{\mathfrak{r}}_s = (\mathfrak{r}_a X_r{}^a) \cdot (\mathfrak{r}_b X_s{}^b) = \mathfrak{r}_a \cdot \mathfrak{r}_b X_r{}^a X_s{}^b \tag{8}$$

$$\bar{g}_{rs} = g_{ab} X_r{}^a X_s{}^b \tag{9}$$

the quantities g_{ab} and $\bar{g}_{rs}$ are tensor components of the type (0,2,0). The first number in the type description bracket is the number of superscripts on the tensor's component symbols and is the contravariant order, the second number is the number of subscripts and is the covariant order, while the last number is the exponent on the Jacobian determinant $\partial x / \partial \bar{x}$ in the expression for the $\bar{x}$-components of the tensor in terms of the x-components, and is called the weight of the tensor. An implication of Eq. (9) and the column-column rule for expressing the product of two Nth order determinants as an Nth-order determinant is if $\bar{G}$, G, and $J(x/\bar{x})$ denote the determinants of $\bar{g}_{rs}$, g_{ab}, $X_r{}^a$, Eq. (10) holds, because $(g_{ab}x_r{}^a)X_s{}^b$ is the element in

$$\overline{G} = G[J(x/\bar{x})]^2 \tag{10}$$

row r, column s of the column-column triple product $(GJ)J$.

It is convenient at times to introduce a second set of base vectors in each coordinate system by means of the definitions: $g^{ab} \equiv$ (the cofactor of g_{ab})/ G, $\bar{g}^{rs} =$ (the cofactor of $\bar{g}_{rs}$)/$\bar{G}$, and so on, $\mathfrak{r}^a = g^{ab}\mathfrak{r}_b$, $\bar{\mathfrak{r}}^r = \bar{g}^{rs}\bar{\mathfrak{r}}_s$. The principal properties of the new base vectors are expressed by Eqs. (11). As a

$$\mathfrak{r}^a \cdot \mathfrak{r}_b = \delta_b{}^a \qquad g^{ab} = \mathfrak{r}^a \cdot \mathfrak{r}^b \qquad \mathfrak{r}_a = g_{ab}\mathfrak{r}^b \tag{11}$$

preliminary to the proof of these relations it may be noted that according to a formula of determinant theory $g_{bc}g^{ac} = \delta_b{}^a$. The symbol δ (called the Kronecker delta) may be defined by the statements $\delta_1{}^1 = \delta_2{}^2 = \cdots = \delta_N{}^N = 1$, $\delta_b{}^a = 0$ for $a \neq b$. It acts as a substitution operator; thus (for $N = 3$) relation (12) holds. Also, the relation (13) holds

$$\delta_2{}^c \mathfrak{r}_c = \delta_2{}^1 \mathfrak{r}_1 + \delta_2{}^2 \mathfrak{r}_2 + \delta_2{}^3 \mathfrak{r}_3 = \mathfrak{r}_2 \tag{12}$$

$$X_s{}^a X_b{}^s = \partial x^a / \partial \bar{x}^s \, \partial \bar{x}^s / \partial x^b = \partial x^a / \partial x^b = \delta_b{}^a \tag{13}$$

since x^a and x^b belong to the same coordinate system. Equations (11) may be established as follows: $\mathfrak{r}^a \cdot \mathfrak{r}_b = g^{ac}\mathfrak{r}_c \cdot \mathfrak{r}_b = g^{ac}g_{bc} = \delta_b{}^a$, $\mathfrak{r}^a \cdot \mathfrak{r}^b = g^{ac}\mathfrak{r}_c \cdot \mathfrak{r}^b = g^{ac}\delta_c{}^b = g^{ab}$, $g_{ab}\mathfrak{r}^b = g_{ab}g^{bc}\mathfrak{r}_c = \delta_a{}^c \mathfrak{r}_c = \mathfrak{r}_a$.
If a given vector $\boldsymbol{\alpha}$ is expressible in the form $\boldsymbol{\alpha} = a^a \mathfrak{r}_a$, the quantities a^a are called the scalar components of $\boldsymbol{\alpha}$ relative to the set $\mathfrak{r}_a$. Since $\boldsymbol{\alpha} \cdot \mathfrak{r}^b = a^a \mathfrak{r}_a \cdot \mathfrak{r}^b = a^a \delta_a{}^b = a^b$, it follows that $\boldsymbol{\alpha} = (\boldsymbol{\alpha} \cdot \mathfrak{r}^a)\mathfrak{r}_a = \boldsymbol{\alpha} \cdot (\mathfrak{r}_b g^{ab})\mathfrak{r}_a = (\boldsymbol{\alpha} \cdot \mathfrak{r}_b)\mathfrak{r}^b$.
In particular, relation (14) holds. Also, relation (15) holds, that is, Eq. (16) is established.

$$\bar{\mathfrak{r}}^r X_r{}^a = [(\bar{\mathfrak{r}}^r X_r{}^a) \cdot \mathfrak{r}_b]\mathfrak{r}^b = (\bar{\mathfrak{r}}^r X_r{}^a) \cdot (\bar{\mathfrak{r}}_s X_b{}^s)\mathfrak{r}^b$$
$$= \delta_s{}^r X_r{}^a X_b{}^s \mathfrak{r}^b = X_s{}^a X_b{}^s \mathfrak{r}^b = \delta_b{}^a \mathfrak{r}^b = \mathfrak{r}^a \tag{14}$$

$$g^{ab} = \mathfrak{r}^a \cdot \mathfrak{r}^b = (\bar{\mathfrak{r}}^r X_r{}^a) \cdot (\bar{\mathfrak{r}}^s X_s{}^b) = \bar{g}^{rs} X_r{}^a X_s{}^b \tag{15}$$

$$\mathfrak{r}^a = \bar{\mathfrak{r}}^r X_r{}^a, g^{ab} = \bar{g}^{rs} X_r{}^a X_s{}^b \tag{16}$$

Thus, the quantities g^{ab} are tensor components of the type (2,0,0), that is, contravariant of order two.

The dual sets $\mathfrak{r}_a$, $\mathfrak{r}^a$ and g_{ab}, g^{ab} provide the basis for a dual representation of various entities. For example, if the contravariant components x'^a of the tangent vector $\mathfrak{r}'(\mathfrak{r}' = x'^a \mathfrak{r}_a)$ for a given curve in an R_N in E_L are denoted by V^a and if $V_b \stackrel{\text{def}}{=} g_{ab}V^a$, then $\mathfrak{r}' = V^a \mathfrak{r}_a = V^a(g_{ab}\mathfrak{r}^b) = V_b \mathfrak{r}^b$. Also the dual sets $\mathfrak{r}_a$ and $\mathfrak{r}^a$ can be used to separate a given vector $\boldsymbol{\alpha}$ associated with a point P of R_N (but not tangent to R_N) into its tangent and perpendicular parts. For example, suppose that $\boldsymbol{\alpha}$ is a vector in an E_3, associated with a point P in a subspace R_2, and that the vectors $\mathfrak{r}_a$, $\mathfrak{r}^a$ are evaluated at P. It then follows that: (1) $(\boldsymbol{\alpha} \cdot \mathfrak{r}^a)\mathfrak{r}_a$ is tangential to R_2 (since $\mathfrak{r}_1$ and $\mathfrak{r}_2$ are); (2) $\mathfrak{r}_b \cdot [\boldsymbol{\alpha} - (\boldsymbol{\alpha} \cdot \mathfrak{r}^a)\mathfrak{r}_a] = 0$; and (3) $\boldsymbol{\alpha} = (\boldsymbol{\alpha} \cdot \mathfrak{r}^a)\mathfrak{r}_a + [\boldsymbol{\alpha} - (\boldsymbol{\alpha} \cdot \mathfrak{r}^a)\mathfrak{r}_a]$. In addition, $(\boldsymbol{\alpha} \cdot \mathfrak{r}^a)\mathfrak{r}_a = (\boldsymbol{\alpha} \cdot \mathfrak{r}_b g^{ab})\mathfrak{r}_a = (\boldsymbol{\alpha} \cdot \mathfrak{r}_b)\mathfrak{r}^b$. In particular, $\mathfrak{r}_{ab}$ ($= \partial^2 \mathfrak{r}/\partial x^a \partial^b$) is not, in general, tangent to the subspace R but its tangential part is given by either $(\mathfrak{r}_{ab} \cdot \mathfrak{r}_c)\mathfrak{r}^c$ or $(\mathfrak{r}_{ab} \cdot \mathfrak{r}^c)\mathfrak{r}_c$. The coefficients $\mathfrak{r}_{ab} \cdot \mathfrak{r}_c$ and $\mathfrak{r}_{ab} \cdot \mathfrak{r}^c$ are important quantities called Christoffel symbols, frequently denoted by $[ab,c]$ and $\{^c_{ab}\}$. This last symbol is involved in the tangential part of the derivative with respect to t of a vector field $V^a(t)\mathfrak{r}_a(t)$ defined along an arc C, $x^a = x^a(t)$. Thus (denoting differentiation with respect to t by a'), $[(V^a \mathfrak{r}_a)' \cdot \mathfrak{r}^c]\mathfrak{r}_c = [V'^a \delta_a{}^c + V^a\{^c_{ab}\}x'^b]\mathfrak{r}_c$. The coefficient of $\mathfrak{r}_c$, frequently denoted by IV^c, is called the intrinsic derivative of V^c, that is, $IV^c = V'^c + V^a\{^c_{ab}\}x'^b$. Similarly, since $\mathfrak{r}^a \cdot \mathfrak{r}_c = \delta_c{}^a$, $\mathfrak{r}^{a'} \cdot \mathfrak{r}_c + \mathfrak{r}^a \cdot \mathfrak{r}_c' = 0$ and therefore $(V_a \mathfrak{r}^a)' \cdot \mathfrak{r}_c = V_a' \delta_c{}^a + V_a \mathfrak{r}^{a'} \cdot \mathfrak{r}_c = V_c' - V_a\{^a_{bc}\}x'^b = IV_c$ (by definition). The process of intrinsic differentiation has been extended to higher-order tensors, for R_N in E_L, and for R_N not imbedded, and also for more general spaces. Another differentiation process which when applied to a tensor yields a tensor and is formally quite similar to the intrinsic derivative is the Lie derivative.

Since $I[V^c(x)]$ and $I[V_c(x)]$ are functions of the sets x, x', it follows that $\partial(IV^c)/\partial x'^b$ and $\partial(IV_c)/\partial x'^b$ are tensors (called the covariant derivative of V^c and V_c with respect to x^b).

Higher abstractions. A higher-order abstraction can be obtained by discarding the enveloping euclidean space E_L and considering the space R_N with a coordinate system (x) and a set of functions $g_{ab}(x)$ to be given from the beginning. If the g_{ab}'s are such that $g_{ab}V^a V^b$ is positive for all choices of the real numbers V^a excepting $V^1 = V^2 = \cdots = V^N = 0$, then the quadratic form $g_{ab}V^a V^b$ and for brevity the g's themselves are said to be positive definite. In the preceding R_N, $V^a \mathfrak{r}_a$ is a vector $\boldsymbol{\alpha}$ in E_L and $g_{ab}V^a V^b = |\boldsymbol{\alpha}|^2 > 0$ and therefore the g_{ab}'s are positive definite. The special and general theories of relativity involve forms $g_{ab}V^a V^b$ which can be positive, negative, or zero for real values of V^a.

In the present abstract case, certain of the preceding formulas may be adopted as definitions. Thus, a set of contravariant quantities V^a will be said to be the components of a vector and for brevity the set V^a itself will be called a vector. The terms magnitude of a vector and the angle A between two vectors $\mathbf{U}^a$ and $\mathbf{V}^a$ are defined by relations (17).

$$|\mathbf{V}^a|^2 \overset{\text{def}}{=} g_{ab}\mathbf{V}^a\mathbf{V}^b$$

$$\cos A \overset{\text{def}}{=} g_{ab}\mathbf{U}^a\mathbf{V}^b / \sqrt{(g_{cd}\mathbf{U}^c\mathbf{U}^d)(g_{ef}\mathbf{V}^e\mathbf{V}^f)} \tag{17}$$

The quantities $\bar{g}_{rs}$ associated with system $(\bar{x})$ may be defined as the coefficients of $\overline{\mathbf{U}^r\mathbf{V}^s}$ in the transform $g_{ab}\overline{\mathbf{U}}{}^r X_r{}^a \overline{\mathbf{V}}{}^s X_s{}^b$ of $g_{ab}\mathbf{U}^a\mathbf{V}^b$. Thus, $\bar{g}_{rs} = g_{ab}X_r{}^a X_s{}^b$ and the quantities g_{ab}, $\bar{g}_{rs}$ are again tensor components of the type (0,2,0). The existence of the quantities g^{ab} requires that $G \neq 0$. In this case the covariant counterparts $\mathbf{V}_a$ of the contravariant $\mathbf{V}^a$ and $\mathbf{V}^a$ are related thus: $\mathbf{V}_a = g_{ab}\mathbf{V}^b$, $\mathbf{V}^a = g^{ab}\mathbf{V}_b$.

Equipollent displacement. Let $E(A)$ and $E(B)$ be the tangent planes at points A and B of a surface R_2 in E_3 with $E(A)$ not $\parallel$ to $E(B)$. Further, let $\boldsymbol{\alpha}$ be a vector in $E(A)$, for example, at A with $\boldsymbol{\alpha}$ not $\parallel$ to the $E(A)$, $E(B)$ intersection. There is then no vector $\boldsymbol{\beta}$ at B which is in the plane $E(B)$ and equipollent (equal and parallel) to $\boldsymbol{\alpha}$ in the ordinary E_3 sense of parallelism. The question arises whether it is possible to broaden the concept of equipollence so that there will be determined a unique vector $\boldsymbol{\beta}$ at B which is equipollent to $\boldsymbol{\alpha}$ relative to R_2 and preserve a substantial core of the cardinal properties of parallelism E_3.

This can be done provided that the new concept be made relative to a curve C in R_2 joining A and B—equipollence (R_2,C). Furthermore, the formulation can be extended readily to the more general cases: R_N in E_L and R_N not imbedded.

Definition. Let C be an arc of class C', expression (18), which lies in an $R_N(G \neq 0)$, and let $V^a(t)$

$$[x^a = x^a(t)] \tag{18}$$

(class C') be a contravariant vector field along C. If $IV^a \underset{t}{\equiv} 0$, then the field V^a will be said to be equipollent relative to R and C. Here I denotes intrinsic differentiation.

In particular, if A and B are points on C corresponding to the values t_1 and t_2 of the parameter t, then the vectors $V^a(t_1)$, $V^a(t_2)$ are equipollent (R_N,C). If the quantities $\{{}^a_{bc}\}x'^c$ are analytic, then the equations $IV^a \overset{\text{def}}{=} V^{a\prime} + V^b\{{}^a_{bc}\}x'^c = 0$, $V^a(t_1) = A^a$ will determine $V^a(t_2)$ uniquely. The Christoffel symbols $\{{}^a_{bc}\}$ may be determined by $\{{}^a_{bc}\} = \mathfrak{r}^a \cdot \mathfrak{r}_{bc}$ in the case of an R_N in E_L and more generally by Eq. (19). The notation $(; \equiv \partial)$ indicates that $;d \equiv$

$$\{{}^a_{bc}\} = g^{ad} 1/2 [-g_{bc;d} + g_{dc;b} + g_{bd;c}] \quad (; \equiv \partial) \tag{19}$$

$\partial/\partial x^d$. Certain properties of intrinsic differentiation $[Ig_{ab} = Ig^{ab} = I\delta_a{}^b = 0$, $(IT_{ab})U^aV^b + T_{ab}(IU^a)V^b + T_{ab}U^a(IV^b) = (T_{ab}U^aV^b)'$ for T, U, V of types (0,2,0), (1,0,0), (1,0,0)] ensure that the magnitudes of and the angles between vectors undergoing equipollent displacement remain constant. Also if E_a is the Euler vector for $F(x,x') \equiv \sqrt{g_{ab}x'^a x'^b}$, $E_a \equiv -\partial F/\partial x^a + (\partial F/\partial x'^a)'$ and the parameter is the arc length s, then $I_s(dx^a/ds) = g^{ab}E_b$. An immediate consequence is that if C satisfies $E_b = 0$ (so that C is an extremal for $\int\sqrt{g_{ab}x'^a x'^b}\,ds$), then C satisfies $I_s(dx^a/ds) = 0$ (dx^a/ds is the unit tangent vector). The shortest arcs are extremal arcs; thus briefly stated the shortest arcs are the straightest.

Subspaces. If, in the generalization from two to N dimensions, the variables y^i are coordinates in an R_L (instead of E_L), then the subset of points

determined by the variables x is an N-dimensional subspace R_N of the Riemannian R_L. R_N is also Riemannian, for the y equations of an R_N-arc $C[x^a = x^a(t)]$ are $y^i = y^i[x^1(t), \dots ,x^N(t)]$ and $g_{ij}y'^iy'^j = g_{ij}\,\partial y^j/\partial x^a\,\partial y^i/\partial x^b\,x'^ax'^b = g_{ab}x'^ax'^b$, $g_{ab} \overset{\text{def}}{=} g_{ij}\,\partial y^i/\partial x^a\,\partial y^j/\partial x^b$. The quantities $\partial y^i/\partial x^a$ for each value of i (1 to L) are covariant relative to transformations of the x's. Hence the g_{ab}'s of R_N are determined by the g_{ij}'s of R_L and are of type (0,2,0). In particular, if $\boldsymbol{\alpha}$ and $\boldsymbol{\beta}$ are two perpendicular vectors at a point P of R_L, and $C(u)$ is the geodesic arc of R_L through P and tangent to $\cos u\boldsymbol{\alpha} + \sin u\boldsymbol{\beta}$, then the set of all such arcs is an $R_2[R_2(\boldsymbol{\alpha},\boldsymbol{\beta})]$ in R_L. The Gaussian curvature of $R_2(\boldsymbol{\alpha},\boldsymbol{\beta})$ is called the curvature $C(R_L,\boldsymbol{\alpha},\boldsymbol{\beta})$ of R_L for the orientation $\boldsymbol{\alpha}$, $\boldsymbol{\beta}$. If $C(R_L,\boldsymbol{\alpha},\boldsymbol{\beta})$ is everywhere independent of the choice of $\boldsymbol{\alpha}$, $\boldsymbol{\beta}$, then C does not vary from point to point.

Riemann-Christoffel tensor. The covariant derivative $V^a{}_{,b}$ of $V^a(x)$ [type (1,0,0) class C''] is given by $V^a{}_{,b} = V^a{}_{;b} + V^c\{{}^a_{bc}\}(; \equiv \partial)$ and is of type (1,1,0). Similarly, the covariant derivative of $V^a{}_{,c}$ with respect to x^d, namely, $V^a{}_{,c,d}$ is given by $V^a{}_{,c,d} = V^a{}_{,c;d} + V^e{}_{,c}\{{}^a_{ed}\} - V^a{}_{,b}\{{}^b_{cd}\}$ and $V^a{}_{,c,d} - V^a{}_{,d,c} = V^bR^a{}_{bcd}$ with $R^a{}_{bcd} = \{{}^a_{bc}\}_{;d} - \{{}^a_{bd}\}_{;c} + \{{}^e_{bc}\}\{{}^a_{de}\} - \{{}^e_{bd}\}\{{}^a_{ce}\}$. The quantities $R^a{}_{bcd}$ are the components of the celebrated Riemann-Christoffel tensor [type (1,3,0)]; some writers use $R^a{}_{bdc}$ for $R^a{}_{bcd}$.

Evidently, $R^a{}_{bcd} = 0$ is a necessary and sufficient condition for the equality of $V^a{}_{,c,d}$ and $V^a{}_{,d,c}$ with V^a of class C'', but otherwise arbitrary. Also, if for a certain coordinate system (x), the g's are constant (intrinsic flatness), then in $(x)\{{}^a_{bc}\}$ and $R^a{}_{bcd}$ vanish identically for all admissible values of the indices. But $R^a{}_{bcd}$ is a tensor and $R^a{}_{bcd} \equiv 0$ implies $\bar{R}^r{}_{stu} \equiv 0$ in $(\bar{x})$. Consequently, if for some coordinate system $(\bar{x})$ and some particular choice of $r,s,t,u\ \bar{R}^r{}_{stu} \neq 0$, then a coordinate system does not exist in the given R_N for which the g's are constant. The conditions $\{{}^a_{bc}\}$ analytic, $R^a{}_{bcd} \equiv 0$ are sufficient for intrinsic flatness. For intrinsically flat R_N equipollence is independent of the arc.

The associated tensors $R_{abcd}\ (\equiv g_{ae}R^e{}_{bcd})$, R_{bd} (the Ricci-Einstein tensor, $R_{bd} \equiv R^a{}_{bad}$), and R (the scalar curvature, $R = g^{bd}R_{bd}$), are all of fundamental importance in Riemannian geometry and relativity. In particular, the Gaussian curvature of an R_2 is determined through the associated R_{1212}, and the Riemann tensor is involved in the fundamental equation for determining the g_{ab} of general relativity and thereby the equations of motion of test particles and the paths of light rays.

Mappings and motions. For each fixed value of t a set of N equations $\bar{x}^r = f^r(x^1, \dots ,x^N,t)$ suitably restricted can be used in two different ways: (1) to determine a coordinate transformation [in which case a given point P has two sets of coordinates (x) and $(\bar{x})$]; or (2) to determine a mating of points of the space without change of coordinate system. In this latter case, if P is a point of the space R with coordinates $x^1, \dots ,x^N$ (in system x) and $t = t_1$ (an admissible value of t), then the point $M(P,t_1)$ with coordinates $x^{*1}, \dots ,x^{*N}$ (system x) determined by $x^{*a} = f^a(x^1, \dots ,x^N,t_1)$ will be regarded as the mate point of P, relative to these equations and the given fixed value of t. This mating of points is herein designated a mapping to emphasize that the (x^*,x,t) equations are not to be interpreted as determining a coordinate transformation. If $x^{*a} = f^a(x,t_1)$ has a unique single-valued inverse $x^a = g^a(x^*,t_1)$ valid for x in a certain domain D and this equation is used to determine the coordinate transformation $\bar{x}^{*r} = g^r(x^*,t_1) = g^r[x^*(x,t_1),t_1]$, then $\bar{x}^{*r}|M(P,t_1) = x^r|P$. In other words, the x-coordinates of the point P (domain D) are the same as the $\bar{x}^*$-coordinates of the mate point $M(P,t_1)$ of the domain $M(D,t_1)$. Thus if (1) $x^a = x^a(u)$ is the equation of a parameterized arc C (in D and of class C') and (2) $P(u)$ denotes the point on C determined by a specified value of u, and $MP(u)$ is the corresponding mate point on MC, then $x^a|P(u) = \bar{x}^{*a}|MP(u)$ and $dx^a/du|P(u),C = d\bar{x}^{*a}/du|MP(u),MC$.

If the functional structure in terms of the coordinates $\bar{x}^*$ (domain MD) of each component $\bar{g}^*_{ab}$ of the metric tensor is the same as the functional structure, in terms of coordinates x, of the corresponding component g_{ab} (domain D), then $\bar{g}^*_{ab}|MP = g_{ab}|P$ and a metric entity in MD equals the corresponding entity in D. In particular, for the curves MC and G, $\bar{g}^*_{ab}d\bar{x}^{*a}/dud\bar{x}^{*b}/du = g_{ab}dx^a/dudx^b/du$ and the magnitudes of tangent vectors and arc lengths are preserved by the mapping. Also, for two class C' arcs in D intersecting at a point P and given by $x^a = x_1{}^a(u)$, $x^a = x_2{}^a(v)$, it follows that $\bar{g}^*_{ab}d\bar{x}^*_1{}^a/dud\bar{x}^*_2{}^b/dv|MP = g_{ab}dx_1{}^a/dudx_2{}^b/dv|P$, and so the magnitude of the intersection angle is unchanged by the mapping.

In particular, if the set of equations for a mapping of a Riemannian space into itself involves one or more parameters and the magnitudes of metric entities are preserved by the mapping, then the mapping is called a motion. Considerable research has been done on groups of motions.

Generalizations. Some of the generalizations are: the geometry of paths (a nonmetric differential geometry based on quantities $\Gamma_{bc}{}^a$ which transform in the manner of Christoffel symbols and provide the definition of parallel displacement); Finsler geometry [a function $F(x,x')$ which is positively homogeneous of degree one in the x''s takes the place of $\sqrt{g_{ab}x'^a x'^b}$]; higher order differential geometries (Kawaguchi spaces, F is a function of the x's, x''s, and higher-order derivatives). *See* CALCULUS OF TENSORS; CALCULUS OF VECTORS; NONEUCLIDEAN GEOMETRY. [HOMER V. CRAIG]

Bibliography: L. P. Eisenhart, *Riemannian Geometry*, 2d ed., 1950; D. F. Lawden, *Introduction to Tensor Calculus and Relativity*, 2d ed., 1975; C. W. Misner, K. S. Thorne, and J. A. Wheeler, *Gravitation*, 1973; J. L. Synge and A. Schild, *Tensor Calculus*, 1969, reprint 1978.

Rigid body

An idealized extended solid whose size and shape are definitely fixed and remain unaltered when forces are applied. Treatment of the motion of a rigid body in terms of Newton's laws of motion leads to an understanding of certain important aspects of the translational and rotational motion of real bodies without the necessity of considering the complications involved when changes in size and shape occur. Many of the principles used to treat the motion of rigid bodies apply in good approximation to the motion of real elastic solids. *See* RIGID-BODY DYNAMICS. [DUDLEY WILLIAMS]

Rigid-body dynamics

A rigid body is defined as an assemblage or system of mass particles that are located rigidly with respect to one another and therefore can have no motion relative to each other. Motion of a rigid body can occur by movement of all points of the body in a parallel direction through equal distances during a given interval of time, called translation, or by movement of all points in circles about a common axis with a common angular velocity, called rotation, or by combined translation and rotation.

Center of mass. The center of mass of a rigid body is a single point located within the body such that any force acting externally on the body along a line of action which passes through this point will result in pure translation, that is, no rotation (Fig. 1). *See* CENTER OF MASS.

To calculate the position of the center of mass, consider a rigid body divided into elemental volumes (or particles) labeled dV in Fig. 2. Each such elemental volume is located from a fixed point O by means of a position vector $\mathbf{r}$ and has a mass density ρ.

The total mass m of the rigid body can then be expressed by the volume integral in Eq. (1).

$$m = \int_V \rho\, dV \tag{1}$$

The center of mass of the rigid body is then defined as the point within the rigid body whose

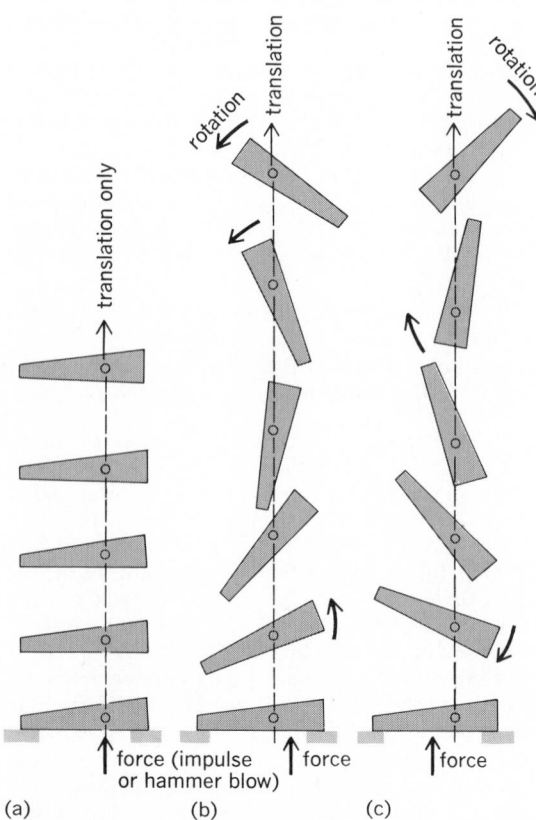

Fig. 1. Resultant motion for (*a*) a force acting at the center of mass, (*b*) a force acting at the right of the center of mass, and (*c*) a force acting at the left of the center of mass. The resultant motion in each case is shown. Force on center of mass does not produce rotation.

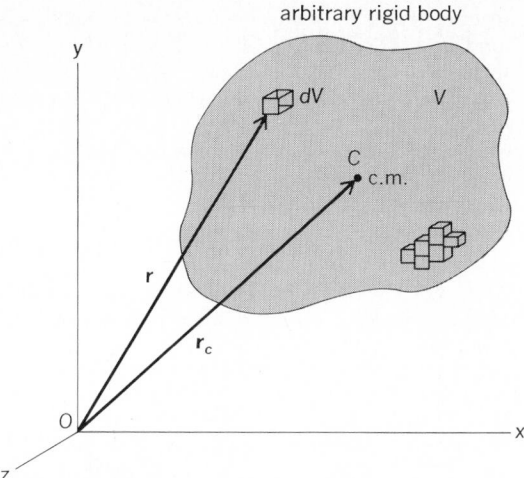

Fig. 2. A rigid body of arbitrary shape may be considered to be composed of a continuous distribution of volume elements (or particles). A typical volume element dV and position vectors are shown.

position vector $\mathbf{r}_c$ is given by Eq. (2).

$$\mathbf{r}_c = \frac{1}{m}\int_V \mathbf{r}\rho\,dV \qquad (2)$$

The coordinates of the mass center are then given by Eqs. (3), where x, y, z are the coordinates of the volume dV.

$$x_c = \frac{1}{m}\int_V x\rho\,dV \qquad y_c = \frac{1}{m}\int_V y\rho\,dV$$
$$z_c = \frac{1}{m}\int_V z\rho\,dV \qquad (3)$$

Translational motion. If a group of forces push or act on a rigid body in such a way that the line of action of these forces passes through the center of mass, pure translational motion results (Fig. 3). If the line of action does not pass through the center of mass, the resulting motion is a combination of both translation and rotation. However, in either

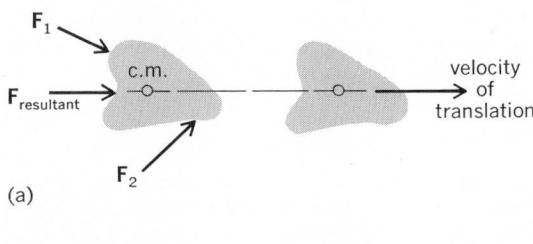

(a)

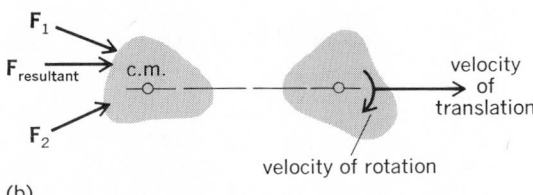

(b)

Fig. 3. Translational motion. (a) Pure translation caused by a resultant force passing through the center of mass of a rigid body. (b) Combined translation and rotation caused by a resultant force which does not pass through the center of mass.

case the motion of the center of mass of the rigid body is governed by Eq. (4), derived directly from

$$\Sigma\mathbf{F} = m\mathbf{a}_c = m\frac{d^2\mathbf{r}_c}{dt^2} \qquad (4)$$

Newton's second law applied to each particle of the rigid body and summed for all such particles. Here $\mathbf{a}_c = d^2\mathbf{r}_c/dt^2$ is the vector acceleration of the center of mass and $\Sigma\mathbf{F}$ is the sum of all external forces acting on the rigid body. All internal forces sum to zero because such forces between particles of the body occur in equal and opposite pairs (Newton's third law).

Equation (4) states that the motion of the center of mass of a rigid body is the same as would be the case if all of the mass of the body were concentrated at this mass center and all external forces were applied there. The motion of translation of the center of mass of a rigid body is therefore determined by the same methods as those used to determine the motion of a particle (Fig. 4). The one significant

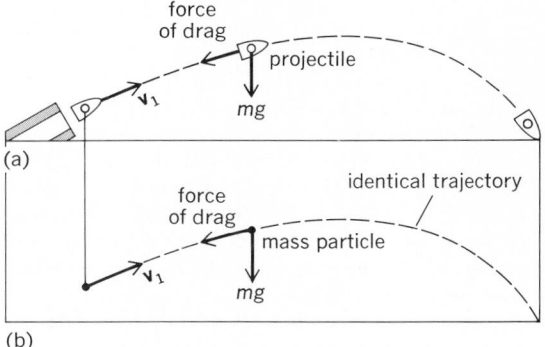

Fig. 4. Translational motion of projectile. (a) The motion of the center of mass of the projectile is the same as (b) the motion in of a mass particle of equal mass and subjected to the same external forces.

difference from particle dynamics, however, is that the forces are those acting on the actual rigid body and as such often depend upon the shape of the body, its orientation in space, and sometimes upon the time rate of change of this orientation in space. Examples are the motion of an airplane or projectile through air or of a submarine through water.

The motion in translation may also be expressed in terms of a vector quantity called linear momentum. Because the velocity of the center of mass is given by Eq. (5), Eq. (4) can be written as Eq. (6),

$$\mathbf{v}_c = \frac{d\mathbf{r}_c}{dt} \qquad (5)$$

$$\Sigma\mathbf{F} = m\frac{d\mathbf{v}_c}{dt} = \frac{d}{dt}(m\mathbf{v}_c) = \frac{d\mathbf{P}}{dt} \qquad (6)$$

where $\mathbf{P} = m\mathbf{v}_c$ is defined as the linear momentum of the rigid body. Thus the resultant external force $\mathbf{F}$ acting on a rigid body is equal to the time rate of change of the linear momentum of the body.

If no external forces are acting, Eq. (6) states that the linear momentum of the body remains constant—a statement of the law of conservation of linear momentum. *See* CONSERVATION OF MOMENTUM; MOMENTUM.

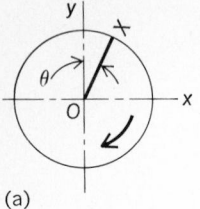

(a)

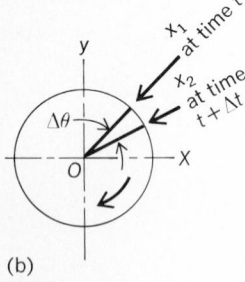

(b)

Fig. 5. Wheel or disk rotating about a perpendicular axis through O, showing (a) angular quantities as function of time and (b) instantaneous angular velocity.

Rotational motion. If a rigid body is rotating about an axis fixed in space, for example, a wheel turning on a shaft, all points on the axis remain fixed, while other points in or on the rigid body move in circular paths concentric to the axis and in planes perpendicular to the axis. If the angular position θ of any given radius OX of the body is specified as a function of time, then the position, velocity, and acceleration of every point of the body are known (Fig. 5).

As an example, if two successive positions of OX are taken over a small time interval Δt then the instantaneous angular velocity ω of the rigid body at time t (Fig. 5b) is specified as in Eq. (7).

$$\omega = \frac{\text{change in angular displacement of } OX}{\text{time interval}}$$

$$= \lim_{\Delta t \to 0} \frac{\Delta \theta}{\Delta t} = \frac{d\theta}{dt} \qquad (7)$$

Similarly, the instantaneous angular acceleration at time t is given by Eq. (8).

$$\alpha = \frac{\text{change in angular velocity}}{\text{time interval}}$$

$$= \lim_{\Delta t \to 0} \frac{\Delta \omega}{\Delta t} = \frac{d\omega}{dt} = \frac{d^2\theta}{dt^2} \qquad (8)$$

As is the case for linear velocity and acceleration, both angular velocity and acceleration are vectors. The vectors representing these quantities are drawn along the axis of rotation in a direction which represents the direction of advance of a right-handed screw which is rotated in a manner designated by the angular velocity and acceleration respectively (Fig. 6).

If now a plane body such as a disk or wheel is pictured rotating about a fixed axis, the tangential velocity of any point B on this body resulting from the rotation is given by the vector product of the angular velocity ω of the body and the vector distance from the axis of rotation to the point in question, as in Eq. (9), or in this case, as in Eq. (10),

$$\mathbf{v}_B = \boldsymbol{\omega} \times \mathbf{r} = \frac{d\mathbf{r}}{dt} \qquad (9)$$

$$\mathbf{v}_B = \omega r \boldsymbol{\theta}_1 \qquad (10)$$

where $\mathbf{v}_B$ is the vector velocity of point B in a direction normal to the radius $\mathbf{r}$ and angular velocity vector $\boldsymbol{\omega}$ (Fig. 7). This direction is designated by

the unit vector $\boldsymbol{\theta}_1$. The acceleration of point B is found by taking the derivative of $\mathbf{v}_B$ in Eq. (9), by Eq. (11a), or in the planar case of Fig. 7, by Eq. (11b). The acceleration of B is therefore composed

$$\mathbf{a}_B = \frac{d\mathbf{v}_B}{dt} = \frac{d}{dt}(\boldsymbol{\omega} \times \mathbf{r}) = \left(\frac{d\boldsymbol{\omega}}{dt} \times \mathbf{r}\right) + \left(\boldsymbol{\omega} \times \frac{d\mathbf{r}}{dt}\right)$$

$$= (\boldsymbol{\alpha} \times \mathbf{r}) + \boldsymbol{\omega} \times (\boldsymbol{\omega} \times \mathbf{r}) \qquad (11a)$$

$$\mathbf{a}_B = \alpha r \boldsymbol{\theta}_1 - \omega^2 r \mathbf{r}_1 \qquad (11b)$$

of two parts—a tangential acceleration αr and a radial acceleration directed inward toward the axis of rotation equal to $\omega^2 r$.

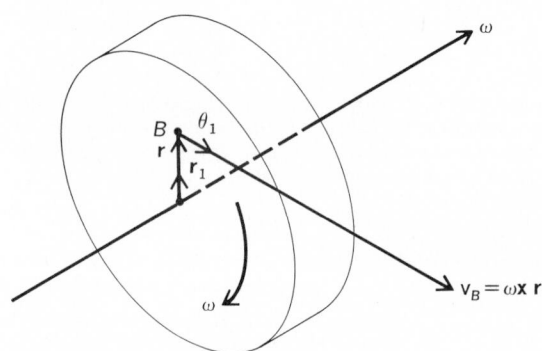

Fig. 7. Velocity of point B for a rotating disk.

For a three-dimensional rigid body, if the vector $\mathbf{r}$ is specified as the radius vector from any point on the axis of rotation to the point B, Eqs. (9) and (11a) are equally valid for this case. *See* ACCELERATION; ROTATIONAL MOTION; VELOCITY.

Rotation about translating axis. If a body rotates about an axis that is translating, for example, an automobile wheel or a spinning missile, the velocity of each point is the vector sum of the velocity of translation of the axis of rotation and the velocity that results from the rotation about the axis.

Angular motion; angular momentum. The equations that describe the rotational or angular motion of a rigid body are again derivable from Newton's second law. In Fig. 2 a rigid body of arbitrary shape is located and moving in an inertial system. This body may be subdivided into mass particles dV of mass m_i and Newton's law applied to each particle, resulting in Eq. (12).

$$m_i \frac{d^2\mathbf{r}}{dt^2} = \mathbf{F} \qquad (12)$$

If both sides of Eq. (12) are multiplied by the position vector $\mathbf{r}$ (by means of a cross product), Eq. (13) is derived. Here the vector product $(\mathbf{r} \times \mathbf{F})$ is

$$m_i \left(\mathbf{r} \times \frac{d^2\mathbf{r}}{dt^2}\right) = (\mathbf{r} \times \mathbf{F}) \qquad (13)$$

defined as the moment $\mathbf{M}$ or torque about the point O of the resultant force $\mathbf{F}$ acting on dV. *See* TORQUE.

Equation (13) may be written as Eq. (14), where

$$m_i \left(\mathbf{r} \times \frac{d^2\mathbf{r}}{dt^2}\right) = m_i \frac{d}{dt}\left(\mathbf{r} \times \frac{d\mathbf{r}}{dt}\right) = \frac{d}{dt}\left(\mathbf{r} \times m_i \frac{d\mathbf{r}}{dt}\right)$$

$$= \frac{d}{dt}(\mathbf{r} \times m_i \mathbf{v}) \qquad (14)$$

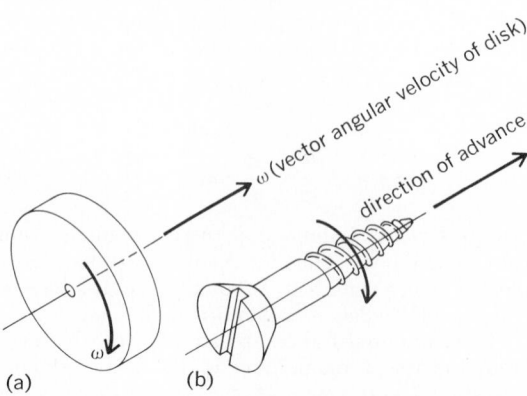

Fig. 6. Vector representation of the angular velocity of (a) a rotating disk and (b) a rotating screw.

$\mathbf{r}$, $\mathbf{v}$, and m_i are the position vector, velocity, and mass, respectively, of the element dV. Equation (14) is valid for each elemental particle of the rigid body, and by summing the equation for all such particles of the body, the equation of angular motion is obtained from Eq. (15a), where m_i has been replaced by $\rho \, dV$ as before. Equation (15b) ex-

$$\Sigma \mathbf{M}_o = (\mathbf{M}_o)_{\text{ext}} = \frac{d}{dt} \int_V (\mathbf{r} \times \mathbf{v}) \rho \, dV = \frac{d\mathbf{H}_o}{dt} \quad (15a)$$

$$\int_V (\mathbf{r} \times m_i \mathbf{v}) = \int_V (\mathbf{r} \times \mathbf{v}) \rho \, dV = \mathbf{H}_o \quad (15b)$$

presses the moment of linear momentum or the angular momentum about a point O of a rigid body. On the left-hand side of Eq. (15a), the summation process results in zero for all equal and opposite collinear forces between particles (internal forces), and only the moment $\mathbf{M}_o$ of the external forces exists.

Equation (15a) expresses the principle of angular momentum, that the time rate of change of angular momentum of a rigid body about any point O fixed in an inertial system is equal to the resultant moment of external forces about the same point O. If the rigid body is not acted upon by any external moment, the angular momentum must remain constant. This is the principle of conservation of angular momentum.

An important extension of Eq. (15a), the angular momentum equation, is as follows. Equation (15a) is precisely valid for the case in which the angular momenta and the moments of external forces are taken about the center of mass of a rigid body, even though the center of mass does not remain at rest in an inertial system. In this case the equation is that given as Eq. (16), where c denotes the center

$$\mathbf{M}_c = \frac{d\mathbf{H}_c}{dt} \quad (16)$$

of mass. Equation (16) is useful, for example, in predicting the motion of a projectile moving through air under accelerated conditions. The angular position of the projectile affects the aerodynamic forces acting on the projectile, and Eq. (16) is necessary for the solution of the problem. The fact that the angular equation can be written about the moving center of mass fixed within the projectile, instead of about some fixed point located exterior to the projectile, results in great simplification in the analysis of the motion.

Moments and products of inertia. To investigate the rotation of a rigid body, either Eq. (15a) or (16) is used, wherein the resultant moment and the angular momentum are taken either about a point fixed in an inertial system or about the center of mass of the body. To handle the equations most easily, however, it is usually essential that the point about which the moments and angular momenta are calculated be fixed in the body itself. Therefore, if the body contains a point fixed in an inertial system, this point is taken as the origin of a coordinate system. If no point within the body is fixed, then the center of mass is taken as the origin.

Because either Eq. (15a) or Eq. (16) describes the rotation of the body in either of these two cases, they are treated together. Consider a set of axes xyz attached to the rigid body with either a fixed point or center of mass as origin O. The ele-

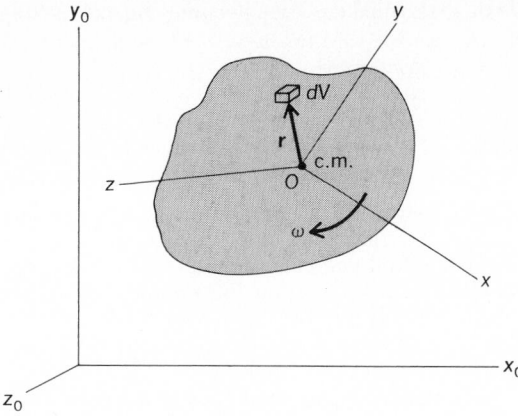

Fig. 8. Rigid body of arbitrary shape with an *xyz* coordinate system attached at origin O (fixed point or center of mass) and moving with the body.

ments or particles of the rigid body are at rest with respect to these axes, and as a result the velocity of any element dV relative to the origin O (Fig. 8) is given by Eq. (17).

$$\frac{d\mathbf{r}}{dt} = \mathbf{v} = \boldsymbol{\omega} \times \mathbf{r} \quad (17)$$

Then the angular momentum of the body about O, Eq. (15b) may be defined by Eq. (18).

$$\mathbf{H}_o = \int (\mathbf{r} \times \mathbf{v}) \rho \, dV = \int \mathbf{r} \times (\boldsymbol{\omega} \times \mathbf{r}) \rho \, dV \quad (18)$$

Now $\mathbf{r}$ and $\boldsymbol{\omega}$ are vectors which can be expressed in terms of their components along x, y, z axes, as in Eqs. (19). Substituting Eqs. (19) into (18) gives Eq. (20).

$$\mathbf{r} = \mathbf{i}x + \mathbf{j}y + \mathbf{k}z$$
$$\boldsymbol{\omega} = \mathbf{i}\omega_x + \mathbf{j}\omega_y + \mathbf{k}\omega_z \quad (19)$$

$$\mathbf{H}_o = \mathbf{i} \int [\omega_x(y^2 + z^2) - \omega_y xy - \omega_z xz] \rho \, dV$$
$$+ \mathbf{j} \int [-\omega_x xy + \omega_y(x^2 + z^2) - \omega_z yz] \rho \, dV$$
$$+ \mathbf{k} \int [-\omega_x xz - \omega_y yz + \omega_z(x^2 + y^2)] \rho \, dV \quad (20)$$

The values ω_x, ω_y, ω_z are the same for all volume elements in the rigid body, and Eq. (20) can be rewritten in a particularly useful way as Eq. (21),

$$\mathbf{H}_o = \mathbf{i}(I_x \omega_x - I_{xy} \omega_y - I_{xz} \omega_z)$$
$$+ \mathbf{j}(-I_{yx} \omega_x + I_y \omega_y - I_{yz} \omega_z)$$
$$+ \mathbf{k}(-I_{zx} \omega_x - I_{zy} \omega_y + I_z \omega_z) \quad (21)$$

where Eqs. (22) apply. These are called the mo-

$$I_x = \int (y^2 + z^2) \rho \, dV \qquad I_y = \int (z^2 + x^2) \rho \, dV$$
$$I_z = \int (x^2 + y^2) \rho \, dV \quad (22)$$

ments of inertia of the rigid body about the x, y, z axis through O, and the terms in Eqs. (23) are called the products of inertia.

$$I_{zy} = I_{yz} = \int yz \rho \, dV$$
$$I_{xz} = I_{zx} = \int xz \rho \, dV \quad (23)$$
$$I_{yx} = I_{xy} = \int xy \rho \, dV$$

These six quantities are geometrical constants associated with the body and do not vary with its motion or with time.

If Eq. (21) is substituted into Eq. (15a) or (16) and the time derivatives of the moving unit vectors $\mathbf{i}$, $\mathbf{j}$, k taken, the final equation becomes Eq. (24), where H_x, H_y, and H_z are the x, y, and z components of $\mathbf{H}_o$ in Eq. (21). *See* MOMENT OF INERTIA.

$$\mathbf{M}_o = \mathbf{i}(H_x + \omega_y H_z - \omega_z H_y) \\ + \mathbf{j}(H_y + \omega_z H_x - \omega_x H_z) \\ + \mathbf{k}(H_z + \omega_x H_y - \omega_y H_x) \tag{24}$$

Principal axes. An important theorem exists which states that at any point in a rigid body it is possible to find and construct a set of rectangular coordinate axes such that the products of inertia vanish and only the three moments of inertia exist. This is the principal axis theorem, and such coordinates are called principal axes.

In particular, in any rigid body that has two perpendicular planes of symmetry, the coordinate planes of the principal axes coincide with the planes of symmetry.

If the principal axes are chosen for the x, y, z coordinate system, Eq. (24) simplifies as Eq. (25).

$$\mathbf{M}_o = \mathbf{i}(I_x \dot{\omega}_x + (I_z - I_y)\omega_y \omega_z) \\ + \mathbf{j}(I_y \dot{\omega}_y + (I_x - I_z)\omega_x \omega_z) \\ + \mathbf{k}(I_z \dot{\omega}_z + (I_y - I_x)\omega_x \omega_y) \tag{25}$$

Equation (25) is called Euler's equation, after the famous mathematician of the 18th century, and can be used to solve the majority of rigid-body dynamics problems. *See* EULER'S EQUATIONS OF MOTION.

Work and energy relations. The work done by the forces acting on a single mass particle (labeled i) as the particle moves from point 1 to point 2 in space is defined by the vector equation given here as Eq. (26), where $\mathbf{F}_i$ is the resultant vector force

$$W_i = \int_1^2 \mathbf{F}_i \cdot d\mathbf{r}_i \tag{26}$$

acting on the particle and $d\mathbf{r}_i$ is the vector displacement of the particle as it travels from A to a neighboring point B in time dt (Fig. 9).

For a rigid body considered as a system of such elemental particles, the work done by all of the external forces acting on the rigid body (the inter-

nal forces in a rigid body do no work) from Eqs. (26) and (12) is given by Eq. (27).

$$W_{1,2} = \int_1^2 \sum (\mathbf{F}_i \cdot d\mathbf{r}_i) = \int_1^2 \sum \left(m_i \frac{d^2 \mathbf{r}_i}{dt^2} \right) \cdot d\mathbf{r}_i \\ = \int_{t_1}^{t_2} \sum m_i \frac{d}{dt}\left(\frac{d\mathbf{r}_i}{dt} \cdot \frac{d\mathbf{r}_i}{dt} \right) dt \\ = \int_{t_1}^{t_2} \frac{d}{dt} \sum \frac{m_i v_i^2}{2} dt = \int_1^2 d(T) = T_2 - T_1 \tag{27}$$

Thus, the change in the kinetic energy T of a rigid body in going from locations 1 to 2 in space is equal to the work done on the rigid body during this period. Or, the rate of change of kinetic energy of the body is equal to the rate at which work is done by all external forces acting on the body. If the forces acting are a function of the coordinates only and the work done is therefore independent of the path the body follows (gravitational forces, for example), the system is said to be conservative (frictionless, without energy dissipation). Then the work done on the body by the forces acting on it, when this body moves in a conservative force field from locations 1 to 2 in space, is called the potential energy of the body at 2 with respect to 1, or is equal to the negative of the change in potential energy of the body in passage from 1 to 2. Then Eq. (28) holds. Here $\Delta V_{1,2}$ is the change in potential

$$W_{1,2} = -\Delta V_{1,2} \tag{28}$$

energy of the body. From Eqs. (27) and (28) Eq. (29) is derived.

$$T_1 + V_1 = T_2 + V_2 \tag{29}$$

Equation (29) is called the conservation of energy equation for a conservative system wherein the sum of kinetic and potential energies remains constant throughout the motion. *See* CONSERVATION OF ENERGY; ENERGY; WORK.

Rolling motion. The motion of a cylinder rolling without slipping on a surface offers an interesting and important application of Eqs. (4) and (25). If a moving coordinate system is located at the center of mass, as shown in Fig. 10, then $\omega_x = \omega_y = 0$ and $\omega_z = \omega$. Then the equations for linear and angular motion become Eqs. (30a)–(30c).

$$\Sigma F_x = ma_x = m \frac{dv_x}{dt} = m \frac{d^2 x_o}{dt^2} \tag{30a}$$

$$\Sigma F_y = 0 \tag{30b}$$

$$M_o = I_o \frac{d\omega}{dt} \tag{30c}$$

The energy equation for such a conservative system is given by Eq. (31), where Eq. (32) applies and V is the potential energy of the cylinder.

$$\Delta T = -\Delta V \tag{31}$$

$$T = \frac{1}{2}mv_x^2 + \frac{1}{2}I_o\omega^2 \tag{32}$$

The specification of no slipping requires, first, that the cylinder complete one revolution as the center O advances a distance of one circumference or that Eqs. (33) hold and, second, that a frictional force f, acting at the point of contact, is

$$r\theta = x \qquad r\left(\frac{d\theta}{dt}\right) = r\omega = \frac{dx}{dt} = v_x \tag{33}$$

tional force f, acting at the point of contact, is

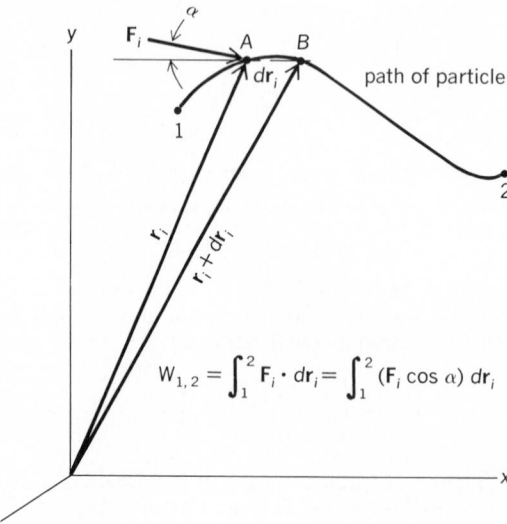

$$W_{1,2} = \int_1^2 \mathbf{F}_i \cdot d\mathbf{r}_i = \int_1^2 (\mathbf{F}_i \cos \alpha) \, d\mathbf{r}_i$$

Fig. 9.   Illustration of the work done by a force moving a mass particle along an arbitrary path.

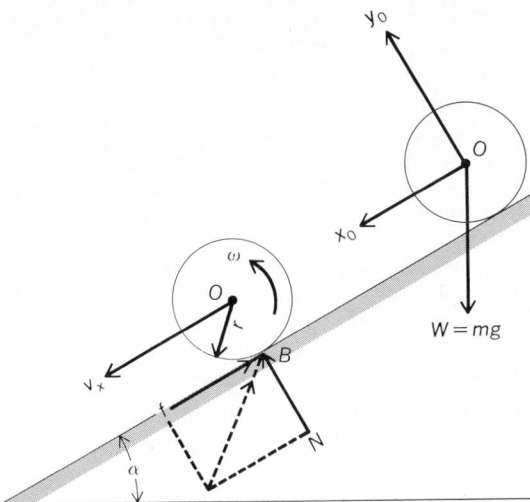

Fig. 10. Diagram of a cylinder rolling down an inclined plane without slipping.

sufficiently large to create a moment M_o that satisfies Eq. (30a).

If the cylinder in Fig. 10, starting from rest, rolls without slipping down the incline under the influence of gravity, the motion can be determined from Eqs. (30) which become for this case Eqs. (34)–(36), where N is the normal force and K is the

$$W \sin \alpha - f = m \frac{d^2 x_o}{dt^2} = m \frac{dv_x}{dt} \qquad (34)$$

$$W \cos \alpha - N = 0 \qquad (35)$$

$$rf = I_o \frac{d\omega}{dt} = (mK^2) \frac{d\omega}{dt} \qquad (36)$$

radius of gyration such that $mK^2 = I_o$. *See* RADIUS OF GYRATION.

Substitution of Eq. (36) into (34) gives Eq. (37),

$$mg \sin \alpha - \frac{mK^2}{r} \frac{d\omega}{dt} = m \frac{dv_x}{dt} \qquad (37)$$

and of Eq. (33) into (37) results in Eq. (38), where g

$$\frac{dv_x}{dt} = \frac{g \sin \alpha}{1 + (K^2/r^2)} = a_x \qquad (38)$$

is the acceleration of gravity. The cylinder therefore rolls with a constant acceleration a_x.

The motion can also be determined by use of Eq. (31), the conservation of energy equation.

If the frictional force f were zero, then ω would be zero [Eq. (36)] and pure translation would take place at an acceleration of $g \sin \alpha$. The value of acceleration is therefore always less for the frictional (no slip) case but approaches the $g \sin \alpha$ value when the mass of the cylinder is so concentrated at the center that $K \to 0$. If the cylinder is a thin-walled tube such that $K \to r$, then Eq. (39) holds.

$$\frac{dv_x}{dt} = \frac{1}{2} g \sin \alpha \qquad (39)$$

If the cylinder is solid and uniform, $K^2 = r^2/2$, and for this case Eq. (40) holds.

$$\frac{dv_x}{dt} = \frac{2}{3} g \sin \alpha \qquad (40)$$

The frictional force required to prevent slipping is given by Eq. (41).

$$f = \frac{mgK^2 \sin \alpha}{r^2 + K^2} \qquad (41)$$

From Eq. (35), the normal force is given by Eq. (42).

$$N = mg \cos \alpha \qquad (42)$$

Since $f = \mu N$, where μ is defined as the coefficient of static friction, Eqs. (41) and (42) show that in order for rolling to take place without slipping the coefficient must be such that Eq. (43) holds.

$$\mu \geqq \frac{K^2 \tan \alpha}{K^2 + r^2} \qquad (43)$$

See STATICS.

Instantaneous axis. Consider a rigid body moving in a completely general manner. A point A is chosen on or within the body as an arbitrary base point and the velocity of A denoted by $\mathbf{v}_A$. The velocity of any other point B on or within the body is given by Eq. (44), where $\mathbf{r}_{AB}$ is the vector distance

$$\mathbf{v}_B = \mathbf{v}_A + (\boldsymbol{\omega} \times \mathbf{r}_{AB}) \qquad (44)$$

between A and B directed toward B. If the base point A is changed, the translational velocity $\mathbf{v}_A$ will be different but the angular velocity $\boldsymbol{\omega}$ will remain the same. The vector $\boldsymbol{\omega}$ pertains to the angular motion of the body as a whole and is to be regarded as a free vector since it does not depend upon a choice of base point.

Now, it is always possible to find a point C in space (not necessarily on or in the body) and an axis through this point parallel to $\boldsymbol{\omega}$ such that the velocity of any point B on or in the body is given at any instant by Eq. (45). Therefore, it will also be true from Eq. (44) that Eq. (46) is valid.

$$\mathbf{v}_B = \boldsymbol{\omega} \times \mathbf{r}_{CB} \qquad (45)$$

$$\mathbf{v}_A = \boldsymbol{\omega} \times \mathbf{r}_{CA} \qquad (46)$$

Such a point C is called an instantaneous center (with an instantaneous velocity of zero) and the axis through C an instantaneous axis. The motion

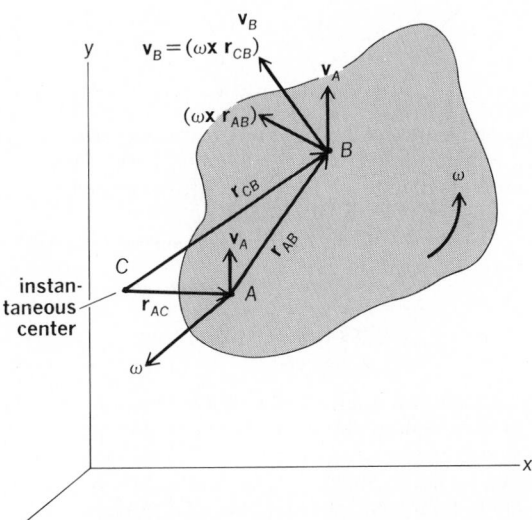

Fig. 11. Sketch showing location of instantaneous center of a rigid body moving in an arbitrary manner.

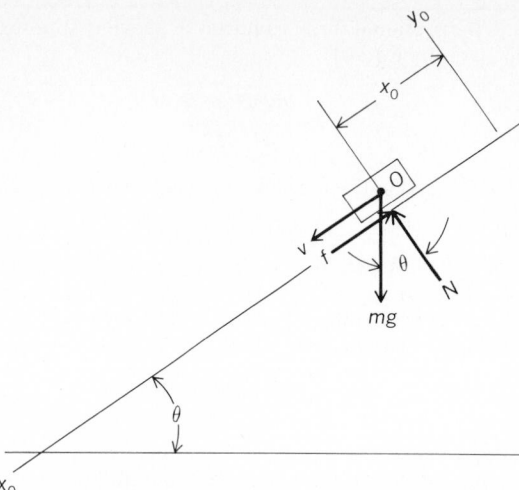

Fig. 12. Block sliding down an inclined plane.

of a rigid body may therefore be considered as a succession of pure rotations about the instantaneous axis.

The instantaneous axis can always be found as follows. If $\mathbf{v}_A$ is the known velocity of a base point A (Fig. 11), then the instantaneous axis is found by shifting the axis of rotation to pass through a new point C located a vector distance $\mathbf{r}_{AC}$ from A such that Eq. (47) holds.

$$-\mathbf{v}_A = \boldsymbol{\omega} \times \mathbf{r}_{AC} \tag{47}$$

For the rolling cylinder case in Fig. 10, the instantaneous center, Eq. (47) is the point of contact B because $v_x = \mathbf{r}_{Bx}\omega$.

Sliding motion. Newton's second law, the principles of friction, and experimental values of μ are needed to determine the motion of sliding rigid bodies. As a simple example, consider the problem of a block sliding down an inclined plane (Fig. 12). The vector equation of motion is Eq. (48), and the

$$\Sigma \mathbf{F} = m \frac{d\mathbf{v}_c}{dt} \tag{48}$$

two scalar equations of motion for this case are Eqs. (49) and (50), where $f/N = \mu'$, and μ', the

$$mg \sin\theta - f = \frac{dv}{dt} = m \frac{d^2x}{dt^2} \tag{49}$$

$$mg \cos\theta - N = 0 \tag{50}$$

coefficient of sliding or kinetic friction, is dependent upon the materials and smoothness of the surfaces and may be estimated from experimental values given in the literature.

Equations (49) and (50) can be solved to give the results defined by Eqs. (51).

$$\begin{aligned} v &= g(\sin\theta - \cos\theta)t + v_0 \\ x &= g(\sin\theta - \cos\theta)t^2/2 + v_0 t \\ f &= \mu'(mg\cos\theta) \end{aligned} \tag{51}$$

Combined rolling and sliding. The problem of a cylinder rolling down an incline is now presented again for the case where Eq. (43) is not fulfilled; that is, if the moment of inertia of the cylinder is made large, the slope of the incline large, or the cylinder radius small, the case easily arises for which (Fig. 10) Eq. (52) holds.

$$\frac{f}{N} = \mu' < \frac{K^2 \tan\alpha}{K^2 + r^2} \tag{52}$$

The cylinder then rolls and slides in a manner governed by the principle of kinetic friction, and if μ' is known as an experimental coefficient, then the simultaneous solution of Eqs. (34)–(36) and (52) gives the equations of motion. *See* POINSOT'S METHOD. [RAY E. BOLZ]

Bibliography: V. D. Barger and M. G. Olsson, *Classical Mechanics: A Modern Perspective*, 1973; H. Goldstein, *Classical Mechanics*, 2d ed., 1980; D. Halliday and R. Resnick, *Physics*, 3d ed., 1977; L. Meirovitch, *Methods of Analytical Dynamics*, 1970; K. Symon, *Mechanics*, 3d ed., 1971.

Ritz's combination principle

The empirical rule, formulated by W. Ritz in 1905, that sums and differences of the frequencies of spectral lines often equal other observed frequencies. The rule is an immediate consequence of the quantum-mechanical formula $hf = E_i - E_f$ relating the energy hf of an emitted photon to the initial energy E_i and final energy E_f of the radiating system; h is Planck's constant and f is the frequency of the emitted light. For example, the figure shows the photon energies hf_{32}, hf_{31}, hf_{30} associated with transitions from level 3 to lower-lying levels, and so

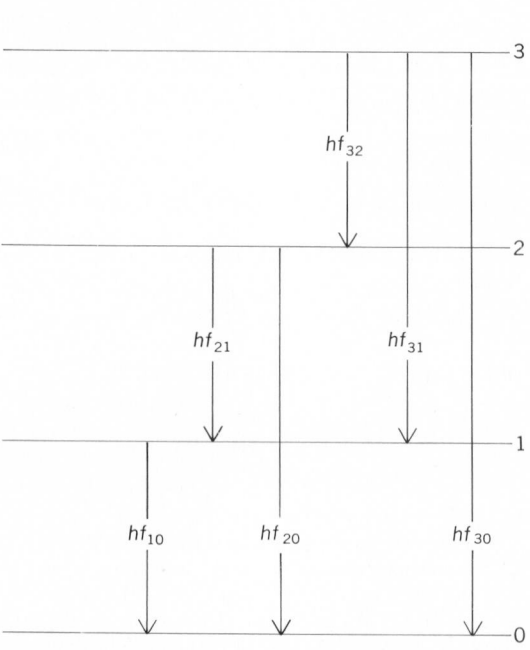

Energy levels and emitted frequencies.

on. Level 3 may radiate directly to the ground state 0, emitting f_{30}, or it may first make a transition to level 2, which subsequently radiates to the ground state, and so on. Since the total energy emitted in these two alternative means of making transitions from 3 to 0 is exactly the same, namely $E_3 - E_0$, it follows that $hf_{30} = hf_{32} + hf_{20}$. Similarly, $hf_{30} = hf_{32} + hf_{21} + hf_{10}$, and so forth. *See* ATOMIC STRUCTURE AND SPECTRA; ENERGY LEVEL; QUANTUM MECHANICS. [EDWARD GERJUOY]

Rotational motion

The motion of a rigid body which takes place in such a way that all of its particles move in circles about an axis with a common angular velocity; also, the rotation of a particle about a fixed point in space. Rotational motion is illustrated by (1) the fixed speed of rotation of the Earth about its axis; (2) the varying speed of rotation of the flywheel of a sewing machine; (3) the rotation of a satellite about a planet, in which both the speed of rotation and the distance from the center of rotation may vary; (4) the motion of an ion in a cyclotron, where the angular speed of rotation remains constant, but the radius of the circular motion increases; and (5) the motion of a pendulum, in which case the particles describe harmonic motion along a circular arc.

This discussion of rotational motion is limited to circular motion such as is exhibited by the first and second examples. For information concerning the other examples *see* HARMONIC MOTION; PARTICLE ACCELERATOR; PENDULUM.

Circular motion is a rotational motion in which each particle of the rotating body moves in a circular path about an axis. The motion may be uniform, that is, with constant angular velocity, or nonuniform, with changing angular velocity.

Uniform circular motion. The speed of rotation, or angular velocity, remains constant in uniform circular motion. In this case, the angular displacement θ experienced by the particle or rotating body in a time t is $\theta = \omega t$, where ω is the constant angular velocity.

Nonuniform circular motion. A special case of circular motion occurs when the rotating body moves with constant angular acceleration. If a body is moving in a circle with an angular acceleration of α radians/sec², and if at a certain instant it has an angular velocity ω_0, then at a time t sec later, the angular velocity may be expressed as $\omega = \omega_0 + \alpha t$, and the angular displacement as $\theta = \omega_0 t + \frac{1}{2}\alpha t^2$. *See* ACCELERATION; VELOCITY.

Banking of curves. When a car travels around a horizontal curve on a highway, the path is a circular arc of radius R, where R is the radius of curvature of the roadway. In order to have the car move in this circular arc, a horizontal external force must be applied to give the car an acceleration perpendicular to its path, that is, toward the center of rotation. This force must equal Mv^2/R, where M is the mass of the car and v its speed. This centripetal force is supplied by the friction between the tires and the road. If the force of friction is not great enough to produce this acceleration, the inertia of the car will tend to make it continue with its speed in a straight line, tangent to the road rather than around the curve, and this will cause the car to slide off the road. *See* CENTRIPETAL FORCE.

To reduce the probability of skidding, roadways are customarily banked as illustrated in Fig. 1, which shows a car of mass M going away from the reader with a speed v and making a right-hand turn along an arc of radius R. The roadway must exert a vertical force F_W upward, equal and opposite to the weight $W = Mg$ of the car (g is the acceleration of gravity), and a horizontal centripetal force $F_C = Mv^2/R$ to make the car move in a circular arc. The net force N of the road on the car is the vector sum of these two forces.

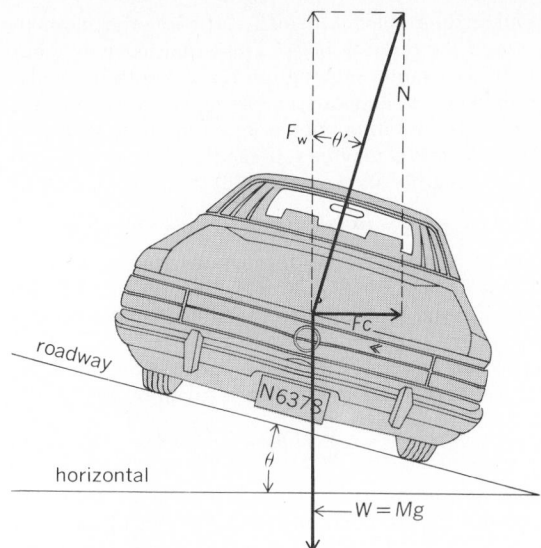

Fig. 1. Banking of a curve.

From the diagram, it can be seen that the angle θ' which N makes with the vertical is given by Eq. (1). If this angle θ' is equal to the bank angle θ of

$$\tan \theta' = \frac{F_C}{F_W} = \frac{Mv^2/R}{Mg} = \frac{v^2}{gR} \qquad (1)$$

the road, the force N of the road on the car is perpendicular to the roadway, and there will be no tendency to skid. Equation (1) shows that the correct bank angle is proportional to the square of the speed and inversely proportional to the radius of the curve. For a given curve, there is no correct bank angle for all speeds; thus roadways are banked for the average speed of traffic. Bank angle enters into the design of railroads and into the banking of an airplane when it executes a turn.

Work and power relations. A rotating body possesses kinetic energy of rotation which may be expressed as $T_{\text{rot.}} = \frac{1}{2}I\omega^2$, where ω is the magnitude of the angular velocity of the rotating body and I is the moment of inertia, which is a measure of the opposition of the body to angular acceleration. The moment of inertia of a body depends on the mass of a body and the distribution of the mass relative to the axis of rotation. For example, the moment of inertia of a solid cylinder of mass M and radius R about its axis of symmetry is $\frac{1}{2}MR^2$.

To impart kinetic energy to a rotating body, work must be done. In Fig. 2 there is represented a solid cylinder of mass M and radius R, capable of rotation without friction about an axis perpendicular to the plane of the page through O. By means of a cord wrapped around the cylinder, a constant force F is applied, thus imparting angular acceleration to the cylinder. If the cylinder is originally at rest and the force F acts through a distance $s = R\theta$ equal to the arc PP', thus rotating the cylinder through the angle θ, the work W done is $W = Fs = FR\theta = L\theta$, where $L = FR$ is called the torque or moment of force. The action of this torque L is to produce an angular acceleration α according to Eq. (2), where $I\omega$, the product of moment of inertia

$$L = I\alpha = I\frac{d\omega}{dt} = \frac{d}{dt}(I\omega) \qquad (2)$$

ROTATIONAL MOTION

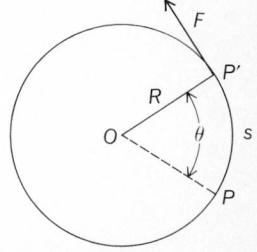

Fig. 2. Work in angular motion.

and angular velocity, is called the angular momentum of the rotating body. This equation points out that the angular momentum $I\omega$ of a rotating body, and hence its angular velocity ω, remains constant unless the rotating body is acted upon by a torque. Both L and $I\omega$ may be represented by vectors.

It is readily shown by Eq. (3) that the work done

$$W = L\theta = I\alpha\theta = I\alpha\tfrac{1}{2}(\alpha t^2) = \tfrac{1}{2}I(\alpha t)^2 = \tfrac{1}{2}I\omega^2 \quad (3)$$

by the torque L acting through an angle θ on a rotating body originally at rest is exactly equal to the kinetic energy of rotation, because, for the case at hand, $\theta = \frac{1}{2}(\alpha t^2)$ and $\omega = \alpha t$.

Power is defined as rate of doing work, and the power P in rotational motion is given by Eq. (4).

$$P = \frac{dW}{dt} = \frac{d}{dt}(L\theta) = L\frac{d\theta}{dt} = L\omega \quad (4)$$

See ANGULAR MOMENTUM; MOMENT OF INERTIA; POWER; RIGID-BODY DYNAMICS; TORQUE; WORK.
[CARL E. HOWE/R. J. STEPHENSON]

Runge vector

The Runge vector describes certain unchanging features of a nonrelativistic two-body interaction for which the potential energy is inversely proportional to the distance r between the bodies or, alternatively, in which each body exerts a force on the other that is directed along the line between them and proportional to r^{-2}. Two basic interactions in nature are of this type: the gravitational interaction between two masses (called the classical Kepler problem), and the Coulomb interaction between like or unlike charges (as in the hydrogen atom). Both at the classical level and the quantum-mechanical level, the existence of a Runge vector is a reflection of the symmetry inherent in the interaction. *See* COULOMB'S LAW; NONRELATIVISTIC QUANTUM THEORY; QUANTUM MECHANICS; SYMMETRY LAWS.

Classical Kepler problem. The nonrelativistic two-body problem can be transformed into an equivalent problem for one body attracted to a fixed center. In the equivalent one-body problem, the motion is governed by the relation in Eq. (1),

$$d\mathbf{p}/dt = -\lambda r^{-2}\hat{\mathbf{r}} \quad (1)$$

where $\hat{\mathbf{r}} = \mathbf{r}/r$, $\mathbf{p} = m\, d\mathbf{r}/dt$, m is the reduced mass, and λ is a constant (positive for attractive force) that characterizes the strength of the interaction. It follows that the energy E and the angular momentum $\mathbf{L}$ are constants of the motion and that Eqs. (2) hold. The Runge vector, defined by Eq. (3),

$$\begin{aligned} E &= p^2/2m - \lambda/r \\ \mathbf{L} &= \mathbf{r} \times \mathbf{p} \end{aligned} \quad (2)$$

$$\mathbf{R} = \mathbf{p} \times \mathbf{L} - \lambda m\hat{\mathbf{r}} \quad (3)$$

is also conserved (that is, does not change with time). Since $\mathbf{R} \cdot \mathbf{L} = 0$, it follows that $\mathbf{R}$ is a fixed vector in the plane of the orbit. If $\mathbf{R}$ is taken to be the fixed direction from which the azimuthal angle θ is measured, Eq. (4) holds. Here $R = |\mathbf{R}| =$

$$1/r = (\lambda m/L^2)[1 + (R/\lambda m)\cos\theta] \quad (4)$$

$[2mEL^2 + \lambda^2 m^2]^{1/2}$. Hence $\mathbf{R}$ points in the apsidal direction and its magnitude determines the eccentricity $\varepsilon = R/\lambda m$. The diagram illustrates the situation for an elliptical orbit. Since the particle in or-

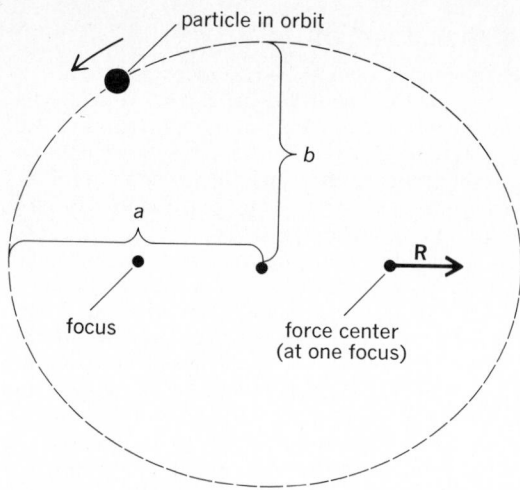

Runge vector for elliptical motion, $(\mathbf{R} = \lambda m\sqrt{a^2 - b^2}/a)$.

bit endlessly retraces the same path (dashed line in diagram), the shape of the elliptical path (described by the eccentricity) and its orientation (described by the direction $\mathbf{R}/R$) are constant. A single-valued constant quantity giving a complete description of the orbit is possible only because of the periodic nature of the motion; that is, the orbit is reentrant. Moreover, a single-valued conserved vector in the plane of motion occurs only in the nonrelativistic problem in which the force varies as r^{-2}. In the corresponding relativistic problem, the ellipse precesses.

This symmetry (constancy of the orbit path in the bound case), which leads to the Runge vector, is reflected in other aspects of the Kepler problem. The fact that the corresponding Hamilton-Jacobi equation is separable not only in spherical coordinates but also in parabolic coordinates implies the existence and form of the Runge vector. The components of the Runge vector may also be related to the generating functions of canonical transformations that transform a given orbit pertaining to a fixed energy into a different orbit of the same energy. The group of transformations comprising those generated by rotation in three-dimensional space (namely, generated by $\mathbf{L}$) as well as by the Runge vector is isomorphic to rotations in four-dimensional space. *See* CANONICAL TRANSFORMATIONS.

Quantum-mechanical problem. The Hamiltonian of the Schrödinger equation analogous to the classical problem is given by Eq. (5), where $\mathbf{p}$ is the

$$H = (\mathbf{p} \cdot \mathbf{p})/2m - \lambda/r \quad (5)$$

operator $-i\hbar\nabla$. In addition to the angular momentum operator $\mathbf{L} = \mathbf{r} \times \mathbf{p}$, there exists a conserved Hermitian Runge vector operator given by Eq. (6). Its time independence is assured by the fact that it

$$\mathbf{R} = \tfrac{1}{2}(\mathbf{p} \times \mathbf{L} - \mathbf{L} \times \mathbf{p}) - \lambda m\hat{\mathbf{r}} \quad (6)$$

commutes with H. Its existence and form can be inferred from the fact that the time-independent Schrödinger equation is separable in both spherical and parabolic coordinates.

Instead of the Runge vector given above, the Runge-Lenz vector operator $\mathbf{A} = (-2mH)^{-1/2}\mathbf{R}$ is usually used. It is well defined on all eigenstates and is Hermitian for bound states ($E < 0$). The square of the Runge-Lenz operator is given by Eq. (7), where $\hbar$ is Planck's constant divided by 2.

$$\mathbf{A} \cdot \mathbf{A} = -\lambda^2 m (2H)^{-1} - (\mathbf{L} \cdot \mathbf{L} + \hbar^2) \qquad (7)$$

See EIGENFUNCTION; EIGENVALUE.

The Runge-Lenz vector operator $\mathbf{A}$ and the angular momentum operator $\mathbf{L}$ have the commutation properties characteristic of the Lie algebra 0(4) (orthogonal group of rotations in four dimensions). Explicitly, Eqs. (8)–(10) hold, where ε_{ijk} is

$$A_i L_j - L_j A_i = i\hbar \varepsilon_{ijk} A_k \qquad (8)$$

$$A_i A_j - A_j A_i = i\hbar \varepsilon_{ijk} L_k \qquad (9)$$

$$L_i L_j - L_j L_i = i\hbar \varepsilon_{ijk} L_k \qquad (10)$$

antisymmetric in its indices with $\varepsilon_{123} = 1$. Moreover, since both A_3 and L_3 commute with each other as well as with the Hamiltonian, a simultaneous eigenfunction of these operators can be found. This corresponds to separating the Schrödinger equation in parabolic coordinates $\xi = r(1 + \cos\theta)$, $\eta = r(1 - \cos\theta)$, and ϕ, instead of the usual spherical coordinates r, the polar angle θ, and the azimuthal angle ϕ. These simultaneous eigenfunctions are not also eigenfunctions of the angular momentum since L^2 does not commute with A_3. For energy $E = k^2(2m)^{-1}$ which is greater than zero, one particular eigenfunction is of special interest. It is the one whose L_3 and A_3 eigenvalues are respectively zero and $(\hbar - im\lambda k^{-1})$. It is defined by Eq. (11), where $_1F_1$ is the confluent hypergeometric

$$\Psi = e^{ikr\cos\theta} \, _1F_1 \left[im\lambda(\hbar k)^{-1}, 1, ik\eta \right] \qquad (11)$$

function. As $r \to \infty$, this wave function behaves as if it were a plane wave plus an outgoing spherical wave, and hence this single eigenfunction has the asymptotic properties of a scattering solution for the Schrödinger equation. The scattering solution is useful in a discussion of scattering of a beam of noninteracting charged nonrelativistic particles off a Coulomb force center since, asymptotically, the plane wave represents that which is scattered by the Coulomb force center. The same eigenfunction is expressible as an infinite sum of angular momentum eigenfunctions, that is, the so-called partial wave expansion of the scattering eigenfunction. *See* DISPERSION RELATIONS.

Symmetry and invariance group. Since the potential for the hydrogen atom is spherically symmetric, it is apparent that both the classical and quantum-mechanical problem enjoy three-dimensional rotational symmetry. The additional symmetry peculiar to the $1/r$ potential extends that symmetry to a four-dimensional symmetry. The space of the four dimensions is that of the variables $2(-2mE)^{1/2} \mathbf{p}/(p^2 - 2mE)$ and $(2mE + p^2)/(p^2 - 2mE)$, where $\mathbf{p}$ is the ordinary three-dimensional momentum. It can be shown that the Schrödinger equation in momentum space can be converted into an integral equation symmetric in these four variables by projecting the momentum space stereographically onto a four-dimensional unit sphere. The abstract Kepler problem in n spatial dimensions has been analyzed with similar results.

The additional internal symmetry is responsible for the degeneracy (that is, number of distinct states of a single energy) characteristic of the hydrogen atom. The group 0(4) is an invariance group

of the hydrogen atom since all the bound states of a given energy constitute the basis for an irreducible representation of that group. In turn, all the bound states constitute an irreducible representation of the noncompact DeSitter group 0(4,1). This larger group (whose generators do not all commute with the Hamiltonian) is called the noninvariance group of the problem. The conceptual analogy between energy levels and mass, hydrogen atom states and elementary particle states, has given rise to the hope that perhaps the spectrum of elementary particles can also be understood in terms of an appropriate invariance and noninvariance group. *See* ELEMENTARY PARTICLE; ENERGY LEVEL.

[D. M. FRADKIN]

Bibliography: I. M. Bander and C. Itzykson, Group theory and the hydrogen atom, *Rev. Mod. Phys.*, 38:330–346, 1966; H. Goldstein, *Classical Mechanics*, 2d ed., 1980; L. D. Landau and E. M. Lifshitz, *Mechanics*, 3d ed., 1976; E. C. G. Sudarshan, N. Mukunda, and L. O'Raifeartaigh, Group theory of the Kepler problem, *Phys. Lett.*, 19:322, 1965.

Rydberg atom

An atom which possesses one valence electron orbiting about an atomic nucleus within an electron shell well outside all the other electrons in the atom. Such an atom approximates the hydrogen atom in that a single electron is interacting with a positively charged core. Early observations of atomic electrons in such Rydberg quantum states involved studies of the Rydberg series in optical spectra. Electrons jumping between Rydberg states with adjacent principal quantum numbers, n and $n - 1$, with n near 80 produce microwave radiation. Microwave spectral lines due to such electronic transitions in Rydberg atoms have been observed both in laboratory experiments and in the emissions originating from certain low-density partially ionized portions of the universe called HII regions. *See* ELECTRON CONFIGURATION.

The valence electron's classical orbit radius increases as n^2, while the orbital velocity decreases as n^{-1}. The binding energy between the electron and the core decreases as n^{-2}. Thus Rydberg atoms with $n = 80$ have diameters of 10^{-6} m and are as large as some bacteria. Yet they are very delicate, being easily distorted by weak electric and magnetic fields. They are also easily destroyed in collisions with other atomic particles. The natural lifetime of an undisturbed Rydberg atom increases as n^3 for a given electronic angular momentum and has millisecond values for n near 80.

The advent of the laser has made possible the production of sizable numbers of Rydberg atoms within a bulb containing gas at low pressures, 10^{-2} torr (1.3 Pa) or less. The rapid energy-resonance absorption of several laser light photons by an atom in its normal or ground state results in a Rydberg atom in a state with a selected principal quantum number. Aggregates of Rydberg atoms have been used as sensitive detectors of infrared radiation, including thermal radiation. They have also been observed to collectively participate in spontaneous photon emission, called superradiance. Such aggregates form the active medium for infrared lasers that operate through the usual laser mechanism of collective stimulated photon emission. All these developments are based upon the great sensitivity of Rydberg atoms to external electromagnetic radiation fields. Atoms with n near 40 can absorb almost instantaneously over a hundred microwave photons and become ionized at easily achievable microwave power levels. Isotope separation techniques have been developed that combine the selectivity of laser excitation of Rydberg states with the ready ionizability of Rydberg atoms. Such applications have been pursued for atoms ranging from deuterium through uranium. *See* INFRARED RADIATION; LASER.

The positively charged core attracting a Rydberg valence electron may be a molecular ion rather than an atomic ion. Even more complex Rydberg-like quantum states exist in solids, in which a valance electron can be weakly bound either to an impurity ion or to a hole vacancy in a semiconductor. Thus many of the phenomena observed in studies with Rydberg atoms also can occur in molecules and in the solid state. The consequences of this for physics, chemistry, biology, and technology largely remain unexplored. *See* CRYSTAL DEFECTS; SEMICONDUCTOR.

In 1979 the first experiments were carried out on doubly excited atoms, in which laser radiation was used to drive simultaneously two atomic electrons out of their normal atomic shells into different Rydberg valence shells. Thus the study of planetary atoms having two or more valence electrons in widely separated classical orbits has become a reality. *See* ATOMIC STRUCTURE AND SPECTRA.

[JAMES E. BAYFIELD]

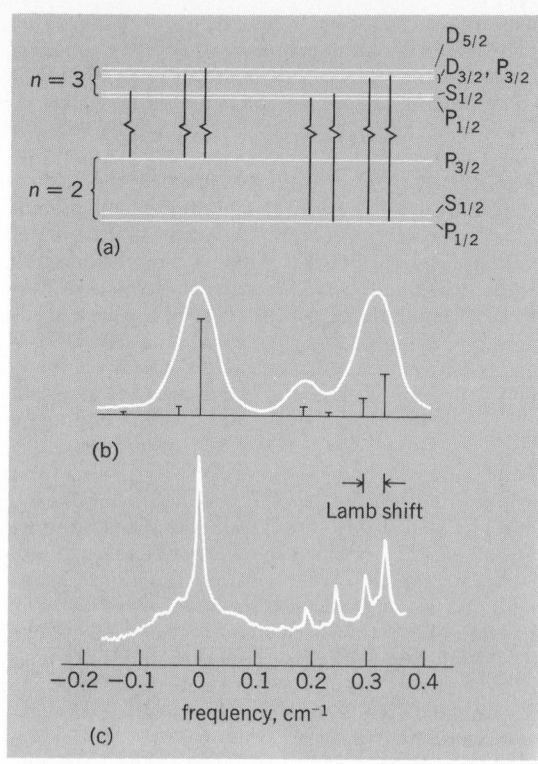

Deuterium Balmer line D_α. (*a*) Energy levels with fine-structure transitions. (*b*) Emission-line profile of a cooled deuterium gas discharge (temperature 50 K) and theoretical fine-structure lines with relative transition probabilities (*from B. P. Kibble et al., J. Phys., B6:1079–1089, 1973*). (*c*) Saturation spectrum with optically resolved Lamb shift (*from T. W. Hansch et al., Precision measurement of the Rydberg constant by laser saturation spectroscopy of the Balmer α line in hydrogen and deuterium, Phys. Rev. Lett., 32:1336–1340, 1974*).

Rydberg constant

An atomic constant describing the binding energy between electron and atomic nucleus. It is connected with the other universal physical constants by the relation, in cgs electrostatic units, $R_\infty = 2\pi^2 me^4/ch^3$. Here, m and e are the mass and charge of the electron, c the velocity of light, and h is Planck's constant. (In SI units it is given by $R_\infty = \mu_0^2 me^4 c^3/8h^3$ where $\mu_0 = 4\pi \times 10^{-7}$ henry/meter is the permeability of vacuum.) The Rydberg constant is determined from measurements of the wavelength of spectral lines of hydrogenlike atoms, which permit accurate theoretical calculations. The relationship between the wavelength and the Rydberg constant is given by Bohr's formula, corrected for finite nuclear mass, relativistic effects, and Lamb shifts. The symbol R_∞ indicates that the constant refers to a hypothetical atom of infinite mass. To correct for the finite nuclear mass M of a real atom, one replaces the electron mass m in the preceding formula by the reduced mass of electron and nucleus, that is, by $mM/(m + M)$, and one obtains a somewhat smaller Rydberg $R = R_\infty/(1 + mM)$. Because the Rydberg constant can be determined with high precision, it has traditionally been a cornerstone in the evaluation of the other fundamental constants; it gives a precise relation between these constants, which must be satisfied by any set of constants that may be adopted.

Conventional measurements. Prior to 1974, the value of the Rydberg constant was based on wavelength measurements of visible emission lines of gas discharges (hydrogen, deuterium, tritium, and singly ionized helium), performed by conventional high-resolution interferometry. Such measurements have always been hampered by the random, high-velocity thermal motion of the light atoms.

Those atoms when moving toward an observer appear to emit or absorb at higher frequency than atoms at rest, and atoms moving away appear to emit at lower frequencies. This Doppler effect badly blurs the intricate fine structure of the spectral lines of hydrogenlike atoms (illustration *a*), even when the discharge tube is cooled to cryogenic temperatures (illustration *b*), and it has prevented any significant improvement of conventional wavelength measurements beyond an accuracy of 1 part in 10^{-7} since the 1940s. A Rydberg value $R\infty = 10,973,731.77 \pm 0.83$ m^{-1}, based on the average of four different conventional measurements, was adopted in the 1973 adjustment of the fundamental constants. *See* DOPPLER EFFECT; FINE STRUCTURE.

Laser spectroscopy. An almost tenfold improvement in the accuracy of the Rydberg constant was achieved in 1974, after the technique of laser saturation spectroscopy had made it possible to overcome the Doppler broadening of the red hydrogen Balmer-α line. An intense, highly monochromatic beam from a pulsed tunable dye laser was used, in effect, to label atoms of slow velocity in a Wood-type glow discharge, by removing them temporarily from the absorbing lower level. The resulting bleaching effect was recorded with a weaker, counterpropagating probe beam from the same laser. In this way, single fine-structure components of the Balmer-α line could be resolved (illustration *c*). A new Rydberg value, $R\infty$ 10,973,731.43 $\pm$ 0.10 m^{-1}, was obtained by measuring the wavelength of the strong $2P_{3/2}-3D_{5/2}$ component in deuterium and hydrogen. The reference standard was a He-Ne gas laser, locked to an absorption line of molecular iodine.

A subsequent study of the hydrogen Balmer-α line by laser polarization spectroscopy in 1978 confirmed the earlier Doppler-free measurement and provided another threefold improvement in accuracy. A new Rydberg value, $R\infty =$ 10,973,731.476 $\pm$ 0.032 m^{-1}, was obtained from the wavelength of the narrow but weak $2S_{1/2}-3P_{3/2}$ fine-structure component. Polarization spectroscopy is related to saturation spectroscopy, but achieves higher sensitivity by monitoring the interaction of two counterpropagating laser beams in an absorbing gas via changes in light polarization rather than intensity. The higher sensitivity permitted measurements at lower discharge current, gas pressure, and laser intensity, so that small systematic line shifts could be accounted for more accurately. An independent Rydberg measurement in 1979, based on saturation spectroscopy of the hydrogen Balmer-α line, yielded a confirming, though somewhat less accurate, result of $R\infty =$ 10,973,731.513 $\pm$ 0.085 m^{-1}.

Research in progress. Efforts have been undertaken to observe hydrogen Balmer lines by high-resolution laser spectroscopy in a beam of metastable atoms undisturbed by collisions. An even more accurate Rydberg should result from these experiments, in particular if it becomes feasible to directly measure the frequency of the laser rather than its wavelength.

Further improvements in accuracy can be expected when the method of Doppler-free two-photon spectroscopy is applied to measure the frequency of the two-photon transition from the 1S hydrogen ground state to the metastable 2S state. In this method the atom absorbs two laser photons of half the transition energy and coming from opposite directions, so that their first-order Doppler shifts cancel. The feasibility of Doppler-free two-photon excitation of the hydrogen 1S-2S transition has already been demonstrated. The long 1/7-s lifetime of the metastable 2S state promises an extremely narrow natural line width, which may permit an ultimate resolution of a few parts in 10^{15}. The continuing role of the Rydberg constant as one of the best-known constants of physics thus seems certain for the foreseeable future. *See* ATOMIC CONSTANTS; ATOMIC STRUCTURE AND SPECTRA; LASER.

[THEO W. HÄNSCH]

Bibliography: E. R. Cohen and N. B. Taylor, The 1973 least squares adjustment of the fundamental constants," *J. Phys. Chem. Ref. Data*, 2: 663–734, 1973; J. E. M. Goldsmith et al., New measurement of the Rydberg constant using polarization spectroscopy of Hα, *Phys. Rev. Lett.*, 41: 1525–1528, 1978; T. W. Hänsch et al., Precision measurement of the Rydberg constant by laser saturation spectroscopy of the Balmer-α line in hydrogen and deuterium, *Phys. Rev. Lett.*, 32: 1336–1340, 1974; T. W. Hänsch et al., The spectrum of atomic hydrogen, *Sci. Amer.*, 240(3): 94–110, March 1979; B. W. Petley and K. Morris, A measurement of the Rydberg constant, *Nature*, 279:141–142, 1979.

Scattering experiments (atoms and molecules)

Experiments in which an incident particle or system of particles, such as an electron, atom, or molecule, is deflected by collision with an atom or a molecule. Such experiments are useful for many reasons: They provide checks on the theory of scattering and yield information on the nature of atomic and molecular forces. They are also important since the experiments can be designed to simulate conditions in planetary atmospheres, to provide information on electric discharges, to assist in the development of gaseous lasers, and to aid in the theory of stellar absorption and of planetary nebulae. *See* LASER.

Classification of collisions. An impact between two atomic systems is said to be elastic if it involves no transfer of energy between the internal motions within the two systems and the motion of relative translation. Otherwise it is inelastic or superelastic according to whether energy is given to, or taken from, internal motion. *See* COLLISION.

If radiation is emitted during the impact, the collision is radiative; otherwise it is nonradiative.

Rearrangement collisions are those in which there is a redistribution of particles between the colliding systems after the impact.

In general, in any type of collision, scattering occurs; that is, the direction of relative motion of the colliding systems before and after impact is rotated to a new direction.

Any number of systems may be involved in an impact. Usually collisions between two initial systems are studied. More than two systems may result from the impact, in which case the directions and energies of motion of all the resultants need to be specified.

Collision rates. Specification of collision rates is best done by first introducing the concept of total collision cross section. Consider a beam of electrons passing through a gas. If an electron is regarded as lost from the beam because of a change in direction or energy by a collision with a gas molecule, the beam current will be reduced by a factor $e^{-\alpha x}$ in passing a distance x through the gas. The quantity α can be written as NQ, where N is the number of gas molecules per unit volume and Q is the total effective cross section for collisions between electrons, of speed v, and the gas molecules.

Effective cross sections for different types of collisions follow by introducing probabilities p_j which give the chance that a collision is of a particular type j. The quantity p_jQ is then the effective cross section Q_j for a collision of this type.

Scattering in a particular type of collision is specified in terms of a differential cross section. If $i_j(\theta,\phi)$ is the chance that, in a collision of type j, the direction of motion of the electron is turned through an angle θ into the solid angle $\sin\theta\,d\theta\,d\phi$, the corresponding differential cross section or scattering intensity is given by notation (1).

$$p_jQi_j(\theta,\phi)\sin\theta\,d\theta\,d\phi \qquad (1)$$

In many experiments the electrons diffuse as a swarm rather than as a directed beam. If $f(E)\,dE$ is the number of electrons of the swarm with energy between E and $E + dE$, the number of collisions of type j occurring per second is given by notation (2),

$$N \int f(E) \, Q_j(E) v \, dE \qquad (2)$$

where v is the speed of an electron of energy E. These definitions and formulas may be extended to cover collisions between more complicated systems.

In general, cross sections defined in this manner have definite values and may be measured with apparatus of sufficient resolving power. For impact between charged particles the total cross section is unbounded, but the scattered intensity remains definite at all angles greater than zero.

Cross-section evaluation. The simplest problem is that of scattering of a beam of structureless particles of mass m and speed v by a center which exerts a force of potential $V(r)$, r being the distance of a particle from the center. Differential cross sections for this case may be calculated without approximation by the methods of quantum theory. If $V(r)$ exceeds the kinetic energy $1/2mv^2$ at distances $r < a$ where $a \gg \hbar/mv$, classical mechanics may be used to calculate the differential cross section for all angles $\theta > \hbar/mva$. If $V(r)$ is much less than $1/2mv^2$ for all r or for $r < a$ where $a \ll \hbar/mv$, a quantum-mechanical perturbation treatment, known as Born's first approximation, is valid.

For the special case $V = A/r$, where A is a constant, the classical theory and Born's first approximation both give the exact value for the differential cross sections for all values of θ.

For collisions between systems with internal structure, no exact theoretical calculation of cross sections is possible. If the interaction between the systems is weak, Born's first approximation may be extended to these cases. In general this will be so if the velocity of relative translation of the colliding systems is much greater than the velocities of the internal motions. When this is not satisfied, there is no general method of approximation, but methods applicable to certain types of collision have been developed. The scope of such methods has been greatly increased by the availability of automatic means for high-speed computation.

Electron-atom collisions. These have been studied intensively, both experimentally and theoretically. Different types of collision that may occur include (1) elastic; (2) inelastic, involving excitation of discrete atomic states; (3) ionization; (4) radiative capture to form negative ions; and (5) radiative collisions in which the electron is not captured.

Radiative collisions are much less probable than nonradiative collisions, whereas the cross sections for inelastic collisions are comparable with those for elastic collisions at electron energies that are large compared with the minimum energy necessary to produce excitation.

The cross section for electron impact with atoms at low velocities varies with electron energy in complicated ways. These can be analyzed into gradual variations, in which a change of electron energy of a few electronvolts (eV) is required to produce much change in the cross section, and much more rapid fluctuations occurring within an electron energy range of less than 0.1 eV.

Experiments in which the energy homogeneity of the incident electrons is of the order of several tenths of an electronvolt reveal only the broad features. Figure 1 shows results of measurement of this kind of total cross section of slow electrons in argon, krypton, and xenon. The observed differential cross sections for elastic scattering in the same velocity range exhibit maxima and minima as functions of the angle of scattering θ, the number increasing with the atomic number and, for a given atom, with electron velocity (Fig. 2).

These effects are due to diffraction of the electrons by the scattering atoms. For low-velocity electrons elastic scattering is the most important, and the full quantum theory of scattering by the undisturbed field of the atom can be used to calculate the cross sections. In a complete theory, allowance must be made for exchange of electrons between the atom and incident beam and for polarization of the atom during the impact. The latter leads to increased scattering of small angles. *See* ELECTRON DIFFRACTION.

The fine structure in the total cross section can be observed by using beams of electrons of high-energy homogeneity. Figure 3 shows the observed variation with electron energy of the fraction of electrons in a beam that pass undeviated through a certain pressure of argon. The transmitted fraction fluctuates rapidly—even in the small electron energy range of 0.5 eV shown.

Fine structure arises from the existence of unstable states in which an electron is bound to an excited atom. The energy E_c of such a state $= E_0 + E$, where E_0 is the energy of the ground state of the atom, so that $E > 0$. After a mean time τ, usually between 10^{-14} and 10^{-15} sec, an electron will be emitted with energy E leaving the normal atom once more. Because of the uncertainty principle, the energy of the state will be uncertain by an amount $\Delta E = \hbar/\tau$, where $\hbar$ is Planck's constant, divided by 2π, that is, typically between 0.05 and 0.5 eV. If an electron of energy within the range $E_c - E_0 \pm \Delta E$ collides with the atom, there is a high

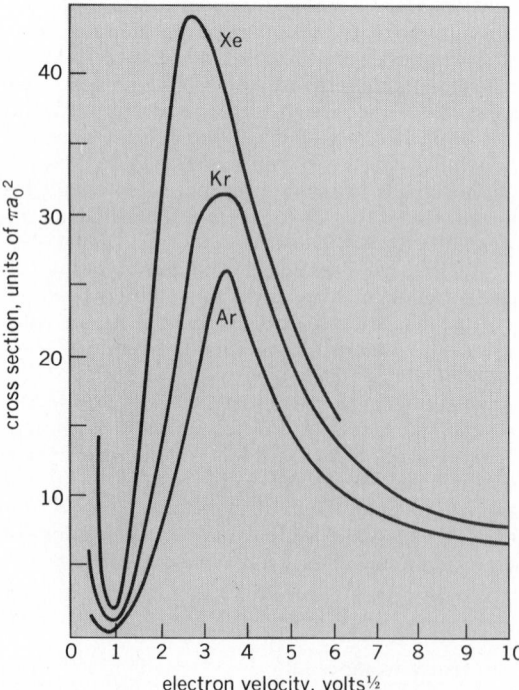

Fig. 1. Observed variation of the total cross sections for collisions of slow electrons in argon, krypton, and xenon; $a_0 = 0.53 \times 10^{-8}$ cm.

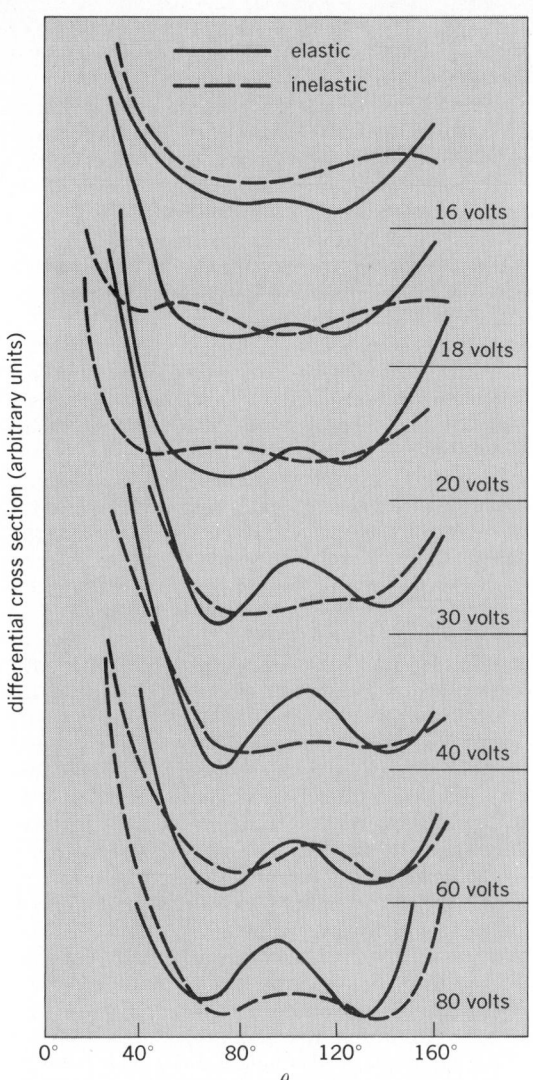

Fig. 2. Observed differential cross sections for elastic and inelastic scattering of electrons in argon for various electron energies. The various electron energies are given in electronvolts.

probability it will be temporarily captured into the unstable state, to be reemitted again in some direction uncorrelated with its initial direction of motion; that is, it will be scattered. Such scattering is only probable when the electron energy is within ΔE of the energy of some unstable state, and this leads to the rapid fluctuations in the cross sections.

Inelastic nonradiative collisions have been studied experimentally by a variety of methods. Excitation of discrete states has been investigated by optical and electrical techniques. The optical method measures the intensity of radiation emitted at a particular wavelength from an electron beam of definite length that is passed through a gas at low pressure. For excitation of metastable states, relative yield at different electron velocities is observed either by use of optical absorption or by observing the electric current emitted from a surface on which the metastable atoms impinge. An important electrical method involves measurement of the fraction of electrons that have lost discrete amounts of energy in diffusing to the walls of

a cylinder from an axial source.

Ionizing collisions have been studied by measuring the number of positive ions produced by an electron beam of definite energy passing for a definite distance through the gas at low pressure. The relative probabilities of collisions in which one, two, or more electrons are removed from the atom have been observed in some cases by performing a mass-spectrographic analysis of the product ions.

The variation of the inelastic collision cross sections with electron velocity for electrons with energies not much greater than the excitation threshold, exhibits fluctuations similar to those observed in the total cross section. Figure 4 illustrates this behavior for the excitation of the 2^1S, 2^3S, 2^1P, and 2^3P states of helium observed with high-resolution energy-analyzing equipment. The explanation of these fluctuations is similar to that for elastic scattering; the only difference is that the unstable negative ion produced by capture of the incident electron breaks up to leave the atom in an excited state and the ejected electron with correspondingly reduced energy.

Fluctuations in the ionization cross section also arise from excitation of unstable doubly excited states of the atom concerned.

The cross section for excitation of an atom by an electron, of angular momentum quantum number ℓ and incident energy E, rises from the threshold as $(E-E_a)^{\ell+1/2}$, where E_a is the excitation energy. For single ionization the cross section rises from the threshold, according to current theory, as $E-E_i$, where E_i is the ionization energy. *See* ANGULAR MOMENTUM.

Among broad features of the variation of inelastic cross sections, those for excitation of an optically allowed transition have the form of curve 1 in Fig. 5. The maximum occurs for energies a few times E_a, and at high electron speeds v the cross section falls off as $v^{-2} \ln (\alpha v)$, where α is a constant.

Excitation of a state with multiplicity different from that of the ground state can take place only through electron exchange (except for heavy atoms such as mercury, for which it is not a good approximation to assign a definite multiplicity to a particu-

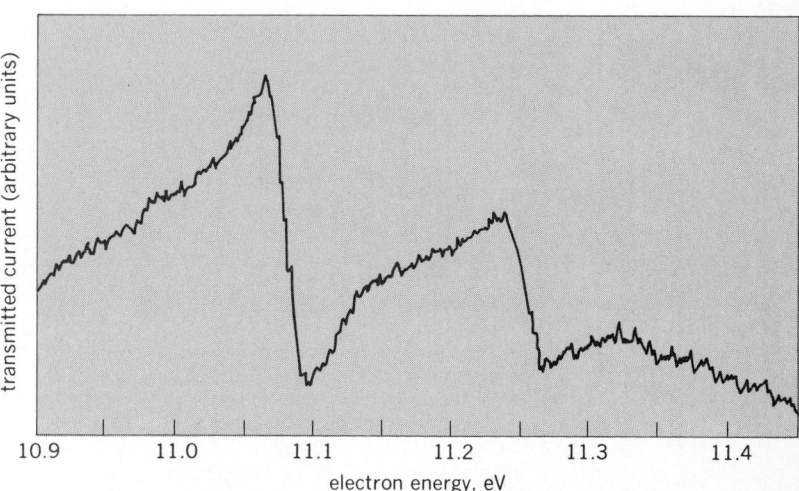

Fig. 3. Observed variation with electron energy of the intensity of an electron beam after transmission through argon; the greater the transmitted intensity, the smaller the scattering cross sections.

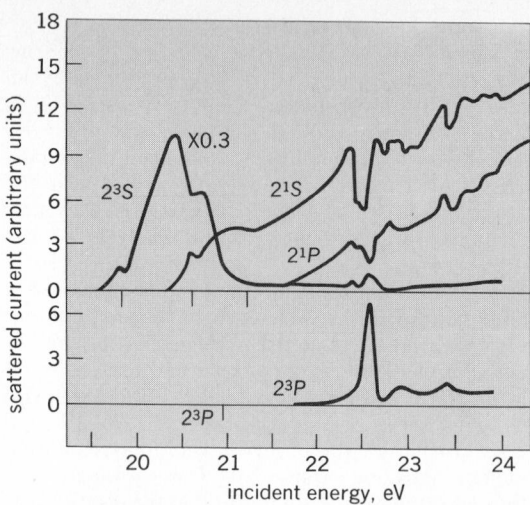

Fig. 4. Observed variation with incident electron energy of the current of electrons scattered without deviation after exciting the 2^1S, 2^1P, 2^3S, and 2^3P states of helium.

lar state). The excitation cross sections are large only for electron energies within a few electron volts of the threshold (Fig. 5, curve 2).

Cross sections for excitation of transitions involving no change of multiplicity but which are optically forbidden (Fig. 5, curve 3) have the same general form as in curve 1 of Fig. 5, but the maximum occurs at a smaller value of E/E_a, and the decrease at high velocity is proportional to v^{-2}.

Differential cross sections for inelastic collisions have been observed extensively, especially for optically allowed transitions. At not too large electron velocities, maxima and minima are found in the scattered intensity for excitation of a particular state. These resemble corresponding results for elastic scattering and are due to the effect of the atomic field in producing strong distortions of the incident and outgoing electron waves from the plane-wave form.

Born's first approximation gives good results for the cross section for electron energies greater than

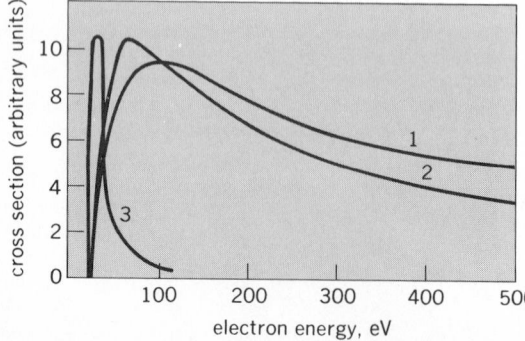

Fig. 5. Observed cross sections for excitation of different states of helium by electron impact. Curve 1, 3^1P, an optically allowed transition; curve 2, 4^1D, an optically disallowed transition involving no change of multiplicity; curve 3, 4^3S, an optically disallowed transition with change of multiplicity. The cross sections are given in arbitrary units that are different for each curve; the emphasis is on illustration of the variation of cross sections with electron energy.

10 times the threshold energy. At lower energies it overestimates the cross section for excitation of optically allowed transitions.

Improved theories have been developed that do not assume the coupling between all atomic states to be weak. Such theories have been successful in predicting the location and characteristics of the "resonance" fluctuations for collisions with simple atoms such as hydrogen and helium and have also had some success in the calculation of cross sections for excitation of low-lying excited states.

Polarization of the light emitted by electron impact has been studied experimentally and theoretically. Account must be taken of the fine and hyperfine structure of the levels concerned. When this is done, there is agreement between theory and the more precise experiments.

If an electron collides with an atom containing a single electron in its outer shell, exchange of the two electrons may occur. Even if they have opposite spin, the exchange does not change the internal energy of the atom. The total scattered amplitude is made up of separate contributions from direct and exchange processes. To determine these amplitudes separately, it is necessary to measure the scattering of polarized electrons by polarized atoms. Such measurements, which provide a very severe test of approximate theories, have become possible through the development of techniques for producing both polarized sources and detectors for measuring the polarization of the scattered electrons. The first measurements were for ionizing collisions with lithium atoms, but measurements are not available for elastic and ionizing collisions with hydrogen atoms.

In an inelastic collision in which an optically allowed transition occurs, it is not possible, from measurements of total and differential cross sections, to determine the amplitudes separately for excitation of transitions between degenerate levels with different components of angular momentum about the direction of the incident beam. Separate information about these amplitudes may be obtained from observing in coincidence the scattered electron and the emitted photon under fixed conditions of angular correlation and photon polarization. Such measurements are made by using sensitive electron and photon detectors.

Coincidence-counting techniques are also being applied to study ionization, the scattered and ejected electron being observed in coincidence. With suitable choice of incident-electron energy and of the coincidence geometry, it is possible, for example, to observe the momentum distribution of electrons in atoms and molecules.

The experimental study of the scattering of electrons by atoms in short-lived excited states is possible through the use of lasers to produce the excited atoms in a specified state. In principle, such experiments are the time-reversed form of the experiments in which photons and scattered electrons from an inelastic collision are observed in coincidence.

The study of the scattering of electrons by atoms in the presence of a strong electromagnetic field has become practicable through the availability of high-power lasers.

Much attention has been paid to the experimental and theoretical study of inelastic collisions in which excitation of an electron or electrons from an inner shell of the target atoms occurs. X-ray emission resulting from such collisions in the gas phase has been observed.

The cross section for radiative capture of an electron by an atom to give a negative ion is of the order 10^{-21} cm^2 or less. At small electron velocities v, the cross section varies as v^2 for capture into an s state, but is independent of v for capture into a p state. A number of observations of the affinity spectrum due to these capture processes have been made under conditions of shock-wave excitation or in high-pressure arcs.

Cross sections for the inverse process, that of photodetachment of an electron from a negative ion, have been measured for many negative ions, using laser light, and from these the radiative capture cross sections may be obtained by using what is known as the theory of detailed balancing.

Photodetachment is of major importance in the solar atmosphere. The H$^-$ ions determine the frequency distribution of the continuous emission from the Sun in the visible region. At longer wavelengths, absorption by free electrons in the neighborhood of hydrogen atoms is important. This is the inverse process to electron scattering by hydrogen atoms, in which the electrons emit radiation without being captured.

Electron-ion collisions. Cross sections for further ionization of positive ions such as He$^+$, Ne$^+$, and N$^+$ by electron impact have been measured by observing the doubly charged ions produced when an electron beam is crossed by an ion beam. The cross section for excitation of He$^+$ (2P) has also been measured.

For collisions with positive ions, the cross sections start from a finite value at the threshold.

Cross sections for detachment of electrons from H$^-$ ions by electron impact have also been measured by crossed-beam methods. *See* MOLECULAR BEAMS.

The elastic scattering of electrons by a positive ion X$^+$, say, can be related to the quantum defects of the bound states of the atom X.

Electron-molecule collisions. If the electrons collide with a molecule instead of an atom, additional possibilities arise. Molecular vibration or rotation may be excited, and dissociation of the molecule into two or more neutral or ionized fragments may occur. Rapid fluctuations with electron energy of the cross sections for impact of slow electrons are more pronounced for molecules than for atoms, but they arise from the same cause — the resonant capture of electrons in certain narrow energy ranges to produce unstable negative molecular ions.

Experiments with electron swarms and with electron beams have shown that, in certain molecules such as N$_2$, CO, O$_2$, and NO, vibrational excitation takes place through the formation of such an unstable negative molecular ion that may break up to leave the molecule in its ground electronic state but with vibrational excitation. If this mechanism operates, the chance of vibrational excitation per collision over the effective electron energy range may be as high as 10%; otherwise it is about 1%. That for excitation of rotation, obtained from analysis of data on the mean energy of electron swarms diffusing through gases in an electron field,

is much higher for molecules with permanent electron dipole or quadrupole moments. *See* MOLECULAR STRUCTURE AND SPECTRA.

Cross sections for dissociation of H$_2$$^+$ by electron impact have been measured by crossed-beam methods.

Extensive studies have been made of the relative probabilities of production of different positively charged fragments due to impact of electrons having energies near 100 ev with various molecules. This has been done by using mass spectrography and is useful for analysis of industrial gases and vapors, especially hydrocarbons.

Negative-ion formation by dissociative attachment may occur in electron impact with a molecule via a process such as in reaction (3). In O$_2$ and

$$AB + e^- \rightarrow A + B^- \qquad (3)$$

CO, these cross sections are of the order 10^{-19} to 10^{-21} cm^2 in certain narrow electron energy ranges. For N$_2$O at 1000 K, it is as high as 10^{-15} cm^2 for thermal electrons. Large cross sections are also found for many halogen-containing molecules.

By using laser selection of the initial vibrational state, it is possible to observe separately the dissociative attachment of electrons to the isotope molecules ^{34}SF$_6$ and ^{32}SF$_6$.

Dissociation of a molecule into positive and negative ions may occur if the electrons are sufficiently energetic.

Recombination. An electron and positive ion may recombine to form a neutral system or system through: (1) radiative capture; (2) dielectronic recombination involving the sequence A$^+$ + e^- → A*, A* → A + $h\nu$, where A* is an unstable, autoionizing state; (3) dissociative recombination, AB$^+$ + e^- → A + B; or (4) three-body recombination, A$^+$ + e^- + B → A + B or A$^+$ + e^- + e^- → A + e^-. Process 1 has a low cross section of order 10^{-19} cm^2. Process 2 is important in stellar atmospheres. Process 3 is important in planetary ionospheres and in molecule formation in interstellar space. It has been studied experimentally for many ions. Cross sections at thermal energies vary from 10^{-15} cm^2 for simple diatomic ions to 10^{-12} cm^2 for large clustered ions. Process 4, involving two electrons, is important in dense plasmas. *See* PLASMA PHYSICS.

Bremsstrahlung. An electron accelerated in the field of an atom or ion may emit radiation in a so-called free-free transition. Such bremsstrahlung is an important sink of energy in a hot plasma, and also may be used to monitor plasma conditions.

The inverse process of light absorption by the electron is important in the long-wave side of the solar spectrum. Direct observation of such absorption has been made by using a laser light source. *See* BREMSSTRAHLUNG.

Positron-atom and molecule collisions. Total cross sections for collisions of slow positrons with atoms have been measured as a function of positron velocity by using time-of-flight and other techniques. The positron source is a suitable moderator which decelerates positrons from a radioactive source to a narrow range of low energies. Positrons are detected from annihilation gamma rays by coincidence-counting techniques. Figure 6 shows observed cross sections for helium, neon, and argon.

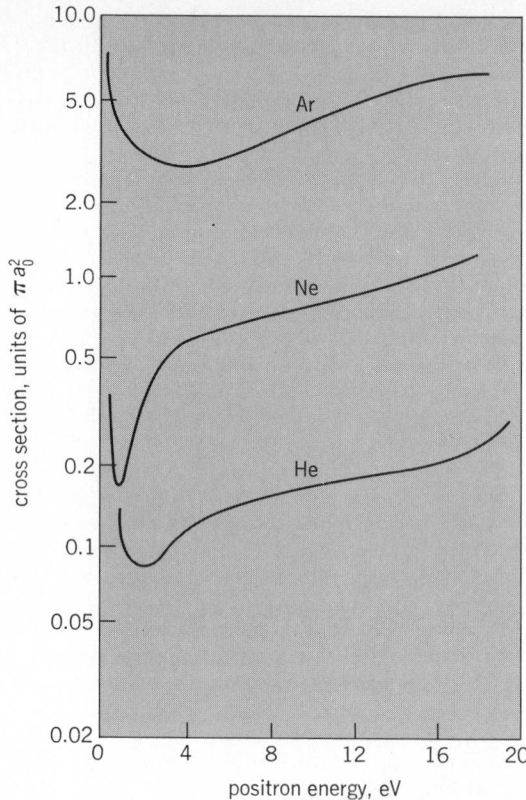

Fig. 6.   Total cross sections for collisions of positrons in helium, neon, and argon as a function of positron energy.

Cross sections for annihilation of positrons in different gases can also be measured from observations of the mean lifetime after its production before a positron annihilates in a dense gas. *See* POSITRON.

Collisions of atomic systems. If both the colliding systems are atoms or molecules, elastic, excitation, or ionization processes can occur. The inelastic processes occur only if the energy of relative motion is great enough. Either or both of the colliding systems may be excited.

There are certain additional possibilities which can be called transfer collisions. They are given by reactions (4) and (5). The last process is often called mutual neutralization.

$$\left. \begin{array}{l} A' + B \rightarrow A + B' \\ A' + B \rightarrow A + B^+ + e^- \end{array} \right\} \begin{array}{l} \text{transfer of} \\ \text{excitation} \end{array} \quad (4)$$

$$\left. \begin{array}{l} A^\pm + B \rightarrow A + B^\pm \\ A^+ + B^- \rightarrow A' + B'' \end{array} \right\} \text{transfer of charge} \quad (5)$$

If either or both of the colliding systems should be molecular, an ionic reaction as in (6) may occur.

$$AB^+ + CD \rightarrow AC^+ + BD \quad (6)$$

Three-body reactions such as (7) may also occur.

$$A^+ + B + C \rightarrow A^+B + C \quad (7)$$

Ionic reactions. Ionic reactions at thermal or near-thermal energies, involving both positive and negative ions, are very important in the upper atmospheres of planets and in electrical discharge and breakdown phenomena generally. Measurement of reaction rates has been greatly facilitated by the use of the flowing afterglow technique, in which the ions in question flow in a buffer gas through a region containing the reactants, after which their concentrations and that of the reaction products are measured. An upper limit to the reaction cross section for (6) is given by $2\pi(\alpha e^2/Mv^2)$, where α is the polarizibility of CD, v is the relative velocity of impact, e the electronic charge, and M the reduced mass.

Elastic scattering. Elastic cross sections for atom-atom impacts may be calculated by treating the interaction between the atoms as static, so that the problem is reducible to that of scattering by a fixed center of force. A typical interaction is strongly repulsive at short distances and attractive at large distances r, falling off as r^{-6} for large r. Collisions in which $2\pi\epsilon r_m/hv$ is > 1 (where ϵ is the magnitude of the maximum attraction energy, r_m is the atomic separation at this maximum, h is Planck's constant, and v is the relative velocity of impact) are referred to as slow collisions. Under these circumstances the total elastic cross section Q takes the form of Eq. (8), where Q varies as

$$Q = \overline{Q} + \Delta Q \quad (8)$$

$v^{-2/5}$, and ΔQ oscillates with v. Figure 7 illustrates the variation of $\Delta Q/Q$ with v for collisions of krypton atoms with argon atoms. When $2\pi\epsilon r_m/hv < 1$, the cross section varies very slowly and gradually with v. *See* INTERMOLECULAR FORCES.

The scattered intensity per unit angle typically exhibits slowly varying undulations on which are superimposed more rapid fluctuations. An example showing calculated and observed angular distributions for collisions of sodium atoms with krypton atoms is given in Fig. 8.

Oscillations also occur in the scattered intensity for low-energy collisions between identical atoms, such as ^{4}He atoms in collisions with ^{4}He. These have been observed at collision energies of the order 0.05 eV.

Cross sections which determine viscosity and diffusion may be calculated by classical mechanics except for light gases such as helium at very low temperatures. Below 10 K the viscosity of the isotope ^{3}He differs considerably from that of ^{4}He due to quantum effects. *See* VISCOSITY.

Many measurements have been made of differential cross sections for elastic scattering of ions by atoms and molecules. For all but very small angles of scattering and very low energies, the re-

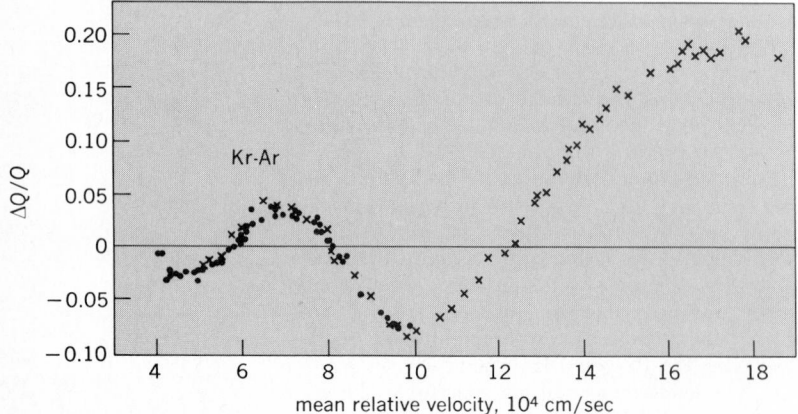

Fig. 7.   Observed variation with relative velocity of the ratio $\Delta Q/Q$ of undulating component to the elastic cross section for collision of krypton atoms with argon atoms.

sults are as would be expected from classical theory.

Transfer collisions. Transfer of charge may be studied by firing an ion beam of homogeneous energy through a gas and measuring the net current of slow positive ions produced for a given path length. Information about excitation transfer is available only from indirect methods involving the quenching of radiation by foreign atoms or other methods.

The magnitude and dependence of the cross section on relative velocity of impact, for a transfer collision, depends very strongly on the amount of energy ΔE transferred between translational and internal motion.

Figure 9 illustrates typical observed cross sections for charge-transfer collisions between positive ions and neutral atoms as functions of the relative impact velocity v. In general, the maximum cross section occurs roughly when Eq. (9) holds,

$$2\pi a \Delta E/(hv) \simeq 1 \qquad (9)$$

where a is a length of the order of atomic dimensions. When $\Delta E = 0$, as in reaction (10), the cross

$$He^+ + He \rightarrow He^+ + He \qquad (10)$$

section falls steadily as the value of v increases.

In all these cases, the magnitude of the cross section never exceeds gas-kinetic values by a large factor. This is because there is no long-range interaction between either the initial or the final system. For charge transfer, in which either the initial or the final system is charged, as in reactions (11),

$$\begin{aligned} A^+ + B^- &\rightarrow A' + B'' \\ \text{or} \qquad A^{++} + B &\rightarrow A^+ + B^+ \end{aligned} \qquad (11)$$

this is no longer true, and cross sections greatly in excess of gas-kinetic values may result. For a given pair of reactants, the minimum cross section need not occur when $\Delta E = 0$, but will usually occur for small values of ΔE.

A similar situation arises with excitation transfer. Cross sections much greater than gas-kinetic values may occur for small ΔE only when both transitions that take place in the reactants are optically allowed.

In general, the cross section for a transfer collision can greatly exceed the gas-kinetic value only when there is a long-range interaction between either the initial or the final reacting systems.

Differential cross sections for charge-transfer collisions have been measured especially for resonant charge transfer such as in Eq. (10). Regular oscillations in the dependence on both the angle of scattering and the impact energy are observed under many conditions, due to the charge-transfer process.

Excitation and ionization. Provided the velocity v of relative motion is large compared with the velocity u of the internal electronic motions concerned, the cross section for excitation of a particular atomic state is nearly the same for impact of singly charged positive ions as for electrons of the same relative velocity v. This is also true for ionization.

When $v < u$, the cross section for excitation by positive ion impact is, in general, small and decreases rapidly as v decreases. Neutral atoms or molecules may be more effective than ions of the same relative velocity in this range, the decrease

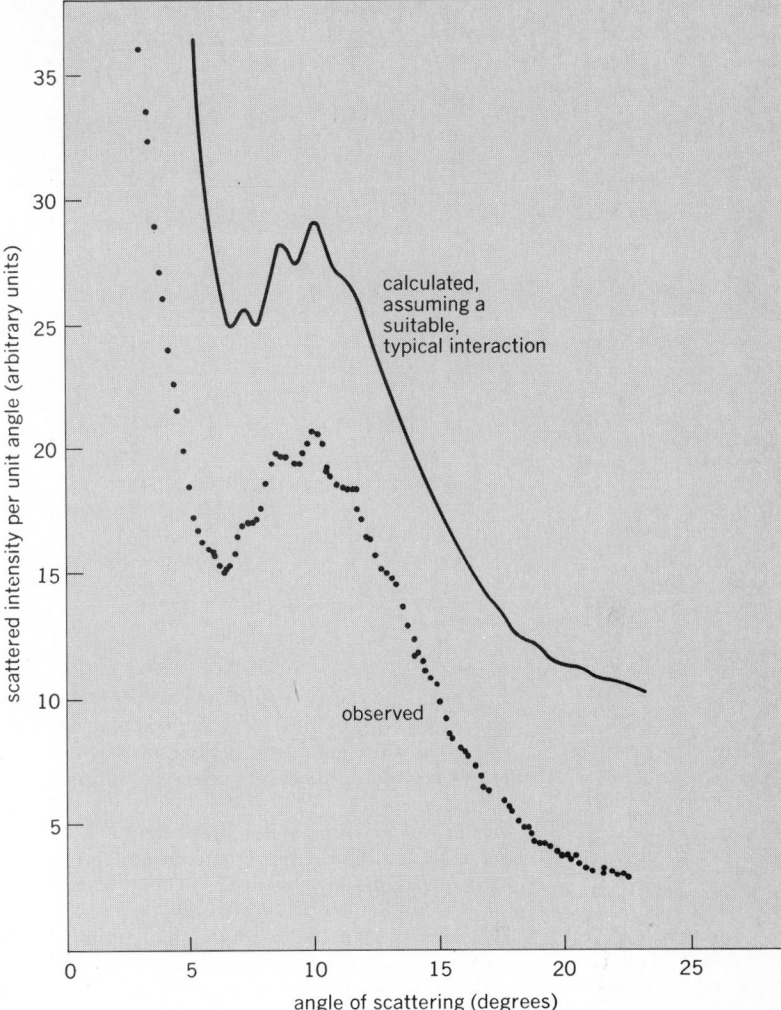

Fig. 8. Distribution per unit angle for scattering of sodium atoms by krypton atoms; the mean relative velocity of collision is 817 m/sec.

of the cross section as v decreases often being more gradual. The cross section for impact of neutral atoms of kinetic energy E is much smaller, at low energies, than for electrons of the same kinetic energy (Fig. 10), just as for ions the cross section is determined mainly by the impact veloc-

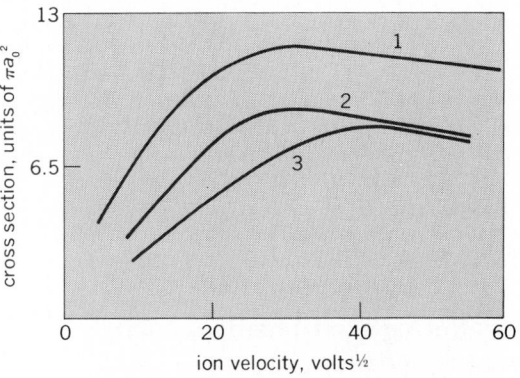

Fig. 9. Some typical observed cross sections for charge transfer processes: (1) $O^+ + Kr \rightarrow Kr^+ + O + 0.43$ eV; (2) $Br^+ + Xe \rightarrow Xe^+ + Br - 0.28$ eV; (3) $C^+ + Xe \rightarrow Xe^+ + C - 0.86$ eV.

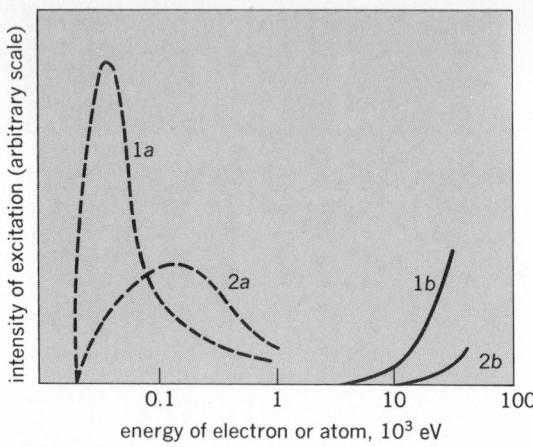

Fig. 10. A pronounced difference exists in the variation with energy of the cross sections for excitation of certain lines by electrons and by hydrogen atoms. Excitation of the line with wavelength 388.8 nm: curve 1*a*, by electrons; curve 1*b*, by hydrogen atoms. Excitation of the line with wavelength 396.4 nm: curve 2*a*, by electrons; curve 2*b*, by hydrogen atoms.

ity. There are many exceptions to these general rules, and a great deal of experimental work is directed toward obtaining further insight into the factors that determine the cross section for a particular process.

Collisions between atoms may result also in spin exchange, depolarization of atoms aligned or oriented in magnetic fields, and reversal of electron spin. Cross sections for these collisions at thermal energies may be measured by optical pumping and magnetic resonance techniques. They are very small for electron spin reversal ($< 10^{-23} cm^2$) but large (of order $10^{-14} cm^2$) for depolarizing collisions, and comparable with gas-kinetic values for spin exchange.

Radiative collisions between atoms and ions. Complex atoms and ions emit radiation when in collision at energies of a few keV. Such collisions have been intensively studied, especially as, under certain circumstances, the radiation is characteristic of transitions between states of the quasi-molecule formed on close approach of the colliding systems. In the limit, these states are those of the superheavy ion which would be formed if the colliding systems coalesced. *See* NONRELATIVISTIC QUANTUM THEORY; QUASIATOM; SCATTERING EXPERIMENTS (NUCLEI).   [HARRIE W. MASSEY]

Bibliography: J. B. Hasted, *Physics of Atomic Collisions*, 2d ed., 1972; E. W. McDaniel, *Collision Phenomena in Ionized Gases*, 1964; E. W. McDaniel, *Ion-Molecule Reactions*, 1970; E. W. McDaniel and E. A. Mason, *The Mobility and Diffusion of Ions in Gases*, 1973; H. S. W. Massey, E. H. S. Burhop, and H. B. Gilbody, *Electronic and Ionic Impact Phenomena*, 2d ed., vols. 1–5, 1969–1974; N. F. Mott and H. S. W. Massey, *The Theory of Atomic Collisions*, 3d ed., 1965.

Scattering experiments (nuclei)

Experiments in which beams of particles such as electrons, nucleons, alpha particles and other atomic nuclei, and mesons are deflected by elastic collisions with atomic nuclei. Much is learned from such experiments about the nature of the scattered particle, the scattering center, and the forces acting between them. Scattering experiments, made possible by the construction of high-energy particle accelerators and the development of specialized techniques for detecting the scattered particles, are one of the main sources of information regarding the structure of matter. *See* PARTICLE ACCELERATOR; PARTICLE DETECTOR.

In the broad sense, any nuclear reaction is an example of scattering. However, this article treats elastic scattering only in the more restricted sense given below, and only insofar as it involves atomic nuclei. *See* COULOMB EXCITATION; GIANT NUCLEAR RESONANCES; NEUTRON SPECTROMETRY; NUCLEAR REACTION; NUCLEAR SPECTRA; NUCLEAR STRUCTURE.

Definitions of elastic scattering. The word "elastic" is used to indicate the absence of energy loss. If particle A collides with particle B of finite mass, there is a loss in the energy of A even if no energy has been transferred to the internal degrees of freedom of either A or B. Sometimes such a collision is referred to as inelastic, in order to distinguish its character from that of a collision with a particle having an infinite mass or its idealization, a fixed center of force. This terminology is not useful in the present context, because in the center-of-mass system of the two particles the sum of kinetic energies after the collision is the same as before. The distinction between elastic and inelastic scattering is made therefore on the basis of whether there are internal energy changes in the colliding particles. The collison is said to be inelastic even if the energy changes of the two particles compensate so as to leave the sum of the kinetic energies in the center-of-mass system unaltered. The treatment of inelastic scattering involves nuclear reaction theory because nuclear reactions markedly influence the scattering. *See* COLLISION.

Cross sections. The results of scattering measurements and calculations are generally expressed in terms of differential cross sections, which furnish a quantitative measure of the probability that the incident particle is scattered through an angle, θ. Cross sections have units of area, and in nuclear physics the convenient measure of area is the barn and its subunits, the millibarn and microbarn (1 barn $= 10^{-28} m^2$).

Coulomb scattering by nuclei. The simplest type of elastic scattering experiment relevant to the study of nuclei involves the deflection of incident electrically charged particles by the Coulomb field of the target nucleus. Such experiments provided the first evidence for atomic nuclei. H. Geiger and E. Marsden observed in 1909 that low-energy alpha particles could be scattered through large angles in collisions with gold and silver targets, and Ernest Rutherford showed in 1911 that these results could be understood if the scattering center consisted of a positively charged region (the nucleus) considerably smaller in size than that occupied by an individual target atom. The yield of scattered alpha particles as a function of angle, called the angular distribution, was measured in detail in a subsequent (1913) experiment by Geiger and Marsden, and the results were entirely consistent with the Rutherford expression for the differential cross section given by Eq. (1), where Z_1

$$\frac{d\sigma}{d\Omega}(\theta)_{\text{point}} = [Z_1 Z_2 \, e^2/16\pi\epsilon_0 E]^2 \sin^{-4}\theta/2 \qquad (1)$$

and Z_2 are the atomic numbers of the target and projectile, e is the electronic charge, ϵ_0 is the permittivity of free space, E is the center-of-mass energy of the alpha particle projectile, and θ is the center-of-mass scattering, or observation angle.

Equation (1) is appropriate only when the classical distance of closest approach, $d = Z_1 Z_2 \, e^2/4\pi\epsilon_0 E$, is greater than the combined size of the colliding pair; that is, only if the two charge distributions never overlap during the course of their interaction. The 1913 experiments involved collisions of 7.68-MeV alpha particles (1 MeV = 1.6×10^{-13} joule) with gold targets, and for this situation $d = 2.96 \times 10^{-14}$ m (29.6 fermis or femtometers in conventional nuclear physics notation, with 1 fm = 10^{-15} m). The detailed agreement between the Rutherford prediction of Eq. (1) and the 7.68-MeV data thus implied that the nuclear charge is contained within a region smaller than 30 fm in radius. Subsequent experiments involving higher-energy projectiles have determined the actual half-density radius of the gold nucleus to be approximately 7 fm.

Although nuclear charge radii can be measured by systematically increasing the energy of the incident alpha particle until Eq. (1) no longer describes the angular distributions, specifically nuclear interactions come into play when the colliding nuclei overlap. In that situation, nuclear size is no longer the only factor being probed. This complexity is avoided when high-energy electrons are used as projectiles, since electrons are not subject to the nuclear forces. At very low energies the elastic scattering of electrons by nuclei conforms to the predictions of Eq. (1), but much more information is revealed at higher energies where the electrons can penetrate into the target nucleus. In the latter case and neglecting magnetic effects for the moment, the finite extent of the nucleus requires that Eq. (1) be modified by the presence of a nuclear form factor, $F(q^2)$, as in Eq. (2), in which $\vec{q}$

$$\frac{d\sigma}{d\Omega}(\theta) = \frac{d\sigma}{d\Omega}(\theta)_{\text{point}} [F(\vec{q}^2)]^2 \qquad (2)$$

is the momentum transfer in the collision. The significance of $F(q^2)$ can be appreciated by noting that at nonrelativistic energies, and assuming the Born approximation to be valid, $F(q^2)$ is the Fourier transform of the nuclear charge distribution.

If sufficiently complete measurements over a wide range of momentum transfer are available, the form factor can be determined in detail, and the nuclear charge density, $\rho_{ch}(r)$, obtained directly through the inverse transform. In practice, a functional form is usually assumed for $\rho_{ch}(r)$, and its parameters are adjusted to fit the electron scattering data. In this manner it is found that the experiments are consistent with nuclear charge distributions specified by Eq. (3). The parameter c is the

$$\rho_{ch}(r) = \rho_0 [1 + e^{(r-c)/a}]^{-1} \qquad (3)$$

radius at which the density falls to one-half its central value. It assumes the typical value $c \approx 1.1 \, A^{1/3}$ fm, where A is the atomic number of the target

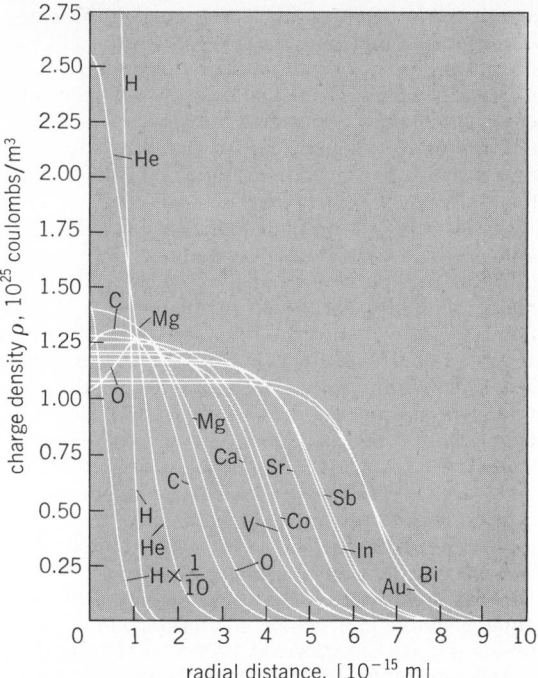

Fig. 1. Summary of the nuclear charge distributions found for various nuclei by electron scattering methods. (*From R. Hofstadter, Nuclear and nucleon scattering of high-energy electrons, Annu. Rev. Nucl. Sci., 7:231–316, 1957*)

nucleus. The quantity a (typically ≈ 0.55 fm) reflects the fact that nuclear surfaces are not sharply defined, so that the density falls off gradually with increasing radius. The so-called central charge density, ρ_0, is found to be approximately 1.1×10^{25} coulombs/m³, which corresponds to 0.07 proton per cubic femtometer. That ρ_0 is approximately the same for all but the lightest nuclei and $c \propto A^{1/3}$ implies that nuclear matter is nearly incompressible. The charge distributions of a variety of nuclei, as determined by electron scattering measurements, are shown in Fig. 1.

Electron scattering measurements involving large angular momentum transfer are sensitive to the distribution of magnetism as well as charge, and Eq. (2) then must be generalized to reflect the contributions from the two distributions. In addition, very precise measurements show that Eq. (3) is only approximately correct and the form of $\rho_{ch}(r)$ varies slightly from nucleus to nucleus as a consequence of variations in the structure of the nuclei in question.

Electron-nucleon scattering. Scattering of electrons by hydrogen gives information regarding the electron-proton interaction. From measurements of the variation of the differential cross section, it has been found necessary to postulate that both the proton charge and its intrinsic magnetic moment are distributed through a finite volume. Existing work favors the assumption of similarity of shape of these distributions. Energies in excess of 200 MeV are required for the detection of these effects, and the anomalous magnetic moment of the proton plays an important part at high energies. The experiments make it probable that the charge density has a root-mean-square radius of approximately 0.8×10^{-13} cm. Measurements on the scattering of electrons by deuterium at large angles and higher energies lead to the conclusion that the magnetic moment of the neutron is not concentrated at a point. If it is assumed that neutron and proton magnetic moments are distributed through nearly the same volumes, a good representation of the scattering measurements is obtained.

Analyses of scattering data at various angles and energies indicate that there is no net charge density within the volume occupied by the neutron. This result is in agreement with measurements of the neutron-electron interaction made by scattering very slow neutrons from atomic electrons and atomic nuclei. While there is an interaction equivalent to a potential energy of approximately -3900 eV through a distance of e^2/mc^2, where m is the electron mass and c is the speed of light, amounting to approximately 2.8×10^{-13} cm, it is accounted for qualitatively as a consequence of what E. Schrödinger called the *Zitterbewegung* (tremblatory motion) expected for the neutron.

Electron scattering by nucleons has also provided valuable information concerning subnuclear processes. *See* ELEMENTARY PARTICLE; QUARKS.

Low-energy np scattering. Since the neutron is electrically neutral, an understanding of how it scatters in an interaction with a proton requires, even for low collision energies, a knowledge of the properties of the nuclear forces. Conversely, the scattering experiments probe these properties. The *np* force is responsible for the binding together of a neutron, *n*, a proton, *p*, to form a deuteron, and the detailed connection between *np* scattering cross sections and the properties of the deuteron were perceived in the early 1930s. For example, the assumption that the spatial extension, or range, of the *np* force is relatively small makes it possible to estimate the magnitude of the force, given the measured binding energy of the deuteron, and Eugene Wigner used this information in 1933 to calculate *np* scattering cross sections. The measured yield was found to be larger than would be expected if the forces between free neutrons and free protons were the same as those in the deuteron. To explain this difference, it was postulated by Wigner in 1935 that the *np* interaction is spin-dependent; that is, it depends on the relative orientation of the spins of the interacting particles. The proton and neutron are known to have a spin of $\frac{1}{2}$; that is, their intrinsic angular momenta are known to be $(h/2\pi)/2 = \hbar/2$, where h is Planck's constant. According to quantum mechanics, when two spins s_1, s_2 combine vectorially, only the values given by Eq. (4) are possible for the resultant s.

$$s = s_1 + s_2, s_1 + s_2 - 1, \ldots, |s_1 - s_2| \qquad (4)$$

For the *np* system, therefore, the resultant spins are 0 or 1. In the first case one speaks of a singlet, and in the second of a triplet. *See* ANGULAR MOMENTUM; SPIN.

The singlet state behaves much like a round and perfectly smooth object which has the same appearance no matter how it is viewed, corresponding to only one possibility of forming a state with $s = 0$. The state with $s = 1$, on the other hand, can have three distinct spin orientations. Measurement of the projection of s on an axis fixed in space can give only the three values (1, 0, −1), again in units $\hbar$. When protons with random spin directions col-

lide with neutrons also having random spin directions, the triplet state is formed three times as often as the singlet. The deuteron, however, is in a triplet state. Thus the hypothesis of spin dependence can account for the difference between the forces in the deuteron and those in *np* scattering.

The neutron-hydrogen scattering experiments on which these conclusions were based were performed with slow neutrons having energies of a few electronvolts or less. The general quantum-mechanical theory of scattering is much simplified in this case. Because of the small range of nuclear forces, only the collisions with zero orbital angular momentum ($L\hbar = 0$) play a role, collisions with higher orbital angular momenta missing the region within which nuclear interactions take place. States with $L = 0$ (called S states) have the property of spherical symmetry, and nuclear forces matter in this case only inasmuch as they modify the spherically symmetric part of the wave function.

The long-wavelength or low-energy scattering cross section can be described completely and simply by a quantity called the scattering length. For interparticle distances r greater than the range of nuclear forces, the spherically symmetric part of the wave function has the form given by expression (5), where a and C are constants. The con-

$$C[1 + (a/r)] \qquad (5)$$

stant a is called the scattering length. It has the following meaning. If $R(r)$ denotes the wave function describing the *np* relative motion, the product $rR(r)$, when plotted against r, is represented by a straight line which cuts the axis of r at a distance a from the origin of coordinates. If the intersection is to the left (right) of the origin, a is counted as positive (negative). The two conditions are illustrated in Figs. 2 and 3. The possibilities $a > 0$ and $a < 0$

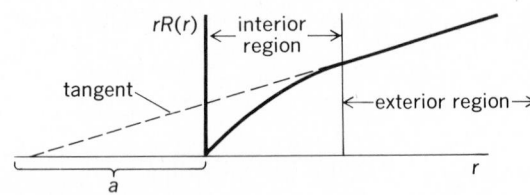

Fig. 2. Scattering length in the case of a virtual level.

are sometimes referred to as those of the virtual and real level, respectively. Sometimes the opposite convention regarding the sign of a is used. The convention adhered to here provides the simplest connection with phase shifts.

The scattering cross section for a state with well-defined a is $4\pi a^2$, and thus low-energy *np* scattering is described by Eq. (6), where σ_{np} is the total

$$\sigma_{np} = 4\pi(\tfrac{1}{4} a_s^2 + \tfrac{3}{4} a_t^2) \qquad (6)$$

cross section and a_s and a_t are the singlet and triplet scattering lengths, respectively. Comparison of the scattering yields with the deuteron data provides the values $a_s = 24$ fm and $a_t = -5.4$ fm.

There are several additional means by which *np* scattering can be calculated. For example, with the techniques outlined below ("Potential scattering"), effective interaction potentials can be found which simulate the *np* nuclear forces to the extent that

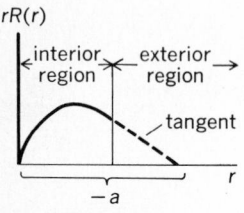

Fig. 3. Scattering length in the case of a real level.

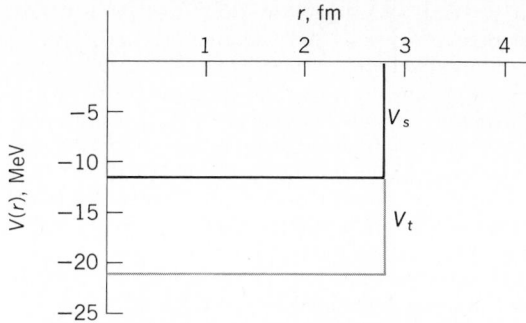

Fig. 4. Examples of singlet (V_s) and triplet (V_t) potential wells suitable for low-energy *np* scattering. The potentials are not uniquely determined; many other equally satisfactory possibilities exist.

calculations using the potentials reproduce the measured cross sections. Two potentials corresponding to singlet and triplet $L = 0$ scattering are required for a description of low-energy *np* data, as is illustrated in Fig. 4. The term "potential well" is often used to describe these potential energy surfaces, because the system can be trapped in the region of space occupied by the potential somewhat similarly to the way in which water is trapped in a well. For nucleon-nucleon scattering, it is frequently useful to express the potential energy in the form $V(r) = V_0 f(r/b)$, where f is a function which determines the shape of the well. The constants V_0 and b are usually referred to as the depth and range parameters. The scattering length determines approximately the product $V_0 b^2$ for a potential well of assigned shape. The variation of the cross section with energy E through an energy range of a few MeV can be used for the determination of b. *See* NONRELATIVISTIC QUANTUM THEORY; QUANTUM MECHANICS.

Low-energy pp scattering. The *pp* and *pn* interactions are believed to be closely equal. Gregory Breit proposed in 1936 the hypothesis of charge independence of nuclear forces, which supposes that nuclear forces, acting in addition to the electrostatic (Coulomb) repulsion, in the *pp*, *pn*, and *nn* cases are equal to each other. Because of the limits of experimental accuracy and uncertainties in the theoretical interpretation, the hypothesis is not established with perfect accuracy, but it is believed to hold within a few percent for the depth parameter V_0 if the range parameter b is specified as the same in the three cases. The concept of charge independence also has important implications for the structure of nuclei. *See* ISOTOPIC SPIN.

Phase shifts. Scattering can be treated by means of phase shifts, which will be illustrated for two spinless particles. The wave function of relative motion will be considered first for a state of definite orbital angular momentum $L\hbar$, with L representing an integer. Outside the range R of nuclear forces, the wave function may be represented by Eq. (7), where k is 2π times the reciprocal of the

$$\psi_L = Y_{LM}(\theta, \phi)\mathscr{F}_L(kr)/(kr) \qquad (7)$$
$$(r > R)$$

wavelength, the so-called wave number, θ and ϕ are the colatitude and azimuthal angles of a polar coordinate system, and Y_{LM} is the spherical har-

monic of order L and azimuthal quantum number M. The form of ψ_L is determined by the Schrödinger wave equation which restricts $\mathscr{F}_L$ by the differential equation, Eq. (8), it being supposed

$$\{d^2/dr^2 + [k^2 - L(L+1)/r^2]\}\mathscr{F}_L = 0 \qquad (8)$$
$$(r > R)$$

that there is no Coulomb field. In the absence of nuclear forces, $\mathscr{F}_L$ satisfies the same equation at all distances and, aside from a constant factor, has its asymptotic form determined by the boundary conditions at $r = 0$ as in notation (9). In the pres-

$$\mathscr{F}_L \underset{r \to \infty}{\sim} \sin(kr - L\pi/2) \qquad (9)$$

ence of nuclear interactions, the asymptotic form is given by notation (10), where δ_L is a constant,

$$\mathscr{F}_L \sim \sin(kr - L\pi/2 + \delta_L) \qquad (10)$$

called the phase shift, which determines the scattering. *See* SPHERICAL HARMONICS.

When elastic scattering is the only process that can take place, as for low- and medium-energy nucleon-nucleon scattering, the phase shifts are real numbers. Above a few hundred MeV, nucleon-nucleon scattering becomes strongly inelastic because of pion production, however, and the phase parameters become complex numbers reflecting the loss of flux from the elastic channel. Nucleon-nucleus and nucleus-nucleus interactions are characterized by the presence of many nonelastic scattering channels, except at very low energies.

The wave packet representing the scattered particle consists of a superposition of all possible partial waves, $\mathscr{F}_L$, and calculation of the scattering cross section requires a knowledge of all phase shifts. In practice, however, only partial waves with orbital angular momenta corresponding to impact parameters within the range of the nuclear forces suffer nuclear phase shifts different from zero. The direct method of varying phase shifts to reproduce experimental data is most useful, therefore, when only a few L values satisfy this condition, as is the case for low- and intermediate-energy nucleon-nucleon scattering and for low-energy nucleon-nucleus and nucleus-nucleus collisions.

For charged spinless particles, such as two alpha particles, the phase shifts caused by specifically nuclear forces add to the asymptotic phase of the functions $\mathscr{F}_L$ for the Coulomb case, which differs from the non-Coulomb case only through the replacement of $kr - L\pi/2$ by $kr - L\pi/2 - \eta \ln(2kr) + \arg \Gamma(L + 1 + i\eta)$, where $\eta = Z_1 Z_2 e^2/4\pi\epsilon_0 \hbar v$, $Z_1 e$ and $Z_2 e$ represent the charges on the colliding particles, e represents the electronic charge, and v represents the relative velocity. *See* GAMMA FUNCTION.

As in the *np* scattering case discussed above, the presence of spin adds additional complications. For example, if two particles with spin $\frac{1}{2}$ collide, it is necessary, in general, to introduce phase shifts for each state with definite total angular momentum $J\hbar$.

Potential scattering. The direct determination of phase shifts through comparison with scattering data becomes difficult in situations where large numbers of partial waves, $\mathscr{F}_L$, are modified by the nuclear scattering center. The usual procedure in this case is to simulate the influence of the nuclear interactions by means of potentials, $V(r)$. For the nonrelativistic scattering of two spinless, uncharged particles, Eq. (8) is supplemented by the Schrödinger equation for $r < R$, Eq. (11), in which

$$\{d^2/dr^2 + [k^2 - L(L+1)/r^2 - 2mV(r)/\hbar^2]\}\mathscr{F}_L(r) = 0$$
$$(r < R) \qquad (11)$$

m is the reduced mass of the colliding pair. Equations (8) and (11) are then solved subject to appropriate boundary conditions to determine the $\mathscr{F}_L(r)$ and hence δ_L for each L. A functional form is generally assumed for $V(r)$, and its parameters are varied and Eqs. (8) and (11) solved iteratively until the measured cross sections, polarizations, and so forth are reproduced.

From the viewpoint of microscopic models of nucleon-nucleon interactions, it appears highly improbable that a description of nucleon-nucleon scattering in terms of two-body energy-indepen-

dent local potentials can have fundamental significance. Nevertheless, in a limited energy range, it is practical and customary to represent scattering by means of such a potential. As noted above, the simple potentials illustrated in Fig. 4 suffice for the description of low-energy np scattering. At somewhat higher energies, but below the threshold of meson production, a real potential may still be used, but different potentials are required for triplet-even, triplet-odd, singlet-even, and singlet-odd states to account for the (even, odd) parity dependence of the interaction. *See* PARITY.

Intermediate energy np and pp scattering. In the intermediate-energy region (10–440 MeV) the analysis of experimental material is more difficult than at low energies because of the necessity of employing many phase shifts and coupling constants. Analysis in terms of phase parameters involves fewer assumptions than that in terms of potentials. Except for approximations connected with the infrared catastrophe and related small inaccuracies in relativistic treatment of Coulomb scattering, it is based on very generally accepted assumptions, such as the validity of time reversal and parity symmetries for strong interactions. With infinite experimental accuracy, it should be possible to extract all the phase parameters from measurements of the differential cross section, the polarization, spin correlation coefficients, and "triple scattering" quantities describing spin orientation which, for unpolarized incident beams and unpolarized targets, require three successive scatterings.

The analysis is usually carried out by assuming that for sufficiently high L and J the phase parameters may be represented by means of the one-pion exchange approximation. The value of the pion-nucleon coupling constant g^2 is often varied in an attempt to improve the fit to experimental data, and values for best fits are compared with those from pion physics. Reasonable agreement usually results.

The consistency of values of g^2 from pp data with those from np measurements indicates approximate validity of charge independence at the larger distances.

Potentials to be used in a nonrelativistic Schrödinger equation and capable of representing pp and np scattering have been devised either on a purely phenomenological or semiphenomenological basis. The former way provides a more accurate representation of the data. Nonrelativistic local potentials required from 0 to 310 MeV are different according to whether the state is even or odd, singlet or triplet. It is necessary to use central, tensor, spin-orbit, and quadratic spin-orbit parts of the potential. Most of the accurately adjusted potentials employ hard cores within which the potential is infinite. The spin-orbit potential suggested by pp scattering data indicated the probable participation of vector-meson exchange in nucleon-nucleon scattering, anticipating the discoveries of the ω- and ρ-mesons, as well as fictitious mesons, the latter partly intended as a representation of simultaneous two-pion exchange. At short distances, the potentials are often modified in order to improve agreement with experiment. Superposition of the single-boson potentials combined with the short-range modifications of the resulting potential gives the so-called one-boson exchange (OBE)

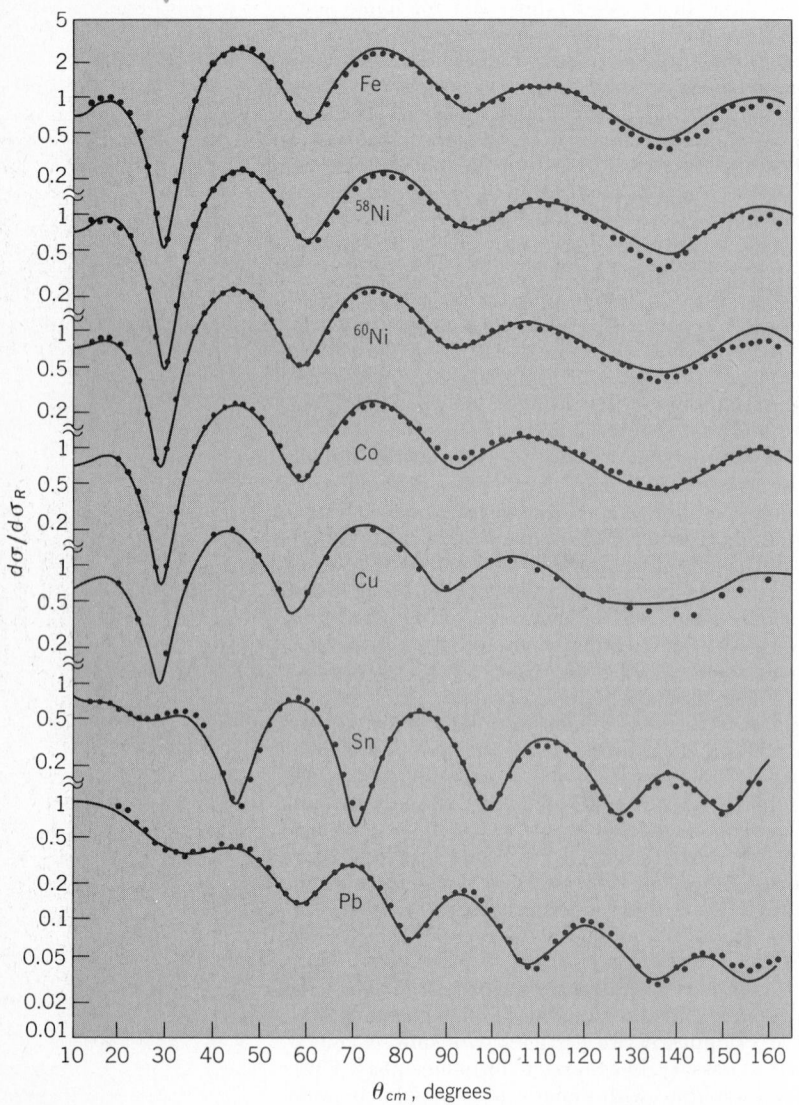

Fig. 5. Measured (points) and calculated (curves) differential cross sections for the scattering of 30-MeV protons from a variety of target nuclei, as a function of scattering angle in the center-of-mass system θ_{cm}. The theoretical calculations made use of optical potential wells. The vertical scale represents the differential cross section $d\sigma$ divided by the Rutherford differential cross section $d\sigma_R$. (From G. R. Satchler, *Optical model for 30 MeV proton scattering, Nucl. Phys., A92:273–305, 1966*)

potentials. These provide a fair but not excellent reproduction of phenomenological fit phase shifts. *See* QUANTUM FIELD THEORY.

Nucleon-nucleus scattering. Nucleon-nucleus scattering at low and intermediate energies ($E <$ 300 MeV) is of interest primarily in connection with nuclear structure studies. The scattering wave functions, $\mathscr{F}_L(r)$, are required for quantitative interpretation of the inelastic scattering and reaction experiments which have been the main source of information regarding the properties of nuclear excited states. These scattering functions are obtained as the solutions of Eq. (11) when elastic data are reproduced by using potential models.

Nucleon-nucleus scattering experiments can be accounted for by a potential-well model with a complex potential (optical model) making use of a spin-orbit interaction term. The details of angular distributions of the cross section and polarization are reproduced remarkably well, and experiments favor some potential well shapes over others. Wells thus determined are wider than those obtained from electron-nucleus scattering experiments, and similar functional shapes work in both cases. The potential energy represented by the wells is added to the electrostatic potential energy in the calculations. The electrostatic potential energy is approximated by a central potential corresponding to the average distribution of nuclear charge. The quantitative success of the potential-well approach to the scattering of 30-MeV protons by a variety of nuclei is illustrated in Fig. 5.

Optical model potential fits to nucleon-nucleus data are in general agreement with the data in a wide energy range from several MeV to about 300 MeV, but the parameters of the potential have to be varied progressively. The spin-orbit potential found to represent the scattering data at the lower energies has the same value of its ratio to $(dV_{ct})/(rdr)$, where V_{ct} is the central potential, as it has in the shell theory of nuclear structure. At higher energies (300 MeV), it has been found that the spin-orbit potential has to be used with a smaller strength than in shell theory. The real part of the central potential decreases with energy and becomes almost zero at 300 MeV. Data at 1 GeV on proton scattering from carbon can be accounted for on the optical model by means of an imaginary central and real spin-orbit potential.

The real part of the volume potential $V(\vec{r}_p)$ can be calculated by folding the interaction potential between the projectile and a target nucleon, $v_{tp}(\vec{r})$, with the nuclear density of the target as in Eq. (12).

$$V(\vec{r}_p) = \int \rho(\vec{r}_t) v(\vec{r}_{tp}) d\vec{r}_t \qquad (12)$$

Effective, rather than free, nucleon-nucleon interactions are used for $v(\vec{r}_{tp})$ because the target nucleon is embedded in a nucleus. The short range of $v(\vec{r}_{tp})$ implies that the resulting central potential has a spatial distribution closely similar to, but somewhat more rounded than, that of the target nuclear density, as is illustrated in Fig. 6. The imaginary part of the optical potential is phenomenologically determined.

Nucleus-nucleus scattering. Composite nuclei ranging from deuterium through uranium have been accelerated to energies as high as several GeV per constituent nucleon. Elastic scattering is an important process only toward the lower end

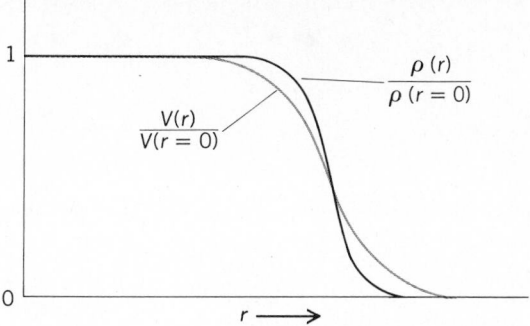

Fig. 6. Comparison between the shape of a nuclear density distribution, $\rho(r)$, and a potential, $V(r)$, obtained from it by folding in an interaction potential, $v(r)$, with a finite but short range as in Eq. (12). (*From G. R. Satchler, Introduction to Nuclear Reactions, copyright © 1980 by John Wiley and Sons, Inc.; used with permission*)

of this energy range, perhaps only below a few tens of MeV per nucleon; at the higher energies, the principal interest is in nonelastic collisions, which are expected to provide information concerning properties of nuclear matter under unusual conditions (like high temperature and high density).

Nucleus-nucleus scattering at low and intermediate energies is characterized by strong absorption associated with the relatively short mean free path for nuclei in nuclear matter. Only peripheral or glancing collisions are likely to lead to direct elastic scattering; more head-on collisions lead to more complicated processes such as compound nucleus formation which absorb flux from the incident beam. Since the de Broglie wavelengths of the incident nuclei are typically comparable to or smaller than the size of the target nuclei, the strong absorption for head-on collisions frequently causes the scattering to be diffractive in nature, and the elastic scattering angular distributions to resemble those observed in the scattering of light by small opaque obstacles. A typical example, corresponding to optical Fraunhöfer diffraction, is illustrated in Fig. 7.

Good fits to nucleus-nucleus scattering data are obtained by using potential wells. For collisions involving relatively light nuclei, the real potentials may be derived by folding an effective nucleon-nucleon interaction with the densities of the two colliding nuclei, in a slight generalization of Eq. (12). For more massive nuclei, analogies with the collisions of liquid droplets, provide guidance in determining the potentials. In many cases, purely phenomenological potentials provide the best fits to the data. There is considerable ambiguity in the potential parameters, because the scattering depends primarily on the values of the potentials at distances corresponding to glancing collisions, and is relatively insensitive to the values at smaller distances. The imaginary potentials are deeper than those used for nucleon-nucleus scattering because of the strong absorption property. Except for collisions involving light nuclei, no very strong evidence for the influence of spin-orbit potentials on elastic scattering yields has been found, but there is evidence suggesting that these may become important for more massive nuclei at energies in excess of 10 MeV per nucleon.

The optical model potential is not a potential in

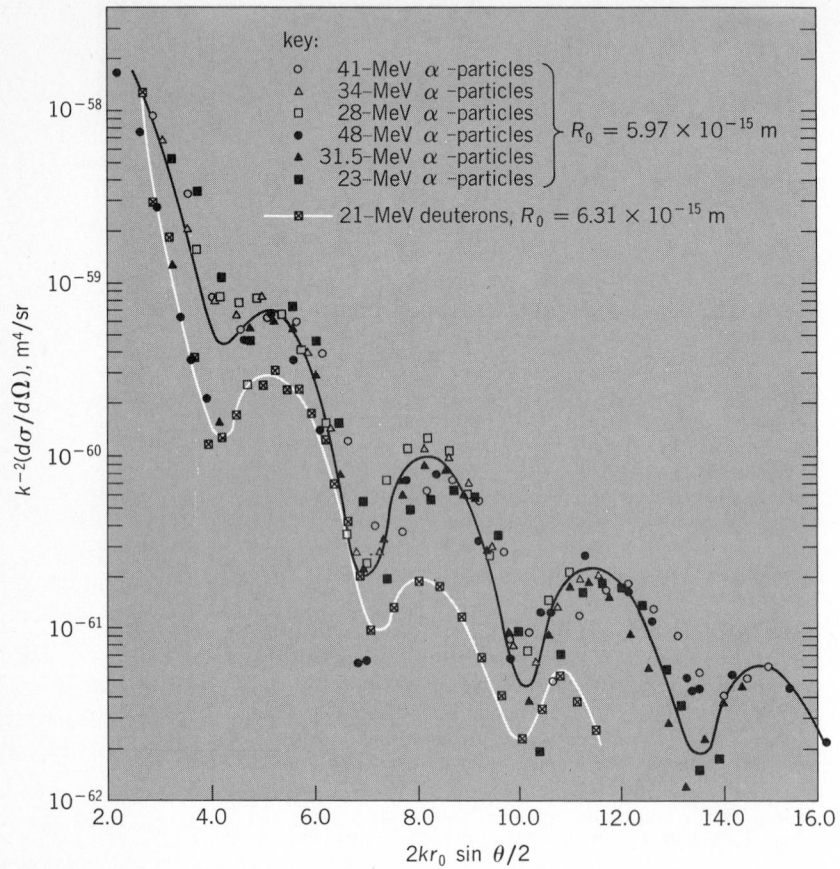

key:
○   41-MeV α-particles
△   34-MeV α-particles
□   28-MeV α-particles
●   48-MeV α-particles
▲   31.5-MeV α-particles
■   23-MeV α-particles
⊠   21-MeV deuterons, $R_0 = 6.31 \times 10^{-15}$ m

$R_0 = 5.97 \times 10^{-15}$ m

Fig. 7. Angular distribution for the elastic scattering of deuterons and alpha-particles from magnesium, showing the typical Fraunhöfer-like diffraction pattern. (*From J. S. Blair et al., Diffraction analysis of elastic and inelastic scattering by magnesium, Nucl. Phys., 17:641–654, 1960*)

the ordinary sense. When inserted in the wave equation (11), it gives agreement with experiment by simulating the complicated interactions between the two colliding many-body systems in terms of a prescribed functional form. Possible variations in this form, associated with the varying proximity of the two nuclei during the collision, are generally neglected. Alternative methods, such as the time-dependent Hartree-Fock approach, are being developed to treat nucleus-nucleus interactions from a more fundamental perspective.

Nucleus-nucleus scattering sometimes exhibits anomalous behavior which cannot be reproduced with potential models and which is believed to reflect the transient formation of moleculelike configurations. These configurations, or resonances, signal their presence by producing a rapid variation in the energy dependence of the scattering yields and by modifying the behavior of the angular distribution data. *See* NUCLEAR MOLECULE.

Meson scattering. The scattering of π-mesons (pions) by nucleons has been studied intensively as an example of a strong interaction between a boson and a baryon, and particularly because of the connection with nucleon-nucleon interactions. Prominent resonances occur in pion-nucleon scattering yields, the best known of which appears for pion bombarding energies of about 200 MeV. Charge independence in pion-nucleon interactions is confirmed by the scattering measurements and their phase-shift analyses.

The scattering of pions by nuclei has been studied over the energy range extending from below 20 MeV to beyond 60 GeV. At low energies the experiments probe nucleon-nucleon correlations within nuclei. Attempts to relate the scattering behavior to the properties of the nuclear wave functions have been partially successful. Comparison of π+ with π− scattering yields provides information concerning differences between proton and neutron distributions within nuclei. Future study of pion-nucleus interactions at high energies is expected to reveal properties of the strong interactions, modified by the nuclear environment, at short distances.

The study of kaon-nucleus interactions, still in its infancy, is currently of interest primarily in connection with the production of nuclei in which, for example, a neutron is replaced by a lambda particle (Λ), and the investigation of Λ-nucleon and related interactions. *See* MESON.   [K. A. ERB]

Bibliography: R. C. Barrett and D. F. Jackson, *Nuclear Sizes and Structure*, 1977; G. Breit and R. D. Haracz, Nucleon-nucleon scattering, in E. H. S. Burhop (ed.), *High-Energy Physics*, 1967; M. Goldberger and K. M. Watson, *Collision Theory*, 1964; G. R. Satchler, *Introduction to Nuclear Reactions*, 1980.

Scattering matrix

A matrix which expresses the initial state in a scattering experiment in terms of the possible final states, and hence enters the calculation of the probabilities that certain reactions will occur in a collision of two or more particles. The scattering matrix was introduced by J. A. Wheeler in 1937 in the discussion of the theory of nuclear reactions. Previous work on scattering theory (applied to atomic collisions) had been based for the most part on the use of the Schrödinger equation for the direct calculation of the scattering amplitude. However, the multiplicity of different reactions in a typical nuclear collision and the uncertainties in the form of the nuclear forces made a more general approach to reaction theory desirable.

The problem is compounded in the relativistic domain, in which particles may be created or destroyed in a collision and the forces between elementary particles are known only approximately. There is, furthermore, no useful relativistic analog of the Schrödinger equation. It was therefore suggested by W. Heisenberg in 1943 that the S matrix should play a fundamental rather than a subsidiary role in relativistic quantum mechanics, and considerable progress has been made toward this goal. *See* ELEMENTARY PARTICLE; MATRIX MECHANICS; NUCLEAR REACTION; QUANTUM MECHANICS; SCATTERING EXPERIMENTS (NUCLEI).

Definition and properties. The initial state of a system of particles specified by the set of quantum numbers γ may be described by an "ingoing" wave function |λ in⟩. Ingoing functions satisfy boundary conditions such that it is possible to construct from them wave functions in which the individual particles are localized and converge toward the region of interaction prior to their collision. The final states of the system are described by a similar set of outgoing wave functions |β, out⟩, from which wave functions can be constructed which describe particles that diverge from the interaction region at times long after the collision. It is convenient to normalize the in- and out-states to unit ingoing or outgoing particle flux in the center-of-mass coordinate system. An in-state $|\lambda, \text{in}\rangle$ may be reexpressed in terms of the out-states as in Eq. (1). The

$$|\gamma, \text{in}\rangle = \Sigma_\beta |\beta, \text{out}\rangle S_{\beta\gamma} \qquad (1)$$

matrix of probability amplitudes $S_{\beta\gamma}$, which relates all possible initial and final states, is the scattering or S matrix.

Conservation of probability in the possible reactions requires that S be a unitary matrix, that is, that $\Sigma_\gamma S_{\gamma\alpha}^* S_{\gamma\alpha} = \Sigma_\gamma S_{\alpha\gamma}^* S_{\beta\gamma}^* = S_{\alpha\beta}$, where $\delta_{\alpha\beta}$ has the value 1 if $\alpha = \beta$ and 0 if $\alpha \neq \beta$. Additional restrictions on S may be deduced if it is assumed that the interactions are invariant under Lorentz transformations, the discrete operations of reflection of the space or time axes, particle-antiparticle interchange (charge conjugation), or such internal symmetries as isotopic-spin and unitary symmetry. *See* RELATIVITY; SYMMETRY LAWS; UNITARY SYMMETRY.

The expansion of the ingoing states in terms of the outgoing states provides a clear physical picture of the results to be expected from a collision of the particles in the initial state. The scattered wave is obtained by subtracting from the final wave those components which did not interact. The resulting probability amplitude for finding the state $|\beta, \text{out}\rangle$ in the scattered wave is given by $S_{\beta\gamma} - \delta_{\beta\gamma}$. The scattering amplitude $f_{\beta\gamma}$ is obtained by multiplying the probability amplitude by the flux in the actual incident state. For a two-particle plane-wave state, this factor is $\hbar/2p$, where p is the momentum of either incident particle in the center-of-mass system. Thus Eq. (2) holds. The

$$f_{\beta\gamma} = (\hbar/2ip)(S_{\beta\gamma} - \delta_{\beta\gamma}) \qquad (2)$$

scattering cross section for a transition to a two-particle final state is then given by $4\pi|f_{\beta\gamma}|^2$.

Perturbation calculations. A Lorentz covariant perturbation theory for the direction calculation of S on the basis of quantum field theory was developed in 1948–1949 by J. Schwinger, R. P. Feynman, and F. J. Dyson. The S-matrix element $S_{\beta\gamma}$ may be written as the matrix element of time-ordered exponential operator between states $|\gamma\rangle$ and $|\beta\rangle$ of the noninteracting system as in Eq. (3). Here

$$S_{\beta\gamma} = \left\langle \beta \left| \exp\left[-\left(\frac{i}{h}\right) \int d^4 x H'(x) \right] \right| \gamma \right\rangle^+ \qquad (3)$$

$H'(x)$ is the interaction Hamiltonian, and the $+$ on the bracket indicates that the operators $H'(x)$, $H'(x')$, . . . in the Taylor series expansion of the exponential are to be arranged from left to right in order of decreasing time variables. The terms in this Taylor series can be represented by means of the diagrammatic technique introduced by Feynman (see illustration). Each vertex and leg in the diagrams is associated with the component of the matrix element corresponding to the process depicted. The set of Feynman diagrams thereby provides a convenient algorithm for the construction of S to any order in H'. *See* PERTURBATION (QUANTUM MECHANICS); QUANTUM ELECTRODYNAMICS; QUANTUM FIELD THEORY.

Dispersion theory. The covariant perturbation techniques have been remarkably successful in quantum electrodynamics; the fine structure constant is small, and provides a natural expansion parameter for the perturbation series. In contrast, the strength of the specifically nuclear forces between elementary particles prevents a meaningful perturbation expansion of S for particle reactions. Much emphasis has consequently been given in recent years to Heisenberg's original conjecture that Lorentz invariance and unitarity could be used in part to determine S. It has in fact been possible to develop at least a partial dynamical theory of the S matrix by supplementing the requirements of Lorentz invariance and unitarity by analyticity conditions derived from perturbation theory. Although this dispersion-relation, or S-matrix, theory of strong interactions has been quite successful in some instances, particularly in the study of low-energy baryon-meson scattering, the theoretical basis of the approach remains obscure. *See* DISPERSION RELATIONS. [LOYAL DURAND, III]

Bibliography: W. O. Amrein et al., *Scattering Theory in Quantum Mechanics*, 1977; G. F. Chew, *S-Matrix Theory of Strong Interactions*, 1962, reprint 1974; J. E. Farina, *Quantum Theory of Scattering Processes*, pt. 1, 1976, pt. 2, 1973; R. G. Newton, *Scattering Theory of Waves and Particles*, 1966; M. Reed and B. Simon, *Methods of Modern Mathematical Physics*, vol. 3: *Scattering Theory*, 1979; J. R. Taylor, *Scattering Theory*, 1972.

Scattering of electromagnetic radiation

The process in which energy is removed from a beam of electromagnetic radiation and reemitted with a change in direction, phase, or wavelength. All electromagnetic radiation is subject to scattering by the medium (gas, liquid, or solid) through which it passes. In the short-wavelength, high-energy regime in which electromagnetic radiation is most easily discussed by means of a particle description, these processes are termed photon scattering. At slightly longer wavelengths, the scattering of x-rays provides the most effective means of determing the structure of crystalline solids. In the visible wavelength region, scattering of light produces the blue sky, red sunsets, and white clouds. At longer wavelengths, scattering of radio waves determines their characteristics as they pass through the atmosphere. *See* LIGHT.

It has been known since the work of J. Maxwell in the 19th century that accelerating electric charges radiate energy and, conversely, that electromagnetic radiation consists of fields which accelerate charged particles. Light in the visible, infrared, or ultraviolet region interacts primarily with the electrons in gases, liquids, and solids — not the nuclei. The scattering process in these wavelength regions consists of acceleration of the electrons by the incident beam, followed by reradiation from the accelerating charges. *See* ELECTROMAGNETIC RADIATION.

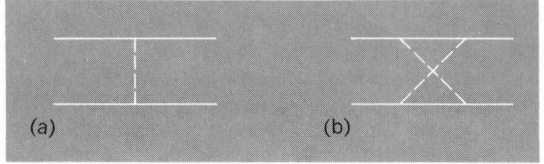

Typical Feynman diagrams for two-particle scattering. (a) Second-order diagram; (b) fourth-order diagram. Solid lines represent scattered particles; broken lines, particles which transmit force between them.

Scattering processes may be divided according to the time between the absorption of energy from the incident beam and the subsequent reradiation. True "scattering" refers only to those processes which are essentially instantaneous. Mechanisms in which there is a measurable delay between absorption and reemission are usually termed luminescence. If the delay is longer than a microsecond or so, the process may be called fluorescence; and mechanisms involving very long delays (seconds) are usually termed phosphorescence. *See* ABSORPTION OF ELECTROMAGNETIC RADIATION; FLUORESCENCE; LUMINESCENCE; PHOSPHORESCENCE.

Inelastic scattering. Instantaneous scattering processes may be further categorized according to the wavelength shifts involved. Some scattering is "elastic"; there is no wavelength change, only a phase shift. In 1928 C. V. Raman discovered the process in which light was inelastically scattered and its energy was shifted by an amount equal to the vibrational energy of a molecule or crystal. Such scattering is usually called the Raman effect. This term has been used in a more general way, however, often to describe inelastic scattering of light by spin waves in magnetic crystals, by plasma waves in semiconductors, or by such exotic excitations as "rotons," the elementary quanta of superfluid helium. *See* RAMAN EFFECT.

Brillouin and Rayleigh scattering. In liquids or gases two distinct processes generate inelastic scattering with small wavelength shifts. The first is Brillouin scattering from pressure waves. When a sound wave propagates through a medium, it produces alternate regions of high compression (high density) and low compression (or rarefaction). A picture of the density distribution in such a medium is shown in Fig. 1. The separation of the high-density regions is equal to one wavelength λ for the sound wave propagating. Brillouin scattering of light to higher (or lower) frequencies occurs because the medium is moving toward (or away from) the light source. This is an optical Doppler effect, in which the frequency of sound from a moving object is shifted up in frequency as the object moves toward the observer. *See* DOPPLER EFFECT.

The second kind of inelastic scattering studied in fluids is due to entropy and temperature fluctuations. In contrast to the pressure fluctuations due to sound waves, these entropy fluctuations do not generate scattering at sharp, well-defined wavelength shifts from the exciting wavelength; rather, they produce a broadening in the scattered radiation centered about the exciting wavelength. This is because entropy fluctuations in a normal fluid are not propagating and do not, therefore, have a characteristic frequency; unlike sound they do not move like waves through a liquid—instead, they diffuse. Under rather special circumstances, however, these entropy or temperature fluctuations can propagate; this has been observed in superfluid helium and in crystalline sodium fluoride at low temperatures, and is called second sound. *See* ENTROPY.

Scattering from entropy or temperature fluctuations is called Rayleigh scattering. In solids this process is obscured by scattering from defects and impurities. Under the assumption that the scattering in fluids is from particles much smaller than the wavelength of the exciting light, Lord Rayleigh derived in 1871 an equation, shown below, for such

$$r^2\,I(\theta)/I_0 = \pi\,d\,\lambda^{-4}v^2(1+\cos^2\theta)\,(n-1)^2$$

scattering; here $I(\theta)$ is the intensity of light scattered from an incident beam of wavelength λ and intensity I_0 at a distance r; d is the number of scattering particles; v is the volume of the disturbing particle; and n is the index of refraction of the fluid. The $\cos\theta$ term is present for unpolarized incident light, where θ is the scattering angle. Measurement of the ratio in Rayleigh's equation allows the determination of either Avogadro's number N or the molecular weight M of the fluid, if the other is known. The dependence of scattering intensity upon the inverse fourth power of the wavelength given in Rayleigh's equation is responsible for the fact that daytime sky looks blue and sunsets red: blue light is scattered out of the sunlight by the air molecules more strongly than red; at sunset, more red light passes directly to the eyes without being scattered.

Large particles. Rayleigh's derivation of his scattering equation relies on the assumption of small, independent particles. Under some circumstances of interest, both of these assumptions fail. Colloidal suspensions provide systems in which the scattering particles are comparable to or larger than the exciting wavelengths. Such scattering is called the Tyndall effect and results in a nearly wavelength-independent (that is, white) scattering spectrum. The Tyndall effect is the reason clouds are white (the water droplets become larger than the wavelengths of visible light). *See* TYNDALL EFFECT.

The breakdown of Rayleigh's second assumption—that of independent particles—occurs in all liquids. There is strong correlation between the motion of neighboring particles. This leads to fixed phase relations and destructive interference for most of the scattered light. The remaining scattering arises from fluctuations in particle density discussed above and was first analyzed theoretically by A. Einstein in 1910 and by M. Smoluchowski in 1908.

Rayleigh's basic theory has been extended by several authors. Rayleigh in 1911 and R. Gans in 1925 derived scattering formulas appropriate for spheres of finite size, and in 1947 P. Debye extended the theory to include random coil polymers. The combined results of these three workers, generally called the Rayleigh-Gans-Debye theory, is valid for any size of particle, provided the refractive index n

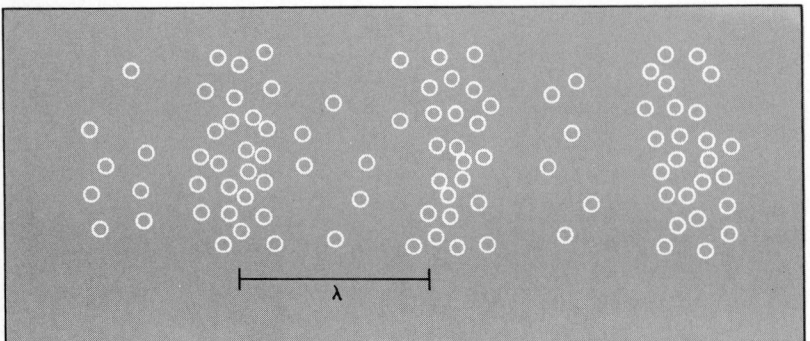

Fig. 1. Pressure-induced density fluctuations in a fluid.

is near unity; whereas the Rayleigh theory is valid for any index, provided the particles are very small. The index *n* of any medium would be unity if there were no scattering.

A more complete theory than Rayleigh-Gans-Debye was actually developed earlier by G. Mie in 1908; however, Mie's theory generally requires numerical solution. Mie's theory is valid even for particles larger than the wavelength of light used. For such particles, large phase shifts occur for the scattered light. Mie scattering exhibits several maxima and minima as a function of scattering angle; the positions of these maxima depend upon particle size, as indicated in Fig. 2. These secondary maxima are essentially higher-order Tyndall scattering.

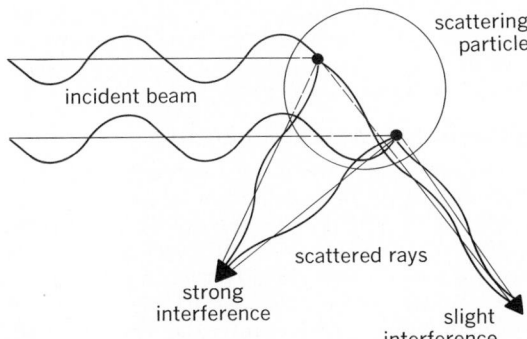

Fig. 2. The role of interference in determining the angular dependence of light scattering from particles that are comparable in size to the wavelength.

Critical opalescence. The most striking example of light scattering is that of critical opalescence, discovered by T. Andrews in 1869. As the critical temperature of a fluid is approached along the critical isochore, fluctuations in the density of the medium become larger than a wavelength and stay together for longer times. The independent particle approximation becomes very bad, and the intensity of light scattered in all directions becomes very large. The fluid, which might have been highly transparent 1°C higher in temperature, suddenly scatters practically all of the light incident upon it. The regions of liquidlike density in the fluid have molecules moving in phase, or "coherently," within them. Such coherent regions in fluids near their critical temperatures behave very much like the large spherical particles diagrammed in Fig. 2, thus establishing the relationship between Mie scattering of colloids and critical opalescence. The theory of scattering from correlated particles was developed primarily by L. S. Ornstein and F. Zernike in 1914–1926. *See* COHERENCE; CRITICAL PHENOMENA.

Nonlinear scattering. At the power densities available with pulsed lasers, a variety of other scattering mechanisms can become important. (For nonlaser excitation these processes exist but are generally too weak to be detected.) Such processes include second-harmonic generation, in which the energy from two quanta (photons) in the incident beam produces one doubly energetic photon. A practical example is the production of green light from the red light of a ruby laser. Third-harmonic

generation (and similar higher-order processes) also occurs as an intense scattering mechanism at sufficiently high power densities. Whereas second-harmonic generation can occur only in certain crystals, for reasons of symmetry, third-harmonic generation occurs in any medium. Other more exotic nonlinear scattering mechanisms include sum generation and difference generation, in which two beams of light are combined to produce scattering having frequencies which are the sum or difference of the frequencies in the two incident beams, and rather weak processes such as the hyper-Raman effect, in which the scattered light has a frequency which is twice that of the incident beam plus (or minus) some vibrational energy. All these effects are called nonlinear because their intensities vary according to the square or higher power of the incident intensity. These nonlinear processes may be very important in astrophysics (for example, in stellar interiors), in addition to laboratory laser experiments. *See* NONLINEAR OPTICS.

[J. F. SCOTT]

Bibliography: W. Hayes and R. Loudon, *Scattering of Light by Crystals*, 1978; B. J. Berne and R. Pecora, *Dynamic Light Scattering*, 1976.

Schlieren photography

An optical technique that detects density gradients occurring in a gas flow. The schlieren system is used particularly in supersonic wind tunnels because it clearly shows the density gradients created by the shock and expansion waves of the airflow around the wind tunnel model.

A simple schlieren system operates as shown in the figure. A source of light is shielded so that only a small rectangular slit emits light. A lens is placed at its focal distance from the slit so that the light is bent into a parallel beam. A second lens collects the parallel beam into an image of the slit and forms an inverted image on the screen or photographic plate. If a knife-edge is moved into the light stream near the slit image, the image at the screen darkens uniformly. Consider the system just described to be oriented so that the parallel light beam crosses a wind tunnel section. A light ray *a*, bent from the parallel path by density gradients in the test section, cannot be brought to focus at the slit image and is interrupted by the knife-edge so that a dark spot occurs at *a'* on the screen. Light ray *b*, deflected the opposite way by a different density gradient, escapes the knife-edge and appears as a light spot at *b'* on the screen. Thus a picture of the density gradients appears on the screen. Sometimes light intensity charts can be correlated with numerical values of

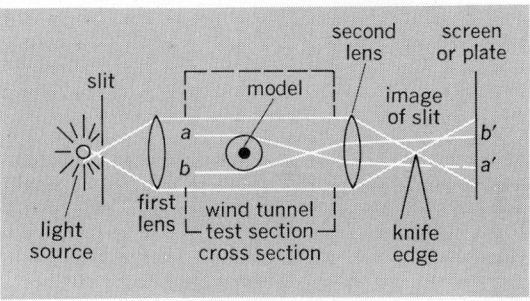

Basic schlieren optical system.

the density gradients. However, numerical values of density can be obtained only from schlieren pictures of airflow about two-dimensional models or about simple axisymmetric models.

For high-speed photographs (3000–4000 frames per second) of unsteady flows, dc light sources are used. Microsecond flash durations from high-intensity gas sources are used in the study of turbulences. To better study specific regions of a flow pattern, the knife-edge is rotated to different positions and moved up, down, left, or right. For sharp images at the center of a test section, a multiple-slot or focusing schlieren system is used.

[D. PHILIP ANKENEY]

Schottky anomaly

A contribution to the heat capacity of a solid arising from the thermal population of discrete energy levels as the temperature is raised. The effect is particularly prominent at low temperatures, where other contributions to the heat capacity are generally small. *See* SPECIFIC HEAT OF SOLIDS.

Discrete energy levels may arise from a variety of causes, including the removal of orbital or spin degeneracy by magnetic fields, crystalline electric fields, and spin orbit coupling, or from the magnetic hyperfine interaction. Such effects commonly occur in paramagnetic ions.

The thermal population of an energy level E_i is proportional to the Boltzmann factor, $\exp(-E_i/k_B T)$, where T is the temperature and k_B is Boltzmann's constant ($k_B = 1.38 \times 10^{-23}$ J K^{-1}). The largest contributions to the Schottky heat capacity, C_{Schottky}, will therefore occur at temperatures where $k_B T$ is comparable with the E_i. Quantitative estimates of C_{Schottky} may be made in terms of the Boltzmann factors. For an assembly of N, identical noninteracting systems, each with n low-lying energy levels E_i ($i = 1 \ldots n$), the total internal energy is given by the equation below. The corresponding heat ca-

$$U = \frac{N \sum_{i=1}^{n} E_i \exp(-E_i/k_B T)}{\sum_{i=1}^{n} \exp(-E_i/k_B T)}$$

pacity is then given by $C_{\text{Schottky}} = dU/dT$. The simplest example is the case $n = 2$, corresponding to a single splitting ϵ between the ground state ($i = 1$) and the first excited state ($i = 2$). The illustration shows the heat capacity for this case plotted as a

function of $k_B T/\epsilon$. For more complex energy-level schemes, the curves are generally similar, with a single broad peak near $k_B T/\bar{\epsilon} \sim 1$, where $\bar{\epsilon}$ is the mean energy above the ground state. However, more structured curves, with more than one peak, are also possible if the energy-level scheme includes excitation energies of very different relative magnitudes. *See* BOLTZMANN STATISTICS.

Corresponding to the Schottky heat capacity, there is a contribution to the entropy, $S = \int (C_{\text{Schottky}}/T)dT$. This can act as a barrier to the attainment of low temperatures if the substance is to be cooled either by adiabatic demagnetization or by contact with another cooled substance. Conversely, a substance with a Schottky anomaly can be used as a heat sink in experiments at low temperatures (generally below 1 K) to reduce temperature changes resulting from the influx or generation of heat. *See* ADIABATIC DEMAGNETIZATION; LOW-TEMPERATURE PHYSICS.   [W. P. WOLF]

Bibliography: H. M. Rosenberg, *Low Temperature Solid State Physics*, 1963.

Schrödinger's wave equation

The differential equation of quantum mechanics (first proposed by Erwin Schrödinger in 1926) whose solution determines the average result, also termed expectation value, of every conceivable experiment on the physical system under examination. When solved, the Schrödinger equation yields the wave function; from the wave function, expectation values are computed. The term wave equation comes from the resemblance of the Schrödinger equation to those differential equations describing acoustic and electromagnetic waves. Also, its consequences and mode of derivation are consistent with the tenet that electrons and other particulate constituents of matter have wavelike properties. *See* NONRELATIVISTIC QUANTUM THEORY; QUANTUM MECHANICS; WAVE EQUATION; WAVE MOTION.

[EDWARD GERJUOY]

Scintillation counter

A particle or radiation detector which operates through emission of light flashes that are detected by a photosensitive device, usually a photomultiplier. The scintillation counter not only can detect the presence of a particle, gamma ray, or x-ray, but can measure the energy, or the energy loss, of the particle or radiation in the scintillating medium. The sensitive medium may be solid, liquid, or gaseous, but is usually one of the first two. The scintillation counter is one of the most versatile particle detectors, and is widely used in industry, scientific research, and radiation monitoring, as well as in exploration for petroleum and radioactive minerals that emit gamma rays. Many low-level radioactivity measurements are made with scintillation counters.

The scintillation phenomenon was used by E. Rutherford and colleagues, who observed the light flashes from screens of powdered zinc sulfide struck by alpha particles. The scattering of alpha particles by various foils was used by Rutherford to establish the modern notion of the nuclear atom. In modern times photomultipliers, which feed into amplifiers, have replaced the eye and have also proved to be very efficient in measuring the time of arrival of a particle, as well as its energy or

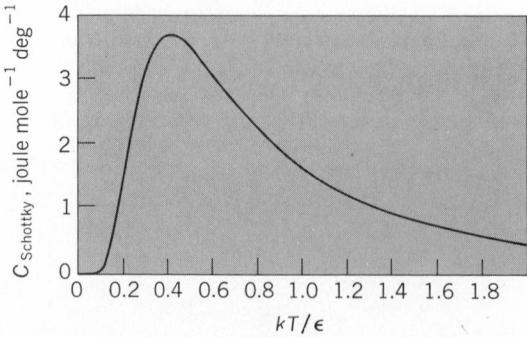

Theoretical curve for the Schottky heat capacity when there are two energy levels separated by a splitting ϵ. (*From H. M. Rosenberg, Low Temperature Solid State Physics, Claredon Press, 1963*)

energy loss. Scintillations are light pulses emitted by atoms which return to their ground states after having been raised to excited states by passing particles. Scintillations are one example of luminescent behavior, and are related to the process called fluorescence. *See* FLUORESCENCE; LUMINESCENCE.

Operation of bulk counter. Following the innovation in 1947 of using bulk material, instead of screens of powdered luminescent materials, scintillation counters have been made of transparent crystalline materials or liquids or plastics. In order to be an efficient detector, the bulk scintillating medium must be transparent to its own luminescent radiation, and since some detectors are quite extensive, covering meters in length, the transparency must be of a high order. One face of the scintillator is placed in optical contact with the photosensitive surface of the photomultiplier, as shown in the illustration. In order to direct as much as possible of the light flash to the photosensitive surface, reflecting material is placed between the scintillator and the inside surface of the container.

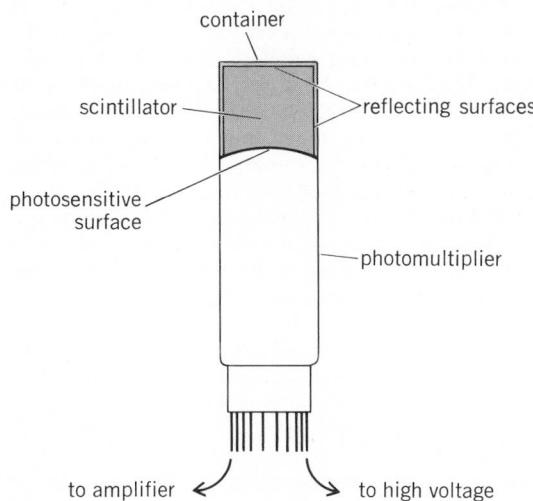

Diagram of a scintillation counter.

In many cases it is necessary to collect the light from a large area and transmit it to the small surface of a photomultiplier. In this case, a "light pipe" leads the light signal from the scintillator surface to the photomultiplier with only small loss. The best light guides and light fibers are made of glass, plastic, or quartz. It is also possible to use lenses and mirrors in conjunction with scintillators and photomultipliers. *See* OPTICAL FIBERS.

A charged particle, moving through the scintillator, leaves a trail of excited atoms which emit the characteristic luminescence of that particular material. When a particle stops in the material, all its energy may be lost in the scintillator, and therefore, after calibration with known sources, the particle's energy can be measured. When a particle passes through a scintillator, the energy loss of the particle is measured. When a gamma ray converts to charged particles in a scintillator, its energy may also be determined. When the scintillator is made of dense material and of very large

dimensions, the entire energy of a very energetic particle or gamma ray may be contained within the scintillator, and again the original energy may be measured. Such is the case for energetic electrons, positrons, or gamma rays which produce electromagnetic showers in the scintillator. When the sizes of the light flashes are measured in the photomultipliers, the results are recorded in pulse-height analyzers, and then one can readily determine the energy spectra of particles in these various cases.

Characteristics. Scintillation counters have several characteristics which make them particularly useful as detectors of x-rays, gamma rays, and other nuclear and high-energy particles.

Efficiency and size. Counting efficiencies close to 100% are often not difficult to achieve, and these detectors can often be made quite thin and small and can then define a particle's position rather accurately. Arrays of small crystals can be used, and "light pipes" can be employed to bring the luminous radiation some distance away to more bulky photomultipliers. When detection of very energetic particles is desired, the scintillator can be made very large and massive, and often requires the use of many photomultipliers "looking" at the scintillator through transparent windows.

Speed. Useful scintillation materials emit light flashes as short as 10^{-9} s, although most inorganic materials have light flashes 10 to 100 times longer. Even so, the flash is very rapid. High speed allows very fast counting and avoids pileup, which is very characteristic of older and slower particle detectors such as the Geiger counter, which has been largely superseded. The Cerenkov counter is similar to a scintillation counter, but depends only on the index of refraction of the transparent medium and the particles' velocity in the medium, and not specifically on the crystalline luminescent processes. *See* CERENKOV RADIATION.

Energy resolution. While the energy resolution is quite good in scintillation counters, their performance in this respect is markedly inferior to that of semiconductor detectors such as silicon, germanium, or lithium-drifted germanium at low temperatures. Semiconductor counters have been rather limited in size, so that they do not usually compete with the large inorganic scintillators needed in high-energy research. Often anticoincidence combinations are made between semiconductor counters and scintillation counters. *See* JUNCTION DETECTOR.

Substances used. Scintillation counter materials are generally classified into inorganic and organic types.

Inorganic scintillators. Inorganic scintillators are generally characterized by the presence of heavy elements. The most useful inorganic scintillator is sodium iodide activated with a small amount of thallium salt. The usual designation of this material is NaI(Tl). Cesium iodide is also useful, and may be activated with thallium or sodium. Other useful crystalline scintillators include calcium fluoride, barium fluoride, and cadmium tungstate. Bismuth germanate is a new and interesting scintillator. NaI(Tl) has a decay time of 2.5×10^{-7} s, and other alkali halide materials are similar. NaI(Tl) is particularly useful for detecting gamma rays because of the presence of iodine, which is a relatively heavy atom of high

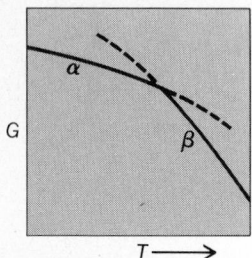

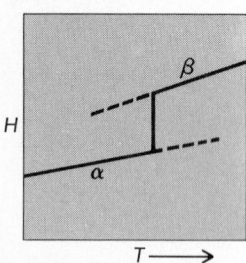

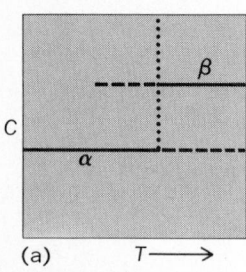

(a)

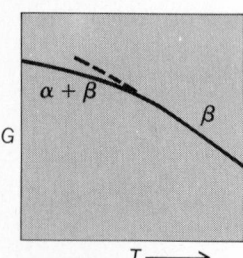

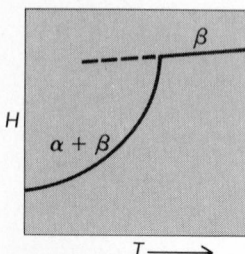

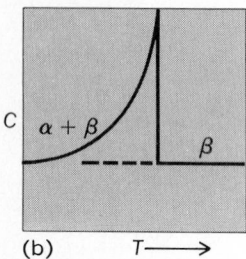

(b)

Diagrams of typical
thermodynamic behavior
at (a) first-order and (b)
second-order transitions.

atomic number (53). The high atomic number is important in both the photoelectric and pair production processes which result from the interaction of a gamma ray with the base material.

Organic scintillators. Materials of this type are typically naphthalene, anthracene, trans-stilbene, terphenyl, $C_{24}H_{16}N_2O_2$, and similar materials. Very successful liquid scintillators have been made by dissolving terphenyl in an organic liquid such as xylene, or in a polymeric material to make a solid plastic scintillator. Organic scintillators find special application when short decay times of the scintillation flashes are required. Scintillation decay times of 10^{-9} s are not unusual, and therefore very high counting rates, avoiding pile-up, may be achieved. Since organic materials contain hydrogen, they are also very useful in detecting neutrons. Neutrons collide with protons (hydrogen nuclei), thereby transferring their kinetic energy to charged particles. Organic scintillators are not as useful as inorganic scintillators for the detection of gamma rays, electrons, or positrons. Large volumes of liquid scintillators have been used to detect the elusive neutrinos. *See* LIQUID SCINTILLATION DETECTOR; NEUTRINO.

Coincidence counting. Scintillation counters are often used in temporal coincidence with other particle or radiation detectors. The coincidence technique eliminates or reduces the problem of local background or false events. For large counting systems recording complicated events, such as the "NaI(Tl) crystal ball" used in high-energy physics, the coincidence technique is used to make a fast trigger for desirable events and to exclude false or uninteresting events. It is also possible to use fast organic scintillators separated by a given distance to measure the time of flight (delay in time between the two pulses) of particles traversing a common pair. In this way the speed of a particle may be measured. Speeds approaching the velocity of light can easily be measured by a fast scintillator pair. Also radioactive decay times, or unstable particle decays, such as those of pions or muons, may be studied in the same scintillator by observing the incident particle pulse and later the pulse produced by the new particle or particles resulting from the decay.

Applications. Fundamental discoveries in physics and astrophysics have been made with scintillators. In particular, the NaI(Tl) detector has been used in nuclear physics, elementary particle physics, atomic physics, solid-state studies, chemistry, nutrition studies, biology, nuclear medicine, astrophysics and space physics, geology, nuclear reaction technology and monitoring, petroleum and mineral resources exploration, and oceanography. *See* PARTICLE DETECTOR.

[ROBERT HOFSTADTER]

Bibliography: J. B. Birks, *The Theory and Practice of Scintillation Counting*, 1964; F. D. Brooks, Development of organic scintillators, *Nucl. Sci.*, pp. 477–505, 1979; R. L. Heath, R. Hofstadter, and E. B. Hughes, Inorganic scintillators in detectors, *Nucl. Sci.*, pp. 431–476, 1979.

Second-order transition

A change of state through which the free energy of a substance and its first derivatives are continuous functions of temperature and pressure, but at which the second derivatives are discontinuous.

For all physical and chemical processes carried out reversibly, the free energy changes continuously. At an ordinary phase transition, such as the boiling of a liquid, the entropy S, enthalpy H, and volume V show sharp discontinuities when plotted as functions of the temperature T or pressure P. Because all these functions are first derivatives of the free energy, as shown in notation (1), such

$$S = -(\partial G/\partial T)_P$$
$$H = [\partial(G/T)/\partial(1/T)]_P \qquad (1)$$
$$V = (\partial G/\partial P)_T$$

phase changes are usually called first-order transitions.

However, for many systems there are points at which the entropy, enthalpy, and volume are continuous, but at which temperature or pressure derivatives, such as the heat capacity $C_p = (\partial H/\partial T)_P$, the coefficient of thermal expansion $\alpha = (\partial \ln V/\partial T)_P$, and the isothermal compressibility $\kappa = (\partial \ln V/\partial P)_T$, show discontinuities. Because these correspond to second derivatives of the free energy, this phenomenon is called a second-order transition.

The illustration shows the typical thermodynamic behavior at first- and second-order transitions. The dotted vertical line for the heat capacity at the first-order transition is a zero-width line of infinite height, representing a finite nonzero area (a Dirac delta function): the heat of transition, absorbed at a single temperature. The dashed lines show metastable phases continued beyond the transition temperature (for example, superheated liquid above the boiling point and supercooled vapor below). Both the low-temperature (α) and high-temperature (β) phases show such extensions beyond a first-order transition, whereas only the β phase shows such an extension at a second-order transition.

Qualitatively all theories of second-order transitions have the following features in common: A system is capable of existing in two forms, one (α) having a lower enthalpy H and a lower entropy S than the other (β). At sufficiently low temperature the enthalpy difference will be the dominant factor and the system will be all α, whereas at sufficiently high temperatures it will be largely or entirely β. If the conditions were such that the α and β forms could not coexist, there would be a first-order transition at the temperature shown by Eq. (2).

$$T = (H_\beta - H_\alpha)/(S_\beta - S_\alpha) \qquad (2)$$

On the other hand, if the change from α to β can take place gradually (that is, if a mixed phase including both forms can exist) and if the energy required to convert an element of the system (a molecule or group of molecules) from α to β decreases as the amount of β increases, a second-order transition will occur. This changeover from α to β in a sense catalyzes itself, so one refers to phenomena of this kind as cooperative. The temperature at which the last trace of α disappears is the λ-point or Curie point; it is, of course, meaningless to extrapolate the $\alpha + \beta$ curve beyond this point. The heat capacity is very large at the λ-point; it is difficult to be sure whether in some systems it may not actually become infinite; in any case the area under the curve (ΔH) is finite. Important examples of second-order transitions are given in the following paragraphs.

Ferromagnetism. In certain metals and alloys (iron and nickel) at low temperatures, the atomic magnets are arranged into ordered groups or domains which can orient in a magnetic field. As the temperature increases, the order within the domains decreases until, at the Curie temperature, all long-range order is gone and only paramagnetic behavior remains. *See* FERROMAGNETISM.

Order-disorder in crystals. In certain solid solutions (such as β-brass, Cu-Zn), the different atoms are distributed regularly in an alternating arrangement. As the temperature increases, the two kinds of atoms exchange positions until all long-range order is lost at the Curie point, above which the arrangement is essentially random. A similar phenomenon occurs in the solid ammonium halides; each $NH_4{}^+$ tetrahedron can have two different orientations. At low temperatures, all have the same orientation; above the Curie temperature, they are distributed randomly between the two.

Liquid helium. Below 2.19 K, helium shows peculiar superfluid properties. At the lowest attainable temperature, all the molecules are in a superfluid state; as the temperature increases, more and more molecules are excited to nonsuperfluid levels until, at the λ-point (2.19 K), the superfluid properties have disappeared and the helium is an ordinary liquid. *See* SUPERFLUIDITY.

[ROBERT L. SCOTT]

Bibliography: J. D. Fast, *Thermodynamics and Phase Relations*, vol. 1, 1965; H. K. Henisch and R. Roy, *Phase Transitions and Their Applications in Materials Science*, 1974; L. D. Landau and E. M. Lifschitz, *Statistical Physics*, 2d ed., 1969; C. N. Rao and K. J. Rao, *Phase Transition in Solids: An Approach to the Study of Physics and Chemistry of Solids*, 1978.

Second sound

A type of wave propagated in the superfluid phase of liquid helium (helium II) and in certain other substances under special conditions. Predicted independently by L. Tisza in 1938 and L. Landau in 1941, such waves were observed first by V. Peshkov in 1944 and have subsequently provided a rich source of information for the study and understanding of the superfluid state. The name is misleading since second sound is not in any sense a sound wave, but a temperature or entropy wave. In ordinary or first sound, pressure and density variations propagate with very small accompanying variations in temperature; in second sound, temperature variations propagate with no appreciable variation in density or pressure.

Two-fluid model. The two-fluid model of helium II provides further insight into the nature of second sound. In this model the liquid can be described as consisting of superfluid and normal components of densities ρ_s and ρ_n, respectively, such that the total density $\rho = \rho_s + \rho_n$. The superfluid component is frictionless and devoid of entropy; the normal component has a normal viscosity and contains the entropy and thermal energy of the system. As the temperature goes from 0 K to T_λ (the superfluid transition temperature), ρ_s goes from ρ to 0, and ρ_n from 0 to ρ. In a temperature or second-sound wave, the normal and superfluid flows are oppositely directed so that $\rho_s \mathbf{V}_s + \rho_n \mathbf{V}_n = 0$, where $\mathbf{V}_s$ and $\mathbf{V}_n$ are the superfluid and normal flow velocities. Thus a variation in relative densities of the two components, and hence a temperature fluctuation, propagates with no change in total density or pressure. In a first-sound wave, the two components move in phase, that is, $\mathbf{V}_n \cong \mathbf{V}_s$.

Second sound can be observed by methods similar to those used with first sound except that a heater and thermometer replace the sound transmitter and receiver. Thus an electrical resistor in the form of a thin metallic film or wire is the most common type of transducer for second sound.

Velocity. The velocity of second sound goes from 0 at T_λ to a value approaching that of first sound at 0 K as seen in the figure. This is in excel-

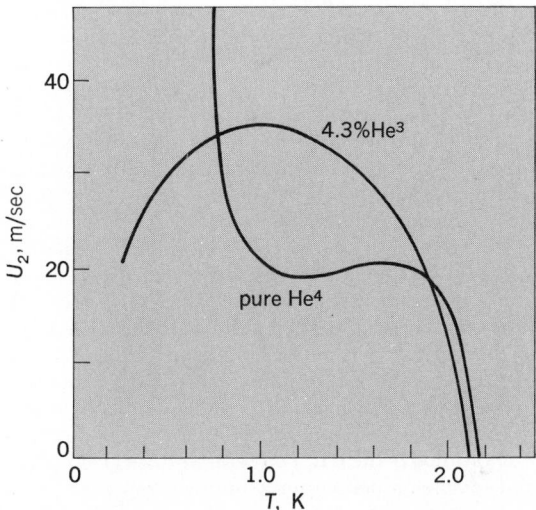

Velocity of second sound as a function of temperature for pure ^{4}He and a 4.3% ^{3}He mixture. (*J. C. King and H. A. Fairbank, Second sound in He³-He⁴ mixtures below 1 K, Phys. Rev., 93:21, 1954*)

lent agreement with the theoretical relation in Eq. (1) deduced from the two-fluid model, where T is

$$u_2{}^2 = \frac{\rho_s}{\rho_n} \frac{T}{C} S^2 \tag{1}$$

the temperature, C the specific heat, and S the entropy of the liquid. At temperatures below about 0.5 K, Landau showed that the only excitations present in superfluid helium are longitudinal phonons (quantized sound waves), and hence heat pulses would be propagated as a fluctuation in the density of these phonons. As the temperature approaches 0, Eq. (1) for second-sound velocity reduces to $u_2 = u_1 \sqrt{3}$, where u_1 is the phonon or first-sound velocity. Experimentally this value of velocity is approached at about 0.5 K, but another effect interferes—the interactions between phonons is very weak and the phonon mean free path goes rapidly toward infinity as the temperature and the density of phonons are reduced. Thus, below 0.5 K second sound is no longer observed in the true sense but the phonons propagate ballistically, interacting only with the walls of the container. This behavior is indeed found, as evidenced by large spreading of temperature pulses and the arrival of the leading edge with the phonon velocity.

In a variety of experiments second sound displays the properties expected in wave phenomena such as diffraction, interference, and shock effects for large amplitude pulses.

³He-⁴He mixtures. Even a small addition of the lighter isotope helium-3 significantly affects many of the superfluid properties including second sound, especially in the very low temperature range. At the lowest temperatures where the ³He excitations dominate the entropy and specific heat, the second-sound velocity is proportional to the square root of the temperature as expected for first sound in an ideal gas. Study of the second sound in this region provides an excellent means of measuring the details of the ³He interactions.

Liquid ³He. Liquid ³He is a superfluid only at very low temperatures (below about 0.002 K). Second sound should exist in the principal phases (A and B) of this superfluid, but the expected velocity is quite small and the attenuation so large as to make direct detection unlikely. However, a new condensed form of second-sound wave and spin density wave can be propagated in a restricted narrow region of the ³He phase diagram, the A_1 phase. This phase lies between the normal phase and the superfluid A phase and exists only in a strong magnetic field.

The superfluid component of the A_1 phase is apparently made up of pairs of ³He atoms with their nuclear magnetic moments lined up parallel to the strong applied magnetic field. Counterflow of the normal and superfluid components, therefore, involves both the temperature and spin density changes in this normal second-sound wave/spin density wave combination. The velocity is directly proportional to the reduced temperature $(1 - T/T_{A_1})$ and reaches a maximum value of about 1.4 m/s at the lower temperature limit of the A_1 phase. Here T_{A_1} is the transition temperature between the A_1 phase and the normal phase. *See* LIQUID HELIUM; SUPERFLUIDITY.

Solid dielectric crystals. Theoretical predictions that second sound should exist in certain solid dielectric crystals under suitable conditions have been confirmed experimentally for solid helium single crystals at temperatures between 0.4 and 1.0 K. In a dielectric solid, heat pulses are carried only by phonons, as for liquid helium II, at very low temperatures. If resistive phonon scattering processes dominate, such as phonon-phonon umklapp processes, scattering by isotope or chemical impurities, or scattering by boundaries or imperfections in the crystal, a first-order diffusive flow equation would govern heat or temperature propagation. However, if the mean free path for such processes is large and the mean free path for nondissipative phonon-phonon normal processes is small, the second-order wave equation is obeyed, and heat pulses are propagated as second sound. That second sound has not yet been seen in most other dielectric crystals is probably due to the diffusive scattering of the phonons by impurities. However, evidence of second-sound propagation has also been found in pure sodium fluoride (NaF) crystals, and fully developed second sound has been found in pure semimetal bismuth crystals. Second sound has provided a good method of measuring directly the mean free path for phonon normal scattering processes. *See* LOW-TEMPERATURE ACOUSTICS.

Liquid crystals. Another quite different class of materials can exhibit second sound. In smectic A liquid crystals, when the wave vector is oblique with respect to the layers of these ordered structures, a modulation of the interlayer spacing can propagate at nearly constant density. Detailed studies of the excitation and detection of such second-sound waves have yet to be made. It is likely that second sound will continue to be a valuable probe for studying the properties of interesting ordered materials. [HENRY A. FAIRBANK]

Bibliography: K. H. Bennemann and J. B. Ketterson, *The Physics of Liquid and Solid Helium*, 1976; L. R. Corruccini and D. D. Osheroff, Observation of second sound in ³He-A_1, *Phys. Rev. Lett.*, 45:2029–2032, 1980; J. Wilks, *The Properties of Liquid and Solid Helium*, 1967.

Secondary emission

The emission of electrons from the surface of a solid into vacuum caused by bombardment with charged particles, in particular with electrons. The mechanism of secondary emission under ion bombardment is quite different from that under electron bombardment; the discussion here is limited to the latter case because it is in this sense that the term secondary emission is generally used.

The bombarding electrons and the emitted electrons are referred to, respectively, as primaries and secondaries. Secondary emission has important practical applications because the secondary yield, that is, the number of secondaries emitted per incident primary, may exceed unity. Thus, secondary emitters are used in electron multipliers, especially in photomultipliers, and in other electronic devices such as television pickup tubes, storage tubes for electronic computers, and so on.

Secondary yield. The most thoroughly investigated property of secondary emission is the yield as a function of the energy of the primaries. The

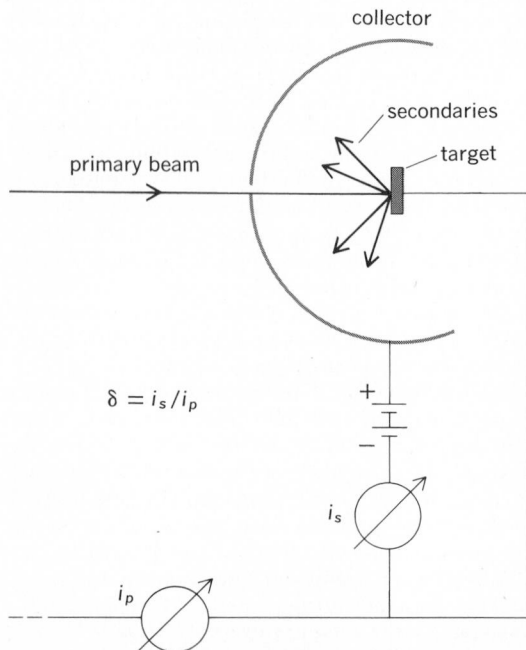

Fig. 1. Schematic circuit for measuring secondary yield; i_p and i_s represent the primary and secondary currents.

yield may be measured by means of the circuit shown schematically in Fig. 1. A beam of primary electrons strikes a target with an energy determined by the potential difference between the target and the cathode. The primary beam passes through a hole in the collector, which has been made positive with respect to the target. The secondaries emitted by the target then flow to the collector, and the yield is obtained as the ratio of the secondary current i_s to the primary current i_p.

Mechanism of the process. The emission of secondary electrons can be described as the result of three processes: (1) excitation of electrons in the solid into high-energy states by the impact of high-energy primary electrons, (2) transport of these secondary electrons to the solid-vacuum interface, and (3) escape of the electrons over the surface barrier into the vacuum. The efficiency of each of these three processes, and hence the magnitude of the secondary emission yield δ, varies greatly for different materials.

Taking into account the material characteristics that, in addition to the value of the primary energy E_p determine the yield δ, one arrives at the equation below for the dependence of δ upon E_p,

$$\delta = \frac{B_1 B_2 E_p}{\epsilon R}\left(1 - e^{-R/L}\right)$$

where ϵ is the energy required to produce a secondary electron; R is the range of primary electrons; B_1 is the coefficient, taking into account that only a fraction of the excited electrons diffuse toward the surface; L is the mean free path of the secondary electron; and B_2 is the probability that an electron reaching the solid-vacuum interface can escape over the surface barrier. Here ϵ and R are associated with process 1 above, B_1 and L with process 2, and B_2 with process 3. Without giving a detailed derivation of the equation above, it is qualitatively plausible that δ increases with increasing B_1, B_2, E_p, and L and decreases with increasing ϵ and R.

On the basis of the equation, a universal curve can be derived (Fig. 2) in which $\delta/\delta_{\max}$ is plotted versus $E_p/E_{p\max}$, where $\delta_{\max}$ is the maximum yield and $E_{p\max}$ is the corresponding primary energy. Whereas curves for the absolute values of δ versus E_p vary over a wide range for different materials, experiments have generally confirmed the validity of the curve in Fig. 2. The peak in the curve can be interpreted as follows: With increasing E_p, the number of secondaries produced within the solid increases, but at the same time the primaries penetrate to a greater depth in the material. Because of energy loss processes, the "escape depth" of the secondaries has a finite value which is determined by some of the parameters entering the equation above. Thus the peak of the curve represents the point beyond which the number of secondaries produced at a depth greater than the escape depth exceeds the number of additional secondaries produced due to the higher primary energy. Because the escape depth is predominantly determined by L and B_2 in the equation, the variations in these two parameters are the main reasons why the δ values for different materials vary over such a wide range.

Experimental yield curves. In a discussion of measured δ versus E_p curves, it is useful to con-

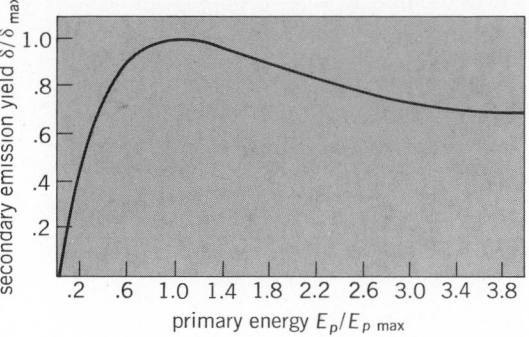

Fig. 2. Theoretical curve of secondary emission yield as a function of primary energy in normalized coordinates.

sider metals and semiconductors (or insulators) separately. In metals the secondaries lose their energy rapidly by electron-electron scattering. As a result, L and B_2 in the equation are small, and the escape depth is of the order of at most nanometers. Hence $\delta_{\max}$, and consequently $E_{p\max}$, have low values, typically well below 2.

For semiconductors and insulators, the situation is more complicated and is best understood in terms of energy-band models. Referring to Fig. 3, the highest δ values that can be obtained depend on the relative position of the top of the valence band (where the secondary electrons originate), the bottom of the conduction band, and the vacuum level. Three typical band models are shown in Fig. 3. The model shown in Fig. 3a is characterized by a small ratio of band-gap energy E_G to electron affinity E_A. In the model shown in Fig. 3b the E_G to E_A ratio is large. In the model shown in Fig. 3c the bands are bent downward to such an extent that the vacuum level lies below the bottom of the conduction band in the bulk. A material with this characteristic is said to have negative effective electron affinity. This concept is also of great importance in photoelectric emission. The differences in secondary emission yields associated with each of the three band models can be qualitatively summarized as follows.

$E_G \ll E_A$ *model.* Secondary electrons excited from the valence band to levels above the vacuum level tend to lose their energy by exciting additional electrons from the valence band into the

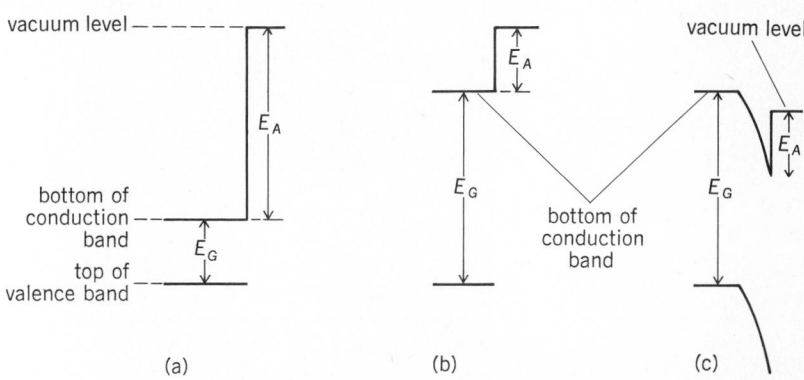

Fig. 3.   Typical energy band models for semiconductors (or insulators). (a) $E_G \ll E_A$ (b) $E_G \gg E_A$ (c) Negative electron affinity.

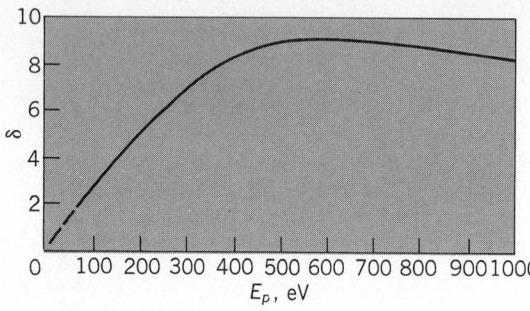

Fig. 4. Secondary emission yield versus primary energy for MgO.

conduction band and thus to arrive at the solid-vacuum interface with insufficient energy to overcome the surface barrier. In other words, the escape depth is very small, and the maximum δ values are below 2, similar to those of metals. Examples of this model are germanium and silicon.

$E_G \gg E_A$ *model.* Whereas secondary electrons excited from the valence band gradually lose energy by phonon-phonon scattering, an appreciable number of secondaries reach the solid-vacuum interface with sufficient energy to overcome the surface barrier. In other words, the escape depth is larger than in the case where $E_G \ll E_A$, of the order of tens of nanometers, and maximum δ values in the 8–15 range are typically obtained. Most of the materials used in practical devices fall into this category. Examples are MgO (see Fig. 4), BeO, Cs_3Sb (cesium antimonide), and KCl.

Negative effective electron affinity. Here the vacuum level is below the bottom of the conduction band. This case differs drastically from that

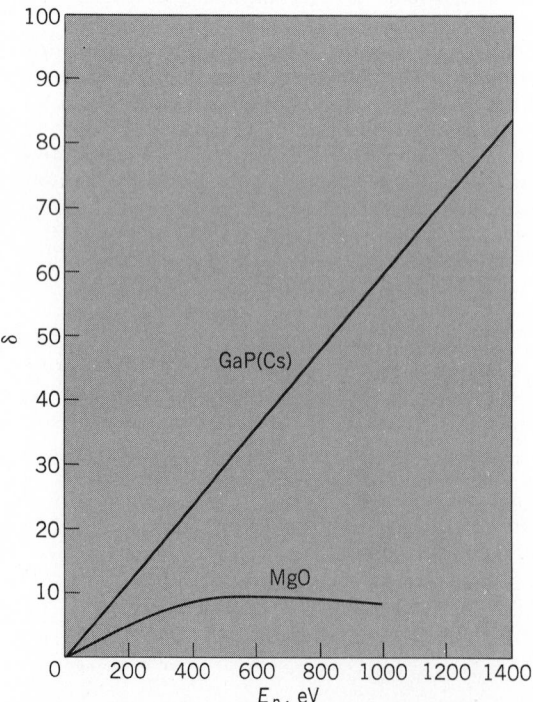

Fig. 5. Secondary emission yield versus primary energy for GaP(Cs). The curve for MgO is shown for comparison.

shown in Fig. 3b because electrons that have dropped to the bottom of the conduction band as a result of phonon-phonon scattering still have enough energy to escape into the vacuum. Because the lifetime of electrons in the bottom of the conduction band is orders of magnitude longer than in states above this level, the escape depth of the secondaries is orders of magnitude greater than in the case represented in Fig. 3b. The most important material in this category is cesium-activated gallium phosphide, GaP(Cs). Figure 5 shows the δ versus E_p curve for GaP(Cs) by comparison with MgO. Because of the much greater escape depth, δ values exceeding 100 are readily obtained. The curve for GaP(Cs) still follows quite closely the universal curve (Fig. 2), but the $E_{p_{max}}$ value is now in the 5–10-kV region compared with several hundred volts for materials represented in Fig. 3a or b.

Materials of the GaP(Cs) type represent a major breakthrough in the use of secondary emission for practical devices. In photomultipliers, for example, GaP(Cs) is superior to the conventional materials represented in Fig. 3b for a number of reasons, the greatest advantage being the improved signal-to-noise ratio. Negative effective electron affinity materials are not in universal use because of the more complex activation procedure and the associated higher cost. *See* BAND THEORY OF SOLIDS; SEMICONDUCTOR.

Dependence on angle of primary beam. The secondary yield for a given primary energy in-

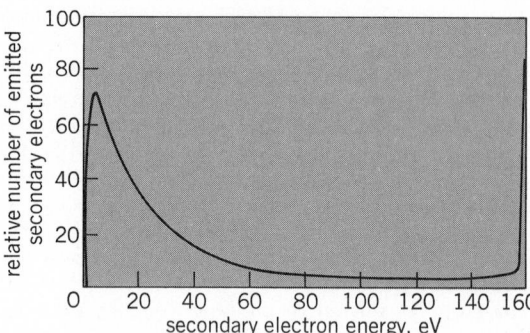

Fig. 6. Relative number of secondary electrons as a function of secondary energy for $E_p = 160$ V.

creases as the angle θ between the primary beam and the normal to the surface increases; the secondaries are then produced closer to the surface and consequently have a larger escape probability. At the same time, the energy for which the yield reaches its maximum value increases with increasing θ.

Secondary energies. A typical energy distribution of secondary electrons emitted by a silver target bombarded with primaries of 160-eV energy is given in Fig. 6. Note that most of the secondaries have relatively low energies. A small fraction of the emitted electrons have the same energy as the incident primaries and are called reflected primaries.

[ALFRED H. SOMMER]

Bibliography: R. L. Bell, *Negative Electron Affinity Devices,* 1973; A. J. Dekker, Secondary electron emission, in *Solid State Physics,* vol. 6, p. 251, 1958; R. U. Martinelli and D. G. Fisher, Application of semiconductors with negative electron affinity surfaces to electron emission devices, *Proc. IEEE*, 62:1339–1360, 1974; R. E. Simon and B. F. Williams, Secondary-electron emission, *IEEE Trans. Nucl. Sci.*, NS-15:167–170, 1968.

Seebeck effect

A thermoelectric phenomenon discovered in 1821 by T. J. Seebeck. He found that near a closed circuit composed of two linear conductors of different metals, a magnetic needle would be deflected if, and only if, the two junctions were at different temperatures; and, further, that if the cooler junction were to become the warmer, the direction of deflection would be reversed. Seebeck advanced the incorrect hypothesis that the conductors were magnetized directly by the temperature difference. Nevertheless, the current and the corresponding electromotive force (emf) in the circuit have been named after him.

The Seebeck emf which gives rise to the current is directly measurable with a voltmeter or potentiometer. It depends only upon the material and temperatures. The current can be calculated from the emf if the resistance of the circuit is known. The circuit (called a thermocouple) is analogous to one with voltaic cells connected in series, the Seebeck emf being the sum of emfs at the junctions (Peltier) and emfs distributed along the conductors (Thomson). *See* PELTIER EFFECT; THERMOELECTRICITY; THOMSON EFFECT.

[JOHN W. STEWART]

Selection rules

General rules concerning the transitions which may occur between the states of a quantum-mechanical physical system. They derive in almost all cases from the symmetry properties of the states and of the interaction which gives rise to the transitions. The system may have a classical (nonquantum) counterpart, and in this case the selection rules may often be related to the classical conserved quantities. A first use of selection rules is in determining the symmetry classes of the states; but in a great variety of ways they may yield other information about the system and the conservation laws. *See* QUANTUM MECHANICS; SYMMETRY LAWS.

Angular momentum and parity rules. For an isolated system the total angular momentum is a conserved quantity; this fact derives from a fundamental fact of nature, namely, that space is isotropic. Each state is then classifiable by angular momentum J and its z component M $(=-J, -J+1, \ldots, +J)$. Angular momenta combine in a vectorial fashion. Thus, if the system makes a particle-emitting transition $J_1, M_1 \rightarrow J_2, M_2$, the emitted particles must carry away angular momentum (j, μ), where $\mathbf{j} = \mathbf{J}_1 - \mathbf{J}_2$. This implies that $\mu = M_1 - M_2$ and that j takes on values $J_1 - J_2, J_1 - J_2 + 1, \ldots, (J_1 + J_2)$. Thus in transitions $(J = 4 \leftrightarrow J = 2)$ the possible j values comprise only 2, 3, 4, 5, 6, and, if it is also specified that $M_1 - M_2 = \pm 4$, only 4, 5, 6. Observe that J_z is additive. *See* ANGULAR MOMENTUM; QUANTUM NUMBERS.

Another fundamental symmetry, the parity, which determines the behavior of a system (or of its description) under inversion of the coordinate axes, is conserved by the strong and electromagnetic interactions, and gives a classification of

systems as even $(\pi = +1)$ or odd $(\pi = -1)$. Under combination the parity combines multiplicatively. Thus, if the transition above is $4^{\pm} \rightarrow 2^{\mp}$, it follows that $j^{\pi} = 2^-, \ldots, 6^-$, while $4^{\pm} \rightarrow 2^{\pm}$ would give $j^{\pi} = 2^+, \ldots, 6^+$. The angular momentum $\mathbf{j}$ may be a combination of intrinsic spin $\mathbf{s}$ and orbital angular momentum $\mathbf{l}$. Scalar, pseudoscalar, vector, and pseudovector particles are respectively characterized by $s^{\pi_s} = 0^+, 0^-, 1^-, 1^+$, where π_s is the "intrinsic" parity, while l always carries $\pi_l = (-1)^l$. *See* NONRELATIVISTIC QUANTUM THEORY; PARITY; SPIN.

Electromagnetic transitions. The photon is a transverse vector particle; this implies that its electric field vector (classically) or its intrinsic spin vector (quantum-mechanically) lies in the plane normal to the propagation direction. Thus, while an ordinary vector particle with total angular momentum j may, for $l > 0$, have $l = j, j-1, j+1$, corresponding to the three internal degrees of freedom, a transverse particle may have $l = j$ or else a particular linear combination of $l = j \pm 1$ (two internal degrees). An $l = j$ photon ("magnetic" type, or Mj) then carries parity $(-1)^{j+1}$, while the $l = j \pm 1$ combination (electric type, or Ej) carries $\pi = (-1)^j$; the magnetic-electric nomenclature is appropriate because, in the long-wavelength (LWL) limit (that is, when the photon wavelength is much greater than the radius of the source) the radiation is generated by a magnetic-type interaction (for example, $\mathbf{L} \cdot \mathbf{H}$ or $\mathbf{S} \cdot \mathbf{H}$ for $M1$, with $\mathbf{H}$ the magnetic field) or electric-type interaction ($\mathbf{P} \cdot \mathbf{E}$ for $E1$, with $\mathbf{P}$ the dipole operator and $\mathbf{E}$ the electric field). For $j = 0$, the only combination of s and l is $(s, l, j) = (1, 1, 0)$, which describes an oscillating $E0$ field, which however vanishes outside the source and hence does not radiate. Thus, for transitions of the form $J_1^{\pm} \leftrightarrow J_2^{\pm}$, photons of type $M1, E2, M3, \ldots$, may be emitted, and, for $J_1^{\pm} \leftrightarrow J_2^{\mp}$, $E1, M2, E3, \ldots$, the j values being restricted as above. If M_1, M_2 are also specified, as in Zeeman transitions, there are further restrictions. For dipole transitions $(j = 1)$, only $\mu = 0$ (π-lines) and $\mu = \pm 1$ (σ-lines) occur, the former not being radiated parallel to the magnetic field because of the photon transversality. If $M_1 = M_2 = 0$, the transition is forbidden unless $J_1 + J_2 + j$ is even; this inversion rule, not restricted to electromagnetic transitions, arises from the behavior of states and operators under a rotation through π radians about an axis in the x-y plane. It is important in discussing molecular symmetries and isospin selection rules. *See* MULTIPOLE RADIATION.

Approximate rules. The "exact" selection rules are to be supplemented by approximate rules which depend on the internal structure of the system. To the extent that L, S, the total orbital and spin angular momenta, are good quantum numbers, as happens in many atoms and a few nuclei, the angular momentum radiated in a long-wavelength $E1$ or $E2$ transition (whose transition operators are spin-independent) must come from the orbital structure; thus $(\Delta \mathbf{L}, \Delta \mathbf{S}) = (1, 0), (2, 0)$, respectively. For the $M1$ operator, which has the form $\alpha \mathbf{L} + \beta \mathbf{S}$, $(\Delta \mathbf{L}, \Delta \mathbf{S})$ can have the values $(1, 0)$ or $(0, 1)$ but not $(0, 0)$ or $(1, 1)$; however, beyond that, since $\mathbf{L}$ and $\mathbf{S}$ can generate no orbital, spin, or radial excitations, $M1$ transitions in LS-coupling occur only between two levels of the same term (although this is not true for Mj with $j > 1$).

Many processes, including photon emission and β-decay, occur via one-body operators [for example, by

$$\mathbf{P} = \sum_i e_i \mathbf{r}(i)$$

for $E1$, where the sum is over particles, with the ith particle having charge e_i and position operator $\mathbf{r}(i)$]. These generate only single-particle transitions, in which at most one particle changes orbit. (The number of such particles is $\delta n = 1$ for odd parity, as in

$$p^2 \xrightarrow{E1} sp, pd$$

$\delta n = 0, 1$ for even, as in

$$d^2 \xrightarrow{E2} d^2, sd, dg$$

$\delta n = 0$ for $M1$ in LS-coupling, but not in jj-coupling, since, for example,

$$f_{5/2} \xrightarrow{M1} f_{7/2}$$

is allowed). Because of this "$\delta n \leq 1$" rule, the $(J\pi)$ rules apply to individual particles (for example, $\Delta l^\pi = 1^-$ for $E1$) and thus greatly facilitate study of the microscopic structures of the states involved. *See* ATOMIC STRUCTURE AND SPECTRA; NUCLEAR STRUCTURE.

Forbidden transitions. Forbidden transitions are those which are hindered by a significant selection rule or combination of them (the term is often applied in a more restricted sense in some domains, for example, in atomic physics, to all photon transitions except $E1$). Several examples will be used to illustrate the remarkable variety of information which they yield.

Parity violation. The α-decay process $A(J^\pi) \rightarrow B(0^+) + \alpha(0^+)$ is allowed only if $\pi = (-1)^J$ since the final angular momentum J is orbital only. The forbidden process, with A a $2-$ state in ^{16}O and B the ^{12}C ground state, has been observed, with a lifetime on the order of 10^{13} larger than that of a nearby $2+$ state in ^{16}O. Since neither of the decay products has a nearby $0-$ companion, the decay can be ascribed to a small $2+$ admixture in the $2-$ state, the admixing "intensity" (square of the $2+$ amplitude) being then on the order of 10^{-13}. This gives a measure of a fundamental symmetry breaking, the violation of parity by the weak interaction. This example, like several of those following, involves symmetry breaking in a two-level substructure, the simple quantum mechanics of which is explained in the illustration.

Two close-lying states Ψ_A and Ψ_F are eigenfunctions of a hamiltonian $H = H^{(0)} + H^{(1)}$ with

energies E_A and E_F (illustration a), where $H^{(0)}$ preserves a symmetry and the small $H^{(1)}$ breaks it. Transitions occur to a distant state Ψ_D (with energy E_D), in which the symmetry admixing is negligible. $\Psi_A^{(0)}$ and $\Psi_F^{(0)}$, eigenfunctions of $H^{(0)}$, have different symmetries. The selection rules are such that the transition $\Psi_A^{(0)} \rightarrow \Psi_D$ is allowed, while $\Psi_F^{(0)} \rightarrow \Psi_D$ is forbidden. The effect of $H^{(1)}$, which has no diagonal matrix elements in the $\Psi^{(0)}$ states, is to "rotate" the states through a small angle ϕ (illustration b), thereby admixing the symmetries, generating a weak transition for Ψ_F, and slightly shifting the energies. The ratio of the transition strengths S determines ϕ, which in turn fixes the admixing intensity and the level shifts. Diagonalizing the 2×2 matrix gives: $S_F/S_A = \tan^2 \phi$; $(\Psi_F^{(0)} H^{(1)} \Psi_A^{(0)}) = (E_F - E_A) \sin \phi \cos \phi$; $(E_F - E_F^{(0)}) = -(E_A - E_A^0) = (E_F - E_A) \sin^2 \phi$. The perturbation solution (small ϕ) is often adequate. *See* PERTURBATION (QUANTUM MECHANICS).

Multiple-photon transitions. Because the $M1$ operator has no radial dependence in the long-wavelength limit, the single-photon transition between the $2S_{1/2}$ and $1S_{1/2}$ states in hydrogen is strongly inhibited; the lifetime should be on the order of 10^5 s. The very-low-energy $E1$ transition to $2P_{1/2}^-$ (which lies below $2S_{1/2}^+$ by the Lamb shift energy), followed by $E1$ to ground, would take on the order of 10^8 s. The dominant process instead, with a lifetime on the order of 10^{-1} s, is the simultaneous emission of two photons, which share the energy to produce a continuous photon spectrum. In nuclei the two-photon $0^+ \rightarrow 0^+$ transition has also been observed, as well as $E0$ deexcitation by ejection of a penetrating atomic electron ("complete internal conversion") and by the closely related emission of an e^+-e^- pair. All of these processes are forbidden for $0^+ \leftrightarrow 0^-$, which has not been observed. *See* NONLINEAR OPTICS; RYDBERG CONSTANT.

LS-forbidden transitions. Early examples discovered in optical astronomy involve $6s6p\ ^3P_0 \leftarrow 6s^2\ ^1S_0 \rightarrow 6s6p\ ^3P_2$ in HgI. The first is absolutely forbidden, while the second would require the highly improbable $M2$, with wave-function corrections (since $s \rightarrow p$ would require $E1$). But the decoupling effect of the hyperfine interaction between valence electrons and the nuclear magnetic moment generates a small $^3P_1$ amplitude in the $^3P_0$ state (the total angular momentum being $F = 1/2$ in each case) and thereby, as in the illustration, an allowed but greatly retarded $^1S_0 \rightarrow ^3P_1$ transition. A similar process occurs for the second case. Astronomical observation of such forbidden lines indicates that particle densities in the source are low; otherwise deexcitation would occur via collisions.

Spherical shell-model orbits in nuclei. The process of adding a neutron in orbit (l, j) to a nucleus, $A + n(l, j) \rightarrow B$, is realizable in (d, p) "direct" reactions, whose angular distribution determines l and whose magnitude gives the intrinsic neutron-capture probability. If one of the nuclear states involved is 0^+, the j value must equal the angular momentum of the other. For example, if A is describable as $(j^{2n})_{J=0^+}$, there should be large cross sections to $(j^{2n+1})_{J=j}$ and $((j^{2n})_0 j')_{J=j'}$, in which the particles in orbit j are themselves coupled to zero. Forbidden transitions to $(j^{2n+1})_{J=j' \neq j}$ are, however, often observed. They arise via admixtures (on the

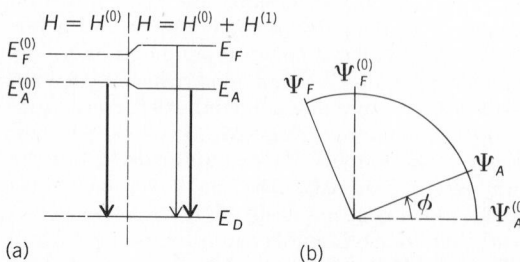

Symmetry admixing in a two-level substructure. (*a*) Energy-level diagram. (*b*) Rotation of states.

order of a few percent) of $((j^{2n})_0 j')$ in the final state which, by the analysis associated with the illustration, give measures of the orbital "purity." *See* Nuclear reaction.

K selection rules in deformed nuclei. There is a J_z splitting of single-particle orbits in many heavy nuclei; this is produced by the large permanent quadrupole deformation of the nucleus which gives rise to a noncentral single-particle potential. K labels the angular-momentum component along the symmetry axis, whose rotation then generates from each intrinsic state a band of levels with $J \geq K$, all with the same parity. If two levels belong to bands with very different K values (specifically if $\Delta = |K_1 - K_2| - j$ is large, where the multiparity j follows from the J^π values), electromagnetic transitions between them will be very highly forbidden. This is so formally because the j-multipole transition operator can only generate $|\Delta K| \leq j$, and more physically because the very different rotations involved in the two states generate large differences between all the corresponding orbits which play a role in the rotation. Empirically one encounters Δ values as high as 8 and finds retardations on the order of $10^{2\Delta}$, the process itself occurring via Coriolis-type admixtures. Similar multiparticle forbiddenness would obtain for transitions between states built on the two potential minima observed in fission studies for many nuclei. *See* Nuclear fission.

Molecular selection rules. The example of K selection rules in deformed nuclei involves both single-particle and rotational motion in these nuclei. Vibrational motion, especially quadrupole, can be found also. The classification of the motions derives from molecular physics and is most developed in that domain. To the extent that the molecular motions can be well separated, the angular momentum transferred in a transition must come from one or another of them. Along with the selection rules implied by this (the rule for harmonic vibrations is $n_v \to n_v \pm 1$), the various coupling possibilities and any special symmetries displayed by the intrinsic structure must be considered. Let N, Λ, Σ be the projection onto the symmetry axis (the line joining the two nuclei for the diatomic molecules to which the discussion is now restricted) of the collective rotational, the electronic orbital, and the electronic spin angular momenta. Λ and Σ may combine to Ω, the total intrinsic angular momentum along the symmetry axis; then the rotation gives $J = \Omega, \Omega + 1, \ldots$. This is Hund's case A. Alternatively (case B), $\Lambda + N = K$, which combines with S to give $\mathbf{J} = \mathbf{S} + \mathbf{K}$, so that $|K - S| \leq J \leq K + S$. In both cases the supplementary rules for $E1$ (with natural extensions to other multipoles) that $\Delta \Lambda = 0, \pm 1$ and $\Delta S = 0$ are valid, while for case A, $\Delta \Sigma = 0$, $\Delta \Omega = 0 \pm 1$, and for case B, $\Delta \mathbf{K} = 1$, these being valid if the spin-orbit interactions are weak.

The special symmetries of the intrinsic structure refer to its behavior under a reflection through the symmetry center and through a plane containing the symmetry axis. The first operation, $\mathscr{P}$, defines the intrinsic parity, labeled as g (for $\pi = +1$) and u ($\pi = -1$). The second, $\mathscr{S}$, defines a symmetry $\pm$; in nuclear physics one commonly uses $\mathscr{R} = \mathscr{P}\mathscr{S}$, with eigenvalue $r = \pm 1$, which generates a rotation through π radians about an axis normal to the symmetry axis. For homonuclear molecules the inter-

change of the two nuclei defines states as symmetric (s) or antisymmetric (a). Then for $E1$ the rules $g \leftrightarrow u$, $+ \leftrightarrow -$ are valid, and with homonuclear molecules $s \to s$, $a \to a$. *See* Molecular structure and spectra.

Further symmetries. The isospin symmetry of the elementary particles is almost conserved, being broken by electromagnetic and weak interactions. It is described by the group SU(2), of unimodular unitary transformations in two dimensions. Since the SU(2) algebra is identical with that of the angular momentum SO(3), isospin behaves like angular momentum with its three generators $\mathbf{T}$ replacing $\mathbf{J}$. For a nucleon $t = 1/2$, the values $t_z = 1/2, -1/2$ may be assigned for protons (p) and neutrons (n), respectively. Then, for nuclear states, $T_z = (Z - N)/2$ (where Z is the atomic number and N is the neutron number) and $T (\geq |T_z|)$ is an almost good quantum number. The argument associated with the illustration can often be used to measure the isospin admixing; for example, with the pair of decays ^{15}N ($T = 1/2$, $3/2) \to {}^{14}$N ($T = 0 + n$), the second of which is forbidden by the selection rule $\Delta T = 1/2$, one finds about 4% admixing. Since the nucleon charge is $|e|\{\frac{1}{2} + t_z\}$, the electromagnetic transition operators split into an isoscalar part (for which $\Delta T = 0$) and an isovector ($\Delta T = 1$). Since the isoscalar part of the (long-wavelength) $E1$ operator $|e|\{\frac{1}{2}\sum_i \mathbf{r}(i) + \sum_i t_z(i)\mathbf{r}(i)\}$ is ineffective, being proportional to the center-of-mass vector, and the inversion rule described earlier forbids isovector transitions in self-conjugate nuclei ($T_z = 0$, $N = Z$), it follows that $E1$ transitions between states of the same T are forbidden in such nuclei. *See* Isotopic spin.

The isospin group is a subgroup of SU(3) which defines a more complex fundamental symmetry of the elementary particles. Two of its eight generators commute, giving two additive quantum numbers, T_z and strangeness S' (or, equivalently, charge and hypercharge). The strangeness is conserved ($\Delta S' = 0$) for strong and electromagnetic, but not for weak, interactions. The selection rules and combination laws for SU(3) and its many extensions, and the quark-structure ideas underlying them, correlate an enormous amount of information and make many predictions about the elementary particles. *See* Baryon; Elementary particle; Meson; Quarks; Unitary symmetry.

The fundamental β-decay transition is $n \to p + e^- + \bar{\nu}$. For the Gamow-Teller decay mode (electron and antineutrino in a triplet state) and the Fermi mode (singlet), the transition operators are respectively components of $\sum_i \mathbf{s}(i) \mathbf{t}(i)$ and of $\mathbf{T}$. When these decays are realized in a nucleus, in which also $p \to n + e^+ + \nu$ may occur, the rules $\Delta \pi = 0$, $\Delta \mathbf{T} = 1$, $\Delta T_z = \pm 1$ clearly hold, with $\Delta \mathbf{J} = 1, 0$ for Gamow-Teller and Fermi, respectively. But for Fermi decay there is also the very strong rule (which includes $\Delta T = 0$) that the decay takes place only to a single state, the isobaric analog, which differs from the parent only in T_z, the transition rate being then easily calculable. This special feature, analogous to that for $M1$ transitions in LS-coupling, arises from the fact that the transition operator is a generator of an (almost) good symmetry. Besides superallowed Fermi transitions between $J = 0$ analog states (forbidden for Gamow-Teller), there are also forbidden Fermi decays between nonanalog states which, as in the illustration, arise from isospin admixing and have been used to give measures for it. Special rules, by no means as well satisfied, arise also for Gamow-Teller decays from the fact that the 15 components of the basic transition operators are the generators of an SU(4) group whose symmetry is fairly good in light nuclei. (For these same nuclei a different SU(3) realization than above gives a model for rotational bands.) Besides the allowed transitions, there is also a hierarchy of forbidden transitions in nuclei which, through recoil (retardation) effects, involve the orbital angular momentum and have therefore less restrictive selection rules. *See* NEUTRINO; RADIO-ACTIVITY.

A great variety of other groups have been introduced to define relevant symmetries for atoms, molecules, nuclei, and elementary particles. They all have their own selection rules, representing one aspect of the symmetries of nature.

[J. B. FRENCH]

Bibliography: A. Bohr and B. R. Mottelson, *Nuclear Structure*, vol. 2, 1975; R. H. Garstang, Forbidden transitions, in D. R. Bates (ed.), *Atomic and Molecular Processes*, pp. 1–46, 1962.

Semiconductor

A solid crystalline material whose electrical conductivity is intermediate between that of a metal and an insulator. Semiconductors exhibit conduction properties that may be temperature-dependent, permitting their use as thermistors (temperature-dependent resistors), or voltage-dependent, as in varistors. By making suitable contacts to a semiconductor or by making the material suitably inhomogeneous, electrical rectification and amplification can be obtained. Semiconductor devices, rectifiers, and transistors have replaced vacuum tubes almost completely in low-power electronics, making it possible to save volume and power consumption by orders of magnitude. In the form of integrated circuits, they are vital for complicated systems. The optical properties of a semiconductor are important for the understanding and the application of the material. Photodiodes, photoconductive detectors of radiation, injection lasers, light-emitting diodes, solar-energy conversion cells, and so forth are examples of the wide variety of optoelectronic devices. *See* LASER; TRANSISTOR.

CONDUCTION IN SEMICONDUCTORS

The electrical conductivity of semiconductors ranges from about 10^3 to 10^{-9} ohm^{-1} cm^{-1}, as compared with a maximum conductivity of 10^7 for good conductors and a minimum conductivity of 10^{-17} ohm^{-1} cm^{-1} for good insulators.

The electric current is usually due only to the motion of electrons, although under some conditions, such as very high temperatures, the motion of ions may be important. The basic distinction between conduction in metals and in semiconductors is made by considering the energy bands occupied by the conduction electrons.

A crystalline solid consists of a large number of atoms brought together into a regular array called a crystal lattice. The electrons of an atom can each have certain energies, so-called energy levels, as predicted by quantum theory. Because the atoms of the crystal are in close proximity, the electron orbits around different atoms overlap to some extent, and the electrons interact with each other; consequently the sharp, well-separated energy levels of the individual electrons actually spread out into energy bands. Each energy band is a quasi-continuous group of closely spaced energy levels. *See* BAND THEORY OF SOLIDS.

At absolute zero temperature, the electrons occupy the lowest possible energy levels, with the restriction that at most two electrons may be in the same energy level. In semiconductors and insulators, there are just enough electrons to fill completely a number of energy bands, leaving the rest of the energy bands empty. The highest filled energy band is called the valence band. The next higher band, which is empty at absolute zero temperature, is called the conduction band. The conduction band is separated from the valence band by an energy gap which is an important characteristic of the semiconductor. In metals, the highest energy band that is occupied by the electrons is only partially filled. This condition exists either because the number of electrons is not just right to fill an integral number of energy bands or because the highest occupied energy band overlaps the next higher band without an intervening energy gap. The electrons in a partially filled band may acquire a small amount of energy from an applied electric field by going to the higher levels in the same band. The electrons are accelerated in a direction opposite to the field and thereby constitute an electric current. In semiconductors and insulators, the electrons are found only in completely filled bands, at low temperatures. In order to increase the energy of the electrons, it is necessary to raise electrons from the valence band to the conduction band across the energy gap. The electric fields normally encountered are not large enough to accomplish this with appreciable probability. At sufficiently high temperatures, depending on the magnitude of the energy gap, a significant number of valence electrons gain enough energy thermally to be raised to the conduction band. These electrons in an unfilled band can easily participate in conduction. Furthermore, there is now a corresponding number of vacancies in the electron population of the valence band. These vacancies, or holes as they are called, have the effect of carriers of positive charge, by means of which the valence band makes a contribution to the conduction of the crystal. *See* HOLES IN SOLIDS.

The type of charge carrier, electron or hole, that is in largest concentration in a material is sometimes called the majority carrier and the type in smallest concentration the minority carrier. The majority carriers are primarily responsible for the conduction properties of the material. Although the minority carriers play a minor role in electrical conductivity, they can be important in rectification and transistor actions in a semiconductor.

Electron distribution. The probability f for an energy level E to be occupied by an electron is given by the Fermi-Dirac distribution function, Eq. (1),

$$f = \left[1 + \exp\left(\frac{E - W}{kT}\right) \right]^{-1} \tag{1}$$

where k is the Boltzmann constant and T is the

absolute temperature. The parameter W is the Fermi energy level; an energy level at W has a probability of 1/2 to be occupied by an electron. The Fermi level is determined by the distribution of energy levels and the total number of electrons. *See* FERMI-DIRAC STATISTICS.

In a semiconductor, the number of conduction electrons is normally small compared with the number of energy levels in the conduction band, and the probability for any energy level to be occupied is small. Under such a condition, the concentration of conduction electrons is given by Eq. (2),

$$N_n = \frac{2}{h^3} (2\pi m_n kT)^{3/2} \exp\left[\frac{(W - E_c)}{kT}\right] \quad (2)$$

where h is Planck's constant, E_c is the lowest energy of the conduction band, and m_n is called the effective mass of conduction electrons. The effective mass is used in place of the actual mass to correct the coefficient in the equation and to bring the results in line with experimental observations. This correction is necessary because the theory leading to these equations is based upon electrons moving in a field free space, which is not the exact picture. The electrostatic Coulomb potential throughout the crystal is varying in a periodic manner, the variation being due to the electric fields around the atomic centers. The concentration of holes in the valence band is given by Eq. (3),

$$N_p = \frac{2}{h^3} (2\pi m_p kT)^{3/2} \exp\left[\frac{(E_v - W)}{kT}\right] \quad (3)$$

where m_p is the effective mass of a hole and E_v is the highest energy of the valence band.

Mobility of carriers. The velocity acquired by charge carriers per unit strength of applied electric field is called the mobility of the carriers. The velocity in question is the so-called drift velocity in the direction of the force exerted on the carriers by the applied field. It is added to the random thermal velocity. In semiconductors the carrier mobility normally ranges from 10^2 to 10^5 cm^2/(s) (volt). A material's conductivity is the product of the charge, the mobility, and the carrier concentration.

Electrons in a perfectly periodic potential field can be accelerated freely. Impurities, physical defects in the structure, and thermal vibrations of the atoms disturb the periodicity of the potential field in the crystal, thereby scattering the moving carriers. It is the resistance produced by this scattering that limits the carriers to only a drift velocity under the steady force of an applied field.

Intrinsic semiconductors. A semiconductor in which the concentration of charge carriers is characteristic of the material itself rather than of the content of impurities and structural defects of the crystal is called an intrinsic semiconductor. Electrons in the conduction band and holes in the valence band are created by thermal excitation of electrons from the valence to the conduction band. Thus an intrinsic semiconductor has equal concentrations of electrons and holes. The intrinsic carrier concentration N_i is determined by Eq. (4),

$$N_i = \frac{2}{h^3} (2\pi kT)^{3/2} (m_n m_p)^{3/4} \exp\left(-\frac{E_g}{2kT}\right) \quad (4)$$

where E_g is the energy gap. The carrier concentration, and hence the conductivity, is very sensitive to temperature and depends strongly on the energy gap. The energy gap ranges from a fraction of 1 ev to several electron volts. A material must have a large energy gap to be an insulator.

Impurity semiconductors. Typical semiconductor crystals such as germanium and silicon are formed by an ordered bonding of the individual atoms to form the crystal structure. The bonding is attributed to the valence electrons which pair up with valence electrons of adjacent atoms to form so-called shared pair or covalent bonds. These materials are all of the quadrivalent type; that is, each atom contains four valence electrons, all of which are used in forming the crystal bonds. *See* CRYSTAL STRUCTURE.

Atoms having a valence of +3 or +5 can be added to a pure or intrinsic semiconductor material with the result that the +3 atoms will give rise to an unsatisfied bond with one of the valence electrons of the semiconductor atoms, and +5 atoms will result in an extra or free electron that is not required in the bond structure. Electrically, the +3 impurities add holes and the +5 impurities add electrons. They are called acceptor and donor impurities, respectively. Typical valence +3 impurities used are boron, aluminum, indium, and gallium. Valence +5 impurities used are arsenic, antimony, and phosphorus.

Semiconductor material "doped" or "poisoned" by valence +3 acceptor impurities is termed *p*-type, whereas material doped by valence +5 donor material is termed *n*-type. The names are derived from the fact that the holes introduced are considered to carry positive charges and the electrons negative charges. The number of electrons in the energy bands of the crystal is increased by the presence of donor impurities and decreased by the presence of acceptor impurities. Let N be the concentration of electrons in the conduction band and let P be the hole concentration in the valence band. For a given semiconductor, the relation $NP = N_i^2$ holds, independent of the presence of impurities. The effect of donor impurities tends to make N larger than P, since the extra electrons given by the donors will be found in the conduction band even in the absence of any holes in the valence band. Acceptor impurities have the opposite effect, making P larger than N.

At sufficiently high temperatures, the intrinsic carrier concentration becomes so large that the effect of a fixed amount of impurity atoms in the crystal is comparatively small and the semiconductor becomes intrinsic. When the carrier concentration is predominantly determined by the impurity content, the conduction of the material is said to be extrinsic. There may be a range of temperature within which the impurity atoms in the material are practically all ionized; that is, they supply a maximum number of carriers. Within this temperature range, the so-called exhaustion range, the carrier concentration remains nearly constant. At sufficiently low temperatures, the electrons or holes that are supplied by the impurities become bound to the impurity atoms. The concentration of conduction carriers will then decrease rapidly with decreasing temperature, according to either exp $(-E_i/kT)$ or exp $(-E_i/2kT)$, where E_i is the ionization energy of the dominant impurity.

Physical defects in the crystal structure may have similar effects as donor or acceptor impurities. They can also give rise to extrinsic conductivity.

An isoelectronic impurity, that is, an atom which has the same number of valence electrons as the host atom, does not bind individual carriers as strongly as a donor or an acceptor impurity. However, an isoelectronic impurity may show an appreciable binding for electron hole pairs, excitons, and thereby have important effects on the properties. An example is nitrogen substituting for phosphorus in gallium phosphide; the impurity affects the luminescence of the material.

Hall effect. Whether a given sample of semiconductor material is *n*- or *p*-type can be determined by observing the Hall effect. If an electric current is caused to flow through a sample of semiconductor material and a magnetic field is applied in a direction perpendicular to the current, the charge carriers are crowded to one side of the sample, giving rise to an electric field perpendicular to both the current and the magnetic field. This development of a transverse electric field is known as the Hall effect. The field is directed in one or the opposite direction depending on the sign of the charge of the carrier. *See* Hall effect.

The magnitude of the Hall effect gives an estimate of the carrier concentration. The ratio of the transverse electric field strength to the product of the current and the magnetic field strength is called the Hall coefficient, and its magnitude is inversely proportional to the carrier concentration. The coefficient of proportionality involves a factor which depends on the energy distribution of the carriers and the way in which the carriers are scattered in their motion. However, the value of this factor normally does not differ from unity by more than a factor of two. The situation is more complicated when more than one type of carrier is important for the conduction. The Hall coefficient then depends on the concentrations of the various types of carriers and their relative mobilities.

The product of the Hall coefficient and the conductivity is proportional to the mobility of the carriers when one type of carrier is dominant. The proportionality involves the same factor which is contained in the relationship between the Hall coefficient and the carrier concentration. The value obtained by taking this factor to be unity is referred to as the Hall mobility.

MATERIALS AND THEIR PREPARATION

The group of chemical elements which are semiconductors includes germanium, silicon, gray (crystalline) tin, selenium, tellurium, and boron.

Elemental semiconductors. Germanium, silicon, and gray tin belong to group IV of the periodic table and have crystal structures similar to that of diamond. Germanium and silicon are two of the best-known semiconductors. They are used extensively in devices such as rectifiers and transistors. Gray tin is a form of tin which is stable below 13°C. White tin, which is stable at higher temperatures, is metallic. Gray tin has a small energy gap and a rather large intrinsic conductivity, about 5×10^3 ohm^{-1} cm^{-1} at room temperature. The *n*-type and *p*-type gray tins can be obtained by adding aluminum and antimony, respectively.

Selenium and tellurium both have a similar structure, consisting of spiral chains located at the corners and centers of hexagons. The structure gives rise to anisotropy of the properties of single crystals; for example, the electrical resistivity of tellurium along the direction of the chains is about one-half the resistivity perpendicular to this direction. Selenium has been widely used in the manufacture of rectifiers and photocells.

Semiconducting compounds. A large number of compounds are known to be semiconductors. Copper(I) oxide (Cu_2O) and mercury(II) indium telluride ($HgIn_2Te_4$) are examples of binary and ternary compounds. The series zinc sulfide (ZnS), zinc selenide (ZnSe), zinc telluride (ZnTe), and the series zinc selenide (ZnSe), cadmium selenide (CdSe), and mercury(II) selenide (HgSe) are examples of binary compounds consisting of a given element in combinations with various elements of another column in the periodic table. The series magnesium antimonide (Mg_2Sb_2), magnesium telluride (MgTe), and magnesium iodide (MgI_2) is an example of compounds formed by a given element with elements of various other columns in the periodic table.

A group of semiconducting compounds of the simple type AB consists of elements from columns symmetrically placed with respect to column IV of the periodic table. Indium antimonide (InSb), cadmium telluride (CdTe), and silver iodide (AgI) are examples of III-V, II-IV, and I-VI compounds, respectively. The various III-V compounds are being studied extensively, and many practical applications have been found for these materials. Some of these compounds have the highest carrier mobilities known for semiconductors. The compounds have zincblende crystal structure which is geometrically similar to the diamond structure possessed by the elemental semiconductors, germanium and silicon, of column IV, except that the four nearest neighbors of each atom are atoms of the other kind. The II-VI compounds, zinc sulfide (ZnS) and cadmium sulfide (CdS), are used in photoconductive devices. Zinc sulfide is also used as a luminescent material. *See* Luminescence; Photoconductivity.

Binary compounds of the group lead sulfide (PbS), lead selenide (PbSe), and lead telluride (PbTe) are sensitive in photoconductivity and are used as detectors of infrared radiation. The compounds, bismuth telluride (Bi_2Te_3) and bismuth selenide (Bi_2Se_3), consisting of heavy atoms, are found to be good materials for thermocouples used for refrigeration or for conversion of heat to electrical energy. *See* Thermoelectricity.

The metal oxides usually have large energy gaps. Thus pure oxides are usually insulators of high resistivity. However, it may be possible to introduce into some of the oxides impurities of low ionization energies and thus obtain relatively good extrinsic conduction. Copper(I) oxide (Cu_2O) was one of the first semiconductors used for rectifiers and photocells; extrinsic *p*-type conduction is obtained by producing an excess of oxygen over the stoichiometric composition, that is, the 2-to-1 ratio of copper atoms to oxygen atoms. A number of oxide semiconductors can be obtained by replacing some of the normal metal atoms with metal atoms of one more or less valency. The method is called controlled valence. An example of such a semiconductor is nickel oxide containing lithium.

Some compounds with rare-earth or transition-metal ions in their composition, such as EuTe and NiS_2, are semiconductors with magnetic properties. Another interesting type of semiconductor is characterized by layered structures. The interaction within a layer is significantly stronger than that between layers. A number of semiconductors of this type are known, such as PbI_2, GaSe, and various transition-metal dichalcogenides such as $SnSe_2$ and MoS_2.

Preparation of materials. The properties of semiconductors are extremely sensitive to the presence of impurities. It is therefore desirable to start with the purest available materials and to introduce a controlled amount of the desired impurity. The zone refining method is often used for further purification of obtainable materials. The floating zone technique can be used, if feasible, to prevent any contamination of molten material by contact with crucible.

For basic studies as well as for many practical applications, it is desirable to use single crystals. Various methods are used for growing crystals of different materials. For many semiconductors, including germanium, silicon, and the III-V compounds, the Czochralski method is commonly used. The method of condensation from the vapor phase is used to grow crystals of a number of semiconductors, for instance, selenium and zinc sulfide. For materials of high melting points, such as various metal oxides, the flame fusion or Vernonil method may be used. *See* CRYSTAL GROWTH.

The introduction of impurities, or doping, can be accomplished by simply adding the desired quantity to the melt from which the crystal is grown. Normally, the impurity has a small segregation coefficient, which is the ratio of equilibrium concentrations in the solid and the liquid phases of the material. In order to obtain a desired impurity content in the crystal, the amount added to the melt must give an appropriately larger concentration in the liquid. When the amount to be added is very small, a preliminary ingot is often made with a larger content of the doping agent; a small slice of the ingot is then used to dope the next melt accurately. Impurities which have large diffusion constants in the material can be introduced directly by holding the solid material at an elevated temperature while this material is in contact with the doping agent in the solid or the vapor phase.

A doping technique, ion implantation, has been developed and used extensively. The impurity is introduced into a layer of semiconductor by causing a controlled dose of highly accelerated impurity ions to impinge on the semiconductor.

A growing subject of scientific and technological interest is amorphous semiconductors. In an amorphous substance the atomic arrangement has some short-range but no long-range order. The representative amorphous semiconductors are selenium, germanium, and silicon in their amorphous states, and arsenic and germanium chalcogenides, including such ternary systems as Ge-As-Te and Si-As-Te. Some amorphous semiconductors can be prepared by a suitable quenching procedure from the melt. Amorphous films can be obtained by vapor deposition.

RECTIFICATION IN SEMICONDUCTORS

In semiconductors, narrow layers can be produced which have abnormally high resistances. The resistance of such a layer is nonohmic; it may depend on the direction of current, thus giving rise to rectification. Rectification can also be obtained by putting a thin layer of semiconductor or insulator material between two conductors of different material.

Barrier layer. A narrow region in a semiconductor which has an abnormally high resistance is called a barrier layer. A barrier may exist at the contact of the semiconductor with another material, at a crystal boundary in the semiconductor, or at a free surface of the semiconductor. In the bulk of a semiconductor, even in a single crystal, barriers may be found as the result of a nonuniform distribution of impurities. The thickness of a barrier layer is small, usually 10^{-3} to 10^{-5} cm.

A barrier is usually associated with the existence of a space charge. In an intrinsic semiconductor, a region is electrically neutral if the concentration n of conduction electrons is equal to the concentration p of holes. Any deviation in the balance gives a space charge equal to $e(p - n)$, where e is the charge on an electron. In an extrinsic semiconductor, ionized donor atoms give a positive space charge and ionized acceptor atoms give a negative space charge. Let N_d and N_a be the concentrations of ionized donors and acceptors, respectively. The space charge is equal to $e(p - n + N_d - N_a)$.

A space charge is associated with a variation of potential. A drop in potential $-\Delta V$ increases the potential energy of an electron by $e\Delta V$; consequently every electronic energy level in the semiconductor is shifted by this amount. With a variation of potential, the electron concentration varies proportionately to $\exp(eV/kT)$ and the hole concentration varies as $\exp(-eV/kT)$. A space charge is obtained if the carriers, mainly the majority carriers, fail to balance the charge of the ionized impurities.

A conduction electron in a region where the potential is higher by ΔV must have an excess energy of $e\Delta V$ in order for it to have the minimum energy on reaching the low potential region. Electrons with less energy cannot pass over to the low potential region. Thus a potential variation presents a barrier to the flow of electrons from high to low potential regions. It also presents a barrier to the flow of holes from low to high potential regions.

Surface barrier. A thin layer of space charge and a resulting variation of potential may be produced at the surface of a semiconductor by the presence of surface states. Electrons in the surface states are bound to the vicinity of the surface, and the energy levels of surface states may lie within the energy gap. Surface states may arise from the adsorption of foreign atoms. Even a clean surface may introduce states which do not exist in the bulk material, simply by virtue of being the boundary of the crystal.

The surface is electrically neutral when the surface states are filled with electrons up to a certain energy level ϵ in the energy gap E_g, which is the energy difference between the bottom of the conduction band E_c and the top of the valence band

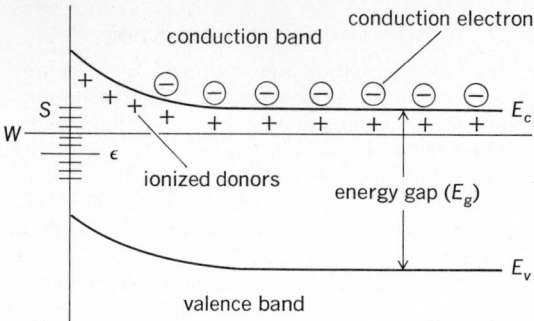

Fig. 1. Energy diagram of a surface barrier as employed in an *n*-type semiconductor.

E_v. If the Fermi level W in the bulk semiconductor lies higher in the energy gap, more surface states would be filled, giving the surface a negative charge. As a result the potential drops near the surface and the energy bands are raised for *n*-type material (Fig. 1). With the rise of the conduction band, the electron concentration is reduced and a positive space charge due to ionized donors is obtained. The amount of positive space charge is equal to the negative surface charge given by the electrons in the surface states between ϵ and the Fermi level.

Contact barrier. The difference between the potential energy E_0 of an electron outside a material and the Fermi level in the material is called the work function of the material. Figure 2 shows the energy diagram for a metal and a semiconductor, the work functions of which differ by eV. Upon connecting the two bodies electrically, charge is transferred between them so that the potential of the semiconductor is raised relative to that of the metal; that is, the electron energy levels in the

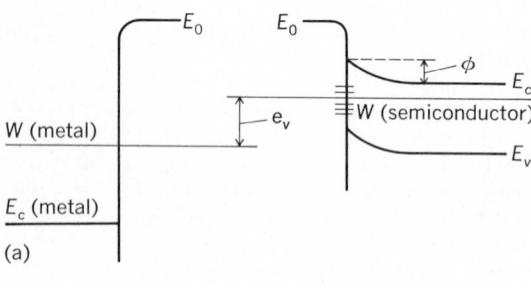

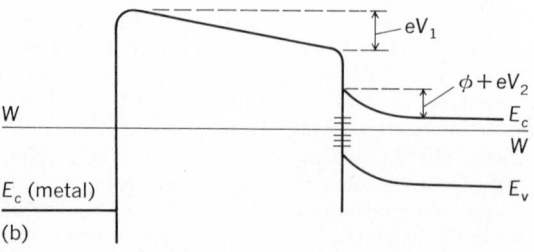

Fig. 2. Energy diagram for a metal (left) and an *n*-type semiconductor (right). E_0 is the potential energy of an electron outside the material, E_c is the energy at the bottom of the conduction band, and E_v is the energy at the top of the valence band. (*a*) Semiconductor and metal isolated. (*b*) Semiconductor and metal in electrical contact, $eV_1 + eV_2 = eV$.

semiconductor are lowered. Equilibrium is established when the Fermi level is the same in the two bodies. In this case, the metal is charged negatively and the semiconductor is charged positively. The negative charge on the metal is concentrated close to the surface, as is expected in good conductors. The positive charge on the semiconductor is divided between the increase of space charge in an extension of the barrier and the depopulation of some of the surface states. The charging of the semiconductor is brought about by a change of eV_2 in the barrier height ϕ. The sum of eV_2 and the potential energy variation eV_1 in the space be-

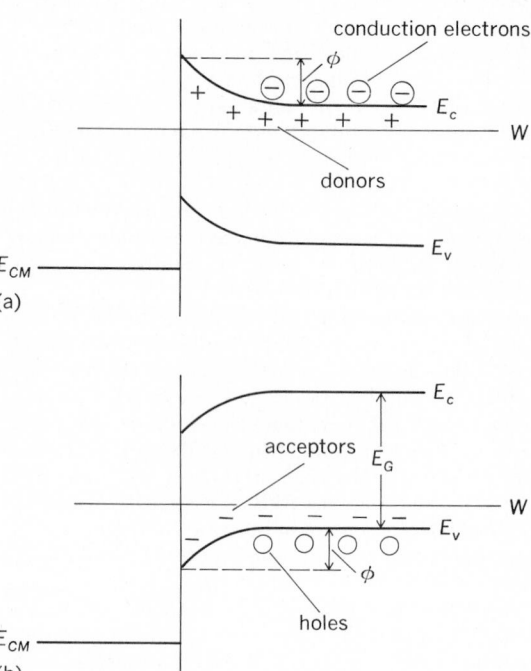

(a)

(b)

Fig. 3. Energy diagrams of a rectifying contact between a metal and a semiconductor: (*a*) *n*-type semiconductor, (*b*) *p*-type semiconductor.

tween the two bodies is equal to the original difference eV between the work functions.

With decreasing separation between the two bodies, the division of eV will be in favor of eV_2. However, if there is a very large density of states, a small eV_2 gives a large surface charge on the semiconductor due to the depopulation of surface states.

It is possible that eV_2 is limited to a small value even at the smallest separation, of the order of an interatomic distance in solids. In such cases, the barrier height remains nearly equal to the value ϕ of the free surface, irrespective of the body in contact. This situation has been found in germanium and silicon rectifiers. Before the explanation was given by J. Bardeen, who postulated the existence of surface states, it had been assumed that the height of a contact barrier was equal to the difference of the work functions.

The understanding and the application of metal-semiconductor contacts have been extended to various kinds of contacts, such as that between different semiconductors, heterojunctions, and metal oxide semiconductor (MOS) junctions.

Single-carrier theory. The phenomenon of rectification at a crystal barrier can be described according to the role played by the carriers. Where the conduction property of the rectifying barrier is determined primarily by the majority carriers, the single-carrier theory is employed. Such cases are likely to be found in semiconductors with large energy gaps, for instance, oxide semiconductors. Figure 3 shows the energy diagrams of metal-semiconductor contact rectifiers under conditions of equilibrium. The potential variation in the semiconductor is such as to reduce the majority carrier concentration near the contact. If the energy bands were to fall in the case of an n-type semiconductor or to rise in the case of a p-type semiconductor, the majority carrier concentration would be enhanced near the contact, and the contact would not present a large and rectifying resistance. It is clear that in the cases shown in Fig. 3, the minority carrier concentration increases near the contact. However, if the energy gap is large, the minority carrier concentration is normally very small, and the role of minority carriers may be still negligible even if the concentration is increased.

Under equilibrium conditions, the number of carriers passing from one body to the other is balanced by the number of carriers crossing the contact in the opposite direction, and there is no net current. The carriers crossing the contact in either direction must have sufficient energies to pass over the peak of the barrier. The situations under applied voltages are shown in Fig. 4 for the case of an n-type semiconductor. When the semiconductor is made positive, its energy bands are depressed and the height of the potential barrier is increased, as shown in Fig. 4a. Fewer electrons in the semiconductor will be able to cross over into the metal, whereas the flow of electrons across the contact from the metal side remains unchanged. Consequently, there is a net flow of electrons from the metal to the semiconductor. The flow of electrons from the metal side is the maximum net flow obtainable. With increasing voltage, the current saturates and the resistance becomes very high. Figure 4b shows the situation when the semiconductor is negative under the applied voltage. The energy bands in the semiconductor are raised. The flow of electrons from the semiconductor to the metal is increased, since electrons of lower energy are able to go over the peak of the barrier. The result is a net flow of electrons from the semiconductor to the metal. There is no limit to the flow in this case. In fact, the electron current increases faster than the applied voltage because there are increasingly more electrons at lower energies. The resistance decreases, therefore, with increasing voltage. The direction of current for which the resistance is low is called the forward direction, while the opposite direction is called the reverse or blocking direction. A general expression for the current can be written in the form of Eq. (5), where

$$j = enC\left(\exp\frac{-\phi}{kT}\right)\left[\exp\left(\frac{eV}{kT}\right) - 1\right] \qquad (5)$$

j is the current density, n is the carrier concentration in the bulk of the semiconductor, ϕ is the barrier height, and V is the applied voltage taken as positive in the forward direction. The factor C

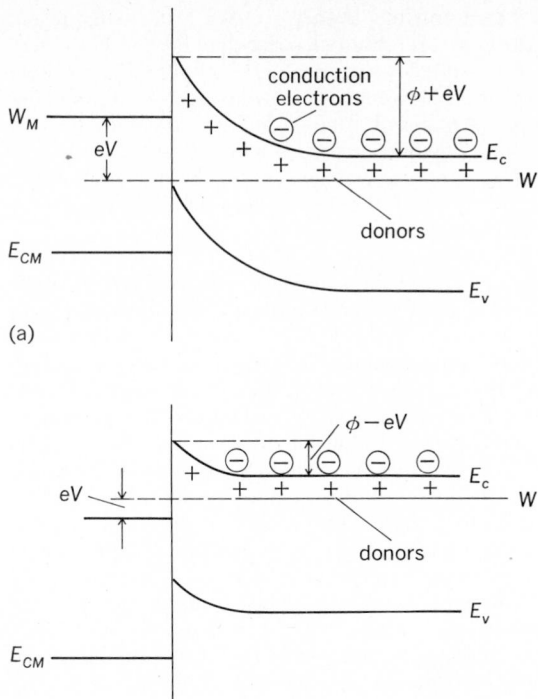

(a)

(b)

Fig. 4. Energy diagrams of a rectifying contact between a metal and an n-type semiconductor under an applied voltage V. (a) Positive semiconductor. There is a net flow of electrons from metal to semiconductor. (b) Negative semiconductor. There is a net flow of electrons from semiconductor to metal.

depends on the theory appropriate for the particular case.

Diffusion theory. When there is a variation of carrier concentration, a motion of the carriers is produced by diffusion in addition to the drift determined by the mobility and the electric field. The transport of carriers by diffusion is proportional to the carrier concentration gradient and the diffusion constant. The diffusion constant is related to the mobility, and both are determined by the scattering suffered by moving carriers. The average distance traveled by a carrier in its random thermal motion between collisions is called the mean free path. If barrier thickness is large compared to mean free path of carriers, motion of carriers in the barrier can be treated as drift and diffusion. This viewpoint is the basis of the diffusion theory of rectification. According to this theory, the factor C in Eq. (5) depends on the mobility and the electric field in the barrier.

Diode theory. When the barrier thickness is comparable to or smaller than the mean free path of the carriers, then the carriers cross the barrier without being scattered, much as in a vacuum tube diode.

According to this theory, the factor C in the rectifier equation is $v/4$, where v is the average thermal velocity of the carriers.

Tunneling theory. Instead of surmounting a potential barrier, carriers have a probability of penetrating through the barrier. The effect, called tunneling, becomes dominant if the barrier thickness is sufficiently small. This effect is important in many applications.

Two-carrier theory. Often the conduction through a rectifying barrier depends on both electron and hole carriers. An important case is the *pn* junction between *p*- and *n*-sections of a semiconductor material. Also, in metal-semiconductor rectifiers, the barrier presents an obstacle for the flow of majority carriers but not for the flow of minority carriers, and the latter may become equally or more important.

Rectification at pn junctions. A *pn* junction is the boundary between a *p*-type region and an *n*-type region of a semiconductor. When the impurity content varies, there is a variation of electron and hole concentrations. A variation of carrier concentrations is related to a shift of the energy bands relative to the constant Fermi level. This is brought about by a variation of the electrostatic potential which requires the existence of a space charge. If the impurity content changes greatly within a short distance, a large space charge is obtained within a narrow region. Such is the situation existing in a rectifying *pn* junction.

When a voltage is applied to make the *n*-region negative relative to the *p*-region, electrons flow from the *n*-region, where they are abundant, into the *p*-region. At the same time, holes flow from the *p*-region, where holes are abundant, into the *n*-region. The resistance is therefore relatively low. The direction of current in this case is forward. Clearly, the resistance will be high for current in the reverse direction.

With a current in the forward direction, electrons in the *n*-region and holes in the *p*-region flow toward the junction and there must be continuous hole-electron recombination in the neighborhood of the junction. The minority carrier concentration in each region is increased near the junction because of the influx of the carriers from the other region. This phenomenon is known as carrier-injection. When there is a current in the reverse direction, there must be a continuous generation of holes and electrons in the neighborhood of the junction, from which electrons flow out into the *n*-region and holes flow out into the *p*-region. Thus current through a *pn* junction is controlled by the hole-electron recombination or generation in the vicinity of the junction.

The transistor consists of two closely spaced *pn* junctions in a semiconductor with an order *pnp* or *npn*.

Contact rectification. If the height of a rectifying contact barrier is high, only a very small fraction of majority carriers can pass over the barrier. The fraction may be so small as to be comparable with the concentration of the minority carriers, provided the energy gap is not too large. The current due to the minority carriers becomes appreciable if the barrier height above the Fermi level approaches the energy difference between the Fermi level and the top of the valence band (Fig. 3).

The concentration of minority carriers is higher at the contact than in the interior of the semiconductor. With a sufficiently high barrier, it is possible to obtain at the contact a minority carrier concentration higher than that of the majority carriers. The small region where this condition occurs is called the inversion layer.

As in the case of a *pn* junction, a forward current produces injection of minority carriers. With the presence of an inversion layer, the injection can be so strong as to increase appreciably the conductivity in the vicinity of the contact. Ordinarily, contact rectifiers consist of a semiconductor in contact with a metal whisker. For large forward currents, the barrier resistance is small, and the resistance of the rectifier is determined by the spreading resistance of the semiconductor for a contact of small area. By increasing the conductivity in the vicinity of the contact where the spreading resistance is concentrated, carrier-injection may reduce considerably the forward resistance of the rectifier.

Surface electronics. The surface of a semiconductor plays an important role technologically, for example, in field-effect transistors and charge-coupled devices. Also, it presents an interesting case of two-dimensional systems where the electric field in the surface layer is strong enough to produce a potential wall which is narrower than the wavelengths of charge carriers. In such a case, the electronic energy levels are grouped into subbands, each of which corresponds to a quantized motion normal to the surface, with a continuum for motion parallel to the surface. Consequently, various properties cannot be trivially deduced from those of the bulk semiconductor. *See* SURFACE PHYSICS.

[H. Y. FAN]

Bibliography: B. L. Crowder (ed.), *Ion Implantation in Semiconductors and Other Materials*, 1973; N. B. Hannay and U. Colombo (eds.), *Electronic Materials*, 1973; *Nuovo Cimento*, vol. 38B, no. 2, 1977; Proceedings of the 8th (1976 International) Conferences on Solid State Devices, *Jap. J. Appl. Phys.*, suppl. 16–1, 1977; F. Seiz and D. Turnbull (eds.), *Solid State Physics*, vol. 1, 1955; W. Shockley, *Electrons and Holes in Semiconductors*, 1950; *Surf. Sci.*, vol. 73, 1978; S. Sze, *Physics of Semiconductor Devices*, 1969; R. K. Willardson and A. C. Beer (eds.), *Semiconductors and Semimetals*, vols. 1–10, 1966–1975.

Series

The indicated sum of a succession of numbers or terms. Series are used to obtain approximate values of infinite repeating decimals, to solve transcendental equations, to obtain values of logarithms or trigonometric functions, to evaluate integrals, and to solve boundary value problems.

For a finite series, with only a limited number of terms, the sum is found by addition. For an infinite series, with an unlimited number of terms, a sum or value can be assigned only by some limiting process. When the simplest such process yields a value, the infinite series is convergent. Many tests for convergence enable one to learn whether a sum can be found without actually finding it.

If each term of an infinite series involves a variable x and the series converges for each value of x in a certain range, the sum will be a function of x. Often the sum is a given function of x, $f(x)$, for which a series having terms of some given form is desired. Thus the Taylor's series expansion

$$f(x) = \sum_{n=0}^{\infty} f^{(n)}(a) \frac{(x-a)^n}{n!}$$

can be found for a large class of functions, the analytic functions, and represents such functions for sufficiently small values of $|x - a|$. For a much less restricted type of function on the interval $-\pi < x < \pi$, a Fourier series expansion of the form

$$\tfrac{1}{2}A_0 + \sum_{n=1}^{\infty} (A_n \cos nx + B_n \sin nx)$$

can be found.

Finite series. Here the problem of interest is to determine the sum of the first n terms,

$$S_n = u_0 + u_1 + u_2 + \cdots + u_{n-1}$$

when u_n is a given function of n. Examples are the arithmetic series, with $u_n = a + nd$ and $S_n = (n/2)[2a + (n-1)d]$, and the geometric series, with $u_n = ar^n$ and $S_n = a(1 - r^n)/(1 - r)$. See PROGRESSION.

If v_n is any function such that $v_{n+1} - v_n = u_n$, then $S_n = v_n - v_0$. For methods of solving this difference equation $\Delta v_n = u_n$ see INTERPOLATION.

For example, if u_n is any polynomial of the nth degree, then v_n will be a polynomial of the $(n+1)$st degree. In particular,

$$1 + 2 + 3 + \cdots + n = \frac{n(n+1)}{2}$$

$$1^2 + 2^2 + 3^2 + \cdots + n^2 = \frac{n(n+1)(2n+1)}{6}$$

$$1^3 + 2^3 + 3^3 + \cdots + n^3 = \frac{n^2(n+1)^2}{4}$$

As an example with u_n a rational function of n, from

$$\frac{-1}{n+2} - \frac{-1}{n+1} = \frac{1}{(n+2)(n+1)}$$

it may be concluded that if

$$u_n = \frac{1}{(n+1)(n+2)} \quad \text{then} \quad S_n = 1 - \frac{1}{n+1} = \frac{n}{n+1}$$

Convergence and divergence. An infinite series is the indicated sum of an unlimited number of terms

$$u_0 + u_1 + u_2 + \cdots + u_n + \cdots$$

or more briefly

$$\sum_{n=0}^{\infty} u_n$$

or simply Σu_n, read "sigma of u_n." The sum S_n of the first n terms is known as the nth partial sum. Thus S_n is the finite sum

$$\sum_{k=0}^{n-1} u_k$$

If, as n increases indefinitely or becomes infinite, the partial sum S_n approaches a limit S, then the infinite series Σu_n is convergent. S denotes the sum or value of the series. For example, if $|r| < 1$,

$$S = \Sigma ar^n = \frac{a}{1-r} \quad \text{since} \quad S_n = a\frac{1-r^n}{1-r}$$

and

$$S = \Sigma \frac{1}{(n+1)(n+2)} = 1 \quad \text{since} \quad S_n = 1 - \frac{1}{n+1}$$

If, as n becomes infinite, the partial sum S_n does not approach a finite limit, then the infinite

series Σu_n is divergent. For example, $\Sigma 1$ diverges, since here $S_n = n$ becomes infinite with n. Also $\Sigma(-1)^n$ diverges, since here $S_n = \tfrac{1}{2}[1 - (-1)^{n+1}]$ which is alternately one and zero.

It follows from the definition of S_n that $S_{n+1} - S_n = u_n$. For a convergent series, one may take limits in this equality and so deduce that

$$\lim_{n \to \infty} u_n = S - S = 0$$

Thus, if as n becomes infinite, either u_n approaches a nonzero limit or u_n fails to approach a limit, the series must diverge. This checks the earlier conclusion about $\Sigma 1$, with $\lim u_n = 1$, and also that about $\Sigma(-1)^n$, where $u_n = (-)^n$ does not approach a limit. A convergent series remains convergent if a finite number of terms is added, removed, or changed either at the beginning or distributed throughout the series. This changes all S_n after a certain point by the same finite constant. Thus the limit S is changed by this same constant, but there is no change in the fact of approach to a limit.

Positive series. These are series each of whose terms is a positive number or zero. For such series, the partial sum S_n increases as n increases. If for some fixed number A, no sum S_n ever exceeds A, the sums are bounded and admit A as an upper bound. In this case, S_n must approach a limit, and the series is convergent. If every fixed number is exceeded by some S_n, the sums are unbounded. In this case, S_n must become positively infinite and the series is divergent. The tests for convergence of positive series are tests for boundedness, and this is shown by a comparison of S_n with the partial sums of another series or with an integral.

The integral test. Let the function $f(x)$ be positive and always decrease as x increases for x greater than m, some fixed positive integer. Then the series $\Sigma f(n)$, with $u_n = f(n)$, converges if the integral

$$\int_m^{\infty} f(x)\, dx$$

converges. The series diverges if the integral diverges. The principal application of this test is that with $f(x) = 1/x^p$, where p is any positive constant. The result is that the series

$$\sum_{n=1}^{\infty} \frac{1}{n^p}$$

converges if p is greater than 1 and diverges if p is less than or equal to 1. Such divergent series as

$$\sum_{n=1}^{\infty} \frac{1}{n} \quad \text{or} \quad \sum_{n=1}^{\infty} \frac{1}{\sqrt{n}}$$

illustrate that series may diverge and still have $\lim u_n = 0$. The slow divergence of

$$\sum_{n=1}^{\infty} \frac{1}{n}$$

may be seen from the fact that over 10^{433} terms must be taken to make the partial sums exceed 1000.

Comparison tests. Let k be any positive constant, Σu_n a positive series to be tested, and Σc_n a positive series known to be convergent. Then if $u_n \leq kc_n$ for all n greater than some fixed integer m, the series Σu_n converges.

If Σd_n is a positive series known to be divergent,

and $u_n \geqq k d_n$ for all n greater than some fixed integer m, then the series Σu_n is divergent.

As corresponding tests involving limits, if

$$\lim_{n \to \infty} \frac{u_n}{c_n} = L$$

where L is a finite limit, then Σu_n converges. But if

$$\lim_{n \to \infty} \frac{u_n}{d_n} = L \quad (L > 0, L = +\infty)$$

then the series Σu_n is divergent.

The ratio test. For positive series, the simple ratio test is based on a consideration of

$$\lim_{n \to \infty} \frac{u_{n+1}}{u_n} = t$$

If t is less than unity, the series converges. If t is greater than unity, or if the ratio becomes positively infinite, the series diverges.

If $t = 1$, no conclusion can be drawn directly. But in many such cases

$$\frac{u_{n+1}}{u_n} = 1 - \frac{b}{n} + \frac{c}{n^2} + \cdots$$

may be written. In this case the series converges if $b > 1$, and diverges if $b \leqq 1$.

Cauchy's test, which is related to the ratio test but depends on a single term, is as follows. If for a positive series

$$\sqrt[n]{u_n} \leqq r < 1 \quad \text{for all } n \text{ greater than } m$$

the series converges. If $\sqrt[n]{u_n} \geqq 1$ for an infinite number of values of n, the series diverges.

Alternating series. These are series whose terms are alternately positive and negative. For such a series, if each term is numerically less than the preceding term, and

$$\lim_{n \to \infty} u_n = 0$$

the series converges. An example is

$$\Sigma (-1)^n \frac{1}{n+1}$$

For such a series, the difference between S_n, the nth partial sum, and S, the sum of the series, is numerically less than the first unused term, u_n.

Absolute convergence. For any series Σu_n which may have both positive and negative terms, the series of absolute values, $\Sigma |u_n|$, is a positive series whose convergence may be proved by one of the tests for positive series. If $\Sigma |u_n|$ converges, then Σu_n necessarily converges and is said to converge absolutely. The sum of an absolutely convergent series is independent of the order of the terms.

Conditional convergence. A series which converges but which does not converge absolutely is said to be conditionally convergent. For such a series, a change in the order of the terms may change the sum or cause divergence. In fact, by a suitable rearrangement, any sum may be obtained. The series $1 - \frac{1}{2} + \frac{1}{3} - \cdots$ is conditionally convergent with sum $\log_e 2$. The rearrangement $1 + \frac{1}{3} - \frac{1}{2} + \frac{1}{5} + \frac{1}{7} - \frac{1}{4} + \cdots$ obtained by taking blocks of two positive terms and then one negative term is conditionally convergent with sum $\frac{3}{2} \log_e 2$.

Operations on series. Two convergent series, if added termwise, give a convergent series. If $\Sigma u_n = S$ and $\Sigma v_n = T$, then $\Sigma(u_n + v_n) = S + T$. If both series are absolutely convergent, then the double series $\Sigma\Sigma u_m v_n$, where m and n each run from 1 to infinity, converges to the product ST absolutely, and so does any rearrangement. In particular, this is true for the Cauchy product, with $u_0 v_n + u_1 v_{n-1} + \cdots + u_{n-1} v_1 + u_n v_0$ as its $(n+1)$st term.

If Σu_n and Σv_n each converge, and if the Cauchy product series converges, its sum is ST. This will necessarily be the case if at least one of the series converges absolutely.

In any convergent series, parentheses may be inserted to form a new convergent series, with the same sum. But the removal of parentheses may convert a convergent series to a divergent one; for example, $\Sigma(1-1) = 0$ becomes $\Sigma(-1)^n$, which diverges.

Power series. These are series with $u_n = a_n x^n$. For such a series, it may happen that

$$\lim_{n \to \infty} \left| \frac{a_{n+1}}{a_n} \right| = A$$

If $A = 0$, the series converges for all values of x. If $A \neq 0$, the series converges for all x of the interval $-1/A < x < 1/A$. It will diverge for all x with $|x| > 1/A$. For any power series, the interval of convergence is related in this way to a number A, which, however, in the general case has to be given by the superior limit of

$$\sqrt[n]{|a_n|}$$

Similar remarks apply to the series with $u_n = a_n(x-c)^n$. Here the interval of convergence is $|x - c| < 1/A$, where A may be

$$\lim \left| \frac{a_{n+1}}{a_n} \right| \quad \text{or} \quad \lim \sqrt[n]{|a_n|}$$

if these limits happen to exist. In any case, A is equal to the superior limit,

$$\overline{\lim} \sqrt[n]{|a_n|}$$

One of the most important power series is the binomial series:

$$1 + mx + \frac{m(m-1)}{1 \cdot 2} x^2 + \cdots$$
$$+ \frac{m(m-1)(m-2) \cdots (m-n+1)}{n!} x^n + \cdots$$

When m is a positive integer, this is a finite sum of $m + 1$ terms which equals $(1 + x)^m$ by the binomial theorem. When m is not a positive integer, the interval of convergence is $-1 < x < 1$, and for x in this interval, the sum of the series is $(1 + x)^m$. See BINOMIAL THEOREM.

The sum function of a power series is continuous inside the interval of convergence. If the series converges at either end of the interval, the function is continuous at this end. Inside the interval of convergence, a power series may be integrated termwise. At any point inside the interval, it may be differentiated termwise. For example, from the series

$$\frac{1}{1-x} = 1 + x + x^2 + \cdots + x^n + \cdots$$

$$\text{for } -1 < x < 1$$

differentiation gives

$$\frac{1}{(1-x)^2}=1+2x+\cdots+nx^{n-1}+\cdots$$

Integration gives

$$\log_e(1-x)=-x-\frac{x^2}{2}-\frac{x^3}{3}-\cdots-\frac{x^{n+1}}{n+1}-\cdots$$

Since this converges when $x=1$, $\log_e 2=1-\frac{1}{2}+\frac{1}{3}-\frac{1}{4}+\cdots$.

Taylor series. Let the power series $\Sigma a_n(x-c)^n$ have the sum function $f(x)$. Then $a_0=f(c)$, $a_n=f^{(n)}(c)/n!$, and the series is the Taylor series of $f(x)$ at $x=c$. Thus, every power series whose interval of convergence has positive length can be put in the form

$$f(x)=f(c)+f'(c)\frac{x-c}{1!}+\cdots$$
$$+f^{(n)}(c)\frac{(x-c)^n}{n!}+\cdots$$

where $f(x)$ is the sum function.

The Maclaurin series is the special case of Taylor series with $c=0$

$$f(x)=f(0)+f'(0)\frac{x}{1!}+\cdots+f^{(n)}(0)\frac{x^n}{n!}+\cdots$$

Remainder term. For any function which is finite together with all of its derivatives for $x=c$, the difference between the function and the first $(n+1)$ terms of its Taylor series is the remainder R_n. For a suitable value x_1 in the interval $c<x_1<x$,

$$R_n=f^{(n+1)}(x_1)\frac{(x-c)^{n+1}}{(n+1)!}$$

which is Lagrange's form for R_n. This gives $x_1=c+\theta h$, where $h=x-c$, and θ is a suitable value between 0 and 1. It is true that, frequently, if $h\to 0$, $\theta\to 1/(n+2)$. Cauchy's form for R_n is

$$R_n=f^{(n+1)}(c+\theta h)(1-\theta)^n\frac{h^{n+1}}{n!}$$

This may be used to prove that the binomial series converges to $(1+x)^m$, by showing that $R_n\to 0$ as $n\to\infty$. Lagrange's form for R_n may be used to show that

$$e^x=1+x+\frac{x^2}{2!}+\cdots+\frac{x^n}{n!}+\cdots$$

$$\sin x=x-\frac{x^3}{3!}+\frac{x^5}{5!}-\cdots$$

$$\cos x=1-\frac{x^2}{2!}+\frac{x^4}{4!}-\cdots$$

are each Maclaurin series valid for all values of x.

Operations on power series. Two power series $\Sigma a_n(x-c)^n$ and $\Sigma A_n(x-c)^n$ may be added or multiplied by the Cauchy product rule. The resultant series will converge in an interval at least as large as the smaller of the intervals of convergence of the two given series. The first series may be divided by the second, provided that the divisor series is not zero at $x=c$, to give a series for the quotient with some nonzero interval of convergence. A Taylor series with constant term c may be substituted in another series about $x=c$. A power series may be inverted; that is, if $y=\Sigma a_n(x-c)^n$, with $a_1\neq 0$, there is an expansion $x=\Sigma b_n(y-a_0)^n$

where $b_0=c$, and the other b_n may be found by substitution.

Complex series. The series Σz_n, in which the general term is the complex number x_n+iy_n, with $i^2=-1$, is said to converge to $A+iB$ if $\Sigma x_n=A$ and $\Sigma y_n=B$. Thus the tests for real series may be made on Σx_n and Σy_n, or the positive series $\Sigma|z_n|$ may be tested. If this converges, the given complex series converges absolutely, and necessarily converges.

Complex power series. The series with $u_n=a_n z^n$, where the $a_n=p_n+iq_n$ are now complex numbers, and the complex variable $z=x+iy$, are of great importance in the theory of analytic functions of a complex variable. More generally, power series occur in $(z-c)$, with $u_n=a_n(z-c)^n$, where c is any complex number. There is always a radius of convergence $R=1/A$, where A may be

$$\lim\left|\frac{a_{n+1}}{a_n}\right| \quad\text{or}\quad \lim\sqrt[n]{|a_n|}$$

if these limits exist. In any case, A is equal to the superior limit

$$\overline{\lim}\sqrt[n]{|a_n|}$$

If A if finite and not zero, so is R, and there is a circle of convergence $|z-c|<R$ within which the series converges. If $A=0$, the series converges for all values of z. If $A=\infty$, $R=0$, and the series converges for no value except $z=c$. For example, $\Sigma n!z^n$ converges for $z=0$, but for no other values of z; $\Sigma(z-3)^n/2^n$, with $A=\frac{1}{2}$, $R=2$, converges [to $2/(5-z)$] for $|z-3|<2$; $\Sigma z^n/n!$, with $A=0$, $R=\infty$, converges (to e^z) for all values of z.

Uniform convergence. Let each term of a series be a function of z, $u_n=g_n(z)$. Let S_n be the sum of the first n terms, and S the sum to which the series converges for a particular value of z. Then $R_n=S-S_n$ is the remainder after n terms, and for the particular value of z, $\lim R_n$ must equal zero. If, for a given range of z, it is possible to make $R_n(z)$ arbitrarily small for sufficiently large n without specifying which z in the range is under consideration, the series converges uniformly. The Weierstrass comparison test may be used to test for uniformity. If ΣU_n is a convergent series of positive constants, or a uniformly convergent series of positive terms, and $|u_n|\leq U_n$ in the range considered, then Σu_n converges uniformly in the range. For a uniformly convergent series, if each $u_n(z)$ is continuous, the sum function $S(z)$ is continuous. A uniformly convergent series may be integrated termwise. If the differentiated series converges uniformly, the series may be differentiated termwise.

Analytic functions. A function $f(z)$ is analytic in a two-dimensional region if, at each point of the region, the derivative $f'(z)$ exists. A uniformly convergent series of analytic functions has a sum function which is analytic and, in particular, may be integrated or differentiated termwise. These results all apply to any power series whose radius of convergence R is not zero, since every power series $\Sigma a_n(z-c)^n$ converges uniformly in any circle $|z-c|<R_1$, where $R_1<R$. Thus the sum function $f(z)=\Sigma a_n(z-c)^n$ is an analytic function of z for any z such that $|z-c|<R$. For such z, the function $f(z)$ possesses derivatives of every order, each expandable in a series obtained by termwise differentiation. Each such series has the same radius of convergence R. The power series is neces-

sarily the Taylor series of the function $f(z)$, so that $a_n = f^{(n)}(c)/n!$. If a function is single-valued and analytic in a two-dimensional region of the complex plane, and c is any point inside this region, then there is a Taylor expansion $f(z) = \Sigma f^{(n)}(c)(z-c)^n/n!$. Its radius of convergence R is at least as great as the largest circle with center at c, all of whose interior points are interior points of the given region of analyticity.

Fourier series. Let $f(x)$ be a periodic function of period T, so that $f(x+T) = f(x)$. Then the Fourier series for $f(x)$ is

$$A + \sum_{n=1}^{\infty} (A_n \cos n\omega x + B_n \sin n\omega x)$$

where $\omega = 2\pi/T$ and

$$A = \frac{1}{T}\int_a^{a+T} f(x)\, dx \quad A_n = \frac{2}{T}\int_a^{a+T} f(x)\cos n\omega x\, dx$$

$$B_n = \frac{2}{T}\int_a^{a+T} f(x)\sin n\omega x\, dx$$

That is, A is the average of $f(x)$, A_n is twice the average of $f(x)\cos n\omega x$, and B_n is twice the average of $f(x)\sin n\omega x$ over any interval of length T. Because of the periodicity, it is immaterial which interval, or value of a, is used. If, in each interval of period T, the graph of $f(x)$, which need not be continuous, is made up of arcs which collectively have finite length, then the Fourier series necessarily converges to $f(x)$ at each point where $f(x)$ is continuous. At each point of discontinuity the series converges to $\frac{1}{2}[f(x-)+f(x+)]$, the average of the right- and left-hand limits. If $f(x)$ is continuous at all points, and has a uniformly bounded derivative, the series will converge uniformly for all values of x.

A Fourier series may always be integrated termwise, but it may not be permissible to differentiate such a series termwise unless $f'(x)$ satisfies some condition for development in a Fourier series.

Fourier sine series. On the interval $0 < x < L$, the Fourier sine series

$$f(x) = \sum_{n=1}^{\infty} B_n \sin n\omega x$$

where $\omega = \pi/L$ and $B_n = \frac{2}{L}\int_0^L f(x)\sin n\omega x\, dx$

may be considered as the Fourier series for the function $F(x)$ which is odd, $F(x) = -F(-x)$, of period $2L$, $F(x+2L) = F(x)$, and which equals $f(x)$ on the interval $0 < x < L$. Thus if $f(x)$ was an odd function for $-L < x < L$, the sine series is valid inside this interval.

Fourier cosine series. On the interval $0 \le x \le L$, the Fourier cosine series

$$f(x) = A + \sum_{n=1}^{\infty} A_n \cos n\omega x$$

where $\omega = \pi/L$ $\quad A = \frac{1}{L}\int_0^L f(x)\, dx$

$$A_n = \frac{2}{L}\int_0^L f(x)\cos n\omega x\, dx$$

may be considered as the Fourier series for the function $F(x)$ which is even, $F(x) = F(-x)$, of period $2L$, $F(x+2L) = F(x)$, and which equals $f(x)$ on the interval $0 < x < L$. Thus if $f(x)$ was an even function for $-L \le x \le L$, the cosine series is valid in-

side this interval. The sum of the sine or cosine series is always $\frac{1}{2}[F(x-)+F(x+)]$ for each value of x, so that the sum of $F(x)$ at each point of continuity.

Cesaro summability. Let S_n be the sum of the first n terms of a series Σu_n, and form the sequence

$$C_1 = S_1/1, C_2 = \frac{1}{2}(S_1 + S_2), \ldots, C_n = \frac{1}{n}\sum_{k=1}^{n} S_k$$

If $\quad \lim_{n \to \infty} C_n = L$

exists, the series is said to be summable in the sense of Cesaro, or $C(1)$ to L. A convergent series with sum S is necessarily summable $C(1)$ to S, but a divergent oscillating series may be summable $C(1)$. For example, if $u_n = (-1)^n$, the series is $1 - 1 + 1 - \cdots$ which diverges. But $S_k = 1$ and 0 alternately, and $L = \frac{1}{2}$. Thus the series is summable $C(1)$ to $\frac{1}{2}$, a value also made plausible from consideration of the limit as $x \to 1$ of the identity

$$\frac{1}{1+x} = 1 - x + x^2 - x^3 + \cdots$$

There are other more elaborate methods of summability by which sums can be assigned to certain divergent series.

The Fejér theorem. There are continuous functions whose Fourier series do not converge. But, for any continuous periodic function, the theorem of Fejér asserts that the Fourier series is always summable $C(1)$ to the function for every value of x.

Convergence in the mean. Let $f(x)$ be periodic with T, and

$$T_n = a + \sum_{k=1}^{n=1} (a_k \cos k\omega x + b_k \sin k\omega x)$$

be any trigonometric sum of order $n-1$. Then T_n may be regarded as an approximation to $f(x)$, and the degree of the approximation may be measured by the average of the square of the error,

$$E_n = \frac{1}{T}\int_a^{a+T} [f(x) - T_n(x)]^2\, dx$$

Then

$$E_n = \frac{1}{T}\int_a^{a+T} [f(x)]^2\, dx - A^2 - \frac{1}{2}\sum_{k=1}^{n-1} (A_k^2 + B_k^2)$$
$$+ (A-a)^2 + \frac{1}{2}\sum_{k=1}^{n-1} (A_k - a_k)^2 + (B_k - b_k)^2$$

This shows that, for all trigonometric sums of given order, S_n, the one formed with Fourier coefficients, $a = A$, $a_k = A_k$, $b_k = B_k$, makes the average error least. Moreover, for any function such that

$$\int_a^{a+T} [f(x)]^2\, dx$$

is finite, as n becomes infinite the limit of E_n, the average squared error for S_n, is zero. Thus

$$\lim_{n \to \infty} \int_a^{a+T} [f(x) - S_n(x)]^2\, dx = 0$$

This is the condition for the trigonometric sums S_n to converse in the mean to $f(x)$.

Integration. If the sequence $S_n(x)$ converges in the mean to $f(x)$, and $g(x)$ is any square-integrable fixed function, it is true that

$$\lim \int S_n(x) g(x)\, dx = \int f(x) g(x)\, dx$$

for the same limits on the integrals in both members. That is, the series with partial sums $S_n(x)$ may be integrated termwise, and the same is true of the series with partial sums $g(x)S_n(x)$. This fact may be used to derive the formulas for the Fourier coefficients in terms of integrals which were given above.

Again, if $T_n(x)$ converges in the mean to $g(x)$, it is true that

$$\lim \int S_n(x)T_n(x)\,dx = \int f(x)g(x)\,dx$$

with the same limits on the integrals in both members. From this Parseval's identity may be deduced,

$$\int_a^{a+T} f(x)g(x)\,dx = AA' + \tfrac{1}{2}\sum_{k=1}^{\infty}(A_kA_k' + B_kB_k')$$

where A', A_k', B_k' are Fourier coefficients for $g(x)$.

Examples of Fourier series. For $-\pi < x < \pi$, there is a series

$\sin ax$

$$= \frac{2\sin a\pi}{\pi}\left(\frac{\sin x}{1^2 - a^2} - \frac{2\sin 2x}{2^2 - a^2} + \frac{3\sin 3x}{3^2 - a^2} - \cdots\right)$$

The analogous expansion for $\cos ax$, with $x = 0$ and $a = z/\pi$ leads to

$$\csc z = \frac{1}{z} - \frac{2z}{z^2 - \pi^2} + \frac{2z}{z^2 - 2^2\pi^2} - \frac{2z}{z^2 - 3^2\pi^2} + \cdots$$

The expansion for $\cos ax$ is valid for $x = \pi$. With this and $a = z/\pi$, it gives

$$\cot z = \frac{1}{z} + \frac{2z}{z^2 - \pi^2} + \frac{2z}{z^2 - 2^2\pi^2} + \frac{2z}{z^2 - 3^2\pi^2} + \cdots$$

The last two expansions hold for any complex z for which no denominator is zero.

Infinite product for the sine. A series for $\log_e[(\sin z)/z]$ may be found from the expansion of $\cot z - 1/z$ by integration. This leads to

$$\sin z = z\left(1 - \frac{z^2}{\pi^2}\right)\left(1 - \frac{z^2}{2^2\pi^2}\right)\left(1 - \frac{z^2}{3^2\pi^2}\right)\cdots$$

the infinite product for the sine. Putting $z = \pi/2$ leads to

$$\frac{\pi}{2} = \frac{2\cdot2\cdot4\cdot4\cdot6\cdot6\cdots}{1\cdot3\cdot3\cdot5\cdot5\cdot7\cdots}$$

which is Wallis's product. Equating the z^3 term on the right with the term $-z^3/3!$ in $\sin z$ gives

$$1 + \frac{1}{2^2} + \frac{1}{3^2} + \frac{1}{4^2} + \cdots = \frac{\pi^2}{6}$$

This is a special case of

$$\sum_{n=1}^{\infty}\frac{1}{n^{2k}} = \frac{2^{2k-1}\pi^{2k}}{(-1)^{k-1}(2k)!}B_{2k}$$

where the B_{2k} are rational fractions, the Bernouilli numbers. $B_2 = \frac{1}{6}$, $B_4 = -\frac{1}{30}$, $B_6 = \frac{1}{42}$. For n odd, B_n is zero unless $n = 1$, and $B_1 = -\frac{1}{2}$. These B_n are the coefficients in

$$\frac{x}{e^x - 1} = \sum_{n=0}^{\infty}B_n\frac{x^n}{n!}$$

They occur in such expansions as

$$\tan x = \sum_{k=1}^{\infty}\frac{(2^{2k}-1)2^{2k}(-1)^{k-1}}{(2k)!}B_{2k}x^{2k-1}$$

$$\cot x = \frac{1}{x} + \sum_{k=1}^{\infty}\frac{(-1)^k 2^{2k}}{(2k)!}B_{2k}x^{2k-1}$$

$$\csc x = \frac{1}{x} + \sum_{k=1}^{\infty}\frac{(2 - 2^{2k})(-1)^k}{(2k)!}B_{2k}x^{2k-1}$$

Stirling's formula for m!. The expansion

$$\log_e(m!) = \log_e\sqrt{2\pi} + \left(m + \frac{1}{2}\right)(\log_e m) - m$$
$$+ \sum_{r=1}^{n}\frac{B_{2r}m^{-2r+1}}{2r(2r-1)} + R_n$$

leads to a divergent series if R_n is omitted and $n \to \infty$. But the series is asymptotic in the sense that R_n is always numerically less than the last term in the sum which involves B_{2n}. Thus for large m, a few terms of the sum give a good approximation. This leads to Stirling's approximation to $m!$, $m! \sim \sqrt{2\pi m}\,m^m e^{-m}$. For $m \to \infty$, the absolute error becomes infinite, but the percentage error is of the order of $e^{1/12m} \sim \frac{1}{12}m$ which is small even for moderate m.

Applications of series. Series sometimes appear in disguised form in arithmetic. Thus the approximation of a rational number by an infinite repeating decimal is really a geometric series, $\frac{1}{3} = 0.3$ being a series with $u_n = 3/(10^{n+1})$ with sum $\frac{1}{3}$.

It is possible to use $1 = 0.9$ to find the fraction represented. For example, $0.285714 = 285714/999999 = \frac{2}{7}$.

Roots are often conveniently found by the binomial series. For instance, for small x, $(1+x)^{1/n} \sim 1 + (x/n)$, $\log_e(1+x) \sim x$.

Solution of equations. Sometimes algebraic or transcendental equations are best solved by reverting series. By translations of x and y, $y = y_0 + Y$, $x = x_0 + X$, it is possible to make $Y = 0$ correspond to $X = 0$ near the points of interest. Also, a change of scale, $X = kX'$, makes the first coefficient one. Then with new notation, $y = x + bx^2 + cx^3 + dx^4 + ex^5 + \cdots$, with x and y small. The reverted series is $x = y - by^2 + (2b^2 - c)y^3 - (5b^3 - 5bc + d)y^4 + (14b^4 - 21b^2c + 6bd + 3c^2 - e)y^5 + \cdots$. For example, with $d = 0$, $e = 0$, $y = -q$, this gives a series solution of the cubic equation $cx^3 + bx^2 + x + q = 0$ for small values of q.

Tables. The values of logarithms for tables may be computed by judicious use of the series

$$\log_e\frac{1+x}{1-x} = 2\left(x + \frac{x^3}{3} + \frac{x^5}{5} + \cdots\right)$$

The expansions

$$\sin x = x - \frac{x^3}{3!} + \frac{x^5}{5!} - \cdots$$

$$\cos x = 1 - \frac{x^2}{2!} + \frac{x^4}{4!} - \cdots$$

$$\sin^{-1}x = x + \frac{1}{2}\frac{x^3}{3} + \frac{1\cdot3}{2\cdot4}\frac{x^5}{5} + \frac{1\cdot3\cdot5}{2\cdot4\cdot6}\frac{x^7}{7} + \cdots$$

or

$$\tan^{-1}x = x - \frac{x^3}{3} + \frac{x^5}{5} - \frac{x^7}{7} + \cdots$$

are useful in computing tables of trigonometric functions. The last series makes it possible to compute π easily by using relations like

$$\frac{\pi}{4} = 2 \tan^{-1} \tfrac{1}{3} + \tan^{-1} \tfrac{1}{7}$$

or

$$\frac{\pi}{4} = 4 \tan^{-1} \tfrac{1}{5} - \tan^{-1} \tfrac{1}{239}$$

Integrals may often be found from series. For example,

$$\int_0^x e^{-x^2}\, dx = x - \frac{x^3}{3} + \frac{x^5}{5\cdot 2!} - \frac{x^7}{7\cdot 3!} + \cdots$$

makes it possible to evaluate the probability integral. For large values of x it is easier to use the divergent, but asymptotic expression

$$\int_0^x e^{-x^2}\, dx =$$

$$\frac{\sqrt{\pi}}{2} - e^{-x^2}\left(\frac{1}{2x} - \frac{1}{2^2 x^3} + \frac{3}{2^3 x^5} - \frac{3\cdot 5}{2^4 x^7} + \cdots\right)$$

Ordinary differential equations. An ordinary point of a linear differential equation is a value of x at which all the coefficients are analytic, with the first coefficient not zero. For x near such a value, the complete solution may be expressed in terms of Taylor series. For example, $x = 0$ is an ordinary point of

$$(1 - x^2)\frac{d^2 y}{dx^2} - 2x\frac{dy}{dx} + n(n+1)y = 0$$

which is Legendre's equation. For n zero or an integer, one solution is

$$P_n(x) = \sum_{k=0}^{\infty} (-1)^k \frac{(2n-2k)!}{2^n k!(n-k)!(n-2k)!}$$

the Legendre polynomial of degree n.

Fourier-Legendre series. If, for $-1 < x < 1$, the function $f(x)$ satisfies the conditions for expansion in a Fourier series, there is an expansion

$$f(x) = \sum_{n=0}^{\infty} a_n P_n(x)$$

where

$$a_n = \frac{2n+1}{2}\int_{-1}^{1} f(x) P_n(x)\, dx$$

The sum to $(n+1)$ terms gives the polynomial of the nth degree best approximating $f(x)$ in the sense of least square error.

Regular singular point. Let a differential equation have the form

$$A_2\frac{d^2 y}{dx^2} + A_1\frac{dy}{dx} + A_0 y = 0$$

where $A_2(x)$ is analytic and not zero at x_0, $(x - x_0)$ $A_1(x)$ and $(x - x_0)^2 A_0(x)$ are each analytic at x_0. Then if x_0 is not an ordinary point, it is a regular singular point. Near such a point, at least one solution may be found in the form $(x - x_0)^s$ times a Taylor series, where s is some real or complex value. For example, for n zero or any positive integer, the Bessel function of order n

$$J_n(x) = \sum_{k=0}^{\infty} \frac{(-1)^k x^{n+2k}}{2^{n+2k} k!(n+k)!}$$

is a solution of Bessel's differential equation

$$\frac{d^2 y}{dx^2} + \frac{1}{x}\frac{dy}{dx} + \left(1 - \frac{n^2}{x^2}\right)y = 0$$

Partial differential equations. Certain boundary-value problems in partial differential equations may be solved by the use of series. Thus, in two dimensions, Laplace's equation in polar coordinates is

$$\frac{\partial^2 U}{\partial r^2} + \frac{1}{r^2}\frac{\partial^2 U}{\partial \theta^2} + \frac{1}{r}\frac{\partial U}{\partial r} = 0$$

Let the values on the boundary of a circular region $r = a$ be given in the form $f(\theta)$, and let the Fourier expansion of this function of period 2π be

$$f(\theta) = A + \sum_{n=1}^{\infty} (A_n \cos n\theta + B_n \sin n\theta)$$

Then the solution of Laplace's equation with $U(a,\theta) = f(\theta)$ is

$$U(r,\theta) = A + \sum_{n=1}^{\infty} \left(\frac{r}{a}\right)^n (A_n \cos n\theta + B_n \sin n\theta)$$

Again, the solution of Laplace's equation

$$\frac{\partial^2 U}{\partial x^2} + \frac{\partial^2 U}{\partial y^2} = 0$$

with $U(0,y) = 0$, $U(L,y) = 0$, $U(x,+\infty) = 0$, $U(x,0) = f(x)$, with Fourier sine expansion

$$f(x) = \sum_{n=1}^{\infty} B_n \sin\frac{n\pi x}{L}$$

is given by

$$U(x,y) = \sum_{n=1}^{\infty} B_n e^{-n\pi y/L} \sin\frac{n\pi x}{L}$$

See FOURIER SERIES AND INTEGRALS.

[PHILIP FRANKLIN/SALOMON BOCHNER]

Bibliography: W. Kaplan, *Advanced Calculus*, 2d ed., 1973; I. S. Sokolnikoff and R. M. Redheffer, *Mathematics of Physics and Modern Engineering*, 2d ed., 1966; S. K. Stein, *Calculus and Analytic Geometry*, 2d ed., 1977.

Set theory

A mathematical term referring to the study of collections or sets. Consider a collection of objects (such as points, dishes, equations, chemicals, numbers, or curves). This set may be denoted by some symbol, such as X. It is useful to know properties that the set X has, irrespective of what the elements of X are. The cardinality of X is such a property.

Cardinality of sets. Two sets A and B are said to have the same cardinal written $C(A) = C(B)$, provided there is a one-to-one correspondence between the elements of A and the elements of B. For finite sets this notion coincides with the phrase "A has the same number of elements as B." However, for infinite sets the above definition yields some interesting consequences. For example, let A denote the set of integers and B the set of odd integers. The function $f(n) = 2n - 1$ shows that $C(A) = C(B)$. Hence, an infinite set may have the same cardinal as a part or subset of itself.

Subset. A is called a subset of B if each element of A is an element of B, and it is expressed as $A \subset B$. The collection of odd integers is a subset of the reals. Also, each set is a subset of itself.

Continuum hypothesis. An infinite set is called uncountable if it cannot be put in a one-to-one correspondence with the positive integers. Here is one of the unsolved problems of set theory. It is of particular interest since so many mathematicians have tried unsuccessfully to solve it. If X is an uncountable subset of the reals R, is $C(X)$ equal to $C(R)$? The conjecture that the answer is in the affirmative is called the continuum hypothesis. It has been shown that the answer cannot be decided by using the ordinary axioms of set theory.

Comparing cardinality of sets. One says that $C(A) \leqq C(B)$ if there is a one-to-one correspondence between the elements of A and a subset of the elements of B. One useful theorem that can be proved states that any two sets A, B are comparable, that is, either $C(A) \leqq C(B)$ or $C(B) \leqq C(A)$ (possibly both). Another theorem states that if $C(A) \leqq C(B)$ and $C(B) \leqq C(A)$, then $C(A) = C(B)$. Each of these results may be proved by using well orderings of A and B.

Ordering. An ordering is one way of setting up a one-to-one correspondence between two sets of the same cardinality. A relation $<$ is an order relation for a set X if it satisfies the following conditions:

1. If x_1, x_2 are two elements of X, either $x_1 < x_2$ or $x_2 < x_1$ (any two elements are related).

2. $x_1 \not< x_1$ (no element is less than itself).

3. If $x_1 < x_2$ and $x_2 < x_3$, then $x_1 < x_3$ (the order relation is transitive).

The following are examples of ordered sets: a horizontal line where $<$ means "is to the left of"; the reals where $<$ means "is less than"; and the collection of words in the dictionary where ordering is alphabetical.

Well ordering. An ordering of a set is called a well ordering if it satisfies the additional condition:

4. Each non-null subset Y of X has a first element; that is, there is an element y_o of Y such that if y' is another element of Y, $y_o < y'$.

The natural ordering of the positive integers is a well ordering, but neither the natural ordering of the integers nor of the reals is a well ordering. A well ordering for the integers is 0, 1, 2, . . . , −1, −2, ... Since a well ordering of the reals cannot be written down, one might guess that there is none. This guess is shown to be false by the theorem that states that any set X has a well ordering. In proving this, one considers the collection Z of all non-null subsets of X, selects a point $x_a = f(z_a)$ from each element z_a of Z, and well orders X so that if S_a is the set of all elements that procedes x_a, then $x_a = f(Z - S_a)$.

Some of the theorems proved by well ordering are so strange that their truths do not seem intuitively obvious. Well ordering is also used to construct pathological examples which serve as counterexamples to various conjectures. These counterexamples are useful since they show that the conjectures are false and it is useless to try to prove them.

Formation of sets. One approved method of forming a set is to consider a property P possessed by certain elements of a given set X. The set of elements of X having property P may be considered as a set Y. The expression $p \epsilon X$ is used to denote the fact that p is an element of X. Then $Y = \{p/p \epsilon X$ and p has property $P\}$. Another approved method is to consider the set Z of all subsets of a given set X. It may be shown in this case that $C(X) < C(Z)$.

Paradoxically, it is not permissible to regard the collection of all sets as a set. If such a collection X were called a set, and Z were used to denote the set of all subsets of X, one would arrive at the absurdity that $C(X) < C(Z)$.

Operations with sets. In set theory, one is interested not only in the properties of sets but also in operations involving sets: addition, subtraction, multiplication, and mapping.

Sum or union. The sum of A and B ($A + B$ or $A \cup B$) is the set of all elements in either A or B: that is, $A + B = \{p/p \epsilon A$ or $p \epsilon B\}$.

Intersection, product, or common part. The intersection of A and B ($A \cdot B$, $A \cap B$, or AB) is the set of all elements in both A and B; that is, $A \cdot B = \{p/p \epsilon A$ and $p \epsilon B\}$. If there is no element which is in both A and B, one says that A does not intersect B and writes $A \cdot B = 0$.

Difference. The expression $A - B$ is used to denote the collection of elements of A that do not belong to B; that is, $A - B = \{p/p \epsilon A$ and $p \notin B\}$. If $A \subset B$, it is expressed as $A - B = 0$.

An example. If a person were to squirt some black ink on a plane, the set A of points in the dark spot would be an example of a point set. Suppose a set B is determined by squirting some red ink on the plane. Then $A + B$ designates the set of points covered by ink, $A \cdot B$ designates the set covered by both kinds of ink, and $A - B$ designates those covered by black but not red ink (Fig. 1).

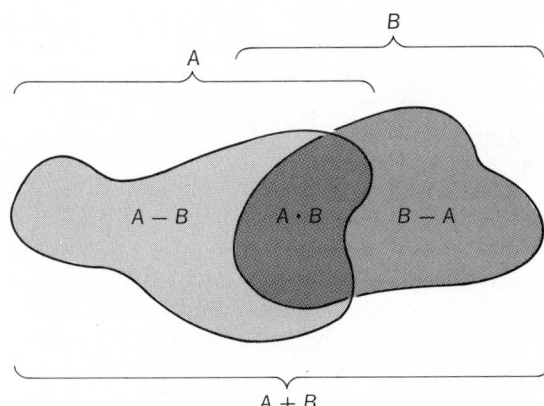

Fig. 1. Sets of points on a plane.

Boolean algebra. By using the previous notation, it follows that the sets of Fig. 1 satisfy some of the familiar laws of algebra as

$$A + B = B + A$$
$$A \cdot (B + C) = A \cdot B + A \cdot C$$

However, other identities are not so familiar:

$$X - (A + B) = (X - A) \cdot (X - B)$$
$$X - A \cdot B = (X - A) + (X - B)$$

See BOOLEAN ALGEBRA.

Transformations. A transformation of a set X into a set Y is a function that assigns a point of Y to each point of X. The transformation shown in Fig. 2 is the vertical projection of a set X onto a segment Y. The point assigned to X under a transformation f is called the image of x and denoted by $f(x)$. Also, the set of all points x sent into a particular point y of Y is called the inverse of y and denoted by $f^{-1}(y)$. The inverse of y shown in Fig. 2 is the sum of three segments. The equation $f(x) = x^2$ rep-

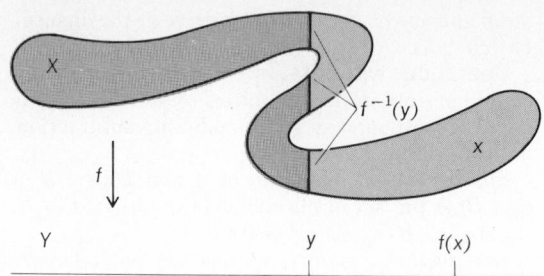

Fig. 2.   Vertical projection of set *X* onto segment *Y*.

resents a transformation that takes each real number into its square, for example, 2 to 4, −5 to 25, and so on. Rotations, congruences, and similarities are examples of transformations from geometry. However, in general, a transformation may change both the size and shape of an object.

Topology. Topology is one of the branches of mathematics that makes extensive use of set theory. Here, not only does one have sets of points for consideration but also collections of interiors of spheres or neighborhoods. These neighborhoods enable one to study limit points and the continuity of transformations. *See* CONFORMAL MAPPING; TOPOLOGY.   [R. H. BING]

Bibliography: P. R. Halmos, *Naive Set Theory*, 1974; E. Kamke, *Theory of Sets*, transl. from 2d German ed., 1950; K. Kuratowski and A. Mostawski, *Set Theory*, 2d ed., 1976; A. Levy, *Basic Set Theory*, 1978.

Shadow

A region of darkness caused by the presence of an opaque object interposed between such a region and a source of light. A shadow can be totally dark only in that part called the umbra, in which all parts of the source are screened off. With a point source, the entire shadow consists of an umbra, since there can be no region in which only part of the source is eclipsed. If the source has an appreciable extent, however, there exists a transition surrounding the umbra, called the penumbra, which is illuminated by only part of the source. Depending on what fraction of the source is exposed, the illumination in the penumbra varies from zero at the edge of the full shadow to the maximum where the entire source is exposed. The edge of the umbra is not perfectly sharp, even with an ideal point source, because of the wave character of light. *See* DIFFRACTION.

For example, the shadow of the Moon shown in the illustration has an umbra which barely touches the Earth during a total eclipse of the Sun. The eclipse is only partial for an observer situated in the penumbra. Astronauts circling the Earth or the Moon observe this shadow pattern during each orbit.

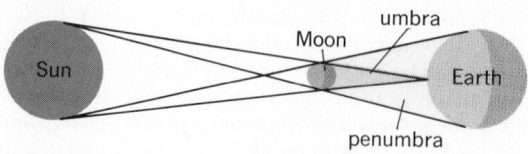

Shadow of the Moon at the time of an eclipse of the Sun. (Relative sizes are not to scale.)

The term shadow is also used with other types of radiation, such as sound or x-rays. In the case of sound, pronounced shadows are formed only for high frequencies. The high frequency of visible light waves makes possible shadows that are nearly black. X-ray photographs are shadowgrams in which the bones and tissues appear by virtue of their different opacities or degrees of absorption. The bending of light rays by reflection or refraction may also produce shadow patterns. An important practical example is the schlieren method of observing flow patterns in wind tunnels. *See* SCHLIEREN PHOTOGRAPHY.

[FRANCIS A. JENKINS/WILLIAM W. WATSON]

Shadowgraph of fluid flow

A simple method of making visible the disturbances that occur in fluid flow at high velocity. The three principal methods of optical fluid flow measurements, schlieren, interferometer, and shadowgraph, depend on the fact that light passing through a flow field of varying density is retarded differently through the field, resulting in a turning of the wavefronts, that is, a refraction of the rays, and in a relative phase shift along different rays. The first of these, the refraction of the rays, is the basis for shadowgraph flow visualization.

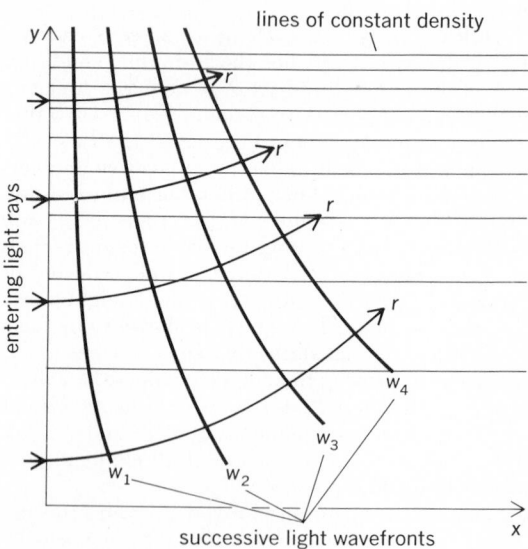

Fig. 1.   Density gradient in fluid flow results in a turning of the wavefronts, that is, a refraction of light rays.

Figure 1 shows light crossing a flow field of varying density in the y direction, as represented by the lines of constant density. The light can be considered to be acting as a wave entering at the left. The lines marked w are the wavefronts at successive times, and the lines orthogonal to them and marked r are the rays of light. Because large density gradients have a greater effect on the velocity of the light, the wavefront turns as shown. The ray of light is turned through the same angle. It can be shown that for plane flow, where conditions are the same along any x direction, the deflection of the ray is proportional to the density gradient.

For three-dimensional flow, as in a wind

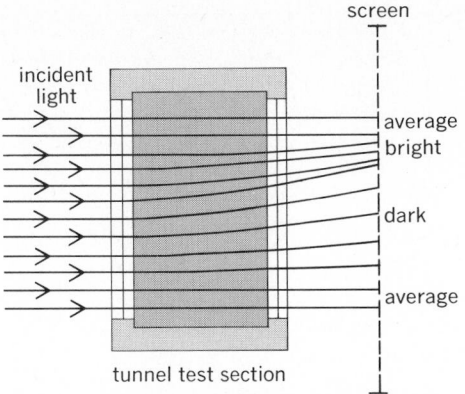

Fig. 2. Shadowgraph produced by light passing through fluid in test section indicates flow pattern.

tunnel, the final deflection depends upon all of the density gradients encountered.

The application of the shadowgraph to a wind tunnel is shown in Fig. 2. The rays enter normal to the sidewall through the window at the left. (The window is optical glass of high quality to reduce refraction and to assure that observed effects are caused by the flow in the test section.) As they pass through the test section, they encounter a change in density of the fluid in the tunnel and are deflected. The light rays then fall on a screen, where they are observed or photographed. Where the rays have crowded together, the screen is brightly illuminated, and where the rays diverge, the screen is dark. Where the spacing is unchanged, the illumination is normal, even though there has been a change of density along the path of the ray.

The light need not be parallel when it enters the test section, so the slit source and lens systems of the other methods are not required. This simplicity makes the shadowgraph system considerably less expensive than other methods; it is often used where the finer resolution of the other systems is not required or desirable. *See* INTERFEROMETRY; SCHLIEREN PHOTOGRAPHY.

[RICHARD F. CHANDLER]

Shock wave

A fully developed compression wave of large amplitude. Shock waves arise from sharp and violent disturbances generated from a lightning stroke, bomb blast, or other form of intense explosion, and from steady supersonic flow over bodies. The abrupt nature of a shock wave can best be visualized from a schlieren photograph or shadowgraph of supersonic flow over objects. Such photographs show well-defined surfaces in the flow field across which the density changes rapidly, in contrast to waves within the range of linear dynamic behavior of the fluid. Measurements of fluid density, pressure, and temperature across the surfaces show that these quantities always increase along the direction of flow, and that the rates of change are usually so rapid as to be beyond the spatial resolution of most instruments. These surfaces of abrupt change in fluid properties are called shock waves or shock fronts. *See* SCHLIEREN PHOTOGRAPHY; SHADOWGRAPH OF FLUID FLOW; WAVE MOTION IN FLUIDS.

Shock waves in supersonic flow may be classified as normal or oblique according to whether the orientation of the surface of abrupt change is perpendicular or at an angle to the direction of flow. A schlieren photograph of a supersonic flow over a blunt object is shown in Fig. 1. Although this photograph was obtained from a supersonic flow over a stationary model in a shock tube, the general shape of the shock wave around the object is quite typical of those observed in a supersonic wind tunnel, or of similar objects (or projectiles) flying at supersonic speeds in a stationary atmosphere. The shock wave in this case assumes an approximately parabolic shape and is clearly detached from the blunt object. The central part of the wave, just in front of the object, may be considered an approximate model of the normal shock; the outer part of the wave is an oblique shock wave of gradually changing obliqueness and strength.

Normal shock wave. The changes in thermodynamic variables and flow velocity across the shock wave are governed by the laws of conservation of mass, momentum, and energy, and also by the equation of state of the fluid. Thus, for the case of normal shock, the sketch in Fig. 2 illustrates a steady flow across a stationary wavefront. The mass flow and momentum equations are the same as for an acoustic wave. However, in a shock wave, changes in pressure and density across the wavefront can no longer be considered small. As a consequence, the velocity of propagation of the shock wave relative to the undisturbed fluid is given by Eq. (1). In addition, conservation of thermal and

$$u_1{}^2 = \frac{\rho_2 \, (p_2 - p_1)}{\rho_1 \, (\rho_2 - \rho_1)} \qquad (1)$$

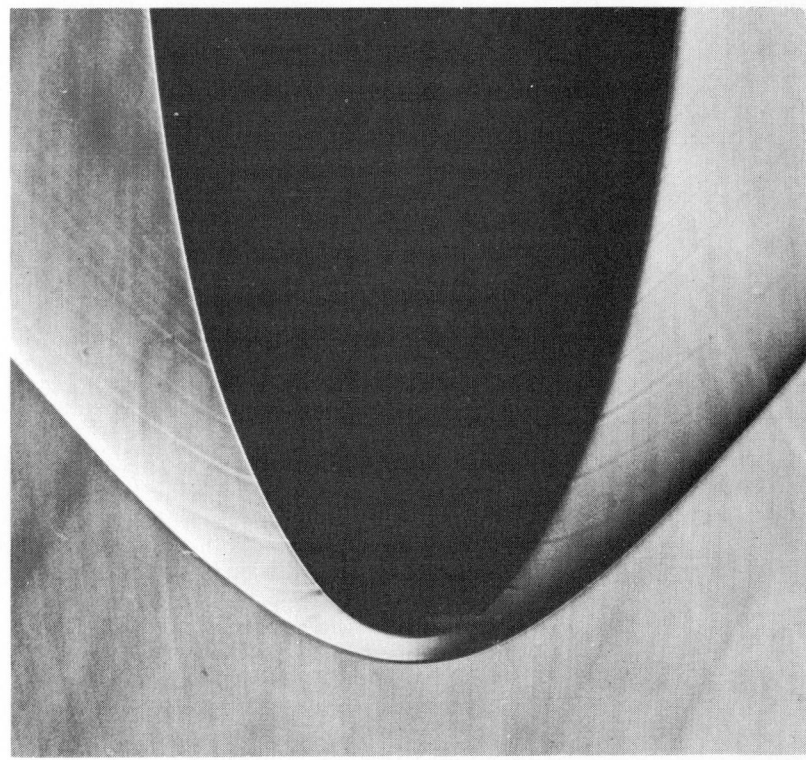

Fig. 1. Schlieren photograph of supersonic flow over blunt object. Shock wave is approximately parabolic, and detached from object. (*Avco Everett Research Laboratory, Inc.*)

kinetic energy across the shock front requires the validity of Eq. (2), where h is the specific enthalpy

$$h_1 + \tfrac{1}{2}u_1^2 = h_2 + \tfrac{1}{2}u_2^2 \qquad (2)$$

(or total heat per unit mass) of the fluid. By eliminating u_2 and u_1 with the aid of Eq. (1) and the law of conservation of mass, the energy equation becomes Eq. (3). If the thermodynamic properties of

$$h_2 - h_1 = \frac{1}{2}\left(\frac{1}{\rho_1} + \frac{1}{\rho_2}\right)(p_2 - p_1) \qquad (3)$$

the fluid are known, specific enthalpy h can be expressed as a function of pressure and density, or of any other pair of thermodynamic variables. Equations (1) and (3), together with the appropriate equation of state of the fluid, are known as the Rankine-Hugoniot equations for normal shock waves. From this set of equations, all thermodynamic variables behind the shock front (denoted by subscript 2) can be expressed as functions of the propagation velocity of the shock wave and the known initial state of the fluid (denoted by subscript 1). For example, if the fluid is a perfect gas of constant specific heats, enthalpy h can be written as Eq. (4), where γ is the ratio of specific heats

$$h = C_p T = \frac{\gamma}{\gamma - 1}\frac{p}{\rho} = \frac{a^2}{\gamma - 1} \qquad (4)$$

and a is the adiabatic speed of sound given by $(\gamma R T)^{1/2}$. For this case, the pressure and density ratios across the shock front are given by Eqs. (5) and (6). The temperature ratio, deduced from the

$$\frac{p_2}{p_1} = \frac{2\gamma M_1^2 - (\gamma - 1)}{\gamma + 1} \qquad (5)$$

$$\frac{\rho_2}{\rho_1} = \frac{u_1}{u_2} = \frac{(\gamma + 1)M_1^2}{2 + (\gamma - 1)M_1^2} \qquad (6)$$

perfect gas law, is given by Eq. (7). These expres-

$$\frac{T_2}{T_1} = \frac{[2\gamma M_1^2 - (\gamma - 1)][2 + (\gamma - 1)M_1^2]}{(\gamma + 1)^2 M_1^2} \qquad (7)$$

sions show that the ratios of all thermodynamic variables and flow velocities across the shock depend on only one parameter for a given gas, which is the Mach number M_1 of the flow relative to the shock front, where $M_1 = u_1/a_1$, or, in other words, M_1 equals the velocity of the shock wave divided by the speed of sound for the gas into which the shock propagates. Because of this, the magnitude of M_1 is often used as a measure of the strength of the shock. For comparison with the amplitude of acoustic waves, sound waves correspond to values of $M_1 \cong 1.001$ or less.

The results of Eqs. (5), (6), and (7) have been derived for gases of constant specific heats. From molecular and atomic physics, it is well known that, when a gas is heated to high temperatures, vibrational excitation, dissociation, and ionization take place, with accompanying changes in heat capacities of the gas. Therefore, for strong shock waves, the appropriate expression for the specific enthalpy h, and the equation of state, which takes into account these phenomena, must be used in place of Eq. (4) to obtain the shock wave solution from Eqs. (1) and (3). The ratios p_2/p_1, ρ_2/ρ_1, and T_2/T_1 for normal shock waves in air at standard atmospheric density are plotted in Figs. 2, 3, and 4.

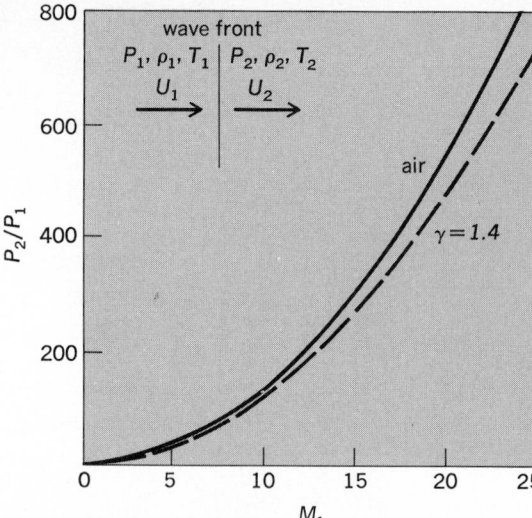

Fig. 2. Pressure ratio across normal shock wave in air at standard atmospheric density.

The approximate solutions, as given by Eqs. (6) and (7), hold only for the weaker shock waves ($M_1 < 6$), even though the pressure ratio is relatively insensitive to the changes in heat capacities of the gas.

Because the Rankine-Hugoniot equations do not impose any limit on the value of M_1, there remains the question of whether a shock wave can propagate into an undisturbed gas at a speed somewhat lower than the speed of sound of this gas, as would be the case for $M_1 < 1$. Although this question cannot be answered by the first law of thermodynamics, an examination of the change in specific entropy across the shock front provides an answer. Thus Eq. (8) holds, which, for gases of constant

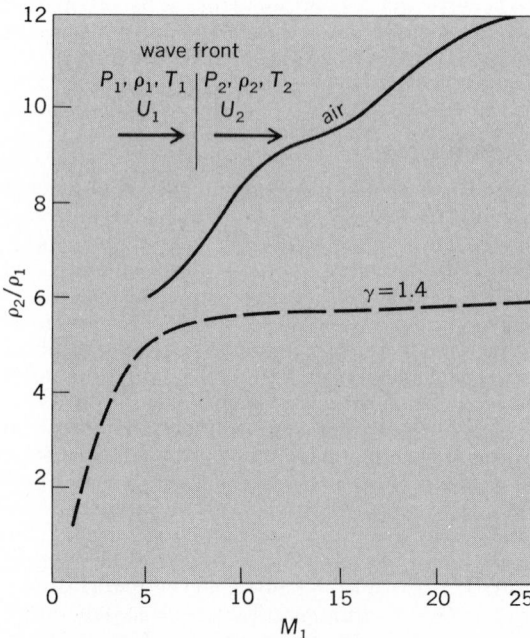

Fig. 3. Density ratio across normal shock wave in air at standard atmospheric density.

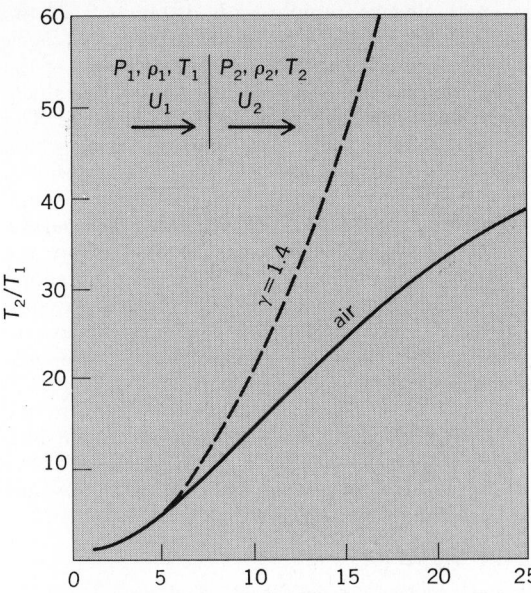

Fig. 4. Temperature ratio across normal shock wave in air at standard atmospheric density.

$$\Delta S_{12} = S_2 - S_1 = \int_1^2 \frac{dE + p \, dv}{T} \qquad (8)$$

specific heats, becomes Eq. (9), showing that en-

$$\frac{\Delta S_{12}}{R} =$$

$$\frac{1}{\gamma - 1} \ln \frac{[2\gamma M_1^2 - (\gamma - 1)] \, [2 + (\gamma - 1)M_1^2]^\gamma}{(\gamma + 1)^{\gamma + 1} M_1^{2\gamma}} \qquad (9)$$

tropy change ΔS_{12} assumes a negative value when M_1 is noticeably less than unity. This violates the second law of thermodynamics, which states that the entropy accompanying any naturally occurring processes always tends to increase. Therefore, one may conclude that shock waves always travel at supersonic speeds relative to the fluids into which they propagate.

Oblique shock wave. The changes in flow variables across an oblique shock wave are also governed by the laws of conservation of mass, momentum, and energy in a coordinate system which is stationary with respect to the shock front. In this case, the problem is slightly complicated by the fact that the flow velocity will experience a sudden change of direction as well as magnitude in crossing the shock front. Thus, if β_1 and β_2 denote the acute angles between the initial and final flow velocity vectors and the shock surface (Fig. 5), then in crossing the oblique shock, the flow will be deflected by a finite amount $\theta = \beta_1 - \beta_2$.

The oblique shock solution can be obtained directly from the complete set of conservation equations. However, the solution already obtained for normal shock waves provides the following simplifying information.

The rate of mass flow per unit area across the shock wave is determined by the normal component of the flow velocity (Fig. 5). Thus, for conservation of mass across the shock, Eq. (10) holds. On

$$\rho_1 u_1 \sin \beta_1 = \rho_2 u_2 \sin \beta_2 \qquad (10)$$

the other hand, conservation of the parallel component of momentum across the shock front requires that Eq. (11) hold. Equations (10) and (11) show that

$$\rho_1 u_1^2 \sin \beta_1 \cos \beta_1 = \rho_2 u_2^2 \sin \beta_2 \cos \beta_2 \qquad (11)$$

Eq. (12) holds. It is equivalent to the statement that

$$u_1 \cos \beta_1 = u_2 \cos \beta_2 \qquad (12)$$

the tangential component of the flow velocity must remain unchanged in crossing the oblique shock wave. Therefore, the resultant flow across the oblique shock shown in Fig. 5 will be identical to what an observer would see if he moved at a uniform velocity $u_1 \cos \beta_1$ along the surface of a normal shock wave propagating at a velocity $u_1 \sin \beta_1$. Such a translation of the frame of reference in the direction parallel to the shock front should not change the strength of the shock wave; thus, the changes in thermodynamic variables across the shock should depend only on the velocity component normal to the shock wave. The substitution of M_1 in Eqs. (5), (6), (7), and (9) with $M_1 \sin \beta_1$ gives the corresponding expressions for oblique shock waves in gases of constant specific heats. Again, thermodynamic considerations show that the normal component of the flow velocity into the oblique shock wave must be at least sonic. Therefore, in a supersonic stream of Mach number $M_1 > 1$, the value of β_1 must lie within the range given by expression (13). The lower limit corresponds to the

$$\sin^{-1} \frac{1}{M_1} \le \beta_1 \le \frac{\pi}{2} \qquad (13)$$

Mach angle of acoustic waves, while the upper limit corresponds to that of the normal shock wave. At both limits, the flow deflection angle $\theta = \beta_1 - \beta_2$ value of β_2 can be obtained from Eqs. (10) and (12). Thus Eq. (14) holds. The flow deflection angle θ for

$$\beta_2 = \tan^{-1} \left(\frac{\rho_1}{\rho_2} \tan \beta_1 \right) \qquad (14)$$

oblique shock waves in air at normal density is plotted in Fig. 6 as a function of M_1 and β_1. For a given Mach number M_1, the flow deflection angle first increases with the wave angle β_1, reaches a maximum value θ_{max}, and then decreases toward

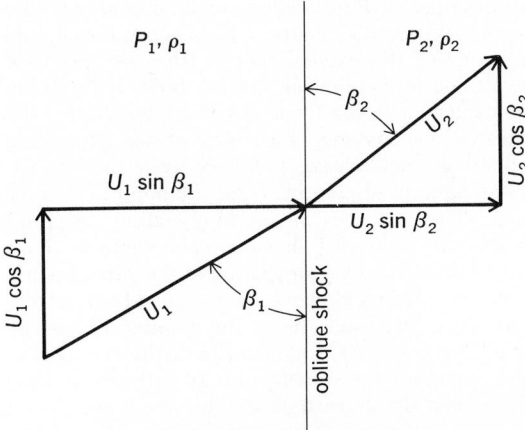

Fig. 5. Flow across an oblique shock wave.

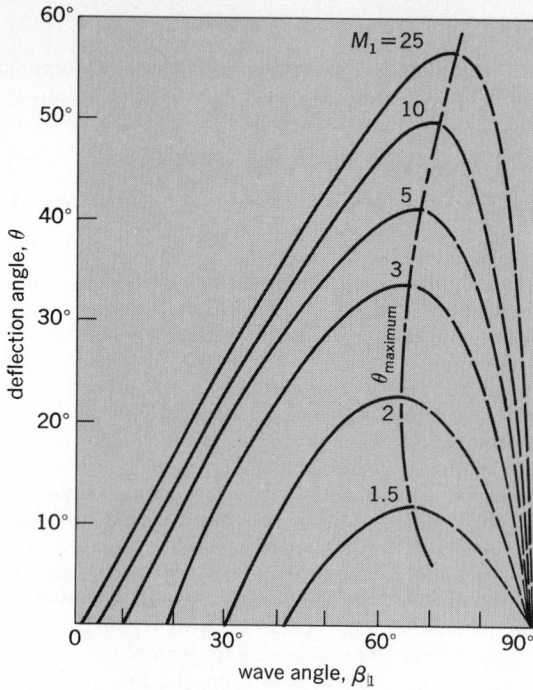

Fig. 6. Flow detection angle for oblique shock waves in air at standard atmospheric density.

zero again as the wave angle approaches that of the normal shock. Conversely, for any given flow deflection angle $\theta < \theta_{max}$ such as would be produced by sudden introduction of a wedge or an inclined plane surface into the initially uniform supersonic stream, there exist two possible values of β_1. The higher value of β_1 corresponds to the stronger shock. If the wedge angle or the inclination of the plane surface so introduced exceeds the value of θ_{max} for the given M_1, the shock wave either will become detached from the obstructing object or will form a more complicated pattern. The question of exactly what shock-wave pattern to expect from a given situation is complicated by interaction of the resultant shock wave and flow pattern on the overall boundary condition as well as on the local state of the flow.

Bomb blast. When energy is suddenly released into a fluid in a concentrated form, such as by a chemical or a nuclear explosion, the local temperature and pressure may rise instantly to such high values that the fluid tends to expand at supersonic speed. When this occurs, a blast wave forms, and propagates the excess energy from the point of explosion to distant parts of the fluid. If the point of explosion is far from any fluid boundary, the blast wave assumes the form of an expanding spherical shock wave followed by a radially expanding fluid originating from the point of detonation. The changes in thermodynamic variables across the spherical shock are the same as those for a normal shock propagating at the same instantaneous velocity. However, because of the continuous expansion and the finite amount of energy available from the explosion, both the strength of the shock and the specific energy of the expanding fluid must decay with time. The decay of a blast wave goes through three principal stages. A strong shock period begins immediately after the formation of the blast wave, during which the shock

strength decays rapidly with distance from the point of detonation. During this stage, the shock velocity decays with the inverse 3/2 power of the distance, and the overpressure behind the shock decays with the inverse cube of the distance. The second stage is a transition period, during which the strong spherical shock gradually changes into an acoustic wave. During the last stage or residual acoustic decay period, the acoustic wave carries the sound of explosion great distances from the point of detonation. As characteristic of sound propagation in three dimensions, the overpressure carried by the spherical acoustic wave decays inversely with distance, and velocity of propagation is constant.

[SHAO-CHI LIN]

Bibliography: H. W. Liepmann and A. Roshko, *Elements of Gasdynamics*, 1957; R. Zucker, *Fundamentals of Gas Dynamics*, 1977; M. J. Zucrow and J. D. Hoffman, *Gas Dynamics*, 2 vols., 1976, 1977.

Sine wave

A wave having a form which, if plotted, would be the same as that of a trigonometric sine or cosine function. It generally results from the solution of a problem having a one-dimensional space coordinate, such as the transverse vibrations of a string, longitudinal vibrations of a bar, or the propagation of plane waves of electromagnetic radiation or sound.

The sine wave may be thought of as the projection on a plane of the path of a point moving around a circle at uniform speed. For example, in illustration *a*, assume that the moving point travels around the circle at constant speed *v*. The projection onto a plane would trace back and forth on the line indicated as the point went around the circle. If the plane is now moved to the left at constant

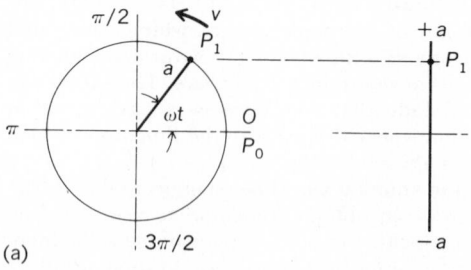

(a)

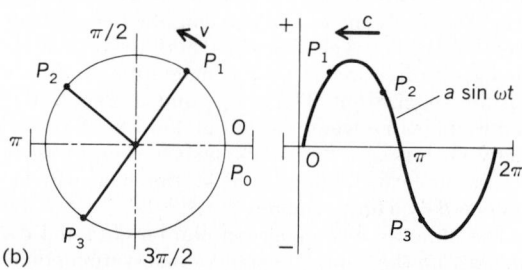

(b)

Simple harmonic motion and sine wave. (*a*) Point *P* moves around the circle at constant tangential speed *v*; its projection moves up and down between limits *+a* and *−a*. (*b*) Same conditions as in preceding, except that the plane on which the projection is plotted is traveling to the left at speed *c*, thereby generating a sine wave. Angular speed is ω rad/sec; the total angle traced out from zero position equals ωt rad.

speed c, as in illustration b, the resulting trace has the form of the graph of the sine function.

The sine wave trace of illustration b describes a body moving in simple harmonic motion. This motion changes in magnitude and time so that it repeats itself exactly, as long as uniform speed is maintained. It is characteristic of one-dimensional vibrations and one-dimensional waves having no dissipation. See HARMONIC MOTION.

The sine wave is the basic function employed in harmonic analysis. It can be shown that any complex motion in a one-dimensional system can be described as the superposition of sine waves having certain amplitude and phase relationships. The technique for determining these relationships is known as Fourier analysis. See FOURIER SERIES AND INTEGRALS; WAVE EQUATION: WAVE MOTION.

[WILLIAM J. GALLOWAY]

Single crystal

In crystalline solids the atoms or molecules are stacked in a regular manner, forming a three-dimensional pattern which may be obtained by a three-dimensional repetition of a certain pattern unit called a unit cell. When the periodicity of the pattern extends throughout a certain piece of material, one speaks of a single crystal. A single crystal is formed by the growth of a crystal nucleus without secondary nucleation or impingement on other crystals. See CRYSTAL STRUCTURE; CRYSTALLOGRAPHY.

Growth techniques. Among the most common methods of growing single crystals are those of P. Bridgman and J. Czochralski. In the Bridgman method the material is melted in a vertical cylindrical vessel which tapers conically to a point at the bottom. The vessel then is lowered slowly into a cold zone. Crystallization begins in the tip and continues usually by growth from the first formed nucleus. In the Czochralski method a small single crystal (seed) is introduced into the surface of the melt and then drawn slowly upward into a cold zone. Single crystals of ultrahigh purity have been grown by zone melting. Single crystals are also often grown by bathing a seed with a supersaturated solution, the supersaturation being kept lower than is necessary for sensible nucleation.

When grown from a melt, single crystals usually take the form of their container. Crystals grown from solution (gas, liquid, or solid) often have a well-defined form which reflects the symmetry of the unit cell. For example, rock salt or ammonium chloride crystals often grow from solutions in the form of cubes with faces parallel to the (100) planes of the crystal, or in the form of octahedrons with faces parallel to the (111) planes. The growth form of crystals is usually dictated by kinetic factors and does not correspond necessarily to the equilibrium form. See CRYSTAL GROWTH.

Physical properties. Ideally, single crystals are free from internal boundaries. They give rise to a characteristic x-ray diffraction pattern. For example, the Laue pattern of a single crystal consists of a single characteristic set of sharp intensity maxima. See X-RAY DIFFRACTION.

Many types of single crystal exhibit anisotropy, that is, a variation of some of their physical properties according to the direction along which they are measured. For example, the electrical resistivity of a randomly oriented aggregate of graphite crystallites is the same in all directions. The resistivity of a graphite single crystal is different, however, when measured along different crystal axes. This anisotropy exists both for structure-sensitive properties, which are strongly affected by crystal imperfections (such as cleavage and crystal growth rate), and structure-insensitive properties, which are not affected by imperfections (such as elastic coefficients).

Anisotropy of a structure-insensitive property is described by a characteristic set of coefficients which can be combined to give the macroscopic property along any particular direction in the crystal. The number of necessary coefficients can often be reduced substantially by consideration of the crystal symmetry; whether anisotropy, with respect to a given property, exists depends on crystal symmetry.

The structure-sensitive properties of crystals (for example, strength and diffusion coefficients) seem governed by internal defects, often on an atomic scale. See CRYSTAL DEFECTS.

[DAVID TURNBULL]

Bibliography: B. R. Pamplin, *Crystal Growth*, 1975; F. Rosenberger, *Fundamentals of Crystal Growth One: Macroscopic Equilibrium and Transport Concepts*, 1979.

Skin effect

The crowding of high-frequency electric current into a thin surface layer of a conductor.

For a steady unidirectional flow of electricity, the current is uniformly distributed over the cross section of a uniform conductor; that is, the current density (current per unit area) is the same at all points in a cross section. For an alternating current, there is no longer this simple uniformity, but the current density is greater near the outer surface than at the center. The magnitude of the nonuniformity increases as the frequency rises. For low frequencies, the effect is very small, but at frequencies for which the wavelength within the conducting material is comparable with the dimensions of the conductor, or smaller, the entire current may be considered to be within a relatively thin surface layer.

The skin effect is largely due to self-induced electromotive forces (emf's) which are different for different paths within the conductor. These emf's increase with frequency, since they depend upon the rate of change of flux. The emf's are smallest for paths that link the smallest flux. The internal linkages decrease as the frequency increases, and for infinite frequency there would be no internal linkages. For this condition, the skin effect may be called complete.

The skin effect results in a resistance for a conductor that is greater for an alternating current than for a direct current, since the effective cross section of the conductor is decreased. Also, the resistance varies with frequency, increasing as the frequency rises. Again, this behavior follows from the decrease in the effective area of the conductor.

[KENNETH V. MANNING]

Bibliography: B. I. Bleaney and B. Bleaney, *Electricity and Magnetism*, 3d ed., 1976; A. E. Fitzgerald et al., *Basic Electrical Engineering*, 5th ed., 1981; W. H. Hayt, Jr., *Engineering Electromagnetics*, 4th ed., 1981; E. M. Pugh, *Principles of Electricity and Magnetism*, 2d ed., 1970.

Solid-state physics

The study of the physical properties of solids, such as electrical, dielectric, elastic, and thermal properties, and their understanding in terms of fundamental physical laws. Most problems in solid-state physics would be called solid-state chemistry if studied by scientists with chemical training, and vice versa. Solid-state physics emphasizes the properties common to large classes of compounds rather than the dependence of properties upon compositions, the latter receiving greater emphasis in solid-state chemistry. In addition, solid-state chemistry tends to be more descriptive, while solid-state physics focuses upon quantitative relationships between properties and the underlying electronic structure.

Many of the scientists who study the physics of liquids identify with solid-state physics, and the term "condensed-matter physics" is increasingly replacing "solid-state physics" as a division of physics. It includes noncrystalline solids such as glass as well as crystalline solids. *See* AMORPHOUS SOLID.

Electronic structure of solids. In solid-state physics it is generally assumed that the electronic states can be described as wavelike. The individual electronic states, called Bloch states, have energies which depend upon the wave number (a vector equal to the momentum divided by Planck's constant $\hbar$), and the wave number is restricted to a domain called the Brillouin zone. This energy

given as a function of the wave number is called the band structure. It is shown for two lines (Γ to X and Γ to L) in the Brillouin zone for metallic copper in the illustration. There are several curves, called bands, for each line. *See* BRILLOUIN ZONE.

These bands can be calculated by using quantum theory; the results of such a calculation are given as the continuous lines in the illustration. It is also possible to measure them directly by using the photoelectric effect, as indicated by the points in the illustration. In any solid the electrons occupy the lowest-energy electronic states. In copper there are 11 electrons per atom to occupy the bands shown, and they fill the states up to the Fermi energy E_F, indicated in the illustration. One of the bands is partly occupied (that is, it crosses the energy E_F), so that this solid is a metal; under an applied field or temperature gradient, the electrons shift from state to state, causing currents to flow and making copper a good conductor of electricity and heat and making it shiny. Light at the blue end of the spectrum can also lift electrons from the nearly horizontal bands to the empty states; the absorption of the blue light gives the reddish color to copper. *See* CRYSTAL ABSORPTION SPECTRA; NONRELATIVISTIC QUANTUM THEORY.

If copper had an additional electron per atom (as does zinc), there would be just enough electrons to fill the lower bands (the bottom five curves). The illustration suggests that there would then be no partially filled bands, and thus an energy gap separating the full bands from the empty ones. Actually in copper and in zinc some of the upper bands dip below the peak energy of the lower bands, as would be the case if the upper band at L dropped below 2 eV in the illustration. Thus copper would still be a metal, as is zinc, even if it had an additional electron.

When such energy gaps between occupied and empty bands do occur, as in silicon, the solid is no longer metallic. Applied fields only shift electrons from state to state within individual bands, and since the same states remain occupied, there is no current generated by, nor any heat flow due to, the electrons. If the gaps are small, as is the 1.13-eV gap of silicon, a small number of electrons are thermally shifted at room temperature to the empty bands, called conduction bands, and electrical conductivity is introduced; it can also be accomplished by substituting impurities, called dopants. Phosphorus, for example, has one more electron than silicon and contributes that electron to the conduction band. Such systems in which conductivity can be introduced are called semiconductors. When the energy gaps are large, as is the 8.5-eV gap for rock salt, there are ordinarily no empty states in the lower bands and no electrons in the conduction bands, and the crystals are insulators. *See* BAND THEORY OF SOLIDS; SEMICONDUCTOR.

Cooperative phenomena. Each state in the copper bands shown in the illustration can accommodate two electrons of opposite spin. The two spin states have the same energy, so that they are both filled if the energy is less than E_F, or both empty if the energy is greater than E_F; there is no net electronic spin in the system and no magnetism. In nickel, less than copper in atomic number by one, the energies of the two spin states shift from each other. Electrons with spin of one sign

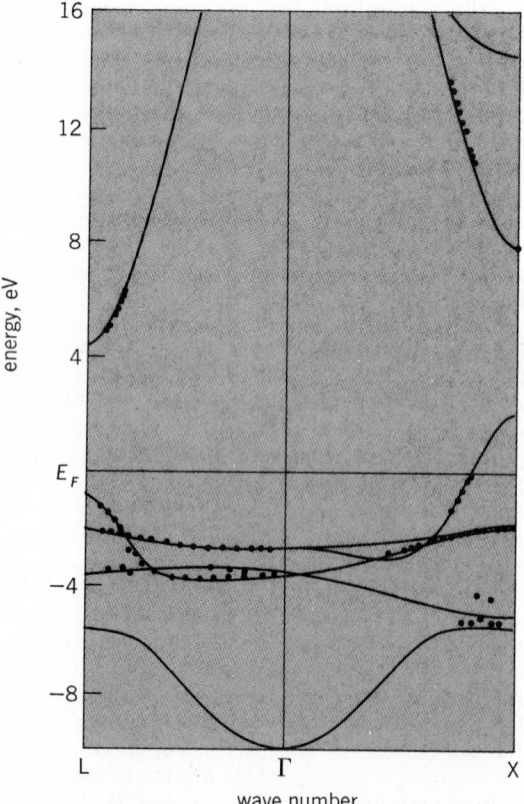

Energy bands of copper. (*From J. A. Knapp, F. J. Himpsel, and D. E. Eastman, Experimental energy band dispersions and lifetimes for valence and conduction bands of copper using angle-resolved photoemission, Phys. Rev., B19: 4952–4964, 1979*)

become more numerous, and the metal becomes ferromagnetic; each of the bands shown in the illustration has become split into two. Such a transition is called a cooperative phenomenon since it is caused by the mutual lowering of each other's energies by electrons of the same spin. The theoretical and experimental study of such cooperative transitions is an active area of solid-state physics research. *See* FERROMAGNETISM; PHASE TRANSITIONS.

Superconductivity. Another cooperative phenomenon is superconductivity, in which pairs of electrons of opposite spin and moving in opposite directions form Cooper pairs that condense into a cooperative electronic state. There is again an energy gap between the occupied Cooper pair states and the empty states from broken pairs, but in this case the entire cooperative state can drift to carry a current without exciting individual electrons from the cooperative state. Such a flow is called supercurrent since it flows without any electrical resistance whatsoever; it occurs only in some metals and only at very low temperatures. *See* SUPERCONDUCTIVITY.

Small supercurrents can even flow through very thin insulators by quantum-mechanical tunneling, an effect predicted by B. D. Josephson. It was predicted and observed after the discovery by I. Giaever that individual electrons can tunnel through thin oxides; Giaever used this tunneling to make a direct measurement of the energy gap arising from the Cooper pairs in superconductors.

Crystal structure and lattice vibrations. The total energy of a solid includes a sum of the energies of the occupied electronic states. Since the energy bands depend upon the positions of the atoms, so does the total energy, and the stable crystal structure is that which minimizes this energy. The theory has not proved adequate to really predict the crystal structure of various solids, but it is possible to predict the changes in energy under various distortions of the lattice. There are in fact three times as many independent distortions, called normal modes, as there are atoms in the solid. Each has a wave number, and the frequencies of the normal vibrational modes, as a function of wave number in the Brillouin zone, form vibrational bands in direct analogy with the electronic energy bands shown in the illustration. These can be directly calculated from quantum theory or measured by using neutron or x-ray diffraction. *See* CRYSTAL STRUCTURE; LATTICE VIBRATIONS; NEUTRON DIFFRACTION; X-RAY DIFFRACTION.

In quantum theory any normal mode can be excited only to discrete quantized vibrational energies. In solids a mode in the nth state of excitation is said to contain n phonons. These phonons interact with the electrons and cause the reduction in conductivity with increasing temperature in metals. This same electron-phonon interaction also provides the attraction that holds electrons together in Cooper pairs to form the cooperative superconducting state. *See* ELECTRICAL RESISTIVITY; PHONON.

Intrinsic and extrinsic properties. Solid-state physics includes the study of all of these properties of ideal crystals, called intrinsic properties. It also includes the study of defects in the structure, impurities, and surfaces, all of which are called extrinsic properties. The study of extrinsic properties can be made with very much the same experimental tools, and they can be understood by using quantum theory with very much the same concepts which are useful for studying intrinsic properties. *See* CRYSTAL DEFECTS.

[WALTER A. HARRISON]

Bibliography: N. W. Ashcroft and N. D. Mermin, *Solid State Physics*, 1976; W. A. Harrison, *Electronic Structure and the Properties of Solids*, 1980; C. Kittel, *Introduction to Solid State Physics*, 5th ed., 1976; F. Seitz et al. (eds.), *Solid State Physics*, vols. 1–34, 1955–1979.

Soliton

An isolated wave which propagates without dispersing its energy over larger and larger regions of space. In most of the scientific literature, the requirement that two solitons emerge unchanged from a collision is also added to the definition; otherwise one speaks of a solitary wave.

There are many equations of mathematical physics which have solutions of the soliton type. Correspondingly, the phenomena which they describe, be it the motion of waves in shallow water or in an ionized plasma, exhibit solitons. The first observation of this kind of wave was made in 1834 by John Scott Russell, who followed on horseback a soliton propagating in the windings of a channel. Scott Russell's account was published in 1845; 50 years later D. J. Korteweg and H. de Vries proposed an equation for the motion of waves in shallow waters which possesses soliton solutions, and thus established a mathematical basis for the study of the phenomenon. Interest in the subject, however, lay dormant for many years, and the major body of investigations began only in the 1950s. Researches done by analytical methods and by numerical methods made possible with the advent of computers gradually led to a complete understanding of solitons. *See* WAVE MOTION IN LIQUIDS.

Eventually, the fact that solitons exhibit particle-like properties, because the energy is at any instant confined to a limited region of space, received attention, and solitons were proposed as models for elementary particles. However, it is difficult to account for all of the properties of known particles in terms of solitons. More recently it has been realized that some of the quantum fields which are used to describe particles and their interactions also have solutions of the soliton type. The solitons would then appear as additional particles, and may have escaped experimental detection because their masses are much larger than those of known particles. In this context the requirement that solitons emerge unchanged from a collision has been found too restrictive, and particle theorists have used the term soliton where traditionally one would speak of a solitary wave. *See* ELEMENTARY PARTICLE; QUANTUM FIELD THEORY.

Sin-ϕ or sine-Gordon model. The sin-ϕ or sine-Gordon model is formulated in terms of a field $\phi(x,t)$ depending on one spatial coordinate x and the time variable t, and describes one-dimensional propagation of waves. The energy density, defined by Eq. (1), leads to Eq. (2) as the equation

$$\mathscr{E} = \frac{\alpha}{2}\left(\frac{\partial \phi}{\partial t}\right)^2 + \frac{\beta}{2}\left(\frac{\partial \phi}{\partial x}\right)^2 + \frac{\gamma}{2}\left(1 - \cos\frac{2\pi\phi}{\phi_0}\right) \quad (1)$$

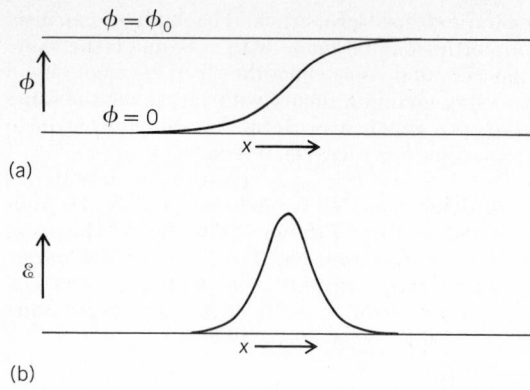

Fig. 1. Static soliton solution of the sin-ϕ model. (a) Field ϕ. (b) Energy density $\mathscr{E}$.

$$\alpha \frac{\partial^2 \phi}{\partial t^2} - \beta \frac{\partial^2 \phi}{\partial x^2} = -\frac{\pi \gamma}{\phi_0} \sin \frac{2\pi \phi}{\phi_0} \qquad (2)$$

of motion, which is one of the simplest equations having soliton solutions. Here α, β, γ, and ϕ_0 are positive constant parameters. The first term in the energy density is sensitive to the rate of variation in time of the field at a definite location and becomes larger as this rate increases. The second term depends on the rate of variation of the field along the x-coordinate axis and also becomes larger as this rate of variation increases. Thus, the first two terms in the energy will be minimal (and indeed will equal zero) if the field is constant in space and time. The third term depends on the actual value of the field. When $\phi = 0$, $\pm\phi_0$, $\pm 2\phi_0$, or any integer multiple of ϕ_0,

$$\cos \frac{2\pi \phi}{\phi_0} \qquad \begin{array}{l} \text{(argument expressed in radians,} \\ \text{otherwise } 2\pi \text{ is replaced by } 360°) \end{array}$$

becomes equal to 1 and the third term in the ener-

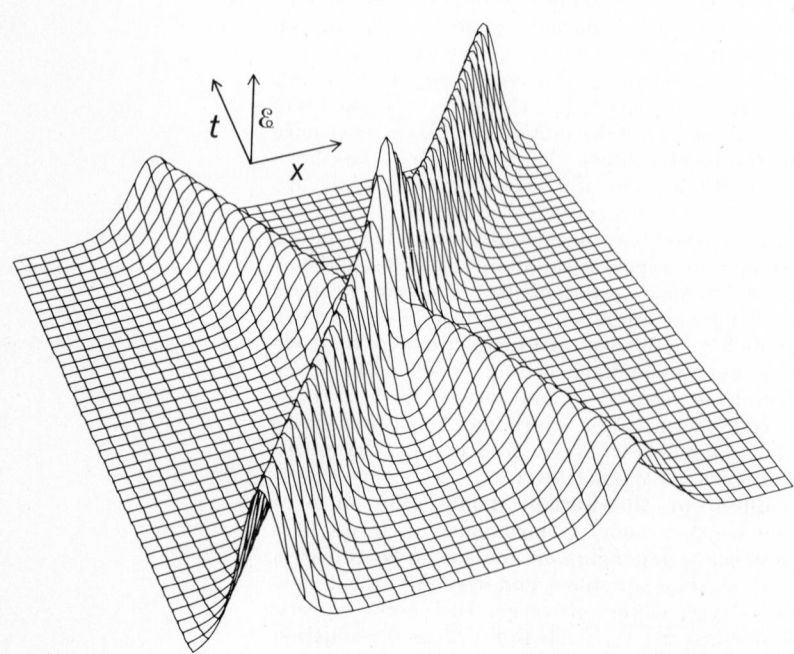

Fig. 2. Space and time variation of the energy density in a collision between two solitons.

gy vanishes. When

$$\phi = \pm \frac{\phi_0}{2}, \pm \frac{3\phi_0}{2}, \pm \frac{5\phi_0}{2}, \ldots$$

the third term in the energy is maximal and equal to γ, and for all other values of ϕ it varies between 0 and γ. The state of minimum energy, or state of rest or vacuum state, is therefore achieved when the field takes a value constant in space and time and equal to 0, $\pm\phi_0$, $\pm 2\phi_0$, ... Thus there is not a single vacuum, but many states of minimal energy (multiple vacua).

Solitons occur when the field approaches different vacuum values in the two opposite directions of the x-coordinate axis. Simple and intuitive reasoning shows that if different vacua are reached as $x \rightarrow +\infty$ and $x \rightarrow -\infty$, a region of transition must exist where the energy density is nonvanishing. If these two vacuum values differ precisely by ϕ_0, the system exhibits a single soliton. The field profile, constant in time, which minimizes the energy gives rise to the static soliton solution (Fig. 1). The energy is thus confined to a definite region of space.

A more gradual passage of ϕ between the two vacuum states (which would spread out the energy) would lead to an increase of the contribution of the last term to the energy not compensated by the decrease in the second term. A more abrupt transition would also increase the total energy. Configurations in which the field undergoes more than one transition between neighboring vacuum states exhibit multiple solitons.

Solutions with moving solitons also exist, and the three-dimensional computer-generated graph of Fig. 2 illustrates the encounter of two solitons moving in opposite directions (the wave with the higher peak in energy is moving faster). It is apparent from the graph that the two waves emerge from the collision with shape and speed unchanged. This remarkable feature of the solitary waves results from special properties of the sin-ϕ equation (and of other equations having soliton solutions) and, as mentioned already, is often considered part of the definition of a soliton.

Other systems with solitons. Many equations which describe one-dimensional propagation of waves give origin to soliton solutions. The equation of Korteweg and de Vries, the nonlinear Schrödinger equation, and the Boussinesq equation belong to this category. The preservation of shape and velocity of the waves after a collision has been related to the existence of conservation laws. Any isolated dynamical system obeys conservation laws: the total energy of the system and its total momentum, for instance, do not change in time. In a perfectly elastic frontal collision of two billiard balls of the same mass, these two conservation laws are sufficient to predict the outcome of the collision: the two balls just exchange their speeds. The systems exhibiting solitons are characterized by obeying a very large number of conservation laws, well beyond what would be expected of an ordinary dynamical system. The existence of these conservation laws implies, and in turn follows from, the fact that the system is integrable; by rather elaborate analytical techniques, a new set of variables may be found for the description of the motion which evolves in time independently and according to very simple rules. *See* CONSERVA-

TION OF ENERGY; CONSERVATION OF MOMENTUM; SYMMETRY LAWS.

While one-dimensional solitons are very interesting for a variety of applications, an isolated wave must be three-dimensional to be interpreted as a particle. The existence of systems exhibiting solitary waves in two and three spatial dimensions has been demonstrated. These confined waves are not expected to preserve their dynamical state in a collision (although at present it cannot be excluded that some have this property), but particle theorists have called them solitons as well. In some cases, the confinement of the wave can be related to the existence of multiple states of minimum energy (multiple vacua), as for the sin-ϕ model; such waves are called topological solitons because of the special geometrical properties of the arrangement of vacua around the wave. The topological features are not universal, however, and nontopological solitons exist as well.

A quantum-mechanical formulation of the dynamics of solitons has been developed, and it has been recognized that many field theories used to describe particles have solutions of the soliton type. These quantum solitons should not be identified with the known atomic, nuclear, or subnuclear particles, which are rather associated with elementary quantum-mechanical excitations of the fields or bound states of these excitations. The solitons, if indeed realized as particles in nature, would have much larger masses, and still await experimental detection.   [CLAUDIO REBBI]

Bibliography: L. D. Faddeev and V. E. Korepin, Quantum theory of solitons, *Phys. Rep.*, 42C(1): 1–87, 1978; C. Rebbi, Solitons, *Sci. Amer.*, 240(2): 92–116, 1979; A. C. Scott, F. Y. F. Chu, and D. W. McLaughlin, The soliton: A new concept in applied science, *Proc. IEEE*, 61(10):1443–1483, 1973.

Sonic barrier

An aeronautical term coined during World War II to symbolize the technical difficulties for manned aircraft in accelerating through the speed of sound. In designing and testing subsonic airplanes to fly at increasingly higher speeds, aeronautical engineers found the following adverse effects as the speed of the airplane approached the speed of sound: a rapid rate of increase in drag, breakdown of lift, and loss of control and maneuverability of the airplane. Such experience suggested that these violent and uncontrollable aerodynamic phenomena might be inevitable in the transonic range of speeds (that is, in the immediate vicinity of the speed of sound). These adverse effects are collectively and loosely termed the sonic barrier. On the other hand, cannoneers had long known that ballistic artillery shells had no difficulty moving through the atmosphere at speeds considerably in excess of the speed of sound.

As a result of intensive research during and after World War II, the source of most of the difficulties has been traced to the formation of local shock waves, which in turn interact with the boundary layer and induce flow separation from the aerodynamic surfaces. By careful choice of geometrical shapes, such as use of thin wings and slim bodies, it has been found possible to design airplanes with relatively smooth flow characteristics through the transonic speed range. Although the increase in drag with flight speed is inevitable, it can be kept at a manageable level and be overcome by more powerful propulsion devices.

Although flight at speeds in excess of the speed of sound has been achieved, the theoretical problem of prescribing the transonic flow field over objects of arbitrary geometric shape remains a formidable one. The difficulty is partly mathematical and partly the result of flow discontinuities, such as shock waves and flow separation, which make it difficult to specify the boundary conditions for the flow field, even though the shapes of the objects themselves are known.

[SHAO-CHI LIN]

Bibliography: C. Ferrari and F. Tricomi, *Transonic Aerodynamics*, 1968; K. Oswaitish and D. Rues (eds.), *Symposium Transsonicum 2*, 1976.

Sonics

A term used to describe the technology of sound, or elastic wave motion, as applied to problems of measurement, control, and processing. It is a branch of acoustics which relates primarily to processes and techniques, as distinguished from those studies which relate to human hearing.

Ultrasonics is the term applied to sound whose frequency is in excess of approximately 20,000 hertz (Hz). *See* ULTRASONICS.

Infrasonics is the term applied to sound whose frequency is less than approximately 15 Hz. These sounds are also referred to as having subsonic frequencies. Many problems in vibration, oceanography, seismology, and the dynamic behavior of elastic materials are analyzed by treating the phenomenon being studied as sound waves of infrasonic frequency. *See* INFRASOUND; SOUND.

[WILLIAM J. GALLOWAY]

Sound

Subjectively, what is heard with the ears; objectively, a mechanical disturbance from equilibrium in an elastic material medium, that is, a material medium which, when disturbed, will return to its original equilibrium state after the disturbing influence has been removed. Most of the solids, liquids, and gases of common experience possess such elasticity to a certain degree and can therefore act as sources, transmitters, and receivers of sound.

Originally sound referred entirely to what is heard. Later, however, interest developed in sounds which human beings cannot hear, either because the frequency is too low (infrasound) or too high (ultrasound). Audible sound is usually described as audio or sonic. The characterization of sound by frequency (associated with pitch in hearing—the higher pitch corresponding to higher frequency) is well known, since most sounds of interest are produced by the mechanical vibration of material media, for which the frequency means the number of complete vibrations or cycles per second, now referred to as hertz (Hz). Another important quantity characterizing sound is the sound pressure, that is, the excess pressure in the medium being traversed by the sound over the normal equilibrium pressure in the undisturbed medium. The pascal (Pa), the SI unit of pressure, is 1 newton/m². The cgs unit of pressure is the dyne/cm², which is equal to one-tenth of a pascal. Normal atmospheric pressure is about 10^5 Pa.

Figure 1 shows the range of sound frequencies and pressures of interest in human experience and to the acoustical scientist and engineer. Acoustical scientists have devised methods of producing sounds over an extended range of pressures and frequencies.

This article describes the physical nature of sound and its propagation through material media, as well as methods for its production and detection.

Nature of sound propagation. Sound is propagated through a medium such as air by means of wave motion. In wave propagation the medium as a whole does not move, but only the disturbance, for example, the hump on the water surface in the case of a water wave. For a sound wave in a fluid like air, the "hump" is the excess pressure p in the medium constituting the disturbance. Such an excess pressure (which for ordinary audible sound may be less than 0.1 Pa; Fig. 1) produced at one point does not remain there but moves through the fluid with velocity c, which is called the velocity of sound; in the case of air at 20°C this is 344 m/s.

To describe sound-wave motion effectively, it is necessary to express it mathematically. For simplicity the discussion will be restricted to a sound wave propagated in one direction, which is taken to be the x axis in a rectangular coordinate system, with velocity c. The excess pressure p in the medium due to the wave motion can then be represented by Eq. (1). That is, p is a function of

$$p = f(x - ct) \qquad (1)$$

the variable $x - ct$, which depends on both space and time through the combination $x - ct$. This does indeed represent a wave, since Eq. (2), which says

$$f(x_1 - ct_1) = f(x_2 - ct_2) \qquad (2)$$

that the disturbance p at the point x_1 at time t_1 moves to point x_2 at time t_2, is valid, provided that Eq. (3) is true; and Eq. (3) says that the distance

$$(x_2 - x_1) = c(t_2 - t_1) \qquad (3)$$

$(x_2 - x_1)$ is traversed in time $(t_2 - t_1)$ with velocity c.

In order to pursue the subject further, it is necessary to set up the differential equation satisfied by wave motion. If the expression for p in Eq. (1) is differentiated twice with respect to x and t, the result is the second-order partial differential equation, Eq. (4), which is called the wave equa-

$$\frac{\partial^2 p}{\partial t^2} = c^2 \frac{\partial^2 p}{\partial x^2} \qquad (4)$$

tion for the propagation of excess pressure p in the x direction with velocity c. The first-order equation obtained by differentiating Eq. (1) once with respect to x and t will not uniquely define the wave velocity independently of the direction of the wave. Hence a second-order equation is necessary (and mathematically sufficient).

The description of sound propagation through material media is essentially the study of the solutions of Eq. (4), when generalized to three space dimensions. *See* WAVE; WAVE MOTION.

Harmonic waves. The most important type of sound wave in practice is that called harmonic. For such a wave in the x direction with the velocity c, the excess pressure is given by Eq. (5). The

$$p = p_0 \cos\left[2\pi\nu\left(t - \frac{x}{c}\right)\right] \qquad (5)$$

quantity p_0 is called the amplitude of the wave; it is the maximum value of the excess pressure as the wave travels along. The minimum value is of course $-p_0$. The quantity ν is the frequency of the wave. At any given point the value of the excess pressure repeats itself ν times a second or in the time $1/\nu$, which is called the period of the wave, designated by T. Another important quantity connected with the wave is the wavelength, designated by λ. At any given time the excess pressure repeats itself after a space interval given by Eq. (6).

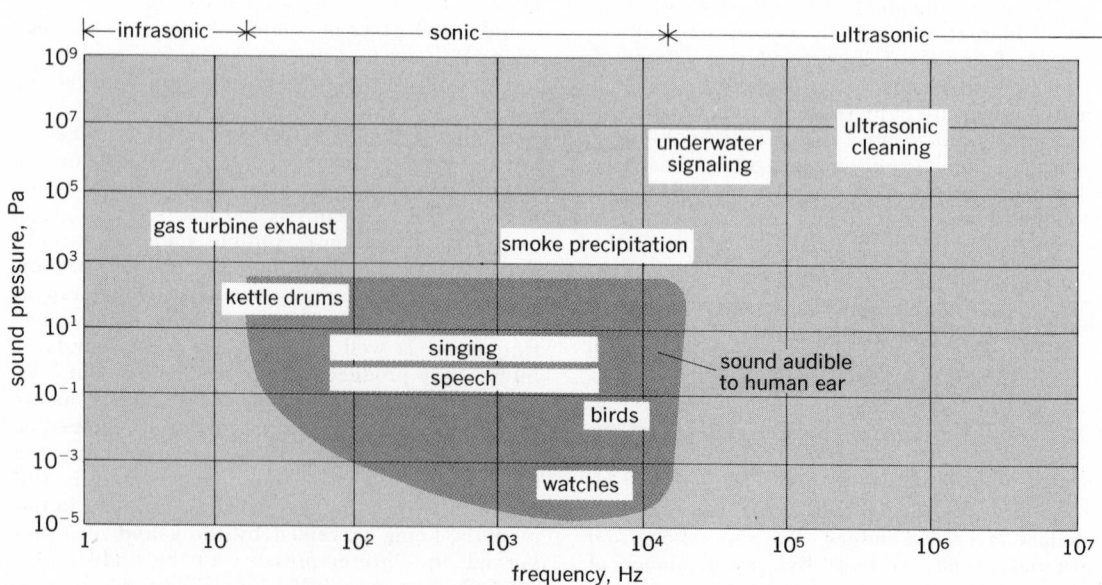

Fig. 1.   Spectrum of sound. The labeled boxes roughly locate the intensity and frequency of some infrasonic, sonic, and ultrasonic phenomena. (*From G. E. Henry, Sci. Amer., 190(5):54–63, 1954*)

$$\lambda = c/\nu = cT \tag{6}$$

This relation between wave length, velocity, and frequency holds for harmonic waves of all kinds and in all material media. Low frequencies are always associated with long wavelengths, and vice versa. For an audible sound wave in air with frequency 10^3 Hz, the wavelength is about 34 cm. On the other hand, for an ultrasonic wave of frequency 10^6 Hz, the wavelength is about 3.4×10^{-2} cm.

Particle displacement and velocity. Associated with the excess pressure p of the sound wave, there is also an actual displacement of the medium from equilibrium in the x direction. This is usually called the particle displacement. If it is denoted by ξ, the corresponding elastic strain is $\partial \xi / \partial x$, and from Hooke's law of elasticity Eq. (7) follows,

$$p = -B \, \partial \xi / \partial x \tag{7}$$

where B is the bulk or volume modulus of elasticity. Substituting from Eq. (5) into Eq. (7) and integrating (with the arbitrary constant of integration set equal to zero) yields Eq. (8). Newton's second

$$\xi = \frac{p_0 c}{2\pi \nu B} \cdot \sin \left[2\pi \nu \left(t - \frac{x}{c} \right) \right] \tag{8}$$

law of motion applies to each moving element in the medium traversed by the sound wave. Writing this for an element of length dx and cross-sectional area S, and using Eq. (7), yields Eq. (9), where ρ_0

$$-S \frac{\partial p}{\partial x} \, dx = \rho_0 S \frac{\partial^2 \xi}{\partial t^2} \, dx = SB \frac{\partial^2 \xi}{\partial x^2} \, dx \tag{9}$$

is the constant equilibrium density of the medium. The result is that ξ satisfies the same partial differential equation as p, namely Eq. (10); that is, ξ

$$\frac{\partial^2 \xi}{\partial t^2} = \frac{B}{\rho_0} \cdot \frac{\partial^2 \xi}{\partial x^2} \tag{10}$$

is propagated as a wave in the x direction with velocity given by Eq. (11). Hence Eq. (8) becomes

$$c = \sqrt{B/\rho_0} \tag{11}$$

Eq. (12). The so-called particle velocity u in the

$$\xi = \frac{p_0}{2\pi \nu \rho_0 c} \cdot \sin \left[2\pi \nu \left(t - \frac{x}{c} \right) \right] \tag{12}$$

medium traversed by the sound wave is the time derivative of ξ, and is thus given by Eq. (13). *See* ELASTICITY; NEWTON'S LAWS OF MOTION.

$$u = \frac{\partial \xi}{\partial t} = \frac{p_0}{\rho_0 c} \cdot \cos \left[2\pi \nu \left(t - \frac{x}{c} \right) \right] \tag{13}$$

Intensity. A very important quantity descriptive of a sound wave is the intensity I, qualitatively a measure of what might be called its strength but precisely defined as the time average rate of flow of energy per unit time through unit area of the medium through which the wave is passing, it being understood that the area in question is perpendicular to the direction of wave propagation. This is the time average power transported by the wave. From the principles of mechanics it is the time average of the product of the particle velocity u and the acoustic pressure p. Hence, Eqs. (5) and (13) yield Eq. (14) for a harmonic wave in the x

$$I = \overline{up} = \frac{p_0{}^2}{\rho_0 c} \cdot \overline{\cos^2 \left[2\pi \nu \left(t - \frac{x}{c} \right) \right]}$$
$$= \frac{1}{2} \frac{p_0{}^2}{\rho_0 c} \tag{14}$$

direction. If p_0 is given in Pa, ρ_0 in kg/m³, and c in m/s (SI units), the intensity is expressed in the units W/m². Alternatively, if p_0 is given in dynes/cm², ρ_0 in g/cm³, and c in cm/s (cgs units), the intensity is expressed in the units ergs/cm² s. A frequently used unit is the watt/cm², where 1 W/cm² = 10^{-4} W/m² = 10^7 ergs/cm² s.

A very important unit of intensity in practical sound problems is the decibel (dB). A sound of intensity I is said to be a number of decibels above or below a standard reference intensity I_0, given by expression (15). Thus, if $I = 2I_0$, I is about 3.01

$$10 \log_{10} I/I_0 \tag{15}$$

dB above I_0, or if $I = I_0/2$, I is about 3.01 dB below I_0. For many problems in audio acoustics I_0 is taken as the threshold of hearing (the minimum audible intensity), which for the average person with so-called normal hearing is approximately 10^{-16} W/cm². In terms of this reference, normal conversational speech directly in front of the mouth has an intensity of about 65 dB, whereas next to an aircraft jet engine the intensity is about 160 dB. For a normal human ear the intensity at which sound ceases to be heard but produces pain in the ear and incipient damage is about 120 dB. *See* DECIBEL.

Impedance. Another important quantity connected with a sound wave is its impedance Z, defined by Eq. (16), where the acoustic volume cur-

$$Z = \frac{\text{acoustic pressure}}{\text{acoustic volume current}} \tag{16}$$

rent is the product of the particle velocity u and the area S through which the sound wave passes. From Eqs. (5) and (13) it follows that the impedance is given by Eq. (17). The impedance for a unit

$$Z = \rho_0 c/S \tag{17}$$

area is $\rho_0 c$, which is called the specific acoustic impedance, represented by Z_s. The important role played by this in sound problems is emphasized by its appearance in Eq. (14) for acoustic intensity. *See* ACOUSTIC IMPEDANCE.

The harmonic sound wave considered so far in this section is called a plane wave since the particle displacement and pressure are uniform over any plane surface perpendicular to the direction of propagation. Such a plane is called a wavefront.

Spherical waves. A type of sound wave of even greater importance than the plane wave is the spherical wave. Such a wave spreads equally in all directions from a point source; it can be realized ideally by a small sphere that pulsates radially. The radiated sound then moves out so that the particle velocity u and the sound pressure p are uniform over a sphere or radius r with center at the source and are, of course, dependent on r. For a harmonic spherical wave the sound pressure can be written in the form of Eq. (18). It has become

$$p = \frac{p_0}{r} \cos \left[2\pi \nu \left(t - \frac{r}{c} \right) \right] \tag{18}$$

customary to represent both plane and spherical waves by means of complex variable notation, in which Eq. (18) becomes Eq. (19), where $i = \sqrt{-1}$.

$$p = \frac{p_0}{r} \cdot e^{i\left[2\pi\nu\left(t-\frac{r}{c}\right)\right]}$$

$$= \frac{p_0}{r}\left\{\cos\left[2\pi\nu\left(t-\frac{r}{c}\right)\right] + i\sin\left[2\pi\nu\left(t-\frac{r}{c}\right)\right]\right\} \quad (19)$$

The expression for p in Eq. (18) is the real part of the complex expression in Eq. (19). Expressed in the same notation, the particle velocity is given by Eq. (20), and its real part is given by Eq. (21). The specific acoustic impedance for a spherical wave is most conveniently expressed in complex notation as Eq. (22). The real part of the specific im-

$$u = \frac{-ip_0}{2\pi\nu\rho_0 r}\left(\frac{1}{r} + \frac{2\pi i\nu}{c}\right)e^{i\left[2\pi\nu\left(t-\frac{r}{c}\right)\right]} \quad (20)$$

$$u_{real} = \frac{p_0}{2\pi\nu\rho_0 r}\left\{\frac{\sin\left[2\pi\nu\left(t-\frac{r}{c}\right)\right]}{r}\right.$$
$$\left. + \frac{2\pi\nu}{c} \cdot \cos\left[2\pi\nu\left(t-\frac{r}{c}\right)\right]\right\} \quad (21)$$

$$Z_s = \frac{\rho_0 c \cdot 4\pi^2\nu^2 r^2/c^2}{1 + 4\pi^2\nu^2 r^2/c^2} + i\frac{\rho_0 c \cdot 2\pi\nu r/c}{1 + 4\pi^2\nu^2 r^2/c^2} \quad (22)$$

pedance is called the specific acoustic resistance and the imaginary part is called the specific acoustic reactance. The former is analogous to electrical resistance and the latter to electrical inductance. The specific acoustic impedance of a plane sound wave is real ($= \rho_0 c$) and is a specific resistance. *See* ELECTRICAL IMPEDANCE; ELECTRICAL RESISTANCE; REACTANCE.

For great distances from the source for which $4\pi^2\nu^2 r^2/c^2 \gg 1$ (strictly this means distances for which $r \gg \lambda$, the wavelength), the specific resistance for the spherical wave approaches that for a plane wave, namely $\rho_0 c$, while the specific reactance becomes vanishingly small. At very large distances from the source compared with the wavelength, any finite portion of the sphere through which the sound wave passes becomes more nearly plane.

The intensity of a spherical sound wave from a source whose constant power output is W is given by Eq. (23). Common sense agrees with this result,

$$I = \frac{W}{4\pi r^2} \quad (23)$$

since the constant power output of the source is distributed over a surface which increases directly with r^2 (the area of a sphere of radius r being $4\pi r^2$).

Velocity of sound. The general theoretical expression for the velocity of sound in a compressible fluid medium in which the instantaneous pressure and density are respectively p and ρ is given by Eq. (24), where the ratio $dp/d\rho$ must be

$$c = \sqrt{\frac{dp}{d\rho}} \quad (24)$$

taken under adiabatic conditions, that is, such that no heat enters or leaves the element of the medium as the sound wave passes through it. For a liquid

for which the adiabatic bulk modulus is B, Eq. (24) becomes Eq. (11). For a gas, Eq. (24) gives Eq. (25),

$$c = \sqrt{\gamma p/\rho} \quad (25)$$

where γ is the ratio of the specific heat at constant pressure to that at constant volume. For air, $\gamma = 1.4$. From the general gas equation of an ideal gas, Eq. (24) takes the form of Eq. (26), where R is the

$$c = \sqrt{\gamma RT} \quad (26)$$

gas constant per unit mass, and T is the absolute temperature. Denoting the velocity of sound in the gas at 0°C by c_0 permits Eq. (26) to be put in the form of Eq. (27), where Θ is temperature in Celsius degrees. *See* GAS; WAVE MOTION IN FLUIDS.

$$c = c_0\sqrt{1 + \Theta/273} \quad (27)$$

Sound is propagated through the bulk of an elastic solid by means of a longitudinal compressional wave (specifically referred to in elasticity theory as an irrotational wave) with velocity given by Eq. (28), where μ is the shear modulus of the solid.

$$c = \sqrt{\frac{B + 4\mu/3}{\rho}} \quad (28)$$

The word longitudinal means that the displacement of the medium through which the sound is passing lies in the direction of propagation. If the solid is in the shape of a rod with cross section much smaller than its length, sound is propagated with velocity given by Eq. (29), where Y is Young's modulus. *See* ELASTICITY.

$$c = \sqrt{\frac{Y}{\rho}} \quad (29)$$

In addition to longitudinal waves, solids can also propagate pure shear waves which are transverse in character, in which the displacement of the medium is at right angles to the direction of propagation (strictly called solenoidal waves in elasticity theory). Though not strictly a sound wave in the usually accepted sense, this type of wave is considered acoustical in character; it is exemplified as a component of seismological or earthquake waves.

Tables 1, 2, and 3 list values of sound velocity in meters per second in selected gases, liquids, and solids at the indicated temperatures. Specific acoustic resistances are also listed.

The velocity of sound varies with the temperature [Eq. (27) for gases], usually increasing as the

Table 1. Sound velocity in selected gases*

Gases	c	ρ_0	$R \times 10^{-4}$
Chlorine, 0°C	206	0.00317	0.0065
Carbon dioxide, 0°C	258	0.00198	0.0050
Oxygen, 0°C	317	0.00143	0.0045
Air (dry, 20°C, 760 mmHg)	344	0.001205	0.0042
Carbon monoxide, 0°C	337	0.00125	0.0030
Methane, 0°C	432	0.00072	0.0030
Steam, 100°C	405	0.00058	0.0023
Hydrogen, 0°C	1270	0.00009	0.0011

*c = velocity of sound in m/s at 20°C, where not otherwise stated; ρ_0 = density in g/cm³; $R = \rho_0 c$, specific acoustic resistance in g/s cm². The substances are arranged in each group in the order of decreasing R.

SOURCE: From G. W. Stewart and R. B. Lindsay, *Acoustics*, Van Nostrand, 1930.

Table 2. Sound velocity in selected liquids*

Liquids	c	ρ_0	$R \times 10^{-4}$
Mercury	1407	13.6	191
Water (sea)	1490	1.025	15.3
Chloroform	1032	1.48	15
Water (air-free)	1461	1.00	14.6
Alcohol, 12.5°C	1241	0.81	10
Benzol, 17°C	1166	0.90	10

*Quantities tabulated are same as in Table 1.
SOURCE: From G. W. Stewart and R. B. Lindsay, *Acoustics*, Van Nostrand, 1930.

Table 3. Sound velocity in selected solids*

Solids	c	ρ_0	$R \times 10^{-4}$
Platinum	2690	21.4	575
Nickel	4973	8.6 – 8.9	427 – 443
Cobalt	4724	8.7	412
Gold (hard)	2100	19.3	406
Steel	5000	7.8	390
Glass (upper limit)	6000	5.9	354
Gold (soft)	1743	19.3	333
Copper	3560	8.9	317
Brass	3500	8.5	298
Silver	2610	10.5	273
Zinc	3700	7.1	262
Cadmium	2307	8.6	199
Tin	2500	7.3	183
Lead	1227	11.4	140
Aluminum	5104	2.6	132
Glass (lower limit)	5000	2.4	120
Marble	3810	2.6	99
Brick	3652	1.4 – 2.2	51 – 80
Upper limit — ash†	4670	0.85	40
Lower limit — elm†	1013	0.54	5
Paraffin	1300	0.90	11
Tallow	390	0.95	3.7
Cork†	500	0.24	1.2
Wax, 17°C	880	0.96	0.84
Wax, 28°C	440	0.96	0.44
Rubber††	31 – 69	0.95	0.29 – 0.66

*Quantities tabulated are same as in Table 1.
†Wood varies with the kind, specimen, and relation of the wave direction to the grain.
‡Variable; dependent on color and manufacture, as well as on temperature.
SOURCE: From G. W. Stewart and R. B. Lindsay, *Acoustics*, Van Nostrand, 1930.

temperature increases, and hence can have different values in different parts of a medium. This is of particular importance in sound propagation through the atmosphere. The same is true for sound transmitted through the ocean. Even in an isothermal medium the velocity will change with the frequency. This is known as sound dispersion. It is a very small effect at low frequencies and becomes perceptible only at very high or ultrasonic frequencies. At ultrasonic frequencies the effective sound velocity in a relatively thin solid rod may vary considerably from the value given by Eq. (29) due to the interaction between the lateral strain and the longitudinal strain associated with the propagation of the wave. This change in velocity with frequency is called configurational dispersion. *See* ULTRASONICS.

Sound rays. Though sound, like light, is propagated by waves, under certain conditions its transmission may be described adequately by means of rays, which are lines perpendicular to the wavefront in each case, giving the direction of sound propagation. The necessary conditions for the use of rays are: the wavefront must not change its direction materially over a distance of one wavelength; the amplitude of the sound pressure must not change materially over the same distance; finally the wavelength itself (in the case of transmission through a nonhomogeneous medium) must change very little over a distance of its own order of magnitude.

The equations describing sound rays are obtained from the solution of the eikonal equation. For a medium in which the sound velocity varies from point to point, as in the atmosphere and in the sea, this is given by Eq. (30), where $c(x,y,z)$ is the

$$\left(\frac{\partial\psi}{\partial x}\right)^2 + \left(\frac{\partial\psi}{\partial y}\right)^2 + \left(\frac{\partial\psi}{\partial z}\right)^2 = \frac{c_0^2}{[c(x,y,z)]^2} \qquad (30)$$

spatially variable sound velocity and c_0 is the constant sound velocity at some arbitrarily chosen reference point. The function $\psi(x,y,z)$ is the eikonal. When multiplied by $2\pi\nu/c_0$, it is the spatially variable phase in the harmonic wave being propagated through the medium in which the pressure is in the general form of Eq. (31). In analogy with op-

$$p = p_0(x,y,z)e^{i[2\pi\nu t - 2\pi\nu/c_0 \cdot \psi(x,y,z)]} \qquad (31)$$

tics, the ratio $c_0/c(x,y,z)$ is the variable index of refraction n of the medium with respect to sound propagation.

An example of the use of the eikonal equation is sound propagation through a stratified medium like the sea when the temperature and sound velocity vary only with depth, measured as z below the surface, which lies along the x axis. The eikonal equation then is Eq. (32), where $n(z)$ is the index

$$\left(\frac{\partial\psi}{\partial x}\right)^2 + \left(\frac{\partial\psi}{\partial z}\right)^2 = [n(z)]^2 \qquad (32)$$

of refraction. The complete integral solution of this partial differential equation is Eq. (33), where C_1

$$\psi(x,z) = C_1 x + \int \sqrt{[n(z)]^2 - C_1^2}\, dz + C_2 \qquad (33)$$

and C_2 are arbitrary constants to be evaluated from the boundary conditions. If the algebraic equation of the sound ray is written in the form of Eq. (34), the corresponding ordinary differential

$$z = f(x) \qquad (34)$$

equation is Eq. (35). For the plausible case in which

$$\frac{dz}{dx} = \frac{\partial\psi/\partial z}{\partial\psi/\partial x} \qquad (35)$$

the index of refraction is given by Eq. (36), where a

$$n = \frac{1}{1 + az} \qquad (36)$$

is some constant, the rays obeying Eq. (35) are circular arcs, concave upward for $a > 0$ and concave downward for $a < 0$.

Ray acoustics has proved very useful in the study of sound propagation in the atmosphere and in the sea.

Reflection and refraction. When a sound wave strikes a surface separating media of different density and elasticity, reflection and refraction can take place. The laws governing these phenomena are the same as those for light. The incident wave normal (or ray, if sound rays apply) makes the same angle with the normal to the surface as the reflected wave normal, and both lie in the same plane. The refracted wave normal obeys Snell's law; that is, if the angle this makes with the normal to the surface is denoted by ϕ_t, whereas the angle of incidence is ϕ_i, then Eq. (37) holds, where c_i is

$$\frac{\sin \phi_t}{c_t} = \frac{\sin \phi_i}{c_i} \qquad (37)$$

the velocity of sound in the medium from which the wave is incident and c_t is the velocity of sound in the medium into which the sound is passing. *See* REFRACTION OF WAVES.

When sound crosses an interface such as is being considered here, some of the energy in it is reflected and some transmitted. In the case of oblique incidence of a plane wave, the ratio of the intensity of the reflected wave I_r to that of the incident wave I_i is given by Eq. (38), where ρ_i and ρ_t

$$\frac{I_r}{I_i} = \frac{(\rho_t/\rho_i - \cot \phi_t/\cot \phi_i)^2}{(\rho_t/\rho_i + \cot \phi_t/\cot \phi_i)^2} \qquad (38)$$

are the densities of the incident and refracting media respectively and ϕ_i and ϕ_t are the angles given in Eq. (37). The ratio of the intensity I_t of the transmitted wave to I_i is given by Eq. (39). For

$$\frac{I_t}{I_i} = \frac{4\rho_t c_i/\rho_i c_t}{(\rho_t/\rho_i + \cot \phi_t/\cot \phi_i)^2} \qquad (39)$$

normal incidence $\cos \phi_i = \cos \phi_t = 1$ and Eqs. (38) and (39) become respectively Eqs. (40) and (41).

$$\frac{I_r}{I_i} = \left(\frac{\rho_t c_t - \rho_i c_i}{\rho_t c_t + \rho_i c_i}\right)^2 \qquad (40)$$

$$\frac{I_t}{I_i} = \frac{4\rho_i \rho_t c_i c_t}{(\rho_t c_t + \rho_i c_i)^2} \qquad (41)$$

The important role of the specific acoustic impedance of a plane wave, namely ρc, in the foregoing formulas, is noteworthy. If the specific impedances of the two media are equal, $I_t = I$ and $I_r = 0$ at normal incidence. The greater the difference between the two specific impedances, the more sound is reflected and the less transmitted. The transmission from air to water, and vice versa, is a good illustration.

The above formulas assume that $c_i > c_t$, or that if $c_i < c_t$, then $\phi_i < $ arc sin (c_i/c_t). Otherwise total reflection occurs as in the analogous case of light waves. *See* REFLECTION AND TRANSMISSION COEFFICIENTS; REFLECTION OF ELECTROMAGNETIC RADIATION.

Diffraction and scattering. Like light waves, sound waves bend around obstacles, that is, experience diffraction. Indeed, sound can be heard around a corner on account of diffraction, which in the case of audible sound is a much more obvious phenomenon than the diffraction of visible light because the wavelength of audible sound is much greater than that of visible light. The bending of sound waves due to diffraction becomes less marked as the wavelength decreases relative to the spatial dimensions of the diffracting obstacles. With ultrasonic radiation it is possible to produce sound shadows. Another common name for sound diffraction is scattering, particularly if many obstacles are involved. The case of a rigid sphere is typical. If a plane harmonic wave of frequency ν is incident on a rigid sphere of radius a fixed in a medium in which the sound velocity is c, let I_0 be

the intensity of the incident wave. The cases for which the sound wavelength is either very long or very short compared with a are the ones of greatest practical interest. The power of the scattered radiation P_s for the two cases in question is given by Eqs. (42) and (43). The case given in Eq. (42) refers

$$P_s = \frac{256\pi^5 a^6}{9\lambda^4} \cdot I_0 \qquad \text{for } 2\pi a/\lambda \ll 1 \qquad (42)$$

$$P_s = 2\pi a^2 I_0 \qquad \text{for } 2\pi a/\lambda \gg 1 \qquad (43)$$

to scattering by a sphere of radius much smaller than the wavelength. An illustration is the scattering of audible sound by small bubbles in sea water. This is usually referred to as Rayleigh scattering since Lord Rayleigh first investigated it for the analogous case of light and applied the theory to the explanation of the blue color of the sky as a scattering phenomenon. Equation (43) corresponds to scattering by spherical obstacles of dimensions very large compared with the wavelength. Here ray acoustics can be applied to a good approximation. A sound shadow develops on the side of the sphere opposite to the incident plane radiation. *See* DIFFRACTION; SCATTERING OF ELECTROMAGNETIC RADIATION.

The intermediate scattering case in which $2\pi a/\lambda$ is of the order of unity presents considerable mathematical difficulty.

Sound pulses. The progressive sound waves described mathematically so far in this article are an idealization of the sound radiation encountered in practice, even though the results derived agree rather well with experience. No physically realizable sound wave can be infinite in space or time. All practical sound waves are really finite wave trains or pulses. Such a harmonic pulse can be represented mathematically by Eqs. (44). For

$$f(x - ct) = A \cos k_0(x - ct) \quad \text{for } |x - ct| < L/2$$
$$f(x - ct) = 0 \quad \text{for } |x - ct| > L/2 \qquad (44)$$

convenience the so-called wave parameter $k_0 = 2\pi/\lambda_0 = 2\pi\nu_0/c$ (λ_0 is the effective wavelength for the pulse and ν_0 the effective frequency) is introduced. Equation (44) represents a harmonic wave pulse of length L moving through the medium with velocity c. The pulse may be analyzed into an infinitely continuous set of harmonic components with continuously variable wave parameter k by means of the Fourier integral theorem. The amplitude of the harmonic wave component of wave parameter k is given by expression (45). This has a

$$\frac{A}{\pi} \cdot \frac{\sin(k_0 - k)L/2}{k_0 - k} \qquad (45)$$

maximum for $k = k_0$, equal to $AL/2\pi$. For larger and smaller values of k the amplitude falls off rather sharply; most of the contribution to the harmonic analysis of the pulses is confined to the range bounded by values of k satisfying Eq. (46). This

$$(k_0 - k)L = \pm 2\pi \qquad (46)$$

means physically that the smaller the length of the wave pulse the greater is the range of the wave parameter k needed for its resolution into harmonic continuous waves, and vice versa. The same statement applies to the frequency. Its practical importance is that if a device is to amplify a sound pulse without distortion, it must have a frequency response sufficiently wide to take care of the whole frequency spread associated with the pulse length. *See* FOURIER SERIES AND INTEGRALS.

The preceding considerations have assumed that the sound wave velocity c remains constant throughout the medium for all frequencies. However, if a sound pulse moves through a dispersive medium—for example, through a gas like carbon dioxide at very high frequency—the pulse is resolved into a group of waves traveling with the velocity (the so-called group velocity) given by Eq. (47), where c_0 is the mean phase velocity and λ_0 is

$$U = c_0 - \lambda_0 \left(\frac{dc}{d\lambda} \right)_0 \qquad (47)$$

the wavelength of the component of maximum amplitude in the group. The derivative in Eq. (47) is to be taken at the value $\lambda = \lambda_0$. Wave groups in acoustics are of importance mainly at very high frequencies and very low pressures in gases. *See* GROUP VELOCITY; PHASE VELOCITY.

Transducers. Sound sources are referred to as transducers, since they transfer energy of various forms into sound. The most important source of sound in human experience is the human voice, which in ordinary conversation and singing produces power outputs of from 30 to 60 microwatts. (Fig. 1). Other important sources are mechanically vibrating solids, liquids, and gases. Solid vibrators include strings, rods, plates, and membranes (as in musical instruments). The organ pipe is an example of a sound source based on vibrating air. The turbulent flow of air in the exhaust of a jet plane produces intense sound. The break of the surf on the shore is a liquid source of sound.

More sophisticated sound sources depend on the ability to produce mechanical vibrations by means of magnetic and electrical effects. The ordinary telephone receiver and the radio loudspeaker are illustrations. Still more elaborate and valuable for many applications are piezoelectric and magnetostrictive oscillators. These electroacoustic transducers have found wide use as both sources and receivers, for example, in underwater sound, in sound recording and reproduction, in radio and

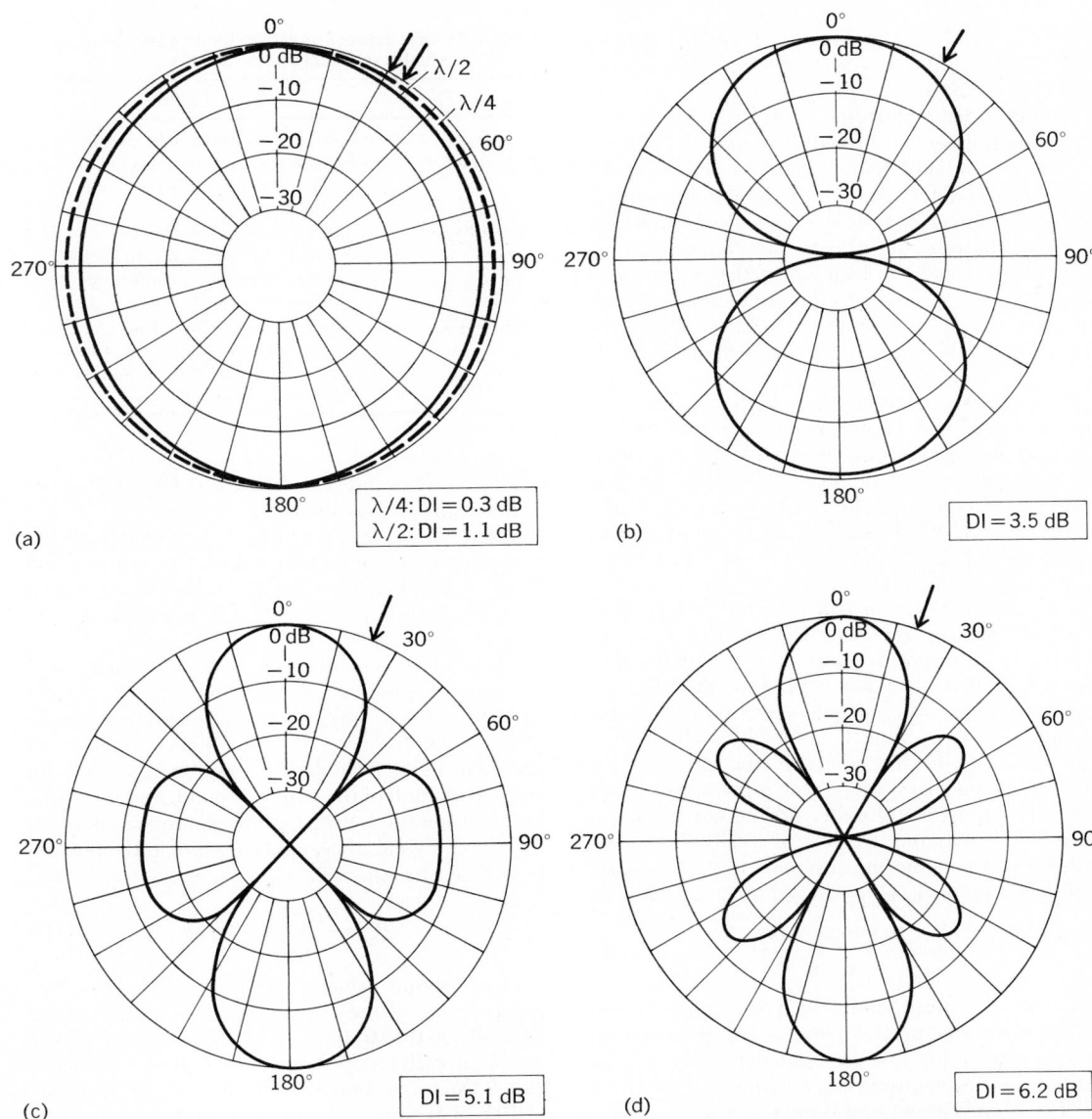

Fig. 2. Directivity patterns for a linear line array radiating uniformly along its length d. The boxes give the directivity index ($DI = 10 \log_{10} q$) at $\theta = 0°$. One angle of zero directivity index is also indicated by the arrow for each length. (a) $d = \lambda/4$ indicated by broken line and $\lambda/2$ indicated by solid line. (b) $d = \lambda$. (c) $d = 3\lambda/2$. (d) $d = 2\lambda$.

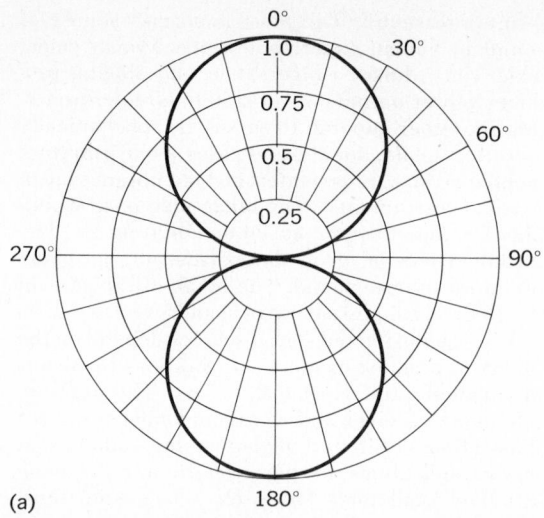

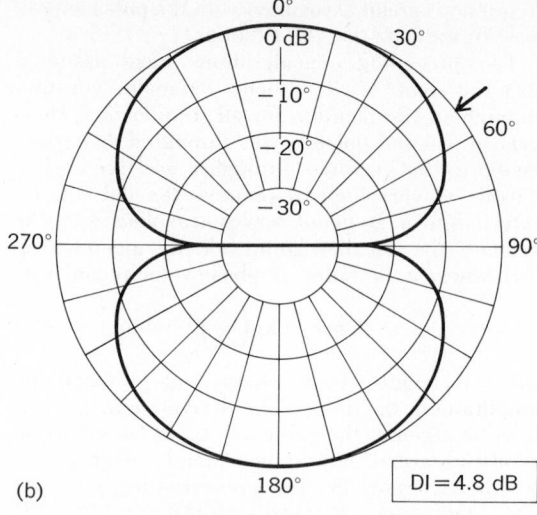

(a)

(b)

DI = 4.8 dB

Fig. 3.   Directivity patterns for a doublet sound source. (a) Sound-pressure ratio p/p_0 versus θ. (b) 20 log p/p_0 versus θ. The box gives directivity index ($DI = 10 \log_{10} q$) at $\theta = 0°$. One angle of zero directivity index is also indicated by the arrow.

television, and in the industrial applications of sound. *See* MAGNETOSTRICTION; PIEZOELECTRICITY; TRANSDUCER.

Radiation from sources. The ideal point source has already been discussed. The sound intensity produced at a point distant from a practical sound transducer is determined by the total power output of the transducer, and by the type of source, which in turn controls its directional characteristics. Other factors are the location of the source with respect to the ground or other reflecting, refracting, diffracting, or absorbing surfaces, as well as the nature of the medium through which the sound passes.

In sectorial radiation, in which the radiated sound is confined to a portion of a sphere—for example, in the case of a constant source mounted on a large plane surface (a so-called half-space)—more sound power will be radiated through a unit area at any given distance from the source than would be radiated through a unit area of a whole sphere; hence the intensity is increased.

The directivity factor q of a source is defined as the multiplicative increase that must be made in the power output of a spherical source (with base power output W) to make it produce the same intensity as a source of the same power output radiating into a spherical sector of less than 360°. For example, the directivity factor for radiation into a half-space is $q = 2$. Table 4 gives the values of the directivity factor under various conditions.

A source may be directive of itself without the influence of neighboring surfaces. In that case the directivity factor is given by Eq. (48), where W is

$$q = \frac{4\pi r^2 [p_D(r,\theta,\phi)]^2}{\rho_0 c W} \qquad (48)$$

the acoustic power radiated by a directional source producing a root-mean-square sound pressure p_D at a distance r and in a direction defined by the colatitude and colongitude angles θ and ϕ respectively. The quantities ρ_0 and c have the usual significance.

Examples of directivity patterns of three common sound sources are given in Figs. 2, 3, and 4.

Table 4. Relation between directivity factor q and the size of space into which a spherical source radiates

Type of space	Description	q
Full-space	Spherical source in free space	1
Half-space	Spherical source radiating to one side of an infinitely large, flat plane	2
Quarter-space	Spherical source radiating outward from the inside of the intersection of two infinitely large, flat planes	4
Eighth-space	Spherical source radiating outward from the inside of the intersection of three infinitely large, flat planes	8

The directivity index is defined as $DI = 10 \log_{10} q$, and is expressed in decibels.

The linear line array (Fig. 2) is common in the use of underwater sound, and may be constructed from a number of identical effective spherical sources closely spaced along a line, all radiating in phase.

The doublet sound source (Fig. 3) is made up of two simple spherical sources operating exactly out of phase; that is, one source is expanding while the other is contracting. If the distance b between the sources of a doublet is small compared with the wavelength of the emitted sound, and if the distance r from the source to the observer satisfies the condition $r \geq \lambda/6$, the sound pressure at the observer is given by Eq. (49), where u_0 is the root-mean-

$$P_D = \frac{\pi \rho_0 \nu^2 u_0 b}{rc} \cdot \cos \theta \qquad (49)$$

square volume velocity (in cubic meters per second) of each of the simple sources making up the doublet, ν the frequency, ρ_0 the density of the medium, c the velocity of sound in the medium, and θ the angle between the line drawn from the center of the doublet to the observer and the line drawn through the two sources of the doublet.

The rigid circular piston (Fig. 4) in an infinite baffle is similar to a circular loudspeaker in the wall of a room or in the side of a large box; none of the sound being radiated to the rear can get around to the front again.

Figure 5 gives the directivity indexes for the principal axes of three kinds of circular piston radiators as a function of the ratio of the circumference $2\pi a$ of the radiator to the wavelength λ. A circular piston in the end of a long tube is equivalent to a circular loudspeaker in a small box. An unbaffled piston is equivalent to a circular loudspeaker, both sides of which are allowed to radiate into the same medium.

Most sound transducers are reversible; that is, they can be used as receivers of sound as well as sources. A familiar example is the megaphone, which, when used for speaking, increases the directivity of the voice and also helps to amplify it because of the better acoustic impedance match with the external air provided by the large open end. The same effect is true of horns in general. When the megaphone or horn is used as a receiver at the ear, it increases the directivity of the ear and also amplifies the received sound. Hence all the directivity patterns given in Figs. 2, 3, and 4

are valid for receivers of the same construction.

For a vibrating plate treated as a plane circular piston of radius a with all points vibrating perpendicular to the surface with the same frequency, amplitude, and phase, in the region for which the distance r from the center of the piston along the axis of the latter (the line perpendicular to its surface) satisfies inequality (50), the intensity for

$$a < r < ka^2/2\pi \qquad (50)$$

given power output passes through a series of maxima and minima (zero) at points given by Eqs. (51) and (52), where $n = 1, 2, 3, \ldots$. If inequality

$$r = ka^2/4\pi n \qquad \text{(maximum)} \qquad (51)$$

$$r = ka^2/2\pi(2n+1) \qquad \text{(minimum)} \qquad (52)$$

(53) holds, that is, at great distances from the

$$r \gg ka^2/2\pi \qquad (53)$$

source, the intensity at any point P on the axis of the source is given approximately by Eq. (54),

$$I_p = \frac{\rho_0 c k^2 a^4 \xi_0^2}{8r^2} \qquad (54)$$

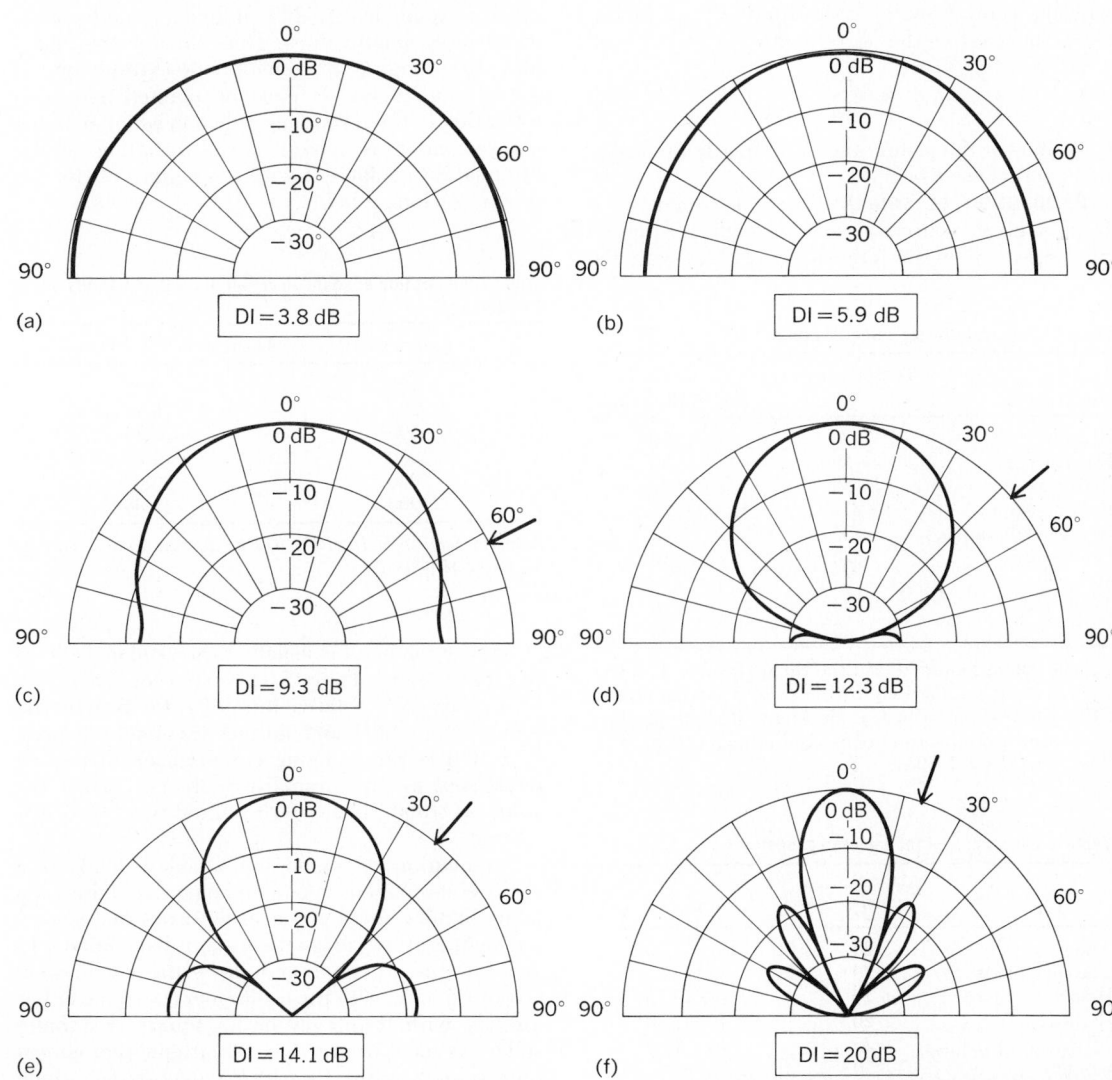

Fig. 4. Directivity patterns for rigid circular piston in an infinite baffle as a function of $2\pi a/\lambda$, where a is the radius of the piston. The boxes give the directivity index at $\theta = 0°$. One angle of zero directivity index is also indicated by the arrow. (a) $2\pi a/\lambda = 1$. (b) $2\pi a/\lambda = 2$. (c) $2\pi a/\lambda = 3$. (d) $2\pi a/\lambda = 4$. (e) $2\pi a/\lambda = 5$. (f) $2\pi a/\lambda = 10$.

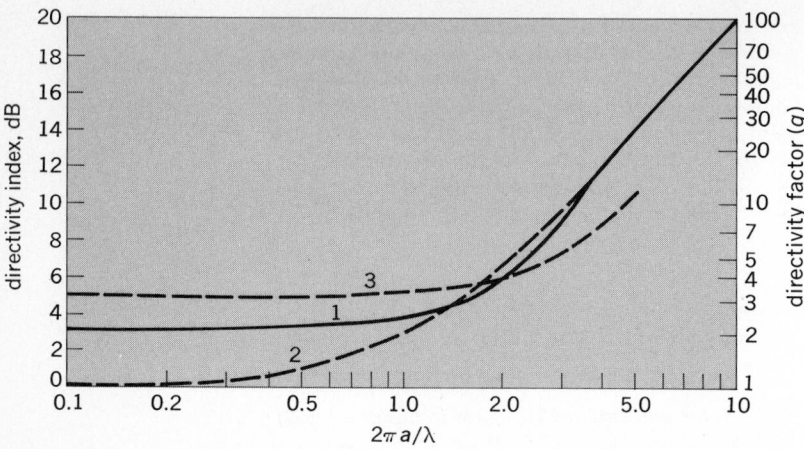

Fig. 5. Directivity indexes for the radiation from one side only of a piston in an infinite plane baffle (curve 1), piston in the end of a long tube (curve 2), and piston in free space without any baffle (curve 3).

where $\dot{\xi}_0$ is the displacement velocity of the piston. In this region the radiation behaves to a good approximation as a spherical wave. Beyond $r = ka^2/2\pi$ the radiation is roughly conically divergent, with the conical angle (in steradians) given by Eq. (55). The shorter the wavelength compared with

$$\theta = \frac{\pi^2 \lambda^2}{a^2} \tag{55}$$

the radius of the piston, the more nearly the radiation approaches a beam.

Attenuation of sound. When a sound wave from a given source passes through any actual physical medium, its intensity decreases with dis-

tance from the source. This is known as attenuation. Part of this is due to the geometrical spreading of the wavefront, unless it is confined in a tube or duct. The rest of the attenuation is due to the influence of the medium and the surfaces over which or through which the sound passes. Reflection, refraction, and scattering can all serve to increase the effective attenuation of sound propagation in a given direction. The medium itself produces attenuation by changing a part of the acoustic energy into heat, a kind of attenuation known as sound absorption. Mechanisms responsible for this are viscosity, heat conduction, and molecular relaxation processes. They all result in intensity variation with distance r along the sound path of the form of Eq. (56), where I_0 is the initial intensity,

$$I = I_0 e^{-\alpha r} \tag{56}$$

$r = 0$, and α is the absorption or attenuation coefficient, which in general for fluids increases with the square of the frequency. It also depends on the physical properties of the medium. Tables 5 and 6 list experimental values of α/ν^2 for some selected liquids and gases. All the values in these tables are based on measurements at frequencies above 1 MHz. Measurements of sound absorption at low frequencies are difficult to carry out because of the very small values. The values in the tables may, to a rough approximation, be extrapolated to lower frequencies. It must be pointed out, however, that in the frequency region in which molecular relaxation is effective, the variation of the attenuation coefficient with the square of the frequency no longer holds.

Table 5. Values of α/ν^2 for selected liquids*

Liquid	Temperature, °C	$\alpha/\nu^2 \times 10^{17}$, s²/cm
Benzene	20	750
Ethyl alcohol	30	50
Glycerine	20	2200
Helium (liquid)	−277	231
Hydrogen (liquid)	−256	5
Lead (liquid)	340	10
Mercury	25	8
Tin (liquid)	240	6
Water	20	25

*The values for liquid lead and tin are from R. T. Beyer and S. V. Letcher, *Physical Ultrasonics*, Academic Press, 1969. The rest are from J. J. Markham, R. T. Beyer, and R. B. Lindsay, Absorption of sound in fluids, *Rev. Mod. Phys.*, 23:353–411, 1951.

Table 6. Values of α/ν^2 for selected gases*

Gas	Temperature, °C	$\alpha/\nu^2 \times 10^{13}$, s²/cm
Argon	20	1.85
Carbon dioxide	20	27.0
Helium	20	0.53
Hydrogen	20	3.60
Nitrogen	25	1.90
Oxygen	50	2.20

*From J. J. Markham, R. T. Beyer, and R. B. Lindsay, Absorption of sound in fluids, *Rev. Mod. Phys.*, 23:353–411, 1951.

Table 7. Absorption of sound in dry air at 20°C, at various frequencies*

Frequency, Hz	Absorption, dB/100 m
125	0.06
250	0.16
500	0.35
1000	1.24
2000	4.20
4000	10.80

From *American Institute of Physics Handbook*, 3d ed., McGraw-Hill, 1957)

Absorption in air is usually expressed in decibels per 100 meters. Table 7 lists experimental values for dry air (10% relative humidity) for various frequencies at 20°C and normal atmospheric pressure. Other things being equal, absorption in air decreases as the humidity increases. Thus the value at 1000 Hz for 90% humidity is 0.38 dB/100 m.

Attenuation of sound in extended solids is a very complicated phenomenon, depending on many factors, including heat flow, internal viscosity, temperature, magnetism, and dislocations. In general, the attenuation follows the exponential law in Eq. (56), with the coefficient α varying either directly with frequency or its square. For finite solids, as used in practice, the attenuation is best represented by the logarithmic decrement, which is the logarithm to the base e of the ratio of successive particle displacement amplitudes at a given point in the solid. Table 8 lists values of the

Table 8. Values of the logarithmic decrement δ for selected solids*

Solid	δ
Aluminum (annealed), 200 K	0.03×10^{-3}
Aluminum (annealed), 275 K	0.1×10^{-3}
Copper (unannealed), 100 K	13.5×10^{-3}
Copper (unannealed), 250 K	0.65×10^{-3}
Lead, 225 K	3.15×10^{-3}
Lead, 250 K	9.5×10^{-3}
Silver (annealed), 250 K	
(little change with temperature)	0.04×10^{-3}
Polystyrene (room temperature)	48×10^{-3}

*Resonant frequencies in the range $10-50$ kHz.
SOURCE: From P. G. Bordoni, Elastic and anelastic behavior of some metals at very low temperatures, *J. Acoust.Soc.Amer.*, 26:495–502, 1954.

logarithmic decrement δ for certain selected solids. *See* SOUND ABSORPTION.

High-intensity sound. As Fig. 1 indicates, sound waves can be produced over a wide intensity range. To the normal ear, sound of intensity 10^{-16} W/cm² is audible. However, it is possible to produce intensities as high as 10^8 W/cm². Sounds of intensity above 1 W/cm² are referred to as macrosonic. Such sound waves, used mainly for underwater transmission and for industrial purposes, are very dangerous to the human ear. They are almost always ultrasonic in character.

The mathematical analysis of Eqs. (1)–(4) applies only to sound of low intensity and is inadequate to describe macrosonic radiation. For a macrosonic wave in the x direction, it is customary to follow the fate of a particle of fluid which was at point $x = a$ at time $t = 0$ and has reached the point x at time t. Then x is a function of the independent variables a and t. Placing $x = a + \xi$, where ξ is the particle displacement due to the wave propagation, leads to the wave equation (57), where A and B are

$$\frac{\partial^2 \xi}{\partial t^2} = \frac{c_0^2}{\left(1 + \dfrac{\partial \xi}{\partial a}\right)^{2+B/A}} \cdot \frac{\partial^2 \xi}{\partial a^2} \tag{57}$$

given by Eqs. (58) and (59). The derivatives in the

$$A = \rho_0 \left(\frac{\partial p}{\partial \rho}\right)_{S, \rho_0} = \rho_0 c_0^2 \tag{58}$$

$$B = \rho_0^2 \left(\frac{\partial^2 p}{\partial \rho^2}\right)_{S, \rho_0} \tag{59}$$

expressions for A and B are taken at constant entropy S and constant density ρ_0. It can be shown that for a gas $B/A = \gamma - 1$, where γ is the ratio of the specific heat at constant pressure to that at constant volume. For liquids, B/A values depart markedly from $\gamma - 1$. For distilled water at 30°C $B/A = 5.2$, and for mercury at the same temperature $B/A = 7.8$.

Since Eq. (57) is a nonlinear partial differential equation, the subject of high-intensity sound propagation governed by it is known as nonlinear acoustics. The various solutions obtained for Eq. (57) indicate that as a harmonic wave of this kind progresses, it will not retain its sinusoidal shape but the crests will tend to overtake the troughs. In the ideal case this leads to the development of a discontinuity in which the wave "breaks," as when water waves break on the seashore. The presence

of viscosity and other dissipative mechanisms prevents this from happening to an intense sound wave. The wave ultimately assumes the sawtooth shape indicated in Fig. 6. The result is known as a shock wave, characterized by a relatively large change in excess pressure and density across a very small region of space, known as the shock front. The sonic boom is an example of a shock-wave. *See* SHOCK WAVE.

Macrosonic radiation produces a number of interesting phenomena in liquids. One is the development of a radiation pressure, that is, a non-oscillating pressure different from the normal pressure in a sound wave. This pressure is similar to the radiation pressure due to the electromagnetic radiation. Its value is a function of the average energy density in the sound radiation. A related effect is the production of direct current streaming in the liquid. Finally, if the intensity and frequency are high enough, cavitation may be caused, that is, the development of bubbles of gas in the liquid. These effects make it possible to use macrosonic radiation at high frequency in the cleaning of small metal parts. They also serve to promote the production of aerosols coagulation in liquids.

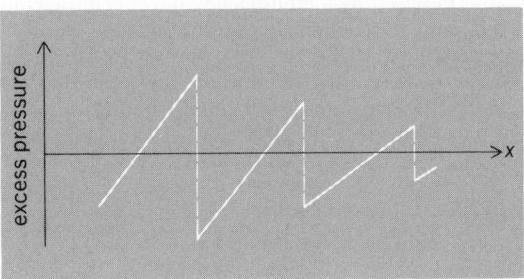

Fig. 6. Excess pressure as a function of distance from the source in a macrosonic wave. (*From R. B. Lindsay, Mechanical Radiation, McGraw-Hill, 1960*)

Atmospheric acoustics. Sound propagation through the atmosphere is complicated by many factors, including wind, spatial and temporal variations in temperature, and the presence of water vapor or fog, rain, and snow. These produce variations in velocity as well as attenuation due to reflection, refraction, and scattering, in addition to normal sound absorption. The presence of the ground and buildings in inhabited areas also affects propagation.

Under normal daytime conditions the air temperature decreases with height above the surface. Since the velocity of sound in air varies as the square root of the absolute temperature [Eq. (26)], refraction of sound wavefronts takes place as they move upward. Since the variation in velocity is generally gradual, ray acoustics may be used to describe propagation. Snell's law of refraction [Eq. (37)] then ordains that under normal conditions rays of sound tend to bend upward, thus decreasing the range from its value in an isothermal atmosphere. On the other hand, when after a clear night there is a temperature inversion in the early morning, with the temperature increasing upward to a certain height, the rays are bent downward and sound carries farther.

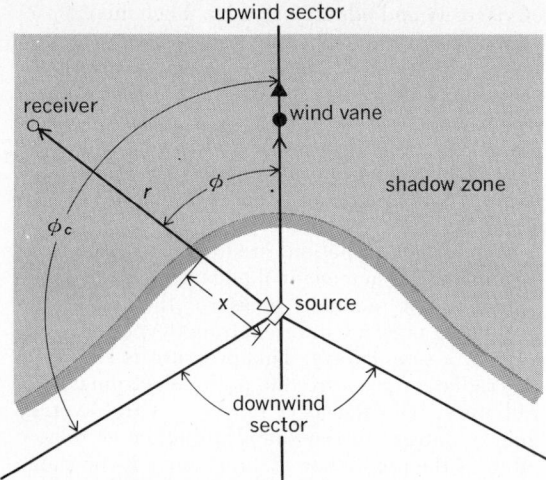

Fig. 7. Geometry of sound propagation over open level terrain (plane view). Average daytime conditions are shown, x = distance from source to shadow zone; ϕ = angle between wind and sound; ϕ_c = critical angle.

Under certain extreme meteorological conditions involving temperature inversion, most of the sound radiated by a land-based source will be confined in the inversion layer, producing considerable increase in intensity at large distances from the source. On the other hand, there are conditions in which a shadow zone is formed into which little sound penetrates. The "silence" zone in the case of sound due to explosions is well known.

When sound is radiated from a source near the surface of the earth and the wind blows with velocity increasing upward, sound rays traveling with the wind are curved downward toward the earth, whereas those traveling against the wind are moved upward away from the earth. When trying to transmit sound against the wind, one should employ an elevated source rather than one on the ground. Shadow zones can also be produced by the influence of wind on sound propagation. This is illustrated in Fig. 7, where a sound source and receiver are a distance r apart. The average direction from which the wind is blowing is indicated by a wind vane, and this direction makes an angle ϕ with the line connecting source and receiver. A shadow zone will generally result on the upward side of the source, since sound rays traveling upwind tend to be bent upward by the wind. If the

temperature decreases upward with height, this tends to accentuate the development of the shadow zone. The wind blowing in the direction indicated in Fig. 7 tends to bend sound rays downward in the downwind direction. At some critical angle ϕ_c, the wind and temperature gradient effects cancel each other and the shadow zone vanishes. As a result, the plane is divided into an upwind sector $2\phi_c$ and a downwind sector $360° - 2\phi_c$.

Experiments have shown that the effective excess attenuation due to temperature variation and wind is frequently radically different upwind and downwind, with a gradual transition at the boundaries $\phi = \pm\phi_c$. On a sunny day, with moderate winds, the excess attenuation upwind, inside the shadow zone, is typically 20–30 dB higher than that for the same distance downwind.

The effect of humidity on the velocity of sound is indicated by the fact that in saturated air at 20°C and normal atmospheric pressure the velocity is about 0.35% greater than that in completely dry air at the same temperature and pressure. When sound is incident normally from dry air on a fog bank in which the density is roughly 1% less than that of dry air, the ratio of the reflected intensity to the incident intensity is about 1/160,000. However, for an angle of incidence greater than 84° total reflection can take place, and a few successive reflections of this kind can largely change the direction of the sound.

The natural echo is another feature of atmospheric sound. In addition to the echo due to reflection from a solid wall like a cliff or the side of a building, the echo from a grove of trees, due to diffractive scattering as well as reflection, presents the interesting feature that the intensity of the octave compared with that of the fundamental in the scattered sound is many times greater than in the original incident sound. This leads to what Lord Rayleigh called harmonic echoes. It is an illustration of his law of scattering [Eq. (42)]. *See* ECHO.

For sound attenuation in the atmosphere see Table 7. *See* ATMOSPHERIC ACOUSTICS.

Bioacoustics. Bioacoustics in its broadest sense treats of the role of sound in the behavior of living things. Hearing and speech of human beings constitute a large segment of this discipline. Yet the acoustical properties of the biological media making up living systems of all kinds wholly aside from the ear and vocal mechanisms have considerable importance. The use of sound (particularly at high

Table 9. Ultrasonic propagation properties of some tissues at 1 MHz*

Tissue	Attenuation coefficient, cm^{-1}	Velocity, m/s	Density, g/cm^3	Specific impedance, 10^5 kg/m^{2}s	Trends
Blood	0.014	1566	1.04	1.63	
Fat	0.07	1478	0.92	1.36	
Nerve	0.10	—	—	—	
Muscle	0.14	1552	1.04	1.62	increasing structural protein content / decreasing water content
Blood vessel	0.20	1530	1.08	1.65	
Skin	0.31	1519	—	1.58	
Tendon	0.56	1750	—	—	
Cartilage	0.58	1665	—	—	
Bone	1.61	3445	1.82	6.27	

*Listed in order of increasing attenuation coefficient. Except for bone, the attenuation coefficient follows the relation $\alpha = \alpha_1 \nu^{1.1}$, where α_1 is the attenuation coefficient at 1 MHz (from table) and ν is frequency in MHz.

SOURCE: From S. A. Goss, R. L. Johnston, and F. Dunn, Comprehensive compilation of empirical ultrasonic properties of mammalian tissues, *J. Acoust. Soc. Amer.*, 64;423–457, 1978.

frequency) in the diagnosis and treatment of disease has necessitated the acquisition of precise knowledge of the acoustical properties of animal tissue. *See* ACOUSTICAL HOLOGRAPHY.

Table 9 presents values of ultrasonic properties of some biological tissues at 1 MHz listed in order of increasing attenuation coefficient.

Since all mammals and many other animal species, including reptiles, fishes, birds, and insects, can generate and detect sound, the study of the mechanisms responsible for these activities and the way they are used in animal behavior has become of increasing importance. Bats and whales (the porpoise is a very conspicuous example) emit sounds which on striking obstacles are reflected; the reflected sound can be detected by the emitting animal and used in the location of the obstacles. This is known as echolocation or animal sonar. It is widely used by the animals in question to avoid obstacles in their motion, to locate prey, and to communicate with other animals of the same species.

Sound in confined spaces. When sound is confined in its propagation to a closed space as through a tube or in a room, the reflection from surfaces leads to the production of standing waves, that is, waves moving in opposite directions and either reinforcing each other or canceling each other out. The reinforcement that takes place at certain frequencies is called resonance, and the corresponding frequencies are known as normal modes. For example, the ideal resonance frequencies for an organ pipe open at both ends are given by Eq. (60),

$$\nu = \frac{nc}{2l} \tag{60}$$

where l is the length of the pipe, c is the velocity of sound in air, and n is any integer. Actually the reflection at each open end is not quite ideal, resulting in so-called end corrections, which effectively lower the ideal resonance frequencies somewhat. *See* RESONANCE (ACOUSTICS AND MECHANICS).

In the case of sound in a closed room as in a concert hall the presence of normal modes accentuates sounds of certain frequencies, thus providing problems in architectural acoustics. For a rectangular enclosure of dimensions l_1, l_2, l_3 in the x, y, z directions, the normal modes due to the reflections at the walls, floor, and ceiling have frequencies given by Eq. (61), where c again is the

$$\nu_{n_1, n_2, n_3} = \pi c \sqrt{\left(\frac{n_1}{l_1}\right)^2 + \left(\frac{n_2}{l_2}\right)^2 + \left(\frac{n_3}{l_3}\right)^2} \tag{61}$$

velocity of sound in the room, and n_1, n_2, n_3 are any three integers. *See* REVERBERATION.

[R. BRUCE LINDSAY]

Bibliography: L. L. Beranek, *Acoustic Measurements*, 1949; L. L. Beranek, *Acoustics*, 1954; L. L. Beranek, *Noise Reduction*, 1960; C. M. Harris (ed.), *Handbook of Noise Control*, 2d ed., 1979; L. E. Kinsler and A. R. Frey, *Fundamentals of Acoustics*, 2d ed., 1962; W. E. Kock, *Sound Waves and Light Waves*, 1965; R. B. Lindsay (ed.), *Acoustics: Historical and Philosophical Development*, 1973; R. B. Lindsay, *Mechanical Radiation*, 1960; P. M. Morse and K. U. Ingard, *Theoretical Acoustics*, 1968; Lord Rayleigh, *Theory of Sound*, reprint 1945; R. W. B. Stephens and A. E. Bate, *Acoustics and Vibrational Physics*, 1966; A. B. Wood, *A Textbook of Sound*, 3d ed., 1955.

Sound absorption

The process by which the intensity of sound is diminished due to the conversion of the energy of the sound wave into heat. The absorption of sound is a special but important case of sound attenuation. No matter what material medium sound passes through, its intensity, measured by the average flow of energy in the wave per unit time per unit area perpendicular to the direction of propagation, decreases with distance from the source. This is called attenuation. In the simple case of a point source of sound radiating into an ideal medium, the intensity decreases inversely as the square of the distance from the source, because the spherical area through which the same amount of energy passes per unit time increases as the square of this distance. This may be called geometrical attenuation.

In addition to this attenuation due to spreading, there is attenuation caused by scattering in sound propagation in the open air and sea. Sound can be reflected and refracted when incident on media of different physical properties, and diffracted as it bends around obstacles. These processes lead to effective attenuation, for example, in fog, which is obvious in practice and can be measured but is difficult to calculate theoretically with precision.

In actual material media the geometrical attenuation is supplemented by absorption due to the interaction between the sound wave and the physical properties of the medium. This dissipates the sound energy by transforming it into heat and hence decreases the intensity of the wave. In all practical cases the attenuation due to such absorption is exponential in character. Thus, if I_0 denotes the intensity of the sound at unit distance from the source, then I, the intensity at distance r in the same units, has the form of Eq. (1), where

$$I = I_0 e^{-\alpha r} \tag{1}$$

e represents the base of napierian logarithms, namely, 2.71828 . . . , and α is called the intensity absorption coefficient. Then α can be expressed by Eq. (2). Hence α has the dimensions of reciprocal

$$\alpha = -\frac{1}{r} \log_e I/I_0 \tag{2}$$

length or distance. However, it is customary to express the logarithm, which is a pure dimensionless number, in nepers, and hence α is in nepers/unit length. Since $10 \log_{10} I/I_0$ is the number of decibels between the intensity I and I_0, the absorption coefficient can also be expressed in decibels/unit length (usually abbreviated to dB/unit length or distance. *See* DECIBEL.

Classical sources of sound absorption. The four so-called classical origins of dissipative sound absorption in material media are: shear viscosity, heat conduction, heat radiation, and diffusion. These will be discussed first as they apply to fluids, namely, liquids and gases.

Viscosity. In the flow of a fluid, viscosity represents the tendency for the fluid flowing in one layer parallel to the direction of motion to retard the motion in adjoining layers. The viscosity of a liquid (indicated qualitatively by the stickiness or sluggishness of its flow) decreases as the temperature rises, whereas that of gases increases with the temperature. *See* VISCOSITY.

Appropriate modifications of the standard wave equation for plane sound waves to take account of the effect of viscosity of a fluid leads to Eq. (3)

$$\alpha_v = \frac{16\pi^2 f^2 \eta}{3\rho_0 c^3} \qquad (3)$$

for the absorption coefficient α_v as defined in Eq. (1), but specifically due to viscosity, where η is the viscosity, f is the frequency of the sound wave in hertz, c is its velocity, and ρ_0 is the equilibrium density of the fluid. The expression for α_v in Eq. (3) holds only for frequencies that are not too high. For example, in the case of air, it fails at frequencies on the order of 10^9 Hz, which can now be realized in ultrasonic research. *See* HYPERSONICS; ULTRASONICS.

Heat conduction. The role of heat conduction can be qualitatively understood as follows. The compression associated with the propagation of a sound wave in a fluid produces a local increase in temperature, which sets up a temperature gradient through which heat will flow by conduction before the succeeding rarefaction cools the gas. This is an irreversible phenomenon, leading to energy dissipation into heat.

The value of the sound absorption coefficient due to heat conduction is given by Eq. (4), where

$$\alpha_h = \frac{\gamma - 1}{\gamma} \cdot \frac{4\pi^2 f^2 M\kappa}{\rho_0 c^3 C_v} \qquad (4)$$

$\gamma = C_p/C_v$, the ratio of the molar specific heat at constant pressure C_p to that at constant volume C_v, M is the mass per mole or the molecular weight, ρ_0 the equilibrium density, and κ the thermal conductivity. Equation (4) is again valid only for frequencies that are not too high. *See* CONDUCTION (HEAT); HEAT CAPACITY.

Substitution of the appropriate numerical values for the physical parameters shows that for gases the value of α_h is somewhat less than α_v for the same frequency. For liquids, except for liquid metals, $\alpha_v \gg \alpha_h$ for the same frequency. It is usually considered appropriate to add α_v and α_h to get the total absorption due to these two processes.

Heat radiation. Thermal radiation results whenever one portion of a fluid acquires a temperature different from that of an adjacent portion; it is an electromagnetic-wave phenomenon different in character from heat transfer by conduction. Calculation shows that thermal radiation plays little actual role in sound absorption except possibly at very high frequencies in very-low-pressure gases. *See* HEAT RADIATION.

Diffusion. Diffusion operates mainly in the case of fluid mixtures, for example, air as a mixture mainly of nitrogen and oxygen, in which the two gases have a slight tendency to diffuse into each other. But for ordinary gases in which the components differ little in molecular weight, the effect of diffusion on absorption is negligible compared with that of viscosity and heat conduction. However, in gaseous mixtures in which the components differ markedly in molecular weight, for example, a mixture of the rare gases helium and krypton, the effect of diffusion can be much in excess of

that due to viscosity and heat conduction combined.

Measurement of absorption in fluids. Sound absorption in fluids can be measured in a variety of ways, referred to as mechanical, optical, electrical, and thermal methods. All these methods reduce essentially to a measurement of sound intensity as a function of distance from the source.

Mechanical. The mechanical method, confined to the estimate of absorption in liquids, uses radiation pressure, proportional to the sound intensity, to "weigh" the sound in a vertical beam.

Optical. The optical method, which is also mainly used for liquids, is based on the diffraction of a beam of light by the sound beam, the latter being treated essentially as a diffraction grating. The greater the intensity of the sound, the more diffraction bands are produced. *See* DIFFRACTION.

Electrical. The electrical methods, the most generally used for both gases and liquids, employ an acoustic transducer like a piezoelectric crystal to receive the sound beam as it traverses the medium whose absorption coefficient is to be evaluated, and to measure the variation in the sound intensity as the distance from the source is varied. An alternative is the use of an acoustic interferometer, which functions essentially in the same way as an optical interferometer. Still another scheme is based on the use of the reverberation time technique. In this method the medium is confined in a closed spherical vessel with highly reflective walls. The reverberation time, that is, the time taken by a sound pulse initially produced at the center of the vessel by a transducer (which can also act as a sound receiver) to decay to one-millionth of its original intensity provides a measure of the intensity loss and leads to an evaluation of the absorption coefficient. This method has been successfully applied to liquids and gases. *See* REVERBERATION.

Thermal. The thermal method undertakes to estimate the loss in intensity of a sound wave by direct measurement of the heat produced in its passage. This demands very careful experimental technique, and the difficulties associated with it have provoked controversy about the usefulness of the method.

Experimental values for gases. Measurements carried out by these methods show that the experimental values for the sound absorption coefficient (as a function of frequency) of a monatomic gas like helium agree rather well with the values calculated on the basis of viscosity and heat conduction, except at very low pressures and very high frequencies. On the other hand, for more complicated gases, like air (though in general α varies roughly directly as the square of the frequency, as predicted by classical theory), the actual numerical values are greater (and in many cases very much greater) than the theoretically predicted ones. For example, for air at 20°C and normal atmospheric pressure, the theoretical value of $(\alpha_v + \alpha_h)/f^2$ is 1.27×10^{-12} neper s²/cm, whereas the experimental value of α/f^2 for air of low humidity under the same conditions is approximately 8×10^{-12} in the same units.

Experimental values for liquids. The situation with respect to liquids is about the same. Mercury and the liquefied gases like argon, oxygen, and nitrogen have absorption coefficients in moderately good agreement (that is, within $10-15\%$) with the classically calculated values. But ordinary liquids like water and aqueous salt solutions show values much in excess of those classically calculated. The observed excess in the case of organic liquids like carbon disulfide, for example, is very large indeed, of the order of several hundred times. A satisfactory explanation for this excess absorption has been found in molecular and related relaxation processes.

Molecular relaxation in gases. Although sound propagation in a gaseous medium is phenomenologically a wave process, it actually takes place through the collisions of the molecules composing the gas. According to the kinetic theory, the molecules of a gas are in ceaseless random motion. When the gas is locally compressed in the passage of a sound wave, the average translational kinetic energy of the molecules is locally increased. This increase is conveyed through collisions to the adjoining group of molecules, which in turn passes it on through the gas in the form of the sound-wave energy.

A sound wave thus involves the momentary local compression of the medium, which by increasing the local density also leads, according to kinetic theory, to an increase in the average translational kinetic energy and hence to an increase in local pressure. In the case of a gas made up of polyatomic molecules, the increase in pressure cannot take place instantaneously, since some of the extra translational kinetic energy is transferred to the internal rotational and vibrational molecular-energy states. After the local density has begun to decrease with the passage of the wave disturbance, the pressure continues to build up as energy is transferred back to the translational form from the internal states. The excess density and excess pressure thus get out of phase, as in the standard hysteresis phenomenon, and this leads to an effective transformation of acoustic-wave energy into heat, corresponding to sound absorption. This process is termed a relaxation process, and the time taken for the density at any point to return to within the fraction $1 - 1/e$ (approximately 63%) of its equilibrium value is termed the relaxation time, denoted by τ. *See* MOLECULAR STRUCTURE AND SPECTRA.

Calculation of the effect. The results of the theoretical calculation of the effect of molecular relaxation will be discussed for the special case in which there are only two energy states of each molecule: (1) that in which it possesses translational kinetic energy only, and (2) that in which it possesses translational kinetic energy plus a single internal energy level. Here the relaxation time is given by Eq. (5), where k_{12}^0 is the average number of

$$\tau = \frac{1}{k_{12}^0 + k_{21}^0} \qquad (5)$$

transitions per second from energy state 1 to energy state 2, and k_{21}^0 is the average number of transitions per second from state 2 to state 1. The superscripts mean that the transition rates refer to the equilibrium condition of the gas as unaffected by the sound wave. Analysis then gives the sound absorption coefficient α due to this type of relaxation process in the form of Eq. (6), where f is the

$$\alpha = K \cdot \frac{4\pi^2 f^2 \tau}{1 + 4\pi^2 f^2 \tau^2} \qquad (6)$$

frequency and K is a constant containing the molar gas constant R, the sound velocity, and the specific heats of the gas. If $2\pi f\tau \ll 1$, α varies as the square of the frequency, as in the case of viscous and heat conduction absorption. However, in the more general case, if $\alpha/2\pi f$ is plotted as a function of frequency, it has a maximum for $f = \frac{1}{2} \pi\tau$, the plot taking the form of an approximate bell-shaped curve. Such a plot, carried out experimentally, provides a method for evaluating the relaxation time τ.

The theoretical evaluation of τ from Eq. (5) involves essentially a calculation in quantum statistics. The agreement of theory with experiment for many gases for the excess absorption over that attributable to viscosity, heat conduction, and diffusion is satisfactory. The relaxation process described here is usually termed thermal relaxation, since the specific heats of the gas enter prominently into the process.

Examples. Examples of gases for which experimental values are in reasonable agreement with theory include hydrogen, for which the excess absorption over classical values arises primarily from rotational relaxation, with a relaxation time on the order of 10^{-8} s, and oxygen, for which the excess low-frequency absorption depends primarily on vibrational relaxation, with a relaxation time on the order of 3×10^{-3} s at room temperature. In polyatomic gases it is rare for a single relaxation process to be solely effective, and rather elaborate analysis is required to take account of multiple relaxations.

Accompanying relaxation absorption in gases at very high frequencies and low pressures, there also occurs sound dispersion, that is, the variation of sound velocity with frequency.

Relaxation processes in liquids. The excess absorption in liquids above the classically predicted values has also been attributed to relaxation processes. The assumption of thermal relaxation has been successful for certain nonpolar liquids like benzene. It has not worked, however, for polar liquids like water. For the latter, so-called structural relaxation has been successfully invoked. This assumes that liquid water can exist in two different states of mutual molecular orientation and that the passage of sound affects the average rate of energy transfer back and forth between the two states. The state of lower energy is assumed to be similar to the structure of ice, in which each water molecule is surrounded on the average (in spite of the random motion of the molecules) by its four nearest neighbors arranged at the corners of a regular tetrahedron with the given molecule in the center. The higher energy state corresponds to a more closely packed configuration of the molecules with an arrangement like that of a face-centered cubic crystal. In water in equilibrium, most of the molecules are in the lower energy state, but the passage of sound causes a transfer of some more molecules into the higher state and upsets the equilibrium. There is the usual lag in the return of the molecules to the lower state, leading to a relaxation time, and this produces absorption, as in the similar case in gases. This theory has been successful in accounting for the excess sound absorption in water as a function of temperature in the range from 10 to 80°C. In this range the appropriate relaxation time has been found to be of the order of 10^{-12} s.

Excess absorption in aqueous solutions has been treated successfully by so-called chemical relaxation. A chemical reaction in the liquid like a dissociation-association process can provide two energy states, the transfer of energy between which will be affected by the compressive stress involved in the passage of a sound wave. The dissociation reaction of magnesium sulfate has been successful in explaining the excess sound absorption in sea water.

Sound absorption in solids. Sound transmission and absorption in solids assume particular importance in view of the applications to solid acoustic delay lines and filters. In the case of solids, it is often convenient to use a temporal intensity absorption coefficient ω, which represents the average loss in intensity per unit time at any point in the solid, and is related to the previously defined spatial absorption coefficient α by Eq. (7), where λ

$$\alpha\lambda = \omega T \tag{7}$$

is the wavelength of the sound wave and T is the period. The use of ω is often convenient in the case of finite solid rods employed in practical applications. In place of ω, one may also use the logarithmic decrement δ, which is the natural logarithm of the ratio of successive vibrational maxima at any given point in the solid traversed by the sound. Equation (8) holds for small damping, that is, for $\delta \ll 1$.

$$\delta = \omega T = \alpha\lambda \tag{8}$$

The decrement is dependent on the nature of the solid, the temperature, and the sound frequency. For a metal like steel at room temperature and not too high frequency, the decrement can be of the order of 10^{-3}, whereas for fused silica at room temperature it can be as low as 2×10^{-5}. In general, δ increases with frequency.

The theory of sound attenuation in solids is complicated because of the presence of many mechanisms responsible for it. These include heat conductivity, sound scattering due to grain boundaries, magnetic domain losses in ferromagnetic materials, interstitial diffusion of atoms, and dislocation relaxation processes in metals. In addition, in metals at very low temperature the interaction between the lattice vibrations (phonons) due to sound propagation and the valence electrons plays an important role, particularly in the superconducting domain. *See* SOUND; SUPERCONDUCTIVITY.

[R. BRUCE LINDSAY]

Bibliography: R. T. Beyer and S. V. Letcher, *Physical Ultrasonics*, 1969; K. F. Herzfeld and T. A. Litovitz, *Absorption and Dispersion of Ultrasonic Waves*, 1959; R. B. Lindsay, *Mechanical Radiation*, 1960; R. B. Lindsay (ed.), *Physical Acoustics*, 1974; W. P. Mason, *Physical Acoustics and the Properties of Solids*, 1958; W. P. Mason (ed.), *Physical Acoustics*, vol. 2, pt. A: Properties of gases, liquids, and solutions, 1965; P. M. Morse and K. U. Ingard, *Theoretical Acoustics*, 1968.

Space-time

A term used to denote the geometry of the physical universe as suggested by the theory of relativity. It is also called space-time continuum. Whereas in Newtonian physics space and time had been considered quite separate entities, A. Einstein and H. Minkowski showed that they are actually intimately intertwined. Isaac Newton's ideas on space and time are summarized in the following list:

1. Given two events, each of which is clearly localized in space and lasts only for an instant in time, such as two strokes of lightning striking small targets, all observers of the events will be in agreement as to which of the two events took place earlier in time, or whether they were actually simultaneous.

2. If the events were not simultaneous, the interval of time between them is an absolute entity, agreed on by all competent observers.

3. The spatial distance between two simultaneous events is an absolute entity, agreed on by all competent observers.

Of these three, the first assumption, concerning the concept of simultaneity of distant events, is the crucial one, the other two depending on it. Simultaneity, however, can be given an unambiguous meaning only if there is available some instantaneous method of signaling over finite distances. Actually, according to the theory of relativity, the greatest speed of transmission of intelligence of any kind is the speed of light c, equaling about 3×10^{10} cm/sec. Moreover, any signal traveling precisely at the speed c appears to travel at that same speed to all conceivable observers, regardless of their own states of motion. This is the only reasonable interpretation of the results of the Michelson-Morley experiment and the effect of aberration. Accordingly, the question of whether two given events are simultaneous or not can be decided only with the help of signals that at best have traveled from the sites of these events to the station of the observer at the speed of light. *See* LIGHT; RELATIVITY.

Under these circumstances, Einstein showed that in general two observers, each using the same techniques of observation but being in motion relative to each other, will disagree concerning the simultaneity of distant events. But if they do disagree, they are also unable to compare unequivocally the rates of clocks moving in different ways, or the lengths of scales and measuring rods. Instead, clock rates and scale lengths of different observers and different frames of reference must be established so as to assure the principal observed fact. Each observer, using his own clocks and scales, must measure the same speed of propagation of light. This requirement leads to a set of relationships known as the Lorentz transformations. *See* LORENTZ TRANSFORMATIONS.

In accordance with the Lorentz transformations, both the time interval and the spatial distance between two events are relative quantities, depending on the state of motion of the observer who carries out the measurements. There is, however, a new absolute quantity that takes the place of the two former quantities. It is known as the invariant, or proper, space-time interval τ and is defined by

Eq. (1), where T is the ordinary time interval, R the

$$\tau^2 = T^2 - \frac{1}{c^2}R^2 \qquad (1)$$

distance between the two events, and c the speed of light in empty space. Whereas T and R are different for different observers, τ has the same value. In the event that Eq. (1) would render τ imaginary, its place may be taken by σ, defined by Eq. (2). If both τ and σ are zero, then a

$$\sigma^2 = R^2 - c^2 T^2 \qquad (2)$$

light signal leaving the location of one event while it is taking place will reach the location of the other event precisely at the instant the signal from the latter is coming forth.

The existence of a single invariant interval led the mathematician Minkowski to conceive of the totality of space and time as a single four-dimensional continuum, which is often referred to as the Minkowski universe. In this universe, the history of a single space point in the course of time must be considered as a curve (or line), whereas an event, limited both in space and time, represents a point. So that these geometric concepts in the Minkowski universe may be distinguished from their analogs in ordinary three-dimensional space, they are referred to as world curves (world lines) and world points, respectively.

Minkowski geometry. The geometry of the Minkowski universe in some respects resembles the geometry of ordinary (euclidean) space but differs from it in others. The Minkowski universe has four dimensions instead of the three dimensions of ordinary space; that is to say, for a complete identification a world point requires four pieces of data, for instance, three space coordinates and a time reading. But there are in the Minkowski universe world points, world lines, two-dimensional surfaces (including planes), three-dimensional surfaces (often called hypersurfaces), and four-dimensional domains. A hypersurface may, for instance, be a spatial domain (volume) at one instant in time, or it may be a two-dimensional (ordinary) surface for an extended period of time. Thus one may form all the geometric figures that are also possible in a four-dimensional euclidean space.

The indefinite metric. In a euclidean space there are the cartesian coordinate systems, those rectilinear systems of coordinates that are mutually perpendicular and whose coordinate values correspond to real lengths. The distance S between two points whose coordinate differences are X, Y, and Z, respectively, is given by Eq. (3). The coordinate

$$S^2 = X^2 + Y^2 + Z^2 \qquad (3)$$

transformations that lead from one cartesian coordinate system to another are called orthogonal coordinate transformations. Formally, they are the coordinate transformations that preserve the precise form of Eq. (3). Likewise, the Lorentz transformations preserve the precise form of Eqs. (1) and (2), respectively. A coordinate system in which these two equations hold is called a Lorentzian frame of reference.

Whereas the form on the right of Eq. (3) is positive definite, that is, always greater than or equal to zero, the right-hand sides of Eqs. (1) and (2) are

indefinite; that is, they may be positive or negative. This fact represents the single but all-important difference between a four-dimensional euclidean space and a Minkowski universe. The forms (1), (2), and (3) are called metrics. Hence the Minkowski universe is said to possess an indefinite metric.

As a result of the indefinite character of the metric, a triangle in the Minkowski universe may possess one side that is longer than the sum of the two others; conversely, one of the three sides may have the length zero. Depending on whether τ or σ is real, or both vanish, an interval is classified as timelike, spacelike, or a null-interval (see illustration).

Improper Lorentz transformations. In the Minkowski universe there are three different types of Lorentz transformations involving some kind of reflection, in addition to the more usual Lorentz transformations (called proper Lorentz transformations). The first type of improper Lorentz transformation changes the sign of all three spatial coordinates but leaves the sense of the time axis unchanged. This transformation changes a right-handed screw into a left-handed screw; it is also called a parity transformation. The second type of improper Lorentz transformation interchanges the future with the past but leaves the space coordinates unchanged. This transformation is called time reversal. The third improper Lorentz transformation reflects both the space and the time coordinates; it bears no special name of its own. The original arguments which led to the formulation of the special theory of relativity all support the proposition that the laws of nature are invariant under proper Lorentz transformations. They are inconclusive as to whether the laws of nature should also be invariant under the improper Lorentz transformations. The laws of mechanics and of electrodynamics have this property; the second

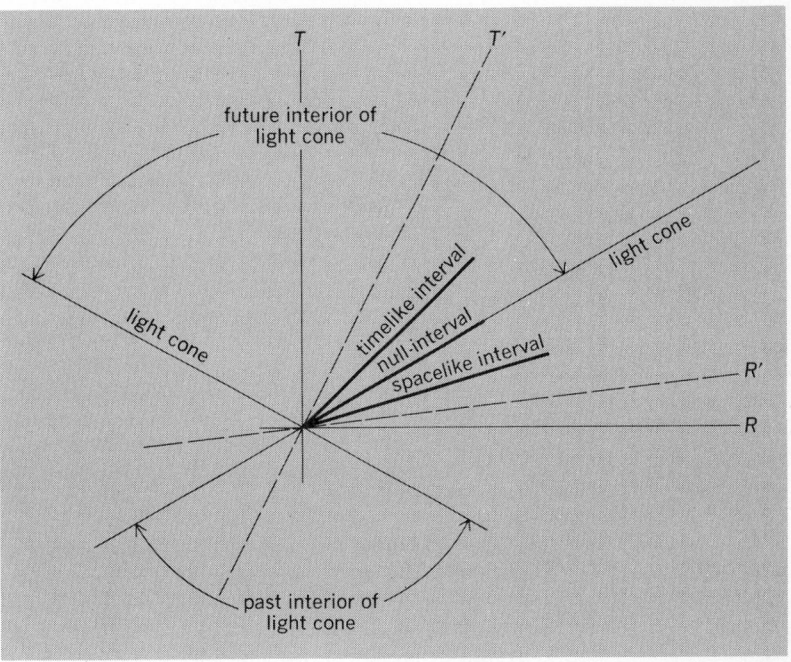

Two-dimensional sketch of three possible classes of intervals. The null-directions actually form a three-dimensional surface having the shape of a double cone, called the light cone. An alternative Lorentz frame is indicated by dashed lines.

law of thermodynamics distinguishes between past and future in a nonsymmetric manner, but it is usually assumed that this feature of thermodynamics is to be explained by its statistical nature and that it has no bearing on the properties of the underlying basic dynamics. *See* SYMMETRY LAWS.

Until recent times, it had therefore been taken for granted that the basic laws of nature should make no distinction between right-handed and left-handed screws, and that they should not discriminate between future and past. Certain difficulties in the interpretation of the decay of *K*-mesons led T. D. Lee and C. N. Yang to suspect that these assumptions might not be tenable, and they suggested some experiments on meson decay and on radioactive β-decay, which showed that in these particle transformations nature certainly discriminates between left-handed and right-handed screws. *See* PARITY.

It appears there are two kinds of neutrinos (electrically neutral particles which travel at the speed of light and are endowed with an intrinsic spin equal to that of electrons). One kind of neutrino always spins clockwise when viewed in the direction of its travel, the other counterclockwise. The exact role of the improper Lorentz transformations in nature is under intensive theoretical and experimental investigation. *See* NEUTRINO.

Curved space-time. Whereas the Minkowski universe is the appropriate geometric model for the special theory of relativity, the general theory of relativity makes use of a further generalization. In the Minkowski universe a particle that is not subject to external forces and which therefore travels along a straight line (in ordinary space) and at a uniform speed is represented by a straight world line. In general relativity, an external gravitational force is indistinguishable from an inertial force, which in special relativity would arise if a non-Lorentzian frame of reference were to be employed. Accordingly one requires a four-dimensional space in which it is impossible to distinguish between a Lorentzian and a non-Lorentzian frame of reference whenever a gravitational field is present. Such a space cannot be flat, as the Minkowski universe is, but must be curved. Such an aggregate is called a Riemannian space. In a general Riemannian space there are no straight lines but only curves. *See* FRAME OF REFERENCE; GRAVITATION; RIEMANNIAN GEOMETRY.

[PETER G. BERGMANN]

Spallation reaction

A nuclear reaction in which the interacting nuclei are disintegrated into a large number of the constituent protons, neutrons, and other light particles. Compared to more conventional reactions involving the simple transfer of nucleons between the colliding nuclei, spallation is a violent process occurring at high incident energy. An extreme case is shown in the illustration, the interaction of an argon nucleus (18 protons and 22 neutrons) with a lead nucleus (82 protons and 126 neutrons) in a streamer chamber at an incident energy of 72,000 MeV. The emitted charged particles appear as tracks of electrical discharge in a gas. In this example the nuclei are apparently completely shattered into individual nucleons. Spallation can also be induced by lighter particles, such as high-energy protons; however, these reactions are usual-

Collision of an argon nucleus with a lead nucleus at an incident energy of 72,000 MeV in a streamer chamber. The nucleus is incident from the left. (*From L. S. Schroeder, Streamer chambers—their use for nuclear science experiments, Nucl. Instr. Meth., 162:395–404, 1979*)

ly less violent and produce only a few particles (typically 3 to 10). Although the most abundant final products are protons, neutrons, and alpha particles, the spallation of larger fragments such as lithium and carbon also occurs. The short, thick tracks in the illustration are probably created by the emission of these larger fragments of nuclear matter.

The detailed mechanism whereby the incident energy is communicated to the nuclei in the spallation process is not well understood, but there may well be analogies with the breakup of other forms of matter. Like the nucleus, a substance such as Silly Putty will deform into new shapes under a gentle collision, but will shatter like a piece of glass when subjected to a hard, rapid blow. There is evidence that the breakup of nuclei also sets in suddenly at energies of a few tens of MeV per colliding nucleon. The process may be used to measure the tensile strength of nuclear matter. It is suggested that the spallation debris of a nuclear collision could contain information about high density and shock waves in nuclei. The spallation reaction also plays an important role in determining the abundance of elements in the universe. In particular, ^{6}Li, ^{9}Be, ^{10}B, and ^{11}B are created from spallation of interstellar gas by galactic cosmic rays, which leads to collisions of hydrogen with abundant nuclei such as carbon, oxygen, and silicon. *See* NUCLEAR REACTION.

[DAVID K. SCOTT]

Bibliography: H. Reeves, W. A. Fowler, and F. Hoyle, Galactic cosmic ray origin of Li, Be and B in stars, *Nature*, 226:727–729, 1970; D. K. Scott, Towards relativistic heavy ion collisions, *Prog. Particle Nucl. Phys.*, 4:5–93, 1980.

Spark chamber

A triggered electronic particle-detecting device whose purpose is to make visible and to locate accurately in space the tracks of charged particles. It is generally classified as wide-gap or narrow-gap; the most common form is a narrow-gap array of parallel-plate condensers, the plates of which are spaced about 3/8 in. apart, filled with a mixture of helium and neon at atmospheric pressure. A spark-chamber system requires an external initiating signal from an auxiliary particle-detection system, which triggers the application of a short-duration high-voltage pulse to the array of plates. The high-voltage pulse produces a spark discharge that follows or marks the path of the ionizing particles.

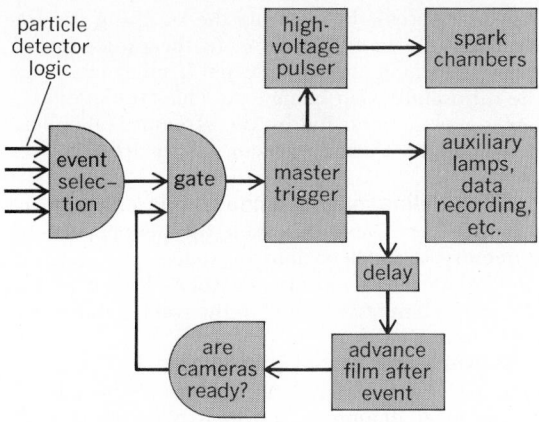

Fig. 1. Spark-chamber system using camera recording.

ber is therefore inherently anisotropic in its response.

In narrow-gap chambers (gaps up to about 3/4 in.), the discharge usually follows the electric field; that is, it is perpendicular to the plates. In wide-gap chambers, it follows the particle trajectory if the trajectory is not too different from the electric field direction—in practice at angles up to about 50° with the field. Whenever the particle path is

If several tracks are present they are all visible, up to at least 20 (although the chamber will not resolve tracks less than about 1 mm apart). Stereo cameras usually provide accurate track location; electronic and digital methods of track location are also used. Spark chambers are characterized by somewhat lower resolution than bubble chambers, accuracy of track location ranging from moderate to excellent (the best performance exceeding that of the bubble chamber), a high rate of data collection, and a unique triggering capability that selects only those events that have passed logical selection tests to be photographed. *See* BUBBLE CHAMBER; PARTICLE DETECTOR; SPARK COUNTER.

GENERAL PROPERTIES

Figure 1 shows a schematic diagram of a spark-chamber system using cameras for data recording.

Characteristic times. The time during which the ionization produced by the particle can produce a good track when the high-voltage pulse is applied is called the memory time of the chamber; it may be decreased by an electric clearing field that removes electrons rapidly, or by electronegative gases like sulfur dioxide, deliberately introduced to capture electrons and provide chemical clearing. Memory times as short as 0.2 μsec or as long as 30 μsec can thus be obtained. If the high-voltage pulse is applied when no particle has produced a track, or after the track is cleared, no spark will occur (spurious sparks may sometimes appear; good design eliminates or minimizes them).

After the spark discharge, all the ionization produced must be removed before another event can be observed. The spark-chamber recovery time (or dead time) depends on the total energy dissipated in the spark; camera recording requires brighter sparks, hence more energy and longer dead times (usually $10-30$ msec). Short dead times (less than 1 msec) can be achieved with digital readout chambers with dim or invisible spark discharges.

Particle tracks and direction. In any spark chamber the appearance of the discharge depends upon the direction the charged particle has taken in the chamber. Because the pulsed electric field has a definite direction, the nature of the discharge is determined by the direction of the electric field as well as by the particle track. The spark cham-

(a)

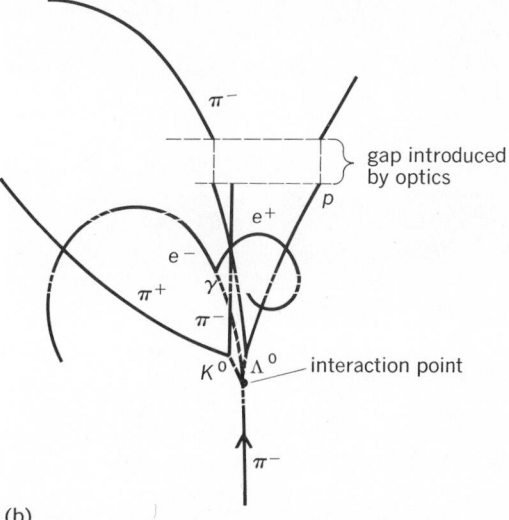

(b)

Fig. 2. A narrow-gap spark-chamber system in a magnetic field. (*a*) The photograph is taken through a pair of cylindrical field lenses that allow the camera to see into the narrow gaps. The large gap just above the center, where the right-hand edge appears bent, is spurious; it is introduced by the optics. (*b*) Interpretation of the event. A 1.2 Bev/c π^--meson enters from below and strikes an invisible target in the gap below the first full-width chamber. The interaction is $\pi^- + p \rightarrow K^0 + \Sigma^0$, where the Σ^0 decays almost instantaneously by the mode $\Sigma^0 \rightarrow \Lambda^0 + \gamma$. The ordinary decay products are neutral and hence leave no tracks until they decay or interact, K^0 decays into $\pi^+ + \pi^-$, Λ^0 into $p + \pi^-$. The gamma ray interacts to produce a positron-electron pair $e^+ - e^-$. (*A. Roberts, Spark chambers, Encyclopaedic Dictionary of Physics, Pergamon Press, 1962–1964*)

more or less parallel to the plates (at right angles to the electric field), the discharge occurs as a "curtain" of streamers randomly distributed along the track with a spacing of a few millimeters.

In practice this lack of isotropy is often not a serious disadvantage. Figure 2, which shows a complex event in a set of narrow-gap chambers, gives a good indication of how well tracks may be followed and located no matter what their direction. A well-designed spark-chamber system can be isotropic in its particle detection efficiency.

TYPES OF SPARK CHAMBER

Spark chambers may be classified according to several different characteristics. There are wide-gap and narrow-gap chambers; sampling, projection, and track-delineating chambers; analog and digital chambers; optical, acoustic, and electronic data-readout chambers.

Sampling chamber. A sampling chamber is one that yields as output data the coordinates of a single point on the track (or tracks) in each gap. All narrow-gap chambers are of this type, including the oldest type of spark chamber, the multiplate chamber with individual gap spacings of 1/2 in. or less.

Projection chamber. A projection chamber is one in which the track of the particle is perpendicular, or nearly so, to the electric field; then the individual electrons of the track each produce a

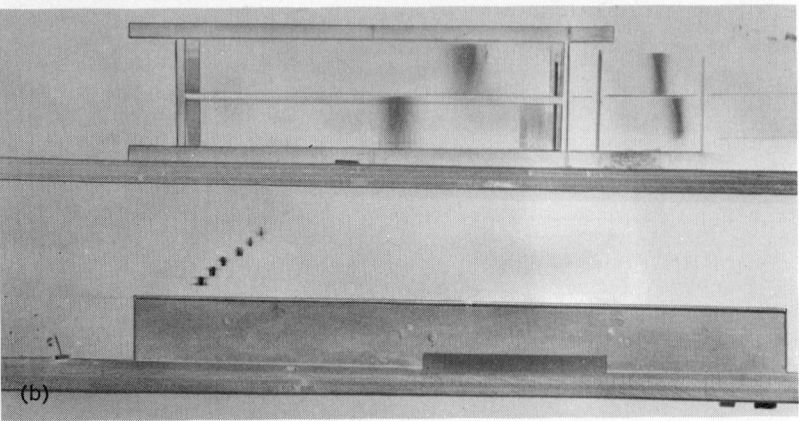

(a)

(b)

Fig. 3. Comparison between a wide-gap and a narrow-gap chamber stacked to observe the same particles. The wide-gap chamber (above) has a mirror showing a 90° stereo view at right. (a) A cosmic-ray particle making a small angle with the electric field direction. (b) A similar track, in this case making a large angle. Note the curtain discharge in the wide-gap chamber. (*A. Roberts, Spark chambers, Encyclopaedic Dictionary of Physics, Pergamon Press, 1962–1964*)

streamer across the gap, and the resulting curtain contains information only as to the projection of the track along the electric field, information on the third dimension being lost. This type of chamber is now superseded by the streamer chamber, in which the third-dimension information is preserved.

Track-delineating chamber. A track-delineating chamber is one in which the actual particle trajectory is made visible in space, just as in a cloud chamber or bubble chamber. Wide-gap and streamer chambers belong in the track-delineating category.

Conventional narrow-gap chamber. The great majority of spark-chamber experiments have used this kind of chamber, in which photographs are taken of the sparks between plates 1/4 to 3/8 in. apart. There may or may not be a magnetic field. Under certain conditions the spark may follow the particle trajectory when it makes an angle with the plates, but in general the spark follows the electric field and coincides with the track in at most one point. Such a chamber is therefore a sampling chamber, since it samples rather than delineates the track.

When the track approaches parallelism with the plates, the number of sparks increases until finally a curtain of streamers marks the track. There is still track location information in such curtains, but frequently the track is simply not measured in those gaps.

Very narrow gaps (down to 2–3 mm) are occasionally used where particle fluxes are high, to achieve very rapid clearing (<0.2 μsec) of unwanted electrons, and thus short memory times. At about 3/4 in. there is a transition to wide-gap chamber operation, which is distinguished by a change from the sampling to the delineating mode. Instead of following the electric field, which is normal to the plates, the spark now follows the track, even if it makes an angle up to 40–50° with the electric field.

Wide-gap chambers. These contain fewer plates, hence less matter than narrow-gap chambers, and offer less particle scattering and absorption. It is easier to "see into" a wide-gap chamber, since one is not peering down a narrow channel. The intrinsic accuracy of track location in wide-gap chambers (at least in the spark mode) is extremely good; it approaches that of the bubble chamber. The chief disadvantages are the need for a high pulse voltage and the loss of information from tracks parallel to the plates, a problem which is more serious than in a narrow-gap chamber, which restricts the loss to a small region for each track.

Figure 3 compares narrow- and wide-gap chambers, with the same particle traversing them. Figure 4 shows a pair of 20-cm gaps with tracks traversing them, both without and with a magnetic field.

The streamer chamber is a development of the wide-gap chamber; the same wide-gap chamber can be operated in either the spark or the streamer mode. If the high-voltage pulse on a wide-gap chamber is made very short (about 20 to 50 nsec) and very intense (20 kv/cm) a streamer discharge begins to grow along the electric field from many different electrons along the particle track. If the track is nearly parallel to the plates, the streamers are approximately at right angles to it, and grow to a length of perhaps 1/4 to 1/2 in. Viewed end-on along the electric field, each streamer appears as a bright dot. The track of the particle then resembles that of a particle in a cloud chamber. The fact that any number of such tracks can be seen simultaneously makes the chamber suitable for looking at complex events.

The tracks in the streamer mode are much dimmer than those of a conventional wide-gap chamber, since much less energy goes into the discharge. Their intensity is also a complicated function of their direction; tracks along the electric field tend to break into a conventional spark mode, very much brighter than the streamers. The resulting wide light-intensity range gives rise to problems in photography. Figure 5 shows a set of tracks in a streamer chamber in a magnetic field.

Heavy spark-chamber plates. Spark-chamber plates usually contain as little matter as possible, to avoid scattering or interacting with the charged particles whose tracks are being measured. Stretched aluminum-foil plates one- or two-thousandths of an inch thick are common; aluminum-coated self-supporting low-density polyurethane foam plates are also used. Chambers in which neu-

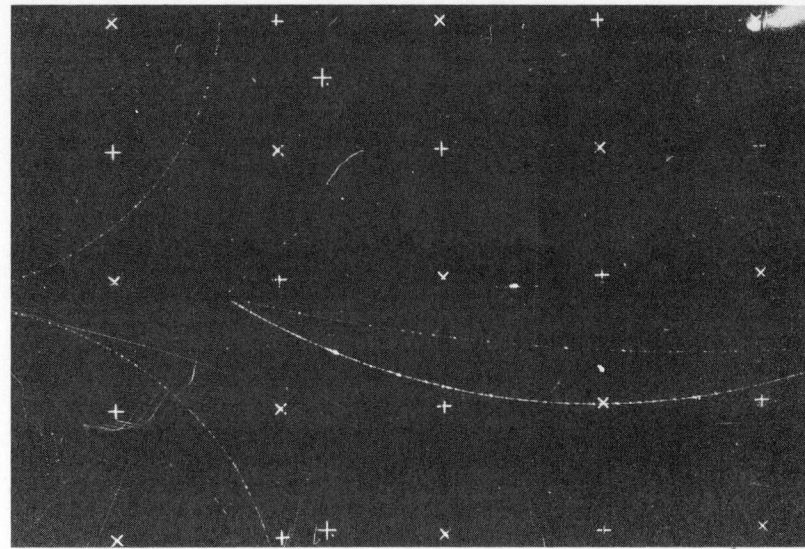

Fig. 5.   Tracks in a wide-gap chamber in a magnetic field, operating in the streamer mode. (*Courtesy of R. F. Mozley*)

tral particles are to be detected use thick plates in which the neutral particles interact to produce charged particles. Thus, for gamma rays, lead plates may be used; then the electron-positron shower produced by the high-energy gamma ray is visible, and the point of interaction, the direction, and the approximate energy of the gamma ray may be measurable. By using separate chambers for measuring charged particle momentum (with thin plates) and for detecting neutral particles (thick plates) optimum conditions for both can be achieved simultaneously.

Heavy plate chambers can also be used for detecting neutrons by their nuclear interactions and for scattering protons (using carbon or aluminum) in order to measure their polarization.

DATA READOUT

Data-readout methods include digitized chambers, sonic chambers, and the use of vidicon tube and magnetic fields.

Figures 6 and 7 illustrate two of the principles described below.

Digitized chambers. In the narrow-gap chamber, the spark samples the track at one point, so that the spark coordinate may be regarded as digitized on the z axis, normal to the plates; in the plane normal to that axis it is still in analog form. The other axis may also be made to yield digitized data if the homogeneous plates of the chamber are instead replaced by arrays of parallel wires, spaced about 1 mm apart. By noting on which wire the spark occurs, the coordinate at right angles to the wire direction in the wire plane is given a digitized value. If the two wire planes are then oriented at right angles to each other (Fig. 6), all three dimensions of the spark location will be in digital form.

Of the two different readout methods illustrated, only one would be used in a real chamber. The x array uses ferrite memory cores, which are set by the spark current. They can be read out directly into a computer. The y array uses a magnetostrictive ribbon AA, in which the current pulse on the wire to which a spark has occurred produces an

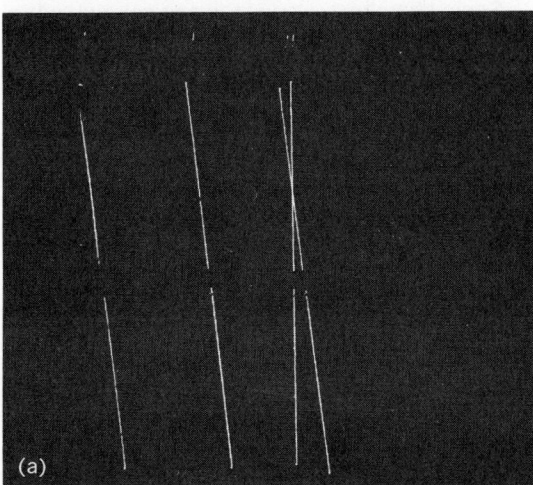

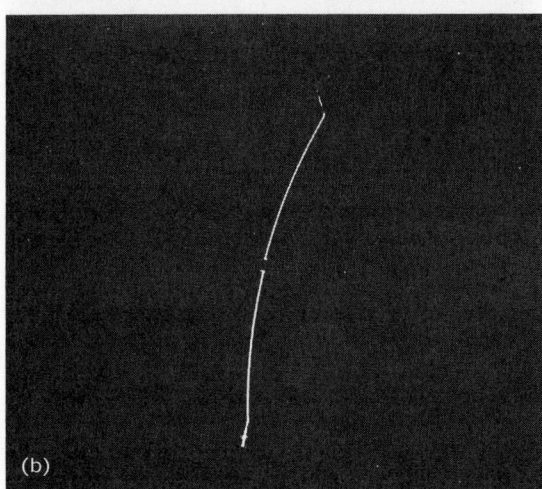

Fig. 4.   Tracks in a pair of wide-gap chambers operating in the spark mode. (*a*) No magnetic field. (*b*) With magnetic field. The kinks at the end are reflections in chamber plates. (*Courtesy of K. Strauch*)

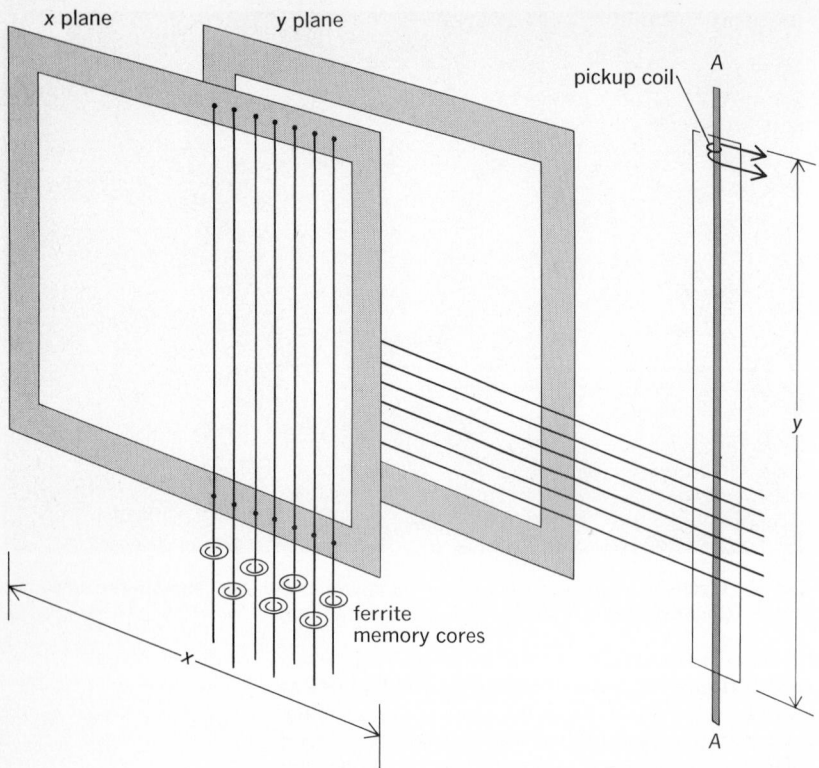

x plane y plane

pickup coil

A

y

ferrite
memory cores

x

A

Fig. 6.   A digitized spark chamber. Each plate is an array of parallel wires. One array determines the x coordinate of a spark, the other the y coordinate.

elastic (sonic) pulse which may be detected by a pickup coil at the end. The time it takes the pulse to arrive determines the wire at which it originated. For use in magnetic fields, the magnetostrictive ribbon is replaced by a nonmagnetic elastic ribbon. An elastic pulse produced by the mechanical shock associated with a spark from the wire to the ribbon across a tiny gap propagates down the ribbon and is detected by a piezoelectric pickup at the end (not shown).

Digitized wire chambers have assumed great importance because of the possibility of introducing their output directly into a computer. Many

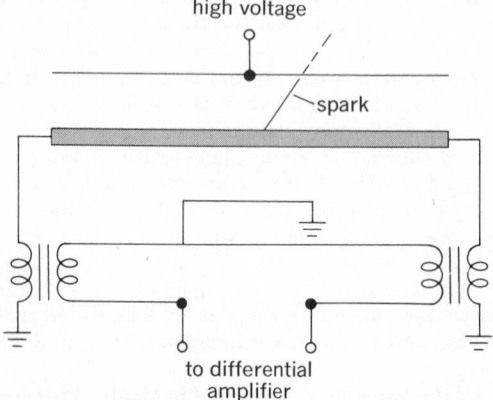

high voltage

spark

to differential
amplifier

Fig. 7.   Electronic determination of spark position in a narrow gap by the Charpak-Massonet current-distribution system. The spark location is determined by measuring the way in which the spark current divides between the two parallel paths to ground.

experiments in which the chambers were connected on-line to a computer have been run, so that the reconstruction and analysis of each event are immediately computed and made available to the experimenter. In an alternative arrangement, the data are recorded on magnetic tape, ready for later introduction into a computer, with slightly delayed analysis; this avoids tying down a computer to the experiment, and allows the use of a larger, more powerful computer.

Use of vidicon tube. An alternative method for digitizing the output of narrow-gap chambers is the use of a vidicon tube. As in all television pickup tubes, the stored image is in the form of a charge distribution on the photocathode, the amount of charge corresponding to intensity of light. The charge distribution is scanned in the conventional television raster, by an electron gun in the tube, and the output amplitude then corresponds to the light intensity. In this case, by using a uniform linear horizontal scan and measuring the position of the spark image along the scan, the position of the spark image can be digitized and recorded on magnetic tape. The chief drawback of the vidicon is its relatively limited resolution and sensitivity, compared to film recording; the convenience of immediate digitization and the elimination of film may well compensate for this disadvantage.

Sonic chambers. One of the earliest methods of locating a spark in a gap was by timing the arrival of the sound from the spark. The instant at which the spark occurs is known; hence, given the location of two microphones in each gap that receive the sound signal, and the time of arrival of the sound from the spark at each microphone, the position of the sound source is readily found. This method is difficult to apply if there is more than one spark present; it has been generally superseded by other techniques.

Electronic readout. Another method applicable to the location of a single spark in a narrow gap is the current-distribution method of G. Charpak and L. Massonet, shown in Fig. 7. The location of the spark (only one is allowed) is determined by observing how the spark current divides between the two available paths to ground. The transformer method shown measures the current ratio (or difference), and the sign and magnitude of the output pulse (properly normalized) determines the location of the spark (in one coordinate) to within 0.5 mm. The other coordinate can be obtained from another gap.

Use of magnetic fields. Any spark chamber operates successfully in magnetic fields up to at least 20 kgauss; the spark discharge is hardly affected. The magnetic field allows the sign and momentum of the particles to be measured by observing the curvature. The magnetic field restricts the data-readout technique somewhat: Certain types of digital readout using magnetic core storage or magnetostrictive delay lines are interfered with. It converts the system into a sort of electronic cloud chamber, with some important advantages, namely, the ability to tolerate relatively large fluxes of particles, the ability to select particular types of events by suitable triggering logic, the absence of background, ready adaptability to automatic data processing, and intrinsically higher precision.

The spark chamber is readily adaptable to complex systems in which several different kinds of chambers are used, including those with heavy plates for inducing interactions with neutral particles.

Spark-chamber photographs, if taken in a well-designed system, have certain advantages over bubble-chamber photographs when it comes to automatic processing. Many spark-chamber experiments have yielded film which has been successfully processed by semiautomatic or fully automatic film-reading data-processing systems; only a beginning in this direction has been made in bubble-chamber film, which generally contains many more background tracks and presents a much more difficult problem. Scanning can also be simpler (it can also be more difficult if the system is poorly designed). *See* CERENKOV RADIATION; SCINTILLATION COUNTER.   [ARTHUR ROBERTS]

Bibliography: O. C. Allkofer, *Spark Chambers*, 1969; G. Charpak, L. Massonet, and J. Favier, The development of spark chamber techniques, *Progr. Nucl. Tech. Instrum.*, 1:323, 1965; J. W. Cronin, Spark chambers, in R. P. Shutt (ed.), *Bubble and Spark Chambers: Principles and Use*, vol. 1, 1967; E. Segré, *Nuclei and Particles*, 2d ed., 1977.

Spark counter

A particle detector which uses the ionization produced in a gas by high-speed charged particles to trigger a spark between two electrodes. Spark counters react in a very short time (about 10^{-9} sec) to the particle, and thus can be used for fast timing. The spark is visible and can be photographed. *See* PARTICLE DETECTOR.

The principal components of a spark counter are two plane, parallel metallic electrodes, with a gas between the electrodes consisting of a mixture of argon and an organic gas such as xylene (see illustration). A potential difference of about 2000 volts is placed across the electrodes, which are spaced about 2 mm apart. The function of the xylene gas is to aid in quenching the discharge which occurs between the plates when ions are produced in the gas by a charged particle passing through the counter. Although the response time of a spark counter is very fast, the counting rate is very low, since additional quenching must be provided by an electronic circuit which has a recovery time of 0.25 sec. *See* SPARK CHAMBER.

[WILLIAM B. FRETTER]

Bibliography: J. W. Keuffel, Parallel-plate counters, *Rev. Sci. Instrum.*, 20:202–208, 1949; E. Segré, *Nuclei and Particles*, 2d ed., 1977.

Special functions

Functions which occur often enough to acquire a name. Some of these, such as the exponential, logarithmic, and the various trigonometric functions, are extensively taught in school and occur so frequently that routines for calculating them are built into many pocket calculators. *See* DIFFERENTIATION; LOGARITHM; TRIGONOMETRY.

The more complicated special functions, or higher transcendental functions as they are often called, have been extensively studied by many mathematicians because they arose in the problems which were being studied. Among the more useful functions are the following: the gamma function defined by Eq. (1), which generalizes the

$$\Gamma(x) = \int_0^\infty t^{x-1} e^{-t} \, dt \tag{1}$$

factorial; the related beta function defined by Eq. (2), which generalizes the binomial coefficient; and

$$B(x,y) = \int_0^1 t^{x-1}(1-t)^{y-1} \, dt \tag{2}$$

elliptic integrals, which arose when mathematicians tried to determine the arc length of an ellipse, and their inverses, the elliptic functions. The hypergeometric function and its generalizations includes many of the special functions which occur in mathematical physics, such as Bessel functions, Legendre functions, error functions, and the classical orthogonal polynomials of Jacobi, Laguerre, and Hermite. The zeta function defined by Eq. (3) has many applications in number theory,

$$\zeta(s) = \sum_{n=1}^\infty n^{-s} \tag{3}$$

and it also arises in M. Planck's work on radiation. *See* BESSEL FUNCTIONS; ELLIPTIC FUNCTION AND INTEGRAL; GAMMA FUNCTION; HEAT RADIATION; HYPERGEOMETRIC FUNCTIONS; LEGENDRE FUNCTIONS; ORTHOGONAL POLYNOMIALS.

Bernoulli polynomials and numbers. Bernoulli polynomials, defined by the generating function in Eq. (4), and Bernoulli numbers, $B_n = B_n(0)$, arise

$$\frac{te^{xt}}{e^t - 1} = \sum_{n=0}^\infty B_n(x) \frac{t^n}{n!} \qquad |t| < 2\pi \tag{4}$$

in many applications. (Different notations are used, especially for Bernoulli numbers.) They occur in the Euler-Maclaurin summation formula (5), where

$$\sum_{j=a}^n f(j) = \int_a^n f(x) \, dx + \tfrac{1}{2} f(a) + \tfrac{1}{2} f(n)$$
$$+ \sum_{k=1}^m \frac{B_{2k}}{(2k)!} \{ f^{(2k-1)}(n) - f^{(2k-1)}(a) \} + R_m(n) \tag{5}$$

a, m, and n are arbitrary integers, $a < n$, $m > 0$, and notation (6) applies. The symbol $[x]$ stands for

$$R_m(n) = -\int_a^n \frac{B_{2m}(x - [x])}{(2m)!} f^{(2m)}(x) \, dx \tag{6}$$

the largest integer less than or equal to x. This formula is very useful in finding numerical values of some slowly convergent infinite series or partial sums of divergent series. An important sum which can be evaluated using Bernoulli numbers is given by Eq. (7).

$$\sum_{k=1}^\infty \frac{1}{k^{2n}} = \frac{(2\pi)^n B_{2n}}{2(2n)!} \tag{7}$$

Bernoulli polynomials and numbers also play an important role in number theory, combinatorial analysis, and the study of spline functions. Splines are piecewise polynomials which are spliced together in a smooth manner, up to a certain degree of differentiability. They are very useful in fitting curves to numerical data and are the solutions to a number of important extremal problems. *See* COMBINATORIAL THEORY.

More complicated functions. All of the special functions mentioned above can be given by explicit representations, either as a series or an integral. There are other functions, among them Lamé functions, spheroidal and ellipsoidal wave functions, and Mathieu functions, which arise as solutions to important differential equations, and for which explicit formulas are lacking. At least the formulas are not nearly as explicit as for the above functions. The ordinary differential equations satisfied by these functions arise from Laplace's equation or the wave equation when it is solved by separation of variables in certain systems of curvilinear coordinates. Instead of having explicit integral representations, these functions satisfy special integral equations. *See* DIFFERENTIAL EQUATION; INTEGRAL EQUATION; LAPLACE'S DIFFERENTIAL EQUATION; WAVE EQUATION.

For example, the prolate spheroidal wave functions of order zero, $\psi_n(x)$, $n=0,1,\ldots$, are the bounded continuous solutions of differential equation (8) and integral equation (9), where χ_n and

$$(1-x^2)\frac{d^2\psi_n}{dx^2} - 2x\frac{d\psi_n}{dx} + (\chi_n - c^2x^2)\psi_n = 0 \quad (8)$$

$$\lambda_n\psi_n(x) = \int_{-1}^{1}\frac{\sin c(x-y)}{\pi(x-y)}\psi_n(y)\,dy \quad (9)$$

λ_n are eigenvalues. Differential equation (8) is the usual starting place for the study of these functions. It arises when solving the wave equation in prolate spheroidal coordinates, and the integral equation has classically been derived as a property of these functions. Integral equation (9) is of interest because of its applications to stochastic processes, the study of lasers, the uncertainty principle, antenna theory, and the statistical theory of energy levels of complex systems. *See* STOCHASTIC PROCESS; UNCERTAINTY PRINCIPLE.

Historical development. Typically, special functions are discovered in the course of working on one problem. Many properties are then discovered, not only to aid in the solution of the problem which gave rise to them, but as a mathematical exercise, and then some of the new properties are used to solve completely different problems.

Each generation of mathematicians has a way of looking at mathematics which is different from previous generations. In the 19th century, complex analysis was being developed, and many important properties of special functions were discovered in the course of this development. In the second half of the 20th century, Lie group and Lie algebra methods have been applied to obtain other important properties of special functions. *See* COMPLEX NUMBERS AND COMPLEX VARIABLES; LIE GROUP.

[RICHARD ASKEY]

Bibliography: M. Abramowitz and I. Stegun, *Handbook of Mathematical Functions*, 1964; R. Askey (ed.), *Theory and Application of Special Functions*, 1975; A. Erdélyi et al., *Higher Transcendental Functions*, 3 vols., 1953–1955; F. W. J. Olver, *Asymptotics and Special Functions*, 1974; G. Szegö, *Orthogonal Polynomials*, 4th ed., 1975.

Specific charge

The ratio of charge to mass expressed as e/m, of a particle. The acceleration of a particle in electromagnetic fields is proportional to its specific charge. Specific charge can be determined by measuring the velocity v which the particle acquires in falling through an electric potential V ($v=\sqrt{2eV/m}$); by measuring the frequency of revolution in a magnetic field H (the so-called cyclotron frequency $\omega=eH/mc$, where c is the velocity of light); or by observing the orbit of the particles in combined electric and magnetic fields. In the mass spectrograph, the fields are arranged so that particles of differing velocities but of the same e/m are focused at a point. *See* ELEMENTARY PARTICLE.   [CHARLES J. GOEBEL]

Specific gravity

The specific gravity of a material is defined as the ratio of its density to the density of some standard material, such as water at a specified temperature, for example, 60°F, or (for gases) air at standard conditions of temperature and pressure. Specific gravity is a convenient concept because it is usually easier to measure than density, and its value is the same in all systems of units. *See* DENSITY.

[LEO NEDELSKY]

Specific heat

The ratio of the amount of heat required to raise unit mass of a material 1 degree in temperature to the amount of heat required to raise the same mass of a reference substance 1 degree in temperature. Both measurements are made at a reference temperature and in nearly all cases at either constant volume or constant pressure. Water is usually the reference substance. Because the heat capacity of water is nearly unity, the value of specific heat for a material is nearly equal to its heat capacity. Specific heat, as defined here, is a ratio without units, although it is often defined differently. For clarity it is recommended that thermodynamic discussion be carried out in terms of heat capacity instead of specific heat. Also, it is desirable to define heat units in electrical terms. *See* HEAT CAPACITY; SPECIFIC HEAT OF SOLIDS.

[HAROLD CHRISTIAN WEBER]

Specific heat of solids

When 1 gram (g) of a material absorbs an amount of heat ΔQ and this causes the temperature of the material to increase an amount ΔT, then the ratio $s=\Delta Q/\Delta T$ is often called the specific heat of the material, although other definitions are also used. The heat capacity C of a body of mass M is the product $C=Ms$. The atomic and molecular heats are the heat capacities of a gram-atomic weight and a gram-molecular weight of material, respectively.

The measured heat capacity of solids is usually made at some constant pressure P, such as atmospheric pressure, and is represented by the symbol C_p. The theoretical heat capacity is most often calculated for constant volume V, and is denoted by C_V. The difference C_P-C_V is essentially the heat per degree required to expand the solid against its internal elastic forces. The difference is given by Eq. (1). Here α_V is the temperature

$$C_P - C_V = \alpha_V^2 VT/\chi \quad (1)$$

coefficient of volume expansion (at constant pressure), V the volume, T the temperature in K, and χ the isothermal compressibility. The quantities represented by the symbols C_P and C_V are often referred to loosely as specific heats, although they are really heat capacities. *See* HEAT CAPACITY.

Dulong-Petit law. P. Dulong and A. Petit observed in 1819 that, although the specific heats of the solid elements at room temperature differ widely from one another, the atomic heats are nearly all the same, the values being about 6.3 cal/°C. A theoretical explanation was given by F. Richarz in 1893. It is an extension of the theory of the specific heat of an ideal gas. According to the kinetic theory of gases, the thermal energy of an ideal monatomic gas is the same as its kinetic energy. From this, it was deduced that the atomic heat of such a gas is $3R/2$, where R, the gas constant, is about 2.0 cal/°C. The thermal energy of a solid, however, is the energy of the harmonic motion of the atoms, and this, on the average, is half kinetic and half potential. Richarz then supposed that $3R/2$ is the atomic heat arising from the mean kinetic energy, and $3R/2$ that arising from the mean potential energy, yielding a total atomic heat of $3R$ or 6.0 cal/°C. *See* KINETIC THEORY OF MATTER.

The Dulong-Petit law is quite accurate at room temperature. To find s for many solid elements, one need only substitute the atomic weight A from a periodic table into the formula $s = 6/A$. However, it was noticed, even in the 19th century, that there are important exceptions to the law, notably diamond, germanium, and silicon, whose atomic heats at room temperature are considerably smaller than $3R$. Furthermore, many solids showed a decrease in C_V as the temperature was lowered to that of liquid nitrogen, which is 77 K or −196°C.

Einstein theory. The quantum hypothesis which M. Planck introduced into the theory of blackbody radiation in 1900 did not become a general principle until Albert Einstein applied it with success to the photoelectric effect in 1905 and to the theory of specific heats in 1907. In his theory of specific heats, Einstein sought to show that the observed failure of the classical theory, which gives $C_V = 3R$ for the atomic heat, could be explained in terms of the quantum hypothesis.

A so-called Planck oscillator can absorb or emit radiation only in integral amounts $nh\nu$, where n is an integer, h is Planck's constant, and ν is the natural frequency of the oscillator. The temperature is introduced by considering the mean value of the energy ϵ of such an oscillator, using the classical Boltzmann statistics. The result is given by Eq. 2,

$$\bar{\epsilon} = h\nu/[\exp(h\nu/kT) - 1] \qquad (2)$$

where k is the Boltzmann constant and T is the absolute temperature. *See* BOLTZMANN STATISTICS; HEAT RADIATION; QUANTUM MECHANICS.

Einstein's theory assumes that each atom of the solid oscillates with the same frequency ν_E and that this is the frequency observed in infrared absorption studies in crystals. Each atom vibrates in three dimensions and therefore has the mean energy $3\bar{\epsilon}$. The energy E of the solid is $3N\bar{\epsilon}$, if it contains Avogadro's number of atoms N. The quantum hypothesis then leads to Eq. (3). The frequency ν_E is called the Einstein frequency. *See* ABSORPTION OF ELECTROMAGNETIC RADIATION.

$$E = 3Nh\nu_E/[\exp(h\nu_E/kT) - 1] \qquad (3)$$

A parameter called the Einstein characteristic

temperature Θ_E is defined by equating one quantum of energy $h\nu_E$ to the classical energy kT of an oscillator and denoting the particular value of T obtained in this manner by Θ_E. According to Einstein, the thermal energy Q of the solid is just the energy E of vibration, so that $C_V = dQ/dT = dE/dT$. This yields the Einstein formula of specific (atomic) heats, Eq. (4). Here $y = h\nu_E/kT = \Theta_E/T$ and $Nk = R$, the gas constant.

$$C_V = 3Ry^2 e^y/(e^y - 1)^2 \qquad (4)$$

A plot of $C_V/3R$ versus T/Θ_E is shown in Fig. 1. At $T = \Theta_E$, the value of $C_V/3R$ is 0.92, which means that, at this temperature, C_V has 92% of the Dulong-Petit value. Above this temperature, C_V approaches $3R$ with increasing temperature. Below this temperature, C_V decreases to zero, practically vanishing at $T < 0.1\Theta_E$. Einstein's theory thus concludes that C_V is temperature-dependent. Furthermore, the observation that $C_V/3R = 0.31$ for diamond at $T = 331$ K is explained by stating that diamond has a value of Θ_E equal to about 1800 K, which corresponds to an infrared wavelength of 11 μm.

The prediction contained in the theory that C_V practically vanishes below $T = 0.1\Theta_E$ stimulated W. Nernst and his assistants to make experimental investigations of C_V down to 16 K. It was found that C_V is still appreciable at $T < 0.1\Theta_E$ for all substances examined; therefore Einstein's theory fails at these low temperatures. However, it appeared from the data that C_V approaches zero at 0 K, in keeping with deductions from Nernst's heat theorem.

Debye theory. The next advance in the theory of specific heats began with the suggestion of E. Madelung and W. Sutherland that the Einstein frequency is equivalent not only to the infrared absorption frequency of the crystal but also to the frequency of the shortest sound wave (or elastic wave) which can propagate through the crystal. This wave travels with the velocity of sound and has a wavelength of about twice the interatomic distance. Since sound waves of longer wavelength can also propagate through the crystal, Madelung made the further suggestion that a whole spectrum of acoustical frequencies should be used in computing C_V rather than just the single frequency ν_E.

In 1912 two theories of the specific heats of solids appeared, incorporating these ideas, one by P. Debye and the other by M. Born and T. von Kármán. Both theories use an acoustical spectrum containing so many frequencies that the spectra can be treated as continuous for purposes of computation. The number of waves (or modes) with frequencies between ν and $\nu + d\nu$ in the solid is thus represented by $g(\nu)\,d\nu$. The energy associated with each of these waves is that of a Planck oscillator, so that one obtains for the total energy E Eq. (5). The two theories differ in the manner of

$$E = \int_0^\infty \frac{g(\nu)h\nu\,d\nu}{e^{h\nu/kT} - 1} \qquad (5)$$

estimating $g(\nu)$. Only the simpler Debye theory is discussed in this article.

In order to estimate $g(\nu)$, Debye made two assumptions. One is that the solid is a continuous medium. With this idea, $g(\nu)$ is computed in a manner analogous to that employed in the theory of blackbody radiation, resulting in Eq. (6). The

$$g(\nu) = 4\pi V \left(\frac{1}{U_l^3} + \frac{2}{U_t^3}\right)\nu^2 \qquad (6)$$

symbol U_l represents the velocity of longitudinal sound waves and U_t that of transverse waves. The volume of the solid is V. The second assumption is that the total number of waves is equal to $3N$, where N is the number of atoms in the crystal. This assumption implies that the solid is not really continuous after all and that the shortest permissible wavelengths are those of about two interatomic distances. The restriction is expressed mathematically by Eq. (7), which serves to define a Debye

$$\int_0^{\nu_D} g(\nu)\, d\nu = 3N \qquad (7)$$

frequency ν_D. The Debye frequency is the maximum allowable frequency. Thus for $\nu > \nu_D$, $g(\nu)$ is zero and the value of the integral above this limiting frequency is zero. This allows the upper limit in Eq. (5) to be replaced by ν_D.

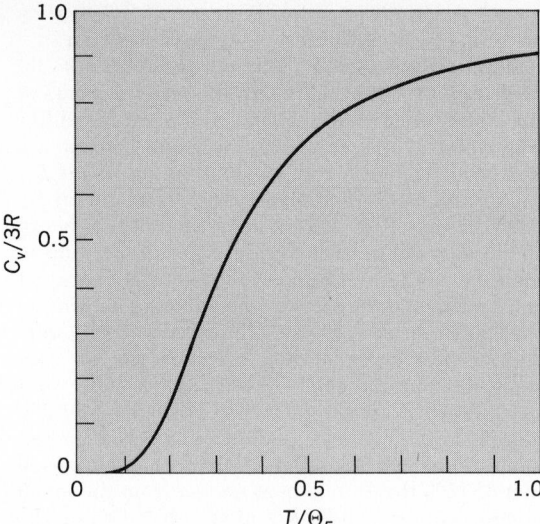

Fig. 1.   Einstein specific heat curve.

Debye characteristic temperature of solid elements, K

Element	Θ	Element	Θ	Element	Θ
Ar	85	Ga	240	Pd	275
Ag	215	Ge	360	Pr	74
Al	394	Gd	152	Pt	230
As	285	Hg	100	Sb	200
Au	170	In	129	Si	625
B	1250	K	100	Sn, gray	260
Be	1000	Li	400	Sn, white	170
Bi	120	La	132	Ta	225
C, diamond	1860	Mg	318	Th	100
Ca	230	Mn	400	Ti	380
Cd	120	Mo	380	Tl	96
Co	385	Na	150	V	390
Cr	460	Ne	63	W	310
Cu	315	Ni	375	Zn	234
Fe	420	Pb	88	Zr	250

Debye temperature. It is customary to replace the Debye frequency ν_D by the Debye characteristic temperature Θ, defined by the relation $k\Theta = h\nu_D$. From this, and from Eqs. (5), (6), and (7), the energy E is defined by Eq. (8), where $z = h\nu/kT$.

$$E = 9R\frac{T^4}{\Theta^3}\int_0^{\Theta/T}\frac{z^3\, dz}{e^z - 1} \qquad (8)$$

Equation (8) can be integrated and then C_V deduced from $C_V = dE/dT$. The result is the infinite series given by Eq. (9)

$$C_V/3R = \frac{4\pi^4}{5}\left(\frac{T}{\Theta}\right)^3 - \frac{3\Theta/T}{e^{\Theta/T} - 1} + 12\log\left(1 - e^{-\Theta/T}\right)$$
$$-36\frac{T}{\Theta}\sum_{n=1}^{\infty}\left\{\left[1 + \frac{2T}{n\Theta} + \frac{2}{n^2}\left(\frac{T}{\Theta}\right)^2\right]\frac{e^{-n\Theta/T}}{n^2}\right\} \qquad (9)$$

As T approaches infinity, the Dulong-Petit value of C_V is obtained. To see this, note that z approaches zero in this limit and that the integrand in Eq. (8) reduces to z^2. Integration then leads to $E = 3RT$, from which $C_V = 3R$. On the other hand, as T approaches 0 K, the Debye T^3 law results, as given by Eq. (10). This law is contained in the first term

$$C_V = \frac{12\pi^4 R}{5}\left(\frac{T}{\Theta}\right)^3 \qquad (10)$$

of Eq. (9). The other terms in Eq. (9) contribute less than 1% to C_V at temperatures below $T = \Theta/12$. Figure 2 shows a plot of $C_V/3R$ versus T/Θ as given by the Debye theory. The table lists the values of Θ required to fit the Debye formula for C_V to the experimental data of solid elements in the region near where C_V is about half the Dulong-Petit value. The corresponding values of Θ_E determined in this manner are smaller and are approximately $3\Theta/4$.

The courses of the two curves shown in Figs. 1 and 2 are quite similar for T above about 0.2Θ. The critical test distinguishing between the two theories must therefore be made at temperatures below about 0.1Θ, where the Debye T^3 law should hold. The T^3 law was first verified by A. Eucken and F. Schwers in 1913 by measuring the heat capacity of a number of insulators. It failed for metals. The reason for this failure is now under-

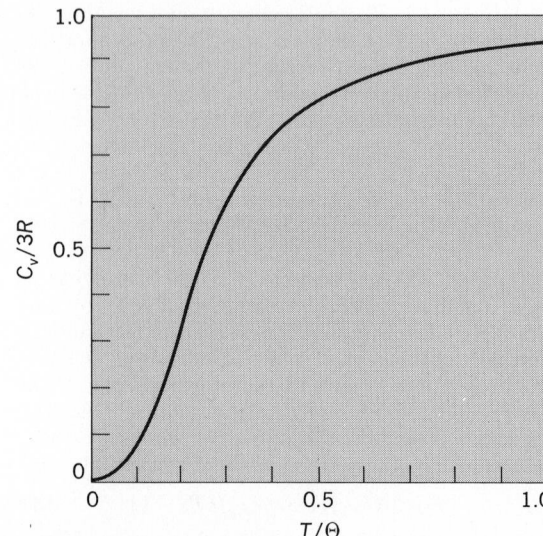

Fig. 2.   Debye specific heat curve.

stood, for A. Sommerfeld's theory of metals (1928) shows that the conduction electrons can make an important contribution to the heat capacity. According to Sommerfeld, there must be a linear term in the temperature included in the expression for C_v in order to account for the electron contribution. Thus Eq. (11) is written. The coefficient γ in

$$C_v = \gamma T + (12\pi^4 R/5)(T/\Theta)^3 \qquad (11)$$

the electron term is sometimes called the Sommerfeld gamma. To analyze low-temperature C_v data for metals, C_v/T versus T^2 is plotted. According to Eq. (11), this should give a straight line of slope $12\pi^4 R/5\Theta^3$ and of intercept γ on the C_v/T axis. *See* FREE-ELECTRON THEORY OF METALS.

Deviations. As experimental measurements became more precise, it was noticed that the data could not be fitted to a Debye curve. One sensitive test is to calculate the Debye Θ for each experimental value of C_v and T, after correcting for the electron contribution. If the data satisfy the Debye formula, then Θ should be independent of the temperature. In most cases, the data do not satisfy this criterion.

A particularly marked deviation from Debye's theory occurs for cadmium. Figure 3 shows a plot of Θ versus T for this element. The data were treated in the following manner. First, C_v was calculated from the measured C_p data, using essentially Eq. (1). Then the 12 items of data below about 3 K were plotted on the basis of Eq. (11) and γ and Θ determined from the straight line graph. Next, all C_v data were corrected for the electron term and then the Θ for each point computed from tables based on Eqs. (8) or (9). The result is plotted in Fig. 3.

Agreement with Debye theory in the case of cadmium exists only below about 3 K, the only part of the curve where Θ is substantially constant. Thus the T^3 law holds (to better than 1%) only below about $T = \Theta/50$, instead of $\Theta/12$ as required by the Debye theory. For most of the solids examined, this limitation of the range of the validity of the T^3 law to the region $T < \Theta/50$ seems to occur. M. Blackman explained this in 1935 on the basis of the lattice dynamics of Born and von Kármán. He gave the estimate $T < \Theta/50$ as the "true" range of the T^3 law for most solids, an estimate which he made when the experimental evidence was still rather meager.

Experimental verification. An important independent check of the theory of specific heats can be made, based on Eq. (6). The velocity of sound is measured for single crystals at temperatures in the "true" T^3 region of specific heats. Since the velocity of sound depends on the direction of propagation through the crystal (an effect known as anisotropy of the velocity of sound), the inverse cube of the velocity must be averaged over all directions. This is done theoretically from the velocity of sound measurements made in several appropriate directions in a single crystal. When this is done, Θ at 0 K can be calculated. The comparison of the values calculated in this manner with those obtained from low-temperature specific heat data has been made for many substances, including copper, silver, and gold. The velocity of sound values of Θ for these last three elements are respectively 345, 226, and 162, which compare well with the corresponding specific heat values 345, 226, and 165 K. *See* CONDUCTION (HEAT); LATTICE VIBRATIONS; THERMAL CONDUCTION IN SOLIDS; ULTRASONICS.

[JULES DE LAUNAY]

Bibliography: N. W. Ashcroft and N. D. Mermin, *Solid State Physics*, 1976; S. Fluegge (ed.), *Handbuch der Physik*, vol. 7, pt. 1, 1955; C. Kittel, *Introduction to Solid State Physics*, 5th ed., 1976; F. Seitz and D. Turnbull (eds.), *Solid State Physics*, vol. 2. 1956.

Speckle

The generation of a random intensity distribution, called a speckle pattern, when light from a highly coherent source, such as a laser, is scattered by a rough surface or inhomogeneous medium. Although the speckle phenomenon has been known since the time of Isaac Newton, the development of the laser is responsible for the present-day interest in speckle. Speckle has proved to be a universal nuisance as far as most laser applications are concerned, and it was only in the mid-1970s that investigators turned from the unwanted aspects of speckle toward the uses of speckle patterns, in a wide variety of applications. *See* LASER.

Basic phenomenon. Objects viewed in coherent light acquire a granular appearance as illustrated in Fig. 1. The detailed irradiance distribution of this granularity appears to have no obvious relationship to the microscopic properties of the illuminated object, but rather it is an irregular pattern that is best described by the methods of probability theory and statistics. Although the mathematical description of the observed granularity is rather complex, the physical origin of the observed speckle pattern is easily described. The surfaces of most materials are extremely rough on the scale of an optical wavelength (approximately 5×10^{-7} m). When nearly monochromatic light is reflected from such a surface, the optical wave resulting at any moderately distant point consists of many coherent wavelets, each arising from a different microscopic element of the surface. Since the distances traveled by these various wavelets may differ by several wavelengths if the surface is truly rough, the interference of the wavelets of various phases results in the granular pattern of intensity called speckle. If a surface is imaged with a per-

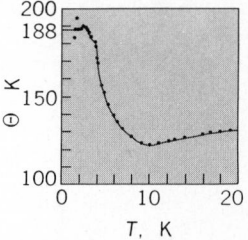

SPECIFIC HEAT OF SOLIDS

Fig. 3. Plot of Θ versus T for cadmium, illustrating a deviation from Debye's theory. (After P. L. Smith and N. M. Wolcott, *Phil. Mag.*, ser. 8, 1:854–865, 1956)

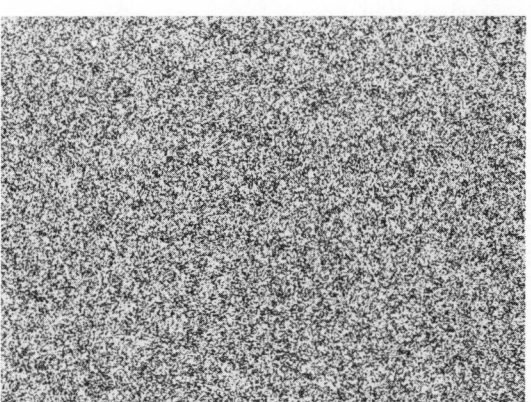

Fig. 1. Photograph of speckle pattern generated by illuminating rough surface with laser radiation.

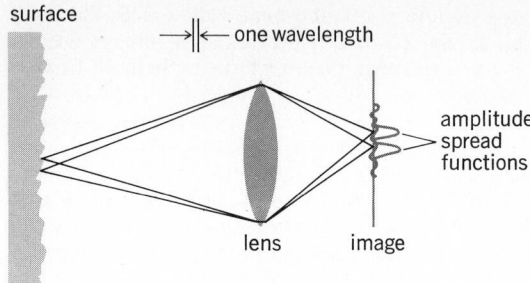

Fig. 2.   Physical origin of speckle for an imaging system.

fectly corrected optical system as shown in Fig. 2, diffraction causes a spread of the light at an image point, so that the intensity at a given image point results from the coherent addition of contributions from many independent surface areas. As long as the diffraction-limited point-spread function of the imaging system is broad by comparison with the microscopic surface variations, many dephased coherent contributions add at each image point to give a speckle pattern.

The basic random interference phenomenon underlying laser speckle exists for sources other than lasers. For example, it explains radar "clutter," results for scattering of x-rays by liquids, and electron scattering by amorphous carbon films. Speckle theory also explains why twinkling may be observed for stars, but not for planets. *See* CO-HERENCE; DIFFRACTION; INTERFERENCE OF WAVES.

Applications. The principle applications for speckle patterns fall into two areas: metrology, and stellar speckle interferometry.

Metrology. In the metrology area, the most obvious application of speckle is to the measurement of surface roughness. If a speckle pattern is produced by coherent light incident on a rough surface, then surely the speckle pattern, or at least the statistics of the speckle pattern, must depend upon the detailed surface properties. If the surface root-mean-square roughness is small compared to one wavelength of the fully coherent radiation used to illuminate the surface, the roughness can be determined by measuring the speckle contrast. If the root-mean-square roughness is large compared to one wavelength, the radiation should be spatially coherent polychromatic, instead of monochromatic; the roughness is again determined by measuring the speckle contrast.

An application of growing importance in engineering is the use of speckle patterns in the study of object displacements, vibration, and distortion that arise in nondestructive testing of mechanical components. The key advantage of speckle methods in this case is that the speckle size can be adjusted to suit the resolution of the most convenient detector, whether it be film or a television camera, while still retaining information about displacements with an accuracy of a small fraction of a micrometer. Although several techniques for using laser speckle in metrology are available, the basic idea can be explained by the following example for measuring vibration. A vibrating surface illuminated with a laser beam is imaged as shown in Fig. 2. A portion of the laser beam is also superimposed on the image of the vibrating surface. If

the object surface is moving in and out with a travel of one-quarter of a wavelength or more, the speckle pattern will become blurred out. But for areas of the surface that are not vibrating (the nodal regions) the eye will be able to distinguish the fully developed high-contrast speckle pattern.

Electronic speckle interferometry (ESPI) is finding much use in industrial applications involving measurement of the vibrational properties of structures. In ESPI a television camera is used to view a speckle pattern. The output of the television camera is processed electronically to emphasize the high spatial frequency structure. The result is viewed on a television monitor (Fig. 3). The bright fringes correspond to the region of the object which are at the nodes of the vibration pattern. By knowing the angles at which the object is illuminated and viewed, it is possible to calculate the amplitude of the vibration from the number of dark fringes present. Thus, simply by looking at the television monitor it is possible to determine the vibrational pattern over the entire object. The amplitude and frequency of the drive source can be changed and instantly the new vibrational pattern can be observed.

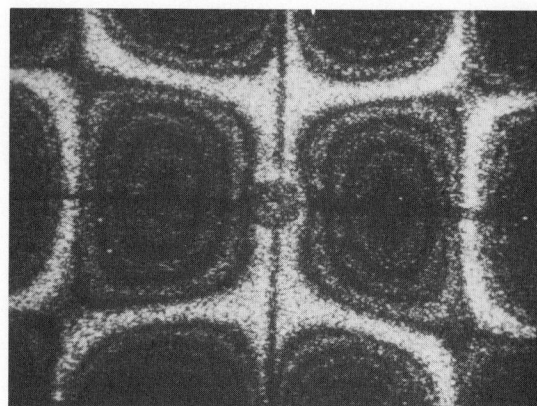

Fig. 3.   Electronic speckle interferogram of vibrating object.

Speckle patterns can also be used to examine the state of refraction of the eye. If a diffusing surface illuminated by laser light moves perpendicular to the line of sight of an observer, the speckles may appear to move with respect to the surface. For a normal eye, movement in a direction opposite to that of the surface indicates underaccommodation, and movement with the surface indicates overaccommodation. If the speckles do not move, but just seem to "boil," the observed surface is imaged on the retina.

Stellar speckle interferometry. Stellar speckle interferometry, which has many similarities with the laser speckle methods used in metrology, is a technique for obtaining diffraction-limited resolution of stellar objects despite the presence of the turbulent atmosphere which limits the resolution of conventional pictures to approximately 1 arcsecond. If a short-exposure photograph is taken of a magnified image of an unresolved star and a narrow-bandwidth spectral filter is used, the picture has a specklelike structure. The size of the speckles is equal to the diffraction-limited resolution limit of the telescope, regardless of the resolution limit determined by the turbulent atmosphere. This means that the short-exposure photograph of a resolvable object, for example a binary star, contains information about the object down to the diffraction limit of the telescope, which is approximately 0.02 arc-second for the 5-m Palomar Mountain telescope, 1/50 the resolution limit set by the atmosphere. Hence, by extracting correctly the information in short-exposure pictures of objects with more than one resolvable element, detail down to the diffraction limit of the telescope can be observed. The technique has proved to be of enormous value in the study of binary stars and centrosymmetric resolvable stars, and research is directed at making the technique useful for observing objects having a more general shape. *See* INTERFEROMETRY.

[JAMES C. WYANT]

Bibliography: J. C. Dainty (ed.), *Laser Speckle and Related Phenomena*, 1975; R. K. Erf (ed.), *Speckle Metrology*, 1978; M. Francon, *Laser Speckle and Applications in Optics*, 1979; G. A. Slettemoen, *Appl. Opt.*, 19:616–623, 1980.

Spectrum

The term spectrum is applied to any class of similar entities or properties strictly arrayed in order of increasing or decreasing magnitude. In general, a spectrum is a display or plot of intensity of radiation (particles, photons, or acoustic radiation) as a function of mass, momentum, wavelength, frequency, or some other related quantity. For example, a β-ray spectrum represents the distribution in energy or momentum of negative electrons emitted spontaneously by certain radioactive nuclides, and when radionuclides emit α-particles, they produce an α-particle spectrum of one or more characteristic energies. A mass spectrum is produced when charged particles (ionized atoms or molecules) are passed through a mass spectrograph in which electric and magnetic fields deflect the particles according to their charge-to-mass ratios. The distribution of sound-wave energy over a given range of frequencies is also called a spectrum. *See* SOUND.

In the domain of electromagnetic radiation, a spectrum is a series of radiant energies arranged in order of wavelength or of frequency. The entire range of frequencies is subdivided into wide intervals in which the waves have some common characteristic of generation or detection, such as the radio-frequency spectrum, infrared spectrum, visible spectrum, ultraviolet spectrum, and x-ray spectrum. Spectra are also classified according to their origin or mechanism of excitation, and are referred to as emission, absorption, continuous, line, and band spectra. *See* ELECTROMAGNETIC RADIATION.

An emission spectrum is produced whenever the radiations from an excited light source are dispersed. Excitation of emission spectra may be by thermal energy, by impacting electrons and ions, or by absorption of photons. Depending upon the nature of the light source, an emission spectrum may be a continuous or a discontinuous spectrum, and in the latter case, it may show a line spectrum, a band spectrum, or both.

An absorption spectrum is produced against a background of continuous radiation by interposing matter that reduces the intensity of radiation at certain wavelengths or spectral regions. The energies removed from the continuous spectrum by the interposed absorbing medium are precisely those that would be emitted by the medium if properly excited. This reciprocity of absorption and emission is known as Kirchhoff's principle; it explains, for example, the absorption spectrum of the

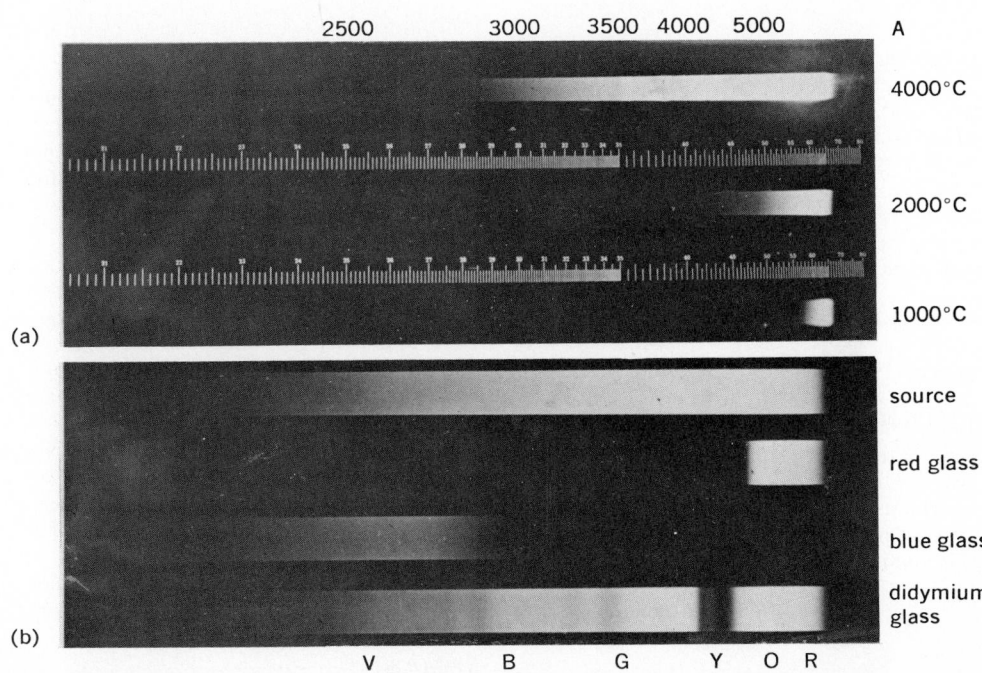

Spectra of a blackbody. (a) Continuous emission spectra of a solid, taken with a quartz spectrograph. (b) Continuous absorption spectra of the source alone and three kinds of colored glass interposed on it. (*From F. A. Jenkins and H. E. White, Fundamentals of Optics, 3d ed., McGraw-Hill, 1957*)

Sun, in which thousands of lines of gaseous elements appear dark against the continuous-spectrum background.

A continuous spectrum contains an unbroken sequence of waves or frequencies over a long range (see illustration).

In illustration *a* the spectra for 1000 and 2000°C were obtained from tungsten filament, that for 4000°C from a positive pole of carbon arc. The upper spectrum in *b* is that of the source alone, extending roughly from 400 to 650 nm. The others show the effect on this spectrum of interposing three kinds of colored glass.

All incandescent solids, liquids, and compressed gases emit continuous spectra, for example, an incandescent lamp filament or a hot furnace. In general, continuous spectra are produced by high temperatures, and under specified conditions the distribution of energy as a function of temperature and wavelength is expressed by Planck's law. *See* HEAT RADIATION; PLANCK'S RADIATION LAW.

Line spectra are discontinuous spectra characteristic of excited atoms and ions, whereas band spectra are characteristic of molecular gases or chemical compounds. *See* ATOMIC STRUCTURE AND SPECTRA; BAND SPECTRUM; LINE SPECTRUM; MOLECULAR STRUCTURE AND SPECTRA.

[W. F. MEGGERS/W. W. WATSON]

Speed

The time rate of change of position of a body without regard to direction. It is the numerical magnitude only of a velocity and hence is a scaler quantity. Linear speed is commonly measured in such units as meters per second, miles per hour, or feet per second. It is the most frequently mentioned attribute of motion.

Average linear speed is the ratio of the length of the path Δs traversed by a body to the elapsed time Δt during which the body moved through that path, as in Eq. (1), where s_0 and t_0 are the initial

$$\text{Speed (average)} = \frac{s_f - s_0}{t_f - t_0} = \frac{\Delta s}{\Delta t} \qquad (1)$$

position and time, respectively, s_f and t_f are the final position and time and Δ stands for "the change in."

Instantaneous speed, defined by Eq. (2), is the limiting value of the foregoing ratio as the elapsed time approaches zero.

$$\text{Speed (instantaneous)} = \lim_{\Delta t \to 0} \frac{\Delta s}{\Delta t} = \frac{ds}{dt} \qquad (2)$$

See VELOCITY.   [ROGERS D. RUSK]

Spherical harmonics

A spherical harmonic or solid spherical harmonic of degree n is a homogeneous function, $R_n(x,y,z)$, of degree n which satisfies Laplace's equation, Eq. (1). Here n is any number and $(x^2 + y^2 + z^2)^{(-n-1)/2}$ $R_n(x,y,z)$ is a spherical harmonic of degree $-n-1$.

$$\Delta R \equiv \frac{\partial^2 R}{\partial x^2} + \frac{\partial^2 R}{\partial y^2} + \frac{\partial^2 R}{\partial z^2} = 0 \qquad (1)$$

There are analogous definitions for spaces of any number of dimensions. In the present article, n

will be a nonnegative integer and R_n a polynomial in x, y, z (polynomial spherical harmonic). In terms of spherical coordinates r, θ, ϕ, $R_n(x,y,z) = r^n S_n(\theta,\phi)$ where S_n, a polynomial in $\cos \theta$, $\sin \theta$, $\cos \phi$, $\sin \phi$ is a spherical surface harmonic of degree n. There are $2n + 1$ linearly independent spherical surface harmonics of degree n; any spherical surface harmonic of degree n is a linear combination of these, and conversely any linear combination of spherical surface harmonics of degree n is again a spherical surface harmonic of degree n.

Applications. Spherical harmonics occur in potential theory. They occur in connection with Laplace's equation not only in spherical coordinates but also in spheroidal coordinates (spheroidal harmonics) and confocal coordinates (ellipsoidal surface harmonics). In the latter case their occurrence is due to the circumstance that the natural affine mapping of an ellipsoid onto a sphere carries the partial differential equation of ellipsoidal surface harmonics into the partial differential equation satisfied by spherical surface harmonics. In spherical coordinates, spherical surface harmonics occur in connection with Laplace's and Poisson's equations, the wave equation, the Schrödinger equation, and generally in connection with partial differential equations of the form $\Delta U + f(r) U = 0$. In the latter case one has special solutions of the form $F(r) S_n(\theta,\phi)$, where F satisfies the ordinary differential equation (2). In geometry,

$$\frac{d^2 F}{dr^2} + \frac{2}{r} \frac{dF}{dr} + \left[f(r) - \frac{n(n+1)}{r^2} \right] F = 0 \qquad (2)$$

spherical surface harmonics are used in the theory of surfaces. In mathematical physics, spherical harmonics appear in the theories of gravitation, electricity and magnetism, hydrodynamics, and in other fields.

Spherical harmonics of degree *n*. Notation (3), with $f(u)$ an integrable function, is a polynomial spherical harmonic of degree n, and every such

$$\int_{-\pi}^{\pi} (x \cos u + y \sin u + iz)^n f(u) \, du \qquad (3)$$

spherical harmonic can be so represented. The representation is not unique. If c_n is a constant, $h_1, h_2, \ldots, h_n$ are n directions (not necessarily distinct), and $\partial/\partial h$ denotes directional differentiation in the direction h, then notation (4) is a polyno-

$$c_n r^{2n+1} \frac{\partial^n}{\partial h_1 \cdots \partial h_n} \frac{1}{r} \qquad (4)$$

mial spherical harmonic of degree n, and every such spherical harmonic can be so represented. The representation is unique. In a zonal spherical harmonic the n directions coincide, and in a sectorial spherical harmonic they are in a plane at angles of π/n. If $n - m$ directions coincide in the axis and the remaining directions are in the plane perpendicular to the axis at angles of π/m, one has a tesseral spherical harmonic of degree n and order m.

Explicit forms. For spherical harmonics whose axis is the z axis, $(m = 0, 1, 2, \ldots, n)$, Eq. (5)

$$S_n^{\pm m}(\theta,\phi) \equiv \frac{(-1)^{n-m} r^{n+1}}{(n-m)!} \frac{\partial^{n-m}}{\partial z^{n-m}} \left(\frac{\partial}{\partial x} \pm i \frac{\partial}{\partial y} \right)^m \frac{1}{r}$$

$$= P_n^m (\cos \theta) \, e^{\pm im\phi} \qquad (5)$$

represents a linearly independent system of spherical surface harmonics of degree n. $S_n{}^m \pm S_n{}^{-m}$ is a zonal, sectorial, tesseral spherical surface harmonic of degree n and order m according as $m = 0$, $m = n$, $1 \leq m \leq n - 1$. The $P_n{}^m(w)$ are associated Legendre functions which satisfy the associated Legendre equation (6). $P_n{}^0 = P_n$ is the Legendre polynomial of degree n.

$$(1 - w^2)\frac{d^2P}{dw^2} - 2w\frac{dP}{dw}$$
$$+ \left[n(n+1) - \frac{m^2}{1 - w^2}\right]P = 0 \quad (6)$$

Properties. A function is said to be harmonic in a region if it is a twice continuously differentiable solution of Laplace's equation there, and if, in addition, it vanishes at infinity in case the point at infinity is an interior point of the region. Every function harmonic inside a sphere about the origin can be expanded in a series $\sum_{n=0}^{\infty} r^n S_n(\theta, \phi)$ convergent inside that sphere. Every function harmonic outside a sphere about the origin can be expanded in a series $\sum_{n=0}^{\infty} r^{-n-1} S_n(\theta, \phi)$ convergent outside that sphere. The reciprocal distance of the points (x, y, z) and $(0, 0, a)$ is harmonic in the regions $r < a$ and $r > a$ and possesses in these regions the expansions of Eqs. (7).

$$\frac{1}{\sqrt{a^2 - 2ar\cos\theta + r^2}} = \begin{cases} \sum_{n=0}^{\infty} \dfrac{r^n}{a^{n+1}} P_n(\cos\theta) & r < a \\[2ex] \sum_{n=0}^{\infty} \dfrac{a^n}{r^{n+1}} P_n(\cos\theta) & r > a \end{cases} \quad (7)$$

θ, ϕ determine a point on the unit sphere. The scalar product of two functions, f and g, on the unit sphere is suitably defined as Eq. (8) where $\bar{g}$ is the

$$(f, g) = \int_0^\pi \int_{-\pi}^\pi f(\theta, \phi)\overline{g(\theta, \phi)}\sin\theta\,d\theta\,d\phi \quad (8)$$

complex conjugate of g. If $(f, g) = 0$, f and g are orthogonal. Spherical surface harmonics are functions on the unit sphere. Any two spherical surface harmonics of different degrees are orthogonal. The spherical surface harmonics of Eq. (9) form an orthogonal system; that is, $S_n{}^m(\theta, \phi)$

$$S_n{}^m(\theta, \phi)\,m = -n, -n+1, \ldots, n \quad (9)$$
$$n = 0, 1, \ldots$$

thogonal system; that is, $(S_n{}^m, S_{n'}{}^{m'}) = 0$ unless $m = m'$ and $n = n'$. This orthogonal system is complete; that is, a continuous function which is orthogonal to all the $S_n{}^m$ vanishes identically. With an integrable function f is associated the Laplace expansion (10)

$$\sum_{n=0}^{\infty} \sum_{m=-n}^{n} C_{mn} S_n{}^m \quad (10)$$

where
$$C_{mn} = \frac{(f, S_n{}^m)}{(S_n{}^m, S_n{}^m)}$$

Under suitable conditions, the Laplace expansion will converge to f. For instance, if f is continuous and continuously differentiable on the unit sphere, then the Laplace expansion converges to f uniformly.

Let (θ_0, ϕ_0) be a fixed point, and let $\cos\gamma = \cos\theta \cos\theta_0 + \sin\theta \sin\theta_0 \cos(\phi - \phi_0)$ be the spherical distance of (θ, ϕ) and (θ_0, ϕ_0). The Laplace expansion of Eq. (11) is the addition theorem of Legendre

$$P_n(\cos\gamma) = P_n(\cos\theta)P_n(\cos\theta_0)$$
$$+ 2\sum_{m=-n}^{n} \frac{(n-m)!}{(n+m)!}$$
$$\cdot P_n{}^m(\cos\theta)P_n{}^m(\cos\theta_0)\cos m(\phi - \phi_0) \quad (11)$$

polynomials and expresses the change to a new axis through the point (θ_0, ϕ_0). Other spherical surface harmonics have corresponding addition theorems.

Let $\cos\gamma$ be the spherical distance of (θ, ϕ) and (θ_0, ϕ_0), and let $K(w)$ be a continuous function for $-1 \leq w \leq 1$. Then for any spherical surface harmonic S_n of degree n, Eq. (12) holds,

$$\int_0^\pi \int_{-\pi}^\pi K(\cos\gamma)S_n(\theta, \phi)\sin\theta\,d\theta\,d\phi$$
$$= \lambda_n S_n(\theta_0, \phi_0) \quad (12)$$

where
$$\lambda_n = 2\pi \int_{-1}^{1} K(w)P_n(w)\,dw$$

See DIFFERENTIAL EQUATION.   [A. ERDÉLYI]

Bibliography: R. Askey, *Orthogonal Polynomials and Special Functions*, 1975; A. Erdélyi et al., *Higher Transcendental Functions*, 3 vols., 1953–1955; E. W. Hobson, *Spherical and Ellipsoidal Harmonics*, 1955; I. N. Sneddon, *Special Functions of Mathematical Physics and Chemistry*, 3d ed., 1980; E. T. Whittaker and G. N. Watson, *A Course of Modern Analysis*, 4th ed., 1927.

Spin

The intrinsic angular momentum of a particle. It is that part of the angular momentum of a particle which exists even when the particle is at rest, as distinguished from the orbital angular momentum. The total angular momentum of a particle is the sum of its spin and its orbital angular momentum resulting from its translational motion. The general properties of angular momentum in quantum mechanics imply that spin is quantized in half integral multiples of $\hbar$ ($\hbar = h/2\pi$, where h is Planck's constant); orbital angular momentum is restricted to half *even* integral multiples of $\hbar$. A particle is said to have spin $\frac{3}{2}$, meaning that its spin angular momentum is $\frac{3}{2}\hbar$. *See* ANGULAR MOMENTUM.

A nucleus, atom, or molecule in a particular energy level, or a particular elementary particle, has a definite spin; for instance, a deuteron has spin 1, a $_3\text{Li}^5$ nucleus in its ground state has spin $\frac{3}{2}$, and an electron has spin $\frac{1}{2}$. The spin is an intrinsic or internal characteristic of a particle, along with its mass, charge, and isotopic spin. *See* ISOTOPIC SPIN; SYMMETRY LAWS.

A particle of spin s has $2s + 1$ spin states, since according to quantum mechanics the projection of an angular momentum of magnitude j along an axis can have the $2j + 1$ integrally spaced values j, $j - 1 \cdots -j + 1, -j$. These spin states represent an internal degree of freedom of a particle, in addition to its external freedom of motion in three-dimensional space.

In field theory, in which particles are regarded as quanta of a field, the spin of the particle is determined by the tensor character of the field. For instance, the quanta of a scalar field have spin 0 and the quanta of a vector field have spin 1. A celebrated theorem of quantum field theory, proved first by W. Pauli, states a connection between spin and statistics: A particle with half even integral spin obeys Bose-Einstein statistics and is called a boson; a particle with half odd integral spin obeys Fermi-Dirac statistics and is called a fermion. *See* ELECTRON SPIN; QUANTUM FIELD THEORY; QUANTUM MECHANICS; QUANTUM STATISTICS.

[CHARLES J. GOEBEL]

Spin-density wave

The ground state of a metal in which the conduction-electron spin density has a sinusoidal variation in space. The periodicity of the wave is unrelated to the lattice periodicity. Instead it is determined by the dimensions of the conduction-electron Fermi surface in momentum space.

Description. In a metal the conduction-electron charge densities for up and down spin are ordinarily equal. Their sum is, say, $\rho_0(\vec{r})$, and they exhibit a dependence on position $\vec{r}$ having the same spatial periodicity as that of the lattice. A metal with a spin-density wave (SDW) has, instead, charge densities given by Eq. (1). The spin density $\vec{\sigma}(\vec{r})$ is the

$$\rho_\pm(\vec{r}) = \tfrac{1}{2}\rho_0(\vec{r})[1 \pm p \cos \vec{Q} \cdot \vec{r}] \qquad (1)$$

difference between $\rho_+(\vec{r})$ and $\rho_-(\vec{r})$, and is therefore given by Eq. (2), where $\hat{\epsilon}$ is a unit vector de-

$$\vec{\sigma}(\vec{r}) = \hat{\epsilon}p\rho_0(\vec{r}) \cos \vec{Q} \cdot \vec{r} \qquad (2)$$

fining the axis of spin quantization, and p is the amplitude of the spin-density wave. The wave vector $\vec{Q}$ is determined by the conduction-electron Fermi surface. The wavelength, $\lambda = 2\pi/\vec{Q}$, of the spin-density wave is not an integral multiple of a lattice periodicity. That is to say, the spin-density wave is incommensurate. *See* FERMI SURFACE.

Origin. The Pauli exclusion principle automatically keeps electrons having parallel spin apart. The reduction in Coulomb interaction that results is called exchange energy. This reduction is increased when the conduction-electron charge densities are modulated according to Eq. (1). However, the modulation requires an increase in conduction-electron kinetic energy. The spin-density wave instability theorem of A. W. Overhauser shows that a new reduction in energy can always be achieved if the quantum theory is treated in an approximation that neglects electron-electron scattering. Were it not for these latter correlations, every metal would have one or more spin-density waves. *See* EXCLUSION PRINCIPLE.

The correlation energy, which contributes to a metal's stability, arises primarily from scattering of up-spin electrons by down-spin electrons. Since a spin-density wave tends to stratify opposite spin densities in alternating layers, as indicated in Eq. (1), the scattering is reduced, thereby tending to suppress spin-density wave instability. This contrasts with a charge-density wave (CDW), where exchange and correlation energy shifts act in unison. Consequently spin-density wave states are less likely to occur than charge-density wave states. The only metallic element known to have a spin-density wave is chromium. *See* CHARGE-DENSITY WAVE.

Detection. Since the total electronic charge density, the sum of ρ_+ and ρ_- in Eq. (1), is just $\rho_0(\vec{r})$, a spin-density wave cannot be detected in an x-ray diffraction experiment. However, the elec-

tronic magnetic moment (1 Bohr magneton), together with the spin-density wave spin density of Eq. (2), leads to a sinusoidal magnetic field. Neutrons, which also have a magnetic moment, can therefore experience Bragg diffraction from a spin-density wave. There will be two magnetic satellites, on opposite sides of each crystallographic Bragg reflection.

The magnetic origin of spin-density wave satellites can be demonstrated in two ways, the first being their absence in an x-ray experiment. The second is by means of neutron diffraction with a spin-polarized beam, since only a magnetic reflection can cause reversal of neutron polarization during diffraction. *See* NEUTRON DIFFRACTION; X-RAY DIFFRACTION.

Q-domains and polarization domains. Chromium metal, which has cubic symmetry above 311 K, acquires a spin-density wave in a first-order transition on cooling below that temperature. The direction of $\vec{Q}$ is along any one of the three cubic axes. Generally a single-crystal sample will subdivide more or less equally into $\vec{Q}$-domains of the three types. However, it is possible to produce a single-$\vec{Q}$ sample by cooling through the transition in a strong magnetic field.

Just below the transition temperature, the spin-density wave has transverse polarization. That is, $\hat{\epsilon}$ of Eq. (2) is along one of the two cubic axes perpendicular to $\vec{Q}$. Even if the crystal is single-$\vec{Q}$, it may divide into polarization domains having either of the two directions of $\hat{\epsilon}$. When chromium is cooled below 123 K, the polarization flips from transverse to longitudinal; $\hat{\epsilon}$ is then parallel to $\vec{Q}$. Polarization domains are no longer possible. *See* DOMAIN.

Other properties. The length λ of the spin-density wave in chromium is larger than the lattice constant by about 3%, an increment that varies continuously with temperature. Despite the smallness of this deviation, the spin-density wave remains incommensurate. However, the spin-density wave can be made commensurate by changing the Fermi-surface dimension. Addition of a few percent rhenium, which increases the election/atom ratio, suffices.

The conduction electrons experience a sinusoidal potential energy proportional to Eq. (2). This changes their energy spectrum and the topology of their Fermi surface. New optical absorption processes become possible, and low-temperature magnetoconductivity properties become anomalous. *See* BAND THEORY OF SOLIDS; CRYSTAL STRUCTURE; MAGNETORESISTANCE.

[ALBERT W. OVERHAUSER]

Bibliography: A. W. Overhauser, Spin density waves in an electron gas, *Phys. Rev.*, 128:1437–1452, 1962.

Spin glass

One of a wide variety of materials which contain interacting atomic magnetic moments and also possess some form of disorder, in which the temperature variation of the magnetic susceptibility undergoes an abrupt change in slope, that is, a cusp, at a temperature generally referred to as the freezing temperature, T_f. At temperatures below T_f the spins have no long-range magnetic order, but instead are found to have static or quasistatic orientations which vary randomly over macroscopic distances. The latter state is referred to as spin-glass magnetic order. Spin-glass ordering is usually detected by means of magnetic susceptibility measurements, although additional data are required to demonstrate the absence of long-range order. Closely related susceptibility cusps can also be observed by using neutron diffraction. It is not generally agreed whether spin glasses undergo a phase transition or not. *See* MAGNETIC SUSCEPTIBILITY; NEUTRON DIFFRACTION; PHASE TRANSITIONS.

Spin-glass transition. An example of a spin-glass ordering transition is shown in Fig. 1, where magnetic susceptibility χ is plotted as a function of absolute temperature T for the case of dilute alloys of iron in gold. The iron solute atoms carry a magnetic moment and occupy sites at random in the gold lattice. Neighboring moments interact with one another via the Ruderman-Kittel-Kasuya-Yosida (RKKY) exchange interaction U, given by Eq. (1), which is mediated by the conduction electrons of the gold. The exchange constant J_{ij}

$$U = -\sum_{ij} J_{ij} \vec{S}_i \cdot \vec{S}_j \qquad (1)$$

oscillates rapidly in sign with increasing distance between spins $\vec{S}_i$ and $\vec{S}_j$. The sharp susceptibility peak in Fig. 1 occurs at a temperature T_f such that $k_B T_f$ has the value of a typical exchange constant J_{ij}, where k_B is Boltzmann's constant. In this case the necessary disorder arises from the random placement of the magnetic moments. The resulting configuration of exchange couplings is such that at low temperatures the minimum energy state is one where the moment orientations are only correlated locally and are effectively random over macroscopic distances.

As seen in Fig. 1, the sharpness of the transition is destroyed by the application of a relatively small magnetic field, $H_0 \sim 100$ oersteds (1 Oe = 79.6 ampere-turns/m). This is evidence that the transition is a cooperative effect involving a large number of moments, since the motion of an isolated

moment would be largely unaffected by a field of 100 Oe at this temperature.

Some of the basic elements of the spin-glass transition are reproduced in a mean-field model calculation. In this model one calculates the thermodynamic behavior of a system of classical spins which interact via exchange couplings having the form given by Eq. (1). The distribution of J_{ij} parameters is taken to be a gaussian random function. Below the temperature T_f, the system is found to be ordered in the sense that each spin $\vec{S}_i$ will retain some memory of its orientation $\vec{S}_i(t_1)$ at time t_1 at a much later time t_2, even though $<\vec{S}_i(t_1)> = 0$. where $<\,>$ denotes an average over all spins. The ordering effect is expressed by the Edwards-Anderson order parameter $q = <\vec{S}_i(t_1) \cdot \vec{S}_i(t_2)>$. Their theory gives Eq. (2) for the magnetic sus-

$$\chi = \chi_c(1 - q) \qquad (2)$$

ceptibility χ. Here $\chi_c = a/T$ is the Curie susceptibility, where a is the Curie constant. For $T > T_f$, $q = 0$; and for $T < T_f$, $q = \frac{1}{2}[(T_f/T)^2 - 1]$. Thus $\chi(T)$ is just the Curie susceptibility for $T > T_f$, and for $T < T_f$ is given by Eq. (3) for $T_f - T \ll T_f$, where b is

$$\chi(T) \cong a/T_f - b(T_f - T)^2 \qquad (3)$$

a constant. Equation (2) therefore exhibits a cusp similar to that shown in Fig. 1. The Edwards-Anderson model also predicts a singularity in the specific heat at T_f. However, this is not observed experimentally.

Occurrence of spin-glass ordering. Spin-glass ordering has been found to occur in a wide variety of disordered magnetic materials, exhibiting several distinct types of disorder. The following list is intended to be illustrative but not exhaustive. The dilute alloy Au:Fe cited above is disordered as a consequence of random placement of iron atoms in the gold lattice. For similar reasons other dilute alloys with noble-metal hosts (for example, Cu:Mn, Ag:Mn, and Au:Mn) are also spin glasses. Some of these systems show spin-glass behavior over a wide range of composition, limited only by the Kondo effect and possible ferromagnetic behavior at low and high concentrations of the magnetic ion, respectively. At low concentrations the exchange interactions of these alloys are well characterized as to their strength and distribution. Another type of metallic spin glass is one which contains a relatively high concentration of magnetic ions and is structurally amorphous. Examples of this are $GdAl_2$ and $(Fe_xMn_{1-x})_{0.75}(P_{16}B_6Al_3)_{0.25}$. In the former case the gadolinium ions are the magnetic constituent. Special techniques are required to prepare these materials in a structurally amorphous state. The second example cited is from a family of materials known as metglass and is a reentrant spin glass, as discussed below. *See* AMORPHOUS SOLID; KONDO EFFECT.

There are also a number of interesting cases of insulating spin glasses. The earliest examples reported are the cobalt and manganese aluminosilicate glasses, which are structurally amorphous, as their name implies, and also contain a sizable fraction (20–40%) of magnetic ions. Both these and the amorphous metallic systems above are rather poorly characterized as to the nature of the interactions between magnetic species.

The best-characterized and most thoroughly studied insulating spin glass is the system

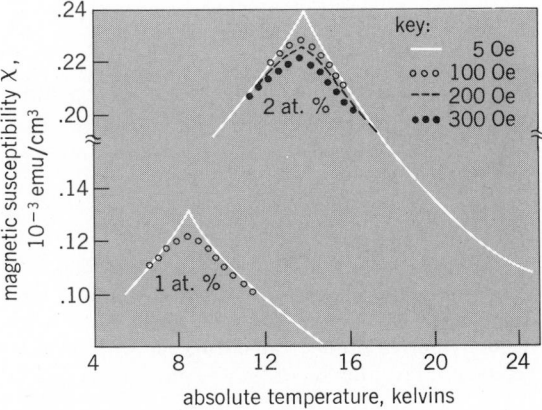

Fig. 1. Magnetic susceptibility versus absolute temperature for two different concentrations of iron alloyed with gold. The data curves indicate measurements in several applied field strengths as shown. 1 Oe = 79.6 ampere-turns/m. Magnetic susceptibility in SI units = magnetic susceptibility in emu multiplied by $4\pi = 12.57$. (*From V. Cannella and J. A. Mydosh, Magnetic ordering in gold-iron alloys, Phys. Rev., B6:4220–4237, 1972*)

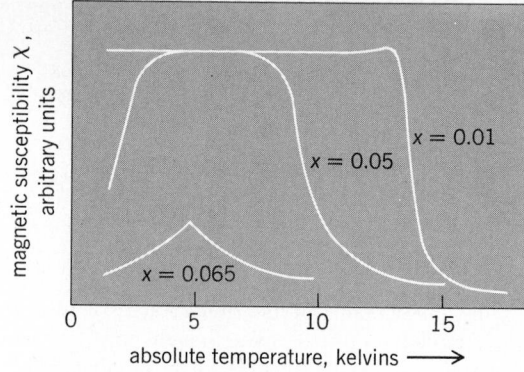

Fig. 2. Magnetic susceptibility versus absolute temperature for the compound $(Pd_{0.9965}Fe_{0.0035})_{1-x}Mn_x$ for three values of the manganese atomic fraction x. The scale for the curve marked $x = 0.065$ has been magnified four times. (*From G. J. Nieuwenhys, B. H. Verbeek, and J. A. Mydosh, Towards a uniform magnetic phase diagram for magnetic alloys with mixed types of order, J. Appl. Phys., 50:1685–1690, 1979*)

$Eu_xSr_{1-x}S$. This compound is a ferromagnet for $x = 1$ as a consequence of the ferromagnetic nearest-neighbor exchange interaction among the europium, while the original crystalline structure is retained. At sufficient dilution ($x < 0.5$) the system becomes a spin glass through the agency of antiferromagnetic second-neighbor exchange couplings. At higher europium concentrations this system also becomes reentrant.

Ferromagnetic to spin-glass transition. A number of compounds are found to form ferromagnetic order at some Curie temperature T_c, then exhibit a collapse of that order as the temperature is lowered below T_c. Such a system is termed a reentrant spin glass, because the low-temperature phase is in some sense ordered without having long-range magnetic order. An example of this behavior is shown in Fig. 2, where magnetic susceptibility versus absolute temperature is plotted for three concentrations of manganese in the compound $(Pd_{0.9965}Fe_{0.0035})_{1-x}Mn_x$. For $x = 0.01$ the compound is a ferromagnet, with χ saturated for all temperatures below 13 K. For $x = 0.05$ the compound becomes ferromagnetic at $T \sim 9K$, but the ferromagnetism collapses into spin-glass behavior for $T < 3$ K. This is the reentrant-spin-glass case. For higher manganese Mn concentrations (for example, $x = 0.065$) the ferromagnetism disappears, and one has a simple spin glass as in Fig. 1. Whether the spin-glass transition of the ferromagnetic to spin-glass transition is a true phase transition is a matter of some controversy. Reentrant-spin-glass behavior has been observed in a variety of other systems as well, including $Eu_xSr_{1-x}S$, $Au_{1-x}Fe_x$, Fe_xCr_{1-x}, and amorphous $(Fe_{1-x}Mn_x)_{0.75}(P_{16}B_6 Al_3)_{0.25}$. The first of these is a nonconductor, and the others are magnetic alloys. See FERROMAGNETISM. [R. E. WALSTEDT]

Spinor

The two-component function $\chi(x)$ which transforms linearly, according to Eq. (1), when the coordinate system is rotated in space, where x stands

$$\chi'(x') = U\chi(x) \tag{1}$$

dinate system is rotated in space, where x stands

collectively for the space coordinates $\vec{x}$ and the time coordinate t. U is a matrix with complex components, as in Eq. (2), where a and b are restricted

$$U = \begin{bmatrix} a & b \\ -b^* & a^* \end{bmatrix} \tag{2}$$

by the condition $aa^* + bb^* = 1$. The parameters a and b and their complex conjugates a^* and b^* form a set of four real numbers, often referred to as the Cayley-Klein parameters. They are chosen so that U is unitary and unimodular (that is, has determinant equal to 1), and therefore belongs to the group $SU(2)$. See CAYLEY-KLEIN PARAMETERS; GROUP THEORY; MATRIX THEORY; NONRELATIVISTIC QUANTUM THEORY.

Relativistic generalization. More generally, the term "spinor" also applies to functions used in the Dirac and other relativistic wave equations which transform linearly under a proper Lorentz transformation according to Eq. (3), where the 2×2 matrix A has the general form shown in Eq. (4). The

$$\chi'(x') = A\chi(x) \tag{3}$$

$$A = e^{i\vec{\tau} \cdot \vec{\sigma}/2} \tag{4}$$

matrices $\vec{\sigma}$ are the Pauli spin matrices shown in Eqs. (5). A is unimodular and thus belongs, in

$$\sigma_1 = \begin{bmatrix} 0 & 1 \\ 1 & 0 \end{bmatrix} \quad \sigma_2 = \begin{bmatrix} 0 & -i \\ i & 0 \end{bmatrix} \quad \sigma_3 = \begin{bmatrix} 1 & 0 \\ 0 & -1 \end{bmatrix} \tag{5}$$

general, to the group $SL(2,C)$. The three complex parameters $\vec{\tau}$ describe the particular proper Lorentz transformation. In detail, if x is taken into x' by the Lorentz transformation $a_{\mu\nu}$ according to Eq. (6) (a repeated index is to be summed upon

$$\chi'_\mu = a_{\mu\nu}x_\nu \quad \mu,\nu = 1,2,3,4 \tag{6}$$

and x_4 is it, the factor of the speed of light c being omitted), then Eq. (7) follows († means Hermitian

$$A^\dagger \sigma_\mu A = a_{\mu\nu}\sigma_\nu \tag{7}$$

conjugate), where the fourth matrix σ_4 is defined as i times the 2×2 identity matrix. The three Pauli matrices $\vec{\sigma}$ and σ_4 form a complete set of 2×2 matrices.

The parameters $\vec{\tau}$ and $a_{\mu\nu}$ can be related through Eq. (7). It is very useful to write A as in Eq. (8),

$$A = \cos\frac{\tau}{2} + i\frac{\vec{\tau}}{\tau} \cdot \vec{\sigma}\sin\frac{\tau}{2} \tag{8}$$

with τ the length of $\vec{\tau}$ such that $\tau^2 = \vec{\tau} \cdot \vec{\tau}$. This form follows from the expansion of A in powers of $i\vec{\tau} \cdot \vec{\sigma}/2$ given by Eq. (9).

$$e^{i\vec{\tau} \cdot \vec{\sigma}/2} = \sum_{n=0}^{\infty} \frac{1}{n!}(i\vec{\tau} \cdot \vec{\sigma}/2)^n \tag{9}$$

One divides the sum into even and odd values of n and uses the relation $(\vec{\sigma} \cdot \vec{\tau})^2 = \tau^2$ to simplify products, the result being Eq. (8). Then, for a pure (velocity only) Lorentz transformation with relative velocity $\vec{v}$, given by Eqs. (10), one finds from Eq. (7)

$$a_{ij} = \delta_{ij} + (\gamma - 1)\frac{v_i v_j}{v^2} \quad i,j = 1,2,3$$

$$a_{i4} = -a_{4i} = i\gamma v_i \tag{10}$$

$$a_{44} = \gamma = \frac{1}{\sqrt{1-v^2}}$$

that $\vec{\tau}$ is given by Eq. (11). A rotation through angle

$$\vec{\tau} = i\frac{\vec{v}}{v}\text{arctanh}\,(v) \qquad (11)$$

θ in the right-handed sense about an axis θ/θ is given by Eqs. (12), where ϵ_{ijk} is the Levi-Civita

$$a_{ij} = \cos\theta\delta_{ij} + (1-\cos\theta)\theta_i\theta_j/\theta^2 + \sin\theta\epsilon_{ijk}\theta_k/\theta \qquad (12)$$

$$a_{i4} = a_{4i} = 0 \qquad a_{44} = 1$$

density, defined to be zero if any two of the indices ijk are equal, and either $+1$ or -1 otherwise, according as ijk is an even or odd permutation of 123. The parameters $\vec{\tau}$ are then given by Eq. (13). In this

$$\vec{\tau} = \vec{\theta} \qquad (13)$$

case, A is also unitary and has the form of Eq. (2). *See* LORENTZ TRANSFORMATIONS; RELATIVISTIC QUANTUM THEORY; RELATIVITY; WAVE EQUATION.

Dotted and undotted indices. Components of spinors that transform under proper Lorentz transformations according to Eq. (3) with A given by Eq. (4) are by convention labeled by lower dotted indices. Thus the detailed expression of Eq. (3) is given by Eq. (14). A spinor φ with a lower-undotted

$$\chi'_\alpha(x') = (e^{i\vec{\tau}\cdot\vec{\sigma}/2})_{\alpha\beta}\chi_{\dot{\beta}}(x) \qquad \alpha,\beta = 1,2 \qquad (14)$$

index by convention transforms with the complex conjugate matrix according to Eq. (15), and so dots

$$\varphi'_\alpha(x') = (e^{i\vec{\tau}\cdot\vec{\sigma}/2})^*_{\alpha\beta}\varphi_\beta(x) \qquad (15)$$

on spinor indices indicate complex conjugation of ordinary functions (c-numbers) and Hermitian conjugation of operator functions (q-numbers). Upper index spinors are defined by employing the raising matrix $i\sigma_2$. The rule is given by Eq. (16),

$$\varphi^\alpha(x) = (i\sigma_2)_{\alpha\beta}\varphi_\beta(x) \qquad (16)$$

and φ^α has the Lorentz transformation property of Eq. (17). Since $i\sigma_2$ is a real matrix, the same rais-

$$\varphi'^\alpha(x') = (e^{i\vec{\tau}^*\cdot\vec{\sigma}/2})_{\alpha\beta}\varphi^\beta(x) \qquad (17)$$

ing rule applies to dotted spinors. Lorentz scalars, functions that are invariant to proper Lorentz transformations, are therefore bilinear functions such as $\chi_{\dot{\alpha}}\varphi^{\dot{\alpha}}$, $\chi^\beta\varphi_\beta$, and so on, made by summing over an upper and lower spinor index of the same dottedness. If only rotations are involved, then the parameters $\vec{\tau}$ are real, and the dottedness does not matter.

Multi-index spinors. Spinors with more than one index are well defined mathematically and useful physically. In fact, the connection between spinors and tensors is made with multi-index spinors. For example, a four-vector V_μ is related to a two-index, mixed-index spinor $V_{\alpha\dot{\beta}}$ by Eq. (18), and the inverse

$$V_\mu = \frac{1}{2}(\sigma_\mu)_{\alpha\beta}V_{\alpha\dot{\beta}} \qquad (18)$$

connection is given by Eq. (19). Furthermore, the

$$V^{\alpha\dot{\beta}} = -(\sigma_\mu)_{\alpha\beta}V_\mu \qquad (19)$$

tensor $T_{\mu\nu}$ with real ij and 44 components and pure imaginary $i4$ and $4i$ components for $i,j = 1,2,3$, is related to the spinor $T_{\alpha\dot{\beta}}$ and its complex conjugate by Eq. (20). (T means the transpose of a matrix.) If

$$T_{\mu\nu} = \frac{i}{4}\{(\sigma_\mu{}^T\sigma_2{}^T\sigma_\nu)_{\alpha\beta}T_{\alpha\dot{\beta}} + (\sigma_\mu\sigma_2{}^T\sigma_\nu{}^T)_{\alpha\beta}T_{\alpha\beta}\} \qquad (20)$$

$T_{\alpha\dot{\beta}}$ is symmetric ($T_{\dot{\beta}\alpha} = T_{\alpha\dot{\beta}}$), then the tensor $T_{\mu\nu}$ is antisymmetric ($T_{\nu\mu} = -T_{\mu\nu}$).

Representation of particles. Particles with integer or half-integer spin $s = 0,\frac{1}{2},1,\ldots$ may be represented by the spinor $\chi_{\dot{\alpha}_1\cdots\dot{\alpha}_{2s}}$, completely symmetric on all $2s$ spinor indices. The symmetry reduces the number of independent components to $2s+1$, the correct number for describing a spin s particle. Spin $\frac{1}{2}$ particles, the electron, neutrino, and others, may have one index representation with two independent components. The photon and other spin 1 particles have a two-index representation with three independent components. The $2s+1$ independent components for spin s are thus organized into a wave function ψ to represent that spin. Then, in place of A in Eq. (4), one has the $(2s+1)$-square matrices $e^{i\vec{\tau}\cdot\vec{s}}$ with spin s matrices $\vec{s}$, and the Lorentz transformation rule is given by Eq. (21). The matrices $e^{i\vec{\tau}\cdot\vec{s}}$ also form a representa-

$$\psi'(x') = e^{i\vec{\tau}\cdot\vec{s}}\psi(x) \qquad (21)$$

tion of the Lorentz group. This follows from a proof which uses the fact that products of such matrices depend only on the commutation relations of the spin matrices and so give the same results for higher spin as for spin $\frac{1}{2}$. To make detailed calculations, expansions analogous to Eq. (8) are available for higher spin. For example, the spin 1 expansion is given by Eq. (22), and relations analo-

$$e^{i\vec{\tau}\cdot\vec{s}} = 1 + (\cos\tau - 1)\left(\frac{\vec{\tau}}{\tau}\cdot\vec{s}\right)^2 + i\frac{\vec{\tau}}{\tau}\cdot\vec{s}\sin\tau \qquad (22)$$

gous to Eq. (7) may also be found. *See* SPIN.

Considering the inversion of space coordinates ($\vec{x}$ replaced by $-\vec{x}$), in addition to the proper Lorentz transformations, leads to the introduction of the conjugate spinor $\varphi^{\alpha_1\cdots\alpha_{2s}}$ into the physical theory via the space inversion (parity transformation) as shown in Eqs. (23). Particles with mass,

$$\chi'_{\alpha_1\cdots\alpha_{2s}}(-\vec{x},t) = \eta_p\varphi^{\alpha_1\cdots\alpha_{2s}}(\vec{x},t)$$

$$|\eta_p|^2 = 1 \qquad (23)$$

such as the electron, are usually described by two spinors, one of each type. On the other hand, particles with zero mass, such as the neutrino, are described by only one type of spinor and can cause processes in which they take part to be parity nonconserving in a maximal way. An example of such a process is the beta decay of a nucleus. *See* ELEMENTARY PARTICLE; PARITY.

[DAVID L. WEAVER]

Bibliography: C. Itzykson and J. B. Zuber, *Quantum Field Theory*, 1980; L. D. Landau and E. M. Lifshitz, *Quantum Mechanics*: *Nonrelativistic Theory*, 3d ed., 1977.

SQUID

A device which, in its original form, consists of two Josephson tunnel junctions connected in parallel on a superconducting loop (Fig. 1). The term is an acronym for superconducting quantum interference device. A small applied current I flows

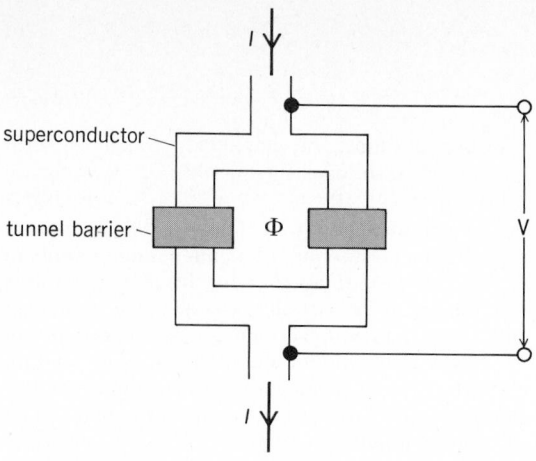

Fig. 1. Direct-current SQUID with enclosed magnetic flux Φ.

through the junctions as a supercurrent, without developing a voltage, by means of Cooper pairs tunneling through the barriers. However, when the applied current exceeds a certain critical value I_c, a voltage V is generated. As shown in Fig. 2, the value of I_c is an oscillatory function of the magnetic flux Φ threading the loop, with a period of one flux quantum, $\Phi_0 = h/2e \approx 2.07 \times 10^{-15}$ weber. The oscillations arise from the interference of the two waves describing the Cooper pairs at the two junctions, in a way that is closely analogous to the interference between two coherent electromagnetic waves. Thus the SQUID is often also called an interferometer. *See* INTERFERENCE OF WAVES; JOSEPHSON EFFECT; SUPERCONDUCTIVITY.

The SQUID has important device applications. When each Josephson tunnel junction is shunted with an external resistance to eliminate hysteresis on the current-voltage characteristic and the SQUID is biased with a constant current greater than the critical value I_c, the voltage across the SQUID is also an oscillatory function of Φ. If one measures the change in voltage produced by the application of a flux equivalent to a small fraction of one flux quantum, one has a very sensitive magnetometer. Since the device in this mode operates with a constant bias current, it is usually referred to as the dc SQUID. Another important potential application is as a logic element or memory cell in high-speed computers. When an unshunted (hysteretic) SQUID is appropriately current-biased, the application of a flux pulse switches it from the zero-voltage to the nonzero-voltage state, a

function that can be used to perform logic; three-junction SQUIDs are also used for this purpose. The SQUID can be used as a dissipation-free memory cell to store a 1 or 0 as a clockwise or anticlockwise circulating persistent supercurrent.

The rf SQUID consists of a single junction interrupting a superconducting loop. It can be operated as a magnetometer by coupling it to the inductor of an LC-tank circuit excited at its resonant frequency by a rf current. The rf voltage across the tank circuit oscillates as a function of the magnetic flux in the loop, with period Φ. The rf SQUID is in fact misnamed, since no interference takes place.

[JOHN CLARKE]

Bibliography: R. C. Jaklevic et al., Quantum interference effects in Josephson tunneling, *Phys. Rev. Lett.*, 12:159–160, 1964; B. D. Josephson, Possible new effects in superconductive tunneling, *Phys. Lett.*, 1:251–253, 1962.

Standing wave

A disturbance which is oscillatory in time and which has an amplitude that varies in space between zero and a maximum value. Standing waves may be formed, for example, near an ideal boundary by the interaction of incident and perfectly reflected traveling waves. The points in space where the amplitudes are zero are called nodes. Antinodes are points where amplitude is a maximum.

For the case of sinusoidal waves, the equations of the standing waves assume a particularly simple form. Consider the two waves q_1 and q_2 of amplitude A given by Eqs. (1) and (2). Here k is the wave

$$q_1 = A \cos k(ct - x) \tag{1}$$

$$q_2 = A \cos k(ct + x) \tag{2}$$

number, c wave velocity, t time coordinate, and x space coordinate. The wave q_1 is traveling in the $+x$ direction, q_2 in the $-x$ direction. The sum of these two waves is then given by Eq. (3). Thus the

$$\begin{aligned} q_1 + q_2 &= A[\cos k(ct - x) + \cos k(ct + x)] \\ &= 2A \cos kx \cos kct \end{aligned} \tag{3}$$

amplitude at any position is the absolute value (without regard to algebraic sign) of $2A \cos kx$. For positions x such that $kx = (n + 1/2)\pi$, where n is an integer, the amplitude is zero; for $kx = n\pi$ the amplitude is a maximum. These positions of x are, respectively, the nodal and antinodal points in the standing wave.

Standing waves are a limiting case of stationary waves. These occur when at least one of the enclosure terminations absorbs a part of the energy of the incident waves, as well as reflecting a portion, resulting in a net power loss from the source.

Acoustic waves can form stationary waves in closed systems with the proper dimensions and boundary conditions. The ratio of the maximum to the minimum effective sound pressure, called the standing-wave ratio, can be used to measure the acoustic impedance of the system. Electromagnetic waves can form stationary waves in waveguides and transmission lines, and the standing-wave ratio can be used to determine the impedance of the waveguide or transmission line. *See* WAVE; WAVE EQUATION; WAVE MOTION.

[WILLIAM J. GALLOWAY]

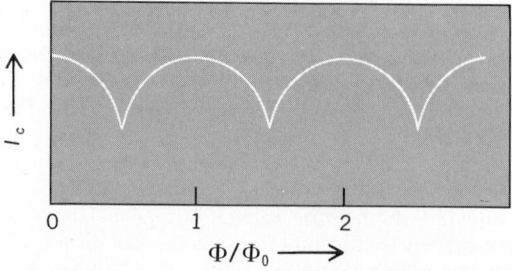

Fig. 2. Graph of maximum supercurrent I_c versus Φ/Φ_0 for a dc SQUID.

Stark effect

The effect of an electric field on spectrum lines. The electric field may be externally applied, but in many cases it is an internal field caused by the presence of neighboring ions or atoms in a gas, liquid, or solid. Discovered in 1913 by J. Stark, the effect is most easily studied in the spectra of hydrogen and helium, by observing the light from the cathode dark space of an electric discharge. Because of the large potential drop across this region, the lines are split into several components. For observation perpendicular to the field, the light of these components is linearly polarized. The splitting can be easily resolved with a spectrograph of moderate dispersion, amounting in the case of the H_γ line at 434 nm to a total spread of 3.2 nm for a field of 104,000 volts/cm.

Linear Stark effect. This effect exhibits large, nearly symmetrical patterns. Examples are shown in the illustration, where the symbols π and σ refer to the two states of polarization. The interpretation of the linear Stark effect was one of the first successes of the quantum theory. According to this theory, the effect of the electric field on the electron orbit is to split each energy level of the principal quantum number n into $2n - 1$ equidistant levels, of separation proportional to the field strength. Thus the higher members of a series show a larger number of components and greater overall splittings. The criterion for the occurrence of the linear Stark effect is that the splitting of the levels shall be large compared to the natural separation of levels of the same n but different L, where L is the quantum number of orbital angular momentum. *See* QUANTUM NUMBERS; ZEEMAN EFFECT.

Quadratic Stark effect. This occurs in lines resulting from the lower energy states of many-electron atoms. Here the large separation of states of different L results from the penetration of the valence electrons into the core of other electrons, with the result that the permanent dipole moment associated with hydrogenlike orbits no longer exists. There is, however, a small induced dipole moment due to polarization of the atom. This moment is proportional to the electric field strength, and since the energy change is proportional to the product of the dipole moment and the field strength, the energy levels shift by an amount depending on the square of the field. Thus all levels have shifts of the same sign and therefore are displaced to lower energies. Each field-free level is also split, as a result of the space quantization of the angular momentum vector J, so that there are $J + 1$ components if J is integral, or $J + 1/2$ components if it is half-integral. Because the lower levels are usually less displaced than the higher ones, the quadratic Stark effect ordinarily shows itself as a shift of the lines toward the red end of the spectrum, with an accompanying separation into several components. The quadratic Stark effect is basic to the explanation of the formation of molecules from atoms, of dielectric constants, and of the broadening of spectral lines.

Intermolecular Stark effect. Produced by the action of the electric field from surrounding atoms or ions on the emitting atom, the intermolecular effect causes a shifting and broadening of spectrum lines. The molecules being in motion, these fields are inhomogeneous in space and also in time. Hence the line is not split into resolved components but is merely widened. Particularly in the electric discharge through gases with high currents, the large ion density may cause very wide lines. The amount of the broadening is found to run parallel to the sensitivity of the line to the Stark effect and thus is greatest for those lines susceptible to the linear effect.

Inverse Stark effect. This is the effect as observed with absorption lines. It has been detected, for example, by applying an electric field to potassium vapor, and measuring a small displacement of the absorption lines toward the red. The displacements are found to be proportional to the square of the field strength, as in the quadratic Stark effect. In addition, certain lines of large n, which are normally forbidden by the selection rule for L, are observed to appear in the presence of the field. This type of transition is said to be produced by "forced dipole radiation." Such forbidden lines also may be obtained in emission. In the illustration, the lines $2^1P - 6^1S$ and $2^1P - 4^1F$ are examples of forced dipole transitions.

[F. A. JENKINS/W. W. WATSON]

Bibliography: G. Herzberg, *Atomic Spectra and Atomic Structure*, 2d ed., 1944; F. K. Richtmyer, E. H. Kennard, and J. W. Cooper, *Introduction to Modern Physics*, 6th ed., 1969; M. R. Wehr, J. A. Richards, and T. W. Adair, *Physics of the Atom*, 3d ed., 1978.

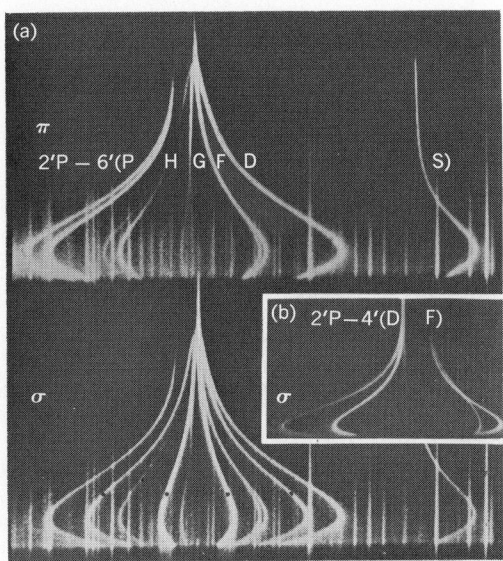

← direction of increasing field strength

Stark effect for helium lines. (*a*) The 414.4- and 416.9-nm lines at field strengths 0–85,000 volts/cm. Note the disappearance of $^1P - ^1H$ line at a certain field strength, and symmetry of the σ components (those with polarization perpendicular to the field), indicated by dots. (*b*) The 492.2-nm line at 0–40,000 volts/cm. Note the crossing-over of two $^1P - ^1F$ components. (*J. S. Foster and C. Foster, McGill University*)

Static electricity

The study of electric charges at rest and the fields they produce. The fundamental fact of static electricity (or electrostatics) is that similarly electrified bodies repel each other, whereas oppositely electrified bodies attract each other. Coulomb's law of force is the basic quantitative law of electrostatics. *See* COULOMB'S LAW; ELECTRIC CHARGE; ELECTROSTATICS.

[RALPH P. WINCH]

Statics

The branch of mechanics that describes bodies which are acted upon by balanced forces and torques so that they remain at rest or in uniform motion. This includes point particles, rigid bodies, fluids, and deformable solids in general. Static point particles, however, are not very interesting, and special branches of mechanics are devoted to fluids and deformable solids. For example, hydrostatics is the study of static fluids, and elasticity and plasticity are two branches devoted to deformable bodies. Therefore this article will be limited to the discussion of the statics of rigid bodies in two- and three-space dimensions. *See* ELASTICITY; HYDROSTATICS; MECHANICS.

Mechanics is the study of motions in terms of mass, length, time, and forces. In statics the bodies being studied are in equilibrium. Positions of points in space, velocities, forces, and torques are all vector quantities, since each has direction, magnitude, and units. Beams, bridges, machine parts, and so on are not really rigid, but whenever the largest change in length of a portion of a body Δl is much smaller than that length l, that is, $l \gg \Delta l$, then it is a satisfactory approximation to treat that body as a rigid body. Many of the objects used in architecture, engineering, and physics can be satisfactorily idealized as rigid bodies, and much of building, machine, bridge, and dam design is based upon the study of statics.

The equilibrium conditions are very similar in the planar, or two-dimensional, and the three-dimensional rigid body statics. These are the vector sum of all forces acting upon the body must be zero; and the resultant of all torques about any point must be zero. Thus it is necessary to understand the vector sums of forces and torques.

Force as a vector. The physical effect of a force is a push or pull on an object at its point of application. The effect of a force is to change the velocity of the body. The SI units of force are newtons (1 N = 1 kg-m-s^{-2}). The vector nature of forces is expressed by their direction in space. This direction is the direction of the push or pull exerted by the force.

The notation for a vector quantity is an overarrow, for example, $\vec{F}$ or $\vec{P}$. In Fig. 1 a two-dimensional vector $\vec{P}$ is shown as a directed line segment $\overline{AB}$ in the plane of the paper. A three-dimensional vector $\vec{P}$ is shown in Fig. 2 as the directed line segment $\overline{OD}$. A complete description of a vector is

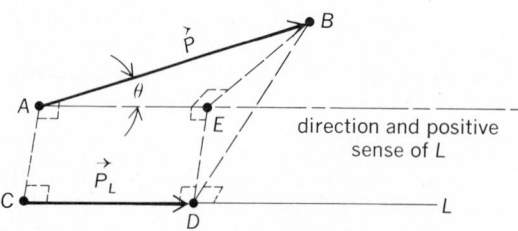

Fig. 1.   Directed line *AB* or $\vec{P}$ is a vector quantity in the plane.

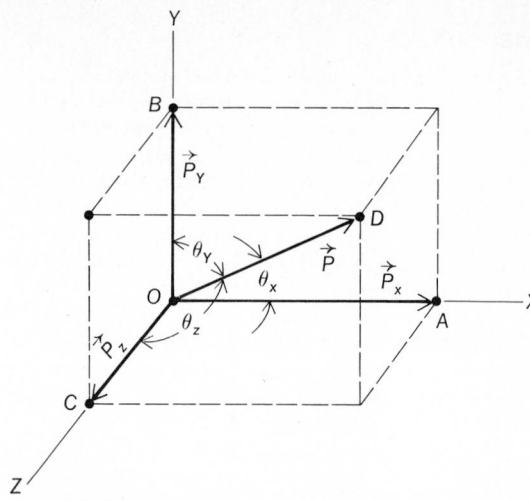

Fig. 2.   Rectangular components of a force vector $\vec{P}$ in three-dimensional space.

obtained by specifying a coordinate system and giving its components in that coordinate system. The sum of any two vectors $\vec{A} + \vec{B}$ is the sum of their components, so that $\vec{A}$ and $\vec{B}$ must have the same number of components.

Superposition and transmissibility. In studying statics problems (and general mechanics problems), two principles, superposition and transmissibility, are used repeatedly on force vectors. They are applicable to all vectors, but specifically to forces and torques. (first moments of forces).

1. The principle of superposition of d-dimensional vectors is that the sum of any two d-vectors is another d-vector. Of course, some or all components can be zero. This principle is illustrated in Fig. 3. The two vectors labeled Q, which can be shown applied to a rigid structural member D, add up to zero, and so do the two labeled R. Superposition applies to all sums of vectors not just sums which vanish.

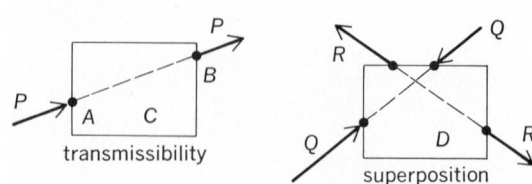

Fig. 3.   Illustration of two theorems of statics.

2. The principle of transmissibility of a force applied to a rigid body is that the same mechanical effect is produced by any shift of the application of the force. The principle of transmissibility is illustrated in Fig. 3. The solid rectangle C represents a solid structural member and the line AB is the line of action for the equivalent forces P. Since these two forces have the same magnitudes and direction and are applied along the same line, by transmissibility, all of their mechanical effects are equivalent. To use the superposition principle to add two vectors, the principle of transmissibility is used to move some vectors along their line of action in order to add to their components.

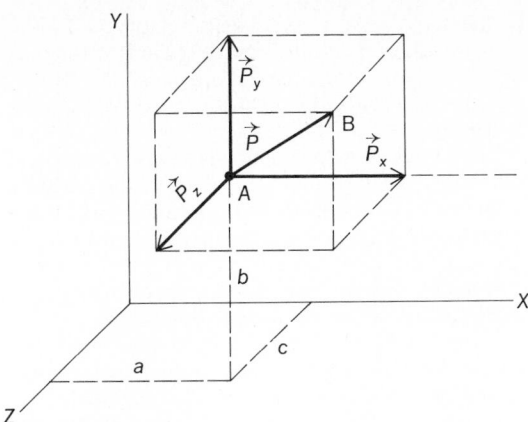

Fig. 4. Moments about axis in rectangular coordinates.

Components of a force. Figure 1, the two-dimensional example, shows the construction by which orthogonal force components are defined. In this figure $\vec{P}$ or $\overline{AB}$ is a force vector, and L a directed line whose positive sense is toward its labeled end. Construction lines AC and BD are in planes (not shown) normal to L, and θ is the direction angle of $\vec{P}$ relative to L; it is a plane angle between the directions in the positive senses of $\vec{P}$ and L; further, $0 \leq \theta \leq 180°$.

The orthogonal vector component of force $\vec{P}$ on directed line L is a force of direction and magnitude given by P_L or CD where P_L is in the direction of L. Its magnitude is given by Eq. (1).

$$P_L = \overline{CD} = \overline{AE} = \overline{AB}|\cos\theta| = P|\cos\theta| \qquad (1)$$

The component of $\vec{P}$ on L is $P_L = P\cos\theta$, where P_L is positive if $0° \geq \theta < 90°$ and negative if $90° < \theta \leq 180°$. The absolute magnitude of P_L is designated $|P_L|$.

The rectangular components of a force are its components on mutually perpendicular lines.

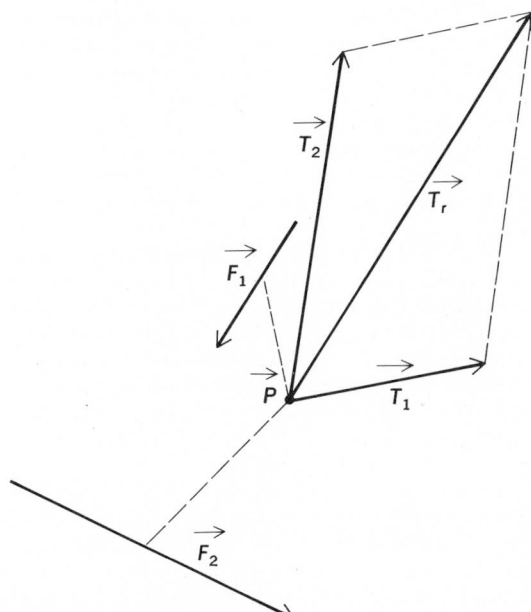

Fig. 5. Resultant of two torques.

In Fig. 2 $\vec{P}_X$ or $\overline{OA}$, $\vec{P}_Y$ or $\overline{OB}$, and $\vec{P}_Z$ or $\overline{OC}$ are the rectangular vector components of $\vec{P}$ or $\overline{OD}$ in the directions of lines (axes) X, Y, and Z, respectively.

The corresponding components have the magnitudes $P_X = P\cos\theta_X$, $P_Y = P\cos\theta_Y$, and $P_Z = P\cos\theta_Z$, and are in the X, Y, Z directions, respectively. The magnitude of the three-dimensional vector $\vec{P}$ is given by Eq. (2).

$$P = \sqrt{P_X^2 + P_Y^2 + P_Z^2} \qquad (2)$$

Moment of a force. The moment of a force about a directed line is a signed number whose value can be obtained by applying these two rules:

1. The moment of a force about a line parallel to the force is zero.

2. The moment of a force about a line normal to a plane containing the force is the product of the magnitude of the force and the least distance from the line to the line of the force. Conventionally, the moment is positive if the force points

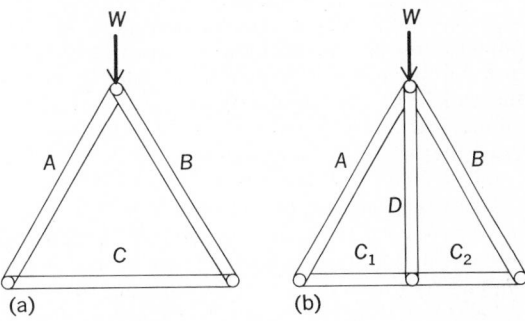

Fig. 6. Statically determinate and indeterminate structures. (a) The forces in members A and B are independent of their strengths (for small deformations). (b) The relative strength of the added member D will determine how much of the load W it supports.

counterclockwise about the line as viewed from the positive end of the line.

In Fig. 4 the moments of $\vec{P}$ about the X, Y, and Z coordinate axes are $M_X = bP_Z - cP_Y$ about the X axis; $M_Y = cP_X = aP_Z$ about the Y axis; and $M_Z = aP_Y - bP_X$ about the Z axis. The resultant of the two torques or moments of force, $\vec{T}_1$ and $\vec{T}_2$ about a point P is shown as $\vec{T}_r$ in Fig. 5. The components of these torques are due to forces $\vec{F}_1$ and $\vec{F}_2$ oriented with respect to $\vec{P}$ as shown.

Figure 6 illustrates cases where the forces in a structure may or may not depend on the relative strength of its parts. Case 6a is called statically determinate, and case 6b statically indeterminate. The terms are actually misnomers, for the forces can be found in both cases. Better terminology would be rigidly determinable and indeterminable. In case 6b, if each member were rigid (infinitely strong), the forces in each could not be determined.

In summary, the statics of rigid bodies in statically determinate structures is carried out by summing vector forces and vector torques and setting the resultants equal to zero. *See* FORCE; TORQUE.

[BRIAN DE FACIO]

Bibliography: F. J. Beer and E. Russell Johnston, *Mechanics for Engineers*: *Statics*, S. Timoshenko and D. H. Young, *Engineering Mechanics*: *Statics*, 1937.

Stationary state

In quantum mechanics, an energy state for which the probability of any observation is independent of time, that is, stationary. Because stationary states endure in time, they conveniently characterize physical systems. For example, at any instant, it is customary and useful to think that each atom in a discharge tube is in one of its stationary states, with most atoms in the ground state, and the radiation from the tube a measure of the populations of the excited states.

A stationary state may be bound or unbound; bound states are localized, unbound states are not. Usually, the potential energy of the particles composing the system is zero at infinite separation, in which event the energies of bound states are negative, of unbound states, nonnegative. For instance, in the physical system consisting of a proton and an electron, the bound states are the negative energy states of neutral atomic hydrogen, the unbound states are the states of the hydrogen ion H^+ in which the electron no longer is found mainly in the vicinity of the proton. Whether bound or unbound, stationary states are states of definite energy; that is, when several atoms are in the same stationary state, their measured energies always will be equal. This property need not hold for other observables; that is, the measured momentum (relative to the proton) of an electron in the ground state of hydrogen does not always have the same value. *See* ENERGY LEVEL; EXCITED STATE; GROUND STATE; NONRELATIVISTIC QUANTUM THEORY; QUANTUM MECHANICS.

[EDWARD GERJUOY]

Statistical mechanics

That branch of physics which endeavors to explain the macroscopic properties of a system on the basis of the properties of the microscopic constituents of the system. Usually the number of constituents is very large. All the characteristics of the constituents and their interactions are presumed known; it is the task of statistical mechanics (often called statistical physics) to deduce from this information the behavior of the system as a whole.

SCOPE OF STATISTICAL MECHANICS

Elements of statistical mechanical methods are present in many widely separated areas in physics. For instance, in the classical Boltzmann problem one attempts to explain the thermodynamic behavior of gases on the basis of classical mechanics applied to the system of molecules. Historically, this was the first systematic investigation of a statistical problem, and many of the procedures and methods originate from this investigation. It is important to realize that statistical mechanics gives more than just an explanation of already known phenomena. By using statistical methods, it often becomes possible to obtain expressions for empirically observed parameters, such as viscosity coefficients, heat conduction coefficients, and virial coefficients, in terms of the forces between molecules. This kind of result, a direct relation between an observed macroscopic entity and an intermolecular potential, constitutes one of the main achievements of statistical physics. *See* BOLTZMANN STATISTICS; INTERMOLECULAR FORCES; KINETIC THEORY OF MATTER.

Statistical considerations also play a significant role in the description of the electric and magnetic properties of materials. In this case one would hope to deduce such characteristics as the dielectric constant, electrical conductivity, and magnetic permeability from the known properties of the atom. In this connection one should observe that Maxwell's equations are macroscopic equations, for matter in the bulk. To obtain the Maxwell equations from the Lorentz (electron) equations is typically a statistical question; new information which can be obtained from this approach is, for example, the temperature dependence of the dielectric constant.

If the problem of molecular structure is attacked by statistical methods, the contributions of internal rotation and vibration to thermodynamic properties, such as heat capacity and entropy, can be calculated for models of various proposed structures. Comparison with the known properties often permits the selection of the correct molecular structure. The statistical description of the activated complex in reaction mechanisms has enabled chemists to formulate a theory of absolute reaction rates.

Perhaps the most dramatic examples of phenomena requiring statistical treatment are the cooperative phenomena or phase transitions. In these processes, such as the condensation of a gas, the transition from a paramagnetic to a ferromagnetic state, or the change from one crystallographic form to another, a sudden and marked change of the whole system takes place. The appropriate description of such processes is one of the most difficult but most interesting problems of statistical physics.

Statistical considerations of quite a different kind occur in the discussion of problems such as the diffusion of neutrons through matter. In this case, one knows the probability of the various events which affect the neutron, such as the capture probability and scattering cross section. The problem here is to describe the physical situation after a large number of these individual events. The procedures used in the solution of these problems are very similar to, and in some instances taken over from, kinetic considerations. Similar problems occur in the theory of cosmic-ray showers. Here the probabilities of the basic processes—pair production, annihilation, ionization, meson production—are all presumed known. The major task is the computation of the combined effect of a large number of these individual events. Special techniques of statistical physics, the use of distribution functions and transport-type equations, find extensive applications in these areas. *See* BOLTZMANN TRANSPORT EQUATION.

It happens in both low-energy and high-energy nuclear physics that a considerable amount of energy is suddenly liberated. An incident particle may be captured by a nucleus, or a high-energy proton may collide with another proton. In either case, there is a large number of ways (a large number of degrees of freedom) in which this energy may be utilized. In the nuclear case, there are usually many decay and excitation modes; in the case of the proton-proton collision, there is enough energy available for the creation of a number of mesons. To survey the resulting processes, one can again invoke statistical considerations. The statistical problem here is to calculate the probability that, given a total amount of energy, a particular process will occur. Of course, the statistical method cannot help in the investigation of the actual mechanisms of these processes. However, the statistical factors must be considered in the interpretation of experiments. *See* SCATTERING EXPERIMENTS (NUCLEI).

Of considerable importance in statistical physics are the random processes, also called stochastic processes or sometimes fluctuation phenomena. The Brownian motion, the motion of a particle moving in an irregular manner under the influence of molecular bombardment, affords a typical example. The process may be described in terms of a fluctuating force acting on a particle, perhaps in addition to a systematic force. The interest in this problem centers around a calculation of the position and velocity of a particle after it has experienced many collisions or after the fluctuating force has acted for a long time. The stochastic processes are in a sense intermediate between purely statistical processes, where the existence of fluctuations may safely be neglected, and the purely atomistic phenomena, where each particle requires its individual description. All statistical considerations involve, directly or indirectly, ideas from the theory of probability of widely different levels of sophistication. The use of probability notions is, in fact, the distinguishing feature of all statistical considerations. *See* BROWNIAN MOVEMENT; PROBABILITY; PROBABILITY (PHYSICS); STATISTICS; STOCHASTIC PROCESS.

METHODS OF STATISTICAL MECHANICS

In the following sections procedures used in statistical mechanics are discussed.

Phase space. Consider a system of N particles, each of mass m, contained in a volume V. Call the positions of the particles $x_1, y_1, z_1, \ldots, x_N, y_N, z_N$, their cartesian velocities $v_x, \ldots, v_{z_N}$, and their momenta $p_{x_1}, \ldots, p_{z_N}$. The simplest statistical description concentrates on a discussion of the distribution function $f(x,y,z;v_x,v_y,v_z;t)$. The quantity $f(x,y,z;v_x,v_y,v_z;t)$ $(dxdydzdv_xdv_ydv_z)$ gives the (probable) number of particles of the system in those positional and velocity ranges where x lies between x and $x + dx$; v_x between v_x and $v_x + dv_x$, and so on. These ranges are finite. The six-dimensional space defined by x, y, z, v_x, v_y, v_z is called the μ space, and the behavior of the gas is represented geometrically by the motion of N points in the μ space. It was shown by L. Boltzmann that in the course of time f approaches an equilibrium distribution f^0 as in Eq. (1). Here $d^3x = dxdydz$, A and

$$f^0(\mathbf{x},\mathbf{v})\, d^3x d^3v$$
$$= A \exp\left[-\tfrac{1}{2}\beta m v^2 - \beta U(x,y,z) \right] d^3x d^3v \quad (1)$$

β are parameters, and $U(x,y,z)$ is the potential energy at the point x, y, z. Boltzmann also could interpret Eq. (1) as the most probable distribution.

Actually, important as the use of the distribution function is in practice, there are in principle serious limitations and difficulties associated with it. The limitations refer to the fact that the description by means of a single distribution function is correct only for noninteracting particles; also, the neglect of triple and higher collisions restricts the applicability to very dilute systems. The indiscriminate use of the ideas of Boltzmann leads to paradoxical results.

In the further study of these questions, the phase space of a dynamical system plays an important role. Consider the $6N$-dimensional euclidean space, whose $6N$ axes are given in notation (2). The

$$x_1, y_1, z_1, \ldots, x_N, y_N, z_N; p_{x_1}, \ldots, p_{z_N} \quad (2)$$

state of the system at a given time is completely specified by these $6N$ numbers. These numbers define a point (the representative point, or phase point) in the $6N$-dimensional phase space. In the course of time the variables $x_1, \ldots, p_{z_N}$ change, hence the phase point moves. If the system is conservative, Eq. (3) defines, for the energy E, a

$$\sum_i \frac{\mathbf{p}_i^2}{2m_i} + U(x_1, \ldots, z_N) = E \quad (3)$$

$(6N-1)$-dimensional surface in the phase space called the energy surface. For a conservative system, the phase point in the course of time wanders over the energy surface. Intuitive as this geometrical representation of an involved mechanical system might seem, it is well to keep in mind that the actual trajectory of the phase point is extremely complicated. Also, the use of the phase space does not by itself imply any statistical considerations. A knowledge of the trajectory is precisely equivalent to a solution of the complete mechanical problem, for one would know each p and x as a function of time. It should be clear, however, that many different microscopic states all correspond to the same macroscopic situation.

Observations made on a system always require a finite time; during this time the microscopic details of the system will generally change considerably as the phase point moves. The result of a measurement of a quantity Q will therefore yield the time average, as in Eq. (4). The integral is along the

$$\overline{Q_t} = \frac{1}{t} \int_0^t Q\, dt \quad (4)$$

trajectory in phase space; Q depends on the variables $x_1, \ldots, p_{z_N}$ and t. To evaluate the integral, one must know the trajectory, which, as already mentioned, requires the solution of the complete mechanical problem.

For many years one of the major problems in statistical mechanics was precisely to recast the expression for a time average in a more tractable form. Generally one attempted to replace time averages by phase space averages, by trying to show that the time spent in a region was proportional to the size of the region. The legitimacy of this procedure was studied extensively by mathematicians, but although these studies opened up significant areas in mathematics, they did not really help elucidate problems of physical interest.

To illustrate the subtlety of the problems involved, attention may be drawn to the so-called ergodic theorem of G. D. Birkhoff, which asserts that the limit $\lim\limits_{t \to \infty} \overline{Q}_t$ exists for almost all trajectories, although this limit varies discontinuously from trajectory to trajectory. The Poincaré recurrence theorem, which states that a system within a finite time will return as closely as one pleases to its initial state, is of special interest in statistical mechanics. Apart from providing the mathematical framework for these theorems, the phase space provides also the appropriate framework for the formulation of ensemble theory, which is the basis of modern statistical mechanics.

Ensembles; Liouville's theorem. J. Willard Gibbs first suggested that instead of calculating a time average for a single dynamical system, one should instead consider a collection of systems, all similar to the original one. Such an ensemble of systems is to be constructed in harmony with the available knowledge of the single system, and may be represented by an assembly of points in the phase space, each point representing a single system. If, for example, one knows the energy of a system precisely, but nothing else, the appropriate representative example would be a uniform distribution of ensemble points over the energy surface, and no ensemble points elsewhere. An ensemble is characterized by a density function $\rho(x_1, \ldots, z_N; p_{x_1}, \ldots, p_{z_N}; t) \equiv \rho(x,p,t)$. The significance of this function is that the number of ensemble systems dN_e contained in the volume element $dx_1 \cdots dz_N; dp_x \cdots dp_{z_N}$ of the phase space (this volume element will be called $d\Gamma$), at time t is as given in Eq. (5).

$$\rho(x,p,t)\,d\Gamma = dN_e \qquad (5)$$

The ensemble average of any quantity Q is given by Eq. (6). The basic idea now is to replace the

$$Q_{\text{ens}} = \frac{\int Q\rho\,d\Gamma}{\int \rho\,d\Gamma} \qquad (6)$$

time average of an individual system by the ensemble average, at a fixed time, of the representative ensemble. Stated formally, one identifies $\overline{Q}_t$ defined by Eq. (4), in which no statistics is involved, with $\overline{Q}_{\text{ens}}$ defined by Eq. (6), in which probability assumptions are explicitly made. Another form of this same connection between the behavior of the individual system and the ensemble is that the probability that the individual system at time t will be in a region R of the phase space is given by the fraction of ensemble systems contained in R at time t, Eq. (7).

$$P(R,t) = \frac{\int_R \rho\,d\Gamma}{\int \rho\,d\Gamma} \qquad (7)$$

It is clear, on the basis of these relations, that the complete statistical behavior of a system is known once its representative ensemble has been obtained. In the construction of such ensembles, one always assumes that accessible parts of the phase space should be weighted equally. There is one other general requirement that the density function must satisfy. Inasmuch as the number of ensemble members remains constant (no ensemble members are created or destroyed in the course of time), there is a continuity equation, Eq. (8), for the

$$\frac{\partial \rho}{\partial t} + \sum_{i=1}^{3N}\left[\frac{\partial}{\partial x_i}\left(\rho\,\frac{dx_i}{dt}\right) + \frac{\partial}{\partial p_i}\left(\rho\,\frac{dp_i}{dt}\right)\right] = 0 \qquad (8)$$

density function ρ which simply states that the change in the number of systems in a volume of phase space per unit time equals the difference of the number of systems flowing in and flowing out of that volume. If one now expresses the time derivatives dx_i/dt and dp_i/dt through the Hamiltonian equations, one obtains Eq. (9). Here $d\rho/dt$ is the

$$\frac{\partial \rho}{\partial t} + \sum_{i=1}^{3N}\left(\frac{\partial \rho}{\partial x_i}\frac{dx_i}{dt} + \frac{\partial \rho}{\partial p_i}\frac{dp_i}{dt}\right) \equiv \frac{d\rho}{dt} = 0 \qquad (9)$$

substantial, or total derivative. *See* HAMILTON'S EQUATIONS OF MOTION.

Equation (9), called the Liouville theorem, asserts that the time rate of change of ρ is zero along a streamline in phase space. One can also deduce from Eq. (9) that the ensemble systems move in the course of time in a volume-preserving fashion through the phase space. This means that if at some initial time one considers a set of ensemble points which occupy a volume V_0, then in the course of time each representative point will move in a complicated fashion. At time t one can consider the volume made up by the ensemble points which initially were in V_0. This volume V_t equals V_0. The shape of the volume changes, but the actual volume is constant. In the attempts which have been made to justify the postulate that $\overline{Q}_t = \overline{Q}_{\text{ens}}$, this volume-preserving property plays a dominant role. The proof of the Poincaré recurrence theorem also depends crucially on this property. Yet in spite of the general significance of the Liouville theorem, Eq. (9) generally does not help in the task of setting up an appropriate representative ensemble. (See, however, the subsequent discussion on nonequilibrium theory of liquids.) The choice of the ensemble is usually determined by physical considerations and by the already mentioned postulate of equal a priori probabilities in phase space. Perhaps most important is the a posteriori justification of the statistical procedures, obtained through a comparison with the experimental facts. It is important that inasmuch as the methods are statistical, one can compute not only average values but also fluctuations around these average values. It turns out that usually these fluctuations are extremely small; however, in those instances in which they are appreciable, they can be compared with experiments, providing additional support for the statistical procedures. Hence, despite the fact that no completely rigorous mathematical justification of the ensemble methods exists, these methods are physically so plausible and lead to such agreement with experiments for a variety of systems that they reasonably must be considered as a well-established part of statistical physics.

Relation to thermodynamics. It is certainly reasonable to assume that the appropriate ensemble for a thermodynamic equilibrium state must be described by a density function which is independent of the time, since all the macroscopic averages which are to be computed as ensemble averages are time-independent. Thus one has an equilibrium ensemble when $\partial\rho/\partial t = 0$. In that case it follows from Liouville's theorem that Eq. (10) holds. It is

$$\sum_{i=1}^{3N}\left(\frac{\partial \rho}{\partial x_i}\frac{dx_i}{dt} + \frac{\partial \rho}{\partial p_i}\frac{dp_i}{dt}\right) = 0 \qquad (10)$$

easy to see that if ρ is a function of a quantity α (which is a function of x and p and which is a constant of the motion, such as the energy, so that $\partial\alpha/\partial t = 0$), then any $\rho(\alpha)$ satisfies Eq. (10). Hence any density function ρ which is a function of x and p through the dependence of the energy E on p and x satisfies the Liouville equation, Eq. (9). The functional form of $\rho(E)$ is left completely unspecified. If one deals with an isolated system, where the energy is specified within a well-defined range, the most obvious example to use would be the so-called microcanonical ensemble defined by Eq. (11a), where c is a constant, for the energy E between E_0 and $E_0 + \Delta E$; for other energies Eq. (11b) holds. By using Eq. (6), one may calculate any

$$\rho(p,x) = c \qquad (11a)$$

$$\rho(p,x) = 0 \qquad (11b)$$

microcanonical average. The calculations, which involve integrations over volumes bounded by two energy surfaces, are not trivial. Still, one may obtain in this way many of the results of classical Boltzmann statistics.

For applications and for the interpretation of thermodynamics, the canonical ensemble is much more preferable. In this case one describes a system which is not isolated but which is in thermal contact with a heat reservoir. By describing the complete system (system plus reservoir) by a microcanonical ensemble, it may be shown that the system itself may be represented by Eq. (12a). Here N_e is the total number of ensemble systems, $E(x,p)$ is the mechanical energy of the system, and ψ and θ are parameters independent of x and p characterizing the canonical ensemble. From the fact that Eq. (12b) is valid, one deduces that the

$$\rho(x,p) = N_e \exp\left[\frac{\psi - E(x,p)}{\theta}\right] \qquad (12a)$$

$$\int \rho(x,p)\, d\Gamma = N_e \qquad (12b)$$

two parameters ψ and θ are not independent, but that Eq. (13) applies. The quantity Z is usually

$$e^{-(\psi/\theta)} = \int d\Gamma \exp\left[-\frac{E(x,p)}{\theta}\right] \equiv Z \qquad (13)$$

called the partition function. It is customary to define Z for a system of N identical particles with a different multiplicative constant by Eq. (14). Here

$$Z' = \frac{1}{N! h^{3N}} \int \exp\left[-\frac{E(p,x)}{\theta}\right] d\Gamma \qquad (14)$$

h is the Planck constant. It is now very important that the parameters θ and ψ be identified with definite thermodynamic functions. In this connection, consider some of the thermodynamic relations. Call the entropy η, the thermodynamic energy (internal energy) ϵ, and the free energy (Helmholtz free energy) ψ'. Then some examples of thermodynamic relations are, for gases, described by pressure, volume, and temperature (P, V, T) as in Eqs. (15). *See* THERMODYNAMIC PRINCIPLES.

$$\psi' = \epsilon - T\eta \qquad (15a)$$

$$d\psi' = -\eta\, dT - P\, dV \qquad (15b)$$

$$\frac{\partial\psi'}{\partial T} = -\eta \qquad \frac{\partial\psi'}{\partial V} = -P \qquad (15c)$$

$$\epsilon = \psi' - T(\partial\psi'/\partial T) \qquad (15d)$$

From Eqs. (12a), (6), and (13) one may calculate

the average energy $\bar{E}$, Eq. (16). For an ideal gas, in

$$\bar{E} = \theta^2 \frac{\partial}{\partial\theta} \log Z \qquad (16)$$

which the energy of a molecule is $\mathbf{p}_i{}^2/2m$, Eq. (17)

$$Z = \int \cdots \int dx_1 \ldots dp_{zN} \exp\left(-\frac{1}{2m\theta}\sum_{i=1}^N \mathbf{p}_i{}^2\right)$$

$$= V^N (2\pi m\theta)^{3N/2} \qquad (17)$$

holds. (The evaluation of the integrals is elementary; the space integrations merely contribute V^N, and the momentum integrals are Gaussian.) One then obtains Eq. (18a) as the energy of an ideal gas, having N molecules in a vessel of any volume V. Now one knows by experiment that the thermodynamic energy of 1 mole of an ideal gas is given by Eq. (18b). Here R is the ideal gas con-

$$\bar{E} = \frac{3N}{2}\theta \qquad (18a)$$

$$\epsilon = \tfrac{3}{2}NkT \equiv \tfrac{3}{2}RT \qquad (18b)$$

stant, T is the absolute temperature, N is Avogadro's number, and k is the Boltzmann constant. Comparison of Eqs. (18a) and (18b) leads to the identification of $\theta = kT$. It is also easy to deduce the Maxwell-Boltzmann distribution. Applying Eq. (7), one finds for the probability that molecule 1 will have a given range of positions and momenta as defined by Eq. (19). The canceled symbols in the $P(\mathbf{p}_1\mathbf{x}_1)d^3p_1\,d^3x_1$

$$= \frac{\int \cdots \int \cancel{d^3x_1}\, \cancel{d^3p_1}\, d^3x_2 \cdots d^3p_N\, e^{-E/\theta}}{\int \cdots \int d^3x_1 \cdots d^3p_N\, e^{-E/\theta}}$$

$$= \frac{1}{V} e^{-\mathbf{p}^2/(2m\theta)} (2m\theta)^{-3/2}\, d^3p_1\, d^3x_1 \qquad (19)$$

numerator mean that one does not integrate over the coordinates and momenta of the molecule 1. The result given by Eq. (19) is again true for an ideal gas. The identification of $\theta = kT$, however, is made in general. By differentiating Eq. (13) with respect to θ, one may establish from Eq. (18a) the general connection (using $\theta = kT$) written as Eq. (20). When this general relation, deduced from the

$$\bar{E} = \psi - T\frac{\partial\psi}{T} = \epsilon \qquad (20)$$

canonical ensemble, is compared with the general thermodynamic relation (15d), one sees indeed that the parameter ψ and the free energy ψ' play identical roles. Again, for an ideal gas one can compute ψ, and it is the same as the ideal gas free energy. Hence one identifies ψ and ψ' in general. Equation (13) relates the thermodynamic free energy to an integral containing the mechanics of the microscopic problem. It is also interesting to observe that a quantity η' defined by Eq. (21) plays in

$$\eta' = \frac{\int \rho \log \rho\, d\Gamma}{\int \rho\, d\Gamma} \qquad (21)$$

all instances the role of the entropy. In fact, one may show that $\psi = \epsilon - T\eta'$, in harmony with Eq. (15a). One can follow these thermodynamic analogies through in all detail to see that the canonical ensemble combined with the relevant definitions does indeed reproduce the formal aspects of thermodynamics. The average energy $\bar{E}$ for a canonical ensemble has already been calculated. It is important to know just how often appreciable deviations from this average energy can be expected. For this, one needs to calculate the fractional fluctuations in energy given by notation (22). The defini-

$$\frac{\overline{E^2} - (\overline{E})^2}{(\overline{E})^2} \qquad (22)$$

tion of $\overline{E^2}$, in Eq. (23), follows again from Eq. (6).

$$\overline{E^2} = \frac{\int E^2 e^{-E/\theta} \, d\Gamma}{\int e^{-E/\theta} \, d\Gamma} \qquad (23)$$

From Eq. (23) one may directly express $\overline{E^2}$ in terms of Z, the partition function. For an ideal gas one can then obtain, using Eq. (17), an explicit expression, Eq. (24), for the fluctuations in energy.

$$\left(\frac{\overline{E^2} - (\overline{E})^2}{(\overline{E})^2}\right)^{1/2} = \sqrt{\frac{2}{3N}} \qquad (24)$$

Since one deals with systems where the number of particles N is about 10^{22}, the chance of observing a sizable deviation from the average energy is extremely small. Using these same methods, defining in addition the specific heat $C_v = \partial \overline{E}/\partial T$, one may show generally that the fractional fluctuation in energy is given by Eq. (25). Even though the energy

$$\left(\frac{\overline{E^2} - (\overline{E})^2}{(\overline{E})^2}\right)^{1/2} = \left(\frac{kT^2 C_v}{(\overline{E})^2}\right)^{1/2} \qquad (25)$$

fluctuations are negligible for an ideal gas, this is not always so for other systems. For solids at low temperatures, the fluctuations may be appreciable. When phase transitions (of the first order) take place, C_v becomes infinite, indicating via Eq. (25) that the fluctuations become very large. It is clear that the main problem of the applications is reduced to the calculation of the partition function Z; all thermodynamic entities follow from Eqs. (16) and (13).

There is yet another ensemble which is extremely useful and which is particularly suitable for quantum-mechanical applications. Much work in statistical mechanics is now based on the use of this so-called grand canonical ensemble. In the grand ensemble one has a collection of systems; the number of particles in each system is no longer the same, but varies from system to system. The density function $\rho(N,p,x) \, d\Gamma_N$ gives the probability that there will be in the ensemble a system having N particles, and that this system, in its $6N$-dimensional phase space Γ_N, will be in the region of phase space $d\Gamma_N$. The function ρ is given by Eq. (26). Here θ, Ω, and μ are the parameters charac-

$$\rho(N,p,x) = \exp\left(\frac{\Omega + \mu N - E(p,x)}{\theta}\right) \qquad (26)$$

terizing the ensemble, just as ψ and θ characterize the canonical ensemble. It is again true that these parameters are directly related to thermodynamic state functions. The detailed argument follows the identical pattern indicated in the discussion of the canonical ensemble. The normalization condition is now given by Eq. (27). The grand canonical aver-

$$\sum_N \int d\Gamma_N \rho(N,p,x)$$
$$= \sum_N \int d\Gamma_N \exp\left(\frac{\Omega + \mu N - E(p,x)}{\theta}\right) = 1 \qquad (27)$$

age of any quantity Q is given by Eq. (28) and the

$$\overline{Q}_{\text{gr}} = \sum_N \int d\Gamma_N \rho(N,p,x) \, Q_N(p,x) \qquad (28)$$

grand partition function by Eq. (29a). Again one customarily uses for a system consisting of N identical particles Eq. (29b). From Eq. (27) one sees

$$Z_{\text{gr}} = \sum_N e^{\mu N/\theta} \int d\Gamma_N \, e^{-(E/\theta)} \qquad (29a)$$

$$Z_{\text{gr}} = \sum_N \frac{1}{N!} \frac{1}{h^{3N}} e^{\mu N/\theta} \int d\Gamma_N \, e^{-(E/\theta)} \qquad (29b)$$

immediately that Eq. (30) holds. From Ω, some-

$$Z_{\text{gr}} = e^{-(\Omega/\theta)} \qquad (30)$$

times called the grand potential, all other thermodynamic functions may be computed. The parameter μ is the chemical potential. It is defined by $\mu = (\partial \xi/\partial N)_{P,T}$ where ξ is the Gibbs free energy or thermodynamic potential; $\xi = \psi + PV$. Formally, these results are written as Eqs. (31). From a knowledge

$$P = -\left(\frac{\partial \Omega}{\partial V}\right)_{\mu,T} \qquad (31a)$$

$$\eta = -\left(\frac{\partial \Omega}{\partial T}\right)_{V,\mu} \qquad (31b)$$

$$\overline{N}_{\text{gr}} = -\left(\frac{\partial \Omega}{\partial \mu}\right)_{V,T} \qquad (31c)$$

of the grand partition function as a function of V, T, and μ, all thermodynamic functions follow. For an ideal gas, for example, Eq. (29b) in conjunction with Eq. (17) yields Eq. (32a). Here λ is the thermal de Broglie wavelength, defined by Eq. (32b). From

$$Z_{\text{gr}} = \sum_{N=1}^{\infty} \frac{1}{N!} \left(\frac{2\pi m\theta}{h^2}\right)^{(3/2)N} V^N e^{\mu N/\theta}$$
$$= \exp\left(e^{\mu/\theta} \frac{V}{\lambda^3}\right) \qquad (32a)$$

$$\lambda = \frac{\hbar}{\sqrt{2\pi mkT}} \qquad (32b)$$

Eqs. (32a) and (30) one sees that Eq. (33) holds.

$$\Omega = -\theta e^{\mu/\theta} \frac{V}{\lambda^3} \qquad (33)$$

Application of Eqs. (31) yields the usual results for ideal gases, such as $PV = \overline{N}kT$; however, the grand average of $\overline{N}$ enters the relations now, rather than just $\overline{N}$. From the definitions (28) and (29b) one may show that the fractional fluctuation in the number N is given by Eq. (34). Thus for gases, the fluctua-

$$\left(\frac{\overline{N_{\text{gr}}^2} - (\overline{N}_{\text{gr}})^2}{\overline{N}_{\text{gr}}^2}\right)^{1/2} = \frac{1}{\sqrt{\overline{N}_{\text{gr}}}} \qquad (34)$$

tions in number are negligibly small, showing that the number of particles in the grand canonical ensemble is sharply peaked around the average value. One therefore expects that the physical results deduced from the canonical and grand canonical ensemble will be the same; the calculations, however, are frequently (especially in quantum problems) simpler in the grand canonical than in the canonical scheme.

Applications. Two of the newest and most important applications of statistical physics involve the theory of nonideal gases and the theory of non-equilibrium states of dense gases and liquids.

Theory of nonideal gases. The importance of relations such as Eq. (13), which connects the thermodynamic free energy to the partition function, lies in the general validity of these connections. Specifically, when dealing with a gas of interacting molecules, where the energy is given by Eq. (35)

$$E(x_1, \ldots ,z_N;p_{x_1}, \ldots ,p_{z_N})$$

$$= \sum_{i=1}^{N} \frac{p_{x_i}{}^2 + p_{y_i}{}^2 + p_{z_i}{}^2}{2m} + U(x_1, \ldots ,z_N) \quad (35)$$

with the potential U a known function, the relation is still valid. In that case, one may obtain the partition function directly in terms of U by performing the integrations over the momenta as in Eq. (36).

$$Z_N = \frac{1}{N!} \frac{1}{\lambda^{3N}} \int \cdots \int d^3x_1 \cdots d^3x_N$$

$$\exp\left[-\frac{U(x_1, \ldots ,z_N)}{kT}\right] \quad (36)$$

Since from Z_N all thermodynamic relations follow, Eq. (36) is the appropriate starting point for all studies of nonideal gases. The canonical ensemble has been used in obtaining Z_N, which depends on V through the integration limits, on T through λ (see 32b), and on N through the number of integrations. Since Z_N gives the free energy immediately, one can get the equation of state (the relation between P, V, and T) by computing $P = -\partial\psi/\partial V$. The experimental results on the equations of state may be expressed in terms of the so-called virial expansion, Eq. (37). Here $B(T)$ and $C(T)$ are the second

$$P = \frac{RT}{V}\left(1 + \frac{B(T)}{V} + \frac{C(T)}{V^2} + \cdots\right) \quad (37)$$

and third virial coefficients; they are determined experimentally. It is clear that Eq. (37) is a development starting from the ideal gas equation. A first task for statistical mechanics would be to deduce the form (37), as well as an explicit expression for $B(T)$, and so on. Let relation (38a) be set up, and assume now that the potential $U(x_1, \ldots ,z_N)$ is such that it is possible to make an unambiguous distinction between interacting and noninteracting particles. Any potential which vanishes for a separation larger than a critical amount has this property, and any short-range potential approximates it. Call a separated configuration one in which none of the molecules of one group interacts with any of another. If two such groups are called α and β, one has, for such a configuration, Eq. (38b). Hence, by

$$\exp\left[-\frac{U(x_1, \ldots ,z_N)}{kT}\right] \equiv W(1,2, \ldots ,N) \quad (38a)$$

$$U = U(\alpha) + U(\beta) \quad (38b)$$

$$W = W(\alpha)W(\beta) \quad (38c)$$

$$W(1) = S(1) = 1$$
$$W(1,2) = S(1,2) + S(1)S(2)$$
$$W(1,2,3) = S(1,2,3) + S(1)S(2,3) + S(2)S(1,3)$$
$$+ S(3)S(1,2) + S(1)S(2)S(3) \quad (38d)$$

$$W(1, \ldots ,N) = S(1, \ldots ,N)$$
$$+ S(1)S(2, \ldots ,N) + \cdots$$
$$+ S(1,2)S(3, \ldots ,N) + \cdots S(1) \cdots S(N)$$

relation (38a), Eq. (38c) holds. Consider now a set of functions S defined by Eqs. (38d). From Eqs. (38c) and (38d) one now proves the important property: S functions vanish for a separated configuration. For instance $S(1,2, \ldots ,l)$ is zero, unless the molecules 1, 2, $\ldots$, l form a nonseparated configuration. This is an important property, for it allows the integration of the S functions in a simple fashion. Consider integral (39a). Imagine that molecule 1 is fixed somewhere in the middle of the vessel. Since the S functions vanish for a separated configuration, the l molecules must all be quite near to molecule 1. Roughly speaking, if a is the range of the molecular forces, integral (39a) will get contributions only from a range of about la around molecule 1. If la is smaller than $V^{1/3}$, the first $l-1$ integrations will be independent of V. The last integrations over the coordinates of 1 will just contribute V, for molecule 1 can be placed anywhere in V. (Wall effects are ignored.) Thus Eq. (39b) holds. The $l!$ is a normalization factor and b_l

$$\int \cdots \int d^3x_1 \cdots d^3x_l S(1, \ldots ,l) \quad (39a)$$

$$\int \cdots \int dx_1 \cdots dz_l S(1, \ldots ,l)$$
$$= Vl!b_l(T) \quad (39b)$$

is independent of V. The b_l are called cluster integrals. Using Eqs. (36), (38a), (38d), and (39b), one may obtain the pressure in terms of the cluster integrals as in Eq. (40). One therefore has succeeded

$$P = \frac{RT}{B}\left(1 - \frac{Nb_2}{V}\right.$$

$$\left. + 4(N^2b_2{}^2 - 2N^3b_3)\frac{1}{V^2} + \cdots\right) \quad (40)$$

in deducing the experimental form of the virial development, while the cluster integrals are related to the experimental virial coefficients by Eqs. (41a) and (41b). Using Eq. (39b) for $l=2$ gives a direct relation between B and the intermolecular force potential $U(r)$ as in Eq. (41c). This is a "per-

$$B(T) = -Nb_2 \quad (41a)$$

$$C(T) = N^2(4b_2{}^2 - 2b_3) \quad (41b)$$

$$B(T) = 2\pi N \int_0^\infty r^2\,dr(1 - e^{-[U(r)/kT]}) \quad (41c)$$

fect" statistical formula. In the case of helium, where both $U(r)$ and B are known, Eq. (41c) may in fact be checked. [At low temperatures, quantum effects, not included in Eq. (41c), begin to play an important role.]

One can obtain the complete equation of state using a similar procedure. The result comes out in an implicit form, as in Eqs. (42a) and (42b). These

$$P = kT\Sigma b_l z^l \quad (42a)$$

$$\frac{N}{V} = \sum_{l=1}^{N} lb_l z^l \quad (42b)$$

equations are from J. E. Mayer. In principle, one can eliminate the auxiliary variable z between Eqs. (42a) and (42b) to obtain a relation between P, V, and T. To have a useful relation one must, of course, know the general character of the cluster integrals which, except for special molecular models, is not known. A question which has concerned physicists for some time is whether or not the system of equations (42) actually predicts a condensation phenomenon. It is well known that every real gas at a low enough temperature will condense when the volume is decreased. During the condensation process the pressure remains constant, the onset of condensation being marked by a discontinuity in the slope of the isotherm. The problem is now whether this information can be obtained from Eqs. (42). An interesting clue was obtained by Mayer. He showed that for low enough temperatures, $b_l(T) \cong \text{constant} \times b_0{}^l$. Substituting in Eq. (42b), one obtains Eq. (43). Since the last power in the series is $N \cong 10^{24}$, one sees that the series given by Eq. (43) is radically different

$$\frac{N}{V} = \sum_{l=1}^{N} l\,(\text{constant})\,(b_0 z)^N \qquad (43)$$

for $b_0 z < 1$ and $b_0 z \gtrdot 1$. If $b_0 z > 1$, $(b_0 z)^N$ is a very large number and a change in N/V will cause a very slight change in z, hence a very slight change in P by Eq. (42a). Therefore, $b_0 z = 1$ separates two ranges: If $b_0 z < 1$, changes in N/V cause reasonable changes in P; if $b_0 z > 1$, P becomes quite insensitive to changes in N/V. This qualitative idea has been refined in many ways, by rigorously studying limit (44). These studies have clarified some as-

$$\lim_{\substack{N \to \infty \\ V \to \infty}} (Z_N)^{1/N} \qquad (44)$$

pects of the situation, but even so the problem is still not completely settled.

Nonequilibrium theory of liquids. It was observed in the earlier discussion of Liouville's theorem that the Liouville equation usually does not help in setting up an appropriate ensemble. Yet the Liouville equation has become the starting point for an important development aimed at an understanding of the nonequilibrium states of dense gases and liquids. This development starts from the Liouville equation, Eq. (45). The interpre-

$$\frac{\partial \rho}{\partial t} + \sum_{i=1}^{3N} \left(\frac{\partial \rho}{\partial x_i} \frac{dx_i}{dt} + \frac{\partial \rho}{\partial p_i} \frac{dp_i}{dt} \right) = 0 \qquad (45)$$

tation that $\rho(x_1, \ldots, p_{z_N}; t)\,d\Gamma$ is the probability of finding the molecules of the system in their prescribed momentum and position ranges may now be used. Define a set of probability functions (or distribution functions) by Eqs. (46). One observes

$$f_1(\mathbf{x}_1, \mathbf{p}_1, t)$$
$$= \int \cdots \int d\cancel{(1)}\, d(2) \cdots dN\, \rho(x, p, t) \qquad (46a)$$

$$f_l(\mathbf{x}_1, \ldots, \mathbf{p}_l; t) = \int \cdots \int d\cancel{(1)} \cdots$$
$$d\cancel{(l)}\, d(l+1) \cdots d(N)\, \rho(x, p, t) \qquad (46b)$$

that f_1 is (apart from constants) the Boltzmann distribution function. The canceled symbol $d(1)$ means that one is not to integrate over $x_1, y_1, z_1,$ p_{x1}, p_{y1}, p_{z1}. The higher distribution functions become important for dense systems. Assume that the potential energy of the system consists of an external potential V_0, and a potential U which is additive. Then by integrating Eq. (45) over all coordinates except those of molecule 1, one obtains an equation for f_1, Eq. (47). The equation for f_l involves f_{l+1}, and so on.

$$\frac{\partial f_1}{\partial t} + \frac{p_\alpha}{m_1} \frac{\partial f_1}{\partial x_\alpha} - \sum_{\alpha=1}^{3} \frac{\partial V_0}{\partial x_\alpha} \frac{\partial f_1}{\partial p_\alpha}$$
$$= \int \int d^3 p_2\, d^3 x_2 \sum_\alpha \frac{\partial U(x_1 x_2)}{\partial x_\alpha} \frac{\partial f_2(p_1 p_2, x_1 x_2)}{\partial p_\alpha} \qquad (47)$$

The discussion of this hierarchy forms the basis of the study of the nonequilibrium phenomena in dense gases. The basic difficulty is that the equations for f_l always involve a function f_{l+1}. To obtain an equation for a single f_l function, one needs to make a guess or assumption about the way in which a higher f function can be expressed in terms of lower ones, if indeed this can be done at all. A frequently discussed possibility is the so-called superposition approximation relation (48).

$$f_2(p_1 x_1, p_2 x_2) \cong f_1(p_1 x_1) f_1(p_2 x_2) \qquad (48)$$

An appeal to probabilities of independent events would make relation (48) appear reasonably plausible. It should be stressed, however, that nowhere does one prove that the approximation of relation (48) is indeed consistent with the system of Eq. (47). The use of relation (48) in Eq. (47) will lead to (nonlinear) integral equations. In spite of the considerable amount of work done with these equations, the results still have not led to great advances in the theory of dense gases. It is of interest to point out, however, that for additive central potentials, defined by $U(1, \ldots, N) = \Sigma U(r_{ij})$, it is possible to express the equation of state in terms of the pair distribution function. This function is defined by Eq. (49a). Here $W(1, \ldots, N)$ is defined by Eq. (38). The function n_2 gives the positional distribution of molecular pairs. As such it is less detailed than f_2, which gives the distribution of pairs both in positions and momenta. Apart from combinatorial factors, one has Eq. (49b). For cen-

$$n_2(\mathbf{x}_1, \mathbf{x}_2) = N(N-1) \qquad (49a)$$

$$\cdot \frac{\int \cdots \int d\cancel{(1)}\, d\cancel{(2)}\, d(3) \cdots d(N)\, W(1, \ldots, N)}{\int \cdots \int d(1) \cdots d(N)\, W(1, \ldots, N)}$$

$$n_2(\mathbf{x}_1, \mathbf{x}_2) = \int \int d^3 p_1\, d^3 p_2\, f_2(\mathbf{x}_1 \mathbf{p}_1, \mathbf{x}_2 \mathbf{p}_2) \qquad (49b)$$

tral additive potentials, it may now be shown that n_2 is in fact a function of the distance r_{12} between the molecules; $n_2 \propto g(r)$, where $g(r)$ is defined as giving the number of molecules between r and $r + dr$.

An analysis similar to the one given in the preceding discussion on nonideal gases now leads to an equation of state, Eq. (50), in terms of $g(r)$ alone.

$$PV = NkT - \frac{2\pi N(N-1)}{3V} \int g(r) \frac{dU}{dr} r^3\, dr \qquad (50)$$

The integral equation referred to for the f_2 functions (once the superposition approximation is made) may now be expressed as an approximate integral equation for $g(r)$. This equation, obtained by M. Born, H. S. Green, and J. Yvon, leads to approximations for the virial coefficient, which are fairly good but not excellent. It is important to recall that the pair distribution function $g(r)$ may be obtained directly from experimental x-ray scattering data. In principle one could use experimental x-ray data to obtain via Eq. (50) data about the equation of state. For this to be a feasible procedure would demand accuracies much beyond the present limits, as well as extensive measurements over a range of temperatures and densities. Unfortunately, therefore, this is not a possible procedure at present (1969), although the formal relation (50) is generally valid. In a current theory of liquid helium, one makes use of the relation between the pair distribution function and neutron-scattering data. *See* LIQUID HELIUM.

Quantum statistical mechanics. The ensemble techniques can also be applied to systems which are described in terms of quantum mechanics. Consider an ensemble, where each system is described by a Hamiltonian operator H. Let $\psi^\alpha(x,t)$ be a wave function (x stands for $\mathbf{x}_1, \ldots, \mathbf{x}_N$, and α characterizes an ensemble member). The Schrödinger equation is written as Eq. (51) *See* NONRELATIVISTIC QUANTUM THEORY.

$$H\psi^\alpha(x,t) \doteq -\frac{\hbar}{i}\frac{\partial \psi^\alpha(x,t)}{\partial t} \tag{51}$$

If $\varphi_n(x)$ is a complete orthogonal set, Eq. (52),

$$\int \varphi_n{}^*\varphi_m = \delta_{nm} \tag{52}$$

the function ψ^α may be developed in terms of the set $\{\varphi\}$ as in Eq. (53). Here $|a_n{}^\alpha(t)|^2$ is the probability that ensemble member α, at time t, is in state n. One has, of course, the fact that Eq. (54) is valid.

$$\psi^\alpha(x,t) = \sum_n a_n{}^\alpha(t)\varphi_n(x) \tag{53}$$

$$\sum_n |a_n{}^\alpha(t)|^2 = 1 \tag{54}$$

Suppose the number of ensemble members is N_e. One defines now an ensemble average of a quantity G which is defined for each ensemble system by G^α as in Eq. (55a). It is important to distinguish the ensemble average from the ordinary quantum-mechanical average, which is defined for a single system as in Eq. (55b). The notation, double

$$\overline{\overline{G}} = \frac{1}{N_e}\Sigma(G^\alpha) \tag{55a}$$

$$\overline{G^\alpha} = \int (\psi^\alpha)^* G_{\text{op}} \psi^\alpha \tag{55b}$$

ble bar for ensemble average, single bar for quantum-mechanical average, stresses the difference. Of special importance in the calculation of averages is the density matrix. The matrix elements of ρ, the density matrix, relative to the set of functions φ are defined by Eq. (56a). One sees that the sum of the diagonal elements of ρ, called the trace of ρ, is given by Eq. (56b). The

$$\rho_{mn} = \frac{1}{N_e}\sum_\alpha (a_n{}^\alpha)^* a_m{}^\alpha = \overline{\overline{a_n{}^* a_m}} \tag{56a}$$

$$\text{Tr}(\rho) = \sum_n \rho_{mn} = \frac{1}{N_e}\sum_\alpha \sum_n |a_n{}^\alpha|^2 = 1 \tag{56b}$$

density matrix ρ is the counterpart of the classical density function. From the Schrödinger equation (53) one deduces immediately the analog of the Liouville theorem, Eq. (57). Here $[a,b] \equiv ab - ba$ is

$$i\hbar\frac{d\rho_{mn}}{dt} = [H,\rho]_{mn} \tag{57}$$

the commutator. Finally one obtains the basic relation, giving the ensemble average of any operator as in Eq. (58). The fact that observable entities, such as ensemble averages, come out in the form

$$\overline{\overline{G}} = \text{Tr}(\rho G) \tag{58}$$

of a trace, implies that the calculated results are independent of the set of functions φ. This is true because a change in this basic set will induce a similarity transformation on both ρ and G, and the trace operation is invariant under such a transformation.

If one has an ensemble in which all ensemble members are in the identical state (ψ^α is independent of α), one has a pure case. If the ψ^α do depend on α, one has a statistical mixture. A necessary and sufficient condition for a pure case is that $\rho^2 = \rho$.

At equilibrium, one has the counterpart of the canonical ensemble. Then the density operator, (see 12a) is written as Eq. (59a). The free energy is again directly related to the partition function as in Eq. (59b). In Eq. (60), Q is an operator e^Q defined by

$$\rho = e^{(\psi - H)/kT} \tag{59a}$$

$$Z = e^{-(\psi/kT)} = \text{Tr}(e^{-(H/kT)}) \tag{59b}$$

$$e^Q = \sum_n \frac{Q^n}{n!} \tag{60}$$

the series expansion. The thermodynamic relations follow as before.

The construction of an ensemble in quantum mechanics must be performed in harmony with the knowledge of the system. If the system is known to be in either one of two quantum states, the ensemble members must be evenly distributed over these states with, in addition, random phases. It is important to include the requirement of random phases; if this were not done, a uniform stationary ensemble would not even remain stationary.

[MAX DRESDEN]

Bibliography: N. Davidson, *Statistical Mechanics*, 1962; H. Eyring et al., *Statistical Mechanics and Dynamics*, 2d ed., 1979; J. O. Hirschfelder, C. F. Curtiss, and R. B. Bird, *Molecular Theory of Gases and Liquids*, 1964; A. Khinchin, *Mathematical Foundations of Statistical Mechanics*, 1949; L. D. Landau and E. Lifshitz, *Statistical Physics*, 2d ed., 1969; T. D. Lee, *Statistical Mechanics*, 1980; J. E. Mayer and M. G. Mayer, *Statistical Mechanics*, 2d ed., 1977; R. C. Tolman, *The Principles of Statistical Mechanics*, 1938, reprint 1980.

Statistics

The field of knowledge concerned with collecting, analyzing, and presenting data. Not only workers in the physical, biological, and social sciences, but also engineers, business managers, government officials, market analysts, and many others regularly use statistical methods in their work. The methods range from simple counting to complex mathematical systems designed to extract the maximum amount of information from very expensive data.

In an important sense statistics may be regarded as a field of application of probability theory. The common problem faced by a physicist reading a meter, an engineer testing a material, an agronomist measuring the yield of a hybrid corn, a chemist determining the concentration of ascorbic acid, and an interviewer studying public opinion is the problem of random variation which prevents repetition of exactly the same result when a measurement is repeated. Statistical methods are employed to assess the magnitude of random variation, to minimize it, to balance it out, to remove it by calculation procedures, and to analyze it by suitably arranged patterns of observation. The theory of probability is concerned with the properties of random variables and hence furnishes the basis for developing techniques for controlling them. *See* PROBABILITY.

Viewing statistics from another direction, it is the science of deriving information about populations by observing only samples of those populations. A population is any well-specified collection of elements. Thus, one may refer to the population of adults in the continental United States viewing television screens at 8:14 P.M. on Aug. 6, 1970; the population of automobiles less than 2 years old registered in Los Angeles County on a certain date; the population of vineyards in France; the hypothetical population of outcomes of tossing a given coin endlessly. Populations may be finite or infinite. An element of a univariate population is characterized by the value of a random variable which measures some single attribute of interest in the population. Thus, one may be interested in whether or not individuals of the television audience were or were not viewing program A; with each individual one may associate a random variable, let it be X, which takes on the value of 1 if the individual is watching A and 0 if he is not. If one were interested in a second characteristic of the elements of the television audience (such as age), he would be said to be dealing with a bivariate population; a third characteristic (such as economic status) would make it a trivariate or, less specifically, a multivariate population.

Random variables are either continuous, which means they can take on any numerical value (the length of a room), or discrete, which means they can take on only a restricted set of values (number of windows in a room).

Distributions. In a univariate population, the population distribution is a curve (function of the random variable which characterizes the elements of the population) from which one can determine the proportion of the population which has elements in a certain range of the random variable. For example, the curve of Fig. 1 provides the distribution of annual incomes of family units in the United States in 1954. The total area under the curve is 1. The area under the curve between any two vertical lines gives the proportion of the families having annual incomes between the two values

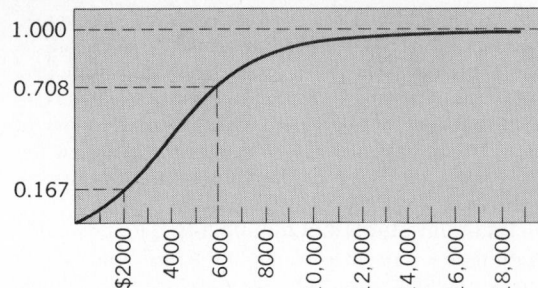

Fig. 2. Cumulative distribution of incomes.

marked on the horizontal scale by the two vertical lines. Thus, the fact that the area under the curve between $2000 and $6000 is 0.541 means that 54.1% of United States family units had incomes in that range in 1954.

The distribution is also referred to as the distribution function, the density function, the frequency function, or the probability density.

The total area under the distribution curve to the left of each point can also be plotted to give a curve which starts at zero and reaches unity as the variable becomes large; the resulting curve is sometimes called the cumulative distribution function, the probability distribution, or simply the distribution. The cumulative form of the curve of Fig. 1 is shown in Fig. 2; the height of the curve at any point on the horizontal scale equals the area to the left of that point under the curve of Fig. 1 and is the proportion of the population having incomes less than the value at that point. The distribution (in either frequency or cumulative form) gives complete information about the way the characterizing variable is spread through the population.

Population parameters. Populations (or population distributions) are often specified incompletely by certain population parameters. Some of these parameters are location parameters or measures of central tendency; a second class of important parameters consists of measures of dispersion or scale parameters.

The most widely used location parameters are the mean, the median, and the mode. The mean is the average over all the population of the values of the random variable. It is often represented by the Greek letter μ. In mathematical terms, if x is the random variable, $f(x)$ the frequency function for a given population, and $F(x)$ its cumulative form, then the mean is as shown in Eq. (1).

$$\mu = \int_{-\infty}^{\infty} xf(x)\,dx = \int_{-\infty}^{\infty} x\,dF(x) \qquad (1)$$

The median, often designated by Med, M, or $X_{.50}$, is a number such that, at most, one-half the values of the variable associated with the elements of the population fall above or below it, as in Eq. (2). The mode is the most frequent value of the ran-

$$\int_{-\infty}^{M} dF(x) \geqq \tfrac{1}{2} \leqq \int_{M}^{\infty} dF(x) \qquad (2)$$

dom variable; if the frequency function has a unique maximum value, the mode is the value of the random variable at which the frequency function reaches its maximum. Location parameters are numbers near the center of the range over which the random variable of the population

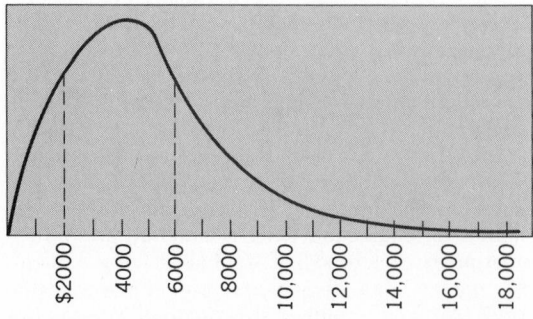

Fig. 1. Distribution of incomes.

varies; different ones arise from different definitions of center; generally the mean is used unless special circumstances make some other location parameter more appropriate.

The extent to which a population is scattered on either side of its center is roughly indicated by measures of dispersion such as the standard deviation, the mean deviation, the interquartile range, the range, and sometimes others. The standard deviation is the square root of the mean square of the deviations from the mean; it is usually denoted by the Greek letter σ; σ^2 is called the variance and is expressed by Eq. (3). The mean deviation, shown

$$\sigma^2 = \int_{-\infty}^{\infty} (x-\mu)^2 f(x)\, dx$$

$$= \int_{-\infty}^{\infty} (x-\mu)^2\, dF(x) \qquad (3)$$

in Eq. (4), is the average over the population of the

$$\text{Mean deviation} = \int_{-\infty}^{\infty} |x-\mu| f(x)\, dx$$

$$= \int_{-\infty}^{\infty} |x-\mu|\, dF(x) \qquad (4)$$

deviations from the mean, all taken to be positive. The interquartile range (often denoted by Q) is the difference $X_{.75} - X_{.25}$, where $X_{.75}$ is the value of the random variable such that one-quarter of the population has values larger than $X_{.75}$, and $X_{.25}$ is the number such that one-quarter of the population has values smaller than $X_{.25}$. The three numbers, $X_{.25}, X_{.50}, X_{.75}$, are called quartiles; these divide the population into quarters. The range is the difference between the largest and the smallest of the population elements.

Samples. If one examines every element of a population and records the value of the random variable for each, complete information is obtained about the distribution of the random variable in the population, and there is no statistical problem. It is usually impossible or uneconomical to make a complete enumeration (or census) of a population, and one must therefore be content to examine only a part or sample of the population. On the basis of the sample, one draws conclusions about the entire population; the conclusions thus drawn are not certain in the sense that they would likely have been somewhat different if a different sample of the population had been examined. The problem of drawing valid conclusions from samples and of specifying their range of uncertainty is known as the problem of statistical inference.

Statisticians distinguish two kinds of samples. A survey sample is one chosen from a population, all the elements of which actually exist. An experiment is a sample chosen from a hypothetical population. Thus if one were interested in the total number of pigs being fattened in a given season by the farmers of the state of Iowa, he would survey-sample the existing population of pigs in Iowa at that time. If one were interested in the effect of a certain hormone on the growth rate of pigs, he would do an experiment by giving a group of pigs the hormone for a period of time; using the sample of observations thus obtained from the experiment, he would draw inferences about the hypothetical population of observations that would have resulted had all pigs been given the hormone.

Random sampling. In planning a sample survey, the manner in which the data to be gathered will fulfill the purpose should be clearly stated. The population to be sampled must be explicitly defined. The method of sampling should be efficient and lead to a straightforward analysis. The question of what elements should be included in the population depends on the purpose of the survey. Thus the population of vineyards in France might include as a vineyard a dozen vines in the backyard of a man living in the heart of Paris if the purpose were to estimate total grape production of France; but the population might be defined to exclude such a small vineyard if the purpose were to estimate the size of the harvest to be available to commercial wineries.

The type of sampling of interest in statistics is probability sampling, because it eliminates subjective aspects from the selection of the sample. In probability sampling, all possible distinct samples are known, the selection of the sample is done randomly according to a preassigned probability, and the method of analysis is predetermined and unambiguous. Only from such samples can inferences about populations be made with measurable precision.

Simple random sampling is a method of selecting a sample of n elements out of a population of N elements so that all such samples have an equal probability of being drawn. This may be done by selecting a first element at random from the population, then a second element at random from the remaining population, and so on until the n elements are selected. Because an element cannot appear more than once in the sample, this is a form of sampling without replacement. The sampling ratio or sampling fraction is n/N.

Often the purpose of drawing a sample is to estimate the mean or average of a characteristic of the population. If y_i is the value of the characteristic of the ith unit, then the population mean is as shown in Eq. (5). The sample mean is the average of the n

$$\mu = \frac{1}{N}\sum_{i=1}^{N} y_i = \frac{1}{N}(y_1 + y_2 + y_3 + \cdots + y_N) \qquad (5)$$

units in the sample and is expressed in Eq. (6).

$$\bar{x} = \frac{1}{n}\sum_{i=1}^{n} x_i = \frac{1}{n}(x_1 + x_2 + \cdots + x_n) \qquad (6)$$

Here the n x's are the n y's selected from the population as the sample. The population total is $N\mu$ and is estimated by $N\bar{x}$. The population variance is defined as shown in Eq. (7), and the sample vari-

$$\sigma^2 = \frac{1}{N}\sum_{i=1}^{N} (y_i - \mu)^2 \qquad (7)$$

ance is usually defined as in Eq. (8), although some

$$s^2 = \frac{1}{n-1}\sum_{i=1}^{n} (x_i - \bar{x})^2 \qquad (8)$$

authors use n instead of $n-1$. The quantities $\bar{x}$ and s^2, the sample mean, and the sample variance are called sample statistics; they are the first and second sample moments and are also estimators of the corresponding population parameters, μ and σ^2.

The estimators are themselves random variables. If one repeatedly drew samples of size n from the population and computed $\bar{x}$ from each sample, he would obtain a population of $\bar{x}$'s with its own distribution which would differ from the distribution of x. It can be shown that the $\bar{x}$ population has exactly the same mean μ as the x population. Further, the variance of the $\bar{x}$ population is as shown in notation (9), where σ^2 is the variance of the x

$$\frac{(N-n)}{(N-1)} \frac{\sigma^2}{n} \qquad (9)$$

population. The first fraction is ordinarily nearly unity so that the variance of $\bar{x}$ is approximately the fraction $1/n$ of the original population variance. As n, the sample size, becomes large, the $\bar{x}$ population becomes more concentrated about μ and the reliability of a particular value of $\bar{x}$ as an estimate of the population mean increases.

The observations of a sample, besides providing estimates of population parameters, can also be used to obtain an estimate of the population's frequency function. This estimate is determined by dividing the range of the sample observations into several intervals of equal length L and counting the number of observations occurring in each interval; these numbers are then divided by nL to determine fractions giving the relative density of the sample occurring in each interval; then on a sheet of graph paper one lays out the intervals on a horizontal axis and plots horizontal lines above each interval at a height equal to the fraction corresponding to the interval; finally the successive plotted horizontal lines are connected by vertical lines to form a broken line curve known as a histogram (Fig. 3). The area under the curve is unity, and the area between any two points gives the fraction of the sample observations lying between those two points. If one takes larger and larger samples, the chosen intervals can be made smaller and the broken line curve will come closer and closer to the underlying population frequency function. Often one does not trouble to divide the interval frequencies by nL to normalize the area of the histogram, but merely plots the frequencies themselves; the resulting broken line curve is still referred to as a histogram.

Sampling techniques. When a population can be regarded as being made up of several nonoverlapping subpopulations, one may draw a sample from it by drawing a simple random sample from each subpopulation (or stratum); this procedure is called stratified random sampling. The method is employed when it is desired to have specific information about each stratum individually, when it is administratively convenient to subdivide the population, when there are natural strata, or when a gain in precision would be realized because each stratum is more homogeneous than the whole population.

Let n be the number of units sampled in the entire population of N units, and let n_h be the number of units sampled in stratum h containing N_h units. Therefore, the sum of all the n_h is n, and of all the N_h is N. Let σ_h be the true standard deviation within stratum h. When $n_h/n = N_h/N_n$, the n_h are said to be proportionately allocated. When $n_h/n = N_h\sigma_h/\Sigma N_h\sigma_h$, the n_h are said to be optimally allocated because the variance of the estimated mean is then smallest for fixed total size of sample. Ordinarily, if the σ_h are well estimated, optimum allocation gives greater precision than proportional allocation, and proportional allocation gives

greater precision than simple random sampling.

Systematic sampling may be regarded as a sampling of units at regular intervals in the population, for example, every tenth unit. Usually this is done with a random start; that is, the first unit in the sample is selected at random from a small group at the beginning. A systematic sample is usually relatively easy and fast to execute, and, if the analogy to stratified sampling is valid, a systematic sample will be more accurate. However, because of the essentially nonrandom nature of a systematic sample, it is difficult to estimate the sample variance unless certain assumptions are made about randomness in the order of the population. If the population has any periodicity, then the estimates made from a systematic sample can be quite poor.

When the elements of a population are in mutually exclusive groups called the primary units of the population, then a sample of these primary units might be made, and then, from those selected, a sample of the individual elements would be made. This procedure is called two-stage sampling or subsampling. Multistage sampling can involve more than two stages of sampling. Although more complicated and difficult to apply and to analyze, multistage sampling offers considerable flexibility for balancing between statistical precision and the cost of sampling.

When too little is known about a population to plan a sample, then two samples are made. The manner in which the second sample is taken is determined from the results of the first sample. In Stein's method of two-stage sampling, the size of the additional sample is decided by using the estimate of the variance from the first sample. In double sampling or two-phase sampling, the first sample is used to gather information on a variate associated with the variate of interest. This information is used to establish strata for drawing a stratified random sample involving the variate of interest.

Ratio and regression estimates are often used when observations are made on one or more variates related to the variate of primary interest.

Sequential sampling and sequential analysis refer to a method of sampling in which the size of the sample is not specified in advance. Observations are drawn one by one until a specified degree of confidence in the information to be obtained has been achieved.

Sampling distributions. Many important sampling distributions are derived for random samples drawn from a normal or Gaussian distribution, which is a bell-shaped symmetrical distribution centered at its mean μ. The distribution is illustrated in Fig. 4 for three different values of σ, the standard deviation. Equation (10) is the mathemat-

$$f(x) = \frac{1}{\sqrt{2\pi}\sigma} e^{-(x-u)^2/\sigma^2} \qquad (10)$$

ical equation for these curves. The area under the curves for the indicated limits is as follows:

Area	Limits
0.50	$\mu - 0.675\sigma$ to $\mu + 0.675\sigma$
0.683	$\mu - \sigma$ to $\mu + \sigma$
0.90	$\mu - 1.645\sigma$ to $\mu + 1.645\sigma$
0.95	$\mu - 1.960\sigma$ to $\mu + 1.960\sigma$
0.99	$\mu - 2.326\sigma$ to $\mu + 2.326\sigma$

STATISTICS

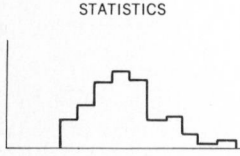

Fig. 3. Histogram.

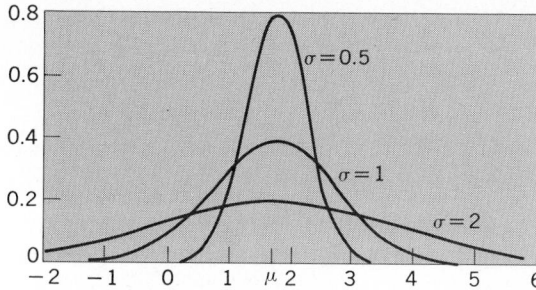

Fig. 4. Normal distribution.

Referring to the above limits, 0.675σ is called the probable error inasmuch as the odds are even that a randomly drawn observation will lie within 0.675σ of the mean.

If samples of size n are drawn from a normal population and the sample mean $\bar{x}$ is computed for each, the $\bar{x}$'s will have a normal distribution with the same mean and with variance σ^2/n. Thus, if samples of size 4 were drawn from a population distributed by the curve marked $\sigma=1$ in Fig. 4, then the means of those samples would be distributed by the curve marked $\sigma=0.5$ (that is, $1/\sqrt{4}$). Further, the probability would be 0.95 that a particular $\bar{x}$ so drawn would lie within $(1.960)(0.5)=0.98$ of the population mean μ.

The central limit theorem of the theory of probability states that under very general conditions the sample mean $\bar{x}$ is approximately normally distributed whatever may be the distribution function for the underlying population. This powerful theorem enables one to make probability statements such as the one at the end of the preceding paragraph in ignorance of the actual population distribution.

Besides the distribution of the mean, two other sampling distributions derived from the normal distribution have wide application. One is the chi-square distribution, which provides the distribution of the sample variance, as shown in Eq. (11).

$$s^2 = \sum_{i=1}^{n} (x_i - \bar{x})^2/(n-1) \qquad (11)$$

Here $x_1, x_2, \ldots, x_n$ are the observations of a random sample of size n drawn from a normal population.

The quantity given in Eq. (12) has a distribution

$$\chi^2 = (n-1)s^2/\sigma^2 \qquad (12)$$

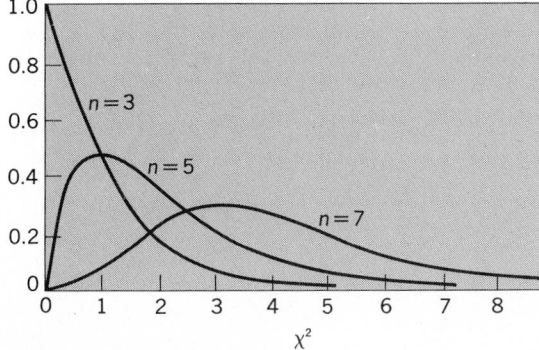

Fig. 5. Chi-square distributions.

illustrated in Fig. 5 for three values of n. The mathematical form for the distribution curve is shown in notation (13) and is often referred to as

$$\frac{1}{\left(\dfrac{n-3}{2}\right)!}\left(\frac{n-1}{2\sigma^2}\right)^{\frac{n-1}{2}}(\chi^2)^{\frac{n-3}{2}}e^{-(n-1)\chi^2/2\sigma^2} \qquad (13)$$

the chi-square distribution with $n-1$ degrees of freedom.

The most useful random variable for interval estimation of a population mean is shown in Eq. (14),

$$t = \frac{\sqrt{n}\,(\bar{x}-\mu)}{s} \qquad (14)$$

which has a symmetrical distribution very similar in appearance to the curves plotted in Fig. 4. The mathematical form for the distribution is notation (15). This is referred to as the t distribution or the

$$\frac{[(n-2)/2]!}{\sqrt{(n-1)\pi}\,[(n-3)/2]!}\bigg/\left(1+\frac{t^2}{n-1}\right)^{n/2} \qquad (15)$$

Student distribution with $n-1$ degrees of freedom. The following gives limits which include 95% of the area under the curve for a few values of n:

n	Limits
2	-12.71 to $+12.71$
3	-4.30 to $+4.30$
4	-3.18 to $+3.18$
10	-2.20 to $+2.20$
25	-2.06 to $+2.06$
120	-1.98 to $+1.98$

For very large sample sizes, the limits become -1.96 to $+1.96$, as for the normal distribution. An illustration of the use of these limits is given below.

Estimation. In making an estimate of the value of a parameter of a population from a sample, a function (called the estimator) of the observations is used. For example, for estimating the mean of a normal population the mean of the sample observations is usually taken as the estimator. Another estimator is the average of the two most extreme observations. In fact, there is an infinitude of estimators. The problem of estimation is to find a "good" estimator.

A good estimator may be regarded as one which results in a distribution of estimates concentrated near the true value of the parameter and which can be applied without excessive effort. There is no single way of deciding how good an estimator is, but there are several criteria by which an estimator may be judged.

An unbiased estimator is one which results in a distribution of estimates which has a mean exactly equal to the value of the parameter being estimated. Otherwise the estimator is called biased. The bias is the mean of the distribution of the estimator minus the value of the parameter it estimates.

An estimator is said to be consistent if the probability that an estimate will differ from the value of the parameter by more than any fixed amount can be made arbitrarily small by increasing the number of observations.

The variance of an estimator is the mean squared deviation of the estimates from the value of the parameter. The estimator with the smallest variance is called most efficient. The relative efficiency of two estimators is the ratio of the variances. When the numerator of this ratio is the variance of a most efficient estimator, this ratio is simply called the efficiency of the other estimator.

An estimator is said to be sufficient if it contains all the information in the sample regarding the parameter. This is so when the conditional distribution of the sample for a given estimate is independent of the parameter.

There are several methods of constructing estimators of parameters. The method of moments is applied by assuming that the first few sample moments are equivalent to the moments of some distribution, and then solving for the parameters of that distribution. The methods known as least squares, minimum variance, and minimum chi-square all have as their basis the estimation of the values of the parameters which minimize some linear function of the squares of deviations of the observations from the values of the parameters. In applying Bayes' method, the distribution of possible values of the parameter before the sample is taken, called the a priori distribution, is used in conjunction with the observations in the sample to yield an estimator. The method of maximum likelihood uses as the estimate that value of the parameter for which the probability of the sample is highest.

A confidence interval is an interval constructed in such a way that the true parameter value is within this interval with a predetermined probability in repeated sampling; this probability is called the confidence level of the interval. For example, a sample mean $\bar{x}$ known to be normally distributed with variance σ^2 will differ less than 1.96σ from the true but unknown mean μ with probability 0.95. Thus expression (16) can be formulated.

Probability that
$$(-1.96\sigma < \bar{x} - \mu < 1.96\sigma) = 0.95 \quad (16)$$

On solving the two inequalities for μ, expression (16) may be written as expression (17).

Probability that
$$(\bar{x} - 1.96\sigma < \mu < \bar{x} + 1.96\sigma) = 0.95 \quad (17)$$

When a particular value of $\bar{x}$ is computed from the observations, then $\bar{x} \pm 1.96\sigma$ is a pair of numbers which define a particular interval called the confidence interval. Because μ is a fixed number, it is within this particular interval with probability zero or one. However, μ is unknown, so that the confidence to be associated with the interval is stated in terms of the proportion of all intervals constructed in this manner which would include μ, rather than this particular interval. A probability of this sort, valid for a population of outcomes, when used in reference to a particular outcome is called a fiducial probability.

Ordinarily σ is unknown and the estimate s derived from the sample must be used to form a confidence interval; in this case, the t distribution rather than the normal distribution must be used. As an example, suppose a chemist has made four determinations of the atomic weight μ of hydrogen as follows: 1.0066, 1.0090, 1.0084, and 1.0086. The average of these values is given in Eq. (18), and the

$$\bar{x} = \frac{1}{4}(1.0066 + 1.0090 + 1.0084 + 1.0086)$$
$$= 1.0082 \quad (18)$$

sample estimate of the variance of his technique is as shown in Eq. (19), the estimate of σ being

$$s^2 = \frac{1}{3}[(1.0066 - 1.0082)^2 + (1.0090 - 1.0082)^2$$
$$+ (1.0084 - 1.0082)^2 + (1.0086 - 1.0082)^2]$$
$$= .00000123 \quad (19)$$

therefore Eq. (20). It follows from the definition of t

$$s = \sqrt{0.00000123} = 0.0011 \quad (20)$$

in the preceding section that in this case Eq. (21)

$$t = 2(1.0082 - \mu)/0.0011 \quad (21)$$

holds. Using the short t table given in the preceding section, one finds that expression (22) can be

Probability that
$$\left(-3.18 < \frac{2(1.0082 - \mu)}{0.0011} < 3.18\right) = 0.95 \quad (22)$$

formulated. Solving these inequalities for μ gives expression (23), and the chemist can assert with

Probability that
$$(1.00645 < \mu < 1.00995) = 0.95 \quad (23)$$

95% confidence that the atomic weight of hydrogen lies between 1.00645 and 1.00995. For greater precision, more observations are required.

In a similar fashion one may obtain a confidence interval for σ by using s^2 in connection with the chi-square distribution.

A confidence region is the generalization of a confidence interval and refers to the simultaneous estimation of several population parameters. The confidence level is the proportion of the time that the region actually includes the true values of the parameters.

In general, the most desirable confidence interval or region is the smallest one which can be constructed for the selected confidence level.

Tests of hypotheses. Besides estimation of parameters, another major area of statistical inference is the testing of hypotheses. A hypothesis is merely an assertion that a population has a specific property. The test consists of drawing a sample from the population and determining whether or not it is consistent with the assertion. Very often the hypothesis is a statement about the mean of a population; that it has a given value, that it is the same as that of another population, that it exceeds that of another population by at least 10 units, and the like. Thus, one may be comparing a new blend of gasoline with a current blend, a new manufacturing process with an existing one, and so on.

For a very simple illustration of the basic ideas involved, suppose a population is known to have either of two distributions which differ mainly in location. Let the random variable be x and let the two functions of x which determine the two distributions be represented by the symbols $f(x)$ and $g(x)$ as indicated in Fig. 6; that is, $f(x)$ represents the height of the left curve at x, and $g(x)$ represents the height of the right curve at x. Assume that the

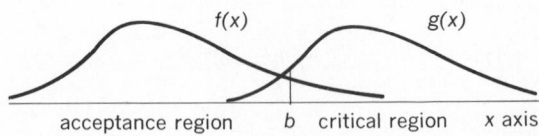

Fig. 6. Simple test of a hypothesis.

population has the distribution $f(x)$; there is in this case only a single alternative, $g(x)$, so that rejection of the hypothesis implies acceptance of the alternative. Suppose that for testing the hypothesis one can afford only a single observation, that is, a sample of size one.

It is obvious in this instance that one should choose some number b in advance, then accept the hypothesis if the observation falls to the left of b and reject it if the observation falls to the right. Values of x to the left of b are said to constitute the acceptance region of the test; values to the right of b constitute the critical region. The area under $f(x)$ to the left of b, which is denoted mathematically by notation (24), is the probability that the hypothe-

$$\int_{-\infty}^{b} f(x) \, dx \qquad (24)$$

sis will be accepted when it is true. The area to the right of b under $f(x)$ is the probability that the hypothesis will be rejected when it is true; this probability is called the type I error of the test. The area to the left of b under $g(x)$ is the probability that the hypothesis will be accepted when it is false and is called the type II error of the test. The area to the right of b under $g(x)$ is called the power of test and is, of course, one minus the type II error.

By moving the point b to the right one can make the type I error as small as he likes, but in doing so the type II error is increased. This dilemma is characteristic of the construction of tests of hypotheses. Ordinarily one arbitrarily chooses the type I error to be some small number such as 0.05 or 0.01 and then chooses the critical region so as to minimize the type II error.

The fraction $f(x)/g(x)$ is called the likelihood ratio or simply the likelihood function of x. From Fig. 6 it is evident that the likelihood is relatively large when the hypothesis is true and small when it is false. For a sample of size n with observations x_1, x_2, . . . , x_n the sample likelihood is defined to be the product of the individual likelihoods given in expression (25). Let all possible samples of size n

$$\frac{f(x_1)}{g(x_1)} \frac{f(x_2)}{g(x_2)} \cdots \frac{f(x_n)}{g(x_n)} \qquad (25)$$

be divided into two sets, with one set (the acceptance region) containing those for which the likelihood is larger than some number b and with the other set (the critical region) containing those samples for which the likelihood is less than b. It can be proved mathematically that this is the best critical region for testing the hypothesis in the sense that it minimizes the type II error for given type I error. The likelihood criterion therefore furnishes a procedure for constructing specific tests of hypotheses. Other approaches to the testing problem, such as Bayes' method or the minimax principle of game theory, lead to the same criterion.

If, for example, one applies the likelihood criterion to the problem of testing whether a normal population has a given mean, perhaps 10, and specifies that the type I error shall be 0.01, then the following procedure results: Draw a random sample of size n; construct an 0.99 confidence interval for the population mean μ; if the number 10 lies within the interval, accept the hypothesis; otherwise, reject it.

An important class of hypotheses has to do with tests of independence in multivariate populations. As an example, one may consider the population of registered voters in the United States as a bivariate population with one variable being their opinions (yes, no, undecided) on some political proposition of the moment, and the other variable being geographical location. Elements of the population may be classified into a so-called contingency table as shown in Fig. 7. The hypothesis of independ-

location	yes	no	undecided
east	n_{11}	n_{12}	n_{13}
south	n_{21}	n_{22}	n_{23}
midwest	n_{31}	n_{32}	n_{33}
west	n_{41}	n_{42}	n_{43}

Fig. 7. Contingency table.

ence asserts that the division of political opinion is unaffected by location. To test that hypothesis, a random sample of the population may be interviewed and classified into the 12 categories or cells of the contingency table. Let n_{ij} be the number of individuals falling in the cell in the ith row and the jth column of the table. The sum of all the n_{ij} is, of course, n, the sample size. The row sums may be denoted by r_1, r_2, r_3, r_4, and the column sums by c_1, c_2, c_3; the sum of each of the sets is n. On applying the likelihood criterion to the test of the hypothesis, one finds that it rests on expression (26), which

$$2(n \log n + \sum_{ij} n_{ij} \log n_{ij}$$
$$- \sum_i r_i \log r_i - \sum_j c_j \log c_j) \qquad (26)$$

is approximately distributed according to the chi-square distribution with six degrees of freedom. To test at the 0.05 level for the type I error, one would compute the above expression and compare it with 12.6, which is the value that marks 95% of the area under the chi-square distribution curve. If the expression turned out to be less than 12.6, one would accept the hypothesis of independence; otherwise, he would reject it.

Had there been R rows and C columns instead of four and three, one would have used the chi-square distribution with $(R-1)(C-1)$ degrees of freedom. For a trivariate population there would have been a three-way contingency table and four hypotheses of independence that could have been tested; three of them assert that a given criterion of classification is independent of the other two, and the fourth asserts that all three are mutually independent.

Design of experiments. An experiment is performed to obtain information about the relations between several variables. For example, one may study the effect of storage temperature and duration of storage on the flavor of a frozen food. Three variables (flavor, temperature, and duration) are involved; one (flavor) is called the subject of the experiment; the other two are called factors which influence the subject. Sometimes the factors have intrinsic value in themselves; sometimes they are merely nuisance variables which must be taken into account because it is impossible to perform the experiment without them.

There exist in the statistical literature great numbers of specific experimental designs. These are patterns for making experimental observations; the actual construction of the designs requires quite advanced mathematics based on group theory, finite geometries, and combinatorial analysis. The mathematical problem is to find a pattern from which it is possible to extract the desired information and yet minimize the number of observations.

Experimental designs are most important to the experimenter when observations are expensive to make and when more than one factor is involved in the experiment. In the past it was believed that the best experimental procedure was to vary factors one at a time. Thus, in the frozen food experiment one might have held storage temperature constant and studied the effect of duration of storage only on flavor. Having determined that relationship, one would then hold duration constant and study the temperature effect. This procedure is not only wasteful of time and resources but may very well lead to erroneous conclusions because in all likelihood there is interaction between the two factors; that is, duration effect probably changes in a not obvious way when temperature is changed. Even if one believes that he can extrapolate accurately the duration effect to other temperatures, there is much to be said for checking the extrapolation procedure in the experiment because it can probably be done at no additional cost if the experiment is well designed.

As an illustration of the use of a design, consider a very simple and useful one called the randomized block design. Suppose a manufacturer contemplates purchasing a machine which can be obtained from one of three sources, X, Y, or Z. He obtains one of each on trial in order to compare their performance using several operators from his own plant. Perhaps five operators, A, B, C, D, and E, are to be used in the proposed experiment, which is a two-factor experiment with performance being the subject and the two factors being men and machines. The operators are not really a factor of interest in evaluating the machines but are, of course, necessary to the experiment.

It would be a mistake to use perhaps 15 operators, putting 5 on each machine, because operators differ in ability and a machine that turned out well might have been fortunate in having a good set of operators assigned to it. It is necessary that the machines be compared in as nearly equivalent circumstances as possible; in this case, that is done by having each operator operate every machine. The machines may then be compared in blocks (operators) which are individually homogeneous although they may be quite different among themselves.

The data are obtained by measuring production of each operator on each machine for a specified period of time and filling in the table of Fig. 8. It is essential that the order in which a man works on the three machines be randomized (by tossing dice, for example) so that minor factors not taken into account in the design will not bias the results. If, for example, learning is an important factor in operating the machines, then the experiment must

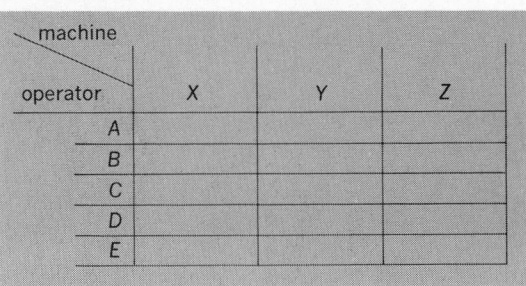

Fig. 8. Data form for randomized block experiment.

be three-factor and a more elaborate design is required.

In the resulting experimental observations the effects of men and of machines are entangled, but there exists a computational technique for well-designed experiments known as the analysis of variance which will disentangle them. This enables the total variation to be broken down into parts, a variance attributable to machines, one to men, and one to interactions; if the experiment were duplicated, one could also break out a variance corresponding to experimental error. One can estimate the individual main effects of machines, the main effects of men, and individual interaction effects. In experiments involving more factors, one may be able (depending on the experimental design selected) to estimate higher-order interactions such as the second-order interactions between the main effects of one factor and the first-order interactions of two other factors.

A common hypothesis in an analysis of variance is the null hypothesis that the variance associated with some factor or interaction is not larger than the experimental error variance: the test criterion is essentially the ratio of the two variances and has the so-called F distribution when the null hypothesis is true. Rejection of the null hypothesis implies that the factor or interaction in question had a significant effect on the subject of the experiment.

Regression and correlation. The regression problem is that of estimating certain unknown constants or parameters occurring in a function which relates several variables; the variables may be random or not. By far the most easily handled cases are those in which the function is linear in the unknown parameters, and it is worth considerable effort to transform the function to that form if at all possible.

The eventual adult height H of a 5-year-old boy may be quite well predicted by a linear function of three variables: his own present height B, his father's height F, and his mother's height M. The linear function is given in Eq. (27), where a, b, c,

$$H = a + bB + cF + dM \qquad (27)$$

and d are the unknown parameters. The three variables B, F, and M are called independent variables, the variable H is called the dependent variable, and the parameters b, c, and d are often referred to as regression coefficients.

To estimate the parameters, one might draw a random sample of 5-year-old boys, measure their heights and those of their parents, then some years later measure their adult heights. One would then have sufficient data from which the parameters could be estimated by procedures entirely analogous to the methods for estimating location and scale parameters described above. In practice, one would not take so long to get the data but would use men whose childhood records were available. This method involves subtle sampling problems, however; for example, persons whose records are available may have been better cared for on the average and hence taller on the average.

The data which supply the estimates of regression coefficients can also be used to estimate the standard deviation, or standard error of regression, σ. Using that estimate and a table of the t distribution, one can compute a prediction interval (analogous to a confidence interval) which will have the desired probability of including the correct adult height of a given boy.

Tall fathers sometimes have short sons and short fathers sometimes have tall sons, but generally a father's height is a good indicator of his son's height; that is, fathers' heights and their sons' heights are positively correlated. To employ a different example: The price of a commodity in a free economy is negatively correlated with the supply of that commodity; there is not a fixed relation between price and supply, but as a general proposition one goes up when the other goes down.

Statisticians have developed measures of the degree of such imprecise relationships called coefficients of correlation. The most widely used one is the Pearson, or product moment, correlation, which is generally denoted by ρ. It measures the degree of linear association or correlation between two random variables of a bivariate or multivariate population and is defined as the mean over all the population of the product of the deviations of the two variables from their means divided by the product of their standard deviations. Mathematically, it is as shown in Eq. (28), where x and y

$$\rho = \int (x-\mu)(y-\nu) f(x,y) \, dx \, dy / \sigma_x \sigma_y \quad (28)$$

are the random variables, μ and ν their means, σ_x and σ_y their standard deviations, and $f(x,y)$ their frequency distribution.

If the two variables are completely unrelated, then $\rho = 0$; if they have a fixed linear relation so that one can be calculated directly from the other, then $\rho = +1$ or $\rho = -1$ depending upon whether their relation is direct or reverse. Otherwise ρ will be some fraction between -1 and $+1$ with the fraction being near zero if there is poor correlation between them.

When the independent variables of a regression function are random variables, there is an equivalence between correlation coefficients and regression coefficients; after one set has been defined, the definition of the other follows automatically; mathematical formulas connecting them may be found in any statistics textbook.

Nonparametric inference. Most techniques of statistical inference rely on the central limit theorem or on sampling distributions derived from normal populations. They are practically always valid for large samples and are often valid for samples of intermediate size. However, there are occasions when one cannot rely on these techniques, particularly when samples are small or when some evident peculiarity of the population (such as marked asymmetry) makes ordinarily

used sampling distributions suspect. In such instances one uses nonparametric methods, which are valid whatever form the population distribution might take.

Nonparametric techniques use the so-called order statistics, which are merely the sample observations arranged in ascending order of magnitude. One may let x_1 be the smallest sample observation, x_2 the next smallest, and so on with x_n being the largest. A basic theorem states that on the average the sample divides the population into $n+1$ equal parts, that is, that $1/(n+1)$ of the population lies between any two successive order statistics. Many nonparametric methods rest on this fact and its consequences.

As a simple illustration of a nonparametric estimate and confidence interval, one may consider the estimation of the population median, M, given an ordered sample of size 5 consisting of x_1, x_2, x_3, x_4, and x_5. The estimate of M among the observations is simply the central sample observation x_3. The extreme observations x_1 and x_5 provide limits for a confidence interval for M; the probability level for the interval is calculated as follows. One-half of the population lies to the right of M, hence the probability is $1/2$ that a randomly drawn observation will lie to its right. The probability is $(1/2)^5$ that all five observations will lie to its right. Similarly, the probability is $(1/2)^5$ that all five will lie to its left. In all other cases the sample will have at least one observation on each side of M; therefore, expression (29) can be formulated. *See*

$$\text{Probability that } (x_1 < M < x_5) = 1 - (1/2)^5 \\ - (1/2)^5 = 30/32 = \text{about } 0.94 \quad (29)$$

ANALYSIS OF VARIANCE; DISTRIBUTION (PROBABILITY).

[ALEXANDER M. MOOD]

Bibliography: K. A. Brownlee, *Statistical Theory and Methodology in Science and Engineering*, 2d ed., 1965; W. J. Dixon and F. J. Massey, *Introduction to Statistical Analysis*, 3d ed., 1969; M. Hamburg, *Basic Statistics: A Modern Approach*, 2d ed., 1979; M. G. Kendall and A. Stuart, *The Advanced Theory of Statistics*, vol. 1, 4th ed., 1977, vol. 2, 4th ed., 1979, vol. 3, 3d ed., 1976; A. M. Mood, *Introduction to the Theory of Statistics*, 3d ed., 1973; D. H. Sanders, *Statistics: A Fresh Approach*, 2d ed., 1979.

Stochastic process

A physical stochastic process is any process governed by probabilistic laws. Examples are (1) development of a population as controlled by Mendelian genetics; (2) Brownian motion of microscopic particles subjected to molecular impacts or, on a different scale, the motion of stars in space; (3) succession of plays in a gambling house; and (4) passage of cars by a specified highway point.

In each case, a probabilistic system is evolving; that is, its state is changing with time. Thus the state at time t depends on chance: It is a random variable $x(t)$. The parameter set of values of t involved is usually (and will always be in this article) either an interval (continuous parameter stochastic process) or a set of integers (discrete parameter stochastic process). Some authors, however, apply the term stochastic process only to the continuous parameter case.

If the state of the system is described by a single

number, $x(t)$ is numerical-valued. In other cases, $x(t)$ may be vector-valued or even more complicated. The discussion in this article will usually be restricted to the numerical case. As the state changes, its values determine a function of time, the sample function, and the probability laws governing the process determine the probabilities assigned to the various possible properties of sample functions.

A mathematical stochastic process is a mathematical structure inspired by the concept of a physical stochastic process, and studied because it is a mathematical model of a physical stochastic process or because of its intrinsic mathematical interest and its applications both in and outside the field of probability. The mathematical stochastic process is defined simply as a family of random variables. That is, a parameter set is specified, and to each parameter point t a random variable $x(t)$ is specified. If one recalls that a random variable is itself a function, if one denotes a point of the domain of the random variable $x(t)$ by ω, and if one denotes the value of this random variable at ω by $x(t,\omega)$, it results that the stochastic process is completely specified by the function of the pair (t,ω) just defined, together with the assignment of probabilities. If t is fixed, this function of two variables defines a function of ω, namely, the random variable denoted by $x(t)$. If ω is fixed, this function of two variables defines a function of t, a sample function of the process.

Probabilities are ordinarily assigned to a stochastic process by assigning joint probability distributions to its random variables. These joint distributions, together with the probabilities derived from them, can be interpreted as probabilities of properties of sample functions. For example, if t_0 is a parameter value, the probability that a sample function is positive at time t_0 is the probability that the random variable $x(t_0)$ has a positive value. The fundamental theorem at this level is that, to any self-consistent assignment of joint probability distributions, there corresponds a stochastic process.

The concept of a stochastic process is so general that the study of stochastic processes includes all of probability theory. Although it would be impossible to make a useful but more restrictive definition, what the probabilist usually has in mind in using the term stochastic process (unless he is interested in the mathematical foundations of the general theory) is a stochastic process whose random variables have some sort of interesting mutual relations. For example, one such relation is that of independence, and one type of stochastic process which was studied long before the term was invented is a sequence of independent random variables. For historical reasons, such a sequence is not commonly thought of as a stochastic process (although its continuous parameter analog, a process with independent increments, to be discussed below, is). The rest of this article is devoted to a general discussion of specific types of stochastic processes which have received the most attention, because they are important in mathematical and nonmathematical applications.

Stationary processes. These are the stochastic processes for which the joint distribution of any finite number of the random variables is unaffected by translations of the parameter; that is, the distribution of $x(t_1 + h), \ldots, x(t_n + h)$ does not depend on h. For a more complete discussion *see* PROBABILITY.

Markov processes. A Markov process is a process for which, if the present is given, the future and past are independent of each other. More precisely, if $t_1 < \cdots < t_n$ are parameter values, and if $1 < j < n$, then the sets of random variables $[x(t_1), \ldots, x(t_{j-1})]$ and $[x(t_{j+1}), \ldots, x(t_n)]$ are mutually independent for given $x(t_j)$. Equivalently, the conditional probability distribution of $x(t_n)$ for given $x(t_1), \ldots, x(t_{n-1})$ depends only on the specified value of $x(t_{n-1})$ and is in fact the conditional probability distribution of $x(t_n)$, given $x(t_{n-1})$. An important and simple example is the Markov chain, in which the number of states is finite or denumerably infinite. (The terminology varies somewhat here.) One simple type of discrete-parameter Markov chain is the following. Let (p_{ij}) be a set of numbers, where i and j range over a finite or infinite set of integers. (In physical language, the number p_{ij} will be the probability that some system has a transition from state i to state j in one step.) The numbers p_{ij} are to satisfy relation (1). The

$$p_{ij} \geq 0 \qquad \sum_j p_{ij} = 1 \qquad (1)$$

random variables of the associated Markov process are integral-valued, denoted by $x(0)$, $x(1)$, $\ldots$. If i_0 is prescribed as the initial state, that is, if $x(0)$ is assigned the value i_0 identically, the probability that $x(k)$ has the value i_k, for $k = 1, \ldots, N$, is product (2). For this process, p_{ij} is

$$p_{i_0 i_1} \cdots p_{i_{N-1} i_N} \qquad (2)$$

the probability that $x(n+1)$ has the value j if $x(n)$ has the value i. The number p_{ij} is also (and this is the characteristic property of Markov processes) the probability that $x(n+1)$ has the value j if $x(n)$ has the value i and if also $x(n-1)$ has any prescribed value a_1, $x(n-2)$ the value a_2, and so on. What makes this a special Markov chain, aside from the fact that the states are denoted by integers, is that the conditional probability just described does not depend on n. The chain is therefore described as having stationary transition probabilities. If the initial state is given a distribution, say by prescribing that $x(0)$ have the value i with probability p_i, evaluation (2) becomes sum (3).

$$\sum_i p_i p_{i i_1} \cdots p_{i_{N-1} i_N} \qquad (3)$$

Here the p_i's are any nonnegative numbers with sum 1. If the initial distribution is chosen, as is not always possible, in such a way that the probability that $x(n)$ has the value j is p_j, not only for $n = 0$ but for all values of n, the resulting process is stationary.

In constructing the corresponding continuous-parameter Markov chain, it is supposed that, for each pair (i,j), there is a function $p_{ij}(.)$, defined for strictly positive t, satisfying relations (4). The

$$p_{ij}(t) \geq 0 \qquad \sum_j p_{ij}(t) = 1$$
$$p_{ij}(s+t) = \sum_k p_{ik}(s)\, p_{kj}(t) \qquad (4)$$

equations of the system in the last line are known as the Chapman-Kolmogorov equations. A Markov stochastic process with continuous parameter ranging from 0 to ∞ can be constructed for which, if again p_j is the probability that $x(0)$ has the value j, and if $0 < t_1 < \cdots < t_n$, the probability that $x(t_k)$ has the value i_k for $k = 1, \ldots, N$ is given by sum (5). For this process, if $s > 0$, the probabil-

$$\sum_i p_i p_{ii_1}(t_1) p_{i_1 i_2}(t_2 - t_1) \cdots$$
$$p_{i_{N-1} i_N}(t_N - t_{N-1}) \quad (5)$$

ity that $x(u + s)$ has the value j, if $x(u)$ has the value i, is $p_{ij}(s)$. The number $p_{ij}(s)$ is also the probability (and this is the characteristic property of Markov processes) that $x(u + s)$ has the value j if $x(u)$ has the value i; and if also $x(u_1)$ has any specified value $a_1, x(u_2)$, the value a_2, and so on, where $u_1, u_2, \ldots$, are any positive numbers less than u. This example is not the general continuous-parameter Markov chain because the transition probability just described does not depend on u, that is, because the chain has stationary transition probabilities. The process is stationary if the probability that $x(u)$ has the value j does not depend on u, for all j. The second line of relations (4) has a simple interpretation: The probability, if $x(u)$ has the value i, that $x(u + s + t)$ has the value j, is the sum over k of the probability that $x(u + s)$ has the value k multiplied by the probability that, if $x(u)$ has the value i, and if $x(u + s)$ has the value k, then $x(u + s + t)$ has the value j. Without the Markov property, the second factor might depend on i. If the number of states is not finite or denumerably infinite, the preceding discussion is modified by replacing sums in notations (1), (3), (4), and (5) by integrals.

Typical applications. Typical questions that have been raised, and solved to a varying degree, about Markov processes with stationary transition probabilities are the following. They are phrased in the continuous-parameter case, for the Markov chain just described, and it is assumed that relation (6) holds. For convenience one defines $p_{ij}(0)$ as 1 if $i = j$ and as 0 otherwise.

$$\lim_{t \to 0} p_{ii}(t) = 1 \quad \text{for all } i \quad (6)$$

1. Does $p_{ij}(t)$ have a limit when $t \to \infty$? In other words, for each j is there a limiting probability that the system is in state j as time passes? The answer is yes, and the limiting probability depends only on the end state j, not on the initial state i, if transitions between all pairs of states are possible.

2. What are the asymptotic properties of $p_{ij}(t)$ as $t \to 0$? The answer is that $p'_{ij}(0)$ always exists and is finite except possibly when $i = j$. Under further hypotheses on the transition probability functions, always satisfied if there are only finitely many states, $p'_{ii}(0)$ is finite and relations (7) hold. These

$$p'_{ij}(t) = \sum_k p_{ik}(t) p'_{kj}(0)$$
$$p'_{ij}(t) = \sum_k p'_{ik}(0) p_{kj}(t) \quad (7)$$

equations can be used to determine the transition probability functions in terms of assigned derivatives when $t = 0$. For example, if c is a strictly positive constant, and if $p'_{ij}(0)$ is specified as c if $j = i + 1$, as $-c$ if $j = i$, and as 0 otherwise, it is shown that $p_{ij}(t) = 0$ if $j < i$, and that otherwise relation (8) is

$$p_{ij}(t) = \frac{(ct)^{j-i} e^{-ct}}{(j-i)!} \quad (8)$$

true. The process with these transition probabilities is known as the Poisson process.

3. What are the properties of the sample functions? Under further restrictions on the process, the sample functions are constant on intervals, changing in jumps from one state to the next, and $-p'_{ij}(0)/p'_{ii}(0)$ is the probability that, if the system is in state i, its next jump will be into state j. If it is in the ith state, the time the system remains in this state thereafter is a random variable with density qe^{-tq}, where $q = -p'_{ii}(0)$. For example, in the case of the Poisson process described above, it is shown that, under proper normalization, and if $x(0) = 0$, the sample functions are integral-valued and monotone, increasing in unit jumps. This process is a mathematical model for the physical process of radioactive decay. That is, $x(t)$ can be interpreted as the number of radioactive disintegrations of a substance by time t. In other interpretations, $x(t)$ is taken as the number of telephone calls initiated by time t, or the number of cars that have passed a given highway point by time t. The constant c is the rate at which these various events occur. In fact, the expected value of $x(u + h) - x(u)$, that is, the expected number of events in a time interval of length h (h here is of course positive) is ch, and the probability that an event will occur in an interval of length h, regardless of the past history of the process, is ch up to higher powers of h.

There are many special types of Markov chains, for which more detailed questions become important. For example, consider the branching processes. In a system of particles, all of the same type, a particle will occasionally split, independently of its past history and of the other particles, into $j \geq 0$ particles with probability q_j. If a particle is observed at time t, the probability that it will split by time $t + h$ is ch, up to higher powers of h, where c is a strictly positive constant. Then the number of particles at time t is a random variable $x(t)$. The $x(t)$ process is a Markov process, and $p'_{ij}(0)$ is easily determined in terms of q_j and c. A more general branching process would permit particles of several types, and each particle would be allowed to split into particles of the various types. The rate c would depend on the particle type. In this case, $x(t)$ is defined as a vector whose ith component is the number of particles of type i at time t. The $x(t)$ stochastic process is a vector-valued Markov process. In studying branching processes, the most natural questions to ask are: What is the probability that the population will die out? If it does not die out, what is the asymptotic distribution of population as time passes? The answers are too technical to be given here.

Transition probabilities. If the states of a Markov process comprise all real numbers, the character of the process may be similar to that of a chain but may also be quite different. For example, the sample functions of the process may be continuous. The most important examples of this type are the diffusion processes. Simplifying somewhat, but not assuming stationary transition probabilities, consider a Markov process for which the probability distribution of the state at time t, given state ξ at time $s < t$, has density $p(s, \xi, t, \eta)$. Then the basic conditions satisfied by the transition density, corresponding to relation (1), are relations (9). Now

$$p(s,\xi,t,\eta) \geq 0 \qquad \int_{-\infty}^{\infty} p(s,\xi,t,\eta)\,d\eta = 1 \tag{9}$$

$$p(s,\xi,t,\eta) = \int_{-\infty}^{\infty} p(s,\xi,u,\varsigma)p(u,\varsigma,t,\eta)\,d\xi \text{ if } s < u < t$$

suppose that limits (10) and (11) exist and have the

$$\lim_{h\downarrow 0}\int_{-\infty}^{\infty} p(s,\xi,s+h,\eta)(\eta-\xi)\,d\eta/h = m(s,\xi) \tag{10}$$

$$\lim_{h\downarrow 0}\int_{-\infty}^{\infty} p(s,\xi,s+h,\eta)(\eta-\xi)^2\,d\eta/h = \sigma^2(s,\xi) \tag{11}$$

indicated values. These limit relations make m and σ^2 the instantaneous rates of change of the displacement and its variance, given a specified time and state. If m and σ are sufficiently regular, and if a further condition is imposed which, roughly, makes improbable significant sample function changes in short times, the corresponding Markov process, when properly normalized, will have continuous sample functions. Moreover, the transition density will then satisfy backward-diffusion equation (12) and forward diffusion equation (13), also

$$\frac{\partial p(s,\xi,t,\eta)}{\partial s} + m(s,\xi)\frac{\partial p}{\partial\xi} + \frac{1}{2}\sigma^2(s,\xi)\frac{\partial^2 p}{\partial\xi^2} = 0 \tag{12}$$

$$\frac{\partial p(s,\xi,t,\eta)}{\partial t} + \frac{\partial}{\partial\eta}[m(t,\eta)p]$$
$$-\frac{1}{2}\frac{\partial^2}{\partial\eta^2}[\sigma^2(t,\eta)p] = 0 \tag{13}$$

known as the Fokker-Planck equation. Conversely, given a pair of coefficient functions m and σ, these second-order parabolic equations can be used to derive the corresponding transition densities.

The simplest nontrivial example of a diffusion process corresponds to the specification $m = 0$ and σ a constant function. In this case, the diffusion process is the Brownian motion process, or Wiener process: The increment $x(t) - x(s)$ has a Gaussian distribution with mean value 0 and variance $\sigma^2|t-s|$. This is a mathematical model for the physical Brownian motion. That is, if $x(t) - x(s)$ represents the displacement in a given direction of a Brownian particle between times s and t, this process is a good model for the actual motion.

Martingales. A martingale is a stochastic process with the property that, if $t_1 < \cdots < t_n$ are parameter values, the expected value of $x(t_n)$, for given $x(t_1), \ldots, x(t_{n-1})$ is equal to $x(t_{n-1})$. That is, the expected future value, given present and past values, is equal to the present value. The interpretation that a martingale can be thought of as the fortune of a player after the successive plays of a fair gambling game is obvious.

Typical results on martingales are the following. If a sequence of random variables is a martingale, it converges under weak conditions which impose certain "bounds" on the random variables, for example, if the expectation of the absolute value of the nth random variable is bounded independently of n. The sample functions of a properly normalized continuous-parameter martingale do not have oscillatory discontinuities.

The applications of martingale theory are too technical to be given here, but one suggestive example will be given, which indicates at least how the theory can be usefully applied to information theory. Let $y, x_1, x_2, \ldots$ be random variables, and let y_n be the expected value of y knowing $x_1, \ldots, x_n$. Then y_n is a random variable which is a function of $x_1, \ldots, x_n$ and the sequence $y_1, y_2, \ldots$ is a martingale. That is, the expected value of a random variable, if one knows more and more, defines a sequence of random variables which is a martingale.

Processes with independent increments. Such a process is a continuous-parameter process with the property that, if $t_1 < \cdots < t_n$ are parameter values, the successive increments in notation (14)

$$x(t_2) - x(t_1), \ldots, x(t_n) - x(t_{n-1}) \tag{14}$$

are mutually independent. If $y(t) = x(t) - x(t_0)$, where t_0 is fixed, the $y(t)$ process is then a Markov process. Both the Poisson and the Brownian motion processes described above have independent increments.

Typical results on these processes include the following. If such a process is properly normalized, its sample functions do not have oscillatory discontinuities. Moreover, the distribution of any increment $x(t) - x(s)$ is infinitely divisible, and there is a standard form for the characteristic function of any such distribution.

[JOSEPH L. DOOB]

Bibliography: T. Assefi, *Stochastic Processes and Estimation Theory with Applications*, 1979; M. S. Bartlett, *An Introduction to Stochastic Processes*, 3d ed., 1978; J. L. Doob, *Stochastic Processes*, 1953; S. Karlin, *A First Course in Stochastic Processes*, 2d ed., 1975; E. Parzen, *Stochastic Processes*, 1962; M. Rosenblatt, *Random Processes*, 1974; S. K. Srinivasan and K. M. Mehata, *Stochastic Processes*, 1978.

Stokes stream function

A degenerate (one-component) vector potential used in analyzing and describing axially symmetric fluid-flow fields. In a steady axially symmetric flow, the rotation of a streamline about the axis of symmetry generates a stream surface. A certain mass rate of flow exists inside this stream surface which is the same at every axial station because, by definition, there is no flow through the stream surface. The value of the Stokes stream function ψ at a point in the flow is equal to $1/2\pi$ times mass rate of flow inside the stream surface passing through that point. If r is the radial coordinate of a point, and x the axial coordinate, then $\psi = \psi(x,r) = $ constant is the equation of a stream surface as shown in the illustration. At a station where $x = $ constant, the differential amount of mass flow between two stream surfaces with radial distance dr between them and of density ρ is given by Eq. (1). Similarly, on an $r = $ constant cylinder, the mass

$$2\pi\,d\psi = 2\pi\frac{\partial\psi}{\partial r}\,dr = (2\pi r\,dr)\,\rho v_x \tag{1}$$

flow between two stream surfaces with axial distance dx between them is given by Eq. (2). Thus

$$2\pi\,d\psi = 2\pi\frac{\partial\psi}{\partial x}\,dx = -(2\pi r\,dx)\,\rho v_r \tag{2}$$

the radial and axial velocity components at any

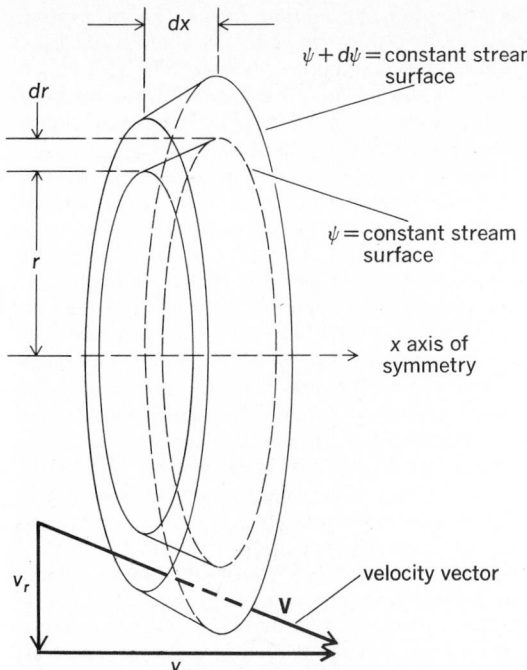

Section of axially symmetric fluid-flow field.

point are given by Eqs. (3), where ρ is fluid density.

$$v_r = -\frac{1}{\rho}\frac{1}{r}\frac{\partial\psi}{\partial x} \qquad v_x = \frac{1}{\rho}\frac{1}{r}\frac{\partial\psi}{\partial r} \qquad (3)$$

If the fluid motion is slow enough to neglect compressibility, ρ = constant and may be dropped from the above relations; ψ then measures the volume rate of flow inside a stream surface. *See* FLUID-FLOW PRINCIPLES.

[ARTHUR E. BRYSON, JR.]

Strange particles

Particles possessing the attribute of strangeness, a quantum number associated with one of the several quarks which are thought to constitute their structure. The nomenclature arose from the fact that the production rates for strange particles, which are compatible with strong interaction times of the order of 10^{-23} s, appeared to be inconsistent with their relatively long lifetimes (in the range 10^{-8} to 10^{-10} s). This inconsistency was resolved by experimental observations that strange particles are invariably produced in pairs (associated production) of equal and opposite strangeness, while their decay channels are restricted to those in which the strangeness quantum number is not conserved. This nonconservation of strangeness greatly suppresses the decay rates relative to the production rates. *See* QUARKS.

Strange particles with baryon number 0 are designated K mesons; those with baryon number 1 are designated hyperons. When characterized by a new quantum number called the hypercharge Y (related to the previously used strangeness quantum number S by $Y = S + B$, where B is the baryon number), strange particles are seen as members of multiplets which include the smaller nonstrange particle multiplets of mesons and nucleons. Additional quantum numbers, designated as charm and beauty (and perhaps there are others), de-

tected since 1974 are expected to expand this multiplet structure still further. Families of particles possessing both strangeness and one or more of these new quantum numbers are also predicted. For discussions of strange particle production processes, reactions, and decay properties *see* BARYON; MESON. *See also* ELEMENTARY PARTICLE; SYMMETRY LAWS.   [HORACE D. TAFT]

Bibliography: R. K. Adair and E. C. Fowler, *Strange Particles*, 1963; F. E. Close, *An Introduction to Quarks and Partons*, 1979.

Strong nuclear interactions

One of the fundamental physical interactions, which acts between a pair of hadrons. Hadrons include the nucleons, that is, neutrons and protons; the strange baryons, such as lambda (Λ) and sigma (Σ); the mesons, such as pion (π) and rho (ρ); and the strange meson, kaon (K). The nature of the interaction is determined principally through observation of the collision of a hadron pair. From this one learns that the interaction has a short range of about 10^{-15} m and is by far the dominant force within this range, being much larger than the electromagnetic interaction, which is next in magnitude. The strong interaction conserves parity and is time-reversal-invariant. *See* BARYON; HADRON; MESON; PARITY; STRANGE PARTICLES; SYMMETRY LAWS.

Meson exchange. The interaction between the baryons, including the nucleons and the strange baryons, is thought to arise from the exchange of mesons. The interaction for relatively large distances between nucleons is generated by the exchange of single pions (Fig. 1a). At shorter separation distances the exchange of two-pion systems, such as the ρ (Fig. 1b), dominates. An important

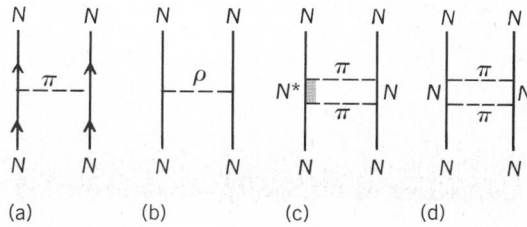

Fig. 1. Interaction between nucleons (a) from exchange of single pion, (b) from exchange of ρ-meson, a two-pion system, (c) from exchange of two separate pions with formation of excited state of nucleon, and (d) without formation of excited state.

contribution is furnished by the exchange of two separate pions with (Fig. 1c) or without (Fig. 1d) the formation of an excited state of the nucleon. The interaction between the strange baryons, and between the strange baryons and the nucleons, also arises in part from the exchange of pions, but kaon exchange can be equally important (Fig. 2).

Range. The range of the interaction generated by the exchanges illustrated in Figs. 1 and 2 can be calculated by using the simple formula below,

$$\text{Range} = \hbar/mc$$

where m is the mass of the exchanged particles, $\hbar$ is Planck's constant divided by 2π, and c is the speed of light. According to the above equation,

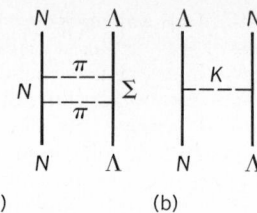

Fig. 2. Interaction between nucleon and strange (Λ) baryon from (a) exchange of pions and (b) exchange of kaon.

the range of the interaction developed when a single pion is exchanged (Fig. 1a) is equal to 1.4×10^{-15} m, while that due to Fig. 1c is 0.7×10^{-15} m.

The internal structure and finite size of the baryons (they consist of three quarks enclosed in a "bag," the quarks interacting through the exchange of gluons) must be considered when the baryons are very close together, that is, for separation distances less than approximately 0.7×10^{-15} m. The consequences for nuclear forces are being actively investigated. *See* GLUONS; QUARKS.

Nucleon-nucleon interaction. The nucleon-nucleon interaction is the most thoroughly investigated strong interaction. This strong interaction is found to be approximately charge-independent; that is, the interactions between two protons and between a neutron and proton are found to be equal when their configurations are identical. The strong interaction is found to be charge-symmetric; that is, the strong interaction between protons is identical to the strong interaction between neutrons, again in similar states. *See* ISOTOPIC SPIN.

The nucleon-nucleon interaction is most precisely known when the nucleons are separated by more than 0.7×10^{-15} m. For much smaller separation the interaction can be described empirically as strongly repulsive, this region being referred to as the hard core. For larger separations the interaction is attractive, falling off exponentially with distances in a manner dictated by the range of the interaction as given by the equation above.

Spin dependence. The nucleons have a spin and therefore their interaction is spin-dependent, that is, dependent on the spatial orientation of their spins. Because of this spin dependence, the simplest atomic nucleus, the deuteron, consisting of a neutron and proton, does not have a spherical shape. The spin dependence of the nucleon-nucleon interaction can be investigated by collisions between polarized nucleons, that is, nucleons whose spin have a definite spatial orientation. A resonance in the proton-proton collision has been discovered when energetic protons polarized in

their direction of motion interact with protons polarized in either the same or opposite direction (Fig. 3). This resonance is thought to be a consequence of the possible excitation of one (or both) nucleons to an excited state (Fig. 1d). *See* SCATTERING EXPERIMENTS (NUCLEI); SPIN.

Nuclear structure. The properties of nuclear forces are reflected in the structure of atomic nuclei. Qualitative aspects such as the short range and strength were first discovered in this manner. Another property is that of saturation; in other words, their general character is such that inside a nucleus each nucleon cannot interact with more than a few neighboring nucleons. The quantitative connection between nuclear forces and nuclear structure is not completely known and continues to be under active investigation. The fundamental goal is the determination of how the forces between a nucleon pair inside the nucleus are modified by the presence of the other nucleons in the nucleus. *See* NUCLEAR STRUCTURE.

Hypernuclei. The force between a nucleon and a Λ or Σ are being determined through the study of the properties of hypernuclei. The hypernuclei under current investigation consist of neutrons and protons and one Λ or Σ. The interaction between a Λ and a nucleon is not strong enough to form a stable two-body system like the deuteron. Another new feature is the existence of a substantial breakdown in charge symmetry, since it is found that the interaction between a Λ and a neutron is not the same as that between a Λ and a proton.

Exotic atoms. Further information about these systems is obtained from the exotic atoms, atoms formed by capturing negatively charged Σs in the attractive electric field of the nucleus. The radius of these Σ atomic orbits is much smaller than that of orbits of electrons in the electric field of the same nucleus, and as a consequence the Σ can interact via the strong interactions with the nucleons in the nucleus. This interaction is revealed by the properties of the x-rays emitted when the Σ changes its orbit.

Exotic atoms can also be formed by an atomic nucleus plus an antiproton. An antiproton has the same mass as the proton but carries a negative charge, whereas the proton carries a positive charge. Experiments involving the collision between antiprotons and protons promise to provide important insights into the strong interactions. *See* HADRONIC ATOM.

Interactions of mesons. Strong interactions exist between the pion and kaon with baryons and between pions and kaons. The interaction most thoroughly studied is that between pions and nucleons. It is found that when these interact they form a system which lives for about 6×10^{-22} s. It is referred to as the delta (Δ). The dynamics of the strong interaction of a pion with a nucleus depends largely on the behavior of the Δ formed by the pion with one of the nucleons in the nucleus. *See* ELEMENTARY PARTICLE; FUNDAMENTAL INTERACTIONS.

[HERMAN FESHBACH]

Bibliography: L. Cohen, *Concepts of Nuclear Physics*, K. Hidaka and A. Yokosawa, *Survey in High Energy Physics 1*, 1980; B. Povh, *Annu. Rev. Nucl. Particle Sci.*, 28:1, 1978.

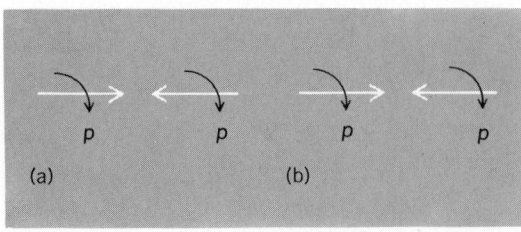

Fig. 3. Collision of two protons polarized in (a) same direction and (b) opposite direction.

Superaerodynamics

That branch of gas dynamics dealing with the flow of gases at such low density that the molecular mean free path is not negligibly small. Under these conditions the gas no longer behaves as a continuous fluid. Important modifications in flow phenomena occur which are ascribable to the discrete molecular structure of the gas. The subject is also often called rarefied gas dynamics. *See* GAS DYNAMICS.

Early work in the field was done by the kinetic theory physicists of the 19th century. This work was mostly concerned with low-speed flow of gas at low pressures through orifices and ducts. Following World War II there has been a revived interest in the field due to problems arising from high-speed flight at extremely high altitudes. This more recent work has thus been more concerned with the supersonic flow of low-pressure gas past aerodynamic objects.

Flow regimes. It is convenient to divide superaerodynamics into three flow regimes. These are called free molecule flow, transition flow, and slip flow, corresponding respectively to highly rarefied, moderately rarefied, and only slightly rarefied flow conditions. Phenomena in the three regimes are quite dissimilar, so the subdivision is useful. The term "rarefied" is relative, but can be made quantitative in terms of a dimensionless number, called the Knudsen number K, defined as the ratio of the mean free path λ divided by a characteristic dimension L of the flow field, for example, the diameter of a duct or the length of a high-altitude rocket ($K = \lambda/L$). Free molecule flow corresponds to $K \gg 1$, slip flow to $K \ll 1$, and transition to intermediate values of K.

The Knudsen number is related to two other basic parameters of fluid mechanics, namely the Mach number $M = V/a$ and the Reynolds number $Re = VL/\nu$, where V is a characteristic velocity of the flow, a is the speed of sound, and ν the kinematic viscosity. Kinetic theory shows that $\nu \sim a\lambda$; it follows that $K \sim M/Re$ and, in particular, that Eq. (1) holds, where γ is the isentropic exponent.

$$K = \sqrt{\frac{\pi}{2}} \gamma \frac{M}{Re} \qquad (1)$$

Free molecule flow. The regime of highly rarefied flow corresponds in the aerodynamic case to flight at altitudes of 100 mi or more. The mean free path is long compared to dimension L. It follows that molecules, which impinge on a surface and are then reemitted, travel very far before colliding with another molecule. The essential simplification of free molecule flow is that such collisions can be neglected. For the aerodynamic case of flow past a convex body the incident molecules are thus not disturbed by the presence of the body. For many applications it is correct to assume that the distribution of molecular velocities for these incident molecules is in Maxwellian equilibrium but with a mean streaming velocity component with respect to the aerodynamic body. For this case the incident flux of mass, momentum, and energy on a surface element can be easily calculated. The net flux then is determined by the nature of the distribution of velocities for the reemit-

ted molecules. Detailed knowledge of this sort of molecule-surface interaction has not yet been obtained. Under some circumstances the reemitted molecules emerge from the surface in Maxwellian equilibrium with it. Such reflection is called diffuse. However, under other circumstances this does not happen. The departure from diffuse reflection depends on the physical and chemical nature of the surface material and of the gas, upon their temperatures, upon the length of time the surface has been exposed to the gas at some particular pressure, and upon the nature of any film of adsorbed gas adhering to the surface. For engineering applications it has been customary to define three surface interaction parameters, also known as accommodation coefficients. These are given by Eqs. (2), where p_i, τ_i, and e_i are the fluxes

$$\alpha = \frac{e_i - e_r}{e_i - e_w}$$

$$\sigma = \frac{\tau_i - \tau_r}{\tau_i} \qquad (2)$$

$$\sigma' = \frac{p_i - p_r}{p_i - p_w}$$

of normal momentum, tangential momentum, and energy incident on the surface; p_r, τ_r, and e_r are the fluxes of these quantities reemitted; while p_w, τ_w, and e_w are the fluxes that would be reemitted for the case of diffuse reflection. Measurements of these quantities for air in contact with typical engineering surfaces are generally in the range $0.8 - 1.0$; however, very low values for α have also been reported, particularly for light gases in contact with very clean crystal surfaces. Utilizing the accommodation coefficient concept and carrying out the indicated calculations leads to basic formulas for net pressure p, shear stress τ, and heat flux q for a surface element at temperature T_w inclined at an angle of attack θ, to an incident free molecule flow of velocity V, temperature T, and density ρ. These are defined by Eqs. (3)–(5), where $S = V/\sqrt{2RT}$ is known as the molecular speed ratio, R is the specific gas constant, and Eq. (6) defines the error integral. By straightforward integration these results can be used to predict overall aerodynamic and heat transfer characteristics for arbitrary convex configurations.

At somewhat higher densities the effect of molecular collisions begins to be manifest. Molecules emitted from the surface begin to shield it from the incident flux, leading to a decrease in drag and heat transfer coefficients. The shielding effect can be shown to be of order L/λ and is also sensitive to the Mach number and the temperature conditions.

Slip flow. This is the flow regime of only slight rarefaction, corresponding in the aerodynamic case to flight at altitudes of the order of $20 - 50$ mi. Noncontinuum effects can be thought of in terms of small corrections to ordinary continuum flow. The characteristic dimension for many cases is the boundary layer thickness δ rather than the body dimension L. For low-speed flow $\delta/L \sim 1/\sqrt{Re_L}$;

$$p=\frac{\rho V^2}{2S^2}\left\{\left(\frac{2-\sigma'}{\sqrt{\pi}}S\sin\theta+\frac{\sigma'}{2}\sqrt{\frac{T_w}{T}}\right)e^{-S^2\sin^2\theta}+\left[(2-\sigma')(S^2\sin^2\theta-\tfrac{1}{2})+\frac{\sigma'}{2}\sqrt{\frac{\pi T_w}{T}}S\sin\theta\right]\left[1+\operatorname{erf}(S\sin\theta)\right]\right\}\quad(3)$$

$$\tau=\frac{\sigma\rho V^2\cos\theta}{2\sqrt{\pi}S}\left\{e^{-S^2\sin^2\theta}+\sqrt{\pi}S\sin\theta[1+\operatorname{erf}(S\sin\theta)]\right\}\quad(4)$$

$$q=\alpha\rho RT\sqrt{\frac{RT}{2\pi}}\left\{\left[S^2+\frac{\gamma}{\gamma-1}-\frac{\gamma+1}{2(\gamma-1)}\frac{T_w}{T}\right]\left(e^{-S^2\sin^2\theta}+\sqrt{\pi}S\sin\theta[1+\operatorname{erf}(S\sin\theta)]\right)-\tfrac{1}{2}e^{-S^2\sin^2\theta}\right\}\quad(5)$$

$$\operatorname{erf}(S\sin\theta)=\frac{2}{\sqrt{\pi}}\int_0^{S\sin\theta}e^{-x^2}dx\quad(6)$$

the significant Knudsen number is given by Eq. (7).

$$K=\lambda/\delta\sim M/\sqrt{Re}\quad(7)$$

For high-speed flow δ becomes dependent on the Mach number, the local heat transfer conditions, and the location on the aerodynamic body, so that a variety of different Knudsen numbers may become relevant.

The term slip flow arises from the phenomenon of slip, according to which a rarefied gas adjacent to a surface does not adhere rigidly to it but rather has a finite velocity, known as the slip velocity, determined by the local stress and temperature gradients. Physically, this phenomenon arises because the gas layer immediately adjacent to the surface is composed of molecules of which one-half are originating in the gas exterior to the surface and one-half are just being emitted from the surface. There is thus a discontinuity in the velocity and temperature. Mathematical analysis leads to a formulation for the velocity in the form of Eq. (8), where the subscript 0 refers to conditions at

$$u_0=\frac{2-\sigma}{\sigma}\lambda_0\left(\frac{\partial u}{\partial y}\right)_0+\frac{3}{4}\frac{\nu_0}{T_0}\left(\frac{\partial T}{\partial x}\right)_0\quad(8)$$

the surface, x is the coordinate along the surface in the direction of the flow, y is the coordinate normal to the surface, and u is the gas velocity in the x direction. The first term on the right side of this equation arises from the applied shear stress, while the second term, called thermal creep, arises from the temperature gradient in the direction of flow. The quantity $(2-\sigma)/\sigma$ is actually only a first approximation to a more complicated function of σ. The entire expression is applicable only for infinitesimal λ, $\partial u/\partial y$, and $\partial T/\partial x$. The corresponding thermal condition for the temperature jump is given by Eq. (9), where T_S is the actual surface

$$T_0-T_S=\frac{2-\alpha}{\alpha}\frac{2\gamma}{\gamma+1}\frac{\lambda_0}{Pr_0}\left(\frac{\partial T}{\partial y}\right)_0\quad(9)$$

temperature and Pr is the Prandtl number. This relation is subject to the same restrictions as the slip velocity condition. These relations express true noncontinuum effects which, by themselves, have the effect of reducing skin friction and heat transfer. Since they are of importance only at appreciable values of $M/\sqrt{Re}$ (or some other relevant Knudsen number), they are often obscured by continuum effects, for example, interaction effects arising from a combination of low Re and high M.

The situation with respect to the relevant flow equations is not quite so simple. Various modifications to the usual continuum Navier-Stokes equations have been proposed. However, in a series of experimental and theoretical investigations, none of these suggested modifications has so far turned out to be superior to the Navier-Stokes equations. Thus most present analytical work in the slip flow regime is being based on the slip velocity and temperature jump boundary conditions and on the Navier-Stokes equations. *See* NAVIER-STOKES EQUATIONS.

Transition flow. No simple formulation for the transition regime has yet been developed, except for the fundamental Maxwell-Boltzmann equation itself. Under some situations this equation can be solved for arbitrary values of λ/L, including those corresponding to transition flow conditions. These special solutions, together with a large number of experimental results, indicate that simple interpolation between slip flow on the one hand, and free molecule flow on the other hand, will usually suffice for the transition flow regime.

[SAMUEL A. SCHAAF]

Superconductivity

A phenomenon occurring at very low temperatures in many electrical conductors, in which the electrons responsible for conduction undergo a collective transition to an ordered state with many unique and remarkable properties. These include the vanishing of resistance to the flow of electric current, the appearance of a large diamagnetism and other unusual magnetic effects, substantial alteration of many thermal properties, and the occurrence of quantum effects otherwise observable only at the atomic and subatomic level.

Superconductivity was discovered by H. Kamerlingh Onnes in Leiden in 1911, while studying the variation with temperature of the electrical resistance of mercury within a few degrees of absolute zero. He observed that the resistance dropped sharply to an unmeasurably small value at a temperature of 4.2 Kamerlingh Onnes's original data are shown in Fig. 1. The temperature at which the transition occurs is called the transition or critical temperature, T_c. The vanishingly small resistance (very high conductivity) below T_c suggested the name given the phenomenon.

In 1933 W. Meissner and R. Ochsenfeld discovered that a metal cooled into the superconducting state in a not-too-large magnetic field expels the

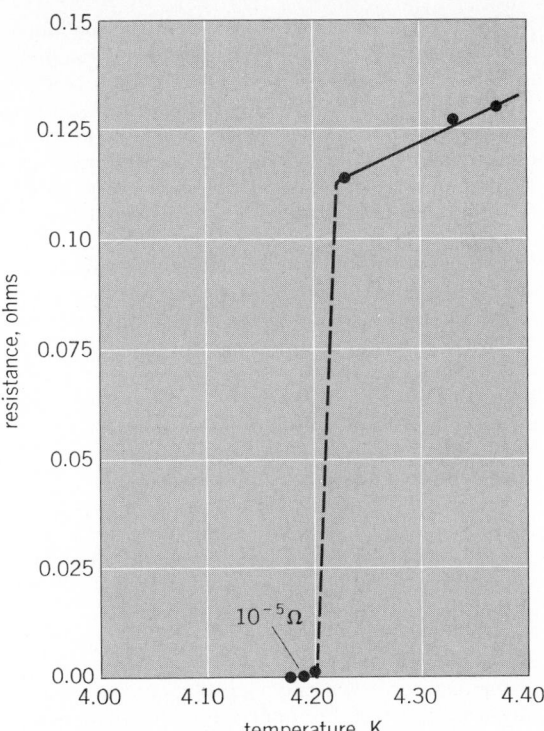

Fig. 1. Resistance in ohms of a specimen of mercury versus absolute temperature. (*From H. Kamerlingh Onnes, Akad. van Wetenschappen, Amsterdam, 14: 113, 818, 1911*)

field from its interior. This discovery demonstrated that superconductivity involves more than simply very high or infinite electrical conductivity, remarkable as that alone is. *See* MEISSNER EFFECT.

Superconductivity remained a much studied but puzzling phenomenon for nearly half a century after its discovery. A great deal of experimental information was amassed on its occurrence and its properties, and several useful phenomenological theories were developed. Then, in 1957, J. Bardeen, L. N. Cooper, and J. R. Schrieffer reported the first successful microscopic theory of superconductivity. It describes how and why the electrons in a conductor may form an ordered superconducting state, and makes predictions about many properties of superconductors which are in good agreement with experimental information.

Since the appearance of the Bardeen-Cooper-Schrieffer (BCS) theory, theoretical and experimental understanding of superconductivity has continued to expand. New superconducting materials and effects have continued to be discovered. Practical applications of the phenomenon are becoming common, ranging from powerful electromagnets and machinery to ultrasensitive electronic instruments and computer elements.

BASIC EXPERIMENTAL PROPERTIES

Soon after its discovery in mercury, superconductivity was found to occur also in such common metals as lead and tin. Initially, the number of known superconductors was quite small. This was so in part because experiments were then confined to temperatures above about 1 K, the minimum temperature readily available using liquid helium (^{4}He) as a refrigerant, so that only superconductors

with transition temperatures above 1 K could be discovered. It was therefore thought that superconductivity might be a relatively rare phenomenon. Scientists now know that it is not, but in fact occurs quite generally. Advances in the technology for achieving low temperatures have pushed the minimum available temperature down to about 0.001 K, and progress in the preparation of materials has greatly expanded the number and variety of materials which have been tested for superconductivity. Some 26 of the metallic elements are known to be superconductors in their normal forms, and another 10 become superconducting under pressure or when prepared in the form of highly disordered thin films (Fig. 2). The number of known superconducting compounds and alloys runs into the thousands. Superconductivity is thus a rather common characteristic of metallic conductors, so much so that its absence is often more unusual and striking than its presence.

Despite the existence of a successful microscopic theory of superconductivity, there are no completely reliable rules for predicting whether a metal will be a superconductor. Certain trends and correlations are apparent among the known superconductors, however—some with obvious bases in the theory—and these provide empirical guidelines in the search for new superconductors. Superconductors with relatively high transition temperatures tend to be rather poor conductors in the normal state. For many years, no superconductors were known among the noble metals, the alkali metals, and the alkaline earth metals. However, cesium, beryllium, and barium have been found to be superconducting under high pressure or in disordered films, and there is some evidence that at least one of the noble metals may be superconducting at extremely low temperatures.

The ordered superconducting state appears to be incompatible with any long-range-ordered magnetic state: None of the ferromagnetic or antiferro-

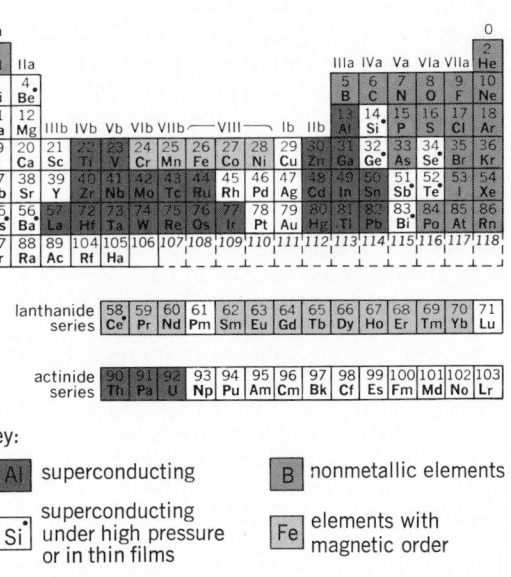

Fig. 2. Superconducting elements in the periodic table. (*From N. W. Ashcroft and N. D. Mermin, Solid State Physics, Holt, Rinehart and Winston, 1976*)

magnetic metals are also superconducting. (Cerium displays antiferromagnetic order but is not superconducting in the phase which exists at ordinary pressures; it is superconducting but not magnetically ordered in a different phase which occurs at pressures above about 5 GPa.) This distaste of superconductors for magnetism extends to the effects of impurities. The presence of nonmagnetic impurities in a superconductor usually has very little effect on the superconductivity, but the presence of impurity atoms which have localized magnetic moments can markedly depress the transition temperature even in concentrations as low as a few parts per million.

Some semiconductors with very high densities of charge carriers are superconducting, and others such as silicon and germanium have high-pressure metallic phases which are superconducting. Many elements which are not themselves superconducting form compounds which are. Examples are CuS and a polymeric form of Sn. Although nearly all the classes of crystal structure are represented among superconductors, certain structures appear to be especially conducive to superconductivity. An example is the so-called A-15 structure shared by a series of intermetallic compounds based on niobium, for example, Nb_3Sn with $T_c = 18.1$ K. High values of transition temperature occur most frequently in elements, compounds, or alloys having three, five, or seven valence electrons per atom.

The highest transition temperature observed is 23 K for a specially prepared alloy of niobium, aluminum, and germanium. There is no widely accepted theoretical proof that superconductivity is necessarily restricted to low temperatures. Encouraged by this and motivated by visions of the immense practical and fundamental implications of a "high-temperature" superconductor, many investigators have searched for materials with higher transition temperatures.

Electrical resistance. It is, of course, not possible to establish that the dc (zero-frequency) electrical resistance of a superconductor is identically zero, but a rather stringent upper limit on the resistance can be established by inducing an electrical current in a superconducting loop or coil and observing whether it dies away in time. The decay time of such a current in a nonsuperconducting coil at low temperatures is on the order of 1 sec or less. Induced currents have been observed to persist in superconducting loops for several years. Very precise measurements of the magnetic field produced by a persistent current, using nuclear magnetic resonance over shorter periods of time, have established that the supercurrent decay time is at least 100,000 years. This implies that the resistance in the superconducting state is at least 10^{12} times less than in the normal state.

It can be shown theoretically that the persistent current state is not really absolutely stable, but only metastable. In superconducting materials such as those used to build superconducting magnets, finite persistent current decay times are often observed, due to processes which cause irreversible redistribution of magnetic flux in the material. But under many conditions the lifetime of the metastable persistent current state is so long that it is not unreasonable to say that the lifetime is infinite, and the electrical resistance is zero.

Magnetic properties. The existence of the Meissner-Ochsenfeld effect, the exclusion of a magnetic field from the interior of a superconductor, is direct evidence that the superconducting state is not simply one of infinite electrical conductivity. If this were so, a superconductor cooled in a magnetic field through its transition temperature would trap the field in its interior. If the external source of the field were subsequently removed, persistent eddy currents would be induced in the superconductor which would preserve the interior field even in the absence of the external source. Instead, the Meissner-Ochsenfeld effect implies that the superconducting state is a true thermodynamic equilibrium state, a new phase which has lower free energy than the normal state at temperatures below the transition temperature and which somehow requires the absence of magnetic flux.

The exclusion of magnetic flux by a superconductor costs some magnetic energy. So long as this cost is less than the condensation energy gained by going from the normal to the superconducting phase, the superconductor will remain completely superconducting in an applied magnetic field. If the applied field becomes too large, the cost in magnetic energy will outweigh the gain in condensation energy, and the superconductor will become partially or totally normal. The manner in which this occurs depends on the geometry and the material of the superconductor. Consider first the simplest geometry, a very long cylinder with field applied parallel to its axis. Two distinct types of be-

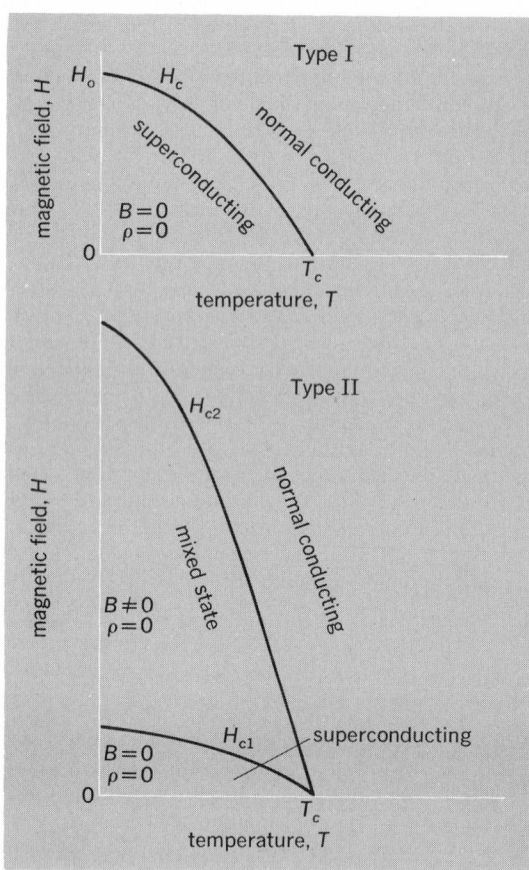

Fig. 3. The superconductive state in the magnetic-field temperature plane.

havior may then occur, depending on the type of superconductor.

Type I superconductors. Below a "critical field" H_c which increases as the temperature decreases below T_c, the magnetic flux is excluded from a type I superconductor, which is said to be perfectly diamagnetic. If the applied field is increased above H_c, the entire superconductor reverts to the normal state and the field penetrates completely. The curve of H_c versus temperature T in Fig. 3 is thus a phase boundary in the magnetic field-temperature plane separating a region where the superconducting phase is thermodynamically stable from the region where the normal phase is stable. The curve of H_c versus T for any type I superconductor is approximately parabolic: To within a few percent, $H_c = H_0[1 - (T/T_c)^2]$, where H_0 is the value of H_c at absolute zero. Values of T_c and H_0 for some typical superconductors are given in the table. All of the known elemental superconductors except niobium are of type I. *See* DIAMAGNETISM.

Type II superconductors. For a type II superconductor, there are two critical fields, the lower critical field H_{c1} and the upper critical field H_{c2}. In applied fields less than H_{c1}, the superconductor completely excludes the field, just as a type I superconductor does below H_c. At fields just above H_{c1}, however, flux begins to penetrate the superconductor in microscopic filaments called fluxoids or vortices. Each fluxoid consists of a normal core in which the magnetic field is large, surrounded by a superconducting region in which flows a vortex of persistent supercurrent which maintains the field in the core. The total magnetic flux in each fluxoid is exactly equal to a fundamental quantum of magnetic flux, $\phi_\Phi = 2.07 \times 10^{-7}$ gauss cm^2 = 2.07×10^{-15} Wb. The diameter of the fluxoid is typically 10^{-7} m. In a sufficiently pure and defect-free type II superconductor, the fluxoids tend to arrange themselves in a regular lattice. This vortex state of the superconductor is known as the mixed state. It exists for applied fields between H_{c1} and H_{c2}. At H_{c2}, the superconductor becomes normal, and the field penetrates completely. (Actually, a superconducting surface sheath may persist up to an even higher critical field H_{c3}, which is approximately 1.5 H_{c2}.) *See* QUANTIZED VORTICES.

In contrast to critical fields in type I superconductors, which tend to be less than 1000 oersteds (1 Oe = 79.6 A/m), H_{c2} for type II superconductors may be several hundred thousand oersteds or more. (The maximum known H_{c2} is about 600,000 Oe). Since a zero-resistance supercurrent can flow in the mixed state in the superconducting regions around the fluxoids, a type II superconductor can carry a lossless current even in the presence of a very large magnetic field. Such superconductors are of practical importance in high-field magnets.

A type II superconductor in the mixed state is not necessarily completely lossless, however. The presence of an electric current creates a force on the fluxoids. They therefore tend to move. Moving magnetic flux creates voltages by electromagnetic induction, and the presence of nonzero voltages together with the current implies power dissipation. This loss mechanism can often be suppressed by introducing defects into the crystal structure of the superconductor which tend to pin down the fluxoids and prevent them from moving.

Penetration depth. The way in which a super-

Values of T_c and H_0 for the superconducting elements*

Element	Phase	T_c (K)	H_0 (oersteds)†
Al		1.196	99
Cd		0.56	30
Ga		1.091	51
Hf		0.09	—
Hg	α(rhomb)	4.15	411
	β	3.95	339
In		3.40	293
Ir		0.14	19
La	α(hcp)	4.9	798
	β(fcc)	6.06	1096
Mo		0.92	98
Nb		9.26	1980†
Os		0.655	65
Pa		1.4	—
Pb		7.19	803
Re		1.698	198
Ru		0.49	66
Sn		3.72	305
Ta		4.48	830
Tc		7.77	1410
Th		1.368	162
Ti		0.39	100
Tl		2.39	171
U	α	0.68	—
	γ	1.80	—
V		5.30	1020
W		0.012	1
Zn		0.875	53
Zr		0.65	47

*From N. W. Ashcroft and N. D. Mermin, *Solid State Physics*, Holt, Rinehart, and Winston, 1976.

†At $T = 0$ K. 1 Oe = 76.9 A/m.

‡For Nb, a type II superconductor, the zero-temperature critical field quoted is obtained from an equal-area construction: The low-field ($H < H_{c1}$) magnetization is extrapolated linearly to a field H_c chosen to give an enclosed area equal to the area under the actual magnetization curve.

conductor excludes from its anterior an applied magnetic field smaller than H_c (type I) or H_{c1} (type II) is by establishing a persistent supercurrent on its surface which exactly cancels the applied field inside the superconductor. This surface current flows in a very thin layer of thickness λ, which is called the penetration depth. The external field also actually penetrates the superconductor within the penetration depth. Lambda depends on the material and on the temperature, the latter variation being given approximately by $\lambda = \lambda_0[1 - (T/T_c)^4]^{-1}$ (λ_0 is the penetration depth at zero temperature for the particular material, and is typically of order 5×10^{-8} m).

Intermediate state. Another kind of magnetic effect occurs in type I superconductors for all but the simplest geometries. The exclusion of magnetic flux by the superconductor distorts the field in its vicinity. As a result, the magnetic field may reach H_c at some points on the surface of the superconductor while remaining below H_c elsewhere. The superconductor near the points of highest field will tend to go normal. The magnetic flux then begins to penetrate the superconductor in nonsuperconducting lamellae separated by regions where superconductivity remains. If the applied field is further increased, the fraction of the specimen occupied by the normal lamellae increases, while the fraction occupied by the intervening superconducting regions decreases. With increasing applied field, this continues until the specimen becomes completely normal and the field penetrates everywhere. A type I superconductor with such alternating superconducting and normal lamellae is in the intermediate state.

There are superficial similarities between the intermediate state in a type I superconductor and the mixed state in a type II superconductor, but the two states are really quite different. In the intermediate state, the lamellae are of macroscopic size, sometimes an appreciable fraction of a millimeter. This size is determined by two energies, a magnetic energy which increases as the size of the lamellae increases, and a positive surface energy associated with the interface between the superconducting and normal regions, rather like the surface energy associated with the surface tension at a liquid-gas interface. This surface energy increases with the supernormal interface area and hence with the number of lamellae, that is, it decreases as the size of the lamellae increases. The magnetic energy favors small lamellae, and the surface energy favors large lamellae. The equilibrium lamellar structure is determined by a compromise between the two and is rather sensitive to the geometry and degree of perfection of the specimen. In type II superconductors, on the other hand, the supernormal interface energy turns out to be negative. It is therefore energetically favorable for the regions of flux penetration to be as small as possible. Their minimum size is limited only by a fundamental quantum constraint on the nature of the superconducting state itself. It is this which sets the level of magnetic flux contained within a fluxoid at the flux quantum mentioned above and determines the microscopic scale of the fluxoid structure in the mixed state.

Critical current. The existence of the critical field leads to another property of superconductors which is of some practical importance. In his Nobel prize lecture, Kamerlingh Onnes referred to the possibility of constructing powerful electromagnets which would consume no electrical power, using superconductors. However, a supercurrent flowing in a superconducting wire will itself create a magnetic field, and this field will drive the superconductor normal at some critical value of the current (the Silsbee critical current). Unfortunately, the critical currents which accompany typical critical fields for type I superconducting wires of macroscopic size are so small that Kamerlingh Onnes's dream remained unrealized until the exploitation of type II superconductors nearly 40 years later.

Thermal properties. The appearance of the superconducting state is accompanied by quite drastic changes in both the thermodynamic equilibrium and thermal transport properties of a superconductor.

Heat capacity. Figure 4 shows the heat capacity of an aluminum specimen in both the normal and superconducting states. In the normal state (produced at temperatures below the transition temperature by applying a magnetic field greater than the critical field), the heat capacity is determined primarily by the normal electrons (with a small contribution from the thermal vibrations of the crystal lattice) and is nearly proportional to the

temperature. In zero applied magnetic field, there appears a discontinuity in the heat capacity at the transition temperature. At temperatures just below the transition temperature, the heat capacity is larger than in the normal state. It decreases more rapidly with decreasing temperature, however, and at temperatures well below the transition temperature varies exponentially as $e^{-\Delta/kT}$, where Δ is a constant and k is Boltzmann's constant. Such an exponential temperature dependence is a hallmark of a system with a gap Δ in the spectrum of allowed energy states. Heat capacity measurements provided the first indications of such a gap in superconductors, and one of the key features of the macroscopic BCS theory is its prediction of just such a gap. *See* SPECIFIC HEAT OF SOLIDS.

Thermal conductivity. Ordinarily a large electrical conductivity is accompanied by a large thermal conductivity, as in the case of copper, used in electrical wiring and cooking pans. However, the thermal conductivity of a pure superconductor is less in the superconducting state than in the normal state, and at very low temperatures approaches zero. This property is applied in "heat switches" for use at low temperatures, in which the thermal contact between two bodies connected by a superconducting wire can be switched on and off simply by application of a magnetic field which switches the superconductivity on and off. Crudely speaking, the explanation for the association of infinite electrical conductivity with vanishing thermal conductivity is that the transport of heat requires the transport of disorder (entropy). The superconducting state is one of perfect order (zero entropy), and so there is no disorder to transport and therefore no thermal conductivity. *See* ENTROPY.

Thermoelectric properties. A combined thermal and electrical effect of interest and practical importance is the Peltier effect, which is the basis of operation of thermocouples used for temperature measurement. If the two junction regions of a loop made of two different metals are maintained at different temperatures, an electrical potential gradient is produced which will drive a current

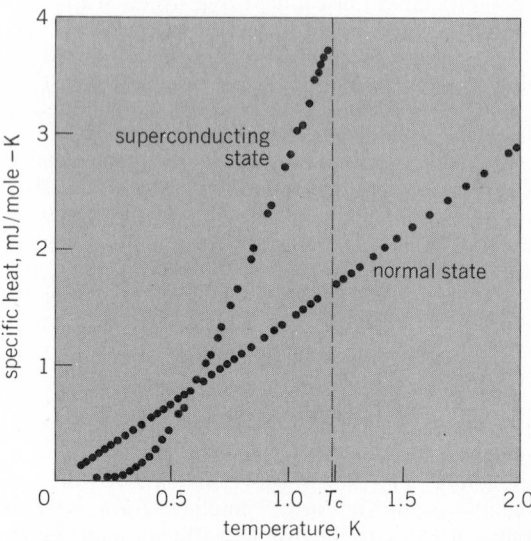

Fig. 4. Low-temperature specific heat of normal and superconducting aluminum. (*From N. E. Phillips, Heat Capacity of Aluminum between 0.1 K and 4 K, Phys. Rev., 114:676–685, 1959*)

around the loop. This effect vanishes in the superconducting state. (That, at least, is the conventional view. Claims have been made for observation of some sort of thermoelectric effects in superconductors.) *See* THERMOELECTRICITY.

High frequency electromagnetic properties. The electrical and magnetic behavior of superconductors at high frequencies differs from the zero frequency behavior described above. In the radio-frequency (up to about 10^8 Hz) and microwave-frequency (from 10^8 to about 10^{11} Hz) regions of the electromagnetic spectrum, it is found that superconductors do not have zero resistance to the flow of current. The resistance and the accompanying electrical energy loss are still much smaller than in the normal state, but they are not zero, and they increase with increasing frequency. On the other hand, in the optical region of the spectrum (about 10^{15} Hz), the electromagnetic response in the superconducting state is indistinguishable from that in the normal state. This may be confirmed simply by looking at a superconductor as it transforms from the normal state to the superconducting state; there is no change in its appearance. Clearly something interesting happens somewhere between 10^{11} and 10^{15} Hz. It is found that in the region of 10^{11} to 10^{12} Hz, depending on the material, the absorption of electromagnetic radiation by a superconductor rises quite sharply from a small value to the value characteristic of the normal state. This behavior provides another clear indication of the presence of a gap in the electronic energy spectrum of a superconductor, rather like the gaps which occur in semiconductors. (In semiconductors, however, gaps tend to be on the order of 1 eV, whereas the gaps in superconductors are typically a thousand times smaller.) The sharp rise in electromagnetic absorption occurs at the frequency for which the energy of a single photon (equal to Planck's constant times the frequency) becomes just sufficient to produce an excitation of some sort (consisting, in fact, of two "electrons") out of the superconducting state across the gap. *See* SEMICONDUCTOR.

Isotope effect. The availability of substantial quantities of separated isotopes of the elements after World War II made it possible to test whether the superconducting state depended in some way on atomic mass. The result provided a crucial key to the development of the BCS microscopic theory of superconductivity. It was found that the superconducting transition temperature for a given element was proportional to $M^{-\frac{1}{2}}$, where M is the isotopic mass. The vibration frequency of a mass M on a spring is proportional to $M^{-\frac{1}{2}}$, and the same relation holds for the characteristic vibrational frequencies of the atoms in a crystal lattice. Thus, the existence of the isotope effect indicated that, although superconductivity is an electronic phenomenon, it nevertheless depends in an important way on the vibrations of the crystal lattice in which the electrons move. Fortunately, not until after the development of the BCS theory was it discovered that the situation is more complicated than it had appeared. For some superconductors, the exponent of M is not $-\frac{1}{2}$, but near zero, and for at least one it is positive. *See* HARMONIC MOTION; ISOTOPE; LATTICE VIBRATIONS.

Absence of effects. While most of the electronic properties of a superconductor are profoundly affected by the transition to the superconducting state, many properties are changed very little if at all. These include the mechanical and elastic properties, tensile strength, sound velocity, and density, among others.

THEORY

The principal theories which have been constructed to explain the basic experimental properties of superconductors are discussed in this section. Those which preceded the BCS microscopic theory are no less useful for being phenomenological and incomplete. The original BCS theory was based on an idealized model, but nevertheless has been broadly successful in explaining the properties of real superconductors. It has been extended and elaborated to cover ever more complex and realistic situations.

Two-fluid model. C. J. Gorter and H. B. G. Casimir introduced in 1934 a phenomenological theory of superconductivity based on the assumption that in the superconducting state there are two components of the conduction electron "fluid" (hence the name given this theory, the "two-fluid model"). One, called the superfluid component, is an ordered condensed state with zero entropy, hence is incapable of transporting heat. It does not interact with the background crystal lattice, its imperfections, or the other conduction electron component and exhibits no resistance to flow. The other component, the normal component, is composed of electrons which behave exactly as they do in the normal state. It is further assumed that the superconducting transition is a reversible thermodynamic phase transition between two thermodynamically stable phases, the normal state and the superconducting state, similar to the transition between the liquid and vapor phases of any substance. The validity of this assumption is strongly supported by the existence of the Meissner-Ochsenfeld effect and by other experimental evidence. This assumption permits the application of all the powerful and general machinery of the theory of equilibrium thermodynamics. The results tie together the observed thermodynamic properties of superconductors in a very satisfying way.

Thermodynamic relations. From a thermodynamic point of view, the superconducting phase appears below the transition temperature because the free energy of the superconducting phase becomes less than the free energy of the normal phase for all temperatures below T_c. The exclusion of magnetic flux by the superconductor in an applied field H increases the free energy per unit volume of superconductor by $\mu_0 H^2/2$ (in SI units); it costs energy to push the flux lines out of the superconducting region. When this increase in free energy becomes equal to the decrease in free energy associated with the normal-to-superconducting transition, it no longer pays the superconductor to remain superconducting, and the superconductor goes normal. Hence, the superconducting condensation energy must equal $\mu_0 H_c^2/2$. This can be verified experimentally by comparing the results of critical field measurements with direct measurements of the heat capacity in the normal and superconducting states. *See* FREE ENERGY.

The zero-field heat capacity in Fig. 4 shows a finite discontinuity at the transition temperature, not an infinite singularity. This means that there is no latent heat at the transition, that is, the transition is "second-order," and therefore the entropy is the same in the superconducting and normal states at the transition temperature. At all but the lowest temperatures, the electronic heat capacity in the superconducting state is nearly proportional to T^3. In the normal state, it is proportional to T, $C_n = \gamma T$, where γ is a constant. All of these facts may be combined to yield the results that the heat capacity in the superconducting state $C_s = 3\gamma T^3/T_c^2$ and that the entropy in the superconducting state is less than that in the normal state for all temperatures below the transition temperature, that is, the superconducting phase is more ordered than the normal state. Further, consideration of the free energy yields the results $\gamma T_c^2 = 2\mu_0 H_0^2$, where H_0 is the critical field at $T = 0$ K (in SI units), and $H_c = (\gamma/2\mu_0)^{\frac{1}{2}} T_c [1 - (T/T_c)^2]$. There is thus a direct thermodynamic connection between the T^3 heat capacity and the parabolic critical field curve; one implies the other. *See* SECOND-ORDER TRANSITION.

More general thermodynamic relations exist between the superconducting parameters H_c and T_c and the normal state heat capacity constant γ, which do not depend on any assumption about the temperature dependence of the heat capacity in the superconducting state or the form of the critical field curve. The good agreement of these relations with experimental data is the basis of the conviction that the superconducting state is indeed a thermodynamic equilibrium phase.

Another result of the two-fluid model is that the normal fluid forms a fraction $x = (T/T_c)^4$ of the total electron fluid. The superfluid fraction $1 - x = 1 - (T/T_c)^4$ therefore rises rapidly from 0 to 1 as T falls below T_c. *See* THERMODYNAMIC PRINCIPLES.

London equation. In order to understand the electromagnetic behavior of superconductors within the two-fluid model, it is necessary to postulate something about how the superfluid responds to electric and magnetic fields. This was done by the brothers F. London and H. London in 1935. It is natural to suppose that any electric field in a superconductor will give rise to a force on the superfluid electrons and hence will accelerate the superfluid. Since the supercurrent $\mathbf{J}_s$ is proportional to the superfluid velocity, this implies a rate of change with time of $\mathbf{J}_s$ which is proportional to the electric field. This is simply equivalent to an assumption that the superfluid has infinite conductivity. Straightforward combination of this assumption with the classical Maxwell equations relating the electric and magnetic fields yields a relation between the supercurrent and the magnetic field which does not lead to the Meissner-Ochsenfeld effect. (This indicates that the Meissner effect is not simply a consequence of infinite conductivity.) The Londons therefore postulated a modification of this relation, now called the London equation, which in one simple form is $\mathbf{J}_s = -K\mathbf{A}$. Here $\mathbf{A}$ is the so-called vector potential from which the magnetic field can be derived, and K is a constant which contains the density of superfluid electrons. This equation does lead to the Meissner effect. *See* MAXWELL'S EQUATIONS.

With the addition of the London equation to describe the electrodynamics, the two-fluid model predicts magnetic-field penetration depths and finite conductivity at nonzero frequencies in reasonable accord with experiment. The source of the latter is qualitatively apparent: At zero frequency all of the current will be carried by the superfluid, which will "short out" the normal fluid. At nonzero frequencies, however, the electromagnetic fields induced in the surface of the superconductor will tend to drive the normal fluid as well as the superfluid, and this will cause some dissipation.

Nonlocal theory. In the course of experiments on the electromagnetic response of superconductors at very high frequencies, A. B. Pippard in 1952 observed significant discrepancies between his results and the predictions of London electrodynamics. He traced these to a failure of the London equation $\mathbf{J}_s = -K\mathbf{A}$. This equation is a "local" one, relating as it does the supercurrent at a given point in the superconductor to the vector potential at that point. Pippard found that the correct relation is nonlocal; the supercurrent at a point actually depends on the vector potential throughout a region around the point of size ξ. The parameter ξ is called the coherence length and may be as large as 10^{-6} m. Nonlocality turns out to be a very fundamental feature of the superconducting state, and ξ is a fundamental length of great importance in a variety of superconducting phenomena. For example, the distinction between type I and type II superconductors depends entirely on the relative sizes of the penetration depth and the coherence length: If λ is greater than ξ, the surface energy is negative and the superconductor is type II. The coherence length can be decreased from its value in a pure superconductor by adding impurities which scatter electrons, and this can change a superconductor from type I to type II.

Ginzburg-Landau theory. V. L. Ginzburg and L. D. Landau proposed in 1950 a highly innovative phenomenological theory of superconductivity which now bears their names. Their objective was to understand the situation at supernormal interfaces like those which occur in the intermediate state of type I superconductors. Since the superconducting state varies from something to nothing at such an interface, they needed a theory which describes a spatial variation of the superconducting state. They began by supposing that the "strength" of the superconducting state can be described by an "order parameter" ψ which may be spatially varying. In the normal state, ψ is zero. Then they assumed that sufficiently near T_c, where superconductivity is weak and ψ is small, the free energy of a superconductor can be expressed as a sum of a series of terms in increasing powers of $|\psi|^2$; $|\psi|^2$ is interpreted as proportional to the superfluid density. The absolute value is used here because ψ is allowed to be a complex number with an amplitude ψ_0 and a phase φ, $\psi = \psi_0 e^{i\varphi}$. This is a crucial if initially inexplicable feature of the order parameter. Thus ψ has something of the character of a quantum-mechanical wave function. This similarity is reinforced by the addition to the free-energy sum of a term involving the spatial gradient of ψ and the magnetic vector potential. This term has the same form as the standard representation of the kinetic energy in the Schrödinger wave equation of quantum mechanics, and here represents the kinetic energy of the superfluid electrons. The equilibrium form of ψ is then assumed to be that which minimizes the free energy. This leads to a rather complicated nonlinear differential equation for ψ. Together with an electrodynamic equation which relates the supercurrent $\mathbf{J}_s$ to ψ, this equation forms a pair of simultaneous differential equations called the Ginzburg-Landau equations. *See* NONRELATIVISTIC QUANTUM THEORY; QUANTUM MECHANICS.

Obtaining solutions of the Ginzburg-Landau equations is in general rather difficult. Nevertheless, they have been a remarkably powerful tool in superconductivity. They are intrinsically nonlocal because of the presence of the spatial gradient of ψ in the original free-energy equation. This has the consequence that a local disturbance (such as a sharp boundary between a normal and a superconducting metal) causes ψ to vary on the scale of a characteristic length which can be identified with the coherence length ξ. The other characteristic length in superconductivity, the penetration depth λ, also appears in a natural way. In 1957 A. A. Abrikosov reported a solution of the Ginzburg-Landau equations for the case $\lambda > \xi$ which constituted the first theoretical explanation of the mixed state in type II superconductors. Although the Ginzburg-Landau equations have many limitations, they continue to provide the basis for most understanding of the spatially varying superconducting state.

Microscopic (BCS) theory. The key to the basic interaction between electrons which gives rise to superconductivity was provided by the isotope effect. It is an interaction mediated by the background crystal lattice and can crudely be pictured as follows: An electron tends to create a slight distortion of the elastic lattice as it moves, because of the Coulomb attraction between the negatively charged electron and the positively charged lattice. If the distortion persists for a brief time (the lattice may ring like a struck bell), a second passing electron will see the distortion and be affected by it. Under certain circumstances, this can give rise to a weak indirect attractive interaction between the two electrons which may more than compensate their Coulomb repulsion.

H. Frölich recognized in 1950 that such an interaction might be responsible for superconductivity. However, all initial attempts to develop a theory based on the interaction failed. The average energy per electron associated with the superconducting condensation is tiny compared with typical electron kinetic or Coulomb interaction energies. It was natural to try to exploit this by trying to arrive at the quantum-mechanical description of electrons in the superconducting state by small perturbations of the description in the normal state. In retrospect, it is clear why this approach failed. The superconducting quantum wave function is qualitatively different from any normal state wave function.

The first forward step was taken by Cooper in 1956, when he showed that two electrons with an attractive interaction can bind together to form a "bound pair" (often called a Cooper pair) if they are in the presence of a high-density fluid of other electrons, no matter how weak the interaction is. The two partners of a Cooper pair have opposite momenta and spin angular momenta. Then, in 1957, Bardeen, Cooper, and Schrieffer showed how to construct a wave function in which all of the electrons (at least, all of the important ones)

are paired. Once this wave function is adjusted to minimize the free energy, it can be used as the basis for a complete microscopic theory of superconductivity.

With the discovery of the BCS superconducting-ground-state wave function, the fundamental reason for the remarkable properties of the superconducting state became clear. An analogy is useful in understanding it. Consider an enormous ballroom packed with dancers, shoulder to shoulder. Suppose each dancer is vigorously doing his or her own individual dance. The dancers will collide with each other and with any other objects which may be scattered about the dance floor. If there is some pressure on the whole group to move toward one side of the ballroom, dancing all the while, the collective motion will be random and chaotic, and a lot of energy will be lost in collisions. This represents the electrons in a normal metal, colliding with each other and with irregularities or impurities in the crystal lattice. If an electric field is applied and a current induced, collisions will dissipate energy and cause a finite conductivity.

Now suppose the dancers are paired in couples, each pair dancing together. This represents Cooper pairs, but now an important feature of the pairs enters. The pairing interaction is so weak that the two members of a pair are separated on the average by a distance which turns out to be just the coherence length. The average distance between two paired electrons not involved in the same pair, however, is about a hundred times smaller. The partners comprising each couple are not dancing cheek to cheek, but are separated by a hundred other dancers. Consequently, if every couple is going to dance together, it is clear that everybody must dance together. The result is a single coherent motion, with order extending all the way across the ballroom. The superconducting state is something like that. A localized perturbation which might deflect a single electron in the normal state, and thus give rise to some resistance, cannot do so in the superconducting state without affecting all the electrons participating in the superconducting ground state at once. That is not impossible, but extremely unlikely, so that a collective drift of the coherent superconducting electrons, corresponding to a current, will be dissipationless.

The successes of the BCS theory and its subsequent elaborations are manifold. One of its key features is the prediction of an energy gap. Excitations called quasiparticles (which are something like normal electrons) can be created out of the superconducting ground state by breaking up pairs, but only at the expense of a minimum energy of Δ per excitation; Δ is called the gap parameter. The original BCS theory predicted that Δ is related to T_c by $\Delta = 1.76 \, kT_c$ at $T = 0$ for all superconductors. This turns out to be nearly true, and where deviations occur they are understood in terms of modifications of the BCS theory. The manifestations of the energy gap in the low-temperature heat capacity and in electromagnetic absorption provide strong confirmation of the theory. The theory accounts for all the thermodynamic properties of superconductors, including such details as deviations from parabolicity of the critical field curve. The theory is intrinsically nonlocal. The Ginzburg-Landau theory and the Pippard nonlocal electrodynamics can be derived from it. The Ginzburg-

Landau order parameter ψ can be associated with the BCS ground-state wave function, and the coherence length with the size or range of a Cooper pair.

FURTHER EXPERIMENTAL PROPERTIES

Some experiments which illustrate some special features of the superconducting wave function are developed in the modern theory of superconductivity discussed in this section.

Flux quantization. Consider the complex order parameter or wave function ψ for a superconductor in the form of a hollow cylinder. If ψ is to be a well-defined object, its phase φ must change by an integral multiple of 2π along any closed path which lies entirely within the superconductor, including any path which surrounds the hole in the cylinder. There is a fundamental relation between the spatial gradient of ψ, hence of its phase, and the magnetic vector potential **A**. A consequence of this relation and the constraint on the phase change around a closed path is that the hole in the cylinder cannot contain an arbitrary amount of magnetic flux, but only integral multiples of a fundamental quantum of flux, $\Phi_0 = h/2e = 2.07 \times 10^{-15}$ Wb in SI units, or $\Phi_0 = hc/2e = 2.07 \times 10^{-7}$ gauss cm^2 in cgs units; h is Planck's constant, c is the velocity of light, and e is the electron charge. This is the reason for the quantization of flux in fluxoids in type II superconductors. There the hole in the cylinder is the normal core of the fluxoid. Flux quantization has been observed in macroscopic hollow cylinders. The factor of 2 in the flux quantum is related to the two electrons of a Cooper pair, so that observation of the expected magnitude of the flux quantum may be interpreted as experimental evidence for electron pairing in the superconducting state.

Quasiparticle tunneling. The phenomenon of tunneling is a direct consequence of the wave nature of material particles and was recognized very early in the development of quantum or wave mechanics. A particle, such as an electron, can pass into a region which classically would be forbidden to it and, if the region is not infinitely thick, can pass (tunnel) through it. Consider, for example, two metals separated by an insulating barrier which prevents electrons from passing between the metals. The resulting structure, called a junction, will be nonconducting. However, if the barrier is thin enough (10^{-9} to 10^{-8} m), electrons can tunnel from one metal to the other and this "tunnel junction" will become conducting. In 1960 I. Giaever discovered that if one or both of the metals in such a junction are superconducting, the dependence of the current on the voltage across the junction becomes highly nonlinear. The lower curve in Fig. 5 shows an example for a junction composed of two thin films of tin separated by a tin oxide barrier. The junction passes very little current until the voltage reaches a value of about 1.2 mV, where it rises sharply. This behavior is a direct consequence of the existence of the superconducting energy gap. The voltage V_g at which the current rises is related to the gap parameter by the simple relation $eV_g = 2\Delta$. Giaever's discovery made it possible to measure the gap parameter with nothing much more than an ammeter and a voltmeter. Previous determinations of the gap had come from much more difficult very-low-temperature heat capacity experiments or far-infrared

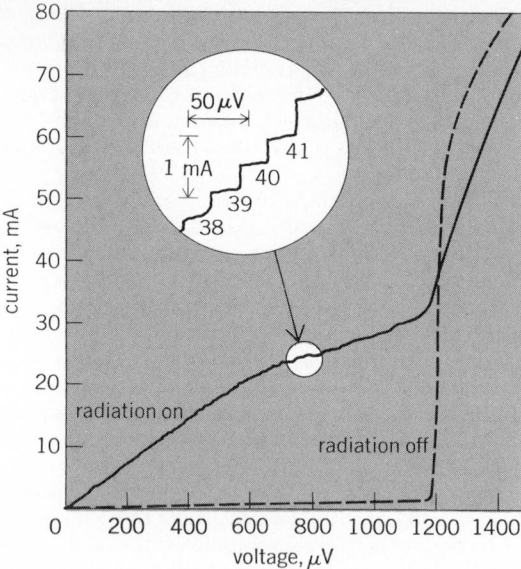

Fig. 5. Current-voltage curves of a Sn-Sn tunnel junction at 1.2 K displaying radiation-induced current steps. (*From W. H. Parker et al., Determination of e/h, Using Macroscopic Quantum Phase Coherence in Superconductors, I, Phys. Rev., 177, 639–664, 1969*)

absorption experiments. The tunnel junction has become the single most valuable tool for the study of the superconducting state. For example, the current-voltage characteristic at voltages above the gap voltage contains small structures which can be unraveled to yield the complete energy dependence of the electron-lattice interaction function responsible for the superconductivity.

Pair tunneling. In 1962 B. D. Josephson made a remarkable theoretical discovery: Not only can quasiparticles ("normal electrons") tunnel through an insulating barrier between two superconductors, but so can Cooper pairs. This implies a coupling between the superconducting wave functions of the two superconductors which leads to the existence of a tunnel supercurrent which depends on the difference between the phases of the two wave functions. Josephson's predictions were soon verified experimentally. They include the following: It is possible for a lossless supercurrent to pass through the insulating barrier; this is called the dc Josephson effect. If a voltage V is applied between the two superconductors, an oscillating supercurrent with frequency $2eV/h$ exists in the junction; this is the ac Josephson effect. The factor of 2 again comes from the pairing. The upper curve in Fig. 5 shows one way to detect the ac supercurrent. A microwave field of frequency ν is applied to the junction. This field frequency modulates the ac supercurrent, producing zero-frequency increments in the supercurrent when the frequency of the ac supercurrent is equal to an integer multiple of the microwave frequency. These appear as steps in the current-voltage characteristic of the junction at voltage intervals of $h\nu/2e$. This effect provides by far the most accurate way to measure the fundamental physical constant h/e, as well as the basis for the United States national standard of voltage. *See* ATOMIC CONSTANTS; ELECTRICAL UNITS AND STANDARDS.

A Josephson tunnel junction can be thought of

as a quantum phase meter. The Josephson effects have been used to confirm experimentally that the superconducting order parameter has all the quantum-phase coherence characteristics of an atomic wave function, but on a scale 10^{10} times larger. The fundamental connection between the quantum phase and the magnetic vector potential makes the Josephson effects incredibly sensitive to magnetic fields. Josephson junctions have been used to detect fields 10^{10} times smaller than the Earth's magnetic field and the tiny magnetic fields produced by neural currents in the human brain. *See* JOSEPHSON EFFECT.

Research on superconductivity has continued actively on a variety of fronts. New superconducting materials have been sought, with higher transition temperatures and more favorable properties for practical applications. There have been theoretical conjectures that much higher transition temperatures, perhaps higher than room temperature, might be achieved with exotic materials in which conduction occurs in one- or two-dimensional molecular structures, or through the exploitation of some electron-electron attractive interaction mechanism which is stronger than the electron-lattice interaction which acts in all known superconductors. These possibilities have been pursued. Superconductivity may have a cosmic role, for the interiors of neutron stars and of the planet Jupiter may be superconducting. There has been much interest in fluctuations in the properties of superconductors near the superconducting phase transition, and in the behavior of superconductors forced into states far from thermal equilibrium. An enormous range of practical applications have been investigated, from superconducting electric power transmission lines to superconducting computers.   [D. N. LANGENBERG]

Bibliography: A. C. Rose-Innes and E. H. Roderick, *Introduction to Superconductivity*, 2d ed., 1977; D. R. Tilley and J. Tilley, *Superfluidity and Superconductivity*, 1975; M. Tinkham, *Introduction to Superconductivity*, 1975, reprint 1980.

Superfluidity

The frictionless flow of liquid helium at low temperature; also, the flow of electric current without resistance in certain solids at low temperature (superconductivity).

Both helium isotopes have a superfluid transition, but the detailed properties of their superfluid states differ considerably because they obey different statistics. He^4, with an intrinsic spin of 0, is subject to Bose-Einstein statistics, and He^3, with a spin of 1/2, to Fermi-Dirac statistics.

Superfluidity in He^4 occurs at temperatures below 2.172 K (lambda point, T_λ) and is a form of Bose-Einstein condensation. Properties of the superfluid state (He II) that are absent in the normal state (He I) include (1) frictionless flow through holes as small as 2 nm in diameter for particle velocities up to approximately 50 cm/s; (2) formation of a highly mobile film over all surfaces touching the bulk liquid; (3) transmission of temperature waves (second sound) in addition to pressure waves (ordinary or first sound); (4) joint occurrence of temperature and pressure differences in connected regions of fluid (thermomechanical effect); (5) inability to carry heat when flowing through small capillaries (mechanocaloric effect); (6) inability to

fully rotate in the manner of a solid body (irrotational flow); and (7) the failure of evaporation and condensation of helium vapor to produce any change in the superfluid velocity.

The more recently discovered superfluid transition in liquid He^3 occurs at approximately .00093 K at zero pressure and increases to .0026 K at the solidification pressure of 34 atm. (1 atm = 1.01×10^5 Pa.) Because the spin of a He^3 atom is the same as that of an electron, the superfluid transition of He^3 is closely related to the common superconducting transition of the electrons in many solids. Of course, the He^3 system does not have the anomalous electrical conductivity property of superconductors since it is neutral, but the parallel spin state of the Cooper pairs in He^3 gives rise to anisotropic properties possible neither with the antiparallel spin state of Cooper pairs in superconductors nor in He^4. There are two distinct superfluid states in He^3 called A and B. The A phase occurs only at a pressure above 21 atm and at higher temperature than the B phase, except in the presence of a magnetic field in which case the A phase extends to lower pressures and a variant of the A phase, called A1, appears near the transition temperature. In addition to its unique magnetic properties, superfluid He^3 is expected to qualitatively display all the properties of superfluid He^4 that are listed above.

Superfluidity in both helium isotopes is a macroscopic quantum effect since a single wave function can describe the superfluid part of the entire system. In mixtures of He^3 and He^4, only the He^4 component is superfluid at currently accessible temperatures.

The term "superfluidity" usually implies He II or the A and B phases of He^3, but the basic similarity between these and the "fluid" consisting of pairs of electrons in superconductors is sufficiently strong to designate the latter as a charged superfluid. Besides flow without resistance, superfluid helium and superconducting electrons display quantized circulating flow patterns in the form of microscopic vortices. *See* BOSE-EINSTEIN STATISTICS; LIQUID HELIUM; QUANTIZED VORTICES; SECOND SOUND; SUPERCONDUCTIVITY.

[LAURENCE J. CAMPBELL]

Bibliography: K. H. Bennemann and J. B. Ketterson, *The Physics of Liquid and Solid Helium*, pt. 1, 1976, pt. 2, 1978; S. J. Putterman, *Superfluid Hydrodynamics*, 1974; D. R. Tilley and J. Tilley, *Superfluidity and Superconductivity*, 1975; J. Wilks, *Introduction to Liquid Helium*, 1970.

Supermultiplet

A generalization of the concept of a multiplet. A multiplet is a set of quantum-mechanical states each of which has the same value of some fundamental quantum numbers and differs from the other members of the set by other quantum numbers, which take values from a range of numbers dictated by the fundamental quantum numbers. There are as many states in the set as there are different values of the varying quantum numbers. The number of states in the set is the multiplicity or dimension of the multiplet. For example, there are $2S + 1$ orientations or projections of a spin S, and these are labeled by M_S with values from $-S$ to $+S$ in integer steps. Each of the $2S + 1$ members of this multiplet, which has the dimension $2S + 1$, is labeled by S and M_S, where S labels the multiplet, and M_S a state within the multiplet. Suppose that T denotes another spinlike quantity but refers to a different physical property. Then the $2T+1$ states also constitute a multiplet labeled by T with members labeled by, say, T_3. If sets of these multiplets are grouped together, then one speaks of a supermultiplet. For example, one supermultiplet might be denoted by a set of fundamental quantum numbers $(\lambda_1 \lambda_2 \lambda_3)$, and S and T would range over a set of values dictated by the values of λ_1, λ_2, and λ_3. For a fixed value of S, M_S would have its usual $2S+1$ values, and for fixed value of T, T_3 would have $2T+1$ different values, ranging from $-T$ to $+T$. The dimension of a supermultiplet in this case would be the sum over all values of S and T of the product $(2S+1)(2T+1)$. A supermultiplet then is a multiplet of multiplets. *See* QUANTUM MECHANICS; QUANTUM NUMBERS.

Wigner supermultiplets. The term supermultiplet was first used by Eugene P. Wigner in a nuclear theory in which protons and neutrons are treated equivalently, and the dependence of the nuclear force on their differences and intrinsic spins is ignored. When used in this connection, one speaks of Wigner supermultiplets.

Each proton and neutron has an intrinsic spin of $\frac{1}{2}$ in units of $\hbar$, which is Planck's constant divided by 2π. The total spin of a collection of protons and neutrons is then obtained by adding their individual spins vectorially, according to the rules of quantum mechanics. This total spin is denoted by S, and the $2S+1$ states corresponding to different M_S constitute a spin multiplet. In addition to having the same spin, the proton and the neutron have essentially the same mass, but differ in that the proton is charged whereas the neutron is not. They are thus regarded as different charge states of the same particle, a nucleon.

In formal mathematics this distinction is made by defining a quantum number called isotopic spin t, which for the nucleon has the value $\frac{1}{2}$. The two charge orientations are taken to be $\frac{1}{2}$ for the proton and $-\frac{1}{2}$ for the neutron, so the nucleon charge in units of electronic charge is $\frac{1}{2}+t_3$, where t_3 is a projection of the isotopic spin I. Isotopic spin is formulated in direct analogy with ordinary spin; thus the total isotopic spin of a collection of nucleons is determined from the individual isotopic spins in the same way that S was determined from the individual spins. This total isotopic spin T labels an isotopic spin (isospin) multiplet consisting of $2T+1$ states labeled by T_3. Since projections of spins or spinlike quantities add algebraically rather than vectorially, $T_3 = \frac{1}{2}(Z-N)$ for a system of A nucleons, where Z is the number of protons, N the number of neutrons, and $N+Z=A$. From this it follows that while a spin multiplet may be associated with a given nucleus (fixed N and Z), an isospin multiplet is distributed over $2T+1$ nuclei, all having the same value of A (isobars) and having $Z = A/2 + T_3$, which is the analog in the nucleus of $\frac{1}{2}+t_3$ for the nucleon. *See* ISOTOPIC SPIN; NEUTRON; PROTON.

Although the force between a pair of nucleons is known to depend at least upon their space, spin, and isospin coordinates, it is sometimes useful to ignore one or more of these dependencies as a first approximation. The Wigner supermultiplet theory is based upon taking into account only the space

dependence. In this case the wave function for a system of A nucleons may be written as a product of wave functions: one involving space coordinates, one involving spin coordinates, and one involving the isospin. In this approximation the energy depends upon the properties of the space function only, including its symmetry under particle exchange. The overall wave function must be antisymmetric under the exchange of a pair of nucleons. To achieve this, the spin-isospin part of the wave function must have a symmetry related to that of the space part. This is called the conjugate symmetry; the quantum numbers which specify this symmetry label the Wigner supermultiplet and dictate the values and combinations of the spin and isospin multiplets belonging to the supermultiplet.

As a specific example, consider two nucleons in a space state of relative orbital angular momentum zero which is symmetric. The spin and isospin functions must then have opposite symmetries, which means that $S=1$ and $T=0$ go together, and $S=0$ goes with $T=1$. All these states are members of the same supermultiplet and in this approximation have the same energy. The energies of the states versus T_3 are illustrated in the figure. The state at $T_3=1$ belongs to $S=0$ and $T=1$ and corresponds to the diproton; similarly, the state at $T_3=-1$ belongs to $S=0$ and $T=1$ and is the dineutron. In figure a all states of the supermultiplet are shown at the same energy so that there are two at $T_3=0$; they have the values $S=0$, $T=1$ and $S=1$, $T=0$. Both of these belong to the deuteron. When the spin dependence of the nucleon interaction is considered the $S=1$, $T=0$ states are lowered in energy relative to the rest of the supermultiplet, and this corresponds to the deuteron ground state. This is the lower level in figure b at $T_3=0$. The $S=0$, $T=1$ state of the deuteron is known to be slightly unbound by about 70 keV. (A bound state of two particles is a state in which energy must be supplied to separate the two in contrast to an unbound state, in which they may separate spontaneously and become free particles. They might also make a transition to a bound state, which would be at lower energy, and release the energy difference as electromagnetic radiation.) Thus one predicts that the diproton and dineutron should not exist; indeed, neither is seen experimentally.

Connections with group theory. It is not possible to completely understand and appreciate the simplification and unification that the supermultiplets bring to the understanding of nuclei without some knowledge of the theory of continuous groups. A few general features may be noted, however, which serve mainly to introduce the vocabulary. This will be useful in connection with the use of other supermultiplets applied to elementary particles in order to attempt a similar unification of their diverse properties. *See* GROUP THEORY.

Let p^+ and p^- denote the spin functions of a proton with spin projections $\frac{1}{2}$ and $-\frac{1}{2}$, respectively. For a collection of protons the spin functions of the jth proton would be $p_j^{\pm}$. If the situation in which these occur does not depend upon spin, then any other sets of pairs of spin functions obtained from these via the linear transformations $ap_j^+ + bp_j^-$, $-b^*p_j^+ + a^*p_j^-$ will suffice. In these expressions, a and b satisfy $a^*a + b^*b = 1$ but are otherwise arbitrary complex numbers; $*$ denotes

SUPERMULTIPLET

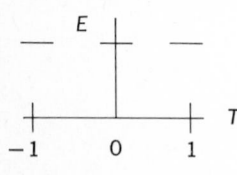

(a)

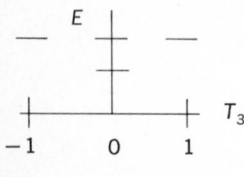

(b)

Energies of states versus T_3. (a) At the same level. (b) At different levels.

complex conjugation. Note that the same transformation is applied simultaneously to each pair $p_j^\pm$. If these conditions are met, then the $2S+1$ members of a spin multiplet formed from the $p_j^\pm$ transform among themselves; that is, if χ_{S,M_S} denotes a member of the spin multiplet S, then under the simultaneous a, b, $-b^*$, a^* transformation, the χ_{S,M_S} become χ'_{S,M_S} given by the equation below. Here the $U^{(S)}_{M'_S,M_S}$ are the analogs for the χ_{S,M_S} of the a, b, a^*, b^*, for the $p_j^\pm$.

$$\chi'_{S,M_S} = \Sigma_{M'_S} U^{(S)}_{M'_S,M_S} \chi_{S,M'_S}$$

These transformations have the necessary mathematical properties to form a group called the special unitary group in two dimensions, $SU(2)$.

The word dimensions refers to the size of the space of the fundamental constituents, the two $p^\pm$. The set of $(2S+1)^2$ $U^{(S)}_{M'_S,M_S}$ is conveniently regarded as a matrix, where M'_S labels the rows, and M_S the columns. The matrices for various values of S are said to be representations of $SU(2)$ because they satisfy the same group properties. Furthermore, the members of the multiplet S are said to be the basis of the representation $U^{(S)}$ since the formation of the χ_{S,M_S} from the $p_j^\pm$ determines how the $U^{(S)}_{M'_S,M_S}$ are formed from the a and b. In general, no smaller subset of the χ_{S,M_S} will transform among themselves, and the representation is said to be irreducible. The dimension of an irreducible representation is the number of basis functions required to define it, here $2S+1$. Characterizing the wave functions of a collection of particles by the χ_{S,M_S} is classifying them by the way in which they transform under the spin $SU(2)$, and ignoring any spin dependence in the nuclear interaction is equivalent to saying that this interaction is invariant under the spin $SU(2)$. Since the isospin formalism is exactly the same as ordinary spin, the isospin multiplets are the bases for irreducible representations of the isospin $SU(2)$. If both isospin and spin dependence of the nuclear interaction are ignored, it is invariant under the group of transformations $SU(2) \otimes SU(2)$, which is a direct product of the spin $SU(2)$ and the isospin $SU(2)$. But in this approximation the interaction is also invariant to a larger group $SU(4)$, which transforms the four fundamental spin-isospin states (n^+, n^-, p^+, p^-) among themselves. A basis of an irreducible representation of $SU(4)$ is a Wigner supermultiplet and is labeled by ($\lambda_1 \lambda_2 \lambda_3$). Within a given ($\lambda_1 \lambda_2 \lambda_3$) several combinations of spin and isospin multiplets occur in general since the $SU(4)$ mixes the fundamental states. Determining the values and combinations of S and T that occur in a given ($\lambda_1 \lambda_2 \lambda_3$) is a well-defined group theory problem called the reduction of $SU(4)$ with respect to $SU(2) \otimes SU(2)$.

The group $SU(3)$ has been used to classify the elementary particles, and each $SU(3)$ irreducible representation consists of several isospin multiplets, together with corresponding values of the hypercharge which corresponds to the group $U(1)$. In this sense an $SU(3)$ irreducible representation is nearly a supermultiplet, except that the multiplicity of a hypercharge state is a trivial value, 1. The members of a given $SU(3)$ multiplet all have the same spin and parity and approximately the same mass. The same $SU(3)$ multiplet but associated with different spin and parity may also occur. For example, the eight dimensional representation of $SU(3)$ is called an octet. The lightest mesons form an octet with spin 0, negative parity. The vector mesons (spin 1, negative parity) also form an octet, as do the baryons (spin $\frac{1}{2}$, positive parity). This existence of several $SU(3)$ multiplets with different spins suggests a higher symmetry.

B. Sakita has proposed a scheme which closely parallels the Wigner supermultiplets. The existence of a fundamental triplet of particles called quarks is presumed. Each has spin $\frac{1}{2}$, and the three play the role here that the proton and neutron did in the nuclear case. There are six fundamental states, and the presumed symmetry is $SU(6)$. Thus sets of elementary particles belong to a given $SU(6)$ supermultiplet, which contains several $SU(3)$ multiplets and spin $SU(2)$ multiplets in different combinations. As an example, the pseudoscalar and vector mesons belong to the same $SU(6)$ supermultiplet. The corresponding group theory problem here is the reduction of $SU(6)$ with respect to $SU(3) \otimes SU(2)$. However (as of 1969), the quarks have not been discovered, and it remains to be seen whether or not these elementary particle supermultiplets or others that could be conceived will yield as much understanding of elementary particles as Wigner supermultiplets have of nuclear structure. *See* BARYON; ELEMENTARY PARTICLE; HYPERON; MESON; QUARKS; SYMMETRY LAWS; UNITARY SYMMETRY.

[S. A. WILLIAMS]

Bibliography: F. J. Dyson, *Symmetry Groups in Nuclear and Particle Physics*, 1966; J. P. Elliot and P. G. Dawber, *Symmetry in Physics*, 2 vols., 1979; J. D. Fox and D. Robson (eds.), *Isobaric Spin in Nuclear Physics*, 1966; W. M. Gibson and R. B. Pollard, *Symmetry Principles in Elementary Particle Physics*, 1980.

Superposition principle

In classical wave theories (optics, acoustics), as in all theories characterized by linear homogeneous differential equations, the sum of any number of solutions to the equations is another solution. Thus (assuming one-dimensional waves for simplicity), the amplitude of the resultant wave at a point x and time t is the linear superposition of the amplitudes of all waves reaching x at time t. This fact, known as the principle of superposition, often also is taken to mean that if each of $\psi_1(x)$ and $\psi_2(x)$ are possible waveforms at $t=0$, then any linear combination $\psi(x) = c_1 \psi_1 + c_2 \psi_2$ is a possible waveform at $t=0$. This latter version is not a consequence of the linearity of the differential equations, but rather is an affirmation of the belief that the waveform can be chosen arbitrarily at any initial instant. Conversely, if $u_n(x,t)$ is the solution which at $t=0$ equals $u_n(x) \equiv u_n(x,0)$, and if $u_n(x)$ are a complete set of functions, then any wave $\psi(x,t)$ which at $t=0$ equals $\psi(x)$ can be written in the form of the equation below, where the constants c_n are determined

$$\psi(x,t) = \sum_n c_n u_n(x,t)$$

from the equation at $t=0$. *See* WAVE MOTION.

The preceding paragraph holds equally well for quantum theory, where the wave function $\psi(x,t)$ obeys the linear homogeneous Schrödinger equation. In quantum theory, however, because the wave function represents physical states, the principle of superposition has a profound significance. In particular, if the observable A is certain to have value α_1 in the state represented by u_1, and is certain to have value α_2 in the state represented by u_2, then $c_1 u_1 + c_2 u_2$ represents a state in which measurement of the observable A is certain to yield either precisely α_1, or else precisely α_2. Thus, because probabilities are proportional to $|\psi|^2$, superposition accounts for phenomena which are difficult to understand from a classical viewpoint. For example, suppose u_1 represents a beam which passes through only one of a pair of slits, and u_2 represents a beam emitted from the same source, but directed at the second slit. Then $\psi = c_1 u_1 + c_2 u_2$ is the wave function in a double slit experiment, with $|c_1|^2$, $|c_2|^2$ the relative intensities of the two beams, that is, $|c_1|^2$, $|c_2|^2$ are the probabilities that particles (photons in light beams, electrons in electron beams) reach the viewing screen via slits 1 and 2, respectively. The beam intensities can be made so low that only one particle reaches the screen each second, which particle (according to the classical viewpoint) must have come either through slit 1 or slit 2. Nonetheless, the probability of observing a particle at a point y on the viewing screen is $|\psi(y)|^2 = |c_1 u_1(y) + c_2 u_2(y)|^2$, wherein the two beams interfere. *See* QUANTUM MECHANICS.

[EDWARD GERJUOY]

Supertransuranics

A group of predicted elements beyond the present periodic table of known elements, with atomic numbers around 114, expected to possess half-lives of the order of a year or longer. These are also referred to as the superheavy elements. Although they have not been discovered experimentally, it is generally believed that the difficulty lies in the synthesis of these elements, rather than in their stability once they are made.

Nuclear stability. The underlying physics responsible for the limited extent of the periodic table is the competition between attractive "nuclear forces" among the nucleons (that is, protons and neutrons) and the repulsive electrostatic forces among all the positively charged protons. The limit of the periodic table at an atomic number Z of approximately 106 is then set by the process of nuclear fission, which takes place when the disruptive effect of electrostatic forces overcomes the cohesive effect of the nuclear forces. *See* NUCLEAR FISSION.

The picture began to change when, in 1964, a clearer understanding was reached on the relation between fission half-lives and a well-known property of the nucleus called the magic numbers. If the number of protons (or neutrons) in a nucleus is equal to the proton magic number (or neutron magic number), then such a nucleus displays some special features; one example is that it is spherical in shape. Work of W. D. Myers and W. J. Swiatecki showed that such a nucleus also displays an extra stability against fission. This means that this nucleus would have a longer half-life than would be expected otherwise. From experiments it has been deduced that the proton magic numbers are 8, 14, 28, 50, and 82. Many people have tried to predict the next magic number, and it is believed by many workers in this field that the next proton magic number is 114. The neutron magic numbers are the same as those for protons at lower numbers, but are different at larger numbers. The neutron magic numbers are 8, 14, 28, 50, 82, and 126, with the next predicted number at 184. *See* MAGIC NUMBERS; NUCLEAR STRUCTURE.

The figure illustrates known nuclei by a peninsula surrounded by the sea of instability. Mountains and ridges on the peninsula, representing regions of extra stability against fission, occur at the proton or neutron magic numbers, shown by grid lines. It can be seen from this figure that nuclei exist with a certain ratio of protons and neutrons. If a nucleus has too many protons or too many neutrons, it will decay by emitting a positron or an electron. The superheavy elements occur in this picture as an island of stability a little distance beyond the known peninsula with the center of the island at the proton magic number 114 and the neutron magic number 184. *See* RADIOACTIVITY.

At about the same time as Myers and Swiatecki's work, V. M. Strutinsky developed a method, now named after him, by which a quantitative estimate can be made of the stability of such superheavy nuclei. This method combines two well-known approaches in nuclear physics: the liquid drop model and the shell model of the nucleus. The first systematic calculation of the half-lives of both the known heavy nuclei and the predicted superheavy nuclei was made by S. G. Nilsson, C. F. Tsang, and coworkers in 1969, using the Strutinsky method. Several more refined calculations were made during the period up to 1974, perhaps the most complete version being that by J. R. Nix and coworkers. To a large extent, these different calculations were in agreement.

Possible natural occurrence. One of the surprising results from the calculations of Nilsson, Tsang, Nix, and others is that these superheavy elements may live as long as the age of the solar system. After considering all the major decay mechanisms, they predicted that several of these elements, in particular the element with $Z = 110$, have half-lives of about 10^8 yr. Thus an interesting possibility arises that if these elements were made in nature along with the other known elements

Scheme illustrating nuclear stability. (*From S. G. Thompson and C. F. Tsang, Superheavy elements, Science, 178:1047–1055, 1972*)

during the formation of the Earth, small fractions could have survived the period of time (approximately 4.5×10^9 yr) since the Earth was formed.

It was realized by workers in the field that great uncertainties were involved in these predictions. If the prediction of 10^8 yr is off by two or three orders of magnitude downward, which is quite possible in the present state of the art in theory, no naturally occurring superheavy elements are expected to be found. Also it is an open question whether such elements would have been made at the formation of the solar system. Nevertheless, extensive effort has been expended in looking for these elements in nature.

A search for new elements on the Earth depends on suitable choices of the most promising minerals and ores containing known elements having chemical properties most resembling those of the elements being sought. Using the time-honored method of D. I. Mendelyeev, it is seen from the periodic table that superheavy elements 110, 111, 112, 113, 114, and 115 should have chemical properties similar to those of Pt, Au, Hg, Tl, Pb, and Bi, respectively. Calculations made by using much more sophisticated methods developed in recent years have also been performed for these elements, giving detailed predictions of their chemical properties.

G. N. Flerov's group at Dubna (Soviet Union), S. G. Thompson's group at Berkeley, G. Herrmann's group at Mainz (Germany), and others looked at a variety of ores and minerals, including natural platinum ores, old lead glasses, moon rocks, and manganese nodules collected from the ocean which are found to be particularly rich in metallic minerals. Typically one tries to detect fission events or accompanying neutron emission characteristic of these superheavy elements. However, in some cases, alpha decay detection is performed, since it has been predicted that alpha emission energies of these elements are exceptionally larger than those of known elements. A number of other searches for superheavy elements in nature have also been made, including a search in cosmic rays which may include elements produced in explosions in distant stars or supernovae. None, so far, has given conclusive evidence of their presence. Their absence may be due to one or both of two reasons. First, it is still an open question whether these superheavy elements were synthesized at the formation of the solar system or in the supernovae. Calculations attempting to answer this question involve great uncertainties, since they are very sensitive to the estimated half-lives of nuclei between the known elements and the island of stability. Second, the half-lives of the superheavy nuclei may be a few orders lower than predicted, and they would have disappeared by radioactive decay during the 4.5×10^9 yr since the Earth was formed. These short half-lives are predicted by a study made in 1974 (see table).

Assuming that the superheavy elements were produced in nature and have half-lives too short to survive until today, attempts have been made to detect fission products of these elements. Exhaustive experiments have been made to find the tracks caused by these fission fragments in, for instance, lead glasses and moon rocks. No clear evidence for

A table of predicted half-lives of several superheavy nuclei

Atomic number (Z)	Neutron number (N)	Mass number (A)	Mean life (τ)
108	180	288	10^{-3} sec
110	184	294	10^5 years
112	184	296	100 years
114	184	298	0.1 year
116	188	304	0.5 min
118	192	310	0.1 sec

superheavy elements was found. In 1975 Edward Anders and his group at Chicago made a careful radiochemical neutron activation analysis of the Allende meteorite. They concluded that the isotopic distribution of the element Xe in a rare chromium mineral sample of the meteorite indicates the existence of an unknown fissioning element, which may be an extinct superheavy element with $Z = 115$, 114, or 113. *See* PARTICLE TRACK ETCHING.

Artificial synthesis. Perhaps it is too much to expect that these superheavy elements should have such long half-lives, given the uncertainties associated with theories. However, nearly all theories predict these superheavy elements to have half-lives of 1 year, 1 day, or at least 1 minute, which would still be extremely interesting and would allow study with available techniques if these elements could be synthesized. Obviously in this case a big jump must be made over the sea of instability to the island. In other words, a target such as U, Pu, or Cm must be bombarded with a heavy ion such as Ar, Ca, or Kr to make the nuclei fuse together to form a superheavy nucleus. Extensive experiments have been made at both Dubna and Orsay (France), but the results have been negative. Work on further experiments has been undertaken, notably at Berkeley and at Darmstadt (Germany).

Numerous problems may hinder the fusing of two nuclei into a superheavy nucleus. These problems have been given intensive theoretical and experimental study. A better understanding has been achieved in some areas, which will affect the conditions under which one would attempt to produce superheavy elements.　　[C. F. TSANG]

Bibliography: E. Anders et al., Extinct superheavy elements in the Allende meteorite, *Science*, 190:1262–1271, 1975; T. Johansson, S. G. Milsson, Z. Szymanski, Theoretical predictions concerning superheavy elements, *Ann. Phys.* (Paris), 5:377–416, 1970; J. R. Nix, Predictions for superheavy nuclei, *Phys. Today*, 25(4):30–38, 1972; *Physica Scripta*, vol. 10A, pp. 1–184, 1974; Superheavy elements: Theoretical predictions and experimental generation, *Proceedings of the 27th Nobel Symposium*, Ronneby, Sweden, June 1974; S. G. Thompson and C. F. Tsang, Superheavy elements, *Science*, 178:1047–1055, 1972.

Surface physics

The study of the structure and dynamics of atoms and their associated electron clouds in the vicinity of a surface, usually at the boundary between a solid and a low-density gas. Thus, surface physics may be regarded as a branch of solid-state physics which deals with those regions of large and rapid variations of atomic and electron density that occur in the vicinity of an interface between the two "bulk" components of a two-phase system. In conventional usage, surface physics is distinguished from interface physics by the restriction of the scope of the former to interfaces between a solid (or liquid) and a low-density gas, often at ultra-high-vacuum pressures of $p = 10^{-10}$ torr (1.33×10^{-8} N/m² or 10^{-13} atm). *See* SOLID-STATE PHYSICS.

More specifically, surface physics is concerned with two separate but complementary areas of investigation into the properties of such solid-"vacuum" interfaces. Ultimately, interest centers on the specification and theoretical prediction of surface composition and structure (that is, the masses, charges, and positions of surface species), of the dynamics of surface atoms (such as surface diffusion and vibrational motion), and of the energetics and dynamics of electrons in the vicinity of a surface (such as electron density profiles and localized electronic surface states). As a practical matter, however, the nature and dynamics of surface species must be determined experimentally by scattering and emission measurements involving particles or electromagentic fields (or both) external to the surface itself. Thus, a second major interest in surface physics is the study of the interaction of external entities (that is, atoms, ions, electrons, and electromagnetic fields) with solids at their vacuum interfaces. It is this aspect of surface physics which most clearly distinguishes it from conventional solid-state physics, because quite different scattering and emission experiments are utilized to examine surface as opposed to bulk properties of a given sample.

Physical principles of measurements.

Since the mid-1960s, surface physics has enjoyed a renaissance by virtue of the development of a host of techniques for characterizing the solid-vacuum interface. While they might appear complicated in detail, all of these techniques are based on one of two simple physical mechanisms for achieving surface sensitivity. The first, which is the basis for field-emission and field-ionization microscopy, is the achievement of surface sensitivity by utilizing electron tunneling through the potential-energy barrier at a surface. This concept reached its apex of application in direct determinations of the energies of individual electronic orbitals of adsorbed complexes via the measurement of the energy distributions either of emitted electrons or of Auger electrons emitted in the process of neutralizing a slow (energy $E \sim 10$ eV) external ion.

The second mechanism for achieving surface sensitivity is the examination of the elastic scattering or emission of particles which interact strongly with the constituents of matter, for example, "low-energy" ($E \lesssim 10^3$ eV) electrons, thermal atoms and molecules, or "slow" (300 eV $\leq E \leq 10^3$ eV) ions. Since such entities lose appreciable ($\Delta E \sim 10$ eV) energy in distances of the order of tenths of nanometers (nm), typical electron analyzers with resolutions of tenths of an electron volt are readily capable of identifying scattering and emission processes which occur in the upper few atomic layers of a solid. This second mechanism is responsible for the surface sensitivity of photoemission, Auger-electron, electron-characteristic-loss, low-energy-

electron-diffraction (LEED), and ion-scattering spectroscopy techniques. The strong particle-solid interaction criterion which renders these measurements surface-sensitive is precisely the opposite of that used in selecting bulk solid-state spectroscopies. In this case, weak particle-solid interactions (that is, penetrating radiations) are desired in order to sample the bulk of the specimen via, for example, x-rays, thermal neutrons, or fast ($E \gtrsim 10^4$ eV) electrons. *See* ELECTRON DIFFRACTION; X-RAY CRYSTALLOGRAPHY.

Surface preparation.

An atomically flat surface, labeled by $M(hkl)$, may be visualized as being obtained by cutting an otherwise ideal, single-crystal solid M along a lattice plane specified by the Miller indices (hkl), and removing all atoms whose centers lie on one side of this plane. On such a surface the formation of a "selvedge" layer can also be envisaged. Such a layer might be created, for example, by the adsorption of atoms from a contiguous gas phase. It is characterized by the fact that its atomic geometry differs from that of the periodic bulk "substrate." From the perspective of atomic structure, this selvedge layer constitutes the "surface" of a solid. In principle, its thickness is a thermodynamic variable determined from the equations of state of the solid and of the contiguous gas phase. In practice, almost all solid surfaces are far from equilibrium, containing extensive regions (micrometers thick) of surface material damaged by sample processing and handling.

Another reason for the renaissance in surface physics is the capacity to generate in a vacuum chamber special surfaces which approximate the ideal of being atomically flat. These surfaces are prepared by cycles of fast-ion bombardment, thermal outgassing, and thermal annealing for bulk samples (for example, platelets with sizes of the order of 1 cm $\times$ 1 cm $\times$ 1 mm) or field evaporation of etched tips for field-ion microscopes. In this fashion, reasonable facsimiles of uncontaminated, atomically flat solid-vacuum interfaces of many simple metals with fcc (face-centered cubic) or bcc (body-centered cubic) structure as well as zincblende and wurtzite semiconductors have been prepared and subsequently characterized by various spectroscopic techniques. Such characterizations must be carried out in an ultra-high vacuum ($p \lesssim 10^{-8}$ N/m²), however, so that the surface composition and structure are not altered by gas adsorption during the course of the measurements.

Experimental apparatus.

Presuming that atomically flat surfaces can be prepared, the bulk of modern experimental surface physics is devoted to the determination of the chemical composition, atomic geometry, and electronic structure of such surfaces. Since different measurements are required to assess each of these three aspects of a surface, the typical surface-characterization instrument consists of equipment for carrying out a combination of several measurements in a single ultra-high-vacuum chamber. Two types of sample geometry are common. Platelet samples are studied using scattering and emission experiments. A typical modern apparatus, such as that shown in Fig. 1, contains an electron gun, an ion gun, an electron energy analyzer, a source of ultraviolet or x-ray electromagnetic radiation, and a

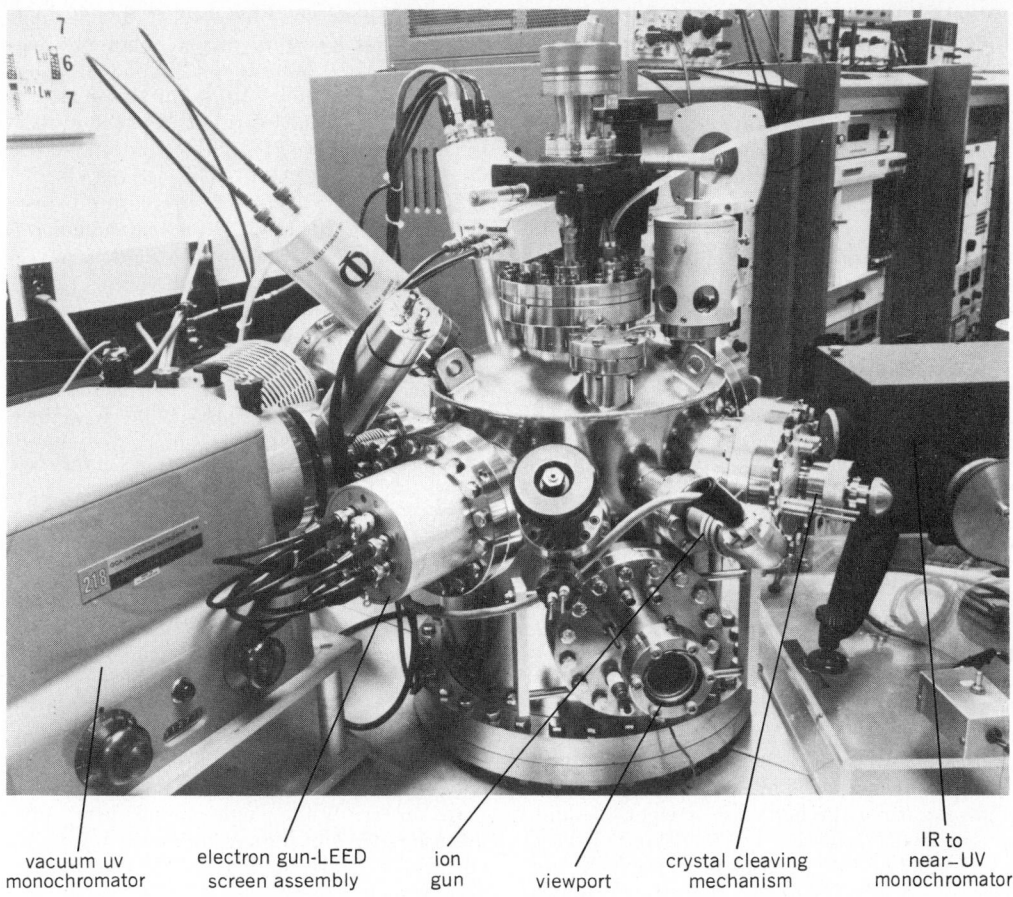

vacuum uv
monochromator

electron gun-LEED
screen assembly

ion
gun

viewport

crystal cleaving
mechanism

IR to
near–UV
monochromator

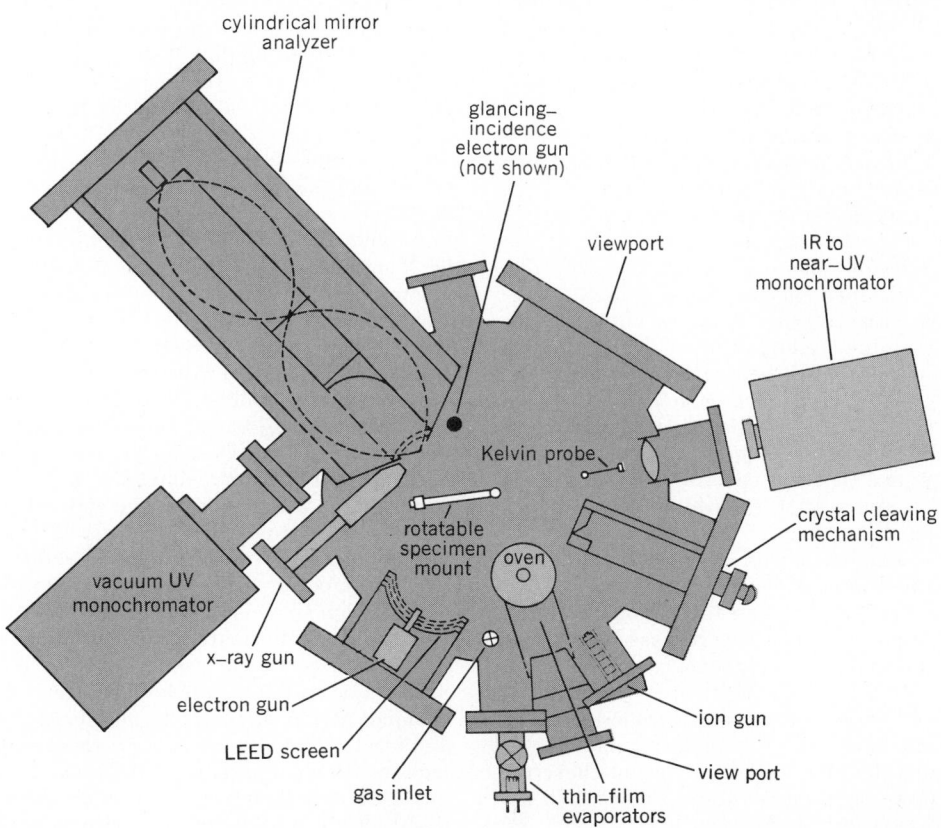

cylindrical mirror
analyzer

glancing–
incidence
electron gun
(not shown)

viewport

IR to
near–UV
monochromator

Kelvin probe

rotatable
specimen
mount

oven

crystal cleaving
mechanism

vacuum UV
monochromator

x–ray gun

electron gun

LEED screen

ion gun

view port

gas inlet

thin–film
evaporators

Fig. 1. Photograph and schematic diagram of modern
multiple-technique ultra-high-vacuum surface characterization instrument for study of insulator surfaces.
(L. J. Brillson, Xerox Corporation)

sample holder permitting precise control of both its orientation and temperature. Occasionally other features (such as a mass spectrometer) also are incorporated for special purposes. For specific applications in which less than a complete characterization of the surface is required, commercial instruments designed to embody only one or two measurements often are available. Such limited-capability instruments commonly are utilized to determine the chemical composition of surfaces by, for example, ion scattering, secondary ion mass spectrometry, x-ray photoemission, or Auger-electron emission. Obviously, the utility of such instruments is not limited to atomically flat or even crystalline surfaces, so that they find widespread applications in metallurgy and polymer science.

The second common sample geometry is an etched tip, about a hundred nanometers in radius. Such specimens are studied by field emission and ionization experiments, which provide a direct magnified image of the surface structure in contrast to the statistical description of platelet surfaces afforded by instruments like that shown in Fig. 1.

Data acquisition, analysis, and theory. Given the ability to perform surface-sensitive spectroscopic measurements, questions naturally arise concerning analysis of the raw spectra to extract parameters characterizing the structure of a given surface and the synthesis of such data to form a coherent picture of the behavior of electrons and atomic species at surfaces. Thus, surface physics may be divided into three types of activity: the acquisition of surface-sensitive spectroscopic data, the analysis of these data using physical models of the appropriate scattering or emission spectroscopy, and the construction of broad theoretical models of surface structure and properties to be tested via critical comparison of their predictions with the results of such data analyses.

Ground- and excited-state properties. Theoretical models have been proposed for the description of two distinct types of surface properties. The stability of surface structures is examined by calculations of ground-state properties, such as surface energies or effective potential-energy diagrams for adsorbed species. These quantities are difficult to measure experimentally, although they are the most direct manifestations of the intrinsic behavior of an undisturbed surface. The interactions of external projectiles or fields with a solid create excited states of the electrons or atoms within the solid. Consequently, the associated scattering and emission spectra indicate the nature and energies of these excited states (called excitations) rather than of the ground state. Two kinds of excitations occur. Electronic excitations are generated when a disturbing force causes the electrons in the solid to alter their quantum states, whereas atomistic excitations are associated with the vibration or diffusion of atomic species (such as adsorbed atoms or molecules). It is important to distinguish between ground-state properties, electronic excitations, and atomistic excitations because rather different models are used to describe each of these three types of phenomena. *See* EXCITED STATE; GROUND STATE.

Quantum theory of surfaces. The structure of the theory of the properties of solid surfaces does not differ in any fundamental way from that of the quantum theory of bulk solids. Specifically, the conventional quantum theory of interacting electron systems is thought to be applicable, although technical refinements are required because of the loss of translational symmetry and the presence of large electron density gradients normal to the surface. *See* NONRELATIVISTIC QUANTUM THEORY.

Macroscopic models. It is premature to speak of an embracing theory of surface phenomena. Rather, a diverse array of specific models have been proposed for the description of various properties. In the case of macroscopic models, the presence of a surface is treated as a boundary condition on an otherwise continuum theory of bulk behavior. Such models have found widespread use in semiconductor and insulator physics because the penetration depth of electrostatic fields associated with surface charges usually is large ($\lambda_e \sim 10^4 - 10^5$ nm) relative to the spatial extend of the charges themselves ($d \lesssim 1$ nm). To describe the atomistic and electronic properties associated with the upper few atomic layers at a surface, however, one must make use of a description of surfaces at the atomistic or electronic level. *See* SEMICONDUCTOR.

Microscopic models. Four major classes of microscopic models of surface properties have been explored. The simplest of these consists of models in which consideration of the electronic motion is suppressed entirely, and the solid is visualized as composed of atomic species interacting via two-body forces. While such models may suffice to describe the vibrational motion of atoms near a surface, they are inadequate to describe ground-state properties such as adsorbate potential-energy curves, although they have been utilized for such calculations. The next more sophisticated models are empirical quantum chemical models (such as "tight-binding" or "pseudopotential" models in solid-state terminology), in which electronic motions are considered explicitly but electron-electron interactions are not. Such models have proved useful in solid-state physics, although their value for surface physics is limited because the large charge rearrangements (relative to the bulk) which occur at surfaces require an accurate, self-consistent treatment of electron-electron and electron-ion interactions. The simplest model in which these interactions are treated explicitly is the "jellium" model of metals, in which the positive charge associated with the ion cores immersed in the sea of conduction electrons is replaced by a uniform positive "background" charge terminating along a plane. This model permits the most accurate (but still approximate) treatment of electron-electron interactions at the expense of losing the effects of atomic lattice structure because of the uniform-positive-background hypothesis. Electronic computers have permitted, however, the construction of semi-empirical models in which both the electron-electron and electron-ion interactions can be treated in a self-consistent, if approximate, fashion. Such models have been applied to examine ground-state electronic charge densities and localized electronic surface states at the surfaces of simple (that is, *s-p* bonded) and transition metals, as well as homopolar semiconductors. The major tests of their adequacy arise from comparisons of their predictions with measured work functions, photoemission spectra, and characteristic electron-loss spectra. The only new result (relative to comparable bulk analyses) emanating from these computations is the recognition that since the local electronic density of states in the upper few layers depends on the geometrical structure of these layers, comparison of the calculated local density of states with observed valence-electron emission spectra provides a qualitative means of assessing the adequacy of proposed models for the atomic geometry of surface species.

Theoretical models for data analysis. Another distinct but important group of theoretical models in surface physics consists of those utilized to analyze observed scattering and emission spectra in order to extract therefrom quantitative assessments of the atomic and electronic structure of surfaces. These models differ substantially from their bulk counterparts because of the necessity of strong particle-solid interactions to achieve surface-sensitive spectroscopies. Consequently, the fundamental assumption underlying the linear-response theory of bulk solid-state spectroscopies—that is, the appropriate particle-solid interaction is weak and hence can be treated by low-order (usually first) perturbation theory—is invalid. This fact results in collision theories of surface-sensitive particle-solid scattering exhibiting a considerably more complicated analytical structure in order to accommodate the strong elastic as well as inelastic scattering of the particle by the various constituents of the solid.

Applications to LEED. While the above considerations are quite general, the special case in which they have been developed in most detail is the coherent scattering (that is, diffraction) of low-energy electrons from the surfaces of crystalline solids. This is an important case because elastic low-energy electron diffraction (ELEED) is the analog of x-ray diffraction for surfaces—that is, it is the vehicle for the achievement of a quantitative surface crystallography. Since 1968 quite complete quantum-field-theory models of the ELEED process have been developed, tested, and reduced to computational algorithms suitable for the routine analysis of ELEED intensity data. From such analyses the surface atomic geometries of the low-index faces of a host of simple metals and polar semiconductors have been determined, as have the geometries of a few simple overlayer structures on the low-index faces of fcc metals. Similar quantitative analyses of inelastic low-energy electron diffraction (ILEED) intensities have yielded the energy-momentum relations of collective surface electronic excitations (surface plasmons). Therefore, adequate quantum field theories of both the elastic and inelastic diffraction of electrons have been constructed and applied to the quantitative characterization of the structure of the low-index faces of crystalline solids via the analysis of LEED intensities. Development of analogous theories of photoelectric emission has been undertaken. *See* QUANTUM FIELD THEORY.

Data acquisition. It is the development of a host of novel surface-sensitive spectroscopic techniques, however, which has provided the foundation for the renaissance in surface physics. Having recognized that low-energy electrons, thermal atoms, and slow ions all constitute surface-sensitive incoming or exit entities in a particle-solid collision experiment, one can envisage a wide variety of surface spectroscopies based on these plus quanta of electromagnetic radiation (photons) as possible incident or detected species. Most of these possibilities actually have been realized in some form. The selection of which technique to use in a particular application depends both upon what one wishes to learn about a surface and upon the relative convenience and destructiveness of the various measurements.

Typically, one wishes to determine the composition of a surface region, and often to ascertain its atomic geometry and electronic structure as well. For planar surfaces of crystalline solids, measurement of the rms surface-atom displacements is an ancillary to that of the atomic geometry because the values of these displacements emerge from an examination of the temperature dependence of the same ELEED intensities whose analysis yields the atomic structure of the surface. In the measurement of any of these quantities, however, important issues are the spatial and depth resolution of the possible techniques. Typically the depth resolution is determined by the particle-solid force law of the incident and exit particles, higher resolution being associated with stronger inelastic collision processes. The lateral spatial resolution depends on the ability to focus the incident beam. It is of the order of 1 cm^2 for photon beams, 10^{-8} to 10^{-12} cm^2 for electron beams, and 10^{-8} cm^2 for ion beams. Thus, scanning microscopies are both feasible and common with electron and ion beams but not with photon beams. Depth resolution is a single monolayer for thermal-atom and slow-ion scattering, and a few monolayers for slow-electron scattering. It can become 1000 atomic layers or more, however, for fast (MeV) ions and (10 keV) electrons.

Surface composition. The elemental composition of surfaces is specified by measuring the masses or atomic numbers, or both, of resident species. Their masses may be ascertained either by the elastic backscattering of slow incident ions (ion scattering spectrometry, or ISS) or by using such ions to erode the surface, detecting the ejected surface species in a mass spectrometer (secondary ion mass spectrometry, or SIMS).

The atomic numbers of surface species are determined by measuring the energy of tightly bound "core" electrons. A schematic diagram illustrating the nature and labeling of the various physical processes which can be utilized to accomplish this task is shown in Fig. 2. An electron, photon, or chemical species incident on a surface excites a low-energy core electron. The binding energy of this electron commonly is determined directly by measuring the energy loss of the incident electron (characteristic loss spectroscopy, or CLS), the energy of the core electron ejected by an incident x-ray photon (x-ray photoelectron spectroscopy, or XPS, sometimes referred to as electron spectroscopy for chemical analysis, or ESCA), or the threshold energy of an incident particle necessary to generate a threshold in the secondary x-ray yield (soft-x-ray appearance potential spectroscopy, or SXAPS). Alternatively, the binding energy of the core electron may be ascertained by "secondary" processes in which an initially empty core state (generated by a direct process) is filled by an electron in a higher energy state. If the filling process is radiative recombination, then the energy of the emitted x-ray yields the binding energy (soft

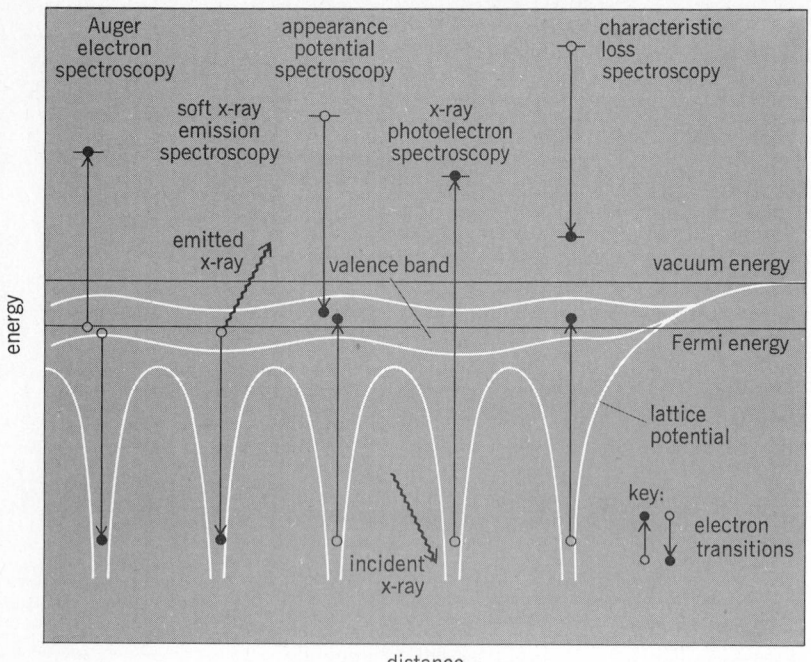

Fig. 2. Schematic diagram of the core-electron transitions utilized to ascertain the atomic number of surface species. (*From R. L. Park, Inner-shell spectroscopy, Phys. Today, 28(4):52–59, April 1975*)

x-ray emission spectroscopy). If this process is radiationless, however, then the energy of the electron excited by the Auger process indicates the binding energy of the initially empty core state (Auger electron spectroscopy, or AES).

Essentially all of these techniques operate on the dictionary premise: that is, calibration spectra are obtained on surfaces of independently known composition, with elemental analysis on unknown samples being performed by comparison of their spectra with the reference calibration spectra. Consequently, although the detailed interpretation of observed line shapes has eluded surface physicists, the use of these spectroscopies for elemental analysis has proved both practical and eminently useful. Difficulties in interpretation have precluded the use of these techniques for quantitative chemical analysis (for example, the determination of whether C and O are adsorbed on aluminum as CO, CO_2, or C on Al_2O_3, and so on). Progress has been made, however, in developing this aspect of the core-electron spectroscopies.

Surface atomic geometry. The atomic geometry of planar surfaces of crystalline solids usually is obtained by electron diffraction, although in certain simple cases slow-ion backscattering or valence-electron photoemission spectroscopy also may be employed. Two experimental configurations commonly are used, as indicated in Fig. 3. The reflection high-energy electron diffraction (RHEED) configuration embodies glancing incidence electrons at keV energies. It yields only the space-group symmetry of the surface, and is quite sensitive to surface topography. The ELEED experiment consists of measuring the backscattering intensities of electrons in the energy range 50 eV $\leq E \leq$ 500 eV. The configuration of diffracted beams reveals the space group symmetry of the surface structure, whereas analysis of their inten-

sities permits determination of their atomic geometry. In the case of tip sample geometries, field-ion microscopy permits the direct imaging of atoms on the tip surface.

Surface atomic motion. The vibrational motion of surface species may be examined either by analysis of the temperature dependence of ELEED intensities or by direct observation of small ($\Delta E \sim 0.01$ eV) electron energy losses caused by the excitation of a normal mode of vibration. The first approach provides the rms vibrational amplitudes of surface species, whereas the second yields the frequencies of localized "surface" normal modes of vibration. Nonstandard equipment is required in both cases: an ultra-high-vacuum goniometer embodying precise temperature control in the former, and a high resolution ($\Delta E \leq 0.01$ eV) electron spectrometer in the latter.

Surface electronic structure. The electronic structure of a solid-vacuum interface is studied by measuring the emission of valence electrons (induced by external fields, electrons, ions, or photons) or the inelastic scattering of an incident electron. A special situation arises when an emitted Auger electron is a valence electron. In this case, the initially empty core state is highly localized in space. Consequently, the emission line shape is a measure of the local electronic structure in the vicinity of this core state. The shifts in energy of core-level photoemission and Auger transitions (called chemical shifts) caused by the nearby electronic charge densities also yield an indication of the local electronic structure around a particular kind of surface atom.

In contrast to these emission processes involving localized core electrons, the photoemission, field-emission, ion-neutralization, and characteristic-loss spectroscopies of valence electrons pro-

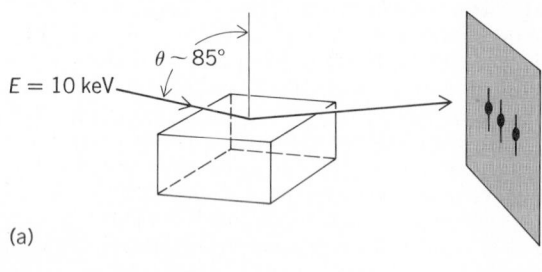

(a)

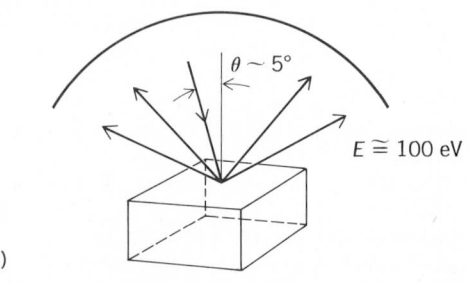

(b)

Fig. 3. Schematic diagram of the two electron diffraction techniques used to determine the geometry of single-crystal surfaces. (a) Reflection high-energy electron diffraction (RHEED). (b) Elastic low-energy electron diffraction (ELEED). (*From C. B. Duke, Determination of the structure and properties of solid surfaces by electron diffraction and emission, in I. Prigogine and S. A. Rice, eds., Adv. Chem. Phys., 27(1):1–209, Wiley-Interscience, 1974*)

vide measures of their average behavior in the vicinity of a surface. Indeed, for precisely this reason such spectra from clean surfaces are quite difficult to interpret because the distinction between "bulk" and "surface" features is often vague. Thus, their major use has occurred in the arena of chemisorption, in which case the changes in spectra upon adsorption can be monitored, and qualitative features of the electronic structure of the chemisorbed complexes inferred therefrom. Since quantitative theoretical models of these emission processes are not available, however, ambiguities in interpretation are common. Although the advent of variable-energy synchrotron radiation sources and measurements of the angular dependence of the emission intensities have reduced the probability of such ambiguities, a proper theoretical analysis of these processes is needed to convert the observed spectra into quantitative indicators of surface structure. The only case for which such a model has been constructed is that of ILEED. The availability of this theoretical model permits the analysis of ILEED intensities to extract the energy-momentum relationship of surface plasma oscillations; this is the only quantitative application developed for valence-electron surface spectroscopy. *See* Synchrotron radiation.

[C. B. Duke]

Bibliography: J. M. Blakely (ed.), *Surface Physics of Materials*, vols. 1 and 2, 1975; H. L. Davis (ed.), Surface physics, *Phys. Today*, special issue, 28(4):23–71, April 1975; H. L. Davis (ed.), Vacuum: A special report, *Phys. Today*, 25(8):23–58, August 1972; F. Garcia-Moliner and F. Flores, *Introduction to the Theory of Solid Surfaces*, 1979; F. O. Goodman (ed.), *Dynamic Aspects of Surface Physics*: *Proceedings of the International School of Physics 'Enrico Fermi,'* course 58, 1975; T. S. Jayadevaiah and R. Vanslow (eds.), Surface science: Recent progress and perspectives, *Crit. Rev. Solid State Sci.*, vol. 4, issues 2 and 3, 1974; H. Kumagai and T. Toya (eds.), Proceedings of the 2d International Conference on Solid Surfaces, *Japan. J. Appl. Phys.*, suppl. 2, pt. 2, 1974; S. R. Morrison, *The Chemical Physics of Surfaces*, 1977; I. Prigogine and S. A. Rice (eds.), Aspects of the study of surfaces, *Adv. Chem. Phys.*, vol. 27, 1974; Proceedings of the Annual Surface Science Symposia of the American Vacuum Society, *J. Vacuum Sci. Tech*, January/February issues; M. Prutton, *Surface Physics*, 1975; G. A. Somorjai, Surface science, *Science*, 201:489–497, 1978; R. Vanselow and S. Y. Tong (eds.), *Chemistry and Physics of Solid Surfaces*, 1977.

Surface tension

The force acting in the surface of a liquid, tending to minimize the area of the surface. Surface forces, or more generally, interfacial forces, govern such phenomena as the wetting or nonwetting of solids by liquids, the capillary rise of liquids in fine tubes and wicks, and the curvature of free-liquid surfaces. The action of detergents and antifrothing agents and the flotation separation of minerals depend upon the surface tensions of liquids.

Surface energy. In the body of a liquid, the time-averaged force exerted on any given molecule by its neighbors is zero. Even though such a molecule may undergo diffusive displacements because of random collisions with other molecules, there

exist no directed forces upon it of long duration. It is equally likely to be momentarily displaced in one direction as in any other. In the surface of a liquid, the situation is quite different; beyond the free surface, there exist no molecules to counteract the forces of attraction exerted by molecules in the interior for molecules in the surface. In consequence, molecules in the surface of a liquid experience a net attraction toward the interior of a drop. These centrally directed forces cause the droplet to assume a spherical shape, thereby minimizing both the free energy and surface area.

From the macroscopic point of view, surface tension may be regarded either as a force exerted normally to a unit length in the surface, or as the work which must be expended upon the liquid to increase its area by unity. Accordingly, surface tension is expressed in centimeter-gram-second (cgs) units of dynes/cm or ergs/cm². From the microscopic point of view, the surface tension (or its equivalent, surface energy) is the reversible isothermal work which must be done in bringing molecules from the interior of the liquid to the surface and creating 1 cm² of new surface thereby.

Most liquids have surface tensions of 20–40 dynes/cm at room temperature, but water has the exceptionally high value of 72.75 dynes/cm at 20°C. Condensed gases such as helium and nitrogen have quite low surface tensions (0.098 dynes/cm at 4.3 K and 6.2 dynes/cm at 90.2 K, respectively). Liquid metals have large surface tensions by comparison: mercury, 470 dynes/cm; and liquid copper at 1131°C has a surface tension of 1103 dynes/cm in hydrogen gas. Small but significant differences in the surface tensions of liquids depend upon the composition of the vapor phase.

In the wetting or nonwetting of solids by liquids, the criterion employed is the contact angle between the solid and the liquid (measured through the liquid) (Fig. 1). A liquid is said to wet a solid if

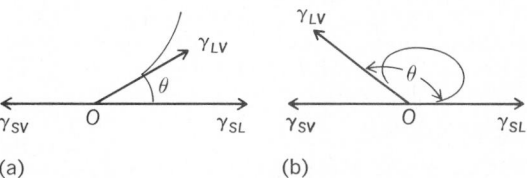

Fig. 1. Contact angle. (*a*) Liquid wets solid. (*b*) Liquid does not wet solid.

the contact angle θ lies between 0 and 90°, and not to wet the solid if the contact angle lies between 90 and 180°. Three interfaces exist when a droplet of liquid contacts a solid, and three corresponding interfacial tensions exist: γ_{SL}, γ_{SV}, and γ_{LV}. The subscripts S, L, and V refer to solid, liquid, and vapor. At equilibrium, a balance of interfacial tensions exists at the line of common contact, which intersects the figures at point O. For the case of a liquid which wets the solid ($\theta < 90°$), this equilibrium is expressed by relation (1).

$$\gamma_{SV} = \gamma_{SL} + \gamma_{LV} \cos \theta \qquad (1)$$

Capillarity. Liquids which wet the walls of fine capillary tubes rise to a height which depends upon the tube radius, the surface tension, the liq-

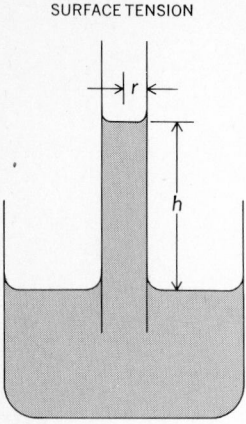

SURFACE TENSION

Fig. 2. Rise of liquid in capillary tube.

uid density, and the contact angle. In Fig. 2, a liquid of density ρ is shown as having risen to a height h in a capillary whose radius is r. A balance exists between the force exerted by gravity on the mass of liquid raised in the capillary and the opposing force caused by surface tension. The former is $\pi r^2 h \rho g$, whereas the latter is $2\pi r \gamma$, assuming the contact angle to be zero. It is clear that $h = 2\gamma/r\rho g$, and that the capillary rise varies inversely with the tube radius and the liquid density. Liquids which do not wet the capillary walls are depressed in height according to the same equation.

The shape of the free surface of a liquid in a vessel is only an approximation to a plane. In narrow tubes the meniscus of a liquid is concave upward if the liquid wets the tube and, conversely, convex upward if it does not wet the tube. A pressure difference exists between the concave and convex sides of the surface, the excess pressure on the concave side over the convex side being given by relation (2), where r_1 and r_2 are the principal radii

$$p = \gamma \frac{1}{r_1} + \frac{1}{r_2} \tag{2}$$

of curvature of the surface. The same equation applies for a bubble of gas within a liquid, with the consequence that the vapor pressure p is larger for small bubbles according to relation (3), where p_0 is

$$\ln \frac{p}{p_0} = \frac{2\gamma}{r\rho} \frac{M}{RT} \tag{3}$$

the vapor pressure over a liquid surface of infinite radius, R is the gas constant, and M is the molecular weight. *See* VAPOR PRESSURE.

Detergents, soaps, and flotation agents owe their usefulness to their ability to lower the surface tension of water, thereby stabilizing the formation of small bubbles of air. At the same time, the interfacial tension between solid particles and the liquid phase is lowered, so that the particles are more readily wetted and floated after attachment to air bubbles.

[NORMAN H. NACHTRIEB]

Bibliography: American Chemical Society, *Contact Angle, Wettability and Adhesion*, 1964; G. M. Barrow, *Physical Chemistry*, 4th ed., 1979.

Susceptance

The imaginary part of the complex representation for the admittance Y, defined by Eq. (1), of a circuit, where G is the real part, called the conduct-

$$Y = G \pm jB \tag{1}$$

ance, and B is the susceptance. Since $Y = 1/Z = 1/(R + jX)$ where X is the total reactance, $X_L - X_C$, and R is the resistance, then Eq. (2) holds and sus-

$$Y = \frac{R}{R^2 + X^2} - j\frac{X}{R^2 + X^2} \tag{2}$$

ceptance $B = X/(R^2 + X^2)$. This is the general expression for susceptance which shows that susceptance is a function involving both resistance and reactance.

If resistance is negligible, then $B = X/X^2 = 1/X$, or the reciprocal of the reactance. This is called simple susceptance and is correct only where the impedance contains no resistance. For simple susceptances Eqs. (3) hold.

Inductive susceptance $B_L = 1/jX_L = -jB_L$ mhos

Capacitive susceptance $B_C = 1/-jX_C = jB_C$ mhos

$$\tag{3}$$

These functions find application chiefly in computation of parallel circuits. *See* ADMITTANCE.

[BURTIS L. ROBERTSON]

Symmetry laws

The physical laws which are the expressions of the existing symmetries. A conservation law results from each such symmetry; that is, from each symmetry the existence of a quantity which is conserved (a constant of the motion) can be deduced. Selection rules result from conservation laws.

Space-time symmetries. A symmetry (or invariance) of the world exists whenever the description of the laws of physics is unaffected by a change in the frame of reference. For instance, the position of the origin of a space coordinate system is quite arbitrary; changing it makes no difference in the description of the motion of bodies because the forces between bodies depend only on their relative positions and not on any absolute position. Equivalently, a system of bodies behaves the same if translated to another place. This symmetry of space to translation implies the conservation of momentum.

Other symmetries of space-time are the irrelevance of (1) the origin of the time coordinate, (2) the orientation of a coordinate system in space, and (3) the velocity of a coordinate system (Lorentz invariance). Each of these implies a conservation law, as shown in the table. All these symmetries are termed continuous because the changes can be arbitrarily small; that is, a finite change can be made bit by bit. The resulting constants of the motion are classical quantities and are additive. *See* FRAME OF REFERENCE; LORENTZ TRANSFORMATION; RELATIVITY; SPACE-TIME.

Discrete symmetries (reflections) also exist, for which the irrelevant change is not arbitrarily small. They imply constants of the motion (parities) in quantum mechanics. These parities are multiplicative. For instance, the direction of increasing time is irrelevant; the world is invariant to time reversal (microscopic reversibility). Although, macroscopically, future and past seem distinct, this is merely a result of the disposition of matter (a state of anomalously small entropy at some time in the past) in the same way that a point of space seems distinct by having a particular piece of matter there. Space is also symmetrical to reflection of space or to inversion, the reflection of all three directions of space; it is irrelevant which is the positive direction of a space axis or of all three axes. This amounts to the irrelevance of whether a right-handed or a left-handed coordinate system is used. The resulting conserved quantity (eigenvalue of space inversion) is (space) parity. Actually, the preceding statements about space inversion symmetry must be qualified. Although inversion symmetry is observed by the strong interactions (such as nuclear forces) and electromagnetic interactions, it is not observed by the weak interactions, such as decays of quasistable elementary particles, including β-decay. The description of a β-decay event depends on the handedness of the coordinate system. *See* FUNDAMENTAL INTERACTIONS; PARITY; STRONG NUCLEAR INTERACTIONS; WEAK NUCLEAR INTERACTIONS.

Invariances (symmetries) and conservation laws

Invariance to	Conserved quantity	Range of validity
Homogeneity of space-time		
Translation of space	Momentum, **p**	
Translation of time	Energy, E	
Isotropy of space-time		
Rotation of space	Angular momentum, **J**	
Lorentz transformation	Velocity of the center of energy (center of mass)	
Interchange of identical particles	Symmetry of the wave function (statistics)	Exact
Inversion of time, space, and charge	CPT	
Gauge invariance		
Charge U_1	Net charge	
Color SU_3	Color[a]	
?	Net number of baryons, e-leptons, μ-leptons	} [b]
Reversal of time	Time parity, T	
Reflection of space and charge	Product of parity and charge parity, CP	} [c]
Reflection (or inversion) of space	Parity, P	
Reflection of charge (charge conjugation)	Charge parity, C	[e]
?	Net number of up quarks, down quarks, strange quarks, charm quarks	} [d]
Flavor SU_2	Isotopic spin, I isotopic parity, G	
Flavor SU_3	"Unitary spin"	[f] [g]

[a]A hypothesized hidden symmetry.

[b]Conserved at the present level of detection, but possibly not exactly conserved.

[c]Violation seen only in K_L decay.

[d]Conservation of these three is equivalent to the conservation of baryon number, I_3 (the charge axis component of isotopic spin) and hypercharge.

[e]Violated by the weak interactions.

[f]Violated by electromagnetic interactions and by the up-down quark mass difference, $m_u - m_d \approx -5$ MeV.

[g]Violated by the strange-nonstrange quark mass difference, $m_s - m_u \approx m_s - m_d \approx 150$ MeV.

Symmetry in quantum mechanics. In quantum mechanics a change in the frame of reference is described by the multiplication of the state vectors by a unitary operator $\psi \rightarrow A\psi$. This has no effect on an observable O if its operator commutes with A, $OA = AO$. In particular, energy eigenvalues are unchanged if the hamiltonian H commutes with A. The totality of operators which commute with H form a group, the symmetry group of H. Starting with an energy eigenstate ψ, the states $A\psi$, $B\psi$, . . . , where A, B, . . . are the members of the symmetry group of H, all have the same energy eigenvalue (are degenerate). A linearly independent set of states which spans all the degenerate states formed from ψ is called a multiplet; its number is called the multiplicity of the multiplet. If ψ_n are the members of a multiplet, the matrices whose elements are the matrix elements of the symmetry operators $A_{mn} = (\psi_m, A\psi_n)$ form a matrix representation of the symmetry group. The importance of this is that all the possible matrix representations of a group can be determined abstractly from the multiplication table of the group, and therefore independently of the physical significance of the transformations. (Further, the matrix elements of other operators are determined by their commutation relations with the symmetry operators; this is the Wigner-Eckart theorem.) For example, if the symmetry group of the hamiltonian is the rotation group (rotations of the spatial coordinate frame about a point), the possible multiplets have multiplicities 1, 2, 3, . . . , corresponding to angular momentum quantum number $j = 0, \frac{1}{2}, 1, ...$ The multiplets of the internal symmetry group SU_2 are the same; for SU_3 they are 1, 3, 6, 8, ... *See* ANGULAR MOMENTUM; DEGENERACY (QUANTUM MECHANICS); GROUP THEORY; NONRELATIVISTIC QUANTUM THEORY.

Internal symmetries. Further symmetries, which are not space-time symmetries, are called internal symmetries. For example, the zero of both scalar and vector electromagnetic potentials is irrelevant; the addition of a constant to an electromagnetic potential is of no consequence (so-called gauge invariance of the first kind). In quantum mechanics this symmetry implies the conservation of charge. There is also a discrete symmetry: It is (nearly) irrelevant which sign of charge is called positive and which is called negative. The qualification "nearly" is necessary here, just as in space inversion, because the weak interactions do not observe the symmetry. Thus the world is (nearly) invariant to charge reversal, the interchange of positive and negative charge. At first sight this symmetry appears not to exist because one can distinguish positive charge as that which is carried by the heavy constituent of matter, the proton, whereas negative charge is that carried by the light constituent, the electron. However, antiprotons (negatively charged protons) and antielectrons (positrons) exist, and a world with (nearly) the same properties as Earth's would result if all electrons were replaced with positrons and all protons by antiprotons and, in general, if all particles were replaced by their antiparticles (the operation of charge conjugation). The resulting (nearly) conserved quantity is termed charge conjugation parity or charge parity, C. More precisely, charge parity is the eigenvalue of the operator of charge reversal. A system can be in an eigenstate of charge reversal only if it goes into itself under the operation, in other words, if it is self-charge conjugate. Such a system must be completely neutral, having no electric or magnetic moments; in fact, it must have no internal quantum numbers of any kind that change sign under charge conjugation. The neutron and the K^0 meson are examples of neutral but non-self-charge conjugate systems ($\bar{n} \neq n$; $\overline{K^0} \neq K^0$). The neutral π-meson, the photon, and the graviton are self-charge-conjugate; their charge parities are the charge parities of their sources, +1, −1, and +1, respectively. A self-charge conjugate system of some interest is positronium, a bound state of an electron and a positron; it has $C = (-)^{s+l}$ in a state or orbital angular momentum l and spin $s(s = 0$ or $1)$. *See* ELEMENTARY PARTICLE; POSITRONIUM; POTENTIALS.

Although the weak interactions are not invariant under charge reversal or space inversion, they are invariant (with one observed exception, K^0 decay) to the combination of these reflections; they are also invariant under time reversal. Equivalently stated, all interactions appear to very nearly conserve CP and also T (C = charge conjugation, P = space inversion, T = time reversal). For instance, consider the decay of a muon into an electron and a pair of neutrinos. The probability that the electron's momentum $\vec{p}_e$ makes an angle θ with the direction of the muon's spin $\vec{s}$ is observed to be of the form given by Eq. (1), where the upper sign

$$P_{\pm} = a \pm b \cos \theta \qquad (1)$$

holds for a positive muon and the lower for a negative muon. In an inverted coordinate system, $\vec{p}_e$ would be reversed in sign, but $\vec{s}$ would not (spin being an axial vector) and so $\cos \theta$ would reverse sign, resulting in Eq. (2). Hence the description of

$$P_{\pm} = a \mp b \cos \theta \qquad (2)$$

the decay is not invariant to space inversion. $P_{\pm}$ would also change from Eq. (1) to Eq. (2) under charge reversal, that is, interchanging what is called positive and negative charge. Thus under both reflections together, $P_{\pm}$ is unchanged and so decay "conserves CP."

A sensitive test of T invariance is the size of a possible electric dipole moment of the neutron. The only direction provided by the neutron to direct such a moment is its spin; under time reversal this would reverse, and so the interaction energy of the moment with an electric field would also reverse, in contrast to all other energies. The upper limit as of 1980 is 2×10^{-24} cm times e, the charge of the electron.

A very sensitive test of CP conservation is provided by K^0 decay. Neutral K mesons are produced by strong interactions, and so are produced as either K^0 or $\overline{K^0}$ mesons. But they decay differently, through weak interactions, and so it is linear combinations of K^0 and $\overline{K^0}$ which decay with definite lifetimes, namely K_L and K_S (L = long life, S = short life). [As an analogy, a light source might produce only right-hand circularly polarized photons (just as in the process $\pi^- p \rightarrow \Lambda K^0$, only K^0, not $\overline{K^0}$, mesons are produced); but if the light passes through Polaroid, it decomposes into two components with different decay (absorption) lengths, namely, photons which are linearly polarized, perpendicular or transverse to the optic axis.] If all interactions, including the weak ones, conserved CP, then K_L and K_S would be eigenstates of CP, and only one of them would decay into two pions, which is a CP eigenstate. In fact, K_S decays almost exclusively into two pions, but K_L is found to also have a nonvanishing, though small, probability of this decay. Thus CP is not exactly conserved in K^0 decay. The smallness of the violation contrasts with the maximal violation in weak interactions of C and P separately. No other system is as sensitive to CP violation as the neutral K mesons (they are a degenerate doublet, which is easily mixed by a

small perturbation); as of 1980 no violation of either CP or T has been observed elsewhere. This makes it difficult to test the various detailed mechanisms which have been suggested for the violation.

CPT theorem. The reflection symmetries are correlated by the *CPT* theorem of G. Lüders. This theorem states that a Lorentz invariant field theory is necessarily invariant to the product of the three reflections: charge conjugation C, space inversion P, and time reversal T. For example, this would mean that K^0 decay, which does not obey CP invariance, also should not obey T, time reversal invariance. In principle, the theorem can be tested experimentally, particularly in K-meson decay, but so far no stringent results have been obtained. Elementary particle theory would be enormously challenged if CPT invariance were found not to hold. *See* MESON.

Invariances of strong interactions. The nuclear force between two protons is found to be identical to the force between two neutrons. This is called the charge symmetry of the nuclear force. More generally, the motion of a system composed of nucleous and pions (the lowest mass quanta of the nuclear force field) is invariant to the operation is $p \leftrightarrow n$, $\pi^+ \leftrightarrow \pi^-$, $\pi^0 \leftrightarrow \pi^0$. The combination of charge conjugation and the charge symmetry operation is called isotopic inversion, G, and carries each π-meson into itself. The π-meson thus has a G parity.

Further, the nuclear force between a neutron and a proton is identical to the force between two protons or two neutrons in the same orbital and spin state (charge independence). *See* NUCLEAR STRUCTURE; SCATTERING EXPERIMENTS (NUCLEI).

Isotopic spin. The foregoing symmetry can be expressed as the isotropy of a three-dimensional "isotopic space," which implies the conservation of "angular momentum," or isotopic spin $\mathbf{I}$, in this space. The component of a particle's isotopic spin along the "third" axis I_3 s related linearly to the charge of the particle. The consequences of charge independence are formally very similar to the consequences of the conservation of angular momentum, except that nothing here corresponds to orbital angular momentum.

In terms of isotopic spin, the charge symmetry operation is a special rotation in isotopic spin space, namely, one which reverses the direction of the third axis. It follows from angular momentum calculus that a system having $I_3 = 0$ has a charge symmetry parity which is $(-)^i$, where i is the magnitude of $\mathbf{I}$ (total isotopic spin) of the system. Thus the charge symmetry parity of the π^0-meson is -1. The π-meson, therefore, has a G parity of -1; nucleonium, a system of nucleon and antinucleon, has $G = (-)^{l+s+i}$ in a state of orbital angular momentum l, spin s (0 or 1), and isotopic spin i (0 or 1).

All the strong interactions are isotropic in isotopic space and conserve isotopic spin; all the strongly interacting elementary particles carry isotopic spin. The anomalous stability of the heavier (the so-called strange) particles is explained by the fact that charge is conserved, and in both the strong and electromagnetic interactions I_3 is conserved (Gell-Mann–Nishijima scheme). *See* ISOTOPIC SPIN.

Unitary symmetry. Charge independence, described above as the isotropy of a three-dimensional space, can also be described as symmetry with respect to arbitrary unimodular unitary transformations of the proton and neutron, that is, invariance of the strong interactions to transformations of the form $p \to \sum \cos \theta e^{i\psi} p + i \sin \theta e^{i\phi} n$, $n \to \cos \theta e^{-i\psi} n + i \sin \theta e^{-i\phi} p$, where p and n stand for the state vectors of a proton and a neutron, respectively. This transformation is called mixing p and n. This group of transformations is called SU_2. The analogous group of transformations on n particles is SU_n. *See* SPINOR; UNITARY SYMMETRY.

It appears that hadrons are well described as compounds of so-called quarks, of which there are n kinds (flavors): $u, d, s, c, b, \ldots$. The net number of quarks of each flavor is changed only in weak interactions. For example, in the strong reaction in Eq. (3) the net number of u, d, and s quarks is 2, 0,

$$\overline{K^0} \, p \to \Sigma^+ \pi^0 \qquad (3)$$

and 1 respectively in both the initial and the final state. Further, the quarks of different flavors interact in the same way with the fundamental strong interaction between them (the "glue" which binds them together to make hadrons). This means that the strong interactions would have a flavor SU_n internal symmetry, were it not for the fact that the quarks of different flavors have different masses; their masses seem to form roughly a geometrical progression. Aside from the effect of this on the internal symmetry of the strong interactions, this has the practical effect that the number of flavors n which are known increases as the maximum energy of particle accelerators increases; at present (1980) n is 5.

The masses of the lightest two quarks, u and d, are small compared to the zero-point energy associated with the binding of the quarks into hadrons, and so the properties of hadrons are rather insensitive to the mixing of u and d quarks; consequently the corresponding SU_2 symmetry is rather good. This is the isotopic spin symmetry described above; there it was described as an insensitivity to the mixing of p and n nucleons, which is equivalent to the mixing of u and d quarks. The next higher mass quark s has a mass which is not much smaller than the zero-point energy, and therefore the mixing of s with u or d quarks affects the properties of a hadron much more than does the mixing of u and d quarks. Hence the corresponding SU_3 symmetry is not as good a symmetry as the SU_2 (isotopic spin) symmetry. The masses of the remaining quarks are so large that it is useful, not to describe the relationship of hadrons which contain various numbers of these heavy quarks in terms of SU_4 and SU_5 symmetries, but rather to calculate the properties of these hadrons by using directly the symmetry principle that the glue couples equally to all flavors of quarks.

In addition to this flavor SU_n symmetry of strong interactions, it is believed that there is also a color SU_3 symmetry which is exact but hidden. *See* COLOR (QUANTUM MECHANICS); FLAVOR; QUARKS.

Chiral symmetry. To describe this symmetry, one must first define helicity, η; η is the projection of the spin of a particle on the direction of its motion, that is, $\eta = \vec{S} \cdot \hat{p}$, where $\vec{S}$ is the particle's spin (in units of $\hbar$, Planck's constant divided by 2π) and $\hat{p} = \vec{p}/|p|$ is its unit momentum vector. In quantum mechanics, η has the quantized values s, $s - 1, \ldots, -s$, where s is the magnitude of the spin of the particle. Thus a spin-$\frac{1}{2}$ particle, such

as e, p, or n, can have just two eigenvalues of η, namely $\frac{1}{2}$ and $-\frac{1}{2}$; these polarization states are called right-hand and left-hand, respectively. Usually, the value of η is not an intrinsic property of a particle because it is not a Lorentz invariant; that is, a particle which is, say, right-handed as observed in one space-time frame will not be as observed in another. But, exceptionally, the helicity of a massless particle is an intrinsic property, independent of the choice of space-time frame (reflections of space excluded), just as much as properties such as charge or i-spin. Hence, a system of massless particles could obey laws such as "the number of left-hand and right-hand particles are each conserved" or "the i-spins carried by the left-hand particles and by the right-hand particles are each conserved." (Relativistic invariance requires that the category "left-hand particles" must also include their charge conjugates, namely, right-hand antiparticles.) Such conservation laws follow from symmetries of the form of a product of an internal symmetry group for the left-hand particles (that is, their field) and a similar one for the right-hand particles. Such a symmetry group is called chiral (meaning "handed"). It is well known that the electromagnetic and weak interactions are chiral-symmetric. But it is obvious that the world as a whole is not chiral-symmetric, because not all (in fact very few) particles are massless. Many scientists believe that the world is fundamentally chiral-symmetric, but that this symmetry is "spontaneously broken." *See* HELICITY.

Gauge invariance. A gauge invariance is an internal symmetry which is local; that is, the symmetry operation is a rotation of the internal coordinates independently at each point of space and time. This symmetry is possible only if there exist "gauge fields," massless vector fields (that is, boson fields whose quanta are massless particles with spin-parity $J^P = 1^-$), one for each independent internal symmetry "rotation." The internal symmetry whose conserved quantum number is charge has long been known to be gauge-invariant; its vector field is the electromagnetic field, whose quanta are photons. The theory of a gauge-invariant SU_2 internal symmetry was first considered by C. N. Yang and R. Mills; the Yang-Mills vector field is an isovector ($i = 1$) field.

There is great interest in gauge-invariant field theories, both for strong and weak interactions: (1) They are the only renormalizable field theories (with one uninteresting exception) which contain vector fields; they can therefore describe weak interactions, which are believed to be mediated by the exchange of intermediate vector bosons. (2) They are the only theories with asymptotic freedom, a term which means roughly that interactions become weaker for larger momentum transfers. As far as known, the "scaling" observed in deep inelastic electron scattering requires the strong interactions to be asymptotically free.

Spontaneous symmetry breaking. At first sight, however, the masslessness of the gauge fields would appear to rule out gauge-invariant theories of the weak and strong interactions, because the only observed massless $J^P = 1^-$ boson is the photon. But a phenomenon called spontaneous symmetry breaking can occur, in which the states of the system do not have all of the symmetry of the equations of motion. A particular consequence can be that some or all of the gauge field quanta are massive (this is known as the Higgs mechanism). Remarkably, the good properties of renormalizability and asymptotic freedom remain true. Further, the algebraic (commutation) relations between the operators which generate the group of internal symmetries remain true, despite the lack of symmetry in the states. Such relations are called a current algebra. An example is the chiral *i*-spin current algebra; although the chiral symmetry is broken and particle states are not chiral eigenstates, the current algebra appears to be valid and a matrix element yields correctly the ratio of the axial vector (Teller) to vector (Fermi) beta-decay constants of the nucleon, the Adler-Weisberger relation. *See* QUANTUM FIELD THEORY.

Baryonic and leptonic charges. An important feature of a gauge field with unbroken gauge invariance is that the internal symmetry, and its corresponding conservation laws, is forced to hold. Electromagnetism provides an example: Charge conservation is a consequence of Maxwell's equations. According to experiment, not only is electric charge conserved, but also three other "charges," namely, baryon number, electron number, and muon number. (These are net numbers; for example, electron number, or better, *e*-lepton number, means the number of electrons and electron neutrinos minus the number of positrons and antielectron neutrinos.) However, these three charges are not the source of any known massless (that is, unbroken) gauge field, and thus there is no known reason for these numbers to be exactly conserved. It is speculated that they are in fact not conserved, and that processes as in Eqs. (4) can occur, though

$$\mu^- \rightarrow e^- \gamma \qquad p \rightarrow e^+ \pi^0 \qquad (4)$$

at a rate too slow to have been yet detected. These speculations are based on so-called grand unified theories in which all particles (quarks and leptons) are treated as basically equivalent. The following is a simple qualitative reason for the existence of such processes: It is known to a very high accuracy that the charge of the proton and the positron are equal, or equivalently that the proton and electron have equal but opposite charges, or that the hydrogen atom is neutral. Why should this be? If baryons and electrons are separately conserved, there is no reason for there to be any relation between the charge of the proton and the electron (although the existence of β-decays like $n \rightarrow pe\bar{\nu}_e$ requires that the charge differences $Q_p - Q_n$ and $Q_{e^+} - Q_{\bar{\nu}_e}$ must be equal). It is only the existence of a reaction like $p \rightarrow e^+ \pi^0$ (where π^0 is certainly neutral, because it is self-charge-conjugate) which forces Q_p to equal Q_{e^+}. *See* BARYON; LEPTON.

Quantum chromodynamics. In the quark theory, which seems highly successful in describing hadrons and their interactions, the glue which binds quarks together to form hadrons is a gauge field of Yang-Mills type in which the internal symmetry is SU_3, and each flavor quark, $u, d, \ldots,$ is a triplet, the defining representation of the SU_3 ("each flavor of quark comes in three colors"). This symmetry is called color SU_3 to distinguish it from the flavor SU_n groups. The glue field couples equally to quarks of all flavors, and this is responsible for the flavor SU_n symmetry. There is a close parallel to electromagnetism: The electromagnetic field is a gauge field which binds electrons to nuclei to form atoms; the electromagnetic field is coupled to electric charge, which is consequently forced to be conserved, and is indifferent to any other properties of the particle which carries the charge. This gauge field theory of the glue is called quantum chromodynamics (QCD).

It is believed that color SU_3 is an exact, unbroken symmetry, despite the fact that hadrons do not occur in color multiplets, nor is there a massless quantum (gluon) of the glue field observed or a long-range (inverse-square) force, analogous to the photon and the Coulomb force. But the analogy to electromagnetism is imperfect, because the glue field is coupled to itself: The glue field is coupled to color, but it itself carries color since it is an octet representation of color SU_3. A consequence is asymptotic freedom, one aspect of which is that the force between two color charges brought close together rises more slowly than inverse square, in contrast to the electric force between two charges which becomes stronger than inverse square, due to vacuum polarization. Conversely, as two color charges are moved farther apart, the force between them falls more slowly than inverse square. At a large enough separation, the well-understood method of calculation, perturbation theory, fails; it is conjectured that the force falls asymptotically to a nonvanishing constant value (roughly 10^5 newtons = 10 tons weight). This is known as the confinement conjecture, since the consequence is that color is confined: Two bodies can separate freely to large distances only if they do not carry color charge, that is, are singlets of color SU_3; hence all hadrons must be colorless, that is, color singlets. Since the gluon is colored, it cannot appear as a free particle. Thus the consequence of confinement—that is, the color force is super-long-range—is that color SU_3, although exact, is a hidden symmetry, not directly observed in hadrons the way that the flavor SU_n symmetries are.

But there are a number of ways that the hidden color does reveal itself. The most dramatic is in very-high-energy collisions involving large momentum transfers, that is, short-distance collisions; these create high-energy quarks and gluons as though these were physical particles (since at short distances confinement is irrelevant). As these colored particles separate from one another, confinement becomes important and each quark or gluon becomes converted into a narrow spray of hadrons, a jet. From the properties of the jets, properties of their progenitors can be deduced. *See* GLUONS; QUANTUM CHROMODYNAMICS.

Selection rules. As stated earlier, selection rules are an important result of conservation laws; They express whether or not particular reactions can satisfy the conservation laws. A few examples follow.

The conservation of angular momentum, parity, and statistics implies, for example, that unless a level of Be^8 has even angular momentum and positive parity, it cannot decay into two α-particles since these two identical spinless bosons can only be in such states.

The conservation of angular momentum and parity implies the selection rules for the emission of radiation. For instance, the selection rules for the emission of electric dipole radiation [$\Delta J = 0, \pm 1$

(but not $J = 0$ to another state with $J = 0$) and parity change] are a consequence of the vector addition rules of angular momentum plus the fact that the electric dipole field is 1^-; that is, its angular momentum is one unit of $\hbar$ and its parity is -1.

The conservation of charge parity C implies that a given state of positronium cannot decay into both an even number and an odd number of photons. Similarly, the conservation of isotopic spin parity G implies that a state of nucleonium cannot decay into both an even number and an odd number of π^- mesons. *See* Nuclear reaction; Selection rules.

[CHARLES J. GOEBEL]

Bibliography: J. W. Cronin, CP symmetry violation, *Rev. Mod. Phys.*, 53(3):378–383, 1981; J. J. Sakurai, *Invariance Principles and Elementary Particles*, 1964; G. 't Hooft, Gauge theories of the forces between elementary particles, *Sci. Amer.*, 242(6):104–138, 1980; C. N. Yang, Law of parity conservation and other symmetry laws, *Science*, 127(3298):565–569, 1958.

Sympathetic vibration

The driving of a mechanical or acoustical system at its resonant frequency by energy from an adjacent system vibrating at this same frequency. Examples include the vibration of wall panels by sounds issuing from a loudspeaker, vibration of machinery components at specific frequencies as the speed of a motor increases, and the use of tuned air resonators under the bars of a xylophone to enhance the acoustic output. Increasing the damping of a vibrating system will decrease the amplitude of its sympathetic vibration but at the same time widen the band of frequencies over which it will partake of sympathetic vibration. *See* Resonance (acoustics and mechanics); Vibration.

[LAWRENCE E. KINSLER]

Synchrotron radiation

Electromagnetic radiation emitted by charged particles in circular motion at relativistic energies. Such emission of light by electrons, guided by celestial magnetic fields, has long been known to astronomers and is responsible, for example, for the beautiful background light in the Crab Nebula. With the construction of high-energy electron synchrotrons and storage rings for high-energy physics research, very powerful sources of synchrotron radiation in the ultraviolet and x-ray parts of the spectrum became available in many countries and, starting in the 1950s, synchrotron radiation research programs were begun on a secondary basis, parasitic to the primary high-energy programs. *See* Particle accelerator; Synchrotron.

These programs have grown in importance, so that in many cases the rings are now partly or completely dedicated to synchrotron radiation research. In addition, new storage rings completely dedicated to synchrotron radiation research are under construction or in operation in several coun-

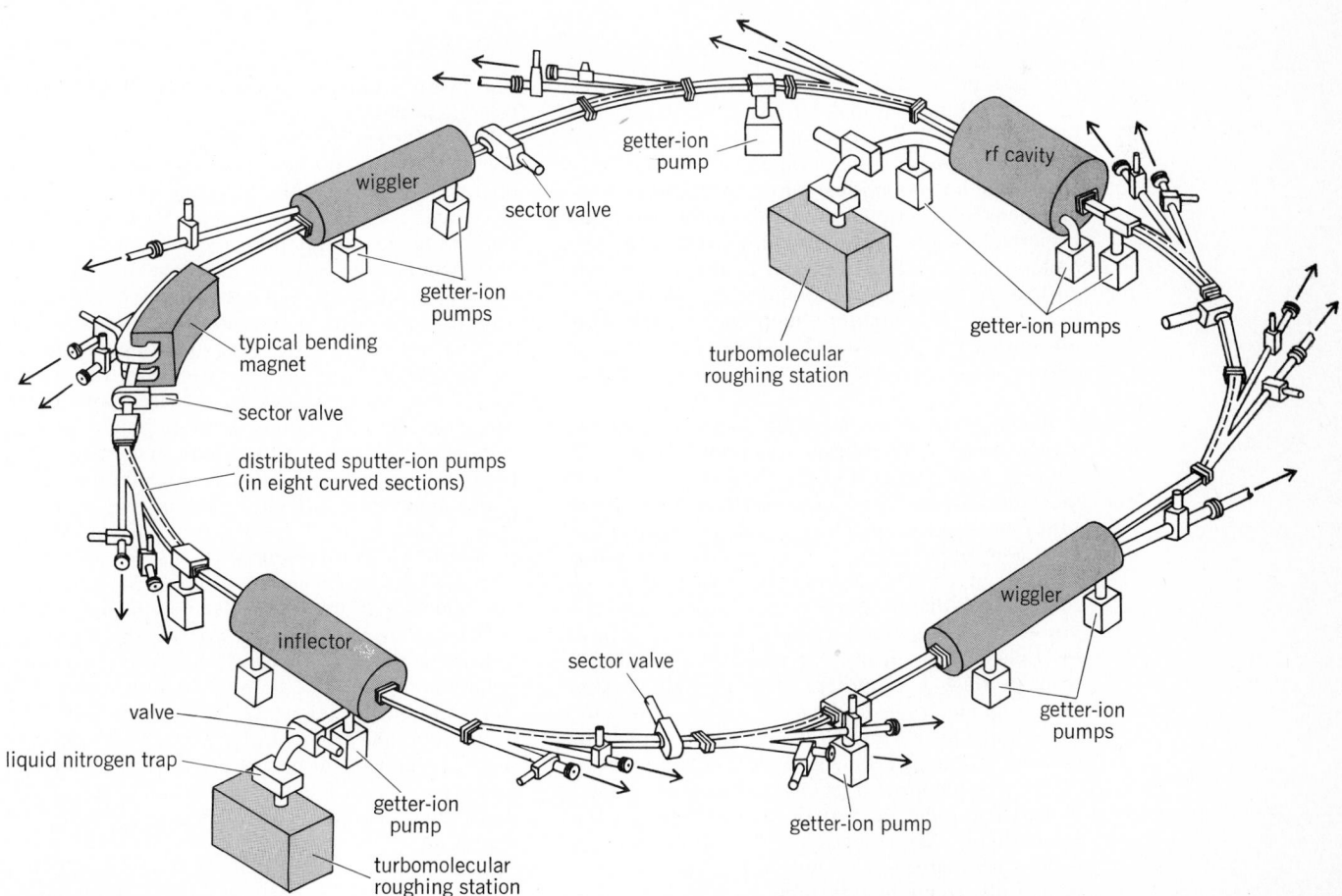

Fig. 1. Electron storage ring designed as a synchrotron radiation source. Only one bending magnet is shown. Not shown is an injector accelerator. (*Courtesy of J. Godell, Brookhaven National Laboratory*)

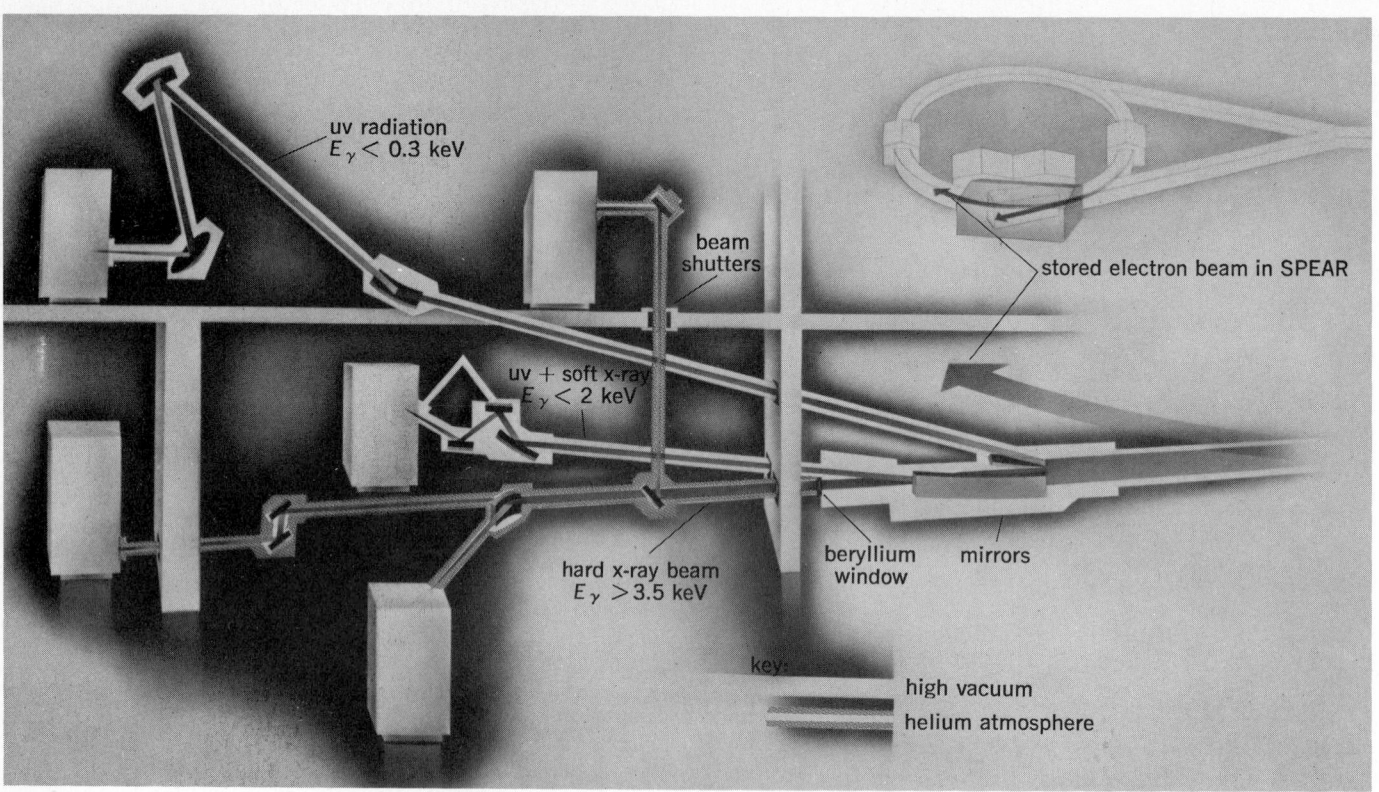

Fig. 2.   First beam line at Stanford Synchrotron Radiation Laboratory. This beam is shared among five simultaneous users. (*Drawing by Walter Zawojski*)

tries (see table). The radiation has many features (natural collimation, high intensity, broad spectral bandwidth, high polarization, pulsed time structure, small source size, and high-vacuum environment) which make it ideal for a wide variety of applications in experimental science and technology. Important results have been achieved in many disciplines such as biochemistry, materials science, surface science, and crystallography through the use of various techniques such as extended x-ray absorption fine structure (EXAFS), photoemission spectroscopy, x-ray scattering, and x-ray microscopy.

Radiation by electrons in circular orbits in early synchrotrons and betatrons was predicted by J. P. Blewett and others before 1945 and was first (accidentally) observed on the General Electric 70-MeV synchrotron in 1947. A detailed theoretical treatment of the properties of synchrotron radiation was given by J. Schwinger in 1949, showing, for example, that for particles of different mass the power radiated varies inversely as the fourth power of mass. This means that synchrotron radiation from protons is 13 orders of magnitude weaker than from electrons (at the same energy and bending radius). In a multi-GeV electron storage ring the rate of energy loss is so high that a circulating electron would lose all its energy in a few milliseconds. To store beams with decay lifetimes of a few hours (limited by collisions with the residual gas even at pressures at 10^{-9} torr or 10^{-7} pascal), powerful radio-frequency systems replenish the lost energy. By contrast, the synchrotron radiation from the 28-GeV proton storage rings at CERN in Geneva is so weak that protons could circulate for

years with negligible loss of energy. *See* RELATIVISTIC ELECTRODYNAMICS.

Laboratories. Research with synchrotron radiation began with the work of D. Tomboulian and Hartman at the 300-MeV Cornell synchrotron in 1955, and was continued by R. Madden and coworkers at the National Bureau of Standards 180-MeV synchrotron in 1962. The constant spectrum and stable intensity characteristic of storage rings were first provided by the 240-MeV storage ring at the University of Wisconsin, which began operating in 1968 as a dedicated synchrotron radiation source. The vastly increased research potential of energetic, intense synchrotron radiation from a high-energy storage ring was first demonstrated in 1972 at the Cambridge Electron Accelerator (CEA) in Massachusetts, which stored beams up to 3.5 GeV. Figure 1 shows a storage ring designed as a synchrotron radiation source.

In 1974 a major research program was begun at the Stanford Synchrotron Radiation Laboratory (SSRL), using the Stanford Positron-Electron Accelerating Ring (SPEAR) at the Stanford Linear Accelerator Center. This was the first time that a large number of scientists had access to the intense radiation produced by a multi-GeV storage ring. The first beam line at SSRL (Fig. 2) was designed so that five simultaneous experiments could share the radiation, each using a specially designed monochromator. The facility has grown with the construction of four additional beam lines and a capability of performing 15 simultaneous experiments. An aerial view of SPEAR and SSRL is shown in Fig. 3. Similar major facilities have developed in Hamburg in Germany, Orsay in

Synchrotron radiation sources

Machine	Location	Energy, GeV	Current, mA	Bending radius, m	Critical energy, keV	Remarks
Storage rings in operation as of September 1981					
PETRA	Hamburg, Germany	15	50	192	39.0	Possible future use for synchrotron radiation research
PEP	Stanford, CA	15	50	165.5	45.2	Synchrotron radiation facility planned
		12	45	(23.6)	(163)	(From 17-kG wiggler)
CESR (Cornell)	Ithaca, NY	8	50	32.5	35.0	Used parasitically
VEPP-4	Novosibirsk, Soviet Union	7	10	16.5	46.1	Initial operation at 4.5 GeV
		4.5		(18.6)	(10.9)	(From 8-kG wiggler)
DORIS	Hamburg, Germany	5	50	12.1	22.9	Partly dedicated
		2.5	300		2.9	
SPEAR	Stanford, CA	4.0	50	12.7	11.1	50% dedicated
		3.0	100		4.7	
		3.0		(5.5)	(10.8)	(From 18-kG wiggler)
SRS	Daresbury, England	2.0	500	5.55	3.2	Dedicated
				(1.33)	(13.3)	(For 50-kG wiggler)
VEPP-3	Novosibirsk, Soviet Union	2.25	100	6.15	4.2	Partly dedicated
				(2.14)	(11.8)	(From 35-kG wiggler)
DCI	Orsay, France	1.8	500	4.0	3.63	Partly dedicated
ADONE	Frascati, Italy	1.5	60	5.0	1.5	Partly dedicated
				(2.8)	(2.7)	(From 18-kG wiggler)
VEPP-2M	Novosibirsk, Soviet Union	0.67	100	1.22	0.54	Partly dedicated
ACO	Orsay, France	0.54	100	1.1	0.32	Dedicated
SOR Ring	Tokyo, Japan	0.40	250	1.1	0.13	Dedicated
SURF II	Washington, DC	0.25	25	0.84	0.041	Dedicated
TANTALUS I	Wisconsin	0.24	200	0.64	0.048	Dedicated
PTB	Braunschweig, Germany	0.14	150	0.46	0.013	Dedicated
N-100	Karkhov, Soviet Union	0.10	25	0.50	0.004	

Storage rings under construction as of September 1981

Machine	Location	Completion:	Energy, GeV	Current, mA	Bending radius, m	Critical energy, keV	Remarks
PHOTON FACTORY	Tsukuba, Japan	1982	2.5	500	8.33	4.16	Dedicated
					(1.67)	(20.5)	(For 50-kG wiggler)
NSLS	Brookhaven National Lab, NY	1981	2.5	500	6.88	5.01	Dedicated
					(1.67)	(20.5)	(For 50-kG wiggler)
ALADDIN	Wisconsin	1981	1.0	500	2.08	1.07	Dedicated
BESSY	West Berlin, Germany	1982	0.80	500	1.83	0.62	Dedicated; industrial use planned
NSLS	Brookhaven National Lab, NY	1981	0.70	500	1.90	0.40	Dedicated
ETL	Electrotechnical Lab, Tsukuba, Japan	1981	0.66	100	2	0.32	Dedicated
UVSOR	Institute of Molecular Science, Okatabi, Japan	1982	0.60	500	2.2	0.22	Dedicated
MAX	Lund, Sweden	1981	0.50	100	1.2	0.23	Dedicated
KURCHATOV	Moscow, Soviet Union		0.45		1.0	0.21	Dedicated

Synchrotrons in operation as of September 1981

Machine	Location	Energy, GeV	Current, mA	Bending radius, m	Critical energy, keV	Remarks
DESY	Hamburg, Germany	7.5	10−30	31.7	29.5	
ARUS	Erevan, Soviet Union	4.5	1.5	24.6	8.22	
BONN I	Germany	2.5	30	7.6	4.6	
SIRIUS	Tomsk, Soviet Union	1.36	15	4.23	1.32	
INS-ES	Tokyo, Japan	1.3	30	4.0	1.22	
PAKHRA	Moscow, Soviet Union	1.3	300	4.0	1.22	
LUSY	Lund, Sweden	1.2	40	3.6	1.06	
FIAN C-60	Moscow, Soviet Union	0.68	10	1.6	0.44	
BONN II	Germany	0.5	30	1.7	0.16	

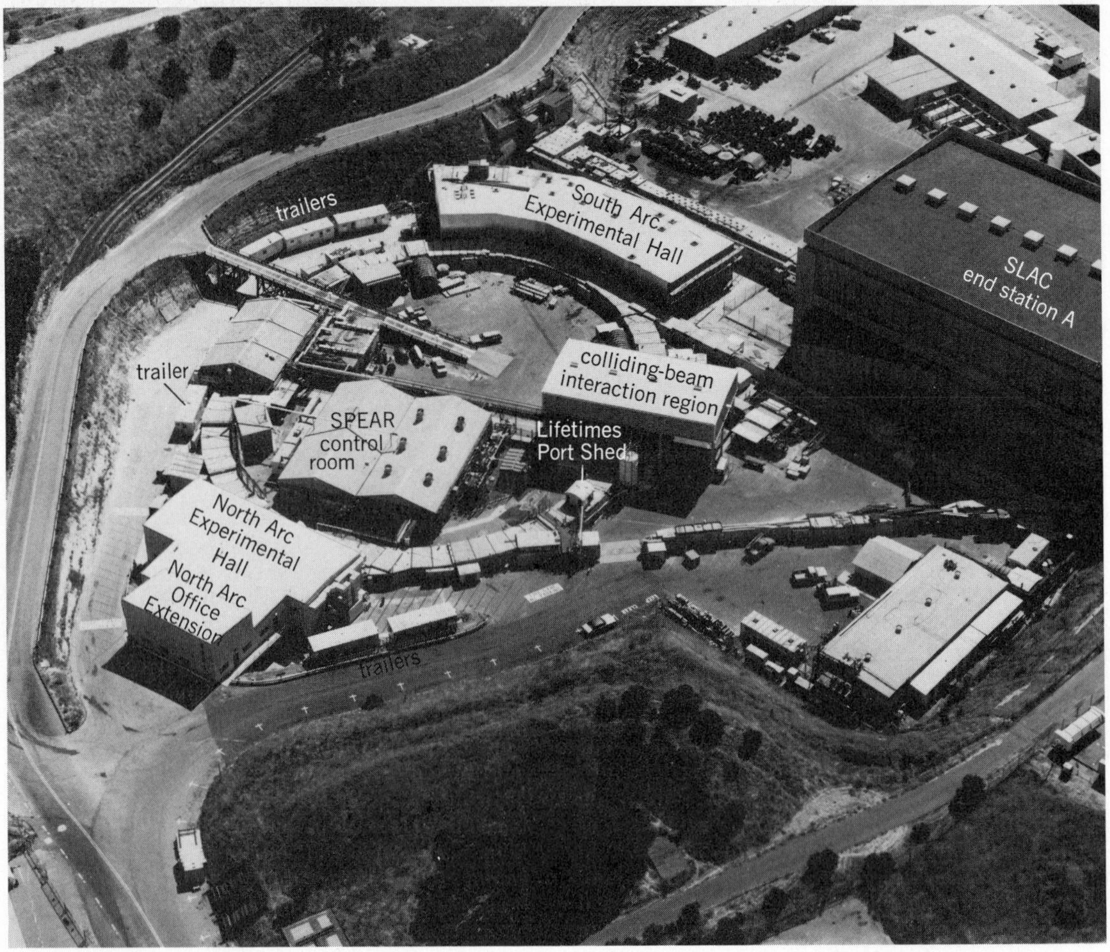

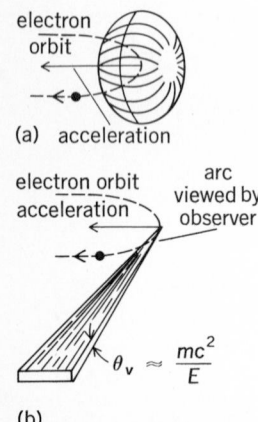

Fig. 3. Aerial photograph of Stanford Synchrotron Radiation Laboratory (SSRL) and Stanford Positron-Electron Accelerating Ring (SPEAR). Trailers are used for office and storage space. (*Photo by J. Faust*)

SYNCHROTRON RADIATION

(a) acceleration

(b)

$$\theta_v \approx \frac{mc^2}{E}$$

Fig. 4. Radiation emission pattern by electrons in circular motion (*a*) at low energy (speed much less than speed of light) and (*b*) at relativistic energy (speed close to speed of light).

France, and Novosibirsk in the Soviet Union. Construction of new facilities has been undertaken in the United States, Western Europe, Japan, and the Soviet Union (see table).

Properties of the radiation. The basic properties of synchrotron radiation, compared with other sources of electromagnetic radiation, may be summarized as follows.

Natural collimation. The normal radiation pattern of an electron in circular motion at low energy is rather nondirectional (Fig. 4*a*). At relativistic energies this pattern is folded forward so that the radiation is emitted in a flat pancake in the plane of the electron orbit (Fig. 4*b*). The vertical opening angle of emission θ_v is given approximately by mc^2/E, where mc^2 is the rest mass energy of the electron (0.51 MeV) and E is the total energy. Thus an electron energy of 2.5 GeV, the opening angle is only 2×10^{-4} radian. This results in very high flux densities on small targets, even at distances $10-20$ m from the source.

High intensity. As a continuum source, synchrotron radiation produces five or more orders of magnitude higher intensity on an experimental sample than powerful rotating-anode x-ray tubes. The total power radiated (in kilowatts) is given by $88E^4I/R$, where E is the electron energy in GeV, I is the electron current in amperes, and R is the radius of curvature in meters. The largest storage rings now used routinely for synchrotron radiation research (SPEAR and DORIS) radiate about 100 kW of synchrotron radiation. The 15-GeV colliding-beam rings PETRA and PEP radiate several megawatts. In storage rings the intensity is particularly stable, decaying slowly over a period of many hours as the stored electron beam current decays.

Broad spectral bandwidth. The power is radiated in a smooth, featureless continuum (Fig. 5) without the spikes and structure associated with other sources. The spectrum is characterized by a single parameter, the critical energy ϵ_c, given (in keV) by $2.2E^3/R$. Typically, useful flux is available out to four or five times the critical energy. For most regions of the ultraviolet and x-ray spectrum, no other source of intense continuum radiation exists, and experiments must work only at the characteristic lines of gas-discharge lamps and x-ray tubes. In use, the broadband radiation is monochromatized by a grating or crystal so that the bandwidth ($\Delta\epsilon/\epsilon$ or $\Delta\lambda/\lambda$, where ϵ and λ are the energy and wavelength of the radiation) is 10^{-3} or smaller. Thus highly monochromatic radiation at any energy within the range of the synchrotron radiation source is readily available.

High polarization. In the plane of the electron orbit and in the vicinity of the critical energy, the radiation is nearly 100% linearly polarized. The polarization drops slightly at locations out of the plane and at longer wavelengths. *See* POLARIZATION OF WAVES.

Pulsed time structure. The radiation is produced in a train of pulses which is different for each source. The pulse duration can be as short as 50 picoseconds, and the interval between pulses can be 1 microsecond or longer.

Small source size. The radiation is emitted from the electron beam, which typically has a cross-sectional area of the order of 1 mm².

High-vacuum environment. Storage rings operate with a pressure in the 10^{-9} torr (10^{-7} Pa) range, and synchrotrons at about 10^{-6} torr (10^{-4} Pa). Thus, if comparable pressure is maintained in the beam lines and experimental equipment, no windows are necessary.

Research applications. Ultraviolet and x-radiation from conventional sources have been employed extensively for decades in research utilizing a great variety of experimental techniques in many disciplines, such as physics, chemistry, and biology, as well as in a variety of technological processes. Many of these techniques and processes have experienced considerable improvements as a result of one or more of the special properties of synchrotron radiation described above. In addition, the availability of synchrotron radiation has made possible major scientific and technological advances which are described below.

Extended x-ray absorption fine structure (EXAFS). A plot of x-ray absorption versus photon energy (Fig. 6) shows the steep rise in the absorption at the absorption-edge energy, where the incident photon is just able to excite core electrons into empty states. Just above this energy, the observed oscillatory structure on the absorption is due to interference between the outgoing photoelectron waves and the electron waves backscattered from atoms adjacent to the absorbing atoms. This interference modulates the probability of exciting the electron, as explained by R. de L. Kronig in 1931. Because the structure arises from an interference phenomenon and the absorption-edge energies of different atomic species are usually quite different, analysis of this structure provides considerable information about the average atomic environment of each atomic species in complex, polyatomic solids, liquids, and gases.

The highly intense, continuous synchrotron radiation spectrum from multi-GeV storage rings is ideally suited for such measurements as a function of photon energy. Consequently, it has been used extensively for structural studies of amorphous materials, heterogeneous catalysts, and noncrystalline metalloproteins whose atomic arrangements are not ordinarily determinable by other structural techniques. These structural studies have led, in turn, to increased understanding of the properties of these materials. EXAFS patterns have also been measured in fractions of a second, so that time-resolved studies of structural changes may be anticipated.

Similar modulations of the x-ray fluorescent yield and the Auger electron yield versus photon energy are caused by the same interference phenomenon. The former is now used extensively to study the atomic environments of extremely dilute (less than 100 parts per million) constituents of

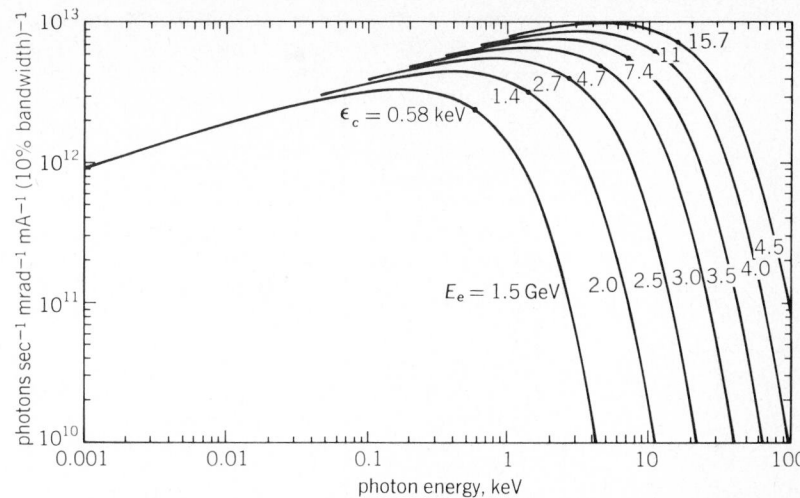

Fig. 5. Spectral distribution of synchrotron radiation from SPEAR; ϵ_c = critical energy; E_e = stored electron beam energy; bending radius = 12.7 m.

complex systems such as alloys, semiconductors, and proteins in solution. The latter is used for the determination of the atomic environments of atoms at the surfaces of materials. *See* AUGER EFFECT; SURFACE PHYSICS.

Photoemission spectroscopy. Energy and angular distributions of electrons ejected from solids by photons of varying incident energy provide basic information about band structure and core levels of both the surface and bulk of the sample. The extremely high intensity and tunability of synchrotron radiation have made possible high-resolution studies which delineate the band structures of crystalline solids. In addition, electrons of energy between 30 and 100 eV have such a large probability for interaction with other atoms that they have an escape depth of only about 0.5 nm in most solids. As a result, the detected electrons reveal the properties of surface electronic energy levels of clean materials and the extrinsic surface states of chemisorbed material, opening a route to the study of the phenomena of oxidation, corrosion, and catalysis. Such studies have led to major revisions of

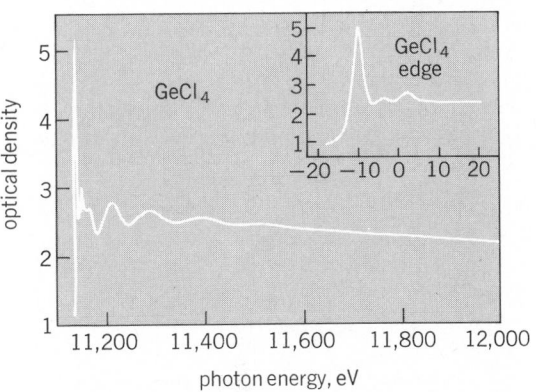

Fig. 6. X-ray absorption spectrum above the Ge K-edge in GeCl₄ vapor. The inset shows the edge region with a shifted energy scale. (*From B. M. Kincaid and P. Eisenberger, Synchrotron radiation studies of the K-edge photoabsorption spectra of Kr, Br₂ and GeCl₂: A comparison of theory and experiment, Phys. Rev. Lett., 34:1361–1364, 1975.*)

models of the metal-semiconductor interface which determines many properties of solid-state electron devices. Again, the tunability, when coupled with the high intensity of synchrotron radiation, makes possible the separate delineation of surface and bulk electronic states. *See* BAND THEORY OF SOLIDS; SEMICONDUCTOR.

In addition, the radiation facilitates photoemission studies in two ways: the high-vacuum environment of the storage ring minimizes contamination of clean surfaces prepared in place; and the sharply pulsed time structure of the radiation makes it possible to measure photoelectron energies by time of flight.

Figure 7 shows photoemission spectra from clean and heavily oxidized gallium arsenide surfaces, each taken in less than 10 min by using synchrotron radiation from SPEAR. The curves taken at 100 eV incident photon energy show clear differences due to oxidation. The new shifted peak in the spectrum of the oxidized surface, to the left of the arsenic peak, shows that the oxygen primarily attaches to the arsenic rather than the gallium (although the reverse was expected from thermodynamic arguments based on bulk chemistry). At 100 eV photon energy, essentially only two atomic layers at the surface are being probed. At higher energies, contributions from the bulk become dominant, as can be seen from the change in relative heights of the shifted and unshifted arsenic peaks. The strength of emission from valence

states is also seen to depend strongly on the state of oxidation and photon energy.

X-ray diffraction and scattering. As time for protracted experiments on multi-GeV storage rings becomes available, synchrotron radiation is having very significant effects on x-ray diffraction structural techniques. The high intensity and natural collimation of synchrotron radiation make it ideal for x-ray small-angle scattering and diffraction experiments in which long-wavelength (5–200 nm) modulations of electron densities are analyzed. Several time-resolved studies of the changes of long spacings in muscles during the contraction cycle have been performed in efforts to understand the associated molecular rearrangements. Methods for the small-angle scattering study of noncyclical dynamic phenomena, such as phase separation in glasses, are also being developed and utilized.

The phenomenon of anomalous scattering has also been utilized in small-angle scattering. When the photon energy is very close to an atom's absorption edge, the atom's scattering cross section (atomic scattering factor) is markedly different from its normal value. This anomalous scattering is used to decrease the contribution of that species to the scattering amplitude, and therefore obtain information about long-wavelength modulations of chemical composition which are not readily observable normally. The tunability of highly monochromatized synchrotron radiation makes such experiments possible. *See* X-RAY CRYSTALLOGRAPHY; X-RAY DIFFRACTION.

Protein structure analysis. The structural analysis of many protein crystals by x-ray diffraction is performed in considerably more detail by using synchrotron radiation because of an unanticipated phenomenon. Frequently the limitation on the number of Bragg diffraction peaks whose intensities can be measured is the x-ray damage of the protein. It has been observed a number of times that a given dose of x-rays does considerably less damage when applied over a short time period than over a long one. As a result, the extremely high intensity of synchrotron radiation makes possible the determination of many more peak intensities before degradation than do conventional sources, so that higher resolution of the protein electron density and atomic arrangements can be obtained.

In addition, initial experiments indicate that anomalous scattering may be used to eliminate the need for the growth of single crystals of three different isomorphic chemical forms of a protein in order to determine the atomic arrangement. Instead, the photon energy may be varied near an absorption edge so that a single chemical form appears to behave like three different forms. This advance is important because it is frequently impossible to grow crystals of the three different forms.

Anomalous scattering has also been utilized to determine atomic arrangements in partially disordered alloys of atoms whose atomic numbers differ by only one. In such cases the atomic scattering factors are almost identical, so that a characterization of the state of order is extremely difficult. By tuning the photon energy close to the absorption edge of one of the elements, its scattering factor may be altered markedly and the characterization performed.

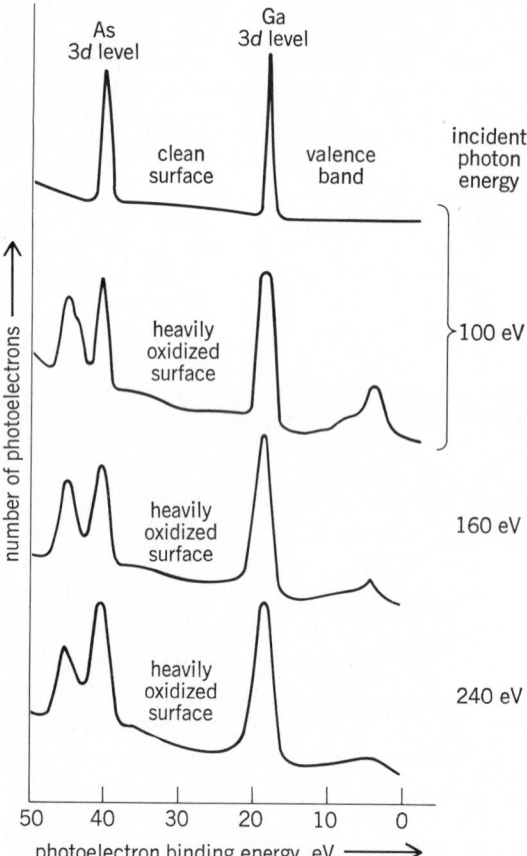

Fig. 7. Photoemission spectra from gallium arsenide surfaces. (*I. Lindau et al., Determination of the oxygen binding site on GaAs (110) using soft x-ray photoemission spectroscopy, Phys. Rev. Lett., 35:1356, 1976*)

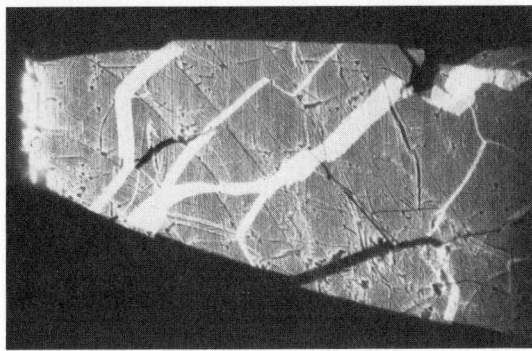

Fig. 8. Diffraction topograph of a lithium fluoride crystal taken in 5 min with SPEAR operating at 1.89 GeV and 7.2 mA. The smallest features are 1 μm. (*Courtesy of W. Parrish, IBM, San Jose, CA*)

Similarly, initial results indicate that detailed information about atomic arrangements in polyatomic amorphous materials can be obtained through techniques which take advantage of this phenomenon.

X-ray diffraction topography. The availability of highly intense white radiation has also accelerated markedly the development of time-resolved studies of defects in nearly perfect crystals through x-ray topography. Such studies include the effects of magnetic fields on domains and of temperature on defects. Studies may be anticipated of defect motion due to stress and of corrosion due to toxic environments on the highly perfect single crystals required in solid-state electronics. The highly collimated monochromatic radiation has also been used to study the relationship between defects in a substrate and those in an epitaxial layer, which are important in many solid-state devices. Figure 8 is an example of an x-ray diffraction topograph taken with synchrotron radiation. *See* CRYSTAL DEFECTS.

Time-resolved fluorescence spectroscopy. Synchrotron radiation is utilized extensively for time-resolved spectroscopy in those temporal domains in which 50 to 200-ps pulses at 1-μs intervals are appropriate. These include studies of electron-hole recombination in semiconductors and rotational times of tryptophan side chains on proteins. The high photon energies available make possible experiments which cannot be performed in the more restricted spectral domain of pulsed lasers.

X-ray fluorescence trace-element analysis. The high intensity, tunability, high collimation, and high linear polarization of synchrotron radiation make it valuable for trace-element analysis through x-ray fluorescence. In a search for superheavy elements, a sensitivity sufficient to observe about 5×10^8 atoms of superheavy elements per sample was demonstrated. With improvements in technique, concentrations as low as 10^6 atoms per sample may be observable. *See* SUPERTRANSURANICS.

X-ray microscopy and lithography. A number of groups have experimented with synchrotron radiation for high-resolution x-ray lithography and microscopy. The features which make it attractive for such applications are: (1) high intensity, which provides for high throughput in lithography and short exposures in microscopy; (2) high collima-

tion, which makes possible high resolution in both techniques and a relative insensitivity to wafer and mask distortion in lithographic production of integrated circuits of micrometer and submicrometer feature size; (3) the ability to obtain high aspect ratios in the lithographic production of microstructures. Figure 9 shows examples of microstructures made by using synchrotron radiation x-ray lithography.

Prospects. With the universal recognition that synchrotron radiation offers extraordinary research opportunities, the field is developing at a rapid pace. The partial or complete dedication of several existing storage rings to synchrotron radiation research resulted in a major increase in photon flux and research capabilities. The dedicated rings in construction will add significantly to these capabilities. In addition, it is becoming clear that

(a)

├─────────── 1 μm ───────────┤

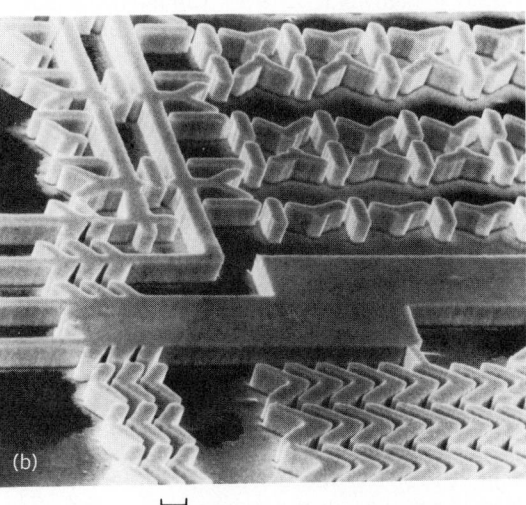

(b)

├── 1 μm ──┤

Fig. 9. Examples of patterns replicated by x-ray lithography by using synchrotron radiation. (*a*) Zone plate pattern. (*b*) Bubble memory pattern. (*Courtesy of E. Spiller, IBM, Yorktown Heights, NY*)

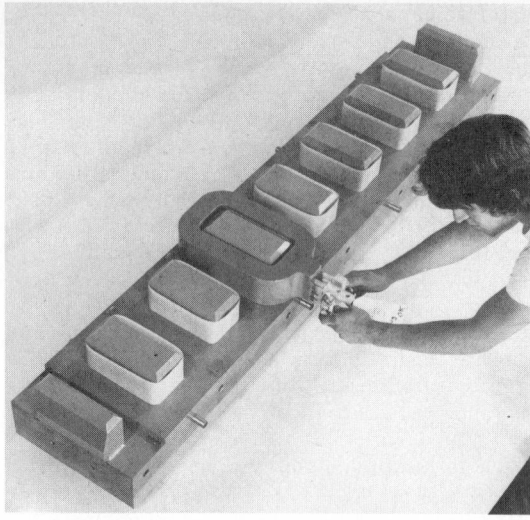

Fig. 10. Wiggler magnet. Lower half of an 18-kilogauss (1.8-tesla) magnet used in SPEAR. The magnet has seven full poles and two end half-poles. Each pole is powered by a coil, one of which is shown. The higher field of the magnet shifts the spectrum to harder x-ray energies, and the overall intensity is enhanced by the number of poles. (*Photo by J. Faust*)

special magnetic structures inserted into straight sections of storage rings can provide greatly enhanced or modified synchrotron radiation. For example, high-field, multipole wiggler magnets (Fig. 10), which cause the beam to oscillate several times in magnetic fields that can be higher than the ring bending magnet field, have been in use at SSRL since 1979. They provide considerably higher intensity, particularly in the 10- to 30-keV region than is otherwise available. Construction of superconducting devices of this type with fields up to 6 teslas has been undertaken at several laboratories. Such devices make it possible for a relatively low-energy machine (about 1 GeV) to produce intense x-ray flux up to about 10 keV.

When the number of oscillations is large (for example, more than about 10), it is possible for interference effects to enhance the intensity at particular wavelengths. Such devices, called undulators, can produce quasimonochromatic peaks whose brightness is two to four orders of magnitude greater than can be produced by normal ring bending magnets. Initial experience with undulators in France and the Soviet Union confirms expectations. Thus it is entirely possible that there will be an increasing trend toward the use of insertion devices to generate synchrotron radiation, rather than bending magnets. It is possible that future machines will be designed with weak bending magnets and a large number of straight sections to accommodate wigglers and undulators, which will be the primary radiation sources.

Major advances are also occurring in the development of specialized instrumentation to collect, focus, monochromatize, and detect the radiation. The vastly increased capabilities that will result from the higher fluxes and improved experimental instrumentation are likely to result in many important scientific breakthroughs.

[HERMAN WINICK; ARTHUR BIENENSTOCK]

Bibliography: C. Kunz (ed.), *Topics in Current Physics—Synchrotron Radiation: Techniques and Applications*, 1979; H. Winick and A. Bienenstock, Synchrotron radiation research, *Annu. Rev. Nucl. Particle Sci.*, 28:33–113, 1978; H. Winick and S. Doniach (eds.), *Synchrotron Radiation Research*, 1980.

Tachyon

A hypothetical faster-than-light particle consistent with the special theory of relativity. According to this theory, a free particle has an energy E and a momentum $\mathbf{p}$ which form a Lorentz four-vector. The length of this vector is a scalar, having the same value in all inertial reference frames. One writes Eq. (1), where c is the speed of light and the

$$E^2 - c^2\mathbf{p}^2 = m^2 c^4 \tag{1}$$

parameter m^2 is a property of the particle, independent of its momentum and energy. Three cases may be considered: m^2 may be positive, zero, or negative. The case $m^2 > 0$ applies for atoms, nuclei, and the macroscopic objects of everyday experience. The positive root m is called the rest-mass. If $m^2 = 0$, the particle is called massless. A few of these are known: the electron neutrino, the muon neutrino, the photon, and the graviton. The third case, $m^2 < 0$, was studied originally by S. Tanaka and by O. M. P. Bilaniuk, V. K. Deshpande, and E. C. G. Sudarshan. Further contributions were made by G. Feinberg, who gave the name tachyons (after a Greek word for swift) to the particles with $m^2 < 0$. Whether such particles exist is an interesting speculation but, as of 1970, there was no experimental evidence for them.

In general, the particle speed is given by Eq. (2).

$$v = \frac{c\mathbf{p}}{E} c \tag{2}$$

If $m^2 > 0$, Eq. (1) implies $E > c\mathbf{p}$ and Eq. (2) gives $v < c$. If $m^2 = 0$, then $E = c\mathbf{p}$ and $v = c$. In case $m^2 < 0$, one finds $E < c\mathbf{p}$ and $v > c$. Tachyons exist only at faster-than-light speeds.

One can eliminate $\mathbf{p}$ or E from Eqs. (1) and (2) to get expressions for E and $\mathbf{p}$ in terms of the speed. In case $m^2 > 0$, the familiar results are given by Eqs. (3), where the radical signs imply the

$$E = \frac{mc^2}{\sqrt{1 - \dfrac{v^2}{c^2}}} \qquad \mathbf{p} = \frac{mv}{\sqrt{1 - \dfrac{v^2}{c^2}}} \tag{3}$$

positive roots. The way to make the analogous formulas for tachyons is to introduce the positive number μ such that $m^2 = -\mu^2$ and Eq. (4) holds.

$$E^2 - c^2\mathbf{p}^2 = -\mu^2 c^4 \tag{4}$$

Then Eqs. (2) and (4) give Eqs. (5). It is seen that

$$E = \frac{\mu c^2}{\sqrt{\dfrac{v^2}{c^2} - 1}} \qquad \mathbf{p} = \frac{\mu v}{\sqrt{\dfrac{v^2}{c^2} - 1}} \tag{5}$$

for ordinary particles as v increases, E increases, but to speed them up to $v = c$ would involve an infinite amount of energy. In contrast for tachyons, as v decreases, E increases, but to slow them down to $v = c$ would involve an infinite amount of energy.

For an ordinary free particle with $m^2 > 0$, there is always a special reference frame, called the rest frame, in which $\mathbf{p} = 0$. From Eq. (1) it is seen that the energy there has the minimum value mc^2. No such special frame exists for massless particles. For tachyons also the situation is quite different.

For a free tachyon a special frame can be found in which $E = 0$ and, according to Eq. (4), **p** has the minimum value μc. Since the tachyons may exist at zero energy, there is no energy obstacle in creating them in elementary particle reactions.

According to electromagnetic theory, a charged particle moving at a speed greater than the speed of light in a medium emits light, the Cerenkov radiation. If charged tachyons existed, they would spontaneously radiate light even in a vacuum. *See* CERENKOV RADIATION.

Attempts to detect tachyons have been made by looking for the Cerenkov radiation (T. Alväger and M. N. Kreisler) and by analyzing for negative m^2 values in elementary particle reactions (C. Baltay, R. Linsker, N. K. Yeh, and G. Feinberg). The conclusions were negative, and present indications are that this type of particle does not exist. *See* ELEMENTARY PARTICLE; RELATIVITY.

[ROLAND H. GOOD, JR.]

Bibliography: O. M. P. Bilaniuk, V. K. Deshpande, and E. C. G. Sudarshan, Meta relativity, *Amer. J. Phys.*, 30:718–723, 1962; G. Feinberg, Particles that go faster than light, *Sci. Amer.*, February, 1970.

Temperature

A concept related to the flow of heat from one object or region of space to another. The term refers not only to the senses of hot and cold but to numerical scales and thermometers as well. Fundamental to the concept of temperature are the absolute scale and absolute zero and the relation of absolute temperatures to atomic and molecular motions.

Numbers for temperatures, such as 100°C and −15°F, have been used for only about 300 years. By the 17th century, science had developed to the point that, to fully describe the properties of matter, a numerical, quantitative scale of temperature differences was needed. For example, in 1756 Joseph Black in Scotland discovered that ice does not change temperature when it melts. Almost all substances behave this way; also, the melting temperature depends on the purity of the substance. Thus one reason for devising a thermometer (literally, a meter for temperature) was that

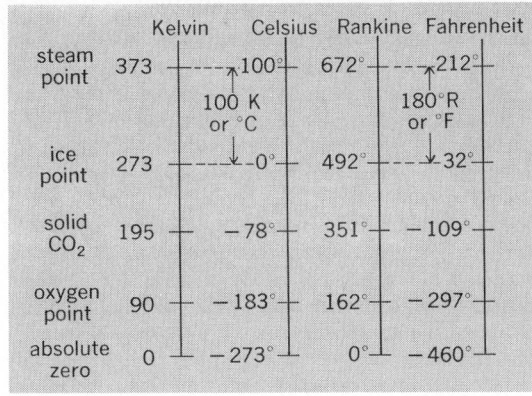

Fig. 2. Comparisons of Kelvin, Celsius, Rankine, and Fahrenheit temperature scales. Temperatures are rounded off to nearest degree. (*M. W. Zemansky, Temperatures Very Low and Very High, Van Nostrand, 1964*)

with it the composition of matter could be studied.

Temperature measurements are useful for studying molecular motions in material. Figure 1 shows how the temperature of 1 g of H_2O, starting as ultracold ice, changes as heat is added at the constant rate of 1 cal/s (4.18 J/s), assuming that no heat is lost to the surroundings. The plateaus on this graph illustrate Black's discovery that the temperature is constant during a phase change solid-to-liquid or liquid-to-gas.

Thermometers do not measure a special physical quantity. They measure length (as of a mercury column) or pressure or volume (with the gas thermometer at the National Bureau of Standards) or electrical voltage (with a thermocouple). The basic fact is that, if a mercury column has the same length when touching two different, separated objects, when the objects are placed in contact no heat will flow from one to the other.

Empirical scales. The numbers on the thermometer scales are merely historical choices; they are not scientifically fundamental. The most widely used scales are the Fahrenheit (°F) and the Celsius (°C). The Centigrade scale with 0° assigned to ice water (ice point) and 100° assigned to water boiling under one atmosphere pressure (steam point) was formerly used, but it has been succeeded by the Celsius scale, defined in a different way than the centigrade scale. However, on the Celsius scale the temperatures of the ice and steam points differ by only a few hundredths of a degree from 0° and 100°, respectively. Figure 2 shows how the Celsius and Fahrenheit scales compare and how they fit onto the absolute scales.

These scales have one common value: −40°C = −40°F. This fact can be used to change a temperature from one scale to the other. Given a temperature in °C or in °F, add 40, multiply by 9/5 if converting from °C to °F or by 5/9 if from °F to °C, then subtract 40. Example: Normal human body temperature is 98.6°F. To convert to °C, 98.6 + 40 = 138.6; 138.6 × 5/9 = 77.0; 77.0 − 40 = 37.0°C.

Absolute temperature scale. In 1848 William Thompson (Lord Kelvin), following ideas of Sadi Carnot, stated the concept of an absolute scale of temperature in terms of measuring amounts of heat flowing between objects. Most important, Kelvin conceived of a body which would not give up any heat and which was at an absolute zero of

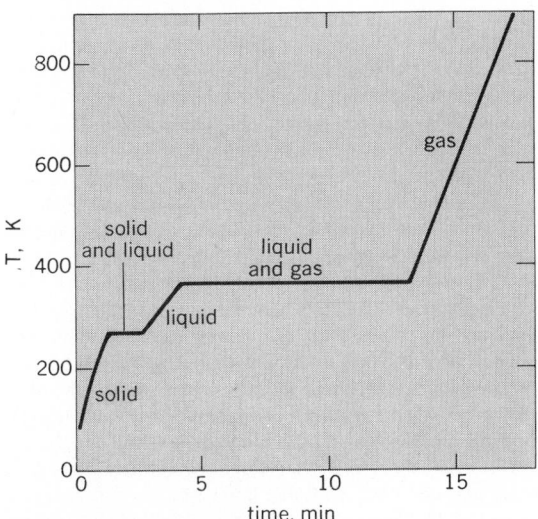

Fig. 1. Temperature of 1 g of H_2O, starting at 0 K, with a constant heat input of 1 cal/s (4.18 J/s).

temperature. Experiments have shown that absolute zero corresponds to $-273.15°C$ or $-459.7°F$. Two absolute scales, shown in Fig. 2, are the Kelvin (K) and the Rankine (°R). *See* ABSOLUTE ZERO.

Interest in temperature and heat flow was stimulated in the early 19th century by efforts to improve the efficiency of steam engines. Out of this came the concept of a Carnot engine. This is not a real machine, but an imagined, ideal, frictionless system. A Carnot engine takes in heat Q_h from a higher temperature source at T_h (kelvins), does work W, and exhausts heat Q_l into a lower temperature source T_l. Two important deductions are: (1) The efficiency of the engine is $W/Q_h = 1 - T_l/T_h$. For example, a Carnot engine operated between the boiling point ($T_h = 373$ K) and the ice point ($T_l = 273$ K) of water has an efficiency of 0.268. A real engine would have an efficiency less than this, but the concept is nevertheless of great importance in engineering. (2) The ratio of temperatures equals the ratio of heats, namely, $T_h/T_l = Q_h/Q_l$. Lord Kelvin suggested that this be the basis for the absolute temperature scale. Define one special system (water at its triple point with ice, liquid, and water vapor present) to have a particular value of absolute temperature (273.16 K). To measure any unknown temperature T_u, operate a Carnot engine between 273.16 K and T_u, measure the heats $Q_{273.16}$ and Q_u absorbed and rejected, and calculate $T_u = 273.16$ K $(Q_u/Q_{273.16})$. *See* THERMODYNAMIC PRINCIPLES; TRIPLE POINT.

In practice, absolute temperatures are not measured this way. Instead low-density helium gas and dilute paramagnetic crystals, the most nearly ideal of real materials, allow measurement of temperatures virtually identical with those defined by a Carnot process. The advantages are that gas pressures and volumes and magnetic fields and magnetizations can be measured more conveniently and accurately than heat flows.

The measurement of a single temperature with a gas or magnetic thermometer is a major scientific event done at a national standards laboratory. Only a few temperatures have been measured, including the freezing point of gold (1337.91 K), and the boiling points of sulfur (717.85 K), oxygen (90.18 K), and helium (4.22 K). Various other types of thermometers (platinum, carbon, and doped germanium resistors; thermocouples) are calibrated at these temperatures and used to measure intermediate temperatures.

Kinetic temperature. An important aspect of the absolute temperature scale is its relation to the motions of atoms and molecules, whether vibrations as in solids and liquids, or straight path flights with collisions as in gases. There are two important facts here: (1) There is a definite distribution of motions. For example, in a gas, even though the motions are chaotic and a particular molecule changes velocity after each collision with another molecule, at any instant a definite number of molecules have a particular velocity. It cannot be said that the gas is at a definite temperature unless the molecules have this definite distribution of velocities, although different small portions of the gas may have definite, though different, temperatures. The same idea holds for the distribution of vibration frequencies in solids and liquids. (2) A body has a minimum amount of motion energy. It was supposed in the 19th century that this minimum was zero energy, but modern theories and experiments show that the minimum is greater than zero. A body in its lowest energy state cannot give out heat and is at absolute zero.

A system may have several degrees of freedom. The molecules of a gas, besides having straight-line motions, may rotate and vibrate and their electrons may be in different energy levels. When a system is in equilibrium, the energies stored in these different degrees of freedom are related to a common absolute temperature. In several cases, one can measure something related to a particular degree of freedom of a system. Then if the system is in equilibrium, its absolute temperature can be inferred.

Examples of this are measurements of temperatures in the Sun's corona and in the remote regions of the Milky Way Galaxy. The Sun's corona is so hot that atoms lose several electrons, that is, are multiply ionized. This greatly affects the wavelengths of light that these atoms emit. From the measured wavelengths that have been identified, the corona's temperature has been estimated at 2×10^6 K.

Besides light, atoms and molecules can emit radio waves. The straight-line motions of atoms toward or away from Earth-based radio telescopes slightly shifts the received wavelenths. From these Doppler shifts, temperatures from 1 to 100 K have been found in the vast hydrogen atom clouds in the galaxy spiral arms. Rotations of OH molecules influence the populations of energy levels and thus affect the intensities of emitted radio signals. From these also, deep-space absolute temperatures are inferred. *See* KINETIC THEORY OF MATTER.

Negative absolute temperatures. Negative Celsius and Fahrenheit temperatures are readily accepted because 0° on these scales is arbitrarily set above absolute zero (Fig. 2.) But the idea that a system could be at, say -50 K, was introduced only in the early 1950s. The concept applies only to systems with a finite number of energy levels; that is, those that can store a finite amount of energy. Thus the translational energy of a gas or the vibration energy of a crystal cannot be at negative absolute temperatures. However, the energies of electron and nuclear magnetic moments (spins) in a magnetic field do have upper limits. Normally there are more spins in lower energy levels, in which case they are at positive absolute temperature. With special techniques one can arrange equal numbers of spins in the energy levels. Then the temperature is infinite K. This does not mean infinite heat can be extracted from such a system. One can go further and cause there to be more spins in the higher energy levels than in the lower, and this situation is described as being at a negative absolute temperature.

The spin system is not in equilibrium with the crystal lattice vibrations which remain at positive temperatures. But at temperatures around a few K, the spins need minutes or hours to exchange heat with the lattice. Negative absolute temperatures are hotter than positive temperatures; this is reasonable since heat flows from the negative temperature spins to the positive temperature lattice. *See* NEGATIVE TEMPERATURE.

Extreme temperatures. For reasons involving both pure and applied science, researchers endeavor to achieve extremes of both low and high temperatures. At very low temperatures, phenomena that are well understood become frozen out and new predicted and unpredicted effects are sought. Experimenters have reported holding 2 kg of copper at about 50 microkelvins for a couple of days. Achieving and measuring such temperatures involves using the nuclear magnetic spin system. In the procedure to reach the 50-μK lattice temperature, the copper spins were brought down to 50 nanokelvins. One object of such research is to see if nuclear spins will spontaneously align as electron spins do in ferromagnetic materials like iron. *See* LOW-TEMPERATURE PHYSICS.

The highest equilibrium temperatures reached on Earth are around 10^7 K in experiments to achieve fusion of hydrogen nuclei. These temperatures have been sustained in very-low-density gases like the Sun's corona for a few seconds. At these temperatures hydrogen nuclei have speeds of about 1.3×10^6 m/s. Such speeds are necessary if the positively charged nuclei are to overcome their electric repulsion force. When the nuclei get close enough together, nuclear forces attract them to fuse and there is a net energy release. The goal of the research is to convert this energy into conventional electrical energy. *See* NUCLEAR FUSION.

[ROLAND A. HULTSCH]

Bibliography: American Society for Testing and Materials, *Evolution of the International Practical Temperature Scale of 1968*, 1974; C. M. Herzfeld (ed.), *Temperature: Its Measurement and Control in Science and Industry*, vol. 3, 1962; K. Mendelssohn, *The Quest for Absolute Zero*, 1966; M. W. Zemansky, *Temperatures Very Low and Very High*, 1964.

Theoretical physics

The description of natural phenomena in mathematical form. It is impossible to separate theoretical physics from experimental physics, since a complete understanding of nature can be obtained only by the application of both theory and experiment. *See* PHYSICS.

Purposes. There are two main purposes of theoretical physics: the discovery of the fundamental laws of nature and the derivation of conclusions from these fundamental laws.

Discovery of fundamental laws. Physicists aim to reduce the number of laws to a minimum to have as far as possible a unified theory. When the laws are known, it is possible from any given initial conditions of a physical system to derive the subsequent events in the system. Sometimes, especially in quantum theory, only the probability of various events can be predicted. *See* NONRELATIVISTIC QUANTUM THEORY; QUANTUM MECHANICS.

Conclusions from fundamental laws. The conclusions to be derived from the fundamental laws of nature may be of several different types.

1. Conclusions may be derived in order to test a given theory, particularly a new theory. An example is the derivation of the spectrum of the hydrogen atom from quantum mechanics; the verification of the predictions by accurate measurements is a good test of quantum mechanics. On rather rare occasions experiment has been found to contradict the predictions of an existing theory, and this has then led to the discovery of important new physical laws. An example is the Michelson-Morley experiment on the constancy of the velocity of light, an experiment which led to special relativity theory. *See* ATOMIC STRUCTURE AND SPECTRA.

2. Theory may be required for experiments designed to determine physical constants. Most fundamental physical constants cannot be accurately measured directly. Elaborate theories may be required to deduce the constant from indirect experiments. An example is the Millikan oil-drop determination of the electron charge, which requires the knowledge of the motion of small droplets in air as deduced from hydrodynamic theory. *See* ATOMIC CONSTANTS.

3. Predictions of physical phenomena may be made in order to gain understanding of the structure of the physical world. In this category fall theories of the structure of the atom leading to an understanding of the periodic system of elements, or of the structure of the nucleus in which various models are tested (for example, shell model or collective model). In the same category fall applications of theoretical physics to other sciences, for example, to chemistry (theory of the chemical bond and of the rate of chemical reactions), astronomy (theory of planetary motion, internal constitution, and energy production of stars), or biology.

4. Engineering applications may be drawn from fundamental laws. All of engineering may be considered an application of physics, and much of it is an application of mathematical physics, such as elasticity theory, aerodynamics, electricity, and magnetism. The generation and propagation of radio waves of all frequencies is an example of application of theoretical physics to direct practice. *See* ELECTRICITY; MAGNETISM.

Content. Apart from the classification of the fields of theoretical physics according to purpose, a classification can also be made according to content. Here one may perhaps distinguish three classification principles: type of force, scale of physical phenomena, and type of phenomena.

Type of force. At present four different types of force are known to physics. The best-understood type of force is electricity and magnetism. Here the fundamental laws, Maxwell's equations, are completely known. Corrections due to quantum theory exist but can be calculated. For practical purposes electromagnetic fields can be calculated with confidence and precision, from dc fields to the shortest γ-ray wavelength. *See* ELECTROMAGNETIC RADIATION; FORCE; MAXWELL'S EQUATIONS.

The second type of force is the gravitational force. For practical purposes, Newton's inverse-square law is usually sufficient. The most complete theory of gravitation, however, is Einstein's general theory of relativity, which has great beauty but only limited experimental confirmation; therefore rival theories are sometimes proposed. *See* GRAVITATION; RELATIVITY.

The strong force which holds atomic nuclei together is the third type. In contrast to the first two types, only some of the general features of the nuclear force are known at present. No exact quantitative predictions from first principles, similar to the very successful ones in electrodynamics, can be made in nuclear theory. It is known that nuclear forces are related to various unstable particles (mesons), but this relation is only partially understood. The nuclear force is the strongest force known to physics but extends only over very small distances. *See* NUCLEAR STRUCTURE.

Distinct from the nuclear force are the weak forces responsible for beta radioactivity and similar phenomena. They are probably closer to being understood than the strong nuclear force.

Scale of physical phenomena. The motion of bodies on the scale of everyday life can be described by the classical mechanics of Isaac Newton. Phenomena in very small dimensions, especially inside atoms or atomic nuclei, must be described by quantum mechanics. The latter theory contains Newton's mechanics as a special case. *See* CLASSICAL MECHANICS.

The description of physical phenomena is also different according to the velocities of the bodies involved. When the velocity is a substantial fraction of that of light, the special theory of relativity must be used to describe the motion. (The special theory of relativity has hardly anything in common with general relativity theory except the name, and in contrast to the general theory is established beyond doubt by an enormous number of experiments.) Newton's classical mechanics again is a special case of the mechanics of special relativity. *See* RELATIVISTIC ELECTRODYNAMICS; RELATIVISTIC MECHANICS.

Special relativity and quantum mechanics are examples of the development of physical theory. Neither of them has made classical mechanics wrong or obsolete, but they have extended classical mechanics into domains which were outside the experience of man until 1900. When a physical law is discovered, it can be expected to hold as long as the general conditions are not radically changed from those holding in the experiments from which the law was originally derived; for example, classical mechanics holds for objects of not too small size moving with moderate velocities. In a completely new area (for example, where there is very small size or very high speed) it cannot be expected a priori that the same laws will continue to hold; but if the laws do change under the new conditions, this does not invalidate the old laws in the domain for which they were originally formulated, except for minor corrections.

The most general theory of motion now known is quantum field theory, which combines both quantum mechanics and relativity theory and at the same time embodies the observed fact that particles can be created and annihilated. This theory may thus be called a unified field theory. Attempts have also been made toward other unification, in particular to unify the theories of two types of forces, gravitational and electromagnetic; this is commonly called unified field theory in the literature. These attempts have not been very successful; moreover, they leave out quantum theory, as well as the nuclear forces, both strong and weak. *See* QUANTUM FIELD THEORY; UNIFIED FIELD THEORY.

Type of phenomena. The most customary classification of theoretical physics is according to the type of phenomena described. The following are the main fields under this heading:

1. Mechanics is the theory of motion of bodies under given forces. It is normally understood to involve classical mechanics only, and includes particle mechanics and mechanics of rigid bodies. In particle mechanics, celestial mechanics is an important subdivision; this includes planetary motion, the motion of artificial satellites, and the complicated motions resulting when three bodies interact (the classical three-body problem). The field of rigid-body mechanics includes the complicated theory of gyroscopic motion with and without external fields of force. *See* MECHANICS; RIGID-BODY DYNAMICS.

2. Continuum mechanics is the theory of motion of bodies, taking into account their internal properties. One branch of this is the theory of elasticity, which is basic for structural engineering design. Another branch is hydro- and aerodynamics. Here a number of problems can be solved approximately by potential theory, but most of modern aerodynamics requires a more physical approach. Knowledge of physical properties, such as those given by the equation of state of a gas, is essential; these properties can be explained only on a molecular scale. Acoustics is a classical branch of continuum mechanics. A combination of aerodynamics and electrodynamics is required for the modern field of magnetohydrodynamics. *See* ACOUSTICS; AERODYNAMICS; CLASSICAL FIELD THEORY; ELASTICITY; HYDRODYNAMICS; MAGNETOHYDRODYNAMICS; POTENTIALS.

3. Heat presents a problem that can be treated on a phenomenological level by thermodynamics, which is the basis of heat engineering, as well as of the theory of chemical equilibrium. On the molecular level, heat is described by statistical mechanics, which may be considered the physical foundation of thermodynamics. Beyond this, statistical mechanics permits the calculation of the properties of bulk substances (gases, liquids, and solids) in terms of their atomic properties. *See* HEAT; STATISTICAL MECHANICS; THERMODYNAMIC PRINCIPLES.

4. Electrodynamics is well understood. Subdivisions are electrostatics; the theory of stationary currents (the basis of electrical generating machinery); the theory of oscillating electrical circuits (the basis of the technology of ordinary radio); the theory of electromagnetic waves, including their propagation in air as well as in waveguides and similar devices (the basis of radar); and finally the electromagnetic theory of light. *See* ELECTRODYNAMICS.

5. Optics is customarily treated as a special field, although, strictly speaking, it is a branch of electrodynamics. Geometrical optics and the theory of diffraction phenomena are two of the principal topics. Emission and absorption of light can be understood only on the basis of atomic physics. The same is true of dispersion, that is, the behavior of the refractive index as a function of frequency. *See* OPTICS.

6. Atomic physics includes the theory of the structure of the atom; the motion of the electrons in the atom; the periodic system; the energy levels and spectral lines of atoms and molecules; the behavior of atoms and molecules in external fields; and collisions of atoms with each other, with electrons, and with other particles. Atomic physics is the basis of the calculation of properties of matter in bulk and of the emission and absorption of light. Related is the theory of molecular structure, which is the basis of theoretical chemistry. Collisions between molecules explain the rate of chemical reactions. *See* ATOMIC PHYSICS; MOLECULAR PHYSICS.

7. Nuclear and particle physics includes the theory of nuclear forces and of the structure of atomic nuclei. A complete theory would predict all energy levels of any nucleus and thus the electromagnetic radiations which can be emitted by the nucleus. The topic also includes the theory of nuclear reactions, which is the basis of the technology of nuclear reactors. In an effort to understand the origin of nuclear forces, theoretical physicists have investigated the production and properties of mesons and the so-called strange particles. Radioactive decay, and particularly beta decay, is another branch of nuclear physics involving weak rather than strong nuclear forces. High-energy nuclear physics aims at understanding the properties of particles—nucleons as well as unstable particles of various kinds. *See* ELEMENTARY PARTICLE; MATHEMATICAL PHYSICS; NUCLEAR PHYSICS.

[HANS A. BETHE]

Thermal conduction in solids

Thermal conduction in a solid is generally measured by stating the thermal conductivity K, which is the ratio of the steady-state heat flow (heat transfer per unit area per unit time) along a long rod to the temperature gradient along the rod. Thermal conductivity varies widely among different types of solids, and depends markedly on temperature and on the purity and physical state of the solids, particularly at low temperatures.

From the kinetic theory of gases the thermal conductivity can be written as in Eq. (1), where S is

$$K = (\text{constant})\, S\, v\, l \qquad (1)$$

the specific heat per unit volume, v is the average particle velocity, and l is the mean free path. In solids, thermal conduction results from conduction by lattice vibrations and from conduction by electrons. In insulating materials, the conduction is by lattice waves; in pure metals, the lattice contribution is negligible and the heat conduction is primarily due to electrons. In many alloys, impure metals, and semiconductors, both conduction mechanisms contribute. *See* CONDUCTION (HEAT); KINETIC THEORY OF MATTER; LATTICE VIBRATIONS; SPECIFIC HEAT OF SOLIDS.

Insulating solids. In insulators, heat is transported by lattice vibrations. The vibrations of the crystal lattice can be resolved into traveling elastic waves (lattice waves). Each lattice wave can be specified by a wave vector **k** and by a mode of polarization (either longitudinal or transverse). The energy of the lattice vibrations is taken as the energy of lattice waves of frequency ν, where ν is of the order of $k_B T/h$; k_B is Boltzmann's constant, T is the absolute temperature, and h is Planck's constant. The energy of each wave is described as consisting of an integral number of phonons, each of energy $h\nu$ and of momentum $h k_B/2\pi$. *See* PHONON.

At equilibrium, the phonons are distributed isotropically in momentum space; when a temperature gradient is established along the crystal, an asymmetry in the phonon distribution is set up. An interchange of energy among the lattice waves, which can result from the anharmonicity of the lattice forces and from lattice imperfections, tends to restore equilibrium; the rate at which equilibrium is restored is associated with a finite mean free path l for the phonons. At low temperatures K varies markedly with temperature and is dependent

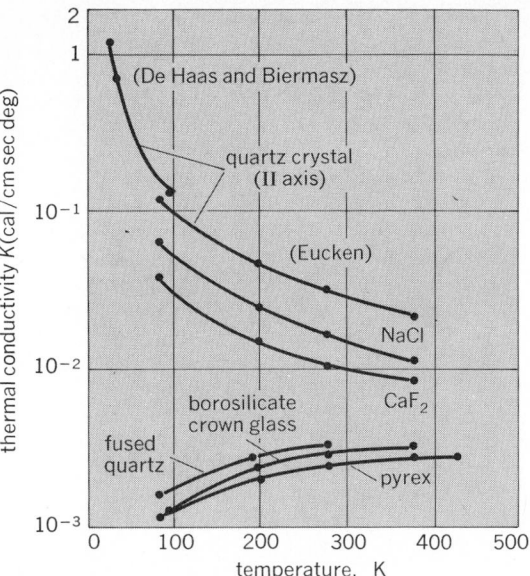

Fig. 1. Thermal conductivity of some insulating and some amorphous materials as a function of temperature. (*After C. Kittel, Solid State Physics, 3d ed., 1966*)

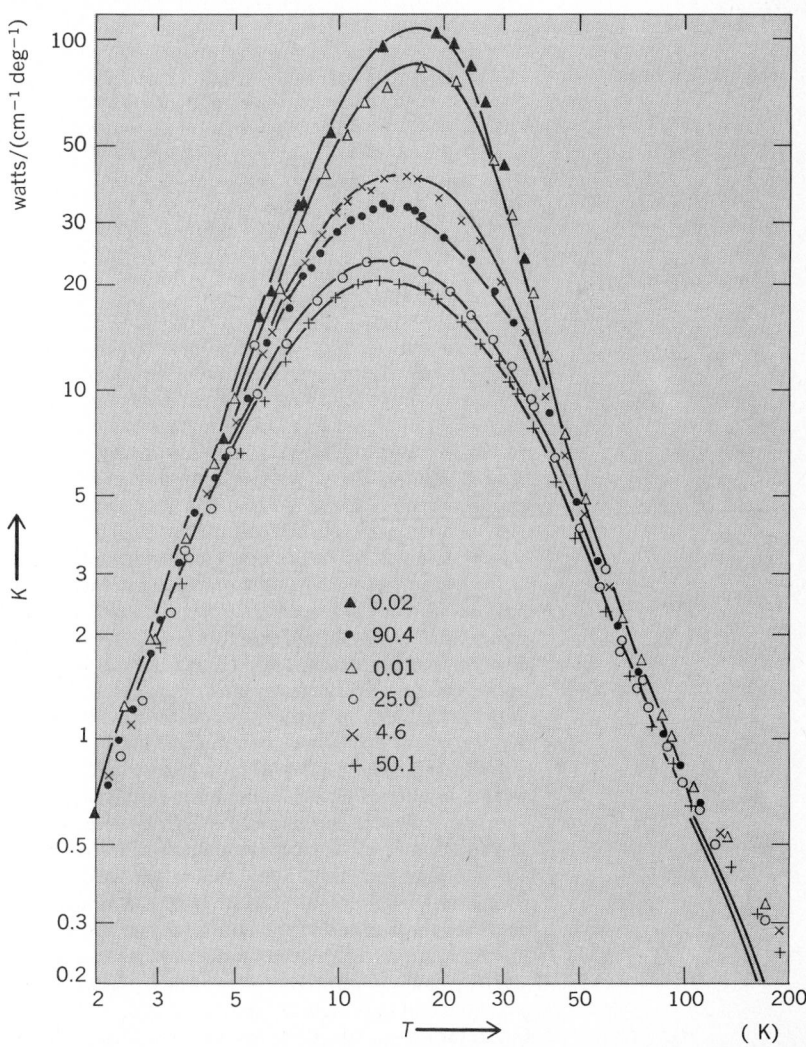

Fig. 2. Thermal conductivity of lithium fluoride as a function of temperature for various percentage concentrations of Li⁶. In the peak the highest value is approximately five times that of the lowest peak. (*After R. Berman, Heat conductivity of non-metallic crystals at low temperatures, Cryogenics, 5:297–305, 1965*)

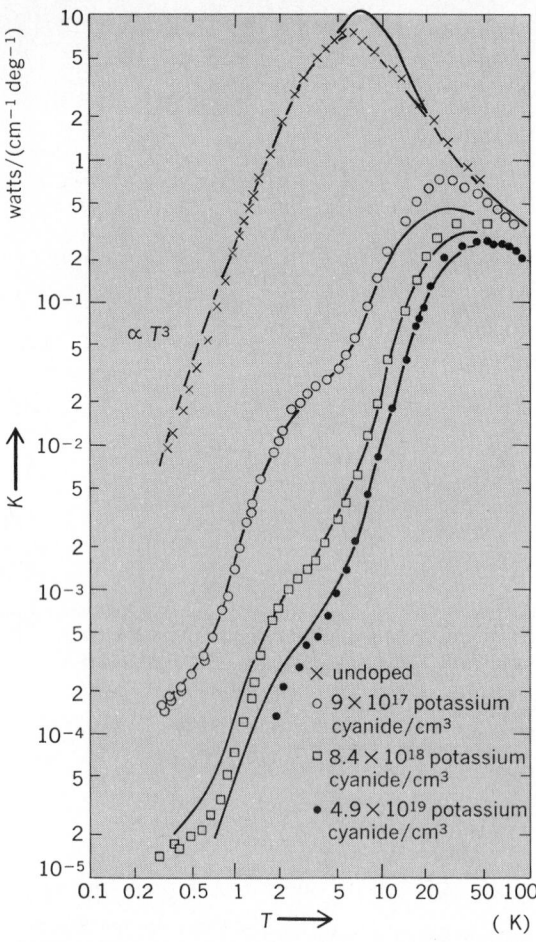

Fig. 3. Thermal conductivity of pure potassium chloride and of potassium chloride containing cyanide ions as a function of temperature. (*After W. D. Seward, in R. Berman, Heat conductivity of non-metallic crystals at low temperatures, Cryogenics, 5:297–305, 1965*)

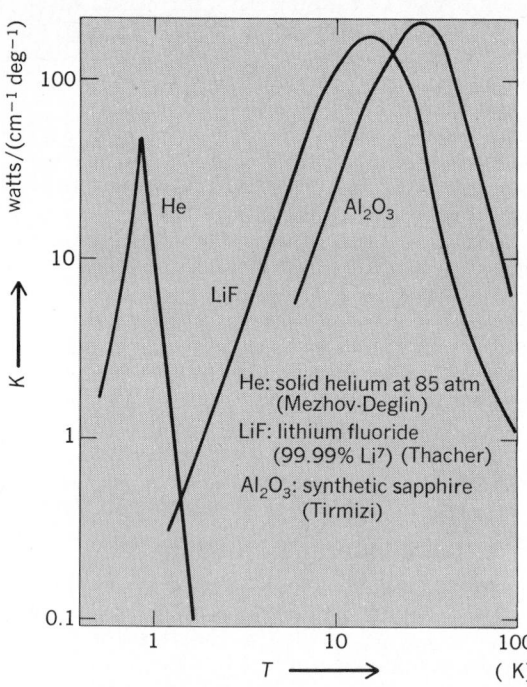

Fig. 4. Thermal conductivity of solid helium as a function of temperature. Thermal conductivities of very pure lithium fluoride and sapphire are given for comparison. (*After R. Berman, Heat conductivity of non-metallic crystals at low temperatures, Cryogenics, 5:297–305, 1965*)

on the variation of l with the phonon frequency ν. Figure 1 shows the conductivity of some insulating materials.

The form of l at low temperatures is closely related to the type of phonon scattering and hence to the type or types of imperfections present in the particular solid. The thermal conductivity at low temperatures is a sensitive function of these imperfections. For example, in Fig. 2, the effect of the isotope of lithium Li^6 on the conductivity of lithium fluoride crystals is shown; here the differences in conductivity are due principally to the mass difference between Li^6 and Li^7.

The introduction of molecular impurities into dielectric crystals produces resonance scattering, as shown in Fig. 3 for the potassium chloride crystals containing cyanide ions; here the center of the resonance (the dip in the thermal conductivity curves) corresponds to the frequency of the particular phonons carrying heat at that temperature. The thermal conductivity of solid helium has been measured at low temperatures, and is shown in Fig. 4; for comparison; the thermal conductivities of very pure lithium fluoride and sapphire are given. At less than 1 K, the conductivity of solid helium increases very rapidly with temperature and is characterized as Poiseuille flow, in which the

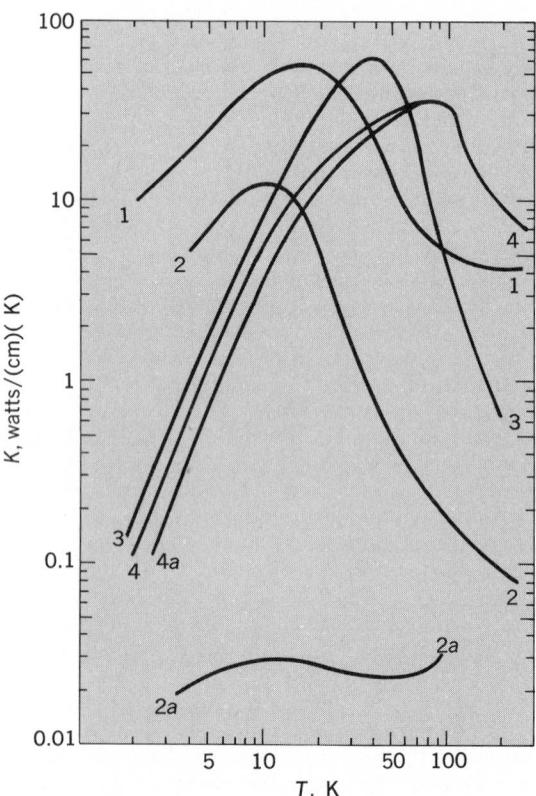

Fig. 5. Thermal conductivity of some insulators as a function of temperature. Curve 1, pure copper (for comparison). Curve 2, quartz crystal. Curve 2a, quartz crystal after intense neutron irradiation. Curve 3, aluminum oxide (sintered alumina). Curve 4, diamond. Curve 4a, diamond, reduced diameter. (*After R. Berman, Heat conductivity of non-metallic crystals at low temperatures, Cryogenics, 5:297–305, 1965*)

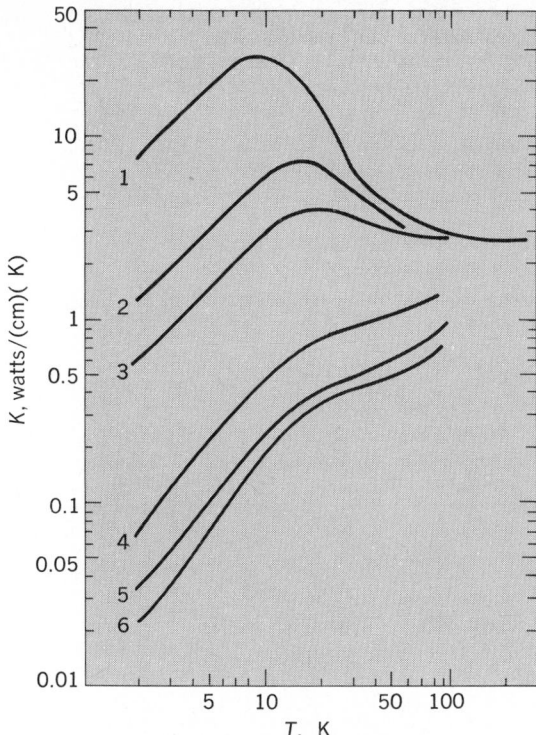

Fig. 6. Thermal conductivity of gold and some alloys as a function of temperature. Curve 1, 99.999% gold, annealed. Curve 2, same, cold-drawn. Curve 3, 99.9% gold, cold-drawn. Curve 4, gold with 0.7% platinum. Curve 5, gold with 1.7% platinum. Curve 6, gold with 0.2% chromium. (*After P. G. Klemens, Thermal conductivity and lattice vibrational modes, in F. Seitz and D. Turnbull, eds., Solid State Physics, vol. 7, 1958*)

phonons are described as flowing as a viscous fluid. *See* FLUID MECHANICS; HYDRODYNAMICS.

Glass and polycrystalline materials. For amorphous materials, glasses, and polycrystalline materials, the mean free path for the phonons is generally small, of the order of the distance between atoms, and is independent of temperature. The thermal conductivity for these materials is much less than that for the insulating materials above; for example, quartz (fused) glass can be compared with crystal quartz in Fig. 1, and sintered alumina (aluminum oxide) in Fig. 5 can be compared with sapphire (single-crystal aluminum oxide) in Fig. 4.

Metals, alloys, and semiconductors. The thermal conductivity of a metal can be written as the sum of an electronic component K_e and a lattice component K_g; K_e and K_g are each limited by various scattering mechanisms. In real crystals, as compared to the ideal crystal with a perfect lattice, the electrons undergo scattering caused by the thermal vibrations of the lattice and by imperfections (such as impurities, point defects, dislocations, and grain boundaries).

In pure metals, the electronic component accounts for nearly all the heat conducted, while the lattice component, in most cases, is negligible. The electronic thermal conductivity is related to the electrical conductivity through the mean free path. In certain temperature regions the value of the mean free path for both thermal and electrical conduction can be assumed to be the same; for these cases, the Wiedemann-Franz law is ap-

plicable, as in Eq. (2), where L (the Lorentz

$$K_e/\sigma T = L = \pi k_B^2/3e^2 \qquad (2)$$

constant) $= 2.45 \times 10^{-8}$ (watt)(ohm)(K^{-2}), σ is the electrical conductivity, and e is the electronic charge. Here the free-electron theory of metals is assumed, in which the conduction electrons form a gas which obeys Fermi-Dirac statistics. *See* FREE-ELECTRON THEORY OF METALS.

The electrical resistivity ρ (the reciprocal of the electrical conductivity σ) can be separated into two parts, one called the residual resistivity ρ_0, which results from the elastic scattering of electrons by imperfections and which is independent of temperature, and the other called the ideal resistivity $\rho_i(T)$, which results from scattering due to lattice vibrations and which is temperature-dependent. Thus the electrical resistivity is written $\rho = 1/\sigma = \rho_0 + \rho_i(T)$. The thermal resistivity for metals W (the reciprocal of the thermal conductivity K_e) can be written analogously as the sum of two terms, one, W_0, due to scattering by imperfections, and the other, $W_i(T)$, due to scattering by lattice vibrations. Thus the thermal conductivity is written $W = 1/K_e = W_0 + W_i(T)$. In Fig. 6 the thermal conductivities of three samples of gold are given; since the purity of the gold differs, the samples differ in both ρ_0 and W_0. In the low-temperature region, the thermal conductivity is proportional to T. At higher temperatures, the thermal resistivity due to lattice vibrations exceeds W_0, and the curves for the three different samples converge.

In the case of alloys, W_0 is much larger than in pure metals and a lattice thermal conductivity must also be included. The thermal conductivities of several alloys of gold are also shown in Fig. 6. It is possible from such curves to obtain K_g as a function of T by also measuring ρ; ρ_0 can be determined from W_0. In well-annealed alloys, K_g is proportional to T^2 at low temperatures; at higher temperatures, K_g decreases with T.

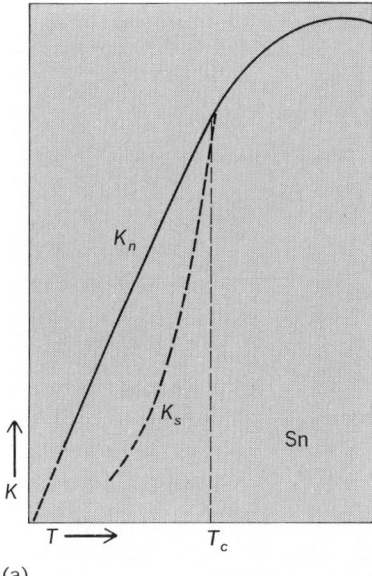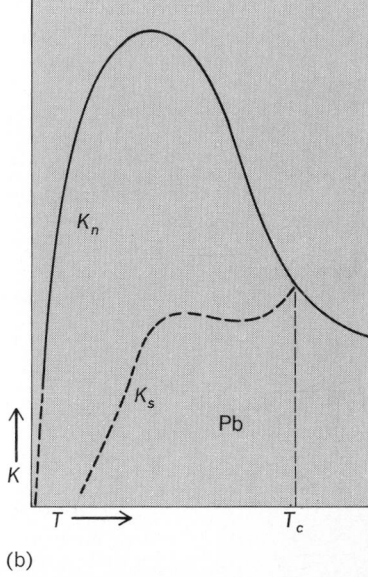

(a)   (b)

Fig. 7. Temperature dependences of thermal conductivity of (a) tin and (b) lead in the normal and superconducting states. (*After K. Mendelssohn and H. M. Rosenberg, The thermal conductivity of metals at low temperatures, in F. Seitz and D. Turnbull, eds., Solid State Physics, vol. 12, 1961*)

In the case of semiconductors, such as high-purity germanium and silicon, phonons are primarily responsible for the thermal conduction. The form of the thermal conductivity versus temperature curves is similar to that for single crystals of dielectric materials.

Superconductors. In superconductors at temperatures below the critical temperature T_c, the electronic conduction is reduced; at sufficiently low temperatures, the thermal conductivity becomes entirely due to lattice waves and is similar to the form of the thermal conductivity of an insulating material. In Fig. 7 the temperature dependences of the thermal conductivity of tin and lead are given; here K_n is the thermal conductivity with the material in the normal state, and K_s is the thermal conductivity with the material in the superconducting state. *See* SUPERCONDUCTIVITY.

[KATHRYN A. MC CARTHY]

Bibliography: R. Berman, *Thermal Conduction in Solids*, 1976; G. E. Childs, L. J. Erick, and R. L. Powell, *Thermal Conductivity of Solids at Room Temperature and Below: A Review and Compilation of the Literature*, Nat. Bur. Stand. Monogr. 131, 1973; P. G. Klemens and T. Kilhus, *Thermal Conductivity*, 1976.

Thermal expansion

Solids, liquids, and gases all exhibit dimensional changes for changes in temperature while pressure is held constant. The molecular mechanisms at work and the methods of data presentation are quite different for the three cases and are therefore discussed separately in this article.

Expansion of solids. The temperature coefficient of linear expansion α_l is defined by Eq. (1),

$$\alpha_l = \frac{1}{l}\left(\frac{\partial l}{\partial t}\right)_{p=\text{const}} \tag{1}$$

where l is the length of the specimen, t is the temperature, and p is the pressure. For each solid there is a Debye characteristic temperature Θ, below which α_l is strongly dependent upon temperature and above which α_l is practically constant. Many common substances are near or above Θ at room temperature and follow the approximate equation Eq. (2), where l_0 is the length at 0°C and

$$l = l_0(1 + \alpha_l t) \tag{2}$$

t is the temperature in °C. The total change in length from absolute zero to the melting point has a range of approximately 2% for most substances. Typical room temperature values of α_l are given in Table 1.

Linear, harmonic vibration of the atoms in a solid cannot account for changes in volume, hence this must result from nonlinearity of the thermally excited vibration. The theory of E. Grüneisen takes this into account and shows the coefficient of expansion to be proportional to the constant-volume specific heat of the solid. At low temperatures (small amplitude vibration), the coefficient of expansion approaches zero.

Pure crystals may have different values of α_l along different axes, but substances such as structural steel have many crystals randomly oriented and are almost free from this effect. At certain temperatures, crystalline substances may change in lattice arrangement, and a sudden change of volume occurs at constant temperature, making α_l

Table 1. Temperature coefficients of linear expansion for typical substances at room temperature

Substance	Coefficient of linear expansion per °C $\times 10^6$
Aluminum, commercial	24
Copper	17
Diamond	1
Glass, commercial	11
Glass, pyrex	3
Granite	8.3
Ice	50
Iron	12
Invar alloy	0.9
Quartz, crystalline	5
Quartz, fused	0.5
Oak, along fiber	5
Oak, across fiber	54
Rubber, hard	80

momentarily infinity. *See* LATTICE VIBRATIONS; SPECIFIC HEAT OF SOLIDS.

Expansion of gases. So-called perfect gases follow the relation in Eq. (3), where p is absolute

$$\frac{pv}{T} = \frac{R}{\text{molecular weight}} \tag{3}$$

pressure, v is specific volume, T is absolute temperature, and R is a constant. The magnitude of R, the so-called gas constant, is 1544 ft-lb/(°R)(lb-mole) in the English system, or 8.3144×10^7 ergs/(K)(g-mole) in the metric system. Real gases often follow this equation closely; for example, Table 2 shows values of R at atmospheric pressure and 0°C. *See* GAS CONSTANT.

Table 2. Values for gas constant R at atmospheric pressure and 0°C

Gas	R
Air	1545
Hydrogen	1546
Nitrogen	1543
Oxygen	1544
Methane	1539

The coefficient of cubic expansion α_v is defined by Eq. (4), and for a perfect gas this is found to be $1/T$.

$$\alpha_v = \frac{1}{v}\left(\frac{\partial v}{\partial t}\right)_{p=\text{const}} \tag{4}$$

The behavior of real gases is largely accounted for by van der Waals' equation, Eq. (5), where a

$$p = \frac{RT}{v-b} - \frac{a}{v^2} \tag{5}$$

and b are constant for a given gas. When the specific volume is large, the effects of these constants are unimportant, and the real gas behaves as a perfect gas. In the regions where a and b have a dominant effect it is usually found desirable to use experimentally determined graphs or charts of properties. *See* GAS; KINETIC THEORY OF MATTER.

Expansion of liquids. For liquids, α_v is somewhat a function of pressure but is largely determined by temperature. Though α_v may often be taken as constant over a sizable range of temperature (as in the liquid expansion thermometer), generally some variation must be accounted for. For example, water contracts with temperature rise from 0 to 4°C, above which it expands at an increasing rate, as shown by the data in Table 3,

Table 3. Behavior of water at different temperatures

t,°C	Volume expansion, ml/g
−10	1.00186
0	1.00013
4	1.00000
10	1.00027
100	1.007

which were taken at atmospheric pressure. One approach to this variation is to evaluate the constants α, β, and γ in Eq. (6), where v_0 is the volume

$$v = v_0(1 + \alpha t + \beta t^2 + \gamma t^3) \qquad (6)$$

at 0°C, and v is the volume at temperature t. Typical values of the coefficients appear in Table 4.

Table 4. Coefficients of volume expansion of gases

Liquid	$\alpha \times 10^3$	$\beta \times 10^6$	$\gamma \times 10^8$
Ethyl alcohol (99.3% by volume)	1.012	2.20	
500 atm	0.866		
3000 atm	0.524		
Carbon tetrachloride	1.184	0.899	1.351
Mercury	1.182	0.0078	
Petroleum	0.8994	1.396	
Water	−0.06427	8.5053	−6.7900

Thermal stresses. When a homogeneous body is subject to constant boundary loads and is raised uniformly in temperature, the stress pattern in it will not change unless its elastic properties change. In general, stresses arise if (1) the body is made up of substances having different coefficients of expansion, (2) changes of boundary dimensions are restrained, or (3) temperature distribution is not uniform. A simple example of the first case is shown in the figure, where the aluminum bar, if heated, would tend to expand faster than the iron bars, thereby putting the iron in tension and the aluminum in compression. Considering the aluminum alone, its change of length would be restrained, and therefore stresses would arise in it. If the aluminum bar were replaced by an iron one and if this one alone were heated, again the other bars would be in tension and the center one in compression. More complex stress patterns may arise in continuous bodies; for example, if the bars were joined along the sides rather than at the ends, shear stresses would arise in the seams. In iron, 360 lb/in.² tensile stress would produce the same elongation as would a temperature rise of 1°C.

Since one source of temperature variation is the gradient necessary for heat transfer, thermal conductivity and heat capacity may both play a role in determining the stress pattern. *See* CONDUCTION (HEAT); HEAT CAPACITY; THERMAL CONDUCTION IN SOLIDS.

[RALPH A. BURTON]

Thermal hysteresis

A phenomenon in which a physical quantity depends not only on the temperature but also on the preceding thermal history. It is usual to compare the behavior of the physical quantity while heating and the behavior while cooling through the same temperature range. The illustration shows the thermal hysteresis which has been observed in the behavior of the dielectric constant of single crystals of barium titanate. On heating, the dielectric constant was observed to follow the path *ABCD*, and on cooling the path *DCEFG*. *See* DIELECTRIC CONSTANT; FERROELECTRICS.

Perhaps the most common example of thermal hysteresis involves a phase change such as solidification from the liquid phase. In many cases these liquids can be dramatically supercooled. Elaborate precautions to eliminate impurities and outside disturbances can be instrumental in supercooling 60 to 80°C. On raising the temperature after freezing, however, the system follows a completely different path, with melting coming at the prescribed temperature for the phase change.

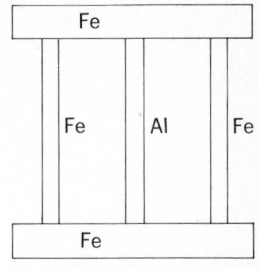

Body composed of substances having different coefficients of expansion. Thermal stresses would arise if it were subjected to heat.

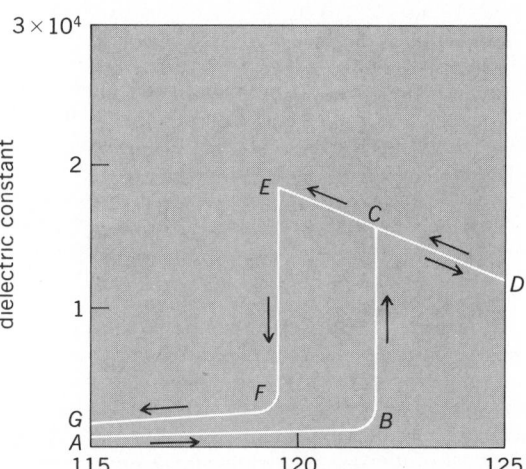

Plot of dielectric constant versus temperature for a single crystal of barium titanate. (*After M. E. Drougard and D. R. Young, Phys. Rev.*, 95:1152–1153, 1954)

Solidification (or appropriate phase change) occurs by a nucleation mechanism, while melting (the reversal of phase) does not. The viable nucleus only forms below the melting temperature, being held up by the competing demands of surface and bulk free energy. Once formed, the nucleus grows rapidly to bring on the sudden precipitation of the solid phase. This rapid release of latent heat and the catastrophic nature of the solidification process illustrate the basic irreversibility and hysteresis of thermally induced phase transformations. *See* CRYSTAL GROWTH; PHASE TRANSITIONS.

[H. B. HUNTINGTON; R. K. MAC CRONE]

Thermodynamic principles

Laws governing the conversion of energy from one form to another. Among the many consequences of these laws are relationships between the properties of matter and the effects of changes in pressure, temperature, electric field, magnetic field, and composition. The great practicality of the science arises from the foundations of the subject. Thermodynamics is based upon observations of common experience that have been formulated into the thermodynamic laws. From these few laws all of the remaining laws of the science are deducible by purely logical reasoning. There is a choice as to which few are considered independent laws, from which the remainder may be derived. A modern tendency is to choose basic laws or postulates that are different from those first discovered. Some of these choices are most useful in that the derivation of the remainder may be accomplished very efficiently. However, those laws that arose from the historical development will be discussed here since they are less abstract and lend themselves to a clearer physical interpretation.

One may say that the whole development of thermodynamic principles was completed when three state functions, the absolute temperature T, the internal energy U, and the entropy S, were defined. The zeroth law formalizes the concept of temperature, the first law defines the internal energy, and the second law brings in the concept of entropy as well as the absolute scale of temperature. Finally, the third law describes the behavior of entropy and internal energy as the absolute temperature approaches zero.

For exposition, it is necessary to define a few terms. A *system* is that part of the physical world under consideration. The rest of the world is the *surroundings*. An *open system* may exchange mass, heat, and work with the surroundings. A *closed system* may exchange heat and work but not mass with the surroundings. An *isolated system* has no exchange with the surroundings. A closed or isolated system is sometimes referred to as a body. Those parts of a system spatially uniform and homogeneous are called *phases*. For example, a liquid together with its vapor may be considered a two-phase system. Systems may be made quite elaborate when required, but since focus is on the thermal properties, single-phase isotropic systems not acted upon by electric or magnetic fields are considered so that the only force allowed is that generated by a uniform normal pressure. Such a restriction is not a basic limitation on the generality of thermodynamics but is simply a pedagogical device.

Specification of equilibrium state. The material properties of concern to thermodynamics are the macroscopic properties such as temperature, pressure, volume, concentration, surface tension, and viscosity. Molecular properties such as interatomic distances are not used. The state of a system is specified by all of the macroscopic properties together with their spatial variation. It is a fact of experience, however, that an isolated system approaches a particularly simple terminal state such that the properties are constant and spatially uniform. This simple state is called an equilibrium state. If one confines attention to a given quantity of a single-phase system, the equilibrum state is completely specified by $r+1$ of its properties, where r is the number of components. For a single-component, single-phase system not subject to magnetic or electric fields, one may fix two properties such as pressure and volume; all the remaining properties such as viscosity, surface tension,

and so forth then assume fixed values. In other words, any macroscopic property of the system may be expressed as a function of the pressure and volume.

Temperature. It is within the scope of thermodynamics to refine the primitive notion of hotness and coldness into an operational and precise concept of temperature. The equilibrium states of a single-component, single-phase fluid provide a starting point. For such a fluid the equilibrium state is defined by fixing two of its properties. For example, one could construct a mercury-in-glass thermometer that has its pressure held constant; then only one other property could be varied independently. If the volume (height of mercury) is observed at any equilibrium state, there is a 1:1 correspondence between it and any other property excepting the pressure. The degree of hotness is one of these properties.

Also note that thermal equilibrium between different systems exists. For example, if the mercury thermometer is placed in contact with a body of quiet water, the mercury will either expand or contract. The volume change of the mercury will eventually stop, and the properties of the mercury will be constant, indicating an equilibrium state; moreover, the water will also have the constant properties of an equilibrium state. Bodies in thermal equilibrium are said to have the same temperature. Thus, one arrives at a method of measuring the temperature of a body of water.

Now suppose there is a large body of water in thermal contact through a wall with a large body of another fluid such as alcohol, both at equilibrium. It is an experimental fact that the mercury-in-glass thermometer will register the same volume when placed in either of the fluids. This fact is most important if one is to attach meaningful numbers to temperature. This fact of experience is designated as the zeroth law of thermodynamics: If two bodies A and B are separately in thermal equilibrium with C, then A and B are in equilibrium with each other. Thus, a useful empirical temperature measurement based upon the volume of mercury under constant pressure is established.

But the empirical temperature scale is unfortunately unique to the choice of the fluid. The mercury-in-glass thermometer is calibrated by bringing it into equilibrium with an ice-water mixture and then with boiling water, all under a pressure of 1 atm. The two mercury levels are marked 0° and 100°, respectively, and linear interpolation is used to assign numbers between the two fixed points. If one constructs and calibrates another thermometer but uses another fluid instead of mercury, one would find that the numerical values between the two fixed points do not agree. Indeed, water cannot be used as a working fluid at all since it has a minimum value of volume at 4°C and yields anomalous readings in this region. It is necessary to choose some fluid as a standard and calibrate all other thermometers by comparison with the standard. The second law of thermodynamics removes this dependence upon a particular material by defining an absolute temperature scale that is independent of the working fluid. Meanwhile the empirical temperature serves a useful operational purpose.

One could have used a low-pressure gas as the thermometric fluid. The volume could have conveniently been held constant, and the pressure of the fluid would have had a 1:1 correspondence with temperature. Here one would have found that all gases at low pressures yield the same temperature scale. This ideal gas temperature scale proves to be identical with the thermodynamic scale of the second law.

In summary, the relationship between the properties of the equilibrium state, the notion of thermal equilibrium, and the zeroth law have been used to establish the property of temperature. It may be noted in passing that temperature is a state property, and for a given mass of a single-phase, single-component fluid the temperature is a function of pressure and volume as in Eq. (1). This can be inverted as in Eqs. (2).

$$t = t(p,V) \qquad (1)$$

$$\begin{aligned} p &= p(t,V) \\ V &= V(p,t) \end{aligned} \qquad (2)$$

Internal energy. Thermodynamics does not define the concepts of energy or work but adopts them from the other macroscopic sciences of mechanics and electromagnetism. Also, the conservation of energy is taken as axiomatic. Therefore, if an isolated system is formed from any part of the world, a definite amount of energy will be trapped in the system. The energy resides in the kinetic and potential energy of the trapped molecules. The trapped energy is of a definite quantity because the isolated system cannot gain or lose energy to the surroundings and remains constant because of the conservation principle. This trapped energy is called the internal energy U.

Because of the conservation of energy, the internal energy of a closed system can be altered only by an exchange of energy with the surroundings. There are only three modes by which the exchange can occur: by mass transfer, heat transfer, or work exchange. So for a closed (no mass transfer), adiabatic (no heat transfer) system the change in internal energy ΔU is equal to the work done by the surroundings on the system, as defined by Eq. (3).

$$\Delta U = W_{AD} \qquad (3)$$

Here a convention has been adopted that work done on the system is positive. There is a great mass of experimental information where work has been done on a closed system enclosed within adiabatic walls. Among these are experiments performed by J. Joule more than a century ago. He caused work to be done on an adiabatically enclosed mass of water in several different ways. A measured amount of work was used to drive an agitator in the water, to create an electric current which was then passed through a coil in the water, to compress a gas in a cylinder immersed in the water, and to rub metal blocks together in the water. In Joule's experiments the same temperature increase was always obtained with the same expenditure of work. It may be concluded from Joule's experiments that the expenditure of a given quantity of work always causes the same change of state regardless of how the work is carried out. Both W_{AD} and ΔU are independent of the path. It is concluded that U is a state function. So for a single-phase, single-component fluid Eq. (4) is

$$W_{AD} = U_2 - U_1 = \Delta U \qquad (4)$$

written, where U_2 and U_1 depend only on the final and initial state, respectively. Also one may write Eqs. (5).

$$U = U(p,V) \qquad U = (t,p) \qquad (5)$$

It is known from experience that the same change in state of a system can be effected by either supplying work to the system in an adiabatic enclosure or by contacting the system through a conducting wall with a higher temperature system. The latter method is a different means of transferring energy than work and is termed heat and given the symbol Q. Measuring the amount of work required to cause the same change in state as an amount of heat enables one to express heat quantities in terms of work quantities. For example, 1 calorie is taken to be the amount of heat necessary to raise 1 g of water 1°C at 15°C and 1 atm. The same change in state can be effected by 4.186×10^7 ergs of work. Therefore, Eq. (6) holds.

$$1 \text{ calorie} = 4.186 \times 10^7 \text{ ergs} \qquad (6)$$

The first law of thermodynamics may now be derived by the useful device of a composite system. Imagine a very large system of water that transfers neither heat nor work to the surroundings. Within this large system there is a small system of a cylinder of gas in thermal contact with the water and having a piston connected to the outside. Work can be done on the small system, and it can in turn interchange heat with the large system called a reservoir. Using subscripts s for the small system and r for the reservoir, the process of doing work can be described as in Eq. (7). But $\Delta U_r = $

$$W_s = \Delta U_s + \Delta U_r \qquad (7)$$

$Q_r = -Q_s$. Therefore, for the small system that is exchanging both heat and work with its surroundings, one writes Eq. (8), omitting the subscript.

$$\Delta U = Q + W \qquad (8)$$

This is the first law of thermodynamics and states that the algebraic sum of heat and work during a process is equal to the change in the state function U. The term $(Q + W)$ is therefore independent of the path taken between the two states. One could, for example, cause 1 g of water to undergo the change in state as in notation (9) by

$$15°C, 1 \text{ atm} \rightarrow 16°C, 1 \text{ atm} \qquad (9)$$

supplying 1 calorie of heat and no work or by doing 4.186×10^7 ergs of work alone, or one could do a great deal of work and abstract all of this energy in the form of heat excepting 1 calorie. Thus, although $(Q + W)$ is independent of the path, neither Q nor W by itself is independent of the path.

It is important to realize that U is a state function and a property of the system whereas W and Q are not. The work, as well as the heat, simply represents energy in transit. Once the energy is in the system, it is not possible to determine whether it came from heat transfer or work transfer; it is simply internal energy.

The differential form of the first law is given by Eq. (10), where q and w represent small quantities.

$$dU = q + w \qquad (10)$$

In general, one may not treat q or w as well-behaved differential coefficients dQ and dW. However, if the change is such that either q or w depends only on the initial and final states and not on the path, Eq. (11) may be correctly written. If two

$$dU = dQ + dW \qquad (11)$$

terms are independent of the path, the third must be also. Obviously, if either q or w is zero, one may properly write Eqs. (12). There is a third case

$$\begin{aligned} dU &= dW \\ dU &= dQ \end{aligned} \qquad (12)$$

where neither q nor $w = 0$, but nevertheless $dU = dQ + dW$ is still proper. Before treating this interesting case one needs to develop the notion of a reversible process.

Reversible and irreversible processes. Any process that occurs in nature is in agreement with the first law, but many processes permissible by the first law never occur. It has already been noted that systems approach an equilibrium state if left to themselves. There is an overwhelming preference for processes to proceed in one direction. Consider Joule's experiment. A falling weight caused a paddle to do work on an adiabatically enclosed body of water. The total effect of the experiment was to increase the internal energy of the water and to lower the weight. The surroundings remained unchanged. The water temperature increased and the volume increased slightly. There is no way one can reverse this process, that is, restore the water to its original state and raise the weight to its original height without also making some additional change in the surroundings. The process is irreversible.

Consider some other processes occurring within an adiabatic enclosure by examining only the initial and final states. Two blocks of copper are initially at different temperatures and finally at the same temperature which is intermediate between the two initial temperatures. A gas is initially filling just half of a container and finally the whole of the container. Again these processes are irreversible; that is, they cannot be reversed without causing some permanent change in the surroundings. The reverses of these processes do not violate the first law; therefore, there must be some condition other than the conservation of energy which is obeyed by those processes which actually take place.

If one were presented with a description of only the initial and final states of these irreversible processes, as was done in the last two examples, one could unerringly decide from the description which state was initial and which was final. The direction of the process is entirely determined by the nature of the states. It may be expected, therefore, that there is some state function that shows which state precedes the other. The function which tells whether a process is possible or not is the entropy S and will be derived from the information that some adiabatic processes are impossible.

Thermodynamics makes use of an idealization, called a reversible process, that is a limiting case of the natural or irreversible process. The reversible process may be defined as one which can be completely reversed without leaving more than a vanishingly small change in the surroundings. It is a consequence of the definition that a reversible process proceeds through a succession of equilibrium states and may be reversed by an infinitesimal change in the external conditions.

Imagine having a cylinder of gas fitted with a frictionless piston. If the piston is moved so slowly that pressure gradients are absent, the gas will be in an equilibrium state at all times. The difference between the gas pressure and the external pressure needs only to be infinitesimal in order to move the frictionless piston. Under the rather restrictive conditions of a reversible process, Eq. (13) may be

$$dW = -p\,dV \qquad (13)$$

quite properly written, where p is the gas pressure and V is the gas volume. The first law now may be written as Eq. (14).

$$dQ = dU + p\,dV \qquad (14)$$

Entropy. The discussion on irreversible processes has led to the second law of thermodynamics, which is just a general statement of the idea that there is a preferred direction for a given process. There are many physical statements of the second law, all being equivalent and leading to the same mathematical statement. The statement of R. Clausius is: "It is *not* possible that, at the end of a cycle of changes, heat has been transferred from a colder to a hotter body without producing some other effect." Lord Kelvin's statement is: "It is *not* possible that, at the end of a cycle of changes, heat has been extracted from a reservoir and an equal amount of work has been produced without producing some other effect."

A specific example of Kelvin's statement may be useful. Work can be converted continuously and completely into heat. For example, work could be expended on rubbing blocks in a large mass of water. The blocks would become infinitesimally hotter than the water and transfer energy to the water by heat flow. The process could be continued indefinitely with the only effect being a complete conversion of work into heat. If, however, heat is converted from the large water reservoir completely into work, some other effect occurs. For example, a gas within a cylinder can be expanded reversibly causing a transfer of heat from the bath to the gas. All of the heat extracted from the bath is converted into work. However, the gas, in this process, has changed its state since its volume is larger. The gas cannot be returned to its original state without undoing the conversion of heat into work already accomplished.

The most efficient way of developing the mathematical consequences of the second law is to proceed from Caratheodory's principle, which can be either taken as another physical expression of the second law or derived from the Clausius or Kelvin statement. Caratheodory's principle is: "In the neighborhood of any equilibrium state of a system there are states which are not accessible by an adiabatic process."

Caratheodory used this principle together with a mathematical theorem that he developed to infer the existence of a state function S and an integrating factor $1/T$, where T is the thermodynamic temperature such that Eq. (15) holds for a reversible

$$dQ_{\text{REV}} = T\,dS \qquad (15)$$

change. The state function S is called the entropy. It can also be shown that the entropy in an adiabatic system increases for an irreversible change and remains constant for a reversible change as in Eq. (16). The implication is that entropy increases for a

$$\Delta S_{AD} \geq 0 \qquad (16)$$

natural change until equilibrium is reached, and then it remains constant at its maximum value.

The first part of the mathematical statement of the second law allows one to write one of the most important thermodynamic equations, Eq. (17). Although this equation was derived for reversible

$$dU = TdS - pdV \qquad (17)$$

changes, it is valid for all changes. All the quantities are functions of state. Therefore, for a change between two states the integral of the equation will be valid even if the path is not reversible. In other words, for a change from a state characterized by (p_1, T_1) to a state characterized by (p_2, T_2), the values of ΔU, ΔV, and ΔS will all have definite values dependent only upon the two states and independent of how the change came about. From this equation are obtained some of the most fruitful applications of thermodynamics to physical problems.

The second part of the mathematical statement is a concise summary of physical statements on the direction of processes. As a simple example, consider pure heat transfer to a body. The heat transfer causes a definite change of state such that $dQ = dU$. The definite change of entropy is then given by Eq. (18). If heat dQ is transferred from a

$$dS = \frac{dQ}{T} \qquad (18)$$

body at temperature T_2 to a body at temperature T_1, the change in entropy is given by Eq. (19).

$$\begin{aligned} dS &= dS_1 + dS_2 \\ &= \frac{dQ}{T_1} - \frac{dQ}{T_2} \\ &= \frac{dQ(T_2 - T_1)}{T_1 T_2} \end{aligned} \qquad (19)$$

Since dS must be positive or zero, $T_2 > T_1$. Therefore heat flows from the hotter body to the colder body.

Also contained in the second part of the entropy statement is the key idea of equilibrium. The equilibrium state of an adiabatic or isolated system is characterized by entropy being at its maximum value consistent with the physical constraints. Therefore, equilibrium states can be determined by setting $dS = 0$. Also, for a maximum in entropy, $d^2S < 0$. This latter condition leads to the notion of stability that is important in the study of phase equilibrium.

When a system is in thermal contact with its surroundings, the entropy of the system may decrease. For example, a gas being compressed isothermally decreases its entropy, but a greater increase of entropy occurs in the surroundings. The total entropy change is always positive. Clausius stated the first and second laws of thermodynamics as: "The energy of the world is constant. The entropy of the world tends toward a maximum."

By way of completeness the third law of thermodynamics needs comments. In the main body of thermodynamics one is mostly interested in changes of entropy and internal energy between states. However, the third law defines an absolute scale for entropy: The entropy of all perfect crystalline solids is zero at absolute zero temperature. The third law is used primarily in classical thermodynamics for the calculation of absolute entropies which combined with thermochemical data permits the calculation of chemical equilibrium. The foundations of the third law, however, are to be found in molecular theory and require a statistical mechanical treatment. *See* STATISTICAL MECHANICS.

Summary. By way of summary, for a closed system all of the fundamentals of thermodynamics are contained in notation (20).

$$\begin{aligned} dU &= q + w \\ dS &= \frac{dQ}{T} \qquad \text{reversible change} \\ dS &\geq 0 \qquad \text{for an isolated system} \\ dU &= TdS - pdV \end{aligned} \qquad (20)$$

The equations are applicable when work is restricted to volume changes only. But the generalization to include changes of polarization, magnetization, surface area, and so forth is quite straightforward. Also, the equations are not applicable to systems that involve irreversible chemical changes, but here too the extension to include these situations presents no difficulty. In actual application it is convenient to define other state functions in terms of those already introduced, but no additional basic principles are needed. *See* ADIABATIC PROCESS; ENTHALPY.

[WILLIAM F. JAEP]

Bibliography: J. Adkins, *Equilibrium Thermodynamics*, 2d ed., 1975; K. Denbigh, *Chemical Equilibrium*, 3d ed., 1971; R. W. Haywood, *Equilibrium Thermodynamics*, 1980; J. S. Hsieh, *Principles of Thermodynamics*, 1974; A. B. Pippard, *Elements of Classical Thermodynamics*, 1966; K. A. Rolle, *Introduction to Thermodynamics*, 2d ed., 1980; M. W. Zemansky and R. Dittman, *Heat and Thermodynamics*, 6th ed., 1981.

Thermodynamic processes

Changes of any property of an aggregation of matter and energy, accompanied by thermal effects. The participants in a process are first identified as a system to be studied; the boundaries of the system are established; the initial state of the system is determined; the path of the changing states is laid out; and, finally, supplementary data are stated to establish the thermodynamic process. These steps will be explained in the following paragraphs. At all times it must be remembered that the only processes which are allowed are those compatible with the first and second laws of thermodynamics: Energy is neither created nor destroyed and the entropy of the system plus its surroundings always increases.

A system and its boundaries. To evaluate the results of a process, it is necessary to know the participants that undergo the process, and their mass and energy. A region, or a system, is selected for study, and its contents determined. This region may have both mass and energy entering or leaving during a particular change of conditions, and these mass and energy transfers may result in changes both within the system and within the surroundings which envelop the system.

As the system undergoes a particular change of

condition, such as a balloon collapsing due to the escape of gas or a liquid solution brought to a boil in a nuclear reactor, the transfers of mass and energy which occur can be evaluated at the boundaries of the arbitrarily defined system under analysis.

A question that immediately arises is whether a system such as a tank of compressed air should have boundaries which include or exclude the metal walls of the tank. The answer depends upon the aim of the analysis. If its aim is to establish a relationship among the physical properties of the gas, such as to determine how the pressure of the gas varies with the gas temperature at constant volume, then only the behavior of the gas is involved; the metal walls do not belong within the system. However, if the problem is to determine how much externally applied heat would be required to raise the temperature of the enclosed gas a given amount, then the specific heat of the metal walls, as well as that of the gas, must be considered, and the system boundaries should include the walls through which the heat flows to reach the gaseous contents. In the laboratory, regardless of where the system boundaries are taken, the walls will always play a role and must be reckoned with.

State of a system. To establish the exact path of a process, the initial state of the system must be determined, specifying the values of variables such as temperature, pressure, volume, and quantity of material. If a number of chemicals are present in the system, the number of variables needed is usually equal to the number of independently variable substances present plus two such as temperature and pressure; exceptions to this rule occur in variable electric or magnetic fields and in some other well-defined cases. Thus, the number of properties required to specify the state of a system depends upon the complexity of the system. Whenever a system changes from one state to another, a process occurs.

Whenever an unbalance occurs in an intensive property such as temperature, pressure, or density, either within the system or between the system and its surroundings, the force of the unbalance can initiate a process that causes a change of state. Examples are the unequal molecular concentration of different gases within a single rigid enclosure, a difference of temperature across the system boundary, a difference of pressure normal to a nonrigid system boundary, or a difference of electrical potential across an electrically conducting system boundary. The direction of the change of state caused by the unbalanced force is such as to reduce the unbalanced driving potential. Rates of changes of state tend to decelerate as this driving potential is decreased.

Equilibrium. The decelerating rate of change implies that all states move toward new conditions of equilibrium. When there are no longer any balanced forces acting within the boundaries of a system or between the system and its surroundings, then no mechanical changes can take place, and the system is said to be in mechanical equilibrium. A system in mechanical equilibrium, such as a mixture of hydrogen and oxygen, under certain conditions might undergo a chemical change. However, if there is no net change in the chemical constituents, then the mixture is said to be in chemical as well as in mechanical equilibrium.

If all parts of a system in chemical and mechanical equilibrium attain a uniform temperature and if, in addition, the system and its surroundings either are at the same temperature or are separated by a thermally nonconducting boundary, then the system has also reached a condition of thermal equilibrium.

Whenever a system is in mechanical, chemical, and thermal equilibrium, so that no mechanical, chemical, or thermal changes can occur, the system is in thermodynamic equilibrium. The state of equilibrium is at a point where the tendency of the system to minimize its energy is balanced by the tendency toward a condition of maximum randomness. In thermodynamics, the state of a system can be defined only when it is in equilibrium. The static state on a macroscopic level is nevertheless underlaid by rapid molecular changes; thermodynamic equilibrium is a condition where the forward and reverse rates of the various changes are all equal to one another. In general, those systems considered in thermodynamics can include not only mixtures of material substances but also mixtures of matter and all forms of energy. For example, one could consider the equilibrium between a gas of charged particles and electromagnetic radiation contained in an oven.

Process path. If under the influence of an unbalanced intensive factor the state of a system is altered, then the change of state of the system is described in terms of the end states or difference between the initial and final properties.

The path of a change of state is the locus of the whole series of states through which the system passes when going from an initial to a final state. For example, suppose a gas expands to twice its volume and that its initial and final temperatures are the same. Various paths connect these initial and final states: isothermal expansion, with temperature held constant at all times, or adiabatic expansion which results in cooling followed by heating back to the initial temperature while holding volume fixed.

Each of these paths can be altered by making the gas do varying amounts of work by pushing out a piston during the expansion, so that an extremely large number of paths can be followed even for such a simple example. The detailed path must be specified if the heat or work is to be a known quantity; however, changes in the thermodynamic properties depend only on the initial and final states and not upon the path.

There are several corollaries from the above descriptions of systems, boundaries, states, and processes. First, all thermodynamic properties are identical for identical states. Second, the change in a property between initial and final states is independent of path or processes. The third corollary is that a quantity whose change is fixed by the end states and is independent of the path is a point function or a property. However, it must be remembered that by the second law of thermodynamics not all states are available (possible final states) from a given initial state and not all conceivable paths are possible in going toward an available state.

Pressure-volume-temperature diagram. Whereas the state of a system is a point function, the change of state of a system, or a process, is a path function. Various processes or methods of change of a system from one state to another may be depicted graphically as a path on a plot using thermodynamic properties as coordinates.

The variable properties most frequently and conveniently measured are pressure, volume, and temperature. If any two of these are held fixed (independent variables), the third is determined (dependent variable). To depict the relationship among these physical properties of the particular working substance, these three variables may be used as the coordinates of a three-dimensional space. The resulting surface is a graphic presentation of the equation of state for this working substance, and all possible equilibrium states of the substance lie on this *P-V-T* surface. The *P-V-T* surface may be extensive enough to include all three phases of the working substance: solid, liquid, and vapor.

Because a *P-V-T* surface represents all equilibrium conditions of the working substance, any line on the surface represents a possible reversible process, or a succession of equilibrium states.

The portion of the *P-V-T* surface shown in Fig. 1 typifies most real substances; it is characterized by contraction of the substance on freezing. Going from the liquid surface to the liquid-solid surface onto the solid surface involves a decrease in both temperature and volume. Water is one of the few exceptions to this condition; it expands upon freezing, and its resultant *P-V-T* surface is somewhat modified where the solid and liquid phases abut.

Gibbs' phase rule is defined in Eq. (1). Here *f* is

$$f = c - p + 2 \tag{1}$$

the degree of freedom; this integer states the number of intensive properties (such as temperature, pressure, and mole fractions or chemical potentials of the components) which can be varied independently of each other and thereby fix the particular equilibrium state of the system (see discussion under Temperature-entropy diagram, below). Also, *p* indicates the number of phases (gas, liquid, or solid) and *c* the number of component substances in the system. Consider a one-component system (a pure substance) which is either in the liquid, gaseous, or solid phase. In equilibrium the system has two degrees of freedom; that is, two independent thermodynamic properties must be chosen to specify the state. Among the thermodynamic properties of a substance which can be quantitatively evaluated are the pressure, temperature, specific volume, internal energy, enthalpy, and entropy. From among these properties, any two may be selected. If these two prove to be independent of each other, when the values of these two properties are fixed, the state is determined and the values of all the other properties are also fixed. A one-component system with two phases in equilibrium (such as liquid in equilibrium with its vapor in a closed vessel) has *f* = 1; that is, only one intensive property can be independently specified. Also, a one-component system with three phases in equilibrium has no degree of freedom. Examination of Fig. 1 shows that the three surfaces (solid-liquid, solid-vapor, and liquid-vapor) are generated by lines parallel to the volume axis. Moving the system along such lines (constant pressure and temperature) involves a heat exchange and a

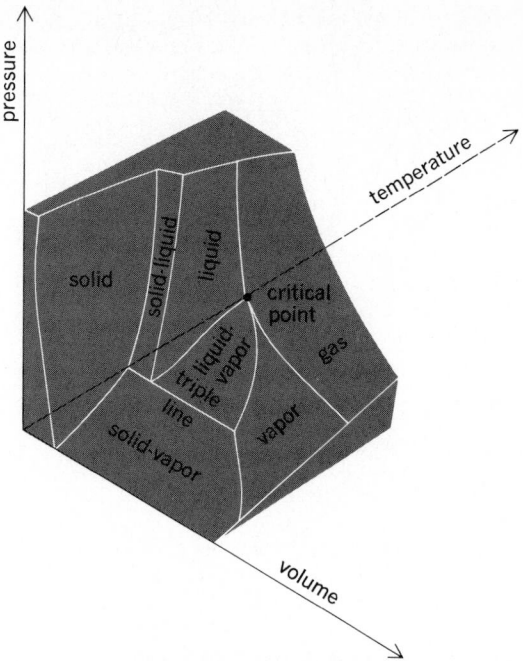

Fig. 1. Portion of pressure-volume-temperature (*P-V-T*) surface for a typical substance.

change in the relative proportion of the two phases. Note that there is an entropy increment associated with this change.

One can project the three-dimensional surface onto the *P-T* plane as in Fig. 2. The triple point is the point where the three phases are in equilibrium. When the temperature exceeds the critical temperature (at the critical point), only the gaseous phase is possible. The gas is called a vapor when it can coexist with another phase (at temperatures below the critical point). The *P-T* diagram for water would have the solid-liquid curve going upward from the triple point to the left (contrary to the ordinary substance pictured in Fig. 2). Then the property so well known to ice skaters would be evident. As the solid-liquid line is crossed from the low-pressure side to the high-pressure side, the

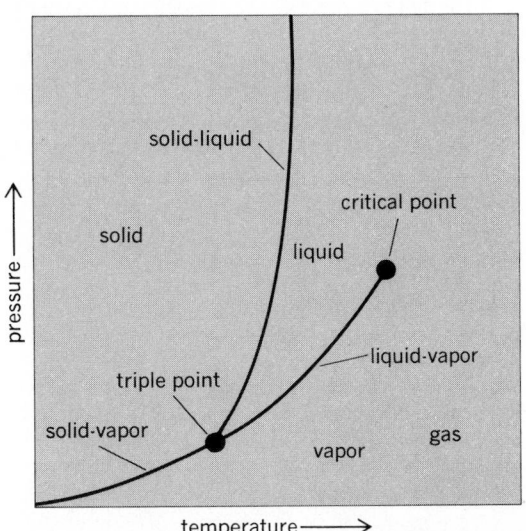

Fig. 2. Portion of equilibrium surface projected on pressure-temperature (*P-T*) plane.

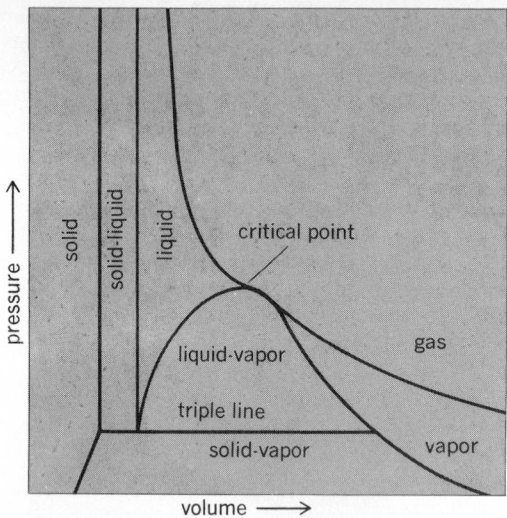

Fig. 3. Portion of equilibrium surface projected on pressure-volume (*P-V*) plane.

water changes from solid to liquid: Ice melts upon application of pressure.

Work of a process. The three-dimensional surface can also be projected onto the *P-V* plane to get Fig. 3. This plot has a special significance: The area under any reversible path on this plane represents the work done during the process. The fact that this *P-V* area represents useful work can be demonstrated by the following example.

Let a gas undergo an infinitesimal expansion in a cylinder equipped with a frictionless piston, and let this expansion perform useful work on the surroundings. The work done during this infinitesimal expansion is the force multiplied by the distance through which it acts, as in Eq. (2), wherein *dW* is

$$dW = F \, dl \tag{2}$$

an infinitesimally small work quantity, F is the force, and dl is the infinitesimal distance through which F acts.

But force F is equal to the pressure P of the fluid times the area A of the piston, or PA. However, the product of the area of the piston times the infinitesimal displacement is really the infinitesimal volume swept by the piston, or $A \, dl = dV$, with dV equal to an infinitesimal volume. Thus Eq. (3) is valid. The work term is found by integration, as in Eq. (4).

$$dW = P \, A \, dl = P \, dV \tag{3}$$

$$_1W_2 = \int_1^2 P \, dV \tag{4}$$

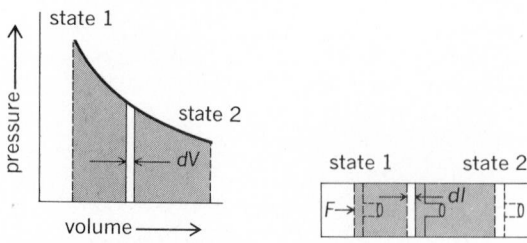

Fig. 4. Area under path in *P-V* plane is work done by expanding gas against piston.

Figure 4 shows that the integral represents the area under the path described by the expansion from state 1 to state 2 on the *P-V* plane. Thus, the area on the *P-V* plane represents work done during this expansion process.

Temperature-entropy diagram. Energy quantities may be depicted as the product of two factors: an intensive property and an extensive one. Examples of intensive properties are pressure, temperature, and magnetic field; extensive ones are volume, magnetization, and mass. Thus, in differential form, work has been presented as the product of a pressure exerted against an area which sweeps through an infinitesimal volume, as in Eq. (5). Note that as a gas expands, it is doing work on

$$dW = P \, dV \tag{5}$$

its environment. However, a number of different kinds of work are known. For example, one could have work of polarization of a dielectric, of magnetization, of stretching a wire, or of making new surface area. In all cases, the infinitesimal work is given by Eq. (6), where X is a generalized applied

$$dW = X dx \tag{6}$$

force which is an intensive quantity such as voltage, magnetic field, or surface tension; and dx is a generalized displacement of the system and is thus extensive. Examples of dx include changes in electric polarization, magnetization, length of a stretched wire, or surface area.

By extending this approach, one can depict transferred heat as the product of an intensive property, temperature, and a distributed or extensive property defined as entropy, for which the symbol is S. See ENTROPY.

If an infinitesimal quantity of heat dQ is transferred during a reversible process, this process may be expressed mathematically as in Eq. (7),

$$dQ = T \, dS \tag{7}$$

with T being the absolute temperature and dS the infinitesimal entropy quantity.

Furthermore, a plot of the change of state of the system undergoing this reversible heat transfer can be drawn on a plane in which the coordinates are absolute temperature and entropy (Fig. 5). The total heat transferred during this process equals the area between this plotted line and the horizontal axis.

Reversible processes. Not all energy contained in or associated with a mass can be converted into useful work. Under ideal conditions only a fraction of the total energy present can be converted into work. The ideal conversions which retain the maximum available useful energy are reversible processes.

Characteristics of a reversible process are that the working substance is always in thermodynamic equilibrium and the process involves no dissipative effects such as viscosity, friction, inelasticity, electrical resistance, or magnetic hysteresis. Thus, reversible processes proceed quasistatically so that the system passes through a series of states of thermodynamic equilibrium, both internally and with its surroundings. This series of states may be traversed just as well in one direction as in the other.

If there are no dissipative effects, all useful work done by the system during a process in one direc-

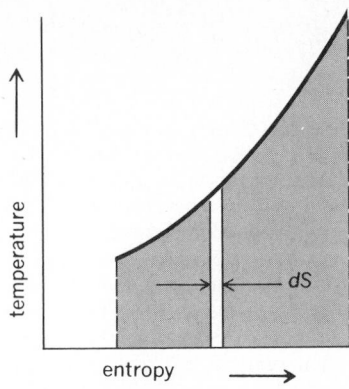

Fig. 5. Heat transferred during a reversible process is area under path in temperature-entropy (*T-S*) plane.

tion can be returned to the system during the reverse process. When such a process is reversed so that the system returns to its starting state, it must leave an effect on the surroundings since, by the second law of thermodynamics, in energy conversion processes the form of energy is always degraded. Part of the energy of the system (including heat source) is transferred as heat from a higher temperature to a lower temperature. The energy rejected to a lower-temperature heat sink cannot be recovered. To return the system (including heat source and sink) to its original state, then, requires more energy than the useful work done by the system during a process in one direction. Of course, if the process were purely a mechanical one with no thermal effects, then both the surroundings and system could be returned to their initial states.

It is impossible to satisfy the conditions of a quasistatic process with no dissipative effects; a reversible process is an ideal abstraction which is not realizable in practice but is useful for theoretical calculations. An ideal reversible engine operating between hotter and cooler bodies at the temperatures T_1 and T_2, respectively, can put out $(T_1 - T_2)/T_1$ of the transferred heat energy as useful work.

There are four reversible processes wherein one of the common thermodynamic parameters is kept constant. The general reversible process for a closed or nonflow system is described as a polytropic process. *See* ISENTROPIC PROCESS; ISOBARIC PROCESS; ISOMETRIC PROCESS; ISOTHERMAL PROCESS; POLYTROPIC PROCESS.

Irreversible processes. Actual changes of a system deviate from the idealized situation of a quasistatic process devoid of dissipative effects. The extent of the deviation from ideality is correspondingly the extent of the irreversibility of the process.

Real expansions take place in finite time, not infinitely slowly, and these expansions occur with friction of rubbing parts, turbulence of the fluid, pressure waves sweeping across and rebounding through the cylinder, and finite temperature gradients driving the transferred heat. These dissipative effects, the kind of effects that make a pendulum or yo-yo slow down and stop, also make the work output of actual irreversible expansions less than the maximum ideal work of a corresponding reversible process. For a reversible process, as stated earlier, the entropy change is given by $dS =$

dQ/T. For an irreversible process even more entropy is produced (turbulence and loss of information) and there is the inequality $dS > dQ/T$.

[PHILIP E. BLOOMFIELD; WILLIAM A. STEELE]

Bibliography: H. A. Bent, *The Second Law,* 1965; J. P. Holman, *Thermodynamics,* 3d ed., 1980; M. Mott-Smith, *The Concept of Energy Simply Explained,* 1934; W. C. Reynolds and H. C. Perkins, *Engineering Thermodynamics,* 2d ed., 1977; F. W. Sears and G. L. Salinger, *Thermodynamics, the Kinetic Theory of Gases and Statistical Mechanics,* 3d ed., 1975; K. Wark, *Thermodynamics,* 3d ed., 1977.

Thermoelectricity

The direct conversion of heat into electrical energy, or the reverse, in solid or liquid conductors by means of three interrelated phenomena—the Seebeck effect, the Peltier effect, and the Thomson effect—including the influence of magnetic fields upon each. The Seebeck effect concerns the electromotive force (emf) generated in a circuit composed of two different conductors whose junctions are maintained at different temperatures. The Peltier effect refers to the reversible heat generated at the junction between two different conductors when a current passes through the junction. The Thomson effect involves the reversible generation of heat in a single current-carrying conductor along which a temperature gradient is maintained. Specifically excluded from the definition of thermoelectricity are the phenomena of Joule heating and thermionic emission. *See* ELECTROMOTIVE FORCE (EMF); JOULE'S LAW.

The three thermoelectric effects are described in terms of three coefficients: the absolute thermoelectric power (or thermopower) S, the Peltier coefficient Π, and the Thomson coefficient μ, each of which is defined for a homogeneous conductor at constant temperature. These coefficients are connected by the Kelvin relations, which convert complete information about one into complete information about all three. It is therefore necessary to measure only one of the three coefficients; usually the thermopower S is chosen. The combination of electrical resistivity, thermal conductivity, and thermopower is sufficient to provide a complete description of the electronic transport properties of conductors for which the electric current and heat current are linear functions of both the applied electric field and the temperature gradient.

Thermoelectric effects have significant applications in both science and technology and show promise of more importance in the future. Studies of thermoelectricity in metals and semiconductors yield information about electronic structure and about the interactions between electrons and both lattice vibrations and impurities. Practical applications include the measurement of temperature, generation of power, cooling, and heating. Thermocouples are widely used for temperature measurement, providing both accuracy and sensitivity. Research has been undertaken concerning the direct thermoelectric generation of electricity using the heat produced by nuclear reactors. Cooling units using the Peltier effect have been constructed in sizes up to those of home refrigerators. Development of thermoelectric heating has also been undertaken.

SEEBECK EFFECT

In 1821 T. J. Seebeck discovered that when two different conductors are joined into a loop, and a temperature difference is maintained between the two junctions, an emf will be generated. Such a loop is called a thermocouple, and the emf generated is called a thermoelectric (or Seebeck) emf.

Measurements. The magnitude of the emf generated by a thermocouple is standardly measured by using the system shown in Fig. 1. Here the contact points between conductors A and B are called junctions. Each junction is maintained at a well-controlled temperature (either T_0 or T_1) by immersion in a bath or connection to a heat reservoir. This bath or reservoir is indicated by the dashed rectangles. From each junction, conductor A is brought to a measuring device, usually a potentiometer. When the potentimeter is balanced, no current flows, thereby allowing direct measurement of the open-circuit emf, undiminished by resistive losses and unperturbed by spurious effects arising from Joule heating or from Peltier heating and cooling at the junctions. This open-circuit emf is the thermoelectric emf.

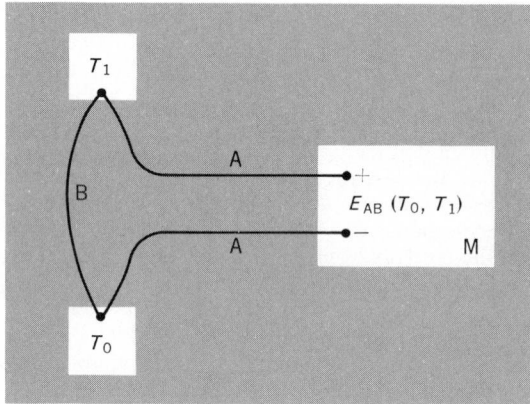

Fig. 1. Diagram of apparatus usually used for measuring thermoelectric (Seebeck) emf $E_{AB}(T_0, T_1)$. M is an instrument for measuring potential.

Equations. According to the experimentally established law of Magnus, for homogeneous conductors A and B the thermoelectric emf depends only upon the temperatures of the two junctions and not upon either the shapes of the samples or the detailed forms of the temperature distributions along them. This emf can thus be described by the symbol $E_{AB}(T_0,T_1)$. According to both theory and experiment, if one of the conductors, say B, is a superconductor in its superconducting state, it makes no contribution to E_{AB}. That is, when B is superconducting, $E_{AB}(T_0,T_1)$ is determined solely by conductor A and can be written as $E_A(T_0,T_1)$. It is convenient to express this emf in terms of a property which depends upon only a single temperature. Such a property is the absolute thermoelectric power (or, simply, thermopower) $S_A(T)$, defined so that Eq. (1) is valid. If $E_A(T,T+\Delta T)$ is known—for example, from measurements involving a superconductor—then $S_A(T)$ can be determined from Eq. (2). If Eq. (1) is valid for any homogeneous

$$E_A(T_0,T_1) = \int_{T_0}^{T_1} S_A(T)\, dt \qquad (1)$$

$$S_A(T) = \lim_{\Delta T \to 0} \frac{E_A(T,T+\Delta T)}{\Delta T} \qquad (2)$$

conductor, then it ought to apply to both sides of the thermocouple shown in Fig. 1. Indeed, it has been verified experimentally that the emf $E_{AB}(T_0,T_1)$ produced by a thermocouple is just the difference between the emfs, calculated using Eq. (1), produced by its two arms. This result can be derived as follows. Employing the usual sign convention, to calculate $E_{AB}(T_0,T_1)$, begin at the cooler bath, labeled T_0, integrate $S_A(T)dT$ along conductor A up to the warmer bath labeled T_1; and then return to T_0 along conductor B by integrating $S_B(T)dT$. This circular excursion produces $E_{AB}(T_0,T_1)$, given by Eq. (3). Inverting the last integral in Eq. (3) gives Eq. (4), which, from Eq. (1), can be rewritten as Eq. (5). Alternatively, combining the two integrals in Eq. (4) gives Eq. (6). Defining S_{AB} according to Eq. (7) then yields Eq. (8).

$$E_{AB}(T_0,T_1) = \int_{T_0}^{T_1} S_A(T)\, dT + \int_{T_1}^{T_0} S_B(T)\, dT \qquad (3)$$

$$E_{AB}(T_0,T_1) = \int_{T_0}^{T_1} S_A(T)\, dT - \int_{T_0}^{T_1} S_B(T)\, dT \qquad (4)$$

$$E_{AB}(T_0,T_1) = E_A(T_0,T_1) - E_B(T_0,T_1) \qquad (5)$$

$$E_{AB}(T_0,T_1) = \int_{T_0}^{T_1} [S_A(T) - S_B(T)]\, dT \qquad (6)$$

$$S_{AB}(T) = S_A(T) - S_B(T) \qquad (7)$$

$$E_{AB}(T_0,T_1) = \int_{T_0}^{T_1} S_{AB}(T)\, dT \qquad (8)$$

Equation (6) shows that $E_{AB}(T_0,T_1)$ can be calculated for a given thermocouple whenever the thermopowers $S_A(T)$ and $S_B(T)$ are known for its two constituents over the temperature range T_0 to T_1. By convention, the signs of $S_A(T)$ and $S_B(T)$ are chosen so that, if the temperature difference $T_1 - T_0$ is taken small enough so that $S_A(T)$ and $S_B(T)$ can be presumed constant, then $S_A(T) > S_B(T)$ when the emf $E_{AB}(T_0,T_1)$ has the polarity indicated in Fig. 1.

Results of equations. These equations lead directly to the following experimentally and theoretically verified results.

Uniform temperature. In a circuit kept at a uniform temperature throughout, $E = 0$ even though the circuit may consist of a number of different conductors. This follows directly from Eq. (8), since $dT = 0$ everywhere throughout the circuit. It follows also from thermodynamic reasoning. If E did not equal 0, the circuit could drive an electric motor and make it do work. But the only source of energy would be heat from the surroundings, which, by assumption, are at the same uniform temperature as the circuit. Thus, a contradiction with the second law of thermodynamics would result.

Homogeneous conductor. A circuit composed of a single, homogeneous conductor cannot produce a thermoelectric emf. This follows from Eq. (6) when $S_B(T)$ is set equal to $S_A(T)$. It is important to emphasize that, in this context, homogeneous means perfectly uniform throughout. A sample

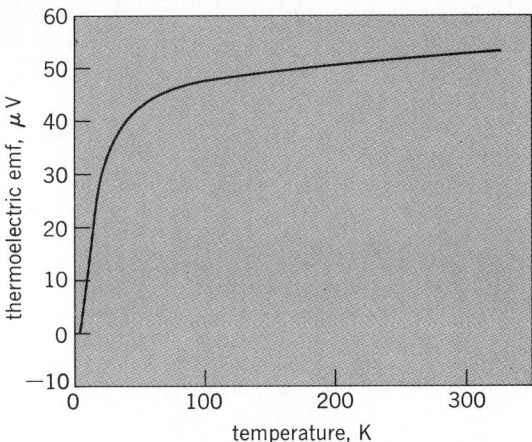

Fig. 2. The thermoelectric emf of a thermocouple formed from pure annealed and pure cold-worked copper. The cold junction reference temperature is 4.2 K. (*From R. H. Kropschot and F. J. Blatt, Thermoelectric power of cold-rolled pure copper, Phys. Rev., 116:617–620, 1959*)

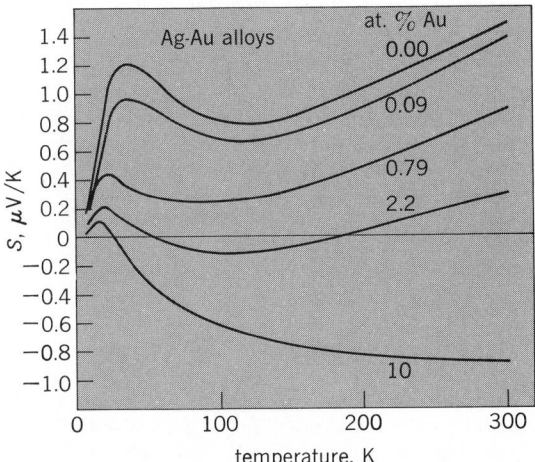

Fig. 3. The thermopower S from 0 to 300 K for pure silver (Ag) and a series of dilute silver-gold (Au) alloys. (*From R. S. Crisp and J. Rungis, Thermoelectric power and thermal conductivity in the silver-gold alloy system from 3–300°K, Phil. Mag., 22:217–236, 1970*)

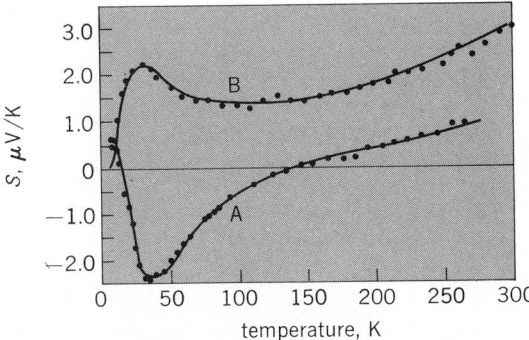

Fig. 4. The thermopower S of zinc (Zn) parallel (A) and perpendicular (B) to the hexagonal axis. (*From V. A. Rowe and P. A. Schroeder, Thermopower of Mg, Cd and Zn between 1.2° and 300°K, J. Phys. Chem. Sol., 31:1–8, 1970*)

made of an isotropic material can be inhomogeneous either because of small variations in chemical composition or because of strain. Figure 2 shows the thermoelectric emf generated by a thermocouple in which one arm is a cold-rolled copper (Cu) sample, and the other arm is the same material after annealing to remove the effects of the strain introduced by the cold-rolling. Figure 3 shows how the addition of impurities can change the thermopower of a pure metal. An additional effect can occur in a noncubic material. As illustrated in Fig. 4, two samples cut in different directions from a noncubic single crystal may be thermoelectrically different even if each sample is highly homogeneous. A thermocouple formed from these two samples will generate a thermoelectric emf. *See* CRYSTAL STRUCTURE.

If material B is superconducting, so that $S_B = 0$, Eq. (5) reduces to $E_{AB}(T_0, T_1) = E_A(T_0, T_1)$, as assumed above. *See* SUPERCONDUCTIVITY.

Source of emf. Finally, Eq. (6) makes it clear that the source of the thermoelectric emf in a thermocouple lies in the bodies of the two materials of which it is composed, rather than at the junctions. This serves to emphasize that thermoelectric emfs are not related to the contact potential or Volta effect, which is a potential difference across the junction between two different metals arising from the difference between their Fermi energies. The contact potential is present even in the absence of temperature gradients or electric currents.

PELTIER EFFECT

In 1834 J. C. A. Peltier discovered that, when an electric current passes through two different conductors connected in a loop, one of the two junctions between the conductors cools, and the other warms. If the direction of the current is reversed, the effect also reverses: the first junction warms, and the second cools. In 1853 Quintus Icilius showed that the rate of heat output or intake at each junction is directly proportional to the current i. The Peltier coefficient Π_{AB} is defined as the heat generated per second per unit current flow through the junction between materials A and B. By convention, Π_{AB} is taken to be positive when cooling occurs at the junction through which current flows from conductor A to conductor B. Quintus Icilius's result guarantees that the Peltier coefficient is independent of the magnitude of the current i. Additional experiments have shown that it is also independent of the shapes of the conductors. It therefore depends only upon the two materials and the temperature of the junction, and can be written as $\Pi_{AB}(T)$ or, alternatively, $\Pi_A(T) - \Pi_B(T)$, where Π_A and Π_B are the Peltier coefficients for materials A and B respectively. The second form emphasizes that the Peltier coefficient is a bulk property which can be defined for a single conductor.

Because of the small amount of heat transfer associated with the Peltier effect, as well as complications resulting from the simultaneous presence of Joule heating and the Thomson effect, $\Pi_{AB}(T)$ is usually difficult to measure accurately, and has therefore rarely been carefully studied. Rather, the value of the Peltier effect is usually determined from the Kelvin relations, using experimental values for S_{AB}.

THOMSON EFFECT AND KELVIN RELATIONS

When an electric current passes through a conductor which is maintained at a constant temperature, heat is generated at a rate proportional to the square of the current. This is called Joule heat, and its magnitude for any given material is determined by the electrical resistivity of the material. In 1854 William Thomson (Lord Kelvin), in an attempt to explain discrepancies between experimental results and a relationship between Π_{AB} and S_{AB} which he had derived from thermodynamic analysis of a thermocouple, postulated the existence of an additional reversible generation of heat when a temperature gradient is applied to a current-carrying conductor. This heat, called Thomson heat, is proportional to the product of the current and the temperature gradient. It is reversible, in the sense that the conductor changes from a generator of Thomson heat to an absorber of Thomson heat when the direction of either the current or the temperature gradient (but not both at once) is reversed. By contrast, Joule heating is irreversible, in that heat is generated for both directions of current flow.

The magnitude of Thomson heat generated (or absorbed) is determined by the Thomson coefficient μ. Using reasoning based upon equilibrium thermodynamics, Thomson derived results equivalent to Eqs. (9) and (10), called the Kelvin (or Kelvin-Onsager) relations.

$$\frac{\Pi_A}{T} = S_A \tag{9}$$

$$\frac{\mu_A}{T} = \frac{dS_A}{dT} \tag{10}$$

Here, μ_A is the Thomson coefficient, defined as the heat generated per second per unit current flow per unit temperature gradient when current flows through conductor A in the presence of a temperature gradient. Equation (10) can be integrated to give Eq. (11), in which the third law of thermodynamics has been invoked to set $S_A(0) = 0$.

$$S_A(T) = \int_0^T \frac{\mu_A(T')}{T'} dT' \tag{11}$$

thermodynamics has been invoked to set $S_A(0) = 0$. By using Eq. (11), $S_A(T)$ can be determined from measurements on a single conductor. In practice, however, accurate measurements of μ_A are very difficult to make; therefore, they have been carried out for only a few metals—most notably lead (Pb)—which then serve as standards for determining $S_B(T)$ by using measurements of $S_{AB}(T)$ in conjunction with Eq. (7).

Long after the Thomson heat was observed and the Kelvin relations were verified experimentally, debate raged over the validity of the derivation employed by Thomson. However, the theory of irreversible processes, developed by L. Onsager in 1931, and by others, yields the same equations and therefore provides them with a relatively firm foundation.

THERMOPOWERS OF METALS AND SEMICONDUCTORS

Since the Kelvin relations provide recipes for calculating any two of the thermoelectric coefficients, S, Π, and μ, from the third, only one of the three coefficients need be measured to determine the thermoelectric properties of any given material. Although there are some circumstances under which one of the other two coefficients may be preferred, because of ease and accuracy of measurement it is almost always the thermopower S which is measured.

Reference materials. Because S must be measured by using a thermocouple, the quantity determined experimentally is $S_A - S_B$, the difference between the thermopowers of the two conductors constituting the couple. Only when one of the arms of the thermocouple is superconducting, and therefore has zero thermopower, can the absolute thermopower of the other arm be directly measured. At temperatures up to about 18 K it is possible to use the superconducting alloy Nb_3Sn for conductor B and thereby determine S_A. For higher temperatures no superconducting wires are generally available. It is thus necessary to have a standard thermoelectric material to use above 18 K. For historical reasons, the reference material for temperatures up to 293 K has been chosen to be Pb. Based upon Thomson coefficient measurements made in the early 1930s, the thermopower of Pb has been calculated from Eq. (11) to have the values given in Table 1. All accepted values for S in this temperature range are ultimately traceable to this table. A redetermination of the Thomson coefficient of Pb has been undertaken. For temperatures above 293 K, no standard exists for which S is as well known as for Pb, but platinum (Pt) is often used because of its high melting temperature, resistance to chemical attack, and availability in high purity.

Temperature variation. Figure 5 shows the variation with temperature of the thermopowers of four different pure metals. The data for the three metals gold (Au), aluminum (Al), and platinum (Pt) are typical of those for most simple metals and for some transition metals as well. The thermopower S consists of a slowly varying portion which increases approximately linearly with absolute tem-

Table 1. The absolute thermoelectric power S of pure lead between 0 and 300 K*

T (K)	S (μV/K)†	T (K)	S (μV/K)†
0	0	60	-0.77_9
5	0	70	-0.78_4
7.5	-0.22_1	80	-0.79_4
8	-0.25_7	90	-0.82_4
8.5	-0.29_7	100	-0.86_5
9	-0.34_3	113.2	-0.91
10	-0.43_4	133.2	-0.96
11	-0.51_6	153.2	-1.02
12	-0.59_3	173.2	-1.06
14	-0.70_6	193.2	-1.10_5
16	-0.77_1	213.2	-1.15
18	-0.78_{45}	233.2	-1.18
20	-0.78_4	253.2	-1.21
30	-0.77_4	273.2	-1.25
40	-0.76_4	293.2	-1.27_5
50	-0.77_4		

*From J. W. Christian et al., Thermoelectricity at low temperatures: VI. A redetermination of the absolute scale of thermo-electric power of lead, *Proc. Roy. Soc.*, A245:213–221, 1958.

†Subscripts indicate figures which are uncertain.

perature, upon which a "hump" is superimposed at lower temperatures. In analyzing these results, S is written as the sum of two terms, as in Eq. (12), where S_d, called the electron-diffusion com-

$$S = S_d + S_g \qquad (12)$$

ponent, is the slowly varying portion, and S_g, called the phonon-drag component, is the hump. For some transition metals, on the other hand, the behavior of S is more complicated, as illustrated by the data for rhodium (Rh) in Fig. 5. Figure 6 shows comparable data for a simple p-type semiconductor, illustrating that the separation of S into S_d and S_g is still valid.

Theory. When a small temperature difference ΔT is established across a conductor, heat is carried from its hot end to its cold end by the flow of both electrons and phonons (quantized lattice vibrations). If the electron current is constrained to be zero, for example, by the insertion of a high resistance measuring device in series with the conductor, the electrons will redistribute themselves in space so as to produce an emf along the conduc-

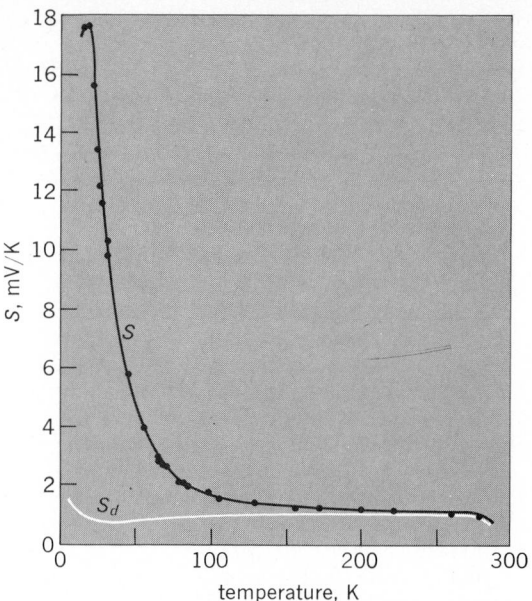

Fig. 6. The thermopower S of p-type germanium (Ge) (1.5×10^{14} acceptors per cubic centimeter) and calculated value for the electron-diffusion thermopower S_d. (*From C. Herring, The role of low-frequency phonons in thermoelectricity and thermal conductivity, Proc. Int. Coll. 1956, Garmisch-Partenkirchen, Vieweg. Braunschweig, p. 184, 1958*)

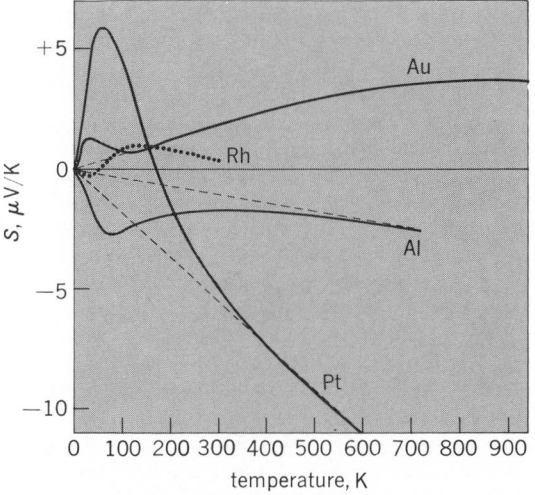

Fig. 5. The thermopower S of the metals gold (Au), aluminum (Al), platinum (Pt), and rhodium (Rh) as a function of temperature. The differences between the solid curves for Pt, Al, and Au and the broken lines indicate the magnitude of the phonon-drag component S_g.

tor. This is the thermoelectric emf. If the phonon flow could somehow be turned off, this emf would be just $S_d \Delta T$. However, the phonon flow cannot be turned off, and as the phonons move down the sample, they interact with the electrons and "drag" them along. This produces an additional contribution to the emf, $S_g \Delta T$. *See* CONDUCTION (HEAT); LATTICE VIBRATIONS; PHONON; THERMAL CONDUCTION IN SOLIDS.

Source of S_d. The conduction electrons in a metal are those having energies near the Fermi energy η. Only these electrons are important for thermoelectricity. As illustrated in Fig. 7, the energy distribution of these electrons varies with the temperature of the metal. When it is at high temperatures, a metal has more high-energy electrons, and less low-energy electrons, than when it is at low temperatures. This means that if a temperature gradient is established along a metal sample,

the total number of electrons will remain constant, but the hot end will have more high-energy electrons than the cold end, and the cold end will have more low-energy electrons. The high-energy electrons will diffuse toward the cold end, and the low-energy electrons will diffuse toward the hot end. However, in general, the diffusion rate is a function of electron energy, and thus a net electron current will result. This current will cause electrons to pile up at one end of the metal (usually the cold end) and thereby produce an emf which opposes the further flow of electrons. When the emf becomes large enough, the current will be reduced to zero.

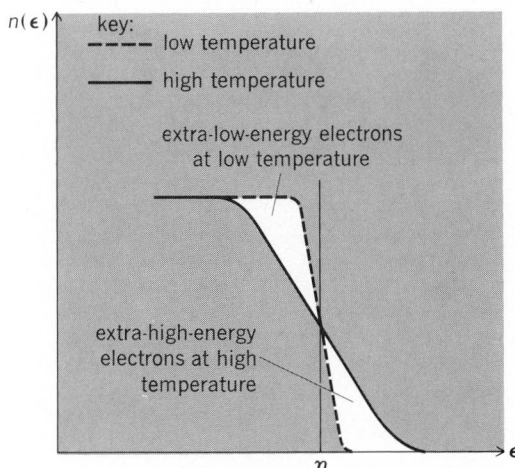

Fig. 7. The variation with energy ϵ of the number of conduction electrons $n(\epsilon)$ in a metal in the vicinity of the Fermi energy η for two different temperatures. A small variation of η with temperature has been neglected.

Table 2. Comparison between theoretical values for S and experimental data

	Thermopower S (μV/K)	
Metal	Theoretical values at 0°C according to Eq. (13)	Experimental data at approximately 0°C
Lithium (Li)	−2	+11
Sodium (Na)	−3	− 6
Potassium (K)	−5	−12
Copper (Cu)	−1.5	+ 1.4
Gold (Au)	−2	+ 1.7
Aluminum (Al)	−0.7	− 1.7

This is the thermoelectric emf arising from electron diffusion. Essentially the same argument applies also to semiconductors, except that in this case the conduction electrons (or holes) are those just above (or just below) the band gap. *See* Free-electron theory of metals; Holes in solids; Semiconductor.

S_d *for a metal.* For a completely free-electron metal, S_d would be given by Eq. (13), where k is

$$S_d = \frac{\pi^2}{2}\frac{k}{e}\left(\frac{kT}{\eta}\right) \qquad (13)$$

Boltzmann's constant, e is the charge on an electron, T is the absolute (Kelvin) temperature, and η is the Fermi energy of the metal. According to Eq. (13), S_d should be negative—since e is a negative quantity—and should increase linearly with T. In Table 2 the predictions of Eq. (13) are compared with experiment for a number of the most free-electron-like metals. Equation (13) is not very satisfactory, in several cases even predicting the wrong sign. To understand the thermopowers of real metals, it is necessary to use a more sophisticated

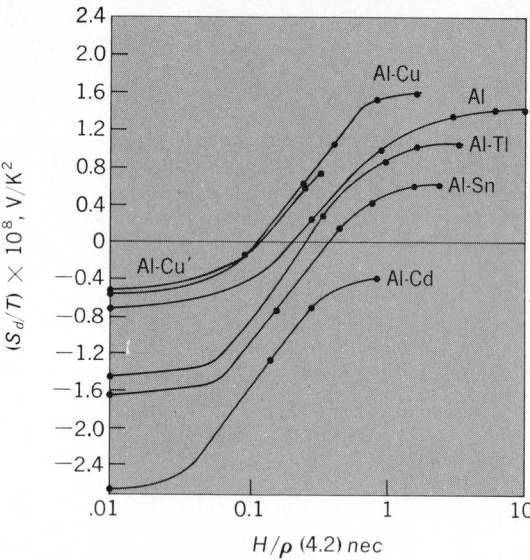

Fig. 9. The variation with magnetic field H of the low-temperature electron-diffusion thermopower S_d of aluminum (Al) and various dilute aluminum-based alloys. Sample labeled Al-Cu' is a second sample of Al-Cu. (*From R. S. Averback C. H. Stephan, and J. Bass, Magnetic field dependence of the thermopower of dilute aluminum alloys, J. Low Temp. Phys., 12:319–346, 1973*)

model which takes into account interactions between the electrons in the metal and the crystal lattice, as well as scattering of the electrons by impurities and phonons. The proper generalization of Eq. (13) is Eq. (14), where $\sigma(\epsilon)$ is a generalized

$$S_d = \frac{\pi^2 k^2 T}{3e}\left[\frac{\delta \ln \sigma(\epsilon)}{\delta \epsilon}\right]_\eta \qquad (14)$$

conductivity defined so that $\sigma(\eta)$ is the experimental electrical conductivity of the metal, and the logarithmic derivative with respect to the energy ϵ is to be evaluated at $\epsilon = \eta$. Equation (14) is able to account, at least in principle, for all the deviations of experiment from Eq. (13). If the logarithmic derivative is negative, S_d will be positive; S_d will differ in magnitude from Eq. (13) if the logarithmic derivative does not have the value $(3/2)\eta^{-1}$; and $S_d(T)$ will deviate from a linear dependence on T if the logarithmic derivative is temperature-dependent. Research interest in S_d in metals centers upon understanding changes in S_d resulting from alloying with both magnetic and nonmagnetic impurities, strain, application of pressure, and application of magnetic fields. In some cases the changes can be dramatic. Figure 8 shows that the addition of very small amounts of the magnetic impurity iron (Fe) can produce enormous changes in S_d for copper (Cu). Sample 1 (in which the deviation of the thermopower from zero is too small to be seen with the chosen scales) is most representative of pure copper because the iron is present as an oxide and is thus not in "magnetic form." Figure 9 shows that, at low temperatures, application of a magnetic field H to Al can cause S_d to change sign. (To obtain a temperature-independent quantity, S_d has been divided by the absolute temperature T. In order to remove the effects of varying impurity concentrations, H has been divided by $\rho(4.2)nec$, where $\rho(4.2)$ is the sample resistivity

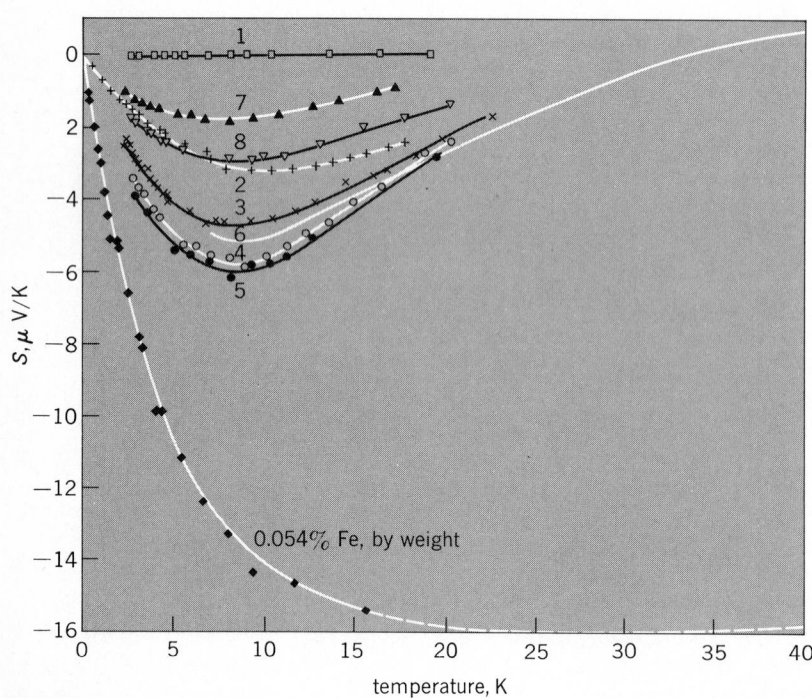

Fig. 8. The low-temperature thermopowers of various samples of copper (Cu) containing very small concentrations of iron (Fe). Specific compositions of samples 1–8 are unknown. (*From A. V. Gold et al., The thermoelectric power of pure copper, Phil. Mag., 5:765–786, 1960*)

at 4.2 K, n is the number of electrons per unit volume in the sample, and c is the speed of light.) Figure 10 illustrates the significant changes which occur in S when a metal melts. Substantial effort has been devoted to the study of thermoelectricity in liquid metals and liquid metal alloys.

S_d *for a semiconductor.* Equation (13) is appropriate for a free-electron gas which obeys Fermi-Dirac statistics. The conduction electrons in a metal obey these statistics. However, there are so few conduction electrons in a semiconductor that, to a good approximation, they can be treated as though they obey a different statistics—Maxwell-Boltzmann statistics. For electrons obeying these statistics, S_d is given by Eq. (15), which predicts

$$S_d = \frac{3}{2}\frac{k}{e} \qquad (15)$$

that S_d will be temperature-independent and will have the value $S_d = -130 \times 10^{-6}$ V/K. For a p-type extrinsic semiconductor, in which the carriers are approximated as free holes, S_d would be just the negative of this value. An examination of the data of Fig. 6 reveals that S_d is very nearly independent of temperature, but is considerably larger than predicted by Eq. (15). Again, a complete understanding of the thermopowers of semiconductors requires the generalization of Eq. (15). The appropriate generalizations are different for single-band and multiband semiconductors, the latter being considerably more complicated. For a single-band (extrinsic) semiconductor, the generalization is relatively straightforward and yields predictions for S_d which, in agreement with experiment, vary slowly with temperature and are several times larger than the prediction of Eq. (15). (The curve for S_d in Fig. 6 is calculated from this generalization.). Experimental interest in the thermopower of semiconductors concerns topics similar to those for metals. In addition, the large magnitudes of the thermopowers of semiconductors continue to spur efforts to develop materials better suited for electrical power generation and thermoelectric cooling. *See* BAND THEORY OF SOLIDS; BOLTZMANN STATISTICS; FERMI-DIRAC STATISTICS.

Source of S_g. Unlike the behavior of S_d, which is determined in both metals and semiconductors primarily by the properties of the charge carriers, the behavior of S_g is determined in both cases primarily by the properties of the phonons. At low temperatures, phonons scatter mainly from electrons or impurities rather than from other phonons. The increase in S_g with increasing temperature shown in Figs. 5 and 6 results from an increasing number of phonons being available to drag the electrons along. However, at higher temperatures the phonons begin to scatter more frequently from each other. At sufficiently high temperatures, phonon-phonon scattering becomes dominant, the electrons are no longer dragged along, and S_g falls off in magnitude. Interest in phonon drag is based on such questions as whether it is the sole source of the humps shown in Figs. 5 and 6, how it changes upon the addition of impurities, and how it varies in the presence of a magnetic field.

APPLICATIONS

The most important practical application of thermoelectric phenomena is in the accurate mea-

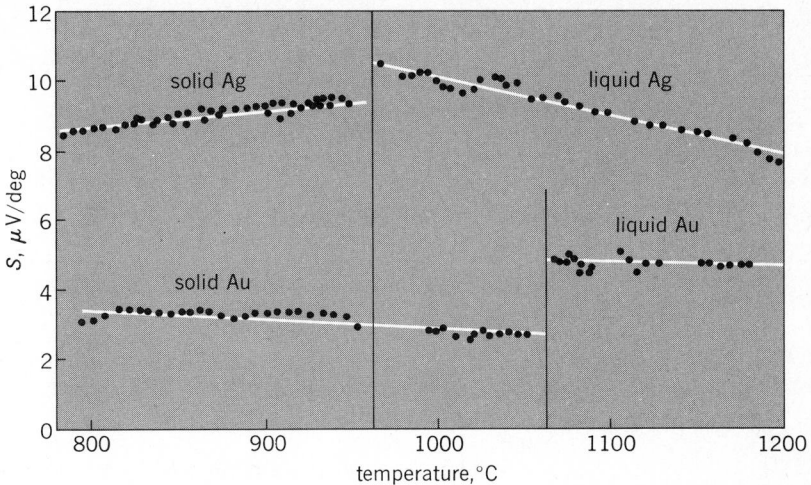

Fig. 10. The changes in the thermopowers of gold (Au) and silver (Ag) upon melting. (*From R. A. Howe and J. E. Enderby, The thermoelectric power of liquid Ag-Au, Phil. Mag., 16:467–476, 1967*)

surement of temperature. The phenomenon involved is the Seebeck effect. Of lesser importance are the direct generation of electric power by application of heat (also involving the Seebeck effect) and thermoelectric cooling and heating (involving the Peltier effect).

A basic system suitable for all four applications is illustrated schematically in Fig. 11. Several thermocouples are connected in series to form a thermopile, a device with increased output (for power generation or cooling and heating) or sensitivity (for temperature measurement) relative to a single thermocouple. The junctions forming one end of the thermopile are all at the same low temperature T_L, and the junctions forming the other end are at the high temperature T_H. The thermopile is connected to a device D which is different for each application. For temperature measurement, the temperature T_L is fixed, for example, by means of a bath; the temperature T_H becomes the running temperature T, which is to be measured; and the device is a potentiometer for measuring the thermoelectric emf generated by the thermopile. For

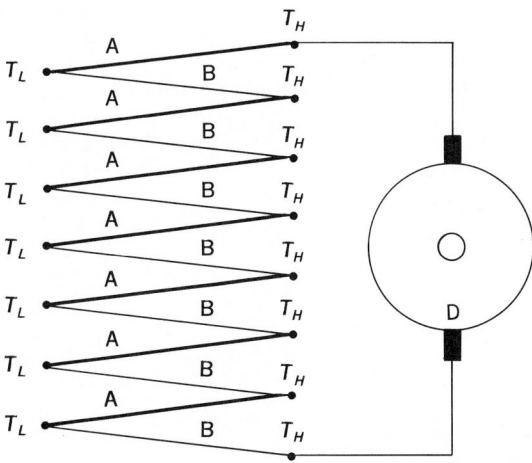

Fig. 11. Thermopile, a battery of thermocouples connected in series; *D* is a device appropriate to the particular application.

power generation, the temperature T_L is fixed by connection to a heat sink; the temperature T_H is fixed at a value determined by the output of the heat source and the thermal conductivity of the thermopile; and the device is whatever is to be run by the electricity which is generated. For heating or cooling, the device is a current generator which passes current through the thermopile. If the current flows in the proper direction, the junctions at T_H will heat up, and those at T_L will cool down. If T_H is fixed by connection to a heat sink, thermoelectric cooling will be provided at T_L. Alternatively, if T_L is fixed, thermoelectric heating will be provided at T_H. Such a system has the advantage that at any given location it can be converted from a cooler to a heater merely by reversing the direction of the current.

Temperature measurement. In principle, any material property which varies with temperature can serve as the basis for a thermometer. In practice, the two properties most often used for precision thermometry are electrical resistance and thermoelectric emf. Thermocouples are widely employed to measure temperature in both scientific research and industrial processes. In the United States alone, several hundred tons of thermocouple materials are produced annually.

Construction of instruments. In spite of their smaller thermopowers, metals are usually preferred to semiconductors for precision temperature measurement because they are cheaper, are easier to fabricate into convenient forms such as thin wires, and have more reproducible thermoelectric properties. With modern potentiometric systems, standard metallic thermocouples provide temperature sensitivities adequate for most needs; small fractions of a degree Celsius are routinely obtained. If greater sensitivity is required, several thermocouples can be connected in series to form a thermopile (Fig. 11). A 10-element thermopile provides a temperature sensitivity 10 times as great as that of each of its constituent thermocouples. However, the effects of any inhomogeneities are also enhanced 10 times.

The thermocouple system standardly used to measure temperature is shown in Fig. 12. It consists of wires of three metals A, B, and C, where C is usually the metal copper. The junction between the wires of metals A and B is located at the temperature to be measured T. Each of these two wires is joined to a wire of metal C at the reference temperature T_0. The other ends of the two wires of metal C are connected to the potentiometer at room temperature T_r. Integrating the appropriate thermopowers around the circuit of Fig. 12 yields the total thermoelectric emf E in terms of the separate emfs generated by each of the four pieces of wire, as given in Eq. (16).

$$E = E_A(T_0,T_1) - E_B(T_0,T_1)$$
$$+ E_{C1}(T_0,T_r) - E_{C2}(T_0,T_r) \quad (16)$$

If the two wires C_1 and C_2 have identical thermoelectric characteristics, the last two terms in this expression cancel, and, with the use of Eq. (5), Eq. (17) results. That is, two matched pieces of metal C

$$E = E_{AB}(T_0,T_1) \quad (17)$$

produce no contribution to the thermoelectric emf of the circuit shown in Fig. 12, provided their ends

are maintained at exactly the same two temperatures. This means that it is not necessary to use either of the sometimes expensive metals making up the thermocouple to go from the reference-temperature bath to the potentiometer. This portion of the circuit can be constructed of any uniform, homogeneous metal. Copper is often used because it is inexpensive, is available in adequate purity to ensure uniform, homogeneous samples when handled with care, can be obtained in a wide variety of wire diameters, and can be either spot-welded or soldered to the ends of the thermocouple wires. Special low-thermal emf alloys are available for making solder connections in thermocouple circuits.

Choice of materials. Characteristics which make a thermocouple suitable as a general-purpose thermometer include adequate sensitivity over a wide temperature range, stability against physical and chemical change under various conditions of use and over extended periods of time, availability in a selection of wire diameters, and moderate cost. No single pair of thermocouple materials satisfies all needs. Platinum versus platinum–10% rhodium can be used up to 1700°C. A combination of the two alloys chromel versus alumel gives greater sensitivity and an emf which is very closely linear with temperature, but this thermocouple cannot be used to so high a temperature. A combination of copper versus the alloy constantan also has high sensitivity above room temperature, and maintains adequate sensitivity down to as low as 15 K. For temperatures of 4 K or lower, special gold-cobalt alloys versus copper or gold-iron alloys versus chromel are used.

Thermocouple tables. To use a thermocouple composed of metals A and B as a thermometer, it is necessary to know how $E_{AB}(T_0,T_1)$ varies with temperature T for some reference temperature T_0. According to Eq. (6), $E_{AB}(T_0,T_1)$ can be determined for any two temperatures T_0 and T_1 if both $S_A(T)$ and $S_B(T)$ are known for all temperatures between T_0 and T_1. Once $S_A(T)$ and $S_B(T)$ are known, it is possible to construct a table of values for $E_{AB}(T_0,T)$ using any arbitrary reference temperature T_0. Such tables are available for the thermocouples mentioned above, and for some others as well, usually with a reference temperature of 0°C. A table of $E_{AB}(T_0,T)$ for one reference temperature T_0 can be converted into a table for any other reference temperature T_1 merely by subtracting a constant value $E_{AB}(T_0,T_1)$ from each entry in the table to give Eq. (18). Here $E_{AB}(T_0,T_1)$ is a positive

$$E_{AB}(T_1,T) = E_{AB}(T_0,T) - E_{AB}(T_0,T_1) \quad (18)$$

quantity when T_1 is greater than T_0 and when $S_{AB}(T)$ is positive between T_0 and T_1.

Other uses. Thermoelectric systems made from even the best available materials have the disadvantages of relatively low efficiencies and concomitant high cost per unit of output. Their use in power generation, heating, and cooling has therefore been largely restricted to situations in which these disadvantages are outweighed by such advantages as small size, low maintenance due to lack of moving parts, quiet performance, light weight, and long life.

Figure of merit. A measure of the utility of a given thermoelectric material for power generation,

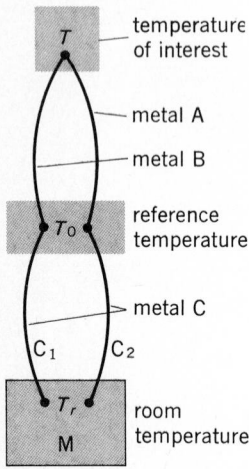

THERMOELECTRICITY

Fig. 12. The thermocouple system standardly used to measure temperature; M is a measuring device, usually a potentiometer, which is at room temperature.

cooling, or heating at a given temperature T is provided by a dimensionless parameter zT, where z is called the figure of merit. The dimensionless parameter zT is given by Eq. (19), where S is the

$$zT = \frac{S^2 \sigma T}{\kappa} \qquad (19)$$

thermopower of the material, σ is its electrical conductivity, and κ is its thermal conductivity. The largest values for zT are attained in semimetals and highly doped semiconductors, which are therefore the materials normally used in practical thermoelectric devices. As illustrated in Fig. 13, for most materials z varies substantially with temperature. The best available thermoelectric materials, such as lead-telluride (Pb-Te) and bismuth-telluride (Bi-Te), have values of z as large as 2 to 4×10^{-3} K^{-1} at their best temperatures. Unfortunately, however, z falls off substantially at both higher and lower temperatures. Combining these materials into thermocouples therefore results in values of z which average less than 1×10^{-3} K^{-1} over a temperature range sufficiently wide to be useful. Such values of z yield conversion efficiencies of only a few percent. A commercially competitive efficiency of about 30% would require a constant value of $z = 5 \times 10^{-3}$ K^{-1} over the temperature range of 300–1000 K.

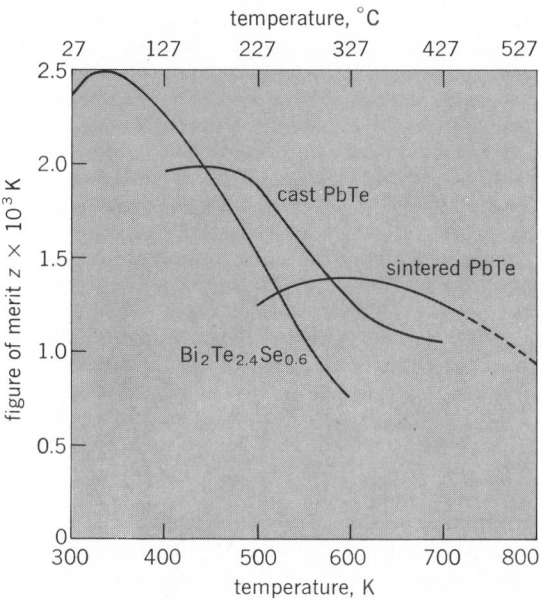

Fig. 13. Temperature variation of the figure of merit for some *n*-type semiconductors. (*From R. R. Heikes and R. W. Ure, Thermoelectricity: Science and Engineering, p. 538, Interscience Publishers, 1961*)

Just as the figures of merit for single materials vary with temperature, so do the figures of merit for thermocouples formed from two such materials. This means that one thermocouple can be better than another at one temperature but less effective at a second temperature. To take maximum advantage of the different properties of different couples, thermocouples are often cascaded as shown in Fig. 14. Cascading produces power generation in stages, the higher temperature of each

stage being determined by the heat rejected from the stage above. Thus, in Fig. 14 the highest and lowest temperatures T_4 and T_1 are fixed by connection to external reservoirs, whereas the middle temperatures T_3 and T_2 are determined by the properties of the materials. By cascading, a series of thermocouples can be used simultaneously in the temperature ranges where their figures of merit are highest. Cascaded thermocouple systems have achieved conversion efficiencies as high as 10–15%.

The quantity of importance in power generation is the figure of merit of a thermocouple rather than the separate figures of merit of its constituents. Although at least one constituent should have a high figure of merit, two constituents with high figures of merit do not necessarily guarantee that the figure of merit of the thermocouple will be high. For example, if the thermopowers of the two constituents are the same, the figure of merit of the couple will be zero.

Thermoelectric generators. A thermoelectric generator requires a heat source and a thermocouple. Kerosine lamps and firewood have been used as heat sources in producing a few watts of electricity in locations where electricity was otherwise unavailable. In the future, sunlight may also be used. Radioactive sources, especially strontium-90, have provided the heat to activate small, rugged thermoelectric batteries for use in lighthouses, in navigation buoys, at isolated weather stations or oil platforms, in spaceships, and in heart pacemakers. Small nuclear batteries have been operating pacemakers implanted in humans since 1970. One such battery, powered by Pu[238] and using a bismuth telluride thermopile module, supplies a few tenths of a volt over a design lifetime of more than 10 years. Nuclear-powered batteries for medical use must be designed to remain intact following the maximum credible accident. Capabilities such as retention of integrity after crushing by 1 ton, or impact at 50 m/s, or saltwater corrosion for centuries, or cremation at temperatures up to 1300°C for half an hour are required. Investigation of the feasibility of thermoelectric generation using the copious heat generated by nuclear reactors has also been undertaken. Here, one major problem lies in the development of efficient thermoelectric materials capable of operating for a long time at the high temperatures which are encountered. *See* RADIOACTIVITY.

Peltier cooling. With available materials, thermoelectric refrigerators suitable for use in homes are more expensive and less efficient than standard vapor-compression-cycle refrigerators. Their use is thus largely restricted to situations in which lower maintenance, increased life, or quiet performance are essential, or in situations (such as in space vehicles or artificial satellites) in which the compressor type of refrigerator is impractical. A number are in use in hotels and other large facilities. A typical unit having about 50-liter capacity requires a dc power input of 40 W, has a refrigerative capacity of 20 kcal/hr (23 W), and a cooling time of 4–5 hr. *See* REFRIGERATION.

For lower temperatures, the proper choice of thermoelectric materials and the use of cascading can result in a reduction in temperature at the coldest junctions of as much as 150°C. Temperature drops of 100°C have been obtained in single

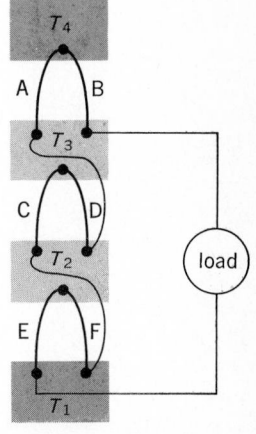

Fig. 14. Three-level cascade consisting of three different thermocouples (A versus B, C versus D, and E versus F) at four temperatures (*T*).

crystals of the semimetal bismuth through use of the thermomagnetic Ettingshausen effect. *See* THERMOMAGNETIC EFFECTS.

Small cooling units with capacities of 10 W or less have been developed for miscellaneous applications such as cold traps for vacuum systems, cooling controls for thermocouple reference junctions, cooling devices for scientific equipment such as infrared detectors, and cold stages on microscopes or on microtomes used for sectioning cooled tissues. However, the real commercial success of thermoelectric refrigeration appears to await development of thermocouple materials with higher figures of merit.

Thermoelectric heating. As noted earlier, a thermoelectric heater is nothing more than a thermoelectric refrigerator with the current reversed. No large heaters have been marketed. However, various small household convenience devices have been developed, such as a baby-bottle cooler-warmer which cools the bottle until just before feeding time and then automatically switches to a heating cycle to warm it, and a thermoelectric hostess cart. *See* ELECTRICITY.   [JACK BASS]

Bibliography: American Institute of Physics, *Temperature, Its Measurement and Control in Science and Industry,* vol. 1, 1941, and vol. 2, 1955; R. D. Barnard, *Thermoelectricity in Metals and Alloys,* 1972; F. J. Blatt et al., *Thermoelectric Power of Metals,* 1976; F. J. Blatt and P. A. Schroeder (eds.), *Thermoelectricity in Metallic Conductors,* 1978; T. C. Harman and H. M. Honig, *Thermoelectric and Thermomagnetic Effects and Applications,* 1967; R. R. Heikes and R. W. Ure, Jr., *Thermoelectricity: Science and Engineering,* 1961; D. K. C. MacDonald, *Thermoelectricity: An Introduction to the Principles,* 1962; A. C. Smith, J. F. Janak, and R. B. Adler, *Electronic Conduction in Solids,* 1967.

Thermoluminescence

A term sometimes used broadly to mean any luminescence appearing in a material due to the application of heat. More frequently the term refers specifically to the luminescence appearing as the temperature of the material is steadily increased.

Many solids that contain luminescent centers often also contain one or more types of centers that can trap electrons or holes. If the solid is exposed to light of sufficiently short wavelength or to x-rays or other high-energy radiation, free electrons and holes are produced in the solid, and some of these charge carriers may be trapped. Very commonly the hole is trapped at the luminescent center itself or at some trapping site nearby while the electron is trapped at a different site, but the reverse situation is also quite possible. If, in the former case, the depth of the electron trap (that is, the amount of energy required to release the electron from the trap) is large and the temperature low, the electron will remain trapped for a long time. If, however, the temperature of the sample is raised slowly, the electron will receive increasing amounts of thermal energy and will eventually escape the trap. An electron thus freed from a trap may go over to a luminescent center and recombine with the hole trapped at or near the center. The energy liberated by the recombination excites the center, causing it to luminesce. For a single type of trap, the glow curve (plot of thermoluminescence intensity as a

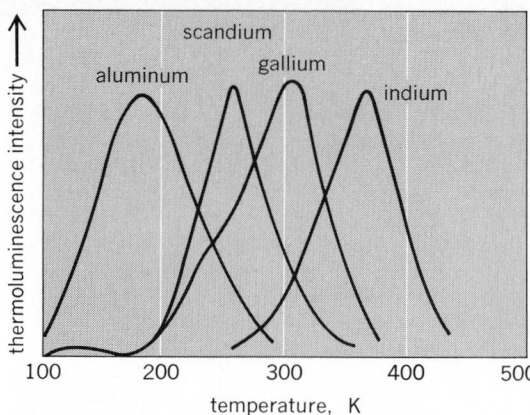

Glow curves for several zinc sulfide phosphors, each of which contains traces of copper and different trivalent ions. Luminescent center is due to presence of copper, the activator in each case. Traps are put in by various trivalent coactivators as shown.

function of temperature) first rises, reaches a maximum, and then decreases to zero when all the traps become emptied. The depth of the trap E (in ergs or other energy units) is given to a good approximation by the equation below, where k is

$$E = 1.51\, kT^*T'/(T^* - T')$$

Boltzmann's constant, T^* is the temperature (in kelvins) of the solid at the peak of the curve, and T' is the temperature on the low-temperature side at which the emission is one-half its peak value. The illustration shows glow curves for several zinc sulfide phosphors. The incorporation of impurities (activators, coactivators) into "host" compounds, such as zinc sulfide, is one of the most important processes for preparing solid materials. When more than one type of trap is present, the glow curve consists of a corresponding number of peaks, which often may be resolved and analyzed as described. Thus thermoluminescence may be used to secure information about the properties of traps in solids. *See* HOLES IN SOLIDS; LUMINESCENCE; TRAPS IN SOLIDS.

[JAMES H. SCHULMAN; CLIFFORD C. KLICK]

Thermomagnetic effects

Electrical and thermal phenomena occurring when a conductor or semiconductor which is carrying a thermal current (that is, is in a temperature gradient) is placed in a magnetic field. *See* SEMICONDUCTOR.

Let the temperature gradient be transverse to the magnetic field H_z, for example, along x. Then the following transverse-transverse effects are observed:

1. Ettingshausen-Nernst effect, an electric field along y, as in Eq. (1), where Q is known as the

$$E_y = Q(\partial T/\partial x)H_z \qquad (1)$$

Ettingshausen-Nernst coefficient. This coefficient is related to the Ettingshausen coefficient P by Eq. (2), where σ is the thermal conductivity in a trans-

$$P = QT\sigma \qquad (2)$$

verse magnetic field. This relation was discovered by P. W. Bridgman; it has been shown to be an example of the Onsager reciprocity relations of irreversible thermodynamics. *See* GALVANOMAGNETIC EFFECTS; THERMOELECTRICITY.

2. Righi-Leduc effect, a temperature gradient along y, as in Eq. (3), where S is known as the Righi-Leduc coefficient.

$$(\partial T/\partial y) = S(\partial T/\partial x)H_z \qquad (3)$$

Also, the following transverse-longitudinal effects are observed:

3. An electric potential change along x, amounting to a change of thermoelectric power.

4. A temperature gradient change along x, amounting to a change of thermal resistance.

Let the temperature gradient be along H. Then changes in thermoelectric power and in thermal conductivity are observed in the direction of H.

For related phenomena *see* HALL EFFECT; MAGNETORESISTANCE.

[ELIHU ABRAHAMS; FREDERIC KEFFER]
Bibliography: *See* GALVANOMAGNETIC EFFECTS.

Thermonuclear reaction

A nuclear fusion reaction which occurs between various nuclei of the light elements when they are constituents of a gas at very high temperatures. Thermonuclear reactions, the source of energy generation in the Sun and the stable stars, are utilized in the fusion bomb. *See* NUCLEAR FUSION.

Thermonuclear reactions occur most readily between isotopes of hydrogen (deuterium and tritium) and less readily among a few other nuclei of higher atomic number. At the temperatures and densities required to produce an appreciable rate of thermonuclear reactions, all matter is completely ionized; that is, it exists only in the plasma state. Thermonuclear fusion reactions may then occur within such an ionized gas when the agitation energy of the stripped nuclei is sufficient to overcome their mutual electrostatic repulsions, allowing the colliding nuclei to approach each other closely enough to react. For this reason, reactions tend to occur much more readily between energy-rich nuclei of low atomic number (small charge) and particularly between those nuclei of the hot gas which have the greatest relative kinetic energy. This latter fact leads to the result that, at the lower fringe of temperatures where thermonuclear reactions may take place, the rate of reactions varies exceedingly rapidly with temperature. *See* PLASMA PHYSICS.

The reaction rate may be calculated as follows: Consider a hot gas composed of a mixture of two energy-rich nuclei, for example, tritons and deuterons. The rate of reactions will be proportional to the rate of mutual collisions between the nuclei. This will in turn be proportional to the product of their individual particle densities. It will also be proportional to their mutual reaction cross section σ and relative velocity v. Thus Eq. (1) gives the

$$R_{12} = n_1 n_2 \langle \sigma v \rangle_{12} \text{ reactions/(cm}^3)(\text{sec}) \qquad (1)$$

rate of reaction. The quantity $\langle \sigma v \rangle_{12}$ indicates an average value of σ and v obtained by integration of these quantities over the velocity distribution of the nuclei (usually assumed to be maxwellian). Since the total density $n = n_1 + n_2$, then if the relative proportions of n_1 and n_2 are maintained, R_{12} varies as the square of the total nuclear particle density.

The thermonuclear energy release per unit volume is proportional to the reaction rate and the energy release per reaction, as in Eq. (2).

$$P_{12} = R_{12}W_{12}\text{ergs/(cm}^3)(\text{sec}) \qquad (2)$$

If this energy release, on the average, exceeds the energy losses from the system, the reaction can become self perpetuating. *See* KINETIC THEORY OF MATTER; MAGNETOHYDRODYNAMICS; NUCLEAR REACTION; PINCH EFFECT.

[RICHARD F. POST]
Bibliography: S. Glasstone and R. H. Lovberg, *Controlled Thermonuclear Reactions*, 1960, reprint 1975; K. Miyamoto, *Plasma Physics for Nuclear Fusion*, 1979; G. Schmidt, *Physics of High-Temperature Plasmas*, 2d ed., 1979.

Thomson effect

A phenomenon discovered in 1854 by William Thomson, later Lord Kelvin. He found that there occurs a reversible transverse heat flow into or out of a conductor of a particular metal, the direction depending upon whether a longitudinal electric current flows from colder to warmer metal or from warmer to colder. Any temperature gradient previously existing in the conductor is thus modified if a current is turned on. The Thomson effect does not occur in a current-carrying conductor which is initially at uniform temperature.

From these observations it may be shown that for copper there is a heat output where positive charge flows down a temperature gradient and a heat input where positive charge flows up a temperature gradient; whereas for iron the reverse is true. All metals may be divided into two classes with respect to the direction of the Thomson effect. These flows of heat require that a distributed seat of electromotive force act at all points in the conductor. The total Thomson emf along the length of a conductor is given by $\int_{T_1}^{T_2}\sigma\,dT$, where σ is the Thomson coefficient for the metal in question, and T_1 and T_2 are the temperatures at the two ends of the conductor. With the discovery of the Thomson effect a complete thermodynamical theory of thermoelectricity became possible. *See* THERMOELECTRICITY.   [JOHN W. STEWART]

Throttled flow

Flow which is forced to pass through a restricted area, where the velocity must be increased. This is also known as choked flow. Most of the kinetic energy produced by reduction of pressure in passing through the constriction is generally converted into thermal energy by turbulent eddying. The net result is a loss in mechanical energy in the system. When a gas is throttled, as by a globe valve, the velocity a short distance downstream from the valve is only a little higher than before the throttling section in most cases, the process being one of constant enthalpy. By introducing mechanical energy losses into a flow system by a throttling valve, the amount of flow may be controlled. *See* ISENTROPIC PROCESS.

A special throttling effect is produced when gas flows through a constriction, such as a nozzle, at sonic velocity. When this occurs, further reduction of downstream pressure does not alter upstream conditions and the flow remains constant.

[VICTOR L. STREETER]

Time-of-flight spectrometers

A general class of instruments in which the speed of a particle is determined directly by measuring the time that it takes to travel a measured distance. By knowing the particle's mass, its energy can be calculated. If the particles are uncharged (for example, neutrons), difficulties arise because standard methods of measurement (such as deflection in electric and magnetic fields) are not possible. The time-of-flight method is a powerful alternative, suitable for both uncharged and charged particles, that involves the measurement of the time t that a particle takes to travel a distance l. If the rest mass of the particle is m_0, its kinetic energy E_T can be calculated from its measured speed, $v = l/t$, using the equation below, where c is the speed of light.

$$E_T = m_0 c^2 \{ [1 - (v/c)^2]^{-1/2} - 1 \}$$

$$\approx m_0 v^2 / 2 \qquad \text{if } v \ll c$$

Some idea of the time scales involved in measuring the energies of nuclear particles can be gained by noting that a slow neutron of kinetic energy $E_T = 1$ eV takes 72.3 μs to travel 1 m. Its flight time along a 10-m path (typical of those found in practice) is therefore 723 μs, whereas a 4-MeV neutron takes only 361.5 ns.

The time intervals are best measured by counting the number of oscillations of a stable oscillator that occur between the instants that the particle begins and ends its journey (see illustration). Oscillators operating at 100 MHz are in common use. If the particles from a pulsed source have different energies, those with the highest energies arrive at the detector first. Digital information from the "gated" oscillator consists of a series of pulses whose number $N(t)$ is proportional to the time-of-flight t. These pulses can be counted and stored in an on-line computer that provides many thousands of sequential "time channels," t_0, $t_0 + \Delta t$, $t_0 + 2\Delta t$, $t_0 + 3\Delta t$, . . . , where t_0 is the time at which the particles are produced and Δt is the period of the oscillator. To store an event in channel $N(t)$, the contents of memory address $N(t)$ are updated by "adding 1."

Time-of-flight spectrometers have been used increasingly for energy measurements of uncharged and charged elementary particles, electrons, atoms, and molecules. Their popularity is due to the broad energy range that can be covered, their high resolution ($\Delta E_T / E_T \approx 2\Delta t / t$, where ΔE_T and Δt are the uncertainties in the energy and time measurements, respectively), their adaptability for studying different kinds of particles, and their relative simplicity. See NEUTRON SPECTROMETRY.

[FRANK W. K. FIRK]

Bibliography: J. A. Harvey (ed.), *Experimental Neutron Resonance Spectroscopy*, 1970.

Topological dynamics

The study and application of topological transformation groups. Topological dynamics originated in the late-19th-century investigations of the qualitative behavior of the solutions to the differential equations of classical mechanics by Henri Poincaré. The emergence of topological dynamics as a mathematical discipline occurred in the 20th century with an abstract formulation of certain qualitative features of the classical systems by G. D. Birkhoff.

Topological transformation groups. A topological transformation group, or simply transformation group, is a triple (G,X,π), where G is a topological group called the phase group, X a topological space called the phase space, and π a continuous mapping of $G \times X$ onto X satisfying the homomorphism rule $\pi(gh,x) = \pi(g,\pi(h,x))$, $g,h \in G$, $x \in X$. The set $\{\pi(g,x)|g \in G\}$ is called the orbit through x. When G is the group $\mathbf{R}$ of real numbers, the transformation group is called a flow; when G is the group $\mathbf{Z}$ of integers, the transformation group is called a cascade. See ABSTRACT ALGEBRA; GROUP THEORY; TOPOLOGY.

Homomorphism and isomorphism. Let (G,X,π) and (G,X',π') be transformation groups. A continuous mapping φ of X onto X' is called a homomorphism if it is equivariant, that is $\varphi(\pi(g,x)) = \pi'(g,\varphi(x))$. If φ is one-to-one with a continuous inverse, it is called an isomorphism.

Invariant measure. The Borel subsets of a topological space are the elements of the smallest σ-algebra of subsets of the space containing the open sets; a Borel measure is a measure on the Borel sets. A Borel measure μ is invariant for a transformation group (G,X,π) if for all Borel sets A, $\mu(A_g) = \mu(A)$, where $A_g = \{\pi(g,x)|x \in A\}$. The use of invariant measures in the study of dynamical systems is original with Poincaré.

Classical dynamical systems. A classical (autonomous) dynamical system is a first-order system of ordinary differential equations, Eq. (1).

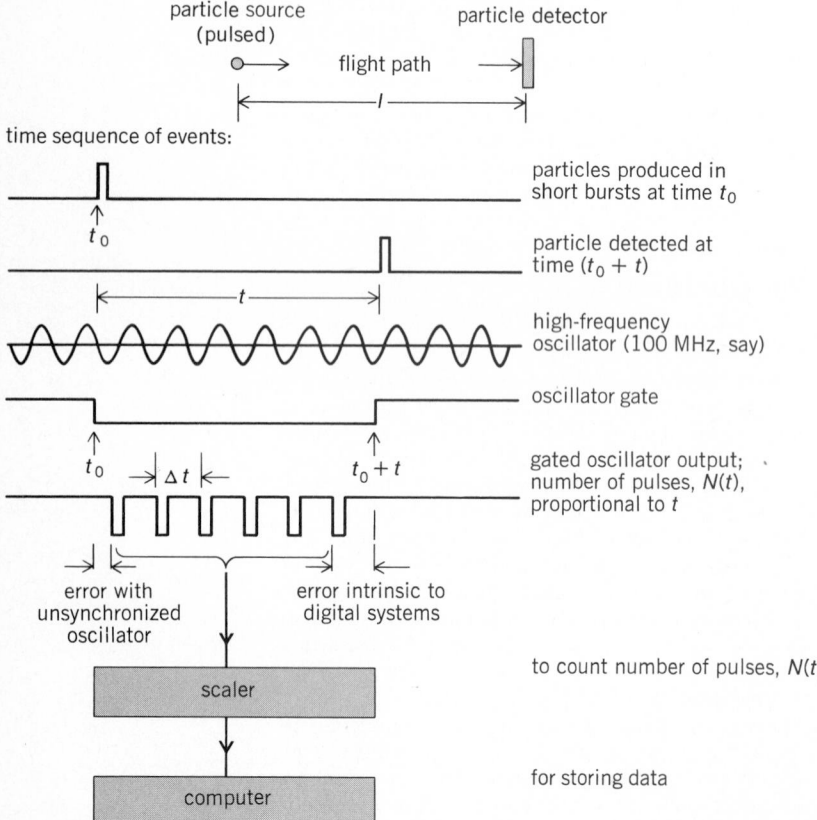

particle source (pulsed)　　　　particle detector

flight path

l

time sequence of events:

particles produced in short bursts at time t_0

t_0

particle detected at time $(t_0 + t)$

t

high-frequency oscillator (100 MHz, say)

oscillator gate

t_0　　Δt　　$t_0 + t$

gated oscillator output; number of pulses, $N(t)$, proportional to t

error with unsynchronized oscillator　　error intrinsic to digital systems

to count number of pulses, $N(t)$

scaler

for storing data

computer

A schematic diagram of a time-of-flight spectrometer.

$$\frac{dx_i}{dt} = F_i(x_1, \ldots, x_n) \qquad (1 \leq i \leq n) \qquad (1)$$

The functions F_i, defined on an open set X in n-dimensional space ($\mathbf{R}^n$), are assumed to be such that (a) for each $y \in X$ there is a unique solution to (1), $x(t)$, $-\infty < t < \infty$, with $x(0) = y$; and (b) the solution in (a) varies continuously in y, t. A flow $(\mathbf{R}, X, \pi)$ may be associated to Eq. (1) by defining $\pi(t, y)$ to be the value at time t of the solution in (a). The homomorphism rule $\pi(s + t, y) = \pi(s, \pi(t, y))$ is a consequence of the uniqueness statement in (a). *See* DIFFERENTIAL EQUATION.

Hamiltonian systems. Let $X = U \times \mathbf{R}^m$, where U is an open subset of $\mathbf{R}^m$. Denote coordinates in X by $(q, p) = (q_1, \ldots, q_m, p_1, \ldots, p_m)$. A continuously differentiable (C^1) function H on X determines a Hamiltonian system, given by Eqs. (2). In

$$\frac{dq_i}{dt} = \frac{\partial H}{\partial p_i}$$
$$\qquad (1 \leq i \leq m) \qquad (2)$$
$$\frac{dp_i}{dt} = -\frac{\partial H}{\partial q_i}$$

terms of the Lagrange equations for a conservative system with m degrees of freedom, $q_1, \ldots, q_m$ represent position coordinates, $p_1, \ldots, p_m$ represent impulses or momenta, and $H = T + V$ is total energy (kinetic plus potential). H is called the Hamiltonian of Eq. (2). *See* HAMILTON'S EQUATIONS OF MOTION; LAGRANGE'S EQUATIONS.

Liouville's theorem. Assume the functions F_i in Eq. (1) are C^1 on X. Lebesgue measure on X is an invariant measure for the associated flow if and only if the vector $F = (F_1, \ldots, F_n)$ has divergence 0, Eq. (3). An immediate consequence of

$$\operatorname{div} F = \sum_{j=1}^{n} \frac{\partial F_j}{\partial x_j} = 0 \qquad (3)$$

this theorem is that the flow associated to a Hamiltonian system with a C^2 Hamiltonian function preserves Lebesgue measure.

Poincaré recurrence theorem. This theorem, probably the first theorem in "ergodic theory," asserts that if a flow has a finite invariant measure μ, then with respect to μ almost every point $x \in X$ is Poisson-stable; for any neighborhood U of x there exist arbitrarily large (positive and negative) values of t such that $\pi(t, x) \in U$. Every Poisson-stable point belongs to the nonwandering set, the closed flow-invariant set of points x such that for any neighborhood U, $U \cap U_t \neq \phi$ for arbitrarily large values of t. In the case of Eq. (1), if X has finite volume and $\operatorname{div} F = 0$, it follows that every point belongs to the nonwandering set. *See* STATISTICAL MECHANICS.

Flows on manifolds. Let X be a differentiable manifold. A vector field F on X generates the flow $(\mathbf{R}, X, \pi)$ if $(d/dt)\pi(t, x) = F(\pi(t, x))$. That is, for every C^1 function f on X, $(d/dt)f(\pi(t, x))|_{t=0} = F(x)f(x)$. In the case of a classical system of Eq. (1), the vector field is given by Eq. (4). If X is a compact manifold,

$$F(x) = \sum_{j=1}^{n} F_j(x) \frac{\partial}{\partial x_j} \qquad (4)$$

every C^1 vector field on X generates a C^1 flow.

Integrals. Flows on manifolds occur naturally in the study of classical systems of Eq. (1). A C^1 function H is an integral of the system defined by Eq. (1) if its gradient is everywhere perpendicular to the vector field defined by Eq. (4); that is, Eq. (5) is

$$\sum_{j=1}^{n} F_j(x) \frac{\partial H}{\partial x_j} = 0 \qquad (5)$$

satisfied. For example, the Hamiltonian of a Hamiltonian system defined by Eq. (2) is automatically an integral of that system. By the chain rule, if H is an integral, the level surface $H = c$ is flow-invariant. For almost all real numbers c, this level surface is empty or else an $(n - 1)$-dimensional differentiable manifold (Sard's theorem), and in the latter case $(\mathbf{R}, \{H = c\}, \pi)$ is a flow on a manifold.

Geodesic flow. Let M be a compact Riemannian manifold with tangent bundle TM, and let T_1M be the bundle of tangent vectors of unit length. If v_p is a unit tangent vector at p, there exists a unique geodesic $x(t)$, $-\infty < t < \infty$, parametrized by arc length "in the direction v_p" (so that negative arc length makes sense) with $x(0) = p$, $(dx/ds)|_{s=0} = v_p$. Define $\pi(t, v_p) = (dx/ds)|_{s=t}$. Since arc length is the parameter, $\pi(t, v_p)$ is a unit tangent vector [at $x(t)$], and $(\mathbf{R}, T_1M, \pi)$ is a flow, called geodesic flow. *See* RIEMANNIAN GEOMETRY.

Principle of least action. Let U be an open set in $\mathbf{R}^m$. Given a positive definite kinetic energy, defined by Eq. (6), and a potential energy $V(q)$, the

$$T(q, \dot{q}) = \frac{1}{2} \sum_{i, j=1}^{m} a_{ij}(q) \dot{q}_i \dot{q}_j \qquad (6)$$

Lagrange equations of motion are given by Eq. (7)

$$\frac{d}{dt} \frac{\partial L}{\partial \dot{q}_i} - \frac{\partial L}{\partial q_i} = 0 \qquad (1 \leq i \leq m) \qquad (7)$$

where $L = T - V$. [The impulses p_i in Eq. (2) are defined by Eq. (8).] The principle of least action of

$$p_i = \sum_{j=1}^{m} a_{ij}(q) \dot{q}_j \qquad (8)$$

Maupertuis-Euler-Lagrange-Jacobi states that among all the curves C on which the total energy $H = T + V$ is constant (h), the one joining a pair of points x_0, x_1 in U and representing a true motion of the system is the one which minimizes the "action integral" $_C\int 2T$. Using Eq. (9), one can

$$T = h - V = (h - V)^{1/2} \left(\sum_{i, j=1}^{m} a_{ij}(q) \dot{q}_i \dot{q}_j \right)^{1/2} \qquad (9)$$

define a metric by Eq. (10). The principle of least

$$ds^2 = 2(h - V(q)) \sum_{i, j=1}^{m} a_{ij}(q) dq_i dq_j \qquad (10)$$

action now says that geodesics for ds^2 are true motions of the system. The speed of travel along a geodesic is $ds/dt = h - V$, but by a "change of clock" this speed can be made to be 1, and the geodesic flow results. *See* LEAST-ACTION PRINCIPLE.

Whitney sums. Let X be a differentiable manifold with tangent bundle TX. TX is the Whitney sum of continuous bundles A and B over X, $TX = A \oplus B$, if for each x the fibers of A and B at x are complementary subspaces of the tangent space at x. One can speak similarly of a Whitney sum of three or more bundles.

Hyperbolic structure. Let X be a compact Riemannian manifold on which the length of a tangent vector is denoted $\|v\|$. A hyperbolic structure for a flow $(\mathbf{R},X,\pi)$ is a decomposition of TX into a Whitney sum of three bundles, E,E^u, and E^s, each of which is flow-invariant (for the derivative of the flow) and such that (i) E is tangent to the flow (that is, to the orbits) and (ii) there exist constants $c < \infty$ and $\lambda > 0$ such that inequalities (11) and (12)

$$\|d\pi(t,x)v_x\| \le ce^{-\lambda t}\|v_x\| \qquad (v_x \in E_x^s, t \ge 0) \tag{11}$$

$$\|d\pi(t,x)v_x\| \le ce^{\lambda t}\|v_x\| \qquad (v_x \in E_x^u, t \le 0) \tag{12}$$

are satisfied. E^s is called the stable bundle and E^u the unstable bundle. The concept of a hyperbolic structure is also used for a diffeomorphism (or for the cascade generated by its repeated application); it differs from the hyperbolic structure of a flow only in the requirement that $TX = E^s \oplus E^u$, necessary because one cannot speak of a tangent to an orbit of a cascade.

Anosov flow. A differentiable flow on a manifold of dimension greater than 1 is said to be an Anosov flow if there exists a hyperbolic structure for the flow. The most important example of an Anosov flow is the geodesic flow on a compact Riemannian manifold of negative sectional curvature.

Anosov diffeomorphism. A diffeomorphism is an Anosov diffeomorphism if there exists a hyperbolic structure for the diffeomorphism. The known examples are algebraic in nature. For example, let A be an $n \times n$ matrix with integer entries and determinant 1. As a linear transformation of $\mathbf{R}^n$, A sends the integer lattice $\mathbf{Z}^n$ to itself and therefore defines an automorphism of the quotient $\mathbf{R}^n/\mathbf{Z}^n = T^n$, the n-dimensional torus. If A has no eigenvalues of absolute value 1, the automorphism ("hyperbolic automorphism") is an Anosov diffeomorphism. J. Franks and A. Manning have shown that any Anosov diffeomorphism of the torus is isomorphic to some hyperbolic automorphism of the same torus; that is, the corresponding cascades are isomorphic.

Structural stability. A differentiable flow $(\mathbf{R},X,\pi)$ on a compact manifold X is "structurally stable" if the phase portrait (or orbit structure) is insensitive to small perturbations in the equations (vector field) governing the flow. It is possible to give the set of all C^1 vector fields, which has a natural vector space structure, a norm (the C^1 norm) with respect to which it is a Banach space. A C^1 vector field F is structurally stable if there exists $\epsilon > 0$ such that whenever G is a C^1 vector field with $\|F - G\| < \epsilon$, G and F generate flows with isomorphic orbit structure; that is, there is a homeomorphism of X which takes G orbits into F orbits, although "time" may not be preserved. The equations governing any physical system can in practice only be approximated, but when the "true" system is structurally stable, a sufficiently good approximation to it will have the same orbit structure. One can also define structural stability for diffeomorphisms (which also have a natural C^1 topology).

Anosov's theorem. Every Anosov flow or diffeomorphism is structurally stable. This implies the geodesic flow on a compact, negatively curved Riemannian manifold is structurally stable; similarly, a hyperbolic automorphism of the torus is structurally stable.

Symbolic dynamics. Let Λ_r be a set with $r > 1$ elements, say $\Lambda_r = \{1,2, \ldots ,r\}$. Provided with a natural metric, the set X_r of all bisequences $x = \{x_n\}_{n=-\infty}^\infty$ of elements from Λ_r is a compact metric space, and the left shift, σ, $(\sigma x)_n = x_{n+1}$, a homeomorphism of this space. This defines a cascade called the shift (on r symbols), written (σ,X_r) [rather than $(\mathbf{Z},X_r,\pi)$]. Symbolic dynamics is the study of the shift, for its own sake and for application to other systems.

J. Hadamard (1898) set up a correspondence between certain symbolic sequences and geodesics on a negatively curved surface. Utilizing this correspondence, M. Morse (1922) proved the corresponding geodesic flow has minimal sets not consisting of a single closed orbit. (A minimal set is a nonempty closed invariant set containing no proper subset with the same property.) Generalizing this approach, Y. G. Sinai and R. Bowen have constructed an elaborate symbolic dynamics for flows and diffeomorphisms which are suitably "hyperbolic" (for example, Anosov). The theory has numerous applications, among them (a) a generalization of Morse's result to Anosov flows with smooth invariant measure, in particular the geodesic flow on a compact negatively curved manifold; (b) the statement that every minimal set for an Anosov flow is one-dimensional (that is, "small"); and (c) remarkable asymptotic formulas for the distribution of periodic orbits in Anosov flows (with application to the distribution of closed geodesics on a negatively curved manifold).

Ergodic theory. Ergodic theory is the study of measure-preserving transformations. Let X be a set, $\mathscr{B}$ a σ-field of subsets of X, and μ a measure on $\mathscr{B}$. A transformation T of X to itself is measure-preserving if $T^{-1}B \in \mathscr{B}$ and $\mu(T^{-1}B) = \mu(B)$ for all $B \in \mathscr{B}$. If T is one-to-one onto, and if both T and T^{-1} are measure-preserving, T is said to be an automorphism of $(X,\mathscr{B},\mu)$. If G is a topological group, a representation of G on (X,B,μ) is a homomorphism $(g \to T^g)$ from G to the automorphism group of $(X,\mathscr{B},\mu)$ with the property that $T^g x$ is measurable on the product space $G \times X$. Representations T^g and T_0^g on $(X,\mathscr{B},\mu)$ and $(X_0,\mathscr{B}_0,\mu_0)$ are isomorphic if there exists a one-to-one equivariant map φ of X onto X_0 such that $\mu_0(\varphi B) = \mu(B)$ for all $B \in \mathscr{B}$, $\mu(\varphi^{-1}B_0) = \mu_0(B_0)$ for all $B_0 \in \mathscr{B}_0$.

Birkhoff ergodic theorem. Let T^t be a representation of $\mathbf{R}$ on a probability space $(X,\mathscr{B},\mu)$. Birkhoff's ergodic theorem says that if f is integrable on X, then for almost all $x \in X$ the limit given by Eq. (13)

$$\lim_{T \to \infty} \frac{1}{T} \int_0^T f(T^t x)\,dt = \bar{f}(x) \tag{13}$$

exists; furthermore, $\bar{f}$ is integrable, and Eq. (14)

$$\int_X f(x)\mu(dx) = \int_X \bar{f}(x)\mu(dx) \tag{14}$$

holds. A similar statement holds for a representation of $\mathbf{Z}$, the integral in Eq. (13) being replaced by

a sum $\dfrac{1}{T}\displaystyle\sum_{t=0}^{T-1} f(T^t x)$.

Ergodicity and mixing. A group of automorphisms of a probability space is ergodic (or metrically transitive) if every measurable invariant set has measure 0 or 1. In the presence of ergodicity the function $\bar{f}$ in Eq. (13) is almost everywhere equal to $_x\!\int f(x)\mu(dx)$, and hence "time averages equal space averages." Because of this interpretation, ergodicity is important for flows with invariant measure which are unstable, that is, sensitive to perturbations of the initial conditions (for example, Anosov flows).

An equivalent formulation of ergodicity (for $\mathbf{R}$) is the condition that Eq. (15) is satisfied for all

$$\lim_{T\to\infty} \frac{1}{T} \int_0^T \mu(A \cap T^t B)\, dt = \mu(A)\mu(B) \qquad (15)$$

$A, B \in \mathscr{B}$. The condition for mixing is an "un-averaged" analog of Eq. (15) given by Eq. (16).

$$\lim_{t\to\infty} \mu(A \cap T^t B) = \mu(A)\mu(B) \qquad (16)$$

Between the two notions is the notion of weak mixing, defined by Eq. (17). If A and $T^t B$ are independent sets, then by definition $\mu(A \cap T^t B) =$

$$\lim_{T\to\infty} \frac{1}{T} \int_0^T |\mu(A \cap T^t B) - \mu(A)\mu(B)|\, dt = 0 \qquad (17)$$

$\mu(A)\mu(T^t B) = \mu(A)\mu(B)$. Ergodicity and the two kinds of mixing thus express, to varying degrees, the notion of asymptotic independence. D. V. Anosov and Sinai have proved an Anosov diffeomorphism with smooth invariant measure is mixing; an Anosov flow with smooth invariant measure is ergodic and, if it is weak-mixing, mixing; the geodesic flow on a compact, negatively curved Riemannian manifold is mixing (in dimension two, a classical theorem of E. Hopf and G. A. Hedlund). Stronger statements have since been made.

Entropy. Let T be a measure-preserving transformation of a probability space $(X, \mathscr{B}, \mu)$. Define $\eta(t)$, $0 \le t \le 1$, to be 0 if $t = 0$ and $-t \log t$ otherwise. If $P = \{A_1, \ldots, A_r\}$ is a partition of X into pairwise disjoint measurable sets, let P_N, $N \ge 1$, be the partition whose elements are sets of the form $A_{j_0} \cap T^{-1} A_{j_1} \cap \ldots \cap T^{-(N-1)} A_{j_{N-1}}$, $1 \le j_0, \ldots j_{N-1} \le r$. Define $H(P_N) = \sum_{A \in P_N} \eta(\mu(A))$. The limit $\lim_{N\to\infty} H(P_N)/N = h(T, P)$ exists and is called the entropy of T with respect to P. One has $0 \le h(T, P) \le \log r$; $h(T, P) = 0$ if and only if P is measurable with respect to the σ-algebra generated by $T^{-1}P$, $T^{-2}P$, ... Thus entropy is a measure of "determinism." Finally $h(T)$, the entropy of T, is the supremum of $h(T, P)$ over all P. In some cases $h(T) = \infty$; but, for example, if T is a diffeomorphism of a compact differentiable manifold X, and if μ is smooth, then Kushnirenko's theorem states that $h(T) < \infty$. *See* ENTROPY.

Ornstein's theorem. If $p_1, \ldots, p_r$ are probabilities, it is possible to define a Borel probability measure μ on the space X_r of symbol sequences which governs the process of selecting elements from Λ_r independently and according to the given probabilities. The shift σ preserves μ, and is called a Bernoulli shift. A. N. Kolmogorov and Sinai showed that $h(\sigma) = \sum_{j=1}^r \eta(p_j)$. Since entropy is an isomorphism invariant, this settled the long-standing question of whether the two shift with probabilities $(1/2, 1/2)$ is isomorphic to the three shift with probabilities $(1/3, 1/3, 1/3)$. (They are not isomorphic because $\log 2 \ne \log 3$.) D. Ornstein (1969) established the deep result that entropy is a complete invariant among the Bernoulli shifts;

two Bernoulli shifts with the same entropy are isomorphic. *See* PROBABILITY.

Bernoulli shifts as models. Ornstein's results have led to the measure theoretic classification of a number of cascades and flows. N. Friedman, Ornstein, Sinai, and R. Azencott showed that a C^2 Anosov diffeomorphism with smooth invariant measure is Bernoulli; Y. Katznelson showed that an ergodic automorphism of the n-dimensional torus is Bernoulli; Ornstein and B. Weiss showed that the geodesic flow on a compact Riemannian manifold of negative curvature is Bernoulli where a flow is defined to be Bernoulli if each of its time t maps is Bernoulli); and M. Ratner showed that an Anosov flow with smooth invariant measure is Bernoulli.

Minimal transformation groups. A transformation group is minimal if every orbit is dense in the phase space. (The concept is due to Birkhoff.) It is of interest to classify minimal transformation groups, one reason being that in an arbitrary compact phase space there exist nonempty closed invariant sets which are minimal. In what follows, X is a compact metric space with metric $d(x, y)$, (G, X, π) and $\pi(g, x)$ will be written (G, X) and gx, and transformation groups (arbitrary G) will be called flows.

Equicontinuous flows. A flow is equicontinuous if for each $\epsilon > 0$ there exists $\delta > 0$ such that $d(gx, gy) < \epsilon$, for all $g \in G$, whenever $d(x, y) < \delta$. $D(x, y) = \sup_g d(gx, gy)$ is a compatible G invariant metric for X when (G, X) is equicontinuous. Since the group of isometries of (X, D) is, with a natural topology, compact, G maps homomorphically into a compact metric group acting on X. If (G, X) is minimal, the closure of the image of G in this group acts transitively on X. The study of minimal equicontinuous flows is therefore essentially the study of transitive compact group actions.

Distal flows. A point $x \in X$ is distal for (G, X) if $\inf_g d(gx, gy) = 0$ implies $y = x$. The flow is distal if every point is distal. Every equicontinuous flow is distal, but there exist minimal distal flows which are not equicontinuous. (However, R. Ellis has shown that if G is finitely generated and X totally disconnected, distality implies equicontinuity. For example, a minimal distal subset of the shift is finite.) Minimal distal flows have been characterized by H. Furstenberg in terms of "isometric extensions." A homomorphism $\varphi\colon (G, X) \to (G, X')$ defines an isometric extension if there is a continuous function R on $\Delta = \{(x, y) \in X \times X \mid \varphi(x) = \varphi(y)\}$ such that (a) $R(gx, gy) = R(x, y)$ for all $g \in G$, and (b) for each $x' \in X'$, R defines a metric on $\varphi^{-1}x'$. Furstenberg's structure theorem states that a minimal distal flow can be "built up" by a transfinite sequence of isometric extensions and "inverse limits" beginning with the trivial flow (one-point phase space).

Point distal flows. If there exists a distal point with dense orbit, the flow (G, X) is point-distal. Point-distal flows are minimal and not necessarily distal. For example, there exist nonequicontinuous, hence nondistal, cascades with totally disconnected phase space. Furstenberg's structure theorem is generalized to point-distal flows using the notion of an almost one-to-one extension, a homomorphism $\varphi\colon (G, X) \to (G, X')$ such that for some $x' \in X'$, $\varphi^{-1}x'$ is a single point. W. Veech and Ellis have shown that every point-distal flow has an almost one-to-one extension which can be built up from the one-point flow by a transfinite sequence of isometric extensions, almost one-to-one extensions, and inverse limits. A general structure theory for minimal flows has been established.   [WILLIAM A. VEECH]

Bibliography: V. I. Arnold and A. Avez, *Ergodic Problems of Classical Mechanics*, 1968; J. R. Brown, *Ergadic Theory and Topological Dynamics*, 1976; R. Ellis, *Lectures on Topological Dynamics*, 1970; Z. Nitecki, *Differential Dynamics*, 1971; D. Ornstein, *Ergodic Theory, Randomness, and Dynamical Systems*, 1974; C. L. Siegel and J. K. Moser, *Lectures on Celestial Mechanics*, 1971.

Topology

The study of topological spaces and continuous maps. Precise mathematical definitions of topological space and continuous map require no more than elementary set theory.

Sets. A set is any collection of things or objects. If X is a set and x is a member of the set X, one writes $x \in X$, which reads, in words, x belongs to X. The symbol ϵ denotes membership. An element or member of a set X is also called a point of X. The rational numbers form a set frequently denoted by Q, and $x \in Q$ means that x is a rational number.

If X and Y are sets, then $X \cup Y$ is the set of elements which belong to either X or Y (including those elements which belong to both X and Y). The notation $X \cup Y$ is read X union Y. Similarly if X and Y are sets, then $X \cap Y$ is the set of elements which belong to both X and Y, and the notation $X \cap Y$ is read X intersect Y.

Suppose, for example, that Q is the set of rational numbers, X is the set of rational numbers greater than or equal to zero, and Y is the set of rational numbers less than or equal to zero. Under these conditions $X \cup Y = Q$, and $X \cap Y$ is the set whose only element is the number zero.

Suppose that for each element i of a set I there is given a set A_i, the set consisting of all the sets A_i is denoted by $\{A_i\}_{i \in I}$, and is said to be a collection of sets indexed on the set I. If $\{A_i\}_{i \in I}$ is a collection of sets indexed on the set I, then $\bigcup_{i \in I} A_i$ is the set consisting of those elements which belong to at least one of the sets A_i, and is called the union of $\{A_i\}_{i \in I}$. Similarly $\bigcap_{i \in I} A_i$ is the set consisting of those elements which belong to every one of the sets A_i, and is called the intersection of $\{A_i\}_{i \in I}$. In case I is the set with exactly two elements 1 and 2, $\{A_i\}_{i \in I}$ has two members A_1 and A_2. Moreover, in this case $\bigcup_{i \in I} A_i = A_1 \cup A_2$, and $\bigcap_{i \in I} A_i = A_1 \cap A_2$.

If X is a set, then a set A is a subset of X if every element of A is also an element of X. For example, if X is the set of automobiles built during any one year, and A is the set of green automobiles built during that year, then A is a subset of X. Every set has a subset called the empty set. The empty set is the set with no elements whatsoever, and is frequently denoted by ϕ. The notation $A \cap B = \phi$ means that the intersection of the set A and B is empty, in other words, that the sets A and B are disjoint. *See* SET THEORY.

Topological space. This is a set of points X, together with a collection of subsets of X called open subsets of X where the following assumptions are made:

1. ϕ and X are open subsets of X.
2. If A and B are open subsets of X, then $A \cap B$ is an open subset of X.
3. If $\{A_i\}_{i \in I}$ is a collection of open subsets of X, then $\bigcup_{i \in I} A_i$ is an open subset of X.

Suppose that R is the set of real numbers. If $r \in R$, in other words, if r is a real number, let $|r|$ denote the absolute value of r. This means that, if r is greater than or equal to zero, then $|r| = r$, but if r is less than zero then $|r| = -r$. In order to make the real numbers into a topological space, open subsets of R are defined using the notion of absolute value. Precisely, a subset U of R is open if for every $x \in U$ there is a real number ε_x greater than zero having the property that if for some real number y, $|y - x| < \varepsilon_x$, then $y \in U$. In other words, if $x \in U$, then any real number sufficiently close to x belongs to U also. With this definition of open subset of the real numbers it is not difficult to verify that the axioms for a topological space are satisfied. When one talks of the real numbers in mathematics, one usually means the real numbers as a topological space, that is to say, the set of real numbers together with the collection of open subsets that are defined above.

Continuous map. In order to continue the discussion of topology, it is necessary to introduce some further notions of set theory. First, suppose that X and Y are sets. Define the product of the sets X and Y to be the set consisting of pairs of elements (x,y) such that $x \in X$ and $y \in Y$. The product of X and Y is denoted by $X \times Y$. Now a function f from X to Y is defined to be a subset f of the product $X \times Y$ such that, if $x \in X$, there exists a unique $y \in Y$ such that $(x,y) \in f$. In this case y is said to be the value of f at x, and is denoted by $f(x)$. Intuitively a function from X to Y is thought of as a rule which assigns to each element of the set X an element of the set Y. The standard mathematical notation for a function f from X to Y is $f: X \to Y$.

Let $f: X \to Y$ be a function and suppose U is a subset of Y, then $f^{-1}(U)$ is the subset of X consisting of those points x such that $f(x) \in U$. If X and Y are topological spaces, then $f: X \to Y$ is a continuous map, or continuous function, (1) if f is a function from X to Y, and (2) if, whenever U is an open subset of Y, the set $f^{-1}(U)$ is an open subset of X. If X is any topological space, then the identity function $i: X \to X$ defined by $i(x) = x$ is a continuous map called the identity map of X. Suppose X, Y, and Z are topological spaces, $f: X \to Y$ is a continuous map, and $g: Y \to Z$ is a continuous map. In this case the function $g \circ f: X \to Z$ defined by $(g \circ f)(x) = g(f(x))$ is a continuous map.

The topological spaces X and Y are said to be homeomorphic if there exist continuous maps $f: X \to Y$ and $g: Y \to X$ such that $f \circ g: Y \to Y$ is the identity map of Y and $g \circ f: X \to X$ is the identity map of X. In this case the maps f and g are said to be homeomorphisms, and g is frequently denoted by f^{-1}. The symbol f^{-1} is read f inverse.

Again, assume that R is the topological space of real numbers. Define a function $f: R \to R$ by $f(x) = 0$ if x is less than zero, and $f(x) = 1$ if x is greater than or equal to zero. The function $f: R \to R$ is not a continuous map. This is because $f(0) = 1$, but there are points x arbitrarily close to 0 such that $f(x) = 0$. In other words the function f is not smooth; it jumps at zero. Intuitively a continuous map $f: X \to Y$ is a function such that if x and x' are close together then $f(x)$ and $f(x')$ are also close together where the notion of closeness is determined by the open subsets of X and Y respectively.

Metrics. When the topology on the real numbers was defined, this was a special case of defining a topology by using a metric. Let X be a set. A metric on X is a function $\rho : X \times X \to R$ such that the following axioms obtain:

1. If x and x' belong to X, then $\rho(x,x')$ is greater than or equal to zero, and $\rho(x,x') = 0$ if and only if $x = x'$.

2. If x and x' belong to X, then $\rho(x,x') = \rho(x',x)$.

3. If x, x', and x'' belong to X, then $\rho(x,x'')$ is less than, or equal to, $\rho(x,x') + \rho(x',x'')$.

If $\rho : X \times X \to X$ is a metric, then $\rho(x,x')$ is called the distance from x to x'. Axiom 1 says that the distance between any two points of X is greater than or equal to zero, and is different from zero if the points are different. Axiom 2 says that the distance from x to x' is the same as the distance from x' to x. Axiom 3, the so-called triangle axiom, says intuitively that it is shorter to proceed from x to x'' along a straight line than it is first to proceed from x to x' along a straight line and then proceed from x' to x'' along another straight line.

If X is a set, and $\rho : X \times X \to R$ is a metric on X, one says that a subset U of X is open if, whenever $x \in U$, there exists a real number ε_x greater than zero such that if x' is another point of X and $\rho(x,x')$ is less than ε_x then x' also belongs to U. In other words, if $x \in U$ and x' is a point which is very close to x, then x' also belongs to U. One verifies easily that, if open subset of X is defined as above, then axioms for the open subsets of a topological space are verified. Thus any set X with a metric $\rho : X \times X \to R$ defines a topological space.

When R is the set of real numbers and $\rho : R \times R \to R$ is defined by $\rho(x,x') = |x - x'|$, it may be proved that ρ is a metric on R. Then clearly the notion of open subset defined from this metric is exactly the notion of open subset defined earlier.

Suppose that X is a set, $\rho : X \times X \to X$ is a metric on X, Y is a set, and $\rho' : Y \times Y \to Y$ is a metric on Y. Further assume that the notion of open subset of X is defined by using the metric ρ, and that the notion of open subset of Y is defined by using the metric ρ'. In this case it may be proved that $f : X \to Y$ is a continuous map if and only if for every $x \in X$ and ε_x greater than zero, there exists δ_x greater than zero such that if $\rho(x,x')$ is less than δ_x then $\rho'(f(x),f(x'))$ is less than ϵ_x. Thus the idea that $f : X \to Y$ is continuous whenever x and x' are close together implies that $f(x)$ and $f(x')$ are close together; the idea is precise when the open subsets of X and Y are defined by metrics.

Construction of topological spaces. One of the most important processes in topology is the construction of new topological spaces from old ones. Two particularly common and useful methods of doing this will be illustrated here. First, suppose X is a topological space and A is a subset of X. A subset U of A is open if there exists an open subset V of X such that $A \cap V = U$. When A is considered as a topological space with the notion of open subset defined in this manner, A is said to be a subspace of X. Second, suppose that X and Y are topological spaces. A subset U of $X \times Y$ is open if for every point $(x,y) \in U$ there exists an open subset V of X and an open subset W of Y such that $V \times W$ is a subset of U. When $X \times Y$ is considered as a topo-

logical space with open subset defined in this manner, it is called the product space of X and Y.

Having defined the concept of product space, it is now easy to define that of euclidean n-space. However, before doing this it is well to remark that, if X, Y, and Z are topological spaces, then the space $(X \times Y) \times Z$ and $X \times (Y \times Z)$ are the same. In other words, the operation of taking the product space of several topological spaces is associative. Now euclidean 1-space is just the space of real numbers R. Euclidean 1-space also called the real line or just the line is denoted by R or R^1. The notion of euclidean n-space for any positive integer n may be defined by induction. If euclidean n-space is defined and denoted by R^n, euclidean $(n+1)$ space is defined to be $R^n \times R$ and denoted by R^{n+1}.

The preceding inductive definition of R^n is convenient for some purposes, but it is useful to have a direct description also. This may be accomplished by first letting R^n denote the set of n-tuples $(x_1, \ldots, x_n)$ such that each x_i is a real number, and then defining a metric on R^n by letting the distance

$$\rho\big((x_1, \ldots, x_n),(y_1, \ldots, y_n)\big)$$

from point $(x_1, \ldots, x_n)$ to point $(y_1, \ldots, y_n)$ be the square root of $\sum_{i=1}^{n}(x_i - y_1)^2$. Thus, if $n = 2$, then $R^2 = R \times R$ and the distance $\rho\big((x_1,x_2),(y_1,y_2)\big)$ from the point (x_1,x_2) to the point (y_1,y_2) is the square root of $(x_1 - y_1)^2 + (x_2 - y_2)^2$. The space R^2 is frequently called the plane or the euclidean plane and corresponds to the intuitive notion of a flat or plane surface. The space R^3 is sometimes called just space and corresponds to the intuitive idea of space. The distance between two points (x_1,x_2,x_3) and (y_1,y_2,y_3) in R^3 is the square root of $(x_1 - y_1)^2 + (x_2 - y_2)^2 + (x_3 - y_3)^2$.

The n-dimensional sphere, denoted by S^n, is the subspace of R^{n+1} consisting of those points whose distance from the origin $(0, \ldots, 0)$ of R^{n+1} is exactly 1. Thus the 1-sphere also called the circle consists of the points in the plane whose distance from the point $(0,0)$ is exactly 1. Consequently the circle is the boundary of the disk consisting of those points in the plane whose distance from the origin is less than or equal to 1. Similarly the 2-sphere is the boundary or surface of the solid ball consisting of those points of R^3 whose distance from the origin is less than or equal to 1. It is pictured as the surface of an ordinary solid ball such as a croquet ball. The 2-sphere is an example of the general notion of surface. There are many other examples of surfaces, a common one being the torus, which is just the topological space $S^1 \times S^1$. It may be thought of as the surface of a doughnut or as the inner tube of an automobile tire.

Two topological spaces which are homeomorphic cannot be distinguished by the methods of topology. Any topological property of one is also a topological property of the other. Consequently one frequently calls any topological space homeomorphic to the n-sphere S^n, an n-dimensional sphere. For example, any circle in the plane is homeomorphic to the standard circle S^1. Suppose that X is the topological space which is the boundary of a square in the plane. Let Y be a circle in the plane which completely contains X in its interior. Choose a point * in the plane inside the square. Now, for every point $x \in X$, let $f(x)$ be the point of Y obtained by proceeding in a straight line from * to x and then continuing along the same straight line from x to the point $f(x)$ of Y. The function $f:X \to Y$ thus obtained is a homeomorphism. Consequently X, the boundary of a square, is homeomorphic to a circle. Similarly it may be proved that a square in the plane is homeomorphic to the disk bounded by a circle.

Since homeomorphic spaces cannot be distinguished by topological methods, one of the important problems of topology is to determine whether two topological spaces are homeomorphic or not. For example, is the sphere S^1 homeomorphic to the sphere S^2? Intuitively it does not seem so, but one would like a proof. One proof may be obtained by giving topological significance to the idea of dimension.

Suppose that X is a topological space. An open covering of X is a collection of sets $\{A_i\}_{i \in I}$ such that each A_i is an open subset of X, and $X = \bigcup_{i \in I} A_i$. If $\{A_i\}_{i \in I}$ and $\{B_j\}_{j \in J}$ are open coverings of X, the covering $\{A_i\}_{i \in I}$ is said to be finer than the covering $\{B_j\}_{j \in J}$ if for each index $i \in I$ there exists an index $j \in J$ such that A_i is a subset of B_j. Let $\{A_i\}_{i \in I}$ be an open covering of the space X. This covering of X is said to have dimension less than or equal to n if the intersection $A_{i_0} \cap \cdots \cap A_{i_n}$ of any $(n+1)$ distinct sets of the covering is empty. Thus an open covering $\{A_i\}_{i \in I}$ has dimension less than or equal to zero if $A_i \cap A_j$ is empty for A_i different from A_j. It has dimension less than or equal to 1 if for every three sets A_{i_1}, A_{i_2}, A_{i_3} such that no two are equal the intersection $A_{i_1} \cap A_{i_2} \cap A_{i_3}$ is empty, and so forth. The topological space X has dimension n if n is the least integer with the property that for every open covering of X there is a finer open covering of X which has dimension less than or equal to n. From the definition of dimension just given it is clear that, if X is an n-dimensional topological space and Y is homeomorphic with X, then Y is an n-dimensional topological space. In other words, the notion of dimension is a topological invariant. Moreover, it is possible to prove that euclidean n-space R^n is an n-dimensional topological space, and that the n-sphere S^n is an n-dimensional topological space. This means, among other things, that the circle S^1 is not homeomorphic to the 2-sphere S^2. Further it means that there is topological significance to the n of S^n or R^n.

There are other important properties of topological spaces which are defined by using the notion of open covering. Probably the most important of these is compactness. A topological space X is compact if and only if for any open covering $\{a_i\}_{i \in I}$ of X there is a finite number of indices $i_1, \ldots,$ $i_n \in I$ such that $X = \bigcup_{j=1}^{n} A_{i_j}$. For any positive integer n the sphere S^n is a compact topological space. A general theorem asserts that, if X and Y are compact topological spaces, then the product space $X \times Y$ is also compact. This implies that the torus is a compact topological space, for it is the product space $S^1 \times S^1$ and S^1 is compact.

The general notion of topological space is not sufficiently restrictive for most purposes. Therefore some additional axioms are almost always assumed—in particular the Hausdorff separation axiom. A topological space X satisfies the Haus-

dorff separation axiom, or is a Hausdorff space, if for every two distinct points x and x' of X there are open subsets U and V of X such that $x \in U$, $x' \in V$, and $U \cap V$ is empty. If X and Y are compact Hausdorff spaces and $f:X \to Y$ is a continuous map such that $f(x) = f(x')$ implies $x = x'$ and for every $y \in Y$ there exists $x \in X$ such that $f(x) = y$, then f is a homeomorphism.

All the topological spaces so far considered, the spheres, the euclidean spaces, and the subspaces or products of these spaces, are Hausdorff spaces. In fact, any subspace of a Hausdorff space, and any product of Hausdorff spaces, is again a Hausdorff space. Further, if X is a topological space such that the topology is derived from a metric on X, then X is a Hausdorff space.

Manifolds. Though Hausdorff spaces form a much more interesting class of spaces than general topological spaces, the most important class of topological spaces, which is still much smaller, consists of the manifolds. An n-manifold is a Hausdorff space X such that the following conditions exist:

1. For every point $x \in X$ there is an open subset U_x of X such that $x \in U_x$ and such that U_x is homeomorphic with an open subset of R^n.
2. There exists a metric $\rho : X \times X \to R$ such that the topology on X is induced by the metric.

In defining n-manifold one always assumes condition 1, which states that within short distances of some fixed point it seems as if one is in euclidean n-space. Having assumed condition 1, there are several other conditions which are sometimes assumed instead of condition 2. For example, it is sometimes assumed that there exist compact subsets $X_1, X_2, \ldots, X_k, \ldots$ of X such that $X = \bigcup_{n=1}^{\infty} X_i$. Another frequent assumption is that there exists a countable number of points $x_1, x_2, \ldots,$ $x_k, \ldots$ of X such that if U is any nonempty open subset of X, then there is some integer k such that $x_k \in U$. All these possible variations of condition 2 are essentially equivalent once condition 1 is assumed.

The spheres and the euclidean spaces are examples of manifolds. Further products of spheres and euclidean spaces are manifolds, for if X is an m-manifold and Y is an n-manifold, then $X \times Y$ is an $(m+n)$-manifold. This implies in particular that the torus $T = S^1 \times S^1$ is a 2-manifold.

A manifold is a topological space which is an n-manifold for some positive integer n. The dimension of this topological space is n. Moreover, the dimension of any open subset is also n.

Before proceeding further with the discussion of manifolds, it is necessary to introduce another general topological notion. A topological space X is connected if, whenever U and V are nonempty open subsets of X such that $X = U \cup V$, the set $U \cap V$ is nonempty. In other words, X is connected if it cannot be expressed as the union of two disjoint nonempty open subsets.

One of the most important problems of topology is the problem of classification of connected n-manifolds. If X is a connected 1-manifold then either X is compact and is homeomorphic with S^1, or X is not compact and is homeomorphic with R^1. Thus from the point of view of topology these are just two connected 1-manifolds, namely the circle and the line.

The problem of classifying compact connected 2-manifolds has also been solved. These manifolds, also called surfaces, are classified by giving a list of standard surfaces, and then proving that any compact connected 2-manifold is homeomorphic to one and only one in the list. The list just mentioned is long and will not be given here. The 2-sphere, euclidean 2-space, and the torus are all examples of 2-manifolds. However, as mentioned previously, R^2 is not compact.

Though no such neat classification exists for higher-dimensional manifolds, there has been much recent progress in studying manifolds of dimension 3 and of dimension greater than 4, as will be described below.

Homotopy theory. If X and Y are topological spaces, then two continuous maps $f_0: X \to Y$ and $f_1: X \to Y$ are said to be homotopic if there exists a continuous map $F: I \times X \to Y$ such that $F(0,x) = f_0(x)$ and $F(1,x) = f_1(x)$, where I is the subspace of the real numbers consisting of those numbers which are greater than or equal to 0 and less than or equal to 1. A map $f: X \to Y$ is said to be a trivial map if $f(x) = f(x')$ for every pair of points x and x' belonging to X. A map $f_0: X \to Y$ is homotopically trivial if there exists a trivial map f_1 which is homotopic if one can be smoothly deformed into the other. Thus a map is homotopically trivial if it can be deformed into a map which sends everything into a single point.

Suppose that X is a connected 2-manifold and every map $f: S^1 \to X$ is homotopically trivial; then either X is compact and is homeomorphic with S^2, or X is not compact and is homeomorphic with R^2.

Two topological spaces X and Y have the same homotopy type if there exist maps $f: X \to Y$ and $g: Y \to X$ such that $f \circ g: Y \to Y$ is homotopic to the identity map of Y and $g \circ f: X \to X$ is homotopic to the identity map of X. Consequently homeomorphic spaces always have the same homotopy type, but the converse is far from true. For example, for any positive integer n, the space R^n has the same homotopy type as the space consisting of a single point.

If X is a connected n-manifold and, for every positive integer q less than n, every map $f: S^q \to X$ is homotopically trivial, then either X is compact and has the same homotopy type as the sphere S^n or X is not compact and has the same homotopy type as R^n. The fact that at present the preceding result cannot be replaced by one saying that under the same conditions X is either homeomorphic with S^n or R^n unless n is less than or equal to 2, seems to be one of the main obstacles to proving classification theorems for manifolds of dimension greater than 2.

In the study of topology, instead of attacking problems directly, algebraic invariants frequently are attached to topological spaces, and then these invariants are studied. The main impetus to starting to work in this direction was given by the French mathematician Henri Poincaré at the turn of the century. The work of Poincaré was done chiefly in connection with the problem of classifying manifolds. He conjectured that if X is a compact connected 3-manifold and every map $f: S^1 \to X$ is homotopically trivial, then X is homeomorphic with a 3-sphere. As mentioned earlier, there is as yet no proof of this result, though it still seems

possible that one may be found.

The study of topology by means of algebraic invariants of topological spaces is a comparatively new field of mathematics. Since most of these invariants are rather difficult to describe, no attempt to do so will be made here, but the definition of one set of these invariants, the homotopy groups, will now be outlined. These invariants are the easiest to describe, but among the most difficult to compute.

A topological space X is pathwise connected if for every pair of points x and x' belonging to X there is a continuous map $f: I \to X$ such that $f(0) = x$ and $f(1) = x'$, in other words, if one can draw an arc between any two points of the space. A manifold is connected if, and only if, it is pathwise connected.

Suppose that X and Y are topological spaces and $x_0 \in X$ and $y_0 \in Y$ are chosen and called base points. A map $f: X \to Y$ is said to preserve base points if $f(x_0) = y_0$. Two maps $f: X \to Y$ and $g: X \to Y$ which preserve base points are homotopic relative to the base points x_0 and y_0 if there exists a map $F: I \times X \to Y$ such that the following conditions obtain:

1. $F(0,x) = f(x)$ for any $x \in X$.
2. $F(1,x) = g(x)$ for any $x \in X$.
3. $F(t,x_0) = y_0$ for any $t \in I$.

Now one says that two maps $f,g: X \to Y$ which preserve base point are equivalent if they are homotopic relative to the base points. The set of such equivalence classes of maps is denoted by $\pi((X,x_0), (Y,y_0))$. Such an equivalence class is called a homotopy class of maps.

For the sphere S^n choose a base point $e_n \in S^n$ once and for all. Let $\pi_n(X,x_0)$ be $\pi((S^n,e_n),(X,x_0))$ for every space X with base point x_0. Let $S^n \vee S^n$ be the space obtained by taking two copies of S^n and identifying the point e_n in one copy with the point e_n in the other copy. This space may be thought of as two tangent n-spheres. Define a map $\theta: S^n \to S^n \vee S^n$ by collapsing an equator of S^n through the base point to obtain two tangent spheres. Suppose that $f,g: S^n \to X$ are maps which preserve base points. Define $f \vee g: S^n \vee S^n \to X$, by mapping points of the first tangent sphere by means of f and of the second by means of g. This definition is legitimate since $f(e_n) = x_0 = g(e_n)$. Now $(f \vee g) \circ \theta: S^n \to X$ and preserves base point. One verifies that, if f' is homotopic to f and g' is homotopic to g, then $(f' \vee g') \circ \theta$ is homotopic to $(f \vee g) \circ \theta$ where all homotopies are relative to the base points. If for any such map f one denotes the homotopy class of f by $[f]$, then $[(f \vee g) \circ \theta]$ depends only on $[f]$ and $[g]$. Then mapping $\varphi: \pi_n(X,x_0) \times \pi_n(X,x_0) \to \pi_n(X,x_0)$ defined by $\varphi([f],[g]) = [(f \vee g) \circ \theta]$ determines a group operation in the set $\pi_n(X,x_0)$ which is now called the n-dimensional homotopy group of X relative to the base point x_0.

If X is pathwise connected, then the group $\pi_n(X,x_0)$ is independent of the choice of base point x_0. Further, if X and Y are pathwise connected spaces having the same homotopy type, then for any $x_0 \in X$ and any $y_0 \in Y$ the groups $\pi_n(X,x_0)$ and $\pi_n(Y,y_0)$ are isomorphic.

The group $\pi_1(X,x_0)$ is also called the fundamental group or Poincaré group of the space X based at the point x_0. This group was discovered and investigated by Poincaré. The groups $\pi_n(X,x_0)$ were discovered some 30 years later by Witold Hurewicz. It is not difficult to prove that, if X is a space having the homotopy type of a point, then $\pi_n(X,x_0)$ has a single element. Further it may be proved that the group $\pi_q(S^n,e_n)$ has a single element if q is less than n, but is isomorphic with the group of integers if $q=n$. This proves that S^n does not have the homotopy of a point.

Even though the groups $\pi_q(X,x_0)$ are abelian for $q>1$, they are difficult to compute. It may be shown that if n is greater than 1, then the group $\pi_q(S^n, e_n)$ has more than one element for an infinite number of integers q. *See* CONFORMAL MAPPING.

[JOHN C. MOORE]

Differential and geometric topology. In euclidean space R^n, one may do differential calculus, and in a manifold M, for an open subset U_i of M, the homeomorphism $h_i:V \to U_i$ where V is open in R^n, can be used to do differential calculus in U_i (transferring the situation to V using h_i). If a set of such homeomorphisms h_i (called charts) are given, so that the U's are an open cover of M, and so that differential calculus on U_i and U_j is compatible for different i and j, this determines what is called a differential structure on M, and M is called a differential manifold.

For differential manifolds M and N, a smooth map $f: M \to N$ is a map which is a differentiable function; that is, it is differentiable for each chart in the two manifolds. Such smooth maps are continuous, but there are many continuous maps which are not smooth. A homeomorphism $F: M \to N$ is called a diffeomorphism if f and f^{-1} are both smooth maps.

One may try to compare the three types of equivalence (homotopy equivalence, homeomorphism, and diffeomorphism) among differential manifolds. Examples where homotopy equivalence did not imply homeomorphism were found in the 1930s, but many believed that homeomorphism might imply diffeomorphism. But in 1956, J. Milnor showed the existence of differential manifolds, homeomorphic to the sphere S^7 but not diffeomorphic to it, so that these three types of equivalence are definitely distinct. In the 1960s and 1970s one of the main concerns of this subject became the detailed comparison of these notions, together with the notion of piece-wise linear (PL) equivalence, a concept between homeomorphism and diffeomorphism.

If instead of the compatibility of the differential calculus on different charts U_i and U_j defined at some point $x \in M$, it is required that the linear structures near x be the same, so that the homeomorphisms in the expression below are piece-wise

$$V_j \xleftarrow{h_j} U_i \cap U_j \xrightarrow{h_i} V_i$$

linear, in a technical sense that will not be made explicit, then this defines what is called a PL linear structure on M. The notions of PL map and PL homeomorphism are defined analogously. It was shown by S. S. Cairns and J. H. C. Whitehead that a differential manifold admits a unique PL structure, in which the linear structure is smooth.

The first generally positive result, obtained by S. Smale in 1960, showed that for compact manifolds of sufficiently high dimension ($m \geq 5$ in subsequent versions), if M^m is homotopy-equivalent to S^m, then it is homeomorphic to it.

From the results of Milnor, there developed a theory classifying up to diffeomorphism the possible compact differential manifolds that are homotopy-equivalent to S^m ($m \geq 5$), and later from this developed a theory (called surgery theory) for classifying the compact differential manifolds of dimensions greater than 5 with a given homotopy type. This classification became mainly reduced to problems in homotopy theory, and to questions in algebra on classification of quadratic forms and their automorphisms. These algebraic problems differed from problems studied earlier in that the forms were defined over more complicated rings.

In one striking success, the theory led to great progress on two old questions: When can a given topological manifold be given the structure of a PL manifold; or when can it be triangulated as a simplicial complex, that is, be made homeomorphic to a polyhedron in R^q. Examples were given to show that the first construction is not always possible, and the problem was reduced to calculating an algebraic invariant. A comprehensive theory was developed for the second problem, but some key steps are still lacking for a complete solution, analogous to the solution of the first problem.

In dimension 3, a powerful geometric technique has been introduced by W. Thurston. He has shown that many 3-dimensional manifolds admit a noneuclidean (lobachevskian) geometry. This has made it possible to attack many topological problems in dimension 3 by using geometrical and analytical arguments. For example, it was shown that any homeomorphism $f:S^3 \to S^3$ of finite period which keeps a simple closed curve fixed is topologically equivalent to a rotation.

Such theorems are not true in dimensions $n > 4$. Surgery theory has shown the existence of infinite families of such periodic diffeomorphisms on S^n, all of period p, which are all topologically distinct from each other, and from any rotation. Much other strange behavior is now known and well understood for these higher dimensions (greater than 4).

Manifolds of dimension 4 remain the realm of greatest mystery in this area of topology. Here the techniques of surgery theory lose their force, while no powerful geometrical techniques exist, such as are being developed in dimension 3. There are two outstanding questions concerning dimensions 3 and 4:

1. Can any 4-dimensional manifold be given a smooth structure? Any 3-manifold can, but there are 5-manifolds which cannot.

2. If M^m is m-manifold ($m = 3$ or 4) and M^m is homotopy equivalent to S^m, is M homeomorphic to S^m? (This is the Poincaré conjecture.) As mentioned, this is known to be true for $m \neq 3$ or 4.

[WILLIAM BROWDER]

Bibliography: M. A. Armstrong, *Basic Topology*, 1980; W. Browder, *Surgery on Simply-Connected Manifolds*, 1972; S. Eilenberg and N. Steenrod, *Foundations of Algebraic Topology*, 1952; J. Hempel, *3-Manifolds*, 1976; M. W. Hirsch, *Differential Topology*, 1976; R. C. Kirby and L. C. Siebenman, *Foundational Essays on Topological Manifolds, Smoothings and Triangulations*, 1977; S. Lefschetz, *Introduction to Topology*, 1949; M. G. Murdeshwar, *General Topology*, 1981; C. T. Wall, *Surgery on Compact Manifolds*, 1970; L. Wilder, *Topology of Manifolds*, 2d ed., 1963, reprint 1979.

Torque

The product of a force and its perpendicular distance to a point of turning, also called the moment of the force. Torque produces torsion and tends to produce rotation. Torque arises from a force or forces acting tangentially to a cylinder or from any force or force system acting about a point. A couple, consisting of two equal, parallel, and oppositely directed forces, produces a torque or moment about the central point. A prime mover such as a turbine exerts a twisting effort on its output shaft, measured as torque. In structures, torque appears as the sum of moments of torsional shear forces acting on a transverse action of a shaft or beam.

[NELSON S. FISK]

Torricelli's theorem

The speed of efflux of a liquid from an opening in a reservoir equals the speed that the liquid would acquire if allowed to fall from rest from the surface of the reservoir to the opening.

Torricelli, a student of Galileo, observed this relationship in 1643. In equation form, $v^2 = 2gh$, in which v is the speed of efflux, h the head (or elevation difference between reservoir surface and center line of opening if in a vertical plane), and g the acceleration due to gravity. (The equation is the same as that for a solid particle dropped a distance h in a vacuum.) The relationship can be derived from the energy equation for flow along a streamline, if energy losses are neglected.

An orifice (opening in the wall or bottom of a reservoir) is used as a flow-measuring device. From Torricelli's theorem, by solving for v and multiplying by the flow area, an expression for discharge Q, in volume per unit time, is obtained. In equation form, $Q = C_d A \sqrt{2gh}$, in which A is the area of opening and C_d is a dimensionless coefficient, determined experimentally, that corrects for contraction of the jet as it leaves the orifice and for energy loss due to viscosity. When h is measured, Q may be determined from the formula.

[VICTOR L. STREETER]

Trajectory

The curve described by a body moving through space, as of a meteor through the atmosphere, a planet around the Sun, a projectile fired from a gun, or a rocket in flight. In general, the trajectory of a body in a gravitational field is a conic section—ellipse, hyperbola, or parabola—depending on the energy of motion. The trajectory of a shell or rocket fired from the ground is a portion of an ellipse with the Earth's center as one focus; however, if the altitude reached is not great, the effect of gravity is essentially constant, and the parabola is a good approximation.

[JOHN P. HAGEN]

Transducer

Any device or element which converts an input signal into an output signal of a different form. An example is the microphone, which converts vibrations caused by an impinging sound wave into an electrical signal. This electrical signal can be measured to determine the magnitude of the sound wave; it can be recorded (through the use of another transducer); or it can be used to control some instrument. Although the most common transducers are designed to transform periodic signals (such as sound waves or alternating-current electrical signals), the word also applies to devices which convert static signals from one form to another. An example of this is the barometer, which produces a signal proportional to the atmospheric pressure. The input signal is the atmospheric pressure. The output signal can either be a mechanical displacement (a dial reading or a liquid level) or it can be a direct-current electrical signal. A different type of transducer is the photoelectric cell, which produces an electrical signal in response to incident light. The most widely used class of transducers is the electromechanical transducer, which converts an electrical signal into a mechanical signal (a vibration or a displacement) or vice versa. Aside from the microphone mentioned above, this class includes phonograph pickups, loudspeakers, automobile horns, doorbells, and underwater transducers.

[M. A. BREAZEALE]

Bibliography: D. E. Gray (ed.), *Amer. Inst. Phys. Handb.*, 3d ed., 1972; J. D. Lenk, *Handbook of Controls and Instrumentation*, 1980; W. P. Mason (ed.), *Phys. Acoust.*, vol. 1, pts. A and B, 1964; P. H. Sydenham, *Transducers in Measurement and Control*, 2d ed., 1980.

Transistor

An active component of an electronic circuit which may be used as an amplifier, detector, or switch. A transistor consists of a small block of semiconducting material to which at least three electrical contacts are made. Transistors are of two general types, bipolar and field-effect. The bipolar type involves excess minority current carrier injection. The field-effect type involves only majority current carriers. Historically the bipolar type was developed before the field-effect type. Today both are widely used. The unmodified term transistor usually refers to the bipolar type.

In a bipolar-type transistor, at least one contact is ohmic (that is, nonrectifying), and at least one contact is rectifying. Usually there are two closely spaced rectifying contacts and one ohmic contact.

The operation of a simple transistor consists of the control of the current flowing in the high-resistance direction through one rectifying contact (called the collector) by the current flowing in the low-resistance direction in the other rectifying contact (called the emitter). The third contact, which is ohmic, is called the base contact.

These contacts usually consist of two or more regions. The regions in which the actual rectification processes take place are called the emitter barrier and collector barrier. The region between these two barriers is called the base region, or simply the base. The regions outside of these barriers are called the emitter and collector regions.

Transistors are used in radio receivers, in electronic computers, in electronic instrumentation and control equipment, and in almost any electronic circuit where vacuum tubes are useful and the required voltages are not too high. Transistors have the advantages over their vacuum-tube counterparts of being much smaller, consuming less power, and having no filament to burn out. They are at a disadvantage in that they do not yet operate at as high voltages as some vacuum tubes and their action is degraded at high temperatures.

Classification of transistors. Transistors are classified chiefly by four criteria: (1) by the type and number of structural regions of the semiconductor crystal; (2) by the technology used in fabrication; (3) by the semiconductor material used; and (4) by the intended use of the device. A typical designation following this scheme would be *npn* double-diffused silicon switching transistor. It is not necessary to include all of the above criteria in a single designation nor to rigidly follow this order.

A modern transistor type is the *npn* double-diffused silicon planar passivated transistor (Fig. 1). The term double-diffused refers to the fabrication technique in which the base region is formed by diffusion through a mask into the body of the silicon wafer which forms the collector region.

In turn, the emitter region is formed by diffusion through a second mask into the previously formed base region. The term planar refers to the fact that all three electrical connections are found on a single surface of the device. The term passivated means that the surface to which all junctions return is protected by a layer of naturally grown silicon oxide which, together with an overcoating of glass or other inert material, passivates the surface, electrically minimizing leakage currents. The dou-

ble-diffusion process allows very close control of narrow base widths. The base diffusion provides a resistivity gradient in the base region which has an associated electric field. In this field charge transport is by drift. Such transistors have been called drift transistors to distinguish them from most other transistors in which the charge transport is by a diffusion process. Silicon planar transistors have power ratings in the 100 mW to 50 W range with characteristic frequencies between 50 and 2000 MHz, usually of the *npn* type. The designation *npn* stands for the conductivity type of the emitter, base, and collector regions, respectively. The *n* stands for negative since the charge on an electron is negative and electrons carry most of the current in a region of *n*-type conductivity. In a region of *p*-type conductivity most of the current is carried by electron vacancies, called holes, which behave as if they were positively charged. For a detailed discussion of conductivity type *see* SEMICONDUCTOR.

A historically important type was the *pnp* alloy-junction germanium transistor. This type was very widely used in the first decade of the solid-state electronics era. The term alloy-junction in this transistor designation refers to the fabrication method. The emitter and collector regions were produced by recrystallization from an alloy of some suitable metal doped with a *p*-type impurity. The alloy had previously been fused in contact with the opposite surfaces of the original *n*-type semiconductor body and had dissolved some of the semiconductor material. Fused-junction is equivalent terminology. This type of transistor was made in power ratings from 50 mW to 200 W, and in frequency ranges up to about 20 MHz.

Transistor action. To explain transistor action in more detail, some of the basic properties of a semiconductor material are first presented. An *n*-type semiconductor contains electrons, and a *p*-type semiconductor contains holes. These are called the majority carriers of the two types. Actually there are always present a small number of holes in an *n*-type semiconductor and a small number of electrons in a *p*-type semiconductor. These are called the minority carriers of the two types. At a given temperature with a given material the product of the densities of the majority and minority carriers is a constant. This means that if there is present a very high density of majority carriers (low-resistivity material), there will be a correspondingly low density of minority carriers.

The emitter current controls the collector current in a simple transistor. To understand this, first consider the magnitude of the collector current in the absence of emitter current. In normal operation the collector barrier is biased in the high-resistance (reverse) direction. Under this condition of bias the majority carriers are stopped by the barrier, and only the minority carriers are free to flow. If the collector barrier is a silicon *pn* junction, the minority-carrier diffusion current is negligible and the reverse-bias leakage current will consist of thermally generated carriers and be in the nanoampere range. If emitter current is present, the portion consisting of carriers entering the base will continue across the collector barrier and thus control the collector current.

Injection. The emitter controls the density of minority carriers by injecting extra minority car-

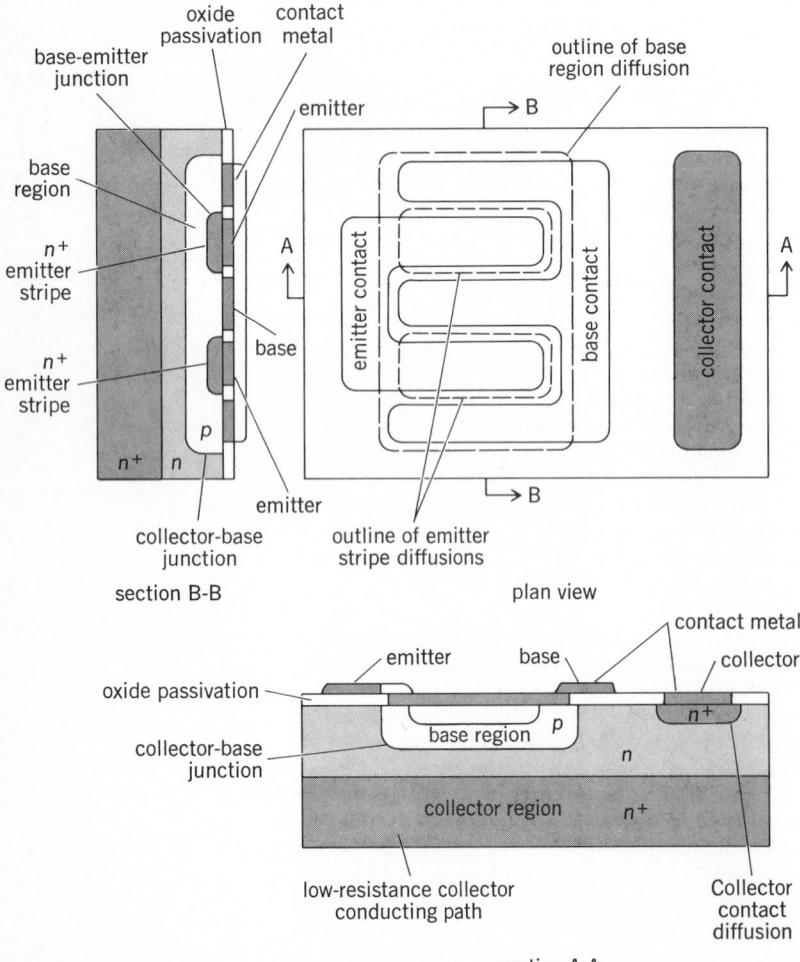

Fig. 1. Plan and sections of a planar *npn* double-diffused silicon transistor.

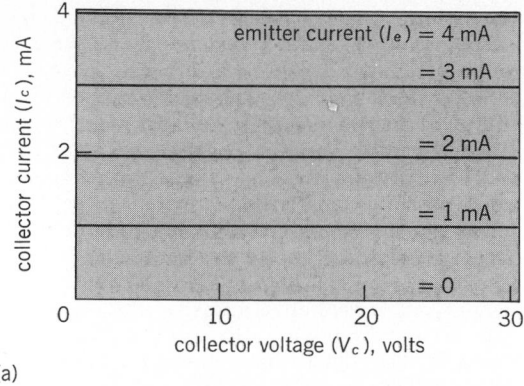

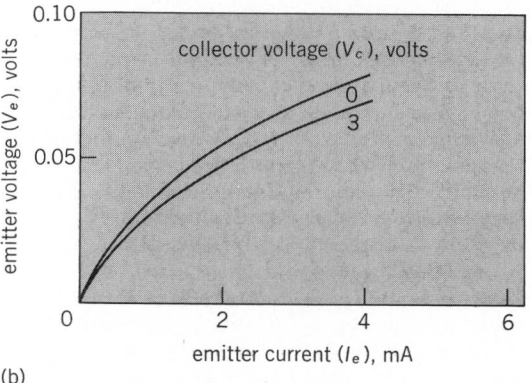

Fig. 2. Typical transistor dc characteristics. (a) Collector characteristics. (b) Emitter characteristics.

riers into the base region when the emitter is biased in the low-resistance (forward) direction. This is the fundamental process of simple transistor action. Whenever a rectifying barrier is forward-biased, extra minority carriers are added to the semiconductor near the barrier. Since the source of these minority carriers is the majority-carrier density on the other side of the barrier, it is clear that the largest part of the forward current will be carried by those carriers which come from the largest majority density. A *pn* junction will have a high injection efficiency for electrons if the *n* region has a much larger density of carriers (lower resistivity) than the *p* region. Therefore, in a *npn* transistor the emitter *n* region should have a low resistivity compared to the *p*-type base region. The phenomenon of minority-carrier injection is observed also in rectifying metal-semiconductor contacts, and such contacts may be used as emitters as well as *pn* junctions.

Current gain. The current gain α of a simple transistor may be expressed as the product of three factors: the fraction γ of the emitter current carried by the injected carriers, and fraction β of the injected carriers which arrive at the collector barrier, and the current multiplication factor α^* of the collector. For a double-diffused transistor, typical values of these factors are $\gamma = 0.985$, $\beta = 0.999$, and $\alpha^* = 1.000$, giving $\alpha = 0.984$. From this it can be seen that most of the current which flows into the emitter flows right on through the base region and out the collector, while only a small fraction (here 0.016) flows out the base connection.

For a fixed value of emitter current I_e there is a fixed value of collector current αI_3 added to the

collector-barrier leakage current I_{co}, giving a total collector current, $I_c = I_{co} + \alpha I_e$. This means that the slope of the dc characteristics should be the same as the slope of the collector-barrier leakage current curve for $I_e = 0$. The typical characteristics in Fig. 2 illustrate this. The slope of the collector leakage curve is very low since the collector voltage does not influence the relatively fixed number of minority carriers carrying the current.

High-frequency effects. These originate in three distinct properties of transistors: the transit time of injected carriers across the base region, the charging time of the collector- or emitter-barrier capacitance through the base-region and collector-region resistances in series, and the time required to build up the proper density of injected carriers in the base region (called storage-capacity effect). In alloy-junction transistors with a base region of uniform resistivity, the transport of injected carriers across the base is usually the limiting factor. Of course, base transit time alone introduces only a phase shift between the emitter and collector signals, but this time also gives a chance for injected carriers, bunched by the emitter signal, to d.f-fuse apart and therefore degrade the signal (Fig. 3).

In double-diffused (drift) transistors the base transit time is usually negligible compared to the charging time of the collector or emitter capacitance, and in some units the storage capacity (often called diffusion capacity) seems to be an appreciable limitation.

Storage capacity also shows up in another way in transistors used as switches. Here it introduces a time delay both in turning on and in turning off the transistor. The turn-off delay is usually longer than the turn-on delay, because the density of injected carriers in the base region has had time to

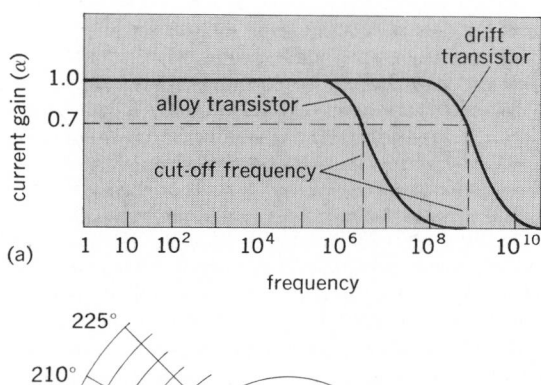

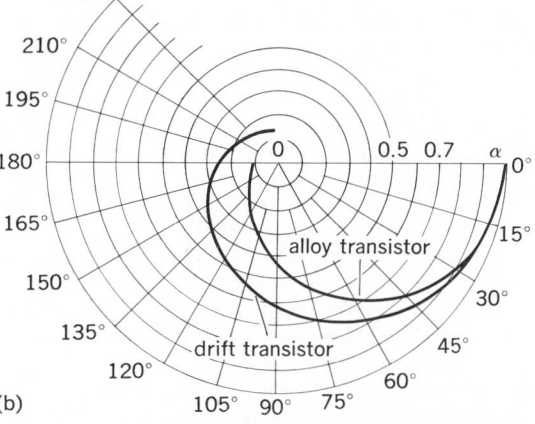

Fig. 3. Typical transistor frequency characteristics. (a) Frequency dependence of α. (b) Phase of collector current I_c versus emitter current I_e.

build up to large values during the time the transistor was on, and therefore takes a long time to subside to the level where the transistor can turn off. These delays are only slightly related to the actual time of rise or fall of the collector level, which is determined primarily by collector-capacitance—base-resistance time constant.

A fabrication technique, called epitaxial growth, is used to minimize the storage capacity effects in high-speed transistors. In this process a transistor structure is formed entirely in a very thin skin of good semiconductor material grown upon the surface of a wafer of heavily doped material. The heavily doped material has very low lifetime for excess carriers and, therefore, a very low storage effect, as well as a low series resistance. The collector junction of such a transistor is close to this low-lifetime material but is formed in the high-quality, epitaxially grown skin so that its properties are not degraded by the heavily doped material. Such transistors are called epitaxial transistors.

Close control of the injection ratio γ, defined above, is afforded by the fabrication technique of ion implantation. In this technique a beam of ions composed of the desired dopant material is accelerated to a specific kinetic energy and caused to strike the surface of the region to be doped. The ions penetrate the surface and remain embedded in the semiconductor material. By controlling the ion-beam current and the time of bombardment, a very accurate control of the total number of dopant ions in the region is achieved. After heating the semiconductor to diffusion temperature, the ions move on into the material, creating the emitter and base regions of the double-diffused structure. These regions now have precisely controlled doping and hence show a γ-factor within $\pm1\%$ of the design value.

Transistor noise. Noise is quite low if a low source impedance is used. With source impedances of about 1000 ohms, a good junction transistor will have a noise factor of about 4 dB. The noise factor is independent of the connection but rises with source impedances above 10,000 ohms and with frequencies below 1000 Hz.

Temperature effects. These are most marked in connection with the collector-barrier leakage current with no emitter current flowing I_{co}. This current increases exponentially with temperature and leads to a phenomenon called thermal runaway. If a transistor is operated at a given ambient temperature and a given initial power dissipation, this power will soon raise the temperature of the collector barrier, which then draws more current and in turn increases the dissipation. The process is cumulative, and precautions must be taken to stabilize against it. Current gain increases slightly with increasing temperature in most *npn* transistors, but this is a small effect unless the current gain is unusually close to unity.

Power switching. There are several transistor structures which are used for power switching and make use of current gains greater than unity to achieve a thyraton-like characteristic. These devices are often called four-layer devices since they usually contain four regions of alternating *n*- and *p*-type semiconductor material. Connections are made to the end regions and to one of the interior regions. The end regions are oppositely biased so

that the center junction is reverse-biased. The connection to the interior region is then the control and is usually called the gate. When the gate is biased to cause injection of excess carriers across the junction between it and the nearest end connection, the device is triggered on and a saturation current is drawn between the two end connections normally called anode and cathode. Such devices are normally classified as rectifiers but in reality are a form of transistor.

Field-effect transistor (FET). There are two major types of field-effect transistors, the junction-gate FET (JFET) and the insulated-gate FET (IGFET). The IGFET is commonly called MOSFET or MOS transistor. The acronym MOS stands for metal-oxide-semiconductor which describes, in order, the structure of the device from the gate toward the channel. The JFET was developed first, since it involved no technology beyond that of the planar bipolar silicon transistor. The development of the MOSFET was delayed while the technology was extended to stable control of silicon surface potential. The MOSFET is very widely applied in large-scale integration, particularly in implementing large random-access high-speed memories for computers.

JFET. Figure 4*a* shows a section of a JFET. The channel consists of relatively low-conductivity semiconductor material sandwiched between two regions of high-conductivity material of opposite type. When these junctions are reverse-biased, the junction depletion regions encroach upon the channel and finally, at a high reverse bias, pinch it off entirely. The thickness of the channel, and hence its conductivity, is controlled by the voltage on the two gates. This device is therefore normally on and may be switched off. It is called a depletion-mode FET. In practice this FET has an input impedance several orders of magnitude greater than that of a silicon bipolar transistor. JFETs are made in both *n*-channel and *p*-channel types. They are used in amplifiers, oscillators, mixers, and switch-

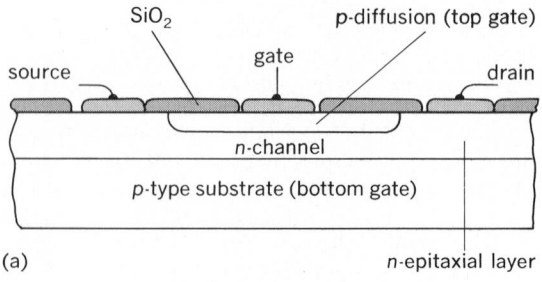

(a)

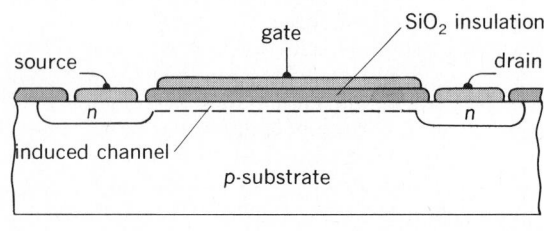

(b)

Fig. 4. Field effect transistors. (*a*) Junction-gate FET (JFET). (*b*) Insulated-gate FET (IGFET or MOSFET).

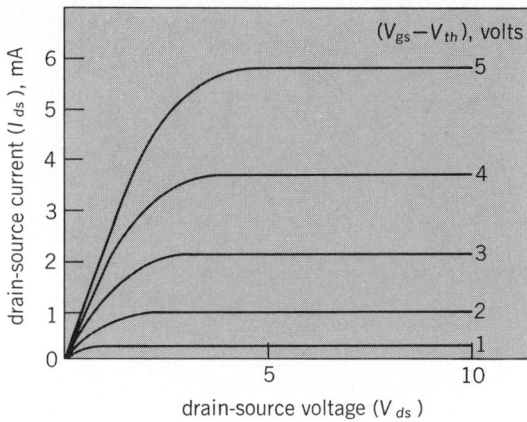

Fig. 5. Typical MOSFET drain characteristics.

es. The general performance limits are about 500 MHz, 1 W, 100 V, and 100 mA (saturation drain current). They also find application in integrated circuits employing bipolar transistors since their technology is compatible.

MOSFET. Figure 4*b* shows a section of a MOSFET. Here the source and drain regions consist of *n* diffusion in a *p*-type substrate. The gate is a metal film evaporated on a thin SiO insulator spanning the separation between the source and drain. With no voltage on the gate, the source and drain are insulated from each other by their surrounding junctions. When a positive voltage is applied to the gate, electrons are induced to move to the surface of the *p*-type substrate immediately beneath the gate, producing a thin surface of induced *n*-type material which now forms a channel connecting the source and drain. Such a surface layer is called an inversion layer since it is of opposite conductivity type to the substrate. The number of induced electrons is directly proportional to the gate voltage, so that the conductivity of the channel increases with gate voltage. This device is called an *n*-channel enhancement-mode MOSFET. It is normally off at zero gate voltage.

Because of the quality of the silicon dioxide gate insulator, the input impedance of a MOSFET is several orders of magnitude greater than that of a JFET. Typical MOSFET dc characteristics are shown in Fig. 5. The low-drain voltage channel resistance is inversely proportional to $(V_{gs} - V_{th})$, where V_{gs} is the gate-source voltage and V_{th} is the threshold voltage, and the saturation drain current is proportional to $(V_{gs} - V_{th})^2$.

MOSFET devices are fabricated in both *p*-channel and *n*-channel types, as well as for both depletion (normally on) and enhancement (normally off) modes of operation. In a MOSFET the mode of operation is determined by a threshold voltage of the gate at which the device changes from off to on, or vice versa. In modern technology this threshold voltage can be set for a wide range of values by the use of ion implantation through the gate oxide.

MOSFET discrete devices are used for ultra-high-input impedance amplifiers such as electrometers where the input leakage current is less than 10^{-14} A. Dual-gate depletion types can be used as mixers up to 1000 MHz, and power-switching types (the VMOS discussed below) are good to

25 W, 2 A, or 100 V. Most integrated circuits using MOSFETs are called CMOS integrated circuits, where the C stands for complementary. These circuits use *n*-channel and *p*-channel types together to achieve digital logic. Typical propagation delay times through small-scale integrated building-block circuits such as three-input NAND or NOR gates is about 20 nanoseconds for a 20-picofarad load. At a 10-MHz clock rate, the power dissipation for such a gate is about 10 mW. For large-scale integration, a typical 16 kilobit random-access memory has an access time of 200 ns, an active power of 500 mW, and a standby power of 20 mW.

VMOS and SOS. There are a number of variations of the MOS technology. Two of particular interest are VMOS (V for vertical) and SOS (silicon on sapphire). The VMOS device is fabricated by etching a notch down through a planar double-diffused structure similar to that of an *npn* bipolar transistor. The surface of the notch is first oxidized and then covered with the gate metallization. The source contact bridges the n^+-*p* junction near the surface, and the drain connection corresponds to the collector contact of the bipolar structure. The channel length is now determined by the thickness of the *p* region. This allows controlled short channels and gives both high current and high voltage capability.

The SOS device is fabricated in a very small silicon body grown epitaxially on a sapphire substrate. An experimental MOS/SOS 1000-bit memory has shown a standby power of only 1 μW.

Double-base diode. Also called a unijunction transistor, this consists of a single rectifying contact situated approximately midway along a semiconductor bar which carries two ohmic contacts at its ends. If a steady bias is applied between the ends of the bar, a negative-resistance diode characteristic is observed between the rectifying contact and one end of the bar. This device is used primarily for switching.

Transistor manufacture. The manufacture of transistors has required a whole new field of exacting technology. Good semiconductor material requires the maintenance of chemical purities far beyond the spectroscopic range. A purity of 1 part in 10^8 is not unusual. Most devices must be made from oriented single crystals of semiconductor material which can have only very low densities of structural defects. *See* SINGLE CRYSTAL.

Physical tolerances of the high-frequency transistor structures are microscopic; the separation of emitter and collector junctions must be of the order of a few micrometers in these units.

To solve these problems new techniques have appeared. Purity is achieved by melting a small zone of a bar, or ingot, and gradually passing this molten zone from one end of the bar to the other. Impurities in the material remain in the liquid phase and are carried along with the molten zone, leaving high-purity material behind.

Tolerances are achieved by a collection of new techniques, such as epitaxial growth, solid-state diffusion, ion implantation, and the photolithographic delineation of diffusion masks.

[LLOYD P. HUNTER]

Bibliography: J. Millman, *Micro-Electronics*, 1979; E. S. Yang, *Fundamentals of Semiconductor Devices*, 1978.

Transition point

The point at which a substance changes from one state of aggregation to another. This general definition would include the melting point (transition from solid to liquid), boiling point (liquid to gas), or sublimation point (solid to gas); but in practice the term transition point is usually restricted to the transition from one solid phase to another, that is, to the temperature (for a fixed pressure, usually 1 atm) at which a substance changes from one crystal structure to another.

Some typical examples of transition points are:

$$\beta\text{-Fe} \xrightarrow{\text{at 1180 K}} \gamma\text{FE}$$
(body-centered cubic)   (face-centered cubic)

$$S_8 \xrightarrow{\text{at 369 K}} S_8$$
(rhombic)   (monoclinic)

$$CCl_4 \xrightarrow{\text{at 225.5 K}} CCl_4$$
(monoclinic)   (tectragonal)

$$NH_4NO_3 \xrightarrow{\text{at 305.3 K}} NH_4NO_3$$
(β-rhombic)   (α-rhombic)

$$NH_4NO_3 \xrightarrow{\text{at 357.4 K}} NH_4NO_3$$
(α-rhombic)   (trigonal)

Another kind of transition point is the culmination of a gradual change (for example, the loss of ferromagnetism in iron or nickel) at the lambda point, or Curie point. This behavior is typical of second-order transitions. *See* SECOND-ORDER TRANSITION; TRIPLE POINT.

[ROBERT L. SCOTT]

Transition radiation detectors

Detectors of energetic charged particles that make use of radiation emitted as the particle crosses boundaries between regions with different indices of refraction. An energetic charged particle moving through matter momentarily polarizes the material nearby. If the particle crosses a boundary where the index of refraction changes, the change in polarization gives rise to the emission of electromagnetic transition radiation. About one photon is emitted for every 100 boundaries crossed, for transitions between air and matter of ordinary density. Transition radiation is emitted even if the velocity of the particle is less than the light velocity of a given wavelength, in contrast to Cerenkov radiation. Consequently, this radiation can take place in the x-ray region of the spectrum where there is no Cerenkov radiation, because the index of refraction is less than one. *See* CERENKOV RADIATION; REFRACTION OF WAVES.

The radiation extends to frequencies greater than the plasma frequency by the factor $\gamma =$ particle energy divided by particle mass. The production of x-rays requires γ equal to or greater than 1000. A threshold as high as this is difficult to achieve by other means. This fact has led to the application of this effect for the identification of high-energy particles. For example, electrons of about 10^9 eV will produce x-rays of a few kiloelectronvolts, while the threshold for pions is on the order of 10^{11} eV. The solid material should be of low atomic number, carbon or lighter, to minimize absorption of x-rays, which are emitted close to the particle direction. The material is often in the form of foils, which must be of the order of 1/100 mm thick to avoid destructive interference of the radiation from the two surfaces, and similarly, the foil spacing is typically 1/10 mm. Random assemblies of fibers or foams are almost as effective as periodic arrays of foils. Effective electron detectors have been made with several hundred foils followed by a xenon proportional chamber for x-ray detection. *See* PARTICLE DETECTOR; PLASMA PHYSICS; RELATIVISTIC MECHANICS. [WILLIAM J. WILLIS]

Bibliography: X. Artru, G. B. Yodh, and G. Menessier, Practical theory of the multilayered transition radiation detector, *Phys. Rev. D*, 12: 1289–1306, 1975; J. Cobb et al., Transition radiators for electron identification at the CERN ISR, *Nucl. Instrum. Method.*, 140:413–427, 1977; G. M. Garibian, Transition radiation effects in particle energy losses, *Sov. Phys. JETP*, 10:372–376, 1960.

Translucent medium

A medium which transmits rays of light so diffused that objects cannot be seen distinctly; that is, the medium is only partially transparent. Familiar examples are various forms of glass which admit considerable light but impede vision. Inasmuch as the term translucent seems to imply seeing, usage of the term is ordinarily limited to the visible region of the spectrum. *See* TRANSPARENT MEDIUM.

[M. G. MELLON]

Transparent medium

Ordinarily, a medium which has the property of transmitting rays of light in such a way that the human eye may see through the medium distinctly. It is pervious to light, that is, to the visible region of the electromagnetic spectrum. By extension, a medium may be described as transparent to other regions of the spectrum, such as x-rays and microwaves. Just as a blue filter passes blue rays, an ultraviolet filter might be considered as passing ultraviolet rays, or being transparent to them. *See* ABSORPTION OF ELECTROMAGNETIC RADIATION; TRANSLUCENT MEDIUM. [M. G. MELLON]

Traps in solids

Electron traps are defects or chemical impurities in semiconductors and in insulators which capture mobile electrons in a special way. The electrons are immobilized, prevented from recombining with (annihilating) positive holes, and are released some time later as mobile electrons again. Analogous traps act on mobile positive holes. A defect which captures a mobile electron and aids recombination with a hole (or vice versa) is called a recombination center. A defect (nickel in germanium, for example) may be a recombination center at one temperature and a trap at a different one.

Electrons or holes may remain in traps for short times or for as long as months or years. They may be released by heating the solid or by irradiating it with infrared. Traps have an important influence in photoconduction, luminescence, and the photographic process. *See* BAND THEORY OF SOLIDS; HOLES IN SOLIDS; LUMINESCENCE; PHOTOCONDUCTIVITY. [L. APKER]

Trichroism

When certain optically anisotropic transparent crystals are subjected to white light, a cube of the material is found to transmit a different color through each of the three pairs of parallel faces. Such crystals are sometimes termed trichroic, and the phenomenon is called trichroism. This expression is used only rarely today since the colors in a particular crystal can appear quite different if the cube is cut with a different orientation with respect to the crystal axes. Accordingly, the term is frequently replaced by the more general term pleochroism. Even this term is being replaced by the phrase linear dichroism or circular dichroism to correspond with linear birefringence or circular birefringence. *See* BIREFRINGENCE; DICHROISM; PLEOCHROISM.

Cordierite is a typical trichroic crystal. In light with a vibration direction parallel to the X axis of the index ellipsoid the crystal appears yellow. With the vibration direction parallel to the Y axis the crystal is dark violet. In the Z direction the crystal is clear.

The phenomena of trichroism can be explained crudely as follows. Classically, one can consider an electron in a biaxial crystal as having three different force constants associated with a displacement directed along each of the principal axes. Linear polarized light traveling along the X axis with its electric vector parallel to the Y axis will displace the electron against the Y force constant and will experience a certain absorption and retardation. It will be unaffected by the force constants in the X and Z directions. Similarly, polarized light traveling in the Y direction will experience absorption and retardation. Unpolarized light will also be absorbed in a different fashion depending on the direction of propagation. In this case light traveling in the X direction can be considered as composed of an equal mixture of light polarized parallel to the Y axis and the Z axis. The absorption will be intermediate between the two polarization directions. *See* CRYSTAL OPTICS; POLARIZED LIGHT.

[BRUCE H. BILLINGS]

Bibliography: N. H. Hartshorne and A. Stuart, *Practical Optical Crystallography*, 2d ed., 1970; E. A. Wood, *Crystals and Light: An Introduction to Optical Crystallography*, 1977.

Trigonometry

The study of triangles and the trigonometric functions. The functions of trigonometry are extremely important because they may be used to represent ranges of values which are repeated again and again. They represent periodic phenomena such as the motions of pendulums or the analysis of alternating-current electricity. In fact, they play a basic role in the theories of light, sound, radio, television, and generally in all phenomena of a vibratory character.

Also, trigonometry is used to find a vast network of lengths that cannot be measured directly. Surveyors use it to find heights of mountains, distances across lakes and countries, and positions of places; engineers use it in the design of large structures and roads; astronomers use it in accurate measurements of time, and in locating the positions of bodies in the sky; and navigators on the sea and in the air use it to find latitudes, longitudes, and directions.

The subject of trigonometry may be divided into plane trigonometry and spherical trigonometry.

PLANE TRIGONOMETRY

Plane trigonometry, for the most part, deals with triangles with straight lines in a plane as sides. Unless otherwise specified, it is understood that all figures considered in this section lie in a plane.

Angles. If a half-line, or ray, having end point O (see Fig. 1) rotates about O from an initial position OA to terminal position OB, it is said to generate the angle AOB. A rotation of 1/360 part of a complete rotation about a point is called a degree and written $1°$. One-sixtieth of a degree is called a minute and written $1'$. Generally angles generated by rotation in a clockwise direction are called negative angles, and those generated by counterclockwise turning are called positive angles. Figure 2, indicating various angles, illustrates this concept.

A radian is an angle subtended at the center of a circle by an arc of the circle equal in length to its radius (see Fig. 3). It is used in purely theoretical discussions because it avoids cumbersome constants.

Trigonometric functions. Figure 4 shows an angle θ with vertex O at the origin of a system of rectangular coordinates and with initial line OA directed in the positive direction of the x axis. Let P, any point except O, on the terminal ray of the angle θ, have coordinates x and y and be at a distance r from O. Then the six functions, sine, cosine, tangent, cotangent, secant, and cosecant of θ, respectively abbreviated by $\sin \theta$, $\cos \theta$, $\tan \theta$, $\cot \theta$, $\sec \theta$, $\csc \theta$, are defined by Eqs. (1) and (2),

$$\sin \theta = y/r, \cos \theta = x/r, \tan \theta = y/x \qquad (1)$$

$$\csc \theta = r/y, \sec \theta = r/x, \cot \theta = x/y \qquad (2)$$

where r is positive. These ratios are independent of the position of the point (x,y) on the terminal ray. *See* COORDINATE SYSTEMS.

Figure 4 shows that any angle with initial side OA obtained by adding an integral multiple of $360°$ to θ has the same terminal ray as θ, and therefore that Eq. (3) can be written, where fn stands for sin,

$$\text{fn} (\theta + k360°) = \text{fn} (\theta) \qquad (3)$$

cos, tan, cot, sec, or csc, and k is an integer. In other words the trigonometric functions are periodic, and a period is $360°$ or 2π radians. This is the property that makes them so useful in theoretical considerations involving periodic values.

Figure 5 shows four angles $60°$, $120°$, $240°$, and $300°$ having terminal lines containing respective points $(1,\sqrt{3})$, $(-1,\sqrt{3})$, $(-1,-\sqrt{3})$, and $(1,-\sqrt{3})$.

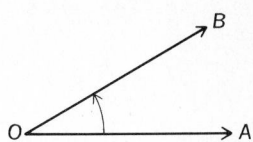

TRIGONOMETRY

Fig. 1. Generation of angle *AOB*.

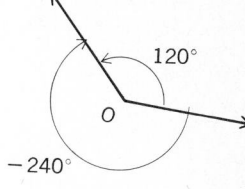
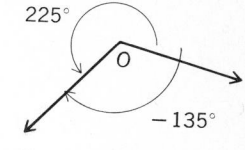
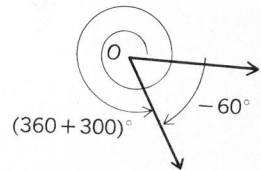

TRIGONOMETRY

Fig. 2. Generation of positive and negative angles.

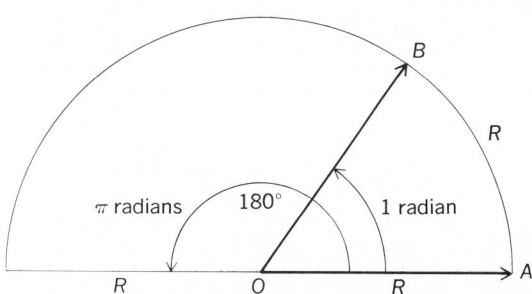

Fig. 3. Radian measurement.

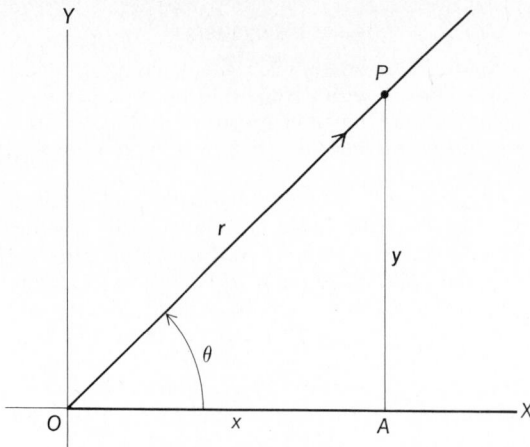

Fig. 4. Angle *AOP* in the rectangular coordinate system.

Since $r = \sqrt{1^2 + (\sqrt{3})^2} = 2$ in all cases the values from Eq. (1) can be tabulated as in notation (4).

	sin	cos	tan	cot	sec	csc	
60°	$\frac{1}{2}\sqrt{3}$	$\frac{1}{2}$	$\sqrt{3}$	$\frac{1}{3}\sqrt{3}$	2	$2/\sqrt{3}$	
120°	$\frac{1}{2}\sqrt{3}$	$-\frac{1}{2}$	$-\sqrt{3}$	$-\frac{1}{3}\sqrt{3}$	-2	$2/\sqrt{3}$	(4)
140°	$-\frac{1}{2}\sqrt{3}$	$-\frac{1}{2}$	$\sqrt{3}$	$\frac{1}{3}\sqrt{3}$	-2	$-2/\sqrt{3}$	
300°	$-\frac{1}{2}\sqrt{3}$	$\frac{1}{2}$	$-\sqrt{3}$	$-\frac{1}{3}\sqrt{3}$	2	$-2/\sqrt{3}$	

In all cases $|x| \leqq r$, $|y| \leqq r$, and therefore the range of the sine and of the cosine is -1 to $+1$ inclusive. Figure 5 shows the angles 0°, 90°, 180°, and 270° containing on their terminal rays the respective points (1,0), (0,1), (−1,0) and (0,−1). Using Fig. 5 and Eqs. (1) and (3), the values may be tabulated as in notation (5), where k is an integer.

	0° + k360°	90° + k360°	180° + k360°	270° + k360°	
sin	0	1	0	−1	
cos	1	0	−1	0	(5)
tan	0	Undefined	0	Undefined	

If the angle θ is near to, but less than 90°, x is small compared to y and therefore the tangent is large; in fact tan θ ranges through all positive numbers as angle θ increases from 0° to 90°. The graphs of the trigonometric functions give a clear picture of the ranges and general behavior of these functions. The three functions shown in Eq. (2) are comparatively unimportant. The cotangent is the reciprocal of the tangent, cot $\theta = 1/\tan \theta$, and it takes on all positive values as θ varies from 0° to 90°. Also, $r \geqq |x|$ and $r \geqq |y|$; therefore, $|\sec \theta| \geqq 1$ and $|\csc \theta| \geqq 1$.

From Eqs. (1) and (2), it follows that Eqs. (6) and (7) can be written. Also from Fig. 4, $y^2 + x^2 = r^2$.

$$\csc \theta = \frac{1}{\sin \theta} \quad \sec \theta = \frac{1}{\cos \theta} \quad \cot \theta = \frac{1}{\tan \theta} \quad (6)$$

$$\tan \theta = \frac{\sin \theta}{\cos \theta} \quad (7)$$

Dividing this through by r^2 and using Eq. (1),

$(\sin \theta)^2 + (\cos \theta)^2 = 1$, which is written as in Eq. (8).

$$\sin^2 \theta + \cos^2 \theta = 1 \quad (8)$$

Using Eqs. (6), (7), (8), and others mentioned below, many complicated expressions can be transformed to simple ones.

Addition formulas. The addition formulas of trigonometry, namely, Eqs. (9) and (10), lead to practically all of the relations between the trigonometric functions.

$$\cos (\phi + \theta) = \cos \phi \cos \theta - \sin \phi \sin \theta \quad (9)$$

$$\sin (\phi + \theta) = \sin \phi \cos \theta + \cos \phi \sin \theta \quad (10)$$

Figure 6 is used in the proof of Eqs. (9) and (10). It represents a set of coordinate axes and a circle O of radius 1 with center at (0,0). Points A (1,0), B, C, D are so placed on circle O that angle AOB is an angle α, and COD is angle AOB revolved through angle θ about (0,0). Hence B is the point (cos α, sin α), C is (cos θ, sin θ), and D is [cos $(\alpha + \theta)$, sin $(\alpha + \theta)$]. Use of the distance formula to express the fact that chord CD equals chord AB gives Eq. (11). *See* ANALYTIC GEOMETRY.

$$\cos^2 (\alpha + \theta) - 2 \cos (\alpha + \theta) \cos \theta + \cos^2 \theta$$
$$+ \sin^2 (\alpha + \theta) - 2 \sin (\alpha + \theta) \sin \theta + \sin^2 \theta$$
$$= 1 - 2 \cos \alpha + \cos^2 \alpha + \sin^2 \alpha \quad (11)$$

Substituting Eq. (8) into Eq. (11) three times, once for $\theta = \alpha + \theta$, once for $\theta = \theta$, and once for $\theta = \alpha$, and simplifying slightly, one obtains Eq. (12).

$$\cos (\alpha + \theta) \cos \theta + \sin (\alpha + \theta) \sin \theta = \cos \alpha \quad (12)$$

Formula (12) holds for any values of angles α and θ. In Eq. (12), α may be replaced by $(\phi - \theta)$ to obtain Eq. (13). It is surprising and pleasing to derive with

$$\cos (\phi - \theta) = \cos \phi \cos \theta + \sin \phi \sin \theta \quad (13)$$

ease a long list of the useful formulas of trigonometry from Eqs. (13) and (5). In Eq. (13), ϕ may be replaced by 90°, and Eq. (5) used to obtain Eq. (14).

$$\cos (90° - \theta) = \sin \theta \quad (14)$$

Equation (14) shows that cos [90° − (−θ)] = sin

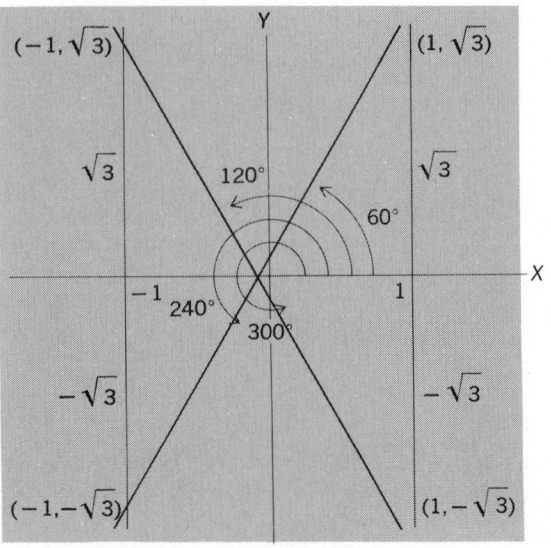

Fig. 5. Four angles in rectangular coordinate.

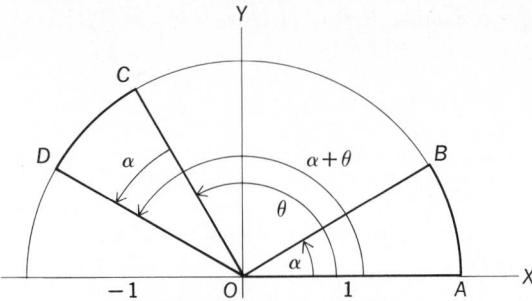

Fig. 6. Diagram for Eqs. (9) and (10).

($-\theta$) and in combination with Eqs. (13) and (5) shows that cos $[\theta + (-90°)] = -\sin \theta$; therefore Eq. (15) holds.

$$\sin (-\theta) = -\sin \theta \qquad (15)$$

In Eq. (13), ϕ may be replaced by 0 and, making use of Eq. (3), Eq. (16) may be written.

$$\cos (-\theta) = \cos \theta \qquad (16)$$

In Eq. (13) θ may be replaced by $-\theta$, and, using Eqs. (15) and (16), Eq. (9) is obtained. In Eq. (14) θ may be replaced by $(90° - \phi)$ to get Eq. (17).

$$\cos \phi = \sin (90° - \phi) \qquad (17)$$

In Eq. (13) ϕ may be replaced by $(90° - \phi)$, and, using Eqs. (14) and (17), Eq. (10) is obtained. In Eq. (10) θ may be replaced by $-\theta$, and using Eqs. (15) and (17), one may write Eq. (18).

$$\sin (\phi - \theta) = \sin \phi \cos \theta - \cos \phi \sin \theta \qquad (18)$$

Reduction formulas. Equations (19)–(24), together with Eq. (5), make it possible to express a trigonometric function of any angle as a function of a positive angle no greater than 45°. Tables giving the values of the trigonometric functions at convenient intervals have been computed. Because of these reduction formulas, it is necessary to compute the value of the functions only for the interval from 0 to 45°.

Using Eqs. (9), (10), (13), and (18), together with Eq. (5), the reduction formulas given in Eqs. (19)–(24) may be verified. For example, by Eqs. (3) and (24), tan (1745°) = tan (1745° − 4 · 360°) = tan 305° = tan (270° + 35°) = −cot 35°.

$$\sin (90° \pm \theta) = \cos \theta, \cos (90° \pm \theta) = \mp \sin \theta \qquad (19)$$

$$\tan (90° \pm \theta) = \mp \cot \theta, \cot (90° \pm \theta) = \mp \tan \theta \qquad (20)$$

$$\begin{aligned}\sin (180° \pm \theta) &= \mp \sin \theta \\ \cos (180° \pm \theta) &= -\cos \theta\end{aligned} \qquad (21)$$

$$\begin{aligned}\tan (180° \pm \theta) &= \pm \tan \theta \\ \cot (180° \pm \theta) &= \pm \cot \theta\end{aligned} \qquad (22)$$

$$\begin{aligned}\sin (270° \pm \theta) &= -\cos \theta \\ \cos (270° \pm \theta) &= \pm \sin \theta\end{aligned} \qquad (23)$$

$$\begin{aligned}\tan (270° \pm \theta) &= \mp \cot \theta \\ \cot (270° \pm \theta) &= \mp \tan \theta\end{aligned} \qquad (24)$$

Double angle and half-angle formulas. Substituting θ for ϕ in Eqs. (9) and (10), one gets Eqs. (25) and (26).

$$\sin 2\theta = 2 \sin \theta \cos \theta \qquad (25)$$

$$\cos 2\theta = \cos^2 \theta - \sin^2 \theta \qquad (26)$$

Solving Eq. (26) and $\cos^2 \theta + \sin^2 \theta = 1$ for $\sin \theta$

and cos θ, and replacing θ by $\tfrac{1}{2}\theta$ in the result, one obtains Eqs. (27) and (28).

$$\sin \tfrac{1}{2}\theta = \pm \sqrt{(1 - \cos \theta)/2} \qquad (27)$$

$$\cos \tfrac{1}{2}\theta = \pm \sqrt{(1 + \cos \theta)/2} \qquad (28)$$

Dividing Eq. (27) by Eq. (28), member by member, one obtains Eq. (29).

$$\tan \tfrac{1}{2}\theta = \pm \sqrt{\frac{1 - \cos \theta}{1 + \cos \theta}} \qquad (29)$$

Conversion formulas. Equations (30)–(33) below are used to transform a product into a sum; Eqs. (35)–(38) are used to transform a sum into a product. This is generally done for some purpose such as evaluation by logarithms or by a computing machine.

First adding and then subtracting Eqs. (9) and (13), member by member, one gets Eqs. (30) and (31).

$$\cos (\phi + \theta) + \cos (\phi - \theta) = 2 \cos \phi \cos \theta \qquad (30)$$

$$\cos (\phi + \theta) - \cos (\phi - \theta) = -2 \sin \phi \sin \theta \qquad (31)$$

Similarly, from Eqs. (10) and (18), one gets Eqs. (32) and (33).

$$\sin (\phi + \theta) + \sin (\phi - \theta) = 2 \sin \phi \cos \theta \qquad (32)$$

$$\sin (\phi + \theta) - \sin (\phi - \theta) = 2 \cos \phi \sin \theta \qquad (33)$$

In Eqs. (30)–(33), by making the substitution shown as Eq. (34), the relationships labeled (35)–(38) are obtained.

$$\phi = \tfrac{1}{2}(A + B), \theta = \tfrac{1}{2}(A - B) \qquad (34)$$

$$\cos A + \cos B = 2 \cos \tfrac{1}{2}(A + B) \cos \tfrac{1}{2}(A - B) \qquad (35)$$

$$\cos A - \cos B = -2 \sin \tfrac{1}{2}(A + B) \sin \tfrac{1}{2}(A - B) \qquad (36)$$

$$\sin A + \sin B = 2 \sin \tfrac{1}{2}(A + B) \cos \tfrac{1}{2}(A - B) \qquad (37)$$

$$\sin A - \sin B = 2 \cos \tfrac{1}{2}(A + B) \sin \tfrac{1}{2}(A - B) \qquad (38)$$

Inverse trigonometric functions. In the equation $x = \sin y$, y is an angle having x as sine. The symbol $\sin^{-1} x$ (or arcsin x) means the angle having x as sine, or the inverse sine of x. Evidently arcsin $\tfrac{1}{2}$ has the values 30° + $k360°$ and 150° + $k360°$, where k is an integer. A like notation applies to the other functions. For example, arccos $\tfrac{1}{2}$ has the values ±60° + $k360°$. Generally a particular value, called the principal value, is chosen and designated by Arcsin x, Arccos x, Arctan x, etc. Arcsin x and Arctan x are chosen in the range from −90° to +90°, and Arccos x and Arccot x are chosen in the range 0° to 180°. Thus Arcsin $\tfrac{1}{2}$ = 30°, Arcsin $(-\tfrac{1}{2})$ = −30°, Arccos $(\tfrac{1}{2})$ = 60°, Arccos $(-\tfrac{1}{2})$ = 120°, Arctan (-1) = −45°, and Arccot (-1) = 135°.

Solution of rectilinear figures. If a side and another part of a right triangle are known, the triangle can be solved. Consider, for example, the right triangle ABC of Fig. 7 having AB = 25.00 ft, angle B = 90°, and angle A = 60°. From a table of trigonometric functions, tan 60 = 1.732. Then from

TRIGONOMETRY

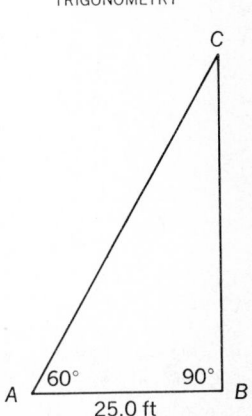

Fig. 7. A right triangle.

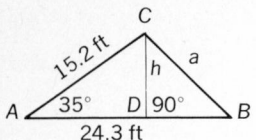

Fig. 8. Triangle for
problem solving.

Fig. 7, $BC/AB = BC/25.00 = \tan 60° = 1.732$ (nearly),
and $BC = 25.00 (1.732) = 43.30$ ft. Similarly, from
Fig. 7, $AB/AC = 25.00/AC = \cos 60° = 0.5000$, and
$AC = 50.00$ ft. Note that the process of finding an
unknown part consists in writing a formula containing the two knowns and the unknown and solving it for the unknown.

A rectilinear figure can often be solved by drawing perpendiculars to sides of the figure to divide it
into right triangles and then solving the set of right
triangles in order. For example, the triangle ABC
of Fig. 8, having $AB = 24.3$ ft, $AC = 15.2$ ft, angle
$A = 35°$, can be solved by drawing the altitude CD
from C to side AB, solving first the right triangle
ACD, and then the right triangle CDB.

However, various formulas are used for solving
rectilinear figures. For convenience, the vertices of
a triangle are denoted by capital letters A, B, C,
and the respective opposite sides by small letters
a, b, c, as indicated in Fig. 9.

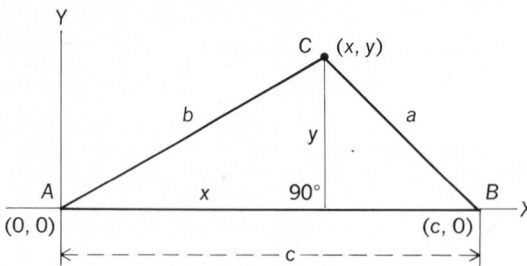

Fig. 9. Generalized triangle.

Law of sines. Figure 9 shows a coordinate system with vertex A of a triangle at the origin, side
AB on the positive side of the x axis, and vertex C
above the x axis. By the law of sines Eq. (39) can be
written.

$$\frac{a}{\sin A} = \frac{b}{\sin B} = \frac{c}{\sin C} \tag{39}$$

Let (x,y) designate the point C in Fig. 9. Then,
Eqs. (40a) hold for all angles A and B less than
180°. Division of the first equation of Eq. (40a) by
the second gives Eq. (40b).

$$\frac{y}{b} = \sin A \qquad \frac{y}{a} = \sin B \tag{40a}$$

$$\frac{a}{b} = \frac{\sin A}{\sin B} \quad \text{or} \quad \frac{a}{\sin A} = \frac{b}{\sin B} \tag{40b}$$

A like argument would show that $b/\sin B =
c/\sin C$. The law of sines is used in solving triangles
in which the given parts are two angles and a side,
or two sides and the angle opposite one of them.
Thus, if a triangle has $A = 65°$, $B = 40°$, $a =
5.0 \times 10^2$, then $b/\sin 40° = 500/\sin 65°$. Hence
$b = 500 \sin 40°/\sin 65° = 350$ nearly. Also from
$c/\sin 75° = 500/\sin 65°$, $c = 530$ nearly.

Law of cosines. The law of cosines may be stated
for the general triangle of Fig. 9 as in Eq. (41).

$$a^2 = b^2 + c^2 - 2bc \cos A \tag{41}$$

From Fig. 9 also, $x^2 + y^2 = b^2$ and $(x-c)^2 + y^2 =
a^2$. Subtracting the first of these equations from
the second, member by member, and replacing x
in the result by its equal $b \cos A$, Eq. (41) is obtained after a slight simplification. Similarly, by

interchanging letters in the argument just made,
Eqs. (42) can be written.

$$b^2 = a^2 + c^2 - 2ac \cos B \\ c^2 = a^2 + b^2 - 2ab \cos C \tag{42}$$

Law of tangents. Other formulas relating to
triangles can be derived from the law of sines and
the law of cosines. The law of tangents may be
stated, relative to the triangle of Fig. 9, as in
Eq. (43).

$$\frac{\tan \frac{1}{2}(A-B)}{\tan \frac{1}{2}(A+B)} = \frac{a-b}{a+b} \tag{43}$$

From the law of sines Eq. (44) is obtained. Two

$$\frac{\sin A}{\sin B} = \frac{a}{b} \tag{44}$$

equations are formed, one by subtracting 1 from
both members of Eq. (44) and the other by adding 1
to both members; then the first of these equations
is divided by the second and simplified to get Eq.
(45). The numerator and the denominator of the

$$\frac{\sin A - \sin B}{\sin A + \sin B} = \frac{a-b}{a+b} \tag{45}$$

left member of Eq. (45) may be replaced by their
values from Eqs. (37) and (38), and the result
simplified to Eq. (43) by using Eq. (7). Two other
forms of the law of tangents are obtained, one by
replacing B by C and b by c in Eq. (43) and the
other by replacing A by C and a by c.

The law of tangents is frequently used in solving
a triangle when two sides and the included angle
are given. Thus, for a triangle having $a = 46.0$, $b =
31.0$, $C = 60°$, $(A + B) = 180° - 60° = 120°$, and from
Eq. (43), $\tan \frac{1}{2}(A-B) = \tan \frac{1}{2}(120°) (46-31)/
(46+31)$. From this equation, $\frac{1}{2}(A-B) = 18.6°$
nearly. Solving this with $\frac{1}{2}(A+B) = 60°$, $A =
78.6°$, $B = 41.4°$. Side c can now be found by the
law of sines.

Half-angle formulas. Two half-angle formulas
for the general triangle of Fig. 9 are given in Eqs.
(46), where Eqs. (47) apply.

$$\sin \frac{1}{2}A = \sqrt{\frac{(s-b)(s-c)}{bc}}$$
$$\cos \frac{1}{2}A = \sqrt{\frac{s(s-a)}{bc}} \tag{46}$$

$$s = \frac{1}{2}(a+b+c) \\ s - a = \frac{1}{2}(-a+b+c) \tag{47}$$

Solving Eq. (41) for $\cos A$ and adding 1 to both
sides of the resulting equation, Eq. (48a) is obtained.

Substituting in Eq. (48) $2 \cos^2 \frac{1}{2}A$ for $1 + \cos A$,
from Eq. (28), and using Eqs. (47), one of Eqs. (48b)
is obtained.

$$1 + \cos A = 1 + \frac{c^2 + b^2 - a^2}{2cb} = \frac{(c+b)^2 - a^2}{2cb}$$
$$= \frac{(a+b+c)(c+b-a)}{2cb} \tag{48a}$$

$$2 \cos^2 \frac{1}{2}A = \frac{2s(2)(s-a)}{2cb}$$
$$\cos \frac{1}{2}A = \sqrt{\frac{s(s-a)}{cb}} \tag{48b}$$

The formula for $\sin \frac{1}{2}A$ is proved by a like procedure. Also, from Eq. (46) by division one gets Eq. (49). By interchanging the letters applied to the

$$\tan \tfrac{1}{2}A = \sqrt{\frac{(s-b)(s-c)}{s(s-a)}}$$

$$= \frac{1}{s-a}\sqrt{\frac{(s-a)(s-b)(s-c)}{s}} \qquad (49)$$

different parts in the proof of Eq. (49), formulas can be derived for $\tan \frac{1}{2}B$ and for $\tan \frac{1}{2}C$. The half-angle formulas, Eqs. (50), can be written, where Eq. (51) applies. Formulas (50) are used to solve a triangle for which three sides are known.

$$\tan \tfrac{1}{2}A = \frac{r}{s-a} \qquad \tan \tfrac{1}{2}B = \frac{r}{s-b}$$

$$\tan \tfrac{1}{2}C = \frac{r}{s-c} \qquad (50)$$

$$r = \sqrt{\frac{(s-a)(s-b)(s-c)}{s}} \qquad (51)$$

Using the fact from Eq. (25) that $\sin 2(\tfrac{1}{2}A) = 2 \sin \tfrac{1}{2}A \cos \tfrac{1}{2}A$, the fact that the area K of triangle ABC in Fig. 9 is $\tfrac{1}{2}cy = \tfrac{1}{2}cb \sin A$, and using Eq. (46), Heron's famous formula for area K of a triangle, Eq. (52), can be derived.

$$K = \sqrt{s(s-a)(s-b)(s-c)} \qquad (52)$$

Also, it is rather easy to derive the formulas for the radii r and R of respective inscribed and circumscribed circles of a triangle, Eqs. (53).

$$r = \sqrt{\frac{(s-a)(s-b)(s-c)}{s}} \qquad R = \frac{abc}{4K} \qquad (53)$$

Values of trigonometric functions. If θ is an angle in radians and $n! = 1 \cdot 2 \cdot 3 \cdots n$, then $\sin \theta$ and $\cos \theta$ are defined for all values of θ by the endless series shown as Eqs. (54) and (55).

$$\sin \theta = \theta - \frac{\theta^3}{3!} + \frac{\theta^5}{5!} - \cdots$$

$$+ (-1)^{n+1}\frac{\theta^{2n+1}}{(2n+1)!} + \cdots \qquad (54)$$

$$\cos \theta = 1 - \frac{\theta^2}{2!} + \frac{\theta^4}{4!} - \cdots$$

$$+ (-1)^n \frac{\theta^{2n}}{2n!} + \cdots \qquad (55)$$

To find a value of the sine or the cosine of an angle, only as many terms of the series are used as are necessary to ensure required accuracy. The error from cutting off the series at any term after the first is less than the numerical value of the first term not used. Thus to find $\sin 0.1$ accurate to six decimal places, only the first two terms of the series in Eq. (54) are used to get $\sin 0.1 = 0.1 - (0.1)^3/6 = 0.099833$ (nearly). The first unused term was $(0.1)^5/5!$ and $(0.1)^5/5! < 0.0000001$. Hence the error in $\sin 0.1$ due to neglecting all terms of Eq. (54) except the first two is less than 0.0000001. Taking advantage of Eqs. (3) and (19)–(24), it is necessary to compute only values of functions for angles $\tfrac{1}{4}\pi$ or less. For five decimal place accuracy, four terms of Eq. (54) and four terms of Eq. (55) are sufficient. Obviously the series in Eqs. (54) and (55) furnish a powerful and convenient method of finding the sine or the cosine of any angle with any required degree of accuracy.

Generalizations of trigonometry. Many interesting generalizations of trigonometry have been made. One of them, because of its importance in advanced mathematics and science, will be considered here. Assume that the complex numbers $a + bi$, where $i^2 = -1$, obey the basic laws of algebra and that the exponential function has been suitably defined. The trigonometric functions can then be introduced by Eqs. (56) and (57), where θ is

$$\cos \theta = \frac{e^{i\theta} + e^{-i\theta}}{2} \qquad \sin \theta = \frac{e^{i\theta} - e^{-i\theta}}{2i} \qquad (56)$$

$$\tan \theta = \frac{1}{\cot \theta} = \frac{\sin \theta}{\cos \theta} \qquad \sec \theta = \frac{1}{\cos \theta} \qquad (57)$$

$$\csc \theta = \frac{1}{\sin \theta}$$

angle measure in radians. By multiplying the second equation of (56) by i, and adding the result to the first, Eq. (58) is obtained.

$$e^{i\theta} = \cos \theta + i \sin \theta \qquad (58)$$

Equating the nth powers of the sides of Eq. (58), De Moivre's theorem, Eq. (59), is obtained. The

$$e^{in\theta} = \cos n\theta + i \sin n\theta = (\cos \theta + i \sin \theta)^n \qquad (59)$$

laws of trigonometry are easily derived from Eqs. (56) and (57). To prove that $\cos(-\theta) = \cos \theta$, replace θ by $-\theta$ in the first equation of (56). This produces no change in the value of the cosine. To prove that $\sin 2\theta = 2 \sin \theta \cos \theta$, replace each function by its value from Eq. (56) to get Eq. (60).

$$\frac{e^{i2\theta} + e^{-i2\theta}}{2i} = \frac{2(e^{i\theta} - e^{-i\theta})}{2i}\frac{(e^{i\theta} + e^{-i\theta})}{2} \qquad (60)$$

The two sides of Eq. (60) are equal. Similar proofs apply for all the other laws of trigonometry. However this new trigonometry includes more than the old. For example, in the first equation of Eqs. (56), replacing θ by $5i$ gives Eq. (61). When "imaginary

$$\cos 5i = \frac{e^{-5} + e^5}{2} = 74.21 \text{ (nearly)} \qquad (61)$$

angles" are introduced, the relation $|\cos \theta| \leqq 1$ no longer applies. This theory involving the imaginary variable θ has many important applications in the theory of electricity

SPHERICAL TRIGONOMETRY

A great circle on a sphere is the intersection of the sphere with a plane through the center. Spherical trigonometry treats, for the most part, of spherical triangles having as sides arcs of great circles on a sphere. Two cases of these triangles are highly important. In one case the vertices are points on the Earth, and in the other on celestial bodies such as the Sun, planets, and stars. Astronomers, surveyors, and navigators on ships and airplanes apply the formulas of spherical trigonometry to find such values as the time of day, directions of motion, and positions of ships, airplanes, and reference points. Thus spherical trigonometry is basic in astronomy, in certain kinds of surveying, and in navigation. It is also used in mathematics and its applications.

Spherical triangle. Figure 10 represents parts of three planes which form a trihedral angle at the center O of a sphere and which intersect the sphere in a spherical triangle ABC. Each side of

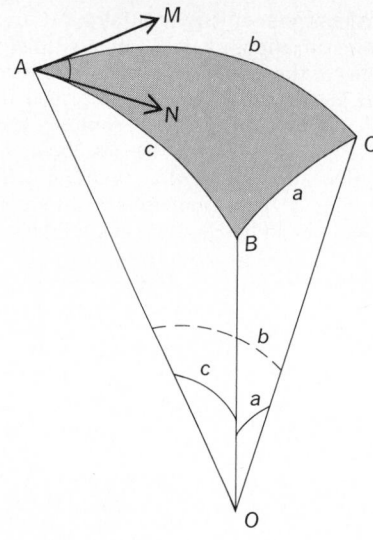

Fig. 10. Spherical triangle.

the triangle is measured by the corresponding face angle subtended at the center of the sphere and is expressed by the same number of angular units. The tangents AN and AM to the arcs AB and AC at A, being perpendicular to the radius OA, are the sides of a plane angle of dihedral angle $M\text{-}AO\text{-}N$. The angle NAM is, by definition, the angle A of spherical triangle ABC. The sides and the angles of spherical triangle ABC are stated in the same angular units. This article will deal only with spherical triangles having sides and angles less than $180°$.

Formulas of right spherical triangles. All spherical triangles can be solved by means of ten formulas obtainable from Napier's rules.

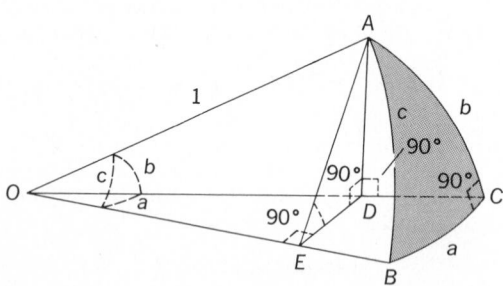

Fig. 11. Right spherical triangle.

Fig. 11 shows a right spherical triangle on a sphere of radius 1 unit and center O, and with angle $C = 90°$. Therefore the dihedral angle $A\text{-}OC\text{-}B$ is a right dihedral angle. Assume that sides a, b, c are all less than $90°$. Pass a plane through A perpendicular to edge OB meeting OB in E and OC in D. Then, from geometry, angles ADO, ADE, DEO, and AEO are right angles. Hence angle $AED =$ angle B since it is the plane angle of dihedral angle $A\text{-}OB\text{-}C$. Figure 11 shows that $OE = \cos c$, $EA = \sin c$, $OD = \cos b$, and $DA = \sin b$. Also from Fig. 11 Eqs. (62)–(65) can be obtained.

$$\cos c = OE = OD\cos a = \cos b \cos a \qquad (62)$$

$$\sin b = AD = \frac{AD}{EA}EA = \sin B \sin c \qquad (63)$$

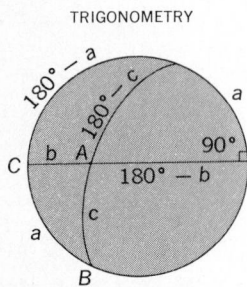

TRIGONOMETRY

Fig. 12. Orthogonal projection of four triangles.

$$\cos B = \frac{ED}{EA} = \frac{ED}{OE}\frac{OE}{EA} = \tan a \cot c \qquad (64)$$

$$\sin a = \frac{ED}{OD} = \frac{ED}{DA}\frac{DA}{OD} = \cot B \tan b \qquad (65)$$

Similarly, by passing a plane through B perpendicular to OA, Eqs. (66)–(68) can be derived.

$$\sin a = \sin A \sin c \qquad (66)$$
$$\cos A = \tan b \cot c \qquad (67)$$
$$\sin b = \tan a \cot A \qquad (68)$$

Using the value of cot A obtained from Eq. (68) and that of cot B from Eq. (65), using Eq. (62), Eq. (69) is found.

$$\cot A \cot B = \frac{\sin b}{\tan a}\frac{\sin a}{\tan b}$$

$$= \cos b \cos a = \cos c \qquad (69)$$

Just as Eq. (69) may be derived from Eqs. (62)–(68), Eqs. (70) and (71) may be derived.

$$\cos B = \cos b \sin A \qquad (70)$$
$$\cos A = \cos a \sin B \qquad (71)$$

In deriving Eqs. (62)–(71), it was assumed that a, b, and c were less than $90°$. If one or more of them were $90°$, or greater, at least four parts of triangle ABC would be $90°$ or greater. Then the triangle would be solved by inspection, and it could be shown by substitution that Eqs. (62)–(71) hold in such cases. Figure 12 represents the orthogonal projection of four triangles composing a hemisphere on the base plane of the hemisphere. In Fig. 12, a, b, c denote sides less than $90°$. Of four right triangles composing a hemisphere and each having only one part $90°$, one must have three sides each less than $90°$. Equations (62)–(71) hold for this spherical triangle, and by substitution, it could then be shown that they hold for the three others. Hence Eqs. (62)–(71) hold for all spherical triangles having parts each less than $180°$.

Napier's rules. Equations (62)–(71) can easily be written by two rules. Figure 13 shows the so-called circular parts a, b, Co-A, Co-c, Co-B of a right spherical triangle; the prefix Co- indicates "complement of." Thus Co-B means $(90° - B)$. Each part has two contiguous parts and two noncontiguous parts; speaking of any part as the middle part, the contiguous parts are called adjacent parts, and the noncontiguous parts are called opposite parts. Napier's rules follow.

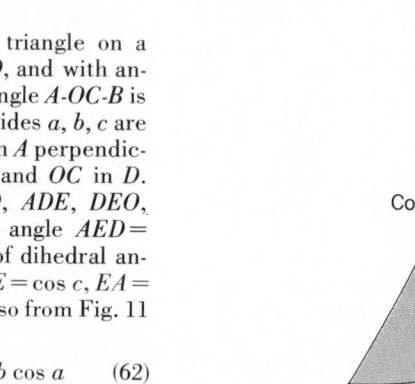

Fig. 13. Circular parts of a right spherical triangle.

Rule I. The sine of any middle part is equal to the product of the cosines of the opposite parts.

Rule II. The sine of any middle part is equal to the product of the tangents of the adjacent parts.

The 10 equations read by Napier's rules are just Eqs. (62)–(71). Thus, from Fig. 13, using c as middle part, by rule I, sin (Co-c) = cos c = cos a cos b. And by rule II, cos c = tan (Co-A) tan (Co-B) = cot A cot B. These are Eqs. (62) and (69), respectively.

If two parts of a right spherical triangle are given, the remaining parts can be found since a formula read by Napier's rules applied for two given parts and a desired part has only one unknown and can be solved for that unknown.

For example, if the given parts are $a = 20°$, $c = 150°$, Napier's rules may be used to obtain the following relations: cos 150° = cos 20° cos b; cos B = tan 20° cot 150°; and sin 20° = sin 150° sin A. Solving these gives $b = 157°$, $B = 129°$, $A = 43°$ nearly. Angle A was taken as 43°, instead of 180° − 43° = 137°, by rule III, stated below.

Whenever the formula used contains the sine of the unknown part, an angle and its supplement are obtained. In this case the following rules are used.

Rule III. A nonright angle of a spherical right triangle and the side opposite are both less than 90° or both greater than 90°.

Rule IV. If the legs of a right spherical triangle are both less than 90° or both greater than 90°, the hypotenuse is less than 90°; otherwise it is not less than 90°.

Rule III follows from Eqs. (70) and (71) and rule IV from Eq. (62). There may be two solutions, one solution, or no solution of a right spherical triangle when the given parts are an angle and the side opposite. In this case rules III and IV are used to assign computed parts to the solutions.

Polar triangle. If a spherical triangle has parts A, B, C, a, b, c and its polar parts A', B', C', a', b', c', then Eqs. (72) may be written. Often it is easy to

$$
\begin{aligned}
A &= 180° - a' \\
A' &= 180° - a \\
B &= 180° - b' \\
B' &= 180° - b \\
C &= 180° - c' \\
C' &= 180° - c
\end{aligned}
\qquad (72)
$$

solve the polar triangle of a given triangle. In this case rule V is used. *See* POLAR TRIANGLE.

Rule V. To solve a triangle Eq. (72) may be used to find the parts of the polar triangle, next the polar triangle is solved, and then the parts of the original triangle are found by using Eq. (72).

A quadrantal triangle is one having a side 90°. A polar or a quadrantal triangle for which three parts are known may be solved by using rule V.

Solution of oblique spherical triangles. Spherical triangles are usually classified on the basis of given parts. Six cases are referred to as follows:

1. Given two sides and the included angle
2. Given two angles and the included side
3. Given three sides
4. Given three angles
5. Given two sides and an angle opposite one of them
6. Given two angles and a side opposite one of them

Essentially rule V reduces these six cases to the

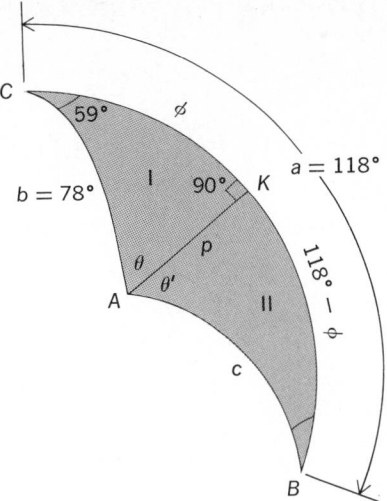

Fig. 14. The triangle and given parts.

three, 1, 3, and 5, since the polar triangles associated with cases 2, 4, and 6, are those of cases 1, 3, and 5.

To solve an oblique spherical triangle, divide it into two right triangles which may be solved in succession by Napier's rules. For example, take the case in which $a = 118°$, $b = 78°$, $C = 59°$. Figure 14 represents the triangle and the given parts. From A an arc perpendicular to side a is drawn meeting it in K. Applying Napier's rules first solve right triangle AKC, then obtain two parts of triangle AKB, and finally solve triangle AKB. From the results the required solution is obtained.

Law of sines. Corresponding to the law of sines and the law of cosines in plane trigonometry, there are two basic laws having the same names in spherical trigonometry. The law of sines for any spherical triangle having angles A, B, C and respective opposite sides a, b, c is expressed by Eq. (73). Figure 15 represents any spherical triangle

$$
\frac{\sin a}{\sin A} = \frac{\sin b}{\sin B} = \frac{\sin c}{\sin C}
\qquad (73)
$$

ABC lettered in the conventional way with arc h drawn from C perpendicular to side AB or AB prolonged and meeting it in D. Applying Napier's rules first to right triangle ACD and then to triangle CDB: sin h = sin b sin A; sin h = sin a sin B. For these equal values of sin h, a/sin A = b/sin B. A like procedure proves that b/sin B = c/sin C.

Law of cosines. Assume in Fig. 15 that D lies on arc AB, denote arc AD by ϕ and arc DB by $(c - \phi)$. Having applied Napier's rules to right triangles

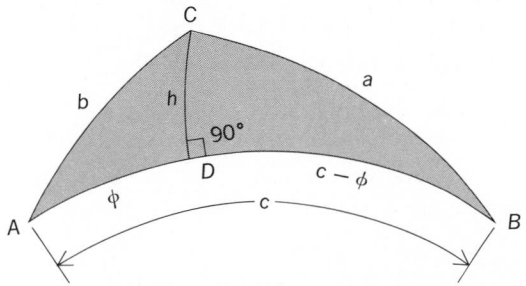

Fig. 15. Spherical triangle.

ADC and *DBC*, then Eqs. (74), (75), and (76) can be written.

$$\cos b = \cos h \cos \phi \qquad (74)$$

$$\cos a = \cos h \cos (c - \phi)$$
$$= \cos h \, (\cos c \cos \phi + \sin c \sin \phi) \qquad (75)$$

$$\cos A = \tan \phi \cot b \qquad (76)$$

Dividing Eq. (75) by Eq. (74) member by member, replacing $\tan \phi$ in the result by $\cos A \tan b$ from Eq. (76), and finally multiplying through by $\cos b$ gives Eq. (77). It is not necessary to assume

$$\cos a = \cos b \cos c + \sin b \sin c \cos A \qquad (77)$$

that *D* is on arc *AB*. If it is assumed that the positive direction of the great circle *AB* is from *A* to *B* and the negative direction opposite and that ϕ or $(c - \phi)$ may be negative, the argument applies generally, unless $b = 90°$. In this case Eq. (77) becomes $\cos a = \sin c \cos A$, and this may be proved by applying Napier's rule to the polar of the given triangle. By repeating the argument after interchanging the names of parts on Fig. 15, Eqs. (78) and (79) can be written.

$$\cos b = \cos a \cos c + \sin a \sin c \cos B \qquad (78)$$

$$\cos c = \cos a \cos b + \sin a \sin b \cos C \qquad (79)$$

If Eqs. (77), (78), and (79) are applied to the polar triangle of *ABC* by using formulas of Eq. (72), the law of cosines for angles is obtained, Eqs. (80), (81), and (82).

$$\cos A = -\cos B \cos C + \sin B \sin C \cos a \qquad (80)$$

$$\cos B = -\cos A \cos C + \sin A \sin C \cos b \qquad (81)$$

$$\cos C = -\cos A \cos B + \sin A \sin B \cos c \qquad (82)$$

The law of cosines is very useful in the derivation of other formulas. Also, it could be used to solve triangles. Evidently, if three sides of a triangle are known, Eqs. (77), (78) and (79) could be used to find the angles, and if three angles are known, Eqs. (80), (81), and (82) could be used to find the sides. Also, if two sides and the included angle are given, one of Eqs. (77), (78), or (79) would give the third side. Then the other two formulas could be used to get the other two angles. Similarly Eqs. (80), (81), and (82) could be used to solve a triangle for which two angles and the included sides are given.

Half-angle formulas. The half-angle formulas of spherical trigonometry are Eqs. (83).

$$\tan \tfrac{1}{2}A = \frac{r}{\sin(s - a)}$$

$$\tan \tfrac{1}{2}B = \frac{r}{\sin(s - b)} \qquad (83)$$

$$\tan \tfrac{1}{2}C = \frac{r}{\sin(s - c)}$$

For Eqs. (83), Eqs. (84) apply.

$$s = \tfrac{1}{2}(a + b + c)$$
$$r = \sqrt{\sin(s - a)\sin(s - b)\sin(s - c)/\sin s} \qquad (84)$$

Equations (83) are derived by a method analogous to that used to get the half-angle formulas of plane trigonometry. Evidently these formulas are used to solve a spherical triangle with three sides given, or, by rule V, one for which three angles are given. Other formulas could be obtained by using

Eq. (72) to apply Eqs. (83) and (84) to the polar triangle and dropping primes. One formula thus obtained involving $S = \frac{1}{2}(A + B + C)$ is Eq. (85).

$$\cot \tfrac{1}{2}a = \frac{R}{\cos(S - A)} \qquad (85)$$
$$R = \sqrt{\frac{\cos(S - A)\cos(S - B)\cos(S - C)}{-\cos S}}$$

Napier's analogies. The following formulas, Eqs. (86) and (87), called Napier's analogies, are analogous to the law of tangents of plane trigonometry.

$$\frac{\sin \tfrac{1}{2}(A - B)}{\sin \tfrac{1}{2}(A + B)} = \frac{\tan \tfrac{1}{2}(a - b)}{\tan \tfrac{1}{2}c} \qquad (86)$$

$$\frac{\cos \tfrac{1}{2}(A - B)}{\cos \tfrac{1}{2}(A + B)} = \frac{\tan \tfrac{1}{2}(a + b)}{\tan \tfrac{1}{2}c} \qquad (87)$$

Four others are obtained from these by interchange of letters. Also formulas may be obtained by using Eq. (72) to apply Eqs. (86) and (87) to the polar triangle. The formulas thus obtained from Eqs. (86) and (87) are Eq. (88). Observe that, if *A*,

$$\frac{\sin \tfrac{1}{2}(a - b)}{\sin \tfrac{1}{2}(a + b)} = \frac{\tan \tfrac{1}{2}(A - B)}{\cot \tfrac{1}{2}C}$$
$$\frac{\cos \tfrac{1}{2}(a - b)}{\cos \tfrac{1}{2}(a + b)} = \frac{\tan \tfrac{1}{2}(A + B)}{\cot \tfrac{1}{2}C} \qquad (88)$$

B, and *C* are given, Eqs. (86) and (87) may be used to find *a* and *b*. The polar formulas and rule V or Eq. (88) may be used to solve a triangle when two sides and the included angle are given.

When two sides and the angle opposite one of them (case 5) are given, the law of sines is used to get one angle and then Napier's analogies are used to find the other parts. In this case two solutions, one solution, or no solutions are possibilities. The solution of case 6 is carried out by the same method applied to the polar triangle. Also the law of sines and Eq. (88) may be used for this case.

[LYMAN M. KELLS]

Bibliography: I. Drooyan et al., *Trigonometry: An Analytic Approach*, 3d ed., 1979; H. S. Engelsohn, *Trigonometry*, 1980; G. Fuller, *Plane Trigonometry with Tables*, 5th ed., 1978; E. R. Heinemann, *Plane Trigonometry*, 5th ed., 1980; K. J. Smith, *Trigonometry for College Students*, 2d ed., 1980.

Triple point

A particular temperature and pressure at which three different phases of one substance can coexist in equilibrium. In common usage these three phases are normally solid, liquid, and gas, although triple points can also occur with two solid phases and one liquid phase, with two solid phases and one gas phase, or with three solid phases.

According to the Gibbs phase rule, a three-phase situation in a one-component system has no degrees of freedom (that is, it is invariant). Consequently, a triple point occurs at a unique temperature and pressure, because any change in either variable will result in the disappearance of at least one of the three phases.

Triple points are shown in the illustration of part of the phase diagram for water. Point *A* is the well-known triple point for Ice I (the ordinary low-pressure solid form) + liquid water + water vapor at

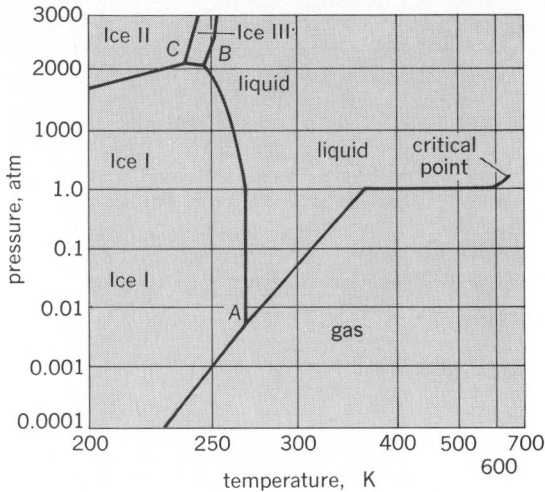

Phase diagram for water, showing gas, liquid, and several solid (ice) phases; triple points at *A, B,* and *C.* The pressure scale changes at 1 atm from logarithmic scale at low pressure to linear at high pressure. 1 atm = 6.895 kPa.

0.0099°C (273.16 K) and a pressure of 0.00603 atm (4.58 mm Hg or 611 Pa). In 1954 the thermodynamic temperature scale (the absolute of Kelvin scale) was redefined by setting this triple-point temperature for water equal to exactly 273.16 K. Point *B,* at 251.1 K and 2047 atm (207.4 MPa) pressure, is the triple point for liquid water + Ice I + Ice III; and point *C,* at 238.4 and 2100 atm (212.8 MPa) pressure, is the triple point for Ice I + Ice II + Ice III. At least four other triple points are known at higher pressures, involving other crystalline forms of ice.

For most substances the solid-liquid-vapor triple point has a pressure less than 1 atm (101.325 kPa); such substances then have a liquid-vapor transition at 1 atm (normal boiling point). However, if this triple point has a pressure above 1 atm, the substance passes directly from solid to vapor at 1 atm.

For a two-component system, the invariant point in a phase diagram is a quadruple point at which four phases coexist. The three-phase situation is then represented by a line in the three-dimensional pressure-temperature-composition diagram. *See* TRANSITION POINT.

[ROBERT L. SCOTT]

Triton

The nucleus of $_1H^3$ (tritium); it is the only known radioactive nuclide belonging to hydrogen. The triton is produced in nuclear reactors by neutron absorption in deuterium ($_1H^2 + _0n^1 \rightarrow _1H^3 + \gamma$), and decays by β^- emission to $_2He^3$ with a half-life of 12.4 years. The spin of the triton is $\frac{1}{2}$, its magnetic moment is 2.9788 nuclear magnetons, and its mass is 3.01700 atomic mass units. Much of the interest in producing $_1H^3$ arises from the fact that the fusion reaction $_1H^3 + _1H^1 \rightarrow _2He^4$ releases about 20 Mev of energy. Tritons are also used as projectiles in nuclear bombardment experiments. *See* NUCLEAR REACTION. [HENRY E. DUCKWORTH]

Turbulent flow

Motion of fluids in which local velocities and pressures fluctuate irregularly. Most flows observed in nature, such as rivers and winds, are turbulent.

Such flows occur at high Reynolds numbers. In turbulent flow, motion of the fluids is steady only insofar as the temporal mean values of velocities and pressures are concerned. The velocity and the pressure distributions in turbulent flows as well as the energy losses are determined mainly by the turbulent fluctuations.

Random nature. The essential characteristic of turbulent flow is that the fluctuations are random. Hence, the solution of the turbulence problem requires the application of methods of statistical mechanics. In turbulent flow, the most important phenomenon is the transfer of forces by such random motion. The rate of turbulent transfer, such as heat transfer, shearing stress, and diffusion, is much higher than that due to molecular mechanism in laminar flow.

In turbulent motion, even though the fluid is regarded as a continuum with an average overall molecular motion, turbulent velocity fluctuations must be superimposed on the mean motion. The separation of mean motion and turbulent fluctuation depends mainly on the scale of turbulence. Different scales give different descriptions of turbulent flow. Once the scale of turbulence is chosen, the instantaneous velocity component u_i for instance, is given by Eq. (1), where u_i is the *i*th

$$u_i = \bar{u}_i + u'_i \qquad (1)$$

component of the total fluid velocity, $\bar{u}_i$ is the *i*th mean velocity component, and u'_i is the *i*th component of the turbulent fluctuating velocity.

Turbulent stresses. The instantaneous velocity component u_i satisfies the Navier-Stokes equations of motion of a viscous fluid. The substitution of the expression for the instantaneous velocity components into the Navier-Stokes equations and the use of the mean values of the equations give the Reynolds equations for turbulent flow. The difference of the Reynolds equation from the Navier-Stokes equations is due to the additional terms of turbulent stresses. The turbulent normal stresses are $-\rho\overline{u'^2_i}$, and the turbulent shearing stresses are $-\rho\overline{u'_i u'_j}$, where $i \neq j$ and ρ is the density of the fluid.

These stresses represent the rate of transfer of momentum across the corresponding surfaces because of turbulent velocity fluctuations.

Semiempirical theories of turbulence. To illustrate various semiempirical theories of turbulent flow, consider a simple parallel mean flow $\bar{u} = \bar{u}(y)$, $\bar{v} = 0$, $\bar{w} = 0$, with fluctuating velocity components u', v', and w', where u, v, and w are the x, y, and z components of velocity, respectively. Semiempirical theories of turbulence are formulated based on various hypotheses about the turbulent stresses.

J. Boussinesq introduced the turbulent exchange coefficient ϵ such that Eq. (2) holds. In ac-

$$\overline{u'v'} = -\epsilon \frac{d\bar{u}}{dy} \qquad (2)$$

tual analysis, further hypotheses are necessary about the variations of ϵ, which are different for different flow.

L. Prandtl originated the mixing-length theory, in which the fluctuating value of a transferable quantity q of the fluid may be written as Eq. (3),

$$|q'| = l \frac{d\bar{q}}{dy} \qquad (3)$$

where l is the mixing length. For simple parallel flow, the exchange coefficient becomes Eq. (4).

$$\epsilon = -\overline{lv'} \tag{4}$$

There are many different mixing-length theories based on different quantities being transferred. In Prandtl's momentum transfer theory, the momentum of the fluid elements is assumed to be preserved in the mixing process. Prandtl obtained his famous formula for shearing stress τ of nearly parallel turbulent flow as defined by Eq. (5). This

$$\tau = \rho l^2 \left| \frac{d\bar{u}}{dy} \right| \frac{d\bar{u}}{dy} \tag{5}$$

formula has been used successfully in the calculation of many turbulent flow problems such as flow along a flat plate and in jet and wakes.

In Taylor's vorticity transfer theory, the transferable quantity is vorticity. Theodor von Kármán made a similarity hypothesis to determine the mixing length so that no special model for a transferable quantity is required.

Logarithmic velocity profile. The most important deduction from von Kármán's similarity hypothesis is the universal velocity distribution for the flow in circular pipes or between parallel plane walls. Von Kármán first pointed out that the ratio between the velocity defect $U_m - \bar{u}$ and the quantity $\sqrt{\tau_0/\rho}$ is a universal function of the ratio $(y_0 - y)/y_0$, as in Eq. (6), where y_0 is the radius of

$$\frac{U_m - \bar{u}}{\sqrt{\tau_0/\rho}} = f\left(\frac{y_0 - y}{y_0}\right) \tag{6}$$

the circular pipe or the half-width between the two plates, y is the distance from the wall, $\bar{u}$ is axial velocity, U_m is the maximum axial velocity occurring at $y = y_0$, which is the center of the channel, and τ_0 is the shearing stress at the wall $y = 0$.

The universal velocity distribution is given by Eq. (7), where Δ is of the same order of magnitude

$$\bar{u} = 2.5 \sqrt{\frac{\tau_0}{\rho}} \log \frac{y + \Delta}{\delta} \tag{7}$$

as the thickness of the laminar sublayer, which is negligible. For smooth wall, length δ is determined by a physical parameter such as density and viscosity of the fluid; for rough surface, it is determined by the roughness of the wall.

For engineering application, a nondimensional pipe-resistance coefficient λ is used such that Eq. (8) holds. Here $p_1 - p_2$ is the pressure drop along a

$$\frac{dp}{dx} = \frac{p_1 - p_2}{L} = \frac{\lambda}{d} \frac{\rho}{2} u_0^2 \tag{8}$$

pipe of length L and diameter d, and u_0 is the average mean velocity over a section of the pipe. *See* DIMENSIONLESS GROUPS.

For smooth pipe, Eq. (9) holds.

$$\frac{1}{\sqrt{\lambda}} = 2.0 \log_{10}\left(\frac{u_0 d}{\nu} \sqrt{\lambda}\right) - 0.80 \tag{9}$$

For pipe of rough surface, the resistance depends on size, shape, and spacing of the roughness elements. Only for closely packed roughness can linear dimension h of the roughness alone be used to describe the roughness. For a completely rough pipe in which $(\sqrt{\tau_0/\rho})h/\nu > 100$, where ν is the coefficient of kinematic viscosity, the resistance

law is given by Eq. (10).

$$\frac{1}{\sqrt{\lambda}} = 2.00 \log_{10}\left(\frac{y_0}{h}\right) + 1.74 \tag{10}$$

In the noncircular pipe, the characteristic length is often represented by the hydraulic mean length L_h, which is $L_h = 2A/L_w$, where A is the cross-sectional area of the pipe and L_w is the wetted circumferential length. If the hydraulic mean length is used instead of the radius of the circular pipe, resistance law (9) may be used for noncircular pipes with an accuracy within a few percent.

Turbulent jet mixing. Another type of turbulent flow without a solid wall in the flow field is known as the free-turbulence problem. Jet mixing and wakes are in this class. *See* WAKE FLOW.

In free-turbulence problems, the application of Prandtl's mixing length theory is more successful than that for the turbulent boundary-layer flow along a solid wall. In the free-turbulence problem, simple and plausible assumptions on the variation of mixing length in the flow field are possible. The mean velocity distribution calculated on the basis of these assumptions agrees well with the experimental results over a major portion of the flow field. *See* BOUNDARY-LAYER FLOW.

For a turbulent jet in a medium at rest, the jet spreads linearly. For a wake of a body of revolution, the width of the wake increases with $(C_D S_b x)^{1/3}$, where C_D is the drag coefficient of the body, S_b is the reference area of the body, and x is the distance from the body.

For a first approximation, velocity distributions in a jet mixing region and in a wake may be represented by error functions.

For turbulent jet mixing of fluids of different temperatures or of different densities, the spread of temperature and of concentration are about the same, and they are usually wider than the spread of velocity profile.

Statistical theory of turbulence. Even though the semiempirical theory has had successfully predicted mean velocity distributions in many practical problems, it has serious limitations and inconsistencies. For an understanding of turbulent flow in general, a study of the mean velocity distribution is insufficient. The fields of turbulent fluctuations must be studied in detail. Because turbulent-velocity fluctuations of a fluid are much too complicated, changing too rapidly in time and location to be known in all their details, a study of only some mean values is feasible. These mean quantities include the intensity of turbulent fluctuations, the correlation functions, and the spectrum of turbulence.

Modern statistical theory of turbulence was developed by G. I. Taylor, who introduced the correlation function, the spectrum, and the concept of statistically isotropic turbulence. Great simplification can be obtained by the isotropic property. Hence, most results from statistical theory of turbulence are concerned with isotropic turbulence.

Correlation function. Consider the fluctuating variables u_A and u_B between stations A and B, and assume that there exists a certain correlation between them. The correlation function is then given by Eq. (11), where the bar means taking the aver-

$$\rho_{AB} = \overline{u_A u_B} \tag{11}$$

age. The correlation coefficient R_{AB} is given by Eq.

(12). The correlation coefficient lies within the limits of -1 and $+1$.

$$R_{AB} = \frac{\overline{u_A u_B}}{\sqrt{\overline{u_A^2}} \sqrt{\overline{u_B^2}}} \qquad (12)$$

Von Kármán first pointed out the tensor character of the correlation function. Both von Kármán and Taylor studied extensively the correlation functions between the components of fluctuating velocity at the same time at two different points of the fluid for isotropic turbulence. These correlation functions had been measured by hot-wire anemometer. Experimental results check well with theory.

For isotropic turbulence, the correlation function is a function of time t and distance r between two points. The curvature of the double correlation curve at $r = 0$ determines a microscale of turbulence, which is a measure of the size of the smallest eddies in the turbulent flow, these eddies being responsible for the dissipation of turbulent energy. The integration of the correlation coefficient over $0 < r < \infty$ gives the scale of turbulence; the scale of turbulence is a measure of the large eddies in the turbulent flow.

Spectrum. A more detailed description of turbulence can be obtained by considering the distribution of energy among eddies of different sizes. This description can be put into precise mathematical form by considering the distribution of energy with frequency or with wave number, which is known as the spectrum of turbulence. Spectral density is a Fourier transform of correlation coefficient. Spectrum of turbulence can be measured by hot-wire anemometer.

Local isotropy. The most significant idea contributed to the problem of turbulent shear flow in recent years is the hypothesis of local isotropy proposed by A. N. Kolmogoroff. He suggested that the fine structure in turbulent shear flow may be isotropic. Turbulent motion is considered to be a mixture of eddies of all sizes from the largest, whose dimensions are comparable with those of the main flow or of the turbulence-producing mechanism such as a grid of bars in a wind tunnel, down to the smallest eddies. When turbulent motion starts, the mean flow breaks up, or the eddies produced by the grid break up into smaller eddies, their motions being unstable; these in turn break up into smaller eddies and so on, until eddies are produced of a small enough size to be stable; this gives a lower bound to the eddy size. Kolmogoroff's idea may be expressed by saying that there is something universal about small eddies; below a certain eddy size the nature of the motion is unaffected by the origin of the turbulence, and it is expected that eddies, small compared with the dimensions of the mean flow, will be statistically isotropic. Kolmogoroff's idea of local isotropy has been verified experimentally by many research workers.

Except for the concept of local isotropy, little has been accomplished for the statistical theory of maintained shear turbulence. However, the statistical theory of isotropic turbulence shows which quantities are important in describing the fluctuating field; they include turbulent intensities, correlation function, spectrum, and probability distribution. In the experimental investigations of

shear flow, such as flow in circular pipe, channel, boundary layer over a flat plate, jets, and wakes, these are the quantities to be measured.

Turbulent diffusion. Diffusion is a fundamental process of turbulence. There is an essential difference between molecular and turbulent diffusion. In molecular diffusion, the medium consists of discrete particles, while in turbulent diffusion, the medium is continuous. The old method of investigating turbulent diffusion is semi-empirical; it uses Boussinesq's turbulent exchange coefficient. The new method of solution for turbulent diffusion uses the statistical theory of turbulence. In statistical theory, two different approaches have been used. One is the continuous stochastic process in which the diffusion equation is obtained from a probabilistic integral equation. The other approach is the random walk method.

Compressible fluid flow. For the turbulent flow of an incompressible fluid, the effect of variation of density in the expression of turbulent stresses is neglected. This effect is no longer negligible for the turbulent flow of a compressible fluid and cannot be neglected for high-speed flow, flow with large variation of temperature, or both. The study of the turbulent flow of a compressible fluid requires the correlation of velocity components, of velocity and density, and of pressure and velocity. To obtain these three correlations is a complicated procedure.

Mixing-length theories may be extended to compressible fluid. For two-dimensional parallel flow with mean flow field Eqs. (13) hold. The fluctua-

$$\begin{aligned} \bar{u} &= \bar{u}(y) & \bar{v} &= 0 \\ \bar{\rho} &= \bar{\rho}(y) & \overline{T} &= \overline{T}(y) \end{aligned} \qquad (13)$$

tions of velocity component u, density ρ, and temperature T of the fluid may be written as Eqs. (14), where l, l_ρ, and l_T are the corresponding mix-

$$|u'| = l \frac{d\bar{u}}{dy} \qquad |\rho'| = l_\rho \frac{d\bar{\rho}}{dy} \qquad |T'| = l_T \frac{d\overline{T}}{dy} \qquad (14)$$

ing lengths for the velocity, density, and temperature. It is customary to assume that these mixing lengths are equal to simplify the analysis and to aid in solving practical problems. Experimental evidences indicate, however, that they are not equal. For instance, in jet mixing of a compressible fluid, the spread of temperature is wider than that of velocity. One way to explain this phenomenon is to assume that l_T is larger than l.

The statistical theory of isotropic turbulence has been extended to the case of compressible fluids.

Electrically conducting fluid flow. Magnetohydrodynamics deals with flow in electrically conducting fluids in which electromagnetic forces are of the same order of magnitude as gas-dynamic forces such as pressure and viscosity. Magnetohydrodynamics is important in problems of astrophysics, geophysics, and the behavior of interstellar gas masses, as well as in such engineering problems as reentry of intercontinental ballistic missiles, controlled fusion, and plasma jets. Because of the large dimensions, it seems probable that the normal state of motion in the cosmos should be turbulent. The high speed of intercontinental ballistic missiles also causes the flow to be turbulent. Controlled fusion research shows that the turbulent dissipation in magnetohydrodynamics is a main difficulty to be overcome before controlled fusion becomes successful. In the study of turbulence in magnetohydrodynamics, correlations between magnetic field and velocity components are important. *See* MAGNETOHYDRODYNAMICS; PLASMA PHYSICS. [SHIH I. PAI]

Bibliography: P. Bradshaw, *An Introduction to Turbulence and Its Measurements*, 1975; W. Frost and T. H. Moulden (eds.), *Handbook of Turbulence: Fundamentals and Applications*, 1977; J. O. Hines, *Turbulence*, 2d ed., 1975; A. J. Reynolds, *Turbulent Flows in Engineering*, 1974; H. Tennekes and J. L. Lumley, *First Course in Turbulence*, 1972.

Twinning

A process in which two or more crystals, or parts of crystals, assume orientations such that one may be brought to coincidence with the other by reflection across a plane or by rotation about an axis. Crystal twins represent a particularly symmetric kind of grain boundary; however, the energy of the twin boundary is much lower than that of the general grain boundary because some of the atoms in the twin interface are in the correct positions relative to each other. *See* GRAIN BOUNDARIES.

In the general grain boundary, all the neighbors of the atoms of the interface are in distorted positions. The usual definition of a twin relationship between two crystals states that there exists a set of parallel equivalent crystal planes of atoms which is common to both twins, but that rows of atoms are discontinuous across the interface. Quite commonly, twins are mirror images of each other, as in the figure. Also, it is common for a twin in a crystal to leave the nearest neighbors of the atoms in the interface unchanged in orientation, but to place the atoms in the second neighbor shell in altered positions. This feature is also true of the twin in the figure. It is sometimes possible to create a twin in a crystal by putting an external stress on the crystal; in other cases, twins are found "grown in." *See* CRYSTAL GROWTH.

[ROBB M. THOMSON]

Bibliography: G. A. Chadwick and D. A. Smith (eds.), *Grain Boundary Structure and Properties*, 1976; M. V. Klassen-Nekludova, *Mechanical Twinning of Crystals*, 1964; B. R. Pamplin, *Crystal Growth*, 1975.

Tyndall effect

Visible scattering of light along the path of a beam of light as it passes through a system containing discontinuities. The luminous path of the beam of light is called a Tyndall cone. An example is shown in the illustration. In colloidal systems the brilliance of the Tyndall cone is directly dependent on the magnitude of the difference in refractive index between the particle and the medium. In aqueous gold sols, where the difference in refractive index is high, strong Tyndall cones are observed.

For systems of particles with diameters less than one-twentieth the wavelength of light, the light scattered from a polychromatic beam is predominantly blue in color and is polarized to a degree which depends on the angle between the observer and the incident beam. The blue color of tobacco smoke is an example of Tyndall blue. As particles are increased in size, the blue color of

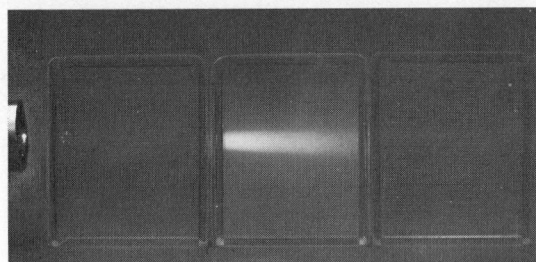

The luminous light path known as the Tyndall cone or Tyndall effect. (*H. Steeves and R. G. Babcock*)

scattered light disappears and the scattered radiation appears white. If this scattered light is received through a nicol prism which is oriented to extinguish the vertically polarized scattered light, the blue color appears again in increased brilliance. This is called residual blue, and its intensity varies as the inverse eighth power of the wavelength. *See* SCATTERING OF ELECTROMAGNETIC RADIATION. [QUENTIN VAN WINKLE]

Ultrasonics

The science of sound waves having frequencies above the audible range, that is, above about 20,000 hertz (Hz = cycle per second). Original workers in this field adopted the term supersonics. However, this name was also used in the study of airflow for velocities faster than the speed of sound. The present convention is to use the term ultrasonics as defined above. The term silent sound also has been used to denote ultrasonic waves. The term pretersonics is now used to refer to frequencies about 10^{10} Hz. Since there is no marked distinction between the propagation and the uses of sound waves above and below 20,000 Hz, the division is rather artificial. In this article the emphasis is on instrumentation, engineering applications, and analytical uses. *See* SOUND.

ULTRASONIC PROJECTORS AND DETECTORS

The earliest instruments for producing ultrasonic waves in air were the Galton whistle and the Hartmann generator. These devices produce sound waves by blowing a jet of high-pressure air from a narrow slit against a sharp metal edge. The Hartmann generator raises the velocity of the jet above that of the sound waves and in effect generates standing shock waves.

Piezoelectricity and magnetostriction. The usual types of generators for air, liquids, and solids are the piezoelectric and magnetostrictive generators. X-cut quartz crystals are used to produce longitudinal waves in gases, liquids, and solids. Y-cut and AC-cut quartz crystals are used to produce shear or transverse waves in solids. Crystals of these types are utilized in instruments such as the acoustic interferometer, which operates by sending an ultrasonic beam through a gas or liquid to be measured and by obtaining a standing wave system between the driving crystal and a reflector whose distance from the crystal can be varied by a screw system. Such systems provide accurate velocity and attenuation data for gases and liquids.

Pulse systems. These have been used to measure properties of liquids and solids. A short burst of ultrasonic waves is sent into the medium and is

TWINNING

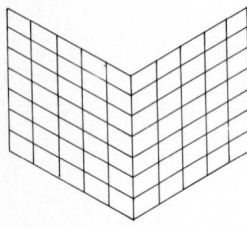

An example of a twinned crystal. One atom plane is common to each half of the crystal, but the other lines of atoms suffer a discontinuity at the twin boundary.

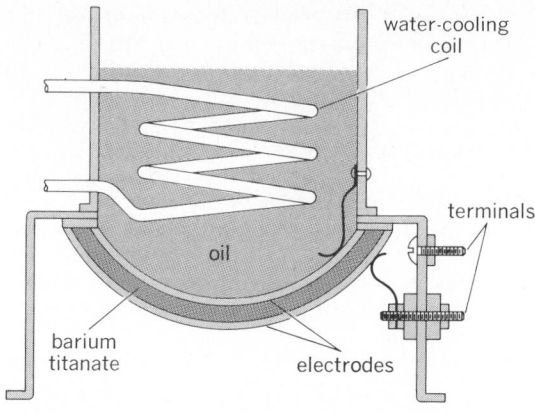

Fig. 1. Cross-sectional view of a barium titanate ultrasonic generator. Barium titanate is made to generate sound waves. Material to be treated is placed in a glass container and lowered into the oil. (*From H. F. Olson, Acoustical Engineering, 3d ed., Van Nostrand, 1957*)

reflected back. By timing the received pulse with respect to the transmitted pulse, or by a phasing technique, accurate velocity measurements can be made. Such techniques have been used widely in measuring the elastic constants of small specimens. The attenuation also can be measured by the rate at which pulses decrease with distance transmitted, but careful consideration must be given to spreading loss and to the losses in the seals connecting the transducers to the specimens.

High-power devices. For higher powers, ferroelectric ceramics, such as barium titanate (Fig. 1), PZT (lead titanate zirconate), and $NaKNbO_3$, or magnetostrictive materials, such as nickel or ferrites, commonly are employed. They generally are used for ultrasonic cutting, wear or fatigue testing, and ultrasonic welding.

Shear waves in liquids. A number of shearwave transducers, most of them employing torsion-

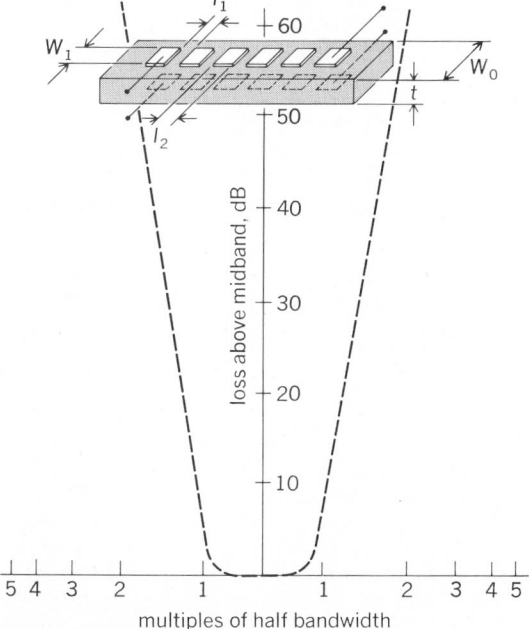

Fig. 2. Monolithic quartz crystal mechanical filter and characteristic. (*After R. A. Sykes and W. D. Beaver*)

al or shear-wave generators of quartz, have been used to measure the shear viscosity and shear stiffness of liquids. By this means it has been proved that moderately viscous liquids have elastic, as well as viscous properties. For pretersonic frequencies (above 500 MHz), all liquids are found to have shear elastic properties. The viscoelastic properties of lubricating oils have been shown to contribute to the load-carrying capacity of spur gears operating at high speeds.

ENGINEERING APPLICATIONS

The engineering applications of ultrasonics can be divided into those dealing with low-amplitude sound waves and those dealing with high-amplitude (now usually called macrosonic) waves.

Low-amplitude applications. Low-amplitude applications are in sonar (an underwater-detection apparatus), in the measurement of the elastic constants of gases, liquids, and solids by a determination of the velocity of propagation of sound waves, in the measurement of the attenuation of sound waves, and in a number of ultrasonic devices such as delay lines, mechanical filters, inspectoscopes, and thickness gages.

All these applications depend on the modifications that boundaries and imperfections in the materials cause in wave-propagation properties. The attenuation and scattering of the sound in the mediums are important factors in determining the frequencies used and the sizes of the pieces that can be utilized or investigated.

Delay lines. These are useful for storing information for a certain period of time. Such lines are used in moving-target-indicator radar systems, in pulse-decoding systems, and in computers.

Dispersive-type delay lines have become of considerable importance. In such devices use is made of the fact that the group velocity of acoustic waves in circular wires or thin strips drops materially when the wavelength approaches the diameter or thickness of the line. This drop in group velocity, which is a function of the frequency, has the effect of transferring a short amplitude-modulated pulse into a long frequency-modulated pulse which does not overload the output amplifier. This makes possible a longer-range radar without an increase in the limiting power. By using a delay line with the inverse delay-frequency characteristic, the properties of the amplitude-modulated pulse are restored on the receiving end and a good distance-time discrimination is obtained.

Mechanical filters. These are used for separating telephone communications sent simultaneously over one transmission line. Early forms used torsional or flexural vibrations to provide filtering action. A later and more useful form is called the monolithic filter. It consists of a series of vacuum-deposited electrodes with a ratio of total electrode mass to the mass of the crystal plate between electrodes of $R = 0.02$ to 0.04. The resonant frequencies of the electroded sections are lower than that of the unplated crystal, and it has been shown that the energy in the plated part is trapped in the plated section. Only a small amount is transmitted to the next plated section. This amount can be adjusted by controlling the length l_1 of the unplated section between plated sections of length l_2. The ratio of l_1 to l_2 controls the bandwidth. By using as many as six sections of platings, this device acts as a six-section filter and has characteristics similar to the filter shown in Fig. 2. Attenuation characteristics obtained in this way are sufficiently good for the device to be used as a channel filter in a long-distance carrier or submarine cable system. The filter can be made from a single quartz crystal, and hence considerable economies can be obtained.

Ultrasonic inspectoscopes. These transmit sound waves into a metal casting or other solid piece and determine the presence of flaws by reflections or by an interruption of the sound-wave transmission through the piece. Frequencies ranging from 500 kHz to 15 MHz are used. Such devices are among the best means for determining defects in metals, glasses, and ceramics, and they have also been applied in the inspection of automobile tires.

Ultrasonic thickness gages. These have been used in measuring the thickness of pieces when one side is not accessible, such as in boilers. Pulsing systems and continuous-frequency systems based on the resonance principle are used.

Surface acoustic-wave devices. One application of ultrasonics involves the use of Rayleigh surface waves which were first studied in the case of earthquake waves on the surface of the Earth. In this mode the motion is along the surface with a small penetration into the interior. Such devices may be used in obtaining high-frequency wave filters (Fig. 3a). Surface waves are generated by equally spaced interdigital electrode transducers. These generate a band of frequencies whose midfrequency is determined when the half Rayleigh wavelength equals the spacing of the transducer electrodes. Since the transducer sends out waves in both directions, the wave toward the back is absorbed by a resistive termination. The theoretical

and measured attenuations (insertion losses) and the nearly linear phase shift are shown in Fig. 3b.

One of the largest uses of these devices is in obtaining dispersive delay lines used in chirp radars. The dispersion is generated by progressively larger spacings of the driving electrodes. The initial pulse is spread over the whole time space between pulses and is transmitted as a frequency-modulated radio wave which does not overload the final amplifier stage. The received reflections are sent through a reverse-type transducer which restores the accuracy of distance and directional determinations.

Most of the surface-wave devices use lithium niobate ($LiNbO_3$), which has a high electromechanical coupling and a low acoustic attenuation. For low-temperature coefficients, lithium tantalate and S-cut quartz are sometimes used.

Other uses for surface acoustic-wave devices are for oscillators and for various processing devices.

Acoustic emission. An effect related to internal friction, motion of dislocations, and fatigue in materials is the noise in the specimens produced by strain. This is called acoustic emission. The first research on this phenomenon was done for rocks. This work started in the 1930s and was undertaken to test mine areas which were near danger regions for slides. Since the attenuation of sound waves in rocks is quite large at high frequencies, most of the measured frequencies were in the 150–10,000-Hz range. Later work has attempted to relate acoustic emission to earthquake properties. A relation found by C. H. Scholz indicates that acoustic emission and microfracturing in rocks could be directly related to the inelastic part of the stress-strain behavior, that is, to the internal friction. This suggests that the acoustic emission in rocks is connected with dislocation motion as has been established for metals.

For metals, one of the first measurements—which involved a burst-type acoustic emission—emerged from the work of W. P. Mason, H. J. McSkimin, and W. Shockley on the effect of a stress on a tin polycrystal with large grain sizes. The experimental arrangement and the type of response measured are shown in Fig. 4. The response is in a direction to indicate a relief of the force applied by the turning of the screw thread. The fine structure shown on the curves indicates that the process is one first proposed by F. C. Frank for the dynamic generation of dislocations. In this process a dislocation starts across the glide plane with the shear velocity and reaches the edge of the crystal, and a return dislocation is generated on an adjacent slip plane by the momentum associated with the motion of the first dislocation. This process keeps up until the dissipation associated with the motion brings it to a halt. This appears to be after about 10 oscillations for the curve marked A and 15 for B. *See* CRYSTAL DEFECTS.

J. Kaiser applied a steadily increasing stress to a metallic specimen and observed the acoustic emission which occurred in the form of pulses which could be counted. He also observed that emissions are not generated during the reloading of a material until the stress exceeds its previous high value, an effect known as the Kaiser effect. This effect applies to most metals but generally not to other materials.

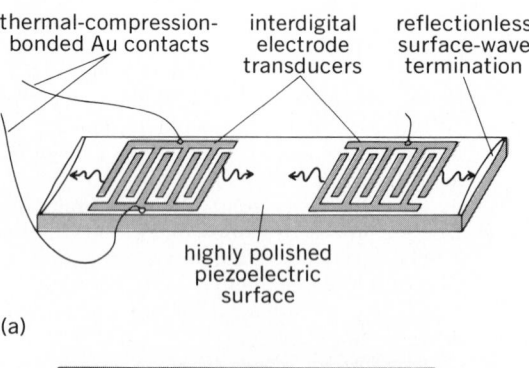

thermal-compression-bonded Au contacts   interdigital electrode transducers   reflectionless surface-wave termination

highly polished piezoelectric surface

(a)

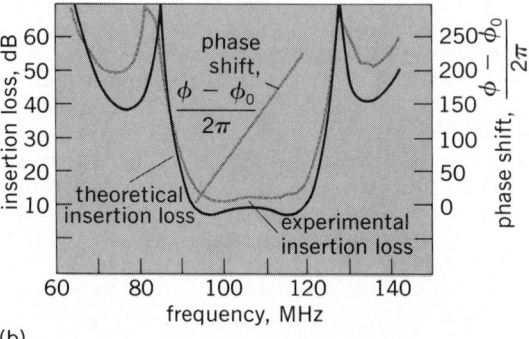

(b)

Fig. 3. Surface-wave transducers used as a bandpass filter (*a*) Configuration. (*b*) Characteristics. (*From E. A. Kraut, ed., in T. Kallard, Acoustic surface wave and acoustooptic devices, Optosonic Press, 1971*)

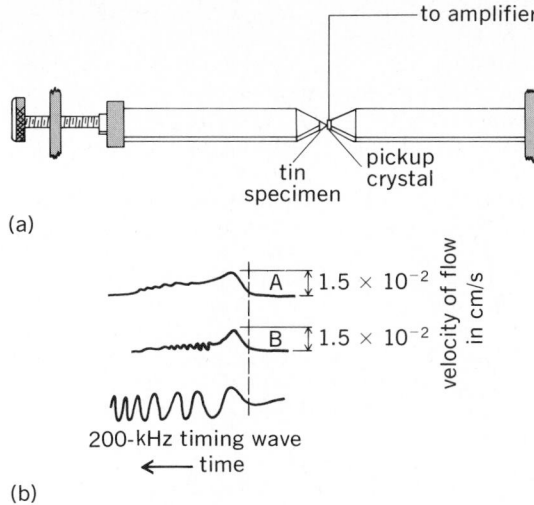

(a)

(b)

Fig. 4. Measurement of acoustic emission resulting from stress of a tin polycrystal with large grains. (*a*) Arrangement of specimen and pickup crystal. (*b*) Response. (*From W. P. Mason, H. J. McSkimin, and W. Shockley, Ultrasonic observation of twinning in tin, Phys. Rev., 73:1213–1214, 1948*)

B. H. Schofield demonstrated that single crystals were important sources of emission by showing that grain boundaries are not the only sources of emission. In this work, Schofield was the first to make a distinction between burst-type and continuous-type emissions. His later work showed that the major contributions to acoustic emissions was from volume effects, not surface effects.

This fundamental work has had wide application in such subjects as structural integrity of metals and rocks, flaws in metals, composite materials, concrete, ceramics, ice, soils, wood, integrity of welds, martensite transformations, nuclear power reactors, and leaks in pressure systems.

Acoustic emission is used to test the goodness of spot welds on relay contacts (Fig. 5*a*). An electrical pulse is used to weld the contact to the relay. The strain associated with the weld sets up an acoustic emission which is picked up by a transducer mounted on the frame. This output is amplified

and filtered, after which it is displayed on a cathode-ray tube with the waveform shown in Fig. 5*b*. A count threshold is established, and the number of times the waveform exceeds the threshold is counted. It is found that the strength of the weld is nearly proportional to the emission count. If the count is below the required value, a buzzer is activated, and the weld is rejected.

High-amplitude applications. High-amplitude acoustic waves (macrosonic) have been used in a variety of applications involving gases, liquids, and solids. Some common applications are mentioned below.

Effects due to cavitation. Holes (gas-bubble cavities) can be created in a liquid by high-intensity sound waves. When such a cavity collapses, extremely high pressures are produced. The process, called cavitation, is the origin of a number of mechanical, chemical, and biological effects.

The cavitation effect can be used to disperse metals and sulfur in solutions, to produce extrafine grain photographic emulsions, and to achieve a finer texture (smaller grain size) and more uniform alloying of a molten metal. In chemistry, cavitation can be used to break long-chain polymers into shorter chains, affording a polymer of more uniform chain length than is possible with other depolymerizing methods. Cavitation forces also can be used to sterilize milk.

Other high-amplitude effects. Ultrasound is used widely in the cleaning of metal parts, such as in watches. The large acoustic forces actually break off particles and contaminants from metal surfaces. Ultrasound has been investigated for washing textiles.

One of the principal applications of ultrasonics to gases is particle agglomeration. This depends upon the fact that light particles can follow the rapid motion of the sound waves, whereas heavy ones cannot. Hence light particles will strike and stick to heavy ones, reducing the number of small particles in the gas. The heavy particles eventually will fall to a collecting plate or can be drawn there by means of an electric field. This technique has been used in industry to collect fumes, dust, sulfuric acid mist, carbon black, and other substances.

Another industrial use of ultrasonics has been to

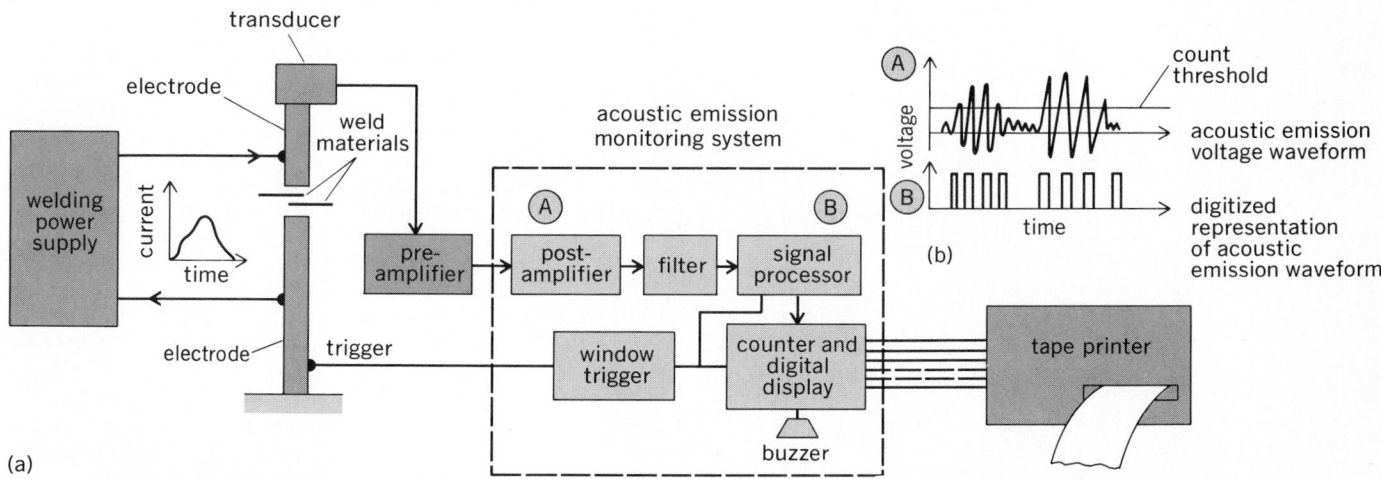

(a)

Fig. 5. Weld-monitoring system used for testing relay welds. (*a*) Block diagram of system. (*b*) Waveforms obtained at A and B in *a*. (*From S. A. Gahr and C. H. Payne,* Acoustic emission detection, real time monitoring of the quality of resistance welds, West Elec. Eng., 23(4):2–29, October 1979)

produce alloys, such as lead-aluminum and lead-tin-zinc, that could not be produced by conventional metallurgical techniques. Shaking by ultrasonic means causes lead, tin, and zinc to mix.

Ultrasound has been used in medical therapy, but there is little agreement as to the benefits. Location of cancers and other growths by ultrasonic pulsing methods has been reported and general use of ultrasonics in medical diagnosis is becoming common.

ANALYTICAL USES

In addition to their engineering applications, high-frequency sound waves have been used to determine the specific types of motions that can occur in gaseous, liquid, and solid mediums. Both the velocity and attenuation of a sound wave are functions of the sound frequency. By studying the changes in these properties with changes of frequency, temperature, and pressure, indications of the motions taking place can be obtained.

Sound attenuation in fluids. In monatomic gases and monatomic liquids such as mercury, the sound attenuation can be explained as absorption due to viscosity and heat conduction. For such fluids, the attenuation A satisfies Eq. (1), where f is

$$A = \frac{2\pi^2 f^2}{\rho v^3}\left[\frac{4}{3}\eta + \frac{(\nu-1)K}{C_p}\right] \qquad (1)$$

the frequency, ρ the density, v the sound velocity, η the coefficient of viscosity, ν the ratio of specific heats, K the thermal conductivity, C_p the specific heat at constant pressure, and A the attenuation in nepers per centimeter.

Polyatomic liquids show additional attenuation due to relaxations of two types. Thermal relaxations, which have been demonstrated for gases and nonassociated liquids, that is, liquids which contain nonpolar molecules, occur by an interchange of energy between the longitudinal sound wave and the rotational and internal modes of motion of the gas or liquid molecules.

Structural relaxations occur for associated liquids, for polymer liquids, and also for solids. These relaxations take place when one part of the molecule moves from one position to another under the combined effect of the thermal- and sound-wave energy. A definite structure, such as that which occurs in associated liquids and polymer liquids, is required. *See* SOUND ABSORPTION.

Effects in solids. For solids, a variety of effects cause attenuation and velocity dispersion. Probably the simplest of these are thermal effects.

Thermal effects. When a solid body is compressed by an acoustic wave, the compressed part becomes hotter and the expanded part cooler. Thermal energy is transmitted from the hot part to the cool part. Since this energy comes from the acoustic wave, a loss or attenuation of the wave results which is proportional to the square of its frequency. For bars in flexural vibration, the thermal path is quite short, and the effect produced is large. Below a frequency f_0, determined by Eq. (2a), such a source produces an internal friction $1/Q$ given by Eq. (2b). Here K is the thermal conductivi-

$$f_0 = \pi K / 2 C_p \rho W^2 \qquad (2a)$$

$$\frac{1}{Q} = \frac{Y_0^\sigma - Y_0^\theta}{Y_0^\theta}\left(\frac{ff_0}{f^2 + f_0^2}\right) \qquad (2b)$$

ty, C_p the specific heat per gram at constant pressure, f the frequency of the sound wave, ρ the density, and W the width of the bar in centimeters and Y_0^σ and Y_0^θ are the adiabatic and isothermal values of Young's modulus, respectively. The velocity increases as a function of frequency, as shown by Fig. 6, while a corresponding internal friction occurs, as shown by the solid line.

The thermal path l for a longitudinal wave becomes smaller as the frequency of vibration increases. It is given by Eq. (3), where v is the velocity of propagation. Inserting this expression in Eq. (2a), with $l = W$, the relaxation frequency for a longitudinal wave is then given by Eq. (4). Above this

$$l = v / 2f \qquad (3)$$

$$f_0 = C_p \rho v^2 / 2\pi K \qquad (4)$$

frequency, the material is isothermal, whereas below f_0 it is adiabatic. This frequency is usually above 10^{10} Hz for most materials. The attenuation for a longitudinal wave for this thermoelastic effect is given by Eq. (5), where γ is the Grüneisen con-

$$A_{\text{(nepers/cm)}} = \frac{\omega^2}{2\rho V_l^3}\left[\frac{\gamma^2 KT}{V_l^2}\right] \qquad \gamma = \frac{3B\alpha}{C} \qquad (5)$$

stant, B the bulk modulus, α the thermal expansion coefficient, C the specific heat per unit volume, ω the frequency of measurement times 2π, ρ the density of the medium, V_l the longitudinal velocity, and T the absolute temperature in degrees Kelvin. This source of attenuation is quite large for metals but provides only about 4% of the thermal attenuation for insulators.

The main thermal attenuation is provided by the Akheiser effect. This loss is determined by the thermal conductivity and the nonlinear third-order elastic moduli. *See* HYPERSONICS.

Other relaxations. A number of relaxation phenomena are associated with the motion of impurity atoms, grain boundaries, domain boundaries, and

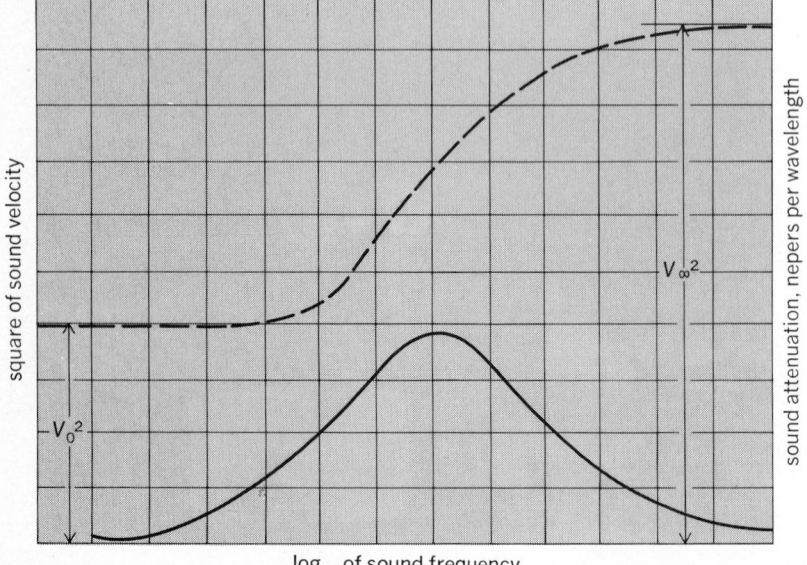

Fig. 6. Velocity dispersion (broken line) and corresponding attenuation per wavelength peak (solid line) for a medium with a single relaxation. The velocity increases as a function of frequency. One neper equals 8.686 dB.

other motions occurring in a solid. Interstitial atoms, such as nitrogen and carbon in iron, can cause an appreciable acoustic loss. These impurity atoms have preferred positions between the iron atoms in the crystal lattice. When a sound wave stretches the lattice in one direction and compresses it in a direction perpendicular to the first, the interstitial atoms, actuated by thermal energy, tend to go to the most open regions. When a compression due to the sound wave occurs, the reverse motion takes place. Since it requires a thermal activation energy H to move the impeding atoms aside, the frequency of jumping f follows Eq. (6),

$$f = f_0 e^{-H/RT} \qquad (6)$$

where f_0 is frequency of vibration of nitrogen atom due to thermal motion ($\cong 10^{13}$ Hz), R the energy necessary to increase the temperature of 1 mole of atoms (6.025×10^{23} atoms) by 1°C, and T the temperature in kelvins (K). Since H is about 16,400 cal/mole (68,600 J/mole) for nitrogen and R is 2 cal (8.31 J) the relaxation frequency for this process is about 1 Hz at room temperature.

Other relaxations involving substitutional atoms have been observed at higher temperatures, since the substitutional atoms in this case have higher activation energies. Relaxations involving the rotation of grains in polycrystalline samples have been observed at high temperatures and low frequencies.

Much faster relaxations occur in magnetic processes involving the motion of domain walls in magnetic materials. A demagnetized specimen is made up of a number of domains within which the direction of magnetism is the same. Domains with directions of magnetism at right angles to or at 180° from the original direction are separated by regions called Bloch walls, in which the direction of magnetism changes from one domain to the other by small steps in the orientation of magnetism. A compressive stress in the same direction as the magnetic flux—for a positive magnetostrictive material—causes the domain to shrink, whereas it causes domains directed at 90° to expand. Hence, the domain wall moves as the stress changes from compressive to extensional. For a discussion of Bloch walls *see* FERROMAGNETISM.

The domain walls can be held up by dislocations and other imperfections in the magnetic material, and a definite magnetic field or stress is required before the domain wall moves at all. On the reverse cycle the domain wall lags behind the applied stress (magnetic or elastic). The effect produces a hysteresis loop in the material and an acoustic loss called the microhysteresis effect. As the direction of magnetism changes, eddy currents are generated. These limit the velocity with which a domain wall can move and produce an acoustic loss for alternating stresses called the microeddy current effect. For a given size domain, there is some frequency for which the velocity is only half as large as that for low frequencies for the same applied magnetic field. The loss at this frequency is a maximum, and hence this frequency is a relaxation frequency. It can be shown that this frequency is determined by Eq. (7), where R is the

$$f_0 = R/96\chi_0 l^2 \qquad (7)$$

electrical resistivity of the material, χ_0 the initial magnetic susceptibility for a demagnetized mate-

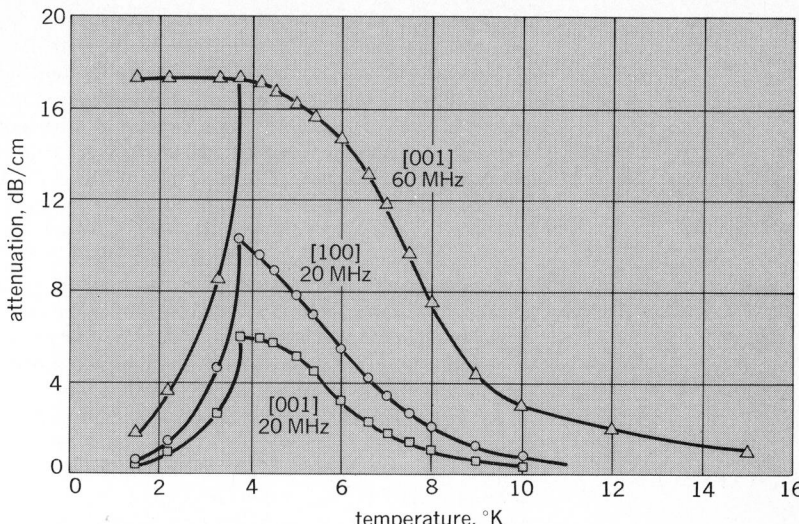

Fig. 7. Longitudinal sound-wave attenuation measurements for a single crystal of tin along the (001) axis and along the (100) axis.

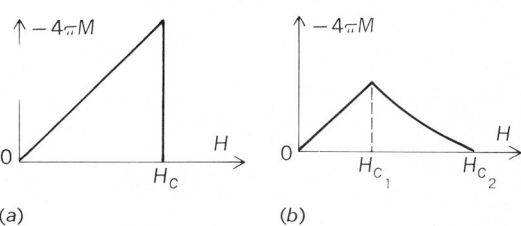

Fig. 8. Magnetization curves of long cylinders of (a) type I and (b) type II superconductors. The applied field H is directed along the axis of the cylinder.

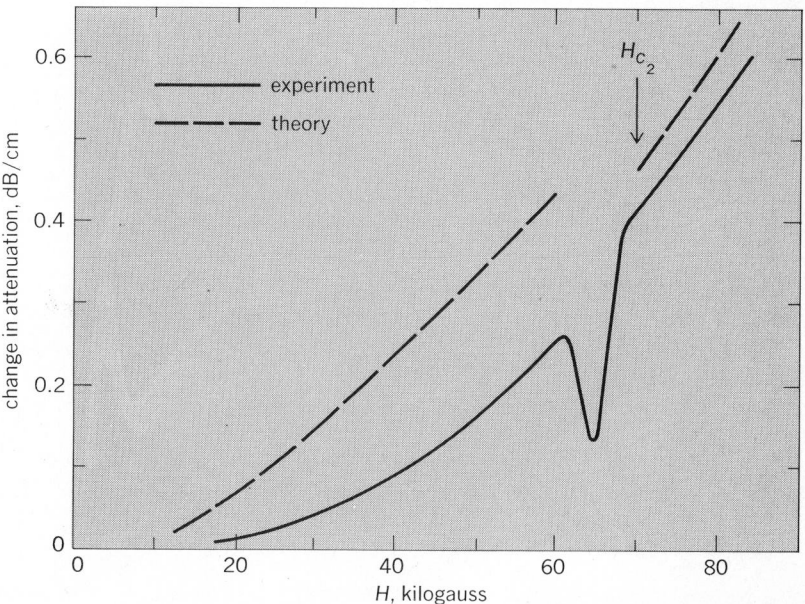

Fig. 9. Magnetic field variation of the attenuation of 9.1-MHz shear waves parallel to H in annealed Nb–25% Zr with temperature at 4.2 K. The broken curves are calculated values. 1 kilogauss = 0.1 tesla (*After Y. Shapira*)

rial, and l the thickness of a domain. For nickel, for example, this frequency is in the order of 10^5 Hz.

Many other relaxations occur, depending on the nature of the solid-state motion that can occur in the material. Ultrasonic measurements carried over wide frequency and temperature range are powerful tools for investigating such motions.

Low-temperature data. Ultrasonic waves have provided significant information on processes occurring at temperatures near absolute zero. In liquids the most important results have been obtained for liquid helium, while for solids results have been obtained with metals at low temperatures which reveal a considerable amount of information about the mechanism of superconductivity.

Liquid helium. When helium is liquefied at its boiling point (4.2 K) and cooled further, at the so-called lambda (λ) point (2.2 K) there is a transformation of helium I into helium II. At the λ point there is an ambiguity in the sound velocity and also a high attenuation. Helium II has a vanishing small viscosity and a high thermal conductivity. The former leads to a small acoustic attenuation for normal sound, while the latter leads to the capabil-

ity of transmitting thermal waves, the so-called second sound. Second sound can be initiated and detected by thermal means, and it has been found that the velocity is zero at 2.2 K, rises to a maximum of 20 m/sec at 1.7 K, and decreases thereafter at lower temperatures. The velocity of normal sound varies from 230 m/sec near absolute zero to 180 m/sec near 2.2 K. *See* LIQUID HELIUM; LOW-TEMPERATURE ACOUSTICS; SECOND SOUND.

Attenuation at low temperatures. At very low temperatures the ultrasonic attenuation of pure normally conducting metals becomes high. Figure 7 shows measurements of pure tin for two directions in the crystal and for two frequencies. Above 10 K the ultrasonic attenuation is relatively small and increases as the square of the frequency. At 4 K, at which temperature tin is still in the normal state, the attenuation is high and increases in proportion to the frequency. It has been shown that the added attenuation in the normal state is due to the transfer of momentum and energy from the acoustic wave to the free electrons in the metal. If the acoustic wavelength is greater than the electronic mean free path, this transfer determines an effective viscosity, and the attenuation increases in proportion to the square of the frequency. When the mean free path becomes longer than the acoustic wavelength, as it does at low temperatures, the energy communicated to the electrons is not returned to the acoustic wave and a high attenuation results. The attenuation is proportional to the number of times the crystal lattice vibrates and hence to the frequency.

As the temperature drops below the temperature at which tin becomes superconductive (3.71 K), this source of attenuation drops rapidly to zero. The form of the curve has been used to confirm the Bardeen-Cooper-Schrieffer energy-gap theory of superconductivity. However, at lower frequencies, that is, from 10 to 100 MHz, losses due to dislocations can occur. These are different for the normal and superconducting states, and this difference has to be taken account of in order to determine the form of the energy-gap relation. For frequencies above 100 MHz, the attenuation due to dislocations is small compared to the electron-phonon loss, and direct measurements give the shape of the energy-gap curves.

Acoustic measurements are also useful for type II or high-field superconductors (HFS). For these types of superconductors, which are used for superconducting magnets, there are two critical fields, rather than the single field of type I superconductors. Figure 8 compares the magnetization curves of type I and type II superconductors when the magnetic field H is directed along the axis of the cylinder. In type I the magnetic flux is completely excluded from the interior of the material below H_c. For type II superconductors the magnetic flux is completely excluded from the interior only below H_{c_1}. Bètween H_{c_1} and H_{c_2} the magnetic flux consists of flux vortices in the form of filaments directed along H, embedded in a superconducting material. When a dc electric current flows in a direction normal to H, each vortex experiences a force normal to its length, which causes it to move. The vortices are pinned by defects, and a finite current density is required before the vortices move. An alternating current or alternating stress causes motions of the pinned vortices which

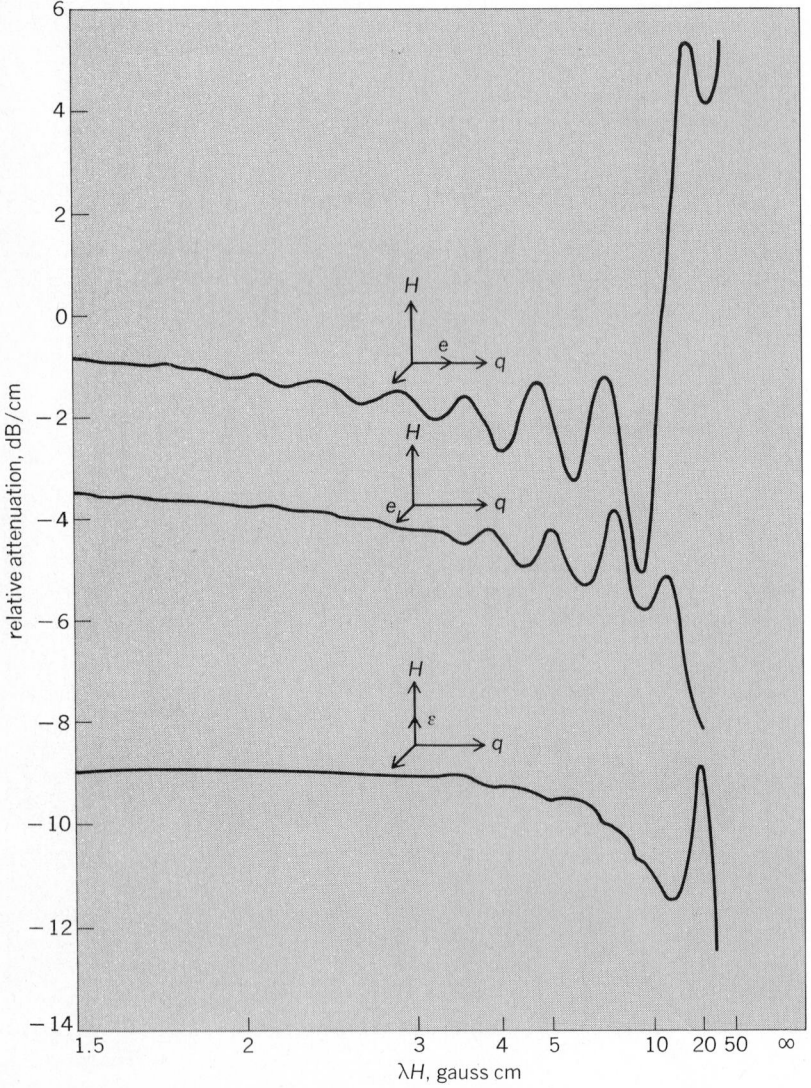

Fig. 10. Relative attenuation in pure single-crystal copper as a function of the product of the wavelength times the magnetic field for several orientations of magnetic field and wave direction. 1 gauss cm = 10^{-6} tesla · m. (*After R. W. Morse*)

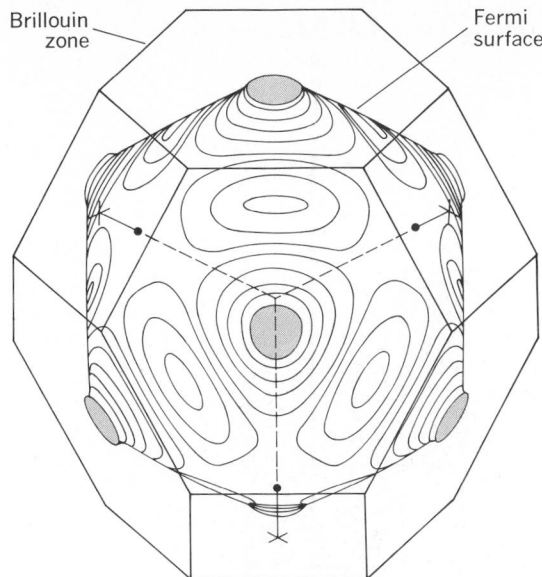

Fig. 11. The probable Fermi surfaces for monovalent copper, gold, and silver and their relation to the Brillouin zone. (*After A. B. Pippard*)

lag behind the applied forces. The result is an acoustic attenuation. Figure 9 shows the change in attenuation of a 9.1-MHz shear wave plotted as a function of the magnetic field. A sharp dip occurs near the superconducting field H_{c_2}. Above H_{c_2} the material is in the normal state and the attenuation rises rapidly with the field.

Magnetoacoustics and Fermi surface. In the presence of a magnetic field, the attenuation in metals in the normal state shows variations which are cyclic when plotted as a function of λH, where H is the magnetic field. Figure 10 shows measurements in a very pure copper single crystal at 4.2 K. These cyclic variations can be related to the shape of the Fermi surface, which is a constant-energy surface that bounds the occupied states of electrons in momentum space. The electrical effects in a metal are primarily determined by the electrons whose energy is near the Fermi surface, since these are the only ones free to move. For free electrons, such surfaces are spherical with a radius determined by the Fermi energy.

The effect of the periodic crystal potential in the band theory approximation is to distort the Fermi-surface from a spherical surface. Electrons of the same energy (which all lie on the Fermi surface) will then have different momenta. Figure 11 shows the probable Fermi surfaces for monovalent copper, gold, and silver and their relation to the Brillouin zone. If the orbit of an electron in momentum space carries it to the Brillouin-zone face, the electron will be refracted to the opposite Brillouin-zone face. In momentum space this has the effect of repeating the zone over and over in an extended zone scheme. The effect of a magnetic field is to localize the electrons that can move onto a plane perpendicular to the magnetic field in momentum space. It can be shown that the periodicity of the attenuation-λH curves can be related to the linear dimension of the Fermi surface perpendicular to the magnetic field and perpendicular in momentum space to the direction of wave propagation in real space. The various measurements of Fig. 10 give details of the Fermi surface for different directions in momentum space. *See* FERMI SURFACE.

Several other types of oscillations in the attenuation occur. These are the de Haas–van Alphen oscillations of the attenuation, the giant quantum oscillations, acoustic cyclotron resonance, and open orbit resonances. Reviews by B. W. Roberts (1968) and Y. Shapira (1968) have surveyed these effects and have applied them to 21 metals and semiconductors. *See* DE HAAS–VAN ALPHEN EFFECT.

Ultrasonic imaging. Ultrasonic methods have been used to produce light images of ultrasonic beams, outputs from ultrasonic transducers, and visual pictures of transmission through live tissues and have been used for the reproduction of pictures through sound and ultrasound holographic methods. The first process used was the Debye-Sears ultrasonic cell, for which the acoustic wavefronts act as optical gratings to diffract the light on either side of the central spot. By focusing the undiffracted light by a lens and absorbing it at a small black pinhead, the diffracted light can be made to pass the pinhead and be the only light present from the cell. By focusing on the center of the cell, a picture of the details of the ultrasonic sound transmission is obtained.

A modification of this technique, suitable for very high frequencies, is to reflect a monochromatic light beam, from a laser, off the wavefront at the Bragg angle in a fused silica delay line. This angle

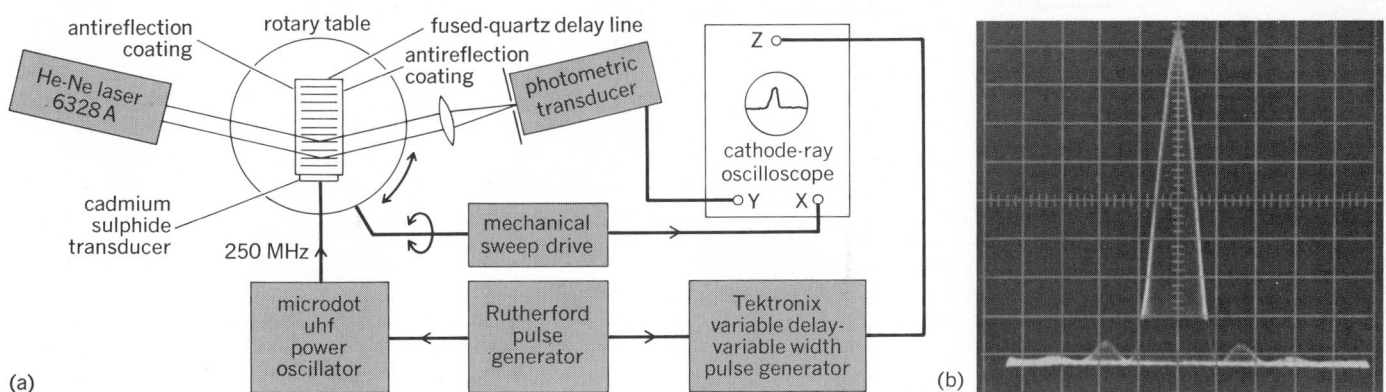

Fig. 12. Light imaging of ultrasonic beam. (*a*) Arrangement for observing angular dependence of deflected light intensity. (*b*) The angular dependence for a rectangular beam which follows closely the theoretical $(\sin x/x)^2$ behavior. (*After H. G. Cohen and E. Gordon*)

is related to the frequency by Eq. (8) for the first

$$2\lambda_s \sin \theta = \lambda_l \qquad (8)$$

diffracted ray. Here λ_s and λ_l are the sound and light wavelengths, and θ is the angle of observation. By rotating the delay line slightly, a complete picture of angular response width of the acoustic transducer can be obtained, as shown by Fig. 12. The figure shows also the equipment necessary to produce such measurements. By putting the end of the fused silica rod into a liquid and determining the width of the Bragg line from the acoustic waves set up in the liquid, accurate velocity and attenuation for longitudinal waves in the liquid can be obtained. This produces results intermediate in frequency between the ultrasonic pulsing method and the Brillouin scattering method.

A direct picture display of ultrasonic waves sent through a sample to be inspected or through live tissues has been achieved by converting the ultrasonic sound wave to a voltage difference by means of a piezoelectric crystal. This voltage pattern on the crystal is scanned by an electronic beam, and it modulates the intensity of this beam. The modulated intensity is amplified by an electron multiplier and controls the intensity of the beam in a television tube, scanned at the same rate as the beam scan of the piezoelectric plate.

Finally, acoustic means have been used to reproduce pictures by holographic methods. They are not yet as useful as light methods. *See* ACOUSTICAL HOLOGRAPHY.

[WARREN P. MASON]

Bibliography: R. T. Beyer and S. V. Letcher, *Physical Ultrasonics*, 1969; S. A. Gahr and C. H. Payne, Acoustic emission detection, real time monitoring of the quality of resistance welds, *West. Elec. Eng.*, 23(4):21–29, October 1979; W. A. Harrison and M. B. Webb (eds.), The Fermi surface, *Proceedings of the International Conference on Fermi-Metal Surfaces*, 1960; W. P. Mason (ed.), *Physical Acoustics*, vols. 5 and 4B, 1968, and vol. 11, 1975; W. P. Mason, *Physical Acoustics and the Properties of Solids*, 1958; R. A. Sykes and W. D. Beaver, High frequency monolithic crystal filter, *Proceedings of the Symposium on Frequency Control*, Bell Syst. Monogr. no. 5214, 1966; P. Vigoureux, *Ultrasonics*, 1950.

Ultraviolet radiation

Electromagnetic radiation in the wavelength range 4–400 nanometers. The ultraviolet region begins at the short wavelength (violet) limit of visibility and extends to the wavelength of long x-rays. It is loosely divided into the near (400–300 nm), far (300–200 nm), and extreme (below 200 nm) ultraviolet regions (see illustration). In the extreme ultraviolet, strong absorption of the radiation by air requires the use of evacuated apparatus; hence this region is called the vacuum ultraviolet. Important phenomena associated with ultraviolet radiation include biological effects and applications, the generation of fluorescence, and chemical analysis through characteristic absorption or fluorescence.

Biological effects of ultraviolet radiation include erythema or sunburn, pigmentation or tanning, and germicidal action. The wavelength regions responsible for these effects are indicated in the figure. Important biological uses of ultraviolet radiation include therapy, the production of vitamin

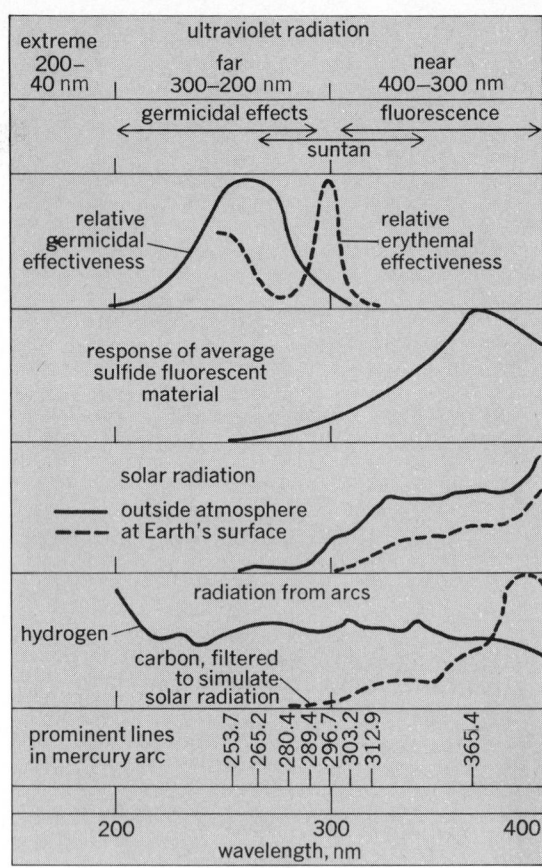

Phenomena associated with ultraviolet radiation. (*After L. R. Koller and General Electric*)

D, the prevention and cure of rickets, and the disinfection of air, water, and other substances.

Fluorescence and phosphorescence are phenomena often generated as a result of the absorption of ultraviolet radiation. These phenomena are utilized in fluorescent lamps, in fluorescent dyes and pigments, in ultraviolet photography, and in phosphors. The effectiveness of ultraviolet radiation in generating fluorescence is shown in the figure. *See* FLUORESCENCE; PHOSPHORESCENCE.

Chemical analysis may be based on characteristic absorption of ultraviolet radiation. Alternatively, the fluorescence arising from absorption in the ultraviolet region may itself be analyzed or observed.

Sources of ultraviolet radiation include the Sun (although much solar ultraviolet radiation is absorbed in the atmosphere); arcs of elements such as carbon, hydrogen, and mercury; and incandescent bodies. The wavelengths produced by some sources of ultraviolet radiation are indicated in the figure.

Artificial sources of ultraviolet light are often used to simulate the effects of solar ultraviolet radiation in the study of the deterioration of materials on exposure to sunlight. Trace amounts of chemicals which strongly absorb ultraviolet radiation may effectively stabilize materials against such degradation.

Detectors of ultraviolet radiation include biological and chemical systems (the skin, the eye of an infant, or eye without a lens, and photographic materials are sensitive to this radiation), but more useful are physical detectors such as phototubes, photovoltaic or photoconductive cells, or radiometric devices.   [FRED W. BILLMEYER]

Bibliography: W. Harm, Biological Effects of Ultraviolet Radiation, 1980; C. N. Rao, Ultra-Violet and Visible Spectroscopy: Chemical Applications, 3d ed., 1975; B. Vodar and J. Romand, Some Aspects of Vacuum Ultraviolet Radiation in Physics, 1974.

Umklapp process

A concept in the theory of transport properties of solids which has to do with the interaction of three or more waves in the solid, such as lattice waves or electron waves. In a continuum, such interactions occur only among waves described by wave vectors k_1, k_2, and so on, such that the interference condition, given by Eq. (1), is satisfied. The sign of

$$k_1 + k_2 + k_3 = 0 \qquad (1)$$

k depends on whether the wave absorbs or emits energy. Since $\hbar k$ is the momentum of a quantum (or particle) described by the wave, Eq. (1) corresponds to conservation of momentum. In a crystal lattice further interactions occur, satisfying Eq. (2),

$$k_1 + k_2 + k_3 = b \qquad (2)$$

where b is any integral combination of the three inverse lattice vectors b_i, defined by $a \cdot b_j = 2\pi\delta_{ij}$, the a's being the periodicity vectors. The group of processes described by Eq. (2) are the Umklapp processes or flip-over processes, so called because the total momentum of the initial particles or quanta is reversed. See CRYSTAL STRUCTURE.

Examples of Umklapp processes are (1) interactions of three lattice waves due to anharmonic lattice forces; of these, only processes described by Eq. (2) produce intrinsic thermal resistance in nonmetals; the exponential variation of the thermal resistance observed in dielectric crystals at low temperatures confirms the concept of Umklapp processes. (2) Scattering of electrons by lattice waves, causing electrical and thermal resistance in metals; it has become clear that the observed properties cannot be accounted for in terms of processes described only by Eq. (1), but Umklapp processes must also be considered. (3) Bragg reflection, which can be regarded as an Umklapp process involving only two waves. See CONDUCTION (HEAT); THERMAL CONDUCTION IN SOLIDS; X-RAY DIFFRACTION.

[PAUL G. KLEMENS]

Bibliography: C. Kittel, Introduction to Solid State Physics, 5th ed., 1976; J. M. Ziman, Principles of the Theory of Solids, 2d ed., 1979.

Uncertainty principle

In quantum mechanics the precept (W. Heisenberg, 1927) that accurate measurement of an observable quantity necessarily produces uncertainties in one's knowledge of the values of other observables. It is also called the indeterminacy principle.

In particular, for a single particle, Eqs. (1) hold.

$$\Delta x \, \Delta p_x \gtrsim h/2\pi \qquad (1a)$$

$$\Delta t \, \Delta E \gtrsim h/2\pi \qquad (1b)$$

In Eq. (1a), Δx represents the uncertainty (error) in

the location of the x coordinate of the particle at any instant, and Δp_x is the simultaneous uncertainty in its x component of momentum; $h =$ Planck's constant $= 6.61 \times 10^{-27}$ erg-sec. Equation (1b) is more subtle than Eq. (1a), and its interpretation depends somewhat on the circumstances. For instance, Eq. (1b) relates the uncertainty ΔE in an energy measurement to the time interval Δt during which the measurement was performed; however, Eq. (1b) also relates the uncertainty ΔE in the energy radiated by a system to the uncertainty Δt in the time at which it radiates, that is, the uncertainty in its lifetime. For derivation of Eqs. (1) from the formal postulates of quantum theory see NONRELATIVISTIC QUANTUM THEORY.

For a discussion of the physical significance of Eqs. (1), especially their relation to the dual wave-particle properties of matter and radiation (summarized in the de Broglie relations $E = hf$, $p = h/\lambda$, where f is the frequency and λ the wavelength), see QUANTUM MECHANICS.

The present article deduces Eqs. (1) from analyses of specific experiments. Only two simple experiments are examined here. Although the sources of error are sometimes arcane, no actual or imagined (Gedanken) experiment has failed to agree with the restrictions of Eqs. (1).

Position measurement. In the illustration an electron is viewed with light of wavelength λ. A photon scatters from the electron, through the lens, and is observed at Q. If the microscope had infinite resolution, observing a photon at Q would precisely locate the electron at P; with the actual lens of finite aperture AB, a photon has an appreciable probability of reaching Q if the electron is as far from P as P', where $P'A - P'B \cong \lambda/2$. Thus, the uncertainty in position of the electron is $\Delta x \cong 2PP' \cong \lambda/\sin \phi$, where ϕ is the aperture angle of the lens (at P). The photon, which originally had momentum h/λ in the x direction, may have x momentum anywhere between $\pm(h/\lambda) \sin (\phi/2)$ after scattering, depending on the path the photon takes through the lens. Thus, the uncertainty in x momentum transferred to the electron is given by Eq. (2). Therefore, $\Delta x \, \Delta p_x \cong h$, which is consistent with Eq. (1a). See RESOLVING POWER.

$$\Delta p_x \cong 2(h/\lambda) \sin (\phi/2) \cong (h/\lambda) \sin \phi \qquad (2)$$

Energy measurement. Suppose the kinetic energy $E = \frac{1}{2}mv^2$ of a particle moving along x is found by measuring its speed v. Then $\Delta E = mv \, \Delta v = v \, \Delta p_x$. If the measurement takes a time Δt, the position of the particle is uncertain by an amount $\Delta x = v \, \Delta t$. Then $\Delta E \, \Delta t = \Delta p_x \, \Delta x$ which, accepting Eq. (1a), demonstrates Eq. (1b).

The uncertainty principle refers only to uncertainties at the instant of measurement, which uncertainties prevent complete certainty about the future course of the system under observation. Measurements made at time t_0 may reveal the precise position and momentum a particle had at a time $t < t_0$; this information is not usable for predictions at times $t > t_0$, however, once the past-revealing but future-obscuring measurement at t_0 has been performed. Equations (1), which are deduced nonrelativistically, are modified when the special theory of relativity is taken into account. For example, relativity requires the recoil photon in the figure to have a longer wavelength than the incident photon, with the result that Δx cannot be

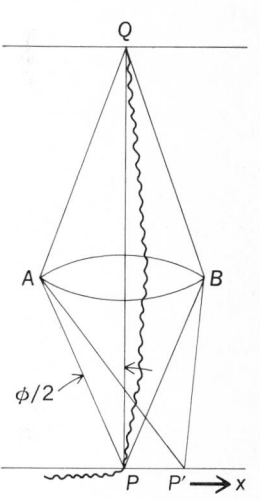

Microscope for locating electron.

less than $\sim h/mc$; Eq. (1a) permits precise position measurements, $\Delta x \rightarrow 0$, provided $\Delta p_x \rightarrow \infty$ is acceptable. *See* COMPTON EFFECT; QUANTUM ELECTRODYNAMICS; RELATIVISTIC MECHANICS; RELATIVITY. [EDWARD GERJUOY]

Bibliography: S. G. Gasiorowicz, *Quantum Physics*, 1974; W. Heisenberg, *The Physical Principles of Quantum Theory*, 1949; J. L. Powell and B. Crasemann, *Quantum Mechanics*, 1961; D. Park, *Introduction to the Quantum Theory*, 2d ed., 1974.

Unified field theory

Any theory which attempts to express gravitational theory and electromagnetic theory within a single unified framework. Usually the term is used in a somewhat narrower sense to refer to attempts to generalize Albert Einstein's general theory of relativity from a theory of gravitation alone to a theory of gravitation and classical electromagnetism. *See* ELECTROMAGNETISM; GRAVITATION; RELATIVITY.

Motivation. General relativity theory, which dates from 1916, achieves a geometric interpretation of gravity that is mathematically elegant and experimentally successful. It ascribes the force of gravitation to a distortion of the four-dimensional space-time manifold, in which test bodies move on generalized straight lines or geodesics. The projections of these paths into three-dimensional space are the familiar trajectories of freely falling bodies. The second universal force of classical physics, the electromagnetic field, is not explained by classical general relativity as a geometric entity. Thus there have been numerous attempts to expand the geometric structure of general relativity to include the electromagnetic field and interactions of charged particles. *See* CALCULUS OF TENSORS.

Any such attempt must overcome three basic difficulties. First, the gravitational field is described by the 10 independent components of the symmetrical metric tensor. Classical electromagnetism is based on a vector potential with four components; thus a combined theory must provide for at least 14 geometric entities. Second, in gravitational theory it is natural to attempt a geometric interpretation since the inertial and gravitational mass of a body are the same, and since the motion of a test body is independent of its mass. This is the famous principle of equivalence; it is not true in electromagnetic theory in that the motion of a test body in an electromagnetic field is strongly dependent on its charge-to-mass ratio. Third, the enormous difference in the characteristic strengths of the electromagnetic and gravitational fields has to be reconciled. For example, the approximate ratio of electromagnetic to gravitational force between two free electrons is the large number 4×10^{42}.

It should be noted that general relativity and classical electromagnetism are entirely consistent and compatible in their original form. Moreover, there is no experimental evidence against either theory on a macroscopic level. Thus the motivation to interpret the two forces within a single theoretical framework is an esthetic desire for mathematical elegance and simplicity, and is not due to an inconsistency or inability of either theory to predict experimental results.

Classical attempts. One of the earliest and most mathematically interesting attempts at a unified field theory was made by Hermann Weyl in 1918. Weyl generalized the differential geometry used in general relativity theory to allow the introduction of precisely four new quantities to be associated with the electromagnetic vector potential. These were introduced by assuming that, under a generalized parallel displacement of a four-vector over a coordinate change dx^β, the four-dimensional length l of the four-vector is not constant but changes according to the simple linear relation given below. The four quantities ϕ^β were then

$$dl = \phi_\beta dx^\beta l$$

taken to be proportional to the electromagnetic vector potential. Although the above assumption does lead to an interesting mathematical structure encompassing the electromagnetic field, there were serious objections raised by Einstein: The Weyl theory implies a dependence of atomic phenomena on the history of individual atoms, which is difficult to reconcile with experiment. Subsequently, because of the above difficulty and its lack of predictive power, the theory was abandoned by Weyl. *See* SPACE-TIME.

Another interesting approach was made by T. Kaluza. The essence of Kaluza's theory was to increase the number of dimensions to five. In five dimensions the symmetric metric tensor has 15 components, which is one more than necessary to accommodate the gravitational and electromagnetic fields. To recover the four-dimensional nature of normal space-time, the metric tensor was subjected to somewhat artificial constraints: It was assumed to be independent of the fifth coordinate, and the fifth diagonal component was taken to be unity; that is, the four-dimensional world that one observes was taken to be the projection of a more basic underlying five-dimensional manifold. Despite a certain degree of mathematical beauty, the Kaluza theory and various extensions of it have not been successful in making any new predictions, and it has now been bypassed by the mainstream of theoretical physics.

In his later years Einstein devoted considerable effort to the consideration of nonsymmetric metric tensors in four dimensions. Such a tensor has 16 components. The geometrical aspects of the theory proposed and begun by Einstein have been developed by V. Hlavaty, but the basic physical problems were unsolved at the time of Einstein's death and have remained so.

Geometrodynamics. It can be safely stated that none of the early attempts at a unified field theory has succeeded in combining gravitational and electromagnetic theory with as much mathematical elegance and physical coherence as classical general relativity or classical electromagnetism. More recent attempts by J. A. Wheeler and C. W. Misner follow a somewhat less pretentious course but have achieved interesting results. Instead of reformulating the theories of gravitation and electromagnetism, Wheeler and Misner retained the original equations of general relativity and classical electromagnetism in the absence of charge and investigated the mathematical consequences. The gravitational and electromagnetic theories both involve second-order differential equations. By a mathematical process which has become known as geometrization, the two sets of second-order equations can be combined in a single set of fourth-order differential equations in which the electromagnetic field does not explicitly appear. However, the metric retains sufficient information about the energy and momentum density associated with the electromagnetic field that the electromagnetic field may be recovered from it; that is, the electromagnetic field may be considered as being completely specified by the gravitational field produced by its energy density. Since no new equations or physical assumptions enter this formulation, it is called the already-unified field theory.

As mentioned above, the already-unified field theory does not allow the a priori existence of charges; therefore the difficulty associated with the very large electromagnetic force between charges (see the preceding section on motivation) is bypassed. The problem is then to reintroduce charge into the theory, since a universe with no charge is not very interesting. This is accomplished by allowing the space-time manifold to be multiply connected, so that electric-field lines can enter a region of space and apparently disappear, only to reappear in another region of space. This structure in the manifold is descriptively referred to as a wormhole and is shown in two dimensions in the illustration.

Since one end of the wormhole is a region into which field lines enter, it therefore appears to be a negative charge, whereas the other end appears to be a positive charge from which field lines emerge. The effect of charge has therefore been obtained, even though no singular point charges are present and all electric-field lines are continuous.

The already-unified field theory in multiply connected space involves only geometry and is thus referred to by its inventors as a geometrodynamical theory. Because of its inherent mathematical difficulty, progress has been slow; many problems, such as the motion of the ends of the wormhole, remain to be solved before any definitive predictions of the theory can be made.

Quantum theory. In the early days of general relativity theory, from 1915 to about 1925, there was intense interest and activity in the formulation of unified field theories. The discovery of viable theories of quantum mechanics by E. Schrödinger and W. Heisenberg in 1925 diverted much of this interest to the newly accessible and exciting world of atomic and nuclear behavior. Thus a large part of the motivation to construct a classical unified field theory has given way to interest in the new vistas in the world of the very small. In this context, moreover, many physicists no longer believe that a useful and consistent unified field theory can be constructed without taking cognizance of the quantized nature of the world. At one time Einstein believed that an approach in five dimensions, like that of Kaluza, might allow for the inclusion of quantum effects, but the hope was short-lived. Similarly, in 1927 F. London attempted to relate the Weyl theory to the early ad hoc and provisionary quantum theory of N. Bohr, but this also proved unsuccessful. More recently the proponents of geometrodynamics have attempted to quantize that theory, but with very limited success. *See* NONRELATIVISTIC QUANTUM THEORY; QUANTUM FIELD THEORY; RELATIVISTIC QUANTUM THEORY.

It seems likely that nature observes a close unity between the world of very large-scale and very small-scale phenomena. Thus a deep understanding of the large-scale behavior described by general relativity, the small-scale behavior described by quantum mechanics, and the omnipresent electromagnetic force may well demand a thorough revision of the ideas of space and time that will go far beyond the original goal of the classical unified field theories.   [R. J. ADLER]

Bibliography: R. Adler, M. Bazin, and M. Schiffer, *Introduction to General Relativity*, 2d ed., 1975; P. G. Bergmann, *Introduction to the Theory of Relativity*, 1942, reprint 1976; A. Einstein, *The Meaning of Relativity*, 5th ed., 1956; C. W. Misner et al., *Gravitation*, 1973.

Unitary symmetry

One of the approximate internal symmetry laws obeyed by the strong interactions of elementary particles. A system of particles has an SU_n internal symmetry if all of the particles can be described as compounds of a fundamental multiplet of n particles, and if all physical properties of the system are unchanged by an arbitrary unitary transformation of the fundamental multiplet. *See* SYMMETRY LAWS.

An analog is the approximate spin independence of electrons under electrostatic forces (as in an atom): There is a fundamental doublet, namely the spin-up electron and the spin-down electron. Denoting these two states by $|u\rangle$ and $|d\rangle$, all physical properties (energy eigenvalues, charge density, and so on) are unchanged by the replacements shown in Eqs. (1), where α and β are complex

$$|u\rangle \rightarrow \alpha|u\rangle + \beta|d\rangle \qquad |d\rangle \rightarrow -\beta^*|u\rangle + \alpha^*|d\rangle$$
$$|\alpha|^2 + |\beta|^2 = 1 \tag{1}$$

numbers. This transformation corresponds to a rotation of space (α and β can be expressed in terms of the three numbers which describe the rotation). It is easily seen that states of several electrons decompose under the rotation, Eqs. (1); for example, the two-electron state $(|u,d\rangle - |d,u\rangle)/\sqrt{2}$ is unchanged by the rotation, and the three remaining two-electron states, $|u,u\rangle$, $|d,d\rangle$, and $(|u,d\rangle + |d,u\rangle)/\sqrt{2}$, transform to linear combinations of themselves. This is the decomposition into singlet and triplet spin states, that is, into total spin $S = 0$ and 1 respectively; the nonmixing between them is equivalent to the invariance of S to rotation. The group of all the transformations of two states which preserves their scalar products $[\langle u|d\rangle = 0, \langle u|u\rangle = \langle d|d\rangle = 1]$ is known as the two-dimensional unitary group, U_2; the transformations of Eqs. (1) form a subgroup known as SU_2 which merely lacks the uninteresting transformations of the form $|u\rangle \rightarrow e^{i\varphi}|u\rangle$ and $|d\rangle \rightarrow e^{i\varphi}|d\rangle$, that is, an equal change of phase of the two states.

Charge independence. The strong interactions are approximately invariant to such a group; the fundamental doublet can be taken to be the nucleon, with the up and down states proton and neutron. This SU_2 symmetry is known as charge independence, or, loosely, as i-spin conservation, the analog to the electron spin being known as i-spin $\mathbf{I}$. The nucleon has i-spin $\frac{1}{2}$; the pion has i-spin 1. Although the transformation of the pion states under i-spin rotations follows correctly from regarding the pion as the triplet state of a pair of nucleons, a further symmetry, nucleon conservation, requires the pion to be regarded instead

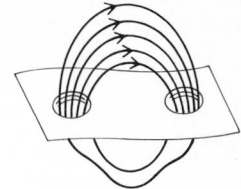

Electric-field lines entering and leaving wormhole, shown here in two dimensions.

as a compound of a nucleon and antinucleon. However, the statement "$|\pi\rangle$ can be regarded as $|N\overline{N}\rangle$" is only a statement about the transformation of pion states under i-spin rotations; it is not a statement about the physical structure of the pion. It is like saying that a triplet state of a calcium atom transforms under rotations like the triplet state of two electrons; this does not mean that the atom is a compound of two electrons. *See* ISOTOPIC SPIN.

SU$_3$ symmetry. The existence of strange particles shows that the symmetry group of the strong interactions is yet larger; an additional quantum number strangeness is conserved equally well as I_3. This can be achieved by adding a third fundamental particle, an i-spin singlet $|\lambda\rangle$ to carry strangeness; the additional symmetry is invariance to a relative phase change of $|N\rangle$ and $|\lambda\rangle$, that is, $|N\rangle \rightarrow e^{i\psi}|N\rangle$ and $|\lambda\rangle \rightarrow e^{-i\psi}|\lambda\rangle$, a U$_1$ group. When a sufficient number of strange particles had been observed, it was seen that they, together with the old nonstrange particles, were grouped into multiplets whose members had the same space-time quantum numbers (except for mass; the masses of the members are only similar not equal). This suggested the existence of a yet larger symmetry; it has turned out that this symmetry is the group of unitary transformations of a triplet of fundamental particles, SU$_3$. (The i-spin plus strangeness symmetry of the triplet described above, that is, SU$_2 \times$ U$_1$, is a subgroup of SU$_3$.) This symmetry is often loosely called unitary symmetry.

A striking difference in the manifestations of SU$_2$ and SU$_3$ is that whereas all possible multiplets of the former appear in nature, only those multiplets of the latter appear which can be regarded as compounds of the fundamental triplet in which the net number of component fundamental particles (number of particles minus number of antiparticles) is an integral multiple of 3. (An analog in SU$_2$ would be compounds of an even number of the fundamental doublet particles; these states would have integral, never half-odd-integral, i-spin.) In particular, no particle which could be regarded as the fundamental triplet is found. Despite this nonappearance, it turns out that a great deal about the strongly interacting particles (hadrons) is at least qualitatively explained if they are regarded as physical compounds of a fundamental triplet of particles, to which the name quark has been given. There are interesting theories to explain why single quarks are never observed. *See* HADRON; QUARKS.

The illustration shows the values of the additive quantum numbers I_3 and Y of the states of some SU$_3$ multiplets. Y (hypercharge) is essentially the same as strangeness. The states of the triplet labeled d, u, s correspond to the triplet n, p, λ discussed above [d = down, u = up (referring to I_3), and s = strange]. By assigning appropriate values of charge Q to the quarks, one gets the correct charges of hadrons, which obey $Q = I_3 + \frac{1}{2}Y$.

Unitary symmetry is only approximate. Its "breaking" appears to obey octet dominance. This means, for example, that the part of a mass operator which is not invariant under SU$_3$ (so that its matrix elements are the mass deviations of the members of an SU$_3$ multiplet) transforms as an octet. The consequence is the mass formula of

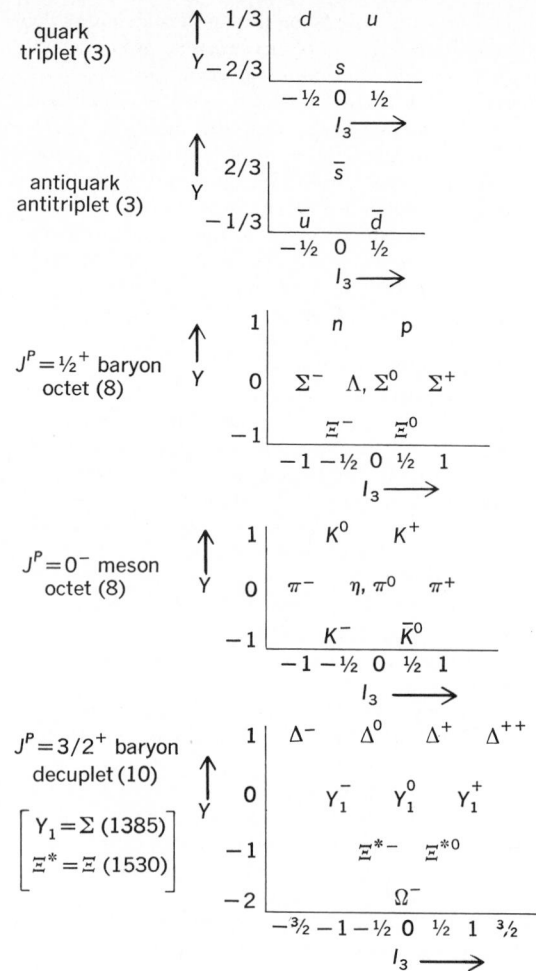

Weight diagrams of some SU$_3$ multiplets showing the values of I_3 and Y (hypercharge) of the members.

M. Gell-Mann and S. Okubo, given by Eq. (2),

$$m(I, Y) = a + bY + c[I(I+1) - \tfrac{1}{4}Y^2] \quad (2)$$

where a, b, and c are constants, different for each multiplet, generally. (An analogy to this would be that the energy operator of atoms in a weak magnetic field transforms as a vector, and therefore the energy of an atom has the form $a + bJ_3$, where J_3 is the projection of the spin of the atom along the direction of the magnetic field.) This implies that the masses of the four i-spin multiplets of an octet have one relation between them, given by Eq. (3). This holds well for the $\frac{1}{2}^+$ baryon multiplet

$$m(\tfrac{1}{2}, +1) + m(\tfrac{1}{2}, -1)$$
$$= \tfrac{3}{2}m(0,0) + \tfrac{1}{2}m(1,0) \quad (3)$$

(i-spin multiplets N, Ξ, Λ, Σ) and also for the 0^- meson octet (K, $\overline{K}$, η, π) if m in the formula is replaced by squares of masses (field theory gives some reason for this, for mesons). For a decuplet, such as the $\frac{3}{2}^+$ baryons [Δ, $\Sigma(1385)$, $\Xi(1530)$, Ω], the GM-O formula, Eq. (2), predicts equal spacing of the masses; this served as a prediction of the Ω, at its correct mass. The GM-O formula does not hold for the eight 1^- mesons [$K(890)$, $\overline{K}(890)$, ω, ρ]. This is explained by the existence of a ninth 1^- meson of similar mass, the ϕ; the SU$_3$ breaking interaction has strongly mixed the original I, $Y = 0$,

0 "ϕ_0" member of the octet with the original singlet "ω_0," and pushed their masses apart. This is also true of the nine 2^+ mesons. In both cases, the lower I, $Y = 0$, 0 meson is nearly degenerate in mass with the I, $Y = 1$, 0 meson (that is, ω with ρ; f' with A_2), and the upper one is strongly coupled to $K\overline{K}$ (the major decay mode of both ϕ and f') but nearly decoupled from pions.

SU_3 symmetry, of course, implies relations between strong coupling constants, for example, $\pm\sqrt{3}\,g_{KN\Lambda} \pm g_{KN\Sigma} = 2g_{\pi NN}$, and between cross sections; these, when testable experimentally, seem to agree. With assumptions about the SU_3 transformation behavior, namely, as octets, of electromagnetic and weak current operators, there are many relations between their matrix elements, which agree well. For instance, the magnetic moments of the $\frac{1}{2}^+$ baryon octet are predicted to have the relations given by Eqs. (4). The relations

$$\mu(\Sigma^+) = \mu(p)$$
$$2\mu(\Lambda) = \mu(\Xi^0) = -2\mu(\Sigma^0) = \mu(n) \qquad (4)$$
$$\mu(\Sigma^-) = \mu(\Xi^-) = -[\mu(p) + \mu(n)]$$

among the weak couplings are described by the Cabibbo theory. *See* BARYON; MESON; WEAK NUCLEAR INTERACTIONS.

SU_N symmetry. According to the argument given above, hadrons have the approximate symmetry SU_N, where N is the number of kinds of quarks, or flavors. Five flavors of quark are known; in addition to the quarks with the flavors up, down, and strange described above, two more quarks, charm and bottom, have been found. There is reason to think that the number of flavors is even, and that a top partner to the bottom quark will be found, having a mass which puts it just out of reach of present-day (1980) accelerators. But it is totally unknown how many yet heavier quarks there are. Parallel to the quarks, there are six kinds of leptons currently known. The resulting SU_6 symmetry plays a role in their weak interactions. *See* CHARM; FLAVOR; J PARTICLE; LEPTON; UPSILON PARTICLES.

Color SU_3. In the quark model of hadrons, baryons are correctly described only if quarks carry, in addition to spin and flavor, another quantum number, color, which can take on three values. The resulting symmetry, color SU_3, is thought to be exact; that is, the differently colored quarks of a given flavor are thought to be absolutely equivalent. However, this exact symmetry is rather well hidden, because apparently all free particles belong to the singlet representation of the symmetry. *See* COLOR (QUANTUM MECHANICS).

Nonrelativistic symmetries. There are unitary symmetries which are not purely internal, in the strict sense, namely those in which it is supposed that the interaction between the fundamental particles is spin-independent (as for nonrelativistic electrons interacting with electrostatic forces). This was suggested by E. Wigner as an approximate symmetry in light nuclei; there the fundamental particle is a quartet, with states $|pu\rangle$, $|pd\rangle$, $|nu\rangle$, $|nd\rangle$ (where pu means a proton with spin up, and so on), and the symmetry group is therefore SU_4. Similarly, if the interactions of the SU_3 quarks are supposed to be spin-independent, the symmetry group is SU_6. It is known that such unitary symmetries can never be exact in a relativistic theory,

but that fact does not prevent them from being approximately true, at least for low-lying states. *See* ELEMENTARY PARTICLE.

[CHARLES J. GOEBEL]

Bibliography: M. Gell-Mann and Y. N'eman (eds.), *The Eightfold Way*, 1964; N. P. Samios, M. Goldberg, and B. T. Meadows, Hadrons and *SU*(3): A critical review, *Rev. Mod. Phys.*, 46:49–81, 1974.

Units of measurement

Values, quantities, or magnitudes in terms of which other such are expressed. Units are grouped into systems, suitable for use in the measurement of physical quantities and in the convenient statement of laws relating physical quantities. A quantity is a measureable attribute of phenomena or matter.

A given physical quantity A, such as length, time, or energy, is the product of a numerical value or measure $\{A\}$ and a unit $[A]$. Thus Eq. (1) holds.

$$A = \{A\}[A] \qquad (1)$$

The unit $[A]$ can be chosen arbitrarily, but it is desirable to define units in such a way that they are derived from a few base units by equations without numerical factors other than unity, and that the equations between numerical values of quantities have exactly the same form as the equations between the quantities. For example, the kinetic energy E of a body is given in terms of its mass M and speed V by Eq. (2), where $E = \{E\}[E]$,

$$E = \tfrac{1}{2}MV^2 \qquad (2)$$

$M = \{M\}[M]$, $V = \{V\}[V]$, and $\frac{1}{2}$ is called a definitional factor and is dimensionless. If the units of E, M, and V are defined in such a way that Eq. (3)

$$[E] = [M][V]^2 \qquad (3)$$

holds, then the equation between the numerical values is Eq. (4). A system of units defined in this

$$\{E\} = \tfrac{1}{2}\{M\}\{V\}^2 \qquad (4)$$

way is called a coherent system. It is constructed by defining the units of a few base quantities independently; these are called base units. The units of all other quantities are defined by equations similar to Eq. (3) with no numerical factors other than unity, and are called derived units. It is a matter of choice which quantities are to be considered as the base quantities and also how many. It is desirable to have as few base quantities as possible, provided no serious inconveniences are encountered. *See* DIMENSIONAL ANALYSIS.

In 1960 the General Conference on Weights and Measures (CGPM) gave official status to a single practical system, the International System of Units, abbreviated SI in all languages. The system is a modernized version of the metric system. The SI, as subsequently extended, includes seven base units, two supplementary units, and nineteen derived units with special names. These derived units, and others without special names, are derived from the base and supplementary units in a coherent manner. A set of prefixes is used to form decimal multiples and submultiples of the SI units.

The CGPM and its subsidiary body, the International Committee for Weights and Measures (CIPM), have accepted certain units which are not part of the SI but which are widely used or are useful in specialized fields, for use with the SI or for temporary use in those fields. For definitions of the seven SI base units, discussion of the physical standards for realizing these units, and tables of the SI base, supplementary, and derived units and units accepted for use with the SI *see* PHYSICAL MEASUREMENT.

This article discusses how the various derived units are defined in terms of the base units, both in the SI and in other coherent systems, and how units in different systems, as well as arbitrary units which are not part of any system, are related. Various units which are not accepted for use with the SI, but whose use is nevertheless common, are discussed. However, the use of all units which are not SI units or specifically accepted for use with the SI has been discouraged, and whenever possible, units outside the SI should be replaced by SI units or by their multiples and submultiples formed by attaching SI prefixes.

GEOMETRICAL UNITS

Units of plane angle and solid angle are purely geometrical. The SI units of plane and solid angle are called supplementary units, and may be regarded either as base units or as derived units.

Plane angle units. The radian (rad), the SI unit of plane angle, is the plane angle between two radii of a circle which cut off on the circumference an arc equal in length to the radius. Since the circumference of a circle is 2π times the radius, the complete angle about a point is 2π rad.

The degree and its decimal submultiples can be used with the SI when the radian is not a convenient unit. By definition, 2π rad $= 360°$. The minute $[1' = (1/60)°]$ and the second $[1'' = (1/60)']$ can also be used.

Steradian. The steradian (sr), the SI unit of solid angle, is the solid angle which, having its vertex at the center of a sphere, cuts off an area on the surface of the sphere equal to that of a square with sides of length equal to the radius of the sphere. Since the total area of a sphere is 4π times the square of its radius, the complete solid angle about a point is 4π sr.

MECHANICAL UNITS

In mechanics, it is convenient to have three base quantities, and two of these are generally chosen to be length and time. Systems of mechanical units may be classified as absolute systems, in which the third base quantity is mass, and gravitational systems, in which the third base quantity is force.

Absolute systems. Two absolute systems of metric units are commonly employed, each named for its base units of length, mass, and time: the mks (meter-kilogram-second) absolute system, and the cgs (centimeter-gram-second) absolute system. The mks absolute system is the mechanical portion of the SI. Once the length, mass, and time units are selected, Newton's second law in the form $F = ma$ is used to define the unit of force F in terms of mass (m) units and acceleration (a) units, which involve length and time units. The unit of force derived in this way through Newton's second law is called the newton in the mks system and the dyne in the cgs system. Units of work W are defined by the relation $W = FL$, where L represents a displacement in the direction of a force F. Units of power P in turn are defined by the relation $P = W/t$ and are therefore the ratio of work units to units of time t. A coherent absolute system of British units is based on the foot, the pound (1 lb $\cong$ 0.4536 kg), and the second; and the derived unit of force is called the poundal (pdl), a unit which is not often used. Various quantities expressed in the three absolute systems are listed in Table 1. *See* KINETICS; NEWTON'S LAWS OF MOTION.

Gravitational systems. Gravitational systems, in which the base quantities are length, force, and time, have been frequently employed by engineers, and are therefore sometimes called technical systems.

In the British gravitational system the standard force, called the pound force (lbf), is the gravitational force exerted on a 1-lb mass at a location on the Earth's surface where the acceleration of a freely falling body due to gravity has the standard value $g_n = 9.80665$ m/s^2 $\cong$ 32.174 ft/s^2. More simply stated, 1 lbf is the force which imparts an acceleration of magnitude g_n to a 1-lb mass. A coherent system is obtained by using Newton's second law, $F = ma$, to define a mass unit called the slug (1 lbf = 1 slug·ft/s^2). The work unit in this system is the foot-pound force (ft·lbf). *See* FREE FALL.

In the mks gravitational system the unit of acceleration is the meter per second squared (m/s^2), and the unit of force is taken to be the Earth's gravitational force on a 1-kg mass at a location where this force imparts an acceleration of magnitude g_n. Unfortunately, this unit of force has also been called a kilogram, but to distinguish it from the unit of mass it should be called kilogram force

Table 1. Absolute systems of units

Quantity	Defining relation	mks (SI)	cgs	British
Length (L)	Fundamental	m	cm	ft
Mass (m)	Fundamental	kg	g	lb
Time (t)	Fundamental	s	s	s
Velocity (v)	$v = L/t$	m/s	cm/s	ft/s
Acceleration (a)	$a = v/t$	m/s^2	cm/s^2	ft/s^2
Force (F)	$F = ma$	1 N = 1 kg·m/s^2	1 dyne = 1 g·cm/s^2	1 pdl = 1 lb·ft/s^2
Work (W)	$W = FL$	1 J = 1 N·m	1 erg = 1 dyne·cm	ft·pdl
Power (P)	$P = W/t$	1 W = 1 J/s	erg/s	ft·pdl/s
Momentum (p)	$p = mv$	kg·m/s	g·cm/s	lb·ft/s

Table 2. Gravitational systems of units

Quantity	Defining relation	Units		
		mks	cgs	British
Length (L)	Fundamental	m	cm	ft
Force (F)	Fundamental	kilogram force (kgf)	gram force (gf)	pound force (lbf)
Time (t)	Fundamental	s	s	s
Mass (m)	$m = F/a$	kgf·s^2/m	gf·s^2/cm	1 slug = 1 lbf·s^2/ft
Velocity (v)	$v = L/t$	m/s	cm/s	ft/s
Acceleration (a)	$a = v/t$	m/s^2	cm/s^2	ft/s^2
Work (W)	$W = F L$	m·kgf	cm·gf	foot-pound force (ft·lbf)
Power (P)	$P = W/t$	m·kgf/s	cm·gf/s	ft·lbf/s
Momentum (p)	$p = m v$	kgf·s	gf·s	slug·ft/s

(kgf). There is no generally accepted name for the mks gravitational unit of mass, although the name metric slug has been used. The cgs gravitational system is constructed in a similar manner. Various quantities expressed in the three gravitational systems are listed in Table 2.

Noncoherent gravitational systems in which force is expressed in pounds force (lbf) or a comparable kilogram force (kgf) or gram force (gf), and mass in pounds, kilograms, or grams, can be set up by writing Newton's second law in the form $g_n F = ma$.

Length units. The meter (m) is the SI base unit of length. The use of special names for decimal submultiples of the meter should be avoided, and units formed by attaching appropriate SI prefixes to the meter should be used instead. Thus the micron (μ), which was defined as 10^{-6} m, should be replaced by the micrometer (μm), which has the same value; and the millimicron (mμ), which was defined as 10^{-9} m, should be replaced by nanometer (nm). The fermi, which was defined as 10^{-15} m, and was used to measure nuclear distances, should be replaced by the femtometer (fm), which has the same value.

The angstrom (A) is equal to 10^{-10} m. Although it has been accepted for temporary use with the SI, it is preferable to replace this unit with the nanometer, using the relation 1 A = 0.1 nm.

The nautical mile (nmi), equal to 1852 m, has been accepted for temporary use with the SI in navigation.

The foot (ft) is, as discussed above, the unit of length in the British systems of units, and it is also in customary use in the United States. Since 1959 the foot has been defined as exactly 0.3048 m. The yard (yd) is defined as exactly 3 ft or 0.9144 m.

Units used to measure x-ray wavelengths. Relative measurements of x-ray wavelengths can be made to a higher accuracy than absolute measurements. That is, the ratio of two x-ray wavelengths can be determined with a higher accuracy than the ratio of either of them to the meter. The same situation holds for dimensions of crystal lattices, which are derived from x-ray wavelengths by x-ray diffraction experiments. For this reason, x-ray wavelengths and dimensions of crystal lattices have been expressed in units that are defined in terms of a standard x-ray wavelength or crystal lattice dimension. *See* X-RAY CRYSTALLOGRAPHY; X-RAYS.

Before 1965, most x-ray wavelengths were ex-

pressed in terms of the X-unit, which is approximately 10^{-13} m. The grating constant of calcite was defined to be exactly 3029.04 X-units. Subsequent absolute measurements of x-ray wavelengths with ruled gratings indicated that the X-unit exceeds 10^{-13} m by about 2 parts per thousand. Furthermore, in practice, workers in the field began using definitions of the X-unit based on various x-ray wavelengths instead of the calcite grating definition, and subsequent precise measurements indicated that these wavelength standards differed from each other and from the calcite grating standard by as much as 20 parts per million.

The X-unit has been superseded by the A* unit, introduced by J. A. Bearden in 1965, which is based on the tungsten $K\alpha_1$ line as a standard. The peak of this line is defined as exactly 0.2090100 A*. X-ray wavelength tables have been published in terms of this unit. At the time the A* unit was defined, it was thought to equal 10^{-10} m (the angstrom unit, A) to within 5 parts per million, but the A* unit is now believed to be 20±5 parts per million larger than 10^{-10} m.

Units used in astronomy. Special units whose values are obtained experimentally are used in astronomy. The astronomical unit and parsec are accepted for use with the SI. The parsec rather than the light-year is used in technical literature.

Area units. The square meter (m^2), the SI unit of area, is the area of a square with sides of length 1 m. Other area units are defined by forming squares of various length units in the same manner. The hectare (ha) is equal to 1 square hectometer (1 hm^2) or equivalently to 10^4 m^2. Its use with the SI is permitted for expressing land or water areas.

Cross sections, which measure the probability of interaction between an atomic nucleus, atom, or molecule and an incident particle, have the dimensions of area, and the appropriate SI unit for expressing them is therefore the square meter. The barn (b), a unit of cross section equal to 10^{-28} m^2, has been accepted for temporary use with the SI. Typical nuclear reactions have cross sections ranging from millibarns to several thousand barns. A related quantity, which is connected with the probability that a reaction will emit radiations in a particular direction, is the differential cross section, which has the dimensions of barns per steradian. *See* NUCLEAR REACTION.

Units of volume. The cubic meter (m^3), the SI unit of volume, is the volume of a cube with sides of length 1 m. Other units of volume are defined by forming cubes of various length units in the same manner. The liter (symbol L in the United States) is equal to 1 cubic decimeter (1 dm³), or equivalently to 10^{-3} m³. It has been accepted for use with the SI for measuring volumes of liquids and gases.

Time units. The second (s) is the SI base unit of time. However, other units of time in customary use, such as the minute (1 min = 60 s), hour (1 h = 60 min), and day (1 d = 24 h), are used in the SI.

Frequency units. The hertz (Hz), the SI unit of frequency, is equal to 1 cycle per second. A periodic oscillation has a frequency of n hertz if it goes through n cycles in 1 s. Other units of frequency are defined by forming reciprocals of time units in the same manner. *See* FREQUENCY.

Speed and velocity units. The meter per second (m/s), the SI unit of speed or velocity, is the magnitude of the constant velocity at which a body traverses 1 m in 1 s. Other speed and velocity units are defined by dividing a unit of length by a unit of time in the same manner. *See* SPEED; VELOCITY.

The knot (kn) is equal to 1 nautical mile per hour (1 nmi/h); it has been accepted for temporary use with the SI.

Acceleration units. The meter per second squared (m/s²), the SI unit of acceleration, is the magnitude of the constant acceleration of a body whose velocity changes by 1 m/s in 1 s. Other units of acceleration are defined by dividing a unit of velocity by a unit of time in the same manner. *See* ACCELERATION.

The gal or galileo (symbol Gal) is equal to 1 cm/s², or equivalently to 10^{-2} m/s². This unit and its decimal submultiple the milligal (1 mGal = 10^{-3} Gal = 10^{-5} m/s²) are employed in geodesy and geophysics to express the acceleration of gravity, and have been accepted for temporary use with the SI.

Mass units. The kilogram (kg), the SI base unit of mass, is the only SI unit whose name, for historical reasons, contains a prefix. Names of decimal multiples and submultiples of the kilogram are formed by attaching prefixes to the word gram (g). Since 1 kg = 10^3 g, 1 g = 10^{-3} kg. The metric ton (t), which is equal to 10^3 kg or 1 megagram (Mg), is permitted in commercial usage of the SI.

The pound (lb), is, as discussed above, the unit of mass in the British absolute system, and is also in customary use in the United States. In 1959 the pound was defined to be exactly 0.45359237 kg.

The slug is, as discussed above, the unit of mass in the British gravitational system. By definition 1 pound force (lbf) acting on a body of mass 1 slug produces an acceleration of 1 foot per second squared (1 ft/s²). The slug is equal to approximately 32.174 lb or 14.594 kg. *See* MASS.

Force units. The newton (N), the SI unit of force, is the force which imparts an acceleration of 1 meter per second squared (1 m/s²) to a body having a mass of 1 kg.

The dyne, the cgs absolute unit of force, is the force which imparts an acceleration of 1 centimeter per second squared (1 cm/s²) to a body having a mass of 1 g. Since 1 cm/s² = 10^{-2} m/s², and 1 g = 10^{-3} kg, it follows that 1 dyne = 10^{-5} N.

The unit of force in the British absolute system is, as discussed above, the poundal (pdl), the force which imparts an acceleration of 1 foot per second squared (1 ft/s²) when applied to a body of mass 1 lb. One poundal is approximately 0.13825 N.

As discussed above, the units of force in the mks gravitational, cgs gravitational, and British gravitational systems are the forces which impart an acceleration equal to the standard acceleration of gravity, g_n = 9.80665 m/s² $\cong$ 32.174 ft/s², when applied to bodies having masses of 1 kg, 1 g, and 1 lb, respectively. These units are named the kilogram force (kgf), gram force (gf), and pound force (lbf) respectively. Unfortunately, these units have also been called simply the kilogram, gram, and pound, giving rise to confusion with the mass units of the same name; in proper usage the suffix "force" should always be employed. The pound has also been called the pound mass (lbm) to further distinguish mass and force units. One pound force is approximately 4.4482 N. *See* FORCE.

Pressure and stress units. The pascal (Pa), the SI unit of pressure and stress, is the pressure or stress of 1 newton per square meter (N/m²). This is a rather small unit for most practical purposes; for example, atmospheric pressure is approximately 10^5 Pa. Thus, most pressures are most readily expressed in decimal multiples of the pascal formed by attaching the appropriate SI prefix.

The dyne per square centimeter (dyne/cm²), the cgs absolute unit of pressure, has sometimes been called the barye, but this name is uncommon. Other units of pressure can also be formed by dividing various units of force by various units of area, such as the pound force per square inch (lbf/in.², frequently abbreviated psi).

Pressure has been frequently expressed in terms of the bar and its decimal submultiples, where 1 bar = 10^6 dynes/cm² = 10^5 Pa. A reference level of 1 microbar (1 μbar = 10^{-6} bar = 1 dyne/cm² = 0.1 Pa) is commonly used in the calibration of microphones, hydrophones, and loudspeakers. The millibar (1 mbar = 10^{-3} bar = 10^2 dynes/cm² = 10^2 Pa = 0.1 kPa) is commonly used in meteorology. The temporary use of the millibar with the SI has been allowed in order to permit meteorologists to communicate easily within their profession, but the kilopascal should be used in presenting meteorological data to the public.

Pressures are also frequently expressed in terms of the height of a column of either mercury or water which the pressure will support. This practice is very convenient in conjunction with the use of barometers or other instruments in which pressure is determined from such column heights. However, the pressure which supports a liquid column of given height depends on the acceleration due to gravity and on the density of the liquid (and, in turn, on its temperature), and these must therefore be specified for accurate work.

Two other units which have been frequently used for measuring pressure are the standard atmosphere and the torr. The standard atmosphere (atm) is exactly 101,325 Pa, which is approximately the average value of atmospheric pressure at sea level. The torr is exactly 1/760 atmosphere, or approximately 133.322 Pa. To within 1 part per million, it is equal to the pressure of a column of mercury of height 1 millimeter (1 mmHg) at a temperature of 0°C when the acceleration due to gravity has the standard value g_n = 9.80665 m/s². *See* PRESSURE.

Energy and work units. The joule (J), the SI unit of energy or work, is the work done by a force of magnitude 1 newton when the point at which the force is applied is displaced 1 m in the direction of the force. Thus, joule is a short name for newton-meter (N·m) of energy or work. *See* ENERGY; WORK.

Units of energy or work in other systems are defined by forming the product of a unit of force and a unit of length in precisely the same manner as in the definition of the joule. Thus, the erg, the cgs absolute unit of energy or work, is the product of 1 dyne and 1 cm. Erg is a short name for dyne-centimeter of energy or work. Since 1 dyne $= 10^{-5}$ N and 1 cm $= 10^{-2}$ m, it follows that 1 erg $= 10^{-7}$ J.

The foot-poundal (ft·pdl), the British absolute unit of energy or work, is the product of 1 poundal and 1 foot. The foot-pound, or, more properly, the foot-pound force (ft·lbf), the British gravitational unit of energy or work, is the product of 1 lbf and 1 ft. Foot-poundal and foot-pound are also names of units of torque, discussed below.

Power-time products. Sometimes energy is measured in units which are products of a unit of power and a unit of time. Since 1 watt (W) of power equals 1 joule per second (1 J/s), as discussed below, the joule is equivalent to 1 watt-second (1 W·s). In electrical power applications, energy is frequently measured in kilowatt-hours (kWh), where 1 kWh $= (10^3 \text{W}) (3600 \text{ s}) = 3.6 \times 10^6$ J.

Calorie and British thermal unit. The calorie and the British thermal unit were originally defined as the quantities of heat required to raise the temperature of a specified mass of water by a specified temperature. Usually these units have been used in connection with energy as heat, but they can also be used when referring to energy as work or energy in any other form. *See* HEAT.

The calorie was originally defined as the quantity of heat required to raise the temperature of 1 g of air-free water 1°C under a constant pressure of 1 atm. However, the magnitude of the calorie, so defined, depends on the place on the Celsius temperature scale at which the measurement is made. The 15°C calorie is based on the temperature interval from 14.5 to 15.5°C, but other temperature intervals have been used. At the Fifth International Conference on the Properties of Steam in London in 1956, the International (Steam) Table calorie was defined as exactly 4.1868 J; this is the type of calorie most frequently used in mechanical engineering. The thermochemical calorie, which has been used in thermochemistry in preference to the other types of calorie, is exactly 4.184 J.

Since the calorie is a relatively small unit, the kilocalorie (kcal), also called the large calorie and the kilogram-calorie, is used to designate 10^3 calories. The energy intake of the human body has been commonly expressed as the number of large calories which the eaten food will liberate as it passes through the body. Unfortunately, confusion arises because it is usually not indicated that kilocalories rather than calories are being used.

The British thermal unit (Btu) was originally defined as the quantity of heat required to raise the temperature of 1 lb of air-free water 1°F under a constant pressure of 1 atm. Again, the temperature interval must be specified; the 60° Btu, based on the interval from 59.5 to 60.5°F, is frequently used.

The International Table Btu and the thermochemical Btu are both defined to be of such magnitude that the values of specific heat capacity of any substance are equal in size, whether expressed in Btu per pound per degree Fahrenheit [Btu/(lb·°F)] or in calories per gram per degree Celsius [cal/(g·°C)], when the corresponding type of calorie is used. From this it follows that each type of Btu is equal to approximately 251.996 times the corresponding type of calorie. Then the International Table Btu is approximately 1055.056 J, and the thermochemical Btu is approximately 1054.350 J.

Electronvolt. The electronvolt (eV), whose value is experimentally determined, is frequently used to express the energies of atomic systems. It has been accepted for use with the SI. *See* ELECTRONVOLT.

Power units. The watt (W), the SI unit of power, is the power which gives rise to the production of energy at the rate of 1 joule per second (1 J/s). Other units of power can be defined by forming the ratio of a unit of energy to a unit of time in the same manner. *See* POWER.

The horsepower (hp) is equal to exactly 550 ft·lbf/s, or approximately 745.700 W. It frequently has been employed to express the power generated by engines and machinery.

Torque units. The newton-meter (N·m), the SI unit of torque, is the magnitude of the torque produced by a force of 1 newton acting at a perpendicular distance of 1 m from a specified axis of rotation. The joule should never be used as a synonym for this unit; although the two units are both products of 1 newton and 1 m, the orientation of force and length is quite different in the two cases.

Units of torque in other systems are defined by forming the product of a unit of force and a unit of length in precisely the same manner as in the definition of the newton-meter. Thus the dyne-centimeter (dyne·cm), the cgs absolute unit of torque, is the product of 1 dyne and 1 cm; the erg should never be used as a synonym for this unit.

The foot-poundal (ft·pdl), the British absolute unit of torque, is the product of 1 poundal and 1 foot. The foot-pound (ft·lbf), the British gravitational unit of torque, is the product of 1 lbf and 1 ft. These units are sometimes called the poundal-foot (pdl·ft) and pound-foot (lbf·ft) to distinguish them from the units of energy or work.

The SI unit of torque does work of 1 N·m on a body which rotates through 1 rad in the direction of the torque. Some standards of metric practice therefore suggest that the SI unit of torque can be designated newton-meter per radian (N·m/rad) in order to further distinguish it from the unit of energy or work. The names of other torque units should then be similarly modified. *See* ROTATIONAL MOTION; TORQUE.

Viscosity units. For units of dynamic viscosity and kinematic viscosity *see* VISCOSITY.

Permeability units. The darcy is a commonly used unit of permeability to fluid flow. The permeability of a rock, brick, or other porous substance is 1 darcy if 1 cm³ of a fluid of 1-centipoise viscosity will flow through a section 1 cm thick and of 1-cm² cross section in 1 s at a pressure difference of 1 atm. In core analysis and other measurements of petroleum-bearing rock, the unit used is the millidarcy, which is 10^{-3} darcy.

The SI unit of permeability is the permeability of a porous substance such that 1 m³ of a fluid of viscosity 1 Pa·s will flow through a section 1 m thick and of 1-m² cross section in 1 s at a pressure difference of 1 Pa. This unit has no special name, but has the dimensions of 1 m². The SI unit of permeability is exactly 1.01325×10^{12} darcy, and 1 darcy $\cong 0.987 \times 10^{-12}$ m².

ELECTRICAL UNITS

For a general discussion of electrical units, including the SI or mks system, three cgs systems [electrostatic system of units (esu), electromagnetic system of units (emu), and gaussian system], and definitions of the SI units ampere (A), volt (V), ohm (Ω), coulomb (C), farad (F), henry (H), weber (Wb), and tesla (T), *see* ELECTRICAL UNITS AND STANDARDS.

This section discusses some additional SI units and some units in the cgs electromagnetic system which are frequently encountered in scientific literature in spite of the fact that their use has been discouraged.

Siemens. The siemens (S), the SI unit of electrical conductance, is the electrical conductance of a conductor in which a current of 1 ampere is produced by an electric potential difference of 1 volt. The conductance G is defined by the equation $I = GV$, where I is the current in amperes, V is the potential difference in volts, and G the conductance in siemens. The conductance of an electrical conductor in siemens is the reciprocal of its resistance in ohms. *See* CONDUCTANCE; ELECTRICAL RESISTANCE.

The siemens was formerly called the mho (℧) to illustrate the fact that the unit is the reciprocal of the ohm.

Abampere. The abampere (abA), the cgs electromagnetic unit of current, is that current which, if maintained in two straight, parallel conductors of infinite length, of negligible circular cross section, and placed 1 cm apart in vacuum, would produce between these conductors a force equal to 2 dynes per centimeter of length.

The abampere is equal to exactly 10 A. That a current of 10 A satisfies the above definition can be seen from the following argument: If two straight, parallel conductors of infinite length, separated by distance r, carry currents I_1 and I_2, then the force F on segment of one of the conductors of length l is given by Eq. (5). The constant μ_0 is the permea-

$$F = \frac{\mu_0 I_1 I_2 l}{2\pi r} \qquad (5)$$

bility of vacuum, $4\pi \times 10^{-7}$ newton per ampere squared. Thus, when the current in each wire is 1 A and the distance between the conductors is $r = 1$ m, the force between them is 2×10^{-7} newton per meter of length, as in the definition of the ampere. When $I_1 = I_2 = 10$ A, and $l = r = 1$ cm $= 10^{-2}$ m, the force between the conductors, given by Eq. (6), is the force specified in the definition of the abampere. *See* AMPERE'S LAW; BIOT-SAVART LAW.

$$F = \frac{(4\pi \times 10^{-7})(10)(10)(10^{-2})}{2\pi \, (10^{-2})} \text{ newton}$$
$$= 2 \times 10^{-5} \text{ newton} = 2 \text{ dynes} \qquad (6)$$

Abvolt. The abvolt (abV), the cgs electromagnetic unit of electrical potential difference and electromotive force, is the difference of electrical potential between two points of a conductor carrying a constant current of 1 abA, when the power dissipated between these points is equal to 1 erg per second. Then 1 abV $= 10^{-8}$ V, as can be seen from Eq. (7).

$$1 \text{ abV} = \frac{1 \text{ erg/s}}{1 \text{ abA}} = \frac{10^{-7} \text{ J/s}}{10 \text{ A}} = 10^{-8} \frac{\text{W}}{\text{A}} = 10^{-8} \text{ V} \qquad (7)$$

The abampere and abvolt are not frequently used, but they figure in the definitions of the maxwell, gauss, oersted, and gilbert, given below.

Maxwell. The maxwell (Mx), the cgs electromagnetic unit of magnetic flux, is the magnetic flux which, linking a circuit of one turn, produces in it an electromotive force of 1 abV as it is reduced to zero in 1 s. Then 1 maxwell $= 10^{-8}$ weber, as can be seen from Eq. (8). *See* MAGNETIC FLUX.

$$1 \text{ Mx} = 1 \text{ abV·s} = 10^{-8} \text{ V·s} = 10^{-8} \text{ Wb} \qquad (8)$$

Units of magnetic flux density. The gauss (Gs), the cgs electromagnetic unit of magnetic flux density (also called magnetic induction), is a magnetic flux density of 1 maxwell per square centimeter (1 Mx/cm²). Then 1 gauss $= 10^{-4}$ tesla, as can be seen from Eq. (9).

$$1 \text{ Gs} = \frac{1 \text{ Mx}}{(1 \text{ cm})^2} = \frac{10^{-8} \text{ Wb}}{(10^{-2} \text{ m})^2} = 10^{-4} \frac{\text{Wb}}{\text{m}^2} = 10^{-4} \text{ T} \qquad (9)$$

Alternative equivalent definitions for the tesla and gauss follow from Eq. (10) for the magnetic

$$B = \frac{F}{Il \sin \theta} \qquad (10)$$

induction B, where F is the force exerted on a current element of length l and current I making angle θ with the magnetic induction vector. This equation is valid whether quantities are expressed in SI or in cgs electromagnetic units. Thus, an alternative definition for the tesla is the constant magnetic induction which exerts a force of 1 newton on a straight wire of length 1 m perpendicular to the magnetic induction vector and carrying a current of 1 ampere; that is, 1 T $= 1$ N/(A·m). The gauss can be defined, in precisely the same manner, as 1 dyne per abampere-centimeter [1 Gs $= 1$ dyne/(abA·cm)]. *See* MAGNETIC INDUCTION.

Units of magnetic field strength. The SI unit of magnetic field strength is 1 ampere per meter (1 A/m), which is the magnetic field strength at a distance of 1 m from a straight conductor of infinite length and negligible circular cross section which carries a current of 2π A. (Other geometric configurations can be used to define this unit; the choice is arbitrary.) This definition is based on the definition, in the SI, of the magnetic field strength H_{SI}. At a distance r from a long straight conductor carrying current I, H_{SI} is given by Eq. (11). The

$$2\pi r H_{SI} = I \qquad (11)$$

left-hand side of this equation is the line integral of H_{SI} around a circular path, all of whose points are at distance r from the conductor. Substituting $r = 1$ m and $I = 2\pi$ A in this equation gives $H_{SI} = 1$ A/m. The SI unit has sometimes been called the ampere-turn per meter because Eq. (11) can be modified to Eq. (12), where i is the current in each turn

$$2\pi r H_{SI} = ni \qquad (12)$$

of a coil, part of which forms the long straight conductor in question, and n is the number of turns in the coil. Substituting $r = 1$ m, $i = 2\pi$ A, and $n = 1$ turn in Eq. (12) gives $H_{SI} = 1$ A·turn/m. However, the name ampere per meter has been adopted by the CGPM.

The oersted (Oe), the cgs electromagnetic unit of magnetic field strength, is the magnetic field strength at a distance of 1 cm from a straight conductor of infinite length and negligible circular cross section which carries a current of 0.5 abA. Again, this definition is based on the definition of magnetic field strength. In the cgs electromagnetic system the magnetic field strength H_{emu} at a distance r from a long straight conductor carrying current I is given by Eq. (13), which differs from Eq. (11) by the factor 4π on the right-hand side. Substituting $r = 1$ cm and $I = 0.5$ abA in this equation gives $H_{emu} = 1$ abA/cm. Thus, 1 Oe = 1 abA/cm in the cgs electromagnetic system, but the method by which the quotient is formed is different from that in the SI unit, being based on Eq. (13) rather than Eq. (11).

$$2\pi r H_{emu} = 4\pi I \qquad (13)$$

When the magnetic field strength in cgs electromagnetic units is 1 Oe, the magnetic field strength in SI units is found by substituting in Eq. (11) $I = 0.5$ abA and $r = 1$ cm, which gives Eq. (14). Thus 1 Oe corresponds to $(10^3/4\pi)$ A/m $\cong 79.577$ A/m. *See* MAGNETIC FIELD.

$$H_{SI} = \frac{0.5 \text{ abA}}{2\pi (1 \text{ cm})} = \frac{0.5 (10 \text{ A})}{2\pi (10^{-2} \text{ m})} = \frac{10^3}{4\pi} \frac{\text{A}}{\text{m}} \qquad (14)$$

Units of magnetic potential and mmf. The ampere serves as the SI unit of magnetic potential difference and magnetomotive force (mmf), as well as the unit of current. In the SI system, the magnetomotive force around a closed path equals the current passing through a surface enclosed by the path. Thus, 1 A is the magnetomotive force around a closed path when a current of 1 A passes through an enclosed surface. The SI unit has also been called the ampere-turn, because the magnetomotive force around a path that loops around a current coil equals the product of the number of turns in the coil and the current in each turn; however, the name ampere has been adopted by the CGPM.

The gilbert (Gb), the cgs electromagnetic unit of magnetic potential difference and magnetomotive force, is the magnetomotive force around a closed path enclosing a surface through which flows a current of $(1/4\pi)$ abA. This definition is based on the fact that in the cgs electromagnetic system the magnetomotive force around a closed path is 4π times the current passing through an enclosed surface. When this current is $(1/4\pi)$ abA, the magnetomotive force is therefore 1 abA in cgs electromagnetic units, but the unit is given the special name gilbert. When the magnetomotive force in cgs electromagnetic units is 1 gilbert, the magnetomotive force in SI units is $(1/4\pi)$ abA = $(10/4\pi)$ A. Thus, 1 Gb corresponds to $(10/4\pi)$ A $\cong 0.79577$ A. *See* MAGNETOMOTIVE FORCE.

Photometric units involve a new base quantity, luminous intensity. For the definition of the candela (cd), the SI unit of luminous intensity, *see* PHOTOMETRY; PHYSICAL MEASUREMENT.

For a discussion of photometric units, including units of illuminance (illumination), luminance, and in particular the SI units lux (lx) and candela per square meter (cd/m²), *see* LUMINANCE.

This section gives an explicit definition of the lumen and discusses units of luminous energy.

Lumen. The lumen (lm), the SI unit of luminous flux, is the luminous flux emitted within a unit solid angle (1 steradian) by a point source having a uniform intensity of 1 candela. It follows, therefore, that a light source having an intensity of 1 candela in every direction will be emitting a total luminous flux of 4π lumens.

The lumen is also equal to the luminous flux received on a unit surface, all points of which are at a unit distance from a point source having a uniform intensity of 1 candela.

The output of light sources is given in lumens. *See* LUMINOUS FLUX.

Luminous energy units. The lumen-second (lm·s), the SI unit of luminous energy (also called quantity of light), is the luminous energy radiated or received over a period of 1 s by a luminous flux of 1 lumen. This unit is also called the talbot. Other units of luminous energy include the lumen-hour (1 lm·h = 3600 lm·s) and the million-lumen-hour. *See* LUMINOUS ENERGY.

Certain quantities and units are used particularly in the area of ionizing radiation. The special units curie, roentgen, rad, and rem, which were previously adopted for use in this area, are not coherent with the SI, but their temporary use with the SI has been approved while the transition to SI units takes place. The CGPM, acting on proposals of the International Commission on Radiation Units and Measurements (ICRU) and the International Commission on Radiological Protection (ICRP), has adopted the special names becquerel, gray, and sievert for the SI derived units of activity, absorbed dose, and dose equivalent.

Activity units. The becquerel (Bq), the SI unit of activity (radioactive disintegration rate), is the activity of a radionuclide decaying at the rate of one spontaneous nuclear transition per second; thus 1 Bq = 1 s^{-1}. The curie (Ci), the special unit of activity, is equal to 3.7×10^{10} Bq (this unit was originally chosen to approximate the activity of 1 g of radium-226). *See* RADIOACTIVITY.

Exposure units. The SI unit of exposure to ionizing radiation, 1 coulomb per kilogram (1 C/kg), is the amount of electromagnetic radiation (x-radiation or gamma radiation) which in 1 kg of pure dry air produces ion pairs carrying 1 coulomb of charge of either sign. (The ionization arising from the absorption of bremsstrahlung emitted by electrons is not to be included in measuring the charge). *See* BREMSSTRAHLUNG.

The roentgen (R), the special unit of exposure, is equal to 2.58×10^{-4} C/kg. (This is equivalent to 1 cgs electrostatic unit of charge in 1.293 mg of air, which is the mass of 1 cm³ of air at temperature 0°C and pressure 1 atm.)

Absorbed dose and kerma units. The gray (Gy), the SI unit of absorbed dose, is the absorbed dose when the energy per unit mass imparted to matter by ionizing radiation is 1 joule per kilogram (1 J/kg).

The rad (rd), the special unit of absorbed dose, is equal to 10^{-2} Gy. In air, the number of rad per roentgen is 0.877, and this is also approximately true in soft tissue.

Kerma (an acronym for kinetic energy released in matter) is related to absorbed dose and is measured in the same units. The kerma is the sum of the initial kinetic energies of all the charged ionizing particles liberated by uncharged ionizing particles in an element of matter divided by the mass of that element. Except for a correction due to energy lost to bremsstrahlung, the exposure is the ionization equivalent of the air kerma.

Dose equivalent units. Different types of radiation cause slightly different effects in biological tissue. For this reason, a weighted absorbed dose called the dose equivalent is used in comparing the effects of radiation on living systems. The dose equivalent is the product of the absorbed dose and various dimensionless modifying factors stipulated by the ICRP. The chief such factor, the quality factor, depends on the linear energy transfer, and this, in turn, depends on the kind of incident radiation and its energy. For electromagnetic radiation and electrons the quality factor equals 1; for heavier charged particles it is greater than 1.

The sievert (Sv), the SI unit of dose equivalent, is the dose equivalent when the absorbed dose of ionizing radiation multiplied by the stipulated dimensionless factors is 1 joule per kilogram (1 J/kg).

The rem, the special unit of dose equivalent, is equal to 10^{-2} Sv.

OTHER UNITS

Logarithmic measures may be used with the SI. *See* BEL; DECIBEL.

For quantities and units pertaining to sound transmission *see* SOUND.

For temperature scales *see* TEMPERATURE.

For units measuring quantities in chemistry *see* ATOMIC MASS UNIT; ATOMIC WEIGHT; MOLE; RELATIVE ATOMIC MASS; RELATIVE MOLECULAR MASS.

[JONATHAN F. WEIL]

Bibliography: *Guidelines for Use of the Modernized Metric System*, NBS Lett. Circ. LC 1120, 1979; Institute of Electrical and Electronics Engineers, *IEEE Standard Metric Practice*, IEEE Std. 268–1979, 1979; International Commission on Radiation Units and Measurements, *Radiation Quantities and Units*, ICRU Rep. 33, 1980; International Organization for Standardization, *Units of Measurement*, ISO Stand. Handb. 2, 1979; *The International System of Units (SI)*, Nat. Bur. Stand. Spec. Publ. 330, 1977; L. V. Judson and L. E. Barbrow, *Units and Systems of Weights and Measures: Their Origin, Development and Present Status*, Nat. Bur. Stand. Lett. Circ. LC 1035, 1960, amended 1976.

Upsilon particles

A family of subnuclear particles having about 10 times the mass of the proton and generally described as the atomlike combination of a new quark with its antiquark. The new quark, designated *b* quark (for bottom or beauty), is about five times as massive as a proton and has a charge one-third that of the electron.

Discovery. In 1970 a group of physicists (Fermi National Accelerator Laboratory or Fermilab). proposed to search for new phenomena by bombarding nuclei with an intense stream of the highest-energy protons available (400 GeV). They would then study the deexcitation of a struck particle by measuring the quanta of electromagnetic radiation that emerged. Simple theory predicted that quanta of long wavelengths (low energy) would be plentiful, but that quanta of short wavelengths (high energy) would be rare. In the detailed description of the experiment, the direct object of study would not be the quanta; rather the experimenters preferred to study one of the products which can be substituted for a quantum of radiation: a pair of mu mesons (muons). Muons have a unique ability to penetrate large amounts of solid matter with only small perturbations of their motion. A pair of muons is equivalent to a photon born in a very high-energy collision, but the muons are easier to measure. Calculations of the speeds and directions of the muons yield the mass or energy equivalent of the produced radiation. This then acts as a probe of any special structures that may have been reached in the initial, catastrophic collision.

The experiment took a long time to reach fruition. It was run many times with slowly improving sensitivity until late 1976. At that time, several new ideas were developed, and an intense effort was mounted to prepare a new version of the experiment. The new run started on May 1, 1977. As the data accumulated, it became clear that the

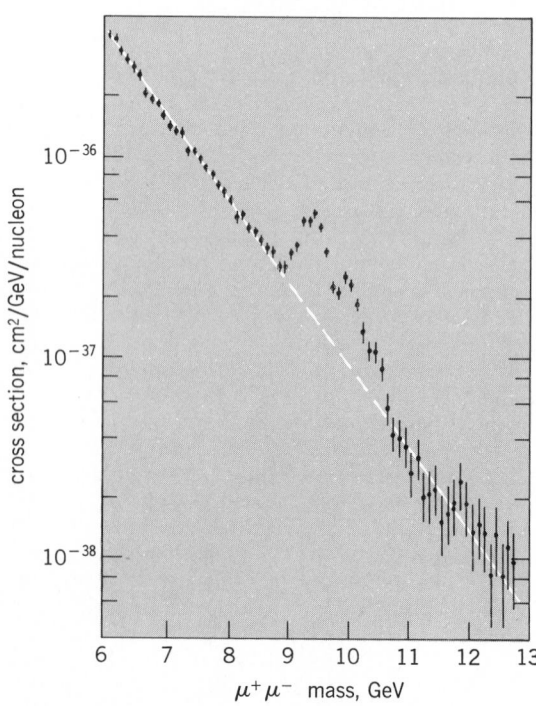

Fig. 1. Cross section for producing muon ($\mu^+\mu^-$) pairs in proton-nucleus collisions, at an angle of 90° with the incident beam in the center-of-mass frame, plotted as a function of $\mu^+\mu^-$ mass. (*From W. R. Innes et al., Observation of structure in the* Υ *region, Phys. Rev. Lett.,* 39: 1240–1242, 1977*)*

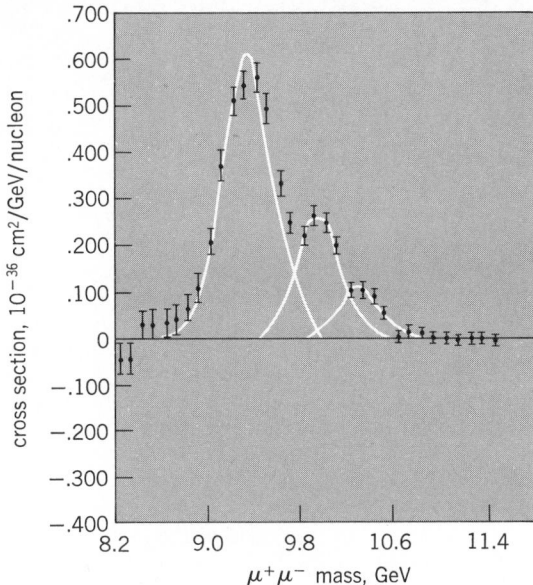

Fig. 2. Cross section for producing muon ($\mu^+\mu^-$) pairs, as in Fig. 1, in the upsilon region, with background subtracted from cross section.

Properties of upsilon family

	State	Mass difference MeV*	Relative width to e^+e^-
Y	1S	0	1.0
Y'	2S	560.7	0.44
Y"	3S	891.1	0.35
Y'''	4S	1114	0.25†

*Best Y mass: $9.4340 \pm .0002$ GeV.
†Absolute Y''' width: 10.8 ± 0.9 MeV.

new arrangement had been highly successful. Data accumulated at 100 times the rate achieved in any previous experiment. The experimenters scanned computer output showing a steeply falling graph of yield versus mass of the muon pair. But soon a new and startling effect appeared. A reversal of the downward trend appeared near 10 GeV (Fig. 1).

The experiment was run with various improvements until January 1978. The final result is shown in Fig. 2, and an analysis of this indicated the existence of three distinct peaks, with properties defined in the table. The peaks were named upsilon (Y), upsilon prime (Y'), and upsilon double-prime (Y"). The interpretation of these peaks followed from an experimental proof that the lifetime of the

particles was long by normal criteria. This meant that the decay of Y into hadrons containing old quarks was forbidden. The prohibition against decay extended the lifetime and, by a basic law of quantum physics, made the peaks look narrow.

Thus the upsilon had to be composed of a new quark. The Fermilab data indicated that the charge of the new quark, b, was probably one-third that of the electron.

Confirmation and properties. In early 1978 a group in Hamburg, at the Deutsches Elektronen Synchrotron (DESY) laboratory, succeeded in producing Y via the collisions of positrons and electrons: $e^+ + e^- \rightarrow Y$. The data confirmed the interpretation that Y is a bound state of a b quark and its antiquark: $Y = b\bar{b}$. The experiment also confirmed the charge of b to one-third. The natural interpretation of the Y' and Y" is that these are excited states of the $b\bar{b}$ system, designated by the atomic structure notation: Y (1S), Y' (2S), Y" (3S).

The upsilon family has now been studied in detail at the Cornell Electron Storage Ring (CESR) in Ithaca, NY, where two independent experiments obtained the spectrum shown in Fig. 3. In these colliding-beam experiments, the energy of the e^+ and e^- are totally consumed in producing the Y, and the experiment detects this via an increase in the collision rate when the sum of the energies is precisely equal to the mass of the upsilon. In this way, the resolution is much higher than in the Fermilab experiment. These experiments indicat-

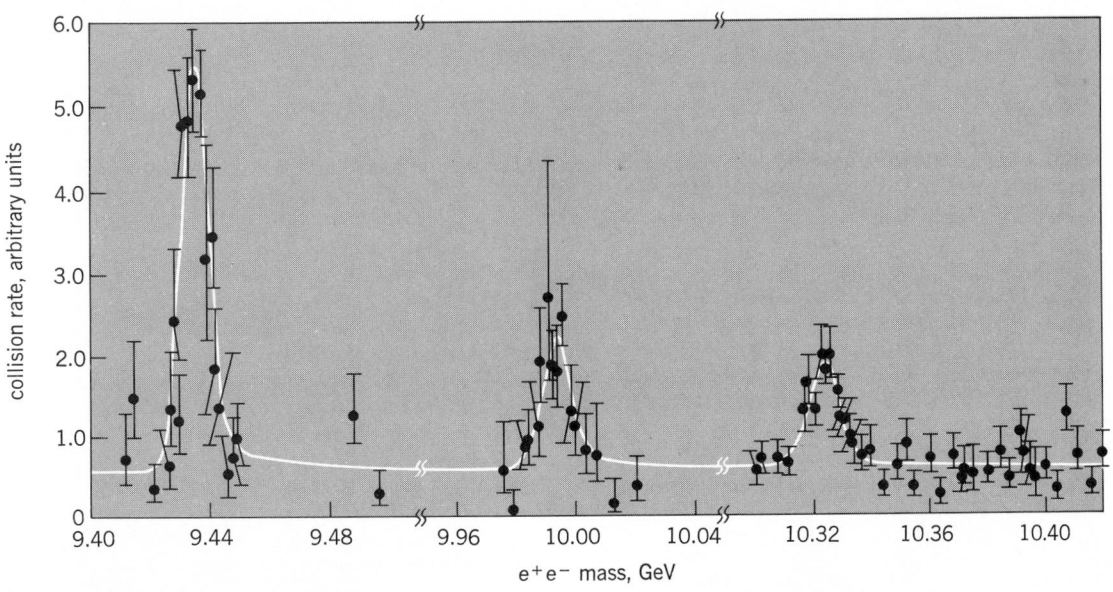

Fig. 3. Collision rate for electron-positron (e^+e^-) colliding beams, plotted as a function of e^+e^- mass in the upsilon region. (*From T. Böhringer et al., Observation Y, Y',* and Y" at the Cornell Electron Storage Ring, Phys. Rev. Lett., 44:1111–1114, 1980)

ed the existence of a fourth particle, presumably Y‴ (4S). The newest state has a much shorter lifetime (indicated by its width) because it is heavy enough to permit the decay into objects which contain a b quark and one of the older quarks, for example, $\bar{b}u + b\bar{u}$.

The study of the spectra of the b-quark system reveals much detailed information on the forces between quarks.

The discovery of the b quark led to a revision of the ideas associated with quarks and leptons and an ordering into three "generations" of quark-lepton groupings.

Generation			
I	up quark	electron	
	down quark	electron neutrino	
II	charmed quark	muon	
	strange quark	muon neutrino	
III	top quark	tau	
	bottom quark	tau neutrino	

The tau lepton was discovered in 1977. A vigorous search for the top quark has been unsuccessful so far. It is now frequently assumed that all matter is composed of various combinations of these objects.

The production of upsilons is studied in both proton machines, where the collisions of quarks and gluons gives rise to Y, Y′, and so forth, and also in electromagnetic collisions of electrons and positrons. The upsilon is known to decay into various states in addition to the state which led to its discovery, $Y \rightarrow \mu^+ + \mu^-$. Other modes of decay have been observed, including $Y' \rightarrow Y + \pi^+ + \pi^-$ and Y, Y′, Y″ → hadrons. See ELEMENTARY PARTICLE; LEPTON; QUARKS.

[LEON M. LEDERMAN]

Bibliography: D. Andrews et al., Observation of a fourth upsilon state in $e^+ e^-$ annihilations, *Phys. Rev. Lett.*, 45:219–221, 1980; G. Finocchiaro et al., Observation of the Y‴ at the Cornell Electron Storage Ring, *Phys. Rev. Lett.*, 45:222–225; 1980; S. W. Herb, et al., Observation of a dimuon resonance at 9.5 GeV in 400-GeV proton-nucleus collisions, *Phys. Rev. Lett.*, 39:252–255, 1977; L. M. Lederman, The upsilon particle, *Sci. Amer.*, 239(4):72–80, 1978.

Van der Waals equation

An empirical equation of state, presented by J. D. van der Waals in 1873, which takes into account the finite size of the molecules and the attractive forces between them. For a homogeneous gas at pressure p, molar volume v_N, and absolute temperature T, the following equation holds.

$$p = \frac{R_u T}{v_N - b} - \frac{a}{(v_N)^2}$$

or

$$\left(p + \frac{a}{(v_N)^2}\right)(v_N - b) = R_u T$$

The terms a and b are constants and evaluated from experimental data. The molar specific volume v_N is equal to the total volume V divided by the number of moles N. R_u is the universal gas constant. If the units for a and b are $(atm)(ft^6)/lbmole^2$ and $ft^3/lbmole$ °R. If the units for a and b are $(psia)(ft^6)/lbmole^2$ and $ft^3/lbmole$, the value and units for R_u are 10.73 $(psia)(ft^3)/lbmole$ °R. If the units for a and b are $(Pa)(m^6)/mole^2$ and $m^3/mole$ respectively, the value and units for R_u are 8.31 $(Pa)(m^3)/(mole)(K)$. Values for a and b for a few gases are presented in the table. See THERMODYNAMIC PRINCIPLES.

[GEORGE A. HAWKINS]

Bibliography: J. B. Jones and G. A. Hawkins, *Engineering Thermodynamics*, 1960; S. L. Kittsley, *Physical Chemistry*, 3d ed., 1969; W. C. Reynolds and H. C. Perkins, *Engineering Thermodynamics*, 2d ed., 1977; K. Wark, *Thermodynamics*, 3d ed., 1977.

Velocity

The time rate of change of position of a body in a particular direction. Linear velocity is velocity along a straight line, and its magnitude is commonly measured in such units as meters per second (m/sec), feet per second (ft/sec), and miles per hour (mph). Since both a magnitude and a direction are implied in a measurement of velocity, velocity is a directed or vector quantity, and to specify a velocity completely, the direction must always be given. The magnitude only is called the speed. See SPEED.

Linear velocity. A body need not move in a straight line path to possess linear velocity. The instantaneous velocity of any point of a body un-

Approximate values for the constants a and b

	a			b	
Gas	$(atm)(ft^6)/lbmole^2$	$(psia)(ft^6)/lbmole^2$	$(Pa)(m^6)/mole^2$	$ft^3/lbmole$	$m^3/mole$
Air	344	5,052	0.136	0.587	3.66×10^{-5}
Carbon dioxide	926	13,600	0.366	0.686	4.28×10^{-5}
Helium	8.57	126	0.00338	0.372	2.32×10^{-5}
Neon	55.4	814	0.0219	0.282	1.76×10^{-5}
Oxygen	350	5,140	0.138	0.510	3.18×10^{-5}

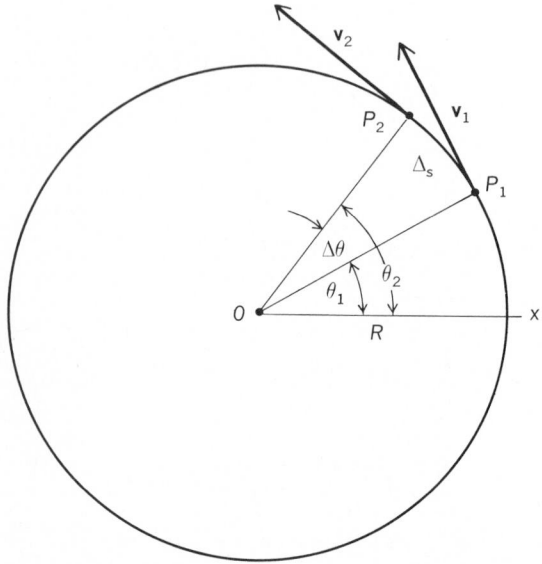

Fig. 1. Illustration of angular displacement, angular speed, and tangential velocity.

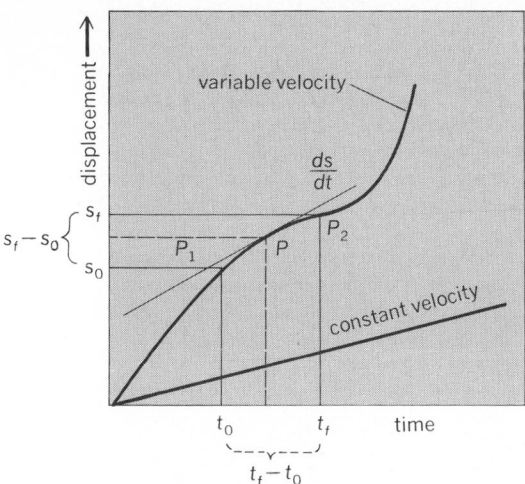

Fig. 2. The average velocity from P_1 to P_2 is $(s_f - s_0)/(t_f - t_0)$. The instantaneous velocity at point P is the limit of the ratio representing the average velocity as the interval approaches zero.

dergoing circular motion is a vector quantity, such as v_1 or v_2 in Fig. 1. When a body is constrained to move along a curved path (Fig. 2), it possesses at any point an instantaneous linear velocity in the direction of the tangent to the curve at that point. The average value of the linear velocity is defined as the ratio of the displacement to the elapsed time interval during which the displacement took place. The displacement of a body from an initial position s_0 to a final position s_f after time t is equal to $s_f - s_0$. The corresponding time interval is $t_f - t_0$. The magnitude of the average velocity is then given by Eq. (1), where Δs is displacement and Δt is the corresponding elapsed time.

$$\bar{v} = \frac{\text{displacement}}{\text{elapsed time}} = \frac{s_f - s_0}{t_f - t_0} = \frac{\Delta s}{\Delta t} \qquad (1)$$

The magnitude of the instantaneous velocity v of a body is the limiting value of the foregoing ratio as the interval approaches zero. In the notation of calculus, Eq. (2), ds/dt is the instantaneous time rate of change of displacement (Fig. 2).

$$v = \lim_{\Delta t \to \Delta} \frac{\Delta s}{\Delta t} = \frac{ds}{dt} \qquad (2)$$

The velocity of a body, like its position, can only be specified relative to a particular frame of reference. Consequently, all velocities are relative. *See* RELATIVE MOTION.

Angular velocity. The representation of angular velocity ω as a vector is shown in Fig. 3. The vector is taken along the axis of spin. Its length is proportional to the angular speed and its direction is that in which a right-hand screw would move. If a body rotates simultaneously about two or more rectangular axes, the resultant angular velocity is the vector sum of the individual angular velocities. Thus, if a body rotates about an x axis with an angular velocity $\boldsymbol{\omega}_x$, and simultaneously about a y axis with an angular velocity $\boldsymbol{\omega}_y$, the resultant angular velocity $\boldsymbol{\omega}$ is the vector sum given by Eq. (3).

$$\boldsymbol{\omega} = \boldsymbol{\omega}_x + \boldsymbol{\omega}_y \qquad (3)$$

It should be emphasized that whereas angular velocities are commutative in addition, that is, they may be added in any order, angular displacements are not commutative. *See* ROTATIONAL MOTION.

Angular displacement. Figure 1 represents a body rotating with circular motion about an axis through O perpendicular to the figure. Line OP_1 is the position of some radius in the body at a time t_1, with θ_1 being the angular displacement from a reference line. Line OP_2 is the position of the same radius at a later time t_2, with the angular displacement θ_2. Angular displacement may be measured in degrees, radians, or revolutions.

Angular speed. From Fig. 1, it is seen that the body has rotated through the angle $\Delta\theta = \theta_2 - \theta_1$ in the time $\Delta t = t_2 - t_1$. The average angular speed $\bar{\omega}$ is defined by $\bar{\omega} = \Delta\theta/\Delta t$, the instantaneous angular speed ω being $\omega = d\theta/dt$. Although it is customary in most scientific work to express angular speed in radians per second, it is common in engineering practice to use the units of revolutions per minute (rpm) or revolutions per second (rps).

Tangential velocity. When a particle rotates in a circular path of radius R through an angular distance $\Delta\theta$ in a time Δt, as in Fig. 1, it traverses a linear distance Δs.

The average linear speed $\bar{v}$ is given by Eq. (4),

$$\bar{v} = \frac{\Delta s}{\Delta t} = \frac{R\,\Delta\theta}{\Delta t} = \bar{\omega}R \qquad (4)$$

since $\Delta s = R\Delta\theta$. Similarly, the instantaneous speed v is given by $v = \omega R$. The direction of this instantaneous speed is tangential to the circular path at the point in question. Any vector $\mathbf{v}$ drawn in this direction represents the tangential velocity.

Combined velocities. A body may have combined linear and angular motions, as is the case when the wheel of a moving automobile rolls along the ground with an angular velocity about its axle which moves with a linear velocity parallel to the pavement. In this case, a point on the rim of the tire describes a curved path called a cycloid. If a circular body rolls on the surface of a sphere, a point on the periphery of the rotating body describes a curve called an epicycloid. *See* EPICYCLOID. [ROGERS D. RUSK]

Bibliography: H. Goldstein, *Classical Mechanics,* 2d ed., 1980; C. Kittel, W. D. Knight, and M. A. Ruderman, *Mechanics,* vol. 1, 2d ed., 1973; R. Resnick and D. Halliday, *Physics,* pt. 1, 3d ed., 1977.

VELOCITY

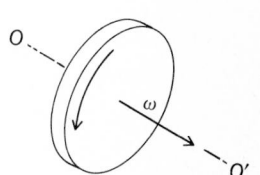

Fig. 3. Angular velocity shown as an axial vector.

Vibration

The term used to describe a continuing periodic change in the magnitude of a displacement with respect to a specified central reference. The periodic motion may range from the simple to-and-fro oscillations of a pendulum, through the more complicated vibrations of a steel plate when struck with a hammer, to the extremely complicated vibrations of large structures such as an automobile on a rough road. Vibrations are also experienced by atoms, molecules, and nuclei.

A mechanical system must possess the properties of mass and stiffness or their equivalents in order to be capable of self-supported free vibration. Stiffness implies that an alteration in the normal configuration of the system will result in a restoring force tending to return it to this configuration. Mass or inertia implies that the velocity imparted to the system in being restored to its normal configuration will cause it to overshoot this configuration. It is in consequence of the interplay of mass and stiffness that periodic vibrations in mechanical systems are possible.

Mechanical vibration. This is the term used to describe the continuing periodic motion of a solid body at any frequency. When the rate of vibration of the solid body ranges between 20 and 20,000 hertz (Hz), it may also be referred to as an acoustic vibration, for if these vibrations are transmitted to a human ear they will produce the sensation of sound. The vibration of such a solid body in contact with a fluid medium such as air or water induces the molecules of the medium to vibrate in a similar fashion and thereby transmit energy in the form of an acoustic wave. Finally, when such an acoustic wave impinges on a material body, it forces the latter into a similar acoustic vibration. In the case of the human ear it produces the sensation of sound. Thus, vibrations in solid and fluid bodies are essential to production, transmission, and reception of sound. *See* MECHANICAL VIBRATION.

One degree of freedom. Systems with one degree of freedom are those for which one space coordinate alone is sufficient to specify the system's displacement from its normal configuration. It is the simplest yet the most fundamental type of vibration system. An idealized example known as a simple oscillator consists of a point mass m fastened to one end of a massless spring and constrained to move back and forth in a line about its undisturbed position (Fig. 1). Although no actual acoustic vibrator is identical with this idealized example, the actual behavior of many vibrating systems when vibrating at low frequencies is similar and may be specified by giving values of a single space coordinate. They include loudspeaker cones, telephone diaphragms, microphone diaphragms, and drum membranes.

When the restoring force of the spring of a simple oscillator on its mass m is directly proportional to the displacement x of the latter from its normal position, the system vibrates in a sinusoidal manner called simple harmonic motion. This motion is identical with the projection of uniform circular motion on a diameter of a circle and is represented by the equation $x = A \sin 2\pi f t$. The frequency of vibration f is given by Eq. (1), where s is the constant of proportionality between force and stretch or compression of the movable end of the spring.

$$f = (1/2\pi) \sqrt{s/m} \qquad (1)$$

The constant A represents the amplitude of the vibration, that is, the maximum displacement of the mass on either side of its rest position. The magnitude of A is determined by the manner in which the motion is initially started. Note that the frequency of vibration of a simple oscillator is independent of the amplitude of its vibration. *See* HARMONIC MOTION.

The variation in velocity v of the point mass of a simple oscillator is given by Eq. (2) and acceleration a is given by Eq. (3). Note that for a given

$$v = dx/dt = 2\pi f A \cos 2\pi f t \qquad (2)$$

$$a = d^2x/dt^2 = -4\pi^2 f^2 A \sin 2\pi f t \qquad (3)$$

displacement amplitude A, the velocity amplitude $2\pi f A$ is directly proportional to the frequency, and the acceleration amplitude $4\pi^2 f^2 A$ is directly proportional to the square of the frequency. Consequently, large velocity and acceleration amplitudes of a simple oscillator are more readily obtained at high than at low frequencies.

Energy considerations. The total mechanical energy of a simple oscillator equals the sum of the kinetic energy associated with the moving mass and the potential energy stored in the distorted spring. When no frictional forces are present, these two types of energy are continuously being converted from one to the other and back again as the vibration cycle progresses. Their sum remains constant at a value given by Eq. (4).

$$E = sA^2/2 = 2m(\pi f A)^2 \qquad (4)$$

When frictional or other dissipative forces are present, the simple oscillator gradually changes its mechanical energy of vibration into heat energy with an attendant reduction in its amplitude of vibration. No one simple law is capable of describing the variation of frictional forces as a material body vibrates. However, in many important cases the frictional force is opposite to the direction of motion and proportional to the velocity of the vibrating mass. For this type of damping force, the amplitude of vibration of the simple oscillator decreases exponentially with time in accordance with the equation $A = A_0 e^{-\alpha t}$, where α is a constant directly proportional to the frictional force and inversely proportional to the mass of the oscillator (Fig. 2). The natural frequency of free oscillation of

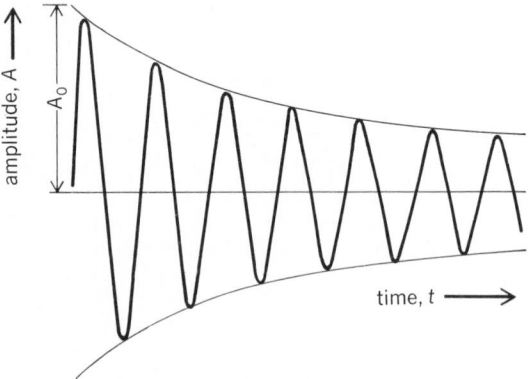

Fig. 2. Damped vibration of simple oscillator.

a damped oscillator is slightly less than that of the same oscillator system without damping. The amount by which the frequency is lowered increases with increased damping. *See* DAMPING.

External driving force. A simple oscillator, or some equivalent system, is often maintained in a condition of steady vibration by the application of a periodic external driving force. After a sufficient time has elapsed, any natural free-oscillation frequency of the oscillator dies out and it vibrates solely at the frequency of the impressed driving force. When the applied driving force is sinusoidal, as represented, for example, by the function $F(\cos 2\pi f t)$, the forced vibration ultimately reaches a

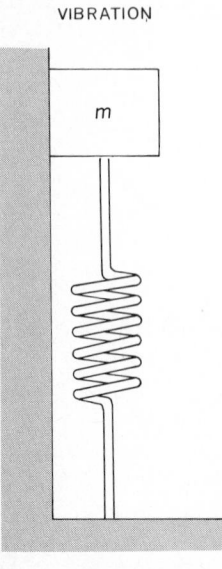

Fig. 1. Simple oscillator.

steady-state amplitude of $A = F/2\pi fZ$ where Eq. (5)

$$Z = [R^2 + (2\pi fm - s/2\pi f)^2]^{1/2} \qquad (5)$$

applies. Z is known as the mechanical impedance of the oscillator. From this equation it is apparent that the driven oscillator will have a maximum amplitude when the forcing frequency is near $(1/2\pi)\sqrt{s/m}$, the free oscillation frequency of the oscillator. *See* FORCED OSCILLATION.

Nonlinear systems. When the amplitude of vibration of any real oscillator system becomes large, the elastic restoring force no longer is proportional to the displacement x. The term s relating force and stretch in the spring is not independent of x and may either increase or decrease as x increases. In either case, the motion no longer is sinusoidal and the frequency is not independent of the displacement amplitude A. The frequency becomes higher when s increases with increasing x and lower for a decreasing s as x increases. The latter characterizes the large-amplitude oscillations of a pendulum bob. *See* PENDULUM.

Two degrees of freedom. When two simple vibrating systems are interconnected by a flexible connection, the combined system has two degrees of freedom. Consider the simple oscillator of mass m_1 and spring s_1 connected to a second oscillator of mass m_2 and spring s_2 by means of the spring s (Fig. 3). The motion of m_1 is completely described by the displacement x_1 and that of m_2 by x_2. The system is said to have two degrees of freedom since two independent space coordinates are required to specify the motion.

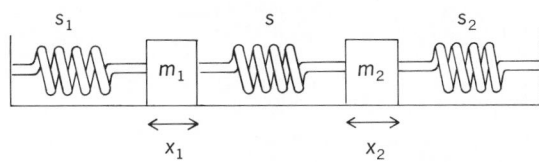

Fig. 3. Simple oscillator with two degrees of freedom.

A system having two degrees of freedom has two normal modes of vibration of respective frequencies f' and f''. Both of these frequencies differ from the respective natural frequencies f_1 and f_2 of the individual uncoupled oscillators. The larger the force constant s of the coupling spring, the larger becomes the frequency difference $(f' - f'')$ relative to $(f_1 - f_2)$; that is, f' increases relative to f_1, and f'' decreases relative to f_2.

Transverse vibrations are defined as those which occur when the vibrations of the medium in question are perpendicular to the direction of propagation of the exciting wave; longitudinal vibrations occur when the vibrations of the medium are lengthwise, along the direction of propagation of the wave. An example of a vibrating system with two degrees of freedom is given by the transverse vibrations of two masses m_1 and m_2 fastened at intermediate points on a tightly stretched wire rigidly supported at each end. Here, the higher frequency mode f' corresponds to a method of vibration in which the individual motions of the two masses are oppositely directed at all times. The lower frequency mode f'' corresponds to one in which the two masses move together in phase, that is, in the same transverse direction at all times.

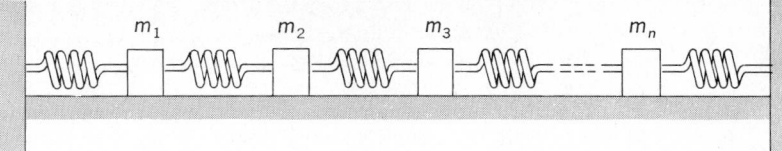

Fig. 4. An example of a vibrating system with several degrees of freedom.

Several degrees of freedom. A vibrating system is said to have several degrees of freedom if many space coordinates are required to describe its motion. One example is n masses $m_1, m_2, \ldots, m_n$ constrained to move in a line and interconnected by $(n-1)$ coupling springs with additional terminal springs leading from m_1 and m_n to rigid supports (Fig. 4). This system has n normal modes of vibration, each of a distinct frequency.

Vibrations of elastic bodies. The primary sources of energy for producing sound waves in fluids are vibrations of elastic bodies at frequencies ranging from 20 to 20,000 Hz. In turn, the ultimate detection of sound waves by such devices as the human ear or a microphone requires the presence of an elastic body being forced into vibration by the impinging sound waves.

Vibrations of elastic bodies are not usually a simple harmonic motion of just one frequency but instead are of a complex nature, having many natural frequencies and modes of vibration. This complexity arises from numerous factors. Not all the parts of a solid body move together in phase with each other. For instance, a loudspeaker diaphragm has its mass spread over a considerable surface area whose individual sectors may vibrate with different phases and amplitudes with respect to adjacent sectors. Consequently, it often is necessary to give the displacement at each point on the diaphragm as a function of time in order to describe the motion adequately. Most solid bodies are capable of displacements in any direction in space, so that three independent space coordinates are required to specify their vibration. The vibration of a solid body is influenced by interaction with its surrounding medium. The natural frequencies and modes of vibration of a solid body depend upon its shape and dimensions.

It is sometimes possible to confine the mode of vibration to one type having a limited number of natural frequencies. The vibrations of a stretched wire are predominately transverse, although it also is possible to excite weak longitudinal vibrations along the wire. As the diameter of a wire increases relative to its length, the importance of the longitudinal mode of vibration increases until in a thin rod both transverse and longitudinal modes are readily excited. As the diameter of a rod is further increased relative to its length, it ultimately may be regarded as a thin plate capable of vibrating in a thickness mode along its axis or in transverse modes along its surface. In each of these examples it is possible to choose a method of excitation which will encourage one mode of vibration and discourage others.

In any consideration of the vibration of solid bodies, characteristics of major acoustical significance include (1) the natural free vibration frequencies of the body, (2) the segmental vibration pattern for each mode of vibration, and (3) the

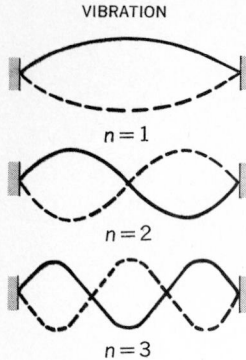

Fig. 5. First, second, and third normal modes of a vibrating string.

efficiency with which the vibrations are coupled to the surrounding medium.

Vibrations of strings. The transverse vibrations of thin strings result from tension forces in the string tending to restore any displaced portion of the string to its equilibrium position. The natural frequencies f of a thin string of length l rigidly supported at its ends are given by the equation $f = nc/2l$, where n may have any integral value 1, 2, 3, . . . and c is the velocity with which transverse waves are propagated along the string. In turn, $c = \sqrt{T/m}$, where T is the tension to which the string is stretched and m is the mass per unit length of string. The fundamental, or lowest frequency mode of vibration, of a string is $c/2l$. In addition, the string is capable of vibrating at harmonic overtone frequencies of $2\times$, $3\times$, $4\times$, . . . this frequency (Fig. 5). The simple integral number relationship between the various frequencies of vibration of a string leads to a pleasant harmonic tonal structure which accounts for the use of vibrating strings as the primary source of sound for such musical instruments as the violin, piano, and harp.

The relative amplitudes of the various harmonic modes of vibration of a string depend upon the particular manner in which the string is initially excited. For instance, if a string is plucked at its center, the even integral (2, 4, 6, . . .) harmonic frequencies will be weak compared to the odd integral frequencies. As a string freely vibrates, nodal positions of no displacement are spaced at intervals of l/n along its length. For a string plucked at its center, the even harmonic frequencies are weak since all have a nodal position at the midpoint of the string.

Vibrations of membranes. The transverse vibrations of stretched membranes result from tension forces in the membrane tending to restore any displaced portion of the membrane to its equilibrium position. The theory of vibrations of stretched membranes of either square or rectangular shape is primarily of mathematical interest. However, the vibrations of a circular membrane have certain physical characteristics which find practical applications. These applications include parchment membranes used for drumheads and the stretched thin steel or aluminum diaphragms of condenser microphones.

When a circular membrane is rigidly clamped along its outer radius, it will vibrate at a fundamental frequency given by Eq. (6), where T is the ten-

$$f = (0.384/a)\sqrt{Tt/\rho} \qquad (6)$$

sion to which the membrane is stretched, t its thickness, ρ its surface density, and a its radius. When vibrating in this mode, the entire surface of the membrane moves in phase with a maximum amplitude at its center. The other possible modes of vibration are not integral multiples of f, but have frequency ratios of 2.295, 3.60, 4.90, and so forth, to the fundamental. When vibrating in these latter modes, one or more nodal circles are present in which the membrane is oppositely displaced on adjacent sides of the circles. The average displacement of the membrane vibrating in this manner is small, because of cancellations between the oppositely phased circular sectors. This condition leads to low acoustical efficiency and limited practical significance for either the production or reception

of sound. For example, the response of a condenser microphone at first gradually falls off and then becomes very irregular at frequencies well above the fundamental of its diaphragm.

Vibrations of plates and diaphragms. The transverse vibrations of thin plates or diaphragms result from elastic restoring forces produced when their surfaces are deformed. The natural modes of vibration of such plates are involved functions of boundary shape and dimensions, whether the boundaries are free to vibrate or are rigidly clamped, and the elastic properties of its material. When a thin circular plate is rigidly clamped at its rim, its fundamental frequency is given by Eq. (7), where t is the thickness of the plate, a its radius, ρ

$$f = 0.47 \frac{t}{a^2}\sqrt{\frac{Y}{\rho(1-\sigma^2)}} \qquad (7)$$

its density, Y is Young's modulus, and σ is Poisson's ratio for the material of the plate. When vibrating in this mode, the entire surface of the plate moves in phase with a maximum amplitude at its center. The overtone modes of vibration have frequency ratios of 3.88, 8.70, and so on, to the fundamental. When the plate is vibrating in these latter modes, one or more nodal circles are present in which the plate is oppositely displaced on adjacent sides of the circles. The average displacement of the plate's surface under these conditions is small because of cancellations between the oppositely phased circular sectors. As in the case of circular membranes, this condition leads to low acoustical efficiency.

The theory of transverse vibration of thin plates finds application in the design of such devices as telephone receiver diaphragms, horn-type loudspeaker diaphragms, underwater sound projectors used in sonar systems, and in understanding the vibrations of wall panels, floors, automobile body panels, and hull plating of ships.

The fundamental mode of vibration of a circular plate free at its rim has a frequency some 13% lower than when rigidly clamped. The relatively involved problem of transverse vibrations of a thin flat plate becomes still further complicated when the plate has a simple curved surface, and a theoretical solution is impossible when the plate is curved into the shape of a bell.

Vibrations of rods. Rod vibrations primarily consist of two simple types: (1) longitudinal vibrations along the long axis of the rod, and (2) transverse vibrations at right angles to this axis.

Longitudinal vibrations. When any plane cross section in a long thin rod is displaced longitudinally relative to adjacent planes, elastic restoring forces caused by either compression or tension in the rod tend to restore the plane to its normal position. These forces result in the propagation of longitudinal waves along the axis of the rod. The interference of two such waves traveling in opposite directions sets up a pattern of standing waves in the rod having certain discrete frequencies. The magnitude of these natural frequencies depends upon the length and material of the rod and upon the particular constraints existing at the two ends of the rod.

If the rod is either free at both ends or rigidly clamped at both ends, its fundamental frequency is given by $f = c/2l$, where l is its length and c is the velocity with which longitudinal waves are propagated in the rod. When vibrating in its fundamental longitudinal mode, the length of the rod is one-half the wavelength of the waves being propagated along the axis of the rod. Overtone frequencies are given by integral multiples ($n = 2, 3, 4, \ldots$) of the fundamental frequency. In the case of these latter modes of vibration, nodal positions having no longitudinal displacements are spaced at intervals of $l/2n$ along the length of the rod.

When the rod is rigidly clamped at one end and free at the other, its fundamental frequency is given by $f = c/4l$. Its overtone frequencies are the odd integral ($n = 3, 5, 7, \ldots$) multiples of this frequency. Numerous types of constraints may be placed on the ends of a rod, for example, stiffness of elastic springs, inertia of masses, and so forth. When a rod free at one end is mass-loaded with a mass m at the opposite end, its fundamental frequency gradually decreases from $c/2l$ to $c/4l$ as the magnitude of the mass increases from zero to infinity.

Whenever a given rod is vibrating longitudinally in one of its natural modes, it may be supported or clamped at a nodal position without interfering with this particular mode of vibration. Since only a few modes of vibration, or in some cases only one, will have a nodal position at a given location, a judicious choice of support position may reduce or eliminate unwanted modes of vibration.

An important practical application for longitudinal vibrators is their use in sonar transducers, where both the magnetostrictive vibrations of nickel tubes and the piezoelectric vibrations of crystals are utilized.

Transverse vibrations. A long rod is capable of vibrating transversely as well as longitudinally, and it is often difficult to produce one motion without exciting the other. When a circular rod is rigidly clamped at one end and free at the other (Fig. 6) its fundamental frequency is $f = 0.28\ ac/l^2$ where c is the velocity of longitudinal waves in the material of the rod, a is its radius, and l its length. The overtone modes of vibration have frequency ratios of 6.27, 17.5, 34.4, and so on, to that of the fundamental.

If the rod were to have a rectangular instead of a circular cross section, its fundamental frequency would be $f = 0.16\ tc/l^2$, where t is the lateral thickness of the rod in the direction of transverse vibration. Thin rods or reeds of rectangular cross section clamped at one end have numerous applications, including use as a vibrating mouthpiece in the woodwind family of musical instruments and use as resonance vibrators in vibration frequency indicators. The latter type of instrument contains a large number of thin reeds, each having a different resonance frequency. When the base of the indicator is placed in contact with a vibrating object, the free end of that reed having the same natural frequency will be set into vigorous vibration, and its visual motion can be used to indicate on an appropriate scale the numerical value of the vibration frequency.

Another important type of transverse vibration is that of a rod free at both ends. The fundamental frequency of such a rod of rectangular cross section is $f = 0.33\ tc/l^2$, and there are overtones in ratio to this frequency of 2.75, 5.4, 8.9, and so on. The bars of a xylophone and similar musical instruments vibrate in this manner.

Vibration measuring equipment. Vibrations of solid bodies may result in the generation and transmission of unwanted noise. To determine the source of such noise and to devise methods for its elimination, it frequently becomes necessary to measure these vibrations.

When the vibration of a point in a material body is to be measured, a device must be provided with a sensing element which indicates either the displacements, velocities, or accelerations of the point. Such a device is usually either some mechanical system which indicates the characteristics of the vibration by means of a mechanical pointer, or a pickup capable of converting mechanical energy to electrical or some other form of energy. In conjunction with associated equipment, these pickups may be used to measure solely vibration amplitudes or to give a detailed picture of the entire vibration pattern. A vibrometer, or vibration meter, is a device indicating solely amplitude of vibration, while a vibrograph provides a complete oscillographic record of the vibration.

Vibration meter. A typical electrical vibration meter consists of an electromechanical pickup, adjustable attenuator, amplifier, integrating network, and a direct reading meter capable of being calibrated to read displacement, velocity, or acceleration amplitudes. Connections are also provided for oscillographic presentation, for a pair of headphones for listening to the vibration being measured, for connection to a vibration analyzer, or for connection to an electronic frequency counter.

Mechanical and mechanooptical vibrometers and vibrographs employ gear trains or mechanical or optical lever arms to magnify the vibratory motions before they are indicated or recorded. These devices are usually held by hand, with a projecting probe contacting the vibrating surface (Fig. 7). Mechanical vibrometers are normally used at frequencies up to 500 Hz.

Mechanical vibrographs are instruments containing a moving paper or film on which a scribing device records the amplitude of the motion being measured. In one type, shown in Fig. 8, the case and recording paper themselves move in proportion to the motion being measured and, for

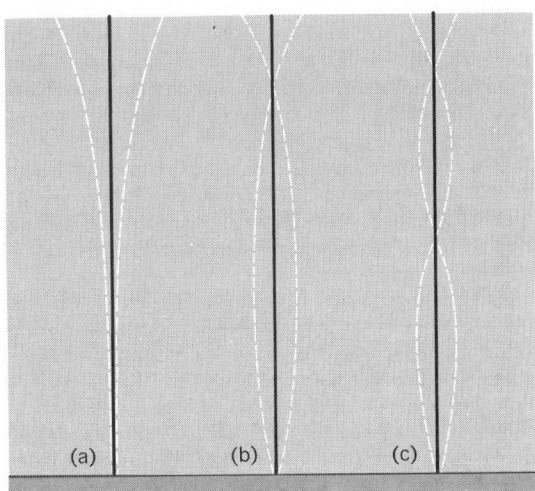

Fig. 6. Transverse vibrations of a long circular rod. (*a*) The fundamental mode. (*b*) The first-overtone mode. (*c*) The second-overtone mode.

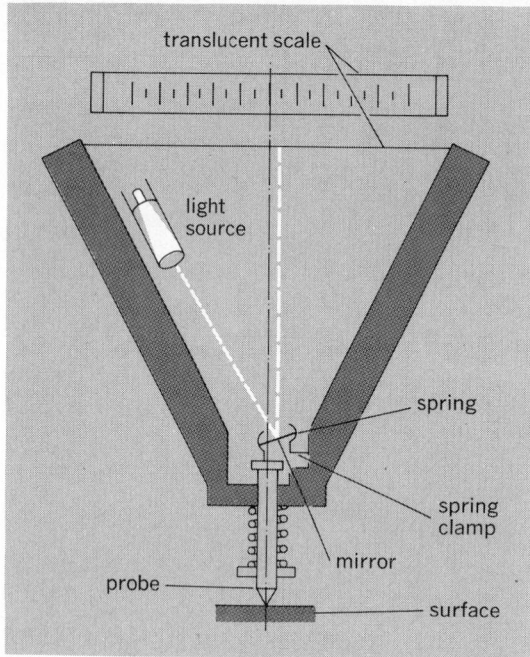

Fig. 7. Mechanooptical vibrometer. The motion given to the probe by the vibrating surface is used to rock a mirror and thereby actuate an optical lever arm. A light beam reflected from the mirror and focused onto the scale provides an indication of the vibration amplitude. (*General Electric Co.*)

sufficiently high frequencies, a mass supported on a soft spring with an attached scribing stylus serves as a stationary reference. In a second type, the case and recording paper are rigidly fixed, and the vibrations are transmitted to the stylus by means of a probe in contact with the vibrating body.

Resonant vibrators. The simple measurement of frequency of vibrating mechanical systems is often made with an instrument containing a series of resonant mechanical vibrators. These sensing elements are a series of cantilever-mounted reeds weighted at their free ends; their natural transverse frequencies are selected to cover a frequency spectrum of interest. Such instruments usually have a relatively small frequency range. By use of a series of these instruments, each covering a

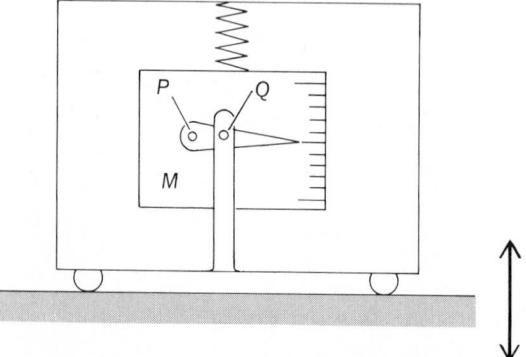

Fig. 8. Mechanical vibrograph. The scribing stylus pivots on point P, which is attached to the mass M. The case is connected to the stylus at point Q. (*From C. M. Harris, ed., Handbook of Noise Control, McGraw-Hill, 1957*)

different frequency range, an entire range from about 20 to 500 Hz may be covered.

Vibration pickups. These are electromechanical transducers capable of converting mechanical vibrations into electrical voltages. Depending upon their sensing element and output characteristics, such pickups are referred to as accelerometers, velocity pickups, or displacement pickups. *See* SOUND; WAVE MOTION.

[LAWRENCE E. KINSLER]

Bibliography: R. E. Bishop, *Vibration*, 2d ed., 1979; J. P. Den Hartog, *Mechanical Vibrations*, 4th ed., 1956; B. Pippard, *The Physics of Vibration*, vol. 1, 1978; W. Thomson, *Theory of Vibrations with Applications*, 2d ed., 1981.

Virial equation

In thermodynamics, an empirical equation of state with additional terms beyond those for an ideal gas. The additional terms account for some of the differences between real gases and ideal gases. For an ideal gas Eq. (1) holds, where p is the pres-

$$pV = NR_uT \quad \text{or} \quad pv_N = R_uT \qquad (1)$$

sure, V the total volume, $v_N = V/N$ the molar volume, N the number of moles, R_u the universal gas constant, and T the absolute temperature. To account for departures by real gases from the idealized relation, Eq. (1) may be written as Eq. (2).

$$\frac{pv_N}{R_uT} = B_0 + \frac{B_1(T)}{v_N} + \frac{B_2(T)}{(v_N)^2} + \frac{B_3(T)}{(v_N)^3} + \cdots \qquad (2)$$

The Bs are functions of temperature and are known as the virial coefficients. *See* VAN DER WAALS EQUATION. [GEORGE A. HAWKINS]

Bibliography: G. N. Hatsopoulos and J. H. Keenan, *Principles of General Thermodynamics*, 1965; J. B. Jones and G. A. Hawkins, *Engineering Thermodynamics*, 1960; K. Wark, *Thermodynamics*, 1966.

Virtual work principle

The principle stating that the total virtual work done by all the forces acting on a system in static equilibrium is zero for a set of infinitesimal virtual displacements from equilibrium. The infinitesimal displacements are called virtual because they need not be obtained by a displacement that actually occurs in the system. The virtual work is the work done by the virtual displacements, which can be arbitrary, provided they are consistent with the constraints of the system. *See* CONSTRAINT.

The principle of virtual work is equivalent to the conditions for static equilibrium of a rigid body expressed in terms of the total forces and torques. That is, the principle of virtual work can be derived from these conditions, and conversely. *See* STATICS.

One advantage of the principle of virtual work is that it can serve as a basis for all of statics. In the solution of problems the principle of virtual work is often useful for eliminating the need for consideration of the forces of constraint, since these forces often are perpendicular to the virtual displacements and consequently do no work.

The following problem provides an application of the principle of virtual work (see illustration). A uniform plank of weight W rests on a smooth floor and leans against a smooth wall. The angle between the plank and the floor is ϕ. A weightless,

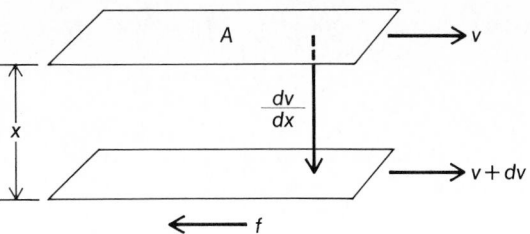

Fig. 1. Viscous shear in fluids.

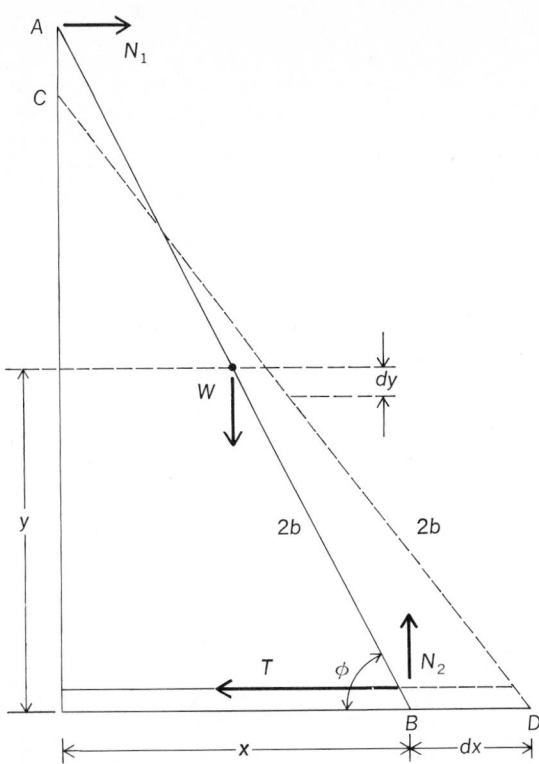

Illustration of principle of virtual work. (*From R. A. Becker, Introduction to Theoretical Mechanics, McGraw-Hill, 1954*)

inextensible string connects the lower end of the plank to the wall. The value of the tension T in the string is desired. Let $2b$ be the length of the plank. Then, as can be seen from the sketch, Eqs. (1)

$$y = b \sin \phi \qquad x = 2b \cos \phi \qquad (1)$$

hold. When ϕ increases by $d\phi$, the resulting changes dx and dy in x and y, respectively, are given by Eqs. (2). According to the principle of virtual work, the total work done in the virtual displacement produced by $d\phi$ must be zero. The

$$dy = b \cos \phi \, d\phi \qquad dx = -2b \sin \phi \, d\phi \qquad (2)$$

tual work, the total work done in the virtual displacement produced by $d\phi$ must be zero. The forces of reaction N_1 and N_2 are normal to the wall, since the wall is smooth. Consequently they do no work. The total work, therefore, is the work done by the forces W and T moving through their respective virtual displacements dy and dx. This work is given by Eq. (3). Solving for the tension T gives Eq. (4). The same result is obtained from the equilib-

$$0 = -2bT \sin \phi \, d\phi + Wb \cos \phi \, d\phi \qquad (3)$$

$$T = (W/2) \cot \phi \qquad (4)$$

rium conditions expressed in terms of forces and torques. When the principle of virtual work is used, the forces of reaction need not be considered. [PAUL W. SCHMIDT]

Bibliography: R. A. Becker, *Introduction to Theoretical Mechanics*, 1954; H. Goldstein, *Classical Mechanics*, 2d ed., 1980.

Viscosity

The resistance that a gaseous or liquid system offers to flow when it is subjected to a shear stress. Viscosity is a measure of the internal friction that arises when there are velocity gradients within the system. For fluids (gases and liquids) its mean-

ing is conceptually and operationally well defined. In the regime of laminar or streamline flow the force required to maintain a velocity gradient, (dv/dx), between planes of fluid of area A is described by Newton's equation (1) and Fig. 1. The

$$f = \eta \, A\left(\frac{dv}{dx}\right) \qquad (1)$$

proportionally constant η is called the viscosity coefficient. Its dimensions are (mass)(length)$^{-1}$(time)$^{-1}$, and in the cgs system the unit of viscosity is the poise (1 g · cm^{-1} s^{-1}). It is the force per unit area (dynes cm^{-2}) required to sustain a unit velocity gradient (cm s^{-1} cm^{-1}) normal to the flow direction. In the International System (SI) the unit of viscosity is kg · m^{-1} s^{-1}, and is hence larger than the poise by a factor of 10; conversely 1 kg · m^{-1} s^{-1} = 10 poise.

Simple gases typically have viscosities in the range of 100 to 200 micropoise at standard temperature and pressure (273 K, 1 atm or 101,325 Pa), whereas simple liquids under the same conditions have coefficients of viscosity about two orders of magnitude larger. The table lists values for the coefficients of viscosity of selected gases and liquids. The flow characteristics of gases and simple liquids such as water, carbon tetrachloride, and ethyl alcohol are accurately described by Eq. (1), and such fluids are called newtonian fluids. Aqueous suspensions, such as clays, gelatin, and agar, are termed non-newtonian fluids because their viscosities may depend upon the rate of shear and prior treatment. Hydrophilic sols often form extended networks involving water, and their non-newtonian behavior is believed to be due to the breakdown of their structure under shear. *See* FLUID FLOW; NEWTONIAN FLUID.

Molecular basis of viscosity in gases. The origin of internal friction (viscosity) at the molecular

Coefficients of viscosity of selected gases and liquids

Substance	Temperature, °C	η, poise*
Hydrogen	0	84.2×10^{-6}
Helium	0	186×10^{-6}
Nitrogen	0	167×10^{-6}
Oxygen	0	181×10^{-6}
Water (liquid)	20	10.1×10^{-3}
Ethyl alcohol	20	12.0×10^{-3}
Diethyl ether	20	2.5×10^{-3}
Carbon tetrachloride	20	9.8×10^{-3}
Mercury	20	15.5×10^{-3}
Glycerin	20	10.69
Glass	400	10^{13}
Glass	800	10^7

*1 poise = 0.1 kg · m^{-1} s^{-1}.

level is the net transfer of momentum between layers of fluid moving with different velocities in parallel flow by the mechanism of molecular collisions. In this process the directed energy of fluid flow is degraded to random thermal energy (heat).

It was one of the early triumphs of the kinetic theory of gases that established the relationship between the viscosity of a hard-sphere gas and the mean speed of its molecules. If x in Fig. 1 represents the mean free path λ of molecules in hypothetical planes that move with velocities equal to v and v', respectively, an exchange of molecules between the planes will result in the net transfer of momentum per unit time equal to $\frac{1}{3}An\bar{c}(mv - mv')$, where n is the number of molecules per unit volume, m is the mass of a molecule, mv and mv' are the additional momenta of molecules in the planes in consequence of their shear velocities, and $\bar{c}$ is the mean speed of molecules, $(8kT/\pi m)^{1/2}$, where k is Boltzmann's constant and T is the absolute temperature. Since the gas density is $\rho = nm$, the retarding force that resists the shear is given by Eq. (2). Combining Eqs. (1) and (2) gives Eq. (3).

$$f = \frac{1}{3}A\rho\bar{c}\left(\frac{dv}{dx}\right)\lambda \qquad (2)$$

$$\eta = \frac{1}{3}\bar{c}\rho\lambda \qquad (3)$$

Substitution for the mean free path $[\lambda = (n\sqrt{2\pi d^2})^{-1}$, where d is the average diameter of a molecule] permits restatement of Eq. (3) as Eq. (4).

$$\eta = \frac{m\bar{c}}{3\sqrt{2\pi d^2}} \qquad (4)$$

More refined calculations for hard-sphere gases replace the factor $\frac{1}{3}$ in Eq. (4) by 0.499, but the functional form is correct. It predicts that the viscosity of gases should be independent of pressure because the mean free path and gas density are affected in opposite ways by pressure, and this prediction is in accord with experiment up to moderately high pressures. Equation (4) also predicts a $T^{1/2}$ dependence of the gas viscosity, which is in fair agreement with experiment. Real gases show a somewhat stronger temperature dependence because their molecules are not ideal hard spheres. Equation (4) has had widespread application to the determination of the diameters of molecules, and the agreement is good but not perfect. Such minor discrepancies as do exist with molecular diameters determined by other methods (such as molar refraction, equation of state, and electron diffraction) are due to the fact that each method probes a somewhat different region of the potential surface of real molecules. *See* KINETIC THEORY OF MATTER; MEAN FREE PATH.

Viscosity of liquids. Momentum transfer between shearing layers also underlies the viscous behavior of simple liquids, but since the mean free path has little meaning for liquids, no simple relation such as Eq. (4) exists for them. In contrast to the behavior of gases, temperature decreases the viscosities of simple liquids and its effect is much larger. The temperature dependence of the viscosity of simple liquids bears no simple relationship to gas kinetic theory, but instead generally follows an exponential law of the form of Eq. (5),

$$\eta = A \exp (B/RT) \qquad (5)$$

where A and B are parameters characteristic of the liquid and are reasonably constant over finite ranges of temperature; R is the gas constant, $8.314 \text{ J} \cdot \text{mol}^{-1} \cdot \text{K}^{-1}$. The form of Eq. (5) is the same as that typically found for transport properties in the defect crystalline state, where the concept of a simple thermally activated process is generally accepted. This similarity in temperature dependence has in the past led to various "hole" theories of the liquid state. These have been based upon analogy with the vacancy model of defect crystals, and the parameter B has been identified with the energy required to create a void of molecular dimensions in the liquid and to move a nearby molecule into it. Such hole theories are now generally thought to be oversimplified, and viscous flow, like diffusion in liquids, is thought to be a highly complex process in which many molecules participate. *See* CRYSTAL DEFECTS.

Measurements of viscosity as a function of pressure likewise show completely different behavior for gases and liquids. Whereas the former show little dependence of viscosity on pressure in the low-density region, very high hydrostatic pressure generally increases the viscosity of liquids, sometimes quite markedly. In the region of laminar flow, nevertheless, Newton's equation accurately describes the viscous behavior of simple liquids, and the presumption is that the transfer of momentum between shearing layers involves a high degree of correlated molecular motions.

When the shear velocity exceeds a critical value in vessels of a given radius, streamline flow is replaced by turbulent flow. The criterion for the onset of turbulence in a tube of radius r is that the Reynolds number $2r\rho v/\eta$ exceed a certain value (approximately 2000 for normal liquids). *See* REYNOLDS NUMBER; TURBULENT FLOW.

Measurement of viscosity. The laminar flow of both gases and liquids in long narrow tubes is described by Poiseuille's equation (6), where η is the

$$\eta = \frac{\pi(p_1 - p_2)r^4 t}{8Vl} \qquad (6)$$

fluid viscosity (poise or $\text{kg} \cdot \text{m}^{-1} \text{s}^{-1}$), where r is the radius of the tube (cm or m), l is its length (cm or m), $(p_1 - p_2)$ is the pressure drop (dynes cm^{-2} or Pa) across the tube, and V is the volume of fluid (cm^3 or m^3) that flows through the tube in time t (s). Poiseuille's equation is based on the assumption that the layer of fluid in contact with the tube wall is stationary. It provides a basis for the absolute measurement of the viscosity of both gases and liquids. *See* BOUNDARY-LAYER FLOW.

More commonly for liquids, relative measurements of viscosity are made by use of either the Ostwald viscometer or the falling-sphere viscometer. The former (Fig. 2) consists of two glass bulbs separated by a length of capillary tubing. Liquid is drawn up into the upper bulb, and the time required for its meniscus to fall between calibration marks above and below the upper bulb is accurately measured. A similar measurement is made with a liquid of known viscosity. From Eq. (6), Eq. (7)

$$\frac{\eta_1}{\eta_2} = \frac{\rho_1 t_1}{\rho_2 t_2} \qquad (7)$$

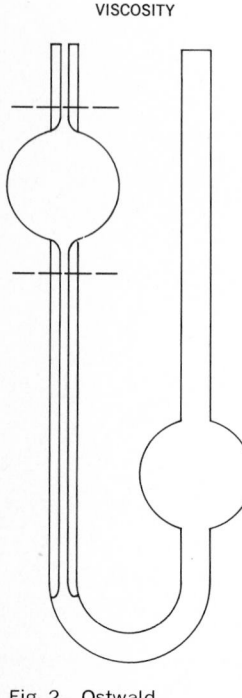

VISCOSITY

Fig. 2. Ostwald viscometer.

follows. Here η_1 and η_2 are the viscosities of the two liquids, ρ_1 and ρ_2 are their densities, and t_1 and t_2 are the corresponding flow times. Equation (7) takes an even simpler form if the viscosity is divided by the density of the liquid. This quantity, called the kinematic viscosity, is measured in units termed stokes (cm² s⁻¹) in cgs units, and in units of m² s⁻¹ in SI.

The falling-sphere viscometer is based upon Stokes' law for the frictional force on a spherical body of radius r falling with constant velocity in a fluid of viscosity η in an unbounded space, Eq. (8).

$$f = 6\pi\eta vr \tag{8}$$

This force is equal and opposite to the net force of gravity acting on the sphere, as in Eq. (9), where ρ

$$f = {}^4/_3\pi r^3(\rho - \rho')g \tag{9}$$

and ρ' are the densities of a metal sphere and the fluid, and g is the acceleration of gravity.

Equations (8) and (9) lead to the absolute viscosity of the fluid, Eq. (10).

$$\eta = \frac{2gr^2(\rho - \rho')}{9v} \tag{10}$$

As with the Ostwald viscometer, it is simpler to compare the times of fall of the sphere in fluids of known and unknown viscosity, and to use Eq. (11).

$$\frac{\eta_1}{\eta_2} = \frac{t_1(\rho - \rho'_1)}{t_2(\rho - \rho'_2)} \tag{11}$$

Other methods for the absolute or relative measurement of fluid viscosities are based upon the determination of the torque exerted upon a cylinder immersed in a fluid when a coaxial cylinder is rotated with constant velocity, or the damping of the amplitude of an oscillating disk suspended in the fluid by a torsion fiber.

Flow behavior of complex fluids. Many fluids display flow behavior that deviates profoundly from that of simple gases and liquids. This subject is normally treated by the field of rheology. A few examples will serve to indicate the complexity of flow behavior in some fluids. Monoclinic sulfur, whose molecules consist of puckered rings of eight sulfur atoms, melts at 95.5°C to form a simple liquid (S_λ) of the same molecularity. Its viscosity is low enough to classify it as a normal liquid, and its viscosity decreases with temperature in the normal manner. Between 160 and 180°C, however, the viscosity increases dramatically by many orders of magnitude, and it appears that ring opening occurs followed by the formation of long-chain polymers. Above this temperature interval, the viscosity again decreases as thermal energy breaks up the long chains into smaller units. The process is highly irreversible, and crystalline sulfur may be recovered only by condensation from sulfur vapor.

Various colloidal dispersions of solids in oil or aqueous media decrease their viscosity when stirred at constant temperature, and revert to their former state of higher viscosity when the shear stresses are reduced. This phenomenon of thixotropy is an essential property of paints that contain solid pigments.

The flow of blood in mammalian vascular systems is non-newtonian, and Poiseuille's law is not obeyed. In part this behavior is attributable to the presence of red corpuscles and other suspended bodies, but the phenomenon is very complex.

Glasses are amorphous solids, structurally much closer to liquids than to crystals. Even at ordinary temperatures they deform under stress over long periods of time, and their viscosity varies over tens of orders of magnitude as the temperature is raised to the softening point. Profound structural changes in the random three-dimensional network and of the dynamical modes of local structural elements take place as the temperature of a glass is increased.

Some adhesives exhibit flow along directions that are not parallel to the direction of stress. Such fluids are anisotropic, and their flow properties are tensors. *See* LIQUID; NON-NEWTONIAN FLUID; RHEOLOGY.

[NORMAN H. NACHTRIEB]

Bibliography: P. W. Atkins, *Physical Chemistry*, 1978; J. O. Hirschfelder, C. F. Curtiss, and R. B. Bird, *The Molecular Theory of Gases and Liquids*, 1954; E. A. Moelwyn-Hughes, *Physical Chemistry*, 2d ed., 1964; W. J. Moore, *Physical Chemistry*, 4th ed., 1972.

Vortex

A line vortex in two-dimensional fluid flow produces a flow or circulation around the line.

Free vortex. Consider the effect of rotating a right-circular cylinder of radius r_0 about its axis with a peripheral velocity v_0 in a fluid otherwise at rest. The fluid in contact with the surface of the cylinder rotates with the cylinder. Fluid at greater radius is also set in motion in concentric circles with velocity diminishing as the radius increases. This type of fluid motion, in which the velocity varies inversely as the radius, is referred to as a free vortex.

If the cylinder is reduced to zero radius in such a manner that $v_0 r_0$ remains constant in the limit as r_0 approaches zero, a line vortex results. The velocity at the line is infinite, so the line itself must be considered as a singular line, to be excluded from the actual fluid.

Examples of vortices occur frequently in nature. The tornado is an example of a free vortex, with high velocities near its center, and correspondingly low pressure intensities. The waterspout is its counterpart over water.

The fluid motion in the case of a line vortex in an ideal (frictionless and incompressible) fluid is irrotational; that is, its motion may be described in terms of a velocity potential.

Vortex tube. If a small spherical particle of a frictionless fluid could be considered as suddenly solidified, its resulting rotation could be expressed by a vector parallel to the axis of rotation, with its length proportional to the angular velocity, and with its direction indicating the sense of rotation by the right-hand rule. When the rotation vector is everywhere zero throughout a region of fluid, the motion of that fluid is irrotational. When some finite fluid regions have nonzero values of the rotation vector, then this fluid has vorticity. A vortex line is a line drawn through the fluid such that it is everywhere tangent to the rotation vector. A collection of vortex lines through a small closed curve defines a vortex tube, which has certain special properties.

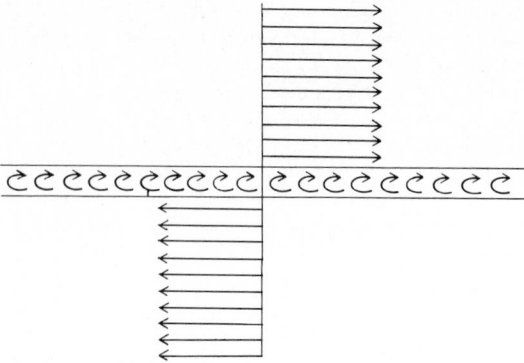

Fig. 1. Vortex sheet which is formed between oppositely directed streams.

1. The circulation about a vortex tube is everywhere the same along its length. Circulation is defined as the line integral of the velocity vector around a closed path.

2. The vortex tube cannot end in the fluid. It must either extend to a boundary or close upon itself.

3. Vortex lines move with the fluid. Vorticity of a fluid is a property of the fluid itself and not the space it occupies.

A smoke ring is a practical example of a closed vortex tube. A circular vortex tube in otherwise still fluid will translate perpendicular to the plane of the ring without change in size.

Vortex arrays. A discontinuity in fluid velocity along a surface, such as slippage of one layer of fluid over another, may be handled as a vortex sheet in an otherwise continuous flow. In this case all vortex lines are in the surface (Fig. 1). A practical case of a vortex sheet is the flow downstream from an airfoil when the velocity leaving the upper surface is higher than the velocity leaving the under surface. *See* AIRFOIL.

When real fluid flows around a body, such as wind blowing across a cable, the fluid rotation in the boundary layer causes vortices to form along the downstream side of the body. For certain conditions they remain near the body and are referred to as bound vortices.

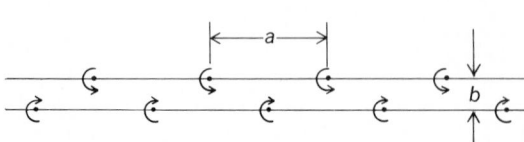

Fig. 2. Kármán vortex street consists of alternate vortices of opposite rotational sense.

For higher velocities the vortices form, grow, and are then systematically shed from the downstream side of the body, forming a vortex street (Fig. 2). The unsymmetrical case shown is called the Kármán vortex street after T. von Kármán, who first identified them and showed that the motion is stable when $b/a = 0.281$. *See* FLUID-FLOW PRINCIPLES; KARMAN VORTEX STREET.

[VICTOR L. STREETER]

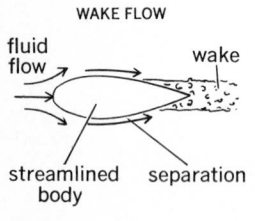

WAKE FLOW

fluid flow   wake

streamlined   separation
body

Wake formed downstream from a streamlined body.

Wake flow

Turbulent eddying flow that occurs downstream from bluff bodies. When fluid flows along the boundary of a solid body, the fluid near the boundary is slowed down by viscous shear stresses exerted at the boundary. This action is progressive along the body and, under certain conditions, fluid near the boundary is brought to rest, which causes the fluid moving near the body to separate from the body (see illustration). A wake develops and produces additional form drag on the body.

When the flow does not separate from the boundary, as in ideal fluid flow situations, the high fluid velocity at the largest cross section of the body is reduced along the downstream portion of the body with recovery of pressure. Formation of a wake stops the pressure recovery process, leaving a low pressure intensity in the wake and a resultant pressure force on the body acting in the downstream direction. The fluid in the wake is highly turbulent, containing vortices that are shed from the body. The wake continues downstream with the flow.

A wake is formed downstream from bluff bodies such as bridge piers, smoke stacks, buildings, or trees. For unsteady flow cases, such as the motion of a train or a ship, the fluid behind the moving object has great turbulence remaining in it and, in the case of a ship, is easily discerned for a great distance. *See* FLUID-FLOW PRINCIPLES.

[VICTOR L. STREETER]

Wave

The general term applied to the description of a disturbance which propagates from one point in a medium to other points without giving the medium as a whole any permanent displacement.

Waves are generally described in terms of their amplitude, and how the amplitude varies with both space and time. The actual description of the wave amplitude involves a solution of the wave equation and the particular boundary conditions for the case being studied. In the cases most often considered, the wave equation is simplified to a second-order, linear, partial differential equation. This equation for a one-dimensional space coordinate x is written as Eq. (1), where ϕ is the amplitude, c is

$$\frac{\partial^2 \phi}{\partial x^2} = \frac{1}{c^2} \frac{\partial^2 \phi}{\partial t^2} \qquad (1)$$

the wave velocity, and t is the time coordinate. The generalized solutions of this equation are of the form given by Eq. (2). The first term indicates a

$$\phi = F(x - ct) + G(x + ct) \qquad (2)$$

wave traveling in the positive x direction at a velocity c, and the second term a wave traveling in the negative x direction. The functions F and G are determined by the particular properties of the boundary conditions of the problem. In the one-dimensional case these are usually sine or cosine waves. The velocity of propagation c is proportional to the square root of the ratio of the elastic to the inertial constants of the medium. *See* SINE WAVE; WAVE EQUATION; WAVE MOTION.

Acoustic waves, or sound waves, are a particular kind of the general class of elastic waves. Elastic waves are propagated in media having two properties, inertia and elasticity. Elasticity of the medium is required in order to provide a force which tends to restore a displaced particle of the medium to its original position. Inertia is required to enable the displaced particle to transfer momentum to an adjoining particle. A shear wave, or rotational wave, is a wave in an elastic medium which causes an element of the medium to change its shape without a change of volume.

Electromagnetic waves (for example, light waves and radio waves) are not elastic waves and therefore can travel through a vacuum. The velocity of the wave depends on the medium through which the wave travels, but in a vacuum it is a constant, c, approximately equal to 3×10^8 m/sec. *See* ELECTROMAGNETIC WAVE; SHOCK WAVE; STANDING WAVE. [WILLIAM J. GALLOWAY]

Wave equation

The name given to certain partial differential equations in classical and quantum physics which relate the spatial and time dependence of physical functions. In this article the classical and quantum wave equations are discussed separately, with the classical equations first for historical reasons. *See* NONRELATIVISTIC QUANTUM THEORY.

Classical wave equation. In classical physics the name wave equation is given to the linear, homogeneous partial differential equations which have the form of Eq. (1). Here v is a parameter with

$$\left[\nabla^2 - \frac{1}{v^2}\frac{\partial^2}{\partial t^2}\right]f(\mathbf{r},t) = 0 \tag{1}$$

the dimensions of velocity; $\mathbf{r}$ represents the space coordinates x, y, z; t is the time; and ∇^2 is Laplace's operator defined by Eq. (2). The function $f(\mathbf{r},t)$ is a

$$\nabla^2 \equiv \frac{\partial^2}{\partial x^2} + \frac{\partial^2}{\partial y^2} + \frac{\partial^2}{\partial z^2} \tag{2}$$

physical observable; that is, it can be measured and consequently must be a real function. Laplace's operator may also be written in polar coordinates $\mathbf{r}$, θ, and ϕ; if f depends only on the radial distance r, the wave equation has the simple form shown by Eq. (3).

$$\left[\frac{\partial^2}{\partial r^2} + \frac{2}{r}\frac{\partial}{\partial r} - \frac{1}{v^2}\frac{\partial^2}{\partial t^2}\right]f(\mathbf{r},t) = 0 \tag{3}$$

The simplest example of a wave equation in classical physics is that governing the transverse motion of a string under tension and constrained to move in a plane. In this case the wave equation is given by Eq. (4), where $y(x,t)$ is the transverse dis-

$$\frac{\partial^2 y(x,t)}{\partial x^2} = \frac{1}{v^2}\frac{\partial^2 y(x,t)}{\partial t^2} \tag{4}$$

placement of the string from its equilibrium position at time t. If the string is thought of as being made up of many separate particles, x is the label for the particle. The velocity parameter is $v = \sqrt{T/\lambda}$, with λ the linear mass density and T the tension in the string, both uniform. Equation (4) is valid when the slope of the string is small, that is, $\partial y/\partial x \ll 1$. A solution to Eq. (4) is any function of the form $f(x - vt)$, where f is arbitrary except that it must be twice differentiable. Since at $t = 0$ one would have $y(x,0) = f(x)$, $f(x)$ specifies the initial shape of the string. Since $f(x - vt)$ differs from $f(x)$ only by a translation along the x axis of

magnitude vt, it is apparent that the shape of the string moves along the x axis as a wave with shape determined by $f(x)$ and with velocity v, which is called the wave propagation velocity. A point to note is that v is not the transverse velocity $\partial y/\partial t$ of the particles in the string. In fact, since Eq. (5)

$$\frac{\partial y}{\partial t} = -v\frac{\partial y}{\partial x} \tag{5}$$

holds, the transverse velocity for the solution $f(x - vt)$ is always small and opposite in sign to the slope. Another solution to Eq. (4) could be $g(x + ct)$. which corresponds to the shape of the string moving in the negative x direction, that is, with velocity $-v$. The most general solution to Eq. (4) is the linear combination given by Eq. (6). The par-

$$y(x,t) = a + f(x - vt) + a - g(x + vt) \tag{6}$$

ticular choice of constants $a\pm$ is determined by the shape of the string at $t = 0$ and by the transverse velocity at $t = 0$, that is, by $y(x,0)$ and $\partial y(x,t)/\partial t|_{t=0}$.

A second type of classical physical situation in which the wave equation, Eq. (1), supplies a mathematical description of the physical reality is the propagation of pressure waves in a fluid medium. Such waves are called acoustical waves, the propagation of sound being an example. Suppose that x is the undisturbed position of a particle in the fluid and let the displacement at time t be $z(x,t)$ so that those particles that have position x when undisturbed have position $x + z(x,t)$ at time t after being disturbed. If $\partial z/\partial x \ll 1$, the acoustic approximation, then the displacement $z(x,t)$ satisfies Eq. (7),

$$\frac{\partial^2 z(x,t)}{\partial x^2} = \frac{1}{\bar{v}^2}\frac{\partial^2 z(x,t)}{\partial t^2} \tag{7}$$

where $\bar{v}^2 = dP(\rho)/d\rho|_{\rho=\rho_0}$, with ρ the density of the fluid, ρ_0 the undisturbed density, and $P(\rho)$ the pressure. The acoustic approximation must be used to derive Eq. (7) from the density equation, the equation of state for the fluid, and Newton's second law of motion. If the fluid is specified to be air, then $\bar{v}$ is the velocity of sound. Of course, for acoustical waves and vibrating strings there may be three-dimensional motion in the general case, satisfying Eq. (1) with the appropriate parameters.

A third example of a classical physical situation in which Eq. (1) gives a description of the phenomena is afforded by electromagnetic waves. In a region of space in which the charge and current densities are zero, Maxwell's equations for the photon lead to the wave equations, Eqs. (8). Here

$$\left[\nabla^2 - \frac{1}{c^2}\frac{\partial^2}{\partial t^2}\right]\mathbf{E}(\mathbf{r},t) = 0$$
$$\left[\nabla^2 - \frac{1}{c^2}\frac{\partial^2}{\partial t^2}\right]\mathbf{B}(\mathbf{r},t) = 0 \tag{8}$$

$\mathbf{E}$ is the electric field strength and $\mathbf{B}$ is the magnetic flux density; they are both vectors in ordinary space. The parameter c is the speed of light in vacuum. A vector potential $\mathbf{A}(\mathbf{r},t)$ and a scalar potential $\phi(\mathbf{r},t)$ may be introduced by the identifications $\mathbf{B}(\mathbf{r},t) = \nabla x \mathbf{A}(\mathbf{r},t)$ and $\mathbf{E}(\mathbf{r},t) = -\nabla\phi - \frac{1}{c}\frac{\partial \mathbf{A}}{\partial t}$, where ∇x is the curl operator and ∇ the gradient operator. The potentials then also satisfy the wave equation.

If relativistic notation is used so that $A_x = A_1$,

$A_y = A_2$, $A_z = A_3$, and $A_4 = i\phi$ with $i = \sqrt{-1}$, then Eq. (9) can be derived (all Greek indices run from

$$\left[\nabla^2 - \frac{1}{c^2}\frac{\partial^2}{\partial t^2}\right]A_\mu(\mathbf{r},t) = 0 \tag{9}$$

1 to 4). This wave equation may be written in a form that is manifestly covariant (form invariant) with respect to the Lorentz transformations by the further identifications $x = x_1$, $y = x_2$, $z = x_3$, and $x_4 = ict$ so that the operator in Eq. (9) can be written as Eq. (10), where the repeated index in the second

$$\nabla^2 - \frac{1}{c^2}\frac{\partial^2}{\partial t^2} = \frac{\partial^2}{\partial x_1^2} + \frac{\partial^2}{\partial x_2^2}$$
$$+ \frac{\partial^2}{\partial x_3^2} + \frac{\partial^2}{\partial x_4^2} \equiv \frac{\partial}{\partial x_\mu}\frac{\partial}{\partial x_\mu} \tag{10}$$

equation is defined to mean that a sum is to be taken, as indicated above (Einstein summation convention). Then the manifestly covariant form of Eq. (9) is written as Eq. (11). Both $x = (\mathbf{r}, ict)$ and

$$\frac{\partial}{\partial x_\mu}\frac{\partial}{\partial x_\mu}A_r(\mathbf{r},t) = 0 \tag{11}$$

$A = (\mathbf{A}, i\phi)$ are Lorentz four-vectors, and the product of two four-vectors, summed over the repeated index, is a scalar, such as $\frac{\partial}{\partial x_\mu}\frac{\partial}{\partial x_\mu}$. *See* ELECTROMAGNETIC RADIATION; MAXWELL'S EQUATIONS.

Quantum wave equation. The above discussion leads to the consideration of the quantum-mechanical wave equations. The nonrelativistic Schrödinger equation is an example. In this article only the relativistic quantum-mechanical wave equations are considered. The first example may be found from Eq. (11) by subtracting from the operator $\frac{\partial}{\partial x_\mu}\frac{\partial}{\partial x_\mu}$ a constant term $\frac{m^2 c^2}{\hbar^2}$, where m is the mass and $\hbar$ is Planck's constant divided by 2π. The new equation is Eq. (12) and is called the Schrödinger-Klein-Gordon equation. Every free

$$\left[\frac{\partial}{\partial x_\mu}\frac{\partial}{\partial x_\mu} - \frac{m^2 c^2}{\hbar^2}\right]\varphi(\mathbf{r},t) = 0 \tag{12}$$

elementary particle has a wave function $\varphi(\mathbf{r},t)$ which must satisfy this equation although $\varphi(\mathbf{r},t)$ are different types of functions for particles with different spins. If $\varphi(\mathbf{r},t)$ has the form of a plane wave given by Eq. (13) corresponding to a particle

$$\varphi(\mathbf{r},t) = \chi(\mathbf{P})\exp\left[i/\hbar(\mathbf{P}\cdot\mathbf{x} - Et)\right] \tag{13}$$

with momentum $\mathbf{P}$ and total energy E, then insertion of Eq. (13) into Eq. (12) yields Eq. (14). $\chi(\mathbf{P})$

$$\frac{1}{\hbar^2}\left[E^2 - c^2 p^2 - m^2 c^4\right]\chi(\mathbf{P}) = 0 \tag{14}$$

in Eqs. (13) and (14) can be any function of the momentum $\mathbf{P}$. Since the term in brackets is the relativistic connection between energy and momentum for a free particle, it is identically zero; therefore $(\mathbf{r},t)$ defined by Eq. (13) satisfies Eq. (12). Historically, Eq. (14) came before Eq. (12).

All the above wave equations are second-order in both space and time derivatives. A different type of relativistic quantum-mechanical wave equation is the Dirac equation, which is first-order

in both space and time derivatives and has the form shown in Eq. (15). Here $\boldsymbol{\alpha}$ is a three-compo-

$$\left[\frac{c\hbar}{i}\boldsymbol{\alpha}\cdot\nabla + mc^2\beta\right]\psi(\mathbf{r},t) = i\hbar\frac{\partial\psi(\mathbf{r},t)}{\partial t} \tag{15}$$

nent object whose components are each a 4×4 matrix, β is a 4×4 matrix, and the 4×1 column matrix $\psi(\mathbf{r},t)$ is a wave function corresponding to a particle of intrinsic spin 1/2.

If ψ_1, ψ_2, ψ_3, ψ_4 are the four components of $\psi(\mathbf{r},t)$, then Eq. (15) is the matrix form of the system of Eq. (16), where $\alpha_i^{\sigma\tau}$ is the $\sigma\tau$ element of the 4×4

$$\frac{c\hbar}{i}\sum_{\tau=1}^{4}\left(\alpha_1^{\sigma\tau}\frac{\partial}{\partial x_1} + \alpha_2^{\sigma\tau}\frac{\partial}{\partial x_2} + \alpha_3^{\sigma\tau}\frac{\partial}{\partial x_3}\right)\psi_\tau$$
$$+ \sum_{\tau=1}^{4}\beta_{\sigma\tau}mc^2\Psi_\tau = i\hbar\frac{\partial\psi_\sigma}{\partial t} \tag{16}$$

matrix which forms the ith component of $\boldsymbol{\alpha}$ in Eq. (15). Because it is necessary that each component of the solution $\psi(\mathbf{r},t)$ of Eq. (15) also satisfy Eq. (12), $\boldsymbol{\alpha}$ and β have certain anticommutation relations with each other which imply that the product of any of these matrices with itself is the 4×4 unit matrix. Equation (15) may be written in a manifestly covariant form by making the definitions $\gamma \equiv -i\beta\boldsymbol{\alpha}$, $\gamma_4 \equiv \beta$ and by rearranging terms in Eq. (15). The result is (in index notation) Eq. (17). The

$$\left[\gamma_\mu\frac{\partial}{\partial x_\mu} + \frac{mc}{\hbar}\right]\psi(\mathbf{r},t) = 0 \tag{17}$$

term $\beta\boldsymbol{\alpha}$ is interpreted as a three-component object with matrices $\beta\alpha_1$, $\beta\alpha_2$, and $\beta\alpha_3$ as components. The quantum-mechanical functions such as $\psi(\mathbf{r},t)$ and $\varphi(\mathbf{r},t)$ differ from their classical counterparts $y(\mathbf{r},t)$ and so forth, in that the quantum functions are not directly observable quantities and are, in fact, complex. In general, only bilinear combinations of the quantum-mechanical wave functions, suitably averaged with certain operators, and over some region of space, are physically observable.

The Dirac equation is the most prominent example of the type of wave equation (relativistic quantum mechanical) that is linear in the space and time derivatives.

A second example is the Rarita-Schwinger equation, Eq. (18). This has the same form as the Dirac

$$\left[\gamma_\mu\frac{\partial}{\partial x_\mu} + \frac{mc}{\hbar}\right]\psi_v(\mathbf{r},t) = 0 \tag{18}$$

equation except that the wave function $\psi_v(\mathbf{r},t)$ is now a 4×1 column matrix with each of the four components being independently a four-vector so that $\psi_v(\mathbf{r},t)$ has 16 components. Equation (18) is intended to describe a free relativistic particle (and antiparticle) with intrinsic spin 3/2. For this only eight independent complex components are needed; in addition to Eq. (18), the auxiliary condition defined by Eq. (19) is required.

$$\gamma_v\psi_v(\mathbf{r},t) = 0 \tag{19}$$

Equations (18) and (19) together imply a third equation, Eq. (20). Exactly as in the Dirac equa-

$$\frac{\partial}{\partial x_v}\psi_v(\mathbf{r},t) = 0 \tag{20}$$

tion, each component of the Rarita-Schwinger wave function satisfies Eq. (12). The necessity of auxiliary conditions, such as Eqs. (19) and (20), to eliminate redundant components in the wave function is a characteristic of the approach to relativistic wave equations which involve only first derivatives of space and time and certain matrices. These considerations for spins 1/2 and 3/2 may be extended to higher half-integral spins by including more four-vector indices on the wave function.

Another way of writing such a first-order matrix–differential wave equation is to make all the components of the wave function matrix components. The wave function $\bar{\psi}(\mathbf{r},t)$ is then a matrix satisfying a wave equation of the form of Eq. (21),

$$\left[\beta_\mu \frac{\partial}{\partial x_\mu} + \frac{mc}{\hbar}\right]\psi(\mathbf{r},t) = 0 \qquad (21)$$

where the four β_μ are a set of singular matrices whose structure depends on the intrinsic spin of the particle to be described. For spin zero the β_μ are 5×5 matrices and for spin 1 they are 10×10. Of course, each component of $\bar{\psi}(\mathbf{r},t)$ must also satisfy Eq. (12). Equation (20) is called the Duffin-Kemmer-Petiau equation.

Free, relativistic quantum-mechanical particles with tensor equations may also be described. The best-known examples of this type are the Proca equations for spin 1 which have the form shown in Eqs. (22), and so are differential relations between

$$F_{\alpha\beta} = \frac{\partial \omega_\beta}{\partial x_\alpha} - \frac{\partial \omega_\alpha}{\partial x_\beta}$$

$$\frac{\partial F_{\alpha\beta}}{\partial x_\alpha} = \frac{m^2 c^2}{\hbar^2 \omega_\beta} \qquad (22)$$

a four-vector field $\omega_\beta(\mathbf{r},t)$ and a second-rank tensor field $F_{\alpha\beta}(\mathbf{r},t)$. By differentiating the second of Eqs. (22) with respect to x_β, the auxiliary condition shown in Eq. (23) can be obtained; by inserting the

$$\frac{\partial}{\partial x_\beta} \omega_\beta(\mathbf{r},t) = 0 \qquad (23)$$

first of Eqs. (22) into the second and using Eq. (23), Eq. (12) can be satisfied by each of the components of ω_ν. Thus it is again apparent that auxiliary conditions are needed to eliminate redundant components from the quantum-mechanical equations of motion.

The most general set of wave equations and auxiliary conditions capable of describing quantum-mechanically a free relativistic particle with mass and internal integral or half-integral spin are the Dirac-Fierz-Pauli spinor wave equations.

A characteristic of all the above quantum-mechanical formulations is that the function satisfying the wave equation has more components than necessary to describe a free particle and antiparticle with the given spin. A description receiving much attention is that involving only irreducible representations of the homogeneous Lorentz group.

With this description a system with spin s is described by a wave function with $2s + 1$ components for the particle and $2s + 1$ components for the antiparticle, the minimum number for describing such a system. For the wave equation in this description, there may be only first-order time derivatives and complicated spatial derivatives, or there may be uniform dependence of the wave equation on both space and time derivatives, the order depending on the spin. Such a description shows the greatest promise for future developments in the field of relativistic quantum-mechanical wave equations. *See* QUANTUM MECHANICS.

[DAVID L. WEAVER]

Bibliography: J. D. Bjorken and S. D. Drell, *Relativistic Quantum Mechanics*, 1964; C. A. Coulson and A. Jeffrey, *Waves: A Mathematical Approach to the Common Types of Wave Motion*, 2d ed., 1978; C. Itzykson and J. B. Zuber, *Quantum Field Theory*, 1980; G. A. Korn and T. M. Korn, *Mathematical Handbook for Scientists and Engineers*, 2d ed., 1968; P. M. Morse and H. Feshbach, *Methods of Theoretical Physics*, 2 vols., 1953.

Wave mechanics

The modern theory of matter holding that elementary particles (such as electrons, protons, and neutrons) have wavelike properties. In 1924 Louis de Broglie postulated that the same wave-corpuscle duality which was then known to exist in the case of light might also occur in matter; this hypothesis was subsequently verified experimentally. With contributions by the mathematical physicists Erwin Schrödinger, Max Born, Werner Heisenberg, P. A. M. Dirac, and others, this theory of matter has become the highly successful quantum mechanics of the present day. For a discussion of wave mechanics and wave-particle duality *see* QUANTUM MECHANICS. *See also* DE BROGLIE WAVELENGTH.

[EDWARD GERJUOY]

Wave motion

The process by which a disturbance at one point in space is propagated to another point more remote from the source with no net transport of the material of the medium itself. For example, sound is a form of wave motion; wind is not. Wave motion can occur only in a medium in which energy can be stored in both kinetic and potential form. In a mechanical medium, kinetic energy results from inertia and is stored in the velocity of the molecules, while potential energy results from elasticity and is stored in the displacement of the molecules.

In a free traveling wave (as distinguished from a stationary or standing wave) one part of the medium disturbs an adjacent part, thereby imparting energy to it. This portion of the medium, in turn, disturbs another part, thereby causing a flow of energy in a given direction away from the source (see illustration). More technically, wave propagation is the result of kinetic energy at one point being transferred into potential energy at an adjacent point, and vice versa. The rate of travel of the disturbance, or velocity of propagation, is determined by the constants of the medium. A stationary wave is the combination of two waves of the same frequency and strength traveling in opposite directions so that no net transfer of energy away from the source takes place. A standing wave is the same but with the returning wave (toward the source) being of lesser intensity than the outwardly traveling wave so that a net transfer of energy away from the source does take place. *See* STANDING WAVE.

Wave motion can occur in a vacuum (electro-

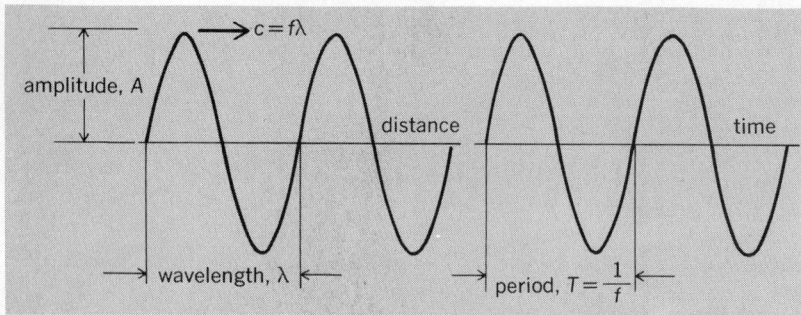

Relation between frequency, wavelength, and velocity in wave propagation.

magnetic waves), in gases (sound waves), in liquids (hydrodynamic waves), and in solids (vibration waves). Electromagnetic waves can also travel in gases, liquids, and solids provided that the electrical conductivity of the medium is not perfect or that the imaginary part of the dielectric constant is not infinitely great. By current usage, elastic waves propagated in gases, liquids, and solids, regardless of whether one can hear them or not, are called acoustic waves.

In this article, it is shown that wave motion can exist in any continuous medium in which potential and kinetic energy can be stored. The wave equation is derived, and solutions for electromagnetic and mechanical media are given. It is shown that longitudinal waves may exist in all mechanical media, and that transverse waves may exist in electromagnetic and solid mechanical media. In all cases, viscosity or heat losses in mechanical media, and resistance and magnetic losses in electromagnetic media are neglected. Only homogeneous and isotropic media are considered. For information which is related to, and supplements, the present discussion *see* DIFFRACTION; ELECTROMAGNETIC RADIATION; HARMONIC MOTION; HUYGENS' PRINCIPLE; INTERFERENCE OF WAVES; LIGHT; MAXWELL'S EQUATIONS; REFRACTION OF WAVES; SUPERPOSITION PRINCIPLE; VIBRATION; WAVE EQUATION; WAVE MOTION IN FLUIDS; WAVE MOTION IN LIQUIDS.

Fundamental relations. A wave is commonly referred to in terms of either its wavelength or its frequency. In any type of wave motion, these two quantities are related to a third quantity, velocity of propagation, by the simple relation in Eq. (1),

$$f\lambda = c \tag{1}$$

where f = frequency, λ = wavelength, and c = velocity of propagation. The period T is the reciprocal of the frequency, and the amplitude A is the maximum magnitude taken on by the variable of the wave at a given point in space. The physical significance of these quantities is illustrated in the figure. It is a basic property of wave motion that the frequency of a wave remains constant under all circumstances except for a relative motion between the source of the wave and the observer. The case of a frequency shift due to relative motion, known as the Doppler effect, has many interesting applications. On the other hand, the velocity of propagation is dependent on the properties of the medium (and, sometimes, also on the frequency) and the wavelength will vary with the velocity in accordance with Eq. (1). *See* DOPPLER EFFECT.

Electromagnetic waves. The media in which electromagnetic waves travel possess no elasticity or inertia, but rather the ability to store energy in the electric and magnetic fields. The electric field corresponds in every respect to the field of an irrotational fluid motion, and the mathematical formulations of the motions of acoustic waves and electromagnetic waves are similar.

In the customary manner, in an electromagnetic medium one defines the vectorial factors **E** as the electric field strength and **H** as the magnetic field strength, with both magnitudes and directions at every point in space.

Next, it is assumed that the medium is homogeneous and isotropic (that is, that the dielectric constant κ and the permeability μ are constants and scalars), that there are no applied electromotive forces in the portion of the medium being dealt with, and that the electrical conductivity σ of the medium is zero. (Waves can be propagated in media where σ is greater than zero but less than infinity.) Then the field equations can be written as Eqs. (2)–(5), where c will be shown to be the ve-

$$\frac{\kappa}{c}\frac{\partial \mathbf{E}}{\partial t} = \text{curl } \mathbf{H} \tag{2}$$

$$\frac{-\mu}{c}\frac{\partial \mathbf{H}}{\partial t} = \text{curl } \mathbf{E} \tag{3}$$

$$\text{div } \mathbf{H} = 0 \tag{4}$$

$$\text{div } \mathbf{E} = 0 \tag{5}$$

locity of propagation of the electromagnetic wave in a vacuum, where $\kappa = \mu = 1$.

Wave equation. J. C. Maxwell recognized about 1863 that these basic equations could be combined to yield an equation resembling the wave equation for mechanical wave motion. Thus he predicted the existence of electromagnetic waves which had not been suspected theretofore. Later, electromagnetic waves proved to be identical with light waves.

The combination of Eqs. (2)–(5) yields the wave equation, Eq. (6). In a vacuum, the wave equation

$$\frac{\kappa\mu}{c^2}\frac{\partial^2 \mathbf{E}}{\partial t^2} = \nabla^2 \mathbf{E} \tag{6}$$

becomes Eq. (7) and also Eq. (8), where ∇^2 is a sca-

$$\frac{\partial^2 \mathbf{E}}{\partial t^2} = c^2 \nabla^2 \mathbf{E} \tag{7}$$

$$\frac{\partial^2 \mathbf{H}}{\partial t^2} = c^2 \nabla^2 \mathbf{H} \tag{8}$$

lar operator called the Laplacian. *See* LAPLACIAN.

Plane wave propagation. For simplicity, particular solutions of the wave equation for the case of plane wave propagation will now be sought. A wave is called plane when a family of parallel planes can be taken in the field such that the electric and magnetic field strengths are constant in magnitude and direction at all points of any given member of the family; the planes are called wavefronts, the direction perpendicular to them the wave normal. If the axis of x is taken in the direction of the wave normal, then the wavefronts are parallel to the plane of yz.

Since, for this case, **E** and **H** are to be constant in any one wavefront, the partial derivatives with respect to y or z must vanish. Then the field equations reveal that Eq. (9) is valid. Therefore, insofar

$$\frac{\partial E_x}{\partial t} = \frac{\partial E_x}{\partial x} = \frac{\partial H_x}{\partial t} = \frac{\partial H_x}{\partial x} = 0 \qquad (9)$$

as wave propagation is concerned, Eq. (10) holds.

$$E_x = H_x = 0 \qquad (10)$$

The meaning of Eq. (10) is that neither **E** nor **H** can have a periodically changing component in the direction in which the wave is traveling. That is to say, the waves are not longitudinal but transverse.

Of the remaining four equations developing out of the field equations, two of them connect E_y and H_z, and the other two E_z and H_y. Dealing with the pairs, one obtains the one-dimensional wave equation, Eq. (11), where E_y, E_z, H_y, or H_z may (any one) be inserted within the parentheses of Eq. (11).

$$\frac{\partial^2(\ \ )}{\partial t^2} = c^2 \frac{\partial^2(\ \ )}{\partial x^2} \qquad (11)$$

The general solution of Eq. (11) can be written in the form of Eq. (12). The first term of Eq. (12) is

$$E_y = F_1(x - ct) + F_2(x + ct) \qquad (12)$$

associated with a wave traveling in the positive x direction and the second term with a wave traveling in the minus x direction. This is true because the argument of F_1 is the same whenever $x = ct$, so that as t becomes greater, the position of a wavefront moves in the positive direction of x. Since for F_2, $-x = ct$, the opposite is true.

If only the outward traveling wave is considered, Eq. (13) holds. Such a wave is transmitted without

$$E_y = F_1(x - ct) \qquad (13)$$

change of shape, because its position x can always equal ct. That is, at any given elapsed time t, E_y will have the same value at $x = ct$ as it had at $x = t = 0$. Also, because $c = x/t$, c has the dimensions of a velocity and is the speed at which the wave travels.

Light is also an electromagnetic process so that one should be able to state the optical properties of a substance once its electrical constants are given. However, at the high frequencies of visible light, there is generally dispersion in the medium so that κ and μ vary with frequency, and this variation must be taken into account if Maxwell's equations are to give a description of optical phenomena which fits measured data.

Acoustic waves in gases. In gases, the existence of inertia (resulting in kinetic energy) and elasticity (resulting in potential energy) is obvious. For example, imagine taking from the medium a small packet of gas enclosed by a weightless deformable membrane. When this packet is squeezed, it is found to have stiffness (or elasticity). Its mass equals the mass of the air molecules contained therein. Also, the deformable membrane keeps the mass inside constant.

From this description, it can be seen that it is a relatively simple matter to describe the motion of such a packet as a function of time. The mass element is labeled in any convenient manner, the most common being by its location at any convenient time. This method is called the Lagrangian description of motion.

To derive the equations for wave motion in a gas, express the concepts of inertia by Newton's second law, of elasticity by the perfect gas law, and of

the deformable packet by the conservation of mass law. Assume that the packet is rectangular in shape. To find the equation of motion, suppose that the box is situated in a medium in which the pressure p changes in space at a space rate given by Eq. (14), where **i**, **j**, and **k** are unit vectors in the

$$\text{grad } p \equiv \mathbf{i}\frac{\partial p}{\partial x} + \mathbf{j}\frac{\partial p}{\partial y} + \mathbf{k}\frac{\partial p}{\partial z} \qquad (14)$$

x, y, and z directions, respectively, and p is the pressure at a point.

The net force **f** acting to move the box in some direction is equal to the vector summation of the gradients in force across the three pairs of faces of the packet times the respective separations of these faces; in the positive direction, Eq. (15)

$$\mathbf{f} = -\left[\mathbf{i}\left(\frac{\partial p}{\partial x}\Delta x\right)\Delta y\,\Delta z + \mathbf{j}\left(\frac{\partial p}{\partial y}\Delta y\right)\Delta x\,\Delta z \right.$$
$$\left. + \mathbf{k}\left(\frac{\partial p}{\partial z}\Delta z\right)\Delta x\,\Delta y \right] = -V\,\text{grad } p \qquad (15)$$

holds. Here V is the average volume of the packet. The positive gradient causes an acceleration of the box in the negative direction of x.

By Newton's second law, the force acting to accelerate the packet is given by Eq. (16), where **q**

$$\mathbf{f} = \rho'V\frac{\partial \mathbf{q}}{\partial t} \qquad (16)$$

is the average vector velocity of the gas in the packet, ρ' the average density of the gas in the packet, and $\rho'V$ the total (constant) mass of the gas in the packet. In writing Eq. (16), it has been assumed that the packet is never displaced from its equilibrium position by a significant part of a wavelength of sound.

Combining Eqs. (15) and (16) yields Eq. (17). In keeping with the approximation just stated, it has

$$-\text{grad } p = \rho_0\frac{\partial \mathbf{q}}{\partial t} \qquad (17)$$

been assumed that ρ' (the instantaneous density) does not appreciably deviate from the average density ρ_0. These approximations are acceptable, providing the sound pressures are below about 100 dynes/cm² (10 newtons/m²).

Assuming adiabatic expansions and contractions of the gas, and that the gas is perfect, Eq. (18)

$$\frac{dP}{P} = -\frac{\gamma\,dV}{V} \qquad (18)$$

is obtained. Here P is the total pressure in the gas and γ is the ratio of specific heat at constant pressure to specific heat at constant volume.

Let P and V be defined by Eqs. (19), where p and

$$P = P_0 + p$$
$$V = V_0 + \tau \qquad (19)$$

τ are time-varying quantities and P_0 and V_0 are equilibrium values. If the inequalities of Eqs. (20) hold, then Eq. (21) is valid.

$$p \ll P_0$$
$$\tau \ll V_0 \qquad (20)$$

$$\frac{1}{P_0}\frac{\partial p}{\partial t}=\frac{-\gamma}{V_0}\frac{\partial \tau}{\partial t} \qquad (21)$$

To satisfy the law of mass conservation, one writes Eq. (22) where ξ is the average vector displacement of the box.

$$\tau = V_0 \operatorname{div} \xi \qquad (22)$$

Differentiation of Eq. (22) with respect to time, and substitution of it in Eq. (21), yields Eq. (23).

$$\frac{\partial p}{\partial t}=-\gamma P_0 \operatorname{div} \mathbf{q} \qquad (23)$$

The elimination of p from Eqs. (17) and (23) yields the wave equation Eq. (24), where, by definition, Eq. (25) applies.

$$\frac{\partial^2 p}{\partial t^2}=c^2 \nabla^2 p \qquad (24)$$

$$c^2 = \gamma P_0/p_0 \qquad (25)$$

One-dimensional plane waves. An acoustic wave is called plane when a family of parallel planes can be taken in the medium such that the pressures and particle velocities are constant in magnitude and the particle velocities are constant in direction at all points of any given member of the family.

Since p and $\mathbf{q}$ are to be constant in any one wavefront, the partial derivatives with respect to y or z must vanish, and in Eq. (23), Eq. (26) applies.

$$\operatorname{div} \mathbf{q} = \frac{\partial q_x}{\partial x} \qquad (26)$$

Since Eq. (26) reveals that the x component of $\mathbf{q}$ is the only component of $\mathbf{q}$ remaining in the equations basic to the wave equation, the wave is longitudinal.

The one-dimensional wave equation becomes Eq. (27).

$$\frac{\partial^2 p}{\partial t^2}=c^2 \frac{\partial^2 p}{\partial x^2} \qquad (27)$$

The general solution to Eq. (27) is exactly the same as that for electromagnetic waves and is given by Eq. (12). The discussion following Eq. (12) is also valid here.

One-dimensional spherical waves. In free space, it is frequently desired to express mathematically the radiation of sound from a nondirectional source. In this case the sound wave expands as it travels away from the source and the wavefront is always a spherical surface. The operator on the right side of Eq. (24) can be written in a form suitable to spherical coordinates.

Assuming equal radiation in all directions, the wave equation in one-dimensional spherical coordinates then becomes Eq. (28).

$$\frac{\partial^2 p}{\partial r^2}+\frac{2}{r}\frac{\partial p}{\partial r}=\frac{1}{c^2}\frac{\partial^2 p}{\partial t^2} \qquad (28)$$

Differentiation shows that Eq. (28) can also be written as Eq. (29).

$$\frac{\partial^2 (pr)}{\partial t^2}=c^2 \frac{\partial^2 (pr)}{\partial r^2} \qquad (29)$$

Equation (29) has the same form as that of Eq. (27). Hence, the same formal solution applies to either equation, except that the dependent variable is $p(x,t)$ in one case and $pr(r,t)$ in the other case.

The solution to Eq. (29) for the outward traveling wave only (free space) is thus given by Eq. (30).

$$p=\frac{1}{r}F_1(r-ct) \qquad (30)$$

Note that, just as for the plane wave, the wave is propagated without change of shape. However, the magnitude of the sound pressure decreases inversely with distance owing to the spreading of the wave as it propagates.

Acoustic waves in liquids. Acoustic waves in liquids obey the same equations as those in gases. The velocity of propagation is greater in liquids than in gases, and viscous losses are often higher. For information on an important example of wave motion in liquids *see* UNDERWATER SOUND.

Elastic and flexural waves in solids. Several different types of acoustic waves may exist in solids, depending on the different manners in which potential energy is stored in the solid. The wave equations associated with several of these types are reviewed in the following paragraphs.

Waves on flexible stretched strings. The wave equation for a flexible stretched string is given by Eq. (31), where ξ_y is the displacement of the string

$$\frac{\partial^2 \xi_y}{\partial t^2}=c^2 \frac{\partial^2 \xi_y}{\partial x^2} \qquad (31)$$

at a point x along the string in the y direction (perpendicular to the string). The speed of propagation c is equal to the square root of the ratio of the tension (in dynes) to the linear density of the string (in g/cm²).

The solution to this equation is identical to that for Eq. (11). Because the motion of the elements of the string is perpendicular to the string, at least for small displacements, the waves are said to be transverse.

Flexural waves in bars. For the purposes of this article, it is assumed that a string has tension with negligible stiffness, while a bar has stiffness without tension. When a bar is bent, its lower half is compressed and its upper half is stretched, or vice versa. When the bending force is removed, the bar attempts to regain its equilibrium position. The restoring force is due to the moment of the forces about the neutral plane in the bar and is related to the cross-sectional dimensions and the Young's modulus of the material.

The transverse wave equation describing the motion of such a bar is given by Eq. (32), where ξ_y

$$\frac{\partial^2 \xi_y}{\partial t^2}=-\kappa^2 \frac{Y}{\rho}\frac{\partial^4 \xi_y}{\partial x^4} \qquad (32)$$

is the displacement perpendicular to the neutral plane of the bar, Y is Young's modulus, p is the density of the bar, and κ is the radius of gyration of the cross section. Values of κ for some of the simpler cross-sectional shapes are given by Eqs. (33).

Rectangle: Length b parallel to center line, width a perpendicular to center line.

$$\kappa = a/\sqrt{12} \qquad (33a)$$

Circle: Radius a, $\kappa = a/2$ $\qquad (33b)$

Circular ring: Outer radius a, inner radius b,

$$\kappa = 0.5(a^2 + b^2)^{1/2} \qquad (33c)$$

Equation (32) differs from the usual wave equation, Eq. (11), in that it has a fourth derivative with respect to x instead of a second derivative.

The function $F_1(x-ct)$ is not a solution, so that a bar satisfying Eq. (32) cannot have waves traveling along it with a velocity independent of frequency and with an unchanged shape.

If one considers excitation of the bar by one frequency f at a time, a solution to Eq. (32) may be written as Eq. (34), where $\mu = (f^2\rho/4\pi^2 Y\kappa^2)^{1/4}$, $f = 2\pi\mu^2\kappa\sqrt{Y/\rho}$, $A = $ amplitude, and $\phi = $ phase angle.

$$\xi_y = A\cos[2\pi(\mu x - ft) + \phi] \qquad (34)$$

The velocity of propagation of the wave is equal to $(f/\mu) = (4\pi^2 Y\kappa^2/\rho)^{1/4}\sqrt{f}$. It obviously depends on the frequency f of the exciting wave. Such a velocity for a simple harmonic wave is called the phase velocity. For a complex wave with several components differing in frequency, the shape must change with distance of travel. A bar is sometimes said to be a dispersing medium for waves of bending. *See* PHASE VELOCITY.

There is also a possibility for a longitudinal wave in such a bar. In this case the bar is excited at an end in the direction of its longitudinal axis. The wave equation is identical to Eq. (31), except that ξ_x replaces ξ_y. The motions of the particles of the bar are in a line with the longitudinal axis of the bar. Longitudinal and transverse waves may, and generally do, coexist.

Waves in plates. Wave motion in a plate is similar to that in a bar. The bending of a plate compresses the material on the inside of the bend and stretches it on the outside. But when a material is compressed, it tries to spread out in a direction perpendicular to the compressional force, so that when a plate is bent downward in one direction there is a tendency for it to curl up in a direction at right angles to the bend. The ratio of the sideways spreading to the compression is called Poisson's ratio and is designated by the letter σ.

The wave equation for the plate is given by Eq. (35), where η is the displacement of the plate perpendicular to its surface, h is the thickness of the plate, and c_L is the longitudinal plate velocity given by Eq. (36).

$$\Delta^4\eta + \frac{12}{(hc_L)^2}\frac{\partial^2\eta}{\partial t^2} = 0 \qquad (35)$$

pendicular to its surface, h is the thickness of the plate, and c_L is the longitudinal plate velocity given by Eq. (36).

$$c_L = \sqrt{\frac{Y}{\rho(1-\sigma^2)}} \qquad (36)$$

From a solution to the equation it is possible to find the velocity of propagation for the transverse (bending) wave, as in Eq. (37). Just as for the bar,

$$c_B = \sqrt{1.8hc_L}\sqrt{f} \qquad (37)$$

the velocity of propagation is dependent on frequency, and the plate is said to be a dispersing medium.

Steady-state wave motion. Very frequently the source of a wave is steady and the wave produced by it is periodic, at least for a long period of time compared to the buildup and decay times of the wave. When this is true, it is convenient to specify the functions $F_1(x-ct)$ and $F_2(x+ct)$ of Eq. (12) by a summation of sinusoidal functions, as in Eq. (38),

$$F(x-ct) = \sum_\nu A_\nu \cos\left[\omega_\nu\left(t - \frac{x}{c}\right) + \theta_\nu\right] \qquad (38)$$

where $\omega_\nu = 2\pi f_\nu$; f_ν is the frequency of the νth component of the wave, and θ_ν is the phase angle of the νth component, which is determined when $x = t = 0$.

From the well-known theory of Fourier series, any function $F_1(x-ct)$, if it repeats on itself periodically, can be represented by a linear summation of sine or cosine wave functions. The component with the lowest frequency f_1 is the fundamental component or the first harmonic; each higher component, called the $(\nu - 1)$th overtone, or the νth harmonic, has a frequency equal to νf_1. In other words, ν is an integer and the waveform $F_1(x-ct)$ is represented by a linear summation of cosine terms with frequency components f_ν that are harmonically related (f, $2f$, $3f$, $4f$, and so forth) and at $t = x = 0$ have phase angles θ_1, θ_2, θ_3, and so forth, that may differ from zero. *See* FOURIER SERIES AND INTEGRALS.

As a further simplification, each component of Eq. (38) is usually written as Eq. (39), where $k = $

$$P = A\cos(\omega t - kx + \theta) \qquad (39)$$

ω/c is called the wave number. The period of this component is T and equals $1/f$. As before, θ is the phase angle in radians.

Note that A is the peak amplitude of the component wave. Generally, in acoustics, measuring devices read the root-mean-square value of a wave, so that the intensity of the wave is designated by $A_{rms} = A/\sqrt{2}$, and the strength of a wave represented by Eq. (38) is given by Eq. (40).

$$A_{rms} = \sqrt{A^2_{1\,rms} + A^2_{2\,rms} + \cdots} \qquad (40)$$

[LEO L. BERANEK]

Bibliography: L. L. Beranek, *Acoustics*, 1954; F. R. Connor, *Wave Transmission*, 1971; I. G. Main, *Vibrations and Waves in Physics*, 1978; P. M. Morse and K. U. Ingard, *Theoretical Acoustics*, 1968; Lord Rayleigh, *The Theory of Sound*, 2d ed., reprint 1945; A. Shadowitz, *The Electromagnetic Field*, 1974; H. D. Young, *Fundamentals of Waves, Optics and Modern Physics*, 2d ed., 1975.

Wave motion in fluids

Wave motion is the basic mechanism by which local disturbances are propagated from one part of a fluid to another. As characteristic of all wave motions, this mechanism allows local disturbances to be passed from one part of a fluid onto the next without net mass motion. The direction along which the local disturbances are transmitted is called the direction of propagation, and the speed of the disturbances relative to the fluid is called the wave speed, or the speed of propagation.

Applications. Wave phenomena have widespread applications. Marine equipment such as the fathometer and sofar rely upon wave propagation. Shock waves used in sofar propagate long distances underwater, being refracted by the isothermal layers in the oceans. The study of waves is directly applicable to supersonic aircraft, wind tunnels, shock tubes, rocket combustion oscillation, nuclear-bomb blasts, controlled fusion processes in plasma, and ultrasonic processes such as cleaning and inspection.

Waves in fluids are diffracted and refracted so that they can be focused to produce intense concentration of energy. For example, in shock tubes, energy is focused into a sharp pulse, the fluid there being caused to glow by the intense excitation so produced. Ultrasonic waves are focused and directed through tanks of mercury to provide short-term memory for large-scale computers.

Fluid waves caused by successive firing of charges from a vertical sounding rocket, such as the Aerobee as it climbs above the altitude in which ballons are effective, are recorded on the ground to provide data for the determination of wind velocity and air density profiles in the upper atmosphere. The technique is analogous to seismic exploration for petroleum. Waves in the upper atmosphere have been found to influence weather, and their Fourier analysis has greatly advanced the techniques of long-range weather forecasting. Shock waves are used to prepare free radicals and in forcing certain chemical processes.

Wave classification. In contrast to surface waves, which are transverse oscillations caused by gravity acting on a liquid having a free surface, wave motion within a fluid is generated by successive compression and expansion of adjacent volume elements of a compressible fluid. Because compression and expansion of an ordinary fluid can only proceed along the direction of propagation of the disturbance, waves within a fluid are mostly longitudinal waves.

Waves in a fluid can be classified as compression waves and expansion waves, according to whether the disturbance is a compression or an expansion. They can further be classified according to the amplitude of the disturbance and the chemical nature of the fluid. For example, waves of small amplitude are called acoustic (or sound) waves; compression waves propagating in chemically inert fluids are called shock waves; waves propagating in the Earth are seismic waves; and waves of large amplitude generated by rapid chemical reactions in explosive fluids are called detonation waves and can propagate much faster than sound waves. Waves in an electrically conducting fluid in the presence of strong magnetic fields are called magnetohydrodynamic waves. *See* MAGNETOHYDRODYNAMICS.

Acoustic waves. The acoustical wave equation can be derived formally through linearization of the equations of motion. A more straightforward approach is to consider a plane wavefront that moves from right to left at a constant speed u_1 in a fluid which is initially at rest and of density ρ_1. If an observer fixes his attention on the wavefront by also moving from right to left at the same speed u_1, he will witness a steady flow of fluid from left to right across the wavefront (Fig. 1). Represent the flow velocity and the fluid density to the right of the wavefront by u_2 and ρ_2, respectively; then conservation of mass requires the validity of Eq. (1)

$$\rho_1 u_1 = \rho_2 u_2 \tag{1}$$

because no fluid mass can accumulate at the wavefront in a steady state. Furthermore, an increase in fluid momentum across the wavefront can be supported only by a corresponding drop in pressure from p_1 to p_2 in the fluid. Therefore Eq. (2) holds. If

$$\rho_2 u_2{}^2 - \rho_1 u_1{}^2 = p_1 - p_2 \tag{2}$$

the disturbance is so weak that the fractional changes in flow velocity, fluid density, and pressure across the wavefront are much smaller than unity, the changes can be written as Eqs. (3). By

$$\begin{aligned} u_2 &= u_1 + du \\ \rho_2 &= \rho_1 + d\rho \\ p_2 &= p_1 + dp \end{aligned} \tag{3}$$

substituting Eqs. (3) and neglecting the product terms of the differential quantities, Eqs. (1) and (2) become Eqs. (4) and (5). An expression

$$\rho_1 du + u_1 d\rho = 0 \tag{4}$$

$$2\rho_1 u_1 du + u_1{}^2 d\rho = -dp \tag{5}$$

for u_1 is obtained in Eq. (6) by eliminating du from

$$u_1{}^2 = dp/d\rho \tag{6}$$

these two equations. Because this derivation began with a wavefront moving in a fluid initially at rest, the above result shows that any small disturbance, if propagated by wave motion at all, must propagate in relation to the fluid at the speed of sound a, defined by Eq. (7). If the disturbance is periodic in

$$a = u_1 = \sqrt{dp/d\rho} \tag{7}$$

time with a fundamental frequency ν, as in a musical note, then there will be a wavelength λ associated with the wave motion $\lambda = a/\nu$. Furthermore, Eq. (7) shows that a depends only on the variation of pressure with density as caused by a small mechanical disturbance in the fluid, and that a is real as long as $dp/d\rho$ is positive. Historically, Isaac Newton first attempted to derive the speed of sound in air from a somewhat different approach. He arrived at a result which was equivalent to assuming an isothermal compression process. Thus, from the equation of state for a perfect gas (Boyle's law), Eq. (8), he obtained Eq. (9). However, experi-

$$p = \rho RT \tag{8}$$

$$a = \sqrt{RT} = \sqrt{p/\rho} \tag{9}$$

mental measurements of the speed of sound in air turned out to be consistently higher than his prediction. This led P. S. Laplace to suspect that compression and expansion processes associated with acoustic waves should obey the adiabatic law $p\rho^{-\gamma} = \text{constant}$ ($\gamma = C_p/C_v$, being the ratio of specific heats), instead. If this pressure-density relationship is assumed, the speed of sound is then given by Eq. (10). The above result has been found

$$a = \sqrt{\gamma RT} = \sqrt{\gamma p/\rho} \tag{10}$$

to agree so well with experimental observations under ordinary conditions that measurement of the speed of sound has become a standard method for determining the value of γ for various gases.

For liquids, and for gases at extreme temperatures and densities, or for disturbances of very high frequencies, the adiabatic law loses its usual significance, so that the speed of sound (or the compressibility $d\rho/dp$) has to be obtained from direct measurement or from more exact theories.

The speed of sound in several common fluids at room temperature is given in the table together with the speed of sound in some common solids for comparison.

At this point, it may be of value to give some numerical examples of sound amplitude. The

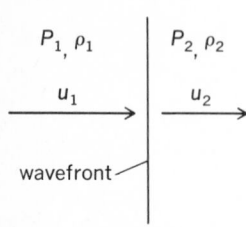

WAVE MOTION IN FLUIDS

P_1, ρ_1　　P_2, ρ_2

u_1　　u_2

wavefront

Fig. 1. Flow across a plane wavefront.

Speed of sound in common substances

Substance	Speed of sound, ft/sec*
Air	1,130
Hydrogen	4,320
Water	4,800
Mercury	4,600
Paraffin	4,300
Lead	4,030
Aluminum	16,700
Iron	16,800

*1 ft/sec = 0.3 m/sec.

smallest periodic pressure amplitude detectable by the human ear is in the order of 10^{-3} dyne/cm², or 10^{-9} atm (at about 2000 Hz). On the other hand, the threshold of feeling corresponds to a pressure amplitude roughly 10^6 times greater. Thus, these limits of what is ordinarily called sound correspond to pressure fluctuations of 10^{-9} to 10^{-3} atm, a fact which justifies the assumption of small disturbance in the foregoing derivation of the speed of sound.

Zone of action and zone of silence. Small disturbances can propagate in a fluid only at a finite speed. The same is true for disturbances of large amplitude. Therefore, when an object moves through a stationary fluid at a speed in excess of the speed at which disturbances can be propagated, there is a boundary that divides the fluid into two distinct regions: the zone of action that has, and the zone of silence that has not yet, been affected by the motion of the object at any given instant. To understand this behavior, consider Fig. 2. *See* SHOCK WAVE.

Let u denote the speed of the object and a denote the speed of propagation of the fluid disturbance. To simplify the discussion, assume that the object is in uniform rectilinear motion and that the amplitude of the disturbance is so weak that a can be identified with the speed of sound in the stationary fluid. At any instant t, when the object is at a certain point P, the object will generate disturbances that will propagate away from P in all directions with velocity a. After time interval Δt, the object would have traveled a distance $u\Delta t$ and would have moved to new position P' along its

trajectory, while the disturbances generated at point P would still be confined to a spherical surface of radius $a\Delta t$. Because u is greater than a, all disturbances generated by the object up to the time $t + \Delta t$ will be confined within the conical surface of the half-vertex angle given by Eq. (11), with

$$\beta = \sin^{-1}(a/u) \qquad (11)$$

vertex at point P' and axis along PP'. The region inside such a conical surface is called the zone of action at the time $t + \Delta t$; correspondingly, the region outside the conical surface, which is still out of reach of the disturbances, is called the zone of silence at the same instant.

The ratio between the speed of the object and the speed of sound u/a is known as the Mach number of the moving object, and is usually denoted by the symbol M. The motion of the object is accordingly called subsonic when $M < 1$, transonic when $M \cong 1$, supersonic when $M > 1$, and hypersonic when $M \gg 1$. From Eq. (11) and from Fig. 2, which has been depicted for supersonic motion, it can be deduced that the zone of action extends to all parts of the fluid for subsonic motion. For transonic motion, the zone of action covers approximately the rear half of the fluid, up to the plane tangent to the nose of the moving object; for hypersonic motion, the zone of action is confined to a relatively slender cone about the trajectory of the object.

Waves of larger amplitude. In physical optics, it is well known that wavelets from a distributed light source in space can be superposed according to Huygens' principle. For wave motion in fluids, however, the analogy holds only as long as the amplitude of the resultant disturbance remains small enough for the acoustic approximation to apply. For waves of larger amplitude, the nonlinear behavior of the fluid dynamic equations must be taken into account. To visualize these nonlinear effects, consider the following one-dimensional problem.

A semi-infinite tube is filled with a gas initially at rest. At one end of this tube is a movable piston. The piston is suddenly given a small velocity toward the gas. As a result, a compression wave is generated, and propagates along the tube at the speed of sound of the undisturbed gas. Suppose that, soon after this wave has started, the piston is given an additional increment in velocity. A second wave is formed and propagates along the tube behind the first. Because the pressure has increased slightly across the first wave, so has the temperature and the speed of sound. Furthermore, the gas behind the first wave is already moving along the tube at the piston velocity after the first impulse. Thus, the second wave, which propagates with respect to a moving gas ahead of itself and at a sound speed slightly that is higher than that of the first wave, will soon catch up with the first wave.

If the accelerating motion of the piston continues, the succeeding wavelets will propagate along the tube at increasingly higher velocity, so that the shape of the resultant compression pulse from the piston motion will appear steeper as it progresses. Similarly, a decelerating piston motion produces an expansion pulse that flattens out as it progresses along the tube. It is this asymmetry between the two processes that makes the occurrence of large-

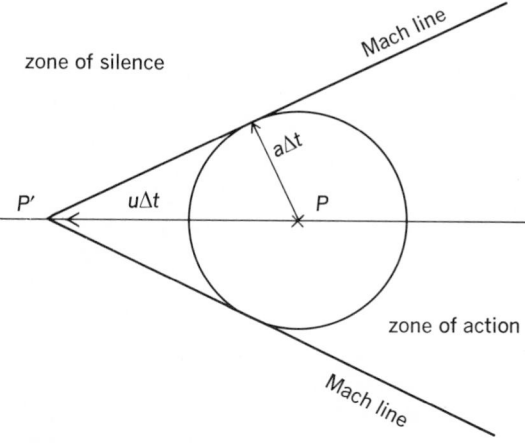

Fig. 2. A point source of disturbance in uniform rectilinear motion, depicted for supersonic motion.

amplitude compression waves, called shock waves, a more noticeable phenomenon in nature than are expansion waves.

Shock waves are characterized by rapid changes in fluid density, pressure, and temperature along the direction of flow. Bomb blasts start as shock waves.

Seismic waves. Seismic waves pass through the ground; they arise either from natural readjustment of faults in the Earth's crust or by man-made explosions. According to their modes of propagation, seismic waves may be divided into two main groups, namely, body and surface waves.

Body waves, which propagate through the inside of the Earth, may further be subdivided into dilation (longitudinal) waves, which are similar to acoustic waves in compressible fluids, and shear (transverse) waves, which arise on account of the large shear resistance of most elastic solids. For any given medium, dilation waves usually have approximately twice the velocity of propagation of shear waves.

Surface waves from any distant earthquake always arrive after both the dilation waves and the shear waves, and they normally register a much larger amplitude signal on the seismogram. From the known relationship between the propagation velocities and the mechanical properties of various substances, seismologists extract valuable information about the structure of the Earth's interior from the seismograms and apply the results to such purposes as mine prospecting. *See* SONIC BARRIER.   [SHAO-CHI LIN]

Bibliography: R. Courant and K. O. Friedrichs, *Supersonic Flow and Shock Waves*, 1948, reprint 1977; J. Lighthill, *Waves in Fluids*, 1977; Lord Rayleigh, *The Theory of Sound*, 2d ed., reprint, 1945; F. W. Sears, et al., *University Physics*, 5th ed., 1976; G. B. Whitham, *Linear and Nonlinear Waves*, 1974.

Wave motion in liquids

A temporal variation in fluid velocity which is propagated through a fluid medium. The speed of propagation of the disturbance, or change in fluid velocity relative to the initial velocity of the fluid medium, is known as the wave celerity. The dynamic behavior of the wave depends upon the method of generation of the disturbance, the boundary conditions of the fluid medium, and the fluid properties. This article treats those aspects of wave motion appropriate to liquids in which disturbances are propagated at a gas-liquid interface and are primarily dependent upon the gravitational fluid property (surface tension and viscosity being of secondary importance). Wave motions which occur in confined fluids (either liquid or gaseous) are primarily dependent upon the elastic property of the medium. *See* WAVE MOTION IN FLUIDS.

The fundamental concepts of gravity waves in liquids are presented in the one-dimensional form.

Oscillatory waves. The term oscillatory implies a periodicity in the form of a disturbance moving past a fixed point. Figure 1 is a definition sketch for an oscillatory wave propagating in a liquid of constant density ρ and depth h measured from the bottom to the still-water level (SWL). Wavelength L is the horizontal distance between successive crests of the wave. Wave height H is the vertical

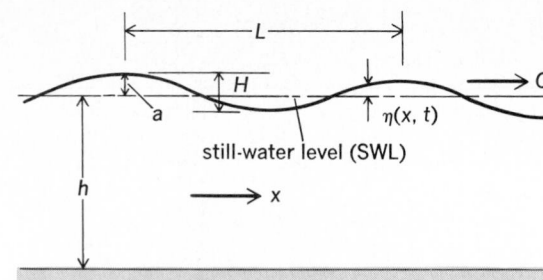

Fig. 1. Definition sketch for an oscillatory wave.

distance from crest to trough; amplitude a is the distance from the still-water level to the crest, and η is the elevation of the free surface with respect to the still-water level at any position x and instant of time t. In the linearized theory of small-amplitude waves ($H/L < 0.03$) the wave profile is sinusoidal and is given by Eq. (1), where T is the wave

$$\eta = a \sin 2\pi \left(\frac{t}{T} - \frac{x}{L} \right) \tag{1}$$

period. By definition, the celerity or speed of propagation $C = L/T$ and is given by Eq. (2). The first

$$C = \sqrt{\left(\frac{\sigma}{\rho} \frac{2\pi}{L} + \frac{gL}{2\pi} \right) \tanh 2\pi \frac{h}{L}} \tag{2}$$

term on the right expresses the influence of surface tension σ and need be considered only for waves of very small length (of the order of magnitude of 1 in.). In the remaining development, only gravity waves will be considered, as expressed by the second term of the celerity equation.

Oscillatory waves may be generated in a rectangular channel by a simple harmonic translation of a vertical wall forming one end of the flume. The wave amplitude will be determined by the displacement (stroke) of the wall, and the wavelength will be a function of the period of oscillation. The two major classes of oscillatory waves, deep-water and shallow-water waves, are determined by the magnitude of the ratio of liquid depth to wavelength h/L. An inspection of the celerity equation for gravity waves shows that as the depth becomes large in comparison with wavelength, Eqs. (3) hold.

$$\tanh 2\pi \frac{h}{L} \to 1$$
$$C = \sqrt{\frac{gL}{2\pi}} \tag{3}$$

Hence, in a deep-water wave the celerity is a function only of the wavelength. This approximation is close if the depth is greater than one-half the wavelength. On the other hand, as the depth becomes small in comparison with wavelength, Eqs. (4) hold. Hence, the celerity depends only on the

$$\tanh 2\pi \frac{h}{L} \to 2\pi \frac{h}{L}$$
$$C = \sqrt{gh} \tag{4}$$

depth in a shallow-water wave. Somewhat arbitrarily, the limit $h/L = 1/10$ is generally applied to this type of wave motion. In deep-water waves, individual fluid particles tend to move in circular orbits. The radius of the surface particle orbit is equal to

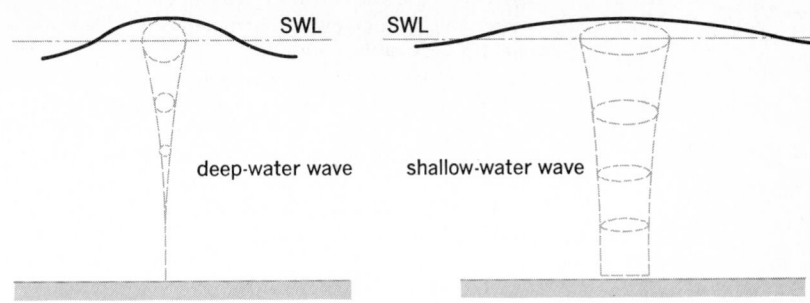

Fig. 2. Fluid particle orbital motions in deep- and shallow-water oscillatory waves:

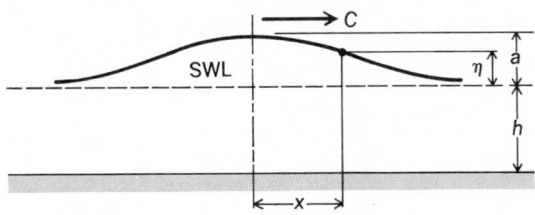

Fig. 3. Definition sketch for a solitary wave.

the wave amplitude and the radius decreases exponentially with depth (Fig. 2). At a depth of one-half the wavelength, the orbital radius is about 1/20 of the amplitude. A zone of essentially zero fluid motion is rapidly approached and the character of the wave is therefore not affected by the total depth of the liquid.

In shallow water, no vertical particle motion can exist at the bottom; thus the wave characteristics are modified. The particle orbits are flat ellipses in which the minor axis is depressed to zero at the bottom (Fig. 2).

The energy of a wave consists of equal amounts of potential energy (due to particle position above or below the still water level) and kinetic energy (due to the motion of particles in their orbits). The rate of propagation of energy in the direction of wave travel is known as the group velocity to distinguish it from phase velocity C. In deep-water waves the group velocity is one-half the phase velocity; in shallow-water waves the two propagation velocities are equal.

Standing waves. A standing wave can be considered to be composed of two equal oscillatory wave trains traveling in opposite directions. The phase velocity of the resulting wave is zero; nevertheless the velocity of propagation of the component waves retains its usual meaning. In the notation of the previous section, the equation for the profile of a standing wave is obtained by adding the elevations of waves moving in the positive and negative x directions, given by Eqs. (5) and (6),

$$\overrightarrow{\eta}_1 = a \sin 2\pi \left(\frac{t}{T} - \frac{x}{L}\right) \qquad (5)$$

$$\overleftarrow{\eta}_2 = a \sin 2\pi \left(\frac{t}{T} + \frac{x}{L}\right) \qquad (6)$$

respectively. Hence Eq. (7) holds. If the length of

$$\eta = \eta_1 + \eta_2 = H \sin 2\pi \frac{t}{T} \cos 2\pi \frac{x}{L} \qquad (7)$$

the basin l in which a disturbance occurs is an integral number n of half wavelengths, a self-perpetuating (except for frictional dissipation) standing wave will result. Therefore, if $l = nL/2$, Eq. (8) is

$$\eta = H \sin 2\pi \frac{t}{T} \cos \frac{\pi n x}{l} \qquad (8)$$

valid. For long waves, as with shallow water, in a basin of uniform depth, the period of oscillation T is defined by Eq. (9). Standing waves frequently

$$T = 2l/n \sqrt{gh} \qquad (9)$$

occur in canal locks as a result of filling disturbances and in large lakes, bays, and estuaries as a result of wind or tidal action.

Solitary waves. A solitary wave consists of a single crest above the original liquid surface which is neither preceded nor followed by another elevation or depression of the surface. Such a wave is generated by the translation of a vertical wall starting from an initial position at rest and coming to rest again some distance downstream. In practice, solitary waves are generated by a motion of barges in narrow waterways or by a sudden change in the rate of inflow into a river; they are therefore related to a form of flood wave. The amplitude of the wave is not necessarily small compared to the depth, and the wavelength is theoretically infinite

because the elevation of the surface approaches the still water level asymptotically with distance as shown in Fig. 3. The profile of the solitary wave is given by Eq. (10) and the celerity by Eq. (11). When

$$\eta = a \operatorname{sech}^2 \left[\frac{x}{h} \sqrt{\frac{3}{4} \frac{a}{h}}\right] \qquad (10)$$

$$C = \sqrt{g(h+a)} \qquad (11)$$

the solitary wave amplitude becomes approximately equal to the depth, the wave profile becomes unstable and a breaking wave results.

Surges. A surge is generated by the forward motion of a vertical wall, at a constant speed, as shown schematically in Fig. 4. Surges in open channels are analogous to shock waves produced in a tube by the continuous motion of a piston. A zone of violent eddy motion occurs at the wavefront and the analysis of such motions must take into account the appreciable energy dissipation in this region. The velocity of propagation of a surge is given by Eq. (12). If a velocity V_1 equal and op-

$$C = \sqrt{gh_1} \left[\frac{1}{2} \frac{h_2}{h_1} \left(\frac{h_2}{h_1} + 1\right)\right]^{1/2} \qquad (12)$$

posite to C is imposed on the fluid upstream of the disturbance, the absolute velocity of the surge front will become zero. In this form the surge is known as a hydraulic jump and it is frequently

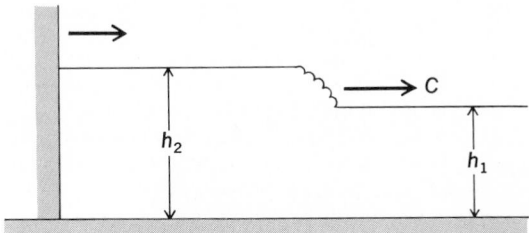

Fig. 4. Definition sketch for a surge wave.

employed as a means of dissipating flow energy at the bottom of dam spillways.

[DONALD R. F. HARLEMAN]

Bibliography: J. Roberts, *Internal Gravity Waves in the Ocean*, 1975; R. C. H. Russell and D. H. Macmillan, *Waves and Tides*, 1953, reprint 1971; R. Silvester, *Coastal Engineering One: General Propagation and Influence of Waves*, 1974; J. J. Stoker, *Water Waves*, 1957.

Wave optics

The branch of optics which treats of light (or electromagnetic radiation in general) with explicit recognition of its wave nature. The counterpart to wave optics is ray optics or geometrical optics, which does not assume any wave character but treats the propagation of light as a straight-line phenomenon except for changes of direction induced by reflection or refraction. *See* GEOMETRICAL OPTICS; OPTICS.

Any optical phenomenon which is correctly describable in terms of geometrical optics can also be correctly described in terms of wave optics. However, the many phenomena of interference, diffraction, and polarization are incontrovertible evidence of the wave nature of light, and geometrical optics often gives an incomplete or incorrect description of the behavior of light in an optical system. This is especially true if changes of refractive index occur within a space which is of the order of several wavelengths of the light. *See* DIFFRACTION; INTERFERENCE OF WAVES; POLARIZED LIGHT.

[RICHARD C. LORD]

Wave packet

In wave phenomena, a superposition of waves of differing lengths, so phased that the resultant amplitude is negligibly small except in a limited portion of space whose dimensions are the dimensions of the packet. If the reciprocal wavelengths $k = \lambda^{-1}$ forming a one-dimensional packet lie in a band Δk, the minimum dimension of the packet is $\Delta x \cong (2\pi\Delta k)^{-1}$. When all the component waves move in the same direction, the packet speed is the group velocity $v_g = df/dk$ evaluated at the mean k; f is the frequency. When the phase velocity c depends on λ, $v_g \neq c$, and Δx changes with time. *See* GROUP VELOCITY; NONRELATIVISTIC QUANTUM THEORY; PHASE VELOCITY; QUANTUM MECHANICS.

[EDWARD GERJUOY]

Wavelength

The distance between two points on a wave which have the same value and the same rate of change of the value of a parameter, for example, electric intensity, characterizing the wave. The wavelength, usually designated by the Greek letter λ, is equal to the speed of propagation c of the wave divided by the frequency of vibration f; that is, $\lambda = c/f$ (see illustration). *See* WAVE.

The wavelength for a sound of a given frequency varies greatly, depending upon the speed of propagation in the medium in which the sound is moving. For example, a sound wave having a frequency of 1000 Hz would have a wavelength of approximately 1 ft (0.3 m) in air, $4\frac{1}{2}$ ft (1.3 m) in water, and 17 ft (5.1 m) in steel. The wavelength of electro-

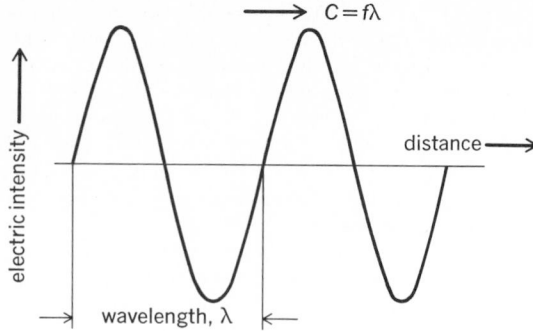

Wavelength λ and related quantities.

magnetic waves depends on the velocity of light in the material in which the waves are traveling. *See* WAVE MOTION.   [WILLIAM J. GALLOWAY]

Waves and instabilities in plasmas

A plasma may be defined as a highly ionized gas composed of a nearly equal number of positive and negative free charges (positive ions and electrons). Because of the mutually coupled nature of electromagnetic fields within a plasma and the motion of the plasma charges themselves, a plasma can support unusual oscillations and wave motions, both stable and unstable. These wave motions are considered in this article. *See* MAGNETOHYDRODYNAMICS; NUCLEAR FUSION.

STABLE WAVE MOTIONS

A rough subdivision of the stable wave motions in a homogeneous plasma is possible in terms of four characteristic frequencies of processes within the plasma, assumed immersed in a magnetic field. In many typical cases these frequencies occur in the following order: (i) interparticle collision frequencies (lowest); (ii) ion cyclotron frequency, ω_{ci}; (iii) electron cyclotron frequency, ω_{ce}; (iv) electron plasma frequency, $\omega_{pe} = \sqrt{n_e e^2 / \epsilon_0 m_e}$ (highest), where n_e is the density of the plasma electrons, and e and m_e are the charge and mass of the electron, respectively (all in SI units).

Alfvén waves. At frequencies below (i) a plasma would behave as an ordinary gas and propagate a simple sound wave. Such waves would be of importance only at high density or at low plasma temperatures, where collision cross sections are large. At frequencies between (i) and (ii) one finds the lowest frequency characteristic wave of the plasma state. This wave is the hydromagnetic or Alfvén wave, dependent on the electrical conductivity of the plasma state. One may recall that an important property of a conducting medium threaded by a magnetic field is that the magnetic flux remains constant within the medium for transverse motions occurring rapidly compared to resistive (collisional) times. In effect, the field lines become loaded with the mass of the plasma, thus slowing the propagation of low-frequency electromagnetic waves along the field, much as attaching closely spaced weights on a string slows the propagation of waves along the string. Here the "weights" are the ions (and electrons) of the plasma; the "strings" are the magnetic field lines.

Ion and electron cyclotron waves. As the wave frequency is increased toward (ii), the ion cyclotron frequency, the Alfvén wave splits into two circularly polarized waves, one rotating in the same sense as the ions and the other in the opposite sense, that is, in the sense of rotation of the electrons. The first of these, called the ion cyclotron wave, propagates at frequencies below the ion cyclotron frequency. This wave becomes highly dispersive, its group velocity (energy propagation velocity) approaching zero as the driving frequency approaches the ion cyclotron frequency. The second of the two waves, called the electron cyclotron wave, is of special interest, since it provides an explanation of the whistler phenomenon observed in ionospheric research. Whistlers are intermittent trains of low-frequency radio waves of natural origin, observed at the Earth's surface as audio tones of descending pitch. These waves, launched by lightning bolts, are guided up into the ionosphere and then back to another point on the Earth's surface far removed from the point of origin, where those magnetic lines of forces leaving the launching position again return to earth. That a descending audio tone (whistle) is detected is explained by the fact that the energy propagation velocity (group velocity) of the wave trains is a function of frequency, with higher frequencies having the greater group velocities and thus arriving earlier.

All of the salient properties of the Alfvén waves, and the ion and electron cyclotron waves just described are deducible from a general wave-propagation equation for these modes. For propagation parallel to the field lines, such as described above, the situation is described through the theoretical expression for the effective dielectric constant of the plasma K at the frequency in question. The wave phase velocity, $v_p = \omega/k$, is related to K through $\sqrt{K} = kc/\omega = c/v_p$. Propagation requires that v_p be a real number, that is, K be positive. The group velocity, that is, the velocity of propagation of energy by the wave, is given by $v_g = d\omega/dk$. The expression giving the dielectric constant for the two waves is shown in Eq. (1). The upper

$$K_{i,e} = 1 - 2\frac{\omega_p{}^2}{(\omega \mp \omega^+)(\omega \pm \omega^-)} \qquad (1)$$

sign choice corresponds to ion cyclotron and the lower to electron cyclotron waves. Here ω^+ and ω^- are the numerical values of the ion and electron cyclotron frequencies, respectively. Also $\omega_p{}^2 = (\omega_{pe}{}^2 + \omega_{pi}{}^2)/2$, the average squared plasma frequency, which is $\approx \omega_{pe}{}^2/2$, since the ion mass M_i is much greater than m_e. As can be seen from Eq. (1), propagation of the ion cyclotron wave ($K_i > 0$) is only possible if $\omega < \omega^+$ (excluding very high frequencies where $\omega > \omega_{pe}$). However, the electron cyclotron branch (the whistler mode) can be seen to propagate for all frequencies for which $\omega < \omega^-$, even when $\omega_{pe} \gg \omega^-$. In this case and when $\omega^- \gg \omega \gg \omega^+$, Eq. (1) reduces to Eq. (2), yielding the group velocity in Eq. (3). Thus $v_g \propto$

$$K_e \approx \frac{\omega_{pe}{}^2}{\omega\omega^-} = \frac{k^2c^2}{\omega^2} \qquad (2)$$

$$v_g = \frac{d\omega}{dk} = 2\frac{(\omega\omega^-)^{1/2}}{\omega_{pe}} \cdot c \qquad (3)$$

$\omega^{1/2}$; that is, the higher frequency components of

such waves travel faster and therefore will arrive first at a distant receiver, explaining the descending tone character of whistlers. *See* GROUP VELOCITY; PHASE VELOCITY.

At very low frequencies, $\omega \ll \omega^+$. When inserted in Eq. (1), this limit shows (as noted earlier) that the cyclotron waves merge to become Alfvén waves, for which the plasma exhibits the so-called hydromagnetic dielectric constant independent of frequency. In this limit Eq. (1) takes the form of Eq. (4), where ρ is the mass density of the plasma.

$$K_e = K_i = K_H = 1 + \frac{\mu_0 \rho c^2}{B^2} \qquad (4)$$

In the limit $\rho c^2 \gg B^2/\mu_0$, that is, $K_H \gg 1$, this expression yields the result for the ordinary Alfvén wave velocity, $v_A = c\sqrt{B^2/\mu_0\rho c^2}$.

At very high frequencies, such that $\omega > \omega_{pe}$, and at densities sufficiently high to ensure that $\omega_{pe} \gg \omega^-$, the wave properties are not appreciably influenced by the magnetic field. Under these circumstances Eq. (1) again predicts propagation, characterized by a dielectric constant shown in expression (5). The wave phase velocity found

$$K \approx [1 - (\omega_{pe}{}^2/\omega^2)] \qquad (5)$$

from the above expression is given by Eq. (6). Note that the phase velocity is greater than the velocity

$$v_p = c[1 - (\omega_{pe}{}^2/\omega^2)]^{-1/2} \qquad (6)$$

of light, as in a microwave waveguide. The indicated dependence of phase velocity on plasma density can be put to practical use to measure the density itself by propagating microwave or laser beams through the plasma and by measuring the resultant phase shifts (relative to vacuum) by interferometric techniques. Since $v_g = d\omega/dk$ and $\sqrt{K} = kc/\omega$, the group velocity of these waves is given by Eq. (7) and is therefore always less than the velocity of

$$v_g = c[1 - \omega^2/\omega_{pe}^2]^{1/2} \qquad (7)$$

light, as required by relativity.

When the waves are not launched parallel to the magnetic field lines or when the magnetic field is not uniform, the situation, except in some special cases, becomes much more complicated than that implied by Eq. (1). For example, if the waves are launched in a direction perpendicular to the field and are polarized with the electric vector parallel to the magnetic field lines, then the dielectric constant is given by Eq. (8); that is, it reduces to the

$$K_\parallel = [1 - (\omega_{pe}{}^2/\omega^2)] \qquad (8)$$

result previously obtained for no magnetic field. The physical reason is that in this case the electrons of the plasma are free to move back and forth along the lines of the field in response to the wave electric fields.

In the case where a perpendicular-launched wave has its electric vector polarized in a direction perpendicular to the field lines, a different situation obtains. Here, below their cyclotron frequency, the electrons can only respond to the wave at their $\mathbf{E} \times \mathbf{B}$ drift velocity and a new situation arises. The dielectric constant now takes the form of Eq. (9), where K_e and K_i are the electron and ion

$$K_\perp = \frac{2K_e K_i}{K_e + K_i} \qquad (9)$$

cyclotron wave dielectric constants previously defined by Eq. (1). In this case the propagation takes on a somewhat complicated "hybrid-wave" character, propagation only being possible for certain discrete bands of frequencies, those where K_e and K_i have signs such that $K_\perp > 0$. An example is the case of propagation at the frequency $\omega = (\omega^+ \omega^-)^{1/2}$, the geometric mean of the ion and electron cyclotron frequencies. $K_\perp$ is here defined by expression (10). Propagation is thus possible in this case provided that the density is low enough to satisfy the inequality which is given in expression (11).

$$K_\perp \approx 1 - (\omega_{pe}/\Omega_e)^4 \, (\omega^+ \omega^-) \qquad (10)$$

$$(\omega_{pe}/\omega^-)^2 < (M_i/m_e)^{1/2} \qquad (11)$$

The density implied by Eq. (11) can be far higher than the one associated with the cutoff point for propagation of the parallel polarized wave, so that this wave can penetrate the plasma when the other would not. *See* DIELECTRIC CONSTANT.

Plasma oscillation wave. The foregoing discusses waves where effects associated with the random thermal motions of the plasma particles are relatively unimportant. These waves, transverse and electromagnetic in nature, are related to ordinary electromagnetic waves. In plasma another type of wave can exist that has no analogy in ordinary media. This wave, the electrostatic plasma oscillation wave, is longitudinal rather than transverse, and the thermal motions of its particles play an essential role. One can understand this wave in terms of a two-component picture of plasma: Plasma is after all a mixture of two compressible gases, the electron gas and the ion gas, coupled to each other through electrostatic forces. Displacement or compression of the electron or ion gas relative to the other clearly can lead to local departure from the average state of charge neutrality which is characteristic of the plasma state. Such a state of charge separation will result in the setting up of electrostatic fields, and thus restoring forces, opposing the change. The displaced charges will thus tend to oscillate about the equilibrium (charge-neutralized) position. The electrons, being much less massive than the ions, will be the most active members of the mutual electron-ion oscillations, so that the ions remain nearly fixed, similar to raisins in a pudding. The random thermal motions of the electrons can strongly influence the propagation of these oscillations, tending to damp the waves by resonant coupling of energy out of the wave into particle kinetic energy. *See* KINETIC THEORY OF MATTER.

The simplest illustration of the longitudinal plasma wave is provided by considering a slab of plasma. Consider first the limiting case where the electrons and ions are so "cold" that their thermal motions can be ignored. If the electron gas is now displaced to the right relative to the ion gas by an infinitesimal amount δx, a surface charge density of magnitude $\sigma = (n_e e)\delta x$ will be set up on each surface, positive on the left surface and negative on the right surface. These charge layers will produce

an electric field within the plasma, of magnitude given by Eq. (12). The restoring force on each dis-

$$E = \sigma/\epsilon_0 = (n_e e/\epsilon_0)\delta x \quad \text{(SI)} \qquad (12)$$

placed electron will thus be defined by Eq. (13).

$$F = Ee = -(n_e e^2/\epsilon_0)\delta x \qquad (13)$$

Such a force will therefore result in an acceleration given by Eq. (14).

$$a = \frac{F}{m} = -\left(\frac{n_e e^2}{\epsilon_0 m_e}\right)\delta x \qquad (14)$$

Acceleration being opposite in direction to and proportional to force, there results a simple harmonic motion of the displaced electrons at the frequency given by Eq. (15), that is, at the ubiquitous

$$\omega = \omega_{pe} = \left(\frac{n_e e^2}{\epsilon_0 m_e}\right)^{1/2} \qquad (15)$$

plasma frequency. I. Langmuir was among the first to identify the fundamental importance of the plasma oscillations, but he did not treat the effect of thermal notions. In the case above, the plasma waves described are not dispersive; that is, their frequency is the same for all wavelengths. In this case ω is not a function of k, so that the group velocity $d\omega/dk$ is identically zero (the waves do not propagate).

The first rigorously correct treatment of the longitudinal plasma oscillations was given by L. Landau. He not only deduced the correct dispersion relation for these waves but showed that there exists a damping mechanism for the waves not dependent on collisions, the so-called Landau damping. This damping occurs in situations where the particles of the plasma are able to increase their average energy at the expense of the plasma wave and thus damp it out, even in cases where the dissipative effects of collisions are unimportant. This same wave-particle coupling mechanism, operating in reverse, is also the source of some important unstable plasma behavior to be discussed later.

A physical picture of Landau damping can be constructed along the following lines: A longitudinal electrostatic wave propagating in a plasma will accelerate or decelerate charged particles of the plasma depending on their instantaneous phase with respect to the wave (Fig. 1). However, only

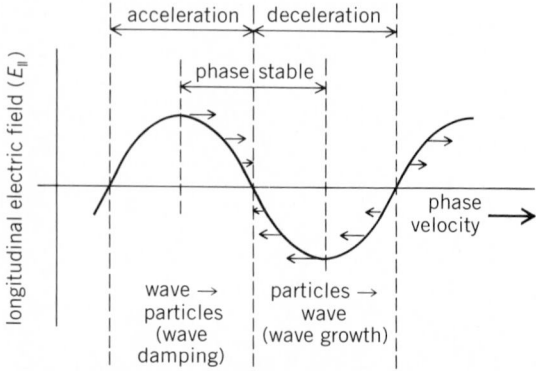

Fig. 1. Schematic of mechanisms of wave-growth and wave-damping processes arising from wave-particle interactions in a plasma.

those particles that have velocity components in the direction of the wave that are comparable to its phase velocity will undergo a significant energy exchange. Particles with velocities v that are not close to the phase velocity will undergo little net energy exchange. This result follows since such particles will average their exchanges over accelerating and decelerating phases of the wave, and thus will undergo little energy exchange with the wave. Considering now those electrons that do have velocity components closely matching that of the wave, it can be seen from Fig. 1 that electrons moving slightly slower than the wave will tend to be speeded up to match the wave phase velocity, while those moving slightly faster will be slowed down, both classes therefore tending to be trapped (phase-bunched) at a stable phase. If, in the vicinity of $v = v_p$ there are more particles with slower velocity than there are with higher velocity (that is, if $\partial f / \partial v|_{v_p} < 0$, where f is the velocity distribution function of the particles), a net energy transfer from the wave to the particles will occur; that is, the wave will be damped.

Computation of the correct dispersion relation for electrostatic plasma waves involves integrals of the wave-particle interactions over the velocity distribution functions of the particles, electrons, and ions. [Dispersion relation refers to the relationship between wave frequency ω and wave number $k = 2\pi / \lambda$, written in the form $D(\omega, k) = 0$ from which the wave propagation and damping can be determined.] The energy transfer between particles and waves is thus calculated, as weighted by the number of particles in each velocity range of the velocity distribution.

Note that whether damping or growth is found by solving the dispersion relation for ω and k will depend on the sign of slope of the distribution function $(\partial f_0 / \partial v_\parallel)$ as evaluated near the resonance phase velocity $v_\parallel = \omega / k$. If $(\partial f_0 / \partial v_\parallel) < 0$, as in a Maxwellian velocity distribution, damping will be found. But if $(\partial f_0 / \partial v_\parallel) > 0$, wave growth may occur, depending on the overall value of the integrals. Wave growth corresponds to instability, since a test wave initially launched into the plasma would grow exponentially as it propagates. Such phenomena will be discussed at greater length in connection with plasma instabilities. *See* DISPERSION RELATIONS.

For the case of a Maxwellian distribution function for the electrons, $f_0 \sim \exp[-(v_\perp^2 + v_\parallel^2) / v_0^2]$, with mean square velocity $\langle v^2 \rangle = \frac{3}{2} v_0^2$. Here the dispersion relation takes the form of Eq. (16), where $\sigma_{i,e} = (\omega / k v_{0\,i,e})$.

$$D(\omega, k) = 0$$

$$= k^2 + \sum_{i,e} \left(\frac{\omega_p^2}{v_0^2} \right)_{i,e} [2(1 + \sigma_{1,e} Z(\sigma_{i,e}))] \quad (16)$$

The function $Z(\sigma)$, given in Eq. (17) is the so-called plasma dispersion function. This function has been tabulated (see B. D. Fried and S. D. Conte in the bibliography). For the present purpose it will be sufficient to state its asymptotic form in the limit $\sigma \gg 1$ and for $Im\,\sigma < 0$ (damped waves) as in Eq. (18). Neglecting the ion terms, an

$$Z(\sigma) = \frac{1}{\pi^{1/2}} \int_{-\infty}^{\infty} \frac{e^{-u}\, du}{u - \sigma} \quad (17)$$

approximate form of the dispersion relation is then given by Eq. (19).

$$Z(\sigma) \approx \left\{ i 4\pi^{1/2}\, \sigma e^{-\sigma 2} \quad - \left[\frac{1}{\sigma^2} + \frac{3}{2\sigma^4} + \cdots \right] \right\} \quad (18)$$

$$D(k, \omega) = 0$$

$$= k^2 - \frac{\omega_{pe}^2}{v_{oe}^2} \left\{ \left[\frac{1}{\sigma^2} + \frac{3}{2\sigma^4} \right] - i [4\pi^{1/2}\, \sigma e^{-\sigma 2}] \right\} \quad (19)$$

The existence of the imaginary component means that the dispersion relation cannot be satisfied for real ω and real k, that is, the wave will be damped.

The damping constant per wavelength can be stated in terms of the ratio of the phase velocity v_p of the wave to the electron thermal velocity as in Eq. (20).

$$\alpha = -2\pi^{1/2} \left(\frac{v_p}{v_{oe}} \right)^2 \left(\frac{\omega_{pe}}{\omega} \right) \exp \left[-\left(\frac{v_p}{v_{oe}} \right)^2 \right] \quad (20)$$

Thus the damping depends exponentially on the square of the ratio of the phase velocity to the mean thermal electron velocity. This follows from the fact that the number of electrons with velocities such as to promote absorption of energy from the wave decreases exponentially with the phase velocity at which this absorption takes place. This important theoretical result has been verified experimentally in one of the classic experiments of modern plasma physics.

Ion acoustic wave. To conclude the subject of longitudinal plasma waves, mention should be made of what has been called ion acoustic waves. If the dispersion relationship above is evaluated at much lower frequencies, a wave is found for which the phase velocity is as given by Eq. (21). Here,

$$v_p = \frac{\omega}{k} = \sqrt{\frac{3kT_e}{2M_i}} \frac{1}{(1 + \frac{3}{2} k^2 \lambda_D^2)^{1/2}} \quad (21)$$

T_e is the kinetic temperature of the plasma electrons and λ_D is the Debye length, $\lambda_D = (\epsilon_0 k T_e / n_e e^2)^{1/2}$ (SI units).

For long wavelengths such that $k\lambda_D \ll 1$, this wave has a phase velocity equal to the thermal velocity the ions would have if they were at the electron temperature—hence ion acoustic waves, since acoustic velocities are also equal to thermal velocities.

To recapitulate the discussion of plasma waves, it has been shown that a homogeneous plasma can support many types of waves, both transverse (electromagnetic) and longitudinal (electrostatic). The importance of these waves in plasma physics, both in space and in the laboratory, is that they provide the means for propagating almost all of the important plasma effects, whether it be the whistlers of ionospheric research or the unstable phenomena encountered in magnetically confined plasma.

Drift waves. One important class of waves has not been discussed. This class is composed of drift waves arising in the presence of density gradients in a magnetically confined plasma, for example, at its surface. In the appropriate limit these waves resemble the waves that propagate at the interface between two fluids of different density (such as water and air) in the presence of a gravity field. Here the magnetic field external to the plasma plays the role of the lighter fluid, and the centrifugal forces arising from particle motions along curved lines of force provide the "gravity" force.

In plasmas of high density, such that the hydromagnetic dielectric constant $K_H \gg 1$, the expression for the phase velocity of these surface waves takes the same form as that found in the case of a classical fluid with variable density, and it is written as Eq. (22). Here $\epsilon = (1/n)(dn/dx)$, the

$$v_H = \frac{\omega}{k} = \sqrt{\frac{\epsilon g}{k^2}} \qquad (22)$$

logarithmic density gradient. Note now that when ϵ and g are in the same direction (lighter fluid on top of heavier fluid) the phase velocity is real, corresponding to stable waves. But when (ϵg) is negative, v_H, and therefore ω, is imaginary, corresponding to unstable growth of the wave. Both examples have their analogy in plasma behavior.

Solitons. Thus far, plasma waves have been discussed from the standpoint of linear theory. That is, situations have been discussed where the local properties of the plasma itself (density, velocity distributions, and so forth) are tacitly assumed to be essentially uninfluenced by the presence of the waves, which are in this sense of "small amplitude." From the description of the phenomenon of Landau damping, however, it can be seen that physical mechanisms for strong wave-particle interactions exist, suggesting that in the case of large-amplitude waves entirely new phenomena could arise. Among these is the phenomenon of the solitary plasma wave, the soliton.

The name soliton was coined to describe a localized propagating plasma disturbance that retains its form over time. Solitons can exist when the amplitude of the disturbance is sufficiently large that nonlinear interactions between elementary waves can give rise to a propagating self-perpetuating localized electrostatic potential well within which the departures from charge neutrality creating that well are maintained. Solitons are met in many areas of physics besides plasma physics. In hydrodynamics, for example, they occur in the propagation of water waves. In plasma physics, solitons are found, for example, in the case of ion acoustic waves. If the analysis of such waves is carried out so as to include finite-amplitude wave interaction effects, and if the results are given in a frame moving at the ion acoustic speed C_s, there results a nonlinear partial differential equation for the potential ϕ, Eq. (23), where A and B are positive constants calculated from the plasma parameters.

$$\frac{\partial \phi}{\partial t} + A \frac{\partial^3 \phi}{\partial x^3} + B \frac{\partial \phi^2}{\partial x} = 0 \qquad (23)$$

tive constants calculated from the plasma parameters.

This equation is the Korteweg–de Vries equation, encountered first (in 1895) in the analysis of another type of soliton (solitary water waves in shallow water channels). The existence of propagating, highly localized disturbances with remarkable stability of form can be shown directly from the Korteweg–de Vries equation. Assuming the existence of a traveling-wave solution, that is, substituting the new variable $u = x - Vt$ into the Korteweg–de Vries equation, where V is the velocity of the traveling wave relative to C_s, there results an

ordinary differential equation (24). This equation may be integrated to yield Eq. (25), where E is a constant of integration.

$$-V \frac{d\phi}{du} + A \frac{d^3\phi}{du^3} + B \frac{d(\phi^2)}{du} = 0 \qquad (24)$$

$$\frac{-V}{2} \phi^2 + \frac{B}{3} \phi^3 + \frac{A}{2} \left(\frac{d\phi}{du}\right)^2 = E \qquad (25)$$

The form of this solution suggests viewing it as an energy conservation equation for a "particle" of mass A moving in the cubic "potential" $U = [-\frac{1}{2} V \phi^2 + \frac{1}{3} B \phi^3]$. When $V > 0$ (that is, when the soliton velocity exceeds the ion acoustic-wave velocity) this equivalent potential possesses a localized negative minimum permitting a nonoscillatory (that is, constant-form) traveling disturbance. Using this line of reasoning, it therefore is apparent that there exists a unique relationship between the amplitude and the velocity of the soliton. Setting $d\phi/du = 0$ (peak of disturbance) in Eq. (25) gives directly Eq. (26).

$$\phi_{max} = 3V/2B \qquad (26)$$

In other words, the velocity of the soliton varies linearly with its peak amplitude. (Higher-order nonlinearities, not included in the Korteweg–de Vries equation, limit the value of the peak amplitude ϕ_{max} for a solution. This in turn limits the maximum soliton velocity to about 1.4 times the ion acoustic-wave velocity, that is, $V < 0.4\, C_s$, approximately.)

The geometrical form of the solitons defined by the Korteweg–de Vries equation may be determined by solving for $d\phi/du$ in Eq. (25) and integrating. Analytic solutions representing one or more solitons, having the general asymptotic single-peaked form $\phi \sim \mathrm{sech}^2[k(x - Vt)]$, where k is the wave number for the wave, can be obtained in this way (Fig. 2). In the case of multiple-soliton solutions, the relationship between amplitude and velocity of solitons cited above permits the analysis of interacting solitons. The remarkable property exhibited in such encounters, as seen both in the solutions (analytic or computer-generated) and in actual laboratory experiments with plasma soli-

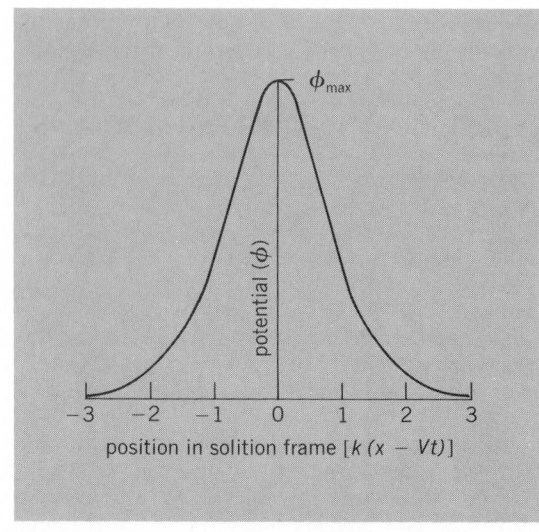

Fig. 2.   General asymptotic single-peaked form of a soliton.

tons, is that two solitons may "collide," pass through each other, and emerge unscathed (that is, still possessing the same asymptotic form, amplitude, and velocity). This property is traceable to the combination of dispersive and nonlinear characteristics of the defining equations, reflecting in turn the operation of basic cooperative phenomena inherent to the plasma state. *See* SOLITON.

PLASMA INSTABILITIES

Instability of plasma, particularly magnetically confined plasma, is probably the most recondite property of this state of matter. An isotropic, isothermal, homogeneous plasma with a Maxwellian particle distribution could not exhibit any instability, any more than could an ordinary gas under the same equilibrium conditions. However all plasmas created in the laboratory or existent in space differ in some respect from the above equilibrium state, possessing a higher degree of order and thus a lower state of entropy. This difference in entropy represents free energy which makes it thermodynamically possible for instability or turbulence to occur. In hot plasma such unstable behavior may provide a much more rapid path toward a state of uniformity and randomness than that provided by collisional processes. Since the achievement of such a state implies the disappearance of magnetic confinement (the plasma would now be in physical contact with the surrounding walls, and at room temperature), any process which hastens the approach to equilibrium reduces the effectiveness of magnetic confinement. In some cases the result may be to shorten the confinement time by factors of a million or more over that which would obtain with collisions alone operative. A simple rule for the time for growth to destructive amplitude of the most catastrophic plasma instabilities is to divide the mean overall dimension of the plasma by the thermal velocity of the ions. This time interval is usually a very short one for a hot plasma, being of the order of a few microseconds for even a large laboratory plasma, whereas the theoretical time scale for the relaxation of magnetic confinement through collisions alone might be many seconds under the same plasma conditions.

Two general classes of plasma instabilities can be distinguished, differing markedly in their nature and origin: (i) gross instabilities; these might also be called configuration-space instabilities since they arise because of order in configuration space (pressure gradients, magnetic field curvature, and so on); and (ii) velocity-space instabilities, arising from ordering in velocity space (streaming motions, anisotropy in velocity distributions, departures from a Maxwellian distribution, and so forth).

Hydromagnetic instabilities. The hydromagnetic instabilities represent the main class of configuration-space instabilities. This type of instability has already figured in the discussion of unstable drift waves. Instabilities of this type can occur if there exists a possible motion for the plasma compatible with the constant flux condition imposed by its electrical conductivity, such that it can expand while moving as a whole across the field. In this way the plasma can tap the free-energy reservoir of expansion to drive an unstable motion, expelling it from the field. An important necessary condition for the possibility of such hy-

dromagnetic instabilities follows from a simple model: In Fig. 3 a cross-section view is shown of a column of magnetically confined plasma, centered on the axis of the confinement chamber. Also shown are two possible displacements of the column, one toward a region where the field is stronger than at the center and the other toward a position of weaker field. For both displacements the total magnetic flux through the column is constrained to remain constant, owing to the electrical conductivity of the plasma. In the first displacement the plasma must compress, since $\phi = B \cdot$ (Area) and B increases; this is a stable displacement, since work must be done to perform the displacement. In the second displacement the plasma will expand, releasing internal energy; this is an unstable displacement, leading to further motion and eventual expulsion of the plasma. Putting this idea in mathematical form as relation (27), the mag-

$$V \sim \int \frac{dl}{B} \tag{27}$$

nitude of infinitesimal volume element V of the plasma that will be involved in the expansion, if there is to be instability, is found to be proportional to the integral of $1/B$ along an infinitesimal flux tube of the confining field, that is, along field lines. It follows that if the integral does not exhibit a maximum at the undisplaced position of the plasma, the plasma may be subject to hydromagnetic instabilities. An example of this is the simple mirror field (Fig. 4). Here the field lines are concave to the plasma (bulge outward), corresponding to weakening of the field radially outward from the plasma. Here the integral has a minimum value on the central axis so that instability is predicted. Also consistent with this prediction, recalling the drift wave picture, this situation corresponds to a case where the density gradient and the effective gravity are oppositely directed, leading to unstable drift waves.

An important case of a field configuration that does not permit this type of instability is the magnetic well. A simple example of a coil producing such a field and the configuration of the lines of magnetic force in such a field is the "baseball" or "tennis-ball" coil, shown in Fig. 5. This coil is so named because its windings are shaped like the seam on a baseball or tennis ball. Note that as seen from the inside the field lines are everywhere convex toward the center, corresponding to the fact that the field increases in every direction from the

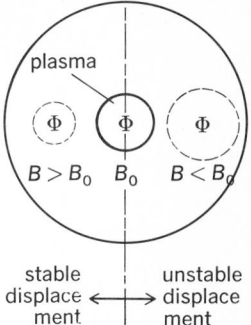

Fig. 3. Schematic of hydromagnetically stable and unstable displacements of a plasma column.

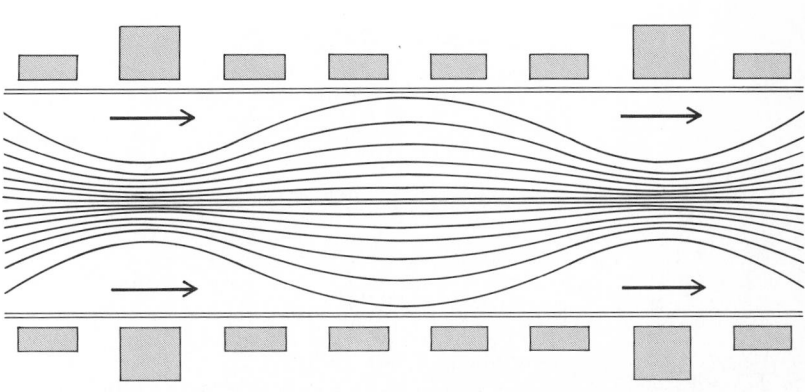

Fig. 4. Illustration of magnetic mirror. Field lines bulge outward.

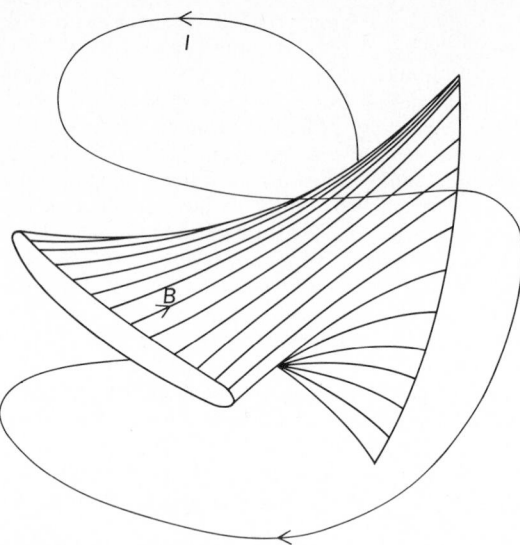

Fig. 5. Coil configuration and field line pattern in "base-ball" magnetic well field.

center of the spherical containment volume. The first experimental demonstration of the efficacy of magnetic wells in suppressing hydromagnetic instability in mirror-confined plasma was performed by A. F. Ioffe in 1961 in the Soviet Union.

While the magnetic-well principle can be readily applied in open-ended systems, it cannot be utilized in its pristine form in closed, or toroidal, systems. The reason is a topological one: In a system where the lines return on themselves, it is not possible for these same field lines to be everywhere convex toward the plasma; somewhere they must bend back on themselves. These regions of "bad" curvature, places where the field weakens in a direction outward from the plasma, are prone to hydromagnetic instability.

However, ways have been found for shaping toroidal confining fields so that they are less subject to hydromagnetic instabilities, either by extension of the magnetic-well idea or by the use of sheared magnetic fields. To understand shear, consider a tubular section of a confining field, as shown in Fig. 6. Here successive "layers" of the flux lines differ in helical pitch angle, so that the overall pattern resembles a basket weave.

Even in toroidal systems that employ shear stabilization, the combination of the effects of bad curvature regions, taken together with the mirror trapping of particles can tend to produce instabilities of the so-called trapped-particle type. The mechanism involved is the following: In a toroidal

confinement field with helical lines (for example, the tokamak), any given field line as it passes around the torus will alternately come closer to the inner wall and, a half-turn of the helix later, closer to the outer wall. But in any toroidal field the main toroidal field is always strongest nearest the inner wall, falling off in the outward direction inversely with radius. Thus along each helical line the field intensity will alternately strengthen and weaken as that line makes its closest approach to the inner wall or to the outer wall. In this way, along any given line there will be formed mirror-trapping regions, namely those regions lying closest to the outer wall. The existence of these regions divides the population of contained particles into two distinct classes—trapped particles, those that are repeatedly reflected back and forth between two adjacent mirrors, and passing particles, those whose pitch angles are such that they are not trapped but pass entirely around the torus. Collisional effects will populate both classes; how long a given particle remains in one class will depend on the rate of collisions compared to the "bounce period" (time for a trapped particle to make a round trip between the middle of a trapping region and a mirror region). As the plasma temperature increases, collisions become less frequent and particles may remain trapped in a given mirror cell for many bounce periods. In such a situation these particles will experience the destabilizing effects of the bad curvature regions for a long enough time to allow the growth of hydromagnetic instability feeding from this effect; conversely, at low plasma temperatures (or very high plasma density) collisions would be frequent enough to wash out the unstable tendency, since particles would exchange rapidly between the trapped and passing classes.

Trapped-particle modes represent a potential threat to toroidal confinement, one that would tend to worsen as the plasma became hot enough to approach fusion reactor conditions. Whether the existence of these modes would cause excessive losses has not been established, as neither experiment nor theory has gone far enough to provide definitive answers.

Two additional instabilities of the hydromagnetic variety are the "firehose" and the "mirror" instabilities. As has been discussed, in connection with collision processes in a plasma, in a hot plasma the kinetic pressure exerted by the plasma often may be anisotropic; that is, the component of pressure perpendicular to the magnetic lines may differ from that parallel to the lines. The concept of pressure components has figured in the discussion of pressure balance in a confined plasma, where the relationship involved only the perpendicular pressure component $p_\perp$. *See* PLASMA PHYSICS.

The firehose instability is so named because of its similarity to the form assumed by a flexible hose lying on a suitably flat surface while carrying a high flow of water. Any local curvature in the line grows as a result of the centrifugal force generated by the water as it flows around the bend. In this way an originally nearly straight length of hose assumes a snakelike pattern on the ground.

In the very same way, in a plasma where $p_\parallel \gg p_\perp$, the flow of plasma along field lines will, once a critical pressure value is exceeded, break into an unstable behavior that will tend to ripple

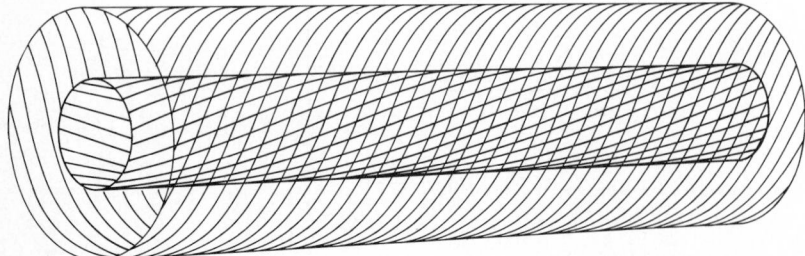

Fig. 6. Schematic of pattern of field lines in magnetic field with shear.

the tubes of magnetic flux just as the firehose develops ripples. Theory gives the critical condition for the firehose instability for plasma contained in a region of uniform field of original value (without the plasma) B_o, as in expression (28). As expected,

$$(p_\parallel - p_\perp) > B_o{}^2/2\mu_0 \qquad (28)$$

in an isotropic plasma ($p_\parallel = p_\perp$), the firehose instability cannot occur. In terms of $\beta = p_\perp/(p_\perp + B_o{}^2/2\mu_0)$ the critical condition is defined by expression (29), consistent with the limiting value of $\beta = 1$

$$\beta > 2p_\perp/(p_\perp + p_\parallel) \qquad (29)$$

for magnetostatic equilibrium in an isotropic plasma.

The mirror instability has nothing to do directly with magnetic mirror confinement. The appellation comes from the physical mechanism involved in the instability. Consider an anisotropic plasma that either an incipient kinking or constriction of the plasma leads to local enhancement of the confining field in such a way as to worsen the situation. In kinking, the field on the convex side of the in which $p_\perp \gg p_\parallel$: that is, the averaged random particle velocities along the field lines are much less than those associated with cyclotron rotations around the lines. Suppose now that the originally uniform field is momentarily weakened infinitesimally locally, for example, by an upward statistical fluctuation in the local particle density. This weakening will create a local mirror trapping zone within which more particles can be trapped in transit through the region, leading to more trapping, and so forth, the end result being production of unstable perturbations in the magnetic field. In the case where electron and ion temperature components are all equal or where $T_e \ll T_i$, a simple expression for the critical pressure is given by relations (30) and (31), again consistent with the limit-

$$(p_\perp{}^2/p_\parallel) > (p_\perp + B^2/2\mu_0) \quad \text{unstable} \quad (30)$$

that is, $\qquad \beta > (p_\parallel/p_\perp) \qquad \text{unstable} \quad (31)$

ing value $\beta = 1$ for equilibrium in an isotropic plasma.

Other instabilities of the hydromagnetic variety include two of the first predicted ones, the "kink" instability and the "pinch-off" instability of a plasma confined solely by the magnetic field of currents flowing through it (Fig. 7). Again the idea is bend is weakened, leading to further sideways displacement, bending, and so forth. In pinch-off instability, a tendency to neck down enhances the field at the plasma surface (same current flowing through a smaller area), locally constricting the plasma still further, and so forth.

By now virtually every theoretically predicted hydromagnetic instability has been experimentally observed. Also, the theoretically predicted means for stabilizing these instabilities, when properly applied, have in all cases proved effective.

While hydromagnetic instabilities are the most catastrophic in their effects on the plasma of all instabilities, their characteristic growth times and frequencies lie roughly midway in the time scale of plasma instabilities. In descending order of frequency one can discern three categories of instabilities, when considering magnetically confined plasmas: (i) the ion cyclotron (or electron cyclo-

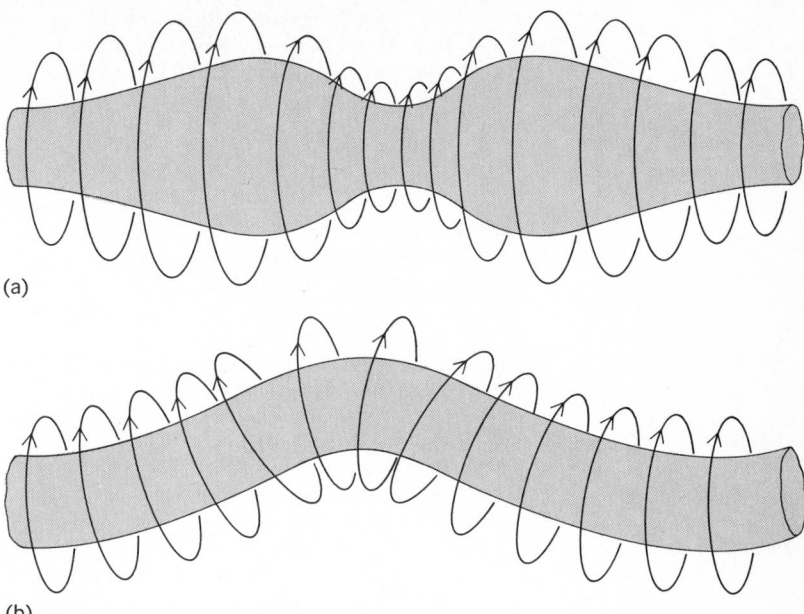

(a)

(b)

Fig. 7. Two kinds of hydromagnetic instabilities. (a) "Pinch-off" and (b) "kink" in a plasma column carrying a current.

tron) instabilities, yet to be discussed; (ii) the hydromagnetic instabilities, with wavelengths long compared to the mean ion orbit radii a_i, and corresponding frequencies roughly a factor ka_i lower than cyclotron frequencies ($k = 2\pi/\lambda$); these have just been discussed; and (iii) the "universal" instability and related types, involving resistive effects and with growth rates a factor of order $k^2a_i{}^2$ lower than that of the ion cyclotron types. This latter type will only be briefly discussed here, though it plays an important role in the behavior of plasmas confined in closed systems.

The universal instability. The universal instability has been given this name because it presumably can occur in any magnetically confined plasma, even in regions where the confining field is uniform and characterized by straight lines of force. As with the hydromagnetic drift instability, the free-energy reservoir is the expansion energy of the plasma, coupling to a slow drift wave associated with the presence of a density gradient. This instability has been observed experimentally, and has been shown to lead to relatively rapid destruction of magnetic confinement when it occurs. Fortunately for the future of magnetic confinement, theory (verified experimentally) has pointed the way to control of the instability. The idea is to take advantage of the fact that the mechanism of the instability, involving the necessity for a small component of electric field along the plasma, requires a very long wavelength along the field lines. If the characteristic length of the confining fields (length between mirrors in a magnetic well, or field periodicity length in an average well) is shorter than this critical wavelength, the instability cannot "fit" in the system and will be suppressed. A rough stability criterion at high densities ($\lambda_D \ll a_i$) is given by relation (32), where L is the critical length

$$L < \frac{10}{\epsilon}\left(1 + \frac{T_i}{T_e}\right) \qquad (32)$$

and $\epsilon = (1/n)(dn/dr)$ is the reciprocal of the scale length for the transverse density gradient.

The instabilities discussed to this point are ones that do not depend on the details of the distribution functions of the plasma particles, but only on general properties or averaged quantities, such as density gradients or differences in pressure components. An important class of instabilities involves the details of the distribution functions in an essential way.

In the discussion of Landau damping of electrostatic waves it was mentioned that non-Maxwellian particle distributions of types for which $(\partial f_o / \partial v_x) > 0$ for some range of particle velocity components in a direction x in the plasma can lead to unstably growing waves, namely those whose phase velocities and directions match with this direction and velocity range.

High-frequency stream instability. One of the earliest predicted and earliest studied of instabilities of this type is the two-stream instability. One common situation where this instability can arise is where two plasmas counterstream through each other at a sufficient relative speed. Even if in their own frames of reference each plasma is characterized by an isotropic Maxwellian distribution, when viewed from the other frame the displaced Maxwellian distribution possesses a region of positive slope (Fig. 8).

For this case one writes the dispersion equation, Eq. (33), where $v = \omega / \omega_{pe}$ and $v_0 = k_{||} v_0 / \omega_{pe}$. Note

$$1 - \tfrac{1}{2}\left[\frac{1}{(v - v_0)^2} + \frac{1}{(v + v_0)^2}\right] = 0 \qquad (33)$$

first that if $v_o \geq 1$ all of the four roots of the above equation are real, corresponding to stability, so that a sufficient condition for stability is given by relation (34). Thus instability can only occur if the

$$k_{||} > \omega_{pe}/v_s \qquad (34)$$

wavelength exceeds a critical value; not all wavelengths are unstable.

Not only does there exist a critical wavelength, but there is also a critical streaming velocity below which the system is stable for any wavelength. This critical value may be found from the dispersion equation, from which an approximate expression, relation (35), is found; that is, $v_s < 0.7\, v_0$, stable.

$$\left(\frac{v_s}{v_0}\right)^2 < \tfrac{1}{2} \quad \text{stable} \qquad (35)$$

Thus the counterstreaming velocity must exceed a critical value compared to the thermal spread of velocities before instability can occur. This circumstance illustrates an important point in the theory of wave-particle instabilities. It is not enough for a free-energy reservoir to exist to ensure instability. The free-energy reservoir must satisfy some minimum quantitative condition, and the instability wavelength must also conform to restrictions imposed by the physical mechanisms involved.

Loss-cone instabilities. An important class of instabilities closely related to the streaming instabilities is predicted to occur in plasmas contained between magnetic mirrors. As noted earlier, such plasmas are necessarily deficient in particles moving too nearly parallel to the direction of the mag-

netic field, since such particles are not reflected by the mirrors. In many ways the loss-cone instabilities resemble the unstable behavior in those atomic systems which are associated with masers and lasers. In these devices the amplification mechanism that is inherent in their operation depends upon the existence of an "inverted population" of atomic states, where through optical or other pumping techniques some of the higher states of the atom system are caused to be more heavily populated than the lower states, contrary to the usual equilibrium situation. In such a circumstance a photon wave train passing through the system will induce atomic transitions both upward, extracting energy from the waves, and downward, giving up energy to the wave. In a system with an inverted population the downward transitions exceed the upward, leading to net amplification. If the inversion is sufficient, if the system is long enough, or if it has good enough mirrors at its ends, laser action will take place, leading to an exponential growth of the light intensity. *See* MASER; OPTICAL PUMPING.

In the streaming instabilities or their cousins, the loss-cone instabilities, the regions of positive slope in the velocity distribution functions play the same role as the inverted population of the laser. Plasma waves moving at plasma velocities lying within these regions can grow exponentially in amplitude at the expense of the free energy represented by these departures from an isotropic Maxwellian state.

The details of the particle distribution functions in mirror systems of course depend on the way in which the plasma has been formed and upon its subsequent history. One important case is that of collisional equilibrium, that state achieved when collisions, operating to drive the distributions toward a Maxwellian state, compete with the mirror losses, which continually deplete the distribution of particles whose velocity vectors lie in the loss cone. It is possible to compute the shape of such distributions by use of the Fokker-Planck equation for collisional processes. The results of such calculations can then be represented in terms of particle velocity components perpendicular to and parallel to the direction of the magnetic field. When evaluated at the lowest field point in a mirror system with a mirror ratio of 1.5, the collisional distribution appears as shown in Fig. 8b. For purpose of comparison, a Maxwellian distribution is shown on the left.

Theoretical calculation of the critical conditions for loss-cone instabilities proceeds along lines similar to those used above to calculate the two-stream instability. The situation is, however, made complicated by the necessity to include the cyclotronic motion of the particles in the analysis. As a result, in the general case many unstable modes are found. All share one feature in common, however. Each possesses a threshold density below which it will not occur, and this threshold density increases as one considers better- and better-randomized distributions. In fact many of the more esoteric modes cannot occur at all in well-randomized plasmas, for example, those plasmas contained in collisional equilibrium in deep magnetic wells.

An additional and very important stabilizing effect is being explored theoretically and experi-

WAVES AND INSTABILITIES
IN PLASMAS

$f(v_\perp{}^2, v_{||}{}^2)$

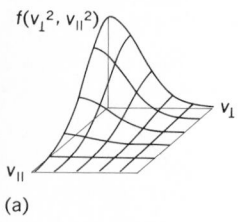

(a)

$f(v_\perp{}^2, v_{||}{}^2)$

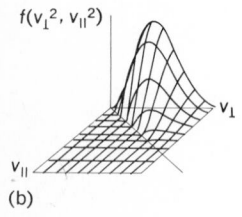

(b)

Fig. 8. Velocity distribution. (a) Maxwellian distribution. (b) Loss cone in a mirror machine, mirror ratio = 1.5.

$$0 = 1 - \sum_{j(\text{species})} \frac{\omega_{pj}^2}{k^2} \sum_{n=-\infty}^{\infty} 2\pi \int_{-\infty}^{\infty} dv_{\parallel} \int_{0}^{\infty} v_{\perp} \, dv_{\perp} J_n^2 \left(\frac{k_{\perp} v_{\perp}}{\omega_{cj}} \right) \frac{1}{(\omega + n\omega_{cj} + k_{\parallel} v_{\parallel})} \left\{ \frac{n\omega_{cj}}{v_{\perp}} \frac{\partial f_o}{\partial v_{\perp}} + k_{\parallel} \frac{\partial f_o}{\partial v_{\parallel}} \right\} \quad (36)$$

mentally. This is the effect of finite plasma length and the effects of the spatial variation of the confining fields on those modes that depend on sharp cyclotronic resonances with the ions for their growth. The finite length effect is related to the earlier-cited analogy with the laser: An instability that grows by wave amplification along the field lines may not be able to grow, from the fluctuational levels launching it, to disruptive amplitude if the plasma length is short enough, and if wave reflections at the end are not strong. The second stabilizing effect, detuning of resonances, is physically obvious, but very difficult to analyze theoretically.

All theoretical analyses of loss-cone modes start from a basic dispersion equation, first derived by I. Bernstein, and employed first by E. Harris. This imposing equation, Eq. (36), is then simplified or specialized to the case at hand.

To illustrate the results of the application of this equation to the loss-cone instability problem, some important examples are cited below.

The high-frequency convective loss-cone (HFCLC) mode is a laserlike mode involving the growth of high frequency ($\omega \gg \omega_{ci}$) waves that propagate along the field lines. In this context the word "convective" refers to an instability that grows in its moving frame of reference, that is, as it convects along the field. This type is to be contrasted with the nonconvective or standing-wave types of instabilities (sometimes called absolute instabilities) that grow without propagating along the plasma. Convective instabilities are generally less dangerous than nonconvective ones, as they can in principle be made innocuous through reducing their growth rates by better randomization, by restricting the plasma length, and by creating situations that minimize wave reflections at the ends of the plasma. (Strong reflections can convert a convective instability into a standing-wave type.) From the dispersion equation one can show that the growth length (length for e-folding of the HFCLC instability) is given by Eq. (37). The term

$$L = 2a_i (1 + \omega_{ce}^2 / \omega_{pe}^2)^{1/2} [y \, \text{Im} \, F(y)]^{-1} \quad (37)$$

$[yF(y)]$ is a function determined from the dispersion equation, with $y = \omega / k \bar{v}$, a dimensionless frequency parameter.

Since at least 10 e-folding lengths would be required to amplify the normal fluctuations up to an important level, one can take $L_c = 10 L_o$ as the critical length. For well-randomized distributions $|y I_m F(y)| \ll 1$ for all y and the critical lengths turn out to 200 to 500 $a_i (1 + \omega_{ce}^2 / \omega_{pi}^2)^{1/2}$, that is, quite long, as measured in units of the ion orbit radii a_i, even at very high densities, where $\omega_{ce}^2 / \omega_{pe}^2 \ll 1$. On the other hand, narrow, δ-function-like distributions can exhibit very short growth lengths, on the order of 1 or 2 orbit radii.

A second example of instabilities of loss-cone origin derivable from Eq. (36) relates to a very common situation in plasma confinement experiments. Suppose that a hot plasma characterized by

a loss-cone distribution is confined between mirrors, while a "cold" plasma of much lower density, composed either of ions of the same type or of heavier-mass impurity ions, is introduced. The dispersion equation shows that this combination can lead to "double-hump" instabilities, in which case the circularly moving hot ions, in streaming through the slow-moving colder ions, stimulate a type of two-stream instability. This instability can take place in the limit $k_{\parallel} \rightarrow 0$, that is, very long wavelength along the field lines, a limit in which the stabilizing Landau damping effects of the electrons are inoperative, so that only ion contributions need be considered.

The general type of analysis described here can be applied to other instabilities of related types. One such instability is the drift cyclotron loss-cone (DCLC) instability. This instability is predicted to occur in plasmas where the combination of a density gradient and a loss-cone distribution is present. At high densities the critical gradient for this instability to occur in a very long plasma (where field curvature stabilization can be neglected) is rather small, corresponding to required plasma diameters of the order of 100 ion orbit radii. There is some evidence, however, that the DCLC mode is stabilized by magnetic wells or by finite length effects.

Many other examples of loss-cone-like instabilities could be cited. In general, however, these instabilities tend either to be unimportant in well-randomized plasmas, or to be so sharply resonant that magnetic gradient detuning effects should stabilize them. In fact the belief is growing that it should be possible to stabilize all loss-cone modes in mirror-confined plasma by controlling the distribution functions and by the use of deep magnetic-well systems of properly scaled length.

In connection with the whole class of instabilities that have just been discussed, it is important to note that their critical conditions can be expressed in terms of dimensionless frequency ratios and dimensionless lengths, so that the results have much wider applicability than at first appears. For example, the loss-cone instabilities have thresholds that scale as $\omega_{pi}^2 / \omega_{ci}^2$, that is, as n/B^2; critical lengths that scale as L/a_i, that is, in units of ion orbit radii. Results found experimentally for one set of conditions can therefore be expected to scale to others, provided the critical ratios are preserved. This fact has important consequences for the experimental study of these instabilities and for the extrapolation of the results.

Nonlinear phenomena. The instability processes just described have been concerned with threshold criteria and the initial growth of instabilities. An important aspect of instability phenomena and of cooperative effects in plasma remains to be discussed. This aspect has to do with nonlinear processes, an example of which is the onset of anomalous losses caused by the plasma instabilities just discussed. Linear theory predicts exponential growth of an instability; exponential growth cannot continue indefinitely but must saturate at a level determined by other, nonlinear processes. An example from the loss-cone instabilities would be the fact that the instability leads to the loss of particles. But escaping particles correspond to a partial filling of the loss cone, a process weakening the instability mechanism itself. In fact the general effect of instabilities is a tendency to destroy the condition that gave rise to them in the first place, by depletion of the free-energy reservoir that drives them. If the free-energy reservoir is weak, the saturation amplitude of the instability will be weak. No simple rules can be given for calculating nonlinear processes of this kind, but extensive theoretical work has been directed at special problems tractable enough for analysis.

Another type of nonlinear process in plasma has to do with the nonlinear coupling of plasma waves. Owing to nonlinear effects, two finite-amplitude plasma waves, say of frequencies ω_1 and ω_2, may under the proper conditions couple to produce a "beat" wave of frequency $(\omega_1 + \omega_2)$. Even if the original waves are not damped, the beat wave may be, because of its frequency domain, leading to so-called nonlinear Landau damping of the primary waves. Also, through the same mechanism two linearly stable waves can be rendered unstable by nonlinear coupling if their beat wave fits in a frequency-wavelength domain that permits unstable growth.

Closely related to the above are the so-called parametric instabilities of a plasma. These instabilities can occur when a driven oscillation of the plasma exists which can couple nonlinearly to other plasma waves in such a way as to cause the unstable growth of those waves. The driving oscillation could either be an external source of radio-frequency energy or even conceivably another plasma instability. Similar phenomena can give rise to so-called explosive instabilities, where parametric coupling effects regenerate on themselves to cause extremely rapid growth of a plasma disturbance.

The field of nonlinear effects in plasma is a difficult one for analysis, but it is clear that some of the most important problems of plasma physics lie in this area.

[RICHARD F. POST]

Bibliography: F. Chen, *Introduction to Plasma Physics*, 1974; R. C. Davidson, *Methods in Nonlinear Plasma Theory*, 1972; B. D. Fried and S. D. Conte, *The Plasma Dispersion Function*, 1961; N. Krall and A. Trivelpiece, *Principles of Plasma Physics*, 1973; R. F. Post, Controlled fusion research and high temperature plasmas. *Annu. Rev. Nucl. Sci.*, 20:509–558, 1970; G. Schmidt, *Physics of High Temperature Plasmas*, 2d ed., 1979; L. Spitzer, *Physics of Fully Ionized Gases*, 1962; T. H. Stix, *The Theory of Plasma Waves*, 1962.

Weak nuclear interactions

Fundamental interactions of nature that play a significant role in elementary-particle and nuclear physics, and are distinguished from other such interactions by special properties such as participation of all the fundamental fermions and failure to conserve parity. Of the four fundamental interactions of nature (gravitational, strong, electromagnetic, and weak), only the strong, electromagnetic, and weak forces are significant for elementary-particle and nuclear physics, given present understanding and foreseeable methods of observation.

The weak force has very short range (less than 10^{-17} m) and is extremely feeble compared to strong and electromagnetic forces, but can be distinguished from these two by its special character. For example, according to the present view, all of matter consists of certain fundamental spin-1/2 constituents, the quarks and leptons, collectively called the fundamental fermions (Table 1). While only the quarks participate in strong interactions, and only the quarks and charged leptons e, μ, and τ participate in electromagnetic interactions, all of the fundamental fermions, including neutrinos, engage in weak interactions. Also, the strong and electromagnetic interactions respect spatial inversion symmetry (they conserve parity) and are also particle-antiparticle (charge conjugation) symmetric, whereas the weak interaction violates these two symmetries. *See* FUNDAMENTAL INTERACTIONS; LEPTON; PARITY; QUARKS; SYMMETRY LAWS.

Table 1. Fundamental constituents of matter*

Constituent	Electric charge	Generation†		
		1	2	3
Quarks	$+2/3\ \lvert e\rvert$ $-1/3\ \lvert e\rvert$	$\begin{pmatrix} u \\ d \end{pmatrix}$	$\begin{pmatrix} c \\ s \end{pmatrix}$	$\begin{pmatrix} t(?)\ddagger \\ b \end{pmatrix}$
Leptons	0 $-\lvert e\rvert$	$\begin{pmatrix} \nu_e \\ e^- \end{pmatrix}$	$\begin{pmatrix} \nu_\mu \\ \mu^- \end{pmatrix}$	$\begin{pmatrix} \nu_\tau(?)\S \\ \tau^- \end{pmatrix}$

*For each quark and lepton there exists a corresponding antiparticle with the same mass but opposite charge.
†More quark and lepton generations may exist.
‡Evidence for t has not yet been obtained.
§Presumed to exist.

Weak interactions are classified as "charged" or "neutral," depending on whether or not a particle participating in a weak reaction suffers a change of electric charge of one electronic unit. For example, the weak neutrino-nucleon scattering reaction of Eq. (1) is charged since the neutrino ν_μ with zero

$$\nu_\mu + n \rightarrow \mu^- + p \qquad (1)$$

charge transforms into a muon μ^- with negative charge, while the neutron n (zero charge) becomes a proton p (positive charge). (Here the neutron may be considered as being composed of two d quarks and one u quark, and the proton of one d quark and two u quarks. In Eq. (1) a d becomes a u.) On the other hand, the weak elastic scattering reaction of Eq. (2) is neutral. Observed charged weak interac-

$$\nu_\mu + p \rightarrow \nu_\mu + p \qquad (2)$$

tions include nuclear beta decay and electron capture, muon capture on nuclei, and the slow decays of unstable elementary particles such as the μ and τ leptons, π, K, and charmed mesons, and hyperons and charmed baryons (Table 2). Also, there are the charged neutrino-nucleon and neutrino-lepton scattering reactions. Neutral weak interactions were first observed in 1973, and include neutrino-nucleon and neutrino-lepton scattering as well as the electron-nucleon reaction of Eq. (3),

$$e^- + N \rightarrow e^- + N \qquad (3)$$

which can also occur by electromagnetic interac-

Table 2. Classification of weak processes, with some examples

Charged weak interactions

    Purely leptonic

$$\mu^- \to e^- \bar{\nu}_e \nu_\mu$$ Muon decay
$$\nu_\mu e^- \to \mu^- \nu_e$$
$$\nu_e e^- \to \nu_e e^-$$

    Semileptonic

$$n \to p e^- \bar{\nu}_e$$ Neutron decay
$$\nu_e n \to p e^-$$ Inverse beta decay
$$\nu_\mu n \to p \mu^-$$
$$\pi^+ \to \mu^+ \nu_\mu$$ Pion decay
$$\mu^- p \to n \nu_\mu$$ Muon capture
$$\Sigma^- \to n e^- \bar{\nu}_e$$ Hyperon decay

    Purely hadronic

$$K^+ \to \pi^+ \pi^0$$ Kaon decay
$$K^- \to \pi^+ \pi^- \pi^0$$ Kaon decay
$$\Lambda^0 \to p \pi^-$$ Hyperon decay

Neutral weak interactions

    Neutrino-lepton scattering
$$\nu_\mu + e \to \nu_\mu + e$$
    Neutrino-nucleon scattering
$$\nu_\mu + N \to \nu_\mu + N$$
    Electron-nucleon scattering
$$e + N \to e + N$$
    Lepton-lepton scattering   Neutral weak
$$e^+ + e^- \to \mu^+ + \mu^-$$ contribution not
                                yet observed

    Hadron-hadron interaction   Competes with
$$N + N \to N + N$$ electromagnetic
                                 and strong
                                 interactions

tion. *See* BARYON; ELEMENTARY PARTICLE; HYPERON; MESON.

In recent years the most important development in the study of weak interactions has been the creation by many workers, but principally by S. Weinberg (1967) and A. Salam (1968), of a successful theory based on the principles of local gauge invariance and spontaneous symmetry breaking. This theory proposes a single basis for the weak and electromagnetic interactions, and indeed, despite striking differences in the observed characteristics of strong, electromagnetic, and weak interactions, important theoretical ideas of a similar type suggest that all these interactions possess a common origin.

Early study. The development of understanding about weak interactions is inseparably linked with other major developments of 20th-century physics: relativity, quantum mechanics and quantum field theory, and nuclear and elementary-particle physics in general. The study of weak interactions began with the discovery of radioactivity by H. Becquerel in 1896, and the recognition shortly thereafter that in one form of radioactivity the decaying nucleus emits "beta rays" (electrons). Nuclear beta decay was thus the first known weak process. In 1914 J. Chadwick observed that the electrons in beta decay are emitted with a continuous spectrum of energies. This result and subsequent observations in the 1920s led to a crisis, for since the energy available to the electron in beta decay is essentially the difference in rest energies of the initial and final nuclei, which is a definite quantity, it appeared as though the principle of conservation of energy was violated. In order to rescue that fundamental law as well as those of conservation of linear and angular momentum, also in jeopardy, W. Pauli proposed in 1930 and again in 1933 that a

neutral particle of small or vanishing rest mass (later called the neutrino) is emitted along with the electron in nuclear beta decay and that it escapes observation because of its feeble interactions with surrounding matter. *See* NEUTRINO; RADIOACTIVITY.

Fermi theory. In 1934 Enrico Fermi proposed a theory of beta decay based on Pauli's neutrino hypothesis and constructed by analogy with quantum electrodynamics, which had been developed a few years before by P. A. M. Dirac. In quantum electrodynamics the interaction between an electron and the electromagnetic field is described by the interaction lagrangian density of Eq. (4), where $A^\lambda(\mathbf{x},t)$ is the four-vector potential of

$$\mathscr{L}_{EM} = -ej_\lambda(\mathbf{x},t) A^\lambda(\mathbf{x},t) \tag{4}$$

the electromagnetic field, and ej_λ is the electromagnetic current density of the electron, another four-vector. Fermi proposed that the beta-decay interaction is also described by the coupling of two four-vectors. In the simplest of nuclear beta decays, namely neutron beta decay, Eq. (5), one has a

$$n \to p + e^- + \bar{\nu}_e \tag{5}$$

four-vector current density describing the neutron-proton transformation: $\bar{\psi}_p \gamma^\lambda \psi_n$; and another four-vector current density describing the creation of e^- and $\bar{\nu}_e$: $\bar{\psi}_e \gamma^\lambda \psi_{ve}$, where ψ_p, ψ_n, ψ_e, and ψ_{ve} are Dirac field operators and the γ^λ are 4×4 matrices appearing in Dirac's relativistic quantum theory. Thus Fermi obtained the beta-decay lagrangian density of Eq. (6), where the hermitian conjugate

$$\mathscr{L}_\beta = \frac{G}{\sqrt{2}} \bar{\psi}_p \gamma_\lambda \psi_n \cdot \bar{\psi}_e \gamma^\lambda \psi_{ve} + \text{h.c.} \tag{6}$$

term (h.c.) was intended to account for nuclear β^+-decay and electron capture. G, the so-called weak interaction, or Fermi coupling constant, must be determined by experiment and is found to be given by Eqs. (7) or in units $\hbar = c = 1$ by Eq. (8), where m_p

$$\begin{aligned} G &= 1.43506 \pm 0.00026 \times 10^{-49} \text{ erg} \cdot \text{cm}^3 \\ &= 1.43506 \pm 0.00026 \times 10^{-62} \text{ J} \cdot \text{m}^3 \end{aligned} \tag{7}$$

$$G = 1.03 \times 10^{-5} \, m_p^{-2} \tag{8}$$

is the proton mass. *See* POTENTIALS; QUANTUM ELECTRODYNAMICS; QUANTUM FIELD THEORY; RELATIVISTIC QUANTUM THEORY; RELATIVITY.

Fermi's theory gave a good account of many aspects of nuclear beta decay, especially when generalized somewhat to include other bilinear covariants than vector $\times$ vector, by G. Gamow and E. Teller in 1936. Also, it contained the essence of many future developments. However, it was recognized almost immediately (by W. Heisenberg in 1936, among others) that the Fermi theory cannot be fundamental, since when it is applied to high-energy processes such as neutrino-electron scattering, it leads to a failure of unitarity (nonconservation of probability). Also, the Fermi theory contains incurable divergences which occur in the calculation of higher-order corrections: the theory is not renormalizable.

Parity violation. During the 25 years following Fermi's proposal, many weak processes were uncovered in addition to nuclear beta decay, with the discovery of new elementary particles and elucidation of their decay schemes. Gradually it became clear that these bear many similarities to nuclear

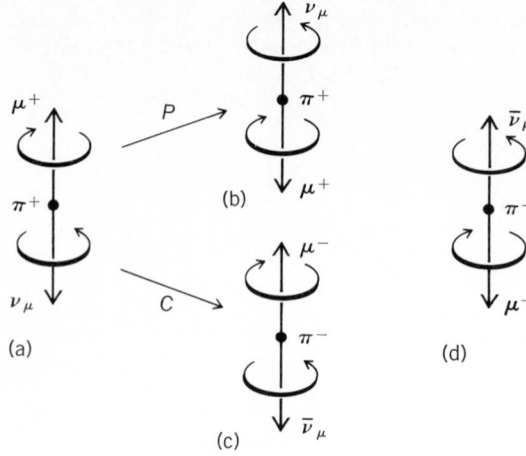

Fig. 1.   Schematic diagram of weak decay of the pion, $\pi^+ \to \mu^+ \nu_\mu$, illustrating parity (P) and charge conjugation (C) violation in weak interactions. (a) Decay of π^+. (b) Result of P transformation not observed. (c) Result of C transformation not observed. (d) Result of CP transformation: observed $\pi^- \to \mu^- \nu_\mu$ decay.

beta decay and all are but different manifestations of a universal weak interaction. Perhaps the most dramatic achievement of this period was the important discovery in 1956 by T. D. Lee and C. N. Yang that parity is violated in the weak interaction. An example of parity violation is the decay of the pion, $\pi^+ \to \mu^+ \nu_\mu$, as seen in the pion rest frame (Fig. 1). In observed $\pi^+ \to \mu^+ \nu_\mu$ decay (Fig. 1a), the μ^+ spin is found experimentally to be opposite to its motion [helicity $h(\mu^+) = -1$]. Since the π^+ spin is zero and angular momentum is conserved, this implies $h(\nu_\mu) = -1$ as well. A parity (P) transformation results in reversal of μ^+ and ν_μ momenta but leaves spins invariant. Thus under P, $h(\mu^+)$ is reversed (Fig. 1b); but this is never observed, implying maximal P violation. *See* HELICITY.

A charge conguation (C) transformation on $\pi^+ \to \mu^+ \nu_\mu$ decay changes the signs of all charges, but leaves spins and momenta unchanged yielding a μ^- with $h(\mu^-) = -1$ (Fig. 1c). Again, this is never observed; that is, the weak interactions violate charge conjugation maximally. On the other hand, a CP transformation on $\pi^+ \to \mu + \nu_\mu$ decay results in observed $\pi^- \to \mu^- \nu_\mu$ decay (Fig. 1d).

Feynman–Gell-Mann theory. The discovery of parity violation led, through a spurt in experimen-

tal and theoretical work, to the ultimate generalization and refinement of Fermi's theory, proposed by R. Feynman and M. Gell-Mann in 1958, and independently by R. Marshak and E. Sudarshan, and S. Gershtein and Y. Zel'dovich in the same year. In this formulation, the charged weak interactions are described by the lagrangian density of Eq. (9), where J_λ, the "universal" charged weak cur-

$$\mathscr{L}_w = \frac{G}{\sqrt{2}} J_\lambda{}^\dagger J^\lambda \tag{9}$$

rent, is a generalization of the original four-vector currents proposed by Fermi. The new current J_λ contains not only a vector (V) but also an axial vector (A) portion. Thus in the lagrangian of Eq. (9) one obtains terms of the form $V \times V$ and $A \times A$ (which are scalars and do not change sign under spatial inversion) and also terms $V \times A$ and $A \times V$ (which are pseudoscalars and change sign under spatial inversion). The appearance of scalar and pseudoscalar terms in $\mathscr{L}_w$ yields parity-violating effects such as the emission of particles with definite helicity in weak decays, the asymmetric angular distribution of beta electrons emitted in the decay of polarized nuclei, and so forth.

The Feynman–Gell-Mann scheme provides an excellent account of the observed features of charged weak interactions, but like the Fermi theory it cannot be fundamental, since it is not renormalizable, and leads to absurdities at sufficiently high energies.

Weinberg-Salam model. One may modify the Fermi–Feynman–Gell-Mann scheme by assuming that the weak interaction does not occur at a single space-time point, but proceeds by exchange of an intermediate boson. This idea, originally suggested by J. Schwinger in 1957, arises naturally by analogy with electromagnetism, where the Coulomb force between two charged particles (say electron and proton) occurs by exchange of a photon, the zero-mass, spin-1 quantum of the electromagnetic field (Fig. 2a). However, in the case of the weak interaction, the intermediate bosons (again vector or spin-1 quanta) must be massive, since the weak interaction has short range. Moreover there must exist at least three such bosons: a charged W^- and its charge-conjugate W^+ to transmit the charged weak interaction (Fig. 2b) and a neutral boson Z^0 for the neutral weak interaction (Fig. 2c). The problem was how to construct a renormalizable theory of weak interactions in which massive vector bosons are exchanged between fundamental fermions.

The solution was indeed found in a theory uniting weak and electromagnetic interactions and based on a combination of subtle ideas. First is the notion that the theory must be invariant under local gauge transformations, that is, gauge transformations which can vary in an arbitrary manner from one space-time point to another. A theory with this property, first considered by Yang and R. L. Mills in 1954, represents a generalization of electrodynamics, and introduces the intermediate bosons, charged and neutral, in a natural way. At this stage the theory is renormalizable, but the bosons are massless, which cannot correspond to reality.

The second important idea, which solves the problem of massless vector bosons, is spontaneous symmetry breaking. This means that one has a

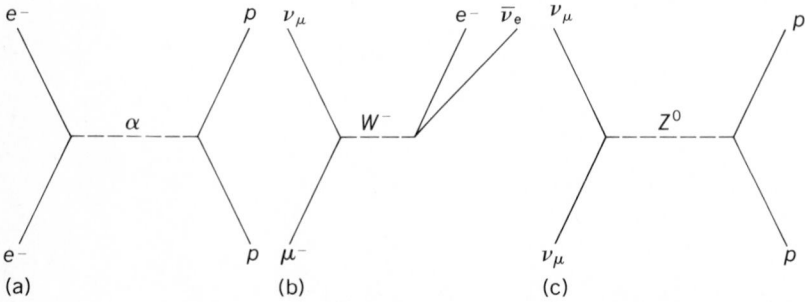

Fig. 2.   Schematic (Feynman) diagrams of electromagnetic and weak interactions, occurring by exchange of intermediate bosons. In each diagram time flows upward. (a) Coulomb interaction (electromagnetic), occurring by photon exchange. (b) Example of charged weak interactions: muon decay ($\mu^- \to e^- \nu_\mu \bar{\nu}_e$), occurring by $W-$ exchange. (c) Example of neutral weak interaction: $\nu_\mu p$ scattering, occurring by Z^0 exchange.

quantum field theory in which the lagrangian possesses a certain symmetry not shared by a particular state of the system (that is, the ground state). When this situation occurs in an ordinary non-gauge field theory, it was shown by J. Goldstone in 1961 that there is a massless spin-0 excitation—the so-called Goldstone boson—corresponding to each degree of freedom in which the symmetry is broken. There is no experimental evidence for such bosons, so it would appear that spontaneous symmetry breaking cannot occur. However, the proof of the Goldstone theorem is based on two assumptions, namely, a positive metric in Hilbert space, and manifest covariance. In a gauge theory (for example electrodynamics, or the Yang-Mills theory) these assumptions cannot both be valid. Thus gauge theories evade the Goldstone theorem, as was first noted by P. Higgs in 1964. A most extraordinary result of these arguments is that the difficulties associated with massless vector bosons and Goldstone bosons neutralize one another: by a suitable transformation the Goldstone bosons disappear from the theory and the vector bosons acquire mass. Weinberg (1967) and independently Salam (1968) thus showed how to construct in a natural way a gauge theory of the Yang-Mills type combining weak and electromagnetic interactions. Essential earlier ideas had also been contributed by S. Glashow. Finally, as was proved by G. 't Hooft in 1971, the Weinberg-Salam theory is renormalizable even after the $W^\pm$ and Z^0 acquire mass.

The Weinberg-Salam model is constructed to conform to known and valid results of quantum electrodynamics and the older charged weak interactions scheme of Feynman and Gell-Mann. The startling new predictions arising from this theory are in the domain of neutral weak interactions, where experiments carried out between 1973 and 1980 confirmed the correctness of the theory. These included neutrino-electron scattering, neutrino-nucleon scattering, parity violation in atoms, and most notably scattering of polarized electrons on nucleons. In the latter two cases one deals with interference between the weak and electromagnetic interactions.

Present and future problems. Although the predictions of the Weinberg-Salam model must be regarded as correct in the low-energy domain (that is, for experiments where the energy momentum transmitted by the intermediate boson is small compared to rest energies of $W^\pm$ or Z^0), many important questions remain unanswered.

Existence of intermediate bosons. It has not been determined whether the $W^\pm$ and Z^0 bosons really exist, and whether their masses are as predicted by Weinberg and Salam, namely, $m_W \sim 70$ GeV/c^2 and $m_Z \sim 90$ GeV/c^2. Much experimental effort will be devoted to direct observations of $W^\pm$ and Z^0.

Fermion masses. The mechanism by which the fundamental fermions acquire mass has not been determined. In the simplest version of the theory, they have zero mass, but additional mass-generating mechanisms can be invoked without spoiling gauge invariance. However, there is then no longer any particular reason to assume that the neutrino masses are zero or even equal. This leads to the important question of coupling between the various types of neutrinos, or as it is called, to neutrino oscillations, which were suggested by B. Pontecorvo. Experiments to detect finite neutrino mass

and the existence of neutrino oscillations have been very actively discussed, but by 1980 no definite results had been obtained. This question has important implications for astrophysics and cosmology.

CP violation. All experimental evidence is consistent with *CPT* (combined charge conjugation, parity, and time reversal) invariance of strong, electromagnetic, and weak interactions. Violation of *CP* invariance manifests itself in the appearance of certain "forbidden" weak decays of the neutral K mesons. The K mesons belong to the $J^P = 0^-$ (spin = 0, negative parity) $SU(3)$ meson octet: (K^+, K^0) and $(\overline{K}^0, K^-)$ form two isodoublets in this octet with strangeness $S = +1$ and $S = -1$, respectively. K^+ and K^- are in fact charge conjugates, as are K^0 and $\overline{K}^0$. The states $|K^0\rangle$, $|\overline{K}^0\rangle$ are by definition eigenstates of the strong and electromagnetic hamiltonians and simultaneously eigenstates of strangeness with eigenvalues ± 1, respectively.

Since the strong and electromagnetic interactions conserve strangeness, the relative phase of $|K^0\rangle$, $|\overline{K}^0\rangle$ is arbitrary and one may define $|\overline{K}^0\rangle \equiv CP|K^0\rangle$. Neither K^0 nor $\overline{K}^0$ has a definite lifetime for weak decay, but one can form two independent linear superpositions of $|K^0\rangle$ and $|\overline{K}^0\rangle$, namely, the states $|K_L^0\rangle$ and $|K_S^0\rangle$ (L and S for "long" and "short," respectively); the particles K_L^0 and K_S^0 do have definite lifetimes but no definite strangeness or isospin. The short-lived K_S^0 decays in only two significant modes: $\pi^+\pi^-$ and $\pi^0\pi^0$ with total inverse lifetime $\gamma_S = 1.134 \times 10^{10}$ s^{-1}. The long-lived component K_L^0 decays in many modes, semileptonic and nonleptonic, and has the total inverse lifetime $\gamma_L = 1.93 \times 10^7$ s^{-1}. Before the discovery of *CP* violation in 1964, it was thought that the states $|K_L^0\rangle$ and $|K_S^0\rangle$ could be expressed by Eq. (10). Since the final states $\pi^+\pi^-$ and $\pi^0\pi^0$ each have

$$|K_S^0\rangle = \frac{1}{\sqrt{2}}\left[|K^0\rangle - |\overline{K}^0\rangle\right]$$

$$|K_L^0\rangle = \frac{1}{\sqrt{2}}\left[|K^0\rangle + |\overline{K}^0\rangle\right] \tag{10}$$

CP eigenvalue +1, this "explained" why $K_S^0 \rightarrow \pi^+\pi^-$ and $K_S^0 \rightarrow \pi^0\pi^0$ are allowed $(CP|K_S^0\rangle = +|K_S^0\rangle)$ and why the decays $K_L^0 \rightarrow \pi^+\pi^-$, $\pi^0\pi^0$ are forbidden $(CP|K_L^0\rangle = -|K_L^0\rangle)$ within the framework of *CP* invariance. However, it has been shown experimentally that the decays $K_L^0 \rightarrow \pi^+\pi^-$, $\pi^0\pi^0$ actually occur (albeit with low probability). Thus Eq. (10) must be replaced by Eq. (11),

$$|K_S^0\rangle = \frac{1}{\sqrt{2(1 + |\epsilon|^2)}}\left[(1 + \epsilon)|K^0\rangle - (1 - \epsilon)|\overline{K}^0\rangle\right]$$

$$|K_L^0\rangle = \frac{1}{\sqrt{2(1 + |\epsilon|^2)}}\left[(1 + \epsilon)|K^0\rangle + (1 - \epsilon)|\overline{K}^0\rangle\right] \tag{11}$$

where Re$(\epsilon) = 1.42 \times 10^{-3}$, Arg $\epsilon = 42.5^0$. It is not known whether *CP* violation (as manifested by nonzero ϵ) originates in the strong, electromagnetic, or weak interactions, or possibly even a new "superweak" interaction. It has been proposed that *CP* violation is connected with Cabibbo rotations, discussed below. *See* EIGENFUNCTION; EIGENVALUE; UNITARY SYMMETRY.

Cabibbo's hypothesis. Although the fundamental quark states for strong interactions are as shown in

Table 1, extensive experimental data on weak interactions of strongly interacting particles show that there exists a "rotation" of quark states d and s through a Cabibbo angle $\theta \simeq 13°$ to a new basis suitable for weak interactions, given by Eqs. (12).

$$d_\theta = d \cos \theta + s \sin \theta$$
$$s_\theta = -d \sin \theta + s \cos \theta \tag{12}$$

Thus the first two quark doublets suitable for weak interactions are shown in notation (13), in-

$$\begin{pmatrix} u \\ d_\theta \end{pmatrix} \quad \begin{pmatrix} c \\ s_\theta \end{pmatrix} \tag{13}$$

stead of those of Table 1. More generally, it has been suggested that a transformation of the three quark states d, s, and b to a new basis may be effected by means of four parameters, three Cabibbo angles and a phase factor δ, and that the phase factor describes CP violation. The physical significance of these transformations has not been determined.

Cause of parity violation. There is as yet no satisfactory explanation for parity violation in weak interactions, perhaps the most striking experimental fact about the weak force.

[EUGENE D. COMMINS]

Bibliography: E. D. Commins, *Weak Interactions*, 1973; R. E. Marshak, Riazuddin, and C. P. Ryan, *Theory of Weak Interactions in Particle Physics*, 1969; L. B. Okun, *Weak Interactions of Elementary Particles*, 1965; A. Salam, Weak and electromagnetic interactions, *Nobel Symposium, Gothenburg*, pp. 367–377, 1968; S. Weinberg, A model of leptons, *Phys. Rev. Lett.*, 19:1264–1266, 1967; C. S. Wu and S. A. Moszkowski, *Beta Decay*, 1966.

Weight

The gravitational weight of a body is the force with which the Earth attracts the body. By extension, the term is also used for the attraction of the Sun or a planet on a nearby body. This force is proportional to the body's mass and depends on the location. Because the distance from the surface to the center of the Earth decreases at higher latitudes, and because the centrifugal force of the Earth's rotation is greatest at the Equator, the observed weight of a body is smallest at the Equator and largest at the poles. The difference is sizable, about 1 part in 300. At a given location, the weight of a body is highest at the surface of the Earth; it diminishes with altitude and with the depth below the surface. For example, the weight of a body diminishes by about 0.1% if it is raised 2 mi above the Earth's surface or taken 4 mi below the surface. Weight also depends to a smaller but measurable degree on the density of the Earth's crust below the body. Weight is measured by several procedures.

Since weight is a force, it is expressed in force units. In the United States, the commonest unit of weight is the pound, sometimes written pound force or pound weight, to distinguish it from the mass unit, pound. Pound weight is the weight of a 1-pound mass at a location where the acceleration of gravity is 32.174 ft/sec². Where the acceleration of gravity is g, the weight of a 1-pound mass is $g/32.174$ pound weight.

In terms of the international kilogram the pound avoirdupois is now defined as being exactly equivalent to 0.45359237 kilogram. In relation to three smaller avoirdupois weight units, 1 pound equals 16 ounces, 256 drams, and 7000 grains. Since the relation of the pound to the kilogram is divisible by 7, 1 grain equals 0.06479891 gram, exactly.

Besides the avoirdupois pound there is the troy or apothecary pound, in which there are 5760 grains, so that this pound is equal to 576/700 avoirdupois pound. In general, the term pound is taken to mean the avoirdupois pound unless definitely stated otherwise.

The carat is a unit of weight used in evaluating precious stones, and 1 carat equals 200 milligrams.

[HOWARD S. BEAN]

Bibliography: A. V. Astin, *Refinement of Values for the Yard and Pound*, Fed. Regist. Doc. no. 59-5442, July 1, 1959; U.S. Department of Commerce, National Bureau of Standards, *Units of Weight and Measure: Definitions and Tables of Equivalents*, Misc. Publ. no. 286, May, 1967.

Wentzel-Kramers-Brillouin method

A special technique for obtaining an approximation to the solutions of the one-dimensional time-independent Schrödinger equation, valid when the wavelength of the solution varies slowly with position; also known as the WKB method. It is named after G. Wentzel, H. A. Kramers, and L. Brillouin, who independently in 1926 contributed to its understanding in the quantum-mechanical application. It was, however, studied earlier by J. Liouville (1837), Lord Rayleigh (1912), and H. Jeffreys (1923). It is also called the BWK method, and the JWKB method, the classical approximation, the quasi-classical approximation, and the phase integral method.

The system to be considered is a particle of mass m moving nonrelativistically in a potential $V(x)$. The Schrödinger equation for stationary states of energy E is Eq. (1), where $p^2(x)$ is defined

$$\frac{d^2\psi}{dx^2} + \frac{p^2(x)\psi}{\hbar^2} = 0 \tag{1}$$

by Eq. (2) and $\hbar$ is Planck's constant divided by 2π.

$$p^2(x) = 2m\,[E - V(x)] \tag{2}$$

Two types of regions are to be considered as illustrated in Fig. 1: (I) The classically allowed type in which $p^2(x)$ is positive. Here $p(x)$ is defined to be the positive root, and it is the magnitude of the classical momentum. (II) The classically unallowed type in which $p^2(x)$ is negative. Here $p(x)$ is defined to be $i|p(x)|$. Typically $p^2(x)$ goes linearly through zero at the point where the region changes from classically allowed to unallowed. Such a point x_1, where $p^2(x_1) = 0$, is a turning point of the classical motion.

The solutions of differential equation (1) can be written, in a formal way, as Eq. (3). In this expres-

$$\psi = \exp\left[\pm \frac{i}{\hbar} \int dx \, \sqrt{p^2(x)} \right. $$
$$\left. \pm i\hbar \frac{d}{dx} \sqrt{p^2(x)} \pm i\hbar \frac{d}{dx} \sqrt{\cdots} \right] \tag{3}$$

sion one uses all the upper signs uniformly or all the lower signs uniformly to obtain the two independent solutions. The constants in the indefinite integrals are the integration constants for the differential equation. The symbol $\sqrt{}$ indicates that the root of all that follows is to be taken, choosing the sign as suggested by the convention given above for $p(x)$. One verifies this formal expression of the solutions through Eqs. (4). Thus if the

$$\frac{d}{dx}\exp\left[\,\right]=\left[\pm\frac{i}{\hbar}\sqrt{p^2(x)\pm i\hbar\frac{d}{dx}\sqrt{\cdots}}\,\right]\exp\left[\,\right] \quad (4a)$$

$$\frac{d^2}{dx^2}\exp\left[\,\right]=\left[\pm\frac{i}{\hbar}\frac{d}{dx}\sqrt{p^2(x)\pm\cdots}\,\right]\exp\left[\,\right]$$

$$+\left[-\frac{p^2(x)}{\hbar^2}\mp\frac{i}{\hbar}\frac{d}{dx}\sqrt{\cdots}\,\right]\exp\left[\,\right]$$

$$=-\frac{p^2(x)}{\hbar^2}\exp\left[\,\right] \quad (4b)$$

process indicated in Eq. (3) converges, it produces the solutions of the differential equation.

The WKB approximation to the solutions is found by evaluating the process to two orders only, Eq. (5). In a classically unallowed region the

$$\psi\simeq\exp\left[\pm\frac{i}{\hbar}\int dx\,\sqrt{p^2(x)\pm i\hbar\frac{dp}{dx}(x)}\,\right]$$

$$\simeq\exp\left[\pm\frac{i}{\hbar}\int dx\left\{p\pm\frac{i\hbar}{2p}\frac{dp}{dx}\right\}\right]$$

$$=\frac{1}{\sqrt{p}}\exp\left[\pm\frac{i}{\hbar}\int dx\,p(x)\right] \quad (5)$$

approximate solutions are usually written as Eq. (6).

$$\psi=\frac{1}{\sqrt{|p|}}\exp\left[\mp\frac{1}{\hbar}\int dx\,|p(x)|\right] \quad (6)$$

Evidently the requirement for the applicability of the approximation is that $\hbar\,dp/dx$ should be small compared to p^2. In terms of the local de Broglie wavelength $\lambda=h/p$, this requirement is that $d\lambda/dx$ should be small compared to 2π. The WKB approximation therefore applies in the case of slowly varying wavelength.

Connection formulas. The approximation ordinarily applies in any region which does not include turning points, but it breaks down as $p^2(x)$ approaches zero, and in fact the approximate solution diverges as $p^{-1/2}$ at a turning point. There are connection formulas which relate the approximate solutions in a classically allowed region to the approximate solutions in a neighboring classically unallowed region. For the case illustrated in Fig. 1 the formulas are given by Eqs. (7). The integrals

$$\frac{1}{\sqrt{p}}\sin\left[\frac{1}{\hbar}\int_x^{x_1}p(\xi)\,d\xi+\frac{\pi}{4}\right]\leftrightarrow$$

$$\frac{1}{2}\frac{1}{\sqrt{|p|}}\exp\left[-\frac{1}{\hbar}\int_{x_1}^x|p(\xi)|\,d\xi\right] \quad (7a)$$

$$\frac{1}{\sqrt{p}}\cos\left[\frac{1}{\hbar}\int_x^{x_1}p(\xi)\,d\xi+\frac{\pi}{4}\right]\leftrightarrow$$

$$\frac{1}{\sqrt{|p|}}\exp\left[\frac{1}{\hbar}\int_{x_1}^x|p(\xi)|\,d\xi\right] \quad (7b)$$

are always written in these formulas so that they

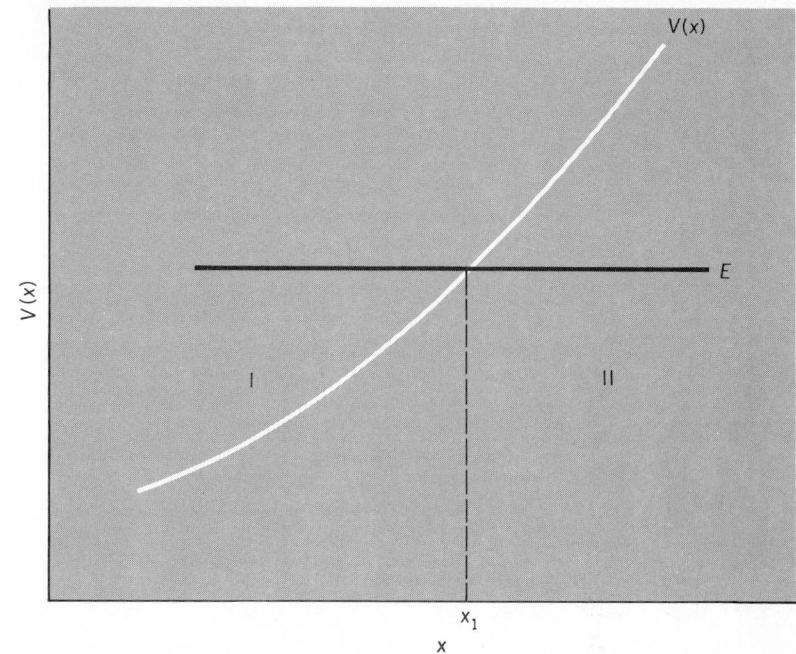

Fig. 1.   A classically allowed region I to the left of a classically unallowed region II.

are increasing functions of x away from the turning point. Similar formulas, given in Eqs. (8),

$$\frac{1}{2}\frac{1}{\sqrt{|p|}}\exp\left[-\frac{1}{\hbar}\int_x^{x_2}|p(\xi)|\,d\xi\right]\leftrightarrow$$

$$\frac{1}{\sqrt{p}}\,\sin\left[\frac{1}{\hbar}\int_{x_2}^x p(\xi)\,d\xi+\frac{\pi}{4}\right] \quad (8a)$$

$$\frac{1}{\sqrt{|p|}}\exp\left[\frac{1}{\hbar}\int_x^{x_2}|p(\xi)|\,d\xi\right]\leftrightarrow$$

$$\frac{1}{\sqrt{p}}\,\cos\left[\frac{1}{\hbar}\int_{x_2}^x p(\xi)\,d\xi+\frac{\pi}{4}\right] \quad (8b)$$

apply when the allowed region is to the right, as illustrated in Fig. 2. The proof of these formulas is too long to be given in this article.

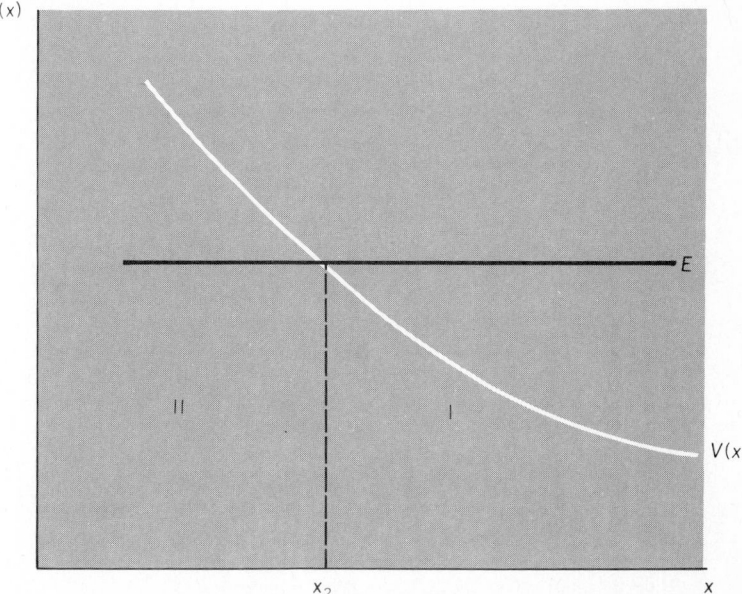

Fig. 2.   Classically allowed region I to the right of classically unallowed region II.

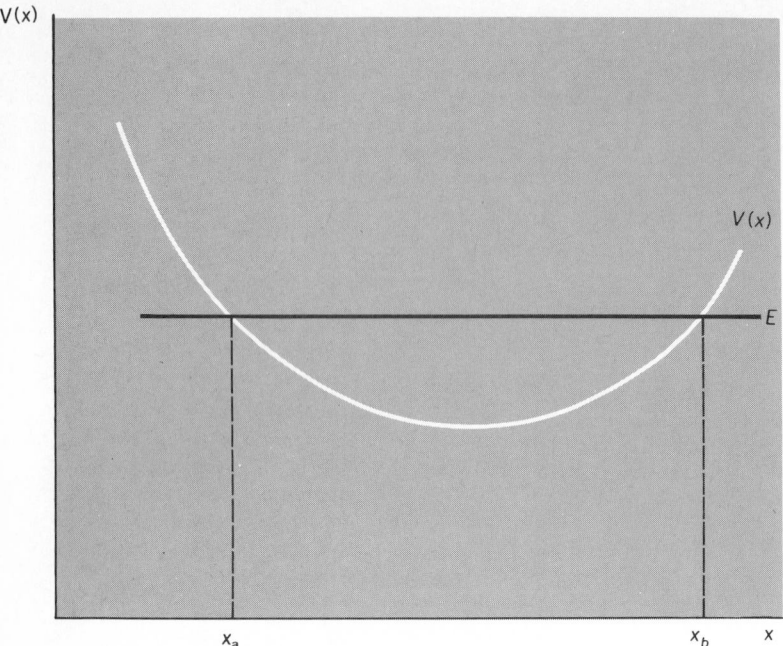

Fig. 3. A potential well with turning points at x_a and x_b.

$$\frac{A}{\sqrt{p}}\sin\left[\frac{1}{\hbar}\int_{x_a}^{x}p(\xi)\,d\xi+\frac{\pi}{4}\right]$$
$$=\frac{B}{\sqrt{p}}\sin\left[\frac{1}{\hbar}\int_{x}^{x_b}p(\xi)\,d\xi+\frac{\pi}{4}\right] \quad (9)$$

$$\frac{1}{\hbar}\int_{x_a}^{x_b}p(\xi)\,d\xi=\left\{n+\frac{1}{2}\right\}\pi \quad (10)$$

0,1,2,3, ... This condition alternatively can be written as in Eq. (11). For each allowed value of n there is an energy E_n that satisfies this condition;

$$2\int_{x_a}^{x_b}\sqrt{2m\,[E-V(\xi)]}\,d\xi=\left\{n+\frac{1}{2}\right\}h \quad (11)$$

these are the allowed energy levels of the potential well, in the WKB approximation.

It is at this point that one can see the connection between modern quantum mechanics and the old quantum theory of N. Bohr and A. Sommerfeld. In the old quantum theory the allowed energy levels were found by setting the integral of the momentum, over a complete cycle, equal to an integer times Planck's constant. This condition coincides with Eq. (11), except that half-integer numbers are called for.

Barrier penetration. Another important application of the WKB approximation is in problems of barrier penetration. Consider a one-dimensional potential barrier as shown in Fig. 4, and suppose there is a particle beam of energy E incident on the barrier from the left. In the classical problem the particles of energy less than the peak potential energy would simply be reflected, but in the quantum-mechanical problem the particle beam is partly reflected and partly transmitted through the classically unallowed region. The transmission coefficient D is defined as the ratio of the transmitted particle current to the incident current. By applying the connection formulas straightforwardly at the turning points and approximating for small D, one finds Eq. (12). This result has wide

$$D=\exp\left[-\frac{2}{\hbar}\int_{a}^{b}|p(\xi)|\,d\xi\right] \quad (12)$$

applicability. It was used, for example, in the Fowler-Nordheim theory of field emission and in G. Gamow's explanation of α-decay. *See* FIELD EMISSION; POTENTIAL BARRIER; RADIOACTIVITY.

Generalization. The WKB approximation is based on the exponential function, the function that occurs in the exact solution of the free-particle Schrödinger equation. S. C. Miller and R. H. Good, Jr., developed a generalization of the WKB approximation, based on the functions that occur in the solution of an arbitrary comparison equation. The comparison equation is chosen so as to be similar to the equation under study and so as to be soluble, in the same spirit as perturbation theory. Qualitatively better approximations can be set up this way; for example, in the potential well problem, approximate solutions finite through both turning points can be found. *See* NONRELATIVISTIC QUANTUM THEORY; PERTURBATION (QUANTUM MECHANICS); QUANTUM MECHANICS.

[ROLAND H. GOOD, JR.]

Bibliography: S. C. Miller and R. H. Good, Jr., A WKB-type approximation for the Schrödinger

Potential well. An important application of the WKB approximation is in obtaining an approximation to the allowed energy levels in a potential well. Suppose the well is as shown in Fig. 3. Only integrable functions, with decreasing exponential dependence, may occur in the classically unallowed regions to the left of x_a and to the right of x_b. Applying connection formulas (8a) and (7a) at the turning points x_a and x_b, one obtains two expressions for the approximate solution in the allowed region, Eq. (9). In order for these two expressions to coincide, it is necessary that Eq. (10) be satisfied, where n indicates one of the values

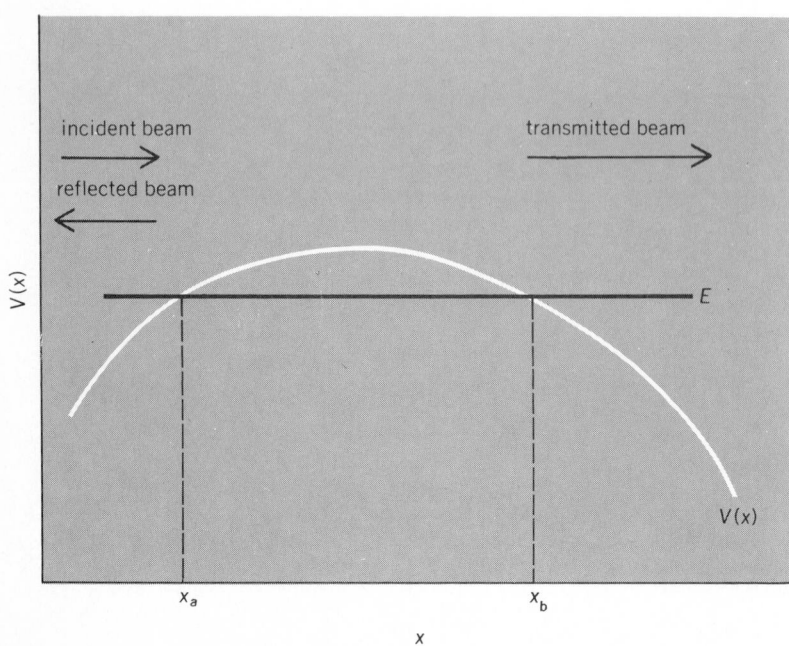

Fig. 4. A potential barrier with turning points at x_a and x_b.

equation, *Phys. Rev.*, 91, 174–179, 1953; E. L. Murphy and R. H. Good, Jr., WKB connection formulas, *J. Math. Phys.*, 43:251–254, 1964; D. Park, *Introduction to the Quantum Theory*, 2d ed., 1974; L. I. Schiff, *Quantum Mechanics*, 3d ed., 1968.

Wiedemann-Franz law

An empirical law of physics which states that the ratio of the thermal conductivity of a metal to its electrical conductivity is a constant times the absolute temperature, as given by Eq. (1). Here K_c is

$$K_c = L_0 \sigma T \qquad (1)$$

the thermal conductivity due to the conduction electrons, σ is the electrical conductivity, T is the absolute temperature, and L_0 is known as the Lorentz number. For the case of a degenerate electron gas, the value of L_0 is given by Eq. (2), where k

$$L_0 = (\pi^2/3)(k/e)^2 \qquad (2)$$

is the Boltzmann constant and e is the electronic charge. The Wiedemann-Franz law provides an important check on theories of electrical and thermal conductivity. *See* CONDUCTION (HEAT); FREE-ELECTRON THEORY OF METALS; THERMAL CONDUCTION IN SOLIDS.

[FRANK J. BLATT]

Bibliography: R. Berman, *Thermal Conduction in Solids*, 1976; C. Kittel, *Introduction to Solid-State Physics*, 5th ed., 1976; P. G. Klemens, Thermal conductivity of solids at low temperatures, *Encyclopedia of Physics*, vol. 19, 1956; J. M. Ziman, *Principle of the Theory of Solids*, 2d ed., 1979.

Work

In physics, the term work refers to the transference of energy that occurs when a force is applied to a body that is moving in such a way that the force has a component in the direction of the body's motion. Thus work is done on a weight that is being lifted, or on a spring that is being stretched or compressed, or on a gas that is undergoing compression in a cylinder.

When the force acting on a moving body is constant in magnitude and direction, the amount of work done is defined as the product of just two factors: the component of the force in the direction of motion, and the distance moved by the point of application of the force. Thus the defining equation for work W is Eq. (1), where f and s are the

$$W = f \cos \phi \cdot s \qquad (1)$$

magnitudes of the force and displacement, respectively, and ϕ is the angle between these two vector quantities (Fig. 1). Because $f \cos \phi \cdot s = f \cdot s \cos \phi$, work may be defined alternatively as the product of the force and the component of the displacement in the direction of the force. In Fig. 2 the work of the constant force f when the application point moves along the curved path from P to P', and therefore undergoes the displacement $\overline{PP'}$, is $f \cdot \overline{PP'} \cos \phi$, or $f' \overline{PE}$.

Work is a scalar quantity. Consequently, to find the total work done on a moving body by several

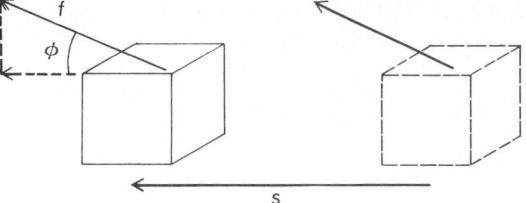

Fig. 1. Work of constant force f is $fs \cos \phi$.

different forces, the work of each may be computed separately and the ordinary algebraic sum taken.

Examples and sign conventions. Suppose that a car slowly rolls forward a distance of 10 m along a straight driveway while a man pushes on it with a constant magnitude of 200 newtons of force (200 N) and let Eq. (1) be used to compute the work W done under each of the following circumstances: (1) If the man pushes straight forward, in the direction of the car's displacement, then $\phi = 0°$, $\cos \phi = 1$, and $W = 200$ N $\times 1 \times 10$ m $= 2000$ N $\cdot$ m $= 2000$ joules; (2) if he pushes in a sideways direction making an angle ϕ of 60° with the displacement, then $\cos 60° = 0.50$ and $W = 1000$ joules; (3) if he pushes against the side of the car and therefore at right angles to the displacement, $\phi = 90°$, $\cos \phi = 0$, and $W = 0$; (4) if he pushes or pulls backward, in the direction opposite to the car's displacement, $\phi = 180°$, $\cos \phi = -1$, and $W = -2000$ joule.

Notice that the work done is positive in sign whenever the force or any component of it is in the same direction as the displacement; one then says that work is being done *by* the agent exerting the force (in the example, the man) and *on* the moving body (the car). The work is said to be negative whenever the direction of the force or force component is opposite to that of the displacement; then work is said to be done *on* the agent (the man) and *by* the moving body (the car). From the point of view of energy, an agent doing positive work is losing energy to the body on which the work is done, and one doing negative work is gaining energy from that body.

Units of work and energy. These consist of the product of any force unit and any distance unit. Units in common use are the foot-pound, the foot-poundal, the erg, and the joule. The product of any

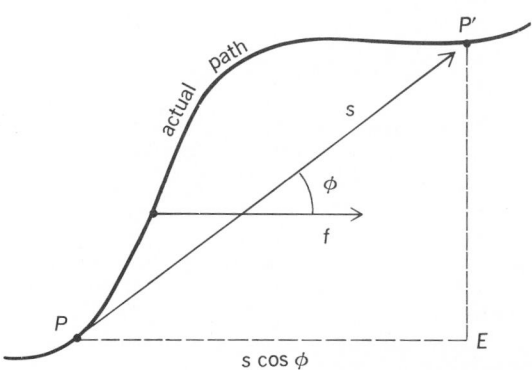

Fig. 2. The work done in traversing any path connecting points P and P' is $f \cdot \overline{PE}$, assuming the force f to be constant in magnitude and direction.

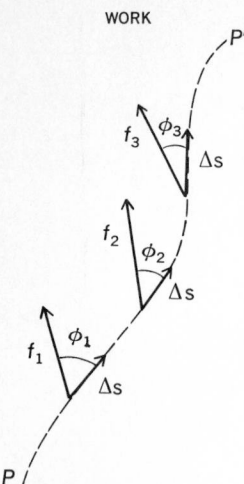

Fig. 3. Work done by a variable force.

power unit and any time unit is also a unit of work or energy. Thus the horsepower-hour (hp-hr) is equivalent, in view of the definition of the horsepower, to 550 ft-lbf/sec × 3600 sec, or 1,980,000 ft-lbf, or (1,980,000)(0.3048 m)(4.45 N) = 2,684,520 joules. Similarly, the watt-hour is 1 joule/sec × 3600 sec, or 3600 joule; and the kilowatt-hour is 3,600,000 joule.

Work of a torque. When a body which is mounted on a fixed axis is acted upon by a constant torque of magnitude τ and turns through an angle θ (radians), the work done by the torque is $\tau\theta$.

Work principle. This principle, which is a generalization from experiments on many types of machines, asserts that, during any given time, the work of the forces applied to the machine is equal to the work of the forces resisting the motion of the machine, whether these resisting forces arise from gravity, friction, molecular interactions, or inertia. When the resisting force is gravity, the work of this force is mgh, where mg is the weight of the body and h is the vertical distance through which the body's center of gravity is raised. Note that if a body is moving in a horizontal direction, h is zero and no work is done by or against the gravitational force of the Earth. If a person holds an object or carries it across level ground, he does no net work against gravity; yet he becomes fatigued because his tensed muscles continually contract and relax in minute motions, and in walking he alternately raises and lowers the object and himself.

The resisting force may be due to molecular forces, as when a coiled elastic spring is being compressed or stretched. From Hooke's law, the average resisting force in the spring is $-\frac{1}{2}ks$, where k is the force constant of the spring and s is the displacement of the end of the spring from its normal position; hence the work of this elastic force is $-\frac{1}{2}ks^2$.

If a machine has any part of mass m that is undergoing an acceleration of magnitude a, the resisting force $-ma$ which the part offers because of its inertia involves work that must be taken into account; the same principle applies to the resisting torque $-I\alpha$ if any rotating part of moment of inertia I undergoes an angular acceleration α.

When the resisting force arises from friction between solid surfaces, the work of the frictional force is $-\mu f_n s$, where μ is the coefficient of friction for the pair of surfaces, f_n is the normal force pressing the two surfaces together, and s is the displacement of the one surface relative to the other during the time under consideration. The frictional force μf_n and the displacement s giving rise to it are always opposite in direction ($\phi = 180°$).

The work done by any conservative force, such as a gravitational, elastic, or electrostatic force, during a displacement of a body from one point to another has the important property of being path-independent: Its value depends only on the initial and final positions of the body, not upon the path traversed between these two positions. On the other hand, the work done by any nonconservative force, such as friction due to air, depends on the path followed and not alone on the initial and final positions, for the direction of such a force varies with the path, being at every point of the path tangential to it. *See* FORCE.

Since work is a measure of energy transfer, it can be calculated from gains and losses of energy. It is useful, however, to define work in terms of forces and distances or torques and angles because these quantities are often easier to measure than energy changes, especially if energy changes are produced by nonconservative forces.

Work of a variable force. If the force varies in magnitude and direction along the path $\overline{PP'}$ of its point of application, one must first divide the whole path into parts of length Δs, each so short that the force component $f\cos\phi$ may be regarded as constant while the point of application traverses it (Fig. 3). Equation (1) can then be applied to each small part and the resulting increments of work added to find the total work done. Various devices are available for measuring the force component as a function of position along the path. Then a work diagram can be plotted (Fig. 4). The total work done between positions s_1 and s_2 is represented by the area under the resulting curve between s_1 and s_2 and can be computed by measuring this area, due allowance being made for the scale in which the diagram is drawn.

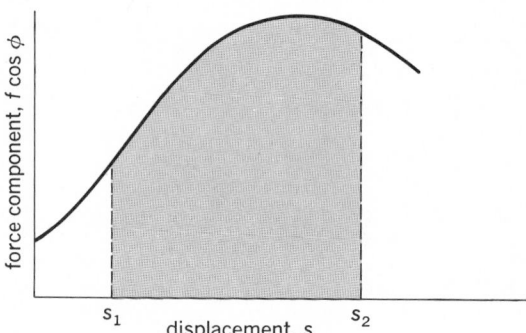

Fig. 4. A work diagram.

For an infinitely small displacement ds of the point of application of the force, the increment of work dW is given by Eq. (2), a differential expression

$$dW = f\cos\phi\,ds \qquad (2)$$

sion that provides the most general definition of the concept of work. In the language of vector analysis, dW is the scalar product of the vector quantities $\mathbf{f}$ and $d\mathbf{s}$; Eq. (2) then takes the form $dW = \mathbf{f} \cdot d\mathbf{s}$. If the force is a known continuous function of the displacement, the total work done in a finite displacement from point P to point P' of the path is obtained by evaluating the line integral in Eq. (3).

$$W = \int_P^{P'} f\cos\phi\,ds = \int_P^{P'} \mathbf{f} \cdot d\mathbf{s} \qquad (3)$$

When a variable torque of magnitude τ acts on a body mounted on a fixed axis, the work done is given by $W = \int_{\theta_1}^{\theta_2} \tau\,d\theta$, where $\theta_2 - \theta_1$ is the total angular displacement expressed in radians. *See* ENERGY. [LEO NEDELSKY]

Bibliography: F. Bueche, *Understanding the World of Physics*, 1981; D. Halliday and R. Resnick, *Fundamentals of Physics*, 2d ed., 1981; E. M. Rogers, *Physics for the Inquiring Mind*, 1960.

Work function (electronics)

A quantity with the dimensions of energy which determines the thermionic emission of a solid at a given temperature. The thermionic electron current density J emitted by the surface of a hot conductor at a temperature T is given by the Richardson formula, $J = AT^2 e^{-\phi/kT}$, where A is a constant, k is Boltzmann's constant ($= 1.38 \times 10^{-23}$ joule per degree Celsius) and ϕ is the work function; the last may be determined from a plot of log (J/T^2) versus $1/T$. For metals, ϕ may also be determined by measuring the photoemission as a function of the frequency of the incident electromagnetic radiation; ϕ is then equal to the minimum (threshold) frequency for which electon emission is observed times Planck's constant h ($= 6.63 \times 10^{-34}$ joule sec). The work function of a solid is usually expressed in electronvolts (1 eV is the energy gained by an electron as it passes through a potential difference of 1 volt, and is equal to 1.60×10^{-19} joule). A list of average values of work functions (in electronvolts) for metals is given in the table.

Average values of work functions for metals, in electronvolts

Metal	Value	Metal	Value	Metal	Value
Al	4.20	Cs	1.93	Na	2.28
Ag	4.46	Cu	4.45	Ni	4.96
Au	4.89	Fe	4.44	Pd	4.98
Ba	2.51	K	2.22	Pt	5.36
Cd	4.10	Li	2.48	Ta	4.13
Co	4.41	Mg	3.67	W	4.54
Cr	4.60	Mo	4.24	Zn	4.29

The work function of metals varies from one crystal plane to another and also varies slightly with temperature (approximately 10^{-4} eV/degree). For a metal, the work function has a simple interpretation. At absolute zero, the energy of the most energetic electrons in a metal is referred to as the Fermi energy; the work function of a metal is then equal to the energy required to raise an electron with the Fermi energy to the energy level corresponding to an electron at rest in vacuum. The work function of a semiconductor or an insulator has the same interpretation, but in these materials the Fermi level is in general not occupied by electrons and thus has a more abstract meaning. *See* FIELD EMISSION. [ADRIANUS J. DEKKER]

Work function (thermodynamics)

A thermodynamic function, also called the work content, Helmholtz free energy, or by the European school, simply the free energy. It is defined as the internal energy E of a system minus the temperature-entropy product, TS, and has a characteristic value for each state of a system. In an isothermal process, the maximum work which can be done by a system is equal to the decrease in its work function. When a process such as a chemical reaction occurs spontaneously at constant temperature and volume, it is characterized by a decrease in the work function. Consequently, the criterion for equilibrium under these conditions is that the work function for the system should be at a minimum. *See* FREE ENERGY.

[PAUL BENDER; WILLIAM A. STEELE]

X-ray crystallography

The study of crystal structure by x-ray diffraction techniques. The prediction in 1912 by the German physicist Max von Laue that crystals might be employed as natural diffraction gratings in the study of x-rays was experimentally verified in the same year by W. Friedrich and P. Knipping, who obtained diffraction patterns photographically by the so-called Laue method. Almost immediately after (1913), W. Lawrence Bragg not only successfully analyzed the structures of NaCl and KCl by Laue photographs but also developed a simple treatment of x-ray scattering by a crystal (the Bragg law) which proved much easier to apply than the more complicated but equivalent Laue theory of diffraction. The availability of the first x-ray spectrometer, constructed by his father, William H. Bragg, as well as the substitution of monochromatic (single wavelength) rather than polychromatic x-ray radiation, enabled W. Lawrence Bragg to determine a number of simple crystal structures, including those of diamond; zincblende, ZnS; fluorspar, CaF_2; and pyrites, FeF_2. Since the inauguration of x-ray crystallography as a science, x-ray diffraction has become a powerful tool for the investigation of both the structure of crystals and the nature of x-rays. This article is concerned with the former application. For the theoretical and experimental aspects of x-ray diffraction *see* X-RAY DIFFRACTION.

Structurally, a crystal is a three-dimensional periodic arrangement in space of atoms, groups of atoms, or molecules. If the periodicity of this pattern extends throughout a given piece of material, one speaks of a single crystal. The exact structure of any given crystal is determined if the locations of all atoms making up the three-dimensional periodic pattern called the unit cell are known. The very close and periodic arrangement of the atoms in a crystal permits it to act as a diffraction grating for x-rays. W. Lawrence Bragg treated the phenomenon of the interference of x-rays with crystals as if the x-rays were being reflected by successive parallel equidistant planes of atoms in the crystal. His important equation relating the perpendicular spacing d of lattice planes in a crystal, the glancing angel θ of the reflected beam, and the x-ray wavelength λ is $n\lambda = 2d \sin \theta$. This expression provides the basic condition that the difference in path length for waves reflected from successive planes must be an integral number of wavelengths $n\lambda$ in order for the waves reflected from a given set of lattice planes to be in phase with one another. Instead of referring to the nth order of a reflected beam, modern crystallographers customarily redefine Bragg's equation as $\lambda = 2d_{hkl} \sin \theta$, where d_{hkl} represents the perpendicular interplanar distance between adjacent lattice planes having the Miller indices (hkl). According to this viewpoint, any nth-order diffraction maxima for waves reflected from a set of planes (hkl) with spacing d_{hkl} is equivalent to a first-order reflection due to a parallel set of planes (nh, nk, nl) with the perpendicular distance $d_{nh,nk,nl} = d_{hkl}/n$.

For a crystal with known unit-cell size and shape and with arbitrary orientation in a parallel beam of x-rays of known wavelength, several questions must be answered. First, it must be ascertained

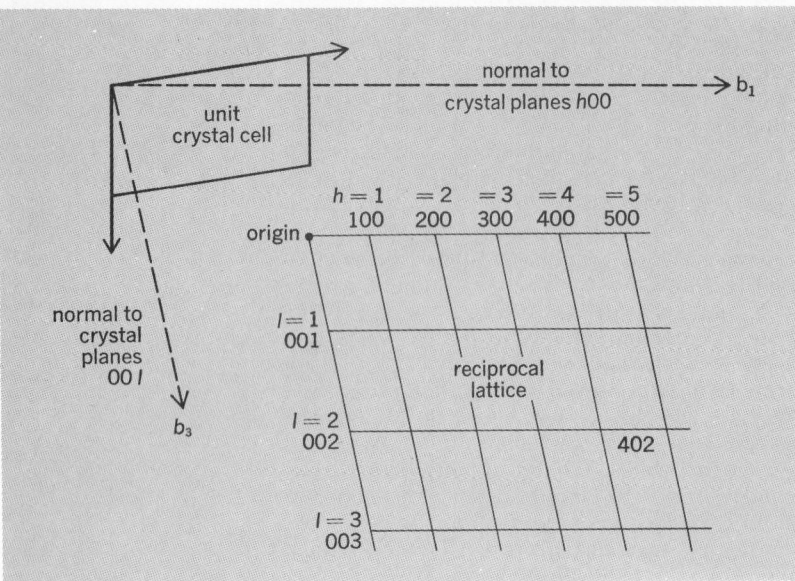

Fig. 1. Reciprocal lattice.

what plane, if any, is obeying Bragg's law and is reflecting the rays. Second, the direction of the reflected x-ray must be accurately measured. The Bragg law treatment does not permit these questions to be answered readily, and for this reason the concept of a reciprocal lattice model was introduced. *See* CRYSTAL STRUCTURE; CRYSTALLOGRAPHY.

Reciprocal lattice. The development and application of the reciprocal lattice to x-ray crystallography have been primarily the results of work by P. P. Ewald (1913 and 1921), J. D. Bernal (1926), and M. J. Buerger (1935). In general, a reciprocal lattice consists of a three-dimensional array of points which is related to the crystal lattice (commonly called the direct lattice) in that each set of planes (hkl) in the crystal lattice is represented in reciprocal space by a point denoted by the coordinates hkl (without parentheses). Figure 1 illustrates in two dimensions the reciprocal lattice geometrically produced in the following way from a unit crystal cell. An origin is chosen and given the coordinates 000. To every set of parallel planes of Miller index (hkl) in direct space a reciprocal lattice vector B_{hkl}, which is perpendicular to this set

of planes, is constructed from the reciprocal lattice origin at a distance inversely proportional to the interplanar spacing d_{hkl}. Thus, the set of parallel direct-lattice planes (402) with an interplanar spacing d_{402} is represented in reciprocal space in Fig. 1 by the point 402. The normal vector B_{402}, which is directed along the perpendicular of the (402) direct-lattice planes from the origin of the reciprocal lattice to the point with coordinates 402, is of length $|B_{402}| = K/d_{402}$, where K is an arbitrary constant. The value of K which simply scales the reciprocal lattice is normally taken as unity or as the wavelength λ; for purposes of this discussion it is taken as 1. All the mathematical definitions of the reciprocal lattice vectors given in the section on reciprocal lattice in the article "X-ray diffraction" in this encyclopedia are based on a K value of unity. It can be shown that this procedure applied to a three-dimensional periodic crystal lattice will result in a three-dimensional reciprocal lattice. Each point of coordinates hkl in this reciprocal lattice will be reciprocal to a set of planes (hkl) in the direct crystal lattice.

Sphere of reflection. The geometrical interpretation of x-ray diffraction from a crystal can be best interpreted by the use of a "sphere of reflection" in reciprocal space as illustrated in Fig. 2. A sphere of radius $1/\lambda$ (corresponding to $K = 1$) is drawn with the direction of the primary x-ray beam (that is, both the incident and transmitted rays) denoted by the unit vector s_0 assumed to travel along a diameter. The crystal C is imagined to be at the center of this sphere of reflection. The origin 000 of the reciprocal lattice is placed at the point 0, where the transmitted beam emerges from the sphere of reflection. A diffracted beam will be formed if the surface of the sphere of reflection intercepts any reciprocal lattice point (other than the origin 0). In Fig. 2 the reciprocal lattice point P lies on the surface of the sphere of reflection. Hence, two important results follow: (1) The set of crystal planes (hkl) to which this point is the reciprocal obeys Bragg's law and reflects the incident x-ray beam; (2) the direction of the diffracted rays will be from the point C at the center of the sphere to where the point P lies on the surface of the sphere. The angle of diffraction between the diffracted beam s and the primary beam s_0 is 2θ.

The reciprocal lattice, by virtue of its definition, is tied to the actual lattice insofar as orientation is concerned. As the crystal (and therefore the crystal lattice) is turned about an axis of rotation, the reciprocal lattice, in turning a similar angle about a parallel axis through its origin 0, passes through the sphere of reflection (which is fixed for the primary x-ray beam that is stationary relative to the movement of the crystal). In the rotating crystal method all diffraction maxima that can possibly be recorded by any chosen x-ray wavelength λ must be represented by those points hkl of the reciprocal lattice that cut the sphere of reflection. These points of possible diffraction maxima all lie within a sphere of radius $2/\lambda$ called the limiting sphere. This geometrical interpretation of Bragg's law enables one to understand readily both the geometry of reflection of x-rays and the manipulation of all the single-crystal cameras and diffractometers now in use.

Recording techniques. Since x-rays are scattered by the electrons of the atoms, the intensity of

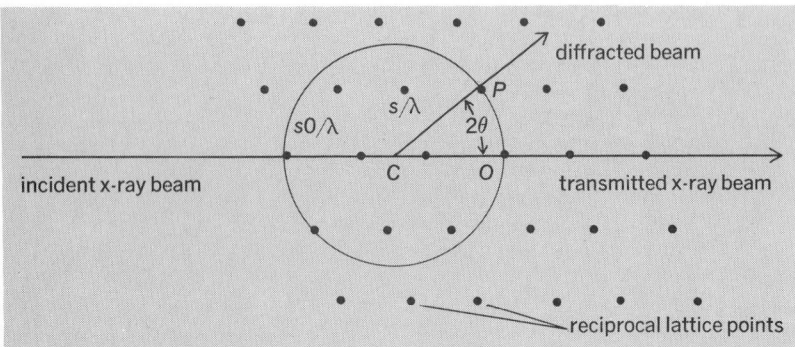

Fig. 2. Sphere of reflection in reciprocal space. Black dots represent reciprocal lattice points. Diffraction occurs when any reciprocal lattice point hkl cuts sphere of reflection, as at point P.

each diffracted beam depends on the positions of the atoms in the unit cell. Any alteration in atomic coordinates would result in changes of the intensities relative to one another. Hence, the first important step in a crystal-structure determination is concerned with the collection of the intensity data by a recording technique which effectively measures the intensity at each reciprocal lattice point.

In the early days of x-ray crystallography, the classical Bragg x-ray spectrometer was used with an ionization chamber to collect data on the intensity of x-ray reflections. It was soon found possible and more convenient to use photographic recording for such data collection. Until the early 1960s the great majority of structural determinations were based on photographically recorded intensities which were usually visually estimated. Nevertheless, this process of photographic data collection is time-consuming, and normally several months are needed to record, judge, average, and scale the intensities of three-dimensional film data. The estimated level of accuracy of the relative intensities obtained photographically is 15–20%, which is generally sufficient for the solution of stereochemical and conformational problems but not necessarily for a reliable determination of atomic thermal motion or bonding. Since about 1953 highly stabilized x-ray diffraction units and sensitive, reliable detectors such as Geiger, scintillation, and proportional counters have been developed. This modern instrumentation, coupled with the demand for making the intensity collection from crystals both more rapid and more accurate (especially that from proteins because of their instability and large number of data to be recorded), has resulted in a gradual return to direct recording by counter techniques. The commercial availability of automatic diffractometers which use computed circuitry to synchronize the movement of the crystal and the detector has effectively revolutionized data collection from crystals such that upward of 200–1000 diffracted beams can be obtained daily with an error factor within 5%. *See* SCINTILLATION COUNTER.

Structural analysis. Although the size and shape of the unit cell determine the geometry of the diffraction maxima, the intensity of each reflection is determined by the number, character, and distribution of the atoms within the unit cell. The second stage in a structural analysis is the solution of the phase relations among the diffracted beams (that is, the phase problem) from which a correct trial structure can be obtained. The intensity of each diffracted beam is related to the square of its amplitude (that is, $I_{hkl} = k|F_{hkl}|^2$, where $|F_{hkl}|$ is the observed structure factor amplitude). Each diffracted beam has not only a characteristic intensity but also a characteristic phase angle α_{hkl} associated with it which expresses the degree to which the diffracted beam is in plane with the other diffracted beams.

Because the electron density is a real, positive quantity which varies continuously and periodically in a crystal, the electron scattering density $\rho(xyz)$ is derivable from the three-dimensional Fourier series as in Eq. (1), where $\rho(xyz)$ represents the electron density at any point with fractional coordinates x, y, z in the unit cell of volume V. Hence, if the characteristic amplitude $|F_{hkl}|$ and its phase α_{hkl} for each diffracted beam are known, the electron density can be calculated at fractional grid points in the unit cell. A properly phased three-dimensional electron-density map effectively provides peaks characteristic of direct images of atoms in the unit cell. The phase problem in x-ray crystallography arises because experimental measurements yield only the magnitudes of the structure factors $|F_{hkl}|$ but not the phases α_{hkl}. Hence, a structural analysis involves a search for the characteristic phases to be utilized together with the observed amplitudes in order to obtain a three-dimensional electron-density map and thereby to determine the crystal structure.

The structure factor itself is related to the scattering by the atoms in the unit cell by Eq. (2),

$$F_{hkl} = \Sigma_n f_n T_n \exp[2\pi i(hx_m + ky_n + lz_n)] \quad (2)$$

where x_n, y_n, z_n are the fractional coordinates of atom n along the three crystallographic axes. The f_n's are the individual atomic scattering factors which are known for an atom at rest. At zero 2θ angle of scattering for which all the electrons of an atom scatter in phase, f_n is equal to the atomic number. The T_n's are the individual modifications of the f_n's as a result of thermal motion. If the positions of the atoms in the unit cell are known, the complex structure factor for each diffracted beam hkl can be calculated and both its magnitude and phase obtained by Eq. (3).

$$F_{hkl} = A_{hkl} + iB_{hkl} \quad (3)$$
$$A_{hkl} = \Sigma_n f_n T_n \cos 2\pi(hx_n + ky_n + lz_n)$$
$$B_{hkl} = \Sigma_n f_n T_n \sin 2\pi(hx_n + ky_n + lz_n)$$
$$i = \sqrt{-1}$$

Consequently Eqs. (4) hold.

$$|f_{hkl}| = \sqrt{A^2_{hkl} + B^2_{hkl}} \quad (4)$$
$$\tan \alpha_{hkl} = B_{hkl}/A_{hkl}$$

For centrosymmetric crystals which have a center of symmetry located at the corner of each unit cell, the structure factor f_{hkl} simplifies to a real rather than a complex number; that is, since for each atom at x, y, z there must exist a centrosymmetrically related atom at $-x$, $-y$, $-z$, it follows that $B_{hkl} = 0$ and $F_{hkl} = \pm A_{hkl}$. The phases of the diffracted beams then are restricted to being either completely in phase (that is, $\alpha_{hkl} = 0°$ corresponding to $F_{hkl} = +A_{hkl}$) or completely out of phase (that is, $\alpha_{hkl} = 180°$ corresponding to $F_{hkl} = -A_{hkl}$) with one another.

Patterson map. Various means of deducing the phase angle of each diffracted beam have been used. One of the most useful general approaches is based on the classic discovery by Patterson in the 1930s that when the experimentally known quantities $|F_{hkl}|^2$ instead of F_{hkl} are used as coefficients in the Fourier series, the maxima in this summation then correspond not to atomic centers but rather to a map of interatomic vectors. This vector map represents a superposition of all interatomic vectors between pairs of atoms translated to the origin of the unit cell. Since the peak height for a given vector between two atoms is approximately

$$\rho(xyz) = \frac{1}{V}\Sigma_h \Sigma_k \Sigma_l |F_{hkl}| \cos\{2\pi(hx + ky + lz) - \alpha_{hkl}\} \quad (1)$$

proportional to the product of their atomic numbers, the vectors between heavy atoms (that is, those of high atomic number) usually stand out strongly against the background of heavy-light and light-light atom vectors. Consequently, the Patterson map is especially applicable to the structure determination of compounds containing a small number of relatively heavy atoms because approximate coordinates of these heavy atoms can be obtained provided the heavy-atom vectors are correctly recognized in the Patterson map. Computers have enabled the development of a number of powerful techniques in unraveling the Patterson vector map in terms of atomic coordinates; these include multiple superpositions of parts of the Patterson map and "image-seeking" with known vectors.

Fourier electron-density map. Any resulting trial structure consisting of initial coordinates for some of the atoms is often sufficient for location of the other atoms by the application of the method of successive Fourier electron-density maps. The phases calculated from the initial parameters of the presumably known atoms, together with the observed $|F_{hkl}|$, are utilized to compute a density map. New coordinates are obtained not only for the peaks corresponding to these known atoms but also for the other peaks which it is hoped can be interpreted from stereochemical considerations as being due to additional atoms in the structure. The Fourier process is then reiterated, with the new phases calculated from the modified coordinates of the previous set of atoms plus the coordinates of newly located atoms. If a correct distinction between the "true" and the "false" peaks is made, the electron-density function usually converges to give the entire crystal structure.

Statistical method. Another powerful and eminently successful approach to the phase problem, known as the direct or statistical method, makes use of probability theory to generate an adequate set of phases by consideration solely of the structure amplitudes. The mathematical fundamentals of this procedure for direct phase determination are primarily due to the extensive work of J. Karle and H. Hauptman. Although the use of direct methods has been limited almost entirely to centrosymmetric crystals, statistical methods are also meeting with success in the solution of complex noncentrosymmetric crystal structures.

Refinement of parameters. Once the phase problem is solved and the approximately correct trial structure is known, the last step of the structural analysis involves the refinement of the positional and thermal parameters of the atoms. Normally this refinement is carried out analytically by the application of a nonlinear least-squares procedure in which a weighted quantity such as $\sum w [|F(hkl)|_{\text{obs}} - |F(hkl)|_{\text{calc}}]^2$ (where the weights w are appropriate to the experiment) is minimized with respect to the parameters. This method of refinement not only gives the best values for the parameters but also provides a means of obtaining an estimation of the statistical errors in the atomic parameters including standard deviations of bond lengths and angles. Although there is no single reliable method for directly assessing the accuracy of a structural determination, a criterion commonly used is the unweighted reliability

factor or discrepancy index R_1, defined by Eq. (5)

$$R_1 = \frac{\sum\limits_{hkl} \left| |F(hkl)|_{\text{obs}} - |F(hkl)|_{\text{calc}} \right|}{\sum\limits_{hkl} |F(hkl)|_{\text{obs}}} \tag{5}$$

as the summation of the absolute difference in the observed and calculated structure amplitudes divided by the summation of the observed amplitudes. The better the structure, including the atomic coordinates and thermal parameters, is known, the more nearly will the calculated amplitudes agree with the observed ones, and hence the lower will be the R_1 value. Discrepancy values of "finished" modern structural analyses found in the literature vary from approximately 0.15 to less than 0.05, depending upon a number of factors, such as the complexity of the structure and the number and quality of the data obtained.

Before the development of large electronic computers the determination of a crystal structure of more than about 20 atoms was not feasible. Nowadays, single-crystal analyses of uncomplicated structures may require the location of as many as 100 atoms, but there is no high correlation between the complexity of the structural determination and the number of atoms involved.

X-ray crystallography provides the quantitative foundation on which much of modern structural chemistry is based. The hundreds of crystal structures analyzed by x-ray diffraction include those of vitamin B_{12} by D. Hodgkin and coworkers (1957), for which she received a Nobel prize in medicine in 1964, and proteins such as myoglobin and hemoglobin, for which J. C. Kendrew and M. Perutz received Nobel prizes in 1962. X-ray crystallography also has been widely used in inorganic and organic chemistry not only to determine the structure of the compound but also in many cases to directly obtain the chemical composition of the compound. The so-called anomalous dispersion technique, first utilized by J. M. Bijvoet in 1951, has enabled x-ray determination of the absolute configuration of a large number of molecules. *See* ABSORPTION OF ELECTROMAGNETIC RADIATION.

[LAWRENCE F. DAHL]

Bibliography: M. J. Buerger, *Crystal Structure Analysis*, 1960, reprint 1979; J. P. Glusker and K. N. Trueblood, *Crystal Structure: A Primer*, 1972; G. H. Stout and L. H. Jensen, *X-Ray Structure Determination*, 1968; A. J. Wilson, *Elements of Crystallography*, 1970; M. M. Woolfson, *Introduction to X-Ray Crystallography*, 1970.

X-ray diffraction

The scattering of x-rays by matter with accompanying variation in intensity in different directions due to interference effects. X-ray diffraction is one of the most important tools of solid-state chemistry, since it constitutes a powerful and readily available method for determining atomic arrangements in matter. X-ray diffraction methods depend upon the fact that x-ray wavelengths of the order of 1 nanometer are readily available and that this is the order of magnitude of atomic dimensions. When an x-ray beam falls on matter, scattered x-radiation is produced by all the atoms. These scattered waves spread out spherically from all the atoms in the sample, and the interference effects of the scattered radiation from the different atoms cause the intensity of the scattered radiation to exhibit maxima and minima in various directions. *See* DIFFRACTION.

Some of the uses of x-ray diffraction are: (1) differentiation between crystalline and amorphous materials; (2) determination of the structure of crystalline materials (crystal axes, size and shape of the unit cell, positions of the atoms in the unit cell); (3) determination of electron distribution within the atoms, and throughout the unit cell; (4) determination of the orientation of single crystals; (5) determination of the texture of polygrained materials; (6) identification of crystalline phases and measurement of the relative proportions; (7) measurement of limits of solid solubility, and determination of phase diagrams; (8) measurement of strain and small grain size; (9) measurement of various kinds of randomness, disorder, and imperfections in crystals; and (10) determination of radial distribution functions for amorphous solids and liquids.

For the study of crystal structure by x-ray diffraction techniques *see* X-RAY CRYSTALLOGRAPHY.

DIFFRACTION THEORY

When x-rays fall on the atoms of a substance, the scattered radiation is of two kinds: Compton modified scattering of increased wavelength which is incoherent with respect to the primary beam, and unmodified scattering coherent with the primary beam. Because of interference effects from the unmodified scattering by the different atoms of the sample, the intensity of unmodified scattering varies in different directions. A diagram of this variation in direction of intensity of unmodified scattering is called the diffraction pattern of the substance. This pattern is determined by the kinds of atoms and their arrangement in the sample; for simple structures the atomic arrangement is readily deduced from the diffraction pattern. *See* COMPTON EFFECT.

The atomic scattering factor f is defined as the ratio of the amplitude of unmodified scattering by an atom to the amplitude of scattering by a free electron, which scatters according to classical theory. In general f is a real number which decreases with $(\sin \theta)/\lambda$, where θ is the grazing angle and λ the wavelength, from an initial value $f = Z$, where Z is the number of electrons in the atom. However, if the x-ray wavelength is close to an absorption edge of the atom, f becomes complex.

If the electron density in the atom has spherical symmetry and if the x-ray wavelength is small compared to all the absorption-edge wavelengths, Eq. (1) holds. Here $k = 4\pi(\sin \theta)/\lambda$ and $\rho(r)$ is the electron density (electrons per unit volume).

$$f = \int_0^\infty 4\pi r^2 \rho(r) \frac{\sin kr}{kr} dr \qquad (1)$$

A crystalline structure is one in which a unit of structure called the unit cell repeats at regular intervals in three dimensions. The repetition in space is determined by three noncoplanar vectors $\mathbf{a}_1\mathbf{a}_2\mathbf{a}_3$, called the crystal axes. The positions of the atoms in the unit cell are expressed by a set of base vectors $\mathbf{r}_n$. The position of atom n in the unit

cell $q_1q_2q_3$ is given by Eq. (2). *See* CRYSTAL STRUCTURE; CRYSTALLOGRAPHY.

$$\mathbf{R}_{nq} = q_1\mathbf{a}_1 + q_2\mathbf{a}_2 + q_3\mathbf{a}_3 + \mathbf{r}_n \qquad (2)$$

For a crystal containing $N_1N_2N_3$ repetitions in the $\mathbf{a}_1\mathbf{a}_2\mathbf{a}_3$ directions, the intensity of unmodified scattering is given by Eq. (3). Here I_e is the inten-

$$\begin{aligned} I = I_e \sum_{nq} f_n \exp\left[\frac{2\pi i}{\lambda}(\mathbf{s}-\mathbf{s}_0)\cdot\mathbf{R}_{nq}\right] \\ \cdot \sum_{n'q'} f_{n'} \exp\left[\frac{-2\pi i}{\lambda}(\mathbf{s}-\mathbf{s}_0)\cdot\mathbf{R}_{n'q'}\right] \\ = I_e FF^* \frac{\sin^2\left[(\pi/\lambda)(\mathbf{s}-\mathbf{s}_0)\cdot N_1\mathbf{a}_1\right]}{\sin^2\left[(\pi/\lambda)(\mathbf{s}-\mathbf{s}_0)\cdot\mathbf{a}_1\right]} \\ \cdot \frac{\sin^2\left[(\pi/\lambda)(\mathbf{s}-\mathbf{s}_0)\cdot N_2\mathbf{a}_2\right]}{\sin^2\left[(\pi/\lambda)(\mathbf{s}-\mathbf{s}_0)\cdot\mathbf{a}_2\right]} \\ \cdot \frac{\sin^2\left[(\pi/\lambda)(\mathbf{s}-\mathbf{s}_0)\cdot N_3\mathbf{a}_3\right]}{\sin^2\left[(\pi/\lambda)(\mathbf{s}-\mathbf{s}_0)\cdot\mathbf{a}_3\right]} \qquad (3) \end{aligned}$$

sity, at a distance R and angle 2θ, scattered by a free electron according to classical theory, and $FF^* = |F|^2$. For an unpolarized primary beam of intensity I_o, Eq. (4) holds, and $e^4/(m^2c^4) = 7.94$

$$I_e = I_o \frac{e^4}{m^2c^4R^2}\left(\frac{1+\cos^2 2\theta}{2}\right) \qquad (4)$$

$\times 10^{-26}$ cm^2 if R is expressed in centimeters. Here m is the mass of the electron, c is the velocity of light, and F is the structure factor, a complex quantity given by a summation over all the atoms of the unit cell as in Eq. (5), where $\mathbf{s}_0$ and $\mathbf{s}$ are

$$F = \sum_n f_n \exp\left[\frac{2\pi i}{\lambda}(\mathbf{s}-\mathbf{s}_0)\cdot\mathbf{r}_n\right] \qquad (5)$$

unit vectors in the directions of the primary and diffracted beams.

Laue equations and Bragg's law. The condition for a crystalline reflection is that the three quotients of Eq. (3) exhibit maxima, and this occurs if all three denominators vanish. Expressing the denominators in terms of three integers α, β, and γ, the three Laue equations, Eqs. (6), are obtained. These express the condition for a diffracted beam.

$$\begin{aligned} (\mathbf{s}-\mathbf{s}_0)\cdot\mathbf{a}_1 &= \alpha\lambda \\ (\mathbf{s}-\mathbf{s}_0)\cdot\mathbf{a}_2 &= \beta\lambda \\ (\mathbf{s}-\mathbf{s}_0)\cdot\mathbf{a}_3 &= \gamma\lambda \end{aligned} \qquad (6)$$

It is convenient to introduce the concept of sets of crystallographic planes. As illustrated by Fig. 1, the set of planes with Miller indices hkl is a set of parallel equidistant planes, one of which passes through the origin, and the next nearest makes intercepts $\mathbf{a}_1/h$, $\mathbf{a}_2/k$, $\mathbf{a}_3/l$ on the three crystallographic axes.

In terms of sets of planes hkl, the diffraction conditions are expressed by the Bragg law, Eq. (7),

$$\lambda = 2d_{hkl} \sin \theta \qquad (7)$$

where θ is the angle which the primary and diffracted beams make with the planes hkl and d_{hkl} is the spacing of the set. As seen from Fig. 2, the Bragg law is simply the condition that the path difference for rays diffracted from two successive

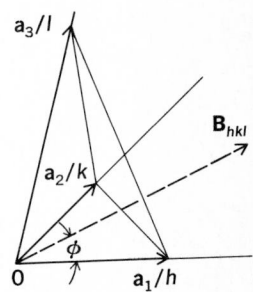

X-RAY DIFFRACTION

Fig. 1. Crystallographic planes with Miller indices hkl.

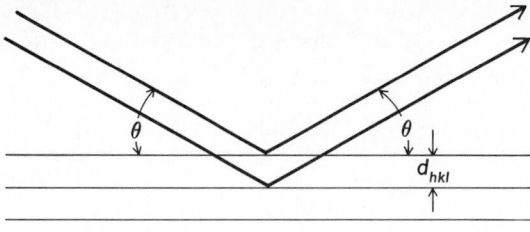

Fig. 2.   Interference conditions involved in Bragg law.

hkl planes be one wavelength. In the early days of x-ray diffraction, the Bragg law was written $n\lambda = 2d \sin\theta$, and $n = 1, 2, 3$ corresponded to first-, second-, and third-order diffraction from the planes of spacing d. That notation has been largely dropped, and instead of being called second-order diffraction from planes *hkl*, it is called diffraction from the planes $2h, 2k, 2l$.

Reciprocal lattice. The understanding and interpretation of x-ray diffraction in crystals is greatly facilitated by the concept of a reciprocal lattice. In terms of the crystal axes $\mathbf{a}_1\mathbf{a}_2\mathbf{a}_3$, three reciprocal vectors are defined by Eqs. (8). From these defini-

$$\mathbf{b}_1 = \frac{\mathbf{a}_2 \times \mathbf{a}_3}{\mathbf{a}_1 \cdot \mathbf{a}_2 \times \mathbf{a}_3} \qquad \mathbf{b}_2 = \frac{\mathbf{a}_3 \times \mathbf{a}_1}{\mathbf{a}_1 \cdot \mathbf{a}_2 \times \mathbf{a}_3}$$
$$\mathbf{b}_3 = \frac{\mathbf{a}_1 \times \mathbf{a}_2}{\mathbf{a}_1 \cdot \mathbf{a}_2 \times \mathbf{a}_3} \qquad (8)$$

tions it follows that Eq. (9) is valid. In terms of

$$\mathbf{a}_i \cdot \mathbf{b}_j = \begin{cases} 1 & i = j \\ 0 & i \neq j \end{cases} \qquad (9)$$

integers *hkl*, the terminal points of the vectors in Eq. (10) generate a lattice of points called the

$$\mathbf{B}_{hkl} = h\mathbf{b}_1 + k\mathbf{b}_2 + l\mathbf{b}_3 \qquad (10)$$

reciprocal lattice. Each point in the lattice is specified by the integers *hkl*, and the vectors $\mathbf{B}_{hkl}$ represent two important properties of the sets of *hkl* planes: (1) $\mathbf{B}_{hkl}$ is perpendicular to the *hkl* planes, and (2) $|\mathbf{B}_{hkl}| = 1/d_{hkl}$. These two relations are readily proved from the geometry of Fig. 1. As seen, $\mathbf{a}_2/k - \mathbf{a}_1/h$ and $\mathbf{a}_3/l - \mathbf{a}_2/k$ are vectors lying in the *hkl* plane. From Eqs. (9) and (10), Eqs. (11) are

$$\left(\frac{\mathbf{a}_2}{k} - \frac{\mathbf{a}_1}{h}\right) \cdot \mathbf{B}_{hkl} = 0 \qquad \left(\frac{\mathbf{a}_3}{l} - \frac{\mathbf{a}_2}{k}\right) \cdot \mathbf{B}_{hkl} = 0 \quad (11)$$

obtained, and hence $\mathbf{B}_{hkl}$ is perpendicular to the planes *hkl*. The spacing of the planes *hkl* is given by Eq. (12).

$$d_{hkl} = \left|\frac{\mathbf{a}_1}{h}\right| \cos\phi = \frac{\mathbf{a}_1}{h} \cdot \frac{\mathbf{B}_{hkl}}{|\mathbf{B}_{hkl}|} = \frac{1}{|\mathbf{B}_{hkl}|} \quad (12)$$

Equivalence of the three Laue equations and the Bragg law can be shown as follows: Any vector $\mathbf{r}$ can be expressed by Eq. (13). Let $\mathbf{r}$ be the vector

$$\mathbf{r} = (\mathbf{r} \cdot \mathbf{a}_1)\mathbf{b}_1 + (\mathbf{r} \cdot \mathbf{a}_2)\mathbf{b}_2 + (\mathbf{r} \cdot \mathbf{a}_3)\mathbf{b}_3 \quad (13)$$

$(\mathbf{s} - \mathbf{s}_0)$ and combine it with the three Laue equations and Eq. (13) to obtain Eq. (14). The Bragg law

$$\mathbf{s} - \mathbf{s}_0 = \lambda(\alpha\mathbf{b}_1 + \beta\mathbf{b}_2 + \gamma\mathbf{b}_3) \qquad (14)$$

can be written in vector form as Eq. (15) since the

$$\mathbf{s} - \mathbf{s}_0 = \lambda\mathbf{B}_{hkl} = \lambda(h\mathbf{b}_1 + k\mathbf{b}_2 + l\mathbf{b}_3) \qquad (15)$$

usual form of the Bragg law is simply an equality in the magnitudes of the vectors: $|\mathbf{s} - \mathbf{s}_0| = 2\sin\theta$ and $|\mathbf{B}_{hkl}| = 1/d_{hkl}$. Comparison of Eqs. (14) and (15) shows that the integers α, β, γ of the three Laue equations are simply the Miller indices *hkl* of the Bragg law.

The positions of the atoms in the unit cell are represented by a set of atomic coordinates x_n, y_n, z_n such that for atom n Eq. (16) holds. For a Bragg

$$\mathbf{r}_n = \mathbf{a}_1 x_n + \mathbf{a}_2 y_n + \mathbf{a}_3 z_n \qquad (16)$$

law reflection *hkl*, the structure factor takes the simple form of Eq. (17).

$$F_{hkl} = \sum_n f_n \exp\left[2\pi i(hx_n + ky_n + lz_n)\right] \quad (17)$$

Integrated intensity. In general, the intensity of a Bragg reflection, as expressed by Eq. (3), is not an experimentally measurable quantity. Other factors, such as the degree of mosaic structure in the crystal and the degree of parallelism of the primary beam, have a profound influence on the measured diffracted intensity for any setting of the crystal. To obtain measurements characteristic of the crystalline structure, it is necessary to adopt a more useful concept, the integrated intensity. For a small single crystal, it is postulated that the crystal is to be turned at constant angular velocity ω through the Bragg law position, and that the total diffracted energy of the reflection is to be measured. The integrated intensity E is then given by Eq. (18), where $d\alpha$ is a change in orientation of the

$$E = \int \int I\frac{d\alpha}{\omega} dA \qquad (18)$$

crystal and dA is an element of area at the point of observation

Most of the equations used in x-ray diffraction studies are derived on the assumption that the intensity of the diffracted beam is so small that any interaction with the primary beam can be neglected. These are classed as the equations for the ideally imperfect crystal. For powder samples, in which the individual crystals are extremely small, and for highly deformed single crystals, if the intensity of the diffracted beam is small, the ideally imperfect crystal is usually a good approximation. For the ideally perfect crystal, it is necessary to use a more elaborate theory which allows for the interaction of diffracted radiation with the primary beam. In general, it is the integrated intensity which is measured, and theory shows that the integrated intensity for an ideally imperfect crystal is larger than that for the ideally perfect crystal. Many of the crystalline samples used for x-ray diffraction studies are not ideally imperfect, and the measured integrated intensity is accordingly less than that predicted by the ideally imperfect crystal formulas which are used in the interpretation. The situation is usually handled by adding a correction factor called the extinction correction to the formulas for the integrated intensity from the ideally imperfect crystal.

Atomic coordinates. To have complete information about a crystalline structure, it is necessary to know all the atomic coordinates $x_n y_n z_n$ of the n

atoms making up the unit cell. The atomic coordinates appear in the structure factor as given by Eq. (17), and sometimes the coordinates are obtained directly from structure factor values. Another way is to plot the electron density in the unit cell and infer the atomic positions from peaks in the electron density function. The electron density in the unit cell is given by the triple Fourier series shown in Eq. (19) for which the coefficients are simply the

$$\rho(xyz)$$

$$= \frac{1}{V} \sum_h \sum_k \sum_l F_{hkl} \exp\left[-2\pi i\left(\frac{hx}{a} + \frac{ky}{b} + \frac{lz}{c}\right)\right] \quad (19)$$

structure factors F_{hkl}. However, from experimental measurements of either an intensity or an integrated intensity, values for $|F_{hkl}|^2$ can be obtained. These yield the magnitude of F_{hkl} but not the phase. This is the most serious limitation to a straightforward determination of crystalline structures by x-ray diffraction methods. The ambiguity in the phase of F_{hkl} prevents the use of the Fourier plot of Eq. (19) as a general method for determining any crystalline structure.

Simple structures are uniquely determined by combining the x-ray intensity results with space group theory. The space group of a crystal is the repeating spatial arrangement of symmetry elements which the structure displays. Considering all the possible symmetry elements which can exist in a crystalline structure, group theory shows that there are only 230 essentially different possible combinations, and these constitute the 230 space groups. A knowledge of the macroscopic symmetry of the crystal, coupled with the systematically vanishing x-ray reflections, usually determines the space group. The limitations imposed by the space group on the possible atomic positions, coupled with the limitations imposed by the measured $|F_{hkl}|^2$, often allow a complete and unique structure determination for not too complicated structures. For highly complex structures, it is never certain that x-ray diffraction analysis can yield a complete structure determination. Additional techniques such as the isomorphous replacement by heavy atoms, the use of Patterson plots, and the determination of phase relations from inequalities are used with success on some of the complex structures. *See* GROUP THEORY.

Many structures of interest in solid-state chemistry exhibit various kinds of randomness and imperfections. The precise nature of these is sometimes of more interest than the ideal average structure. Randomness and imperfections in a structure show themselves by producing a diffuse intensity in addition to the sharp Bragg reflections. The temperature vibration of the atoms produces a diffuse intensity called temperature diffuse scattering. Quantitative measurements of this scattering lead to values for the velocity of high-frequency elastic waves and to a complete experimental determination of the spectrum of the elastic waves which constitute the thermal vibrations of the crystal. In alloys showing order-disorder changes, the short-range order parameters are obtained from quantitative measurement of the diffuse intensity which results from randomness in the atomic arrangement.

CRYSTALLINE DIFFRACTION

The techniques employed in the study of crystalline substances are discussed below.

Laue method. The Laue pattern uses polychromatic x-rays provided by the continuous spectrum from an x-ray tube operated at $35-50$ kV. The transmission Laue pattern is obtained by passing a finely collimated beam through a thin single crystal and recording the diffracted beams on a photographic film placed several centimeters beyond the crystal. For each set of planes hkl, θ is fixed, and the Bragg law is satisfied by selecting the proper λ from the primary beam. In a Laue pattern, the different diffracted beams have different wavelengths, and their directions are determined solely by the orientations of the hkl planes. Transmission Laue patterns were once used for structure determinations, but their many disadvantages have made them practically obsolete.

On the other hand, the back-reflection Laue pattern is used a great deal in the study of the orientation of crystals. The back-reflection Laue camera is shown schematically in Fig. 3. The polychromatic beam enters through a hole in the x-ray film and falls on a single crystal whose orientation can be

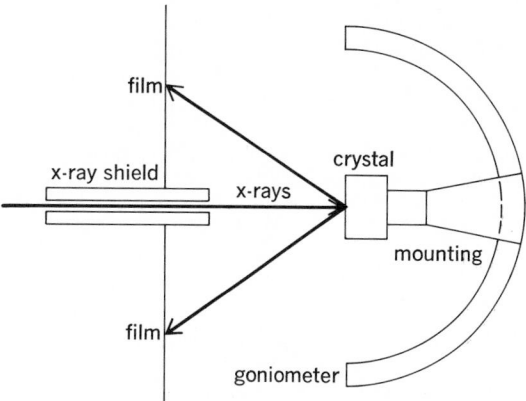

Fig. 3. Schematic of back-reflection Laue camera.

set as desired by a system of goniometer circles. Diffracted beams bent through angles 2θ approaching 180° are registered on the photographic film. For cubic crystals it is very easy to read the crystal orientation from a back-reflection Laue pattern, and the patterns find considerable use in the cutting of single-crystal metal ingots.

Rotating crystal method. The original rotating crystal method was employed in the Bragg spectrometer. A sufficiently monochromatic beam, of wavelength of the order of 1 A, is obtained by using the strong $K\alpha_1\alpha_2$ doublet with a filter which suppresses the $K\beta$ line and much of the continuous spectrum. The beam is collimated by a system of slits and then falls on the large extended face of a single crystal as shown by Fig. 4. Originally the diffracted beam was measured with an ionization chamber, but Geiger counters and proportional counters have largely replaced the ionization chamber. Both the crystal and the chamber turn about the spectrometer axis.

The Bragg spectrometer has been used extensively in obtaining quantitative measurements of the integrated intensity from planes parallel to the

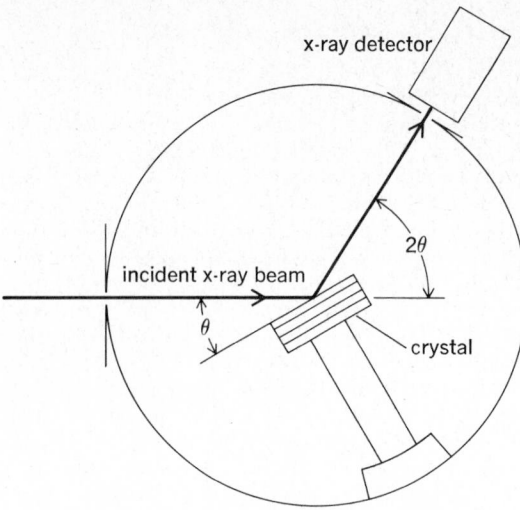

Fig. 4. Schematic of Bragg spectrometer.

face of the crystal. The chamber is set at the correct 2θ-angle with a slit so wide that all of the radiation reflected from the crystal can enter and be measured. The crystal is turned at constant angular speed ω through the Bragg law position, and the total diffracted energy E received by the ionization chamber during this process is measured. Similar readings with the chamber set on either side of the peak give a background correction. For this type of measurement, the integrated intensity E is given by Eq. (20), where P_0 is the power of the primary

$$E = \frac{P_0}{2\mu\omega}\,\frac{e^4}{m^2c^4}\,\frac{\lambda^3 F^2}{v^2}\left(\frac{1+\cos^2 2\theta}{2\sin 2\theta}\right)\exp\left[-2M\right] \quad (20)$$

beam, μ is the linear absorption coefficient in the crystal, ω is the angular velocity of the crystal, λ is the x-ray wavelength, F is the structure factor, v is the volume of the unit cell, and $\exp\left[-2M\right]$ is the so-called Debye factor allowing for temperature vibration. When more than one kind of atom is present, this factor must be incorporated in F^2.

Measurements of the integrated intensity E give quantitative values of F^2 directly. When a Geiger counter is used in place of an ionization chamber for this type of measurement, it is necessary to employ a narrow counterslit and traverse the

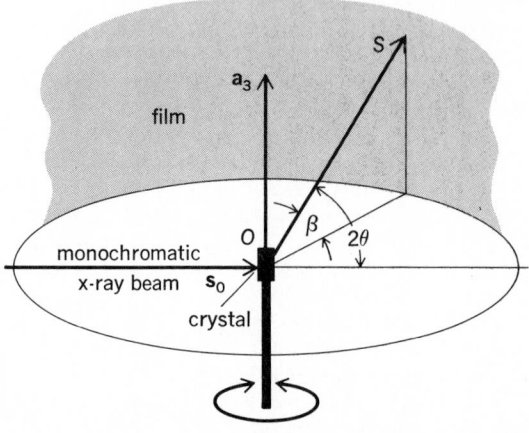

Fig. 5. Schematic of rotation camera.

counter through the reflected beam, since the sensitivity of a Geiger counter is not constant over a large window opening.

The rotation camera, which is frequently used for strucutre determinations, is illustrated in Fig. 5. The monochromatic primary beam $\mathbf{s}_0$ falls on a small single crystal at O. The crystal is mounted with one of its axes (say, $\mathbf{a}_3$) vertical, and it rotates with constant velocity about the vertical axis during the exposure. The various diffracted beams are registered on a cylindrical film concentric with the axis of rotation. For a rotation about $\mathbf{a}_3$ it follows that $\mathbf{s}_0 \cdot \mathbf{a}_3 = 0$, and the third Laue equation gives Eq. (21). The diffracted beams form the ele-

$$\sin\beta = \frac{l\lambda}{|\mathbf{a}_3|} \quad (21)$$

ments of a set of cones, and the intersection of these cones with the cylindrical film gives a set of horizontal lines of diffraction spots. This type of pattern is called a rotation pattern, and the horizontal rows of spots are called layer lines. As seen from Eq. (21), the measured values of $\sin\beta$ give directly the length of the axis about which the crystal was rotated, and the layer line in which a spot occurs gives the l index of the reflection. Similar rotations about the other two axes give corresponding information. More elaborate variations of the rotation method, such as those of the Weissenberg and the precession cameras, involve a motion of the film in addition to the rotation of the crystal.

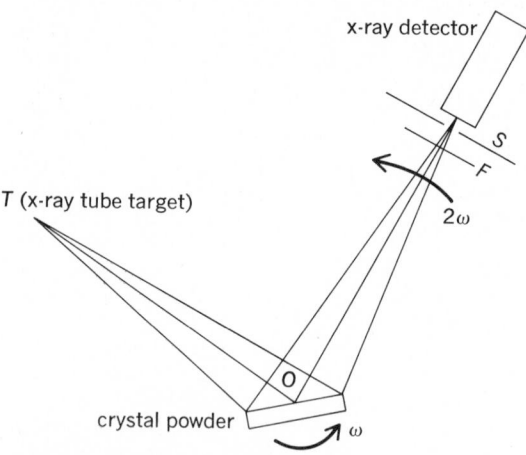

Fig. 6. Schematic representation of the Geiger counter diffractometer for powder samples.

Powder method. The powder method involves the diffraction of a collimated monochromatic beam from a sample containing an enormous number of tiny crystals having random orientation. Since about 1950 an increasing number of powder pattern studies have been made with Geiger counter, or porportional counter, diffractometers. The apparatus is shown schematically in Fig. 6. X-rays diverging from a target at T fall on the sample at O, the sample being a flat-faced briquet of powder. Diffracted radiation from the sample passes through the receiving slit at s and enters the Geiger counter. During the operation the sample turns at angular velocity ω and the counter at 2ω. The distances TO and OS are made equal to satisfy approximate focusing conditions. A filter F before the receiving slit gives the effect of a sufficiently monochromatic beam. A chart recording of the amplified output of the Geiger counter gives directly a plot of intensity versus scattering angle 2θ.

NONCRYSTALLINE DIFFRACTION

For a noncrystalline substance such as a glass or a liquid, a more general expression for the intensity of diffracted radiation is required. If the instantaneous position of each atom in the sample is represented by a vector $\mathbf{r}_n$, the diffracted intensity is given by Eq. (22). A particularly useful variation

$$I = I_e \sum_q \sum_n f_q f_n \exp\left[\frac{2\pi i}{\lambda}(\mathbf{s} - \mathbf{s}_0)\cdot(\mathbf{r}_q - \mathbf{r}_n)\right] \quad (22)$$

of Eq. (22) given as Eq. (23), is obtained by computing the average intensity $\langle I \rangle$ when the sample as a rigid array is allowed to take with equal probability all orientations in space. In Eq. (23) $k = 4\pi(\sin\theta)/\lambda$ and $r_{qn} = |\mathbf{r}_q - \mathbf{r}_n|$.

$$\langle I \rangle = I_e \sum_q \sum_n f_q f_n \frac{\sin(kr_{qn})}{kr_{qn}} \quad (23)$$

The fact that there are fairly definite nearest-neighbor and second-neighbor distances in a glass or liquid means that Eq. (23) will show peaks and dips when the intensity is plotted against $(\sin\theta)/\lambda$. Peaks and dips in an x-ray diffraction pattern merely indicate the existence of preferred interatomic distances, not that the material is necessarily crystalline. X-ray patterns of noncrystalline materials are usually analyzed by a Fourier inversion of Eq. (23), which yields a radial distribution function giving the probability of finding neighboring atoms at any distance from an average atom.

[BERTRAM E. WARREN]

Gases. Gases and liquids are found to give rise to x-ray diffraction patterns characterized by one or more halos or interference rings which are usually somewhat diffuse. These diffraction patterns which are similar to those for glasses and amorphous solids, are due to interference effects depending both upon the electronic distribution of each of the individual atoms or molecules and upon their relative positions in the system.

For monatomic gases the only appreciable interference effects giving rise to a distribution of scattered intensities are those produced by the electronic distribution about each nucleus. These interference effects giving rise to so-called coherent intensities are the result of the interference of the individual waves scattered by electrons in different parts of the atom. The electronic distribution of an atom is described in terms of a characteristic atomic scattering factor which is defined as the ratio of the resultant amplitude scattered by an atom to the amplitude that a free electron would scatter under the same conditions. At zero-angle scattering the atomic scattering factor is equal to the atomic number of the atom. The coherent intensity in a given direction is proportional to the square of the atomic scattering factor. If it is assumed that the electronic distribution is spherically symmetrical, the atomic scattering factors can be readily obtained from the observed intensities. For molecular gases the interference effects depend not only on the scattering factors of the atoms but also on their relative positions in the mole-

cule. One can observe only an average intensity scattered over a period of time during which the molecules have taken innumerable positions with respect to the incident beam. Interference effects due to the relative packing of the atoms or molecules can be neglected for dilute gases but not for dense gases.

As in the case of x-ray diffraction by crystals, light atoms such as hydrogen are difficult to detect in the presence of heavy atoms. Because of the shorter exposure times required, electron diffraction rather than x-ray diffraction has been utilized in studies of the structures of gaseous molecules. Both methods appear to be comparable in view of the accuracy of the intensity measurements and the technical difficulties involved.

Liquids. One cannot, as in the cases of dilute gases and crystalline solids, derive unambiguous, detailed descriptions of liquid structures from diffraction data. Nevertheless, diffraction studies of liquids do provide most useful information. Instead of comparing the experimental intensity distributions with theoretical distributions computed for various models, the experimental results are usually provided in the form of a radial distribution function which specifies the density of atoms or electrons as a function of the radial distance from any reference atom or electron in the system without any prior assumptions about the structure. From the radial distribution function one can obtain (1) the average interatomic distances most frequently occurring in the structure corresponding to the positions of the first, second, and possibly third nearest neighbors; (2) the distribution of distances; and (3) the average coordination number for each interatomic distance. The interpretation of these diffraction patterns given by the radial distribution function usually is not straightforward and in general it can be said only that a certain assumed structural model and arrangement is not inconsistent with the observed diffraction data. The models considered represent only a description of the time-average environment about any given atom or molecule within the liquid.

There are great experimental difficulties in obtaining accurate intensity data. The sources of error are many; for a detailed treatment the reader is referred to publications of C. Finback. A brief description of some results obtained by x-ray diffraction of liquids is given below.

Liquid elements. The radial distribution function, first used in a study of liquid mercury, has been applied to a considerable number of liquid elements mainly to compare their physical properties in the liquid and crystalline states. In most cases a lower first mean coordination number is found in the liquid state, exceptions are liquid gallium, bismuth, germanium, and lithium. The radial distribution curves give direct evidence for the existence of molecules in some liquid elements (for example, N_2, O_2, Cl_2, and P_4) and imply the existence of more complicated atomic aggregates in a few cases. Argon and helium have been extensively studied in the liquid and vapor states over wide ranges of temperature and pressure.

Liquid water and solutions. A prime example illustrating the considerable structural information made available from modern x-ray liquid diffractometry investigations is the detailed analysis of liquid water, which revealed the following significant features: (1) There are distinct structural deviations of water molecules from a uniform distribution of distances to about 0.8 nm at room temperature; (2) the first prominent maximum, corresponding to near-neighbor interactions, shifts gradually from 0.282 nm at 4°C to 0.294 nm at 200°C; (3) the average coordination number in liquid water from 4 to 200°C is approximately constant and slightly larger than four; and (4) the radial distribution of oxygen atoms in water at 4°C is not significantly different from that in deuterium oxide at the same temperature. Comparison of calculated radial distribution functions for various proposed liquid water models (which are sufficiently defined at the molecular level) based on those derived from patterns of liquid water have shown that the only realistic model which gives agreement with data from both large- and small-angle x-ray scattering is related to a modification of the ordinary hexagonal ice structure. This solid-state structure is similar to that of the hexagonal form of silicon dioxide, tridymite, with each oxygen atom tetrahedrally surrounded by neighboring oxygen atoms to give layers of puckered six-membered rings with dodecahedral cavities large enough (radius 0.295 nm) to accommodate a water molecule.

In terms of an average configuration, the liquid water phase may be regarded as a "mixture" model comprising network water molecules forming a slightly expanded ordinary ice structure (each oxygen atom forming nearly four hydrogen bonds with neighboring oxygen atoms) and the cavity water molecules interacting with the network by less specific but by no means negligible forces. It must be emphasized that both kinds of water molecules instantaneously exist in environments which are distorted from the average, as implied by sizable root-mean-square variations in interatomic distance.

Radial distribution curves for concentrated $FeCl_3$ solutions indicate a large degree of local ordering of the ions with formation of $Fe^{3+} - Cl^-$ complexes. Studies on metal-metal solutions, colloidal solutions, and molecular solutions have been made. More definite results have been obtained for concentrated solutions of strongly scattering solutes in weakly scattering solvents. Examples are the proof of the existence of a polymeric species in aqueous $Bi(ClO_4)_3$, evidence that in aqueous solution the HgX_4^{2-} anions (X = Cl, Br, and I) are tetrahedral, and definite evidence of ion-pair formation in aqueous BaI_2.

Molten salts. A molten salt is considered to be a loose and expanded imitation of the solid with the same coordination scheme and short-range order. Careful x-ray diffraction studies of a number of molten salts have indicated that melts do not possess such quasi-crystalline structures but instead have quite open structures with a wide variety of individual ion coordinations. Interpretations of radial distribution functions for several other molten salts have been made. Liquid $AlCl_3$ appears to consist mainly of Al_2Cl_6 molecules; liquid SnI_4 is composed of independent tetrahedral molecules. The results for other molten salts are not as conclusive. *See* ELECTRON DIFFRACTION; NEUTRON DIFFRACTION.

[LAWRENCE F. DAHL]

Bibliography: C. S. Barrett and T. B. Massalski, *Structure of Metals*, 3d ed., 1980; A. G. Brown, *X-Rays and Their Applications*, 1975; M. J. Buerger, *Crystal Structure Analysis*, 1960, reprint 1979; N. A. Dyson, *X-Rays in Atomic and Nuclear Physics*, 1973; G. H. Stout and L. H. Jensen, *X-Ray Structure Determination*, 1968.

X-ray optics

A title-by-analogy of those phases of x-ray physics in which x-rays demonstrate properties similar to those of light waves. X-ray optics is also called roentgen optics.

Optics is that branch of physical science which deals with the nature and properties of light and the laws of its modification by opaque and transparent bodies. Essentially, then, it involves the electromagnetic wave phenomena of reflection from mirrors and of refraction or change in direction of a beam in passing the boundary between two media, as from air into a glass prism, thus spreading white light out into a rainbow or spectrum of colors. It also includes the diffraction or bending around opaque obstacles such as slits or the lines in ruled gratings, the spreading of light into a spectrum, and the polarization or constraint of vibrations in all directions transverse to the direction of wave propagation into one direction, as in certain crystals, or by grazing reflection from surfaces. Optics thus is concerned with the control and focusing of the paths of light rays, with image formation, and with instrumentation, such as mirrors, lenses in microscopes and telescopes, refractometers, polarimeters, interferometers, and spectrometers with prisms and ruled gratings. *See* GEOMETRICAL OPTICS; OPTICS; PHYSICAL OPTICS.

X-rays, when first discovered by W. K. Röntgen, seemed to possess none of the optical properties mentioned. Later, however, polarization, diffraction, reflection, and refraction were all detected by using crystals having refractive indices that were slightly less than 1. Total reflection or diffraction of x-rays from ruled gratings could occur only with incident beams at very small grazing angles of the order of a few minutes. Until this discovery by A. Compton in 1922 all attempts to concentrate or focus x-rays by lenses and mirrors had failed. *See* X-RAYS.

X-ray microscopes. In 1948, accepting the fact of total reflection at very small grazing angles, P. Kirkpatrick devised the first true x-ray microscope. A concave spherical mirror receiving x-rays at grazing incidence images a point into a line in accordance with the focal length $f = Ri/2$, where R is the radius of curvature and i the grazing angle. The image is subject to an aberration such that a ray reflected at the periphery of the mirror misses the focal point of central rays by a distance $S = 1.5Mr^2/R$, where M is the magnification of the mirror and r the radius of the mirror face. The possible resolving power actually achieved by Kirkpatrick was such as to resolve points separated by 7 nm, independent of wavelength.

Point images of points, and therefore extended images of extended objects, may be produced by causing x-rays to reflect from two concave mirrors in series crossed at right angles to each other. Elliptical gold mirrors were found superior to spherical ones. This gave the impetus to a number of other varieties of x-ray microscopy, including contact and projection microradiography.

An outstanding success of this reflecting optical system is in the isolation and focusing of very soft (low-energy) x-rays with wavelengths up to 4.5 nm (the $K\alpha$ line of copper) in the remarkable work of B. Henke on microradiography of single blood cells.

Beam formation. Bent crystals can be used as diffraction gratings to focus diffracted beams and to form enlarged images of a specimen, itself serving as the source of x-rays or bathed by a beam of x-rays, in accordance with the same principles and equations that apply to light optics.

In x-ray fluorescent emission spectrometers where it is necessary to concentrate beams of low intensities, bent reflecting or transmitting crystals (for example, mica) are so arranged that the theoretical curvature required can be varied with the diffraction angle of a spectrum line. Focusing powder diffraction cameras, in order to gain maximum intensities, having the specimen and the photographic film on the same circumference so that a larger area of sample can be irradiated with focusing of all diffracted wavelets into a sharp interference. Optical principles are involved in assuring sharp-edged shadows in the projection x-ray microscope, which uses x-rays from a point source. Inherent enlargement without loss of detail is gained on images recorded on film at large distances from the specimen. *See* X-RAY DIFFRACTION.

Continuing effort is being made to improve the collimation of x-ray beams into finer pencils while retaining the maximum intensity of the radiation. Reflection from walls under grazing-angle conditions is one such method used. Thus calcite-faced slits, very fine lead-glass capillary tubes, converging polished walls, bent optical flats, and similar devices have been used to produce beams smaller than 1 μm in width with 200 times the intensity of a slit system having the same definition at the same working distance; at the same time these beams are partly monochromatized. Conversely, the total reflection of x-rays may be used to investigate the structure of optical surfaces from the widening of images; even the best-polished surfaces consist of hills and valleys about 1 nm in height and 1 μm in width. *See* COMPTON EFFECT; X-RAY CRYSTALLOGRAPHY.

[GEORGE L. CLARK]

X-ray tube

An electronic device used for the generation of x-rays. X-rays are produced in the x-ray tube by accelerating electrons to a high velocity by an electrostatic field and then suddenly stopping them by collision with a solid body, the so-called target, introstatic field and then suddenly stopping them by directions from the spot on the target where the collisions take place. The x-rays are due to the mutual interaction of the fast-moving electrons with the electrons and positively charged nuclei which constitute the atoms of the target. Depending upon the method used in generating the electrons, x-ray tubes may all be classified in two general groups, gas tubes and high-vacuum tubes. *See* X-RAYS.

Gas x-ray tubes. In gas tubes electrons are freed from a cold cathode by positive ion bombardment. For the existence of the positive ions a certain gas

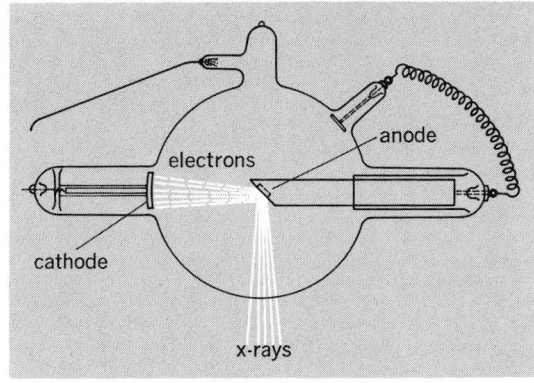

Fig. 1. A commercial model of gas x-ray tube.

pressure is required without which the tube will allow no current to pass. In the earliest gas tube the electrons liberated by positive ion bombardment from a flat aluminum cathode were emitted in a direction perpendicular to the cathode surface. The electrons traveled in straight lines until they impinged upon the glass end wall of the tube, where x-rays were generated.

Many designs of gas tubes have been built for useful application, particularly in the medical field. Metals, such as platinum and tungsten, have been placed in the path of the electron beam to replace glass as the target. Concave metal cathodes are used to focus the electrons on a small area of the metal target and increase the sharpness of the resulting shadows on the fluorescent screen or the photographic film. Figure 1 shows one form of commercial gas x-ray tube.

The useful life of a gas x-ray tube is dependent on maintaining a constant gas pressure. In operation, the gas pressure in the tube changes, either rising or falling, depending on the particular operating conditions and on the past history of the tube. The pressure usually is reduced by operating the tube intermittently with small currents. When the pressure becomes too low for satisfactory operation, it can be raised again by various methods, prolonging the useful life of the tube. The methods for raising the pressure are heating the glass bulb; heating a suitable chemical placed in a side tube; or diffusing gas through a thin-walled tube of platinum, palladium, or unglazed porcelain joined to the tube envelope. The pressure may often be raised when operating the tube continuously, as in therapeutic work. The gas pressure is often regulated automatically. Many forms of automatic osmoregulators were designed and used to admit or remove gas to maintain the desired operating pressure in the gas x-ray tube.

The size of the bulb in commercial tubes was increased as the power input was increased, to reduce the pressure change in operation and to reduce the local heating of the glass envelope resulting from bombardment of electrons reflected from the focal spot. The aluminum cathodes were made heavier to withstand the increased positive-ion bombardment. The thin metal targets were also replaced by a heavier mass of metal. Targets now consist of two main parts: a refractory metal face, such as platinum, to take the direct impact of the electron beam, and a heavy backplate of a good heat-conducting material, such as copper, to con-

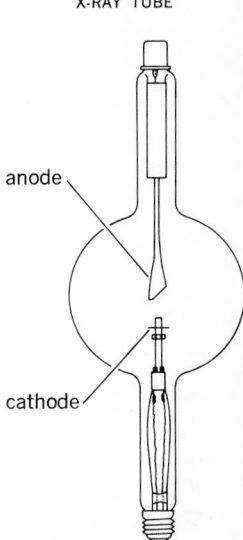

Fig. 2. Early commercial model of single-section, hot-cathode, high-vacuum x-ray tube.

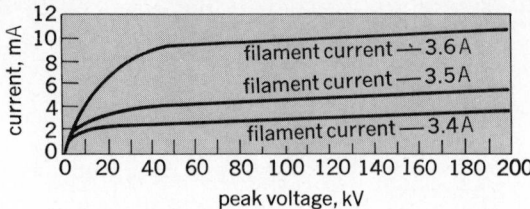

Fig. 3. Curves showing relation of current to voltage in hot-cathode, high-vacuum x-ray tube.

duct the heat away from the focal spot and store it temporarily.

Early in the development of x-ray tubes metals of high atomic weight were known to be the most efficient x-ray generators. W. K. Röntgen had used both aluminum and platinum as targets. W. D. Coolidge and others later used thorium and uranium with increased efficiency. The principal properties desired for a suitable target material are (1) high atomic number to give best x-ray efficiency, (2) high melting point and high thermal conductivity to permit maximum energies for a given size of focal spot, and (3) low vapor pressure to reduce the rate of evaporation of the metal on the walls of the glass envelope. Ductile tungsten as developed by Coolidge was found to combine these desired properties of an x-ray target to the greatest degree.

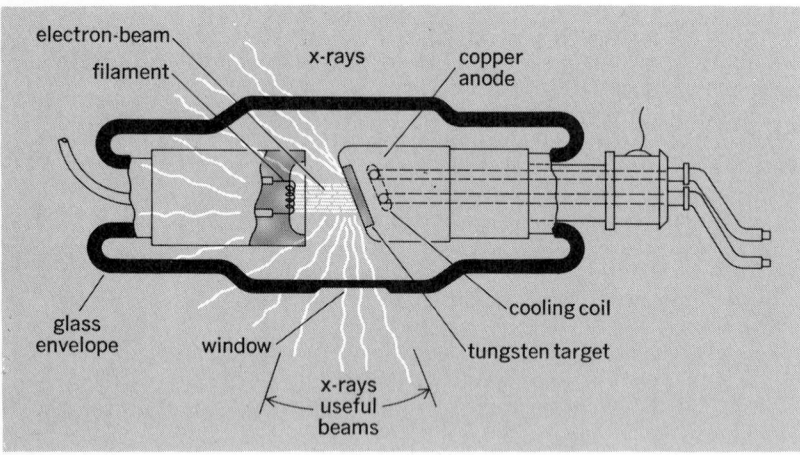

Fig. 4. Single-section, hot-cathode, high-vacuum x-ray tube.

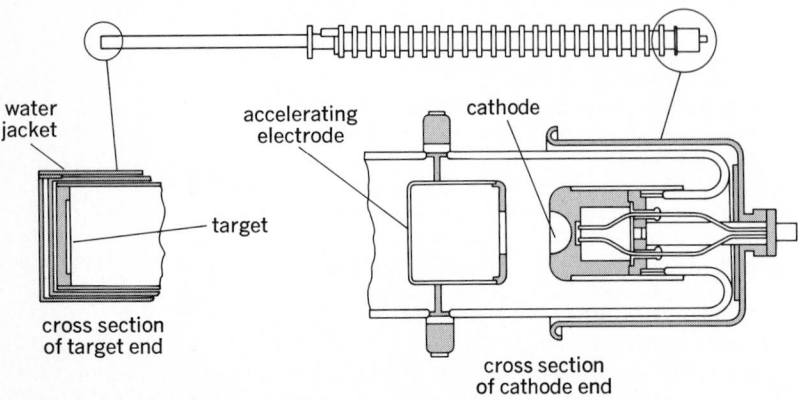

Fig. 5. A 2-MeV multisection, hot-cathode, high-vacuum x-ray tube.

High-vacuum x-ray tubes. The operational difficulties and erratic behavior of gas x-ray tubes are inherently associated with the gas itself and the positive ion bombardment that takes place during operation. The high-vacuum x-ray tube eliminates these difficulties by using other means of emitting electrons from the cathode.

Coolidge made a high-vacuum x-ray tube with a hot tungsten-filament cathode and a solid tungsten target. This hot-cathode high-vacuum type of x-ray tube permitted stable and reproducible operation with relatively high voltages and large masses of metals. The vacuum was so good that positive ions did not play either an essential or a harmful role in the tube operation. The earliest commercial form of the hot-cathode high-vacuum type of x-ray tube (Fig. 2) utilizes a solid tungsten target. The independent relation of the current to the impressed voltage in a tube of this type is shown in Fig. 3. The different curves are for different filament temperatures and show that, over the operating range of x-ray voltages, the discharge current is practically independent of voltage. One typical form of a modern commercial hot-cathode high-vacuum x-ray tube (Fig. 4) is built with a liquid-cooled, copper-backed tungsten target, which operates over a wide range of energy ratings and is capable of rectifying its own current.

The inherent advantages of the hot-cathode high-vacuum tube over the gas tube are (1) flexibility—the voltage and current may be varied independently; (2) stability—this permits more accurate reproducibility of results; (3) small size; (4) operation—it can be operated directly from a transformer, making possible a very simple unit; and (5) long life.

High-voltage x-ray tubes. The operating voltage of a single-section hot-cathode x-ray tube is limited by the field current pulled from the cathode by the electrostatic field. If produced in an x-ray tube, field currents may cause erratic and uncontrollable operating behavior. Single-section x-ray tubes must preferably be operated at voltages below which field currents cannot be produced. *See* FIELD EMISSION.

There are, however, some highly specialized applications of x-ray tubes for which the tube is designed specifically to produce and use field currents to generate x-rays. To build x-ray tubes that operate stably and continuously at very high voltages, the multisection principle is usually employed. These tubes are made with many intermediate sections between the cathode and anode sections. The voltage applied across each section of this multisection tube is always less than that at which field currents will be produced. The electrons emitted from the cathode are accelerated by the voltage applied across each intermediate section. The sum of all of these sectional voltages determines the voltage rating of the multisection x-ray tube. By this procedure, x-ray tubes can be built to generate x-rays at many megaelectron-volts. Figure 5 is a commercial 2-MeV x-ray tube built according to this principle.

The toroidal electromagnetic type of hot-cathode high-vacuum x-ray tube, first successfully built by D. W. Kerst, permits the generation of x-rays at energy levels of many million electron volts without encountering the high-voltage insulation problem that is present in more conventional types of x-ray tubes. This x-ray tube consists of a circular "doughnut" type of high-vacuum envelope housing the hot-cathode and the x-ray target. This tube is designed to guide and accelerate electrons in a circular orbit in a machine called the betatron. In this toroidal-shaped betatron tube (Fig. 6) the electrons are injected and energized to many millions of volts before they strike the target to produce x-rays. The average electron beam current is very small, but x-ray tubes of this type have been built to generate x-rays at energies ranging from 1 to over 100 MeV. The synchrotron is a circular accelerator related to the betatron and contains a similar doughnut-shaped x-ray tube. *See* PARTICLE ACCELERATOR.

The multisection linear accelerator type of x-ray tube is another hot-cathode high-vacuum device that is capable of generating intense x-rays over a wide range of energy levels ranging from 1 to many MeV. It consists (Fig. 7) of an electron gun and a copper pipe waveguide in which the electrons are accelerated by the axial electric field produced by a high-frequency oscillator. At the output end of the waveguide, the electrons strike a target and generate x-rays. X-ray tubes of this type have been built for operation at 5–15 MeV and higher, and with average currents of the order of 25 μA or more.

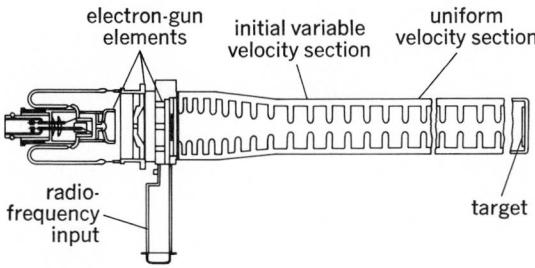

Fig. 7. A multisection linear accelerator type of hot-cathode, high-vacuum x-ray tube.

Applications. Many special forms of x-ray tubes, following the basic patterns already outlined, have been built for application in medicine, industry, and fundamental science. They vary from a few inches to many feet in length. X-ray tubes are used as an aid in medical diagnosis and in therapeutic treatment. They are used in the nondestructive testing of materials throughout all industry. In both fundamental and applied science they are widely used in crystal-diffraction work, chemical analysis by x-ray spectra and absorption, and research on atomic structure.

To provide maximum use in these wide fields of application, particular design features are built into the tubes to meet the special requirements. Some of these special requirements are low or high currents, low- or high-voltage x-rays, large or small focal spot, high x-ray beam intensity for short or long intervals, and choice of x-ray target material to generate a particular quality of x-ray spectra. X-ray and electrical protection must be ensured.

The overall capacity of an x-ray tube is determined principally by the target material, the area of the focal spot, the duration of the energy applied, and the temperature of the target during the

time the electron energy is applied. The modern x-ray tube is a precision tool of great stability and flexibility, capable of controlled operation with currents and voltages of any desired magnitude, and generating x-rays over a wide range of energy and intensity. *See* NUCLEAR RADIATION.

[ERNEST E. CHARLTON]

Bibliography: W. L. Bloom et al., *Medical Radiographic Technic*, 3d ed., 1979; R. Halmshaw, *Industrial Radiology Techniques*, 1971; P. A. Myers, *An Introduction to Radiographic Technique*, 1980.

X-rays

X-rays, or roentgen rays, are electromagnetic waves in which periodically variable electric and magnetic fields are perpendicular to each other and to the direction of propagation. Thus they are identical in nature with visible light and all the other types of radiation that constitute the electromagnetic spectrum (ultraviolet, infrared, γ-rays from radioactive atomic disintegrations, microwaves, and radio or Hertzian waves). In general, x-rays are generated as the result of energy transitions of atomic electrons caused by the bombardment of a material of high atomic weight by high-energy electrons. *See* ELECTROMAGNETIC RADIATION.

Röntgen's findings. The unequivocal establishment of the nature of these rays was not made in W. R. Röntgen's experiments following the discovery of "a new kind of ray" in 1895. In his first communication, Röntgen described the properties of these rays as follows: They were invisible; moved in straight lines; were unaffected by electric or magnetic fields, and hence not electrically charged; passed through matter opaque to ordinary light (since they penetrated through the black cardboard around his cathode-ray tube); were differentially absorbed by matter of different densities or of different atomic weights; affected photographic plates; produced fluorescence in certain chemicals, such as in the barium platinocyanide screen with which the initial discovery was made and in the wall of his glass tube opposite the cathode; produced ionization in gases; and were evidently produced by the stoppage at the anode of the beam of rays (identified by J. J. Thomson in 1897 as electrons) issuing from the cathode in his vacuum tube.

Along with all these definitive characteristics of the rays, however, other crucial experiments designed to establish similarity or differences from ordinary light were clearly called for. The fundamental optical properties of light were well established in 1895: reflection from mirrors; refraction in prisms (change in direction in passing from air into glass, for example), by means of which a beam of white light could be spread out into a rainbow or spectrum of colors; diffraction by narrow slits or ruled gratings, also a method of producing spectra; and polarization, or constraint of the transverse vibrations to a single direction. In spite of the best efforts of Röntgen, no indubitable evidence of any of these four optical phenomena could be found. Hence the designation "x"—unknown—was assigned by Röntgen. Many theories were proposed to account for the apparently unique quality of x-rays, which seemed to be so closely similar and yet so greatly different from light; some suggested that

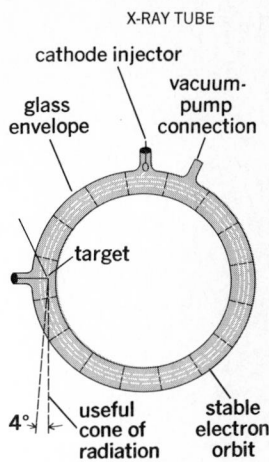

X-RAY TUBE

Fig. 6. Betatron "doughnut" high-vacuum x-ray tube. (*General Electric Co.*)

they were vortex rings in the ether, and waves with longitudinal vibrations, that is, vibrations parallel to the direction of propagation as in sound waves, instead of transverse as with light.

Later discoveries. Inevitably, other scientists studying the enigma found the essential experimental conditions to prove that x-rays can be polarized (C. Barkla, 1905, by scattering from carbon); diffracted by crystals (M. von Laue, W. Friedrich, and P. Knipping, 1912); refracted in prisms and in crystals; reflected by mirrors; and diffracted by ruled gratings (A. Compton, 1921–1922). Instead of being refracted in passing from a less dense medium (air) to a more dense medium (a glass prism or a crystal) in the same direction as light so that the index of refraction is always greater than 1, x-rays are deviated in the opposite direction by a very small amount, so that the index of refraction is less than 1 by an amount as small as 10^{-6}. Thus total reflection from mirrors is observed only when the beam impinges at a very small grazing angle, a necessary condition understandably missed by Röntgen. Similarly, the beam must graze a ruled diffraction grating if a spectrum is to be observed. *See* X-RAY OPTICS.

From 1895 to 1912 there seemed to be no analyzer capable of dispersing an x-ray beam into a spectrum. The spectacular Laue diffraction pattern of a zinc sulfide crystal in 1912 proved the electromagnetic wave nature of x-rays and the ordered structure of crystals with atoms lying on families of planes to constitute three-dimensional diffraction gratings, all governed by the simple Bragg law $n\lambda = 2d \sin \theta$ (which must be corrected for refraction in extremely accurate work). Here n is an integer indicating the order of the spectrum, λ the wavelength, d the crystal lattice spacing of one set of planes, and θ the angle between the incident ray and this set of planes.

The range of x-rays in the electromagnetic spectrum, as excited in x-ray tubes by the bombardment of anode targets by cathode electrons under a high accelerating potential, overlaps the ultraviolet range on the order of 100 nm on the long-wavelength side, and the shortest-wavelength limit moves downward as voltages increase. An accelerating potential of 10^9 volts, now readily generated, produces a λ of 0.00001×10^{-8} cm (10^{-6} nm). An average wavelength used in research is 0.1 nm, or about 1/6000 the wavelength of yellow light.

Quantum theory. In the consideration of roentgen rays as continuous electromagnetic waves, it must not be dismissed that they also appear to be propagated in discontinuous bundles, or quanta, in accordance with the laws first enunciated by M. Planck and extended by A. Einstein early in the 20th century. In diffraction, refraction, polarization, and interference phenomena, x-rays, together with all other related radiations, appear to act as waves and λ has a real significance. Beams of corpuscular electrons and neutrons are diffracted so that they too have wavelengths. In other phenomena—such as the appearance of sharp spectral lines, a definite short-wavelength limit λ_0 of the continuous "white" spectrum [defined by $\lambda_0 = hc/eV$, where h is Planck's constant, c the velocity of electromagnetic radiation (including light and x-rays), e the charge of the electron, and V the accelerating voltage], the shift in wavelength of x-rays scattered by electrons in atoms (Compton effect),

and the photoelectric effect—the energy seems to be propagated and transferred in quanta defined by values of $h\nu$, where the frequency ν is c/λ. These quanta are called photons. *See* COMPTON EFFECT; ELECTRON DIFFRACTION; NEUTRON DIFFRACTION; QUANTUM MECHANICS.

Applications. Important uses have been found for x-rays in many fields of scientific endeavor. For example, roentgen spectrometry is the science of measuring λ values with a known crystal of lattice spacing d; roentgen diffractometry is the science of determining unknown values of d, and thereby crystal structures, with x-ray beams of known λ. In both cases, the experimental measurement is that of the angle θ. Extensive tables of the wavelengths of x-ray emission lines in series (K, L, M, and so on) and so-called absorption edges, characteristic of the chemical elements, afford the necessary information for chemical analyses, exactly as in the case of optical emission spectra and for derivation of theories of atomic structure to account for the origin of spectra. *See* HISTORADIOGRAPHY; MICRORADIOGRAPHY; RADIATION BIOLOGY; RADIOGRAPHY; RADIOLOGY; X-RAY CRYSTALLOGRAPHY; X-RAY DIFFRACTION; X-RAY TUBE.

[GEORGE L. CLARK]

Bibliography: A. G. Brown, *X-rays and Their Applications*, 1975; N. A. Dyson, *X-rays in Atomic and Nuclear Physics*, 1973; F. K. Richtmyer, E. H. Kennard, and J. N. Cooper, *Introduction to Modern Physics*, 6th ed., 1969.

Zeeman effect

A splitting of spectral lines when the light source being studied is placed in a magnetic field. First discovered by P. Zeeman in 1896, the Zeeman effect furnishes information of prime importance in the analysis of spectra. Each kind of spectral term has its characteristic mode of splitting, and the types of terms are most definitely identified by this property. Furthermore, the effect allows an evaluation of the ratio of charge to mass of the electron and an evaluation of its precise magnetic moment.

Normal Zeeman effect. This is a splitting into two or three lines, depending on the direction of observation, as shown in Fig. 1. The light of these components is polarized in ways indicated in the figure. The normal effect is observed for all lines belonging to singlet systems, those for which the spin quantum number $S = 0$. The change of frequency $\Delta\nu_n$ of the shifted components can be evaluated on classical electromagnetic principles as follows. Assume that the electron of charge e revolves in a circular orbit of radius r and circular frequency ω radians per second (Fig. 2). If a mag-

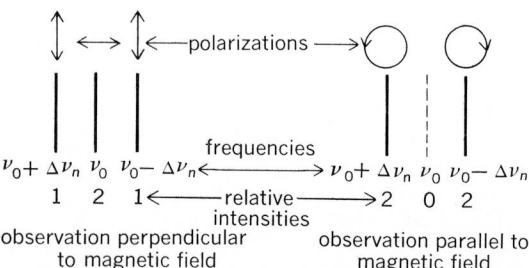

Fig.1. Triplet observed in normal Zeeman effect.

netic field H is applied perpendicular to the plane of the orbit, the electron will be speeded up or slowed down because of the changing flux through its orbit, just as in the electron accelerator known as the betatron.

Denoting the centripetal force holding the electron on its orbit before application of the field by f_0 (Fig. 2b), and the additional force due to the motion of the electron across the field by f_H (Fig. 2a and c), one has Eqs. (1). In Fig. 2a, where the two forces

$$f_0 = m\omega^2 r$$
$$f_H = He(\omega \pm \Delta\omega)r \qquad (1)$$

are in the same direction, one may equate their

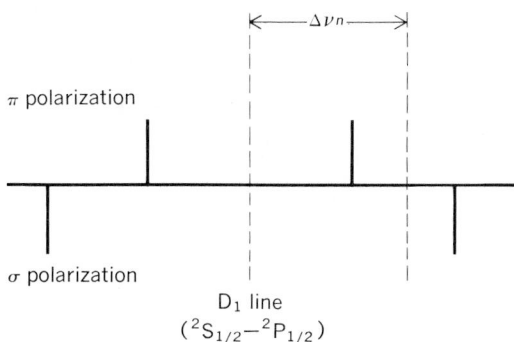

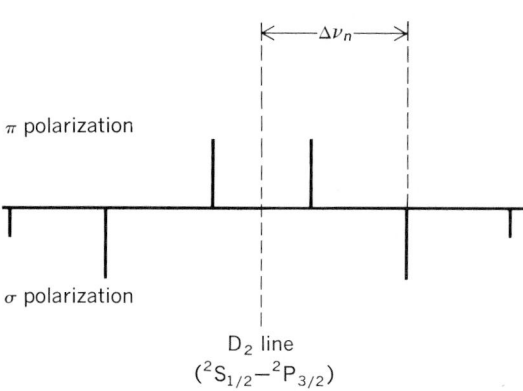

Fig. 3. Anomalous Zeeman effect of the sodium lines. $\Delta\nu_n$ denotes the normal Zeeman splitting, while π and σ refer to polarizations like those of the central and outer components in the normal effect illustrated in Fig. 1. The heights of the lines indicate their relative intensities.

sum to the centripetal force on an electron of frequency $\omega + \Delta\omega$, obtaining Eq. (2). Solution of this

$$f_0 + f_H = m\omega^2 r + He(\omega + \Delta\omega)r$$
$$= m(\omega + \Delta\omega)^2 r \qquad (2)$$

equation for $\Delta\omega$, under the assumption that it is small compared to ω itself, yields Eqs. (3). the

$$\Delta\omega = eH/2m \qquad \Delta\nu_n = eH/4\pi m \qquad (3)$$

latter expression following from the fact that $\nu = \omega/2\pi$. This relation, although derived for a special case, is generally valid for any system of particles having a particular value of e/m and moving under

the action of a central force. *See* LARMOR PRECESSION.

On substitution of the ratio of charge to mass of an electron, one obtains Eq. (4). Conversely,

$$\Delta\nu_n = 1.3996 \times 10^6 H \text{ sec}^{-1} \qquad (4)$$

from the observed spectroscopic splitting $\Delta\nu_n$ and measurement of the field strength, the value of e/m for the electron has been evaluated as 1.7572 ± 0.0007 emu/g. This is in good agreement with the figure determined by other methods.

Anomalous Zeeman effect. This effect is a more complicated type of line splitting, so named because it did not agree with the predictions of classical theory. It occurs for any spectral line arising from a combination of terms of multiplicity greater than one. As examples, Fig. 3 gives diagrams of the theoretical patterns for the yellow lines of sodium, belonging to a doublet system, while Fig. 4 shows some actual patterns observed for doublets and quartets in rhodium.

Since multiplicity in spectral lines is caused by the presence of a resultant spin vector S of the electrons, the anomalous effect must be attributed to a nonclassical magnetic behavior of the electron spin. While classical theory associates with the vector L of the orbital angular momentum a magnetic moment as in Eq. (5), it is necessary, in explaining the anomalous Zeeman effect, that the

$$\mu_L = (eh/4\pi mc)L \qquad (5)$$

magnetic moment corresponding to S be as in Eq. (6). Thus the spin generates twice as much magnetic moment, relative to its angular momentum,

$$\mu_S = (eh/2\pi mc)S \qquad (6)$$

as does the orbital motion. In an atom for which

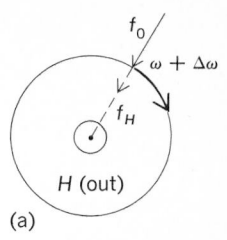

(a)

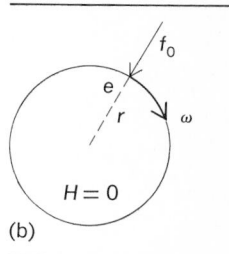

(b)

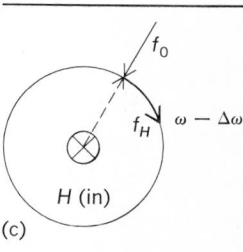

(c)

Fig. 2. Effect of a magnetic field H applied perpendicular to a circular electron orbit. Diagrams a–c are explained in the text.

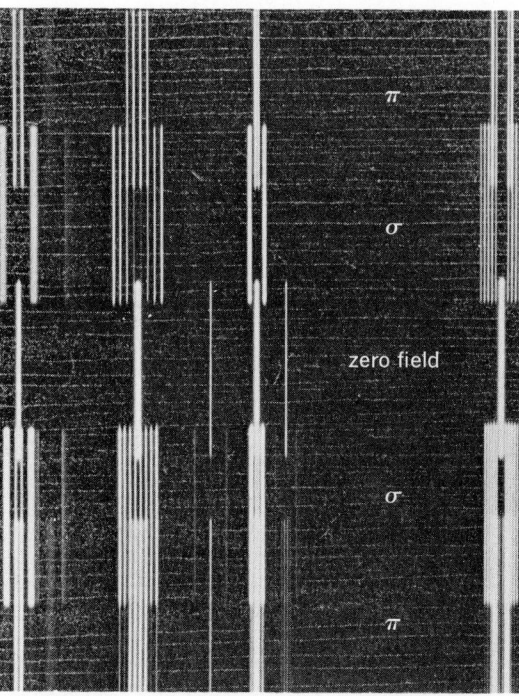

Fig. 4. Zeeman effect of the rhodium spectrum in the wavelength range 3479–3462 A. Field strengths, 70,000 oersteds (lower exposure) and 90,500 (upper exposure). (G. R. Harrison and F. Bitter, Massachusetts Institute of Technology)

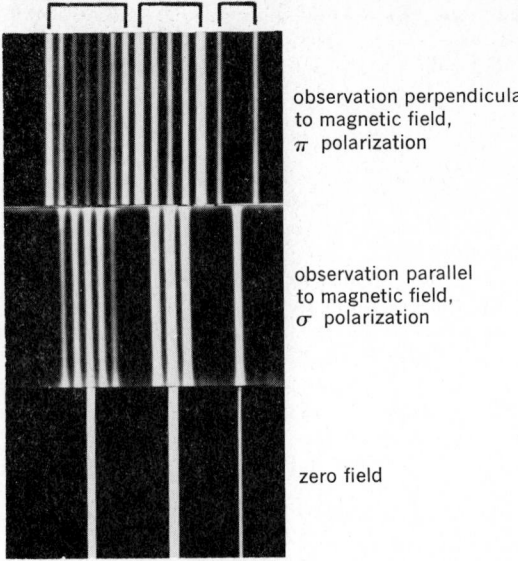

Fig. 5. Zeeman effect of three adjacent rotational lines of the hydrogen molecule. (*Johns Hopkins University*)

both L and S are finite, the effective magnetic moment may be written as Eq. (7), where J mea-

$$\mu_J = g(eh/4\pi mc)J = g\mu_0 J \qquad (7)$$

sures the total angular momentum (in LS coupling, the resultant of L and S), and μ_0 is the Bohr magneton, $eh/4\pi mc$. *See* PARAMAGNETISM.

Theory gives values of g, the Landé g factor, which are characteristic of the type of spectral term. In LS coupling, the value is given by Eq. (8).

$$g = 1 + \frac{J(J+1) + S(S+1) - L(L+1)}{2J(J+1)} \qquad (8)$$

For a classical electron orbit $g = 1$, which yields the normal Zeeman effect. When the spin S is present, however, the changes in energy produced by the magnetic field, which are proportional to μ_J, are just g times as great, and this fact is responsible for the anomalous Zeeman effect. It also should be mentioned that both theory and experiment now show that the g factor for the electron is not exactly 2, but 2.00229.

The component of μ_J in the field direction is $g\mu_0 M$, where M is the quantized component of J in this direction. The energy terms become $T = T_0 + g\mu_0 M$, where T_0 is the term value with no field. This magnetic quantum number M has only the $2J + 1$ values, $J, J-1, J-2, \ldots, -J$, and the allowed transitions between energy terms must obey the selection rule $\Delta M = 0, \pm 1$. To explain the statement that the normal Zeeman effect is observed for all lines in singlet systems, recall that S is then zero, $J = L$, and hence $g = 1$.

Quadratic Zeeman effect. The quadratic effect, which depends on the square of the field strength, is of two kinds. The first results from the second-order terms that were neglected in the preceding derivation, and the second, from the diamagnetic reaction of the electron when revolving in large orbits.

Inverse Zeeman effect. This is the Zeeman effect of absorption lines. It is closely related to the Faraday effect, the rotation of plane-polarized light

by matter situated in a magnetic field. *See* ATOMIC STRUCTURE AND SPECTRA; FARADAY EFFECT.

Zeeman effect in molecules. This effect is, in general, so small as to be unobservable, even for molecules which have a permanent magnetic moment. Each level with a total angular momentum J splits into $2J + 1$ components, as in the case of atoms.

The component of the magnetic moment along the direction of the external field is small, however, because the rotation of the molecule, which carries the magnetic moment along with it, causes the principal part of the magnetic moment to average out to zero. The consequence is that the magnetic levels have an extremely narrow spacing except for cases where the molecule has either very little rotation or none at all. An exception occurs for some light molecules where the magnetic moment is coupled so lightly to the frame of the molecule that it can orient itself freely in the magnetic field just as for atoms (Fig. 5).

Zeeman effect in crystals. A clear Zeeman effect also can be observed in many crystals with sharp spectrum lines in absorption or fluorescence. Such crystals are found particularly among the salts of the rare earths. In these cases the internal electric field in the crystal splits and shifts the level of the free ion. When the number of electrons is even and the crystal symmetry low, this electric splitting is complete. No degeneracy remains, and there can be no further splitting by a magnetic field. If the number of electrons is odd, or if for an even number there is high crystal symmetry, the levels occur in degenerate pairs which are split by a magnetic field. Each line is then split

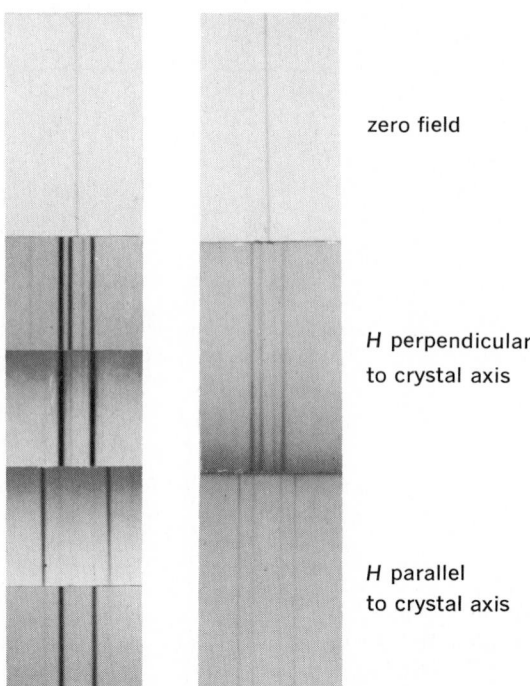

Fig. 6. Zeeman effect of an absorption line of neodymium chloride, ($NdCl_3$), left with polarization and right without polarization. The splitting is quite different, depending on whether the trigonal crystal axis is parallel or perpendicular to the magnetic field. $H = 35,000$ oersteds. (*Johns Hopkins University*)

into four components (Fig. 6). For cubic crystal symmetry and when angular momentum is due only to electron spin, splitting into more than four components may occur.

Nuclear Zeeman effect. The magnetic moment of the nucleus causes a Zeeman splitting in atomic spectra which is of an order of magnitude a thousand times smaller than the ordinary Zeeman effect. This Zeeman effect of the hyperfine structure usually is modified by a nuclear Paschen-Back effect, first studied by E. Back and S. A. Goudsmit for the spectral lines of bismuth. *See* PASCHEN-BACK EFFECT.

A strong magnetic field actually may modify the intensity and selection rules so that usually absent lines may appear. For example, the *J* selection rule is no longer valid in a magnetic field. *See* SELECTION RULES.

[F. A. JENKINS; G. H. DIEKE/W. W. WATSON]

Bibliography: G. Herzberg, *Atomic Spectra and Atomic Structure*, 1944; F. A. Jenkins and H. E. White, *Fundamentals of Optics*, 4th ed., 1976; J. H. Van Vleck, *The Theory of Electric and Magnetic Susceptibilities*, 1932; M. R. Wehr, J. A. Richards, and T. W. Adair, *Physics of the Atom*, 3d ed., 1978.

Appendix

Scientific Notation in the Encyclopedia

Since the Encyclopedia is a work of science, it was necessary to fol-
low in the articles the scientific style of using symbols, abbreviations,
and exact names. This section discusses the most frequently used
conventions in the Encyclopedia and includes tables for convenient
reference. The relation between the three primary measurement
systems is also clarified.

U.S. Customary System and the metric system

Scientists and engineers have been using two major systems of units in measurement. These are commonly called the U.S. Customary System (inherited from the British Imperial System) and the metric system.

In the U.S. Customary System the units yard and pound with their divisions, such as the inch, and multiples, such as the ton, are basic. The metric system was evolved during the 18th century and has been adopted for general use by most countries. Nearly everywhere it is used for precise measurements in science. The meter and kilogram with their multiples, such as the kilometer, and fractions, such as the gram, are basic to the metric system.

In the U.S. Customary System, units of the same kind are related almost at random. For example, there are the units of length, the inch, yard, and mile. In the metric system the relationships between units of the same kind are strictly decimal (millimeter, meter, and kilometer).

However, to complicate matters in scientific writing, there is no uniformity within each of these two systems as to the choice of units for the same quantities. For example, the hour or the second, the foot or the inch, and the centimeter or the millimeter could be chosen by a scientist as the unit of measurement for the quantities time and length.

Introduction of the International System, or SI

To simplify matters and to make communication more understandable, an internationally accepted system of units is coming into use. This is termed the International System of Units, which is abbreviated SI in all languages.

Fundamentally the system is metric with the base units derived from scientific formulas or natural constants. For example, the meter in the SI is defined as the length equal to 1 650 763.73 wavelengths in vacuum of the radiation corresponding to the transition between the electronic energy levels $2p_{10}$ and $5d_5$ of the krypton-86 atom. Previously, in the metric system, the meter was

Introduction of the International System, or SI (cont.)

defined as the distance between two marks on a specific metal bar.

In a similar way the second in the SI is defined as the duration of 9 192 631 770 periods of the radiation corresponding to the transition between two hyperfine levels of the ground state of the cesium-133 atom.

Interestingly, the kilogram, the SI unit of mass, is still the mass of the kilogram kept at Sèvres, France. However, it is possible that eventually the unit will be redefined in terms of atomic mass.

Although the SI is increasing in usage by scientists and engineers, there are some units in everyday use which will probably remain, for example, minute, hour, day, degree (angle), and liter. The point should be made, however, that these terms will not be employed in a scientific context if the SI is fully adopted.

Because of their extremely common use among scientists, several units are still permitted in conjunction with SI units, for example, the electron volt, rad, roentgen, barn, and curie. In time their usage might be phased out.

One further point is that in October, 1967, the Thirteenth General Conference of Weights and Measures decided to name the SI unit of thermodynamic temperature "kelvin" (symbol K) instead of "degree Kelvin" (symbol °K). For example, the notation is 273 K and not 273°K.

The base units and derived units of the SI are shown in **Table 1** and **Table 2.**

In the SI the prefixes differ from a unit in steps of 10^3. A list of prefix terms, symbols, and their factors is given in **Table 3**. Some examples of the use of these prefixes follow:

$$1000 \text{ m} = 1 \text{ kilometer} = 1 \text{ km}$$

$$1000 \text{ V} = 1 \text{ kilovolt} = 1 \text{ kV}$$

$$1\,000\,000 \ \Omega = 1 \text{ megohm} = 1 \text{ M}\Omega$$

$$0.000\,000\,001 \text{ s} = 1 \text{ nanosecond} = 1 \text{ ns}$$

Only one prefix is to be employed for a unit. For example:

$$1000 \text{ kg} = 1 \text{ Mg} \qquad \text{not 1 kkg}$$

$$10^{-9} \text{ s} = 1 \text{ ns} \qquad \text{not 1 m}\mu\text{s}$$

$$1\,000\,000 \text{ m} = 1 \text{ Mm} \qquad \text{not 1 kkm}$$

Also, when a unit is raised to a power, the power applies to the whole unit including the prefix. For example:

$$\text{km}^2 = (\text{km})^2 = (1000 \text{ m})^2 = 10^6 \text{ m}^2 \qquad \text{not 1000 m}^2$$

Table 1. Base units of the International System

Quantity	Name of unit	Unit symbol
length	meter	m
mass	kilogram	kg
time	second	s
electric current	ampere	A
temperature	kelvin	K
luminous intensity	candela	cd
amount of substance	mole	mol

Table 2. Derived units of the International System

Quantity	Name of unit	Unit symbol, where differing from basic form	Unit expressed in terms of base or supplementary units*
area	square meter		m^2
volume	cubic meter		m^3
frequency	hertz	Hz	s^{-1}
density	kilogram per cubic meter		kg/m^3
velocity	meter per second		m/s
angular velocity	radian per second		rad/s
acceleration	meter per second squared		m/s^2
angular acceleration	radian per second squared		rad/s^2
volumetric flow rate	cubic meter per second		m^3/s
force	newton	N	$kg \cdot m/s^2$
surface tension	newton per meter, joule per square meter	$N/m, J/m^2$	kg/s^2
pressure	newton per square meter, pascal	$N/m^2, Pa$	$kg/m \cdot s^2$
viscosity, dynamic	newton-second per square meter, pascal-second	$N \cdot s/m^2, Pa \cdot s$	$kg/m \cdot s$
viscosity, kinematic	meter squared per second		m^2/s
work, torque, energy, quantity of heat	joule, newton-meter, watt-second	$J, N \cdot m, W \cdot s$	$kg \cdot m^2/s^2$
power, heat flux	watt, joule per second	$W, J/s$	$kg \cdot m^2/s^3$
heat flux density	watt per square meter	W/m^2	kg/s^3
volumetric heat release rate	watt per cubic meter	W/m^3	$kg/m \cdot s^3$
heat transfer coefficient	watt per square meter kelvin	$W/m^2 \cdot K$	$kg/s^3 \cdot K$
heat capacity (specific)	joule per kilogram kelvin	$J/kg \cdot K$	$m^2/s^2 \cdot K$
capacity rate	watt per kelvin	W/K	$kg \cdot m^2/s^3 \cdot K$
thermal conductivity	watt per meter kelvin	$W/m \cdot K, \dfrac{J \cdot m}{s \cdot m^2 \cdot K}$	$kg \cdot m/s^3 \cdot K$
quantity of electricity	coulomb	C	$A \cdot s$
electromotive force	volt	$V, W/A$	$kg \cdot m^2/A \cdot s^3$
electric field strength	volt per meter	V/m	$kg \cdot m/A \cdot s^3$
electric resistance	ohm	$\Omega, V/A$	$kg \cdot m^2/A^2 \cdot s^3$
electric conductivity	ampere per volt meter	$A/V \cdot m$	$A^2 \cdot s^3/kg \cdot m^3$
electric capacitance	farad	$F, A \cdot s/V$	$A^3 \cdot s^4/kg \cdot m^2$
magnetic flux	weber	$Wb, V \cdot s$	$kg \cdot m^2/A \cdot s^2$
inductance	henry	$H, V \cdot s/A$	$kg \cdot m^2/A^2 \cdot s^2$
magnetic permeability	henry per meter	H/m	$kg \cdot m/A^2 \cdot s^2$
magnetic flux density	tesla, weber per square meter	$T, Wb/m^2$	$kg/A \cdot s^2$
magnetic field strength	ampere per meter		A/m
magnetomotive force	ampere		A
luminous flux	lumen	lm	$cd \cdot sr$
luminance	candela per square meter		cd/m^2
illumination	lux, lumen per square meter	$lx, lm/m^2$	$cd \cdot sr/m^2$
activity (of radionuclides)	becquerel	Bq	s^{-1}
absorbed dose	gray	$Gy, J/kg$	$m^2 \cdot s^{-2}$
dose equivalent	sievert	Sv	$m^2 \cdot s^{-2}$

*Supplementary units are: plane angle, radian (rad): solid angle, steradian (sr).

Table 3. Prefixes for units in the International System

Prefix	Symbol	Power	Example	Prefix	Symbol	Power	Example
exa	E	10^{18}		deci	d	10^{-1}	
peta	P	10^{15}		centi	c	10^{-2}	
tera	T	10^{12}		milli	m	10^{-3}	milligram (mg)
giga	G	10^{9}		micro	μ	10^{-6}	microgram (μg)
mega	M	10^{6}	megahertz (MHz)	nano	n	10^{-9}	nanosecond (ns)
kilo	k	10^{3}	kilometer (km)	pico	p	10^{-12}	picofarad (pf)
hecto	h	10^{2}		femto	f	10^{-15}	
deka	da	10^{1}		atto	a	10^{-18}	

Introduction of the International System, or SI (cont.)

Some common units defined in terms of SI units are given in **Table 4** (the definitions in the fourth column are exact).

Table 4. Some common units defined in terms of SI units

Quantity	Name of unit	Unit symbol	Definition of unit
length	inch	in.	2.54×10^{-2} m
mass	pound (avoirdupois)	lb	0.45359237 kg
force	kilogram-force	kgf	9.80665 N
pressure	atmosphere	atm	101325 Pa
pressure	torr	torr	(101325/760) Pa
pressure	conventional millimeter of mercury*	mmHg	$13.5951 \times 980.665 \times 10^{-2}$ Pa
energy	kilowatt-hour	kWh	3.6×10^{6} J
energy	thermochemical calorie	cal	4.184 J
energy	international steam table calorie	cal_{IT}	4.1868 J
thermodynamic temperature (T)	degree Rankine	°R	(5/9) K
customary temperature (t)	degree Celsius	°C	$t(°C) = T(K) - 273.15$
customary temperature (t)	degree Fahrenheit	°F	$t(°F) = T(°R) - 459.67$
radioactivity	curie	Ci	3.7×10^{10} Bq
energy†	electron volt	eV	$eV \approx 1.60219 \times 10^{-19}$ J
mass†	unified atomic mass unit	u	$u \approx 1.66057 \times 10^{-27}$ kg

*The conventional millimeter of mercury, symbol mmHg (not mm Hg), is the pressure exerted by a column exactly 1 mm high of a fluid of density exactly 13.5951 g · cm⁻³ in a place where the gravitational acceleration is exactly 980.665 cm · s⁻². The mmHg differs from the torr by less than 2×10^{-7} torr.
†These units defined in terms of the best available experimental values of certain physical constants may be converted to SI units. The factors for conversion of these units are subject to change in the light of new experimental measurements of the constants involved.

Conversion factors for the measurement systems

Because it will take some years for all scientists and engineers to convert to the SI, the Encyclopedia has retained the U.S. Customary and metric systems, but has incorporated SI units when preparation of the text permitted. Conversion factors between the three measurement systems are given in **Table 5** for some prevalent units; in each of the subtables the user proceeds as follows:

To convert a quantity expressed in a unit in the left-hand column to the equivalent in a unit in the top row of a subtable, multiply the quantity by the factor common to both units.

The factors have been carried out to seven significant figures, as derived from the fundamental constants and the definitions of the units. However, this does not mean that the factors are always known to that accuracy. Numbers followed by ellipses are to be continued indefinitely with repetition of the same pattern of digits. Factors written with fewer than seven significant digits are exact values. Numbers followed by an asterisk are definitions of the relation between the two units.

Table 5. Conversion factors for the U.S. Customary System, metric system, and International System

A. UNITS OF LENGTH

Units	cm	m	in.	ft	yd	mile
1 cm	= 1	0.01^*	0.3937008	0.03280840	0.01093613	6.213712×10^{-6}
1 m	= 100.	1	39.37008	3.280840	1.093613	6.213712×10^{-4}
1 in.	$= 2.54^*$	0.0254	1	0.08333333...	0.02777777...	1.578283×10^{-5}
1 ft	= 30.48	0.3048	$12.^*$	1	0.3333333...	$1.893939... \times 10^{-4}$
1 yd	= 91.44	0.9144	36.	$3.^*$	1	$5.681818... \times 10^{-4}$
1 mile	$= 1.609344 \times 10^5$	1.609344×10^3	6.336×10^4	$5280.^*$	1760.	1

B. UNITS OF AREA

Units	cm^2	m^2	in.^{2}	ft^2	yd^2	mile^2
1 cm^2	= 1	10^{-4*}	0.1550003	1.076391×10^{-3}	1.195990×10^{-4}	3.861022×10^{-11}
1 m^2	$= 10^4$	1	1550.003	10.76391	1.195990	3.861022×10^{-7}
1 in.^{2}	$= 6.4516^*$	6.4516×10^{-4}	1	$6.944444 \times 10^{-3}...$	7.716049×10^{-4}	2.490977×10^{-10}
1 ft^2	= 929.0304	0.09290304	$144.^*$	1	0.1111111...	3.587007×10^{-8}
1 yd^2	= 8361.273	0.8361273	1296.	$9.^*$	1	3.228306×10^{-7}
1 mile^2	$= 2.589988 \times 10^{10}$	2.589988×10^6	4.014490×10^9	$2.78784 \times 10^{7*}$	3.0976×10^6	1

continued

Conversion factors for the measurement systems (cont.)

Table 5. Conversion factors for the U.S. Customary System, metric system, and International System (cont.)

C. UNITS OF VOLUME

Units	m³	cm³	liter	in.³	ft³	qt	gal
1 m³	= 1	10^6	10^8	6.102374×10^4	35.31467	1.056688×10^3	264.1721
1 cm³	= 10^{-6}	1	10^{-3}	0.06102374	3.531467×10^{-5}	1.056688×10^{-3}	2.641721×10^{-4}
1 liter	= 10^{-3}	1000.*	1	61.02374	0.03531467	1.056688	0.2641721
1 in.³	= 1.638706×10^{-5}	16.38706*	0.01638706	1	5.787037×10^{-4}	0.01731602	4.329004×10^{-3}
1 ft³	= 2.831685×10^{-2}	28316.85	28.31685	1728.*	1	2.992208	7.480520
1 qt	= 9.46353×10^{-4}	946.353	0.946353	57.75	0.0342014	1	0.25
1 gal (U.S.)	= 3.785412×10^{-3}	3785.412	3.785412	231.*	0.1336806	4.*	1

D. UNITS OF MASS

Units	g	kg	oz	lb	metric ton	ton
1 g	= 1	10^{-3}	0.03527396	2.204623×10^{-3}	10^{-6}	1.102311×10^{-6}
1 kg	= 1000.	1	35.27396	2.204623	10^{-3}	1.102311×10^{-3}
1 oz (avdp)	= 28.34952	0.02834952	1	0.0625	2.834952×10^{-5}	$5. \times 10^{-4}$
1 lb (avdp)	= 453.5924	0.4535924	16.*	1	4.535924×10^{-4}	0.0005
1 metric ton	= 10^6	1000.*	35273.96	2204.623	1	1.102311
1 ton	= 907184.7	907.1847	32000.	2000.*	0.9071847	1

E. UNITS OF DENSITY

Units	g · cm⁻³	g · L⁻¹, kg · m⁻³	oz · in.⁻³	lb · in.⁻³	lb · ft⁻³	lb · gal⁻¹
1 g · cm⁻³	= 1	1000.	0.5780365	0.03612728	62.42795	8.345403
1 g · L⁻¹, kg · m⁻³	= 10^{-3}	1	5.780365×10^{-4}	3.612728×10^{-5}	0.06242795	8.345403×10^{-3}
1 oz · in.⁻³	= 1.729994	1729.994	1	0.0625	108.	14.4375
1 lb · in.⁻³	= 27.67991	27679.91	16.	1	1728.	231.
1 lb · ft⁻³	= 0.01601847	16.01847	9.259259×10^{-3}	5.7870370×10^{-4}	1	0.1336806
1 lb · gal⁻¹	= 0.1198264	119.8264	4.749536×10^{-3}	4.3290043×10^{-3}	7.480519	1

Table 5. Conversion factors for the U.S. Customary System, metric system, and International System (cont.)

F. UNITS OF PRESSURE

Units	Pa, N·m⁻²	dyn·cm⁻²	bar	atm	kgf·cm⁻²	mmHg (Torr)	in. Hg	lbf·in.⁻²
1 Pa, 1 N·m⁻²	=1	10	10^{-5}	9.869233×10^{-6}	1.019716×10^{-5}	7.500617×10^{-3}	2.952999×10^{-4}	1.450377×10^{-4}
1 dyn·cm⁻²	=0.1	1	10^{-6}	9.869233×10^{-7}	1.019716×10^{-6}	7.500617×10^{-4}	2.952999×10^{-5}	1.450377×10^{-5}
1 bar	$=10^{5}$*	10^{6}	1	0.9869233	1.019716	750.0617	29.52999	14.50377
1 atm	=101325.0*	1013250.	1.013250	1	1.033227	760.	29.92126	14.69595
1 kgf·cm⁻²	=98066.5	980665.	0.980665	0.9678411	1	735.5592	28.95903	14.22334
1 mmHg (Torr)	=133.3224	1333.224	1.333224×10^{-3}	1.3157895×10^{-3}	1.3595099×10^{-3}	1	0.03937008	0.01933678
1 in. Hg	=3386.388	33863.88	0.03386388	0.03342105	0.03453155	25.4	1	0.4911541
1 lbf·in.⁻²	=6894.757	68947.57	0.06894757	0.06804596	0.07030696	51.71493	2.036021	1

G. UNITS OF ENERGY†

Units	g mass (energy equiv)	J	int J	cal	cal$_{IT}$	Btu$_{IT}$	kW hr	hp hr	ft-lbf	cu ft-lbf in.⁻²	liter-atm
1 g mass (energy equiv)	=1	8.987552×10^{13}	8.986069×10^{13}	2.148076×10^{13}	2.146640×10^{13}	8.518555×10^{10}	2.496542×10^{7}	3.347918×10^{7}	6.628878×10^{13}	4.603388×10^{11}	8.870024×10^{11}
1 J	$=1.112650 \times 10^{-14}$	1	0.999835	0.2390057	0.2388459	9.478172×10^{-4}	$2.777777... \times 10^{-7}$	3.725062×10^{-7}	0.7375622	5.121960×10^{-3}	9.869233×10^{-3}
1 int J	$=1.112834 \times 10^{-14}$	1.000165	1	0.2390452	0.2388853	9.479735×10^{-4}	2.778236×10^{-7}	3.725676×10^{-7}	0.7376839	5.122805×10^{-3}	9.870862×10^{-3}
1 cal	$=4.655328 \times 10^{-14}$	4.184*	4.183310	1	0.9993312	3.965667×10^{-3}	$1.1622222... \times 10^{-6}$	1.558562×10^{-6}	3.085960	2.143028×10^{-2}	0.04129287
1 cal$_{IT}$	$=4.658443 \times 10^{-14}$	4.1868*	4.186109	1.000669	1	3.968321×10^{-3}	1.163000×10^{-6}	1.559609×10^{-6}	3.088025	2.144462×10^{-2}	0.04132050
1 Btu$_{IT}$	$=1.173908 \times 10^{-11}$	1055.056	1054.882	252.1644	251.9958*	1	2.930711×10^{-4}	3.930148×10^{-4}	778.1693	5.403953	10.41259
1 kW hr	$=4.005540 \times 10^{-8}$	3600000.*	3599406.	860420.7	859845.2	3412.142	1	1.341022	2655224.	18439.06	35529.24
1 hp hr	$=2.986931 \times 10^{-8}$	2684519.	2684077.	641615.6	641186.5	2544.33	0.7456998	1	1980000.*	13750.	26494.15
1 ft-lbf	$=1.508551 \times 10^{-14}$	1.355818	1.355594	0.3240483	0.3238315	1.285067×10^{-3}	3.766161×10^{-7}	$5.050505... \times 10^{-7}$	1	$6.944444... \times 10^{-3}$	0.01338088
1 cu ft-lbf in.⁻²	$=2.172313 \times 10^{-12}$	195.2378	195.2056	46.66295	46.63174	0.1850497	5.423272×10^{-5}	$7.272727... \times 10^{-5}$	144.*	1	1.926847
1 liter-atm	$=1.127393 \times 10^{-12}$	101.3250	101.3083	24.21726	24.20106	0.09603757	2.814583×10^{-5}	3.774419×10^{-5}	74.73349	0.5189825	1

†The electrical units are those in terms of which certification of standard cells, standard resistances, and so forth, is made by the National Bureau of Standards. Unless otherwise indicated, all electrical units are absolute.

Units of temperature in measurement systems

Temperature is a basic physical quantity. It is a measure of the thermal energy of random motion of particles in a system. As such it has been chosen as one of the base quantities in the SI. It is to be treated as are the units length, mass, time, electric current, and luminous intensity. In the SI the unit of length is the meter, the unit of time the second, and so on. The question arises as to the choice of the unit of temperature in the SI.

In the past it was customary to refer to scales of temperature, for example, the Celsius and Fahrenheit scales. On the Celsius scale, 0 designates the freezing point (ice point) and 100 the boiling point (steam point) of water. Corresponding numbers on the Fahrenheit scale are 32 and 212. There are 100 units between the ice point and steam point on the Celsius scale, and 180 units between these points in the Fahrenheit system.

By measuring the volume changes of a gas within the 100-unit interval of the ice point and steam point of water on the Celsius scale, it was found that a numerical value could be assigned for a basic unit of temperature. Careful measurement of this ice-steam interval in a gas thermometer determined that the ice point of water should be assigned the value of 273.15 kelvins. The unit of temperature was thus called the kelvin with the symbol K. Further experiments led to the decision to define the kelvin in the SI along the same lines but in terms of the triple point of water. This is the temperature and pressure at which ice, liquid water, and water vapor coexist at equilibrium. The triple point was chosen because it was a more reproducible value than the ice point.

This change led to the SI definition of temperature in terms of the triple point of water, which contains exactly 273.16 kelvins.

It follows that the Celsius temperature (°C) is an intermediate scale. It is useful in defining Kelvin temperature in the SI. Celsius temperature (t) is related to Kelvin temperature (K) as follows:

$$t_{\text{ice point}} = 0°\text{C}$$

$$t_{\text{steam point}} = 100°\text{C}$$

$$0\,\text{K} = -273.15°\text{C}$$

A summary of the conventions in the SI as proposed in the Thirteenth General Conference of Weights and Measures pertaining to temperature units is given below.

1. The unit of SI temperature is the kelvin, symbol K.

2. The word "scale" is not to be used except in terms of measurement of temperature between certain fixed points on the Celsius scale.

3. The terms "thermodynamic scale" or "absolute scale" are not to be used to describe temperature. The degree sign is to be eliminated with the symbol K.

4. When Celsius temperatures are used (°C), it is understood that the temperature unit is the kelvin.

Not all scientists and engineers have adopted the SI of temperature terminology. For this reason the contributors to the Encyclopedia have retained the term "scale" in relation to thermodynamic temperature. Furthermore, many engineers in the United States still use the Fahrenheit system in discussing practical engineering systems.

In converting Fahrenheit (°F) to Celsius (°C) the following formula applies.

$$°\text{C} = \frac{°\text{F} - 32°}{1.8}$$

Units of temperature in measurement systems (cont.)

In converting Celsius to Fahrenheit the following formula can be used.

$$°F = (°C \times 1.8) + 32°$$

In changing from Celsius terminology (t) to kelvin units (K) the following formula can be used.

$$K = t + 273.15$$

Symbols for the chemical elements

The mass number, atomic number, number of atoms, and ionic charge of an element are indicated in the Encyclopedia by means of four indices placed around the symbol. The positions occupied are left upper index, mass number (the right upper index was formerly used); left lower index, atomic number; right upper index, ionic charge; and right lower index, number of atoms; for example, $^{12}_{6}C$ and Ca^{2+}. The atomic number, which is redundant, is omitted in most cases; that is, the former example is written as ^{12}C.

Ionic charge is indicated by a plus or minus superscript following the symbol of the ion; for multiple charges an Arabic superscript numeral precedes the plus or minus sign, for example, Na^+, NO_3^-, Ca^{2+}, PO_4^{3-} (an alternative sometimes used is Ca^{++}, O^{--}, and so on).

An alphabetical list of the elements, their symbols, and their atomic numbers is shown in **Table 6**. (The symbol Lw was originally designated for element 103, but in 1963 the symbol Lr was proposed. Elements 104, 105, and 106 have been reported, but no official names or symbols have yet been assigned.)

Table 6. The 103 chemical elements

Name	Symbol	At. no.	Name	Symbol	At. no.	Name	Symbol	At. no.	Name	Symbol	At. no.
Actinium	Ac	89	Erbium	Er	68	Mercury	Hg	80	Samarium	Sm	62
Aluminum	Al	13	Europium	Eu	63	Molybdenum	Mo	42	Scandium	Sc	21
Americium	Am	95	Fermium	Fm	100	Neodymium	Nd	60	Selenium	Se	34
Antimony	Sb	51	Fluorine	F	9	Neon	Ne	10	Silicon	Si	14
Argon	Ar	18	Francium	Fr	87	Neptunium	Np	93	Silver	Ag	47
Arsenic	As	33	Gadolinium	Gd	64	Nickel	Ni	28	Sodium	Na	11
Astatine	At	85	Gallium	Ga	31	Niobium	Nb	41	Strontium	Sr	38
Barium	Ba	56	Germanium	Ge	32	Nitrogen	N	7	Sulfur	S	16
Berkelium	Bk	97	Gold	Au	79	Nobelium	No	102	Tantalum	Ta	73
Beryllium	Be	4	Hafnium	Hf	72	Osmium	Os	76	Technetium	Tc	43
Bismuth	Bi	83	Helium	He	2	Oxygen	O	8	Tellurium	Te	52
Boron	B	5	Holmium	Ho	67	Palladium	Pd	46	Terbium	Tb	65
Bromine	Br	35	Hydrogen	H	1	Phosphorus	P	15	Thallium	Tl	81
Cadmium	Cd	48	Indium	In	49	Platinum	Pt	78	Thorium	Th	90
Calcium	Ca	20	Iodine	I	53	Plutonium	Pu	94	Thulium	Tm	69
Californium	Cf	98	Iridium	Ir	77	Polonium	Po	84	Tin	Sn	50
Carbon	C	6	Iron	Fe	26	Potassium	K	19	Titanium	Ti	22
Cerium	Ce	58	Krypton	Kr	36	Praseodymium	Pr	59	Tungsten	W	74
Cesium	Cs	55	Lanthanum	La	57	Promethium	Pm	61	Uranium	U	92
Chlorine	Cl	17	Lawrencium	Lr (Lw)	103	Protactinium	Pa	91	Vanadium	V	23
Chromium	Cr	24	Lead	Pb	82	Radium	Ra	88	Xenon	Xe	54
Cobalt	Co	27	Lithium	Li	3	Radon	Rn	86	Ytterbium	Yb	70
Copper	Cu	29	Lutetium	Lu	71	Rhenium	Re	75	Yttrium	Y	39
Curium	Cm	96	Magnesium	Mg	12	Rhodium	Rh	45	Zinc	Zn	30
Dysprosium	Dy	66	Manganese	Mn	25	Rubidium	Rb	37	Zirconium	Zr	40
Einsteinium	Es	99	Mendelevium	Md	101	Ruthenium	Ru	44			

Symbols and abbreviations in scientific writing

For convenience, commonly used symbols encountered in scientific writings are listed in **Table 7**. Symbols following the ellipses and separated by commas are alternatives that are used only when there is some reason for not using the symbol given first.

Symbols for particles and quanta are as follows:

neutron	n	pion	π
proton	p	muon	μ
deuteron	d	electron	e
triton	t	neutrino	ν
α-particle	α	photon	γ

Table 7. Commonly used symbols in scientific literature

SPACE, TIME, MASS, AND RELATED QUANTITIES

length l
height h
radius r
diameter d
path, length of arc s
plane angle $\alpha, \beta, \gamma, \theta, \phi, \psi$
solid angle ω
area A, S
volume $V...v$
specific volume v
wavelength λ
wavenumber σ, ν
time t
period or other characteristic interval T, τ
frequency ν, f
angular frequency ($2\pi\nu$) ω
velocity $v...u, w$
angular velocity ω
acceleration a
acceleration of free fall g
mass m
moment of inertia I
density ρ
relative density d

MOLECULAR AND RELATED QUANTITIES

molecular mass m
molar mass M
Avogadro's number N_0, L, N
number of molecules N
number of moles n
mole fraction $x...X, y$
molality m
concentration c
molar concentration of substance B $c_B, [B], c(B)$
molecular concentration C
partition function Q
statistical weight $g...p$
symmetry number σ
characteristic temperature Θ
diameter of molecule $\sigma...D$
mean free path l
diffusion coefficient D
osmotic pressure Π
surface concentration Γ

MECHANICAL AND RELATED QUANTITIES

force F
force due to gravity (weight) $G...W$
moment of force M
power P
pressure p, P
traction σ
shear stress τ
modulus of elasticity E
shear modulus G
compressibility κ
compression modulus ($1/\kappa$)   K
viscosity η
fluidity ϕ
kinematic viscosity ν
friction coefficient f
surface tension $\gamma...\sigma$
angle of contact θ

THERMODYNAMIC AND RELATED QUANTITIES

temperature $\theta...t$
temperature, absolute T
gas constant $R, \boldsymbol{R}$
Boltzmann constant $k, \boldsymbol{k}$
heat q, Q
work w, A
energy (Gibbs ϵ) $E...U$
entropy (Gibbs η) S
*Helmholtz free energy (Gibbs ψ) A
enthalpy (Gibbs χ) H
*Gibbs function (ζ) $G...F$
heat capacity C
specific heats c_p, c_v
ratio c_p/c_v γ, κ
chemical potential μ
activity, absolute λ
activity, relative a
activity coefficient f, γ
osmotic coefficient g, ϕ
thermal conductivity λ
Joule-Thomson coefficient μ

*The terms for the Helmholtz and Gibbs energies were modified by action of the IUPAC Council, Montreal, August, 1961, as follows:
Helmholtz energy (Gibbs $\psi = E - TS$) A Gibbs energy (Gibbs $\zeta = H - TS$) G

Symbols and abbreviations in scientific writing (cont.)

The meaning of abbreviated notations for nuclear reactions should be the following:

$$\text{initial nuclide} \left(\begin{array}{cc} \text{incoming} & \text{outgoing} \\ \text{particle(s)} & \text{particle(s)} \\ \text{or quanta,} & \text{or quanta} \end{array} \right) \text{final nuclide}$$

Some examples are:

$$^{14}\text{N}(\alpha, p)^{17}\text{O} \qquad\qquad ^{59}\text{Co}(n, \gamma)^{60}\text{Co}$$
$$^{23}\text{Na}(\gamma, 3n)^{20}\text{Na} \qquad\qquad ^{31}\text{P}(\gamma, pn)^{29}\text{Si}$$

The Greek alphabet is frequently used to represent terms. A listing of the Greek alphabet is shown in **Table 8**.

Table 7. Commonly used symbols in scientific literature (cont.)

CHEMICAL REACTIONS

stoichiometric number of molecules (negative for reactants, positive for products) ν

standard equation of chemical reaction $\Sigma \nu_B B = 0$

affinity $(-\Sigma \nu_B \mu_B)$ of a reaction A

equilibrium constant K

equilibrium quotient or equilibrium product (of molalities) Q

extent of reaction $(dn_B = \nu_B d\xi)$ ξ

degree of reaction (*e.g.*, degree of dissociation) α

rate constant k

collision number (collisions per unit volume and unit time) Z

rate constant corresponding to the rate Z z

rate of reaction $v...r, s, J$

LIGHT

Planck's constant $h, \boldsymbol{h}$

Planck's constant divided by 2π $\hbar$

quantity of light Q

radiant power, flux of light (dQ/dt) Φ

luminous intensity $(d\Phi/d\omega)$ I

illumination $(d\Phi/dS)$ E

luminance L, B

luminous emittance H

absorption factor (fraction of incident radiant power which is absorbed) α

reflection factor (fraction of incident radiant power which is reflected) ρ

transmission factor (fraction of incident radiant power which is transmitted) τ

transmittance $(T = I/I_0)$ T

absorption (extinction) coefficient $[\kappa lc = \ln(1/T)]$ κ

absorbance (extinction) $[A = \log(1/T)]$ $A...E$

absorptivity (specific absorbance) (decadic absorption or extinction coefficient) a

molar absorptivity (molar decadic absorption or extinction coefficient) $(\epsilon lc = A)$ ϵ

refraction index n

refractivity r

angle of optical rotation α

ELECTRICITY AND MAGNETISM

elementary charge $e, \boldsymbol{e}$

quantity of electricity Q

charge density ρ

surface charge density σ

electric current $I...i$

electric current density J

electric potential V

electric field strength E

electric displacement D

electrokinetic potential ζ

capacity C

permittivity (dielectric constant) ϵ

dielectric polarization P

dipole moment μ

electric polarizability of a molecule α, γ

magnetic field strength H

magnetic induction B

magnetic permeability μ

magnetization M

magnetic susceptibility χ

resistance R

resistivity ρ

self inductance L

mutual inductance M, L_{12}

reactance X

impedance Z

admittance Y

ELECTROCHEMISTRY

Faraday's constant (the faraday) $F, \boldsymbol{F}$

charge number of an ion, plus or minus z

degree of electrolytic dissociation α

ionic strength $I...\mu$

electrolytic conductivity (specific conductance) κ

equivalent or molar conductance of electrolyte or ion Λ

transport number t, T

electromotive force E

overpotential η

Table 8
Greek alphabet

Upper and lower cases	Name
A α	Alpha
B β	Beta
Γ γ	Gamma
Δ δ	Delta
E ϵ	Epsilon
Z ζ	Zeta
H η	Eta
Θ θ	Theta
I ι	Iota
K κ	Kappa
Λ λ	Lambda
M μ	Mu
N ν	Nu
Ξ ξ	Xi
O o	Omicron
Π π	Pi
P ρ	Rho
Σ σ ς	Sigma
T τ	Tau
Υ υ	Upsilon
Φ ϕ	Phi
X χ	Chi
Ψ ψ	Psi
Ω ω	Omega

Mathematical signs and symbols

Symbol	Definition	Symbol	Definition	Symbol	Definition
$+$	plus (sign of addition)	$\propto$	varies as	$\oint$	line integral around a closed path
$+$	positive	∞	infinity	Σ	(sigma) summation of
$-$	minus (sign of subtraction)	$\sqrt{}$	square root of	$f(x), F(x)$	functions of x
$-$	negative	$\sqrt[3]{}$	cube root of	$\exp x = e^x$	(e = naperian log base) (abbreviation for e^x)
$\pm (\mp)$	plus or minus (minus or plus)	$\therefore$	therefore		
$\times$	times, by (multiplication sign)	$\parallel$	parallel to	∇	del or nabla, vector differential operator
$\cdot$	multiplied by	$()[]\{\}$	parentheses, brackets and braces; quantities enclosed by them to be taken together in multiplying, dividing, etc.	∇^2	Laplacian operator
$\div$	sign of division			$\mathcal{L}$	Laplace operational symbol
$/$	divided by			$4!$	factorial $4 = 1 \times 2 \times 3 \times 4$
$:$	ratio sign, divided by, is to	$\overline{AB}$	length of line from A to B	$\lvert x \rvert$	absolute value of x
$::$	equals, as (proportion)	π	(pi), $= 3.14159+$	$\dot{x}$	first derivative of x with respect to time
$<$	less than	$°$	degrees		
$>$	greater than	$'$	minutes	$\ddot{x}$	second derivative of x with respect to time
$\ll$	much less than	$''$	seconds		
$\gg$	much greater than	$\angle$	angle	$\mathbf{A} \times \mathbf{B}$	vector product; magnitude of $\mathbf{A}$ times magnitude of $\mathbf{B}$ times sine of the angle from $\mathbf{A}$ to $\mathbf{B}$; $AB \sin \overline{AB}$
$=$	equals	dx	differential of x		
$\equiv$	identical with	Δ	(delta) difference		
$\sim$	similar to	Δx	increment of x		
$\approx$	approximately equals	$\partial u/\partial x$	partial derivative of u with respect to x	$\mathbf{A} \cdot \mathbf{B}$	scalar product of $\mathbf{A}$ and $\mathbf{B}$; magnitude of $\mathbf{A}$ times magnitude of $\mathbf{B}$ times cosine of the angle from $\mathbf{A}$ to $\mathbf{B}$; $AB \cos \overline{AB}$
$\cong$	approximately equals, congruent	$\int$	integral of		
$\leq$	equal to or less than	$\int_b^a$	integral of, between limits a and b		
$\geq$	equal to or greater than				
$\neq$	not equal to				
$\to \doteq$	approaches				

Mathematical notation

Mathematical logic.

$p, q, P(x)$	Sentences, propositional functions, propositions	$0, 1$	Truth, falsity (values)
		$=$	Identity
$\neg p, \sim p, \text{non } p, Np$	Negation, read "not p" ($\neq$: read "not equal")	$\overset{Df}{=}, \overset{df}{=}, \underset{df}{=}, \equiv$	Definitional identity
$p \lor q, p + q, Apq$	Disjunction, read "p or q," "p, q," or both	$\blacksquare$	"End of proof"; "QED"
$p \land q, p \cdot q, p \& q, Kpq$	Conjunction, read "p and q"		
$p \to q, p \supset q, p \Rightarrow q,$ Cpq	Implication, read "p implies q" or "if p then q"	**Set theory, relations, functions.**	
		X, Y	Sets
		$x \in X$	x is a member of the set X
$p \leftrightarrow q, p \equiv q, p \Longleftrightarrow q,$ $Epq, p \text{ iff } q$	Equivalence, read "p is equivalent to q" or "p if and only if q"	$x \notin X$	x is not a member of X
		$A \subset X, A \subseteq X$	Set A is contained in set X
n.a.s.c.	Read "necessary and sufficient condition"	$A \not\subset X, A \not\subseteq X$	A is not contained in X
$(), [], \{\}, \cdots, \cdot\cdot$	Parentheses	$X \cup Y, X + Y$	Union of sets X and Y
$\forall, \forall, \Sigma$	Universal quantifier, read "for all" or "for every"	$X \cap Y, X \cdot Y$	Intersection of sets X and Y
		$+, \dotplus, \bigcirc$	Symmetric difference of sets
$\exists, \exists, \Pi$	Existential quantifier, read "there is a" or "there exists"	$\bigcup X_i, \Sigma X_i$	Union of all the sets X_i
		$\bigcap X_i, \Pi X_i$	Intersection of all the sets X_i
$\vdash$	Assertion sign ($p \vdash q$: read "q follows from p"; $\vdash p$: read "p is or follows from an axiom," or "p is a tautology"	$\varnothing, 0, \Lambda$	Null set, empty set
		$X', \mathbf{C}X, CX$	Complement of the set X
		$X - Y, X \backslash Y$	Difference of sets X and Y
		$\hat{x}(P(x)), \{x \mid P(x)\},$ $\{x : P(x)\}$	The set of all x with the property P

(x,y,z), $\langle x,y,z \rangle$	Ordered set of elements x, y, and z; to be distinguished from (x,z,y), for example
$\{x,y,z\}$	Unordered set, the set whose elements are x, y, z, and no others
$\{a_1, a_2, \ldots, a_n\}$, $\{a_i\}_{i=1,2,\ldots,n}$, $\{a_i\}_{i=1}^n$	The set whose members are a_i, where i is any whole number from 1 to n
$\{a_1, a_2, \ldots\}$, $\{a_i\}_{i=1,2,\ldots}$, $\{a_i\}_{i=1}^\infty$	The set whose members are a_i, where i is any positive whole number
$X \times Y$	Cartesian product, set of all (x,y) such that $x \in X$, $y \in Y$
$\{a_i\}_{i \in I}$	The set whose elements are a_i, where $i \in I$
xRy, $R\{x,y\}$	Relation
$\equiv$, $\cong$, $\sim$, $\simeq$	Equivalence relations, for example, congruence
$\geqq$, $\geq$, $>$, $\&$, $\gg$, $\leqq$, $\leq$, $<$	Transitive relations, for example, numerical order
$f: X \to Y$, $X \xrightarrow{f} Y$, $X \to Y$, $f \in Y^X$	Function, mapping, transformation
f^{-1}, $\overset{-1}{f}$, $X \xleftarrow{f^{-1}} Y$	Inverse mapping
$g \circ f$	Composite functions: $(g \circ f)(x) = g(f(x))$
$f(X)$	Image of X by f
$f^{-1}(X)$	Inverse-image set, counter image
1-1, one-one	Read "one-to-one correspondence"
$\begin{array}{ccc} X & \xrightarrow{f} & Y \\ \phi \downarrow & & \downarrow \psi \\ W & \xrightarrow{g} & Z \end{array}$	Diagram: the diagram is commutative in case $\psi \circ f = g \ \phi$
$f\|A$	Partial mapping, restriction of function f to set A
X, card X, $|X|$	Cardinal of the set A
$\aleph_0$, d	Denumerable infinity
$\mathfrak{c}$, c, $2^{\aleph_0}$	Power of continuum
ω	Order type of the set of positive integers
σ-	Read "countably"

Number, numerical functions.

1.4; 1,4; 1·4	Read "one and four-tenths"
1(1)20(10)100	Read "from 1 to 20 in intervals of 1, and from 20 to 100 in intervals of 10"
const	Constant
$A \geqq 0$	The number A is nonnegative, or, the matrix A is positive definite, or, the matrix A has nonnegative entries
$x\|y$	Read "x divides y"
$x \equiv y \bmod p$	Read "x congruent to y modulo p"
$a_0 + \dfrac{1}{a_1+} \dfrac{1}{a_2+} \cdots$, $\quad a_0 + \dfrac{1|}{|a_1} + \cdots$	Continued fractions
$[a,b]$	Closed interval
$[a,b)$, $[a,b[$	Half-open interval (open at the right)
(a,b), $]a,b[$	Open interval
$[a,\infty)$, $[a,\to[$	Interval closed at the left, infinite to the right
$(-\infty, \infty)$, $]\leftarrow,\to[$	Set of all real numbers
$\max_{x \in X} f(x)$, $\max\{f(x) | x \in X\}$	Maximum of $f(x)$ when x is in the set X
min	Minimum
sup, l.u.b.	Supremum, least upper bound
inf, g.l.b.	Infimum, greatest lower bound
$\lim_{x \to a} f(x) = b$, $\lim_{x=a} f(x) = b$, $f(x) \to b$ as $x \to a$	b is the limit of $f(x)$ as x approaches a
$\lim_{x \to a-} f(x)$, $\lim_{x=a-0} f(x)$, $f(a-)$	Limit of $f(x)$ as x approaches a from the left
lim sup, $\overline{\lim}$	Limit superior
lim inf, $\underline{\lim}$	Limit inferior
l.i.m.	Limit in the mean
$z = x + iy = re^{i\theta}$, $\zeta = \xi + i\eta$, $w = u + iv = \rho e^{i\phi}$	Complex variables
z^*	Complex conjugate
Re, $\Re$	Real part
Im, $\Im$	Imaginary part
arg	Argument
$\dfrac{\partial(u,v)}{\partial(x,y)}$, $\dfrac{D(u,v)}{D(x,y)}$	Jacobian, functional determinant
$\displaystyle\int_E f(x) \, d\mu(x)$	Integral (for example, Lebesgue integral) of function f over set E with respect to measure μ
$f(n) \sim \log n$ as $n \to \infty$	$f(n)/\log n$ approaches 1 as $n \to \infty$
$f(n) = O(\log n)$ as $n \to \infty$	$f(n)/\log n$ is bounded as $n \to \infty$
$f(n) = o(\log n)$	$f(n)/\log n$ approaches zero
$f(x) \nearrow b$, $f(x) \uparrow b$	$f(x)$ increases, approaching the limit b
$f(x) \downarrow b$, $f(x) \searrow b$	$f(x)$ decreases, approaching the limit b
a.e., p.p.	Almost everywhere
ess sup	Essential supremum
C^0, $C^0(X)$, $C(X)$	Space of continuous functions
C^k, $C^k[a,b]$	The class of functions having continuous kth derivative (on $[a,b]$)

C'	Same as C^1	$A^*, \tilde{A}$	Adjoint, Hermitian conjugate of A
$\mathrm{Lip}_\alpha, \mathrm{Lip}\,\alpha$	Lipschitz class of functions	$\mathrm{tr}\,A, \mathrm{Sp}\,A$	Trace of the matrix A
$L^p, L_p, L^p[a,b]$	Space of functions having integrable absolute pth power (on $[a,b]$)	$\det A, |A|$	Determinant of the matrix A
		$\Delta^n f(x), \Delta_h{}^n f, \underset{h}{\Delta}{}^n f(x)$	Finite differences
L'	Same as L^1	$[x_0, x_1], [x_0, x_1, x_2],$	Divided differences
$(C,\alpha), (C,p)$	Cesàro summability	$\underset{x_1}{\Delta} u_{x_0}, [x_0, x_1]_f$	

Special functions.

$[x]$	The integral part of x	$\nabla f, \mathrm{grad}\,f$	Read "gradient of f"
$\binom{n}{k}, {}^nC_k, {}_nC_k$	Binomial coefficient $n!/k!(n-k)!$	$\nabla \cdot \mathbf{v}, \mathrm{div}\,\mathbf{v}$	Read "divergence of $\mathbf{v}$"
		$\nabla \times \mathbf{v}, \mathrm{curl}\,\mathbf{v}, \mathrm{rot}\,\mathbf{v}$	Read "curl of $\mathbf{v}$"
$\left(\dfrac{n}{p}\right)$	Legendre symbol	$\nabla^2, \Delta, \mathrm{div}\,\mathrm{grad}$	Laplacian
		$[X,Y]$	Poisson bracket, or commutator, or Lie product
$e^x, \exp x$	Exponential function		
$\sinh x, \cosh x, \tanh x$	Hyperbolic functions	$\mathrm{GL}(n,R)$	Full linear group of degree n over field R
$\mathrm{sn}\,x, \mathrm{cn}\,x, \mathrm{dn}\,x$	Jacobi elliptic functions	$\mathrm{O}(n,R)$	Full orthogonal group
$\wp(x)$	Weierstrass elliptic function	$\mathrm{SO}(n,R), \mathrm{O}^+(n,R)$	Special orthogonal group
$\Gamma(x)$	Gamma function		
$J_\nu(x)$	Bessel function	**Topology.**	
$\chi_X(x)$	Characteristic function of the set X: $\chi_X(x)=1$ in case $x \in X$, otherwise $\chi_X(x)=0$	E^n	Euclidean n space
		S^n	n sphere
		$\rho(p,q), d(p,q)$	Metric, distance (between points p and q)
$\mathrm{sgn}\,x$	Signum: $\mathrm{sgn}\,0=0$, while $\mathrm{sgn}\,x = x/|x|$ for $x \neq 0$	$\overline{X}, X^-, \mathrm{cl}\,X, X^c$	Closure of the set X
		$\mathrm{Fr}X, \mathrm{fr}X, \partial X, \mathrm{bdry}\,X$	Frontier, boundary of X
$\delta(x)$	Dirac delta function	$\mathrm{int}\,X, \mathring{X}$	Interior of X
		T_2 space	Hausdorff space
		F_σ	Union of countably many closed sets

Algebra, tensors, operators.

$+, \cdot, \times, \circ, \mathsf{T}, \tau$	Laws of composition in algebraic systems	G_δ	Intersection of countably many open sets
$e, 0$	Identity, unit, neutral element (of an additive system)	$\dim X$	Dimensionality, dimension of X
$e, 1, I$	Identity, unit, neutral element (of a general algebraic system)	$\pi_1(X)$	Fundamental group of the space X
		$\pi_n(X), \pi_n(X,A)$	Homotopy groups
$e, \mathfrak{e}, E, P$	Idempotent	$H_n(X), H_n(X,A;G), H_*(X)$	Homology groups
a^{-1}	Inverse of a		
$\mathrm{Hom}(M,N)$	Group of all homomorphisms of M into N	$H^n(X), H^n(X,A;G), H^*(X)$	Cohomology groups
G/H	Factor group, group of cosets		

Probability and statistics.

$[K:k]$	Dimension of K over k	X, Y	Random variables
$\oplus, \dotplus$	Direct sum	$P(X \leqq 2), \mathrm{Pr}\{X \leqq 2\}$	Probability that $X \leqq 2$
$\otimes$	Tensor product, Kronecker product	$P(X \leqq 2 | Y \geqq 1)$	Conditional probability
		$E(X), \mathsf{E}(X)$	Expectation of X
$\wedge$	Exterior product, Grassmann product	$E(X | Y \geqq 1)$	Conditional expectation
$\vec{x}, \mathbf{x}, \mathfrak{x}, \underline{x}$	Vector	c.d.f.	Cumulative distribution function
$\vec{x} \cdot \vec{y}, \mathbf{x} \cdot \mathbf{y}, (\mathfrak{x}, \mathfrak{h})$	Inner product, scalar product, dot product	p.d.f.	Probability density function
$\mathbf{x} \times \mathbf{y}, [\mathfrak{x}, \mathfrak{h}], \mathbf{x} \wedge \mathbf{y}$	Outer product, vector product, cross product	c.f.	Characteristic function
		$\bar{x}$	Mean (especially, sample mean)
$|x|, \|x\|, \|x\|, \|x\|_p$	Norm of the vector x	σ, s.d.	Standard deviation
Ax, xA	The image of x under the transformation A	$\sigma^2, \mathrm{Var}, \mathrm{var}$	Variance
		$\mu_1, \mu_2, \mu_3, \mu_i, \mu_{ij}$	Moments of a distribution
δ_{ij}	Kronecker delta: $\delta_{ii}=1$, while $\delta_{ij}=0$ for $i \neq j$	ρ	Coefficient of correlation
$A', {}^tA, A^t, {}^tA$	Transpose of the matrix A	$\rho_{12 \cdot 34}$	Partial correlation coefficient

Fundamental constants

Compiled by E. R. Cohen and B. N. Taylor under the auspices of the CODATA Task Group on Fundamental Constants. This set has been officially adopted by CODATA and is taken from J. Phys. Chem. Ref. Data, Vol. 2, No. 4, p. 663 (1973) and CODATA Bulletin No. 11 (December 1973).

Quantity	Symbol	Numerical Value *	Uncert. (ppm)	SI † ← Units →	cgs ‡
Speed of light in vacuum	c	299792458(1.2)	0.004	$m \cdot s^{-1}$	10^2 $cm \cdot s^{-1}$
Permeability of vacuum	μ_0	4π =12.5663706144		10^{-7} $H \cdot m^{-1}$ 10^{-7} $H \cdot m^{-1}$	
Permittivity of vacuum, $1/\mu_0 c^2$	ϵ_0	8.854187818(71)	0.008	10^{-12} $F \cdot m^{-1}$	
Fine-structure constant, $[\mu_0 c^2/4\pi](e^2\hbar c)$	α α^{-1}	7.2973506(60) 137.03604(11)	0.82 0.82	10^{-3}	10^{-3}
Elementary charge	e	1.6021892(46) 4.803242(14)	2.9 2.9	10^{-19} C	10^{-20} emu 10^{-10} esu
Planck constant	h $\hbar = h/2\pi$	6.626176(36) 1.0545887(57)	5.4 5.4	10^{-34} $J \cdot s$ 10^{-34} $J \cdot s$	10^{-27} $erg \cdot s$ 10^{-27} $erg \cdot s$
Avogadro constant	N_A	6.022045(31)	5.1	10^{23} mol^{-1}	10^{23} mol^{-1}
Atomic mass unit, $10^{-3} kg \cdot mol^{-1} N_A^{-1}$	u	1.6605655(86)	5.1	10^{-27} kg	10^{-24} g
Electron rest mass	m_e	9.109534(47) 5.4858026(21)	5.1 0.38	10^{-31} kg 10^{-4} u	10^{-28} g 10^{-4} u
Proton rest mass	m_p	1.6726485(86) 1.007276470(11)	5.1 0.011	10^{-27} kg u	10^{-24} g u
Ratio of proton mass to electron mass	m_p/m_e	1836.15152(70)	0.38		
Neutron rest mass	m_n	1.6749543(86) 1.008665012(37)	5.1 0.037	10^{-27} kg u	10^{-24} g u
Electron charge to mass ratio	e/m_e	1.7588047(49) 5.272764(15)	2.8 2.8	10^{11} $C \cdot kg^{-1}$	10^7 $emu \cdot g^{-1}$ 10^{17} $esu \cdot g^{-1}$
Magnetic flux quantum, $[c]^{-1}(hc/2e)$	Φ_0 h/e	2.0678506(54) 4.135701(11) 1.3795215(36)	2.6 2.6 2.6	10^{-15} Wb 10^{-15} $J \cdot s \cdot C^{-1}$	10^{-7} $G \cdot cm^2$ 10^{-7} $erg \cdot s \cdot emu^{-1}$ 10^{-17} $erg \cdot s \cdot esu^{-1}$
Josephson frequency-voltage ratio	$2e/h$	4.835939(13)	2.6	10^{14} $Hz \cdot V^{-1}$	
Quantum of circulation	$h/2m_e$ h/m_e	3.6369455(60) 7.273891(12)	1.6 1.6	10^{-4} $J \cdot s \cdot kg^{-1}$ 10^{-4} $J \cdot s \cdot kg^{-1}$	$erg \cdot s \cdot g^{-1}$ $erg \cdot s \cdot g^{-1}$
Faraday constant, $N_A e$	F	9.648456(27) 2.8925342(82)	2.8 2.8	10^4 $C \cdot mol^{-1}$	10^3 $emu \cdot mol^{-1}$ 10^{14} $esu \cdot mol^{-1}$
Rydberg constant, $[\mu_0 c^2/4\pi]^2(m_e e^4/4\pi\hbar^3 c)$	R_∞	1.097373177(83)	0.075	10^7 m^{-1}	10^5 cm^{-1}
Bohr radius, $[\mu_0 c^2/4\pi]^{-1}(\hbar^2/m_e e^2) = \alpha/4\pi R_\infty$	a_0	5.2917706(44)	0.82	10^{-11} m	10^{-9} cm
Classical electron radius, $[\mu_0 c^2/4\pi](e^2/m_e c^2) = \alpha^3/4\pi R_\infty$	$r_e = \alpha \lambdabar_C$	2.8179380(70)	2.5	10^{-15} m	10^{-13} cm
Thomson cross section, $(8/3)\pi r_e^2$	σ_e	0.6652448(33)	4.9	10^{-28} m^2	10^{-24} cm^2
Free electron g-factor, or electron magnetic moment in Bohr magnetons	$g_e/2 = \mu_e/\mu_B$	1.0011596567(35)	0.0035		
Free muon g-factor, or muon magnetic moment in units of $[c](e\hbar/2m_\mu c)$	$g_\mu/2$	1.00116616(31)	0.31		
Bohr magneton, $[c](e\hbar/2m_e c)$	μ_B	9.274078(36)	3.9	10^{-24} $J \cdot T^{-1}$	10^{-21} $erg \cdot G^{-1}$
Electron magnetic moment	μ_e	9.284832(36)	3.9	10^{-24} $J \cdot T^{-1}$	10^{-21} $erg \cdot G^{-1}$
Gyromagnetic ratio of protons in H_2O	γ'_p $\gamma'_p/2\pi$	2.6751301(75) 4.257602(12)	2.8 2.8	10^8 $s^{-1} \cdot T^{-1}$ 10^7 $Hz \cdot T^{-1}$	10^4 $s^{-1} \cdot G^{-1}$ 10^3 $Hz \cdot G^{-1}$
γ'_p corrected for diamagnetism of H_2O	γ_p $\gamma_p/2\pi$	2.6751987(75) 4.257711(12)	2.8 2.8	10^8 $s^{-1} \cdot T^{-1}$ 10^7 $Hz \cdot T^{-1}$	10^4 $s^{-1} \cdot G^{-1}$ 10^3 $Hz \cdot G^{-1}$
Magnetic moment of protons in H_2O in Bohr magnetons	μ'_p/μ_B	1.52099322(10)	0.066	10^{-3}	10^{-3}
Proton magnetic moment in Bohr magnetons	μ_p/μ_B	1.521032209(16)	0.011	10^{-3}	10^{-3}
Ratio of electron and proton magnetic moments	μ_e/μ_p	658.2106880(66)	0.010		
Proton magnetic moment	μ_p	1.4106171(55)	3.9	10^{-26} $J \cdot T^{-1}$	10^{-23} $erg \cdot G^{-1}$
Magnetic moment of protons in H_2O in nuclear magnetons	μ'_p/μ_N	2.7927740(11)	0.38		
μ'_p/μ_N corrected for diamagnetism of H_2O	μ_p/μ_N	2.7928456(11)	0.38		
Nuclear magneton, $[c](e\hbar/2m_p c)$	μ_N	5.050824(20)	3.9	10^{-27} $J \cdot T^{-1}$	10^{-24} $erg \cdot G^{-1}$
Ratio of muon and proton magnetic moments	μ_μ/μ_p	3.1833402(72)	2.3		
Muon magnetic moment	μ_μ	4.490474(18)	3.9	10^{-26} $J \cdot T^{-1}$	10^{-23} $erg \cdot G^{-1}$
Ratio of muon mass to electron mass	m_μ/m_e	206.76865(47)	2.3		

Fundamental constants (cont.)

Quantity	Symbol	Numerical Value *	Uncert. (ppm)	SI †	← Units → cgs ‡
Muon rest mass	m_μ	1.883566(11)	5.6	10^{-28} kg	10^{-25} g
		0.11342920(26)	2.3	u	u
Compton wavelength of the electron, $h/m_e c = \alpha^2/2R_\infty$	λ_C	2.4263089(40)	1.6	10^{-12} m	10^{-10} cm
	$\lambda_C = \lambda_C/2\pi = \alpha a_0$	3.8615905(64)	1.6	10^{-13} m	10^{-11} cm
Compton wavelength of the proton, $h/m_p c$	$\lambda_{C,p}$	1.3214099(22)	1.7	10^{-15} m	10^{-13} cm
	$\lambda_{C,p} = \lambda_{C,p}/2\pi$	2.1030892(36)	1.7	10^{-16} m	10^{-14} cm
Compton wavelength of the neutron, $h/m_n c$	$\lambda_{C,n}$	1.3195909(22)	1.7	10^{-15} m	10^{-13} cm
	$\lambda_{C,n} = \lambda_{C,n}/2\pi$	2.1001941(35)	1.7	10^{-16} m	10^{-14} cm
Molar volume of ideal gas at s.t.p.	V_m	22.41383(70)	31	10^{-3} $m^3 \cdot mol^{-1}$	10^3 $cm^3 \cdot mol^{-1}$
Molar gas constant, $V_m p_0/T_0$	R	8.31441(26)	31	$J \cdot mol^{-1} \cdot K^{-1}$	10^7 $erg \cdot mol^{-1} \cdot K^{-1}$
($T_0 \equiv 273.15$ K; $p_0 \equiv 101325$ Pa$\equiv$1atm)		8.20568(26)	31	10^{-5} $m^3 \cdot atm \cdot mol^{-1} \cdot K^{-1}$	10 $cm^3 \cdot atm \cdot mol^{-1} \cdot K^{-1}$
Boltzmann constant, R/N_A	k	1.380662(44)	32	10^{-23} $J \cdot K^{-1}$	10^{-16} $erg \cdot K^{-1}$
Stefan-Boltzmann constant, $\pi^2 k^4/60\hbar^3 c^2$	σ	5.67032(71)	125	10^{-8} $W \cdot m^{-2} \cdot K^{-4}$	10^{-5} $erg \cdot s^{-1} \cdot cm^{-2} \cdot K^{-4}$
First radiation constant, $2\pi hc^2$	c_1	3.741832(20)	5.4	10^{-16} $W \cdot m^2$	10^{-5} $erg \cdot cm^2 \cdot s^{-1}$
Second radiation constant, hc/k	c_2	1.438786(45)	31	10^{-2} $m \cdot K$	$cm \cdot K$
Gravitational constant	G	6.6720(41)	615	10^{-11} $m^3 \cdot s^{-2} \cdot kg^{-1}$	10^{-8} $cm^3 \cdot s^{-2} \cdot g^{-1}$
Ratio, kx-unit to ångström, $\Lambda = \lambda(\text{Å})/\lambda(\text{kxu})$; $\lambda(\text{CuK}\alpha_1) \equiv 1.537400$ kxu	Λ	1.0020772(54)	5.3		
Ratio, Å* to ångström, $\Lambda^* = \lambda(\text{Å})/\lambda(\text{Å}^*)$; $\lambda(\text{WK}\alpha_1) \equiv 0.2090100$ Å*	Λ^*	1.0000205(56)	5.6		

ENERGY CONVERSION FACTORS AND EQUIVALENTS

Quantity	Symbol	Numerical Value *	Units	Uncert. (ppm)
1 kilogram ($kg \cdot c^2$)		8.987551786(72)	10^{16} J	0.008
		5.609545(16)	10^{29} MeV	2.9
1 Atomic mass unit ($u \cdot c^2$)		1.4924418(77)	10^{-10} J	5.1
		931.5016(26)	MeV	2.8
1 Electron mass $m_e \cdot c^2$)		8.187241(42)	10^{-14} J	5.1
		0.5110034(14)	MeV	2.8
1 Muon mass ($m_\mu \cdot c^2$)		1.6928648(96)	10^{-11} J	5.6
		105.65948(35)	MeV	3.3
1 Proton mass ($m_p \cdot c^2$)		1.5033015(77)	10^{-10} J	5.1
		938.2796(27)	MeV	2.8
1 Neutron mass ($m_n \cdot c^2$)		1.5053738(78)	10^{-10} J	5.1
		939.5731(27)	MeV	2.8
1 Electron volt		1.6021892(46)	10^{-19} J	2.9
			10^{-12} erg	2.9
	1 eV/h	2.4179696(63)	10^{14} Hz	2.6
	1 eV/hc	8.065479(21)	10^5 m^{-1}	2.6
			10^3 cm^{-1}	2.6
	1 eV/k	1.160450(36)	10^4 K	31
Voltage-wavelength conversion, hc		1.986478(11)	10^{-25} $J \cdot m$	5.4
		1.2398520(32)	10^{-6} $eV \cdot m$	2.6
			10^{-4} $eV \cdot cm$	2.6
Rydberg constant	$R_\infty hc$	2.179907(12)	10^{-18} J	5.4
			10^{-11} erg	5.4
		13.605804(36)	eV	2.6
	$R_\infty c$	3.28984200(25)	10^{15} Hz	0.075
	$R_\infty hc/k$	1.578885(49)	10^5 K	31
Bohr magneton	μ_B	9.274078(36)	10^{-24} $J \cdot T^{-1}$	3.9
		5.7883785(95)	10^{-5} $eV \cdot T^{-1}$	1.6
	μ_B/h	1.3996123(39)	10^{10} $Hz \cdot T^{-1}$	2.8
	μ_B/hc	46.68604(13)	$m^{-1} \cdot T^{-1}$	2.8
			10^{-2} $cm^{-1} \cdot T^{-1}$	2.8
	μ_B/k	0.671712(21)	$K \cdot T^{-1}$	31
Nuclear magneton	μ_N	5.505824(20)	10^{-27} $J \cdot T^{-1}$	3.9
		3.1524515(53)	10^{-8} $eV \cdot T^{-1}$	1.7
	μ_N/h	7.622532(22)	10^6 $Hz \cdot T^{-1}$	2.8
	μ_N/hc	2.5426030(72)	10^{-2} $m^{-1} \cdot T^{-1}$	2.8
			10^{-4} $cm^{-1} \cdot T^{-1}$	2.8
	μ_N/k	3.65826(12)	10^{-4} $K \cdot T^{-1}$	31

* Note that the numbers in parentheses are the one standard-deviation uncertainties in the last digits of the quoted value computed on the basis of internal consistency, that the unified atomic mass scale $^{12}C \triangleq 12$ has been used throughout, that u=atomic mass unit, C=coulomb, F=farad, G=gauss, H=henry, Hz=hertz=cycle/s, J=joule, K=kelvin (degree Kelvin), Pa=pascal=$N \cdot m^{-2}$, T=tesla (10^4 G), V=volt, Wb=weber=$T \cdot m^2$, and W=watt. In cases where formulas for constants are given (e.g., R_∞), the relations are written as the product of two factors. The first factor, in brackets, is to be included only if all quantities are expressed in SI units. We remind the reader that with the exception of the auxiliary constants which have been taken to be exact, the uncertainties of these constants are correlated, and therefore the general law of error propagation must be used in calculating additional quantities requiring two or more of these constants.

† Quantities given in u and atm are for the convenience of the reader; these units are not part of the International System of Units (SI).

‡ In order to avoid separate columns for "electromagnetic" and "electrostatic" units, both are given under the single heading "cgs Units." When using these units, the elementary charge e in the second column should be understood to be replaced by e_m or e_s, respectively.

PERIODIC TABLE OF THE ELEMENTS

KEY

Oxidation States →
+1
+3

79 ← Atomic Number

Au ← Symbol
Gold ← Name

Atomic Weight → 196.9665

Electron Configuration → -32-18-1

Transition Elements

Noble Gases

Group	I	II													III	IV	V	VI	VII	0
																				2 He Helium 4.00260 — 2
+1	**1** H Hydrogen 1.0079														**5** B Boron 10.81 2-3	**6** C Carbon 12.011 2-4	**7** N Nitrogen 14.0067 2-5	**8** O Oxygen 15.9994 2-6	**9** F Fluorine 18.99840 2-7	**10** Ne Neon 20.179 2-8
	3 Li Lithium 6.941 2-1	**4** Be Beryllium 9.01218 2-2													**13** Al Aluminum 26.98154 2-8-3	**14** Si Silicon 28.086 2-8-4	**15** P Phosphorus 30.97376 2-8-5	**16** S Sulfur 32.06 2-8-6	**17** Cl Chlorine 35.453 2-8-7	**18** Ar Argon 39.948 2-8-8

Filled Shells

	21 Sc Scandium 44.9559 -8-9-2	22 Ti Titanium 47.90 -8-10-2	23 V Vanadium 50.9414 -8-11-2	24 Cr Chromium 51.996 -8-13-1	25 Mn Manganese 54.9380 -8-13-2	26 Fe Iron 55.847 -8-14-2	27 Co Cobalt 58.9332 -8-15-2	28 Ni Nickel 58.70 -8-16-2	29 Cu Copper 63.546 -18-18-1	30 Zn Zinc 65.38 -8-18-2

19 K Potassium 39.098 -8-8-1
20 Ca Calcium 40.08 -8-8-2
31 Ga Gallium 69.72 -8-18-3
32 Ge Germanium 72.59 -8-18-4
33 As Arsenic 74.9216 -8-18-5
34 Se Selenium 78.96 -8-18-6
35 Br Bromine 79.904 -8-18-7
36 Kr Krypton 83.80 -8-18-8

37 Rb Rubidium 85.4678 -18-8-1
38 Sr Strontium 87.62 -18-8-2
39 Y Yttrium 88.9059 -18-9-2
40 Zr Zirconium 91.22 -18-10-2
41 Nb Niobium 92.9064 -18-12-1
42 Mo Molybdenum 95.94 -18-13-1
43 Tc Technetium (97) -18-13-2
44 Ru Ruthenium 101.07 -18-15-1
45 Rh Rhodium 102.9055 -18-16-1
46 Pd Palladium 106.4 -18-18-0
47 Ag Silver 107.868 -18-18-1
48 Cd Cadmium 112.40 -18-18-2
49 In Indium 114.82 -18-18-3
50 Sn Tin 118.69 -18-18-4
51 Sb Antimony 121.75 -18-18-5
52 Te Tellurium 127.60 -18-18-6
53 I Iodine 126.9045 -18-18-7
54 Xe Xenon 131.30 -18-18-8

55 Cs Cesium 132.9054 -18-8-1
56 Ba Barium 137.34 -18-8-2
57-71 See Lanthanides
72 Hf Hafnium 178.49 -32-10-2
73 Ta Tantalum 180.9479 -32-11-2
74 W Wolfram 183.85 -32-12-2
75 Re Rhenium 186.207 -32-13-2
76 Os Osmium 190.2 -32-14-2
77 Ir Iridium 192.22 -32-15-2
78 Pt Platinum 195.09 -32-17-1
79 Au Gold 196.9665 -32-18-1
80 Hg Mercury 200.59 -32-18-2
81 Tl Thallium 204.37 -32-18-3
82 Pb Lead 207.2 -32-18-4
83 Bi Bismuth 208.9804 -32-18-5
84 Po Polonium (209) -32-18-6
85 At Astatine (210) -32-18-7
86 Rn Radon (222) -32-18-8

87 Fr Francium (223) -18-8-1
88 Ra Radium 226.0254 -18-8-2
89-102 See Actinides
104 Rf-Ku Rutherfordium (Kurchatovium) (261) -32-10-2
105 Ha Hahnium (262) -32-11-2
106 (263) -32-12-2

Lanthanides

57 La Lanthanum 138.9055 -18-9-2	58 Ce Cerium 140.12 -19-9-2	59 Pr Praseodymium 140.9077 -21-8-2	60 Nd Neodymium 144.24 -22-8-2	61 Pm Promethium (145) -23-8-2	62 Sm Samarium 150.4 -24-8-2	63 Eu Europium 151.96 -25-8-2	64 Gd Gadolinium 157.25 -25-9-2	65 Tb Terbium 158.9254 -26-9-2	66 Dy Dysprosium 162.50 -28-8-2	67 Ho Holmium 164.9304 -29-8-2	68 Er Erbium 167.26 -30-8-2	69 Tm Thulium 168.9342 -31-8-2	70 Yb Ytterbium 173.04 -32-8-2	71 Lu Lutetium 174.97 -32-9-2

Actinides

89 Ac Actinium (227) -18-9-2	90 Th Thorium 232.0381 -18-10-2	91 Pa Protactinium 231.0359 -20-9-2	92 U Uranium 238.029 -21-9-2	93 Np Neptunium 237.0482 -22-9-2	94 Pu Plutonium (244) -24-8-2	95 Am Americium (243) -25-8-2	96 Cm Curium (247) -25-9-2	97 Bk Berkelium (247) -27-8-2	98 Cf Californium (251) -28-8-2	99 Es Einsteinium (254) -29-8-2	100 Fm Fermium (257) -30-8-2	101 Md Mendelevium (258) -31-8-2	102 No Nobelium (255) -32-8-2	103 Lr Lawrencium (260) -32-9-2

Filled Shells: 2, 2-8, 2-8-18, 2-8-18-32

Note:
Atomic weights are those of the most commonly available long-lived isotopes on the 1973 IUPAC Atomic Weights of the Elements. A value given in parentheses denotes the mass number of the longest-lived isotope. Adapted from *Merck Index: An Encyclopedia of Chemicals and Drugs*, Merck and Co., Inc.; 9th ed.; 1976.

Contributors

Contributors

A

Abrahams, Prof. Elihu. *Department of Physics, Rutgers University.* ANTIFERROMAGNETISM; CURIE-WEISS LAW; DOMAIN (CRYSTALLOGRAPHY); LANGEVIN FUNCTION; THERMOMAGNETIC EFFECTS; other articles—all coauthored.

Adair, Dr. Robert K. *Department of Physics, Yale University.* ISOTOPIC SPIN.

Adams, Dr. E. Dwight. *Department of Physics, University of Florida.* QUANTUM SOLIDS.

Adler, Dr. Ronald J. *Lockheed Palo Alto Research Laboratories, Palo Alto, CA.* RELATIVITY; UNIFIED FIELD THEORY.

Alburger, Dr. David E. *Brookhaven National Laboratory, Upton, NY.* MULTIPOLE RADIATION.

Allan, William. *Dean (retired), School of Engineering, City College of City University of New York.* HYDROSTATICS.

Ankeney, D. Philip. *Naval Weapons Center, China Lake, CA.* SCHLIEREN PHOTOGRAPHY.

Apker, Dr. L. *General Electric Research Laboratory, Schenectady, NY.* PHOTOVOLTAIC EFFECT; TRAPS IN SOLIDS.

Appelquist, Dr. Thomas. *Sloane Laboratory, Department of Physics, Yale University.* COLOR (QUANTUM MECHANICS).

Applegate, Charles E. *Consulting Engineer, Weston, MA.* ELECTRICAL RESISTANCE; ELECTRICAL RESISTIVITY; OHM'S LAW.

Arnstein, Dr. Karl. *Vice President in Charge of Engineering (retired), Goodyear Aircraft Corporation, Akron, OH.* PASCAL'S LAW—coauthored.

Aron, Dr. Walter. *Scientist, Science Applications, Inc., Palo Alto, CA.* ELECTRICITY.

Askey, Dr. Richard. *Mathematics Research Center, University of Wisconsin.* HYPERGEOMETRIC FUNCTIONS; LEGENDRE FUNCTIONS; SPECIAL FUNCTIONS.

Aspnes, Dr. David. *Bell Telephone Laboratories, Murray Hill, NJ.* CRYSTAL ABSORPTION SPECTRA.

B

Bäckström, Prof. Gunnar. *Institute of Physics, University of Umea, Sweden.* ELECTRON-POSITRON PAIR PRODUCTION.

Bagley, Dr. Brian G. *Bell Laboratories, Murray Hill, NJ.* AMORPHOUS SOLID.

Baltay, Dr. Charles. *Department of Physics, Columbia University.* NEUTRINO.

Bars, Dr. Itzhak. *Department of Physics, Yale University.* RENORMALIZATION.

Barton, Dr. Mark. *Brookhaven National Laboratory, Upton, NY.* PARTICLE ACCELERATOR—in part.

Bashkin, Dr. Stanley. *Department of Physics, University of Arizona.* BEAM-FOIL SPECTROSCOPY.

Bass, Prof. Jack. *Department of Physics, Michigan State University.* THERMOELECTRICITY.

Bauer, Prof. Charles L. *Department of Metallurgical Engineering and Materials Science, Carnegie-Mellon University.* GRAIN BOUNDARIES.

Bauer, Dr. E. *Physikalisches Institut der Technischen Universitat Clausthal, West Germany.* ELECTRON DIFFRACTION—in part.

Bayfield, Dr. James E. *Department of Physics, University of Pittsburgh.* ELECTRON CONFIGURATION; MOLECULAR BEAMS; RYDBERG ATOM.

Beams, Dr. Jesse W. *Deceased; formerly, Department of Physics, University of Virginia.* ELECTROMAGNETIC FIELD.

Bean, Howard S. *Deceased; formerly, Consultant on Fluid Metering, Liquids and Gases, Sedona, AZ.* WEAK NUCLEAR INTERACTIONS.

Beaumont, Prof. Ross A. *Department of Mathematics, University of Washington.* DISCRIMINANT; EQUATIONS, THEORY OF; MATRIX THEORY; POLYNOMIAL SYSTEMS OF EQUATIONS.

Bederson, Dr. Benjamin. *Department of Physics, New York University.* ATOMIC PHYSICS; MOLECULAR PHYSICS.

Bender, Prof. Paul J. *Professor of Physical Chemistry, University of Wisconsin.* FREE ENERGY; INTERNAL ENERGY; WORK FUNCTION (THERMODYNAMICS)—coauthored.

Beranek, Dr. Leo L. *Chief Scientist (retired), Bolt Beranek and Newman, Inc., Cambridge, MA.* WAVE MOTION.

Bergmann, Prof. Peter G. *Department of Physics, Syracuse University.* SPACE-TIME.

Beringer, Prof. Robert. *Department of Physics, Yale University.* PARTICLE ACCELERATOR—in part.

Bethe, Prof. Hans A. *Laboratory of Nuclear Studies, Cornell University.* MATHEMATICAL PHYSICS; THEORETICAL PHYSICS.

Betts, Dr. Russell. *Argonne National Laboratory, Argonne, IL.* NUCLEAR ISOMERISM.

Bichsel, Dr. Hans. *Department of Radiology, University of Washington.* CHARGED PARTICLE BEAMS.

Biedenharn, Prof. L. C. *Department of Physics, Duke University.* ANGULAR MOMENTUM.

Bienenstock, Dr. Arthur. *Professor of Applied Physics, Stanford University; and Director, Stanford Synchrotron Radiation Laboratory.* SYNCHROTRON RADIATION.

Billings, Dr. Bruce H. *Special Assistant to the Ambassador for Science and Technology, Embassy of the United States of America, Taipei.* BIREFRINGENCE; CRYSTAL OPTICS; DICHROISM; PLEOCHROISM; POLARIZATION OF WAVES; other articles.

Billmeyer, Prof. Fred W., Jr. *Department of Chemistry, Rensselaer Polytechnic Institute.* ULTRAVIOLET RADIATION.

Bing, Prof. R. H. *Department of Mathematics, University of Wisconsin.* SET THEORY.

Birkhoff, Prof. Garrett. *Department of Mathematics, Harvard University.* BOOLEAN ALGEBRA.

Bjorken, Prof. J. D. *Stanford Linear Acceleration Center, Stanford University.* QUANTUM ELECTRODYNAMICS; QUANTUM FIELD THEORY.

Blatt, Prof. Frank J. *Department of Physics, Michigan State University.* FREE ELECTRON THEORY OF METALS; HALL EFFECT; MATTHIESSEN'S RULE; WIEDEMANN-FRANZ LAW.

Bloomfield, Prof. Philip E. *Deceased; formerly, Department of Physics, University of Pennsylvania; and City College, City University of New York.* ADIABATIC PROCESS; ISENTROPIC FLOW; ISOMETRIC PROCESS; THERMODYNAMIC PROCESSES—coauthored; other articles.

Blumenthal, Prof. Leonard M. *Defoe Distinguished Professor of Mathematics, University of Missouri.* ANALYTIC GEOMETRY; EPICYCLOID.

Boast, Dr. Warren B. *Anson Marston Distinguished Professor Emeritus of Electrical Engineering, Iowa State University.* PHOTOMETRY.

Bochner, Dr. Salomon. *Department of Mathematics, Rice University.* LOGARITHM; SERIES—both validated.

Bolz, Dr. Ray E. *Leonard S. Case Professor and Dean of Engineering, Case Western Reserve University.* NUTATION; POINSOT'S METHOD; PRECESSION; RIGID-BODY DYNAMICS.

Boynton, Dr. Robert M. *Department of Psychology, University of California, San Diego.* COLOR.

Brand, Dr. Louis. *Deceased; formerly, M. D. Anderson Professor of Mathematics, Department of Mathematics, University of Houston.* MATRIX CALCULUS; PROGRESSION.

Breazeale, Prof. M A. *Department of Physics and Astronomy, University of Tennessee.* TRANSDUCER.

Brickley, Robert L. *Marketing Manager, Westinghouse Electric Corporation, Springfield, MA.* ANALYSIS OF VARIANCE.

Brienza, Dr. Michael J. *United Technologies Corporation, Norden Division, Norwalk, CT.* ACOUSTOOPTICS.

Brockway, Prof. Lawrence O. *Department of Chemistry, University of Michigan.* ELECTRON DIFFRACTION.

Bromley, Prof. D. Allan. *Henry Ford II Professor, and Director, A. W. Wright Nuclear Structure Laboratory, Yale University.* COMPTON WAVELENGTH; HELICITY; HYPERCHARGE; MAGIC NUMBERS; NUCLEAR PHYSICS; other articles.

Brooks, Dr. Frank D. *Department of Physics, University of Capetown, Rondebosch, South Africa.* LIQUID SCINTILLATION DETECTOR.

Brouwer, Prof. Dirk *Deceased; formerly, Director, Observatory, Yale University.* GRAVITY.

Browder, Dr. William. *Department of Mathematics, Princeton University.* TOPOLOGY—in part.

Brown, Prof. Frederick C. *Department of Physics, University of Illinois.* COLOR CENTERS.

Brylawski, Prof. Thomas. *Department of Mathematics, University of North Carolina.* COMBINATORIAL THEORY.

Bryson, Prof. Arthur E., Jr. *Chairman, Department of Aeronautics and Astronautics, Stanford University.* BERNOULLI'S THEOREM; EULER'S MOMENTUM THEOREM; FLUID-FLOW PRINCIPLES; LAPLACE'S IRROTATIONAL MOTION; NAVIER-STOKES EQUATIONS; other articles.

Bube, Dr. Richard H. *Professor of Electrical Engineering, Department of Material Sciences, Stanford University.* PHOTOCONDUCTIVITY.

Burton, Prof. Ralph A. *Chairman, Department of Mechanical Engineering and Astronautical Sciences, Northwestern University.* THERMAL EXPANSION.

C

Callaway, Prof. Joseph. *Department of Physics and Astronomy, Louisiana State University.* BAND THEORY OF SOLIDS; BLOCH THEOREM; BRILLOUIN ZONE; HOLES IN SOLIDS; KRONIG-PENNEY MODEL.

Cambel, Dr. Ali B. *Executive Vice President for Academic Affairs, Wayne State University.* GAS DYNAMICS—coauthored.

Cameron, Dr. A. E. *Analytical Chemistry Division, Oak Ridge National Laboratory, Oak Ridge, TN.* ISOTOPE.

Campbell, Dr. Laurence J. *Low Temperature Physics Group, Los Alamos Scientific Laboratory, Los Alamos, NM.* MICROSCOPIC QUANTIZATION EFFECTS; SUPERFLUIDITY.

Chan, Dr. Moses H. W. *Department of Physics, Pennsylvania State University.* CRITICAL PHENOMENA.

Chandler, Richard F. *Mechanical Engineer, Air Force Missile Development Center, Holloman Air Force Base, NM.* SHADOWGRAPH OF FLUID FLOW.

Charlton, Dr. Ernest E. *Research Engineer (retired), Schenectady, NY.* X-RAY TUBE.

Clark, Dr. David J. *Lawrence Berkeley Laboratory, University of California, Berkeley.* PARTICLE ACCELERATOR—in part.

Clark, Prof. George L. *Emeritus Research Professor of Analytical Chemistry, University of Illinois.* X-RAY OPTICS; X-RAYS.

Clarke, Prof. John. *Department of Physics, University of California, Berkeley.* SQUID.

Clemence, Dr. G. M. *Deceased; formerly, Observatory, Yale University.* GRAVITY—validated.

Collins, Dr. Dean R. *Director, CCD Technology Laboratory, Central Research Laboratories, Texas Instruments Inc., Dallas, TX.* LIGHT AMPLIFIER.

Conwell, Dr. Esther M. *Xerox Webster Research Center, Rochester NY.* INTEGRATED OPTICS.

Cook, Dr. Richard K. *Formerly, National Bureau of Standards.* INFRASOUND.

Cooley, Prof. Hollis R. *Professor Emeritus of Mathematics, New York University.* ALGEBRA.

Corben, Dr. Herbert C. *Deceased; formerly, Ramo-Wooldridge Corporation, Los Angeles, CA.* DYNAMICS; KINEMATICS; MOTION.

Cornish, Dr. Joseph J., III. *Chief Engineer, Technology, Lockheed-Georgia Company, Marietta, GA.* BOUNDARY-LAYER FLOW.

Craig, Prof. Homer V. *Department of Mathematics, University of Texas, Austin.* RIEMANNIAN GEOMETRY.

Curtiss, Prof. C. F. *Department of Chemistry, University of Wisconsin.* GAS—coauthored.

D

Dahl, Dr. Lawrence F. *Department of Chemistry, University of Wisconsin.* X-RAY CRYSTALLOGRAPHY; X-RAY DIFFRACTION—in part.

Dalitz, Prof. Richard H. *Department of Theoretical Physics, Oxford University, England.* BARYON; DALITZ PLOT; HYPERON; MESON.

Danby, Dr. J. M. A. *Department of Mathematics, North Carolina State University.* GRAVITATION.

Daniels, Dr. Farrington. *Deceased; formerly, Professor Emeritus, Solar Energy Laboratory, University of Wisconsin.* HALF-LIFE—in part.

Daniels, Prof. James M. *Department of Physics, University of Toronto, Canada.* DYNAMIC NUCLEAR POLARIZATION.

Daugherty, Prof. Robert L. *Division of Engineering, California Institute of Technology.* FLUID FLOW.

de Boor, Dr. Carl. *Mathematical Research Center, University of Wisconsin.* NUMERICAL ANALYSIS.

DeCarli, Dr. P. S. *Poulter Laboratory, Stanford Research Institute.* HIGH-PRESSURE PHYSICS—coauthored.

De Cicco, John. *Department of Mathematics, Illinois Institute of Technology.* NONEUCLIDEAN GEOMETRY.

DeFacio, Prof. Brian. *Ames Laboratory, Iowa State University.* BEAT; ENERGY; STATICS.

Dehmelt, Dr. Hans. *Department of Physics, University of Washington.* NUCLEAR QUADRUPOLE RESONANCE.

Dekeyser, Prof. Willy C. *Laboratory for Crystallography, Ghent, Belgium.* COORDINATION NUMBER; CRYSTAL STRUCTURE; CRYSTALLOGRAPHY; ISOMORPHISM (CRYSTALLOGRAPHY); POLYMORPHISM.

Dekker, Dr. Adrianus J. *Professor of Solid State Physics, University of Groningen, Netherlands.* WORK FUNCTION (ELECTRONICS).

de Launay, Jules. *Consultant, Solid State Division, U.S. Naval Research Laboratory.* LATTICE VIBRATIONS; PHONON; SPECIFIC HEAT OF SOLIDS.

Den Hartog, Prof. J. P. *(Retired) Department of Mechanical Engineering, Massachusetts Institute of Technology.* MECHANICAL VIBRATION—coauthored.

Dick, Prof. B. Gale. *Department of Physics, University of Utah.* MADELUNG CONSTANT.

Dieke, Prof. G. H. *Deceased; formerly, Chairman, Department of Physics, Johns Hopkins University.* FARADAY EFFECT; MAGNETOOPTICS; ZEEMAN EFFECT—in part.

Dillon, J. F., Jr. *Bell Telephone Laboratories, Murray Hill, NJ.* CURIE TEMPERATURE.

Doob, Prof. Joseph L. *Department of Mathematics, University of Illinois.* STOCHASTIC PROCESS.

Dresden, Prof. Max. *Institute for Theoretical Physics, State University of New York, Stony Brook.* BOLTZMANN CONSTANT; BROWNIAN MOVEMENT; FERMI-DIRAC STATISTICS; KINETIC THEORY OF MATTER; QUANTUM STATISTICS; other articles.

Duckworth, Dr. Henry E. *Department of Physics, University of Manitoba, Canada.* ATOMIC NUCLEUS; DEUTERON; MASS NUMBER; RADIOISOTOPE; TRITON; other articles.

Duguay, Dr. Michel A. *Laser Development, Sandia Laboratories, Albuquerque, NM.* ELECTROOPTICS; KERR EFFECT.

Duke, Dr. C. B. *Manager, Molecular and Organic Materials Area, Xerox Corporation, Rochester, NY.* SURFACE PHYSICS.

Durand, Prof. Loyal, III. *Department of Physics, University of Wisconsin.* SCATTERING MATRIX.

Dym, Prof. Clive L. *Department of Civil Engineering, University of Massachusetts.* MECHANICAL VIBRATION—coauthored.

E

Edeskuty, Dr. F. J. *Los Alamos Scientific Laboratory, Los Alamos, NM.* CRYOGENICS—coauthored.

Egan, Patrick O. *Department of Physics, Yale University.* MUONIUM.

Emin, David. *Sandia Laboratories, Albuquerque, NM.* POLARON—in part.

Erb, Dr. K. A. *Oak Ridge National Laboratory, Oak Ridge, TN.* SCATTERING EXPERIMENTS (NUCLEI).

Erdélyi, A. *Professor of Mathematics, University of Edinburgh, Scotland.* ELLIPTIC FUNCTION AND INTEGRAL; SPHERICAL HARMONICS.

Evans, Prof. Robley D. *(Retired) Department of Physics, Massachusetts Institute of Technology.* HALF-LIFE—in part.

Ewing, Prof. George E. *Department of Chemistry, Indiana University.* INTERMOLECULAR FORCES.

Eyring, Prof. Henry. *Institute for the Study of Rate Processes, University of Utah.* RHEOLOGY.

F

Fairbank, Prof. Henry A. *Department of Physics, Duke University.* SECOND SOUND.

Fan, Prof. H. Y. *Department of Physics, Purdue University.* SEMICONDUCTOR.

Feldman, Dr. Barry J. *Los Alamos Scientific Laboratory, Los Alamos, NM.* OPTICAL PHASE CONJUGATION—coauthored.

Feller, Prof. William. *Deceased; formerly, Department of Mathematics, Princeton University.* DISTRIBUTION (PROBABILITY); PROBABILITY.

Feshbach, Dr. Herman. *Department of Physics, Massachusetts Institute of Technology.* PERTURBATION (MATHEMATICS); STRONG NUCLEAR INTERACTIONS.

Firk, Prof. Frank W. K. *Electron Accelerator Laboratory, Yale University.* TIME-OF-FLIGHT SPECTROMETERS.

Fisher, Dr. Robert A. *Los Alamos Scientific Laboratory, Los Alamos, NM.* OPTICAL PHASE CONJUGATION—coauthored.

Fisher, Prof. Russell A. *Department of Physics, Northwestern University.* DEGREE OF FREEDOM; EULER'S EQUATIONS OF MOTIONS; KINETICS.

Fisk, Prof. Nelson S. *Department of Civil Engineering, Columbia University.* CENTER OF GRAVITY; MOMENT OF INERTIA; RADIUS OF GYRATION; TORQUE; other articles.

Fitch, Prof. Val. *Department of Physics, Princeton University.* FLAVOR.

Foster, Dr. Mark G. *Department of Electrical Engineering, School of Engineering and Applied Science, University of Virginia.* MAGNETIC CIRCUITS.

Fowler, Prof. Michael. *Department of Physics, University of Virginia.* KONDO EFFECT.

Fowler, Prof. W. Beall. *Department of Physics, Lehigh University.* CRYSTAL DEFECTS.

Fradkin, Dr. David M. *Department of Physics, Wayne State University.* PERTURBATION (QUANTUM MECHANICS); RUNGE VECTOR.

Frame, Dr. J. Sutherland. *Professor Emeritus, Department of Mathematics, Michigan State University.* GEOMETRY.

Franklin, Dr. Philip. *Deceased; formerly, Professor of Mathematics, Massachusetts Institute of Technology.* SERIES.

French, Dr. J. B. *Department of Physics, University of Rochester.* SELECTION RULES.

Fretter, Prof. William B. *Department of Physics, University of California, Berkeley.* CERENKOV RADIATION; SPARK COUNTER.

Fuchs, Prof. Ronald. *Department of Physics, Iowa State University.* BOLTZMANN TRANSPORT EQUATION.

G

Galloway, Dr. William J. *Bolt, Beranek and Newman, Inc., Canoga Park, CA.* ACOUSTIC IMPEDANCE; FREQUENCY; PERIOD; PHASE; WAVE; other articles.

Gelatt, Dr. C. D., Jr. *Watson Research Center, IBM, Yorktown Heights, NY.* COHESION.

Gerjuoy, Dr. Edward. *Department of Physics, University of Pittsburgh.* DE BROGLIE WAVELENGTH; EXCITED STATE; MATRIX MECHANICS; NONRELATIVISTIC QUANTUM THEORY; RITZ'S COMBINATION PRINCIPLE; other articles.

Giedt, Prof. Warren H. *Department of Mechanical Engineering, University of California, Davis.* CONDUCTION (HEAT); CONVECTION (HEAT).

Gingrich, Prof. Newell S. *Department of Physics, University of Missouri.* CLASSICAL MECHANICS.

Gluckstern, Prof. Robert L. *Head, Department of Physics and Astronomy, University of Massachusetts.* RELATIVISTIC ELECTRODYNAMICS.

Goebel, Prof. Charles J. *Department of Physics, University of Wisconsin.* BREMSSTRAHLUNG; DISPERSION RELATIONS; ELEMENTARY PARTICLE; GRAVITON; PARITY; other articles.

Goldhaber, Dr. A. S. *Department of Physics, State University of New York, Stony Brook.* MAGNETIC MONOPOLES.

Goldhaber, Dr. Gerson. *Lawrence Berkeley Laboratory, University of California, Berkeley.* GOLDHABER TRIANGLE.

Gomer, Prof. Robert. *James Franck Institute, University of Chicago.* FIELD EMISSION.

Good, Dr. Irving J. *Department of Statistics and Statistical Laboratory, College of Arts and Sciences, Virginia Polytechnic Institute.* MONTE CARLO METHOD.

Good, Prof. Roland H., Jr. *Head, Department of Physics, Pennsylvania State University.* TACHYON; WENTZEL-KRAMERS-BRILLOUIN METHOD.

Goodheart, Prof. Clarence F. *Department of Electrical Engineering, Union College.* CIRCUIT (ELECTRICITY).

Goodman, Prof. Bernard. *Department of Physics, University of Cincinnati.* ACTION; CAYLEY-KLEIN PARAMETERS; DYNAMICS; EULER ANGLES; MECHANICS; other articles.

Goodman, Dr. Joseph W. *Stanford Electronics Laboratories, Stanford University.* HOLOGRAPHY.

Gränicher, Prof. H. *Laboratory of Solid State Physics, Swiss Federal Institute of Technology, Zurich.* PIEZOELECTRICITY; PYROELECTRICITY.

Graves, Prof. Lawrence M. *Professor Emeritus of Mathematics, University of Chicago.* INTEGRATION.

Greenberg, Prof. Jack S. *Wright Nuclear Structure Laboratory, Yale University.* QUASIATOM.

Greenspan, Martin. *National Bureau of Standards.* DAMPING; FORCED OSCILLATION; MECHANICAL IMPEDANCE.

Gross, Prof. Jonathan L. *Department of Mathematics, Columbia University.* GRAPH THEORY.

Gunning, Dr. Robert C. *Department of Mathematics, Princeton University.* COMPLEX NUMBERS AND COMPLEX VARIABLES.

H

Hagen, Dr. John P. *Department of Astronomy, Pennsylvania State University.* CENTER OF MASS; TRAJECTORY.

Hamilton, Dr. Joseph H. *Department of Physics-Astronomy, Vanderbilt University.* RADIOACTIVITY.

Hanna, Prof. Stanley S. *Department of Physics, Stanford University.* GIANT NUCLEAR RESONANCES.

Hansch, Dr. Theo W. *Department of Physics, Stanford University.* RYDBERG CONSTANT.

Hansen, Arthur G. *President, Georgia Institute of Technology.* DYNAMIC SIMILARITY.

Harleman, Dr. Donald R. F. *Department of Civil Engineering, Massachusetts Institute of Technology.* WAVE MOTION IN LIQUIDS.

Harris, Dr. Cyril M. *Professor of Electrical Engineering and Architecture, Department of Electrical Engineering, Columbia University.* ACOUSTICAL IMAGE; REVERBERATION.

Harris, Dr. J. *Postgraduate School of Studies in Chemical Engineering, University of Bradford, England.* FLUIDS; NON-NEWTONIAN FLUID—both coauthored.

Harrison, George R. *Deceased; formerly, Dean Emeritus, School of Science, Massachusetts Institute of Technology.* DIFFRACTION GRATING; LINE SPECTRUM. INCANDESCENCE; RADIOMETRY; RESOLVING POWER (OPTICS)—all three validated.

Harrison, Dr. Walter A. *Department of Applied Physics, Stanford University.* SOLID-STATE PHYSICS.

Harvey, Dr. John A. *Oak Ridge National Labora-*

tory, Oak Ridge, TN. NEUTRON SPECTROMETRY.

Hatch, Dr. Eastman N. *Dean of Graduate Studies, Utah State University.* COMPTON EFFECT.

Hawkings, Prof. George A. *Vice President for Academic Affairs, Purdue University.* DALTON'S LAW; VAN DER WAALS EQUATION; VIRIAL EQUATION.

Hearmon, R. F. S. *Formerly, Timber Mechanics Section, Forest Products Research Laboratory, Princes Risborough, Bucks, England.* ELASTICITY.

Herber, Prof. Rolfe H. *Department of Chemistry, Rutgers University.* MÖSSBAUER EFFECT.

Herz, Prof. Carl S. *Department of Mathematics, McGill University.* FOURIER SERIES AND INTEGRALS; ORTHOGONAL POLYNOMIALS.

Herzberger, Dr. Max J. *Consulting Professor, Department of Physics, Louisiana State University.* ABERRATION (OPTICS); GEOMETRICAL OPTICS; LENS; MAGNIFICATION; OPTICAL PRISM; other articles.

Hess, Dr. George B. *Department of Physics, University of Virginia.* LIQUID HELIUM.

Higgins, Prof. Richard J. *Department of Physics, University of Oregon.* RELAXATION TIME OF ELECTRONS.

Hill, Prof. Edward L. *Professor of Physics and Mathematics, School of Physics, University of Minnesota.* GALILEAN TRANSFORMATIONS; INERTIA OF ENERGY; LORENTZ TRANSFORMATIONS; REST MASS.

Hirschfelder, Dr. J. O. *Department of Chemistry, University of Wisconsin.* GAS—coauthored.

Hobbs, Dr. Herman H. *Department of Physics, George Washington University.* CRYSTAL.

Hofstadter, Dr. Robert. *Department of Physics, Stanford University.* SCINTILLATION COUNTER.

Holdeman, Dr. Louis B. *National Bureau of Standards.* JOSEPHSON EFFECT.

Horen, Dr. D. J. *Nuclear Division, Oak Ridge National Laboratory, Oak Ridge, TN.* NUCLEAR SPECTRA.

Horton, G. A. *Manager, Engineering Laboratories, Westinghouse Electric Corporation, Cleveland, OH.* LUMINOUS EFFICACY; LUMINOUS EFFICIENCY.

Howe, Prof. Carl E. *Deceased; formerly, Professor Emeritus of Physics, Oberlin College.* ACCELERATION—in part; CENTRIFUGAL FORCE; ROTATIONAL MOTION.

Hoxton, Prof. Llewellyn G. *Deceased; formerly, Professor Emeritus of Physics, University of Virginia.* JOULE'S LAW.

Hudson, Dr. Ralph P. *Bureau International des Poids et Mesures, Sèvres, France.* ADIABATIC DEMAGNETIZATION; LOW-TEMPERATURE PHYSICS.

Hughes, Prof. Vernon. *Department of Physics, Yale University.* POSITRONIUM.

Huizenga, Dr. John R. *Nuclear Structure Research Laboratory, University of Rochester.* DEEP INELASTIC COLLISIONS; NUCLEAR FISSION.

Hull, Dr. McAllister H., Jr. *Department of Physics and Astronomy, State University of New York, Buffalo.* ELECTRON CAPTURE; LINE INTEGRAL; MAGNETON; POINT SOURCE; RADIATION.

Hultsch, Dr. Roland A. *Department of Physics, University of Missouri.* TEMPERATURE.

Hunter, Prof. Lloyd P. *Department of Electrical Engineering, University of Rochester.* TRANSISTOR.

Huntington, Prof. H. B. *Department of Physics, Rensselaer Polytechnic Institute.* HYSTERESIS; THERMAL HYSTERESIS—both coauthored.

Hurst, Dr. G. S. *Department of Physics, Oak Ridge National Laboratory, Oak Ridge, TN.* RESONANCE IONIZATION SPECTROSCOPY.

I

Iachello, Dr. F. *Department of Physics, Yale University.* NUCLEAR STRUCTURE.

J

Jacobs, Prof. Stephen F. *Optical Sciences Center, University of Arizona.* LASER—coauthored.

Jacobson, Prof. Nathan. *Department of Mathematics, Yale University.* ABSTRACT ALGEBRA; LINEAR ALGEBRA.

Jaep, William F. *Central Research Department, Experimental Station, E. I. du Pont de Nemours and Company, Wilmington, DE.* ENTROPY—in part; GIBBS FUNCTION; THERMODYNAMIC PRINCIPLES.

Jafarey, Prof. S. *Behlen Laboratory of Physics, University of Nebraska.* PHASE TRANSITIONS—coauthored.

Jenkins, Prof. Francis A. *Deceased; formerly, Department of Physics, University of California, Berkeley.* ATOM; HUYGENS' PRINCIPLE; ISOELECTRONIC SEQUENCE; PASCHEN-BACK EFFECT; RESOLVING POWER (OPTICS); other articles.

Jones, Prof. Lawrence W. *Harrison M. Randall Laboratory of Physics, University of Michigan.* QUARKS.

K

Kabler, Dr. Milton N. *Naval Research Laboratory, Washington, DC.* EXCITON.

Kanzig, Prof. Werner. *Department of Physics, Massachusetts Institute of Technology.* FERROELECTRICS.

Keating, Richard E. *Time Service Division, U.S. Naval Observatory, Washington, DC.* CLOCK PARADOX.

Keffer, Prof. Frederic. *Department of Physics, University of Pittsburgh.* ANTIFERROMAGNETISM; CURIE-WEISS LAW; DOMAIN (CRYSTALLOGRAPHY); LARMOR PRECESSION; MAGNETORESISTANCE—all coauthored; other articles.

Keller, Dr. Joseph M. *Department of Physics, Iowa State University.* ANHARMONIC OSCILLATOR; HARMONIC MOTION; PENDULUM; PERIODIC MOTION; other articles.

Kells, Prof. Lyman M. *Deceased; formerly, Professor Emeritus, U.S. Naval Academy.* TRIGONOMETRY.

Ketterson, Dr. J. B. *Department of Physics, Northwestern University.* DE HAAS–VAN ALPHEN EFFECT.

Kinch, Dr. Michael A. *Central Research Laboratories, Texas Instruments Inc., Dallas, TX.* PHOTOELECTRICITY.

Kinsler, Prof. Lawrence E. *Professor of Physics, U.S. Naval Postgraduate School, Monterey, CA.* RESONANCE (ACOUSTICS AND MECHANICS); SYMPATHETIC VIBRATION; VIBRATION.

Klemens, Dr. Paul G. *Department of Physics, University of Connecticut.* UMKLAPP PROCESS.

Klick, Dr. Clifford C. *Superintendent, Solid State Division, U.S. Naval Research Laboratory.* FLUORESCENCE; FRANCK-CONDON PRINCIPLE; LUMINESCENCE; PHOSPHORESCENCE; THERMOLUMINESCENCE—all coauthored.

Kliewer, Prof. K. L. *Department of Physics, Iowa State University.* ANGULAR FREQUENCY.

Knapp, Prof. Anthony W. *Department of Mathematics, Cornell University.* LIE GROUP.

Knebelman, Prof. Morris S. *Professor of Mathematics, Bucknell University.* COORDINATE SYSTEMS.

Koch, Dr. Peter M. *Department of Physics, Yale University.* ISOTOPE SHIFT.

Kock, Prof. Winston E. *Director, Herman Schneider Laboratory, College of Engineering, University of Cincinnati.* ACOUSTICAL HOLOGRAPHY.

Kogelnik, Dr. Herwig. *Crawford Hill Laboratory, Bell Laboratories, Holmdel, NJ.* REFLECTION OF ELECTROMAGNETIC RADIATION.

Koller, Noémie. *Department of Physics, Rutgers University.* NUCLEAR MOMENTS.

Kovar, Dr. Dennis G. *Hahn-Meitner-Institut für Kernforschung, Berlin, West Germany.* NUCLEAR RADIATION; NUCLEAR REACTION.

Kryter, Dr. Karl D. *Director, Sensory Sciences Research Center, Stanford Research Institute, Menlo Park, CA.* BEL; DECIBEL.

Kunz, Dr. Kaiser S. *Research Professor of Physics, New Mexico State University.* EXTRAPOLATION.

Kusch, Prof. Polykarp. *Department of Physics, Columbia Radiation Laboratory, Columbia University.* ATOMIC BEAMS.

L

Lach, Dr. Joseph. *Fermi National Accelerator Laboratories, Batavia, IL.* ANTIMATTER.

Landes, Dr. Hugh S. *Department of Electrical Engineering, University of Virginia.* RECIPROCITY PRINCIPLE.

Landshoff, Dr. Rolf. *Consulting Scientist, Lockheed Missiles and Space Company, Palo Alto, CA.* MAGNETOHYDRODYNAMICS.

Landweber, Dr. Louis. *Iowa Institute of Hydraulic Research, University of Iowa.* HYDRODYNAMICS.

Lanford, Prof. William A. *Department of Physics, State University of New York, Albany.* IONIZATION CHAMBER.

Langenberg, Prof. D. N. *Department of Physics, University of Pennsylvania.* SUPERCONDUCTIVITY.

Larsen, Dr. David M. *Francis Bitter National Magnet Laboratory, Massachusetts Institute of Technology.* POLARON—in part.

Lass, Dr. Harry. *Research Specialist, Jet Propulsion Laboratory, California Institute of Technology.* CALCULUS OF TENSORS; CALCULUS OF VECTORS.

Lassila, Prof. Kenneth E. *Department of Physics, Iowa State University.* QUANTIZATION; QUANTUM; QUANTUM NUMBERS.

Lederman, Leon M. *Director, Fermi National Accelerator Laboratory, Batavia, IL.* UPSILON PARTICLES.

Lin, Dr. Shao-Chi. *Department of Applied Mechanics and Engineering Sciences, University of California, San Diego.* SHOCK WAVE; SONIC BARRIER; WAVE MOTION IN FLUIDS.

Linde, Dr. R. K. *President, Envirodyne, Inc., Los Angeles, CA.* HIGH-PRESSURE PHYSICS—coauthored.

Lindsay, Prof. R. Bruce. *Hazard Professor of Physics, Emeritus, Brown University.* ACOUSTICS; SOUND; SOUND ABSORPTION.

Lineberger, Dr. W. C. *Department of Chemistry, University of Colorado.* ELECTRON AFFINITY.

Liu, Dr. S. H. *Department of Physics, Iowa State University.* DENSITY MATRIX.

Ljung, Dr. Donovan A. *Fermi National Accelerator Laboratory, Batavia, IL.* BUBBLE CHAMBER.

Lord, Prof. Richard C. *Department of Chemistry, Massachusetts Institute of Technology.* OPTICS; PHYSICAL OPTICS; RAMAN EFFECT; WAVE OPTICS.

Lowan, Dr. Arnold N. *Deceased; formerly, Chairman, Department of Physics, Yeshiva University.* LOGARITHM.

Luebbers, Dr. Ralph H. *Department of Chemical Engineering, University of Missouri.* HEAT TRANSFER.

M

McCarthy, Prof. Kathryn A. *Professor of Physics, and Dean of Graduate School of Arts and Sciences, Tufts University.* THERMAL CONDUCTION IN SOLIDS.

McCoy, Prof. Barry M. *Division of Engineering and Applied Physics, Harvard University.* ISING MODEL—coauthored.

MacCrone, Prof. R. K. *Department of Physics, Rensselaer Polytechnic Institute.* HYSTERESIS; THERMAL HYSTERESIS—both coauthored.

McKenzie, Dr. James M. *Sandia Laboratories, Albuquerque, NM.* CRYSTAL COUNTER; JUNCTION DETECTOR.

Mackey, Prof. George W. *Department of Mathematics, Harvard University.* OPERATOR THEORY.

McNish, Alvin G. *National Bureau of Standards, Chevy Chase, MD.* ELECTRICAL UNITS AND STANDARDS.

Mann, Prof. A. K. *Department of Physics, University of Pennsylvania.* HADRON.

Manning, Dr. Kenneth V. *Professor Emeritus, Pennsylvania State University.* AMPÈRE'S LAW; DEMAGNETIZATION; EDDY CURRENT; RELUCTANCE; SKIN EFFECT; other articles.

Margenau, Prof. Henry. *Department of Physics, Yale University.* PROBABILITY (PHYSICS).

Markus, John. *Consultant, Sunnyvale, CA.* CONDUCTIVITY.

Marshak, Dr. Harvey. *National Bureau of Standards.* NUCLEAR ORIENTATION.

Mason, Dr. Warren P. *Associate Editor, "Journal of the Acoustical Society of America," West Orange, NJ.* HYPERSONICS; ULTRASONICS.

Massey, Sir Harrie W. *Department of Physics, University College, London, England.* SCATTERING EXPERIMENTS (ATOMS AND MOLECULES).

May, James E. *Lockheed-California Company, Burbank, CA.* INCOMPRESSIBLE FLOW.

Meggers, Dr. W. F. *Deceased; formerly, National Bureau of Standards.* BAND SPECTRUM; SPECTRUM.

Meissner, Dr. Hans W. *Department of Physics, Stevens Institute of Technology, Hoboken, NJ.* MEISSNER EFFECT.

Meixner, Dr. Josef. *Director, Institute of Theoretical Physics, Aachen, West Germany.* BESSEL FUNCTIONS; GAMMA FUNCTION.

Mellon, Prof. M. G. *Department of Chemistry, Purdue University.* OPAQUE MEDIUM; TRANSLUCENT MEDIUM; TRANSPARENT MEDIUM.

Menkes, Joshua. *Science and Technology Division, Institute for Defense Analyses, Arlington, VA.* GAS DYNAMICS—coauthored.

Miller, Dr. Glenn H. *Weapons Effects Divisions, Sandia Laboratories, Albuquerque, NM.* ELECTRONVOLT; IONIZATION POTENTIAL.

Moffett, Dr. Mark B. *Naval Underwater Systems Center, New London Laboratory, New London, CT.* PARAMETRIC ARRAYS.

Mood, Dr. Alexander M. *Director, Public Policy Research Organization, University of California, Irvine.* STATISTICS.

Moore, Prof. John C. *Department of Mathematics, Princeton University.* TOPOLOGY—in part.

Mulliken, Prof. Robert S. *Institute of Molecular Biophysics, Florida State University.* MOLECULAR STRUCTURE AND SPECTRA; MOLECULE.

Murphy, Dr. Glenn. *Department of Nuclear Engineering, Iowa State University.* MACH NUMBER; REYNOLDS NUMBER.

Murray, Dr. Thomas P. *Applied Research Laboratory, United States Steel Corporation, Monroeville, PA.* PYROMETER.

N

Nachtrieb, Prof. Norman H. *Chairman, Department of Chemistry, University of Chicago.* LIQUID; SURFACE TENSION; VISCOSITY.

Nagel, Suzanne R. *Bell Laboratories, Murray Hill, NJ.* OPTICAL FIBERS.

Nedelsky, Prof. Leo. *Department of Physical Science, University of Chicago.* CONSERVATION OF ENERGY—validated; DENSITY; INERTIA; SPECIFIC GRAVITY; WORK; other articles.

Nicodemus, Dr. Fred E. *Physicist, Optical Radiation Section, Heat Division, Institute for Basic Standards, National Bureau of Standards.* RADIANCE.

O

Overhauser, Dr. Albert W. *Department of Physics, Purdue University.* CHARGE-DENSITY WAVE; SPIN-DENSITY WAVE.

P

Packard, Prof. Richard E. *Department of Physics, University of California, Berkeley.* QUANTIZED VORTICES.

Pai, Dr. Shih-I. *Institute of Physical Science and Technology, College of Engineering, University of Maryland.* CREEPING FLOW; LAMINAR FLOW; POTENTIAL FLOW; TURBULENT FLOW.

Pake, Dr. George E. *Vice President, Xerox Corporation; General Manager, Xerox Palo Alto Research Center, Palo Alto, CA.* FORCE.

Panofsky, Prof. Wolfgang K. H. *Professor of Physics and Director, Stanford Linear Accelerator Center, Stanford University.* CYCLOTRON.

Park, Prof. David. *Department of Physics, Williams College.* EIGENFUNCTION; EIGENVALUE (QUANTUM MECHANICS); ENERGY LEVEL.

Percus, Prof. J. K. *Courant Institute, New York University.* MANY-BODY THEORY.

Perl, Prof. Martin L. *Stanford Linear Accelerator Center, Stanford, CA.* LEPTON.

Phillips, Dr. James A. *CTR-Division Office, Los Alamos Scientific Laboratory, Los Alamos, NM.* PINCH EFFECT.

Pierce, Prof. Allan D. *Department of Transportation, Transportation Systems Center, Cambridge, MA.* ATMOSPHERIC ACOUSTICS.

Polking, Dr. John C. *Department of Mathematics, Rice University.* DIFFERENTIAL EQUATION; INTEGRAL EQUATION.

Pollard, Dr. William G. *Oak Ridge Associated Universities.* PHYSICS.

Post, Dr. Richard F. *Lawrence Livermore Laboratory, Livermore, CA.* LAWSON CRITERION; NUCLEAR FUSION; PLASMA PHYSICS; THERMONUCLEAR REACTION; WAVES AND INSTABILITIES IN PLASMAS.

Price, Dr. P. Buford. *Lawrence Berkeley Laboratory, University of California, Berkeley.* PARTICLE TRACK ETCHING.

Putnam, Prof. Russell C. *Professor Emeritus, Case Institute of Technology.* CANDLEPOWER; ILLUMINANCE; LUMINANCE; LUMINOUS FLUX; other articles.

Q

Quigg, Dr. C. *Fermi National Accelerator Laboratory, Batavia, IL.* GLUONS; QUANTUM CHROMODYNAMICS.

R

Radebaugh, Dr. R. *National Bureau of Standards.* KAPITZA RESISTANCE.

Ramsey, Dr. Norman F. *Department of Physics, Harvard University.* NEGATIVE TEMPERATURE.

Rasmussen, Dr. Norman C. *Department of Nuclear Engineering, Massachusetts Institute of Technology.* CHAIN REACTION.

Rebbi, Dr. Claudio. *Department of Physics, Brookhaven National Laboratory, Upton, NY.* SOLITON.

Reinmuth, Prof. William H. *Department of Chemistry, Columbia University.* DIELECTRIC CONSTANT—in part.

Reintjes, John F. *Naval Research Laboratory, Washington, DC.* NONLINEAR OPTICS.

Rich, Prof. Arthur. *Department of Physics, University of Michigan.* ELECTRON MAGNETIC MOMENT; ELECTRON SPIN.

Robb, D. D. *D. D. Robb and Associates, Consulting Engineers, Electric Power Systems, Salina, KS.* DIRECT CURRENT.

Roberts, Dr. Arthur. *National Accelerator Laboratory, Batavia, IL.* SPARK CHAMBER.

Roberts, Prof. Louis D. *Department of Physics, University of North Carolina.* HYPERFINE STRUCTURE.

Robertson, Prof. Burtis L. *Professor of Electrical Engineering (retired), University of California, Berkeley.* ADMITTANCE; POWER FACTOR; Q (ELECTRICITY); REACTANCE; SUSCEPTANCE; other articles.

Rockett, Frank H. *Engineering Consultant, Charlottesville, VA.* BOYLE'S LAW; ENTROPY; FOUCAULT PENDULUM; KNUDSEN NUMBER; PRESSURE; other articles.

Rohrlich, Prof. F. *Department of Physics, Syracuse University.* POTENTIALS.

Roller, Dr. Duane E. *Deceased; formerly, Harvey Mudd College.* CONSERVATION OF ENERGY.

Ross, Dr. Robert S. *Goodyear Aerospace Corporation, Akron, OH.* PASCAL'S LAW—coauthored.

Rusk, Dr. Rogers D. *Mount Holyoke College.* ACCELERATION—in part; DISPLACEMENT; FREE FALL; RECTILINEAR MOTION; RELATIVE MOTION; other articles.

Russell, Dr. H. W. *Deceased; formerly, Technical Director, Battelle Memorial Institute, Columbus, OH.* INCANDESCENCE; RADIOMETRY.

S

Salam, Prof. Abdus. *Director, International Centre for Theoretical Physics, Trieste, Italy.* FUNDAMENTAL INTERACTIONS.

Samios, Dr. Nicholas P. *Brookhaven National Laboratory, Upton, NY.* CHARM.

Sandler, Prof. Stanley I. *Department of Chemical Engineering, University of Delaware.* ISOTHERMAL PROCESS; PHASE RULE; POLYTROPIC PROCESS.

Sanford, Dr. James R. *Associate Director, Brookhaven National Laboratory, Upton, NY.* PARTICLE ACCELERATOR—in part.

Sargent, Dr. Murray, III. *Max Planck Institut für Festkoperforschung, Stuttgart, West Germany.* PHOTON.

Schaaf, Prof. Samuel A. *Professor of Engineering Science, and Chairman, Division of Aeronautical Sciences, University of California, Berkeley.* SUPERAERODYNAMICS.

Schawlow, Prof. Arthur L. *Department of Physics, Stanford University.* LASER.

Schmidt, Dr. Paul W. *Department of Physics, University of Missouri.* COLLISION; D'ALEMBERT'S PRINCIPLE; IMPULSE (MECHANICS); MOMENTUM; VIRTUAL WORK PRINCIPLE; other articles.

Schulman, Dr. James H. *U.S. Naval Research Laboratory.* FRANCK-CONDON PRINCIPLE. FLUORESCENCE; LUMINESCENCE; PHOSPHORESCENCE; THERMOLUMINESCENCE—all four coauthored.

Schumacher, Dr. Uwe. *Max Planck Institut für Plasmaphysik, Munich, West Germany.* PARTICLE ACCELERATOR—in part.

Schwartz, Dr. H. M. *Department of Physics, University of Arkansas.* RELATIVISTIC MECHANICS; RELATIVITY—in part.

Scott, Dr. David K. *National Superconducting Cyclotron Laboratory, Michigan State University.* SPALLATION REACTION.

Scott, Prof. J. F. *Department of Physics, University of Colorado.* SCATTERING OF ELECTROMAGNETIC RADIATION.

Scott, Dr. John E., Jr. *Department of Aerospace Engineering and Engineering Physics, University of Virginia.* COMPRESSIBLE FLOW.

Scott, Prof. Robert L. *Department of Chemistry, University of California, Los Angeles.* FIRST-ORDER TRANSITION; SECOND-ORDER TRANSITION; TRANSITION POINT; TRIPLE POINT.

Segré, Prof. Emilio G. *Department of Physics, University of California, Berkeley.* PROTON.

Sell, Dr. Heinz G. *Metals Development Section, Westinghouse Lamp Divisions, Bloomfield, NJ.* BLACKBODY; EMISSIVITY; GRAYBODY; HEAT RADIATION; PLANCK'S CONSTANT—all coauthored; other articles.

Sellars, Dr. John R. *Manager, Engineering Mechanics Operations, TRW, Inc., Redondo Beach, CA.* AEROSTATICS.

Sellin, Dr. Ivan A. *Department of Physics, University of Tennessee.* ATOMIC STRUCTURE AND SPECTRA.

Sellmyer, Prof. D. J. *Behlen Laboratory of Physics, University of Nebraska.* PHASE TRANSITIONS—coauthored.

Sertorio, Dr. Luigi. *Department of Physics, University of Rome, Italy.* REGGE POLE.

Shank, Dr. C. V. *Bell Telephone Laboratories, Holmdel, NJ.* OPTICAL PULSES.

Shankar, Dr. R. *Department of Physics, Yale University.* INSTANTON.

Shannon, Prof. R. R. *Optical Sciences Center, University of Arizona.* MIRROR OPTICS.

Shaw, Prof. Gordon L. *Department of Physics, University of California, Irvine.* CAUSALITY.

Simon, Dr. Barry. *Department of Mathematics, Princeton University.* FEYNMAN INTEGRAL.

Skilling, Prof. Hugh Hildreth. *Department of Electrical Engineering, Stanford University.* ALTERNATING CURRENT; RESONANCE (ALTERNATING-CURRENT CIRCUITS).

Slichter, Dr. Charles P. *Department of Physics, University of Illinois.* MAGNETIC RELAXATION; MAGNETIC RESONANCE.

Smythe, Dr. William R. *Department of Physics, California Institute of Technology.* DISPLACEMENT CURRENT; ELECTROMAGNETIC RADIATION; ETHER HYPOTHESIS; MAXWELL'S EQUATIONS; REFLECTION AND TRANSMISSION COEFFICIENTS; other articles.

Snell, Dr. Arthur H. *Associate Director, Oak Ridge National Laboratory, Oak Ridge, TN.* NEUTRON; NEUTRON OPTICS.

Sommer, Dr. Alfred H. *Thermo Electron Corporation, Waltham, MA.* SECONDARY EMISSION.

Souder, Dr. Paul A. *Department of Physics, Yale University.* PIONIUM.

Squire, Dr. Charles F. *Fellow, American Physical Society, College Station, TX.* ABSOLUTE ZERO.

Steele, Dr. William A. *Department of Chemistry, Pennsylvania State University.* ISOBARIC PROCESS—validated. ENTHALPY HEAT; THERMODYNAMIC PROCESSES; WORK FUNCTION (THERMODYNAMICS)—all four coauthored.

Stehle, Dr. Philip M. *Department of Physics, University of Pittsburgh.* CANONICAL TRANSFORMATIONS; HAMILTON-JACOBI THEORY; HAMILTON'S EQUATIONS OF MOTION; LAGRANGE'S EQUATIONS; LEAST-ACTION PRINCIPLE; other articles.

Stein, Dr. Nelson. *Physics Division, Los Alamos Scientific Laboratory, Los Alamos, NM.* ANALOG STATES.

Stelson, Dr. Paul H. *Physics Division, Oak Ridge National Laboratory, Oak Ridge, TN.* COULOMB EXCITATION.

Stephenson, Dr. R. J. *Department of Physics, Wooster College.* CENTRIPETAL FORCE. ACCELERATION; CENTRIFUGAL FORCE; ROTATIONAL MOTION—all three validated.

Stewart, Dr. John W. *Department of Physics, University of Virginia.* CONDUCTION (ELECTRICITY); DIMENSIONAL ANALYSIS; JOULE'S LAW—validated; REFRACTION OF WAVES; SEEBECK EFFECT; other articles.

Stratton, Dr. Robert. *Central Research Laboratories, Texas Instruments, Inc., Dallas.* POTENTIAL BARRIER.

Strauch, Dr. Karl. *Physics Laboratories, Harvard University.* PARTICLE ACCELERATOR—in part.

Streeter, Prof. Victor L. *Department of Civil Engineering, College of Engineering, University of Michigan.* ARCHIMEDES' PRINCIPLE; FLUID MECHANICS; THROTTLED FLOW; TORRICELLI'S THEOREM; VORTEX; other articles.

Stroke, Dr. George W. *Department of Electrical Sciences, and Head, Electro-Optical Sciences Center, State University of New York, Stony Brook.* DOPPLER EFFECT; LIGHT.

Sturge, Dr. M. D. *Bell Laboratories, Murray Hill, NJ.* JAHN-TELLER EFFECT.

T

Taft, Prof. Horace D. *Department of Physics, Yale University.* STRANGE PARTICLES.

Tang, Prof. K. Y. *Deceased; formerly, Department of Electrical Engineering, Ohio State University.* KIRCHHOFF'S LAWS OF ELECTRIC CIRCUITS.

Taylor, Prof. Angus E. *President's Office, University of California, Berkeley.* CALCULUS; DIFFERENTIATION; PARTIAL DIFFERENTIATION.

Taylor, Dr. Barry N. *National Bureau of Standards.* ATOMIC CONSTANTS.

Thomson, Prof. Robb M. *Chairman, Department of Materials Science, State University of New York, Stony Brook.* TWINNING.

Tigner, Dr. M. *Laboratory of Nuclear Studies, Cornell University.* PARTICLE ACCELERATOR—in part.

Ting, Prof. Samuel C. C. *Department of Physics, Massachusetts Institute of Technology.* J PARTICLE.

Tompkins, Prof. Charles B. *Department of Mathematics, University of California, Los Angeles.* CALCULUS OF VARIATIONS.

Townes, Prof. Charles H. *Department of Physics, University of California, Berkeley.* MASER—coauthored.

Tsang, Dr. C. F. *Lawrence Berkeley Laboratory, University of California, Berkeley.* SUPERTRANSURANICS.

Turnbull, Prof. David. *Department of Applied Physics, Harvard University.* ANISOTROPY; CRYSTAL GROWTH; CRYSTAL WHISKERS; ISOTROPY; SINGLE CRYSTAL.

V

Van Winkle, Prof. Quentin. *Department of Chemistry, Ohio State University.* TYNDALL EFFECT.

Veech, Dr. William A. *Department of Mathematics, Rice University.* TOPOLOGICAL DYNAMICS.

W

Waddington, Prof. Thomas C. *Department of Chemistry, University of Durham.* AVOGADRO NUMBER; GAS CONSTANT; MOLE; RELATIVE ATOMIC MASS; RELATIVE MOLECULAR MASS; other articles.

Waldron, Dr. Robert D. *Director, Research Enterprises, Scottsdale, AZ.* DIELECTRIC CONSTANT—in part; DIPOLE MOMENT; ELECTRET; ELECTROSTRICTION; POLAR MOLECULE; other articles.

Walker, Lawrence R. *Bell Laboratories, Murray Hill, NJ.* FERRIMAGNETISM.

Walsh, Prof. Joseph L. *Department of Mathematics, University of Maryland.* CONFORMAL MAPPING; LAPLACE'S DIFFERENTIAL EQUATION.

Walsh, Dr. Peter J. *Department of Physics, Fairleigh Dickinson University.* BLACKBODY; EMISSIVITY; GRAYBODY; HEAT RADIATION; PLANCK'S CONSTANT—all coauthored; other articles.

Walsh, Dr. Walter M. *Bell Laboratories, Murray Hill, NJ.* CYCLOTRON RESONANCE EXPERIMENTS.

Walstedt, Dr. R. E. *Bell Laboratories, Murray Hill, NJ.* SPIN GLASS.

Walter, William C. *President, Frontier Systems, Rolling Hills, CA.* AERODYNAMICS; AEROMECHANICS.

Warren, Prof. Bertram E. *Department of Physics, Massachusetts Institute of Technology.* X-RAY DIFFRACTION—in part.

Watson, William W. *Professor Emeritus of Physics, Yale University.* AUGER EFFECT. ATOM; MAGNETOOPTICS; PASCHEN-BACK EFFECT; ZEEMAN EFFECT—all four validated; other articles.

Weaver, Prof. David L. *Department of Physics, Tufts University.* SPINOR; WAVE EQUATION.

Weber, Harold Christian. *Chemical Engineer, Boston, MA.* ENTHALPY—coauthored; HEAT—coauthored; HEAT CAPACITY; SPECIFIC HEAT.

Wegner, Dr. Harvey E. *Physics Division, Brookhaven National Laboratory, Upton, NY.* PARTICLE ACCELERATOR—in part.

Weil, Jonathan F. *Staff Editor, "McGraw-Hill Encyclopedia of Science and Technology," McGraw-Hill Book Company, New York, NY.* ATOMIC MASS UNIT; UNITS OF MEASUREMENT.

Weil, Robert T., Jr. *Deceased; formerly, Dean, School of Engineering, Manhattan College.* KIRCHHOFF'S LAWS OF ELECTRIC CIRCUITS—validated.

Welch, Arthur A. *Engineering Consultant, Purcellville, VA.* CONDUCTANCE.

West, Dr. William. *Eastman Kodak Company, Rochester, NY.* ABSORPTION OF ELECTROMAGNETIC RADIATION; DISPERSION (RADIATION); OPTICAL PUMPING.

Whitehead, Prof. W. Dexter. *Department of Physics, Center for Advanced Studies, University of Virginia.* MEAN FREE PATH.

Widder, Prof. David V. *Department of Mathematics, Harvard University.* INTERGRAL TRANSFORM; LAPLACE TRANSFORM.

Wiegand, Dr. Clyde E. *Lawrence Berkeley Laboratory, University of California, Berkeley.* HADRONIC ATOM.

Wightman, Prof. Arthur S. *Department of Physics, Joseph Henry Laboratories, Princeton University.* GROUP THEORY.

Wildhack, William A. *Consultant, formerly, Associate Director, Institute for Basic Standards, National Bureau of Standards.* PHYSICAL MEASUREMENT.

Wilkinson, Prof. D. H. *Department of Nuclear Physics, Oxford University.* NUCLEAR BINDING ENERGY—in part.

Wilkinson, Dr. Michael K. *Associate Director, Solid State Division, Oak Ridge National Laboratory, Oak Ridge, TN.* NEUTRON DIFFRACTION.

Wilkinson, Dr. W. L. *Postgraduate School of Studies in Chemical Engineering, University of Bradford, England.* FLUIDS; NON-NEWTONIAN FLUID—both coauthored.

Williams, Prof. Dudley. *Department of Physics, Kansas State University.* DIMENSIONS (MECHANICS); MINIMAL PRINCIPLES; NEWTON'S LAWS OF MOTION; PARTICLE; RIGID BODY; other articles.

Williams, Prof. Gary A. *Department of Physics, University of California, Los Angeles.* LOW-TEMPERATURE ACOUSTICS.

Williams, S. A. *Department of Physics, Iowa State University.* AMPLITUDE (WAVE MOTION); EXCLUSION PRINCIPLE; GROUP VELOCITY; PHASE VELOCITY; SUPERMULTIPLET.

Williamson, Dr. K. D. *Los Alamos Scientific Laboratory, Los Alamos, NM.* CRYOGENICS—coauthored.

Willis, Dr. William J. *Department of Physics, Brookhaven National Laboratory, Upton, NY.* TRANSITION RADIATION DETECTORS.

Winch, Prof. Ralph P. *Department of Physics, Williams College.* CAPACITANCE; COULOMB'S LAW; ELECTROSCOPE; MAGNET; PERMITTIVITY; other articles.

Winick, Dr. Herman. *Deputy Director, Stanford Linear Accelerator Center, Stanford, CA.* SYNCHROTRON RADIATION—coauthored.

Winter, Prof. Rolf G. *Department of Physics, College of William and Mary.* COHERENCE.

Wolf, Dr. W. P. *Department of Physics, Yale University.* SCHOTTKY ANOMALY.

Wolfe, Dr. William L. *Optical Sciences Center, University of Arizona.* INFRARED RADIATION; OPTICAL MATERIALS.

Wu, Prof. Tai Tsun. *Division of Engineering and Applied Physics, Harvard University.* ISING MODEL—coauthored.

Wyant, Prof. James C. *Optical Sciences Center, University of Arizona.* INTERFEROMETRY; SPECKLE.

Y

Yost, Prof. Don M. *Formerly, California Institute of Technology.* QUATERNIONS.

Young, Dr. Robert W. *Associate Editor, "Journal of the Acoustical Society of America," San Diego, CA.* FUNDAMENTAL FREQUENCY; LEVEL; MODE OF VIBRATION.

Index

Index

Asterisks indicate page references to article titles.